P9-EEH-412

CHILTON®

ASIAN
SERVICE MANUAL
2010 EDITION
VOLUME III
INFINITI
NISSAN

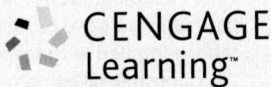

CENGAGE
Learning™

Australia • Brazil • Japan • Korea • Mexico • Singapore • Spain • United Kingdom • United States

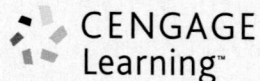
CENGAGE
Learning™

CHILTON®
Asian Service Manual
2010 Edition
Volume III
Infiniti, Nissan

**Vice President,
Technology Professional
Business Unit:**
Gregory L. Clayton

**Publisher,
Technology Professional
Business Unit:**
David Koontz

Director of Marketing:
Beth A. Lutz

Director Education Production:
Carolyn Miller

Marketing Manager:
Jennifer Barbic

Marketing Coordinator:
Rachael Torres

Chilton Content Specialist:
Paula Baillie

Graphical Designer:
Melinda Possinger

Art Director:
Benjamin Gleeksman

Sr. Content Project Manager:
Elizabeth C. Hough

Senior Editor:
Christine L. Sheeky

Editors:
Dennis L. Bailey

Jim Bailey

Sherry Burdette

Nick D'Andrea

Maureen Lazarz

Printed in the United States of America
1 2 3 4 5 6 7 15 14 13 12 11

For product information and technology
assistance, contact us at
**Professional & Career Group customer
Support, 1-800-648-7450.**
For permission to use material from
this text or product,
submit all requests online at
www.cengage.com/permissions.
Further permissions questions can be e-mailed to
permissionrequest@cengage.com

ISBN-13: 978-1-1110-3766-6
ISBN-10: 1-1110-3766-3
ISSN: 1939-621X

Chilton
5 Maxwell Drive
Clifton Park, NY 12065-2919
USA

Chilton products are represented in Canada by Nelson Education, Ltd.

NOTICE TO THE READER

Contents

Sections

Model Index

USING THIS INFORMATION

Organization

To find where a particular model section or procedure is located, look in the Table of Contents. Main topics are listed with the page number on which they may be found. Following the main topics is an alphabetical listing of all of the procedures within the section and their page numbers.

Manufacturer and Model Coverage

This product covers 2009–2010 Asian models that are produced in sufficient quantities to warrant coverage, and which have technical content available from the vehicle manufacturers before our publication date. Although this information is as complete as possible at the time of publication, some manufacturers may make changes which cannot be included here. While striving for total accuracy, the publisher cannot assume responsibility for any errors, changes, or omissions that may occur in the compilation of this data.

Part Numbers and Special Tools

Part numbers and special tools are recommended by the publisher and vehicle manufacturer to perform specific jobs. Before substituting any part or tool for the one recommended, you must be completely satisfied that neither your personal safety, nor the performance of the vehicle will be endangered.

ACKNOWLEDGEMENT

The publisher would like to express appreciation to the following vehicle manufacturers for their assistance in producing this manual: Nissan North America, including Infiniti and Nissan Divisions. No further reproduction or distribution of the material in this manual is allowed without the expressed written permission of the vehicle manufacturers and the publisher.

PRECAUTIONS

Before servicing any vehicle, please be sure to read all of the following precautions, which deal with personal safety, prevention of component damage, and important points to take into consideration when servicing a motor vehicle:

• Always wear safety glasses or goggles when drilling, cutting, grinding or prying.

• Steel-toed work shoes should be worn when working with heavy parts. Pockets should not be used for carrying tools. A slip or fall can drive a screwdriver into your body.

• Work surfaces, including tools and the floor should be kept clean of grease, oil or other slippery material.

• When working around moving parts, don't wear loose clothing. Long hair should be tied back under a hat or cap, or in a hair net.

• Always use tools only for the purpose for which they were designed. Never pry with a screwdriver.

• Keep a fire extinguisher and first aid kit handy.

• Always properly support the vehicle with approved stands or lift.

• Always have adequate ventilation when working with chemicals or hazardous material.

• Carbon monoxide is colorless, odorless and dangerous. If it is necessary to operate the engine with vehicle in a closed area such as a garage, always use an exhaust collector to vent the exhaust gases outside the closed area.

• When draining coolant, keep in mind that small children and some pets are attracted by ethylene glycol antifreeze, and are quite likely to drink any left in an open container, or in puddles on the ground. This will prove fatal in sufficient quantity. Always drain the coolant into a sealable container.

• To avoid personal injury, do not remove the coolant pressure relief cap while the engine is operating or hot. The cooling system is under pressure; steam and hot liquid can come out forcefully when the cap is loosened slightly. Failure to follow these instructions may result in personal injury. The coolant must be recovered in a suitable, clean container for reuse. If the coolant is contaminated it must be recycled or disposed of correctly.

• When carrying out maintenance on the starting system be aware that heavy gauge leads are connected directly to the battery. Make sure the protective caps are in place when maintenance is completed. Failure to follow these instructions may result in personal injury.

• Do not remove any part of the engine emission control system. Operating the engine without the engine emission control system will reduce fuel economy and engine ventilation. This will weaken engine performance and shorten engine life. It is also a violation of Federal law.

• Due to environmental concerns, when the air conditioning system is drained, the refrigerant must be collected using refrigerant recovery/recycling equipment. Federal law requires that refrigerant be recovered into appropriate recovery equipment and the process be conducted by qualified technicians who have been certified by an approved organization, such as MACS, ASI, etc. Use of a recovery machine dedicated to the appropriate refrigerant is necessary to reduce the possibility of oil and refrigerant incompatibility concerns. Refer to the instructions provided by the equipment manufacturer when removing refrigerant from or charging the air conditioning system.

• Always disconnect the battery ground when working on or around the electrical system.

• Batteries contain sulfuric acid. Avoid contact with skin, eyes, or clothing. Also, shield your eyes when working near batteries to protect against possible splashing of the acid solution. In case of acid contact with skin or eyes, flush immediately with water for a minimum of 15 minutes and get prompt medical attention. If acid is swallowed, call a physician immediately. Failure to follow these instructions may result in personal injury.

• Batteries normally produce explosive gases. Therefore, do not allow flames, sparks or lighted substances to come near the battery. When charging or working near a battery, always shield your face and protect your eyes. Always provide ventilation. Failure to follow these instructions may result in personal injury.

• When lifting a battery, excessive pressure on the end walls could cause acid to spew through the vent caps, resulting in personal injury, damage to the vehicle or battery. Lift with a battery carrier or with your hands on opposite corners. Failure to follow these instructions may result in personal injury.

• Observe all applicable safety precautions when working around fuel. Whenever servicing the fuel system, always work in a well-ventilated area. Do not allow fuel spray or vapors to come in contact with a spark, open flame, or excessive heat (a hot drop light, for example). Keep a dry chemical fire extinguisher near the work area. Always keep fuel in a container specifically designed for fuel storage; also, always properly seal fuel containers to avoid the possibility of fire or explosion. Do not smoke or carry lighted tobacco or open flame of any type when working on or near any fuel-related components.

• Fuel injection systems often remain pressurized, even after the engine has been turned OFF. The fuel system pressure must be relieved before disconnecting any fuel lines. Failure to do so may result in fire and/or personal injury.

• The evaporative emissions system contains fuel vapor and condensed fuel vapor. Although not present in large quantities, it still presents the danger of explosion or fire. Disconnect the battery ground cable from the battery to minimize the possibility of an electrical spark occurring, possibly causing a fire or explosion if fuel vapor or liquid fuel is present in the area. Failure to follow these instructions can result in personal injury.

• The EPA warns that prolonged contact with used engine oil may cause a number of skin disorders, including cancer! You should make every effort to minimize your exposure to used engine oil. Protective gloves should be worn when changing oil. Wash your hands and any other exposed skin areas as soon as possible after exposure to used engine oil. Soap and water, or waterless hand cleaner should be used.

• Some vehicles are equipped with an air bag system, often referred to as a Supplemental Restraint System (SRS) or Supplemental Inflatable Restraint (SIR) system. The system must be disabled before performing service on or around system components, steering column, instrument panel components, wiring and sensors. Failure to follow safety and disabling procedures could result in accidental air bag deployment, possible personal injury and unnecessary system repairs.

• Always wear safety goggles when working with, or around, the air bag system. When carrying a non-deployed air bag, be sure the bag and trim cover are pointed away from your body. When placing a non-deployed air bag on a work surface, always face the bag and trim cover upward, away from the surface. This will reduce the motion of the module if it is accidentally deployed.

• Electronic modules are sensitive to electrical charges. The ABS module can be damaged if exposed to these charges.

• Brake pads and shoes may contain asbestos, which has been determined to be a cancer-causing agent. Never clean brake surfaces with compressed air. Avoid inhaling brake dust. Clean all brake surfaces with a commercially available brake cleaning fluid.

• When replacing brake pads, shoes, discs or drums, replace them as complete axle sets.

• When servicing drum brakes, disassemble and assemble one side at a time, leaving the remaining side intact for reference.

• Brake fluid often contains polyglycol ethers and polyglycols. Avoid contact with the eyes and wash your hands thoroughly after handling brake fluid. If you do get brake fluid in your eyes, flush your eyes with clean, running water for 15 minutes. If eye irritation persists, or if you have taken brake fluid internally, immediately seek medical assistance.

• Clean, high quality brake fluid from a sealed container is essential to the safe and proper operation of the brake system. You should always buy the correct type of brake fluid for your vehicle. If the brake fluid becomes contaminated, completely flush the system with new fluid. Never reuse any brake fluid. Any brake fluid that is removed from the system should be discarded. Also, do not allow any brake fluid to come in contact with a painted or plastic surface; it will damage the paint.

• Never operate the engine without the proper amount and type of engine oil; doing so will result in severe engine damage.

• Timing belt maintenance is extremely important! Many models utilize an interference- type, non freewheeling engine. If the timing belt breaks, the valves in the cylinder head may strike the pistons, causing potentially serious (also time-consuming and expensive) engine damage.

• Disconnecting the negative battery cable on some vehicles may interfere with the functions of the on-board computer system(s) and may require the computer to undergo a relearning process once the negative battery cable is reconnected.

• Steering and suspension fasteners are critical parts because they affect performance of vital components and systems and their failure can result in major service expense. They must be replaced with the same grade or part number or an equivalent part if replacement is necessary. Do not use a replacement part of lesser quality or substitute design. Torque values must be used as specified during reassembly.

SPECIFICATIONS AND MAINTENANCE CHARTS

ENGINE AND VEHICLE IDENTIFICATION

	Engine						Model Year	
Code ①	Liters (cc)	Cu. In.	Cyl.	Fuel Sys.	Engine Type	Eng. Mfg.	Code ②	Year
VQ35HR	3.5 (3498)	274.2	8	SFI	DOHC	Nissan	8	2008
							9	2009
							A	2010

SFI: Sequential Fuel Injection

DOHC: Double Overhead Camshaft

① Stamped on the upper rear of the engine block, just behind one of the cylinder heads

② 10th digit of the Vehicle Identification Number (VIN)

37663_EX35_C0001

GENERAL ENGINE SPECIFICATIONS

All measurements are given in inches.

Year	Model	Engine Displacement Liters	Engine Series ID/VIN	Net Horsepower @ rpm	Net Torque @ rpm (ft. lbs.)	Bore x Stroke (in.)	Com-pression Ratio	Oil Pressure @ rpm
2008	EX35	3.5	VQ35HR	297@6800	253@4800	3.76 x 3.21	10.6:0	43@2000
2009	EX35	3.5	VQ35HR	297@6800	253@4800	3.76 x 3.21	10.6:0	43@2000
2010	EX35	3.5	VQ35HR	297@6800	253@4800	3.76 x 3.21	10.6:0	43@2000

37663_EX35_C0002

GASOLINE ENGINE TUNE-UP SPECIFICATIONS

Year	Engine Displacement Liters	Engine Series ID/VIN	Spark Plug Gap (in.)	Ignition Timing (deg.) ① MT	Ignition Timing (deg.) ① AT	Fuel Pump (psi)	Idle Speed (rpm) MT	Idle Speed (rpm) AT	Valve Clearance (in.) ② Intake	Valve Clearance (in.) ② Exhaust
2008	3.5	VQ35HR	0.043	—	11-21B	51	—	600-700	0.010-0.013	0.011-0.015
2009	3.5	VQ35HR	0.043	—	11-21B	51	—	600-700	0.010-0.013	0.011-0.015
2010	3.5	VQ35HR	0.043	—	11-21B	51	—	600-700	0.010-0.013	0.011-0.015

NOTE: The Vehicle Emission Control Information label reflects specification changes made during production.

Follow the figures on the label if they differ from those in this chart.

B: Before top dead center

① With terminals TC and CG of DLC3 connected

② Engine cold - approximately 68°F (20°C)

37663_EX35_C0003

CAPACITIES

Year	Model	Engine Displacement Liters	Engine Series ID/VIN	Engine Oil with Filter (qts.)	Transmission (pts.)		Transfer Case (pts.)	Drive Axle		Fuel Tank (gal.)	Cooling System (qts.)
					Manual	Auto. ①		Front (pts.)	Rear (pts.)		
2008	EX35	3.5	VQ35HR	5.0	—	21.8	2.6	1.4	3.0	20.0	9.1
2009	EX35	3.5	VQ35HR	5.0	—	21.8	2.6	1.4	3.0	20.0	9.1
2010	EX35	3.5	VQ35HR	5.0	—	21.8	2.6	1.4	3.0	20.0	9.1

NOTE: All capacities are approximate. Add fluid gradually and check to be sure a proper fluid level is obtained.

① Drain and refill

37663_EX35_C0004

FLUID SPECIFICATIONS

Year	Model	Engine Displacement Liters	Engine Series ID/VIN	Engine Oil	Auto. Trans.	Drive Axle	Transfer Case	Power Steering Fluid	Brake Master Cylinder	Cooling System
2008	EX35	3.5	VQ35HR	5W-30	①	②	③	④	⑤	⑥
2009	EX35	3.5	VQ35HR	5W-30	①	②	③	④	⑤	⑥
2010	EX35	3.5	VQ35HR	5W-30	①	②	③	④	⑤	⑥

DOT: Department Of Transportation

① Genuine NISSAN Matic J ATF

② Genuine NISSAN Differential Oil Hypoid Super GL-5 80W-90 or API GL-5 Viscosity SAE 80W-90

③ Genuine NISSAN Matic D ATF (Continental U.S. and Alaska) or Canada NISSAN Automatic Transmission Fluid or equivalent (if available)

④ Genuine NISSAN PSF or equivalent

⑤ Genuine NISSAN Super Heavy Duty Brake Fluid or equivalent DOT 3

⑥ Genuine NISSAN Long Life Antifreeze/Coolant or equivalent

37663_EX35_C0005

VALVE SPECIFICATIONS

Year	Engine Displacement Liters	Engine Series ID/VIN	Seat Angle (deg.)	Face Angle (deg.)	Spring Test Pressure (lbs. @ in.)	Spring Installed Height (in.)	Stem-to-Guide Clearance (in.)		Stem Diameter (in.)	
							Intake	Exhaust	Intake	Exhaust
2008	3.5	VQ35HR	45.15-45.45	44.23-45.08	84-95 @1.071	1.457	0.0008-0.0021	0.0012-0.0022	0.2348-0.2354	0.2347-0.2350
2009	3.5	VQ35HR	45.15-45.45	44.23-45.08	84-95 @1.071	1.457	0.0008-0.0021	0.0012-0.0022	0.2348-0.2354	0.2347-0.2350
2010	3.5	VQ35HR	45.15-45.45	44.23-45.08	84-95 @1.071	1.457	0.0008-0.0021	0.0012-0.0022	0.2348-0.2354	0.2347-0.2350

37663_EX35_C0006

CAMSHAFT AND BEARING SPECIFICATIONS CHART

All measurements are given in inches.

Year	Engine Displ. Liters	Engine Series ID/VIN	Journal Dia.	Brg. Oil Clearance	Shaft End-play	Runout	Journal Bore	Lobe Height Intake	Lobe Height Exhaust
2008	3.5	VQ35HR	①	②	0.0045-0.0074	0.0020 max.	NA	1.8057-1.8132	1.8061-1.8136
2009	3.5	VQ35HR	①	②	0.0045-0.0074	0.0020 max.	NA	1.8057-1.8132	1.8061-1.8136
2010	3.5	VQ35HR	①	②	0.0045-0.0074	0.0020 max.	NA	1.8057-1.8132	1.8061-1.8136

NA: Not Available

① No. 1: 1.0211-1.0218 in.
   No. 2, 3, 4: 0.9230-0.9238 in.

② No. 1: 0.0018-0.0034 in.
   No. 2, 3, 4: 0.0014-0.0030 in.

37663_EX35_C0007

CRANKSHAFT AND CONNECTING ROD SPECIFICATIONS

All measurements are given in inches.

Year	Engine Displacement Liters	Engine Series ID/VIN	Crankshaft Main Brg. Journal Dia.	Crankshaft Main Brg. Oil Clearance	Crankshaft Shaft End-play	Crankshaft Thrust on No.	Connecting Rod Journal Diameter	Connecting Rod Oil Clearance	Connecting Rod Side Clearance
2008	3.5	VQ35HR	①	0.0014-0.0018	0.0039-0.0098	3	②	0.0013-0.0023	0.0079-0.0138
2009	3.5	VQ35HR	①	0.0014-0.0018	0.0039-0.0098	3	②	0.0013-0.0023	0.0079-0.0138
2010	3.5	VQ35HR	①	0.0014-0.0018	0.0039-0.0098	3	②	0.0013-0.0023	0.0079-0.0138

① Depends on the grade of the crankshaft. The nominal range is: 2.3603-2.3612 in.
② Grade 0: 2.0460-2.0462 in.
   Grade 1: 2.0457-2.0460 in.
   Grade 2: 2.0455-2.0457 in.

37663_EX35_C0008

PISTON AND RING SPECIFICATIONS

All measurements are given in inches.

Year	Engine Displ. Liters	Engine Series ID/VIN	Piston Clearance	Ring Gap Top Compression	Ring Gap Bottom Compression	Ring Gap Oil Control	Ring Side Clearance Top Compression	Ring Side Clearance Bottom Compression	Ring Side Clearance Oil Control
2008	3.5	VQ35HR	0.0004-0.0012	0.0091-0.0130	0.0130-0.0189	0.0067-0.0185	0.0016-0.0031	0.0012-0.0028	0.0026-0.0053
2009	3.5	VQ35HR	0.0004-0.0012	0.0091-0.0130	0.0130-0.0189	0.0067-0.0185	0.0016-0.0031	0.0012-0.0028	0.0026-0.0053
2010	3.5	VQ35HR	0.0004-0.0012	0.0091-0.0130	0.0130-0.0189	0.0067-0.0185	0.0016-0.0031	0.0012-0.0028	0.0026-0.0053

37663_EX35_C0009

TORQUE SPECIFICATIONS
All readings in ft. lbs.

Year	Engine Displacement Liters	Engine Series ID/VIN	Cylinder Head Bolts	Main Bearing Bolts	Rod Bearing Bolts	Crankshaft Damper Bolts	Flywheel Bolts	Manifold		Spark Plugs	Oil Pan Drain Plug
								Intake	Exhaust		
2008	3.5	VQ35HR	①	②	③	④	65	⑤	22	14	25
2009	3.5	VQ35HR	①	②	③	④	65	⑤	22	14	25
2010	3.5	VQ35HR	①	②	③	④	65	⑤	22	14	25

NOTE: Dip main bearing bolts, crankshaft damper bolt, and flywheel bolts in clean engine oil prior to tightening.

① Step 1: Tighten in sequence to 77 ft. lbs.
　Step 2: Completely loosen all in reverse sequence
　Step 3: Tighten in sequence to 30 ft. lbs.
　Step 4: Tighten 90 degrees
　Step 5: Tighten another 90 degrees

② Step 1: Tighten in sequence to 10 ft. lbs.
　Step 2: Tighten in sequence to 26 ft. lbs.
　Step 3: Tighten 90 degrees

③ Step 1: Tighten to 14 ft. lbs.
　Step 2: Tighten 90 degrees

④ Step 1: Tighten to 69 ft. lbs.
　Step 2: Tighten 90 degrees

⑤ Step 1: Tighten to 60 inch lbs.
　Step 2: Tighten to 19 ft. lbs.

37663_EX35_C0010

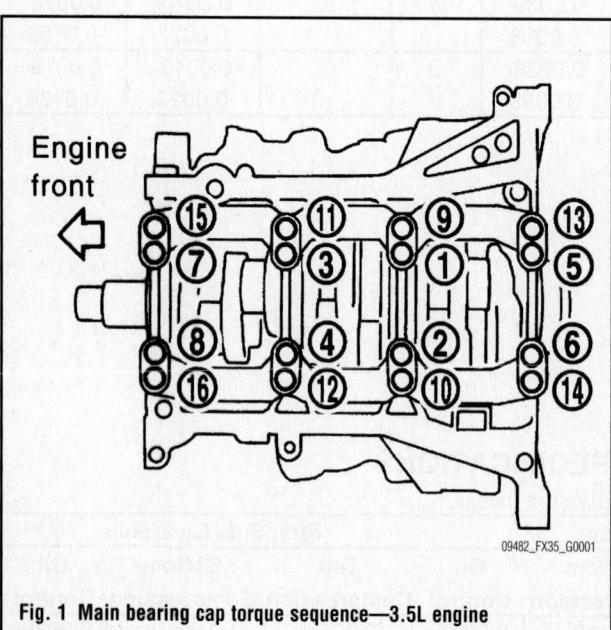

09482_FX35_G0001

Fig. 1 Main bearing cap torque sequence—3.5L engine

WHEEL ALIGNMENT

Year	Model		Caster Range (+/-Deg.)	Caster Preferred Setting (Deg.)	Camber Range (+/-Deg.)	Camber Preferred Setting (Deg.)	Toe-in (Deg.)
2008	EX35	Front	0.75	+3.78	0.75	-0.73	0.060+/-0.040
		Rear	—	—	0.50	-0.80	0.170+/-0.080
2009	EX35	Front	0.75	+3.78	0.75	-0.73	0.060+/-0.040
		Rear	—	—	0.50	-0.80	0.170+/-0.080
2010	EX35	Front	0.75	+3.78	0.75	-0.73	0.060+/-0.040
		Rear	—	—	0.50	-0.80	0.170+/-0.080

NOTE: Measurements are given for unladen vehicle: fuel, engine coolant, and fluid levels are full. Spare tire, jack, hand tools, and mats are in designated positions

37663_EX35_C0011

TIRE, WHEEL AND BALL JOINT SPECIFICATIONS

Year	Model	OEM Tires Standard	OEM Tires Optional	Tire Pressures (psi) Front	Tire Pressures (psi) Rear	Wheel Size	Ball Joint Inspection	Lug Nut Torque (ft. lbs.)
2008	EX35	P225/60R17	P225/55R18	32	32	7.5J x 17 8.0J x 18	①	80
2009	EX35	P225/60R17	P225/55R18	32	32	7.5J x 17 8.0J x 18	①	80
2010	EX35	P225/60R17	P225/55R18	32	32	7.5J x 17 8.0J x 18	①	80

OEM: Original Equipment Manufacturer

PSI: Pounds Per Square Inch

① Replace if any measurable axial end play is found.

37663_EX35_C0012

BRAKE SPECIFICATIONS
All measurements in inches unless noted

Year	Model		Brake Disc Original Thickness	Brake Disc Minimum Thickness	Brake Disc Maximum Runout	Brake Drum Diameter Original Inside Diameter	Max. Wear Limit	Brake Drum Diameter Maximum Machine Diameter	Minimum Lining Thickness	Brake Caliper Bracket Bolts (ft. lbs.)	Brake Caliper Mounting Bolts (ft. lbs.)
2008	EX35	F	1.339	1.260	0.0014	—	—	—	0.079	34	122
		R	0.630	0.551	0.0022	—	—	—	0.079	32	62
2009	EX35	F	1.339	1.260	0.0014	—	—	—	0.079	34	122
		R	0.630	0.551	0.0022	—	—	—	0.079	32	62
2010	EX35	F	1.339	1.260	0.0014	—	—	—	0.079	34	122
		R	0.630	0.551	0.0022	—	—	—	0.079	32	62

F: Front

R: Rear

37663_EX35_C0013

SCHEDULED MAINTENANCE INTERVALS
INFINITI—EX35 & EX35 Journey

TO BE SERVICED	TYPE OF SERVIC	VEHICLE MILEAGE INTERVAL (x1000)														
		7.5	15	22.5	30	37.5	45	52.5	60	67.5	75	82.5	90	97.5	105	120
Accessory drive belts ①	S/I								✓							✓
Air cleaner element (engine)	R			✓					✓				✓			✓
Air conditioner system	S/I	Inspect system operation annually														
Automatic transaxle fluid	S/I		✓		✓		✓		✓		✓		✓		✓	✓
Brake lines, hoses, cables, and connections	S/I		✓		✓		✓		✓		✓		✓		✓	✓
Brake pads, calipers, & rotors	S/I		✓		✓		✓		✓		✓		✓		✓	✓
Differential gear oil	S/I		✓		✓		✓		✓		✓		✓		✓	✓
Driveshafts and CV-boots	S/I		✓		✓		✓		✓		✓		✓		✓	✓
Engine coolant	R								✓				✓			
Engine oil and filter	R	✓	✓	✓	✓	✓	✓	✓	✓	✓	✓	✓	✓	✓	✓	✓
Exhaust pipe connections, muffler, and suspension bolts	S/I				✓				✓				✓			✓
EVAP vapor lines	S/I				✓				✓				✓			✓
Fuel lines and connections	S/I				✓				✓				✓			✓
In-cabin microfilter	S/I		✓		✓		✓		✓		✓		✓		✓	✓
Spark plugs (Platinum-tipped)	R	105,000 miles (under normal usage)														
Steering system	S/I				✓				✓				✓			✓
Suspension system	S/I				✓				✓				✓			✓
Transfer case fluid	S/I		✓		✓		✓		✓		✓		✓		✓	✓
Valve clearance	S/I	Whenever valve noise increases														

R: Replace   S/I: Service or Inspect

① Replace if worn or damaged or if the auto-tensioner has reached its limit (V8)

FREQUENT OPERATION MAINTENANCE (SEVERE SERVICE)

If a vehicle is operated under any of the following conditions it is considered severe service:

- Extremely dusty areas.

- 50% or more of the vehicle operation is in 90°F (32°C) or higher temperatures, or constant operation in temperatures below 32°F (0°C).

- Prolonged idling (vehicle operation in stop and go traffic).

- Frequent short running periods (engine does not warm to normal operating temperatures).

- Police, taxi, delivery usage, or trailer towing usage.

Automatic transaxle fluid (and filter), transfer case fluid, and differential gear oil: check every 15,000 miles, replace every 30,000 miles

Brake pads, calipers & rotors: service or inspect every 7,500 miles

Driveshafts and CV-boots inspect every 7,500 miles

Exhaust system inspect every 7,500 miles

Oil and oil filter: change every 3,750 miles

Steering system and suspension components inspect for looseness and damage every 7,500 miles

37663_EX35_C0014

BRAKES — INFORMATION AND PRECAUTIONS

ANTI-LOCK SYSTEMS

- Certain components within the ABS system are not intended to be serviced or repaired individually.
- Do not use rubber hoses or other parts not specifically specified for and ABS system. When using repair kits, replace all parts included in the kit. Partial or incorrect repair may lead to functional problems and require the replacement of components.
- Lubricate rubber parts with clean, fresh brake fluid to ease assembly. Do not use shop air to clean parts; damage to rubber components may result.
- Use only DOT 3 brake fluid from an unopened container.
- If any hydraulic component or line is removed or replaced, it may be necessary to bleed the entire system.
- A clean repair area is essential. Always clean the reservoir and cap thoroughly before removing the cap. The slightest amount of dirt in the fluid may plug an orifice and impair the system function. Perform repairs after components have been thoroughly cleaned; use only denatured alcohol to clean components. Do not allow ABS components to come into contact with any substance containing mineral oil; this includes used shop rags.
- The Anti-Lock control unit is a microprocessor similar to other computer units in the vehicle. Ensure that the ignition switch is **OFF** before removing or installing controller harnesses. Avoid static electricity discharge at or near the controller.
- If any arc welding is to be done on the vehicle, the control unit should be unplugged before welding operations begin.

DISC AND DRUM SYSTEMS

> **✳✳ CAUTION**
>
> **Dust and dirt accumulating on brake parts during normal use may contain asbestos fibers from production or aftermarket brake linings.**

Breathing excessive concentrations of asbestos fibers can cause serious bodily harm. Exercise care when servicing brake parts. Do not sand or grind brake lining unless equipment used is designed to contain the dust residue. Do not clean brake parts with compressed air or by dry brushing. Cleaning should be done by dampening the brake components with a fine mist of water, then wiping the brake components clean with a dampened cloth. Dispose of cloth and all residue containing asbestos fibers in an impermeable container with the appropriate label. Follow practices prescribed by the Occupational Safety and Health Administration (OSHA) and the Environmental Protection Agency (EPA) for the handling, processing, and disposing of dust or debris that may contain asbestos fibers.

BRAKES — BLEEDING THE BRAKE SYSTEM

BLEEDING PROCEDURE

BLEEDING PROCEDURE

When any part of the hydraulic system has been disconnected for repair or replacement, air may get into the lines and cause spongy pedal action (because air can be compressed and brake fluid cannot). To correct this condition, it is necessary to bleed the hydraulic system so to be sure all air is purged.

When bleeding the brake system, bleed one brake cylinder at a time, beginning at the cylinder with the longest hydraulic line (farthest from the master cylinder) first. ALWAYS keep the master cylinder reservoir filled with brake fluid during the bleeding operation. Never use brake fluid that has been drained from the hydraulic system, no matter how clean it is.

The primary and secondary hydraulic brake systems are separate and are bled independently. During the bleeding operation, do not allow the reservoir to run dry. Keep the master cylinder reservoir filled with brake fluid.

1. Clean all dirt from around the master cylinder fill cap, remove the cap and fill the master cylinder with brake fluid until the level is within ¼ in. (6mm) of the top edge of the reservoir.

2. Clean the bleeder screws at all 4 wheels. The bleeder screws are located on the top of the brake calipers.

3. Attach a length of rubber hose over the bleeder screw and place the other end of the hose in a glass jar, submerged in brake fluid.

4. Open the bleeder screw ½–¾ turn. Have an assistant slowly depress the brake pedal.

> **✳✳ CAUTION**
>
> **Brake fluid contains polyglycol ethers and polyglycols. Avoid contact with the eyes and wash your hands thoroughly after handling brake fluid. If you do get brake fluid in your eyes, flush your eyes with clean, running water for 15 minutes. If eye irritation persists, or if you have taken brake fluid internally, IMMEDIATELY seek medical assistance.**

5. Close the bleeder screw and tell your assistant to allow the brake pedal to return slowly. Continue this process to purge all air from the system.

6. When bubbles cease to appear at the end of the bleeder hose, close the bleeder screw and remove the hose. Tighten the bleeder screw to the proper torque.

7. Check the master cylinder fluid level and add fluid accordingly. Do this after bleeding each wheel.

8. Repeat the bleeding operation at the remaining 3 wheels, ending with the one closet to the master cylinder.

9. Fill the master cylinder reservoir to the proper level.

MASTER CYLINDER BLEEDING

1. Before servicing the vehicle, refer to the Precautions Section.

2. If removed from the vehicle, clamp the master cylinder in a vise with soft-jaw caps.

3. Attach the special tools for bleeding the master cylinder in the following fashion:

   a. Thread the bleeder tube adapters into the primary and secondary outlet ports of the master cylinder and tighten the adapters.

   b. Thread a bleeder tube into each adapter and tighten the tube nuts.

   c. Flex each bleeder tube and place the open ends into the neck of the master cylinder reservoir. Position the open ends of the tubes into the reservoir so their outlets are below the surface of the brake fluid in the reservoir when filled.

➡**Make sure the ends of the bleeder tubes stay below the surface of the brake fluid in the reservoir at all times during the bleeding procedure.**

4. Fill the brake fluid reservoir with fresh brake fluid (DOT 3).

5. Using an appropriately sized wooden dowel as a pushrod, slowly press the pistons inward discharging brake fluid through the bleeder tubes, then release the pressure, allowing the pistons to return to the released position. Repeat this several times until all air bubbles are expelled from the master cylinder bore and bleeder tubes.

6. Remove the bleeder tubes and adapters from the master cylinder and plug the master cylinder outlet ports.

7. Install the fill cap on the reservoir.

8. Remove the master cylinder from the vise.

9. Install the master cylinder on the vehicle.

BLEEDING THE ABS SYSTEM

The bleeding procedure for the ABS System is the same as the conventional bleeding procedure. Refer to Bleeding the Brake System, Bleeding Procedure.

BRAKES | ANTI-LOCK BRAKE SYSTEM (ABS)

WHEEL SPEED SENSORS

REMOVAL & INSTALLATION

Front Wheel

See Figure 2.

1. Before servicing the vehicle, refer to the Precautions Section.
2. Disconnect the negative battery cable.
3. Raise and support the vehicle safely.

4. Remove the tire and wheel assembly.

5. Remove the wheel speed sensor mounting bolts, grommets, and clip.

6. Pull the sensor out, being careful to turn it as little as possible. Do not pull on the sensor harness.

To install:

7. Before installing the sensor be sure no foreign materials such as iron fragments are adhered to the pick-up part of the sensor or to the inside of the sensor mounting hole or on the rotor mounting surface.

8. Tighten the sensor retaining bolt to 57 inch lbs. (6 Nm).

9. Be sure to press the rubber grommets in until they lock at the three locations shown in the illustration.

10. Be sure the harness is not twisted when installed. The white line on the harness must be visible from the front.

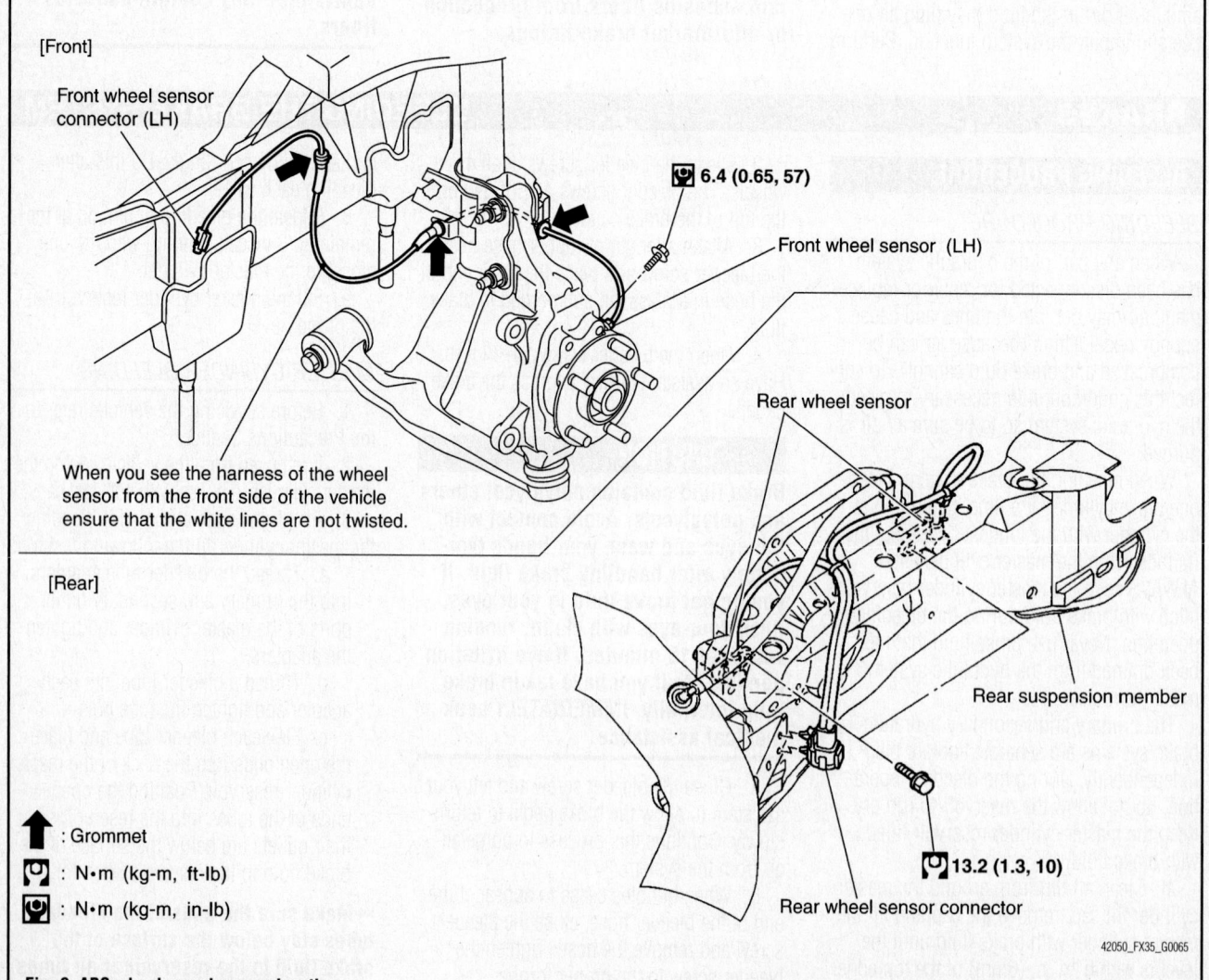

[Front]

Front wheel sensor connector (LH)

6.4 (0.65, 57)

Front wheel sensor (LH)

When you see the harness of the wheel sensor from the front side of the vehicle ensure that the white lines are not twisted.

Rear wheel sensor

Rear suspension member

[Rear]

⬆ : Grommet

⬛ : N•m (kg-m, ft-lb)

⬛ : N•m (kg-m, in-lb)

13.2 (1.3, 10)

Rear wheel sensor connector

Fig. 2 ABS wheel speed sensor location

42050_FX35_G0065

11. Continue the installation in the reverse order of the removal procedure.

Rear Wheel

See Figure 2.

1. Before servicing the vehicle, refer to the Precautions Section.
2. Disconnect the negative battery cable.
3. Raise and support the vehicle safely.
4. Remove the tire and wheel assembly.
5. Remove the wheel speed sensor mounting bolts and clip.
6. Pull the sensor out, being careful to turn it as little as possible. Do not pull on the sensor harness.

To install:

7. Before installing the sensor, be sure no foreign materials (such as iron fragments) are adhered to the pick-up part of the sensor or to the inside of the sensor mounting hole or on the rotor mounting surface.
8. Tighten the sensor retaining bolt to 117 inch lbs. (13 Nm).

9. Be sure to securely attach the harness at the rear suspension member. Be sure the harness is not twisted when installed.
10. Continue the installation in the reverse order of the removal procedure.

WHEEL SPEED SENSOR RINGS (TOOTHED RINGS)

REMOVAL & INSTALLATION

Front Wheel

The wheel speed sensor ring is an integral part of the wheel hub and bearing assembly and is not serviced separately.

Rear Wheel

See Figure 3.

1. Before servicing the vehicle, refer to the Precautions Section.
2. Remove the halfshaft(s). Refer to Halfshafts, removal & installation.
3. Remove the side flange.
4. Using a bearing puller, remove the sensor rotor (ring) from the side flange.

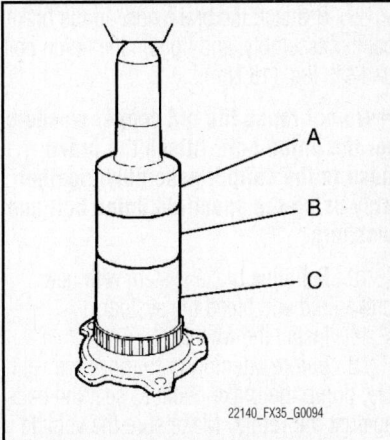

22140_FX35_G0094

Fig. 3 Use a drift set to press the rear sensor rotor (ring) onto the side flange

To install:

5. Using a drift set, as illustrated, press the rear sensor rotor (ring) onto the side flange.
6. Special tools in the drift set:
   • A: ST30720000 (J-25405)
   • B: ST27863000
   • C: KV40104710

BRAKES   FRONT DISC BRAKES

BRAKE CALIPER

REMOVAL & INSTALLATION

See Figure 4.

1. Before servicing the vehicle, refer to the Precautions Section.
2. Raise and safely support the vehicle.
3. Remove the front wheels from the vehicle.
4. Drain the brake fluid.
5. Remove the union bolts and torque member bolts, and remove the brake caliper assembly from the vehicle.
6. If necessary, remove the rotor.

To install:

➡**Use Only new DOT-3 brake fluid. Do not reuse drained brake fluid.**

7. If removed, install the rotor.
8. Install the caliper assembly onto the vehicle. Tighten the sliding pin bolts to 34 ft. lbs. (46 Nm) and the torque member bolts to 122 ft. lbs. (165 Nm).

➡**When attaching the caliper assembly to the vehicle, wipe any oil off the knuckle spindle, washers, and caliper assembly attachment surfaces.**

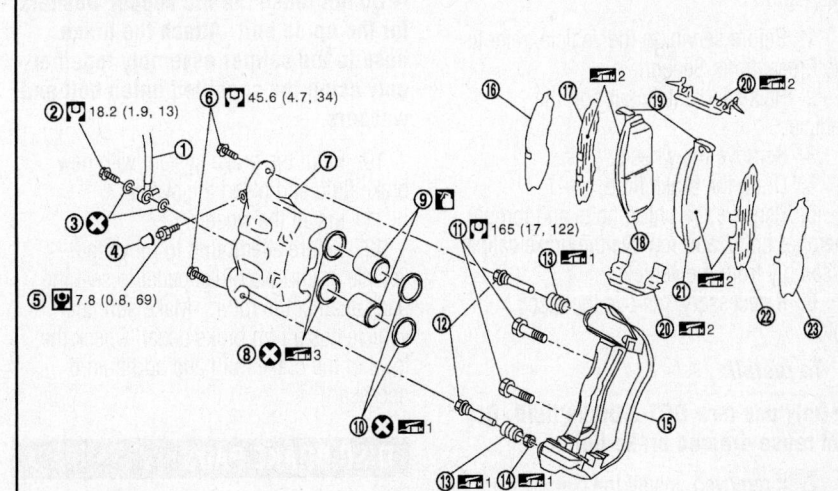

1.	Brake hose	2.	Union bolt	3.	Copper washer
4.	Cap	5.	Bleed valve	6.	Sliding pin bolt
7.	Cylinder body	8.	Piston seal	9.	Piston
10.	Piston boot	11.	Torque member mounting bolt	12.	Sliding pin
13.	Sliding pin boot	14.	Bushing	15.	Torque member
16.	Inner shim cover	17.	Inner shim	18.	Inner pad
19.	Pad wear sensor	20.	Pad retainer	21.	Outer pad

22140_FX35_G0001

Fig. 4 Exploded view of the front disc brake mounting and components

9. Reattach the brake hose to the brake caliper assembly, and tighten the union bolt to 13 ft. lbs. (18 Nm).

➡ **Do not reuse the old copper washers for the union bolt. Attach the brake hose to the caliper assembly together only using the specified union bolt and washers.**

10. Refill the brake system with new brake fluid and bleed the system.

11. Install the wheels.

12. Before attempting to move the vehicle, pump the brake pedal to seat the pads against the rotors. Make sure the vehicle has a firm brake pedal. Check the level of the brake fluid and add fluid if necessary.

DISC BRAKE PADS

REMOVAL & INSTALLATION

See Figure 4.

1. Before servicing the vehicle, refer to the Precautions Section.

BRAKES

BRAKE CALIPER

REMOVAL & INSTALLATION

See Figure 5.

1. Before servicing the vehicle, refer to the Precautions Section.

2. Raise and safely support the vehicle.

3. Remove the wheels.

4. Drain the brake fluid.

5. Remove the union bolts and torque member bolts, and remove the brake caliper assembly from the vehicle.

6. If necessary, remove the disc rotor.

To install:

➡ **Only use new DOT-3 brake fluid. Do not reuse drained brake fluid.**

7. If removed, install the disc rotor.

8. Install the caliper assembly onto the vehicle. Tighten the sliding pin bolts to 32 ft. lbs. (43 Nm) and the torque member bolts to 62 ft. lbs. (84 Nm).

➡ **When attaching the caliper assembly to the vehicle, wipe any oil off the knuckle spindle, washers, and caliper assembly attachment surfaces.**

2. Raise and safely support the vehicle.

3. Remove the front wheels.

4. Remove the lower sliding pin bolt.

5. Suspend the cylinder body with strong cord or wire, then remove the pad and shim from the torque member.

To install:

6. Position the inner shim and shim cover onto the inner pad, and outer shim onto the outer pad.

7. Push the caliper piston in so that the brake pad is firmly installed.

8. Position the cylinder body on the torque member.

➡ **The use of a disc brake piston tool can make it easier to push in the piston.**

❄ WARNING

By pushing in the piston, brake fluid returns to the mas-

9. Reattach the brake hose to the brake caliper assembly, and tighten the union bolt to 13 ft. lbs. (18 Nm).

➡ **Do not reuse the old copper washers for the union bolt. Attach the brake hose to the caliper assembly together only using the specified union bolt and washers.**

10. Refill the brake system with new brake fluid and bleed the system.

11. Install the wheels.

12. Before attempting to move the vehicle, pump the brake pedal to seat the pads against the rotors. Make sure the vehicle has a firm brake pedal. Check the level of the brake fluid and add fluid if necessary.

DISC BRAKE PADS

REMOVAL & INSTALLATION

See Figure 5.

1. Before servicing the vehicle, refer to the Precautions Section.

2. Raise and safely support the vehicle.

3. Remove the wheels.

4. Remove the upper sliding pin bolt.

ter cylinder reservoir tank. Watch the level of the surface of reservoir tank. Brake fluid is corrosive to painted surfaces.

9. Reattach the pad retainer to the torque member. When attaching the pad retainer, attach it firmly so that it does not float up higher than the torque member.

10. Install the lower sliding pin bolt and tighten it to 34 ft. lbs. (46 Nm).

11. Check the brake assembly for drag.

12. Install the wheels.

13. Before attempting to move the vehicle, pump the brake pedal to seat the pads against the rotors. Make sure the vehicle has a firm brake pedal. Check the level of the brake fluid and add fluid if necessary.

REAR DISC BRAKES

5. Suspend the cylinder body with strong cord or wire, then remove the pad and shim from the torque member.

To install:

6. Apply silicon-based grease to the backside of the pad and to both sides of the shim, then attach the inner shim and shim cover to the inner pad. Attach the outer shim and outer shim cover to the outer pad.

7. Install the pad retainer and mount the pad onto the torque member.

8. Position the cylinder body on the torque member.

➡ **Using a disc brake piston tool can make it easier to push in the piston.**

❄ WARNING

By pushing in the piston, brake fluid returns to the master cylinder reservoir tank. Watch the level of the surface of the reservoir tank. Brake fluid is corrosive to painted surfaces.

9. Install the top sliding pin bolt and tighten it to 32 ft. lbs. (43 Nm).

10. Check the brake assembly for drag.

11. Install the wheels

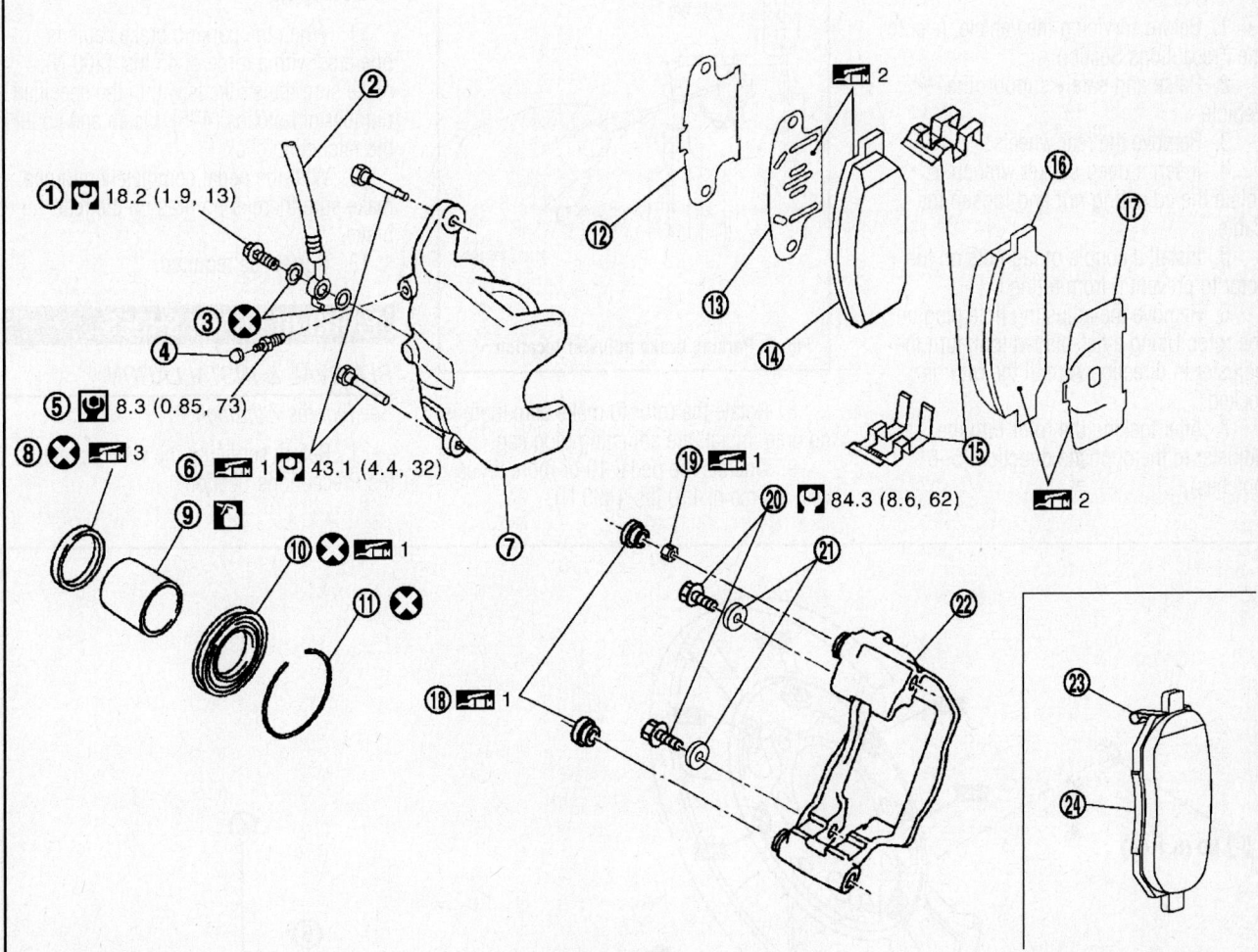

① 18.2 (1.9, 13)

⑤ 8.3 (0.85, 73)

⑥ 1 43.1 (4.4, 32)

⑳ 84.3 (8.6, 62)

1.	Union bolt	2.	Brake hose	3.	Copper washer
4.	Cap	5.	Bleed valve	6.	Sliding pin bolt
7.	Cylinder body	8.	Piston seal	9.	Piston
10.	Piston boot	11.	Retaining ring	12.	Inner shim cover
13.	Inner shim	14.	Inner pad	15.	Pad retainer
16.	Outer pad	17.	Outer shim	18.	Sliding pin boot
19.	Bushing	20.	Torque member mounting bolt	21.	Washer
22.	Torque member	23.	Pad wear sensor	24.	Inner pad (RH)

22140_FX35_G0002

Fig. 5 Exploded view of the rear disc brake mounting and components

BRAKES

PARKING BRAKE

PARKING BRAKE CABLES

ADJUSTMENT

See Figure 6.

1. Before servicing the vehicle, refer to the Precautions Section.
2. Raise and safely support the vehicle.
3. Remove the rear wheels.
4. Insert a deep socket wrench to rotate the adjusting nut and loosen the cable.
5. Install a couple of lug nuts on the rotor to prevent it from tilting.
6. Remove the adjusting hole plug on the rotor. Using a flat-bladed tool, turn the adjuster in direction **A** until the rotor is locked.
7. After locking the rotor turn the adjuster in the opposite direction (5–6 notches).

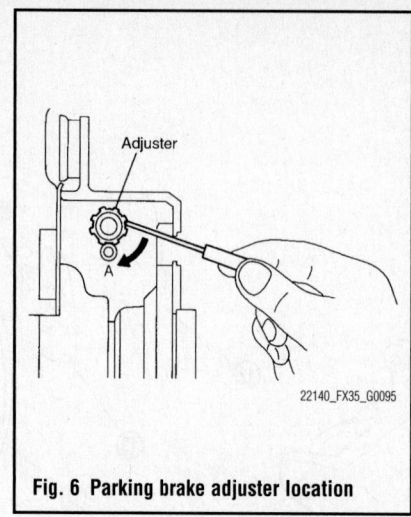

Fig. 6 Parking brake adjuster location

8. Rotate the rotor to make sure there is no drag. Install the adjusting plug cap.
9. Operate the pedal 10 or more times with a force of 110 lbs. (490 N).

10. Rotate the adjusting nut with a deep socket to adjust the pedal stroke.

➡**Do not reuse the adjusting nut after removing it.**

11. When the parking brake cable is operated with a force of 45 lbs. (200 N), make sure the stroke is within the specified number of notches (4–5). Listen and count the ratcheting click.
12. With the pedal completely returned, make sure there is no drag on the rear brake.
13. Correct as required.

PARKING BRAKE SHOES

REMOVAL & INSTALLATION

See Figures 7 through 11.

1. Before servicing the vehicle, refer to the Precautions Section.

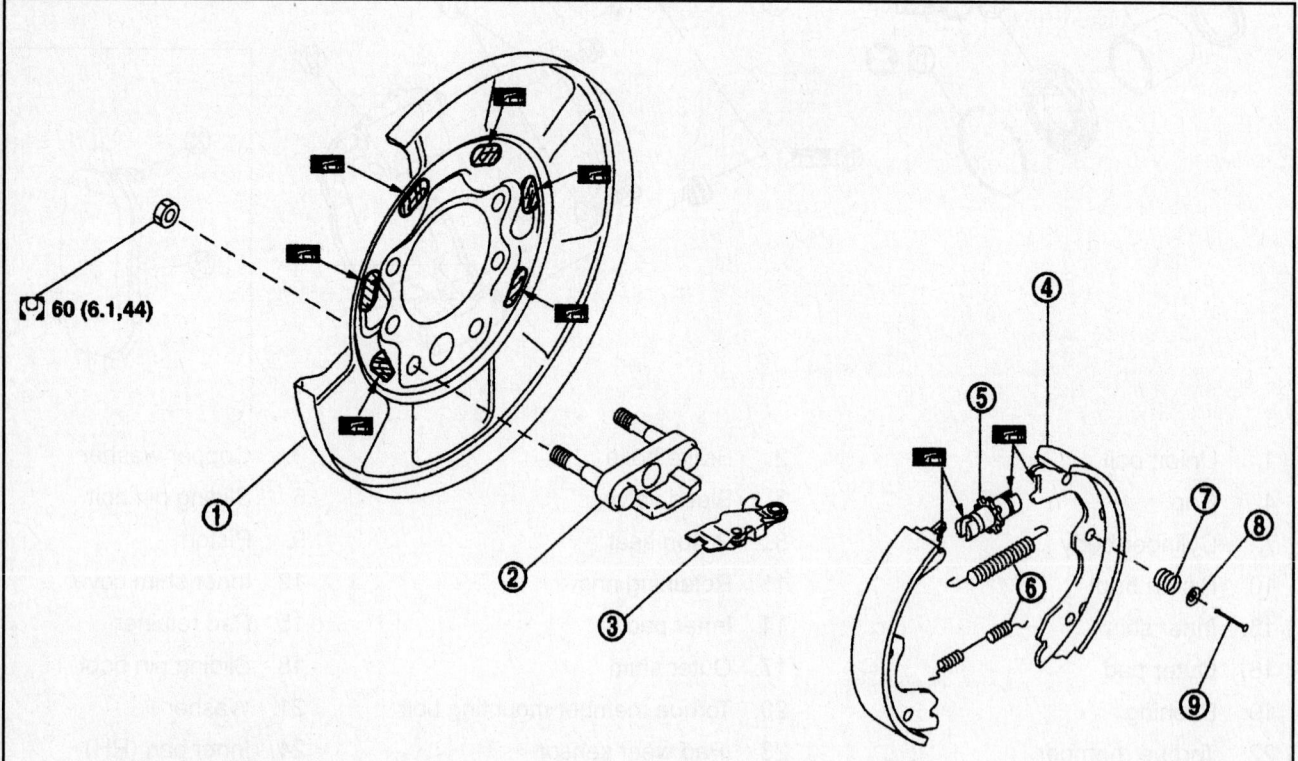

🔩 : PBC (Poly Butyl Cuprysil) grease or silicone-based grease point

🔧 : N·m (kg-m, ft-lb)

1. Back plate	2. Anchor block	3. Toggle lever
4. Shoe	5. Adjuster	6. Return spring
7. Anti-rattle spring	8. Retainer	9. Anti-rattle pin

Fig. 7 Parking brake shoes and related components

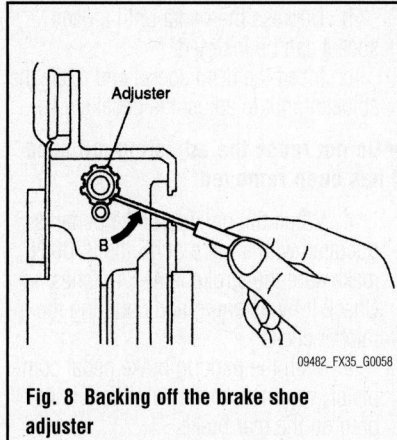

Fig. 8 Backing off the brake shoe adjuster

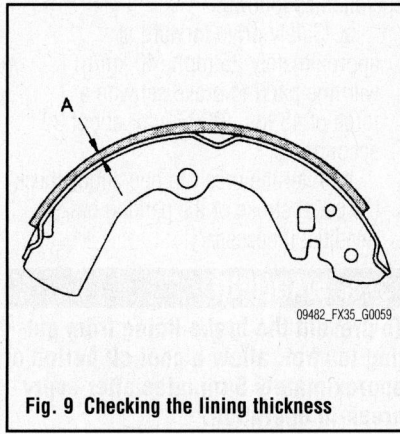

Fig. 9 Checking the lining thickness

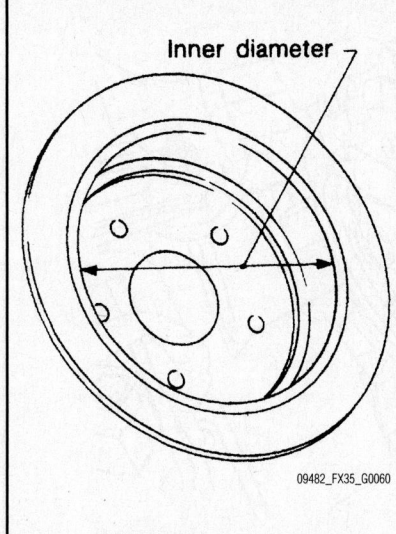

Fig. 10 Checking the drum inner diameter

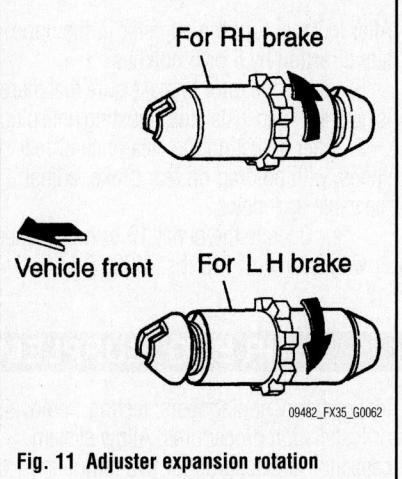

Fig. 11 Adjuster expansion rotation

2. Raise and safely support the vehicle.
3. Release the parking brake.
4. Remove the rear wheels.
5. Remove the rotor. If the rotor cannot be removed:

   a. Secure the rotor in place with lug nuts and remove the adjuster hole plug.

   b. Using a flat-bladed screwdriver, rotate the adjuster in direction **B** to retract and loosen the brake shoes.

6. Remove the anti-rattle pins, retainers, anti-rattle springs, and return springs.

7. Remove the parking brake shoes, adjuster assembly, adjuster spring, and toggle lever.

To install:

8. Check the thickness of the lining. Standard thickness **A**: 0.126 in. (3.2mm). Repair limit thickness **A**: 0.059 in. (1.5mm).

9. Check the drum inner diameter. Standard inner diameter: 7.48 in. (190mm). Maximum inner diameter: 7.52 in. (191mm).

10. Check the following:
    - Shoes for excessive wear, damage, and peeling
    - Shoe sliding surfaces for excessive wear and damage
    - Anti-rattle pins for excessive wear and corrosion
    - Return springs for sagging
    - Check that the adjuster moves smoothly
    - Visually check the inside of the drum for excessive wear, cracks, and damage. Check the inside diameter of the drum

11. Replace any suspect parts.

➡**When disassembling the adjuster, apply silicone grease or equivalent to the threads.**

12. Continue the installation in the reverse order of the removal procedure. Note the following:
    - Apply brake grease to the brake mechanism, being sure to keep the shoes and drum clean
    - Assemble the adjuster so that the threaded part expands when rotating it in the direction shown by the arrow
    - When disassembling the adjuster, apply silicone-based grease to the threads

13. Adjust the parking brake shoe tension:

    a. Adjust the parking brake pedal stroke to 4–5 clicks fully depressed. Insert a deep socket wrench to rotate the adjusting nut and loosen the cable sufficiently. Then, return the pedal.

    b. With the wheels removed, use a lug nut and secure the rotor to hub to prevent it from tilting.

    c. Remove the adjusting hole plug. Using a flat-bladed screwdriver, turn the adjuster clockwise until the rotor is locked. After locking, turn the adjuster in the opposite direction by 5–6 notches.

    d. Rotate the rotor to make sure that there is no drag. Install the adjusting hole plug.

14. After adjusting the clearance of the rear shoes, with no drag on rear brake, adjust the cable as follows:

    a. Operate the pedal 10 or more times with a force of 110 lbs. (490 N).

    b. Depress the pedal until a deep socket can be inserted. Insert the deep socket, and rotate the adjusting nut to adjust the pedal stroke.

➡**Do not reuse the adjusting nut.**

    c. When the parking brake pedal is operated with a force of 45 lbs. (200 N), make sure the stroke is 4–5 notches. (Check it by listening and counting the ratchet clicks).

    d. With the parking brake pedal completely returned, make sure there is no drag on the rear brake.

15. Perform the parking brake break-in operation as follows: Safely, drive forward at approximately 25 mph (40 km/h) with the parking brake set with a force of approximately 45 lbs. (200 N) for about 30 seconds.

16. After the break-in operation, check the pedal stroke of parking brake. Readjust if necessary.

➡**To prevent the brake lining from getting too hot, allow a cool off period of approximately 5 minutes after every break-in operation.**

ADJUSTMENT

See Figure 12.

1. Before servicing the vehicle, refer to the Precautions Section.

2. Adjust the parking brake pedal stroke to 4–5 clicks fully depressed. Insert a deep

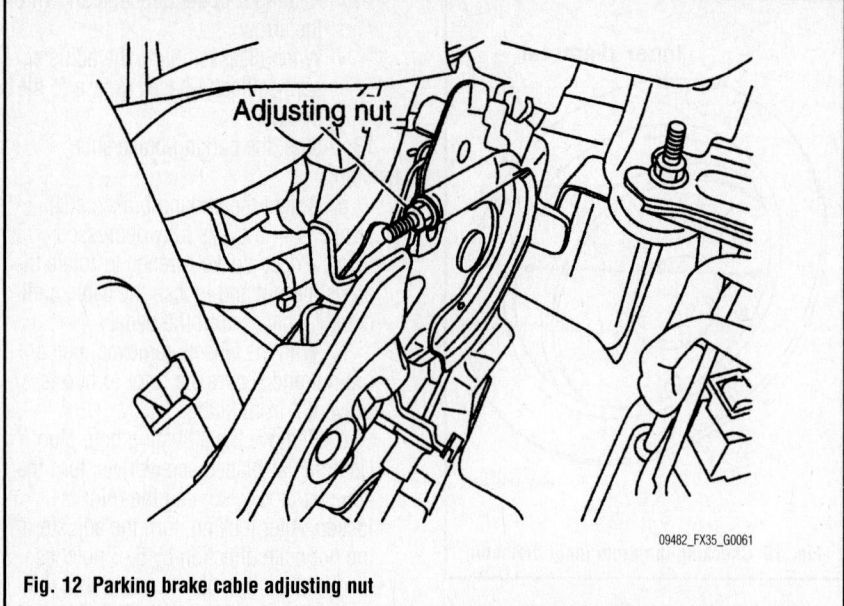

Fig. 12 Parking brake cable adjusting nut

socket wrench to rotate the adjusting nut and loosen the cable sufficiently. Then, return the pedal.

3. Remove the rear wheels.

4. Using a couple of lug nuts, secure the rotor to hub to prevent it from tilting.

5. Remove the adjusting hole plug. Using a flat-bladed screwdriver, turn the adjuster clockwise until the rotor is locked.

After locking, turn the adjuster in the opposite direction by 5 or 6 notches.

6. Rotate the rotor to make sure that there is no drag. Then install the adjusting hole plug.

7. After adjusting the clearance of rear shoes, with no drag on rear brake, adjust the cable as follows:

   a. Operate the pedal 10 or more times with a force of 110 lbs. (490 N).

   b. Depress the pedal until a deep socket can be inserted.

   c. Insert the deep socket and rotate the adjusting nut to adjust the pedal stroke.

➥**Do not reuse the adjusting nut once it has been removed.**

   d. When the parking brake pedal is operated with a force of 45 lbs. (200 N), make sure the stroke is 4–5 notches. Check it by listening and counting the ratchet clicks.

   e. With the parking brake pedal completely returned, make sure there is no drag on the rear brake.

8. Perform the parking brake break-in operation as follows:

   a. Safely drive forward at approximately 25 mph (40 km/h) with the parking brake set with a force of 45 lbs. (200 N) for about 30 seconds.

   b. After the break-in operation, check the pedal stroke of the parking brake. Readjust if necessary.

❋❋ WARNING

To prevent the brake lining from getting too hot, allow a cool off period of approximately 5 minutes after every break-in operation.

CHASSIS ELECTRICAL

GENERAL INFORMATION

❋❋ CAUTION

These vehicles are equipped with an air bag system. The system must be disarmed before performing service on, or around, system components, the steering column, instrument panel components, wiring and sensors. Failure to follow the safety precautions and the disarming procedure could result in accidental air bag deployment, possible injury and unnecessary system repairs.

SERVICE PRECAUTIONS

Disconnect and isolate the battery negative cable before beginning any airbag system component diagnosis, testing, removal, or installation procedures. Allow system capacitor to discharge for two minutes before beginning any component service. This will disable the airbag system. Failure to disable the airbag system may result in accidental airbag deployment, personal injury, or death.

DISARMING THE SYSTEM

1. Before servicing the vehicle, refer to the Precautions Section.

2. Turn the ignition switch to **OFF**.

3. Disconnect the negative battery cable and isolate it from accidental reconnection. Insulate the cable end with high-quality electrical tape or a similar non-conductive wrapping.

AIR BAG (SUPPLEMENTAL RESTRAINT SYSTEM)

4. Wait at least 3 minutes for the system capacitor to discharge before performing any service. The air bag system is designed to retain enough voltage to deploy the air bag for a short period of time after the battery has been disconnected.

ARMING THE SYSTEM

1. Before servicing the vehicle, refer to the Precautions Section.

2. Reconnect the negative battery cable.

3. To confirm proper system operation, turn the ignition switch to the **ON** position. The SRS indicator light should light for at least 7 seconds and then go off.

DRIVE TRAIN

FRONT HALFSHAFT

REMOVAL & INSTALLATION

AWD Models

See Figures 13 through 15.

1. Before servicing the vehicle, refer to the Precautions Section.
2. Raise and safely support the vehicle.
3. Remove the wheels.
4. Remove the undercover.

5. Remove the cotter pin. Then remove the locknut from the driveshaft.
6. Remove the wheel sensor wiring harness from the strut assembly.

> ✳✳ **CAUTION**
>
> **Do not pull on the wheel sensor wiring harness.**

7. Remove the brake hose lock plate. Then remove the brake hose from the strut assembly.

8. Remove the mounting bolts and nuts between the strut assembly and the steering knuckle.
9. Separate the halfshaft from the steering knuckle.

> ✳✳ **CAUTION**
>
> **When removing the halfshaft, do not apply an excessive angle to the halfshaft joint. Also, be careful not to excessively extend the slide joint.**

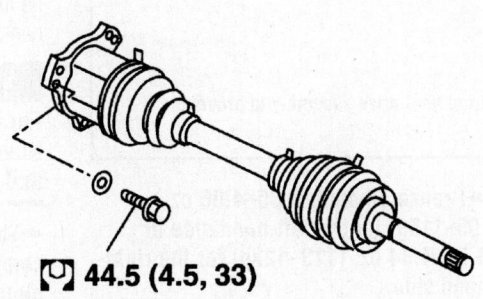
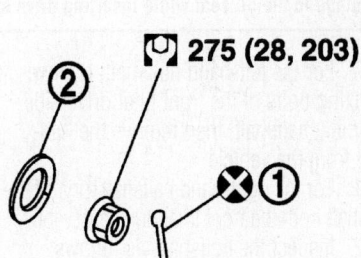

🔧 44.5 (4.5, 33)

🔧 275 (28, 203)

❌ : Always replace after every disassembly

🔧 : N·m(kg-m,ft-lb)

1.   Cotter pin

2.   Washer

67162-FX35-G199

Fig. 13 View of left-hand front halfshaft

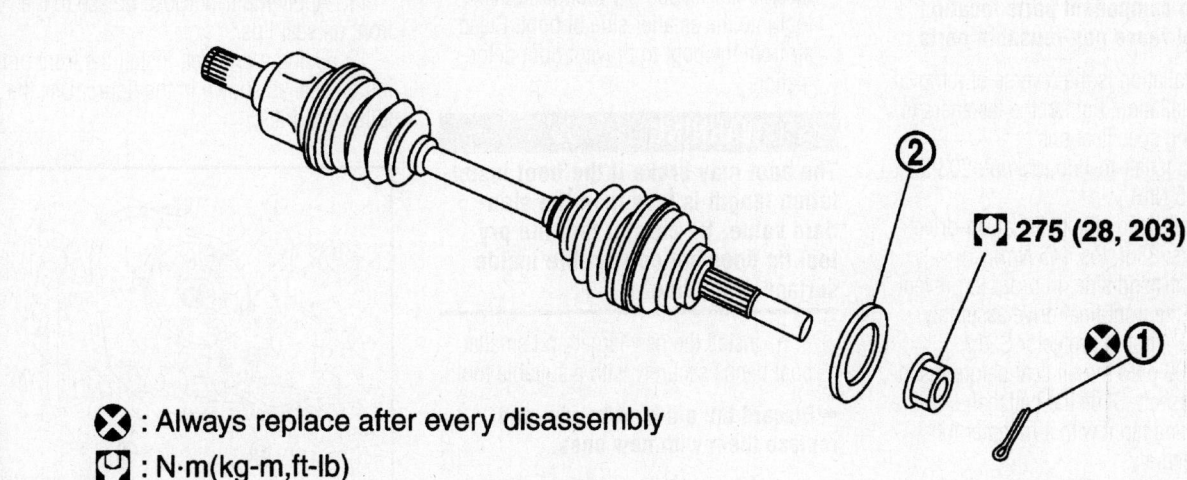

🔧 275 (28, 203)

❌ : Always replace after every disassembly

🔧 : N·m(kg-m,ft-lb)

1.   Cotter pin

2.   Washer

67162-FX35-G200

Fig. 14 View of right-hand front halfshaft

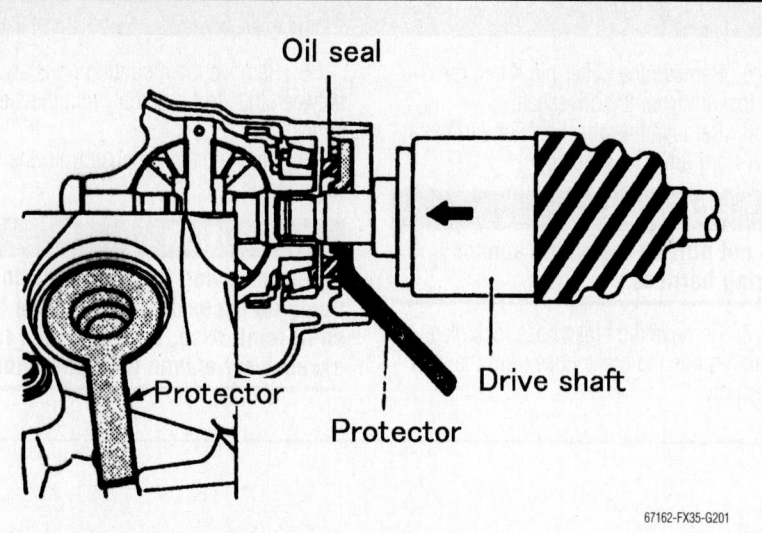

Fig. 15 Place the protector SST: KV38107900 onto front final drive assembly to prevent damage to the oil seal while inserting drive shaft

10. For the left-hand halfshaft, remove the fixing bolts of the front final drive side assembly halfshaft, then remove the halfshaft from the vehicle.

11. For the right-hand halfshaft, pry off the halfshaft from the front final drive assembly.

12. Inspect the halfshaft, as follows:

   a. Move the joint up/down, left /right, and in the axial direction. Check for any rough movement or significant looseness.

   b. Check the boot for cracks or other damage, and also for grease leakage.

   c. If damage is found, disassemble the halfshaft and replace the defective part(s) with new one(s).

To install:

➡**Refer to component parts location and do not reuse non-reusable parts.**

13. Installation is the reverse of removal. During installation, tighten the fasteners to the following specifications:

   a. Halfshaft-to-hub locknut: 203 ft. lbs. (275 Nm).

   b. Left-hand side halfshaft-to-drive unit bolts: 33 ft. lbs. (45 Nm).

14. Right-hand side, in order to prevent damage to the front final drive assembly side oil seal, first fit protector SST: KV38107900 onto the oil seal before inserting the halfshaft. Slide the halfshaft into the slide joint and tap it with a hammer to install it securely.

➡**Be sure to check that the circular clip is securely fastened after inserting the halfshaft.**

15. Check the condition of the wheel sensor wiring harness. Repair if necessary.

➡**Grease amount: 3.35–4.06 oz. (95–115g) for the left-hand side or 3.99–4.34 oz. (113–123g) for the right-hand side.**

   f. Install the boot securely into the grooves (indicated by * marks).

✳✳ CAUTION

If there is grease on the boot mounting surfaces (indicated by * marks) of the shaft and housing of the joint sub assembly, the boot may come off. Remove all grease from the surfaces.

   g. Make sure the boot installation length is according to specification. Insert a flat-bladed pry tool or similar tool into the smaller side of boot. Bleed air from the boot to prevent boot deformation.

✳✳ WARNING

The boot may brake if the boot installation length is less than the standard value. Be careful that the pry tool tip does not contact the inside surface of the boot.

   h. Install the new larger and smaller boot bands securely with a suitable tool.

➡**Discard the old boot bands, and replace them with new ones.**

   i. After installing the joint sub-assembly and shaft, rotate the boot to check whether or not the actual position is correct. If the boot position is not correct, secure the boot with new boot bands again.

FRONT PINION SEAL

REMOVAL & INSTALLATION

See Figures 16 & 17.

1. Before servicing the vehicle, refer to the Precautions Section.

2. Raise and safely support the vehicle.

3. Drain the gear oil.

4. Remove the front propeller shaft.

5. Remove the front halfshafts.

6. Remove the side shaft assembly.

7. Remove the drive pinion lock nut using a flange wrench.

8. Put a matching mark on the end of the drive pinion. The matching mark should be in line with the matching mark on companion flange.

✳✳ WARNING

For the matching marks, use paint. Never damage the companion flange and drive pinion.

➡**The matching mark on the final drive companion flange indicates the maximum vertical runout position.**

9. Remove the companion flange using a puller.

10. Remove the front pinion oil seal using puller KV381054S0 (J-34286), or equivalent.

✳✳ WARNING

Be careful not to damage the gear carrier when removing the pinion seal.

To install:

11. Apply multi-purpose grease to the front oil seal lips.

12. Using a drift set, install the front pinion oil seal as shown in the figure. Use the following drifts:

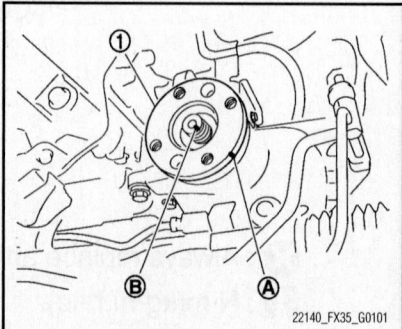

Fig. 16 Put a matching mark (B) on the end of the drive pinion. The matching mark (B) should be in line with the matching mark (A) on companion flange (1)

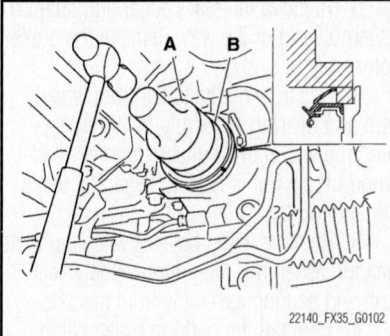

Fig. 17 Using a drift set, install the front pinion oil seal

- A: ST33400001 (J-26082)
- B: KV38102510

➡**Never reuse the oil seal. When installing the seal, never incline the oil seal or install it crooked.**

13. Align the matching mark (B) of drive pinion with the matching mark (A) of the companion flange, and then install the companion flange (1).

14. Apply anti-corrosion oil to the thread and seat of the new drive pinion lock nut, and temporarily tighten the drive pinion lock nut to the drive pinion.

※※ WARNING

Never reuse the drive pinion lock nut.

15. Tighten the drive pinion lock nut using ST3127S000 (J-25765-A), or equivalent, while adjusting the total preload torque.

   a. Drive pinion lock nut tightening torque: 94–181 ft. lbs. (127–245 Nm).

   b. Total preload torque: 14–23 inch lbs. (2–3 Nm).

- Adjust to the lower limit of the drive pinion lock nut tightening torque first
- After adjustment, rotate the drive pinion back and forth 2 to 3 times to check for unusual noise, rotation malfunction, and other malfunctions

- If the measured value is out of the specification, remove the final drive assembly and disassemble the drive pinion parts to check and adjust each part
16. Install the front propeller shaft.
17. Install the side shaft assembly.
18. Install the front halfshafts.
19. Refill the gear oil in the final drive and check the oil level.
20. Check the final drive for oil leakage.

REAR AXLE HOUSING

REMOVAL & INSTALLATION

See Figures 18 and 19.

1. Before servicing the vehicle, refer to the Precautions Section.
2. Raise and safely support the vehicle.
3. Remove the center muffler.
4. Remove the rear stabilizer bar.
5. Remove the driveshaft from the final drive.

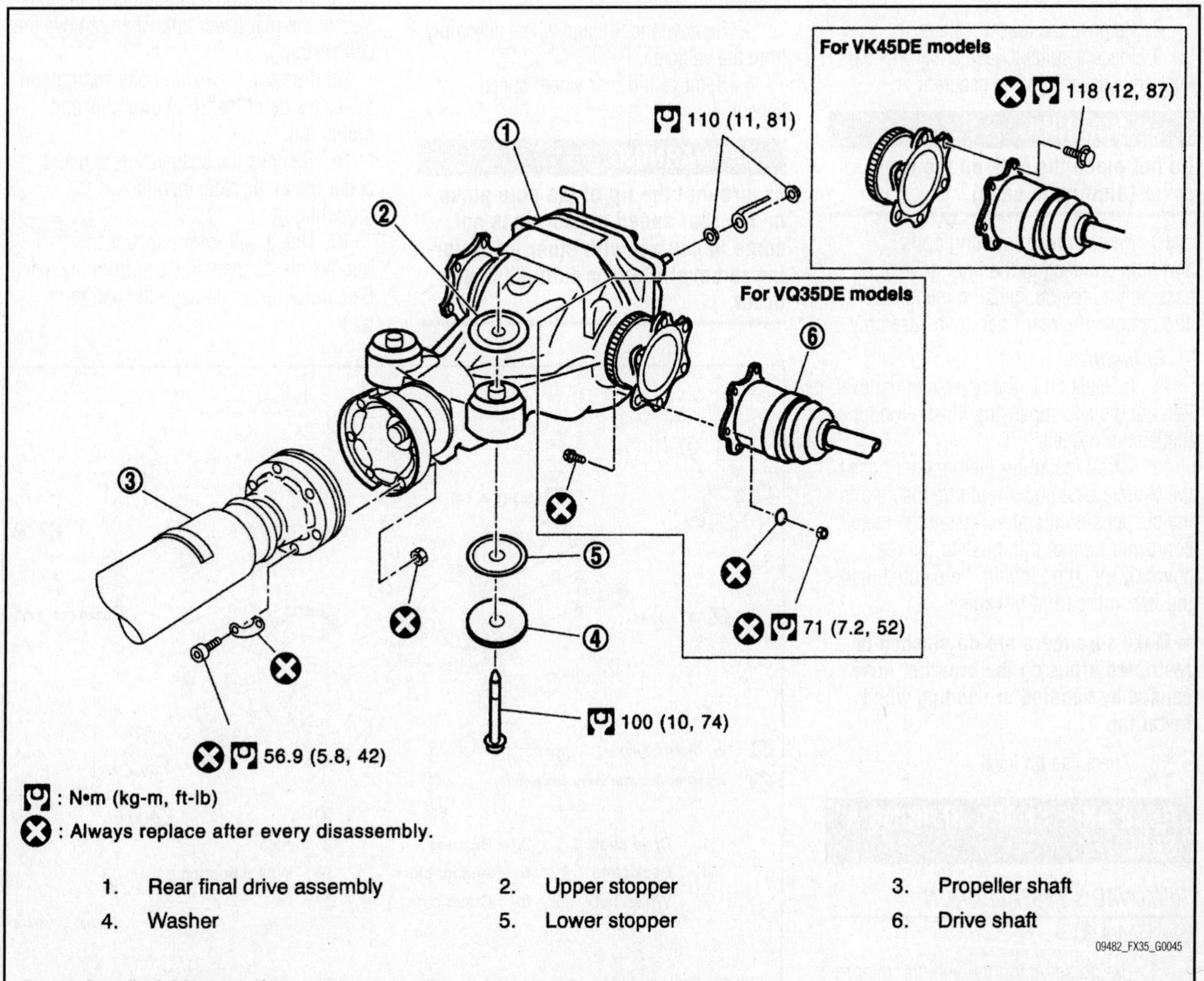

For VK45DE models

✕ 🔧 118 (12, 87)

For VQ35DE models

🔧 110 (11, 81)

🔧 71 (7.2, 52)

🔧 100 (10, 74)

✕ 🔧 56.9 (5.8, 42)

🔧 : N•m (kg-m, ft-lb)

✕ : Always replace after every disassembly.

1. Rear final drive assembly	2. Upper stopper	3. Propeller shaft
4. Washer	5. Lower stopper	6. Drive shaft

Fig. 18 Rear final drive mounting

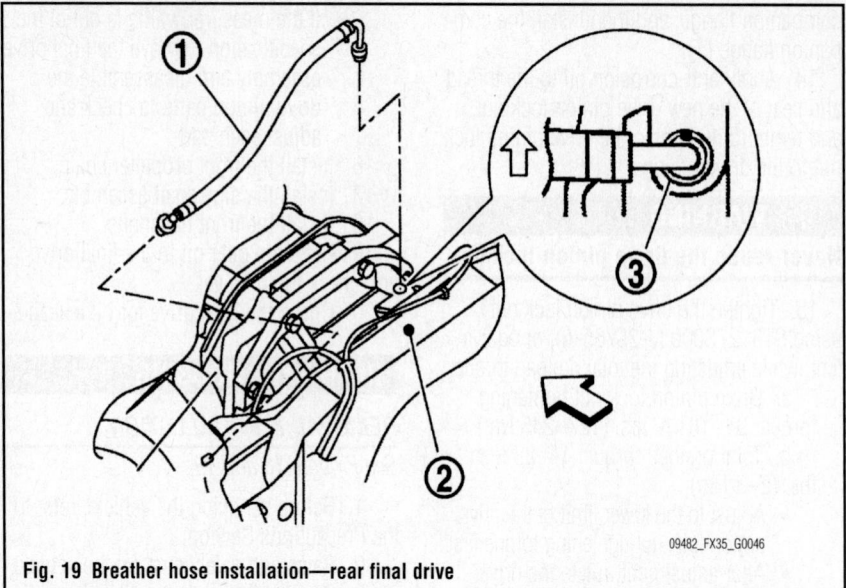

Fig. 19 Breather hose installation—rear final drive

09482_FX35_G0046

6. Remove the halfshaft from final drive. Suspend it out of the way with wire.

7. Remove the breather hose from the final drive.

8. Remove the rear wheel sensor.

9. Place a suitable jack under the rear final drive assembly and secure it.

❋❋ WARNING

Do not place the jack on the rear cover (aluminum case).

10. Remove the mounting bolts and nuts connecting the final drive assembly to the suspension member, and remove the rear final drive assembly.

To install:

11. Installation is the reverse of removal. Refer to the accompanying illustration for tightening torques.

12. When installing the breather hose, the vehicle side end should be inserted in the suspension member. Install the metal connector side of this hose to the rear cover by inserting it with the painted marking facing the front of vehicle.

➡**Make sure there are no pinched or restricted areas on the breather hose caused by bending or winding when installing it.**

13. Check the oil level.

REAR AXLE SHAFT, BEARING & SEAL

REMOVAL & INSTALLATION

See Figure 20.

1. Before servicing the vehicle, refer to the Precautions Section.

2. Disconnect the negative battery cable.

3. Raise and support the vehicle safely.

4. Remove the tire and wheel assembly from the vehicle.

5. Remove the rear wheel speed sensor.

❋❋ WARNING

Ensure that the tip of the pole piece on the rear speed sensor does not come in contact with other parts during removal. Sensor damage could occur.

6. Remove the rear caliper and support assembly out of the way. Remove the brake rotor.

7. Separate the halfshaft from wheel hub and bearing assembly by lightly tapping the end with a suitable hammer and wood block. If it is hard to separate, use a suitable puller.

8. Remove fixing bolts of wheel hub and bearing assembly, then remove the wheel hub and bearing assembly from the axle.

9. Remove the parking brake cable and parking brake shoe from the back plate.

10. Remove the fixing nuts of the anchor block, then remove the anchor block and back plate from the axle.

11. Loosen the fixing bolts and nuts of the front lower link, radius rod, and rear lower link in the side of the suspension member.

12. Set a jack under the rear lower link. Then remove the fixing bolt in the front lower link side of the shock absorber.

13. Remove the bolt and nut in the axle side of the rear lower link. Then remove the coil spring.

14. Remove the fixing bolts and nuts in the axle side of the front lower link and radius rod.

15. Remove the suspension arm and cotter pin at the axle, then loosen the mounting nut.

16. Use a ball joint remover to remove the suspension arm from the axle. Be careful not to damage the ball joint boot.

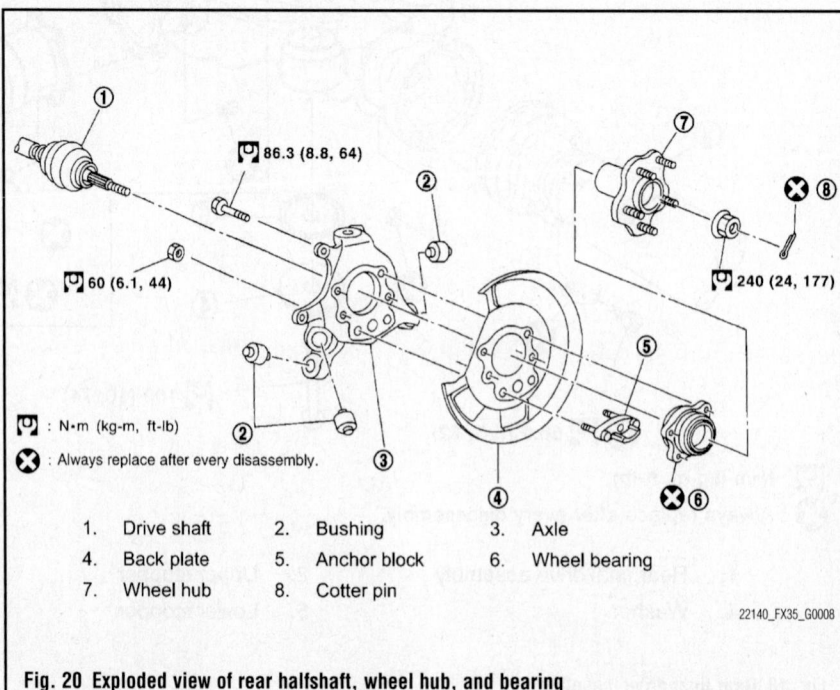

| : N·m (kg-m, ft-lb)
❌ : Always replace after every disassembly.

1. Drive shaft
2. Bushing
3. Axle
4. Back plate
5. Anchor block
6. Wheel bearing
7. Wheel hub
8. Cotter pin

22140_FX35_G0008

Fig. 20 Exploded view of rear halfshaft, wheel hub, and bearing

➡ **Temporarily tighten the mounting nut to prevent damage to the threads and to prevent the ball joint remover from coming off.**

17. Remove the axle/halfshaft from the vehicle.

➡ **When removing the halfshaft, do not apply an excessive angle to the half-shaft joint. Also be careful not to excessively extend the slide joint.**

18. Remove wheel bearing fixing bolts and anchor block fixing nuts, and remove the wheel hub and bearing assembly, back plate, and anchor block from the axle/halfshaft.

19. Using a drift and a puller, press the wheel hub out to remove from wheel bearing.

20. Using a drift and a puller, press the wheel bearing outer side inner race out to remove from wheel hub.

21. Using a suitable drift, remove each bushing from the axle.

To install:

22. Press fit a wheel hub into wheel bearing with a drift.

➡ **Press fit a drift while holding it against the wheel bearing inner side inner race. The wheel bearing cannot be reused.**

23. Install the back plate and wheel hub and bearing assembly.

24. Install the anchor block onto axle.

25. Install the rear axle shaft. Install the companion flange to the rear axle shaft, then install a new self-locking nut.

26. While holding the rear axle shaft in position, tighten a new self-locking nut to 177 ft. lbs. (240 Nm).

27. Install the rear brake disc, caliper assembly and parking brake.

28. Install the tire and wheel assembly and lower the vehicle. Check the parking brake stroke and adjust as required.

29. Before moving the vehicle, pump the brakes until a firm pedal is achieved.

REAR PINION SEAL

REMOVAL & INSTALLATION

See Figures 21 through 25.

1. Before servicing the vehicle, refer to the Precautions Section.

2. Raise and safely support the vehicle.

3. Drain the gear oil.

4. Make a judgment if a collapsible spacer replacement is required.

5. Remove the center muffler.

6. Remove the rear wheel sensor.

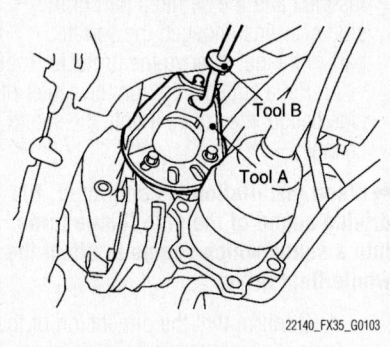

Fig. 21 Attachment installed for use with a sliding hammer

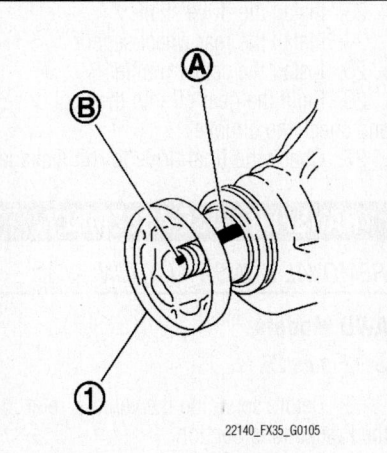

Fig. 22 Put a matching mark (B) on the end of the drive pinion. The matching mark (B) should be in line with the matching mark (A) on the companion flange (1)

7. Remove the drive shaft from the final drive. Suspend it out of the way with wire.

8. Install an attachment to the side flange and then pull out the side flange with a sliding hammer.

   a. For VQ35DE (3.5L) models, use tools:
- A: KV40104100
- B: ST36230000 (J-25840-A)

➡ **Take note of the circular clip installation position on the final drive side.**

   b. For VK45DE (4.5L) models, use tools:
- Tool number A: KV40101000
- B: ST36230000 (J-25840-A)

➡ **Take note of the circular clip installation position on the final drive side.**

9. Remove the propeller shaft.

10. Measure the total preload with the preload gauge using ST3127S000 (J-25765-A), or equivalent.

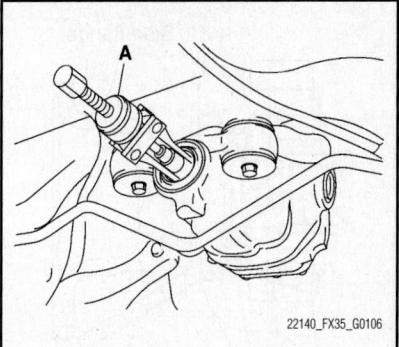

Fig. 23 Remove the pinion seal using puller KV381054S0 (J-34286)

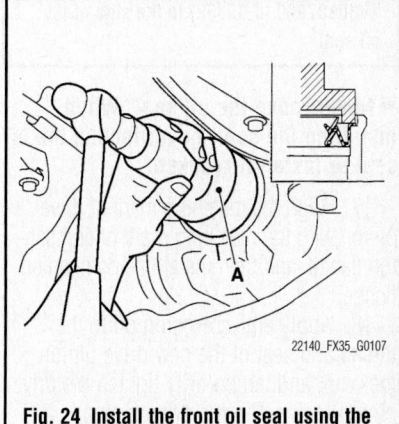

Fig. 24 Install the front oil seal using the drift A: ST30720000 (J-25405)

➡ **Record the preload measurement.**

11. Put a matching mark on the end of the drive pinion. The matching mark should be in line with the matching mark on the companion flange.

✵✵ WARNING

For matching mark, use paint. Never damage the companion flange and drive pinion.

➡ **The matching mark on the final drive companion flange indicates the maximum vertical runout position.**

12. Remove the drive pinion lock nut using the flange wrench.

13. Remove the companion flange using a puller.

14. Remove the front pinion oil seal using puller A: KV381054S0 (J-34286), or equivalent.

To install:

15. Apply multi-purpose grease to front oil seal lips.

16. Install the front oil seal using the drift A: ST30720000 (J-25405) as shown in the figure.

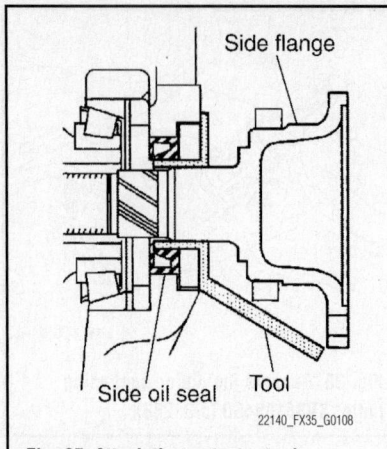

Fig. 25 Attach the protector tool KV38107900 (J-39352) to the side of the oil seal

➡**Never reuse the oil seal. When installing the seal, never incline the oil seal or install it crooked.**

17. Align the matching mark of drive pinion with the matching mark of companion flange, and then install the companion flange.

18. Apply anti-corrosion oil to the thread and seat of the new drive pinion lock nut, and temporarily tighten the drive pinion lock nut to the drive pinion.

✳✳ WARNING

Never reuse drive pinion lock nut.

19. Tighten the drive pinion lock nut using ST3127S000 (J-25765-A), or equivalent, while adjusting the total preload torque.

a. Drive pinion lock nut tightening torque: 109–238 ft. lbs. (147–323 Nm).

b. Total preload torque: should equal the measurement taken during removal plus an additional 1–3 inch lbs. (0.1–0.4 Nm).

- Adjust to the lower limit of the drive pinion lock nut tightening torque first.
- If the preload torque exceeds the specified value, replace the collapsible spacer and tighten it again to adjust. Never loosen the drive pinion lock nut to adjust the preload torque.

20. Make a stamping for identification of the front oil seal replacement frequency.

➡**Be sure to make a stamping after replacing the front oil seal.**

21. Install the propeller shaft.

22. Install the side flange with the following procedure.

a. Attach the protector tool KV38107900 (J-39352) to the side of the oil seal.

b. After the side flange is

inserted and the serrated part of the side gear has engaged the serrated part of flange, remove the protector tool.

c. Put a suitable drift on the center of side flange, then drive it until the sound changes.

➡**When installation is completed, the driving sound of the side flange turns into a sound which seems to affect the whole final drive.**

d. Confirm that the dimension of the side flange installation (measurement A in the figure) comes into the following specification: 12.83–12.91 in. (326–328mm).

23. Install the drive shaft.

24. Install the rear wheel sensor.

25. Install the center muffler.

26. Refill the gear oil into the final drive and check the oil level.

27. Check the final drive for oil leakage.

TRANSFER CASE ASSEMBLY

REMOVAL & INSTALLATION

AWD Models

See Figure 26.

1. Before servicing the vehicle, refer to the Precautions Section.

2. Raise and safely support the vehicle.

3. Remove fixing bolts and nuts of the tunnel stay and the member stay and remove from the vehicle.

4. Remove the exhaust front tube.

5. Remove the front and rear driveshafts.

6. Disconnect the transfer assembly wiring harness connector and separate the wiring harness from the transfer assembly.

7. Remove the air breather hose.

8. Support the transfer assembly with a jack.

9. Remove the engine rear mounting.

10. Remove the transfer mounting bolts and separate the transfer from the transmission.

✳✳ CAUTION

Secure the transfer assembly to a jack or similar support.

To install:

11. Install the transfer mounting bolts and mount the transfer onto the transmission. Tighten the bolts in the proper order and torque specification as illustrated.

12. Install the engine rear mounting.

13. Install the air breather hose.

14. Connect the transfer assembly wiring harness connector and attach the wiring harness to the transfer assembly.

15. Install the front and rear driveshafts.

16. Install the exhaust front tube.

17. Install the fixing bolts and nuts of the tunnel stay and the member stay.

18. Check the fluid level and for fluid leakage.

Bolt No.	1	2	3	4
Quantity	4	3	2	1
Bolt length "ℓ" mm (in)	75 (2.95)	45 (1.77)	40 (1.57)	30 (1.18)
Tightening torque N·m (kg-m, ft-lb)	37 (3.8, 27)			

⊙ : Transfer to Transmission
⊗ : Transmission to transfer

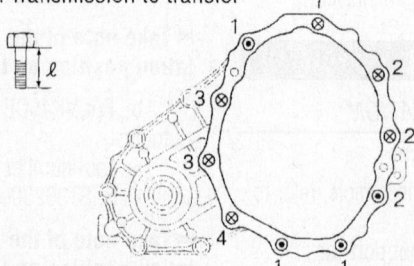

Fig. 26 Transfer case-to-transmission mounting bolt length and tightening specifications illustrated

ENGINE COOLING

ENGINE FAN

REMOVAL & INSTALLATION

See Figure 27.

1. Before servicing the vehicle, refer to the Precautions Section.
2. Disconnect the negative battery cable.
3. Drain and recycle the engine coolant.

✳✳ CAUTION

Never open, service or drain the radiator or cooling system when hot; serious burns can occur from the steam and hot coolant. Also, when draining engine coolant, keep in mind that cats and dogs are attracted to ethylene glycol antifreeze and could drink any that is left in an uncovered container or in puddles on the ground. This will prove fatal in sufficient quantities. Always drain coolant into a sealable container. Coolant should be reused unless

it is contaminated or is several years old.

4. Remove the air duct (inlet), power duct, and air cleaner case assembly.
5. Disconnect the harness connector from the fan motors. Move the harness aside.
6. Remove the radiator cooling fan assembly retaining bolts.
7. Remove the assembly from the vehicle.

To install:

8. Installation is the reverse of the removal procedure. Refer to illustration for torque specifications.
9. Be sure to fill the radiator with the proper grade and type of engine coolant.
10. Start the engine and check for leaks. Correct as required.

RADIATOR

REMOVAL & INSTALLATION

See Figure 28.

1. Before servicing the vehicle, refer to the Precautions Section.
2. Disconnect the negative battery cable.
3. Remove the engine undercover.
4. Drain and recycle the engine coolant.

➡**Be sure the engine is cold before draining the radiator. Do not allow coolant to spill on the drive belts.**

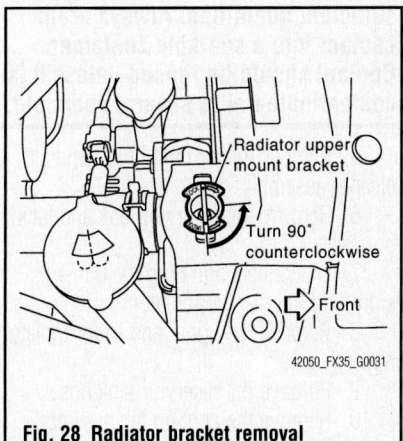

Fig. 28 Radiator bracket removal

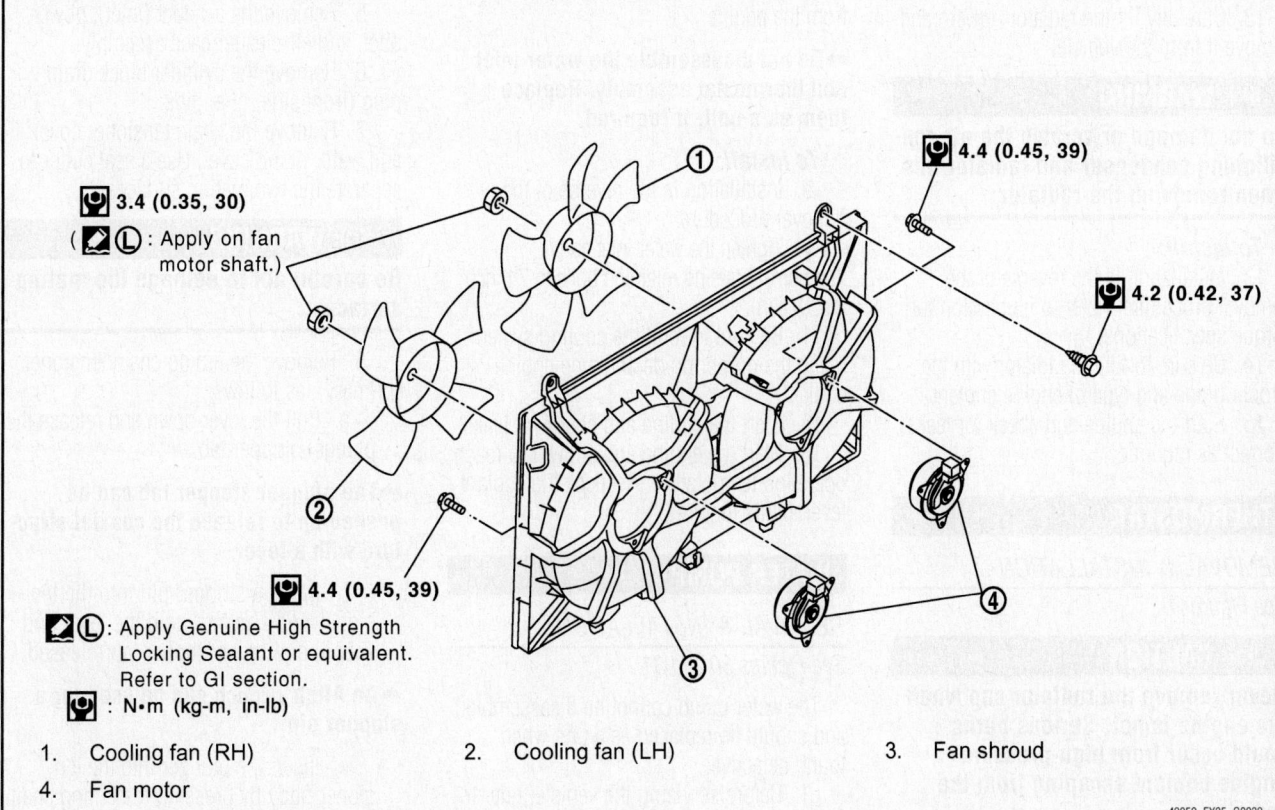

3.4 (0.35, 30)
(L : Apply on fan motor shaft.)

4.4 (0.45, 39)

4.2 (0.42, 37)

4.4 (0.45, 39)

L: Apply Genuine High Strength Locking Sealant or equivalent. Refer to GI section.

: N•m (kg-m, in-lb)

1. Cooling fan (RH)
4. Fan motor
2. Cooling fan (LH)
3. Fan shroud

42050_FX35_G0032

Fig. 27 Cooling fans and related components

✳✳ CAUTION

Never open, service or drain the radiator or cooling system when hot; serious burns can occur from the steam and hot coolant. Also, when draining engine coolant, keep in mind that cats and dogs are attracted to ethylene glycol antifreeze and could drink any that is left in an uncovered container or in puddles on the ground. This will prove fatal in sufficient quantities. Always drain coolant into a sealable container. Coolant should be reused unless it is contaminated or is several years old.

5. Remove the air duct (inlet) and air cleaner assembly.
6. Remove the reservoir tank and tank bracket.
7. Disconnect and plug the transmission fluid lines at the radiator.
8. Remove the upper and lower radiator hoses.
9. Remove the reservoir tank hose.
10. Remove the cooling fan assembly.
11. Rotate the 2 upper radiator mount brackets 90° in the direction shown in the illustration and remove them.
12. Carefully lift the radiator upward and remove it from the vehicle.

✳✳ WARNING

Do not damage or scratch the air conditioning condenser and radiator fins when removing the radiator.

To install:
13. Installation is the reverse of the removal procedure. Refer to illustration for torque specifications.
14. Be sure to fill the radiator with the proper grade and type of engine coolant.
15. Start the engine and check for leaks. Correct as required.

THERMOSTAT

REMOVAL & INSTALLATION
See Figure 29.

✳✳ CAUTION

Never remove the radiator cap when the engine is hot. Serious burns could occur from high-pressure engine coolant escaping from the radiator.

1. Before servicing the vehicle, refer to the Precautions Section.
2. Be sure the engine is cold.

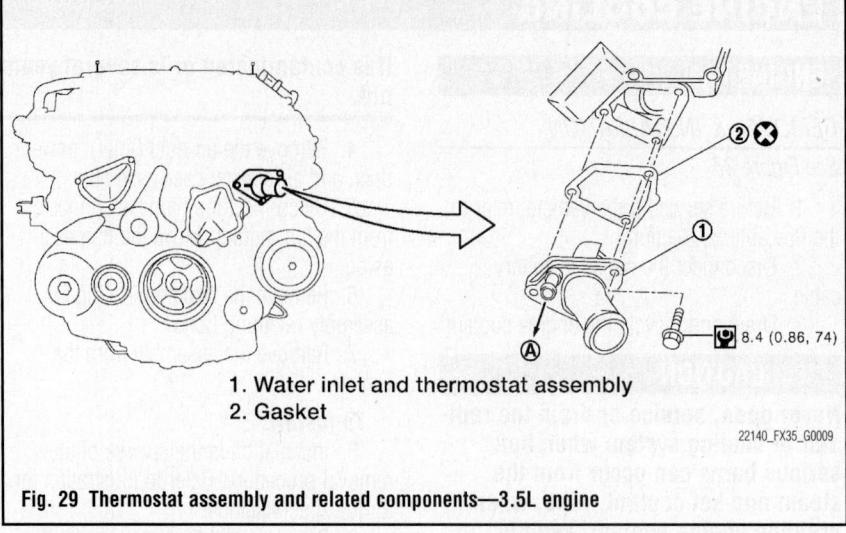

1. Water inlet and thermostat assembly
2. Gasket

⬛ 8.4 (0.86, 74)

22140_FX35_G0009

Fig. 29 Thermostat assembly and related components—3.5L engine

3. Disconnect the negative battery cable.
4. Remove the front engine cover.
5. Drain the engine coolant using the radiator drain plug and the water drain plug at the front of the cylinder block. Properly dispose of used coolant.
6. Remove the air duct (inlet).
7. Remove the water inlet and thermostat housing retaining bolts.
8. Remove the thermostat assembly from the engine.

➡ Do not disassemble the water inlet and thermostat assembly. Replace them as a unit, if required.

To install:
9. Installation is the reverse of the removal procedure.
10. Tighten the water inlet and thermostat housing retaining bolts to 74 inch lbs. (8 Nm).
11. Be sure to refill the cooling system using the proper grade and type engine coolant.
12. Start the engine and check for leaks.
13. Start the engine and allow it to reach operation temperature. Recheck the coolant level, fill as required.

WATER PUMP

REMOVAL & INSTALLATION
See Figures 30 and 31.

The water pump cannot be disassembled and should be replaced as a unit when found defective.

1. Before servicing the vehicle, refer to the Precautions Section.

➡ During service, be sure to prevent engine coolant from contacting the accessory drive belt.

2. Remove the front engine under-cover.
3. Remove the drive belts.
4. Drain the engine coolant from the radiator.

✳✳ CAUTION

Make sure the engine is cold before draining the coolant.

5. Remove the air duct (inlet), power duct, and air cleaner case assembly.
6. Remove the cylinder block drain plug (front side) of engine.
7. Remove the chain tensioner cover and water pump cover. Use a seal cutter to separate the two mating surfaces.

✳✳ WARNING

Be careful not to damage the mating surfaces.

8. Remove the timing chain tensioner (primary), as follows:
   a. Pull the lever down and release the plunger stopper tab.

➡ The plunger stopper tab can be pushed up to release the coaxial structure with a lever.

   b. Insert a stopper pin into the tensioner body hole to hold the lever and keep the plunger stopper tab released.

➡ An Allen wrench can be used for a stopper pin.

   c. Insert the plunger into the tensioner body by pressing the timing chain slack guide.
   d. Keep the slack guide depressed and hold the plunger in by pushing the stopper pin deeper through the lever and into the tensioner body hole.

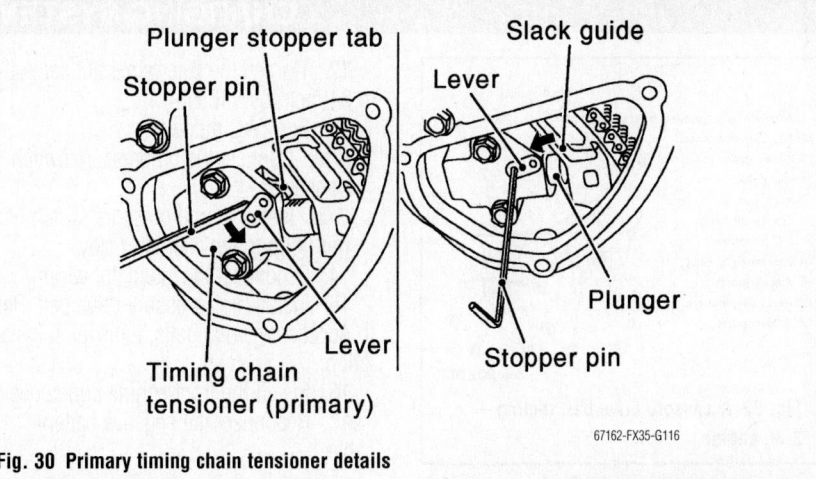

Fig. 30 Primary timing chain tensioner details

e. Turn the crankshaft pulley approximately 20° clockwise, so that the timing chain on the timing chain tensioner (primary) side is loose.

f. Remove the mounting bolts and timing chain tensioner (primary).

✵✵ WARNING

Be careful not to drop mounting bolts inside the chain case.

9. Remove the 3 water pump fixing bolts. Secure a gap between the water pump gear and timing chain, by turning the crankshaft pulley counterclockwise until timing chain slack on the water pump sprocket is at its maximum.

10. Insert M8 bolts into the water pump upper and lower mounting bolt holes until they contact the timing chain case. Then, alternately tighten each bolt ½ turn, and pull out the water pump.

➡**The M8 bolts should be 0.0492 x 1.97 in. (1.25mm x 50mm).**

✵✵ WARNING

Pull the water pump straight out while preventing the vane from contacting the socket in the installation area. Remove the water pump without allowing the sprocket to contact the timing chain.

11. Remove the M8 bolts and O-rings from the water pump.

12. Do not disassemble the water pump.

13. Check for badly rusted or corroded water pump body assembly.

14. Check for rough operation due to excessive end play.

15. If any defects are found, replace the water pump.

To install:

16. Install new O-rings on the water pump.

17. Apply engine oil and engine coolant to the water pump O-rings as illustrated.

18. Position the O-ring with the white paint mark toward the engine front side.

19. Install the water pump.

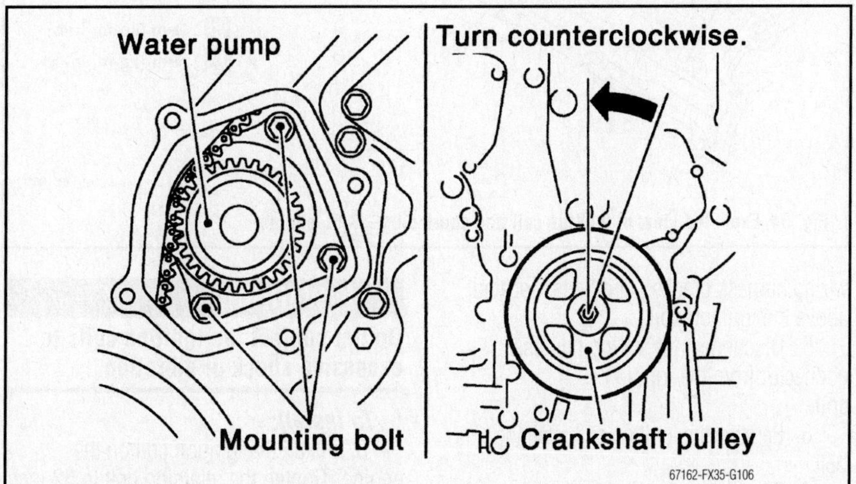

✵✵ CAUTION

Do not allow the cylinder block to damage the O-rings during installation.

20. Make sure the timing chain and water pump sprocket are engaged properly.

21. Install the water pump by alternately and evenly tightening the mounting bolts.

22. Install the timing chain tensioner (primary), as follows:

a. Remove all dust and foreign material completely from the backside of the chain tensioner and from the installation area of the rear timing chain case.

b. Turn the crankshaft pulley clockwise so that the timing chain on the timing chain tensioner (primary) side is loose.

c. Apply engine oil to the oil hole and tensioner when installing the timing chain tensioner.

d. Install the timing chain tensioner (primary).

e. Remove the stopper pin.

23. Install the chain tensioner cover and water pump cover.

➡**Before installing, remove all traces of liquid gasket from the mating surface of the water pump cover and chain tensioner cover using a scraper. Also, remove traces of liquid gasket from the mating surface of the front timing chain case.**

24. Apply a continuous bead of liquid gasket 0.091–0.130 in. (2–3mm) thick to the mating surface of the chain tensioner cover and water pump cover.

➡**Use RTV Silicone Sealant or equivalent.**

25. Install the cylinder block drain plug (front side).

26. Apply thread sealant to the thread of cylinder block drain plug.

➡**Use Genuine Thread Sealant or equivalent.**

27. Install the air duct (inlet), power duct and air cleaner case assembly.

28. Fill the engine with coolant.

29. Install the drive belts.

30. Install the front engine undercover.

31. Check for engine coolant leaks using a radiator cap tester.

32. Start the engine and let it idle for 3 minutes, then raise the engine RPM up to 3,000 RPM under no load to purge air from the high-pressure chamber of the chain tensioner. The engine may produce a rattling noise. This indicates that air remains in the chamber, but it is not a matter of concern.

33. With the engine idling, visually make sure that there are no leaks of engine coolant.

Fig. 31 Water pump mounting bolt locations

ENGINE ELECTRICAL   CHARGING SYSTEM

ALTERNATOR

REMOVAL & INSTALLATION

See Figure 32.

1. Before servicing the vehicle, refer to the Precautions Section.
2. Disconnect the negative battery cable.
3. Remove engine front undercover.
4. Remove the alternator and power steering oil pump belt.
5. Disconnect the alternator connector. Remove the **B** terminal nut.
6. Remove the harness clip and water hose bracket from the alternator.
7. For 2WD models:
    a. Remove the oil pressure switch harness clip from the alternator stay.
    b. Disconnect the oil pressure switch connector.
8. Remove the alternator stay mounting bolts and alternator stay.
9. Remove the alternator mounting bolt.
10. Remove the alternator assembly, by

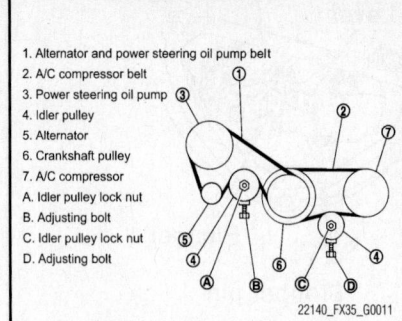

1. Alternator and power steering oil pump belt
2. A/C compressor belt
3. Power steering oil pump
4. Idler pulley
5. Alternator
6. Crankshaft pulley
7. A/C compressor
A. Idler pulley lock nut
B. Adjusting bolt
C. Idler pulley lock nut
D. Adjusting bolt

22140_FX35_G0011

Fig. 32 Accessory drive belt routing—3.5L engine

lowering it out of the bottom of the engine compartment.

To install:

11. Reposition the alternator in place on the engine and tighten the mounting bolts. Tighten the long alternator bolt 48 ft. lbs. (65 Nm) and the short alternator bracket bolts to 21 ft. lbs. (28 Nm).

12. Tighten the **B** terminal nut carefully to 84 inch lbs. (10 Nm).
13. For 2WD models:
    a. Connect the oil pressure switch connector.
    b. Install the oil pressure switch harness clip to the alternator stay.
14. Reconnect the alternator wiring.
15. Install the accessory drive belt. Refer to Accessory Drive Belts, removal & installation.
16. Install the front engine undercover.
17. Reconnect the negative battery cable.

VOLTAGE REGULATOR

REMOVAL & INSTALLATION

The voltage regulator is an internal component of the alternator. In order to replace the voltage regulator, the entire alternator assembly must be replaced. Refer to Alternator, installation & removal.

ENGINE ELECTRICAL   IGNITION SYSTEM

FIRING ORDER

See Figure 33.

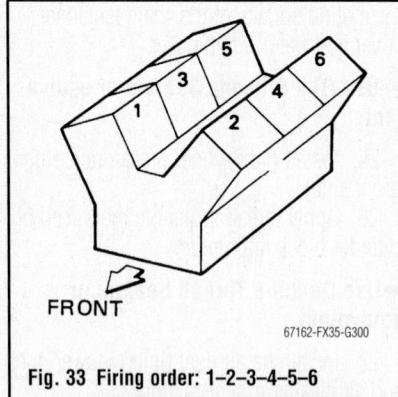

FRONT

67162-FX35-G300

Fig. 33 Firing order: 1–2–3–4–5–6 Distributorless ignition system

IGNITION COIL

REMOVAL & INSTALLATION

See Figures 34 and 35.

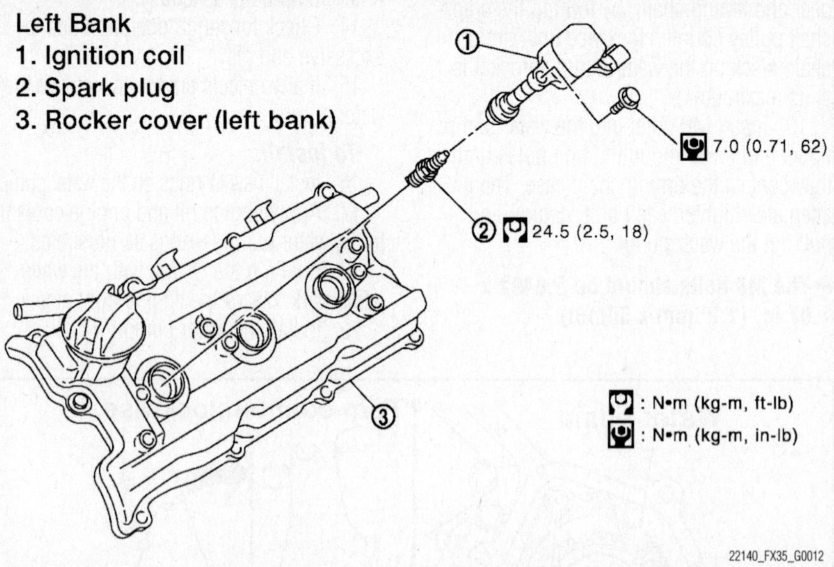

1. Before servicing the vehicle, refer to the Precautions Section.
2. Remove the engine cover.
3. Remove the air duct (for ignition coil of left bank side).
4. Move aside the wiring harness,

Left Bank
1. Ignition coil
2. Spark plug
3. Rocker cover (left bank)

① 7.0 (0.71, 62)

② 24.5 (2.5, 18)

: N•m (kg-m, ft-lb)

: N•m (kg-m, in-lb)

22140_FX35_G0012

Fig. 34 Exploded view of ignition coil and spark plug—3.5L engine

wiring harness bracket, and hoses located above the ignition coil.
5. Disconnect the wiring harness connector from the ignition coil.
6. Remove the ignition coil retaining bolt.
7. Remove the ignition coil.

❊❊ WARNING

Do not subject the ignition coils to excessive shock or vibration.

To install:

8. Install the ignition coil on the engine. Tighten the retaining bolt to 62 inch lbs. (7 Nm).

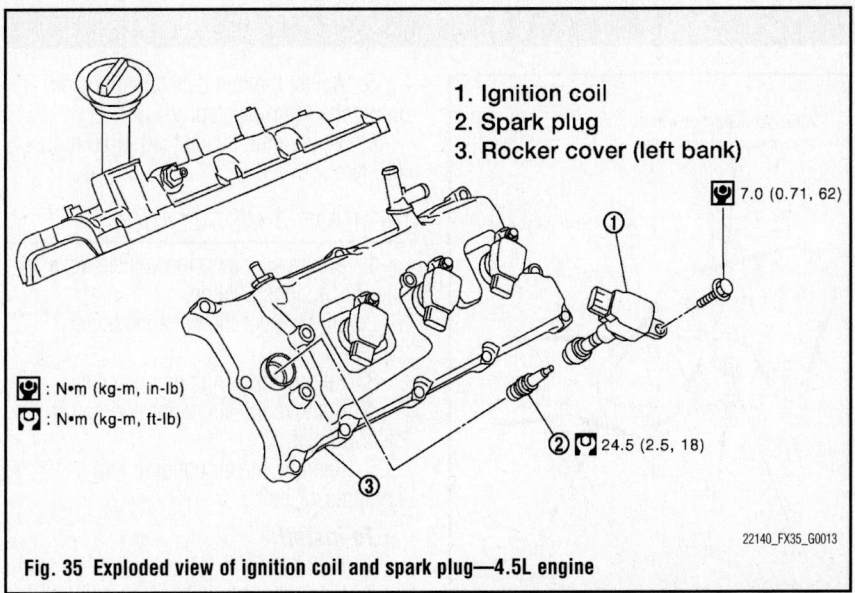

1. Ignition coil
2. Spark plug
3. Rocker cover (left bank)

7.0 (0.71, 62)

: N•m (kg-m, in-lb)

: N•m (kg-m, ft-lb)

24.5 (2.5, 18)

22140_FX35_G0013

Fig. 35 Exploded view of ignition coil and spark plug—4.5L engine

9. Reconnect the wiring harness to the coil.

10. Reposition the wiring harness, bracket and hoses.

11. Install the air duct and the engine cover.

IGNITION TIMING

ADJUSTMENT

The ignition timing is controlled by the Electronic Control Module (ECM). No adjustment is necessary or possible.

SPARK PLUGS

REMOVAL & INSTALLATION

See Figures 34 and 35.

➡ Disconnecting the negative battery cable on some vehicles may interfere with the functions of the on board computer system. The computer may undergo a relearning process once the negative battery cable is reconnected.

1. Disconnect the negative battery cable.
2. Remove the engine cover.
3. Remove the ignition coil retaining bolt.
4. Remove the ignition coil.
5. Remove the spark plug using a spark plug socket and wrench.

To install:

6. Be sure the spark plug gap is to specification: 0.043 in.
7. Carefully install the spark plug and torque to specification: 18 ft. lbs. (25 Nm).
8. Install the ignition coil, torque the retaining bolt to 62 inch lbs. (7 Nm).
9. Install the engine cover.
10. Connect the negative battery cable.

ENGINE ELECTRICAL

STARTER

REMOVAL & INSTALLATION

1. Before servicing the vehicle, refer to the Precautions Section.
2. Disconnect the negative battery cable.
3. Remove the engine rear undercover.
4. Remove the starter electrical wires.
5. Remove the starter retaining bolts.
6. On 2WD vehicles, remove the harness clip bracket.

7. Remove the starter from its mounting.

To install:

8. Installation is the reverse of the removal procedure.
9. Tighten the retaining bolts to 41 ft. lbs. (55 Nm).
10. Tighten the terminal nut to 87 inch lbs. (10 Nm).

SOLENOID OR RELAY REPLACEMENT

1. Before servicing the vehicle, refer to the Precautions Section.

STARTING SYSTEM

2. Disconnect the negative battery cable.
3. Remove the starter.
4. Remove the solenoid attaching screws.
5. Remove the solenoid.

To install:

6. Install the solenoid.
7. Install the solenoid attaching screws.
8. Install the starter.
9. Connect the negative battery cable.

ENGINE MECHANICAL

ACCESSORY DRIVE BELTS

ACCESSORY BELT ROUTING

See Figure 36.

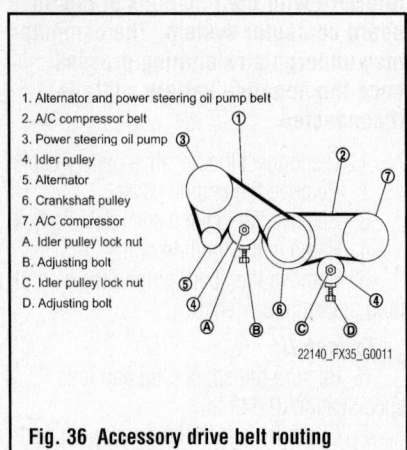

1. Alternator and power steering oil pump belt
2. A/C compressor belt
3. Power steering oil pump
4. Idler pulley
5. Alternator
6. Crankshaft pulley
7. A/C compressor
A. Idler pulley lock nut
B. Adjusting bolt
C. Idler pulley lock nut
D. Adjusting bolt

22140_FX35_G0011

Fig. 36 Accessory drive belt routing

Refer to the accompanying illustration for belt routing.

INSPECTION

Inspect the drive belt for signs of glazing or cracking. A glazed belt will be perfectly smooth from slippage, while a good belt will have a slight texture of fabric visible. Cracks will usually start at the inner edge of the belt and run outward. All worn or damaged drive belts should be replaced immediately.

ADJUSTMENT

Air Conditioning Compressor Belt

See Figures 37 and 38.

1. Before servicing the vehicle, refer to the Precautions Section.

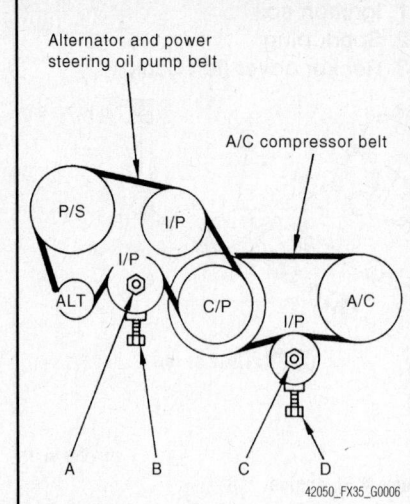

42050_FX35_G0006

Fig. 37 Drive belt tension check—3.5L engine

2. Disconnect the negative battery cable.
3. Remove the engine undercover.
4. Loosen the idler pulley locknut and adjust the tension by turning the adjusting bolt.
5. Adjust the belt deflection/tension using the following table.
6. Tighten the locknut to 26 ft. lbs. (35 Nm).

Alternator And Power Steering Belt

See Figures 37 and 38.

1. Before servicing the vehicle, refer to the Precautions Section.
2. Disconnect the negative battery cable.
3. Remove the engine undercover.
4. Loosen the idler pulley locknut (A) and adjust the tension by turning the adjusting bolt (B).

5. Adjust the belt deflection/tension using the following table.
6. Tighten the locknut (A) to 26 ft. lbs. (35 Nm).

REMOVAL & INSTALLATION

1. Before servicing the vehicle, refer to the Precautions Section.
2. Disconnect the negative battery cable.
3. Remove the engine undercover.
4. Remove the alternator and power steering belt.
5. Remove the air conditioning compressor belt.

To install:

6. Installation is the reverse of removal.
7. Be sure not to get grease or oil on the belts.
8. Make sure the drive belts are correctly engaged with the pulley groove.
9. Torque and adjust the drive belts to specification. Refer to Accessory Drive Belts, Adjustment.

CAMSHAFT AND VALVE LIFTERS

REMOVAL & INSTALLATION

See Figures 39 through 44.

1. Before servicing the vehicle, refer to the Precautions Section.
2. Remove the front timing chain case, camshaft sprocket, timing chain, and rear timing chain case.
3. Remove the Camshaft Position sensor (PHASE) (right and left banks) from the cylinder head back side.

Belt Deflection and Tension

Items	Deflection adjustment		Unit: mm (in)	Tension adjustment		Unit: N (kg, lb)
	Used belt		New belt	Used belt		New belt
	Limit	After adjustment		Limit	After adjustment	
Alternator and power steering oil pump belt	12 (0.47)	7 - 8 (0.28 - 0.31)	6 - 7 (0.24 - 0.28)	294 (30, 66)	730 - 818 (74.5 - 83.4, 164 - 184)	838 - 926 (85.5 - 94.5, 188 - 208)
A/C compressor belt	12 (0.47)	9 - 10 (0.35 - 0.39)	8 - 9 (0.31 - 0.35)	196 (20, 44)	348 - 436 (35.5 - 44.5, 78 - 98)	470 - 559 (47.9 - 57.0, 106 - 126)
Applied pushing force	98 N (10 kg, 22 lb)			—		

22140_FX35_G0014

Fig. 38 Belt deflection and tension table—3.5L engine

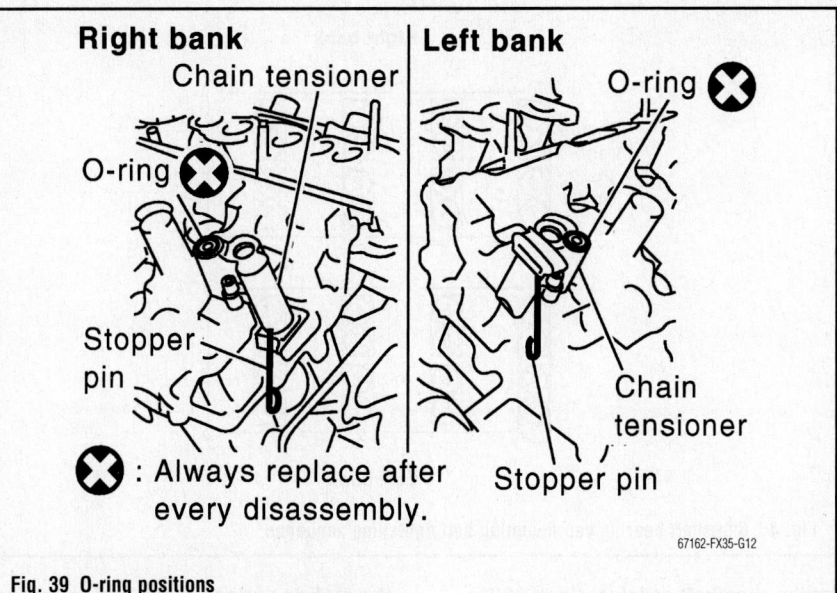

Right bank
Chain tensioner
O-ring ✗
Stopper pin

Left bank
O-ring ✗
Chain tensioner
Stopper pin

✗ : Always replace after every disassembly.

67162-FX35-G12

Fig. 39 O-ring positions

installation positions and store them without mixing them up.

9. Remove the secondary timing chain tensioner from the cylinder head.

➡**Remove the secondary timing chain tensioner with its stopper pin attached. The stopper pin was attached when the secondary timing chain was removed.**

To install:

10. Install the secondary timing chain tensioners on both sides of the cylinder head. Install the timing chain tensioner with its stopper pin attached and with the timing chain tensioner sliding part facing downward on the right-side cylinder head while the sliding part is facing upward on the left-side cylinder head.

11. Install new O-rings as shown in the figure.

✳✳ WARNING

Handle the camshaft position sensor carefully to avoid dropping and shocks. Do not disassemble and do not allow metal powder to adhere to magnetic part at the sensor tip. Do not place sensors in a location where they are exposed to magnetism.

4. Remove the intake valve timing control solenoid valves.

5. Discard the intake valve timing control solenoid valve gaskets.

6. Remove the camshaft brackets. Equally loosen the camshaft bracket bolts in several steps in reverse order of camshaft bearing cap mounting bolt tightening sequence shown in the figure.

➡**Mark the camshafts, camshaft brackets, and bolts so they are placed in the same position and direction at installation.**

7. Remove the camshaft(s).
8. Remove the valve lifters. Identify

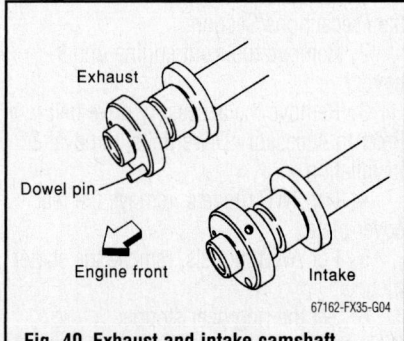

Exhaust
Dowel pin
Engine front
Intake

67162-FX35-G04

Fig. 40 Exhaust and intake camshaft differences

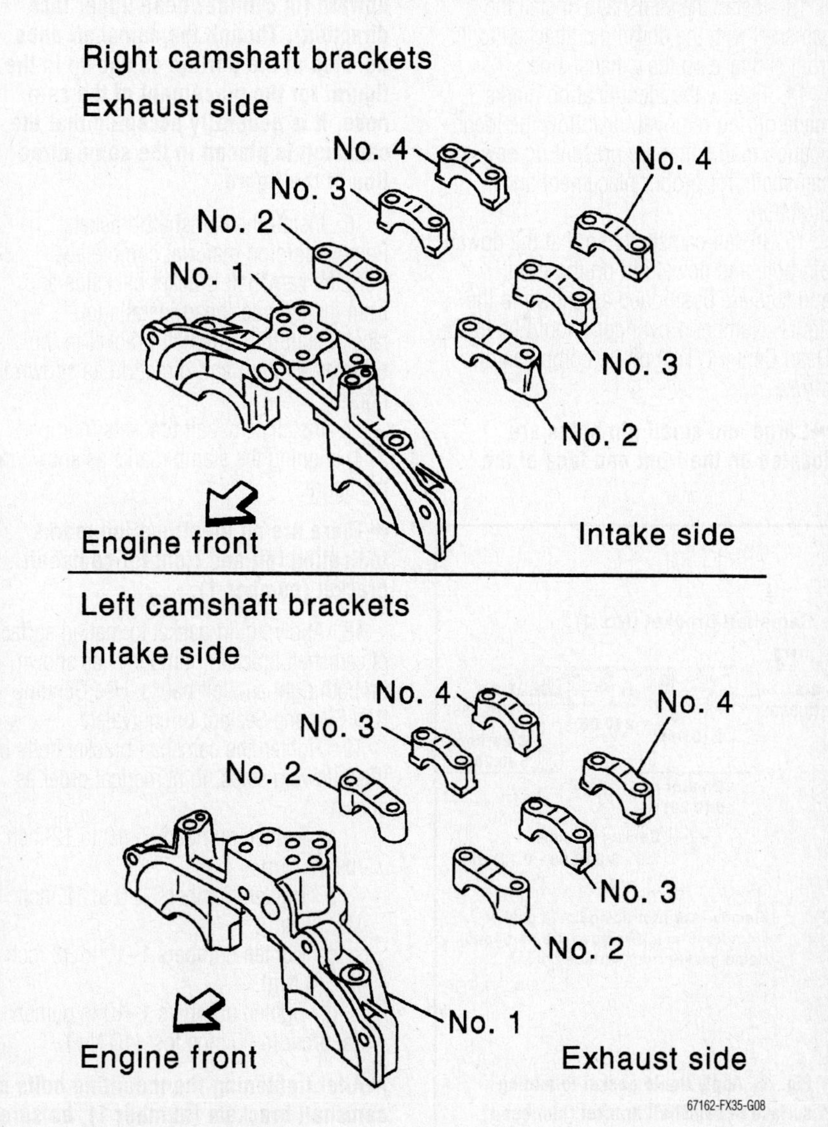

Right camshaft brackets
Exhaust side
No. 4
No. 3
No. 2
No. 1
No. 4
No. 3
No. 2
Engine front
Intake side

Left camshaft brackets
Intake side
No. 4
No. 3
No. 2
No. 4
No. 3
No. 2
No. 1
Engine front
Exhaust side

67162-FX35-G08

Fig. 41 Camshaft bracket positions

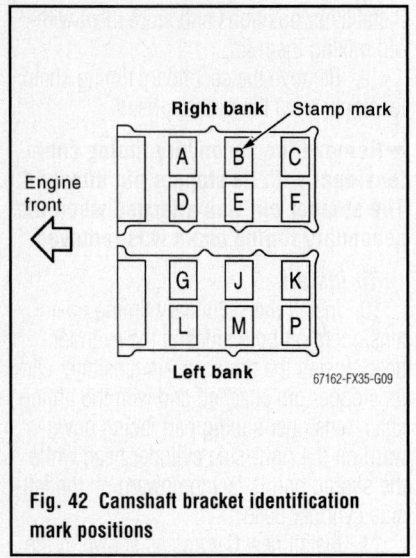

Fig. 42 Camshaft bracket identification mark positions

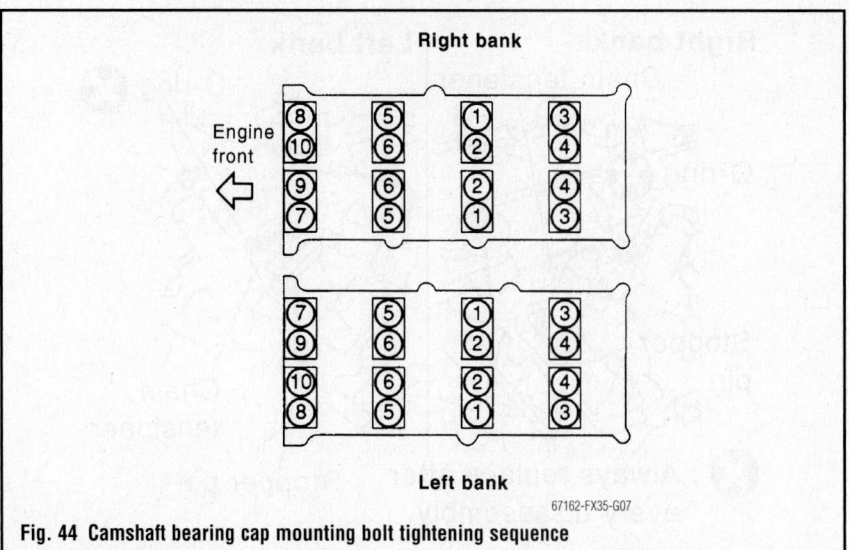

Fig. 44 Camshaft bearing cap mounting bolt tightening sequence

12. Install valve lifters in the original positions.

13. Install the camshafts. Install the camshaft with the dowel pin attached to its front end face on the exhaust side.

14. Follow the identification marks made during removal, or follow the identification marks that are present on new camshafts for proper placement and direction.

15. Install camshafts so that the dowel pin hole and dowel pin on the front end face are positioned as shown in the figure. Number 1 cylinder should be in Top Dead Center (TDC) on its compression stroke.

➡**Large and small pin holes are located on the front end face of the**

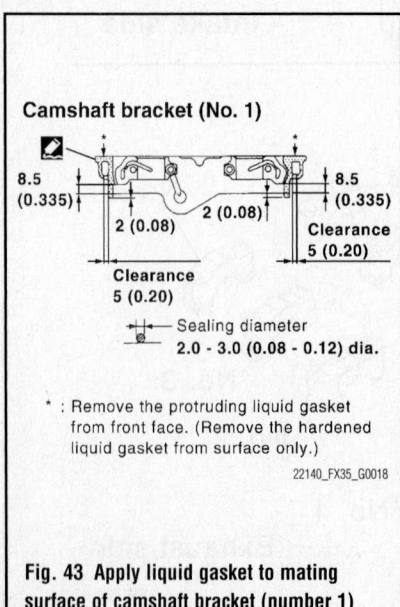

Fig. 43 Apply liquid gasket to mating surface of camshaft bracket (number 1) as shown

intake camshaft at intervals of 180°. Face the small diameter side pin hole upward (in cylinder head upper face direction). Though the camshaft does not stop at the portion as shown in the figure, for the placement of the cam nose, it is generally accepted that the camshaft is placed in the same direction of the figure.

16. Install the camshaft brackets. Remove foreign material completely from the camshaft bracket backside and from the cylinder head installation face. Install the camshaft bracket in the original position and direction as shown in figure.

17. Install camshaft brackets (numbers 2–4) aligning the stamp marks as shown in the figure.

➡**There are no identification marks indicating left and right for camshaft bracket (number 1).**

18. Apply liquid gasket to mating surface of camshaft bracket (number 1) as shown on both right and left banks. Use Genuine RTV Silicone Sealant or equivalent.

19. Tighten the camshaft bracket bolts in the following steps, in numerical order as shown.

 a. Tighten numbers 7–10 to 12 inch lbs. (2 Nm).

 b. Tighten numbers 1–6 to 12 inch lbs. (2 Nm).

 c. Tighten numbers 1–10 to 48 inch lbs. (6 Nm).

 d. Tighten numbers 1–10 in numerical order to 96 inch lbs. (10 Nm).

➡**After tightening the mounting bolts of camshaft brackets (number 1), be sure to wipe off excessive liquid gasket from**

the mating surface of the rocker cover and the mating surface of the rear timing chain case.

20. Measure difference in levels between the front end faces of the number 1 camshaft bracket and the cylinder head. If the measurement is outside the specified range, re-install the camshaft and camshaft brackets.

➡**Camshaft bracket and cylinder head standard is: -0.0055–0.0055 in. (-0.14–0.14mm).**

 a. Measure 2 positions (both intake and exhaust side) for a single bank.

 b. If the measured value is out of the standard, re-install the camshaft bracket (number 1).

21. Inspect and adjust the valve clearance.

22. The remainder of installation is the reverse of removal.

CRANKSHAFT DAMPER

REMOVAL & INSTALLATION

See Figure 45.

1. Before servicing the vehicle, refer to the Precautions Section.

2. Remove the front engine undercover.

3. Remove the accessory drive belt. Refer to Accessory Drive Belts, removal & installation.

4. For 2WD models, remove the rear cover plate.

5. For AWD models, remove the starter motor.

6. Set the ring gear stopper SST: KV10117700 (J44716), or equivalent, to hold the crankshaft in position.

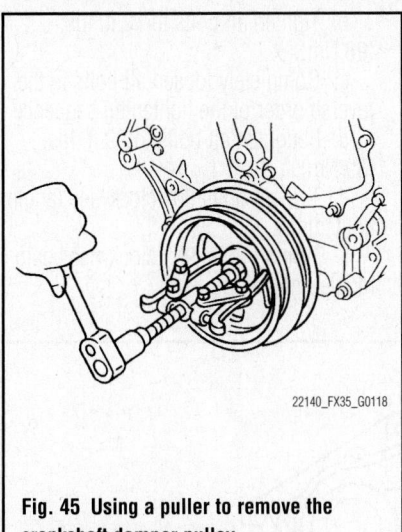

Fig. 45 Using a puller to remove the crankshaft damper pulley

7. Loosen the crankshaft damper pulley bolt until the bolt seating surface is approximately 0.39 in. (10mm) from its original position.

➡️Do not completely remove the crankshaft damper pulley bolt, since it will be used as a supporting point for a suitable puller.

8. Place a suitable puller tab on the holes of the crankshaft damper pulley and pull the crankshaft damper pulley until it releases.

❊❊ WARNING

Do not position the puller on the outer edges of the crankshaft pulley since this can damage the internal damper.

To install:

9. Fix the crankshaft in position using ring gear stopper SST: KV10117700 (J44716) or equivalent.

10. Install the crankshaft pulley, taking care not to damage the front oil seal.

❊❊ WARNING

When press-fitting the crankshaft pulley with a plastic hammer, tap on its center portion, not on the circumference.

11. Tighten the crankshaft bolt to 33 ft. lbs. (44 Nm).

12. Put a paint mark on the crankshaft pulley aligned with the angle mark on the crankshaft pulley bolt. Then, further tighten the bolt 90°.

13. Rotate the crankshaft pulley in the normal direction (clockwise when viewed from front) to confirm that it turns smoothly.

CRANKSHAFT FRONT SEAL

REMOVAL & INSTALLATION

See Figure 46.

1. Before servicing the vehicle, refer to the Precautions Section.

2. Disconnect the negative battery cable.

3. Remove the engine undercover.

4. Remove the accessory drive belts. Refer to Accessory Drive Belts, removal & installation.

5. Remove the crankshaft damper. Refer to Crankshaft Damper, removal & installation.

6. Remove the crankshaft front oil seal using a suitable tool.

❊❊ WARNING

Be careful not to damage the front timing chain case or the crankshaft.

To install:

7. Apply new engine oil to the oil seal lip and the dust seal lip of new crankshaft front oil seal.

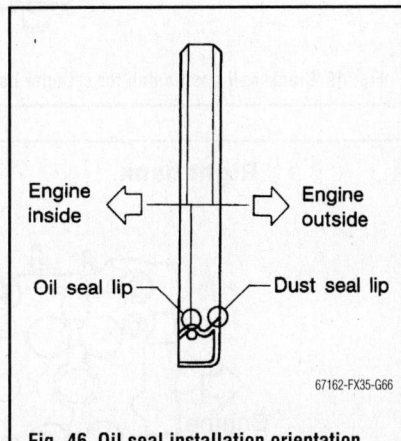

Fig. 46 Oil seal installation orientation

8. Install the crankshaft front oil seal.

➡️The oil seal must be oriented as shown in the figure.

9. Using a suitable drift, press-fit until the height of front oil seal is level with the mounting surface.

➡️A suitable drift should have an outer diameter of 2.36 in. (60mm), and an inner diameter of 1.97 in. (50mm). Make sure the garter spring is in position and the seal lips are not inverted.

❊❊ WARNING

Be careful not to damage front timing chain case or the crankshaft. Press-fit straight and avoid causing burrs or tilting the oil seal during installation.

10. Installation continues in the reverse order of the removal procedure.

CYLINDER HEAD

REMOVAL & INSTALLATION

See Figures 47 through 50.

1. Before servicing the vehicle, refer to the Precautions Section.

2. Remove the camshafts. Refer to Camshaft and Valve Lifters, removal & installation.

3. Temporarily support the front suspension member to support the engine.

➡️Temporary support means that the engine is adequately stable although the weight supported by the hoist may be released. The front suspension member is removed and the cylinder head is hung by the hoist with an engine slinger installed.

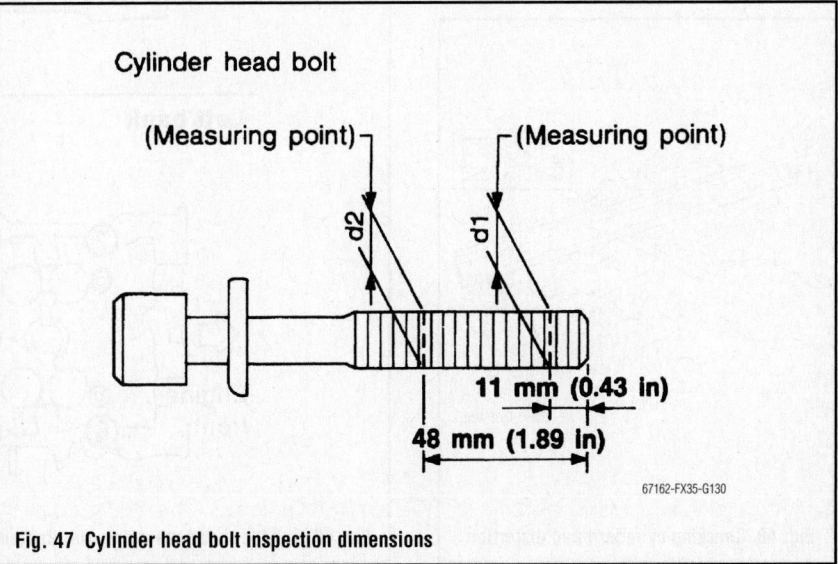

Cylinder head bolt

(Measuring point)   (Measuring point)

d_2 d_1

11 mm (0.43 in)

48 mm (1.89 in)

Fig. 47 Cylinder head bolt inspection dimensions

4. Release the hoist from hanging, then remove the engine slinger.

5. Remove the fuel tube and fuel injector assembly.

6. Remove the intake manifold.

7. Remove the exhaust manifold.

8. Remove the water inlet and thermostat housing.

9. Remove the water outlet and water piping.

10. Loosen the cylinder head bolts in the reverse of the tightening order.

11. Remove the cylinder head.

12. Remove the cylinder head gaskets.

To install:

13. Inspect the cylinder head bolt diameters. The cylinder head bolts are tightened by the plastic zone tightening method. Whenever the size difference between d1 and d2 exceeds the limit, replace the bolt with a new one. The specification for d1 minus d2 for the cylinder head bolts is 0.0043 in. (0.11mm).

➡**If the reduction of the outer diameter appears in a position other than at d2, use it as the d2 point value.**

14. Check the cylinder head for distortion. Using a scraper, wipe off oil, scale, gasket, sealant, and carbon deposits from the surface of the cylinder head. At each of several locations on the bottom surface of the cylinder head, measure distortion in 6 directions. If cylinder head distortion exceeds the recommended limit of 0.004 in. (0.1mm), replace the cylinder head.

✳✳ WARNING

Do not allow gasket fragments to enter engine oil or engine coolant passages.

15. Install a new cylinder head gasket.

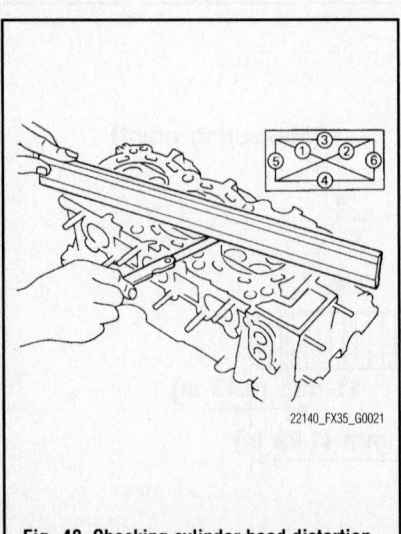

Fig. 48 Checking cylinder head distortion

16. Turn the crankshaft until number 1 piston is set at Top Dead Center (TDC) on the compression stroke. The crankshaft key should line up with the right bank cylinder center line.

17. Install the cylinder head.

18. Install and tighten the cylinder head bolts in the proper order:

a. Apply new engine oil to the threads and seat surfaces of the cylinder head bolts.

b. Tighten all bolts to 72 ft. lbs. (98 Nm).

c. Completely loosen all bolts in the reverse order of the tightening sequence.

d. Retighten all bolts to 29 ft. lbs. (39 Nm).

e. Turn all bolts 90° clockwise (angle tightening).

f. Turn all bolts 90° clockwise again (angle tightening).

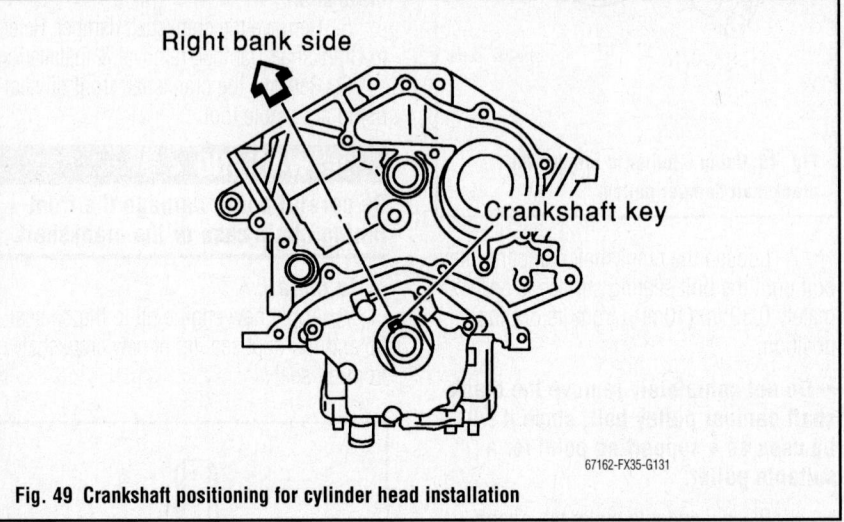

Fig. 49 Crankshaft positioning for cylinder head installation

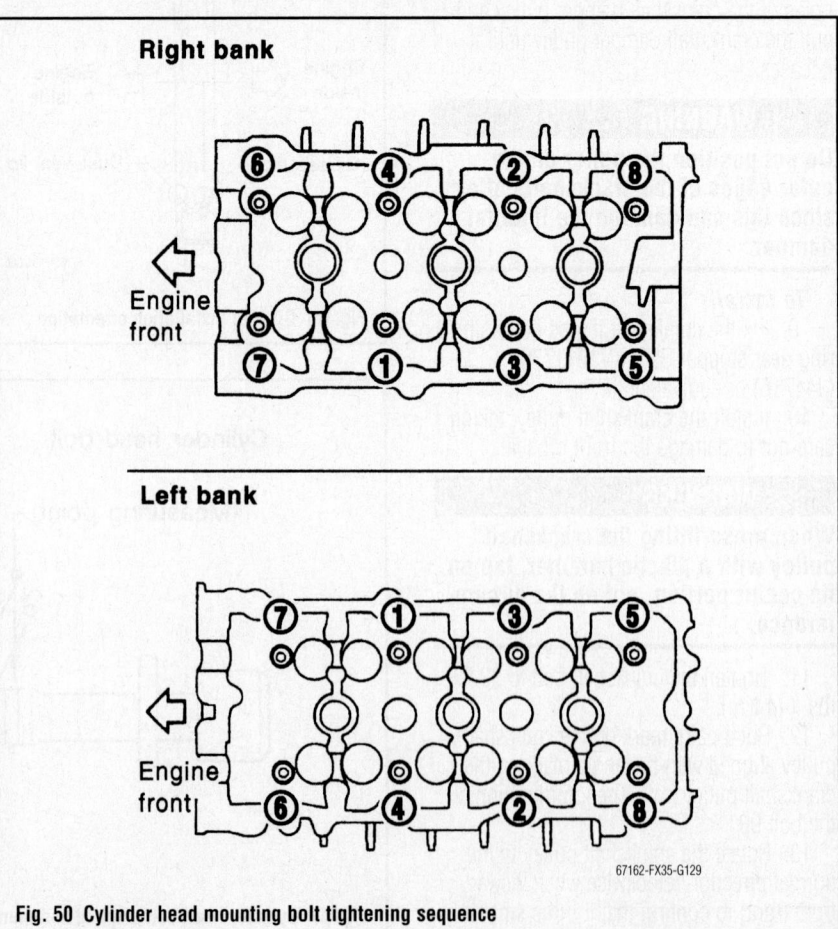

Fig. 50 Cylinder head mounting bolt tightening sequence

Check and confirm the tightening angle by using an angle wrench, or equivalent, and a cylinder head bolt wrench (commercial service tool). Avoid tightening the bolts with a visual inspection only.

19. After installing the cylinder head, measure the distance between the front end faces of the cylinder block and the cylinder head (left and right banks). If the measurement is outside the specified range, re-install the cylinder head.

➡**The specified range measurement is 0.555–0.587 in. (14.1–14.9mm).**

20. Install the water outlet and water piping.
21. Install the water inlet and thermostat housing.
22. Install the exhaust manifold.
23. Install the intake manifold.
24. Install the fuel tube and fuel injector assembly.
25. Install the camshafts. Refer to Camshaft and Valve Lifters, removal & installation.

EXHAUST MANIFOLD

REMOVAL & INSTALLATION

See Figures 51 and 52.

1. Before servicing the vehicle, refer to the Precautions Section.

Perform the work when the exhaust and cooling system have completely cooled down.

2. Remove the engine cover.
3. Remove the air cleaner case and air duct.
4. Remove the front and rear engine undercover and front cross bar.
5. Disconnect the heated oxygen sensor 2 wiring harness connectors (bank 1 and bank 2).
6. Using a heated oxygen sensor wrench, remove the heated oxygen sensors (bank 1 and bank 2).

Be careful not to damage heated oxygen sensor. Discard any heated oxygen sensor which has been dropped from a height of more than 20 in. (51cm) onto a hard surface such as a concrete floor; replace with a new sensor.

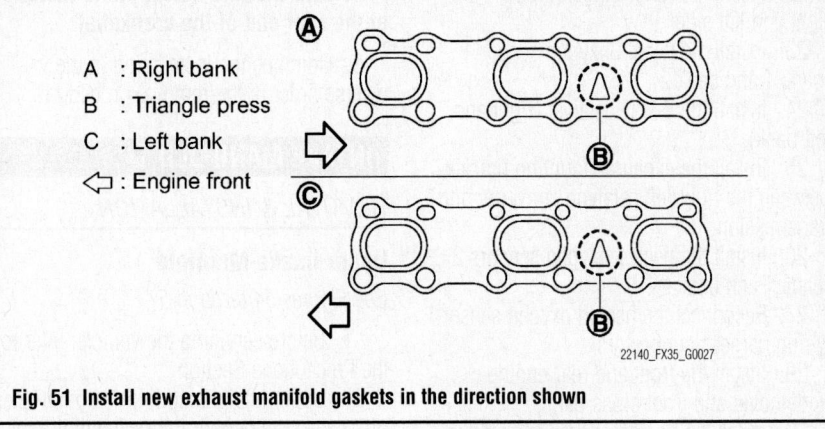

A : Right bank
B : Triangle press
C : Left bank
◁ : Engine front

22140_FX35_G0027

Fig. 51 Install new exhaust manifold gaskets in the direction shown

7. Remove the exhaust mounting bracket between the right/left catalytic converter and transmission.
8. Remove the 3-way catalyst (right and left bank).
9. Disconnect the heated oxygen sensor 1 (bank 1 and bank 2) wiring harness connectors and remove the wiring harness clip.
10. Using the heated oxygen sensor wrench, remove the heated oxygen sensor 1 (bank 1 and bank 2).
11. Remove the water pipes on both the right and left side.
12. Remove the exhaust manifold cover (right and left bank).
13. Loosen the mounting nuts in the reverse order of the tightening sequence shown in the illustration.
14. Remove the exhaust manifold.
15. Remove the exhaust manifold gaskets.

➡**Cover all engine openings to avoid entry of foreign materials.**

To install:

16. Check the surface distortion of the exhaust manifold mating surface with a straightedge and feeler gauge. If it exceeds the limit, replace the exhaust manifold. Limit of surface distortion: 0.012 in. (0.3mm).
17. Install new exhaust manifold gaskets in the direction shown in the figure with the triangle press mark in the correct position.
18. Install the manifold and tighten the mounting nuts in the order shown. If the stud bolts were removed, install them and tighten them to 11 ft. lbs. (15 Nm).
19. Tighten nuts number 1 and 2 in two steps.
20. Tighten all exhaust manifold nuts-to-engine nuts to 22 ft. lbs. (31 Nm).
21. Install the exhaust manifold cover (right and left bank).

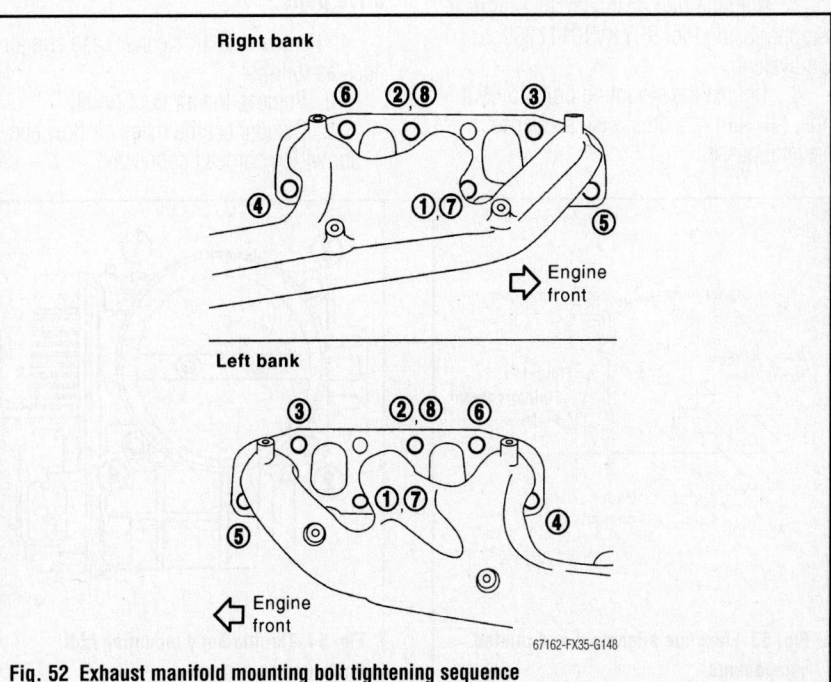

Right bank

Engine front

Left bank

Engine front

67162-FX35-G148

Fig. 52 Exhaust manifold mounting bolt tightening sequence

22. Install the water pipes on both the right and left side.

23. Install the heated oxygen sensor 1 (bank 1 and bank 2).

24. Install the 3-way catalyst (right and left bank).

25. Install the exhaust mounting bracket between the right/left catalytic converter and transmission.

26. Install the heated oxygen sensors 2 (bank 1 and bank 2).

27. Reconnect the heated oxygen sensor wiring harness connectors.

28. Install the front and rear engine undercover and front cross bar.

29. Install the air cleaner case and air duct.

30. Install the engine cover.

FLEXPLATE

REMOVAL & INSTALLATION

See Figure 53.

1. Before servicing the vehicle, refer to the Precautions Section.

2. Remove the transmission.

3. Remove the flexplate retaining bolts.

4. Remove the flexplate from the engine.

To install:

5. When installing the flexplate, be sure to correctly align the crankshaft side guide pin and drive side guide pin hole.

➡ **If not correctly aligned the engine will run rough and turn on the MIL light.**

6. Install the flexplate and reinforcement plate, as shown in the illustration.

7. Hold the ring gear with the ring gear stopper using tool SST:KV10117700, or equivalent.

8. Tighten the mounting bolts to 65 ft. lbs. (88 Nm) in a crosswise sequence in several passes.

➡ **Be sure that the dowel pin is installed at the rear end of the crankshaft.**

9. Continue the installation in the reverse order of the removal procedure.

INTAKE MANIFOLD

REMOVAL & INSTALLATION

Upper Intake Manifold

See Figures 54 through 57.

1. Before servicing the vehicle, refer to the Precautions Section.

The upper intake manifold is constructed of 2 halves: the upper intake manifold collector and the lower intake manifold collector. Together these two components create the upper intake manifold. The 3.5L engine also uses a lower intake manifold plenum.

✳✳ CAUTION

To avoid the danger of being scalded, never drain engine coolant when the engine is hot.

➡ **The gasket for the intake manifold collector (upper) is secured together with the mounting bolt for the intake manifold collector (lower). Thus, even when only the gasket for the upper side is replaced, the gasket for lower side must also be replaced.**

2. Remove the engine cover.

3. Disconnect and plug the water hoses from the intake manifold collector (upper).

➡ **Do not spill engine coolant on the drive belts.**

4. Remove the air cleaner case and air duct, as follows:

a. Remove the air duct (inlet).

b. Disconnect the mass air flow sensor wiring harness connector.

c. Remove the air cleaner case/mass air flow sensor assembly and the air duct/resonator assembly disconnecting them at the joints.

➡ **Add match marks as necessary for easier installation.**

d. Remove the mass air flow sensor from air cleaner case.

✳✳ WARNING

Handle the mass air flow sensor with care. Do not expose it to harsh vibration or shock. Do not disassemble it or touch its sensor.

e. Remove the resonator in the fender, lifting the left fender protector.

5. Remove the electric throttle control actuator, as follows:

a. Disconnect the wiring harness connector.

b. Loosen the bolts in the reverse order as shown in the figure.

✳✳ WARNING

Handle the throttle body carefully to avoid any shock to the electric throttle control actuator. Do not disassemble.

6. Remove the fuel sub-tube mounting bolt to disconnect it from the rear of the intake manifold collector (lower).

7. Disconnect the vacuum hose and water hose from the intake manifold collector (upper).

8. Disconnect the EVAP canister purge volume control solenoid valve bracket mounting bolt from the intake manifold collector (upper).

9. Loosen the bolts in the reverse order of the illustration to remove the intake manifold collector (upper).

10. Remove the PCV hose between the

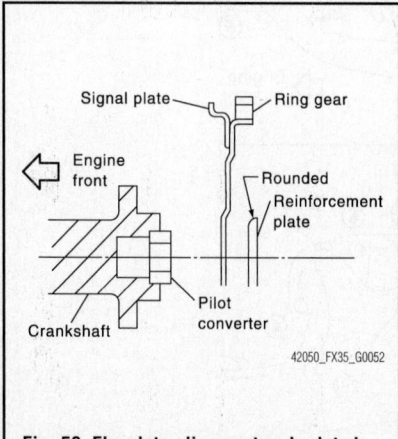

Fig. 53 Flexplate alignment and related components

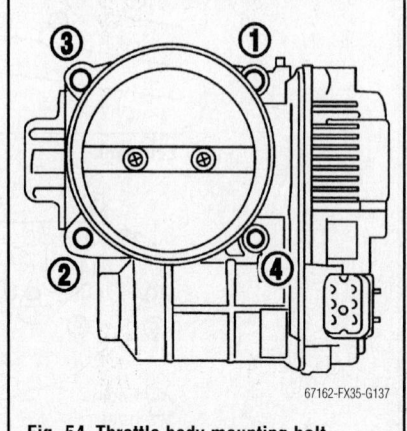

Fig. 54 Throttle body mounting bolt tightening sequence

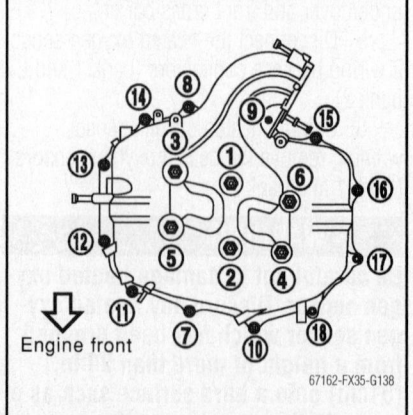

Fig. 55 Tightening sequence for the upper half of the upper intake manifold

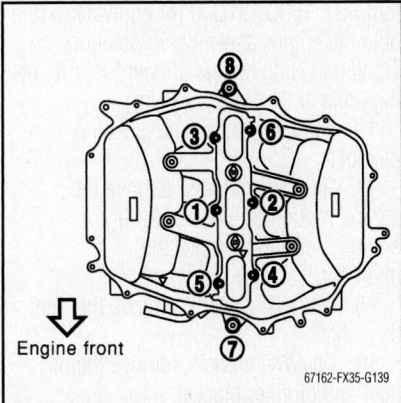

Fig. 56 Tightening sequence for the lower half of the upper intake manifold

intake manifold collector and the right-hand rocker cover.

11. Loosen the bolts in the reverse order of the illustration, and remove the intake manifold collector cover, gasket, intake manifold collector (lower) and gasket.

※※ CAUTION

Cover all engine openings to avoid entry of foreign materials.

To install:

12. Check the surface distortion of both the intake manifold collector (upper and lower) mating surfaces with a straightedge and feeler gauge. If it exceeds 0.004 in. (0.1mm), replace the intake manifold collector (upper and/or lower).

13. Use the accompanying illustration referring to the gasket front marks, as a guide to installing the parts.

14. Install the intake manifold collector (lower). Tighten the mounting bolts in numerical order as shown in the figure.

➡ **Tighten mounting bolts to secure gasket (lower), intake manifold collector (lower), gasket (upper), and intake manifold collector cover.**

15. Reconnect the PCV hose between the intake manifold collector and the right-hand rocker cover.

16. Install the lower manifold. If the stud bolts were removed, install them and tighten them to the specified torque of 52 inch lbs. (6 Nm).

17. The shank length from under the bolt head varies with bolt location. Install the bolts while referring to the numbers shown below and in the illustration of the tightening sequence for the upper half of the upper intake manifold.

➡ **The bolt length does not include pilot portion. Make sure to tighten each bolt in the numerical order as shown in the figure.**

- M6 bolt (length _25mm): Positions 7, 8, 10, 11, 13, 14, 15, 16, and 18
- M6 bolt (length _45mm): Positions 2, 4, 5
- M6 bolt (length _60mm): Positions 1, 3, 6, and 9
- M6 nut: Positions 12 and 17

18. Reattach the EVAP canister purge volume control solenoid valve bracket to the intake manifold collector (upper).

Intake manifold collector (upper)

Intake manifold collector cover

Front mark

Front mark

Gasket

Front mark

Engine front

Gasket

Intake manifold collector (lower)

Front mark

67162-FX35-G140

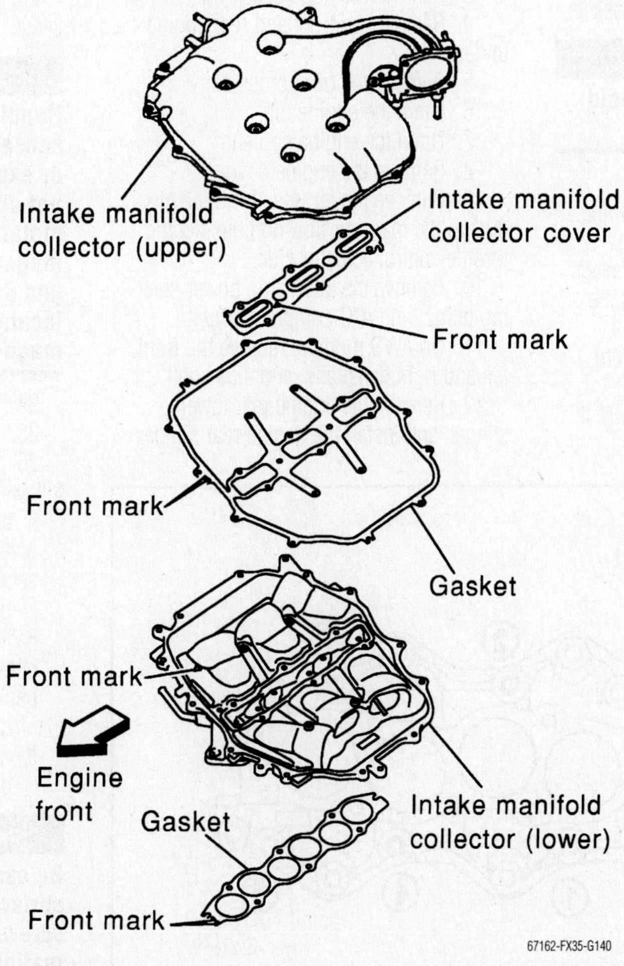

Fig. 57 Upper intake manifold gasket positioning

19. Reattach the vacuum hose and water hose to the intake manifold collector (upper).

20. Reattach the fuel sub-tube to the rear of the intake manifold collector (lower).

21. Install the electric throttle control actuator.

22. Install the mass air flow sensor in the air cleaner case.

23. Install the air cleaner case and air duct.

24. Reconnect the hoses to the intake manifold collector (upper).

25. Install the engine cover.

Lower Intake Manifold

See Figure 58.

1. Before servicing the vehicle, refer to the Precautions Section.

2. Relieve the fuel pressure.

3. Remove the intake manifold collector (upper) and (lower). Refer to Upper Intake Manifold procedure.

4. Remove the fuel tube and fuel injector assembly.

5. Loosen the mounting bolts and nuts in the reverse order of the illustration to remove the lower intake manifold.

6. Remove the intake manifold gaskets.

❋❋ CAUTION

Cover all engine openings to avoid entry of foreign materials.

To install:

7. Check the surface distortion of the intake manifold mating surface with a straightedge and feeler gauge. If it exceeds the limit of 0.04 in. (0.1mm), replace the lower intake manifold.

8. Install new lower intake manifold gaskets.

9. Install the lower intake manifold. If the stud bolts were removed, install them and tighten to 8 ft. lbs. (11 Nm). Tighten all mounting bolts and nuts to the specified torque in 2 or more steps in numerical order shown in the figure, as follows:
   a. Step 1: 60 inch lbs. (7 Nm).
   b. Step 2: 21 ft. lbs. (29 Nm).

10. The remainder of installation is the reverse of removal.

OIL PAN

REMOVAL & INSTALLATION

See Figures 59 through 65.

1. Before servicing the vehicle, refer to the Precautions Section.

❋❋ CAUTION

To avoid the danger of being scalded, never drain the engine oil or engine coolant when the engine is hot.

➡**To remove only the lower oil pan, drain the engine oil and skip to step 26.**

2. Remove the front wheels and tires.

3. Remove the hood assembly.

4. Remove the front and rear engine undercover.

5. Remove the front cross bar.

6. Drain the engine oil.

7. Drain the engine coolant.

8. Remove the engine cover.

9. Remove the air hose from the air duct to the mass air flow and the electric throttle control actuator side.

10. Remove the alternator, power steering pump, and A/C compressor belt.

11. On AWD models, remove the front left and right halfshafts, and side shaft.

12. Remove the engine rear lower slinger, and install the engine rear slinger tool SST: 10006 31U00 (or equivalent) to hold the engine assembly in position. Tighten the engine rear slinger tool mounting bolts to 21 ft. lbs. (28 Nm).

13. Remove the front suspension member.

14. On AWD models, remove the engine mounting bracket, lower engine mounting bracket and insulator.

15. On AWD models, remove the front driveshaft.

16. On AWD models, remove the oil filter and oil filter bracket.

17. Remove the alternator stay.

18. Remove the starter motor.

19. Remove the alternator and power steering pump and A/C compressor idler pulley and bracket assembly.

20. Disconnect the A/T fluid cooler hoses, and remove the oil cooler water pipe mounting bolt.

21. Disconnect the A/T fluid cooler tube.

22. On AWD models, remove the front final drive assembly. .

23. Remove the crankshaft position sensor.

❋❋ WARNING

Handle the crankshaft position sensor carefully to avoid dropping it or exposing it to abrupt shocks. Do not disassemble it, do not allow metal powder to adhere to the magnetic part at the sensor tip, and do not place the sensor in a location where it may be exposed to magnetism.

24. Remove the oil filter, as necessary.

25. Remove the oil cooler, as necessary.

26. Remove the oil pan (lower), as follows:
   a. Loosen the mounting bolts in the reverse order of the tightening sequence.
   b. Insert a seal cutter SST: KV10111100 (J37228) or equivalent, between the upper oil pan and lower oil pan.
   c. Slide the seal cutter by tapping on the side of the tool with a hammer.
   d. Remove the lower oil pan.

❋❋ WARNING

Be careful not to damage the mating surface. Do not use a flat-bladed screwdriver as this could damage the mating surfaces.

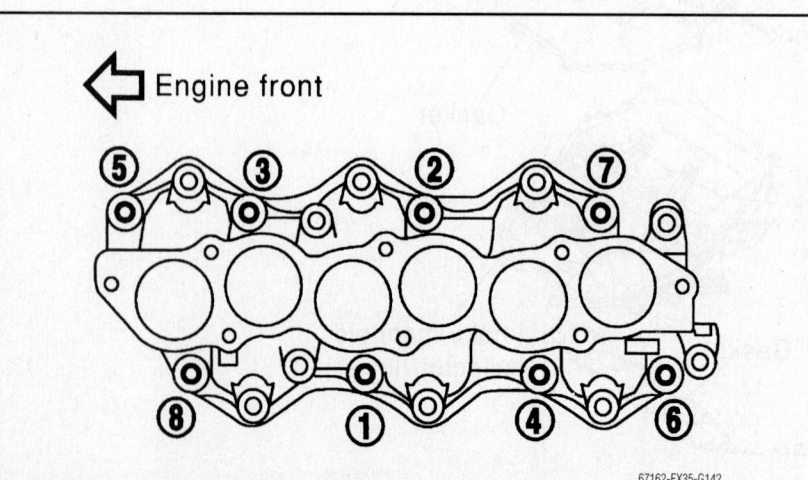

← Engine front

67162-FX35-G142

Fig. 58 Lower intake manifold mounting bolt tightening sequence—3.5L engine

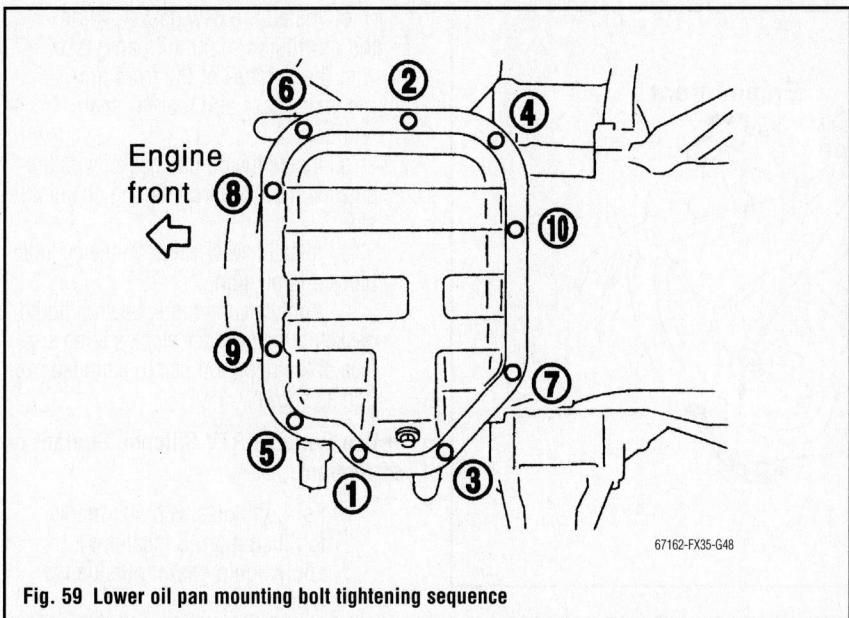

Fig. 59 Lower oil pan mounting bolt tightening sequence

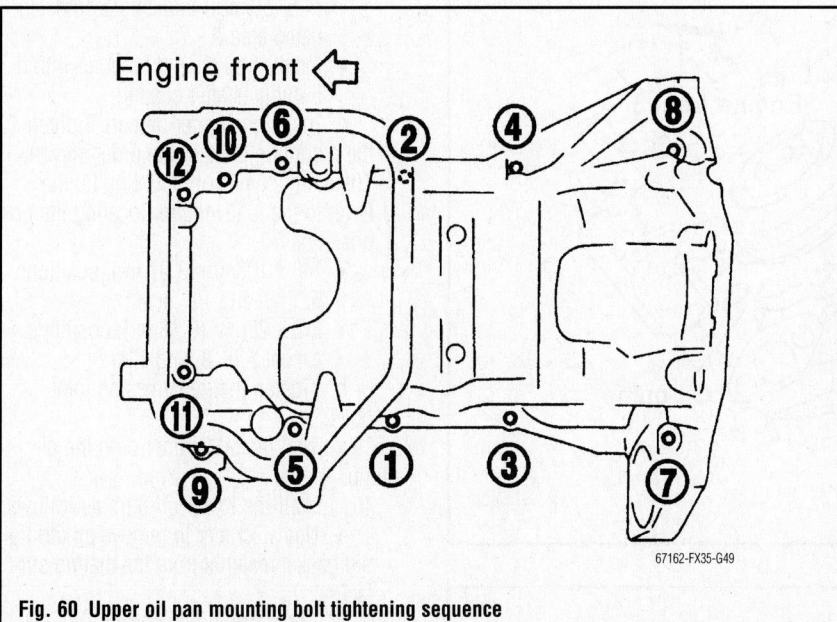

Fig. 60 Upper oil pan mounting bolt tightening sequence

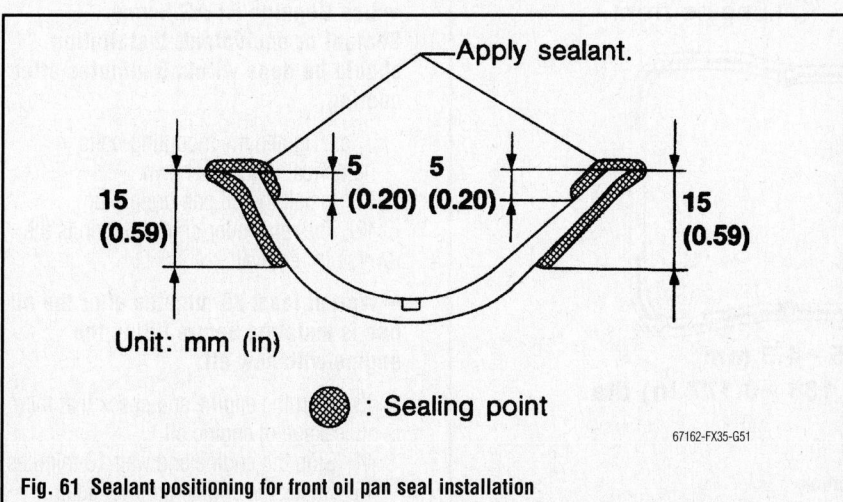

Fig. 61 Sealant positioning for front oil pan seal installation

27. Remove the oil strainer.

28. Remove the transmission joint bolts which pass through the upper oil pan.

29. On 2WD models, remove the rear cover plate.

30. Loosen the upper oil pan bolts in the reverse order of the tightening sequence.

31. Insert a seal cutter SST: KV10111100 (J37228) between the upper oil pan and cylinder block. Slide the seal cutter by tapping on the side of the tool with a hammer.

32. Remove the upper oil pan.

✳✳ WARNING

Be careful not to damage the mating surface. Do not use a flat-bladed screwdriver as this could damage the mating surfaces.

33. Remove the O-rings from the bottom of the cylinder block and oil pump.

34. Remove the oil pan gaskets.

35. For AWD models, remove the axle pipe from the upper oil pan using a suitable drift, if necessary.

36. Clean the oil strainer, if necessary.

To install:

37. On AWD models, install the axle pipe to the oil pan, if removed:

   a. Lubricate the O-ring groove of the axle pipe, O-ring, and O-ring joint of the oil pan with new engine oil.

   b. Install the axle pipe to the oil pan (upper) from the axle pipe flange side (left side) using a suitable drift with an outer diameter of 1.7 to 2.2 in. (43 to 57mm).

➡**Insert the axle pipe with care to prevent the O-ring from sliding.**

38. Install the upper oil pan, as follows:

   a. Use a scraper to remove the old liquid gasket from all mating surfaces. Remove old liquid gasket from the mating surface of the cylinder block, and the bolt holes and threads.

✳✳ WARNING

Do not scratch or damage the mating surfaces when cleaning off the old liquid gasket material.

   b. Apply liquid gasket to the oil pan gaskets as shown.

➡**Use Genuine RTV Silicone Sealant or equivalent.**

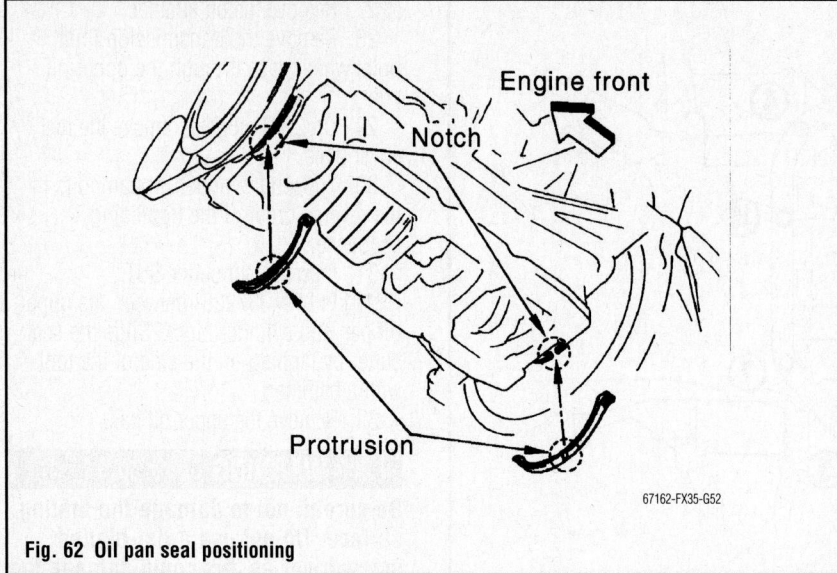

Fig. 62 Oil pan seal positioning

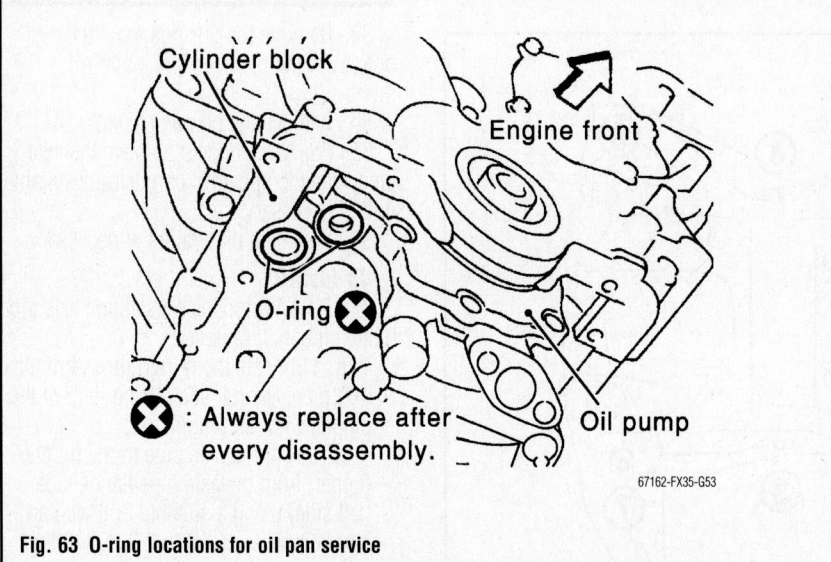

Fig. 63 O-ring locations for oil pan service

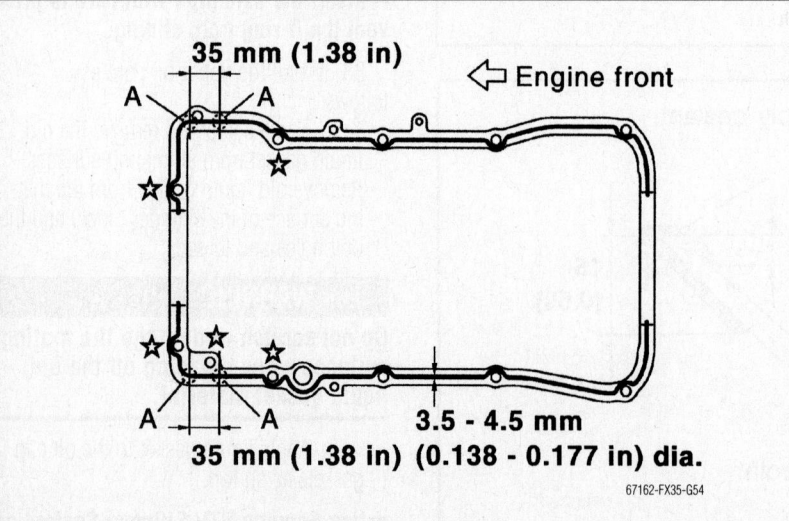

Fig. 64 Liquid gasket positioning for oil pan installation

c. Install the new gasket. Align the protrusion of the oil pan gasket with the notches of the front timing chain case and rear oil seal retainer.

d. Install the oil pan gasket with the smaller arc to the front timing chain case side.

e. Install new O-rings on the cylinder block and oil pump.

f. Apply a continuous bead of liquid gasket to the cylinder block mating surface of the upper oil pan to a limited portion as shown.

➡ **Use Genuine RTV Silicone Sealant or equivalent.**

- For bolt holes with star marks in illustration (5 locations), apply liquid gasket outside the holes.
- Apply a bead of 0.18–0.22 in. (4.5–5.5mm) in diameter to designated area **A**.
- Installation should be done within 5 minutes after coating.

g. Install the upper oil pan. Tighten the mounting bolts in the order shown. There are 2 types of mounting bolts. Refer to the following for locating the bolt positions:

- M8 x 100mm (3.97 in.): positions 5, 7, 8, and 11
- M8 x 25mm (0.98 in.): positions except 5, 7, 8, and 11

h. Tighten the transmission joint bolts.

39. Install the oil strainer onto the oil pump.

40. Install the lower oil pan, as follows:

a. Use a scraper to remove all old liquid gasket material from the mating surfaces.

b. Apply new liquid gasket.

➡ **Use Genuine RTV Silicone Sealant or equivalent. Installation should be done within 5 minutes after coating.**

c. Tighten the mounting bolts in numerical order as shown.

41. Install the oil pan drain plug.

42. The remainder of installation is the reverse of removal.

➡ **Wait at least 30 minutes after the oil pan is installed before filling the engine with new oil.**

43. Start the engine and check that there is no leakage of engine oil.

44. Stop the engine and wait 10 minutes.

45. Check the engine oil level again.

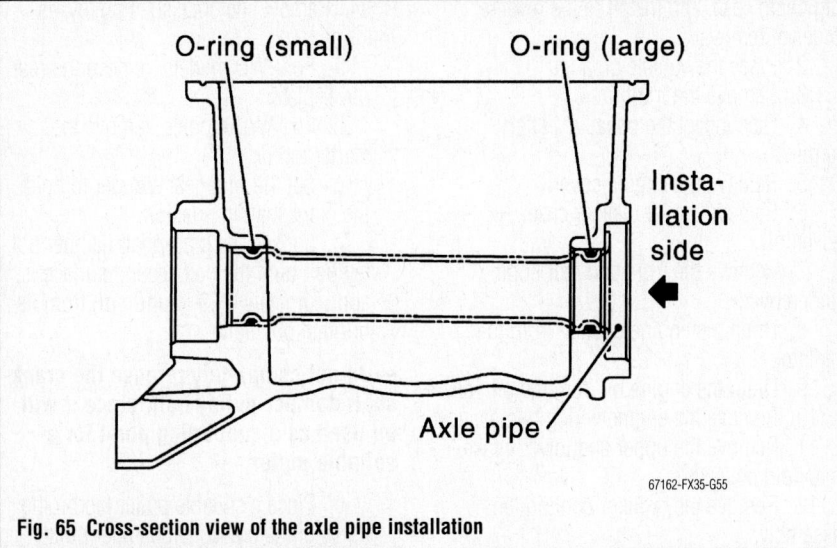

Fig. 65 Cross-section view of the axle pipe installation

OIL PUMP

REMOVAL & INSTALLATION

See Figure 66.

1. Before servicing the vehicle, refer to the Precautions Section.
2. Remove the oil pan (lower and upper) and the oil strainer.
3. Remove the front timing chain case and the timing chain (primary).
4. Remove the oil pump assembly.

To install:

5. Before installation, apply new engine oil to the parts as illustrated in the figure.
6. For pump installation, align the crankshaft flat faces with the oil pump inner rotor flat faces.
7. Installation is the reverse of the removal procedure.
8. After warming up the engine, check for engine oil leakage.
9. Check the engine oil level and add engine oil, as needed.

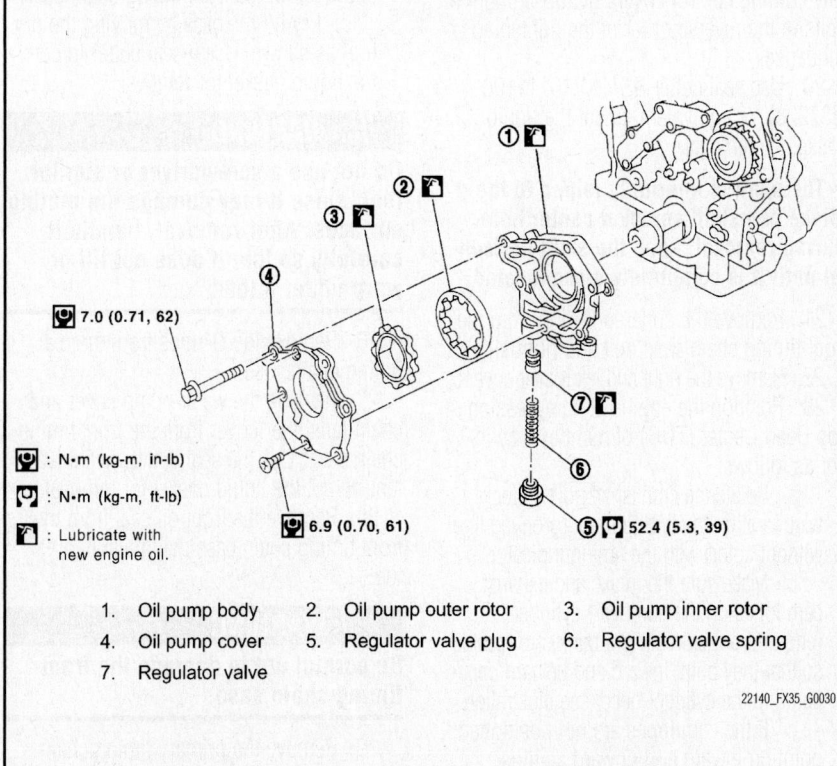

Fig. 66 Exploded view of the oil pump assembly and torque specifications—3.5L engine

REAR MAIN SEAL

REMOVAL & INSTALLATION

See Figures 67 and 68.

1. Before servicing the vehicle, refer to the Precautions Section.
2. Remove the upper oil pan.
3. Remove the transmission assembly.
4. Remove the drive plate.
   a. Install ring gear stopper SST: KV1011770 (J44716), or equivalent, and remove the mounting bolts in a diagonal order.
   b. Carefully remove the drive plate.

❊❊ WARNING

Do not disassemble the drive plate. Never place the drive plate with the signal plate facing down. When handling the signal plate, take care not to damage or scratch it. Handle the signal plate in a manner that prevents it from becoming magnetized.

5. Use seal cutter SST: KV10111100 (J37228), or equivalent, to cut away the old liquid gasket material and remove the rear oil seal retainer.

❊❊ WARNING

Be careful not to damage the mounting surfaces.

➡The rear oil seal and retainer form a single part and are handled as one assembly.

To install:

6. Remove the old liquid gasket from the mating surface of the cylinder block and oil pan using a scraper.
7. Apply new engine oil to the oil and dust seal lips.

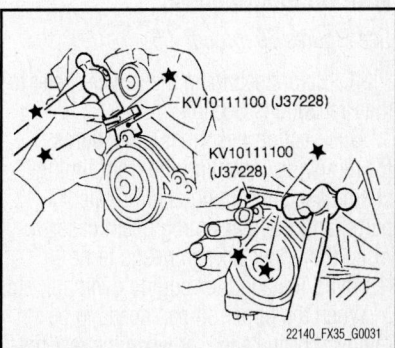

Fig. 67 Use seal cutter SST: KV10111100 (J37228) to cut away the old gasket and remove the rear oil seal retainer

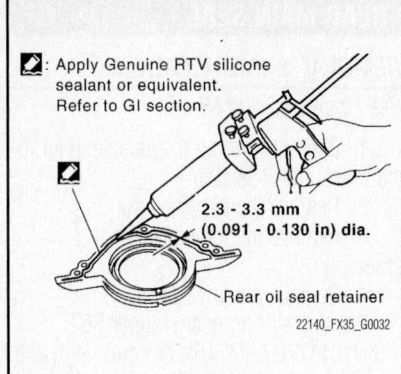

: Apply Genuine RTV silicone sealant or equivalent. Refer to GI section.

2.3 - 3.3 mm (0.091 - 0.130 in) dia.

Rear oil seal retainer

22140_FX35_G0032

Fig. 68 Apply liquid gasket to the rear oil seal retainer as illustrated

8. Apply liquid gasket to the rear oil seal retainer as illustrated.

➡**Use Genuine RTV Silicone Sealant or equivalent. Installation should be done within 5 minutes after coating, otherwise the liquid gasket may not seal properly.**

9. Install the rear oil seal retainer onto the cylinder block.

➡**Make sure the garter spring is in position and the seal lips are not inverted.**

10. The remainder of installation is the reverse of the removal procedure.

ROCKER ARMS/SHAFTS

REMOVAL & INSTALLATION

These engines are not equipped with rocker arms. The camshaft acts directly on the valves through hydraulic/mechanical lash adjusters.

TIMING CHAIN & SPROCKETS

REMOVAL & INSTALLATION

With Oil Pan Removal

See Figures 69 through 95.

1. Before servicing the vehicle, refer to the Precautions Section.

This section describes procedures for removing/installing the front timing chain case and timing chain related parts, and the rear timing chain case, when the upper oil pan needs to be removed/installed for engine overhaul, etc.

When the upper oil pan needs to be removed or installed, or when the rear timing chain case is removed or installed, remove the oil pans (upper and lower) first. Then, remove the front timing chain case, timing chain related parts, and the rear tim-

ing chain case, and install in the reverse order of removal.

2. Place the vehicle on a lift.
3. Remove the front tires.
4. Disconnect the negative battery terminal.
5. Remove the engine cover.
6. Remove the air cleaner case assembly.
7. Remove the front and rear engine undercovers.
8. Drain the engine coolant from the radiator.
9. Drain the engine oil from the oil pan.
10. Remove the engine wiring harnesses.
11. Remove the upper and lower intake manifold collectors.
12. Remove the radiator cooling fan assembly.
13. Remove the A/C compressor from the bracket with its piping connected, and temporarily secure it aside.
14. Remove the power steering oil pump from the bracket with its piping connected, and temporarily secure it aside.
15. Remove the power steering oil pump bracket.
16. Remove the alternator.
17. Remove the water bypass hose, water hose clamp, and idler pulley bracket from the front timing chain case.
18. Remove the upper and lower oil pan.
19. Remove the right and left intake valve timing control covers by loosening the bolts in the reverse order of the tightening sequence.
20. Use seal cutter SST: KV10111100 (J37228), or equivalent, to cut the liquid gasket for removal.

➡**The shaft is internally joined to the intake camshaft sprocket center hole. During removal, keep the shaft horizontal until it is completely disconnected.**

21. Remove the collared O-ring from the front timing chain case (left and right side).
22. Remove the right and left rocker covers.
23. Position the engine at compression Top Dead Center (TDC) of number 1 cylinder as follows:

a. Rotate the crankshaft pulley clockwise to align the timing mark (grooved line without color) with the timing indicator.

b. Make sure the intake and exhaust cam lobes on the number 1 cylinder (engine front side of right bank) are located so that they point inward and upward compared to the cylinder head, see illustration.

c. If the cam lobes are not positioned pointing inward and upward as illustrated, rotate the crankshaft 1 full revolution (360°) until they are.

24. Remove the crankshaft pulley, as follows:

a. For 2WD models, remove the rear cover plate.

b. For AWD models, remove the starter motor.

c. Set the ring gear stopper to hold the crankshaft in position.

d. Loosen the crankshaft damper pulley bolt until the bolt seating surface is approximately 0.39 in. (10mm) from its original position.

➡**Do not completely remove the crankshaft damper pulley bolt, since it will be used as a supporting point for a suitable puller.**

e. Place a suitable puller tab on the holes of the crankshaft damper pulley and pull the crankshaft damper pulley until it releases.

✳✳ WARNING

Do not position the suitable puller tab on the outer edges of the crankshaft pulley since this can damage the internal damper.

25. Remove the front timing chain case, as follows:

a. Loosen the mounting bolts in the reverse order of the tightening sequence.

b. Insert a suitable tool into the notch at the top of the front timing chain case.

c. Pry off the case by moving the pry tool as shown. Use a seal cutter to cut the liquid gasket for removal.

✳✳ WARNING

Do not use a screwdriver or similar tool, since it may damage the mating surfaces. After removal, handle it carefully so that it does not tilt or warp under a load.

26. Remove the O-rings from the rear timing chain case.

27. Remove the water pump cover and chain tensioner cover from the front timing chain case. Use the seal cutter, or equivalent, to cut the liquid gasket for removal.

28. Remove the front oil seal from the front timing chain case using a suitable tool.

✳✳ WARNING

Be careful not to damage the front timing chain case.

29. Remove the primary timing chain tensioner, as follows:

a. Pull the lever down and release the

plunger stopper tab. The plunger stopper tab can be pushed up to release.

b. Insert a stopper pin into the tensioner body hole to hold the lever and keep the tab released.

➡**A 0.098 in. (2.5mm) Allen wrench can be used for a stopper pin.**

c. Insert the plunger into the tensioner body by pressing on the slack guide.

d. Keep the slack guide depressed and hold it by pushing the stopper pin through the lever hole and body hole.

e. Remove the mounting bolts and remove the primary timing chain tensioner.

30. Remove the internal chain guide, tension guide, and slack guide.

➡**The tension guide can be removed after removing the primary timing chain.**

31. Remove the primary timing chain, tension guide, and crankshaft sprocket.

❊❊ WARNING

After removing the timing chain, do not turn the crankshaft and camshaft separately, or the valves may strike piston heads and cause damage.

32. Remove the secondary timing chain and camshaft sprockets, as follows:

a. Attach a suitable stopper pin to the right and left secondary timing chain camshaft chain tensioners.

b. Remove the intake and exhaust camshaft sprocket bolts. Apply paint to the timing chain and camshaft sprockets for alignment during installation. Secure the hexagonal portion of the camshaft using an open-end wrench to hold the camshaft steady while loosening the mounting bolts.

c. Remove the secondary timing chain together with camshaft sprockets. Turn the camshaft slightly to create slack in the timing chain on the secondary timing chain tensioner side. Insert a 0.020 in. (0.5mm) thick metal or resin plate between the timing chain and timing chain tensioner plunger guide. Remove the secondary timing chain together with the camshaft sprockets with the timing chain loose from the guide groove.

➡**Be careful of the plunger coming-off when removing the secondary timing chain. This is because the plunger of the secondary timing chain tensioner moves during operation, which can result in the fixed stopper pin coming off.**

➡**The intake camshaft sprocket is a two-for-one assembly of the primary and secondary sprockets.**

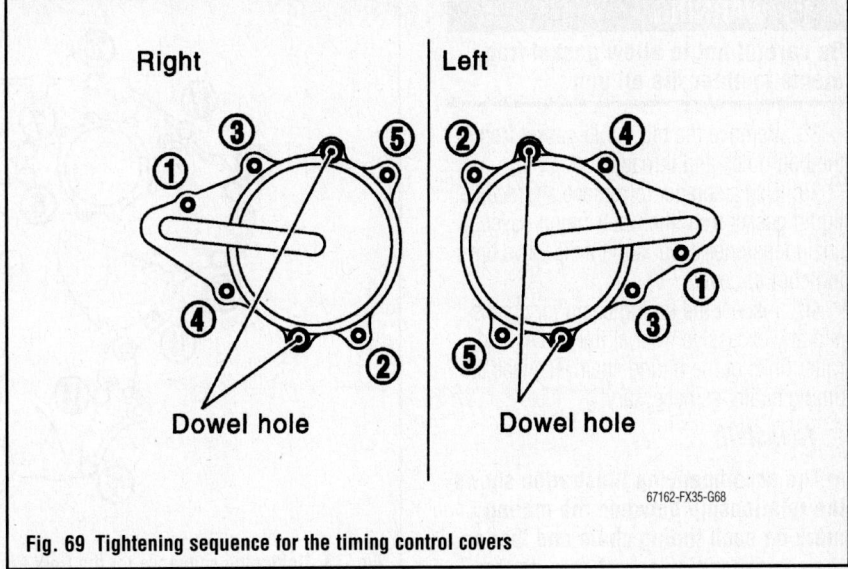

Fig. 69 Tightening sequence for the timing control covers

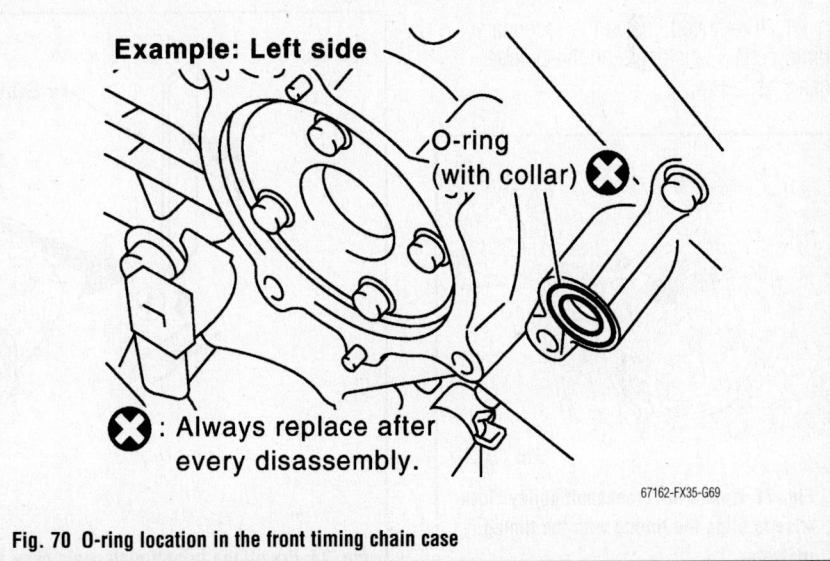

Fig. 70 O-ring location in the front timing chain case

❊❊ WARNING

When handling the intake camshaft sprocket, be careful to handle it carefully to avoid any shock to the camshaft sprocket. Do not disassemble. Do not loosen bolt A as shown in the figure.

33. Remove the rear timing chain case, as follows:

a. Loosen and remove the mounting bolts in the reverse order of the tightening sequence.

b. Cut the sealant using a seal cutter and remove the rear timing chain case.

❊❊ WARNING

Do not remove plate metal cover of engine oil passage. After removing the chain case, do not apply any load on the case which could cause warping.

34. Remove the O-rings from the cylinder head.

35. Remove the O-rings from the cylinder block.

36. If necessary, remove the secondary timing chain tensioners from the cylinder head, as follows:

a. Remove the number 1 camshaft brackets.

b. Remove the secondary timing chain tensioners with the stopper pins attached.

37. Use a scraper to remove all traces of liquid gasket from the front and rear timing chain cases and opposite mating surfaces.

※※ **WARNING**

Be careful not to allow gasket fragments to enter the oil pan.

38. Remove the old liquid gasket from the bolt holes and threads.

39. Use a scraper to remove all traces of liquid gasket from the water pump cover, chain tensioner cover and intake valve timing control covers.

40. Inspect the timing chain for cracks and any excessive wear at link plates and roller links of the timing chain. Replace the timing chain, as necessary.

To install:

➡The accompanying illustration shows the relationship between the mating mark on each timing chain and that on the corresponding sprocket, with the components installed.

41. If removed, install the secondary timing chain tensioners on the cylinder head, as follows:

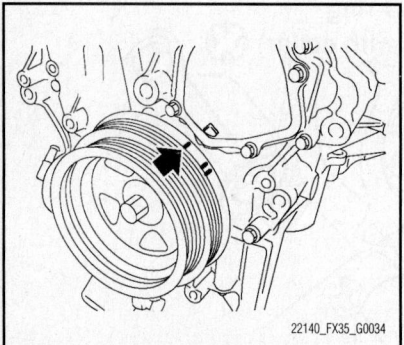

Fig. 71 Rotate the crankshaft pulley clockwise to align the timing with the timing indicator

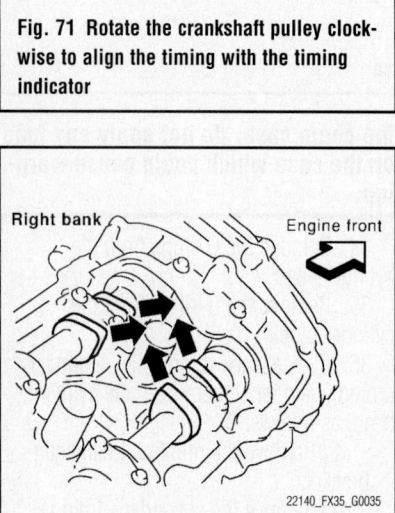

Fig. 72 Make sure the intake and exhaust cam lobes on the number 1 cylinder are located so that they point inward and upward compared to the cylinder head

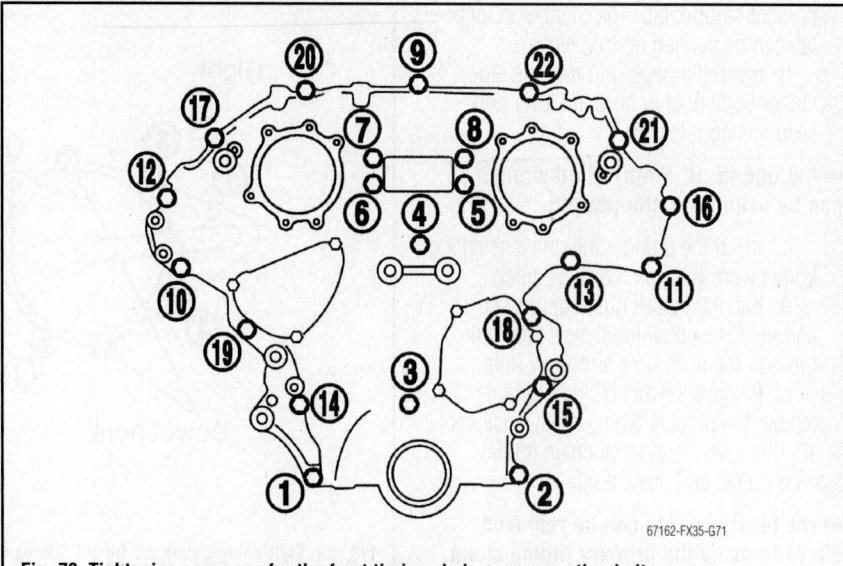

Fig. 73 Tightening sequence for the front timing chain case mounting bolts

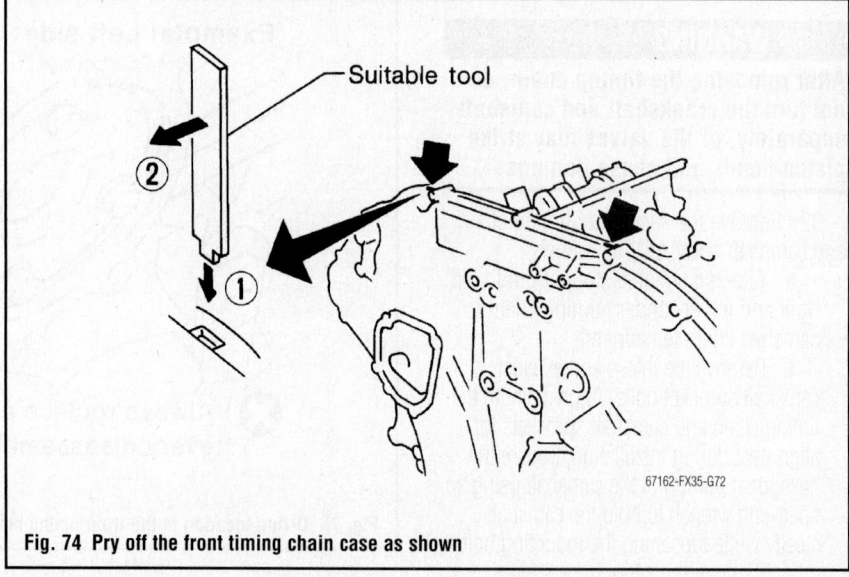

Fig. 74 Pry off the front timing chain case as shown

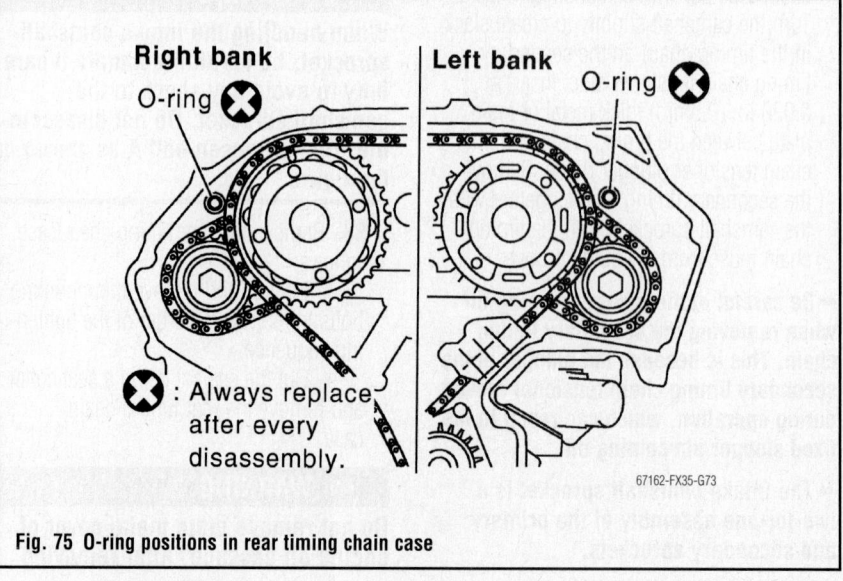

Fig. 75 O-ring positions in rear timing chain case

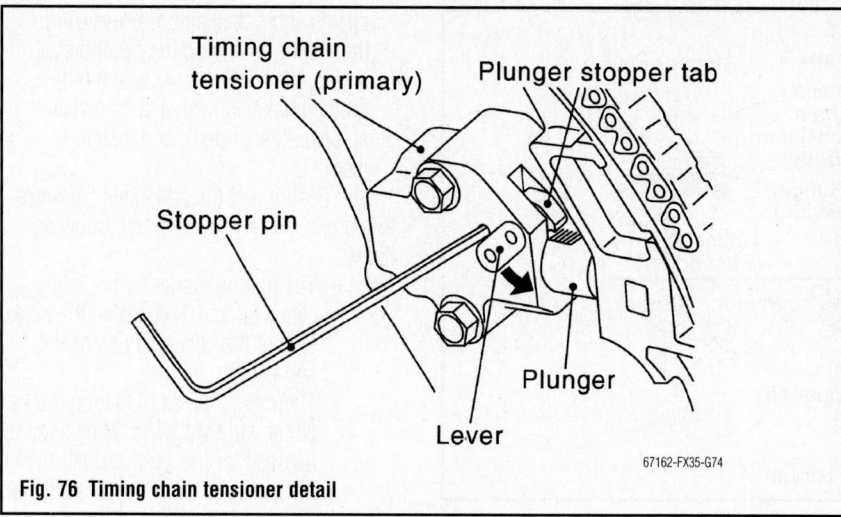

Fig. 76 Timing chain tensioner detail

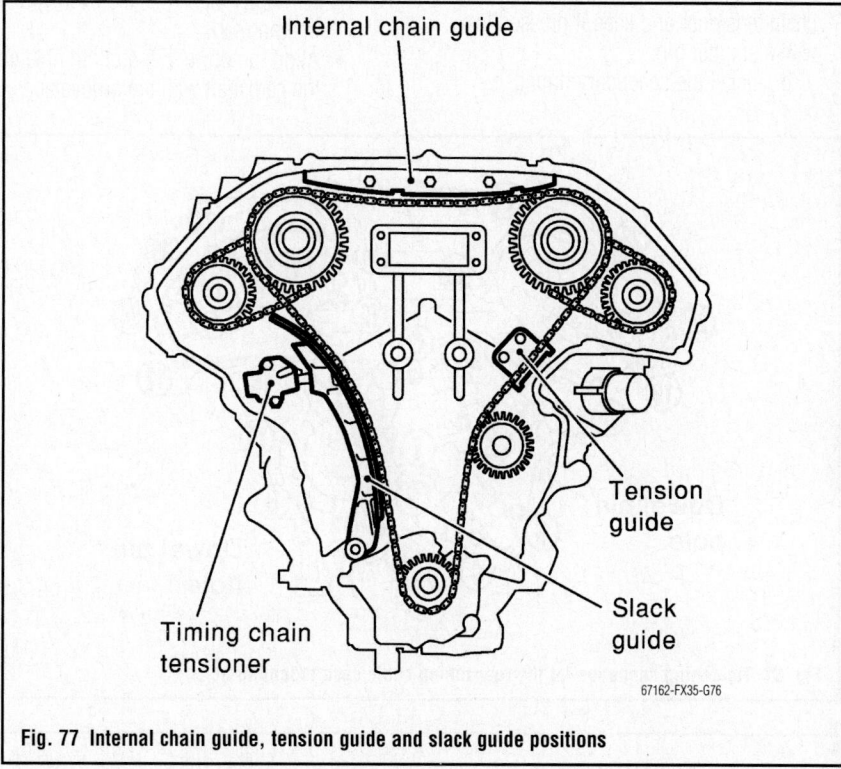

Fig. 77 Internal chain guide, tension guide and slack guide positions

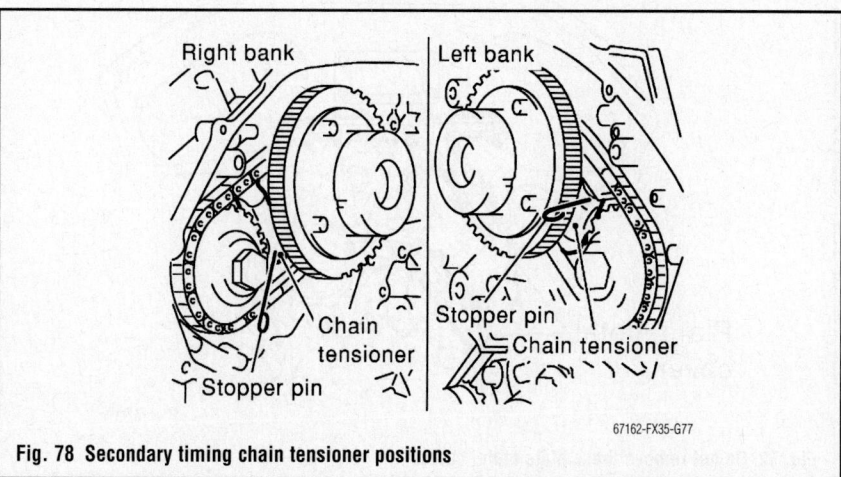

Fig. 78 Secondary timing chain tensioner positions

a. Install the chain tensioners with stopper pins attached and new O-rings.

b. Install the number 1 camshaft brackets. Refer to Camshaft and Valve Lifters, removal & installation.

42. Install new O-rings onto the cylinder block.

43. Install new O-rings on the cylinder head.

44. Apply liquid gasket to the rear timing chain case backside, as shown.

➡**Use Genuine RTV Silicone Sealant or equivalent.**

45. For **A** in the illustration, completely wipe out the liquid gasket extended on a portion touching at engine coolant.

46. Apply liquid gasket on the installation position of the water pump and cylinder head completely.

47. Align the rear timing chain case and water pump assembly with right and left dowel pins on cylinder block and install the case.

➡**Make sure the O-rings stay in place during installation on the cylinder block and cylinder head.**

48. Tighten the mounting bolts in the sequence shown. After all bolts are temporarily tightened, retighten them to 108 inch lbs. (13 Nm) in the tightening sequence. There are 2 bolt lengths used for the timing chain case:

a. 20mm length: Bolt positions 1, 2, 3, 6, 7, 8, 9, and 10.

b. 16mm length: All except bolt positions 1, 2, 3, 6, 7, 8, 9, and 10.

➡**If RTV Silicone Sealant protrudes beyond the sealing surfaces, wipe it off immediately.**

49. After installing the rear timing chain case, check the surface height difference between the rear timing chain case to the cylinder block at the oil pan mounting surface. If the difference is not within -0.009– 0.006 in. (-0.24–0.14mm), repeat the installation procedure.

50. Position the crankshaft so the number 1 piston is set at TDC on the compression stroke. Make sure that the dowel pin hole, dowel pin, and crankshaft key are located as shown.

➡**Though the camshaft does not stop at the position as shown in the figure, for the placement of the cam nose, it is generally accepted the camshaft is placed in the same direction of the figure and as follows:**

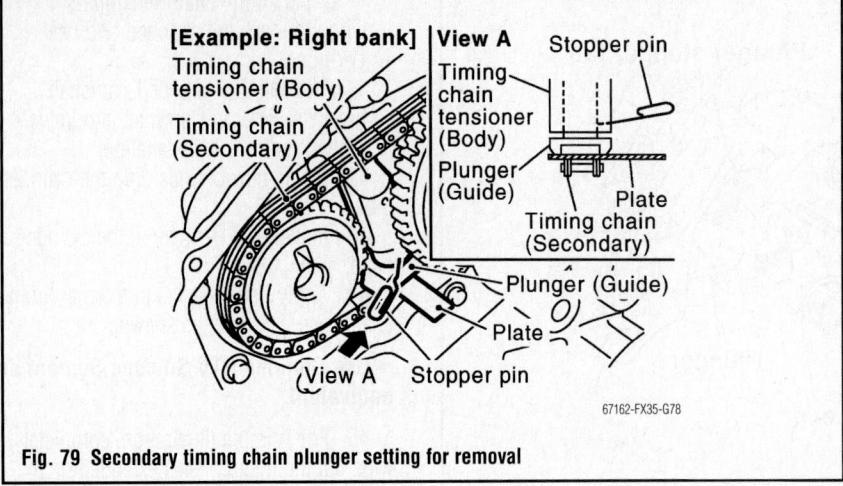

Fig. 79 Secondary timing chain plunger setting for removal

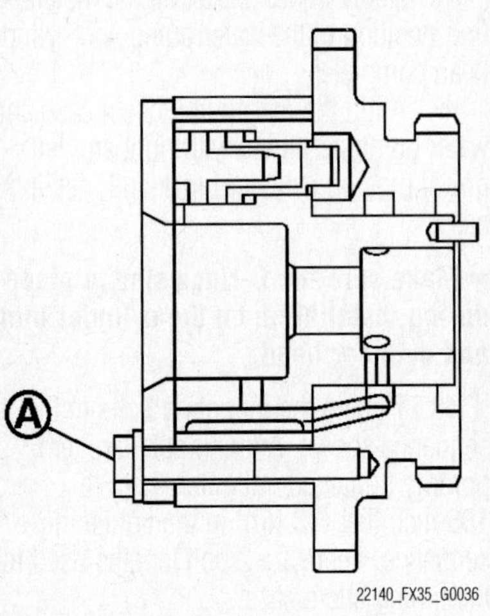

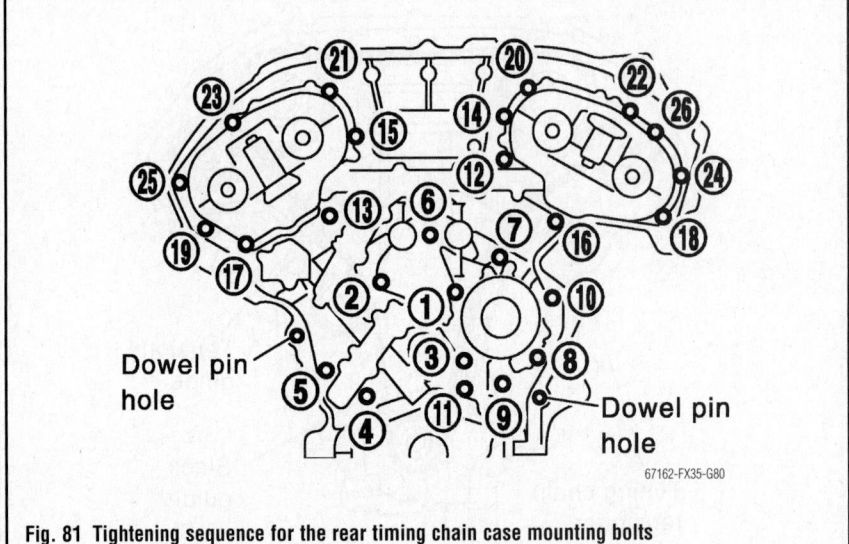

Fig. 80 Cross-section of intake camshaft sprocket—Do NOT loosen bolt A

- Camshaft dowel pin hole (intake side): At the cylinder head upper face side in each bank
- Camshaft dowel pin (exhaust side): At the cylinder head upper face side in each bank
- Crankshaft key: At the cylinder head side of the right bank

✳✳ WARNING

The hole on the small diameter side must be used for the intake side dowel pin hole.

51. Install the secondary timing chains and camshaft sprockets, as follows:

✳✳ WARNING

The matching marks between the timing chain and sprockets slip easily. Confirm all matching mark positions repeatedly during the installation process.

a. Push the plunger of the secondary chain tensioner and keep it pressed in with a stopper pin.

b. Install the secondary timing chains and camshaft sprockets. Align the mating marks on the secondary timing chain (gold link) with the ones on the intake and exhaust camshaft sprockets (stamped), and install them.

52. Ensure that the sprockets are properly positioned by heeding the following items:

- The mating marks for the intake camshaft sprocket are on the back side of the secondary camshaft sprocket.
- There are 2 types of mating marks, circle and oval types. They should be used for the right and left banks, respectively. For the right bank, use the circle type of mating mark, and for the left bank use the oval type of mating mark.
- Align the dowel pin and pin hole on the camshaft with the groove and

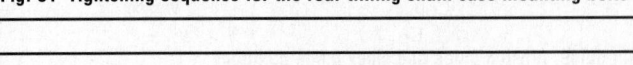

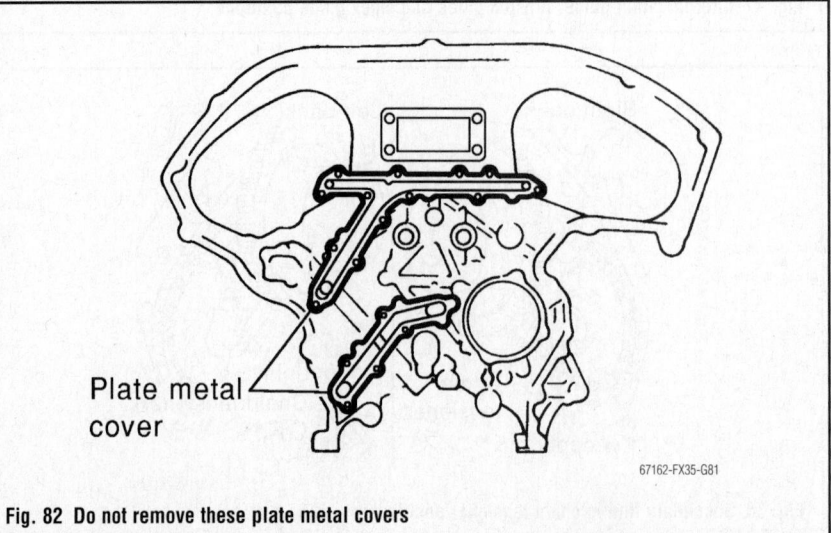

Fig. 81 Tightening sequence for the rear timing chain case mounting bolts

Fig. 82 Do not remove these plate metal covers

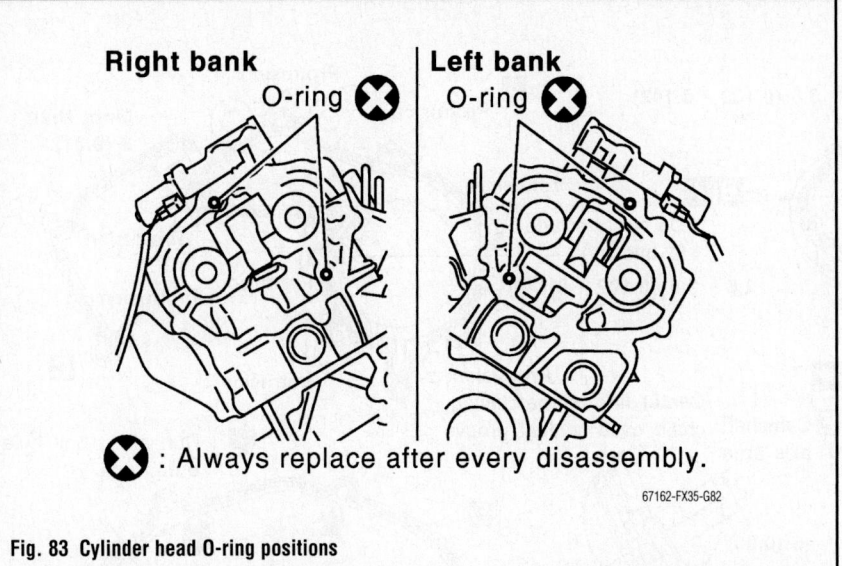

: Always replace after every disassembly.

67162-FX35-G82

Fig. 83 Cylinder head O-ring positions

dowel pin on the sprocket, and install them.
- On the intake side, align the pin hole on the small diameter side of the camshaft front end with the dowel pin on the back side of camshaft sprocket, and install them.
- On the exhaust side, align the dowel pin on the camshaft front end with the pin groove on the camshaft sprocket, and install them.

- In the case that positions of each mating mark and each dowel pin are not fit to the mating parts, make fine adjustments to the position holding the hexagonal portion of the camshaft with a wrench or equivalent.
- Mounting bolts for the camshaft sprockets must be tightened in the next step. Tightening them by hand is enough to prevent the dislocation of dowel pins.
- It may be difficult to visually check

the dislocation of the mating marks during and after installation. To make the matching easier, make a mating mark on the top of the sprocket teeth and its extended line in advance with paint.

53. After confirming the mating marks are aligned, tighten the camshaft sprocket mounting bolts, while securing the camshaft using a wrench on a hexagonal portion of the camshaft.

54. Pull the stopper pins out from the secondary timing chain tensioners.

55. Install the primary timing chain, as follows:

➡**During alignment, be careful to prevent dislocation of the mating mark alignments of the secondary timing chains.**

a. Install the crankshaft sprocket, making sure the mating marks on the crankshaft sprocket face the front of the engine.

b. Install the primary timing chain, so that the mating mark (punched) on the camshaft sprocket is aligned with the pink link on the timing chain, while the mating mark (notched) on the crankshaft sprocket is aligned with the orange one on the timing chain, as shown.

➡**If it is difficult to align mating marks of the primary timing chain with each**

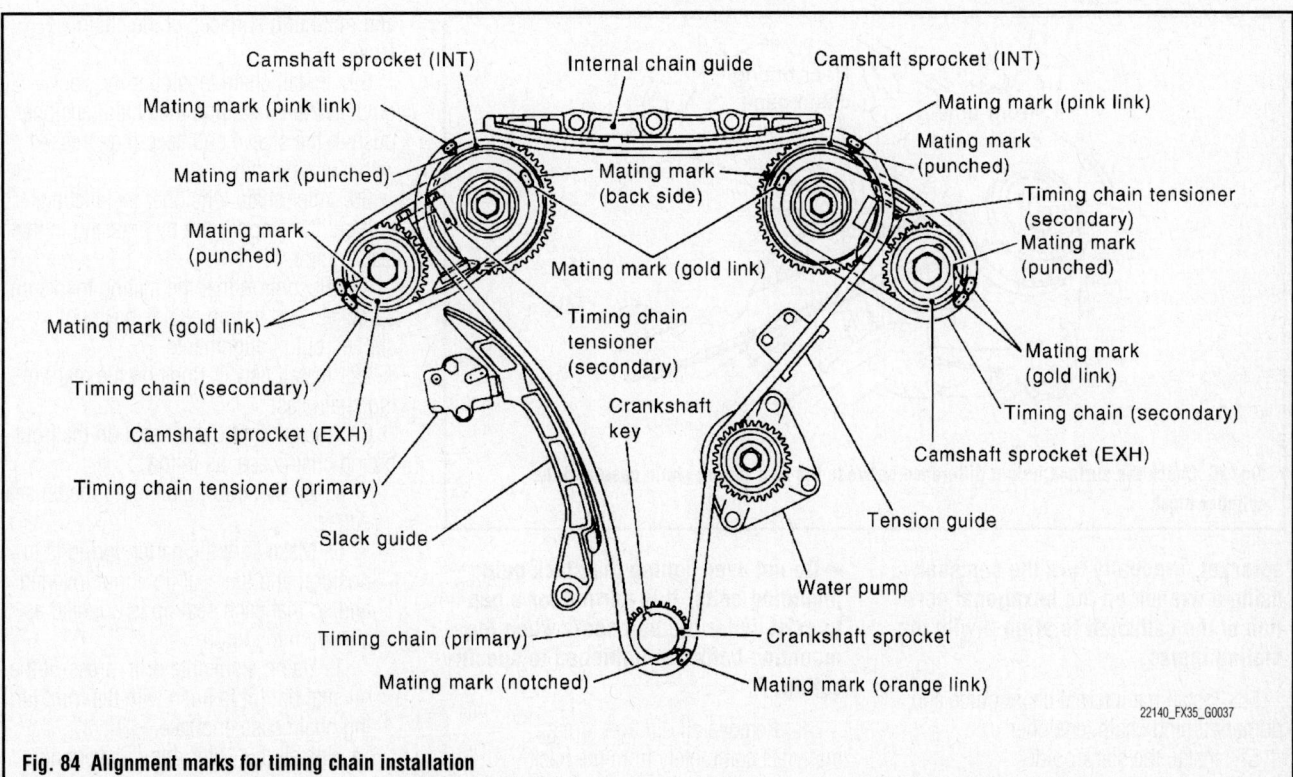

22140_FX35_G0037

Fig. 84 Alignment marks for timing chain installation

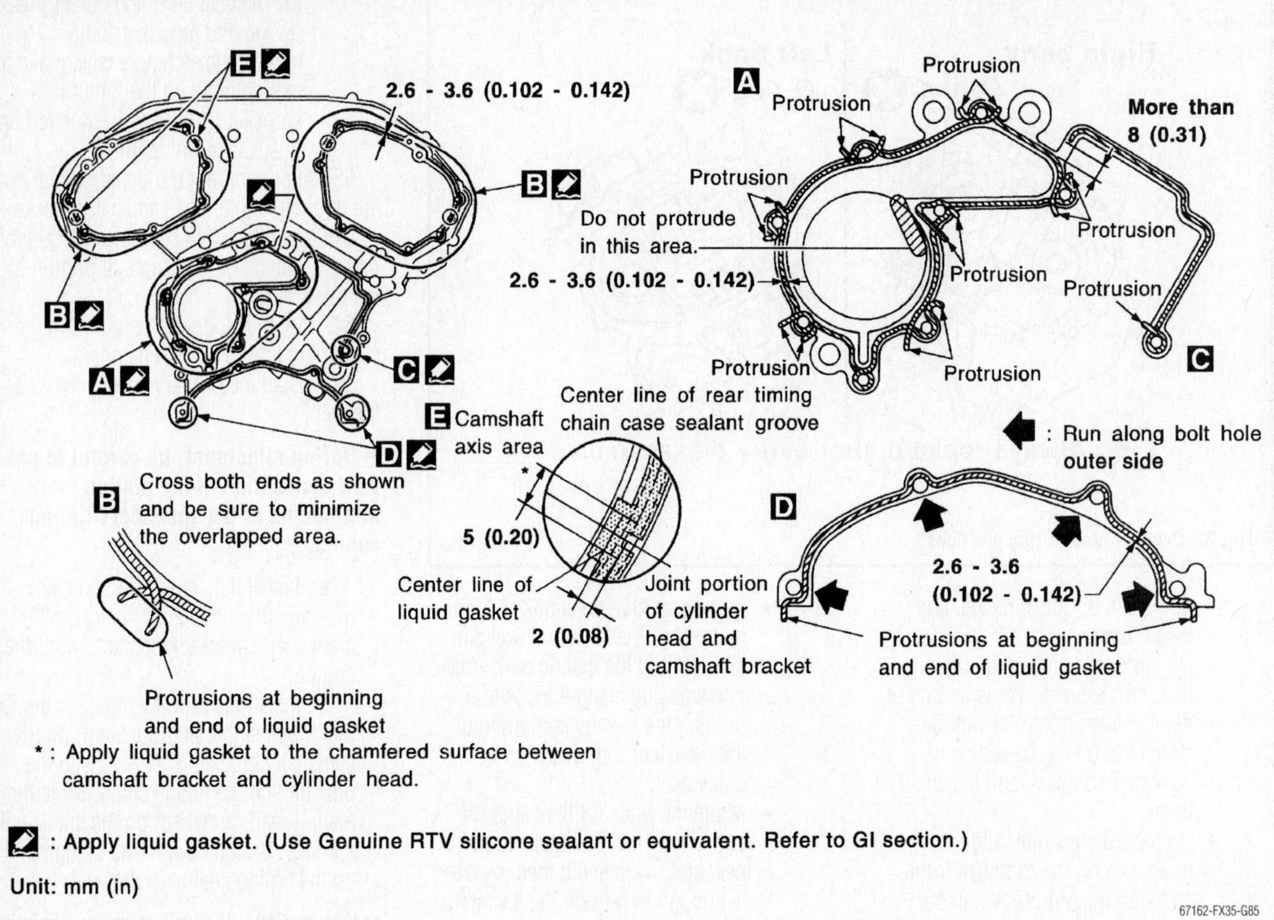

2.6 - 3.6 (0.102 - 0.142)

Protrusion
Protrusion
Protrusion

Do not protrude in this area.

2.6 - 3.6 (0.102 - 0.142)

Protrusion
Protrusion
Protrusion
Protrusion
Protrusion

More than 8 (0.31)

Protrusion
Protrusion

: Run along bolt hole outer side

E Camshaft axis area

Center line of rear timing chain case sealant groove

5 (0.20)

Center line of liquid gasket

2 (0.08)

Joint portion of cylinder head and camshaft bracket

2.6 - 3.6 (0.102 - 0.142)

Protrusions at beginning and end of liquid gasket

B Cross both ends as shown and be sure to minimize the overlapped area.

Protrusions at beginning and end of liquid gasket

* : Apply liquid gasket to the chamfered surface between camshaft bracket and cylinder head.

: Apply liquid gasket. (Use Genuine RTV silicone sealant or equivalent. Refer to GI section.)

Unit: mm (in)

67162-FX35-G85

Fig. 85 Liquid gasket positions for timing chain case installation

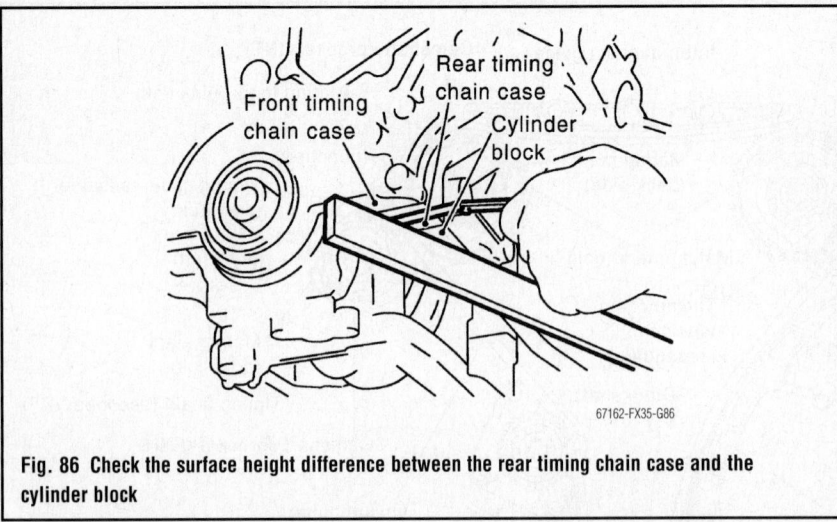

Front timing chain case
Rear timing chain case
Cylinder block

67162-FX35-G86

Fig. 86 Check the surface height difference between the rear timing chain case and the cylinder block

sprocket, gradually turn the camshaft using a wrench on the hexagonal portion of the camshaft to align it with the mating marks.

56. Install the internal chain guide and primary timing chain tensioner.

57. Install the slack guide.

➡ Do not over tighten the slack guide mounting bolts. It is normal for a gap to exist under the bolt seats when the mounting bolts are tightened to specification.

58. Remove all dirt and foreign materials completely from the back and mounting surfaces of the chain tensioner.

59. Install chain tensioner for slack guide. When installing the chain tensioner, push in the sleeve and keep it depressed with a stopper pin.

60. After chain tensioner installation, pull out the stopper pin by pressing in the slack guide.

61. Reconfirm that the mating marks on sprockets and timing chains have not slipped out of alignment.

62. Install new O-rings on the rear timing chain case.

63. Install the front oil seal on the front timing chain case, as follows:

   a. Apply new engine oil to the oil seal edges.

   b. Make sure the garter spring is in position and the seal lip is not inverted and so that each seal lip is oriented as shown in the figure.

   c. Using a suitable drift, press-fit the oil seal until it is flush with the front timing chain case end face.

64. Install the water pump cover and

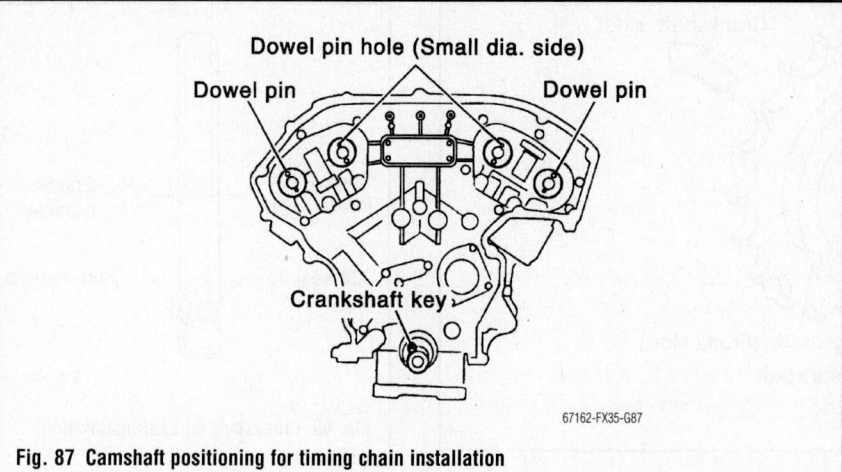

Fig. 87 Camshaft positioning for timing chain installation

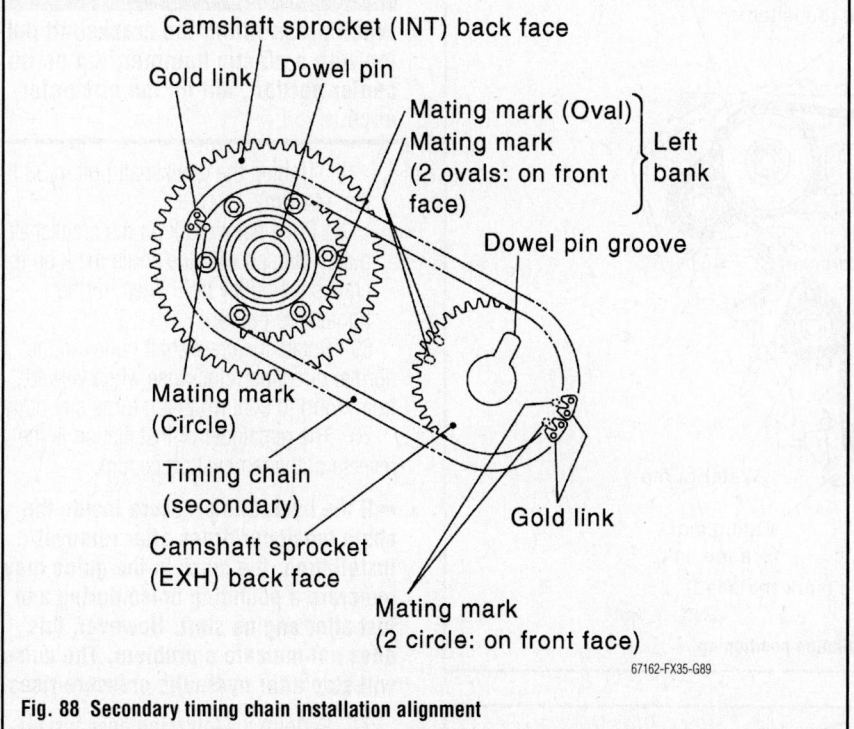

Fig. 88 Secondary timing chain installation alignment

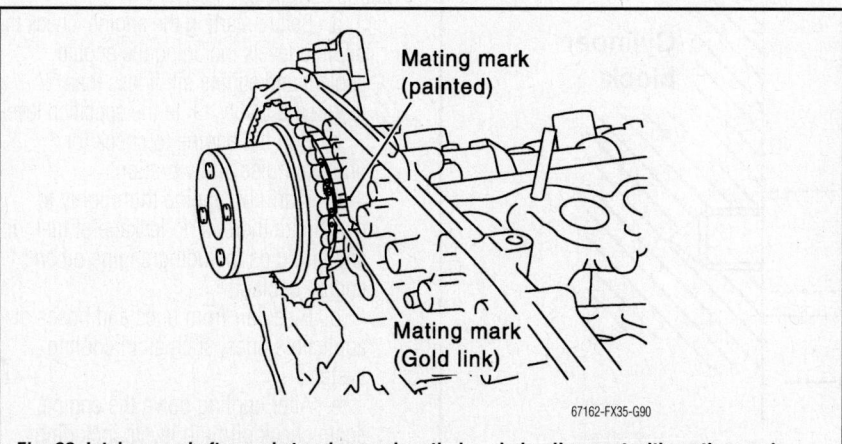

Fig. 89 Intake camshaft sprocket and secondary timing chain alignment with mating mark painted

chain tensioner cover to the front timing chain case. Apply liquid gasket to the front timing chain case front side as shown.

➡**Use Genuine RTV Silicone Sealant or equivalent.**

65. Install the front timing chain case as follows:

a. Apply liquid gasket to front timing chain case back side as shown.

➡**Use Genuine RTV Silicone Sealant or equivalent.**

b. Install the dowel pin on the rear timing chain case into the dowel pin hole on the front timing chain case.

c. Tighten the bolts to the specified torque in the order shown. Refer to the following for locating the bolts.

- 8mm bolts (positions 1 and 2): 21 ft. lbs. (28 Nm)
- 6mm bolts (except positions 1 and 2): 108 inch lbs. (13 Nm)

d. After tightening, retighten them again to the specified torque in the numerical order shown.

66. After installing the front timing chain case, check the surface height difference between the front timing chain case to rear timing chain case on the oil pan mounting surface. The allowable height difference is -0.006–0.006 in. (-0.14–0.14mm). If not within specification, repeat the installation procedure.

67. Install the right and left intake valve timing control covers, as follows:

a. Install the seal rings in the shaft grooves.

b. Apply a continuous bead of liquid gasket to the intake valve timing control covers.

➡**Use Genuine RTV Silicone Sealant or equivalent.**

c. Install the collared O-ring in the front cover engine oil hole (left and right sides).

d. Being careful not to move the seal ring from the installation groove, align the dowel pins on the chain case with the holes to install the intake valve timing control covers. Tighten the bolts in the order shown.

68. Install the crankshaft damper pulley, as follows:

a. Fix the crankshaft in position using ring gear stopper SST: KV10117700 (J44716) or equivalent.

b. Install the crankshaft pulley, taking care not to damage the front oil seal.

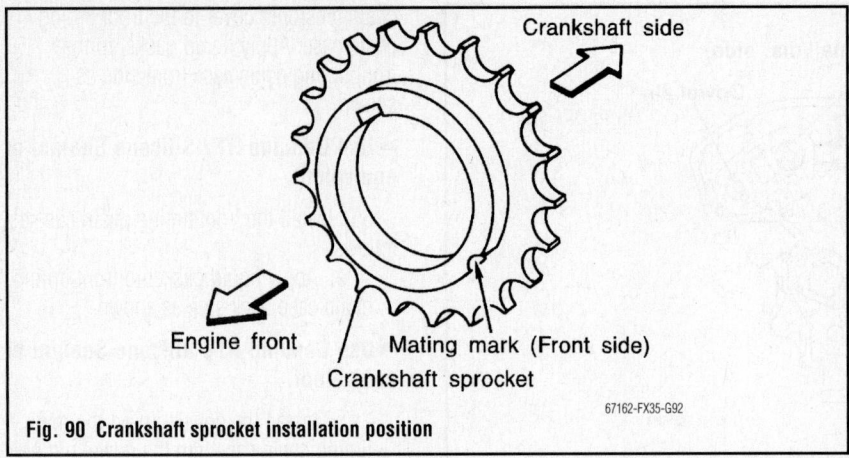

Fig. 90 Crankshaft sprocket installation position

67162-FX35-G92

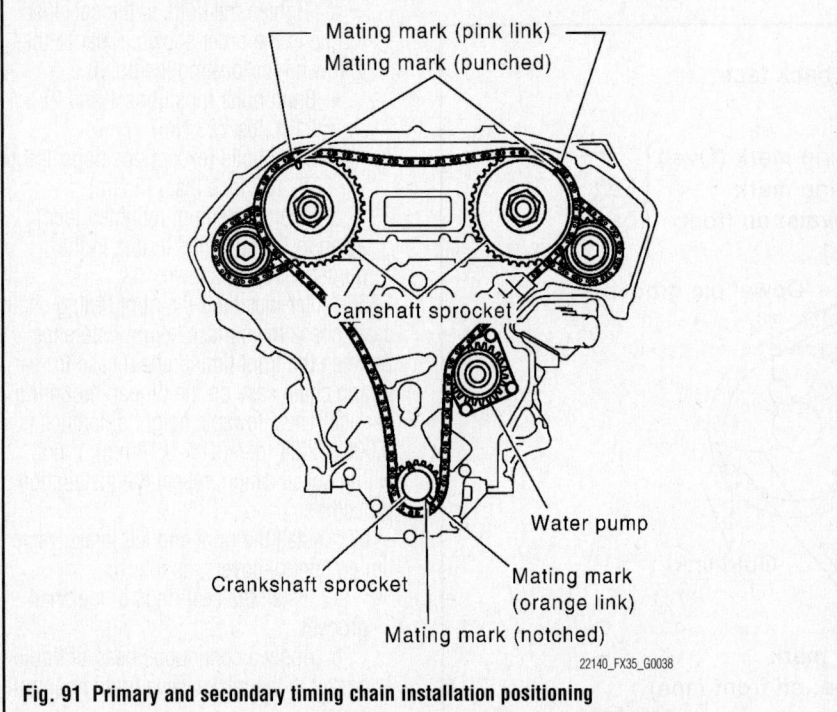

Fig. 91 Primary and secondary timing chain installation positioning

22140_FX35_G0038

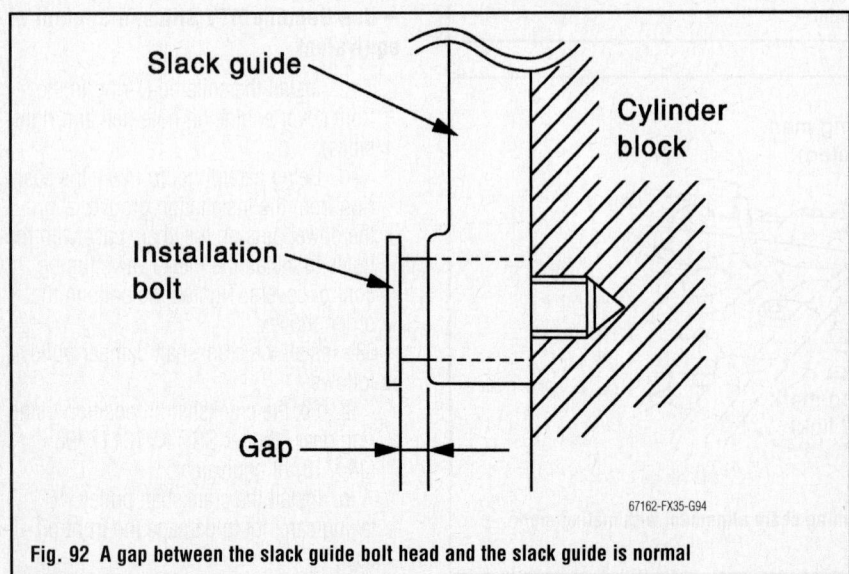

Fig. 92 A gap between the slack guide bolt head and the slack guide is normal

67162-FX35-G94

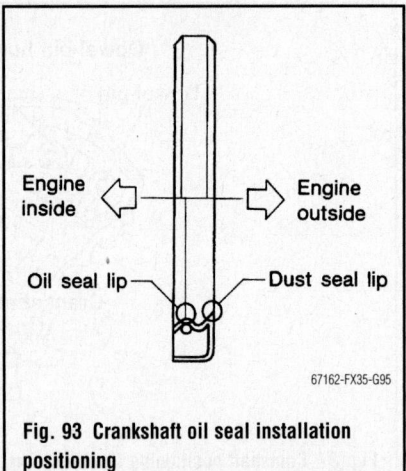

Fig. 93 Crankshaft oil seal installation positioning

67162-FX35-G95

❄❄ WARNING

When press-fitting the crankshaft pulley with a plastic hammer, tap on its center portion, not on the circumference.

c. Tighten the crankshaft bolt to 33 ft. lbs. (44 Nm).

d. Put a paint mark on the crankshaft pulley aligned with the angle mark on the crankshaft pulley bolt. Then, further tighten the bolt 90°.

69. Rotate the crankshaft pulley in the normal direction (clockwise when viewed from front) to confirm that it turns smoothly.

70. The remainder of installation is the reverse of the removal procedure.

➡**If the hydraulic pressure inside the chain tensioner drops after removal/installation, the slack in the guide may generate a pounding noise during and just after engine start. However, this does not indicate a problem. The noise will stop after hydraulic pressure rises.**

71. Perform the following once installation is complete:

a. Before starting the engine, check the oil/fluid levels including the engine coolant and engine oil. If less than required quantity, fill to the specified level.

b. Run the engine to check for unusual noise and vibration.

c. Warm up engine thoroughly to make sure there is no leakage of fuel, or any oil/fluids including engine oil and engine coolant.

d. Bleed air from lines and hoses of applicable lines, such as in cooling system.

e. After cooling down the engine, again check oil/fluid levels including engine oil and engine coolant. Refill to the specified level, if necessary.

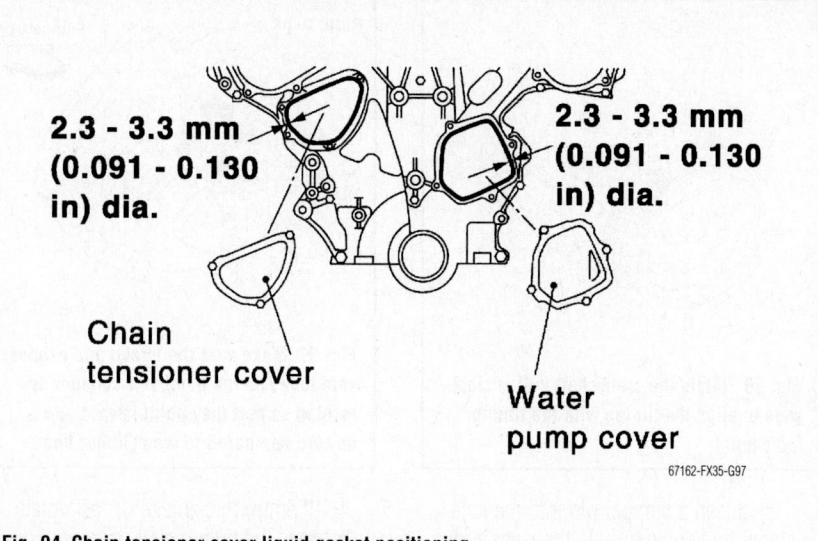

2.3 - 3.3 mm (0.091 - 0.130 in) dia.

2.3 - 3.3 mm (0.091 - 0.130 in) dia.

Chain tensioner cover

Water pump cover

67162-FX35-G97

Fig. 94 Chain tensioner cover liquid gasket positioning

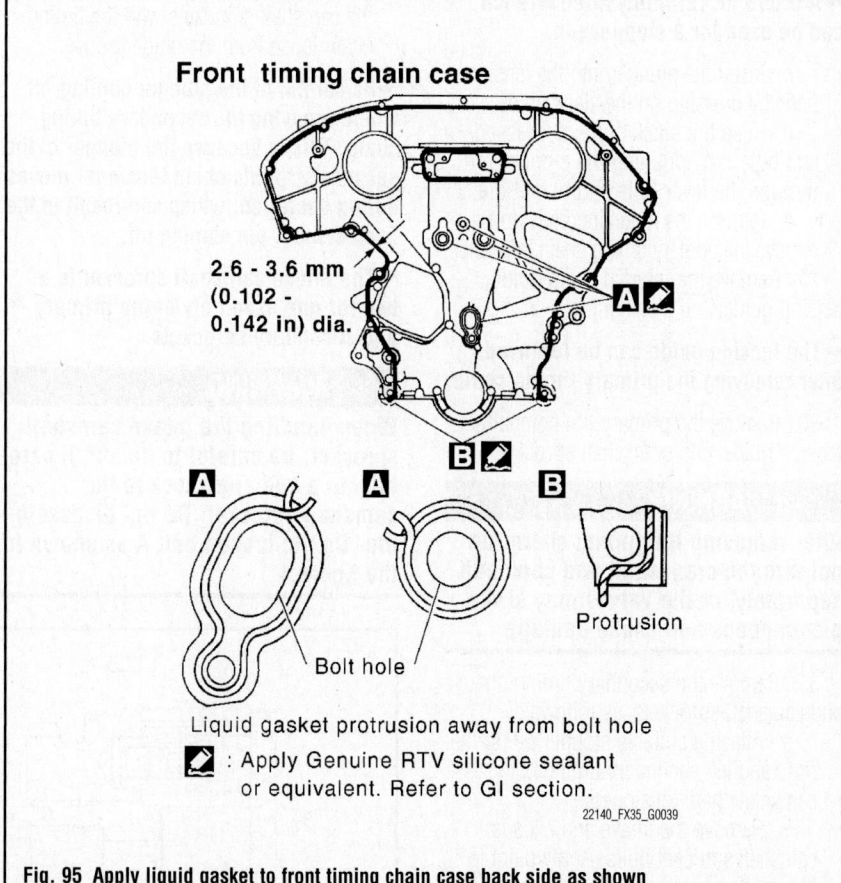

Front timing chain case

2.6 - 3.6 mm (0.102 - 0.142 in) dia.

A

A   A   B

B

Protrusion

Bolt hole

Liquid gasket protrusion away from bolt hole

🖊 : Apply Genuine RTV silicone sealant or equivalent. Refer to GI section.

22140_FX35_G0039

Fig. 95 Apply liquid gasket to front timing chain case back side as shown

Without Oil Pan Removal

See Figures 96 through 123.

1. Before servicing the vehicle, refer to the Precautions Section.

This section describes the removal/installation procedure of the front timing chain case and timing chain related parts without removing the upper oil pan.

2. Position the vehicle onto a lift or support it in a safe manner so that work can be performed on the underside of the vehicle.

3. Disconnect the negative battery terminal.

4. Remove the engine cover.

5. Remove the air cleaner case assembly.

6. Remove the front and rear engine undercover.

7. Drain the engine coolant from the radiator.

8. Drain the engine oil from oil pan.

9. Remove and label the engine wiring harnesses.

10. Remove the upper and lower intake manifold collectors.

11. Remove the power steering oil pump from the bracket with the piping connected, then temporarily secure it aside.

12. Remove the power steering oil pump bracket.

13. Remove the alternator.

14. Remove the water bypass hose, water hose clamp, idler pulley bracket, and accessory drive belt tensioner from the front timing chain case.

15. Remove the right and left intake valve timing control covers, by loosening the bolts in the reverse order as shown.

16. Use seal cutter SST: KV10111100 (J37228), or equivalent, to cut the liquid gasket for removal.

➡ **The shaft is internally joined to the intake camshaft sprocket center hole. During removal, keep the shaft horizontal until it is completely disconnected.**

17. Remove the collared O-ring from the front timing chain case left and right sides.

18. Remove the right and left rocker covers.

➡ **When the secondary timing chain is not removed/installed, the following step and associated sub-steps are not required.**

19. Position the engine with cylinder number 1 on Top Dead Center (TDC) of its compression stroke, as follows:

   a. Rotate the crankshaft pulley clockwise to align the timing mark (grooved line without color) with the timing indicator.

   b. Make sure the intake and exhaust cam lobes on the number 1 cylinder (engine front side of right bank) are located so that they point inward and upward compared to the cylinder head, see illustration.

   c. If the cam lobes are not positioned pointing inward and upward as illustrated, rotate the crankshaft 1 full revolution (360°) until they are.

➡ **When only the primary timing chain is removed, the rocker cover does not need to be removed. To confirm that the number 1 cylinder is at its compression TDC, remove the front timing chain case first. Then, check the mating marks on the camshaft sprockets.**

20. Remove the crankshaft pulley, as follows:

a. Remove the rear cover plate (2WD models) or the starter motor (AWD models) and install the ring gear stopper or equivalent.

b. Loosen the crankshaft pulley bolt until the bolt seating surface is 0.39 in. (10mm) from its original position.

➡ **Do not completely remove the crankshaft damper pulley bolt, since it will be used as a supporting point for a suitable puller.**

c. Place a suitable puller tab on the holes of the crankshaft damper pulley and pull the crankshaft damper pulley until it releases.

✳✳ WARNING

Do not position the suitable puller tab on the outer edges of the crankshaft pulley since this can damage the internal damper.

21. Remove the lower oil pan.

22. Loosen the 2 mounting bolts in the front of the upper oil pan in the reverse order shown.

23. Remove the front timing chain case, as follows:

a. Loosen the mounting bolts in the reverse order of the tightening sequence.

b. Insert a suitable tool into the notch at the top of the front timing chain case, as shown.

✳✳ WARNING

Do not use a screwdriver or similar tool, since it may damage the mating surfaces. After removal, handle it carefully so that it does not tilt or warp under a load.

c. Pry off the case.

24. Remove the O-rings from the rear timing chain case.

25. Remove the water pump cover and the chain tensioner cover from the front timing chain case.

26. Use a seal cutter to cut the liquid gasket.

27. Remove front oil seal from front timing chain case using a suitable pry tool.

✳✳ WARNING

Be careful not to damage the front timing chain case.

28. Remove the primary timing chain tensioner, as follows:

a. Pull the lever down and release the plunger stopper tab. The plunger stopper tab can be pushed up to release.

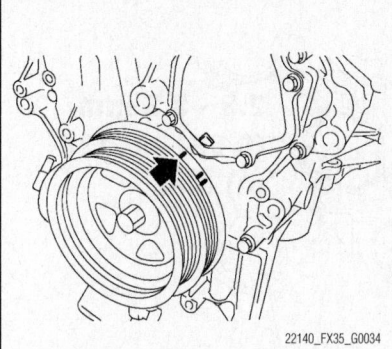

Fig. 96 Rotate the crankshaft pulley clockwise to align the timing with the timing indicator

b. Insert a stopper pin into the tensioner body hole to hold the lever and keep the tab released.

➡ **A 0.098 in. (2.5mm) Allen wrench can be used for a stopper pin.**

c. Insert the plunger into the tensioner body by pressing on the slack guide.

d. Keep the slack guide depressed and hold it by pushing the stopper pin through the lever hole and body hole.

e. Remove the mounting bolts and remove the primary timing chain tensioner.

29. Remove the internal chain guide, tension guide, and slack guide.

➡ **The tension guide can be removed after removing the primary timing chain.**

30. Remove the primary timing chain, tension guide, and crankshaft sprocket.

✳✳ WARNING

After removing the timing chain, do not turn the crankshaft and camshaft separately, or the valves may strike piston heads and cause damage.

31. Remove the secondary timing chain and camshaft sprockets, as follows:

a. Attach a suitable stopper pin to the right and left secondary timing chain camshaft chain tensioners.

b. Remove the intake and exhaust camshaft sprocket bolts. Apply paint to the timing chain and camshaft sprockets for alignment during installation. Secure the hexagonal portion of the camshaft using an open-end wrench to hold the camshaft steady while loosening the mounting bolts.

c. Remove the secondary timing chain together with camshaft sprockets. Turn the camshaft slightly to create slack in the timing chain on the secondary timing chain tensioner side. Insert a 0.020

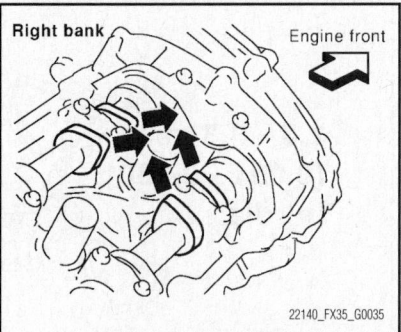

Fig. 97 Make sure the intake and exhaust cam lobes on the number 1 cylinder are located so that they point inward and upward compared to the cylinder head

in. (0.5mm) thick metal or resin plate between the timing chain and timing chain tensioner plunger guide. Remove the secondary timing chain together with the camshaft sprockets with the timing chain loose from the guide groove.

➡ **Be careful of the plunger coming-off when removing the secondary timing chain. This is because the plunger of the secondary timing chain tensioner moves during operation, which can result in the fixed stopper pin coming off.**

➡ **The intake camshaft sprocket is a two-for-one assembly of the primary and secondary sprockets.**

✳✳ WARNING

When handling the intake camshaft sprocket, be careful to handle it carefully to avoid any shock to the camshaft sprocket. Do not disassemble. Do not loosen bolt A as shown in the figure.

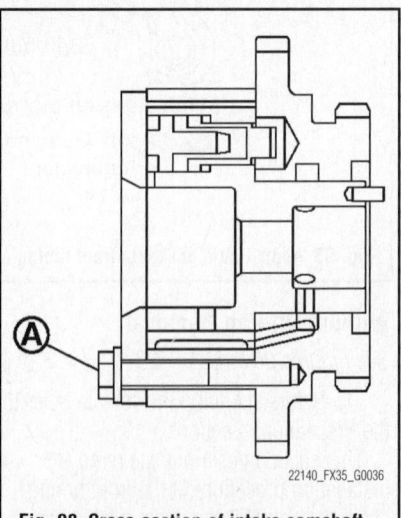

Fig. 98 Cross-section of intake camshaft sprocket—Do NOT loosen bolt A

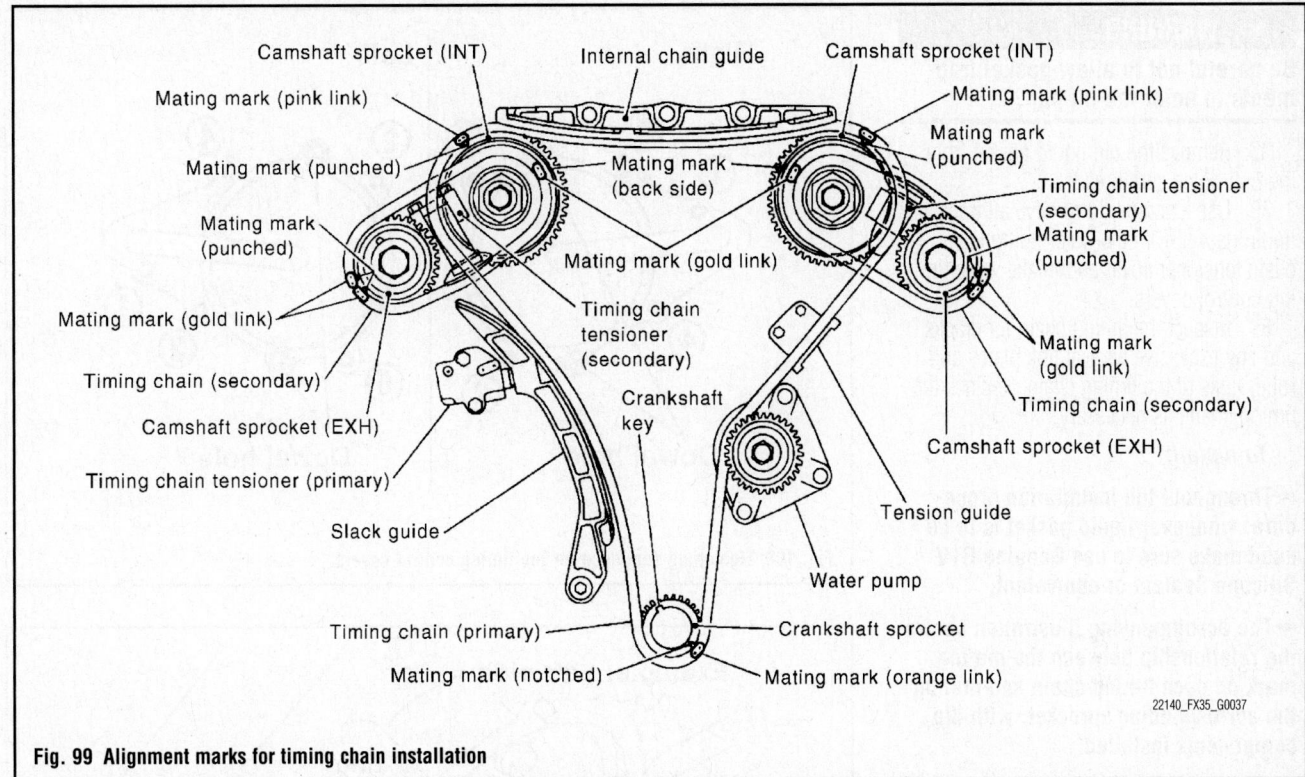

Fig. 99 Alignment marks for timing chain installation

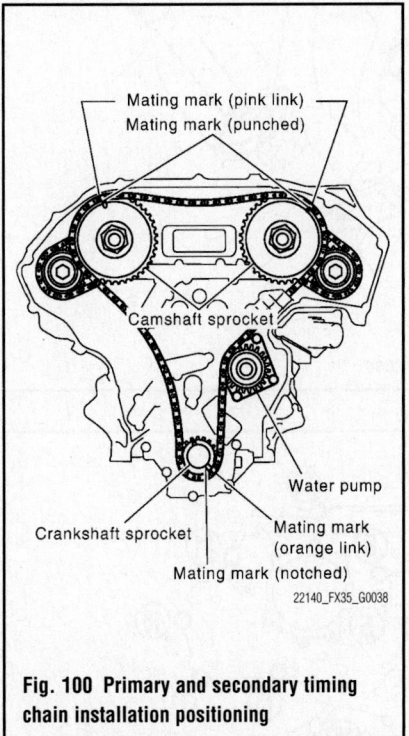

Fig. 100 Primary and secondary timing chain installation positioning

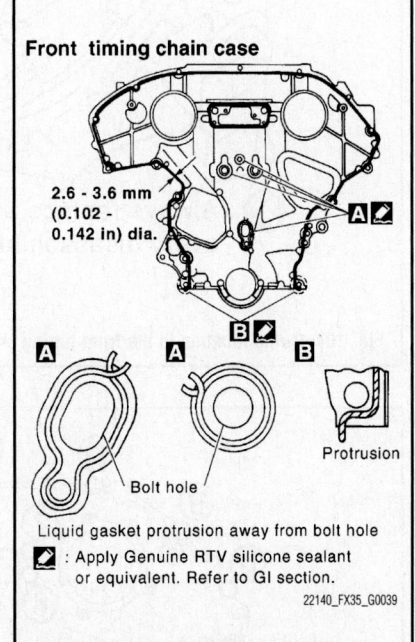

Fig. 101 Apply liquid gasket to front timing chain case back side as shown

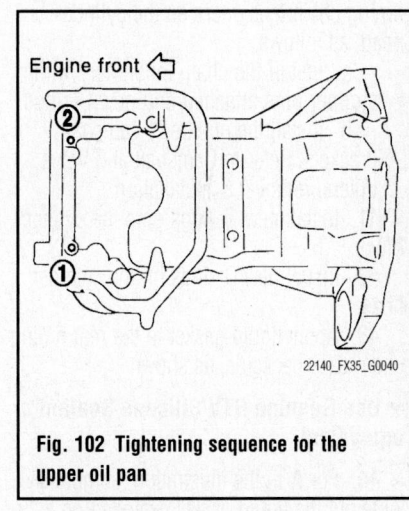

Fig. 102 Tightening sequence for the upper oil pan

32. Remove the rear timing chain case, as follows:

    a. Loosen and remove the mounting bolts in the reverse order of the tightening sequence.

    b. Cut the sealant using a seal cutter and remove the rear timing chain case.

※※ WARNING

Do not remove plate metal cover of engine oil passage. After removing the chain case, do not apply any load on the case which could cause warping.

33. Remove the O-rings from the cylinder head.

34. Remove the O-rings from the cylinder block.

35. If necessary, remove the secondary timing chain tensioners from the cylinder head, as follows:

    a. Remove the number 1 camshaft brackets.

    b. Remove the secondary timing chain tensioners with the stopper pins attached.

36. Use a scraper to remove all traces of liquid gasket from the front and rear timing chain cases and opposite mating surfaces.

※※ WARNING

Be careful not to allow gasket fragments to enter the oil pan.

37. Remove the old liquid gasket from the bolt holes and threads.

38. Use a scraper to remove all traces of liquid gasket from the water pump cover, chain tensioner cover and intake valve timing control covers.

39. Inspect the timing chain for cracks and any excessive wear at link plates and roller links of the timing chain. Replace the timing chain, as necessary.

To install:

➡Throughout the installation procedure, whenever liquid gasket is to be used make sure to use Genuine RTV Silicone Sealant or equivalent.

➡The accompanying illustration shows the relationship between the mating mark on each timing chain and that on the corresponding sprocket, with the components installed.

40. If removed, install the secondary timing chain tensioners on the cylinder head, as follows:

   a. Install the chain tensioners with stopper pins attached and new O-rings.

   b. Install the number 1 camshaft brackets. Refer to Camshaft and Valve Lifters, removal & installation.

41. Install new O-rings onto the cylinder block.

42. Install new O-rings on the cylinder head.

43. Apply liquid gasket to the rear timing chain case backside, as shown.

➡Use Genuine RTV Silicone Sealant or equivalent.

44. For **A** in the illustration, completely wipe out the liquid gasket extended on a portion touching at engine coolant.

45. Apply liquid gasket on the installation position of the water pump and cylinder head completely.

46. Align the rear timing chain case and water pump assembly with right and left dowel pins on cylinder block and install the case.

➡Make sure the O-rings stay in place during installation on the cylinder block and cylinder head.

47. Tighten the mounting bolts in the sequence shown. After all bolts are temporarily tightened, retighten them to 108 inch lbs. (13 Nm) in the tightening sequence. There are 2 bolt lengths used for the timing chain case:

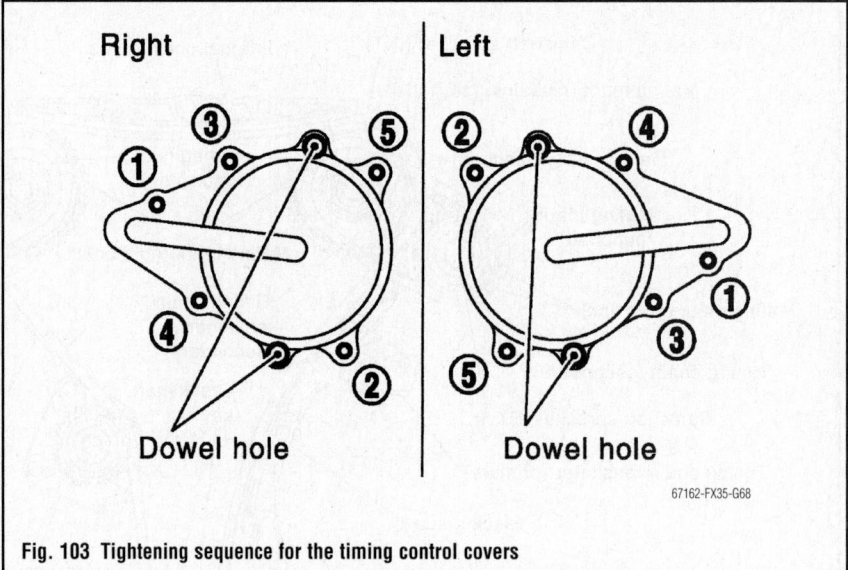

Fig. 103 Tightening sequence for the timing control covers

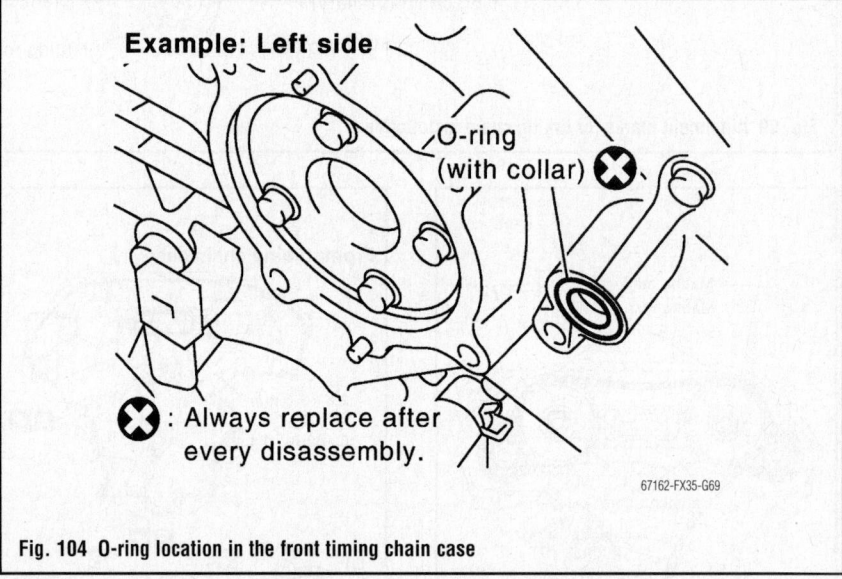

Fig. 104 O-ring location in the front timing chain case

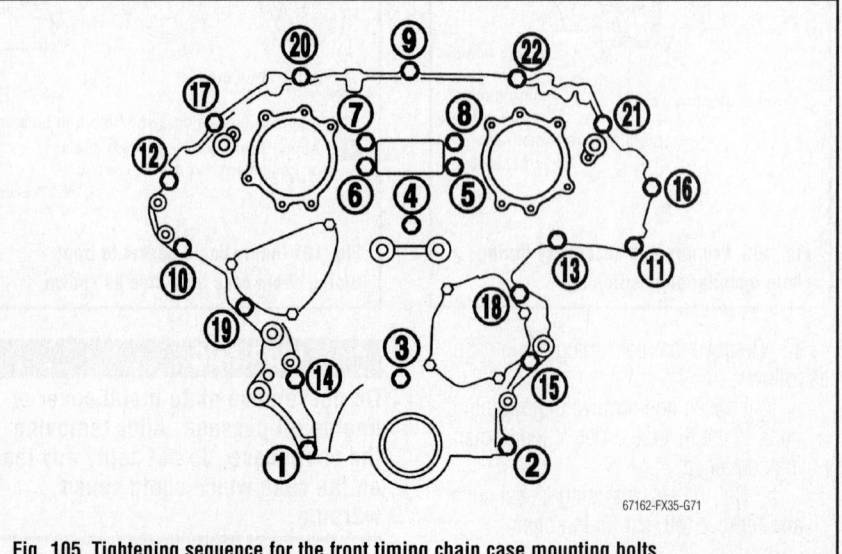

Fig. 105 Tightening sequence for the front timing chain case mounting bolts

a. 20mm length: Bolt positions 1, 2, 3, 6, 7, 8, 9, and 10.

b. 16mm length: All except bolt positions 1, 2, 3, 6, 7, 8, 9, and 10.

➡ **If RTV Silicone Sealant protrudes beyond the sealing surfaces, wipe it off immediately.**

48. After installing the rear timing chain case, check the surface height difference between the rear timing chain case to the cylinder block at the oil pan mounting surface. If the difference is not within -0.009–0.006 in. (-0.24–0.14mm), repeat the installation procedure.

49. Position the crankshaft so the number 1 piston is set at TDC on the compression stroke. Make sure that the dowel pin hole, dowel pin, and crankshaft key are located as shown.

➡ **Though the camshaft does not stop at the position as shown in the figure, for the placement of the cam nose, it is generally accepted the camshaft is placed in the same direction of the figure and as follows:**

- Camshaft dowel pin hole (intake side): At the cylinder head upper face side in each bank
- Camshaft dowel pin (exhaust side): At the cylinder head upper face side in each bank
- Crankshaft key: At the cylinder head side of the right bank

✳✳ WARNING

The hole on the small diameter side must be used for the intake side dowel pin hole.

50. Install the secondary timing chains and camshaft sprockets, as follows:

✳✳ WARNING

The matching marks between the timing chain and sprockets slip easily. Confirm all matching mark positions repeatedly during the installation process.

a. Push the plunger of the secondary chain tensioner and keep it pressed in with a stopper pin.

b. Install the secondary timing chains and camshaft sprockets. Align the mating marks on the secondary timing chain (gold link) with the ones on the intake and exhaust camshaft sprockets (stamped), and install them.

51. Ensure that the sprockets are properly positioned by heeding the following items:

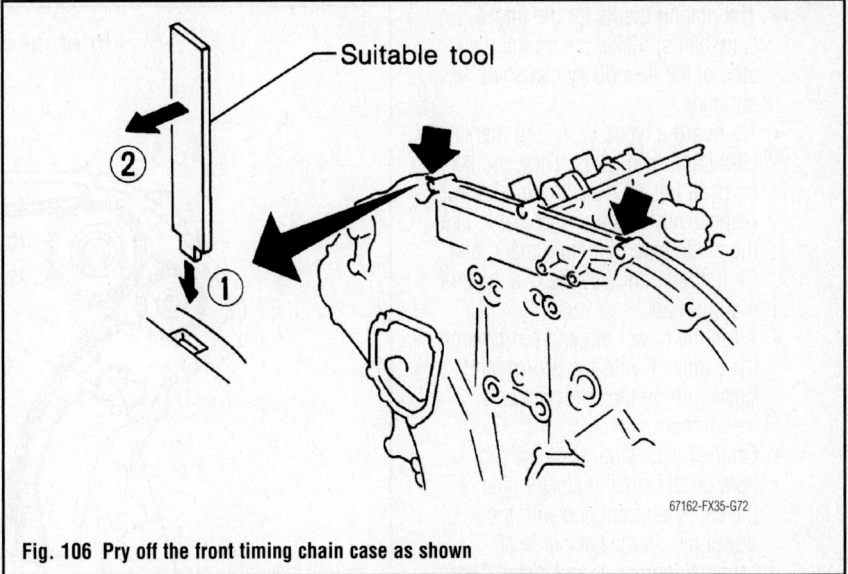
67162-FX35-G72
Fig. 106 Pry off the front timing chain case as shown

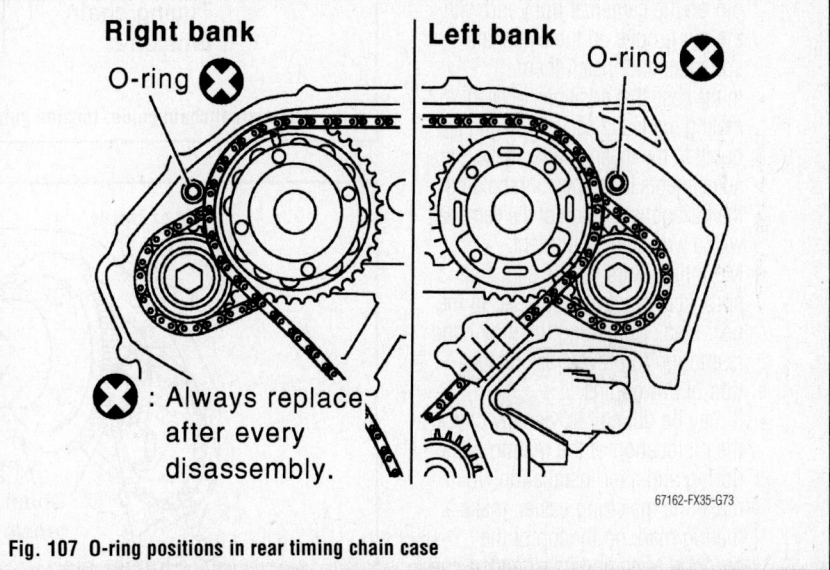
67162-FX35-G73
Fig. 107 O-ring positions in rear timing chain case

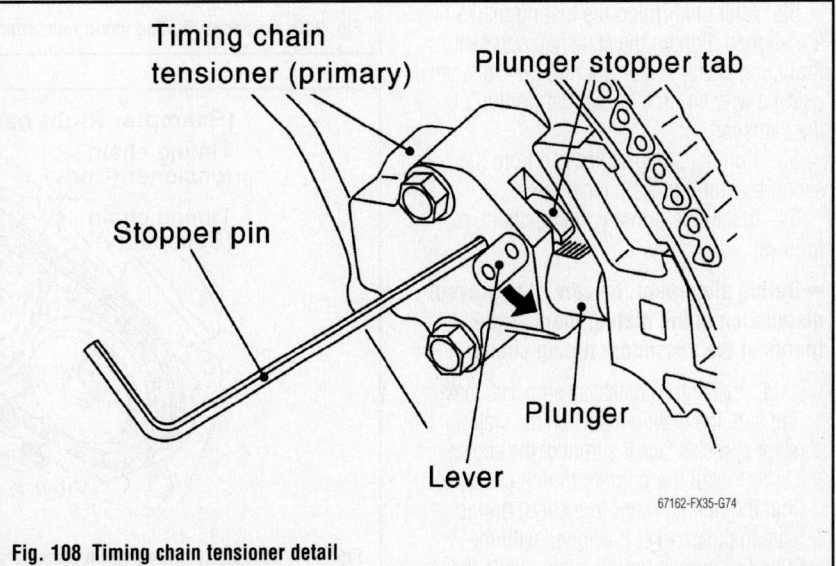
67162-FX35-G74
Fig. 108 Timing chain tensioner detail

- The mating marks for the intake camshaft sprocket are on the back side of the secondary camshaft sprocket.
- There are 2 types of mating marks, circle and oval types. They should be used for the right and left banks, respectively. For the right bank, use the circle type of mating mark, and for the left bank use the oval type of mating mark.
- Align the dowel pin and pin hole on the camshaft with the groove and dowel pin on the sprocket, and install them.
- On the intake side, align the pin hole on the small diameter side of the camshaft front end with the dowel pin on the back side of camshaft sprocket, and install them.
- On the exhaust side, align the dowel pin on the camshaft front end with the pin groove on the camshaft sprocket, and install them.
- In the case that positions of each mating mark and each dowel pin are not fit to the mating parts, make fine adjustments to the position holding the hexagonal portion of the camshaft with a wrench or equivalent.
- Mounting bolts for the camshaft sprockets must be tightened in the next step. Tightening them by hand is enough to prevent the dislocation of dowel pins.
- It may be difficult to visually check the dislocation of the mating marks during and after installation. To make the matching easier, make a mating mark on the top of the sprocket teeth and its extended line in advance with paint.

52. After confirming the mating marks are aligned, tighten the camshaft sprocket mounting bolts, while securing the camshaft using a wrench on a hexagonal portion of the camshaft.

53. Pull the stopper pins out from the secondary timing chain tensioners.

54. Install the primary timing chain, as follows:

➡**During alignment, be careful to prevent dislocation of the mating mark alignments of the secondary timing chains.**

a. Install the crankshaft sprocket, making sure the mating marks on the crankshaft sprocket face the front of the engine.

b. Install the primary timing chain, so that the mating mark (punched) on the camshaft sprocket is aligned with the pink link on the timing chain, while the

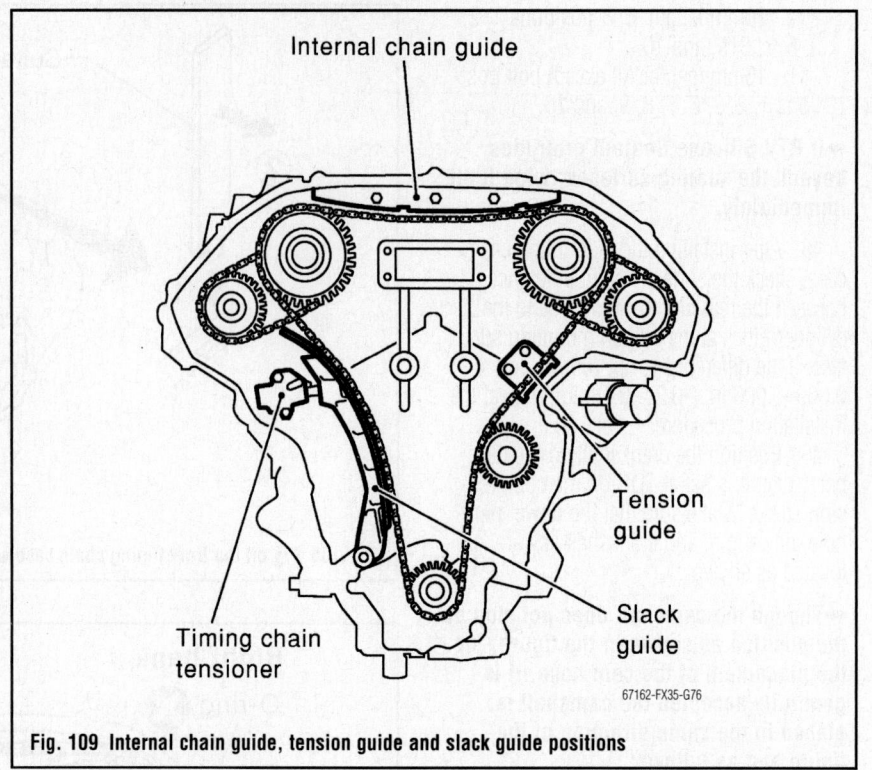
Fig. 109 Internal chain guide, tension guide and slack guide positions

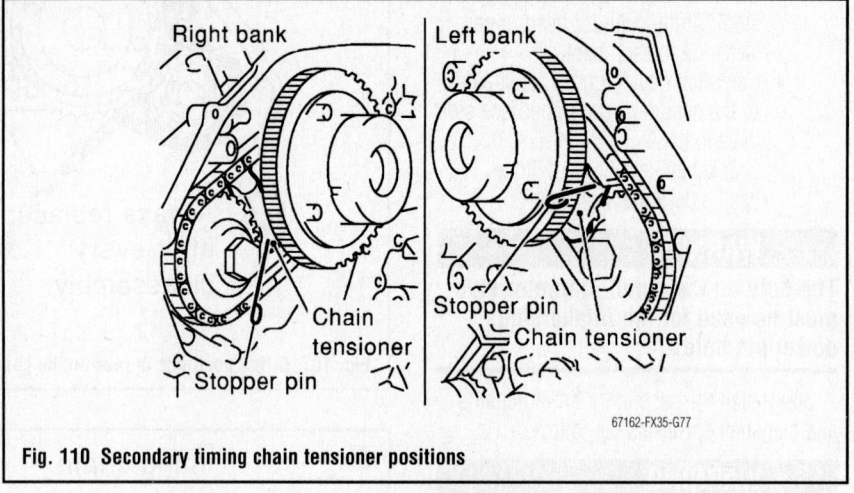
Fig. 110 Secondary timing chain tensioner positions

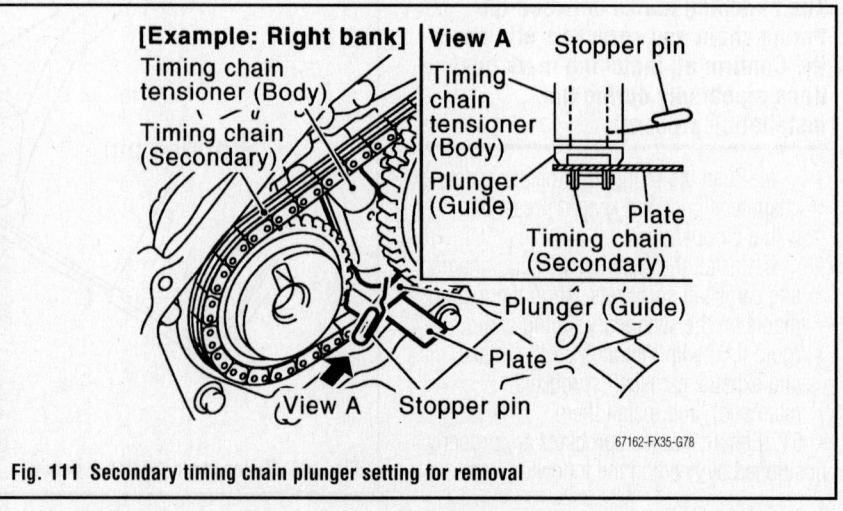
Fig. 111 Secondary timing chain plunger setting for removal

mating mark (notched) on the crankshaft sprocket is aligned with the orange one on the timing chain, as shown.

➡**If it is difficult to align mating marks of the primary timing chain with each sprocket, gradually turn the camshaft using a wrench on the hexagonal portion of the camshaft to align it with the mating marks.**

55. Install the internal chain guide and primary timing chain tensioner.

56. Install the slack guide.

➡**Do not over tighten the slack guide mounting bolts. It is normal for a gap to exist under the bolt seats when the mounting bolts are tightened to specification.**

57. Remove all dirt and foreign materials completely from the back and mounting surfaces of the chain tensioner.

58. Install chain tensioner for slack guide. When installing the chain tensioner, push in the sleeve and keep it depressed with a stopper pin.

59. After chain tensioner installation, pull out the stopper pin by pressing in the slack guide.

60. Reconfirm that the mating marks on sprockets and timing chains have not slipped out of alignment.

61. Install new O-rings on the rear timing chain case.

62. Hammer the dowel pins (right and left) into the front timing chain case up to a point close to the taper in order to shorten the protrusion length.

63. Install the front oil seal on the front timing chain case, as follows:

  a. Apply new engine oil to the oil seal edges.

  b. Make sure the garter spring is in position and the seal lip is not inverted and so that each seal lip is oriented as shown in the figure.

  c. Using a suitable drift, press-fit the oil seal until it is flush with the front timing chain case end face.

64. Install the water pump cover and chain tensioner cover to the front timing chain case. Apply liquid gasket to the front timing chain case front side as shown.

➡**Use Genuine RTV Silicone Sealant or equivalent.**

65. Install the front timing chain case, as follows:

  a. Apply liquid gasket to the front timing chain case back side as shown.

  b. Apply liquid gasket to the oil pan gasket as shown.

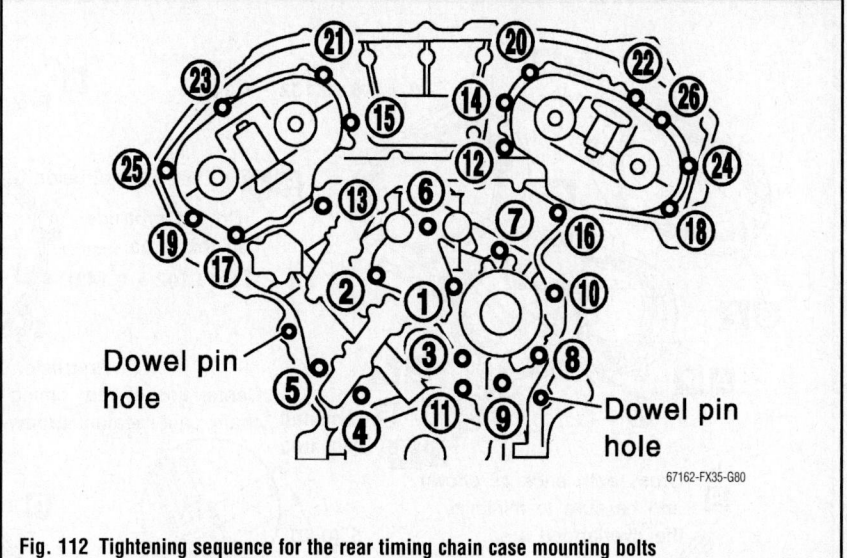

Fig. 112 Tightening sequence for the rear timing chain case mounting bolts

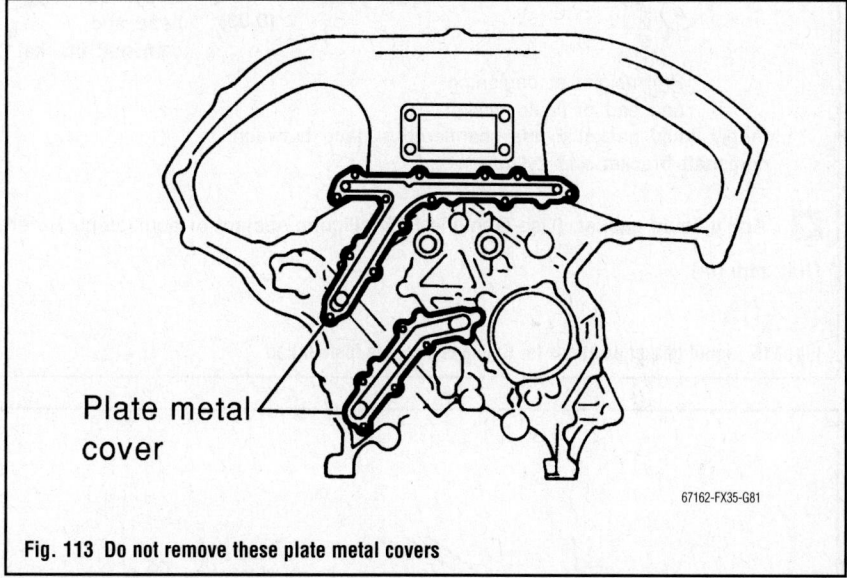

Fig. 113 Do not remove these plate metal covers

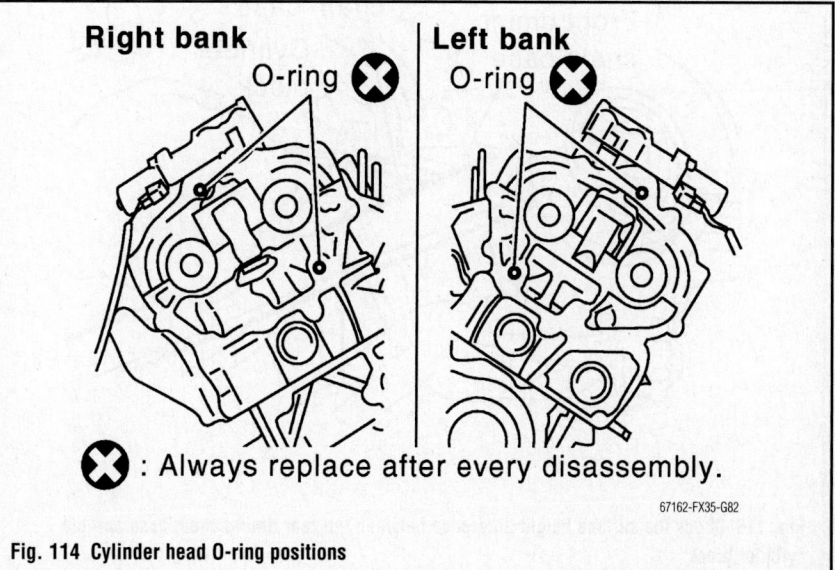

Fig. 114 Cylinder head O-ring positions

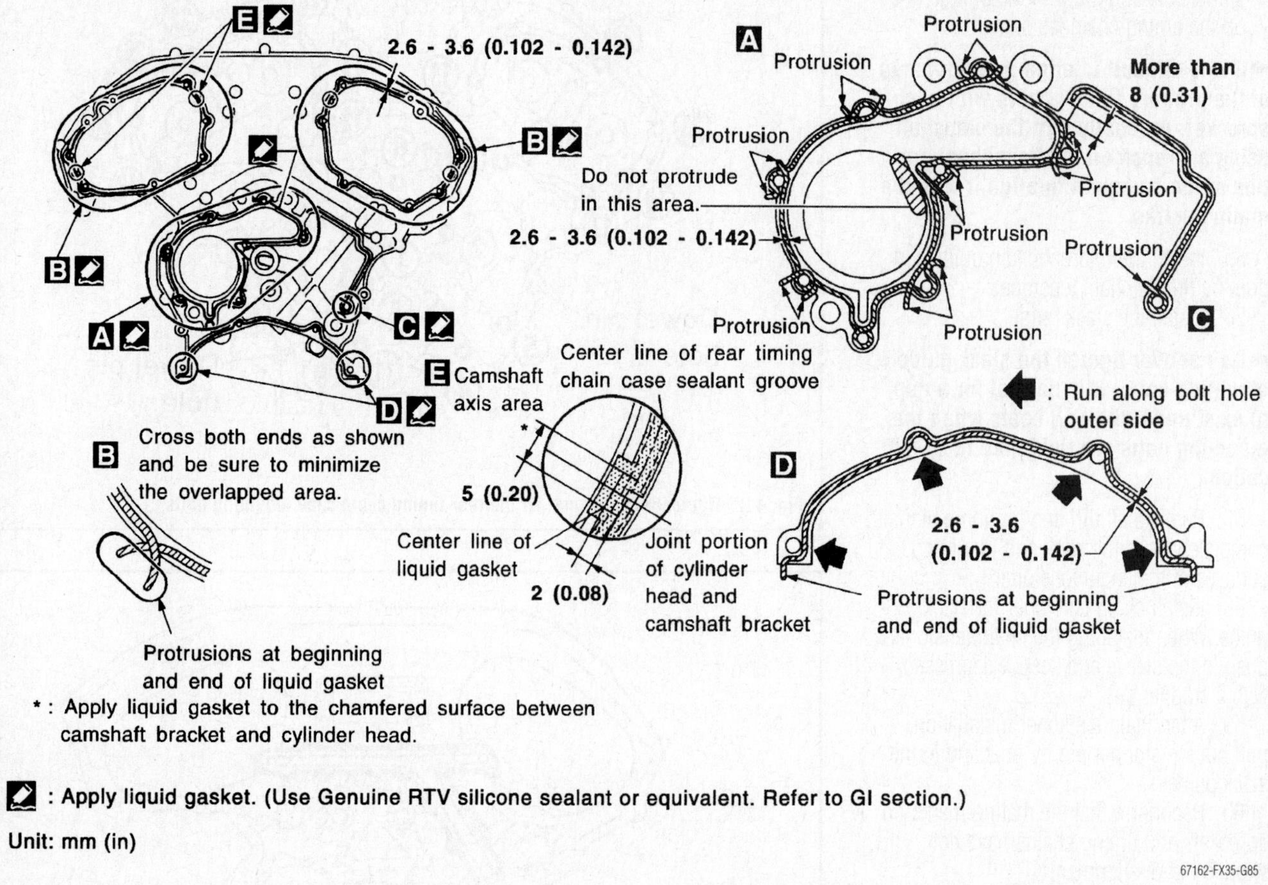

2.6 - 3.6 (0.102 - 0.142)

A

Protrusion
Protrusion

More than
8 (0.31)

Protrusion

B

Do not protrude
in this area.

2.6 - 3.6 (0.102 - 0.142)

Protrusion

Protrusion

Protrusion

C

E

Protrusion

Protrusion

B

A

C

D

Center line of rear timing
chain case sealant groove

E Camshaft
axis area

5 (0.20)

Center line of
liquid gasket

2 (0.08)

Joint portion
of cylinder
head and
camshaft bracket

◄ : Run along bolt hole
outer side

D

2.6 - 3.6
(0.102 - 0.142)

Protrusions at beginning
and end of liquid gasket

B Cross both ends as shown
and be sure to minimize
the overlapped area.

Protrusions at beginning
and end of liquid gasket

* : Apply liquid gasket to the chamfered surface between
camshaft bracket and cylinder head.

✎ : Apply liquid gasket. (Use Genuine RTV silicone sealant or equivalent. Refer to GI section.)

Unit: mm (in)

67162-FX35-G85

Fig. 115 Liquid gasket positions for timing chain case installation

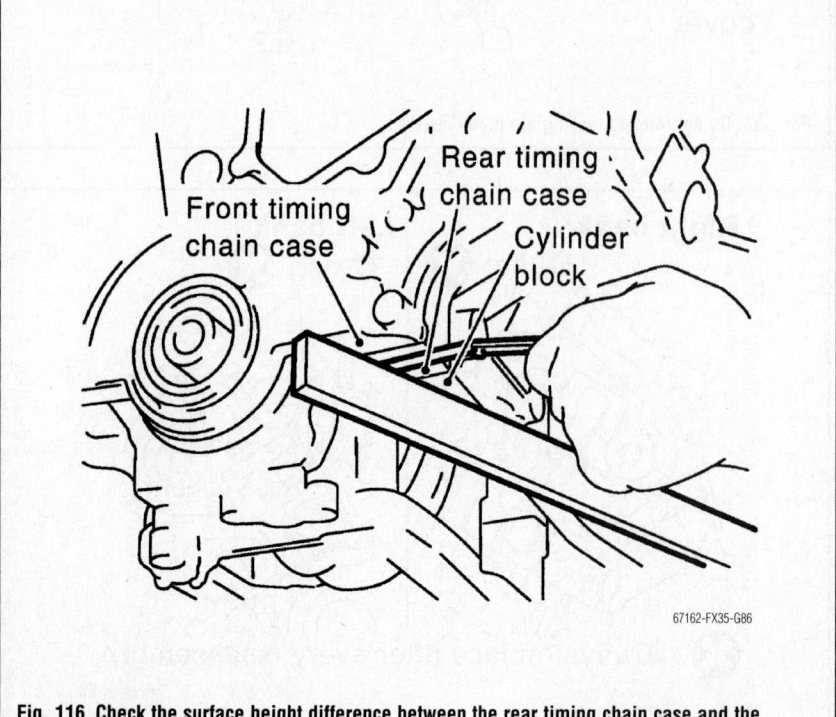

67162-FX35-G86

Fig. 116 Check the surface height difference between the rear timing chain case and the cylinder block

c. Install the new oil pan gasket, and heed the following:
- Align the notch of the front timing chain case with the protrusion of the oil pan gasket
- Apply liquid gasket to the top surface of the upper oil pan as shown

➡**Be careful that oil pan gasket is in place.**

d. Assemble the front timing chain case by fitting the lower end of the front timing chain case tightly onto the top face of the upper oil pan. From the fitting point, make the entire front timing chain case contact the rear timing chain case completely. Then, while pressing the front timing chain case from its front and top as shown in figure, install the bolts and temporarily tighten them by hand. Hammer the dowel pin until the outer end becomes flush with the surface.

e. After hand tightening the mounting bolts, retighten them to specified torque in numerical order shown. Refer to the following for locating the bolts.

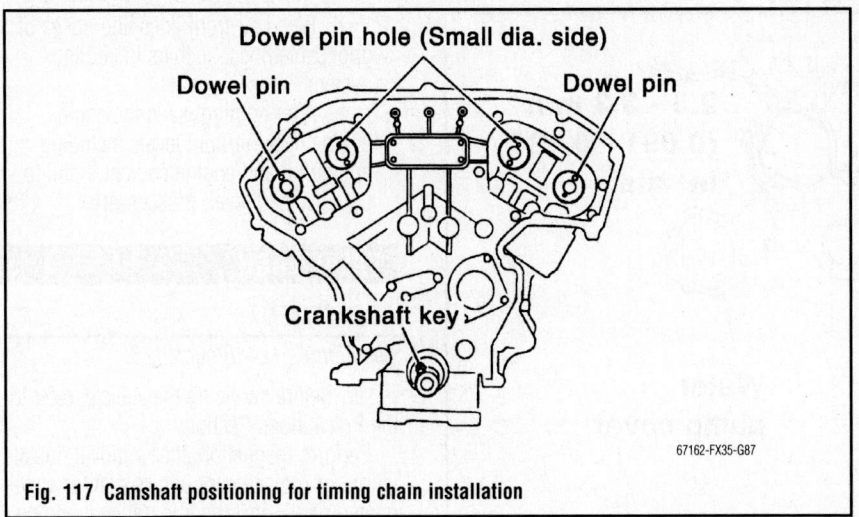

Fig. 117 Camshaft positioning for timing chain installation

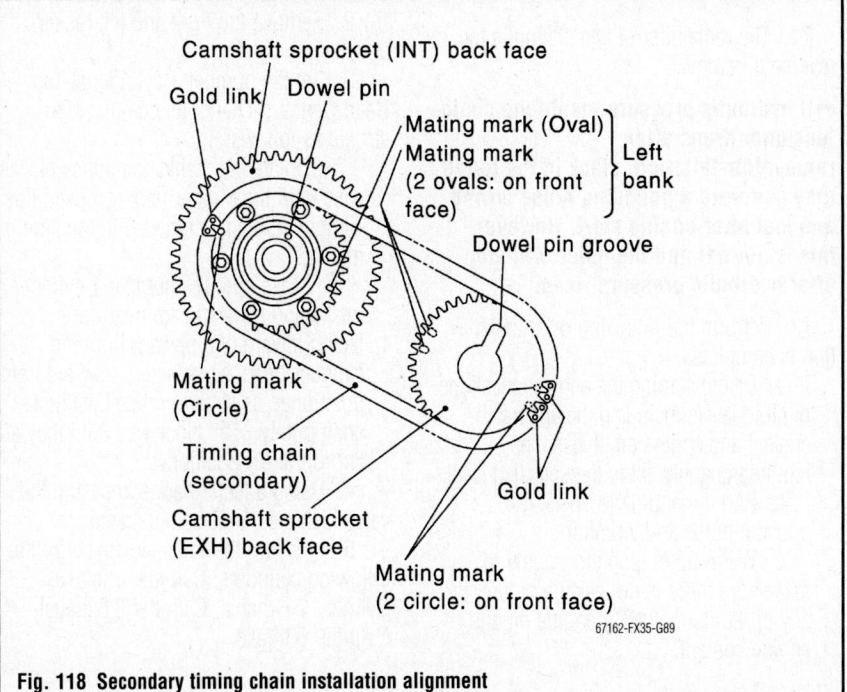

Fig. 118 Secondary timing chain installation alignment

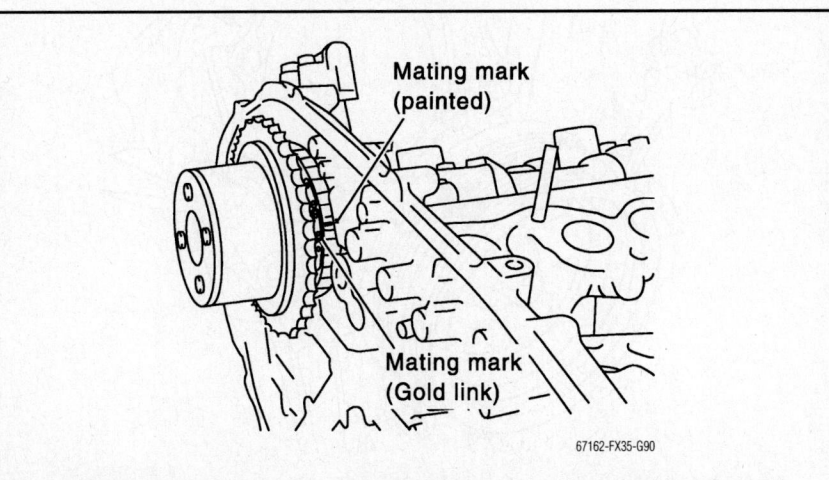

Fig. 119 Intake camshaft sprocket and secondary timing chain alignment with mating mark painted

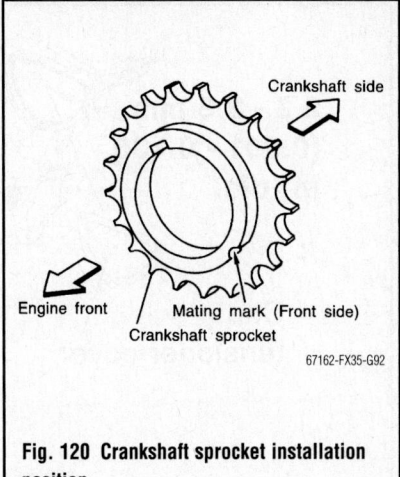

Fig. 120 Crankshaft sprocket installation position

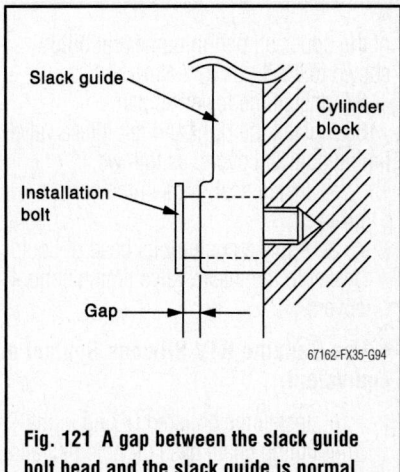

Fig. 121 A gap between the slack guide bolt head and the slack guide is normal

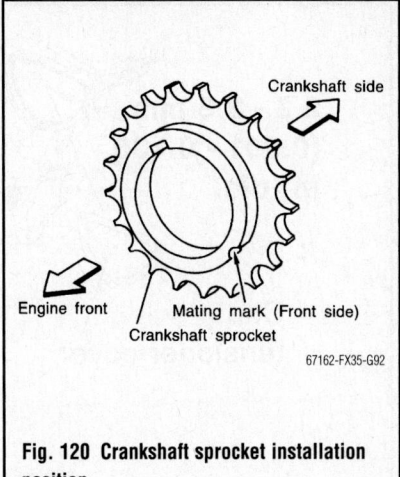

Fig. 122 Crankshaft oil seal installation positioning

- 8mm bolts (positions 1 and 2): 21 ft. lbs. (28 Nm)
- 6mm bolts (except positions 1 and 2): 108 inch lbs. (13 Nm)

f. After tightening, retighten them again to the specified torque in the numerical order shown.

66. Install 2 mounting bolts in the front

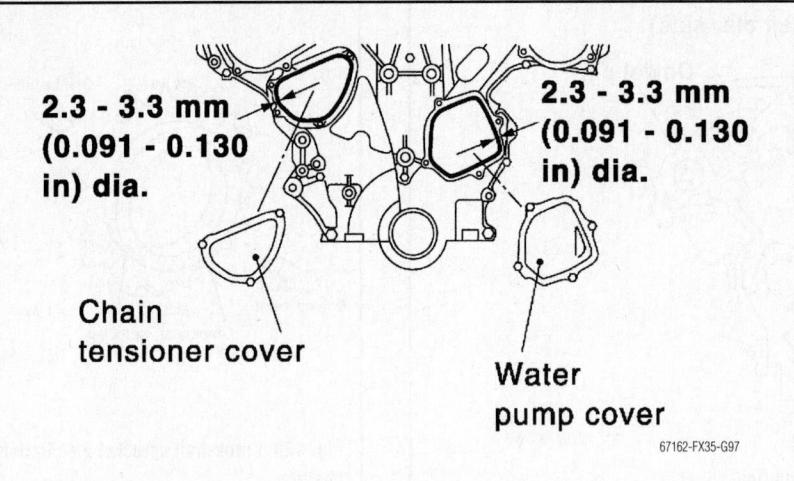

2.3 - 3.3 mm (0.091 - 0.130 in) dia.

2.3 - 3.3 mm (0.091 - 0.130 in) dia.

Chain tensioner cover

Water pump cover

67162-FX35-G97

Fig. 123 Chain tensioner cover liquid gasket positioning

of the upper oil pan in numerical order shown to 13 ft. lbs. (17 Nm).

67. Install the lower oil pan.

68. Install the right and left intake valve timing control covers, as follows:

a. Install seal rings in the shaft grooves.

b. Apply a continuous bead of liquid gasket to the intake valve timing control covers.

➡ **Use Genuine RTV Silicone Sealant or equivalent.**

c. Install the collared O-ring in the front timing chain case oil hole (left and right sides).

d. Being careful not to move the seal ring from the installation groove, align the dowel pins on the chain case with the holes in the intake valve timing control covers.

e. Tighten the bolts in the numerical order shown to 96 inch lbs. (11 Nm).

69. Install the crankshaft damper pulley, as follows:

a. Fix the crankshaft in position using ring gear stopper SST: KV10117700 (J44716) or equivalent.

b. Install the crankshaft pulley, taking care not to damage the front oil seal. When press-fitting crankshaft pulley with a plastic hammer, tap on its center portion (not the circumference).

c. Tighten the crankshaft pulley bolt to 33 ft. lbs. (44 Nm).

d. Put a paint mark on the crankshaft pulley aligned with the angle mark on the crankshaft pulley bolt. Then, further tighten the bolt by 90°.

70. Rotate the crankshaft pulley in the normal direction (clockwise when viewed from the front of the engine) to confirm it turns smoothly.

71. The remainder of installation is the reverse of removal.

➡ **If hydraulic pressure inside the chain tensioner drops after removal/installation, slack in the guide may generate a pounding noise during and just after engine start. However, this is normal and the noise will stop after hydraulic pressure rises.**

72. Perform the following once installation is complete:

a. Before starting the engine, check the oil/fluid levels including the engine coolant and engine oil. If less than required quantity, fill to the specified level.

b. Run the engine to check for unusual noise and vibration.

c. Warm up engine thoroughly to make sure there is no leakage of fuel, or any oil/fluids including engine oil and engine coolant.

d. Bleed air from lines and hoses of applicable lines, such as in cooling system.

e. After cooling down the engine, again check oil/fluid levels including engine oil and engine coolant. Refill to the specified level, if necessary.

VALVE LASH

ADJUSTMENT

See Figures 124 through 128.

1. Before servicing the vehicle, refer to the Precautions Section.

Perform inspection after removal, installation, or replacement of camshaft or valve-related parts, or if there is unusual engine condition regarding valve clearance.

2. Remove the right and left rocker covers.

3. Set the number 1 cylinder at Top Dead Center (TDC) of its compression stroke, as follows:

a. Rotate the crankshaft pulley clockwise until the timing mark (grooved line without color) is aligned with the timing indicator.

b. Make sure the number 1 cylinder intake and exhaust cam noses are facing inward and upward from the cylinder head, as shown. If they are not positioned as shown, rotate the crankshaft pulley 360° clockwise until they are appropriately positioned.

4. Using a feeler gauge, measure the valve lash clearance as illustrated.

5. Measure the valve clearance for the following cylinders: Cylinder 1 Intake, Cylinder 2 Exhaust, Cylinder 3 Exhaust, Cylinder 6 Intake:

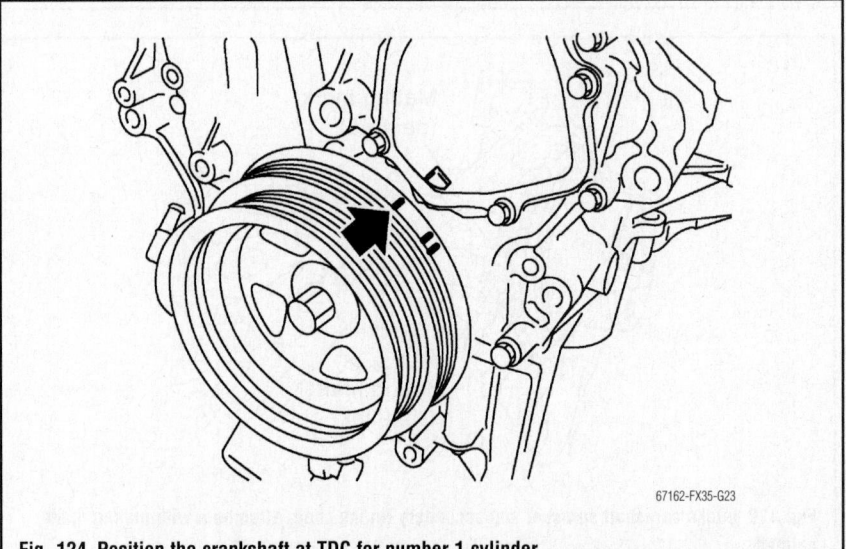

67162-FX35-G23

Fig. 124 Position the crankshaft at TDC for number 1 cylinder

a. Valve clearance cold (68°F/20°C):
- Intake: 0.010–0.013 in.
  (0.26–0.34mm)
- Exhaust: 0.011–0.015 in.
  (0.29–0.37mm)

b. Valve clearance hot (176°F/80°C):
- Intake: 0.012–0.016 in.
  (0.304–0.416mm)
- Exhaust: 0.012–0.017 in.
  (0.308–0.432mm)

6. Rotate the crankshaft by 240° clockwise (when viewed from front) to align the number 3 cylinder at TDC of its compression stroke.

➡**The crankshaft damper pulley mounting bolt flange has a stamped line every 60°. This can be used as a guide to the rotation angle.**

7. Using a feeler gauge, measure the valve clearance for the following cylinders: Cylinder 2 Intake, Cylinder 3 Intake, Cylinder 4 Exhaust, Cylinder 5 Exhaust:

a. Valve clearance standard cold (68°F/20°C):
- Intake: 0.010–0.013 in.
  (0.26–0.34mm)
- Exhaust: 0.011–0.015 in.
  (0.29–0.37mm)

b. Valve clearance hot (176°F/80°C):
- Intake: 0.012–0.016 in.
  (0.304–0.416mm)
- Exhaust: 0.012–0.017 in.
  (0.308–0.432mm)

8. Rotate the crankshaft by 240° clockwise (when viewed from front) to align the number 5 cylinder at TDC of its compression stroke.

9. Using a feeler gauge, measure the valve clearance for the following cylinders: Cylinder 1 Exhaust, Cylinder 4 Intake, Cylinder 5 Intake, Cylinder 6 Exhaust:

a. Valve clearance standard cold (68°F/20°C):
- Intake: 0.010–0.013 in.
  (0.26–0.34mm)

- Exhaust: 0.011–0.015 in.
  (0.29–0.37mm)

b. Valve clearance hot (176°F/80°C):
- Intake: 0.012–0.016 in.
  (0.304–0.416mm)
- Exhaust: 0.012–0.017 in.
  (0.308–0.432mm)

➡**If the inspection was carried out with a cold engine, make sure the values with a fully warmed up engine are still within specifications.**

10. For all valve lifters that are found to be outside the specified range, perform the following steps.

a. Perform adjustment depending on selected head thickness of valve lifter.

b. The specified valve lifter thickness is the dimension at normal temperatures. Ignore dimensional differences caused by temperature. Use the specifications for hot engine condition to adjust.

11. Remove the camshaft.

12. Remove the valve lifters at the locations that are outside the standard.

13. Measure the center thickness of the removed valve lifters with a micrometer.

14. Use the following equation to calculate valve lifter thickness for the replacement lifters.

➡**Valve lifter thickness calculation: thickness of replacement valve lifter = t1 + (C1 - C2). t1 = Thickness of removed valve lifter, C1 = measured valve clearance, C2 = standard valve clearance.**

The thickness of a new valve lifter can be identified by stamp marks on the reverse side (inside the cylinder). Stamp mark 788U or 788R indicates 0.3102 in. (7.88mm) in thickness.

➡**Two types of stamp marks are used for parallel setting and for manufacturer identification. Available thicknesses of valve lifters include 27 sizes**

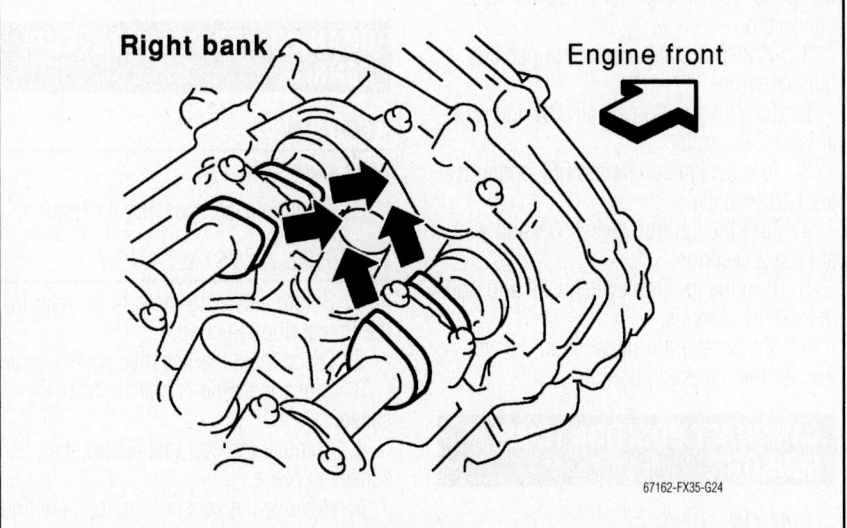

Fig. 125 When at TDC with number 1 cylinder, the camshaft lobes should point as shown

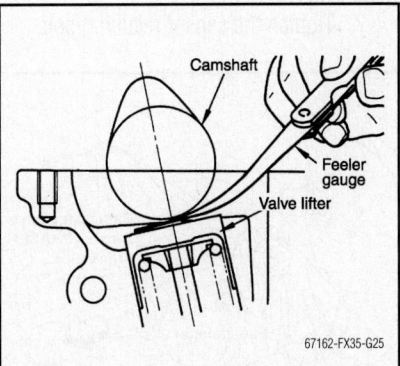

Fig. 126 Measure the valve lash with the camshaft lobe positioned as shown

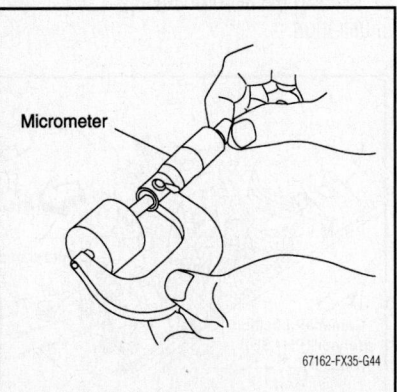

Fig. 127 Measure the valve lifter height as shown

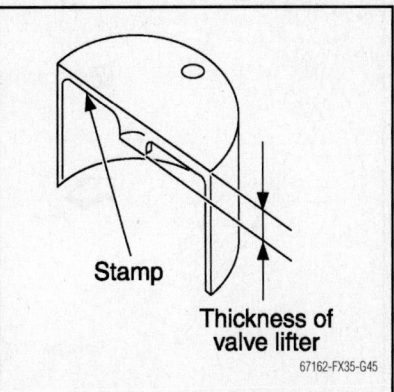

Fig. 128 Valve lifter identification stamp location

covering a range of 0.3102–0.3307 in. (7.88–8.40mm) in steps of 0.0008 in. (0.02mm).

15. Install the selected valve lifter(s).
16. Install the camshaft.

17. Manually turn crankshaft pulley a few turns.
18. Make sure the valve clearances for the cold engine are within specifications by referring to the specified values.

19. After completing the repair, check valve clearances again with the specifications for a warmed engine. Make sure the values are within specifications.

ENGINE PERFORMANCE & EMISSION CONTROLS

ACCELERATOR PEDAL POSITION (APP) SENSOR

LOCATION

See Figure 129.

The Accelerator Pedal Position (APP) sensor is located inside the vehicle. It is part of the accelerator pedal assembly.

REMOVAL & INSTALLATION

1. Before servicing the vehicle, refer to the Precautions Section.
2. Disconnect the accelerator pedal position sensor harness connector.
3. Loosen the mounting nuts and remove the accelerator pedal assembly.

✴ WARNING

Do not disassemble the accelerator pedal assembly or attempt to remove the accelerator pedal position sensor from accelerator pedal assembly. Avoid impact from dropping the sensor during handling. Be careful to keep the accelerator pedal assembly away from water.

To install:
4. Installation is the reverse order of removal.

5. Tighten the mounting nuts to 45 inch lbs. (5 Nm).
6. When the harness connector of the accelerator pedal position sensor is disconnected, perform the Accelerator Pedal Released Position Learning.

Accelerator Pedal Released Position Learning

Accelerator Pedal Released Position Learning is an operation to learn the fully released position of the accelerator pedal by monitoring the accelerator pedal position sensor output signal. It must be performed each time the harness connector of accelerator pedal position sensor or ECM is disconnected.

1. Make sure that accelerator pedal is fully released.
2. Turn the ignition switch ON and wait at least 2 seconds.
3. Turn the ignition switch OFF and wait at least 10 seconds.
4. Turn the ignition switch ON and wait at least 2 seconds.
5. Turn the ignition switch OFF and wait at least 10 seconds.
6. The accelerator pedal should have learned the released position.

CAMSHAFT POSITION (CMP) SENSOR

LOCATION

See Figure 130.

Refer to the accompanying illustration.

REMOVAL & INSTALLATION

1. Before servicing the vehicle, refer to the Precautions Section.
2. Disconnect the negative battery cable.
3. Disconnect the connector from the Camshaft Position (CMP) sensor.
4. Remove the bolt that retains the CMP sensor.
5. Remove the CMP sensor.

To install:
6. Installation is the reverse of the removal procedure.
7. Tighten the bolt that retains the CMP sensor.
8. Connect the sensor connector.

CRANKSHAFT POSITION (CKP) SENSOR

LOCATION

See Figure 131.

Refer to the accompanying illustration.

REMOVAL & INSTALLATION

1. Before servicing the vehicle, refer to the Precautions Section.
2. Disconnect the negative battery cable.
3. Disconnect the connector from the sensor.
4. Remove the bolt that retains the sensor in place.
5. Remove the sensor from its mounting.

To install:
6. Installation is the reverse of the removal procedure.
7. Tighten the sensor retaining bolt.

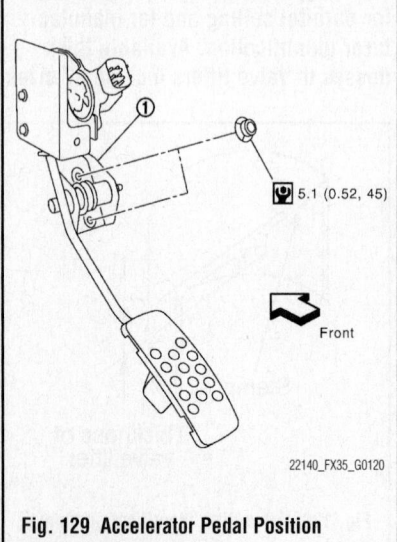

5.1 (0.52, 45)

Front

22140_FX35_G0120

Fig. 129 Accelerator Pedal Position (APP) sensor location (1)

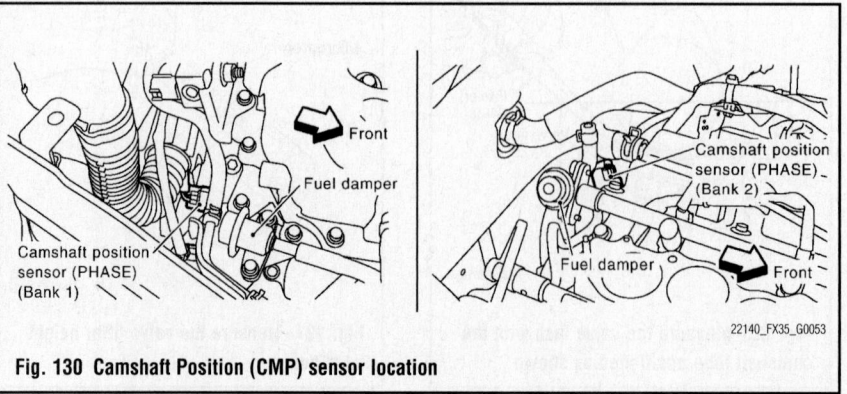

Front
Fuel damper

Camshaft position sensor (PHASE) (Bank 1)

Camshaft position sensor (PHASE) (Bank 2)

Fuel damper
Front

22140_FX35_G0053

Fig. 130 Camshaft Position (CMP) sensor location

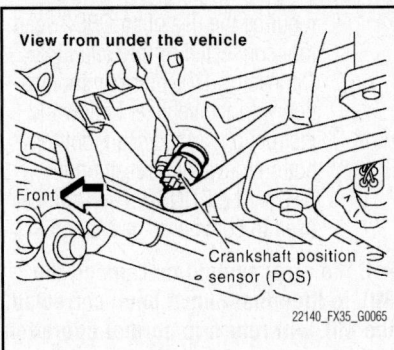

Fig. 131 Crankshaft Position (CKP) sensor location

ELECTRONIC CONTROL MODULE (ECM)

LOCATION

See Figure 132.

Refer to the accompanying illustration.

REMOVAL & INSTALLATION

1. Before servicing the vehicle, refer to the Precautions Section.
2. Turn the ignition switch **OFF**.
3. Disconnect the negative battery cable from the battery.
4. Disconnect the Electronic Control Module (ECM) connectors.
5. Remove the ECM mounting bolts and remove the ECM from the vehicle.

To install:

6. Installation is the reverse of the removal.
7. Tighten the ECM mounting bolts.

✴✴ WARNING

When replacing the ECM, be careful to use the right part number, as damage to the injection system could occur.

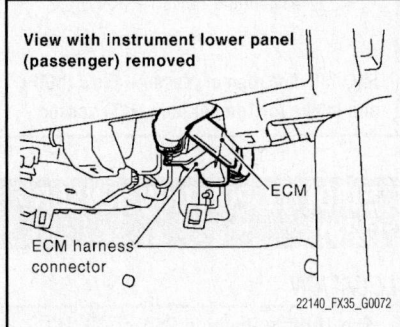

Fig. 132 Electronic Control Module (ECM) location

ENGINE COOLANT TEMPERATURE (ECT) SENSOR

LOCATION

See Figure 133.

Refer to the accompanying illustration.

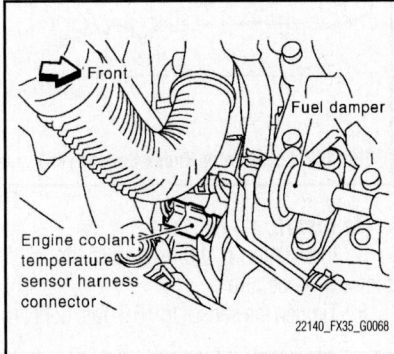

Fig. 133 Engine Coolant Temperature (ECT) sensor location—3.5L engine

REMOVAL & INSTALLATION

1. Before servicing the vehicle, refer to the Precautions Section.
2. Drain the coolant to a level below the bottom of the sensor.
3. Disconnect the ground cable from the battery and then remove the sensor connector.
4. Remove the coolant temperature sensor.

To install:

5. Coat the threads of the sensor with a suitable sealant and thread into the housing.
6. Tighten the sensor.
7. Refill the cooling system to the proper level.
8. Attach the electrical connector to the sensor securely.
9. Connect the negative battery cable.

HEATED OXYGEN (HO2S) SENSOR

LOCATION

See Figures 134 and 135.

Refer to the accompanying illustrations.

REMOVAL & INSTALLATION

✴✴ CAUTION

The temperature of the exhaust system is extremely high after the engine has been run. To prevent personal injury, allow the exhaust system to cool before removing the sensor from the exhaust system.

1. Before servicing the vehicle, refer to the Precautions Section.
2. Disconnect the negative battery cable.
3. Raise and safely support the vehicle, as needed.
4. Detach the electrical connector from the oxygen sensor.
5. Using an oxygen sensor socket, remove the Heated Oxygen Sensor (HO2S).

To install:

6. If installing the old HO2S sensor, coat the threads with anti-seize compound. New sensors are already coated. Take care not to

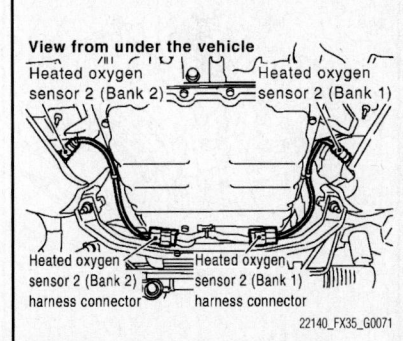

Fig. 135 Location of Heated Oxygen (HO2S) Sensor (bank 1, sensor 2 and bank 2, sensor 2

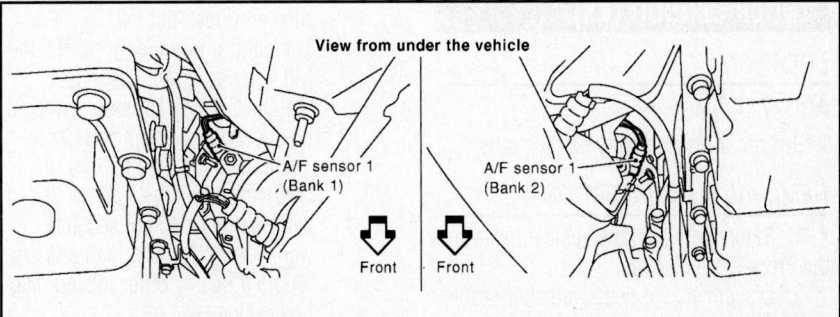

Fig. 134 Location of Heated Oxygen (HO2S) Sensor (bank 1 and bank 2

contaminate the oxygen sensor probe with the anti-seize compound.

7. Install the oxygen sensor. Using the correct tool, tighten the sensor.

8. Attach the wiring to the sensor.

9. Connect the negative battery cable.

INTAKE AIR TEMPERATURE (IAT) SENSOR

LOCATION

See Figure 136.

Refer to the accompanying illustration.

REMOVAL & INSTALLATION

1. Before servicing the vehicle, refer to the Precautions Section.

2. Disconnect the negative battery cable.

3. Disconnect the connector from the sensor.

4. Remove the sensor retaining screws.

5. Remove the sensor from its mounting.

To install:

6. Installation is the reverse of the removal procedure.

7. Handle the sensor assembly carefully, protecting it from impact, extremes of temperature and/or exposure to shop chemicals.

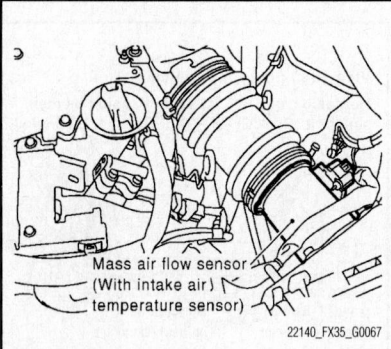

Fig. 136 Location of Mass Air Flow (MAF) and Intake Air Temperature (IAT) sensor— 3.5L engine

KNOCK SENSOR (KS)

LOCATION

See Figure 137.

Refer to the accompanying illustration.

REMOVAL & INSTALLATION

1. Before servicing the vehicle, refer to the Precautions Section.

2. Disconnect the negative battery cable.

3. Disconnect the sensor connector.

4. Remove the sensor from its mounting.

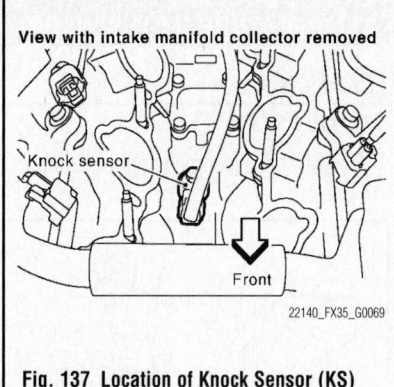

Fig. 137 Location of Knock Sensor (KS)

To install:

5. Installation is the reverse of the removal procedure.

6. Tighten the sensor to 16 ft. lbs. (21 Nm).

MALFUNCTION INDICATOR LIGHT (MIL)

RESET PROCEDURE

1. Proper operation of the Malfunction Indicator Light (MIL):
- The MIL will illuminate with the ignition switch ON and the engine OFF
- The MIL will turn OFF when the engine is started
- The MIL will remain ON if the self-diagnostic system has detected a malfunction
- The MIL may turn OFF if the malfunction is no longer present
- If the MIL is illuminated and then the engine stalls, the MIL will remain illuminated as long as the ignition switch is ON
- If the MIL is not illuminated and the engine stalls, the MIL will not illuminate until the ignition switch is cycled OFF, then ON

2. Resetting the MIL:
- The control module turns OFF the MIL after 3 consecutive ignition cycles that the diagnostic system runs and does not fail
- The control module turns OFF the MIL after a current Diagnostic Trouble Code (DTC) clears when the diagnostic cycle runs and passes
- There may still be a history of DTC's stored in the system. These will clear after 40 consecutive warm-up cycles, if no failures are reported by any other related diagnostic system
- Manual resetting of the MIL and any DTC stored in the system,

requires the use of an OBD2 scan tool connected to the Data Link Connector (DLC) for communication with the vehicle. Follow the instructions of the scan tool for both retrieval and resetting of DTC's. The CONSULT-III® can be used to command the MIL off.

➡If the error symptoms causing the MIL to illuminate have been corrected, the MIL will return to normal operation.

MASS AIR FLOW (MAF) SENSOR

LOCATION

See Figure 138.

Refer to the accompanying illustration.

REMOVAL & INSTALLATION

1. Before servicing the vehicle, refer to the Precautions Section.

2. Disconnect the negative battery cable.

3. Disconnect the connector from the sensor.

4. Remove the air cleaner and air intake assembly, as required.

5. Remove the sensor from its mounting.

To install:

6. Installation is the reverse of the removal procedure.

7. Handle the sensor assembly carefully, protecting it from impact, extremes of temperature, and exposure to shop chemicals.

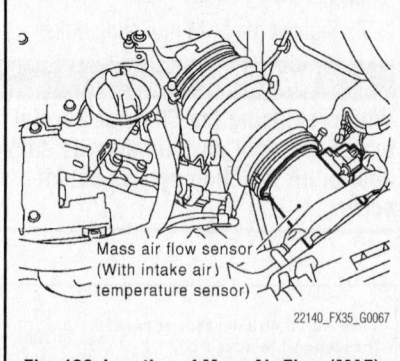

Fig. 138 Location of Mass Air Flow (MAF) and Intake Air Temperature (IAT) sensor

THROTTLE POSITION SENSOR (TPS)°

LOCATION

The Throttle Position Sensor (TPS) is mounted on the throttle body and is incorporated into the throttle body assembly.

REMOVAL & INSTALLATION

The throttle position sensor is an integral part of the throttle body.

1. Before servicing the vehicle, refer to the Precautions Section.

2. Properly relieve the fuel system pressure.

3. Drain the engine coolant.

4. Remove the air intake hose.

5. Remove the battery.

6. Disconnect the throttle position sensor connector.

7. Disconnect the water hose connection.

8. Remove the throttle body retaining bolts.

9. Remove the throttle body from the engine.

10. Discard the gasket.

To install:

✳✳ WARNING

Do not loosen the retaining screws for the resin cover of the throttle body assembly. If the screws are loosened, the sensor incorporated in the resin cover becomes misaligned and the throttle body may not work properly.

11. Align the recess on the intake manifold plenum with the projection of the throttle body gasket.

12. Install the gasket. Install the throttle body to the engine and tighten the retaining bolts to 80 inch lbs. (9 Nm).

➡**Poor idling may result if the throttle body gasket is not installed properly.**

13. Continue the installation in the reverse order of the removal procedure.

14. Connect the negative battery cable.

15. Turn the ignition **ON** and then **OFF**, and keep it off for at least 10 seconds.

16. Complete the vehicle initialization procedure.

VARIABLE CAMSHAFT TIMING OIL CONTROL SOLENOID

LOCATION

See Figure 139.

Refer to the accompanying illustration.

REMOVAL & INSTALLATION

1. Before servicing the vehicle, refer to the Precautions Section.

2. Disconnect the ground cable from the battery.

3. Disconnect the connector from Variable Camshaft Timing Oil Control Solenoid (VCTOCS) on the right-hand bank and/or left-hand bank.

4. Remove the bolt retaining the VCTOCS.

5. Remove the VCTOCS.

To install:

➡**Always use a new gasket/O-ring.**

6. Apply a small amount of engine oil to the new O-ring of the VCTOCS.

7. Install the VCTOCS and tighten the mounting bolt.

8. Connect the electrical connector to the VCTOCS.

VEHICLE SPEED SENSOR (VSS)

LOCATION

The Vehicle Speed Sensor (VSS) is installed on the transmission.

REMOVAL & INSTALLATION

1. Before servicing the vehicle, refer to the Precautions Section.

2. Raise and support the vehicle safely.

3. Place a drip pan below the Vehicle Speed Sensor (VSS) to catch any spilled transmission fluid when it is removed.

4. Disconnect the VSS connector.

5. Remove the sensor from its mounting.

To install:

6. Install the sensor and tighten the attaching bolt.

7. Replace any lost transmission fluid.

8. Connect the sensor electrical connector.

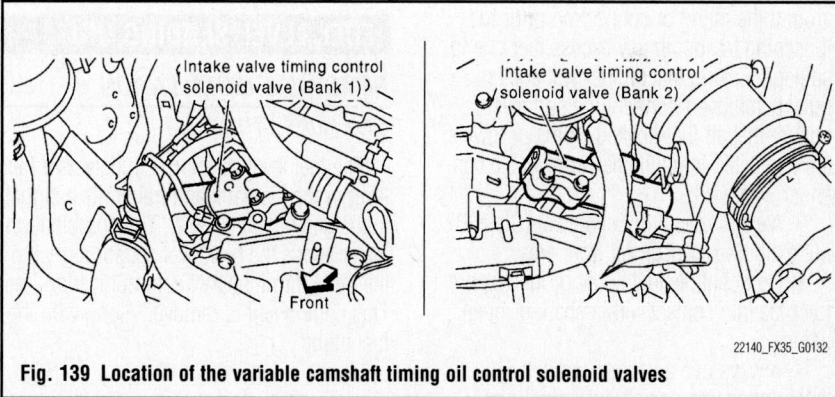

Intake valve timing control solenoid valve (Bank 1)

Front

Intake valve timing control solenoid valve (Bank 2)

22140_FX35_G0132

Fig. 139 Location of the variable camshaft timing oil control solenoid valves

FUEL

FUEL SYSTEM SERVICE PRECAUTIONS

Safety is the most important factor when performing not only fuel system maintenance, but any type of maintenance. Failure to conduct maintenance and repairs in a safe manner may result in serious personal injury or death. Work on a vehicle's fuel system components can be accomplished safely and effectively by adhering to the following rules and guidelines.

• To avoid the possibility of fire and personal injury, always disconnect the negative battery cable unless the repair or test procedure requires that battery voltage be applied.

• Always relieve the fuel system pressure prior to disconnecting any fuel system component (injector, fuel rail, pressure regulator, etc.) fitting or fuel line connection. Exercise extreme caution whenever relieving fuel system pressure to avoid exposing skin, face and eyes to fuel spray. Please be advised that fuel under pressure may penetrate the skin or any part of the body that it contacts.

• Always place a shop towel or cloth around the fitting or connection prior to loosening to absorb any excess fuel due to spillage. Ensure that all fuel spillage is quickly removed from engine surfaces. Ensure that all fuel-soaked cloths or towels are deposited into a flame-proof waste container with a lid.

• Always keep a dry chemical (Class B) fire extinguisher near the work area.

• Do not allow fuel spray or fuel vapors to come into contact with a spark or open flame.

• Always use a second wrench when loosening or tightening fuel line connection fittings. This will prevent unnecessary stress and torsion on fuel piping. Always follow the proper torque specifications.

• Always replace worn fuel fitting O-rings with new ones. Do not substitute fuel hose where rigid pipe is installed.

FUEL SYSTEM PRESSURE

RELIEVING

With The Consult-III® Tool

1. Before servicing the vehicle, refer to the Precautions Section.
2. Turn the ignition switch **ON**.

3. Perform the "FUEL PRESSURE RELEASE" in "WORK SUPPORT" mode with the CONSULT-III®.
4. Start the engine.
5. After engine stalls, crank it over 2–3 times to release all fuel pressure.
6. Turn the ignition switch **OFF**.

Without The Consult-III® Tool

See Figure 140.

1. Before servicing the vehicle, refer to the Precautions Section.
2. Remove the fuel pump fuse located in IPDM E/R.
3. Start the engine.
4. After the engine stalls, crank it over 2–3 times to release all fuel pressure.
5. Turn the ignition switch **OFF**.
6. Reinstall the fuel pump fuse after servicing the fuel system.

FUEL FILTER

REMOVAL & INSTALLATION

The fuel delivery system integrates the fuel filter with the in-tank fuel pump. To service this filter, remove the fuel pump. Refer to Fuel Pump, removal & installation.

FUEL LEVEL SENDING UNIT

REMOVAL & INSTALLATION

See Figures 141 and 142.

The fuel level sending unit detects a fuel level in the fuel tank and transmits a signal to the combination meter. The combination meter sends the fuel level sensor signal to the ECM through CAN communication line. This component is removed along with the fuel pump.

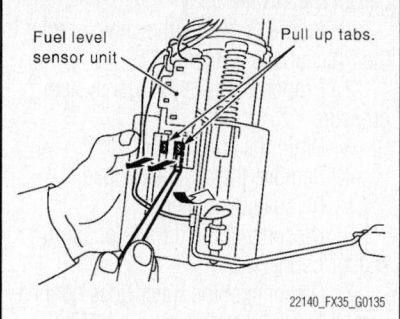

Fig. 141 Using a suitable tool, pull up the tab points on the fuel level sensor unit to release the lock

1. Before servicing the vehicle, refer to the Precautions Section.
2. Remove the fuel pump module. Refer to Fuel Pump, removal & installation.

❊❊ WARNING

Be careful not to bend the float arm during removal, and avoid impacts, such as dropping, when handling the components.

3. Remove the main fuel level sensor unit, fuel filter, and fuel pump assembly, and sub fuel level sensor unit.
4. Disconnect the harness connector. Hold the connector and pull it out (there is no stopper release tab).
5. Using a suitable tool, pull up the tab points, as shown in the figure, to release the lock. Be careful not to damage it.
6. After the fixing tabs are disengaged, slide the fuel level sensor unit out in the direction shown by the arrow.

➡️**Do not disassemble the fuel filter and fuel pump assembly.**

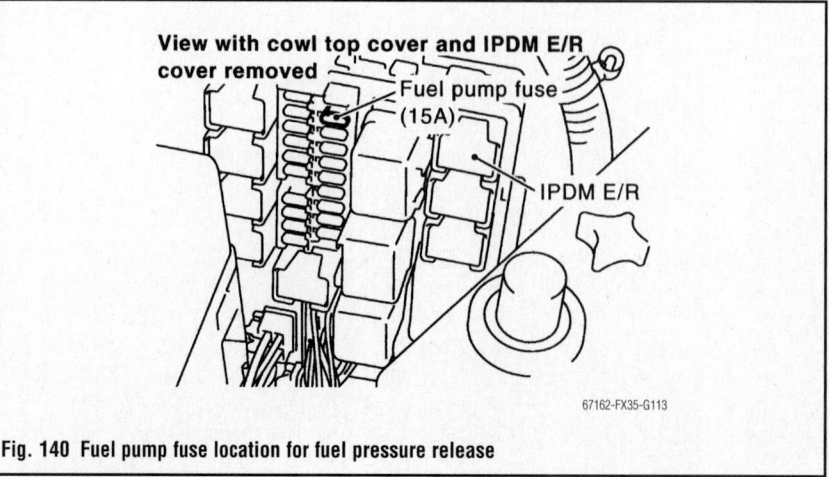

Fig. 140 Fuel pump fuse location for fuel pressure release

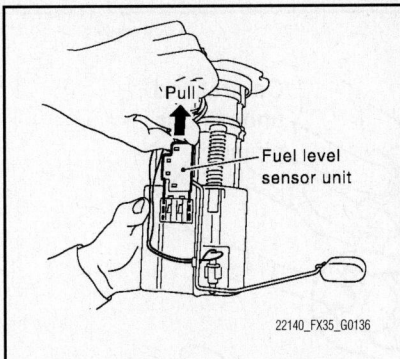

Fig. 142 Slide the fuel level sensor unit out in the direction shown by the arrow

To install:

7. Check for damage to the fuel level sensor unit installation position on the side of fuel filter and fuel pump assembly.

8. Slide the fuel level sensor unit until it aligns into the installation groove, then insert it until it stops.

9. After inserting the fuel level sensor unit, apply force in the reverse direction (removal direction) to ensure it cannot be pulled out.

10. Connect the harness connector securely until the connector stops.

11. Install the fuel pump assembly to the vehicle. Refer to Fuel Pump, removal & installation.

FUEL PUMP

REMOVAL & INSTALLATION

See Figures 143 and 144.

1. Before servicing the vehicle, refer to the Precautions Section.

2. Check the fuel level on the fuel gauge. If the fuel gauge indicates full or almost full, drain the fuel from the fuel tank until the gauge indicates a level near ¾ of a tank.

✳✳ CAUTION

Fuel will be spilled when removing the main and sub fuel level sensor units if the level of the fuel in the tank is higher than about ¾ of a tank.

3. In the case that the fuel pump does not operate, perform the following procedure:

a. Insert a hose of less than 1 in. (25mm) in diameter into the fuel filler tube through the fuel filler opening to draw fuel from the fuel filler tube.

b. Disconnect the fuel filler hose from the fuel filler tube.

c. Insert the fuel tube into the fuel tank through the fuel filler hose to draw the fuel from the fuel tank.

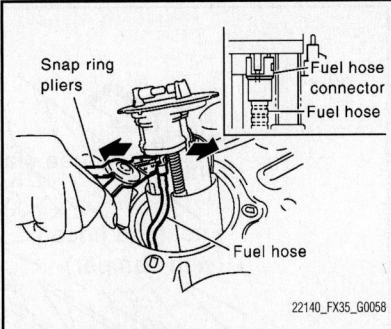

Fig. 143 Raise the fuel pump assembly and using snapring pliers, remove the fuel hose connector

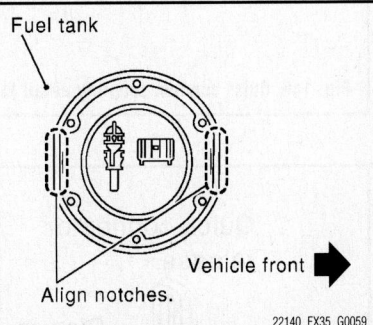

Fig. 144 Install the fuel pump retainer so that its notch becomes parallel with the notch on the fuel tank

4. Release the fuel pressure from the fuel lines.

5. Open the fuel filler lid.

6. Open the filler cap and release the pressure inside the fuel tank.

7. Remove the rear seat cushion, as follows:

a. Pull the lock at the front bottom of the seat cushion forward (1 for each side).

b. Pull the seat cushion upward to release the retaining wire from the plastic hook.

c. Pull the seat cushion forward to remove.

8. Lift up the floor carpet, then remove the inspection hole cover for the main and sub fuel level sensor units by turning the retaining clips clockwise by 90°.

9. Disconnect the wiring harness connector and fuel feed tube.

10. Disconnect the fuel line quick connector, as follows:

a. Hold the sides of the connector, push in the tabs and pull out the tube.

➡**If the quick connector sticks to the tube of the main fuel level sensor unit, push and pull the quick connector several times until they start to move. Then, disconnect them by pulling.**

When dealing with the fuel line quick connector, heed the following:

• The quick connector can be disconnected when the tabs are completely depressed. Do not twist it more than necessary

• Do not use any tools to disconnect the quick connector

• Keep the resin tube away from heat. Be especially careful when welding near the resin tube

• Prevent acidic liquid such as battery electrolyte from getting on the resin tube

• Do not bend or twist the resin tube during connection and disconnection

• Do not remove the remaining retainer on the hard tube (or the equivalent) except when the resin tube or retainer is replaced

• When the resin tube or hard tube (or the equivalent) is replaced, also replace the retainer with a new one

• To keep the connecting portion clean and to avoid damage and foreign materials, cover them completely with plastic bags or a similar material.

✳✳ WARNING

Make sure to not bend the float arm during removal, and avoid impacts, such as falling, when handling components.

11. Remove the main fuel level sensor unit, fuel filter, and fuel pump assembly, and sub fuel level sensor unit, as follows:

a. Remove the main fuel sensor unit retainer.

b. Raise the main fuel level sensor unit, fuel filter and fuel pump assembly, and using snapring pliers, remove the fuel hose connector.

✳✳ WARNING

Be careful not to damage the fuel hose connector by expanding it excessively.

12. Removal of sub fuel level sensor unit:

a. Remove the sub fuel level sensor unit retainer.

b. Raise and release the sub fuel level sensor unit.

To install:

13. Installation is the reverse of removal.

14. When installing the fuel hose connectors insert them fully until a click sound of full stopper engagement is heard.

15. Install the fuel pump retainer so that its notch becomes parallel with the notch on the fuel tank.

16. Tighten the retainer mounting bolts evenly to 20 inch lbs. (2 Nm).

FUEL RAIL AND INJECTORS

REMOVAL & INSTALLATION

See Figures 145 through 150.

1. Before servicing the vehicle, refer to the Precautions Section.
2. Remove the engine cover.
3. Relieve the fuel pressure.
4. Remove the fuel feed hose (with damper) from the fuel sub-tube.

➡**There is no fuel return route.**

✳✳ CAUTION

While the hoses are disconnected, plug them to prevent fuel from draining. Also, do not separate the damper and hose.

5. When separating the fuel feed hose (with damper) and the centralized under-floor piping connection, disconnect the quick connector, as follows:

    a. Remove the quick connector cap from the quick connector connection on the right member side.

    b. Disconnect the fuel feed hose (with damper) from the bracket hose clamp.

➡**Disconnect the quick connector by using quick connector release tool SST: J-45488, or equivalent.**

    c. With the sleeve side of the quick connector release facing the quick connector, install the quick connector release onto the centralized under-floor piping.

    d. Insert the quick connector release into the quick connector until the sleeve

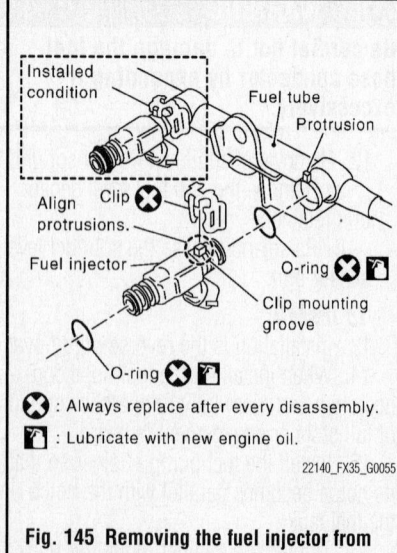

Fig. 145 Removing the fuel injector from the fuel rail (tube)

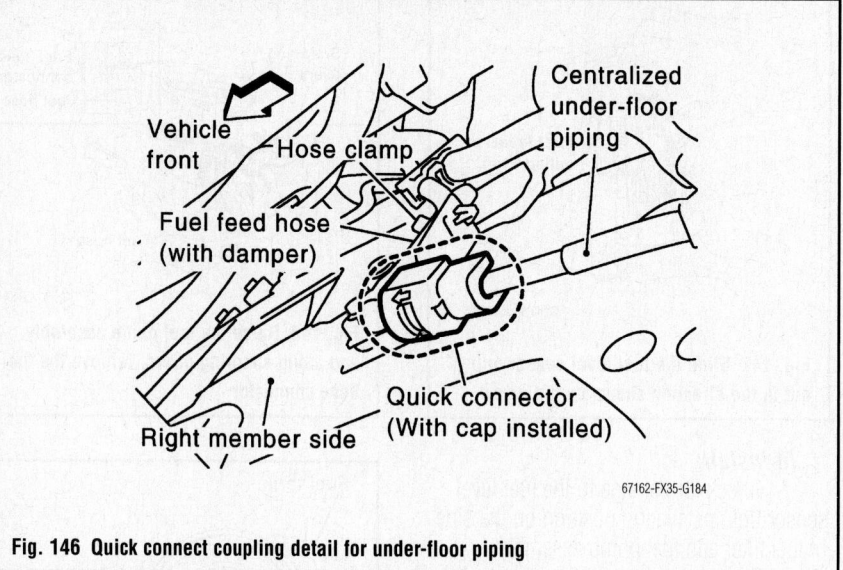

Fig. 146 Quick connect coupling detail for under-floor piping

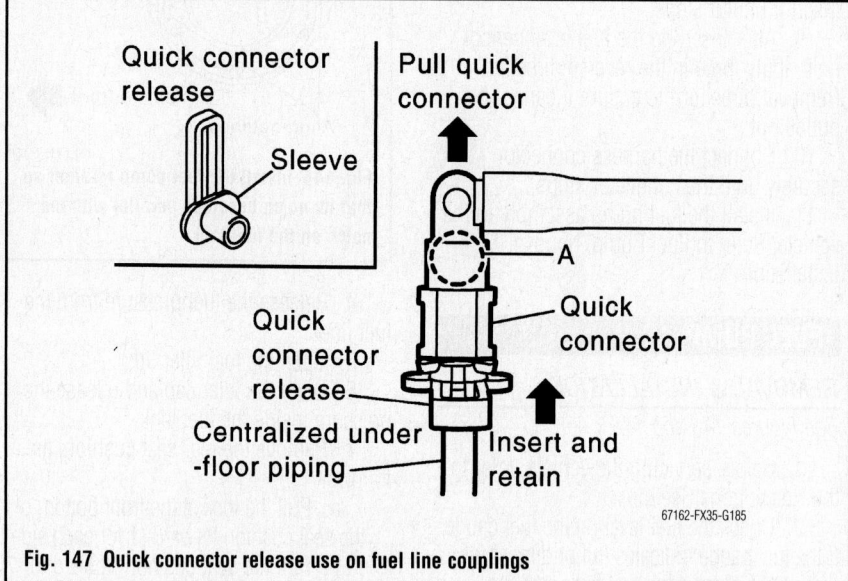

Fig. 147 Quick connector release use on fuel line couplings

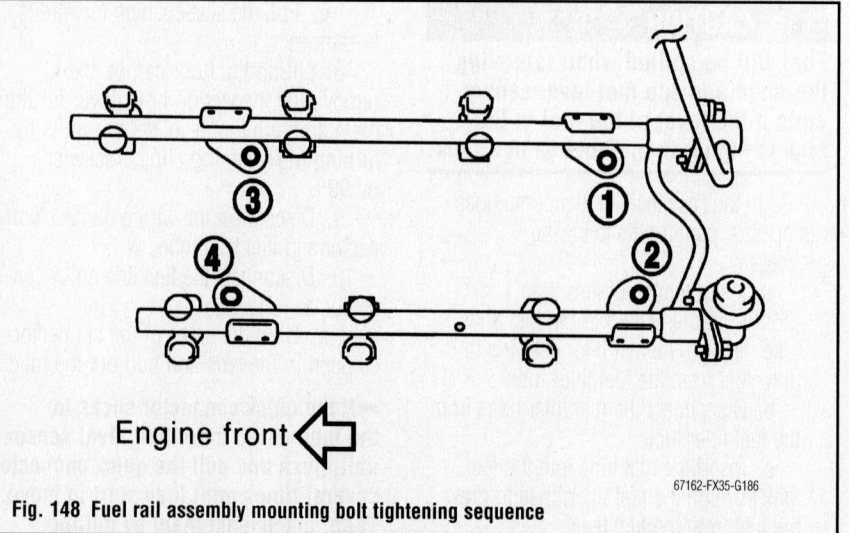

Fig. 148 Fuel rail assembly mounting bolt tightening sequence

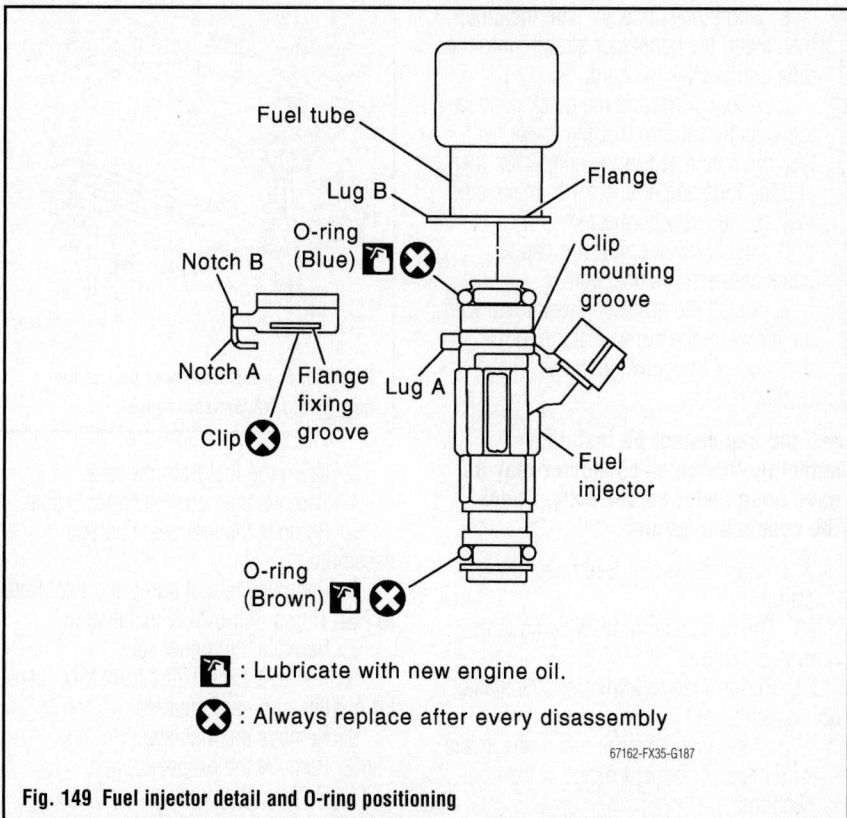

: Lubricate with new engine oil.

: Always replace after every disassembly

67162-FX35-G187

Fig. 149 Fuel injector detail and O-ring positioning

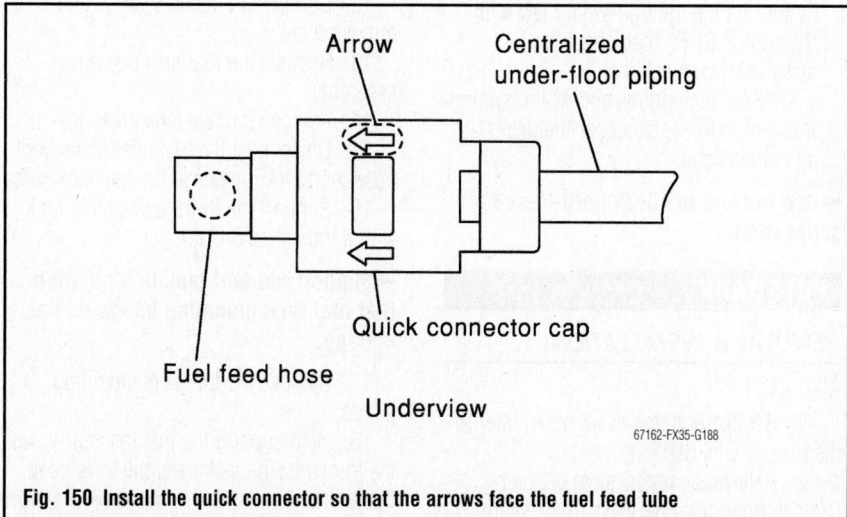

67162-FX35-G188

Fig. 150 Install the quick connector so that the arrows face the fuel feed tube

• Do not bend or twist the connection between the quick connector and the fuel feed hose (with damper) during installation/removal

• To keep the connecting portion clean and to avoid damage from foreign materials, cover them completely with plastic bags or a similar material

6. Remove the upper and lower intake manifold collectors. Refer to Intake Manifold, Upper Intake Manifold, removal & installation.

7. Disconnect the wiring harness connector from the fuel injector.

8. Loosen the mounting bolts in the reverse order shown, and remove the fuel tube and fuel injector assembly.

❋❋ CAUTION

Do not tilt the assembly or the remaining fuel may leak.

9. Remove the fuel injectors from the fuel rail (tube) with the following procedure:

    a. Open and remove the clip.

    b. Remove the fuel injector from the fuel tube by pulling straight.

During injector removal, heed the following items:

• Be careful with the remaining fuel that leaks from the fuel tube

• Be careful not to damage the injector nozzles during removal

• Do not bump or drop the fuel injectors

• Do not disassemble the fuel injectors

10. Remove the fuel sub-tube and fuel damper.

To install:

When handling all O-rings in this procedure, heed the following:

• Handle the O-ring with bare hands. Never wear gloves

• Lubricate the O-ring with new engine oil

• Do not clean the O-ring with solvent

• Make sure that the O-ring and its mating part are free of foreign material

• When installing the O-ring, be careful not to scratch it with a tool or fingernails

• Be careful not to twist or stretch the O-ring

• Insert the O-ring straight into the fuel tube

11. Install the fuel damper and fuel sub-tube.

12. Insert the fuel damper and fuel sub-tube straight into the fuel tube.

13. Tighten the mounting bolts.

14. After tightening the mounting bolts,

contacts and goes no further. Hold the quick connector release at that position.

❋❋ CAUTION

Inserting the quick connector release hard will not disconnect the quick connector. Hold the quick connector release where it contacts and goes no further.

    e. Draw and pull out the quick connector straight from the centralized under-floor piping.

When disconnecting the fuel line, heed the following:

• Pull the quick connector holding **A** position as shown in the figure. Do not pull it with lateral force applied. The O-ring inside the quick connector may be damaged.

• Prepare a container and cloth beforehand as fuel will leak out

• Avoid fire and sparks

• Keep parts away from all heat sources. Especially, be careful when welding is performed

• Do not expose the parts to battery electrolyte or other acids

make sure that there is no gap between the flange and fuel tube.

15. Install O-rings onto the fuel injector—the upper and lower O-rings are different. The O-rings are identified as follows:
- Fuel tube side O-ring: Blue
- Nozzle side O-ring: Brown

16. Install each fuel injector onto the fuel tube, as follows:

a. Insert the clip into the clip mounting groove on the fuel injector.

b. Insert the clip so that lug "A" of the fuel injector matches notch "A" of the clip.

✳✳ CAUTION

Do not reuse old clips. Replace them with new ones. Be careful to keep the clip from interfering with the O-ring. If interference occurs, replace the O-ring with a new one.

c. Insert the fuel injector into the fuel tube, matching it to the axial center, with the clip attached. Insert the fuel injector so that lug "B" of fuel tube matches notch "B" of the clip. Make sure that the fuel tube flange is securely fixed in the groove on the clip.

d. Make sure that installation is complete by checking that the fuel injector does not rotate or come off.

17. Install the fuel tube and fuel injector assembly onto the intake manifold, and tighten the mounting bolts in 2 steps as shown.

a. 1st Step: 84 inch lbs. (10 Nm).

b. 2nd Step: 17 ft. lbs. (24 Nm).

✳✳ WARNING

Be careful not to let the tips of the injector nozzles come into contact with other parts.

18. Connect the injector sub-wiring harness.

19. Install the upper and lower intake manifold collectors. Refer to Intake Manifold, Upper Intake Manifold, removal & installation.

20. Install the fuel sub-tube on the rear end of the lower intake manifold collector.

21. Connect the fuel feed hose and damper. After tightening the mounting bolts, make sure that there is no gap between the flange and the fuel sub-tube.

22. Connect the quick connector between the fuel feed hose and centralized under-floor piping connection with the following procedure:

a. Check the connection for damage and foreign materials.

b. Align the connector with the tube, then insert the connector straight into the tube until a click is heard.

c. After connecting the quick connector, visually confirm that the 2 retainer tabs are connected to the connector, then pull the tube and connector to make sure they are securely connected.

d. Install quick connector cap to quick connector connection.

e. Install the quick connector cap with the arrow on the surface facing in the direction of the quick connector (fuel feed hose side).

➡ **If the cap cannot be installed smoothly, the quick connector may not have been installed correctly. Check the connection again.**

f. Secure the fuel feed hose to the clamp.

23. The remainder of installation is the reverse of removal.

24. Perform the following once installation is complete.

a. After installing the fuel tubes, make sure there is no fuel leakage at the connections.

b. Apply fuel pressure to the fuel lines by turning the ignition switch **ON** with the engine **OFF**. Then check for fuel leaks at all connections.

c. Start the engine, and while holding it at a high RPM, check for fuel leaks at all connections.

➡ **Use mirrors to check hard-to-see connections.**

FUEL TANK

REMOVAL & INSTALLATION

See Figure 151.

1. Before servicing the vehicle, refer to the Precautions Section.

2. Relieve the fuel system pressure. Refer to Relieving Fuel System Pressure.

✳✳ CAUTION

The fuel injection system remains under pressure even after the engine has been turned OFF. Properly relieve fuel pressure before disconnecting any fuel lines. Failure to do so may result in fire or personal injury. Do not allow fuel spray or fuel vapors to come in contact with a spark or an open flame. Keep a dry chemical fire extinguisher nearby. Never store fuel in an open container due to risk of fire or explosion.

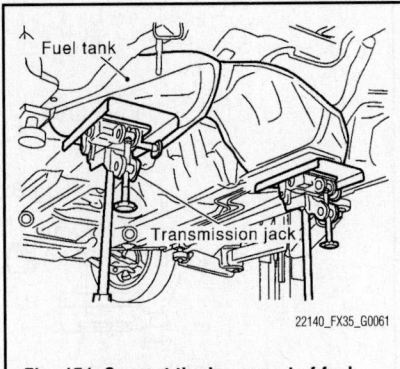

Fig. 151 Support the lower part of fuel tank with a transmission jack

3. Drain the fuel from the tank.

4. Remove the negative battery cable.

5. Remove the rear seat cushion assembly.

6. Remove the fuel pump module. Refer to Fuel Pump, removal & installation.

7. Remove the tunnel stay.

8. Remove the exhaust front tube, center muffler, and main muffler.

9. Remove the insulator.

10. Remove the propeller shaft.

11. Remove the parking rear brake cables.

12. Compress the coil springs and remove them.

13. Remove the rear suspension assembly.

14. Remove the fuel tank protector.

15. Disconnect the fuel filler hose, vent hose, and EVAP hoses at the fuel tank side.

16. Support the lower part of fuel tank with a transmission jack.

➡ **Support the fuel tank in a position that fuel tank mounting bands do not engage.**

17. Remove the fuel tank mounting bands.

18. Help support the fuel tank and lower the transmission jack carefully to remove the fuel tank.

➡ **Make sure that all the connection points have been disconnected. Confirm there is no interference with the vehicle while lowering the fuel tank.**

19. Remove fuel filler tube protector and fuel filler tube, if necessary.

To install:

20. Installation is the reverse of the removal procedure.

21. Tighten the retaining bolts and fasteners to specification as illustrated.

22. Perform the following once installation is complete.

a. After installing the fuel tubes, make sure there is no fuel leakage at the connections.

b. Apply fuel pressure to the fuel lines by turning the ignition switch **ON** with the engine **OFF**. Then check for fuel leaks at all connections.

c. Start the engine, and while holding it at a high RPM, check for fuel leaks at all connections.

➡**Use mirrors to check hard-to-see connections.**

IDLE SPEED

ADJUSTMENT

Idle speed is maintained by the Electronic Control Module (ECM). No adjustment is necessary or possible.

THROTTLE BODY

REMOVAL & INSTALLATION
See Figure 152.

1. Before servicing the vehicle, refer to the Precautions Section.
2. Disconnect the negative battery cable.
3. Remove the engine cover.
4. Disconnect and plug the water hoses from the intake manifold collector (upper).

➡**Do not spill engine coolant on the drive belts.**

5. Remove the air cleaner case and air duct, as follows:
   a. Remove the air duct (inlet).
   b. Disconnect the mass air flow sensor wiring harness connector.
   c. Remove the air cleaner case/mass air flow sensor assembly and the air

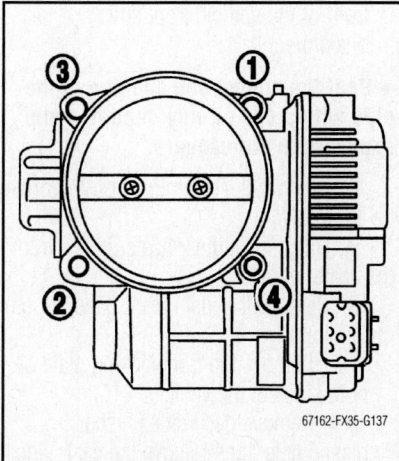

67162-FX35-G137

Fig. 152 Throttle body mounting bolt tightening sequence—3.5L engine

duct/resonator assembly disconnecting them at the joints.

➡**Add match marks as necessary for easier installation.**

d. Remove the mass air flow sensor from air cleaner case.

> ❄❄ **WARNING**
>
> **Handle the mass air flow sensor with care. Do not expose it to harsh vibration or shock. Do not disassemble it or touch its sensor.**

e. Remove the resonator in the fender, lifting the left fender protector.
6. Remove the electric throttle body control actuator, as follows:
   a. Disconnect the wiring harness connector.
   b. Loosen the bolts in the reverse order as shown in the figure.

> ❄❄ **WARNING**
>
> **Handle the throttle body carefully to avoid any shock to the electric throttle control actuator. Do not disassemble.**

To install:

7. Install the electric throttle control actuator. Tighten the bolts to 75 inch lbs. (9 Nm).
8. Install the mass air flow sensor in the air cleaner case.
9. Install the air cleaner case and air duct.
10. Reconnect the hoses to the intake manifold collector (upper).
11. Install the engine cover.
12. Connect the negative battery cable.
13. Perform the following drivability adjustments:
14. Perform the "Throttle Valve Closed Position Learning" procedure (below) when the wiring harness connector of the electric throttle control actuator is disconnected, or perform the "Idle Air Volume Learning" and "Throttle Valve Closed Position Learning" procedures (below) when the electric throttle control actuator is replaced.

Throttle Valve Closed Position Learning

1. Before servicing the vehicle, refer to the Precautions Section.

The Throttle Valve Closed Position Learning procedure is an operation for the ECM to relearn the fully closed position of the throttle valve by monitoring the throttle position sensor output signal. It must be performed each time the wiring harness

connector of the electric throttle control actuator or ECM is disconnected.

2. Make sure that accelerator pedal is fully released.
3. Turn ignition switch **ON**.
4. Turn ignition switch **OFF** wait at least 10 seconds. Make sure that throttle valve moves during the above 10 seconds by confirming the operating sound.

Idle Air Volume Learning

1. Before servicing the vehicle, refer to the Precautions Section.

Idle Air Volume Learning is an operation to learn the idle air volume that keeps each engine within the specific range. It must be performed under any of the following conditions:

- Each time the electric throttle control actuator or ECM is replaced
- Idle speed or ignition timing is out of specification

Before performing the "Idle Air Volume Learning" procedure, make sure that all of the following conditions are satisfied. Learning will be cancelled if any of the following conditions are missed for even a moment.

- Battery voltage: More than 12.9 volts (at idle)
- Engine coolant temperature: 158–212°F (70–100°C)
- PNP switch: ON
- Electric load switch: OFF (air conditioner, headlamp, and rear window defogger)

➡**On vehicles equipped with daytime light systems, if the parking brake is applied before the engine is started, the headlamp will not be illuminated.**

- Steering wheel: Neutral (straight-ahead position)
- Vehicle speed: Stopped
- Transmission: Warmed-up
- For models with CONSULT-III®, drive vehicle until "FLUID TEMP SE 1" in "DATA MONITOR" mode of "A/T" system indicates less than 0.9 volts.
- For models without CONSULT-III®, drive vehicle for 10 minutes.

2. If using the CONSULT-III® tool, perform the following:
   a. Perform the "Accelerator Pedal Released Position Learning" procedure.
   b. Perform the "Throttle Valve Closed Position Learning" procedure.
   c. Start the engine and warm it up to normal operating temperature.
   d. Check that all items listed above are properly set.

e. Select "IDLE AIR VOL LEARN" in "WORK SUPPORT" mode.

f. Touch "START" and wait 20 seconds.

g. Make sure that "CMPLT" is displayed on CONSULT-III® screen. If "CMPLT" is not displayed, the Idle Air Volume Learning procedure will not be carried out successfully.

h. Rev up the engine 2–3 times and make sure that idle speed and ignition timing are within specifications.

3. If NOT using the CONSULT-III® tool, perform the following:

➡ **It is best to keep track of time accurately with a clock.**

➡ **It is impossible to switch the diagnostic mode when an accelerator pedal position sensor circuit has a malfunction.**

a. Perform the "Accelerator Pedal Released Position Learning" procedure.

b. Perform the "Throttle Valve Closed Position Learning" procedure.

c. Start the engine and warm it up to normal operating temperature.

d. Check that all items listed above are properly set.

e. Turn the ignition switch OFF and wait at least 10 seconds.

f. Confirm that the accelerator pedal is fully released, turn the ignition switch ON and wait 3 seconds.

➡ **Repeat the following 2 steps quickly 5 times within 5 seconds.**

g. Fully depress the accelerator pedal.

h. Fully release the accelerator pedal.

i. Wait 7 seconds, fully depress the accelerator pedal and keep it for approx. 20 seconds until the MIL stops blinking and remains ON.

j. Fully release the accelerator pedal within 3 seconds after the MIL turned ON.

k. Start the engine and let it idle.

l. Wait 20 seconds.

m. Rev up the engine 2–3 times and make sure that idle speed and ignition timing are within specifications.

n. If the idle speed and ignition tim-

ing are not within specification, the Idle Air Volume Learning procedure will not be successful.

Accelerator Pedal Released Position Learning

1. Before servicing the vehicle, refer to the Precautions Section.

The "Accelerator Pedal Released Position Learning" procedure is an operation for the ECM to relearn the fully released position of the accelerator pedal by monitoring the accelerator pedal position sensor output signal. It must be performed each time the wiring harness connector of the accelerator pedal position sensor or ECM is disconnected.

2. Make sure that the accelerator pedal is fully released.

3. Turn the ignition switch **ON** and wait at least 2 seconds.

4. Turn the ignition switch **OFF** wait at least 10 seconds.

5. Turn the ignition switch **ON** and wait at least 2 seconds.

6. Turn the ignition switch **OFF** wait at least 10 seconds.

HEATING & AIR CONDITIONING SYSTEM

BLOWER MOTOR

REMOVAL & INSTALLATION
See Figure 153.

✳✳ CAUTION

Before servicing, or working around, the SRS system, turn the ignition switch OFF, disconnect both battery cables and wait at least 3 minutes. When servicing, or working around, the SRS system, do not work directly in front of the air bag module.

1. Before servicing the vehicle, refer to the Precautions Section.

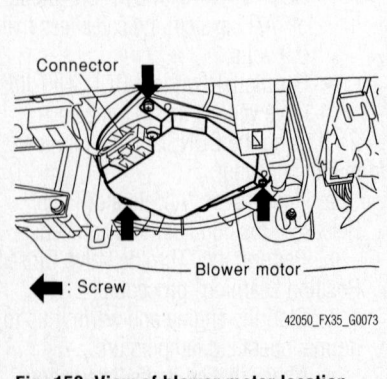

Fig. 153 View of blower motor location

2. Disconnect the negative battery cable.

3. Remove the lower instrument panel, passenger side.

4. Disconnect the blower motor electrical connector.

5. Remove the blower motor retaining screws.

6. Remove the blower motor from its mounting.

To install:

7. Installation is the reverse of the removal procedure.

HEATER CORE

REMOVAL & INSTALLATION
See Figures 154 through 159.

1. Before servicing the vehicle, refer to the Precautions Section.

2. Use approved refrigerant collecting equipment to discharge refrigerant.

✳✳ CAUTION

Make sure the engine is cold before draining the coolant.

3. Drain the coolant from the cooling system.

4. Remove the cowl top cover.

5. Remove the 2 high-pressure pipe mounting clips.

6. Remove the low-pressure flexible hose bracket mounting bolts.

7. Disconnect the evaporator-side one touch joint, as follows:

a. Set disconnect SST: 9253089908 (high-pressure side) and SST: 9253089916 (low-pressure side) on the A/C piping.

b. Slide the disconnect tool toward the front of the vehicle until it clicks.

c. Slide the A/C pipe toward the front of the vehicle front and disconnect it.

➡ **Seal the connection opening of the pipe with a cap or vinyl tape to avoid exposure to atmosphere.**

8. Remove the electronic control throttle assembly.

9. Disconnect the 2 heater hoses from the heater core.

10. Remove the instrument panel assembly, as follows:

a. Remove the front kicking plate on both sides of the vehicle.

b. Remove dash side finisher plastic nuts, then remove the dash side finisher.

c. Pull to the inside of the vehicle,

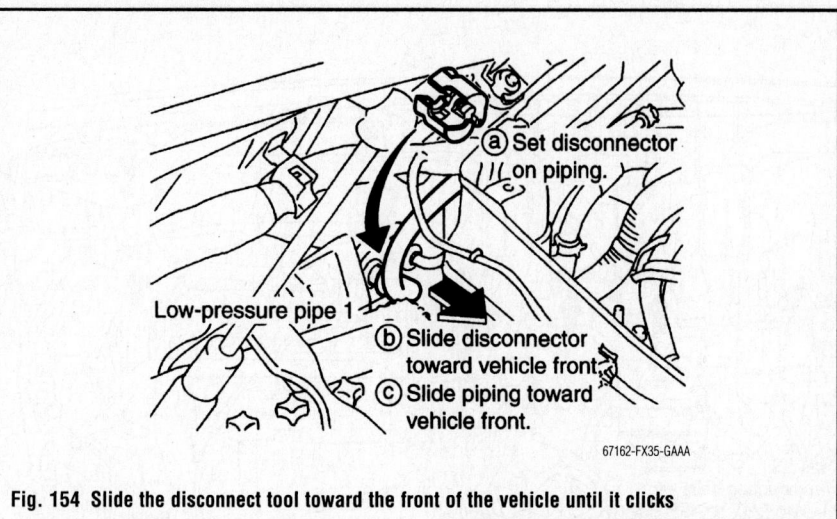

Fig. 154 Slide the disconnect tool toward the front of the vehicle until it clicks

disengage the metal clips and remove the front pillar garnish.

d. To remove the A/T Select Lever Knob, pull down the knob cover. Remove the lock-pin of the select lever knob. Then, lift up the select lever knob and remove it.

e. Insert a remover into the side between the gaps of the instrument clock finisher and pull back to the side.

f. Disconnect the clips and wiring harness connector, then remove the instrument clock finisher.

g. Insert a remover into the side between the gaps of the A/T console finisher and remove it by lifting the A/T console finisher.

h. Disconnect the wiring harness connector.

i. Remove the console finisher screws.

j. Remove the console finishers.

k. To remove the center console, remove the mounting screws, then remove the console sub-wiring harness.

✳✳ WARNING

When removing console, be careful not to pull the wiring harness.

l. Remove the instrument lower cover by pulling down on the front instrument lower cover and disconnecting clips. Pull it horizontally, and remove it from the lower cover pawls.

m. Remove the instrument passenger lower panel screws, disconnect the wiring harness connector, and remove the lower panel.

n. Remove the instrument driver lower panel bolt and screws, detach the data link connector, pull to disengage the clip and pawl by removing panel in a horizontal direction. Then, disconnect the in-vehicle sensor and all electrical parts. Remove the grommet and remove the hood lock cable.

o. Remove the steering column front lower cover screw, disengage the tab, and then remove the steering column

front lower cover. Move the steering column telescopic to the rear most position, and move the steering column tilt to the top position.

p. Remove the steering column lower cover screws, then disengage the tab and remove steering column lower cover.

q. Remove the steering column upper cover.

r. Remove the wiper and washer switch.

s. Remove the lighting and turn signal switch.

t. Pull the steering lock escutcheon back and remove it.

u. Remove the Combination Meter Assembly by removing the bolts and disconnecting the connector bracket. Remove the bolts and disconnect the wiring harness connector.

✳✳ WARNING

To prevent it from being damaged by interference with the combination meter assembly, protect the combination meter assembly with cloths.

v. Remove the instrument panel side panel screws, then pull the panels to the side, disconnect the clip and pawls, and remove the instrument side panels. Perform for both right-hand and left-hand panels.

w. To remove the cluster lid, insert a pry tool into the gap between the instrument panel and pad, pull back towards you, and disconnect the metal clips. Then, disconnect the wiring harness connectors, and remove the cluster lid.

➡ Cover surroundings with cloth to avoid making scratches or causing damage.

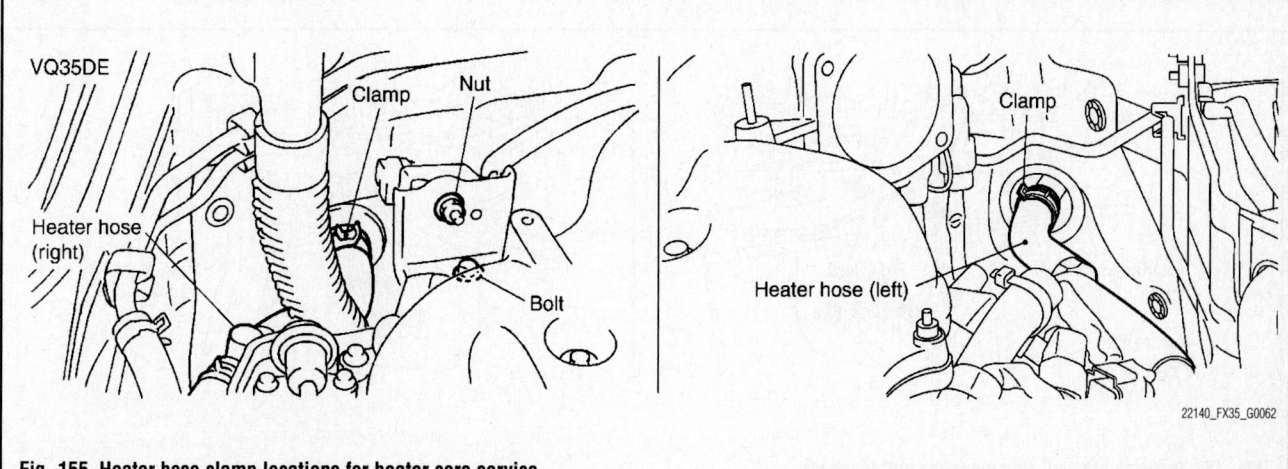

Fig. 155 Heater hose clamp locations for heater core service

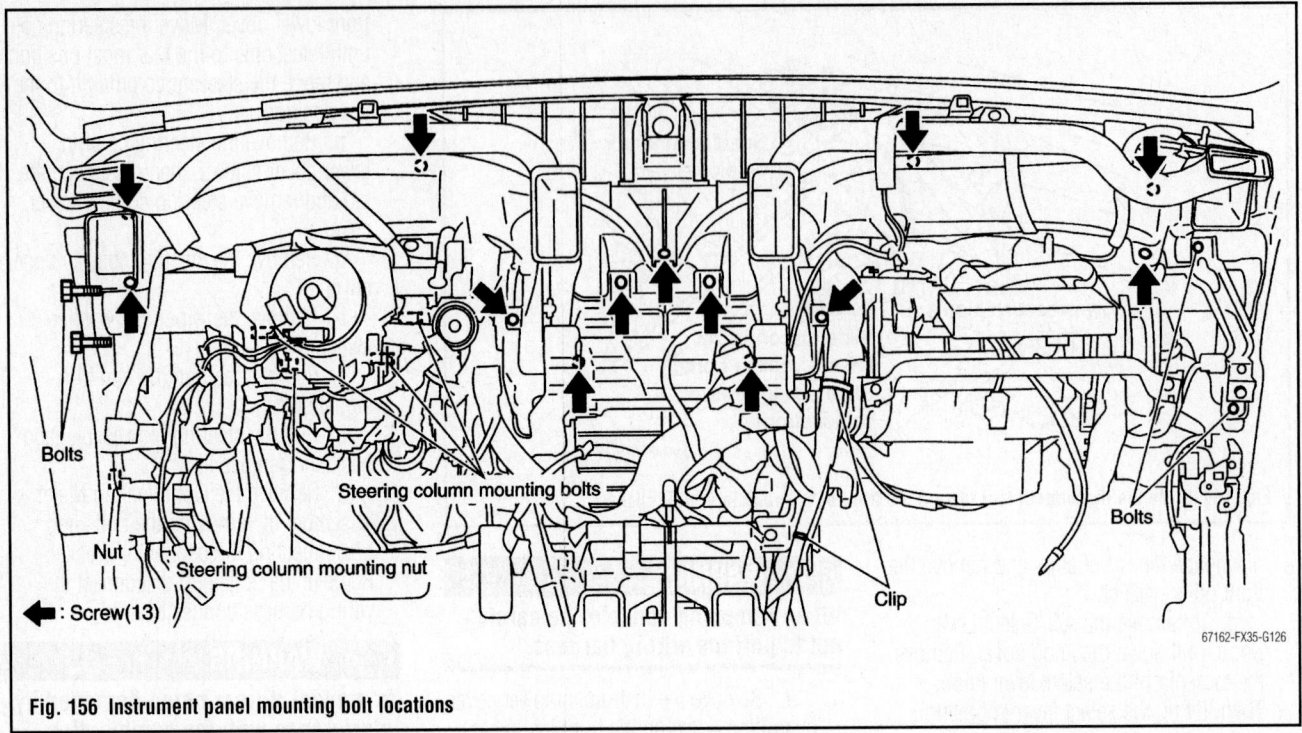

Fig. 156 Instrument panel mounting bolt locations

x. Remove the display unit and audio unit by removing the screws, disconnecting the wiring harness connector, and removing the display unit and audio unit.

✳✳ CAUTION

The unit is heavy, so be careful not to pinch your fingers when working.

y. Insert a thin pry tool into the gaps between the front defroster grille and instrument panel and pad, lift the front defroster grille upward, and remove the front defroster grille. Perform this task for both right-hand and left-hand grilles.

z. Remove the combination meter bracket bolts and remove the bracket from the vehicle.

aa. Once the mounting bolts of the wiring harness clip and steering column assembly are removed, pull the steering column assembly backward, and free the combination meter bracket from the instrument panel and pad.

bb. Remove the side ventilations by inserting a thin pry tool into the gaps between the instrument panel and pad, pull back to disconnect the metal retaining clips. Then, disconnect the door mirror switch wiring harness connectors, and remove the side ventilations.

cc. Remove the instrument panel and pad by removing the bolts and screws. Then, remove the front passenger air bag module, disconnect the wiring harness connectors, and remove the instrument panel and pad from the passenger door opening.

11. Remove the blower unit, as follows:

a. Remove the ECM with the bracket attached.

b. Disconnect the intake door motor connector and blower fan motor connector.

c. Remove the wiring harness clip from the blower unit.

d. Remove the mounting bolt and screws from the blower unit.

✳✳ CAUTION

Move the blower unit rightward, and remove the locating pin and joint. Then, remove the blower unit downward.

e. Remove the blower unit.

12. Remove the instrument stays (driver-side and passenger-side).

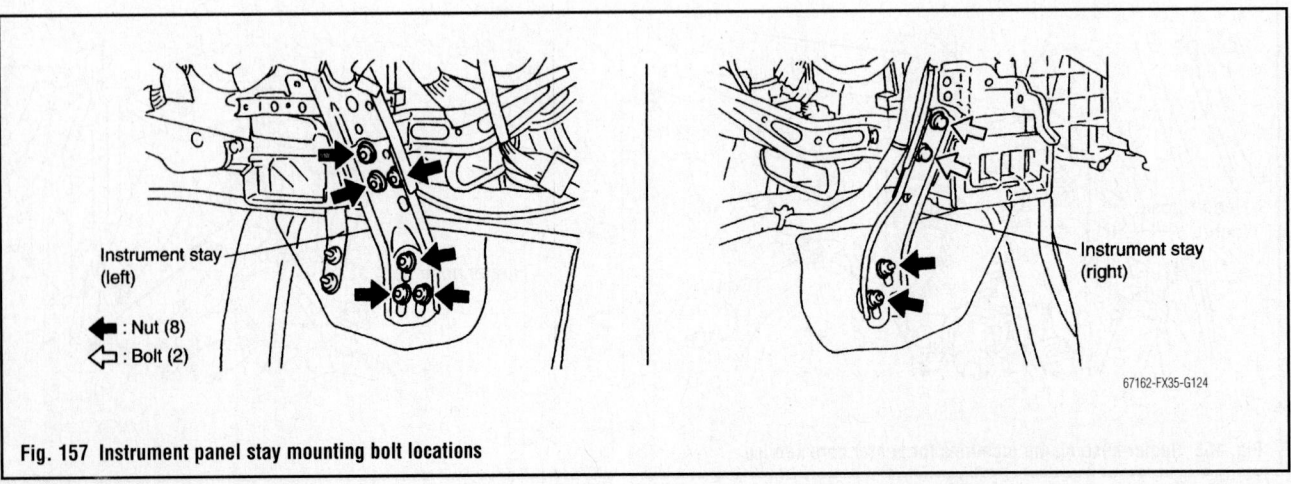

Fig. 157 Instrument panel stay mounting bolt locations

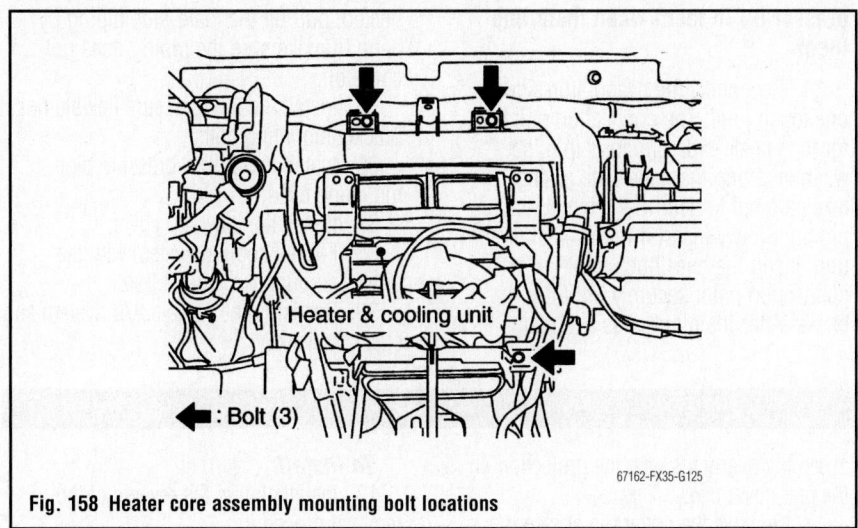

Fig. 158 Heater core assembly mounting bolt locations

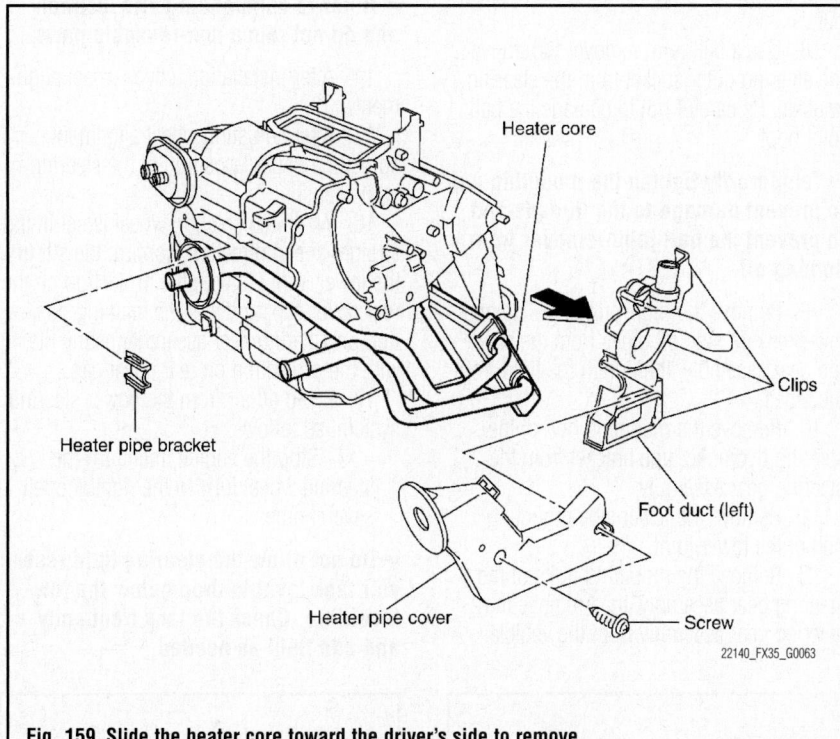

Fig. 159 Slide the heater core toward the driver's side to remove

13. Remove the mounting bolts from the heater and cooling unit.

14. Disconnect the drain hose.

15. Remove the ventilator ducts, defroster nozzle and ducts.

16. Remove the steering member mounting bolts, nut and wiring harness clips.

17. Remove the steering member.

18. Remove the mounting screws and then remove the heater pipe cover.

19. Remove the heater pipe bracket.

20. Slide the heater core (shown in the figure) toward the driver's side.

21. Remove the heater core.

To install:

22. Install the heater core into the heating/cooling unit.

23. Install the heater and cooling unit. Tighten the mounting bolts to 60 inch lbs. (7 Nm).

24. Install the steering member.

25. Install the steering member mounting bolts, nut and wiring harness clips. Tighten the steering member mounting bolts to 108 inch lbs. (12 Nm).

26. Install the ventilator ducts, defroster nozzle and ducts.

27. Reconnect the drain hose.

28. Install the mounting bolts for the heater and cooling unit.

29. Install the instrument stays (driver-side and passenger-side).

✳✳ CAUTION

Make sure the locating pin and joint are securely inserted.

30. Install the blower unit, as follows:
   a. Install the blower unit.
   b. Install the mounting bolt and screws for the blower unit.
   c. Install the wiring harness clip for the blower unit.
   d. Reconnect the intake door motor connector and blower fan motor connector.
   e. Install the ECM with its bracket attached.

31. Install the instrument panel assembly, as follows:
   a. Install the front passenger air bag module and instrument panel and pad.
   b. Install the side ventilations.
   c. Install the combination meter bracket and reinstall the steering column assembly.
   d. Install the front defroster grilles.
   e. Install the display unit and audio unit.
   f. Install the cluster lid.
   g. Install the instrument side panels.
   h. Install the combination meter assembly.
   i. Install the steering lock escutcheon.
   j. Install the lighting and turn signal switch.
   k. Install the wiper and washer switch
   l. Install the steering column upper cover.
   m. Install the steering column lower cover.
   n. Install the steering column front lower cover.
   o. Install the instrument driver lower panel.
   p. Reattach the data link connector.
   q. Reconnect the in-vehicle sensor and all electrical parts.
   r. Install the grommet and hood lock cable.
   s. Install the instrument passenger lower panel.
   t. Install the instrument lower cover.
   u. Connect the center console sub-wiring harness.
   v. Install the console finishers.
   w. Install the console finisher screws.
   x. Reconnect the wiring harness connector.
   y. Install the A/T console finisher.
   z. Install instrument clock finisher.
   aa. Install the A/T select lever knob.

bb. Install the front pillar garnish.

cc. Install the dash side finisher and plastic nuts.

dd. Install the front kicking plate on both sides of the vehicle.

32. Reconnect the two heater hoses to the heater core.

33. Install the electronic control throttle assembly.

➡**Replace the O-rings for A/C piping with new ones, then apply com-** pressor oil to them when installing them.

34. Reconnect the evaporator-side one touch joint. The connection point for the female-side piping is thin, so when inserting the male-side piping, take care not to deform the female-side piping. Slowly insert it in the axial direction. Insert the one-touch joint connection point securely until it clicks. After the piping has been con- nected, pull on the male-side piping by hand to make sure the piping does not come off.

35. Install the low-pressure flexible hose bracket mounting bolts.

36. Install the 2 high-pressure pipe mounting clips.

37. Install the cowl top cover.

38. Fill the cooling system with the proper amount and type of fluid.

39. Recharge the vehicle A/C system and check for leaks.

STEERING

POWER RACK & PINION STEERING GEAR

REMOVAL & INSTALLATION

See Figures 160 through 162.

1. Before servicing the vehicle, refer to the Precautions Section.

※※ CAUTION

The spiral cable may snap due to steering operation if the steering column is separated from the steering gear assembly. Therefore fix the steering wheel with a string to avoid turning it too far.

2. Set the wheels in the straight-ahead position.

3. Raise and safely support the vehicle.

4. Remove the tires from vehicle.

5. Remove the undercover.

6. Confirm the slit of the lower joint fits with the projection on the rear cover cap marking the position on the steering gear assembly nearly fits with the projection on the rear cover cap.

7. Remove the cotter pin at steering outer socket, then loosen the mounting nut.

8. Use a ball joint remover to remove the steering outer socket from the steering knuckle. Be careful not to damage the ball joint boot.

➡**Temporarily tighten the mounting nut to prevent damage to the threads and to prevent the ball joint remover from coming off.**

9. Remove the high-pressure side and low-pressure side oil pipes from the steering gear assembly, then drain the fluid from the pipes.

10. Remove the mounting bolt of the steering hydraulic pipe bracket from the steering gear assembly.

11. Remove the lower side mounting bolt of the lower joint.

12. Remove the mounting bolts of the steering gear assembly, then remove the steering gear assembly from the vehicle.

To install:

13. Installation is the reverse of the removal procedure.

➡**Refer to component parts location and do not reuse non-reusable parts.**

14. After installation, check wheel alignment.

15. After adjusting wheel alignment, adjust the neutral position of the steering angle sensor.

16. When the steering wheel is set in the straight ahead direction, confirm the slit of the lower joint fits with the projection on the rear cover cap, and that the marking position on steering gear assembly nearly fits with the projection on rear cover cap.

17. Bleed all air from the power steering system, as follows:

a. Stop the engine, then turn the steering wheel fully to the right and left several times.

➡**Do not allow the steering fluid reservoir tank level to drop below the low-level line. Check the tank frequently and add fluid as needed.**

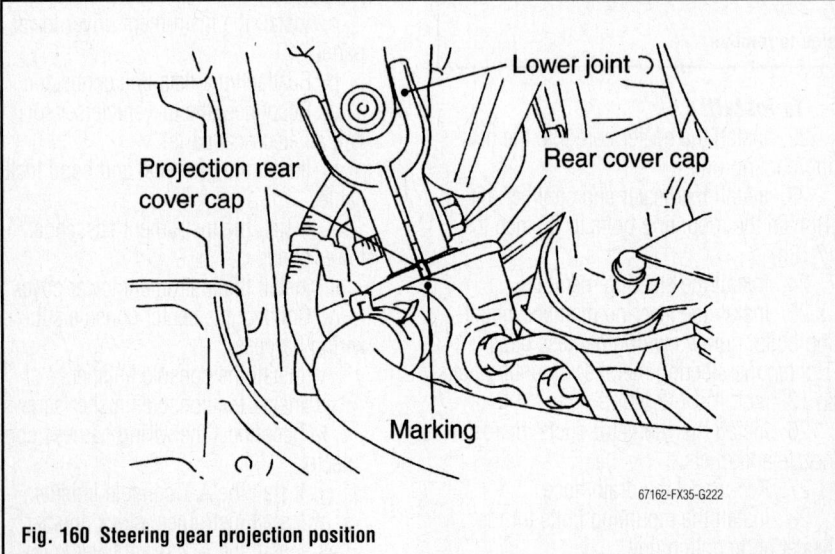

Fig. 160 Steering gear projection position

67162-FX35-G222

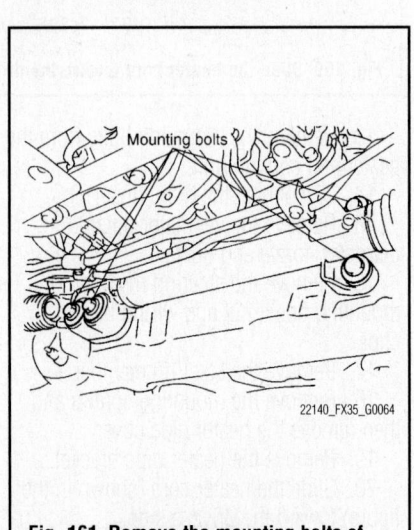

22140_FX35_G0064

Fig. 161 Remove the mounting bolts of the steering gear assembly

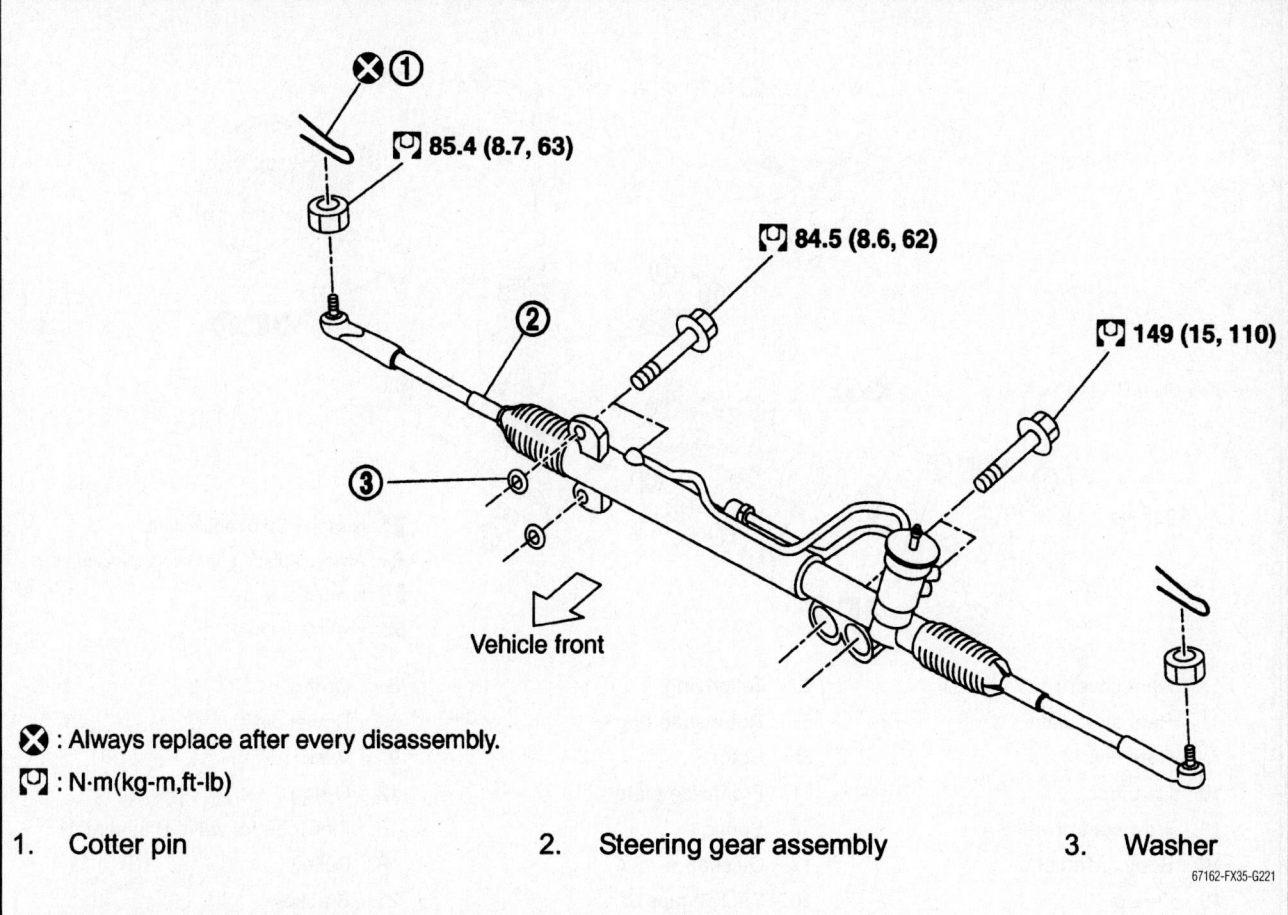

85.4 (8.7, 63)

84.5 (8.6, 62)

149 (15, 110)

Vehicle front

⊗ : Always replace after every disassembly.

⊡ : N·m(kg-m,ft-lb)

1. Cotter pin   2. Steering gear assembly   3. Washer

67162-FX35-G221

Fig. 162 Exploded view of the power steering gear mounting

b. Run the engine at idle speed. Turn the steering wheel fully to the right and then fully to the left, and keep hold for about 3 seconds. Then check whether any fluid leaks have occurred.

c. Repeat sub-step b several times at about 3 second intervals.

✳✳ WARNING

Do not hold the steering wheel in the locked position for more than 10 seconds. There is the possibility that the oil pump may be damaged.

d. Check for air bubbles and cloudiness in the fluid.

e. If air bubbles and/or cloudiness don't fade, stop the engine, and stop air bleeding until the air bubbles and cloudiness fade.

f. Perform until all bubbles and cloudiness are gone.

g. Stop the engine and check the fluid level.

Incomplete air bleeding causes the following. When this happens, bleed the system again:

• Generation of air bubbles in the reservoir tank

• Generation of clicking noise in the oil pump

• Excessive buzzing in the oil pump

➡**When the vehicle is stationary or while the steering wheel is being turned slowly, some noise may be heard from the oil pump or gear. This noise is normal.**

18. Check that the steering wheel turns smoothly when it is turned several times fully to the end of the left and right.

POWER STEERING PUMP

REMOVAL & INSTALLATION

See Figure 163.

1. Before servicing the vehicle, refer to the Precautions Section.

2. Disconnect the negative battery cable.

3. Remove the undercover.

4. Remove the drive belt. Refer to Accessory Drive Belts, removal & installation.

5. Drain the power steering fluid from

the reservoir tank into a suitable container. Properly discard the used fluid.

6. Remove the high pressure and the low pressure lines from the power steering fluid pump.

7. Remove the pump mounting bolts.

8. Remove the pump from the vehicle.

To install:

9. Installation is the reverse of the removal procedure.

10. Bleed the power steering system.

11. Adjust the belt tension.

BLEEDING

1. Before servicing the vehicle, refer to the Precautions Section.

2. Stop the engine.

3. Turn the steering wheel fully to the right and left several times.

➡**Do not allow the fluid level in the reservoir tank to go below the MIN level line. Check and add fluid as needed.**

4. Run the engine at idle speed. Turn the steering wheel fully to the right and then

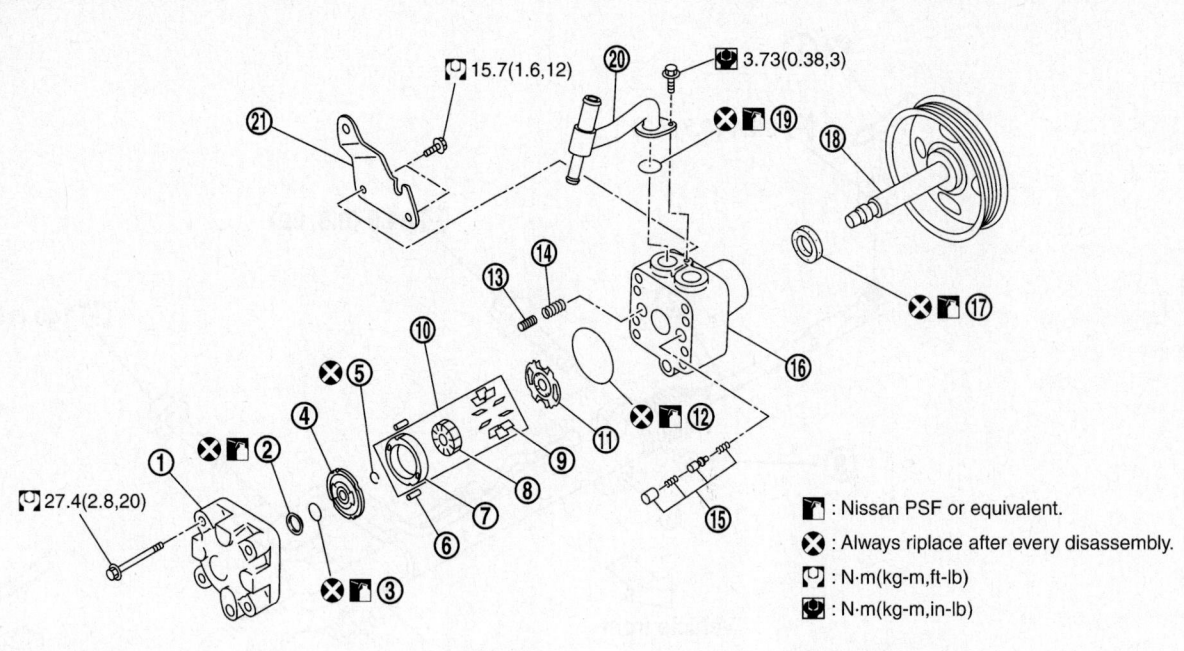

15.7(1.6,12) 3.73(0.38,3)

27.4(2.8,20)

1.	Rear cover	2.	Teflon ring	3.	O-ring
4.	Rear side plate	5.	Rotor snap ring	6.	Dowel pin
7.	Cam ring	8.	Rotor	9.	Vane
10.	Cartridge	11.	Front side plate	12.	O-ring
13.	Flow control valve A	14.	Spring	15.	Flow control valve B assembly
16.	Body assembly	17.	Oil seal	18.	Pulley
19.	O-ring	20.	Suction pipe	21.	Bracket

: Nissan PSF or equivalent.

: Always riplace after every disassembly.

: N·m(kg-m,ft-lb)

: N·m(kg-m,in-lb)

42050_FX35_G0062

Fig. 163 Exploded view of power steering pump

fully to the left. Hold for about 3 seconds. Check for fluid leakage.

5. Repeat the above step several times at 3 second intervals.

❊❊ WARNING

Do not hold the steering wheel in the locked position for more than 10 seconds. Damage to the pump may occur.

6. Check for air bubbles or cloudy fluid. If found, repeat the bleeding procedure.

7. Stop the engine and check the fluid level. Correct as required.

SUSPENSION

FRONT SUSPENSION

CONTROL LINKS

REMOVAL & INSTALLATION

1. Before servicing the vehicle, refer to the Precautions Section.
2. Remove the tires from the vehicle.
3. Remove the undercover.
4. Remove the stabilizer control link lower nut and separate the stabilizer bar and stabilizer control link (the control link is also called the stabilizer connecting rod).
5. Remove the stabilizer control link upper nut.
6. Remove the stabilizer control link.

To install:

7. Check the stabilizer control link for cracks or damage, and replace if necessary.
8. Install in the reverse order of removal.
9. Tighten the stabilizer control link nuts to 75 ft. lbs. (102 Nm).
10. Check the wheel alignment and adjust as necessary.

LOWER BALL JOINT

REMOVAL & INSTALLATION

The front lower control arm ball joints are not separately replaceable from the control arms themselves. If the joints are found

to be defective, the entire assembly must be replaced.

LOWER CONTROL ARM

REMOVAL & INSTALLATION

1. Before servicing the vehicle, refer to the Precautions Section.
2. Raise and safely support the vehicle.
3. Remove the wheels from vehicle.
4. Remove the undercover.
5. Remove the front cross bar.
6. Remove the cotter pin at the lower control arm (also called the transverse link), then loosen the mounting nut.

7. Use a ball joint remover to remove the transverse link from the steering knuckle. Be careful not to damage the ball joint boot.

➡Temporarily tighten the mounting nut to prevent damage to the threads and to prevent the ball joint remover from coming off.

8. Remove the mounting bolts which are at the back of the transverse link (mounting part with body), then separate the transverse link.

9. Remove the mounting bolts which are at the front of the transverse link (mounting part with the front suspension member), then separate the transverse link.

10. Remove the transverse link from the vehicle.

11. Check transverse link and bushing for deformation, cracks, or damage. If any non-standard condition is found, replace it.

12. Check the boot of the ball joint for cracks, or other damage, and also for grease leakage. If any non-standard condition is found, replace it.

13. Manually move ball stud to confirm it moves smoothly with no binding.

➡Before measurement, move ball joint at least ten times by hand to check for smooth movement.

14. Hook a spring scale onto the ball stud tip. Confirm that the spring scale measurement value is within specifications when the ball stud begins to move. If it is outside the specified range, replace the transverse link assembly.

➡Swing torque specification: Less than 5–43 inch lbs. (1–5 Nm), measure value of spring scale: less than 5–43 inch lbs. (1–5 Nm).

15. Attach the mounting nut onto the ball stud. Check that the rotating torque is within specifications with a preload gauge . If it is outside the specified range, replace transverse link assembly.

➡Rotating torque specification: Less than 5–43 inch lbs. (1–5 Nm).

16. Move the tip of ball joint in axial direction to check for looseness. If it is outside the specified range, replace transverse link assembly.

➡Axial end play specification: 0.004 in. (0.1mm).

To install:
17. Install the transverse link on the vehicle.

18. Install and tighten the mounting bolts at the front of the transverse link (mounting part with the front suspension member) to 89 ft. lbs. (120 Nm), then install and tighten the mounting bolts which are at the back of the transverse link (mounting part with body) to 118 ft. lbs. (160 Nm).

19. Reattach the transverse link to the steering knuckle, and tighten the ball joint nut to 105 ft. lbs. (142 Nm). Be careful not to damage the ball joint boot.

20. Install a new cotter pin.

21. Install the front cross bar. Tighten the inner 2 bolts on each end to 33 ft. lbs. (45 Nm) and the outer 2 bolts on each end to 41 ft. lbs. (55 Nm).

22. Install the undercover.

23. Install the tire.

24. Check the wheel alignment.

25. After adjusting wheel alignment, adjust the neutral position of the steering angle sensor.

MACPHERSON STRUT

REMOVAL & INSTALLATION

See Figure 164.

1. Before servicing the vehicle, refer to the Precautions Section.

2. Raise and safely support the vehicle.

3. Make an alignment marking on the camber adjusting bolt and strut for approximate installation alignment later.

4. Remove the wheels from the vehicle.

5. Remove the brake hose lock plate. Then, remove the brake hose from the strut assembly.

6. Remove the wheel sensor wiring harness from the strut assembly.

✳✳ WARNING

Do not pull on the wheel sensor wiring harness.

7. Remove the stabilizer connecting rod upper nut, separate the stabilizer connecting rod and strut assembly.

8. Remove the attaching bolts and nuts between the strut assembly and the steering knuckle.

9. Remove the mounting nuts on the mounting insulator bracket, then remove the strut upper plate, strut spacer and the strut from the vehicle.

To install:

➡Attach strut upper plate as shown in the accompanying illustration.

10. Install the strut upper plate, strut spacer, and the strut onto the vehicle.

11. Install and tighten the mounting nuts on the mounting insulator bracket to 35 ft. lbs. (47 Nm).

12. Install and tighten the attaching bolts and nuts between the strut assembly and the steering knuckle to 134 ft. lbs. (182 Nm).

➡Use the matchmarks made earlier to approximate the front end alignment during installation of the strut assembly.

13. Reattach the stabilizer connecting rod to the strut and tighten the upper nut to 75 ft. lbs. (102 Nm).

14. Install the wheel sensor wiring harness onto the strut assembly.

15. Install the brake hose and the hose lock plate.

16. Install the wheel and tire to the vehicle.

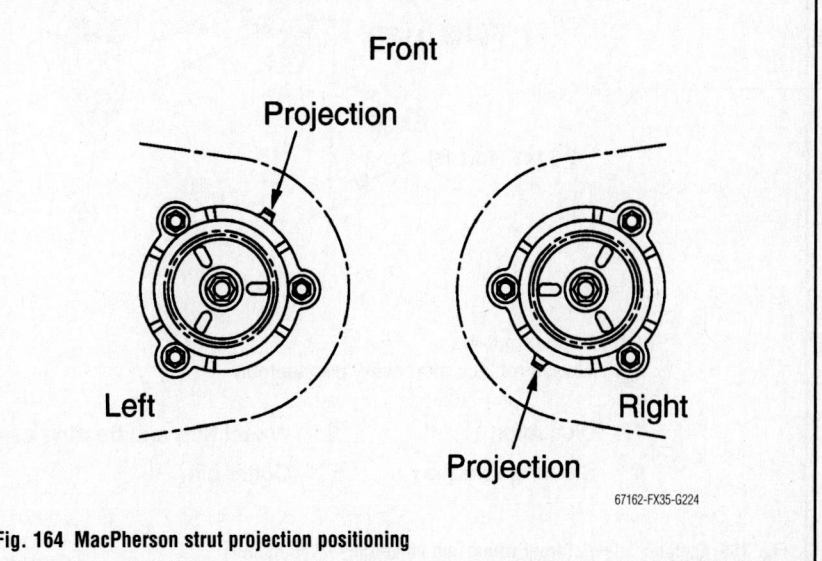

Fig. 164 MacPherson strut projection positioning

67162-FX35-G224

17. After installation, check wheel alignment.

18. After adjusting the wheel alignment, adjust the neutral position of steering angle sensor.

19. Double-check to ensure that the wheel sensor wiring harness is properly routed.

STEERING KNUCKLE

REMOVAL & INSTALLATION

See Figures 165 and 166.

1. Before servicing the vehicle, refer to the Precautions Section.

2. Raise and safely support the vehicle.

3. Remove the appropriate wheel.

4. Remove the brake caliper. Support it in a place where it will not interfere with work.

➡**Avoid depressing brake pedal while brake caliper is removed.**

5. Remove the disc rotor.

6. Remove the wheel sensor from the wheel hub and bearing assembly.

➡**Do not pull on wheel sensor wiring harness.**

7. Remove the cotter pin from the steering outer socket, then loosen the mounting nut.

8. Use a ball joint remover to separate the steering outer socket from the steering knuckle. Be careful not to damage the ball joint boot.

➡**Temporarily tighten the mounting nut to prevent damage to the threads and to prevent the ball joint remover from coming off.**

9. Remove the cotter pin at the transverse link, then loosen the mounting nut.

10. Use a ball joint remover to separate the transverse link from the steering knuckle. Be careful not to damage ball joint boot.

✳✳ CAUTION

Temporarily tighten the mounting nut to prevent damage to the threads and to prevent the ball joint remover from coming off.

11. On AWD models, perform the following:

a. Remove the cotter pin, then remove the lock nut from the halfshaft.

b. Remove the steering knuckle from the halfshaft.

✳✳ WARNING

When removing the steering knuckle, do not apply an excessive angle to the halfshaft joint. Also be careful not to excessively extend the slide joint. Do not hang over the halfshaft without proper support.

12. Remove the mounting bolts and nuts between the strut assembly and the steering knuckle.

13. Remove the steering knuckle from the vehicle.

14. Remove the mounting bolts between the steering knuckle and the wheel hub/bearing assembly.

15. Remove the splash guard and wheel hub/bearing assembly from the steering knuckle.

16. Check for deformities, cracks and damage on all parts and replace if necessary.

17. Inspect the ball joint for boot breakage, axial looseness, and torque of transverse link and steering outer socket ball joint. Maximum of allowable axial end play is 0.002 in. (0.05mm) or less.

To install:

18. Install the splash guard and wheel hub/bearing assembly onto the steering knuckle.

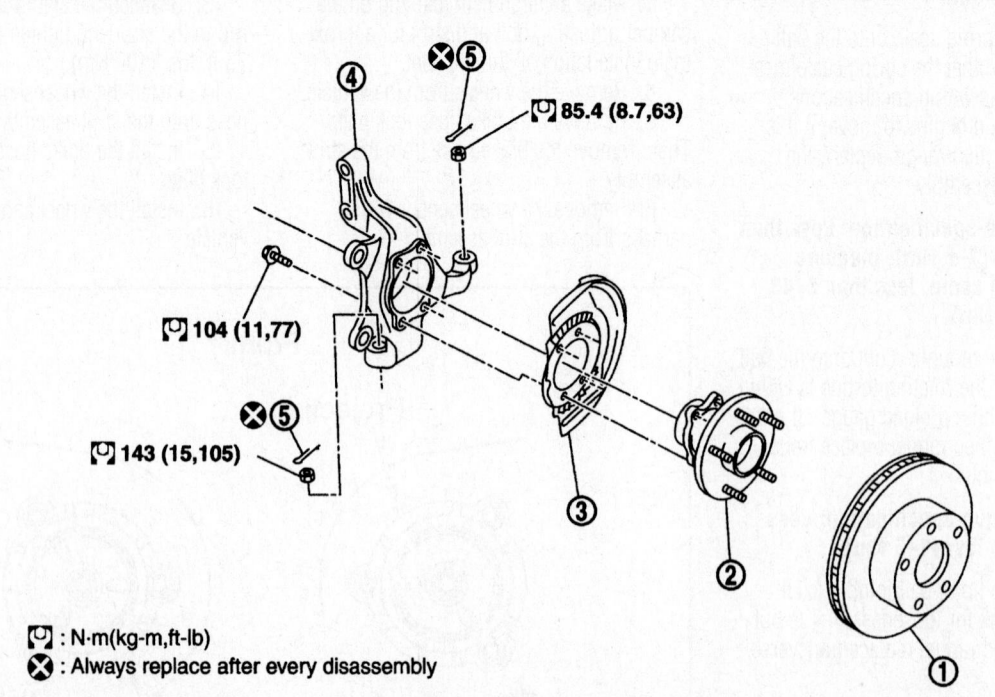

🔧 85.4 (8.7,63)

🔧 104 (11,77)

🔧 143 (15,105)

🔧 : N·m(kg-m,ft-lb)
✖ : Always replace after every disassembly

1. Disc rotor
2. Wheel hub and bearing assembly
3. Splash guard
4. Steering knuckle
5. Cotter pin

67162-FX35-G228

Fig. 165 Exploded view of front wheel hub mounting—2WD models

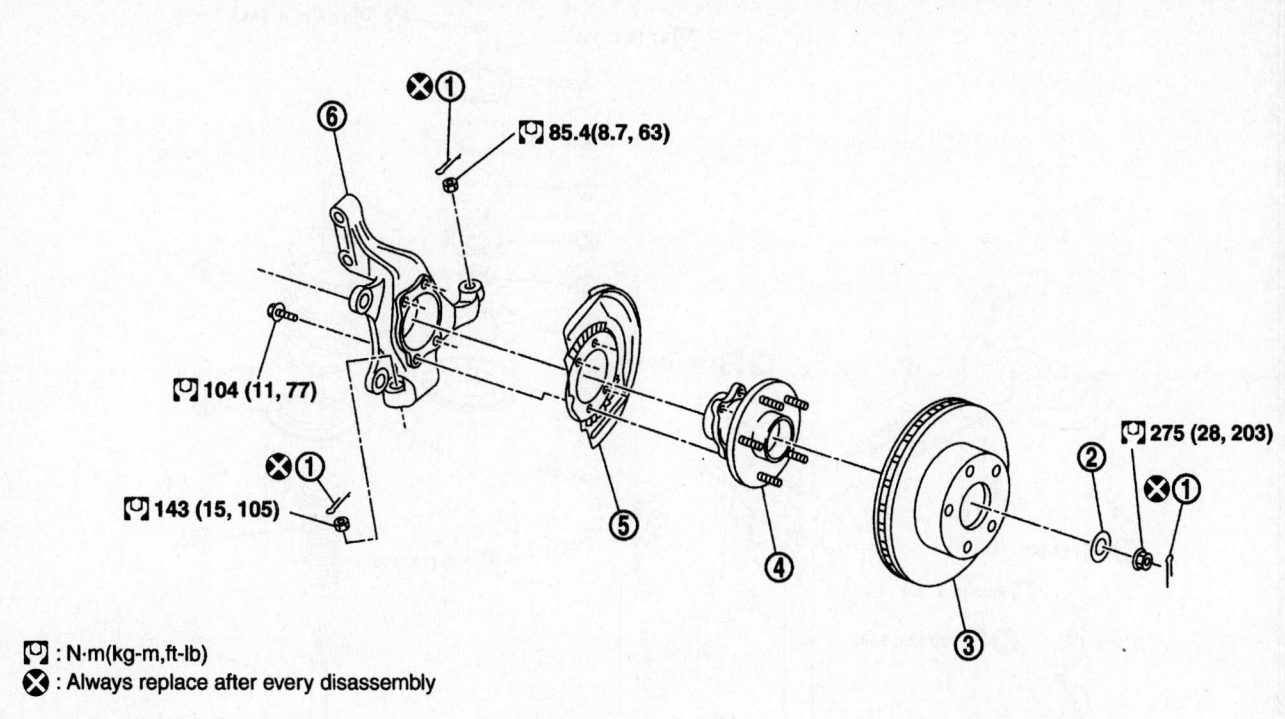

⊡ 85.4(8.7, 63)

⊡ 104 (11, 77)

⊗① 143 (15, 105)

⊡ 275 (28, 203)

⊡ : N·m(kg-m,ft-lb)
⊗ : Always replace after every disassembly

1. Cotter pin
4. Wheel hub and bearing assembly

2. Washer
5. Splash guard

3. Disc rotor
6. Steering knuckle

67162-FX35-G229

Fig. 166 Exploded view of front wheel hub mounting—AWD models

19. Install the mounting bolts between the steering knuckle and the wheel hub/bearing assembly. Tighten the bolts to 77 ft. lbs. (104 Nm).

20. Install the steering knuckle on the vehicle.

21. Install the mounting bolts and nuts to the strut assembly and the steering knuckle. Tighten the bolts to 134 ft. lbs. (182 Nm).

22. On AWD models, perform the following:

    a. Install the steering knuckle onto the halfshaft.

    b. Install and tighten the lock nut on the halfshaft to 203 ft. lbs. (275 Nm). Install a new cotter pin.

23. Reattach the transverse link to the steering knuckle. Tighten the ball joint nut to 105 ft. lbs. (143 Nm).

24. Install a new cotter pin.

25. Reconnect the steering outer socket to the steering knuckle. Tighten the ball joint nut to 63 ft. lbs. (86 Nm).

26. Install a new cotter pin.

27. Install the wheel sensor onto the wheel hub and bearing assembly.

28. Install the disc rotor.

29. Install the brake caliper.

30. Install the wheel.

31. Check wheel alignment.

32. After adjusting wheel alignment, adjust the neutral position of the steering angle sensor.

33. Check the installation condition of the wheel sensor wiring harness.

STABILIZER BAR

REMOVAL & INSTALLATION

See Figure 167.

1. Before servicing the vehicle, refer to the Precautions Section.

2. Raise and safely support the vehicle.

3. Remove the wheels.

4. Remove the fixing bolts and remove the stabilizer connecting rod mount bracket from the suspension arm.

5. Remove the lower side fixing nut on the stabilizer connecting rod and remove the stabilizer connecting rod from the stabilizer bar.

6. Remove the fixing nuts on the stabilizer clamps and remove the stabilizer from the vehicle.

7. Check the stabilizer bar, stabilizer

bushings, stabilizer clamps, stabilizer connecting rod, and stabilizer connecting rod mounting bracket for any deformation, cracks, or damage. Replace if necessary.

To install:

8. Refer to the exploded view for tightening torques. Installation is the reverse of removal.

➡**Do not reuse non-reusable parts during assembly.**

9. The stabilizer bar uses pillow ball type connecting rod, position the ball joint with the case on pillow ball head parallel to the stabilizer bar.

10. When the bushing and clamp are installed to the stabilizer bar, position the bushing and clamp inside of the side slip prevention clamp.

WHEEL HUB & BEARING

REMOVAL & INSTALLATION

See Figures 165 and 166.

1. Before servicing the vehicle, refer to the Precautions Section.

2. Raise and safely support the vehicle.

3. Remove the appropriate wheel.

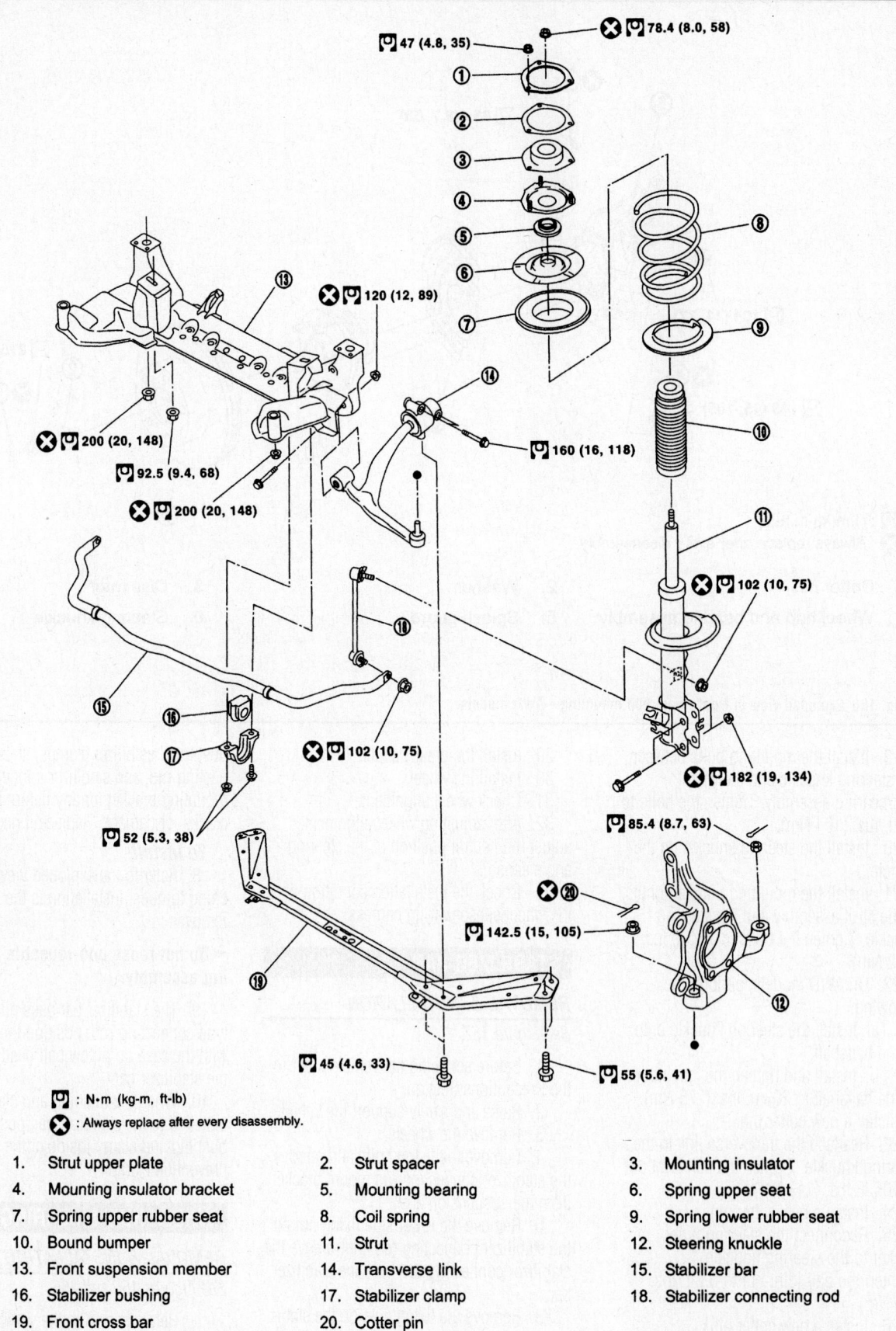

47 (4.8, 35)

78.4 (8.0, 58)

① Strut upper plate
② Strut spacer
③ Mounting insulator
④ Mounting insulator bracket
⑤ Mounting bearing
⑥ Spring upper seat
⑦ Spring upper rubber seat
⑧ Coil spring

120 (12, 89)

200 (20, 148)

92.5 (9.4, 68)

200 (20, 148)

160 (16, 118)

⑨

⑩

⑪

102 (10, 75)

182 (19, 134)

85.4 (8.7, 63)

⑳

142.5 (15, 105)

⑫

102 (10, 75)

52 (5.3, 38)

⑲

45 (4.6, 33)

55 (5.6, 41)

: N·m (kg-m, ft-lb)

: Always replace after every disassembly.

1. Strut upper plate	2. Strut spacer	3. Mounting insulator
4. Mounting insulator bracket	5. Mounting bearing	6. Spring upper seat
7. Spring upper rubber seat	8. Coil spring	9. Spring lower rubber seat
10. Bound bumper	11. Strut	12. Steering knuckle
13. Front suspension member	14. Transverse link	15. Stabilizer bar
16. Stabilizer bushing	17. Stabilizer clamp	18. Stabilizer connecting rod
19. Front cross bar	20. Cotter pin	

67162-FX35-G223

Fig. 167 Exploded view of the front suspension components

4. Remove the brake caliper. Support it in a place where it will not interfere with work.

→**Avoid depressing the brake pedal while brake caliper is removed.**

5. Remove the disc rotor.
6. Remove the wheel sensor from the wheel hub and bearing assembly.

✲✲ WARNING

Do not pull on wheel sensor wiring harness.

7. Remove the cotter pin from the steering outer socket, then loosen the mounting nut.
8. Use a ball joint remover to separate the steering outer socket from the steering knuckle. Be careful not to damage the ball joint boot. Temporarily tighten the mounting nut to prevent damage to the threads and to prevent the ball joint remover from coming off.
9. Remove the cotter pin at the lower control arm (also known as the transverse link), then loosen the mounting nut.
10. Use a ball joint remover to separate the transverse link from the steering knuckle. Be careful not to damage ball joint boot. Temporarily tighten the mounting nut to prevent damage to the threads and to prevent the ball joint remover from coming off.
11. On AWD models, perform the following:
   a. Remove the cotter pin, then remove the lock nut from the halfshaft.

b. Remove the steering knuckle from the halfshaft.

✲✲ WARNING

When removing steering knuckle, do not apply an excessive angle to the halfshaft joint. Also be careful not to excessively extend the slide joint. Do not hang the halfshaft without proper support.

12. Remove the mounting bolts and nuts between the strut assembly and the steering knuckle.
13. Remove the steering knuckle from the vehicle.
14. Remove the mounting bolts between the steering knuckle and the wheel hub/bearing assembly.
15. Remove the splash guard and wheel hub/bearing assembly from the steering knuckle.
16. Check for deformities, cracks, and damage on all parts and replace if necessary.
17. Inspect the ball joint for boot breakage, axial looseness, and torque of transverse link and steering outer socket ball joint. Maximum of allowable axial end play is 0.002 in. (0.05mm) or less.

To install:
18. Install the splash guard and wheel hub/bearing assembly onto the steering knuckle.
19. Install the mounting bolts between the steering knuckle and the wheel hub/bearing assembly. Tighten the bolts to 77 ft. lbs. (104 Nm).

20. Install the steering knuckle on the vehicle.
21. Install the mounting bolts and nuts to the strut assembly and the steering knuckle. Tighten the bolts to 134 ft. lbs. (182 Nm).
22. On AWD models, perform the following:
   a. Install the steering knuckle onto the halfshaft.
   b. Install and tighten the lock nut on the halfshaft to 203 ft. lbs. (275 Nm). Install a new cotter pin.
23. Reattach the transverse link to the steering knuckle. Tighten the ball joint nut to 105 ft. lbs. (143 Nm).
24. Install a new cotter pin.
25. Reconnect the steering outer socket to the steering knuckle. Tighten the ball joint nut to 63 ft. lbs. (86 Nm).
26. Install a new cotter pin.
27. Install the wheel sensor onto the wheel hub and bearing assembly.
28. Install the disc rotor.
29. Install the brake caliper.
30. Install the wheel.
31. Check wheel alignment.
32. After adjusting wheel alignment, adjust the neutral position of the steering angle sensor.
33. Check the installation condition of the wheel sensor wiring harness.

ADJUSTMENT

The front wheel bearings are not adjustable. If the bearings are noisy or become loose, they must be replaced.

SUSPENSION

COIL SPRING

REMOVAL & INSTALLATION

1. Before servicing the vehicle, refer to the Precautions Section.
2. Raise and safely support the vehicle.
3. Remove the rear tire.
4. Position a jack under the rear lower link for support.
5. Loosen the fixing bolt and nut of the rear lower link in the side of the suspension member, and then remove the fixing bolt and nut in the side of the axle.
6. Slowly lower the jack, then remove the upper seat, coil spring and rubber sheet from the rear lower link.
7. Remove the fixing bolt and nut in the

side of the rear suspension member to remove the rear lower link.
8. Check the rear lower link, bushing, and coil spring for deformation, cracks, and damage. Replace the rear lower link and coil spring, if necessary.

To install:
9. Position the rear lower link on the vehicle.
10. Install and tighten the fixing bolt and nut in the side of the rear suspension member to 48 ft. lbs. (65 Nm).
11. Position the upper seat, coil spring and rubber sheet in place, and then slowly raise the jack under the rear lower link.

→**Match up the rubber seat indentions and rear lower link grooves. Also, make sure the spring is not upside**

REAR SUSPENSION

down. The top and bottom are indicated by paint color.

12. Install and tighten the fixing bolt and nut in the side of the axle to 77 ft. lbs. (105 Nm).
13. Slowly lower the jack from under the rear lower link.
14. Install the rear tire.

✲✲ CAUTION

Perform the final tightening of the rear suspension member and axle installation position (rubber bushing) under unladen conditions with the tires on level ground.

15. Check the wheel alignment.
16. After adjusting wheel alignment, adjust the neutral position of the steering angle sensor.

CONTROL ARMS/LINKS

REMOVAL & INSTALLATION

Upper Control Arm

1. Before servicing the vehicle, refer to the Precautions Section.
2. Raise and safely support the vehicle.
3. Remove the rear tire.
4. Remove the stabilizer connecting rod mounting bracket from the suspension arm.
5. Remove the halfshaft from the vehicle.
6. Remove the cotter pin of the suspension arm ball joint, and loosen the nut.
7. Use a ball joint remover or suitable tool to remove the suspension arm from the axle. Be careful not to damage the ball joint boot.

➡**Temporarily tighten the mounting nut to prevent damage to the threads and to prevent the ball joint remover from coming off.**

8. Remove the fixing nuts and bolts between the suspension arm and the rear suspension member.
9. Remove the suspension arm from the vehicle.
10. Check the suspension arm and bushing for deformation, cracks, or damage. If any non-standard condition is found, replace it.
11. Check the boot of the ball joint for cracks or damage and also for grease leakage.
12. Manually move the ball stud to confirm it moves smoothly with no binding.

➡**Before measuring, move ball joint at least 10 times by hand to check for smooth movement.**

13. Hook a spring scale at the cotter pin mounting hole. Confirm the spring scale measurement value is within 2–15 lbs. (10–66 N) when the ball joint stud begins moving. If it is outside the specified range, replace the suspension arm assembly.
14. Attach the mounting nut to the ball stud. Make sure the rotating torque is within 5–30 inch lbs. (1–3 Nm) with a preload gauge . If it is outside the specified range, replace the suspension arm assembly.
15. Move the tip of the ball joint in the axial direction to check for looseness. If it is outside the specified range of 0 in. (0mm), replace the suspension arm assembly.

To install:

16. Install the suspension arm onto the vehicle.

17. Install the fixing nuts and bolts between the suspension arm and the rear suspension member. Tighten them to 53 ft. lbs. (73 Nm).
18. Reattach the suspension arm ball joint to the axle. Tighten the ball joint nut to 96 ft. lbs. (130 Nm).
19. Install a new cotter pin.
20. Install the halfshaft.
21. Install the stabilizer connecting rod mounting bracket onto the suspension arm, and tighten the 2 mounting bolts to 41 ft. lbs. (55 Nm).
22. Install the rear tire.

➡**Do not reuse non-reusable parts.**

23. Perform the final tightening of the rear suspension member installation position (rubber bushing) under unladen conditions with the tires on level ground.
24. Check wheel alignment.
25. After adjusting wheel alignment, adjust the neutral position of the steering angle sensor.

Front Lower Link

1. Before servicing the vehicle, refer to the Precautions Section.
2. Raise and safely support the vehicle.
3. Remove the rear tire.
4. Position a jack under the rear lower link for support.
5. Remove the front lower link protector.
6. Remove the shock absorber assembly from the vehicle.
7. Remove the mounting nut and bolt between the front lower link and the axle.
8. Remove the mounting nut and bolt between the front lower link and the rear suspension member.
9. Remove the front lower link from the vehicle.
10. Check the front lower link and bushing for any deformation, cracks, or damage. Replace it if necessary.

To install:

11. Position the front lower link on the vehicle.
12. Install and tighten the mounting nut and bolt between the front lower link and the rear suspension member to 74 ft. lbs. (101 Nm).
13. Install and tighten the mounting nut and bolt between the front lower link and the axle to 77 ft. lbs. (105 Nm).
14. Install the shock absorber assembly.
15. Install the front lower link protector.
16. Slowly lower the jack from under the rear lower link.
17. Install the rear tire.

✴✴ CAUTION

Perform final tightening of the rear suspension member and axle installation position (rubber bushing) under unladen conditions with the tires on level ground.

18. Check the wheel alignment.
19. After adjusting wheel alignment, adjust the neutral position of the steering angle sensor.

Rear Lower Link

1. Before servicing the vehicle, refer to the Precautions Section.
2. Raise and safely support the vehicle.
3. Remove the rear tire.
4. Position a jack under the rear lower link for support.
5. Loosen the fixing bolt and nut of the rear lower link in the side of the suspension member and then remove the fixing bolt and nut in the side of the axle.
6. Slowly lower the jack, then remove the upper seat, coil spring, and rubber sheet from the rear lower link.
7. Remove the fixing bolt and nut in the side of the rear suspension member to remove the rear lower link.
8. Check the rear lower link, bushing, and coil spring for deformation, cracks, and damage. Replace the rear lower link and coil spring, if necessary.

To install:

9. Position the rear lower link on the vehicle.
10. Install and tighten the fixing bolt and nut in the side of the rear suspension member to 48 ft. lbs. (65 Nm).
11. Position the upper seat, coil spring and rubber sheet in place, and then slowly raise the jack under the rear lower link.

➡**Match up the rubber seat indentions and rear lower link grooves. Also, make sure the spring is not upside down. The top and bottom are indicated by paint color.**

12. Install and tighten the fixing bolt and nut in the side of the axle to 77 ft. lbs. (105 Nm).
13. Slowly lower the jack from under the rear lower link.
14. Install the rear tire.

➡**Perform final tightening of the rear suspension member and axle installation position (rubber bushing) under unladen conditions with the tires on level ground.**

15. Check the wheel alignment.

16. After adjusting wheel alignment, adjust the neutral position of the steering angle sensor.

SHOCK ABSORBER

REMOVAL & INSTALLATION

1. Before servicing the vehicle, refer to the Precautions Section.

2. Raise and safely support the vehicle.

3. Remove the rear tire.

4. Position a jack or equivalent support under the rear lower link.

5. Remove the fixing bolt in the lower side of the shock absorber assembly.

6. Remove the attaching nuts in the upper side of the shock absorber assembly and remove the shock absorber assembly from the vehicle.

7. Check the shock absorber assembly for deformation, cracks, or damage, and replace if necessary.

8. Check the piston rod for damage, uneven wear, or distortion, and replace if necessary.

9. Check the welded and sealed areas for oil leakage, and replace if necessary.

To install:

10. Position the shock absorber assembly on the vehicle.

11. Install and tighten the attaching nuts in the upper side of the shock absorber assembly to 22 ft. lbs. (30 Nm).

12. Install a new upper shock absorber mounting nut and tighten to 33 ft. lbs. (45 Nm).

13. Install the fixing bolt in the lower side of the shock absorber assembly and tighten until snug.

14. Remove the jack or equivalent support from under the rear lower link.

15. Install the tire.

16. With the weight of the vehicle resting on the suspension (empty vehicle), tighten the shock absorber lower fixing bolt to 66 ft. lbs. (89 Nm).

17. Check the wheel alignment.

18. After adjusting wheel alignment, adjust the neutral position of the steering angle sensor.

TESTING

1. Before servicing the vehicle, refer to the Precautions Section.

2. Check for oil leakage around seals and welds.

3. Move the piston rod up and down to check if it operates smoothly without any binding.

WHEEL HUB & BEARING

REMOVAL & INSTALLATION

See Figure 168.

1. Before servicing the vehicle, refer to the Precautions Section.

2. Raise and safely support the vehicle.

3. Remove the rear wheel.

4. Remove the brake caliper. Hang it in a place where it will not interfere with work.

➡**Avoid depressing brake pedal while brake caliper is removed.**

5. Remove the disc rotor.

6. Remove the wheel sensor from the axle.

✳✳ WARNING

Do not pull on the wheel sensor wiring harness.

7. Remove the cotter pin, then remove the locknut from the halfshaft.

8. Separate the halfshaft from the wheel hub and bearing assembly by lightly tapping the end with a suitable hammer and

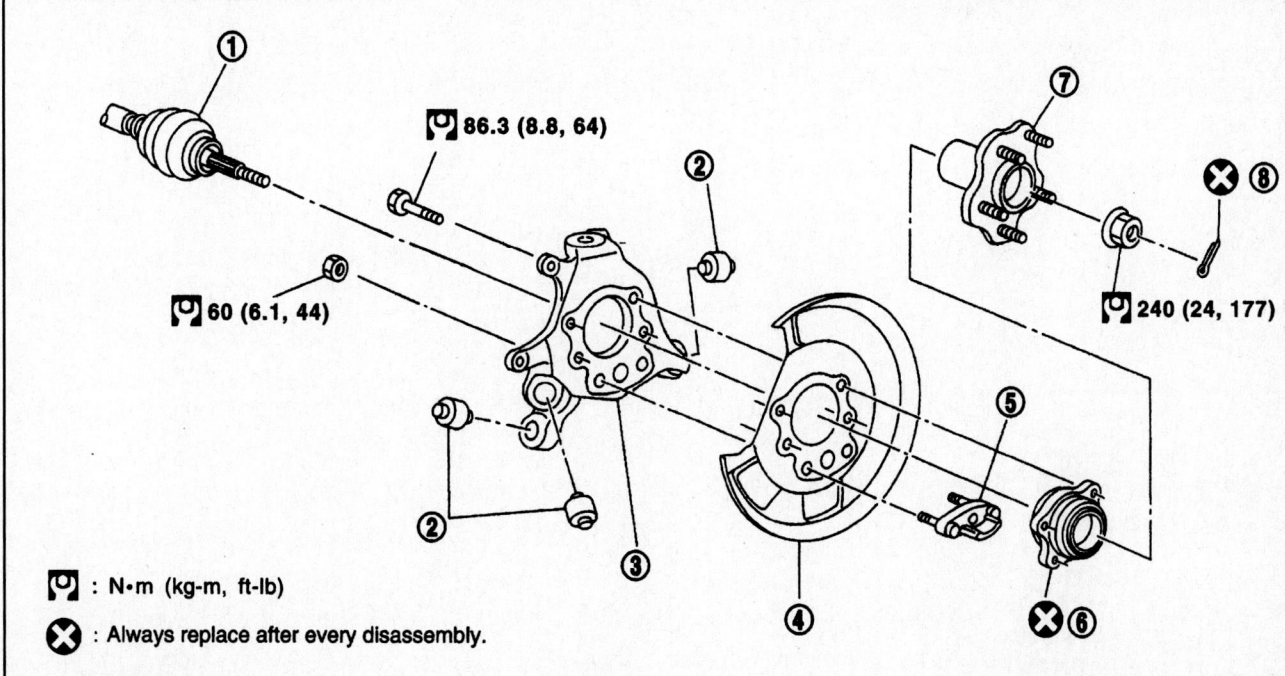

🔧 : N•m (kg-m, ft-lb)

✖ : Always replace after every disassembly.

1.	Drive shaft	2.	Bushing	3.	Axle
4.	Back plate	5.	Anchor block	6.	Wheel bearing
7.	Wheel hub	8.	Cotter pin		

67162-FX35-G230

Fig. 168 Exploded view of rear wheel hub mounting

block of wood. If it is hard to separate, use a suitable puller.

9. Remove the mounting bolts of the wheel hub/bearing assembly, then remove the wheel hub/bearing assembly from the axle.

10. Remove the parking brake cable and parking brake shoe from the brake backing plate.

11. Remove the mounting nuts of the anchor block, then remove the anchor block and backing plate from the axle.

12. Loosen mounting bolts and nuts of the front lower link, radius rod, and rear lower link on the side of the suspension member.

13. Set a jack under the rear lower link, then remove the mounting bolt in the front lower link side of the shock absorber.

14. Remove the bolt and nut in the axle side of the rear lower link, then remove the coil spring.

15. Remove the mounting bolts and nuts in the axle side of the front lower link and radius rod.

16. Remove the suspension arm and cotter pin at the axle, then loosen the mounting nut.

17. Use a ball joint remover (or equivalent) to remove the suspension arm from the axle. Be careful not to damage the ball joint boot.

➡**Temporarily tighten the mounting nut to prevent damage to the threads and to prevent the ball joint remover from coming off.**

18. Remove the axle from the vehicle.

19. Inspect the Ball Joint for boot breakage, axial looseness, and torque of suspension arm ball joint. Maximum of allowable axial end play is 0.00 in. (0.0mm).

To install:

20. Install the axle in the vehicle.

21. Reattach the upper control arm to the axle.

22. Tighten the upper control arm mounting nut to 96 ft. lbs. (130 Nm), and install a new cotter pin at the axle.

23. Install the mounting bolts and nuts in the axle side of the front lower link and radius rod.

24. Install the coil spring, then install the bolt and nut in the axle side of the rear lower link.

25. Install the mounting bolt in the front lower link side of the shock absorber.

26. Tighten the mounting bolts and nuts of the front lower link, radius rod, and rear lower link on the side of the suspension member.

27. Install the anchor block and backing plate onto the axle.

28. Install and tighten the mounting

nuts of the anchor block to 44 ft. lbs. (60 Nm).

29. Install the parking brake cable and parking brake shoe from the brake backing plate.

30. Install the wheel hub/bearing assembly onto the axle, then install the mounting bolts of the wheel hub/bearing assembly and tighten to 64 ft. lbs. (86 Nm).

31. Install the halfshaft to the wheel hub/bearing assembly.

32. Install and tighten the locknut on the halfshaft to 177 ft. lbs. (240 Nm), then install a new cotter pin.

33. Install the wheel sensor on the axle.

34. Install the disc rotor.

35. Install the brake caliper.

36. Install the appropriate wheel.

➡**Perform the final tightening of the rear suspension fasteners under unladen conditions with the tires on level ground.**

37. Check the wheel alignment.

38. After adjusting wheel alignment, adjust the neutral position of the steering angle sensor.

ADJUSTMENT

The rear wheel bearings are not adjustable. If the bearings are noisy or become loose, they must be replaced.

INFINITI

FX35 • FX50

2

SPECIFICATIONS AND MAINTENANCE CHARTS

ENGINE AND VEHICLE IDENTIFICATION

	Engine						Model Year	
Code ①	Liters (cc)	Cu. In.	Cyl.	Fuel Sys.	Engine Type	Eng. Mfg.	Code ②	Year
VQ35HR	3.5 (3498)	213.5	6	SFI	DOHC	Nissan	9	2009
VK50VE	5.0 (5026)	306.7	8	SFI	DOHC	Nissan	A	2010

SFI: Sequential Fuel Injection

DOHC: Double Overhead Camshaft

① Stamped on the upper rear of the engine block, just behind one of the cylinder heads

② 10th digit of the Vehicle Identification Number (VIN)

37663_FX35_C0001

GENERAL ENGINE SPECIFICATIONS

All measurements are given in inches.

Year	Model	Engine Displacement Liters	Engine Series ID/VIN	Net Horsepower @ rpm	Net Torque @ rpm (ft. lbs.)	Bore x Stroke (in.)	Compression Ratio	Oil Pressure @ rpm
2009	FX35	3.5	VQ35HR	303@6800	262@4800	3.76 x 3.21	10.6:1	43@2000
	FX50	5.0	VK50VE	390@6500	369@4400	3.76 x 3.45	10.9:1	
2010	FX35	3.5	VQ35DE	275@6200	268@4800	3.76 x 3.21	10.3:1	43@2000
	FX50	5.0	VK50VE	390@6500	369@4400	3.76 x 3.45	10.9:1	

37663_FX35_C0002

GASOLINE ENGINE TUNE-UP SPECIFICATIONS

Year	Engine Displacement Liters	Engine Series ID/VIN	Spark Plug Gap (in.)	Ignition Timing (deg.) ① MT	Ignition Timing (deg.) ① AT	Fuel Pump (psi)	Idle Speed (rpm) MT	Idle Speed (rpm) AT	Valve Clearance (in.) ① Intake	Valve Clearance (in.) ① Exhaust
2009	3.5	VQ35HR	0.043	—	10-20B	51	—	625-725	0.010-0.013	0.011-0.015
	5.0	VK50VE	0.043	—	10-20B	51	—	600-700	0.010-0.013	0.011-0.015
2010	3.5	VQ35HR	0.043	—	10-20B	51	—	625-725	0.010-0.013	0.011-0.015
	5.0	VK50VE	0.043	—	10-20B	51	—	600-700	0.010-0.013	0.011-0.015

NOTE: The Vehicle Emission Control Information label reflects specification changes made during production.

Follow the figures on the label if they differ from those in this chart.

B: Before top dead center

① Engine cold - approximately 68°F (20°C)

37663_FX35_C0003

CAPACITIES

Year	Model	Engine Displacement Liters	Engine Series ID/VIN	Engine Oil with Filter (qts.)	Transmission (pts.) Manual	Transmission (pts.) Auto. ①	Transfer Case (pts.)	Drive Axle Front (pts.)	Drive Axle Rear (pts.)	Fuel Tank (gal.)	Cooling System (qts.)
2009	FX35	3.5	VQ35HR	5.1	—	19.5	2.1	1.4	3.0	23.8	9.8
	FX50	5.0	VK50VE	7.1	—	24.0	2.1	1.4	3.0	23.8	11.6
2010	FX35	3.5	VQ35HR	5.1	—	19.5	2.1	1.4	3.0	23.8	9.8
	FX50	5.0	VK50VE	7.1	—	24.0	2.1	1.4	3.0	23.8	11.6

NOTE: All capacities are approximate. Add fluid gradually and check to be sure a proper fluid level is obtained.

① Drain and refill

37663_FX35_C0004

FLUID SPECIFICATIONS

Year	Model	Engine Displacement Liters	Engine Series ID/VIN	Engine Oil	Auto. Trans.	Drive Axle	Transfer Case	Power Steering Fluid	Brake Master Cylinder	Cooling System
2009	FX35	3.5	VQ35HR	5W-30	①	②	③	④	⑤	⑥
	FX50	5.0	VK50VE	5W-30	①	②	③	④	⑤	⑥
2010	FX35	3.5	VQ35HR	5W-30	①	②	③	④	⑤	⑥
	FX50	5.0	VK50VE	5W-30	①	②	③	④	⑤	⑥

DOT: Department Of Transportation

① Genuine NISSAN Matic S ATF

② VQ35HR: Genuine NISSAN Differential Oil Hypoid Super GL-5 80W-90 or API GL-5 Viscosity SAE 80W-90

   VK50VE and VQ35HR with towing package: Genuine NISSAN Differential Oil Synthetic 75W-90 or API GL-5 Viscosity SAE 75W-90

③ Genuine NISSAN Matic J ATF

④ Genuine NISSAN PSF or equivalent

⑤ Genuine NISSAN Super Heavy Duty Brake Fluid or equivalent DOT 3

⑥ Genuine NISSAN Long Life Antifreeze/Coolant or equivalent

37663_FX35_C0005

VALVE SPECIFICATIONS

Year	Engine Displacement Liters	Engine Series ID/VIN	Seat Angle (deg.)	Face Angle (deg.)	Spring Test Pressure (lbs. @ in.)	Spring Installed Height (in.)	Stem-to-Guide Clearance (in.) Intake	Stem-to-Guide Clearance (in.) Exhaust	Stem Diameter (in.) Intake	Stem Diameter (in.) Exhaust
2009	3.5	VQ35HR	45.15-45.45	NA	NA	1.457	0.0008-0.0021	0.0012-0.0022	0.2348-0.2354	0.2347-0.2350
	5.0	VK50VE	45.15-45.45	NA	83-96 @1.010	1.396	0.0008-0.0021	0.0012-0.0025	0.2348-0.2354	0.2344-0.2350
2010	3.5	VQ35HR	45.15-45.45	NA	NA	1.457	0.0008-0.0021	0.0012-0.0022	0.2348-0.2354	0.2347-0.2350
	5.0	VK50VE	45.15-45.45	NA	83-96 @1.010	1.396	0.0008-0.0021	0.0012-0.0025	0.2348-0.2354	0.2344-0.2350

NA Not Available

37663_FX35_C0006

CAMSHAFT AND BEARING SPECIFICATIONS CHART

All measurements are given in inches.

Year	Engine Displ. Liters	Engine Series ID/VIN	Journal Dia.	Brg. Oil Clearance	Shaft End-play	Runout	Journal Bore	Lobe Height	
								Intake	Exhaust
2009	3.5	VQ35HR	①	②	0.0045-0.0074	0.0020 max.	NA	1.8057-1.8132	1.8061-1.8136
	5.0	VK50VE	③	④	0.0045-0.0074	0.0020 max.	NA	1.7904-1.7978	1.7904-1.7978
2010	3.5	VQ35HR	①	②	0.0045-0.0074	0.0020 max.	NA	1.8057-1.8132	1.8061-1.8136
	5.0	VK50VE	③	④	0.0045-0.0074	0.0020 max.	NA	1.7904-1.7978	1.7904-1.7978

NA: Not Available

① No. 1: 1.0211-1.0218 in.
   No. 2, 3, 4: 0.9230-0.9238 in.

② No. 1: 0.0018-0.0034 in.
   No. 2, 3, 4: 0.0014-0.0030 in.

③ No. 1: 1.0211-1.0218 in.
   No. 2, 3, 4, 5: 1.0217-1.0224 in.

④ No. 1: 0.0018-0.0034 in.
   No. 2, 3, 4, 5: 0.0012-0.0028 in.

37663_FX35_C0007

CRANKSHAFT AND CONNECTING ROD SPECIFICATIONS

All measurements are given in inches.

Year	Engine Displacement Liters	Engine Series ID/VIN	Crankshaft				Connecting Rod		
			Main Brg. Journal Dia.	Main Brg. Oil Clearance	Shaft End-play	Thrust on No.	Journal Diameter	Oil Clearance	Side Clearance
2009	3.5	VQ35HR	①	0.0014-0.0018	0.0039-0.0098	3	②	0.0016-0.0021	NA
	5.0	VK50VE	③	0.0014-0.0018	0.0039-0.0102	3	2.1242-2.1250	0.0016-0.0021	NA
2010	3.5	VQ35HR	①	0.0014-0.0018	0.0039-0.0098	3	②	0.0016-0.0021	NA
	5.0	VK50VE	③	0.0014-0.0018	0.0039-0.0102	3	2.1242-2.1250	0.0016-0.0021	NA

NA: Not Available

① Depends on the grade of the crankshaft. The nominal range is: 2.5571-2.5581 in.

② Depends on the grade of the crankshaft. The nominal range is: 2.1242-2.1250 in.

③ Journals No 1 and 5: 2.5173-2.5183 in.

   Journals No 2, 3 and 4: 2.5174-2.5182 in.

37663_FX35_C0008

PISTON AND RING SPECIFICATIONS

All measurements are given in inches.

Year	Engine Displ. Liters	Engine Series ID/VIN	Piston Clearance	Ring Gap			Ring Side Clearance		
				Top Compression	Bottom Compression	Oil Control	Top Compression	Bottom Compression	Oil Control
2009	3.5	VQ35HR	0.0004-0.0012	0.0091-0.0130	0.0130-0.0189	0.0067-0.0185	0.0016-0.0031	0.0012-0.0028	0.0022-0.0061
	5.0	VK50VE	0.0004-0.0012	0.0091-0.0130	0.0130-0.0189	0.0067-0.0185	0.0016-0.0031	0.0012-0.0028	0.0022-0.0061
2010	3.5	VQ35HR	0.0004-0.0012	0.0091-0.0130	0.0130-0.0189	0.0067-0.0185	0.0016-0.0031	0.0012-0.0028	0.0022-0.0061
	5.0	VK50VE	0.0004-0.0012	0.0091-0.0130	0.0130-0.0189	0.0067-0.0185	0.0016-0.0031	0.0012-0.0028	0.0022-0.0061

37663_FX35_C0009

TORQUE SPECIFICATIONS

All readings in ft. lbs.

Year	Engine Displacement Liters	Engine Series ID/VIN	Cylinder Head Bolts	Main Bearing Bolts	Rod Bearing Bolts	Crankshaft Damper Bolts	Flywheel Bolts	Manifold		Spark Plugs	Oil Pan Drain Plug
								Intake	Exhaust		
2009	3.5	VQ35HR	①	②	③	④	65	⑤	22	14	25
	5.0	VK50VE	⑥	⑦	⑧	⑨	65	8	21	14	25
2010	3.5	VQ35HR	①	②	③	④	65	⑤	22	14	25
	5.0	VK50VE	⑥	⑦	⑧	⑨	65	8	21	14	25

NOTE: Dip main bearing bolts, crankshaft damper bolt, and flywheel bolts in clean engine oil prior to tightening.

① Step 1: Tighten in sequence to 77 ft. lbs.
   Step 2: Completely loosen all in reverse sequence
   Step 3: Tighten in sequence to 30 ft. lbs.
   Step 4: Tighten 95 degrees
   Step 5: Tighten another 95 degrees

② Step 1: Tighten bolts 17 to 26 in sequence to 18 ft. lbs.
   Step 2: Repeat Step 1.
   Step 3: Tighten bolts 1 to 16 in sequence to 26 ft. lbs.
   Step 4: Tighten 90 degrees

③ Step 1: Tighten to 21 ft. lbs.
   Step 2: Completely loosen bolts
   Step 3: Tighten to 18 ft. lbs.
   Step 4: Tighten 90 degrees

④ Step 1: Tighten to 33 ft. lbs.
   Step 2: Tighten 90 degrees

⑤ Step 1: Tighten to 5 ft. lbs.
   Step 2: Tighten to 19 ft. lbs.

⑥ Step 1: Tighten in sequence to 30 ft. lbs.
   Step 2: Tighten 75 degrees
   Step 3: Completely loosen all in reverse sequence
   Step 4: Tighten in sequence to 30 ft. lbs.
   Step 5: Tighten 65 degrees
   Step 6: Tighten another 65 degrees

⑦ Step 1: Tighten M12 bolts in sequence 1-10 to 40 ft. lbs.
   Step 2: Tighten M9 bolts in sequence 11-20 to 14 ft. lbs.
   Step 3: Tighten M12 bolts in sequence 1-10 another 90 degrees
   Step 4: Tighten M9 bolts in sequence 11-20 another 90 degrees
   Step 5: Tighten M10 bolts in sequence 21-30 to 36 ft. lbs.

⑧ Step 1: Tighten to 21 ft. lbs.
   Step 2: Completely loosen bolts
   Step 3: Tighten to 18 ft. lbs.
   Step 4: Tighten 90 degrees

⑨ Step 1: Tighten to 116 ft. lbs.
   Step 2: Tighten 90 degrees

37663_FX35_C0010

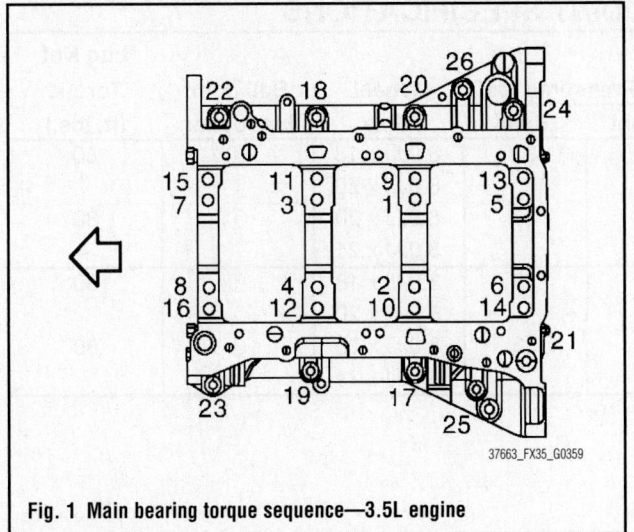

Fig. 1 Main bearing torque sequence—3.5L engine

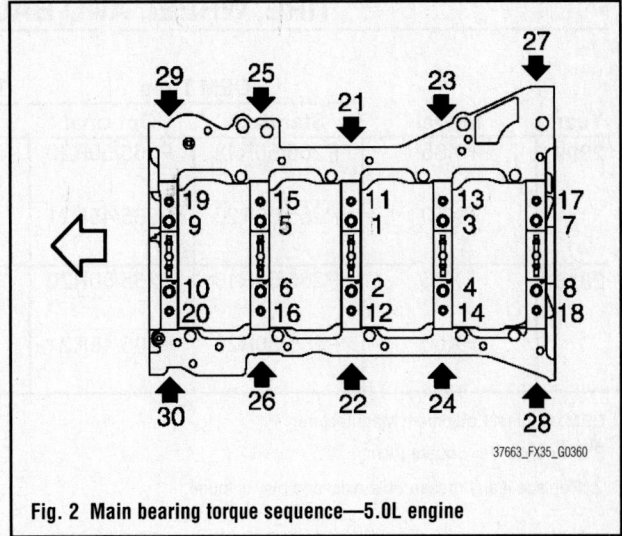

Fig. 2 Main bearing torque sequence—5.0L engine

WHEEL ALIGNMENT

Year	Model		Caster Range (+/-Deg.)	Caster Preferred Setting (Deg.)	Camber Range (+/-Deg.)	Camber Preferred Setting (Deg.)	Toe-in (Deg.)
2009	FX35	Front	0.75	+3.67	0.75	-0.33	0.070+/-0.040
		Rear	—	—	0.50	-1.16	0.120+/-0.120
	FX50	Front	0.75	+3.67	0.75	-0.33	0.070+/-0.040
		Rear	—	—	0.50	-1.16	0.120+/-0.120
2010	FX35	Front	0.75	+3.67	0.75	-0.33	0.070+/-0.040
		Rear	—	—	0.50	-1.16	0.120+/-0.120
	FX50	Front	0.75	+3.67	0.75	-0.33	0.070+/-0.040
		Rear	—	—	0.50	-1.16	0.120+/-0.120

NOTE: Measurements are given for unladen vehicle: fuel, engine coolant, and fluid levels are full. Spare tire, jack, hand tools, and mats are in designated positions

37663_FX35_C0011

TIRE, WHEEL AND BALL JOINT SPECIFICATIONS

Year	Model	OEM Tires		Tire Pressures (psi)		Wheel Size	Ball Joint Inspection	Lug Nut Torque (ft. lbs.)
		Standard	Optional	Front	Rear			
2009	FX35	P265/60R18	P265/50R20	33	33	8.0JJ x 18 8.0JJ x 20	①	80
	FX50	P265/50R20	P265/45R21	33	33	8.0JJ x 20 8.0JJ x 21	①	80
2010	FX35	P265/60R18	P265/50R20	33	33	8.0JJ x 18 8.0JJ x 20	①	80
	FX50	P265/50R20	P265/45R21	33	33	8.0JJ x 20 8.0JJ x 21	①	80

OEM: Original Equipment Manufacturer

PSI: Pounds Per Square Inch

① Replace if any measurable axial end play is found.

37663_FX35_C0012

BRAKE SPECIFICATIONS
All measurements in inches unless noted

Year	Model		Brake Disc			Brake Drum Diameter			Minimum Lining Thickness	Brake Caliper	
			Original Thickness	Minimum Thickness	Maximum Runout	Original Inside Diameter	Max. Wear Limit	Maximum Machine Diameter		Bracket Bolts (ft. lbs.)	Mounting Bolts (ft. lbs.)
2009	FX35	F	1.339	1.260	0.0014	—	—	—	0.079	34	103
		R	0.630	0.551	0.0022	—	—	—	0.079	32	62
	FX50	F	1.260	1.181	0.0014	—	—	—	0.079	34	103
		R	0.787	0.709	0.0022	—	—	—	0.079	32	62
2010	FX35	F	1.339	1.260	0.0014	—	—	—	0.079	34	103
		R	0.630	0.551	0.0022	—	—	—	0.079	32	62
	FX50	F	1.260	1.181	0.0014	—	—	—	0.079	34	103
		R	0.787	0.709	0.0022	—	—	—	0.079	32	62

F: Front

R: Rear

37663_FX35_C0013

SCHEDULED MAINTENANCE INTERVALS
INFINITI—FX35 & FX50

TO BE SERVICED	TYPE OF SERVIC	VEHICLE MILEAGE INTERVAL (x1000)														
		7.5	15	22.5	30	37.5	45	52.5	60	67.5	75	82.5	90	97.5	105	120
Accessory drive belts ①	S/I								✓							✓
Air cleaner element (engine)	R				✓				✓				✓			✓
Air conditioner system	S/I	Inspect system operation annually														
Automatic transaxle fluid	S/I		✓		✓		✓		✓		✓		✓		✓	✓
Brake lines, hoses, cables, and connections	S/I		✓		✓		✓		✓		✓		✓		✓	✓
Brake pads, calipers, & rotors	S/I		✓		✓		✓		✓		✓		✓		✓	✓
Differential gear oil	S/I		✓		✓		✓		✓		✓		✓		✓	✓
Driveshafts and CV-boots	S/I		✓		✓		✓		✓		✓		✓		✓	✓
Engine coolant	R								✓				✓			✓
Engine oil and filter	R	✓	✓	✓	✓	✓	✓	✓	✓	✓	✓	✓	✓	✓	✓	✓
Exhaust pipe connections, muffler, and suspension bolts	S/I				✓				✓				✓			✓
EVAP vapor lines	S/I				✓				✓				✓			✓
Fuel lines and connections	S/I				✓				✓				✓			✓
In-cabin microfilter	S/I		✓		✓		✓		✓		✓		✓		✓	✓
Spark plugs (Platinum-tipped)	R	105,000 miles (under normal usage)														
Steering system	S/I				✓				✓				✓			✓
Suspension system	S/I				✓				✓				✓			✓
Transfer case fluid	S/I		✓		✓		✓		✓		✓		✓		✓	✓
Valve clearance	S/I	Whenever valve noise increases														

R: Replace   S/I: Service or Inspect

① Replace if worn or damaged or if the auto-tensioner has reached its limit (V8)

FREQUENT OPERATION MAINTENANCE (SEVERE SERVICE)

If a vehicle is operated under any of the following conditions it is considered severe service:

- Extremely dusty areas.

- 50% or more of the vehicle operation is in 90°F (32°C) or higher temperatures, or constant operation in temperatures below 32°F (0°C).

- Prolonged idling (vehicle operation in stop and go traffic).

- Frequent short running periods (engine does not warm to normal operating temperatures).

- Police, taxi, delivery usage, or trailer towing usage.

Automatic transaxle fluid (and filter), transfer case fluid, and differential gear oil: check every 15,000 miles, replace every 30,000 miles

Brake pads, calipers & rotors: service or inspect every 7,500 miles

Driveshafts and CV-boots inspect every 7,500 miles

Exhaust system inspect every 7,500 miles

Oil and oil filter: change every 3,750 miles

Steering system and suspension components inspect for looseness and damage every 7,500 miles

37663_FX35_C0014

BRAKES INFORMATION AND PRECAUTIONS

ANTI-LOCK SYSTEMS

- Certain components within the ABS system are not intended to be serviced or repaired individually.
- Do not use rubber hoses or other parts not specifically specified for and ABS system. When using repair kits, replace all parts included in the kit. Partial or incorrect repair may lead to functional problems and require the replacement of components.
- Lubricate rubber parts with clean, fresh brake fluid to ease assembly. Do not use shop air to clean parts; damage to rubber components may result.
- Use only DOT 3 brake fluid from an unopened container.
- If any hydraulic component or line is removed or replaced, it may be necessary to bleed the entire system.
- A clean repair area is essential. Always clean the reservoir and cap thoroughly before removing the cap. The slightest amount of dirt in the fluid may plug an orifice and impair the system function. Perform

repairs after components have been thoroughly cleaned; use only denatured alcohol to clean components. Do not allow ABS components to come into contact with any substance containing mineral oil; this includes used shop rags.
- The Anti-Lock control unit is a microprocessor similar to other computer units in the vehicle. Ensure that the ignition switch is **OFF** before removing or installing controller harnesses. Avoid static electricity discharge at or near the controller.
- If any arc welding is to be done on the vehicle, the control unit should be unplugged before welding operations begin.

DISC AND DRUM SYSTEMS

> **❋❋ CAUTION**
>
> **Dust and dirt accumulating on brake parts during normal use may contain asbestos fibers from production or aftermarket brake linings.**

Breathing excessive concentrations of asbestos fibers can cause serious bodily harm. Exercise care when servicing brake parts. Do not sand or grind brake lining unless equipment used is designed to contain the dust residue. Do not clean brake parts with compressed air or by dry brushing. Cleaning should be done by dampening the brake components with a fine mist of water, then wiping the brake components clean with a dampened cloth. Dispose of cloth and all residue containing asbestos fibers in an impermeable container with the appropriate label. Follow practices prescribed by the Occupational Safety and Health Administration (OSHA) and the Environmental Protection Agency (EPA) for the handling, processing, and disposing of dust or debris that may contain asbestos fibers.

BRAKES BLEEDING THE BRAKE SYSTEM

BLEEDING PROCEDURE

BLEEDING PROCEDURE

> **❋❋ CAUTION**
>
> **Note the following:**

- Turn the ignition switch OFF and disconnect the ABS actuator and electric unit (control unit) connector or the battery negative terminal before performing the work.
- Monitor the fluid level in the reser-

voir tank while performing the air bleeding.
- Always use new brake fluid for refilling. Never reuse the drained brake fluid.

1. Connect a vinyl tube to the bleeder valve of the rear right brake.
2. Fully depress the brake pedal 4 to 5 times.
3. Loosen the bleeder valve and bleed air with the brake pedal depressed, and then quickly tighten the bleeder valve.
4. Repeat steps 2 and 3 until all of the air is out of the brake line.

5. Tighten the bleeder valve to the specified torque.
6. Perform steps 1 to 5 for the right rear brake; left front brake; left rear brake; and right front brake in that order.
7. Check that the fluid level in the reservoir tank is within the specified range after air bleeding.
8. Check each item of brake pedal. Adjust it if the measurement value is not the standard.

BLEEDING THE ABS SYSTEM

Refer to Bleeding the Brake System.

BRAKES ANTI-LOCK BRAKE SYSTEM (ABS)

WHEEL SPEED SENSORS

REMOVAL & INSTALLATION

Front
See Figure 3.

➡**The manufacturer does not provide a specific Removal and Installation procedure for this component. Refer to the graphic(s) when servicing this component.**

1. Be careful with the following when removing sensor.

> **❋❋ CAUTION**
>
> **Note the following:**

- Do not twist sensor harness as much as possible, when removing it. Pull sensors out without pulling sensor harness.
- Be careful to avoid damaging sensor edges or rotor teeth. Remove wheel sensor first before removing front or rear wheel hub. This is to avoid damage to sensor wiring and loss of sensor function.

To install:
2. Tighten installation bolts to the specified torques.
3. Be careful with the following when installing wheel sensor.

> **❋❋ CAUTION**
>
> **Note the following:**

- When installing, make sure there is no foreign material such as iron chips on and in the mounting hole of the wheel sensor. Make sure no foreign material has been caught in

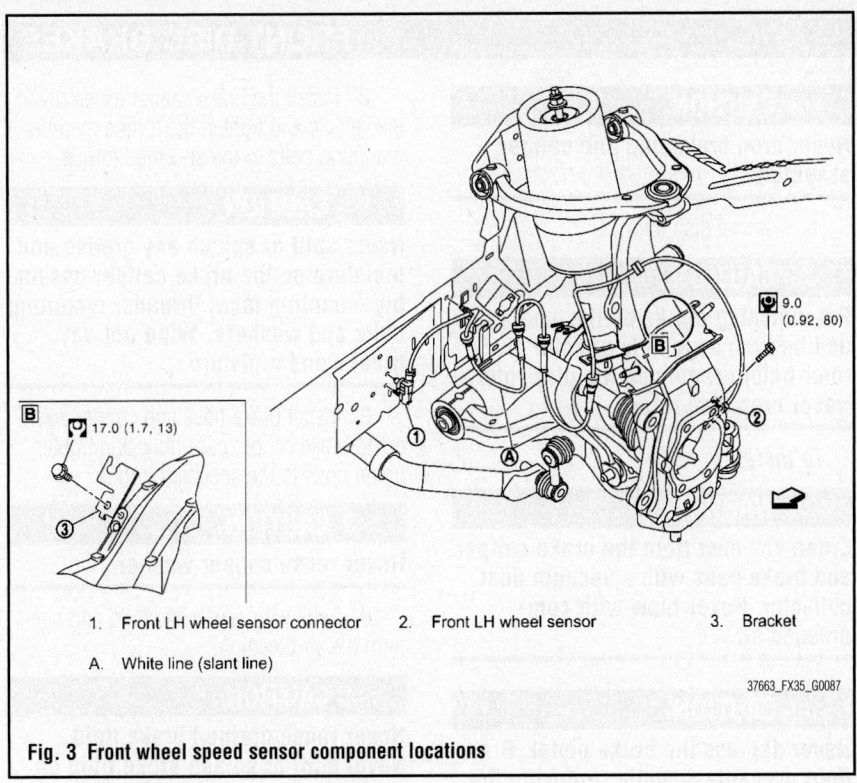

1. Front LH wheel sensor connector   2. Front LH wheel sensor   3. Bracket

A. White line (slant line)

37663_FX35_G0087

Fig. 3 Front wheel speed sensor component locations

• Be careful to avoid damaging sensor edges or rotor teeth. Remove wheel sensor first before removing front or rear wheel hub. This is to avoid damage to sensor wiring and loss of sensor function.

To install:

2. Tighten installation bolts to the specified torques.

3. Be careful with the following when installing wheel sensor.

✳✳ CAUTION

Note the following:

• When installing, make sure there is no foreign material such as iron chips on and in the mounting hole of the wheel sensor. Make sure no foreign material has been caught in the sensor rotor. Remove any foreign material and clean the mount.

• When installing a rear LH wheel sensor, be sure to pass the wheel sensor harness under the breather hose.

the sensor rotor. Remove any foreign material and clean the mount.

• When installing wheel sensor, be sure to press rubber grommets in until they lock at locations shown. When installed, harness must not be twisted.

• When you see the harness of the wheel sensor from the front side of the vehicle ensure that the white lines are not twisted.

Rear

See Figure 4.

➡**The manufacturer does not provide a specific Removal and Installation procedure for this component. Refer to the graphic(s) when servicing this component.**

1. Be careful with the following when removing sensor.

✳✳ CAUTION

Note the following:

• Do not twist sensor harness as much as possible, when removing it. Pull sensors out without pulling sensor harness.

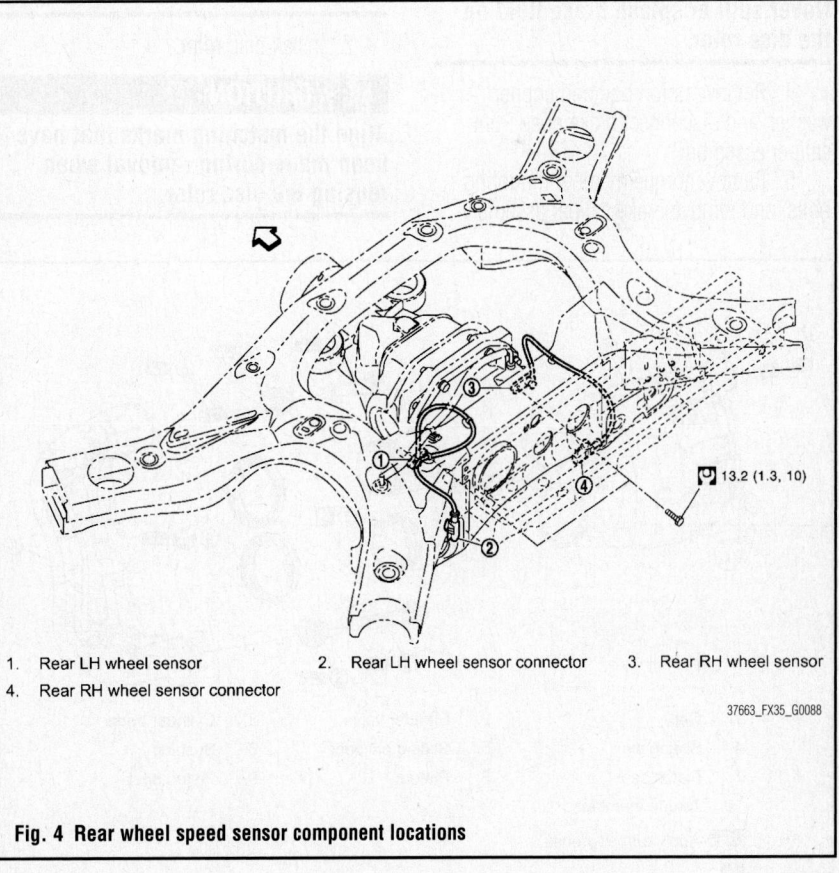

1. Rear LH wheel sensor   2. Rear LH wheel sensor connector   3. Rear RH wheel sensor
4. Rear RH wheel sensor connector

37663_FX35_G0088

Fig. 4 Rear wheel speed sensor component locations

BRAKE CALIPER

REMOVAL & INSTALLATION

2 Piston Type

See Figure 5.

✳ WARNING

Clean any dust from the brake caliper and brake pads with a vacuum dust collector. Never blow with compressed air.

✳ CAUTION

Never depress the brake pedal. Brake fluid may splash while removing the brake hose.

1. Remove tires and wheels.
2. Fix the disc rotor using wheel nuts.
3. Drain brake fluid.

✳ CAUTION

Never spill or splash brake fluid on the disc rotor.

4. Remove union bolt and copper washer, and disconnect brake hose from caliper assembly.
5. Remove torque member mounting bolts, and remove brake caliper assembly.

✳ CAUTION

Never drop brake pad and caliper assembly.

6. Remove disc rotor.

✳ CAUTION

Put matching marks on the wheel hub and bearing assembly and the disc rotor before removing the disc rotor. Never drop disc rotor.

To install:

✳ WARNING

Clean any dust from the brake caliper and brake pads with a vacuum dust collector. Never blow with compressed air.

✳ CAUTION

Never depress the brake pedal. Brake fluid may splash while removing the brake hose.

7. Install disc rotor.

✳ CAUTION

Align the matching marks that have been made during removal when reusing the disc rotor.

8. Install the brake caliper assembly to the vehicle and tighten the torque member mounting bolts to the specified torque.

✳ CAUTION

Never spill or splash any grease and moisture on the brake caliper assembly mounting face, threads, mounting bolts and washers. Wipe out any grease and moisture.

9. Install brake hose and copper washers to brake caliper assembly, and tighten union bolts to the specified torque.

✳ CAUTION

Never reuse copper washer.

10. Refill with new brake fluid and perform the air bleeding.

✳ CAUTION

Never reuse drained brake fluid. Never spill or splash brake fluid on the disc rotor.

11. Check a drag of front disc brake.

4 Piston Type

See Figure 6.

✳ WARNING

Clean any dust from the brake caliper and brake pads with a vacuum dust collector. Never blow with compressed air.

✳ CAUTION

Never depress the brake pedal. Brake fluid may splash while removing the brake hose.

1. Remove tires and wheels.
2. Fix the disc rotor using wheel nuts.
3. Drain brake fluid.

✳ CAUTION

Never spill or splash brake fluid on the disc rotor.

4. Loosen the flare nut with a flare nut wrench and separate the brake tube from caliper.

✳ CAUTION

Note the following:

- Cover flare nut wrench with a cloth as not to damage the caliper.

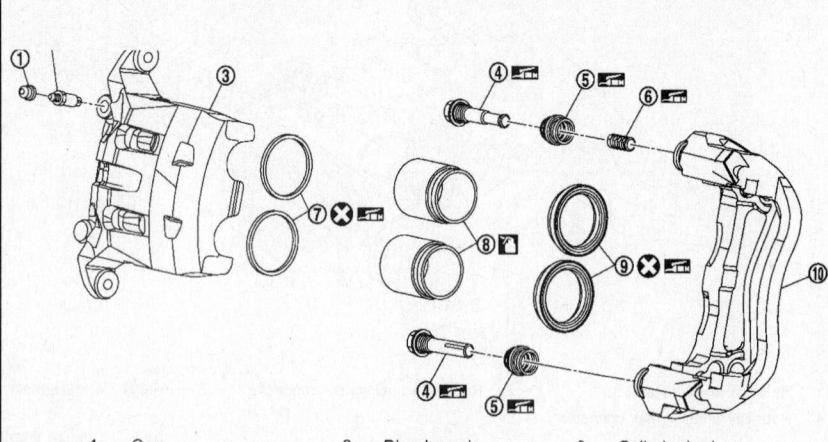

1.	Cap	2.	Bleeder valve	3. Cylinder body
4.	Sliding pin	5.	Sliding pin boot	6. Bushing
7.	Piston seal	8.	Piston	9. Piston boot
10.	Torque member			

🔲 : Apply rubber grease.

🔲 : Apply brake fluid.

37663_FX35_G0093

Fig. 5 Exploded view of brake caliper assembly

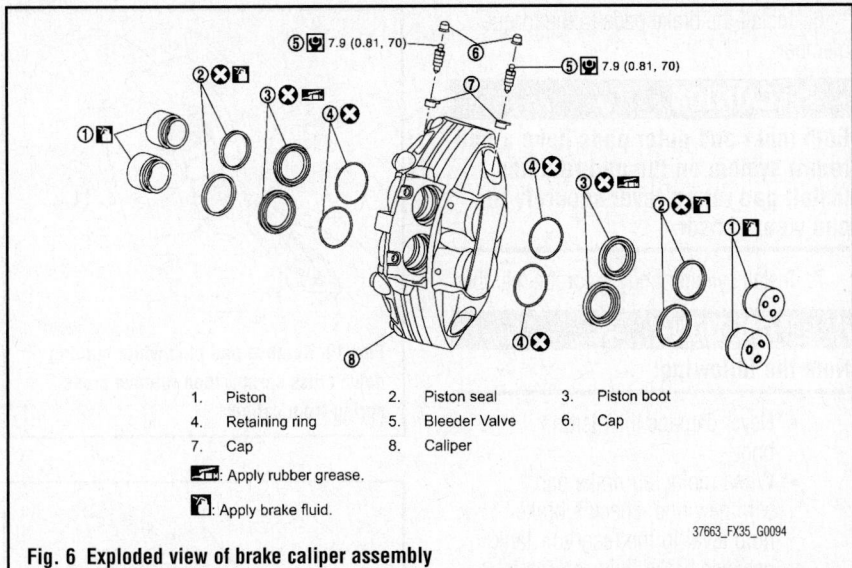

1. Piston
2. Piston seal
3. Piston boot
4. Retaining ring
5. Bleeder Valve
6. Cap
7. Cap
8. Caliper

: Apply rubber grease.

: Apply brake fluid.

37663_FX35_G0094

Fig. 6 Exploded view of brake caliper assembly

- Never scratch the flare nut and the brake tube.
- Never bend sharply, twist or strongly pull out the brake tube.
- Cover open end of brake tube when disconnecting to prevent entrance of dirt.

5. Remove caliper mounting bolts, and remove caliper.

❋❋ CAUTION

Never drop brake pad and caliper.

6. Remove disc rotor.

❋❋ CAUTION

Put matching marks on the wheel hub and bearing assembly and the disc rotor before removing the disc rotor. Never drop disc rotor.

To install:

❋❋ WARNING

Clean any dust from the brake caliper and brake pads with a vacuum dust collector. Never blow with compressed air.

❋❋ CAUTION

Never depress the brake pedal. Brake fluid may splash while removing the brake hose.

7. Install disc rotor.

❋❋ CAUTION

Align the matching marks that have been made during removal when reusing the disc rotor.

8. Install the brake caliper assembly to the vehicle and tighten the torque member mounting bolts to the specified torque.

❋❋ CAUTION

Never spill or splash any grease and moisture on the brake caliper assembly mounting face, threads, mounting bolts and washers. Wipe out any grease and moisture.

9. Tighten the flare nut to the specified torque with a flare nut crowfoot and a torque wrench.

❋❋ CAUTION

Cover crowfoot with a cloth as not to damage the caliper. Never scratch the flare nut and the brake tube.

10. Refill with new brake fluid and perform the air bleeding.

❋❋ CAUTION

Never reuse drained brake fluid. Never spill or splash brake fluid on the disc rotor and caliper.

11. Check a drag of front disc brake.

DISC BRAKE PADS

REMOVAL & INSTALLATION

2 Piston Type
See Figures 7 and 8.

❋❋ WARNING

Clean any dust from the brake caliper and brake pads with a vacuum dust collector. Never blow with compressed air.

❋❋ CAUTION

Never depress the brake pedal while removing the brake pads because the piston may pop out. Never spill or splash brake fluid on the disc rotor.

1. Remove tires and wheels.
2. Remove lower sliding pin bolt.
3. Suspend the cylinder body with suitable wire so that the brake hose will not stretch. Then remove the brake pads, shims, shim covers and pad retainers from the torque member.

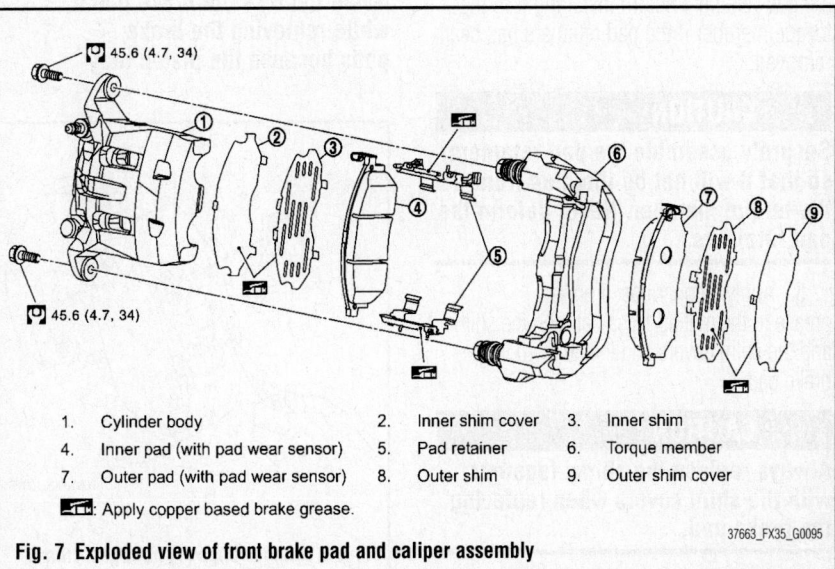

1. Cylinder body
2. Inner shim cover
3. Inner shim
4. Inner pad (with pad wear sensor)
5. Pad retainer
6. Torque member
7. Outer pad (with pad wear sensor)
8. Outer shim
9. Outer shim cover

: Apply copper based brake grease.

37663_FX35_G0095

Fig. 7 Exploded view of front brake pad and caliper assembly

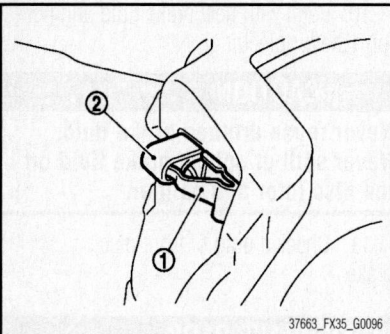

37663_FX35_G0096

Fig. 8 Install pad return lever (1) securely to pad wear sensor (2)

☀ CAUTION

Note the following:

- Never deform the pad retainer when removing the pad retainer from the torque member.
- Never damage the piston boot.
- Never drop the brake pads, shims, and the shim covers.

To install:

☀ WARNING

Clean any dust from the brake caliper and brake pads with a vacuum dust collector. Never blow with compressed air.

☀ CAUTION

Never depress the brake pedal while removing the brake pads or the cylinder body because the piston may pop out. Never spill or splash brake fluid on the disc rotor.

4. Apply Copper based brake grease to the pad retainers before installing it to the torque member if the pad retainers has been removed.

☀ CAUTION

Securely assemble the pad retainers so that it will not be lifted up from the torque member. Never deform the pad retainers.

5. Apply Copper based brake grease to the mating faces between the shims and the shim covers and install them to the brake pad.

☀ CAUTION

Always replace the shims together with the shim covers when replacing the brake pad.

6. Install the brake pads to the torque member.

☀ CAUTION

Both inner and outer pads have a pad return system on the pad retainer. Install pad return lever securely to pad wear sensor.

7. Install cylinder body to torque member.

☀ CAUTION

Note the following:

- Never damage the piston boot.
- When replacing brake pad with new one, check a brake fluid level in the reservoir tank because brake fluid returns to master cylinder reservoir tank when pressing piston in.

➡Use a disc brake piston tool to easily press piston.

8. Install the lower sliding pin bolt and tighten it to the specified torque.

9. Depress the brake pedal several times to check that no drag feel is present for the front disc brake.

4 Piston Type

See Figures 9 through 13.

☀ WARNING

Clean any dust from the brake caliper and brake pads with a vacuum dust collector. Never blow with compressed air.

☀ CAUTION

Never depress the brake pedal while removing the brake pads because the piston may

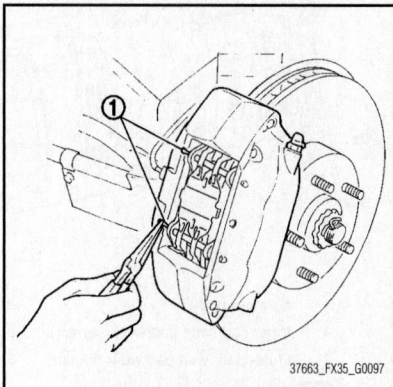

37663_FX35_G0097

Fig. 9 Remove clips (1) from pad pins

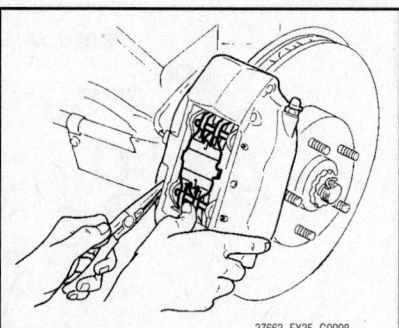

37663_FX35_G0098

Fig. 10 Remove pad pins while holding down cross spring, then remove cross spring from caliper

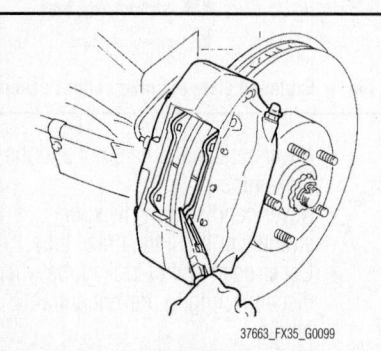

37663_FX35_G0099

Fig. 11 Using pliers, remove brake pads and shims from caliper

pop out. Never spill or splash brake fluid on the disc rotor and caliper.

1. Remove tires and wheels.
2. Remove clips from pad pins.
3. Remove pad pins while holding down cross spring, then remove cross spring from caliper.
4. Using pliers, remove brake pads and shims from caliper.

☀ CAUTION

Never damage the piston boot. Never drop the brake pads, shims.

To install:

☀ WARNING

Clean any dust from the brake caliper and brake pads with a vacuum dust collector. Never blow with compressed air.

☀ CAUTION

Never depress the brake pedal while removing the brake pads because the piston may pop out. Never spill or

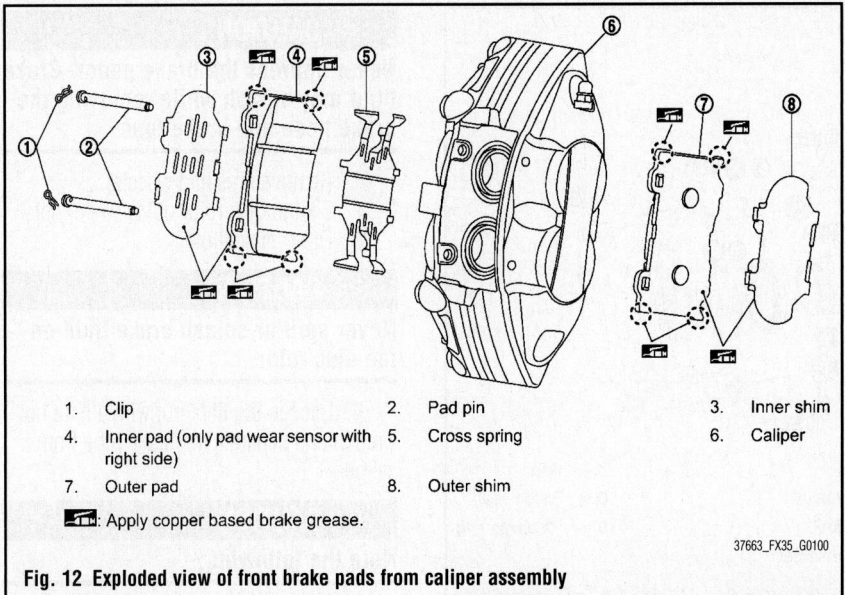

1. Clip
2. Pad pin
3. Inner shim
4. Inner pad (only pad wear sensor with right side)
5. Cross spring
6. Caliper
7. Outer pad
8. Outer shim

🔧: Apply copper based brake grease.

37663_FX35_G0100

Fig. 12 Exploded view of front brake pads from caliper assembly

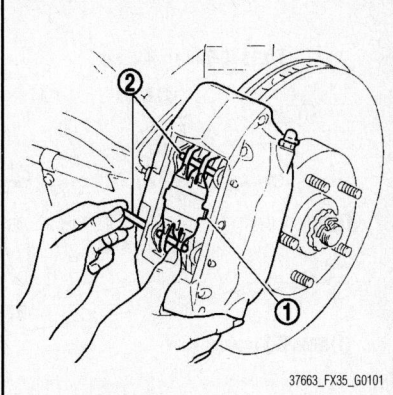

37663_FX35_G0101

Fig. 13 Place the top of cross spring (1) over the upper pad pin (2), press in the cross spring, install lower pad pin from the inner side to the outer side, and secure cross spring

splash brake fluid on the disc rotor and caliper.

5. Apply copper based brake grease to the mating faces between the brake pads and shims, and install shims to the brake pad.

✳✳ CAUTION

Always replace the shims together when replacing the brake pad.

6. Apply copper based brake grease to the mating faces between the brake pads and caliper.

7. Install brake pads to caliper.

✳✳ CAUTION

Note the following:

- Never damage the piston boot.
- In the case of replacing a pad with new one, check a brake fluid level in the reservoir tank because brake fluid returns to master cylinder reservoir tank when pressing piston in.

➡ **Use a disc brake piston tool to easily press piston.**

8. Install upper pad pin from the inner side, then install firmly to the outer side through the hole in the top of brake pad.

9. Place the top of cross spring over the upper pad pin, press in the cross spring, install lower pad pin from the inner side to the outer side, and secure cross spring.

10. Install clips to the pad pins.

✳✳ CAUTION

If clip is not fully attached, pad pin or brake pad could fall out while vehicle is in motion.

11. Depress the brake pedal several times to check that no drag feel is present for the front disc brake.

BRAKES

BRAKE CALIPER

REMOVAL & INSTALLATION

1 Piston Type
See Figures 14 and 15.

✳✳ WARNING

Clean any dust from the brake caliper and brake pads with a vacuum dust collector. Never blow with compressed air.

✳✳ CAUTION

Never depress the brake pedal. Brake fluid may splash while removing the brake hose.

1. Remove tires with power tool.

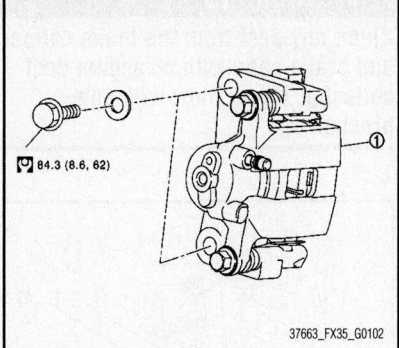

🔩 84.3 (8.6, 62)

37663_FX35_G0102

Fig. 14 Remove torque member mounting bolts, and remove brake caliper assembly (1)

2. Fix the disc rotor using wheel nuts.

3. Drain brake fluid.

REAR DISC BRAKES

✳✳ CAUTION

Never spill or splash brake fluid on the disc rotor.

4. Remove union bolt and copper washers, and disconnect brake hose from caliper assembly.

5. Remove torque member mounting bolts, and remove brake caliper assembly.

✳✳ CAUTION

Never drop brake pad and caliper assembly.

6. Remove disc rotor.

✳✳ CAUTION

Put matching marks on the wheel hub and bearing assembly and the disc rotor before removing the disc rotor. Never drop disc rotor.

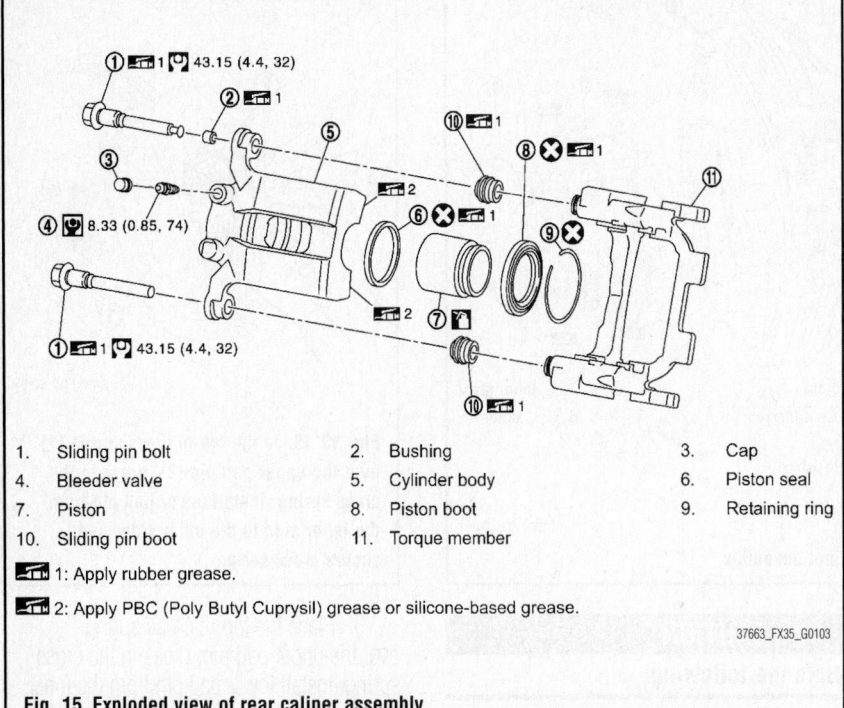

1. Sliding pin bolt
4. Bleeder valve
7. Piston
10. Sliding pin boot

2. Bushing
5. Cylinder body
8. Piston boot
11. Torque member

3. Cap
6. Piston seal
9. Retaining ring

1: Apply rubber grease.

2: Apply PBC (Poly Butyl Cuprysil) grease or silicone-based grease.

37663_FX35_G0103

Fig. 15 Exploded view of rear caliper assembly

To install:

⁜⁜ WARNING

Clean any dust from the brake caliper and brake pads with a vacuum dust collector. Never blow with compressed air.

⁜⁜ CAUTION

Never depress the brake pedal. Brake fluid may splash while removing the brake hose.

7. Install disc rotor.

⁜⁜ CAUTION

Align the matching marks that have been made during removal when reusing the disc rotor.

8. Install the brake caliper assembly to the vehicle and tighten the torque member mounting bolts to the specified torque.

⁜⁜ CAUTION

Never spill or splash any grease and moisture on the brake caliper assembly mounting face, threads, mounting bolts, and washers. Wipe out any grease and moisture.

9. Install brake hose and copper washers to brake caliper assembly, and tighten union bolts to the specified torque.

10. Refill with new brake fluid and perform the air bleeding.

⁜⁜ CAUTION

Never reuse drained brake fluid. Never spill or splash brake fluid on the disc rotor.

11. Check a drag of rear disc brake.

2 Piston Type

See Figures 16 and 17.

⁜⁜ WARNING

Clean any dust from the brake caliper and brake pads with a vacuum dust collector. Never blow with compressed air.

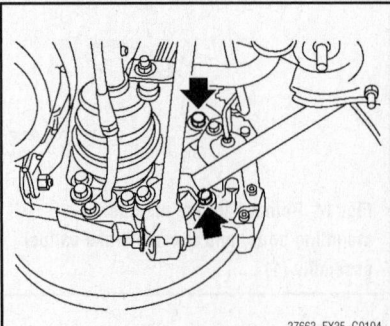

37663_FX35_G0104

Fig. 16 Remove caliper mounting bolts, and remove caliper

⁜⁜ CAUTION

Never depress the brake pedal. Brake fluid may splash while removing the brake hose and brake tube.

1. Remove tires and wheels.
2. Fix the disc rotor using wheel nuts.
3. Drain brake fluid.

⁜⁜ CAUTION

Never spill or splash brake fluid on the disc rotor.

4. Loosen the flare nut with a flare nut wrench and separate the brake tube from caliper.

⁜⁜ CAUTION

Note the following:

- Cover flare nut wrench with a cloth as not to damage the caliper.
- Never scratch the flare nut and the brake tube.
- Never bend sharply, twist or strongly pull out the brake tube.
- Cover open end of brake tube when disconnecting to prevent entrance of dirt.

5. Remove brake hose mounting bolt.

6. Remove caliper mounting bolts, and remove caliper.

⁜⁜ CAUTION

Never drop brake pad and caliper.

7. Remove disc rotor.

⁜⁜ CAUTION

Put matching marks on the wheel hub and bearing assembly and the disc rotor before removing the disc rotor. Never drop disc rotor.

To install:

⁜⁜ WARNING

Clean any dust from the brake caliper and brake pads with a vacuum dust collector. Never blow with compressed air.

⁜⁜ CAUTION

Never depress the brake pedal. Brake fluid may splash while removing the brake hose.

8. Install disc rotor.

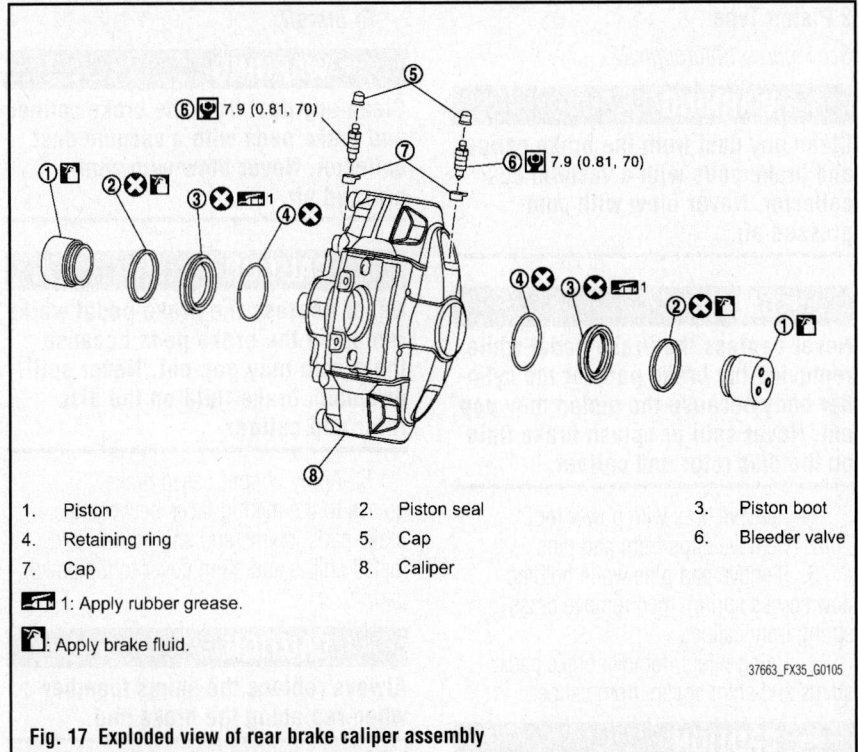

1. Piston
2. Piston seal
3. Piston boot
4. Retaining ring
5. Cap
6. Bleeder valve
7. Cap
8. Caliper

1: Apply rubber grease.

: Apply brake fluid.

37663_FX35_G0105

Fig. 17 Exploded view of rear brake caliper assembly

※※ **CAUTION**

Align the matching marks that have been made during removal when reusing the disc rotor.

9. Install the brake caliper to the vehicle and tighten the caliper mounting bolts to the specified torque.

※※ **CAUTION**

Never spill or splash any grease and moisture on the caliper mounting face, threads, mounting bolts and washers. Wipe out any grease and moisture.

10. Install the brake hose mounting bolt to the specified torque.

11. Tighten the flare nut to the specified torque with a flare nut crowfoot and a torque wrench.

※※ **CAUTION**

Cover crowfoot with a cloth as not to damage the caliper. Never scratch the flare nut and the brake tube.

12. Refill with new brake fluid and perform the air bleeding.

※※ **CAUTION**

Never reuse drained brake fluid. Never spill or splash brake fluid on the disc rotor.

13. Check a drag of rear disc brake.

DISC BRAKE PADS

REMOVAL & INSTALLATION

1 Piston Type

See Figure 18.

※※ **WARNING**

Clean any dust from the brake caliper and brake pads with a vacuum dust collector. Never blow with compressed air.

※※ **CAUTION**

Never depress the brake pedal while removing the brake pads or the cylinder body because the piston may pop out. Never spill or splash brake fluid on the disc rotor.

1. Remove tires with power tool.

2. Remove the upper sliding pin bolt.

3. Suspend the cylinder body with a wire so that the brake hose will not stretch. Remove the brake pads, shims, shim cover and pad retainers from the torque member.

※※ **CAUTION**

Note the following:

- Never deform the pad retainers if removing the pad retainers.
- Never damage the piston boot.
- Never drop the brake pad, shims, and the shim cover.

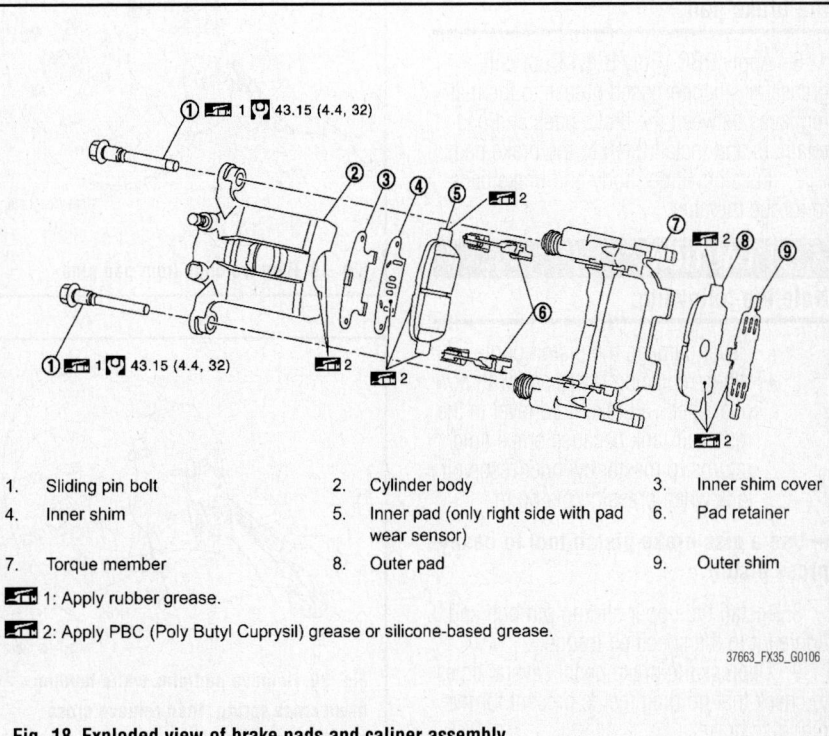

1. Sliding pin bolt
2. Cylinder body
3. Inner shim cover
4. Inner shim
5. Inner pad (only right side with pad wear sensor)
6. Pad retainer
7. Torque member
8. Outer pad
9. Outer shim

1: Apply rubber grease.

2: Apply PBC (Poly Butyl Cuprysil) grease or silicone-based grease.

37663_FX35_G0106

Fig. 18 Exploded view of brake pads and caliper assembly

To install:

※ WARNING

Clean any dust from the brake caliper and brake pads with a vacuum dust collector. Never blow with compressed air.

※ CAUTION

Never depress the brake pedal while removing the brake pads or the cylinder body because the piston may pop out. Never spill or splash brake fluid on the disc rotor.

4. Apply PBC (Poly Butyl Cuprysil) grease or silicone-based grease to the pad retainers before installing it to the torque member if the pad retainers has been removed.

※ CAUTION

Securely assemble the pad retainers so that it will not be lifted up from the torque member. Never deform the pad retainers.

5. Apply PBC (Poly Butyl Cuprysil) grease or silicone-based grease to the mating faces between the shims and the shim cover.

※ CAUTION

Always replace the shims together with the shim cover when replacing the brake pad.

6. Apply PBC (Poly Butyl Cuprysil) grease or silicone-based grease to the mating faces between the brake pads and pad retainers and install them to the brake pads.
7. Install cylinder body and brake pads to torque member.

※ CAUTION

Note the following:

- Never damage the piston boot.
- When replacing brake pad with new one, check a brake fluid level in the reservoir tank because brake fluid returns to master cylinder reservoir tank when pressing piston in.

➡ Use a disc brake piston tool to easily press piston.

8. Install the upper sliding pin bolt and tighten it to the specified torque.
9. Depress the brake pedal several times to check that no drag feel is present for the rear disc brake.

2 Piston Type

See Figures 19 through 22.

※ WARNING

Clean any dust from the brake caliper and brake pads with a vacuum dust collector. Never blow with compressed air.

※ CAUTION

Never depress the brake pedal while removing the brake pads or the cylinder body because the piston may pop out. Never spill or splash brake fluid on the disc rotor and caliper.

1. Remove tires with power tool.
2. Remove clips from pad pins.
3. Remove pad pins while holding down cross spring, then remove cross spring from caliper.
4. Using pliers, remove brake pads, shims and shim covers from caliper.

※ CAUTION

Never damage the piston boot. Never drop the brake pad, shims, and the shim cover.

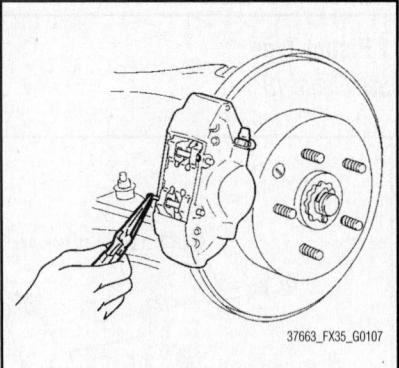

37663_FX35_G0107

Fig. 19 Remove clips from pad pins

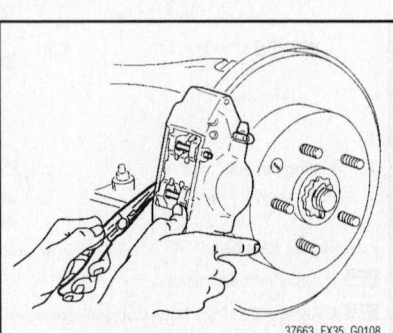

37663_FX35_G0108

Fig. 20 Remove pad pins while holding down cross spring, then remove cross spring from caliper

To install:

※ WARNING

Clean any dust from the brake caliper and brake pads with a vacuum dust collector. Never blow with compressed air.

※ CAUTION

Never depress the brake pedal while removing the brake pads because the piston may pop out. Never spill or splash brake fluid on the disc rotor and caliper.

5. Apply copper based brake grease to the mating faces between the brake pads, shims and shim cover, and install shims and shim cover to the brake pad.

※ CAUTION

Always replace the shims together when replacing the brake pad.

6. Apply copper based brake grease to the mating faces between the brake pads and caliper.
7. Apply copper based brake grease to

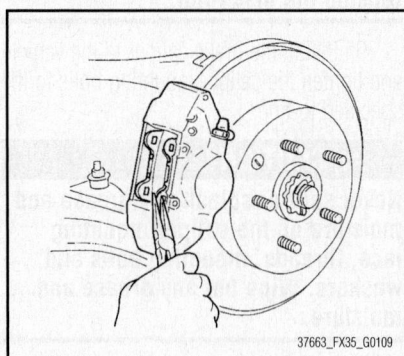

37663_FX35_G0109

Fig. 21 Using pliers, remove brake pads, shims and shim covers from caliper

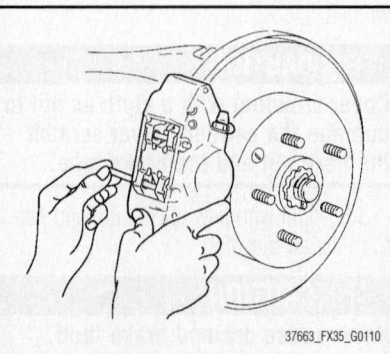

37663_FX35_G0110

Fig. 22 Installing pad pins and cross spring

the mating faces between the brake pads and pad pins.

8. Apply copper based brake grease to the mating faces between the brake pads and cross spring.

9. Install brake pads to caliper.

✳✳ CAUTION

Note the following:

- Never damage the piston boot.
- In the case of replacing a pad with new one, check a brake fluid level

in the reservoir tank because brake fluid returns to master cylinder reservoir tank when pressing piston in.

➡**Use a disc brake piston tool to easily press piston.**

10. Install upper pad pin from the inner side, then install firmly to the outer side through the hole in the top of brake pad.

11. Place the top of cross spring over the upper pad pin, press in the cross

spring, install lower pad pin from the inner side to the outer side, and secure cross spring.

12. Install clips to the pad pins.

✳✳ CAUTION

If clip is not fully attached, pad pin or brake pad could fall out while vehicle is in motion.

13. Depress the brake pedal several times to check that no drag feel is present for the rear disc brake.

BRAKES

PARKING BRAKE CABLES

ADJUSTMENT

See Figure 23.

1. Adjust the cable with the following procedure.

a. Operate the parking brake pedal with a force of 110 lbs. (490 N) for 10 strokes or more.

b. Adjust the parking brake pedal stroke by turning the adjusting nut with a deep socket wrench.

✳✳ CAUTION

Never reuse the adjusting nut if the nut is removed.

c. Operate the parking brake pedal with a force of 44 lbs. (196 N). Check that the pedal stroke is within 2 to 3

notches. (Check it by listening to clicks of ratchet.)

d. Rotate the disc rotor with the parking brake pedal released and check that there is no drag.

PARKING BRAKE SHOES

REMOVAL & INSTALLATION

See Figures 24 through 26.

✳✳ WARNING

Clean any dust from the parking brake shoes and back plates with a vacuum dust collector. Never blow with compressed air.

1. Remove rear tires and wheels.
2. Remove disc rotor.

PARKING BRAKE

✳✳ CAUTION

Parking brake completely in the released position.

3. If disc rotor cannot be removed, remove as follows:

a. Fix the disc rotor with wheel nuts and remove the adjusting hole plug.

b. Using suitable tool, rotate adjuster in the direction to retract and loosen brake shoe.

4. Remove anti-rattle pins, retainers, anti-rattle springs, and return spring, adjuster spring.

✳✳ CAUTION

Never drop the removed parts.

5. Remove parking brake shoes, adjuster assembly, and toggle lever.

✳✳ CAUTION

Note the following:

- The parking brake shoes for the front wheels are made of different materials from those for the rear wheels. Never misidentify them when removing.
- Never drop the removed parts.

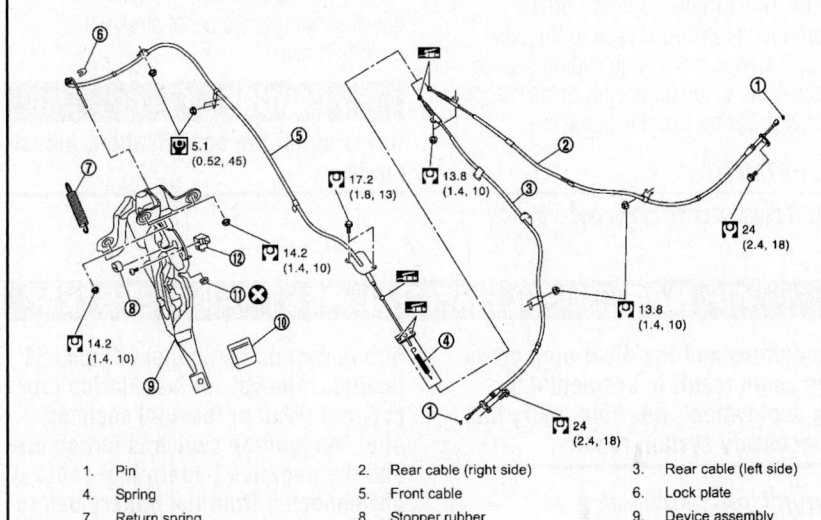

1. Pin
2. Rear cable (right side)
3. Rear cable (left side)
4. Spring
5. Front cable
6. Lock plate
7. Return spring
8. Stopper rubber
9. Device assembly
10. Pedal pad
11. Adjusting nut
12. Parking brake switch

🔧: Apply multi-purpose grease.

37663_FX35_G0111

Fig. 23 Exploded view of parking brake control assembly

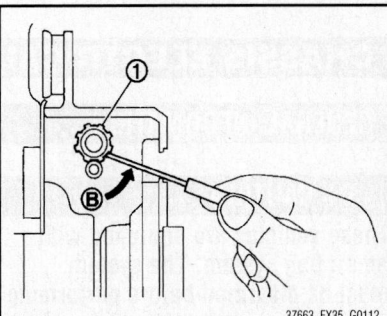

37663_FX35_G0112

Fig. 24 Using suitable tool, rotate adjuster (1) in the direction (B) to retract and loosen brake shoe

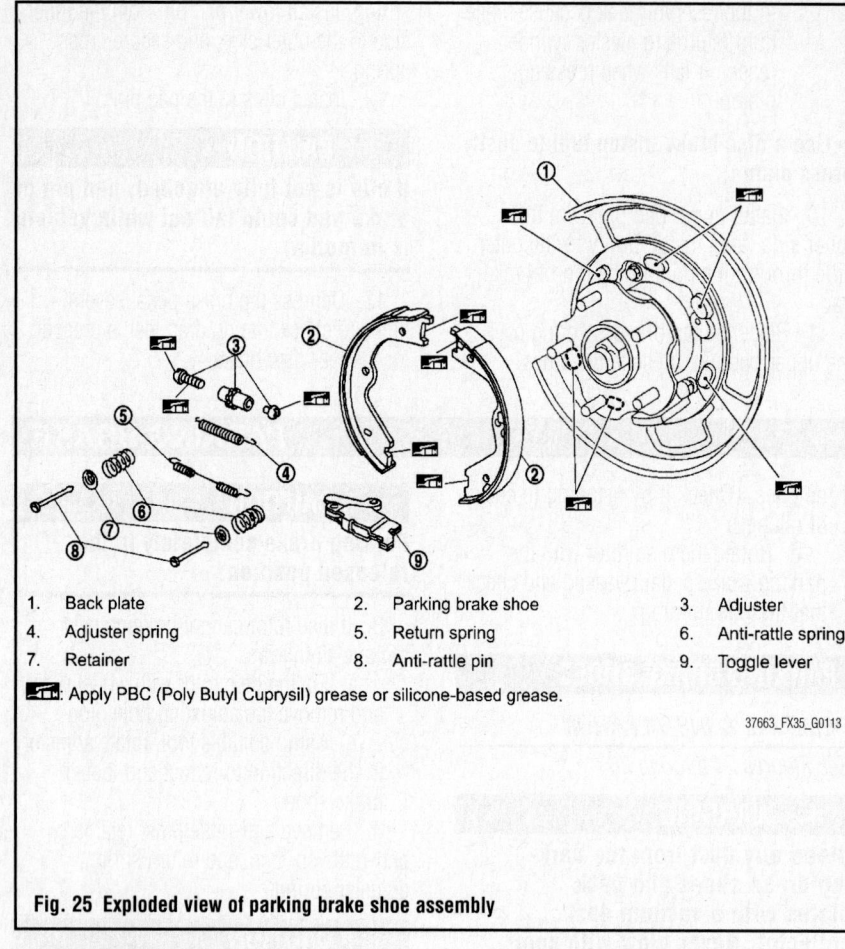

1. Back plate
2. Parking brake shoe
3. Adjuster
4. Adjuster spring
5. Return spring
6. Anti-rattle spring
7. Retainer
8. Anti-rattle pin
9. Toggle lever

: Apply PBC (Poly Butyl Cuprysil) grease or silicone-based grease.

37663_FX35_G0113

Fig. 25 Exploded view of parking brake shoe assembly

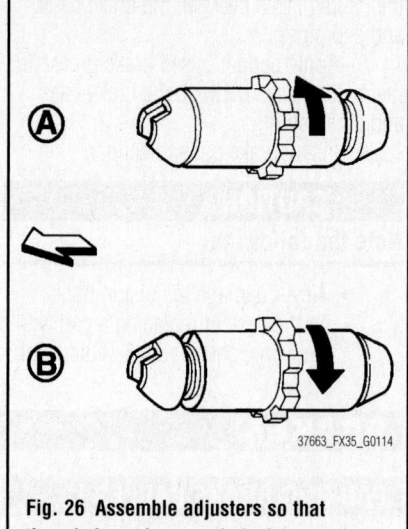

37663_FX35_G0114

Fig. 26 Assemble adjusters so that threaded part is expanded when rotating it in the direction shown by arrow

To install:

6. Note the following, install in the reverse order of removal.

a. Apply PBC (Poly Butyl Cuprysil) grease or silicone-based grease to the back plate and brake shoe.

> ✳✳ **CAUTION**
>
> **The parking brake shoes for the front wheels are made of different materials from those for the rear wheels. Never misidentify them when removing and replacing.**

b. Assemble adjusters so that threaded part is expanded when rotating it in the direction shown by arrow.

c. Shorten adjuster by rotating it.

d. When disassembling apply PBC (Poly Butyl Cuprysil) grease or silicone-based grease to threads.

e. Check brake shoe sliding surface and drum inner surface for grease. Wipe it off if it adhere on the surfaces.

ADJUSTMENT

1. Adjust parking brake pedal stroke.

2. Perform parking brake break-in (drag on) operation by driving vehicle under the following conditions:

- Drive forward
- Vehicle speed: Approximately 19 MPH (30 km/h) set (constant and forward)
- Parking brake operating force: 82 lbs. (365 N) set contact
- Time: Approx. 35 sec.

> ✳✳ **CAUTION**
>
> **To prevent lining from getting too hot, allow a cool off period of approximately 5 minutes after every break-in operation.**

3. After the break-in procedure, check parking brake pedal stroke of parking brake.

> ✳✳ **CAUTION**
>
> **If it is out of the specification, adjust again.**

CHASSIS ELECTRICAL

GENERAL INFORMATION

> ✳✳ **CAUTION**
>
> **These vehicles are equipped with an air bag system. The system must be disarmed before performing service on, or around, system components, the steering column, instrument panel components, wiring and sensors. Failure to follow the safety**

AIR BAG (SUPPLEMENTAL RESTRAINT SYSTEM)

precautions and the disarming procedure could result in accidental air bag deployment, possible injury and unnecessary system repairs.

SERVICE PRECAUTIONS

> ✳✳ **CAUTION**
>
> **Disconnect and isolate the battery negative cable before beginning any**

airbag system component diagnosis, testing, removal, or installation procedures. Wait at least 90 seconds after the ignition switch is turned off and the negative (-) terminal cable is disconnected from the battery before starting the operation. The SRS is equipped with a backup power source, so if work is started within 90 seconds after disconnecting the negative (-) terminal cable from the battery, the SRS may be deployed. Failure to disable the airbag system may result in accidental airbag deployment, personal injury, or death.

DISARMING THE SYSTEM

Before servicing, turn ignition switch OFF, disconnect battery negative terminal and wait 3 minutes or more.

ARMING THE SYSTEM

Connect the negative battery cable.

CLOCKSPRING CENTERING

See Figure 27.

> ❋❋ **CAUTION**
>
> **The spiral cable may snap during steering operation if**

the cable is installed in an improper position.

The neutral position is set as per the following:

1. Carefully turn the spiral cable clockwise to the end position. Then turn it counterclockwise (about 2 and a half turns) and stop turning at the mark on which the stopper insertion holes are in the same position.

2. The service part is installed in the neutral position by the stopper and can be set without adjusting after the stopper is removed.

3. Never over turn the spiral cable or go beyond the number of turns required. (This will cause the cable to snap.)

4. Adjust the spiral cable locating pin to the steering wheel locating pin hole.

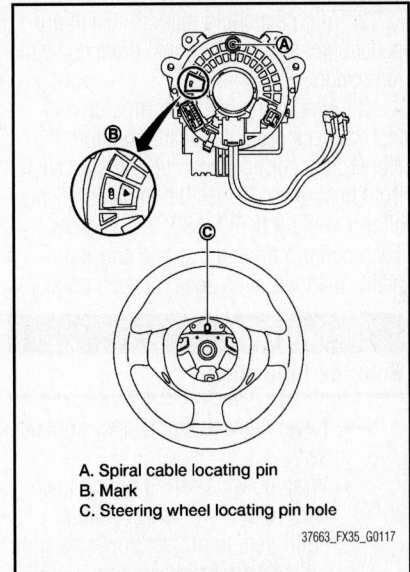

A. Spiral cable locating pin
B. Mark
C. Steering wheel locating pin hole

37663_FX35_G0117

Fig. 27 Spiral cable assembly

DRIVE TRAIN

FRONT DRIVESHAFT

REMOVAL & INSTALLATION

3.5L Engine

See Figures 28 and 29.

1. Shift the transmission to the neutral position, and then release the parking brake.

2. Remove engine undercover.

3. Remove exhaust front tube and three-way catalyst (bank 1).

4. Put matching mark on propeller shaft flange yoke and final drive companion flange.

> ❋❋ **CAUTION**
>
> **For matching mark, use paint. Never damage propeller shaft flange and final drive companion flange.**

5. Remove the propeller shaft assembly fixing bolts.

6. Move steering hydraulic line not to interfere with work.

> ❋❋ **CAUTION**
>
> **Wrap power steering piping interference area with shop cloth or equivalent to protect power steering piping from breakage.**

7. Support transfer assembly with a jack, remove rear engine mounting member.

8. Remove propeller shaft assembly from the front final drive and transfer.

> ❋❋ **CAUTION**
>
> **Note the following:**
>
> • Never damage the transfer front oil seal.
> • Wrap transmission interference area with shop cloth or equivalent to protect propeller shaft from breakage.

9. Remove propeller shaft assembly from O-ring.

10. Remove heat bracket.

To install:

11. Note the following, and install in the reverse order of removal.

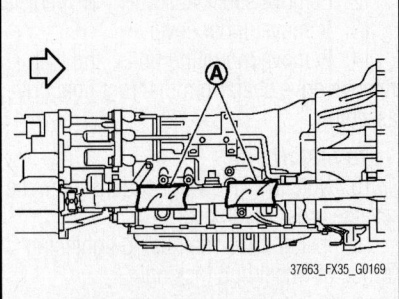

37663_FX35_G0169

Fig. 28 Wrap power steering piping interference area (A) with shop cloth or equivalent to protect power steering piping from breakage

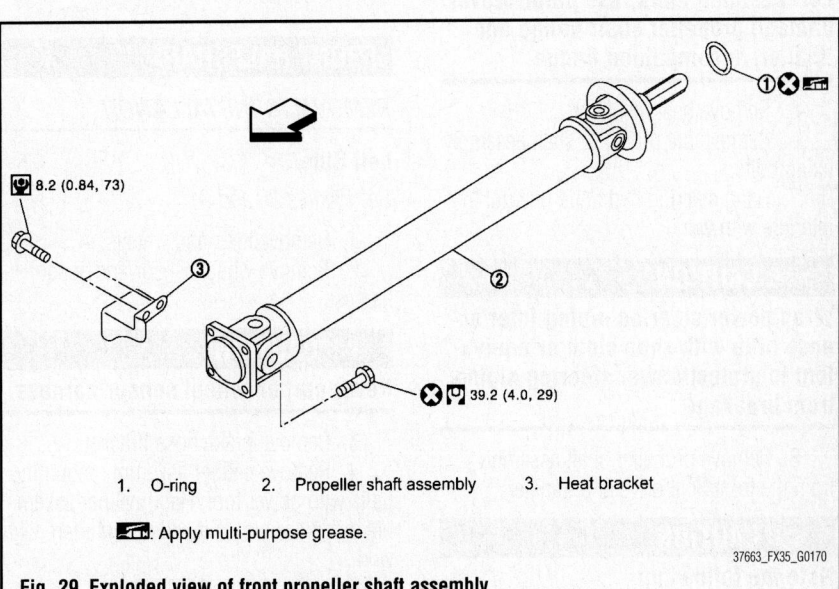

8.2 (0.84, 73)

39.2 (4.0, 29)

1. O-ring   2. Propeller shaft assembly   3. Heat bracket

: Apply multi-purpose grease.

37663_FX35_G0170

Fig. 29 Exploded view of front propeller shaft assembly

12. Align matching mark to install propeller shaft assembly to final drive companion flange.

13. After assembly, perform a driving test to check propeller shaft vibration. If vibration occurred, separate propeller shaft from final drive. Reinstall companion flange after rotating it by 90, 180, 270 degrees. Then perform driving test and check propeller shaft vibration again at each point.

✳✳ CAUTION

Note the following:

- Never damage the transfer front oil seal.
- Wrap power steering piping interference area with shop cloth or equivalent to protect power steering piping from breakage.
- Wrap transmission interference area with shop cloth or equivalent to protect propeller shaft from breakage.
- Never reuse O-ring.
- Apply multi-purpose grease onto O-ring.

5.0L Engine

See Figures 28 and 29.

1. Shift the transmission to the neutral position, and then release the parking brake.
2. Remove engine undercover.
3. Remove exhaust front tube and three-way catalyst.
4. Put matching mark onto propeller shaft flange yoke and final drive companion flange.

✳✳ CAUTION

For matching mark, use paint. Never damage propeller shaft flange and final drive companion flange.

5. Remove heat insulator.
6. Remove the propeller shaft assembly fixing bolts.
7. Hang steering hydraulic line not to interfere with work.

✳✳ CAUTION

Wrap power steering piping interference area with shop cloth or equivalent to protect power steering piping from breakage.

8. Remove propeller shaft assembly from the front final drive and transfer.

✳✳ CAUTION

Note the following:

- Never damage the transfer front oil seal.
- Wrap transmission interference area with shop cloth or equivalent to protect propeller shaft from breakage.

9. Remove propeller shaft assembly from O-ring.
10. Remove heat bracket.

To install:

11. Note the following, and install in the reverse order of removal.
12. Align matching mark to install propeller shaft assembly to final drive companion flange.
13. After assembly, perform a driving test to check propeller shaft vibration. If vibration occurred, separate propeller shaft from final drive. Reinstall companion flange after rotating it by 90, 180, 270 degrees. Then perform driving test and check propeller shaft vibration again at each point.

✳✳ CAUTION

Note the following:

- Never damage the transfer front oil seal.
- Wrap power steering piping interference area with shop cloth or equivalent to protect power steering piping from breakage.
- Wrap transmission interference area with shop cloth or equivalent to protect propeller shaft from breakage.
- Never reuse O-ring.
- Apply multi-purpose grease onto O-ring.

FRONT HALFSHAFT

REMOVAL & INSTALLATION

Left Side

See Figures 30 and 31.

1. Remove tires and wheels.
2. Remove wheel sensor and sensor harness.

✳✳ CAUTION

Never pull on wheel sensor harness.

3. Remove brake hose bracket.
4. Remove caliper assembly mounting bolts with power tool. Hang caliper assembly in a place where it will not interfere with work.

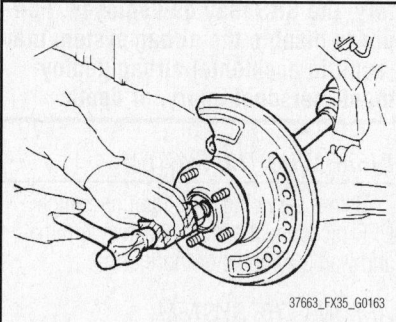

Fig. 30 Hammer the wood to disengage wheel hub and bearing assembly from drive shaft

✳✳ CAUTION

Never depress brake pedal while brake caliper is removed.

5. Remove disc rotor.
6. Remove cotter pin, and then loosen wheel hub lock nut with a power tool.
7. Patch wheel hub lock nut with a piece of wood. Hammer the wood to disengage wheel hub and bearing assembly from drive shaft.

✳✳ CAUTION

Never place drive shaft joint at an extreme angle. Also be careful not to overextend slide joint. Never allow drive shaft to hang down without support for joint sub-assembly, shaft and the other parts.

➡ **Use suitable puller if wheel hub and drive shaft cannot be separated even after performing the above procedure.**

8. Remove wheel hub lock nut.
9. Remove steering outer socket (tie rod end).
10. Separate upper link from steering knuckle.
11. Remove drive shaft from wheel hub and bearing assembly.
12. Remove shock absorber from vehicle.
13. Remove under cover.
14. Remove mounting bolts, and then remove drive shaft from the front final drive assembly.

To install:

15. Note the following, and install in the reverse order of removal.
16. Install drive shaft using tightening torque of wheel hub lock nut.

✳✳ CAUTION

Be sure to use torque wrench to tighten the wheel hub lock nut. Never use a power tool.

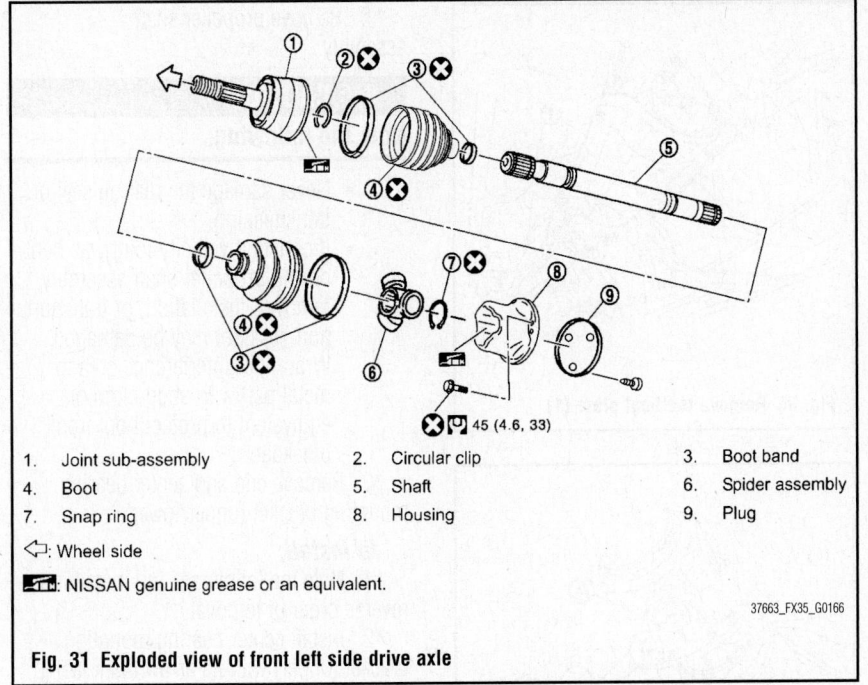

1. Joint sub-assembly
2. Circular clip
3. Boot band
4. Boot
5. Shaft
6. Spider assembly
7. Snap ring
8. Housing
9. Plug

⬅: Wheel side

▨: NISSAN genuine grease or an equivalent.

37663_FX35_G0166

Fig. 31 Exploded view of front left side drive axle

17. Never reuse cotter pin.

Right Side

See Figures 30, 32 through 34.

1. Remove tires and wheels.
2. Remove wheel sensor and sensor harness.

❊❊ CAUTION

Never pull on wheel sensor harness.

3. Remove brake hose bracket.

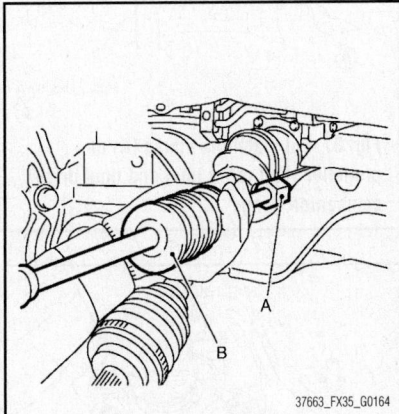

37663_FX35_G0164

Fig. 32 Remove drive shaft from front final drive assembly using the drive shaft attachment (A) KV40107500 and a sliding hammer (B) while inserting tip of the drive shaft attachment between housing and front final drive assembly

4. Remove caliper assembly mounting bolts. Hang caliper assembly in a place where it will not interfere with work.

❊❊ CAUTION

Never depress brake pedal while brake caliper is removed.

5. Remove disc rotor.
6. Remove cotter pin, and then loosen wheel hub lock nut.
7. Patch wheel hub lock nut with a piece of wood. Hammer the wood to disengage wheel hub and bearing assembly from drive shaft.

❊❊ CAUTION

Never place drive shaft joint at an extreme angle. Also be careful not to overextend slide joint. Never allow drive shaft to hang down without support for joint sub-assembly, shaft and the other parts.

➡**Use suitable puller if wheel hub and drive shaft cannot be separated even after performing the above procedure.**

8. Remove wheel hub lock nut.
9. Remove wheel hub and bearing assembly from steering knuckle.
10. Remove fender protector.

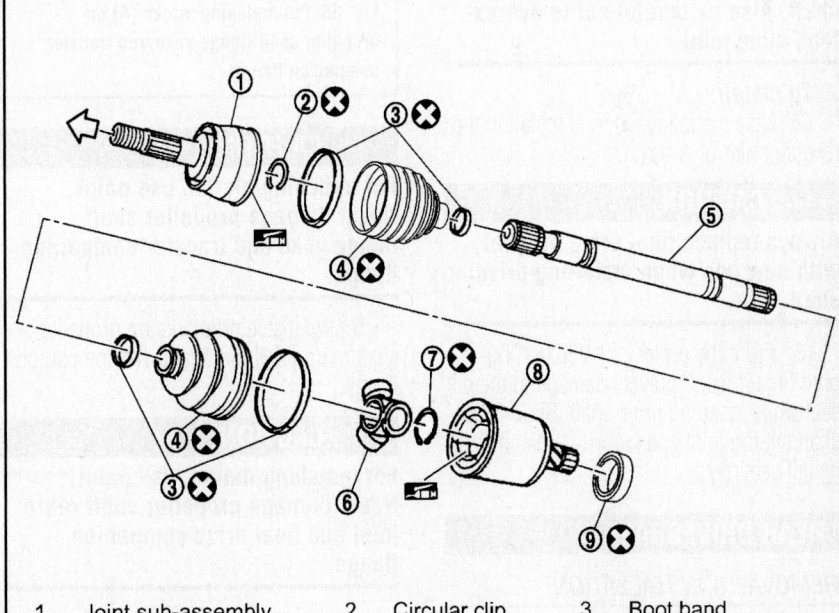

1. Joint sub-assembly
2. Circular clip
3. Boot band
4. Boot
5. Shaft
6. Spider assembly
7. Snap ring
8. Housing
9. Dust shield

⬅: Wheel side

▨: NISSAN genuine grease or an equivalent.

37663_FX35_G0168

Fig. 33 Exploded view of front right side drive axle—AWD

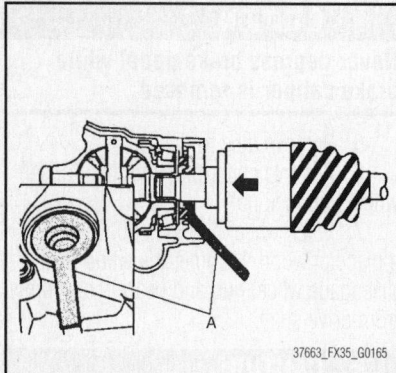

Fig. 34 Place the protector (A) onto final drive to prevent damage to the oil seal while inserting drive shaft

11. Remove drive shaft from front final drive assembly using the drive shaft attachment KV40107500 and a sliding hammer while inserting tip of the drive shaft attachment between housing and front final drive assembly.

⁂ CAUTION

Never place drive shaft joint at an extreme angle when removing drive shaft. Also be careful not to overextend slide joint.

To install:

12. Note the following, and install in the reverse order of removal.

⁂ CAUTION

Always replace final drive oil seal with new one when installing drive shaft.

13. Place the protector KV38107900 onto final drive to prevent damage to the oil seal while inserting drive shaft. Slide drive shaft sliding joint and tap with a hammer to install securely.

REAR DRIVESHAFT

REMOVAL & INSTALLATION

See Figures 35 through 41.

1. Shift the transaxle to the neutral position, and release the parking brake.
2. Remove the center muffler and exhaust front tube.
3. Remove the heat plate.
4. Put matching marks on propeller shaft flange yoke and transfer companion flange.

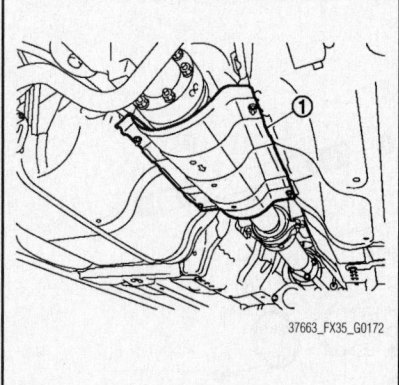

Fig. 35 Remove the heat plate (1)

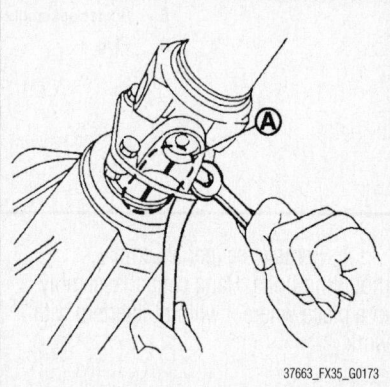

Fig. 36 Put matching marks (A) on propeller shaft flange yoke and transfer companion flange

⁂ CAUTION

For matching marks, use paint. Never damage propeller shaft flange yoke and transfer companion flange.

5. Put matching marks on propeller shaft rebro joint and final drive companion flange.

⁂ CAUTION

For matching marks, use paint. Never damage propeller shaft rebro joint and final drive companion flange.

6. Loosen mounting nuts of center bearing mounting brackets (upper/lower).

⁂ CAUTION

Tighten mounting nuts temporarily.

7. Remove propeller shaft assembly fixing bolts and nuts.
8. Remove center bearing mounting bracket fixing nuts.

9. Remove propeller shaft assembly.

⁂ CAUTION

Note the following:

• Never damage the rear oil seal of transmission.
• If constant velocity joint was bent during propeller shaft assembly removal, installation, or transportation, its boot may be damaged. Wrap boot interference area to metal part with shop cloth or equivalent to protect boot from breakage.

10. Remove clip and center bearing mounting bracket (upper/lower).

To install:

11. Note the following, and install in the reverse order of removal.

12. Install center bearing mounting bracket (upper) with its arrow mark facing forward.

13. Adjust position of center bearing mounting bracket (upper), center bearing mounting bracket (lower) sliding back and forth to prevent play in thrust direction of

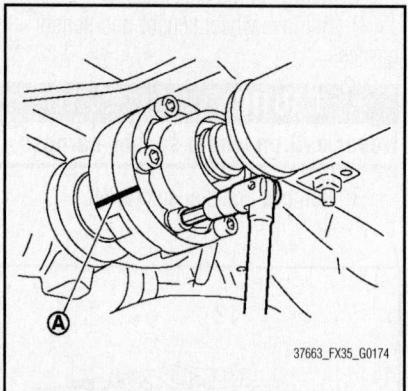

Fig. 37 Put matching marks (A) on propeller shaft rebro joint and final drive companion flange

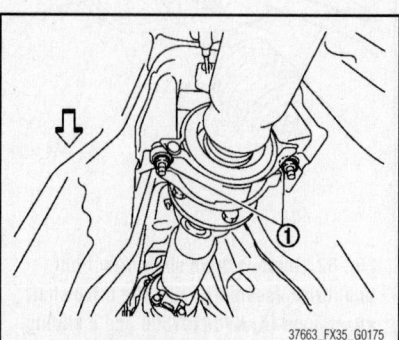

Fig. 38 Loosen mounting nuts (1) of center bearing mounting brackets (upper/lower)

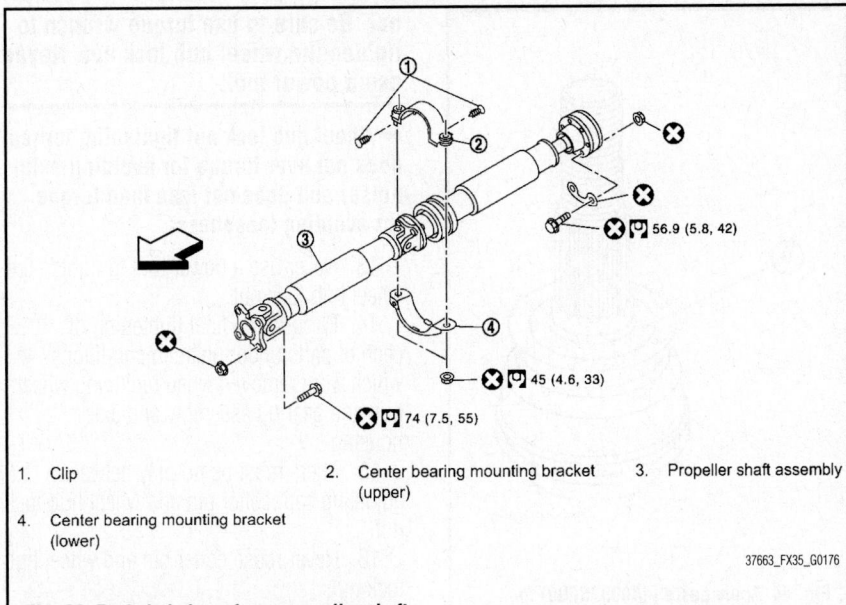

1. Clip
2. Center bearing mounting bracket (upper)
3. Propeller shaft assembly
4. Center bearing mounting bracket (lower)

37663_FX35_G0176

Fig. 39 Exploded view of rear propeller shaft

center bearing insulator. Install center bearing mounting bracket (upper/lower) to vehicle.

14. Align matching marks to install propeller shaft flange yoke and transfer companion flange.

15. Align matching marks to install propeller shaft rebro joint and final drive companion flange.

16. Tighten mounting bolts and nuts in the order shown.

17. After assembly, perform a driving test to check propeller shaft vibra-tion. If vibration occurred, sepa-rate propeller shaft from final drive. Reinstall companion flange after rotating it by 60, 120, 180, 240, 300 degrees. Then perform driving test and check propeller shaft vibration again at each point.

18. If propeller shaft or final drive has been replaced, connect them as follows:

a. Install the propeller shaft while aligning its matching mark with the matching mark on the joint as close as possible.

❄❄ CAUTION

Avoid damaging the rebro joint boot, protect it with a shop cloth or equiva-lent.

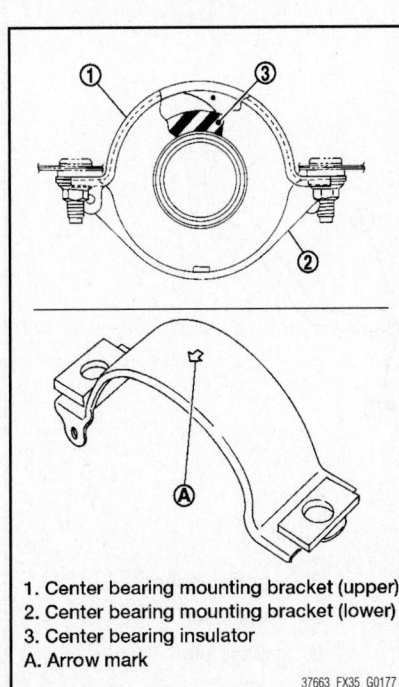

1. Center bearing mounting bracket (upper)
2. Center bearing mounting bracket (lower)
3. Center bearing insulator
A. Arrow mark

37663_FX35_G0177

Fig. 40 Proper installation of center bearing mounting brackets (upper/lower)

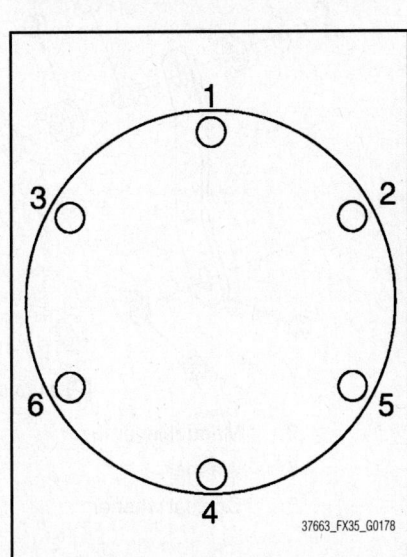

37663_FX35_G0178

Fig. 41 Tighten mounting bolts and nuts in the order shown

REAR HALFSHAFT

REMOVAL & INSTALLATION

See Figures 42 through 44.

1. Remove tires and wheels.
2. Remove cotter pin and adjusting cap, then loosen wheel hub lock nut.
3. Put matching mark on drive shaft and wheel hub and bearing assembly.

❄❄ CAUTION

Use paint or similar substance for matching marks. Never scratch the surface.

4. Remove center muffler.
5. Patch wheel hub lock nut with a piece of wood. Hammer the wood to disen-gage wheel hub and bearing assembly from drive shaft.

❄❄ CAUTION

Never place drive shaft joint at an extreme angle. Also be care-ful not to overextend slide joint. Never allow drive shaft to hang down without support for counterpart such as joint sub-assembly, and other parts.

➡ **Use a suitable puller if wheel hub and bearing assembly and drive shaft cannot be separated even after per-forming the above procedure.**

6. Remove wheel hub lock nut.
7. Remove mounting bolts between side flange and drive shaft.
8. Remove drive shaft.

To install:

9. Note the following, and install in the reverse order of removal.

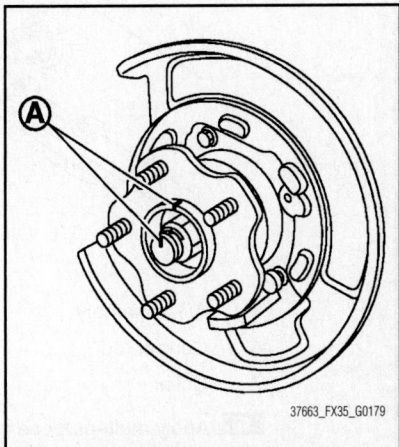

37663_FX35_G0179

Fig. 42 Put matching mark (A) on drive shaft and wheel hub and bearing assembly

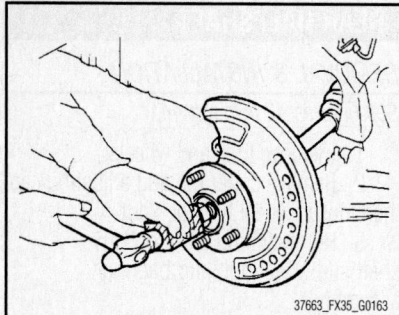

Fig. 43 Hammer the wood to disengage wheel hub and bearing assembly from drive shaft

10. Clean the matching surface of wheel hub lock nut and wheel hub and bearing assembly.

✳✳ CAUTION

Never apply lubricating oil to these matching surface.

11. Clean the matching surface of drive shaft and wheel hub and bearing assembly. And then apply paste (440037S000) to surface (A) of joint sub-assembly of drive shaft.

✳✳ CAUTION

Apply paste to cover entire flat surface of joint sub-assembly of drive shaft.

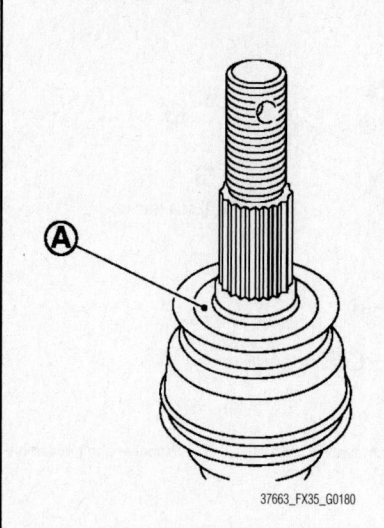

Fig. 44 Apply paste (440037S000) to surface (A) of joint sub-assembly of drive shaft

12. Use the following torque range for tightening the wheel hub lock nut.

✳✳ CAUTION

Since the drive shaft is assembled by press-fitting, use the tightening torque range for the wheel hub lock

nut. Be sure to use torque wrench to tighten the wheel hub lock nut. Never use a power tool.

➡ **Wheel hub lock nut tightening torque does not over torque for avoiding axle noise, and does not less than torque for avoiding looseness.**

13. Never use a power tool to tighten the wheel hub lock nut.

14. Perform the final tightening of each of parts under unladen conditions, which were removed when removing wheel hub and bearing assembly and axle housing.

15. There must be no play between adjusting cap, cotter pin and wheel hub lock nut.

16. Never reuse cotter pin and wheel hub lock nut.

TRANSFER CASE ASSEMBLY

REMOVAL & INSTALLATION

3.5L Engine

See Figures 45 through 51.

1. Remove rear propeller shaft.
2. Remove front propeller shaft.
3. Disconnect AWD solenoid harness connector and separate harness from transfer assembly.

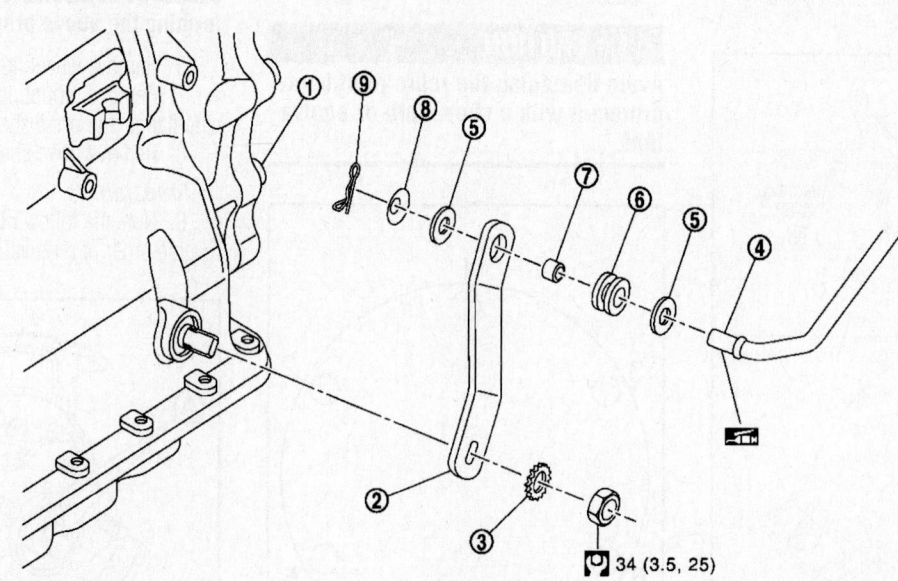

1. A/T assembly	2. Manual lever	3. Lock washer
4. Control rod	5. Washer	6. Insulator
7. Collar	8. Conical washer	9. Snap pin

🔧: Apply multi-purpose grease.

Fig. 45 Exploded view of control rod assembly

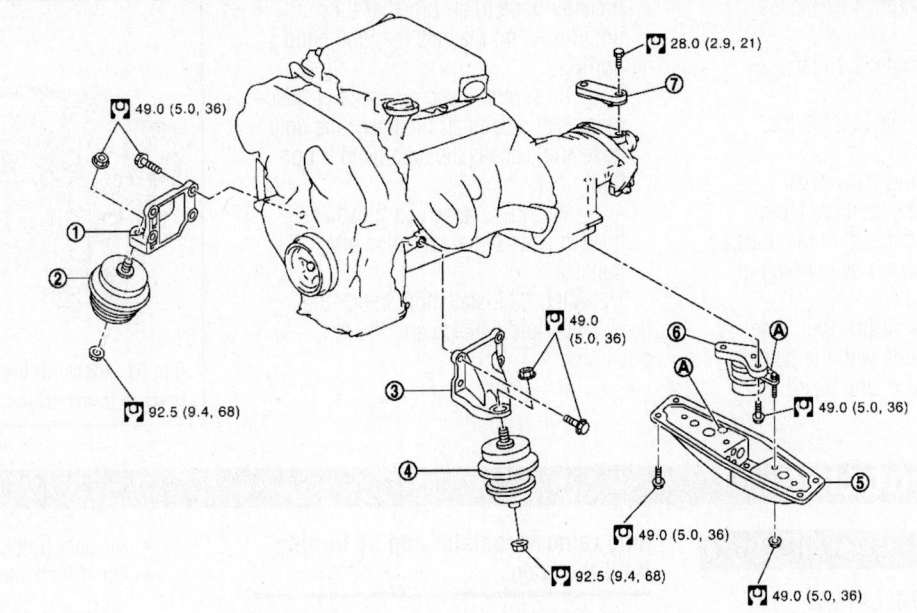

1. Engine mounting bracket (RH)
2. Engine mounting insulator (RH)
3. Engine mounting bracket (LH)
4. Engine mounting insulator (LH)
5. Rear engine mounting member
6. Engine mounting insulator (rear)
7. Dynamic damper
A. Front mark

37663_FX35_G0150

Fig. 46 Exploded view showing engine mounts

4. Remove transfer air breather hose.

5. Remove control rod.

6. Support transfer assembly and transmission assembly with a jack.

7. Remove rear engine mounting member and engine mounting insulator.

8. Lower jack to the position where the top transfer mounting bolts can be removed.

9. Remove transfer mounting bolts and separate transfer from transmission.

�֎֎ CAUTION

Secure transfer assembly and transmission assembly to a jack.

To install:

10. Note the following, and install in the reverse order of removal.

11. When installing the transfer to the transmission, install the mounting bolts following the standard below, tighten bolts to the specified torque.

   a. Bolts A: Length: 2.95 inches (75 mm)

   b. Bolts B: Length: 1.77 inches (45 mm)

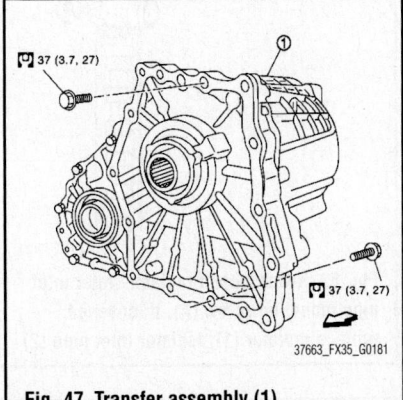

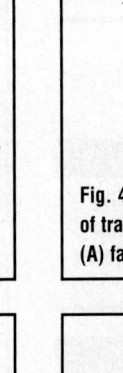

37663_FX35_G0181

Fig. 47 Transfer assembly (1)

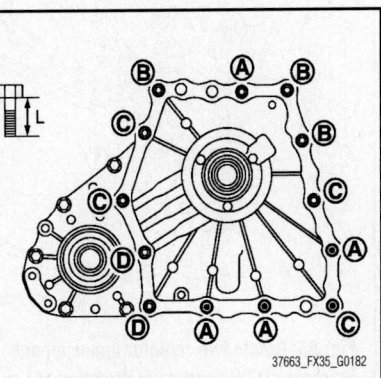

37663_FX35_G0182

Fig. 48 Proper bolt installation locations

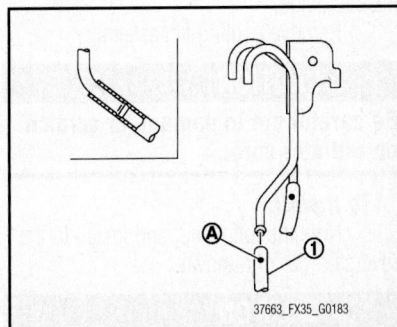

37663_FX35_G0183

Fig. 49 Set transfer air breather hose (1) of transmission side with the paint mark (A) facing upward

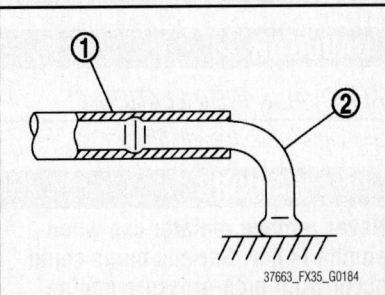

37663_FX35_G0184

Fig. 50 Be sure to insert air breather hose (1) of transfer side to air breather tube (2) until hose end reaches the tube bend portion

c. Bolts C: Length: 1.57 inches (40 mm)

d. Bolts D: Length: 1.18 inches (30 mm)

e. Torque all bolts to 27 ft. lbs. (37 Nm)

12. When installing transfer air breather hose, make sure there are no pinched or restricted areas on the transfer air breather hose caused by bending or winding.

a. Set transfer air breather hose of transmission side with the paint mark facing upward, and insert air

breather hose to air breather tube until hose end reaches the tube bend portion.

b. Be sure to insert air breather hose of transfer side to air breather tube until hose end reaches the tube bend R portion.

c. Be sure to attach air breather hose in parts of transmission and transfer.

13. After the installation, check the fluid level, fluid leakage and the A/T positions.

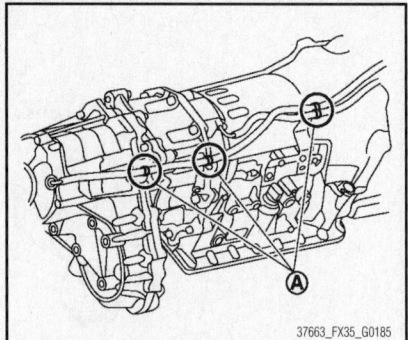

Fig. 51 Attach air breather hose in (A) parts of transmission and transfer

ENGINE COOLING

ENGINE FAN

REMOVAL & INSTALLATION

1. Drain engine coolant.
2. Remove reservoir tank.
3. Remove air cleaner case (bank 1 and bank 2).
4. Remove mounting bolt from high pressure flexible hose bracket.
5. Remove the upper radiator hose.
6. Disconnect harness connector from cooling fan control module, and move harness to aside.
7. Remove cooling fan assembly.

✳✳ CAUTION

Be careful not to damage or scratch on radiator core.

To install:

8. Note the following, and install in the reverse order of removal.

✳✳ CAUTION

Only use genuine parts for cooling fan mounting bolt and observe the specified torque (to prevent radiator from being damaged).

RADIATOR

REMOVAL & INSTALLATION

See Figures 52 through 55.

✳✳ WARNING

Never remove radiator cap when engine is hot. Serious burns could occur from high-pressure engine coolant escaping from water outlet (front). Wrap a thick cloth around the cap. Slowly turn it a quarter of a turn to release built-up pressure. Care-

fully remove radiator cap by turning it all the way.

1. Remove the following parts:
   - Engine under cover
   - Engine cover
   - Air cleaner case

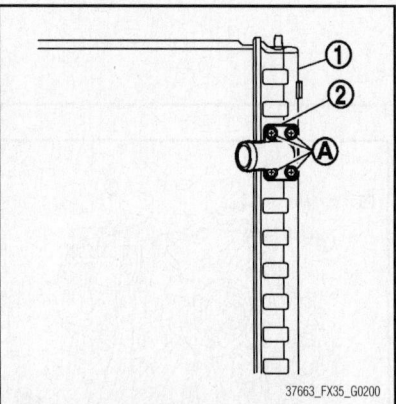

Fig. 52 Never loosen radiator water inlet pipe mounting screw (A). If loosened, replace radiator (1); radiator inlet pipe (2)

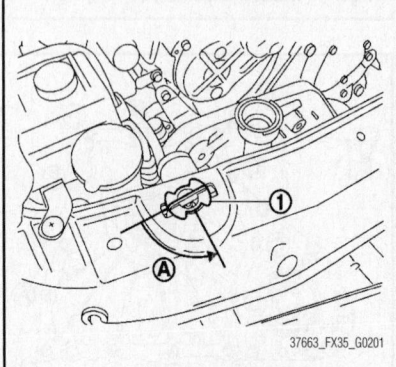

Fig. 53 Rotate two radiator upper mount brackets (1) 90 degrees in direction (A) as shown

- Air duct (inlet)
- Hood lock stay assembly and horn

2. Remove condenser.
3. Drain engine coolant from radiator.

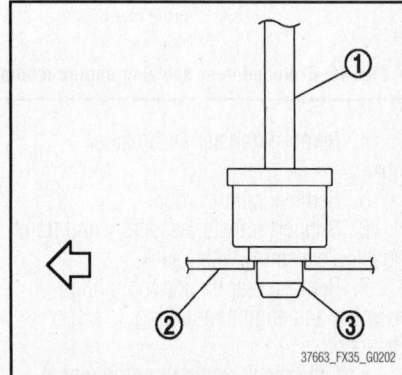

Fig. 54 Lift up and pull the radiator (1) forward, and then remove the mounting rubber (lower) (3) from the radiator core support (2)

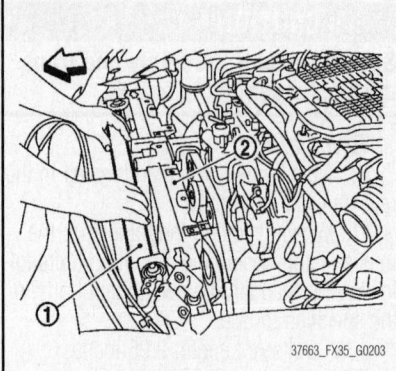

Fig. 55 Remove radiator (1) from front of radiator core support (2)

✳✳ CAUTION

Perform this step when the engine is cold. Never spill engine coolant on drive belt.

4. Disconnect A/T fluid cooler hoses from radiator.

➡ **Install blind plug to avoid leakage of A/T fluid.**

5. Remove radiator hoses (upper and lower) and reservoir tank hose.

✳✳ CAUTION

Note the following:

- Be careful not to allow engine coolant to contact drive belt.
- Never loosen radiator water inlet pipe mounting screw. If loosened, replace radiator.

6. Rotate two radiator upper mount brackets 90 degrees in direction as shown, and remove them.

7. Remove radiator as per the following:

✳✳ CAUTION

Be careful not to damage radiator core.

a. Lift up and pull the radiator forward, and then remove the mounting rubber (lower) from the radiator core support.

b. Remove radiator from front of radiator core support.

To install:

8. Installation is the reverse order of removal.

9. Check for leakage of engine coolant using the radiator cap tester adapter and the radiator cap tester.

10. Start and warm up the engine. Visually check that there is no leakage of engine coolant and A/T fluid.

THERMOSTAT

REMOVAL & INSTALLATION

3.5L Engine

See Figures 56 and 57.

1. Remove engine cover.
2. Remove air duct and air cleaner case assembly (bank 2).
3. Remove reservoir tank.
4. Remove engine undercover with power tool.
5. Drain engine coolant from radiator drain plug at the bottom of radiator.

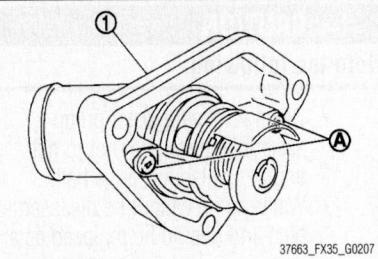

Fig. 56 Remove water inlet and thermostat assembly (1); never loosen these screws (A)

37663_FX35_G0207

✳✳ CAUTION

Perform this step when the engine is cold. Never spill engine coolant on drive belt.

6. Disconnect the lower radiator hose.

7. Disconnect intake valve timing control solenoid valve harness connector (bank 2), and remove intake valve timing control solenoid valve.

8. Remove water inlet and thermostat assembly.

✳✳ CAUTION

Never disassemble water inlet and thermostat assembly. Replace them as a unit, if necessary.

To install:

9. Note the following, and install in the reverse order of removal.

✳✳ CAUTION

Be careful not to spill engine coolant over engine room. Use rag to absorb engine coolant.

10. Check that the reservoir tank cap is tightened.

11. Check for leakage of engine coolant using the radiator cap tester adapter and the radiator cap tester.

12. Start and warm up the engine. Visually check that there is no leakage of engine coolant.

5.0L Engine

See Figures 58 and 59.

1. Remove engine cover and engine room cover (RH and LH).
2. Remove air duct (inlet).
3. Remove reservoir tank.
4. Remove engine undercover.
5. Drain engine coolant from drain plugs on radiator and cylinder block.

✳✳ CAUTION

Perform this step when engine is cold. Never spill engine coolant on drive belts.

6. Disconnect the upper and lower radiator hoses.

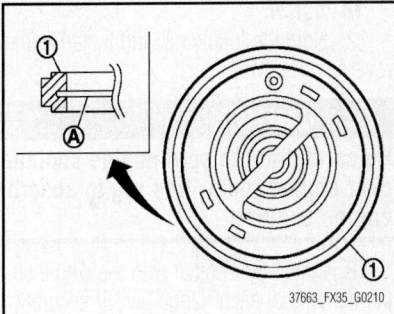

Fig. 58 Install thermostat with the whole circumference of each flange part (A) fit securely inside rubber ring (1)

37663_FX35_G0210

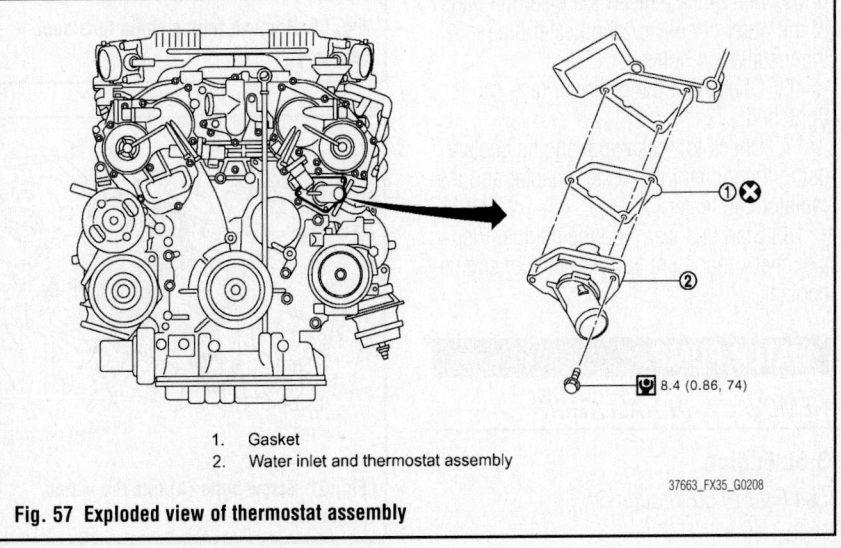

1. Gasket
2. Water inlet and thermostat assembly

8.4 (0.86, 74)

Fig. 57 Exploded view of thermostat assembly

37663_FX35_G0208

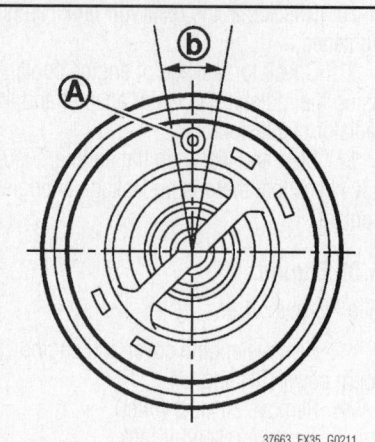

Fig. 59 Install thermostat with jiggle valve (A) facing upwards within the range of 20 degrees (b)

7. Remove intake manifold.
8. Remove water suction pipe and water suction hose.
9. Remove water inlet and thermostat.
10. Remove water connector, heater pipes and heater hoses.
11. Remove thermostat housing.

To install:
12. Note the following, and install in the reverse order of removal.

> ※※ **CAUTION**
>
> **Be careful not to spill engine coolant over engine room. Use rag to absorb engine coolant.**

13. Install thermostat with the whole circumference of each flange part fit securely inside rubber ring.
14. Install thermostat with jiggle valve facing upwards. The position deviation may be within the range of 20 degrees.
15. First apply a neutral detergent to O-rings, then quickly insert the insertion parts of the water connector and heater pipe into the installation holes.
16. Check that the reservoir tank cap is tightened.
17. Check for leakage of engine coolant using the radiator cap tester adapter and the radiator cap tester.
18. Start and warm up the engine. Visually check that there is no leakage of engine coolant.

WATER PUMP

REMOVAL & INSTALLATION

3.5L Engine
See Figures 60 through 64.

> ※※ **CAUTION**
>
> **Note the following:**
>
> - When removing water pump assembly, be careful not to get engine coolant on drive belt.
> - Water pump cannot be disassembled and should be replaced as a unit.
> - After installing water pump, connect hose and clamp securely, then check for leakage using the radiator cap tester and the radiator cap tester adapter.

1. Remove engine cover.
2. Release the fuel pressure.
3. Disconnect the battery cable from the negative terminal.
4. Remove air duct and air cleaner case assembly (bank 1 and bank 2).
5. Remove reservoir tank.
6. Separate engine harness removing their brackets from front timing chain case.
7. Remove engine undercover.
8. Drain engine oil.

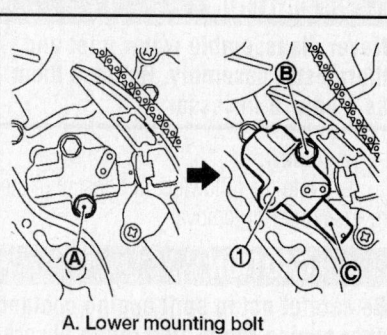

A. Lower mounting bolt
B. Upper mounting bolt
C. Plunger
1. Chain tensioner (primary)

Fig. 60 Remove timing chain tensioner (primary)

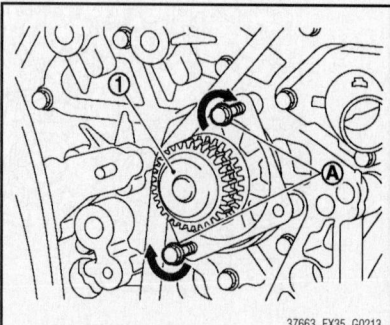

Fig. 61 Screw bolts (A) into the water pump (1)

> ※※ **CAUTION**
>
> **Perform this step when the engine is cold. Never spill engine oil on drive belt.**

9. Drain engine coolant from radiator.

> ※※ **CAUTION**
>
> **Perform this step when the engine is cold. Never spill engine coolant on drive belt.**

10. Remove the lower radiator hose.
11. Remove cooling fan assembly.
12. Remove front timing chain case.
13. Remove timing chain tensioner (primary) as per the following:
   a. Remove lower mounting bolt.
   b. Loosen upper mounting bolt slowly, and then turn chain tensioner (primary) on the upper mounting bolt so that plunger is fully expanded.

➡**Even if plunger is fully expanded, it is not dropped from the body of timing chain tensioner (primary).**

   c. Remove upper mounting bolt, and then remove timing chain tensioner (primary).
14. Remove water pump as per the following:
   a. Remove three water pump mounting bolts. Secure a gap between water pump gear and timing chain, by turning crankshaft counterclockwise until timing chain looseness on water pump sprocket becomes maximum.
   b. Screw M8 bolts (A) length: approximately 2 inches (50 mm) into water pump's upper and lower mounting bolt holes until they reach timing chain case. Then, alternately tighten each bolt for a half turn, and pull out water pump (1).

> ※※ **CAUTION**
>
> **Note the following:**
>
> - Pull straight out while preventing vane from contacting socket in installation area.
> - Remove water pump without causing sprocket to contact timing chain.

   c. Remove M8 bolts and O-rings from water pump.

> ※※ **CAUTION**
>
> **Never disassemble water pump.**

To install:
15. Install new O-rings to water pump.
   a. Apply engine oil to O-ring and engine coolant to O-ring.

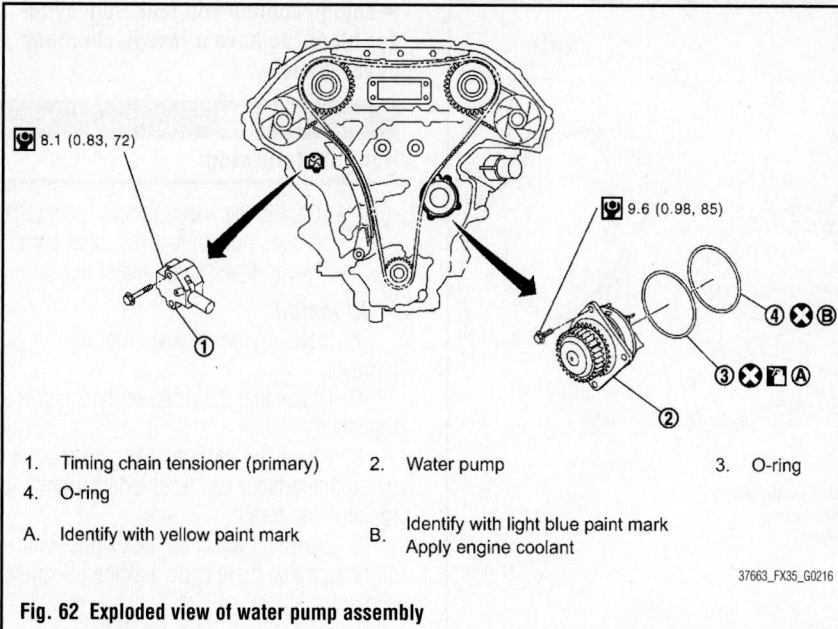

1. Timing chain tensioner (primary) 2. Water pump 3. O-ring
4. O-ring
A. Identify with yellow paint mark B. Identify with light blue paint mark
Apply engine coolant

37663_FX35_G0216

Fig. 62 Exploded view of water pump assembly

b. Locate O-ring with yellow paint mark to front side.

c. Locate O-ring with light blue paint mark to rear side.

16. Install water pump.

✷✷ CAUTION

Never allow cylinder block to nip O-rings when installing water pump.

a. Check timing chain and water pump sprocket are engaged.

b. Insert water pump by tightening mounting bolts alternately and evenly.

17. Install timing chain tensioner (primary) as per the following:

a. Turn crankshaft clockwise so that timing chain on the timing chain tensioner (primary) side is loose.

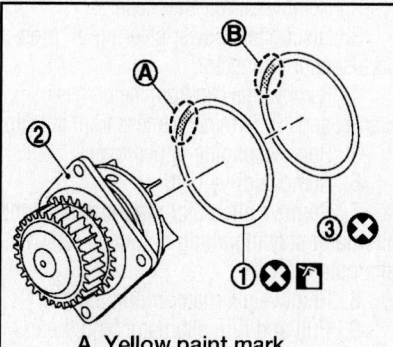

A. Yellow paint mark
B. Light blue paint mark
1. O-ring
2. Water pump
3. O-ring

37663_FX35_G0214

Fig. 63 Install new O-rings to water pump

b. Pull plunger stopper tab up (or turn lever downward) so as to remove plunger stopper tab from the ratchet of plunger.

➥**Plunger stopper tab and lever are synchronized.**

c. Push plunger into the inside of tensioner body.

d. Hold plunger in the fully compressed position by engaging plunger stopper tab with the tip of ratchet.

e. To secure lever, insert stopper pin through hole of lever into tensioner body hole.

➥**The lever parts and the tab are synchronized. Therefore, the plunger will be secured under this condition.**

f. Install timing chain tensioner (primary).

➥**Remove dust and foreign material completely from backside of timing chain**

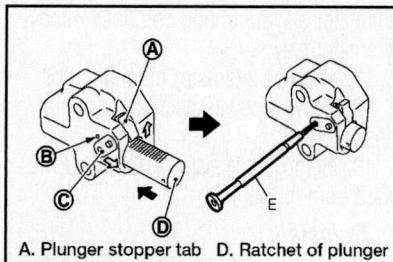

A. Plunger stopper tab D. Ratchet of plunger
B. Tensioner body hole E. Stopper pin
C. Lever

37663_FX35_G0215

Fig. 64 Install timing chain tensioner (primary)

tensioner (primary) and from installation area of rear timing chain case.

g. Remove stopper pin.

h. Check again that timing chain and water pump sprocket are engaged.

18. Install in the reverse order of removal for remaining parts.

a. After starting engine, let idle for three minutes, then rev engine up to 3,000 rpm under no load to purge air from the high-pressure chamber of chain tensioner. Engine may produce a rattling noise. This indicates that air still remains in the chamber and is not a matter of concern.

b. Check that the reservoir tank cap is tightened.

c. Check for leakage of engine coolant using the radiator cap tester adapter and the radiator cap tester.

d. Start and warm up the engine. Visually check that there is no leakage of engine coolant.

5.0L Engine

See Figure 65.

✷✷ CAUTION

Note the following:

- When removing water pump assembly, be careful not to get engine coolant on drive belts.
- Water pump cannot be disassembled and should be replaced as a unit.
- After installing water pump, connect hose and clamp securely, then check for leakage using the radiator cap tester and the radiator cap tester adapter.

1. Remove following parts:
- Engine undercover
- Engine cover, and engine room covers (RH and LH)
- Air duct (inlet)
- Reservoir tank

2. Loosen water pump pulley mounting bolts.

3. Remove alternator, water pump and A/C compressor belt.

4. Remove water pump pulley.

5. Drain engine coolant from drain plugs on radiator and cylinder block.

✷✷ CAUTION

Perform this step when engine is cold. Never spill engine coolant on drive belt.

6. Remove water pump.

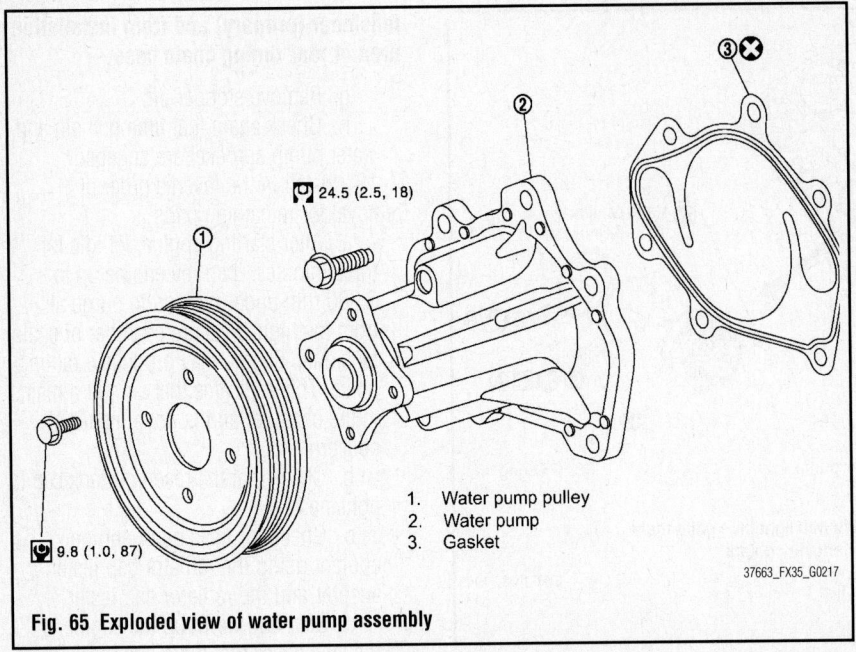

1. Water pump pulley
2. Water pump
3. Gasket

37663_FX35_G0217

Fig. 65 Exploded view of water pump assembly

24.5 (2.5, 18)

9.8 (1.0, 87)

➡Engine coolant will leak from cylinder block, so have a receptacle ready under vehicle.

❊❊ CAUTION
Note the following:

- Handle the water pump vane so that it does not contact any other parts.
- Never disassemble water pump.

To install:

7. Install in the reverse order of removal.

8. Check that the reservoir tank cap is tightened.

9. Check for leakage of engine coolant using the radiator cap tester adapter and the radiator cap tester.

10. Start and warm up the engine. Visually check that there is no leakage of engine coolant.

ENGINE ELECTRICAL

ALTERNATOR

REMOVAL & INSTALLATION

3.5L Engine With 2WD

See Figures 66 and 67.

1. Disconnect the battery cable from the negative terminal.

2. Remove engine front undercover.

3. Remove drive belt.

4. Remove the Splash guard (RH).

5. Disconnect alternator connector.

6. Remove "B" terminal nut.

7. Remove the harness bracket bolts.

8. Remove oil pressure switch harness clip from alternator stay.

9. Disconnect oil pressure switch connector and oil temperature sensor connector.

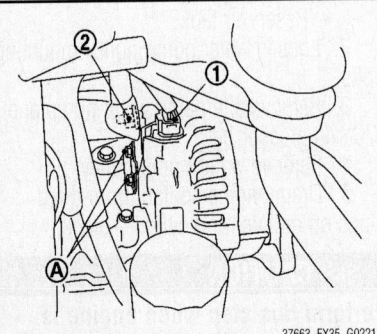

37663_FX35_G0221

Fig. 66 Disconnect alternator connector (1), remove "B" terminal nut (2) and the harness bracket bolts (A)

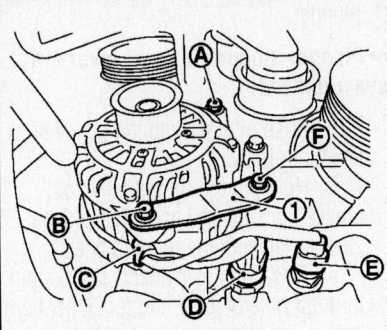

A. Alternator mounting bolt
B. Alternator mounting bolt
C. Oil pressure switch harness clip
D. Oil pressure switch connector
E. Oil temperature sensor connector
F. Alternator stay mounting bolt

37663_FX35_G0222

Fig. 67 Remove oil pressure switch harness clip from alternator stay

10. Remove alternator mounting bolt and alternator stay mounting bolt, then remove alternator stay.

11. Remove alternator mounting bolt.

12. Move a power steering oil pump hose upward.

13. Remove alternator assembly downward from the vehicle.

To install:

14. Install in the reverse order of removal.

❊❊ CAUTION
Be sure to tighten "B" terminal nut carefully.

CHARGING SYSTEM

15. Install alternator, and check tension of belt.

➡For this model, the power generation voltage variable control system that controls the power generation voltage of the alternator has been adopted. Therefore, the power generation voltage variable control system operation inspection should be performed after replacing the alternator, and then make sure that the system operates normally.

3.5L Engine With AWD

See Figures 68 through 71.

1. Disconnect the battery cable from the negative terminal.

2. Remove air cleaner case.

3. Disconnect power steering oil pressure sensor connector.

4. Remove the clip from the harness bracket and "B" terminal harness from the clip.

5. Remove engine undercover.

6. Remove drive belt.

7. Remove alternator mounting bolt and alternator stay mounting bolt, then remove alternator stay.

8. Remove alternator mounting bolt.

9. Pull and turn alternator, and then remove the harness bracket bolts.

10. Disconnect alternator connector.

11. Remove "B" terminal nut.

12. Remove power steering oil pump hose bracket bolts and clamp bolts.

13. Move a power steering oil pump hose upward.

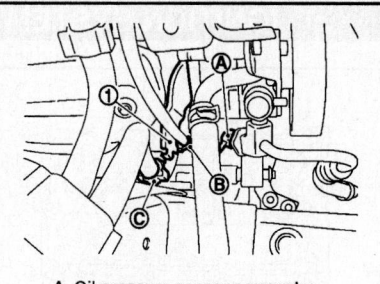

A. Oil pressure sensor connector
B. Clip
C. Clip
1. Harness bracket

37663_FX35_G0223

Fig. 68 Disconnect power steering oil pressure sensor connector

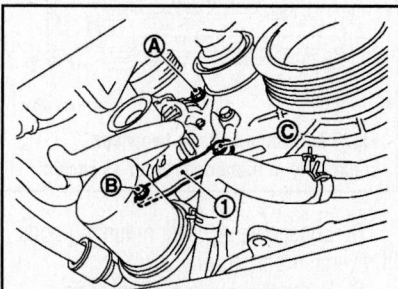

A. Alternator mounting bolt
B. Alternator mounting bolt
C. Alternator stay mounting bolt
1. Alternator stay

37663_FX35_G0224

Fig. 69 Remove alternator mounting bolt and alternator stay mounting bolt, then remove alternator stay

14. Remove alternator assembly downward from the vehicle.

To install:
15. Install in the reverse order of removal.

☀☀ CAUTION

Be sure to tighten "B" terminal nut carefully.

16. Install alternator, and check tension of belt.

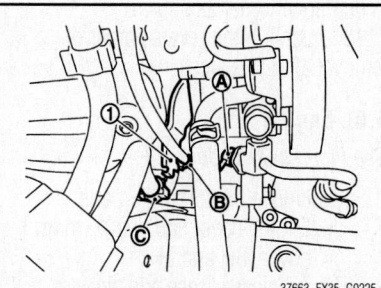

37663_FX35_G0225

Fig. 70 Remove the harness bracket bolts (A); disconnect alternator connector (1); remove "B" terminal nut (2)

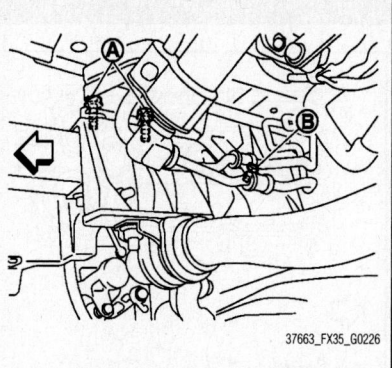

37663_FX35_G0226

Fig. 71 Remove power steering oil pump hose bracket bolts (A) and clamp bolt (B)

➡For this model, the power generation voltage variable control system that controls the power generation voltage of the alternator has been adopted. Therefore, the power generation voltage variable control system operation inspection should be performed after replacing the alternator, and then make sure that the system operates normally.

5.0L Engine

See Figures 72 through 74.

1. Disconnect the battery cable from the negative terminal.
2. Remove drive belt.
3. Remove the air ducts and air cleaner assembly RH.
4. Remove the alternator connector harness bracket.
5. Move a steering hose and harness not to interfere the removal of the alternator.
6. Remove the alternator mounting bolt and alternator mounting bolt.
7. Pull and turn alternator, and then remove the "B" terminal nut and alternator connector.
8. Remove alternator assembly upward from the vehicle.

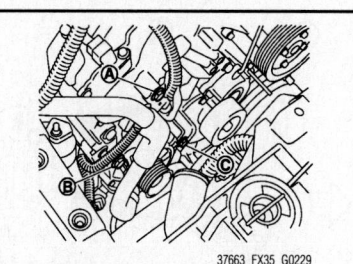

37663_FX35_G0229

Fig. 72 Remove the alternator connector harness bracket (A), the alternator mounting bolt (B) and alternator mounting bolt (C)

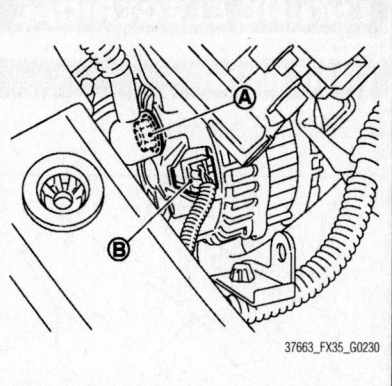

37663_FX35_G0230

Fig. 73 Remove the "B" terminal nut (A) and alternator connector (B)

37663_FX35_G0231

Fig. 74 Remove alternator assembly (1) upward from the vehicle

To install:
9. Install in the reverse order of removal.

☀☀ CAUTION

Be sure to tighten "B" terminal nut carefully.

10. Install alternator, and check tension of belt.

➡For this model, the power generation voltage variable control system that controls the power generation voltage of the alternator has been adopted. Therefore, the power generation voltage variable control system operation inspection should be performed after replacing the alternator, and then make sure that the system operates normally.

VOLTAGE REGULATOR

ADJUSTMENT

The voltage regulator is an integral part of the alternator assembly. No adjustment is possible. If the component is defective, it must be replaced.

FIRING ORDER

See Figures 75 and 76.

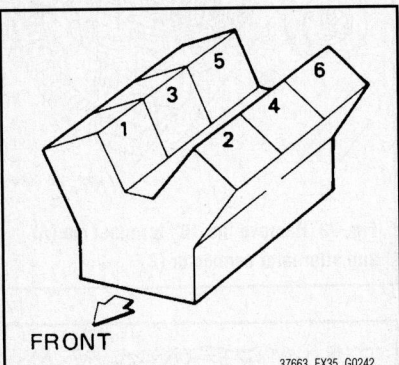

Fig. 75 3.5L engine cylinder number locations

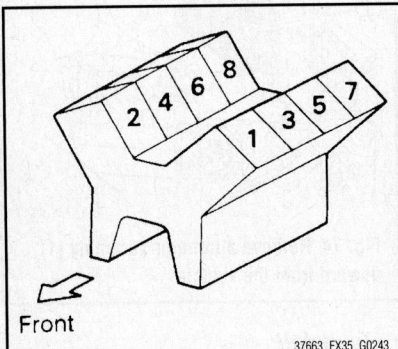

Fig. 76 5.0L engine cylinder number locations

The firing order for the 3.5L engine is 1–2–3–4–5–6.

The firing order for the 5.0L engine is 1–8–7–3–6–5–4–2.

IGNITION COIL & SPARK PLUGS

REMOVAL & INSTALLATION

3.5L Engine

See Figures 77 through 79.

1. Remove the following parts:
   • Engine cover
   • Air cleaner case and air duct
   • Intake manifold collector
2. Disconnect PCV hose from valve cover.
3. Remove camshaft position sensor (PHASE) and exhaust valve timing control position sensor. (bank 1 and bank 2)

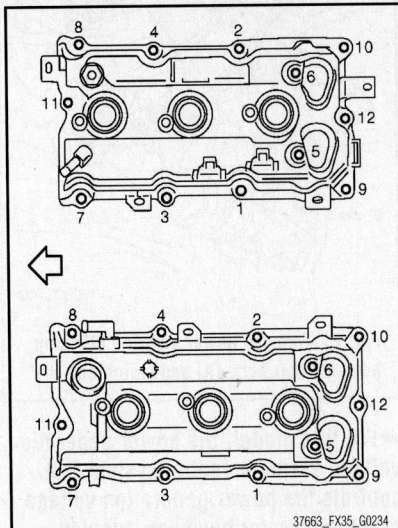

Fig. 77 Loosen valve cover mounting bolts in reverse order shown

✳✳ CAUTION

Note the following:

   • Handle carefully to avoid dropping and shocks.
   • Never disassemble.
   • Never allow metal powder to adhere to magnetic part at sensor tip.
   • Never place sensors in a location where they are exposed to magnetism.
4. Remove PCV valve and O-ring from valve cover, if necessary.
5. Remove oil filler cap from valve cover, if necessary.
6. Remove ignition coil.

✳✳ CAUTION

Never impact ignition coil.

7. Remove harness clips on the valve cover.

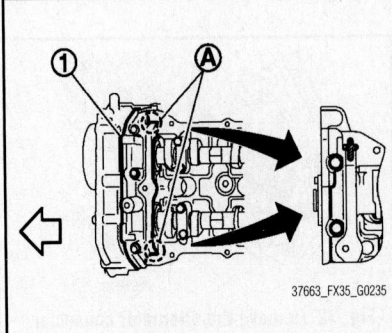

Fig. 78 Apply liquid gasket to the position shown

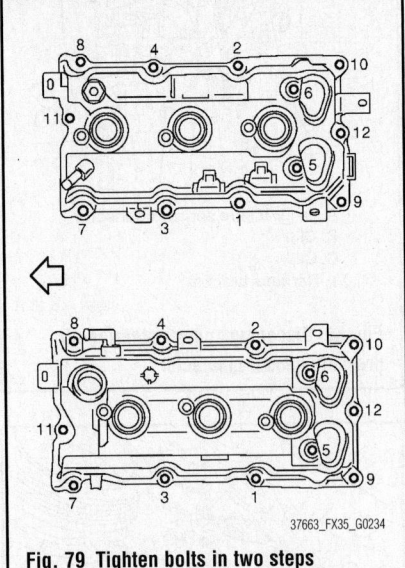

Fig. 79 Tighten bolts in two steps separately in numerical order as shown

8. Loosen valve cover mounting bolts in reverse order shown.
9. Remove valve cover gasket from valve cover.
10. Use scraper to remove all traces of liquid gasket from cylinder head and camshaft bracket (No. 1).

✳✳ CAUTION

Never scratch or damage the mating surface when cleaning off old liquid gasket.

To install:
11. Apply liquid gasket to the position shown.
12. Install valve cover gasket to valve cover.
13. Install valve cover.

➡ **Check if valve cover gasket is not dropped from the installation groove of valve cover.**

14. Tighten bolts in two steps separately in numerical order as shown.
15. Install in the reverse order of removal after this step.

5.0L Engine

See Figures 80 through 84.

1. Remove the following parts:
   • Engine cover and engine room cover (RH and LH)
   • Air cleaner case and air duct
   • Fuel feed hose
2. Disconnect PCV hose from valve cover.
3. Remove ignition coil.

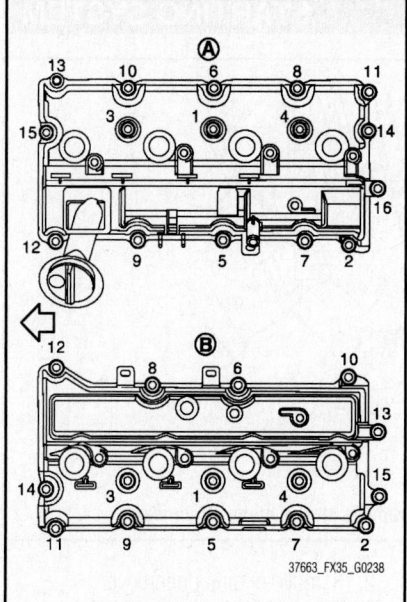

Fig. 80 Remove valve cover; loosen bolts in reverse order shown

➡**Installation position of Ignition coil depends on cylinder position.**

4. Remove spark plugs.

5. Remove valve cover. Loosen bolts in reverse order shown.

6. Remove valve cover gasket from valve cover.

7. Use scraper to remove all traces of liquid gasket from cylinder head & VVEL ladder assembly.

✳✳ CAUTION

Never scratch or damage the mating surface when cleaning off old liquid gasket.

8. Remove PCV valve from valve cover, if necessary.

9. Remove oil filler cap and oil catcher from valve cover, if necessary.

To install:

10. Apply liquid gasket with the tube presser to VVEL ladder assembly and actuator bracket (rear).

➡**The figure shows an example of bank 1 side.**

➡**Apply liquid gasket on the front and rear side of engine first.**

11. Install valve cover gasket to valve cover.

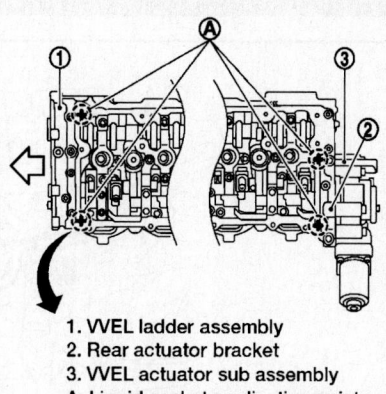

1. VVEL ladder assembly
2. Rear actuator bracket
3. VVEL actuator sub assembly
A. Liquid gasket application point

Fig. 81 Apply liquid gasket to VVEL ladder assembly and actuator bracket (rear)

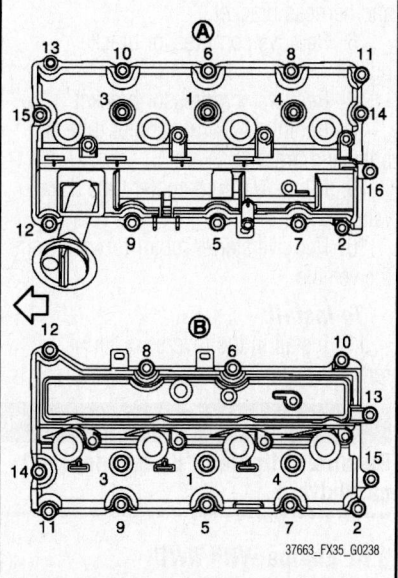

Fig. 82 Tighten bolts in two steps separately in numerical order as shown

12. Install valve cover.

➡**Check that valve cover gasket does not drop from the installation groove of valve cover.**

13. Tighten bolts in two steps separately in numerical order as shown.

➡**Because of the limited working space, use adapter and torque wrench assembly KV10119300 to tighten bolts (on the No.7 and No. 8 cylinders) to the specified torque.**

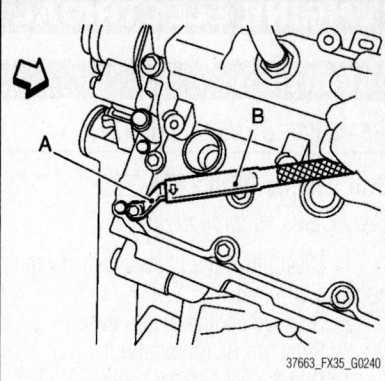

Fig. 83 Because of the limited working space, use adapter (A) and torque wrench (B) assembly KV10119300 to tighten bolts (on the No.7 and No. 8 cylinders) to the specified torque

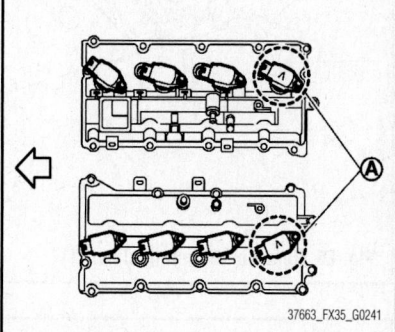

Fig. 84 Install Ignition coil marked with an identification mark (A) on cylinder No. 7 and 8

14. Install spark plug.

15. Install ignition coil.

➡**Install Ignition coil marked with an identification mark on cylinder No. 7 and 8.**

16. Install in the reverse order of removal.

IGNITION TIMING

ADJUSTMENT

Ignition is controlled by the ECM. No adjustment is necessary or possible.

SPARK PLUGS

REMOVAL & INSTALLATION

Refer to Ignition Coil & Spark Plugs.

STARTER

REMOVAL & INSTALLATION

3.5L Engine With 2WD

See Figures 85 through 88.

1. Disconnect the battery cable from the negative terminal.
2. Remove engine undercover.
3. Remove "B" terminal nut.
4. Disconnect "S" connector.

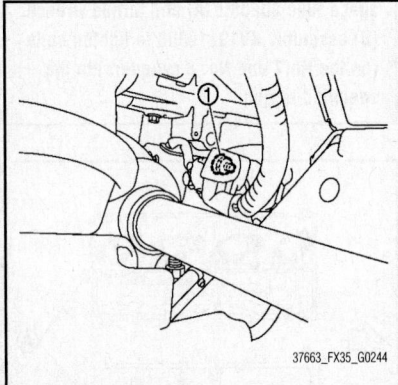

Fig. 85 Remove "B" terminal nut (1)

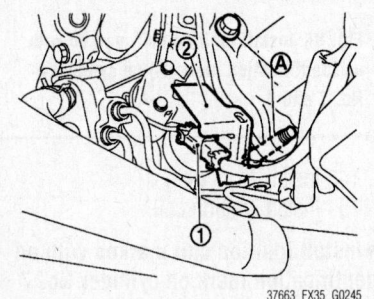

Fig. 86 Disconnect "S" connector (1); remove starter motor mounting bolts (A) and harness bracket (2)

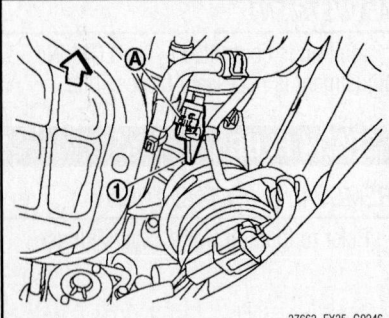

Fig. 87 Remove compressor bracket bolts (A) and compressor bracket (1)

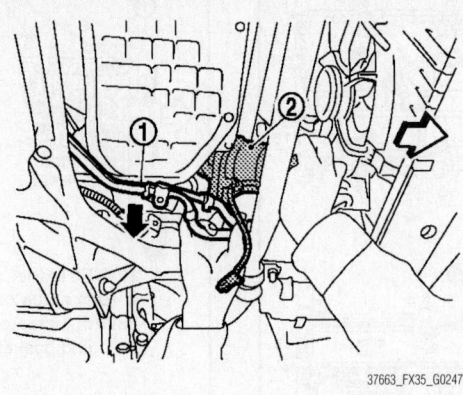

Fig. 88 Move A/T fluid cooler tube (1) downward; remove starter motor (2) forward

5. Remove starter motor mounting bolts and harness bracket.
6. Remove compressor bracket bolts.
7. Remove compressor bracket.
8. Remove A/T fluid cooler tube clip bolts and bracket.
9. Move A/T fluid cooler tube downward.
10. Remove starter motor forward from the vehicle.

To install:
11. Install in the reverse order of removal.

> ✳✳ **CAUTION**
>
> **Be sure to tighten "B" terminal nut carefully.**

3.5L Engine With AWD

See Figures 85 and 86, 89.

1. Disconnect the battery cable from the negative terminal.

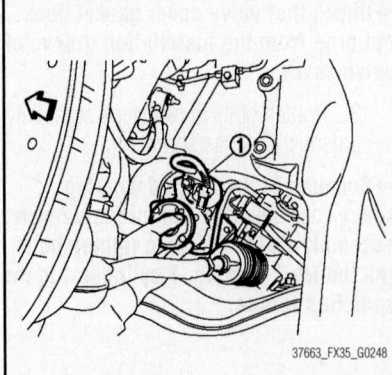

Fig. 89 Remove starter motor (1) to left side from the vehicle

2. Remove engine undercover.
3. Remove "B" terminal nut.
4. Disconnect "S" connector.
5. Remove starter motor mounting bolts and harness bracket.
6. Remove front drive shaft left side housing bolts.
7. Move a front drive shaft left side forward.
8. Remove starter motor to left side from the vehicle.

To install:
9. Install in the reverse order of removal.

> ✳✳ **CAUTION**
>
> **Be sure to tighten "B" terminal nut carefully.**

5.0L Engine

1. Disconnect the battery cable from the negative terminal.
2. Remove engine cover.
3. Remove intake manifold.
4. Remove "B" terminal nut.
5. Disconnect "S" connector.
6. Remove starter motor mounting bolts using power tools.
7. Remove starter motor upward from the vehicle.

To install:
8. Install in the reverse order of removal.

> ✳✳ **CAUTION**
>
> **Be sure to tighten "B" terminal nut carefully.**

ENGINE MECHANICAL

ACCESSORY DRIVE BELTS

ACCESSORY BELT ROUTING

See Figures 90 and 91.

Refer to the accompanying illustrations.

INSPECTION

1. Check that the indicator (notch on fixed side) of each auto-tensioner is within the possible use range.

➡ **Check the each auto-tensioners indication when the engine is cold. When**

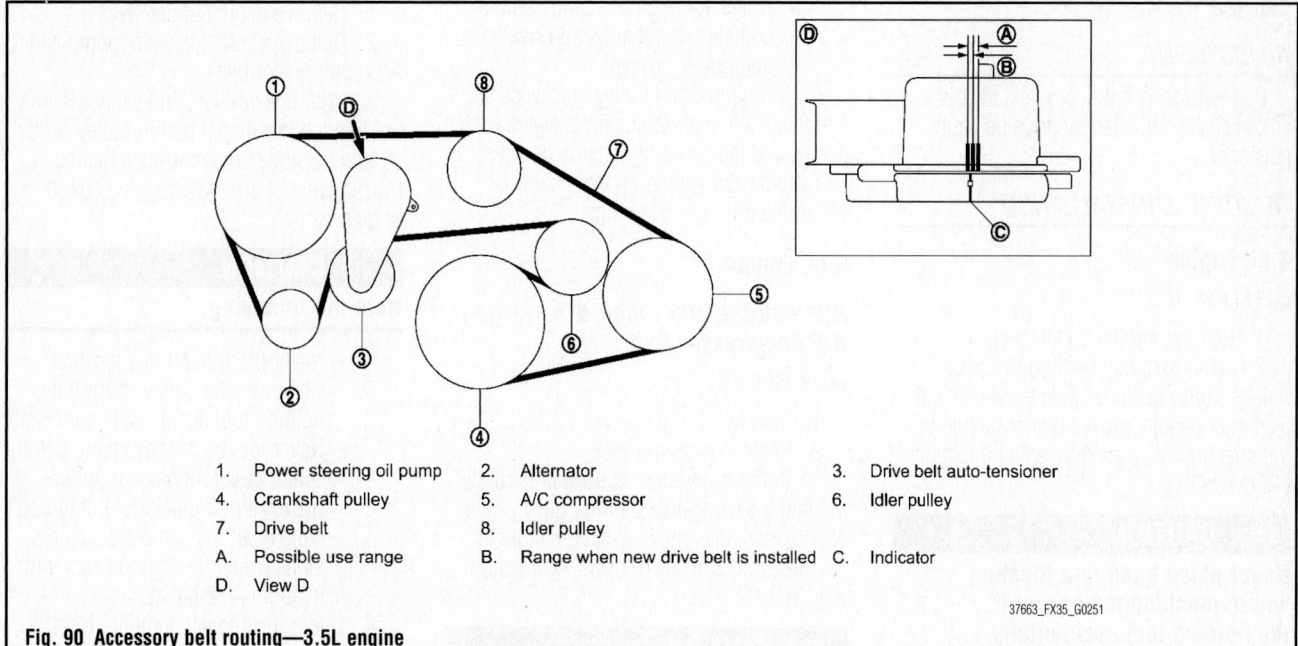

1.	Power steering oil pump	2.	Alternator	3.	Drive belt auto-tensioner
4.	Crankshaft pulley	5.	A/C compressor	6.	Idler pulley
7.	Drive belt	8.	Idler pulley		
A.	Possible use range	B.	Range when new drive belt is installed	C.	Indicator
D.	View D				

37663_FX35_G0251

Fig. 90 Accessory belt routing—3.5L engine

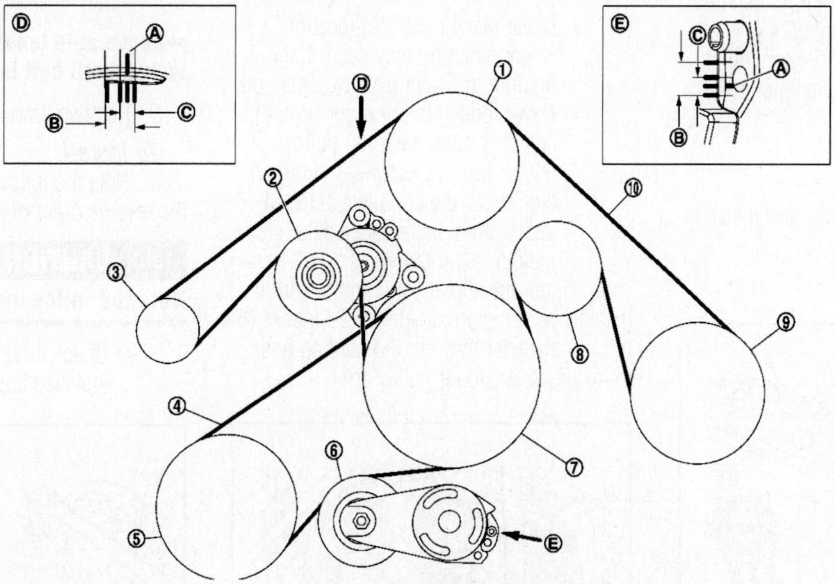

1.	Water pump	2.	Auto-tensioner (for alternator, water pump and A/C compressor belt)	3.	Alternator
4.	Power steering oil pump belt	5.	Power steering oil pump	6.	Auto-tensioner (for power steering oil pump belt)
7.	Crankshaft pulley	8.	Idler pulley	9.	A/C compressor
10.	Alternator, water pump and A/C compressor belt				
A.	Indicator	B.	Possible use range	C.	Range when new drive belt is installed
D.	View D	E.	View E		

37663_FX35_G0252

Fig. 91 Accessory belt routing—5.0L engine

new drive belts is installed, the indicator (notch on fixed side) should be within the range.

2. Visually check entire drive belt for wear, damage or cracks.

3. If the indicator (notch on fixed side) is out of the possible use range or belt is damaged, replace drive belt.

ADJUSTMENT

Belt tension is not necessary, as it is automatically adjusted by drive belt auto-tensioner.

REMOVAL & INSTALLATION

3.5L Engine

See Figure 92.

1. Remove engine undercover.

2. While securely holding the square hole in pulley center of auto tensioner with a spinner handle, move spinner handle in the direction of arrow (loosening direction of drive belt).

> ※※ **CAUTION**
>
> **Never place hand in a location where pinching may occur if the holding tool accidentally comes off.**

3. Under the above condition, insert a metallic bar of approximately 0.24 inches (6 mm) in diameter through the holding boss to lock auto-tensioner pulley arm.

4. Remove drive belt.

To install:

5. Note the following item, and install in the reverse order of removal.

> ※※ **CAUTION**
>
> **Note the following:**
>
> - Check drive belt is securely installed around all pulleys.
> - Check drive belt is correctly engaged with the pulley groove.
> - Check for engine oil and engine coolant are not adhered drive belt and pulley groove.

6. Turn crankshaft pulley clockwise several times to equalize tension between each pulley, and then confirm tension of drive belt at indicator (notch on fixed side) is within the possible use range

5.0L Engine

Alternator, Water Pump & A/C Compressor Belt

See Figure 93.

1. Remove air duct (inlet).
2. Remove reservoir tank.
3. With box wrench, and while securely holding the hexagonal part in pulley center of auto tensioner, move wrench handle in the direction of arrow (loosening direction of belt).

> ※※ **CAUTION**
>
> **Note the following:**
>
> - Never place hand in a location where pinching may occur if the holding tool accidentally comes off.
> - Never loosen the hexagonal part in center of auto tensioner pulley (Never turn it clockwise). If turned clockwise, the complete auto tensioner must be replaced as a unit, including the pulley.

4. Under the above condition, insert a metallic bar of approximately 0.24 inches (6 mm) in diameter through the holding boss to lock auto tensioner pulley arm.

➡Leave auto tensioner pulley arm locked until belt is installed again.

5. Remove alternator, water pump and A/C compressor belt.

Power Steering Oil Pump Belt

See Figure 94.

1. Remove engine undercover.
2. Remove alternator, water pump and A/C compressor belt.
3. With box wrench, and while securely holding the hexagonal part in pulley center of auto tensioner, move wrench handle in the direction of arrow (loosening direction of belt).

> ※※ **CAUTION**
>
> **Note the following:**
>
> - Never place hand in a location where pinching may occur if the holding tool accidentally comes off.
> - Never loosen the hexagonal part in center of auto tensioner pulley (Never turn it clockwise). If turned clockwise, the complete auto tensioner must be replaced as a unit, including the pulley.

4. Under the above condition, insert a metallic bar of approximately 0.24 inches (6 mm) in diameter through the holding boss to lock auto tensioner pulley arm.

➡Leave auto tensioner pulley arm locked until belt is installed again.

5. Remove power steering oil pump belt.

To install:

6. Note the following item, and install in the reverse order of removal.

> ※※ **CAUTION**
>
> **Note the following:**
>
> - Check drive belts are securely installed around all pulleys.

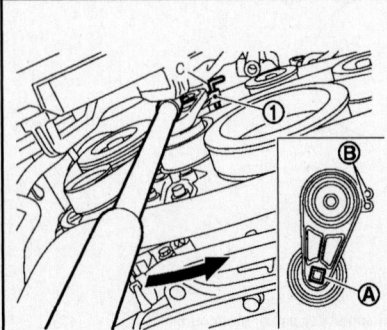

A. Square hole   C. Wrench
B. Holding boss   1. Auto-tensioner

37663_FX35_G0253

Fig. 92 Move spinner handle in the direction of arrow (loosening direction of drive belt)

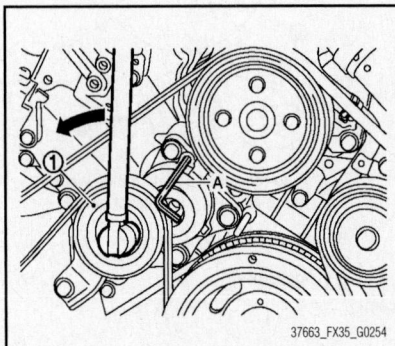

37663_FX35_G0254

Fig. 93 Move wrench handle in the direction of arrow (loosening direction of belt)

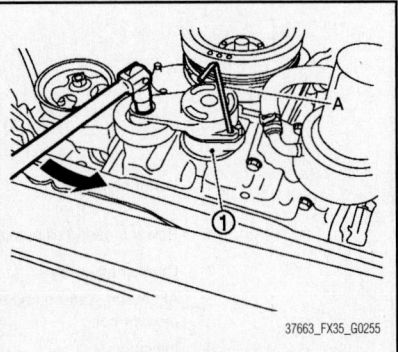

37663_FX35_G0255

Fig. 94 Move wrench handle in the direction of arrow (loosening direction of belt)

- Check drive belts are correctly engaged with the pulley groove.
- Check for engine oil and engine coolant are not adhered drive belts and pulley groove.

7. Turn crankshaft pulley clockwise several times to equalize tension between each pulley, and then confirm tension of drive belts at indicator (notch on fixed side) is within the possible use range.

CAMSHAFT AND VALVE LIFTERS

REMOVAL & INSTALLATION

3.5L Engine

See Figures 95 through 105.

1. Remove front timing chain case, camshaft sprocket and timing chain.
2. Remove fuel sub tube.
3. Remove camshaft sensor bracket. Loosen camshaft sensor bracket bolts in reverse order as shown.

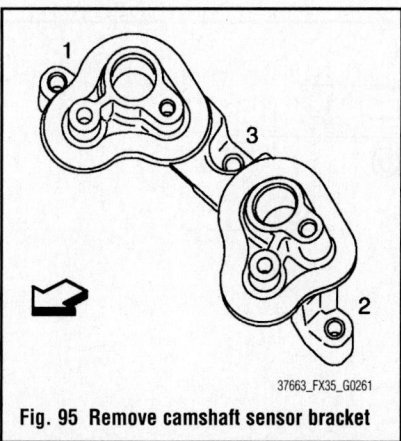

Fig. 95 Remove camshaft sensor bracket

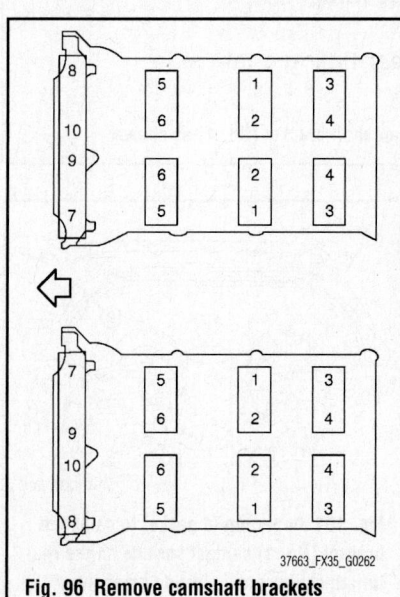

Fig. 96 Remove camshaft brackets

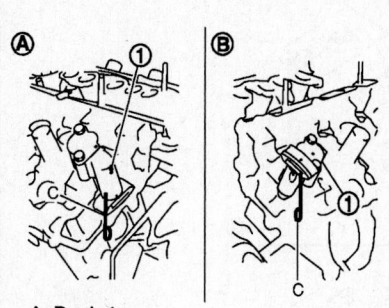

A. Bank 1
B. Bank 2
C. Stopper pin
1. Timing chain tensioners (secondary)

Fig. 97 Remove timing chain tensioners (secondary) from cylinder heads

➡**The order of loosening bolts is the same for bank 1 and bank 2.**

4. Remove camshaft brackets.
   a. Mark camshafts, camshaft brackets and bolts so they are placed in the same position and direction for installation.
   b. Equally loosen camshaft bracket bolts in several steps in reverse order as shown.

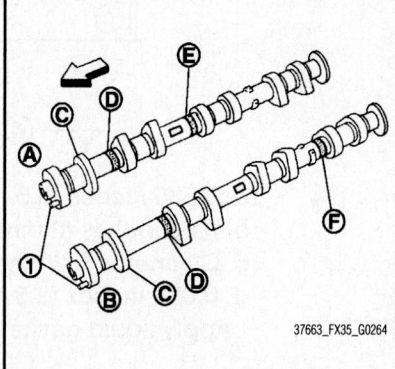

Fig. 98 Camshaft identification marks

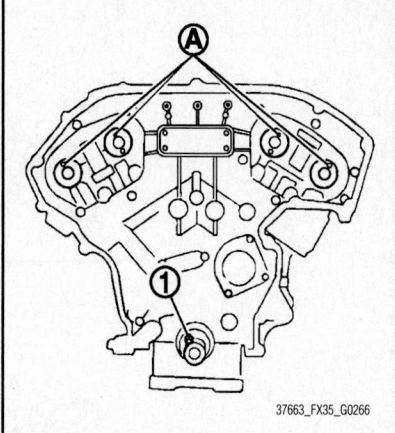

Fig. 100 Install camshaft so that dowel pin (A) on front end face are positioned as shown

5. Remove camshafts.
6. Remove valve lifters.

➡**Identify installation positions, and store them without mixing them up.**

7. Remove timing chain tensioners (secondary) from cylinder heads.
8. Remove timing chain tensioners (secondary) with its stopper pin attached.

➡**Stopper pin should be attached when timing chain (secondary) is removed.**

To install:

9. Install timing chain tensioners (secondary) on both sides of cylinder head.
10. Install timing chain tensioners with its stopper pin attached.
11. Install valve lifters.

➡**Install it in the original position.**

12. Install camshafts.
   a. Follow your identification marks made during removal, or follow the identification marks that are present on new camshafts for proper placement and direction.

Bank	INT/EXH	Dowel pin (1)	Paint marks			Identification mark (C)
			M1 (E)	M2 (F)	M3 (D)	
1	EXH (B)	Yes	No	Green	Light blue	1F
	INT (A)	Yes	Green	No	Light blue	1E
2	INT (A)	Yes	Green	No	Light blue	1G
	EXH (B)	Yes	No	Green	Light blue	1H

Fig. 99 Table describing camshaft identification marks

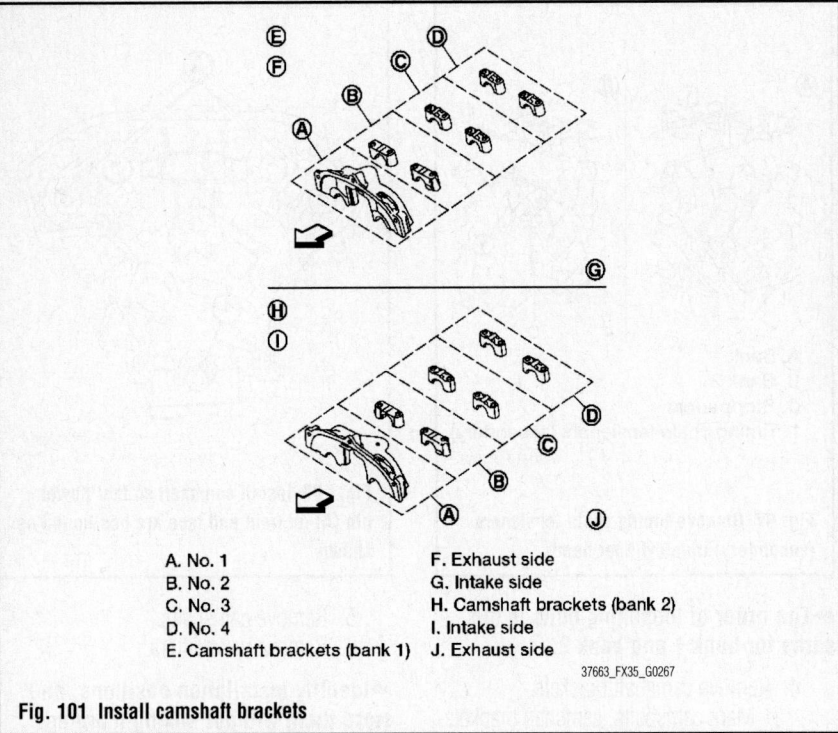

A. No. 1
B. No. 2
C. No. 3
D. No. 4
E. Camshaft brackets (bank 1)
F. Exhaust side
G. Intake side
H. Camshaft brackets (bank 2)
I. Intake side
J. Exhaust side

37663_FX35_G0267

Fig. 101 Install camshaft brackets

rear timing chain case backside as shown on both bank 1 and bank 2.

⁑ CAUTION

For camshaft bracket (No. 1) near installation position, and install it without disturbing the liquid gasket applied to the surfaces.

14. Tighten camshaft bracket bolts in the following steps, in numerical order as shown.
    a. Tighten No. 7 to 10 to 1 ft. lbs. (2 Nm) in numerical order as shown.
    b. Tighten No. 1 to 6 to 1 ft. lbs. (2 Nm) in numerical order as shown.
    c. Tighten No. 1 to 10 to 4 ft. lbs. (6 Nm) in numerical order as shown.
    d. Tighten No. 1 to 10 to 8 ft. lbs. (10 Nm) in numerical order as shown.
15. Install camshaft sensor bracket.
16. Tighten camshaft sensor bracket bolts in numerical order.

➡**The order of tightening bolts is the same for bank 1 and bank 2.**

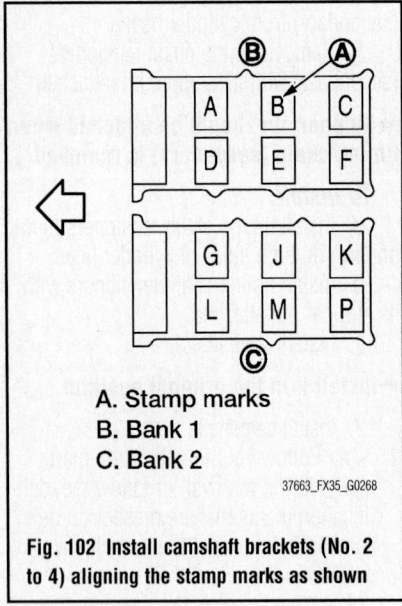

A. Stamp marks
B. Bank 1
C. Bank 2

37663_FX35_G0268

Fig. 102 Install camshaft brackets (No. 2 to 4) aligning the stamp marks as shown

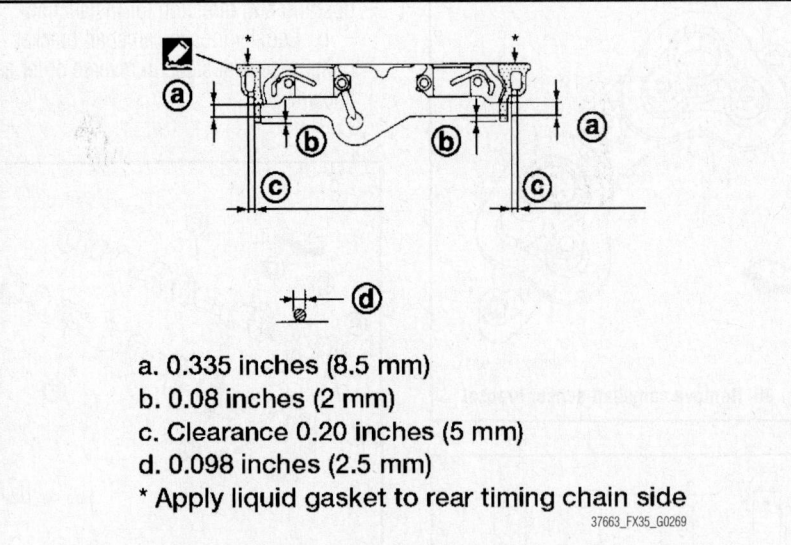

a. 0.335 inches (8.5 mm)
b. 0.08 inches (2 mm)
c. Clearance 0.20 inches (5 mm)
d. 0.098 inches (2.5 mm)
* Apply liquid gasket to rear timing chain side

37663_FX35_G0269

Fig. 103 Apply liquid gasket to mating surface of camshaft bracket (No. 1) as shown

b. Install camshaft so that dowel pin (A) on front end face are positioned as shown. (No. 1 cylinder TDC on its compression stroke)

➡**Though camshaft does not stop at the portion as shown, for the placement of cam nose, it is generally accepted camshaft is placed for the same direction of the figure.**

13. Install camshaft brackets.
    a. Remove foreign material completely from camshaft bracket backside and from cylinder head installation face.

b. Install camshaft bracket in original position and direction as shown.
    c. Install camshaft brackets (No. 2 to 4) aligning the stamp marks as shown.

➡**There are no identification marks indicating bank 1 and bank 2 for camshaft bracket (No. 1).**

d. Apply liquid gasket to mating surface of camshaft bracket (No. 1) as shown on both bank 1 and bank 2.
    e. Apply liquid gasket to camshaft bracket (No. 1) contact surface on the

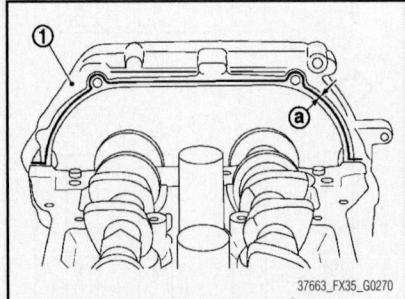

37663_FX35_G0270

Fig. 104 Apply liquid gasket to camshaft bracket (No. 1) contact surface on the rear timing chain case backside as shown

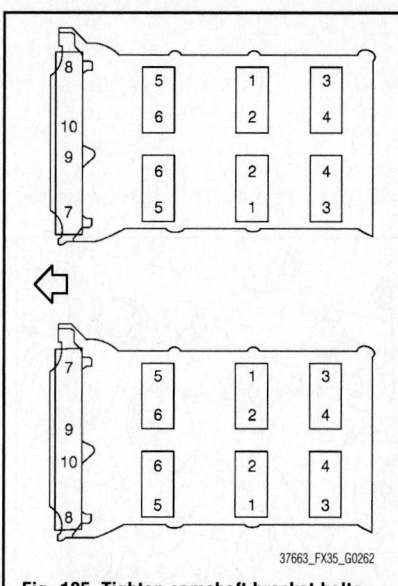

Fig. 105 Tighten camshaft bracket bolts in numerical order

17. Inspect and adjust the valve clearance.

18. Install in the reverse order of removal after this step.

5.0L Engine

See Figures 106 through 124.

✳✳ CAUTION

A high degree of precision is required for a valve on the intake side. Never remove the valve related parts unless necessary.

✳✳ CAUTION

Note the following:

- As for replacement of parts on the intake side as shown in the exploded view, replace VVEL ladder assembly and cylinder head assembly.
- VVEL ladder assembly cannot be replaced as a single part, because it is machined together with cylinder head assembly.

➡**The figure shows an example of bank 1.**

✳✳ CAUTION

Never loosen adjusting bolts and mounting bolts (black color) of VVEL ladder assembly. If loosened, the stroke of cam lift becomes out of adjustment. In such case, replacement of VVEL ladder assembly and cylinder head assembly is required.

1. Remove valve covers (bank 1 and bank 2).

2. Remove VVEL actuator sub assembly as per the following:

✳✳ CAUTION

VVEL actuator sub assembly and VVEL control shaft position sensor are not reusable. Never remove them unless they are required.

a. Remove VVEL control shaft position sensor.

b. Hold the two flat areas of control shaft with a wrench to remove mounting bolts of control shaft.

✳✳ CAUTION

During the operation, never allow a wrench to interfere with other parts. Fix control shaft to prevent the interference of the stopper surface.

c. Remove VVEL actuator sub assembly. Loosen mounting bolts in the reverse order as shown.

✳✳ CAUTION

Note the following:

VVEL Ladder Assembly & Cylinder Head Assembly Features

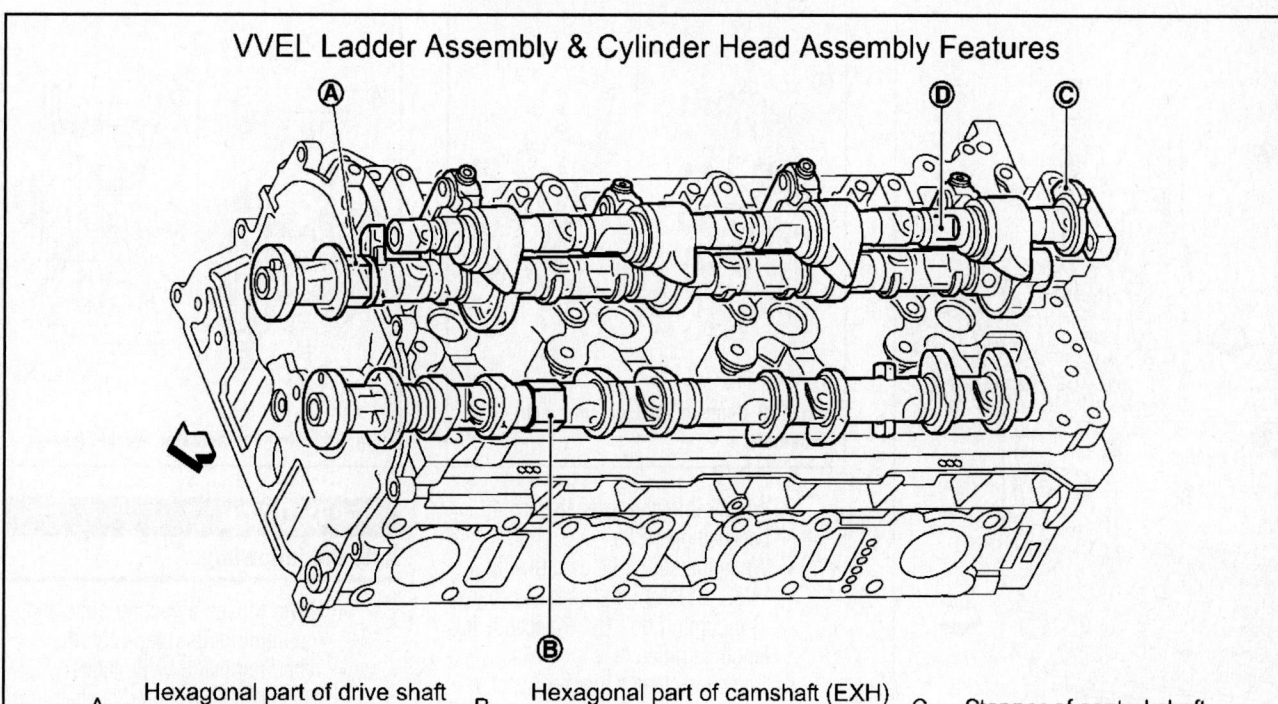

A. Hexagonal part of drive shaft (for holding)

B. Hexagonal part of camshaft (EXH) (for holding)

C. Stopper of control shaft

D. Two flat areas of control shaft (for holding)

⇦ : Engine front

37663_FX35_G0272

Fig. 106 VVEL Ladder Assembly & Cylinder Head Assembly Features

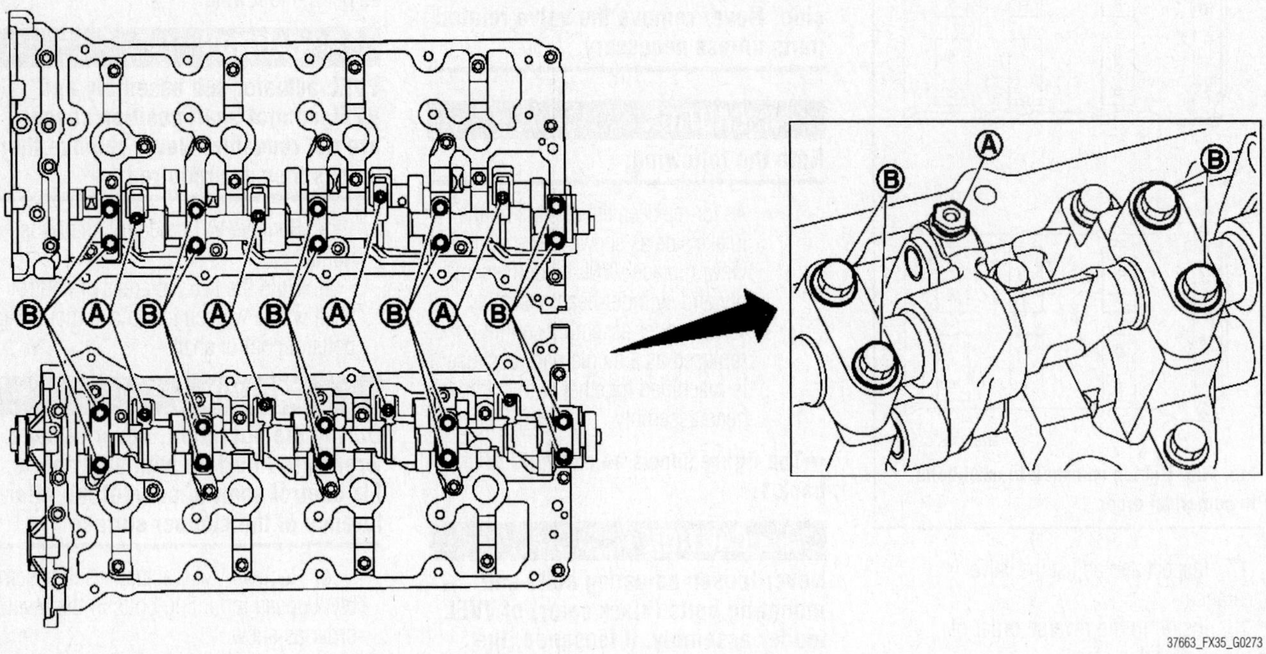

Fig. 107 Never loosen adjusting bolts (A) and mounting bolts (black color) (B) of VVEL ladder assembly

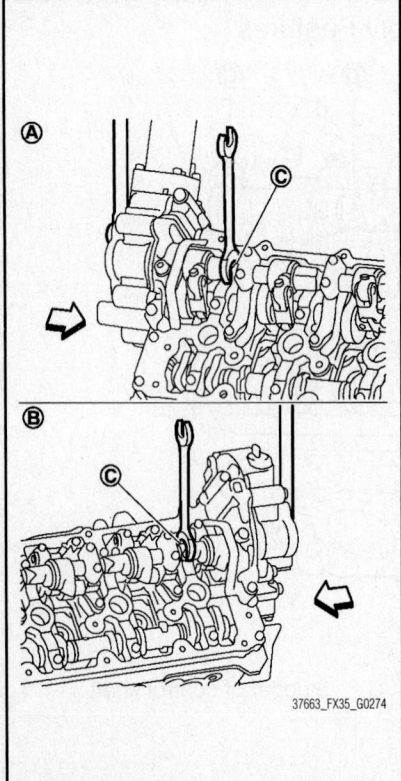

Fig. 108 Hold the two flat areas of control shaft with a wrench to remove mounting bolts of control shaft

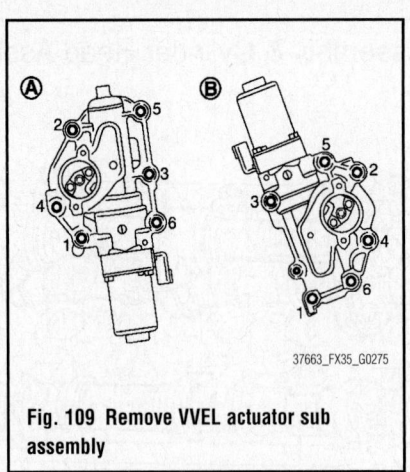

Fig. 109 Remove VVEL actuator sub assembly

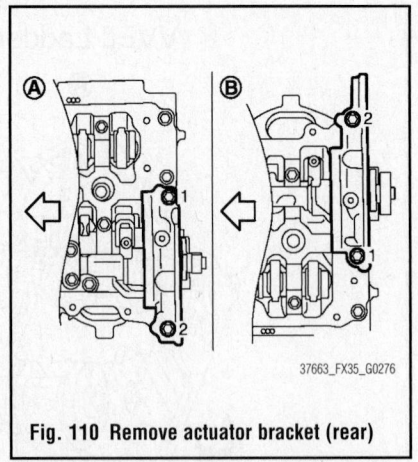

Fig. 110 Remove actuator bracket (rear)

- When removing, prepare wastes because oil spills.
- When installing, be careful with VVEL actuator sub assembly (bank 1) mounting bolt No. 4 because its length is different.

  d. Remove actuator bracket (rear). Loosen mounting bolts in the reverse order as shown.

  3. Remove front cover, camshaft sprockets, and timing chains.

  4. Remove VVEL ladder assembly. Loosen mounting bolts (gold color) in the reverse order as shown.

✳✳ CAUTION

Note the following:

- Never loosen adjusting bolts and mounting bolts (black color).
- When removing VVEL ladder assembly, hold the drive shaft from below so as not to drop it.

  5. Remove camshaft (EXH).

  6. Remove valve lifter, if necessary.

➠Identify installation positions, and store them without mixing them up.

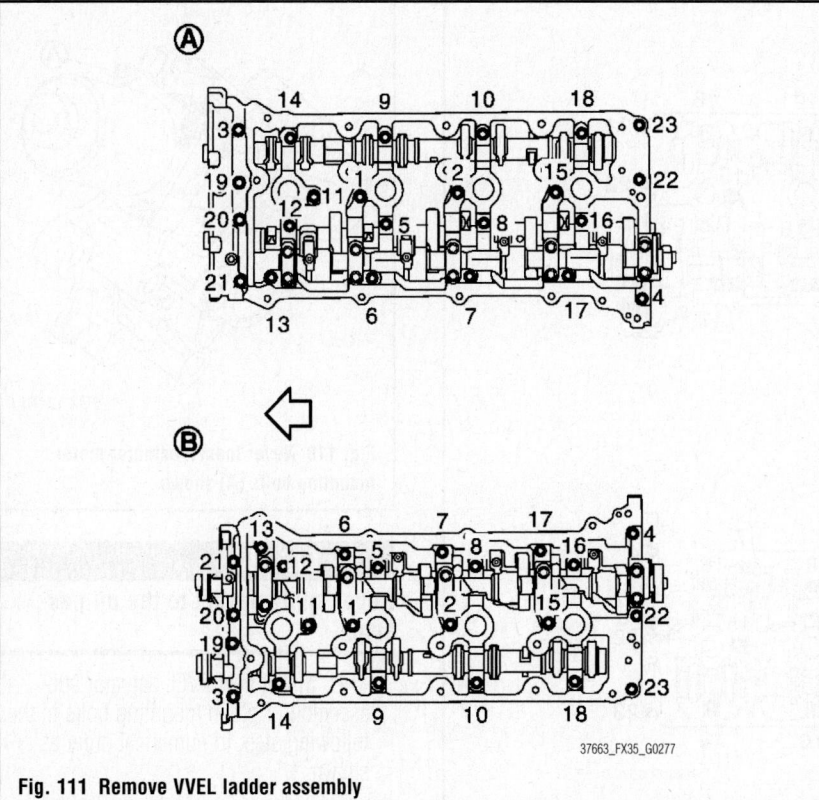

Fig. 111 Remove VVEL ladder assembly

To install:

7. Install valve lifters. Install them in their original positions.

8. Install camshaft (EXH).

➡️**Distinction between camshaft (EXH) is performed with the identification mark.**

9. Install VVEL ladder assembly as per the following:

a. Apply a continuous bead of liquid gasket with tube presser to the cylinder head as shown.

b. Tighten mounting bolts in the following steps, in numerical order as shown.

- Tighten bolts to 1 ft. lbs. (2 Nm) in numerical order as shown.
- Tighten bolts to 4 ft. lbs. (6 Nm) in numerical order as shown.
- Tighten bolts 8 ft. lbs. (10 Nm) in numerical order as shown.

10. Install camshaft sprockets and timing chains.

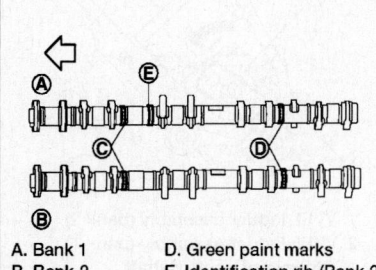

A. Bank 1
B. Bank 2
C. No paint marks
D. Green paint marks
E. Identification rib (Bank 2)

Fig. 112 Exhaust camshaft identification marks

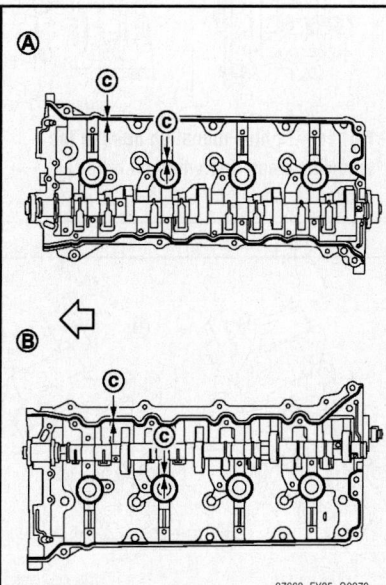

Fig. 113 Apply a continuous bead of liquid gasket (C) to the cylinder heads Bank 1 (A) and Bank 2 (B) as shown

11. Install actuator bracket (rear) as per the following:

a. Refer to the figure to replace new dowel pins, if removed.

b. Apply a continuous bead of liquid gasket with tube presser to the actuator bracket (rear) as shown.

✳✳ CAUTION

Never apply gasket to the oil passage.

c. Tighten mounting bolts in the following steps, in numerical order as shown.

- Tighten bolts to 1 ft. lbs. (2 Nm) in numerical order as shown.
- Tighten bolts to 4 ft. lbs. (6 Nm) in numerical order as shown.
- Tighten bolts to 23 ft. lbs. (31 Nm) in numerical order as shown.

12. Install new VVEL actuator sub assembly as per the following:

✳✳ CAUTION

Regarding replacement, because VVEL actuator sub assembly and VVEL control shaft position sensor are controlled on a one-on-one basis, replace them as a set.

➡️**VVEL actuator arm is factory-fixed at 10 degrees from the small lift with a holding jig. The holding jig is supplied in the new VVEL actuator sub assembly.**

✳✳ CAUTION

Note the following:

- Never disassemble VVEL actuator sub assembly. Never loosen actuator motor mounting bolts shown.
- Never impact VVEL actuator sub assembly.

a. Move control shaft to the position of small lift stopper.

b. The position where a part of the stopper of control shaft contacts VVEL ladder bracket.

✳✳ CAUTION

Be careful not to damage the stopper surface.

c. If control shaft cannot be moved, set crankshaft in position referring to the information below. (To displace cam nose)

- Bank 1: Turn 360 degrees from No. 1 cylinder at TDC
- Bank 2: No. 1 cylinder at TDC

d. Hold two flat areas of control shaft

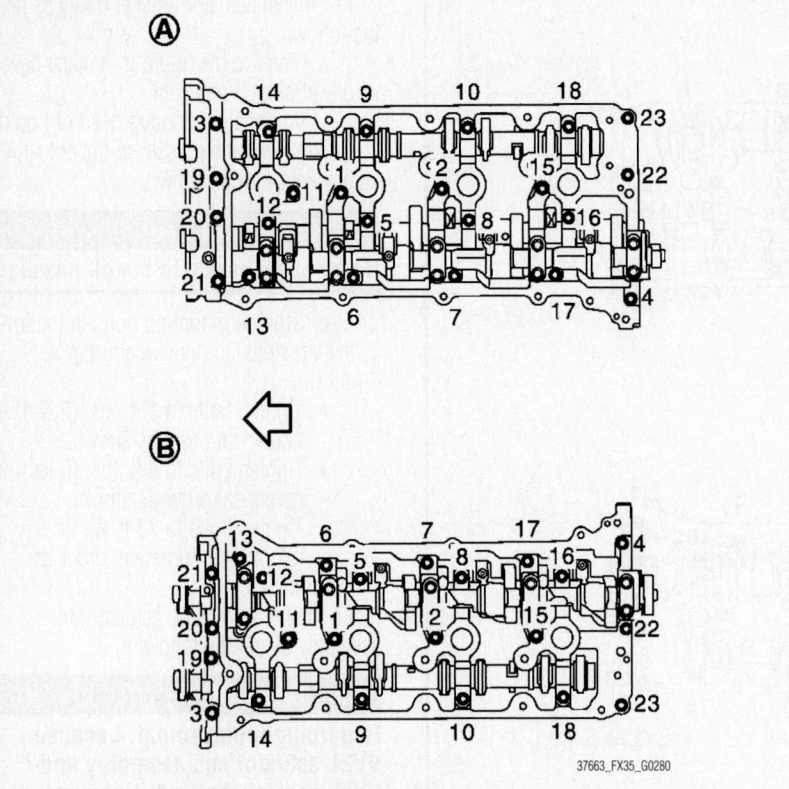

Fig. 114 Tighten mounting bolts in the following steps, in numerical order as shown

with a wrench, and rotate the control shaft (10 degrees from the stopper) to the large lift side. (This is for aligning the bolt hole of control shaft and the hole of VVEL actuator arm.)

➡**The figure shows an example of bank 2.**

e. Apply a continuous bead of liquid gasket with tube presser to the VVEL actuator sub assembly as shown.

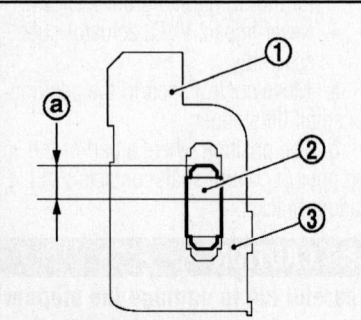

1. Actuator bracket
2. Dowel pins
3. VVEL ladder assembly
a. Bead of liquid gasket
   0.157-0.236 inches (4.0-6.0 mm)

37663_FX35_G0281

Fig. 115 Refer to the figure to replace new dowel pins, if removed

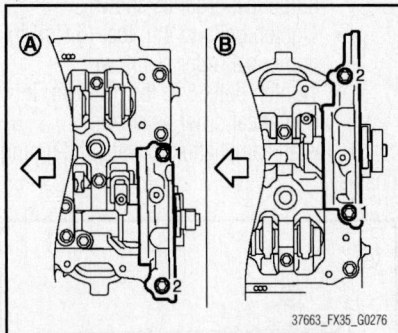

37663_FX35_G0276

Fig. 116 Tighten mounting bolts in the following steps, in numerical order as shown

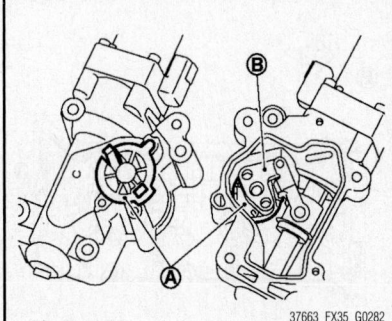

37663_FX35_G0282

Fig. 117 VVEL actuator arm (B) is factory-fixed at 10 degrees from the small lift with a holding jig (A)

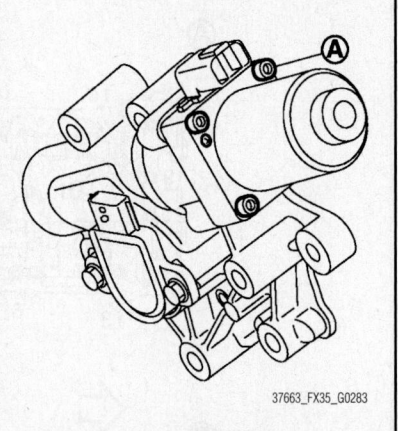

37663_FX35_G0283

Fig. 118 Never loosen actuator motor mounting bolts (A) shown

✳✳ CAUTION

Never apply gasket to the oil passage.

f. Install new VVEL actuator sub assembly. Tighten mounting bolts in the following step, in numerical order as shown.

✳✳ CAUTION

Note the following:

- When installing, be careful with VVEL actuator sub assembly (bank 1)

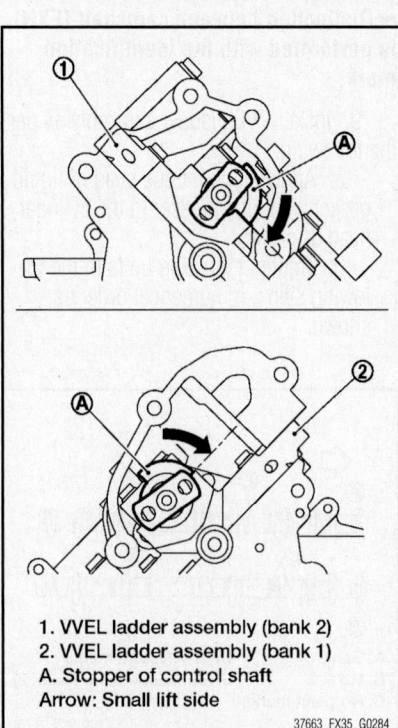

1. VVEL ladder assembly (bank 2)
2. VVEL ladder assembly (bank 1)
A. Stopper of control shaft
Arrow: Small lift side

37663_FX35_G0284

Fig. 119 Move control shaft to the position of small lift stopper

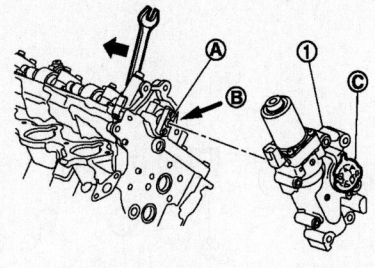

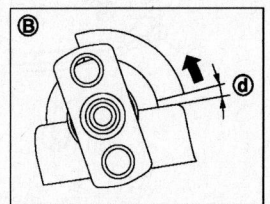

1. VVEL actuator sub assembly (bank 2)
A. Control shaft
B. View B
C. Holding jig
d. 10 degrees
Arrow: Large lift side

37663_FX35_G0285

Fig. 120 Hold two flat areas of control shaft with a wrench, and rotate the control shaft (10 degrees from the stopper) to the large lift side

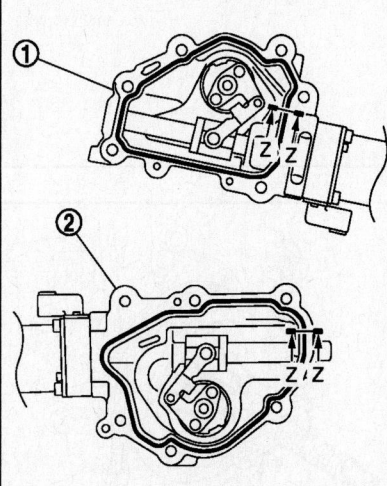

Fig. 121 Apply a continuous bead of liquid gasket to the VVEL actuator sub assembly as shown

mounting bolt No. 4 because its length is different.

- Be sure to check that the VVEL actuator sub assembly is in contact with the cylinder head before tightening the mounting bolts.

g. Remove holding jig.

h. Check that VVEL actuator arm bolt hole is aligned with control shaft tapped

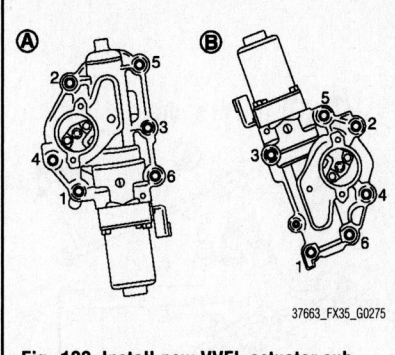

Fig. 122 Install new VVEL actuator sub assembly

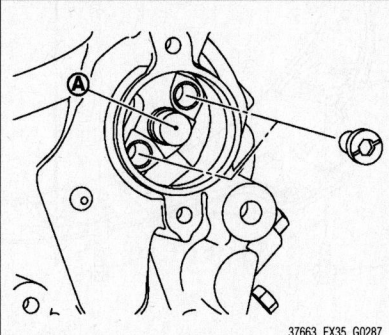

Fig. 123 Never give an impact to the magnet part (A)

hole. If it is not aligned, turn control shaft for alignment.

✳✳ CAUTION

Never give an impact to the magnet part.

i. Hold two flat areas of control shaft with a wrench to tighten mounting bolts of control shaft.

✳✳ CAUTION

During the operation, never allow a wrench to interfere with other parts. Hold control shaft to prevent the interference of the stopper surface.

13. Install new VVEL control shaft position sensor as per the following:

✳✳ CAUTION

Regarding replacement, because VVEL actuator sub assembly and VVEL control shaft position sensor are controlled on a one-on-one basis, replace them as a set.

a. Apply engine oil to O-ring or contact surface of O-ring.

b. Align matching marks B of VVEL

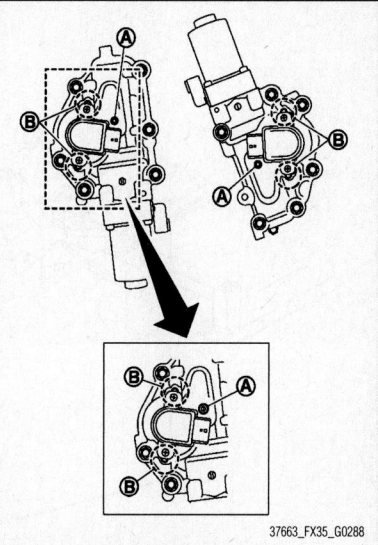

Fig. 124 Align matching marks B of VVEL control shaft position sensor and upper housing

control shaft position sensor and upper housing.

➡**Face connector toward matching mark A.**

c. Temporarily tighten bolt.

d. Adjust VVEL control shaft position sensor after setting the engine assembly in the vehicle.

✳✳ CAUTION

Be sure to adjust VVEL control shaft position sensor.

e. After adjusting VVEL control shaft position sensor, tighten bolts to the specified torque.

14. Install actuator cover.

15. Inspect the valve clearance.

16. Install in the reverse order of removal.

CRANKSHAFT PULLEY DAMPER

REMOVAL & INSTALLATION

3.5L Engine

See Figures 125 through 128.

1. Remove crankshaft pulley as per the following:

a. Remove the accessory drive belt.

b. Remove front cross bar.

c. Remove power steering pipe mounting bolt.

d. Remove rear cover plate and set the ring gear stopper KV10118600 (J-48641) as shown.

e. Loosen crankshaft pulley bolt and rotate bolt seating surface at 0.39 inches (10 mm) from its original position.

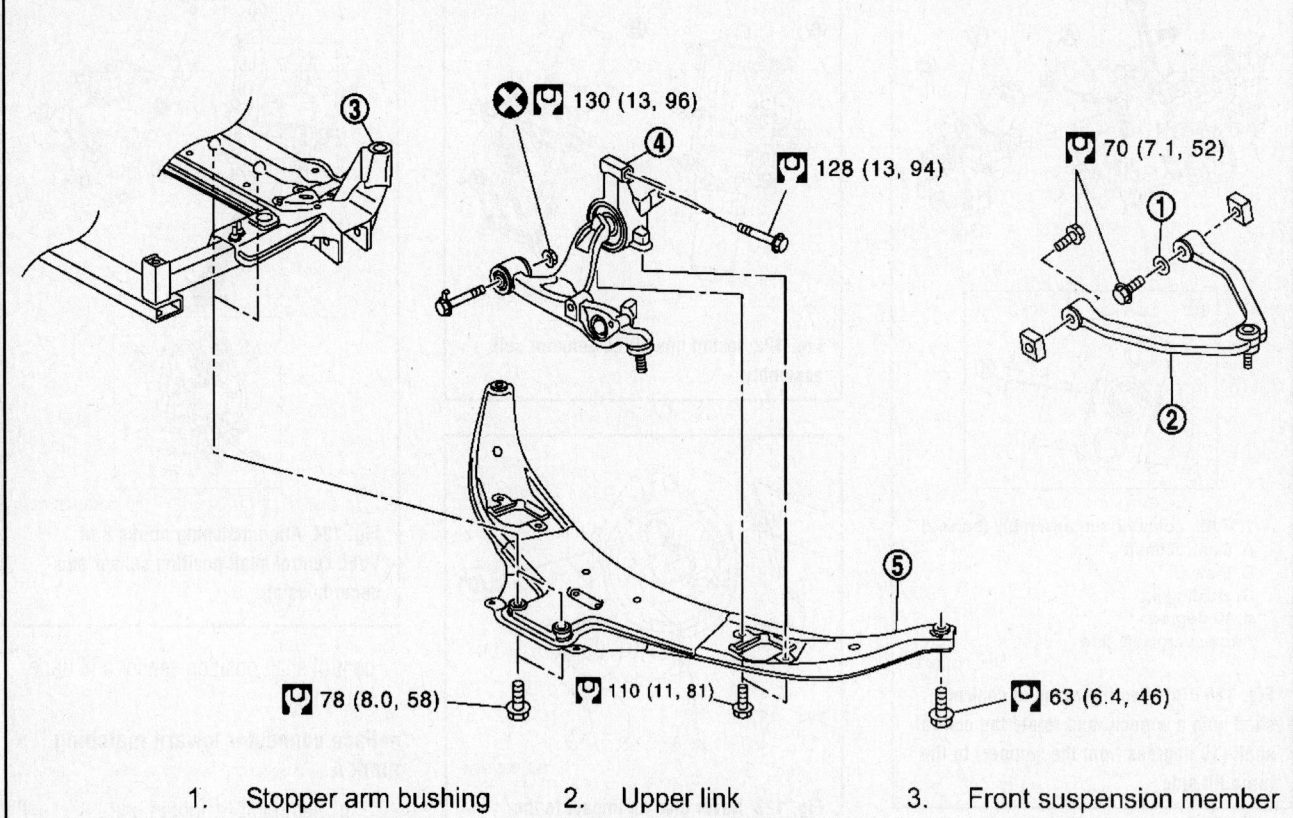

1. Stopper arm bushing
2. Upper link
3. Front suspension member
4. Transverse link
5. Front cross bar

37663_FX35_G0148

Fig. 125 Exploded view showing front cross bar

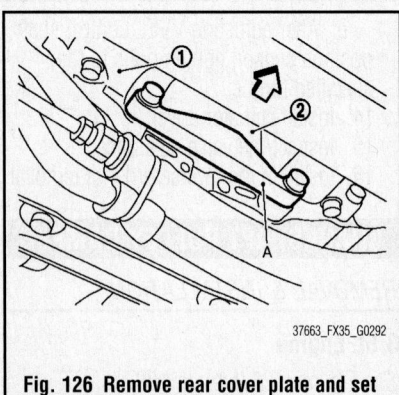

37663_FX35_G0292

Fig. 126 Remove rear cover plate and set the ring gear stopper (A) as shown

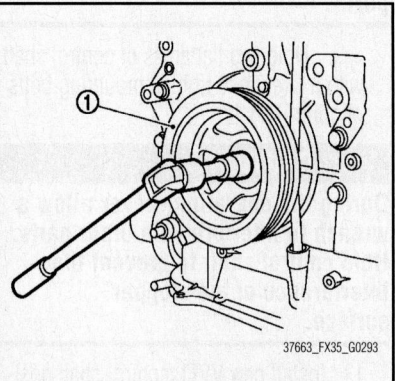

37663_FX35_G0293

Fig. 127 Loosen crankshaft pulley bolt and rotate bolt seating surface at 0.39 inches (10 mm) from its original position

37663_FX35_G0294

Fig. 128 Place suitable puller tab on holes of crankshaft pulley, and pull crankshaft pulley through

✳✳ CAUTION

Never remove crankshaft pulley bolt as it will be used as a supporting point for suitable puller.

f. Place suitable puller tab on holes of crankshaft pulley, and pull crankshaft pulley through.

✳✳ CAUTION

Never put suitable puller tab on crankshaft pulley periphery, as this will damage internal damper.

2. Installation is the reverse of removal.

5.0L Engine

See Figures 125 and 129.

1. Remove the following parts:
- Engine undercover.
- Drive belts
- Cooling fan assembly
- Front cross bar

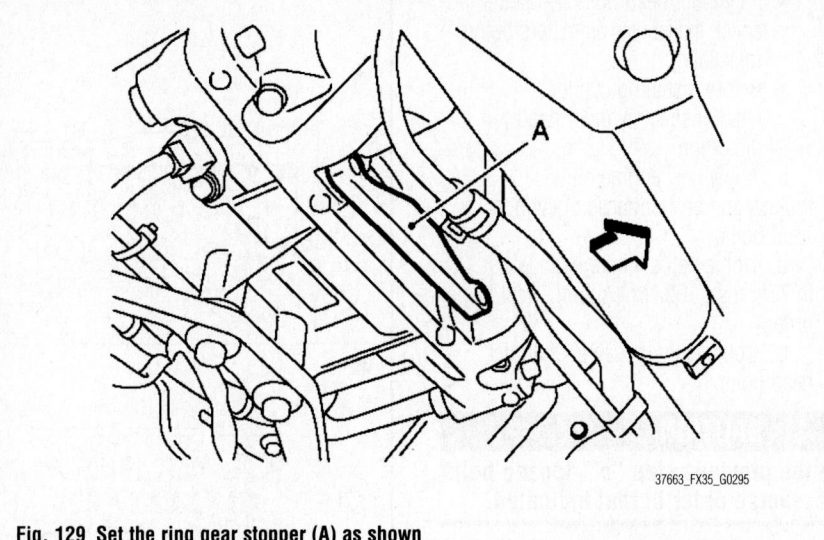

Fig. 129 Set the ring gear stopper (A) as shown

2. Remove crankshaft pulley as per the following:

    a. Remove rear plate cover.

    b. Set the ring gear stopper SST: KV10119200 (J-49277) as shown.

    c. Loosen crankshaft pulley bolt, and then pull crankshaft pulley with both hands to remove it.

✳✳ CAUTION

Never remove crankshaft pulley bolt. Keep loosened crankshaft pulley bolt in place to protect removed crankshaft pulley from dropping.

3. Installation is the reverse of removal.

CRANKSHAFT FRONT SEAL

REMOVAL & INSTALLATION

3.5L Engine

See Figure 130.

1. Remove the following parts:
   • Engine undercover
   • Drive belt
   • Crankshaft pulley

2. Remove front oil seal using a suitable tool.

✳✳ CAUTION

Be careful not to damage front timing chain case and crankshaft.

To install:

3. Apply new engine oil to both oil seal lip and dust seal lip of new front oil seal.

4. Install front oil seal.

    a. Install front oil seal so that each seal lip is oriented as shown.

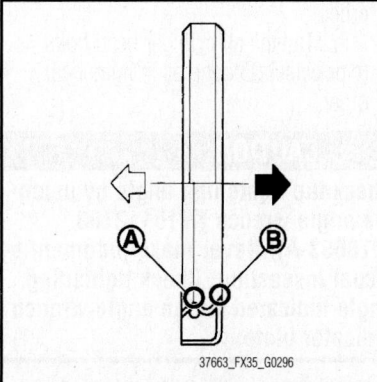

Fig. 130 Install front oil seal so that each seal lip is oriented as shown

    b. Using a suitable drift, press-fit until the height of front oil seal is level with the mounting surface. Suitable drift: outer diameter 2.36 inches (60 mm), inner diameter 2 inches (50 mm).

    c. Check the garter spring is in position and seal lips are not inverted.

✳✳ CAUTION

Note the following:

• Be careful not to damage front timing chain case and crankshaft.

• Press-fit straight and avoid causing burrs or tilting oil seal.

5. Install in the reverse order of removal after this step.

5.0L Engine

See Figures 129 and 131.

1. Remove the following parts:
   • Engine undercover
   • Drive belts

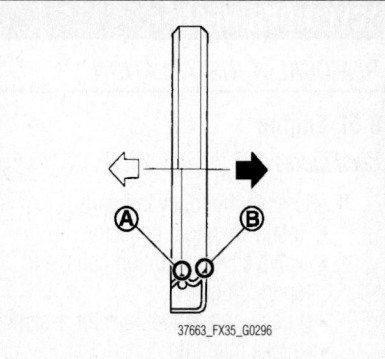

Fig. 131 Apply new engine oil to both oil seal lip (A) and dust seal lip (B)

    • Cooling fan assembly
    • Front cross bar

2. Remove crankshaft pulley as per the following:

    a. Remove rear plate cover.

    b. Set the ring gear stopper KV10119200 (J-49277) as shown.

    c. Loosen crankshaft pulley bolt, and then pull crankshaft pulley with both hands to remove it.

✳✳ CAUTION

Never remove crankshaft pulley bolt. Keep loosened crankshaft pulley bolt in place to protect removed crankshaft pulley from dropping.

3. Remove front oil seal using a suitable tool.

✳✳ CAUTION

Be careful not to damage front cover and crankshaft.

To install:

4. Install front oil seal on front cover.

    a. Apply new engine oil to both oil seal lip and dust seal lip.

    b. Install it so that each seal lip is oriented as shown.

✳✳ CAUTION

Be careful not to scratch or make burrs on circumference of oil seal.

    c. Using a suitable drift with outer diameter: 2.20 inches (56 mm), press-fit oil seal until it becomes flush with front cover end face.

    d. Check the garter spring is in position and seal lips are not inverted.

5. Install in the reverse order of removal.

CYLINDER HEAD

REMOVAL & INSTALLATION

3.5L Engine

See Figures 132 through 136.

1. Remove the following parts:
   - Intake manifold collector
   - valve covers, ignition coils, and spark plugs
   - Fuel tube and fuel injector assembly
   - Intake manifold
   - Exhaust manifold
   - Water inlet and thermostat assembly
   - Water outlets (front and rear), water pipe and heater pipe
   - Timing chain
   - Rear timing chain case
   - Camshafts
2. Remove cylinder head.
   a. Loosen cylinder head bolts in reverse order as shown with cylinder head bolt wrench.
   b. Remove cylinder head gaskets.

To install:

3. Install new cylinder head gaskets.
4. Turn crankshaft until No. 1 piston is set at TDC. Crankshaft key should line up with the bank 1 cylinder center line as shown.
5. Install cylinder head following the steps below to tighten cylinder head bolts in numerical order as shown with cylinder head bolts wrench.

✳✳ CAUTION

Note the following:

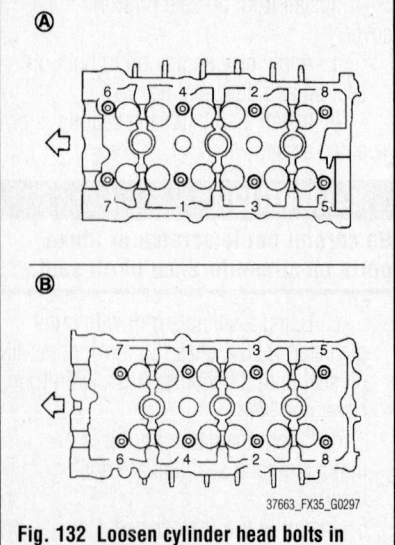

Fig. 132 Loosen cylinder head bolts in reverse order as shown

- If cylinder head bolts re-used, check their outer diameters before installation.
- Before installing cylinder head, inspect cylinder head distortion.
   a. Apply new engine oil to threads and seat surfaces of cylinder head bolts.
   b. Tighten all cylinder head bolts to 77ft. lbs. (105 Nm) in numerical order.
   c. Completely loosen all cylinder head bolts.

✳✳ CAUTION

In the previous step "c", loosen bolts in reverse order of that indicated.

   d. Tighten all cylinder head bolts to 30 ft. lbs. (40 Nm) in numerical order.
   e. Tighten all cylinder head bolts (clockwise) 95 degrees in numerical order.

✳✳ CAUTION

Check the tightening angle by using the angle wrench KV10112100 (BT8653-A). Never make judgment by visual inspection. Check tightening angle indicated on the angle wrench indicator plate.

   f. Tighten all cylinder head bolts again (clockwise) 95 degrees in numerical order.
6. After installing cylinder head, measure distance between front end faces of cylinder block and cylinder head (bank 1 and bank 2). If measured value is out of the standard of 0.555–0.587 inches (14.1–14.9 mm), reinstall cylinder head.

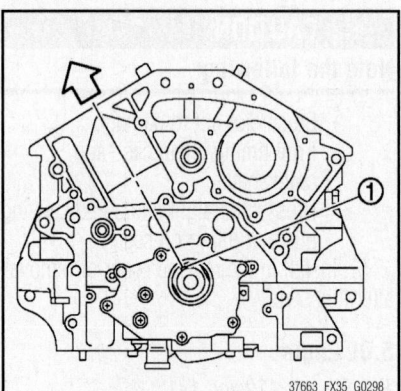

Fig. 133 Turn crankshaft until No. 1 piston is set at TDC

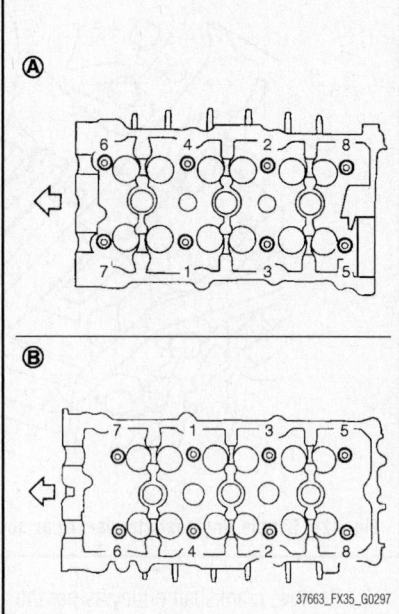

Fig. 134 Install cylinder head following the steps below to tighten cylinder head bolts in numerical order as shown

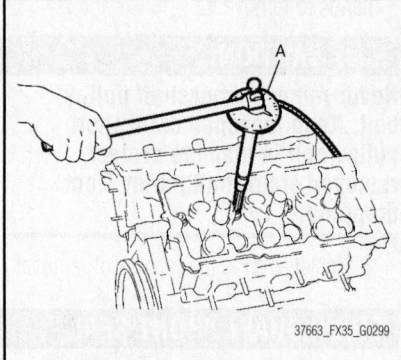

Fig. 135 Check the tightening angle by using the angle wrench KV10112100 (BT8653-A) (A)

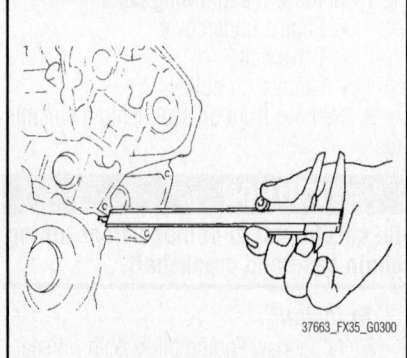

Fig. 136 Measure distance between front end faces of cylinder block and cylinder head (bank 1 and bank 2)

7. Install valve lifters.

✽✽ CAUTION

Install them in their original positions.

8. Install spark plug with spark plug wrench.

9. Install in the reverse order of removal after this step.

10. Before starting engine, check oil/ fluid levels including engine coolant and engine oil. If less than required quantity, fill to the specified level.

11. Use procedure below to check for fuel leakage:

   a. Turn ignition switch "ON" (with engine stopped). With fuel pressure applied to fuel piping, check for fuel leakage at connection points.

   b. Start engine. With engine speed increased, check again for fuel leakage at connection points.

12. Run engine to check for unusual noise and vibration.

13. Warm up engine thoroughly to check there is no leakage of fuel, exhaust gases, or any oil/fluids including engine oil and engine coolant.

14. Bleed air from lines and hoses of applicable lines, such as in cooling system.

15. After cooling down engine, again check oil/fluid levels including engine oil and engine coolant. Refill to the specified level, if necessary.

5.0L Engine

See Figures 137 through 139.

1. Remove the following parts:
- valve covers, ignition coils, and spark plugs
- Intake manifold
- Exhaust manifolds
- Water inlet and thermostat housing
- Water pipe and heater pipe
- Timing chain
- Camshaft (EXH) and VVEL ladder assembly

2. Remove cylinder heads.

   a. Loosen mounting bolts in reverse order as shown.

   b. Use TORX® socket and power tool.

3. Remove cylinder head gaskets.

To install:

4. Install new cylinder head gaskets.

5. Install cylinder head.

✽✽ CAUTION

Note the following:

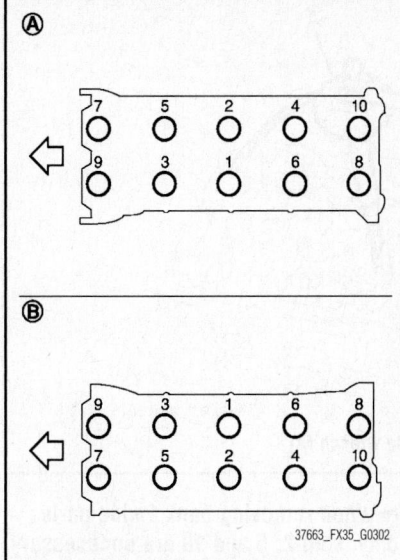

Fig. 137 Loosen mounting bolts in reverse order as shown

- If cylinder head bolts are re-used, check their outer diameters before installation.
- Before installing cylinder head, inspect cylinder head distortion.

6. Tighten cylinder head bolts in numerical order as shown using TORX;rM socket.

   a. Apply new engine oil to threads and seat surfaces of cylinder head bolts.

   b. Tighten all cylinder head bolts to 30 ft. lbs. (40 Nm) in numerical order.

   c. Tighten all cylinder head bolts (clockwise) 75 degrees in numerical order.

   d. Completely loosen all cylinder head bolts.

✽✽ CAUTION

In the previous step "d", loosen bolts in the reverse order of that indicated.

   e. Tighten all cylinder head bolts to 30 ft. lbs. (40 Nm) in numerical order.

   f. Tighten all cylinder head bolts (clockwise) 65 degrees in numerical order.

✽✽ CAUTION

Check the tightening angle using the angle wrench KV10112100 (BT8653-A). Never make judgment by visual inspection. Check tightening angle indicated on the angle wrench indicator plate.

   g. Tighten all cylinder head bolts again (clockwise) 65 degrees in numerical order.

7. Install the valve lifters.

✽✽ CAUTION

Install them in their original positions.

8. Install in the reverse order of removal.

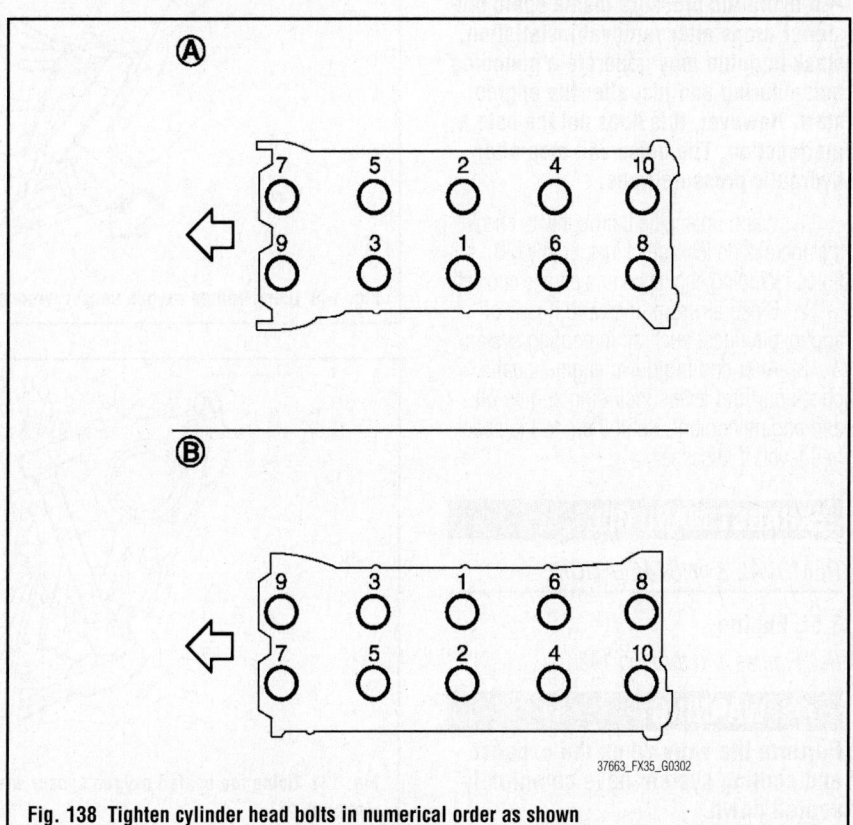

Fig. 138 Tighten cylinder head bolts in numerical order as shown

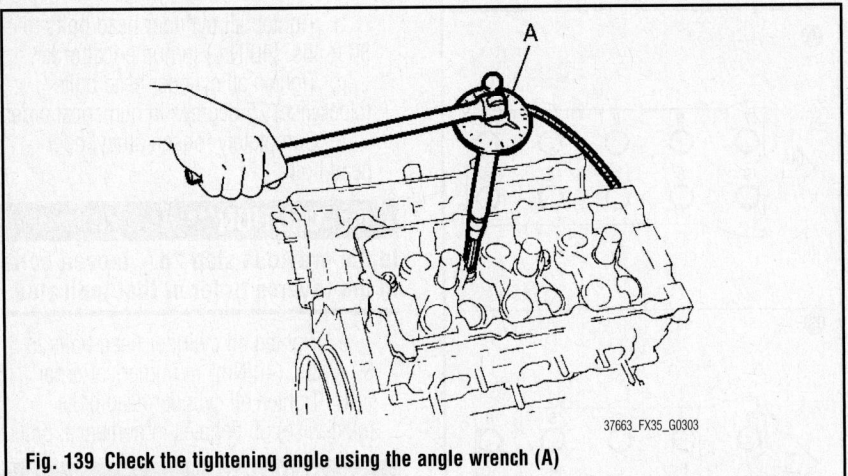

Fig. 139 Check the tightening angle using the angle wrench (A)

9. Before starting engine, check oil/fluid levels including engine coolant and engine oil. If any are less than the required quantity, fill them to the specified level.

10. Follow the procedure below to check for fuel leakage.

a. Turn ignition switch to the "ON" position (with engine stopped). With fuel pressure applied to fuel piping, check for fuel leakage at connection points.

b. Start engine. With engine speed increased, check again for fuel leakage at connection points.

c. Run engine to check for unusual noise and vibration.

➡If hydraulic pressure inside chain tensioner drops after removal/installation, slack in guide may generate a pounding noise during and just after the engine start. However, this does not indicate a malfunction. The noise will stop after hydraulic pressure rises.

11. Warm up engine thoroughly to check that there is no leakage of fuel, or any oil/fluids including engine oil and engine coolant.

12. Bleed air from lines and hoses of applicable lines, such as in cooling system.

13. After cooling down engine, again check oil/fluid levels including engine oil and engine coolant. Refill them to the specified level, if necessary.

EXHAUST MANIFOLD

REMOVAL & INSTALLATION

3.5L Engine

See Figures 140 through 142.

❋❋ WARNING

Perform the work when the exhaust and cooling system have completely cooled down.

➡**When removing bank 1 side parts only, step 2, 5 and 10 are unnecessary.**

1. Remove engine undercover.
2. Drain engine coolant.
3. Remove engine cover.
4. Remove air cleaner case and air duct.
5. Remove heater pipe and water hose.
6. Remove exhaust front tube.
7. Disconnect heated oxygen sensor 2 harness connectors (bank 1 and bank 2) and remove harness clip.

8. Using heated oxygen sensor wrench KV10114400 (J-38365), removal heated oxygen sensor 2.

9. Remove three way catalysts (bank 1 and bank 2).

10. Disconnect steering lower joint at power steering gear assembly side, and release steering lower shaft.

11. Disconnect air fuel ratio sensor 1 harness connectors (bank 1 and bank 2) and remove harness clip.

12. Using the heated oxygen sensor wrench, remove air fuel ratio sensor 1 (bank 1 and bank 2).

❋❋ CAUTION

Be careful not to damage air fuel ratio sensor 1. Discard any air fuel ratio sensor 1 which has been dropped onto a hard surface such as a concrete floor. Replace with a new sensor.

13. Remove exhaust manifold cover (upper) (bank 1 and bank 2).

14. Loosen mounting nuts in the reverse order as shown to remove exhaust manifold.

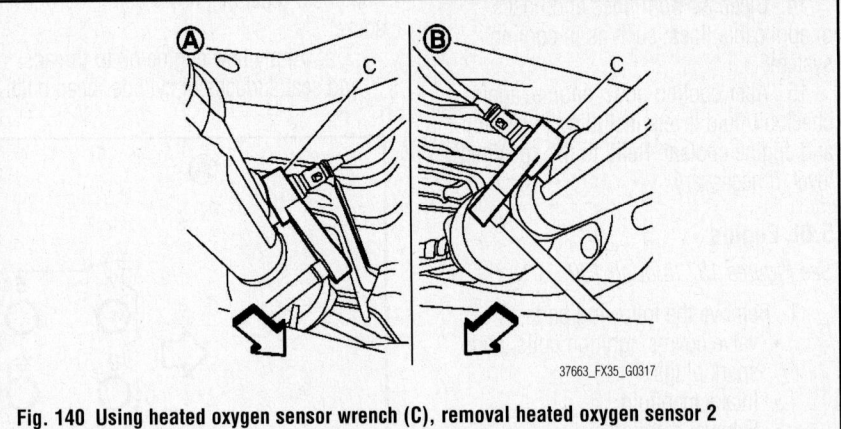

Fig. 140 Using heated oxygen sensor wrench (C), removal heated oxygen sensor 2

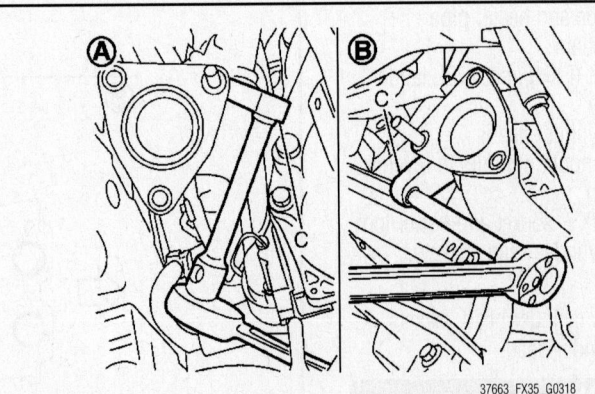

Fig. 141 Using the heated oxygen sensor wrench (C), remove air fuel ratio sensor 1 (bank 1 and bank 2)

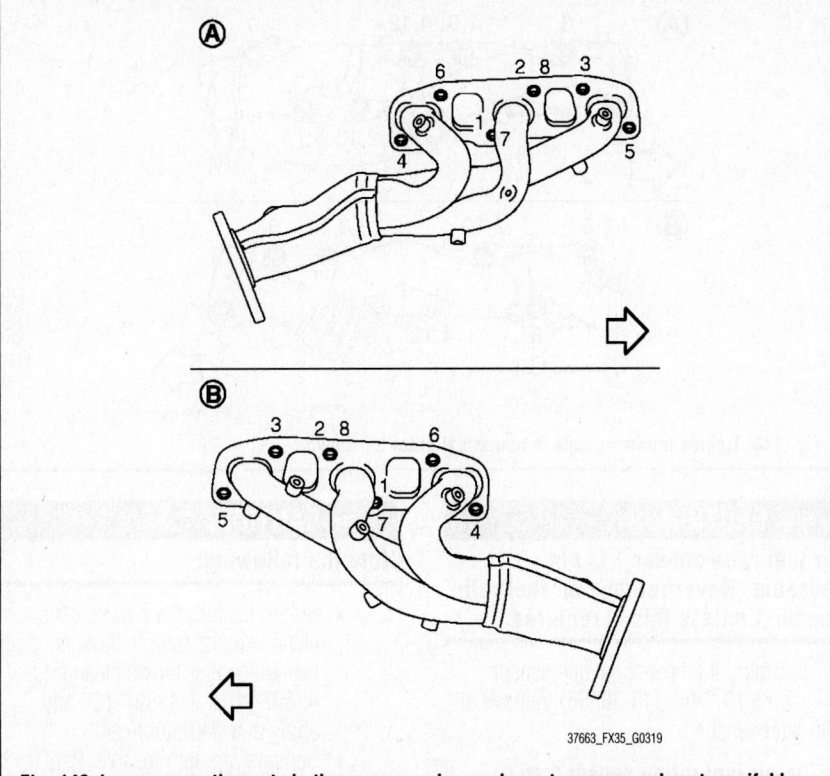

Fig. 142 Loosen mounting nuts in the reverse order as shown to remove exhaust manifold

➡Disregard the numerical order No. 7 and 8 in removal.

15. Remove gaskets.

❄❄ CAUTION

Cover engine openings to avoid entry of foreign materials.

To install:

16. Note the following item, and install in the reverse order of removal.

Exhaust Manifold Gasket

See Figure 143.

1. Install exhaust manifold gasket in direction shown. (Follow same procedure for both banks.)

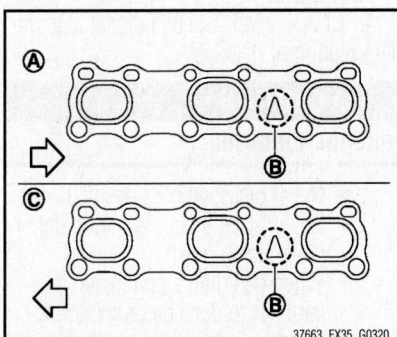

Fig. 143 Install exhaust manifold gasket in direction shown

Exhaust Manifold

See Figure 144.

1. If stud bolts were removed, install them and tighten to 11 ft. lbs. (15 Nm).

2. Install exhaust manifold and tighten mounting nuts in numerical order as shown.

➡**Tighten nuts No. 1 and 2 in two steps. The numerical order No. 7 and 8 shows second step.**

❄❄ CAUTION

Note the following:

- Before installing a new sensors, clean exhaust system threads using heated oxygen sensor thread cleaner tool and apply anti-seize lubricant.
- Never over torque sensors. Doing so may cause damage to sensors, resulting in the "MIL" coming on.

5.0L Engine

See Figures 145 through 148.

1. Remove heated oxygen sensor 2.

❄❄ CAUTION

Heated oxygen sensor 2 is not reusable. Never remove heated oxygen sensor 2 unless this is required.

2. Using the heated oxygen sensor

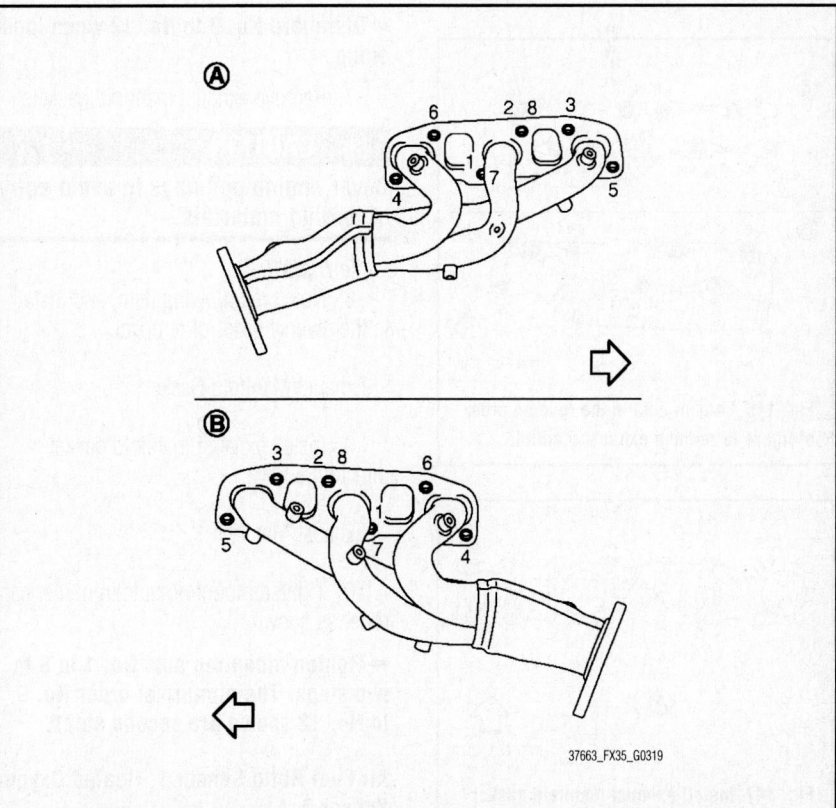

Fig. 144 Install exhaust manifold and tighten mounting nuts in numerical order as shown

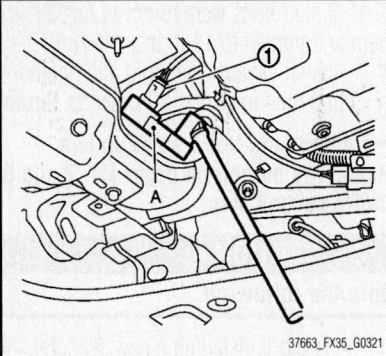

Fig. 145 Using the heated oxygen sensor wrench (A), remove heated oxygen sensor 2 (1)

wrench KV10114400 (J-38365), remove heated oxygen sensor 2.

❋❋ CAUTION
Note the following:

- The heated oxygen sensor 2 is removable under vehicle mounted condition.
- The figure shows an example of bank 1.

3. Remove three way catalyst (bank 1 and bank 2).

4. Remove air fuel ratio sensor 1as per the following:

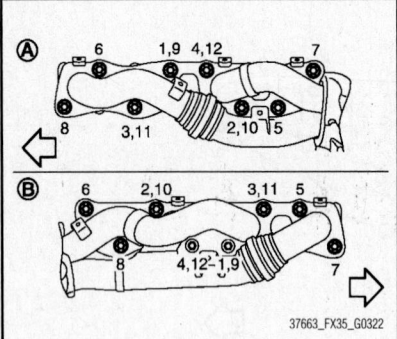

Fig. 146 Loosen nuts in the reverse order of figure to remove exhaust manifold

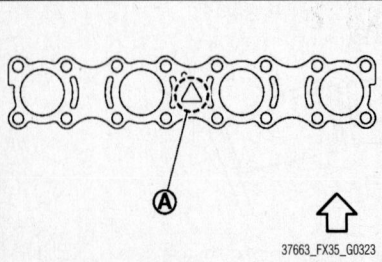

Fig. 147 Install exhaust manifold gasket in direction shown

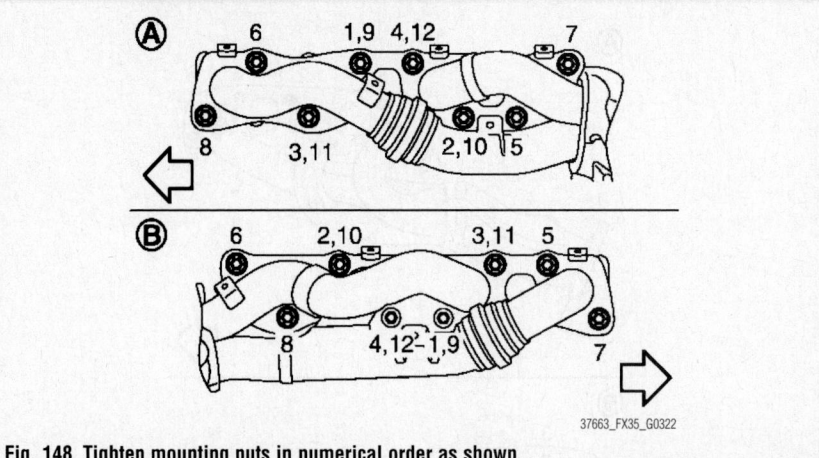

Fig. 148 Tighten mounting nuts in numerical order as shown

❋❋ CAUTION
Air fuel ratio sensor 1 is not reusable. Never remove air fuel ratio sensor 1 unless this is required.

5. Using the heated oxygen sensor wrench KV10114400 (J-38365), remove air fuel ratio sensor 1.

➡**The air fuel ration sensor 1 is removable under vehicle-mounted condition.**

6. Remove exhaust manifold. Loosen nuts in the reverse order of figure to remove exhaust manifold.

➡**Disregard No. 9 to No. 12 when loosening.**

7. Remove exhaust manifold gaskets.

❋❋ CAUTION
Cover engine openings to avoid entry of foreign materials.

To install:

8. Note the following item, and install in the reverse order of removal.

Exhaust Manifold Gasket

9. Install exhaust manifold gasket in direction shown.

Exhaust Manifold

10. Tighten mounting nuts in numerical order as shown.

➡**Tighten mounting nuts No. 1 to 8 in two steps. The numerical order No. 9 to No. 12 shown are second steps.**

Air Fuel Ratio Sensor 1, Heated Oxygen Sensor 2

❋❋ CAUTION
Note the following:

- Before installing a new sensors, clean exhaust system threads using oxygen sensor thread cleaner (J-43897-18 or J-43897-12), and apply anti-seize lubricant.
- Sensors are not reusable. Replace them with a new one after removal. When replacing them, handle with care not to impact on them.
- When installing the new sensors, set the heated oxygen sensor wrench KV10114400(J-38365) (A) in the hexagonal part to tighten the them.
- Never over torque sensors. Doing so may cause damage to the sensors, resulting in "MIL" coming on.

DRIVE PLATE

REMOVAL & INSTALLATION

See Figure 149.

1. Separate engine and transmission.
2. Remove drive plate attaching bolts.
3. Remove reinforcement plate and drive plate.

To install:

4. Check drive plate and signal plate for deformation or damage.

❋❋ CAUTION
Note the following:

- Never disassemble drive plate.
- Never place drive plate with signal plate facing down.
- When handling signal plate, take care not to damage or scratch it.
- Handle signal plate in a manner that prevents it from becoming magnetized.

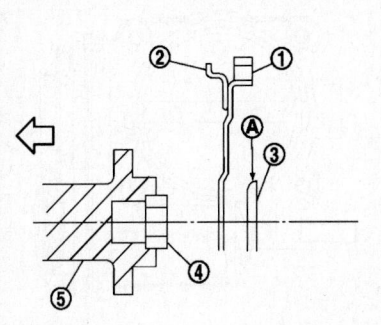

1. Ring gear
2. Drive plate
3. Reinforcement plate
4. Pilot converter
5. Crankshaft
A. Rounded edge

37663_FX35_G0324

Fig. 149 Install drive plate and reinforcement plate as shown

5. If damage is found, replace drive plate.
6. When installing drive plate to crankshaft, be sure to correctly align crankshaft side dowel pin and drive plate side dowel pin hole.

❈❈ CAUTION
If these are not aligned correctly, engine runs roughly and "MIL" turns on.

7. Install drive plate and reinforcement plate as shown.
8. Holding ring gear with the ring gear stopper KV10118600 (J-48641).
9. Tighten the mounting bolts crosswise over several times. Torque to 65 ft. lbs. (88 Nm).

INTAKE MANIFOLD

REMOVAL & INSTALLATION

3.5L Engine
See Figures 150 through 152.

1. Release fuel pressure.
2. Remove intake manifold collector.

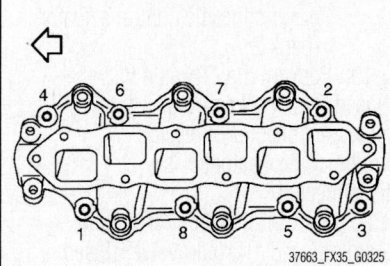

37663_FX35_G0325

Fig. 150 Loosen mounting bolts and nuts in reverse order as shown

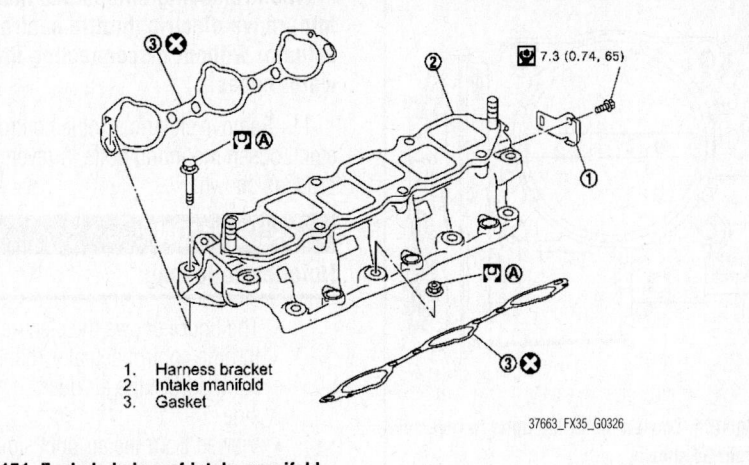

1. Harness bracket
2. Intake manifold
3. Gasket

7.3 (0.74, 65)

37663_FX35_G0326

Fig. 151 Exploded view of intake manifold

3. Remove fuel tube and fuel injector assembly.
4. Remove harness bracket.
5. Loosen mounting bolts and nuts in reverse order as shown to remove intake manifold.

❈❈ CAUTION
Note the following:

- Cover engine openings to avoid entry of foreign materials.
- Put a mark on the intake manifold and the cylinder head with paint before removal because they need installed in the specified direction.
- Remove gaskets.

To install:
7. Note the following item, and install in the reverse order of removal.
8. If stud bolts were removed, install them and tighten to 8 ft. lbs. (11 Nm).
9. Tighten all mounting bolts and nuts to 19 ft. lbs. (25.5 Nm) in two steps in numerical order as shown.

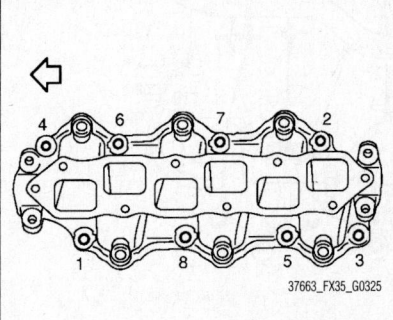

37663_FX35_G0325

Fig. 152 Tighten all mounting bolts and nuts in two steps in numerical order as shown

a. Step 1: 5 ft. lbs. (7.4 Nm)
b. Step 2: 19 ft. lbs. (25.5 Nm)

❈❈ CAUTION
Install intake manifold with the marks (put on the intake manifold and the cylinder head before removal) aligned.

10. Install intake manifold collector.

5.0L Engine
See Figures 153 through 156.

❈❈ WARNING
To avoid the danger of being scalded, never drain the engine coolant when the engine is hot.

1. Remove engine cover and engine room cover (RH and LH).
2. Release fuel pressure.
3. Remove air duct (inlet) and air duct.
4. Remove quick connector cap and disconnect fuel feed hose on engine side.

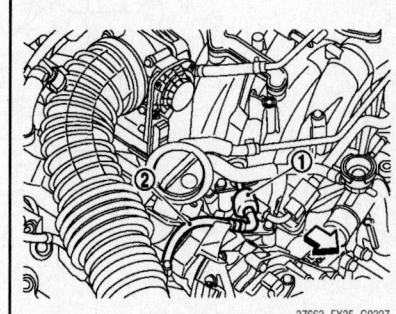

37663_FX35_G0327

Fig. 153 Remove quick connector cap (1) and disconnect fuel feed hose (2) on engine side

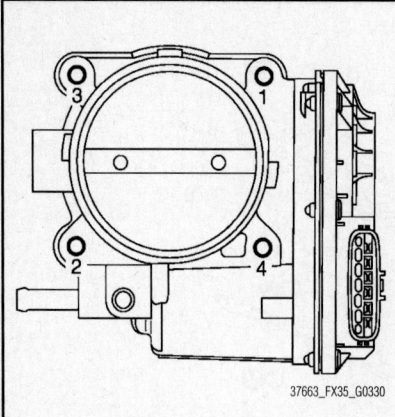

Fig. 154 Loosen mounting bolts in reverse order as shown

5. Remove engine cover bracket.

6. Remove fuel injector and fuel tube assembly.

7. Disconnect Manifold Absolute Pressure (MAP) sensor and air fuel ratio sensor 1 (bank 1) harness connector.

8. Remove vacuum tank, EVAP service port hose and EVAP canister purge control solenoid valve.

9. Disconnect PCV hoses and vacuum hose from intake manifold. Add matching marks as necessary for easier installation.

10. Drain engine coolant from radiator.

❈❈ CAUTION

Perform this step when the engine is cold. Never spill engine coolant on drive belts.

➥ **When removing only intake manifold, move electric throttle control actuator without disconnecting the water hoses.**

11. Remove electric throttle control actuator. Loosen mounting bolts in reverse order as shown.

❈❈ CAUTION
Note the following:

- The figure shows the electric throttle control actuator (bank 1) viewed from the air duct side.
- Viewed from the air duct side, the order of loosening mounting bolts of electric throttle control actuator (bank 1) is the same as that of the electric throttle control actuator (bank 2).

❈❈ CAUTION
Note the following:

- Handle carefully to avoid any impact to electric throttle control actuator.
- Never disassemble.

12. Remove intake manifold. Loosen mounting bolts in reverse order as shown.

13. Remove intake manifold gaskets.

❈❈ CAUTION

Cover engine openings to avoid entry of foreign materials.

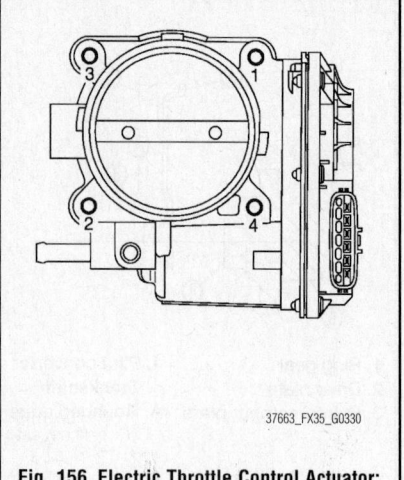

Fig. 156 Electric Throttle Control Actuator: tighten in numerical order as shown

14. Remove Manifold Absolute Pressure (MAP) sensor, if necessary.

❈❈ CAUTION

Handle carefully to avoid any impact to Manifold Absolute Pressure (MAP) sensor.

15. Remove acoustic absorbent.

To install:

16. Note the following item, and install in the reverse order of removal.

17. Intake Manifold: tighten in numerical order as shown.

18. Electric Throttle Control Actuator: tighten in numerical order as shown.

❈❈ CAUTION
Note the following:

- The figure shows the electric throttle control actuator (bank 1) viewed from the air duct side.
- Viewed from the air duct side, the order of tightening mounting bolts of electric throttle control actuator (bank 1) is the same as that of the electric throttle control actuator (bank 2).

19. Perform the "Throttle Valve Closed Position Learning" when harness connector of electric throttle control actuator is disconnected.

20. Perform the "Idle Air Volume Learning" and "Throttle Valve Closed Position Learning" when electric throttle control actuator is replaced.

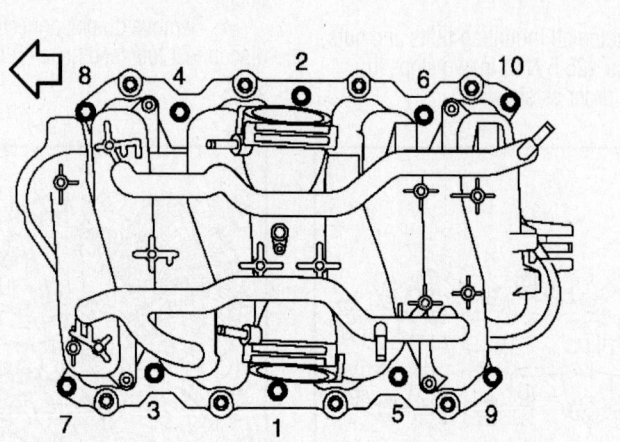

Fig. 155 Intake Manifold: tighten in numerical order as shown; loosen in the reverse order of this sequence

THROTTLE VALVE CLOSED POSITION LEARNING

Throttle Valve Closed Position Learning is a function of ECM to learn the fully closed position of the throttle valve by monitoring the throttle position sensor output signal. It must be performed each time the harness connector of the electric throttle control actuator or ECM is disconnected.

1. Check that accelerator pedal is fully released.
2. Turn ignition switch ON.
3. Turn ignition switch OFF and wait at least 10 seconds.

➡**Check that throttle valve moves during the above 10 seconds by confirming the operating sound.**

IDLE AIR VOLUME LEARNING

Idle Air Volume Learning is a function of ECM to learn the idle air volume that keeps engine idle speed within the specific range. It must be performed under the following conditions:

• Each time the electric throttle control actuator or ECM is replaced.
• Idle speed or ignition timing is out of the specification.

Check that all of the following conditions are satisfied.

✳✳ CAUTION

Learning will be cancelled if any of the following conditions are missed for even a moment.

• Battery voltage: More than 12.9 V (At idle)
• Engine coolant temperature: 158–221°F (70–105°C)
• Selector lever position: P or N
• Electric load switch: OFF (Air conditioner, headlamp, rear window defogger)

➡**On vehicles equipped with daytime light systems, if the parking brake is applied before the engine is started the headlamp will not illuminate.**

• Steering wheel: Neutral (Straight-ahead position)
• Vehicle speed: Stopped
• Transmission: Warmed-up
• With CONSULT-III: Drive vehicle until "ATF TEMP 2" in "DATA MONITOR" mode of "A/T" system indicates less than 0.9 V.
• Without CONSULT-III: Drive vehicle for 10 minutes.

With CONSULT-III

1. Perform Accelerator Pedal Released Position Learning.
2. Perform Throttle Valve Closed Position Learning.
3. Start engine and warm it up to normal operating temperature.
4. Select "IDLE AIR VOL LEARN" in "WORK SUPPORT" mode.
5. Touch "START" and wait 20 seconds.

Without CONSULT-III

✳✳ CAUTION

Note the following:

• It is better to count the time accurately with a clock.
• It is impossible to switch the diagnostic mode when an accelerator pedal position sensor circuit has a malfunction.

6. Perform Accelerator Pedal Released Position Learning.
7. Perform Throttle Valve Closed Position Learning.
8. Start engine and warm it up to normal operating temperature.
9. Turn ignition switch OFF and wait at least 10 seconds.
10. Confirm that accelerator pedal is fully released, turn ignition switch ON and wait 3 seconds.
11. Repeat the following procedure quickly 5 times within 5 seconds.
    a. Fully depress the accelerator pedal.
    b. Fully release the accelerator pedal.
12. Wait 7 seconds, fully depress the accelerator pedal for approx. 20 seconds until the MIL stops blinking and turns ON.
13. Fully release the accelerator pedal within 3 seconds after the MIL turns ON.
14. Start engine and let it idle.
15. Wait 20 seconds.
16. Rev up engine two or three times and check that idle speed and ignition timing are within the specifications.
17. Check the following:
    • Check that throttle valve is fully closed.
    • Check PCV valve operation.
    • Check that downstream of throttle valve is free from air leakage.

▌INTAKE MANIFOLD COLLECTOR

REMOVAL & INSTALLATION

3.5L Engine

See Figures 157 through 159.

✳✳ WARNING

Never drain engine coolant when the engine is hot to avoid the danger of being scalded.

1. Remove engine cover.
2. Remove air cleaner case and air duct.
3. Remove electric throttle control actuator as per the following:
    a. Drain engine coolant, or when water hoses are disconnected, attach plug to prevent engine coolant leakage.

✳✳ CAUTION

Perform this step when engine is cold. Never spill engine coolant on drive belt.

    b. Disconnect water hoses from electric throttle control actuator. When engine coolant is not drained from radiator, attach plug to water hoses to prevent engine coolant leakage.
    c. Disconnect harness connector.
    d. Loosen mounting bolts in reverse order as shown.

✳✳ CAUTION

Note the following:

• When removing only intake manifold collector, move electric throttle control actuator without disconnecting the water hose.
• The figure shows the electric throttle control actuator (bank 1) viewed from the air duct side.
• Viewed from the air duct side, order of loosening mounting bolts of electric throttle control actuator (bank 2) is the same as that of the electric throttle control actuator (bank 1).

✳✳ CAUTION

Handle carefully to avoid any impact to electric throttle control actuator.

4. Disconnect vacuum hose, PCV hose and EVAP hose from intake manifold collector.
5. Remove EVAP canister purge volume control solenoid valve and EVAP tube assembly from intake manifold collector.
6. Loosen mounting bolts and nuts in the reverse order as shown to remove intake manifold collector.

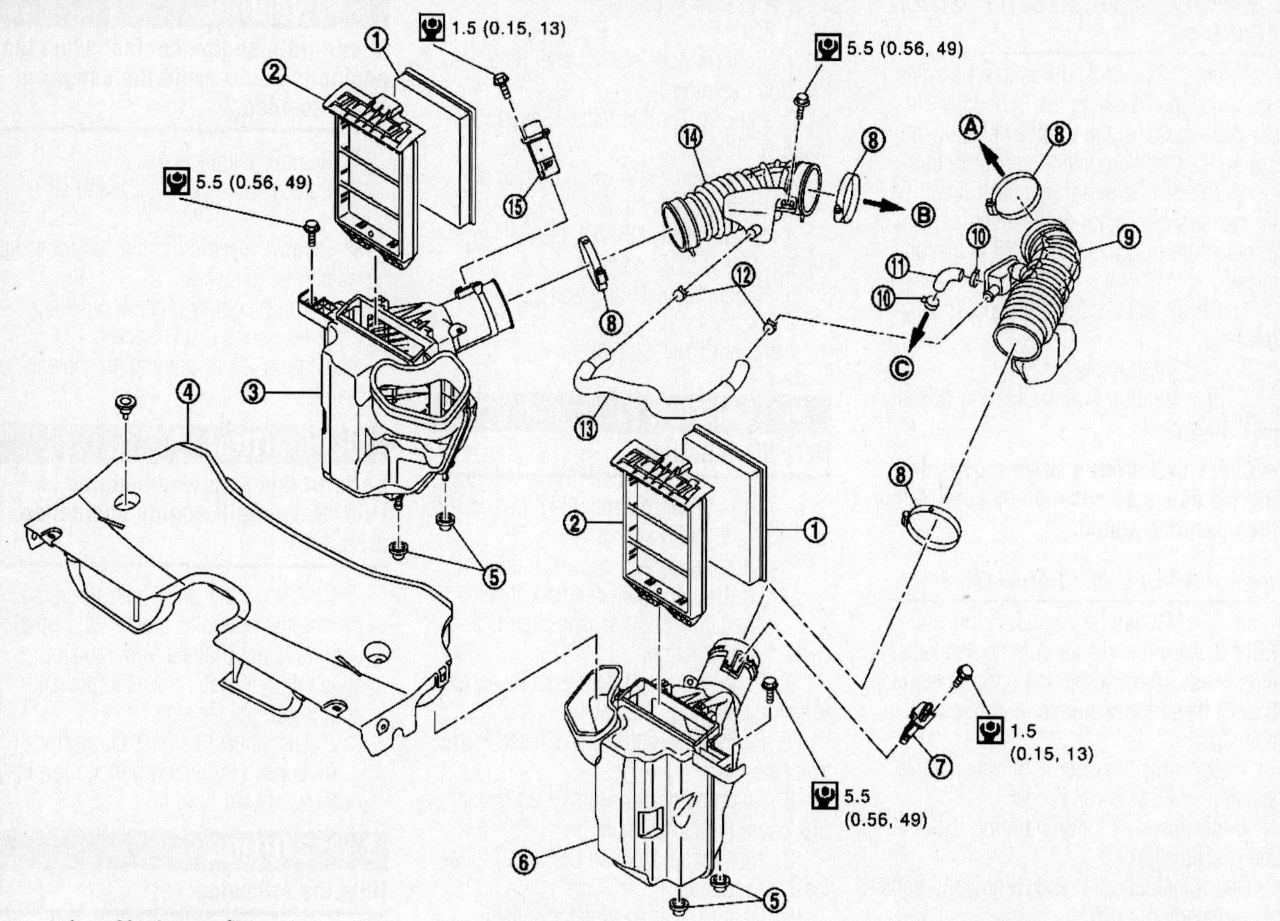

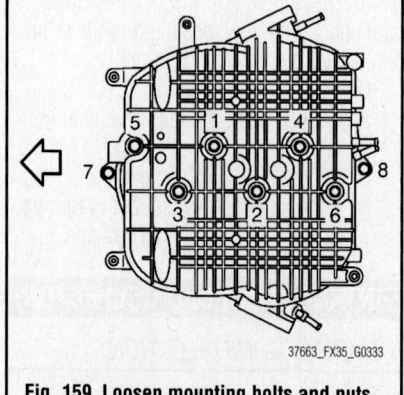

1.5 (0.15, 13)

5.5 (0.56, 49)

5.5 (0.56, 49)

1.5 (0.15, 13)

5.5 (0.56, 49)

1. Air cleaner filter	2. Holder	3. Air cleaner case (bank 1)
4. Air duct (inlet)	5. Grommet	6. Air cleaner case (bank 2)
7. Mass air flow sensor (bank 2)	8. Clamp	9. Air duct (bank 2)
10. Clamp	11. PCV hose	12. Clamp
13. PCV hose	14. Air duct (bank 1)	15. Mass air flow sensor (bank 1)
A. To electric throttle control actuator (bank 2)	B. To electric throttle control actuator (bank 1)	C. To rocker cover (bank 2)

37663_FX35_G0206

Fig. 157 Remove air cleaner case and air duct

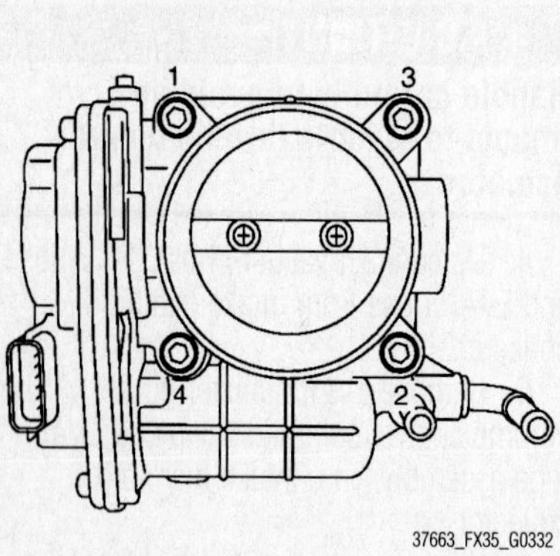

37663_FX35_G0332

Fig. 158 Loosen mounting bolts in reverse order as shown

37663_FX35_G0333

Fig. 159 Loosen mounting bolts and nuts in the reverse order as shown to remove intake manifold collector

To install:

7. Note the following item, and install in the reverse order of removal.

Intake Manifold Collector

See Figure 160.

1. If stud bolts were removed, install them and tighten to 8 ft. lbs. (10.8 Nm).

2. Tighten mounting bolts and nuts in numerical order as shown.

Water Hose

1. Insert hose by 1.06 to 1.26 inches (27 to 32 mm) from connector end.

2. Clamp hose at location of 0.12 to 0.28 inches (3 to 7 mm) from hose end.

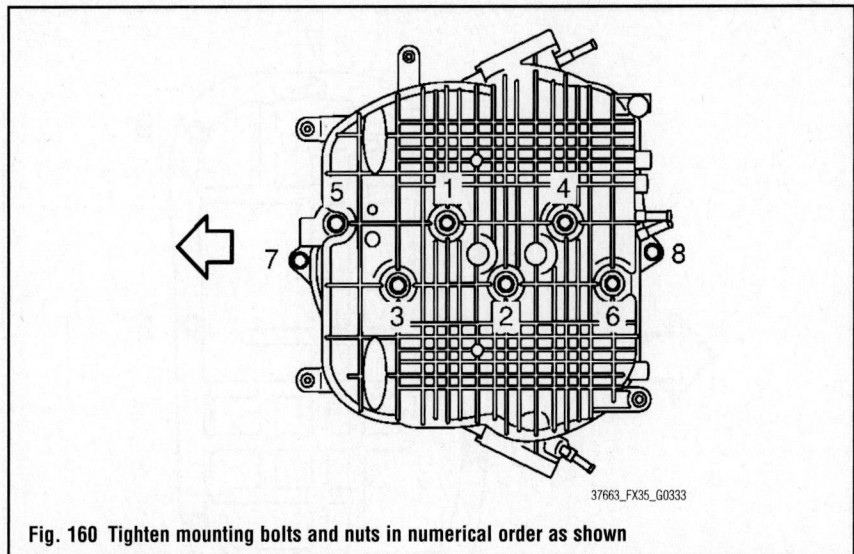

Fig. 160 Tighten mounting bolts and nuts in numerical order as shown

37663_FX35_G0333

Electric Throttle Control Actuator (Bank 1 and Bank 2)

See Figure 161.

1. Tighten in numerical order as shown.

✳✳ CAUTION

Note the following:

- The figure shows the electric throttle control actuator (bank 1) viewed from the air duct side.
- Viewed from the air duct side, order of tightening mounting bolts of electric throttle control actuator (bank 2) is the same as that of the electric throttle control actuator (bank 1).

2. Perform the "Throttle Valve Closed Position Learning" when harness connector of electric throttle control actuator is disconnected.

3. Perform the "Idle Air Volume Learning" and "Throttle Valve Closed Position Learning" when electric throttle control actuator is replaced.

THROTTLE VALVE CLOSED POSITION LEARNING

Throttle Valve Closed Position Learning is a function of ECM to learn the fully closed position of the throttle valve by monitoring the throttle position sensor output signal. It must be performed each time the harness connector of the electric throttle control actuator or ECM is disconnected.

1. Check that accelerator pedal is fully released.
2. Turn ignition switch ON.
3. Turn ignition switch OFF and wait at least 10 seconds.

➡**Check that throttle valve moves during the above 10 seconds by confirming the operating sound.**

IDLE AIR VOLUME LEARNING

Idle Air Volume Learning is a function of ECM to learn the idle air volume that keeps engine idle speed within the specific range. It must be performed under the following conditions:

- Each time the electric throttle control actuator or ECM is replaced.
- Idle speed or ignition timing is out of the specification.

Check that all of the following conditions are satisfied.

✳✳ CAUTION

Learning will be cancelled if any of the following conditions are missed for even a moment.

- Battery voltage: More than 12.9 V (At idle)
- Engine coolant temperature: 158–221°F (70–105°C)
- Selector lever position: P or N
- Electric load switch: OFF
(Air conditioner, headlamp, rear window defogger)

➡**On vehicles equipped with daytime light systems, if the parking brake is applied before the engine is started the headlamp will not illuminate.**

- Steering wheel: Neutral (Straight-ahead position)
- Vehicle speed: Stopped
- Transmission: Warmed-up
- With CONSULT-III: Drive vehicle until "ATF TEMP 2" in "DATA MONITOR" mode of "A/T" system indicates less than 0.9 V.
- Without CONSULT-III: Drive vehicle for 10 minutes.

1. With CONSULT-III:
   a. Perform Accelerator Pedal Released Position Learning.
   b. Perform Throttle Valve Closed Position Learning.
   c. Start engine and warm it up to normal operating temperature.
   d. Select "IDLE AIR VOL LEARN" in "WORK SUPPORT" mode.
2. Touch "START" and wait 20 seconds.
3. Without CONSULT-III:

✳✳ CAUTION

Note the following:

- It is better to count the time accurately with a clock.
- It is impossible to switch the diagnostic mode when an accelerator pedal position sensor circuit has a malfunction.

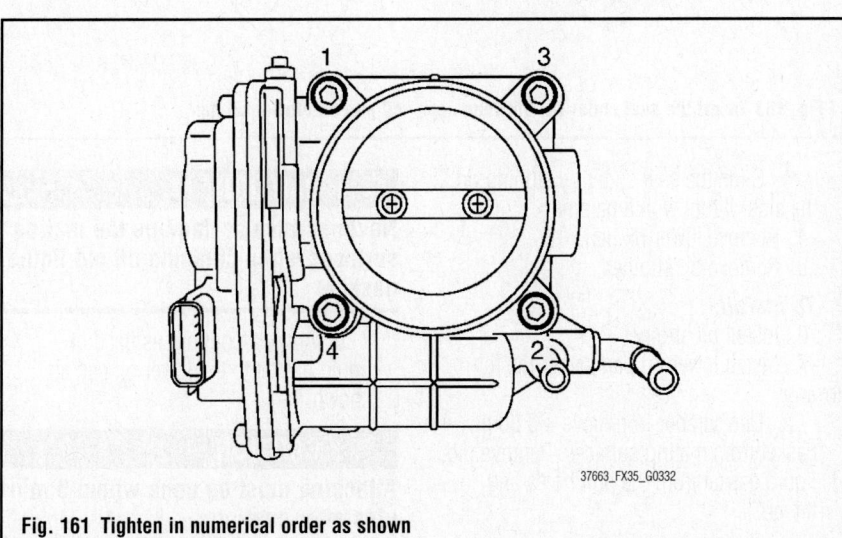

37663_FX35_G0332

Fig. 161 Tighten in numerical order as shown

4. Perform Accelerator Pedal Released Position Learning.

5. Perform Throttle Valve Closed Position Learning.

6. Start engine and warm it up to normal operating temperature.

7. Turn ignition switch OFF and wait at least 10 seconds.

8. Confirm that accelerator pedal is fully released, turn ignition switch ON and wait 3 seconds.

9. Repeat the following procedure quickly 5 times within 5 seconds.

    a. Fully depress the accelerator pedal.

    b. Fully release the accelerator pedal.

10. Wait 7 seconds, fully depress the accelerator pedal for approx. 20 seconds until the MIL stops blinking and turns ON.

11. Fully release the accelerator pedal within 3 seconds after the MIL turns ON.

12. Start engine and let it idle.

13. Wait 20 seconds.

14. Rev up engine two or three times and check that idle speed and ignition timing are within the specifications.

15. Check the following:

- Check that throttle valve is fully closed.
- Check PCV valve operation.
- Check that downstream of throttle valve is free from air leakage.

OIL PAN

REMOVAL & INSTALLATION

3.5L Engine

Lower Oil Pan and Oil Strainer

See Figures 162 through 165.

> ❋❋ **WARNING**
>
> **To avoid the danger of being scalded, never drain engine oil when engine is hot.**

1. Remove engine undercover.
2. Drain engine oil.
3. Remove lower oil pan as per the following:

    a. Loosen mounting bolts in reverse order as shown.

    b. Insert the seal cutter KV10111100 (J-37228) between upper oil pan and lower oil pan.

> ❋❋ **CAUTION**
>
> **Be careful not to damage the mating surfaces. Never insert a screwdriver, this will damage the mating surfaces.**

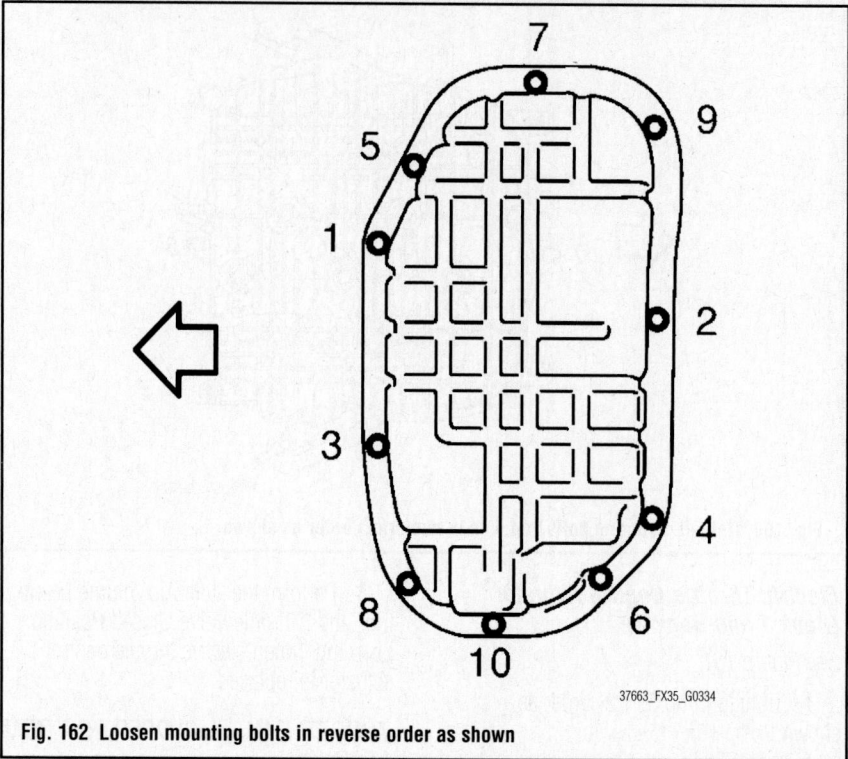

Fig. 162 Loosen mounting bolts in reverse order as shown

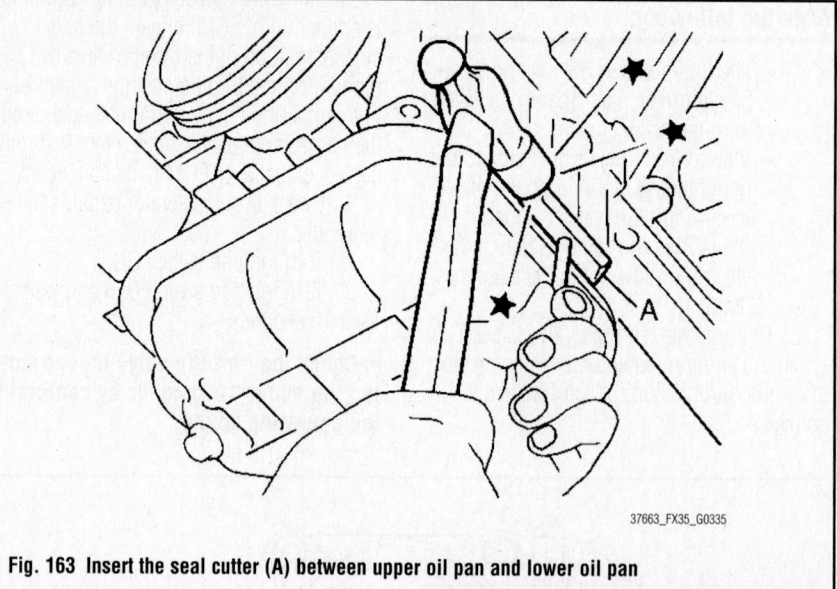

Fig. 163 Insert the seal cutter (A) between upper oil pan and lower oil pan

    c. Slide the seal cutter by tapping on the side of tool with a hammer.

4. Remove lower oil pan.
5. Remove oil strainer.

To install:

6. Install oil strainer.
7. Install lower oil pan as per the following:

    a. Use scraper to remove old liquid gasket from mating surfaces. Remove old liquid gasket from the bolt holes and thread.

> ❋❋ **CAUTION**
>
> **Never scratch or damage the mating surfaces when cleaning off old liquid gasket.**

    b. Apply a continuous bead of liquid gasket to the lower oil pan as shown.

> ❋❋ **CAUTION**
>
> **Attaching must be done within 5 minutes after coating.**

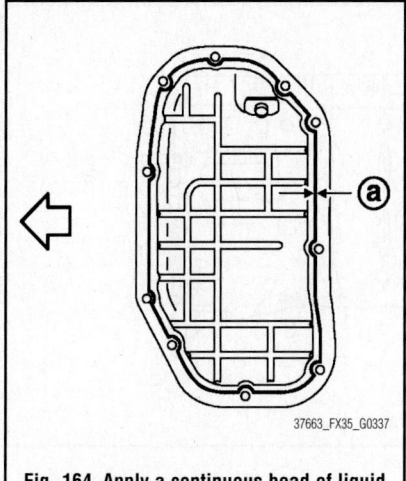

Fig. 164 Apply a continuous bead of liquid gasket to the lower oil pan as shown

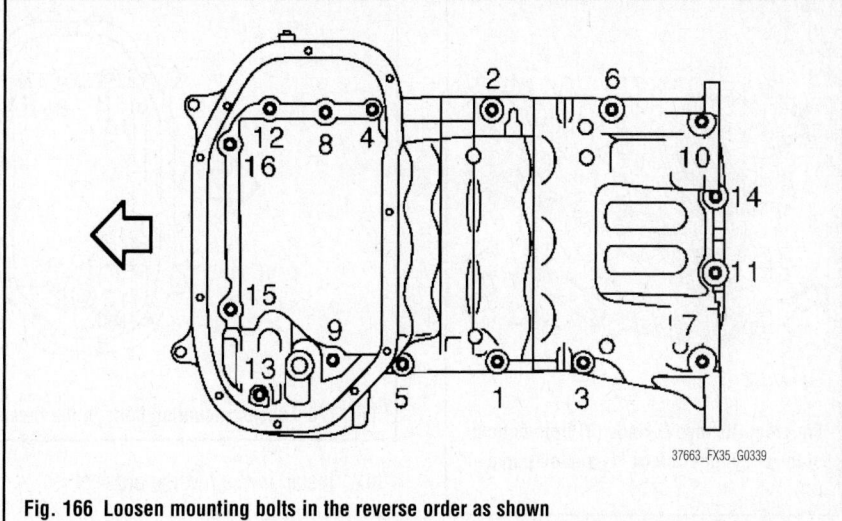

Fig. 166 Loosen mounting bolts in the reverse order as shown

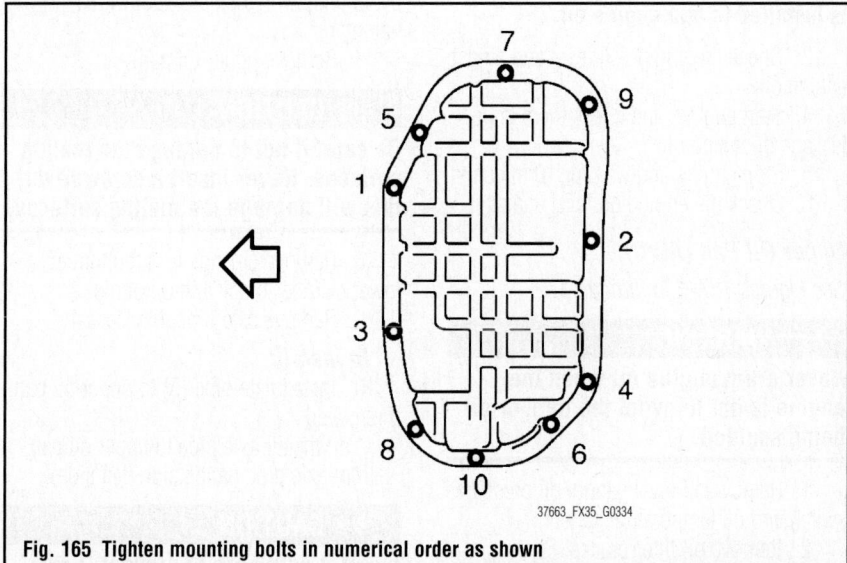

Fig. 165 Tighten mounting bolts in numerical order as shown

c. Install lower oil pan. Tighten mounting bolts in numerical order as shown.

8. Install oil pan drain plug.

9. Install in the reverse order of removal after this step.

➡ **At least 30 minutes after oil pan is installed, pour engine oil.**

10. Check the engine oil level and adjust engine oil.

11. Start engine, and check there is no leakage of engine oil.

12. Stop engine and wait for 10 minutes.

13. Check the engine oil level again.

Upper Oil Pan (2WD)

See Figures 166 through 170.

※※ **WARNING**

Never drain engine oil when the engine is hot to avoid the danger of being scalded.

1. Remove oil level gauge, oil pressure switch and oil temperature sensor.

2. Remove lower oil pan.

3. Remove oil strainer.

4. Loosen mounting bolts in the reverse order as shown.

5. Insert the seal cutter KV10111100 (J-37228) between upper oil pan and lower cylinder block. Slide seal cutter by tapping on the side of tool with a hammer.

6. Remove upper oil pan.

※※ **CAUTION**

Be careful not to damage the mating surfaces. Never insert a

screwdriver, this will damage the mating surfaces.

7. Remove O-rings from bottom of lower cylinder block and oil pump.

To install:

8. Install upper oil pan as per the following:

a. Use a scraper to remove old liquid gasket from mating surfaces.

※※ **CAUTION**

Never scratch or damage the mating surfaces when cleaning off old liquid gasket.

b. Also remove old liquid gasket from mating surface of lower cylinder block.

c. Remove old liquid gasket from the bolt holes and threads.

d. Install new O-rings on the bottom of lower cylinder block and oil pump.

e. Apply a continuous bead of liquid gasket to the cylinder block mating surface of upper oil pan as shown.

※※ **CAUTION**

Note the following:

- For bolt holes with triangle marks (7 locations), apply liquid gasket outside the holes.
- Attaching must be done within 5 minutes after coating.

f. Install upper oil pan.

※※ **CAUTION**

Install avoiding misalignment of both O-rings.

g. Tighten mounting bolts in numerical order as shown.

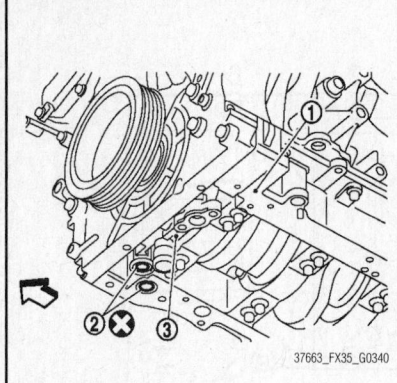

Fig. 167 Remove O-rings (2) from bottom of lower cylinder block (1) and oil pump (3)

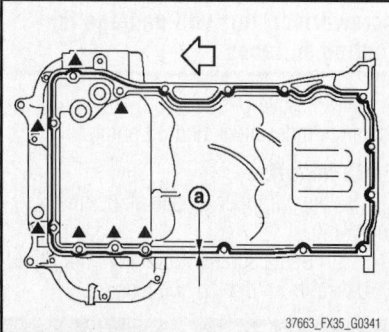

Fig. 168 Apply a continuous bead of liquid gasket (a) to the cylinder block mating surface of upper oil pan as shown

h. There are two types of mounting bolts. 3.5 inches (90 mm) bolts are used in holes 7, 10 and 13. All other holes use 1 inch (25 mm) bolts.
9. Install oil strainer to oil pump.
10. Install lower oil pan.
11. Install oil pan drain plug.

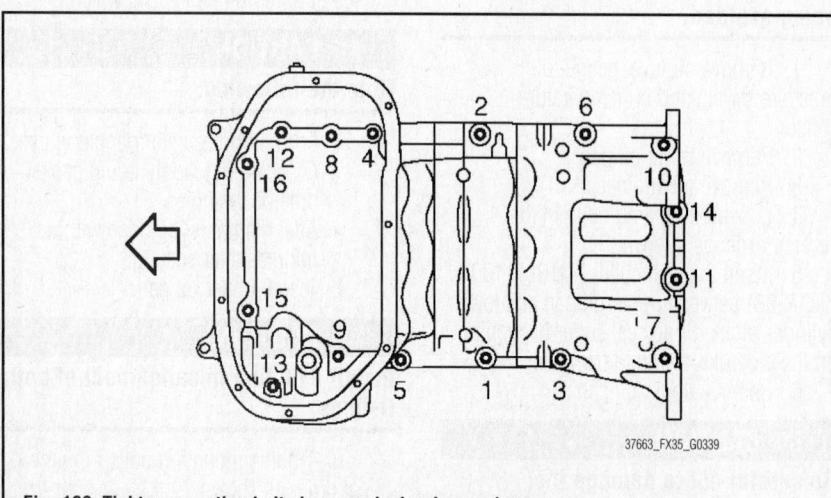

Fig. 169 Tighten mounting bolts in numerical order as shown

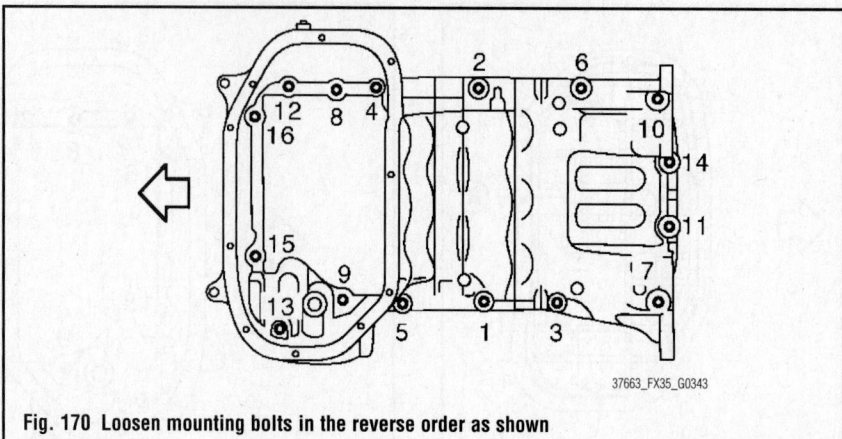

Fig. 170 Loosen mounting bolts in the reverse order as shown

12. Install in the reverse order of removal after this step.

➡ **Wait at least 30 minutes after oil pan is installed to add engine oil.**

13. Check the engine oil level and adjust engine oil.
14. Start engine, and check there is no leakage of engine oil.
15. Stop engine and wait for 10 minutes.
16. Check the engine oil level again.

Upper Oil Pan (AWD)

See Figures 167, 170 through 172.

✳✳ WARNING

Never drain engine oil when the engine is hot to avoid the danger of being scalded.

1. Remove oil level gauge, oil pressure switch and oil temperature sensor.
2. Remove oil filter bracket.
3. Remove lower oil pan.
4. Remove oil strainer.
5. Loosen mounting bolts in reverse order as shown.

6. Insert the seal cutter KV10111100 (J-37228) between upper oil pan and lower cylinder block. Slide seal cutter by tapping on the side of tool with a hammer.
7. Remove upper oil pan.

✳✳ CAUTION

Be careful not to damage the mating surfaces. Never insert a screwdriver, this will damage the mating surfaces.

8. Remove O-rings from bottom of lower cylinder block and oil pump.
9. Remove axle pipe, if necessary.

To install:

10. Install axle pipe (3) to upper oil pan, if removed.
   a. Install axle pipe to upper oil pan from axle pipe flange side (left side).

✳✳ CAUTION

Insert it with care to prevent O-ring from sliding.

11. Install upper oil pan as per the following:
   a. Use a scraper to remove old liquid gasket from mating surfaces.

✳✳ CAUTION

Never scratch or damage the mating surfaces when cleaning off old liquid gasket.

   b. Also remove old liquid gasket from mating surface of lower cylinder block.
   c. Remove old liquid gasket from the bolt holes and threads.
   d. Install new O-rings on the bottom of lower cylinder block and oil pump.
   e. Apply a continuous bead of liquid gasket to the cylinder block mating surface of upper oil pan as shown.

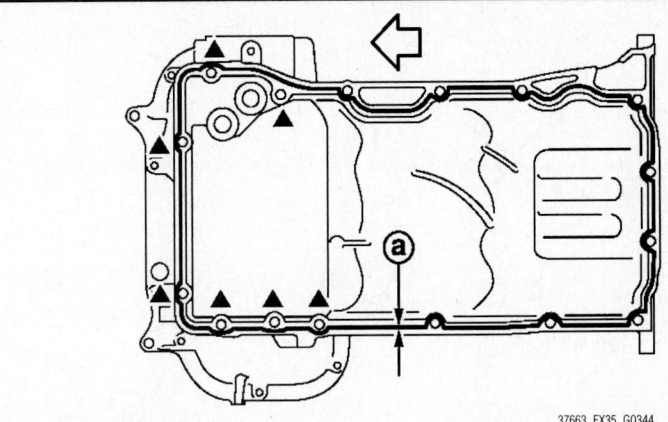

Fig. 171 Apply a continuous bead of liquid gasket to the cylinder block mating surface of upper oil pan as shown

✳✳ CAUTION

Note the following:

- For bolt holes with triangle marks (7 locations), apply liquid gasket outside the holes.
- Attaching must be done within 5 minutes after coating.
- f. Install upper oil pan.

✳✳ CAUTION

Install avoiding misalignment of O-rings.

g. Tighten mounting bolts in numerical order as shown.

h. There are three types of mounting bolts:

- 1 inch (25 mm) bolts: Bolt positions 3, 6, 8, 9, 11, 12, 14, 15, 16
- 2 inches (50 mm) bolt: Bolt position 2
- 3.5 inches (90 mm) bolts: Bolt positions 1, 4, 5, 7, 10, 13

12. Install oil strainer to oil pump.
13. Install lower oil pan.

14. Install oil pan drain plug.
15. Install in the reverse order of removal after this step.

➡ **Wait at least 30 minutes after oil pan is installed to add engine oil.**

16. Check the engine oil level and adjust engine oil.
17. Start engine, and check there is no leakage of engine oil.
18. Stop engine and wait for 10 minutes.
19. Check the engine oil level again.

5.0L Engine

Lower Oil Pan

See Figures 173 through 175.

✳✳ WARNING

To avoid the danger of being scalded, never drain engine oil when engine is hot.

1. Drain engine oil.
2. Remove lower oil pan as per the following:

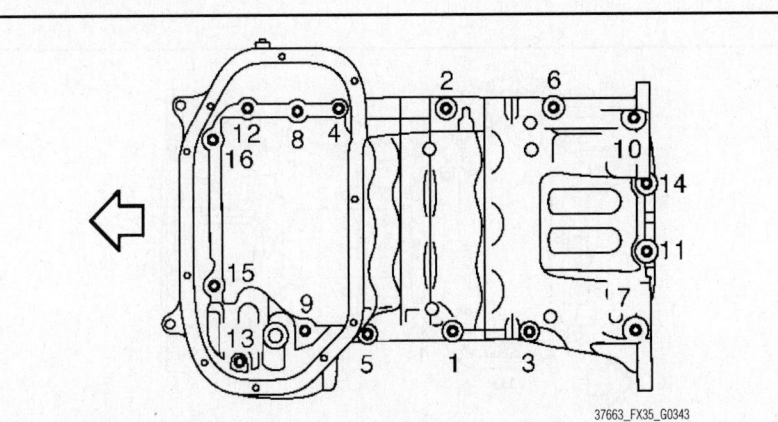

Fig. 172 Tighten mounting bolts in numerical order as shown

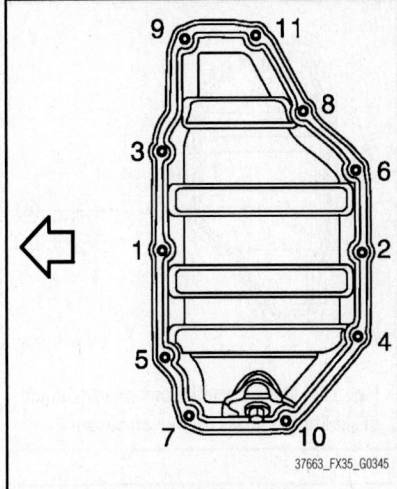

Fig. 173 Loosen mounting bolts in reverse order as shown

a. Loosen mounting bolts in reverse order as shown.
b. Insert the seal cutter KV10111100 (J-37228) between upper oil pan and lower oil pan.

✳✳ CAUTION

Be careful not to damage the mating surfaces. Never insert a screwdriver. This damages the mating surfaces.

c. Slide the seal cutter by tapping on the side of tool with a hammer.
3. Remove lower oil pan.
4. Remove oil strainer.

To install:

5. Install oil strainer.
6. Install lower oil pan as per the following:

a. Use scraper to remove old liquid gasket from mating surfaces.
b. Remove old liquid gasket from the bolt holes and thread.

✳✳ CAUTION

Never scratch or damage the mating surfaces when cleaning off old liquid gasket.

c. Apply a continuous bead of liquid gasket to the lower oil pan as shown.

✳✳ CAUTION

Attaching must be done within 5 minutes after coating.

d. Install lower oil pan.
e. Tighten mounting bolts in numerical order as shown.
7. Install oil pan drain plug.
8. Install in the reverse order of removal after this step.

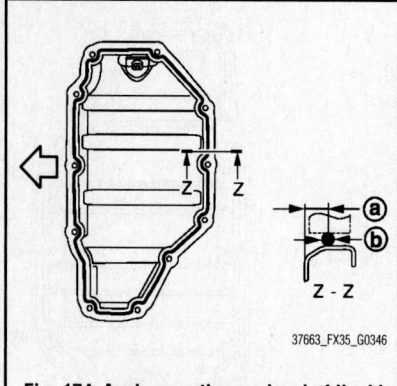

Fig. 174 Apply a continuous bead of liquid gasket to the lower oil pan as shown

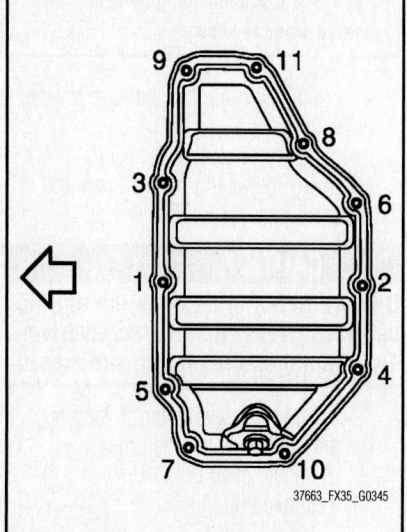

Fig. 175 Tighten mounting bolts in numerical order as shown

→Wait at least 30 minutes after oil pan is installed before adding engine oil.

9. Check the engine oil level and adjust engine oil.

10. Start engine, and check there is no leakage of engine oil.

11. Stop engine and wait for 15 minutes.

12. Check the engine oil level again.

Upper Oil Pan

See Figures 176 through 180.

❊❊ WARNING

To avoid the danger of being scalded, never drain engine oil when engine is hot.

1. Remove oil filter.
2. Remove oil cooler.

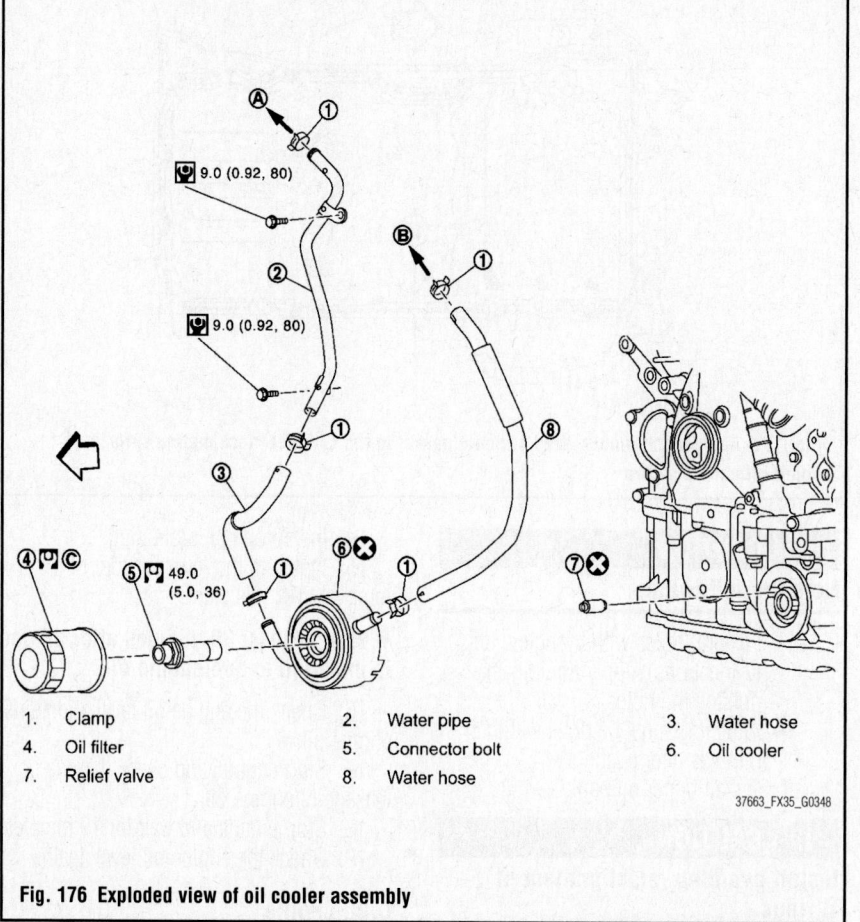

1.	Clamp	2.	Water pipe	3.	Water hose
4.	Oil filter	5.	Connector bolt	6.	Oil cooler
7.	Relief valve	8.	Water hose		

Fig. 176 Exploded view of oil cooler assembly

3. Remove A/C compressor and A/C compressor bracket.

4. Remove oil level gauge and oil level gauge guide.

5. Remove oil pressure switch and oil temperature sensor if necessary.

6. Remove rear plate cover.

7. Remove lower oil pan.

8. Remove oil strainer.

9. Remove upper oil pan as per the following:

   a. Loosen mounting bolts in the reverse order as shown.

→Disregard No. 12, 17 when loosening.

   b. Insert a suitable tool into the notch at upper oil pan as shown.

   c. Pry off case by using a suitable tool.

❊❊ CAUTION

Be careful not to damage the mating surfaces.

10. Remove O-ring from bottom of cylinder block and oil pump.

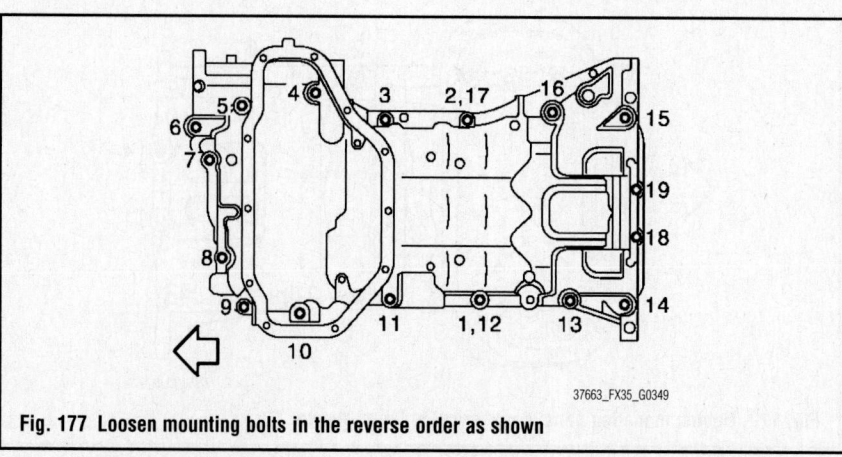

Fig. 177 Loosen mounting bolts in the reverse order as shown

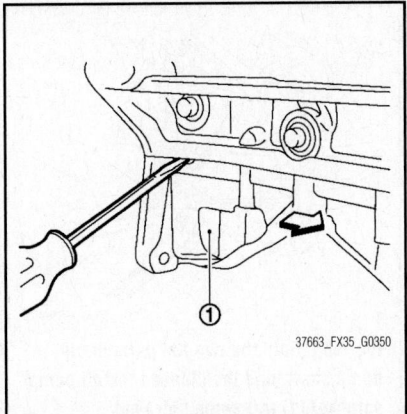

Fig. 178 Insert a suitable tool into the notch at upper oil pan (1) as shown

11. Remove baffle plate, if necessary.

12. Remove axle pipe from upper oil pan, if necessary.

13. Remove axle pipe from upper oil pan using a suitable drift.

To install:

14. Install axle pipe to upper oil pan, if removed.

  a. Lubricate O-ring groove of axle pipe, O-ring, and O-ring joint of oil pan with new engine oil.

  b. Install axle pipe to upper oil pan from drive shaft (LH) side.

❋❋ CAUTION

Insert it with care to prevent O-ring from sliding.

15. Install upper oil pan as per the following:

  a. Use a scraper to remove old liquid gasket from mating surfaces.

  b. Also remove the old liquid gasket from mating surface of cylinder block.

  c. Remove old liquid gasket from the bolt holes and threads.

❋❋ CAUTION

Never scratch or damage the mating surfaces when cleaning off old liquid gasket.

  d. Install new O-rings on the bottom of cylinder block and oil pump.

  e. Apply a continuous bead of liquid gasket to the cylinder block mating surfaces of upper oil pan as shown.

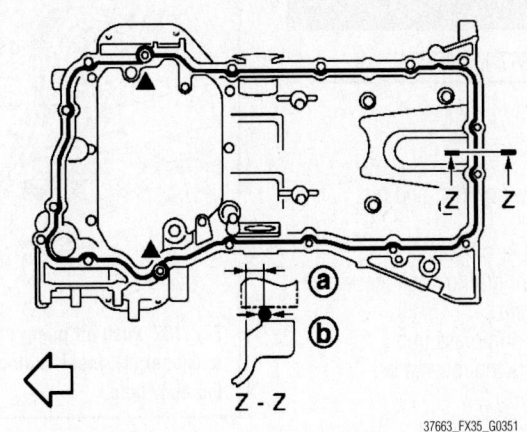

Fig. 179 Apply a continuous bead of liquid gasket to the cylinder block mating surfaces of upper oil pan as shown

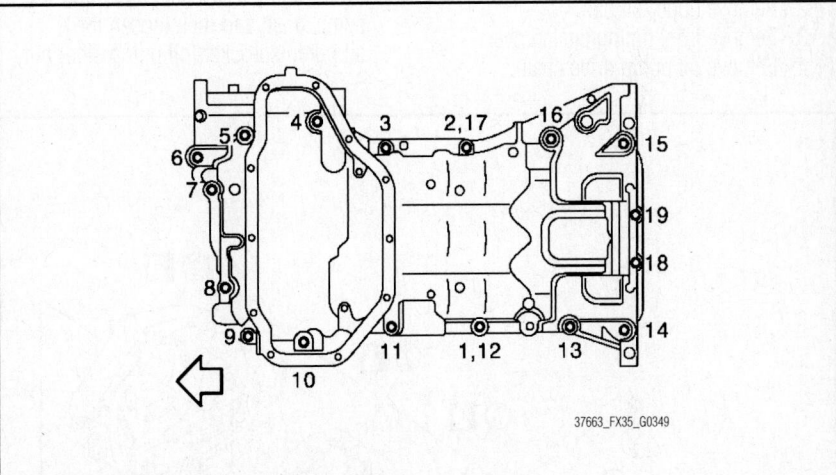

Fig. 180 Tighten mounting bolts in numerical order as shown

❋❋ CAUTION

Attaching must be done within 5 minutes after coating.

  f. Tighten mounting bolts in numerical order as shown.

❋❋ CAUTION

Install avoiding misalignment of O-rings.

➠Tighten mounting bolts No. 1 and 2 in two steps. The numerical order No. 12 and 17 shown as second steps.

  g. There are three types of mounting bolts.

- 1.18 inches (30 mm) bolts: Bolt positions 18, 19
- 3.94 inches (100 mm) bolts: Bolt positions 3, 4, 5, 7, 10, 11, 14, 15

- 1.77 inches (45 mm) bolts: Bolt positions all others except above.

  h. Tighten transmission joint bolts.

  i. Install rear plate cover.

16. Install oil strainer.

17. Install lower oil pan.

18. Install in the reverse order of removal.

➠Wait at least 30 minutes after oil pan is installed before adding engine oil.

19. Check the engine oil level and adjust engine oil.

20. Start engine, and check there is no leakage of engine oil.

21. Stop engine and wait for 15 minutes.

22. Check the engine oil level again.

OIL PUMP

REMOVAL & INSTALLATION

3.5L Engine

See Figure 181.

1. Remove lower oil pan and oil strainer.
2. Remove upper oil pan.
3. Remove front timing chain case and primary timing chain.
4. Remove oil pump assembly.
5. Installation is the reverse of removal.

5.0L Engine

See Figures 182 through 184.

1. Remove lower oil pan and oil strainer.
2. Remove upper oil pan.
3. Remove front timing chain cover.
4. Remove oil pump drive chain.

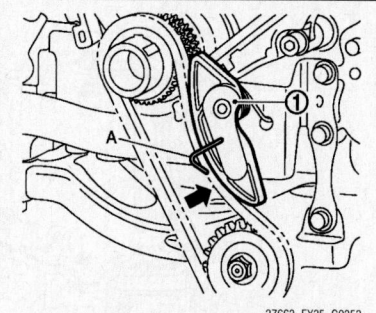

Fig. 182 Push oil pump drive chain tensioner (1); Insert a stopper pin (A) into the body hole

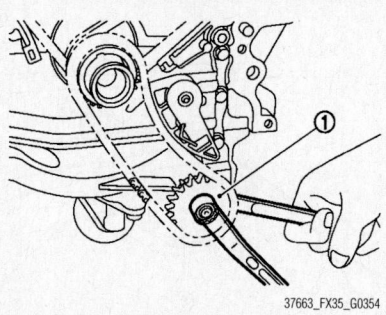

Fig. 183 Hold the two flat parts of oil pump shaft, and then loosen the oil pump sprocket (1) (oil pump side) nut

a. Push oil pump drive chain tensioner.

b. Insert a stopper pin into the body hole.

c. Hold the two flat parts of oil pump shaft, and then loosen the oil pump sprocket (oil pump side) nut.

5. Remove oil pump.

❈❈ CAUTION

Never disassemble oil pump.

6. Installation is the reverse order of removal.

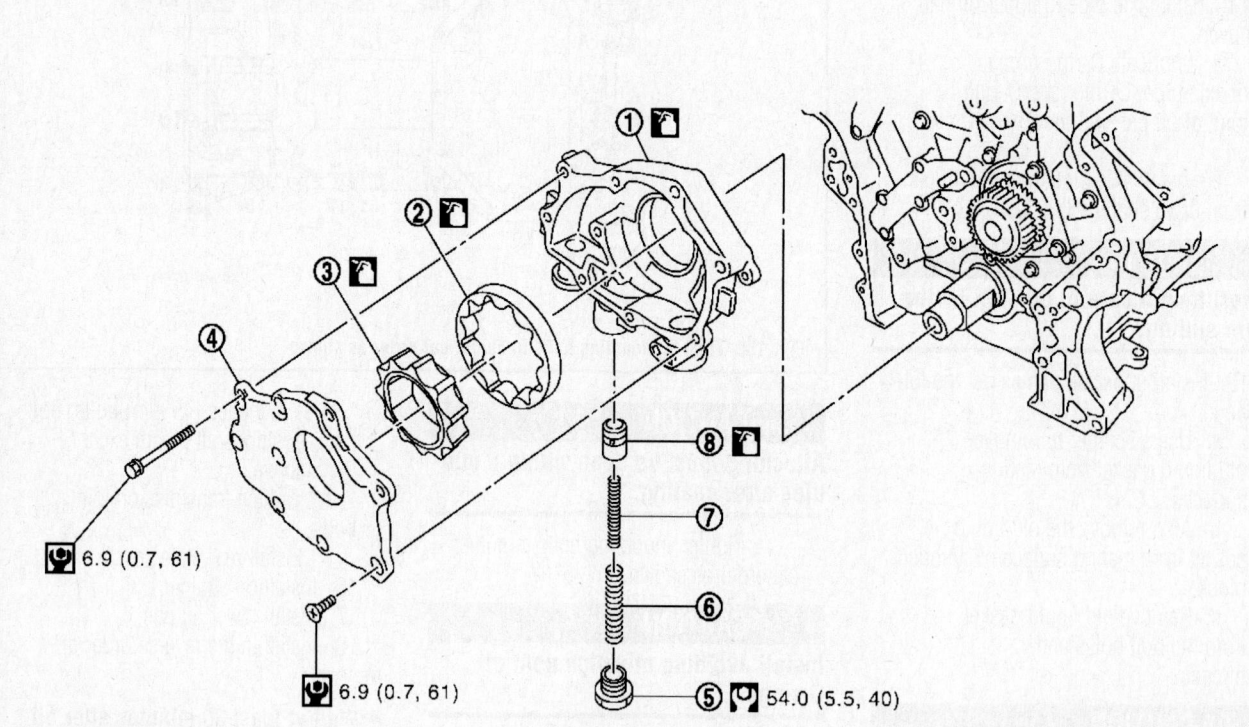

6.9 (0.7, 61)

6.9 (0.7, 61)

5 54.0 (5.5, 40)

1.	Oil pump body	2.	Oil pump outer rotor	3.	Oil pump inner rotor
4.	Oil pump cover	5.	Regulator valve plug	6.	Regulator valve spring
7.	Regulator valve spring	8.	Regulator valve		

Fig. 181 Exploded view of oil pump assembly

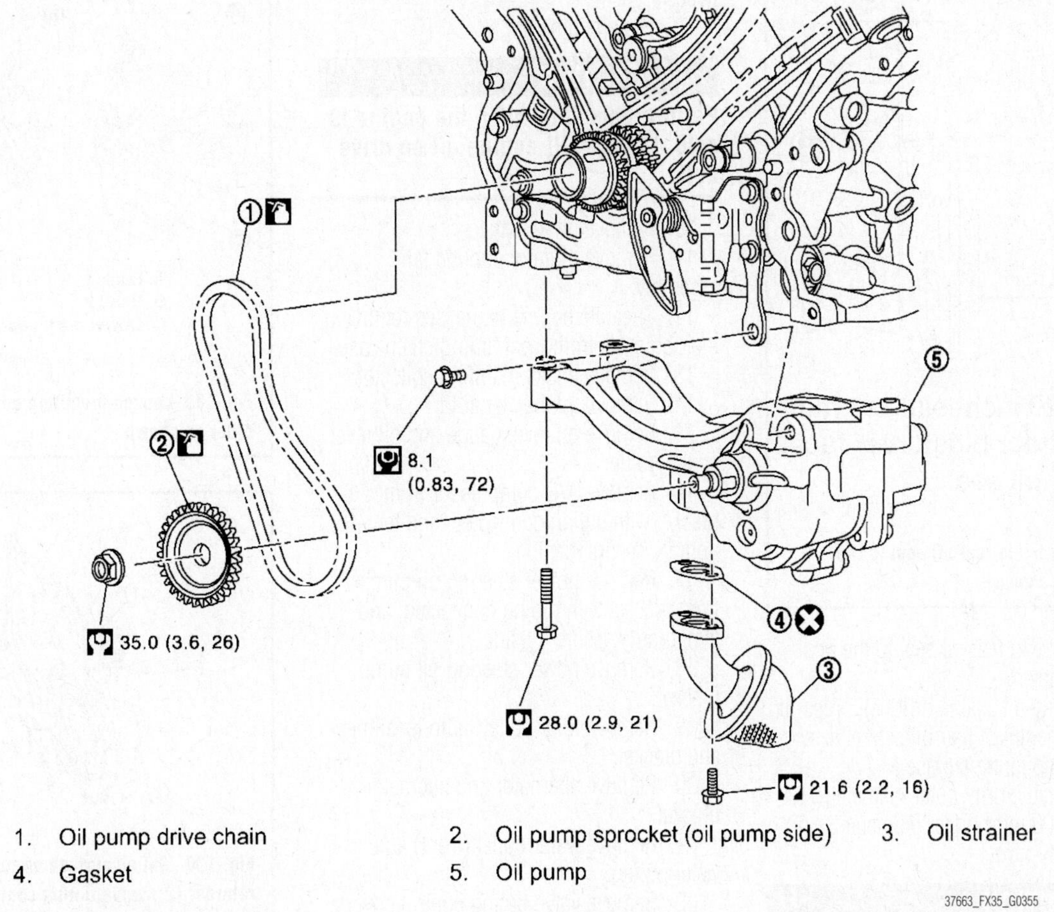

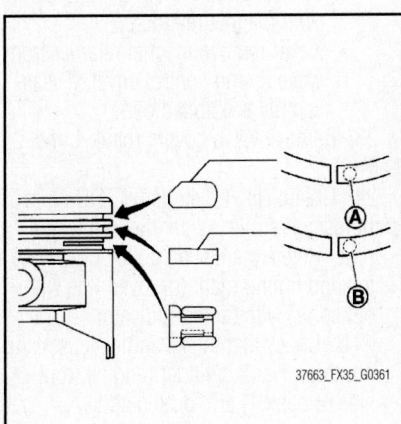

1. Oil pump drive chain
4. Gasket
2. Oil pump sprocket (oil pump side)
5. Oil pump
3. Oil strainer

35.0 (3.6, 26)

8.1 (0.83, 72)

28.0 (2.9, 21)

21.6 (2.2, 16)

37663_FX35_G0355

Fig. 184 Exploded view of oil pump assembly

PISTON AND RING

POSITIONING

See Figures 185 and 186.

Fig. 185 Top ring (A), Second ring (B) and Oil ring locations

37663_FX35_G0361

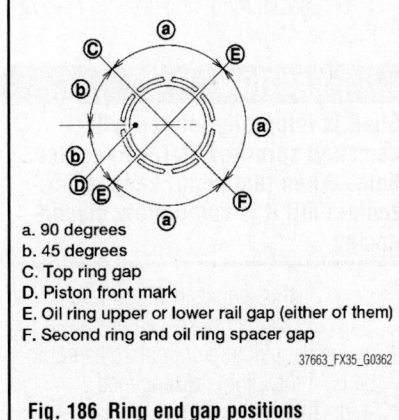

a. 90 degrees
b. 45 degrees
C. Top ring gap
D. Piston front mark
E. Oil ring upper or lower rail gap (either of them)
F. Second ring and oil ring spacer gap

37663_FX35_G0362

Fig. 186 Ring end gap positions

REAR MAIN SEAL

REMOVAL & INSTALLATION

See Figures 187 and 188.

1. Remove transmission assembly.
2. Remove drive plate.
3. Remove rear oil seal with a suitable tool.

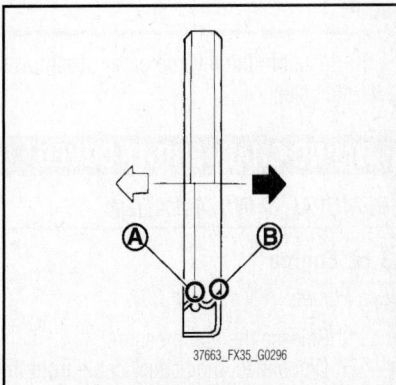

37663_FX35_G0296

Fig. 187 Install rear oil seal so that each seal lip is oriented as shown

✳✳ CAUTION

Be careful not to damage crankshaft and cylinder block.

To install:

4. Install rear oil seal.
    a. Install rear oil seal so that each seal lip is oriented as shown.

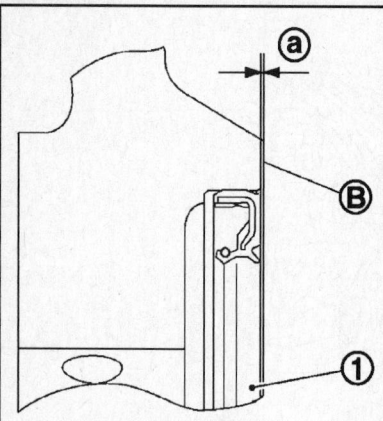

a. 0-0.20 inches (0-0.5 mm)
B. Cylinder block rear face
1. Rear oil seal

37663_FX35_G0363

Fig. 188 Press in rear oil seal to the position as shown

b. Press in rear oil seal to the position as shown.

c. Using a suitable drift (A), press-fit until the height of rear oil seal is level with the mounting surface.

d. Suitable drift: outer diameter 3.94 inches (100 mm), inner diameter 3.35 inches (85 mm).

✳✳ CAUTION

Be careful not to damage crankshaft and cylinder block. Press-fit straight and avoid causing burrs or tilting oil seal.

5. Install in the reverse order of removal after this step.

TIMING CHAIN FRONT COVER

REMOVAL & INSTALLATION

3.5L Engine

See Figures 189 through 201.

1. Release the fuel pressure.
2. Disconnect the battery cable from the negative terminal.
3. Remove engine cover.
4. Remove radiator reservoir tank.
5. Remove air duct and air cleaner case assembly.
6. Remove engine undercover.
7. Drain engine coolant from radiator.

✳✳ CAUTION

Perform this step when the engine is cold. Never spill engine coolant on drive belt.

8. Remove upper and lower radiator hoses.
9. Drain engine oil.

✳✳ CAUTION

Perform this step when the engine is cold. Never spill engine oil on drive belt.

10. Remove drive belt.
11. Remove radiator cooling fan assembly.
12. Separate engine harnesses removing their brackets from front timing chain case.
13. Remove intake manifold collector.
14. Remove intake manifold.
15. Remove oil level gauge and oil level gauge guide.
16. Remove A/C compressor from bracket with piping connected, and temporarily secure it aside.
17. Remove power steering oil pump from bracket with piping connected, and temporarily secure it aside.
18. Remove power steering oil pump bracket.
19. Remove idler pulley, auto tensioner and bracket.
20. Remove alternator and alternator bracket.
21. Remove water outlet (front) and water piping.
22. Remove valve timing control covers (bank 1 and bank 2) and gasket as per the following:
  a. Disconnect valve timing control harness connector
  b. Loosen mounting bolts in reverse order as shown.

✳✳ CAUTION

Shaft is internally jointed with camshaft sprocket (INTAKE) center hole. When removing, keep it horizontal until it is completely disconnected.

c. Shaft is engaged with intake side camshaft sprocket center hole on inside. pull straight out so as not to tilt until the joint is disengaged.

d. The mating surface of magnet retarder may be fitted with the exhaust side camshaft sprocket via the engine oil. Open valve timing control cover carefully.

e. If the mating surface of magnet retarder is fitted with the camshaft sprocket, open the cover within the range that the load is not applied to the harness. And then, remove it so as to prevent magnet retarder from dropping.

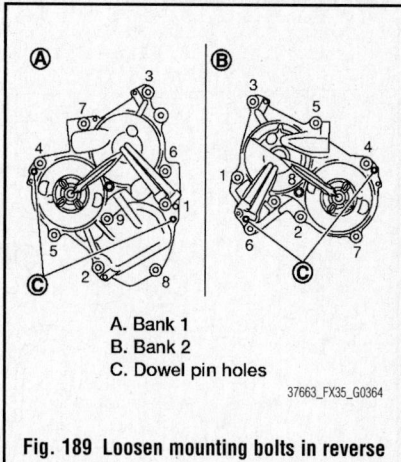

A. Bank 1
B. Bank 2
C. Dowel pin holes

37663_FX35_G0364

Fig. 189 Loosen mounting bolts in reverse order as shown

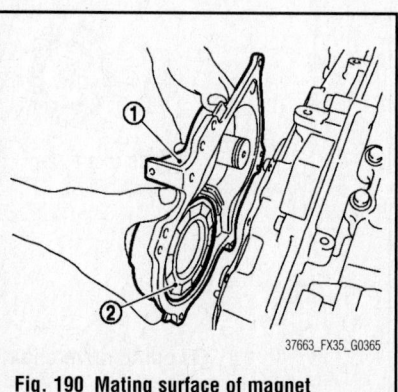

37663_FX35_G0365

Fig. 190 Mating surface of magnet retarder (2); valve timing control cover (1)

✳✳ CAUTION

Note the following:

- Be careful not to damage magnet retarder.
- When carrying valve timing control cover, face the magnet retarder side up to prevent the cover from falling from magnet retarder.
- Never remove magnet retarder from valve timing control cover. (Disassembly prohibited parts)

23. Remove valve covers (bank 1 and bank 2).

24. Obtain No. 1 cylinder at TDC of its compression stroke as per the following:
  a. Rotate crankshaft pulley clockwise to align timing mark (grooved line without color) with timing indicator.
  b. Check that intake and exhaust cam noses on No. 1 cylinder (engine front side of bank 1) are located as shown.
  c. If not, turn crankshaft one revolution (360 degrees) and align as shown.

25. Remove crankshaft pulley as per the following:
  a. Remove front cross bar.

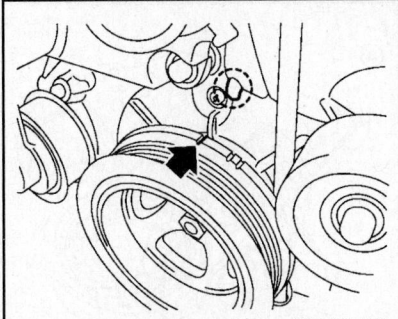

Fig. 191 Rotate crankshaft pulley clockwise to align timing mark (grooved line without color) with timing indicator

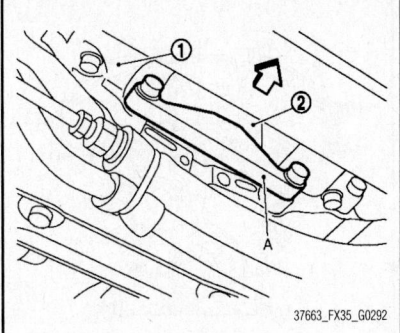

Fig. 193 Remove rear cover plate and set the ring gear stopper (A) as shown

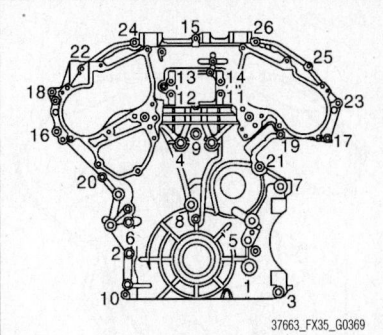

Fig. 195 Loosen mounting bolts in reverse order as shown

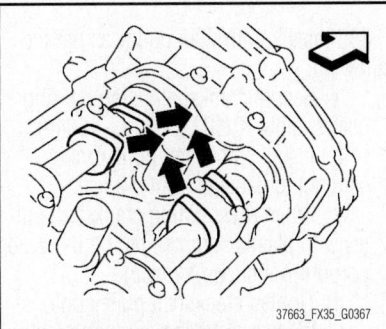

Fig. 192 Check that intake and exhaust cam noses on No. 1 cylinder (engine front side of bank 1) are located as shown

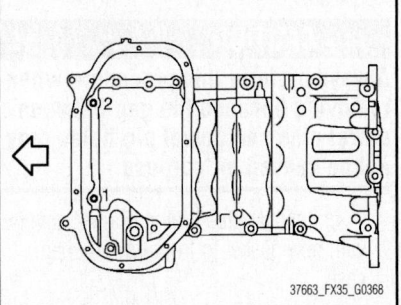

Fig. 194 Loosen two mounting bolts in front of upper oil pan in reverse order as shown

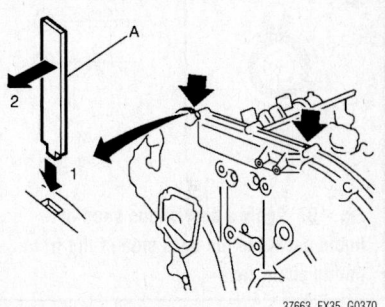

Fig. 196 Insert a suitable tool (A) into the notch at the top of front timing chain case as shown

b. Remove power steering pipe mounting bolt.

c. Remove rear cover plate and set the ring gear stopper KV10118600 (J-48641) as shown.

d. Loosen crankshaft pulley bolt and rotate bolt seating surface at 0.39 inches (10 mm) from its original position.

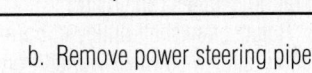 **CAUTION**

Never remove crankshaft pulley bolt as it will be used as a supporting point for suitable puller.

e. Place suitable puller tab on holes of crankshaft pulley, and pull crankshaft pulley through.

⁂ **CAUTION**

Never put suitable puller tab on crankshaft pulley periphery, as this will damage internal damper.

26. Remove lower oil pan.
27. Loosen two mounting bolts in front of upper oil pan in reverse order as shown.
28. Remove front timing chain case as per the following:

a. Loosen mounting bolts in reverse order as shown.

b. Insert a suitable tool into the notch at the top of front timing chain case as shown.

c. Pry off case by moving the suitable tool as shown.

d. Use the seal cutter KV10111100 (J-37228) to cut liquid gasket for removal.

⁂ **CAUTION**

Note the following:

- Never use a screwdriver or something similar.
- After removal, handle front timing chain case carefully so it does not tilt, cant, or warp under a load.

To install:

29. Install front timing chain case as per the following:

a. Check O-rings stay in place during installation to rear timing chain case.

b. Apply a continuous bead of liquid gasket to the back side of the front timing chain case as shown.

c. Apply liquid gasket to top surface of upper oil pan as shown.

d. Install front timing chain case.

⁂ **CAUTION**

Note the following:

- Be careful not to damage front oil seal by interference with front end of crankshaft.
- Attaching must be done within 5 minutes after liquid gasket application.

e. Install front timing chain case as to fit its dowel pin hole together dowel pin on rear timing chain case.

f. Tighten mounting bolts to the specified torque in numerical order as shown.

g. There are two types of mounting bolts.

- M10 bolts: Bolt positions 1, 2, 3, 4, 5, 6, 7: Torque to 41 ft. lbs. (55 Nm)
- M6 bolts: Bolt positions: All except above: Torque to 9 ft. lbs. (13 Nm)

h. After all bolts are tightened, retighten them to the specified torque in numerical order shown.

⁂ **CAUTION**

Be sure to wipe out any excessive liquid gasket leaking on surface mating with upper oil pan.

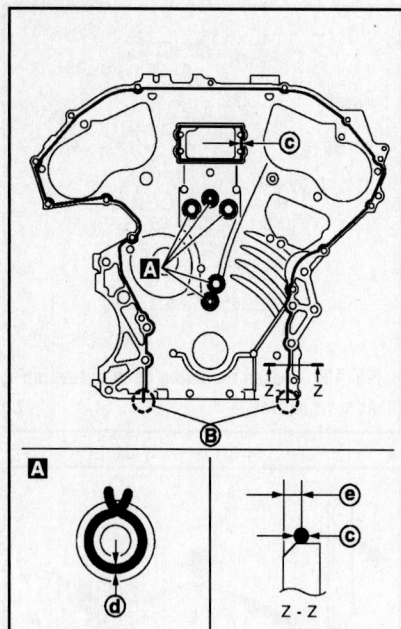

Fig. 197 Apply a continuous bead of liquid gasket to the back side of the front timing chain case

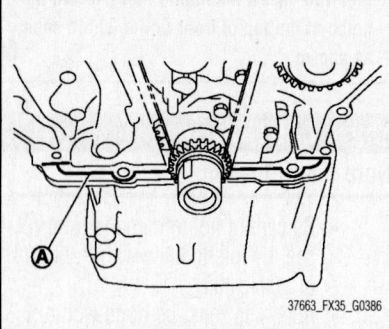

Fig. 198 Apply liquid gasket to top surface of upper oil pan as shown

i. Install two mounting bolts in front of upper oil pan.

30. Install valve timing control covers (bank 1 and bank 2) as per the following:

a. Install new seal rings in shaft grooves for Bank 2.

✳✳ CAUTION

When replacing seal rings, replace all rings with new ones.

b. To check the joint between dowel pins and dowel pin holes, check the looseness in the axle direction by pushing the circumferential looseness (between dowel pins and dowel pin holes) by twisting in the circumferential direction.

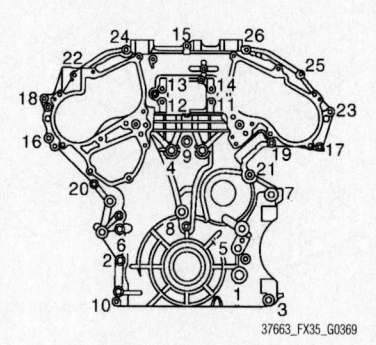

Fig. 199 Tighten mounting bolts in numerical order as shown

✳✳ CAUTION

Always perform this procedure when removing because the gap between dowel pins and dowel pin holes may not be caused on purpose.

c. Install valve timing control cover with new gasket to front timing chain case.

✳✳ CAUTION

Note the following:

- Never face the magnet retarder side down to prevent magnet retarder from dropping.
- Check the mating surface of magnet retarder and the drum of exhaust side camshaft sprocket for foreign materials.
- Align the center of both shaft holes of the shaft and the intake side camshaft sprocket, and then insert them.
- Be careful not to drop the seal ring from the shaft groove.
- When setting the valve timing control cover in position by hand, if valve timing control cover is not contacting with the front timing chain case, the dowel pin of magnet retarder may not be aligned with the dowel pin holes of cover.

d. Being careful not to move seal ring from the installation groove, align dowel pins on front timing chain case with holes to install valve timing control covers.

e. Tighten mounting bolts in numerical order as shown.

f. After all bolts are tightened, tighten No. 1 bolt to the specified torque again.

31. Install lower oil pan.

32. Install valve covers (bank 1 and bank 2).

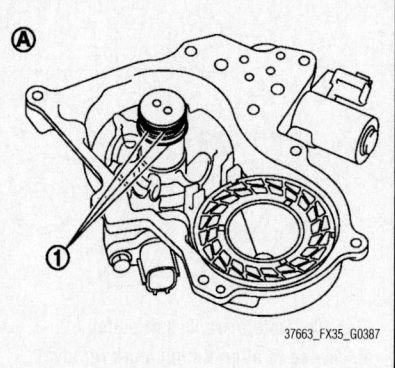

Fig. 200 Install new seal rings (1) in shaft grooves for Bank 2 (A)

33. Install crankshaft pulley as per the following:

a. Secure crankshaft using the ring gear stopper KV10118600 (J-48641).

b. Install crankshaft pulley, taking care not to damage front oil seal.

c. When press-fitting crankshaft pulley with plastic hammer, tap on its center portion (not circumference).

d. Tighten crankshaft pulley bolt.

e. Place a matching mark on crankshaft pulley aligning with the matching mark of crankshaft pulley bolt. Tighten the bolt 90 degrees (one mark).

f. Rotate crankshaft pulley in normal direction (clockwise when viewed from front) to confirm it turns smoothly.

34. Install drive belt auto-tensioner bracket and power steering oil pump bracket as per the following:

a. Install drive belt auto-tensioner bracket, and tighten mounting bolts No. 2, 3. (temporarily)

b. Tighten mounting bolts No. 2, 3. (specified torque)

c. Install power steering oil pump bracket, and tighten mounting bolts No. 1, 4, 5. (temporarily)

d. Tighten mounting bolts No. 1. (specified torque)

e. Tighten mounting bolts No. 4, 5. (specified torque)

35. For the following operations, perform steps in the reverse order of removal.

36. Before starting engine, check oil/fluid levels including engine coolant and engine oil. If less than required quantity, fill to the specified level.

37. Use procedure below to check for fuel leakage.

a. Turn ignition switch "ON" (with engine stopped). With fuel pressure applied to fuel piping, check for fuel leakage at connection points.

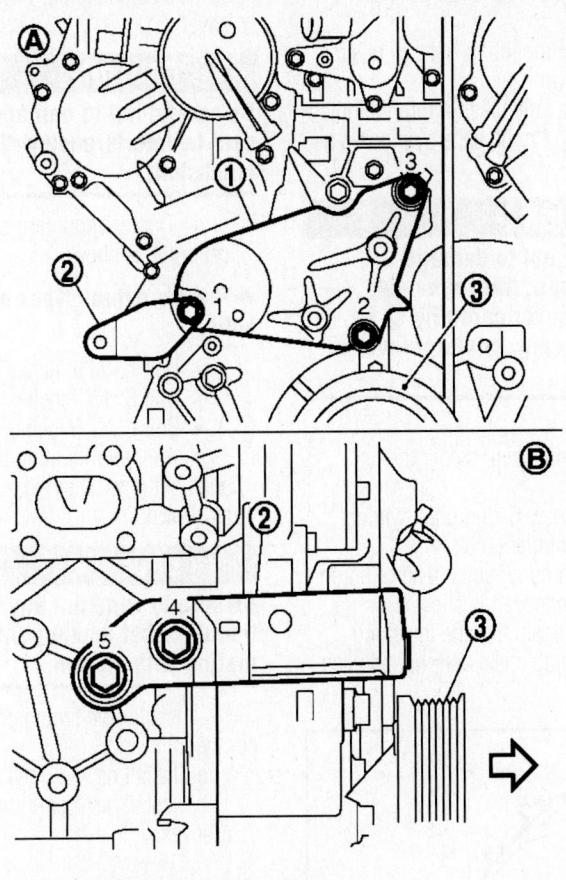

A. Engine front side
B. engine right side
1. Drive belt auto-tensioner
2. Power steering oil pump bracket
3. Crankshaft pulley

37663_FX35_G0388

Fig. 201 Install drive belt auto-tensioner bracket and power steering oil pump bracket

b. Start engine. With engine speed increased, check again for fuel leakage at connection points.

38. Run engine to check for unusual noise and vibration.

➡ **If hydraulic pressure inside chain tensioner drops after removal/installation, slack in guide may generate a pounding noise during and just after the engine start. However, this does not indicate an unusualness. Noise will stop after hydraulic pressure rises.**

39. Warm up engine thoroughly to check there is no leakage of fuel, or any oil/fluids including engine oil and engine coolant.

40. Bleed air from lines and hoses of applicable lines, such as in cooling system.

41. After cooling down engine, again check oil/fluid levels including engine oil and engine coolant. Refill to the specified level, if necessary.

5.0L Engine
See Figures 202 through 210.

1. Remove auto tensioners and idler pulley.

2. Remove oil level gauge and oil level gauge guide.

3. Remove alternator bracket and alternator stay.

4. Remove camshaft position sensors.

✳✳ CAUTION

Note the following:

• Handle carefully to avoid dropping and shocks.

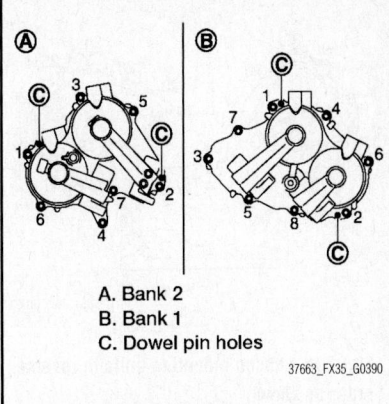

A. Bank 2
B. Bank 1
C. Dowel pin holes

37663_FX35_G0390

Fig. 202 Loosen mounting bolts in the reverse order as shown

• Never disassemble.
• Never allow metal powder to adhere to magnetic part at sensor tip.
• Never place sensors in a location where they are exposed to magnetism.

5. Remove valve timing control cover as per the following:

a. Disconnect valve timing control solenoid valve harness connector.

b. Loosen mounting bolts in the reverse order as shown.

✳✳ CAUTION

Exercise care not to damage mating surfaces. Shaft is internally jointed with camshaft sprocket center hole. When removing, keep it horizontal until it is completely disconnected.

6. Remove valve timing control solenoid valve (INT and EXH), if necessary.

✳✳ CAUTION

Valve timing control solenoid valve is not reusable. Never remove it unless required.

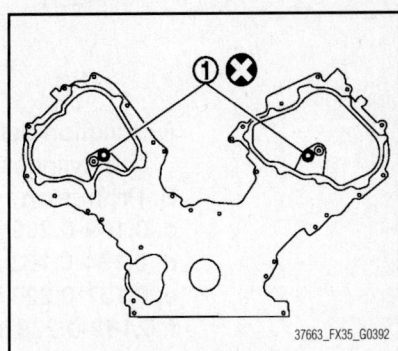

37663_FX35_G0392

Fig. 203 Remove O-rings (1) from front cover

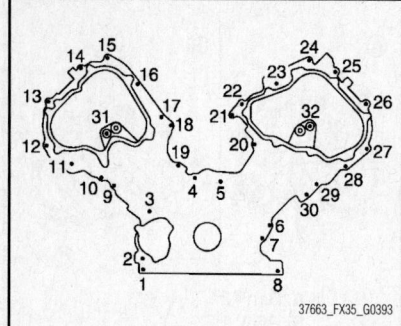

Fig. 204 Loosen mounting bolts in reverse order as shown

7. Remove O-rings from front cover.
8. Remove valve covers.
9. Obtain No. 1 cylinder at TDC of its compression stroke.
10. Remove crankshaft pulley.
11. Remove water pump pulley.
12. Remove lower oil pan and oil strainer.
13. Remove upper oil pan.

14. Remove front cover as per the following:
    a. Loosen mounting bolts in reverse order as shown.
    b. Insert a suitable tool into the notch at front cover. Pry off case by moving a suitable tool.

> ※※ **CAUTION**
>
> **Exercise care not to damage mating surfaces. After removal, handle front cover carefully so it does not tilt, cant, or warp under a load.**

To install:

15. Install front cover as per the following:
    a. Install new O-ring onto cylinder heads and cylinder block.
    b. Apply a continuous bead of liquid gasket to front cover as shown.
    c. Check again that the matching marks on timing chain and that on each

sprocket are aligned. Then, install front cover.

> ※※ **CAUTION**
>
> **Be careful not to damage front oil seal by interference with front end of crankshaft.**

    d. Tighten mounting bolts in numerical order as shown.

➡ **There are three types of mounting bolts.**

- Bolt A: 0.79 inches (20 mm)
- Bolt B: 1.77 inches (45 mm)
- Bolt C: 3.15 inches (80 mm)

    e. After all mounting bolts are tightened, retighten them in numerical order as shown.

> ※※ **CAUTION**
>
> **Be sure to wipe out any excessive liquid gasket leaking onto surface mating with oil pan.**

16. Install valve timing control cover as per the following:
    a. Install new O-rings on front cover.
    b. Install new seal rings in shaft grooves.

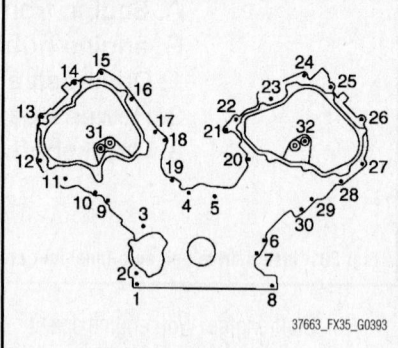

Fig. 206 Tighten mounting bolts in numerical order as shown

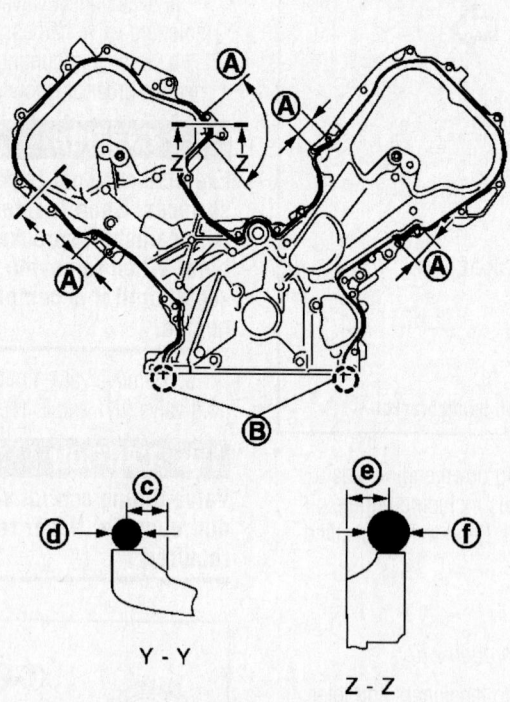

A. Junction between cylinder block and cylinder head
B. Protrusion
c. 0.169-0.209 inches (4.3-5.3 mm)
d. 0.134-0.173 inches (3.4-4.4 mm)
e. 0.157-0.220 inches (4.0-5.6 mm)
f. 0.189-0.228 inches (4.8-5.8 mm)

Fig. 205 Apply a continuous bead of liquid gasket to front cover as shown

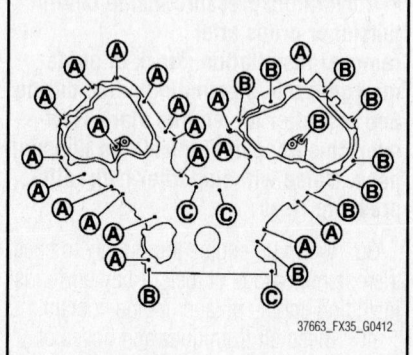

Fig. 207 Proper bolt locations (A, B, C)

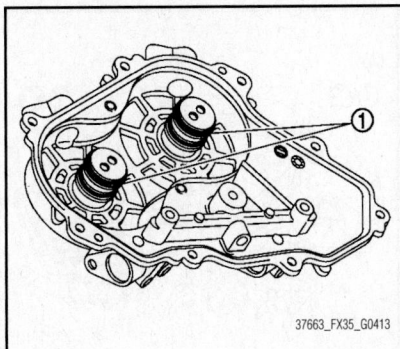

Fig. 208 Install new seal rings (1) in shaft grooves

✳✳ CAUTION

When replacing seal ring, replace all rings with new ones.

c. Apply a continuous bead of liquid gasket to valve timing control covers as shown.

d. Being careful not to move seal ring from the installation groove, align dowel pins on front cover with dowel pin holes to install valve timing control covers.

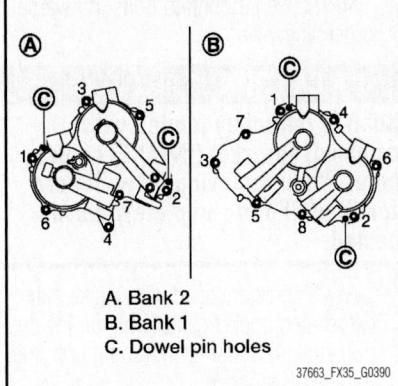

A. Bank 2
B. Bank 1
C. Dowel pin holes

Fig. 210 Tighten mounting bolts in numerical order as shown

e. Tighten mounting bolts in numerical order as shown.

17. Install camshaft position sensor and valve timing control solenoid valve (RH and LH) to valve timing control cover, if removed. Be sure to tighten mounting bolts with flanges completely seated.

18. Install lower oil pan and oil strainer.

19. Install upper oil pan.

20. Install water pump pulley.

21. Install crankshaft pulley.

a. Install crankshaft pulley, taking care not to damage front oil seal.

b. Apply engine oil onto threaded parts of crankshaft pulley bolt and seating area. Lightly tapping its center with plastic hammer, insert crankshaft pulley.

✳✳ CAUTION

Never tap crankshaft pulley on the side surface where belt is installed (outer circumference).

c. Tighten crankshaft pulley bolt.

d. Put a paint mark on crankshaft pulley aligning with angle mark on crankshaft pulley bolt.

e. Tighten crankshaft pulley bolt (clockwise). Check the tightening angle by referencing to the notches. The angle between two notches is 90 degrees.

22. Rotate crankshaft pulley in normal direction (clockwise when viewed from engine front) to confirm it turns smoothly.

23. Install in the reverse order of removal.

24. Before starting engine, check oil/fluid levels including engine coolant and engine oil. If any are less than the required quantity, fill them to the specified level.

25. Follow the procedure below to check for fuel leakage.

a. Turn ignition switch to the "ON" position (with engine stopped). With fuel pressure applied to fuel piping, check for fuel leakage at connection points.

b. Start engine. With engine speed increased, check again for fuel leakage at connection points.

26. Run engine to check for unusual noise and vibration.

➡**If hydraulic pressure inside chain tensioner drops after removal/installation, slack in guide may generate a pounding noise during and just after the engine start. However, this does not indicate a malfunction. The noise will stop after hydraulic pressure rises.**

27. Warm up engine thoroughly to check that there is no leakage of fuel, or any oil/fluids including engine oil and engine coolant.

28. Bleed air from lines and hoses of applicable lines, such as in cooling system.

29. After cooling down engine, again check oil/fluid levels including engine oil and engine coolant. Refill them to the specified level, if necessary.

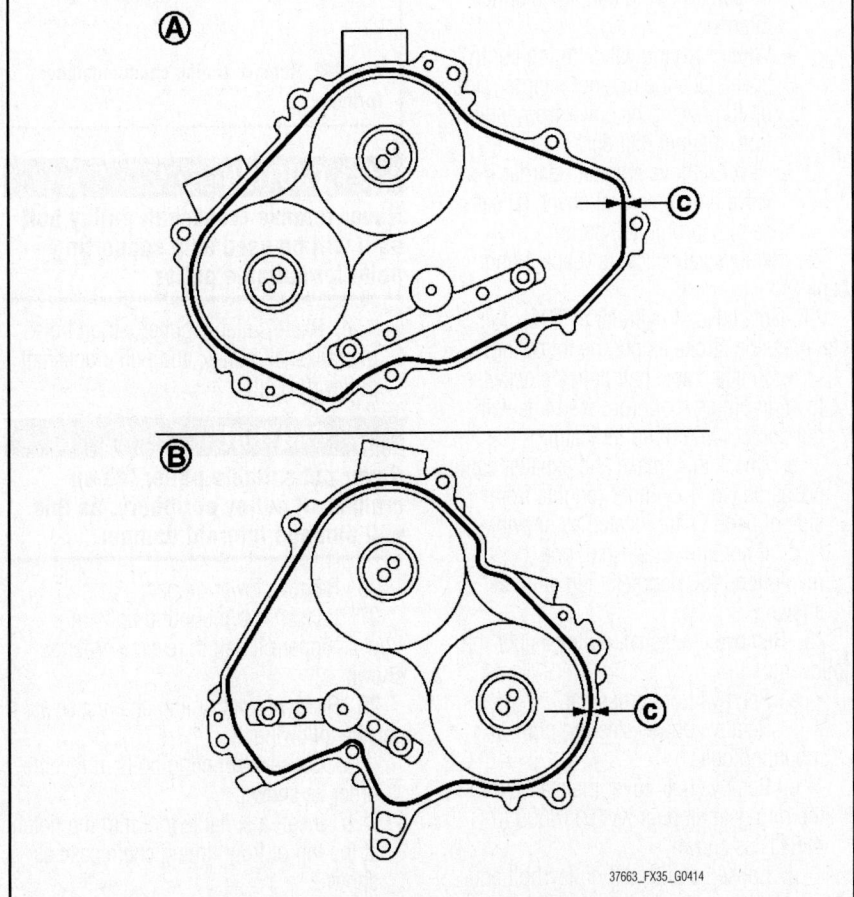

Fig. 209 Apply a continuous bead of liquid gasket to valve timing control covers as shown

TIMING CHAIN & SPROCKETS

REMOVAL & INSTALLATION

3.5L Engine

See Figures 189 through 201, 211 through 225.

1. Release the fuel pressure.
2. Disconnect the battery cable from the negative terminal.
3. Remove engine cover.
4. Remove radiator reservoir tank.
5. Remove air duct and air cleaner case assembly.
6. Remove engine undercover.
7. Drain engine coolant from radiator.

✳✳ CAUTION
Perform this step when the engine is cold. Never spill engine coolant on drive belt.

8. Remove upper and lower radiator hoses.
9. Drain engine oil.

✳✳ CAUTION
Perform this step when the engine is cold. Never spill engine oil on drive belt.

10. Remove drive belt.
11. Remove radiator cooling fan assembly.
12. Separate engine harnesses removing their brackets from front timing chain case.
13. Remove intake manifold collector.
14. Remove intake manifold.
15. Remove oil level gauge and oil level gauge guide.
16. Remove A/C compressor from bracket with piping connected, and temporarily secure it aside.
17. Remove power steering oil pump from bracket with piping connected, and temporarily secure it aside.
18. Remove power steering oil pump bracket.
19. Remove idler pulley, auto tensioner and bracket.
20. Remove alternator and alternator bracket.
21. Remove water outlet (front) and water piping.
22. Remove valve timing control covers (bank 1 and bank 2) and gasket as per the following:
    a. Disconnect valve timing control harness connector

b. Loosen mounting bolts in reverse order as shown.

✳✳ CAUTION
Shaft is internally jointed with camshaft sprocket (INTAKE) center hole. When removing, keep it horizontal until it is completely disconnected.

c. Shaft is engaged with intake side camshaft sprocket center hole on inside. pull straight out so as not to tilt until the joint is disengaged.

d. The mating surface of magnet retarder may be fitted with the exhaust side camshaft sprocket via the engine oil. Open valve timing control cover carefully.

e. If the mating surface of magnet retarder is fitted with the camshaft sprocket, open the cover within the range that the load is not applied to the harness. And then, remove it so as to prevent magnet retarder from dropping.

✳✳ CAUTION
Note the following:

- Be careful not to damage magnet retarder.
- When carrying valve timing control cover, face the magnet retarder side up to prevent the cover from falling from magnet retarder.
- Never remove magnet retarder from valve timing control cover. (Disassembly prohibited parts)

23. Remove valve covers (bank 1 and bank 2).
24. Obtain No. 1 cylinder at TDC of its compression stroke as per the following:
    a. Rotate crankshaft pulley clockwise to align timing mark (grooved line without color) with timing indicator.
    b. Check that intake and exhaust cam noses on No. 1 cylinder (engine front side of bank 1) are located as shown.
    c. If not, turn crankshaft one revolution (360 degrees) and align as shown.
25. Remove crankshaft pulley as per the following:
    a. Remove front cross bar.
    b. Remove power steering pipe mounting bolt.
    c. Remove rear cover plate and set the ring gear stopper KV10118600 (J-48641) as shown.
    d. Loosen crankshaft pulley bolt and rotate bolt seating surface at 0.39 inches (10 mm) from its original position.

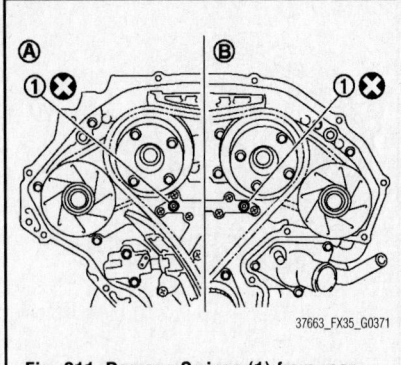

Fig. 211 Remove O-rings (1) from rear timing chain case

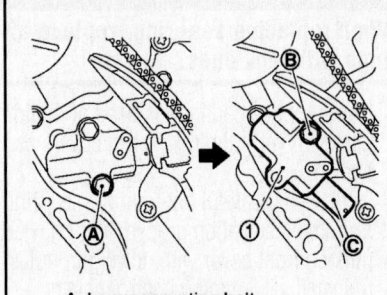

A. Lower mounting bolt
B. Upper mounting bolt
C. Plunger
1. Timing chain tensioner (primary)

Fig. 212 Remove timing chain tensioner (primary)

✳✳ CAUTION
Never remove crankshaft pulley bolt as it will be used as a supporting point for suitable puller.

e. Place suitable puller tab on holes of crankshaft pulley, and pull crankshaft pulley through.

✳✳ CAUTION
Never put suitable puller tab on crankshaft pulley periphery, as this will damage internal damper.

26. Remove lower oil pan.
27. Loosen two mounting bolts in front of upper oil pan in reverse order as shown.
28. Remove front timing chain case as per the following:
    a. Loosen mounting bolts in reverse order as shown.
    b. Insert a suitable tool into the notch at the top of front timing chain case as shown.
    c. Pry off case by moving the suitable tool as shown.

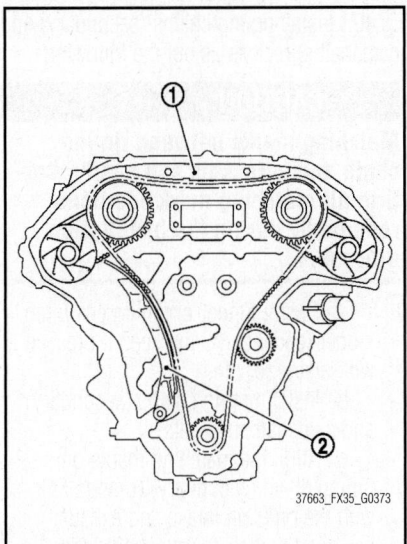

Fig. 213 Remove internal chain guide (1), and slack guide (2)

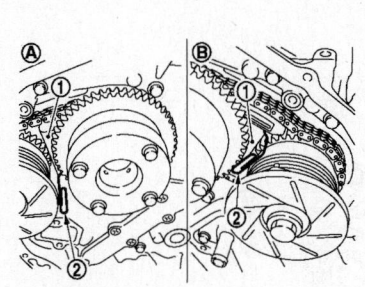

A. Bank 1
B. Bank 2
1. Timing chain tensioners (secondary)
2. Stopper pin

37663_FX35_G0374

Fig. 214 Remove timing chain (secondary) and camshaft sprockets

d. Use the seal cutter KV10111100 (J-37228) to cut liquid gasket for removal.

☀ CAUTION

Note the following:

- Never use a screwdriver or something similar.
- After removal, handle front timing chain case carefully so it does not tilt, cant, or warp under a load.

29. Remove front oil seal from front timing chain case using a suitable tool.

☀ CAUTION

Be careful not to damage front timing chain case.

30. Remove O-rings from rear timing chain case.

31. Remove timing chain tensioner (primary) as per the following:
a. Remove lower mounting bolt.
b. Loosen upper mounting bolt slowly, and then turn timing chain tensioner (primary) on the upper mounting bolt so that plunger is fully expanded.

➡**Even if plunger is fully expanded, it is not dropped from the body of timing chain tensioner (primary).**

c. Remove upper mounting bolt, and then remove timing chain tensioner (primary).

32. Remove internal chain guide (1), and slack guide (2).

33. Remove timing chain (primary) and crankshaft sprocket.

☀☀ CAUTION

After removing timing chain tensioner (primary), never turn crankshaft and camshaft separately, or valves will strike the piston heads.

34. Remove timing chain (secondary) and camshaft sprockets as per the following:
a. Attach suitable stopper pin to the timing chain tensioners (secondary).

☀☀ CAUTION

Note the following:

- Use hard metal pin as a stopper pin.
- For removal of timing chain tensioners (secondary), refer to "Exploded View". Removing camshaft bracket (No. 1) is required.

b. Remove camshaft sprocket mounting bolts (INT and EXH).
c. Secure the hexagonal portion of camshaft using a wrench to loosen mounting bolts.
d. Remove timing chain (secondary) together with camshaft sprockets.

☀☀ CAUTION

Note the following:

- Never loosen the mounting bolts with securing anything other than the camshaft hexagonal portion or with tensioning the timing chain.
- Never disassemble. Never loosen bolts as shown.

35. Remove timing chain tensioners (secondary) from cylinder head as per the following, if necessary.
a. Remove camshaft brackets (No. 1).
b. Remove timing chain tensioners (secondary) with a stopper pin attached.

36. Use a scraper to remove all traces of old liquid gasket from front and rear timing chain cases and upper oil pan, and liquid gasket mating surfaces.

☀☀ CAUTION

Be careful not to allow gasket fragments to enter oil pan.

37. Remove old liquid gasket from bolt hole and thread.

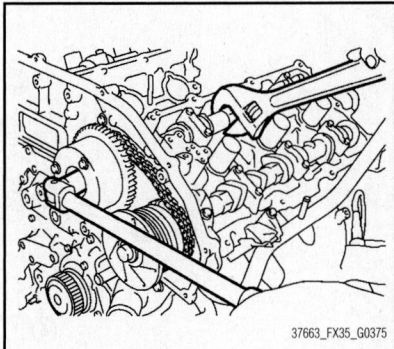

37663_FX35_G0375

Fig. 215 Secure the hexagonal portion of camshaft using a wrench to loosen mounting bolts

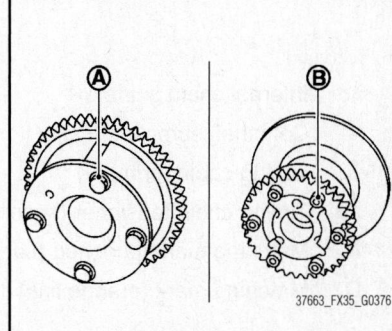

37663_FX35_G0376

Fig. 216 Never loosen bolts (A) and (B) as shown

To install:

➡The below figure shows the relationship between the matching mark on each timing chain and that on the corresponding sprocket, with the components installed.

38. Install timing chain tensioners (secondary) to cylinder head if removed.

39. Check that dowel pin and crankshaft key are located as shown. (No. 1 cylinder at compression TDC)

➡Though camshaft does not stop at the position as shown, for the placement of cam noses, it is generally accepted camshaft is placed the same direction as the figure.

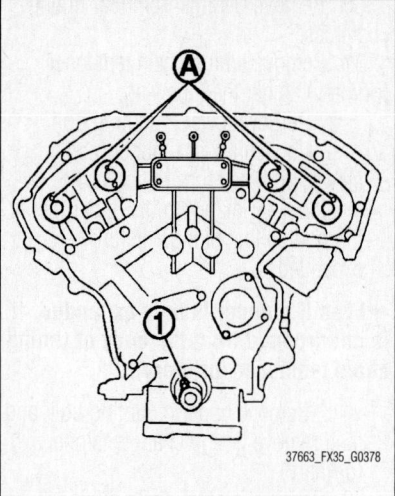

37663_FX35_G0378

Fig. 218 Check that dowel pin (A) and crankshaft key (1) are located as shown

40. Install timing chains (secondary) and camshaft sprockets as per the following:

✳✳ CAUTION

Matching marks between timing chain and sprockets slip easily. Confirm all matching mark positions repeatedly during the installation process.

a. Push plunger of timing chain tensioner (secondary) and keep it pressed in with a stopper pin.

b. Install timing chains (secondary) and camshaft sprockets.

c. Align the matching marks on timing chain (secondary) (orange link) with the ones on intake and exhaust camshaft sprockets (punched), and install them.

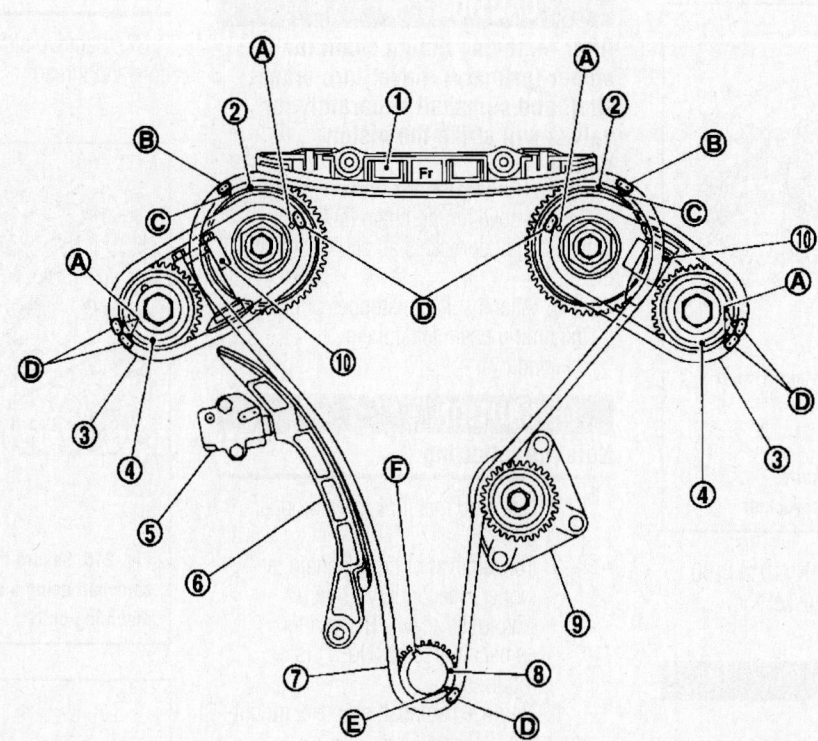

1. Internal chain guide	2. Camshaft sprocket (INT)	3. Timing chain (secondary)
4. Camshaft sprocket (EXH)	5. Timing chain tensioner (primary)	6. Slack guide
7. Timing chain (primary)	8. Crankshaft sprocket	9. Water pump
10. Timing chain tensioner (secondary)		
A. Matching mark [punched (back side)]	B. Matching mark (yellow link)	C. Matching mark (punched)
D. Matching mark (orange link)	E. Matching mark (notched)	F. Crankshaft key

37663_FX35_G0377

Fig. 217 Matching marks on timing chains and sprockets

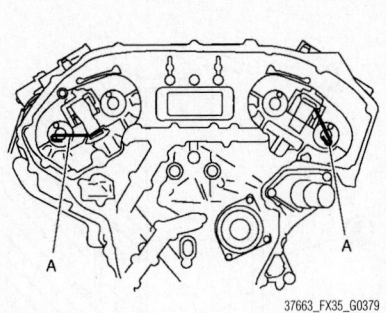

Fig. 219 Push plunger of timing chain tensioner (secondary) and keep it pressed in with a stopper pin (A)

✳✳ CAUTION

Note the following:

- Figure shows bank 1 (rear view).
- Matching marks for camshaft sprockets are on the back side of camshaft sprockets (secondary).
- There are two types of matching marks, circle and oval types. They should be used for the bank 1 and bank 2, respectively.

d. Align dowel pin camshafts with the groove or dowel hole on sprockets, and install them.

e. On the intake side, align dowel pin on camshaft front end with pin groove on the back side of camshaft sprocket, and install them.

f. On the exhaust side, align dowel pin on camshaft front end with pin hole on camshaft sprocket, and install them.

g. In case that positions of each matching mark and each dowel pin are not fit on matching parts, make fine adjustment to the position holding the hexagonal portion on camshaft with wrench or an equivalent.

h. Mounting bolts for camshaft sprockets must be tightened in the next step. Tightening them by hand is enough to prevent the dislocation of dowel pins.

i. Check the matching marks (punched) on each camshaft sprocket are positioned on the matching marks (orange link) on timing chain (secondary).

➡ **Matching mark (punched) is for checking loose at this step.**

j. After confirming the matching marks are aligned, tighten camshaft sprocket mounting bolts.

k. Secure camshaft using a wrench at the hexagonal portion to tighten mounting bolts.

l. Pull stopper pins out from timing chain tensioners (secondary).

41. Install timing chain (primary) as per the following:

a. Install crankshaft sprocket. Check the matching marks on crankshaft sprocket face the front of the engine.

b. Install timing chain (primary).

c. Install timing chain (primary) so the matching mark (punched) on camshaft sprocket (INT) is aligned with the yellow link on timing chain, while the

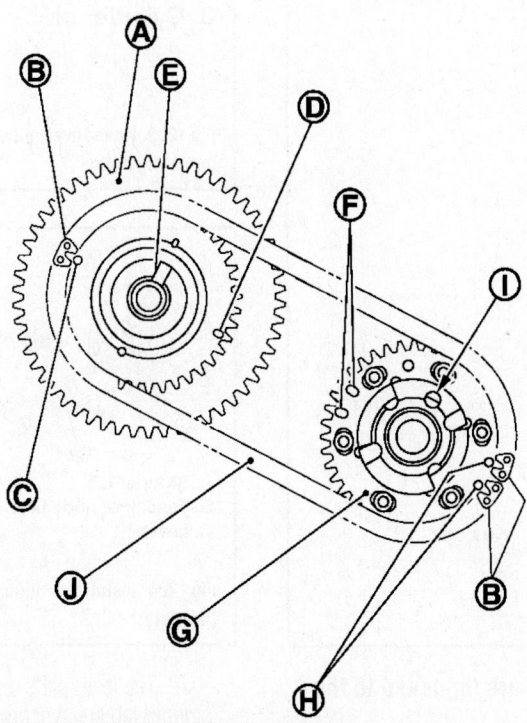

A. Camshaft sprocket (INT) back face
B. Orange link
C. Matching mark (Circle)
D. Matching mark (Oval)
E. Dowel groove
F. Matching mark (2 oval)
G. Camshaft sprocket (EXH) back face
H. Matching mark (2 circle)
I. Dowel hole
J. Timing chain (secondary)

Fig. 220 Align the matching marks on timing chain (secondary) (orange link) with the ones on intake and exhaust camshaft sprockets (punched)

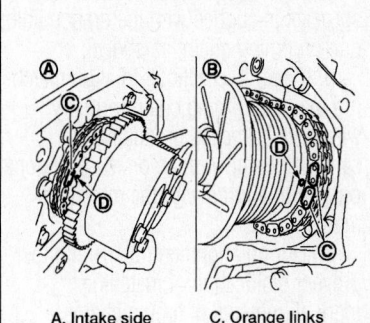

A. Intake side   C. Orange links
B. Exhaust side   D. Punched mark

Fig. 221 Check the matching marks (punched) on each camshaft sprocket are positioned on the matching marks (orange link) on timing chain (secondary)

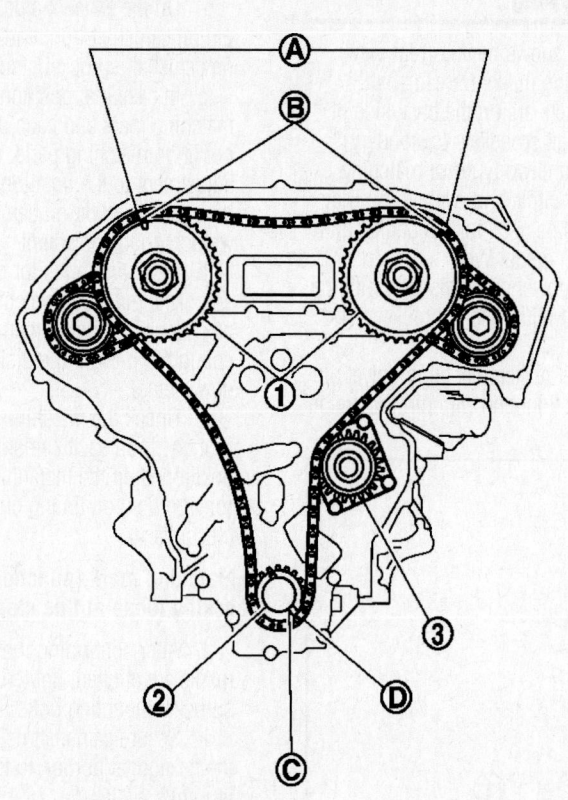

A. Yellow link
B. Punched mark
C. Notched mark
D. Orange link
1. Camshaft sprocket (INTAKE)
2. Crankshaft sprocket

37663_FX35_G0382

Fig. 222 Install timing chain (primary)

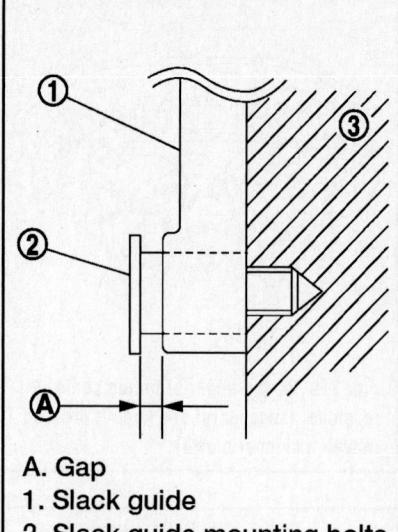

A. Gap
1. Slack guide
2. Slack guide mounting bolts
3. Cylinder block

37663_FX35_G0383

Fig. 223 Never overtighten slack guide mounting bolts

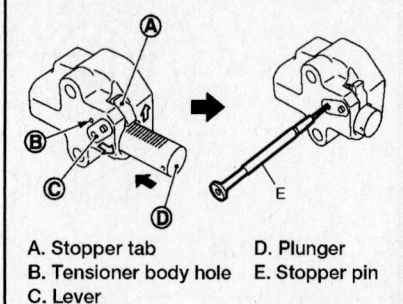

A. Stopper tab   D. Plunger
B. Tensioner body hole E. Stopper pin
C. Lever

37663_FX35_G0384

Fig. 224 Install the timing chain tensioner (primary)

matching mark (notched) on crankshaft sprocket is aligned with the orange link one on timing chain, as shown.

d. When it is difficult to align matching marks of timing chain (primary) with each sprocket, gradually turn camshaft using wrench on the hexagonal portion to align it with the matching marks.

e. During alignment, be careful to prevent dislocation of matching mark alignments of timing chains (secondary).

42. Install internal chain guide and slack guide.

✳✳ CAUTION

Never overtighten slack guide mounting bolts. It is normal for a gap to exist under the bolt seats when mounting bolts are tightened to the specification.

43. Install the timing chain tensioner (primary) with the following procedure:

a. Pull plunger stopper tab up (or turn lever downward) so as to remove plunger stopper tab from the ratchet of plunger.

➡**Plunger stopper tab and lever are synchronized.**

b. Push plunger into the inside of tensioner body.

c. Hold plunger in the fully compressed position by engaging plunger stopper tab with the tip of ratchet.

d. To secure lever, insert stopper pin through hole of lever into tensioner body hole.

e. The lever parts and the plunger stopper tab are synchronized. Therefore, the plunger will be secured under this condition.

➡**Figure shows the example of thin screwdriver being used as the stopper pin.**

f. Install timing chain tensioner (primary).

➡**Remove any dirt and foreign materials completely from the back and the mounting surfaces of timing chain tensioner (primary).**

g. Pull out stopper pin after installing, and then release plunger.

44. Check again that the matching marks on sprockets and timing chain have not slipped out of alignment.

45. Install new O-rings on rear timing chain case.

46. Install new front oil seal on front timing chain case.

a. Apply new engine oil to both oil seal lip and dust seal lip.

b. Install it so that each seal lip is oriented as shown.

c. Using a suitable drift, press-fit oil seal until it becomes flush with front timing chain case end face.

d. Check the garter spring is in position and seal lip is not inverted.

47. Install front timing chain case as per the following:

a. Check O-rings stay in place during installation to rear timing chain case.

b. Apply a continuous bead of liquid gasket to the back side of the front timing chain case as shown.

c. Apply liquid gasket to top surface of upper oil pan as shown.

d. Install front timing chain case.

❋❋ CAUTION

Note the following:

- Be careful not to damage front oil seal by interference with front end of crankshaft.
- Attaching must be done within 5 minutes after liquid gasket application.

e. Install front timing chain case as to fit its dowel pin hole together dowel pin on rear timing chain case.

f. Tighten mounting bolts to the specified torque in numerical order as shown.

g. There are two types of mounting bolts.

- M10 bolts: Bolt positions 1, 2, 3, 4, 5, 6, 7: Torque to 41 ft. lbs. (55 Nm)
- M6 bolts: Bolt positions: All except above: Torque to 9 ft. lbs. (13 Nm)

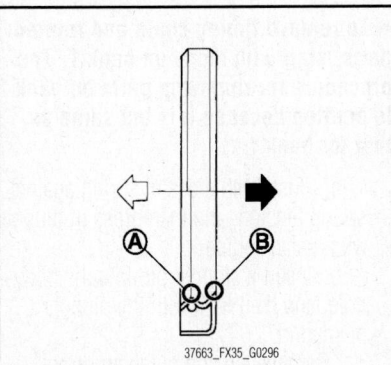

37663_FX35_G0296

Fig. 225 Install it so that each seal lip is oriented as shown

h. After all bolts are tightened, retighten them to the specified torque in numerical order shown.

❋❋ CAUTION

Be sure to wipe out any excessive liquid gasket leaking on surface mating with upper oil pan.

i. Install two mounting bolts in front of upper oil pan.

48. Install valve timing control covers (bank 1 and bank 2) as per the following:

a. Install new seal rings in shaft grooves for Bank 2.

❋❋ CAUTION

When replacing seal rings, replace all rings with new ones.

b. To check the joint between dowel pins and dowel pin holes, check the looseness in the axle direction by pushing the circumferential looseness (between dowel pins and dowel pin holes) by twisting in the circumferential direction.

❋❋ CAUTION

Always perform this procedure when removing because the gap between dowel pins and dowel pin holes may not be caused on purpose.

c. Install valve timing control cover with new gasket to front timing chain case.

❋❋ CAUTION

Note the following:

- Never face the magnet retarder side down to prevent magnet retarder from dropping.
- Check the mating surface of magnet retarder and the drum of exhaust side camshaft sprocket for foreign materials.
- Align the center of both shaft holes of the shaft and the intake side camshaft sprocket, and then insert them.
- Be careful not to drop the seal ring from the shaft groove.
- When setting the valve timing control cover in position by hand, if valve timing control cover is not contacting with the front timing chain case, the dowel pin of magnet retarder may not be aligned with the dowel pin holes of cover.

d. Being careful not to move seal ring from the installation groove, align dowel pins on front timing chain case with holes to install valve timing control covers.

e. Tighten mounting bolts in numerical order as shown.

f. After all bolts are tightened, tighten No. 1 bolt to the specified torque again.

49. Install lower oil pan.

50. Install valve covers (bank 1 and bank 2).

51. Install crankshaft pulley as per the following:

a. Secure crankshaft using the ring gear stopper KV10118600 (J-48641).

b. Install crankshaft pulley, taking care not to damage front oil seal.

c. When press-fitting crankshaft pulley with plastic hammer, tap on its center portion (not circumference).

d. Tighten crankshaft pulley bolt.

e. Place a matching mark on crankshaft pulley aligning with the matching mark of crankshaft pulley bolt. Tighten the bolt 90 degrees (one mark).

f. Rotate crankshaft pulley in normal direction (clockwise when viewed from front) to confirm it turns smoothly.

52. Install drive belt auto-tensioner bracket and power steering oil pump bracket as per the following:

a. Install drive belt auto-tensioner bracket, and tighten mounting bolts No. 2, 3. (temporarily)

b. Tighten mounting bolts No. 2, 3. (specified torque)

c. Install power steering oil pump bracket, and tighten mounting bolts No. 1, 4, 5. (temporarily)

d. Tighten mounting bolts No. 1. (specified torque)

e. Tighten mounting bolts No. 4, 5. (specified torque)

53. For the following operations, perform steps in the reverse order of removal.

54. Before starting engine, check oil/fluid levels including engine coolant and engine oil. If less than required quantity, fill to the specified level.

55. Use procedure below to check for fuel leakage.

a. Turn ignition switch "ON" (with engine stopped). With fuel pressure applied to fuel piping, check for fuel leakage at connection points.

b. Start engine. With engine speed increased, check again for fuel leakage at connection points.

56. Run engine to check for unusual noise and vibration.

➡️**If hydraulic pressure inside chain tensioner drops after removal/installation, slack in guide may generate a pounding noise during and just after the engine start. However, this does not indicate an unusualness. Noise will stop after hydraulic pressure rises.**

57. Warm up engine thoroughly to check there is no leakage of fuel, or any oil/fluids including engine oil and engine coolant.

58. Bleed air from lines and hoses of applicable lines, such as in cooling system.

59. After cooling down engine, again check oil/fluid levels including engine oil and engine coolant. Refill to the specified level, if necessary.

5.0L Engine

See Figures 202 through 206, 208 through 210, 226 through 243.

1. Remove auto tensioners and idler pulley.

2. Remove oil level gauge and oil level gauge guide.

3. Remove alternator bracket and alternator stay.

4. Remove camshaft position sensors.

❄️ **CAUTION**

Note the following:

- Handle carefully to avoid dropping and shocks.
- Never disassemble.
- Never allow metal powder to adhere to magnetic part at sensor tip.
- Never place sensors in a location where they are exposed to magnetism.

5. Remove valve timing control cover as per the following:

   a. Disconnect valve timing control solenoid valve harness connector.

   b. Loosen mounting bolts in the reverse order as shown.

❄️ **CAUTION**

Exercise care not to damage mating surfaces. Shaft is internally jointed with camshaft sprocket center hole. When removing, keep it horizontal until it is completely disconnected.

6. Remove valve timing control solenoid valve (INT and EXH), if necessary.

❄️ **CAUTION**

Valve timing control solenoid valve is not reusable. Never remove it unless required.

7. Remove O-rings from front cover.

8. Remove valve covers.

9. Obtain No. 1 cylinder at TDC of its compression stroke.

10. Remove crankshaft pulley.

11. Remove water pump pulley.

12. Remove lower oil pan and oil strainer.

13. Remove upper oil pan.

14. Remove front cover as per the following:

   a. Loosen mounting bolts in reverse order as shown.

   b. Insert a suitable tool into the notch at front cover. Pry off case by moving a suitable tool.

❄️ **CAUTION**

Exercise care not to damage mating surfaces. After removal, handle front cover carefully so it does not tilt, cant, or warp under a load.

15. Remove front oil seal from front cover using suitable tool.

❄️ **CAUTION**

Be careful not to damage front cover.

16. Remove O-rings from cylinder heads and cylinder block.

17. Remove oil filter (for valve timing control solenoid valve), if necessary.

18. Remove timing chain tensioner cover from front cover, if necessary. Use seal cutter KV10111100 (J-37228) to cut liquid gasket for removal.

19. Remove oil pump drive chain as per the following:

   a. Push oil pump drive chain tensioner.

   b. Insert a stopper pin into the body hole.

   c. Hold the two flat parts of oil pump

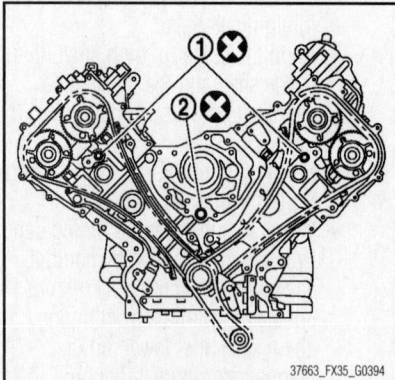

Fig. 226 Remove O-rings (1), (2) from cylinder heads and cylinder block

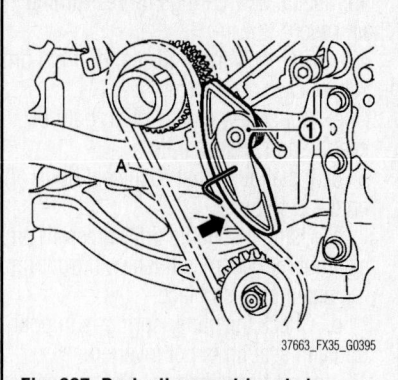

Fig. 227 Push oil pump drive chain tensioner (1); insert a stopper pin (A) into the body hole

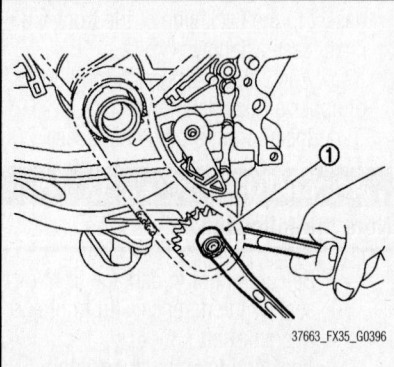

Fig. 228 Hold the two flat parts of oil pump shaft, and then loosen the oil pump sprocket (1) (oil pump side) nut

shaft, and then loosen the oil pump sprocket (oil pump side) nut.

❄️ **CAUTION**

Secure the oil pump unit shaft with the two flat parts.

20. Remove oil pump drive chain tensioner.

21. Remove timing chain tensioner (bank 1) as per the following:

➡️**To remove timing chain and related parts, start with those on bank 1. The procedure for removing parts on bank 2 is omitted because it is the same as that for bank 1.**

   a. Push both sides of spring against spring tension, and then press in plunger with a slack guide.

   b. Insert a stopper pin into the body hole, and then fix it with the plunger pushed in.

22. Remove tension guide and slack guide.

23. Remove timing chain and crankshaft sprocket.

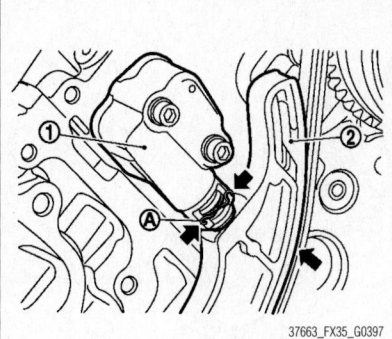

Fig. 229 Timing chain tensioner (bank 1) (1): Push both sides of spring (A) against spring tension, and then press in plunger with a slack guide (2)

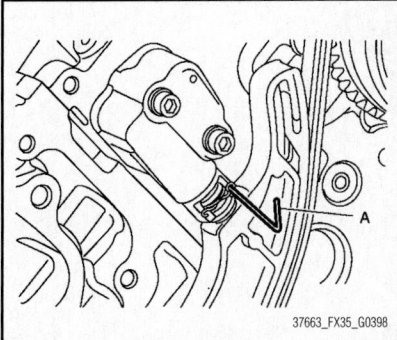

Fig. 230 Insert a stopper pin (A) into the body hole

✳✳ CAUTION

After removing timing chain, never turn crankshaft and camshaft separately, or valves will strike the piston head.

24. Remove camshaft sprocket (INT) and (EXH) as per the following:

   a. Exhaust side: Secure the hexagonal portion of camshaft (EXH) using a wrench to loosen mounting bolt.

   b. Intake side: Secure the hexagonal portion (located in between journal No.1 and journal No. 2) of drive shaft using a wrench to loosen mounting bolt.

➡ **The figure shows an example of bank 2.**

✳✳ CAUTION

Note the following:

• Never loosen the mounting bolt by securing anything other than the camshaft (drive shaft) hexagonal portion or with tensioning the timing chain.

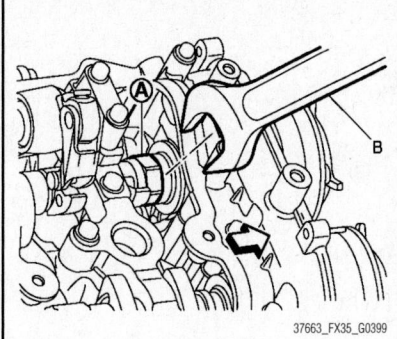

Fig. 231 Secure the hexagonal portion (located in between journal No.1 and journal No. 2) of drive shaft (A) using a wrench (B) to loosen mounting bolt

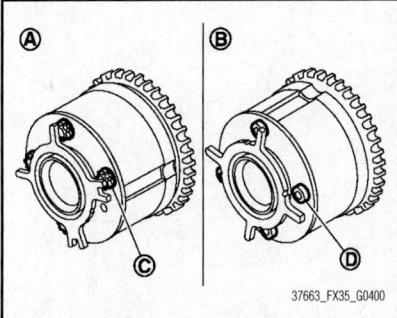

Fig. 232 Never disassemble camshaft sprocket. Never loosen bolts (C), (D) as shown

• When holding the hexagonal part of camshaft (drive shaft) with a wrench, be careful not to allow the wrench to cause interference with other parts.

• Never disassemble camshaft sprocket. Never loosen bolts as shown.

25. Use scraper to remove all traces of old liquid gasket from front cover and opposite mating surfaces. Remove old liquid gasket from bolt hole and thread.

To install:

✳✳ CAUTION

Note the following:

• The below figure shows the relationship between the matching marks on each timing chain and that on the corresponding sprocket, with the components installed.

• Parts with an identification mark (R or L) should be installed on the corresponding bank according to the mark.

• To install timing chain and related parts, start with those on bank 2.

The procedure for installing parts on bank 1 is omitted because it is the same as that for installation on bank 2.

• There is no matching mark in the oil pump related parts.

26. Check that crankshaft key and dowel pin of each camshaft are located as shown.

➡ **Though camshaft does not stop at the position as shown, for the placement of cam nose, it is generally accepted camshaft is placed for the same direction of the figure.**

27. Install camshaft sprockets (INT and EXH).

   a. Install onto correct side by checking with identification mark (A) on surface.

   b. Secure the hexagonal portion of camshaft (EXH) using a wrench to tighten mounting bolt.

   c. Secure the hexagonal portion (located in between journal No.1 and journal No. 2) of drive shaft using a wrench to tighten mounting bolt.

28. Install timing chains as per the following:

   a. Install crankshaft sprockets for both banks. Install each crankshaft sprocket so that its flange side (the larger diameter side without teeth) faces in the direction shown.

➡ **The same parts are used but facing directions are different.**

29. Install timing chains.

   a. Bank 2: Install timing chain so that the matching mark (outer groove) on camshaft sprocket is aligned with the copper link on timing chain, while the matching mark (punched) on crankshaft sprocket is aligned with the yellow link one on timing chain, as shown.

   b. Bank 1: Install timing chain so that the matching mark (outer groove) on camshaft sprocket is aligned with the copper link on timing chain, while the matching mark (notched) on crankshaft sprocket is aligned with the yellow link one on timing chain, as shown.

30. Install slack guides and tension guides onto correct side by checking with identification mark on surface.

✳✳ CAUTION

Never overtighten slack guide mounting bolt. It is normal for a gap to exist under the bolt seats when mounting bolt are tightened to the specification.

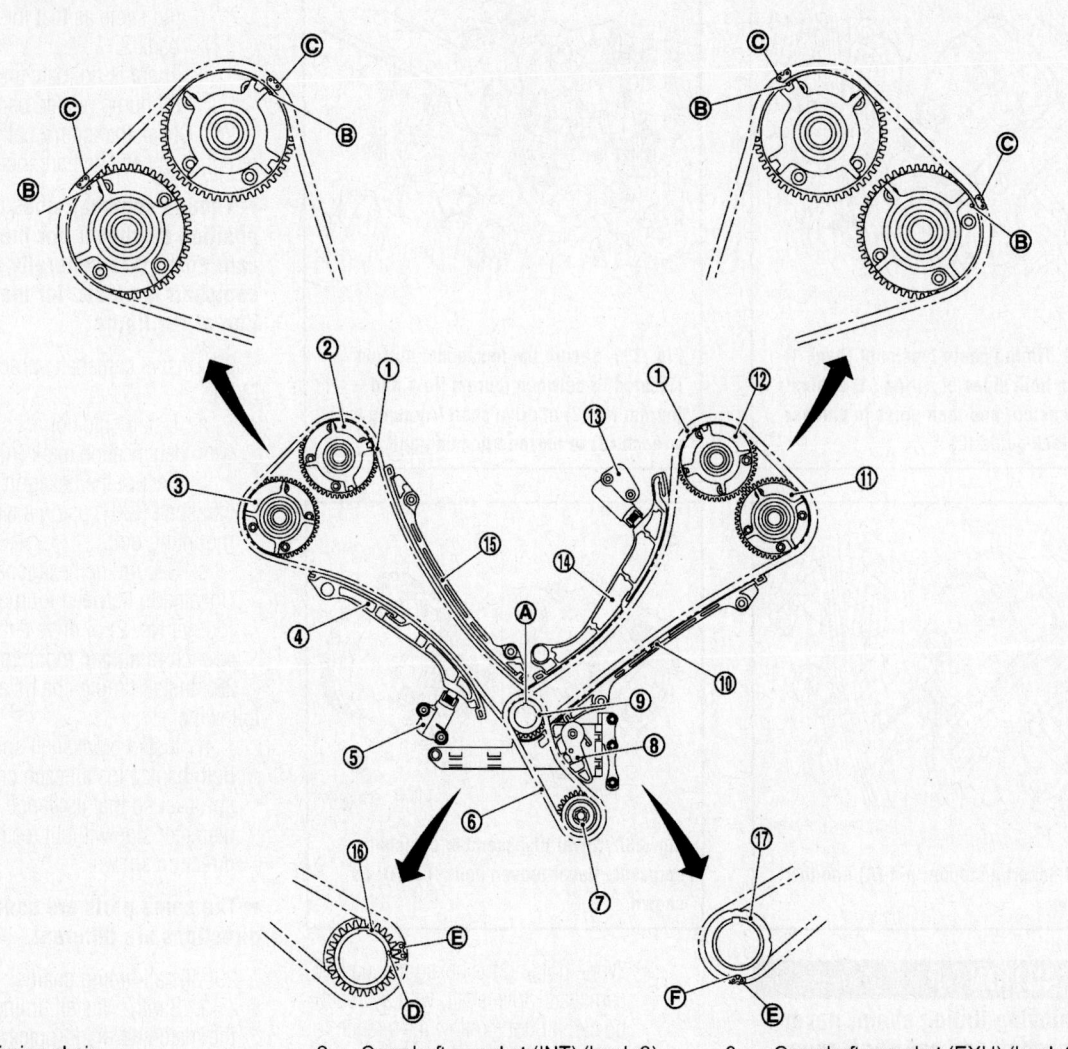

1. Timing chain
2. Camshaft sprocket (INT) (bank 2)
3. Camshaft sprocket (EXH) (bank 2)
4. Slack guide (bank 2)
5. Timing chain tensioner (bank 2)
6. Oil pump drive chain
7. Oil pump sprocket (oil pump side)
8. Oil pump drive chain tensioner
9. Oil pump sprocket (crankshaft side)
10. Tension guide (bank 1)
11. Camshaft sprocket (EXH) (bank 1)
12. Camshaft sprocket (INT) (bank 1)
13. Timing chain tensioner (bank 1)
14. Slack guide (bank 1)
15. Tension guide (bank 2)
16. Crankshaft sprocket (bank 2 side)
17. Crankshaft sprocket (bank 1 side)
A. Crankshaft key
B. Matching mark (outer groove)
C. Matching mark (copper link)
D. Matching mark (punched)
E. Matching mark (yellow link)
F. Matching mark (notched)

37663_FX35_G0401

Fig. 233 This figure shows the relationship between the matching marks on each timing chain and that on the corresponding sprocket, with the components installed

31. Install timing chain tensioner as per the following:

a. Fix the plunger at the most compressed position using a stopper pin (A). Remove any dirt and foreign materials completely from the back and the mounting surfaces of timing chain tensioner.

b. Pull out stopper pin after installing, and then release plunger.

32. Check again that the matching marks on sprockets and timing chain have not slipped out of alignment.

33. Install oil pump drive chain as per the following:

a. Install oil pump drive chain tensioner. Fix the tensioner at the most compressed position using a stopper pin. and then install it.

b. Install the oil pump sprocket (crankshaft side), oil pump sprocket (oil pump side) and oil pump drive chain at the same time. Install each oil pump sprocket so that its flange side (the larger diameter side without teeth) faces in the direction shown.

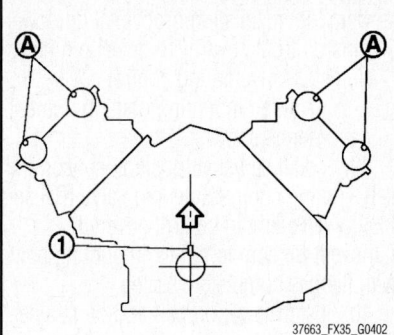

Fig. 234 Check that crankshaft key (1) and dowel pin (A) of each camshaft are located as shown

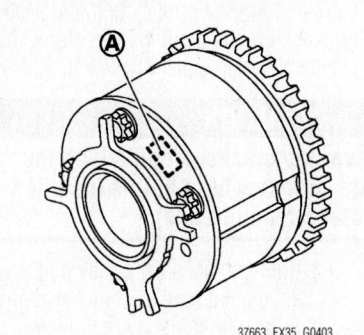

Fig. 235 Install onto correct side by checking with identification mark (A) on surface

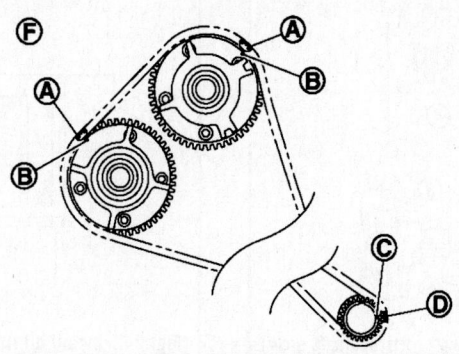

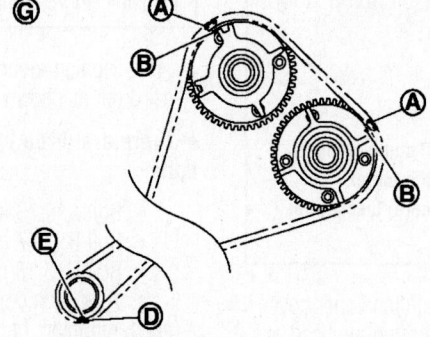

A. Copper link
B. Matching mark (outer groove)
C. Matching mark (punched)
D. Yellow link
E. Matching mark (notched)
F. Bank 2
G. Bank 1

Fig. 237 Install timing chains

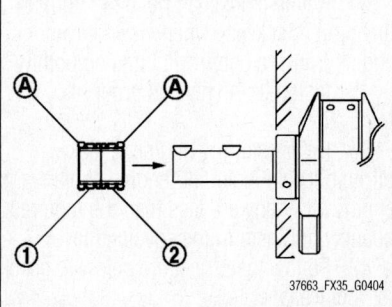

Fig. 236 Install each crankshaft sprocket so that its flange side (the larger diameter side without teeth) (A) faces in the direction shown

➡There are no matching marks on the oil pump related parts.

   c. Hold the two flat parts of oil pump shaft, and then tighten the oil pump sprocket (oil pump side) nut.

❋❋ CAUTION

Secure the oil pump shaft with the two flat parts.

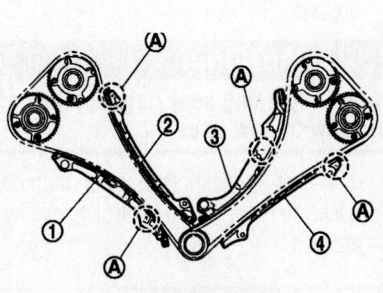

A. Identification mark
1. Slack guide (bank 2)
2. Tension guide (bank 2)
3. Slack guide (bank 1)
4. Tension guide (bank 1)

Fig. 238 Install slack guides and tension guides onto correct side

   d. Pull out the stopper pin after installing the oil pump drive chain. Check that the tension is applied to the oil pump drive chain after installing.
34. Install front oil seal on front cover.

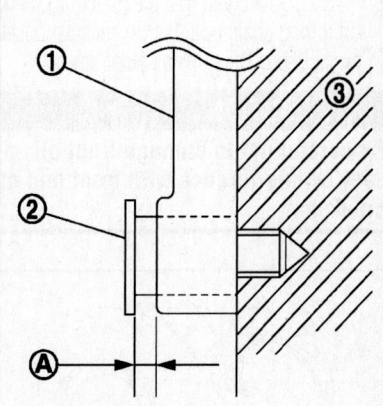

A. Gap
1. Slack guide
2. Slack guide mounting bolt
3. Cylinder block

Fig. 239 Never overtighten slack guide mounting bolt

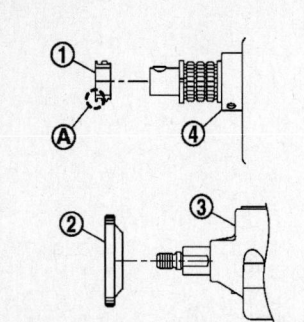

A. Flange side
1. Oil pump sprocket (crankshaft side)
2. Oil pump sprocket (oil pump side)
3. Oil pump
4. Crankshaft

37663_FX35_G0408

Fig. 240 Install each oil pump sprocket so that its flange side (the larger diameter side without teeth) faces in the direction shown

35. Install timing chain tensioner cover to front cover. Apply a continuous bead of liquid gasket to front cover as shown.

36. Install oil filter (for valve timing control solenoid valve) in the direction shown, if removed. Check that the oil filter does not protrude from the upper surface of front cover after installation.

37. Install front cover as per the following:

a. Install new O-ring onto cylinder heads and cylinder block.

b. Apply a continuous bead of liquid gasket to front cover as shown.

c. Check again that the matching marks on timing chain and that on each sprocket are aligned. Then, install front cover.

✳✳ CAUTION

Be careful not to damage front oil seal by interference with front end of crankshaft.

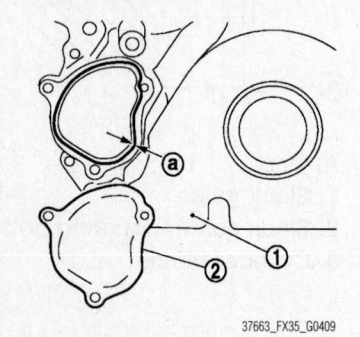

37663_FX35_G0409

Fig. 241 Install timing chain tensioner cover (2) to front cover (1)

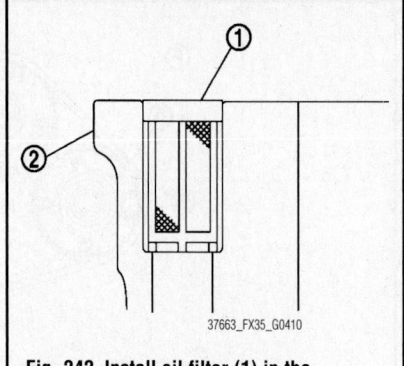

37663_FX35_G0410

Fig. 242 Install oil filter (1) in the direction shown; front cover (2)

d. Tighten mounting bolts in numerical order as shown.

➡**There are three types of mounting bolts.**

- Bolt A: 0.79 inches (20 mm)
- Bolt B: 1.77 inches (45 mm)
- Bolt C: 3.15 inches (80 mm)

e. After all mounting bolts are tightened, retighten them in numerical order as shown.

✳✳ CAUTION

Be sure to wipe out any excessive liquid gasket leaking onto surface mating with oil pan.

38. Install valve timing control cover as per the following:

a. Install new O-rings on front cover.

b. Install new seal rings in shaft grooves.

✳✳ CAUTION

When replacing seal ring, replace all rings with new ones.

c. Apply a continuous bead of liquid gasket to valve timing control covers as shown.

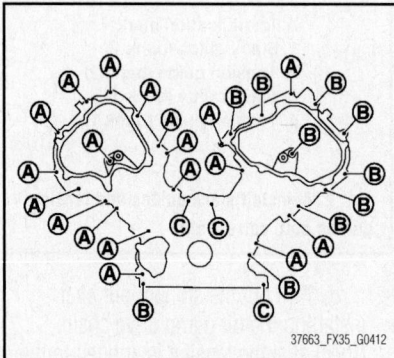

37663_FX35_G0412

Fig. 243 Proper bolt locations (A, B, C)

d. Being careful not to move seal ring from the installation groove, align dowel pins on front cover with dowel pin holes to install valve timing control covers.

e. Tighten mounting bolts in numerical order as shown.

39. Install camshaft position sensor and valve timing control solenoid valve (RH and LH) to valve timing control cover, if removed. Be sure to tighten mounting bolts with flanges completely seated.

40. Install lower oil pan and oil strainer.

41. Install upper oil pan.

42. Install water pump pulley.

43. Install crankshaft pulley.

a. Install crankshaft pulley, taking care not to damage front oil seal.

b. Apply engine oil onto threaded parts of crankshaft pulley bolt and seating area. Lightly tapping its center with plastic hammer, insert crankshaft pulley.

✳✳ CAUTION

Never tap crankshaft pulley on the side surface where belt is installed (outer circumference).

c. Tighten crankshaft pulley bolt.

d. Put a paint mark on crankshaft pulley aligning with angle mark on crankshaft pulley bolt.

e. Tighten crankshaft pulley bolt (clockwise). Check the tightening angle by referencing to the notches. The angle between two notches is 90 degrees.

44. Rotate crankshaft pulley in normal direction (clockwise when viewed from engine front) to confirm it turns smoothly.

45. Install in the reverse order of removal.

46. Before starting engine, check oil/fluid levels including engine coolant and engine oil. If any are less than the required quantity, fill them to the specified level.

47. Follow the procedure below to check for fuel leakage.

a. Turn ignition switch to the "ON" position (with engine stopped). With fuel pressure applied to fuel piping, check for fuel leakage at connection points.

b. Start engine. With engine speed increased, check again for fuel leakage at connection points.

48. Run engine to check for unusual noise and vibration.

➡**If hydraulic pressure inside chain tensioner drops after removal/installation, slack in guide may generate a pounding noise during and just after the engine start. However, this does not indicate a malfunc-**

tion. **The noise will stop after hydraulic pressure rises.**

49. Warm up engine thoroughly to check that there is no leakage of fuel, or any oil/fluids including engine oil and engine coolant.

50. Bleed air from lines and hoses of applicable lines, such as in cooling system.

51. After cooling down engine, again check oil/fluid levels including engine oil and engine coolant. Refill them to the specified level, if necessary.

VALVE COVERS

REMOVAL & INSTALLATION

Refer to Ignition Coil & Spark Plugs procedure in ENGINE ELECTRICAL.

VALVE LASH

ADJUSTMENT

3.5L Engine

See Figures 244 and 245.

➡ **Standard valve clearances:**

  a. Intake:
- Cold: 0.010–0.013 inches (0.26–0.34 mm)
- Hot: 0.012–0.016 inches (0.304–0.416 mm)

  b. Exhaust:
- Cold: 0.011–0.015 inches (0.29–0.37 mm)
- Hot: 0.012–0.017 inches (0.308–0.432 mm)

➡ **Perform adjustment depending on selected head thickness of valve lifter.**

1. Measure the valve clearance.
2. Remove camshaft.
3. Remove valve lifters at the locations that are out of the standard.
4. Measure the center thickness of the removed valve lifters with a micrometer.
5. Use the equation below to calculate valve lifter thickness for replacement.
- Valve lifter thickness calculation: $t = t1 + (C1 - C2)$
- t = Valve lifter thickness to be replaced
- $t1$ = Removed valve lifter thickness
- $C1$ = Measured valve clearance
- $C2$ = Standard valve clearance:

➡ **Thickness of new valve lifter can be identified by stamp marks on the reverse side (inside the cylinder). Stamp mark 788 indicates 7.88 mm (0.3102 in) in thickness. Available thickness of valve lifter: 27 sizes with**

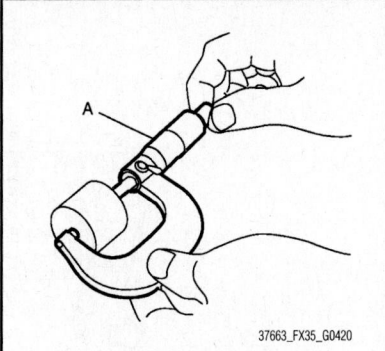
Fig. 244 Measure the center thickness of the removed valve lifters with a micrometer (A)

37663_FX35_G0420

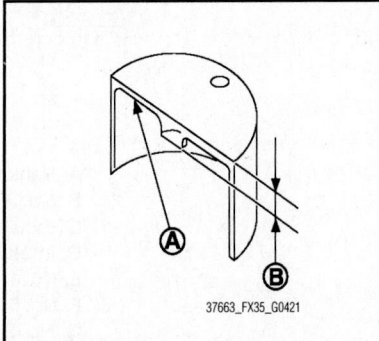
Fig. 245 Valve lifter stamp mark (A) and measured thickness (B)

37663_FX35_G0421

range 7.88 to 8.40 mm (0.3102 to 0.3307 in) in steps of 0.02 mm (0.0008 in) (when manufactured at factory).

6. Install selected valve lifter.
7. Install camshaft.
8. Manually turn crankshaft pulley a few turns.
9. Check that the valve clearances for cold engine are within the specifications by referring to the specified values.
10. Install all removal parts in the reverse order of removal.
11. Warm up the engine, and check for unusual noise and vibration.

5.0L Engine

See Figures 244 and 245.

Perform adjustment depending on selected head thickness of valve lifter (EXH).

1. Measure the valve clearance.
2. Remove VVEL ladder assembly and camshaft (EXH).

✳✳ CAUTION

Never loosen adjusting bolts and mounting bolts (black color) of VVEL ladder assembly.

3. Remove valve lifter (EXH) at the locations that are out of the standard.
4. Measure the center thickness of the removed valve lifters (EXH) with a micrometer (A).
5. Use the equation below to calculate valve lifter (EXH) thickness for replacement.
- Valve lifter (EXH) thickness calculation: $t = t1 + (C1 - C2)$
- t = Valve lifter (EXH) thickness to be replaced
- $t1$ = Removed valve lifter (EXH) thickness
- $C1$ = Measured valve clearance
- $C2$ = Standard valve clearance:

➡ **Thickness of new valve lifter (EXH) can be identified by stamp marks on the reverse side (inside the cylinder). Stamp mark 788 indicates 7.88 mm (0.3102 in) in thickness. Available thickness of valve lifter (EXH): 27 sizes with range 7.88 to 8.40 mm (0.3102 to 0.3307 in) in steps of 0.02 mm (0.0008 in) (when manufactured at factory).**

6. Install selected valve lifter (EXH).
7. Install VVEL ladder assembly and camshaft (EXH).
8. Manually turn crankshaft pulley a few turns.
9. Check that the valve clearances for cold engine are within the specifications by referring to the specified values.
10. Install all removed parts in the reverse order of removal.
11. Warm up the engine, and check for unusual noise and vibration.

INSPECTION

3.5L Engine

See Figures 246 through 251.

Perform inspection as follows after removal, installation or replacement of camshaft or valve-related parts, or if there is unusual engine conditions regarding valve clearance.

In cases of removing/installing or replacing camshaft and valve related parts, or of unusual engine conditions due to changes in valve clearance (found malfunctions during stating, idling or causing noise), perform inspection as follows:

1. Remove valve covers (bank 1 and bank 2).

➡ **Standard valve clearances:**

  a. Intake:
- Cold: 0.010–0.013 inches (0.26–0.34 mm)
- Hot: 0.012–0.016 inches (0.304–0.416 mm)

b. Exhaust:
- Cold: 0.011–0.015 inches (0.29–0.37 mm)
- Hot: 0.012–0.017 inches (0.308–0.432 mm)

2. Measure the valve clearance as per the following:

a. Set No. 1 cylinder at TDC of its compression stroke.

- Rotate crankshaft pulley clockwise to align timing mark (grooved line without color) with timing indicator.
- Check that intake and exhaust cam nose on No. 1 cylinder (engine front side of bank 1) are located as shown.
- If not, turn crankshaft one revolution (360 degrees) and align as shown in the figure.

b. Use a feeler gauge, measure the clearance between valve lifter and camshaft.

- Measure the valve clearances at locations indicated in the figure with No. 1 cylinder at TDC.

c. Rotate crankshaft 240 degrees

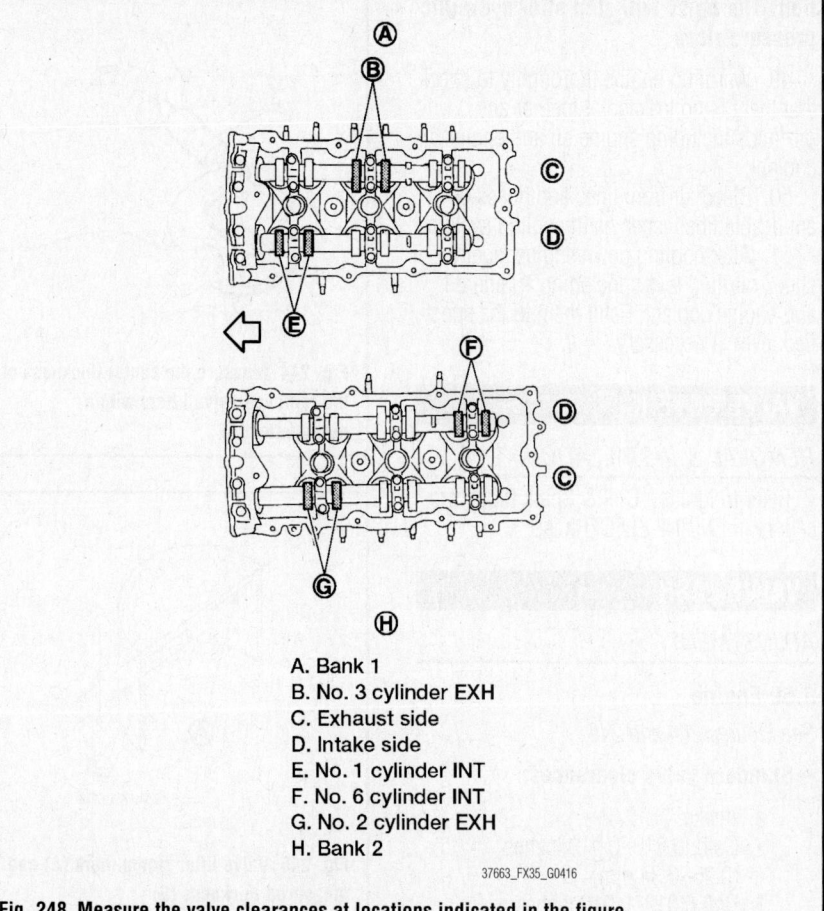

A. Bank 1
B. No. 3 cylinder EXH
C. Exhaust side
D. Intake side
E. No. 1 cylinder INT
F. No. 6 cylinder INT
G. No. 2 cylinder EXH
H. Bank 2

37663_FX35_G0416

Fig. 248 Measure the valve clearances at locations indicated in the figure

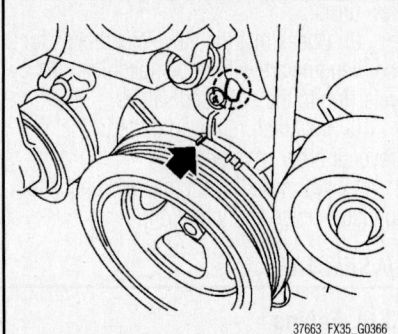

37663_FX35_G0366

Fig. 246 Rotate crankshaft pulley clockwise to align timing mark (grooved line without color) with timing indicator

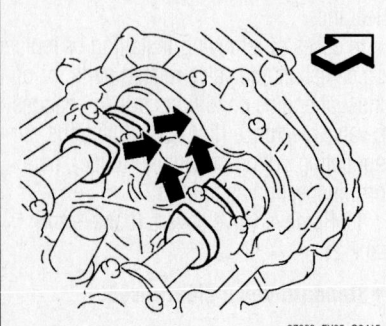

37663_FX35_G0415

Fig. 247 Check that intake and exhaust cam nose on No. 1 cylinder (engine front side of bank 1) are located as shown

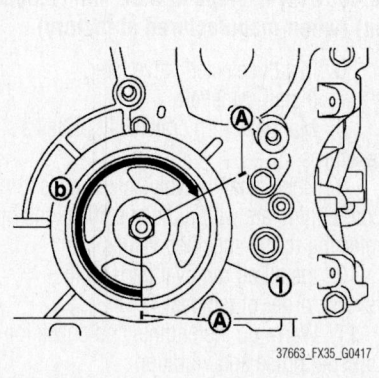

37663_FX35_G0417

Fig. 249 Mark (A) a position 240 degrees (b) from a corner of the hexagonal part of crankshaft pulley (1) mounting bolt as shown

clockwise (when viewed from engine front) to align No. 3 cylinder at TDC its compression stroke.

- Mark a position 240 degrees from a corner of the hexagonal part of crankshaft pulley mounting bolt as shown. Use the hexagonal part as a guide.

- Measure the valve clearances at locations indicated in the figure with No. 3 cylinder at TDC.

d. Rotate crankshaft 240 degrees clockwise (when viewed from engine front) to align No. 5 cylinder at TDC of compression stroke.

- Mark a position 240 degrees from a corner of the hexagonal part of crankshaft pulley mounting bolt. Use the hexagonal part as a guide.

- Measure the valve clearances at locations indicated in the figure with No. 5 cylinder at TDC.

3. Perform adjustment if the measured value is out of the standard.

5.0L Engine

Check valve clearance if applicable to the following cases:

Intake Side:

At the removal and installation of VVEL ladder assembly or valve-related parts, or at the occurrence of malfunction (poor starting, idle malfunction, unusual noise) due to aged deterioration in valve clearance.

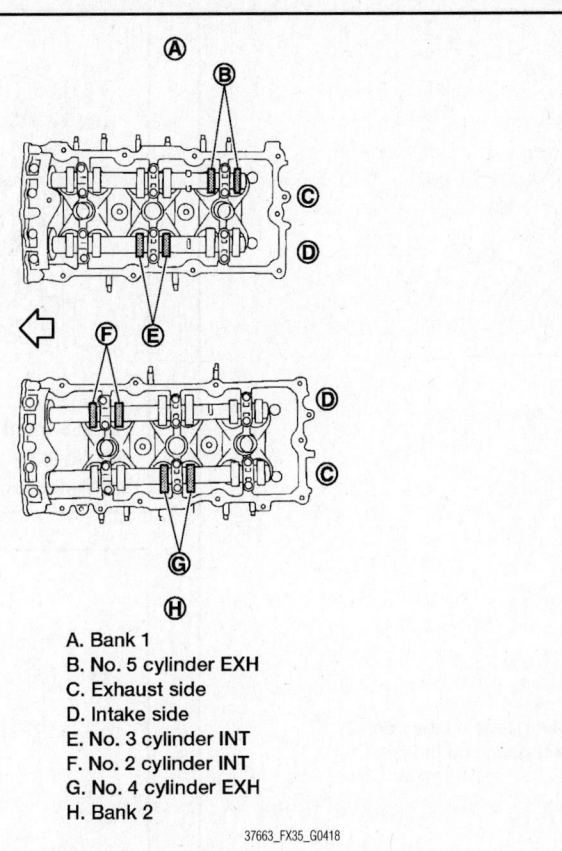

A. Bank 1
B. No. 5 cylinder EXH
C. Exhaust side
D. Intake side
E. No. 3 cylinder INT
F. No. 2 cylinder INT
G. No. 4 cylinder EXH
H. Bank 2

37663_FX35_G0418

Fig. 250 Measure the valve clearances at locations indicated in the figure

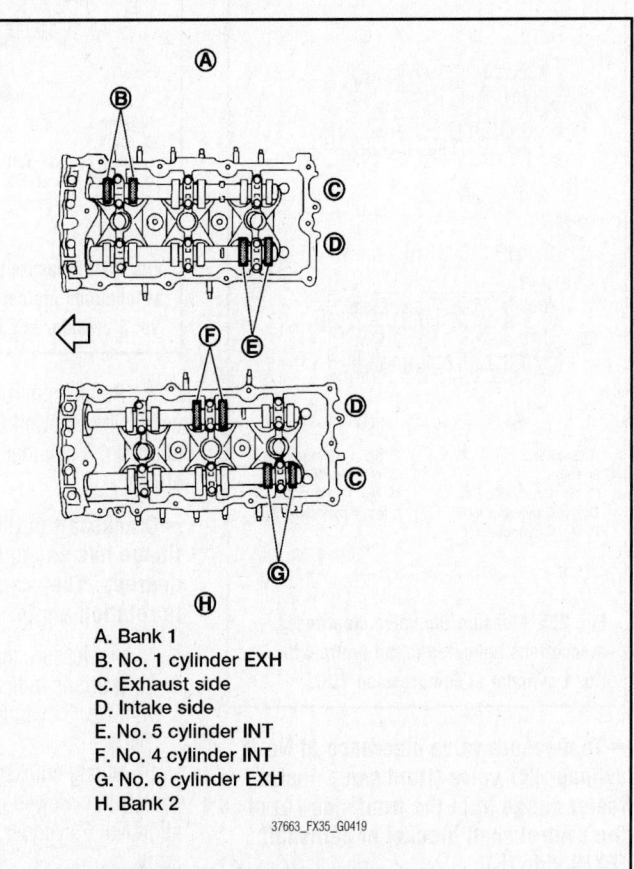

A. Bank 1
B. No. 1 cylinder EXH
C. Exhaust side
D. Intake side
E. No. 5 cylinder INT
F. No. 4 cylinder INT
G. No. 6 cylinder EXH
H. Bank 2

37663_FX35_G0419

Fig. 251 Measure the valve clearances at locations indicated in the figure

✳✳ CAUTION

Valve clearance check on the intake side is not required after replacing the VVEL ladder assembly and cylinder head assembly with a new one. (Install new VVEL ladder assembly and cylinder head assembly in factory-shipped condition because it is factory-adjusted and inspected.)

➡VVEL ladder assembly cannot be replaced as a single part, because it is machined together with cylinder head assembly.

Exhaust Side:

See Figures 252 through 258.

At the removal, installation, and replacement of camshaft (EXH) or valve-related parts, or at the occurrence of malfunction (poor starting, idle malfunction, unusual noise) due to aged deterioration in valve clearance.

1. Remove rocker covers (bank 1 and bank 2).

2. Measure the valve clearance as per the following:

   a. Use the feeler gauge of curved-tip. This allows the feeler gauge to access the clearance between camshaft (drive shaft) nose and valve lifter with ease.

➡Be sure to note the following points when measuring valve clearance on the intake side.

   b. Before measuring, check that the position of drive shaft nose is within the angle shown.

   c. Refer to the figure for the insertion direction of the feeler gauge since the direction depends on the bank.

3. Set No. 1 cylinder at TDC of its compression stroke.

   a. Rotate crankshaft pulley clockwise to align timing mark (grooved line without color) with timing indicator.

   b. Check that exhaust cam nose on No. 1 cylinder (engine front side of bank 1) is located as shown.

   c. If not, turn crankshaft one revolution (360 degrees) and align as shown.

   d. Measure the valve clearances at locations indicated in the figure with No. 1 cylinder at compression TDC

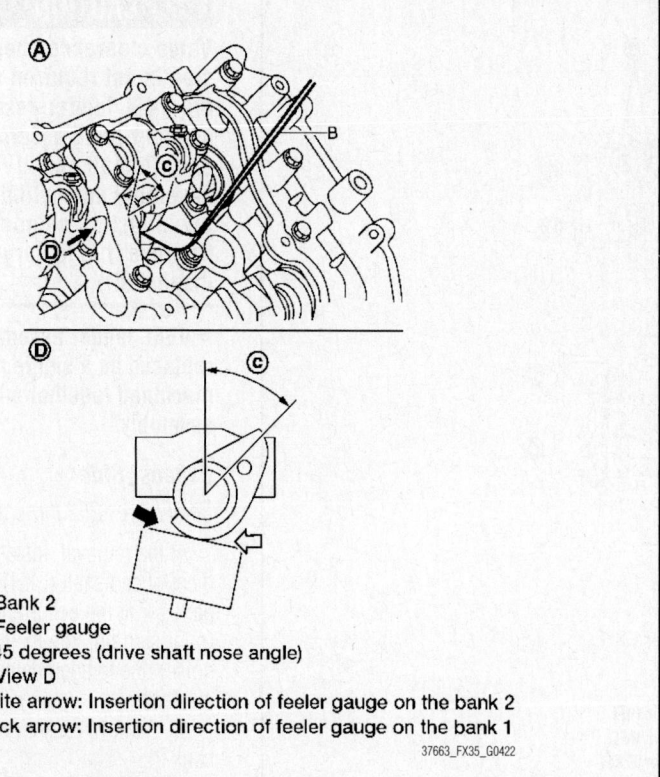

A. Bank 2
B. Feeler gauge
c. 45 degrees (drive shaft nose angle)
D. View D
White arrow: Insertion direction of feeler gauge on the bank 2
Black arrow: Insertion direction of feeler gauge on the bank 1

37663_FX35_G0422

Fig. 252 Measuring valve clearances

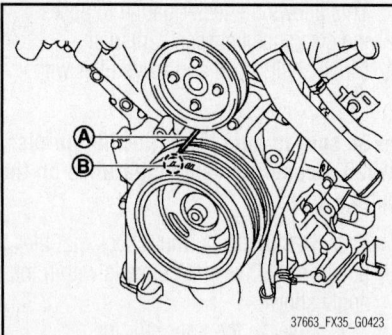

37663_FX35_G0423

Fig. 253 Rotate crankshaft pulley clockwise to align timing mark (grooved line without color) (B) with timing indicator (A)

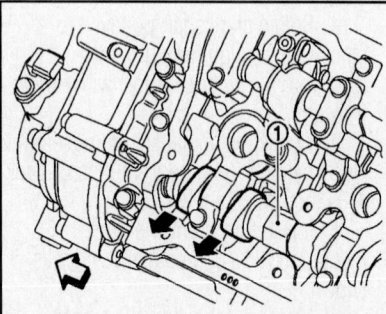

37663_FX35_G0424

Fig. 254 Check that exhaust cam nose on No. 1 cylinder (engine front side of bank 1) is located as shown

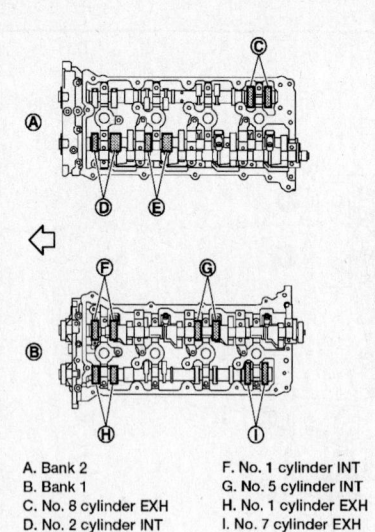

A. Bank 2   F. No. 1 cylinder INT
B. Bank 1   G. No. 5 cylinder INT
C. No. 8 cylinder EXH   H. No. 1 cylinder EXH
D. No. 2 cylinder INT   I. No. 7 cylinder EXH
E. No. 4 cylinder INT

37663_FX35_G0425

Fig. 255 Measure the valve clearances at locations indicated in the figure with No. 1 cylinder at compression TDC

➡To measure valve clearance of No. 1 cylinder INT valve (front side), insert feeler gauge from the front side (A) of the control shaft bracket or camshaft (EXH) side (B).

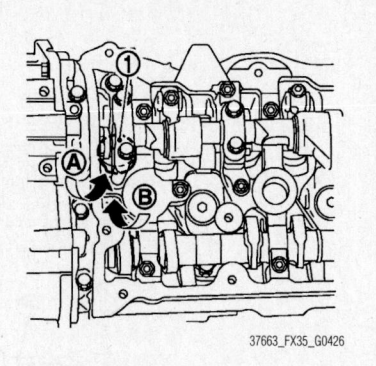

37663_FX35_G0426

Fig. 256 Insert feeler gauge from the front side (A) of the control shaft bracket or camshaft (EXH) side (B)

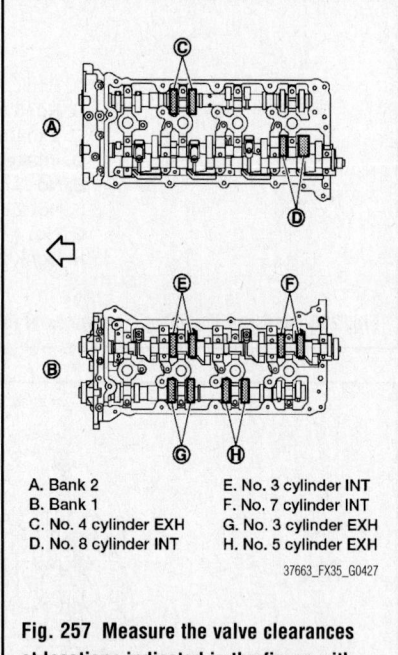

A. Bank 2   E. No. 3 cylinder INT
B. Bank 1   F. No. 7 cylinder INT
C. No. 4 cylinder EXH   G. No. 3 cylinder EXH
D. No. 8 cylinder INT   H. No. 5 cylinder EXH

37663_FX35_G0427

Fig. 257 Measure the valve clearances at locations indicated in the figure with No. 3 cylinder at compression TDC

4. Rotate crankshaft 270 degrees clockwise (when viewed from engine front) to align No. 3 cylinder at TDC its compression stroke.

➡Crankshaft pulley mounting bolt flange has an angle mark every 90 degrees. They can be used as a guide to rotation angle.

a. Measure the valve clearances at locations indicated in the figure with No. 3 cylinder at compression TDC

5. Rotate crankshaft 90 degrees clockwise (when viewed from engine front) to align No. 6 cylinder at TDC of compression stroke.

➡**Crankshaft pulley mounting bolt flange has an angle mark every 90 degrees. They can be used as a guide to rotation angle.**

   a. Measure the valve clearances at locations indicated in the figure with No. 6 cylinder at compression TDC

6. Perform adjustment or replacement if the measured value is out of the standard.

   a. If a valve clearance on the exhaust side is out of specification, adjust the valve clearance.

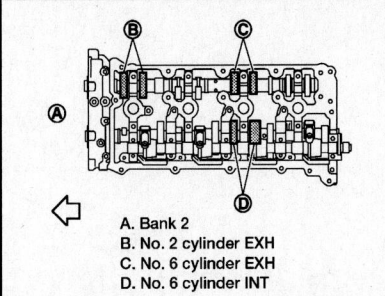

A. Bank 2
B. No. 2 cylinder EXH
C. No. 6 cylinder EXH
D. No. 6 cylinder INT

37663_FX35_G0428

Fig. 258 Measure the valve clearances at locations indicated in the figure with No. 6 cylinder at compression TDC

   b. If a valve clearance on the intake side is out of specification, replace VVEL ladder assembly & cylinder head assembly.

❄❄ CAUTION

Never adjust valve clearance on the intake side.

➡**Since the valve lifter (INT) cannot be replaced by the piece, VVEL ladder assembly & cylinder head assembly replacement are required.**

ENGINE PERFORMANCE & EMISSION CONTROLS

CAMSHAFT POSITION (CMP) SENSOR

LOCATION

The Camshaft Position (CMP) sensor for the 3.5L engine is located on the right rear of the cylinder block.

REMOVAL & INSTALLATION

1. Turn ignition switch OFF.
2. Loosen the mounting bolt of the sensor.
3. Disconnect camshaft position sensor harness connector.
4. Remove the sensor.
5. Installation is the reverse of removal.

CRANKSHAFT POSITION (CKP) SENSOR

REMOVAL & INSTALLATION

1. Turn ignition switch OFF.
2. Loosen the fixing bolt of the sensor.
3. Disconnect crankshaft position sensor harness connector.
4. Remove the sensor.
5. Installation is the reverse of removal.

ELECTRONIC CONTROL MODULE (ECM)

LOCATION

See Figure 259.

Refer to the accompanying illustration.

REMOVAL & INSTALLATION

➡**The manufacturer does not provide a specific Removal and Installation procedure for this component. Refer to the graphic(s) when servicing this component.**

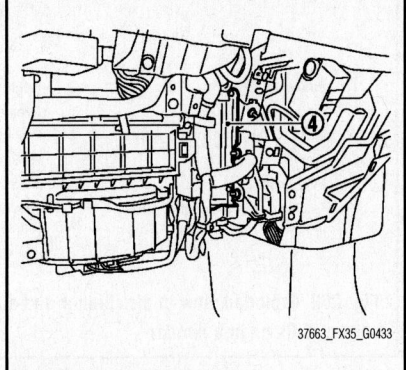

37663_FX35_G0433

Fig. 259 ECM location (4)

ENGINE COOLANT TEMPERATURE (ECT) SENSOR

REMOVAL & INSTALLATION

1. Turn ignition switch OFF.
2. Disconnect engine coolant temperature sensor harness connector.
3. Remove engine coolant temperature sensor.
4. Installation is the reverse of removal.

INTAKE AIR TEMPERATURE (IAT) SENSOR

LOCATION

The Intake Air Temperature (IAT) sensor is an integral part of the Mass Air Flow (MAF) sensor/Intake Air Temperature (IAT) sensor assembly which is mounted on the air intake duct. Refer to the Mass Air Flow section for information regarding servicing this component.

KNOCK SENSOR (KS)

LOCATION

See Figures 260 and 261.

Refer to the accompanying illustrations.

REMOVAL & INSTALLATION

➡**The manufacturer does not provide a specific Removal and Installation procedure for this component. Refer to the graphic(s) when servicing this component.**

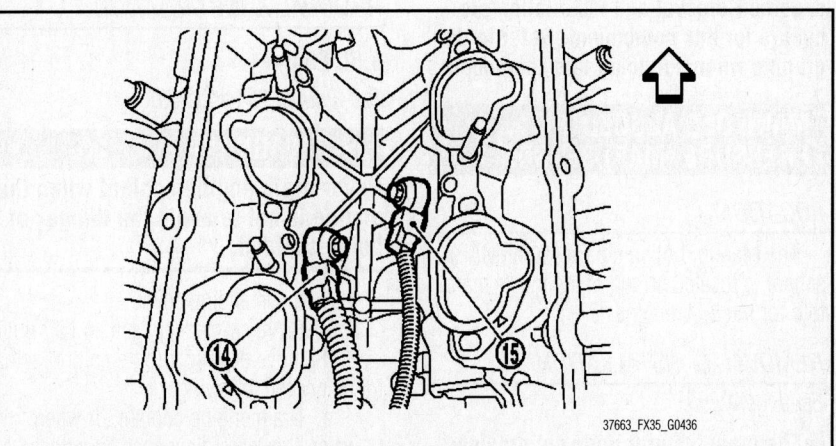

37663_FX35_G0436

Fig. 260 Knock sensor (14, 15) locations—3.5L engine

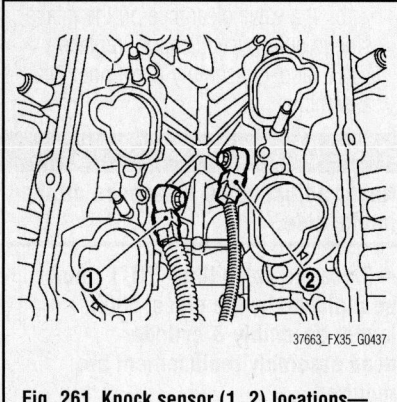

Fig. 261 Knock sensor (1, 2) locations—5.0L engine

MALFUNCTION INDICATOR LIGHT (MIL)

RESET PROCEDURE

The MIL is located on the combination meter.

1. The MIL will illuminate when the ignition switch is turned ON without the engine running. This is a bulb check. If the MIL does not illuminate, check MIL circuit.

2. When the engine is started, the MIL should turn off. If the MIL remains on, the on board diagnostic system has detected an engine system malfunction.

MASS AIR FLOW (MAF) SENSOR

LOCATION

The Mass Air Flow (MAF) sensor is placed in the stream of intake air. It measures the intake flow rate by measuring a part of the entire intake flow.

REMOVAL & INSTALLATION

See Figures 262.

➡The manufacturer does not provide a specific Removal and Installation procedure for this component. Refer to the graphic when servicing this component.

MANIFOLD ABSOLUTE PRESSURE (MAP) SENSOR

LOCATION

The Manifold Absolute Pressure (MAP) sensor is located on top of the intake manifold for the 5.0L engine.

REMOVAL & INSTALLATION

See Figure 263.

➡The manufacturer does not provide a specific Removal and Installation

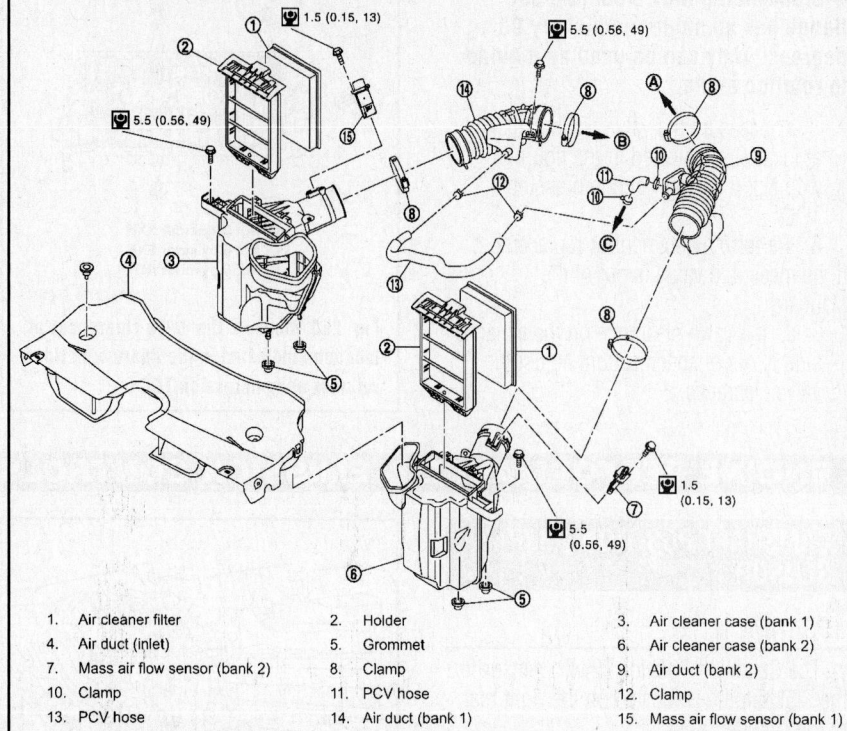

1.	Air cleaner filter	2.	Holder	3.	Air cleaner case (bank 1)
4.	Air duct (inlet)	5.	Grommet	6.	Air cleaner case (bank 2)
7.	Mass air flow sensor (bank 2)	8.	Clamp	9.	Air duct (bank 2)
10.	Clamp	11.	PCV hose	12.	Clamp
13.	PCV hose	14.	Air duct (bank 1)	15.	Mass air flow sensor (bank 1)
A.	To electric throttle control actuator (bank 2)	B.	To electric throttle control actuator (bank 1)	C.	To rocker cover (bank 2)

Fig. 262 Exploded view of air cleaner and air duct assembly showing MAF—3.5L engine shown, 5.0L engine similar

procedure for this component. Refer to the graphic(s) when servicing this component.

THROTTLE CONTROL ACTUATOR

LOCATION

The electric throttle control actuators are located on the intake manifold collector on the 3.5L engine and on the intake manifold of the 5.0L engine.

REMOVAL & INSTALLATION

3.5L Engine

See Figures 264 and 265.

✳✳ WARNING

Never drain engine coolant when the engine is hot to avoid the danger of being scalded.

1. Remove engine cover.
2. Remove air cleaner case and air duct.
3. Remove electric throttle control actuator as per the following:
    a. Drain engine coolant, or when water hoses are disconnected, attach plug to prevent engine coolant leakage.

✳✳ CAUTION

Perform this step when engine is cold. Never spill engine coolant on drive belt.

    b. Disconnect water hoses from electric throttle control actuator. When engine coolant is not drained from radiator, attach plug to water hoses to prevent engine coolant leakage.
    c. Disconnect harness connector.
    d. Loosen mounting bolts in reverse order as shown.

✳✳ CAUTION

Note the following:

- When removing only intake manifold collector, move electric throttle control actuator without disconnecting the water hose.
- The figure shows the electric throttle control actuator (bank 1) viewed from the air duct side.
- Viewed from the air duct side, order of loosening mounting bolts of electric throttle control actuator (bank 2) is the same as that of the electric throttle control actuator (bank 1).

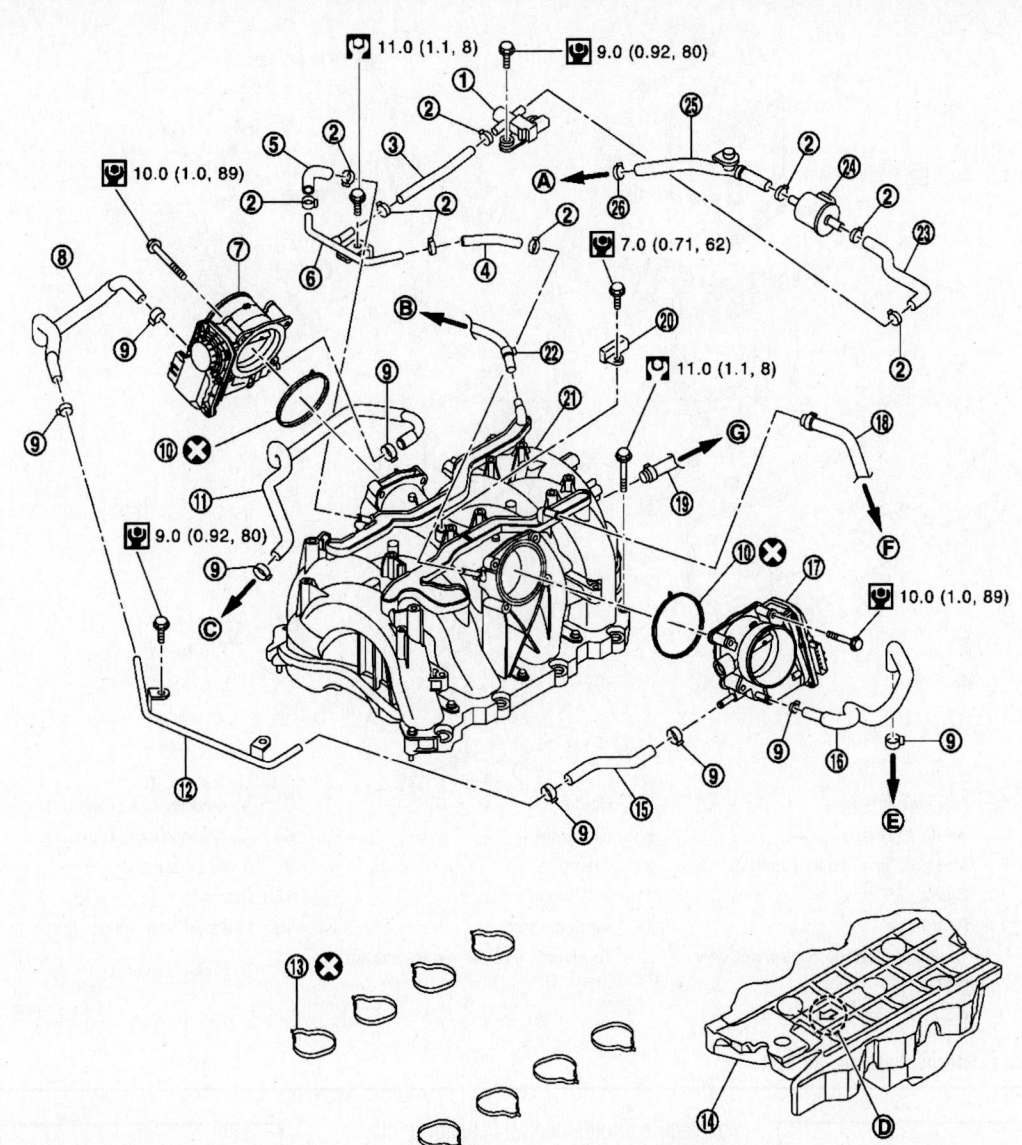

1. EVAP canister purge control solenoid valve
2. Clamp
3. EVAP hose
4. EVAP hose
5. EVAP hose
6. EVAP tube
7. Electric throttle control actuator (bank 2)
8. Water hose
9. Clamp
10. Gasket
11. Water hose
12. Water pipe
13. Gasket
14. Acoustic absorbent
15. Water hose
16. Water hose
17. Electric throttle control actuator (bank 1)
18. PCV hose
19. Vacuum hose
20. Manifold absolute pressure (MAP) sensor
21. Intake manifold
22. PCV hose
23. EVAP hose
24. Vacuum tank
25. EVAP service port hose
26. Clamp
A. To centralized under-floor piping
B. To rocker cover (bank 2)
C. To water inlet
D. Front mark
E. To cylinder head
F. To rocker cover (bank 1)
G. To brake booster

37663_FX35_G0438

Fig. 263 Exploded view of intake manifold assembly—5.0L engine

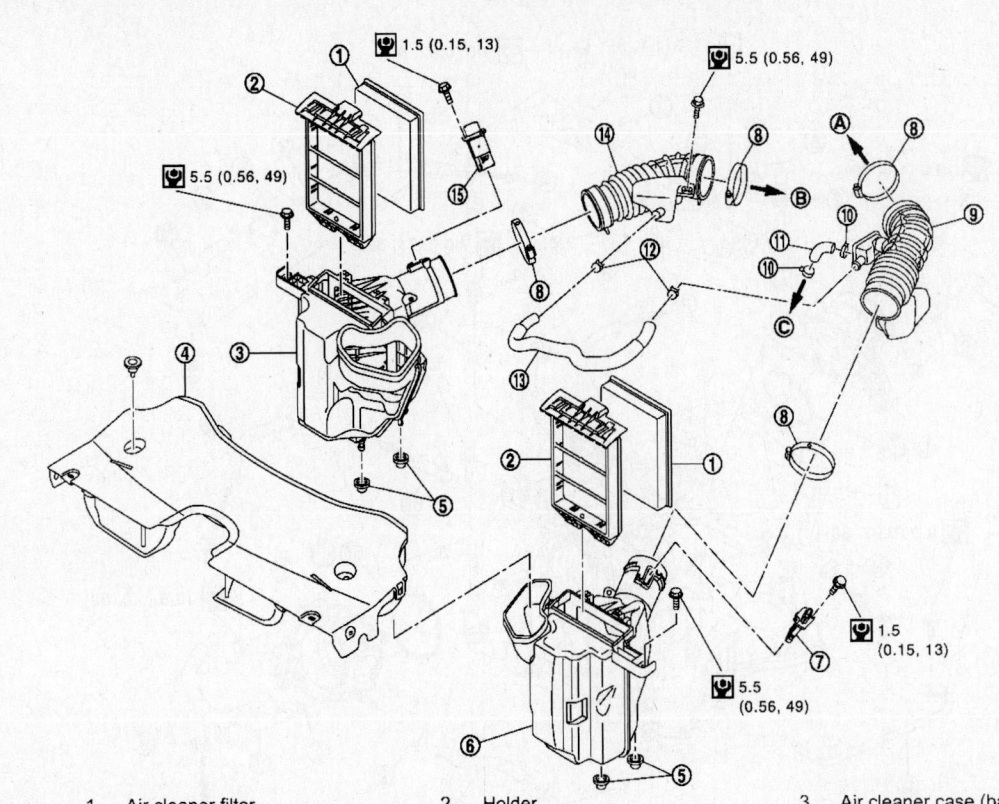

1. Air cleaner filter
2. Holder
3. Air cleaner case (bank 1)
4. Air duct (inlet)
5. Grommet
6. Air cleaner case (bank 2)
7. Mass air flow sensor (bank 2)
8. Clamp
9. Air duct (bank 2)
10. Clamp
11. PCV hose
12. Clamp
13. PCV hose
14. Air duct (bank 1)
15. Mass air flow sensor (bank 1)
A. To electric throttle control actuator (bank 2)
B. To electric throttle control actuator (bank 1)
C. To rocker cover (bank 2)

37663_FX35_G0206

Fig. 264 Remove air cleaner case and air duct

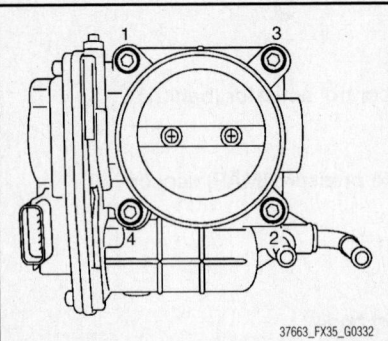

37663_FX35_G0332

Fig. 265 Loosen mounting bolts in reverse order as shown

✴✴ CAUTION

Handle carefully to avoid any impact to electric throttle control actuator.

4. Installation is the reverse of removal.
5. Perform the "Throttle Valve Closed Position Learning" when harness connector

of electric throttle control actuator is disconnected.
6. Perform the "Idle Air Volume Learning" and "Throttle Valve Closed Position Learning" when electric throttle control actuator is replaced.

5.0L Engine

See Figures 266 through 269.

✴✴ WARNING

To avoid the danger of being scalded, never drain the engine coolant when the engine is hot.

1. Remove engine cover and engine room cover (RH and LH).
2. Release fuel pressure.
3. Remove air duct (inlet) and air duct.
4. Remove quick connector cap and disconnect fuel feed hose on engine side.

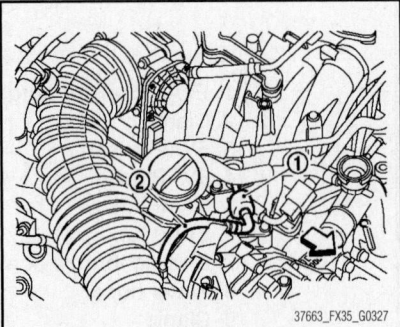

37663_FX35_G0327

Fig. 266 Remove quick connector cap (1) and disconnect fuel feed hose (2) on engine side

5. Remove engine cover bracket.
6. Remove fuel injector and fuel tube assembly.
7. Disconnect Manifold Absolute Pressure (MAP) sensor and air fuel ratio sensor 1 (bank 1) harness connector.

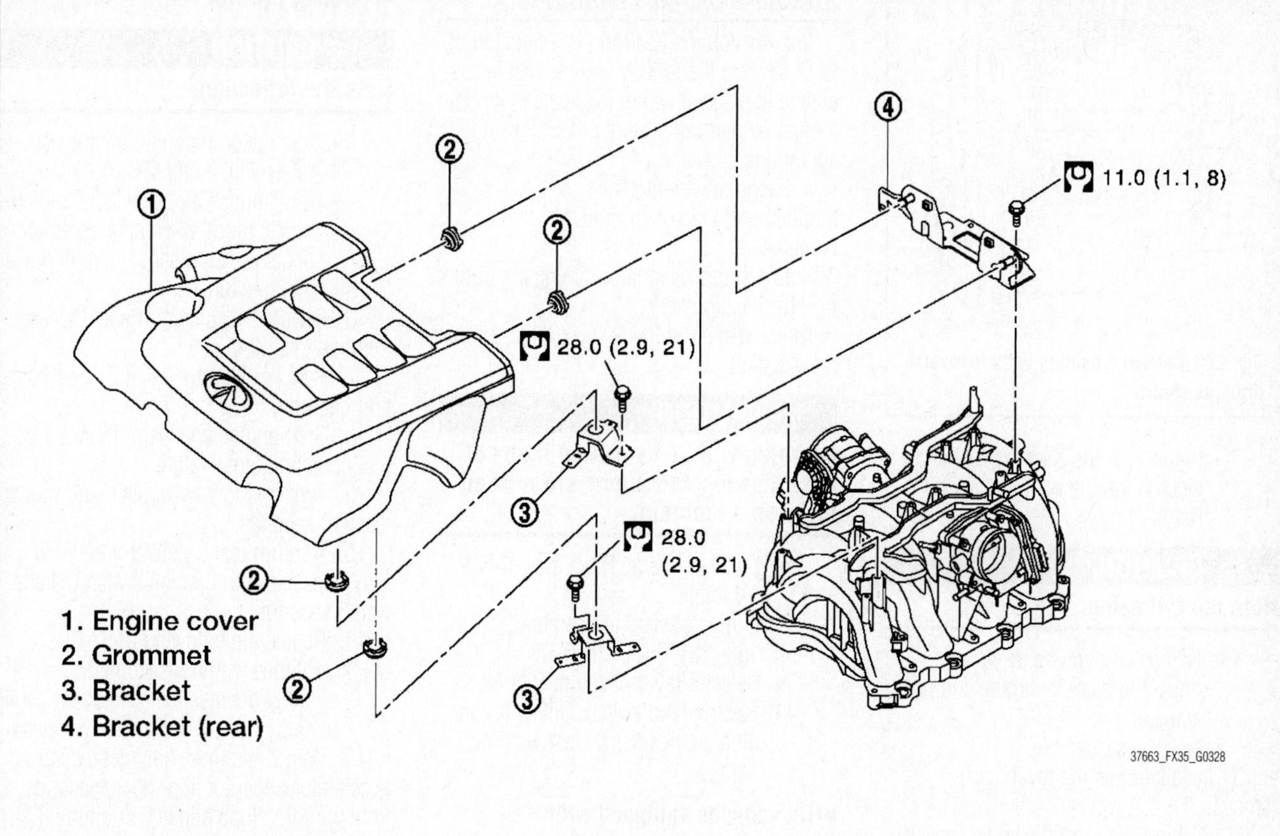

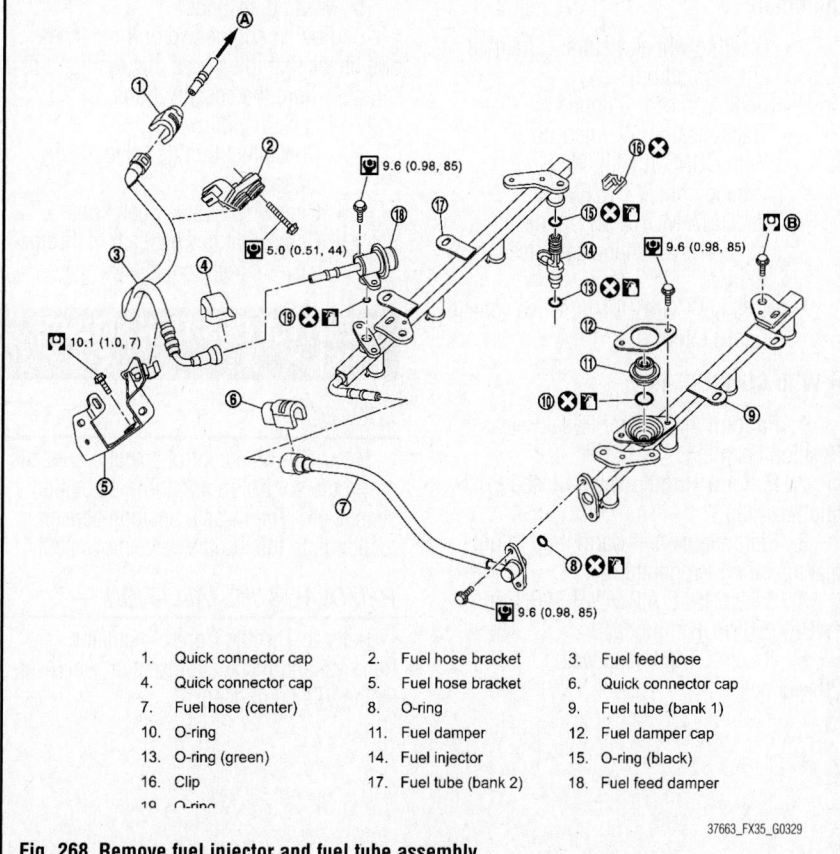

1. Engine cover
2. Grommet
3. Bracket
4. Bracket (rear)

28.0 (2.9, 21)

28.0 (2.9, 21)

11.0 (1.1, 8)

37663_FX35_G0328

Fig. 267 Remove engine cover bracket

9.6 (0.98, 85)

5.0 (0.51, 44)

10.1 (1.0, 7)

9.6 (0.98, 85)

9.6 (0.98, 85)

9.6 (0.98, 85)

1.	Quick connector cap	2.	Fuel hose bracket	3.	Fuel feed hose
4.	Quick connector cap	5.	Fuel hose bracket	6.	Quick connector cap
7.	Fuel hose (center)	8.	O-ring	9.	Fuel tube (bank 1)
10.	O-ring	11.	Fuel damper	12.	Fuel damper cap
13.	O-ring (green)	14.	Fuel injector	15.	O-ring (black)
16.	Clip	17.	Fuel tube (bank 2)	18.	Fuel feed damper
19.	O-ring				

37663_FX35_G0329

Fig. 268 Remove fuel injector and fuel tube assembly

8. Remove vacuum tank, EVAP service port hose and EVAP canister purge control solenoid valve.

9. Disconnect PCV hoses and vacuum hose from intake manifold. Add matching marks as necessary for easier installation.

10. Drain engine coolant from radiator.

❋❋ CAUTION
Perform this step when the engine is cold. Never spill engine coolant on drive belts.

➡**When removing only intake manifold, move electric throttle control actuator without disconnecting the water hoses.**

11. Remove electric throttle control actuator. Loosen mounting bolts in reverse order as shown.

❋❋ CAUTION
Note the following:

- The figure shows the electric throttle control actuator (bank 1) viewed from the air duct side.
- Viewed from the air duct side, the order of loosening mounting bolts of electric throttle control actuator

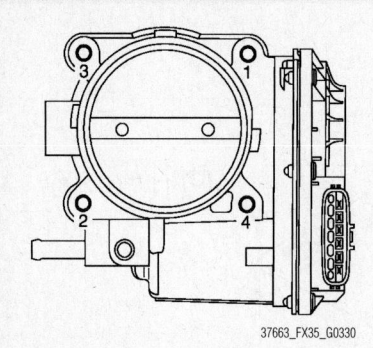

Fig. 269 Loosen mounting bolts in reverse order as shown

(bank 1) is the same as that of the electric throttle control actuator (bank 2).

⁂ **CAUTION**

Note the following:

- Handle carefully to avoid any impact to electric throttle control actuator.
- Never disassemble.

12. Installation is the reverse of removal.

13. Perform the "Throttle Valve Closed Position Learning" when harness connector of electric throttle control actuator is disconnected.

14. Perform the "Idle Air Volume Learning" and "Throttle Valve Closed Position Learning" when electric throttle control actuator is replaced.

THROTTLE VALVE CLOSED POSITION LEARNING

Throttle Valve Closed Position Learning is a function of ECM to learn the fully closed position of the throttle valve by monitoring the throttle position sensor output signal. It must be performed each time the harness connector of the electric throttle control actuator or ECM is disconnected.

1. Check that accelerator pedal is fully released.
2. Turn ignition switch ON.
3. Turn ignition switch OFF and wait at least 10 seconds.

➡**Check that throttle valve moves during the above 10 seconds by confirming the operating sound.**

IDLE AIR VOLUME LEARNING

Idle Air Volume Learning is a function of ECM to learn the idle air volume that keeps engine idle speed within the specific range. It must be performed under the following conditions:

- Each time the electric throttle control actuator or ECM is replaced.
- Idle speed or ignition timing is out of the specification.

Check that all of the following conditions are satisfied.

⁂ **CAUTION**

Learning will be cancelled if any of the following conditions are missed for even a moment.

- Battery voltage: More than 12.9 V (At idle)
- Engine coolant temperature: 158–221°F (70–105°C)
- Selector lever position: P or N
- Electric load switch: OFF (Air conditioner, headlamp, rear window defogger)

➡**On vehicles equipped with daytime light systems, if the parking brake is applied before the engine is started the headlamp will not illuminate.**

- Steering wheel: Neutral (Straight-ahead position)
- Vehicle speed: Stopped
- Transmission: Warmed-up
- With CONSULT-III: Drive vehicle until "ATF TEMP 2" in "DATA MONITOR" mode of "A/T" system indicates less than 0.9 V.
- Without CONSULT-III: Drive vehicle for 10 minutes.

➡**With CONSULT-III:**

1. Perform Accelerator Pedal Released Position Learning.
2. Perform Throttle Valve Closed Position Learning.
3. Start engine and warm it up to normal operating temperature.
4. Select "IDLE AIR VOL LEARN" in "WORK SUPPORT" mode.
5. Touch "START" and wait 20 seconds.

➡**Without CONSULT-III:**

⁂ **CAUTION**

Note the following:

- It is better to count the time accurately with a clock.
- It is impossible to switch the diagnostic mode when an accelerator pedal position sensor circuit has a malfunction.

6. Perform Accelerator Pedal Released Position Learning.
7. Perform Throttle Valve Closed Position Learning.
8. Start engine and warm it up to normal operating temperature.
9. Turn ignition switch OFF and wait at least 10 seconds.
10. Confirm that accelerator pedal is fully released, turn ignition switch ON and wait 3 seconds.
11. Repeat the following procedure quickly 5 times within 5 seconds.
   a. Fully depress the accelerator pedal.
   b. Fully release the accelerator pedal.
12. Wait 7 seconds, fully depress the accelerator pedal for approx. 20 seconds until the MIL stops blinking and turns ON.
13. Fully release the accelerator pedal within 3 seconds after the MIL turns ON.
14. Start engine and let it idle.
15. Wait 20 seconds.
16. Rev up engine two or three times and check that idle speed and ignition timing are within the specifications.
17. Check the following:

- Check that throttle valve is fully closed.
- Check PCV valve operation.
- Check that downstream of throttle valve is free from air leakage.

THROTTLE POSITION SENSOR (TPS)

LOCATION

Electric throttle control actuator consists of throttle control motor, throttle position sensor, etc. The throttle position sensor responds to the throttle valve movement.

REMOVAL & INSTALLATION

Refer to Throttle Control Actuator Removal and Installation section when servicing this component.

FUEL ○ **GASOLINE FUEL INJECTION SYSTEM**

FUEL SYSTEM SERVICE PRECAUTIONS

Safety is the most important factor when performing not only fuel system maintenance, but any type of maintenance. Failure to conduct maintenance and repairs in a safe manner may result in serious personal injury or death. Work on a vehicle's fuel system components can be accomplished safely and effectively by adhering to the following rules and guidelines.

• To avoid the possibility of fire and personal injury, always disconnect the negative battery cable unless the repair or test procedure requires that battery voltage be applied.

• Always relieve the fuel system pressure prior to disconnecting any fuel system component (injector, fuel rail, pressure regulator, etc.) fitting or fuel line connection. Exercise extreme caution whenever relieving fuel system pressure to avoid exposing skin, face and eyes to fuel spray. Please be advised that fuel under pressure may penetrate the skin or any part of the body that it contacts.

• Always place a shop towel or cloth around the fitting or connection prior to loosening to absorb any excess fuel due to spillage. Ensure that all fuel spillage is quickly removed from engine surfaces. Ensure that all fuel-soaked cloths or towels are deposited into a flame-proof waste container with a lid.

• Always keep a dry chemical (Class B) fire extinguisher near the work area.

• Do not allow fuel spray or fuel vapors to come into contact with a spark or open flame.

• Always use a second wrench when loosening or tightening fuel line connection fittings. This will prevent unnecessary stress and torsion on fuel piping. Always follow the proper torque specifications.

• Always replace worn fuel fitting O-rings with new ones. Do not substitute fuel hose where rigid pipe is installed.

FUEL SYSTEM PRESSURE

RELIEVING

With CONSULT-III

1. Turn ignition switch ON.
2. Perform "FUEL PRESSURE RELEASE" in "WORK SUPPORT" mode with CONSULT-III.
3. Start engine.

4. After engine stalls, crank it 2 or 3 times to release all fuel pressure.
5. Turn ignition switch OFF.

Without CONSULT-III

See Figure 270.

Fig. 270 Remove fuel pump fuse (1) located in IPDM E/R (2)

1. Remove fuel pump fuse located in IPDM E/R.
2. Start engine.
3. After engine stalls, crank it 2 or 3 times to release all fuel pressure.
4. Turn ignition switch OFF.
5. Reinstall fuel pump fuse after servicing fuel system.

FUEL FILTER

REMOVAL & INSTALLATION

The fuel filter is an integral part of the Fuel Level Sensor Unit, Fuel Filter, and Fuel Pump Assembly. Refer to the Fuel Level Sensor Unit, Fuel Filter, and Fuel Pump Assembly section when servicing this component.

FUEL LEVEL SENSOR

LOCATION

The fuel level sensor is an integral part of the Fuel Level Sensor Unit, Fuel Filter, and Fuel Pump Assembly. Refer to the Fuel Level Sensor Unit, Fuel Filter, and Fuel Pump Assembly section when servicing this component.

FUEL LEVEL SENSOR UNIT, FUEL FILTER, AND FUEL PUMP ASSEMBLY

REMOVAL & INSTALLATION

See Figures 271 through 278.

1. Check fuel level on fuel gauge. If fuel gauge indicates more than the level as shown (full or almost full), drain fuel from fuel tank until fuel gauge indicates level as shown or below.

➡**Because fuel will be spilled when removing main and sub fuel level sensor units for the top of the fuel is above the main and sub fuel level sensor units installation surface.**

As a guide, fuel level becomes the position as shown or below when approximately 5.25 gals. (20 L) of fuel are drained from fuel tank.

In a case that fuel pump does not operate, perform the following procedure.

a. Insert hose of less than 1 inches (25 mm) in diameter into fuel filler tube through fuel filler opening to draw fuel from fuel filler tube.

b. Disconnect fuel filler hose from fuel filler tube.

c. Insert fuel tube into fuel tank through fuel filler hose to draw fuel from fuel tank.

2. Release the fuel pressure from the fuel lines.
3. Open fuel filler lid.
4. Open filler cap and release the pressure inside fuel tank.
5. Remove rear seat cushion.
6. Peel off floor carpet, then remove inspection hole cover units by turning clips clockwise by 90 degrees.
7. Disconnect harness connector and fuel feed tube.

a. Hold the sides of connector, push in tabs and pull out fuel feed tube.

b. If quick connector sticks to tube of main fuel level sensor unit, push and pull quick connector several times until they start to move. Then disconnect them by pulling.

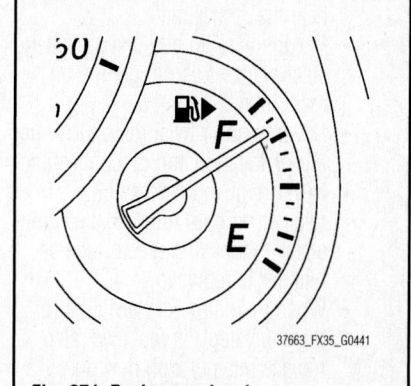

Fig. 271 Fuel gauge level

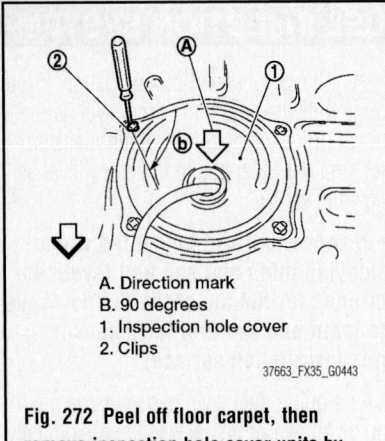

A. Direction mark
B. 90 degrees
1. Inspection hole cover
2. Clips

37663_FX35_G0443

Fig. 272 Peel off floor carpet, then remove inspection hole cover units by turning clips clockwise by 90 degrees

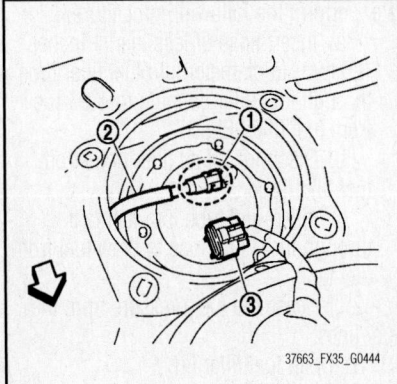

37663_FX35_G0444

Fig. 273 Disconnect harness connector (3) and fuel feed tube (2); quick connector (1)

✳✳ CAUTION

Note the following:

- Quick connector can be disconnected when the tabs are completely depressed. Never twist it more than necessary.
- Never use any tools to disconnected quick connector.
- Keep resin tube away from heat. Be especially careful when welding near the resin tube.
- Prevent acid liquid such as battery electrolyte, etc. from getting on resin tube.
- Never bend or twist resin tube during installation and disconnection.
- Never remove the remaining retainer on hard tube (or the equivalent) except when resin tube or retainer is replaced.
- When resin tube or hard tube (or the equivalent) is replaced, also replace retainer with new one.
- To keep the connecting portion

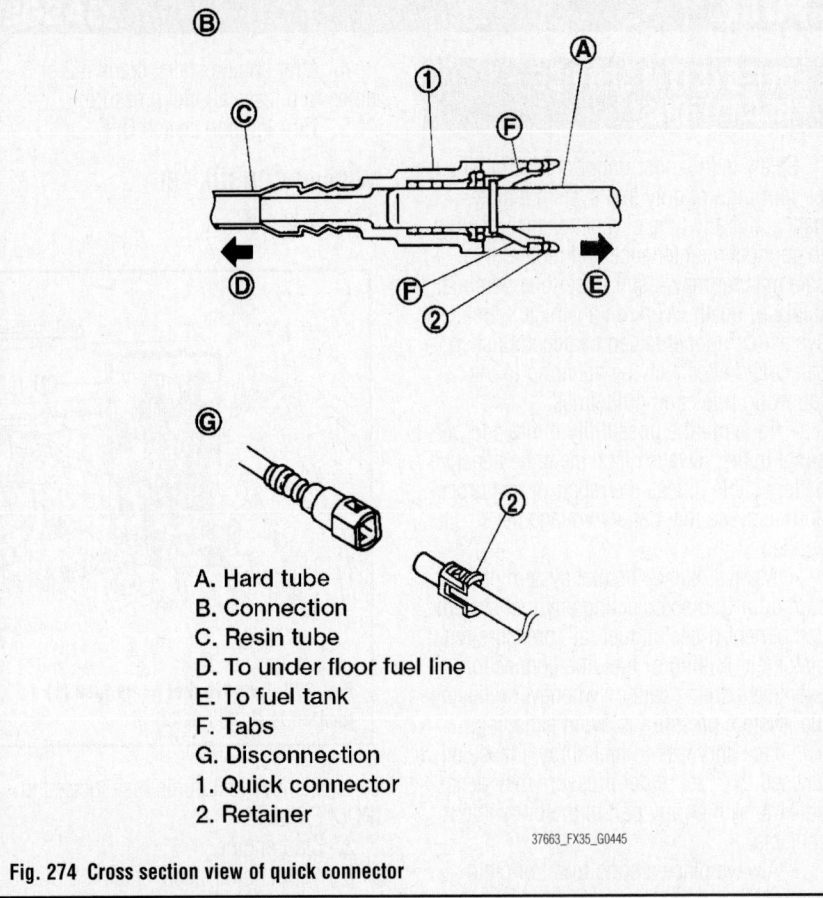

A. Hard tube
B. Connection
C. Resin tube
D. To under floor fuel line
E. To fuel tank
F. Tabs
G. Disconnection
1. Quick connector
2. Retainer

37663_FX35_G0445

Fig. 274 Cross section view of quick connector

clean and to avoid damage and foreign materials, cover them completely with plastic bags or something similar.

8. Remove main fuel level sensor unit, fuel filter and fuel pump assembly, and sub fuel level sensor unit as follows:

✳✳ CAUTION

Never bend float arm during removal. Avoid impacts such as falling when handling components.

a. Removal of main fuel level sensor unit, fuel filter and fuel pump assembly:
b. Remove retainer.
c. Raise main fuel level sensor unit, fuel filter and fuel pump assembly, and disconnect quick connector. Push in tabs and pull out fuel tube.
d. Removal of sub fuel level sensor unit:
e. Remove retainer.
f. Raise and release sub fuel level sensor unit to remove.

To install:

9. Note to the following, and install in the reverse order of removal.

10. Install fuel tube.

11. Face main and sub fuel level sensor units as shown, and install them with the knock pin on back aligned with pin hole on fuel tank.

12. Install retainer so that its notch becomes parallel with the notch on fuel tank.

13. Tighten retainer mounting bolts evenly.

14. Check the connection for damage or any foreign materials.

15. Align the connector with the tube, then insert the connector straight into the tube until a click sound is heard.

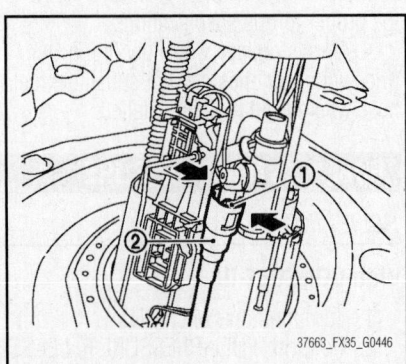

37663_FX35_G0446

Fig. 275 Push in tabs (1) and pull out fuel tube (2)

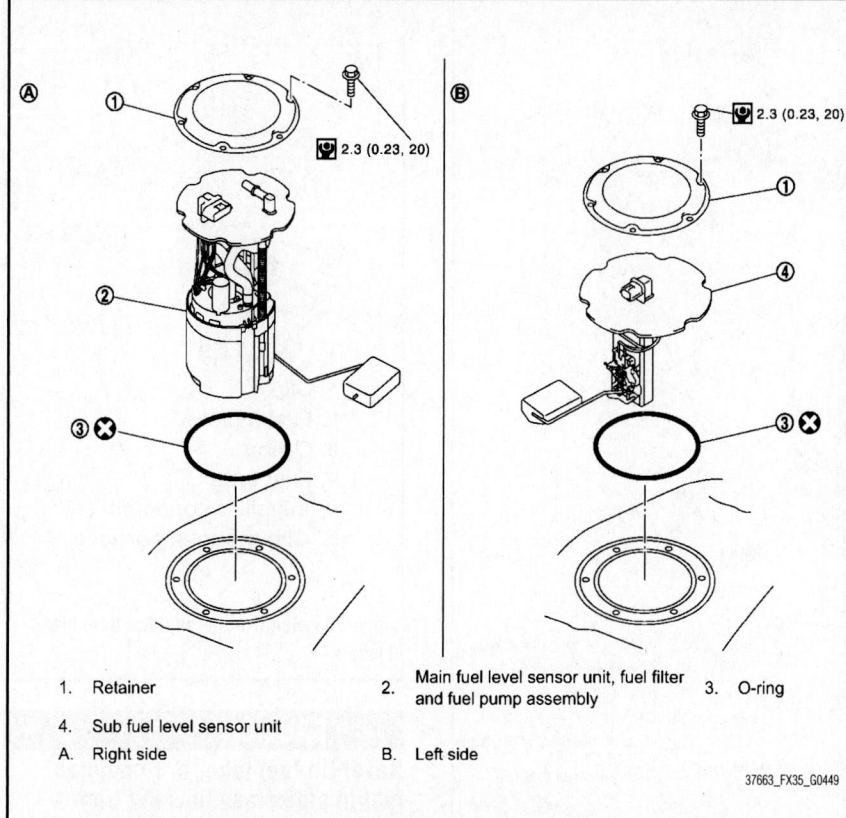

1. Retainer
2. Main fuel level sensor unit, fuel filter and fuel pump assembly
3. O-ring
4. Sub fuel level sensor unit
A. Right side
B. Left side

37663_FX35_G0449

Fig. 276 Exploded view of fuel level sensor unit, fuel filter and fuel pump assembly

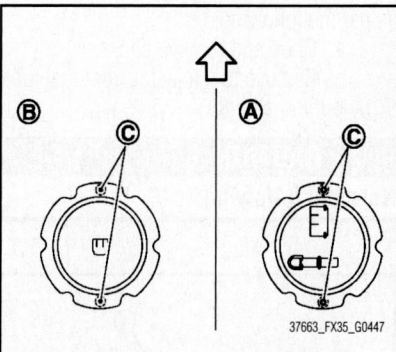

37663_FX35_G0447

Fig. 277 Face main and sub fuel level sensor units as shown, and install them with the knock pin (C) on back aligned with pin hole on fuel tank; right side (A), left side (B)

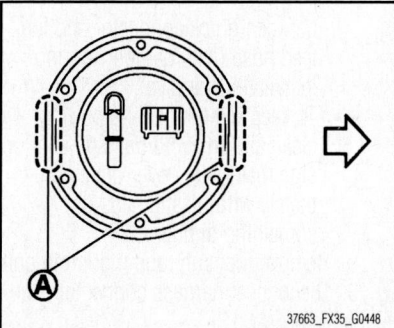

37663_FX35_G0448

Fig. 278 Install retainer so that its notches (A) becomes parallel with the notch on fuel tank

16. After connecting, check that the connection is secure by following method.

  a. Pull the tube and the connector to check they are securely connected.

  b. Visually confirm that the two retainer tabs are connected to the connector.

17. Turn ignition switch "ON" (with engine stopped), then check connections for leakage by applying fuel pressure to fuel piping.

18. Start engine and let it idle and check there are no fuel leakage at the fuel system connections.

FUEL RAIL AND INJECTOR

REMOVAL & INSTALLATION

3.5L Engine

See Figures 279 through 286.

✳✳ WARNING

Note the following:

- Put a "CAUTION: FLAMMABLE" sign in the workshop.
- Be sure to work in a well ventilated area and furnish workshop with a CO2 fire extinguisher.

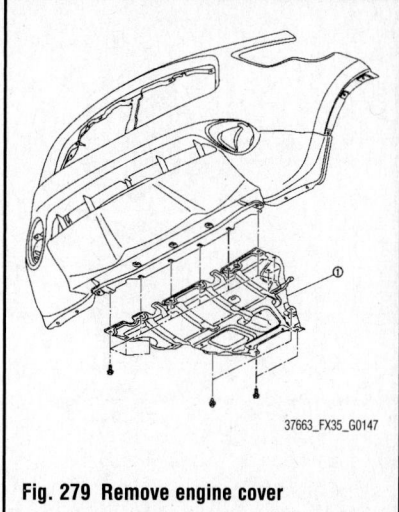

37663_FX35_G0147

Fig. 279 Remove engine cover

- Never smoke while servicing fuel system. Keep open flames and sparks away from the work area.
- Never drain engine coolant when the engine is hot to avoid the danger of being scalded.

1. Release fuel pressure.

2. Disconnect battery cable from the negative terminal.

3. Remove engine cover.

4. Remove air cleaner case and air duct.

5. Remove intake manifold collector.

6. Remove fuel feed hose (with damper) from fuel sub-tube and remove harness bracket.

➡There is no fuel return route.

✳✳ CAUTION

While hoses are disconnected, plug them to prevent fuel from draining. Never separate damper and hose.

7. When separating fuel feed hose (with damper) and centralized under-floor piping connection, disconnect quick connector as per the following:

  a. Remove quick connector cap from quick connector connection on right member side.

  b. Disconnect fuel feed hose (with damper) from bracket hose clamp.

  c. Push in retainer tabs.

  d. Draw and pull out quick connector straight from centralized under-floor piping.

✳✳ CAUTION

Note the following:

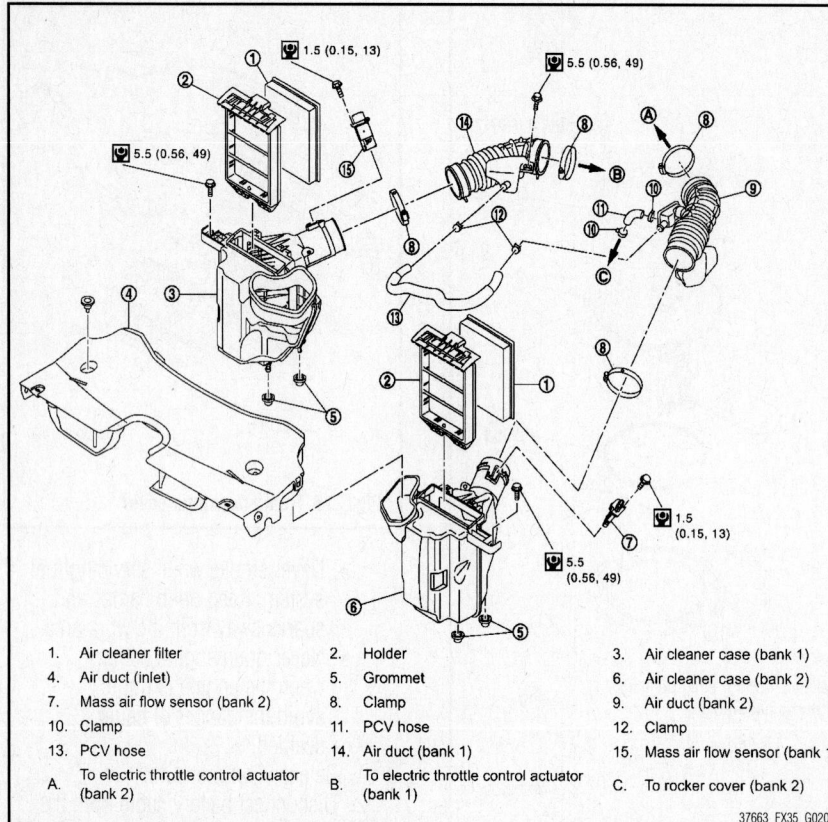

1.	Air cleaner filter	2.	Holder	3.	Air cleaner case (bank 1)
4.	Air duct (inlet)	5.	Grommet	6.	Air cleaner case (bank 2)
7.	Mass air flow sensor (bank 2)	8.	Clamp	9.	Air duct (bank 2)
10.	Clamp	11.	PCV hose	12.	Clamp
13.	PCV hose	14.	Air duct (bank 1)	15.	Mass air flow sensor (bank 1)
A.	To electric throttle control actuator (bank 2)	B.	To electric throttle control actuator (bank 1)	C.	To rocker cover (bank 2)

37663_FX35_G0206

Fig. 280 Remove air cleaner case and air duct

- Pull quick connector holding position (at 90 degrees).
- Never pull with lateral force applied. O-ring inside quick connector may be damaged.
- Prepare container and cloth beforehand as fuel will leak out.
- Avoid fire and sparks.
- Keep parts away from heat source. Especially, be careful when welding is performed around them.
- Never expose parts to battery electrolyte or other acids.
- Never bend or twist connection

between quick connector and fuel feed hose (with damper) during installation/removal.
- To keep the connecting portion clean and to avoid damage and foreign materials, cover them completely with plastic bags or something similar.

8. Remove fuel sub tube mounting bolt.

9. Disconnect harness connector from fuel injector.

10. Loosen mounting bolts in reverse order as shown, and remove fuel tube and fuel injector assembly.

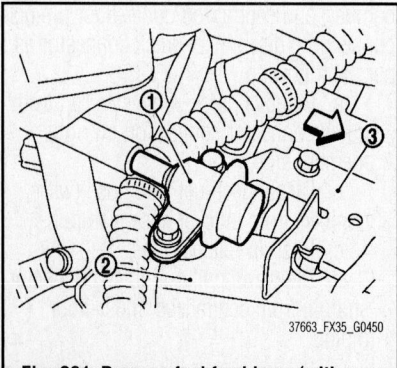

37663_FX35_G0450

Fig. 281 Remove fuel feed hose (with damper) (1) from fuel sub-tube (2) and remove harness bracket (3)

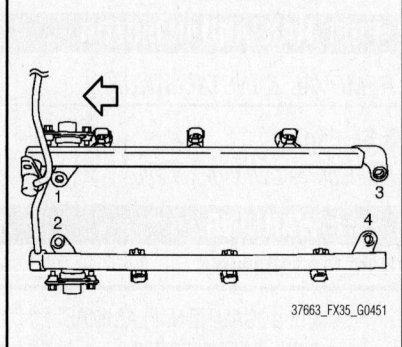

37663_FX35_G0451

Fig. 282 Loosen mounting bolts in reverse order as shown

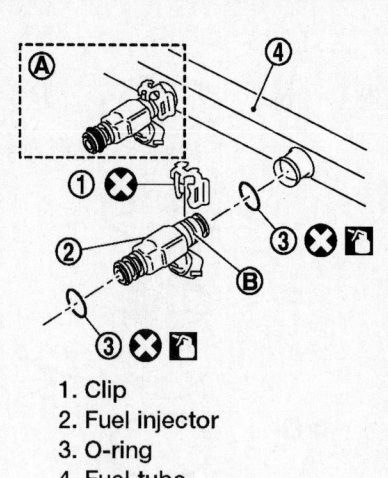

1. Clip
2. Fuel injector
3. O-ring
4. Fuel tube
A. Installed condition
B. Clip mounting groove

37663_FX35_G0452

Fig. 283 Remove fuel injector from fuel tube

✱✱ CAUTION

Never tilt fuel tube, or remaining fuel in pipes may flow out from pipes.

11. Remove fuel injector from fuel tube as per the following:
 a. Open and remove clip.
 b. Remove fuel injector from fuel tube by pulling straight.

✱✱ CAUTION

Note the following:

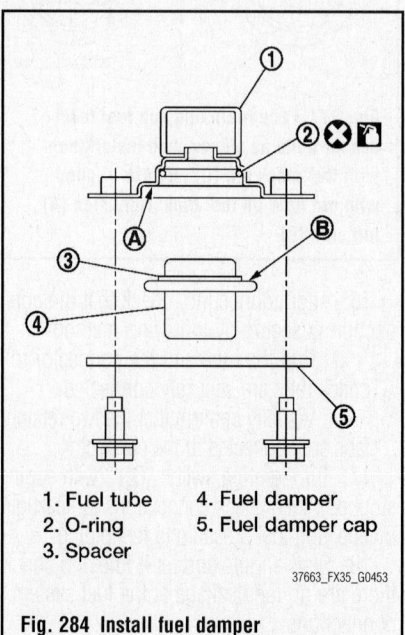

1. Fuel tube 4. Fuel damper
2. O-ring   5. Fuel damper cap
3. Spacer

37663_FX35_G0453

Fig. 284 Install fuel damper

- Be careful with remaining fuel that may go out from fuel tube.
- Be careful not to damage injector nozzles during removal.
- Never bump or drop fuel injector.
- Never disassemble fuel injector.

12. Remove fuel sub-tube and fuel damper, if necessary.

To install:

13. Install fuel damper as per the following:

a. Install new O-ring to fuel tube as shown. When handling new O-ring, be careful of the following caution:

✳✳ CAUTION
Note the following:

- Handle O-ring with bare hands. Never wear gloves.
- Lubricate O-ring with new engine oil.
- Never clean O-ring with solvent.
- Check that O-ring and its mating part are free of foreign material.
- When installing O-ring, be careful not to scratch it with tool or finger-nails. Also be careful not to twist or stretch O-ring. If O-ring was stretched while it was being attached, never insert it quickly into fuel tube.
- Insert new O-ring straight into fuel tube. Never twist it.

b. Install spacer to fuel damper.

c. Insert fuel damper straight into fuel tube.

✳✳ CAUTION
Note the following:

- Insert straight, checking sure that the axis is lined up.
- Insert fuel damper at 29 lbs. (130 N) or less to prevent damage to the parts.
- Insert fuel damper until B is touching A of fuel tube.

d. Tighten bolts evenly in turn. After tightening bolts, check that there is no gap between fuel damper cap and fuel tube.

14. Install fuel sub-tube.

a. When handling new O-rings, be careful of the following caution:

✳✳ CAUTION
Note the following:

- Handle O-ring with bare hands. Never wear gloves.
- Lubricate O-ring with new engine oil.
- Never clean O-ring with solvent.

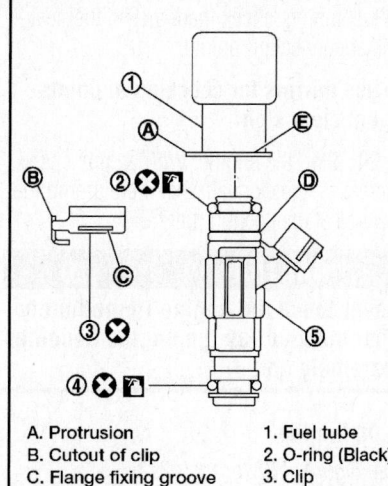

A. Protrusion	1. Fuel tube
B. Cutout of clip	2. O-ring (Black)
C. Flange fixing groove	3. Clip
D. Clip mounting groove	4. O-ring (Green)
E. Fuel tube flange	5. Fuel injector

37663_FX35_G0454

Fig. 285 Install fuel injector to fuel tube

- Check that O-ring and its mating part are free of foreign material.
- When installing O-ring, be careful not to scratch it with tool or finger-nails. Also be careful not to twist or stretch O-ring. If O-ring was stretched while it was being attached, never insert it quickly into fuel tube.
- Insert new O-ring straight into fuel tube. Never decenter or twist it.

b. Insert fuel sub-tube straight into fuel tube.

c. Tighten mounting bolts evenly in turn.

d. After tightening mounting bolts, check that there is no gap between flange and fuel tube.

15. Install new O-rings to fuel injector, paying attention to the following.

✳✳ CAUTION
Note the following:

- Upper and lower O-ring are different. Be careful not to confuse them. Fuel tube side (Black) and Nozzle side (Green)
- Handle O-ring with bare hands. Never wear gloves.
- Lubricate O-ring with new engine oil.
- Never clean O-ring with solvent.
- Check that O-ring and its mating part are free of foreign material.
- When installing O-ring, be careful not to scratch it with tool or finger-nails. Also be careful not to twist or stretch O-ring. If O-ring was

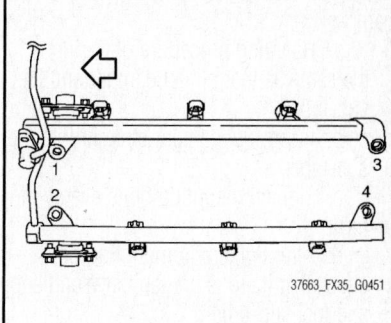

37663_FX35_G0451

Fig. 286 Tighten mounting bolts in two steps in numerical order as shown

stretched while it was being attached, never insert it quickly into fuel tube.

- Insert O-ring straight into fuel injector. Never decenter or twist it.

16. Install fuel injector to fuel tube as per the following:

a. Insert clip into clip mounting groove on fuel injector.

✳✳ CAUTION
Never reuse clip. Replace it with a new one. Be careful to keep clip from interfering with O-ring. If interference occurs, replace O-ring.

b. Insert fuel injector into fuel tube with clip attached.

- Insert it while matching it to the axial center.
- Insert fuel injector so that protrusion of fuel tube matches cutout of clip.
- Check that fuel tube flange is securely fixed in flange fixing groove on clip.

c. Check that installation is complete by checking that fuel injector does not rotate or come off. Check that protrusions of fuel injectors are aligned with cutouts of clips after installation.

17. Install fuel tube and fuel injector assembly to intake manifold. Tighten mounting bolts in two steps in numerical order as shown.

a. Step 1: 7 ft. lbs. (10 Nm)

b. Step 2: 17 ft. lbs. (24 Nm)

✳✳ CAUTION
Be careful not to let tip of injector nozzle come in contact with other parts.

18. Connect injector sub-harness.

19. Install fuel sub tube mounting bolt.

20. Connect fuel feed hose (with damper).

a. Handling procedure of O-ring is the same as that of fuel damper and fuel sub-tube.

b. Insert fuel damper straight into fuel sub-tube.

c. Tighten mounting bolts evenly in turn.

d. After tightening mounting bolts, check that there is no gap between flange and fuel sub-tube.

21. Connect quick connector between fuel feed hose (with damper) and centralized under-floor piping connection as per the following:

a. Check no foreign substances are deposited in and around centralized under-floor piping and quick connector, and no damage on them.

b. Thinly apply new engine oil around centralized under-floor piping from tip end to spool end.

c. Align center to insert quick connector straightly into centralized under-floor piping. Visually confirm that the two retainer tabs are connected to the quick connector.

※※ CAUTION

Note the following:

- Carefully align center to avoid inclined insertion to prevent damage to O-ring inside quick connector.
- Insert until you hear a "click" sound and actually feel the engagement.
- To avoid misidentification of engagement with a similar sound, be sure to perform the next step.

d. Pull quick connector by hand holding position. Check it is completely engaged (connected) so that it does not come out from centralized under-floor piping.

e. Install quick connector cap to quick connector connection. Install quick connector cap with arrow on surface facing the direction of quick connector (fuel feed hose side).

※※ CAUTION

If quick connector cap cannot be installed smoothly, quick connector may have not be installed correctly. Check the connection again.

22. Install in the reverse order of removal after this step.

23. Turn ignition switch "ON" (with the engine stopped). With fuel pressure applied to fuel piping, check there are no fuel leakage at connection points.

➡**Use mirrors for checking at points out of clear sight.**

24. Start the engine. With engine speed increased, check again that there are no fuel leakage at connection points.

※※ CAUTION

Never touch the engine immediately after stopped, as the engine becomes extremely hot.

5.0L Engine

See Figures 287 through 293.

※※ WARNING

Note the following:

- Put a "CAUTION: FLAMMABLE" sign in the workshop.
- Be sure to work in a well ventilated area and furnish workshop with a CO2 fire extinguisher.
- Never smoke while servicing fuel system. Keep open flames and sparks away from the work area.
- To avoid the danger of being scalded, never drain engine coolant when engine is hot.

1. Remove engine cover and engine room cover (RH and LH).

2. Release fuel pressure.

3. Remove the fuel feed hose on the fuel feed damper side with quick connector release J-45488 as per the followings steps.

※※ CAUTION

Use the quick connector release for removing the fuel feed hose on the centralized under-floor piping side as

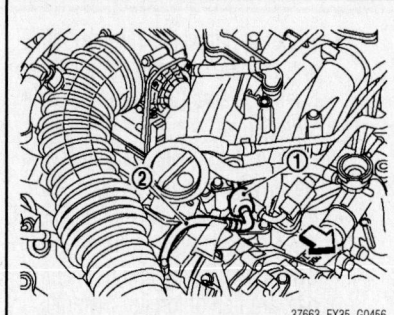

37663_FX35_G0456

Fig. 287 Remove the fuel feed hose (2) on the fuel feed damper side; quick connector cap (1)

well as the fuel feed damper side although the shape of the quick connector is different.

a. Remove quick connector cap from quick connector connection.

b. With the sleeve side of quick connector release facing to quick connector, install quick connector release onto fuel feed hose.

c. Insert quick connector release into quick connector until sleeve contacts and goes no further. Hold quick connector release on that position.

※※ CAUTION

Inserting quick connector release hard will not disconnect quick connector. Hold quick connector release where it contacts and goes no further.

d. Pull out quick connector straight from fuel feed damper.

※※ CAUTION

Note the following:

- Pull quick connector holding position.
- Never pull with lateral force applied. O-ring inside quick connector may be damaged.
- Prepare container and cloth beforehand as fuel will leak out.
- Avoid fire and sparks.
- Keep parts away from heat source. Especially, be careful when welding is performed around them.

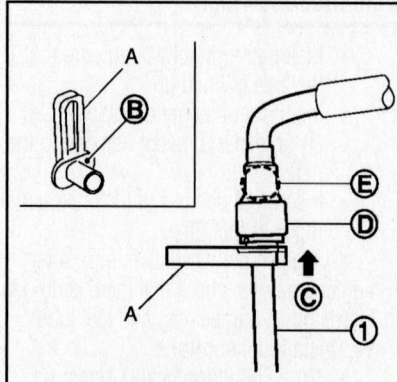

A. Quick connector release
B. Sleeve side
C. Insert and retain
D. Quick connector
E. Holding position
1. Fuel feed damper

37663_FX35_G0457

Fig. 288 Quick connector release

- Never expose parts to battery electrolyte or other acids.
- Never bend or twist connection between quick connector and fuel feed hose during installation/removal.
- To keep the connecting portion clean and to avoid damage and foreign materials, cover them completely with plastic bags or something similar.

4. Remove air duct.

5. Remove electric throttle control actuator.

6. Remove fuel hose (center).

➡**The procedure for removing the quick connector is the same as for removing the fuel feed damper.**

❋❋ **CAUTION**

Disconnect quick connector by using quick connector release J-45488, not by picking out retainer tabs.

7. Remove fuel tube and fuel injector assembly.

❋❋ **CAUTION**

Never tilt it, or remaining fuel in pipes may flow out from pipes.

8. Remove fuel injector from fuel tube as per the following:

   a. Open and remove clip.

   b. Remove fuel injector from fuel tube by pulling straight.

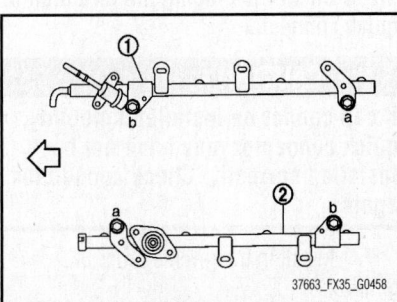

Fig. 289 Loosen mounting bolts (b) first, then loosen mounting bolts (a)

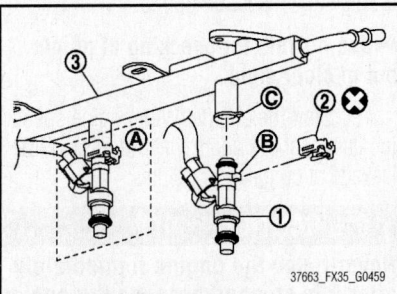

Fig. 290 Remove fuel injector (1) from fuel tube (3), and remove clip (2)

❋❋ **CAUTION**

Note the following:

- Be careful with remaining fuel that may go out from fuel tube.
- Be careful not to damage injector nozzles during removal.
- Never bump or drop fuel injector.
- Never disassemble fuel injector.

9. Disconnect sub harness connector from fuel injectors.

10. Remove fuel damper and fuel feed damper, if necessary.

To install:

11. Install fuel damper as per the following:

   a. Install new O-ring to fuel tube (bank 1). When handling new O-ring, pay attention to the following caution items:

❋❋ **CAUTION**

Note the following:

- Handle O-ring with bare hands. Never wear gloves.
- Lubricate O-ring with new engine oil.
- Never clean O-ring with solvent.
- Check that O-ring and its mating part are free of foreign material.
- When installing O-ring, be careful not to scratch it with tool or fingernails. Also be careful not to twist or stretch O-ring. If O-ring was

stretched while it was being attached, never insert it quickly into fuel tube.

- Insert new O-ring straight into fuel tube. Never decenter or twist it.

   b. Install spacer to fuel damper. Insert fuel damper straight into fuel tube (bank 1).

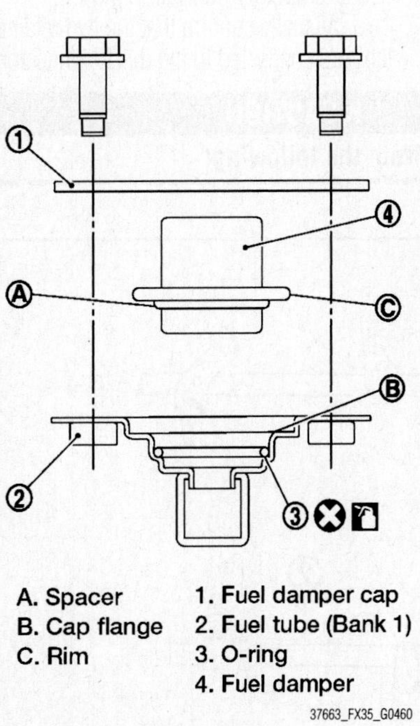

❋❋ **CAUTION**

Note the following:

- Insert straight, check that the axis is lined up.
- Insert fuel damper at 29 lbs. (130 N) or less to prevent damage to the parts.
- Insert fuel damper until the rim reaches the cap flange.

   c. Tighten mounting bolts evenly in turn. After tightening mounting bolts, check that there is no gap between flange and fuel tube (bank 1).

12. Install fuel feed damper.

   a. Handling procedure of O-ring is the same as that of fuel damper.

   b. Insert fuel feed damper straight into fuel tube (bank 2).

❋❋ **CAUTION**

Insert fuel feed damper at 33 lbs. (147 N) or less to prevent damage to the parts

   c. Tighten mounting bolts evenly in turn. After tightening mounting bolts, check that there is no gap between flange and fuel tube (bank 2).

13. Install new O-rings to fuel injector paying attention to the following caution.

❋❋ **CAUTION**

Note the following:

- Upper and lower O-ring are different. Be careful not to confuse them.
- Fuel tube side O-ring (Black); Nozzle side O-ring (Green)
- Handle O-ring with bare hands. Never wear gloves.
- Lubricate O-ring with new engine oil.
- Never clean O-ring with solvent.
- Check that O-ring and its mating part are free of foreign material.
- When installing O-ring, be careful not to scratch it with tool or fingernails. Also be careful not to twist or stretch O-ring. If O-ring was stretched while it was being attached, never insert it quickly into fuel tube.
- Insert O-ring straight into fuel injector. Never decenter or twist it.

14. Install fuel injector to fuel tube as per the following:

A. Spacer
B. Cap flange
C. Rim
1. Fuel damper cap
2. Fuel tube (Bank 1)
3. O-ring
4. Fuel damper

Fig. 291 Install fuel damper

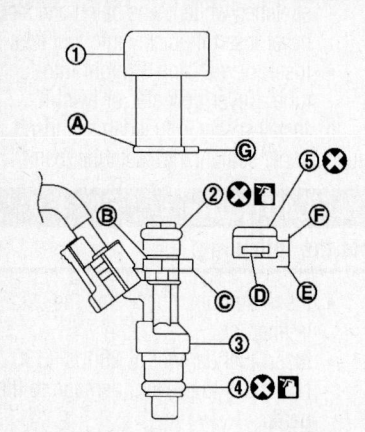

A. Fuel tube flange
B. Clip mounting groove
C. Protrusion
D. Flange fixing groove
E. Cutout of clip
F. Cutout of clip
G. Protrusion

1. Fuel tube
2. O-ring (Black)
3. Fuel injector
4. O-ring (Green)
5. Clip

37663_FX35_G0461

Fig. 292 Install fuel injector to fuel tube

a. Insert clip into clip mounting groove on fuel injector. Insert clip so that protrusion of fuel injector matches cutout of clip.

❋❋ CAUTION

Never reuse clip. Replace it with a new one. Be careful to keep clip from interfering with O-ring. If interference occurs, replace O-ring.

b. Insert fuel injector into fuel tube with clip attached.
- Insert it while matching it to the axial center.
- Insert fuel injector so that protrusion of fuel tube matches cutout of clip.
- Check that fuel tube flange is securely fixed in flange fixing groove on clip.

❋❋ CAUTION

Insert fuel injector at 33 lbs. (147 N) or less to prevent damage to the parts

c. Check that installation is complete by checking that fuel injector does not rotate or come off.

➡**Check that protrusions of fuel injectors and fuel tube are aligned with cutouts of clips after installation.**

15. Install fuel tube and fuel injector assembly. Tighten in two steps.
   a. Step 1: 7 ft. lbs. (10 Nm)
   b. Step 2: 17 ft. lbs. (24 Nm)

❋❋ CAUTION

Be careful not to let tip of injector nozzle come in contact with other parts. Insert fuel injector at 33 lbs. (147 N) or less to prevent damage to the parts

16. Install quick connecters as per the following:

➡**Unless otherwise indicated, the installation to the engine side and centralized under-floor piping side is exactly alike.**

a. Check no foreign substances are deposited in and around fuel piping and quick connector, and no damage on them.
b. Thinly apply new engine oil around fuel piping from tip end to spool end.
c. Align center to insert quick connector straightly into fuel piping.
d. Visually confirm that the two retainer tabs are connected to the quick connector.

❋❋ CAUTION

Note the following:

- Carefully align center to avoid inclined insertion to prevent damage to O-ring inside quick connector.
- Insert until you hear a "click" sound and actually feel the engagement.
- To avoid misidentification of engagement with a similar sound, be sure to perform the next step.

e. Insert quick connector to fuel feed damper piping until top spool is completely inside quick connector and 2nd level spool exposes just below quick connector.

❋❋ CAUTION

Note the following:

- Hold (A) position as shown in the figure when inserting fuel feed hose (1) into quick connector.
- Carefully align center to avoid inclined insertion to prevent damage to O-ring inside quick connector.
- Insert until you hear a "click" sound and actually feel the engagement.
- To avoid misidentification of engagement with a similar sound, be sure to perform the next step.

f. Pull quick connector by hand holding position. Check it is completely engaged (connected) so that it does not come out from fuel piping.
g. Install quick connector cap to quick connector connection.

➡**Install quick connector cap with arrow on surface facing the direction of quick connector.**

❋❋ CAUTION

If cap cannot be installed smoothly, quick connector may have not been installed correctly. Check connection again.

17. Install in the reverse order of removal.
18. Turn ignition switch "ON" (with the engine stopped). With fuel pressure applied to fuel piping, check that there is no fuel leakage at connection points.

➡**Use mirrors for checking at points out of clear sight.**

19. Start the engine. With engine speed increased, check again that there is no fuel leakage at connection points.

❋❋ CAUTION

Never touch the engine immediately after it is stopped because the engine is extremely hot.

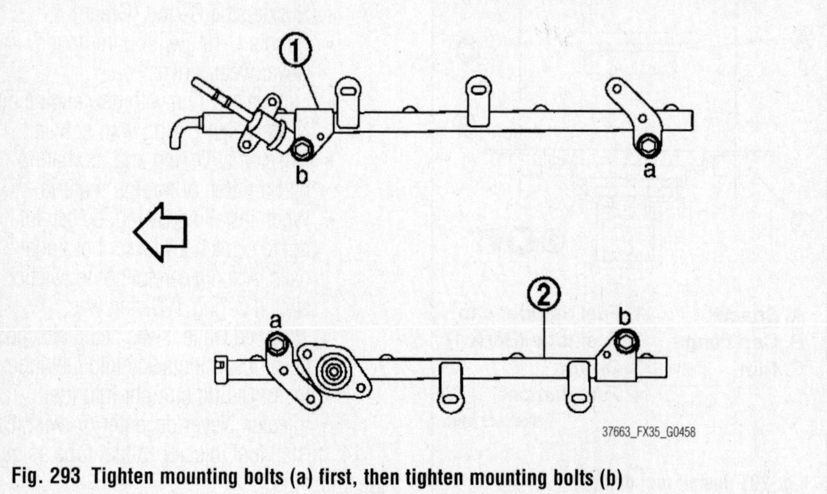

37663_FX35_G0458

Fig. 293 Tighten mounting bolts (a) first, then tighten mounting bolts (b)

FUEL TANK

REMOVAL & INSTALLATION

See Figures 294 through 297.

1. Check fuel level on fuel gauge. If fuel gauge indicates more than the level as shown (full or almost full), drain fuel from fuel tank until fuel gauge indicates level as shown or below.

➡**Because fuel will be spilled when removing main and sub fuel level sensor units for the top of the fuel is above the main and sub fuel level sensor units installation surface.**

As a guide, fuel level becomes the position as shown or below when approximately 5.25 gals. (20 L) of fuel are drained from fuel tank.

In a case that fuel pump does not operate, perform the following procedure.

   a. Insert hose of less than 1 inches (25 mm) in diameter into fuel filler tube through fuel filler opening to draw fuel from fuel filler tube.

   b. Disconnect fuel filler hose from fuel filler tube.

   c. Insert fuel tube into fuel tank through fuel filler hose to draw fuel from fuel tank.

2. Release the fuel pressure from the fuel lines.

3. Open fuel filler lid.

4. Open filler cap and release the pressure inside fuel tank.

5. Remove rear seat cushion.

6. Peel off floor carpet, then remove inspection hole cover units by turning clips clockwise by 90 degrees.

7. Disconnect harness connector and fuel feed tube.

   a. Hold the sides of connector, push in tabs and pull out fuel feed tube.

   b. If quick connector sticks to tube of main fuel level sensor unit, push and pull quick connector several times until they start to move. Then disconnect them by pulling.

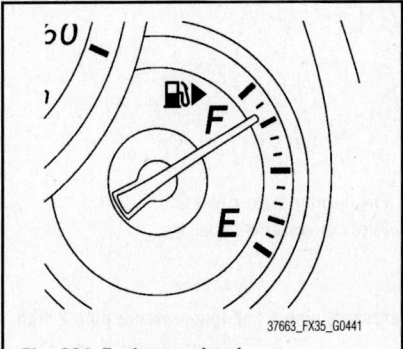

Fig. 294 Fuel gauge level

※※ CAUTION

Note the following:

- Quick connector can be disconnected when the tabs are completely depressed. Never twist it more than necessary.
- Never use any tools to disconnected quick connector.
- Keep resin tube away from heat. Be especially careful when welding near the resin tube.
- Prevent acid liquid such as battery electrolyte, etc. from getting on resin tube.
- Never bend or twist resin tube during installation and disconnection.
- Never remove the remaining retainer on hard tube (or the equivalent) except when resin tube or retainer is replaced.
- When resin tube or hard tube (or the equivalent) is replaced, also replace retainer with new one.
- To keep the connecting portion clean and to avoid damage and foreign materials, cover them completely with plastic bags or something similar.

8. Remove exhaust front tube, center muffler and main muffler.

9. Remove propeller shaft.

10. Remove parking rear brake cables.

11. Remove rear suspension member assembly.

➡**For this service, drive shaft, final drive, and rear suspension member are required not to be separate one another during removal.**

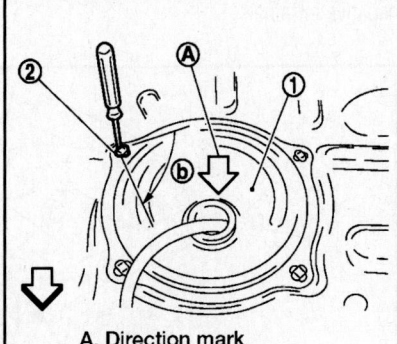

A. Direction mark
B. 90 degrees
1. Inspection hole cover
2. Clips

Fig. 295 Peel off floor carpet, then remove inspection hole cover units by turning clips clockwise by 90 degrees

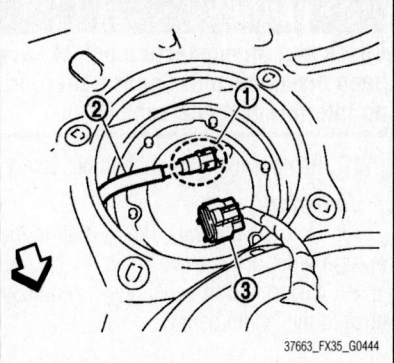

Fig. 296 Disconnect harness connector (3) and fuel feed tube (2); quick connector (1)

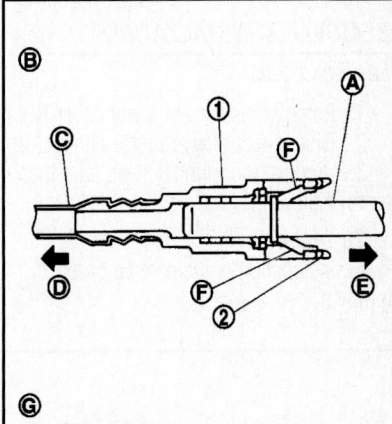

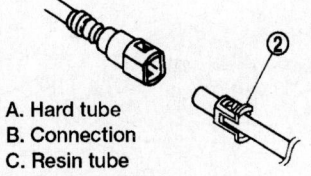

A. Hard tube
B. Connection
C. Resin tube
D. To under floor fuel line
E. To fuel tank
F. Tabs
G. Disconnection
1. Quick connector
2. Retainer

Fig. 297 Cross section view of quick connector

12. Disconnect fuel filler hose (1), EVAP tube (2) and vent hose (3).

13. Remove fuel tank protector.

14. Support the lower part of fuel tank (1) with transmission jack (A).

※※ CAUTION

Support the position that fuel tank mounting bands never engage.

15. Remove fuel tank mounting bands.

16. Supporting with hands, descend transmission jack carefully, and remove fuel tank.

✳✳ CAUTION

Check that all connection points have been disconnected. Confirm there is no interference with vehicle.

17. Remove fuel filler tube, if necessary.

To install:

18. Note the following, and install in the reverse order of removal.

19. Securely clamp fuel hoses and insert hose to the length below.

20. Be sure hose clamp is not placed on swelled area of fuel tube.

21. Tighten the clamp hand with the top mark (A) until the mark is on the bolt head flange.

22. Turn ignition switch "ON" (with engine stopped), and check connections for leakage by applying fuel pressure to fuel piping.

23. Start engine and rev it up and check there are no fuel leakage at the fuel system tube and hose connections.

24. After removing/installing rear suspension assembly, check to adjust wheel alignment.

IDLE SPEED

ADJUSTMENT

Idle speed is controlled by the ECM. No adjustment is necessary or possible.

HEATING & AIR CONDITIONING SYSTEM

BLOWER MOTOR

REMOVAL & INSTALLATION

See Figure 298.

1. Remove instrument lower cover RH.
2. Disconnect blower motor connector.
3. Remove mounting screws, and then remove blower motor.

To install:

4. Installation is the reverse order of removal.

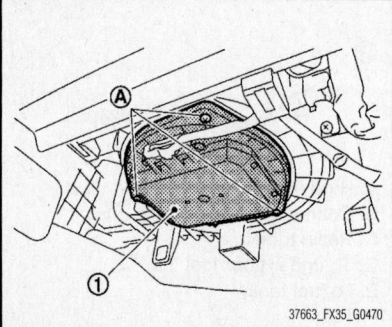

Fig. 298 Remove mounting screws (A), and then remove blower motor (1)

HEATER CORE

REMOVAL & INSTALLATION

See Figure 299.

1. Remove heater & cooling unit assembly.
2. Remove mounting screws, and then remove heater pipe cover.
3. Remove mounting screws, and then remove foot duct (left).
4. Slide heater core to leftward (as shown).

To install:

5. Installation is basically the reverse order of removal.

A. Mounting screws
B. Mounting screws
1. Heater pipe cover
2. Foot duct (left)
3. Heater core

37663_FX35_G0488

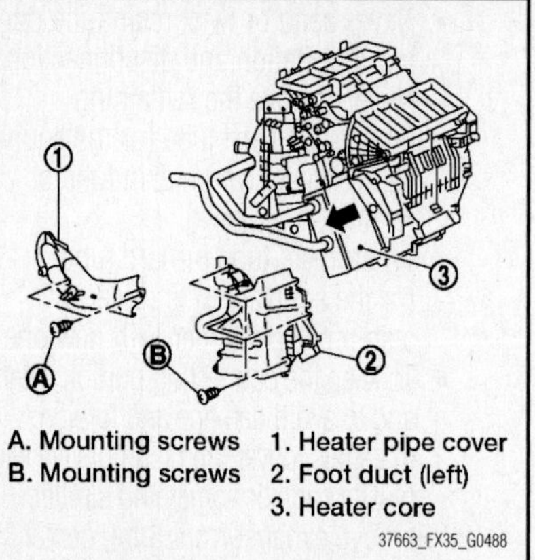

Fig. 299 Exploded view showing heater core removal

HEATER & COOLING UNIT

REMOVAL & INSTALLATION

See Figures 300 through 304.

1. Set the temperature at 60°F (18.0°C).
2. Disconnect the battery cable from the negative terminal.

3. Use a refrigerant collecting equipment (for HFC-134a) to discharge the refrigerant.

4. Drain engine coolant from cooling system.

5. Remove cowl top cover.

6. Remove engine cover.

7. Disconnect one-touch joint between low-pressure pipe 1 and low-pressure pipe 2 with disconnector (SST: 9253089916).

✳✳ CAUTION

Cap or wrap the joint of the A/C piping with suitable material such as vinyl tape to avoid the entry of air.

8. Disconnect one-touch joint between high-pressure pipe 1 and high-pressure pipe 2 with disconnector (SST: 9253089908).

✳✳ CAUTION

Cap or wrap the joint of the A/C piping with suitable material such as vinyl tape to avoid the entry of air.

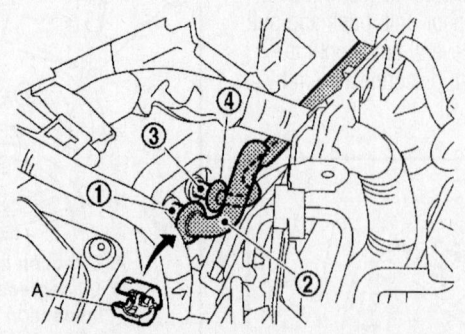

A. Disconnector
1. Low-pressure pipe 1
2. Low-pressure pipe 2
3. High-pressure pipe 2
4. High-pressure pipe 1

37663_FX35_G0489

Fig. 300 Disconnect one-touch joint between low-pressure pipe 1 and low-pressure pipe 2 with disconnector

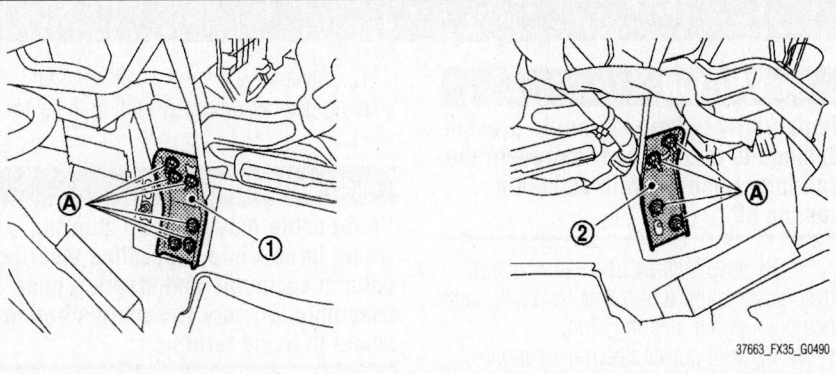

Fig. 301 Remove mounting nuts (A), and then remove instrument stay (left) (1) and instrument stay (right) (2)

9. Remove clamps and then disconnect heater hoses.

10. Remove instrument panel assembly. Refer to Interior section for Instrument Panel Removal and Installation.

11. Remove defroster nozzle and adaptor duct.

12. Remove blower unit.

13. Remove mounting nuts, and then remove instrument stay (left) and instrument stay (right).

14. Disconnect drain hose from heater & cooling unit assembly.

15. Remove mounting bolts from steering member.

16. Remove mounting bolts from steering member.

17. Remove ground bolts from steering member.

18. Remove steering column mounting nuts and bolts.

19. Remove harness connector and clips of vehicle harness from steering member.

20. Remove mounting bolts from steering member.

21. Remove steering member mounting bolt.

22. Remove steering member, and then remove heater & cooling unit assembly.

To install:

23. Installation is basically the reverse order of removal.

❋❋ CAUTION

Note the following:

- Replace O-rings with new ones. Then apply compressor oil to them when installing.
- Female-side piping connection is thin and easy to deform. Slowly insert the male-side piping straight in axial direction.
- Insert piping securely until a clicks is heard.
- After piping connection is completed, pull male-side piping by hand to make sure that connection does not come loose.

24. Recharge the refrigerant.

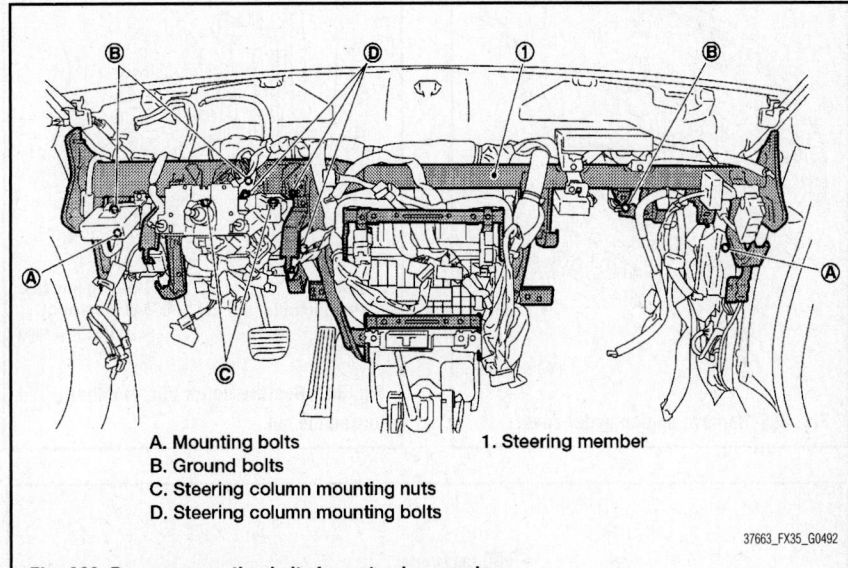

A. Mounting bolts
B. Ground bolts
C. Steering column mounting nuts
D. Steering column mounting bolts

1. Steering member

Fig. 303 Remove mounting bolts from steering member

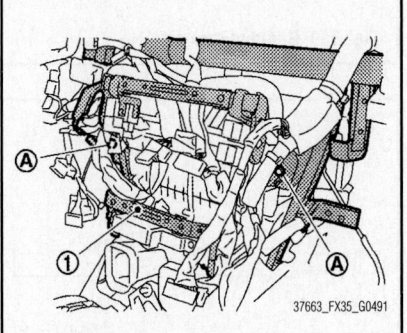

Fig. 302 Remove mounting bolts (A) from steering member (1)

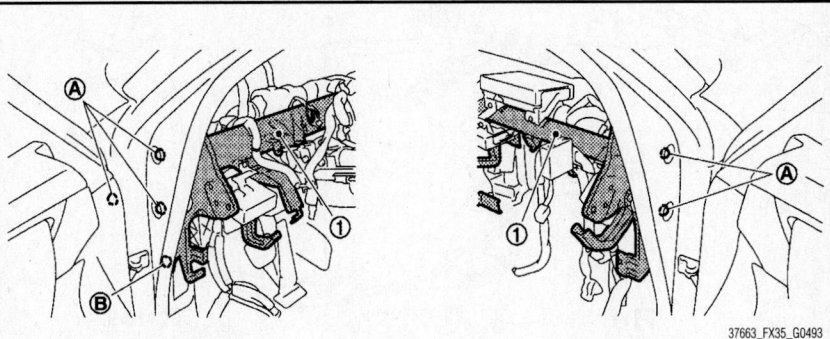

Fig. 304 Remove mounting bolts (A) and steering member mounting bolt (B) from steering member (1)

STEERING

POWER RACK & PINION STEERING GEAR

REMOVAL & INSTALLATION

See Figures 305 through 313.

1. Set the vehicle to the straight-ahead position.
2. Remove tires and wheels.
3. Remove engine under cover.
4. Remove front cross bar.
5. Remove cotter pin, and then loosen the nut.
6. Remove steering outer socket from steering knuckle so as not to damage ball joint boot using suitable ball joint remover.

✳✳ CAUTION

Temporarily tighten the nut to prevent damage to threads and to prevent the ball joint remover from suddenly coming off.

7. Remove high pressure piping and low pressure piping of hydraulic piping, and then drain power steering fluid.
8. Remove power steering solenoid valve harness connector and harness clip.
9. Remove steering hydraulic piping bracket (5.0L engine).
10. Remove rack stay.
11. Remove lower joint mounting bolt (steering gear side).

12. Separate the lower shaft from the steering gear assembly by sliding the side shaft.

✳✳ CAUTION

Spiral cable may be cut if steering wheel turns while separating steering column assembly and steering gear assembly. Be sure to secure steering wheel to avoid turning.

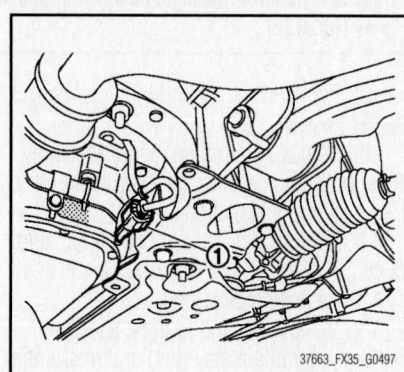

Fig. 308 Remove power steering solenoid valve harness connector (1) and harness clip

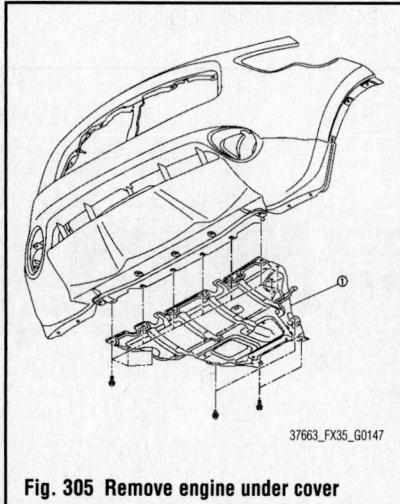

Fig. 305 Remove engine under cover

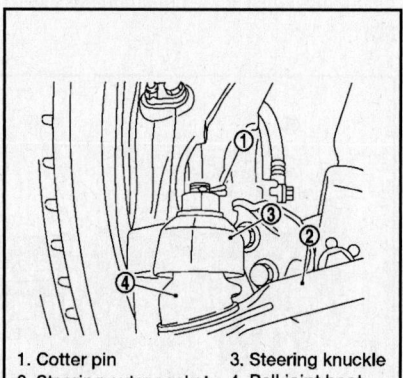

1. Cotter pin
2. Steering outer socket
3. Steering knuckle
4. Ball joint boot

37663_FX35_G0496

Fig. 307 Remove cotter pin, and then loosen the nut

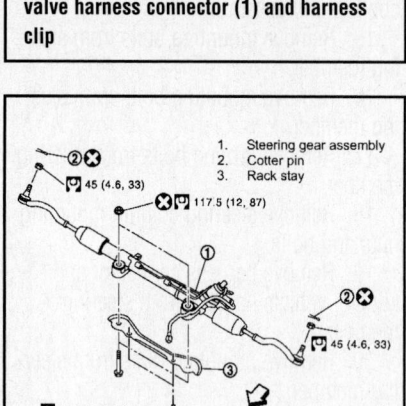
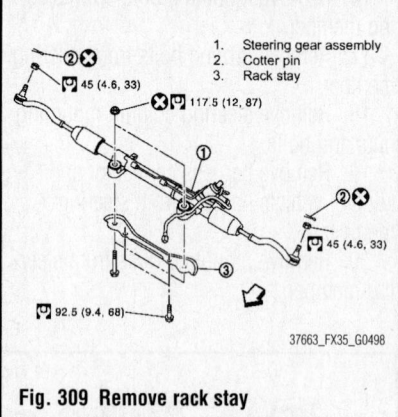

1. Steering gear assembly
2. Cotter pin
3. Rack stay

45 (4.6, 33)
117.5 (12, 87)
45 (4.6, 33)
92.5 (9.4, 68)

37663_FX35_G0498

Fig. 309 Remove rack stay

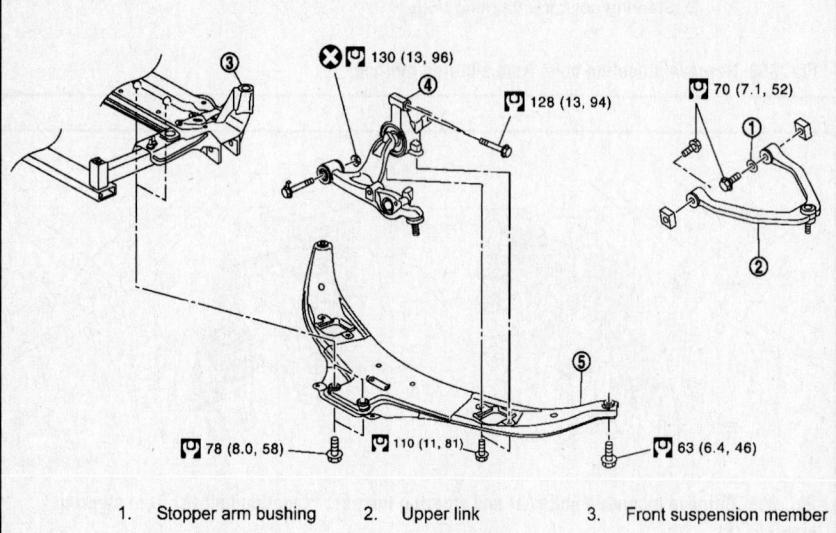

130 (13, 96)
128 (13, 94)
70 (7.1, 52)
78 (8.0, 58)
110 (11, 81)
63 (6.4, 46)

1. Stopper arm bushing
2. Upper link
3. Front suspension member
4. Transverse link
5. Front cross bar

37663_FX35_G0148

Fig. 306 Remove front cross bar

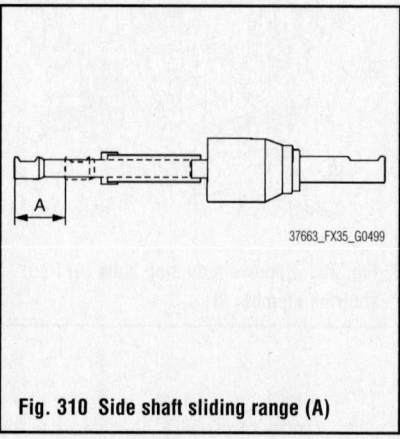

37663_FX35_G0499

Fig. 310 Side shaft sliding range (A)

13. Remove steering gear assembly.

To install:

14. Note the following, and install in the reverse order of removal.

✳✳ CAUTION

Spiral cable may be cut if steering wheel turns while separating steering column assembly and steering gear assembly. Be sure to secure steering wheel using string to avoid turning.

15. Tighten the mounting bolts in the order shown when installing the steering gear assembly.

✳✳ CAUTION

Never reuse the steering gear assembly mounting nut.

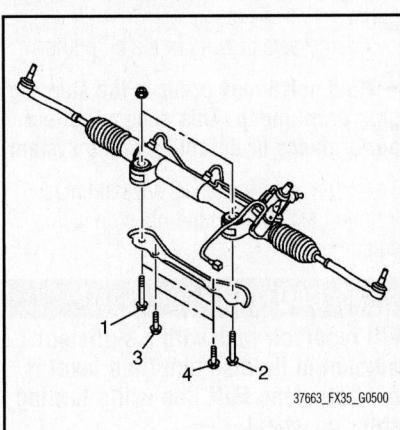

Fig. 311 Tighten the mounting bolts in the order shown

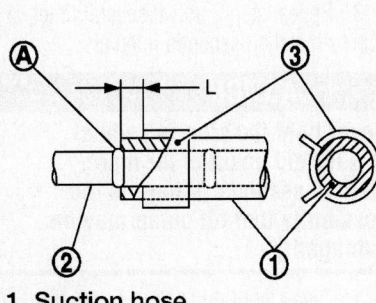

1. Suction hose
2. Tube
3. Clamp
A. Spool of tube
L. Clearance:
   0.12-0.31 inches (3-8 mm)

Fig. 312 Install suction hoses

16. Install suction hoses according to the figure shown.

✳✳ CAUTION

Note the following:

- Never apply fluid to the hose and tube.
- Insert hose securely until it contacts spool of tube.
- Leave clearance when installing clamp.

17. When installing lower joint to steering gear assembly, follow the procedure listed below.

a. Set rack of steering gear in the neutral position.

➡**To get the neutral position of rack, turn gear-sub assembly and measure the distance of inner socket, and then measure the intermediate position of the distance.**

b. Align rear cover cap projection with the marking position of gear housing assembly.

c. Install slit part of lower joint aligning with the rear cover cap projection. Make sure that the slit part of lower joint is aligned with rear cover cap projection and the marking position of gear housing assembly.

18. After installation, bleed air from the steering hydraulic system.

19. Perform final tightening of nuts and bolts on each part under loaded conditions with tires on level ground. Check wheel alignment.

20. Adjust neutral position of steering angle sensor after checking wheel alignment.

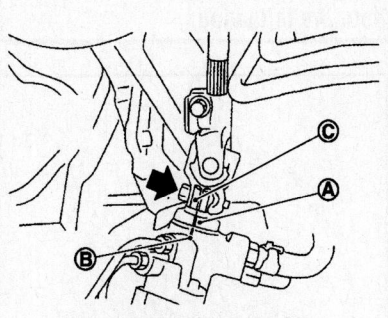

A. Rear cover cap projection
B. Gear housing assembly
C. Lower joint

Fig. 313 Align rear cover cap projection with the marking position of gear housing assembly

ADJUSTMENT OF STEERING ANGLE SENSOR NEUTRAL POSITION

1. Stop the vehicle with front wheels in straight-ahead position.
2. On the CONSULT-III screen, touch "WORK SUPPORT" and "ST ANGLE SENSOR ADJUSTMENT" in order.
3. Touch "START".

✳✳ CAUTION

Do not touch steering wheel while adjusting steering angle sensor.

4. After approximately 10 seconds, touch "END".

➡**After approximately 60 seconds, it ends automatically.**

5. Turn ignition switch OFF, then turn it ON again.

✳✳ CAUTION

Be sure to perform above operation.

6. Run the vehicle with front wheels in straight-ahead position, then stop.
7. Select "STR ANGLE SIG" in "DATA MONITOR" and check steering angle sensor signal.

➡**Steering angle signal: 0+/-2.5°**

8. Erase the self-diagnosis memories of the ABS actuator and electric unit (control unit), ECM and ICC.

POWER STEERING PUMP

REMOVAL & INSTALLATION

3.5L Engine
See Figures 314 and 315.

1. Drain power steering fluid from reservoir tank.
2. Remove the air cleaner and air duct (RH).
3. Loosen drive belt.
4. Remove drive belt from oil pump pulley.
5. Remove pressure sensor connector.
6. Remove copper washers and eye bolt (drain fluid from their piping).
7. Remove suction hose (drain fluid from their piping).
8. Remove oil pump mounting bolts, and then remove oil pump.

To install:

9. Note the following, and install in the reverse order of removal.
10. Install suction hoses according to the figure shown.

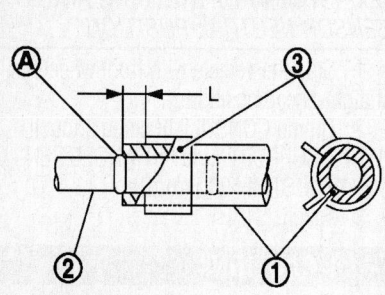

1. Suction hose
2. Tube
3. Clamp
A. Spool of tube
L. Clearance:
   0.12–0.31 inches (3–8 mm)

37663_FX35_G0501

Fig. 314 Install suction hoses

✷✷ CAUTION

Note the following:

- Never apply fluid to the hose and tube.
- Insert hose securely until it contacts spool of tube.
- Leave clearance when installing clamp.

11. Install eye bolt and copper washer to oil pump according to the figure.

✷✷ CAUTION

Note the following:

- Never reuse copper washer.
- Apply power steering fluid to around copper washers, then install eye bolt.

12. Install eye bolt with eye joint (assembled to high pressure hose) protrusion facing with pump side cutout,

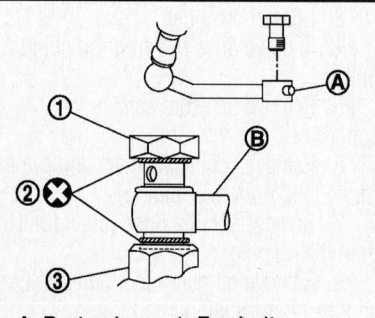

A. Protrusion   1. Eye bolt
B. Eye joint   2. Copper washer
                  3. Oil pump

37663_FX35_G0506

Fig. 315 Install eye bolt and copper washer to oil pump

and then tighten it to 43 ft. lbs. (59 Nm) after tightening by hand.

13. Securely insert harness connector to pressure sensor.

14. Adjust belt tension.

15. Check fluid level, fluid leakage and air bleeding hydraulic system after the installation.

5.0L Engine

See Figures 314 through 316.

1. Drain power steering fluid from reservoir tank.

2. Remove the engine under cover from vehicle.

3. Loosen drive belt.

4. Remove drive belt from oil pump pulley.

5. Remove pressure sensor connector.

6. Remove joint mounting nut.

7. Remove suction hose (drain fluid from their piping).

8. Remove oil pump mounting bolts, and then remove oil pump.

To install:

9. Note the following, and install in the reverse order of removal.

10. Install suction hoses according to the figure shown.

✷✷ CAUTION

Note the following:

- Never apply fluid to the hose and tube.
- Insert hose securely until it contacts spool of tube.
- Leave clearance when installing clamp.

11. Install eye bolt and copper washer to oil pump according to the figure.

✷✷ CAUTION

Note the following:

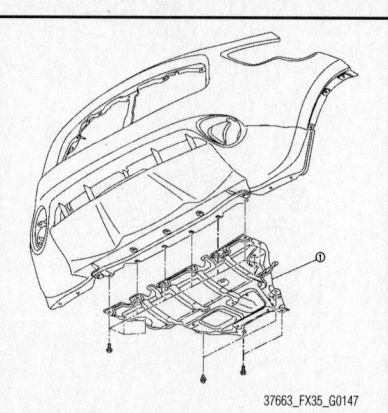

37663_FX35_G0147

Fig. 316 Remove the engine under cover from vehicle

- Never reuse copper washer.
- Apply power steering fluid to around copper washers, then install eye bolt.

12. Install eye bolt with eye joint (assembled to high pressure hose) protrusion facing with pump side cutout, and then tighten it to the specified torque after tightening by hand.

13. Securely insert harness connector to pressure sensor.

14. About the installation of drive belt.

15. Check fluid level, fluid leakage and air bleeding hydraulic system after the installation.

BLEEDING

If air bleeding is not complete, the following symptoms can be observed.

- Bubbles are created in reservoir tank.
- Clicking noise can be heard from oil pump.
- Excessive buzzing in the oil pump.

➡**Fluid noise may occur in the steering gear or oil pump. This does not affect performance or durability of the system.**

1. Turn steering wheel several times from full left stop to full right stop with engine off.

✷✷ CAUTION

Fill reservoir tank with a sufficient amount of fluid so that fluid level is not below the MIN line while turning steering wheel.

2. Start the engine and hold steering wheel at each lock position for 3 seconds at idle to check for fluid leakage.

3. Repeat step 2 above several times at approximately 3 seconds intervals.

✷✷ CAUTION

Never hold the steering wheel in a locked position for more than 10 seconds. (There is the possibility that oil pump may be damaged.)

4. Check fluid for bubbles and white contamination.

5. Stop the engine if bubbles and white contamination do not drain out. Perform step 2 and 3 above after waiting until bubbles and white contamination drain out.

6. Stop the engine, and then check fluid level.

SUSPENSION | **FRONT SUSPENSION**

CONTROL LINKS

REMOVAL & INSTALLATION

Transverse Link

See Figure 317.

1. Remove tires and wheels.
2. Remove shock absorber (strut).
3. Remove stabilizer connecting rod.
4. Remove front cross bar.
5. Remove transverse link from steering knuckle.
6. Set suitable jack under transverse link.
7. Remove transverse link.

➡**If removing transverse link mounting bolt (front side) is difficult, rotating steering wheel and remove steering outer socket.**

To install:

8. Note the following, and install in the reverse order of removal.
9. Never tap on the ball joint cap of the stabilizer connecting rod with a hammer or a similar item when inserting the stabilizer connecting rod into the transverse link.
10. Perform final tightening of bolts and nuts at the front suspension member installation and shock absorber lower side (rubber bushing), under unladen conditions with tires on level ground.
11. Check wheel sensor harness for proper connection.
12. Check wheel alignment.
13. Adjust neutral position of steering angle sensor.

Upper Link

See Figure 318.

1. Remove tires and wheels.
2. Remove shock absorber (strut).
3. Remove upper link and stopper arm bushing.

To install:

4. Note the following, and install in the reverse order of removal.
5. Perform final tightening of bolts and nuts at the vehicle installation position (rubber bushing), under unladen conditions with tires on level ground.

6. Check wheel sensor harness for proper connection.
7. Check wheel alignment.
8. Adjust neutral position of steering angle sensor.

STEERING KNUCKLE

REMOVAL & INSTALLATION

2WD

See Figure 319.

1. Remove tires and wheels.

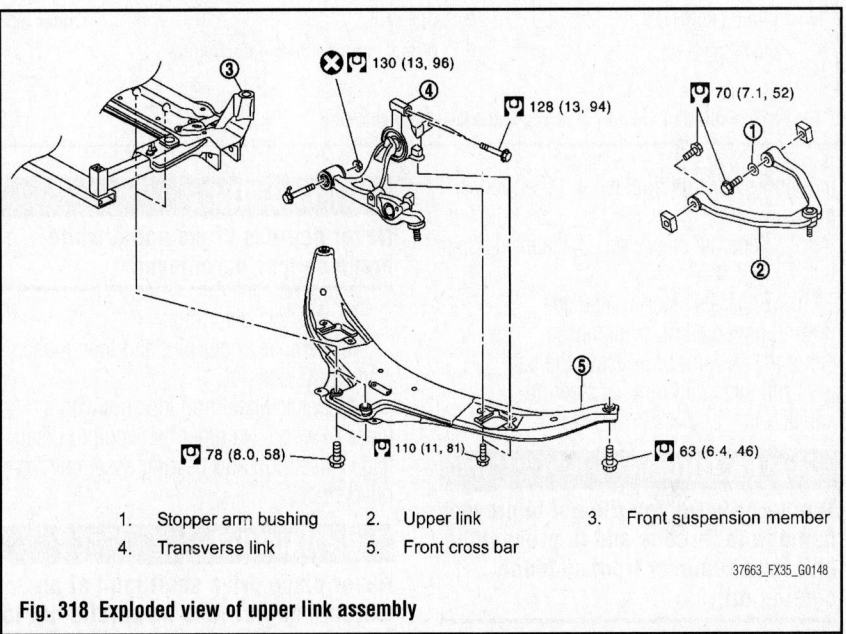

| 1. | Stopper arm bushing | 2. | Upper link | 3. | Front suspension member |
| 4. | Transverse link | 5. | Front cross bar | | |

37663_FX35_G0148

Fig. 318 Exploded view of upper link assembly

2. Remove wheel sensor and sensor harness.

✳✳ CAUTION

Never pull on wheel sensor harness.

3. Remove brake hose bracket.
4. Remove caliper assembly with power tool. Hang caliper assembly in a place where it will not interfere with work.

✳✳ CAUTION

Never depress brake pedal while brake caliper is removed.

5. Remove disc rotor.
6. Remove wheel hub and bearing assembly, and then remove splash guard.
7. Remove steering outer socket.
8. Remove cotter pin of transverse

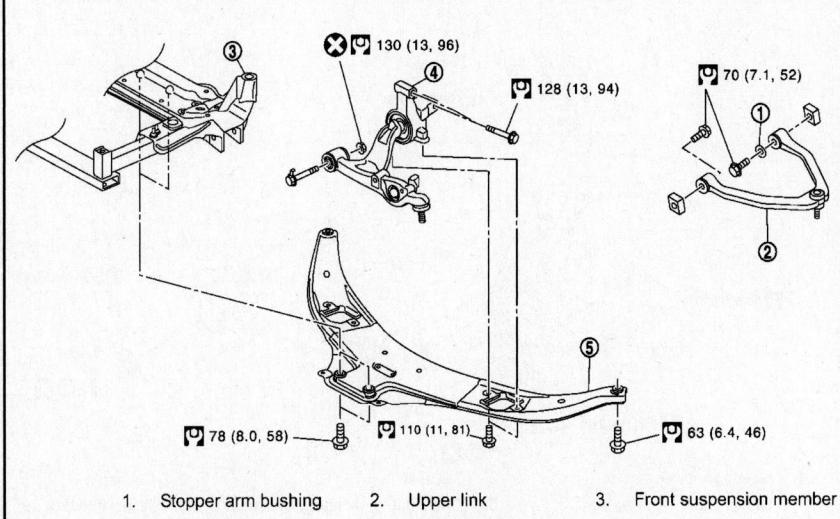

| 1. | Stopper arm bushing | 2. | Upper link | 3. | Front suspension member |
| 4. | Transverse link | 5. | Front cross bar | | |

37663_FX35_G0148

Fig. 317 Exploded view of transverse link assembly

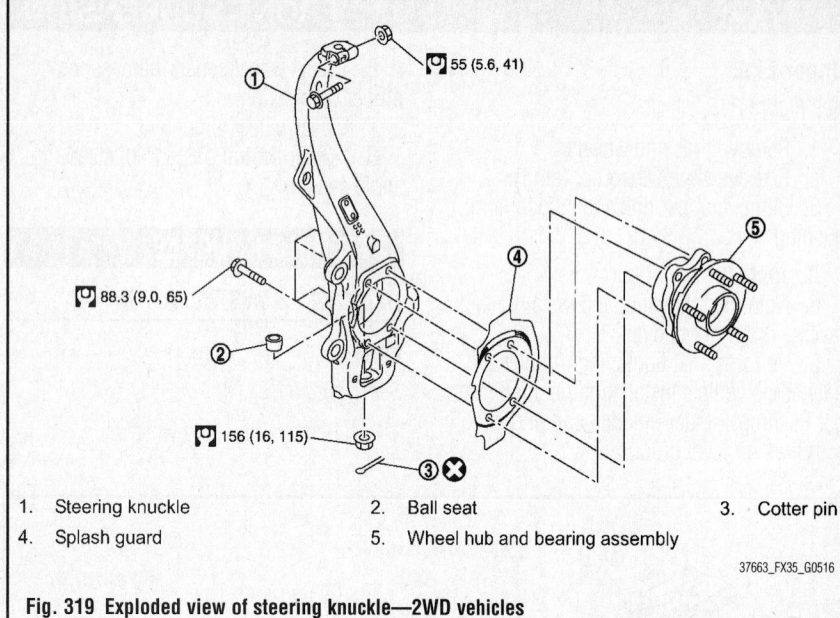

1. Steering knuckle
2. Ball seat
3. Cotter pin
4. Splash guard
5. Wheel hub and bearing assembly

37663_FX35_G0516

Fig. 319 Exploded view of steering knuckle—2WD vehicles

link and steering knuckle, and then loosen nut.

9. Separate steering knuckle from upper link.

10. Separate steering knuckle from transverse link so as not to damage ball joint boot using the ball joint remover, and remove steering knuckle.

> ⁂ **CAUTION**
>
> **Temporarily tighten the nut to prevent damage to threads and to prevent the ball joint remover from suddenly coming off.**

To install:

11. Note the following, and install in the reverse order of the removal.

12. Check wheel sensor harness for proper connection.

13. Check the wheel alignment.

14. Adjust neutral position of steering angle sensor.

AWD

See Figure 320.

1. Remove tires and wheels.
2. Remove wheel sensor and sensor harness.

> ⁂ **CAUTION**
>
> **Never pull on wheel sensor harness.**

3. Remove brake hose bracket.
4. Remove caliper assembly. Hang caliper assembly in a place where it will not interfere with work.

> ⁂ **CAUTION**
>
> **Never depress brake pedal while brake caliper is removed.**

5. Remove disc rotor.
6. Remove cotter pin, and then loosen wheel hub lock nut.
7. Patch wheel hub lock nut with a piece of wood. Hammer the wood to disengage wheel hub and bearing assembly from drive shaft.

> ⁂ **CAUTION**
>
> **Never place drive shaft joint at an extreme angle. Also be careful not to**

overextend slide joint. Never allow drive shaft to hang down without support for or joint sub-assembly, shaft and the other parts.

➡ **Use suitable puller, if wheel hub and bearing assembly and drive shaft cannot be separated even after performing the above procedure.**

8. Remove wheel hub lock nut.
9. Remove wheel hub and bearing assembly, and then remove splash guard.
10. Remove steering outer socket (tie rod end).
11. Remove cotter pin of transverse link and steering knuckle, and then loosen nut.
12. Separate steering knuckle from upper link.
13. Separate steering knuckle link from transverse so as not to damage ball joint boot using the ball joint remover, and remove steering knuckle.

> ⁂ **CAUTION**
>
> **Temporarily tighten the nut to prevent damage to threads and to prevent the ball joint remover from suddenly coming off.**

To install:

14. Note the following, and install in the reverse order of the removal.
15. Perform the final tightening of each of parts under unladen conditions, which were removed when removing wheel hub and bearing assembly and steering knuckle.

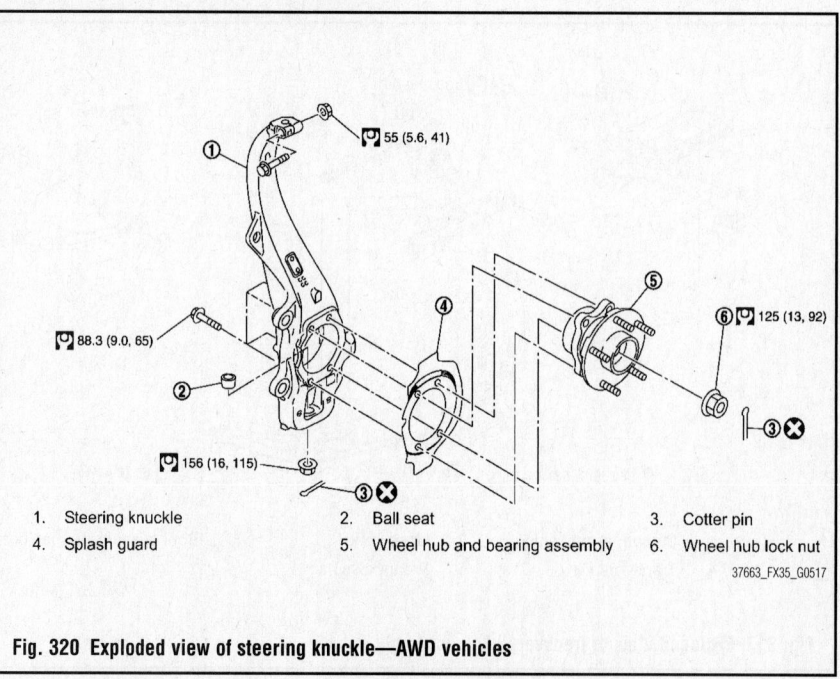

1. Steering knuckle
2. Ball seat
3. Cotter pin
4. Splash guard
5. Wheel hub and bearing assembly
6. Wheel hub lock nut

37663_FX35_G0517

Fig. 320 Exploded view of steering knuckle—AWD vehicles

16. Install drive shaft using tightening torque of wheel hub lock nut.

✳✳ CAUTION

Be sure to use torque wrench to tighten the wheel hub lock nut. Never use a power tool.

17. Check wheel sensor harness for proper connection.
18. Check the wheel alignment.
19. Adjust neutral position of steering angle sensor.

STABILIZER BAR

REMOVAL & INSTALLATION

See Figure 321.

1. Remove under cover.
2. Remove stabilizer connecting rod.

✳✳ CAUTION

Apply a matching mark to identify the installation position.

3. Remove stabilizer clamp and stabilizer bushing.
4. Remove stabilizer bar.

To install:

5. Note the following, and install in the reverse order of removal.
6. Check the matching mark when installing.
7. Tighten the mounting nut to the specified torque while holding a hexagonal part of stabilizer connecting rod side.

STRUT

REMOVAL & INSTALLATION

2WD

See Figure 322.

1. Remove tires and wheels.
2. Remove wheel sensor and harness connector from shock absorber (strut).

✳✳ CAUTION

Never pull on wheel sensor harness.

3. Remove brake hose bracket.

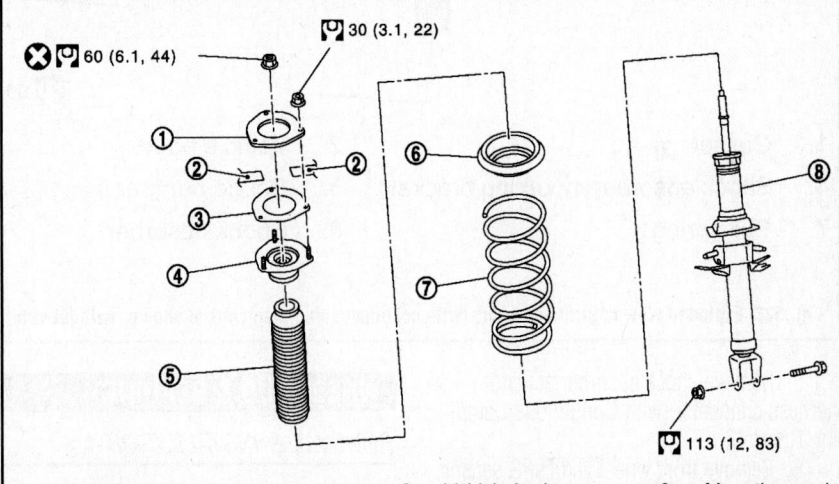

1. Gusset
2. Vehicle body
3. Mounting seal
4. Shock absorber mounting bracket
5. Bound bumper
6. Rubber seat
7. Coil spring
8. Shock absorber

37663_FX35_G0519

Fig. 322 Exploded view of strut assembly

4. Remove shock absorber from transverse link.
5. Separate upper link from steering knuckle.
6. Remove shock absorber mounting bracket mounting nuts, and remove shock absorber (strut) assembly.

➡**If removing shock absorber is difficult, loosen upper link mounting bolts (vehicle side).**

To install:

7. Note the following, and install in the reverse order of removal.
8. Never tap on the ball joint cap of the stabilizer connecting rod with a hammer or a similar item when inserting the stabilizer connecting rod into the transverse link.
9. Perform final tightening of bolts and nuts at the shock absorber lower side (rubber bushing), under unladen conditions with tires on level ground.

AWD

See Figure 323.

1. Remove engine cover.
2. Remove front fender protector.
3. Remove tires and wheels.
4. Remove wheel sensor and harness connector from vehicle.

✳✳ CAUTION

Never pull on wheel sensor harness.

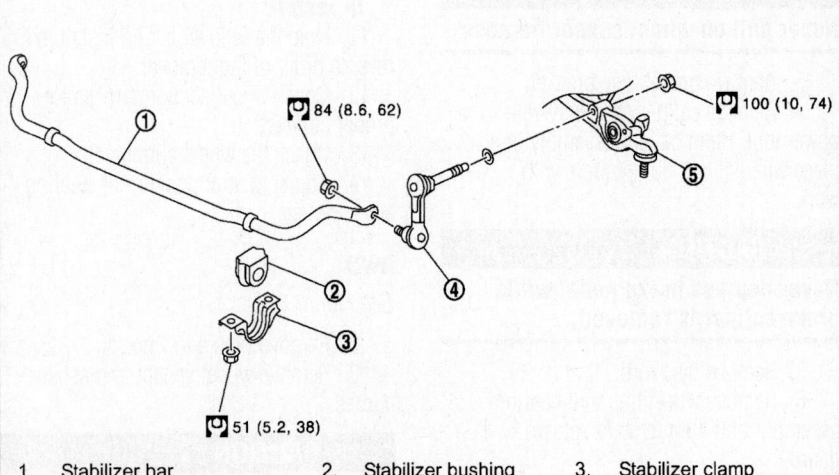

1. Stabilizer bar
2. Stabilizer bushing
3. Stabilizer clamp
4. Stabilizer connecting rod
5. Transverse link

37663_FX35_G0518

Fig. 321 Exploded view of front stabilizer bar

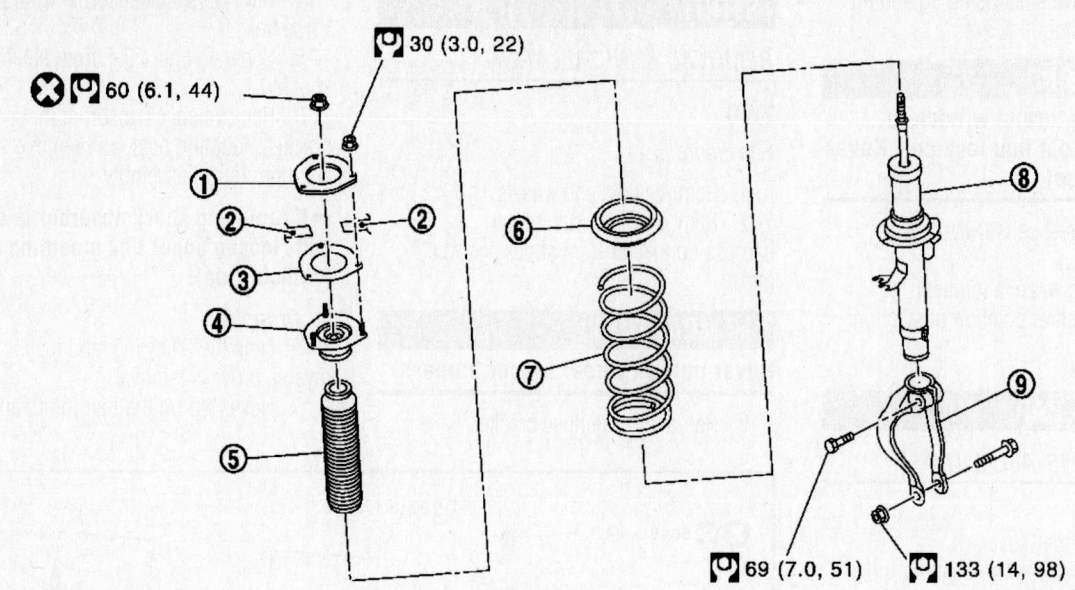

30 (3.0, 22)

60 (6.1, 44)

69 (7.0, 51)   133 (14, 98)

37663_FX35_G0523

1. Gusset	2. Vehicle body	3. Mounting seal
4. Shock absorber mounting bracket	5. Bound bumper	6. Rubber seat
7. Coil spring	8. Shock absorber	9. Shock absorber arm

Fig. 323 Exploded view of strut assembly (with continuous damping control shown, without similar)

5. Remove shock absorber actuator harness connector (with Continuous Damping Control).

6. Remove front wheel vertical G sensor (with Continuous Damping Control).

7. Remove brake hose bracket.

8. Remove stabilizer connecting rod.

9. Remove wheel hub lock nut.

10. Remove shock absorber (strut) from transverse link.

11. Separate upper link from steering knuckle.

12. Separate drive shaft from wheel hub and bearing assembly.

13. Remove shock absorber (strut) assembly.

➡ **If removing shock absorber (strut) is difficult, loosen upper link mounting bolts (vehicle side).**

To install:

14. Note the following, and install in the reverse order of removal.

15. Never tap on the ball joint cap of the stabilizer connecting rod with a hammer or a similar item when inserting the stabilizer connecting rod into the transverse link.

16. Perform final tightening of bolts and nuts at the shock absorber lower side (rubber bushing), under unladen conditions with tires on level ground.

WHEEL HUB & BEARING

REMOVAL & INSTALLATION

2WD

See Figure 324.

1. Remove tires and wheels.

2. Remove wheel sensor and sensor harness.

✳✳ CAUTION
Never pull on wheel sensor harness.

3. Remove brake hose bracket.

4. Remove caliper assembly with power tool. Hang caliper assembly in a place where it will not interfere with work.

✳✳ CAUTION
Never depress brake pedal while brake caliper is removed.

5. Remove disc rotor.

6. Remove wheel hub and bearing assembly, and then remove splash guard.

7. Remove steering outer socket.

8. Remove cotter pin of transverse link and steering knuckle, and then loosen nut.

9. Separate steering knuckle from upper link.

10. Separate steering knuckle from transverse link so as not to damage ball joint boot using the ball joint remover, and remove steering knuckle.

✳✳ CAUTION
Temporarily tighten the nut to prevent damage to threads and to prevent the ball joint remover from suddenly coming off.

To install:

11. Note the following, and install in the reverse order of the removal.

12. Check wheel sensor harness for proper connection.

13. Check the wheel alignment.

14. Adjust neutral position of steering angle sensor.

AWD

See Figure 325.

1. Remove tires and wheels.

2. Remove wheel sensor and sensor harness.

✳✳ CAUTION
Never pull on wheel sensor harness.

3. Remove brake hose bracket.

4. Remove caliper assembly.

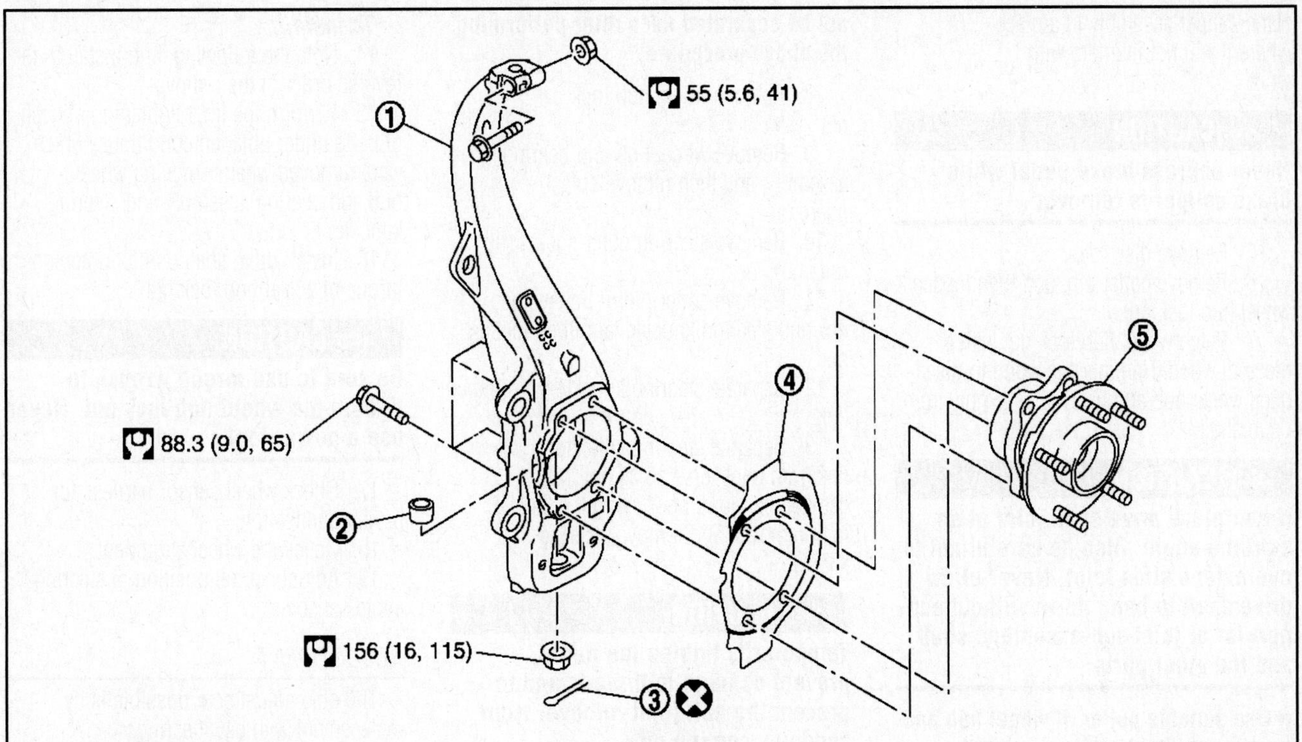

1. Steering knuckle
4. Splash guard

2. Ball seat
5. Wheel hub and bearing assembly

3. Cotter pin

37663_FX35_G0516

Fig. 324 Exploded view of steering knuckle

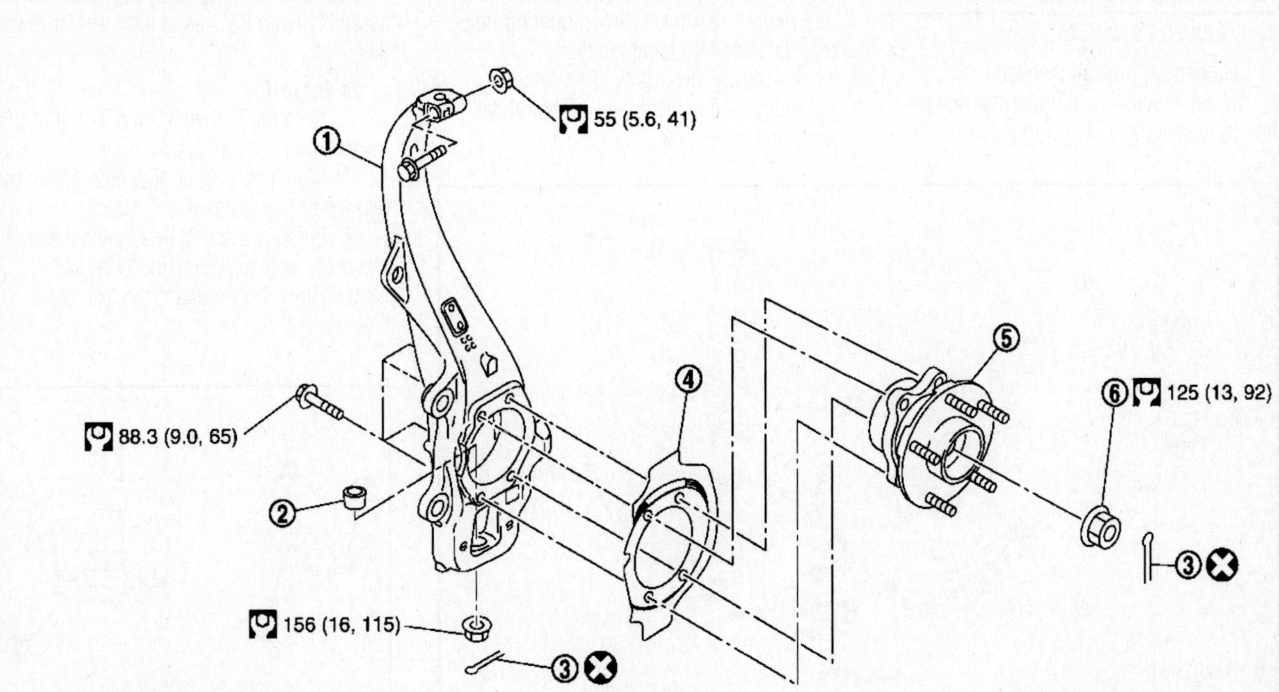

1. Steering knuckle
4. Splash guard

2. Ball seat
5. Wheel hub and bearing assembly

3. Cotter pin
6. Wheel hub lock nut

37663_FX35_G0517

Fig. 325 Exploded view of steering knuckle

Hang caliper assembly in a place where it will not interfere with work.

> ❊❊ **CAUTION**
>
> **Never depress brake pedal while brake caliper is removed.**

5. Remove disc rotor.
6. Remove cotter pin, and then loosen wheel hub lock nut.
7. Patch wheel hub lock nut with a piece of wood. Hammer the wood to disengage wheel hub and bearing assembly from drive shaft.

> ❊❊ **CAUTION**
>
> **Never place drive shaft joint at an extreme angle. Also be careful not to overextend slide joint. Never allow drive shaft to hang down without support for or joint sub-assembly, shaft and the other parts.**

➡ Use suitable puller, if wheel hub and bearing assembly and drive shaft can-not be separated even after performing the above procedure.

8. Remove wheel hub lock nut.
9. Remove wheel hub and bearing assembly, and then remove splash guard.
10. Remove steering outer socket (tie rod end).
11. Remove cotter pin of transverse link and steering knuckle, and then loosen nut.
12. Separate steering knuckle from upper link.
13. Separate steering knuckle link from transverse so as not to damage ball joint boot using the ball joint remover, and remove steering knuckle.

> ❊❊ **CAUTION**
>
> **Temporarily tighten the nut to prevent damage to threads and to prevent the ball joint remover from suddenly coming off.**

To install:

14. Note the following, and install in the reverse order of the removal.
15. Perform the final tightening of each of parts under unladen conditions, which were removed when removing wheel hub and bearing assembly and steering knuckle.
16. Install drive shaft using tightening torque of wheel hub lock nut.

> ❊❊ **CAUTION**
>
> **Be sure to use torque wrench to tighten the wheel hub lock nut. Never use a power tool.**

17. Check wheel sensor harness for proper connection.
18. Check the wheel alignment.
19. Adjust neutral position of steering angle sensor.

ADJUSTMENT

The only adjustment possible is by replacement with new bearing.

SUSPENSION

COIL SPRING

REMOVAL & INSTALLATION

See Figures 326 and 327.

1. Remove tires and wheels.
2. Set suitable jack under rear lower link to relieve the coil spring tension.

3. Loosen rear lower link mounting nuts [rear suspension member side (without RAS) or RAS actuator assembly (with RAS)], and remove rear lower link mounting bolts and nuts (axle housing side).
4. Slowly lower jack, then remove upper seat, coil spring and rubber sheet from rear lower link.

REAR SUSPENSION

5. Remove rear lower link mounting nuts and adjusting bolts [rear suspension member side (without RAS) or RAS actuator assembly (with RAS)] and remove rear lower link.

To install:

6. Note the following, and install in the reverse order of removal.
7. Match up rubber seat indentions and rear lower link grooves and attach.
8. Install coil spring by aligning the lower end of the large diameter side to the step between the rubber seat and the rear lower link.

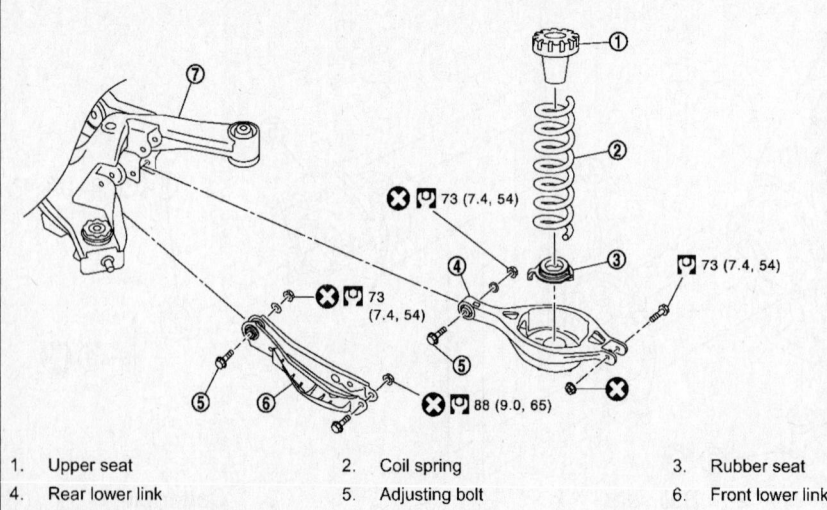

1. Upper seat
2. Coil spring
3. Rubber seat
4. Rear lower link
5. Adjusting bolt
6. Front lower link
7. Rear suspension member

37663_FX35_G0530

Fig. 326 Exploded view of rear lower link and coil spring assembly (without RAS shown, with RAS similar)

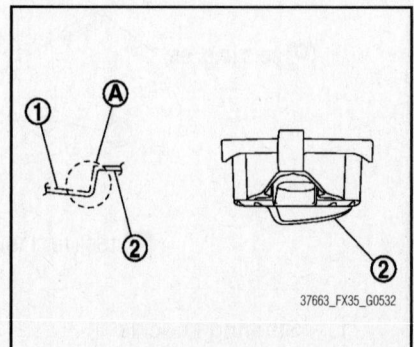

37663_FX35_G0532

Fig. 327 Install coil spring by aligning the lower end of the large diameter side to the step (A) between the rubber seat (1) and the rear lower link (2)

※※ CAUTION

Make sure spring is not upside down.

9. Perform the final tightening of rear suspension member and axle housing rubber bushing position under unladen condition with tires on level ground.

10. Check the wheel alignment.

11. Adjust neutral position of steering angle sensor.

CONTROL ARMS/LINKS

REMOVAL & INSTALLATION

Front Lower Link

See Figures 328.

1. Remove tire and wheels.

2. Set suitable jack under axle assembly to relieve the coil spring tension.

3. Remove shock absorber mounting bolts from front lower link.

4. Remove front lower link mounting bolts and nuts from axle housing.

5. Remove stabilizer clamp and stabilizer bushing.

6. Remove front lower link mounting bolts and nuts from rear suspension member, and remove front lower link.

To install:

7. Note the following, and install in the reverse order of removal.

8. Perform final tightening of rear suspension member and axle installation position (rubber bushing), under unladen conditions with tires on level ground.

Radius Rod

See Figure 329.

1. Remove tire and wheels.

2. Remove radius rod mounting bolt and nut (axle housing side).

3. Remove radius rod mounting bolt (rear suspension member side), and remove radius rod.

To install:

4. Note the following, and install in the reverse order of removal.

5. Perform final tightening of rear suspension member and axle installation posi-

tion (rubber bushing), under unladen conditions with tires on level ground.

Suspension Arm

See Figures 330 and 331.

1. Remove tire and wheels.

2. Remove radius rod.

3. Remove caliper assembly mounting bolts. Hang caliper assembly in a place where it will not interfere with work.

4. Set suitable jack under axle assembly to relieve the coil spring tension.

5. Remove stabilizer connecting rod.

6. Remove drive shaft.

7. Remove height sensor (with xenon head lamp).

8. Remove cotter pin of suspension arm ball joint, and loosen nut.

9. Remove suspension arm mounting bolts and nuts (rear suspension member side).

10. Use the ball joint remover to remove suspension arm from axle housing. Be careful not to damage ball joint boot.

※※ CAUTION

Tighten temporarily mounting nut to prevent damage to threads and to prevent ball joint remover from coming off.

11. Remove suspension arm.

12. Remove stabilizer connecting rod mounting bracket.

To install:

13. Note the following and, install in the reverse order of removal.

14. Perform final tightening of rear suspension member installation position (rubber bussing), under unladen conditions with tires on level ground.

SHOCK ABSORBER

REMOVAL & INSTALLATION

See Figures 332.

1. Remove tires and wheels.

2. Remove shock absorber actuator harness connector (with Continuous Damping Control).

3. Set suitable jack under axle assembly to relieve the coil spring tension.

4. Remove shock absorber (lower side).

5. Gradually lower the jack to remove it from rear lower link.

6. Remove shock absorber assembly mounting nuts (upper side), and then remove shock absorber assembly.

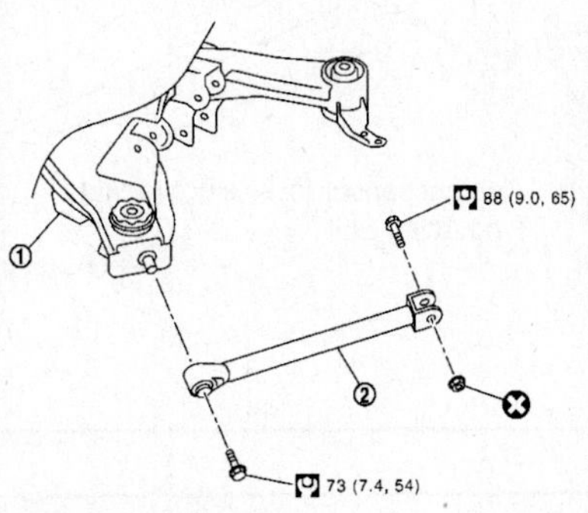

| 1. | Rear suspension member | 2. | Radius rod |

37663_FX35_G0533

Fig. 329 Exploded view of radius rod assembly

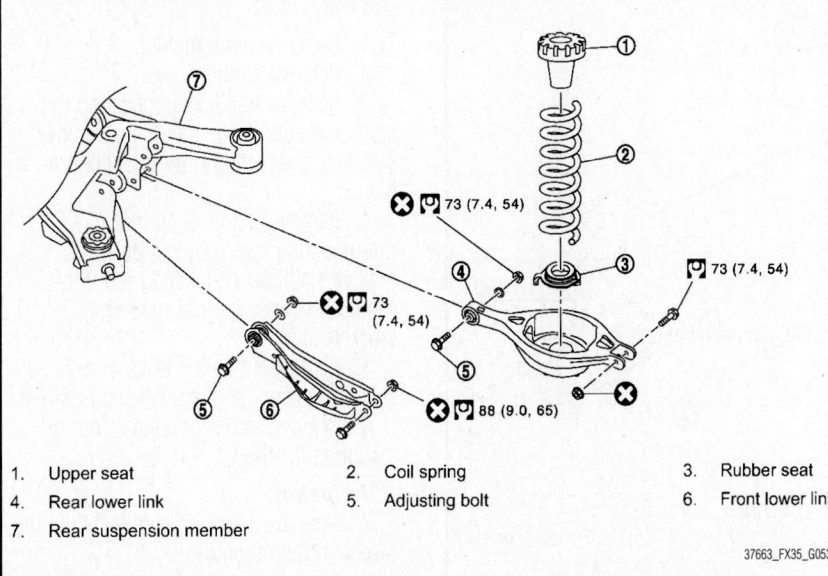

1.	Upper seat	2.	Coil spring	3.	Rubber seat
4.	Rear lower link	5.	Adjusting bolt	6.	Front lower link
7.	Rear suspension member				

37663_FX35_G0530

Fig. 328 Exploded view of front lower link and coil spring assembly (without RAS shown, with RAS similar)

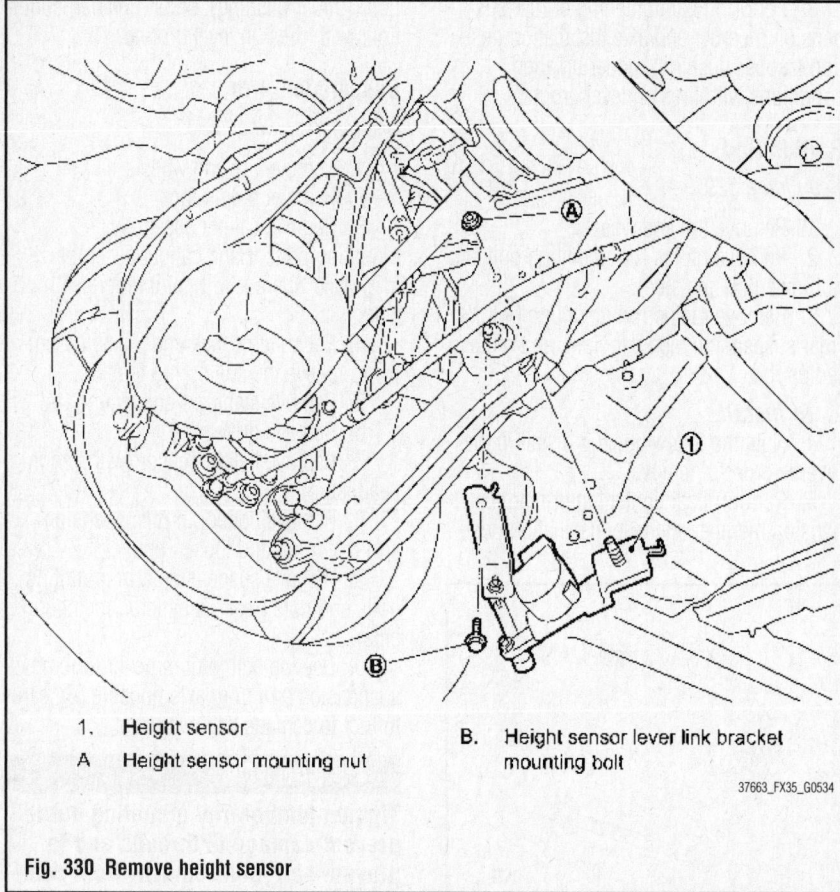

1. Height sensor
A Height sensor mounting nut

B. Height sensor lever link bracket
    mounting bolt

37663_FX35_G0534

Fig. 330 Remove height sensor

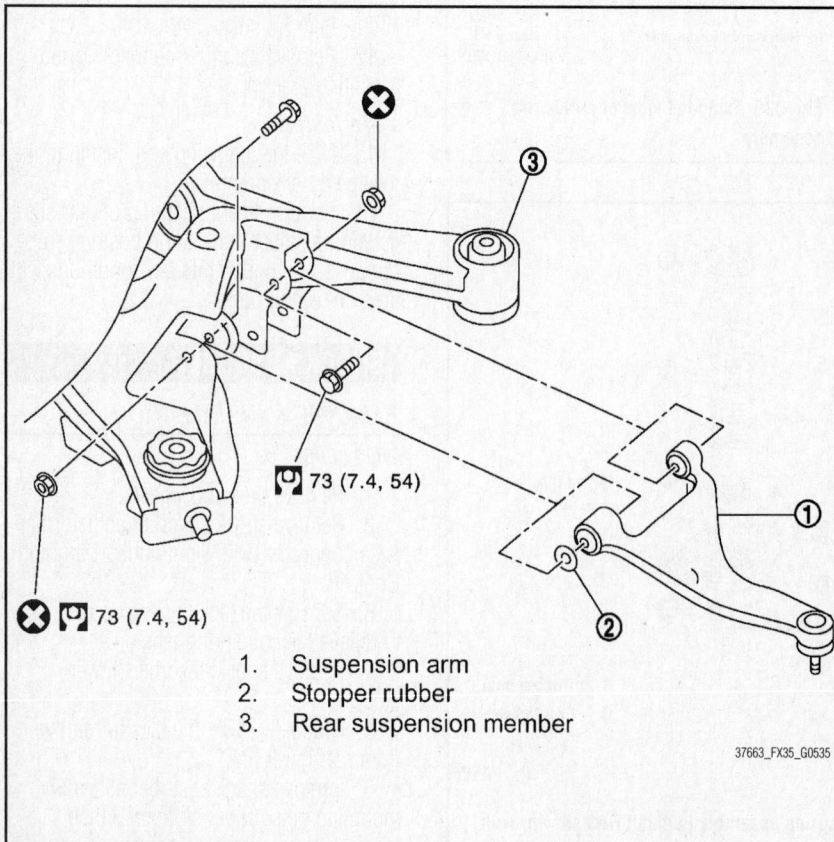

73 (7.4, 54)

73 (7.4, 54)

1. Suspension arm
2. Stopper rubber
3. Rear suspension member

37663_FX35_G0535

Fig. 331 Exploded view of suspension arm assembly

To install:

7. Note the following, and install in the reverse order of removal.

8. Perform final tightening of bolts and nuts at the shock absorber lower side (rubber bushing), under unladen conditions with tires on level ground.

TESTING

Disassembly

※※ CAUTION

Never damage shock absorber piston rod when removing components from shock absorber.

1. Remove cap from mounting bracket
2. Wrap a shop cloth around lower side of shock absorber and fix it with a vise.

※※ CAUTION

Never set the cylindrical part of shock absorber with a vise.

3. Secure the piston rod tip so that piston rod does not turn, and remove piston rod lock nut.

4. Remove mounting seal, mounting bracket and bound bumper cover from shock absorber.

Assembly

1. Install in the reverse order of disassembly.

STABILIZER BAR

REMOVAL & INSTALLATION

See Figure 333.

1. Remove center muffler.
2. Remove under cover.
3. Remove stabilizer connecting rod mounting nuts (lower side), and remove stabilizer connecting rod from stabilizer bar.

4. Remove stabilizer connecting rod mounting nuts (upper side), and remove stabilizer connecting rod from stabilizer connecting rod mounting bracket.

5. Remove mounting bolts or nuts on stabilizer clamp, and remove stabilizer bar.

6. Remove stabilizer connecting rod mounting bracket.

To install:

7. Note the following, and install in the reverse order of removal.

8. Tighten the mounting nut to the specified torque while holding a hexagonal part of stabilizer connecting rod side.

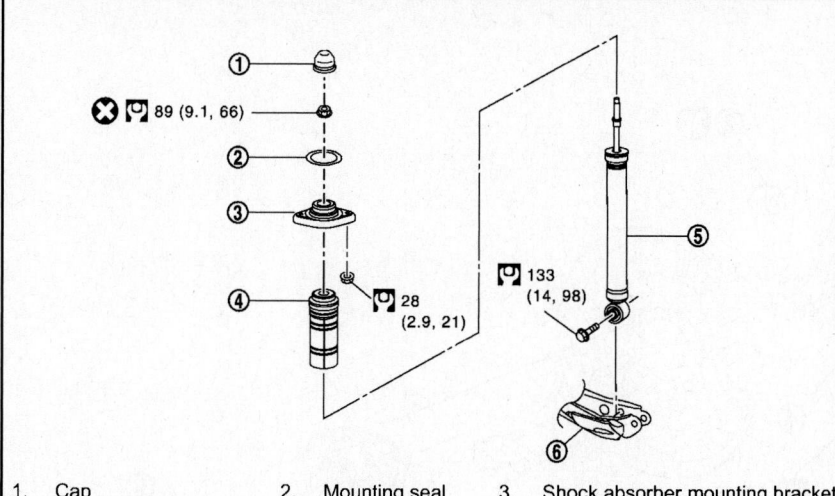

1. Cap
2. Mounting seal
3. Shock absorber mounting bracket
4. Bound bumper cover
5. Shock absorber
6. Front lower link

37663_FX35_G0536

Fig. 332 Exploded view of rear shock absorber assembly (without continuous damping control shown, with similar)

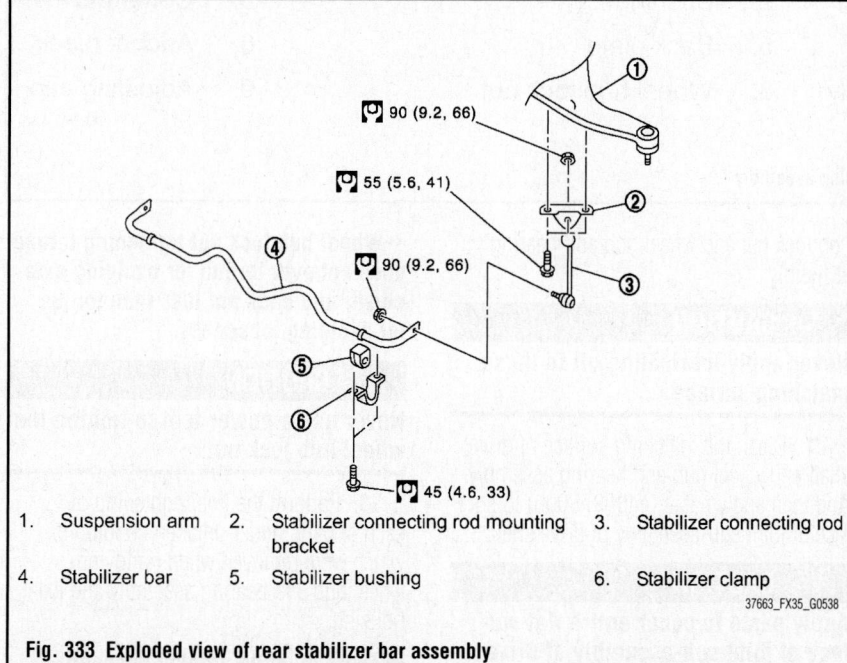

1. Suspension arm
2. Stabilizer connecting rod mounting bracket
3. Stabilizer connecting rod
4. Stabilizer bar
5. Stabilizer bushing
6. Stabilizer clamp

37663_FX35_G0538

Fig. 333 Exploded view of rear stabilizer bar assembly

WHEEL HUB & BEARING

REMOVAL & INSTALLATION

See Figures 334 through 336.

1. Remove tire and wheels.
2. Remove caliper assembly. Hang caliper assembly in a place where it will not interfere with work.

✳✳ CAUTION

Never depress brake pedal while caliper assembly is removed.

3. Remove disc rotor.
4. Remove cotter pin and adjusting cap, then loosen wheel hub lock nut.
5. Put matching mark on drive shaft and wheel hub and bearing assembly.

✳✳ CAUTION

Use paint or similar substance for matching marks. Never scratch the surface.

6. Cover wheel hub lock nut with a piece of wood. Hammer the wood to disen-

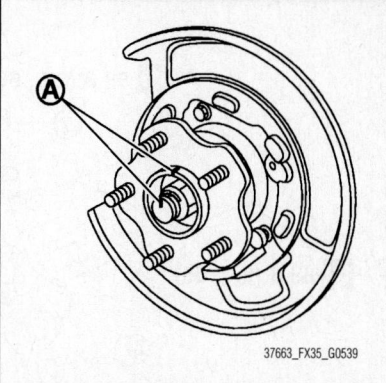

37663_FX35_G0539

Fig. 334 Put matching mark (A) on drive shaft and wheel hub and bearing assembly

gage wheel hub and bearing assembly from drive shaft.

✳✳ CAUTION

Never place drive shaft joint at an extreme angle. Also be careful not to overextend slide joint. Never allow drive shaft to hang down without support for counterpart such as joint sub-assembly, and other parts.

➡Use a suitable puller, if wheel hub and bearing assembly and drive shaft cannot be separated even after performing the above procedure.

7. Remove wheel hub lock nut.
8. Remove parking brake shoe and parking brake cable from back plate.
9. Remove stabilizer connecting rod (upper side).
10. Remove coil spring.
11. Set suitable jack under axle housing.
12. Remove radius rod.
13. Remove shock absorber (lower side).
14. Separate suspension arm from axle housing so as not to damage ball joint boot using ball joint remover, and then remove axle housing from the vehicle.

✳✳ CAUTION

Note the following:

- Temporarily tighten nuts to prevent damage to threads and to prevent the ball joint remover from coming off.
- Never place drive shaft joint at an extreme angle. Also be careful not to overextend slide joint.
- Never allow drive shaft to hang down without support for counterpart such as joint sub-assembly, and other parts.

15. Remove front lower link (axle housing side).

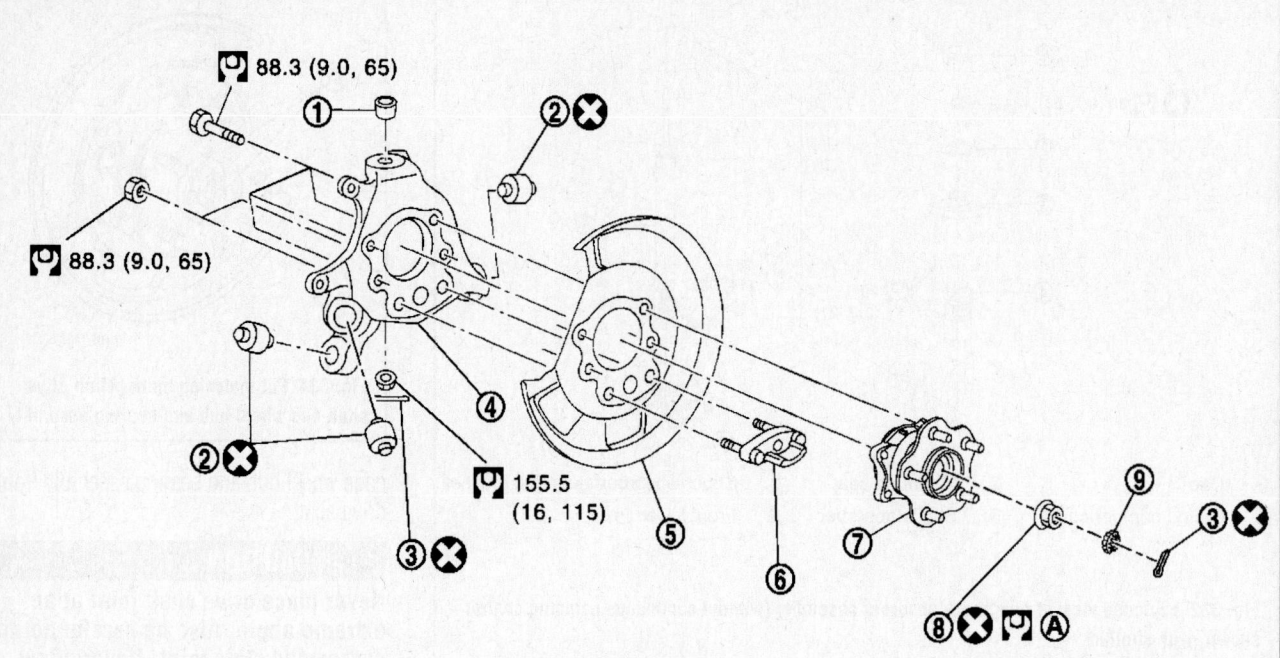

1. Ball seat
4. Axle housing
7. Wheel hub and bearing assembly

2. Bushing
5. Back plate
8. Wheel hub lock nut

3. Cotter pin
6. Anchor block
9. Adjusting cap

37663_FX35_G0540

Fig. 335 Exploded view of rear wheel hub and housing assembly

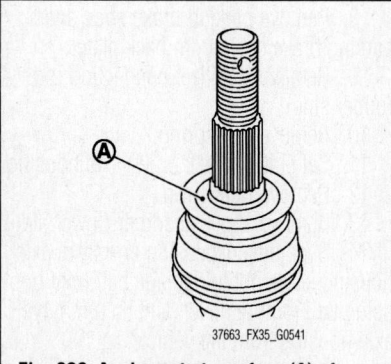

37663_FX35_G0541

Fig. 336 Apply paste to surface (A) of joint sub-assembly of drive shaft

16. Remove rear lower link (axle housing side).

17. Remove wheel hub and bearing assembly.

18. Remove anchor block mounting nuts, and then remove anchor block and back plate from axle housing.

To install:

19. Note the following, and install in the reverse order of removal.

20. Clean the matching surface of wheel hub lock nut and wheel hub and bearing assembly.

> **⁂ CAUTION**
> **Never apply lubricating oil to these matching surface.**

21. Clean the matching surface of drive shaft and wheel hub and bearing assembly. And then apply paste (440037S000) to surface of joint sub-assembly of drive shaft.

> **⁂ CAUTION**
> **Apply paste to cover entire flat surface of joint sub-assembly of drive shaft.**

22. Use the following torque range for tightening the wheel hub lock nut.

> **⁂ CAUTION**
> **Since the drive shaft is assembled by press-fitting, use the tightening torque range for the wheel hub lock nut. Be sure to use torque wrench to tighten the wheel hub lock nut. Never use a power tool.**

➡ **Wheel hub lock nut tightening torque does not over torque for avoiding axle noise, and does not less than torque for avoiding looseness.**

> **⁂ CAUTION**
> **Never use a power tool to tighten the wheel hub lock nut.**

23. Perform the final tightening of each of parts under unladen conditions, which were removed when removing wheel hub and bearing assembly and axle housing.

➡ **There must be no play between adjusting cap, cotter pin and wheel hub lock nut.**

> **⁂ CAUTION**
> **Never reuse cotter pin, wheel hub lock nut and bushing.**

ADJUSTMENT

The only adjustment possible is by replacement with new bearing.

SPECIFICATIONS AND MAINTENANCE CHARTS

ENGINE AND VEHICLE IDENTIFICATION

			Engine					Model Year	
Code ①	Liters (cc)	Cu. In.	Cyl.	Fuel Sys.	Engine Type	Eng. Mfg.		Code ②	Year
VQ37VHR	3.7 (3696)	225	6	MFI	DOHC	Nissan		9	2009
								A	2010

MFI: Multi-port Fuel Injection

DOHC: Double Overhead Camshaft

① The engine VIN "C" is the 4th position of the Vehicle Identification Number (VIN)

② 10th position of the Vehicle Identification Number (VIN)

37663_IG37_C0001

GENERAL ENGINE SPECIFICATIONS

Year	Model	Engine Displacement Liters	Engine ID	Net Horsepower @ rpm	Net Torque @ rpm (ft. lbs.)	Bore x Stroke (in.)	Compression Ratio	Oil Pressure @ rpm
2009	G37	3.7	VQ37VHR	①	②	3.760X3386	11.0:1	43@2000
2010	G37	3.7	VQ37VHR	①	②	3.760X3386	11.0:1	43@2000

① Sedan: 328@7000. Coupe: 330@7000. Convertible: 325@7000.

② Sedan: 269@5200. Coupe: 270@5200. Convertible: 267@5200.

37663_IG37_C0002

ENGINE TUNE-UP SPECIFICATIONS

Year	Model	Engine ID	Spark Plug Gap (in.)	Ignition Timing (deg.) MT	AT	Fuel Pump (psi) ①	Idle Speed (rpm) MT	AT ②	Valve Clearance (in.) Intake ③	Exhaust ③
2009	G37	VQ37VHR	0.043	④	④	51	600-700	600-700	0.010-0.013	0.011-0.015
2010	G37	VQ37VHR	0.043	④	④	51	600-700	600-700	0.010-0.013	0.011-0.015

NOTE: The Vehicle Emission Control Information label often reflects specification changes made during production.

The label figures must be used if they differ from those in this chart.

NA: Not Available

B: Before top dead center

① System pressure at idle with vacuum hose connected; should increase to 43 psi when disconnected

② Automatic transmission in park or neutral

③ Engine cold

④ 5-15 degrees BTDC

37663_IG37_C0003

CAPACITIES

Year	Model	Engine ID	Engine Displacement Liters	Engine Oil with Filter (qts.)	Transmission (pts.) Man	Transmission (pts.) Auto.	Drive Axle Rear (pts.)	Fuel Tank (gal.)	Cooling System (qts.)
2009	G37	VQ37VHR	3.7	5.0	6.0	19.5	①	20.0	②
2010	G37	VQ37VHR	3.7	5.0	6.0	19.5	①	20.0	②

NOTE: All capacities are approximate. Add fluid gradually and check to be sure a proper fluid level is obtained.

① Front: 1 38. Rear: 3.00

② Automatic transmission: 9.00. Manual transmission: 9.18

37663_IG37_C0004

FLUID SPECIFICATIONS

Year	Model	Engine ID	Engine Oil	Auto. Trans.	Manual Trans.	Drive Axle Front	Drive Axle Rear	Transfer Case	Power Steering Fluid	Brake Master Cylinder	Cooling System
2009	G37	VQ37VHR	5W-30	①	②	③	④ ⑤ ⑥	⑦	⑧	DOT 3	⑨
2010	G37	VQ37VHR	5W-30	①	②	③	④ ⑤ ⑥	⑦	⑧	DOT 3	⑨

DOT: Department Of Transportation

① Nissan Matic S ATF. Using non approved fluid will cause deterioration in driveability and transmission damage.

② Nissan (MTF) HQ MULTI 75W-85 or API GL-4, Viscosity 75W-85

③ Nissan differential oil hypoid super GL-5 80W-90 or API GL-5 viscosity SAE 80W-90 gear oil.

④ Sedan and Coupe: Nissan differential oil hypoid super GL-5 80W-90 or API GL-5 viscosity SAE 80W-90 gear oil, except if equipped with 7A/T 2WD.

⑤ Sedan and Coupe: Nissan differential synthetic 75W-90 or API GL-5 synthetic gear oil, viscosity SAE 75W-90 gear oil, if equipped with 7A/T 2WD.

⑥ Convertible: Nissan differential oil hypoid super GL-5 80W-90 or API GL-5 viscosity SAE 80W-90 gear oil, if equipped with M/T. Nissan differential synthetic 75W-90 or API GL-5 synthetic gear oil, viscosity SAE 75W-90 gear oil, if equipped with A/T.

⑦ Nissan Matic J ATF. Using non approved fluid will cause deterioration in driveability and transfer case damage.

⑧ Nissan Power Steering Fluid

⑨ Nissan Long Life antifreeze or equivalent

37663_IG37_C0013

VALVE SPECIFICATIONS

Year	Engine ID	Engine Displacement Liters	Seat Angle (deg.)	Face Angle (deg.)	Spring Test Pressure (lbs. @ in.)	Spring Installed Height (in.)	Stem-to-Guide Clearance (in.) Intake	Stem-to-Guide Clearance (in.) Exhaust	Stem Diameter (in.) Intake	Stem Diameter (in.) Exhaust
2009	VQ37VHR	3.7	45.15-45.45	45	①	②	0.0008-0.0021	0.0012-0.0022	0.2348-0.2354	0.2347-0.2350
2010	VQ37VHR	3.7	45.15-45.45	45	①	②	0.0008-0.0021	0.0012-0.0022	0.2348-0.2354	0.2347-0.2350

① Intake: 43-48 lb at 1.6102 inch. Exhaust: 37-42 lb at 1.4567 inch.

② Intake: 1.7976. Exhaust: 1.7264.

37663_IG37_C0005

CAMSHAFT AND BEARING SPECIFICATIONS CHART

All measurements are given in inches.

Year	Engine Displ. Liters	Engine ID	Journal Dia.	Brg. Oil Clearance	Shaft End-play	Runout	Journal-to-Bore Clearance	Lobe Lift Intake	Lobe Lift Exhaust
2009	3.7	VQ37VHR	① ③	② ③	0.0045-0.0074	0.0008	NA	NA	③ ④
2010	3.7	VQ37VHR	① ③	② ③	0.0045-0.0074	0.0008	NA	NA	③ ④

NA: Not Available

① Front No. 1: 1.0211- 1.0218
  No. 2, 3, 4: 0.9230- 0.9238

② Front No. 1: 0.0018- 0.0034
  No. 2, 3, 4: 0.0014- 0.0030

③ Specification is for exhaust camshaft.

④ 1.7722- 1.7797 bank one. 1.8400- 1.8474 bank two.

37663_IG37_C0014

CRANKSHAFT AND CONNECTING ROD SPECIFICATIONS

All measurements are given in inches.

Year	Engine Displacement Liters	Engine ID	Crankshaft Main Brg. Journal Dia.	Crankshaft Main Brg. Oil Clearance	Crankshaft Shaft End-play	Thrust on No.	Connecting Rod Journal Diameter	Connecting Rod Oil Clearance	Connecting Rod Side Clearance
2009	3.7	VQ37VHR	①	0.0014-0.0018	0.0039-0.0098	3	②	0.0016-0.0021	0.0079-0.0138
2010	3.7	VQ37VHR	①	0.0014-0.0018	0.0039-0.0098	3	②	0.0016-0.0021	0.0079-0.0138

① There are 24 different grades, ranging from 2.5581- 2.5571

② There are 13 different grades, ranging from 2.2441- 2.2446. Specification is for rod big end diameter (without bearing).

37663_IG37_C0008

PISTON AND RING SPECIFICATIONS

All measurements are given in inches.

Year	Engine Displ. Liters	Engine ID	Piston Clearance	Ring Gap Top Compression	Ring Gap Bottom Compression	Ring Gap Oil Control	Ring Side Clearance Top Compression	Ring Side Clearance Bottom Compression	Ring Side Clearance Oil Control
2009	3.7	VQ37VHR	0.0004-0.0012	0.0091-0.0130	0.0091-0.0130	0.0067-0.0185	0.0016-0.0031	0.0012-0.0028	0.0022-0.0061
2010	3.7	VQ37VHR	0.0004-0.0012	0.0091-0.0130	0.0091-0.0130	0.0067-0.0185	0.0016-0.0031	0.0012-0.0028	0.0022-0.0061

37663_IG37_C0007

TORQUE SPECIFICATIONS

All readings in ft. lbs.

Year	Engine Displacement Liters	Engine ID	Cylinder Head Bolts	Main Bearing Bolts	Rod Bearing Bolts	Crankshaft Damper Bolts	Flywheel Bolts	Manifold		Spark Plugs	Oil Drain Plug
								Intake	Exhaust		
2009	3.7	VQ37VHR	①	②	③	④	65	⑤	⑥	18	25
2010	3.7	VQ37VHR	①	②	③	④	65	⑤	⑥	18	25

① Step 1: 77 ft. lbs.

Step 2: Loosen bolts completely

Step 3: 30 ft. lbs.

Step 4: Tighten an additional 95 degrees

Step 5: Tighten an additional 95 degrees

② Step 1: 18 ft. lbs.

Step 2: 26 ft.lbs.

Step 3: Tighten an additional 90 degrees

③ Step 1: 21 ft. lbs.

Step 2: Loosen bolts completely

Step 3: 18 ft.lbs.

Step 4: Tighten an additional 90 degrees

④ Step 1: 33 ft. lbs.

Step 2: Tighten an additional 90 degrees

⑤ Step 1: 5 ft. lbs.

Step 2: 19 ft. lbs.

⑥ Step 1: 11 ft. lbs. in 2 steps

37663_IG37_C0006

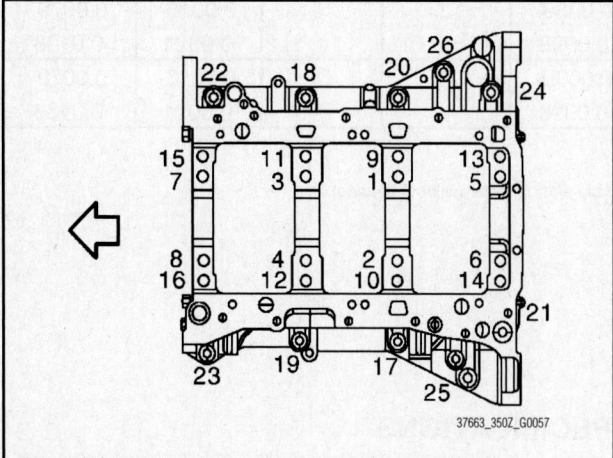

Fig. 1 Main bearing (lower cylinder block bolts) tightening sequence

WHEEL ALIGNMENT

| Year | Model | | Caster | | Camber | | Toe-in |
			Range (+/-Deg.)	Preferred Setting (Deg.)	Range (+/-Deg.)	Preferred Setting (Deg.)	(in.)
2009	G37 (sedan)	F	①	①	②	②	③
		R	—	—	④	④	⑤
	G37 (coupe)	F	⑥	⑥	⑦	⑦	③
		R	—	—	⑧	⑧	⑤
	G37 (convertible)	F	⑨	⑨	⑩	⑩	③
		R	—	—	⑧	⑧	⑤
2010	G37 (sedan)	F	①	①	②	②	③
		R	—	—	④	④	⑤
	G37 (coupe)	F	⑥	⑥	⑦	⑦	③
		R	—	—	⑧	⑧	⑤
	G37 (convertible)	F	⑨	⑨	⑩	⑩	③
		R	—	—	⑧	⑧	⑤

① Min: 3 degrees 50' (3.83 degrees) except sport models. 3 degrees 55' (3.92 degrees) sport models.

　Nominal: 4 degrees 35' (4.58 degrees) except sport models. 4 degrees 40' (4.67 degrees) sport models.

　Max: 5 degrees 20' (5.33 degrees) except sport models. 5 degrees 25' (5.42 degrees) sport models.

② Min: -1 degree 05' (-1.08 degrees)

　Nominal: -0 degree 20' (-0.33 degrees)

　Max: 0 degree 25' (0.42 degrees)

③ Min: 0mm (0 inch) distance

　Nominal: In 1mm (0.04 inch) distance

　Max: In 2mm (0.08 inch) distance

④ Min: -1 degrees 20' (-1.33 degrees) 2WD except sport models and AWD. -1 degree 25' (-1.42 degrees) 2WD sport models and AWD.

　Nominal: -0 degrees 50' (-0.83 degree) 2WD except sport models and AWD. -0 degrees 55' (-0.92 degrees) 2WD sport models and AWD.

　Max: -0 degrees 20' (-0.33 degrees) 2WD except sport models and AWD. -0 degrees 25' (-0.42 degrees) 2WD sport models and AWD.

⑤ Min: 0mm (0 inch) distance

　Nominal: In 2.8mm (0.110 inch) distance

　Max: In 5.6mm (0.220 inch) distance

⑥ Min: 3 degrees 30' (3.50 degrees)

　Nominal: 4 degrees 15' (4.25 degrees)

　Max: 5 degrees 00' (5.00 degrees)

⑦ Min: -1 degree 10' (-1.16 degrees)

　Nominal: -0 degree 25' (-0.42 degrees)

　Max: 0 degree 20' (0.33 degrees)

⑧ Min: -1 degree 45' (-1.75 degrees)

　Nominal: -1 degree 15' (-1.25 degrees)

　Max: -0 degree 45' (-0.75 degrees)

⑨ Min: 4 degrees 05' (4.08 degrees) 18 inch wheel. 4 degrees 10' (4.17 degrees) 19 inch wheel.

　Nominal: 4 degrees 50' (4.83 degrees) 18 inch wheel. 4 degrees 55' (4.92 degrees) 19 inch wheel.

　Max: 5 degrees 35' (5.58 degrees) 18 inch wheel. 5 degrees 40' (5.66 degrees) 19 inch wheel.

⑩ Min: -1 degree 10' (-1.16 degrees)

　Nominal: -0 degree 25' (-0.41 degrees)

　Max: 0 degree 20' (0.33 degrees)

37663_IG37_C0009

TIRE, WHEEL AND BALL JOINT SPECIFICATIONS

Year	Model	OEM Tires		Tire Pressures (psi)		Wheel Size	Lug Nut Torque (ft. lbs.)
		Standard	Optional	Front	Rear		
2009	G37 (sedan)	①	①	33	33	NA	NA
	G37 (coupe)	②	②	③	③	NA	NA
	G37 (convertible)	④	④	⑤	⑤	NA	NA
2010	G37 (sedan)	①	①	33	33	NA	NA
	G37 (coupe)	②	②	③	③	NA	NA
	G37 (convertible)	④	④	⑤	⑤	NA	NA

Note: If specification differes from vehicle placard, use specification given on vehicle placard.

NA: Not Available

OEM: Original Equipment Manufacturer

PSI: Pounds Per Square Inch

① P225/55R17, P225/50R18, 225/50R18, 245/45R18

② P225/50R18, 225/45R19, 245/40R19

③ P225/50R18: 33. 225/45R19 and 245/40R19: 35.

④ P225/50R18, P245/45R18, 225/45R19, 245/40R19. On 225/45R19 and 245/40R19 XL indicates extra load (reinforced) tire.

⑤ P225/50R18 and P245/45R18: 38. 225/45R19 and 245/40R19: 39.

37663_IG37_C0010

BRAKE SPECIFICATIONS
All measurements in inches unless noted

Year	Model		Brake Disc			Minimum Lining Thickness		Brake Caliper	
			Original Thickness	Minimum Thickness	Maximum Run-out	Front	Rear	Bracket Bolts (ft. lbs.)	Mounting Bolts (ft. lbs.)
2009	G37 (sedan)	F	①	②	0.0014	0.079	0.079	NA	NA
		R	③	②	0.0022	0.079	0.079	NA	NA
	G37 (coupe)	F	④	②	0.0014	0.079	0.079	NA	NA
		R	⑤	②	0.0022	0.079	0.079	NA	NA
	G37 (convertible)	F	⑥	②	0.0014	0.079	0.079	NA	NA
		R	⑦	②	0.0022	0.079	0.079	NA	NA
2010	G37 (sedan)	F	①	②	0.0014	0.079	0.079	NA	NA
		R	③	②	0.0022	0.079	0.079	NA	NA
	G37 (coupe)	F	④	②	0.0014	0.079	0.079	NA	NA
		R	⑤	②	0.0022	0.079	0.079	NA	NA
	G37 (convertible)	F	⑥	②	0.0014	0.079	0.079	NA	NA
		R	⑦	②	0.0022	0.079	0.079	NA	NA

NA: Not Available

① Rotor outer diameter x thickness: 2 piston caliper: 12.60x1.102. 4 piston caliper: 13.98x1.260.

② Thickness variation measured at 8 positions: 0.0006

③ Rotor outer diameter x thickness: 1 piston caliper: 12.13x0.630. 2 piston caliper: 13.78x0.787.

④ Rotor outer diameter x thickness: 1 piston caliper: 12.99x1.260. 2 piston caliper: 12.60x1.102. 4 piston caliper: 13.98x1.260.

⑤ Rotor outer diameter x thickness: 1 piston caliper: 12.99x0.630. 1 piston caliper: 12.13x0.630 (if used with 2 piston front caliper). 2 piston caliper: 13.78x0.787.

⑥ Rotor outer diameter x thickness: 1 piston caliper: 12.99x0.630. 4 piston caliper: 13.98x1.260.

⑦ Rotor outer diameter x thickness: 1 piston caliper: 12.99x0.630. 2 piston caliper: 13.78x0.787.

37663_IG37_C0011

SCHEDULED MAINTENANCE INTERVALS
INFINITI G37

TO BE SERVICED	TYPE OF	VEHICLE MILEAGE INTERVAL (x1000)												
		3.75	7.5	15	22.5	30	37.5	45	52.5	60	67.5	75	82.5	90
Engine oil & filter	R	✓	✓	✓	✓	✓	✓	✓	✓	✓	✓	✓	✓	✓
Brake lines & cables	I			✓		✓		✓		✓		✓		✓
Brake pads& rotors	L/I			✓		✓		✓		✓		✓		✓
Driveshaft boots & propeller shaft (4WD)	I					✓				✓				
Automatic transmission, final drive oil & transfer case	I			✓		✓		✓		✓				
LSD gear oil	I			✓		✓		✓		✓		✓		✓
Front wheel bearing grease (4WD)	R					✓				✓				
Air cleaner filter	R					✓				✓				✓
Engine coolant	R									✓				✓
Exhaust system	I						✓	✓	✓	✓				
Spark plugs	R	Replace every 105,000 miles												
Drive belt(s)	I			✓		✓		✓		✓		✓		✓
Cabin air filter	R						✓							✓
Exhaust system	I		✓			✓				✓				✓
Fuel lines	I		✓			✓				✓				
Steering gear (box) & linkage, axle & suspension parts	I					✓				✓				✓
Transfer case	I					✓				✓				✓
Tire rotation			✓	✓	✓	✓	✓	✓	✓	✓	✓	✓	✓	✓
Vapor lines	S/I					✓				✓				✓

R: Replace   S/I: Service or Inspect

FREQUENT OPERATION MAINTENANCE (SEVERE SERVICE)

If a vehicle is operated under any of the following conditions it is considered severe service:

- Extremely dusty areas.

- 50% or more of the vehicle operation is in 32°C (90°F) or higher temperatures, or constant operation in temperatures below 0°C (32°F).

- Prolonged idling (vehicle operation in stop and go traffic).

- Frequent short running periods (engine does not warm to normal operating temperatures).

- Police, taxi, delivery usage or trailer towing usage.

Oil & oil filter: change every 3750 miles.

Brake pads & discs: service or inspect every 7500 miles.

Driveshaft boots: service or inspect every 7500 miles.

Exhaust system: service or inspect every 7500 miles.

Steering gear & linkage, axle & suspension parts: service or inspect every 7500 miles.

Steering linkage ball joints & front suspension ball joints: service or inspect every 7500 miles.

Air cleaner filter: service or inspect every 15,000 miles.

Final drive oil: Change every 30000 miles if towing a trailer.

Transfer case fluid: Change every 30000 miles if towing a trailer.

37663_IG37_C0012

BRAKES · INFORMATION AND PRECAUTIONS

ANTI-LOCK SYSTEMS

• Certain components within the ABS system are not intended to be serviced or repaired individually.

• Do not use rubber hoses or other parts not specifically specified for and ABS system. When using repair kits, replace all parts included in the kit. Partial or incorrect repair may lead to functional problems and require the replacement of components.

• Lubricate rubber parts with clean, fresh brake fluid to ease assembly. Do not use shop air to clean parts; damage to rubber components may result.

• Use only DOT 3 brake fluid from an unopened container.

• If any hydraulic component or line is removed or replaced, it may be necessary to bleed the entire system.

• A clean repair area is essential. Always clean the reservoir and cap thoroughly before removing the cap. The slightest amount of dirt in the fluid may plug an ori-

fice and impair the system function. Perform repairs after components have been thoroughly cleaned; use only denatured alcohol to clean components. Do not allow ABS components to come into contact with any substance containing mineral oil; this includes used shop rags.

• The Anti-Lock control unit is a microprocessor similar to other computer units in the vehicle. Ensure that the ignition switch is **OFF** before removing or installing controller harnesses. Avoid static electricity discharge at or near the controller.

• If any arc welding is to be done on the vehicle, the control unit should be unplugged before welding operations begin.

DISC AND DRUM SYSTEMS

✳✳ CAUTION

Dust and dirt accumulating on brake parts during normal use may contain asbestos fibers from production or aftermarket brake linings. Breathing excessive concentrations of asbestos fibers can cause serious bodily harm. Exercise care when servicing brake parts. Do not sand or grind brake lining unless equipment used is designed to contain the dust residue. Do not clean brake parts with compressed air or by dry brushing. Cleaning should be done by dampening the brake components with a fine mist of water, then wiping the brake components clean with a dampened cloth. Dispose of cloth and all residue containing asbestos fibers in an impermeable container with the appropriate label. Follow practices prescribed by the Occupational Safety and Health Administration (OSHA) and the Environmental Protection Agency (EPA) for the handling, processing, and disposing of dust or debris that may contain asbestos fibers.

BRAKES · BLEEDING THE BRAKE SYSTEM

BLEEDING PROCEDURE

BLEEDING PROCEDURE

➡Whenever the negative battery cable is disconnected the following components will require resetting. The Automatic temperature control system, Automatic drive positioner, Power window control, Sunroof system, Sunshade system, Rear view monitor, Idle Air Volume Learning, Steering Angle Sensor Neutral Position, Audio presets and Navigation. You will need the CONSULT-III diagnostic tool, or equivalent. Follow the directions on the screen of the tool, as needed.

1. Before servicing the vehicle, refer to the Precautions Section.

➡If working near and/or around the SRS system and components, be sure to disable the SRS system. After disabling the system wait three minutes or more before servicing the vehicle.

2. Disconnect the negative battery cable.

➡Turn the ignition switch off and disconnect the ABS actuator and electric control unit connector, or the negative battery cable before performing the work.

➡Monitor the fluid level in the reservoir while performing the work. Always use new brake fluid. Be sure to use the proper grade and type fluid.

➡As required, cover the crowfoot and flare nut wrench with a shop towel to prevent damage to the front four piston type caliper and rear two piston type caliper.

3. Connect a vinyl tube to the bleeder valve of the right rear brake.

4. Fully depress the brake pedal four or five times.

5. Loosen the bleeder valve and bleed the air with the brake pedal depressed, quickly tighten the bleeder valve.

6. Repeat the above step until all air is expelled out of the brake line.

7. Tighten the bleeder valve.

8. Perform the above procedure to the brakes in the following sequence.

9. Right rear brake, left front brake, left rear brake and right front brake.

10. Check and refill the master cylinder, as required.

11. Be sure to perform the reconnect/relearn procedures.

FLUID FILL PROCEDURE

➡Whenever the negative battery cable is disconnected the following compo-

nents will require resetting. The Automatic temperature control system, Automatic drive positioner, Power window control, Sunroof system, Sunshade system, Rear view monitor, Idle Air Volume Learning, Steering Angle Sensor Neutral Position, Audio presets and Navigation. You will need the CONSULT-III diagnostic tool, or equivalent. Follow the directions on the screen of the tool, as needed.

1. Before servicing the vehicle, refer to the Precautions Section.

➡If working near and/or around the SRS system and components, be sure to disable the SRS system. After disabling the system wait three minutes or more before servicing the vehicle.

➡Turn the ignition switch off and disconnect the ABS actuator and electric control unit connector, or the negative battery cable before performing the work.

➡Cover the crowfoot and flare nut wrench with a shop towel to prevent damage to the front four piston type caliper and rear two piston type caliper.

2. Check that there is no foreign material in the reservoir tank or around it. Never reuse used fluid.

3. Loosen the bleeder valve.

4. Slowly depress the brake pedal to the full stroke. Release the pedal.

5. Repeat at intervals of two or three seconds until all brake fluid is discharged.

6. Close the bleeder valve with the brake pedal depressed.

7. Repeat the above on each wheel.

8. Bleed the brake system.

9. Be sure to perform the reconnect/relearn procedures.

BLEEDING THE ABS SYSTEM

➡**Whenever the negative battery cable is disconnected the following components will require resetting. The Automatic temperature control system, Automatic drive positioner, Power window control, Sunroof system, Sunshade system, Rear view monitor, Idle Air Volume Learning, Steering Angle Sensor Neutral Position, Audio presets and Navigation. You will need the CON-**

SULT-III diagnostic tool, or equivalent. Follow the directions on the screen of the tool, as needed.

1. Before servicing the vehicle, refer to the Precautions Section.

➡**If working near and/or around the SRS system and components, be sure to disable the SRS system. After disabling the system wait three minutes or more before servicing the vehicle.**

2. Disconnect the negative battery cable.

➡**Turn the ignition switch off and disconnect the ABS actuator and electric control unit connector, or the negative battery cable before performing the work.**

➡**Monitor the fluid level in the reservoir while performing the work. Always use new brake fluid. Be sure to use the proper grade and type fluid.**

➡**As required, cover the crowfoot and flare nut wrench with a shop towel to prevent damage to the front four piston type caliper and rear two piston type caliper.**

3. Connect a vinyl tube to the bleeder valve of the right rear brake.

4. Fully depress the brake pedal four or five times.

5. Loosen the bleeder valve and bleed the air with the brake pedal depressed, quickly tighten the bleeder valve.

6. Repeat the above step until all air is expelled out of the brake line.

7. Tighten the bleeder valve.

8. Perform the above procedure to the brakes in the following sequence.

9. Right rear brake, left front brake, left rear brake and right front brake.

10. Check and refill the master cylinder, as required.

11. Be sure to perform the reconnect/relearn procedures.

BRAKES | **ANTI-LOCK BRAKE SYSTEM (ABS)**

WHEEL SPEED SENSORS

REMOVAL & INSTALLATION

See Figure 2.

At this time the manufacturer does not provide removal and installation procedures

for this component. The following procedure is a guideline and may differ from the vehicle you are servicing.

➡**Whenever the negative battery cable is disconnected the following components will require resetting. The Auto-**

matic temperature control system, Automatic drive positioner, Power window control, Sunroof system, Sunshade system, Rear view monitor, Idle Air Volume Learning, Steering Angle Sensor Neutral Position, Audio presets

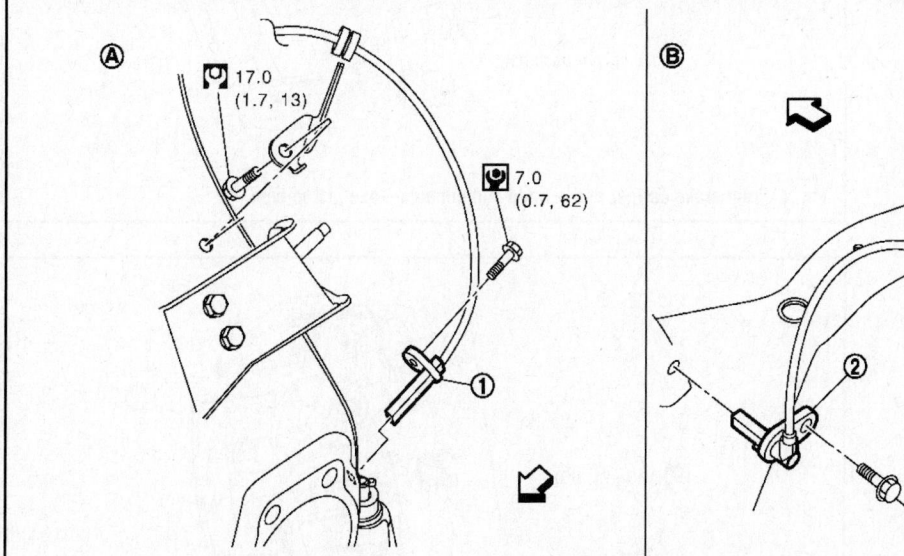

1. Front LH wheel sensor

2. Rear LH wheel sensor

3. Rear RH wheel sensor

A. Front

B. Rear

◁: Vehicle front

37663_IG37_G0192

Fig. 2 Front and rear wheel speed sensors and related components—sedan and coupe

and Navigation. You will need the CON-SULT-III diagnostic tool, or equivalent. Follow the directions on the screen of the tool, as needed.

1. Before servicing the vehicle, refer to the Precautions Section.

➡If working near and/or around the SRS system and components, be sure to disable the SRS system. After disabling the system wait three minutes or more before servicing the vehicle.

2. Disconnect the negative battery cable.

3. Raise and support the vehicle safely.

4. Remove the tire and wheel assembly, as required.

5. Never twist or bend sensor harness when removing it.

6. Pull the wheel sensor out without pulling the harness.

7. Be careful not to damage the sensor edges or rotor teeth.

8. Remove the sensor first, before removing the wheel hub and bearing assembly.

To install:

➡Be sure to use new fasteners, as required.

9. Installation is the reverse of the removal procedure.

WHEEL SPEED SENSOR RINGS (TOOTHED RINGS)

REMOVAL & INSTALLATION
See Figure 3.

The front sensor rotor cannot be disas-

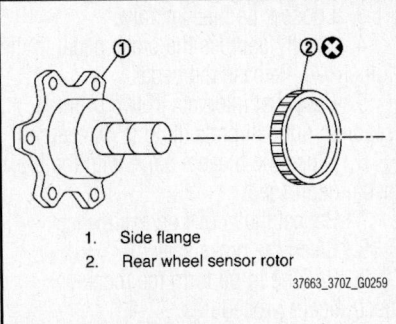

1. Side flange
2. Rear wheel sensor rotor

37663_370Z_G0259

Fig. 3 Rear speed sensor rotor and related components

sembled. To replace the sensor rotor, replace the hub bearing assembly.

At this time the manufacturer does not provide removal and installation procedures for this component, refer to the illustration as required.

BRAKES

FRONT DISC BRAKES

BRAKE CALIPER

REMOVAL & INSTALLATION
See Figures 4 through 6.

➡Whenever the negative battery cable is disconnected the following components will require resetting. The Automatic temperature control system, Automatic drive positioner, Power window control, Sunroof system, Sunshade system, Rear view monitor, Idle Air Volume Learning, Steering Angle Sensor Neutral Position, Audio presets and Navigation. You will need the CON-SULT-III diagnostic tool, or equivalent. Follow the directions on the screen of the tool, as needed.

1. Before servicing the vehicle, refer to the Precautions Section.

➡If working near and/or around the SRS system and components, be sure to disable the SRS system. After disabling the system wait three minutes or more before servicing the vehicle.

2. Disconnect the negative battery cable.
3. Raise and safely support the vehicle.
4. Remove the tire and wheel assembly.
5. Hold the rotor, using the wheel nuts.
6. Drain the brake fluid.
7. On one piston and two piston calipers, remove the union bolt and copper washer. Discard the washer. Disconnect the brake hose from the caliper. Remove the torque member mounting bolts.
8. On four piston caliper, cover the

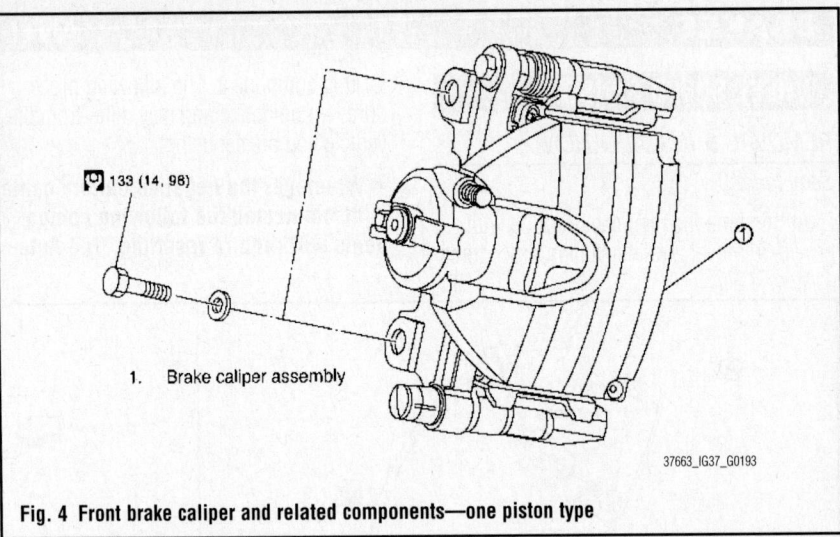

133 (14, 98)

1. Brake caliper assembly

37663_IG37_G0193

Fig. 4 Front brake caliper and related components—one piston type

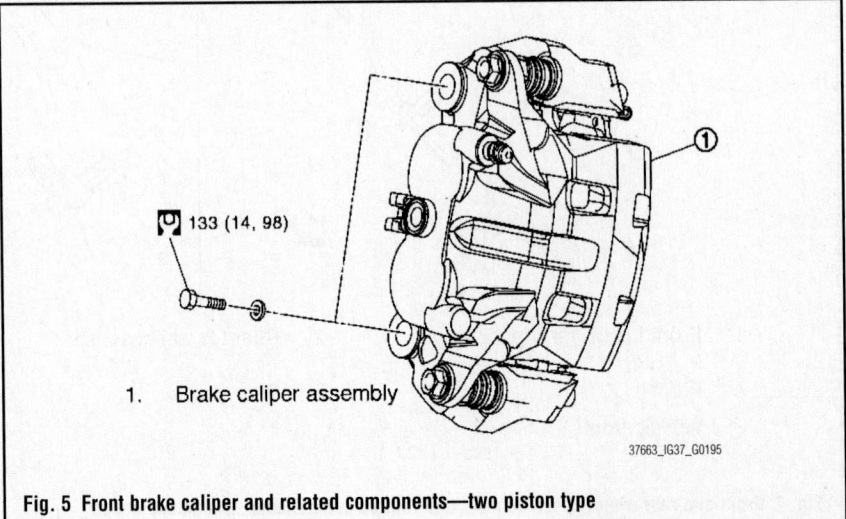

133 (14, 98)

1. Brake caliper assembly

37663_IG37_G0195

Fig. 5 Front brake caliper and related components—two piston type

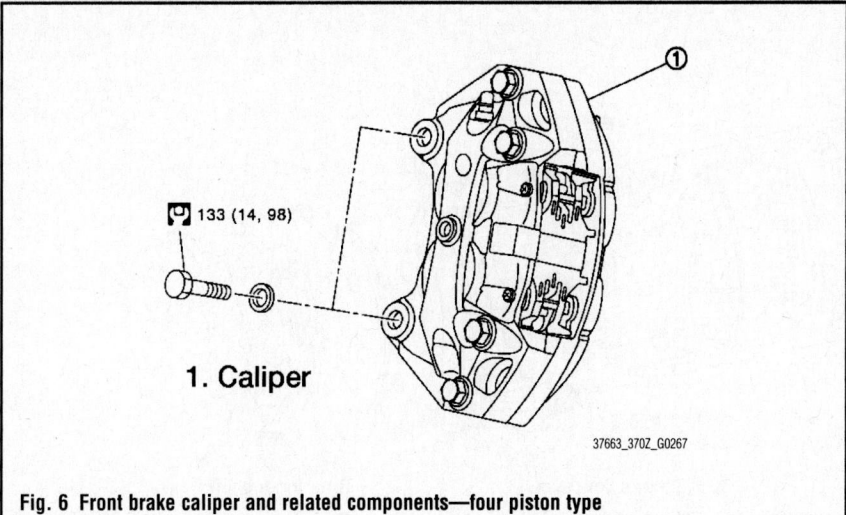

Fig. 6 Front brake caliper and related components—four piston type

flare nut wrench with a shop towel. Never scratch the flare nut and the brake tube. Never bend sharply, twist or strongly pull out the brake tube. Over the open end of the brake tube when disconnecting to prevent the entrance of dirt. Remove the caliper mounting bolts.

9. Remove the caliper assembly.
10. As required, remove the rotor.

To install:

➡Be sure to use new fasteners, as required.

11. Installation is the reverse of the removal procedure.
12. Be sure to use a new cooper washer, one piston and two piston calipers.
13. Bleed the brake system.
14. Be sure to perform the reconnect/relearn procedures.

DISC BRAKE PADS

REMOVAL & INSTALLATION

One and Two Piston Caliper

See Figures 7 and 8.

➡Whenever the negative battery cable is disconnected the following components will require resetting. The Automatic temperature control system, Automatic drive positioner, Power window control, Sunroof system, Sunshade system, Rear view monitor, Idle Air Volume Learning, Steering Angle Sensor Neutral Position, Audio presets and Navigation. You will need the CONSULT-III diagnostic tool, or equivalent. Follow the directions on the screen of the tool, as needed.

1. Before servicing the vehicle, refer to the Precautions Section.

➡If working near and/or around the SRS system and components, be sure to disable the SRS system. After disabling the system wait three minutes or more before servicing the vehicle.

2. Disconnect the negative battery cable.
3. Raise and safely support the vehicle.
4. Remove the tire and wheel assembly.
5. Remove the lower sliding pin bolt.
6. Suspend the caliper using mechanics wire.

➡Do not allow the caliper to hang by the brake line hose.

7. Remove the pads, shims, shim covers and pad retainers from the torque member.

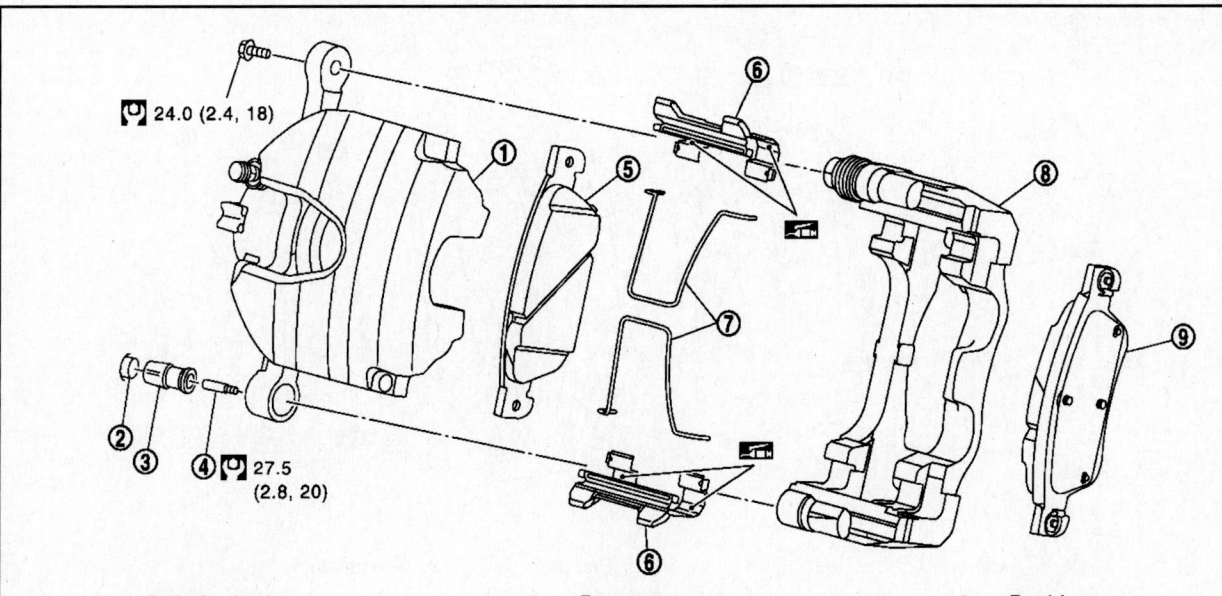

1. Cylinder body
2. Protector
3. Bushing
4. Location pin
5. Inner pad (with pad wear sensor)
6. Pad retainer
7. Pad return spring
8. Torque member
9. Outer pad

: Apply bentonite noise damping brake grease.

Fig. 7 Front brake pads and related components—one piston type

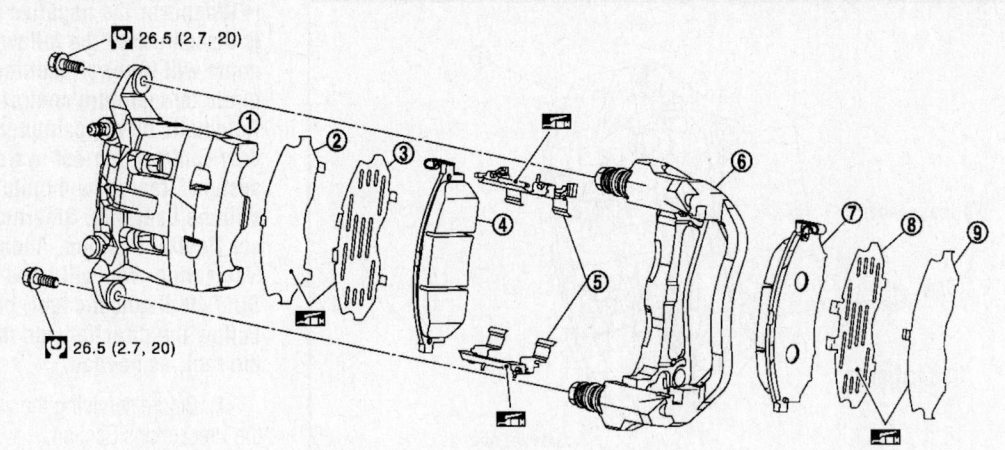

1. Cylinder body
2. Inner shim cover
3. Inner shim
4. Inner pad (with pad wear sensor)
5. Pad retainer
6. Torque member
7. Outer pad (with pad wear sensor)
8. Outer shim
9. Outer shim cover

: Apply copper based brake grease.

37663_IG37_G0196

Fig. 8 Front brake pads and related components—two piston type

To install:

➡**Be sure to use new fasteners, as required.**

8. Installation is the reverse of the removal procedure.

9. Apply copper based brake grease to the pad retainers before installing them to the torque member, if the pad retainers were removed.

➡**Both inner and outer pads have a pad return system on the pad retainer. Install the pad return lever securely to the pad wear sensor.**

10. Depress the brake pedal several times to seat the pads and check that no drag feel is present for the disc rotor.

11. Be sure to perform the reconnect/relearn procedures.

Four Piston Caliper

See Figure 9.

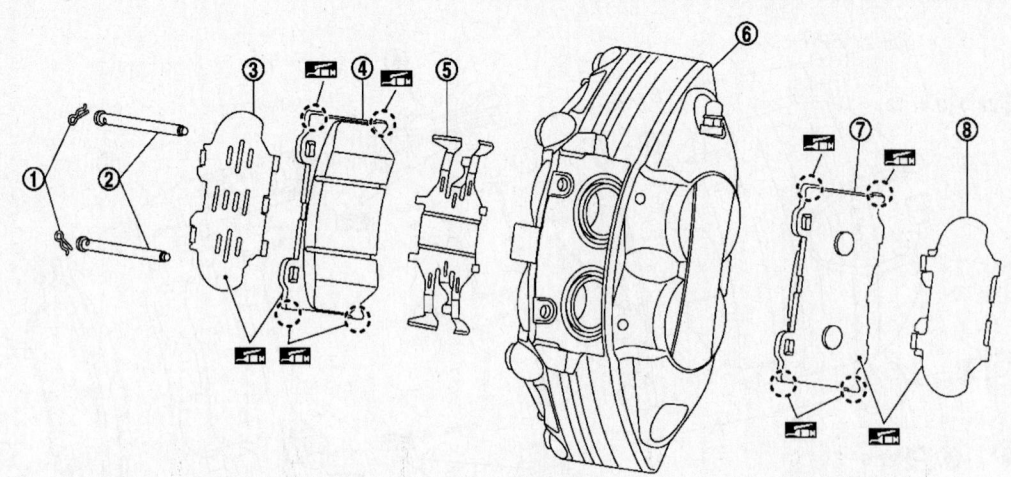

1. Clip
2. Pad pin
3. Inner shim
4. Inner pad (with pad wear sensor)*
5. Cross spring
6. Caliper
7. Outer pad
8. Outer shim

*: Some vehicles has pad wear sensor only for one side.

: Apply copper based brake grease.

37663_370Z_G0265

Fig. 9 Front brake pads and related components—four piston type

→Whenever the negative battery cable is disconnected the following components will require resetting. The Automatic temperature control system, Automatic drive positioner, Power window control, Sunroof system, Sunshade system, Rear view monitor, Idle Air Volume Learning, Steering Angle Sensor Neutral Position, Audio presets and Navigation. You will need the CONSULT-III diagnostic tool, or equivalent. Follow the directions on the screen of the tool, as needed.

1. Before servicing the vehicle, refer to the Precautions Section.

→If working near and/or around the SRS system and components, be sure to disable the SRS system. After disabling the system wait three minutes or more before servicing the vehicle.

2. Disconnect the negative battery cable.

3. Raise and safely support the vehicle.

4. Remove the tire and wheel assembly.

5. Remove the clips from the pad shims.

6. Remove the pad pins while holding down the cross spring. Remove the cross spring from the caliper.

7. Using a pliers, remove the brake pads and shims from the caliper.

To install:

→Be sure to use new fasteners, as required.

8. Installation is the reverse of the removal procedure.

9. Apply copper based brake grease to the mating surfaces between the pads and shims. Install the shims to the brake pads.

10. Install the pads to the caliper.

11. Install the upper pad pin from the inner side, then install firmly to the outer side through the hole in the top of the brake pad.

12. Place the top of the cross spring over the upper pad pin, press the cross spring, install the lower pad pin from the inner side to the outer side and secure the cross spring.

13. Install the clips to the pad pins.

→If the clip is not fully attached, pad pin or brake pad could fall out while the vehicle is in motion.

14. Depress the brake pedal several times to seat the pads and check that no drag feel is present for the disc rotor.

15. Be sure to perform the reconnect/relearn procedures.

BRAKES   REAR DISC BRAKES

BRAKE CALIPER

REMOVAL & INSTALLATION

See Figures 10 and 11.

→Whenever the negative battery cable is disconnected the following components will require resetting. The Automatic temperature control system, Automatic drive positioner, Power window control, Sunroof system, Sunshade system, Rear view monitor, Idle Air Volume Learning, Steering Angle Sensor Neutral Position, Audio presets and Navigation. You will need the CONSULT-III diagnostic tool, or equivalent. Follow the directions on the screen of the tool, as needed.

1. Before servicing the vehicle, refer to the Precautions Section.

→If working near and/or around the SRS system and components, be sure to disable the SRS system. After disabling the system wait three minutes or more before servicing the vehicle.

2. Disconnect the negative battery cable.

3. Raise and safely support the vehicle.

4. Remove the tire and wheel assembly.

5. Hold the rotor, using the wheel nuts.

6. Drain the brake fluid.

7. On one and two piston calipers, remove the union bolt and copper washer. Discard the washer. Disconnect the brake hose from the caliper. Remove the torque member mounting bolts.

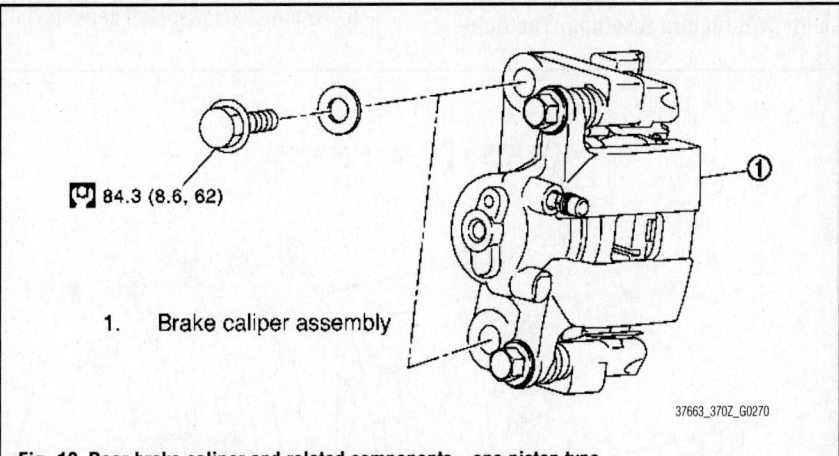

84.3 (8.6, 62)

1. Brake caliper assembly

37663_370Z_G0270

Fig. 10 Rear brake caliper and related components—one piston type

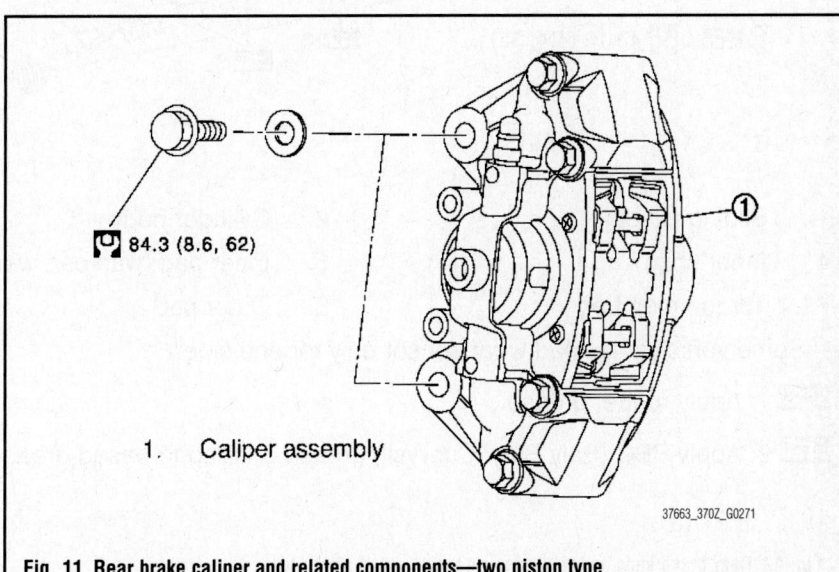

84.3 (8.6, 62)

1. Caliper assembly

37663_370Z_G0271

Fig. 11 Rear brake caliper and related components—two piston type

8. Remove the caliper assembly.

9. As required, remove the rotor.

To install:

➡ Be sure to use new fasteners, as required.

10. Installation is the reverse of the removal procedure.

11. Be sure to use a new cooper washer, two piston caliper.

12. Check for brake drag, correct as required.

13. Bleed the brake system.

14. Be sure to perform the reconnect/relearn procedures.

DISC BRAKE PADS

REMOVAL & INSTALLATION

One Piston Caliper

See Figure 12.

➡ Whenever the negative battery cable is disconnected the following components will require resetting. The Auto-

matic temperature control system, Automatic drive positioner, Power window control, Sunroof system, Sunshade system, Rear view monitor, Idle Air Volume Learning, Steering Angle Sensor Neutral Position, Audio presets and Navigation. You will need the CONSULT-III diagnostic tool, or equivalent. Follow the directions on the screen of the tool, as needed.

1. Before servicing the vehicle, refer to the Precautions Section.

➡ If working near and/or around the SRS system and components, be sure to disable the SRS system. After disabling the system wait three minutes or more before servicing the vehicle.

2. Disconnect the negative battery cable.

3. Raise and safely support the vehicle.

4. Remove the tire and wheel assembly.

5. Remove the lower sliding pin bolt.

6. Suspend the caliper using mechanics wire.

➡ Do not allow the caliper to hang by the brake line hose.

7. Remove the pads, shims, shim covers and pad retainers from the torque member.

To install:

➡ Be sure to use new fasteners, as required.

8. Installation is the reverse of the removal procedure.

9. Apply PBC grease to the mating surfaces between the pads and shims.

10. Apply PBC grease to the mating surfaces between the pad retainers and the pads before installing them to the brake pads.

11. Depress the brake pedal several times to seat the pads and check that no drag feel is present for the disc rotor.

12. Be sure to perform the reconnect/relearn procedures.

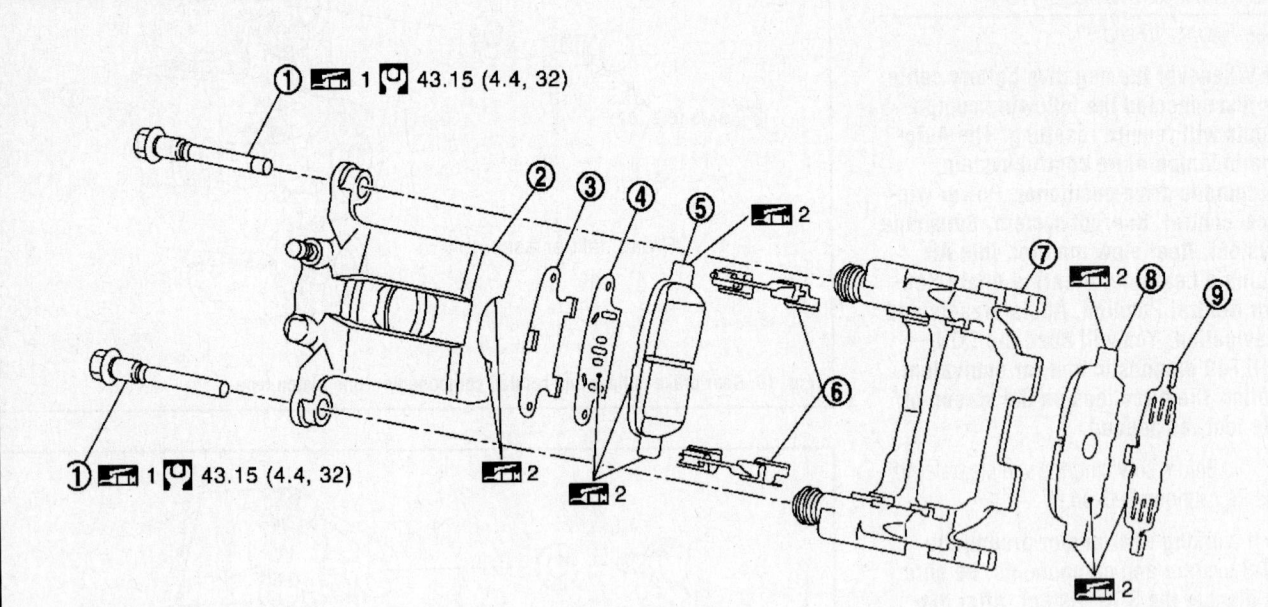

1. Sliding pin bolt
2. Cylinder body
3. Inner shim cover
4. Inner shim
5. Inner pad (with pad wear sensor)*
6. Pad retainer
7. Torque member
8. Outer pad
9. Outer shim

*: Some vehicles has pad wear sensor only for one side.

🔧 1: Apply rubber grease.

🔧 2: Apply PBC (Poly Butyl Cuprysil) grease or silicone-based grease.

37663_370Z_G0268

Fig. 12 Rear brake pads and related components—one piston type

Two Piston Caliper

See Figure 13.

➥Whenever the negative battery cable is disconnected the following components will require resetting. The Automatic temperature control system, Automatic drive positioner, Power window control, Sunroof system, Sunshade system, Rear view monitor, Idle Air Volume Learning, Steering Angle Sensor Neutral Position, Audio presets and Navigation. You will need the CONSULT-III diagnostic tool, or equivalent. Follow the directions on the screen of the tool, as needed.

1. Before servicing the vehicle, refer to the Precautions Section.

➥If working near and/or around the SRS system and components, be sure to disable the SRS system. After disabling the system wait three minutes or more before servicing the vehicle.

2. Disconnect the negative battery cable.

3. Raise and safely support the vehicle.

4. Remove the tire and wheel assembly.

5. Remove the clips from the pad shims.

6. Remove the pad pins while holding down the cross spring. Remove the cross spring from the caliper.

7. Using a pliers, remove the brake pads and shims from the caliper.

To install:

➥Be sure to use new fasteners, as required.

8. Installation is the reverse of the removal procedure.

9. Apply copper based brake grease to the mating surfaces between the pads and caliper, between the pads and pad pins and between the pads and cross spring.

10. Install the pads to the caliper.

11. Install the upper pad pin from the inner side, then install firmly to the outer side through the hole in the top of the brake pad.

12. Place the top of the cross spring over the upper pad pin, press the cross spring, install the lower pad pin from the inner side to the outer side and secure the cross spring.

13. Install the clips to the pad pins.

➥If the clip is not fully attached, pad pin or brake pad could fall out while the vehicle is in motion.

14. Depress the brake pedal several times to seat the pads and check that no drag feel is present for the disc rotor.

15. Be sure to perform the reconnect/relearn procedures.

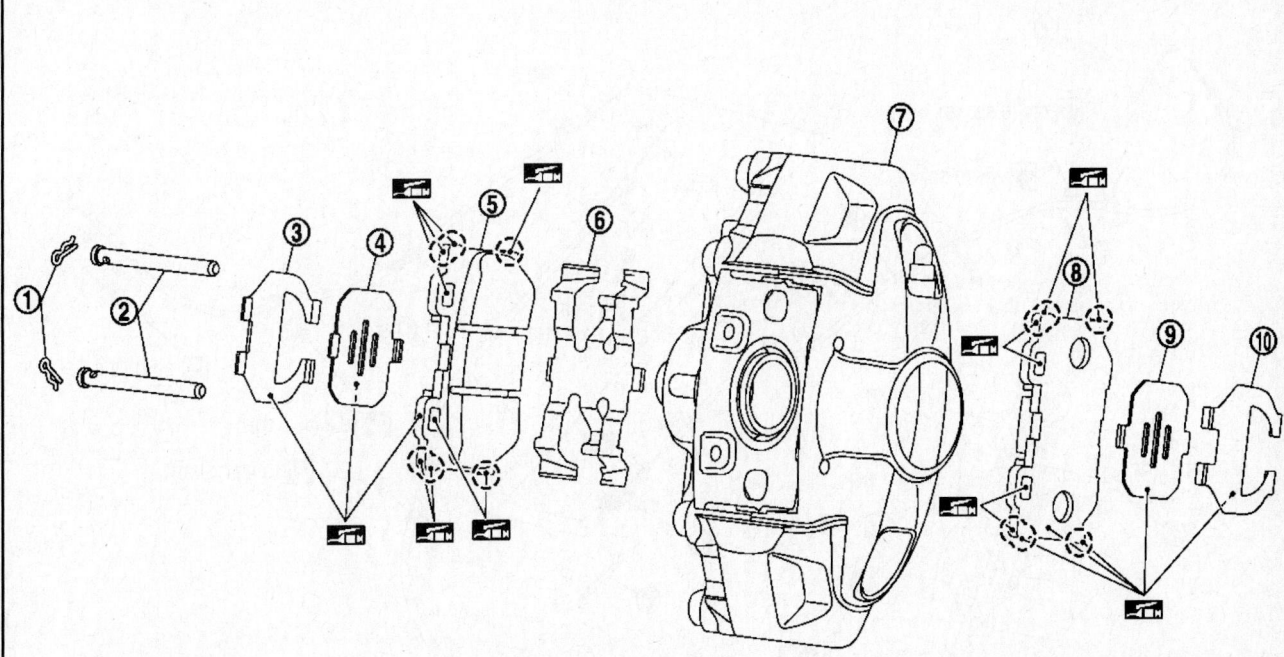

1.	Clip	2.	Pad pin	3.	Inner shim cover
4.	Inner shim	5.	Inner pad (with pad wear sensor)*	6.	Cross spring
7.	Caliper	8.	Outer pad	9.	Outer shim
10.	Outer shim cover				

*: Some vehicles has pad wear sensor only for one side.

⬛: Apply copper based brake grease.

37663_370Z_G0269

Fig. 13 Rear brake pads and related components—two piston type

PARKING BRAKE CABLES

ADJUSTMENT

See Figures 14 through 16.

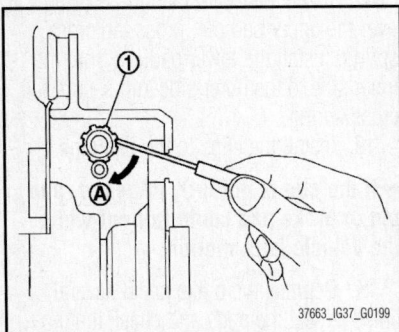

Fig. 14 Parking brake cable adjuster location

1. Before servicing the vehicle, refer to the Precautions Section.

2. To perform adjustment operations, remove tire from the vehicle with power tool.

3. Remove the coin pocket. Insert a deep socket wrench to rotate adjusting nut (1) and loosen the cable sufficiently. Then, return the lever.

4. Using wheel nuts, fix the disc rotor to the hub and prevent it from tilting.

5. Remove adjusting hole plug installed on the disc. Using a flat bladed tool, turn the disc in direction "A" as shown in the figure until the disc is locked. After locking, turn the adjuster in the opposite direction by 5 or 6 notches.

6. Rotate the disc to make sure there is no drag. Install the adjusting hole plug.

7. Adjust cable as follows:

   a. Operate lever/pedal 10 or more times with a force of 66 ft. lbs. for lever type and 110 ft. lbs. for pedal type.

   b. Rotate adjusting nut with deep socket to adjust lever stroke.

   c. When parking brake lever/pedal is operated with a force of 44 ft. lbs. (60 Nm), check that the stroke is 2–3 notches for pedal type and 7–8 notches for lever type (Check it by listening and counting the ratchet clicks).

8. With the lever completely returned, make sure there is no drag on the rear brake.

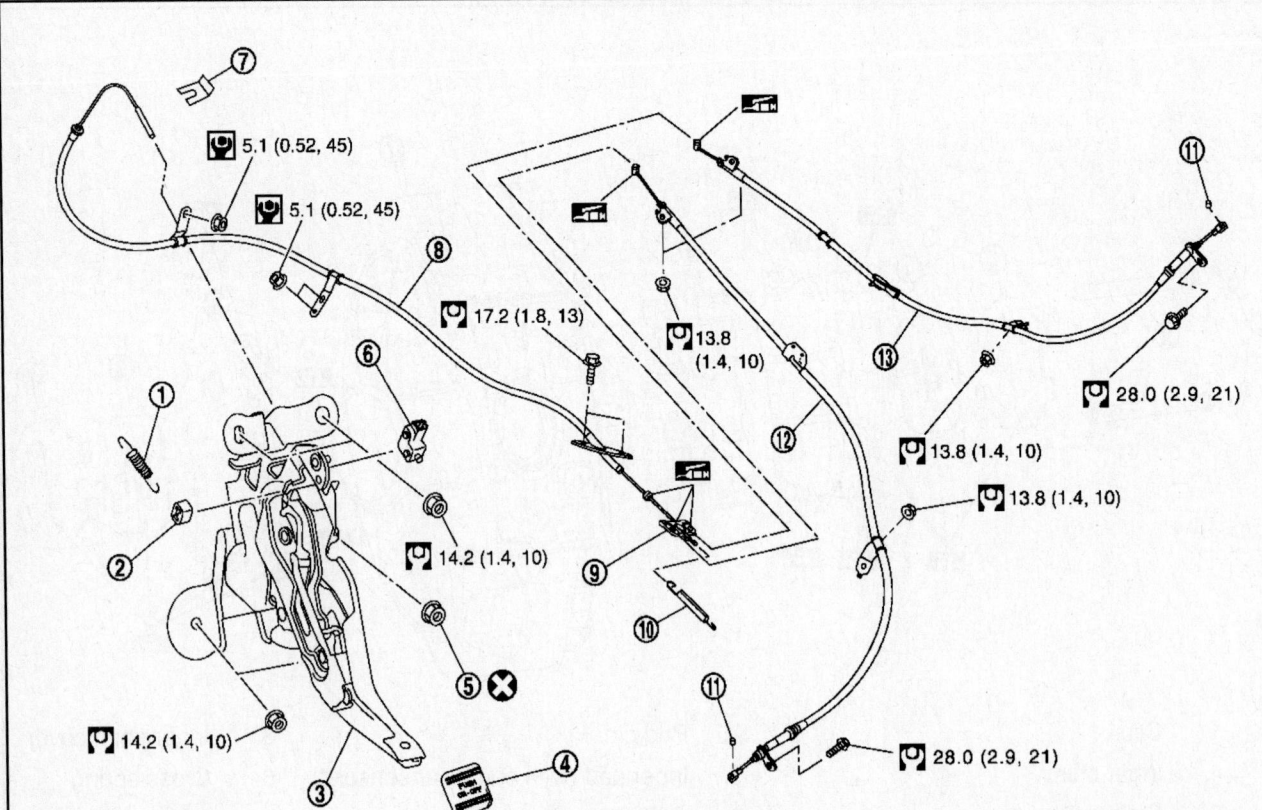

1.	Return spring	2.	Stopper rubber	3.	Device assembly
4.	Pedal pad	5.	Adjusting nut	6.	Parking brake switch
7.	Lock plate	8.	Front cable	9.	Equalizer
10.	Spring	11.	Pin	12.	Rear cable (left side)
13.	Rear cable (right side)				

Fig. 15 Parking brake cables and related components (sedan and coupe)—pedal type

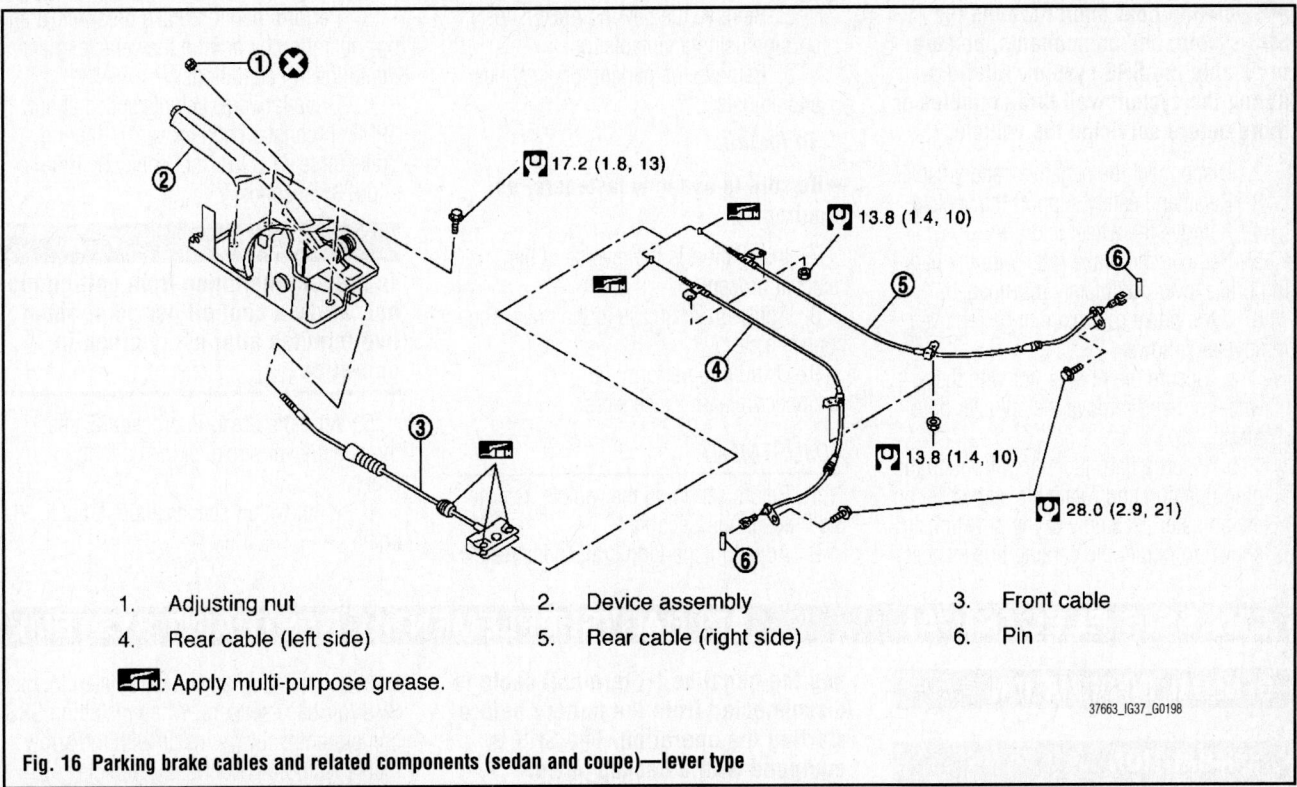

1. Adjusting nut
2. Device assembly
3. Front cable
4. Rear cable (left side)
5. Rear cable (right side)
6. Pin

▣: Apply multi-purpose grease.

37663_IG37_G0198

Fig. 16 Parking brake cables and related components (sedan and coupe)—lever type

PARKING BRAKE SHOES

REMOVAL & INSTALLATION
See Figure 17.

➡Whenever the negative battery cable is disconnected the following components will require resetting. The Automatic temperature control system, Automatic drive positioner, Power window control, Sunroof system, Sunshade system, Rear view monitor, Idle Air Volume Learning, Steering Angle Sensor Neutral Position, Audio presets and Navigation. You will need the CONSULT-III diagnostic tool, or equivalent. Follow the directions on the screen of the tool, as needed.

1. Before servicing the vehicle, refer to the Precautions Section.

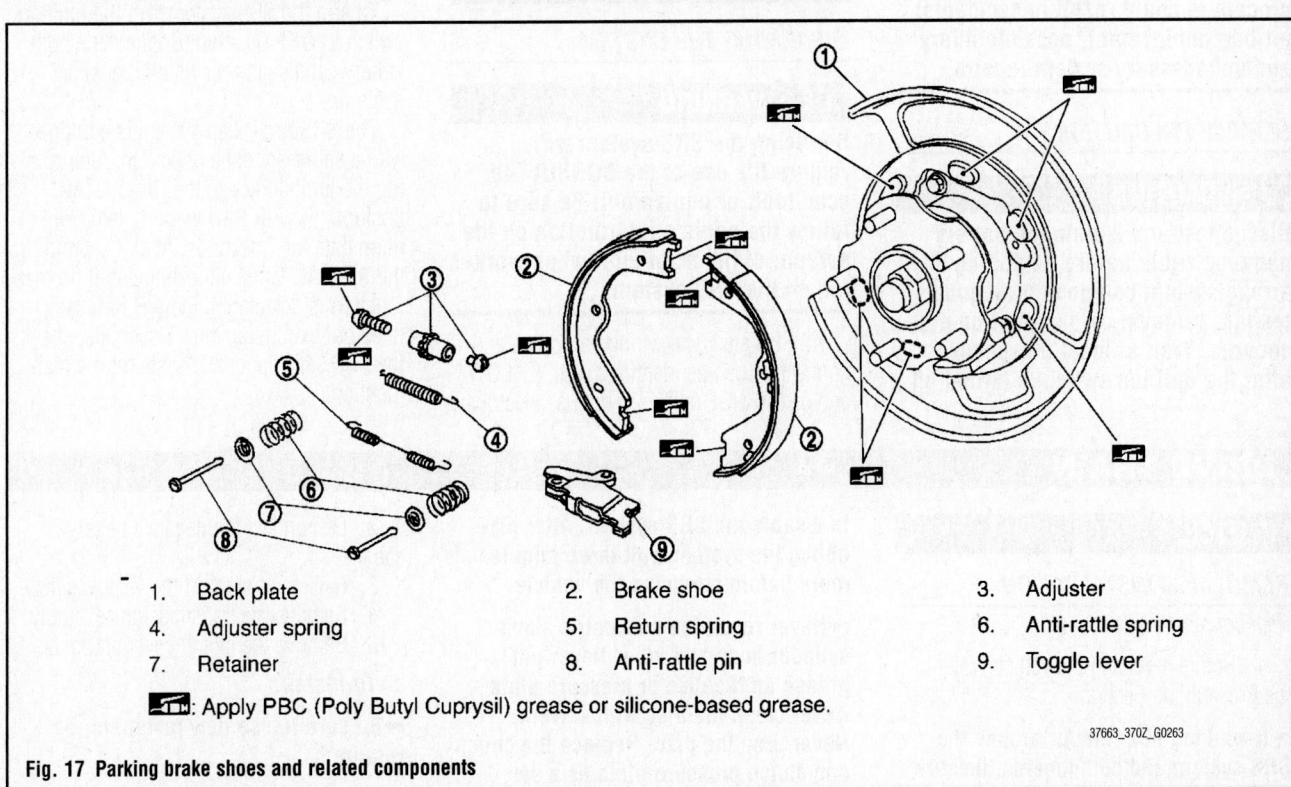

1. Back plate
2. Brake shoe
3. Adjuster
4. Adjuster spring
5. Return spring
6. Anti-rattle spring
7. Retainer
8. Anti-rattle pin
9. Toggle lever

▣: Apply PBC (Poly Butyl Cuprysil) grease or silicone-based grease.

37663_370Z_G0263

Fig. 17 Parking brake shoes and related components

➡If working near and/or around the SRS system and components, be sure to disable the SRS system. After disabling the system wait three minutes or more before servicing the vehicle.

2. Disconnect the negative battery cable.
3. Raise and safely support the vehicle.
4. Remove the wheel and tire.
5. Remove the brake rotor with the parking brake lever completely disengaged.
6. If the brake rotor cannot be removed, remove as follows:
    a. Secure the brake rotor with the wheel nut and remove the adjuster hole plug.
    b. Insert a flat-bladed tool through the plug opening and rotate the star wheel on the adjuster assembly in the direction as shown to retract the parking brake shoes.

c. Remove the parking brake shoe springs using a suitable tool.
    d. Remove the parking brake shoes and adjuster.

To install:

➡Be sure to use new fasteners, as required.

7. Installation is the reverse of the removal procedure.
8. Apply brake grease to the brake shoe contact area.
9. Be sure to perform the reconnect/relearn procedures.

ADJUSTMENT

1. Before servicing the vehicle, refer to the Precautions Section.
2. Adjust the parking brake pedal/lever.

3. Perform parking brake break-in (drag on) operation by driving the vehicle under the following conditions.
4. Drive forward. Vehicle speed about 19MPH (constant and forward). Parking brake force 71.6 lbs. (set contact). Time about 35 seconds.

✳✳ CAUTION

To prevent the lining from getting too hot, allow a cool off period of about five minutes after every break-in operation.

5. After the break-in procedure, check the pedal/lever stroke of the parking brake.
6. If not within specification, adjust again.

CHASSIS ELECTRICAL

GENERAL INFORMATION

✳✳ CAUTION

These vehicles are equipped with an air bag system. The system must be disarmed before performing service on, or around, system components, the steering column, instrument panel components, wiring and sensors. Failure to follow the safety precautions and the disarming procedure could result in accidental air bag deployment, possible injury and unnecessary system repairs.

SERVICE PRECAUTIONS

✳✳ CAUTION

Disconnect and isolate the battery negative cable before beginning any airbag system component diagnosis, testing, removal, or installation procedures. Wait at least 90 seconds after the ignition switch is turned off

AIR BAG (SUPPLEMENTAL RESTRAINT SYSTEM)

and the negative (-) terminal cable is disconnected from the battery before starting the operation. The SRS is equipped with a backup power source, so if work is started within 90 seconds after disconnecting the negative (-) terminal cable from the battery, the SRS may be deployed. Failure to disable the airbag system may result in accidental airbag deployment, personal injury, or death.

DISARMING THE SYSTEM

✳✳ WARNING

Servicing the SRS system will require the use of the CONSULT-III scan tool, or equivalent. Be sure to follow the service information on the screen, of the scan tool, when working on the SRS system.

All SRS electrical wiring harnesses and connectors can be identified with YELLOW and or ORANGE color. Do not use electrical

test equipment on any circuit related to the SRS (air bag) sensors. When installing SRS components, always install with the arrow marks facing the front of the vehicle.

To disarm the SRS system turn the ignition switch to **OFF** position. Then, disconnect the both battery cables starting with the negative cable first and wait at least 3 minutes after the cables are disconnected. Be sure to insulate the battery terminal ends.

ARMING THE SYSTEM

To arm the SRS system turn the ignition switch to **OFF** position. Connect the both battery cables starting with the positive cable first.

The SRS or air bag system is equipped with a self-diagnostic operation. After turning the ignition key to the ON or START position, the AIR BAG warning lamp will illuminate for 7 seconds. After 7 seconds, the AIR BAG lamp will extinguish if no malfunction is detected. If the AIR BAG lamp does not extinguish after 7 seconds, check the SRS self-diagnostic system for a malfunction.

DRIVE TRAIN

CLUTCH

REMOVAL & INSTALLATION

See Figures 18 and 19.

1. Before servicing the vehicle, refer to the Precautions Section.

➡If working near and/or around the SRS system and components, be sure

to disable the SRS system. After disabling the system wait three minutes or more before servicing the vehicle.

➡Never reuse the concentric slave cylinder body and tube. Never put grease on the disc or pressure plate. Never clean the disc with solvent. Never drop the disc. Replace the clutch and clutch pressure plate as a set.

2. Disconnect the negative battery cable.
3. Raise and support the vehicle safely.
4. Remove the transmission assembly.
5. Remove the clutch cover and disc.

To install:

➡Be sure to use new fasteners, as required.

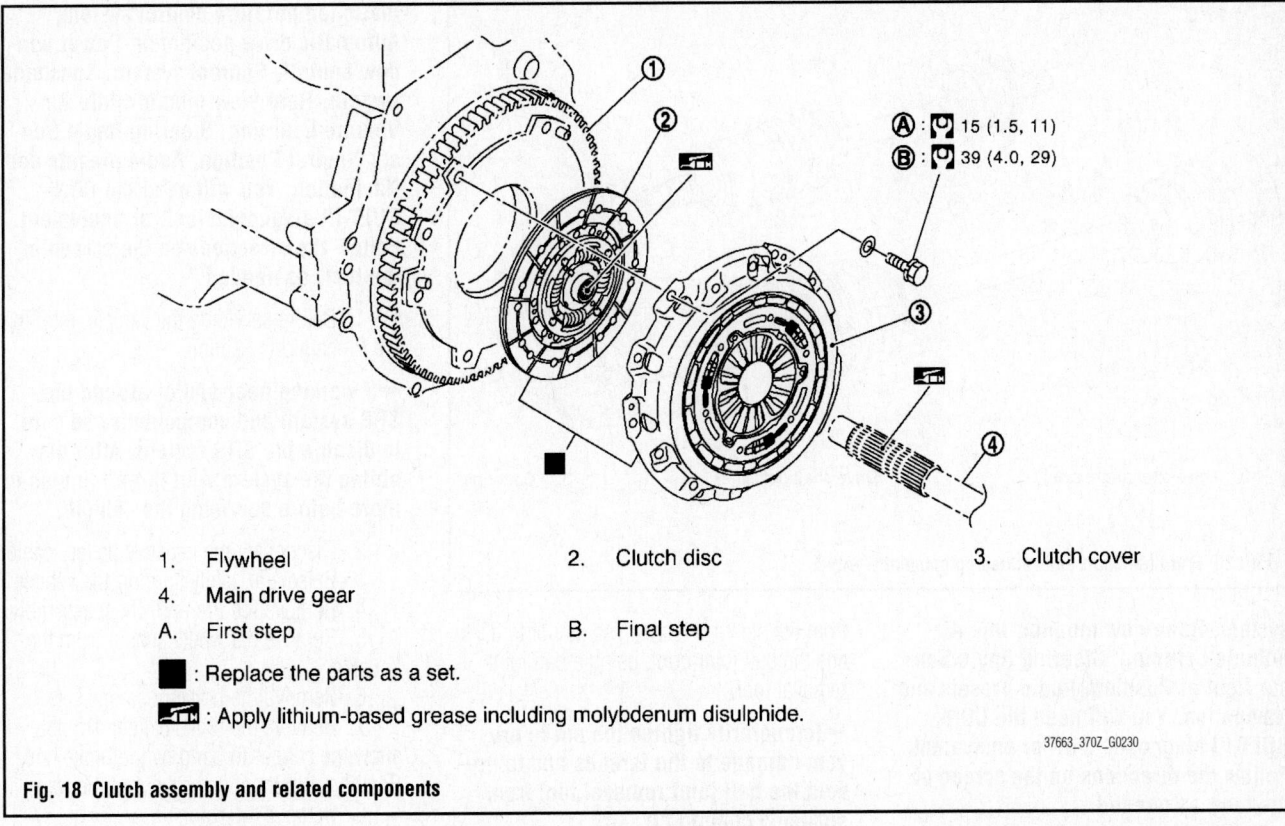

1. Flywheel
2. Clutch disc
3. Clutch cover
4. Main drive gear
A. First step
B. Final step

■ : Replace the parts as a set.

◢▦ : Apply lithium-based grease including molybdenum disulphide.

A : 🔧 15 (1.5, 11)
B : 🔧 39 (4.0, 29)

37663_370Z_G0230

Fig. 18 Clutch assembly and related components

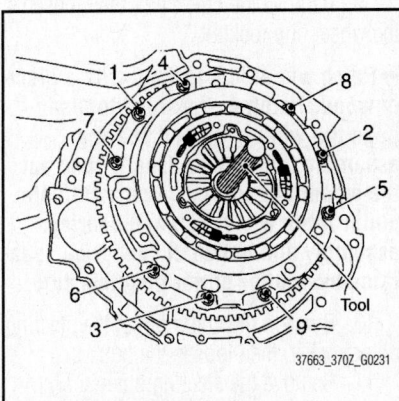

37663_370Z_G0231

Fig. 19 Clutch bolt locations and tightening sequence

6. Installation is the reverse of the removal procedure.

7. Be sure to apply grease to the specified points, see illustration.

8. Be sure to tighten the clutch pressure plate retaining bolts to specification and in the proper sequence. Tighten in two steps.

9. Be sure to perform the reconnect/relearn procedures.

HYDRAULIC SYSTEM BLEEDING

BLEEDING PROCEDURE

At this time the manufacturer does not provide service information for bleeding the system. The following procedure is a guideline and may differ from the vehicle you are servicing.

Bleeding is required to remove air trapped in the hydraulic system. The bleed screw is located on the clutch slave (operating) cylinder.

Some models are also equipped with a clutch damper mechanism. The clutch damper mechanism is bled in exactly the same manner as the operating cylinder. It should be bled along with the operating cylinder.

➡Whenever the negative battery cable is disconnected the following components will require resetting. The Automatic temperature control system, Automatic drive positioner, Power window control, Sunroof system, Sunshade system, Rear view monitor, Idle Air Volume Learning, Steering Angle Sensor Neutral Position, Audio presets and Navigation. You will need the CONSULT-III diagnostic tool, or equivalent. Follow the directions on the screen of the tool, as needed.

1. Before servicing the vehicle, refer to the Precautions Section.

➡If working near and/or around the SRS system and components, be sure to disable the SRS system. After disabling the system wait three minutes or more before servicing the vehicle.

2. Remove the bleed screw dust cap.

3. Attach a transparent vinyl tube to the bleed screw, immersing the free end in a clean container of clean brake fluid.

4. Fill the master cylinder with the proper fluid.

5. Open the bleed screw about ¾ turn.

6. Depress the clutch pedal quickly. Hold it down. Have an assistant tighten the bleed screw. Allow the pedal to return slowly.

7. Repeat the above procedure until no more air bubbles are seen in the fluid container.

8. Remove the bleed tube.

9. Replace the dust cap and refill the master cylinder.

10. Bleed the clutch damper, if equipped.

11. Be sure to perform the reconnect/relearn procedures, as required.

FRONT HALFSHAFT

REMOVAL & INSTALLATION

Left Side

See Figure 20.

➡Whenever the negative battery cable is disconnected the following components will require resetting. The Automatic temperature control system, Automatic drive positioner, Power window control, Sunroof system, Sunshade

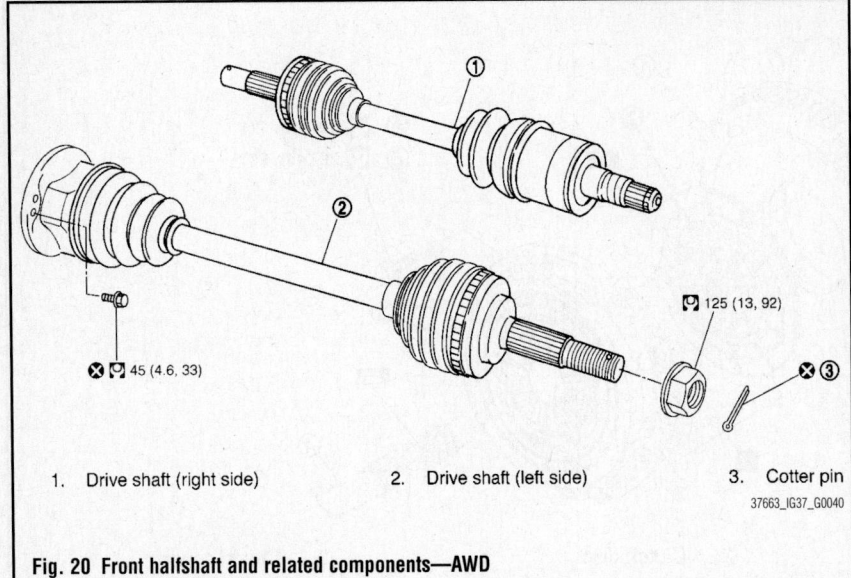

1. Drive shaft (right side)
2. Drive shaft (left side)
3. Cotter pin

37663_IG37_G0040

Fig. 20 Front halfshaft and related components—AWD

system, Rear view monitor, Idle Air Volume Learning, Steering Angle Sensor Neutral Position, Audio presets and Navigation. You will need the CONSULT-III diagnostic tool, or equivalent. Follow the directions on the screen of the tool, as needed.

1. Before servicing the vehicle, refer to the Precautions Section.

➡**If working near and/or around the SRS system and components, be sure to disable the SRS system. After disabling the system wait three minutes or more before servicing the vehicle.**

2. Disconnect the negative battery cable.
3. Raise and safely support the vehicle.
4. Remove the tire and wheel assembly.
5. Remove the wheel sensor from the knuckle.
6. Remove the brake hose bracket.
7. Remove the brake caliper. Do not allow the caliper to hang by the brake hose. Position it to the side and out of the way.
8. Remove the rotor.
9. Remove the cotter pin, then loosen the wheel hub locknut.

➡**Patch wheel hub locknut with a piece of wood. Hammer the wood to disengage the wheel hub and bearing assembly from the halfshaft. Take out the wheel hub locknut. Use a suitable puller if the wheel hub and bearing assembly and halfshaft cannot be separated even after performing this step.**

10. Remove the cotter pin of the steering outer socket, then loosen the nut.
11. Remove the steering outer socket

from the steering knuckle so as not to damage the ball joint boot, using a ball joint removal tool.

➡**Temporarily tighten the nut to prevent damage to the threads and to prevent the ball joint removal tool from suddenly coming off.**

12. Remove the cotter pin of the transverse link and steering knuckle, and then loosen the nut.
13. Separate the transverse link from the steering knuckle so as not to damage the ball joint boot using a ball joint removal too.

➡**Temporarily tighten the nut to prevent damage to the threads and to prevent the ball joint removal tool from suddenly coming off.**

14. Remove the halfshaft from the wheel hub and bearing assembly.
15. Remove the shock from the transverse link
16. Remove the retaining bolts and nuts and remove the halfshaft from the vehicle.

To install:

➡**Be sure to use new fasteners, as required.**

17. Installation is the reverse of the removal procedure.
18. Be sure to perform the reconnect/relearn procedures.

Right Side
See Figure 26.

➡**Whenever the negative battery cable is disconnected the following components will require resetting. The Auto-**

matic temperature control system, Automatic drive positioner, Power window control, Sunroof system, Sunshade system, Rear view monitor, Idle Air Volume Learning, Steering Angle Sensor Neutral Position, Audio presets and Navigation. You will need the CONSULT-III diagnostic tool, or equivalent. Follow the directions on the screen of the tool, as needed.

1. Before servicing the vehicle, refer to the Precautions Section.

➡**If working near and/or around the SRS system and components, be sure to disable the SRS system. After disabling the system wait three minutes or more before servicing the vehicle.**

2. Disconnect the negative battery cable.
3. Raise and safely support the vehicle.
4. Remove the tire and wheel assembly.
5. Remove the wheel sensor from the knuckle.
6. Remove the brake hose bracket.
7. Remove the brake caliper. Do not allow the caliper to hang by the brake hose. Position it to the side and out of the way.
8. Remove the rotor.
9. Remove the cotter pin, then loosen the wheel hub locknut.

➡**Patch wheel hub locknut with a piece of wood. Hammer the wood to disengage the wheel hub and bearing assembly from the halfshaft. Take out the wheel hub locknut. Use a suitable puller if the wheel hub and bearing assembly and halfshaft cannot be separated even after performing this step.**

10. Remove the cotter pin of the steering outer socket, then loosen the nut.
11. Remove the steering outer socket from the steering knuckle so as not to damage the ball joint boot, using a ball joint removal tool.

➡**Temporarily tighten the nut to prevent damage to the threads and to prevent the ball joint removal tool from suddenly coming off.**

12. Remove the cotter pin of the transverse link and steering knuckle, and then loosen the nut.
13. Separate the transverse link from the steering knuckle so as not to damage the ball joint boot using a ball joint removal too.

➡**Temporarily tighten the nut to prevent damage to the threads and to prevent the ball joint removal tool from suddenly coming off.**

14. Remove the halfshaft from the wheel hub and bearing assembly.

15. Remove the shock from the transverse link

16. Using tool SST:KV40107500 and a slide hammer, remove the halfshaft from the final drive unit.

➡ **Never position the halfshaft joint at an extreme angle when removing the halfshaft. Be careful not to overextend the slide joint.**

To install:

➡ **Be sure to use new fasteners, as required.**

17. Installation is the reverse of the removal procedure.

➡ **Always replace the front final drive oil seal when installing the halfshaft.**

18. Be sure to perform the reconnect/relearn procedures.

REAR AXLE HOUSING

REMOVAL & INSTALLATION

See Figures 21 through 25.

➡ **Whenever the negative battery cable is disconnected the following components will require resetting. The Automatic temperature control system, Automatic drive positioner, Power window control, Sunroof system, Sunshade system, Rear view monitor, Idle Air Volume Learning, Steering Angle Sensor Neutral Position, Audio presets and**

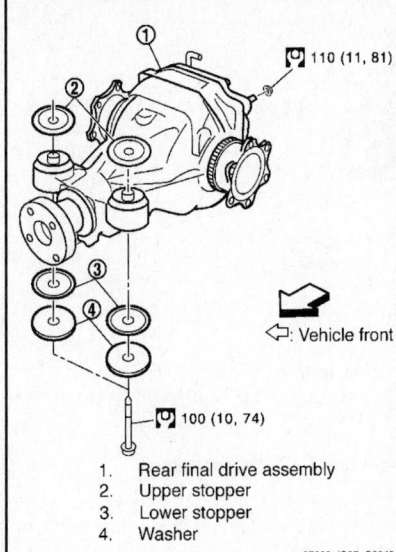

1. Rear final drive assembly
2. Upper stopper
3. Lower stopper
4. Washer

37663_IG37_G0042

Fig. 22 Rear drive axle and related components (AWD)—R200

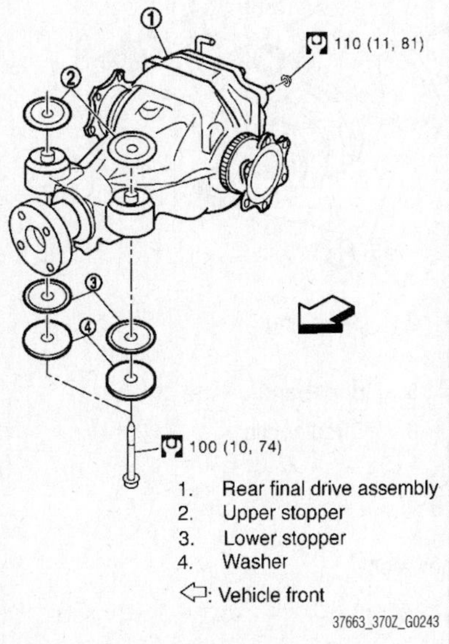

Navigation. You will need the CONSULT-III diagnostic tool, or equivalent. Follow the directions on the screen of the tool, as needed.

1. Before servicing the vehicle, refer to the Precautions Section.

➡ **If working near and/or around the SRS system and components, be sure to disable the SRS system. After disabling the system wait three minutes or more before servicing the vehicle.**

2. Disconnect the negative battery cable.

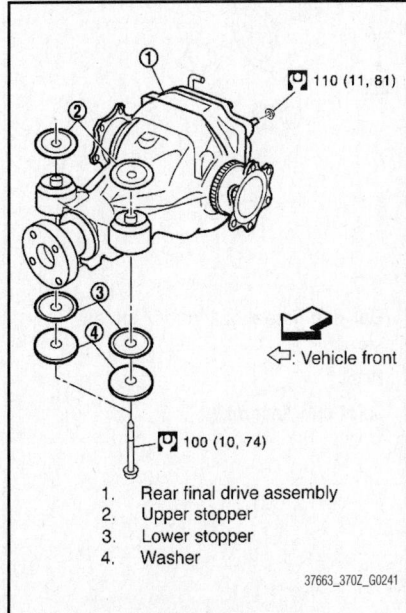

1. Rear final drive assembly
2. Upper stopper
3. Lower stopper
4. Washer

37663_370Z_G0241

Fig. 21 Rear drive axle and related components (except AWD)—R200

110 (11, 81)

100 (10, 74)

⇦: Vehicle front

1. Rear final drive assembly
2. Upper stopper
3. Lower stopper
4. Washer

⇦: Vehicle front

37663_370Z_G0243

Fig. 23 Rear drive axle and related components (manual transmission)—R200V

3. Raise and support the vehicle safely.

4. Remove the center muffler.

5. Remove the stabilizer bar.

6. Remove the driveshaft.

7. Remove the rear halfshafts from the final drive.

8. Remove the breather hose from the final drive. Discard the hose clamp.

9. Remove the rear wheel speed sensors.

10. Position a suitable jack under the assembly.

➡ **Never place the jack on the rear cover.**

11. Remove the mounting bolts and nuts connecting the suspension member and remove the final drive assembly. Be sure the assembly is secured in the jack before removing it.

To install:

➡ **Be sure to use new fasteners, as required.**

12. Installation is the reverse of the removal procedure.

13. Be sure that there are no pinched or restricted areas on the breather hose caused by bending or winding, when installing it.

14. When installing the new hose clamp, install it at the final drive side with the tab facing downward.

15. If the breather connector was removed, install it as shown in the illustration. Never reuse the breather connector and metal connector.

16. Be sure to perform the reconnect/relearn procedures.

REAR HALFSHAFT

REMOVAL & INSTALLATION

See Figure 26.

➡ **Whenever the negative battery cable is disconnected the following components will require resetting. The Automatic temperature control system, Automatic drive positioner, Power window control, Sunroof system, Sunshade system, Rear view monitor, Idle Air Volume Learning, Steering Angle Sensor Neutral Position, Audio presets and Navigation. You will need the CONSULT-III diagnostic tool, or equivalent. Follow the directions on the screen of the tool, as needed.**

1. Before servicing the vehicle, refer to the Precautions Section.

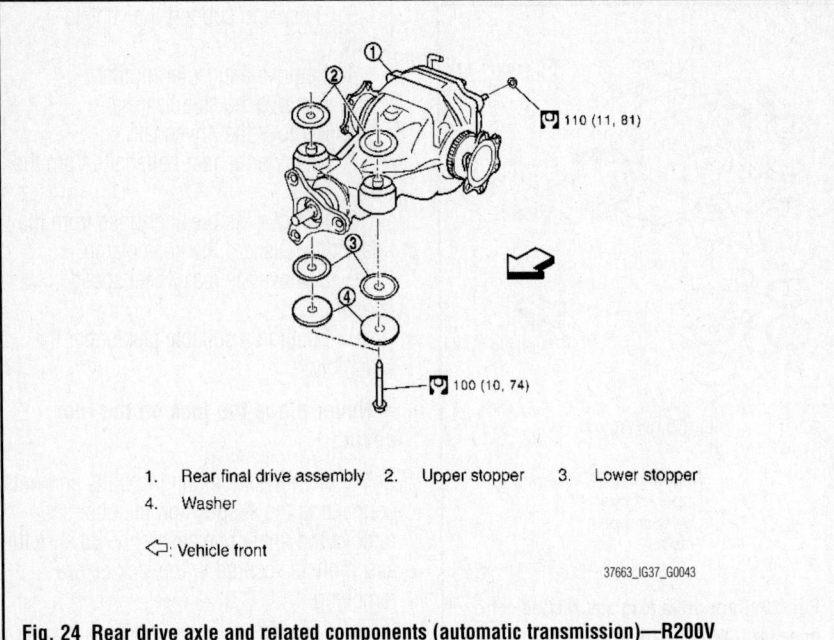

1. Rear final drive assembly   2. Upper stopper   3. Lower stopper
4. Washer

⟵: Vehicle front

37663_IG37_G0043

Fig. 24 Rear drive axle and related components (automatic transmission)—R200V

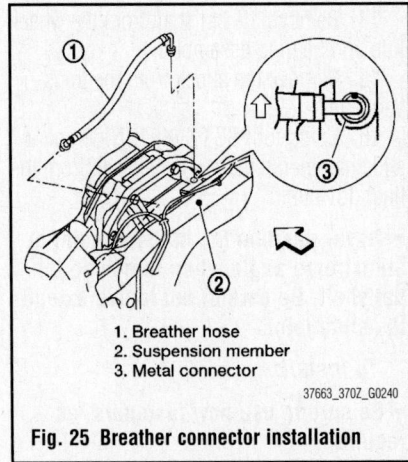

1. Breather hose
2. Suspension member
3. Metal connector

37663_370Z_G0240

Fig. 25 Breather connector installation

➡If working near and/or around the SRS system and components, be sure to disable the SRS system. After disabling the system wait three minutes or more before servicing the vehicle.

2. Disconnect the negative battery cable.

3. Raise and support the vehicle safely.

4. Remove the tire and wheel assemblies.

5. Remove and discard the cotter pin.

6. Loosen the wheel hub locknut.

7. Matchmark the halfshaft and wheel hub and bearing.

8. On convertible remove the diag. brace.

9. Remove the main muffler and center muffler, coupe and sedan.

10. Remove the center muffler, convertible.

11. Patch wheel hub locknut with a piece of wood. Hammer the wood to disengage

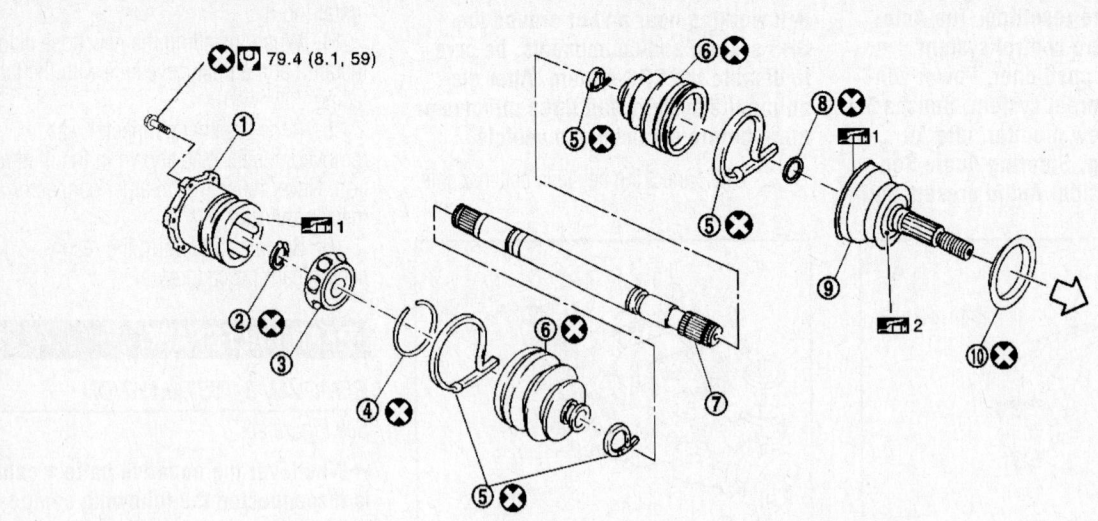

1.	Housing	2.	Snap ring	3.	Ball cage/steel ball/inner race assembly
4.	Stopper ring	5.	Boot band	6.	Boot
7.	Shaft	8.	Circular clip	9.	Joint sub-assembly
10.	Dust shield				

⟵: Wheel side

1: NISSAN genuine grease or an equivalent.

2: Apply paste [service parts (440037S000)].

37663_370Z_G0245

Fig. 26 Rear halfshaft exploded view

the wheel hub and bearing assembly from the halfshaft.

➡**Never position the halfshaft at an extreme angle. Do not overextend the slide joint. Properly support the half-shaft, do not allow it to hang unsupported.**

12. Use a suitable puller if the wheel hub and bearing assembly and halfshaft cannot be separated even after performing the above procedure.

13. Remove the wheel locknut.

14. Remove the mounting bolts between the side flange and the halfshaft.

To install:

➡**Be sure to use new fasteners, as required.**

15. Installation is the reverse of the removal procedure.

16. Clean the matching surface of the halfshaft and wheel hub and bearing assembly.

17. Apply paste part number 440037S000 or equivalent to the surface of the sub joint assembly of the halfshaft.

➡**Apply the paste, about 0.04–0.10 ounce) to cover the entire flat surface of the sub joint assembly of the halfshaft.**

18. The wheel hub locknut tightening specification is 136 ft. lbs. (185 Nm).

19. Perform a final tightening of nuts and bolts of each removed component with the vehicle in an unladen position.

20. Be sure to perform the reconnect/relearn procedures.

REAR PINION SEAL

REMOVAL & INSTALLATION

See Figures 27 through 31.

At this time the manufacturer does not provide removal and installation procedures for this component. The following procedure is a guideline and may differ from the vehicle you are servicing.

➡**Whenever the negative battery cable is disconnected the following components will require resetting. The Automatic temperature control system, Automatic drive positioner, Power window control, Sunroof system, Sunshade system, Rear view monitor, Idle Air Volume Learning, Steering Angle Sensor Neutral Position, Audio presets and Navigation. You will need the CONSULT-III diagnostic tool, or equivalent. Follow the directions on the screen of the tool, as needed.**

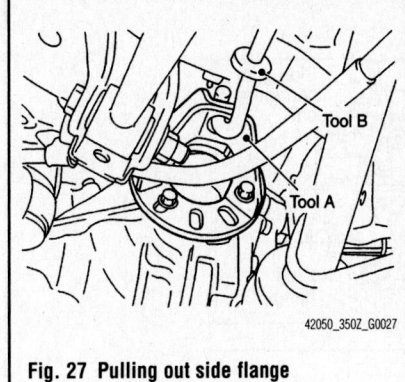

Fig. 27 Pulling out side flange

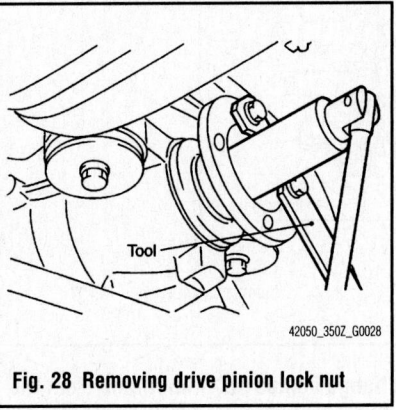

Fig. 28 Removing drive pinion lock nut

1. Before servicing the vehicle, refer to the Precautions Section.

➡**If working near and/or around the SRS system and components, be sure to disable the SRS system. After disabling the system wait three minutes or more before servicing the vehicle.**

2. Disconnect the negative battery cable.

➡**Verify the identification stamp for the replacement frequency, installed on the lower part of gear carrier. Use this to determine the replacement of collapsible spacer when replacing front oil seal. If it is necessary to replace the collapsible spacer, remove final drive assembly and disassemble to replace front oil seal and collapsible spacer.**

3. Drain gear oil.

4. Raise and support the rear of the vehicle safely and remove the wheels.

5. Remove the center muffler.

6. Remove or disconnect the following:
- ABS rear wheel sensor
- Rear halfshaft, and suspend with mechanics wire

7. Install attachments [A: KV40104100 and B: ST36230000 (J-25840-A)] to side flange, and then pull out the side flange with the sliding hammer.

8. Remove drive shaft.

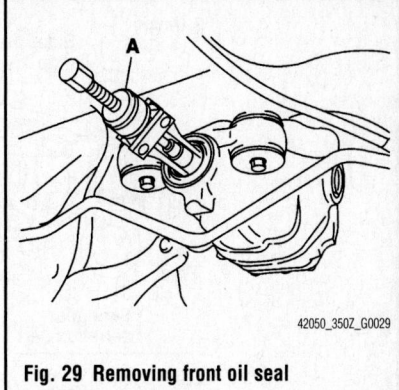

Fig. 29 Removing front oil seal

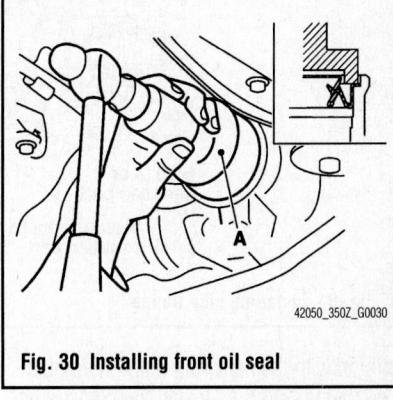

Fig. 30 Installing front oil seal

9. Measure the total preload with the preload gauge tool [ST3127S000 (J-25765-A)] and record the measurement.

10. Remove drive pinion lock nut using the flange wrench KV40104000.

✳ WARNING

For matching mark, use paint. Do not damage drive pinion.

➡**The matching mark "A" on the final drive companion flange indicates the maximum vertical runout position.**

11. Put matching mark on the end of the drive pinion. The matching mark should be in line with the matching mark on the companion flange.

12. Remove companion flange using a puller.

13. Remove front oil seal using the puller [A: KV381054S0 (J-34286)].

To install:

➡**Do not reuse oil seal. Do not incline oil seal when installing.**

14. Apply multi-purpose grease to front oil seal lips.

15. Install front oil seal using the drift [A:ST30720000 (J-25405)] as shown in the illustration.

16. Align the matching mark of drive

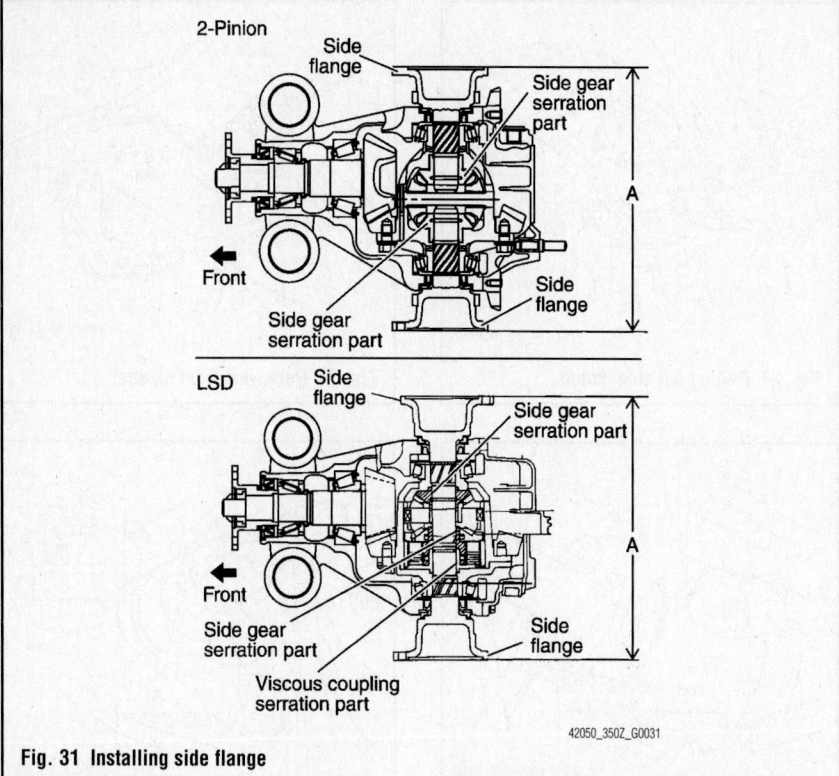

2-Pinion

Side flange

Side gear serration part

Front

Side gear serration part

Side flange

A

LSD

Side flange

Side gear serration part

Front

Side gear serration part

Side flange

Viscous coupling serration part

A

42050_350Z_G0031

Fig. 31 Installing side flange

pinion with the matching mark of companion flange, and then install the companion flange.

➡**Do not reuse drive pinion lock nut.**

17. Apply anti-corrosion oil to the thread and seat of new drive pinion lock nut, and temporarily tighten drive pinion lock nut to drive pinion.

➡**Total preload torque should equal the measurement taken during removal plus an additional 3 inch lbs. (0.1 to 0.4 Nm).**

18. Tighten to drive pinion lock nut to 09–238 ft. lbs. (147–323 Nm), while adjusting total reload torque.

➡**Adjust to the lower limit of the drive pinion lock nut tightening torque first. If the preload torque exceeds the specified value, replace collapsible spacer and tighten it again to adjust. Do not loosen drive pinion lock nut to adjust the preload torque.**

19. Make a stamping for identification of front oil seal replacement frequency. Be sure to make a stamping after replacing front oil seal.

20. Install driveshaft.

➡**Install the RH side flange, then install the LH side flange. If LH side flange is installed first, the RH side**

flange comes out sometimes from the shock of installing the RH side flange [For R200V (with LSD)].

21. Install side flange.
22. Attach the protector to side oil seal.
23. After the side flange is inserted and the serrated part of side gear has engaged the serrated part of flange, remove the protector.
24. Put a suitable drift on the center of side flange, then drive it until sound changes.

➡**When installation is completed, driving sound of the side flange turns into**

a sound which seems to affect the whole final drive.

25. Confirm that the dimension of the side flange is 12.83–12.91 in. (326–328 mm) as shown in the illustration.
26. Install halfshaft.
27. Install rear wheel ABS sensor.
28. Refill gear oil to the final drive and check oil level.
29. Check the final drive for oil leakage.
30. Be sure to perform the reconnect/relearn procedures.

TRANSFER CASE ASSEMBLY

REMOVAL & INSTALLATION

See Figures 32 and 33.

➡**Whenever the negative battery cable is disconnected the following components will require resetting. The Automatic temperature control system, Automatic drive positioner, Power window control, Sunroof system, Sunshade system, Rear view monitor, Idle Air Volume Learning, Steering Angle Sensor Neutral Position, Audio presets and Navigation. You will need the CONSULT-III diagnostic tool, or equivalent. Follow the directions on the screen of the tool, as needed.**

1. Before servicing the vehicle, refer to the Precautions Section.

➡**If working near and/or around the SRS system and components, be sure to disable the SRS system. After disabling the system wait three minutes or more before servicing the vehicle.**

2. Disconnect the negative battery cable.
3. Raise and safely support the vehicle.

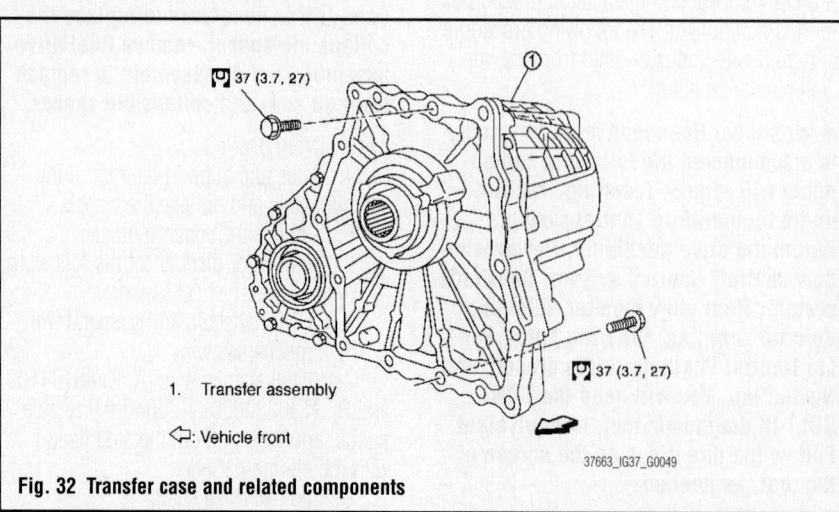

37 (3.7, 27)

①

37 (3.7, 27)

1. Transfer assembly

⬅: Vehicle front

37663_IG37_G0049

Fig. 32 Transfer case and related components

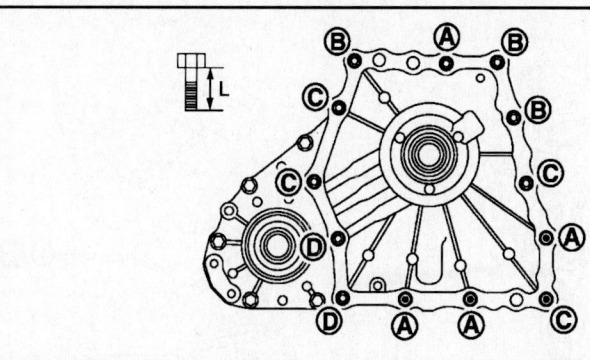

Bolt No.	A	B	C	D
Quantity	4	3	4	2
Bolt length " L " mm (in)	75 (2.95)	45 (1.77)	40 (1.57)	30 (1.18)

●:Transfer to transmission.

◎:Transmission to transfer.

37663_IG37_G0050

Fig. 33 Transfer case bolt tightening sequence

4. Drain the transfer case. Be sure to properly dispose of use fluid.

5. Remove the rear driveshaft.

6. Remove the front driveshaft.

7. Disconnect the AWD solenoid harness connector. Separate the harness from the transfer case assembly.

8. Remove the air breather hose.

9. Remove the control rod.

10. Support the transfer case using a suitable jack.

11. Remove the rear mounting member and engine mounting insulator.

12. Lower the jack to a position where the top transfer case mounting bolts can be removed.

13. Remove the transfer case mounting bolts and separate the transfer case from the transmission.

➡**Be sure to secure the transfer case and transmission assembly to the jack.**

To install:

➡**Be sure to use new fasteners, as required.**

14. Installation is the reverse of the removal procedure.

15. Be sure to fill the transfer case with the proper grade and type fluid.

16. Be sure to perform the reconnect/relearn procedures.

ENGINE COOLING

ENGINE FAN

REMOVAL & INSTALLATION

➡**Whenever the negative battery cable is disconnected the following components will require resetting. The Automatic temperature control system, Automatic drive positioner, Power window control, Sunroof system, Sunshade system, Rear view monitor, Idle Air Volume Learning, Steering Angle Sensor Neutral Position, Audio presets and Navigation. You will need the CONSULT-III diagnostic tool, or equivalent. Follow the directions on the screen of the tool, as needed.**

1. Before servicing the vehicle, refer to the Precautions Section.

➡**If working near and/or around the SRS system and components, be sure to disable the SRS system. After disabling the system wait three minutes or more before servicing the vehicle.**

2. Disconnect the negative battery cable.

3. Remove the reservoir tank.

4. Remove the air cleaner assembly.

5. Disconnect the harness connector from the cooling fan control module, move the harness to the side.

6. Remove the undercover.

7. Disconnect and plug the automatic transmission fluid cooler hose from the fan shroud, if equipped.

8. Remove the cooling fan from under the vehicle.

To install:

➡**Be sure to use new fasteners, as required.**

9. Installation is the reverse of the removal procedure.

10. Be sure to perform the reconnect/relearn procedures.

RADIATOR

REMOVAL & INSTALLATION

See Figure 34.

1. Before servicing the vehicle, refer to the Precautions Section.

➡**If working near and/or around the SRS system and components, be sure to disable the SRS system. After disabling the system wait three minutes or more before servicing the vehicle.**

➡**Never change the engine coolant when the engine is hot. Wrap a thick cloth around the cap and carefully remove it. First turn the cap a quarter**

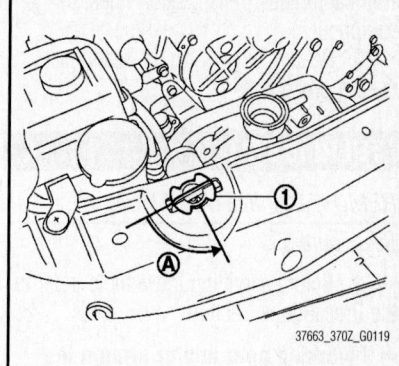

37663_370Z_G0119

Fig. 34 Radiator upper mount bracket removal

of a turn to release any pressure, then turn it all the way. Do not allow coolant to come in contact with the drive belts.

2. Disconnect the negative battery cable.

3. Remove the engine undercover.

4. Drain the engine coolant. Be sure to properly dispose of used coolant.

5. As required, properly discharge the air conditioning system.

6. Remove the engine cover.

7. Remove the air cleaner assembly.

8. Remove the reservoir tank.

9. Remove the radiator core support ornament and radiator core support center. Remove the horn. Remove the hood lock.

10. Remove the condenser pipe assembly.

11. Disconnect and plug the automatic transmission cooler lines at the radiator, if equipped.

12. Remove the upper and lower radiator hoses.

13. Remove the cooling fan assembly.

14. Rotate the two upper radiator mount brackets ninety degrees, in the direction shown in the illustration.

15. To remove the radiator and condenser assembly, lift up and pull the assembly forward and then remove the lower mounting rubber from the radiator support.

16. Remove the assembly from the front of the radiator core support.

To install:

➡**Be sure to use new fasteners, as required.**

17. Installation is the reverse of the removal procedure.

18. Be sure to fill the cooling system with the proper grade and type engine coolant.

19. Check that there is no fluid leakage at the automatic transmission lines, if equipped.

20. Be sure to perform the reconnect/relearn procedures.

THERMOSTAT

REMOVAL & INSTALLATION
See Figure 35.

1. Before servicing the vehicle, refer to the Precautions Section.

➡**If working near and/or around the SRS system and components, be sure to disable the SRS system. After disabling the system wait three minutes or more before servicing the vehicle.**

➡**Never change the engine coolant when the engine is hot. Wrap a thick cloth around the cap and carefully remove it. First turn the cap a quarter of a turn to release any pressure, then turn it all the way. Do not allow coolant to come in contact with the drive belts.**

2. Disconnect the negative battery cable.

3. Remove the engine undercover.

4. Drain the coolant. Be sure to properly dispose of used engine coolant.

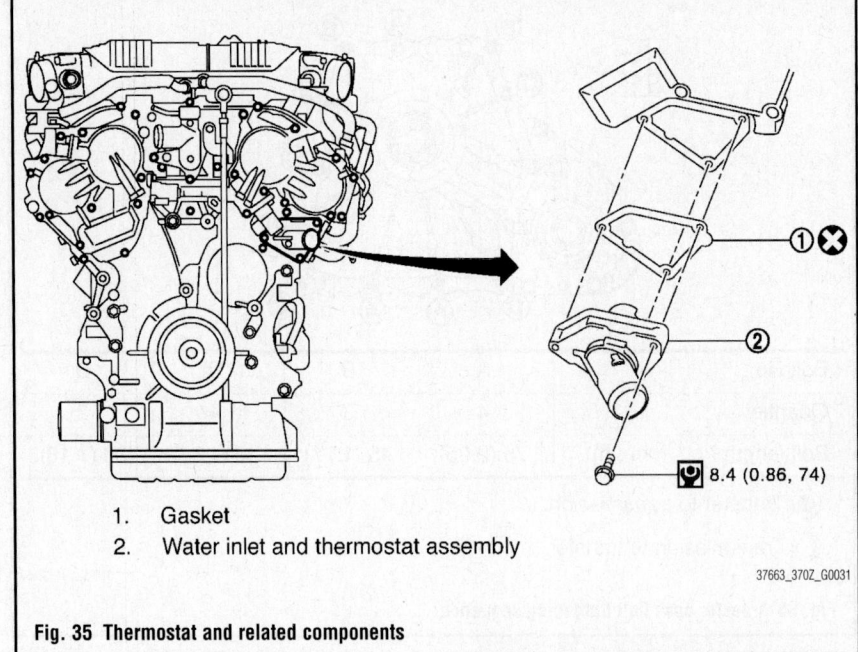

1. Gasket
2. Water inlet and thermostat assembly

37663_370Z_G0031

Fig. 35 Thermostat and related components

5. Remove the engine cover. Remove the air cleaner assembly.

6. Remove the reservoir assembly.

7. Disconnect the radiator hose from the water inlet and thermostat assembly.

8. Disconnect the intake valve timing control solenoid harness connector (left side) and remove the component.

9. Remove the water inlet and thermostat assembly retaining bolts.

10. Remove the component from its mounting. Discard the gasket.

➡**Never disassemble the water inlet and thermostat assembly, replace them as a complete unit.**

To install:

➡**Be sure to use new fasteners, as required.**

11. Installation is the reverse of the removal procedure.

12. Be sure to fill the cooling system with the proper grade and type engine coolant.

13. Be sure to perform the reconnect/relearn procedures.

WATER PUMP

REMOVAL & INSTALLATION
See Figures 36 through 40.

➡**Whenever the negative battery cable is disconnected the following components will require resetting. The Automatic temperature control system,** Automatic drive positioner, Power window control, Sunroof system, Sunshade system, Rear view monitor, Idle Air Volume Learning, Steering Angle Sensor Neutral Position, Audio presets and Navigation. You will need the CONSULT-III diagnostic tool, or equivalent. Follow the directions on the screen of the tool, as needed.

1. Before servicing the vehicle, refer to the Precautions Section.

➡**If working near and/or around the SRS system and components, be sure to disable the SRS system. After disabling the system wait three minutes or more before servicing the vehicle.**

➡**Never change the engine coolant when the engine is hot. Wrap a thick cloth around the cap and carefully**

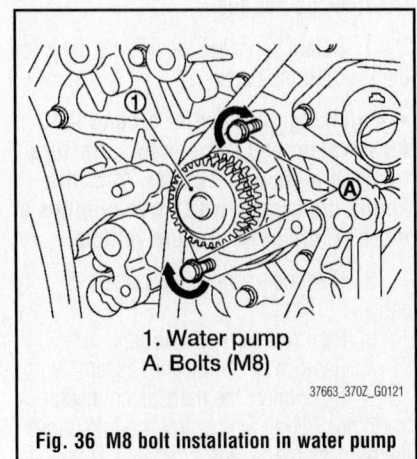

1. Water pump
A. Bolts (M8)

37663_370Z_G0121

Fig. 36 M8 bolt installation in water pump

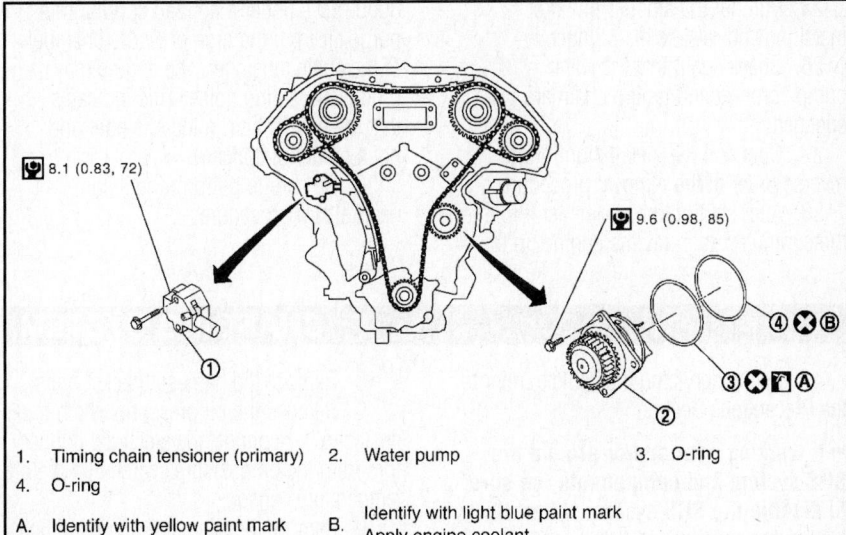

8.1 (0.83, 72)

9.6 (0.98, 85)

1. Timing chain tensioner (primary)
4. O-ring
A. Identify with yellow paint mark

2. Water pump

3. O-ring

B. Identify with light blue paint mark
Apply engine coolant

37663_370Z_G0120

Fig. 37 Water pump and related components

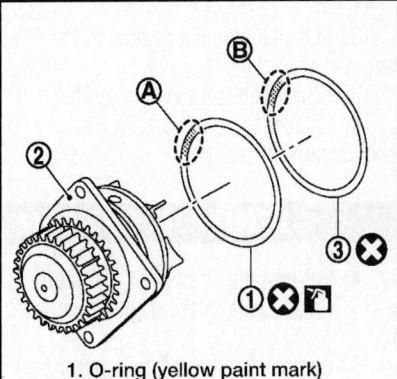

1. O-ring (yellow paint mark)
2. Water pump
3. O-ring (light blue paint mark)
A. Yellow paint mark
B. Light blue paint mark

37663_370Z_G0122

Fig. 38 Water pump and O-ring positioning and color identification

remove it. First turn the cap a quarter of a turn to release any pressure, then turn it all the way. Do not allow coolant to come in contact with the drive belts.

2. Disconnect the negative battery cable.

3. Remove the engine cover.

4. Properly relieve the fuel system pressure.

5. Disconnect the negative battery cable.

6. Remove the air cleaner assembly.

7. Remove the reservoir tank.

8. Separate the engine harness by removing their brackets from the front timing chain case.

9. Remove the engine undercover.

10. Drain the engine oil. Be sure to properly dispose of used oil.

11. Drain the engine coolant. Be sure to properly dispose of used coolant.

12. Remove the radiator hoses.

13. Remove the cooling fan assembly.

14. Remove the front timing chain case cover.

15. To remove the timing chain tensioner (primary), remove the lower mounting bolt. Loosen the upper mounting bolt slowly and then turn the timing chain tensioner (primary) on the upper mounting bolt so that the plunger is fully expanded. Remove the upper mounting bolt and then remove the timing chain tensioner (primary).

➡**Even if the plunger is fully expanded, it does not drop from the body of the timing chain tensioner (primary).**

16. To remove the water pump, remove the three retaining bolts. Secure a gap between the water pump gear and the timing chain, by turning the crankshaft counterclockwise until the timing chain looseness on the water pump sprocket becomes maximum.

17. Screw M8 bolts (pitch 0.0492 inch) in length approximately 1.97 inch into the water pump upper and lower mounting bolt holes until they reach the timing chain case. Then alternately tighten each bolt for a half a turn, and pull out the water pump.

➡**Pull the pump straight out preventing the vane from contacting the socket in the installation area. Remove the pump without causing the sprocket to contact the chain.**

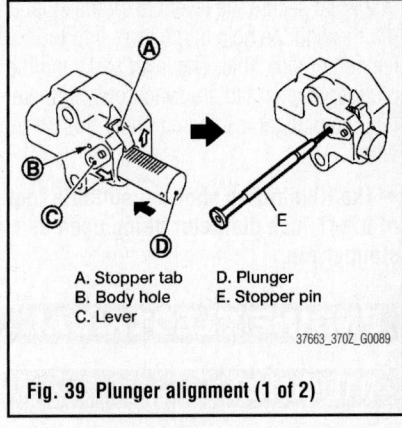

A. Stopper tab
B. Body hole
C. Lever
D. Plunger
E. Stopper pin

37663_370Z_G0089

Fig. 39 Plunger alignment (1 of 2)

18. Remove the M8 bolts and O-rings from the water pump.

To install:

➡**Be sure to use new fasteners, as required.**

19. Apply clean engine oil to the yellow paint marked O-ring. Apply clean coolant to the light blue paint marked O-ring. Install the O-rings on the water pump.

20. Install the water pump. Never allow the cylinder block to nip the O-rings. Check that the chain and pump sprocket are engaged.

21. Tighten the mounting bolts alternately and securely.

22. To install the timing chain tensioner (primary) pull the plunger stopper tab up (or turn lever downward) so as to remove the plunger stopper tab from the ratchet of the plunger. Note that the plunger stopper tab and lever are synchronized.

23. Push the plunger into the inside of the tensioner body. Hold the plunger in the fully compressed position by engaging the plunger stopper tab with the tip of the ratchet.

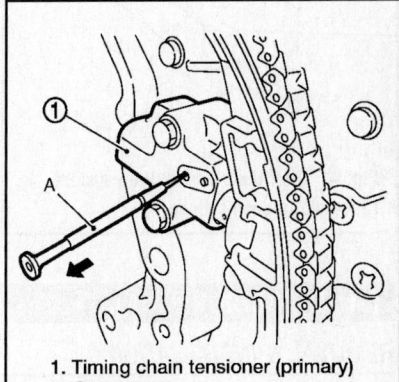

1. Timing chain tensioner (primary)
A. Stopper pin

37663_370Z_G0090

Fig. 40 Plunger alignment (2 of 2)

24. To secure the lever, insert the stopper pin through the hole of the lever into the tensioner body hole. The lever parts and the plunger stopper tab are synchronized, therefore the plunger is secured under this condition.

➡The illustration shows a suitable tool of 0.047 inch diameter being used as a stopper pin.

25. Pull out the stopper pin after installing and release the plunger.

26. Check again that the water pump sprocket and timing chain are engaged.

27. Continue the installation in the reverse order of the removal procedure.

28. After starting the engine let it idle for three minutes, then rev the engine up to 3000 rpm's under a no load condition to purge air from the high pressure chamber of the chain tensioner. The engine may produce a rattling noise. This indicates that there is still air in the chamber and not a matter of concern.

29. Be sure to perform the reconnect/relearn procedures.

ENGINE ELECTRICAL · CHARGING SYSTEM

ALTERNATOR

REMOVAL & INSTALLATION

➡Whenever the negative battery cable is disconnected the following components will require resetting. The Automatic temperature control system, Automatic drive positioner, Power window control, Sunroof system, Sunshade system, Rear view monitor, Idle Air Volume Learning, Steering Angle Sensor Neutral Position, Audio presets and Navigation. You will need the CONSULT-III diagnostic tool, or equivalent. Follow the directions on the screen of the tool, as needed.

1. Before servicing the vehicle, refer to the Precautions Section.

➡If working near and/or around the SRS system and components, be sure to disable the SRS system. After disabling the system wait three minutes or more before servicing the vehicle.

2. Disconnect the negative battery cable.

3. Remove the engine undercover.

4. Remove the air cleaner assembly, AWD.

5. Remove the cooling fan assembly.

6. Remove the drive belt.

7. Disconnect the alternator electrical connectors.

8. Remove the harness bracket bolts.

9. Remove the oil pressure switch harness clip. Disconnect the electrical connectors from the oil pressure switch and the oil temperature sensor.

10. Remove the alternator mounting bolts.

11. Remove the alternator from the vehicle, in the downward direction.

To install:

➡Be sure to use new fasteners, as required.

12. Installation is the reverse of the removal procedure.

13. Check belt tension, as required.

14. Be sure to perform the reconnect/relearn procedures.

ENGINE ELECTRICAL · IGNITION SYSTEM

FIRING ORDER

See Figure 41.

Firing order: 1–2–3–4–5–6

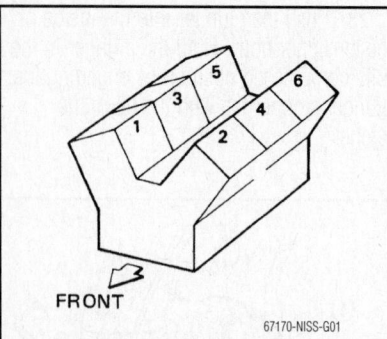

Fig. 41 Distributorless ignition system (one coil on each cylinder)

IGNITION COIL

REMOVAL & INSTALLATION

See Figure 42.

At this time the manufacturer does not provide removal and installation procedures for this component. The following procedure is a guideline and may differ from the vehicle you are servicing.

➡Whenever the negative battery cable is disconnected the following components will require resetting. The Automatic temperature control system, Automatic drive positioner, Power window control, Sunroof system, Sunshade system, Rear view monitor, Idle Air Volume Learning, Steering Angle Sensor Neutral Position, Audio presets and Navigation. You will need the CONSULT-III diagnostic tool, or equivalent. Follow the directions on the screen of the tool, as needed.

1. Before servicing the vehicle, refer to the Precautions Section.

➡If working near and/or around the SRS system and components, be sure to disable the SRS system. After disabling the system wait three minutes or more before servicing the vehicle.

2. Disconnect the negative battery cable.

3. Remove the engine undercover.

4. Remove the air cleaner assembly.

5. Remove the intake manifold collector.

6. Remove the necessary components in order to gain access to the ignition coils.

7. Disconnect the electrical connectors.

8. Remove the retainer bolt.

9. Remove the component from its mounting.

To install:

➡Be sure to use new fasteners, as required.

10. Installation is the reverse of the removal procedure.

11. Be sure to perform the reconnect/relearn procedures.

IGNITION TIMING

INSPECTION

➡The ignition timing is not adjustable. If not within specifications, further diagnostic inspection is required. You will need the CONSULT-III diagnostic tool, or equivalent. Follow the directions on the screen of the tool, as needed.

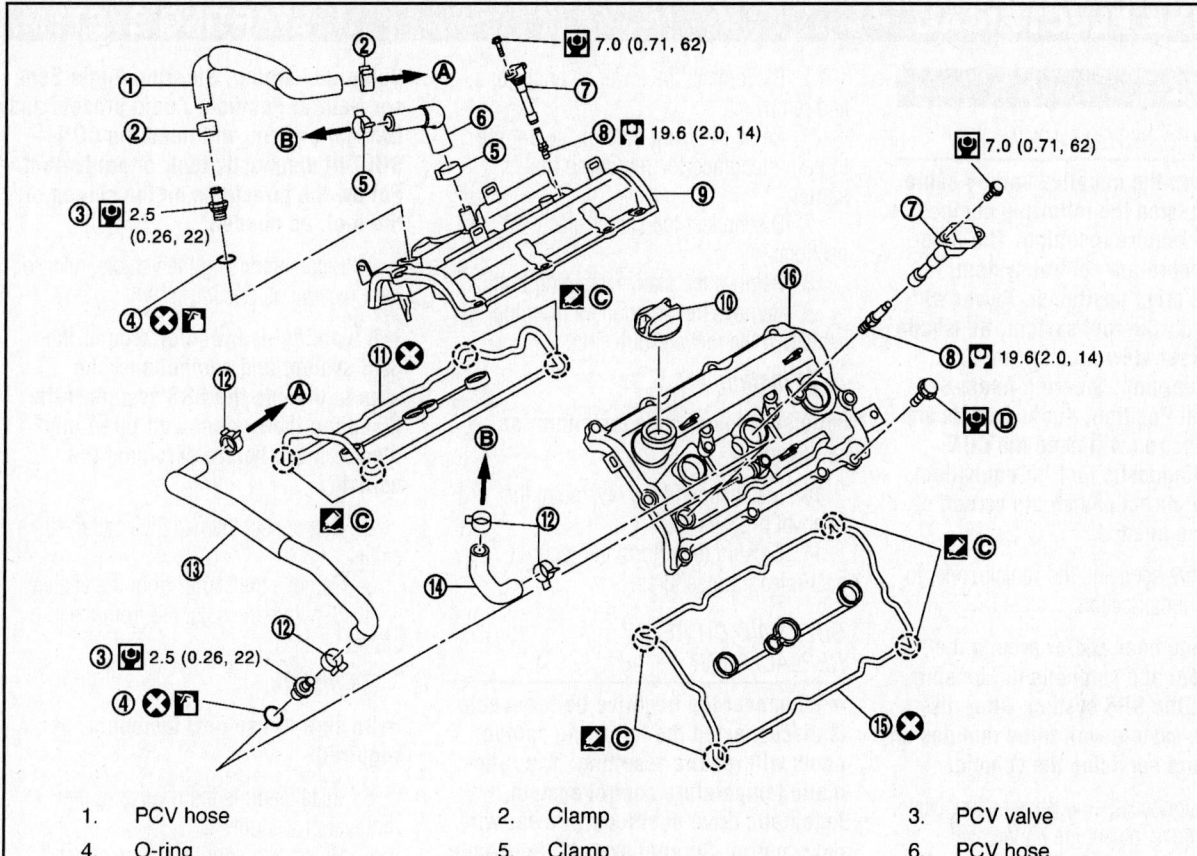

1. PCV hose	2. Clamp	3. PCV valve
4. O-ring	5. Clamp	6. PCV hose
7. Ignition coil	8. Spark plug	9. Rocker cover (bank 1)
10. Oil filler cap	11. Rocker cover gasket (bank 1)	12. Clamp
13. PCV hose	14. PCV hose	15. Rocker cover gasket (bank 2)
16. Rocker cover (bank 2)		
A. To intake manifold collector	B. To air duct	C. VVEL ladder assembly side

37663_370Z_G0007

Fig. 42 Ignition coil and related components

ADJUSTMENT

No timing adjustment is necessary.

SPARK PLUGS

REMOVAL & INSTALLATION

See Figure 42.

➡Whenever the negative battery cable is disconnected the following components will require resetting. The Automatic temperature control system, Automatic drive positioner, Power window control, Sunroof system, Sunshade system, Rear view monitor, Idle Air Volume Learning, Steering Angle Sensor Neutral Position, Audio presets and Navigation. You will need the CONSULT-III diagnostic tool, or equivalent. Follow the directions on the screen of the tool, as needed.

1. Before servicing the vehicle, refer to the Precautions Section.

➡If working near and/or around the SRS system and components, be sure to disable the SRS system. After disabling the system wait three minutes or more before servicing the vehicle.

2. Disconnect the negative battery cable.
3. Remove the engine undercover.
4. Remove the ignition coil.
5. Remove the spark plug from its mounting.

To install:

➡Be sure to use new fasteners, as required.

6. Installation is the reverse of the removal procedure.
7. Be sure to perform the reconnect/relearn procedures.

ENGINE ELECTRICAL   STARTING SYSTEM

STARTER

REMOVAL & INSTALLATION

➡ Whenever the negative battery cable is disconnected the following components will require resetting. The Automatic temperature control system, Automatic drive positioner, Power window control, Sunroof system, Sunshade system, Rear view monitor, Idle Air Volume Learning, Steering Angle Sensor Neutral Position, Audio presets and Navigation. You will need the CONSULT-III diagnostic tool, or equivalent. Follow the directions on the screen of the tool, as needed.

1. Before servicing the vehicle, refer to the Precautions Section.

➡ If working near and/or around the SRS system and components, be sure to disable the SRS system. After disabling the system wait three minutes or more before servicing the vehicle.

2. Disconnect the negative battery cable.
3. Remove the engine undercover.
4. Raise and support the vehicle safely. Remove the tire and wheel assembly.

5. Disconnect the lower steering joint, and remove it.
6. Raise the front side of the engine to gain clearance for removing the starter.
7. Disconnect the starter electrical connections.
8. Remove the starter retaining bolts.
9. Remove the component from the vehicle in the forward direction.

To install:

➡ Be sure to use new fasteners, as required.

10. Installation is the reverse of the removal procedure.
11. Be sure to perform the reconnect/relearn procedures.

SOLENOID OR RELAY REPLACEMENT

➡ Whenever the negative battery cable is disconnected the following components will require resetting. The Automatic temperature control system, Automatic drive positioner, Power window control, Sunroof system, Sunshade system, Rear view monitor, Idle Air Volume Learning, Steering Angle Sensor Neutral Position, Audio presets and Navigation. You will need the CONSULT-III diagnostic tool, or equivalent. Follow the directions on the screen of the tool, as needed.

1. Before servicing the vehicle, refer to the Precautions Section.

➡ If working near and/or around the SRS system and components, be sure to disable the SRS system. After disabling the system wait three minutes or more before servicing the vehicle.

2. Disconnect the negative battery cable.
3. Remove the starter from the vehicle.
4. Separate the solenoid from the starter.

To install:

➡ Be sure to use new fasteners, as required.

5. Installation is the reverse of the removal procedure.
6. Be sure to perform the reconnect/relearn procedures.

ENGINE MECHANICAL

ACCESSORY DRIVE BELTS

ACCESSORY BELT ROUTING

See Figure 43.

Refer to the accompanying illustration.

INSPECTION

1. Inspect belts for cracks, fraying, wear and oil. If necessary, replace.

ADJUSTMENT

No adjustment is necessary.

REMOVAL & INSTALLATION

See Figures 44 and 45.

➡ Whenever the negative battery cable is disconnected the following components will require resetting. The Automatic temperature control system, Automatic drive positioner, Power window control, Sunroof system, Sunshade system, Rear view monitor, Idle Air Volume Learning, Steering Angle Sensor Neutral Position, Audio presets and Navigation. You will need the CON-

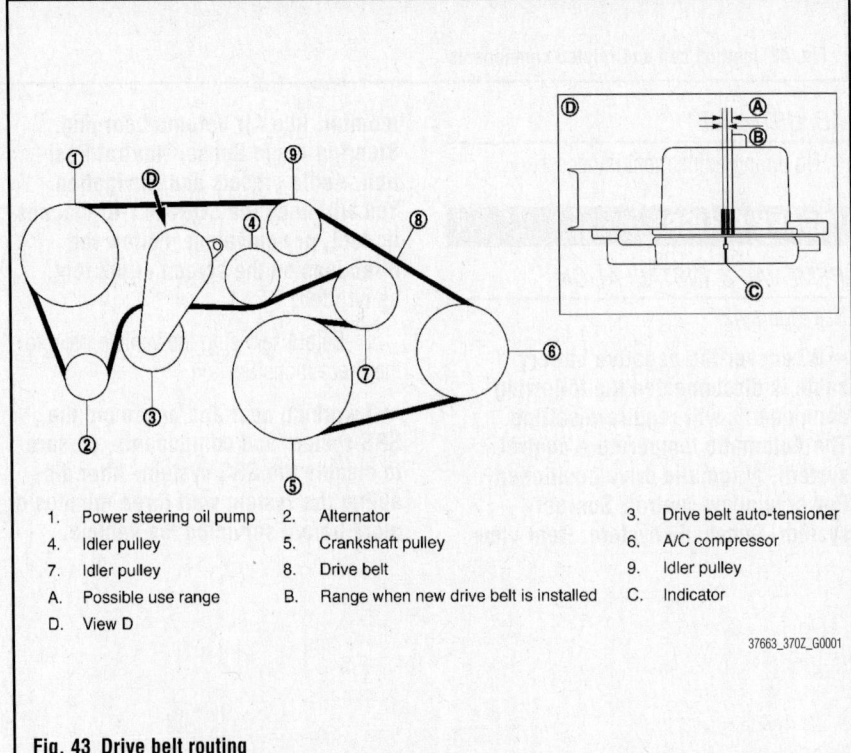

1.	Power steering oil pump	2.	Alternator	3.	Drive belt auto-tensioner
4.	Idler pulley	5.	Crankshaft pulley	6.	A/C compressor
7.	Idler pulley	8.	Drive belt	9.	Idler pulley
A.	Possible use range	B.	Range when new drive belt is installed	C.	Indicator
D.	View D				

37663_370Z_G0001

Fig. 43 Drive belt routing

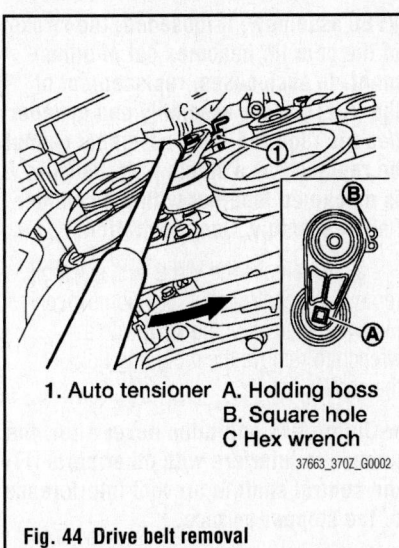

1. Auto tensioner A. Holding boss
                      B. Square hole
                      C Hex wrench

37663_370Z_G0002

Fig. 44 Drive belt removal

SULT-III diagnostic tool, or equivalent. Follow the directions on the screen of the tool, as needed.

1. Before servicing the vehicle, refer to the Precautions Section.

➡**If working near and/or around the SRS system and components, be sure to disable the SRS system. After dis-** abling the system wait three minutes or more before servicing the vehicle.

2. Disconnect the negative battery cable.

3. Remove the engine undercover.

4. On sedan, remove the reservoir tank.

5. On sedan, remove the cooling fan assembly.

6. While securely holding the square hole in the pulley center of the auto tensioner with a spinner handle, move the spinner handle in the direction of the arrow (loosening the drive belt). See illustration.

➡**Never place your hand in a location where pinching may occur if the holding tool accidentally comes off.**

7. Under the above condition insert a metallic bar about 0.24 inch in diameter through the holding boss to lock the auto tensioner pulley arm.

8. Remove the drive belt.

To install:

➡**Be sure to use new fasteners, as required.**

9. Installation is the reverse of the removal procedure.

10. Be sure that the belt is securely installed around the pulleys.

11. Be sure that the belt is correctly engaged with the pulley groove.

12. Turn the crankshaft pulley several times to equalize tension between each pulley, then confirm tension of the drive belt at the indicator (notch on fixed side) is within the possible use range.

13. Be sure to perform the reconnect/relearn procedures.

CAMSHAFT AND VALVE LIFTERS

INSPECTION

Camshaft Lobe Height

1. Measure camshaft cam lobe height at center of lobe.

2. If wear has reduced the lobe height below specifications, replace the camshaft.

Journal Oil Clearance

1. Measure outer diameter of camshaft journal.

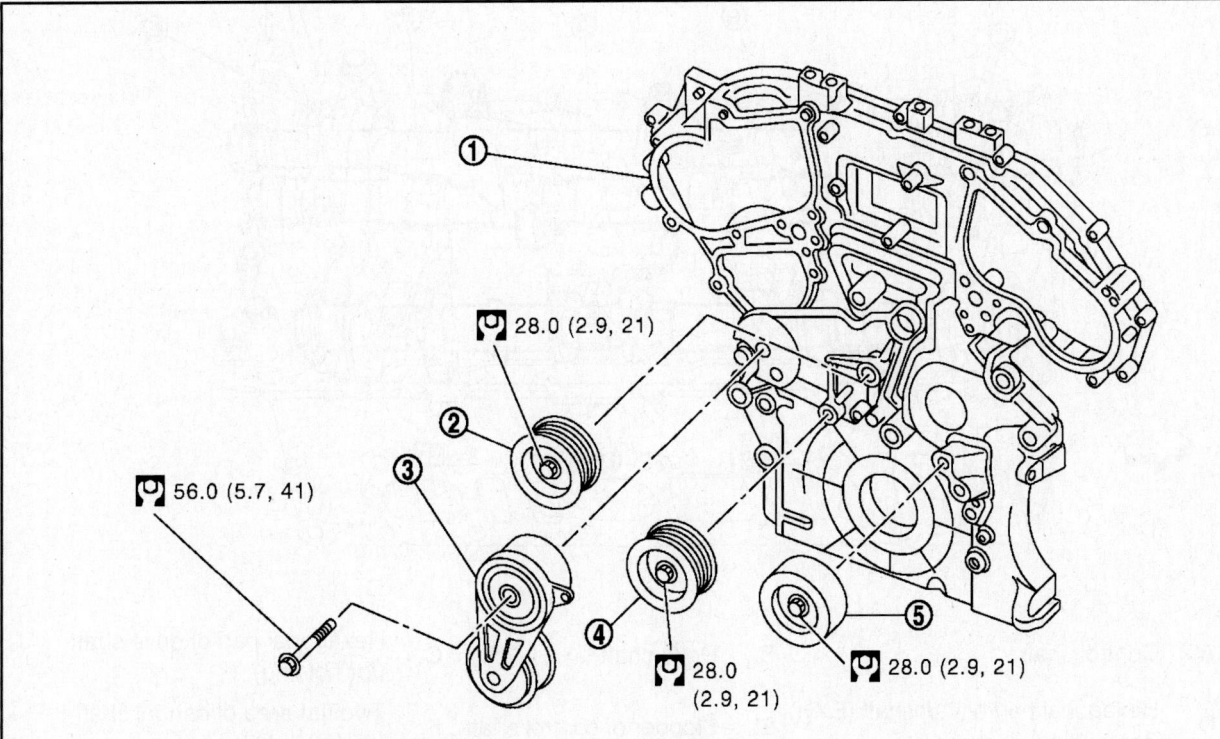

28.0 (2.9, 21)

56.0 (5.7, 41)

28.0 (2.9, 21)

28.0 (2.9, 21)

1.	Front timing chain case	2.	Idler pulley	3.	Drive belt auto-tensioner
4.	Idler pulley	5.	Idler pulley		

37663_370Z_G0063

Fig. 45 Drive belt auto tensioner and related components

2. If wear has reduced the diameter below specifications, replace the camshaft.

Runout

1. Put V-block on precise flat bed and support No. 2 and No. 4 journal of camshaft as shown.

2. Set dial gauges vertically to No. 3.

3. Turn camshaft in one direction slowly by hand, measure the camshaft runout on the dial gauges.

4. If actual runout exceeds the limit, replace the camshaft.

REMOVAL & INSTALLATION

See Figures 46 through 60.

At this time the manufacturer provides service information for this component with the engine removed from the vehicle and positioned in a suitable holding fixture.

➡Whenever the negative battery cable is disconnected the following components will require resetting. The Automatic temperature control system, Automatic drive positioner, Power window control, Sunroof system, Sunshade system, Rear view monitor, Idle Air Volume Learning, Steering Angle Sensor Neutral Position, Audio presets and Navigation. You will need the CONSULT-III diagnostic tool, or equivalent. Follow the directions on the screen of the tool, as needed.

1. Before servicing the vehicle, refer to the Precautions Section.

➡If working near and/or around the SRS system and components, be sure to disable the SRS system. After disabling the system wait three minutes or more before servicing the vehicle.

2. Disconnect the negative battery cable.

3. Remove the engine and position it in a suitable holding fixture.

4. Remove the valve covers.

➡Never loosen the adjusting bolts "A" and black mounting bolts "B" of the VVEL assembly. If loosened, the stroke of the cam lift becomes out of adjustment. In such cases, replacement of the VVEL ladder assembly and cylinder head is required. This assembly cannot be replaced as a single part, because it is machined together with the cylinder head assembly. See illustration.

5. To remove the VVEL sub assembly, remove the control shaft position sensor. Fix two flat areas of the shaft with a wrench to remove the mounting bolts of the shaft.

➡During this operation never allow the wrench to interfere with other parts. Fix the control shaft to prevent interference of the stopper surface.

➡The VVEL actuator sub assembly and control shaft position sensor are not reusable. Never remove them unless required.

6. Remove the VVEL sub actuator assembly by loosening the mounting bolts

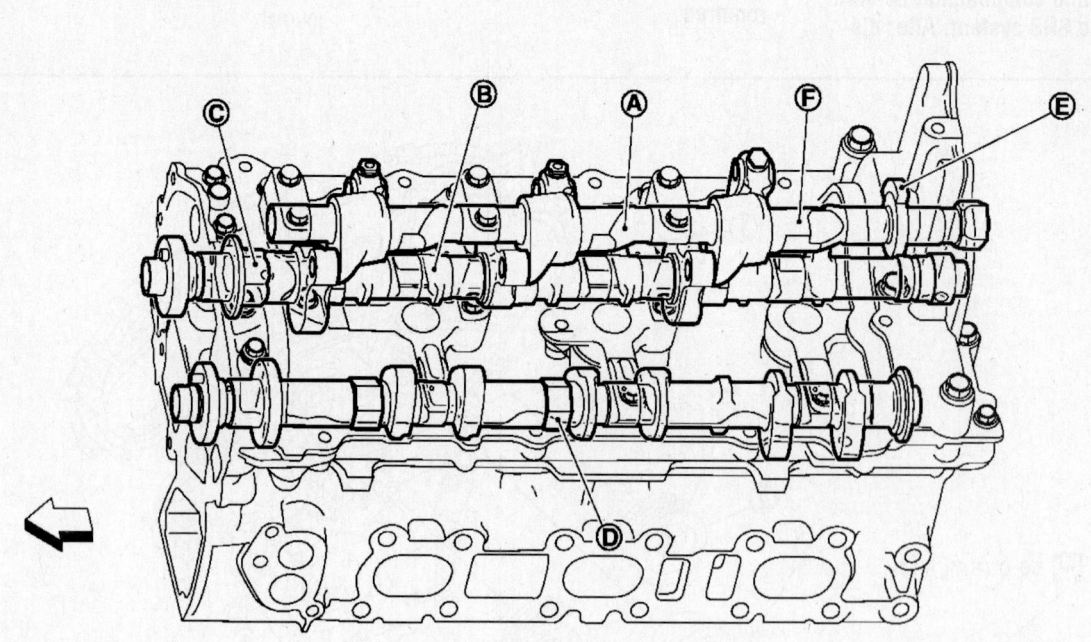

| A. | Control shaft | B. | Drive shaft | C. | Hexagonal part of drive shaft (for holding) |
| D. | Hexagonal part of camshaft (EXH) (for holding) | E. | Stopper of control shaft | F. | Two flat area of control shaft (for holding) |

⇦ : Engine front

37663_370Z_G0005

Fig. 46 VVEL ladder assembly and related components (1 of 2)

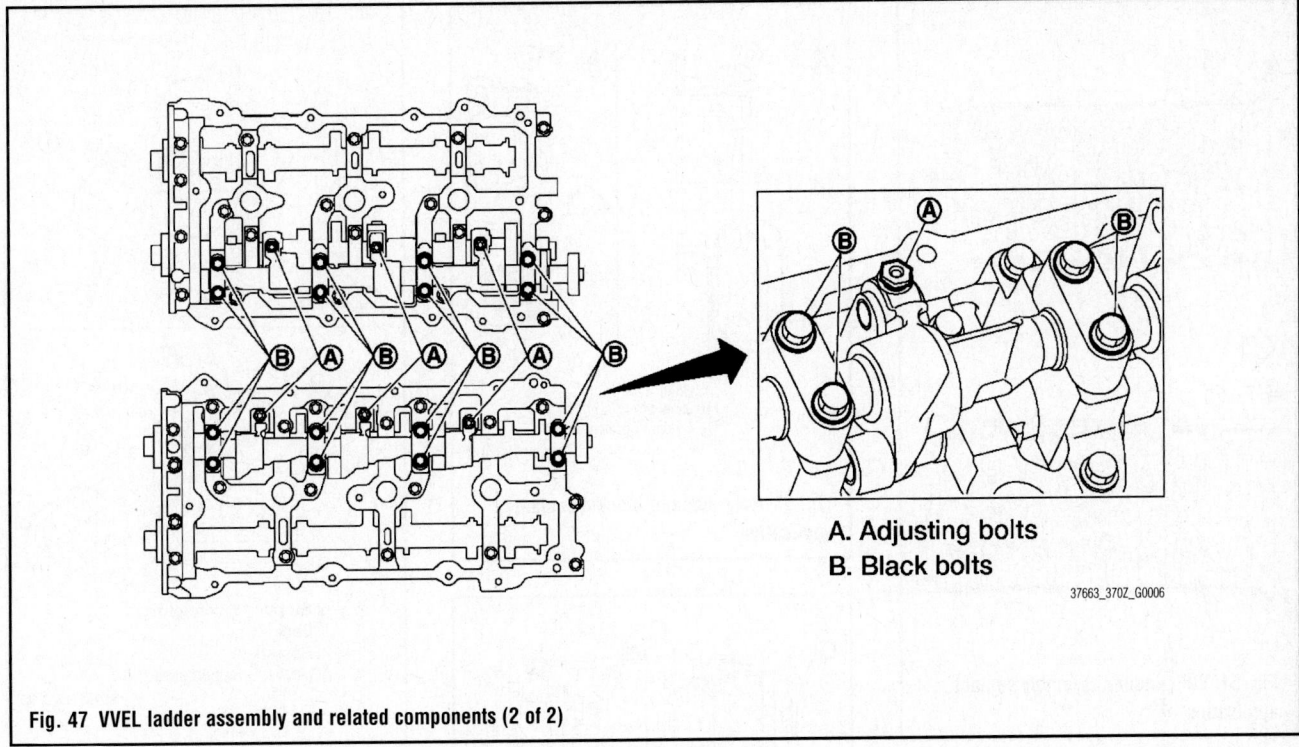

A. Adjusting bolts
B. Black bolts

37663_370Z_G0006

Fig. 47 VVEL ladder assembly and related components (2 of 2)

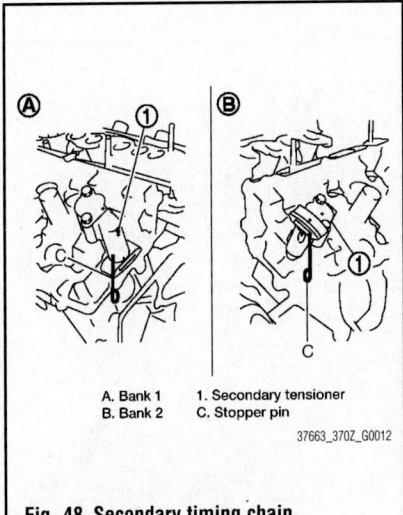

A. Bank 1   1. Secondary tensioner
B. Bank 2   C. Stopper pin

37663_370Z_G0012

Fig. 48 Secondary timing chain tensioners and related components

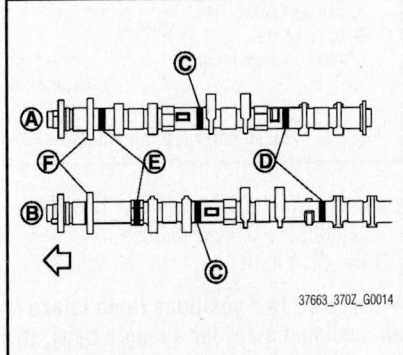

37663_370Z_G0014

Fig. 49 Camshaft (EXH) identification (1 of 2)

Bank	Paint marks			Identification mark (F)
	M1 (C)	M2 (D)	M3 (E)	
Bank 1 (A)	No	Blue	Light blue	1N
Bank 2 (B)	No	Blue	Light blue	1P

37663_370Z_G0015

Fig. 50 Camshaft (EXH) identification (2 of 2)

in the reverse order of the tightening sequence.

7. Remove the actuator bracket by loosening the mounting bolts in the reverse order of the tightening sequence.

➡**When removing, be sure to properly dispose of any oil, as required. When installing be careful with the VVEL actuator subassembly (bank 2) mounting bolt No. 1, because the length is different from the other.**

8. Remove the timing chain case, camshaft sprockets and timing chain.

9. Remove the rear timing chain case.

10. Remove the VVEL ladder assembly.

11. Loosen the mounting bolts (gold color) in the reverse order of the tightening sequence.

➡**Never loosen the adjusting bolts and mounting bolts (black color). When removing the VVEL ladder assembly,**

hold the driveshaft from below as not to drop it.

12. Remove the camshaft (EXH).

13. Remove the valve lifter. Be sure to identify and store properly for reinstallation.

14. Remove the chain tensioners (secondary) from the cylinder head. Remove the component with its stopper pin attached. The stopper pin should be attached when removed.

15. If necessary, remove the oil filter from the cylinder head.

To install:

➡**Be sure to use new fasteners, as required.**

16. Install the secondary timing chain tensioners.

17. Install the oil filter, if removed.

18. Install the valve lifters, in their original position.

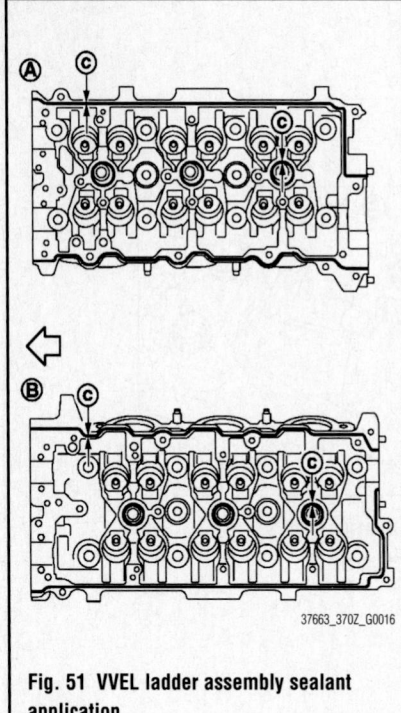

Fig. 51 VVEL ladder assembly sealant application

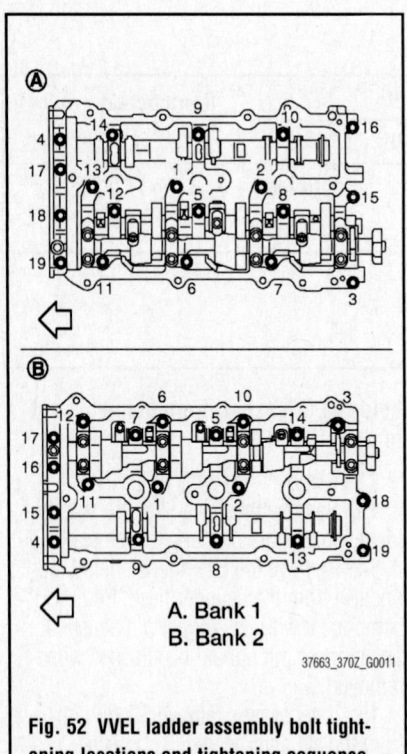

A. Bank 1
B. Bank 2

37663_370Z_G0011

Fig. 52 VVEL ladder assembly bolt tightening locations and tightening sequence

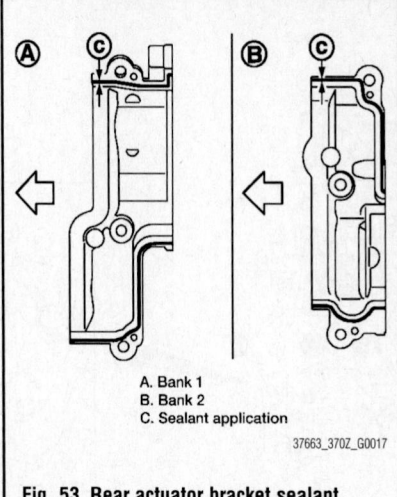

A. Bank 1
B. Bank 2
C. Sealant application

37663_370Z_G0017

Fig. 53 Rear actuator bracket sealant application

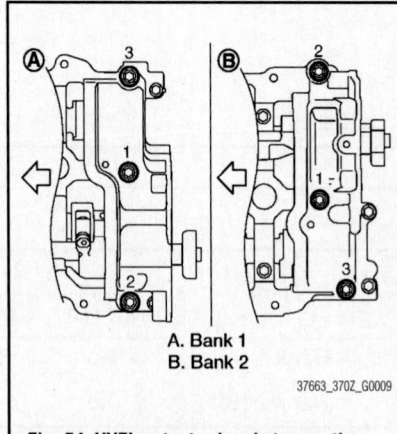

A. Bank 1
B. Bank 2

37663_370Z_G0009

Fig. 54 VVEL actuator bracket mounting bolt tightening locations and tightening sequence

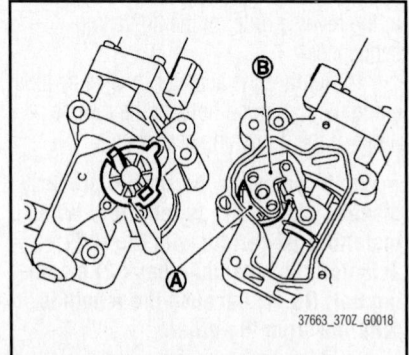

37663_370Z_G0018

Fig. 55 VVEL actuator arm location

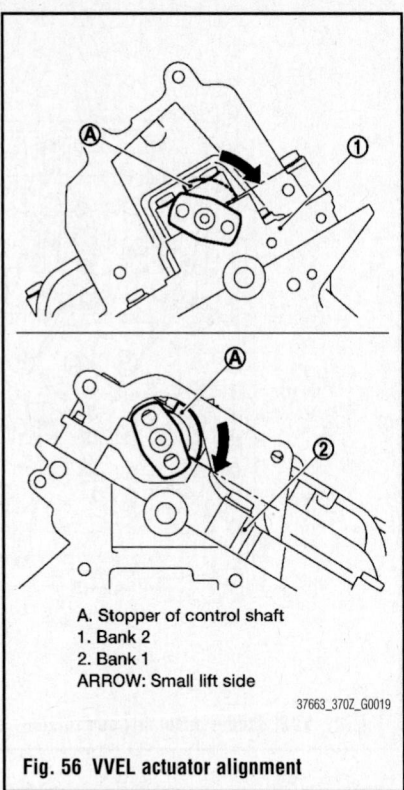

A. Stopper of control shaft
1. Bank 2
2. Bank 1
ARROW: Small lift side

37663_370Z_G0019

Fig. 56 VVEL actuator alignment

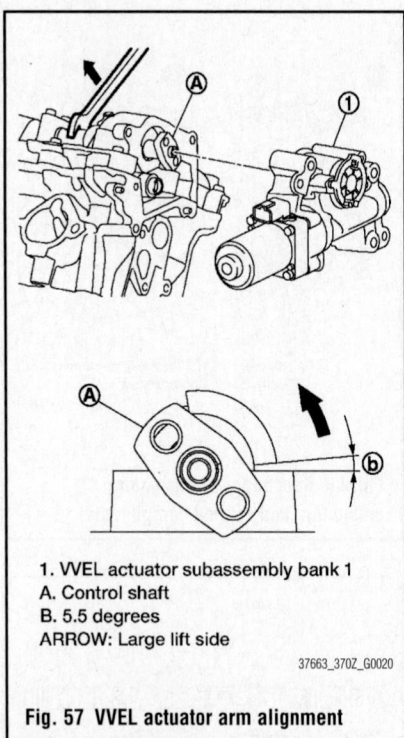

1. VVEL actuator subassembly bank 1
A. Control shaft
B. 5.5 degrees
ARROW: Large lift side

37663_370Z_G0020

Fig. 57 VVEL actuator arm alignment

19. Install the camshaft (EXH).

➡**Note identification mark distinction between camshaft (EXH).**

20. To install the VVEL ladder assembly, apply a continuous bead of liquid gasket (0.134–0.173 inch) to the cylinder head, as shown in the illustration.

21. Tighten the mounting bolts in the proper sequence to specification.

22. Specification is first pass 1 ft. lb. (1.96 Nm). Second pass 4 ft. lbs. (5.88 Nm). Third pass 8 ft. lbs. (10.4 Nm).

23. Measure the difference in levels between the front faces of the VVEL ladder

assembly and the cylinder head. Specification should be -0.0055–0.0055 inch (-0.14 –0.14 mm).

➡**Measure two positions (both intake and exhaust side) for a single bank. If the measured value is out of standard, reinstall the VVEL ladder assembly.**

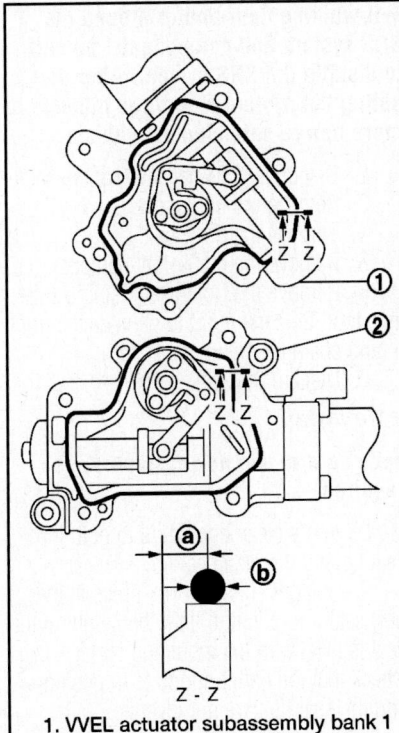

1. VVEL actuator subassembly bank 1
2. VVEL actuator subassembly bank 2
A. 0.157-0.220 inch
B. 0.134-0.173 inch

37663_370Z_G0021

Fig. 58 VVEL actuator subassembly sealant application

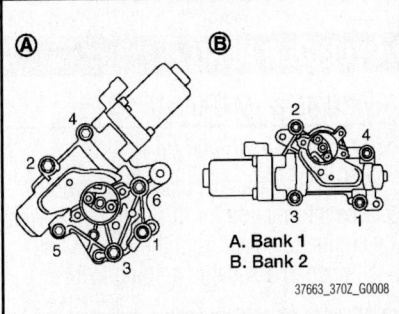

A. Bank 1
B. Bank 2

37663_370Z_G0008

Fig. 59 VVEL actuator sub assembly mounting bolt tightening locations and tightening sequence

24. Install the rear timing case.

25. Install the timing chain sprockets, timing chain and front case.

26. To install the rear actuator bracket, apply a continuous bead of liquid gasket (0.134–0.173 inch) to the rear actuator bracket as shown in the illustration. Never apply gasket material to the oil passage.

27. Tighten the mounting bolts in the proper sequence and to specification, as shown in the illustration.

28. Specification is first pass 1 ft. lb.

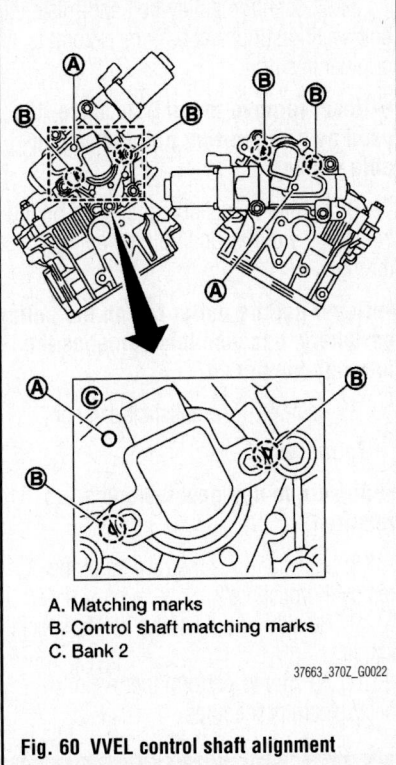

A. Matching marks
B. Control shaft matching marks
C. Bank 2

37663_370Z_G0022

Fig. 60 VVEL control shaft alignment

(1.96 Nm). Second pass 4 ft. lbs. (5.88 Nm). Third pass 23 ft. lbs. (31.4 Nm).

29. Install the new VVEL actuator sub assembly.

➡**Regarding replacement, because the VVEL actuator subassembly and VVEL control shaft position sensor are controlled on a one to one basis, replace them as a set.**

➡**The VVEL actuator arm is factory fixed at 5.5 degrees from the small lift with the holding jig. The holding jig is supplied in the new VVEL actuator subassembly.**

➡**Never disassemble the VVEL actuator sub assembly. Never loosen the actuator mounting motor bolts. Never shock the assembly.**

30. Move the control shaft to the position of small lift stopper. The position where part of the stopper of the control shaft contacts the VVEL ladder bracket. Be careful not to damage the stopper surface. See illustration.

31. If the control shaft cannot be moved, set the crankshaft in the following position. Bank 1: Turn 120 degrees from No. 1 cylinder at TDC. Bank 2: No. 1 cylinder at TDC.

32. Hold two flat areas of the control shaft with a wrench, and rotate the control shaft (5.5 degrees from the stopper) to the large lift side. This is for aligning the bolt

hole of the control shaft and the hole of the VVEL actuator arm.

33. Apply a continuous bead of sealant, as shown in the illustration to the VVEL actuator subassembly.

34. Install the new VVEL actuator subassembly. Tighten the mounting bolts to specification and in the proper sequence.

➡**When installing be careful with the VVEL actuator subassembly (bank 2) mounting No.1 mounting bolt. The length is different. Be sure to check that the VVEL actuator subassembly is in contact with the cylinder head before tightening the mounting bolts.**

35. Remove the jig. Check that the VVEL actuator arm bolt hole is aligned with the control shaft tapped hole. If not turn the control shaft for alignment.

36. Fix the two flat areas of the control shaft with a wrench to install the mounting bolts of the control shaft.

➡**During this operation never allow the wrench to interfere with other parts. Fix the control shaft to prevent interference of the stopper surface.**

37. To install the new VVEL control position sensor. Apply engine oil to the O-ring or contact surface of the O-ring.

38. Align the matching marks of the sensor and upper housing. Face the connector toward the matching mark.

39. Temporarily tighten the bolt.

40. Using the CONSULT III or equivalent, adjust the sensor after installing the engine in the vehicle.

➡**Be sure to adjust this sensor.**

41. Continue the installation in the reverse order of the removal procedure.

CRANKSHAFT DAMPER

REMOVAL & INSTALLATION
See Figures 61 through 63.

➡**Whenever the negative battery cable is disconnected the following components will require resetting. The Automatic temperature control system, Automatic drive positioner, Power window control, Sunroof system, Sunshade system, Rear view monitor, Idle Air Volume Learning, Steering Angle Sensor Neutral Position, Audio presets and Navigation. You will need the CONSULT-III diagnostic tool, or equivalent. Follow the directions on the screen of the tool, as needed.**

1. Before servicing the vehicle, refer to the Precautions Section.

➡️**If working near and/or around the SRS system and components, be sure to disable the SRS system. After disabling the system wait three minutes or more before servicing the vehicle.**

2. Disconnect the negative battery cable.

3. Remove the undercover.

4. Remove the rear cover plate and install the ring gear stopper tool KV10118600 or equivalent.

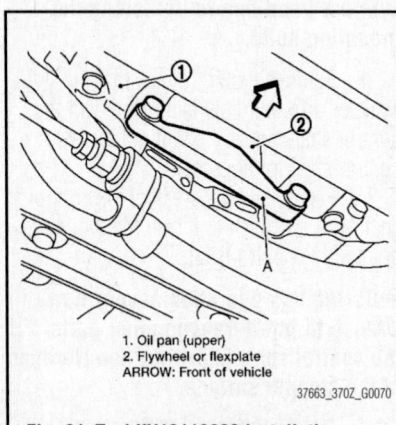

1. Oil pan (upper)
2. Flywheel or flexplate
ARROW: Front of vehicle

37663_370Z_G0070

Fig. 61 Tool KV10118600 installation

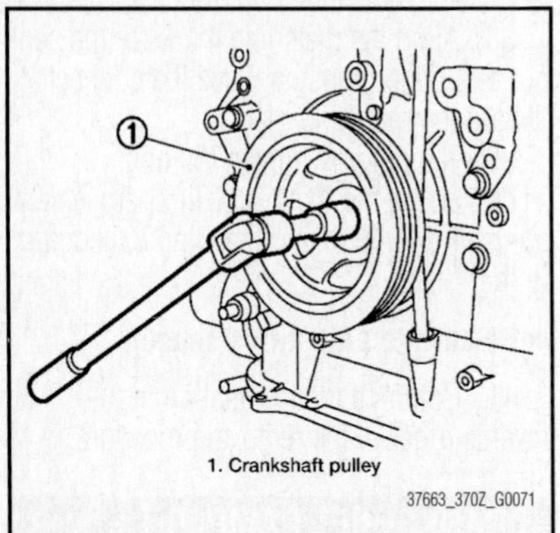

1. Crankshaft pulley

37663_370Z_G0071

Fig. 62 Crankshaft pulley removal (1 of 2)

Fig. 63 Crankshaft pulley removal (2 of 2)

37663_370Z_G0072

5. Loosen the pulley bolt and rotate the bolt seating surface at 0.39 inch from its original position.

➡️**Never remove the bolt because it is used as a supporting point for a suitable puller.**

6. Position a suitable puller tab on the holes of the pulley and pull the pulley through.

➡️**Never put the puller tab on the pulley periphery, because this damages the internal damper.**

7. Remove the crankshaft damper.

To install:

➡️**Be sure to use new fasteners, as required**

8. Installation is the reverse of the removal procedure.

9. Be sure to tighten the bolt to specification.

10. Be sure to perform the reconnect/relearn procedures.

CRANKSHAFT FRONT SEAL

REMOVAL & INSTALLATION

See Figure 64.

➡️**Whenever the negative battery cable is disconnected the following components will require resetting. The Automatic temperature control system, Automatic drive positioner, Power window control, Sunroof system, Sunshade system, Rear view monitor, Idle Air Volume Learning, Steering Angle Sensor Neutral Position, Audio presets and Navigation. You will need the CONSULT-III diagnostic tool, or equivalent. Follow the directions on the screen of the tool, as needed.**

1. Before servicing the vehicle, refer to the Precautions Section.

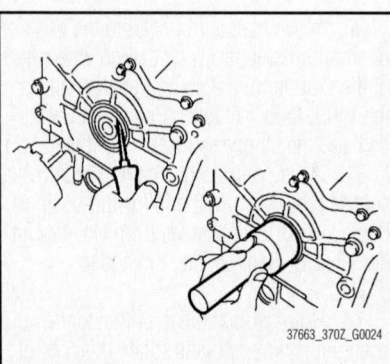

37663_370Z_G0024

Fig. 64 Front oil seal removal and installation

➡️**If working near and/or around the SRS system and components, be sure to disable the SRS system. After disabling the system wait three minutes or more before servicing the vehicle.**

2. Disconnect the negative battery cable.

3. Remove the undercover.

4. Remove the drive belt.

5. Remove the crankshaft damper.

6. Remove the front seal, using a suitable tool. Be careful not to damage the front timing chain case and crankshaft.

7. Discard the seal.

To install:

➡️**Be sure to use new fasteners, as required**

8. Apply clean engine oil to both the seal lip and the dust seal lip.

9. Using a suitable drift, press fit the new seal into position until the height of the seal is level with the mounting surface. Check that the garter spring is in position and the seal lips are not inverted.

➡️ **Be careful not to damage the front timing chain case and crankshaft. Press fit straight ahead and avoid causing burrs or tilting the seal.**

10. Continue the installation is the reverse of the removal procedure.

11. Be sure to perform the reconnect/relearn procedures.

CYLINDER HEAD

REMOVAL & INSTALLATION

See Figures 65 through 67.

At this time the manufacturer provides service information for this component with the engine removed from the vehicle and positioned in a suitable holding fixture.

➡️**Whenever the negative battery cable is disconnected the following components will require resetting. The Automatic temperature control system, Automatic drive positioner, Power window control, Sunroof system, Sunshade system, Rear view monitor, Idle Air Volume Learning, Steering Angle Sensor Neutral Position, Audio presets and Navigation. You will need the CONSULT-III diagnostic tool, or equivalent. Follow the directions on the screen of the tool, as needed.**

1. Before servicing the vehicle, refer to the Precautions Section.

➡️**If working near and/or around the SRS system and components, be**

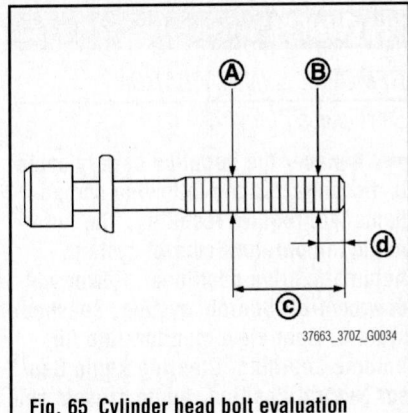

Fig. 65 Cylinder head bolt evaluation

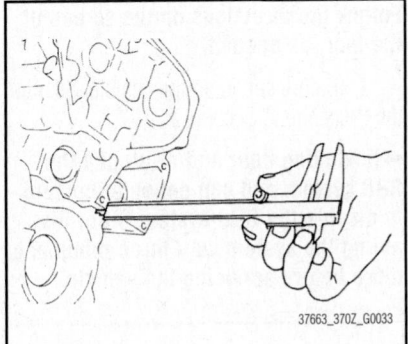

Fig. 66 Cylinder head to cylinder block measurement check

sure to disable the SRS system. After disabling the system wait three minutes or more before servicing the vehicle.

2. Disconnect the negative battery cable.
3. Remove the engine and position it in a suitable holding fixture.
4. Remove the intake manifold collector.
5. Remove the fuel tube and fuel injector assembly.
6. Remove the intake manifold.
7. Remove the valve covers.
8. Remove the exhaust manifold.
9. Remove the water inlet and thermostat assembly. Remove the water pipe and heater pipe assemblies.
10. Remove the timing chain.
11. Remove the Camshaft.
12. Remove the cylinder head retaining bolts in the reverse order of the tightening sequence.
13. Remove the cylinder head gaskets. Discard the gaskets.

To install:

➡Be sure to use new fasteners, as required.

14. Installation is the reverse of the removal procedure.

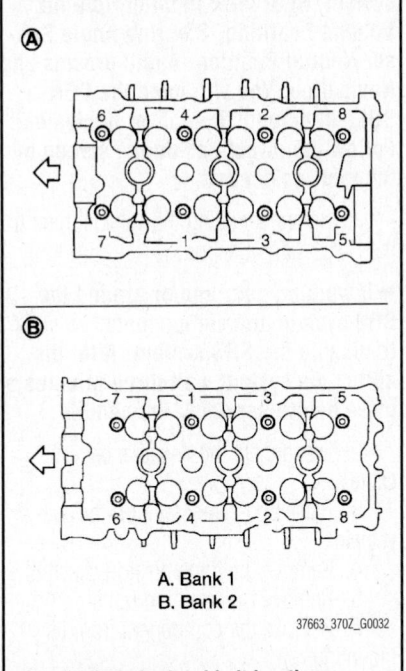

A. Bank 1
B. Bank 2

Fig. 67 Cylinder head bolt locations tightening sequence

15. Be sure to use new gaskets.

➡If the old bolts are being reused check their outer diameter before installation. Out of spec bolts must be replaced. "B" minus "A" should be 0.0071 inch. "C" should be 1.89 inch. "D" should be 0.43 inch.

16. Tighten the cylinder head bolts in the proper sequence and to specification. Coat the bolt threads with clean engine oil before installation.
17. Specification is 77 ft. lbs. (105 NM) first pass. Completely loosen all bolts, in the reverse order of the tightening sequence. Tighten all bolts to 30 ft. lbs. (40.0 Nm) in the proper sequence. Finally turn all bolts 95 degrees clockwise (angle tightening) in the proper sequence.
18. After installing the cylinder head measure the distance between the front end faces of the cylinder block and the cylinder head on both banks.
19. Specification should be 0.555–0.587 inch (14.1–14.9 mm).
20. Continue the installation in the reverse order of the removal procedure.
21. Be sure to perform the reconnect/relearn procedures.

EXHAUST MANIFOLD

REMOVAL & INSTALLATION
See Figures 68 and 69.

➡Whenever the negative battery cable is disconnected the following components will require resetting. The Automatic temperature control system, Automatic drive positioner, Power window control, Sunroof system, Sunshade system, Rear view monitor, Idle Air Volume Learning, Steering Angle Sensor Neutral Position, Audio presets and Navigation. You will need the CONSULT-III diagnostic tool, or equivalent. Follow the directions on the screen of the tool, as needed.

1. Before servicing the vehicle, refer to the Precautions Section.

➡If working near and/or around the SRS system and components, be

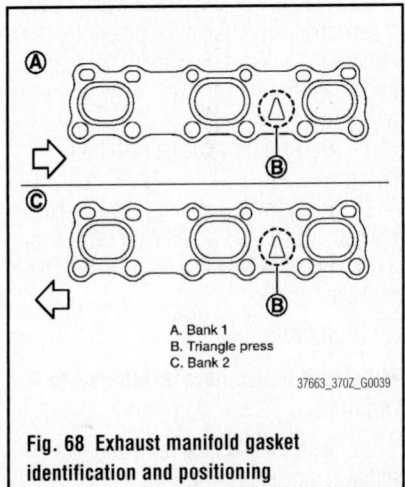

A. Bank 1
B. Triangle press
C. Bank 2

Fig. 68 Exhaust manifold gasket identification and positioning

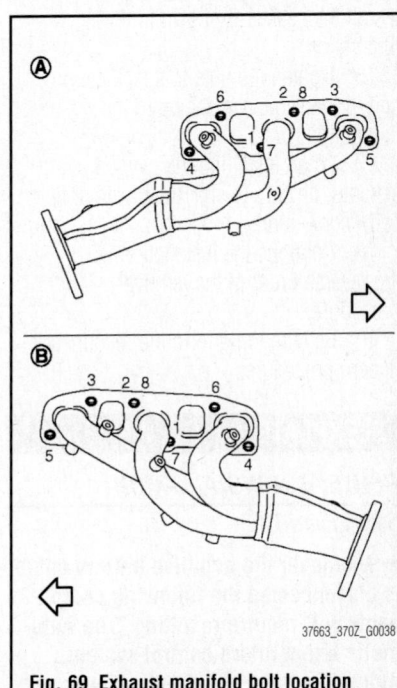

Fig. 69 Exhaust manifold bolt location and tightening sequence

sure to disable the SRS system. After disabling the system wait three minutes or more before servicing the vehicle.

2. Disconnect the negative battery cable.

3. Remove the engine undercover.

4. Drain the engine coolant. Be sure to properly dispose of used coolant.

5. Remove the engine cover.

6. Remove the air cleaner assembly.

7. Remove the water pipe and heater pipe assemblies.

8. Remove the exhaust front tube and three way catalysts.

9. Disconnect the steering lower joint at the power steering gear assembly side, and release the steering lower shaft.

10. Disconnect the air fuel sensor electrical connectors. Using the proper tool remove the sensors. Be careful not to drop or damage the sensors, or they will have to be replaced.

11. Remove the exhaust manifold cover.

12. Remove the retaining nuts in the reverse order of the installation sequence.

13. Remove the exhaust manifold. Discard the gaskets.

To install:

➡Be sure to use new fasteners, as required.

14. Installation is the reverse of the removal procedure.

15. Be sure to use new gaskets. Install the gasket as shown in the illustration.

16. Tighten the nuts to specification and in the sequence shown in the illustration.

17. When installing the oxygen sensors, be sure to coat them with anti-seize compound.

18. Continue the installation in the reverse order of the removal procedure.

19. Be sure to perform the reconnect/relearn procedures.

FLEXPLATE

REMOVAL & INSTALLATION

See Figure 70.

➡Whenever the negative battery cable is disconnected the following components will require resetting. The Automatic temperature control system, Automatic drive positioner, Power window control, Sunroof system, Sunshade

system, Rear view monitor, Idle Air Volume Learning, Steering Angle Sensor Neutral Position, Audio presets and Navigation. You will need the CONSULT-III diagnostic tool, or equivalent. Follow the directions on the screen of the tool, as needed.

1. Before servicing the vehicle, refer to the Precautions Section.

➡If working near and/or around the SRS system and components, be sure to disable the SRS system. After disabling the system wait three minutes or more before servicing the vehicle.

2. Disconnect the negative battery cable.

3. Raise and safely support the vehicle.

4. Remove the transmission assembly.

5. Remove the retaining bolts.

6. Remove the component from its mounting.

To install:

➡Be sure to use new fasteners, as required.

7. Be sure to correctly align the crankshaft side dowel pin and the flexplate side dowel pin hole.

➡If not aligned correctly the MIL light will illuminate.

8. Install the flexplate and reinforcement plate. Be sure to hold the ring gear with the ring gear stopper tool KV10118600 or equivalent.

9. Tighten the retaining bolts to specification in a crisscross pattern.

10. Continue the installation in the reverse order of the removal procedure.

11. Be sure to perform the reconnect/relearn procedures.

FLYWHEEL

REMOVAL & INSTALLATION

See Figures 71 and 72.

➡Whenever the negative battery cable is disconnected the following components will require resetting. The Automatic temperature control system, Automatic drive positioner, Power window control, Sunroof system, Sunshade system, Rear view monitor, Idle Air Volume Learning, Steering Angle Sensor Neutral Position, Audio presets and Navigation. You will need the CONSULT-III diagnostic tool, or equivalent. Follow the directions on the screen of the tool, as needed.

1. Before servicing the vehicle, refer to the Precautions Section.

➡If working near and/or around the SRS system and components, be sure to disable the SRS system. After disabling the system wait three minutes or more before servicing the vehicle.

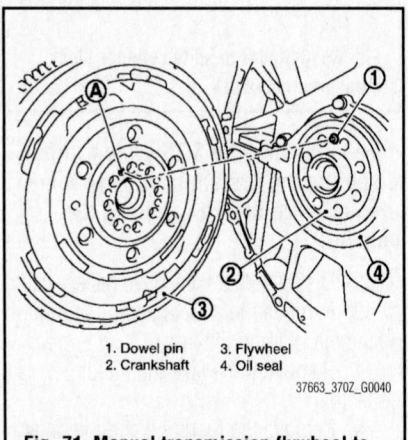

1. Dowel pin
2. Crankshaft
3. Flywheel
4. Oil seal

37663_370Z_G0040

Fig. 71 Manual transmission flywheel to transmission alignment

1. Ring gear
2. Flexplate
3. Reinforcement plate
4. Pilot converter
5. Crankshaft
A. Rounded

37663_370Z_G0042

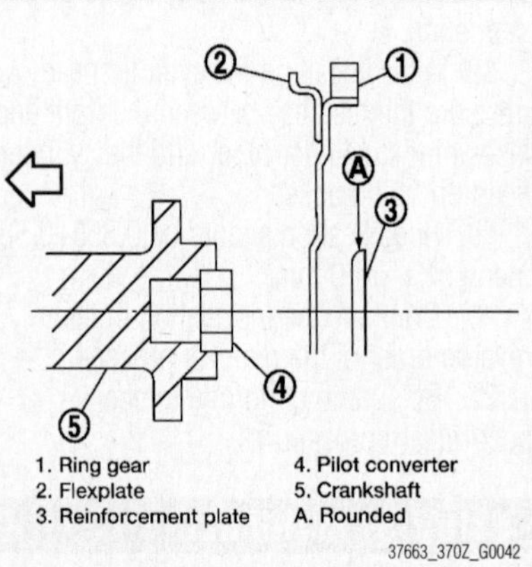

Fig. 70 Automatic transmission flexplate to transmission alignment

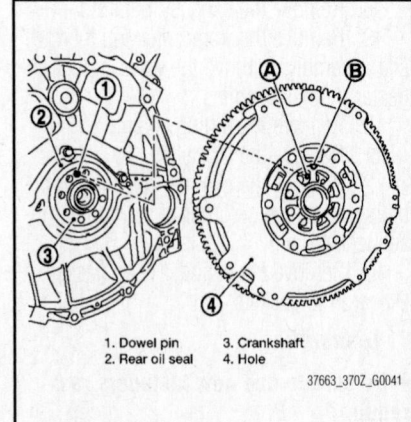

1. Dowel pin
2. Rear oil seal
3. Crankshaft
4. Hole

37663_370Z_G0041

Fig. 72 Manual transmission clutch cover matchmark

2. Disconnect the negative battery cable.

3. Raise and safely support the vehicle.

4. Remove the transmission assembly.

5. Remove the retaining bolts.

6. Remove the component from its mounting.

To install:

➡ **Be sure to use new fasteners, as required.**

7. Be sure that the dowel pin is installed in the crankshaft.

8. Be sure to correctly align the crankshaft side dowel pin and the flywheel side dowel pin hole.

9. There is a matchmark on the clutch cover side of the flywheel.

10. Tighten the retaining bolts to specification in a crisscross pattern.

11. Continue the installation in the reverse order of the removal procedure.

12. Be sure to perform the reconnect/relearn procedures.

INTAKE MANIFOLD

REMOVAL & INSTALLATION

See Figures 73 and 74.

➡ **Whenever the negative battery cable is disconnected the following components will require resetting. The Automatic temperature control system, Automatic drive positioner, Power window control, Sunroof system, Sunshade system, Rear view monitor, Idle Air Volume Learning, Steering Angle Sensor Neutral Position, Audio presets and Navigation. You will need the CONSULT-III diagnostic tool, or equivalent. Follow the directions on the screen of the tool, as needed.**

1. Before servicing the vehicle, refer to the Precautions Section.

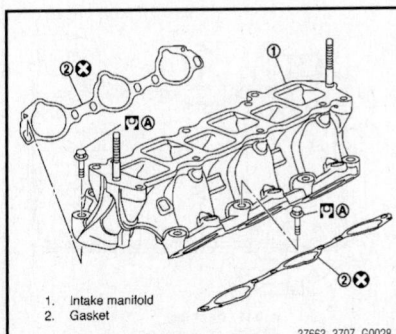

1. Intake manifold
2. Gasket

37663_370Z_G0028

Fig. 73 Intake manifold and related components

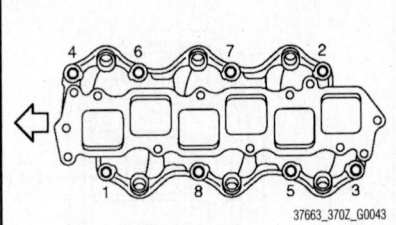

37663_370Z_G0043

Fig. 74 Intake manifold bolt locations and tightening sequence

➡ **If working near and/or around the SRS system and components, be sure to disable the SRS system. After disabling the system wait three minutes or more before servicing the vehicle.**

2. Properly relieve the fuel system pressure.

3. Disconnect the negative battery cable.

4. Remove the intake manifold collector.

5. Remove the fuel tube and fuel injector assembly.

6. Loosen the intake manifold retaining bolts in the reverse order of the installation.

7. Remove the component from its mounting. Discard the gasket.

➡ **Matchmark the manifold and the cylinder head for proper installation, as these components need to be installed in a specified direction.**

To install:

➡ **Be sure to use new fasteners, as required.**

8. Installation is the reverse of the removal procedure.

➡ **Be sure to use the marks made in the removal for proper alignment and installation.**

9. If the stud bolts were removed, install and tighten to 8 ft. lbs. (10.8 Nm).

10. Tighten all mounting bolts to specification and in the proper sequence.

11. Specification is 5 ft. lbs. (7.4 Nm) step one. 19 ft. lbs. (25.5 Nm) step two.

12. Be sure to perform the reconnect/relearn procedures.

INTAKE MANIFOLD COLLECTOR

REMOVAL & INSTALLATION

See Figures 75 and 76.

➡ **Whenever the negative battery cable is disconnected the following components will require resetting. The Auto-**

matic temperature control system, Automatic drive positioner, Power window control, Sunroof system, Sunshade system, Rear view monitor, Idle Air Volume Learning, Steering Angle Sensor Neutral Position, Audio presets and Navigation. You will need the CONSULT-III diagnostic tool, or equivalent. Follow the directions on the screen of the tool, as needed.

1. Before servicing the vehicle, refer to the Precautions Section.

➡ **If working near and/or around the SRS system and components, be sure to disable the SRS system. After disabling the system wait three minutes or more before servicing the vehicle.**

2. Disconnect the negative battery cable.

3. Remove the engine undercover.

4. Drain the engine coolant. Be sure to properly dispose of used coolant.

5. Remove the engine cover.

6. Remove the air cleaner assembly.

7. To remove the electric throttle control actuator, Disconnect the water hoses from the component. Disconnect the harness connector.

8. Loosen the mounting bolts in the reverse order of the tightening sequence.

➡ **When removing only the intake manifold collector, move the electric throttle control actuator without disconnecting the water hose. Handel the component carefully to avoid shock damage.**

9. Disconnect the vacuum hose, PCV hose, and EVAP hose from the intake manifold collector.

10. Remove the EVAP canister purge volume control solenoid valve and EVAP tube assembly from the manifold collector.

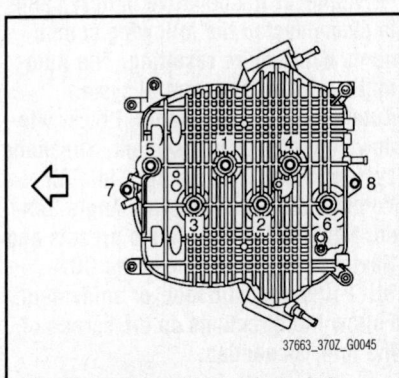

37663_370Z_G0045

Fig. 75 Intake manifold collector bolt locations and tightening sequence

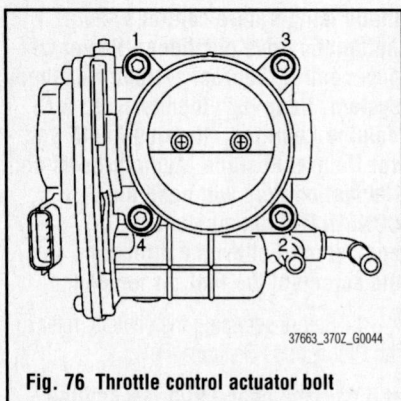

Fig. 76 Throttle control actuator bolt locations and tightening sequence

11. Loosen the retaining bolts and nuts in the reverse order of the tightening sequence and remove the component from its mounting.

To install:

➡ Be sure to use new fasteners, as required.

12. Installation is the reverse of the removal procedure.

13. Tighten the retaining bolts to specification and in the proper sequence.

14. Specification is 8 ft. lbs. (10.8 Nm).

15. Tighten the electric throttle actuator retaining bolts in the proper sequence.

16. Using the CONSULT-III diagnostic tool or equivalent, perform the throttle valve closed position learning and the idle air volume learning procedures.

17. Be sure to perform the reconnect/relearn procedures.

OIL PAN

REMOVAL & INSTALLATION

Lower

See Figures 77 and 78.

➡ Whenever the negative battery cable is disconnected the following components will require resetting. The Automatic temperature control system, Automatic drive positioner, Power window control, Sunroof system, Sunshade system, Rear view monitor, Idle Air Volume Learning, Steering Angle Sensor Neutral Position, Audio presets and Navigation. You will need the CONSULT-III diagnostic tool, or equivalent. Follow the directions on the screen of the tool, as needed.

1. Before servicing the vehicle, refer to the Precautions Section.

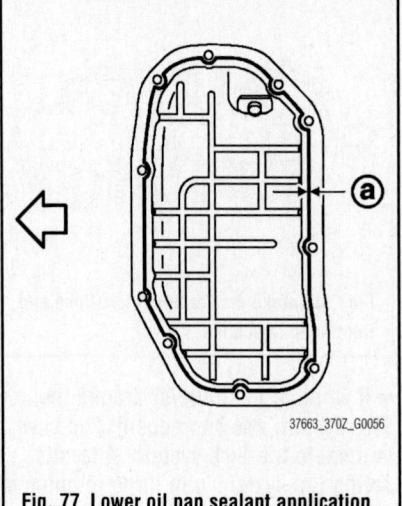

Fig. 77 Lower oil pan sealant application

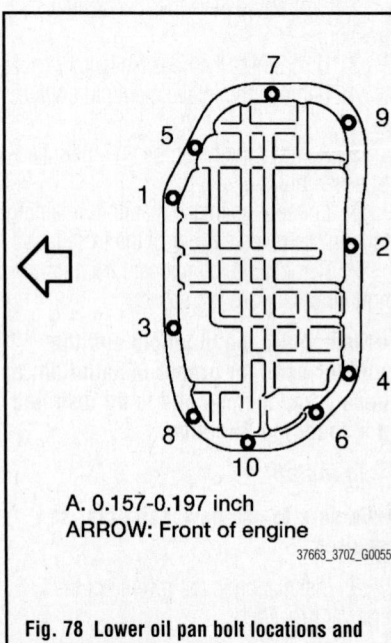

A. 0.157-0.197 inch
ARROW: Front of engine

Fig. 78 Lower oil pan bolt locations and tightening sequence

➡ If working near and/or around the SRS system and components, be sure to disable the SRS system. After disabling the system wait three minutes or more before servicing the vehicle.

2. Disconnect the negative battery cable.

3. Raise and support the vehicle safely.

4. Remove the engine undercover.

5. Drain the engine oil. Be sure to properly dispose of used oil.

6. Loosen the oil pan retaining bolts in the reverse order of the installation.

7. Insert a seal cutter tool between the upper and lower oil pan sealing surfaces. Never use a screwdriver. Slide the tool, by tapping the side of the tool with a hammer. Remove the lower oil pan.

8. Discard the gasket.

To install:

➡ Be sure to use new fasteners, as required.

9. Be sure to use a new gasket.

10. Apply a continuous bead of RTV sealant as shown in the illustration. Install within five minutes.

11. Tighten retaining bolts in the proper sequence and to the proper specification.

12. Continue the installation in the reverse order of the removal procedure.

13. Wait at least thirty minutes after installation before fill the crankcase with clean engine oil.

14. Start the engine and check for leaks. Correct as required.

15. Be sure to perform the reconnect/relearn procedures.

Upper

See Figures 79 and 80.

At this time the manufacturer provides service information for this component with the engine removed from the vehicle and positioned in a suitable holding fixture.

➡ Whenever the negative battery cable is disconnected the following components will require resetting. The Automatic temperature control system, Automatic drive positioner, Power window control, Sunroof system, Sunshade system, Rear view monitor, Idle Air Volume Learning, Steering Angle Sensor Neutral Position, Audio presets and Navigation. You will need the CONSULT-III diagnostic tool, or equivalent. Follow the directions on the screen of the tool, as needed.

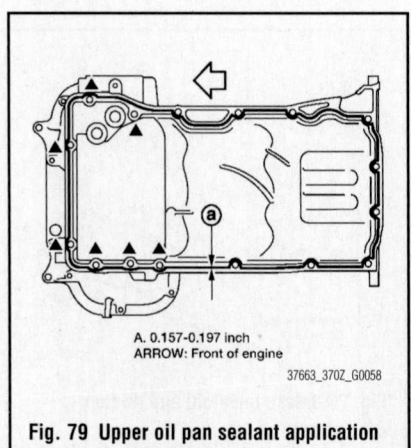

A. 0.157-0.197 inch
ARROW: Front of engine

Fig. 79 Upper oil pan sealant application

1. Before servicing the vehicle, refer to the Precautions Section.

➡ **If working near and/or around the SRS system and components, be sure to disable the SRS system. After disabling the system wait three minutes or more before servicing the vehicle.**

2. Disconnect the negative battery cable.

3. Remove the engine and position it in a suitable holding fixture.

4. Drain the engine oil. Be sure to properly dispose of used oil.

5. Remove the oil level gauge, oil pressure switch and oil temperature sensor.

6. Remove the lower oil pan.

7. Remove the oil strainer.

8. Loosen the retaining bolts in the reverse order of installation.

9. Insert a seal cutter tool between the upper and lower oil pan sealing surfaces. Never use a screwdriver. Slide the tool, by tapping the side of the tool with a hammer. Remove the upper oil pan.

10. Discard the gasket.

11. Remove the O-rings from the bottom of the lower cylinder block and oil pump. Discard the O-rings.

To install:

➡ **Be sure to use new fasteners, as required.**

12. Be sure to use a new gasket and O-rings.

13. Apply a continuous bead of RTV sealant as shown in the illustration. Install within five minutes.

➡ **For bolt holes with the triangle (see illustration) apply liquid gasket outside the holes.**

14. Tighten retaining bolts in the proper sequence and to the proper specification.

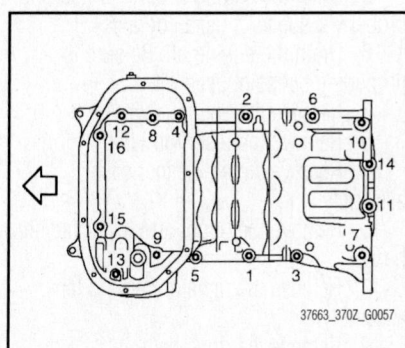

Fig. 80 Upper oil pan bolt locations and tightening sequence

➡ **There are two size bolts. Be careful to install the wrong bolt, in the wrong hole.**

15. Continue the installation in the reverse order of the removal procedure.

16. Wait at least thirty minutes after installation before fill the crankcase with clean engine oil.

17. Start the engine and check for leaks. Correct as required.

18. Be sure to perform the reconnect/relearn procedures.

OIL PUMP

REMOVAL & INSTALLATION

See Figure 81.

➡ **Whenever the negative battery cable is disconnected the following components will require resetting. The Automatic temperature control system, Automatic drive positioner, Power window control, Sunroof system, Sunshade system, Rear view monitor, Idle Air Volume Learning, Steering Angle Sensor Neutral Position, Audio presets and Navigation. You will need the CONSULT-III diagnostic tool, or equivalent. Follow the directions on the screen of the tool, as needed.**

1. Before servicing the vehicle, refer to the Precautions Section.

➡ **If working near and/or around the SRS system and components, be sure to disable the SRS system. After disabling the system wait three minutes or more before servicing the vehicle.**

2. Disconnect the negative battery cable.

3. Remove the upper and lower oil pan assemblies.

4. Remove the timing chain cover and the primary timing chain.

5. Remove the oil pump assembly from its mounting.

To install:

➡ **Be sure to use new fasteners, as required.**

6. Installation is the reverse of the removal procedure.

➡ **When installing, align the crankshaft flat surfaces with the oil pump inner rotor flat surfaces.**

7. Be sure to perform the reconnect/relearn procedures.

INSPECTION

1. Clearance between outer rotor and oil pump body: 0.0045–0.0079 in. (0.114–0.200 mm).

2. Tip clearance between inner rotor and outer rotor: Below 0.0071 in. (0.180 mm).

3. Side clearance with a straight-edge between inner rotor and oil pump body: 0.0012–0.0028 in. (0.030–0.070 mm).

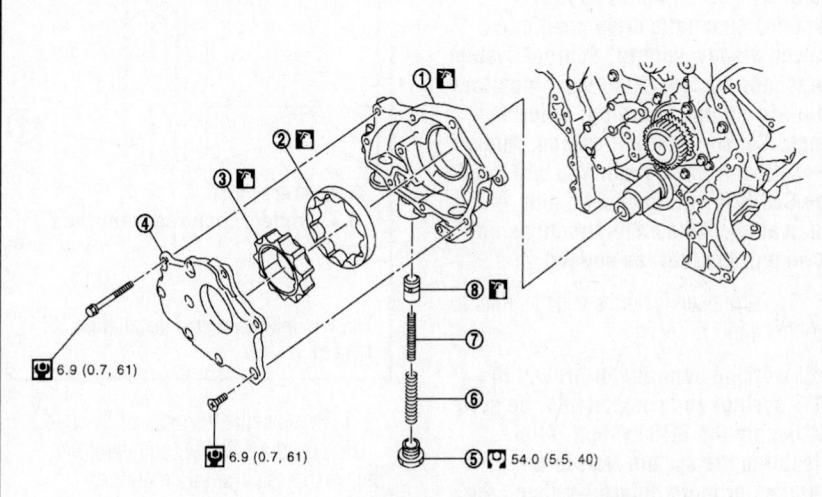

1.	Oil pump body	2.	Oil pump outer rotor	3.	Oil pump inner rotor
4.	Oil pump cover	5.	Regulator valve plug	6.	Regulator valve spring
7.	Regulator valve spring	8.	Regulator valve		

Fig. 81 Oil pump and related components

4. Side clearance with a straight-edge between outer rotor and oil pump body: 0.0020–0.0043 in. (0.050–0.110 mm).

PISTON AND RING

POSITIONING

See Figure 82.

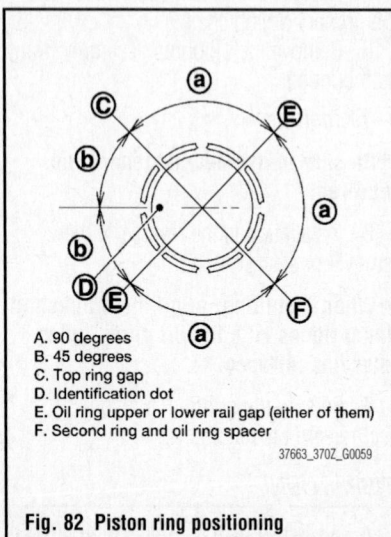

A. 90 degrees
B. 45 degrees
C. Top ring gap
D. Identification dot
E. Oil ring upper or lower rail gap (either of them)
F. Second ring and oil ring spacer

37663_370Z_G0059

Fig. 82 Piston ring positioning

REAR MAIN SEAL

REMOVAL & INSTALLATION

See Figures 83 and 84.

➡Whenever the negative battery cable is disconnected the following components will require resetting. The Automatic temperature control system, Automatic drive positioner, Power window control, Sunroof system, Sunshade system, Rear view monitor, Idle Air Volume Learning, Steering Angle Sensor Neutral Position, Audio presets and Navigation. You will need the CONSULT-III diagnostic tool, or equivalent. Follow the directions on the screen of the tool, as needed.

1. Before servicing the vehicle, refer to the Precautions Section.

➡If working near and/or around the SRS system and components, be sure to disable the SRS system. After disabling the system wait three minutes or more before servicing the vehicle.

2. Disconnect the negative battery cable.
3. Raise and safely support the vehicle.
4. Remove the transmission from the vehicle.

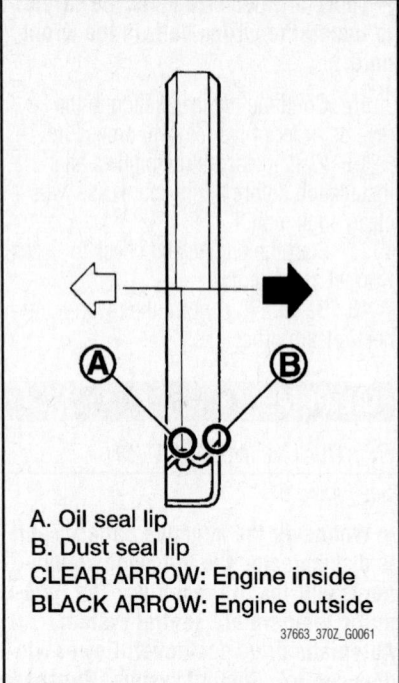

A. Oil seal lip
B. Dust seal lip
CLEAR ARROW: Engine inside
BLACK ARROW: Engine outside

37663_370Z_G0061

Fig. 83 Rear main seal installation (1 of 2)

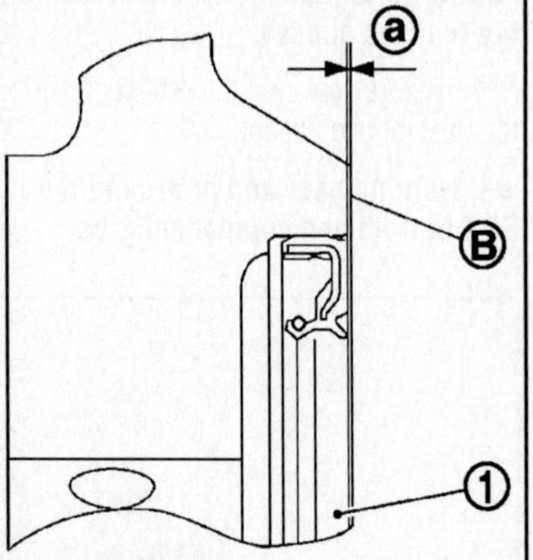

A. 0-0.020 inch
B. Cylinder block rear end face

37663_370Z_G0062

Fig. 84 Rear main seal installation (2 of 2)

5. Remove the flexplate or flywheel.
6. Using a suitable tool, carefully remove the seal from its mounting.

To install:

➡Be sure to use new fasteners, as required.

7. Install the seal so that each seal lip is oriented as shown in the illustration.

8. Press the seal into position as shown in the illustration.
9. Using a suitable drift, press fit the seal until the height of the seal is level with the mounting surface.

➡Be careful to avoid damage to the crankshaft and cylinder block. Press fit straight and avoid causing burrs or tilting the oil seal.

10. Continue the installation in the reverse order of the removal procedure.
11. Be sure to perform the reconnect/relearn procedures.

TIMING CHAIN FRONT COVER

REMOVAL & INSTALLATION

See Figures 85 through 90.

➡Whenever the negative battery cable is disconnected the following components will require resetting. The Automatic temperature control system, Automatic drive positioner, Power window control, Sunroof system, Sunshade system, Rear view monitor, Idle Air Volume Learning, Steering Angle Sensor Neutral Position, Audio presets and Navigation. You will need the CONSULT-III diagnostic tool, or equivalent. Follow the directions on the screen of the tool, as needed.

1. Before servicing the vehicle, refer to the Precautions Section.

➡If working near and/or around the SRS system and components, be sure to disable the SRS system. After disabling the system wait three minutes or more before servicing the vehicle.

2. Relieve the fuel system pressure.
3. Disconnect the negative battery cable.
4. Remove the engine undercover.
5. Drain the engine coolant. Be sure to properly dispose of used coolant.
6. Drain the engine oil. Be sure to properly dispose of used oil.
7. Remove the engine cover.
8. Remove the reservoir tank.
9. Remove the air cleaner case assembly.
10. Remove the upper and lower radiator hoses.
11. Remove the radiator cooling fan assembly.
12. Remove the drive belt.
13. Separate the engine harnesses by removing their brackets from the timing chain cover.

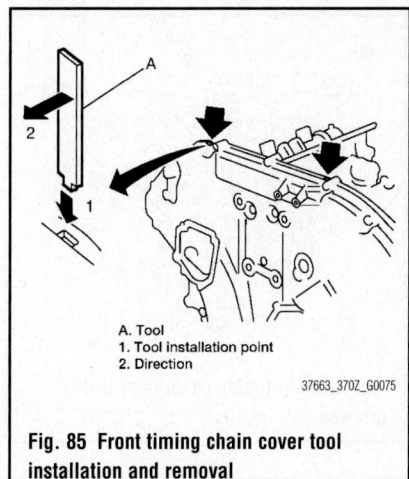

A. Tool
1. Tool installation point
2. Direction

37663_370Z_G0075

Fig. 85 Front timing chain cover tool installation and removal

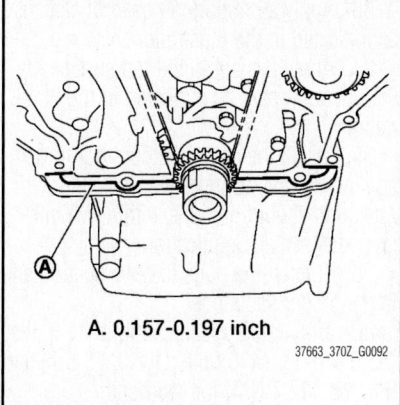

A. 0.157–0.197 inch

37663_370Z_G0092

Fig. 87 Oil pan (upper) sealant application points

14. Remove the intake manifold collector.

15. Remove the fuel sub mounting bolt.

16. Remove the oil level gauge and guide.

17. Remove the air conditioning compressor from the bracket. Secure it to the side. Do not discharge the refrigerant or disconnect the refrigerant hoses.

18. Remove the power steering fluid pump from the bracket with the hoses connected, secure it to the side. Remove the power steering oil pump bracket.

19. Remove the idler pulley, drive belt auto tensioner and bracket.

20. Remove the alternator and alternator bracket.

21. Remove the front water outlet.

22. Remove the Camshaft Position (CMP) sensor.

➡ **Do not drop the sensor. Never disassemble it. Never allow metal powder to adhere to the magnetic portion of the sensor. Never store the sensor where it is exposed to magnetism.**

23. To remove the intake valve timing control covers and gasket, disconnect the intake valve timing control solenoid valve harness connector. Loosen the mounting bolts in the reverse order of the tightening sequence.

➡ **The shaft is internally jointed with the camshaft sprocket (INT) center hole. When removing, keep it horizontal until it is completely disconnected.**

24. The shaft is engaged with the camshaft sprocket (INT) center hole on the inside. Pull straight out so that it does not tilt until the joint is disengaged.

25. Remove the intake valve timing control solenoid valve, if necessary.

➡ **This valve is not reusable. Never remove it unless required.**

26. Remove the valve covers.

27. To position the engine to TDC on the compression stroke, rotate the crankshaft pulley clockwise to align the timing mark (grooved line without color) with the timing indicator. Check that the exhaust cam noses on No. 1 cylinder (engine front side of bank 1) is located as shown in the illustration. If not turn the crankshaft on complete revolution and align.

28. Remove the crankshaft pulley.

29. Remove the lower oil pan.

30. Loosen the mounting bolts in the front of the upper oil pan in the reverse order of the installation sequence.

31. To remove the case cover, loosen the mounting bolts in the reverse order of the installation sequence.

32. Insert a suitable tool (KV10111100) into the notch at the top of the front timing chain cover, as shown in the illustration. Pry off the case by moving the tool, as shown in the illustration.

➡ **Never use a screwdriver or similar item. After removal handle the cover carefully so it does not tilt, cant or warp under load.**

33. Using the proper tool, remove the front case seal, as required.

To install:

➡ **Be sure to use new fasteners, as required.**

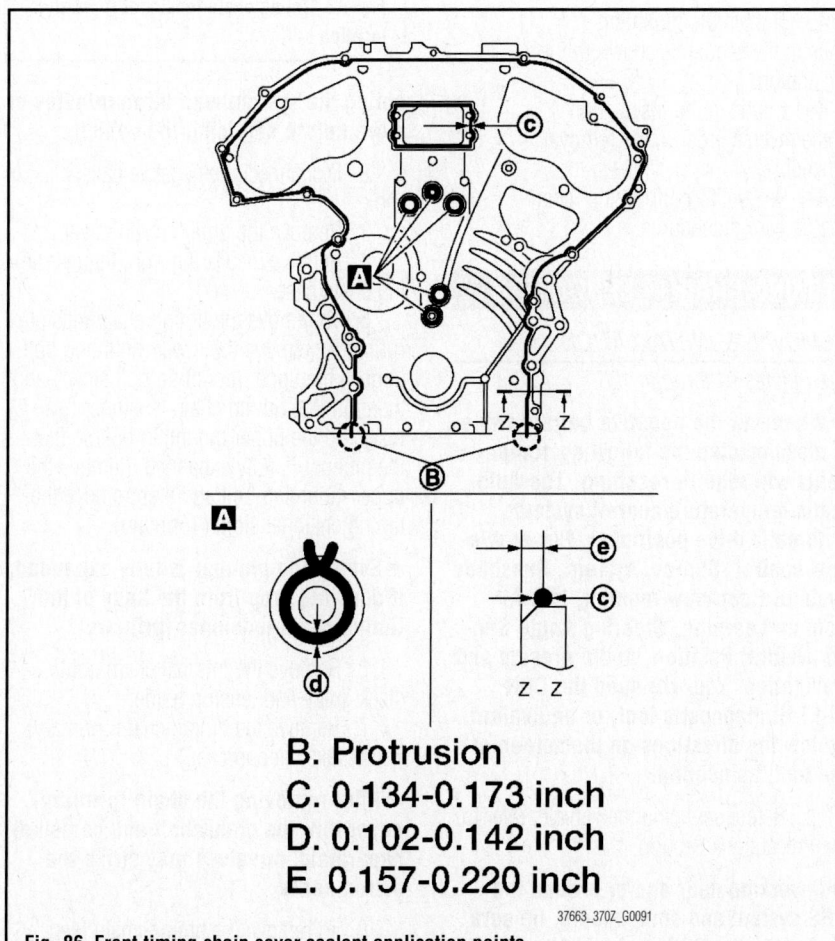

B. Protrusion
C. 0.134–0.173 inch
D. 0.102–0.142 inch
E. 0.157–0.220 inch

37663_370Z_G0091

Fig. 86 Front timing chain cover sealant application points

34. Install new O-rings on the rear timing chain case cover.

➡**Be sure that the O-rings remain in place during installation to the rear case cover.**

35. Install a new oil seal in the front timing chain cover. Coat the seal with clean engine oil before installation. Press fit the seal into position until it becomes flush with the front timing chain case end face. Check that the garter spring is in position and that the seal lip is not inverted.

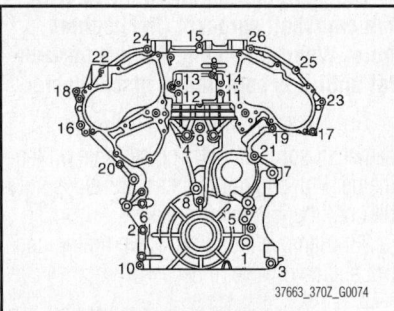

Fig. 88 Front timing chain cover bolt locations and tightening sequence

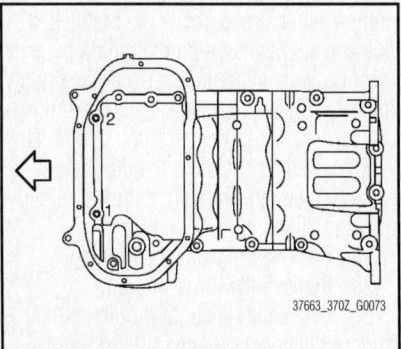

Fig. 89 Upper oil pan front bolt locations and tightening sequence

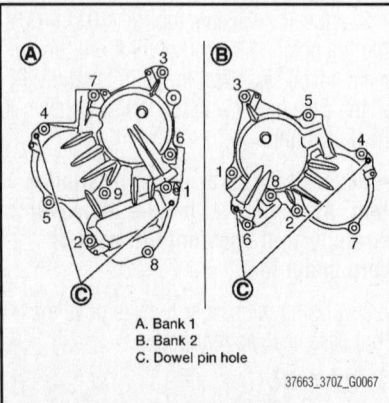

A. Bank 1
B. Bank 2
C. Dowel pin hole

Fig. 90 Intake valve timing control cover bolt locations and tightening sequence

36. Apply a continuous bead of sealant as indicated in the illustration.

37. Apply sealant to the top surface of the upper oil pan, as indicated in the illustration.

38. Install the cover. Be sure not to damage the oil seal during cover installation. Attaching should be done within five minutes after sealant application.

39. Tighten the bolts to specification and in the proper sequence.

40. Tightening specification is 41 ft. lbs. (55.0 Nm) for M10 bolts (1,2,3,4,5,6,7) and 9 ft. lbs. (12.7 NM) for M6 bolts.

41. After all bolts are tightened, retighten them to specification and in the proper sequence.

42. Install the upper oil pan mounting bolts in the proper sequence.

43. To install the valve timing control covers, first install new seal rings in the shaft grooves.

44. Install the covers, using new gaskets.

➡**Align the center of both shaft holes of the camshaft sprocket (INT) and the shaft and then insert them. Be careful not to drop the seal ring from the shaft groove.**

45. Tighten the mounting bolts in the sequence shown in the illustration.

46. Continue the installation in the reverse order of the removal procedure.

47. Be sure to perform the reconnect/relearn procedures.

TIMING CHAIN & SPROCKETS

REMOVAL & INSTALLATION
See Figures 91 through 106.

➡**Whenever the negative battery cable is disconnected the following components will require resetting. The Automatic temperature control system, Automatic drive positioner, Power window control, Sunroof system, Sunshade system, Rear view monitor, Idle Air Volume Learning, Steering Angle Sensor Neutral Position, Audio presets and Navigation. You will need the CONSULT-III diagnostic tool, or equivalent. Follow the directions on the screen of the tool, as needed.**

1. Before servicing the vehicle, refer to the Precautions Section.

➡**If working near and/or around the SRS system and components, be sure to disable the SRS system. After dis-**

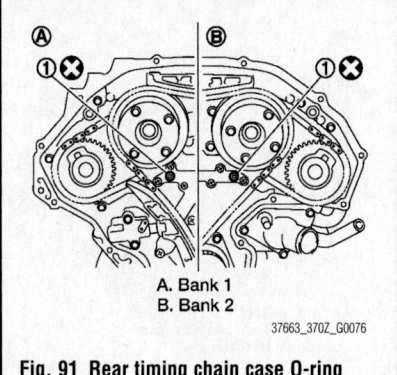

A. Bank 1
B. Bank 2

Fig. 91 Rear timing chain case O-ring location

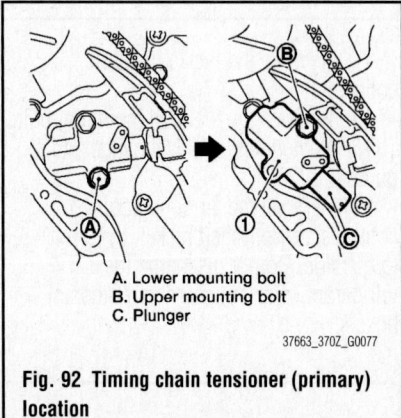

A. Lower mounting bolt
B. Upper mounting bolt
C. Plunger

Fig. 92 Timing chain tensioner (primary) location

abling the system wait three minutes or more before servicing the vehicle.

2. Disconnect the negative battery cable.

3. Remove the timing chain cover.

4. Remove the O-ring from the rear timing chain case.

5. To remove the timing chain tensioner (primary), remove the lower mounting bolt. Loosen the upper mounting bolt slowly and then turn the timing chain tensioner (primary) on the upper mounting bolt so that the plunger is fully expanded. Remove the upper mounting bolt and then remove the timing chain tensioner (primary).

➡**Even if the plunger is fully expanded, it does not drop from the body of the timing chain tensioner (primary).**

6. Remove the internal chain guide, slack guide and tension guide.

7. Remove the timing chain (primary) and crankshaft sprocket.

➡**After removing the chain (primary), never turn the crankshaft and camshaft separately, or valves may strike the piston heads.**

8. To remove the timing chain (secondary) and camshaft sprockets attach a

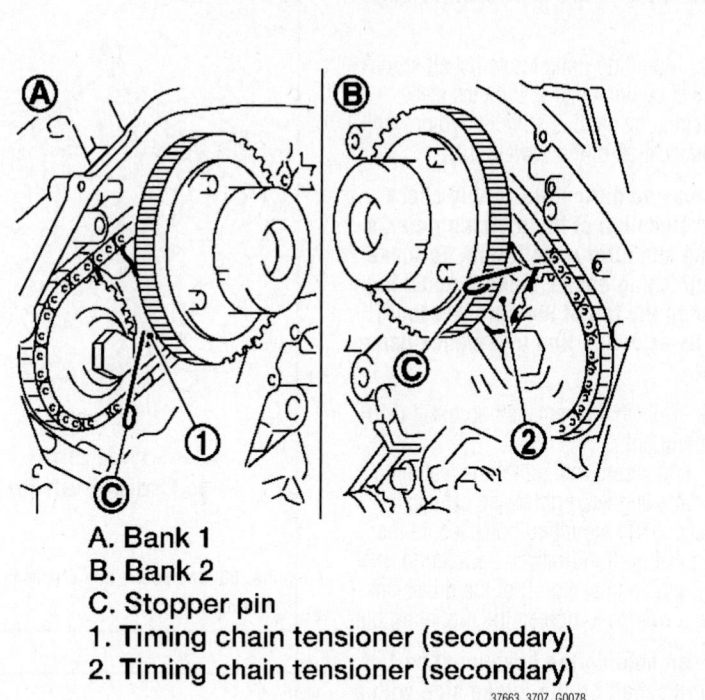

A. Bank 1
B. Bank 2
C. Stopper pin
1. Timing chain tensioner (secondary)
2. Timing chain tensioner (secondary)

37663_370Z_G0078

Fig. 93 Timing chain tensioner (secondary) location

suitable stopper pin (0.020 inch hard metal) to the chain tensioners (secondary).

9. For removal of the chain tensioners (secondary) refer to the illustration.

➡**Removing VVEL ladder assembly is required.**

10. Remove the camshaft sprocket (EXH) mounting bolt.

➡**Secure the hexagonal portion of the camshaft (EXH) using a wrench to loosen mounting bolt. Never loosen the mounting bolt by securing anything other than camshaft (EXH) hexagonal portion or with tensioning the timing chain.**

11. Remove the camshaft sprocket (INT) mounting bolt.

➡**Secure the hexagonal portion (located between journal No. 1 and journal No. 2) of the driveshaft using a wrench to loosen the mounting bolt. Never loosen the mounting bolt by securing anything other than the driveshaft hexagonal portion or with tensioning the timing chain. When holding the hexagonal part of the driveshaft on the intake side with a wrench, be careful not to allow the wrench to cause interference with other parts.**

➡**Never disassemble the camshaft sprocket (INT). Never loosen bolts (A) as shown in the illustration.**

12. Remove the timing chain (secondary) along with the camshaft sprockets.

13. Remove all traces of gasket material from the front and rear timing chain covers. Be sure to remove all material from the bolt holes and threads. See illustration

To install:

➡**Be sure to use new fasteners, as required.**

➡**See illustration that shows the relationship between the matching mark on each timing chain and that on the cor-**

responding sprocket with components installed.

14. Check that the dowel pin and crankshaft key are located as shown in the illustration (engine at TDC on the compression stroke).

➡**Though the camshaft does not stop at the position as shown, for placement of cam noses, it is generally accepted that the camshaft is placed in the same direction as that of the illustration.**

➡**Matching marks between the chain and sprockets slip easily. Confirm all matching mark positions repeatedly during the installation process.**

15. To install the timing chains (secondary) and camshaft sprockets, push the plunger of the timing chain tensioner (secondary) and keep it pressed in with a stopper pin.

16. Install the timing chains (secondary) and camshaft sprockets. See illustration.

➡**Illustration shows bank 1 (rear view)**

17. Align the matching marks on the chain (secondary) (orange link) with the ones on the intake and exhaust camshaft sprockets (punched), and install them.

➡**Matching marks for camshaft sprockets (INT) are on the back side of the camshaft sprockets (secondary). There are two types of matching marks, the circle and the oval. They should be used for bank 1 (circle) and bank 2 (oval) respectively.**

18. Shape (orientation of signal plate) of camshaft sprocket (INT) varies depending on the bank position. See illustration.

19. Align dowel pin camshafts with the pin groove on sprockets and install them.

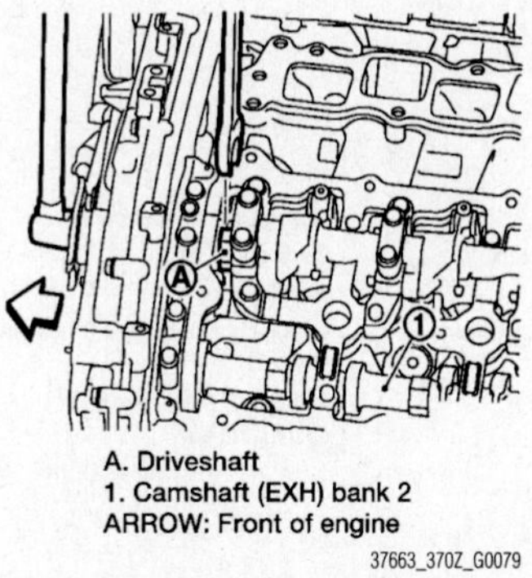

A. Driveshaft
1. Camshaft (EXH) bank 2
ARROW: Front of engine

37663_370Z_G0079

Fig. 94 Camshaft sprocket (INT) removal

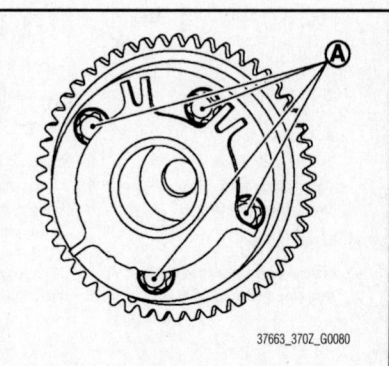

37663_370Z_G0080

Fig. 95 Camshaft sprocket (INT) bolt "A" location

➡In case that positions of each matching mark and each dowel pin do not fit with matching parts, make fine adjustment to the position holding the hexagonal portion on camshaft (EXH) or driveshaft with wrench or equivalent tool.

20. Mounting bolts for camshaft sprockets must be tightened in the next step. Tightening by hand is sufficient to prevent the dislocation of the dowel pins.

➡It may be difficult to visibly check the dislocation of the matching marks during and after installation. To make the matching easier make a matching mark on the top of the sprocket teeth and its extended line in advance using paint.

21. Tighten the camshaft sprocket (EXH) mounting bolt.

22. After confirming that the matching marks are aligned, tighten the camshaft sprocket (INT) mounting bolt. Secure the hexagonal portion (located between journal No. 1 and Journal No. 2) of the driveshaft using a wrench to tighten the mounting bolt.

➡When holding the hexagonal part of the driveshaft on the intake side with a wrench, be careful not to allow the wrench to cause interference with other parts.

23. Pull out the stopper pins from the timing chain tensioners (secondary).

24. To install the timing chain (primary), install the crankshaft sprocket.

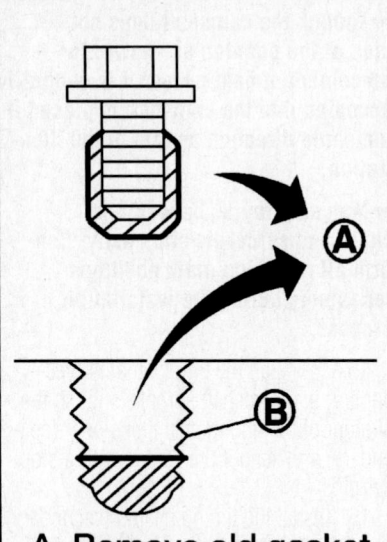

A. Remove old gasket that is stuck
B. Bolt hole

37663_370Z_G0081

Fig. 96 Removing gasket material from bolt hole

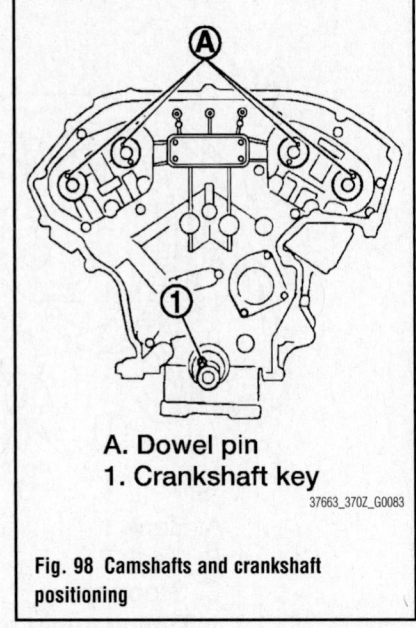

A. Dowel pin
1. Crankshaft key

37663_370Z_G0083

Fig. 98 Camshafts and crankshaft positioning

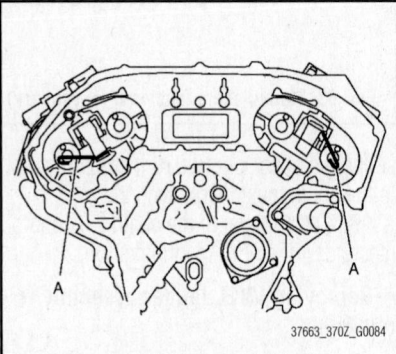

37663_370Z_G0084

Fig. 99 Stopper pin installation and timing chain tensioner (secondary) plunger location

➡Be sure that the matching marks on the crankshaft sprocket face the front of the engine.

25. Install the timing chain (primary) so that the matching mark (punched) on the camshaft sprocket (INT) is aligned with the yellow link on the timing chain while the matching mark (notched) on the crankshaft sprocket is aligned with the orange link on the timing chain. See illustration.

➡When it is difficult to align the matching marks of the timing chain (primary) with each sprocket, gradually turn the driveshaft sing a wrench on the hexagonal portion to align it with the matching marks.

26. Install the internal chain guide, slack guide and tension guide.

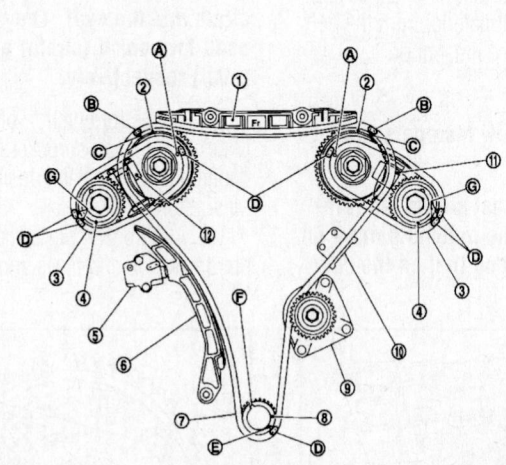

1. Internal chain guide	2. Camshaft sprocket (INT)	3. Timing chain (secondary)
4. Camshaft sprocket (EXH)	5. Timing chain tensioner (primary)	6. Slack guide
7. Timing chain (primary)	8. Crankshaft sprocket	9. Water pump
10. Tension guide	11. Timing chain tensioner (secondary) (bank 2)	12. Timing chain tensioner (secondary) (bank 1)
A. Matching mark [punched (back side)]	B. Matching mark (yellow link)	C. Matching mark (punched)
D. Matching mark (orange link)	E. Matching mark (notched)	F. Crankshaft key

37663_370Z_G0082

Fig. 97 Timing chains and sprockets (relationship) alignment

► **Never overtighten the slack guide mounting bolt. It is normal for a gap to exist under the bolt seats when mounting bolts are tightened to specification.**

27. To install the timing chain tensioner (primary) pull the plunger stopper tab up

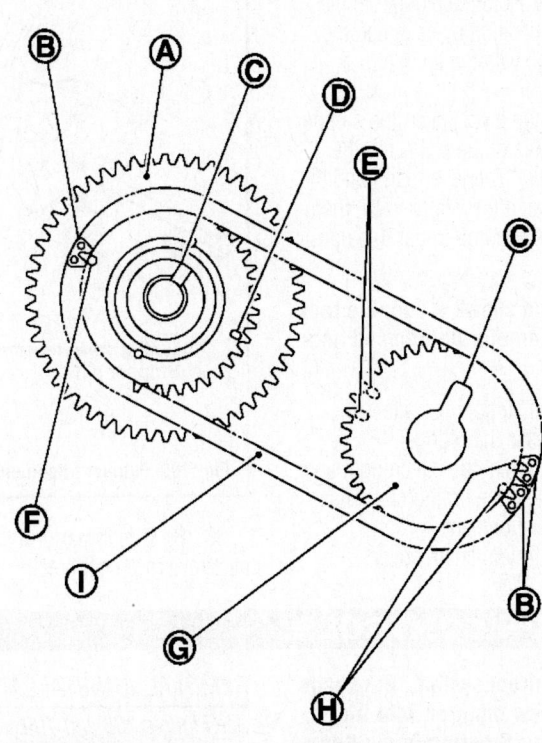

A. Camshaft sprocket (INT) back face
B. Orange link
C. Dowel groove
D. Matching mark (oval)
E. Matching mark (2 oval: on front face)
F. Matching mark (circle)
G. Camshaft sprocket (EXH) back face
H. Matching mark (2 oval: on front face)
I. Timing chain (secondary)

37663_370Z_G0085

Fig. 100 Timing chain tensioner (secondary) and camshaft sprockets—bank 1 rear view

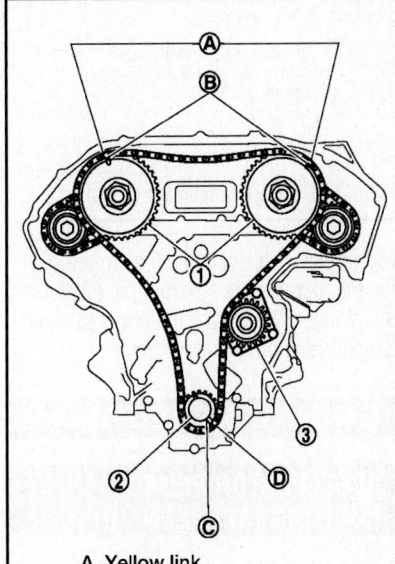

A. Yellow link
B. Punched mark
C. Notched mark
D. Orange link
1. Camshaft sprocket (INT)
2. Crankshaft sprocket
3. Water pump

37663_370Z_G0087

Fig. 103 Camshaft sprocket (INT) signal plate orientation

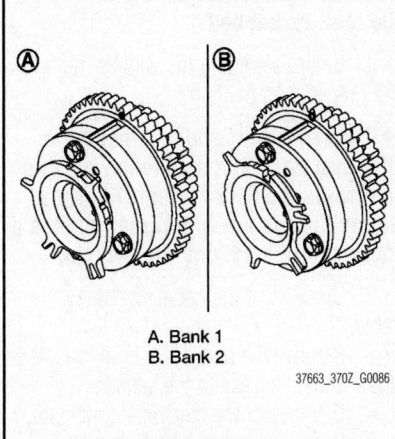

A. Bank 1
B. Bank 2

37663_370Z_G0086

Fig. 101 Camshaft sprocket (INT) signal plate orientation

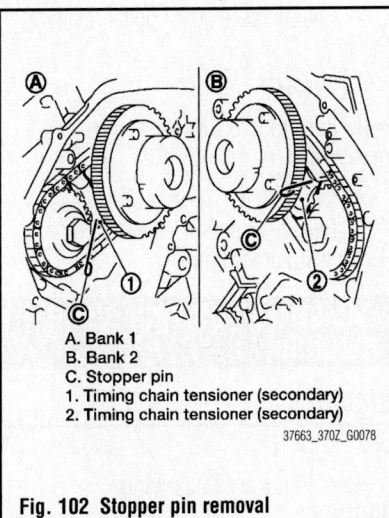

A. Bank 1
B. Bank 2
C. Stopper pin
1. Timing chain tensioner (secondary)
2. Timing chain tensioner (secondary)

37663_370Z_G0078

Fig. 102 Stopper pin removal

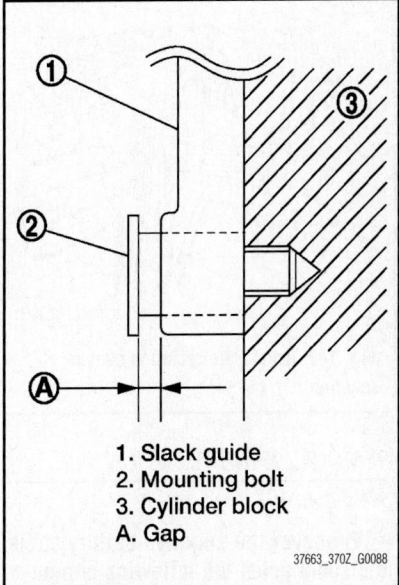

1. Slack guide
2. Mounting bolt
3. Cylinder block
A. Gap

37663_370Z_G0088

Fig. 104 Slack guide mounting bolt installation

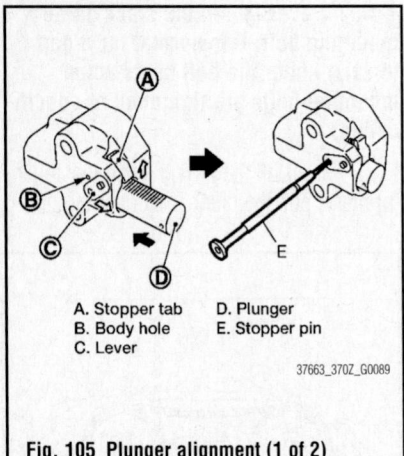

A. Stopper tab   D. Plunger
B. Body hole   E. Stopper pin
C. Lever

37663_370Z_G0089

Fig. 105 Plunger alignment (1 of 2)

(or turn lever downward) so as to remove the plunger stopper tab from the ratchet of the plunger. Note that the plunger stopper tab and lever are synchronized.

28. Push the plunger into the inside of the tensioner body. Hold the plunger in the fully compressed position by engaging the plunger stopper tab with the tip of the ratchet.

29. To secure the lever, insert the stopper pin through the hole of the lever into the tensioner body hole. The lever parts and the plunger stopper tab are synchronized, therefore the plunger is secured under this condition.

➡The illustration shows a suitable tool of 0.047 inch diameter being used as a stopper pin.

30. Pull out the stopper pin after installing and release the plunger.

31. Check again that the matching marks on the sprockets and the timing chain have not slipped out of alignment.

32. Install the timing chain cover.

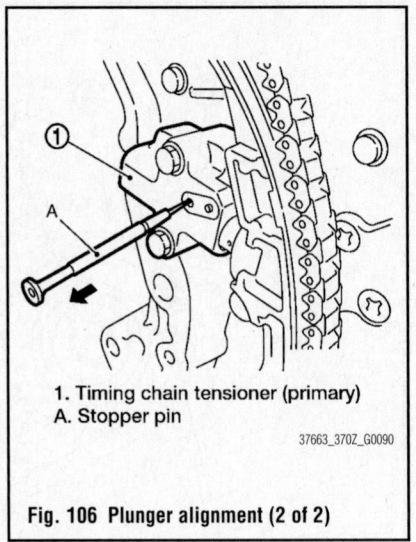

1. Timing chain tensioner (primary)
A. Stopper pin

37663_370Z_G0090

Fig. 106 Plunger alignment (2 of 2)

33. Be sure to perform the reconnect/relearn procedures.

ENGINE PERFORMANCE & EMISSION CONTROLS

ACCELERATOR PEDAL POSITION (APP) SENSOR

LOCATION

See Figure 107.

The sensor detects the accelerator position and sends a signal to the ECM. The Accelerator Pedal Position (APP) sensor has two sensors. These sensors are a kind of potentiometers which transform the accelerator pedal position into output voltage, and emit the voltage signal to the ECM.

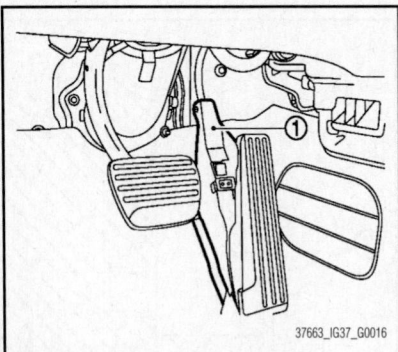

37663_IG37_G0016

Fig. 107 Accelerator position sensor location

REMOVAL & INSTALLATION

See Figure 107.

➡Whenever the negative battery cable is disconnected the following components will require resetting. The Automatic temperature control system, Automatic drive positioner, Power window control, Sunroof system, Sunshade system, Rear view monitor, Idle Air Volume Learning, Steering Angle Sensor Neutral Position, Audio presets and Navigation. You will need the CONSULT-III diagnostic tool, or equivalent. Follow the directions on the screen of the tool, as needed.

1. Before servicing the vehicle, refer to the Precautions Section.

➡If working near and/or around the SRS system and components, be sure to disable the SRS system. After disabling the system wait three minutes or more before servicing the vehicle.

2. Disconnect the negative battery cable.

3. Unplug Accelerator Pedal Position (APP) sensor connector.

4. Remove accelerator pedal retaining bolts.

To install:

➡Be sure to use new fasteners, as required.

5. Installation is reverse of removal.

6. Be sure to perform the reconnect/relearn procedures.

CAMSHAFT POSITION (CMP) SENSOR

LOCATION

See Figure 108.

Refer to the accompanying illustration.

REMOVAL & INSTALLATION

See Figures 108 and 109.

At this time the manufacturer does not provide removal and installation procedures for this component. The following procedure is a guideline and may differ from the vehicle you are servicing.

➡Whenever the negative battery cable is disconnected the following components will require resetting. The Automatic temperature control system, Automatic drive positioner, Power window control, Sunroof system, Sunshade system, Rear view monitor, Idle Air Volume Learning, Steering Angle Sensor Neutral Position, Audio presets and Navigation. You will need the CONSULT-III diagnostic tool, or equivalent. Follow the directions on the screen of the tool, as needed.

1. Before servicing the vehicle, refer to the Precautions Section.

➡If working near and/or around the SRS system and components, be sure to disable the SRS system. After disabling the system wait three minutes or more before servicing the vehicle.

2. Disconnect the negative battery cable.

3. Remove the necessary components in order to gain access to the sensor.

4. Disconnect the electrical connector.

5. Remove the retaining bolt.

6. Remove the component from the vehicle.

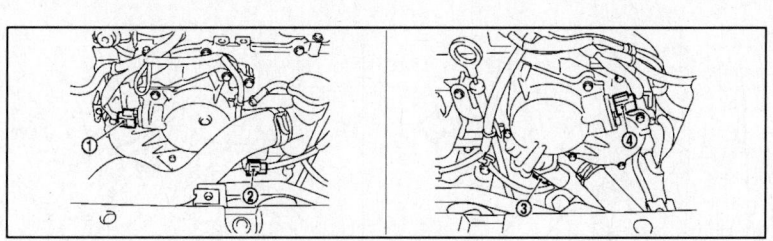

1. Camshaft position sensor (PHASE) (bank 1)
2. Intake valve timing control solenoid valve (bank 1) harness connector
3. Intake valve timing control solenoid valve (bank 2) harness connector
4. Camshaft position sensor (PHASE) (bank 2)

37663_370Z_G0135

Fig. 108 Camshaft Position (CMP) sensor location

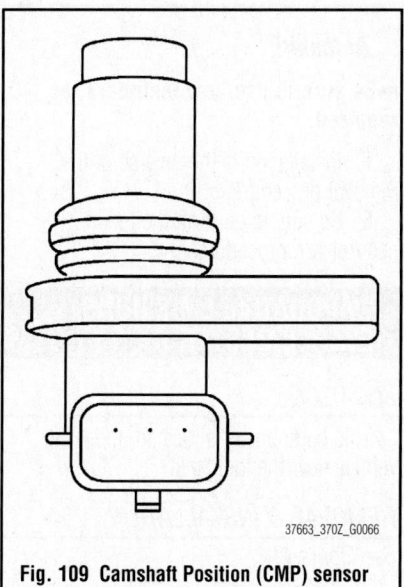

37663_370Z_G0066

Fig. 109 Camshaft Position (CMP) sensor

To install:

➡**Be sure to use new fasteners, as required.**

7. Installation is the reverse of the removal procedure.
8. Be sure to perform the reconnect/relearn procedures.

CRANKSHAFT POSITION (CKP) SENSOR

LOCATION

See Figure 110.

This sensor is located on the cylinder block facing the gear teeth on the signal plate.

REMOVAL & INSTALLATION

See Figures 110 and 111.

At this time the manufacturer does not provide removal and installation procedures

for this component. The following procedure is a guideline and may differ from the vehicle you are servicing.

➡**Whenever the negative battery cable is disconnected the following components will require resetting. The Automatic temperature control system, Automatic drive positioner, Power window control, Sunroof system, Sunshade system, Rear view monitor, Idle Air Volume Learning, Steering Angle Sensor Neutral Position, Audio presets and Navigation. You will need the CONSULT-III diagnostic tool, or equivalent. Follow the directions on the screen of the tool, as needed.**

1. Before servicing the vehicle, refer to the Precautions Section.

➡**If working near and/or around the SRS system and components, be sure to disable the SRS system. After disabling the system wait three minutes or more before servicing the vehicle.**

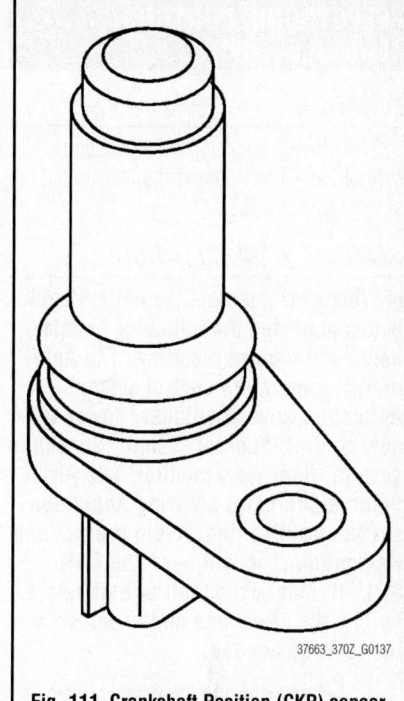

37663_370Z_G0137

Fig. 111 Crankshaft Position (CKP) sensor

2. Disconnect the negative battery cable.
3. Remove the necessary components in order to gain access to the sensor.
4. Disconnect the electrical connector.
5. Remove the retaining bolt.
6. Remove the component from the vehicle.

To install:

➡**Be sure to use new fasteners, as required.**

7. Installation is the reverse of the removal procedure.
8. Be sure to perform the reconnect/relearn procedures.

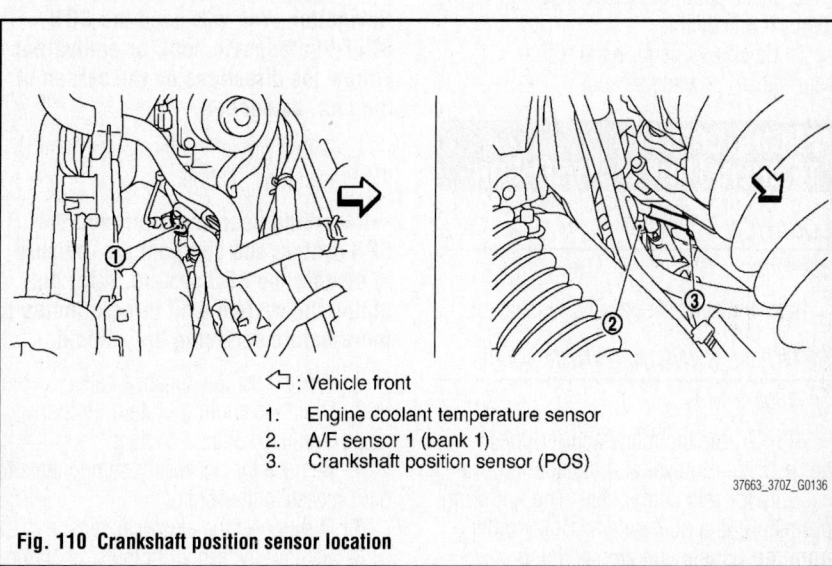

⬅ : Vehicle front

1. Engine coolant temperature sensor
2. A/F sensor 1 (bank 1)
3. Crankshaft position sensor (POS)

37663_370Z_G0136

Fig. 110 Crankshaft position sensor location

ELECTRONIC CONTROL MODULE (ECM)

LOCATION

The ECM is located on the passenger's side of the vehicle behind the instrument assist lower panel.

REMOVAL & INSTALLATION

➡Whenever the negative battery cable is disconnected the following components will require resetting. The Automatic temperature control system, Automatic drive positioner, Power window control, Sunroof system, Sunshade system, Rear view monitor, Idle Air Volume Learning, Steering Angle Sensor Neutral Position, Audio presets and Navigation. You will need the CONSULT-III diagnostic tool, or equivalent. Follow the directions on the screen of the tool, as needed.

1. Before servicing the vehicle, refer to the Precautions Section.

➡If working near and/or around the SRS system and components, be sure to disable the SRS system. After disabling the system wait three minutes or more before servicing the vehicle.

2. Disconnect the negative battery cable.
3. Remove the passenger's side instrument lower panel.
4. Disconnect electrical connectors.
5. Remove the component.

To install:

➡Be sure to use new fasteners, as required.

6. Installation is the reverse of the removal procedure.
7. Be sure to perform the reconnect/relearn procedures.

ENGINE COOLANT TEMPERATURE (ECT) SENSOR

LOCATION

See Figure 112.

Refer to the accompanying illustration.

REMOVAL & INSTALLATION

See Figure 113.

At this time the manufacturer does not provide removal and installation procedures for this component. The following procedure is a guideline and may differ from the vehicle you are servicing.

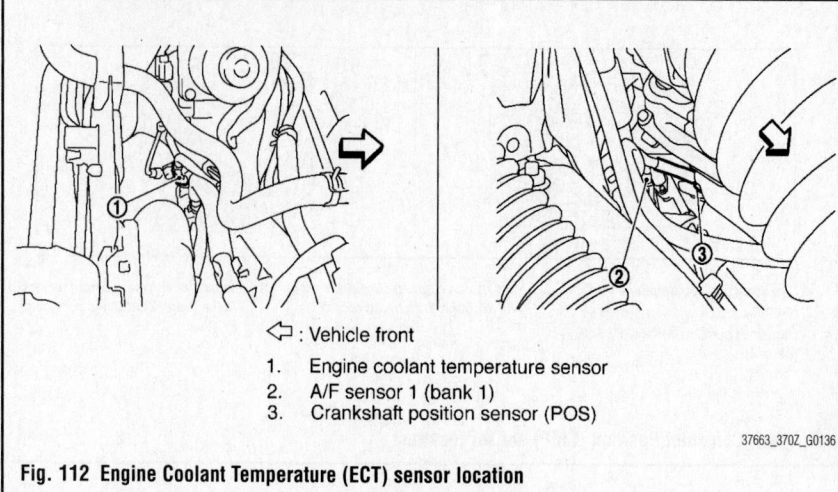

◁ : Vehicle front

1. Engine coolant temperature sensor
2. A/F sensor 1 (bank 1)
3. Crankshaft position sensor (POS)

37663_370Z_G0136

Fig. 112 Engine Coolant Temperature (ECT) sensor location

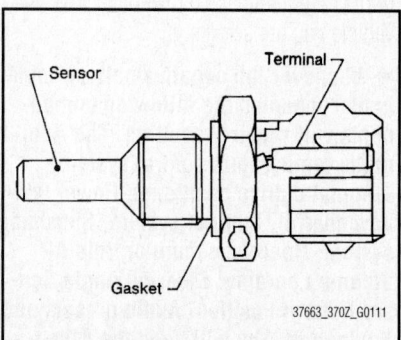

37663_370Z_G0111

Fig. 113 Engine Coolant Temperature (ECT) sensor

➡Whenever the negative battery cable is disconnected the following components will require resetting. The Automatic temperature control system, Automatic drive positioner, Power window control, Sunroof system, Sunshade system, Rear view monitor, Idle Air Volume Learning, Steering Angle Sensor Neutral Position, Audio presets and Navigation. You will need the CONSULT-III diagnostic tool, or equivalent. Follow the directions on the screen of the tool, as needed.

1. Before servicing the vehicle, refer to the Precautions Section.

➡If working near and/or around the SRS system and components, be sure to disable the SRS system. After disabling the system wait three minutes or more before servicing the vehicle.

2. Disconnect the negative battery cable.
3. Drain the cooling system. Properly dispose of used engine coolant.
4. Remove the necessary components to gain access to the sensor.
5. Disconnect the electrical sensor.
6. Remove the sensor from its mounting.

To install:

➡Be sure to use new fasteners, as required.

7. Installation is the reverse of the removal procedure.
8. Be sure to perform the reconnect/relearn procedures.

EVAPORATIVE EMISSIONS (EVAP) CANISTER

LOCATION

This component is located under the vehicle near the fuel tank.

REMOVAL & INSTALLATION

See Figure 114.

➡Whenever the negative battery cable is disconnected the following components will require resetting. The Automatic temperature control system, Automatic drive positioner, Power window control, Sunroof system, Sunshade system, Rear view monitor, Idle Air Volume Learning, Steering Angle Sensor Neutral Position, Audio presets and Navigation. You will need the CONSULT-III diagnostic tool, or equivalent. Follow the directions on the screen of the tool, as needed.

1. Before servicing the vehicle, refer to the Precautions Section.

➡If working near and/or around the SRS system and components, be sure to disable the SRS system. After disabling the system wait three minutes or more before servicing the vehicle.

2. Disconnect the negative battery cable.
3. Raise and support the vehicle safely.
4. Remove the canister retaining bolt.

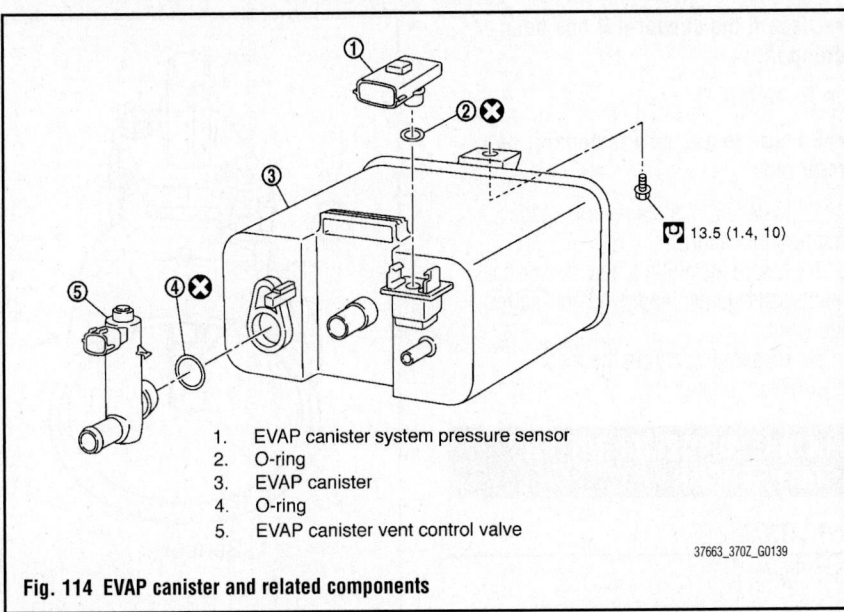

1. EVAP canister system pressure sensor
2. O-ring
3. EVAP canister
4. O-ring
5. EVAP canister vent control valve

37663_370Z_G0139

Fig. 114 EVAP canister and related components

13.5 (1.4, 10)

5. Remove the canister from its mounting.

➡ **The canister vent control valve and system pressure sensor can be removed without removing the canister.**

To install:

➡ **Be sure to use new fasteners, as required.**

6. Installation is the reverse of the removal procedure.
7. Be sure to perform the reconnect/relearn procedures.

HEATED OXYGEN SENSOR (HO2S)

LOCATION

See Figure 115.

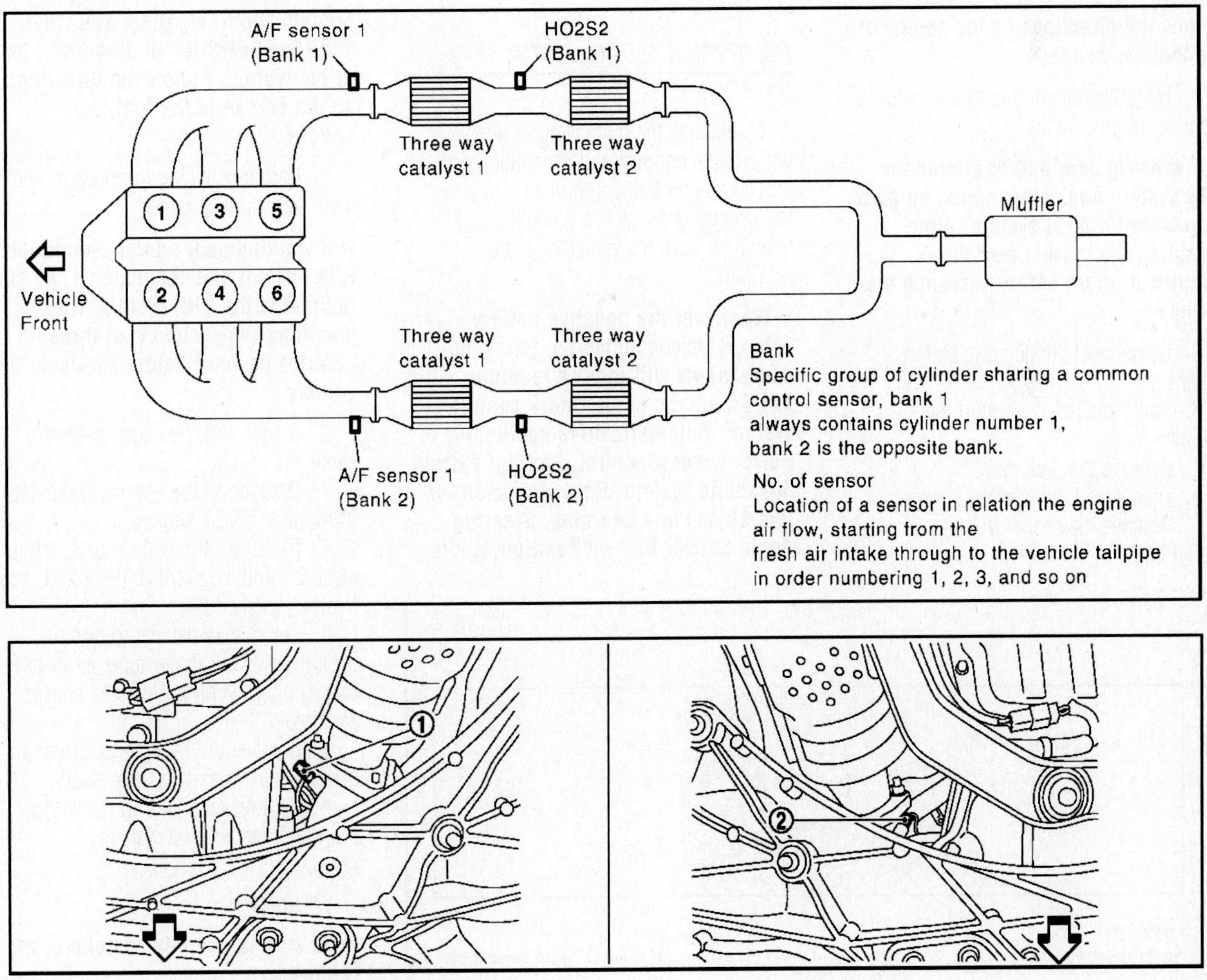

A/F sensor 1 (Bank 1)
HO2S2 (Bank 1)

Three way catalyst 1
Three way catalyst 2

Muffler

Vehicle Front

1 3 5
2 4 6

Three way catalyst 1
Three way catalyst 2

A/F sensor 1 (Bank 2)
HO2S2 (Bank 2)

Bank
Specific group of cylinder sharing a common control sensor, bank 1 always contains cylinder number 1, bank 2 is the opposite bank.

No. of sensor
Location of a sensor in relation the engine air flow, starting from the fresh air intake through to the vehicle tailpipe in order numbering 1, 2, 3, and so on

⇦ : Vehicle front

1. A/F sensor 1 (bank 1)
2. A/F sensor 1 (bank 2)

37663_370Z_G0132

Fig. 115 Oxygen sensors and related components

These sensors are located in the exhaust stream, after the catalytic converter. Two sensors are used, one for each cylinder bank.

REMOVAL & INSTALLATION

See Figure 116.

At this time the manufacturer does not provide removal and installation procedures for this component. The following procedure is a guideline and may differ from the vehicle you are servicing.

→**Whenever the negative battery cable is disconnected the following components will require resetting. The Automatic temperature control system, Automatic drive positioner, Power window control, Sunroof system, Sunshade system, Rear view monitor, Idle Air Volume Learning, Steering Angle Sensor Neutral Position, Audio presets and Navigation. You will need the CONSULT-III diagnostic tool, or equivalent. Follow the directions on the screen of the tool, as needed.**

1. Before servicing the vehicle, refer to the Precautions Section.

→**If working near and/or around the SRS system and components, be sure to disable the SRS system. After disabling the system wait three minutes or more before servicing the vehicle.**

2. Disconnect the negative battery cable.
3. Raise and safely support the vehicle.
4. Remove the undercover.
5. Disconnect the harness connector.
6. Remove the sensor from its mounting.

→**Discard the sensor if it has been dropped.**

To install:

→**Be sure to use new fasteners, as required.**

7. Installation is the reverse of the removal procedure.
8. Before installing a new sensor coat the threads with an approved anti-seize lubricant.
9. Be sure to perform the reconnect/relearn procedures.

INTAKE AIR TEMPERATURE (IAT) SENSOR

LOCATION

See Figure 117.

This sensor is built into the MAF sensor and is services with that component.

REMOVAL & INSTALLATION

See Figure 117.

At this time the manufacturer does not provide removal and installation procedures for this component. The following procedure is a guideline and may differ from the vehicle you are servicing.

→**Whenever the negative battery cable is disconnected the following components will require resetting. The Automatic temperature control system, Automatic drive positioner, Power window control, Sunroof system, Sunshade system, Rear view monitor, Idle Air Volume Learning, Steering Angle Sensor Neutral Position, Audio**

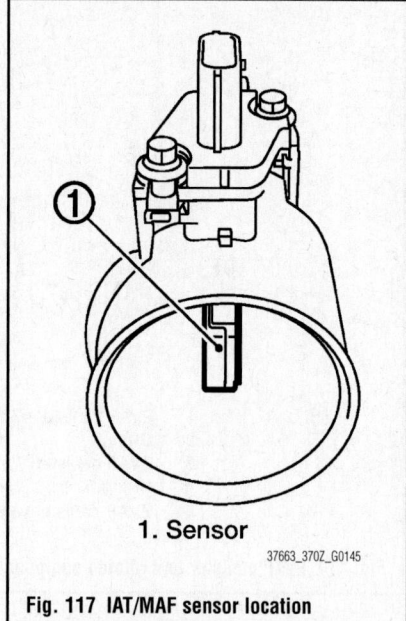

1. Sensor

37663_370Z_G0145

Fig. 117 IAT/MAF sensor location

presets and Navigation. You will need the CONSULT-III diagnostic tool, or equivalent. Follow the directions on the screen of the tool, as needed.

1. Before servicing the vehicle, refer to the Precautions Section.

→**If working near and/or around the SRS system and components, be sure to disable the SRS system. After disabling the system wait three minutes or more before servicing the vehicle.**

2. Disconnect the negative battery cable.
3. Disconnect the harness connector from the IAT/MAF sensor.
4. Disconnect the tube clamp at the electric throttle control actuator and at the fresh air intake tube.
5. Remove air cleaner to electric throttle control actuator tube, air cleaner case (upper) with the IAT/MAF sensor attached.
6. Remove IAT/MAF sensor from air cleaner case (upper), as necessary.
7. Remove resonator in the fender, lifting left fender protector, as necessary.

To install:

→**Be sure to use new fasteners, as required.**

8. Installation is the reverse of the removal procedure.
9. Be sure to perform the reconnect/relearn procedures.

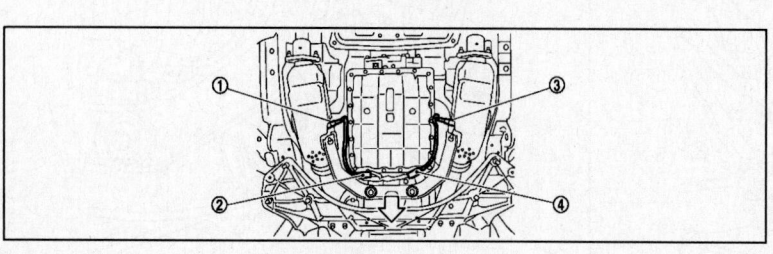

◁ : Vehicle front
1. Heated oxygen sensor 2 (bank 2)
2. Heated oxygen sensor 2 (bank 2) harness connector
3. Heated oxygen sensor 2 (bank 1)
4. Heated oxygen sensor 2 (bank 1) harness connector

37663_370Z_G0134

Fig. 116 Heated oxygen sensor and related components

KNOCK SENSOR (KS)

LOCATION

See Figure 118.

The knock sensors are located under the intake manifold. There are two of them.

REMOVAL & INSTALLATION

See Figure 118.

At this time the manufacturer does not provide removal and installation procedures for this component. The following procedure is a guideline and may differ from the vehicle you are servicing.

➡Whenever the negative battery cable is disconnected the following components will require resetting. The Automatic temperature control system, Automatic drive positioner, Power window control, Sunroof system, Sunshade system, Rear view monitor, Idle Air Volume Learning, Steering Angle Sensor Neutral Position, Audio presets and Navigation. You will need the CONSULT-III diagnostic tool, or equivalent. Follow the directions on the screen of the tool, as needed.

1. Before servicing the vehicle, refer to the Precautions Section.

➡If working near and/or around the SRS system and components, be sure to disable the SRS system. After disabling the system wait three minutes or more before servicing the vehicle.

2. Disconnect the negative battery cable.
3. Remove the intake manifold collector.
4. Remove the intake manifold.
5. Disconnect the sensor electrical connector.

6. Remove the sensor from its mounting.

To install:

➡**Be sure to use new fasteners, as required.**

7. Installation is the reverse of removal procedure.
8. Be sure to perform the reconnect/relearn procedures.

MALFUNCTION INDICATOR LIGHT (MIL)

RESET PROCEDURE

Clearing diagnostic trouble codes resets the MIL.

MASS AIR FLOW (MAF) SENSOR

LOCATION

See Figure 119.

This sensor is placed in the stream of the air intake. It is incorporated along with the IAT sensor.

REMOVAL & INSTALLATION

See Figure 119.

At this time the manufacturer does not provide removal and installation procedures for this component. The following procedure is a guideline and may differ from the vehicle you are servicing.

➡Whenever the negative battery cable is disconnected the following components will require resetting. The Automatic temperature control system, Automatic drive positioner, Power window control, Sunroof system, Sunshade system, Rear view monitor, Idle Air Volume Learning, Steering Angle Sensor Neutral Position, Audio presets and Navigation. You will need the CONSULT-III diagnostic tool, or equivalent. Follow the directions on the screen of the tool, as needed.

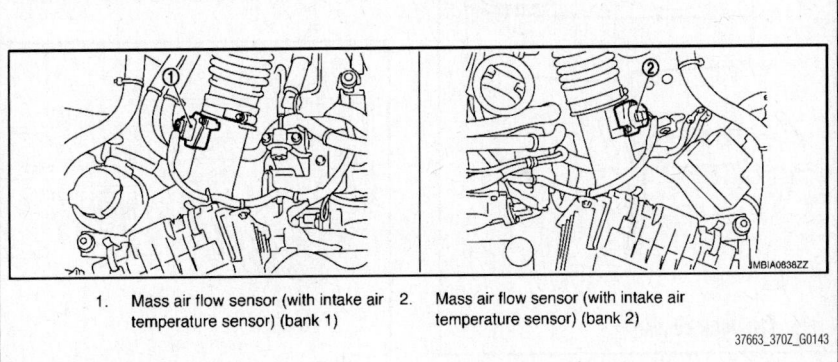

1. Mass air flow sensor (with intake air temperature sensor) (bank 1) 2. Mass air flow sensor (with intake air temperature sensor) (bank 2)

37663_370Z_G0143

Fig. 119 MAF sensor location

1. Before servicing the vehicle, refer to the Precautions Section.

➡If working near and/or around the SRS system and components, be sure to disable the SRS system. After disabling the system wait three minutes or more before servicing the vehicle.

2. Disconnect the negative battery cable.
3. Disconnect the harness connector from the MAF/IAT sensor.
4. Disconnect the tube clamp at the electric throttle control actuator and at the fresh air intake tube.
5. Remove air cleaner to electric throttle control actuator tube, air cleaner case (upper) with the MAF/IAT sensor attached.

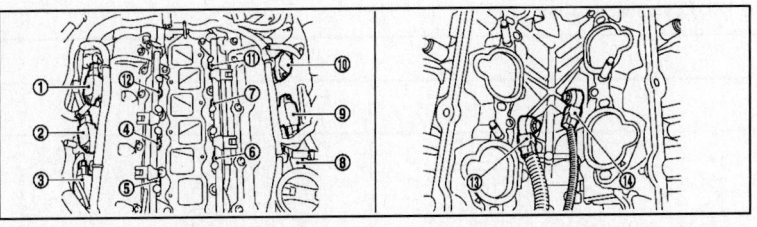

1. Ignition coil No.5 (with power transistor)	2. Ignition coil No.3 (with power transistor)	3. Ignition coil No.1 (with power transistor)
4. Fuel injector No.3	5. Fuel injector No.1	6. Fuel injector No.2
7. Fuel injector No.4	8. Ignition coil No.2 (with power transistor)	9. Ignition coil No.4 (with power transistor)
10. Ignition coil No.6 (with power transistor)	11. Fuel injector No.6	12. Fuel injector No.5
13. Knock sensor (bank 1)	14. Knock sensor (bank 2)	

37663_370Z_G0144

Fig. 118 Knock sensor location

6. Remove MAF/IAT sensor from air cleaner case (upper), as necessary.

7. Remove resonator in the fender, lifting left fender protector, as necessary.

To install:

➡**Be sure to use new fasteners, as required.**

8. Installation is the reverse of the removal procedure.

9. Be sure to perform the reconnect/relearn procedures.

MANIFOLD ABSOLUTE PRESSURE (MAP) SENSOR

LOCATION

See Figure 120.

This sensor is located at the intake manifold collector.

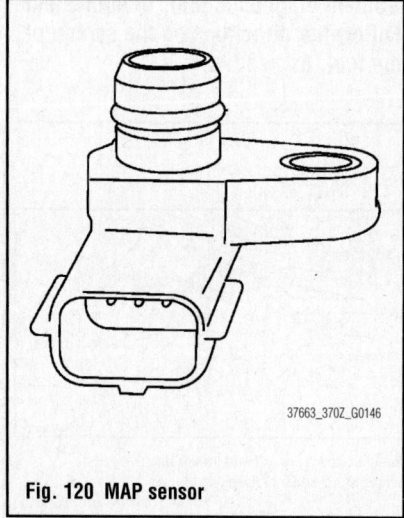

Fig. 120 MAP sensor

REMOVAL & INSTALLATION

See Figure 120.

At this time the manufacturer does not provide removal and installation procedures for this component, refer to the illustration as required.

POSITIVE CRANKCASE VENTILATION (PCV) VALVE

LOCATION

See Figure 121.

Refer to the accompanying illustration.

REMOVAL & INSTALLATION

See Figure 121.

At this time the manufacturer does not provide removal and installation procedures for this component. The following procedure is a guideline and may differ from the vehicle you are servicing.

➡**Whenever the negative battery cable is disconnected the following components will require resetting. The Automatic temperature control system, Automatic drive positioner, Power window control, Sunroof system, Sunshade system, Rear view monitor, Idle Air Volume Learning, Steering Angle Sensor Neutral Position, Audio presets and Navigation. You will need the CONSULT-III diagnostic tool, or equivalent. Follow the directions on the screen of the tool, as needed.**

1. Before servicing the vehicle, refer to the Precautions Section.

➡**If working near and/or around the SRS system and components, be sure to disable the SRS system. After disabling the system wait three minutes or more before servicing the vehicle.**

2. Disconnect the negative battery cable.

3. Remove the necessary components in order to gain access to the component.

4. Disconnect the PCV hose.

5. Remove the valve from its mounting.

To install:

➡**Be sure to use new fasteners, as required.**

6. Installation is the reverse of the removal procedure.

7. Be sure to perform the reconnect/relearn procedures.

THROTTLE POSITION SENSOR (TPS)

LOCATION

See Figure 122.

This sensor is located in the main hose connecting the intake manifold collector . There are two of these sensors.

REMOVAL & INSTALLATION

See Figure 122.

At this time the manufacturer does not provide removal and installation procedures for this component, refer to the illustration as required.

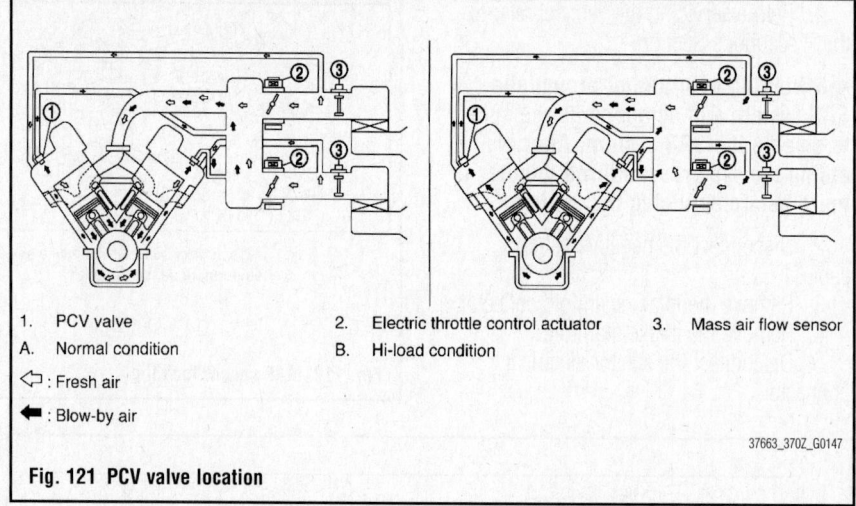

1. PCV valve
A. Normal condition
◁ : Fresh air
◀ : Blow-by air

2. Electric throttle control actuator
B. Hi-load condition

3. Mass air flow sensor

Fig. 121 PCV valve location

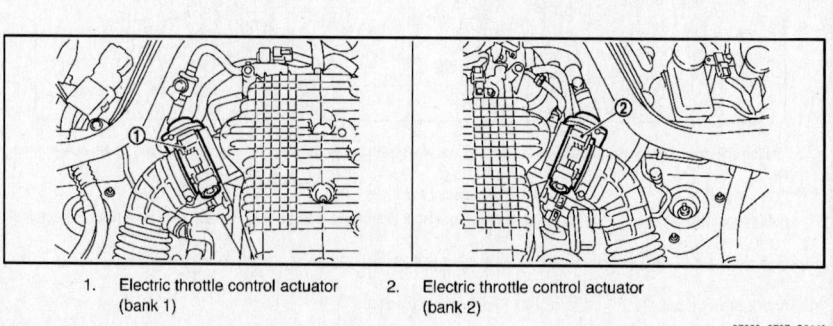

1. Electric throttle control actuator (bank 1)

2. Electric throttle control actuator (bank 2)

Fig. 122 TPS sensor (electronic throttle control) location

FUEL

FUEL SYSTEM SERVICE PRECAUTIONS

Safety is the most important factor when performing not only fuel system maintenance, but any type of maintenance. Failure to conduct maintenance and repairs in a safe manner may result in serious personal injury or death. Work on a vehicle's fuel system components can be accomplished safely and effectively by adhering to the following rules and guidelines.

• To avoid the possibility of fire and personal injury, always disconnect the negative battery cable unless the repair or test procedure requires that battery voltage be applied.

• Always relieve the fuel system pressure prior to disconnecting any fuel system component (injector, fuel rail, pressure regulator, etc.) fitting or fuel line connection. Exercise extreme caution whenever relieving fuel system pressure to avoid exposing skin, face and eyes to fuel spray. Please be advised that fuel under pressure may penetrate the skin or any part of the body that it contacts.

• Always place a shop towel or cloth around the fitting or connection prior to loosening to absorb any excess fuel due to spillage. Ensure that all fuel spillage is quickly removed from engine surfaces. Ensure that all fuel-soaked cloths or towels are deposited into a flame-proof waste container with a lid.

• Always keep a dry chemical (Class B) fire extinguisher near the work area.

• Do not allow fuel spray or fuel vapors to come into contact with a spark or open flame.

• Always use a second wrench when loosening or tightening fuel line connection fittings. This will prevent unnecessary stress and torsion on fuel piping. Always follow the proper torque specifications.

• Always replace worn fuel fitting O-rings with new ones. Do not substitute fuel hose where rigid pipe is installed.

FUEL SYSTEM PRESSURE

RELIEVING

See Figure 123.

➡**Whenever the negative battery cable is disconnected the following components will require resetting. The Automatic temperature control system, Automatic drive positioner, Power win-**

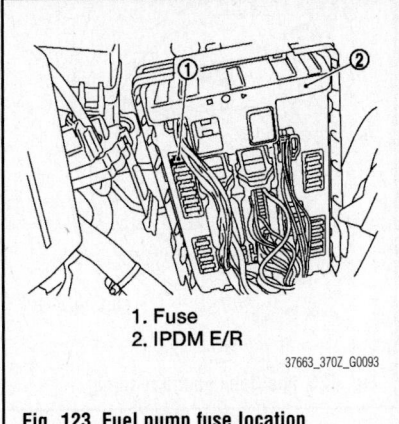

1. Fuse
2. IPDM E/R

37663_370Z_G0093

Fig. 123 Fuel pump fuse location

dow control, Sunroof system, Sunshade system, Rear view monitor, Idle Air Volume Learning, Steering Angle Sensor Neutral Position, Audio presets and Navigation. You will need the CONSULT-III diagnostic tool, or equivalent. Follow the directions on the screen of the tool, as needed.

1. Before servicing the vehicle, refer to the Precautions Section.

➡**If working near and/or around the SRS system and components, be sure to disable the SRS system. After dis-**

abling the system wait three minutes or more before servicing the vehicle.

2. Remove the fuel pump fuse, located in the IPDM E/R.
3. Start the engine.
4. After the engine stalls, crank it two or three times to release all fuel pressure.
5. Turn the ignition switch off.
6. Reinstall the fuel pump fuse after servicing the fuel system.
7. Be sure to perform the reconnect/relearn procedures, as required.

FUEL FILTER

REMOVAL & INSTALLATION

The fuel filter is attached to the fuel pump assembly. The fuel pump must be removed before the filter can be serviced.

FUEL LEVEL SENDING UNIT

LOCATION

The fuel level sending unit is attached to the fuel pump assembly. The fuel level sending unit is located in the fuel tank.

REMOVAL & INSTALLATION

See Figure 124.

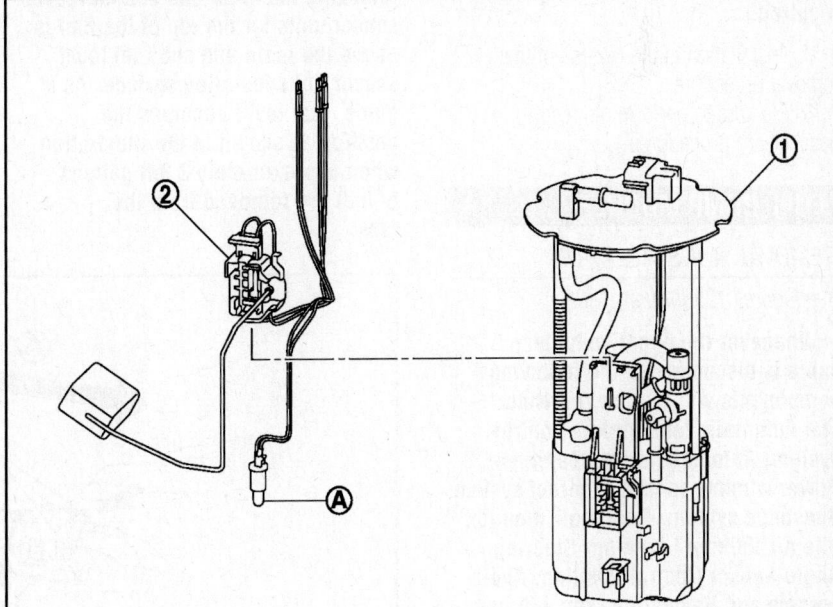

1. Fuel filter and fuel pump assembly
A. Fuel temp sensor
2. Main fuel level sensor unit

37663_370Z_G0095

Fig. 124 Fuel level sending unit and related components

➡Whenever the negative battery cable is disconnected the following components will require resetting. The Automatic temperature control system, Automatic drive positioner, Power window control, Sunroof system, Sunshade system, Rear view monitor, Idle Air Volume Learning, Steering Angle Sensor Neutral Position, Audio presets and Navigation. You will need the CONSULT-III diagnostic tool, or equivalent. Follow the directions on the screen of the tool, as needed.

1. Before servicing the vehicle, refer to the Precautions Section.

➡If working near and/or around the SRS system and components, be sure to disable the SRS system. After disabling the system wait three minutes or more before servicing the vehicle.

2. Properly relieve the fuel system pressure.
3. Disconnect the negative battery cable.
4. Remove the fuel pump module.
5. Discard the gasket.
6. Service the fuel level sending unit, as required.

➡This component cannot be disassembled and should be replaced as a unit.

To install:

➡Be sure to use new fasteners, as required.

7. Installation is the reverse of the removal procedure.
8. Be sure to perform the reconnect/relearn procedures.

FUEL PUMP MODULE

REMOVAL & INSTALLATION

See Figures 125 through 130.

➡Whenever the negative battery cable is disconnected the following components will require resetting. The Automatic temperature control system, Automatic drive positioner, Power window control, Sunroof system, Sunshade system, Rear view monitor, Idle Air Volume Learning, Steering Angle Sensor Neutral Position, Audio presets and Navigation. You will need the CONSULT-III diagnostic tool, or equivalent. Follow the directions on the screen of the tool, as needed.

1. Before servicing the vehicle, refer to the Precautions Section.

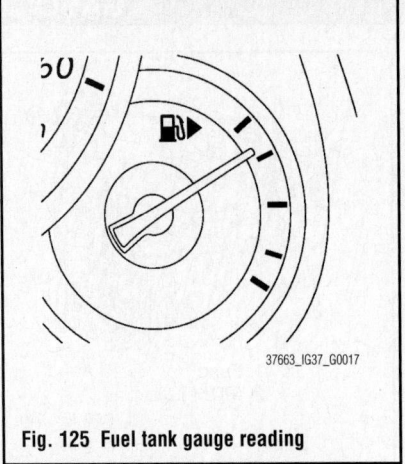

Fig. 125 Fuel tank gauge reading

37663_IG37_G0017

➡If working near and/or around the SRS system and components, be sure to disable the SRS system. After disabling the system wait three minutes or more before servicing the vehicle.

2. Disconnect the negative battery cable.
3. Drain the fuel tank to an acceptable level. If the fuel level indicates more than the level shown in the illustration, full or almost full, drain the fuel from the tank until the fuel gauge indicates a level as shown in the illustration.

➡Because fuel will be spilled when removing the main and sub fuel level sensor units for the top of the fuel is above the main and sub fuel level sensor units installed surface. As a guide, fuel level becomes the position as shown in the illustration when approximately 3 3/8 gallons of fuel are removed from the tank.

4. Properly relieve the fuel system pressure.
5. Remove the fuel tank cap.
6. Remove the rear seat cushion.
7. Peel off the floor carpet. Remove the inspection hole cover.

➡Right side for main fuel level sensor, fuel filter and fuel pump assembly. Left side for sub fuel level sensor unit.

8. Disconnect and fuel feed tube. Disconnect the harness connector.
9. Disconnect the quick connector. Hold the sides of the connector, push in the tabs and pull out the fuel feed tube.

➡The quick connector can be disconnected when the tabs are completely depressed. Never twist it more than necessary. Never use tools to disconnect the quick connector. Cover the fuel line openings to prevent dirt from entering the fuel system.

10. To remove the main fuel level sensor unit, fuel filter and fuel pump assembly, remove the retainer. Raise the unit and disconnect the quick connector.
11. To remove the sub fuel level sensor, remove the retainer. Raise the component and remove it.

To install:

➡Be sure to use new fasteners, as required.

12. Installation is the reverse of the removal procedure.
13. When installing, face the units as shown in the illustration and install them

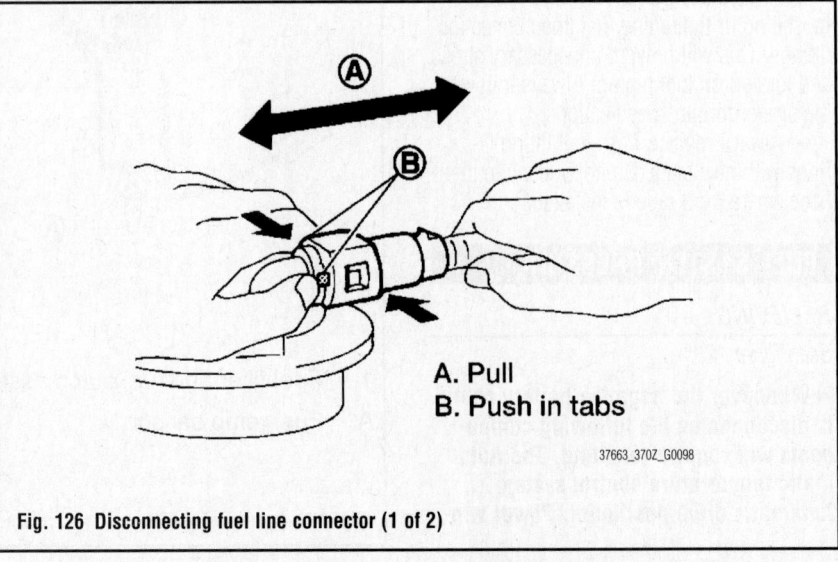

A. Pull
B. Push in tabs

37663_370Z_G0098

Fig. 126 Disconnecting fuel line connector (1 of 2)

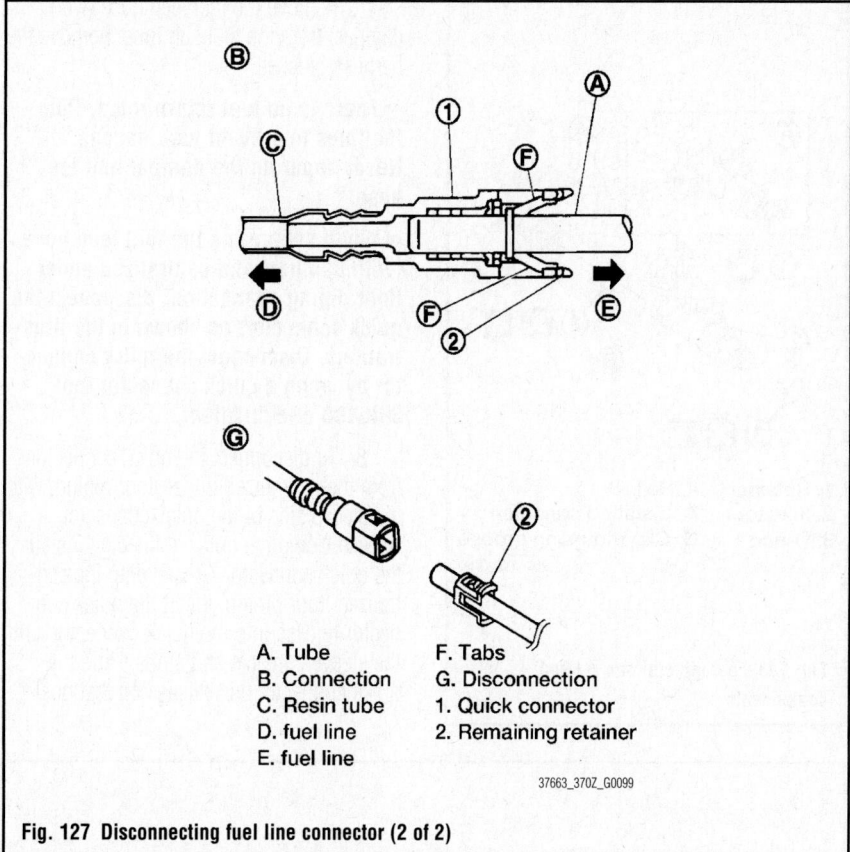

A. Tube   F. Tabs
B. Connection   G. Disconnection
C. Resin tube   1. Quick connector
D. fuel line   2. Remaining retainer
E. fuel line

37663_370Z_G0099

Fig. 127 Disconnecting fuel line connector (2 of 2)

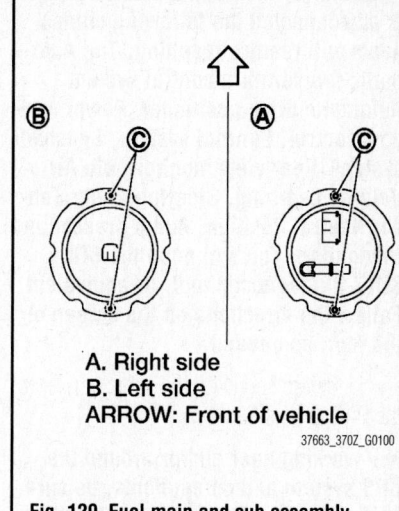

A. Right side
B. Left side
ARROW: Front of vehicle

37663_370Z_G0100

Fig. 129 Fuel main and sub assembly installation alignment

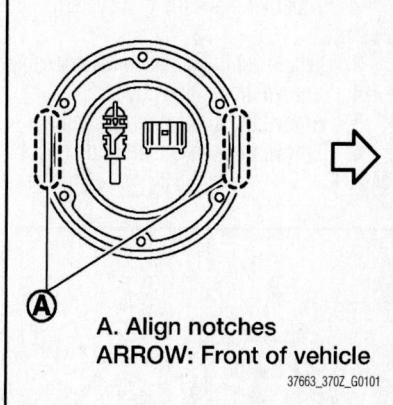

A. Align notches
ARROW: Front of vehicle

37663_370Z_G0101

Fig. 130 Fuel main and sub assembly retainer alignment

with the knock pun on back aligned with the pin hole on the fuel tank.

14. Install the retainer so that its notch becomes parallel with the notch on the fuel tank. Tighten the retainer bolts evenly.

15. Install the fuel pump fuse, if removed.

16. Turn the ignition switch ON (with the engine stopped), check all connections for fuel leakage, correct as required.

17. Start the engine and let it idle, check for fuel leaks. Correct as required.

18. Be sure to perform the reconnect/relearn procedures.

FUEL RAIL AND INJECTOR

REMOVAL & INSTALLATION

See Figures 131 through 136.

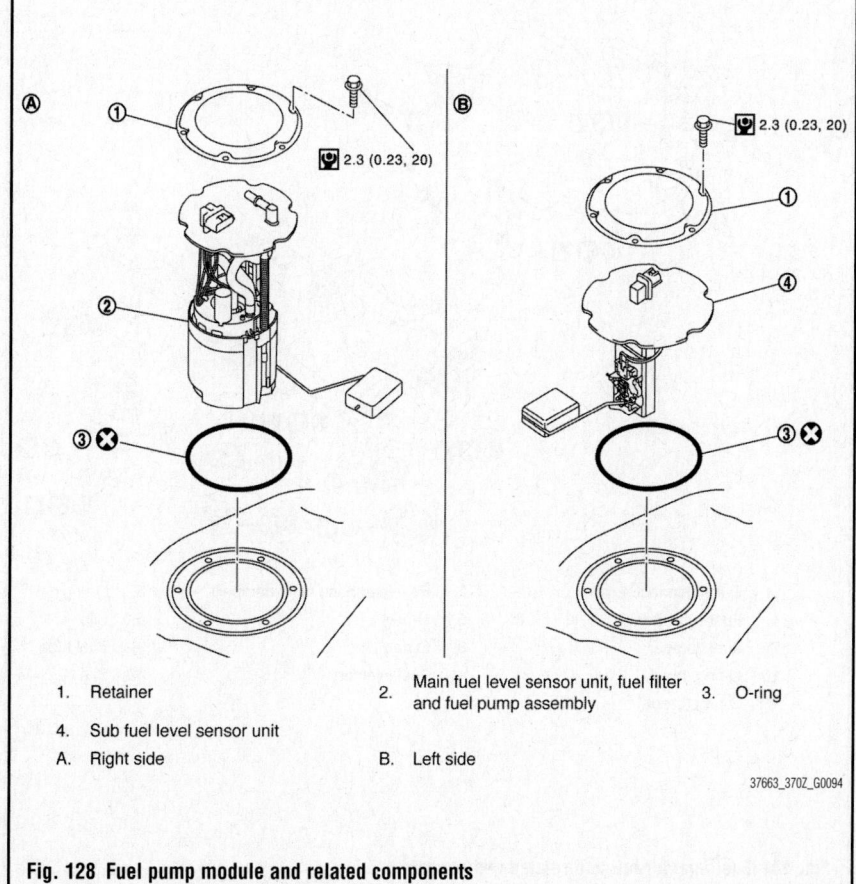

1. Retainer
2. Main fuel level sensor unit, fuel filter and fuel pump assembly
3. O-ring
4. Sub fuel level sensor unit
A. Right side
B. Left side

37663_370Z_G0094

Fig. 128 Fuel pump module and related components

➥Whenever the negative battery cable is disconnected the following components will require resetting. The Automatic temperature control system, Automatic drive positioner, Power window control, Sunroof system, Sunshade system, Rear view monitor, Idle Air Volume Learning, Steering Angle Sensor Neutral Position, Audio presets and Navigation. You will need the CONSULT-III diagnostic tool, or equivalent. Follow the directions on the screen of the tool, as needed.

1. Before servicing the vehicle, refer to the Precautions Section.

➥If working near and/or around the SRS system and components, be sure to disable the SRS system. After disabling the system wait three minutes or more before servicing the vehicle.

2. Properly relieve the fuel system pressure.

3. Disconnect the negative battery cable.

4. Remove the engine cover.

5. Remove the air cleaner assembly.

6. Remove the intake manifold collector.

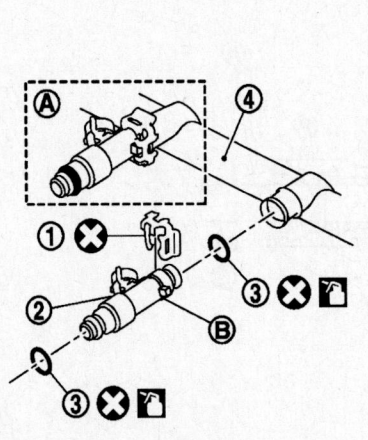

1. Retainer
2. Injector
3. O-ring
4. Rail
A. Installed condition
B. Clip mounting groove

37663_370Z_G0105

Fig. 133 Fuel injector and related components

7. Remove the fuel feed hose (with damper) from the fuel sub tube. Remove the harness bracket.

➥There is no fuel return route. Plug the lines to prevent fuel leakage. Never separate the damper and the hose.

➥When separating the fuel feed hose (with damper) and centralized under floor piping connection, disconnect the quick connector, as shown in the illustrations. Disconnect the quick connector by using a quick connector tool, J-45488 or equivalent.

8. To disconnect the quick connector from the centralized under floor piping, with the sleeve side of the quick connector release facing the quick connector, install the quick connector release onto the centralized floor piping. Insert the quick connector release into the quick connector, until the sleeve contacts and stops. Hold the quick connector and release on that posi-

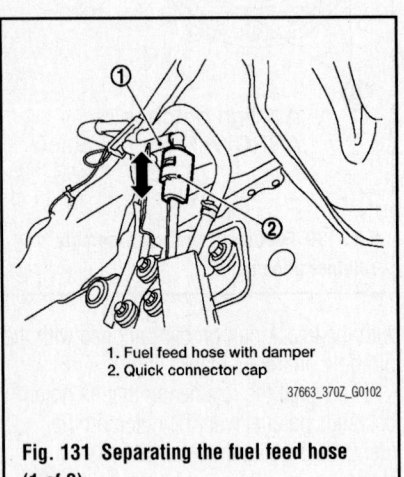

1. Fuel feed hose with damper
2. Quick connector cap

37663_370Z_G0102

Fig. 131 Separating the fuel feed hose (1 of 2)

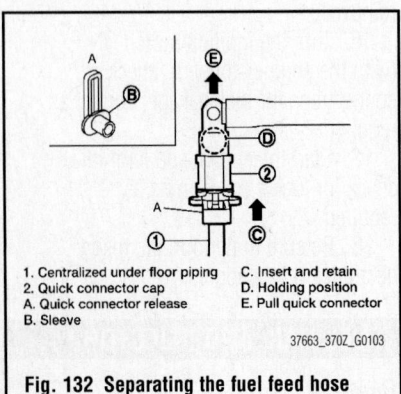

1. Centralized under floor piping
2. Quick connector cap
A. Quick connector release
B. Sleeve
C. Insert and retain
D. Holding position
E. Pull quick connector

37663_370Z_G0103

Fig. 132 Separating the fuel feed hose (2 of 2)

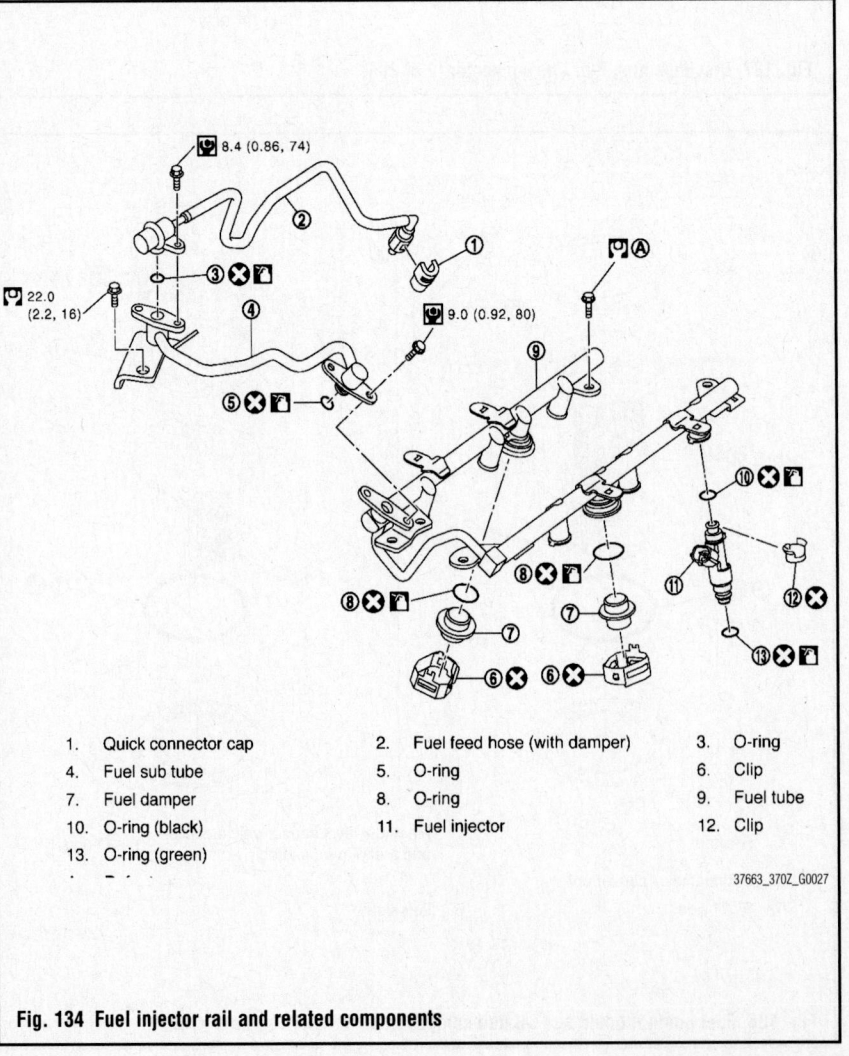

1. Quick connector cap
2. Fuel feed hose (with damper)
3. O-ring
4. Fuel sub tube
5. O-ring
6. Clip
7. Fuel damper
8. O-ring
9. Fuel tube
10. O-ring (black)
11. Fuel injector
12. Clip
13. O-ring (green)

37663_370Z_G0027

Fig. 134 Fuel injector rail and related components

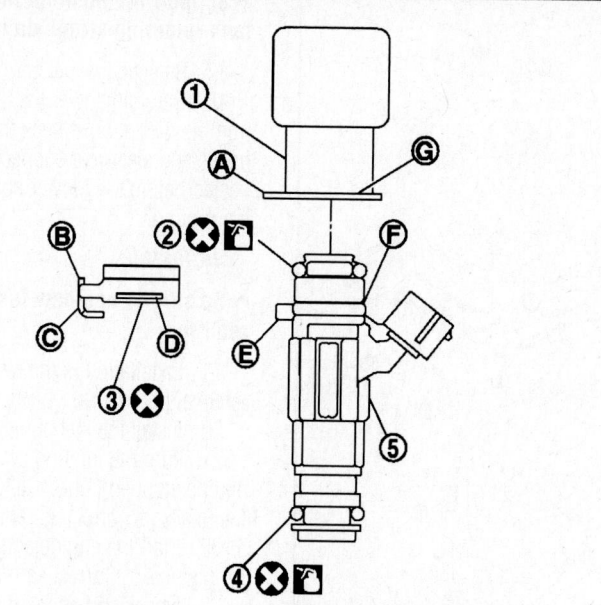

1. Fuel tube
2. Black O-ring
3. Clip
4. Green O-ring
5. Injector

A. Protrusion of fuel tube
B. Cutout of clip
C. Cutout of clip
D. Fixing groove
E. Protrusion of injector
F. Mounting groove
G. Fuel tube flange

37663_370Z_G0106

Fig. 135 Fuel injector installation

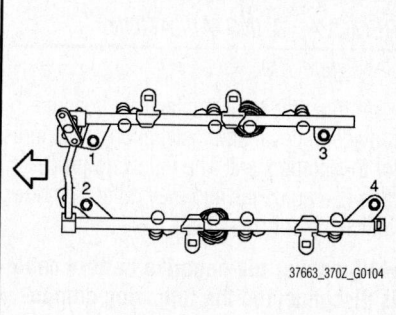

37663_370Z_G0104

Fig. 136 Fuel injector rail bolt locations and tightening sequence

tion. Draw and pull out the quick connector straight from the centralized under floor piping. Never bend or twist the connection between the quick connector and fuel feed hose (with damper) during removal and/or installation. Be sure to have a catch pan available to catch spilled fuel. Cover the fuel line openings to prevent dirt from entering the fuel system.

➡**Inserting the quick connector release hard will not disconnect the quick connector. Hold the quick connector release where it contacts and goes no further.**

➡**Pull the quick connector holding as shown in the removal illustration. Never pull with lateral force. O-ring inside quick connector may be damaged.**

9. Remove the sub tube mounting bolt.

10. Disconnect the harness connector from the fuel injector.

11. Loosen the mounting bolts in the reverse order of the tightening sequence.

12. Remove the assembly from its mounting.

13. Remove the fuel injector from the rail, as required. Discard the O-rings.

14. Do not remove the fuel sub tube and fuel damper.

To install:

➡**Be sure to use new fasteners, as required.**

15. Install new O-rings on the injector, if removed. Lubricate the O-ring with clean engine oil prior to installation.

➡**Fuel tube side O-ring is black, nozzle side is green.**

16. To install the injector, insert a new retaining clip into the mounting groove of the injector. Never reuse the old retaining clip.

17. Do not remove the fuel sub tube and fuel damper.

Insert the injector into the fuel tube, with the clip attached.

➡**Insert it while matching it to the axial center. Insert the injector so that the protrusion of the fuel tube matches the cutout of the clip. Check that the tube flange is securely fixed in the flange fixing groove on the clip. Check that the injector does not rotate. See illustration.**

18. Install the fuel injector rail and injectors. Tighten to specification and in the proper sequence.

19. Tightening specification is 7 ft. lbs. (10.1 Nm), first pass. 17 ft. lbs. (23.6 Nm), second pass.

20. Continue the installation in the reverse order of the removal procedure.

21. Be sure to perform the reconnect/relearn procedures.

FUEL TANK

REMOVAL & INSTALLATION

See Figures 137 and 138.

➡**Whenever the negative battery cable is disconnected the following components will require resetting. The Automatic temperature control system, Automatic drive positioner, Power window control, Sunroof system, Sunshade system, Rear view monitor, Idle Air Volume Learning, Steering Angle Sensor Neutral Position, Audio presets and Navigation. You will need the CONSULT-III diagnostic tool, or equivalent. Follow the directions on the screen of the tool, as needed.**

1. Before servicing the vehicle, refer to the Precautions Section.

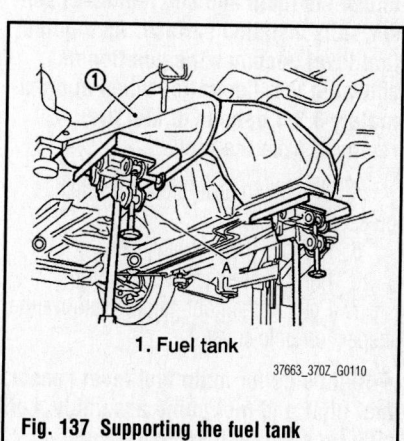

1. Fuel tank

37663_370Z_G0110

Fig. 137 Supporting the fuel tank

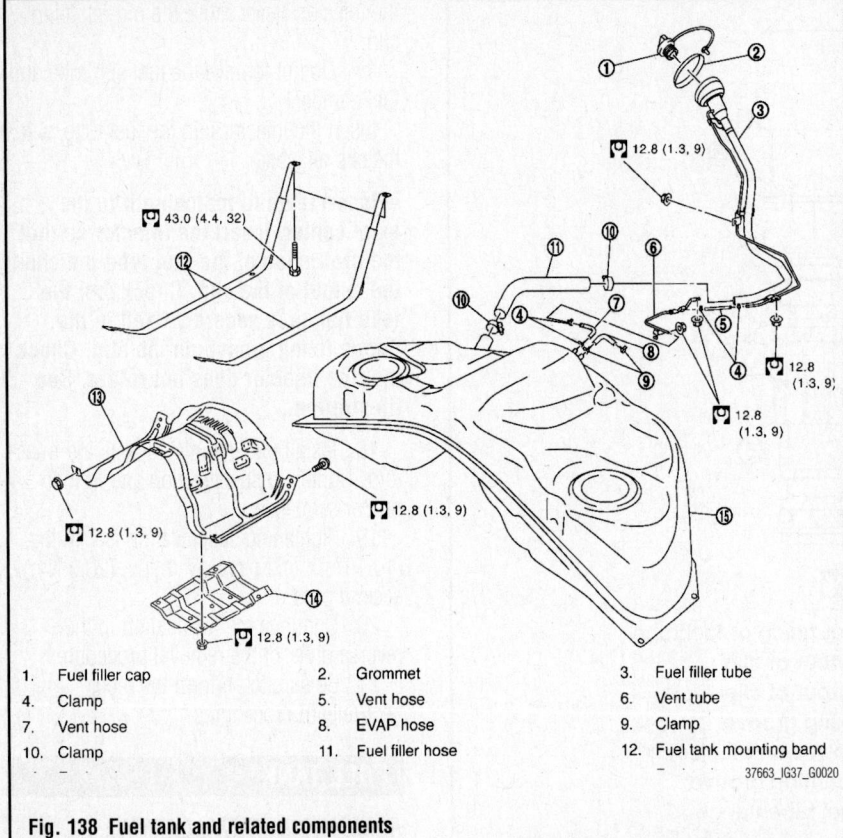

1. Fuel filler cap	2. Grommet	3. Fuel filler tube
4. Clamp	5. Vent hose	6. Vent tube
7. Vent hose	8. EVAP hose	9. Clamp
10. Clamp	11. Fuel filler hose	12. Fuel tank mounting band

37663_IG37_G0020

Fig. 138 Fuel tank and related components

➡If working near and/or around the SRS system and components, be sure to disable the SRS system. After disabling the system wait three minutes or more before servicing the vehicle.

2. Disconnect the negative battery cable.
3. Drain the fuel tank to an acceptable level. If the fuel level indicates more than the level shown in the illustration, full or almost full, drain the fuel from the tank until the fuel gauge indicates a level as shown in the illustration.

➡Because fuel will be spilled when removing the main and sub fuel level sensor units for the top of the fuel is above the main and sub fuel level sensor units installed surface. As a guide, fuel level becomes the position as shown in the illustration when approximately 3 3/8 gallons of fuel are removed from the tank.

4. Properly relieve the fuel system pressure.
5. Remove the fuel tank cap.
6. Remove the rear seat cushion.
7. Peel off the floor carpet. Remove the inspection hole cover.

➡Right side for main fuel level sensor, fuel filter and fuel pump assembly. Left side for sub fuel level sensor unit.

8. Disconnect and fuel feed tube. Disconnect the harness connector.
9. Disconnect the quick connector. Hold the sides of the connector, push in the tabs and pull out the fuel feed tube.

➡The quick connector can be disconnected when the tabs are completely depressed. Never twist it more than necessary. Never use tools to disconnect the quick connector. Cover the fuel line openings to prevent dirt from entering the fuel system.

10. Remove the exhaust front tube, center muffler and main muffler.
11. Remove the driveshaft.
12. Remove the parking brake cables.
13. Remove the rear suspension member assembly.

➡For this service, halfshaft, final drive and rear suspension member are not required to be separated from one another during removal.

14. Disconnect the fuel filler hose, vent hose and EVAP hose at the tube side.
15. Remove the fuel tank protector, if equipped.
16. Properly support the lower part of the fuel tank, with a suitable jack.

➡Support the position that the fuel tank retaining straps do not engage.

17. Remove the fuel tank mounting bands.
18. Carefully lower the tank assembly from its mounting. Check that all required hoses and electrical connectors are disconnected before fully lowering the assembly to the ground.

To install:

➡Be sure to use new fasteners, as required.

19. Installation is the reverse of the removal procedure.
20. Install the fuel pump fuse, if removed.
21. Turn the ignition switch ON (with the engine stopped), check all connections for fuel leakage, correct as required.
22. Start the engine and let it idle, check for fuel leaks. Correct as required.
23. Check and adjust the rear wheel alignment, as required.
24. Be sure to perform the reconnect/relearn procedures.

IDLE SPEED

ADJUSTMENT

Idle speed is not adjustable.

THROTTLE BODY

REMOVAL & INSTALLATION
See Figure 139.

At this time the manufacturer does not provide removal and installation procedures for this component. The following procedure is a guideline and may differ from the vehicle you are servicing.

➡Whenever the negative battery cable is disconnected the following components will require resetting. The Automatic temperature control system, Automatic drive positioner, Power win-

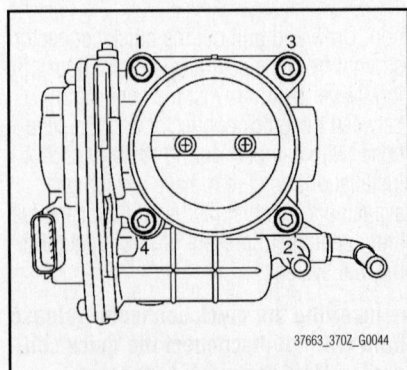

37663_370Z_G0044

Fig. 139 Throttle control actuator bolt locations and tightening sequence

dow control, Sunroof system, Sunshade system, Rear view monitor, Idle Air Volume Learning, Steering Angle Sensor Neutral Position, Audio presets and Navigation. You will need the CONSULT-III diagnostic tool, or equivalent. Follow the directions on the screen of the tool, as needed.

1. Before servicing the vehicle, refer to the Precautions Section.

➡️If working near and/or around the SRS system and components, be sure to disable the SRS system. After disabling the system wait three minutes or more before servicing the vehicle.

2. Disconnect the negative battery cable.
3. Remove the engine undercover.

4. Drain the engine coolant. Be sure to properly dispose of used coolant.
5. Remove the engine cover.
6. Remove the air cleaner assembly.
7. To remove the electric throttle control actuator, Disconnect the water hoses from the component. Disconnect the harness connector.
8. Loosen the mounting bolts in the reverse order of the tightening sequence.
9. Remove the component from its mounting.

➡️When removing only the intake manifold collector, move the electric throttle control actuator without disconnecting the water hose. Handel the component carefully to avoid shock damage.

To install:

➡️Be sure to use new fasteners, as required.

10. Installation is the reverse of the removal procedure.
11. Tighten the intake manifold collector retaining bolts in the proper sequence.
12. Tighten the electric throttle actuator retaining bolts in the proper sequence.
13. Using the CONSULT-III diagnostic tool or equivalent, perform the throttle valve closed position learning and the idle air volume learning procedures.
14. Be sure to perform the reconnect/relearn procedures.

HEATING & AIR CONDITIONING SYSTEM

BLOWER MOTOR

REMOVAL & INSTALLATION
See Figure 140.

➡️Whenever the negative battery cable is disconnected the following components will require resetting. The Automatic temperature control system, Automatic drive positioner, Power window control, Sunroof system, Sunshade system, Rear view monitor, Idle Air Volume Learning, Steering Angle Sensor Neutral Position, Audio presets and Navigation. You will need the CONSULT-III diagnostic tool, or equivalent. Follow the directions on the screen of the tool, as needed.

1. Before servicing the vehicle, refer to the Precautions Section.

➡️If working near and/or around the SRS system and components, be sure to disable the SRS system. After disabling the system wait three minutes or more before servicing the vehicle.

2. Disconnect the negative battery cable.
3. As necessary remove the instrument lower cover.
4. Disconnect the blower motor electrical connector.
5. Remove the blower motor retaining screws.
6. Remove the component from its mounting.

To install:

➡️Be sure to use new fasteners, as required.

7. Installation is the reverse of the removal procedure.
8. Be sure to perform the reconnect/relearn procedures.

HEATER CORE

REMOVAL & INSTALLATION
See Figure 141.

➡️Whenever the negative battery cable is disconnected the following components will require resetting. The Automatic temperature control system, Automatic drive positioner, Power window control, Sunroof system, Sunshade system, Rear view monitor, Idle Air Volume Learning, Steering Angle Sensor Neutral Position, Audio presets and Navigation. You will need the CONSULT-III diagnostic tool, or equivalent. Follow the directions on the screen of the tool, as needed.

1. Before servicing the vehicle, refer to the Precautions Section.

➡️If working near and/or around the SRS system and components, be sure to disable the SRS system. After disabling the system wait three minutes or more before servicing the vehicle.

2. Disconnect the negative battery cable.
3. Remove the heating and cooling unit assembly.
4. Remove the mounting screws and remove the heater pipe cover.
5. Remove the mounting screws and remove the left foot duct.
6. Slide the heater core leftward and remove it.

To install:

➡️Be sure to use new fasteners, as required.

7. Installation is the reverse of the removal procedure.

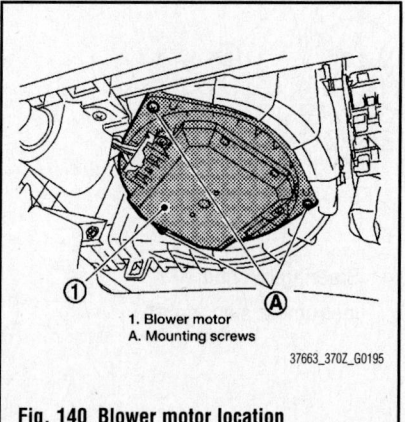

1. Blower motor
A. Mounting screws

37663_370Z_G0195

Fig. 140 Blower motor location

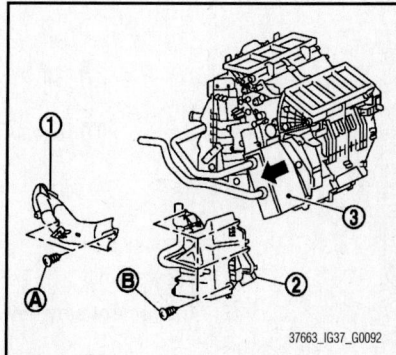

37663_IG37_G0092

Fig. 141 HVAC heater core and related components

8. Be sure to use new O-rings, coated with clean refrigerant oil prior to installation.

9. Properly charge the air conditioning system.

10. Properly refill the cooling system.

11. Start the engine and check the system for proper operation and refrigerant leakage.

12. Using the CONSULT-III diagnostic tool, or equivalent, perform the 4WAS front actuator adjustment if equipped with 4WAS.

13. Be sure to perform the reconnect/relearn procedures.

HEATER AND COOLING UNIT

REMOVAL & INSTALLATION

See Figures 142 and 143.

➡**Whenever the negative battery cable is disconnected the following components will require resetting. The Automatic temperature control system, Automatic drive positioner, Power window control, Sunroof system, Sunshade system, Rear view monitor, Idle Air Volume Learning, Steering Angle Sensor Neutral Position, Audio presets and**

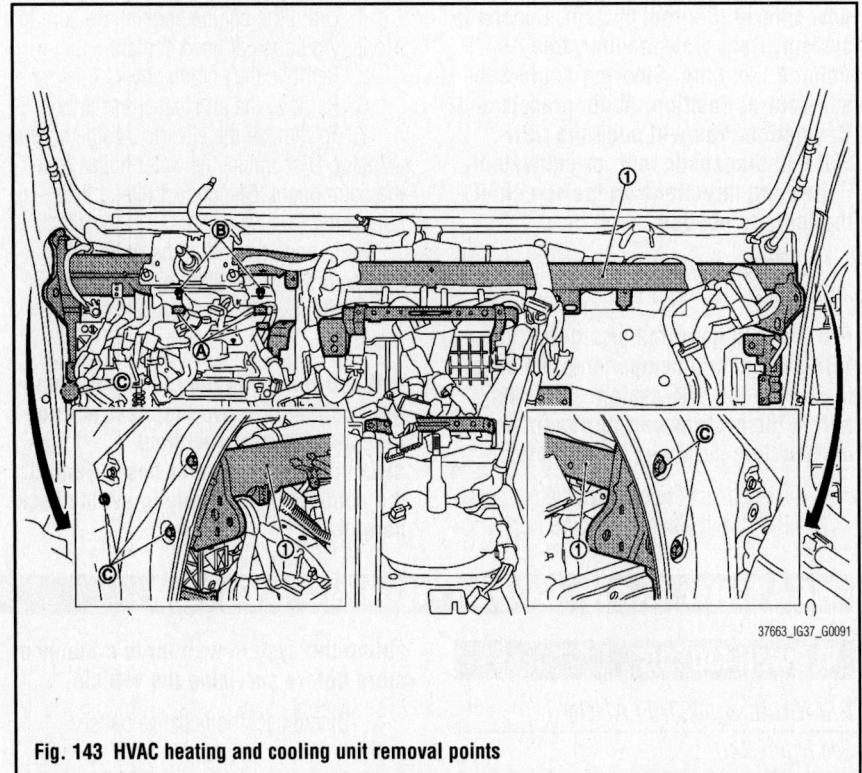

Fig. 143 HVAC heating and cooling unit removal points

37663_IG37_G0091

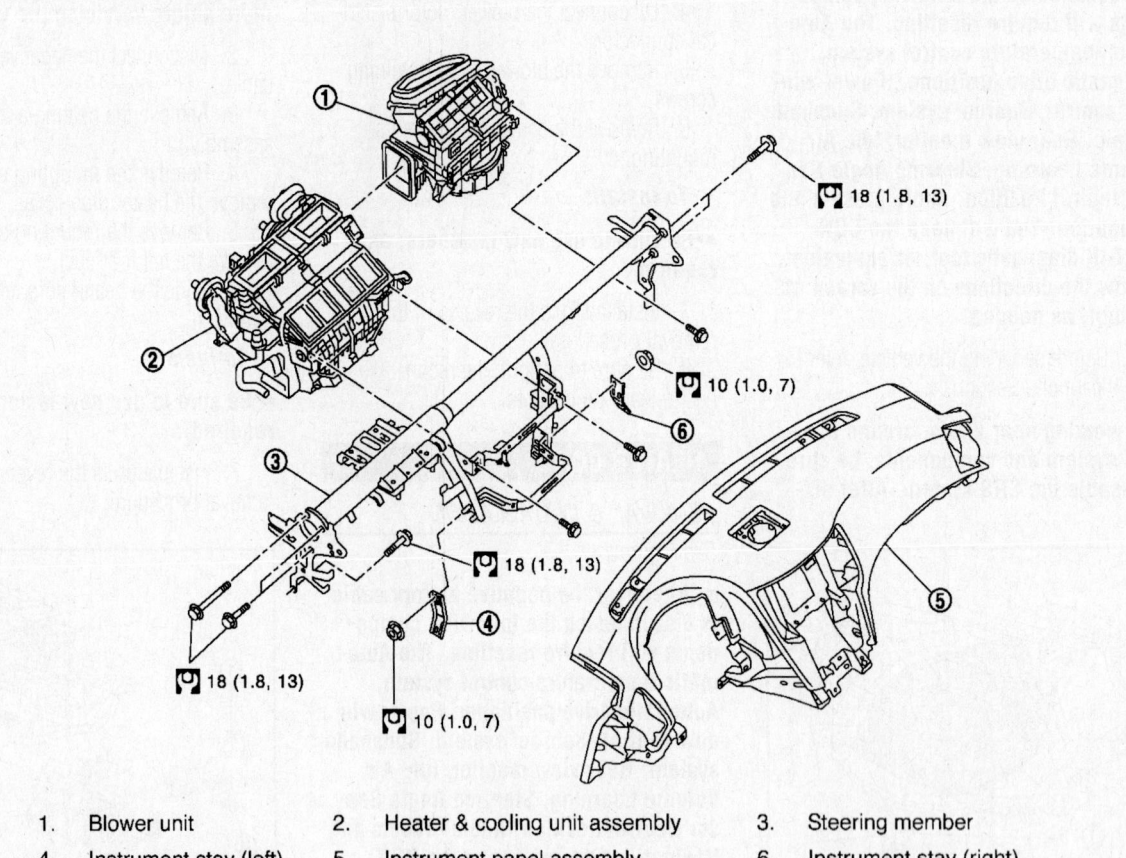

1.	Blower unit	2.	Heater & cooling unit assembly	3.	Steering member
4.	Instrument stay (left)	5.	Instrument panel assembly	6.	Instrument stay (right)

37663_IG37_G0078

Fig. 142 HVAC blower unit and related components

Navigation. You will need the CON-SULT-III diagnostic tool, or equivalent. Follow the directions on the screen of the tool, as needed.

1. Before servicing the vehicle, refer to the Precautions Section.

➡If working near and/or around the SRS system and components, be sure to disable the SRS system. After disabling the system wait three minutes or more before servicing the vehicle.

2. Disconnect the negative battery cable.

3. Properly discharge the air conditioning system

4. Drain the cooling system. Be sure to properly dispose of used coolant.

5. Remove the cowl top cover.

6. Using tool SST: J-45815 disconnect

and plug the one touch connectors at the housing.

7. Remove the clamps and disconnect the heater hoses. Plug the hoses.

8. Remove the instrument panel assembly.

9. Remove the blower unit.

10. Remove the clips of the vehicle harness from the steering member.

11. Remove the instrument left and right stay.

12. Remove the drain hose.

13. Remove the mounting bolts from the heater and cooling unit assembly.

14. Remove the front defroster nozzle, side defroster nozzle and ventilator duct.

15. Remove the steering column mounting bolts and nuts. Remove the steering member mounting bolts. Remove the steering member.

16. Remove the heater and cooling unit assembly.

To install:

➡Be sure to use new fasteners, as required.

17. Installation is the reverse of the removal procedure.

18. Be sure to use new O-rings, coated with clean refrigerant oil prior to installation.

19. Properly charge the air conditioning system.

20. Properly refill the cooling system.

21. Start the engine and check the system for proper operation and refrigerant leakage.

22. Using the CONSULT-III diagnostic tool, or equivalent, perform the 4WAS front actuator adjustment if equipped with 4WAS.

23. Be sure to perform the reconnect/relearn procedures.

STEERING

POWER RACK & PINION STEERING GEAR

REMOVAL & INSTALLATION

See Figures 144 through 150.

➡Whenever the negative battery cable is disconnected the following components will require resetting. The Automatic temperature control system, Automatic drive positioner, Power window control, Sunroof system, Sunshade system, Rear view monitor, Idle Air Volume Learning, Steering Angle Sensor Neutral Position, Audio presets and Navigation. You will need the CONSULT-III diagnostic tool, or equivalent. Follow the directions on the screen of the tool, as needed.

1. Before servicing the vehicle, refer to the Precautions Section.

➡If working near and/or around the SRS system and components, be sure to disable the SRS system. After disabling the system wait three minutes or more before servicing the vehicle.

2. Disconnect the negative battery cable.

3. Position the front tires in the straight ahead position.

4. Perform the 4WAS front actuator neutral position adjustment, using the CONSULT-III diagnostic tool, or equivalent, if equipped with 4WAS.

5. Raise and support the vehicle safely.

6. Remove the tire and wheel assemblies.

7. Remove the front suspension member stay, except AWD.

8. Remove the front cross bar, AWD.

9. Remove the cotter pin and loosen the locknut.

10. Remove the outer steering socket from the steering knuckle, so as not to damage the ball joint boot, using a suitable removal tool.

➡Temporarily tighten the nut to prevent damage to the threads and prevent the ball joint remover from sudden drop.

11. Disconnect and plug the power steering fluid lines. Drain the power steering fluid. Be sure to correctly dispose of used fluid.

12. Remove the power steering solenoid valve harness connector.

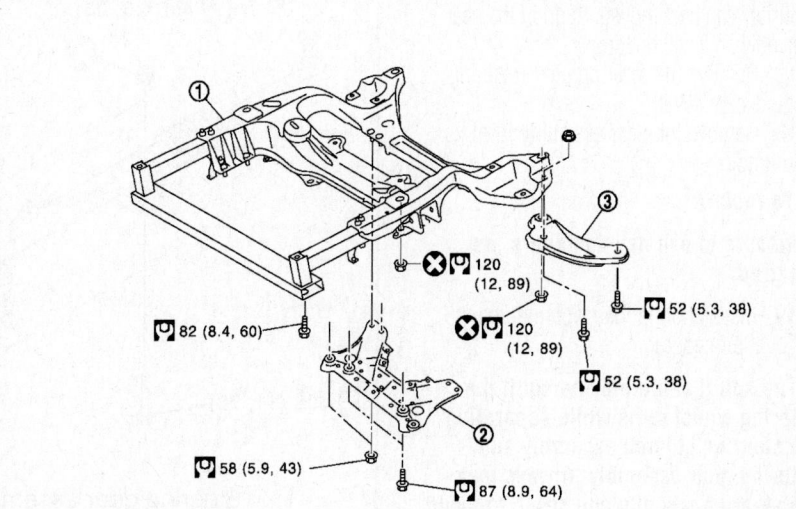

1. Front suspension member 2. Suspension member stay 3. Suspension member sub stay

37663_IG37_G0116

Fig. 144 Front suspension member and related components—convertible

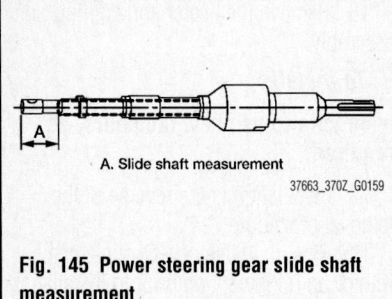

A. Slide shaft measurement

37663_370Z_G0159

Fig. 145 Power steering gear slide shaft measurement

13. Remove the rack stay.

14. Remove the lower joint fixing bolts (steering gear side).

15. Separate the lower shaft from the steering gear assembly by sliding the slide shaft. See illustration.

➡**The spiral cable may be cut if the steering wheel turns while separating the steering column assembly and steering gear assembly. Always lock the steering wheel using string to avoid turning.**

16. On AWD, position a suitable jack to the transmission assembly. Remove the mounting nuts and bolts on the lower side of the shock absorber arm and remove the shock from the transverse link. Position a suitable jack to the front suspension member. Remove the mounting bolts and nuts of the steering gear assembly. Remove the mounting nuts of the engine mounting insulator. Remove the mounting nuts of the front suspension member. Position a suitable jack and slowly lower it to the position where the steering gear assembly can be removed. Support the gear assembly so it will not drop.

17. Remove the steering gear retaining bolts, except AWD.

18. Remove the gear assembly from its mounting.

To install:

➡**Be sure to use new fasteners, as required.**

19. Installation is the reverse of the removal procedure.

➡**The spiral cable may be cut if the steering wheel turns while separating the steering column assembly and steering gear assembly. Always lock the steering wheel using string to avoid turning.**

20. Tighten the mounting bolts in the order shown in the illustration.

21. Tighten step one, temporary and in step two final.

22. When installing the suction hoses, refer to the illustration.

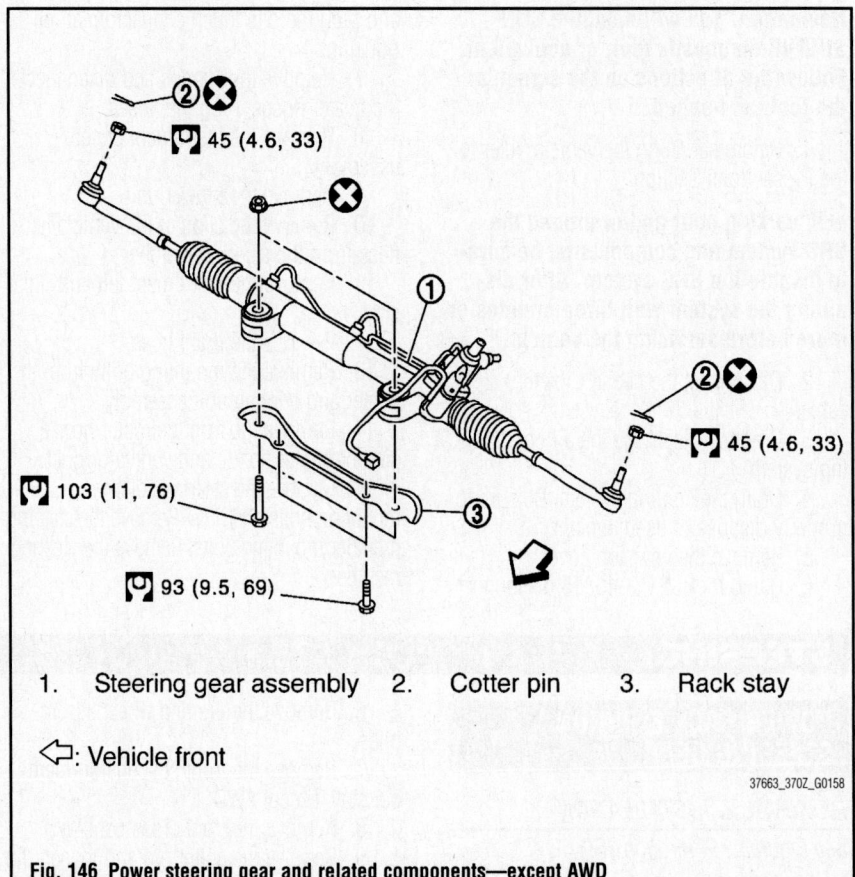

1. Steering gear assembly 2. Cotter pin 3. Rack stay

◁: Vehicle front

37663_370Z_G0158

Fig. 146 Power steering gear and related components—except AWD

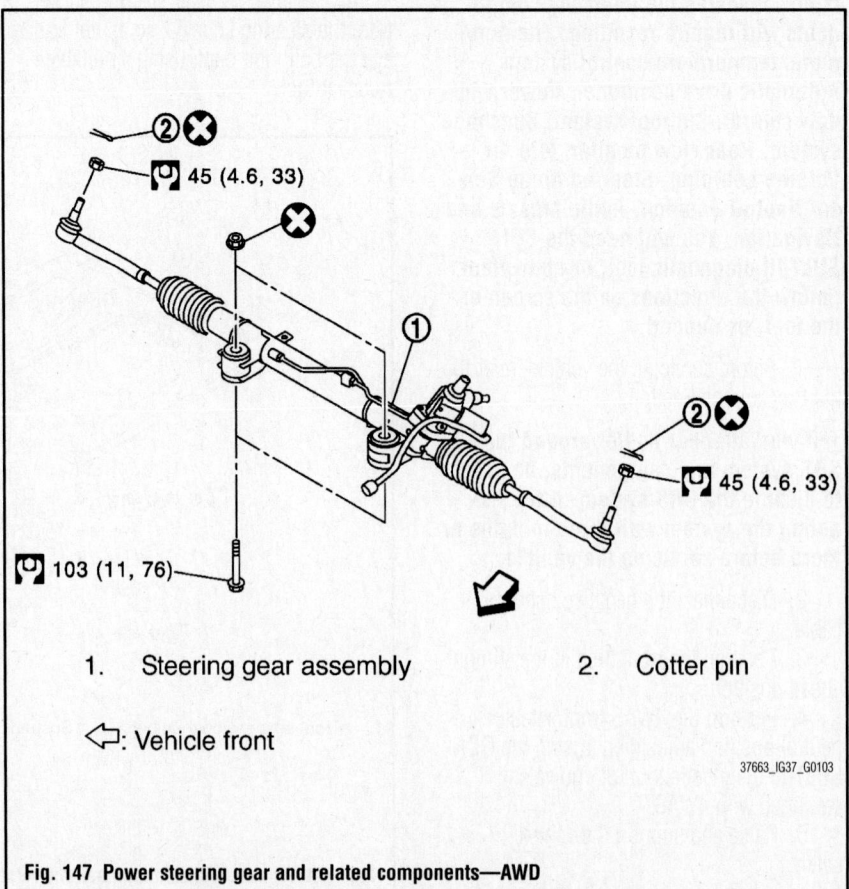

1. Steering gear assembly   2. Cotter pin

◁: Vehicle front

37663_IG37_G0103

Fig. 147 Power steering gear and related components—AWD

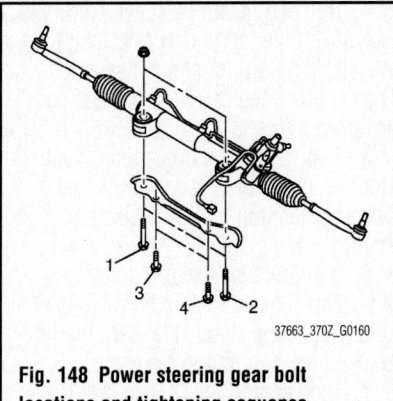

Fig. 148 Power steering gear bolt locations and tightening sequence

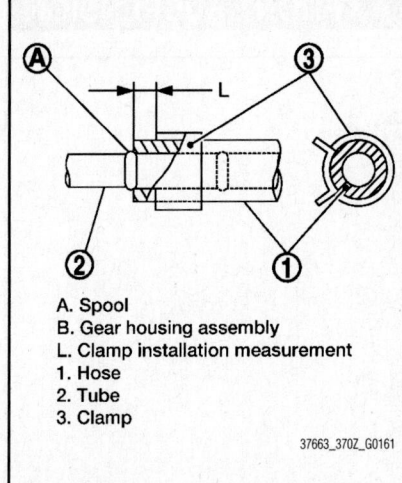

A. Spool
B. Gear housing assembly
L. Clamp installation measurement
1. Hose
2. Tube
3. Clamp

Fig. 149 Power steering suction hose installation

➡**Never apply fluid to the hose and the tube. Insert the hose securely until it contacts the spool of the tube. Install the clamp at the hose (0.12–0.31 inch from the edge of the hose.**

23. To install the lower joint to the steering gear, see the illustration. Position the rack of the steering gear in the neutral position.

➡**To get to the neutral position turn the sub gear assembly and measure the distance of the inner socket, then measure the intermediate position and distance. Align the rear cover cap projection with the marking position of the gear housing assembly. Install the slip part of the lower joint aligning with the rear cover cap projection. Make sure that the slit part of the lower joint is aligned with the rear cover cap protection and the marking position of the gear housing assembly. Make sure that there is no clearance between the lower joint, gear housing assembly and mounting bolt.**

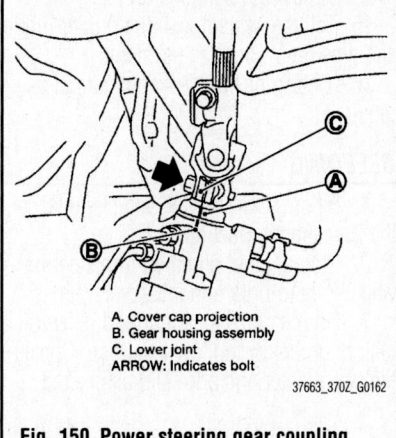

A. Cover cap projection
B. Gear housing assembly
C. Lower joint
ARROW: Indicates bolt

Fig. 150 Power steering gear coupling installation

24. Fill the system. Bleed the system.
25. Perform a final tightening of nuts and bolts of each removed component with the vehicle in an unladen position.
26. Check and adjust the wheel alignment, as required.
27. Adjust the neutral position of the steering angle sensor using the CONSULT-III diagnostic tool, or equivalent, after checking the wheel alignment.
28. Perform the 4WAS front actuator neutral position adjustment, using the CONSULT-III diagnostic tool, or equivalent, if equipped with 4WAS.
29. Be sure to perform the reconnect/relearn procedures.

POWER STEERING PUMP

REMOVAL & INSTALLATION
See Figures 151 and 152.

➡**Whenever the negative battery cable is disconnected the following components will require resetting. The Automatic temperature control system, Automatic drive positioner, Power window control, Sunroof system, Sunshade system, Rear view monitor, Idle Air Volume Learning, Steering Angle Sensor Neutral Position, Audio presets and Navigation. You will need the CONSULT-III diagnostic tool, or equivalent. Follow the directions on the screen of the tool, as needed.**

1. Before servicing the vehicle, refer to the Precautions Section.

➡**If working near and/or around the SRS system and components, be sure to disable the SRS system. After disabling the system wait three minutes or more before servicing the vehicle.**

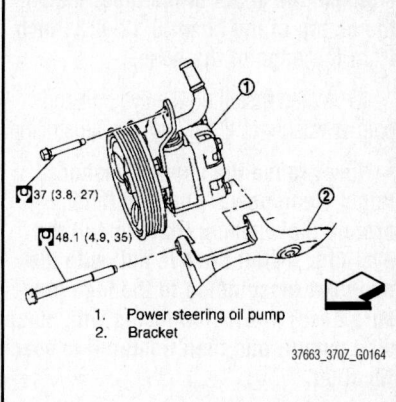

1. Power steering oil pump
2. Bracket

Fig. 151 Power steering pump and related components

2. Disconnect the negative battery cable.
3. Drain the power steering fluid from the reservoir. Be sure to properly dispose of used fluid.
4. Remove the air cleaner assembly.
5. Loosen the drive belt.
6. Remove the belt from the steering pump.
7. Remove the copper washers and eye bolt. Drain fluid from their pipings.
8. Remove the suction hose. Drain fluid from their pipings.
9. Remove the pump retaining bolts.
10. Remove the pump from the vehicle.

To install:

➡**Be sure to use new fasteners, as required.**

11. Installation is the reverse of the removal procedure.
12. When installing the suction hoses see note below.

➡**Never apply fluid to the hose and the tube. Insert the hose securely until it**

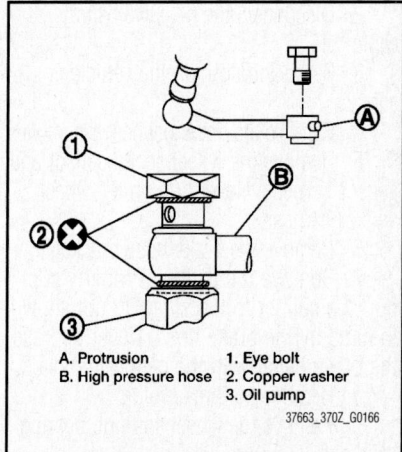

A. Protrusion
B. High pressure hose
1. Eye bolt
2. Copper washer
3. Oil pump

Fig. 152 Power steering pump eye bolt and cooper washer installation

contacts the spool of the tube. Install the clamp at the hose (0.12–0.31 inch from the edge of the hose.

13. When installing the eye bolt and copper washer to the pump, see illustration.

➡Never reuse the cooper washer. Apply clean power steering fluid around the washers, then install the eye bolt. Install the eye bolt with the eye joint (assembled to the high pressure hose) protrusion facing with pump side cutout, and then tighten it to specification.

14. Adjust the belt tension.

15. Fill the system with the proper grade and type fluid. Bleed the system.

16. Check for fluid leakage, correct as required.

BLEEDING

1. Before servicing the vehicle, refer to the Precautions Section.

2. Stop engine, and then turn steering wheel fully to right and left several times.

3. Do not allow steering fluid reservoir tank to go below the low-level line. Check tank frequenter and add fluid as needed.

4. Run engine at idle speed. Turn steering wheel fully to the right and then fully to the left, and keep for about 3 seconds. Then check whether a fluid leakage has occurred.

5. Repeat the 2nd procedure several times at about three seconds interval Check generation of air bubbles and cloud in fluid.

6. If air bubbles and the cloud don't fade, stop engine, hold air bleeding until air bubbles and the cloud fade. Perform the 2nd and the 3rd procedures again.

7. Stop engine, check fluid level.

SUSPENSION   FRONT SUSPENSION

STEERING KNUCKLE

REMOVAL & INSTALLATION

See Figures 153 and 154.

➡Whenever the negative battery cable is disconnected the following components will require resetting. The Automatic temperature control system, Automatic drive positioner, Power window control, Sunroof system, Sunshade system, Rear view monitor, Idle Air Volume Learning, Steering Angle Sensor Neutral Position, Audio presets and Navigation. You will need the CONSULT-III diagnostic tool, or equivalent. Follow the directions on the screen of the tool, as needed.

1. Before servicing the vehicle, refer to the Precautions Section.

➡If working near and/or around the SRS system and components, be sure to disable the SRS system. After disabling the system wait three minutes or more before servicing the vehicle.

2. Disconnect the negative battery cable.

3. Raise and support the vehicle safely.

4. Remove the tire and wheel assemblies.

5. Remove the wheel speed sensor and sensor harness. Never pull on the wheel sensor harness.

6. Remove the brake hose bracket.

7. Remove the caliper. Properly support the caliper to the side. Do not allow it to hang by the brake hose. Never depress the brake pedal with the caliper removed.

8. Remove the brake rotor.

9. Remove the wheel and hub bearing assembly.

10. As required, remove the splash guard shield.

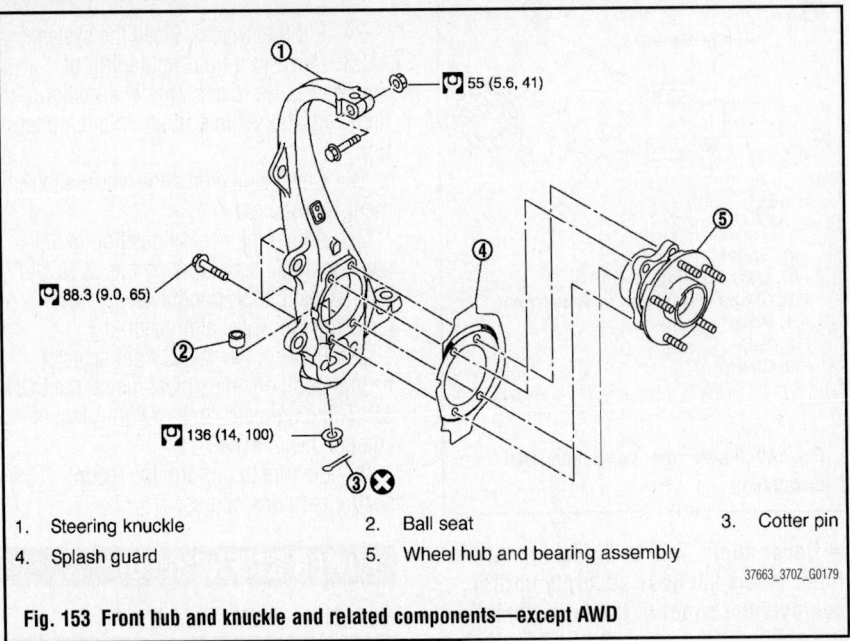

| 1. | Steering knuckle | 2. | Ball seat | 3. | Cotter pin |
| 4. | Splash guard | 5. | Wheel hub and bearing assembly | | |

37663_370Z_G0179

Fig. 153 Front hub and knuckle and related components—except AWD

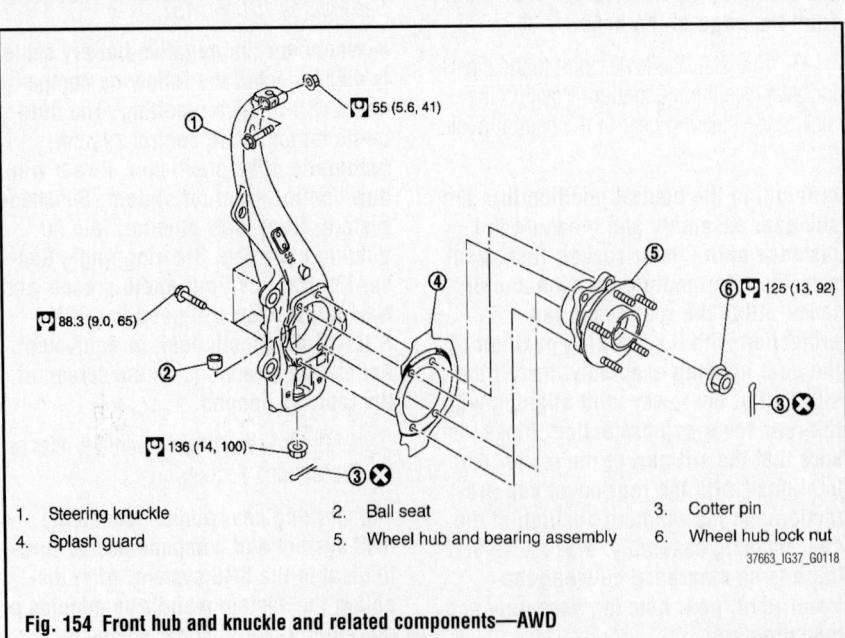

| 1. | Steering knuckle | 2. | Ball seat | 3. | Cotter pin |
| 4. | Splash guard | 5. | Wheel hub and bearing assembly | 6. | Wheel hub lock nut |

37663_IG37_G0118

Fig. 154 Front hub and knuckle and related components—AWD

11. Remove the steering outer socket.

12. Remove the cotter pin of the transverse link and steering knuckle, and then loosen nut.

13. Separate the upper link from the steering knuckle.

14. Separate the transverse link from the steering knuckle, using a ball joint removal tool. Remove the steering knuckle.

➡**Temporarily tighten the nut to prevent damage to the threads and to prevent the removal tool from suddenly coming off.**

To install:

➡**Be sure to use new fasteners, as required.**

15. Installation is the reverse of the removal procedure.

16. Be sure to use new cotter pins.

17. Perform a final tightening of nuts and bolts of each removed component with the vehicle in an unladen position.

18. Check wheel speed sensor for proper operation.

19. Adjust the steering angle sensor neutral position, using the CONSULT-III diagnostic tool, or equivalent.

20. Check and adjust the front alignment, as required.

21. Be sure to perform the reconnect/relearn procedures.

STRUT

REMOVAL & INSTALLATION

See Figures 155 through 157.

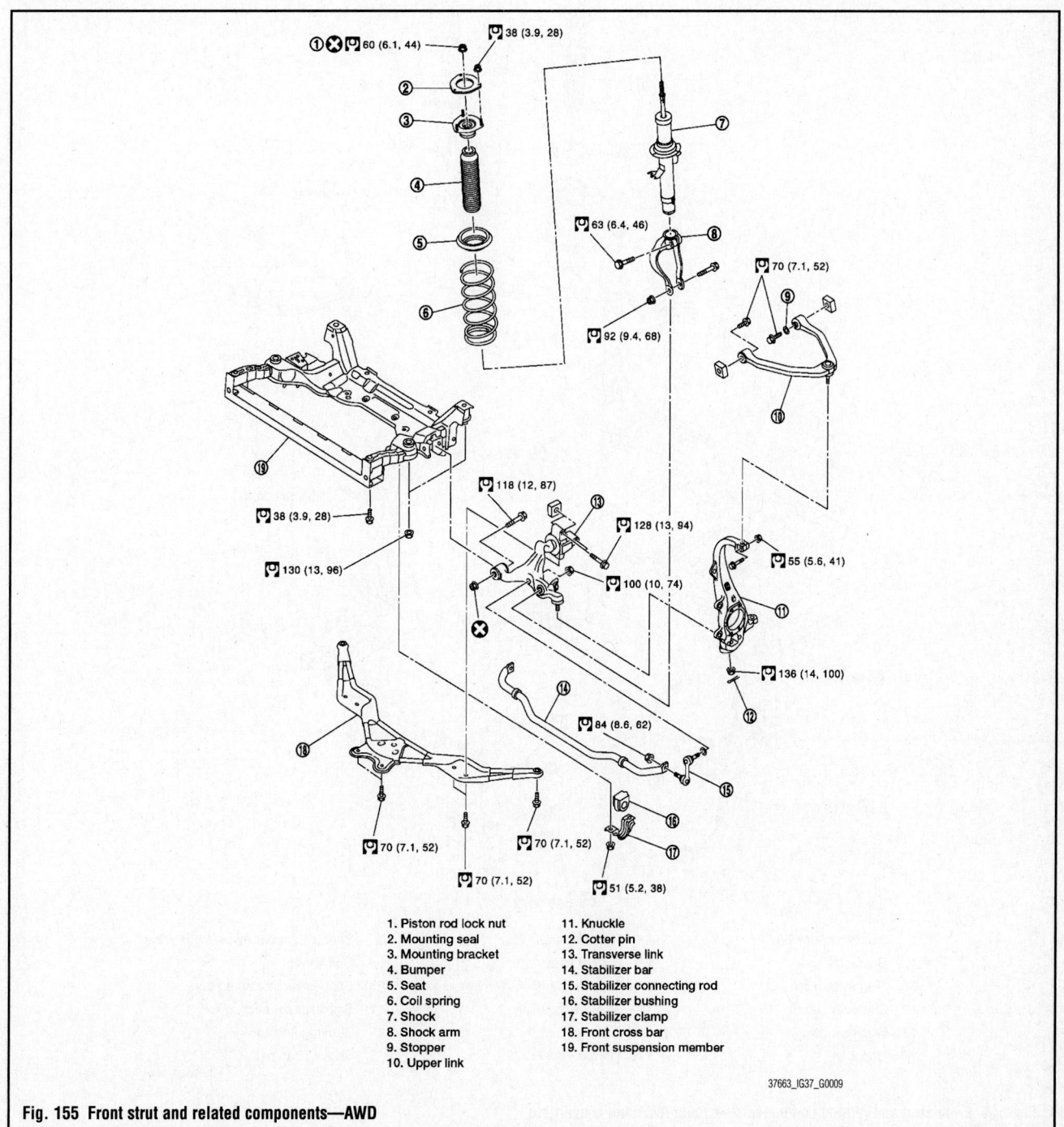

1. Piston rod lock nut
2. Mounting seal
3. Mounting bracket
4. Bumper
5. Seat
6. Coil spring
7. Shock
8. Shock arm
9. Stopper
10. Upper link
11. Knuckle
12. Cotter pin
13. Transverse link
14. Stabilizer bar
15. Stabilizer connecting rod
16. Stabilizer bushing
17. Stabilizer clamp
18. Front cross bar
19. Front suspension member

Fig. 155 Front strut and related components—AWD

➡Whenever the negative battery cable is disconnected the following components will require resetting. The Automatic temperature control system, Automatic drive positioner, Power window control, Sunroof system, Sunshade system, Rear view monitor, Idle Air Volume Learning, Steering Angle Sensor Neutral Position, Audio presets and Navigation. You will need the CONSULT-III diagnostic tool, or equivalent. Follow the directions on the screen of the tool, as needed.

1. Before servicing the vehicle, refer to the Precautions Section.

➡If working near and/or around the SRS system and components, be sure to disable the SRS system. After disabling the system wait three minutes or more before servicing the vehicle.

2. Disconnect the negative battery cable.

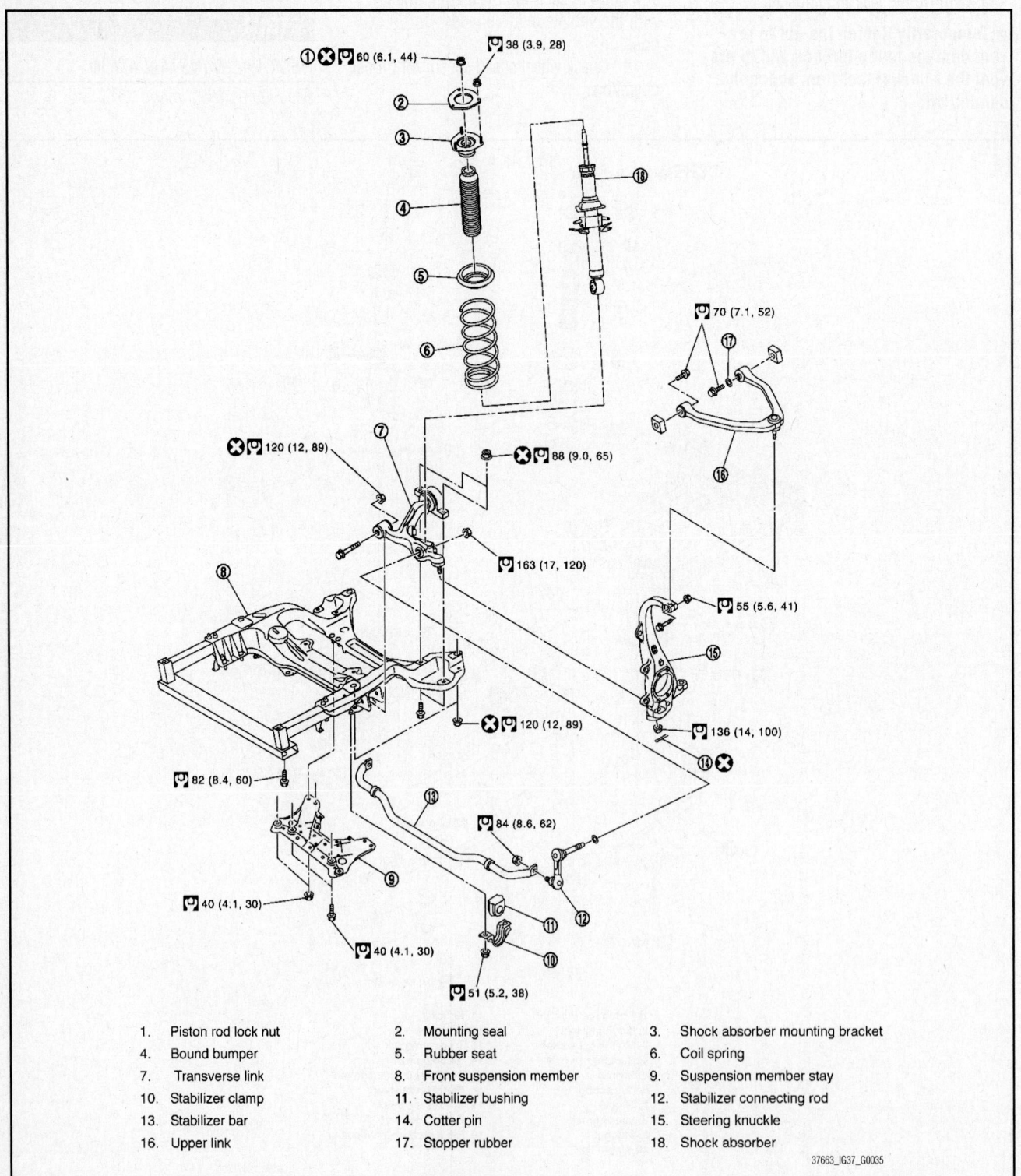

1. Piston rod lock nut
2. Mounting seal
3. Shock absorber mounting bracket
4. Bound bumper
5. Rubber seat
6. Coil spring
7. Transverse link
8. Front suspension member
9. Suspension member stay
10. Stabilizer clamp
11. Stabilizer bushing
12. Stabilizer connecting rod
13. Stabilizer bar
14. Cotter pin
15. Steering knuckle
16. Upper link
17. Stopper rubber
18. Shock absorber

37663_IG37_G0035

Fig. 156 Front strut and related components—except AWD and convertible

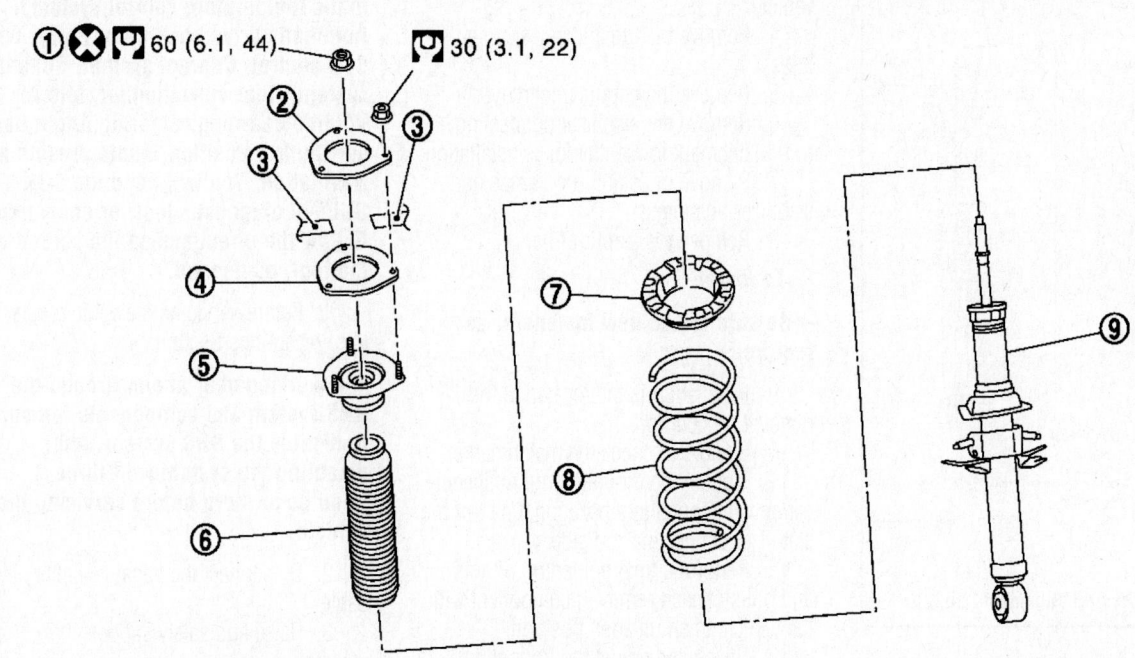

1. Piston rod lock nut	2. Gusset	3. Vehicle body
4. Mounting seal	5. Shock absorber mounting bracket	6. Bound bumper
7. Rubber seat	8. Coil spring	9. Shock absorber

37663_IG37_G0117

Fig. 157 Front strut and related components—convertible

3. Raise and safely support the vehicle.

4. Remove the tire and wheel assemblies, as necessary.

5. Remove the wheel sensor and harness connector from the shock absorber, except AWD.

6. Remove the brake hose bracket, except AWD.

7. Remove the stabilizer connecting rod.

8. Remove the halfshaft, AWD.

9. Separate the upper link from the steering knuckle. Remove the shock absorber assembly mounting nuts.

10. Remove the component from the vehicle.

To install:

➡**Be sure to use new fasteners, as required.**

11. Installation is the reverse of the removal procedure.

➡**Never tap on the ball joint cap of the stabilizer connecting rod with a hammer**
when inserting the stabilizer connecting rod into the transverse link.

12. Perform a final tightening of nuts and bolts of each removed component with the vehicle in an unladen position.

13. Adjust the steering angle sensor neutral position, using the CONSULT-III diagnostic tool, or equivalent.

14. Check and adjust the front alignment, as required.

OVERHAUL

See Figures 158 and 159.

Coil spring removal and installation and any other strut disassembly should go here

At this time the manufacturer does not provide service procedures for this component, refer to the illustration as required.

STABILIZER BAR

REMOVAL & INSTALLATION

See Figures 156, 157 and 160.

➡**Whenever the negative battery cable is disconnected the following compo-**
nents will require resetting. The Automatic temperature control system, Automatic drive positioner, Power window control, Sunroof system, Sunshade system, Rear view monitor, Idle Air Volume Learning, Steering Angle Sensor Neutral Position, Audio presets and Navigation. You will need the CONSULT-III diagnostic tool, or equivalent. Follow the directions on the screen of the tool, as needed.

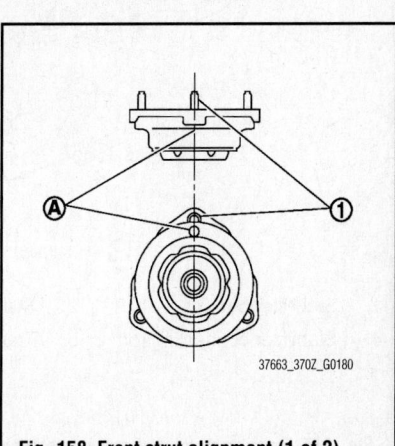

37663_370Z_G0180

Fig. 158 Front strut alignment (1 of 2)

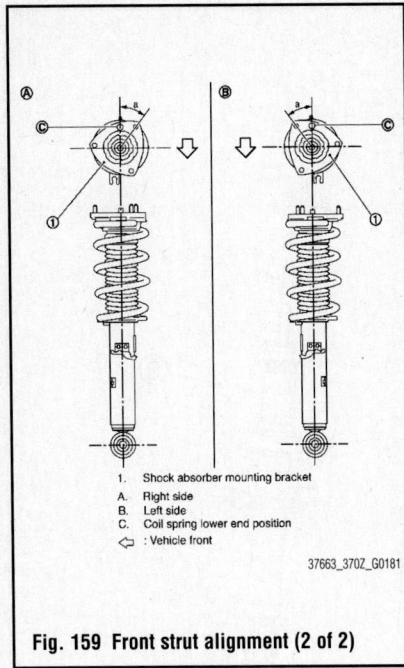

1. Shock absorber mounting bracket
A. Right side
B. Left side
C. Coil spring lower end position
⇦ : Vehicle front

37663_370Z_G0181

Fig. 159 Front strut alignment (2 of 2)

1. Before servicing the vehicle, refer to the Precautions Section.

➡**If working near and/or around the SRS system and components, be sure to disable the SRS system. After disabling the system wait three minutes or more before servicing the vehicle.**

2. Disconnect the negative battery cable.

3. Raise and safely support the vehicle.

4. Remove the tire and wheel assemblies.

5. Remove the engine undercover.

6. Remove the stabilizer connecting rod. Matchmark to identify for reinstallation.

7. Remove the stabilizer clamps and stabilizer bushings.

8. Remove the stabilizer bar.

To install:

➡**Be sure to use new fasteners, as required.**

9. Installation is the reverse of the removal procedure.

10. Be sure to check the matchmarks.

11. Tighten the mounting nut to specification while holding a hexagonal part of the stabilizer connecting rod side.

12. Perform a final tightening of nuts and bolts of each removed component with the vehicle in an unladen position.

13. Check and adjust the front alignment, as required.

14. Be sure to perform the reconnect/relearn procedures.

TRANSVERSE LINK

REMOVAL & INSTALLATION

See Figure 161.

➡**Whenever the negative battery cable is disconnected the following compo-**nents will require resetting. The Automatic temperature control system, Automatic drive positioner, Power window control, Sunroof system, Sunshade system, Rear view monitor, Idle Air Volume Learning, Steering Angle Sensor Neutral Position, Audio presets and Navigation. You will need the CONSULT-III diagnostic tool, or equivalent. Follow the directions on the screen of the tool, as needed.**

1. Before servicing the vehicle, refer to the Precautions Section.

➡**If working near and/or around the SRS system and components, be sure to disable the SRS system. After disabling the system wait three minutes or more before servicing the vehicle.**

2. Disconnect the negative battery cable.

3. Raise and safely support the vehicle.

4. Remove the tire and wheel assemblies.

5. Remove the engine undercover.

6. Remove the stabilizer connecting rod, convertible.

7. Remove the shock absorber, sedan and coupe.

8. Remove the front crossmember, sedan and coupe with AWD.

9. Remove the outer socket from the steering knuckle.

10. Remove the transverse link from the steering knuckle.

11. Position a suitable jack under the transverse link.

12. Remove the transverse link from its mounting.

To install:

➡**Be sure to use new fasteners, as required.**

13. Installation is the reverse of the removal procedure.

➡**Never tap on the ball joint cap of the stabilizer connecting rod with a hammer when inserting the stabilizer connecting rod into the transverse link.**

14. Perform a final tightening of nuts and bolts of each removed component with the vehicle in an unladen position.

15. Check wheel speed sensor for proper operation.

16. Adjust the steering angle sensor neutral position, using the

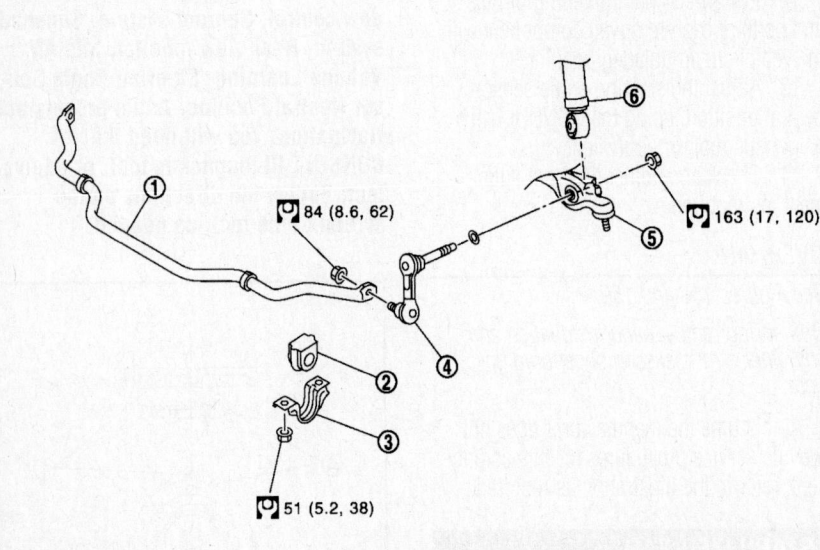

84 (8.6, 62)
163 (17, 120)
51 (5.2, 38)

| 1. | Stabilizer bar | 2. | Stabilizer bushing | 3. | Stabilizer clamp |
| 4. | Stabilizer connecting rod | 5. | Transverse link | 6. | Shock absorber |

37663_370Z_G0178

Fig. 160 Front stabilizer bar and related components—convertible

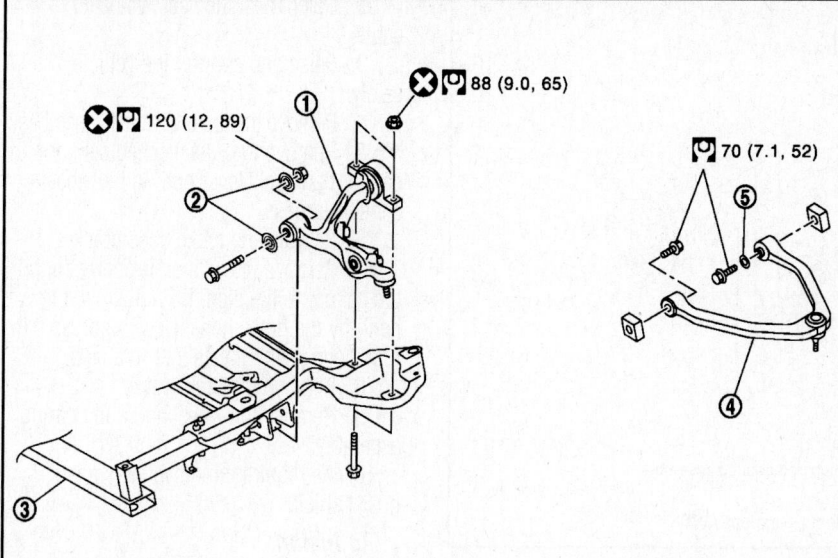

1. Transverse link
2. Stopper bush
3. Front suspension member
4. Upper link
5. Stopper rubber

37663_370Z_G0182

Fig. 161 Front transverse link and related components—convertible

CONSULT-III diagnostic tool, or equivalent.

17. Check and adjust the front alignment, as required.

18. Be sure to perform the reconnect/relearn procedures.

UPPER LINK

REMOVAL & INSTALLATION

See Figure 162.

➡Whenever the negative battery cable is disconnected the following components will require resetting. The Automatic temperature control system, Automatic drive positioner, Power window control, Sunroof system, Sunshade system, Rear view monitor, Idle Air Volume Learning, Steering Angle Sensor Neutral Position, Audio presets and Navigation. You will need the CONSULT-III diagnostic tool, or equivalent. Follow the directions on the screen of the tool, as needed.

1. Before servicing the vehicle, refer to the Precautions Section.

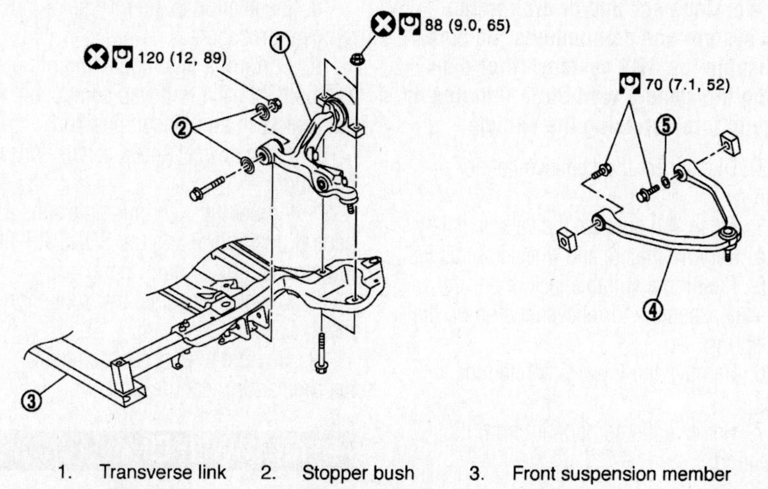

➡If working near and/or around the SRS system and components, be sure to disable the SRS system. After disabling the system wait three minutes or more before servicing the vehicle.

2. Disconnect the negative battery cable.

3. Raise and safely support the vehicle.

4. Remove the tire and wheel assemblies.

5. Remove the shock absorber.

6. Remove the upper link from the steering knuckle.

7. Remove the upper link and stopper rubber.

To install:

➡Be sure to use new fasteners, as required.

8. Perform a final tightening of nuts and bolts of each removed component with the vehicle in an unladen position.

9. Check wheel speed sensor for proper operation.

10. Adjust the steering angle sensor neutral position, using the CONSULT-III diagnostic tool, or equivalent as necessary.

11. Check and adjust the front alignment, as required.

12. Be sure to perform the reconnect/relearn procedures.

WHEEL HUB & BEARING

REMOVAL & INSTALLATION

See Figures 163 and 164.

➡Whenever the negative battery cable is disconnected the following components will require resetting. The Automatic temperature control system, Automatic drive positioner, Power window control, Sunroof system, Sunshade system, Rear view monitor, Idle Air Volume Learning, Steering Angle Sensor Neutral Position, Audio presets and Navigation. You will need the CONSULT-III diagnostic tool, or equivalent. Follow the directions on the screen of the tool, as needed.

1. Before servicing the vehicle, refer to the Precautions Section.

➡If working near and/or around the SRS system and components, be sure to disable the SRS system. After disabling the system wait three minutes or more before servicing the vehicle.

1. Transverse link
2. Stopper bush
3. Front suspension member
4. Upper link
5. Stopper rubber

37663_370Z_G0183

Fig. 162 Front upper link and related components—convertible

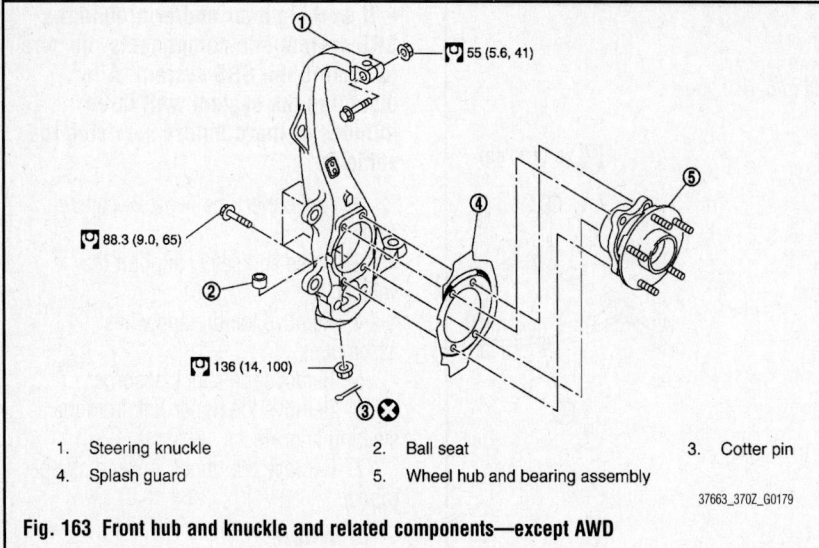

1. Steering knuckle
2. Ball seat
3. Cotter pin
4. Splash guard
5. Wheel hub and bearing assembly

37663_370Z_G0179

Fig. 163 Front hub and knuckle and related components—except AWD

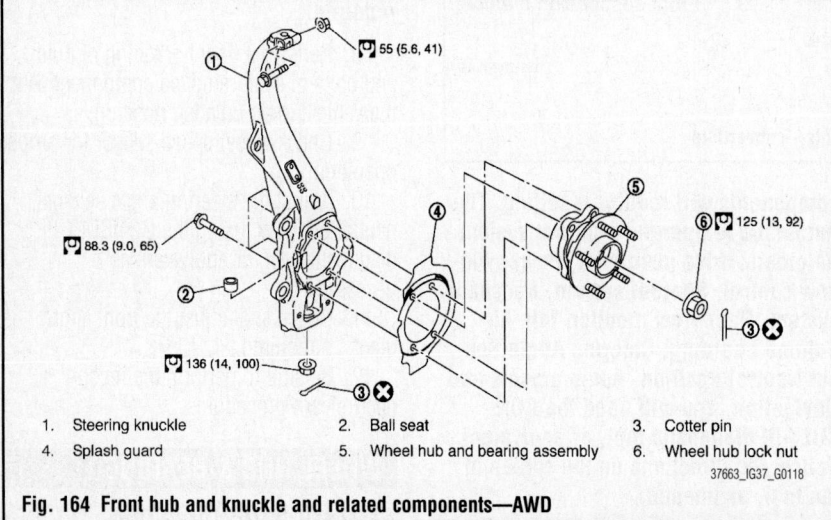

1. Steering knuckle
2. Ball seat
3. Cotter pin
4. Splash guard
5. Wheel hub and bearing assembly
6. Wheel hub lock nut

37663_IG37_G0118

Fig. 164 Front hub and knuckle and related components—AWD

SUSPENSION

FRONT LOWER LINK

REMOVAL & INSTALLATION

See Figure 165.

➡**Whenever the negative battery cable is disconnected the following components will require resetting. The Automatic temperature control system, Automatic drive positioner, Power window control, Sunroof system, Sunshade system, Rear view monitor, Idle Air Volume Learning, Steering Angle Sensor Neutral Position, Audio presets and Navigation. You will need the CONSULT-III diagnostic tool, or equivalent. Follow the directions on the screen of the tool, as needed.**

1. Before servicing the vehicle, refer to the Precautions Section.

➡**If working near and/or around the SRS system and components, be sure to disable the SRS system. After disabling the system wait three minutes or more before servicing the vehicle.**

2. Disconnect the negative battery cable.
3. Raise and support the vehicle safely.
4. Remove the tire and wheel assemblies.
5. Position a suitable jack under the rear axle assembly to relieve tension on the coil spring.
6. Remove the lower link retaining bolts.
7. Remove the component from its mounting.

To install:

➡**Be sure to use new fasteners, as required.**

2. Disconnect the negative battery cable.
3. Raise and support the vehicle safely.
4. Remove the tire and wheel assemblies.
5. Remove the wheel speed sensor and sensor harness. Never pull on the wheel sensor harness.
6. Remove the brake hose bracket.
7. Remove the caliper. Properly support the caliper to the side. Do not allow it to hang by the brake hose. Never depress the brake pedal with the caliper removed.
8. Remove the brake rotor.
9. Remove the wheel and hub bearing assembly.
10. As required, remove the splash guard shield.

To install:

➡**Be sure to use new fasteners, as required.**

11. Installation is the reverse of the removal procedure.
12. Perform a final tightening of nuts and bolts of each removed component with the vehicle in an unladen position.
13. Check wheel speed sensor for proper operation.
14. Adjust the steering angle sensor neutral position, using the CONSULT-III diagnostic tool, or equivalent.
15. Check and adjust the front alignment, as required.

ADJUSTMENT

These bearings are not adjustable. If defective, they must be replaced.

REAR SUSPENSION

8. Installation is the reverse of the removal procedure.
9. Perform a final tightening of nuts and bolts of each removed component with the vehicle in an unladen position.
10. Check wheel speed sensor for proper operation.
11. Adjust the steering angle sensor neutral position, using the CONSULT-III diagnostic tool, or equivalent.
12. Check and adjust the front alignment, as required.
13. Be sure to perform the reconnect/relearn procedures.

REAR LOWER LINK

REMOVAL & INSTALLATION

See Figures 166 through 168.

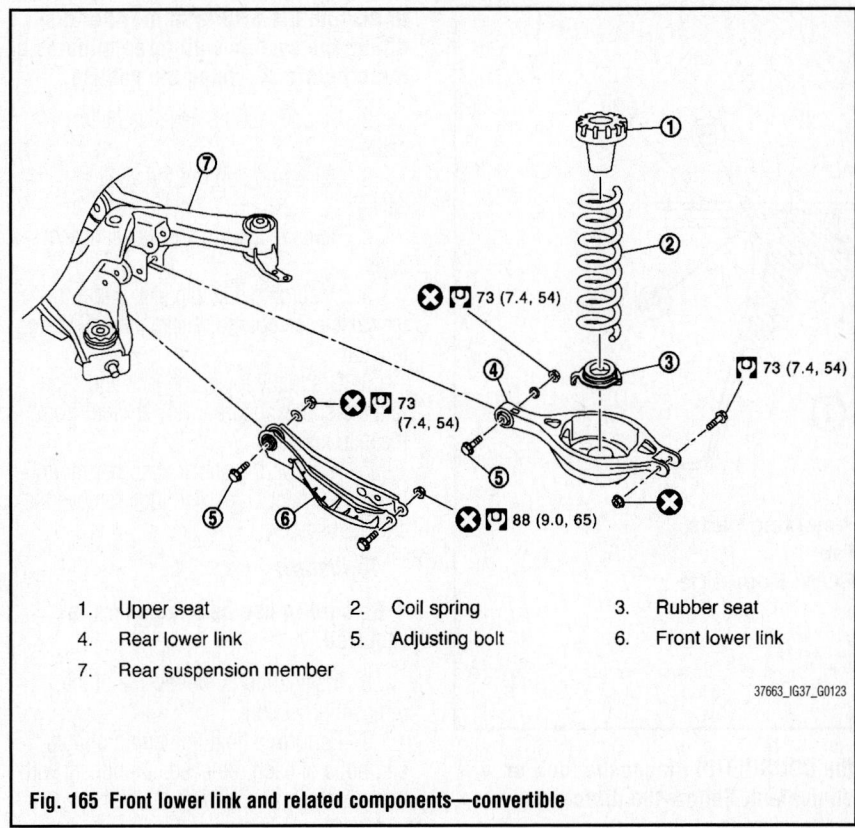

1. Upper seat
2. Coil spring
3. Rubber seat
4. Rear lower link
5. Adjusting bolt
6. Front lower link
7. Rear suspension member

37663_IG37_G0123

Fig. 165 Front lower link and related components—convertible

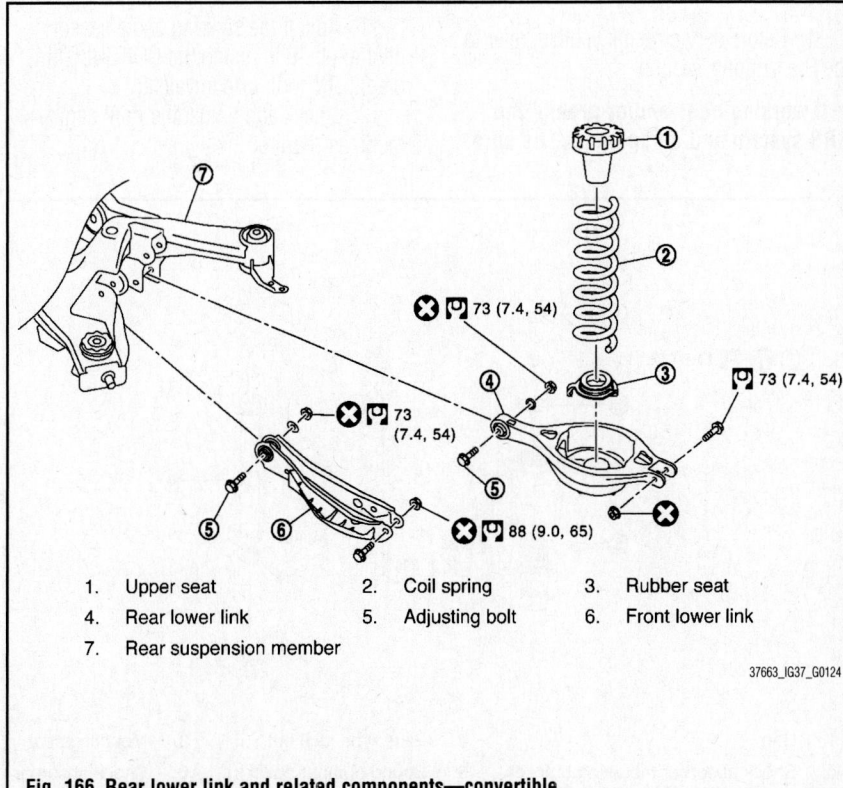

1. Upper seat
2. Coil spring
3. Rubber seat
4. Rear lower link
5. Adjusting bolt
6. Front lower link
7. Rear suspension member

37663_IG37_G0124

Fig. 166 Rear lower link and related components—convertible

➡Whenever the negative battery cable is disconnected the following components will require resetting. The Automatic temperature control system, Automatic drive positioner, Power window control, Sunroof system, Sunshade system, Rear view monitor, Idle Air Volume Learning, Steering Angle Sensor Neutral Position, Audio presets and Navigation. You will need the CONSULT-III diagnostic tool, or equivalent. Follow the directions on the screen of the tool, as needed.

1. Before servicing the vehicle, refer to the Precautions Section.

➡If working near and/or around the SRS system and components, be sure to disable the SRS system. After disabling the system wait three minutes or more before servicing the vehicle.

2. Disconnect the negative battery cable.
3. Raise and support the vehicle safely.
4. Remove the tire and wheel assemblies.
5. Position a suitable jack under the rear lower link to relieve the coil spring tension.
6. Loosen the rear lower link mounting nuts (rear suspension member side).
7. Remove the rear lower link (axle housing side).
8. Slowly lower the jack and remove the upper seat, coil spring and rubber sheet from the rear lower link.
9. Remove the rear lower link.

To install:

➡Be sure to use new fasteners, as required.

10. Installation is the reverse of the removal procedure.
11. Be sure that the upper seat is attached as indicated in the illustration.

➡Make sure that the projecting parts of the floor panel is securely fitted with the upper seat tab.

12. Match up the rubber seat indentations and the rear lower link grooves and attach.
13. Install the coil spring by aligning the lower end of the large diameter side to the step between the rubber seat and the rear lower link.

➡Make sure that the spring is not upside down. The top and bottom are indicated by paint color.

14. Perform a final tightening of nuts and bolts of each removed component with the vehicle in an unladen position.
15. Check wheel speed sensor for proper operation.
16. Adjust the steering angle sensor neutral position, using the CONSULT-III diagnostic tool, or equivalent.

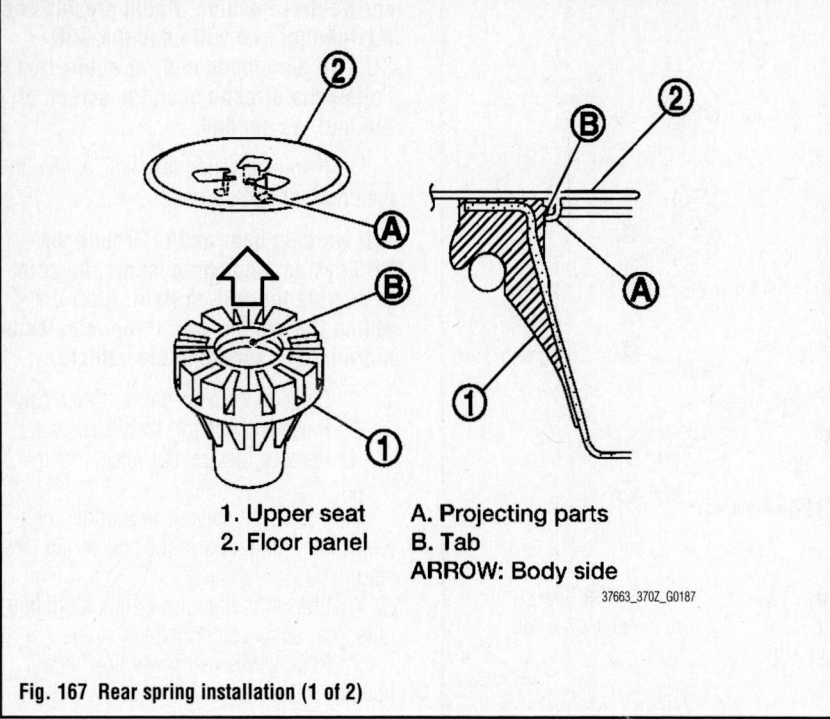

1. Upper seat
2. Floor panel
A. Projecting parts
B. Tab
ARROW: Body side

37663_370Z_G0187

Fig. 167 Rear spring installation (1 of 2)

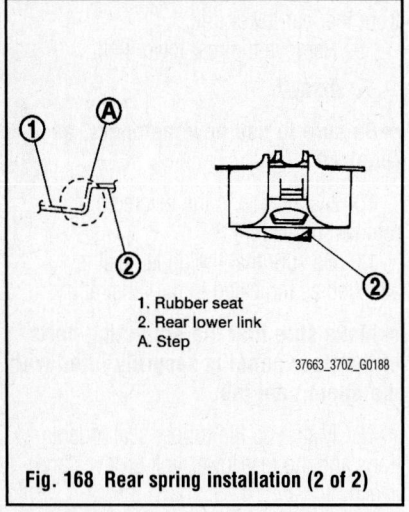

1. Rubber seat
2. Rear lower link
A. Step

37663_370Z_G0188

Fig. 168 Rear spring installation (2 of 2)

17. Check and adjust the front alignment, as required.

SHOCK ABSORBER

REMOVAL & INSTALLATION

See Figure 169.

➡Whenever the negative battery cable is disconnected the following components will require resetting. The Automatic temperature control system, Automatic drive positioner, Power window control, Sunroof system, Sunshade system, Rear view monitor, Idle Air Volume Learning, Steering Angle Sensor Neutral Position, Audio presets and Navigation. You will need

the CONSULT-III diagnostic tool, or equivalent. Follow the directions on the screen of the tool, as needed.

1. Before servicing the vehicle, refer to the Precautions Section.

➡If working near and/or around the SRS system and components, be sure

to disable the SRS system. After disabling the system wait three minutes or more before servicing the vehicle.

2. Disconnect the negative battery cable.

3. Raise and support the vehicle safely.

4. Remove the tire and wheel assemblies.

5. Position a suitable jack under the rear axle assembly to relieve the coil spring tension.

6. Gradually lower the jack and separate the shock absorber (lower side) from the axle housing.

7. Remove the shock absorber mounting nuts (upper side), and then remove the shock absorber.

To install:

➡Be sure to use new fasteners, as required.

8. Installation is the reverse of the removal procedure.

9. Perform a final tightening of nuts and bolts of each removed component with the vehicle in an unladen position.

10. Check wheel speed sensor for proper operation.

11. Adjust the steering angle sensor neutral position, using the CONSULT-III diagnostic tool, or equivalent.

12. Check and adjust the front alignment, as required.

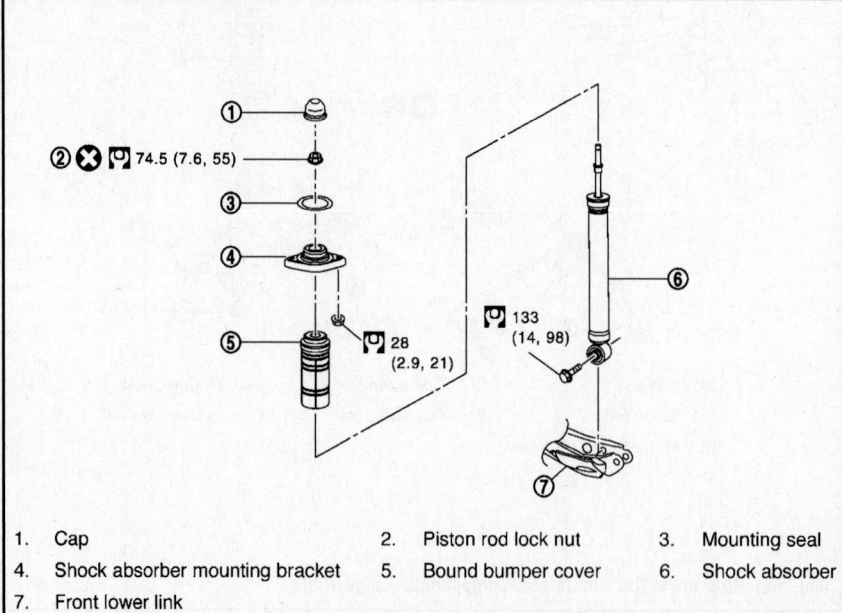

1. Cap
2. Piston rod lock nut
3. Mounting seal
4. Shock absorber mounting bracket
5. Bound bumper cover
6. Shock absorber
7. Front lower link

37663_IG37_G0125

Fig. 169 Rear shock absorber and related components—convertible

13. Be sure to perform the reconnect/relearn procedures.

TESTING

1. Before servicing the vehicle, refer to the Precautions Section.

2. Check the shock for deformation, cracks, damage and replace as required.

3. Check the piston rod for damage, uneven wear or distortion and replace as required.

4. Check the welded and sealed areas for oil leakage and replace as required.

RADIUS ARM

REMOVAL & INSTALLATION

See Figure 170.

➡Whenever the negative battery cable is disconnected the following components will require resetting. The Automatic temperature control system, Automatic drive positioner, Power window control, Sunroof system, Sunshade system, Rear view monitor, Idle Air Volume Learning, Steering Angle Sensor Neutral Position, Audio presets and Navigation. You will need the CONSULT-III diagnostic tool, or equivalent. Follow the directions on the screen of the tool, as needed.

1. Before servicing the vehicle, refer to the Precautions Section.

➡If working near and/or around the SRS system and components, be sure to disable the SRS system. After disabling the system wait three minutes or more before servicing the vehicle.

2. Disconnect the negative battery cable.

3. Raise and support the vehicle safely.

4. Remove the tire and wheel assemblies.

5. Remove the radius rod retaining nuts.

6. Remove the component from the vehicle.

To install:

➡Be sure to use new fasteners, as required.

7. Installation is the reverse of the removal procedure.

8. Perform a final tightening of nuts and bolts of each removed component with the vehicle in an unladen position.

9. Check wheel speed sensor for proper operation.

10. Adjust the steering angle sensor neutral position, using the CONSULT-III diagnostic tool, or equivalent.

11. Check and adjust the front alignment, as required.

12. Be sure to perform the reconnect/relearn procedures.

REAR SUSPENSION ARM

REMOVAL & INSTALLATION

See Figure 171.

➡Whenever the negative battery cable is disconnected the following components will require resetting. The Automatic temperature control system, Automatic drive positioner, Power window control, Sunroof system, Sunshade system, Rear view monitor, Idle Air Volume Learning, Steering Angle Sensor Neutral Position, Audio presets and Navigation. You will need the CONSULT-III diagnostic tool, or equivalent. Follow the directions on the screen of the tool, as needed.

1. Before servicing the vehicle, refer to the Precautions Section.

➡If working near and/or around the SRS system and components, be sure to disable the SRS system. After disabling the system wait three minutes or more before servicing the vehicle.

2. Disconnect the negative battery cable.

3. Raise and support the vehicle safely.

4. Remove the tire and wheel assemblies.

5. Remove the diagonal brace, convertible.

6. On sedan and coupe, remove the caliper and position it to the side with mechanics wire. Do not allow the caliper to hang by the brake hose.

7. Remove the stabilizer connecting rod.

8. Remove the halfshaft.

9. Remove the cotter pin of the suspension arm ball joint, and loosen the nut.

10. Remove the suspension arm (rear suspension member side).

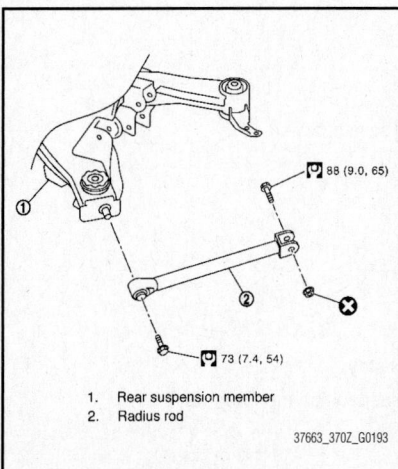

1. Rear suspension member
2. Radius rod

37663_370Z_G0193

Fig. 170 Rear radius arm and related components—convertible

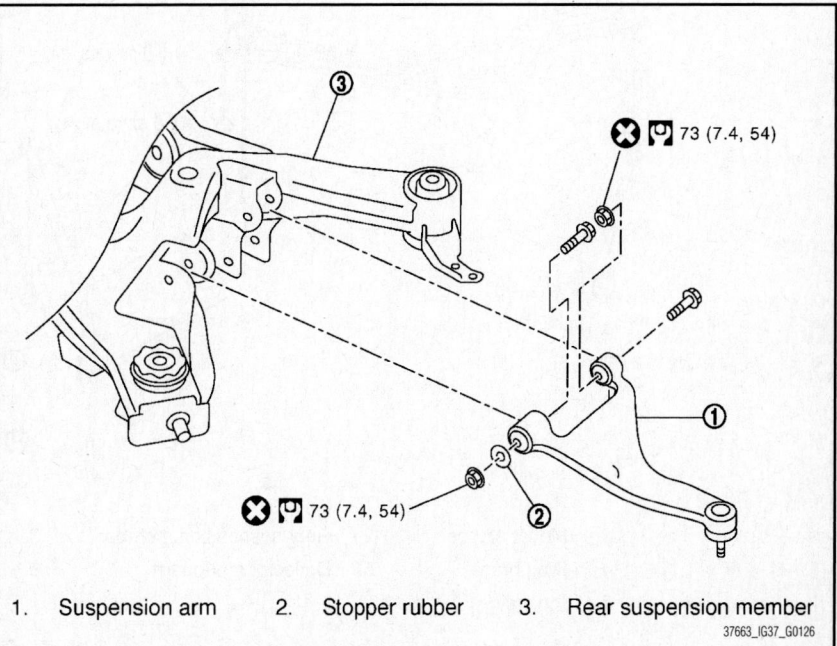

1. Suspension arm   2. Stopper rubber   3. Rear suspension member

37663_IG37_G0126

Fig. 171 Rear suspension arm and related components—convertible

11. Use a ball joint removal tool to remove the suspension arm from the axle housing.

➡**Be careful not to damage the ball joint boot. Temporarily tighten the nut to prevent damage to the threads and to prevent the tool from coming off.**

12. Remove the suspension arm.

13. Remove the stabilizer connecting rod mounting bracket.

To install:

➡**Be sure to use new fasteners, as required.**

14. Installation is the reverse of the removal procedure.

15. Perform a final tightening of nuts and bolts of each removed component with the vehicle in an unladen position.

16. Check wheel speed sensor for proper operation.

17. Adjust the steering angle sensor

neutral position, using the CONSULT-III diagnostic tool, or equivalent.

18. Check and adjust the front alignment, as required.

19. Be sure to perform the reconnect/relearn procedures.

REAR SUSPENSION MEMBER

REMOVAL & INSTALLATION

See Figure 172.

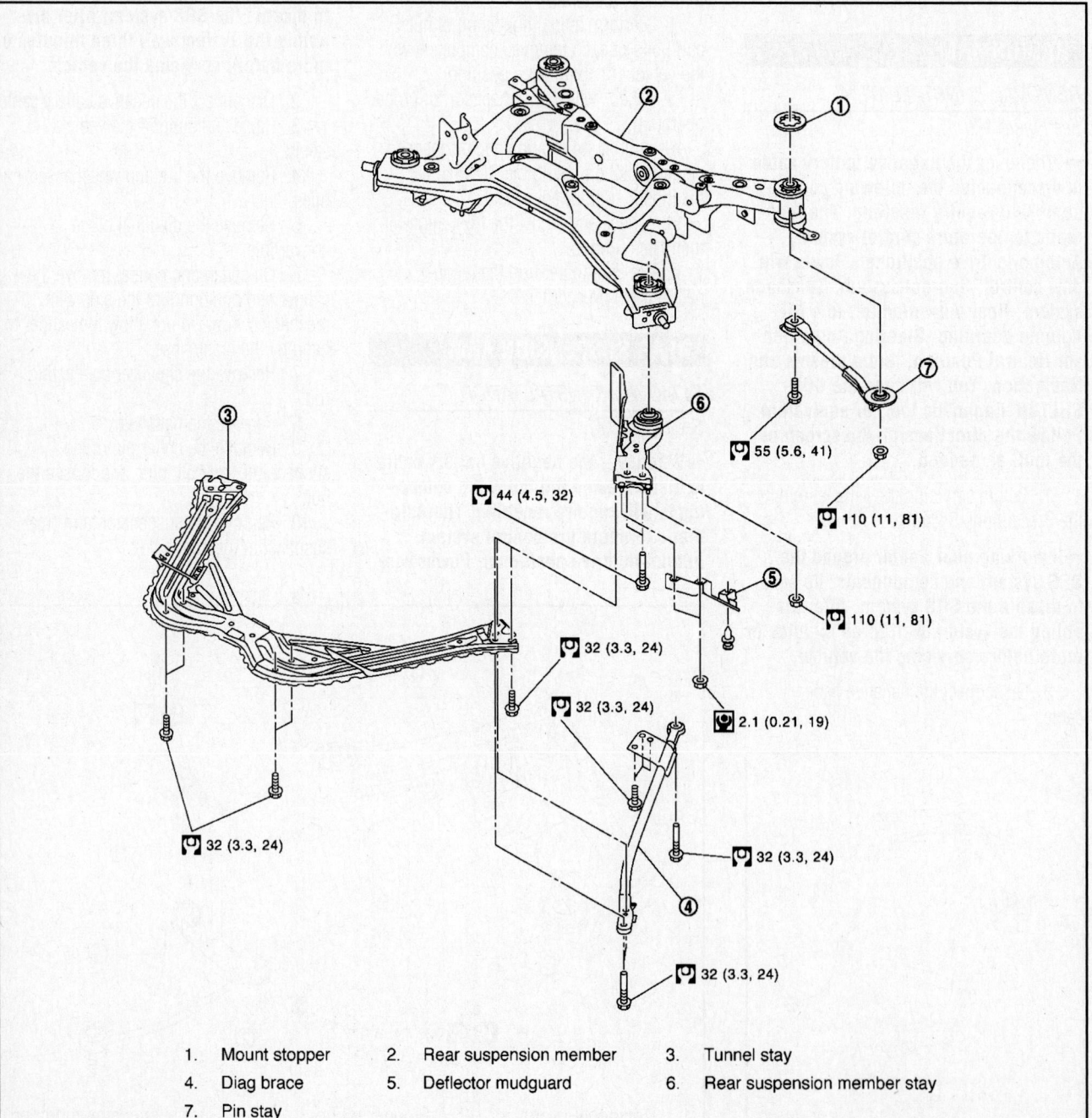

1.	Mount stopper	2.	Rear suspension member	3.	Tunnel stay
4.	Diag brace	5.	Deflector mudguard	6.	Rear suspension member stay
7.	Pin stay				

37663_IG37_G0022

Fig. 172 Rear suspension member assembly and related components—convertible

At this time the manufacturer does not provide removal and installation procedures for this component, refer to the illustration as required.

➡**Whenever the negative battery cable is disconnected the following components will require resetting. The Automatic temperature control system, Automatic drive positioner, Power window control, Sunroof system, Sunshade system, Rear view monitor, Idle Air Volume Learning, Steering Angle Sensor Neutral Position, Audio presets and Navigation. You will need the CONSULT-III diagnostic tool, or equivalent. Follow the directions on the screen of the tool, as needed.**

1. Before servicing the vehicle, refer to the Precautions Section.

➡**If working near and/or around the SRS system and components, be sure to disable the SRS system. After disabling the system wait three minutes or more before servicing the vehicle.**

2. Disconnect the negative battery cable.

3. Be sure to perform the reconnect/relearn procedures.

STABILIZER BAR

REMOVAL & INSTALLATION
See Figure 173.

➡**Whenever the negative battery cable is disconnected the following components will require resetting. The Automatic temperature control system, Automatic drive positioner, Power window control, Sunroof system, Sunshade system, Rear view monitor, Idle Air Volume Learning, Steering Angle Sensor Neutral Position, Audio presets and Navigation. You will need the CONSULT-III diagnostic tool, or equivalent. Follow the directions on the screen of the tool, as needed.**

1. Before servicing the vehicle, refer to the Precautions Section.

➡**If working near and/or around the SRS system and components, be sure to disable the SRS system. After disabling the system wait three minutes or more before servicing the vehicle.**

2. Disconnect the negative battery cable.
3. Raise and support the vehicle safely.
4. Remove the tire and wheel assemblies.
5. Remove the diagonal brace, convertible.
6. Remove the mounting bracket center muffler and remove the mounting rubber of the main muffler, sedan and coupe.
7. Remove the stabilizer connecting rods.
8. Remove the stabilizer clamps. Remove the stabilizer bushings.
9. Remove the stabilizer bar.
10. Remove the stabilizer connecting rod mounting brackets from the suspension arm.

To install:

➡**Be sure to use new fasteners, as required.**

11. Installation is the reverse of the removal procedure.
12. Perform a final tightening of nuts and bolts of each removed component with the vehicle in an unladen position.
13. Check wheel speed sensor for proper operation.
14. Adjust the steering angle sensor neutral position, using the CONSULT-III diagnostic tool, or equivalent.
15. Check and adjust the front alignment, as required.
16. Be sure to perform the reconnect/relearn procedures.

WHEEL HUB & BEARING

REMOVAL & INSTALLATION

Except Convertible
See Figure 174.

➡**Whenever the negative battery cable is disconnected the following components will require resetting. The Automatic temperature control system, Automatic drive positioner, Power window control, Sunroof system, Sunshade system, Rear view monitor, Idle Air Volume Learning, Steering Angle Sensor Neutral Position, Audio presets and Navigation. You will need the CONSULT-III diagnostic tool, or equivalent. Follow the directions on the screen of the tool, as needed.**

1. Before servicing the vehicle, refer to the Precautions Section.

➡**If working near and/or around the SRS system and components, be sure to disable the SRS system. After disabling the system wait three minutes or more before servicing the vehicle.**

2. Disconnect the negative battery cable.
3. Raise and support the vehicle safely.
4. Remove the tire and wheel assembly.
5. Remove the caliper. Properly support the caliper to the side. Do not allow it to hang by the brake hose. Never depress the brake pedal with the caliper removed.
6. Remove the brake rotor.
7. Remove the cotter pin and adjusting cap. Loosen the wheel hub locknut.
8. Matchmark the halfshaft and the wheel hub and bearing assembly.
9. Remove the locknut.
10. Remove the cotter pin, then loosen

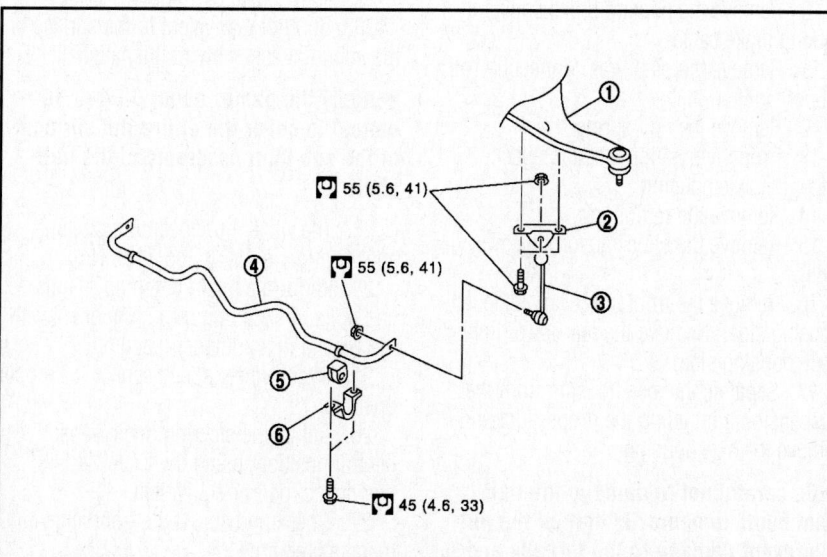

1.	Suspension arm	2.	Stabilizer connecting rod mounting bracket	3.	Stabilizer connecting rod	
4.	Stabilizer bar	5.	Stabilizer bushing	6.	Stabilizer clamp	

37663_370Z_G0191

Fig. 173 Rear stabilizer bar and related components—convertible

the suspension arm mounting nut of the axle housing.

11. Remove the wheel hub and bearing assembly.

To install:

➡**Be sure to use new fasteners, as required.**

12. Installation is the reverse of the removal procedure.

13. Use the matchmarks to align removed components.

14. Clean the matching surface of the halfshaft and wheel hub and bearing assembly.

15. Apply paste part number 440037S000 or equivalent to the surface of the sub joint assembly of the halfshaft.

➡**Apply the paste, about 0.04–0.10 ounce) to cover the entire flat surface of the sub joint assembly of the half-shaft.**

16. The wheel hub locknut tightening specification is 136 ft. lbs. (185 Nm).

17. Perform a final tightening of nuts and bolts of each removed component with the vehicle in an unladen position.

18. Check wheel speed sensor for proper operation.

19. Adjust the steering angle sensor neutral position, using the CONSULT-III diagnostic tool, or equivalent.

20. Check and adjust the front alignment, as required.

Convertible

See Figure 174.

➡**Whenever the negative battery cable is disconnected the following components will require resetting. The Automatic temperature control system, Automatic drive positioner, Power window control, Sunroof system, Sunshade system, Rear view monitor, Idle Air Volume Learning, Steering Angle Sensor Neutral Position, Audio presets and Navigation. You will need the CONSULT-III diagnostic tool, or equivalent. Follow the directions on the screen of the tool, as needed.**

1. Before servicing the vehicle, refer to the Precautions Section.

➡**If working near and/or around the SRS system and components, be sure to disable the SRS system. After disabling the system wait three minutes or more before servicing the vehicle.**

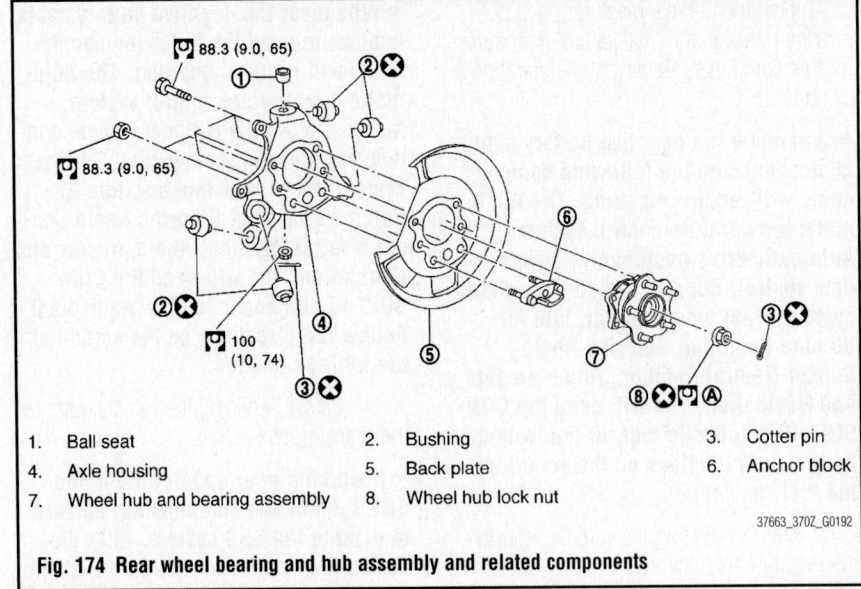

1.	Ball seat	2.	Bushing	3.	Cotter pin
4.	Axle housing	5.	Back plate	6.	Anchor block
7.	Wheel hub and bearing assembly	8.	Wheel hub lock nut		

37663_370Z_G0192

Fig. 174 Rear wheel bearing and hub assembly and related components

2. Disconnect the negative battery cable.

3. Raise and support the vehicle safely.

4. Remove the tire and wheel assembly.

5. Remove the caliper. Properly support the caliper to the side. Do not allow it to hang by the brake hose. Never depress the brake pedal with the caliper removed.

6. Remove the brake rotor.

7. Remove the cotter pin and adjusting cap. Loosen the wheel hub locknut.

8. Matchmark the halfshaft and the wheel hub and bearing assembly.

9. Remove the locknut and spring washer.

10. Remove the parking brake shoe and parking brake cable.

11. Remove the stabilizer connecting rod (upper side).

12. Remove the coil spring.

13. Properly position a suitable jack under the axle housing.

14. Remove the radius rod.

15. Remove the shock absorber (lower side).

16. Remove the front lower link (axle housing side). Remove the rear lower link (axle housing side)

17. Separate the axle housing from the suspension arm, using the proper tool and remove the axle housing.

➡**Be careful not to damage the ball joint boot. Temporarily tighten the nut to prevent damage to the threads and to prevent the tool from coming off.**

➡**Never place the halfshaft at an extreme angle. Be careful not to overextend the slide joint. Never allow**

the halfshaft to hang down with proper support.

18. Remove the wheel hub and bearing assembly.

To install:

➡**Be sure to use new fasteners, as required.**

19. Installation is the reverse of the removal procedure.

20. Use the matchmarks to align removed components.

21. Clean the matching surface of the halfshaft and wheel hub and bearing assembly.

22. Apply paste part number 440037S000 or equivalent to the surface of the sub joint assembly of the halfshaft.

➡**Apply the paste, about 0.04–0.10 ounce) to cover the entire flat surface of the sub joint assembly of the half-shaft.**

23. The wheel hub locknut tightening specification is 136 ft. lbs. (185 Nm).

24. Perform a final tightening of nuts and bolts of each removed component with the vehicle in an unladen position.

25. Check wheel speed sensor for proper operation.

26. Adjust the steering angle sensor neutral position, using the CONSULT-III diagnostic tool, or equivalent.

27. Check and adjust the front alignment, as required.

ADJUSTMENT

These bearings are not adjustable. If defective, they must be replaced.

INFINITI

M35 • M35X • M45 • M45X

SPECIFICATIONS AND MAINTENANCE CHARTS

ENGINE AND VEHICLE IDENTIFICATION

Engine								Model Year	
Code ①	Liters (cc)	Cu. In.	Cyl.	Fuel Sys.	Engine Type	Eng. Mfg.		Code ②	Year
VQ35HR	3.5 (3498)	213	6	MFI	DOHC	Nissan		9	2009
VK45DE	4.5 (4494)	274	8	MFI	DOHC	Nissan		A	2010

MFI: Multi-Port Fuel Injection

DOHC: Double Overhead Camshaft

37663_IM35_C0001

GENERAL ENGINE SPECIFICATIONS

Year	Model	Engine Displacement Liters	Engine ID	Net Horsepower @ rpm	Net Torque @ rpm (ft. lbs.)	Bore x Stroke (in.)	Compression Ratio	Oil Pressure @ rpm
2009	M35/35X	3.5	VQ35HR	303@6800	262@4800	3.76X3.21	10.6:1	43@2000
	M45/45X	4.5	VK45DE	325@6400	336@4000	3.66x3.26	10.5:1	43@2000
2010	M35/35X	3.5	VQ35HR	303@6800	262@4800	3.76X3.21	10.6:1	43@2000
	M45/45X	4.5	VK45DE	325@6400	336@4000	3.66x3.26	10.5:1	43@2000

X denotes AWD

37663_IM35_C0002

ENGINE TUNE-UP SPECIFICATIONS

Year	Engine Displacement Liters	Engine ID	Spark Plug Gap (in.)	Ignition Timing (deg.) MT	Ignition Timing (deg.) AT	Fuel Pump (psi)	Idle Speed (rpm) MT	Idle Speed (rpm) AT	Valve Clearance Intake	Valve Clearance Exhaust
2009	3.5	VQ35HR	0.043	—	15 ①	34	—	600-700	②	③
	4.5	VK45DE	0.043	—	12 ①	34	—	600-700	②	③
2010	3.5	VQ35HR	0.043	—	15 ①	34	—	600-700	②	③
	4.5	VK45DE	0.043	—	12 ①	34	—	600-700	②	③

NOTE: The Vehicle Emission Control Information label often reflects specification changes made during production.

The label figures must be used if they differ from those in this chart.

① +/-5° Before top dead center

② Intake 0.010 - 0.013 Cold

③ Exhaust 0.011 - 0.015 Cold

37663_IM35_C0003

CAPACITIES

Year	Model	Engine Displacement Liters	Engine ID	Engine Oil with Filter (qts.)	Transmission (pts.) Manual	Transmission (pts.) Auto.	Transfer Case (pts.)	Drive Axle (pts.)	Fuel Tank (gal.)	Cooling System (qts.)
2009	M35	3.5	VQ35HR	5.125	—	①	2.6	②	20	9.75
	M45	4.5	VK45DE	5.75	—	①	2.6	②	20	11.0
2010	M35	3.5	VQ35HR	5.125	—	①	2.6	②	20	9.75
	M45	4.5	VK45DE	5.75	—	①	2.6	②	20	11.0

NOTE: All capacities are approximate. Add fluid gradually and check to be sure a proper fluid level is obtained.

① 5A/T: 21-5/8 pts.

   6A/T: 19-1/2 pts.

② Front Axle 1.38 (AWD) -Rear Axle 3.00

37663_IM35_C0005

FLUID SPECIFICATIONS

Year	Model	Engine Displ. Liters	Engine Oil	Auto. Trans.	Drive Axle Front	Drive Axle Rear	Transfer Case	Power Steering Fluid	Brake Master Cylinder	Cooling System
2009	M35	3.5	①	②	GL-5 80W-90	GL-5 80W-90	③	NISSAN PSF	DOT 3	④
	M45	4.5	①	②	GL-5 80W-90	GL-5 80W-90	③	NISSAN PSF	DOT 3	④
2010	M35	3.5	①	②	GL-5 80W-90	GL-5 80W-90	③	NISSAN PSF	DOT 3	④
	M45	4.5	①	②	GL-5 80W-90	GL-5 80W-90	③	NISSAN PSF	DOT 3	④

DOT: Department Of Transpotation

① API Certification Mark 5W-30

② Genuine NISSAN Matic S ATF

③ Genuine NISSAN Matic J ATF

④ NISSAN Long Life Antifreeze/ Coolant or equivalent

37663_IM35_C0004

VALVE SPECIFICATIONS

Year	Engine Displacement Liters	Engine ID	Seat Angle (deg.)	Face Angle (deg.)	Spring Test Pressure (lbs. @ in.)	Spring Free Height (in.)	Stem-to-Guide Clearance (in.) Intake	Stem-to-Guide Clearance (in.) Exhaust	Stem Diameter (in.) Intake	Stem Diameter (in.) Exhaust
2009	3.5	VQ35HR	①	NA	NA	1.7264	0.0008-0.0021	0.0012-0.0022	0.2348-0.2354	0.2347-0.2350
	4.5	VK45DE	①	NA	NA	1.8248-1.8445	0.0008-0.0018	0.0012-0.0022	0.2351-0.2354	0.2347-0.2350
2010	3.5	VQ35HR	①	NA	NA	1.7264	0.0008-0.0021	0.0012-0.0022	0.2348-0.2354	0.2347-0.2350
	4.5	VK45DE	①	NA	NA	1.8248-1.8445	0.0008-0.0018	0.0012-0.0022	0.2351-0.2354	0.2347-0.2350

NA: Not Available

① 45 degrees, 15 minutes to 45 degrees, 45 minutes

37663_IM35_C0006

CAMSHAFT AND BEARING SPECIFICATIONS

All measurements are given in inches.

Year	Engine Displacement Liters	Engine VIN	Journal Diameter	Brg. Oil Clearance	Shaft End-play	Runout	Journal Bore	Lobe Lift Intake	Lobe Lift Exhaust
2009	3.5	VQ35HR	①	②	0.0045-0.0074	0.0045-0.0074	NA	1.8057-1.8132	1.8061-1.8136
	4.5	VK45DE	③	④	0.0045-0.0074	0.0059	1.0236-1.0244	1.7663-1.7738	1.7293-1.7368
2010	3.5	VQ35HR	①	②	0.0045-0.0074	0.0045-0.0074	NA	1.8057-1.8132	1.8061-1.8136
	4.5	VK45DE	③	④	0.0045-0.0074	0.0059	1.0236-1.0244	1.7663-1.7738	1.7293-1.7368

NA: Information not available

Note: (Oil clearance) = (Camshaft bracket inner diameter) – (Camshaft journal diameter).

① No. 1: 1.0211 - 1.0218 inches

   No 2,3 and 4 - 0.9230-0.9238 inches

② No. 1: 0.0018 - 0.0034 inches

   No 2,3,4 and 5 - 0.0014-0.0030 inches

③ No. 1: 1.0212 - 1.0218 inches

   No 2,3 and 4: 1.218-1.0224 inches

④ No. 1: 0.0018 - 0.0033 inches

   No 2,3,4 and 5 - 0.0012-0.0027 inches

37663_IM35_C0007

CRANKSHAFT AND CONNECTING ROD SPECIFICATIONS

All measurements represent standard values and are given in inches.

Year	Engine Displacement Liters	Engine ID	Crankshaft Main Brg. Journal Dia.	Crankshaft Main Brg. Oil Clearance	Crankshaft Shaft End-play	Thrust on No.	Connecting Rod Journal Diameter	Connecting Rod Oil Clearance	Connecting Rod Side Clearance
2009	3.5	VQ35HR	2.5571-2.5581	0.0014-0.0018	0.0039-0.0098	3	NA	0.0016-0.0021	NA
	4.5	VK45DE	2.5173-2.5183	①	0.0039-0.0098	3	NA	0.0008-0.0018	NA
2010	3.5	VQ35HR	2.5571-2.5581	0.0014-0.0018	0.0039-0.0098	3	NA	0.0016-0.0021	NA
	4.5	VK45DE	2.5173-2.5183	①	0.0039-0.0098	3	NA	0.0008-0.0018	NA

① Nos. 1 and 5: 0.00004-0.0004 in.

   Nos. 2, 3 and 4: 0.0003-0.0007 in.

37663_IM35_C0009

PISTON AND RING SPECIFICATIONS

All measurements are given in inches.

Year	Engine Disp. Liters	Engine ID	Piston Clearance	Ring Gap			Ring Side Clearance		
				Top Compression	Bottom Compression	Oil Control	Top Compression	Bottom Compression	Oil Control
2009	3.5	VQ35HR	0.0004-0.0012	0.0091-0.0130	0.0130-0.0189	0.0067-0.0189	0.0016-0.0031	0.0012-0.0028	0.0022-0.0061
	4.5	VK45DE	0.0004-0.0012	0.0087-0.0126	0.0087-0.0126	0.0079-0.0197	0.0018-0.0031	0.0012-0.0028	0.0026-0.0053
2010	3.5	VQ35HR	0.0004-0.0012	0.0091-0.0130	0.0130-0.0189	0.0067-0.0189	0.0016-0.0031	0.0012-0.0028	0.0022-0.0061
	4.5	VK45DE	0.0004-0.0012	0.0087-0.0126	0.0087-0.0126	0.0079-0.0197	0.0018-0.0031	0.0012-0.0028	0.0026-0.0053

37663_IM35_C0008

TORQUE SPECIFICATIONS

All readings in ft. lbs.

Year	Engine Displacement Liters	Engine ID	Cylinder Head Bolts	Main Bearing Bolts	Rod Bearing Bolts	Crankshaft Damper Bolts	Flywheel Bolts	Manifold		Spark Plugs	Oil Pan Drain Plug
								Intake	Exhaust		
2009	3.5	VQ35HR	①	②	③	33	65	④	22	14	25
	4.5	VK45DE	⑤	⑥	⑦	⑧	65	⑨ ⑩	21	18	25
2010	3.5	VQ35HR	①	②	③	33	65	④	22	14	25
	4.5	VK45DE	⑤	⑥	⑦	⑧	65	⑨ ⑩	21	18	25

① Step 1: 77 ft. lbs.
  Step 2: Loosen bolts completely
  Step 3: 30 ft. lbs.
  Step 4: Tighten an additional 95 degrees
  Step 5: Again, tighten an additional 95 degrees.
  Step 6: After installing cylinder head, measure
    distance between front end faces of cylinder
    block and cylinder head (left and right banks)
    Standard : 0.555 - 0.587 in (14.1 - 14.9 mm)

② Step 1: Bolts 17-26: 18 ft. lbs.
  Step 2: Repeat Step 1
  Step 3: Bolts 1-16: 26 ft. lbs.
  Step 4: Bolts 1-16: Additional 90 degrees

③ Step 1: Tighten to 21 ft. lbs.
  Step 2: Completely loosen rod bolts
  Step 3: Tighten to 18 ft. lbs.
  Step 4: Additional 90 degrees

④ Step 1: Tighten stud bolts to 8 ft. lbs.
  Step 2: Tighten to 5 ft. lbs.
  Step 3: Tighten, again, to 19 ft. lbs.

⑤ Step 1: 33 ft.lbs.
  Step 2: Additional 70 degrees
  Step 3: Loosen bolts completely
  Step 3: 33 ft. lbs.
  Step 4: Tighten an additional 60 degrees.

⑥ Step 1: M12 bolts 1-10: 29 ft. lbs.
  Step 2: M9 bolts 11-20: 22 ft. lbs.
  Step 3: M12 bolts: Tighten an additional 40 degrees
  Step 4: M9 bolts: Tighten an additional 30 degrees
  Step 5: M10 side bolts 21-30: 36 ft. lbs.

⑦ Step 1: Tighten to 11 ft. lbs.
  Step 2: Tighten an additional 60 degrees

⑧ Step 1: 69 ft. lbs.
  Step 2: Tighten an additional 90 degrees

⑨ Step 1: Tighten Upper Manifold bolts to 8 ft. lbs.

⑩ Step 1: Tighten stud bolts to 8 ft. lbs.
  Step 2: Tighten to 11 ft. lbs.
  Step 3: Tighten, again, to 21 ft. lbs.

37663_IM35_C0010

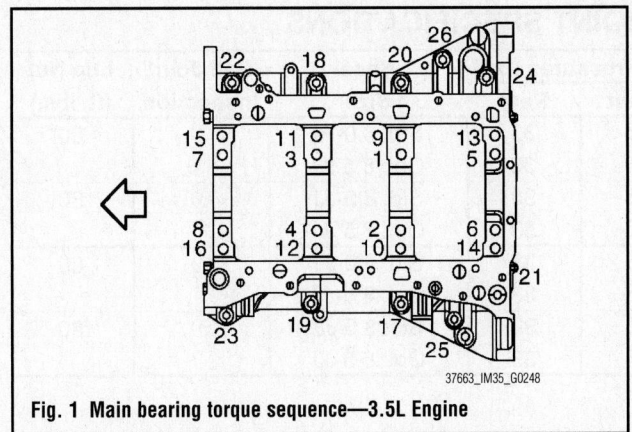

Fig. 1 Main bearing torque sequence—3.5L Engine

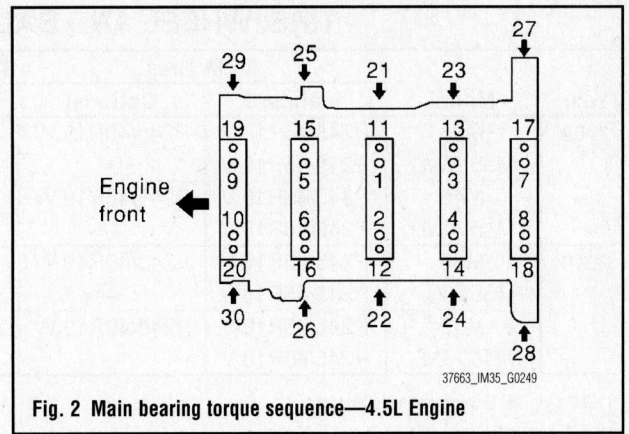

Fig. 2 Main bearing torque sequence—4.5L Engine

WHEEL ALIGNMENT

Year	Model		Caster Range (+/-Deg.)	Caster Preferred Setting (Deg.)	Camber Range (+/-Deg.)	Camber Preferred Setting (Deg.)	Toe-in (Deg.)
2009	M35	F	0.75	①	0.75	-0.25	0.05 +/- 0.05
		R	—	—	②	③	0.12 +0.11/- 0.12
	M35X	F	0.75	3.83	0.75	-0.25	0.04 +/- 0.04
	AWD	R	—	—	0.50	-0.17	0.12 +0.11/- 0.12
	M45	F	0.75	①	0.75	-0.25	0.05 +/- 0.05
		R	—	—	②	③	0.12 +0.11/- 0.12
	M45X	F	0.75	3.83	0.75	-0.25	0.04 +/- 0.04
	AWD	R	—	—	0.50	-0.17	0.12 +0.11/- 0.12
2010	M35	F	0.75	①	0.75	-0.25	0.05 +/- 0.05
		R	—	—	②	③	0.12 +0.11/- 0.12
	M35X	F	0.75	3.83	0.75	-0.25	0.04 +/- 0.04
	AWD	R	—	—	0.50	-0.17	0.12 +0.11/- 0.12
	M45	F	0.75	①	0.75	-0.25	0.05 +/- 0.05
		R	—	—	②	③	0.12 +0.11/- 0.12
	M45X	F	0.75	3.83	0.75	-0.25	0.04 +/- 0.04
	AWD	R	—	—	0.50	-0.17	0.12 +0.11/- 0.12

Note: Measure wheel alignment under unladen conditions.

"Unladen conditions" means that fuel, engine coolant, and lubricant are full. Spare tire, jack, hand tools and mats are in designated position.

① 18" Wheels 4.50°

  19" Wheels 4.58°

② 18" Wheels: 40°

  19" Wheels: 50°

③ Rear 18" Wheels -67°

  Rear 19" Wheels -83°

37663_IM35_C0011

TIRE, WHEEL AND BALL JOINT SPECIFICATIONS

Year	Model	OEM Tires Standard	OEM Tires Optional	Tire Pressures (psi) Front	Tire Pressures (psi) Rear	Wheel Size	Ball Joint Inspection	Lug Nut (ft. lbs.)
2009	M35	P245/45R18 V	P245/40R19 W	33	33	Std: 8.0-JJ	①	80
	M35 AWD	P245/45R18 V	—	33	33	Opt: 8,5-JJ		
	M45	P245/45R18 V	P245/40R19 W	33	33	Std: 8.0-JJ	①	80
	M45 AWD	P245/45R18 V	—	33	33	Opt: 8.5-JJ		
2010	M35	P245/45R18 V	P245/40R19 W	33	33	Std: 8.0-JJ	①	80
	M35 AWD	P245/45R18 V	—	33	33	Opt: 8,5-JJ		
	M45	P245/45R18 V	P245/40R19 W	33	33	Std: 8.0-JJ	①	80
	M45 AWD	P245/45R18 V	—	33	33	Opt: 8.5-JJ		

OEM: Original Equipment Manufacturer

PSI: Pounds Per Square Inch

STD: Standard

OPT: Optional

NA: Not Available

① Replace is any measureable movement is found.

37663_IM35_C0012

BRAKE SPECIFICATIONS
All measurements in inches unless noted

Year	Model	Front Brake Disc Original Thickness	Front Brake Disc Minimum Thickness	Front Brake Disc Maximum Run-out	Rear Brake Disc Original Thickness	Rear Brake Disc Minimum Thickness	Rear Brake Disc Maximum Run-out	Minimum Lining Thickness Front	Minimum Lining Thickness Rear	Brake Caliper Bracket Bolts (ft. lbs.)	Brake Caliper Mounting Bolts (ft. lbs.)
2009	M35	1.102	1.024	0.0014	0.631	0.551	0.0022	0.079	0.079	①	②
	M45	1.102	1.024	0.0014	0.631	0.551	0.0022	0.079	0.079	①	②
2010	M35	1.102	1.024	0.0014	0.631	0.551	0.0022	0.079	0.079	①	②
	M45	1.102	1.024	0.0014	0.631	0.551	0.0022	0.079	0.079	①	②

① Front: 20

Rear: 32

② Front: 98

Rear: 62

37663_IM35_C0013

SCHEDULED MAINTENANCE INTERVALS
Infiniti— M35 & M45

TO BE SERVICED	TYPE OF SERVICE	VEHICLE MILEAGE INTERVAL (x1000)												
		7.5	15	22.5	30	37.5	45	52.5	60	67.5	75	82.5	90	97.5
Engine oil & filter	R	✓	✓	✓	✓	✓	✓	✓	✓	✓	✓	✓	✓	✓
Automatic transaxle fluid ①	S/I		✓		✓		✓		✓		✓		✓	
Brake lines & cables	S/I		✓		✓		✓		✓		✓		✓	
Brake pads & discs	S/I		✓		✓		✓		✓		✓		✓	
Differential gear oil	S/I		✓		✓		✓		✓		✓		✓	
Driveshaft boots	S/I		✓		✓		✓		✓		✓		✓	
Active suspension fluid ②	S/I		✓		✓		✓		✓		✓		✓	
In-cabin microfilter	R		✓		✓		✓		✓		✓		✓	
Air cleaner filter ③	R								✓					
Exhaust system	S/I		✓		✓		✓		✓		✓		✓	
Fuel lines	S/I								✓					
Steering gear & linkage, axle & suspension parts	S/I		✓		✓		✓		✓		✓		✓	
Vapor lines	S/I				✓				✓				✓	
Engine coolant ④	R								✓				✓	
Spark plugs(Platinum-tipped type) ⑤	R	Replace every 105,000 miles												
Drive belts ⑥	S/I								✓					

R: Replace   S/I: Service or Inspect

① If towing a trailer, using a camper or car-top carrier, or driving on rough or muddy roads, CHANGE oil every 30,000 miles or 24 months.

② Replace at 60,000 miles (if not previously replaced).

③ If operating in dusty conditions, more frequent maintenance may be required.

④ After 60,000 miles or 48 months, replace coolant every 30,000 miles or 24 months.

⑤ Platinum-tipped spark plugs should be changed every 105,000 miles.

⑥ After 60,000 miles or 48 months, inspect every 15,000 miles or 12 months. Replace belts if found damaged.

FREQUENT OPERATION MAINTENANCE (SEVERE SERVICE)

If a vehicle is operated under any of the following conditions it is considered severe service:

- Extremely dusty areas.
- 50% or more of the vehicle operation is in 32°C (90°F) or higher temperatures, or constant operation in temperatures below 0°C (32°F).
- Prolonged idling (vehicle operation in stop and go traffic).
- Frequent short running periods (engine does not warm to normal operating temperatures).
- Police, taxi, delivery or trailer towing usage.

Oil & oil filter: change every 3750 miles.

Brake pads & discs: service or inspect every 7500 miles.

Driveshaft boots: service or inspect every 7500 miles

Exhaust system: service or inspect every 7500 miles.

Steering gear, linkage, axle & suspension ball joints: service or inspect every 7500 miles.

Steering linkage, ball joints & front suspension ball joints: service or inspect every 7500 miles.

37663_IM35_C0014

BRAKES — INFORMATION AND PRECAUTIONS

ANTI-LOCK SYSTEMS

• Certain components within the ABS system are not intended to be serviced or repaired individually.

• Do not use rubber hoses or other parts not specifically specified for and ABS system. When using repair kits, replace all parts included in the kit. Partial or incorrect repair may lead to functional problems and require the replacement of components.

• Lubricate rubber parts with clean, fresh brake fluid to ease assembly. Do not use shop air to clean parts; damage to rubber components may result.

• Use only DOT 3 brake fluid from an unopened container.

• If any hydraulic component or line is removed or replaced, it may be necessary to bleed the entire system.

• A clean repair area is essential. Always clean the reservoir and cap thoroughly before removing the cap. The slightest amount of dirt in the fluid may plug an orifice and impair the system function. Perform repairs after components have been thoroughly cleaned; use only denatured alcohol to clean components. Do not allow ABS components to come into contact with any substance containing mineral oil; this includes used shop rags.

• The Anti-Lock control unit is a microprocessor similar to other computer units in the vehicle. Ensure that the ignition switch is **OFF** before removing or installing controller harnesses. Avoid static electricity discharge at or near the controller.

• If any arc welding is to be done on the vehicle, the control unit should be unplugged before welding operations begin.

DISC AND DRUM SYSTEMS

※※ CAUTION

Dust and dirt accumulating on brake parts during normal use may contain asbestos fibers from production or aftermarket brake linings. Breathing excessive concentrations of asbestos fibers can cause serious bodily harm. Exercise care when servicing brake parts. Do not sand or grind brake lining unless equipment used is designed to contain the dust residue. Do not clean brake parts with compressed air or by dry brushing. Cleaning should be done by dampening the brake components with a fine mist of water, then wiping the brake components clean with a dampened cloth. Dispose of cloth and all residue containing asbestos fibers in an impermeable container with the appropriate label. Follow practices prescribed by the Occupational Safety and Health Administration (OSHA) and the Environmental Protection Agency (EPA) for the handling, processing, and disposing of dust or debris that may contain asbestos fibers.

BRAKES — BLEEDING THE BRAKE SYSTEM

BLEEDING PROCEDURE

BLEEDING PROCEDURE

※※ WARNING

Use of any other than the approved DOT 3 brake fluid will cause permanent damage to brake components and will render the brakes inoperative. Failure to follow these instructions may result in personal injury.

※※ CAUTION

Brake fluid contains polyglycol ethers and polyglycols. Avoid contact with eyes. Wash hands thoroughly after handling. If brake fluid contacts eyes, flush eyes with running water for 15 minutes. Get medical attention if irritation persists. If taken internally, drink water and induce vomiting. Get medical attention immediately. Failure to follow these instructions may result in personal injury.

※※ WARNING

Do not allow the brake master cylinder reservoir to run dry during the bleeding operation. Keep the master cylinder reservoir filled with the specified brake fluid. Never reuse the brake fluid that has been drained from the hydraulic system.

※※ WARNING

Brake fluid is harmful to painted and plastic surfaces. If brake fluid is spilled onto a painted or plastic surface, immediately wash it with water.

※※ WARNING

When any part of the hydraulic system has been disconnected or a new component is installed, air may enter the system, causing spongy brake pedal action. This requires the bleeding of the hydraulic system after it has been correctly connected.

1. Connect a vinyl tube to rear right brake caliper bleed valve.

2. Fully depress brake pedal 4 or 5 times.

3. With brake pedal depressed, loosen bleed valve to bleed air in brake line, and then tighten it immediately.

4. Repeat steps 2 and 3 until all of the air is out of the brake line.

5. Tighten the bleed valve to the specified torque.

6. From step 1 to 5, with master cylinder reservoir tank filled at least half way, bleed air from brake hydraulic line bleed valves in the following order:
   • Rear right brake
   • Front left brake
   • Rear left brake
   • Front right brake

BLEEDING THE ABS SYSTEM

Refer to Bleeding the Brake System.

BRAKES

ANTI-LOCK BRAKE SYSTEM (ABS)

WHEEL SPEED SENSORS

REMOVAL & INSTALLATION
See Figure 3.

> ❊❊ **WARNING**
>
> Do not twist sensor harness as much as possible, when removing it. Pull sensors out without pulling on sensor harness.

> ❊❊ **WARNING**
>
> Take care to avoid damaging sensor edges or rotor teeth, remove wheel sensor first before removing front or rear wheel hub. This is to avoid damage to sensor wiring and loss of sensor function.

➥When installing, make sure there is no foreign material such as iron chips on and in the mounting hole of the wheel sensor. Make sure no foreign material has been caught in the sensor rotor. Remove any foreign material and clean the mount.

➥When installing wheel sensor, be sure to press rubber grommets in until they lock at locations shown above.

> ❊❊ **WARNING**
>
> When installed, harness must not be twisted.

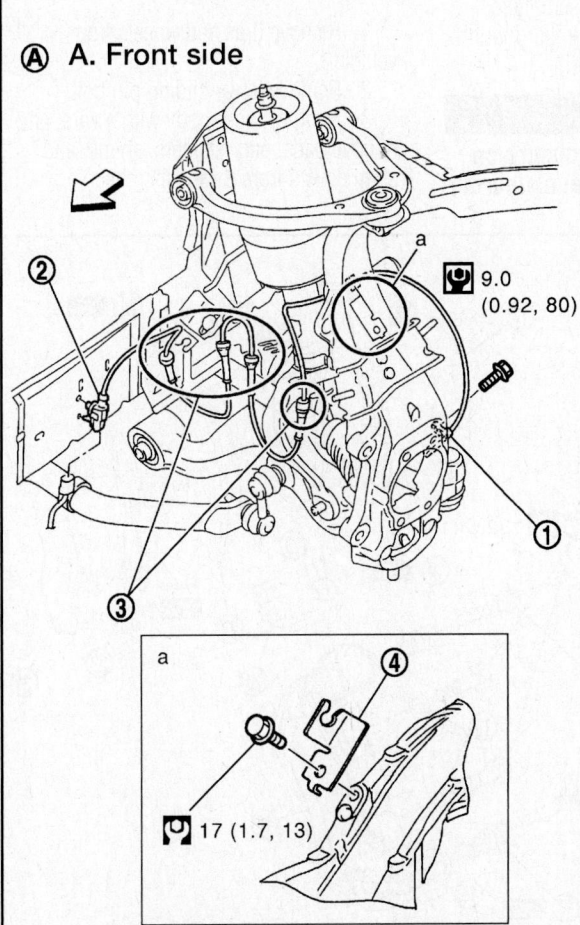

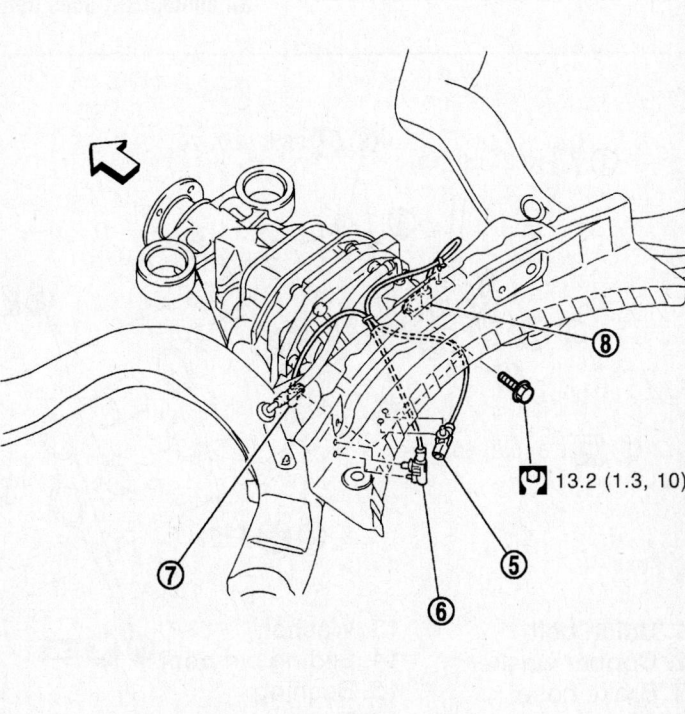

Ⓐ A. Front side

Ⓑ B. Rear side

1. Front LH wheel sensor
2. Front LH wheel sensor connector
3. Clamp
4. Bracket

5. Rear RH wheel sensor connector
6. Rear LH wheel sensor connector
7. Rear LH wheel sensor
8. Rear RH wheel sensor

22140_IM35_G0071

Fig. 3 ABS wheel sensors exploded view

BRAKES   **FRONT DISC BRAKES**

BRAKE CALIPER

REMOVAL & INSTALLATION

See Figures 4 and 5.

1. Remove tires and wheels from vehicle.
2. Fasten disc rotor using wheel nut.
3. Drain brake fluid.
4. Remove union bolt, and then disconnect brake hose from caliper assembly.
5. Remove torque member mounting bolts, and remove brake caliper assembly.

> ※※ **CAUTION**
>
> **Do not drop brake pad.**

To install:

> ※※ **CAUTION**
>
> **Refill with new brake fluid "DOT 3". Never reuse drained brake fluid.**

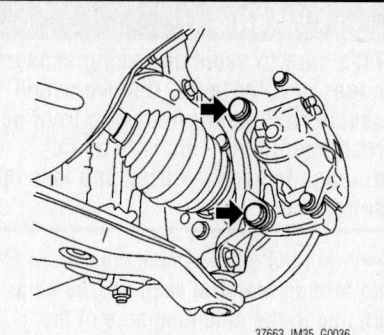

37663_IM35_G0036

Fig. 4 Remove torque member mounting bolts, and remove brake caliper assembly

6. Install brake caliper assembly to vehicle, and tighten torque member mounting bolts to 98 ft. lbs. (133 Nm).

> ※※ **CAUTION**
>
> **Do not allow oil or any moisture on all contact surfaces between steering**

knuckle and caliper assembly, bolts, and washer.

7. Install brake hose to brake caliper assembly, and tighten union bolts to 13 ft. lbs. (18 Nm).
8. Refill with new brake fluid and bleed air.
9. Check front disc brake for drag.
10. Install tires and wheels to vehicle.

DISC BRAKE PADS

REMOVAL & INSTALLATION

See Figures 5 through 6.

1. Remove tires and wheels from vehicle.
2. Remove lower sliding pin bolt.
3. Hang cylinder body with a wire, and remove pads, pad retainers, shims, and shim covers from torque member.

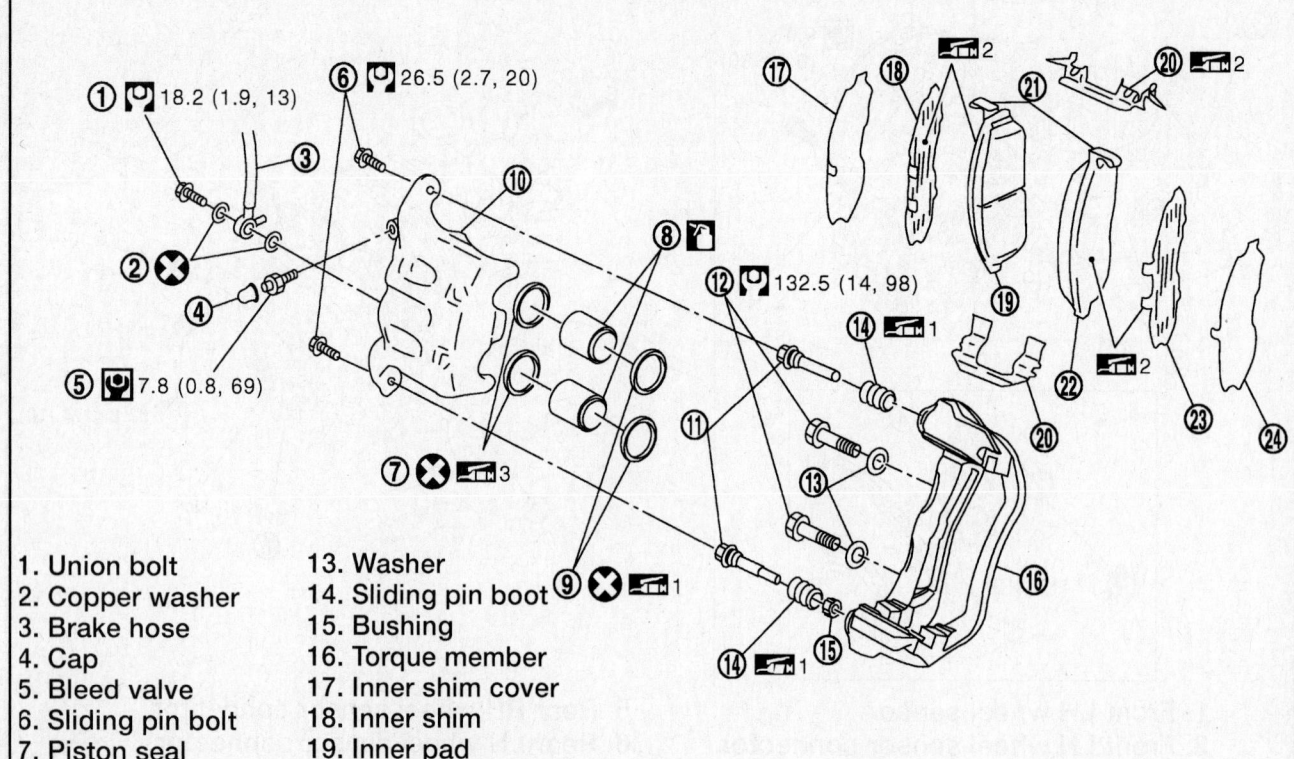

1. Union bolt
2. Copper washer
3. Brake hose
4. Cap
5. Bleed valve
6. Sliding pin bolt
7. Piston seal
8. Piston
9. Piston boot
10. Cylinder body
11. Sliding pin
12. Torque member mounting bolt
13. Washer
14. Sliding pin boot
15. Bushing
16. Torque member
17. Inner shim cover
18. Inner shim
19. Inner pad
20. Pad retainer
21. Pad wear sensor
22. Outer pad
23. Outer shim
24. Outer shim cover

22140_IM35_G0076

Fig. 5 Exploded view of front disc brake assembly

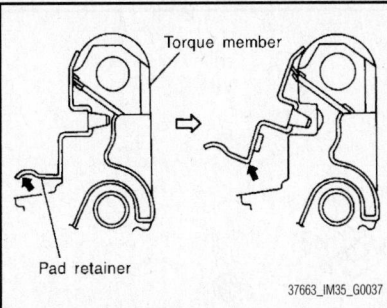

Fig. 6 When removing the pad retainer from the torque member, lift it in the direction indicated by the arrow so that it does not deform

37663_IM35_G0037

⁂ CAUTION

When removing the pad retainer from the torque member, lift it in the direction indicated by the arrow so that it

BRAKES

BRAKE CALIPER

REMOVAL & INSTALLATION

See Figures 7 and 8.

1. Remove tires and wheels from vehicle.
2. Fasten disc rotor using wheel nut.
3. Drain brake fluid.
4. Remove union bolt and then disconnect brake hose from caliper assembly.
5. Remove torque member mounting, bolts, and remove brake caliper assembly.

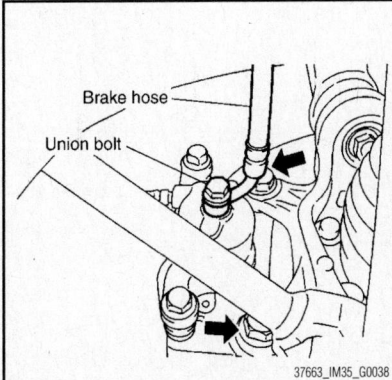

Fig. 7 Remove union bolt and then disconnect brake hose from caliper assembly

37663_IM35_G0038

⁂ CAUTION

Do not drop brake pad.

does not deform. Remember each position of the removed brake pads.

To install:

4. Apply PBC (Poly Butyl Cuprysil) grease or silicone-based grease to between pad and shim. Install shim, shim cover, to pad.
5. Apply PBC (Poly Butyl Cuprysil) grease or silicone-based grease to between pad retainer and pad. Install pad retainers and pads to torque member.

⁂ CAUTION

Securely assemble pad retainers so that they are not being lifted up from torque member. Both inner and outer pads have a pad return system on the pad retainer. Install pad return lever securely to pad wear sensor.

To install:

⁂ CAUTION

Refill with new brake fluid "DOT 3". Never reuse drained brake fluid.

6. Install brake caliper assembly vehicle, and tighten torque member mounting bolts to 62 ft. lbs. (84 Nm).

⁂ CAUTION

Before installing caliper assembly, wipe off oil and moisture on all mounting surfaces of rear axle and caliper assembly and threads, bolts and washers.

7. Install L-shaped pin of brake hose and then tighten union bolt to 13 ft. lbs. (18 Nm).
8. Refill with new brake fluid and bleed air.
9. Check rear disc brake for drag.
10. Install tires and wheels to vehicle.

DISC BRAKE PADS

REMOVAL & INSTALLATION

See Figure 8.

1. Remove tires and wheels from vehicle.
2. Remove lower sliding pin bolt.

6. Install cylinder body to torque member.

⁂ CAUTION

In the case of replacing a pad with new one, check a brake fluid level in the reservoir tank because brake fluid returns to master cylinder reservoir tank when pressing piston in.

➡Use a disc brake piston tool to easily press piston.

7. Install lower sliding pin bolt, and tighten it to 20 ft. lbs. (27 Nm).
8. Check front disc brake for drag.
9. Install tires and wheels to vehicle.

REAR DISC BRAKES

3. Hang cylinder body with a wire, and remove pads, pad retainers, shims, and shim cover from torque member.

⁂ CAUTION

Deform pad retainer when removing pad retainer from torque member. Remember each position of the removed brake pads.

To install:

4. Apply PBC (Poly Butyl Cuprysil) grease or silicone-based grease to between pad and shim. Install inner shim, inner shim cover to inner pad, and outer shim to outer pad.
5. Install pad retainers and pads to torque member.
6. Press in piston until pads can be installed, and then install cylinder body to torque member.

⁂ CAUTION

In the case of replacing a pad with new one, check a brake fluid level in the reservoir tank because brake fluid returns to master cylinder reservoir tank when pressing piston in.

➡Use a disc brake piston tool to easily press piston.

7. Install upper sliding pin bolt and tighten to the specified torque.
8. Check rear disc brake for drag.
9. Install tires to vehicle.

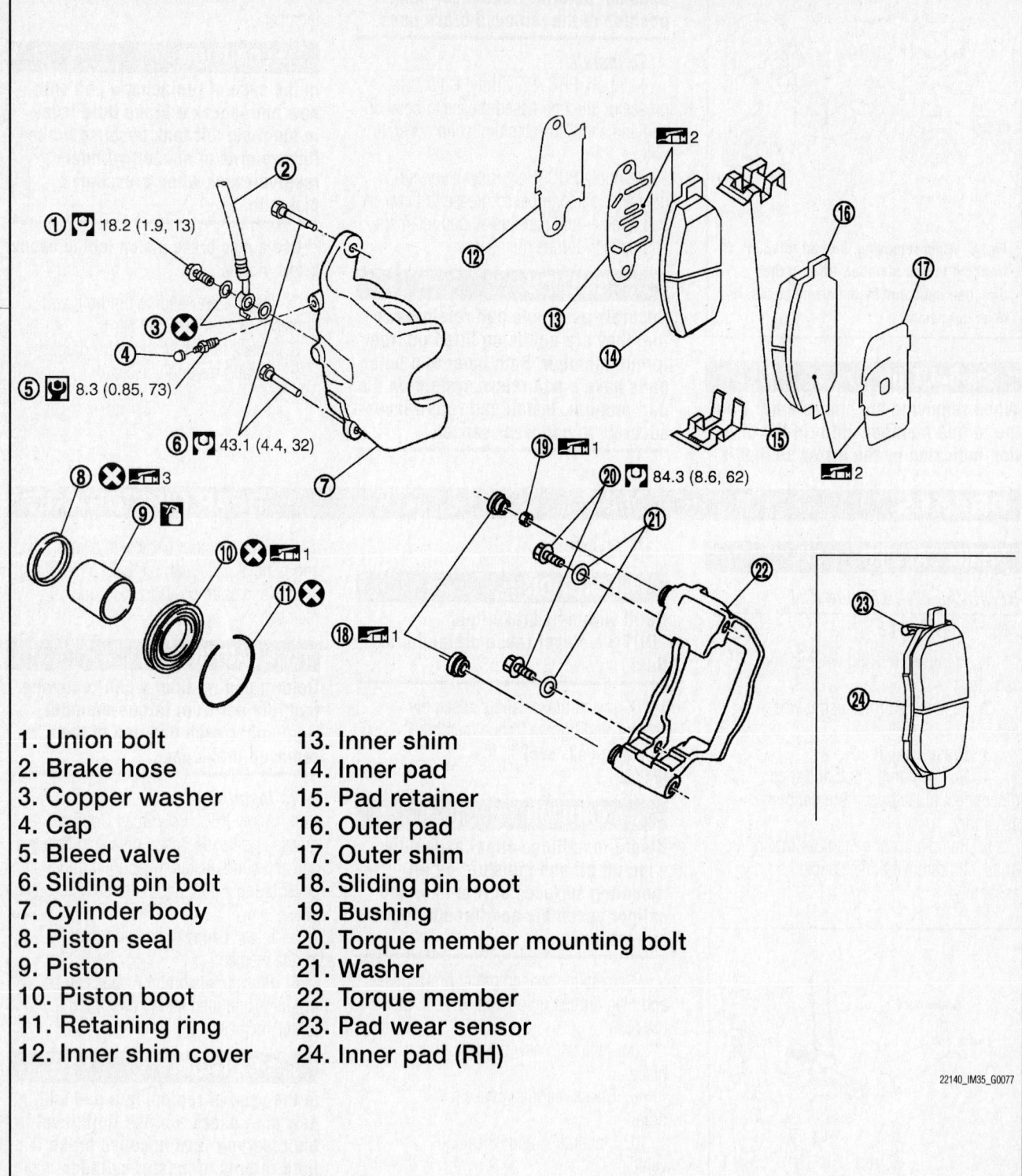

① 🔧 18.2 (1.9, 13)

⑤ 🔧 8.3 (0.85, 73)

⑥ 🔧 43.1 (4.4, 32)

⑳ 🔧 84.3 (8.6, 62)

1. Union bolt
2. Brake hose
3. Copper washer
4. Cap
5. Bleed valve
6. Sliding pin bolt
7. Cylinder body
8. Piston seal
9. Piston
10. Piston boot
11. Retaining ring
12. Inner shim cover
13. Inner shim
14. Inner pad
15. Pad retainer
16. Outer pad
17. Outer shim
18. Sliding pin boot
19. Bushing
20. Torque member mounting bolt
21. Washer
22. Torque member
23. Pad wear sensor
24. Inner pad (RH)

22140_IM35_G0077

Fig. 8 Exploded view of rear brake caliper assembly

BRAKES   **PARKING BRAKE**

PARKING BRAKE CABLES

ADJUSTMENT

1. To perform adjustment operations, remove rear tires from vehicle.

2. Insert a deep socket wrench onto adjusting nut.

3. Rotate adjusting nut to fully loosen cable, and then release parking brake pedal.

4. Secure disc rotor to hub using wheel nut so as not to tilt disc rotor.

5. Remove adjuster hole plug installed on the disc rotor.

6. Turn the adjuster in direction "A" using a flat-bladed screwdriver as shown, until disc rotor is locked.

7. Turn the adjuster in the opposite direction by 5 or 6 notches after locking.

8. Rotate disc rotor to make sure that there is no drag. Install the adjuster hole plug.

9. Adjust parking brake cable with the following procedure.

- Operate parking brake pedal 10 or more times with the force of 110 ft. lbs. (150 Nm).
- Rotate adjusting nut to adjust parking brake pedal stroke using a deep socket wrench.

➡ **Do not reuse adjusting nut after removing it.**

- Operate parking brake pedal with a force of 44 ft.lbs. (60 Nm), make sure the pedal stroke is within the specified number of notches. (Check it by listening and counting ratchet clicks.)

➡ **Pedal stroke 3–4 notches**

➡ **Make sure that there is no drag on rear brake with parking brake pedal completely released.**

PARKING BRAKE SHOES

REMOVAL & INSTALLATION
See Figures 9 and 10.

➡ Clean brakes with a vacuum dust collector to minimize the hazard of air borne particles or other materials.

❋❋ CAUTION

Clean dust on disc rotor and back plate using a vacuum dust collector. Do not blow with compressed air.

➡ Put matching marks on both disc rotor and wheel hub when removing disc rotor.

1. Remove rear tires from vehicle.

2. Remove disc rotor with parking brake pedal completely in the released

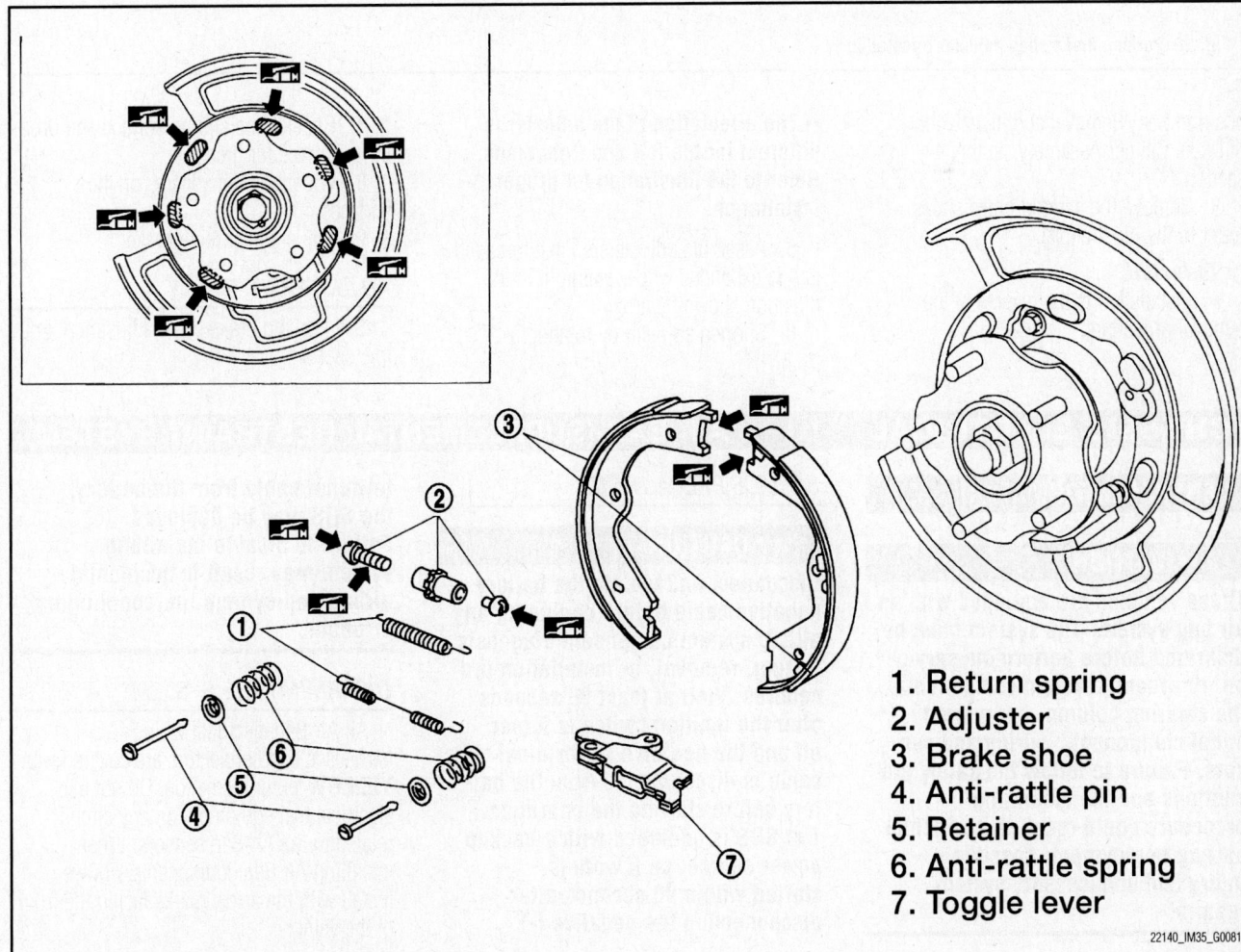

1. Return spring
2. Adjuster
3. Brake shoe
4. Anti-rattle pin
5. Retainer
6. Anti-rattle spring
7. Toggle lever

22140_IM35_G0081

Fig. 9 Parking brake shoe assembly exploded view

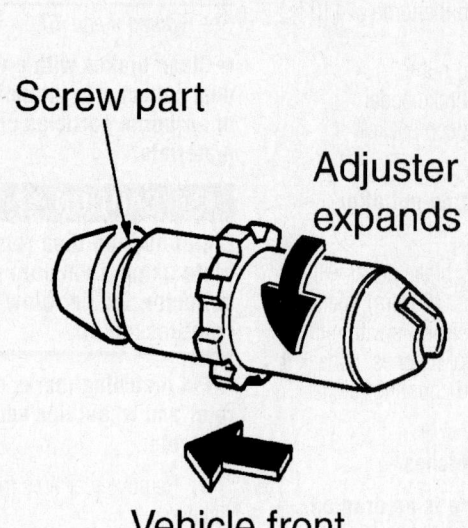

For LH brake

Screw part

Adjuster expands

Vehicle front

For RH brake

Screw part

Adjuster expands

Vehicle front

22140_IM35_G0082

Fig. 10 Parking brake shoe adjuster orientation

position See "Removal and Installation of Brake Caliper Assembly" in this section.

3. Remove the parking brake shoes (refer to the illustration).

To install:

4. Installation is the reverse of the removal procedure.

➡**The orientation of the adjuster is different for the left and right sides. Refer to the illustration for proper installation.**

5. Assemble adjusters so that threaded part is expanded when rotating it in the direction shown by arrow.

6. Shorten adjuster by rotating it.

7. Check shoe sliding surface and drum inner surface for grease.

8. Wipe it off if it adheres on the surfaces.

9. Adjust the parking brake.

ADJUSTMENT

See Parking brake cable adjustment in this section.

CHASSIS ELECTRICAL AIR BAG (SUPPLEMENTAL RESTRAINT SYSTEM)

GENERAL INFORMATION

✳✳ CAUTION

These vehicles are equipped with an air bag system. The system must be disarmed before performing service on, or around, system components, the steering column, instrument panel components, wiring and sensors. Failure to follow the safety precautions and the disarming procedure could result in accidental air bag deployment, possible injury and unnecessary system repairs.

SERVICE PRECAUTIONS

✳✳ CAUTION

Disconnect and isolate the battery negative cable before beginning any airbag system component diagnosis, testing, removal, or installation procedures. Wait at least 90 seconds after the ignition switch is turned off and the negative (-) terminal cable is disconnected from the battery before starting the operation. The SRS is equipped with a backup power source, so if work is started within 90 seconds after disconnecting the negative (-)

terminal cable from the battery, the SRS may be deployed. Failure to disable the airbag system may result in accidental airbag deployment, personal injury, or death.

DISARMING THE SYSTEM

All Air Bag electrical wiring harnesses and connectors are covered with **YELLOW** outer insulation. Do not use electrical test equipment on any circuit related to the Air Bag sensors. When installing Air Bag components, always install with the arrow marks facing the front of the vehicle.

1. Before servicing the vehicle, refer to the precautions in the beginning of this section.

2. Turn the ignition switch to the **OFF** position.

3. Disconnect both battery cables starting with the negative cable first and wait at least 3 minutes after the cables are disconnected. Be sure to insulate the battery terminal ends.

ARMING THE SYSTEM

1. Before servicing the vehicle, refer to the precautions in the beginning of this section.

2. Turn the ignition switch to the **OFF** position.

3. Connect both battery cables starting with the positive cable first.

4. The Air Bag or Air Bag system is equipped with a self-diagnostic operation.

After turning the ignition key to the **ON** or **START** position, the **AIR BAG** warning lamp will illuminate for 7 seconds. After 7 seconds, the **AIR BAG** lamp will extinguish if no malfunction is detected. If the **AIR BAG** lamp does not extinguish after 7 seconds, check the Air Bag self-diagnostic system for a malfunction.

DRIVE TRAIN

FRONT DRIVESHAFT

REMOVAL & INSTALLATION

See Figures 11 and 12.

1. Remove engine undercover.
2. If necessary, remove heat bracket.
3. Remove the three way catalyst (right bank).
4. Put matching marks onto propeller shaft flange yoke and final drive companion flange.

> ### ❄ CAUTION
> **For matching marks, use paint. Never damage propeller shaft flange and companion flange.**

5. Remove the propeller shaft mounting bolts.
6. Remove propeller shaft from the front final drive and transfer.

To install:

7. Note the following, install in the reverse order of removal.

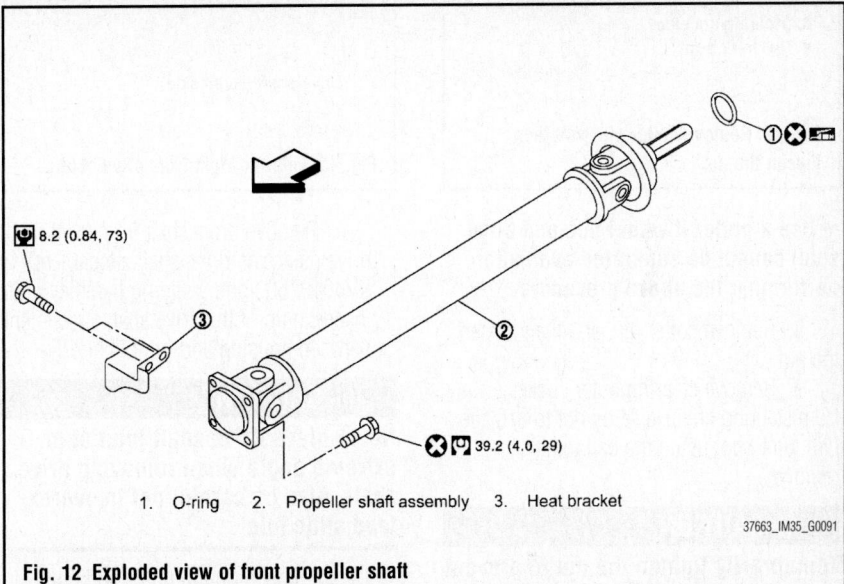

1. O-ring 2. Propeller shaft assembly 3. Heat bracket

8.2 (0.84, 73)
39.2 (4.0, 29)

37663_IM35_G0091

Fig. 12 Exploded view of front propeller shaft

8. Align matching marks to install propeller shaft to final drive companion flange, and then tighten to specified torque.

> ### ❄ CAUTION
> **Never reuse the bolts.**

9. After assembly, perform a driving test to check propeller shaft vibration. If vibration occurred, separate propeller shaft from final drive or transfer. Reinstall companion flange after rotating it by 90, 180, 270 degrees. Then perform driving test and check propeller shaft vibration again at each point.

FRONT HALFSHAFTS

REMOVAL & INSTALLATION

See Figures 13 through 16.

1. Remove tires and wheels from vehicle.
2. Remove wheel sensor from steering knuckle.

> ### ❄ CAUTION
> **Do not pull on wheel sensor harness.**

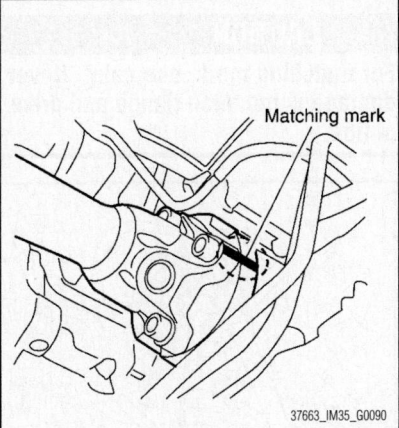

Matching mark

37663_IM35_G0090

Fig. 11 Put matching marks onto propeller shaft flange yoke and final drive companion flange

3. Remove brake hose bracket.
4. Remove torque member mounting bolts. Hang torque member in a place where it will not interfere with work.

> ### ❄ CAUTION
> **Do not depress brake pedal while brake caliper is removed.**

5. Remove disc rotor.
6. Remove cotter pin, then loosen hub lock nut.
7. Separate wheel hub and bearing assembly from drive shaft by lightly tapping the end with a hammer and a wood block, and then remove hub lock nut.

> ### ❄ CAUTION
> **Do not place drive shaft joint at an extreme angle. Also be careful not to overextend slide joint. Do not allow drive shaft to hang down without support for housing (or joint sub-assembly), shaft and the other parts.**

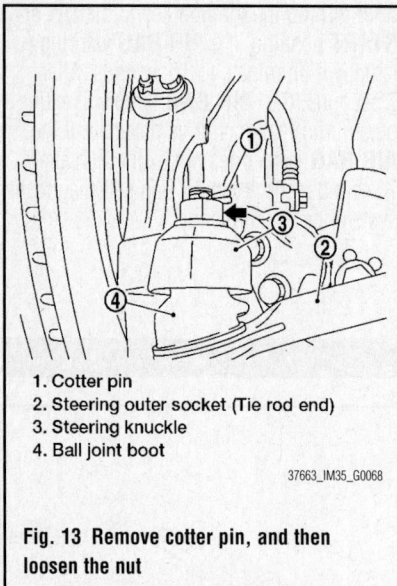

1. Cotter pin
2. Steering outer socket (Tie rod end)
3. Steering knuckle
4. Ball joint boot

37663_IM35_G0068

Fig. 13 Remove cotter pin, and then loosen the nut

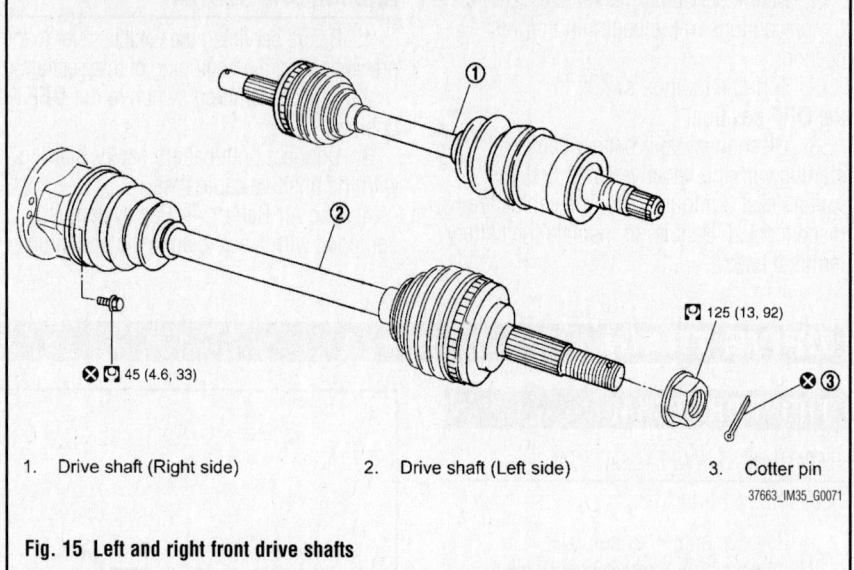

1. Drive shaft (Right side) 2. Drive shaft (Left side) 3. Cotter pin

⊗ ⚙ 45 (4.6, 33)   ⚙ 125 (13, 92)   ⊗ ③

37663_IM35_G0071

Fig. 15 Left and right front drive shafts

➥Use a puller if wheel hub and drive shaft cannot be separated even after performing the above procedure.

8. Remove cotter pin, and then loosen the nut.

9. Remove steering outer socket from steering knuckle so as not to damage ball joint boot using the ball joint remover.

☀ CAUTION

Temporarily tighten the nut to prevent damage to threads and to prevent the ball joint remover from suddenly coming off.

10. Remove drive shaft from wheel hub and bearing assembly.

11. Remove mounting nuts and bolts, and then remove shock absorber arm.

12. Remove drive shaft from front final drive. (Right side)

a. Remove drive shaft from front final drive using the drive shaft attachment KV40107500 and a sliding hammer while inserting tip of the drive shaft attachment between housing and front final drive.

☀ CAUTION

Never place drive shaft joint at an extreme angle when removing drive shaft. Also be careful not to overextend slide joint.

13. Remove mounting nuts and bolts, and then remove drive shaft from vehicle. (Left side)

To install:

☀ CAUTION

Always replace transaxle side oil seal with new one when installing drive shaft.

14. Installation is the reverse order of removal.

15. Place the protector KV38107900

onto front final drive to prevent damage to the oil seal while inserting drive shaft.

16. Slide drive shaft sliding joint and tap with a hammer to install securely. (Right side)

FRONT PINION SEAL

REMOVAL & INSTALLATION

See Figures 17 through 22.

1. Drain gear oil.
2. Remove front propeller shaft.
3. Remove both left and right front drive shafts.
4. Remove side shaft assembly.
5. Remove drive pinion lock nut using a flange wrench.
6. Put matching mark on the end of the drive pinion. The matching mark should be in line with the matching mark on companion flange.

☀ CAUTION

For matching mark, use paint. Never damage companion flange and drive pinion.

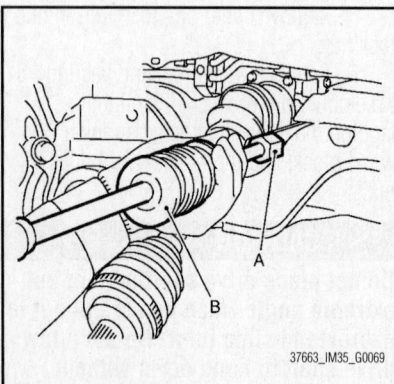

37663_IM35_G0069

Fig. 14 Remove drive shaft from front final drive using the drive shaft attachment (A) and a sliding hammer (B)

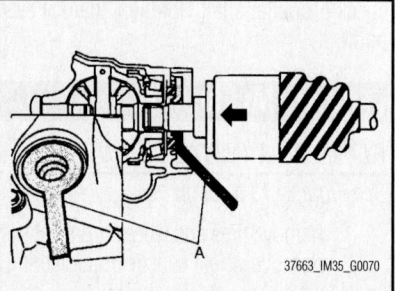

37663_IM35_G0070

Fig. 16 Place the protector (A) onto front final drive to prevent damage to the oil seal while inserting drive shaft

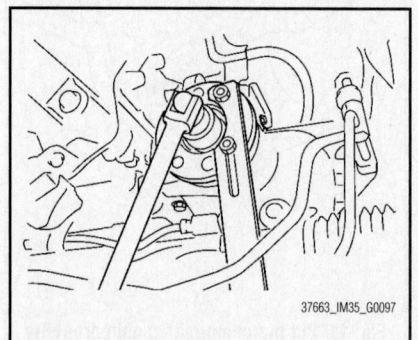

37663_IM35_G0097

Fig. 17 Remove drive pinion lock nut using a flange wrench

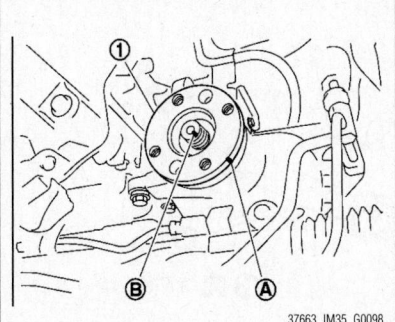

Fig. 18 The matching mark (B) should be in line with the matching mark (A) on companion flange (1)

➡The matching mark on the final drive companion flange indicates the maximum vertical runout position.

7. Remove companion flange using a puller.
8. Remove front oil seal using the puller.

✻✻ CAUTION

Be careful not to damage gear carrier.

To install:

9. Apply multi-purpose grease to front oil seal lips.
10. Using the drifts, install front oil seal as shown.

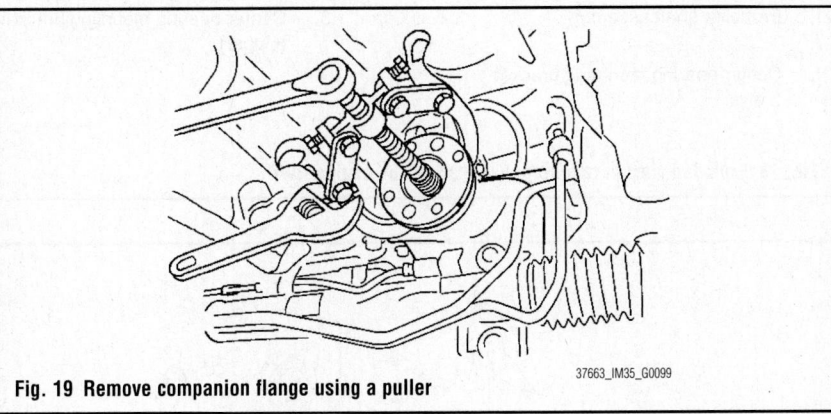

Fig. 19 Remove companion flange using a puller

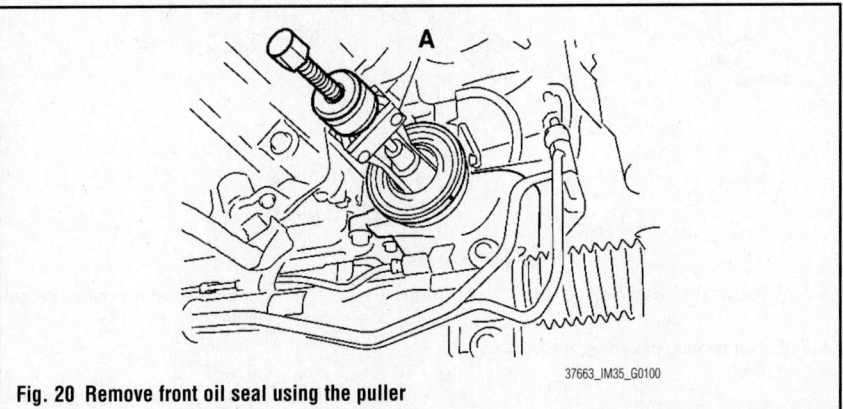

Fig. 20 Remove front oil seal using the puller

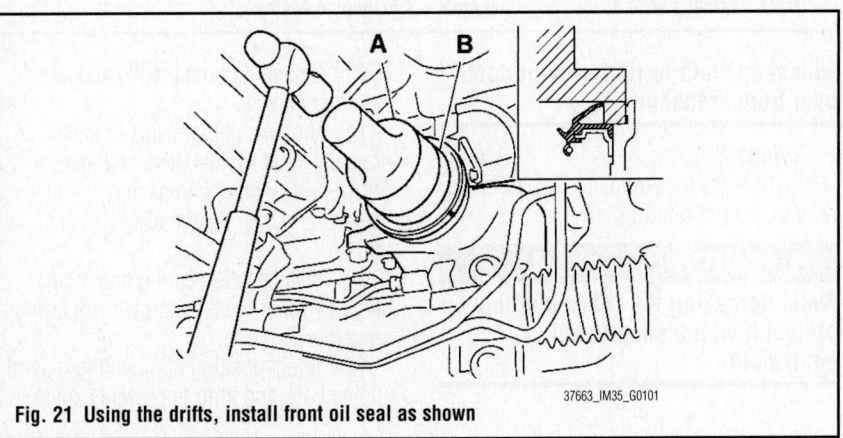

Fig. 21 Using the drifts, install front oil seal as shown

✻✻ CAUTION

Never reuse oil seal. When installing, never incline oil seal.

11. Align the matching mark of drive pinion with the matching mark of companion flange, and then install the companion flange.
12. Apply anti-corrosion oil to the thread and seat of new drive pinion lock nut, and temporarily tighten drive pinion lock nut to drive pinion.

✻✻ CAUTION

Never reuse drive pinion lock nut.

13. Tighten to drive pinion lock nut to 94–181 ft. lbs. (127–245 Nm), while adjust total preload torque to 14–23 inch lbs. (1.56–2.65 Nm).

   a. Adjust to the lower limit of the drive pinion lock nut tightening torque first.

   b. After adjustment, rotate drive pinion back and forth 2 to 3 times to check for unusual noise, rotation malfunction, and other malfunctions.

   c. If measured value is out of the specification, remove final drive assembly and disassemble drive pinion parts to check and adjust each part.
14. Install front propeller shaft.
15. Install side shaft assembly.
16. Install both front drive shafts.

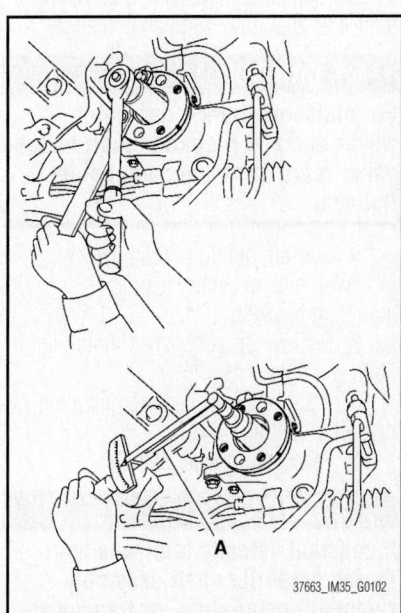

Fig. 22 Tighten to drive pinion lock nut, while adjust total preload torque

17. Refill gear oil to the final drive and check oil level.

18. Check the final drive for oil leakage.

REAR DRIVESHAFT

REMOVAL & INSTALLATION

See Figures 23 through 27.

1. Move the A/T select lever to N position and release the parking brake.

2. Remove the floor reinforcement.

3. Remove the center muffler.

4. Matching marks:

5. For 3.5L engine, put matching marks on propeller shaft rebro joint with final drive companion flange.

> ✳✳ **CAUTION**
>
> **For matching marks, use paint. Never damage propeller shaft rebro joint and companion flange.**

6. For 4.5L engines, put matching marks on propeller shaft rubber coupling with transmission companion flange and on rebro joint with final drive companion flange.

> ✳✳ **CAUTION**
>
> **For matching marks, use paint. Never damage rubber coupling, rebro joint and companion flanges.**

7. For AWD models, put matching marks on propeller shaft flange yoke with transfer companion flange and on rebro joint with final drive companion flange.

> ✳✳ **CAUTION**
>
> **For matching marks, use paint. Never damage propeller shaft flange yoke, rebro joint and companion flanges.**

8. Loosen, but do not remove mounting nuts of center bearing mounting brackets.

9. Remove propeller shaft mounting bolts and nuts.

10. Remove center bearing mounting bracket mounting nuts.

11. Remove propeller shaft.

> ✳✳ **CAUTION**
>
> **If constant velocity joint was bent during propeller shaft assembly removal, installation, or transportation, its boot may be damaged. Wrap boot interference area to metal part**

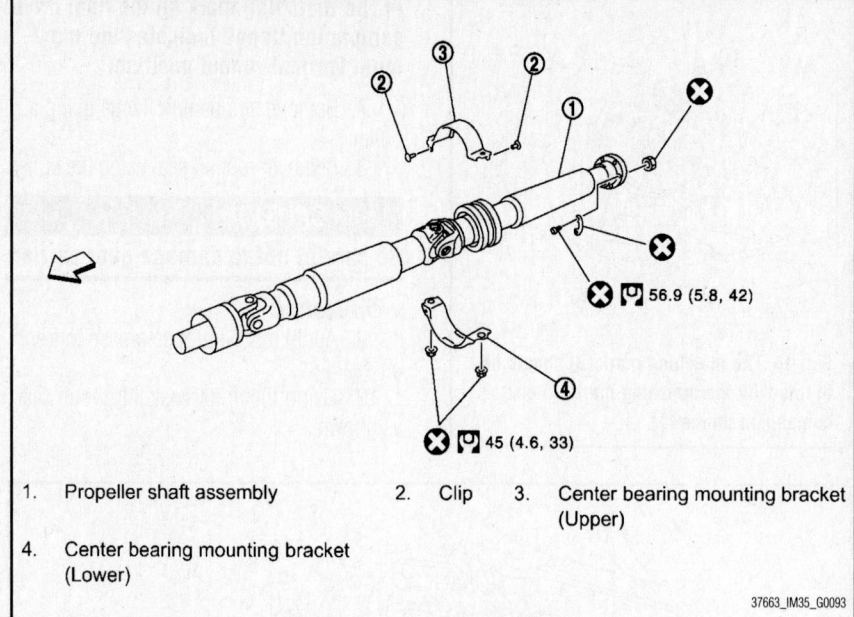

1. Propeller shaft assembly
2. Clip
3. Center bearing mounting bracket (Upper)
4. Center bearing mounting bracket (Lower)

37663_IM35_G0093

Fig. 23 Exploded view of rear propeller shaft—3.5L engine RWD

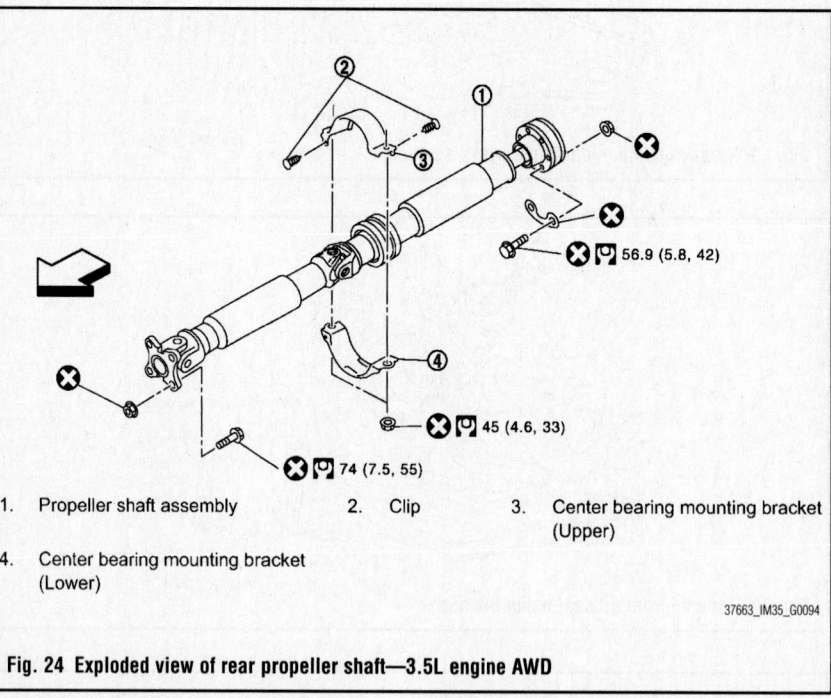

1. Propeller shaft assembly
2. Clip
3. Center bearing mounting bracket (Upper)
4. Center bearing mounting bracket (Lower)

37663_IM35_G0094

Fig. 24 Exploded view of rear propeller shaft—3.5L engine AWD

with shop cloth or rubber to protect boot from breakage.

To install:

12. Note the following, and install in the reverse order of removal.

> ✳✳ **CAUTION**
>
> **Avoid damaging the rebro joint boot, protect it with a shop towel or equivalent.**

13. Tighten mounting bolts and nuts in the order shown.

14. Align matching marks to install propeller shaft to final drive and transfer (AWD models only) companion flanges, and then tighten to specified torque.

15. Install center bearing mounting bracket (Upper) with its arrow mark facing forward.

16. Adjust position of mounting bracket sliding back and forth to prevent play in

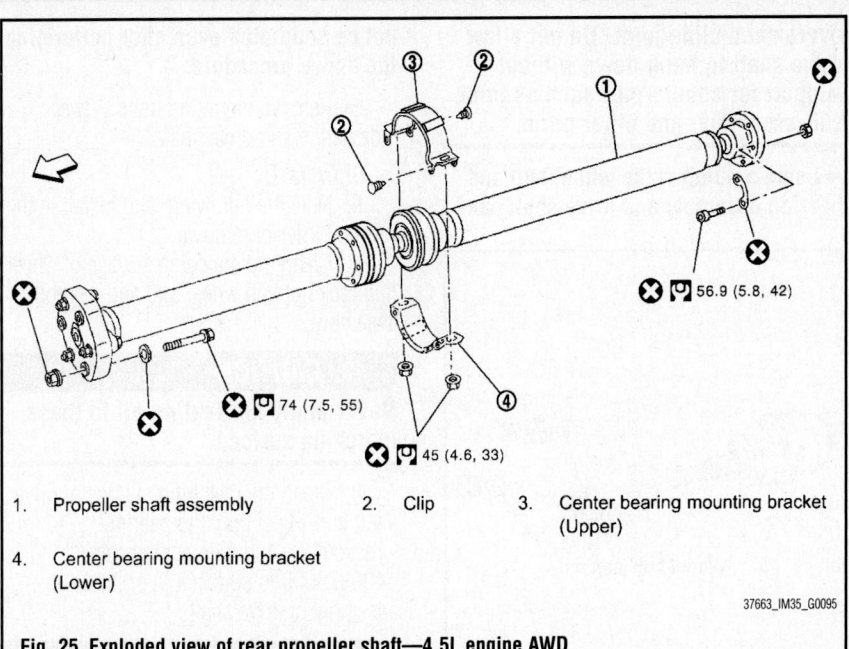

1. Propeller shaft assembly
2. Clip
3. Center bearing mounting bracket (Upper)
4. Center bearing mounting bracket (Lower)

❌ 🔧 56.9 (5.8, 42)
❌ 🔧 74 (7.5, 55)
❌ 🔧 45 (4.6, 33)

37663_IM35_G0095

Fig. 25 Exploded view of rear propeller shaft—4.5L engine AWD

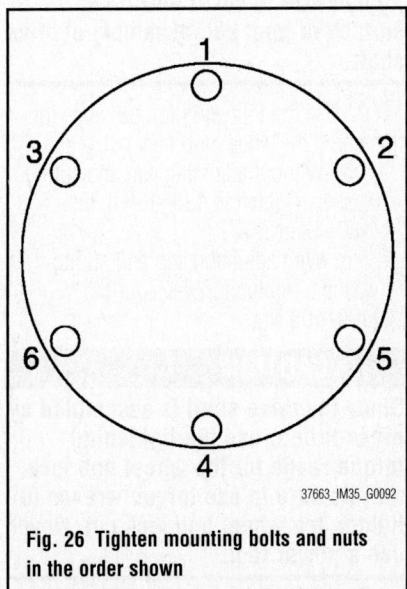

37663_IM35_G0092

Fig. 26 Tighten mounting bolts and nuts in the order shown

thrust direction of center bearing insulator. Install bracket to vehicle.

17. After assembly, perform a driving test to check propeller shaft vibration. If vibration occurred, separate propeller shaft from final drive. Reinstall companion flange after rotating it by 60, 120, 180, 240, 300 degrees. Then perform driving test and check propeller shaft vibration again at each point.

18. If propeller shaft or final drive has been replaced, connect them as follows:

a. Install the propeller shaft while aligning its matching mark A with the matching mark B on the joint as close as possible.

b. Tighten the joint bolts to the specified torque.

❊❊❊ CAUTION

Never reuse the bolts, nuts and washers.

REAR HALFSHAFTS

REMOVAL & INSTALLATION

See Figures 28 through 31.

1. Remove tires and wheels.
2. Remove cotter pin and adjusting cap (if equipped), then loosen wheel hub lock nut.
3. Remove stabilizer connecting rod mounting bracket mounting bolt and free stabilizer connecting rod.
4. Separate the wheel hub and bearing assembly from drive shaft by lightly tapping the end with a hammer and wood block, and

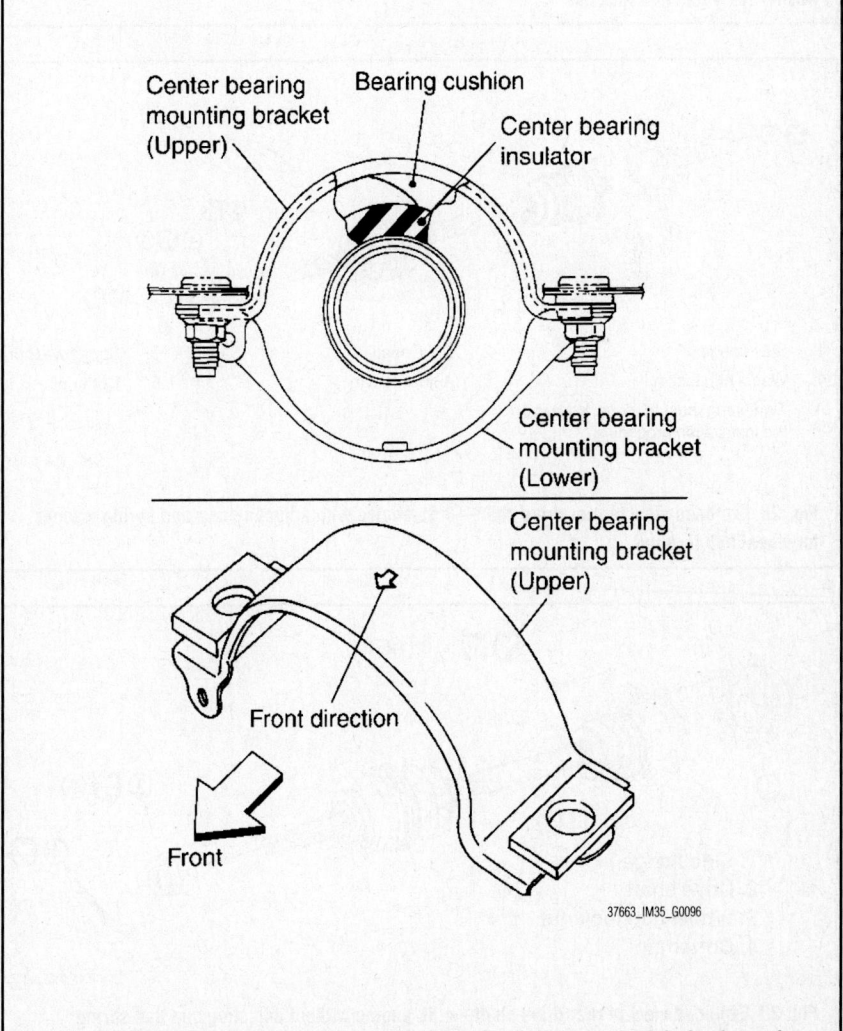

37663_IM35_G0096

Fig. 27 Install center bearing mounting bracket (Upper) with its arrow mark facing forward

then remove wheel hub lock nut and spring washer (if equipped).

✳✳ CAUTION

Do not place drive shaft joint at an extreme angle. Also be careful not to overextend slide joint. **Do not allow drive shaft to hang down without support for counterpart such as joint sub-assembly, and other parts.**

➡ Using a puller if the wheel hub and bearing assembly and drive shaft can- not be separated even after performing the above procedure.

5. Remove mounting bolts between side flange and drive shaft.

To install:

6. Note the following, and install in the reverse order of removal.

7. Clean the matching surface of wheel hub lock nut and wheel hub and bearing assembly.

✳✳ CAUTION

Never apply lubricating oil to these matching surface.

8. Clean the matching surface of drive shaft and wheel hub and bearing assembly. And then apply paste 440037S000 to surface (A) of joint sub-assembly of drive shaft.

✳✳ CAUTION

Apply paste to cover entire flat surface of joint sub-assembly of drive shaft.

9. Use the following torque range for tightening the wheel hub lock nut.
   a. Without adjusting cap and spring washer: Tighten to 133–136 ft. lbs. (180–185 Nm).
   b. With adjusting cap and spring washer: Tighten to 74–77 ft. lbs. (100–105 Nm).

✳✳ CAUTION

Since the drive shaft is assembled by press-fitting, use the tightening torque range for the wheel hub lock nut. Be sure to use torque wrench to tighten the wheel hub lock nut. Never use a power tool.

➡ Wheel hub lock nut tightening torque does not over torque for avoiding axle noise, and does not less than torque for avoiding looseness.

10. Perform the final tightening of each of parts under unladen conditions, which were removed when removing wheel hub and bearing assembly and axle housing.

11. When installing the spring washer, face the identification paint mark to the wheel hub and bearing assembly side. (With adjusting cap and spring washer for wheel hub lock nut)

12. When installing the adjusting cap, check that there must be no play. (With adjusting cap and spring washer for wheel hub lock nut)

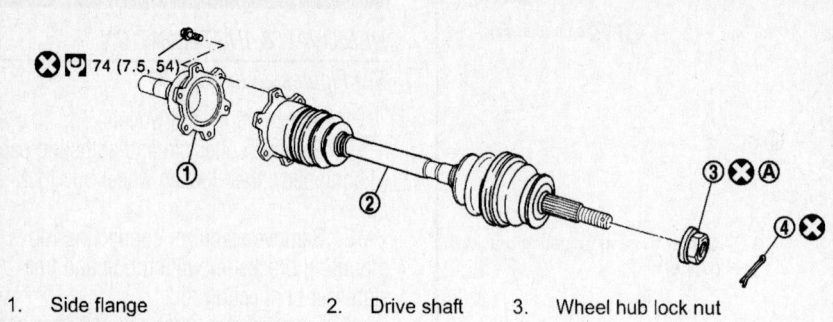

1. Side flange
2. Drive shaft
3. Wheel hub lock nut
4. Cotter pin
A. Tightening must be done following the installation procedure.

37663_IM35_G0103

Fig. 28 Exploded view of rear drive shaft—3.5L engine without adjusting cap and spring washer for wheel hub lock nut

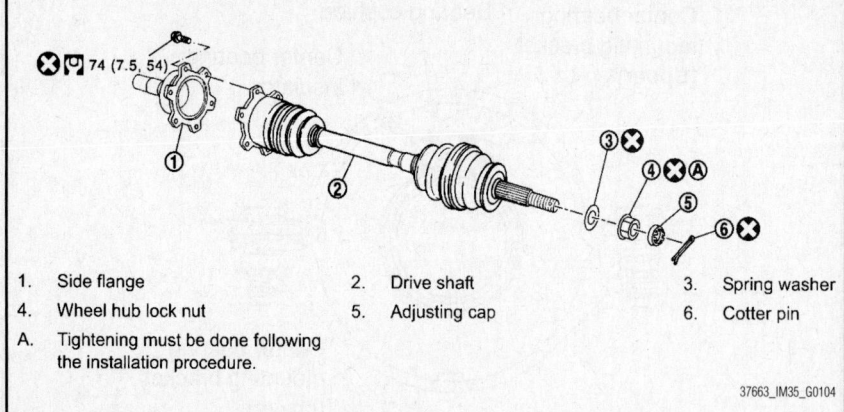

1. Side flange
2. Drive shaft
3. Spring washer
4. Wheel hub lock nut
5. Adjusting cap
6. Cotter pin
A. Tightening must be done following the installation procedure.

37663_IM35_G0104

Fig. 29 Exploded view of rear drive shaft—3.5L engine with adjusting cap and spring washer for wheel hub lock nut

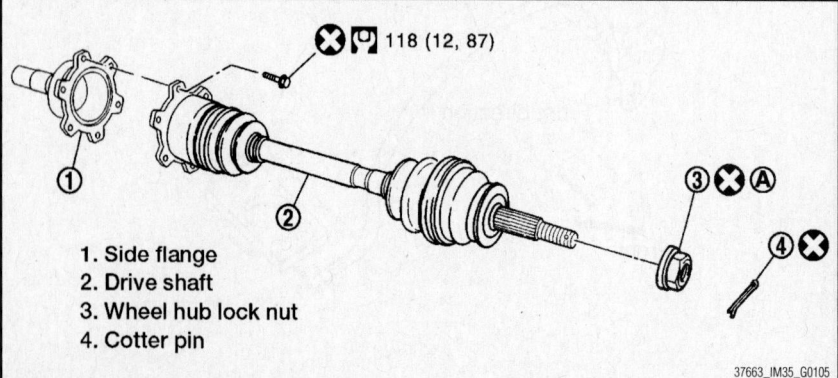

1. Side flange
2. Drive shaft
3. Wheel hub lock nut
4. Cotter pin

37663_IM35_G0105

Fig. 30 Exploded view of rear drive shaft—4.5L engine without adjusting cap and spring washer for wheel hub lock nut

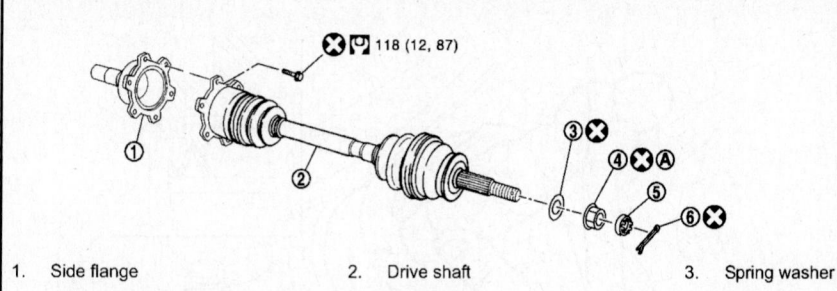

1. Side flange
2. Drive shaft
3. Spring washer
4. Wheel hub lock nut
5. Adjusting cap
6. Cotter pin
A. Tightening must be done following the installation procedure. Refer to "INSTALLATION".

37663_IM35_G0106

Fig. 31 Exploded view of rear drive shaft—4.5L engine with adjusting cap and spring washer for wheel hub lock nut

13. Never reuse cotter pin, wheel hub lock nut, spring washer (if equipped).

REAR PINION SEAL

REMOVAL & INSTALLATION

See Figures 32 through 41.

➡The reuse of collapsible spacer is prohibited in principle. However, it is reusable on a one-time basis only in cases when replacing front oil seal.

➡The diagonally shaded area shows stamping point for replacement frequency of front oil seal.

➡The following shows if a collapsible spacer replacement is needed before replacing front oil seal.

    a. Stamp—collapsible spacer replacement
    b. No stamp—Not required
    c. "0" or "0" on the far right of stamp—Required
    d. "01" or "1" on the far right of stamp—Not required

    e. When collapsible spacer replacement is required, disassemble final drive assembly to replace collapsible spacer and front oil seal.
    1. Drain gear oil.
    2. Make a judgment if a collapsible spacer replacement is required
    3. Remove center muffler.
    4. Remove rear wheel sensor.
    5. Match mark the drive shaft prior to removal.
    6. Remove drive shaft from final drive. Then suspend it by wire.
    7. Install attachment to side flange, on M35 models and then pull out the side flange with the sliding hammer. Tool number A: KV40104100—B: ST36230000 (J- 25840-A)
    8. Install attachment to side flange, on M45 models and then pull out the side flange with the sliding hammer. Tool number A: KV40101000—B: ST36230000 (J-25840-A)

➡Install circle clip in same position it was removed from.

    9. Match mark the propeller shaft prior to removal.
    10. Remove propeller shaft.
    11. Measure the total preload with the preload gauge.
    12. Tool number A: ST3127S000 (J-25765-A)
    13. Record the preload measurement for use on assembly.
    14. Put matching mark on the end of the drive pinion. The matching mark should be in line with the matching mark on companion flange.

✳✳ WARNING

For matching mark, use paint. Never damage companion flange and drive pinion.

➡The matching mark on the final drive companion flange indicates the maximum vertical runout position.

    15. Remove drive pinion lock nut.
    16. Remove companion flange using a puller.
    17. Remove front oil seal using the puller.

To install:

    18. Apply multi-purpose grease to front oil seal lips.
    19. Install front oil seal using the drift as shown.

✳✳ WARNING

Never reuse oil seal.

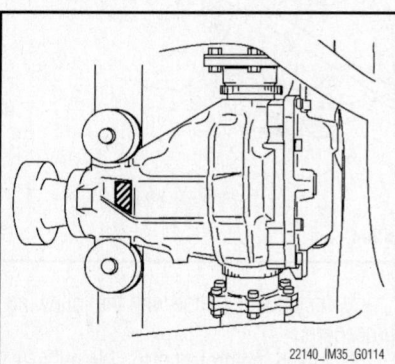

22140_IM35_G0114

Fig. 32 Diagonally shaded area on rear differential showing stamping point for replacement frequency of front oil seal

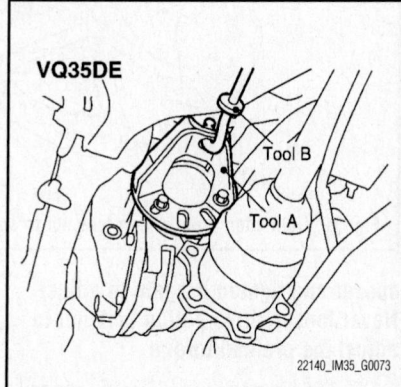

VQ35DE

Tool B

Tool A

22140_IM35_G0073

Fig. 33 Special tool used to remove side flange

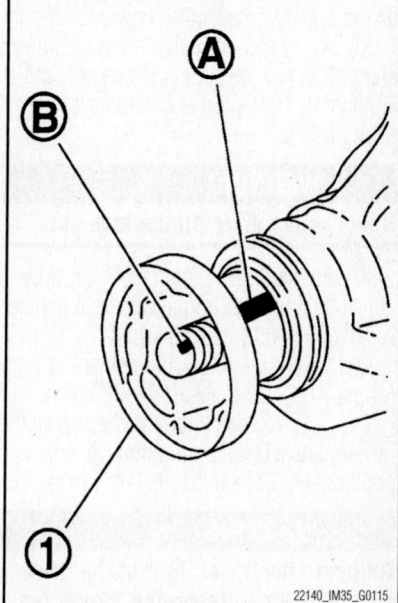

22140_IM35_G0115

Fig. 34 Final drive companion flange (1) Matching mark (A) thru (B)

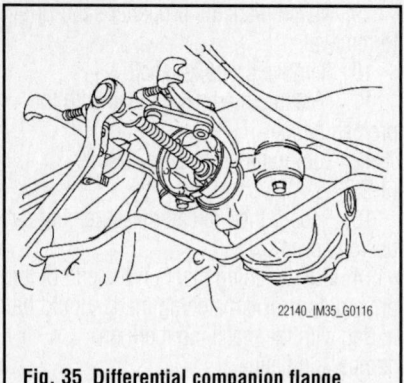

Fig. 35 Differential companion flange puller

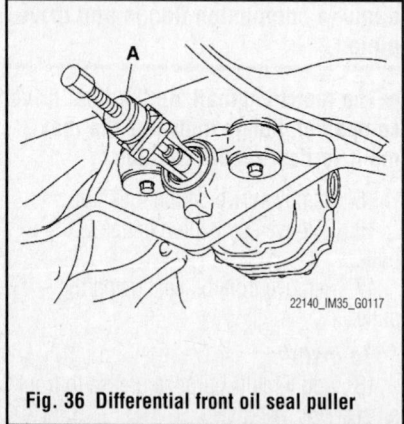

Fig. 36 Differential front oil seal puller

✳✳ WARNING

Never incline oil seal when installing.

20. Align the matching mark of drive pinion with the matching mark of companion flange, and then install the companion flange.

21. Apply anti-corrosion oil to the thread and seat of new drive pinion lock nut, and temporarily tighten drive pinion lock nut to drive pinion.

✳✳ WARNING

Never reuse drive pinion lock nut.

22. Tighten drive pinion lock nut, while adjust total preload torque with Tool number A: ST3127S000 (J-25765-A).

23. Drive pinion lock nut tightening torque: 109–238 ft. lbs. (147–323 Nm).

24. Total preload torque should equal the measurement taken during removal plus an additional 1– 3 inch lbs. (0.1–0.4 Nm).

✳✳ WARNING

Adjust to the lower limit of the drive pinion lock nut tightening torque first.

➡ If the preload torque exceeds the specified value, replace. collapsible

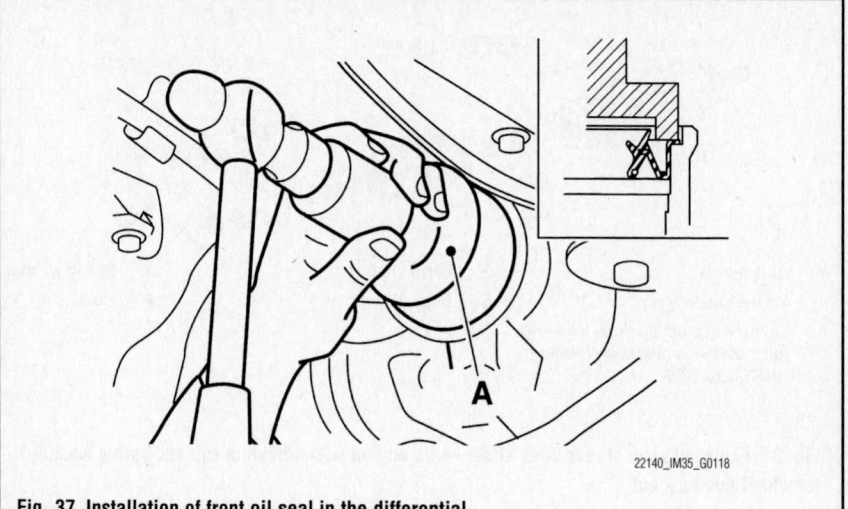

Fig. 37 Installation of front oil seal in the differential

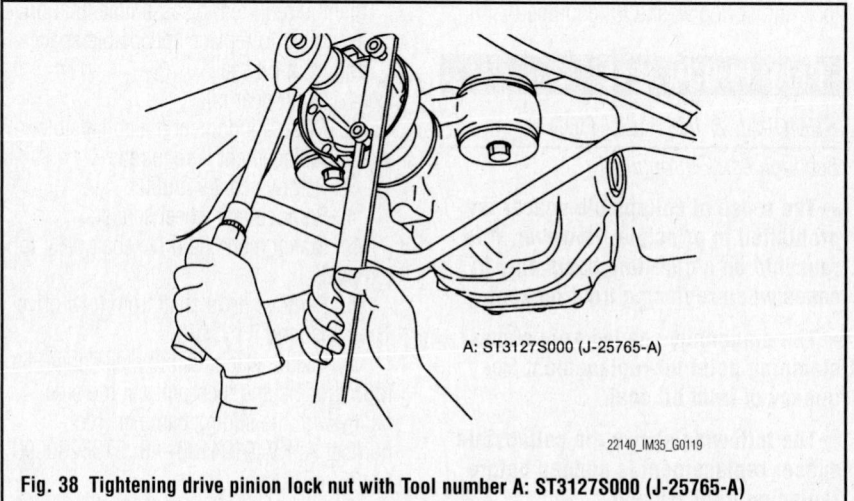

A: ST3127S000 (J-25765-A)

Fig. 38 Tightening drive pinion lock nut with Tool number A: ST3127S000 (J-25765-A)

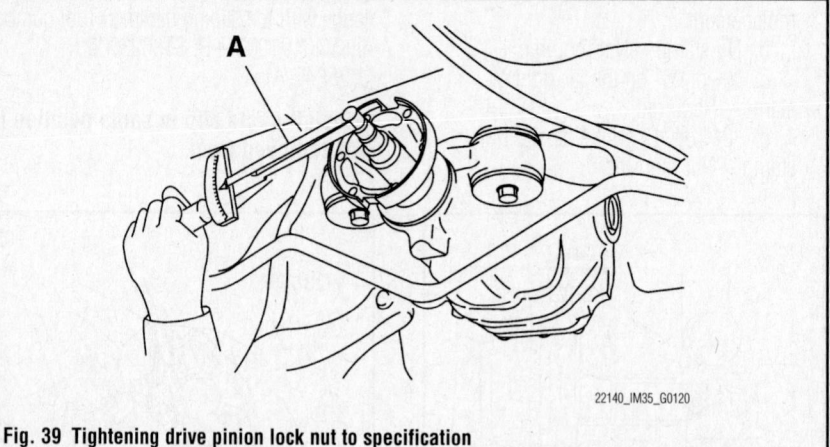

Fig. 39 Tightening drive pinion lock nut to specification

spacer and tighten it again to adjust. **Never loosen drive pinion lock nut to adjust the preload torque.**

25. Make a stamping for identification of front oil seal replacement frequency.

26. Install propeller shaft.

27. Install side flange with the following procedure.

28. Attach the protector to side oil seal.

29. Apply multi-purpose grease to side oil seal lips.

30. Attach the protector to side oil seal. Tool number kv38100200 (J-26233).

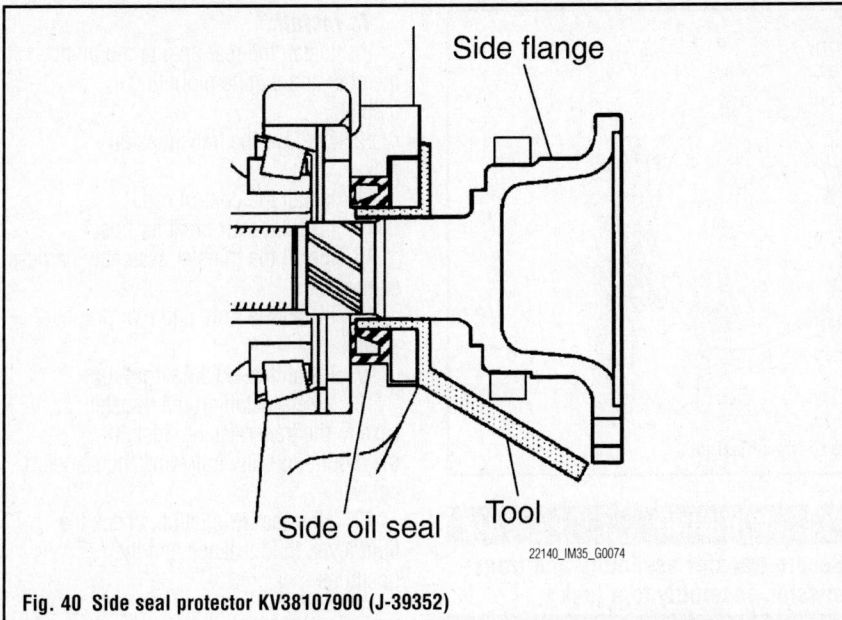

Fig. 40 Side seal protector KV38107900 (J-39352)

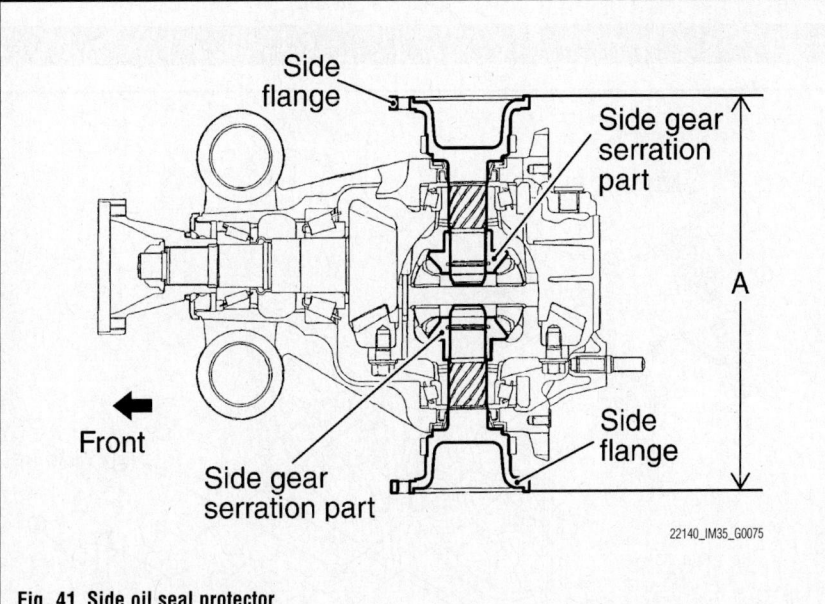

Fig. 41 Side oil seal protector

31. After the side flange is inserted and the serrated part of side gear has engaged the serrated part of flange, remove the protector.

32. Put a suitable drift on the center of side flange, then drive it until sound changes.

➥**When installation is completed, driving sound of the side flange turns into a sound, which seems to affect the whole final drive.**

33. Confirm that the dimension of the side flange installation measurement comes into the following.

34. Measurement A: 12.83–12.91 in (326– 328 mm)

35. Install drive shaft.

36. Install rear wheel sensor.

37. Install center muffler.

38. Refill gear oil to the final drive and check oil level.

39. Check the final drive for oil leakage

TRANSFER CASE ASSEMBLY

REMOVAL & INSTALLATION

See Figures 42 and 43.

1. Remove exhaust front tube.

2. Remove front and rear propeller shaft.

3. Match mark the shaft and companion for reassembly.

4. Disconnect transfer assembly harness connector and separate harness from transfer assembly.

5. Remove air breather hose.

6. Remove control rod.

7. Support transfer assembly and transmission assembly with a jack.

8. Remove rear engine mounting member and engine mounting insulator.

Bolt No.	1	2	3	4
Quantity	4	3	2	1
Bolt length " ℓ " mm (in)	75 (2.95)	45 (1.77)	40 (1.57)	30 (1.18)
Tightening torque N·m (kg-m, ft-lb)	37 (3.8, 27)			

Fig. 42 Bolt installation chart for transfer case assemblies

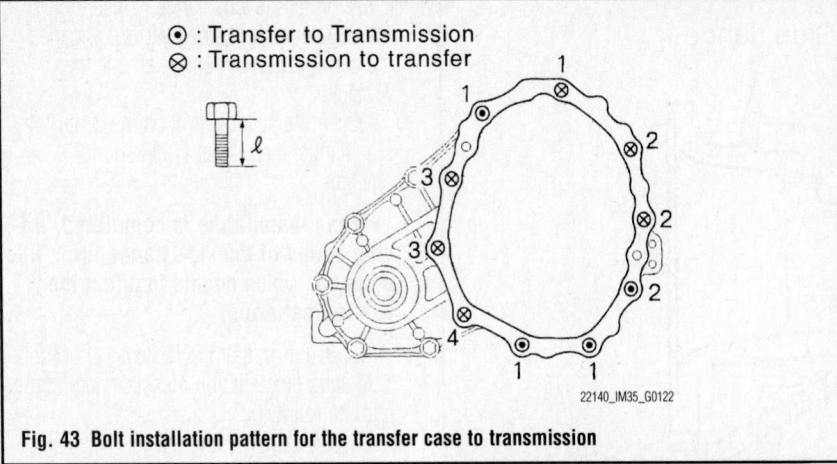

◉ : Transfer to Transmission
⊗ : Transmission to transfer

22140_IM35_G0122

Fig. 43 Bolt installation pattern for the transfer case to transmission

9. Lower jack to the position where the top transfer mounting bolts can be removed.

10. Remove transfer mounting bolts and separate transfer from transmission.

❊❊ WARNING

Secure transfer assembly and transmission assembly to a jack.

To install:

11. Install the rear engine mounting member and engine mounting insulator.

12. Remove the transmission jack.

13. Install the control rod.

14. Install the air breather hose

15. Install the transfer assembly harness connector.

16. Install the front and rear propeller shafts.

17. Install the exhaust front tube.

18. When installing the transfer case to the transmission, install the mounting bolts following the standard below.

19. After the installation, check the fluid level, fluid leakage and the A/T positions

ENGINE COOLING

ENGINE FAN

REMOVAL & INSTALLATION

See Figure 44.

1. Remove engine room cover (RH and LH).

2. Remove air duct (inlet) and air cleaner case assembly.

3. Drain engine coolant from radiator.

4. Remove reservoir tank. (3.5L engine)

5. Disconnect harness connector from cooling fan control module, and move harness to aside.

6. Remove the upper radiator hose.

7. Remove cooling fan assembly.

❊❊ CAUTION

Be careful not to damage or scratch on radiator core.

To install:

8. Note the following, and install in the reverse order of removal.

❊❊ CAUTION

Only use genuine parts for radiator shroud and cooling fan mounting bolt and observe the specified torque (to prevent radiator from being damaged).

9. Check that fan motors operate normally.

➡**Cooling fans are controlled by cooling fan control module.**

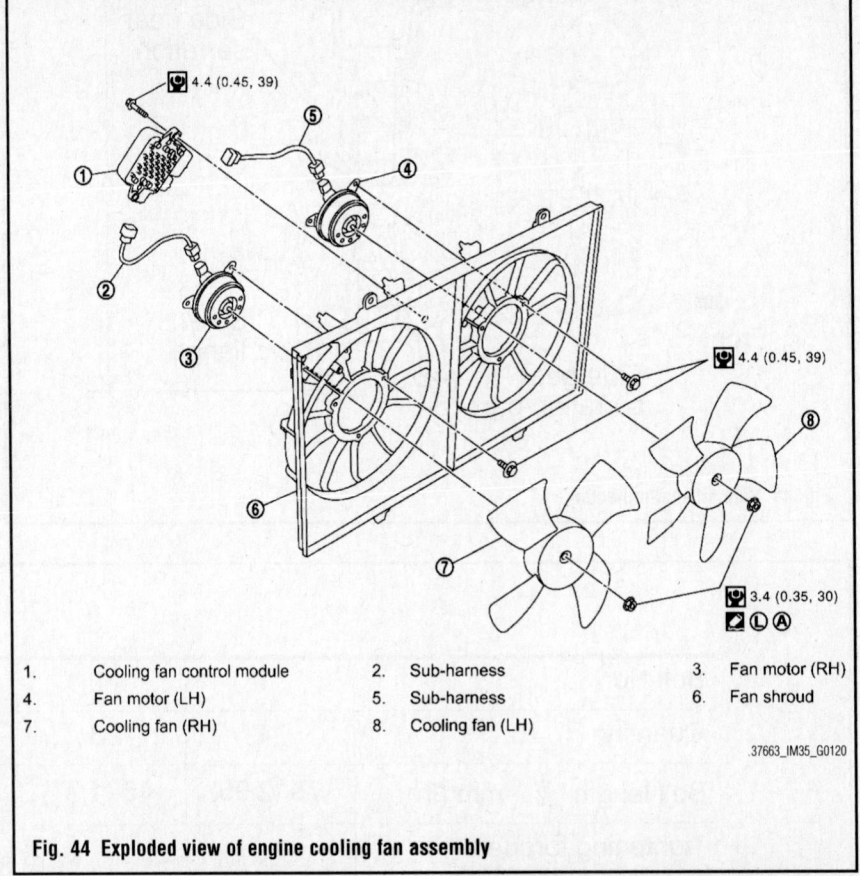

1.	Cooling fan control module	2.	Sub-harness	3.	Fan motor (RH)
4.	Fan motor (LH)	5.	Sub-harness	6.	Fan shroud
7.	Cooling fan (RH)	8.	Cooling fan (LH)		

.37663_IM35_G0120

Fig. 44 Exploded view of engine cooling fan assembly

RADIATOR

REMOVAL & INSTALLATION

See Figures 45 and 46.

❊❊ WARNING

Never remove radiator cap when engine is hot. Serious burns could occur from high-pressure engine

coolant escaping from radiator. **Wrap a thick cloth around the cap. Slowly turn it a quarter of a turn to release built-up pressure. Carefully remove radiator cap by turning it all the way.**

1. Remove the following parts:
   - Front engine undercover
   - Engine room cover (RH and LH)
   - Air duct (inlet) and air cleaner case assembly
2. Remove front grille and front grille support.
3. Drain engine coolant from radiator.

✳✳ CAUTION

Perform this step when the engine is cold. Never spill engine coolant on drive belt.

4. Remove the A/C piping bracket from left side member, and then move the A/C piping out of the way.
5. Disconnect A/T fluid cooler hoses from radiator. Install blind plug to avoid leakage of A/T fluid.
6. Remove the upper and lower radiator hoses and reservoir tank hose.

✳✳ CAUTION

Be careful not to allow engine coolant to contact drive belt.

7. Remove cooling fan assembly.

✳✳ CAUTION

Never damage or scratch radiator core when removing.

8. Remove radiator as follows:
   a. Remove mounting bracket (RH and LH).

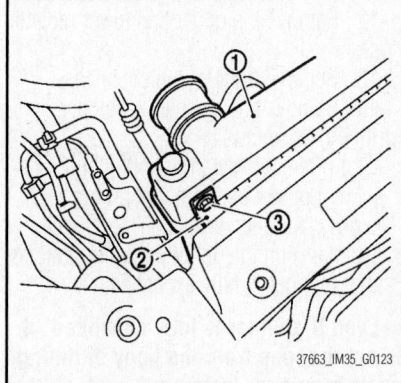

Fig. 45 Pull the radiator (1) rearward from the vehicle, and then remove the mounting bolts (3) and A/C condenser (2)

b. Pull the radiator rearward from the vehicle, and then remove the mounting bolts and A/C condenser.

➡**Figure shows right side.**

c. Lift up and pull the radiator rearward, and then remove the lower mounting rubber from the radiator core support.

✳✳ CAUTION

At this time, A/C condenser is on the lower end of radiator front surface. Minimize the movement to the rear side.

d. Lift up the A/C condenser to disengage the lower end of front surface, and then remove the radiator.

✳✳ CAUTION

Be careful not to damage radiator and A/C condenser core. Minimize the lift of A/C condenser to prevent load from being applied to A/C piping.

e. After removing the radiator, place the A/C condenser on the radiator core support to prevent load from being applied to piping. And then, temporarily secure them using a rope to prevent them from being dropped.

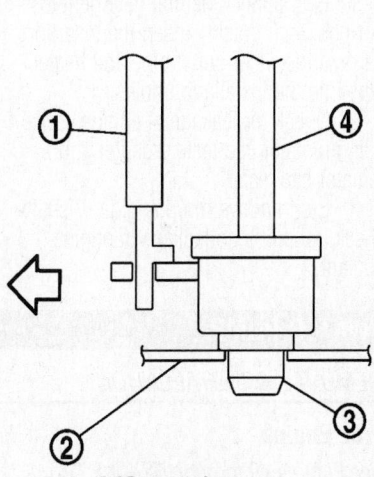

1. A/C condenser
2. Radiator core support
3. Lower mounting rubber
4. Radiator

Fig. 46 Lift up and pull the radiator rearward, and then remove the lower mounting rubber from the radiator core support

To install:

9. Install in the reverse order of removal.
10. Check for leakage of engine coolant using the radiator cap tester adapter and the radiator cap tester.
11. Start and warm up the engine. Visually check that there is no leakage of engine coolant and A/T fluid.

THERMOSTAT

REMOVAL & INSTALLATION

3.5L Engine

1. Remove engine room cover (RH and LH).
2. Remove air duct (inlet).
3. Remove front engine undercover.
4. Drain engine coolant from radiator drain plug at the bottom of radiator, and from water drain plug at the front of cylinder block.

✳✳ CAUTION

Perform this step when the engine is cold. Never spill engine coolant on drive belt.

5. Disconnect the lower radiator hose and oil cooler water hose from water inlet and thermostat assembly.
6. Disconnect intake valve timing control solenoid valve harness connector (bank 2), and remove intake valve timing control solenoid valve (bank 2).
7. Remove water inlet and thermostat assembly.

✳✳ CAUTION

Never disassemble water inlet and thermostat assembly. Replace them as a unit, if necessary.

To install:

8. Note the following, and install in the reverse order of removal.

✳✳ CAUTION

Be careful not to spill engine coolant over engine room. Use rag to absorb engine coolant.

9. Check for leakage of engine coolant using the radiator cap tester adapter and the radiator cap tester.
10. Start and warm up the engine. Visually check that there is no leakage of engine coolant.

4.5L Engine

1. Remove engine room cover (RH and LH).
2. Remove engine cover.
3. Remove air duct (inlet).
4. Drain engine coolant from drain plugs on radiator and both side of cylinder block.

✳✳ CAUTION

Perform this step when engine is cold. Never spill engine coolant on drive belts.

5. Disconnect water suction hose from water inlet.
6. Remove water inlet and thermostat.

✳✳ CAUTION

Never disassemble thermostat.

7. Remove the upper and lower intake manifolds.
8. Disconnect the upper radiator hose from thermostat housing.
9. Disconnect heater hoses from water outlet and heater pipe.
10. Remove thermostat housing, water outlet pipe, water connector, water control valve, water outlet, and heater pipe.

✳✳ CAUTION

Never disassemble water control valve.

To install:
11. Note the following, and install in the reverse order of removal.

✳✳ CAUTION

Be careful not to spill engine coolant over engine room. Use rag to absorb engine coolant.

Thermostat And Water Control Valve
See Figures 47 and 48.

1. Install thermostat and water control valve with the whole circumference of each flange part fit securely inside rubber ring.
2. Install thermostat with jiggle valve facing upwards. (The position deviation may be within the range of +/-10 degrees)
3. Install water control valve with the up-mark facing up and the frame center part facing upwards. (The position deviation may be within the range of +/-10 degrees)

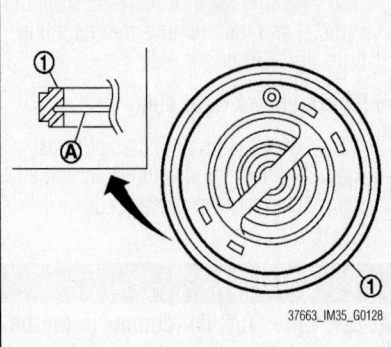

Fig. 47 Install thermostat and water control valve with the whole circumference of each flange (A) part fit securely inside rubber ring (1)

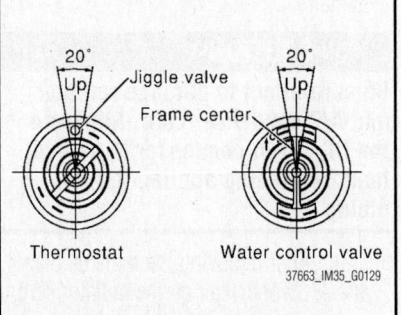

Fig. 48 Installation positions of thermostat and water control valve

Water Outlet Pipe And Heater Pipe

1. First apply a neutral detergent to O-rings, then quickly insert the insertion parts of the water outlet pipe and heater pipe into the installation holes.
2. Check for leakage of engine coolant using radiator cap tester adapter and radiator cap tester.
3. Start and warm up engine. Visually check if there is no leakage of engine coolant.

WATER PUMP

REMOVAL & INSTALLATION

3.5L Engine
See Figures 49 through 52.

✳✳ CAUTION

Note the following:

- When removing water pump assembly, be careful not to get engine coolant on drive belt.
- Water pump cannot be disassembled and should be replaced as a unit.

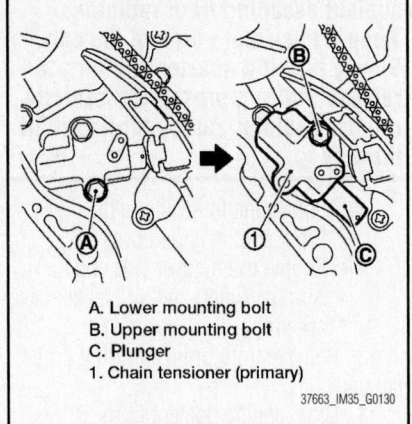

A. Lower mounting bolt
B. Upper mounting bolt
C. Plunger
1. Chain tensioner (primary)

Fig. 49 Remove timing chain tensioner (primary)

1. Remove engine room cover (RH and LH).
2. Remove engine cover.
3. Release the fuel pressure.
4. Disconnect the battery cable from the negative terminal.
5. Remove air duct and air cleaner case assembly (RH and LH).
6. Remove reservoir tank.
7. Separate engine harness removing their brackets from front timing chain case.
8. Remove front engine undercover.
9. Drain engine oil.

✳✳ CAUTION

Perform this step when the engine is cold. Never spill engine oil on drive belt.

10. Drain engine coolant from radiator.

✳✳ CAUTION

Perform this step when the engine is cold. Never spill engine coolant on drive belt.

11. Remove cooling fan assembly.
12. Remove the upper and lower radiator hoses.
13. Remove front timing chain case.
14. Remove timing chain tensioner (primary) as follows:
   a. Remove lower mounting bolt.
   b. Loosen upper mounting bolt slowly, and then turn chain tensioner (primary) on the upper mounting bolt so that plunger is fully expanded.

➡**Even if plunger is fully expanded, it is not dropped from the body of timing chain tensioner (primary).**

   c. Remove upper mounting bolt, and then remove timing chain tensioner (primary).

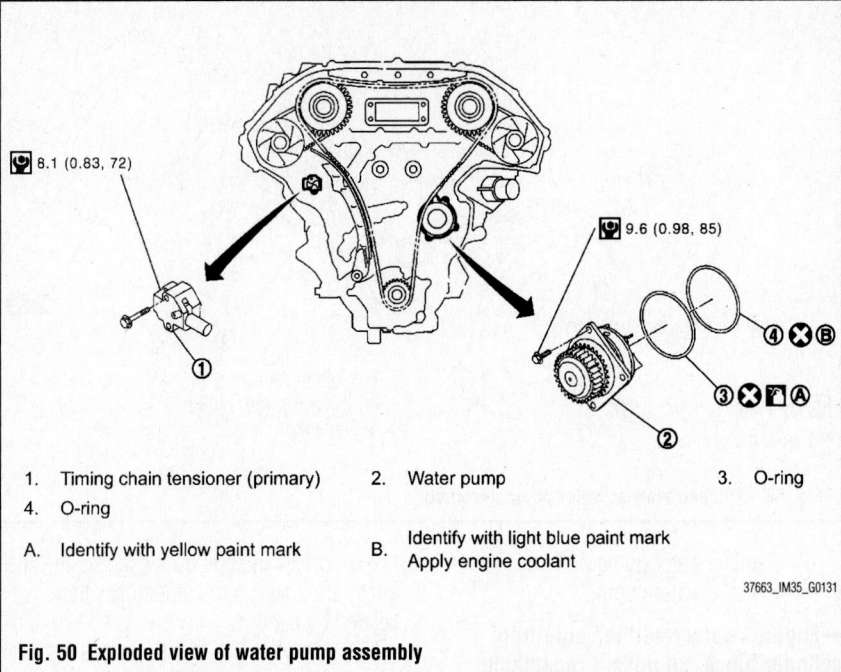

1. Timing chain tensioner (primary)
2. Water pump
3. O-ring
4. O-ring
A. Identify with yellow paint mark
B. Identify with light blue paint mark
   Apply engine coolant

37663_IM35_G0131

Fig. 50 Exploded view of water pump assembly

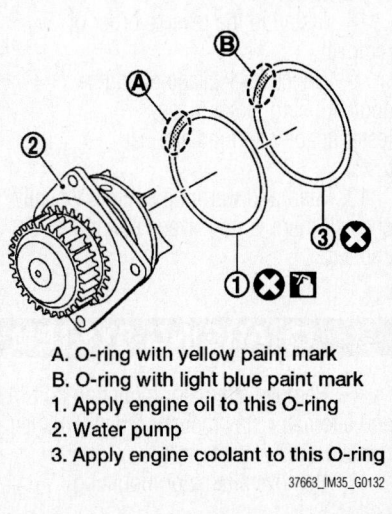

A. O-ring with yellow paint mark
B. O-ring with light blue paint mark
1. Apply engine oil to this O-ring
2. Water pump
3. Apply engine coolant to this O-ring

37663_IM35_G0132

Fig. 51 Install new O-rings to water pump

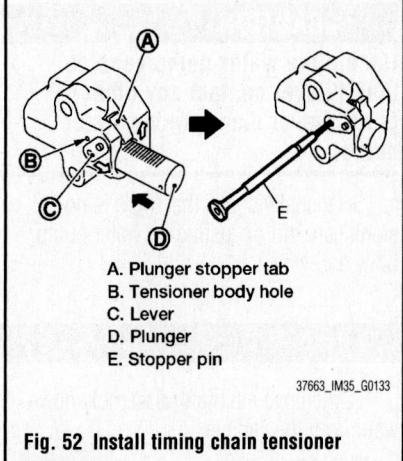

A. Plunger stopper tab
B. Tensioner body hole
C. Lever
D. Plunger
E. Stopper pin

37663_IM35_G0133

Fig. 52 Install timing chain tensioner (primary)

15. Remove water pump as follows:
   a. Remove three water pump mounting bolts. Secure a gap between water pump gear and timing chain, by turning crankshaft counterclockwise until timing chain looseness on water pump sprocket becomes maximum.
   b. Screw bolts into water pumps upper and lower mounting bolt holes until they reach timing chain case. Then, alternately tighten each bolt for a half turn, and pull out water pump.

> ※※ **CAUTION**
> **Pull straight out while preventing vane from contacting socket in installation area. Remove water pump**

without causing sprocket to contact timing chain.

   c. Remove M8 bolts and O-rings from water pump.

> ※※ **CAUTION**
> **Never disassemble water pump.**

To install:
16. Install new O-rings to water pump.
   a. Apply engine oil to O-ring and engine coolant to O-ring as shown.
   b. Locate O-ring with yellow paint mark to front side.
   c. Locate O-ring with light blue paint mark to rear side.
17. Install water pump.

> ※※ **CAUTION**
> **Never allow cylinder block to nip O-rings when installing water pump.**

   a. Check timing chain and water pump sprocket are engaged.
   b. Insert water pump by tightening mounting bolts alternately and evenly.
18. Install timing chain tensioner (primary) as follows:
   a. Turn crankshaft clockwise so that timing chain on the timing chain tensioner (primary) side is loose.
   b. Pull plunger stopper tab up (or turn lever downward) so as to remove plunger stopper tab from the ratchet of plunger.

➡**Plunger stopper tab and lever are synchronized.**

   c. Push plunger into the inside of tensioner body.
   d. Hold plunger in the fully compressed position by engaging plunger stopper tab with the tip of ratchet.
   e. To secure lever, insert stopper pin through hole of lever into tensioner body hole. The lever parts and the tab are synchronized. Therefore, the plunger will be secured under this condition.
   f. Install timing chain tensioner (primary).
   g. Remove dust and foreign material completely from backside of timing chain tensioner (primary) and from installation area of rear timing chain case.
   h. Remove stopper pin.
   i. Check again that timing chain and water pump sprocket are engaged.
19. Install in the reverse order of removal for remaining parts.

➡**After starting engine, let idle for three minutes, then rev engine up to 3,000 rpm under no load to purge air from the high-pressure chamber of chain tensioner. Engine may produce a rattling noise. This indicates that air still remains in the chamber and is not a matter of concern.**

20. Check for leakage of engine coolant using the radiator cap tester adapter and the radiator cap tester.
21. Start and warm up the engine. Visually check that there is no leakage of engine coolant.

4.5L Engine

See Figure 53.

❋❋ CAUTION

Note the following:

- When removing water pump, be careful not to get engine coolant on drive belts.
- Water pump cannot be disassembled and should be replaced as a unit.
- After installing water pump, connect hose and clamp securely, then check for leakage using radiator cap tester and radiator cap tester adapter.
1. Remove following parts:
   - Front engine undercover
   - Engine cover
   - Engine room cover (RH and LH)
   - Air duct (inlet)
   - Alternator, water pump and A/C compressor belt
2. Drain engine coolant from drain plugs on radiator and both side of cylinder block.

❋❋ CAUTION

Perform this step when engine is cold. Never spill engine coolant on drive belts.

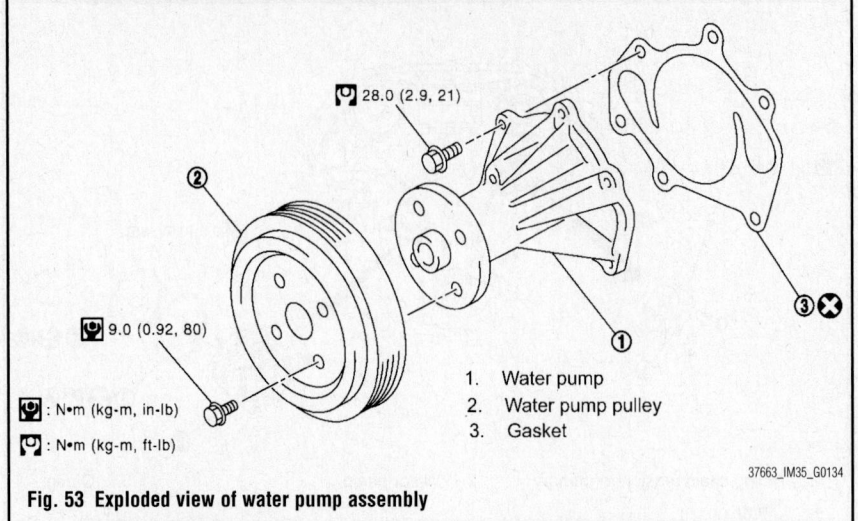

Fig. 53 Exploded view of water pump assembly

28.0 (2.9, 21)
9.0 (0.92, 80)
: N•m (kg-m, in-lb)
: N•m (kg-m, ft-lb)

1. Water pump
2. Water pump pulley
3. Gasket

37663_IM35_G0134

3. Remove water pump pulley.
4. Remove water pump.

➡**Engine coolant will leakage from cylinder block, so have a receptacle ready under vehicle.**

❋❋ CAUTION

Handle the water pump vane so that it never contact any other parts. Never disassemble water pump.

5. Visually check that there is no significant dirt or rusting on water pump body and vane.

6. Check there is no looseness in vane shaft, and that it turns smoothly when rotated by hand.
7. If anything is found, replace water pump.

To install:

8. Install in the reverse order of removal.
9. Check for leakage of engine coolant using radiator cap tester adapter and radiator cap tester.
10. Start and warm up engine. Visually check if there is no leakage of engine coolant.

ENGINE ELECTRICAL   CHARGING SYSTEM

ALTERNATOR

REMOVAL & INSTALLATION

3.5L Engine RWD

1. Before servicing the vehicle, refer to the service precautions.
2. Disconnect the battery cable from the negative terminal.
3. Remove engine front undercover.
4. Remove alternator and power steering oil pump belt.
5. Disconnect alternator connector (1).
6. Remove "B" terminal nut (2).
7. Remove the harness bracket bolts (A).
8. Remove oil pressure switch harness clip (A) from alternator stay.
9. Disconnect oil pressure switch connector (1).
10. Remove alternator mounting bolt (B) and alternator stay mounting bolt (C), then remove alternator stay (2).
11. Remove alternator mounting bolt (D).

12. Remove alternator assembly downward from the vehicle

To install:

13. Installation is the reverse order of removal.
14. Tighten the alternator mounting bolt (C) to 48 ft. lbs. (64.7 Nm).
15. Tighten the alternator stay mounting bolts (A and B) to 21 ft. lbs. (28 Nm).
16. Connect B terminal; and alternator connector.

3.5L Engine AWD

1. Disconnect the battery cable from the negative terminal.
2. Remove power steering oil reservoir tank from the bracket.
3. Remove the clips (A) and the hose clamp (B) from the harness bracket (1).
4. Remove engine front undercover.
5. Remove alternator and power steering oil pump belt.

6. Remove alternator mounting bolt (A) and alternator stay mounting bolt (B), then remove alternator stay (1).
7. Remove alternator mounting bolt (C).
8. Pull and turn alternator, and then remove the harness bracket bolts (A).
9. Disconnect alternator connector (1).
10. Remove "B" terminal nut (2).
11. Remove alternator assembly downward from the vehicle.

To install:

12. Installation is the reverse order of removal.
13. Tighten the alternator mounting bolt (C) to 48 ft. lbs. (64.7 Nm).
14. Tighten the alternator stay mounting bolts (A and B) to 21 ft. lbs. (28 Nm).
15. Connect B terminal; and alternator connector.

4.5L Engine

See Figures 54 through 56.

1. Disconnect the battery cable from the negative terminal.
2. Remove engine front undercover.
3. Remove "B" terminal nut.
4. Disconnect alternator connector.
5. Remove alternator ground harness mounting bolt.
6. Remove the harness bracket bolts.
7. Remove air intake duct.
8. Remove alternator, water pump and A/C compressor belt.
9. Remove power steering oil reservoir tank from the bracket, engine coolant reservoir tank and vacuum tank.
10. Remove the harness clips.
11. Remove alternator mounting bolts.
12. Remove alternator assembly upward.

To install:

13. Installation is the reverse order of removal.

❄❄ CAUTION

Be sure to tighten "B" terminal nut carefully.

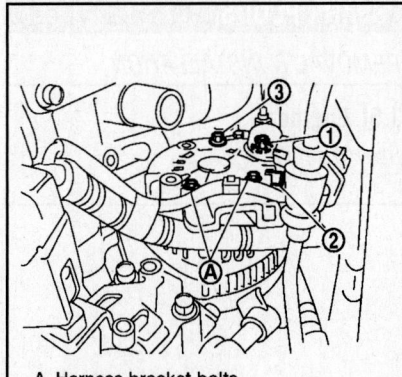

A. Harness bracket bolts
1. "B" terminal nut
2. Alternator connector
3. Alternator ground harness mounting bolt

37663_IM35_G0138

Fig. 54 Disconnecting the alternator

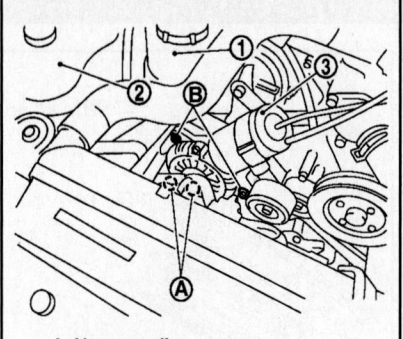

A. Harness clips
B. Alternator mounting bolts
1. Power steering oil reservoir tank
2. Engine coolant reservoir tank
3. Vacuum tank

37663_IM35_G0139

Fig. 55 Remove power steering oil reservoir tank from the bracket, engine coolant reservoir tank and vacuum tank

14. Install alternator, and check tension of belt.

➡ **For this model, the power generation voltage variable control system that controls the power generation voltage of the alternator has been adopted. Therefore, the power generation voltage variable control system operation inspection should be performed after replacing the alternator, and then make sure that the system operates normally.**

VOLTAGE REGULATOR

ADJUSTMENT

The voltage regulator is an integral part of the alternator. No adjustment is possible. If the voltage regulator is found to be defective, it must be replaced.

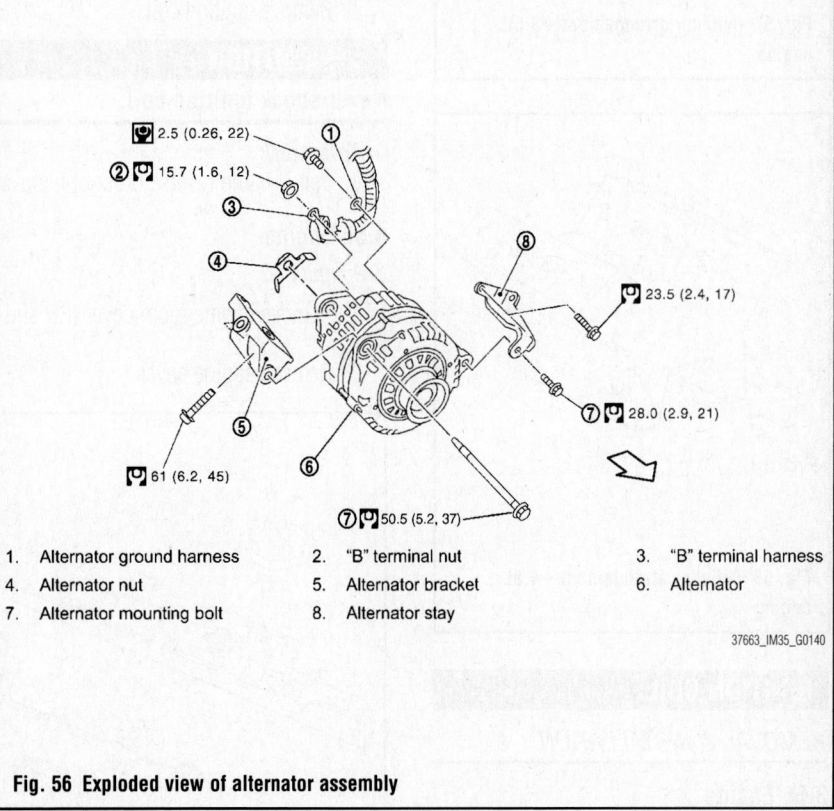

2.5 (0.26, 22)
15.7 (1.6, 12)
23.5 (2.4, 17)
28.0 (2.9, 21)
61 (6.2, 45)
50.5 (5.2, 37)

1. Alternator ground harness
2. "B" terminal nut
3. "B" terminal harness
4. Alternator nut
5. Alternator bracket
6. Alternator
7. Alternator mounting bolt
8. Alternator stay

37663_IM35_G0140

Fig. 56 Exploded view of alternator assembly

FIRING ORDERS

See Figures 57 and 58.

Firing order for the 3.5L engine is
1–2–3–4–5–6.

Firing order for the 4.5L engine is
1–8–7–3–6–5–4–2.

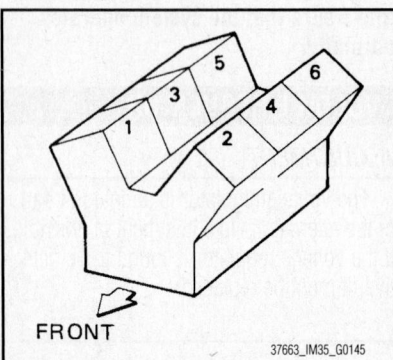

Fig. 57 Cylinder arrangement—3.5L engine

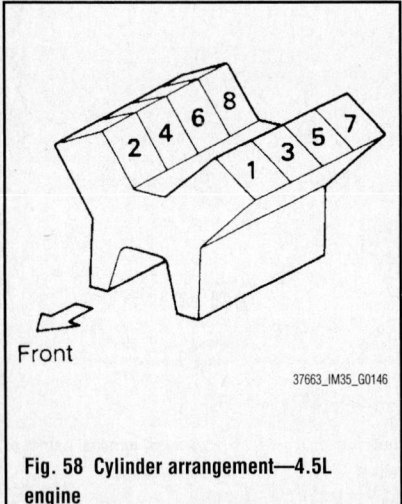

Fig. 58 Cylinder arrangement—4.5L engine

IGNITION COIL

REMOVAL & INSTALLATION

3.5L Engine

See Figure 59.

1. Remove engine room cover (RH and LH).
2. Remove engine cover.
3. Remove air cleaner case and air duct.
4. Move aside harness, harness bracket, and hoses located above ignition coil.
5. Remove electric throttle control actuator.

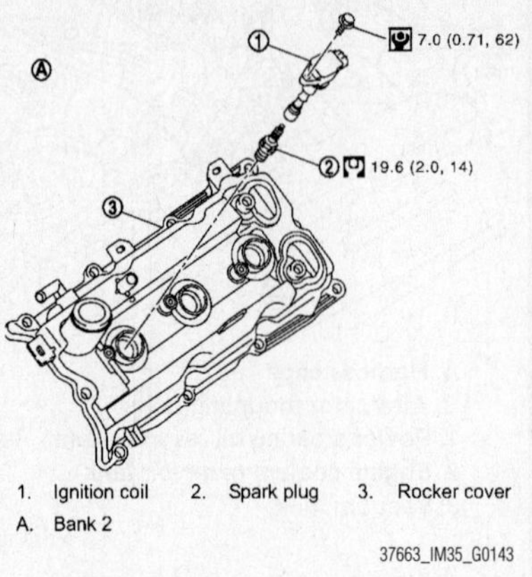

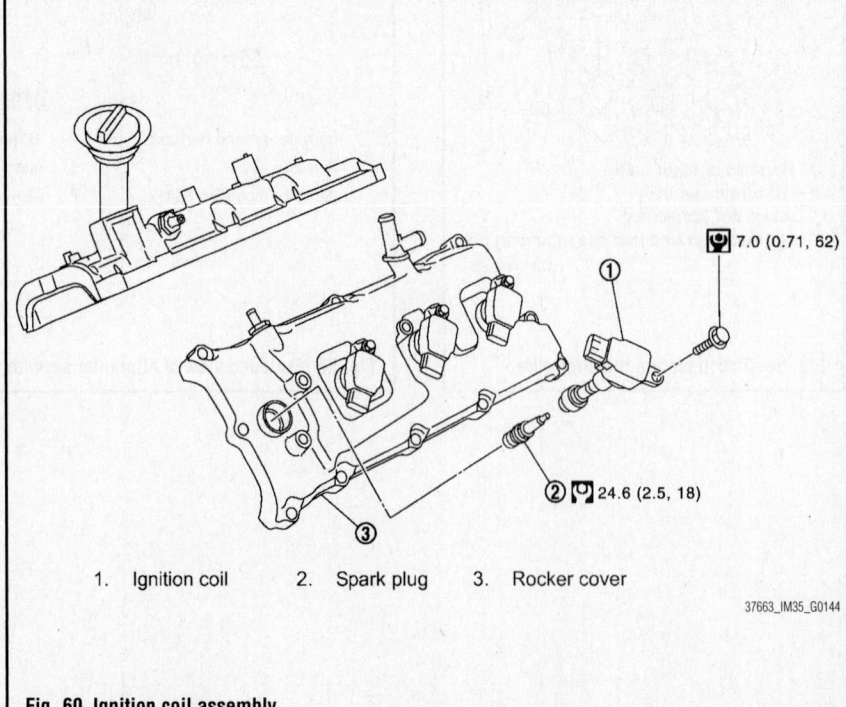

1. Ignition coil
2. Spark plug
3. Rocker cover
A. Bank 2

Fig. 59 Ignition coil assembly

6. Disconnect harness connector from ignition coil.
7. Remove ignition coil.

✳✳ CAUTION
Never shock ignition coil.

To install:
8. Install in the reverse order of removal.

4.5L Engine
See Figure 60.

1. Remove engine room cover (RH and LH).
2. Remove engine cover.

3. Remove air duct (inlet), air cleaner case, and air duct and resonator assembly.
4. Disconnect harness connector from ignition coil.
5. Remove ignition coil.

✳✳ CAUTION
Never shock ignition coil.

To install:
6. Install in the reverse order of removal.

IGNITION TIMING

INSPECTION

The 3.5L engine ignition timing is 15 +/- 5° BTDC.
The 4.5L engine ignition timing is 12 +/- 5° BTDC.

ADJUSTMENT

The ignition timing is controlled by the ECM. No adjustment is possible or necessary.

SPARK PLUGS

REMOVAL & INSTALLATION

3.5L Engine
See Figure 59.

1. Ignition coil
2. Spark plug
3. Rocker cover

Fig. 60 Ignition coil assembly

1. Remove engine room cover (RH and LH).

2. Remove engine cover.

3. Remove air cleaner case and air duct.

4. Move aside harness, harness bracket, and hoses located above ignition coil.

5. Remove electric throttle control actuator.

6. Disconnect harness connector from ignition coil.

7. Remove ignition coil.

※※ CAUTION

Never shock ignition coil.

---

8. Remove the spark plugs.

To install:

9. Install in the reverse order of removal.

4.5L Engine

See Figure 60.

1. Remove engine room cover (RH and LH).

2. Remove engine cover.

3. Remove air duct (inlet), air cleaner case, and air duct and resonator assembly.

4. Disconnect harness connector from ignition coil.

5. Remove ignition coil.

※※ CAUTION

Never shock ignition coil.

---

6. Remove the spark plugs.

To install:

7. Install in the reverse order of removal.

ENGINE ELECTRICAL

STARTING SYSTEM

STARTER

REMOVAL & INSTALLATION

4.5L Engine RWD

See Figure 61.

1. Disconnect the battery cable from the negative terminal.

2. Remove engine front and rear undercover.

3. Remove left engine mounting insulator and left engine mounting bracket.

4. Remove "B" terminal nut.

5. Disconnect "S" connector.

6. Remove the bolt and the harness bracket.

7. Remove starter motor mounting bolts.

8. Remove starter motor forward from the vehicle.

To install:

9. Installation is the reverse order of removal.

※※ CAUTION

Be sure to tighten "B" terminal nut carefully.

4.5L Engine AWD

1. Disconnect the battery cable from the negative terminal.

2. Remove engine front and rear undercover.

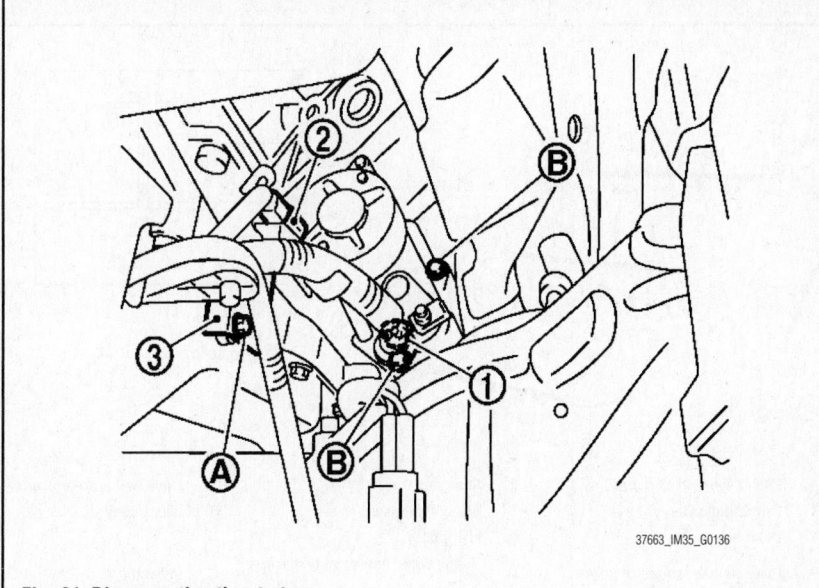

Fig. 61 Disconnecting the starter

37663_IM35_G0136

3. Remove front drive shaft left side housing bolts.

4. Remove left engine mounting insulator and left engine mounting bracket.

5. Remove "B" terminal nut.

6. Disconnect "S" connector.

7. Remove the bolt and the harness bracket.

8. Remove starter motor mounting bolts.

9. Remove starter motor forward from the vehicle.

To install:

10. Installation is the reverse order of removal.

※※ CAUTION

Be sure to tighten "B" terminal nut carefully.

ENGINE MECHANICAL

ACCESSORY DRIVE BELTS

ACCESSORY BELT ROUTING

See Figures 62 and 63.

Refer to the accompanying illustrations.

INSPECTION

3.5L Engine

1. Check that the indicator (notch on mounted side) of drive belt auto-tensioner is within the possible use range (A).

➡**Check the drive belt auto-tensioner indication when the engine is cold.**

2. When new drive belt is installed, the indicator (notch on mounted side) should be within the range (B).

3. Visually check entire drive belt for wear, damage or cracks.

4. If the indicator (notch on mounted side) is out of the possible use range or belt is damaged, replace drive belt.

4.5L Engine

1. Remove air duct (inlet) when inspecting drive belt for alternator, water pump and A/C compressor.

2. Remove front engine undercover when inspecting power steering oil pump belt.

3. Check that indicator (single line notch) of each auto tensioner is within the allowable working range (between three line notches).

➡**Check auto tensioner indication when engine is cold.**

4. When new drive belt is installed, the range should be "A".

5. The indicator notch is located on the moving side of auto tensioner for alternator, water pump and A/C compressor belt, while it is found on the mounted side for power steering oil pump belt.

6. Visually check entire belt for wear, damage or cracks.

7. If the indicator is out of allowable working range or belt is damaged, replace belt.

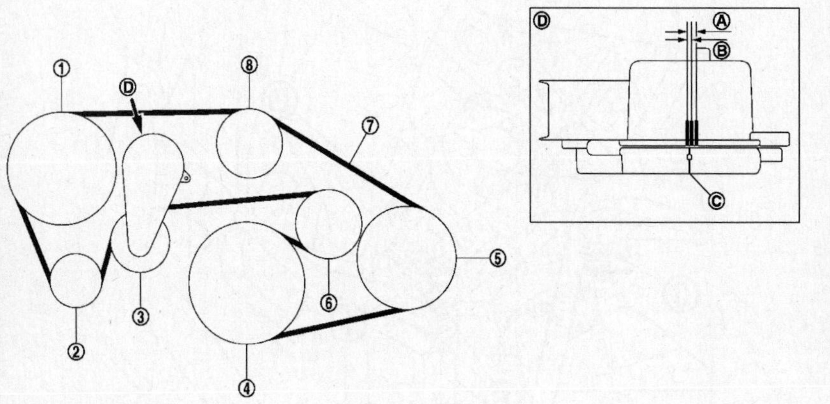

ADJUSTMENT

Belt tension is not necessary, as it is automatically adjusted by drive belt auto-tensioner.

REMOVAL & INSTALLATION

3.5L Engine

See Figure 64.

1. Remove front engine undercover.

2. While securely holding the square hole in pulley center of auto tensioner with a spinner handle, move spinner handle in the direction of arrow (loosening direction of drive belt).

❄❄ CAUTION

Avoid placing hand in a location where pinching may occur if the holding tool accidentally comes off.

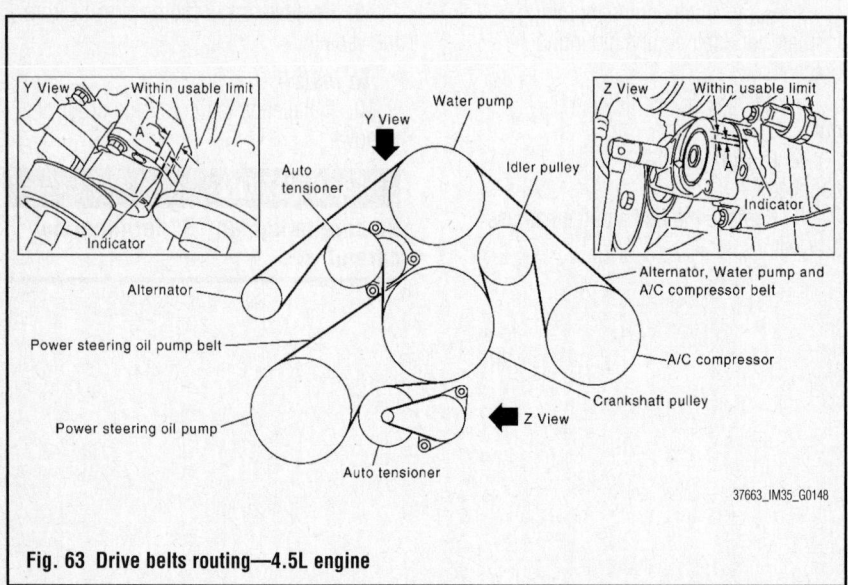

1.	Power steering oil pump	2.	Alternator	3.	Drive belt auto-tensioner
4.	Crankshaft pulley	5.	A/C compressor	6.	Idler pulley
7.	Drive belt	8.	Idler pulley		
A.	Possible use range	B.	Range when new drive belt is installed	C.	Indicator
D.	View D				

37663_IM35_G0147

Fig. 62 Drive belt routing—3.5L engine

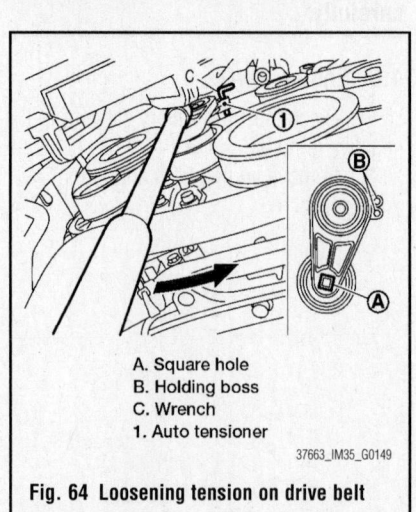

37663_IM35_G0148

Fig. 63 Drive belts routing—4.5L engine

A. Square hole
B. Holding boss
C. Wrench
1. Auto tensioner

37663_IM35_G0149

Fig. 64 Loosening tension on drive belt

3. Under the above condition, insert a metallic bar through the holding boss to lock auto-tensioner pulley arm.

4. Remove drive belt.

To install:

5. Note the following, and install in the reverse order of removal.

✳✳ CAUTION

Note the following:

- Check drive belt is securely installed around all pulleys.
- Check drive belt is correctly engaged with the pulley groove.
- Check for engine oil and engine coolant are not adhered drive belt and pulley groove.

6. Turn crankshaft pulley clockwise several times to equalize tension between each pulley, and then confirm tension of drive belt at indicator (notch on mounted side) is within the possible use range.

4.5L Engine

Alternator, Water Pump And A/C Compressor Belt

See Figure 65.

1. Remove air duct (inlet).

2. With box wrench, and while securely holding the hexagonal part in pulley center of auto tensioner, move wrench handle in the direction of arrow (loosening direction of tensioner).

✳✳ CAUTION

Note the following:

- Avoid placing hand in a location where pinching may occur if the holding tool accidentally comes off.
- Never loosen the hexagonal part in center of drive belt auto tensioner pulley (Never turn it clockwise). If turned clockwise, the complete drive

belt auto tensioner must be replaced as a unit, including the pulley.

3. Under the above condition, insert a metallic bar through the holding boss to lock auto tensioner pulley arm.

➡**Leave auto tensioner pulley arm locked until belt is installed again.**

4. Remove alternator, water pump and A/C compressor belt.

Power Steering Oil Pump Belt

See Figure 66.

1. Remove air duct (inlet).
2. Remove front engine undercover.
3. Remove alternator, water pump and A/C compressor belt.
4. While securely holding the hexagonal protrusion part of auto tensioner pulley with box wrench, move wrench handle in the direction of arrow (loosening direction of tensioner).

✳✳ CAUTION

Avoid placing hand in a location where pinching may occur if holding tool accidentally comes off.

5. Under the above condition, insert a metallic bar through the holding boss to lock auto tensioner pulley arm.

➡**Leave auto tensioner pulley arm locked until belt is installed again.**

6. Remove power steering oil pump belt.

To install:

7. Note the following, and install in the reverse order of removal.

✳✳ CAUTION

Note the following:

- Check belt is securely installed around all pulleys.
- Check belt is correctly engaged with the pulley groove.

- Check for engine oil and engine coolant are not adhered belt and pulley groove.
- Check that belt tension is within the allowable working range, using indicator notch on auto tensioner.

CAMSHAFT AND VALVE LIFTERS

REMOVAL & INSTALLATION

3.5L Engine

See Figures 122 through 130.

1. Remove front timing chain case, camshaft sprocket, and timing chain.
2. Remove fuel sub tube.
3. Remove camshaft sensor bracket bolts in reverse order as shown.

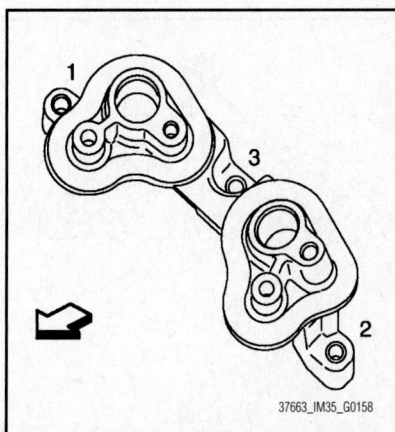

Fig. 67 Remove camshaft sensor bracket bolts in reverse order as shown

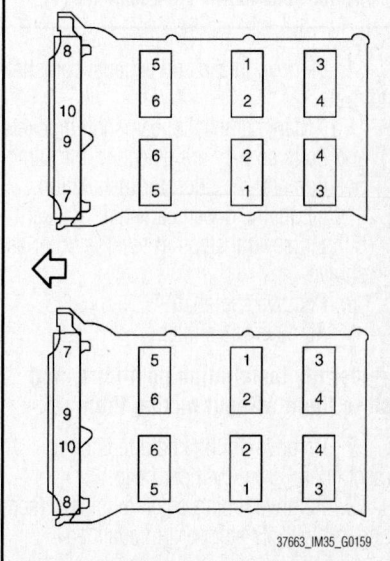

Fig. 68 Equally loosen camshaft bracket bolts in several steps in reverse order as shown

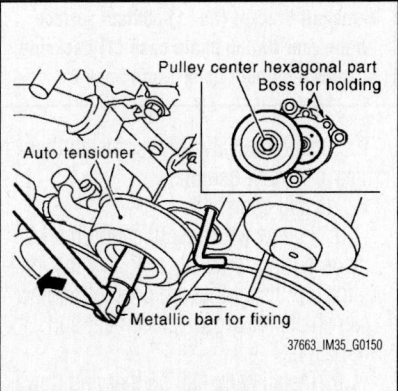

Fig. 65 Loosening tension on drive belt

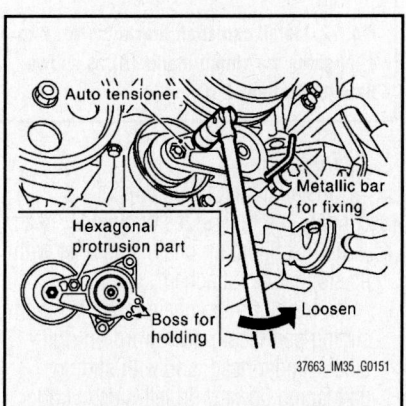

Fig. 66 Loosening tension on drive belt

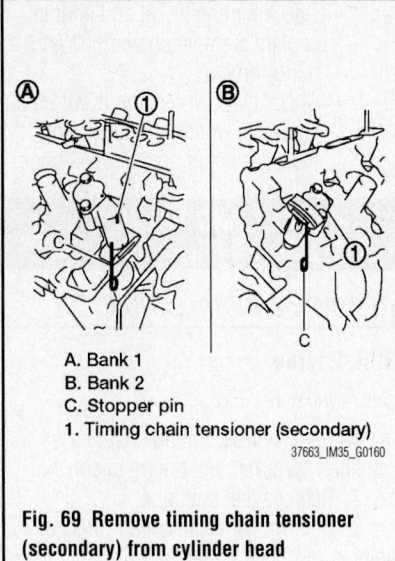

A. Bank 1
B. Bank 2
C. Stopper pin
1. Timing chain tensioner (secondary)

37663_IM35_G0160

Fig. 69 Remove timing chain tensioner (secondary) from cylinder head

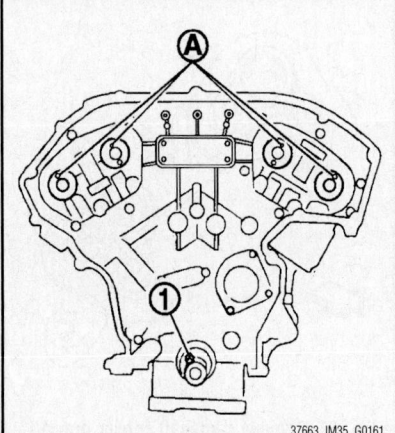

37663_IM35_G0161

Fig. 70 Install camshaft so that and dowel pin (A) on front end face are positioned as shown; crankshaft key (1)

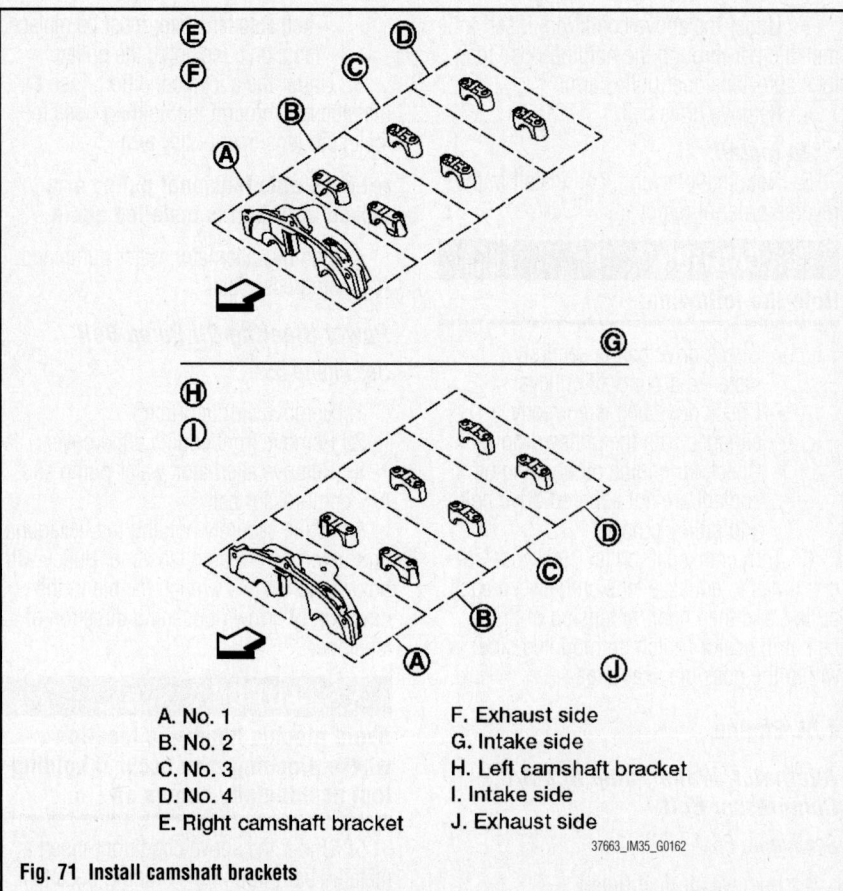

A. No. 1
B. No. 2
C. No. 3
D. No. 4
E. Right camshaft bracket
F. Exhaust side
G. Intake side
H. Left camshaft bracket
I. Intake side
J. Exhaust side

37663_IM35_G0162

Fig. 71 Install camshaft brackets

4. Remove intake and exhaust camshaft brackets.

a. Mark camshafts, camshaft brackets and bolts so they are placed in the same position and direction for installation.

b. Equally loosen camshaft bracket bolts in several steps in reverse order as shown.

5. Remove camshaft.
6. Remove valve lifter.

➡**Identify installation positions, and store them without mixing them up.**

7. Remove timing chain tensioner (secondary) from cylinder head.

8. Remove timing chain tensioner (secondary) with its stopper pin attached.

➡**Stopper pin was attached when timing chain (secondary) was removed.**

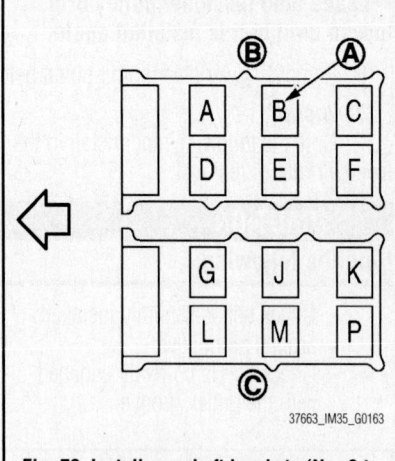

37663_IM35_G0163

Fig. 72 Install camshaft brackets (No. 2 to 4) aligning the stamp marks (A) as shown; Bank 1 (B), Bank 2 (C)

To install:

9. Install timing chain tensioners (secondary) on both sides of cylinder head.

a. Install timing chain tensioner with its stopper pin attached.

b. Install timing chain tensioner with sliding part facing downward on right-side cylinder head, and with sliding part facing upward on left-side cylinder head.

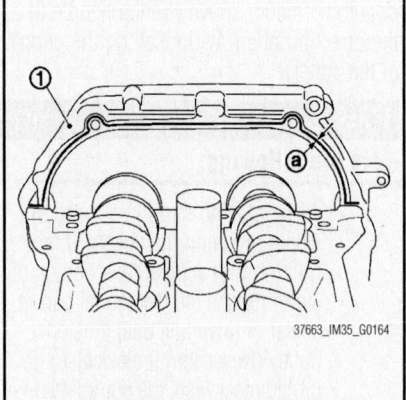

37663_IM35_G0164

Fig. 73 Apply liquid gasket (a) to camshaft bracket (No. 1) contact surface on the rear timing chain case (1) backside as shown on both bank 1 and bank 2

10. Install the valve lifters. Install them in their original positions.

11. Install camshafts.

a. Follow your identification marks made during removal, or follow the identification marks that are present on new camshafts for proper placement and direction.

b. Install camshaft so that and dowel pin on front end face are positioned as

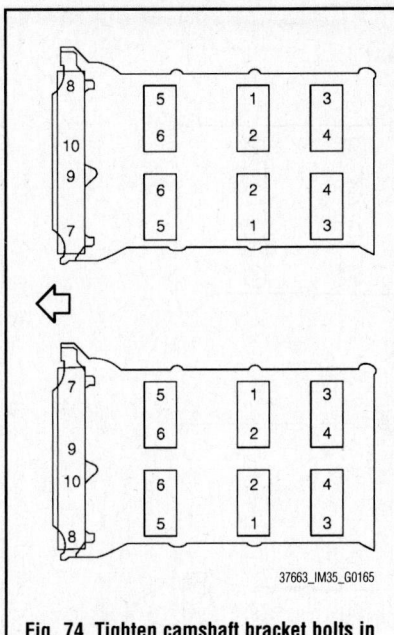

Fig. 74 Tighten camshaft bracket bolts in numerical order as shown

shown. (No. 1 cylinder TDC on its compression stroke)

➡ **Though camshaft does not stop at the position as shown, for the placement of cam nose, it is generally accepted camshaft is placed for the same direction as the figure.**

12. Install camshaft brackets.

a. Remove foreign material completely from camshaft bracket backside and from cylinder head installation face.

b. Install camshaft bracket in original position and direction as shown.

c. Install camshaft brackets (No. 2 to 4) aligning the stamp marks as shown.

➡ **There are no identification marks indicating left and right for camshaft bracket (No. 1).**

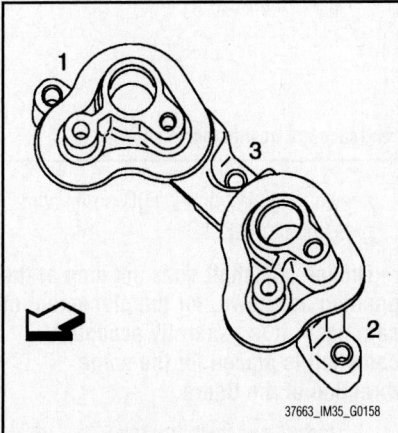

Fig. 75 Tighten camshaft sensor bracket bolts in numerical order as shown

d. Apply liquid gasket to mating surface of camshaft bracket (No. 1).

e. Apply liquid gasket to camshaft bracket (No. 1) contact surface on the rear timing chain case backside as shown on both bank 1 and bank 2.

※ **CAUTION**

For camshaft bracket (No. 1) near installation position, and install it without disturbing the liquid gasket applied to the surfaces.

13. Tighten camshaft bracket bolts in the following steps, in numerical order as shown.

a. Tighten No. 7 to 10 to 1 ft. lb. (2 Nm) in numerical order as shown.

b. Tighten No. 1 to 6 to 1 ft. lb. (2 Nm) in numerical order as shown.

c. Tighten No. 1 to 10 to 4 ft. lbs. (6 Nm) in numerical order as shown.

d. Tighten No. 1 to 10 to 8 ft. lbs. (10 Nm) in numerical order as shown.

※ **CAUTION**

After tightening mounting bolts of camshaft brackets (No. 1), be sure to wipe off excessive liquid gasket from the parts list below.

- Mating surface of rocker cover
- Mating surface of rear timing chain case

14. Inspect and adjust the valve clearance.

15. Install camshaft sensor bracket. Tighten camshaft sensor bracket bolts in numerical order as shown.

➡ **The order of tightening bolts is the same for bank 1 and bank 2.**

16. Install in the reverse order of removal after this step.

※ **CAUTION**

Check when engine is cold so as to prevent burns from any splashing engine oil.

17. Check the engine oil level.

18. Perform the following procedure so as to prevent the engine from being unintentionally started while checking.

a. Release fuel pressure.

b. Disconnect ignition coil and injector harness connectors.

19. Remove intake valve timing control solenoid valve.

20. Crank the engine, and then check that engine oil comes out from valve timing control solenoid valve hole (A). End crank after checking.

※ **WARNING**

Be careful not to touch rotating parts (drive belt, idler pulley, and crankshaft pulley, etc.).

※ **CAUTION**

Engine oil may squirt from intake valve timing control solenoid valve installation hole during cranking. Use a shop cloth to prevent the engine components and the vehicle. Never allow engine oil to get on rubber components such as drive belt or engine mount insulators. Immediately wipe off any splashed engine oil.

➡ **Clean oil groove between oil strainer and intake valve timing control solenoid valve if engine oil does not come out from valve timing control solenoid valve hole.**

21. Remove components between intake valve timing control solenoid valve and camshaft sprocket (INT), and then check each oil groove for clogging.

➡ **Clean oil groove if necessary.**

22. After inspection, install removed parts.

23. The following are procedures for checking fluids leakage, lubricates leakage, and exhaust gases leakage.

a. Before starting engine, check oil/fluid levels including engine coolant and engine oil. If less than required quantity, fill to the specified level.

b. Use procedure below to check for fuel leakage.

- Turn ignition switch "ON" (with engine stopped). With fuel pressure applied to fuel piping, check for fuel leakage at connection points.
- Start engine. With engine speed increased, check again for fuel leakage at connection points.

c. Run engine to check for unusual noise and vibration.

➡ **If hydraulic pressure inside timing chain tensioner drops after removal/ installation, slack in the guide may generate a pounding noise during and just after engine start. However, this is normal. Noise will stop after hydraulic pressure rises.**

d. Warm up engine thoroughly to check there is no leakage of fuel, exhaust gases, or any oil/fluids including engine oil and engine coolant.

e. Bleed air from lines and hoses of applicable lines, such as in cooling system.

f. After cooling down engine, again check oil/fluid levels including engine oil and engine coolant. Refill to the specified level, if necessary.

4.5L Engine

See Figures 76 through 82.

1. Remove engine assembly from vehicle.
2. Remove timing chain.
3. With hexagonal part of camshaft locked with wrench, loosen bolts securing camshaft sprocket to remove camshaft sprocket.

✳✳ CAUTION

Note the following:

- Never loosen mounting bolts with securing anything other than the camshaft hexagonal portion or with tensioning the timing chain.
- After removing timing chain, never turn crankshaft and camshaft separately, or valves will strike the piston head.

4. Remove intake and exhaust camshaft brackets.

a. Mark camshafts, camshaft brackets and bolts so placed in the same position and direction for installation.

b. Equally loosen camshaft brackets and bolts in several steps in reverse order as shown.

c. Lightly tapping with plastic hammer, remove camshaft bracket (No. 1) and camshaft bracket (No. 6).

➡**The bottom surface of each bracket will be stuck to cylinder head because of liquid gasket.**

5. Remove camshaft.

6. Remove valve lifter. Identify installation positions, and store them without mixing them up.

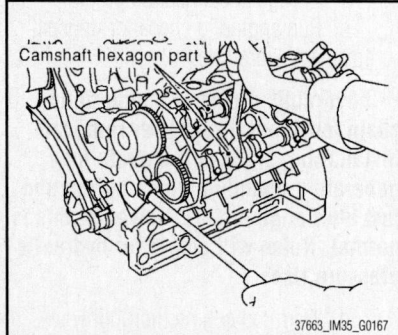

Fig. 76 Loosen bolts securing camshaft sprocket to remove camshaft sprocket

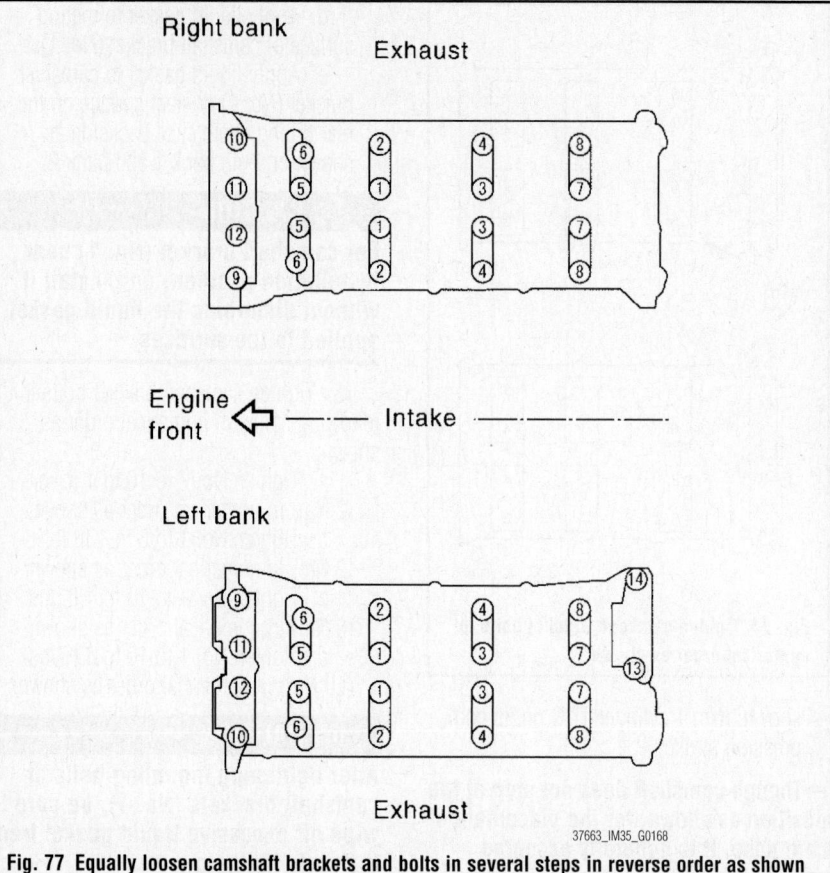

Fig. 77 Equally loosen camshaft brackets and bolts in several steps in reverse order as shown

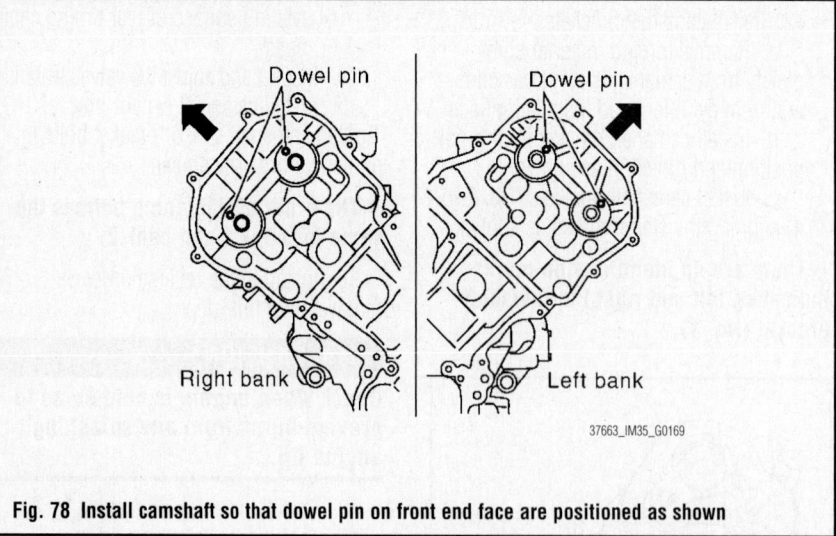

Fig. 78 Install camshaft so that dowel pin on front end face are positioned as shown

To install:

7. Install valve lifters if removed. Install them in their original positions.

8. Install camshafts.

a. Follow your identification marks made during removal, or follow the identification marks that are present on new camshafts for proper placement and direction.

b. Install camshaft so that dowel pin on front end face are positioned as shown. (No. 1 cylinder TDC on its compression stroke)

➡**Though camshaft does not stop at the position as shown, for the placement of cam nose, it is generally accepted camshaft is placed for the same direction of the figure.**

9. Install camshaft brackets.

a. Remove foreign material completely from camshaft bracket backside

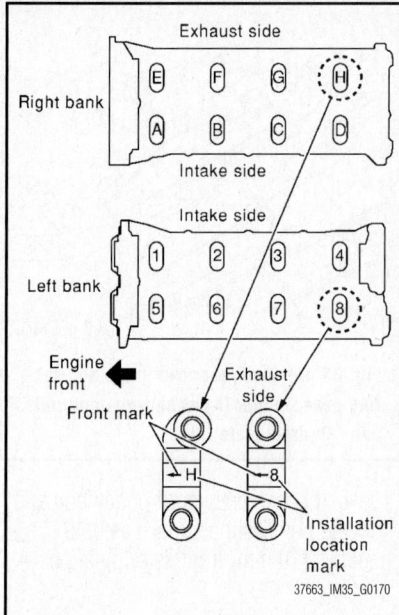

Fig. 79 Install by referring to installation location mark on upper surface and front mark

and from cylinder head installation face.

b. Install by referring to installation location mark on upper surface and front mark.

c. Install so that installation location mark can be correctly read when viewed from the side of left exhaust bank.

d. Apply liquid gasket to mating surface of camshaft bracket (No. 1).

✳✳ CAUTION

After installation, be sure to wipe off any excessive liquid gasket leaking from part "A" and "B" (both on right and left sides). Remove completely any excess of liquid gasket inside bracket.

e. Apply liquid gasket to mating surface of camshaft bracket (No. 6) on left bank intake.

✳✳ CAUTION

After installation, be sure to wipe off any excessive liquid gasket leaking from part "A" and "B" (both on right and left sides). Remove completely any excess of liquid gasket inside bracket.

10. Tighten camshaft bracket bolts in the following steps, in numerical order as shown.
a. Tighten No. 9 to 12 to 1 ft. lbs. (2 Nm) in numerical order as shown.

b. Tighten No. 1 to 8 to 1 ft. lbs. (2 Nm) in numerical order as shown.

c. Tighten No. 13 to 14 to 1 ft. lbs. (2 Nm) in numerical order as shown. (left bank only)

d. Tighten all bolts top 4 ft. lbs. (6 Nm) in numerical order as shown.

e. Tighten No. 1 to 12 to 8 ft. lbs. (10 Nm) in numerical order as shown.

f. Tighten No. 13 to 14 to 23 ft. lbs. (31 Nm) in numerical order as shown. (left bank only)

✳✳ CAUTION

After tightening mounting bolts of camshaft brackets, be sure to wipe off excessive liquid gasket from the parts listed below.

- Mating surface of rocker cover
- Mating surface of front cover

11. Install camshaft sprockets.
a. Install by checking with identification mark on surface.

b. Install camshaft sprocket (EXH) by selectively using the groove of dowel pin according to the bank. (Common part used for both banks.)

c. Lock the hexagonal part of camshaft in the same way as for removal, and tighten mounting bolts.

12. Check and adjust the valve clearance.

13. Install in the reverse order of removal after this step.

✳✳ CAUTION

Note the following:

- Perform this inspection only when DTC P0011 and/or P0021 are detected in self-diagnostic results of CONSULT-III and it is directed according to inspection procedure of EC section.
- Check when the engine is cold so as to prevent burns from any splashing engine oil.

14. Check the engine oil level.

15. Perform the following procedure so

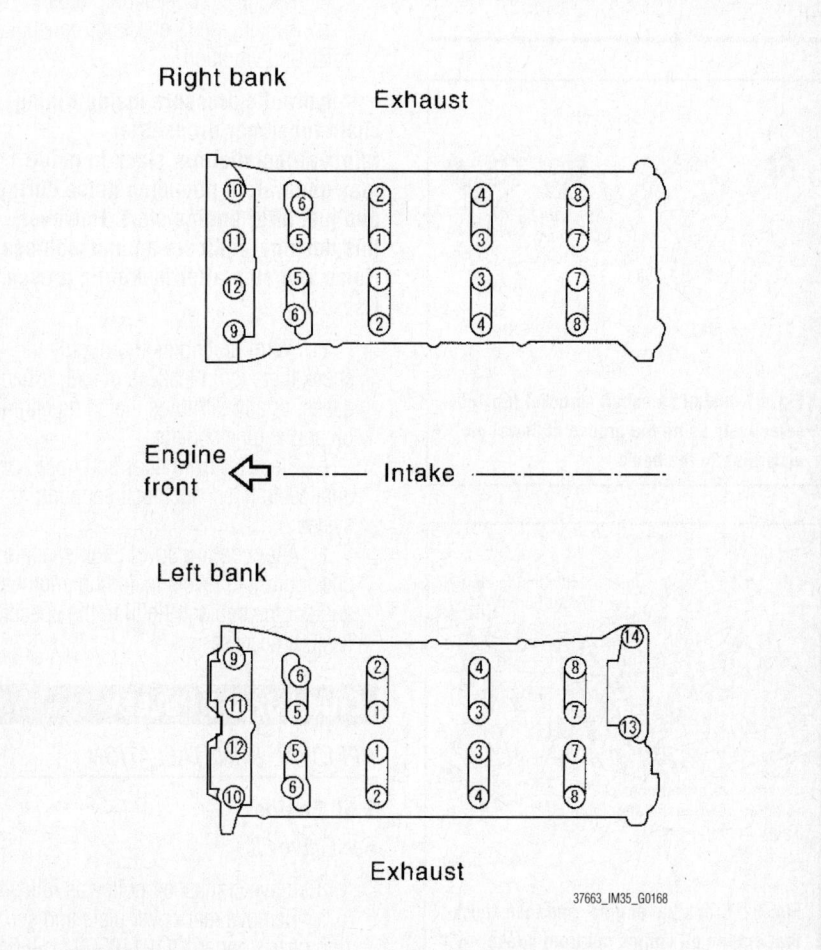

Fig. 80 Tighten camshaft bracket bolts in numerical order as shown

as to prevent the engine from being unintentionally started while checking.

a. Release fuel pressure.

b. Disconnect ignition coil and injector harness connectors.

16. Remove intake valve timing control solenoid valve.

17. Crank the engine, and then check that engine oil comes out from intake valve timing control cover oil hole. End crank after checking.

✴✴ WARNING

Be careful not to touch rotating parts (drive belt, idler pulley, and crankshaft pulley, etc.).

✴✴ CAUTION

Engine oil may squirt from intake valve timing control solenoid valve installation hole during cranking. Use a shop cloth to prevent the engine components and the vehicle. Never allow engine oil to get on rubber components such as drive belt or engine mount insulators. Immediately wipe off any splashed engine oil.

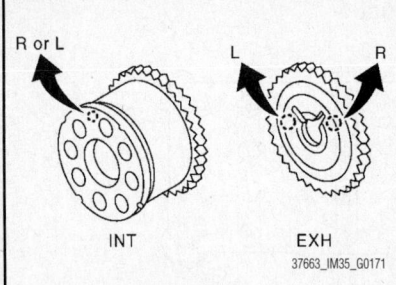

Fig. 81 Install camshaft sprocket (EXH) by selectively using the groove of dowel pin according to the bank

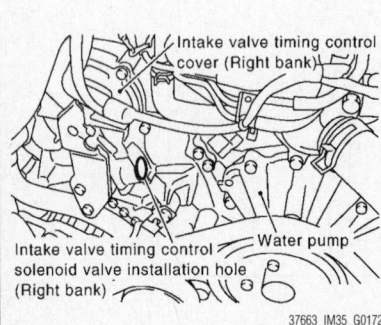

Fig. 82 Crank the engine, and then check that engine oil comes out from intake valve timing control cover oil hole

➡ **Clean oil groove between oil strainer and intake valve timing control solenoid valve if engine oil does not come out from intake valve timing control cover oil hole.**

18. Remove components between intake valve timing control solenoid valve and camshaft sprocket (INT), and then check each oil groove for clogging.

➡ **Clean oil groove if necessary.**

19. After inspection, install removed parts.

20. The following are procedures for checking fluids leakage, lubricates leakage, and exhaust gases leakage.

a. Before starting engine, check oil/fluid levels including engine coolant and engine oil. If less than required quantity, fill to the specified level.

b. Use procedure below to check for fuel leakage.

• Turn ignition switch "ON" (with engine stopped). With fuel pressure applied to fuel piping, check for fuel leakage at connection points.

• Start engine. With engine speed increased, check again for fuel leakage at connection points.

c. Run engine to check for unusual noise and vibration.

➡ **If hydraulic pressure inside timing chain tensioner drops after removal/installation, slack in guide may generate a pounding noise during and just after engine start. However, this does not indicate an unusualness. Noise will stop after hydraulic pressure rises.**

d. Warm up engine thoroughly to check there is no leakage of fuel, exhaust gases, or any oil/fluids including engine oil and engine coolant.

e. Bleed air from lines and hoses of applicable lines, such as in cooling system.

f. After cooling down engine, again check oil/fluid levels including engine oil and engine coolant. Refill to the specified level, if necessary.

CRANKSHAFT PULLEY

REMOVAL & INSTALLATION

3.5L Engine

See Figure 83.

1. Remove crankshaft pulley as follows:

a. Remove rear cover plate and set ring gear stopper KV10118600 (J-48641) as shown.

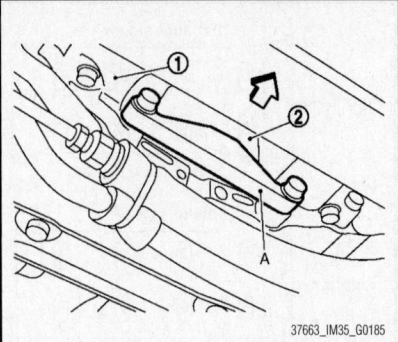

Fig. 83 Remove rear cover plate and set ring gear stopper (A) as shown; upper oil pan (1); drive plate (2)

b. Loosen crankshaft pulley bolt and locate bolt seating surface as 0.39 inches (10 mm) from its original position.

✴✴ CAUTION

Never remove crankshaft pulley bolt as it will be used as a supporting point for suitable puller.

c. Place suitable puller tab on holes of crankshaft pulley, and pull crankshaft pulley through.

✴✴ CAUTION

Never put suitable puller tab on crankshaft pulley periphery, as this will damage internal damper.

To install:

2. Install crankshaft pulley as follows:

a. Secure crankshaft using ring gear stopper KV10118600 (J-48641).

b. Install crankshaft pulley, taking care not to damage front oil seal.

➡ **When press-fitting crankshaft pulley with plastic hammer, tap on its center portion (not circumference).**

c. Tighten crankshaft pulley bolt.

d. Tighten the bolt 90 degrees. Place a matching mark on crankshaft pulley aligning with the matching of crankshaft pulley bolt.

e. Rotate crankshaft pulley in normal direction (clockwise when viewed from front) to confirm it turns smoothly.

4.5L Engine

See Figure 84.

1. Remove crankshaft pulley as follows:

a. Set ring gear stopper.

b. Loosen crankshaft pulley bolt, and then pull crankshaft pulley with both hands to remove it.

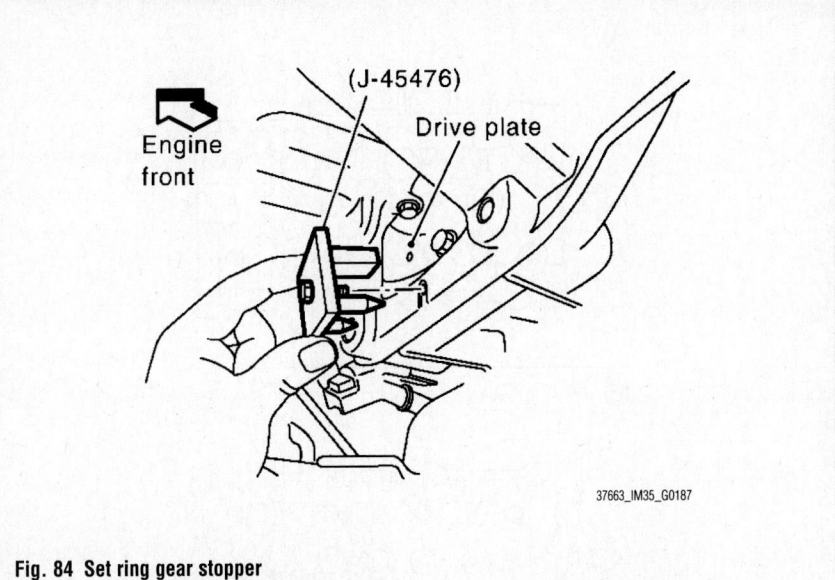

Fig. 84 Set ring gear stopper

⁂ **CAUTION**

Never remove crankshaft pulley bolt. Keep loosened crankshaft pulley bolt in place to protect removed crankshaft pulley from dropping. Never remove balance weight (inner hexagon bolt) at the front of crankshaft pulley.

2. Installation is the reverse of removal.

CRANKSHAFT FRONT SEAL

REMOVAL & INSTALLATION

3.5L Engine

See Figure 85.

1. Remove the following parts:
   • Front engine undercover
   • Drive belt

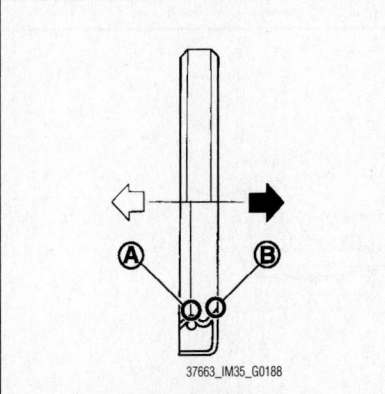

Fig. 85 Install front oil seal so that each seal lip is oriented as shown

• Radiator cooling fan assembly
• Crankshaft pulley
2. Remove front oil seal.

⁂ **CAUTION**

Be careful not to damage front timing chain case and crankshaft.

To install:

3. Apply engine oil to both oil seal lip and dust seal lip of new front oil seal.
4. Install front oil seal.
   a. Install front oil seal so that each seal lip is oriented as shown.
   b. Using suitable drift, press-fit until the height of front oil seal is level with the mounting surface.

⁂ **CAUTION**

Note the following:

• Be careful not to damage front timing chain case and crankshaft.
• Press-fit straight and avoid causing burrs or tilting oil seal.

5. Install in the reverse order of removal after this step.

4.5L Engine

See Figures 84 and 85.

1. Remove the following parts:
   • Front engine undercover
   • Radiator
   • Drive belts
   • Rear plate cover
2. Remove crankshaft pulley as follows:
   a. Set ring gear stopper.

b. Loosen crankshaft pulley bolt, and then pull crankshaft pulley with both hands to remove it.

⁂ **CAUTION**

Note the following:

• Never remove crankshaft pulley bolt. Keep loosened crankshaft pulley bolt in place to protect removed crankshaft pulley from dropping.
• Never remove balance weight (inner hexagon bolt) at the front of crankshaft pulley.

3. Remove front oil seal.

⁂ **CAUTION**

Be careful not to damage front cover and oil pump drive spacer.

To install:

4. Apply new engine oil to both oil seal lip and dust seal lip of new front oil seal.
5. Install front oil seal.
   a. Install front oil seal so that each seal lip is oriented as shown.
   b. Using suitable drift, press fit until the height of front oil seal is level with the mounting surface.
   c. Check the garter spring is in position and seal lips not inverted.

⁂ **CAUTION**

Note the following:

• Be careful not to damage front cover and oil pump drive spacer.
• Press fit straight and avoid causing burrs or tilting oil seal.

6. Install in the reverse order of removal.

CYLINDER HEAD

REMOVAL & INSTALLATION

3.5L Engine

See Figures 86 through 88.

1. Remove engine assembly from vehicle, and separate front suspension member and transmission from engine.
2. Remove the following parts:
   • Fuel tube and fuel injector assembly
   • Intake manifold
   • Exhaust manifolds
   • Water inlet and thermostat assembly
   • Water outlet and water pipe

- Front and rear timing chain case

3. Remove camshafts (INT and EXH).

4. Remove cylinder head bolts in reverse order as shown with cylinder head bolt wrench.

5. Remove cylinder head gaskets.

To install:

6. Install new cylinder head gaskets.

7. Turn crankshaft until No. 1 piston is set at TDC.

➡**Crankshaft key (1) should line up with the bank 1 cylinder center line as shown.**

8. Install cylinder head follow the steps below to tighten cylinder head bolts in numerical order as shown.

※※ CAUTION

If cylinder head bolts reused, check their outer diameters before installation.

a. Apply new engine oil to threads and seat surfaces of cylinder head bolts.

b. Tighten all cylinder head bolts to 77 ft. lbs. (105 Nm).

c. Completely loosen all cylinder head bolts in reverse order.

d. Tighten all cylinder head bolts to 30 ft. lbs. (40 Nm).

e. Turn all bolts 95 degrees clockwise (angle tightening).

※※ CAUTION

Check the tightening angle by using angle wrench KV10112100 (BT8653-A). Avoid judgment by visual inspection without tool. Check tightening angle indicated on angle wrench indicator plate.

f. Turn all bolts 95 degrees clockwise again (angle tightening).

9. After installing cylinder head, measure distance between front end faces of cylinder block and cylinder head (bank 1 and bank 2). If the measured value is out of the standard of 0.555–0.587 inches (14–15 Nm), reinstall cylinder head.

10. Install in the reverse order of removal after this step.

11. Before starting engine, check oil/fluid levels including engine coolant and engine oil. If less than required quantity, fill to the specified level.

12. Use procedure below to check for fuel leakage.

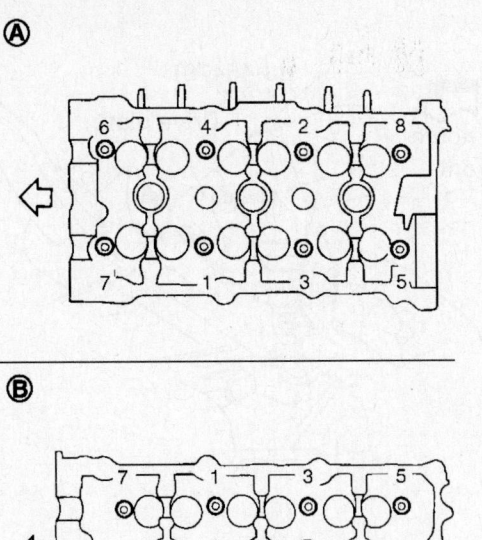

Fig. 86 Cylinder head bolt tightening sequence. Loosen the bolts in the reverse of this sequence

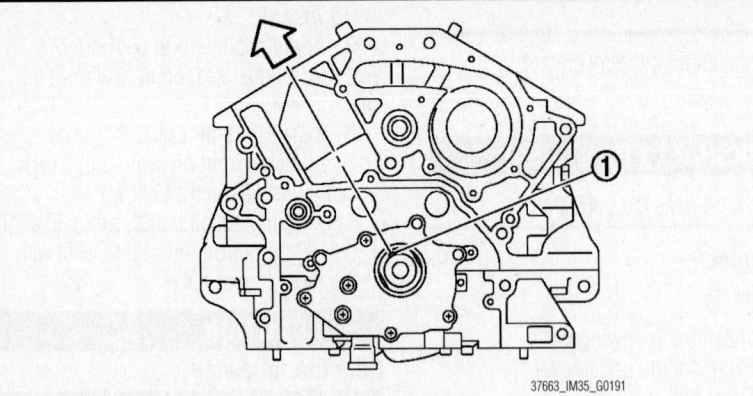

Fig. 87 Turn crankshaft until No. 1 piston is set at TDC

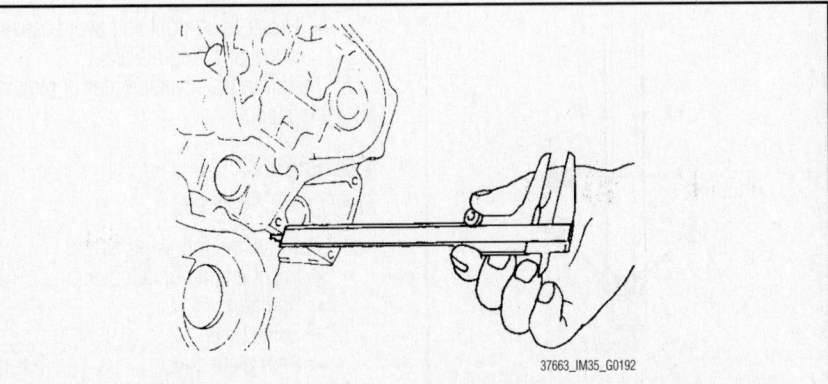

Fig. 88 Measure distance between front end faces of cylinder block and cylinder head (bank 1 and bank 2)

a. Turn ignition switch "ON" (with engine stopped). With fuel pressure applied to fuel piping, check for fuel leakage at connection points.

b. Start engine. With engine speed increased, check again for fuel leakage at connection points.

13. Run engine to check for unusual noise and vibration.

14. Warm up engine thoroughly to check there is no leakage of fuel, exhaust gases, or any oil/fluids including engine oil and engine coolant.

15. Bleed air from lines and hoses of applicable lines, such as in cooling system.

16. After cooling down engine, again check oil/fluid levels including engine oil and engine coolant. Refill to the specified level, if necessary.

4.5L Engine

See Figures 89 and 90.

> ❊❊ **CAUTION**
>
> **Never use washer bolts together with flange bolts, when replacing cylinder head bolts, because there are two kinds of cylinder head bolts.**

1. Remove engine assembly from vehicle.
2. Remove exhaust manifolds.
3. Remove camshafts.
4. Remove cylinder head bolts in reverse order as shown to remove cylinder heads (right bank and left bank).
5. Remove cylinder head gaskets.

To install:

6. Install new cylinder head gaskets.

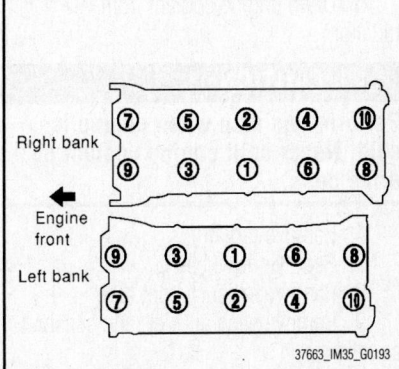

Fig. 89 Cylinder head bolt tightening sequence. Remove cylinder head bolts in the reverse order of this sequence (right bank and left bank)

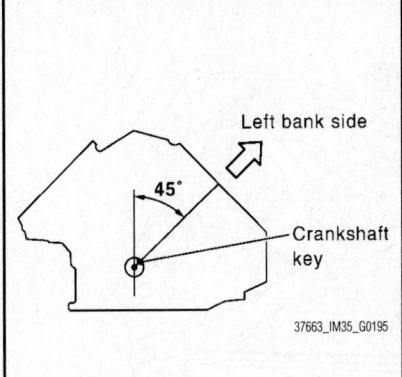

Fig. 90 Turn crankshaft until No. 1 piston is set at TDC

7. Turn crankshaft until No. 1 piston is set at TDC.

➡**Crankshaft key should line up with the left bank cylinder center line as shown.**

8. Install cylinder heads following the steps below to tighten cylinder head bolts in numerical order as shown:

> ❊❊ **CAUTION**
>
> **If cylinder head bolts are reused, check their outer diameters before installation.**

a. Apply new engine oil to threads and seating surface of cylinder head bolts.

b. Tighten all cylinder head bolts to 33 ft. lbs. (44 Nm).

c. Turn all cylinder head bolts 70 degrees clockwise. (angle tightening)

> ❊❊ **CAUTION**
>
> **Check the tightening angle by using angle wrench. Avoid judgment by visual inspection without angle wrench. Check tightening angle indicated on angle wrench indicator plate.**

d. Completely loosen all cylinder head bolts in reverse order.

e. Tighten all cylinder head bolts to 33 ft. lbs. (44 Nm).

f. Turn all cylinder head bolts 60 degrees clockwise. (angle tightening)

> ❊❊ **CAUTION**
>
> **Check the tightening angle by using angle wrench. Avoid judgment by visual inspection without angle**

wrench. **Check tightening angle indicated on angle wrench indicator plate.**

g. Turn all cylinder head bolts 60 degrees clockwise again. (angle tightening)

9. Install remaining parts in the reverse order of removal.

EXHAUST MANIFOLD

REMOVAL & INSTALLATION

3.5L Engine

See Figures 91 through 93.

> ❊❊ **WARNING**
>
> **Perform the work when the exhaust and cooling system have completely cooled down.**

➡**When removing bank 1 side parts only, step 3, 10 and 11 are unnecessary.**

1. Remove engine room cover (RH and LH).
2. Remove engine cover.
3. Drain engine coolant.

> ❊❊ **CAUTION**
>
> **Perform this step when engine is cold. Never spill engine coolant on drive belt.**

4. Remove air cleaner case and air duct.
5. Remove front and rear undercover.
6. Disconnect heated oxygen sensor harness connectors.

> ❊❊ **CAUTION**
>
> **Be careful not to damage heated oxygen sensor 2. Discard any heated oxygen sensor 2 which has been dropped onto a hard surface such as a concrete floor. Replace with a new sensor.**

7. Remove exhaust mounting bracket between three way catalysts (bank 1 and bank 2) and transmission.
8. Remove exhaust front tube and three way catalysts (bank 1 and bank 2).
9. Disconnect harness connector and remove air fuel ratio sensor 1 on both banks using heated oxygen sensor wrench KV10114400 (J38365).
10. Put marks to identify installation positions of each air fuel ratio sensor 1.

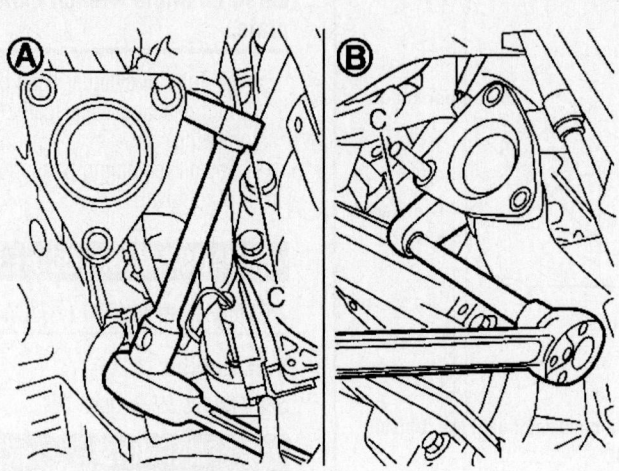

A. Bank 1
B. Bank 2
C. Heated oxygen sensor wrench

37663_IM35_G0177

Fig. 91 Disconnect harness connector and remove air fuel ratio sensor 1 on both banks

✳✳ CAUTION

Be careful not to damage air fuel ratio sensor 1. Discard any air fuel ratio sensor 1 which has been dropped onto a hard surface such as a concrete floor. Replace with a new sensor.

11. Disconnect steering lower joint at power steering gear assembly side, and release steering lower shaft.

12. Remove water bypass pipe and heater pipe.

13. Remove exhaust manifold cover.

14. Loosen mounting nuts in reverse order as shown to remove exhaust manifold.

➡ Disregard the numerical order No. 7 and 8 in removal.

15. Remove gaskets.

✳✳ CAUTION

Cover engine openings to avoid entry of foreign materials.

To install:

16. Note the following, and install in the reverse order of removal.

17. Install exhaust manifold gasket in direction shown.

18. If stud bolts were removed, install them and tighten to 11 ft. lbs. (15 Nm).

19. Install mounting exhaust manifold in numerical order as shown.

➡ Tighten nuts No. 1 and 2 in two steps. The numerical order No. 7 and 8 shown second step.

✳✳ CAUTION

Note the following:

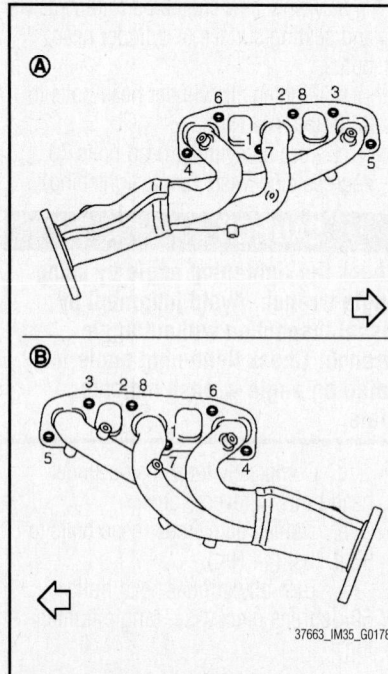

37663_IM35_G0178

Fig. 92 Exhaust manifold bolt tightening sequence shown. Loosen mounting nuts in reverse order of this sequence

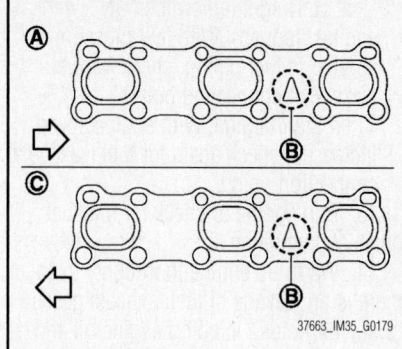

37663_IM35_G0179

Fig. 93 Install exhaust manifold gasket in direction shown

- Before installing a new air fuel ratio sensor 1 and heated oxygen sensor 2, clean exhaust system threads using oxygen sensor thread cleaner (J-43897-18 or J43897-12) and apply anti-seize lubricant.
- Never over torque air fuel ratio sensor 1 and heated oxygen sensor 2. Doing so may cause damage to air fuel ratio sensor 1 and heated oxygen sensor 2, resulting in the "MIL" coming on.

4.5L Engine

See Figures 94 through 96.

✳✳ WARNING

Perform the work when the exhaust and cooling system have completely cooled down.

1. Remove engine room cover (RH and LH).

2. Remove engine cover.

3. Remove air duct (inlet), air cleaner case and air duct and resonator assembly.

4. Remove front and rear engine under covers.

5. Drain engine coolant from radiator.

✳✳ CAUTION

Perform this step when engine is cold. Never spill engine coolant on drive belts.

6. Remove radiator.

7. Remove drive belts.

8. Remove exhaust front tube.

9. Remove each air fuel ratio sensor 1 as follows:

a. Disconnect harness connector of each air fuel ratio sensor 1.

b. Remove each air fuel ratio sensor 1 on both banks with heated oxygen sensor wrench.

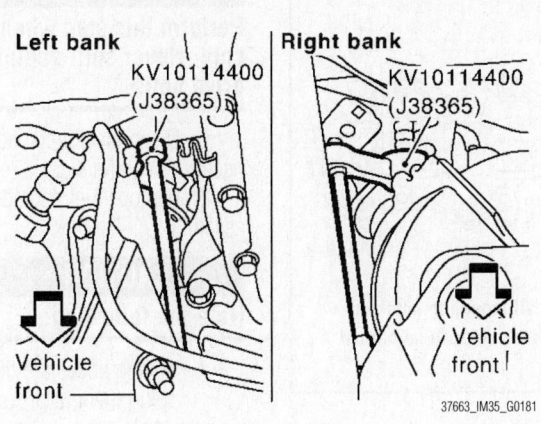

Fig. 94 Remove each air fuel ratio sensor 1 on both banks with heated oxygen sensor wrench

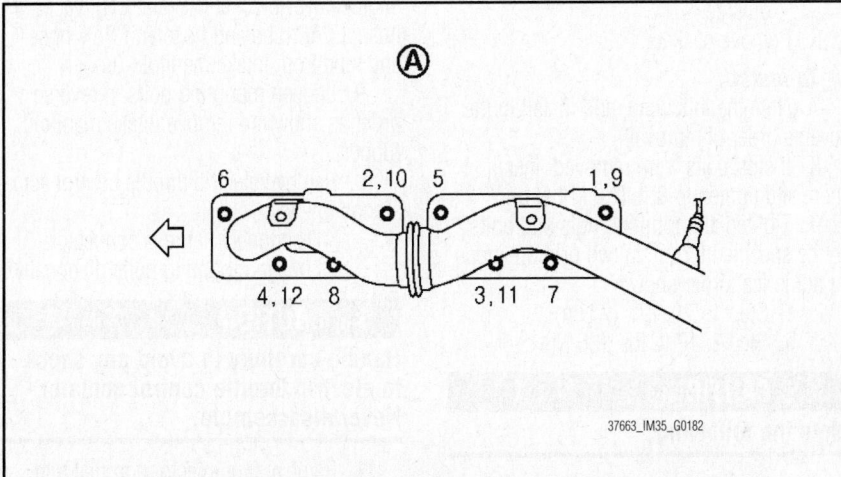

Fig. 95 Left Bank tightening sequence shown; Loosen nuts in the reverse order to remove exhaust manifold and three way catalyst (left bank)

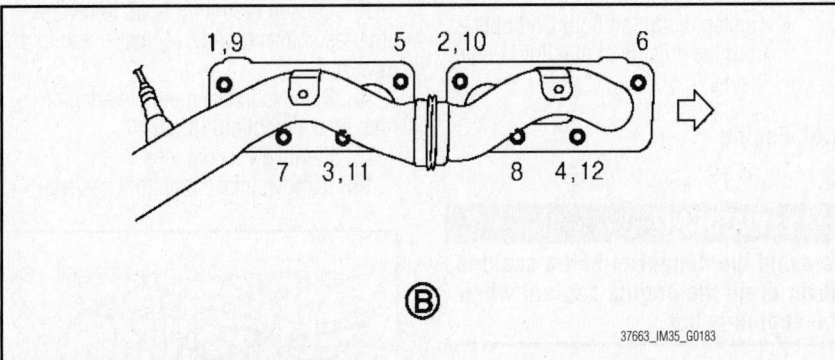

Fig. 96 Right bank tightening sequence shown; loosen nuts in the reverse order to remove exhaust manifold and three way catalyst (right bank)

❋❋ **CAUTION**

Be careful not to damage air fuel ratio sensor 1. Discard any air fuel ratio sensor 1 which has been dropped onto a hard surface such as a concrete floor. Replace with a new one.

10. Remove exhaust manifold and three way catalyst (left bank) as follows:

a. Disconnect A/C piping from A/C compressor, then remove A/C compressor.

b. Remove steering lower joint to enable steering shaft to move freely.

c. Remove starter motor.

d. Remove nuts on bottom of engine mounting insulator (LH), and lift up left side of engine approximately 1.18 inches (3 cm) with transmission jack.

e. Remove exhaust manifold cover (left bank).

f. Loosen nuts in the reverse order to remove exhaust manifold and three way catalyst (left bank).

➡**Disregard No. 9 to No. 12 when loosening.**

11. Remove exhaust manifold and three way catalyst (right bank) as follows:

a. Remove alternator and bracket.

b. Remove nuts on bottom of engine mounting insulator (RH), and lift up right side of engine approximately 1.18 inches (3 cm) with transmission jack.

c. Remove exhaust manifold cover (right bank).

d. Loosen nuts in the reverse order to remove exhaust manifold and three way catalyst (right bank).

➡**Disregard No. 9 to No. 12 when loosening.**

12. Remove exhaust manifold gaskets.

❋❋ **CAUTION**

Cover engine openings to avoid entry of foreign materials.

To install:

13. Note the following, and install in the reverse order of removal.

14. Install exhaust manifold gasket with its directional protrusion set upward.

15. Install exhaust manifold and tighten mounting nuts in numerical order as shown.

➡**Tighten mounting nuts No. 1 to 4 in two steps. The numerical order No. 9 to 12 shown second steps.**

❋❋ **CAUTION**

Note the following:

- Before installing a new air fuel ratio sensor 1, clean exhaust system threads using oxygen sensor thread cleaner (J-43897-18 or J-43897-12), and apply anti-seize lubricant
- Never over torque air fuel ratio sensor 1. Doing so may cause damage to the air fuel ratio sensor 1, resulting in "MIL" coming on.

DRIVE PLATE

REMOVAL & INSTALLATION

1. Secure crankshaft with a ring gear stopper KV10118600 (J-48641), and remove mounting bolts.
2. Loosen mounting bolts in diagonal order.
3. Check for deformation or damage.

✳✳ CAUTION

Note the following:

- Never disassemble drive plate.
- Never place drive plate with signal plate facing down.
- When handling signal plate, take care not to damage or scratch it.
- Handle signal plate in a manner that prevents it from becoming magnetized.

4. Installation is the reverse of removal.

INTAKE MANIFOLD

REMOVAL & INSTALLATION

3.5L Engine

See Figures 97 and 98.

1. Release fuel pressure.
2. Remove intake manifold collector.
3. Remove fuel tube and fuel injector assembly.
4. Remove harness bracket.
5. Loosen mounting nuts and bolts in reverse order as shown to remove intake manifold.

✳✳ CAUTION

Note the following:

- Cover engine openings to avoid entry of foreign materials.
- Put a mark on the intake manifold and the cylinder head with paint

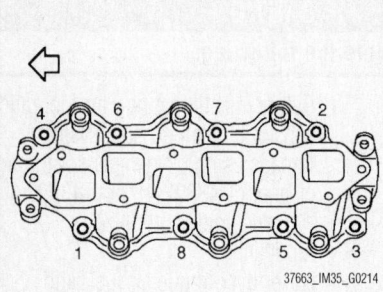

Fig. 97 Loosen mounting nuts and bolts in reverse order as shown

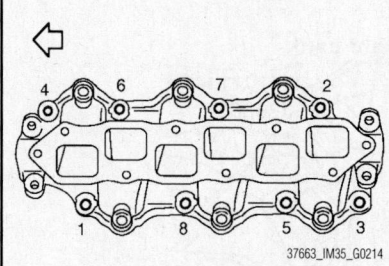

Fig. 98 Tighten all mounting nuts and bolts in two or more steps in numerical order

before removal because they need installed in the specified direction.
- Loosen mounting bolts and nuts from the inside of manifold to the outside.

6. Remove gaskets.

To install:

7. Note the following, and install in the reverse order of removal.
8. If stud bolts were removed, install them and tighten to 8 ft. lbs. (11 Nm).
9. Tighten all mounting nuts and bolts to the specified torque in two or more steps in numerical order shown.
   a. Step 1: 5 ft. lbs. (7 Nm)
   b. Step 2: 19 ft. lbs. (26 Nm)

✳✳ CAUTION

Note the following:

- Install intake manifold with the marks (put on the intake manifold and the cylinder head before removal) aligned.
- Tighten mounting bolts and nuts from the outside of manifold to the inside.

4.5L Engine

See Figures 99 through 101.

✳✳ WARNING

To avoid the danger of being scalded, never drain the engine coolant when the engine is hot.

1. Remove engine room cover (RH and LH).
2. Remove engine cover.
3. Release fuel pressure.
4. Remove air duct (inlet), air cleaner case, and air duct and resonator assembly.
5. Drain engine coolant from radiator.

✳✳ CAUTION

Perform this step when the engine is cold. Never spill engine coolant on drive belts.

6. Disconnect fuel feed hose quick connector on engine side.
7. Remove fuel damper and fuel hose assembly.

✳✳ CAUTION

Note the following:

- While hoses are disconnected, plug them to prevent fuel from draining.
- Never separate fuel damper and fuel hose.

8. Remove or disconnect harnesses, engine cover bracket (RH and LH), vacuum hose, EVAP tube and hose and PCV hose and tube from intake manifold (upper).
9. Loosen mounting bolts in reverse order as shown to remove intake manifold (upper).
10. Remove electric throttle control actuator as follows:
   a. Disconnect harness connector.
   b. Loosen mounting bolts diagonally.

✳✳ CAUTION

Handle carefully to avoid any shock to electric throttle control actuator. Never disassemble.

11. Remove fuel injector and fuel tube assembly.
12. Disconnect water hoses from intake manifold adaptor.
13. Loosen mounting bolts in reverse order as shown to remove intake manifold (lower).
14. Remove intake manifold adaptor from intake manifold (lower).
15. Remove vacuum tank.
16. Remove intake manifold gaskets.

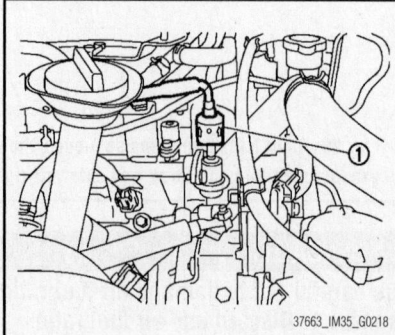

Fig. 99 Disconnect fuel feed hose quick connector (1) on engine side

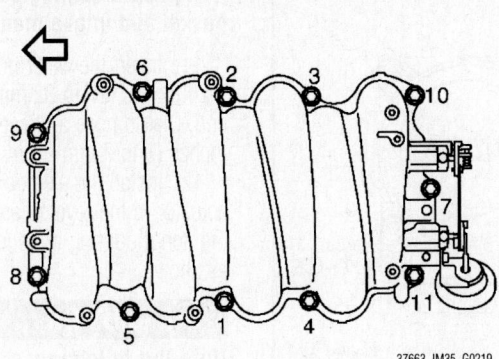

Fig. 100 Upper manifold bolt tightening sequence; Loosen mounting bolts in reverse order as shown

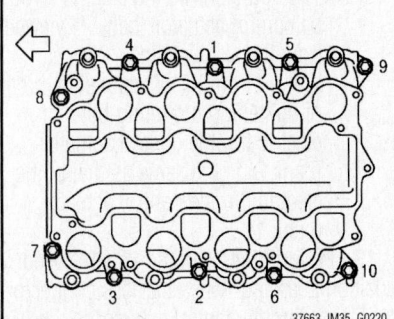

Fig. 101 Intake manifold tightening sequence; Loosen mounting bolts in reverse order as shown

✳✳ CAUTION

Cover engine openings to avoid entry of foreign materials.

To install:

17. Note the following, and install in the reverse order of removal.

18. Tighten the lower and upper manifold bolts in the proper sequence (see illustrations).

19. Install the electric throttle control actuator gasket with its directional protrusion set up/downward. Tighten mounting bolts of electric throttle control actuator equally and diagonally in several steps.

20. Insert water hose by 1.06 to 1.26 inches (27 to 32 mm) from connector end.

21. Perform the "Throttle Valve Closed Position Learning" when harness connector of electric throttle control actuator is disconnected.

22. Perform the "Idle Air Volume Learning" and "Throttle Valve Closed Position Learning" when electric throttle control actuator is replaced.

IDLE AIR VOLUME LEARNING

Idle Air Volume Learning is a function of ECM to learn the idle air volume that keeps engine idle speed within the specific range. It must be performed under the following conditions:
- Each time electric throttle control actuator or ECM is replaced.
- Idle speed or ignition timing is out of specification.

Preconditioning

Check that all of the following conditions are satisfied. Learning will be cancelled if any of the following conditions are missed for even a moment.
- Battery voltage: More than 12.9 V (At idle)
- Engine coolant temperature: 158–221°F (70–105°C)
- Selector lever: P or N
- Electric load switch: OFF (Air conditioner, headlamp, rear window defogger)

➡**On vehicles equipped with daytime light systems, if the parking brake is applied before the engine is started the headlamp will not illuminate.**

- Steering wheel: Neutral (Straight-ahead position)
- Vehicle speed: Stopped
- Transmission: Warmed-up
- With CONSULT-III: Drive vehicle until "ATF TEMP SE 1" in "DATA MONITOR" mode of "A/T" system indicates less than 0.9 V.
- Without CONSULT-III: Drive vehicle for 10 minutes.

Perform Idle Air Volume Learning

With Consult-III

1. Perform Accelerator Pedal Released Position Learning.

2. Perform Throttle Valve Closed Position Learning.

3. Start engine and warm it up to normal operating temperature.

4. Select "IDLE AIR VOL LEARN" in "WORK SUPPORT" mode.

5. Touch "START" and wait 20 seconds.

Without Consult-III

➡**Note the following:**

- It is better to count the time accurately with a clock.
- It is impossible to switch the diagnostic mode when an Accelerator Pedal Position (APP) sensor circuit has a malfunction.

1. Perform Accelerator Pedal Released Position Learning.

2. Perform Throttle Valve Closed Position Learning.

3. Start engine and warm it up to normal operating temperature.

4. Turn ignition switch OFF and wait at least 10 seconds.

5. Confirm that accelerator pedal is fully released, turn ignition switch ON and wait 3 seconds.

6. Repeat the following procedure quickly 5 times within 5 seconds.
   a. Fully depress the accelerator pedal.
   b. Fully release the accelerator pedal.

7. Wait 7 seconds, fully depress the accelerator pedal for approx. 20 seconds until the MIL stops blinking and turns ON.

8. Fully release the accelerator pedal within 3 seconds after the MIL turns ON.

9. Start engine and let it idle.

10. Wait 20 seconds.

11. Rev up the engine 2 or 3 times and check that idle speed and ignition timing are within the specifications.

THROTTLE VALVE CLOSED POSITION LEARNING

Throttle Valve Closed Position Learning is a function of ECM to learn the fully closed position of the throttle valve by monitoring the throttle position sensor output signal. It must be performed each time harness connector of electric throttle control actuator or ECM is disconnected.

1. Check that accelerator pedal is fully released.

2. Turn ignition switch ON.

3. Turn ignition switch OFF and wait at least 10 seconds.

4. Check that throttle valve moves during the above 10 seconds by confirming the operating sound.

INTAKE MANIFOLD COLLECTOR

REMOVAL & INSTALLATION

3.5L Engine

See Figures 102 and 103.

> ❊❊ **WARNING**
>
> **To avoid the danger of being scalded, never drain the engine coolant when the engine is hot.**

1. Remove engine room cover (RH and LH).
2. Remove engine cover.
3. Remove air cleaner case and air duct (RH and LH).
4. Remove electric throttle control actuator (bank 1 and bank 2) as follows:

➡**When removing only intake manifold collector, move electric throttle control actuator without disconnecting water hose.**

  a. Drain engine coolant.

> ❊❊ **CAUTION**
>
> **Perform this step when engine is cold.**

  b. Disconnect water hoses from electric throttle control actuator. When engine coolant is not drained from radiator, attach plug to water hoses to prevent engine coolant leakage.

> ❊❊ **CAUTION**
>
> **Never spill engine coolant on drive belt.**

  c. Disconnect harness connector.
  d. Loosen mounting bolts in reverse order as shown.

> ❊❊ **CAUTION**
>
> **Note the following:**
>
> - Handle carefully to avoid any shock to electric throttle control actuator.
> - Never disassemble.

➡**Note the following:**

- Figure shows electric throttle control actuator (bank 1) viewed from the air duct side.
- Viewed from the air duct side, order of loosening mounting bolts of electric throttle control actuator (bank 2) is the same as that of the electric throttle control actuator (bank 1).

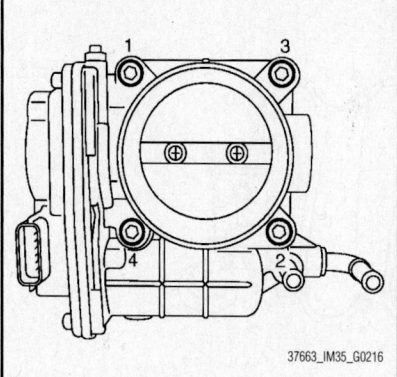

Fig. 102 Electric throttle control actuator tightening sequence; Loosen mounting bolts in reverse order as shown

37663_IM35_G0216

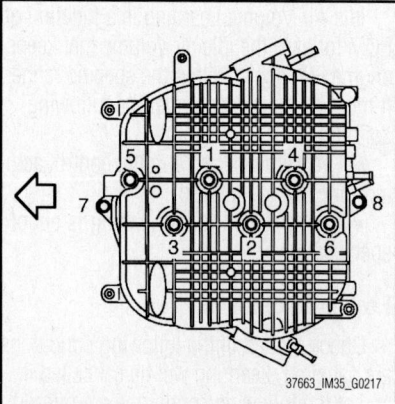

Fig. 103 Intake manifold collector tightening sequence; Loosen mounting bolts in reverse order as shown

37663_IM35_G0217

5. Disconnect vacuum hose, PCV hose and EVAP hose from intake manifold collector.
6. Remove EVAP canister purge volume control solenoid valve and EVAP tube assembly from intake manifold collector.
7. Loosen mounting bolts in reverse order as shown to remove intake manifold collector.

> ❊❊ **CAUTION**
>
> **Cover engine openings to avoid entry of foreign materials.**

8. Remove PCV hose between intake manifold collector and rocker cover (bank 1).

  To install:

9. Note the following, and install in the reverse order of removal.
10. For the intake manifold collector, if stud bolts were removed, install them and tighten to 8 ft. lbs. (11 Nm). Tighten mounting bolts in sequence order as shown.

➡**Tighten mounting bolts to secure gasket and intake manifold collector.**

11. Insert the water hose by 1.06 to 1.26 inches (27 to 32 mm) from connector end. Clamp hose at location of 0.12 to 0.28 inches (3 to 7 mm) from hose end.
12. Install the electric throttle control actuator in the reverse order of removal. Tighten mounting bolts in sequence order as shown.

> ❊❊ **CAUTION**
>
> **Note the following:**
>
> - Handle carefully to avoid any shock to electric throttle control actuator.
> - Never disassemble.
> - The figure shows the electric throttle control actuator (bank 1) viewed from the air duct side.
> - Viewed from the air duct side, order of tightening mounting bolts of electric throttle control actuator (bank 2) is the same as that of the electric throttle control actuator (bank 1).

13. Perform the "Throttle Valve Closed Position Learning" when harness connector of electric throttle control actuator is disconnected.
14. Perform the "Idle Air Volume Learning" and "Throttle Valve Closed Position Learning" when electric throttle control actuator is replaced.

IDLE AIR VOLUME LEARNING

Idle Air Volume Learning is a function of ECM to learn the idle air volume that keeps engine idle speed within the specific range. It must be performed under the following conditions:

- Each time electric throttle control actuator or ECM is replaced.
- Idle speed or ignition timing is out of specification.

Preconditioning

Check that all of the following conditions are satisfied. Learning will be cancelled if any of the following conditions are missed for even a moment.

- Battery voltage: More than 12.9 V (At idle)
- Engine coolant temperature: 158–221°F (70–105°C)
- Selector lever: P or N
- Electric load switch: OFF (Air conditioner, headlamp, rear window defogger)

➡**On vehicles equipped with daytime light systems, if the parking brake is** applied before the engine is started the headlamp will not illuminate.

- Steering wheel: Neutral (Straight-ahead position)
- Vehicle speed: Stopped
- Transmission: Warmed-up
- With CONSULT-III: Drive vehicle until "ATF TEMP SE 1" in "DATA MONITOR" mode of "A/T" system indicates less than 0.9 V.
- Without CONSULT-III: Drive vehicle for 10 minutes.

Perform Idle Air Volume Learning

With Consult-III

1. Perform Accelerator Pedal Released Position Learning.
2. Perform Throttle Valve Closed Position Learning.
3. Start engine and warm it up to normal operating temperature.
4. Select "IDLE AIR VOL LEARN" in "WORK SUPPORT" mode.
5. Touch "START" and wait 20 seconds.

Without Consult-III

➡Note the following:

- It is better to count the time accurately with a clock.
- It is impossible to switch the diagnostic mode when an Accelerator Pedal Position (APP) sensor circuit has a malfunction.

1. Perform Accelerator Pedal Released Position Learning.
2. Perform Throttle Valve Closed Position Learning.
3. Start engine and warm it up to normal operating temperature.
4. Turn ignition switch OFF and wait at least 10 seconds.
5. Confirm that accelerator pedal is fully released, turn ignition switch ON and wait 3 seconds.
6. Repeat the following procedure quickly 5 times within 5 seconds.
    a. Fully depress the accelerator pedal.
    b. Fully release the accelerator pedal.
7. Wait 7 seconds, fully depress the accelerator pedal for approx. 20 seconds until the MIL stops blinking and turns ON.
8. Fully release the accelerator pedal within 3 seconds after the MIL turns ON.
9. Start engine and let it idle.
10. Wait 20 seconds.
11. Rev up the engine 2 or 3 times and check that idle speed and ignition timing are within the specifications.

THROTTLE VALVE CLOSED POSITION LEARNING

Throttle Valve Closed Position Learning is a function of ECM to learn the fully closed position of the throttle valve by monitoring the throttle position sensor output signal. It must be performed each time harness connector of electric throttle control actuator or ECM is disconnected.

1. Check that accelerator pedal is fully released.
2. Turn ignition switch ON.
3. Turn ignition switch OFF and wait at least 10 seconds.
4. Check that throttle valve moves during the above 10 seconds by confirming the operating sound.

OIL PAN

REMOVAL & INSTALLATION

3.5L Engine RWD

See Figures 104 through 108.

※※ **WARNING**

To avoid the danger of being scalded, never drain engine oil when engine is hot.

➡**To remove the upper oil pan, remove engine assembly first. When removing the lower oil pan only, remove engine assembly is not necessary. Perform steps 1, 2 and 10.**

1. Remove front and rear undercover.
2. Drain engine oil.

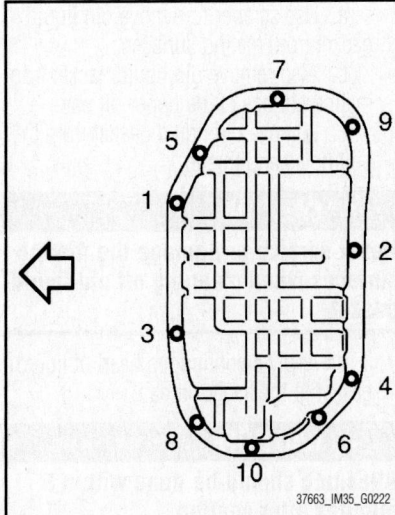

Fig. 104 Lower oil pan tightening sequence shown; Loosen mounting bolts in reverse order of this sequence

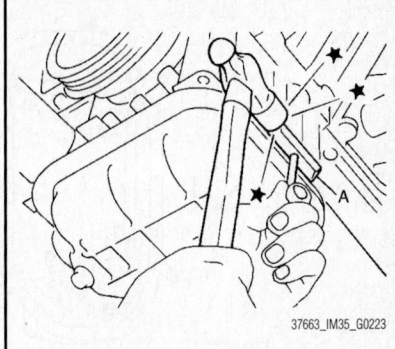

37663_IM35_G0223

Fig. 105 Insert seal cutter (A) between the upper oil pan and the lower oil pan

※※ **CAUTION**

Perform this step when engine is cold. Never spill engine oil on drive belt.

3. Remove engine assembly from the vehicle, and separate front suspension member and transmission from engine.
4. Lift the engine with hoist, and mount it onto widely use engine stand.
5. Remove alternator.
6. Remove starter motor.
7. Remove idler pulley and bracket assembly.
8. Remove oil filter, if necessary.
9. Remove oil temperature sensor, if necessary.
10. Remove the lower oil pan as follows:
    a. Loosen mounting bolts in reverse order as shown to remove the lower oil pan.
    b. Insert seal cutter KV10111100 (J37228) between the upper oil pan and the lower oil pan.

※※ **CAUTION**

Be careful not to damage the mating surfaces. Never insert screwdriver, this will damage the mating surface.

11. Remove oil strainer.
12. Remove rear cover plate.
13. Loosen mounting bolts in reverse order as shown to remove the upper oil pan.
14. Insert seal cutter KV10111100 (J37228) between the upper oil pan and cylinder block. Slide seal cutter by tapping on the side of tool with hammer.

※※ **CAUTION**

Be careful not to damage mating surfaces. Never insert screwdriver, this will damage the mating surface.

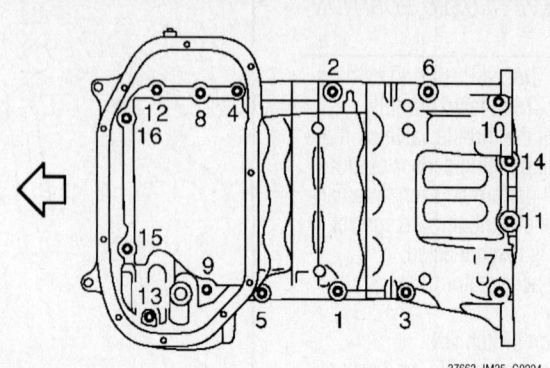

Fig. 106 Upper oil pan bolt tightening sequence shown; Loosen mounting bolts in reverse of this sequence

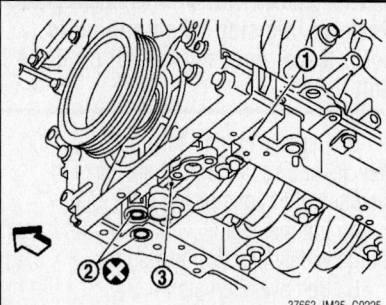

Fig. 107 Remove O-rings (2) from bottom of lower cylinder block (1) and oil pump (3)

15. Remove the upper oil pan.
16. Remove O-rings from bottom of lower cylinder block and oil pump.

To install:

17. Install the upper oil pan as follows:
   a. Use scraper to remove old liquid gasket from mating surfaces.
   b. Also remove the old liquid gasket from mating surface of lower cylinder block.
   c. Remove old liquid gasket from the bolt holes and threads.

✴✴ CAUTION

Never scratch or damage the mating surfaces when cleaning off old liquid gasket.

   d. Install new O-rings on the bottom of lower cylinder block and oil pump.
   e. Apply a continuous bead of liquid gasket to the lower cylinder block mating surface of the upper oil pan as shown.

✴✴ CAUTION

For bolt holes B (7 locations), apply liquid gasket outside the holes.

Attaching should be done within 5 minutes after coating.

   f. Install the upper oil pan.

✴✴ CAUTION

Install avoiding misalignment of both oil pan gasket and O-rings.

   g. Tighten mounting bolts in sequence order as shown.

➡**Bolts measuring 3.62 inches (92 mm) are used at bolt positions 7, 10, and 13. The bolts measuring 1.0 inches (25 mm) are used in all other bolt positions.**

   h. Tighten transmission joint bolts.
18. Install oil strainer to oil pump.

➡**Apply locking sealant to the thread of mounting bolts.**

19. Install the lower oil pan as follows:
   a. Use scraper to remove old liquid gasket from mating surfaces.
   b. Also remove old liquid gasket from mating surface of the upper oil pan.
   c. Remove old liquid gasket from the bolt holes and thread.

✴✴ CAUTION

Never scratch or damage the mating surfaces when cleaning off old liquid gasket.

   d. Apply a continuous bead of liquid gasket to the lower oil pan.

✴✴ CAUTION

Attaching should be done within 5 minutes after coating.

   e. Install the lower oil pan.
   f. Tighten mounting bolts in numerical order as shown.

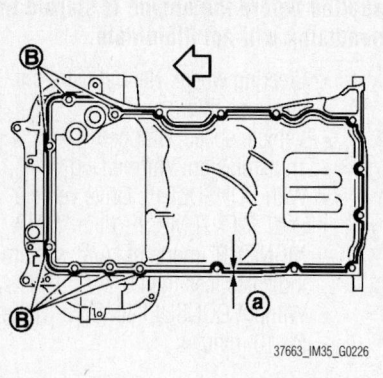

Fig. 108 Apply a continuous bead of liquid gasket to the lower cylinder block mating surface of the upper oil pan as shown

20. Install oil pan drain plug.
21. Install in the reverse order of removal after this step.

➡**At least 30 minutes after oil pan is installed, pour engine oil.**

22. Check engine oil level and adjust engine oil.
23. Start engine, and check there is no leakage of engine oil.
24. Stop engine and wait for 10 minutes.
25. Check engine oil level again.

3.5L Engine AWD

See Figures 104 through 109.

✴✴ WARNING

To avoid the danger of being scalded, never drain engine oil when engine is hot.

➡**To remove the upper oil pan, remove engine assembly first. When removing the lower oil pan only, removal of engine assembly is not necessary. Perform steps 1, 2 and 10.**

1. Remove front and rear undercover.
2. Drain engine oil.

✴✴ CAUTION

Perform this step when engine is cold. Never spill engine oil on drive belt.

3. Remove engine assembly from the vehicle, and separate front suspension member and transmission from engine.
4. Lift the engine with hoist, and mount it onto widely use engine stand.
5. Remove alternator.
6. Remove starter motor.
7. Remove idler pulley and bracket assembly.

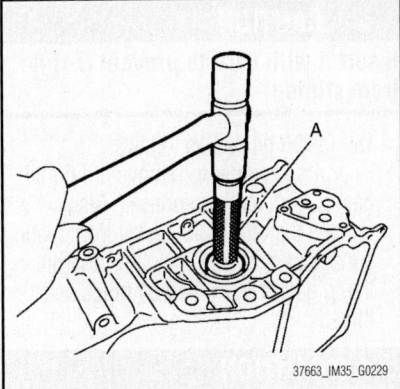

Fig. 109 Remove axle pipe from the upper oil pan using a suitable drift (A)

8. Remove oil filter, if necessary.

9. Remove oil temperature sensor, if necessary.

10. Remove the lower oil pan as follows:

a. Loosen mounting bolts in reverse order as shown to remove the lower oil pan.

b. Insert seal cutter KV10111100 (J37228) between the upper oil pan and the lower oil pan.

✳✳ CAUTION

Be careful not to damage the mating surfaces. Never insert screwdriver, this will damage the mating surface.

11. Remove oil strainer.

12. Remove rear cover plate.

13. Loosen mounting bolts in reverse order as shown to remove the upper oil pan.

14. Insert seal cutter KV10111100 (J37228) between the upper oil pan and cylinder block. Slide seal cutter by tapping on the side of tool with hammer.

15. Remove the upper oil pan.

✳✳ CAUTION

Be careful not to damage mating surfaces. Never insert screwdriver, this will damage the mating surface.

16. Remove O-rings from bottom of lower cylinder block and oil pump.

17. Remove axle pipe, if necessary. Remove axle pipe from the upper oil pan using a suitable drift.

To install:

18. Install axle pipe to the upper oil pan, if removed.

a. Lubricate O-ring groove of axle pipe, O-rings, and O-ring joint of oil pan with new engine oil.

b. Install axle pipe to the upper oil

pan from axle pipe flange side (left side) using a suitable drift.

✳✳ CAUTION

Insert it with care to prevent O-ring from sliding.

19. Install the upper oil pan as follows:

a. Use scraper to remove old liquid gasket from mating surfaces. Also remove the old liquid gasket from mating surface of lower cylinder block. Remove old liquid gasket from the bolt holes and threads.

✳✳ CAUTION

Never scratch or damage the mating surfaces when cleaning off old liquid gasket.

b. Install new O-rings on the bottom of lower cylinder block and oil pump.

c. Apply a continuous bead of liquid gasket to the lower cylinder block mating surface of the upper oil pan as shown.

✳✳ CAUTION

For bolt holes B (7 locations), apply liquid gasket outside the holes. Attaching should be done within 5 minutes after coating.

d. Install the upper oil pan.

✳✳ CAUTION

Install avoiding misalignment of both oil pan gasket and O-rings.

e. Tighten mounting bolts in numerical order as shown.

f. There are three types of mounting bolts. Refer to the following for locating bolts:

- 1 inches (25 mm) bolts used in positions 3, 6, 8, 9, 11, 12, 14, 15, 16
- 2 inches (50 mm) bolt used in position 2
- 3.5 inches (90 mm) bolts used in positions 1, 4, 5, 7, 10, 13

g. Tighten transmission joint bolts.

20. Install oil strainer to oil pump. Apply locking sealant to the thread of mounting bolts.

21. Install the lower oil pan as follows:

a. Use scraper to remove old liquid gasket from mating surfaces. Also remove old liquid gasket from mating surface of the upper oil pan. Remove old liquid gasket from the bolt holes and thread.

✳✳ CAUTION

Never scratch or damage the mating surfaces when cleaning off old liquid gasket.

b. Apply a continuous bead of liquid gasket to the lower oil pan.

✳✳ CAUTION

Attaching should be done within 5 minutes after coating.

c. Install the lower oil pan.

d. Tighten mounting bolts in numerical order as shown.

22. Install oil pan drain plug.

23. Install in the reverse order of removal after this step.

➡**At least 30 minutes after oil pan is installed, pour engine oil.**

24. Check engine oil level and adjust engine oil.

25. Start engine, and check there is no leakage of engine oil.

26. Stop engine and wait for 10 minutes.

27. Check engine oil level again.

4.5L Engine

See Figures 110 through 112.

✳✳ WARNING

To avoid the danger of being scalded, never drain engine oil when engine is hot.

1. Remove front and rear engine under covers.

2. Drain engine oil.

✳✳ CAUTION

Perform this step when engine is cold. Never spill engine oil on drive belts.

3. Remove engine assembly from vehicle.

4. Install engine slingers into front of cylinder head (left bank) and front of cylinder head (right bank).

5. Remove engine mounting insulators (RH and LH) under side nut.

6. Lift with hoist and separate engine and transmission assembly from front suspension member.

✳✳ CAUTION

Avoid damage to and oil/grease smearing or spills onto engine mounting insulator.

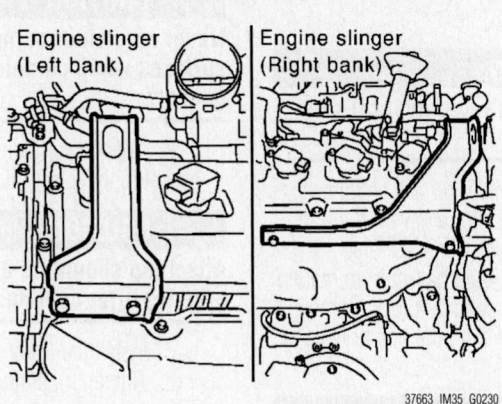

Fig. 110 Install engine slingers into front of cylinder head (left bank) and front of cylinder head (right bank)

7. Remove harness bracket from oil pan. (2WD models)

8. Remove oil filter.

9. Remove oil pan as the follows:

a. Remove rear plate cover.

b. Remove transmission joint bolts which pierce oil pan.

c. Loosen mounting bolts in reverse order as shown.

➡**Disregard the numerical order No. 11 and 17 in removal.**

d. Insert seal cutter between oil pan and cylinder block. Slide seal cutter by tapping on the side of seal cutter with hammer.

e. Remove oil pan.

※※ CAUTION

Be careful not to damage the mating surfaces. Never insert screwdriver, this will damage the mating surface.

f. Remove O-rings from bottom of oil pump and front cover.

10. Remove oil pressure switch if necessary.

11. If necessary, pull axle pipe from oil pan. (AWD models) Hold pipes and pull them out to front drive shaft (left) installing side.

12. Remove oil strainer.

To install:

13. Install oil strainer.

14. Install axle pipe to oil pan, if removed. (AWD models)

a. Lubricate O-ring groove of axle pip, O-ring, and O-ring joint of oil pan with new engine oil.

b. Right/left O-ring diameters differ from each other. O-ring with identifica-

tion paint mark is installed on front drive shaft (left) installing side.

c. Install axle pipe to oil pan from left side.

※※ CAUTION

Insert it with care to prevent O-ring from sliding.

15. Install oil pan as follows:

a. Use scraper to remove old liquid gasket from mating surfaces. Also remove the old liquid gasket from mating surface of cylinder block. Remove old liquid gasket from the bolt holes and threads.

※※ CAUTION

Never scratch or damage the mating surfaces when cleaning off old liquid gasket.

b. Install new O-rings to oil pump and front cover side.

c. Apply a continuous bead of liquid gasket to the cylinder block mating surfaces of oil pan to a limited portion as shown.

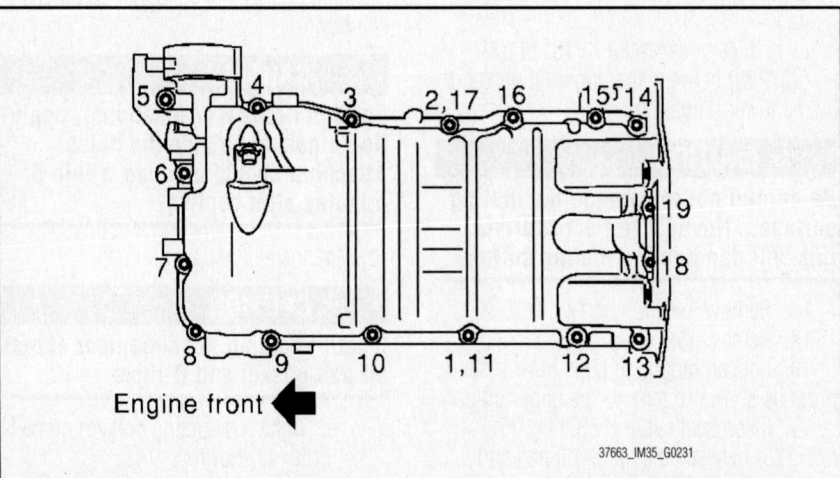

Fig. 111 Oil pan tightening sequence shown; Loosen mounting bolts in reverse order of this sequence

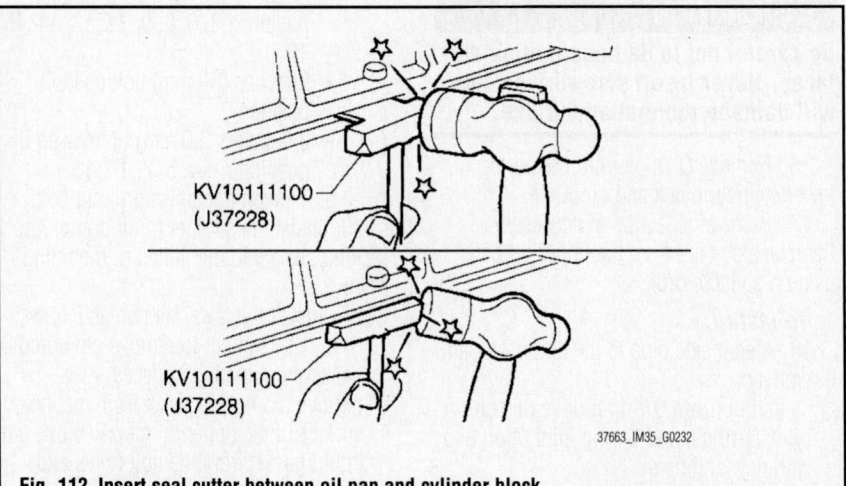

Fig. 112 Insert seal cutter between oil pan and cylinder block

※※ **CAUTION**

Attaching should be done within 5 minutes after application.

    d. Install oil pan.

※※ **CAUTION**

Install avoiding misalignment of O-rings.

    e. Tighten mounting bolts in sequence order as shown.

➡ **Tighten mounting bolts No. 1 and 2 in two steps. The numerical order No. 11 and 17 shown second steps.**

➡ **There are three types of mounting bolts. Refer to the following for locating bolts:**

- 1.18 inches (30 mm) bolts are used in positions 18 and 19
- 4.0 inches (100 mm) bolts are used in positions 5 and 9
- 1.77 inches (45 mm) bolts are used in all other positions
  - f. Tighten transmission joint bolts.
  - g. Install rear plate cover.

16. Install oil pan drain plug with new drain plug washer.

17. Install in the reverse order of removal after this step.

➡ **At least 30 minutes after oil pan is installed, pour engine oil.**

18. Check engine oil level and adjust engine oil.

19. Start engine, and check there is no leakage of engine oil.

20. Stop engine and wait for 15 minutes.

21. Check engine oil level again.

OIL PUMP

REMOVAL & INSTALLATION

3.5L Engine

See Figure 113.

1. Remove the lower oil pan and oil strainer.

2. Remove front timing chain case and timing chain (primary).

3. Remove oil pump assembly.

To install:

※※ **CAUTION**

Before installation, apply new engine oil to the parts as shown.

4. Note the following, and install in the reverse order of removal.

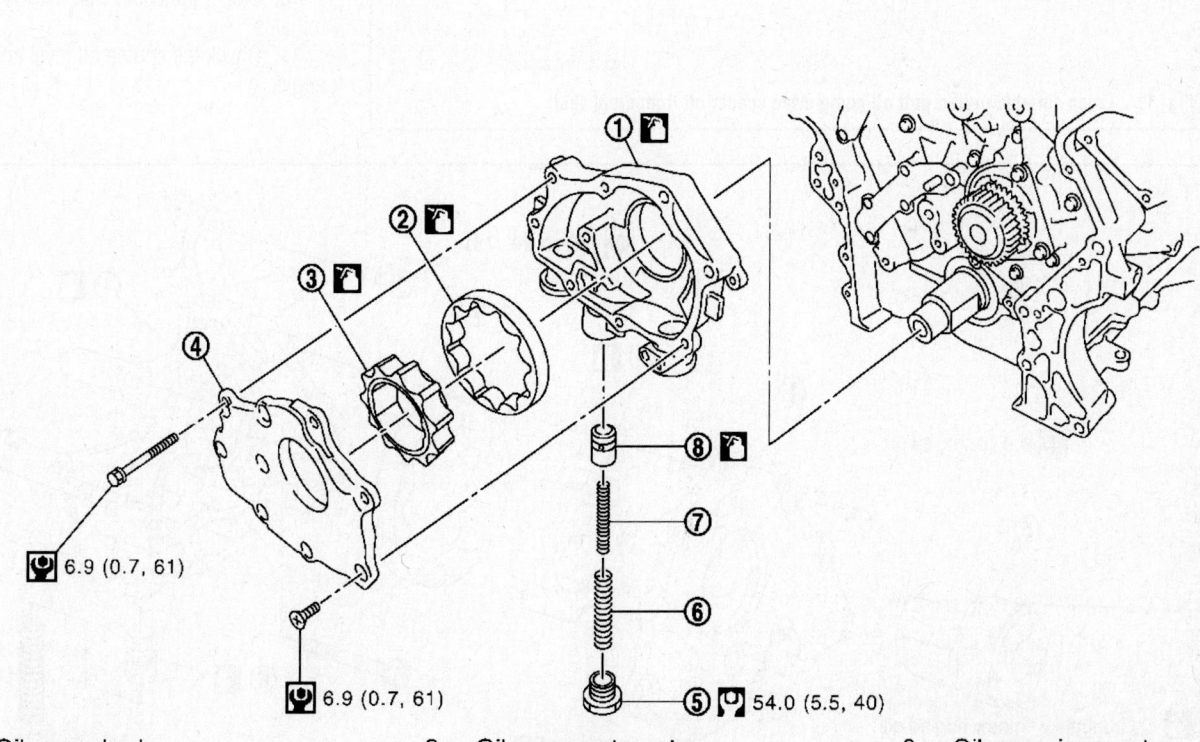

1.	Oil pump body	2.	Oil pump outer rotor	3.	Oil pump inner rotor
4.	Oil pump cover	5.	Regulator valve plug	6.	Regulator valve spring
7.	Regulator valve spring	8.	Regulator valve		

6.9 (0.7, 61)

6.9 (0.7, 61)

54.0 (5.5, 40)

37663_IM35_G0235

Fig. 113 Exploded view of oil pump assembly

➡**When installing, align crankshaft flat faces with oil pump inner rotor flat faces.**

5. Check the engine oil level.

6. Start the engine, and check there is no leakage of engine oil.

7. Stop the engine and wait for 10 minutes.

8. Check the engine oil level and adjust the level.

4.5L Engine

See Figures 114 through 116.

1. Remove engine assembly from vehicle.

2. Remove front cover.

3. Using suitable puller, pull oil pump drive spacer off from crankshaft.

4. Remove oil pump.

To install:

5. Install the oil pump.

6. Install oil pump drive spacer as follows:

    a. Insert oil pump drive spacer according to the directions of crankshaft key and the two flat surfaces of oil pump inner rotor.

➡**If the positional relationship does not allow the insertion, rotate oil pump inner rotor with a finger to allow spacer.**

    b. After confirming that the position of each part is in correct condition to allow for spacer, force fit spacer by lightly tapping with plastic hammer until it contacts and does not go further.

7. Install in the reverse order of removal after this step.

8. Check the engine oil level.

9. Start engine, and check there is no leakage of engine oil.

10. Stop engine and wait for 15 minutes.

11. Check the engine oil level and adjust engine oil.

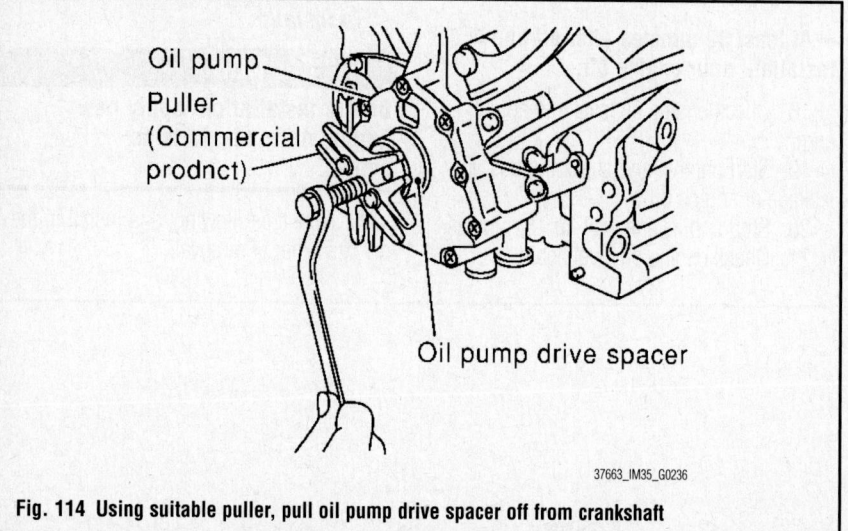

37663_IM35_G0236

Fig. 114 Using suitable puller, pull oil pump drive spacer off from crankshaft

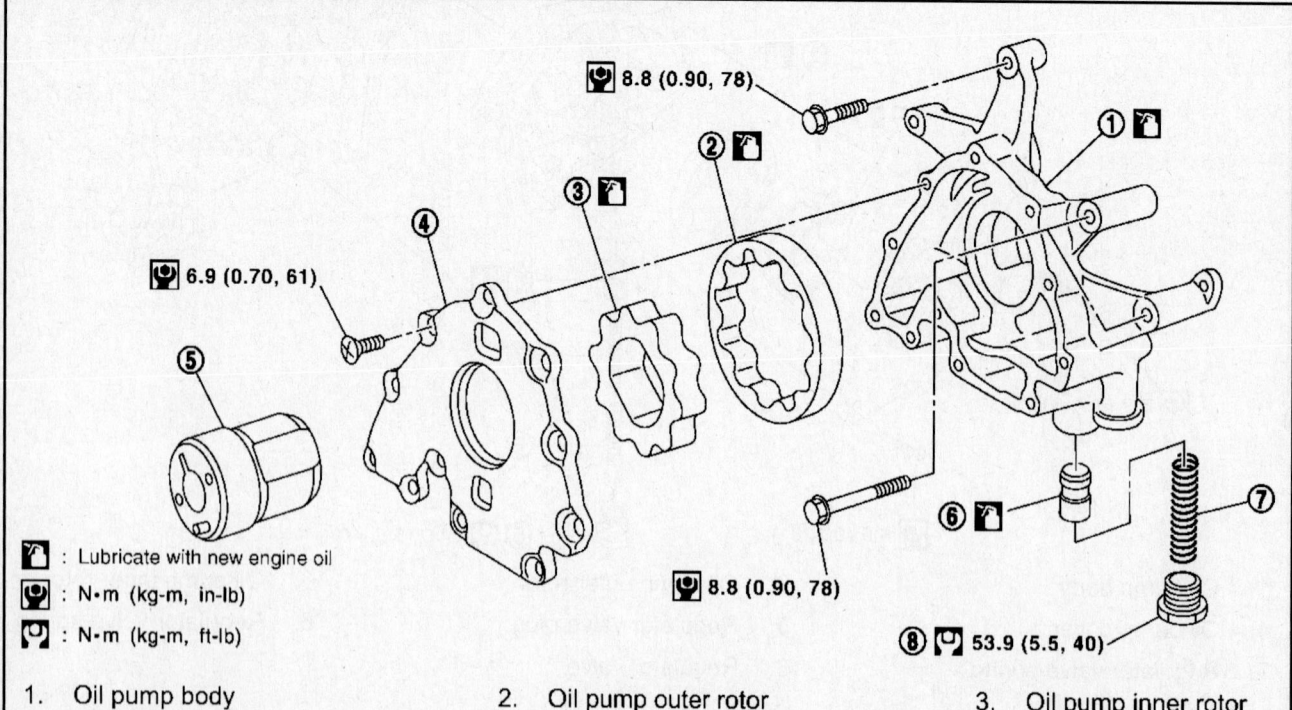

37663_IM35_G0237

	8.8 (0.90, 78)
	6.9 (0.70, 61)
	8.8 (0.90, 78)
	53.9 (5.5, 40)

🛢 : Lubricate with new engine oil

: N·m (kg-m, in-lb)

: N·m (kg-m, ft-lb)

1. Oil pump body
2. Oil pump outer rotor
3. Oil pump inner rotor
4. Oil pump cover
5. Oil pump drive spacer
6. Regulator valve
7. Regulator valve spring
8. Regulator valve plug

Fig. 115 Exploded view of oil pump assembly

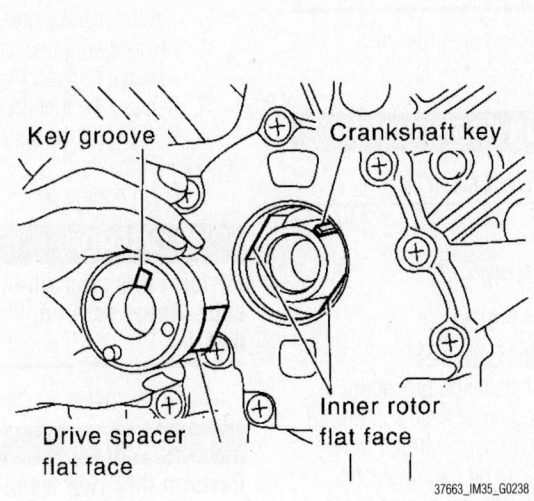

Fig. 116 Insert oil pump drive spacer according to the directions of crankshaft key and the two flat surfaces of oil pump inner rotor

PISTON AND RING

POSITIONING

See Figures 117 and 118.

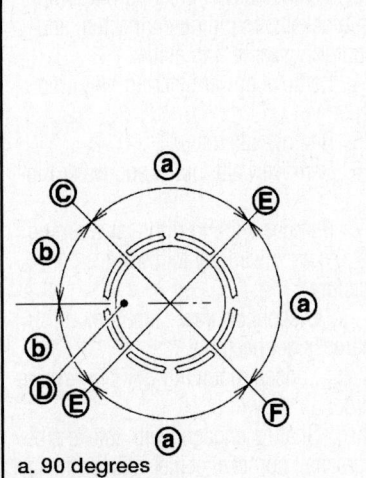

a. 90 degrees
b. 45 degrees
C. Top ring gap
D. Piston front mark
E. Oil ring upper or lower rail gap
F. Second ring and oil ring spacer gap

Fig. 117 Piston ring end gap positions—3.5L engine

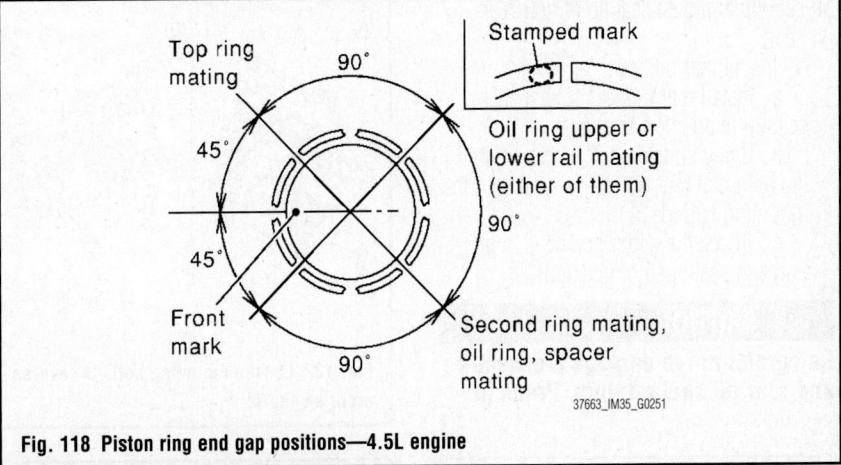

Fig. 118 Piston ring end gap positions—4.5L engine

REAR MAIN SEAL

REMOVAL & INSTALLATION

3.5L Engine

See Figures 119 and 120.

1. Remove transmission assembly.
2. Remove drive plate.
3. Remove rear oil seal.

※※ CAUTION

Be careful not to damage crankshaft and cylinder block.

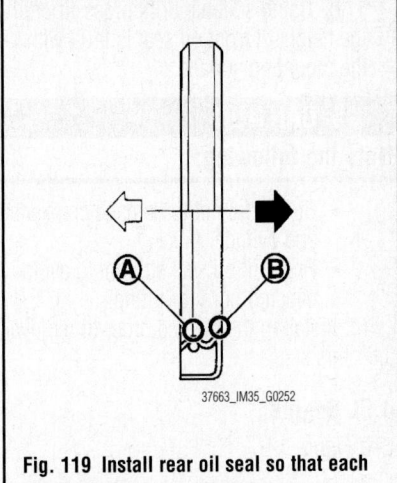

Fig. 119 Install rear oil seal so that each seal lip is oriented as shown

To install:

4. Apply new engine oil to new rear oil seal joint surface and seal lip.
5. Install rear oil seal so that each seal lip is oriented as shown.

　　a. Press in rear oil seal to the position as shown.

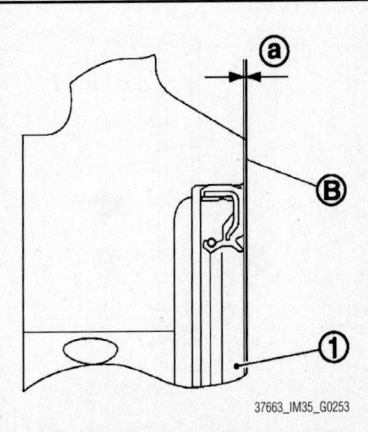

Fig. 120 Press in rear oil seal (1) to the position as shown

b. Using suitable drift, press-fit until the height of front oil seal is level with the mounting surface.

✳✳ CAUTION
Note the following:

- Be careful not to damage crankshaft and cylinder block.
- Press-fit oil seal straight to avoid causing burrs or tilting.

6. Install in the reverse order of removal after this step.

4.5L Engine
See Figure 121.

1. Remove transmission assembly.
   a. Remove drive plate.
   b. Remove rear plate.
2. Remove rear oil seal.

✳✳ CAUTION
Be careful not to damage crankshaft and oil seal retainer surface.

To install:
3. Apply new engine oil to both oil seal lip and dust seal lip of new rear oil seal.
4. Install rear oil seal.
   a. Install rear oil seal so that each seal lip is oriented as shown.
   b. Using suitable drift, press fit until the height of rear oil seal is level with the mounting surface.
   c. Check the garter spring is in position and seal lips not inverted.

✳✳ CAUTION
Be careful not to damage crankshaft and rear oil seal retainer. Press fit

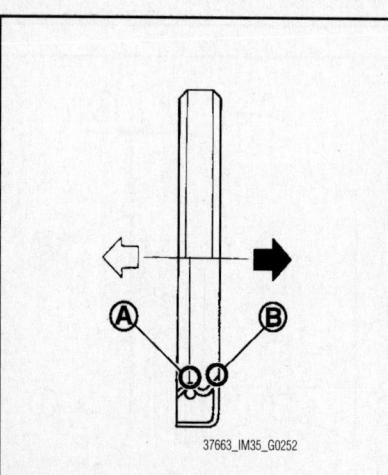

Fig. 121 Install rear oil seal so that each seal lip is oriented as shown

straight and avoid causing burrs or tilting oil seal.

5. Install in the reverse order of removal.

TIMING CHAIN FRONT COVER

REMOVAL & INSTALLATION

3.5L Engine
See Figures 122 through 139.

➡Note the following:

- This section describes removal/installation procedure of front timing chain case and timing chain related parts without removing the upper oil pan on vehicle.
- When the upper oil pan needs to be removed or installed, or when rear timing chain case is removed or installed, remove the upper and lower oil pans first. Then remove front timing chain case, timing

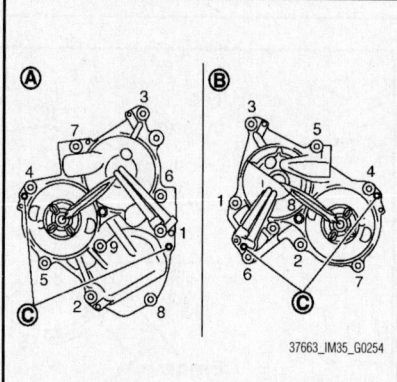

Fig. 122 Loosen mounting bolts in reverse order as shown

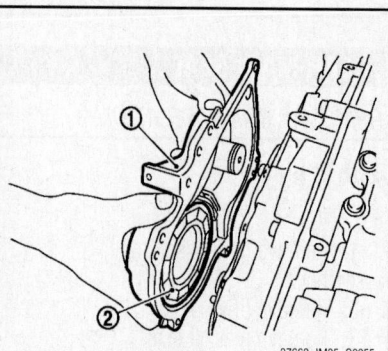

Fig. 123 The mating surface of magnet retarder (2) may be fitted with the exhaust side camshaft sprocket via the engine oil; open valve timing control cover (1) carefully

chain related parts, and rear timing chain case in this order, and install in the reverse order of removal.

1. Remove engine cover.
2. Remove front and rear under covers.
3. Release the fuel pressure.
4. Disconnect the battery cable from the negative terminal.
5. Drain engine oil.

✳✳ CAUTION
Perform this step when engine is cold. Never spill engine oil on drive belt.

6. Drain engine coolant from radiator.

✳✳ CAUTION
Perform this step when engine is cold. Never spill engine coolant on drive belt.

7. Remove radiator cooling fan assembly.
8. Separate engine harnesses removing their brackets from front timing chain case.
9. Remove drive belt.
10. Remove intake manifold collector.
11. Remove harness bracket and fuel sub tube mounting bolt on front timing chain case.
12. Remove oil level gauge and guide.
13. Remove power steering oil pump from bracket with piping connected, and temporarily secure it to aside.
14. Remove power steering oil pump bracket.
15. Remove alternator.
16. Remove water outlet and water piping.
17. Remove left and right valve timing control covers with the following procedure.
   a. Disconnect valve timing control harness connector.
   b. Loosen mounting bolts in reverse order as shown.
   c. Shaft is engaged with intake side camshaft sprocket center hole on inside. Pull straight out so as not to tilt until the joint is disengaged.
   d. The mating surface of magnet retarder may be fitted with the exhaust side camshaft sprocket via the engine oil. Open valve timing control cover carefully.
   e. If the mating surface of magnet retarder is fitted with the camshaft sprocket, open the cover within the range that the load is not applied to the harness. And then, remove it so as to prevent magnet retarder from dropping.

⁕ CAUTION

Note the following:

- Be careful not to damage magnet retarder.
- When carrying valve timing control cover, face the magnet retarder side up to prevent the cover from falling from magnet retarder.
- Never remove magnet retarder from valve timing control cover. (Disassembly prohibited parts)

18. Remove rocker covers (bank 1 and bank 2).

➡**When only timing chain (primary) is removed, rocker cover does not need to be removed.**

19. Obtain No. 1 cylinder at TDC of its compression stroke as follows:

➡**When timing chain is not removed/installed, this step is not required.**

 a. Rotate crankshaft pulley clockwise to align timing mark (grooved line without color) with timing indicator.

 b. Check that intake and exhaust cam noses on No. 1 cylinder (engine front side of bank 1) are located as shown. If not, turn crankshaft one revolution (360 degrees) and align as shown.

➡**When only timing chain (primary) is removed, rocker cover does not need to be removed. To check that No. 1 cylinder is at its compression TDC, remove front timing chain case first. Then check mating marks on camshaft sprockets.**

20. Remove crankshaft pulley as follows:

Fig. 124 Rotate crankshaft pulley clockwise to align timing mark (grooved line without color) with timing indicator

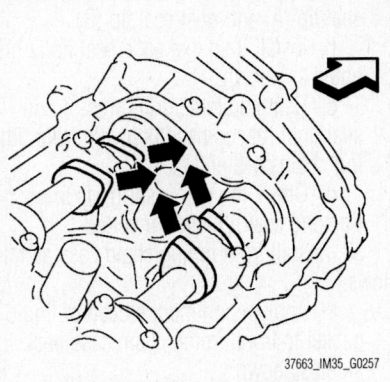

Fig. 125 Check that intake and exhaust cam noses on No. 1 cylinder (engine front side of bank 1) are located as shown

 a. Remove rear cover plate and set ring gear stopper KV10118600 (J-48641) (A) as shown.

 b. Loosen crankshaft pulley bolt and locate bolt seating surface as 0.39 inches (10 mm) from its original position.

⁕ CAUTION

Never remove crankshaft pulley bolt as it will be used as a supporting point for suitable puller.

 c. Place suitable puller tab on holes of crankshaft pulley, and pull crankshaft pulley through.

⁕ CAUTION

Never put suitable puller tab on crankshaft pulley periphery, as this will damage internal damper.

21. Remove the lower oil pan.

Fig. 126 Remove rear cover plate and set ring gear stopper (A) as shown; upper oil pan (1); drive plate (2)

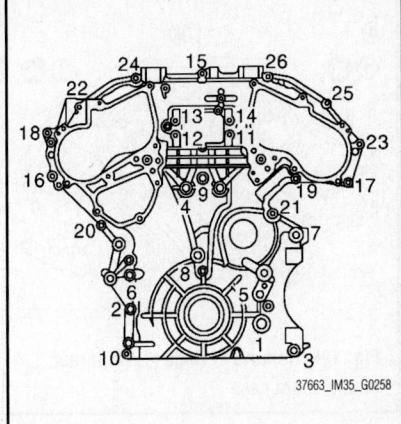

Fig. 127 Loosen mounting bolts in reverse order as shown

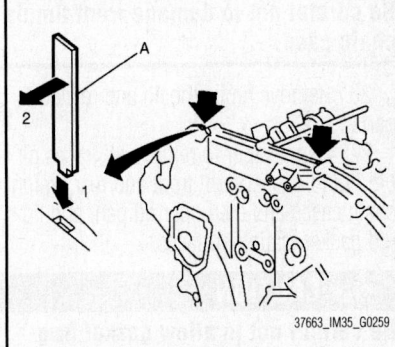

Fig. 128 Insert suitable tool (A) into the notch at the top of the front timing chain case as shown

22. Loosen two mounting bolts in front of the upper oil pan.

23. Remove front timing chain case as follows:

 a. Loosen mounting bolts in reverse order as shown.

 b. Insert suitable tool into the notch at the top of the front timing chain case as shown.

 c. Pry off case by moving tool as shown.

➡**Use seal cutter KV10111100 (J37228) to cut liquid gasket for removal.**

⁕ CAUTION

Note the following:

- Never use screwdriver or something similar.
- After removal, handle front timing chain case carefully so it never tilt, cant, or warp under a load.

24. Remove O-rings from rear timing chain case.

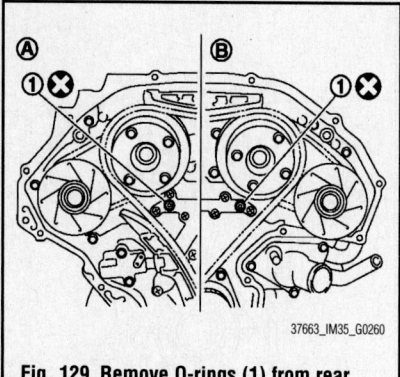

Fig. 129 Remove O-rings (1) from rear timing chain case

25. Remove front oil seal from front timing chain case.

> ❈❈ **CAUTION**
>
> **Be careful not to damage front timing chain case.**

26. Remove timing chain and related parts.

27. Use scraper to remove all traces of old liquid gasket from front and rear timing chain cases and the upper oil pan, and liquid gasket mating surfaces.

> ❈❈ **CAUTION**
>
> **Be careful not to allow gasket fragments to enter oil pan.**

28. Remove old liquid gasket from bolt holes and threads.

To install:

29. Install timing chain and related parts.
30. Install new O-rings on rear timing chain case.
31. Hammer dowel pins (right and left) into front timing chain case up to a point close to taper in order to shorten protrusion length.
32. Install new front oil seal on the front timing chain case.

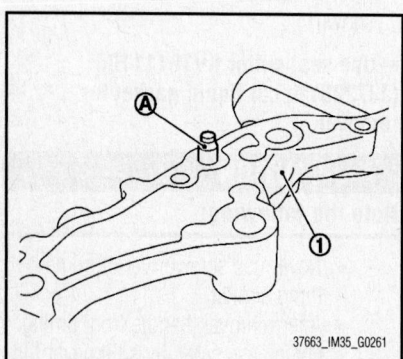

Fig. 130 Hammer dowel pins (right and left) (A) into front timing chain case (1)

a. Apply new engine oil to both oil seal lip (A) and dust seal lip (B).

b. Install it so that each seal lip is oriented as shown.

c. Using suitable drift, press-fit oil seal until it becomes flush with front timing chain case end face.

d. Check the garter spring is in position and seal lip is not inverted.

33. Install front timing chain case as follows:

a. Apply a continuous bead of liquid gasket to front timing chain case back side as shown.

b. Apply liquid gasket to top surface of the upper oil pan as shown.

34. Assemble front timing chain case as follows:

a. Fit lower end of front timing chain case tightly onto top face of the upper oil pan..From the fitting point, make entire front timing chain case contact rear timing chain case completely.

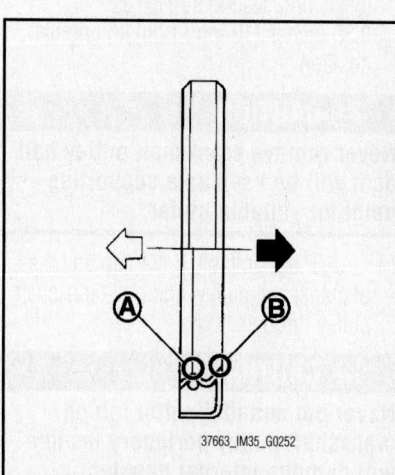

Fig. 131 Install oil seal so that each seal lip is oriented as shown

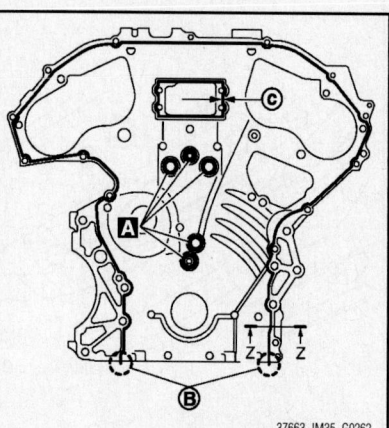

Fig. 132 Apply a continuous bead of liquid gasket to front timing chain case back side as shown

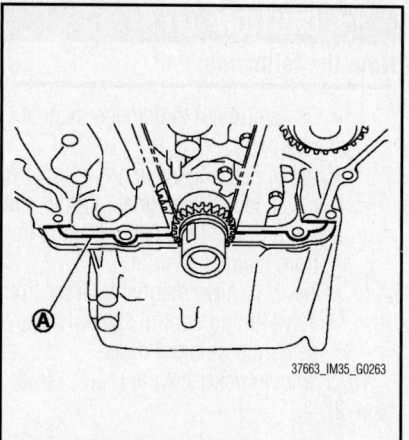

Fig. 133 Apply liquid gasket (A) to top surface of the upper oil pan as shown

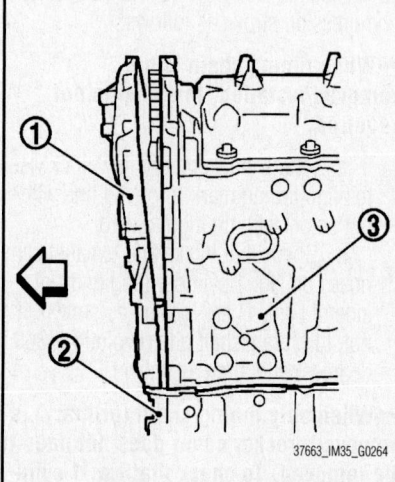

Fig. 134 Fit lower end of front timing chain case (1) tightly onto top face of the upper oil pan (2); cylinder block (3)

> ❈❈ **CAUTION**
>
> **Note the following:**
>
> • Be careful not to damage front oil seal by interference with front end of crankshaft.
> • Attaching should be done within 5 minutes after liquid gasket application.

b. Install front timing chain case as to fit its dowel pin hole together dowel pin on rear timing chain case.

c. Tighten mounting bolts to the specified torque in numerical order as shown.

➥**There are two types of mounting bolts. Refer to the following for locating bolts.**

d. M10 bolts are used in bolt positions 1 through 7. Tighten to 41 ft. lbs. (55 Nm).

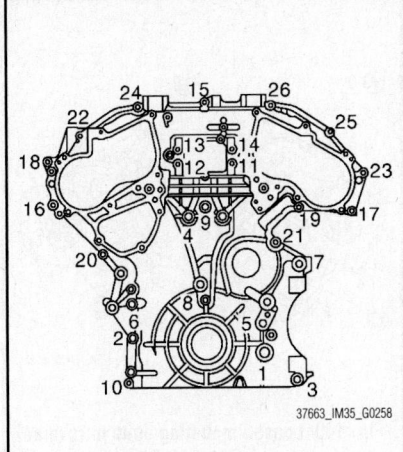

Fig. 135 Tighten mounting bolts to the specified torque in numerical order as shown

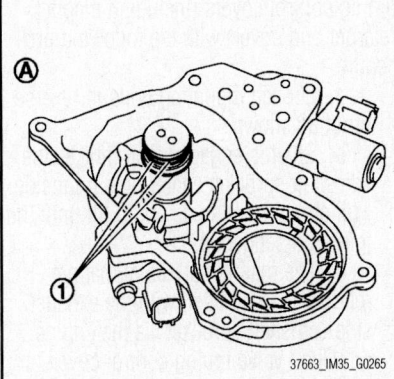

Fig. 136 Install new seal rings (1) in shaft grooves

e. M6 bolts are used in all other bolts positions. Tighten to 9 ft. lbs. (13 Nm).

f. After all bolts tightened, retighten them to the specified torque in numerical order.

35. Install two mounting bolts in front of the upper oil pan.

36. Install the lower oil pan.

37. Install right and left valve timing control covers as follows.

a. Install new seal rings in shaft grooves.

> ❋❋ **CAUTION**

When replacing seal rings, replace all rings with new ones.

b. To check the joint between dowel pins and dowel pin holes, check the looseness in the axle direction by pushing the mating surface of magnet retarder at several places and the circumferential

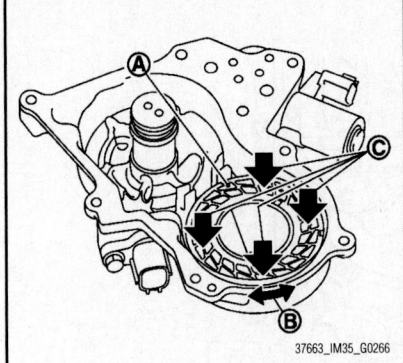

Fig. 137 check the looseness in the axle direction by pushing the mating surface of magnet retarder (A) at several places and the circumferential looseness (between dowel pins and dowel pin holes) by twisting in the circumferential direction

looseness (between dowel pins and dowel pin holes) by twisting in the circumferential direction.

➡ **At position B, it should move slightly. At positions C, it should not be loose.**

> ❋❋ **CAUTION**

Always perform this procedure when removing because the gap between dowel pins and dowel pin holes may not be caused on purpose.

c. Install valve timing control cover with new gasket to front timing chain case.

> ❋❋ **CAUTION**

Note the following:

- Never face the magnet retarder side down to prevent magnet retarder from dropping.
- Check the mating surface of magnet retarder and the drum of exhaust side camshaft sprocket for foreign materials.
- Align the center of both shaft holes of the shaft and the intake side camshaft sprocket, and then insert them.
- Be careful not to drop the seal ring from the shaft groove.
- When setting the valve timing control cover in position by hand, if valve timing control cover is not contacting with the front timing chain case, the dowel pin of magnet retarder may not be aligned with the dowel pin holes of cover.

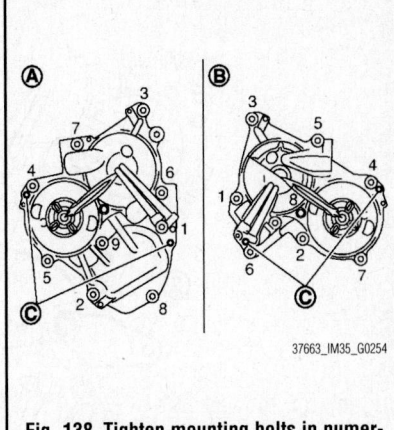

Fig. 138 Tighten mounting bolts in numerical order as shown

d. Tighten mounting bolts in numerical order as shown.

> ❋❋ **CAUTION**

Completely tighten the mounting bolts with the seat surface of valve timing control cover contacting with front timing chain case.

e. After all bolt are tightened, tighten No. 1 bolt to the specified torque again.

38. Install crankshaft pulley as follows:

a. Secure crankshaft using ring gear stopper KV10118600 (J-48641).

b. Install crankshaft pulley, taking care not to damage front oil seal.

➡ **When press-fitting crankshaft pulley with plastic hammer, tap on its center portion (not circumference).**

c. Tighten crankshaft pulley bolt to 33 ft. lbs. (44 Nm).

d. Tighten the bolt 90 degrees (one mark). Place a matching mark on crankshaft pulley aligning with the matching of crankshaft pulley bolt.

e. Rotate crankshaft pulley in normal direction (clockwise when viewed from front) to confirm it turns smoothly.

39. Install power steering oil pump bracket and idler pulley bracket as follows:

a. Tighten mounting bolts in numerical order as shown. (temporarily)

b. Tighten mounting bolts to specified torque in numerical order as shown.

40. For the following operations, perform steps in the reverse order of removal.

41. Before starting engine, check oil/fluid levels including engine coolant and engine oil. If less than required quantity, fill to the specified level.

42. Run engine to check for unusual noise and vibration.

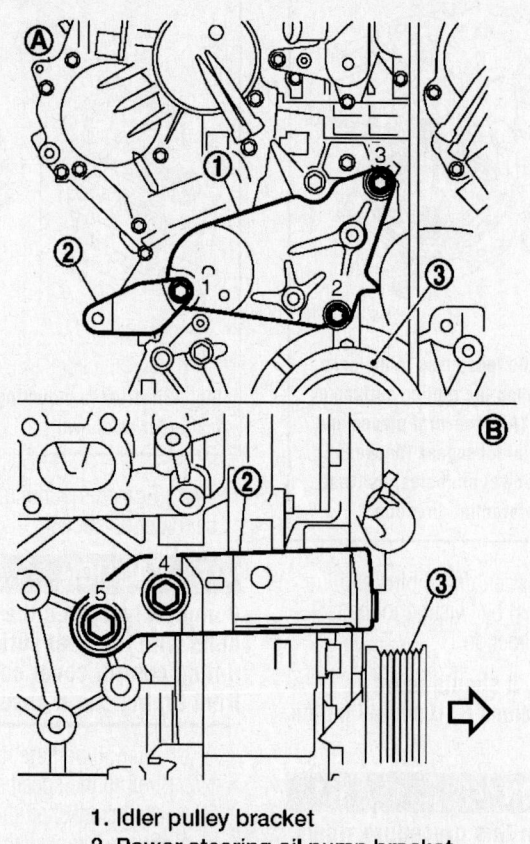

1. Idler pulley bracket
2. Power steering oil pump bracket
3. Crankshaft pulley
A. Engine front side
B. Engine right side

37663_IM35_G0267

Fig. 139 Install power steering oil pump bracket and idler pulley bracket

➡If hydraulic pressure inside timing chain tensioner drops after removal/installation, slack in the guide may generate a pounding noise during and just after engine start. However, this is normal. Noise will stop after hydraulic pressure rises.

43. Warm up engine thoroughly to check there is no leakage of exhaust gases, or any oil/fluids including engine oil and engine coolant.

44. Bleed air from lines and hoses of applicable lines, such as in cooling system.

45. After cooling down engine, again check oil/fluid levels including engine oil and engine coolant. Refill to the specified level, if necessary.

4.5L Engine

For Removal and Installation of Timing Chain Front Cover, refer to Timing Chain & Sprockets Removal and Installation section.

TIMING CHAIN & SPROCKETS

REMOVAL & INSTALLATION

3.5L Engine

See Figures 140 through 175.

➡This section describes procedures for removing/installing front timing chain case and timing chain related parts, and rear timing chain case, when the upper oil pan needs to be removed/installed for engine overhaul, etc.

1. Remove engine assembly from the vehicle, and separate front suspension member and transmission from engine.

2. Lift the engine with hoist and mount it onto widely use engine stand.

3. Remove intake manifold collector.

4. Remove power steering oil pump bracket.

5. Remove alternator.

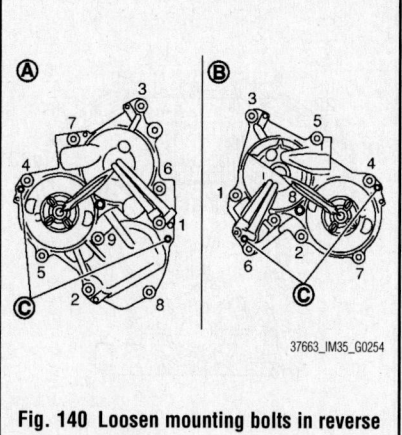

37663_IM35_G0254

Fig. 140 Loosen mounting bolts in reverse order as shown

6. Remove water bypass hose, water hose clamp and idler pulley bracket from front timing chain case.

7. Remove left and right valve timing control covers (including magnet retarder and cover) with the following procedure.

a. Loosen mounting bolts in reverse order as shown.

b. Shaft is engaged with intake side camshaft sprocket center hole on inside. Pull straight out so as not to tilt until the joint is disengaged.

c. The mating surface of magnet retarder may be fitted with the exhaust side camshaft sprocket via the engine oil. Open valve timing control cover carefully.

d. If the mating surface of magnet retarder is fitted with the camshaft sprocket, open the cover within the range that the load is not applied to the harness. And then, remove it so as to prevent magnet retarder from dropping.

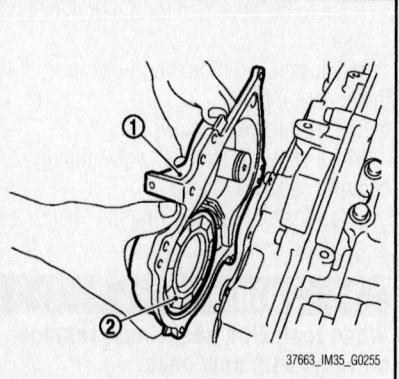

37663_IM35_G0255

Fig. 141 The mating surface of magnet retarder (2) may be fitted with the exhaust side camshaft sprocket via the engine oil; open valve timing control cover (1) carefully

⁕⁕ **CAUTION**

Note the following:

- Be careful not to damage magnet retarder.
- When carrying valve timing control cover, face the magnet retarder side up to prevent the cover from falling from magnet retarder.
- Never remove magnet retarder from valve timing control cover. (Disassembly prohibited parts)

8. Remove rocker covers (bank 1and bank 2).

9. Obtain No. 1 cylinder at TDC of its compression stroke as follows:

a. Rotate crankshaft pulley clockwise to align timing mark (grooved line without color) with timing indicator.

b. Check that intake and exhaust cam noses on No. 1 cylinder (engine front side of bank 1) are located as shown. If not, turn crankshaft one revolution (360 degrees) and align as shown.

Fig. 142 Rotate crankshaft pulley clockwise to align timing mark (grooved line without color) with timing indicator

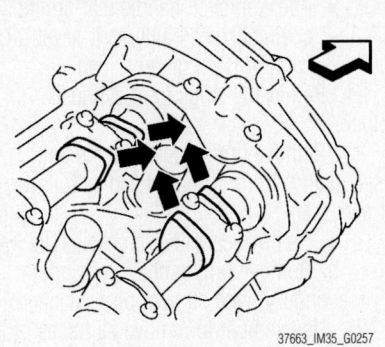

Fig. 143 Check that intake and exhaust cam noses on No. 1 cylinder (engine front side of bank 1) are located as shown

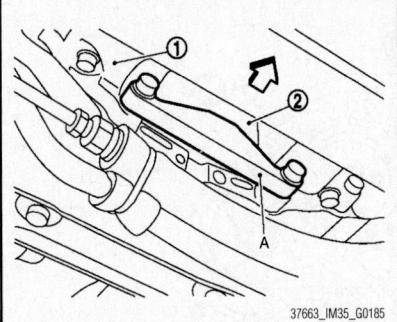

Fig. 144 Remove rear cover plate and set ring gear stopper (A) as shown; upper oil pan (1); drive plate (2)

10. Remove crankshaft pulley as follows:

a. Remove rear cover plate and set ring gear stopper KV10118600 (J-48641) (A) as shown.

b. Loosen crankshaft pulley bolt and rotate bolt seating surface at 0.39 inches (10 mm) from its original position.

⁕⁕ **CAUTION**

Never remove crankshaft pulley bolt as it will be used as a supporting point for suitable puller.

c. Place suitable puller tab on holes of crankshaft pulley, and pull crankshaft pulley through.

⁕⁕ **CAUTION**

Never put suitable puller tab on crankshaft pulley periphery, as this will damage internal damper.

11. Remove the upper and lower oil pans.

12. Remove front timing chain case as follows:

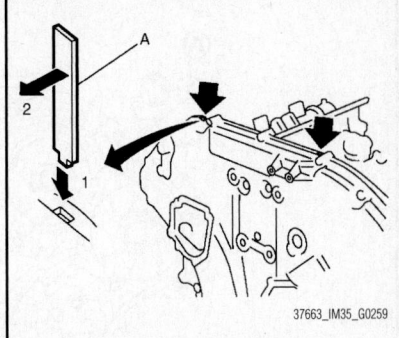

Fig. 146 Insert suitable tool (A) into the notch at the top of the front timing chain case as shown

a. Loosen mounting bolts in reverse order as shown.

b. Insert suitable tool into the notch at the top of the front timing chain case as shown.

c. Pry off case by moving the tool as shown. Use seal cutter KV10111100 (J37228) to cut liquid gasket for removal.

⁕⁕ **CAUTION**

Never use screwdriver or something similar. After removal, handle front timing chain case carefully so it never tilt, cant, or warp under a load.

13. Remove O-ring from rear timing chain case.

14. Remove front oil seal from front timing chain case.

⁕⁕ **CAUTION**

Be careful not to damage front timing chain case.

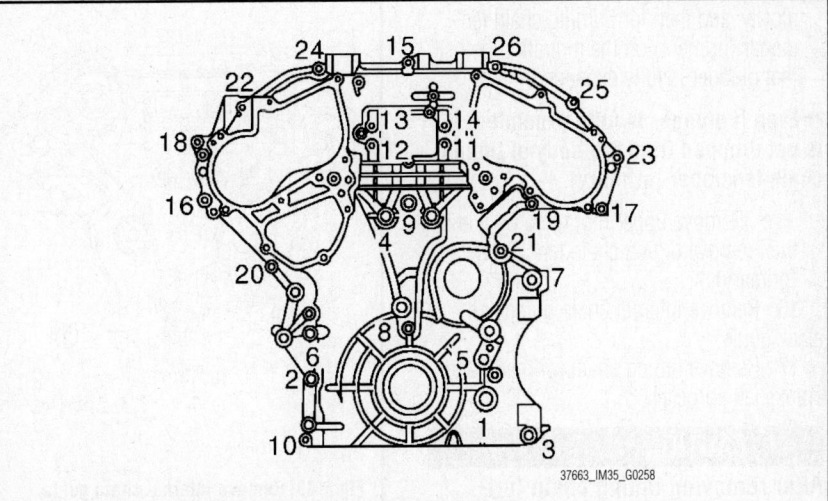

Fig. 145 Loosen mounting bolts in reverse order as shown

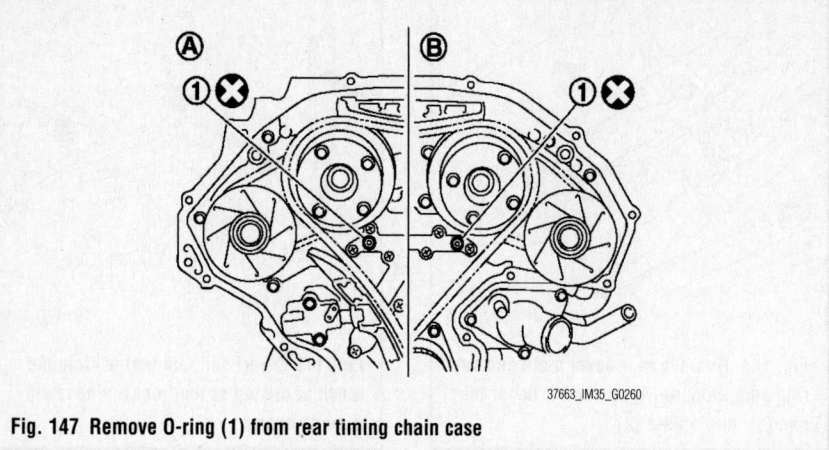

Fig. 147 Remove O-ring (1) from rear timing chain case

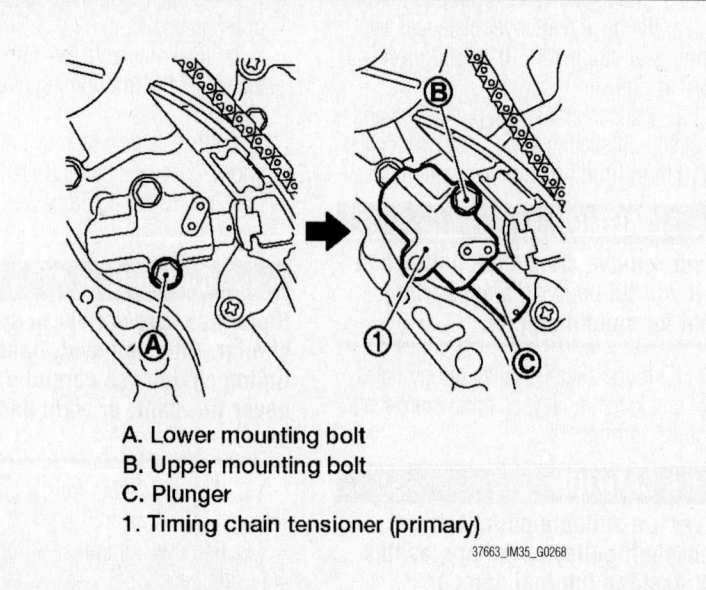

A. Lower mounting bolt
B. Upper mounting bolt
C. Plunger
1. Timing chain tensioner (primary)

37663_IM35_G0268

Fig. 148 Remove timing chain tensioner (primary)

15. Remove timing chain tensioner (primary) as follows:
   a. Remove lower mounting bolt.
   b. Loosen upper mounting bolt slowly, and then turn timing chain tensioner (primary) on the mounting bolt so that plunger is fully expanded.

➡**Even if plunger is fully expanded, it is not dropped from the body of timing chain tensioner (primary).**

   c. Remove upper mounting bolt, and then remove timing chain tensioner (primary).
16. Remove internal chain guide and slack guide.
17. Remove timing chain (primary) and crankshaft sprocket.

❋❋ CAUTION

After removing timing chain (primary), never turn crankshaft and

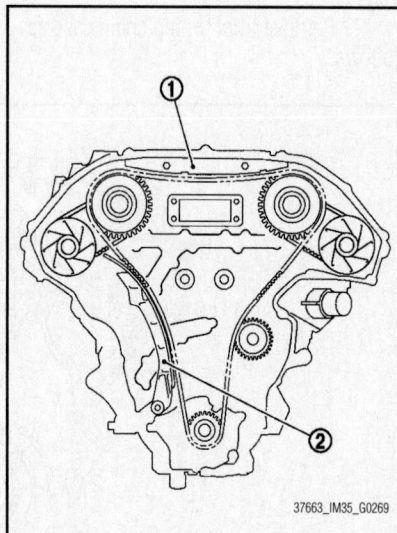

37663_IM35_G0269

Fig. 149 Remove internal chain guide (1) and slack guide (2)

camshaft separately, or valves will strike the piston heads.

18. Remove timing chain (secondary) and camshaft sprockets as follows:
   a. Attach suitable stopper pin to the right and left timing chain tensioners (secondary).

➡**Use a hard metal pin as a stopper pin.**

   b. Remove intake and exhaust camshaft sprocket bolts. Secure the hexagonal portion of camshaft using wrench to loosen mounting bolts.

❋❋ CAUTION

Never loosen mounting bolts with securing anything other than the camshaft hexagonal portion or with tensioning the timing chain.

   c. Remove timing chain (secondary) together with camshaft sprockets.

❋❋ CAUTION

Note the following:

- Handle carefully to avoid any shock to camshaft sprocket.
- Never disassemble.
- Never loosen bolts as shown.
19. Remove water pump.
20. Remove rear timing chain case as follows:
   a. Loosen and remove mounting bolts in reverse order as shown.
   b. Cut liquid gasket using seal cutter KV10111100 (J37228) and remove rear timing chain case.

❋❋ CAUTION

Note the following:

- Never remove plate metal cover of oil passage.
- After removal, handle rear timing chain case carefully so it never tilt, cant, or warp under a load.
21. Remove O-rings from cylinder block.
22. Remove timing chain tensioners (secondary) from cylinder head as follows, if necessary.
   a. Remove camshaft brackets (No. 1).
   b. Remove timing chain tensioners (secondary) with stopper pins attached.
23. Use scraper to remove all traces of old liquid gasket from front and rear timing chain cases, and opposite mating surfaces. Remove old liquid gasket from bolt holes and threads.

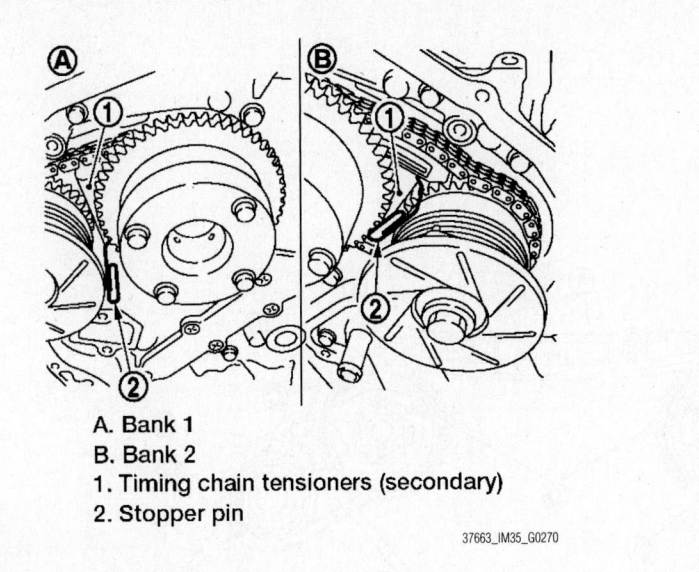

A. Bank 1
B. Bank 2
1. Timing chain tensioners (secondary)
2. Stopper pin

Fig. 150 Remove timing chain (secondary) and camshaft sprockets

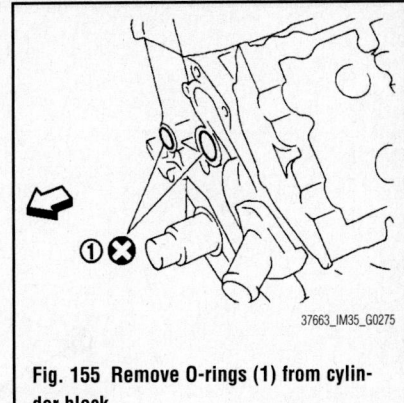

Fig. 155 Remove O-rings (1) from cylinder block

To install:

➡Note the following:

• The below figure shows the relationship between the mating mark on each timing chain and that on

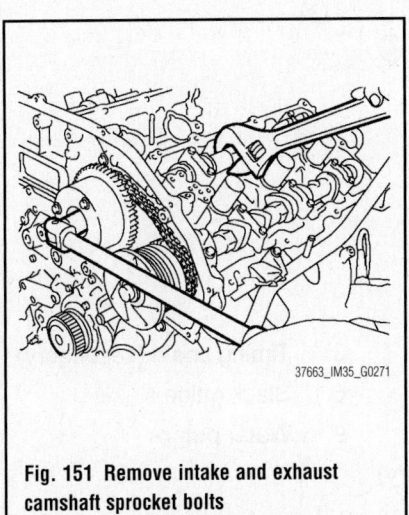

Fig. 151 Remove intake and exhaust camshaft sprocket bolts

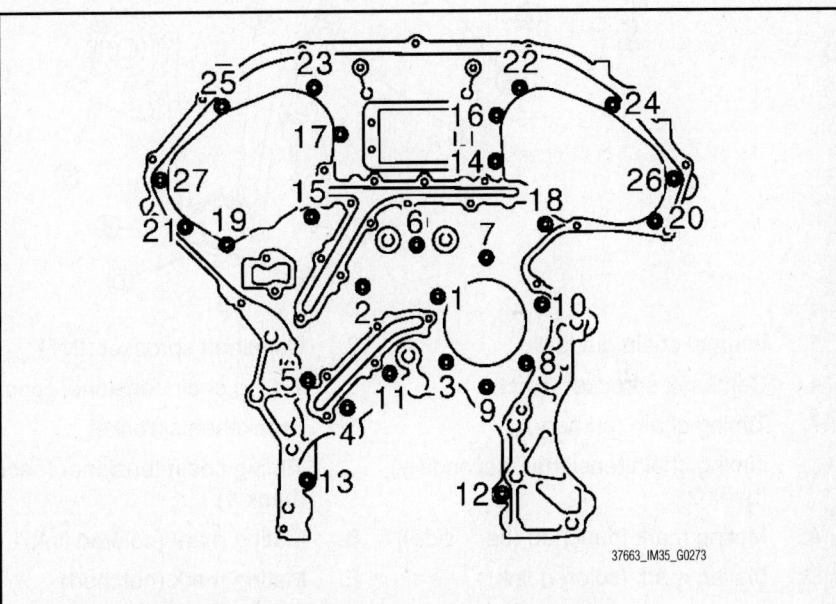

Fig. 153 Loosen and remove mounting bolts in reverse order as shown

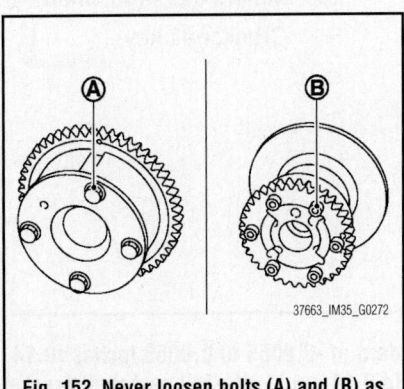

Fig. 152 Never loosen bolts (A) and (B) as shown

24. Check for cracks and any excessive wear at link plates and roller links of timing chain. Replace timing chain if necessary.

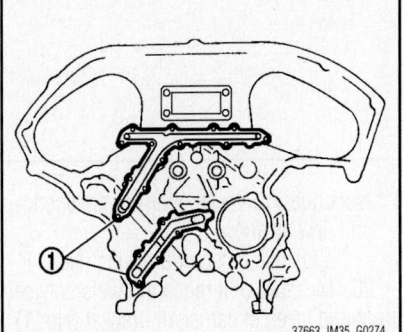

Fig. 154 Never remove plate metal cover (1) of oil passage

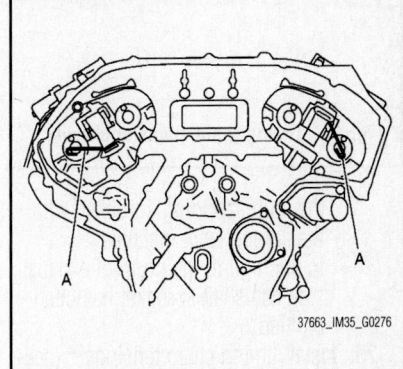

Fig. 156 Remove timing chain tensioners (secondary) with stopper pins (A) attached

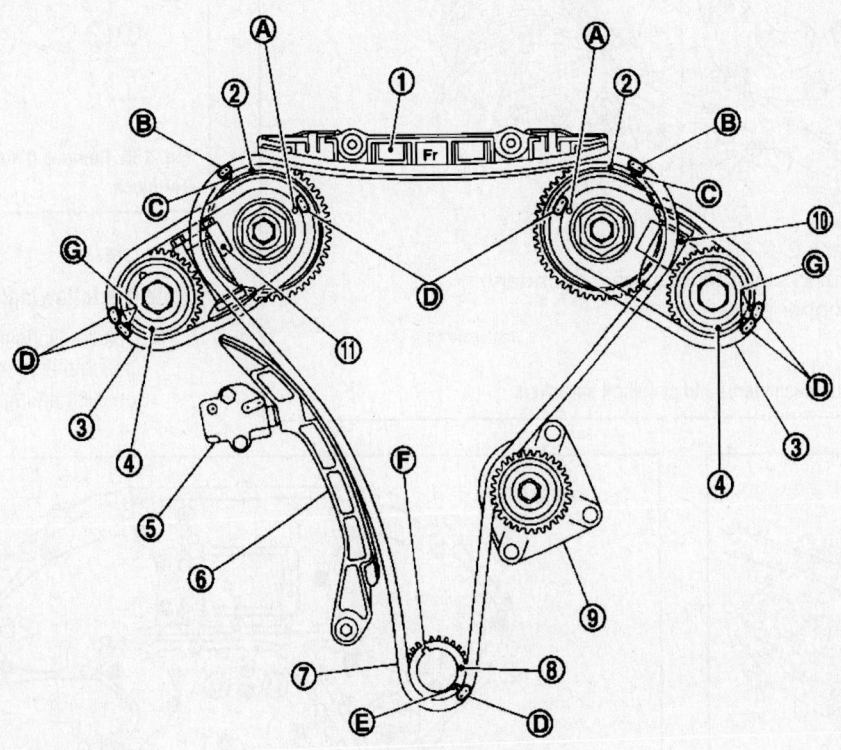

1. Internal chain guide	2. Camshaft sprocket (INT)	3. Timing chain (secondary)
4. Camshaft sprocket (EXH)	5. Timing chain tensioner (primary)	6. Slack guide
7. Timing chain (primary)	8. Crankshaft sprocket	9. Water pump
10. Timing chain tensioner (secondary) (bank 2)	11. Timing chain tensioner (secondary) (bank 1)	
A. Mating mark [punched (back side)]	B. Mating mark (colored link)	C. Mating mark (punched)
D. Mating mark (colored link)	E. Mating mark (notched)	F. Crankshaft key

37663_IM35_G0277

Fig. 157 Timing chain marks orientation

the corresponding sprocket, with the components installed.
- In this figure, the drum of exhaust side camshaft sprocket has been omitted.

25. Install timing chain tensioners (secondary) to cylinder head as follows if removed.

a. Install timing chain tensioners

(secondary) with stopper pin attached and new O-ring.

b. Install camshaft brackets (No. 1).

26. Measure difference in levels between front end faces of camshaft bracket (No. 1) and cylinder head.

➡**Measure two positions (Both intake and exhaust side) for a single bank. If the measured value is out of the stan-**

dard of -0.0055 to 0.0055 inches (0.14 to 0.14 mm), reinstall camshaft bracket (No. 1).

27. Install rear timing chain case as follows:

a. Install new O-rings onto cylinder block.

b. Apply liquid gasket to rear timing chain case back side as shown.

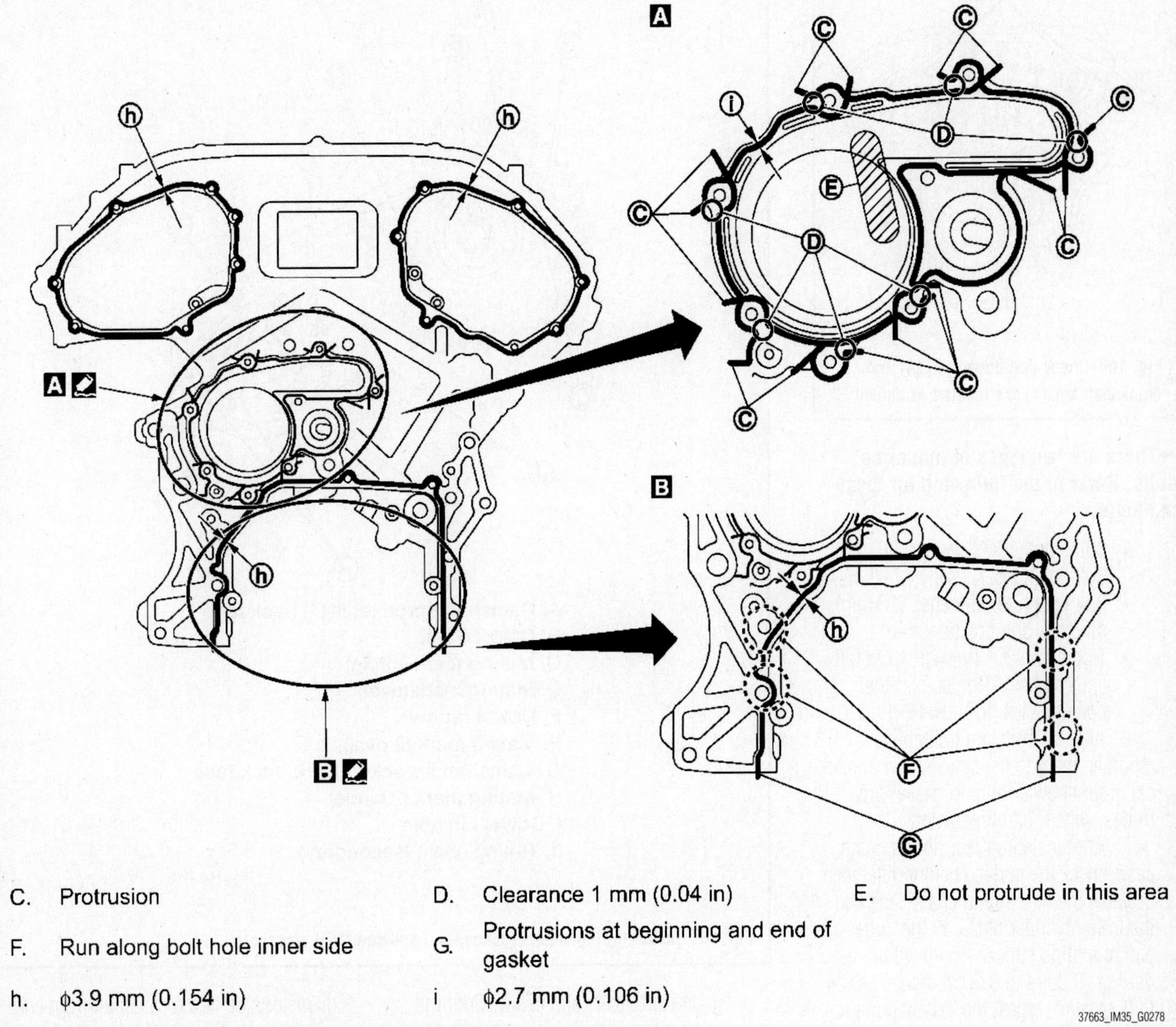

C. Protrusion

D. Clearance 1 mm (0.04 in)

E. Do not protrude in this area

F. Run along bolt hole inner side

G. Protrusions at beginning and end of gasket

h. φ3.9 mm (0.154 in)

i φ2.7 mm (0.106 in)

37663_IM35_G0278

Fig. 158 Apply liquid gasket to rear timing chain case back side as shown

❄❄ CAUTION

Note the following:

- For "A", completely wipe out liquid gasket extended on a portion touching at engine coolant.
- Apply liquid gasket on installation position of water pump and cylinder head very completely.

c. Align rear timing chain case and water pump assembly with dowel pins (right and left) on cylinder block and install rear timing chain case. Check O-rings stay in place during installation to cylinder block.

d. Tighten mounting bolts in numerical order as shown.

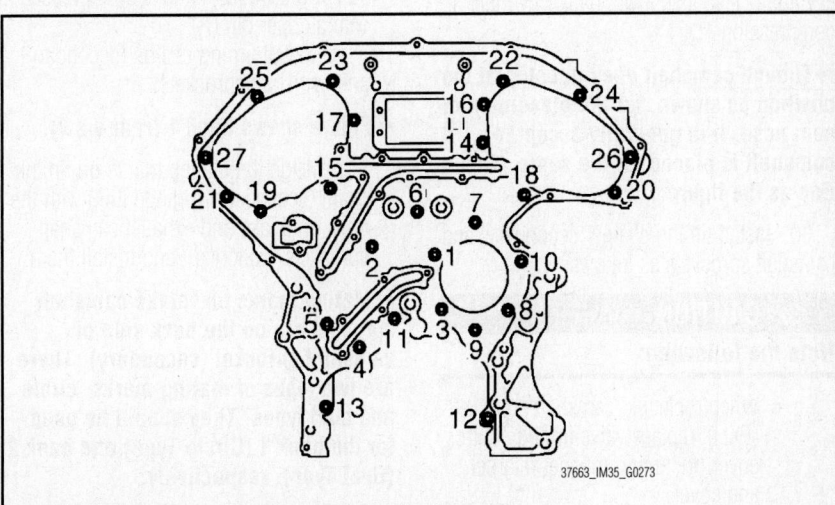

37663_IM35_G0273

Fig. 159 Tighten mounting bolts in numerical order as shown

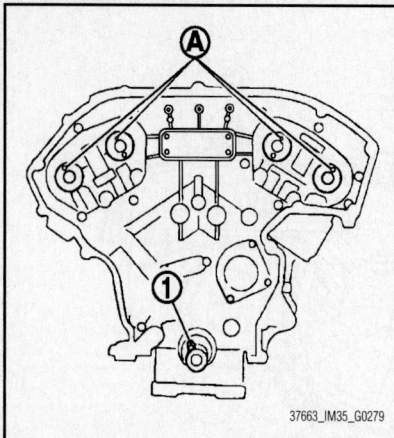

Fig. 160 Check that dowel pin (A) and crankshaft key (1) are located as shown

➡**There are two types of mounting bolts. Refer to the following for locating bolts:**

- Bolt length: 0.79 inches (20 mm): Bolt positions 1, 2, 3, 6, 7, 8, 9, 10
- Bolt length: 0.63 inches (16 mm): All other bolt positions
- Tighten bolts 1 through 13 to 9 ft. lbs. (13 Nm). Tighten all other bolts to 11 ft. lbs. (15 Nm).

e. After all bolts are tightened, retighten them to the specified in numerical order shown. If liquid gasket protrudes, wipe it off immediately.

f. After installing rear timing chain case, check the surface height difference between the rear timing chain case and the lower cylinder block on the upper oil pan mounting surface. If not within standard of -0.0094 to 0.0055 inches (-0.24 to 0.14 mm), repeat the installation procedure.

28. Install water pump with new O-rings.

29. Check that dowel pin and crankshaft key are located as shown. (No. 1 cylinder at compression TDC)

➡**Though camshaft does not stop at the position as shown, for the placement of cam nose, it is generally accepted camshaft is placed for the same direction as the figure.**

30. Install timing chains (secondary) and camshaft sprockets as follows:

❉❉ CAUTION

Note the following:

- When replacing camshaft sprocket (EXH), replace valve timing control cover (including magnet retarder and cover).
- Mating marks between timing chain

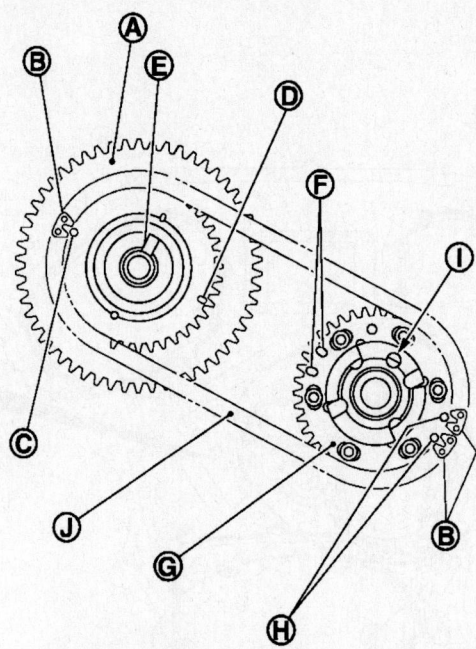

A. Camshaft sprocket (INT) back face
B. Colored link
C. Mating mark (circle)
D. Mating mark (oval)
E. Dowel groove
F. Mating mark (2 oval)
G. Camshaft sprocket (EXH) back face
H. Mating mark (2 circle)
I. Dowel pin hole
J. Timing chain (secondary)

Fig. 161 Install timing chains (secondary) and camshaft sprockets

and sprockets slip easily. Confirm all mating mark positions repeatedly during the installation process.

a. Push plunger of timing chain tensioner (secondary) and keep it pressed in with stopper pin (A).

b. Install timing chains (secondary) and camshaft sprockets.

➡**Figure shows bank 1 (rear view).**

c. Align the mating marks on timing chain (secondary) (orange link) with the ones on intake and exhaust camshaft sprockets (punched), and install them.

➡**Mating marks for intake camshaft sprocket are on the back side of camshaft sprocket (secondary). There are two types of mating marks, circle and oval types. They should be used for the bank 1 (Circle Type) and bank 2 (Oval Type), respectively.**

d. Align dowel pin hole on the small

diameter side of the camshaft front end with dowel pin on the back side of camshaft sprockets, and install them.

e. In case that positions of each mating mark and each dowel pin are not fit on mating parts, make fine adjustment to the position holding the hexagonal portion on camshaft with wrench or equivalent.

f. Mounting bolts for camshaft sprockets must be tightened in the step "d". Tightening them by hand is enough to prevent the dislocation of dowel pins.

g. Check that timing chain (secondary) is not loose from each camshaft sprocket.

h. Check the mating marks (punched) on each camshaft sprocket are positioned on the mating marks (orange link) on timing chain (secondary).

➡**Mating mark (punched) is for checking looseness at this step.**

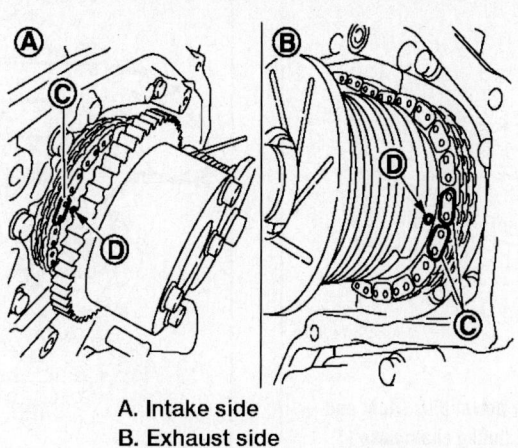

A. Intake side
B. Exhaust side
C. Mating marks (punched)
D. Mating marks (orange link)

37663_IM35_G0281

Fig. 162 Check the mating marks (punched) on each camshaft sprocket are positioned on the mating marks (orange link) on timing chain (secondary)

a. Pull plunger stopper tab up (or turn lever downward) so as to remove plunger stopper tab from the ratchet of plunger.

➡ **Plunger stopper tab and lever are synchronized.**

b. Push plunger into the inside of tensioner body.

c. Hold plunger in the fully compressed position by engaging plunger stopper tab with the tip of ratchet.

d. To secure lever, insert stopper pin through hole of lever into tensioner body hole. The lever parts and the tab are synchronized. Therefore, the plunger will be secured under this condition.

➡ **Figure shows the example of thin screwdriver being used as the stopper pin.**

i. Tighten camshaft sprocket mounting bolts. Secure camshaft using wrench at the hexagonal portion to tighten mounting bolts.

j. Pull stopper pins out from timing chain tensioners (secondary).

31. Install timing chain (primary) as follows:

a. Install crankshaft sprocket. Check the mating marks on crankshaft sprocket face the front of engine.

b. Install timing chain (primary).

c. Install timing chain (primary) so the mating mark (punched) on camshaft sprocket (INT) is aligned with the yellow link on timing chain, while the mating mark (notched) on crankshaft sprocket is aligned with the orange link on timing chain, as shown.

d. When it is difficult to align mating marks of timing chain (primary) with each sprocket, gradually turn camshaft using wrench on the hexagonal portion to align it with the mating marks.

e. During alignment, be careful to prevent dislocation of mating mark alignments of timing chains (secondary).

32. Install internal chain guide and slack guide.

❊❊ CAUTION

Never over tighten slack guide mounting bolts. It is normal for a gap to exist under the bolt seats when mounting bolts are tightened to specification.

33. Install the timing chain tensioner (primary) with the following procedure:

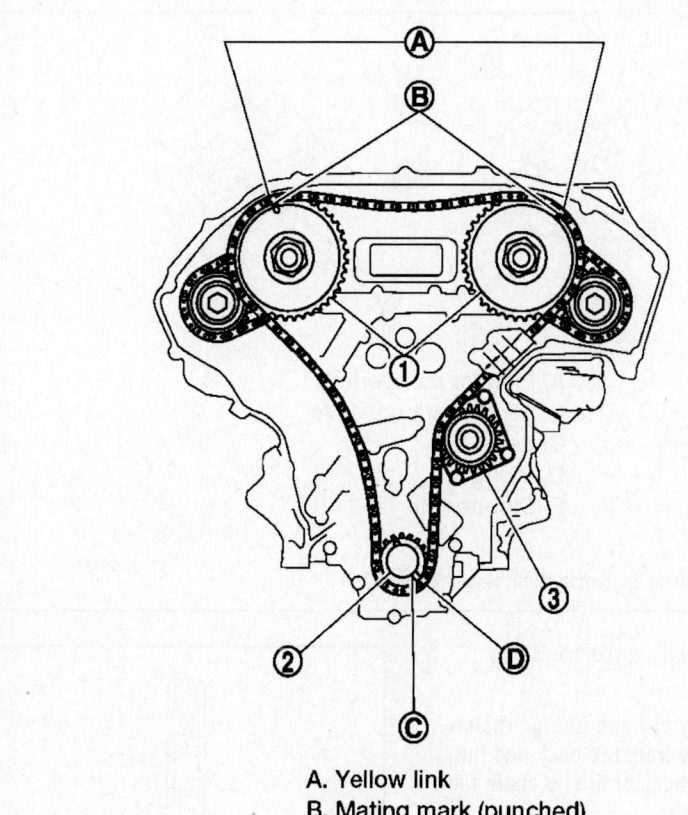

A. Yellow link
B. Mating mark (punched)
C. Mating mark (notched)
D. Orange link
1. Camshaft sprocket INT
2. Crankshaft sprocket
3. Water pump

37663_IM35_G0282

Fig. 163 Install timing chain (primary)

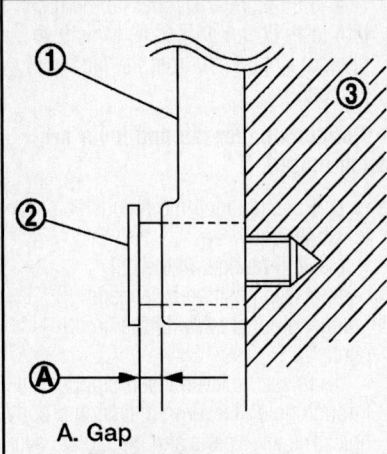

A. Gap
1. Slack guide
2. Slack guide mounting bolts
3. Cylinder block

Fig. 164 It is normal for a gap to exist under the bolt seats when mounting bolts are tightened

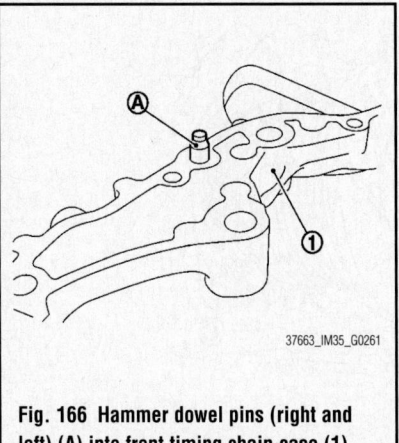

Fig. 166 Hammer dowel pins (right and left) (A) into front timing chain case (1)

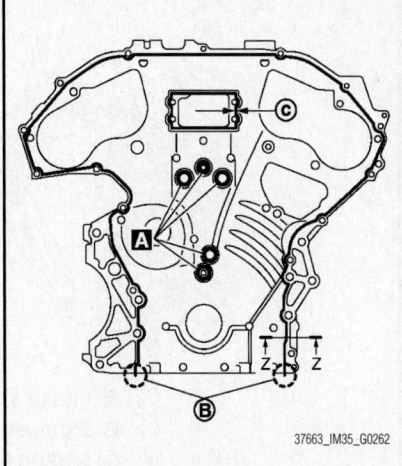

Fig. 168 Apply a continuous bead of liquid gasket to front timing chain case back side as shown

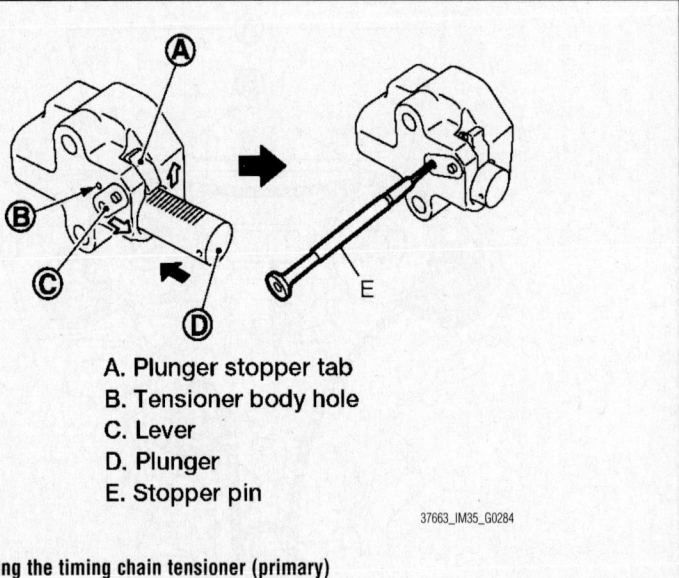

A. Plunger stopper tab
B. Tensioner body hole
C. Lever
D. Plunger
E. Stopper pin

Fig. 165 Installing the timing chain tensioner (primary)

e. Install timing chain tensioner (primary).

➡**Remove any dirt and foreign materials completely from the back and the mounting surfaces of timing chain tensioner (primary).**

f. Pull out stopper pin after installing, and then release plunger.

34. Check again that the mating marks on sprockets and timing chain have not slipped out of alignment.

35. Install new O-ring on rear timing chain case.

36. Hammer dowel pins (right and left) into front timing chain case up to a point

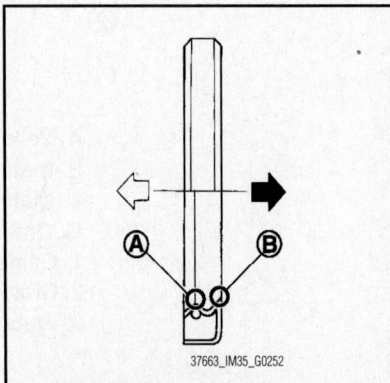

Fig. 167 Install new front oil seal on front timing chain case

close to taper in order to shorten protrusion length.

37. Install new front oil seal on front timing chain case.

a. Apply new engine oil to both oil seal lip and dust seal lip.

b. Install it so that each seal lip is oriented as shown.

c. Using suitable drift, press-fit oil seal until it becomes flush with front timing chain case end face.

d. Check the garter spring is in position and seal lip is not inverted.

38. Install front timing chain case as follows:

a. Apply a continuous bead of liquid gasket to front timing chain case back side as shown.

b. Install front timing chain case as to fit its dowel pin hole together dowel pin on rear timing chain case.

c. Tighten mounting bolts to the specified torque in numerical order as shown.

➡**There are two types of mounting bolts. Refer to the following for locating bolts:**

- M8 bolts: bolt positions 1 through 7. Tighten to 41 ft. lbs. (55 Nm).
- M6 bolts: all other bolt positions. Tighten to 9 ft. lbs. (13 Nm).

d. After all bolts are tightened, retighten them to the specified torque in numerical order shown.

❊❊ **CAUTION**

Be sure to wipe off any excessive liquid gasket leaking on surface mating with the upper oil pan.

e. After installing front timing chain case, check the surface height difference between the following parts on the upper oil pan mounting surface. If not within standard of -0.0094 to 0.0055 inches

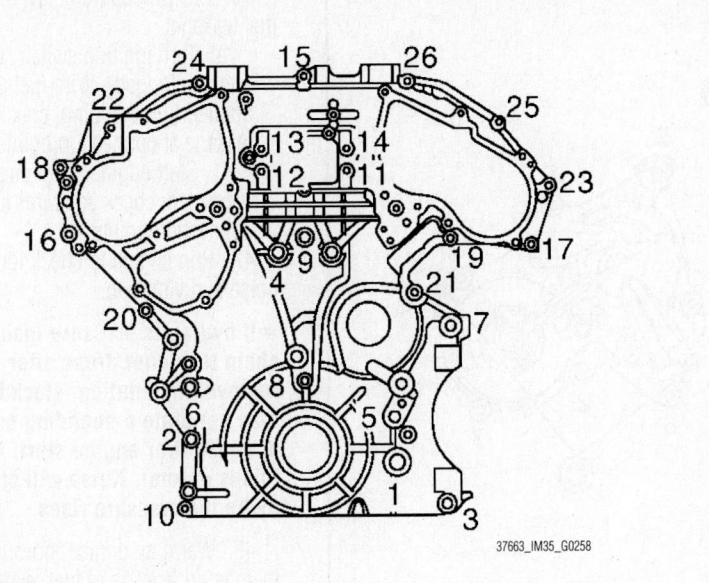

Fig. 169 Tighten mounting bolts in numerical order as shown

(-0.24 to 0.14 mm), repeat the installation procedure.

39. Install right and left valve timing control covers as follows:

a. Install new seal rings in shaft grooves.

> **※ CAUTION**
> **When replacing seal rings, replace all rings with new ones.**

b. To check the joint between dowel pins and dowel pin holes, check the looseness in the axle direction by pushing the mating surface of magnet retarder at several places and the circumferential looseness (between dowel pins and

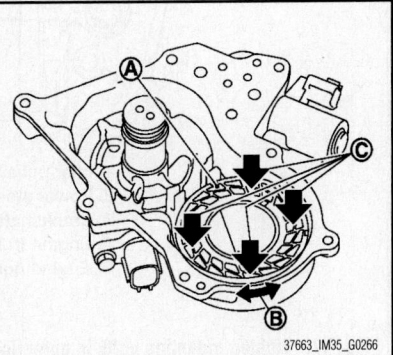

Fig. 171 Install new seal rings (1) in shaft grooves

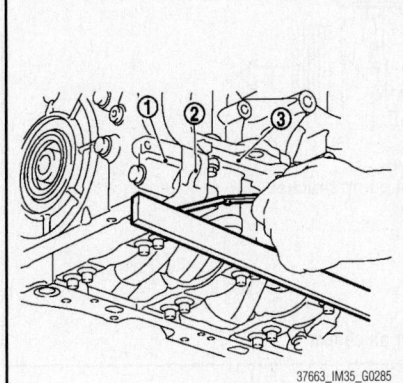

Fig. 170 Front timing chain case (1), rear timing chain case (2), and lower cylinder block (3)

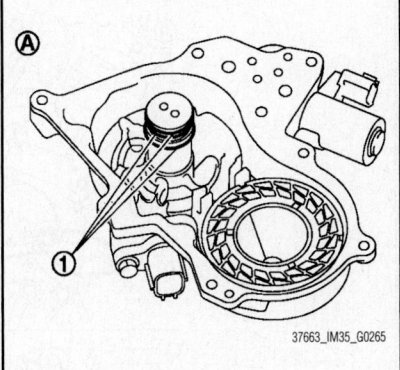

Fig. 172 Check the looseness in the axle direction by pushing the mating surface of magnet retarder (A) at several places and the circumferential looseness (between dowel pins and dowel pin holes) by twisting in the circumferential direction

dowel pin holes) by twisting in the circumferential direction.

➡At position B, it should move slightly. At positions C, it should not be loose.

> **※ CAUTION**
> **Always perform this procedure when removing because the gap between dowel pins and dowel pin holes may not be caused on purpose.**

c. Install valve timing control cover to front timing chain case.

> **※ CAUTION**
> **Note the following:**
> - Never face the magnet retarder side down to prevent magnet retarder from dropping.
> - Check the mating surface of magnet retarder and the drum of exhaust side camshaft sprocket for foreign materials.
> - Align the center of both shaft holes of the shaft and the intake side camshaft sprocket, and then insert them.
> - Be careful not to drop the seal ring from the shaft groove.
> - When setting the valve timing control cover in position by hand, if valve timing control cover is not contacting with the front timing chain case, the dowel pin of magnet retarder may not be aligned with the dowel pin holes of cover. In this case, return to step "b".

d. Tighten mounting bolts in numerical order as shown.

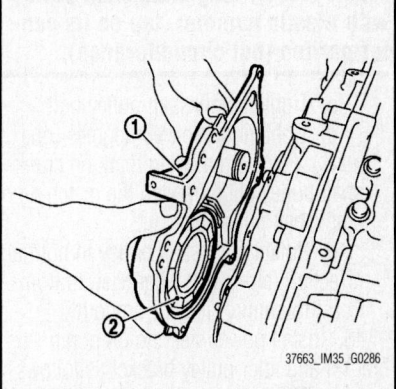

Fig. 173 Install valve timing control cover (1) to front timing chain case; magnet retarder (2)

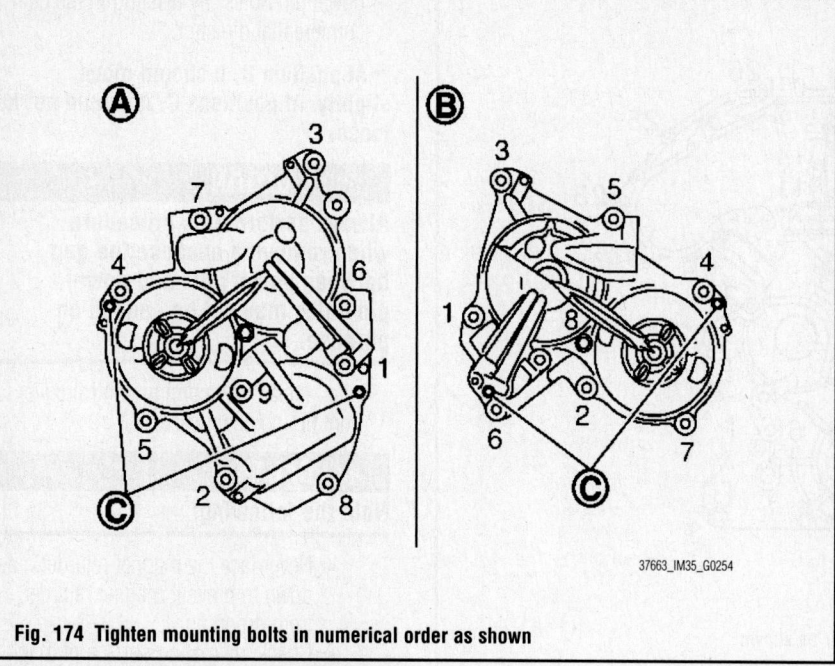

Fig. 174 Tighten mounting bolts in numerical order as shown

❋❋ CAUTION

Completely tighten the mounting bolts with the seat surface of valve timing control cover contacting with front timing chain case.

➡ **After all bolts are tightened, tighten No. 1 bolt to the specified torque again.**

40. Install the upper and lower oil pans.

41. Install rocker covers (bank 1 and bank 2).

42. Install crankshaft pulley as follows:

  a. Secure crankshaft using ring gear stopper KV10118600 (J-48641).

  b. Install crankshaft pulley, taking care not to damage front oil seal.

❋❋ CAUTION

When press-fitting crankshaft pulley with plastic hammer, tap on its center portion (not circumference).

  c. Tighten crankshaft pulley bolt.

  d. Tighten the bolt 90 degrees (one mark). Place a matching mark on crankshaft pulley aligning with the matching of crankshaft pulley bolt.

  e. Rotate crankshaft pulley in normal direction (clockwise when viewed from front) to confirm it turns smoothly.

43. Install power steering oil pump bracket and idler pulley bracket as follows:

  a. Tighten mounting bolts in numerical order as shown. (temporarily)

  b. Tighten mounting bolts to specified torque in numerical order as shown.

44. For the remaining operations, perform steps in the reverse order of removal.

45. Before starting engine, check oil/fluid levels including engine coolant and engine oil. If less than required quantity, fill to the specified level.

46. Use procedure below to check for fuel leakage.

  a. Turn ignition switch "ON" (with engine stopped). With fuel pressure applied to fuel piping, check for fuel leakage at connection points.

  b. Start engine. With engine speed increased, check again for fuel leakage at connection points.

47. Run engine to check for unusual noise and vibration.

➡ **If hydraulic pressure inside timing chain tensioner drops after removal/installation, slack in the guide may generate a pounding noise during and just after engine start. However, this is normal. Noise will stop after hydraulic pressure rises.**

48. Warm up engine thoroughly to check there is no leakage of fuel, exhaust gases, or any oil/fluids including engine oil and engine coolant.

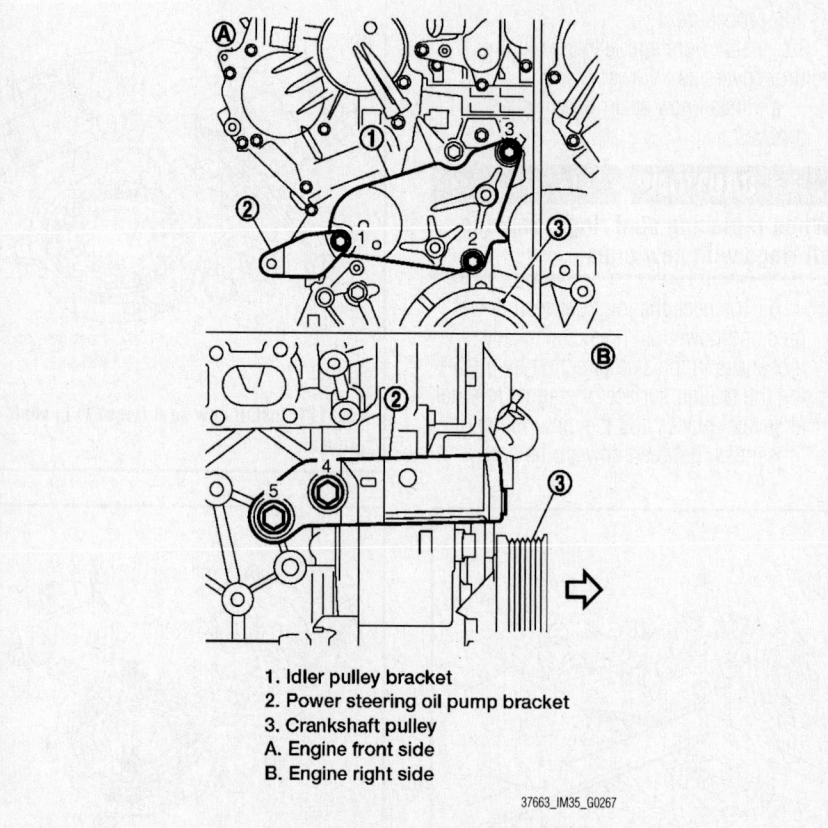

1. Idler pulley bracket
2. Power steering oil pump bracket
3. Crankshaft pulley
A. Engine front side
B. Engine right side

Fig. 175 Tighten mounting bolts in numerical order as shown

49. Bleed air from lines and hoses of applicable lines, such as in cooling system.

50. After cooling down engine, again check oil/fluid levels including engine oil and engine coolant. Refill to the specified level, if necessary.

4.5L Engine

See Figures 176 through 199.

1. Remove engine assembly from vehicle.
2. Remove the following components and related parts:
   - Drive belt auto tensioner and idler pulley
   - Thermostat housing and hoses
   - Ignition coils
   - Rocker covers
3. If necessary, remove intake valve timing control position sensor (right bank and left bank) and Camshaft Position (CMP) sensor (PHASE) from intake valve timing control cover and front cover.

✳✳ CAUTION

Note the following:

- Handle carefully to avoid dropping and shocks.
- Never disassemble.

4. If necessary, remove intake valve timing control solenoid valve from intake valve timing control cover.

✳✳ CAUTION

Note the following:

- Handle components and parts carefully to avoid dropping and shocks.
- Never disassemble.
- Never allow metal powder to adhere to magnetic part at sensor tip.
- Never place sensors in a location where they are exposed to magnetism.

5. Remove intake valve timing control cover as follows:
   a. Loosen and remove mounting

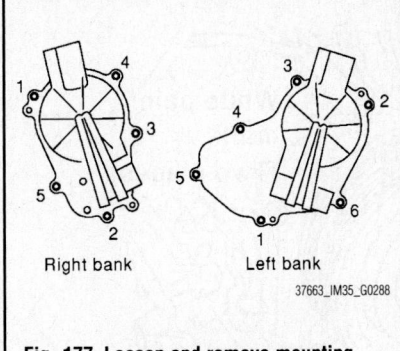

Right bank   Left bank

37663_IM35_G0288

Fig. 177 Loosen and remove mounting bolts in the reverse order as shown

bolts in the reverse order as shown.
   b. Use seal cutter KV10111100 (J-37228) to cut liquid gasket for removal.

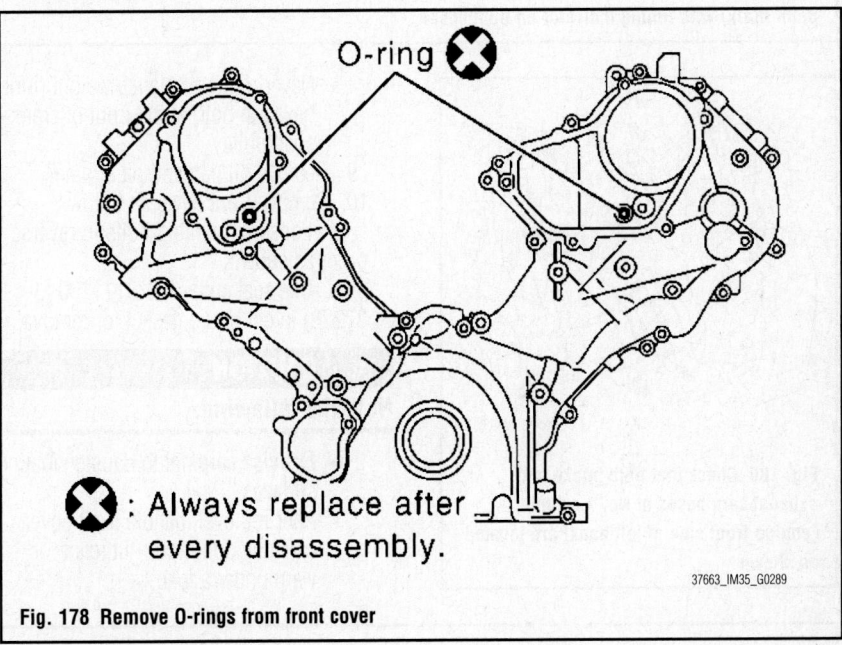

O-ring ✖

✖ : Always replace after every disassembly.

37663_IM35_G0289

Fig. 178 Remove O-rings from front cover

✳✳ CAUTION

Note the following:

- Exercise care not to damage mating surfaces.
- Pull out cover keeping levelness without an angle, as inner part of cover is engaged with the center of camshaft sprocket (INT).

6. Remove O-rings from front cover.
7. Obtain No. 1 cylinder at TDC of its compression stroke as follows:
   a. Rotate crankshaft pulley clockwise to align the TDC identification notch (without paint mark) with timing indicator on front cover.
   b. Check that both intake and exhaust cam noses of No. 1 cylinder (engine front side of left bank) are located as shown. If not, turn crankshaft one revolution (360 degrees) and align as shown.

8. Remove crankshaft pulley as follows:
   a. Remove rear plate cover, and set ring gear stopper (J-45476).
   b. Loosen crankshaft pulley bolt, and then pull crankshaft pulley with both hands to remove it.

✳✳ CAUTION

Note the following:

- Never remove crankshaft pulley bolt. Keep loosened crankshaft pulley bolt in place to protect removed crankshaft pulley from dropping.

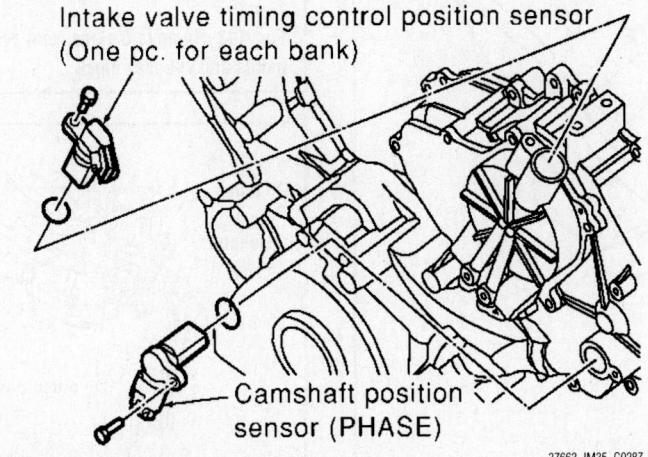

Intake valve timing control position sensor (One pc. for each bank)

Camshaft position sensor (PHASE)

37663_IM35_G0287

Fig. 176 If necessary, remove intake valve timing control solenoid valve from intake valve timing control cover

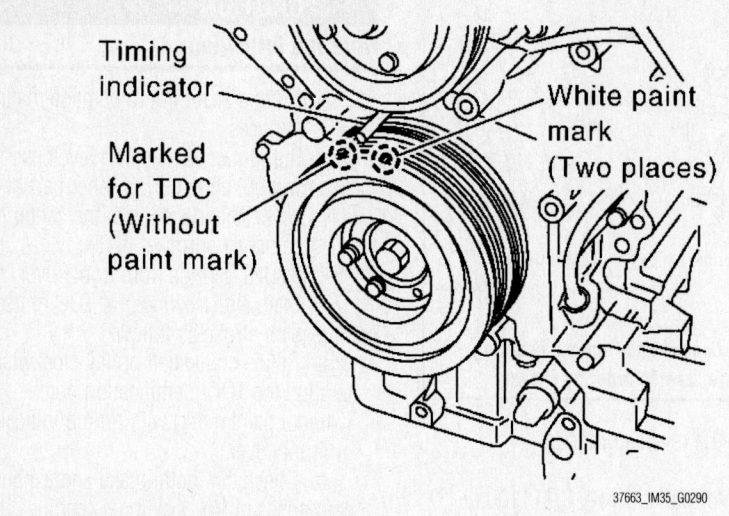

Fig. 179 Rotate crankshaft pulley clockwise to align the TDC identification notch (without paint mark) with timing indicator on front cover

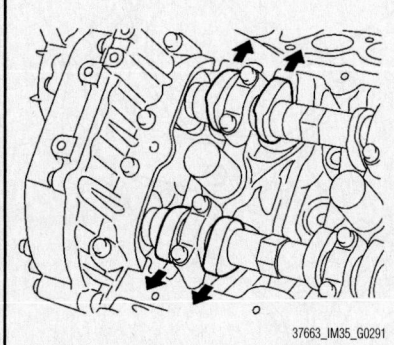

Fig. 180 Check that both intake and exhaust cam noses of No. 1 cylinder (engine front side of left bank) are located as shown

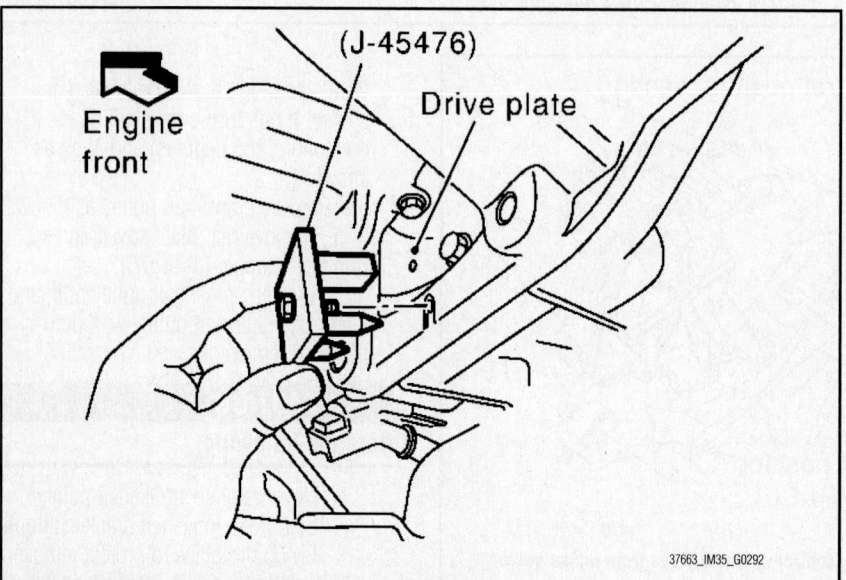

Fig. 181 Remove rear plate cover, and set ring gear stopper

- Never remove balance weight (inner hexagon bolt) at the front of crankshaft pulley.
9. Remove oil pan and oil strainer.
10. Remove front cover as follows:
   a. Loosen mounting bolts in reverse order as shown.
   b. Use seal cutter KV10111100 (J-37228) to cut liquid gasket for removal.

✳✳ CAUTION
Note the following:

- Exercise care not to damage mating surfaces.
- After removal, handle front cover carefully so it never tilt, cant, or warp under a load.

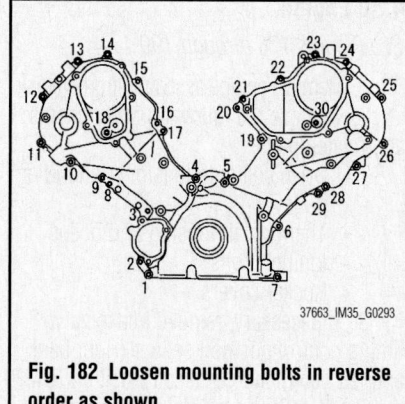

Fig. 182 Loosen mounting bolts in reverse order as shown

11. Remove front oil seal from front cover.

✳✳ CAUTION
Be careful not to damage front cover.

12. Remove O-rings from cylinder heads (right bank and left bank) and cylinder block.
13. Remove chain tensioner cover from front cover. Use seal cutter KV10111100 (J-37228) to cut liquid gasket for remove.
14. Remove oil pump drive spacer.

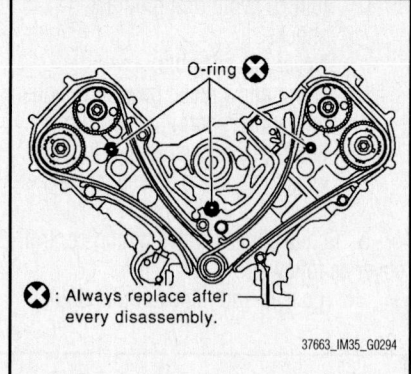

Fig. 183 Remove O-rings from cylinder heads and cylinder block

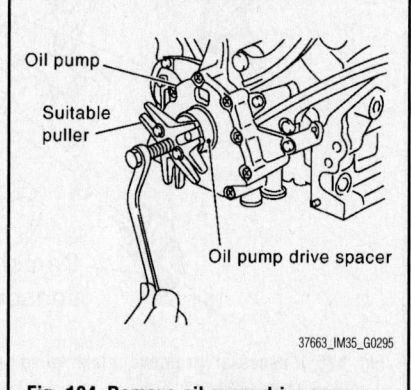

Fig. 184 Remove oil pump drive spacer

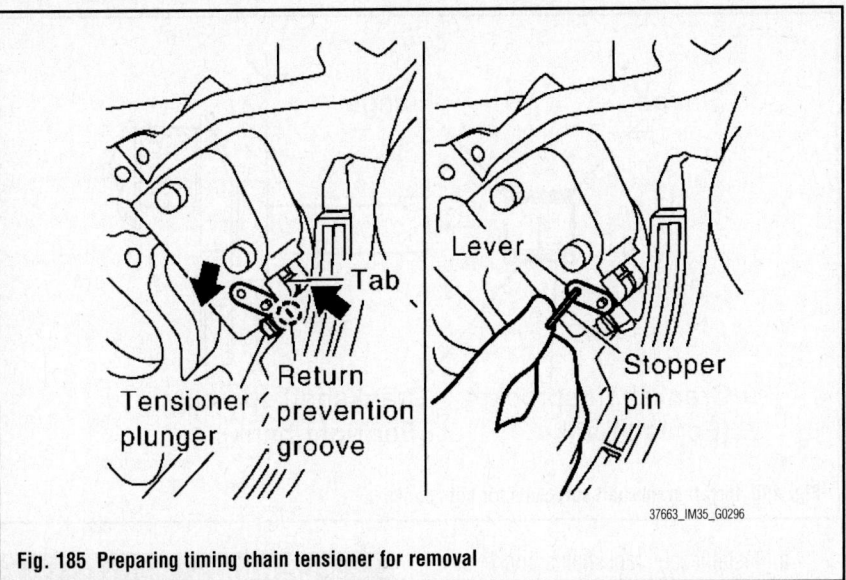

Fig. 185 Preparing timing chain tensioner for removal

➡**The dimension between the centers of the two bolt holes is 1.30 inches (33 mm).**

15. Remove oil pump.

16. Remove chain tensioner (left bank) as follows:

➡**To remove timing chain and related parts, start with those on left bank. The procedure for removing parts on right bank is omitted because it is the same as that for left bank.**

    a. Press tab in the direction of arrow (or turn lever in the direction of arrow) to unlock the locking with the groove that stops tensioner plunger from returning. Lightly press tensioner plunger to release the tension of spring for this operation.

    b. Push in tensioner plunger to align the hole on lever and that on pump main body. Pushing in tensioner too far does not allow the holes to align. Therefore, push in plunger to the degree at which the start of stopper groove and tab engages.

    c. Insert stopper pin to secure the plunger. With plunger secured, remove chain tensioner.

17. Remove chain tension guide and timing chain slack guide.

18. Remove timing chain and crankshaft sprocket.

❋❋ CAUTION

After removing timing chain, never turn crankshaft and camshaft separately, or valves will strike the piston head.

19. With hexagonal part of camshaft locked with wrench, loosen mounting bolts

securing camshaft sprocket to remove camshaft sprocket.

❋❋ CAUTION

Never loosen mounting bolts with securing anything other than the camshaft hexagonal portion or with tensioning the timing chain.

20. Perform same procedure as for left bank, remove timing chain and related parts on right side.

21. Use scraper to remove all traces of old liquid gasket from front cover and opposite mating surfaces. Remove oil liquid gasket from bolt hole and thread.

22. Use scraper to remove all trace of liquid gasket from chain tensioner cover and intake valve timing control covers.

To install:

➡**Note the following:**

- The figure shows the relationship between the mating mark on each timing chain and that on the corresponding sprocket, with the components installed.
- Parts with an identification mark (R or L) should be installed on the corresponding bank according to the mark.

➡**Parts with an identification mark:**

- Camshaft sprocket (INT)
- Dowel pin groove of camshaft sprocket (EXH) (camshaft sprocket is same part both banks)

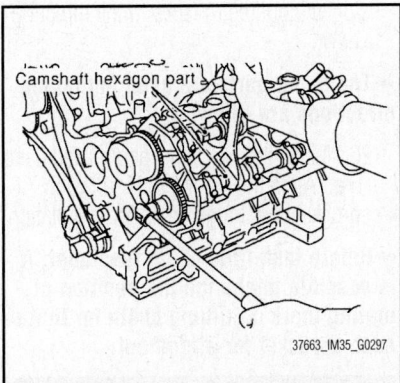

Fig. 186 Loosen mounting bolts securing camshaft sprocket to remove camshaft sprocket

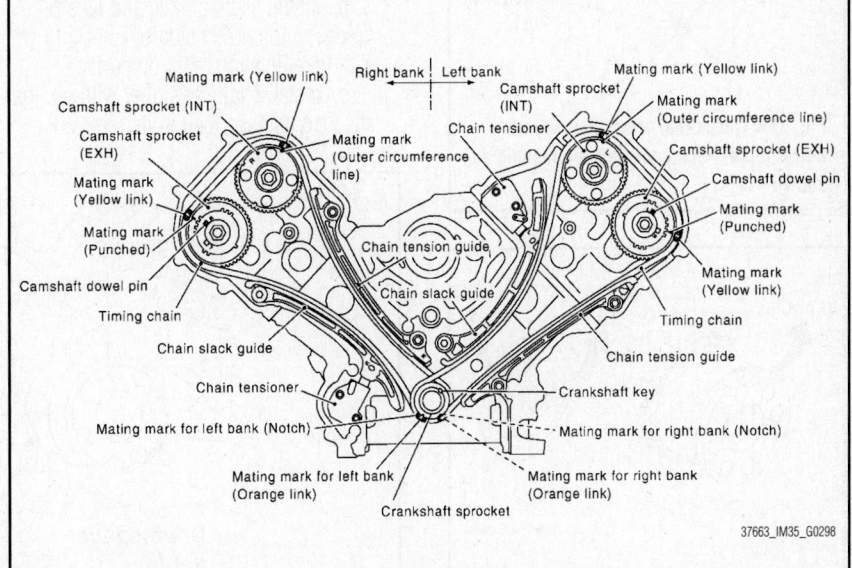

Fig. 187 The relationship between the mating mark on each timing chain and that on the corresponding sprocket

- Chain tension guide
- Chain slack guide

➡**To install timing chain and related parts, start with those on right bank. The procedure for installing parts on left bank is omitted because it is the same as that for installation on right bank.**

23. Check that crankshaft key and dowel pin of each camshaft are located as shown. (No. 1 cylinder at compression TDC)

➡**Though camshaft does not stop at the position as shown, for the placement of cam nose, it is generally accepted camshaft is placed for the same direction of the figure.**

24. Install camshaft sprockets.
   a. Install onto correct side by checking with identification mark on surface.
   b. Install camshaft sprocket (EXH) by selectively using the groove of dowel pin according to the bank. (Common part used for both banks.)
   c. Lock the hexagonal part of camshaft in the same procedure as for removal, and tighten mounting bolts.
25. Install crankshaft sprockets for both banks.

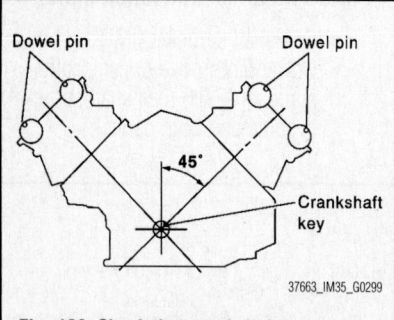

Fig. 188 Check that crankshaft key and dowel pin of each camshaft are located as shown

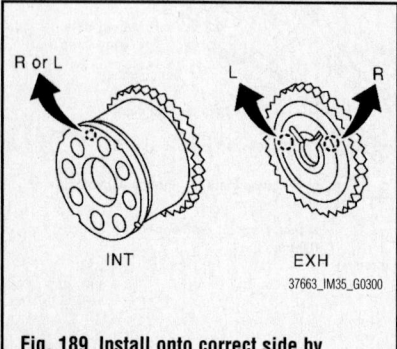

Fig. 189 Install onto correct side by checking with identification mark on surface

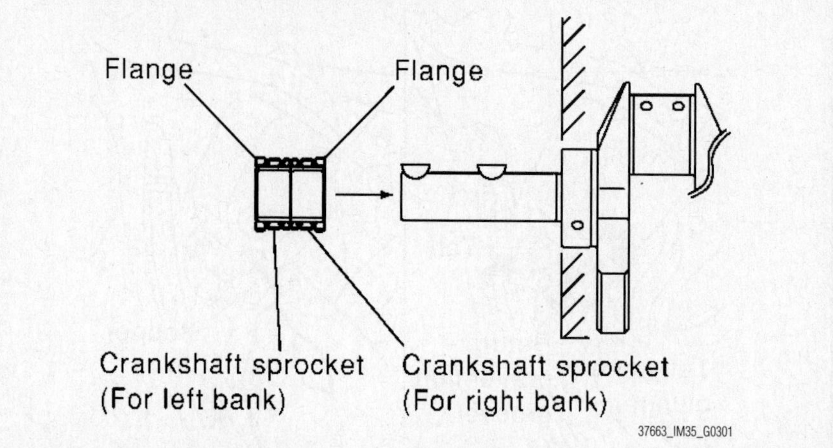

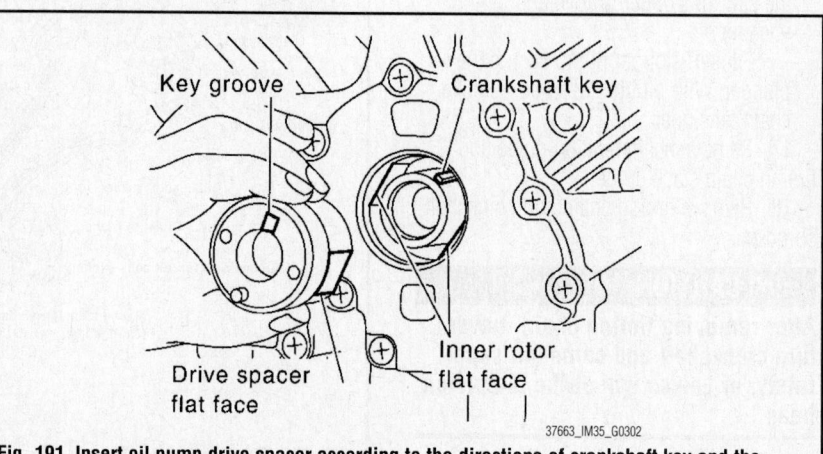

Fig. 190 Install crankshaft sprockets for both banks

   a. Install each crankshaft sprocket so that its flange side (the larger diameter side without teeth) faces in the direction shown.

➡**The same parts are used but facing directions are different.**

26. Install timing chains and related parts.
   a. Align the mating mark on each sprocket and timing chain for installation.

➡**Before installing chain tensioner, it is possible to change the position of mating mark on timing chain for that on each sprocket for alignment.**

❋❋ CAUTION

For the above reason, after the mating marks are aligned, keep them aligned by holding them with a hand.

   b. Install slack guides and tension guides onto correct side by checking with identification mark on surface.
   c. Install chain tensioner with plunger attached as described in its removal.

❋❋ CAUTION

Note the following:

- Before and after the installation of chain tensioner, check that the mating mark on timing chain is not out of alignment.
- After installing chain tensioner, remove stopper pin to release tensioner. Check tensioner is released.
- To avoid chain-link skipping of timing chain, never move crankshaft or camshafts until front cover is installed.

27. Perform the same procedure as for right bank, install timing chain and related parts on left side.
28. Install oil pump.
29. Install oil pump drive spacer as follows:
   a. Insert oil pump drive spacer according to the directions of crankshaft key and the two flat surfaces of oil pump inner rotor. If the positional relationship

Fig. 191 Insert oil pump drive spacer according to the directions of crankshaft key and the two flat surfaces of oil pump inner rotor

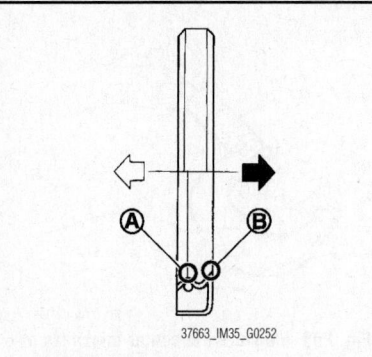

Fig. 192 Install it so that each seal lip is oriented as shown

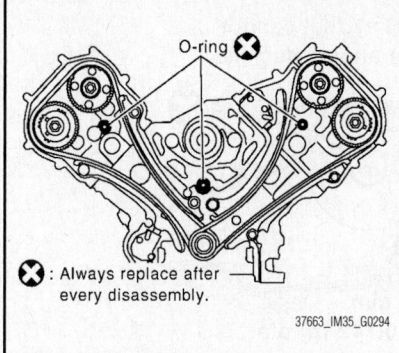

Fig. 194 Install new O-rings from cylinder heads and cylinder block

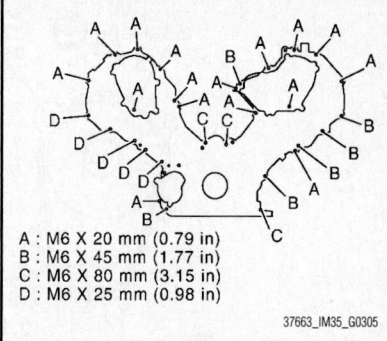

A : M6 X 20 mm (0.79 in)
B : M6 X 45 mm (1.77 in)
C : M6 X 80 mm (3.15 in)
D : M6 X 25 mm (0.98 in)

Fig. 197 Proper bolt positions for the four different types of bolts

does not allow the insertion, rotate oil pump inner rotor with a finger to allow spacer.

b. After confirming that the position of each part is in correct condition to allow for spacer, force fit spacer by lightly tapping with plastic hammer until it contacts and does not go further.

30. Install front oil seal on front cover.

a. Apply new engine oil to both oil seal lip and dust seal lip.

b. Install it so that each seal lip is oriented as shown.

❊❊ CAUTION

Be careful not to scratch or make burrs on circumference of oil seal.

c. Using front oil seal drift, press fit until the height of front oil seal is level with the mounting surface.

d. Check the garter spring is in position and seal lips not inverted.

31. Install chain tensioner cover to front cover.

32. Install front cover as follows:

a. Install new O-rings onto cylinder heads (right bank and left bank) and cylinder block.

b. Apply a continuous bead of liquid gasket to front cover as shown.

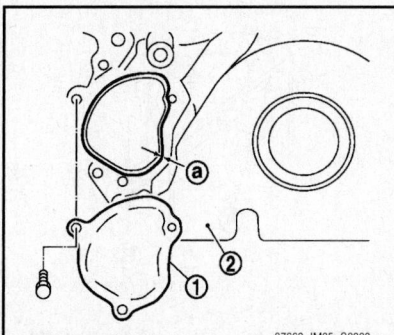

Fig. 193 Install chain tensioner cover to front cover

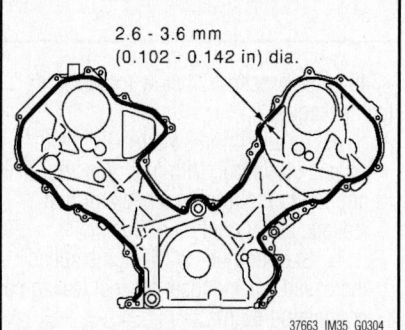

2.6 - 3.6 mm
(0.102 - 0.142 in) dia.

Fig. 195 Apply a continuous bead of liquid gasket to front cover as shown

c. Check again that the mating marks on timing chain and that on each sprocket are aligned. Then, install front cover.

❊❊ CAUTION

Be careful to avoid interference with the front end of oil pump drive spacer. Such interference may damage front oil seal.

d. Tighten mounting bolts in numerical order as shown.

e. There are four types of mounting bolts.

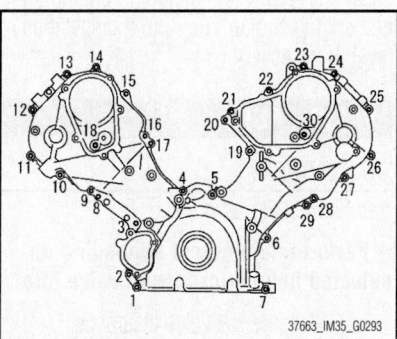

Fig. 196 Tighten mounting bolts in order as shown

f. After all mounting bolts are tightened, retighten them in numerical order as shown.

❊❊ CAUTION

Be sure to wipe off any excessive liquid gasket leaking onto surface mating with oil pan.

33. Install intake valve timing control cover as follows:

a. At the back of intake valve timing control cover, install new seal rings (three for each bank) to the area to be inserted into camshaft sprocket (INT).

❊❊ CAUTION

Never spread seal ring excessively to avoid breaks and deformation.

b. Install new O-rings on front cover.

c. Apply a continuous bead of liquid gasket to intake valve timing control covers as shown.

d. Tighten mounting bolts in numerical order as shown.

34. Install intake valve timing control position sensor, intake valve timing control solenoid valve and Camshaft Position (CMP) sensor (PHASE) to intake valve timing control cover and front cover if removed. Be sure to tighten mounting bolts with flanges completely seated.

35. Install oil pan and oil strainer.

36. Install crankshaft pulley as follows:

a. Secure crankshaft with ring gear stopper (J-45476).

b. Install crankshaft pulley, taking care not to damage front oil seal. Install according to dowel pin of oil pump drive spacer. Lightly tapping its center with plastic hammer, insert pulley.

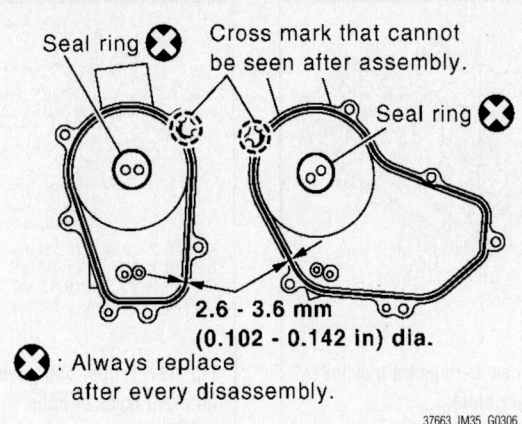

Fig. 198 Apply a continuous bead of liquid gasket to intake valve timing control covers as shown

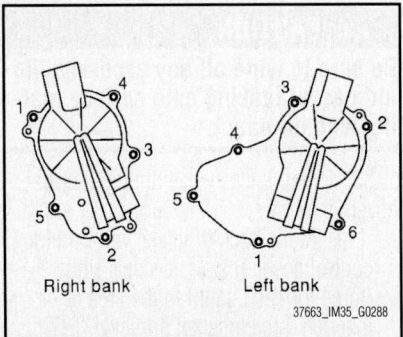

Fig. 199 Tighten mounting bolts in the numerical order as shown

✳✳ CAUTION

Never tap pulley on the side surface where belt is installed (outer circumference).

c. Apply engine oil onto threaded parts of crankshaft pulley bolt and seating area.

d. Tighten crankshaft pulley bolt to 69 ft. lbs. (93 Nm).

e. Put a paint mark on crankshaft pulley aligning with angle mark on crankshaft pulley bolt.

f. Further tighten by 90 degrees. (angle tightening) Check the tightening angle by referencing to the notches. The angle between two notches is 90 degrees.

37. Rotate crankshaft pulley in normal direction (clockwise when viewed from engine front) to confirm it turns smoothly.

38. Install in the reverse order of removal after this step.

39. Before starting engine, check oil/fluid levels including engine coolant and engine oil. If less than required quantity, fill to the specified level.

40. Use procedure below to check for fuel leakage.

a. Turn ignition switch "ON" (with engine stopped). With fuel pressure applied to fuel piping, check for fuel leakage at connection points.

b. Start engine. With engine speed increased, check again for fuel leakage at connection points.

41. Run engine to check for unusual noise and vibration.

➡**If hydraulic pressure inside timing chain tensioner drops after removal/installation, slack in guide may generate a pounding noise during and just after engine start. However, this does not indicate an unusualness. Noise will stop after hydraulic pressure rises.**

42. Warm up engine thoroughly to check there is no leakage of fuel, exhaust gases, or any oil/fluids including engine oil and engine coolant.

43. Bleed air from lines and hoses of applicable lines, such as in cooling system.

44. After cooling down engine, again check oil/fluid levels including engine oil and engine coolant. Refill to the specified level, if necessary.

VALVE LASH

ADJUSTMENT

See Figures 200 and 201.

➡**Perform adjustment depending on selected head thickness of valve lifter.**

1. Measure the valve clearance.
2. Remove camshafts.
3. Remove valve lifters at the locations that are out of the standard.

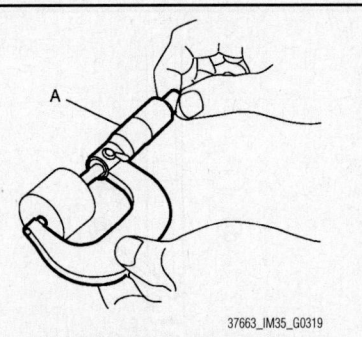

Fig. 200 Measure the center thickness of removed valve lifters with a micrometer (A)

4. Measure the center thickness of removed valve lifters with a micrometer.

5. Use the equation below to calculate valve lifter thickness for replacement.

- Valve lifter thickness calculation: $t = t1 + (C1 - C2)$
- t = Valve lifter thickness to be replaced
- $t1$ = Removed valve lifter thickness
- $C1$ = Measured valve clearance
- $C2$ = Standard valve clearance: Intake: 0.012 inches (0.30 mm); Exhaust: 0.013 inches (0.33 mm)

➡**Thickness of new valve lifter can be identified by stamp marks on the reverse side (inside the cylinder). Stamp mark 788 indicates 0.3102 inches (7.88 mm) in thickness. Available thickness of valve lifter: 27 sizes with range 7.88 to 8.40 mm (0.3102 to 0.3307 inches) in steps of 0.02 mm (0.0008 inches) (when manufactured at factory).**

6. Install selected valve lifter.
7. Install camshaft.
8. Manually turn crankshaft pulley a few turns.
9. Check that the valve clearances for cold engine are within the specifications by referring to the specified values.

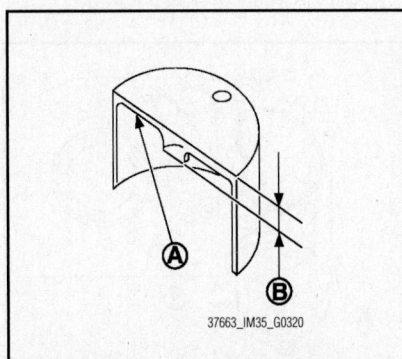

Fig. 201 Thickness of new valve lifter (B) can be identified by stamp marks (A) on the reverse side (inside the cylinder)

10. Install all removal parts in the reverse order of removal.

11. Warm up the engine, and check for unusual noise and vibration.

INSPECTION

3.5L Engine

See Figures 202 through 208.

1. Remove rocker covers (bank 1 and bank 2).

2. Measure the valve clearance as follows:

a. Set No. 1 cylinder at TDC of its compression stroke.

- Rotate crankshaft pulley clockwise to align timing mark (grooved line without color) with timing indicator.
- Check that intake and exhaust cam nose on No. 1 cylinder (engine front side of bank 1) are located as shown.
- If not, turn crankshaft one revolution (360 degrees) and align as shown.

b. Using a feeler gauge, measure the clearance between the valve adjuster and the camshaft. Record any valve clearance

Fig. 202 Rotate crankshaft pulley clockwise to align timing mark (grooved line without color) with timing indicator

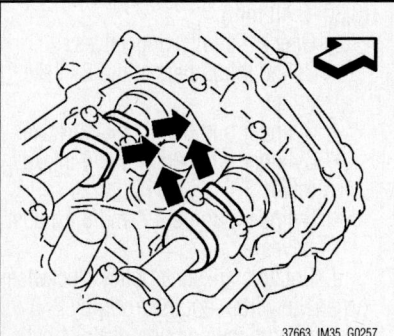

Fig. 203 Check that intake and exhaust cam nose on No. 1 cylinder are located as shown

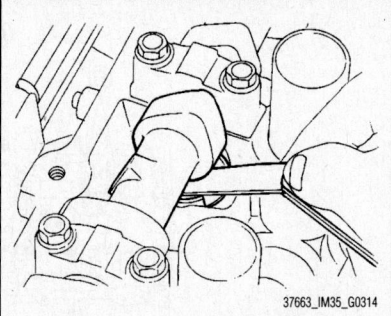

Fig. 204 Use a feeler gauge, measure the clearance between valve lifter and camshaft

measurements that are out of specification. Intake valve clearance (cold) is 0.010–0.013 inches (0.26–0.34 mm) and exhaust valve clearance (cold) is 0.011–0.015 inches (0.29–0.37 mm).

c. Check the following valves:
- Both No. 1 intake valves—right bank
- Both No. 2 exhaust valves—left bank
- Both No. 3 exhaust valves—right bank
- Both No. 6 intake valves—left bank

d. Rotate crankshaft by 240 degrees clockwise (when viewed from engine

front) to align No. 3 cylinder at TDC of its compression stroke.

➡Mark a position 240 degrees from a corner of the hexagonal part of crankshaft pulley mounting bolt as shown.

e. Check the following valves:
- Both No. 2 intake valves—left bank
- Both No. 3 intake valves—right bank
- Both No. 4 exhaust valves—left bank
- Both No. 5 exhaust valves—right bank

f. Rotate crankshaft by 240 degrees clockwise (when viewed from engine front) to align No. 5 cylinder at TDC of its compression stroke.

3. Check the following valves:
- Both No. 1 exhaust valves—right bank
- Both No. 4 intake valves—left bank
- Both No. 5 intake valves—right bank
- Both No. 6 exhaust valves—left bank

4. If the measured values are out of the standard, perform adjustment.

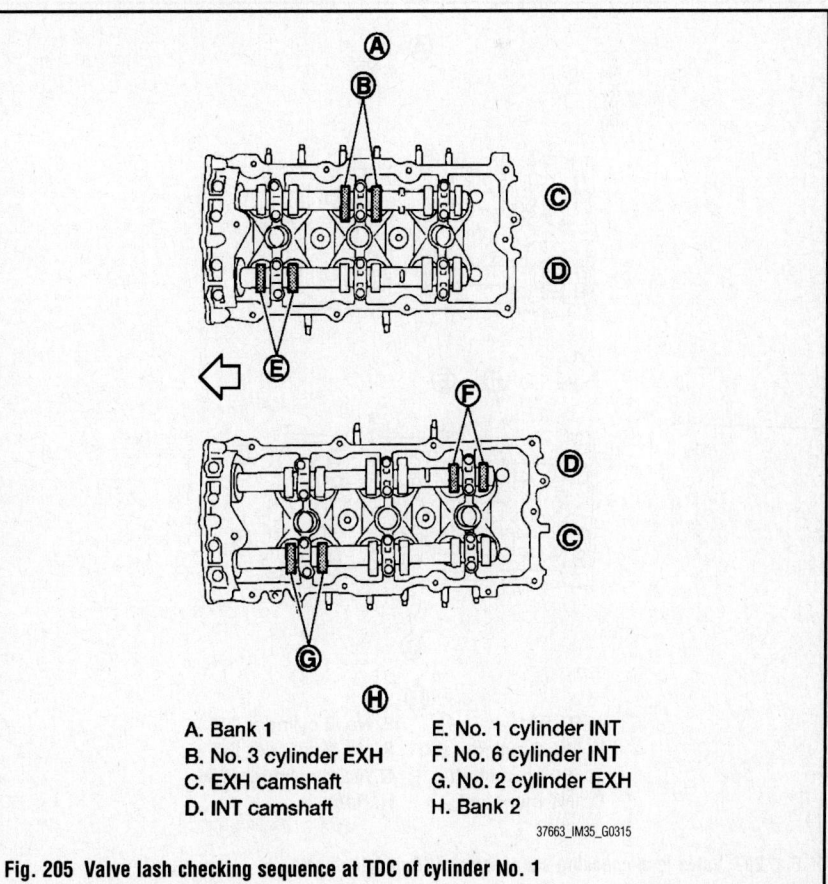

A. Bank 1
B. No. 3 cylinder EXH
C. EXH camshaft
D. INT camshaft
E. No. 1 cylinder INT
F. No. 6 cylinder INT
G. No. 2 cylinder EXH
H. Bank 2

Fig. 205 Valve lash checking sequence at TDC of cylinder No. 1

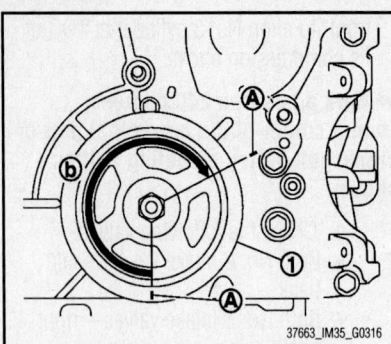

Fig. 206 Mark a position 240 degrees (b) from a corner (A) of the hexagonal part of crankshaft pulley (1) mounting bolt as shown

4.5L Engine

See Figures 209 through 212.

1. Remove rocker covers (right bank and left bank).

2. Measure the valve clearance as follows:

   a. Set No. 1 cylinder at TDC of its compression stroke.

   • Rotate crankshaft pulley in clockwise to align TDC identification notch (without paint mark) with timing indicator on front cover.

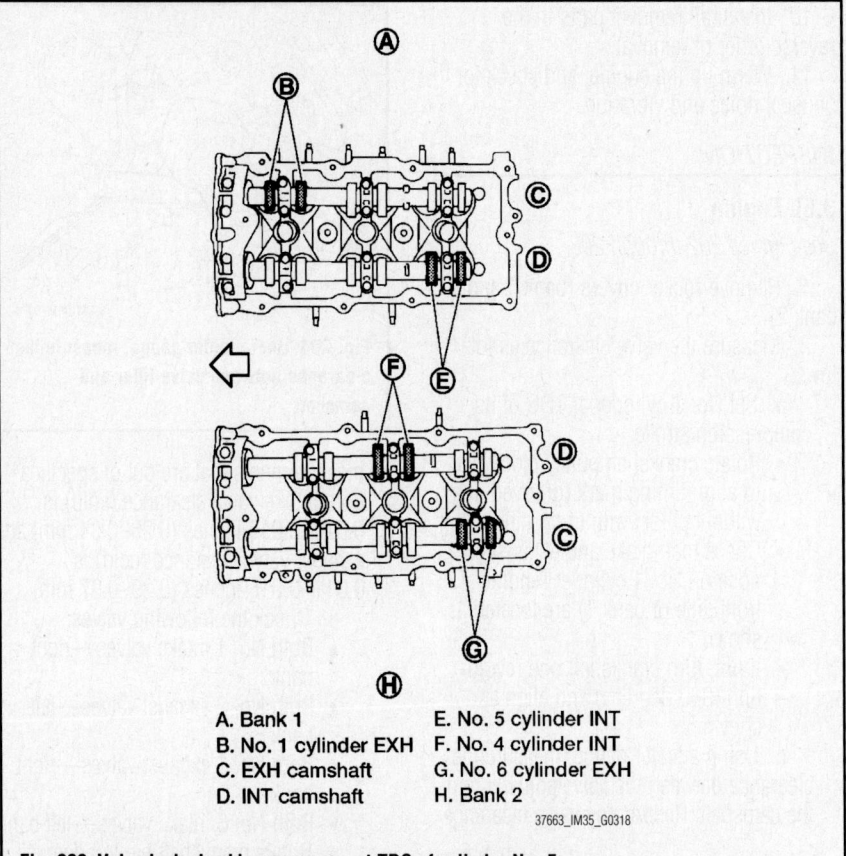

A. Bank 1	E. No. 5 cylinder INT
B. No. 1 cylinder EXH	F. No. 4 cylinder INT
C. EXH camshaft	G. No. 6 cylinder EXH
D. INT camshaft	H. Bank 2

Fig. 208 Valve lash checking sequence at TDC of cylinder No. 5

• Check that both intake and exhaust cam noses of No. 1 cylinder (engine front side of left bank) are located as shown.

• If not, turn crankshaft one revolution (360 degrees) and align as shown.

   b. Using a feeler gauge, measure the clearance between the valve lifter and the camshaft. Record any valve clearance measurements that are out of specification. Intake valve clearance (hot) is 0.012–0.016 inches (0.30–0.41 mm) and exhaust valve clearance (hot) is 0.012–0.017 inches (0.30–0.43 mm).

   c. Check the following valves:
   • Cylinder numbers 1 and 2 intake valves
   • Cylinder number 1 exhaust valves
   • Cylinder numbers 4 and 5 intake valves
   • Cylinder numbers 7 and 8 exhaust valves

   d. Rotate crankshaft pulley clockwise (when view from engine front) by 270 degrees from the position of No. 1 cylinder compression TDC to align No. 3 cylinder at TDC of its compression stroke.

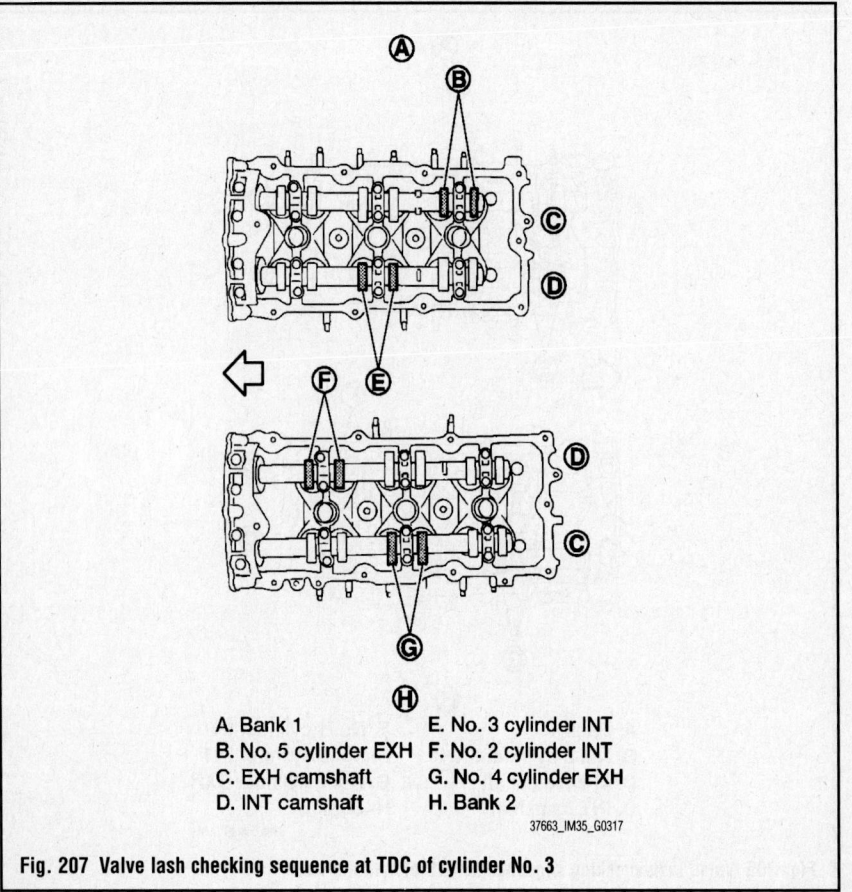

A. Bank 1	E. No. 3 cylinder INT
B. No. 5 cylinder EXH	F. No. 2 cylinder INT
C. EXH camshaft	G. No. 4 cylinder EXH
D. INT camshaft	H. Bank 2

Fig. 207 Valve lash checking sequence at TDC of cylinder No. 3

↑ (filled) : Measurable at No. 1 cylinder compression TDC

↑ (open) : Measurable at No. 3 cylinder compression TDC

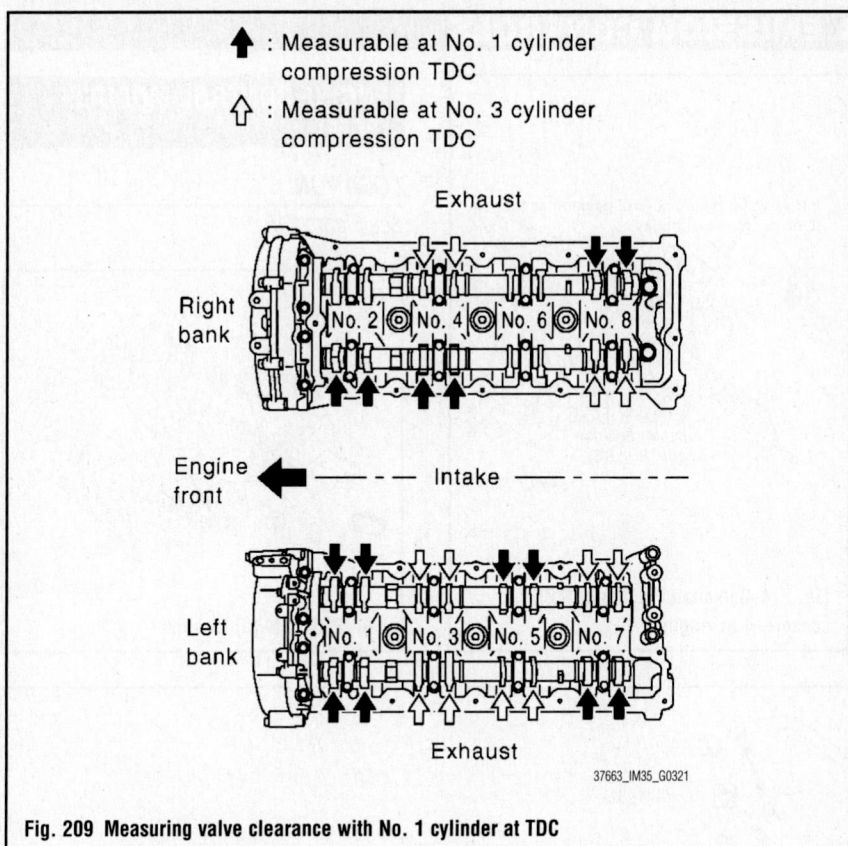

Fig. 209 Measuring valve clearance with No. 1 cylinder at TDC

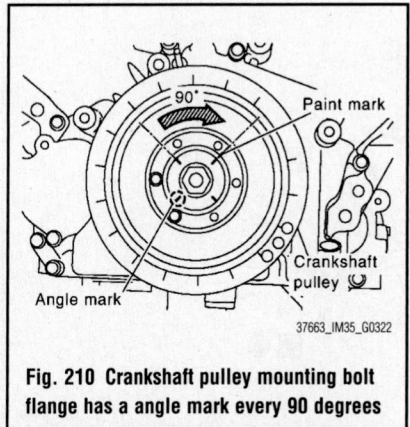

Fig. 210 Crankshaft pulley mounting bolt flange has a angle mark every 90 degrees

➥Crankshaft pulley mounting bolt flange has a angle mark every 90 degrees. They can be used as a guide to rotation angle.

  e. Check the following valves:
- Cylinder numbers 3 and 4 exhaust valves
- Cylinder numbers 3 and 7 intake valves
- Cylinder number 5 exhaust valves

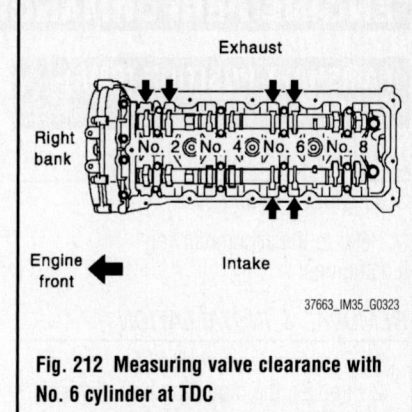

Fig. 212 Measuring valve clearance with No. 6 cylinder at TDC

- Cylinder number 8 intake valves

  f. Rotate crankshaft pulley clockwise (when view from engine front) by 90 degrees from the position of No. 3 cylinder compression TDC to align No. 6 cylinder at TDC of its compression stroke.

  g. Check the following valves:
- Cylinder numbers 2 and 6 exhaust valves
- Cylinder number 6 intake valves

3. Perform adjustment if the measured value is out of the standard.

↑ (filled) : Measurable at No. 1 cylinder compression TDC

↑ (open) : Measurable at No. 3 cylinder compression TDC

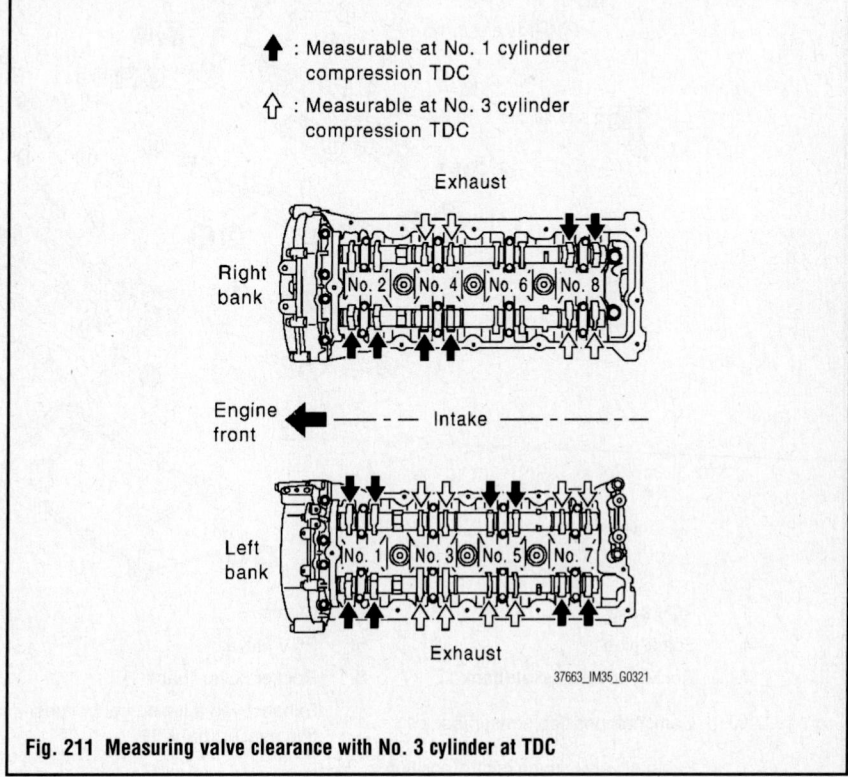

Fig. 211 Measuring valve clearance with No. 3 cylinder at TDC

ENGINE PERFORMANCE & EMISSION CONTROLS

CAMSHAFT POSITION (CMP) SENSOR

LOCATION

See Figures 213 and 214.

Refer to the accompanying illustrations.

REMOVAL & INSTALLATION

1. Turn ignition switch OFF.
2. Loosen the mounting bolt of the sensor.
3. Disconnect Camshaft Position (CMP) sensor harness connector.
4. Remove the sensor.
5. Installation is the reverse of removal.

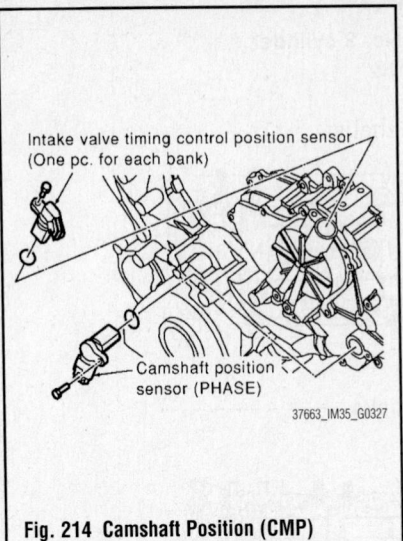

Fig. 214 Camshaft Position (CMP) sensor—4.5L engine

ENGINE CONTROL MODULE (ECM)

LOCATION

See Figure 215.

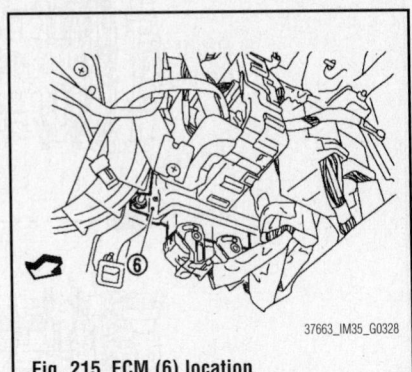

Fig. 215 ECM (6) location

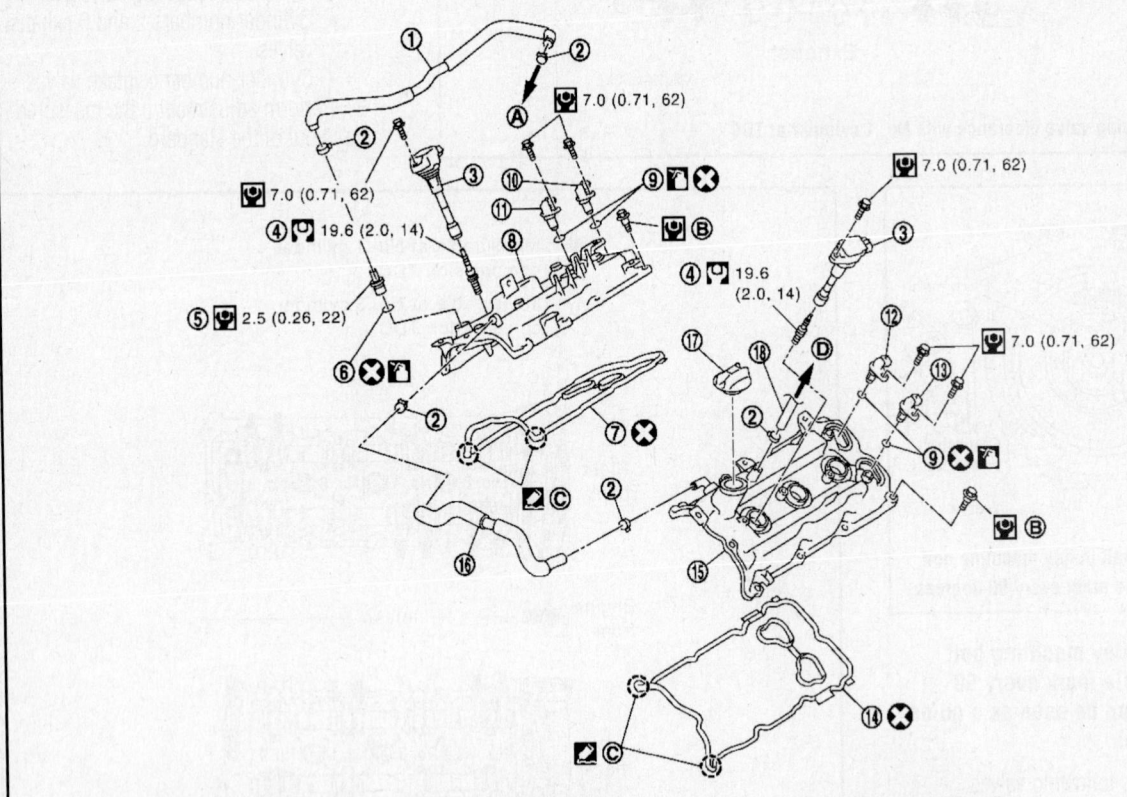

1. PCV hose	2. Clamp	3. Ignition coil
4. Spark plug	5. PCV valve	6. O-ring
7. Rocker cover gasket (bank 1)	8. Rocker cover (bank 1)	9. O-ring
10. Camshaft position sensor (bank 1)	11. Exhaust valve timing control position sensor (bank 1)	12. Camshaft position sensor (bank 2)
13. Exhaust valve timing control position sensor (bank 2)	14. Rocker cover gasket (bank 2)	15. Rocker cover (bank 2)
16. PCV hose	17. Oil filler cap	18. PCV hose

Fig. 213 Exploded view of rocker cover assemblies showing Camshaft Position (CMP) sensor locations—3.5L engine

Refer to the accompanying illustration.

REMOVAL & INSTALLATION

➡**The manufacturer does not provide a specific Removal and Installation procedure for this component. Refer to the graphic(s) when servicing this component.**

ENGINE COOLANT TEMPERATURE (ECT) SENSOR

REMOVAL & INSTALLATION

1. Turn ignition switch OFF.
2. Disconnect engine coolant temperature sensor harness connector.
3. Remove engine coolant temperature sensor.

4. Installation is the reverse of removal.

HEATED OXYGEN SENSOR (HO2S)

LOCATION

See Figures 216 and 217.

Refer to the accompanying illustrations.

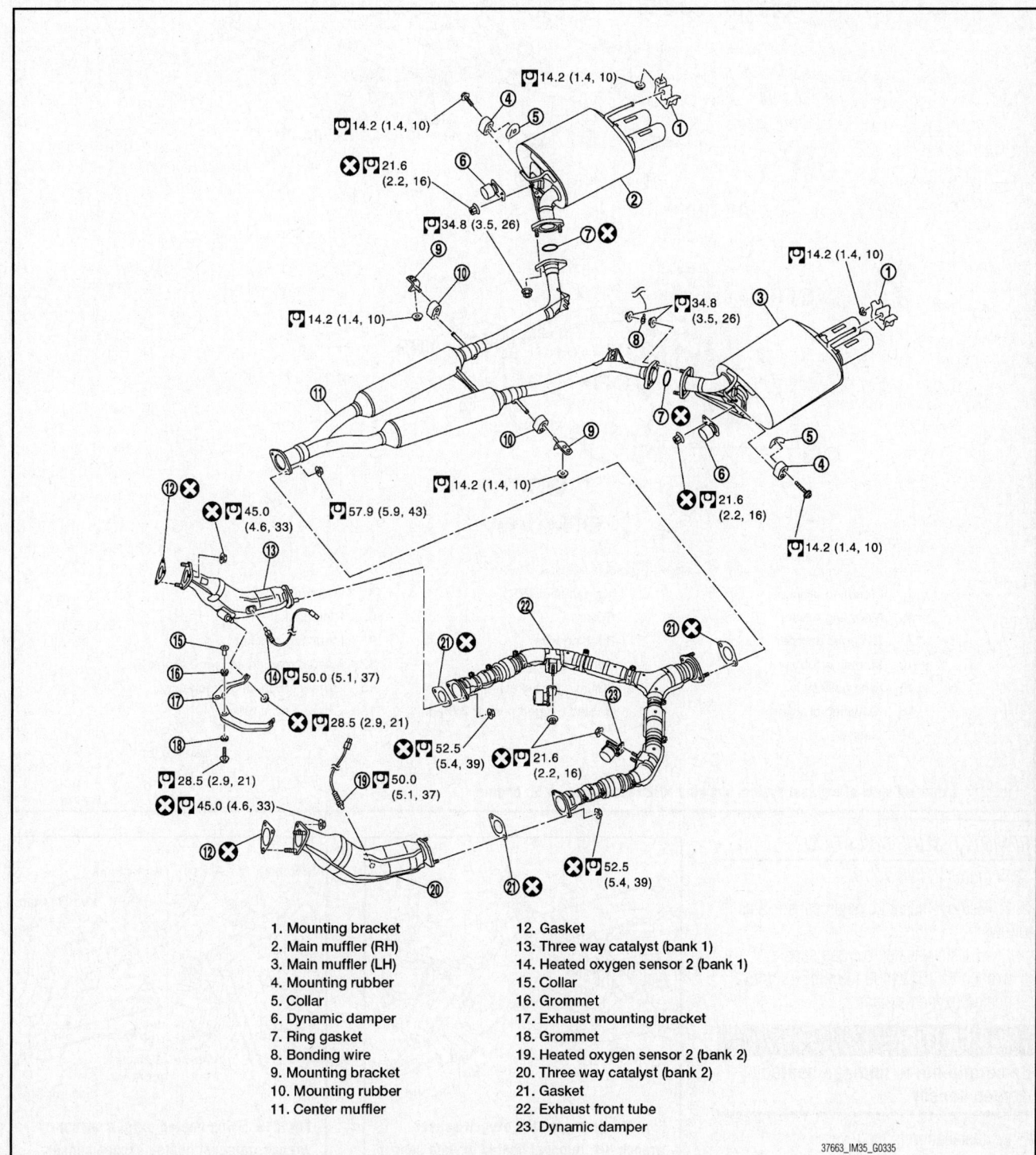

1. Mounting bracket
2. Main muffler (RH)
3. Main muffler (LH)
4. Mounting rubber
5. Collar
6. Dynamic damper
7. Ring gasket
8. Bonding wire
9. Mounting bracket
10. Mounting rubber
11. Center muffler
12. Gasket
13. Three way catalyst (bank 1)
14. Heated oxygen sensor 2 (bank 1)
15. Collar
16. Grommet
17. Exhaust mounting bracket
18. Grommet
19. Heated oxygen sensor 2 (bank 2)
20. Three way catalyst (bank 2)
21. Gasket
22. Exhaust front tube
23. Dynamic damper

37663_IM35_G0335

Fig. 216 Exploded view of exhaust system showing HO2S locations—3.5L engine

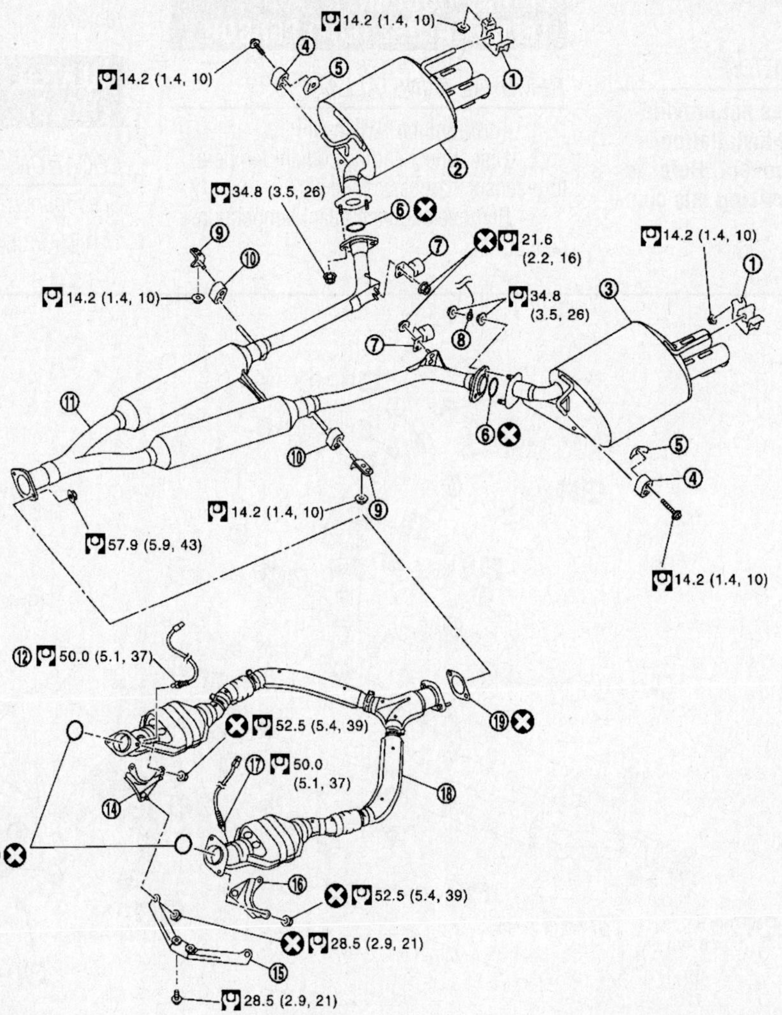

14.2 (1.4, 10)

14.2 (1.4, 10)

14.2 (1.4, 10)

34.8 (3.5, 26)

21.6 (2.2, 16)

14.2 (1.4, 10)

34.8 (3.5, 26)

14.2 (1.4, 10)

14.2 (1.4, 10)

57.9 (5.9, 43)

14.2 (1.4, 10)

50.0 (5.1, 37)

52.5 (5.4, 39)

50.0 (5.1, 37)

52.5 (5.4, 39)

28.5 (2.9, 21)

28.5 (2.9, 21)

1.	Mounting bracket	2.	Main muffler (RH)	3.	Main muffler (LH)
4.	Mounting rubber	5.	Collar	6.	Ring gasket
7.	Dynamic damper	8.	Bonding wire	9.	Mounting bracket
10.	Mounting rubber	11.	Center muffler	12.	Heated oxygen sensor 2 (bank 2)
13.	Ring gasket	14.	Mounting bracket	15.	Exhaust mounting bracket
16.	Mounting bracket	17.	Heated oxygen sensor 2 (bank 1)	18.	Exhaust front tube
19.	Gasket				

37663_IM35_G0336

Fig. 217 Exploded view of exhaust system showing HO2S locations—4.5L engine

REMOVAL & INSTALLATION

See Figures 218 and 219.

1. Remove heated oxygen sensor 2 as follows:

   a. Using heated oxygen sensor wrench KV10114400 (J38365), removal heated oxygen sensor 2.

❉❉ CAUTION

Be careful not to damage heated oxygen sensor 2.

2. Installation is the reverse of removal.

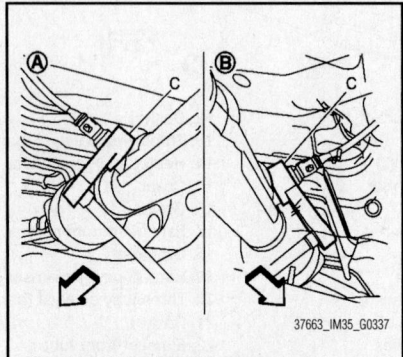

37663_IM35_G0337

Fig. 218 Using heated oxygen sensor wrench (C), removal heated oxygen sensor 2—3.5L engine

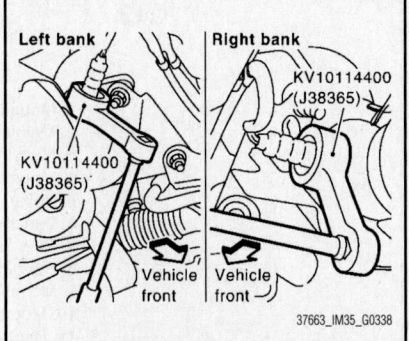

Left bank · Right bank

KV10114400 (J38365)

KV10114400 (J38365)

Vehicle front · Vehicle front

37663_IM35_G0338

Fig. 219 Using heated oxygen sensor wrench, removal heated oxygen sensor 2—4.5L engine

INTAKE AIR TEMPERATURE (IAT) SENSOR

LOCATION

The Intake Air Temperature (IAT) sensor is an integral part of the Mass Air Flow (MAF) sensor/Intake Air Temperature (IAT) sensor assembly which is mounted on the air intake duct. Refer to the Mass Air Flow (MAF) section for information regarding servicing this component.

KNOCK SENSOR (KS)

LOCATION

See Figures 220 and 221.

The knock sensors are located in the valley of the cylinder block beneath the intake manifold.

REMOVAL & INSTALLATION

➡**The manufacturer does not provide a specific Removal and Installation procedure for this component. Refer to the graphic(s) when servicing this component.**

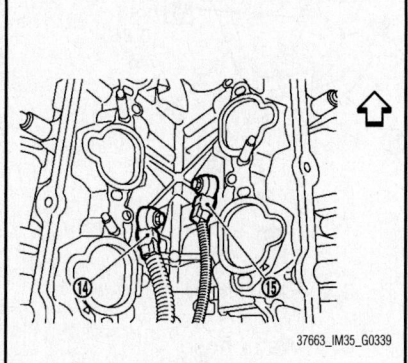

Fig. 220 Knock sensor (14, 15) locations—3.5L engine

Fig. 221 Knock sensor (1, 2) locations—4.5L engine

MALFUNCTION INDICATOR LIGHT (MIL)

RESET PROCEDURE

The Malfunction Indicator Lamp (MIL) is located on the combination meter. The MIL will illuminate when the ignition switch is turned ON without the engine running. This is a bulb check. When the engine is started, the MIL should turn off. If the MIL remains illuminated, the on board diagnostic system has detected an engine system malfunction.

The Malfunction Indicator Lamp (MIL) on the instrument panel illuminates when the same malfunction is detected in two consecutive trips (Two trip detection logic), or when the ECM enters fail-safe mode.

When there is an open circuit on MIL circuit, the ECM cannot warn the driver by illuminating MIL when there is malfunction on engine control system.

Therefore, when electrical controlled throttle and part of ECM related diagnoses are continuously detected as NG for 5 trips, ECM warns the driver that engine control system malfunctions and MIL circuit is open by means of operating the fail-safe function.

The fail-safe function also operates when above diagnoses except MIL circuit are detected and demands the driver to repair the malfunction.

MASS AIR FLOW (MAF) SENSOR

LOCATION

See Figures 222 and 223.

Fig. 223 MAF sensor (1) location—4.5L engine

The Mass Air Flow (MAF)/Intake Air Temperature (IAT) sensors are located on the engine air intake duct.

REMOVAL & INSTALLATION

➡**The manufacturer does not provide a specific Removal and Installation procedure for this component. Refer to the graphic(s) when servicing this component.**

THROTTLE CONTROL ACTUATOR (TAC)

LOCATION

See Figures 224 and 225.

Refer to the accompanying illustrations.

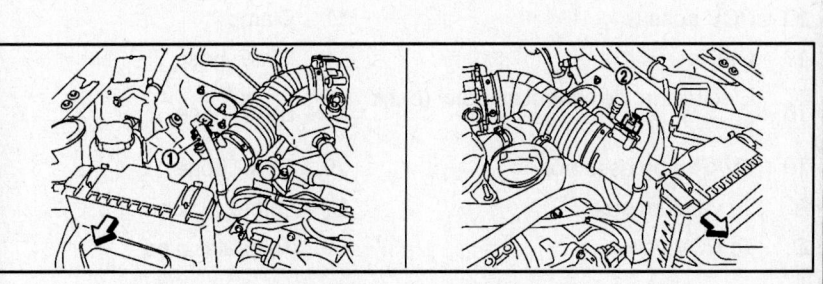

1. Mass air flow sensor (with intake air temperature sensor) (bank 1)
2. Mass air flow sensor (with intake air temperature sensor) (bank 2)

⇦ : Vehicle front

Fig. 222 MAF sensor locations—3.5L engine

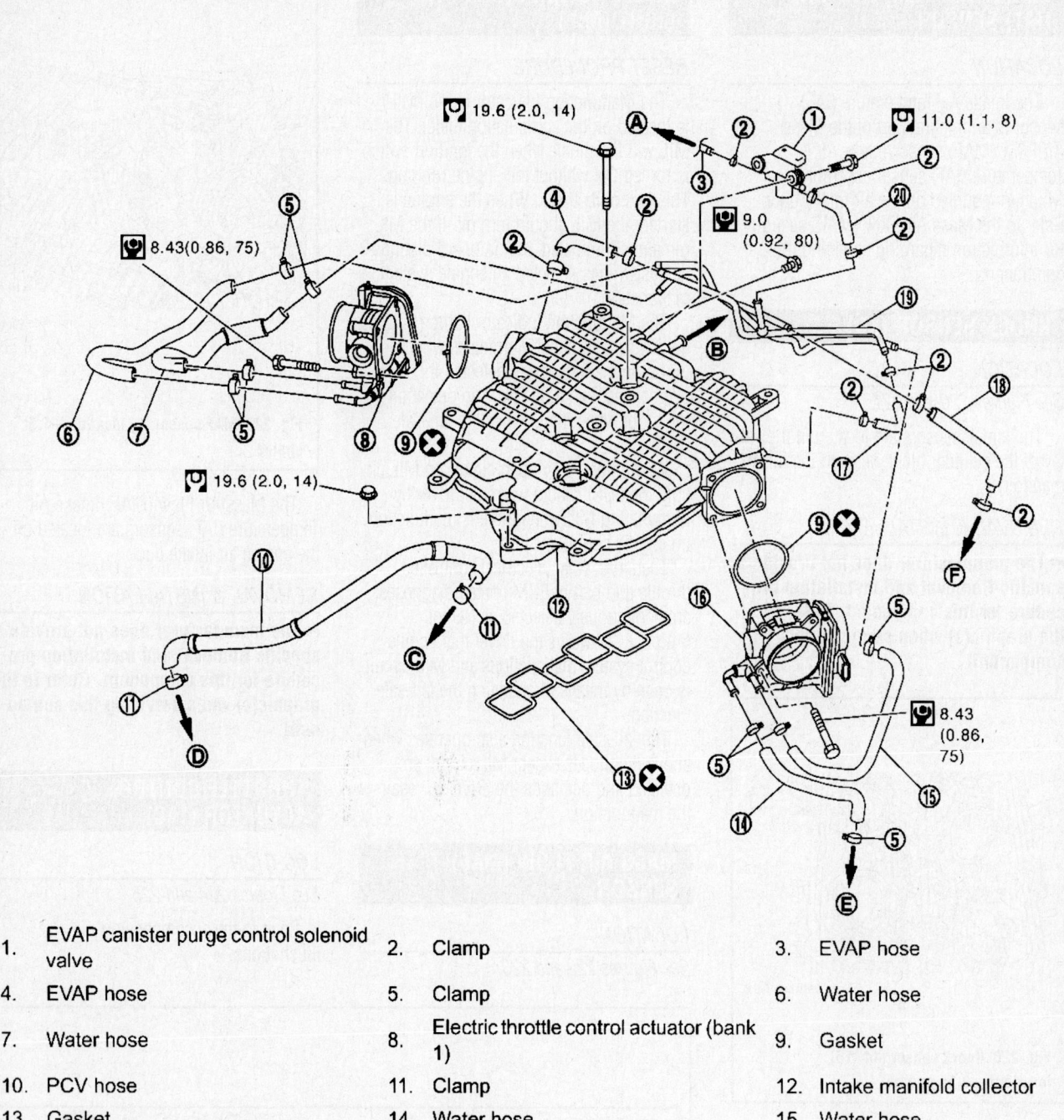

Fig. 224 Exploded view of intake manifold collector showing electric throttle control actuator locations—3.5L engine

1.	EVAP canister purge control solenoid valve	2.	Clamp	3.	EVAP hose
4.	EVAP hose	5.	Clamp	6.	Water hose
7.	Water hose	8.	Electric throttle control actuator (bank 1)	9.	Gasket
10.	PCV hose	11.	Clamp	12.	Intake manifold collector
13.	Gasket	14.	Water hose	15.	Water hose
16.	Electric throttle control actuator (bank 2)	17.	EVAP hose	18.	Water hose
19.	EVAP tube assembly	20.	EVAP hose		
A.	To vacuum pipe	B.	To brake booster	C.	To intake manifold collector
D.	To PCV valve	E.	To heater pipe	F.	To water outlet (rear)

37663_IM35_G0344

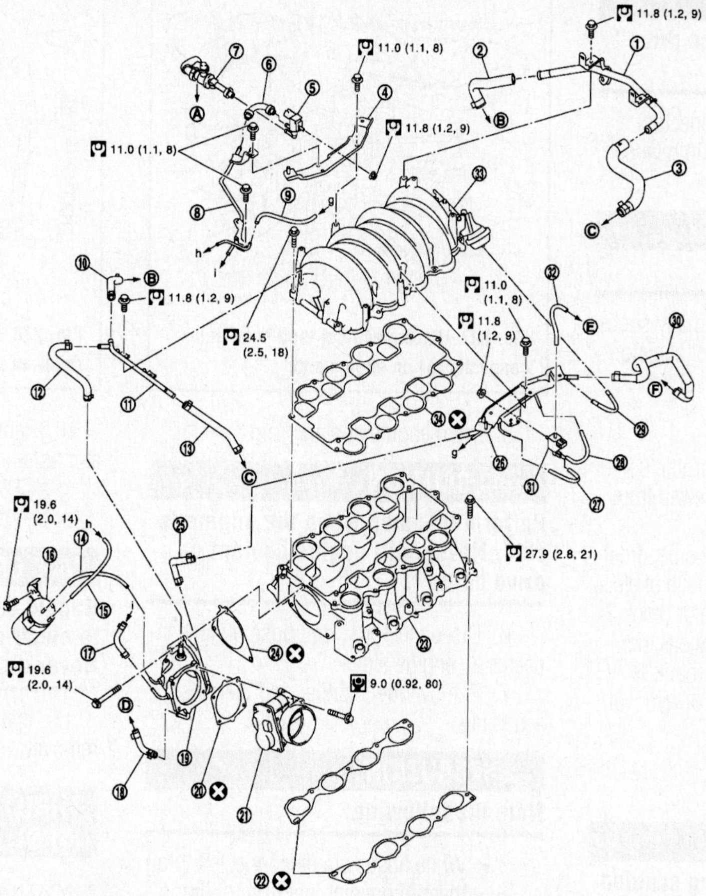

1. PCV tube
2. PCV hose
3. PCV hose
4. Engine cover bracket (RH)
5. EVAP canister purge control solenoid valve
6. EVAP hose
7. EVAP service port
8. EVAP tube
9. Vacuum hose
10. PCV hose
11. PCV tube
12. PCV hose
13. PCV hose
14. Vacuum hose
15. Vacuum hose
16. Vacuum tank
17. EVAP hose
18. Water hose
19. Intake manifold adapter
20. Gasket
21. Electric throttle control actuator
22. Gasket
23. Intake manifold (lower)
24. Gasket
25. Water hose
26. Engine cover bracket (LH)
27. Vacuum hose
28. Vacuum hose
29. Vacuum hose
30. Water hose
31. VIAS control solenoid valve
32. Vacuum hose
33. Intake manifold (upper)
34. Gasket
A. To centralized under-floor piping
B. To rocker cover (right bank)
C. To rocker cover (left bank)
D. To thermostat housing
E. To air duct and resonator assembly
F. To heater pipe

37663_IM35_G0221

Fig. 225 Exploded view of intake manifold showing electric throttle control actuator locations—4.5L engine

REMOVAL & INSTALLATION

3.5L Engine

See Figure 226.

1. Remove engine room cover (RH and LH).
2. Remove engine cover.
3. Remove air cleaner case and air duct (RH and LH).
4. Remove electric throttle control actuator (bank 1 and bank 2) as follows:

➡**When removing only intake manifold collector, move electric throttle control**

actuator without disconnecting water hose.

a. Drain engine coolant.

❋❋ CAUTION

Perform this step when engine is cold.

b. Disconnect water hoses from electric throttle control actuator. When engine coolant is not drained from radiator, attach plug to water hoses to prevent engine coolant leakage.

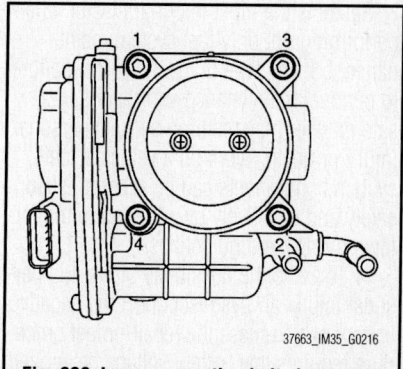

37663_IM35_G0216

Fig. 226 Loosen mounting bolts in reverse order as shown

✳✳ CAUTION
Never spill engine coolant on drive belt.

   c. Disconnect harness connector.
   d. Loosen mounting bolts in reverse order as shown.

✳✳ CAUTION
Note the following:

- Handle carefully to avoid any shock to electric throttle control actuator.
- Never disassemble.

➡**Note the following:**

- Figure shows electric throttle control actuator (bank 1) viewed from the air duct side.
- Viewed from the air duct side, order of loosening mounting bolts of electric throttle control actuator (bank 2) is the same as that of the electric throttle control actuator (bank 1).

5. Installation is the reverse of removal.

4.5L Engine

See Figures 227 and 228.

✳✳ WARNING
To avoid the danger of being scalded, never drain the engine coolant when the engine is hot.

   1. Remove engine room cover (RH and LH).
   2. Remove engine cover.
   3. Release fuel pressure.
   4. Remove air duct (inlet), air cleaner case, and air duct and resonator assembly.

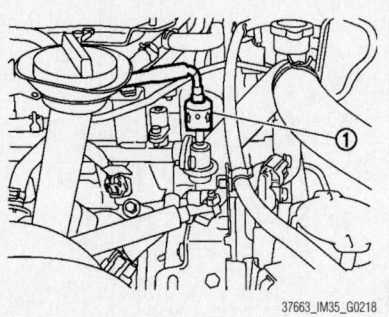

Fig. 227 Disconnect fuel feed hose quick connector (1) on engine side

   5. Drain engine coolant from radiator.

✳✳ CAUTION
Perform this step when the engine is cold. Never spill engine coolant on drive belts.

   6. Disconnect fuel feed hose quick connector on engine side.
   7. Remove fuel damper and fuel hose assembly.

✳✳ CAUTION
Note the following:

- While hoses are disconnected, plug them to prevent fuel from draining.
- Never separate fuel damper and fuel hose.

   8. Remove or disconnect harnesses, engine cover bracket (RH and LH), vacuum hose, EVAP tube and hose and PCV hose and tube from intake manifold (upper).
   9. Loosen mounting bolts in reverse order as shown to remove intake manifold (upper).

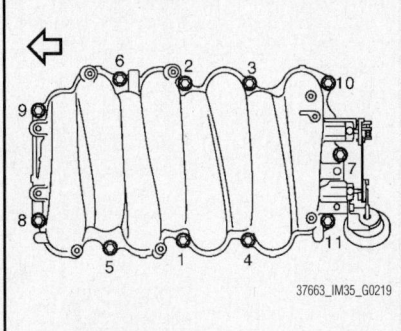

Fig. 228 Loosen mounting bolts in reverse order as shown

   10. Remove electric throttle control actuator as follows:
     a. Disconnect harness connector.
     b. Loosen mounting bolts diagonally.

✳✳ CAUTION
Handle carefully to avoid any shock to electric throttle control actuator. Never disassemble.

   11. Installation is the reverse of removal.

THROTTLE POSITION SENSOR (TPS)

LOCATION

The Throttle Position Sensor is an integral part of the electric throttle control actuator.

REMOVAL & INSTALLATION

Refer to Throttle Control Actuator Removal and Installation when servicing this component.

FUEL   GASOLINE FUEL INJECTION SYSTEM

FUEL SYSTEM SERVICE PRECAUTIONS

Safety is the most important factor when performing not only fuel system maintenance, but any type of maintenance. Failure to conduct maintenance and repairs in a safe manner may result in serious personal injury or death. Work on a vehicle's fuel system components can be accomplished safely and effectively by adhering to the following rules and guidelines.

- To avoid the possibility of fire and personal injury, always disconnect the negative battery cable unless the repair or test procedure requires that battery voltage be applied.
- Always relieve the fuel system pressure prior to disconnecting any fuel system component (injector, fuel rail, pressure regulator, etc.) fitting or fuel line connection. Exercise extreme caution whenever relieving fuel system pressure to avoid exposing skin, face and eyes to fuel spray. Please be advised that fuel under pressure may penetrate the skin or any part of the body that it contacts.
- Always place a shop towel or cloth around the fitting or connection prior to loosening to absorb any excess fuel due to spillage. Ensure that all fuel spillage is quickly removed from engine surfaces. Ensure that all fuel-soaked cloths or towels are deposited into a flame-proof waste container with a lid.
- Always keep a dry chemical (Class B) fire extinguisher near the work area.
- Do not allow fuel spray or fuel vapors to come into contact with a spark or open flame.

- Always use a second wrench when loosening or tightening fuel line connection fittings. This will prevent unnecessary stress and torsion on fuel piping. Always follow the proper torque specifications.
- Always replace worn fuel fitting O-rings with new ones. Do not substitute fuel hose where rigid pipe is installed.

FUEL SYSTEM PRESSURE

RELIEVING

With CONSULT-III

   1. Turn ignition switch ON.
   2. Perform "FUEL PRESSURE RELEASE" in "WORK SUPPORT" mode with CONSULT-III.

3. Start engine.

4. After engine stalls, crank it 2 or 3 times to release all fuel pressure.

5. Turn ignition switch OFF.

Without CONSULT-III

See Figure 229.

1. Remove fuel pump fuse located in IPDM E/R.

2. Start engine.

3. After engine stalls, crank it 2 or 3 times to release all fuel pressure.

4. Turn ignition switch OFF.

5. Reinstall fuel pump fuse after servicing fuel system.

Fig. 229 Remove fuel pump fuse (1) located in IPDM E/R (2)

FUEL FILTER

REMOVAL & INSTALLATION

Refer to Fuel Level Sensor Unit, Filter and Fuel Pump Assembly when servicing this component.

FUEL LEVEL SENSOR UNIT

REMOVAL & INSTALLATION

Refer to Fuel Level Sensor Unit, Filter and Fuel Pump Assembly when servicing this component.

FUEL LEVEL SENSOR UNIT, FUEL FILTER AND FUEL PUMP ASSEMBLY

REMOVAL & INSTALLATION

See Figures 230 through 235.

1. Check fuel level on fuel gauge. If fuel gauge indicates more than the level as shown (full or almost full), drain fuel from fuel tank until fuel gauge indicates level as shown or below.

➡**Because fuel will be spilled when removing main and sub fuel level sensor units for the top of the fuel is above**

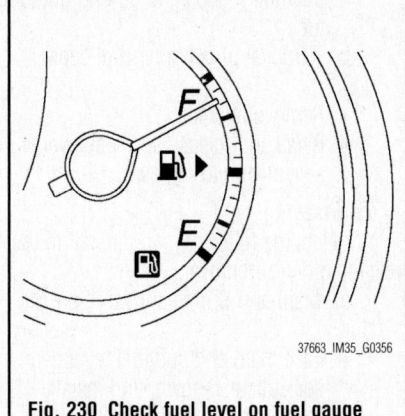

Fig. 230 Check fuel level on fuel gauge

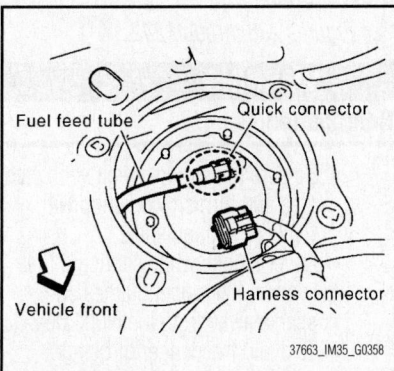

Fig. 231 Remove inspection hole cover units by turning clips clockwise by 90 degrees

the main and sub fuel level sensor units installation surface. As a guide, fuel level becomes the position as shown or below when approximately 5¼ US gal. (20 L) of fuel are drained from fuel tank.

2. In a case that fuel pump does not operate, perform the following procedure.

a. Insert hose of less than 0.98 inches (22 mm) in diameter into fuel filler tube through fuel filler opening to draw fuel from fuel filler tube.

b. Disconnect fuel filler hose from fuel filler tube.

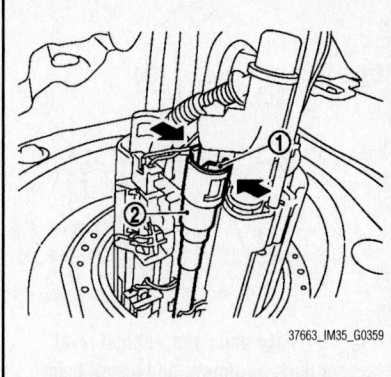

Fig. 232 Disconnect harness connector and fuel feed tube

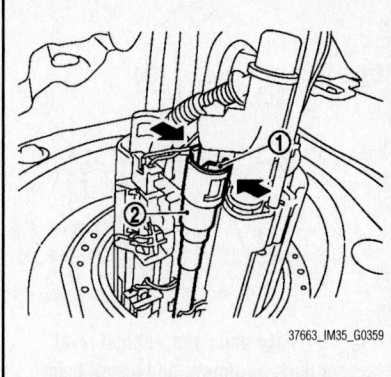

Fig. 233 Raise main fuel level sensor unit, fuel filter and fuel pump assembly, and disconnect quick connector by pushing in tabs (1) and pulling out fuel tube (2)

c. Insert fuel tube into fuel tank through fuel filler hose to draw fuel from fuel tank.

3. Release the fuel pressure from the fuel lines.

4. Open fuel filler lid.

5. Open filler cap and release the pressure inside fuel tank.

6. Remove rear seat cushion.

7. Peel off floor carpet, then remove inspection hole cover units by turning clips clockwise by 90 degrees.

8. Disconnect harness connector and fuel feed tube.

9. Disconnect quick connector as follows:

a. Hold the sides of connector, push in tabs and pull out fuel feed tube.

b. If quick connector sticks to tube of main fuel level sensor unit, push and pull quick connector several times until they start to move. Then disconnect them by pulling.

❋❋ CAUTION

Note the following:

- Quick connector can be disconnected when the tabs are completely depressed. Never twist it more than necessary.
- Never use any tools to disconnected quick connector.
- Keep resin tube away from heat. Be especially careful when welding near the resin tube.
- Prevent acid liquid such as battery electrolyte, etc. from getting on resin tube.
- Never bend or twist resin tube during installation and disconnection.

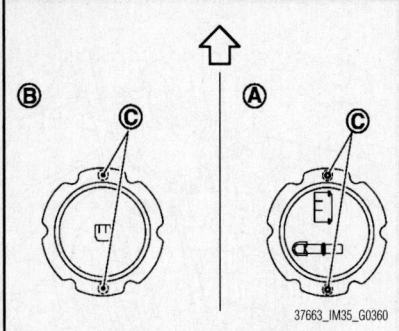

Fig. 234 Face main and sub fuel level sensor units as shown, and install them with the knock pin (C) on back aligned with pin hole on fuel tank; Right side (A), Left side (B)

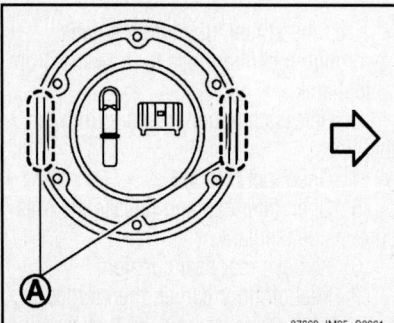

Fig. 235 Install retainer so that its notch becomes parallel with the notch on fuel tank

- Never remove the remaining retainer on hard tube (or the equivalent) except when resin tube or retainer is replaced.
- When resin tube or hard tube (or the equivalent) is replaced, also replace retainer with new one.
- To keep the connecting portion clean and to avoid damage and foreign materials, cover them completely with plastic bags or something similar.

10. Remove main fuel level sensor unit, fuel filter and fuel pump assembly, and sub fuel level sensor unit as follows:

⁎⁎ CAUTION

Never bend float arm during removal. Avoid impacts such as falling when handling components.

a. Removal of main fuel level sensor unit, fuel filter and fuel pump assembly, and sub fuel level sensor unit:
- Remove retainer.
- Raise main fuel level sensor unit, fuel filter and fuel pump assembly, and disconnect quick connector by

pushing in tabs and pulling out fuel tube.
b. Removal of sub fuel level sensor unit:
- Remove retainer.
- Raise and release sub fuel level sensor unit to remove.

To install:

11. Note the following, and install in the reverse order of removal.
a. Main and Sub Fuel Level Sensor Unit:
- Face main and sub fuel level sensor units as shown, and install them with the knock pin on back aligned with pin hole on fuel tank.
- Install retainer so that its notch becomes parallel with the notch on fuel tank.
- Tighten retainer mounting bolts evenly.
b. Connect quick connector as follows:
- Check the connection for damage or any foreign materials.
- Align the connector with the tube, then insert the connector straight into the tube until a click sound is heard.
- After connecting, check that the connection is secure by following method.
- Pull the tube and the connector to check they are securely connected.
- Visually confirm that the two retainer tabs are connected to the connector.

12. Turn ignition switch "ON" (with engine stopped), then check connections for leakage by applying fuel pressure to fuel piping.

13. Start engine and let it idle and check there are no fuel leakage at the fuel system connections.

FUEL RAIL AND INJECTOR

REMOVAL & INSTALLATION

3.5L Engine

See Figures 236 through 242.

⁎⁎ WARNING

Note the following:

- Be sure to work in a well ventilated area and furnish workshop with a CO2 fire extinguisher.
- Never smoke while servicing fuel system. Keep open flames and sparks away from the work area.
- To avoid the danger of being scalded, never drain engine coolant when engine is hot.

1. Remove engine room cover (RH and LH).
2. Remove engine cover.
3. Release fuel pressure.
4. Drain engine coolant, or when water hoses are disconnected, attach plug to prevent engine coolant leakage.

⁎⁎ CAUTION

Perform this step when engine is cold.

5. Remove intake manifold collector.
6. Remove fuel feed hose (with damper) from fuel sub-tube and harness bracket.

➡**There is no fuel return route.**

⁎⁎ CAUTION

While hoses are disconnected, plug them to prevent fuel from draining. Never separate fuel damper and fuel feed hose.

7. When separating fuel feed hose (with damper) and centralized under-floor piping connection, disconnect quick connector as follows:
a. Remove quick connector cap from quick connector connection on right member side.
b. Disconnect fuel feed hose (with damper) from bracket hose clamp.
c. Push in retainer tabs.
d. Draw and pull out quick connector straight from centralized under-floor piping.

⁎⁎ CAUTION

Note the following:

- Never pull with lateral force applied. O-ring inside quick connector may be damaged.

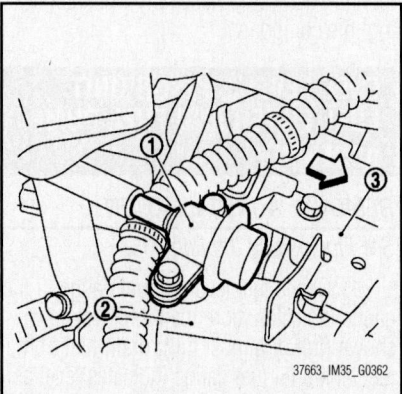

Fig. 236 Remove fuel feed hose (with damper) (1) from fuel sub-tube (2) and harness bracket (3)

- Prepare container and cloth before-hand because fuel will leakage out.
- Avoid fire and sparks.
- Keep parts away from heat source. Especially, be careful when welding is performed around them.
- Never expose parts to battery electrolyte or other acids.
- Never bend or twist connection between quick connector and fuel feed hose (with damper) during installation/removal.
- To keep clean the connecting portion and to avoid damage and foreign materials, cover them completely with plastic bags, etc. or something similar.

8. Disconnect harness connector from fuel injector.

9. Loosen mounting bolts in reverse order as shown, and remove fuel tube and fuel injector assembly.

⁂ CAUTION

Never tilt it, or remaining fuel in pipes may flow out from pipes.

10. Remove fuel injector from fuel tube as follows:
    a. Open and remove clip.
    b. Remove fuel injector from fuel tube by pulling straight.

⁂ CAUTION

Note the following:

- Be careful with remaining fuel that may go out from fuel tube.
- Be careful not to damage injector nozzles during removal.
- Never bump or drop fuel injector.
- Never disassemble fuel injector.

11. Remove fuel sub-tube and fuel damper, if necessary.

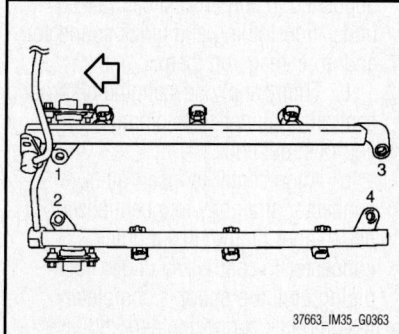

Fig. 237 Loosen mounting bolts in reverse order as shown

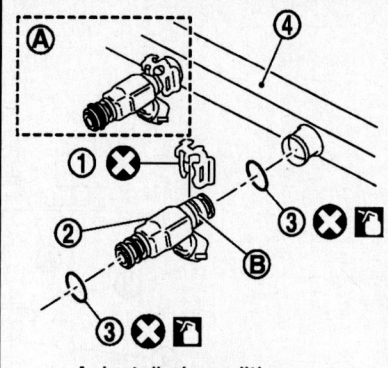

A. Installed condition
B. Clip mounting groove
1. Clip
2. Fuel injector
3. O-ring
4. Fuel tube

37663_IM35_G0364

Fig. 238 Remove fuel injector from fuel tube

To install:

12. Install fuel damper as follows:
    a. Install new O-ring to fuel tube as shown.

⁂ CAUTION

Note the following:

- Handle O-ring with bare hands. Never wear gloves.
- Lubricate O-ring with new engine oil.
- Never clean O-ring with solvent.
- Check that O-ring and its mating part are free of foreign material.
- When installing O-ring, be careful not to scratch it with tool or fingernails. Also be careful not to twist or stretch O-ring. If O-ring was stretched while it was being attached, never insert it quickly into fuel tube.
- Insert O-ring straight into fuel tube. Never decenter or twist it.
  b. Install spacer to fuel damper.
  c. Insert fuel damper straight into fuel tube.

⁂ CAUTION

Insert straight, checking that the axis is lined up. Never pressure-fit with excessive force. Insert fuel damper unit is touching of fuel tube.

d. Tighten bolts evenly in turn.

➡**After tightening bolts, check that there is no gap between fuel damper cap and fuel tube.**

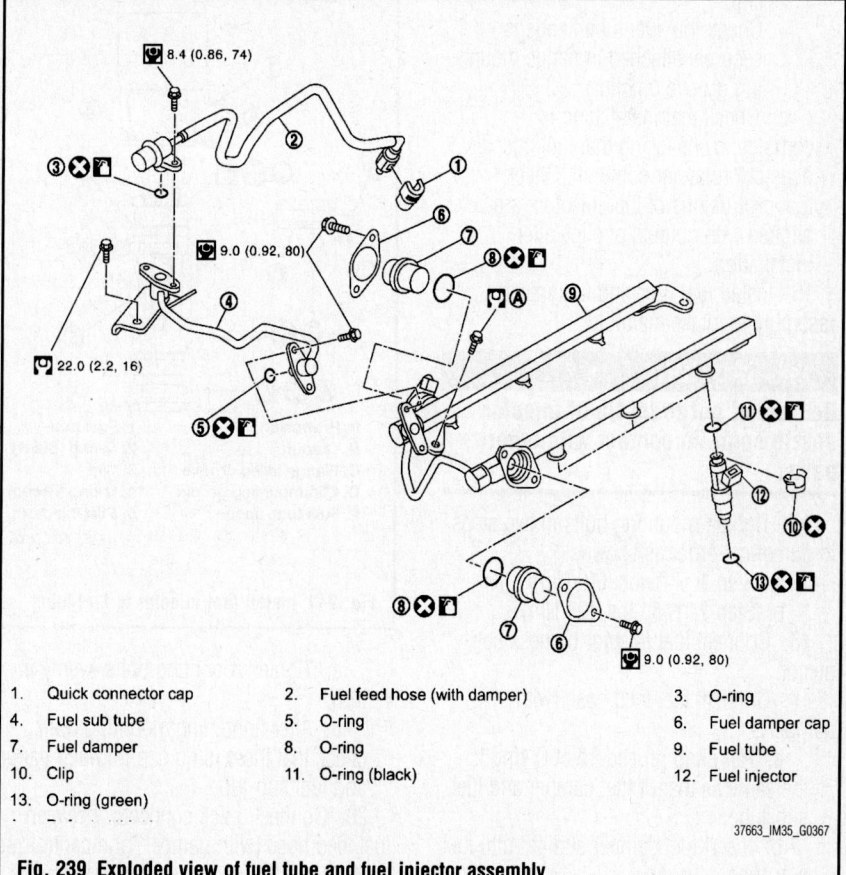

1. Quick connector cap
2. Fuel feed hose (with damper)
3. O-ring
4. Fuel sub tube
5. O-ring
6. Fuel damper cap
7. Fuel damper
8. O-ring
9. Fuel tube
10. Clip
11. O-ring (black)
12. Fuel injector
13. O-ring (green)

37663_IM35_G0367

Fig. 239 Exploded view of fuel tube and fuel injector assembly

13. Install fuel sub-tube.

   a. Insert fuel sub-tube straight into fuel tube.

   b. Tighten mounting bolts evenly in turn.

   c. After tightening mounting bolts, check that there is no gap between flange and fuel sub-tube.

14. Install O-rings to fuel injector, paying attention to the following:

   • Upper and lower O-ring are different. Be careful not to confuse them.
   • Fuel tube side is black
   • Nozzle side is green

15. Install fuel injector to fuel tube as follows:

   a. Insert clip into clip mounting groove on fuel injector.

❊❊ CAUTION

Never reuse clip. Replace it with a new one. Be careful to keep clip from interfering with O-ring. If interference occurs, replace O-ring.

   b. Insert fuel injector into fuel tube with clip attached.

   • Insert it while matching it to the axial center.
   • Insert fuel injector so that protrusion of fuel tube matches cutout of clip.
   • Check that fuel tube flange is securely attached in flange mounting groove on clip.

   c. Check that installation is complete by checking that fuel injector does not rotate or come off. Check that protrusions of fuel injectors are aligned with cutouts of clips after installation.

16. Install fuel tube and fuel injector assembly to intake manifold.

❊❊ CAUTION

Be careful not to let tip of injector nozzle come in contact with other parts.

17. Tighten mounting bolts in two steps in numerical order as shown.

   a. Step 1: 7 ft. lbs. (10 Nm)
   b. Step 2: 17 ft. lbs. (24 Nm)

18. Connect fuel injector harness connector.

19. Connect fuel feed hose (with damper).

   a. Handling procedure of O-ring is the same as that of fuel damper and fuel sub-tube.

   b. Insert fuel damper straight into fuel sub-tube.

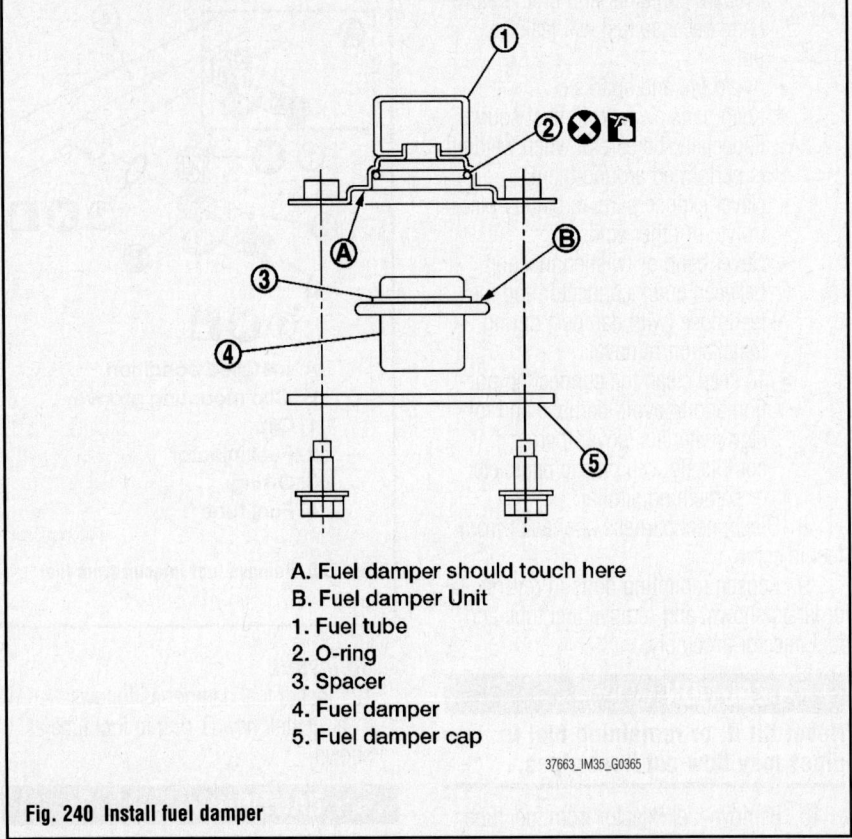

A. Fuel damper should touch here
B. Fuel damper Unit
1. Fuel tube
2. O-ring
3. Spacer
4. Fuel damper
5. Fuel damper cap

37663_IM35_G0365

Fig. 240 Install fuel damper

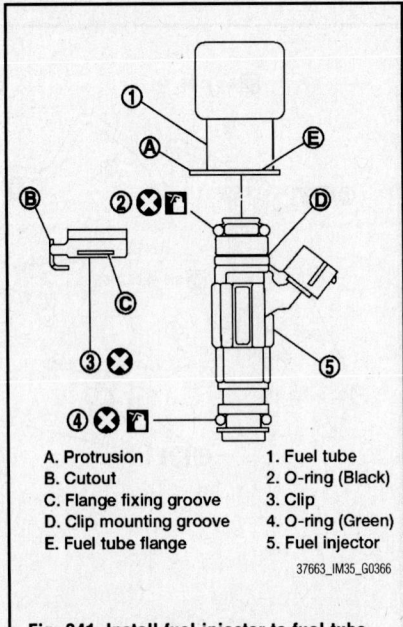

A. Protrusion
B. Cutout
C. Flange fixing groove
D. Clip mounting groove
E. Fuel tube flange
1. Fuel tube
2. O-ring (Black)
3. Clip
4. O-ring (Green)
5. Fuel injector

37663_IM35_G0366

Fig. 241 Install fuel injector to fuel tube

   c. Tighten mounting bolts evenly in turn.

   d. After tightening mounting bolts, check that there is no gap between flange and fuel sub-tube.

20. Connect quick connector between fuel feed hose (with damper) and centralized under-floor piping connection as follows:

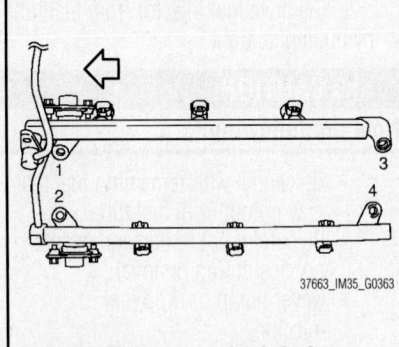

37663_IM35_G0363

Fig. 242 Tighten mounting bolts in two steps in numerical order as shown

   a. Check no foreign substances are deposited in and around centralized under-floor piping and quick connector, and no damage on them.

   b. Thinly apply new engine oil around centralized under-floor piping from tip end to spool end.

   c. Align center to insert quick connector straightly into centralized under-floor piping. Insert quick connector to centralized under-floor piping until top spool is completely inside quick connector, and 2nd level spool exposes right below quick connector.

❋ CAUTION

Note the following:

- Hold align center to avoid inclined insertion to prevent to O-ring inside quick connector.
- Insert until you hear a "click" sound and actually feel the engagement.
- To avoid misidentification of engagement with a similar sound, be sure to perform the next step.

 d. Pull quick connector by hand holding position. Check it is completely engaged (connected) so that it does not come out from centralized under-floor piping.

 e. Install quick connector cap to quick connector connection. Install quick connector cap with arrow on surface facing in direction of quick connector (fuel feed hose side).

❋ CAUTION

If quick connector cap cannot be installed smoothly, quick connector may have not been installed correctly. Check the connection again.

21. Install in the reverse order of removal after this step.
22. Turn ignition switch "ON" (with engine stopped). With fuel pressure applied to fuel piping, check for fuel leakage at connection points.

➡**Use mirrors for checking at points out of clear sight.**

23. Start engine. With engine speed increased, check again for fuel leakage at connection points.

❋ CAUTION

Never touch engine immediately after stopped, as engine becomes extremely hot.

4.5L Engine

See Figures 243 through 247.

❋ WARNING

Note the following:

- Be sure to work in a well ventilated area and furnish workshop with a CO_2 fire extinguisher.
- Never smoke while servicing fuel system. Keep open flames and sparks away from the work area.
- To avoid the danger of being

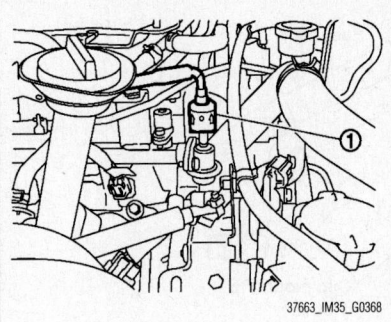

Fig. 243 Disconnect fuel feed hose (1) on engine side

scalded, never drain engine coolant when engine is hot.

1. Remove engine room cover (RH and LH).
2. Remove engine cover.
3. Release fuel pressure.
4. Disconnect fuel feed hose on engine side as follows: (Perform same procedure for the side of centralized under-floor piping as well.)

 a. Remove quick connector cap from quick connector connection.

 b. Disconnect quick connector from fuel feed damper as follows:

❋ CAUTION

Disconnect quick connector by using quick connector release J-45488, not by picking out retainer tabs (centralized under-floor piping side).

- With the sleeve side of quick connector release facing to quick connector, install quick connector release onto fuel tube.
- Insert quick connector release into quick connector until sleeve contacts and goes no further. Hold quick connector release on that position.

❋ CAUTION

Inserting quick connector release hard will not disconnect quick connector. Hold quick connector release where it contacts and goes no further.

- Draw and pull out quick connector straight from fuel feed damper.

❋ CAUTION

Note the following:

- Never pull with lateral force applied. O-ring inside quick connector may be damaged.

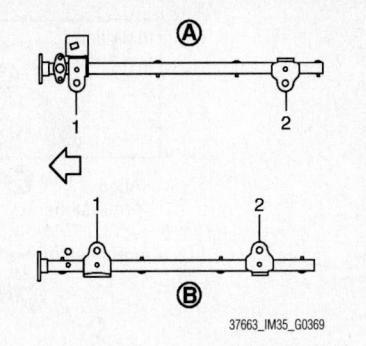

Fig. 244 Loosen mounting bolts in reverse order as shown

- Prepare container and cloth beforehand as fuel will leakage out.
- Avoid fire and sparks.
- Keep parts away from heat source. Especially, be careful when welding is performed around them.
- Never expose parts to battery electrolyte or other acids.
- Never bend or twist connection between quick connector and fuel feed hose during installation/removal.
- To keep clean the connecting portion and to avoid damage and foreign materials, cover them completely with plastic bags or something similar.

5. Disconnect fuel damper and fuel hose assembly from fuel tubes (RH and LH).

❋ CAUTION

While hoses are disconnected, plug them to prevent fuel from draining. Never separate fuel damper and fuel hose.

6. Disconnect harness connector from fuel injector.
7. Loosen mounting bolts in reverse order as shown, and remove fuel tube and fuel injector assembly.

❋ CAUTION

Never tilt it, or remaining fuel in pipes may flow out from pipes.

8. Remove spacers on intake manifold (lower).
9. Remove fuel injector from fuel tube as follows:

 a. Open and remove clip.

 b. Remove fuel injector from fuel tube by pulling straight.

❋ CAUTION

Note the following:

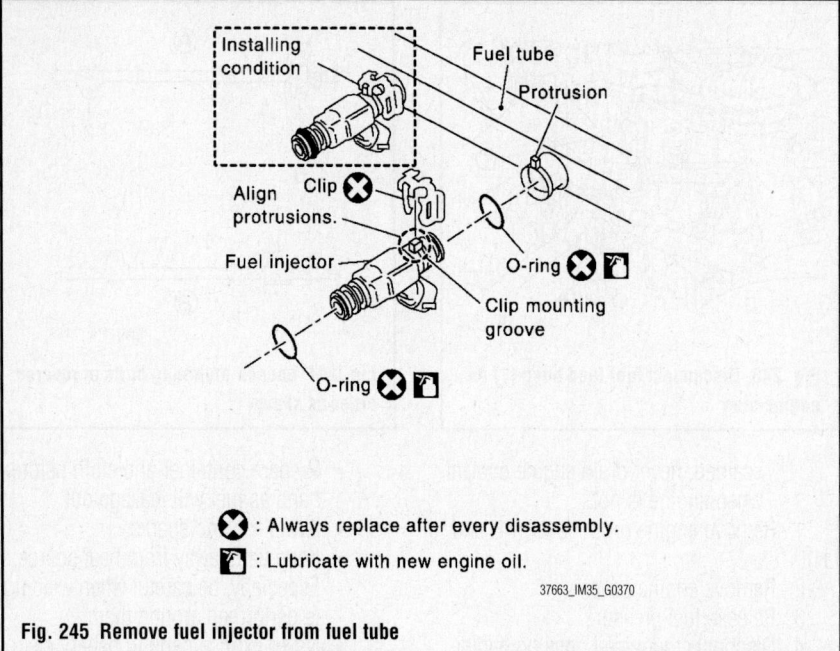

⊗ : Always replace after every disassembly.

🛢 : Lubricate with new engine oil.

37663_IM35_G0370

Fig. 245 Remove fuel injector from fuel tube

- Be careful with remaining fuel that may go out from fuel tube.
- Be careful not to damage injector nozzles during removal.
- Never bump or drop fuel injector.
- Never disassemble fuel injector.

10. Remove fuel feed damper.

To install:

11. Install fuel feed damper.

> **※ CAUTION**
>
> Note the following:
>
> - Handle O-ring with bare hands. Never wear gloves.
> - Lubricate O-ring with new engine oil.
> - Never clean O-ring with solvent.
> - Check that O-ring and its mating part are free of foreign material.
> - When installing O-ring, be careful not to scratch it with tool or fingernails. Also be careful not to twist or stretch O-ring. If O-ring was stretched while it was being attached, never insert it quickly into fuel tube.
> - Insert new O-ring straight into fuel tube. Never decenter or twist it.

a. Insert fuel feed damper straight into fuel tube (RH).

b. Tighten mounting bolts evenly in turn.

c. After tightening mounting bolts, check that there is no gap between flange and fuel tube (RH).

12. Install new O-rings to fuel injector.

- Upper and lower O-ring are different. Be careful not to confuse them.
- Fuel tube side is black
- Nozzle side is green

13. Install fuel injector to fuel tube as follows:

a. Insert clip into clip mounting groove on fuel injector.

➡ **Insert clip so that "protrusion A" of fuel injector matches "cutout A" of clip.**

> **※ CAUTION**
>
> **Never reuse clip. Replace it with a new one. Be careful to keep clip from**

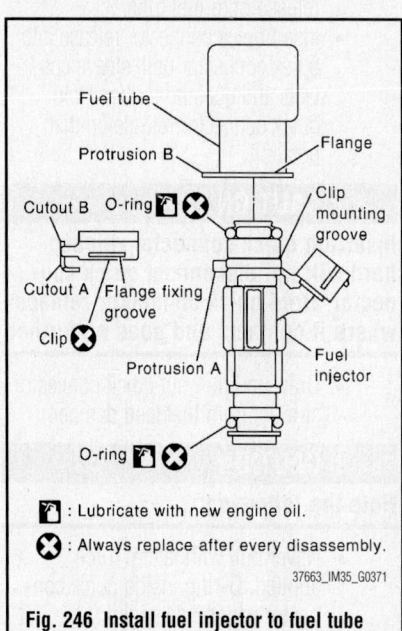

🛢 : Lubricate with new engine oil.

⊗ : Always replace after every disassembly.

37663_IM35_G0371

Fig. 246 Install fuel injector to fuel tube

interfering with O-ring. If interference occurs, replace O-ring.

b. Insert fuel injector into fuel tube with clip attached.
- Insert it while matching it to the axial center.
- Insert fuel injector so that "protrusion B" of fuel tube matches "cutout B" of clip.
- Check that fuel tube flange is securely attached in flange mounting groove on clip.

c. Check that installation is complete by checking that fuel injector does not rotate or come off.

➡ **Check that protrusions of fuel injectors are aligned with cutouts of clips after installation.**

14. Install spacers on intake manifold (lower).

15. Install fuel tube and fuel injector assembly to intake manifold (lower).

> **※ CAUTION**
>
> **Be careful not to let tip of injector nozzle come in contact with other parts.**

16. Tighten mounting bolts in two steps in numerical order as shown.
a. Step 1: 7 ft. lbs. (10 Nm)
b. Step 2: 17 ft. lbs. (24 Nm)

17. Connect fuel feed hose on engine side as follows: (Unless otherwise indicated, the installation to the engine side and centralized under-floor piping side is exactly alike.)

a. Check no foreign substances are deposited in and around fuel tube and quick connector, and no damage on them.

b. Thinly apply new engine oil around fuel tube from tip end to spool end.

c. Align center to insert quick connector straightly into fuel tube.

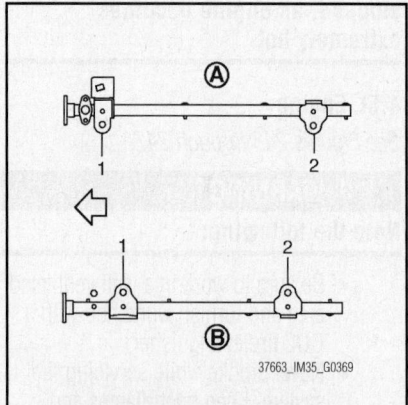

37663_IM35_G0369

Fig. 247 Tighten mounting bolts in two steps in numerical order as shown

- Engine side: Insert fuel tube into quick connector until top spool is completely inside quick connector, and 2nd level spool exposes right below quick connector.
- Centralized under-floor piping side: Visually confirm that the two retainer tabs are connected to the connector.

d. Pull quick connector by hand holding position. Check it is completely engaged (connected) so that it does not come out from fuel tube.

e. Install quick connector cap on quick connector connection.

※※ CAUTION

If cap cannot be installed smoothly, quick connector may have not been installed correctly. Check connection again.

f. Install fuel feed hose to hose clamps.

18. Install in the reverse order of removal after this step.

19. Turn ignition switch "ON" (with engine stopped). With fuel pressure applied to fuel piping, check for fuel leakage at connection points.

➡**Use mirrors for checking at points out of clear sight.**

20. Start engine. With engine speed increased, check again for fuel leakage at connection points.

※※ CAUTION

Never touch engine immediately after stopped, as engine becomes extremely hot.

FUEL TANK

DRAINING

See Figure 248.

1. Check fuel level on fuel gauge. If fuel gauge indicates more than the level as shown (full or almost full), drain fuel from fuel tank until fuel gauge indicates level as shown or below.

➡**Because fuel will be spilled when removing main and sub fuel level sensor units for the top of the fuel is above the main and sub fuel level sensor units installation surface. As a guide, fuel level becomes the position as shown or below when approximately 5¼ US gal. (20 L) of fuel are drained from fuel tank.**

2. In a case that fuel pump does not operate, perform the following procedure.

a. Insert hose of less than 0.98 inches (22 mm) in diameter into fuel

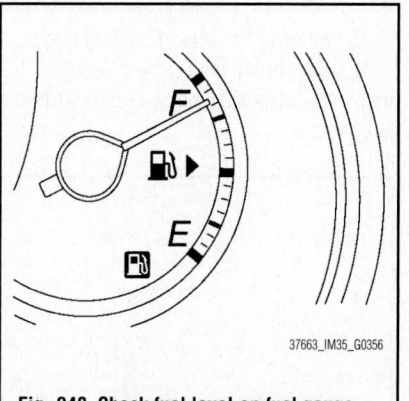

Fig. 248 Check fuel level on fuel gauge

filler tube through fuel filler opening to draw fuel from fuel filler tube.

b. Disconnect fuel filler hose from fuel filler tube.

c. Insert fuel tube into fuel tank through fuel filler hose to draw fuel from fuel tank.

REMOVAL & INSTALLATION

See Figures 249 through 254.

1. Release the fuel pressure from the fuel lines.

2. Open fuel filler lid.

3. Open filler cap and release the pressure inside fuel tank.

4. Remove rear seat cushion.

5. Peel off floor carpet, then remove inspection hole cover units by turning clips clockwise by 90 degrees.

6. Disconnect harness connector and fuel feed tube.

7. Disconnect quick connector as follows:

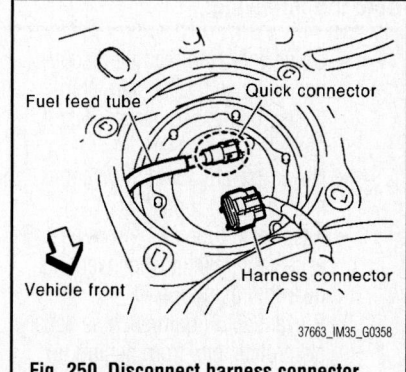

Fig. 250 Disconnect harness connector and fuel feed tube

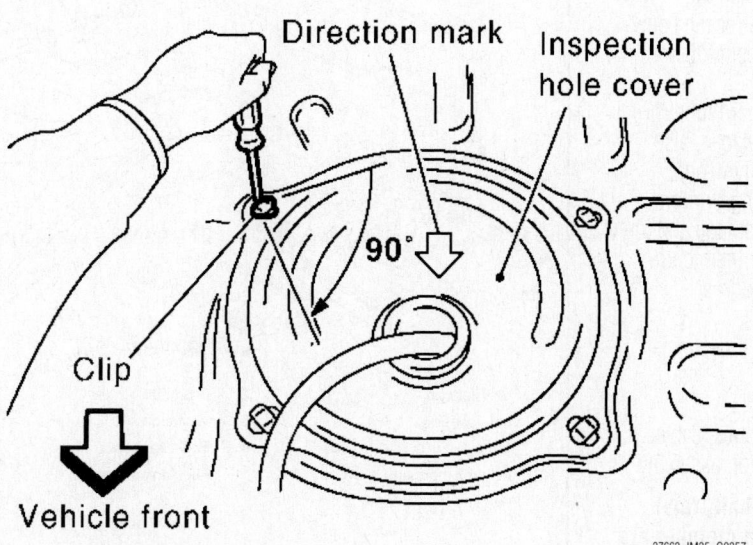

Fig. 249 Remove inspection hole cover units by turning clips clockwise by 90 degrees

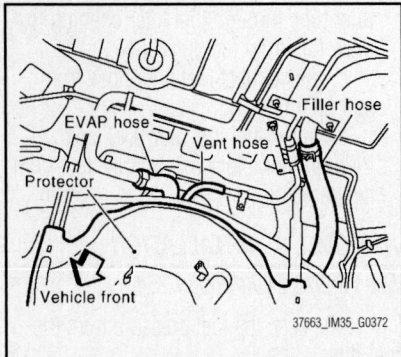

Fig. 251 Disconnect fuel filler hose, vent hose, and EVAP hoses at fuel tank side

a. Hold the sides of connector, push in tabs and pull out fuel feed tube.

b. If quick connector sticks to tube of main fuel level sensor unit, push and pull quick connector several times until they start to move. Then disconnect them by pulling.

✳✳ CAUTION

Note the following:

- Quick connector can be disconnected when the tabs are completely depressed. Never twist it more than necessary.
- Never use any tools to disconnected quick connector.
- Keep resin tube away from heat. Be especially careful when welding near the resin tube.
- Prevent acid liquid such as battery electrolyte, etc. from getting on resin tube.
- Never bend or twist resin tube during installation and disconnection.
- Never remove the remaining retainer on hard tube (or the equivalent) except when resin tube or retainer is replaced.
- When resin tube or hard tube (or the equivalent) is replaced, also replace retainer with new one.
- To keep the connecting portion clean and to avoid damage and foreign materials, cover them completely with plastic bags or something similar.

8. Remove exhaust front tube, center muffler, and main muffler.

9. Remove propeller shaft.

10. Remove parking rear brake cables.

11. Remove rear suspension assembly.

➡**For this service, drive shaft, final drive, and rear suspension member are required not to be separate one another during removal.**

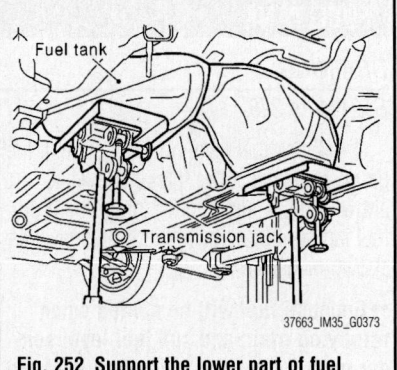

Fig. 252 Support the lower part of fuel tank with transmission jack

12. Disconnect fuel filler hose, vent hose, and EVAP hoses at fuel tank side.

13. Remove fuel tank protector.

14. Support the lower part of fuel tank with transmission jack.

✳✳ CAUTION

Support the position that fuel tank mounting bands never engage.

15. Remove fuel tank mounting bands.

16. Supporting with hands, descend transmission jack carefully, and remove fuel tank.

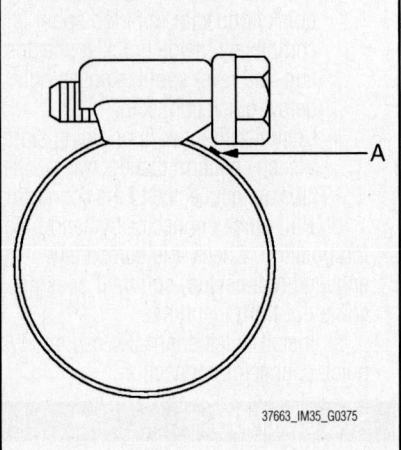

Fig. 254 Tighten the clamp band with the top mark (A) until the mark is on the bolt head flange

✳✳ CAUTION

Check that all connection points have been disconnected. Confirm there is no interference with vehicle.

17. Remove fuel filler tube, if necessary.

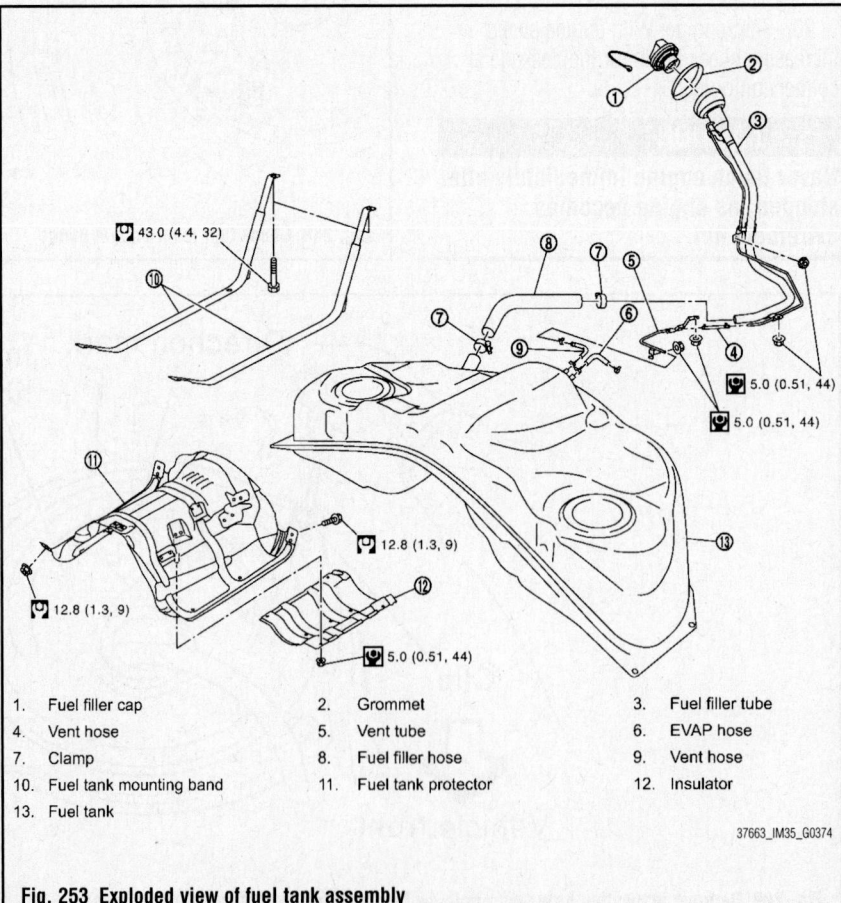

Fig. 253 Exploded view of fuel tank assembly

1. Fuel filler cap
2. Grommet
3. Fuel filler tube
4. Vent hose
5. Vent tube
6. EVAP hose
7. Clamp
8. Fuel filler hose
9. Vent hose
10. Fuel tank mounting band
11. Fuel tank protector
12. Insulator
13. Fuel tank

To install:

18. Note the following, and install in the reverse order of removal.

   a. Surely clamp fuel hoses and insert hose to the length below.

   b. Be sure hose clamp is not placed on swelled area of fuel tube.

   c. Tighten the clamp band with the top mark until the mark is on the bolt head flange.

19. Turn ignition switch "ON" (with engine stopped), and check connections for leakage by applying fuel pressure to fuel piping.

20. Start engine and rev it up and check there are no fuel leakage at the fuel system tube and hose connections.

21. After removing/installing rear suspension assembly, check to adjust wheel alignment and then, adjust neutral position of steering angle sensor.

IDLE SPEED

ADJUSTMENT

Idle speed is controlled by the ECM. No adjustment is necessary or possible.

HEATING & AIR CONDITIONING SYSTEM

BLOWER MOTOR

REMOVAL & INSTALLATION

See Figure 255.

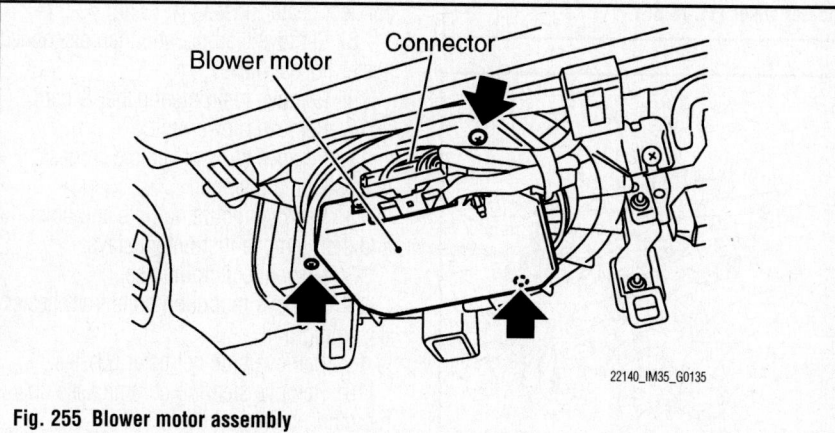

Fig. 255 Blower motor assembly

1. Remove instrument passenger lower cover.

2. Disconnect blower motor connector.

3. Remove mounting screws, and then remove blower motor.

4. Installation is the reverse order of removal.

HEATER CORE

REMOVAL & INSTALLATION

See Figure 256.

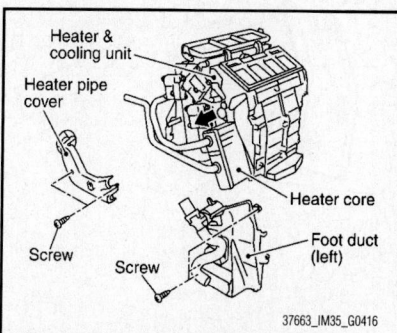

Fig. 256 Remove mounting screws, and then remove foot duct (left)

1. Remove heater & cooling unit assembly.

2. Remove mounting screws, and then remove heater pipe cover.

3. Remove mounting screws, and then remove foot duct (left).

4. Slide heater core to the left.

To install:

5. Installation is the reverse order of removal.

HEATER & COOLING UNIT

REMOVAL & INSTALLATION

See Figures 257 through 263.

1. Use a refrigerant collecting equipment (for HFC-134a) to discharge the refrigerant.

2. Drain coolant from cooling system.

3. Remove cowl top cover.

4. Remove engine cover.

5. Disconnect one-touch joint between low-pressure pipe 1 and low-pressure pipe

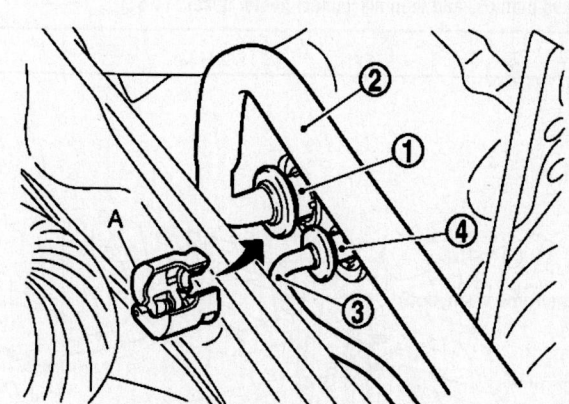

A. Disconnector 9253089916
1. Low-pressure pipe 1
2. Low-pressure pipe 2
3. High-pressure pipe 1
4. High-pressure pipe 2

Fig. 257 Disconnect one-touch joint between low-pressure pipe 1 and low-pressure pipe 2 (3.5L) or low-pressure flexible hose (4.5L) with disconnector 9253089916

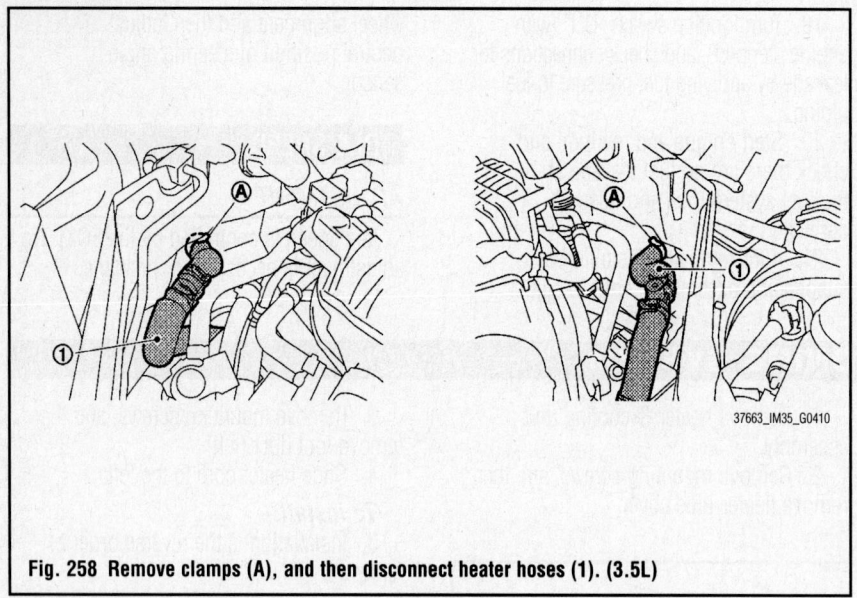

Fig. 258 Remove clamps (A), and then disconnect heater hoses (1). (3.5L)

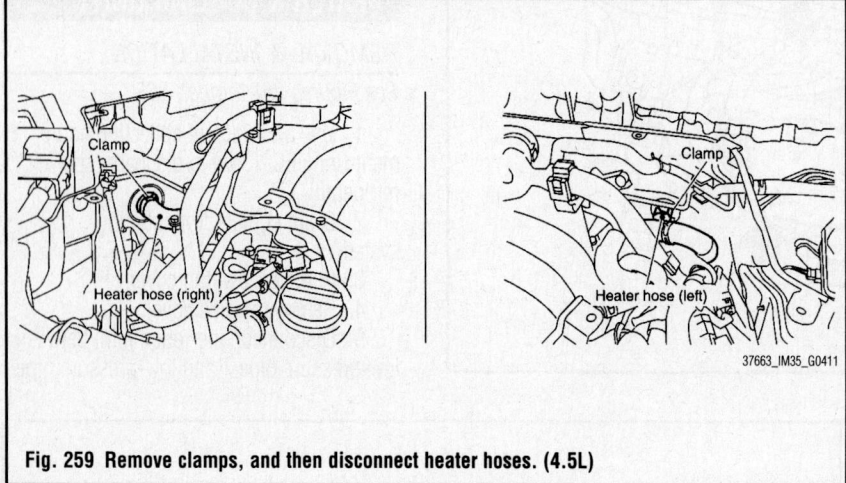

Fig. 259 Remove clamps, and then disconnect heater hoses. (4.5L)

2 (3.5L) or low-pressure flexible hose (4.5L) with disconnector 9253089916.

✳✳ CAUTION

Cap or wrap the joint of the A/C piping with suitable material such as vinyl tape to avoid the entry of air.

6. Disconnect one-touch joint between high-pressure pipe 1 and high-pressure pipe 2 with disconnector 9253089908.

✳✳ CAUTION

Cap or wrap the joint of the A/C piping with suitable material such as vinyl tape to avoid the entry of air.

7. Remove clamps (A), and then disconnect heater hoses (1). (3.5L)
8. Remove clamps, and then disconnect heater hoses. (4.5L)
9. Remove instrument panel & pad.
10. Remove blower unit.
11. Remove clips of vehicle harness from steering member.
12. Remove mounting nuts and bolts, and then remove instrument stays.
13. Disconnect drain hose.
14. Remove mounting bolts from heater & cooling unit.
15. Remove side defroster nozzles.
16. Remove steering column assembly mounting bolts and nut.
17. Remove steering member mounting bolts.
18. Remove steering member, and then remove heater & cooling unit.

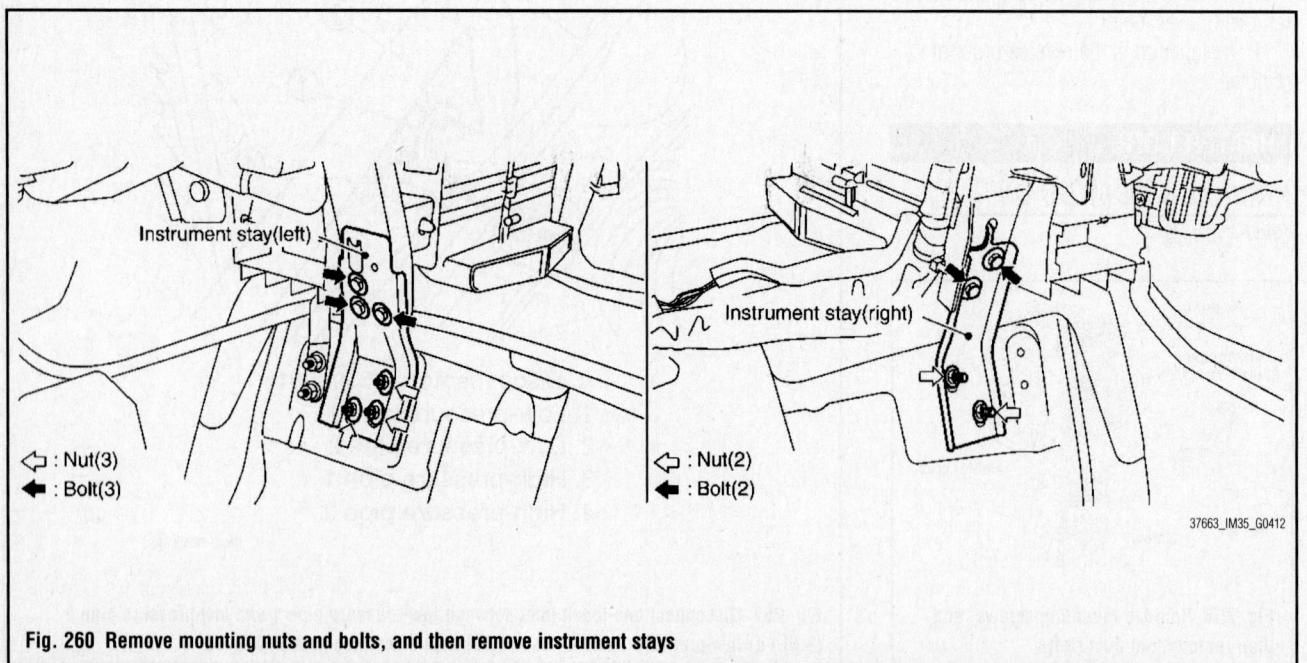

Fig. 260 Remove mounting nuts and bolts, and then remove instrument stays

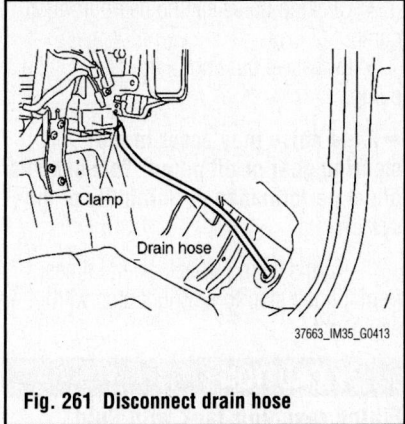

Fig. 261 Disconnect drain hose

Fig. 262 Remove mounting bolts from heater & cooling unit

To install:

19. Installation is basically the reverse order of removal.

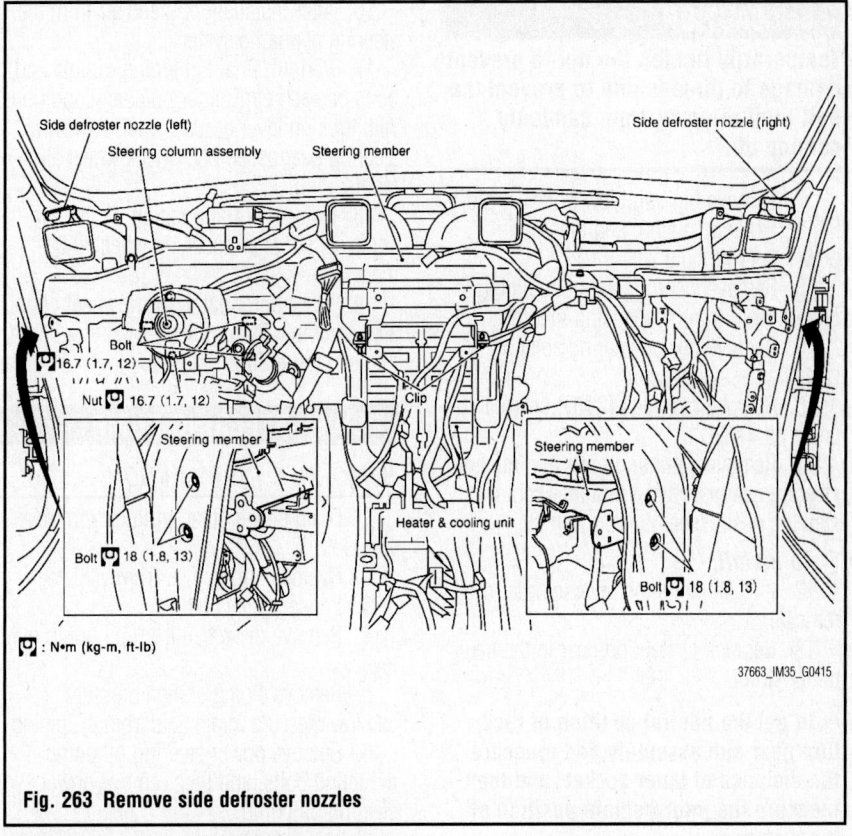

Fig. 263 Remove side defroster nozzles

✺✺ CAUTION
Note the following:

- Replace O-rings with new ones. Then apply compressor oil to them when installing.
- Female-side piping connection is thin and easy to deform. Slowly insert the male-side piping straight in axial direction.

- Insert piping securely until a click is heard.
- After piping connection is completed, pull male-side piping by hand to check that connection does not come loose.
- Check for leakages when recharging refrigerant.
20. Recharge the refrigerant.

STEERING

POWER RACK & PINION STEERING GEAR

REMOVAL & INSTALLATION

See Figures 264 and 265.

1. Set vehicle to the straight-ahead position.
2. Remove tires from vehicle.
3. Remove undercover from vehicle.
4. Remove lower side mounting bolt of lower joint.
5. Remove cotter pin, and then loosen the nut.
6. Remove steering outer socket from steering knuckle so as not to damage ball joint boot using the ball joint remover.

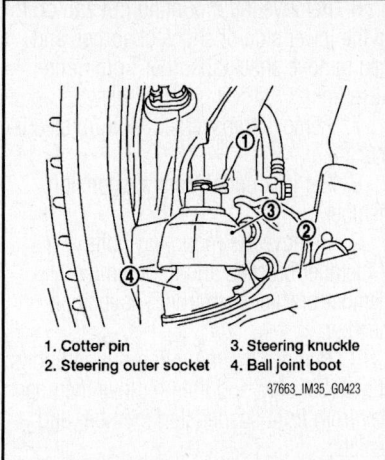

1. Cotter pin
2. Steering outer socket
3. Steering knuckle
4. Ball joint boot

Fig. 264 Remove cotter pin, and then loosen the nut

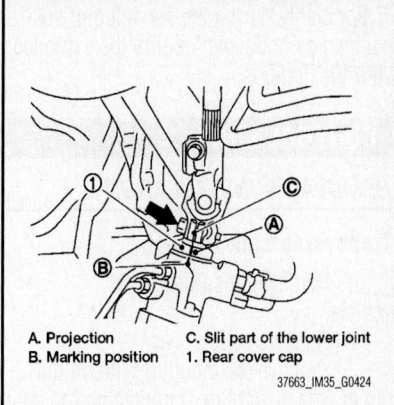

A. Projection
B. Marking position
C. Slit part of the lower joint
1. Rear cover cap

Fig. 265 Align rear cover cap projection with the marking position of gear housing assembly

✳✳ CAUTION

Temporarily tighten the nut to prevent damage to threads and to prevent the ball joint remover from suddenly coming off.

7. Remove high and low pressure piping of hydraulic piping, and then drain power steering fluid.

8. Remove steering hydraulic piping bracket from front suspension member.

9. Remove power steering solenoid valve harness connector.

10. Remove rack stay (2WD) or front cross bar (AWD).

11. Remove mounting bolts and nuts of steering gear assembly, and then remove steering gear assembly from vehicle.

To install:

12. Installation is the reverse order of removal.

13. Set rack of steering gear in the neutral position.

➡**To get the neutral position of rack, turn gear-sub assembly and measure the distance of inner socket, and then measure the intermediate position of the distance.**

14. Align rear cover cap projection with the marking position of gear housing assembly.

15. Install slit part of lower joint aligning with the projection of rear cover cap. Make sure that the slit part of lower joint is aligned with both the projection of rear cover cap and the marking position of gear housing assembly.

16. After installation, bleed air from the steering hydraulic system.

17. Perform final tightening of nuts and bolts on each part under unladen conditions with tires on level ground when removing steering gear assembly. Check wheel alignment.

18. Adjust neutral position of steering angle sensor after checking wheel alignment.

19. Make sure that steering wheel operates smoothly by turning several times from full left stop to full right stop

POWER STEERING PUMP

REMOVAL & INSTALLATION

1. Drain power steering fluid from reservoir tank.

2. Remove undercover from vehicle.

3. Loosen drive belt.

4. Remove drive belt from oil pump pulley.

5. Remove piping of high pressure and low pressure (drain fluid from the piping).

6. Remove power steering oil pump mounting bolts, and then remove power steering oil pump.

To install:

7. Installation is the reverse order of removal.

8. Bleed air.

BLEEDING

If air bleeding is not complete, the following symptoms can be observed.

• Bubbles are created in reservoir tank.

• Clicking noise can be heard from oil pump.

• Excessive buzzing in the oil pump.

➡**Fluid noise may occur in the steering gear or oil pump. This does not affect performance or durability of the system.**

1. Turn steering wheel several times from full left stop to full right stop with engine off.

✳✳ CAUTION

Filling reservoir tank with fluid so as not to lower fluid level below the MIN line while steering wheel turning.

2. Start engine and hold steering wheel at each lock position for 3 seconds at idle to check for fluid leakage.

3. Repeat step 2 above several times at approximately 3 second intervals.

✳✳ CAUTION

Do not hold the steering wheel in a locked position for more than 10 seconds. (There is the possibility that oil pump may be damaged.)

4. Check fluid for bubbles and while contamination.

5. Stop engine if bubbles and white contamination do not drain out. Perform step 2 and 3 above after waiting until bubbles and white contamination drain out.

6. Stop the engine, and then check fluid level.

SUSPENSION

See Figures 266 and 267.

For clarity during the servicing of suspension components, refer to the exploded view illustrations..

CONTROL LINKS

REMOVAL & INSTALLATION

Transverse Link RWD

1. Remove tires from vehicle.

2. Remove undercover.

3. Remove the mounting nut on the upper side of stabilizer connecting rod, and then remove stabilizer connecting rod from transverse link.

4. Separate steering gear assembly and lower joint.

5. Remove rack stay.

6. Remove the mounting nut and bolt on the lower side of shock absorber, and then remove shock absorber from transverse link.

7. Remove transverse link from steering knuckle.

8. Set jack under front suspension member.

9. Remove the mounting bolts of member bracket, and then remove member bracket from front suspension member.

10. Remove the mounting nut and bolts of member stay, and then remove member stay from front suspension member and vehicle.

11. Remove the mounting nut of front suspension member.

FRONT SUSPENSION

12. Gradually lower the suspension member to the position where transverse link mounting bolts is remove.

✳✳ CAUTION

Be careful not to lower it too far. (Do not overload the links)

13. Remove mounting nut and bolts and stopper-arm bush, and then remove transverse link from vehicle.

To install:

14. Installation is the reverse order of removal.

15. Perform final tightening of bolts and nuts at the front suspension member installation position and the shock absorber lower side (rubber bushing) under unladen

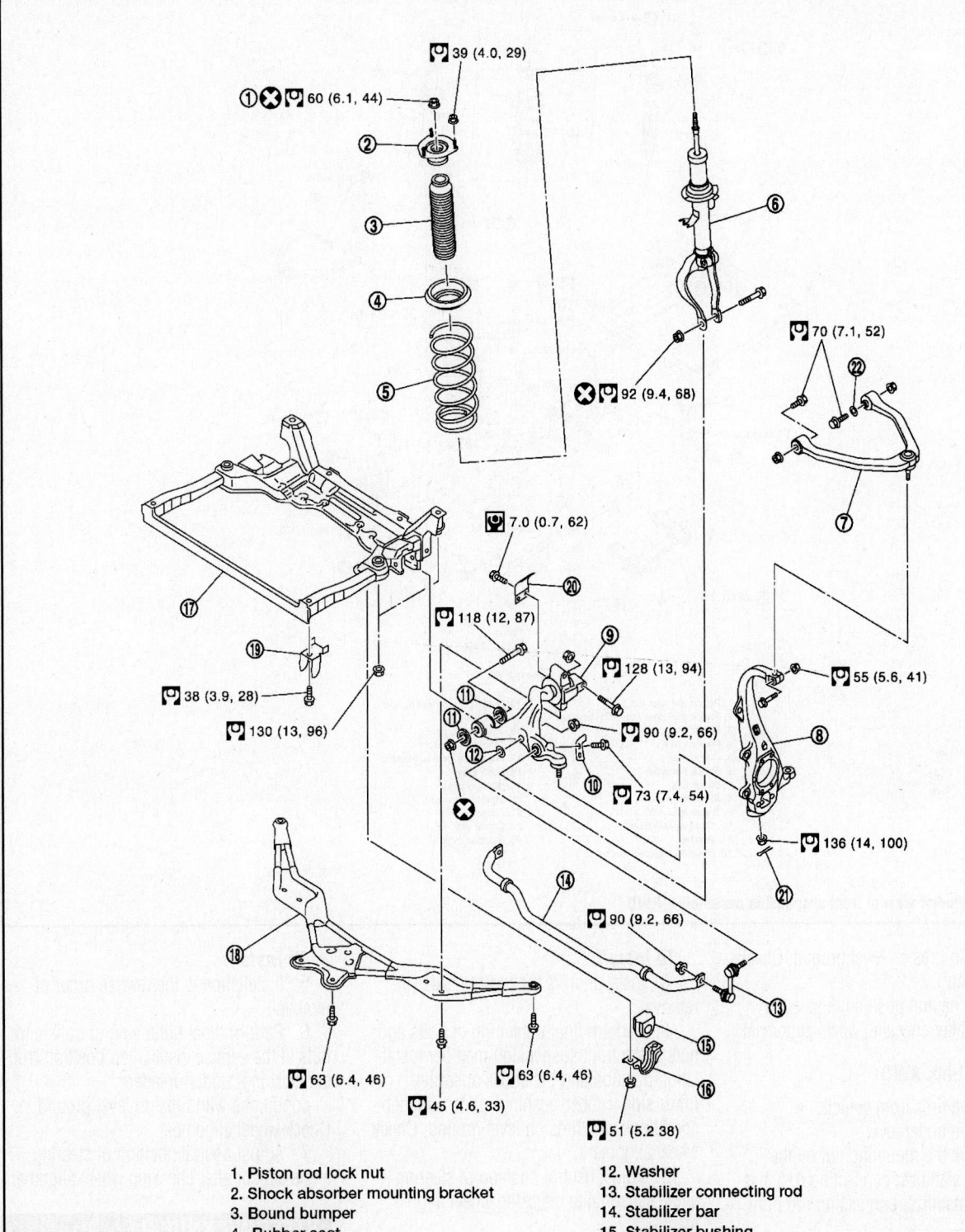

39 (4.0, 29)
60 (6.1, 44)
70 (7.1, 52)
92 (9.4, 68)
7.0 (0.7, 62)
118 (12, 87)
128 (13, 94)
55 (5.6, 41)
38 (3.9, 28)
130 (13, 96)
90 (9.2, 66)
73 (7.4, 54)
136 (14, 100)
90 (9.2, 66)
63 (6.4, 46)
63 (6.4, 46)
45 (4.6, 33)
51 (5.2 38)

1. Piston rod lock nut
2. Shock absorber mounting bracket
3. Bound bumper
4. . Rubber seat
5. Coil spring
6. Shock absorber
7. Upper link
8. Steering knuckle
9. Transverse link
10. Steering stopper bracket
11. Stopper-arm bush

12. Washer
13. Stabilizer connecting rod
14. Stabilizer bar
15. Stabilizer bushing
16. Stabilizer clamp
17. Front suspension member
18. Front cross bar
19. Member bracket
20. Clamp
21. Cotter pin
22. Stopper rubber

37663_IM35_G0436

Fig. 266 Exploded view of front suspension assembly—AWD

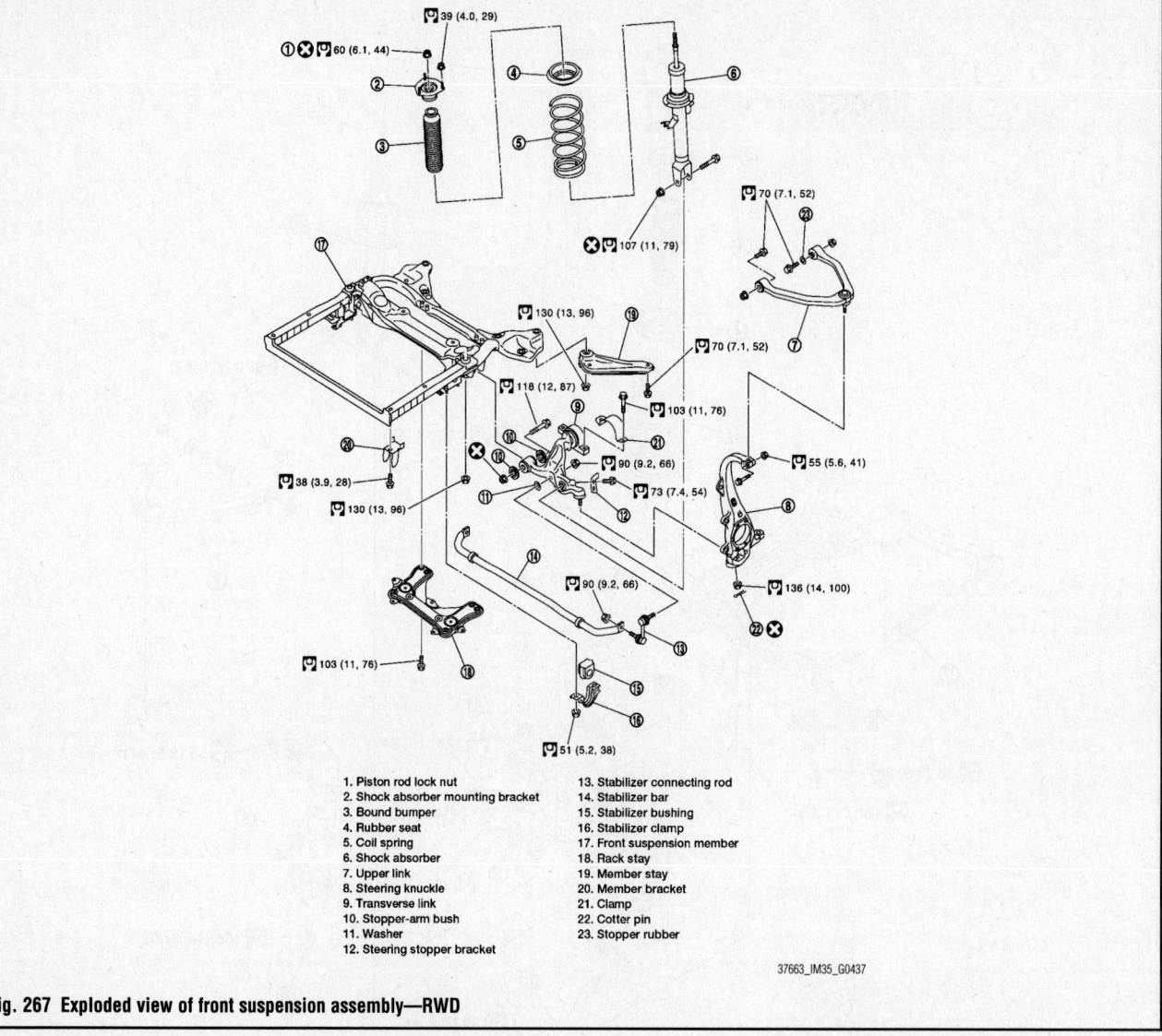

1. Piston rod lock nut
2. Shock absorber mounting bracket
3. Bound bumper
4. Rubber seat
5. Coil spring
6. Shock absorber
7. Upper link
8. Steering knuckle
9. Transverse link
10. Stopper-arm bush
11. Washer
12. Steering stopper bracket
13. Stabilizer connecting rod
14. Stabilizer bar
15. Stabilizer bushing
16. Stabilizer clamp
17. Front suspension member
18. Rack stay
19. Member stay
20. Member bracket
21. Clamp
22. Cotter pin
23. Stopper rubber

37663_IM35_G0437

Fig. 267 Exploded view of front suspension assembly—RWD

conditions with tires on level ground. Check wheel alignment.

16. Adjust neutral position of steering angle sensor after checking wheel alignment.

Transverse Link AWD

1. Remove tires from vehicle.
2. Remove undercover.
3. Remove the mounting nut on the upper side of stabilizer connecting rod, and then remove stabilizer connecting rod from transverse link.
4. Remove the mounting nut and bolt on the lower side of shock absorber arm, and then remove shock absorber arm from transverse link.
5. Remove front cross bar.
6. Remove transverse link from steering knuckle.
7. Remove mounting nuts and bolts and stopper-arm bush, and then remove transverse link from vehicle.

To install:

8. Installation is the reverse order of removal.
9. Perform final tightening of bolts and nuts at the front suspension member installation position and the shock absorber lower side (rubber bushing) under unladen conditions with tires on level ground. Check wheel alignment.
10. Adjust neutral position of steering angle sensor after checking wheel alignment.

Upper Link

1. Remove tires from vehicle.
2. Remove shock absorber.
3. Remove mounting nut and bolt, and then remove upper link from steering knuckle.
4. Remove mounting nuts and bolts, and then remove upper link and stopper rubber from vehicle.

To install:

5. Installation is the reverse order of removal.
6. Perform final tightening of bolts and nuts at the vehicle installation position (rubber bushing) under unladen conditions with tires on level ground. Check wheel alignment.
7. Adjust neutral position of steering angle sensor after checking wheel alignment

STEERING KNUCKLE

REMOVAL & INSTALLATION

See Figures 268 and 269.

1. Remove tires from vehicle.
2. Remove wheel sensor from steering knuckle.

❊❊ CAUTION

Do not pull on wheel sensor harness.

3. Remove brake hose bracket.

4. Remove torque member mounting bolts. Hang torque member in a place where it will not interfere with work.

❋❋ CAUTION

Do not depress brake pedal while brake caliper is removed.

5. Put matching mark on disc rotor and wheel hub and bearing assembly, then remove disc rotor.

6. Remove cotter pin, then loosen hub lock nut. (AWD)

7. Separate wheel hub and bearing assembly from drive shaft by lightly tapping the end with a hammer and a wood block, and then remove hub lock nut. (AWD)

❋❋ CAUTION

Do not place drive shaft joint at an extreme angle. Also be careful not to overextend slide joint. Do not allow drive shaft to hang down without support for housing (or joint sub-assembly), shaft and the other parts.

➡ **Use a puller if wheel hub and bearing assembly and drive shaft cannot be separated even after performing the above procedure.**

8. Remove cotter pin and then loosen the nut.

9. Remove steering outer socket from steering knuckle so as not to damage ball joint boot using the ball joint remover.

❋❋ CAUTION

Temporarily tighten the nut to prevent damage to threads and to prevent the ball joint remover from suddenly coming off.

10. Remove cotter pin of transverse link and steering knuckle, and then loosen nut.

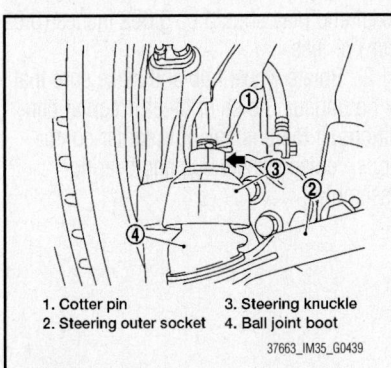

1. Cotter pin
2. Steering outer socket
3. Steering knuckle
4. Ball joint boot

37663_IM35_G0439

Fig. 268 Remove cotter pin and then loosen nut

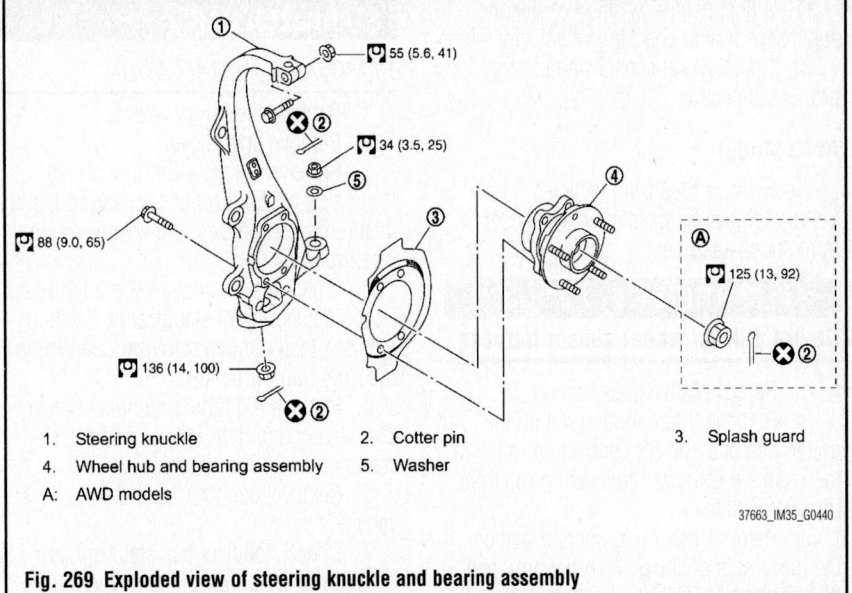

1. Steering knuckle
2. Cotter pin
3. Splash guard
4. Wheel hub and bearing assembly
5. Washer
A: AWD models

37663_IM35_G0440

Fig. 269 Exploded view of steering knuckle and bearing assembly

11. Remove transverse link from steering knuckle so as not to damage ball joint boot using the ball joint remover.

❋❋ CAUTION

Temporarily tighten the nut to prevent damage to threads and to prevent ball joint remover from suddenly coming off.

12. Remove mounting nut and bolt, and then remove steering knuckle from upper link.

13. Remove wheel hub and bearing assembly mounting bolts, and then remove splash guard and wheel hub and bearing assembly from steering knuckle.

To install:

14. Installation is the reverse order of the removal.

15. Perform the final tightening of each of parts under unladen conditions, which were removed when removing wheel hub and bearing assembly and steering knuckle. Check the wheel alignment.

16. Adjust neutral position of steering angle sensor after checking the wheel alignment.

17. Check wheel sensor harness for proper connection.

18. Assemble disc rotor and wheel hub and bearing assembly by aligning each matching mark when installing disc rotor.

STRUT

REMOVAL & INSTALLATION

RWD Model

1. Remove tires from vehicle.

2. Remove harness of wheel sensor from shock absorber.

❋❋ CAUTION

Do not pull on wheel sensor harness.

3. Remove brake hose bracket.

4. Remove the mounting nut on the upper side of stabilizer connecting rod, and then remove stabilizer connecting rod from transverse link.

5. Remove mounting nut and bolt on the lower side of shock absorber, and then remove shock absorber from transverse link.

6. Remove cotter pin of transverse link and steering knuckle, and then loosen nut.

7. Remove transverse link from steering knuckle so as not to damage ball joint boot using the ball joint remover.

❋❋ CAUTION

Temporarily tighten the nut to prevent damage to threads and to prevent ball joint remover from suddenly coming off.

8. Remove the mounting nuts of shock absorber mounting bracket, then remove shock absorber from vehicle.

To install:

9. Installation is the reverse order of removal.

10. Perform final tightening of bolt and nut at the shock absorber lower side (rubber bushing), under unladen conditions with tires on level ground. Check wheel alignment.

11. Adjust neutral position of steering angle sensor after checking wheel alignment.

12. Check wheel sensor harness for proper connection.

AWD Model

1. Remove tires from vehicle.

2. Remove harness of wheel sensor from shock absorber.

> ### ❋❋ CAUTION
> **Do not pull on wheel sensor harness.**

3. Remove brake hose bracket.

4. Remove the mounting nut on the upper side of stabilizer connecting rod, and then remove stabilizer connecting rod from transverse link.

5. Remove mounting nut and bolt on the lower side of shock absorber arm, and then remove shock absorber arm from transverse link.

6. Remove cotter pin of transverse link and steering knuckle, and then loosen nut.

7. Remove transverse link from steering knuckle so as not to damage ball joint boot using the ball joint remover.

> ### ❋❋ CAUTION
> **Temporarily tighten the nut to prevent damage to threads and to prevent ball joint remover from suddenly coming off.**

8. Remove the mounting bolt on the upper side of shock absorber arm, and then remove shock absorber arm from shock absorber.

9. Remove the mounting nuts of shock absorber mounting bracket, then remove shock absorber from vehicle.

To install:

10. Installation is the reverse order of removal.

11. Perform final tightening of bolt and nut at the shock absorber arm lower side (rubber bushing) under unladen conditions with tires on level ground. Check wheel alignment.

12. Adjust neutral position of steering angle sensor after checking wheel alignment.

13. Check wheel sensor harness for proper connection.

STABILIZER BAR

REMOVAL & INSTALLATION

1. Remove tires from vehicle.

2. Remove undercover.

3. Remove the mounting nut on the lower side of stabilizer connecting rod, and then remove stabilizer connecting rod from stabilizer bar.

4. If necessary remove the mounting nut on the upper side of stabilizer connecting rod, and then remove stabilizer connecting rod from transverse link.

5. Remove the mounting nuts of stabilizer clamp, and then remove stabilizer clamp and stabilizer bushing.

6. Remove stabilizer bar from vehicle.

7. Check stabilizer bar, stabilizer connecting rod, stabilizer bushing and stabilizer clamp for deformation, cracks or damage. Replace it if a malfunction is detected.

To install:

8. Installation is the reverse order of removal.

WHEEL HUB & BEARING

REMOVAL & INSTALLATION

See Figure 365.

1. Remove tires from vehicle.

2. Remove wheel sensor from steering knuckle.

> ### ❋❋ CAUTION
> **Do not pull on wheel sensor harness.**

3. Remove brake hose bracket.

4. Remove torque member mounting bolts. Hang torque member in a place where it will not interfere with work.

> ### ❋❋ CAUTION
> **Do not depress brake pedal while brake caliper is removed.**

5. Put matching mark on disc rotor and wheel hub and bearing assembly, then remove disc rotor.

6. Remove cotter pin, then loosen hub lock nut. (AWD)

7. Separate wheel hub and bearing assembly from drive shaft by lightly tapping the end with a hammer and a wood block, and then remove hub lock nut. (AWD)

> ### ❋❋ CAUTION
> **Do not place drive shaft joint at an extreme angle. Also be careful not to overextend slide joint. Do not allow drive shaft to hang down without support for housing (or joint sub-assembly), shaft and the other parts.**

➡**Use a puller if wheel hub and bearing assembly and drive shaft cannot be separated even after performing the above procedure.**

8. Remove wheel hub and bearing assembly mounting bolts, and then remove splash guard and wheel hub and bearing assembly from steering knuckle.

To install:

9. Installation is the reverse order of the removal.

10. Perform the final tightening of each of parts under unladen conditions, which were removed when removing wheel hub and bearing assembly and steering knuckle. Check the wheel alignment.

11. Adjust neutral position of steering angle sensor after checking the wheel alignment.

12. Check wheel sensor harness for proper connection.

13. Assemble disc rotor and wheel hub and bearing assembly by aligning each matching mark when installing disc rotor.

ADJUSTMENT

1. Move wheel hub and bearing assembly in the axial direction by hand. Make sure there is no looseness of wheel bearing. Axial end play should be 0.002 inches (0.05 mm) or less.

2. Rotate wheel hub and make sure that is no unusual noise or other irregular conditions. If there is any of irregular conditions, replace wheel hub and bearing assembly

SUSPENSION   REAR SUSPENSION

See Figure 270.

For clarity during the servicing of suspension components, refer to the exploded view illustration.

COIL SPRING

REMOVAL & INSTALLATION

See Figures 271 and 272.

1. Remove tires and wheels.
2. Set a jack under rear lower link to relieve the coil spring tension.

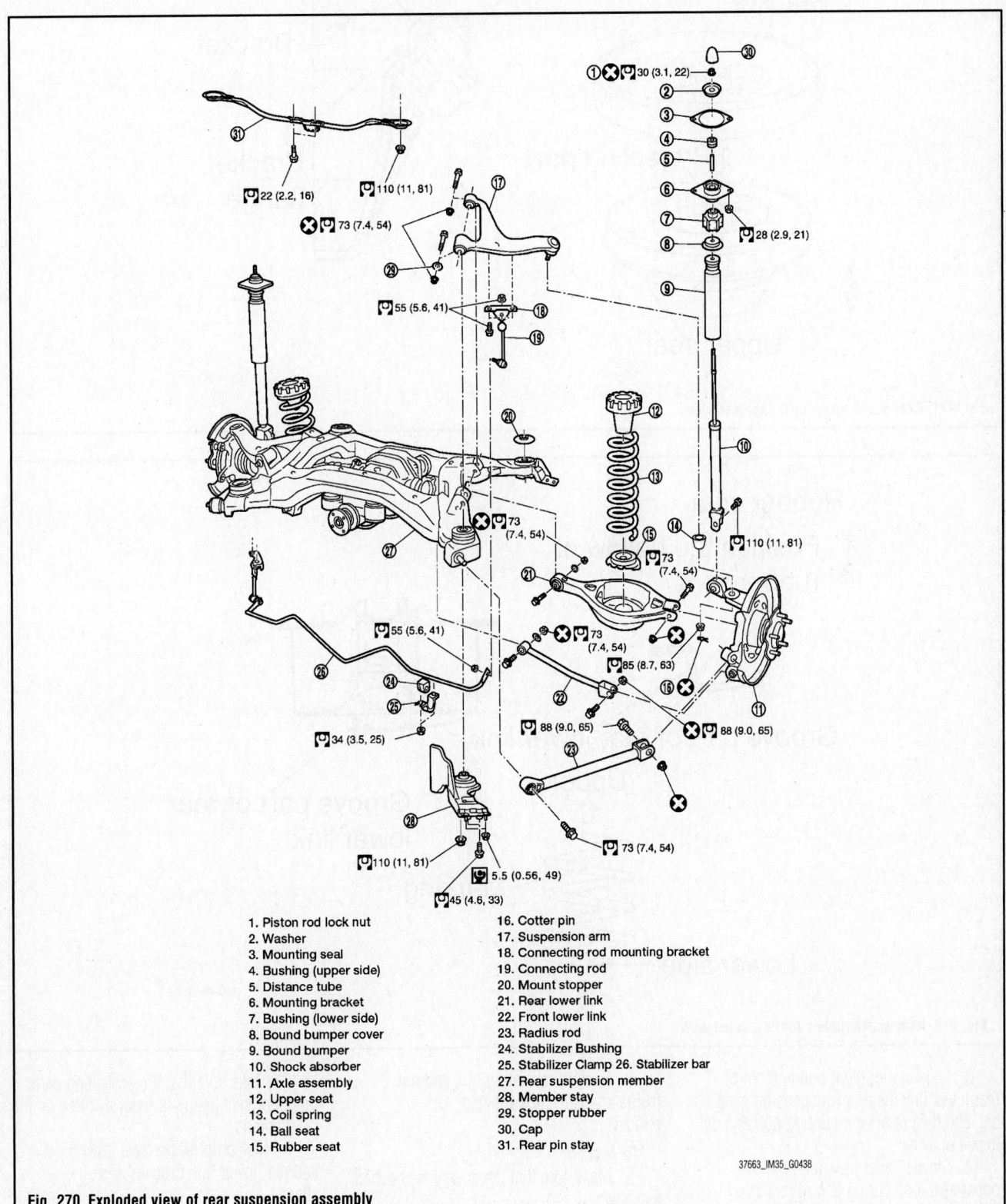

1. Piston rod lock nut
2. Washer
3. Mounting seal
4. Bushing (upper side)
5. Distance tube
6. Mounting bracket
7. Bushing (lower side)
8. Bound bumper cover
9. Bound bumper
10. Shock absorber
11. Axle assembly
12. Upper seat
13. Coil spring
14. Ball seat
15. Rubber seat
16. Cotter pin
17. Suspension arm
18. Connecting rod mounting bracket
19. Connecting rod
20. Mount stopper
21. Rear lower link
22. Front lower link
23. Radius rod
24. Stabilizer Bushing
25. Stabilizer Clamp 26. Stabilizer bar
27. Rear suspension member
28. Member stay
29. Stopper rubber
30. Cap
31. Rear pin stay

37663_IM35_G0438

Fig. 270 Exploded view of rear suspension assembly

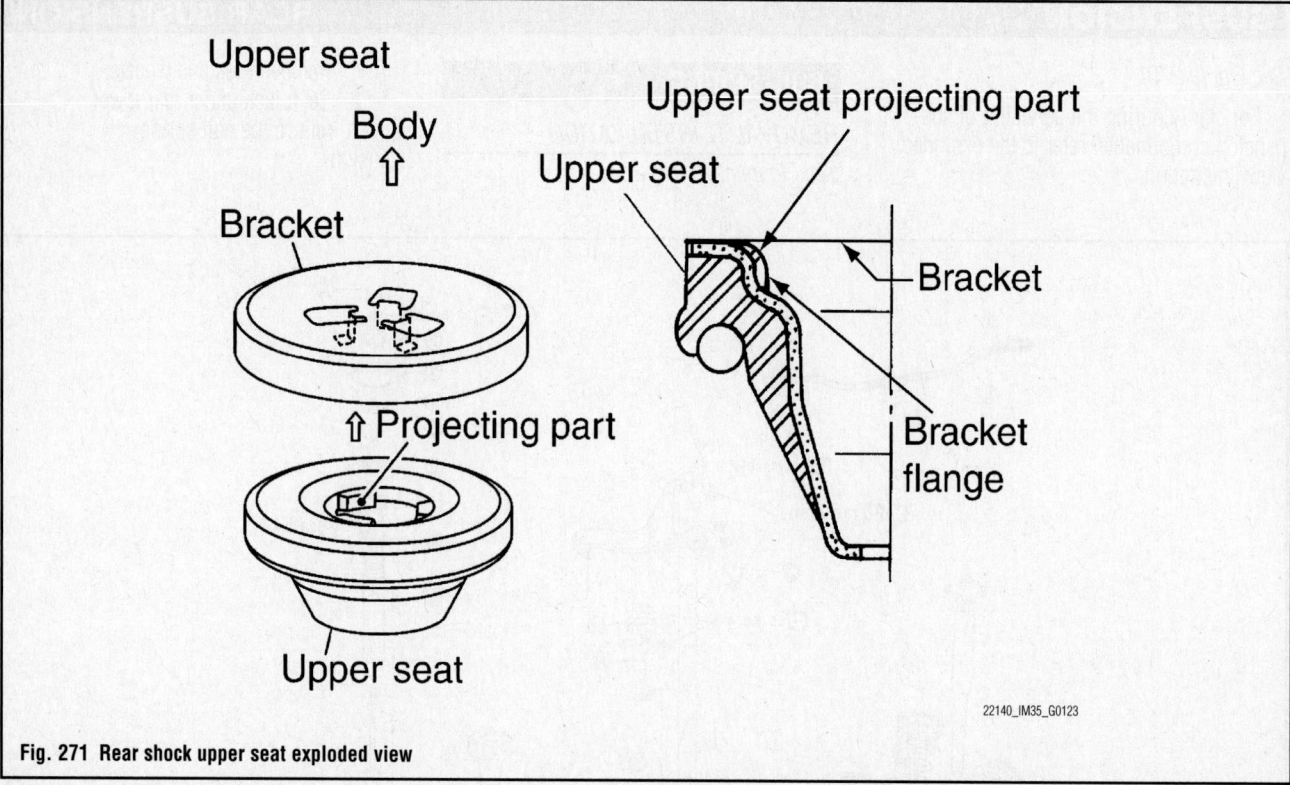

Fig. 271 Rear shock upper seat exploded view

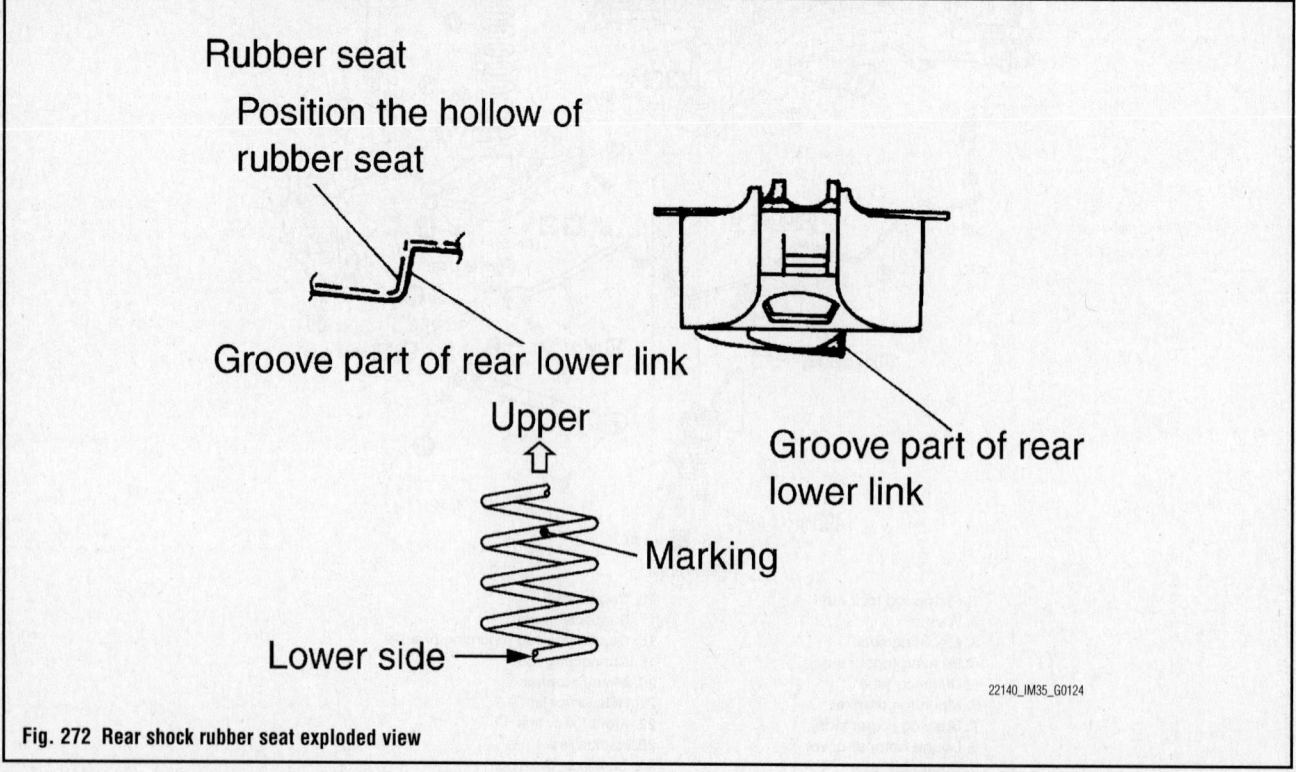

Fig. 272 Rear shock rubber seat exploded view

3. Loosen mounting bolt and nut of rear lower link inside of suspension member, and then remove mounting bolt and nut inside of axle.

4. Slowly lower jack, then remove upper seat, coil spring and rubber sheet from rear lower link.

5. Remove mounting bolt and nut inside of suspension member to remove.

To install:

6. Make sure that upper seat is attached as shown.

7. Make sure that the projecting parts on upper seat inside is securely fitted on the bracket tabs.

8. Match up rubber seat indentions and rear lower link grooves and attach.

✳✳ WARNING

Make sure spring is not upside down. The top and bottom are indicated by paint color.

9. Perform the final tightening of rear suspension member and axle installation position (rubber bushing) under unladen condition with tires on level ground.

10. Tighten of rear suspension member to 53 ft. lbs. (72 Nm).

11. Check wheel alignment.

➡ **Adjust neutral position of steering angle sensor after checking the wheel alignment.**

CONTROL ARMS/LINKS

REMOVAL & INSTALLATION

Suspension Arm

1. Remove tires and wheels.

2. Set a jack under rear lower link to relieve the coil spring tension.

3. Remove connecting rod mounting bracket from suspension arm.

4. Remove mounting nuts and bolts between suspension arm and rear suspension member.

5. Remove cotter pin of suspension arm ball joint, and loosen nut.

6. Use a ball joint remover to remove suspension arm from axle. Be careful not to damage ball joint boot.

✳✳ CAUTION

Tighten temporarily mounting nut to prevent damage to threads and to prevent ball joint remover from coming off.

7. Remove suspension arm and stopper rubber from vehicle.

To install:

8. Installation is the reverse order of removal.

✳✳ CAUTION

Do not reuse non-reusable parts.

9. Perform the final tightening of rear suspension member installation position (rubber bushing) under unladen condition with tires on level ground.

10. Adjust neutral position of steering angle sensor after checking the wheel alignment.

Radius Rod

1. Remove tires and wheels.

2. Remove brake caliper. Hang it in a place where it will not interfere with work.

✳✳ CAUTION

Do not depressing brake pedal while brake caliper is removed.

3. Put matching marks on both disc rotor and the wheel hub and bearing assembly. then remove disc rotor.

4. Remove rear lower link and coil spring.

5. Remove mounting bolt in lower side of shock absorber.

6. Remove mounting bolt and nut in axle side of front lower link.

7. Remove mounting bolt and nut in axle side of radius rod.

8. Remove mounting bolt in rear suspension member side of radius rod, then remove radius rod from vehicle.

9. Check radius rod and bushing for any deformation, cracks, or damage. Replace if there is damage or excessive wear.

To install:

10. Installation is the reverse order of removal.

✳✳ CAUTION

Do not reuse non-reusable parts.

11. Assemble disc rotor and wheel hub and bearing assembly by aligning each matching mark when installing disc rotor.

12. Perform final tightening of rear suspension member and axle installation position (rubber bushing) under unladen condition with tires on level ground. Check wheel alignment.

13. Adjust neutral position of steering angle sensor after checking the wheel alignment.

Front Lower Link

1. Remove tires and wheels.

2. Set a jack under rear lower link to relieve the coil spring tension.

3. Remove mounting nut and bolt between front lower link and rear suspension member.

4. Remove mounting nut and bolt between front lower link and axle.

5. Remove front lower link from vehicle.

6. Check front lower link and bushing for any deformation, cracks, or damage. Replace if there is any damage or excessive wear.

To install:

7. Installation is the reverse order of removal.

✳✳ CAUTION

Do not reuse non-reusable parts.

8. Perform the final tightening of rear suspension member and axle installation position (rubber bushing) under unladen condition with tires on level ground. Check wheel alignment.

9. Adjust neutral position of steering angle sensor after checking the wheel alignment.

Rear Lower Link

See Figures 271 and 272.

1. Remove tires and wheels.

2. Set a jack under rear lower link to relieve the coil spring tension.

3. Loosen mounting bolt and nut of rear lower link inside of suspension member, and then remove mounting bolt and nut inside of axle.

4. Slowly lower jack, then remove upper seat, coil spring and rubber sheet from rear lower link.

5. Remove mounting bolt and nut inside of suspension member to remove.

To install:

6. Make sure that upper seat is attached as shown.

7. Make sure that the projecting parts on upper seat inside is securely fitted on the bracket tabs.

8. Match up rubber seat indentions and rear lower link grooves and attach.

✳✳ WARNING

Make sure spring is not upside down. The top and bottom are indicated by paint color.

9. Perform the final tightening of rear suspension member and axle installation position (rubber bushing) under unladen condition with tires on level ground.

10. Tighten of rear suspension member to 53 ft. lbs. (72 Nm).

11. Check wheel alignment.

➡ **Adjust neutral position of steering angle sensor after checking the wheel alignment.**

SHOCK ABSORBER

REMOVAL & INSTALLATION

1. Remove tires and wheels from vehicle.

2. Set a jack under rear lower link to relieve the coil spring tension.

3. Remove shock absorber lower end bolt.

4. Gradually lower the jack to remove it from rear lower link.

5. Remove shock absorber assembly upper end nuts, and then remove shock absorber assembly from vehicle.

6. Check shock absorber assembly for deformation, cracks, damage, and replace if there is any damage or excessive wear.

7. Check welded and sealed areas for oil leakage, and replace if there are any defects noted.

To install:

8. Installation is the reverse order of removal.

> ❊❊ **CAUTION**
>
> **Do not reuse non-reusable parts.**

9. Perform final tightening of shock absorber assembly lower side (rubber bushing) under unladen condition with tires on level ground. Check wheel alignment.

10. Adjust neutral position of steering angle sensor after checking the wheel alignment.

STABILIZER BAR

REMOVAL & INSTALLATION

1. Remove mounting bracket of center muffler and remove mounting rubber of main muffler.

2. Remove lower side mounting nut on stabilizer connecting rod and remove stabilizer connecting rod from stabilizer bar.

3. Remove mounting nut on stabilizer clamp and remove stabilizer from vehicle.

4. Check stabilizer bar, stabilizer bushings, stabilizer clamp, stabilizer connecting rod and stabilizer connecting rod mounting bracket for any deformation, crack or damage. Replace if there is any damage or excessive wear.

To install:

5. Installation is the reverse order of removal.

> ❊❊ **CAUTION**
>
> **Do not reuse non-reusable parts.**

WHEEL HUB & BEARING

REMOVAL & INSTALLATION

See Figures 273 through 275.

1. Remove tires and wheels from vehicle.

2. Remove rear brake caliper. Hang it in a place where it will not interfere with work.

> ❊❊ **CAUTION**
>
> **Do not depress brake pedal while brake caliper is removed.**

3. Put matching mark on disc rotor and the wheel hub and bearing assembly then removing disc rotor.

4. Remove cotter pin and adjusting cap (if equipped), then loosen wheel hub lock nut.

5. Separate the wheel hub and bearing assembly from drive shaft by lightly tapping the end with a hammer and wood block, and then remove wheel hub lock nut and spring washer (if equipped).

> ❊❊ **CAUTION**
>
> **Do not place drive shaft joint at an extreme angle. Also be careful**

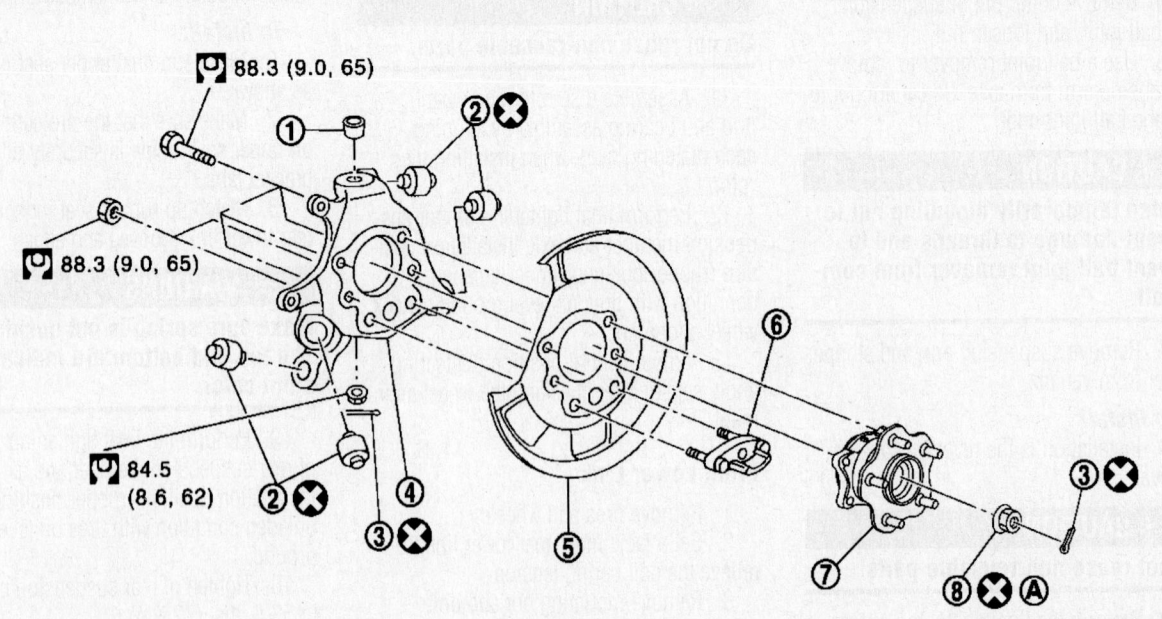

1. Ball seat	2. Bushing	3. Cotter pin
4. Axle housing	5. Back plate	6. Anchor block
7. Wheel hub and bearing assembly	8. Wheel hub lock nut	

37663_IM35_G0444

Fig. 273 Exploded view of wheel hub and bearing assembly (Without adjusting cap and spring washer)

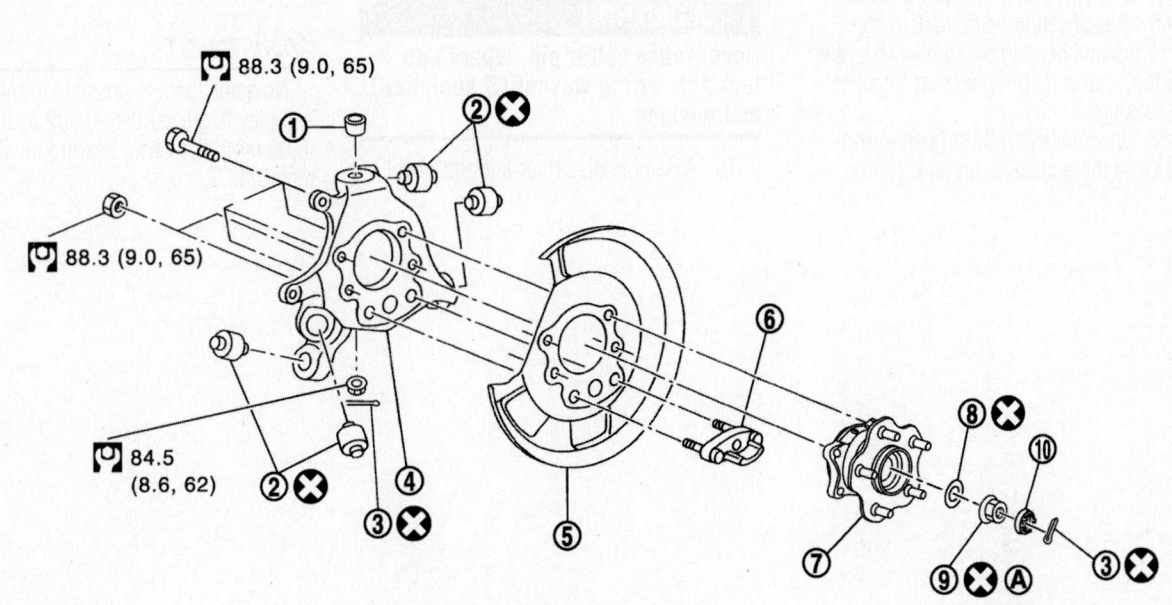

1. Ball seat
4. Axle housing
7. Wheel hub and bearing assembly
10. Adjusting cap

2. Bushing
5. Back plate
8. Spring washer

3. Cotter pin
6. Anchor block
9. Wheel hub lock nut

37663_IM35_G0445

Fig. 274 Exploded view of wheel hub and bearing assembly (With adjusting cap and spring washer)

not to overextend slide joint. Do not allow drive shaft to hang down without support for housing (or joint sub-assembly), shaft and other parts.

➡ Use a puller, if the wheel hub and bearing assembly and drive shaft can-

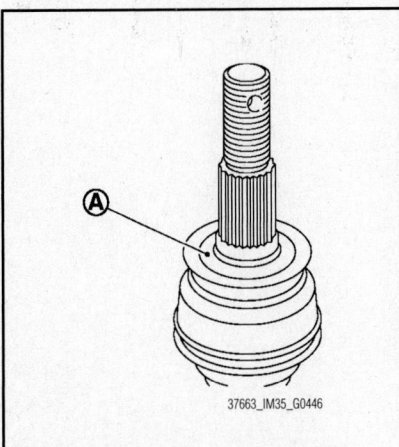

37663_IM35_G0446

Fig. 275 Apply paste to surface (A) of joint sub-assembly of drive shaft

not be separated even after performing the above procedure.

6. Remove the wheel hub and bearing assembly mounting bolts.
7. Remove the wheel hub and bearing assembly.

To install:

8. Note the following, and install in the reverse order of removal.
9. Clean the matching surface of wheel hub lock nut and wheel hub and bearing assembly.

✳✳ **CAUTION**

Never apply lubricating oil to these matching surface.

10. Clean the matching surface of drive shaft and wheel hub and bearing assembly.
11. Apply paste (440037S000) to surface of joint sub-assembly of drive shaft.

✳✳ **CAUTION**

Apply paste to cover entire flat surface of joint sub-assembly of drive shaft.

12. Use the following torque range for tightening the wheel hub lock nut:
- Without adjusting cap and spring washer: 133 to 136 ft. lbs. (180 to 185 Nm)
- With adjusting cap and spring washer: 74 to 77 ft. lbs. (100 to 105 Nm)

✳✳ **CAUTION**

Note the following:

- Since the drive shaft is assembled by press-fitting, use the tightening torque range for the wheel hub lock nut.
- Be sure to use torque wrench to tighten the wheel hub lock nut. Never use a power tool.

➡ Wheel hub lock nut tightening torque does not over torque for avoiding axle noise, and does not less than torque for avoiding looseness.

13. Perform the final tightening of each of parts under unladen conditions, which were removed when removing

wheel hub and bearing assembly and axle housing.

14. When installing the spring washer, face the identification paint mark to the wheel hub and bearing assembly side. (With adjusting cap and spring washer for wheel hub lock nut)

15. When installing the adjusting cap, check that there must be no play. (With

adjusting cap and spring washer for wheel hub lock nut)

✳✳ CAUTION

Never reuse cotter pin, wheel hub lock nut, spring washer (if equipped), and bushing.

16. Assemble disc rotor and the wheel

hub and bearing assembly by aligning each matching mark when installing disc rotor.

ADJUSTMENT

No adjustment is possible. If the axial end play is greater than 0.002 inches (0.05 mm), the wheel bearing must be replaced.

INFINITI

QX56

SPECIFICATIONS AND MAINTENANCE CHARTS

ENGINE AND VEHICLE IDENTIFICATION

Engine							Model Year	
Code ①	Liters (cc)	Cu. In.	Cyl.	Fuel Sys.	Engine	Eng. Mfg.	Code ②	Year
VK56DE	5.6 (5552)	338.8	8	MFI	DOHC	Nissan	9	2009
							A	2010

MFI: Multi-port Fuel Injection

DOHC: Double Overhead Camshafts

① Engine VIN: Z

② 10th digit of the Vehicle Identification Number (VIN)

37663_QX56_C0001

GENERAL ENGINE SPECIFICATIONS

Year	Model	Engine Displacement Liters	Engine ID	Net Horsepower @ rpm	Net Torque @ rpm (ft. lbs.)	Bore x Stroke (in.)	Compression Ratio	Oil Pressure @ rpm
2009	QX56	5.6	VK56DE	320@5200	393@3400	3.86X3.62	9.8:1	43@2000
2010	QX56	5.6	VK56DE	320@5200	393@3400	3.86X3.62	9.8:1	43@2000

37663_QX56_C0002

ENGINE TUNE-UP SPECIFICATIONS

Year	Engine Displacement Liters	Engine ID	Spark Plug Gap (in.)	Ignition Timing	Fuel Pump (psi) ①	Idle Speed	Valve Clearance (in.) In.	Valve Clearance (in.) Ex.
2009	5.6	VK56DE	0.043	②	51	600-700	0.010-0.013	0.011-0.015
2010	5.6	VK56DE	0.043	②	51	600-700	0.010-0.013	0.011-0.015

NOTE: The Vehicle Emission Control Information label often reflects specification changes made during production. The label figures

must be used if they differ from those in this chart.

① At idle

② 15 degrees +/- 5 degrees BTDC

37663_QX56_C0003

CAPACITIES

Year	Model	Engine Displacement Liters	Engine ID	Engine Oil with Filter (qts.)	Transmission (pts.)	Transfer Case (pts.)	Drive Axle Front (pts.)	Drive Axle Rear (pts.)	Fuel Tank (gal.)	Cooling System (qts.)
2009	QX56	5.6	VK56DE	6.50	22.50	6.36	3.380	3.75	28.0	15.25
2010	QX56	5.6	VK56DE	6.50	22.50	6.36	3.380	3.75	28.0	15.25

NOTE: All capacities are approximate. Add fluid gradually and check to be sure a proper fluid level is obtained.

37663_QX56_C0004

FLUID SPECIFICATIONS

Year	Model	Engine Displ. Liters	Engine Oil	Auto. Trans.	Drive Axle Front	Drive Axle Rear	Transfer Case	Power Steering Fluid	Brake Master Cylinder	Cooling System
2009	QX56	5.6	SAE 5W-30	①	②	②	③	④	⑤	⑥
2010	QX56	5.6	SAE 5W-30	①	②	②	③	④	⑤	⑥

① Genuine Nissan Matic S ATF fluid. If not available genuine Nissan Matic J ATF may be used.

② Front: Nissan differential oil hypoid super GL-5 80W-90 or API GL-5 viscosity SAE 80W-90 gear oil.

   Rear: Nissan differential oil API GL-5 synthetic 75W-90 gear oil.

③ Nissan Matic D ATF

④ Nissan Power Steering Fluid

⑤ Nissan Super Heavy Duty DOT 3

⑥ Nissan Long Life antifreeze or equivalent

37663_QX56_C0014

VALVE SPECIFICATIONS

Year	Engine Displacement Liters	Engine ID	Seat Angle (deg.)	Face Angle (deg.)	Spring Test Pressure (lbs. @ in.)	Spring Installed Height (in.)	Stem-to-Guide Clearance (in.) Intake	Stem-to-Guide Clearance (in.) Exhaust	Stem Diameter (in.) Intake	Stem Diameter (in.) Exhaust
2009	5.6	VK56DE	45.15-45.45	45	37.0@1.457	1.9913	0.0008-0.0021	0.0012-0.0025	0.2348-0.2354	0.2344-0.2350
2010	5.6	VK56DE	45.15-45.45	45	37.0@1.457	1.9913	0.0008-0.0021	0.0012-0.0025	0.2348-0.2354	0.2344-0.2350

37663_QX56_C0006

CAMSHAFT SPECIFICATIONS

All measurements are given in inches.

Year	Engine Displ. Liters	Engine ID/VIN	Journal Dia.	Brg. Oil Clearance	Shaft End-play	Runout	Journal Bore	Lobe Height Intake	Exhaust
2009	5.6	VK56DE	1.0217-1.0224	0.0012-0.0028	0.0045-0.0074	0.0008	1.0236-1.0244	1.7663-1.7738	1.7746-1.7821
2010	5.6	VK56DE	1.0278-1.0224	0.0012-0.0028	0.0045-0.0074	0.0008	1.0236-1.0244	1.7663-1.7738	1.7746-1.7821

37663_QX56_C0007

CRANKSHAFT AND CONNECTING ROD SPECIFICATIONS

All measurements are given in inches.

Year	Engine Displ. Liters	Engine ID	Crankshaft Main Brg. Journal Dia.	Main Brg. Oil Clearance	Shaft End-play	Thrust on No.	Connecting Rod Journal Diameter	Oil Clearance	Side Clearance
2009	5.6	VK56DE	①	②	0.0039-0.0102	3	③	0.0008-0.0015	0.0079-0.0157
2010	5.6	VK56DE	①	②	0.0039-0.0102	3	③	0.0008-0.0015	0.0079-0.0157

① There are 24 different grades, ranging from 2.5182- 2.5174

② No. 1 and 5: 0.00004- 0.0004

No. 2, 3 and 4: 0.0003- 0.0007

③ There are 13 different grades, ranging from 2.2441- 2.2446. Specification is for rod bearing housing.

37663_QX56_C0005

PISTON AND RING SPECIFICATIONS

All measurements are given in inches.

Year	Engine Displacement Liters	Engine ID	Piston Clearance	Ring Gap Top Comp.	Bottom Comp.	Oil Control	Ring Side Clearance Top Comp.	Bottom Comp.	Oil Control
2009	5.6	VK56DE	0.0004-0.0012	0.0091-0.0130	0.0098-0.0157	0.0079-0.0236	0.0014-0.0033	0.0012-0.0028	0.0006-0.0073
2010	5.6	VK56DE	0.0004-0.0012	0.0091-0.0130	0.0098-0.0157	0.0079-0.0236	0.0014-0.0033	0.0012-0.0028	0.0006-0.0073

37663_QX56_C0008

TORQUE SPECIFICATIONS
All readings in ft. lbs.

Year	Engine Displacement Liters	Engine ID	Cylinder Head Bolts	Main Bearing Bolts	Rod Bearing Bolts	Crankshaft Damper Bolts	Flywheel Bolts	Manifold		Spark Plugs	Oil Pan Drain Plug
								Intake	Exhaust		
2009	5.6	VK56DE	①	②	③	④	65	NA	25	18	25
2010	5.6	VK56DE	①	②	③	④	65	NA	25	18	25

NA: Not Available

① Step 1: 33 ft. lbs

   Step 2: +70 degrees clockwise

   Step 3: loosen in reverse order of tightening sequence

   Step 4: 33 ft. lbs.

   Step 5: +60 degrees clockwise

   Step 6: +60 degrees clockwise

② Step 1: cap bolts in order 1-10: 29 ft. lbs.

   Step 2: cap sub bolts in order 11-20: 22 ft. lbs.

   Step 3: cap bolts in order 1-10: +40 degrees

   Step 4: cap sub bolts in order 11-20: +30 degrees

   Step 5: side bolts in order 21-30: 36 ft. lbs.

③ Step 1: 11 ft. lbs.

   Step 2: +90 degrees clockwise

④ Step 1: 69 ft. lbs.

   Step 2: +90 degrees

37663_QX56_C0009

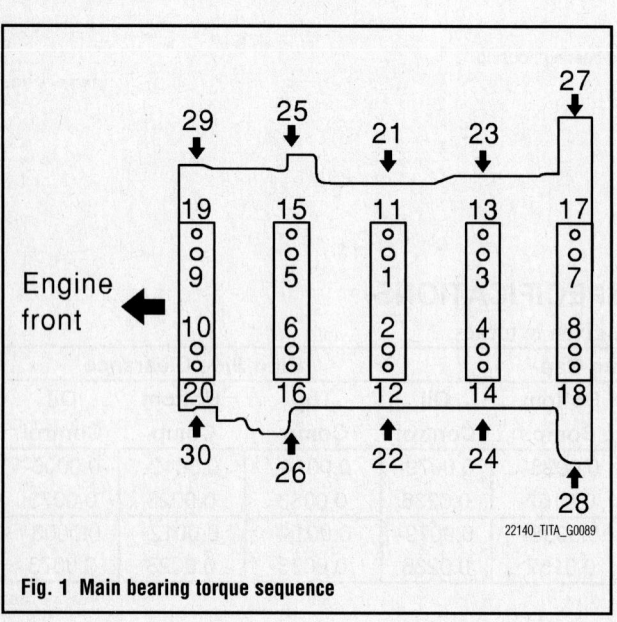

Fig. 1 Main bearing torque sequence

WHEEL ALIGNMENT

Year	Model		Caster Range (+/-Deg.)	Caster Preferred Setting (Deg.)	Camber Range (+/-Deg.)	Camber Preferred Setting (Deg.)	Toe-in (in.)
2009	QX56	2WD	①	①	②	②	NA
		4WD	③	③	④	④	NA
2010	QX56	2WD	①	①	②	②	NA
		4WD	③	③	④	④	NA

NA - not available

① Minimum: 3 degrees 15' (3.25 degrees). Nominal: 4 degrees 0' (4.00 degrees). Maximum: 4 degrees 45' (4.75 degrees).

② Minimum: -0 degrees 51' (-0.85 degrees). Nominal: 0 degrees 6' (0.10 degrees). Maximum: 0 degree 39' (0.65 degree).

③ Minimum: 2 degrees 45' (2.75 degrees). Nominal: 3 degrees 30' (3.50 degrees). Maximum: 4 degrees 15' (4.25 degrees).

④ Minimum: -0 degrees 33' (-0.55 degrees). Nominal: 0 degrees 12' (0.20 degrees). Maximum: 0 degree 57' (0.95 degrees).

37663_QX56_C0010

TIRE, WHEEL AND BALL JOINT SPECIFICATIONS

Year	Model	OEM Tires Standard	OEM Tires Optional	Tire Pressures (psi) Front	Tire Pressures (psi) Rear	Wheel Size	Ball Joint Inspection	Lut Nut Torque (ft. lbs.)
2009	QX56	P275/60R20	None	35	35	①	②	98
2010	QX56	P275/60R20	None	35	35	①	②	98

OEM: Original Equipment Manufacturer

PSI: Pounds Per Square Inch

① 20x8JJ

② Axial play

    Upper: 0

37663_QX56_C0011

BRAKE SPECIFICATIONS

All measurements in inches unless noted

Year	Model		Brake Disc Original Thickness	Brake Disc Minimum Thickness	Brake Disc Maximum Runout	Minimum Pad Thickness	Brake Caliper Bracket Bolts (ft. lbs.)	Brake Caliper Mounting Bolts (ft. lbs.)
2009	QX56	F	1.181	1.102	①	0.039	②	②
		R	0.551	0.472	①	0.039	③	③
2010	QX56	F	1.181	1.102	①	0.039	②	②
		R	0.551	0.472	①	0.039	③	③

① Maximum uneven wear measured at 8 positions: 0.0006. Runout limit, attached to vehicle: Front: 0.001, Rear: 0.002.

② Torque member mounting bolt: 110

    Sliding pin bolt 53

③ Torque member mounting bolt: 76

37663_QX56_C0012

SCHEDULED MAINTENANCE INTERVALS
INFINITI QX56

TO BE SERVICED	SERVICE	VEHICLE MILEAGE INTERVAL (x1000)												
		3.75	7.5	15	22.5	30	37.5	45	52.5	60	67.5	75	82.5	90
Engine oil & filter	R	✓	✓	✓	✓	✓	✓	✓	✓	✓	✓	✓	✓	✓
Brake lines & cables	I			✓		✓		✓		✓		✓		✓
Brake pads& rotors	L/I			✓		✓		✓		✓		✓		✓
Driveshaft boots & propeller shaft (4x4)	I					✓				✓				
Automatic transmission, final drive oil & transfer case	I			✓		✓		✓		✓				
LSD gear oil	I			✓		✓		✓		✓		✓		✓
Front wheel bearing grease (4x4)	R					✓				✓				
Air cleaner filter	R					✓				✓				✓
Engine coolant	R									✓				✓
Exhaust system	I						✓	✓	✓	✓				
Spark plugs	R	Replace every 105,000 miles												
Drive belt(s)	I			✓		✓		✓		✓		✓		✓
Cabin air filter	R							✓						✓
Exhaust system	I		✓			✓				✓				✓
Fuel lines	I		✓			✓				✓				
Steering gear (box) & linkage, axle & suspension parts	I					✓				✓				✓
Transfer case	I					✓				✓				✓
Tire rotation			✓	✓	✓	✓	✓	✓	✓	✓	✓	✓	✓	✓
Vapor lines	S/I					✓				✓				✓

R: Replace   S/I: Service or Inspect   L: Lubricate

FREQUENT OPERATION MAINTENANCE (SEVERE SERVICE)

If a vehicle is operated under any of the following conditions it is considered severe service:

- Extremely dusty areas.

- 50% or more of the vehicle operation is in 32°C (90°F) or higher temperatures, constant operation in temp. below 0°C (32°F).

- Prolonged idling (vehicle operation in stop and go traffic).

- Frequent short running periods (engine does not warm to normal operating temperatures).

- Police, taxi, delivery usage or trailer towing usage.

Oil & oil filter: replace every 3750 miles.

Brake pads, discs, drums & linings: service or inspect every 7500 miles.

Driveshaft boots & propeller shaft: service or inspect every 7500 miles.

Exhaust system: service or inspect every 7500 miles.

Final drive oil: Change every 30000 miles if towing a trailer.

Transfer case fluid: Change every 30000 miles if towing a trailer.

Steering gear (box) & linkage, (steering damper-4x4), axle & suspension parts: service or inspect every 7500 miles.

Steering linkage ball joints & front suspension ball joints: service or inspect every 7500 miles.

37663_QX56_C0013

BRAKES — INFORMATION AND PRECAUTIONS

ANTI-LOCK SYSTEMS

• Certain components within the ABS system are not intended to be serviced or repaired individually.

• Do not use rubber hoses or other parts not specifically specified for and ABS system. When using repair kits, replace all parts included in the kit. Partial or incorrect repair may lead to functional problems and require the replacement of components.

• Lubricate rubber parts with clean, fresh brake fluid to ease assembly. Do not use shop air to clean parts; damage to rubber components may result.

• Use only DOT 3 brake fluid from an unopened container.

• If any hydraulic component or line is removed or replaced, it may be necessary to bleed the entire system.

• A clean repair area is essential. Always clean the reservoir and cap thoroughly before removing the cap. The slightest amount of dirt in the fluid may plug an ori-fice and impair the system function. Perform repairs after components have been thoroughly cleaned; use only denatured alcohol to clean components. Do not allow ABS components to come into contact with any substance containing mineral oil; this includes used shop rags.

• The Anti-Lock control unit is a microprocessor similar to other computer units in the vehicle. Ensure that the ignition switch is **OFF** before removing or installing controller harnesses. Avoid static electricity discharge at or near the controller.

• If any arc welding is to be done on the vehicle, the control unit should be unplugged before welding operations begin.

DISC AND DRUM SYSTEMS

✳✳ CAUTION

Dust and dirt accumulating on brake parts during normal use may contain asbestos fibers from production or aftermarket brake linings. Breathing excessive concentrations of asbestos fibers can cause serious bodily harm. Exercise care when servicing brake parts. Do not sand or grind brake lining unless equipment used is designed to contain the dust residue. Do not clean brake parts with compressed air or by dry brushing. Cleaning should be done by dampening the brake components with a fine mist of water, then wiping the brake components clean with a dampened cloth. Dispose of cloth and all residue containing asbestos fibers in an impermeable container with the appropriate label. Follow practices prescribed by the Occupational Safety and Health Administration (OSHA) and the Environmental Protection Agency (EPA) for the handling, processing, and disposing of dust or debris that may contain asbestos fibers.

BRAKES — BLEEDING THE BRAKE SYSTEM

BLEEDING PROCEDURE

BLEEDING PROCEDURE

➡**Be sure that the master cylinder is full of clean fresh brake fluid before starting the bleeding process. Use only the recommended brake fluid when bleeding the system. Do not allow brake fluid to spill on painted surfaces as damage will occur.**

1. Before servicing the vehicle, refer to the Precautions Section.

➡**If working near and/or around the SRS system and components, be sure to disable the SRS system. After disabling the system wait three minutes or more before servicing the vehicle.**

➡**Whenever the negative battery cable is disconnected the following components will require resetting. The Idle Air Volume Learning, Steering Angle Sensor Neutral Position, Sunroof Memory Reset/Initialization, Automatic Drive Positioner System, Audio presets and Navigation. Use the CONSULT-III diagnostic tool, or equivalent to perform the required resets.**

2. Disconnect the negative battery cable.
3. Turn the ignition switch OFF. Disconnect the ABS actuator and electric control unit connector.
4. Connect a vinyl tube to the rear right bleed valve. Be sure to have a catch pan handy to catch excess brake fluid.
5. Fully depress the brake pedal four or five times.
6. With the brake pedal depressed, loosen the bleed valve to let air out, then tighten it immediately.
7. Repeat the above steps until all air is removed from the system. Be sure to keep watch on the brake fluid level and replenish, as necessary.
8. Tighten the bleed valve.
9. Repeat the above steps at each wheel, with the master cylinder reservoir tank filled at least half way.
10. Bleed the remaining components in the following order: front left, rear left and front right.
11. Be sure to perform the reconnect/relearn procedures.

BLEEDING THE ABS SYSTEM

Refer to "Bleeding the Brake System".

WHEEL SPEED SENSORS

REMOVAL & INSTALLATION

Front Sensor

See Figure 2.

1. Before servicing the vehicle, refer to the Precautions Section.

➡**If working near and/or around the SRS system and components, be sure to disable the SRS system. After disabling the system wait three minutes or more before servicing the vehicle.**

➡**Whenever the negative battery cable is disconnected the following components will require resetting. The Idle Air Volume Learning, Steering Angle Sensor Neutral Position, Sunroof Memory Reset/Initialization, Automatic Drive Positioner System, Audio presets and Navigation. Use the CONSULT-III diagnostic tool, or equivalent to perform the required resets.**

2. Disconnect the negative battery cable.
3. Raise and support the vehicle safely.
4. Remove the tire and wheel assembly.
5. Remove the wheel speed sensor mounting screw.

➡**Remove the rotor to gain access to the wheel sensor mounting bolt.**

6. Pull the sensor out, being careful to turn it as little as possible. Do not pull on the sensor harness.
7. Disconnect the wheel speed sensor electrical connector.
8. Remove the harness from it mount.

To install:

➡**Be sure to use new fasteners, as required.**

9. Inspect the sensor O-ring, replace as required.
10. Before installing the sensor be sure no foreign materials such as iron fragments are adhered to the pick-up part of the sensor or to the inside of the sensor mounting hole or on the rotor mounting surface.
11. Apply a thin coat of a suitable grease to the wheel sensor O-ring and mounting hole.
12. Tighten the sensor retaining bolt to 73 inch lbs.
13. Continue the installation in the reverse order of the removal procedure.
14. Be sure to perform the reconnect/relearn procedures.

Rear Sensor

See Figure 2.

1. Before servicing the vehicle, refer to the Precautions Section.

➡**If working near and/or around the SRS system and components, be sure to disable the SRS system. After disabling the system wait three minutes or more before servicing the vehicle.**

➡**Whenever the negative battery cable is disconnected the following components will require resetting. The Idle Air Volume Learning, Steering Angle Sensor Neutral Position, Sunroof Memory Reset/Initialization, Automatic Drive Positioner System, Audio presets and Navigation. Use the CONSULT-III diagnostic tool, or equivalent to perform the required resets.**

2. Disconnect the negative battery cable.
3. Raise and support the vehicle safely.
4. Remove the tire and wheel assembly.
5. Remove the wheel speed sensor mounting screw.

➡**Remove the rear hub and bearing assembly to gain access to the wheel sensor mounting bolt.**

6. Pull the sensor out, being careful to turn it as little as possible. Do not pull on the sensor harness.

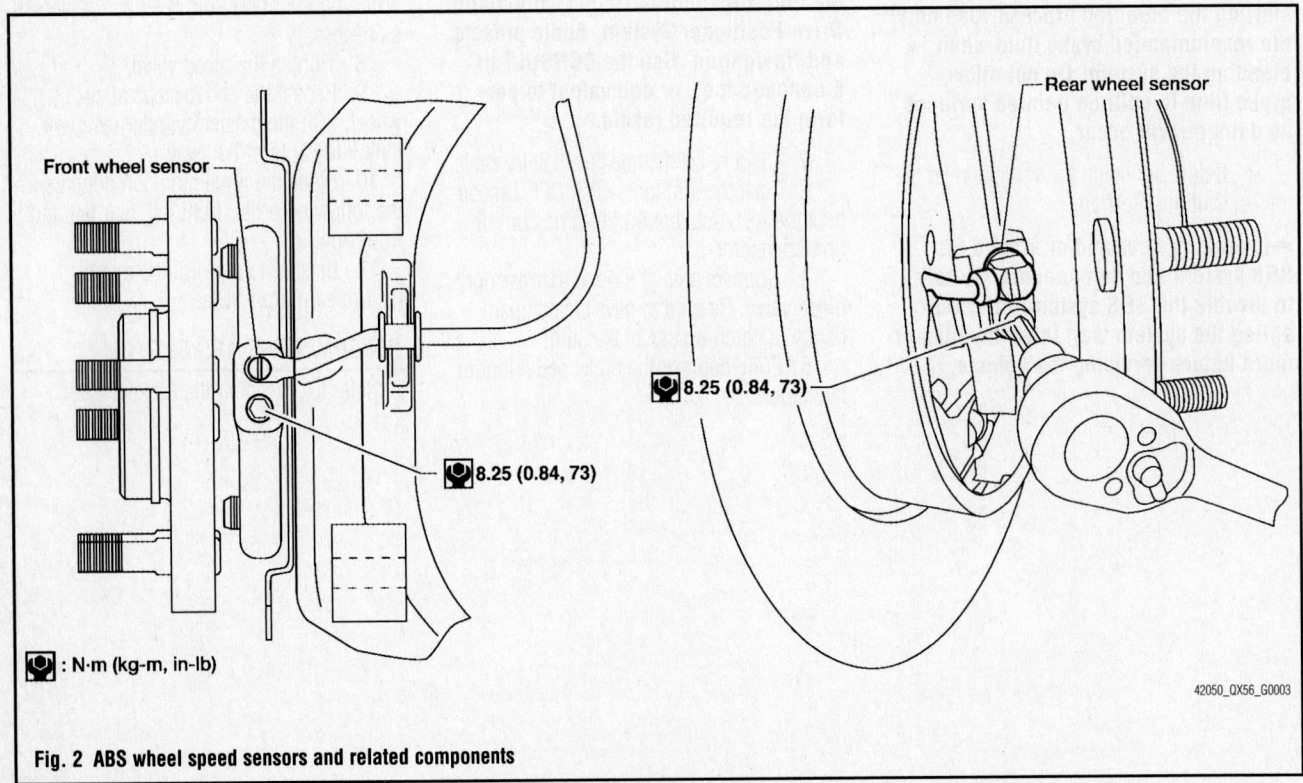

Front wheel sensor

Rear wheel sensor

8.25 (0.84, 73)

8.25 (0.84, 73)

: N·m (kg-m, in-lb)

42050_QX56_G0003

Fig. 2 ABS wheel speed sensors and related components

7. Disconnect the wheel speed sensor electrical connector.

8. Remove the harness from it mount.

To install:

➡**Be sure to use new fasteners, as required.**

9. Inspect the sensor O-ring, replace as required.

10. Before installing the sensor be sure no foreign materials such as iron fragments are adhered to the pick-up part of the sensor or to the inside of the sensor mounting hole or on the rotor mounting surface.

11. Apply a thin coat of a suitable grease to the wheel sensor O-ring and mounting hole.

12. Tighten the sensor retaining bolt to 73 inch lbs.

13. Continue the installation in the reverse order of the removal procedure.

14. Be sure to perform the reconnect/relearn procedures.

BRAKES   FRONT DISC BRAKES

BRAKE CALIPER

REMOVAL & INSTALLATION

See Figure 3.

1. Before servicing the vehicle, refer to the Precautions Section.

➡**If working near and/or around the SRS system and components, be sure to disable the SRS system. After disabling the system wait three minutes or more before servicing the vehicle.**

➡**Whenever the negative battery cable is disconnected the following components will require resetting. The Idle Air Volume Learning, Steering Angle Sensor Neutral Position, Sunroof Memory Reset/Initialization, Automatic Drive Positioner System, Audio presets and Navigation. Use the CONSULT-III diagnostic tool, or equivalent to perform the required resets.**

2. Disconnect the negative battery cable.
3. Drain brake fluid as necessary.
4. Raise and safely support the vehicle.
5. Remove or disconnect the following:
   - Wheel and tire assembly
   - Union bolt
   - Disconnect and plug brake hose. Discard the washer.
   - Caliper-to-torque member slide pins, or remove the caliper and torque member as an assembly
   - Brake caliper

To install:

➡**Be sure to use new fasteners, as required.**

6. Install or connect the following:
   - Brake caliper, tighten torque member bolts to specification and the caliper slide pin to specification
   - Union bolt and tighten to 13 ft. lbs. (18 Nm)

7. Fill the master cylinder and bleed the brake system.

8. Install the wheel and tire assemblies.

9. Be sure to perform the reconnect/relearn procedures.

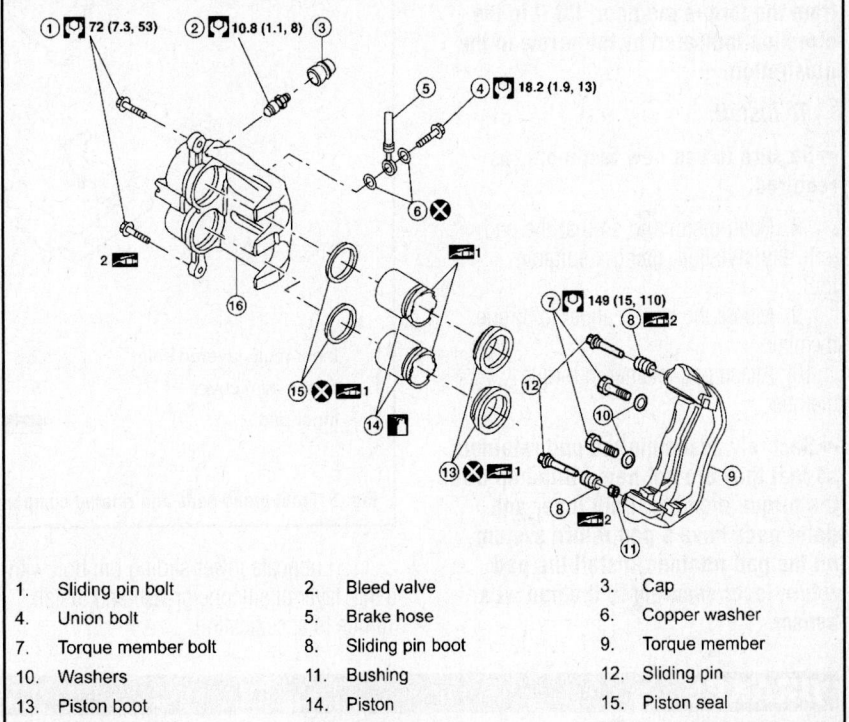

1.	Sliding pin bolt	2.	Bleed valve	3.	Cap
4.	Union bolt	5.	Brake hose	6.	Copper washer
7.	Torque member bolt	8.	Sliding pin boot	9.	Torque member
10.	Washers	11.	Bushing	12.	Sliding pin
13.	Piston boot	14.	Piston	15.	Piston seal
16.	Cylinder body		Brake fluid		1: Molykote M-77 grease

37663_QX56_G0018

Fig. 3 Front brake caliper and related components

DISC BRAKE PADS

REMOVAL & INSTALLATION

See Figures 4 and 5.

1. Before servicing the vehicle, refer to the Precautions Section.

➡**If working near and/or around the SRS system and components, be sure to disable the SRS system. After disabling the system wait three minutes or more before servicing the vehicle.**

➡**Whenever the negative battery cable is disconnected the following components will require resetting. The Idle Air Volume Learning, Steering Angle Sensor Neutral Position, Sunroof Memory Reset/Initialization, Automatic**

Drive Positioner System, Audio presets and Navigation. Use the CONSULT-III diagnostic tool, or equivalent to perform the required resets.

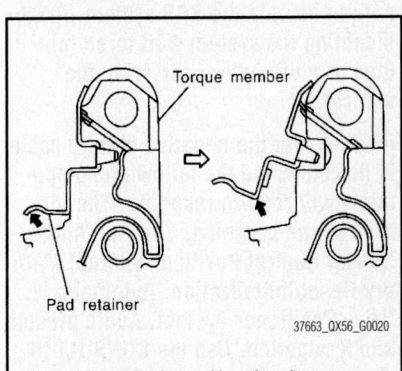

37663_QX56_G0020

Fig. 4 Front brake pad/retainer/torque member removal

2. Disconnect the negative battery cable.

3. Drain brake fluid as necessary.

4. Raise and safely support the vehicle.

5. Remove the wheel and tire assembly.

6. Remove lower sliding pin bolt.

7. Suspend brake caliper with a remove and remove brake pad and shim from torque member.

➡When removing the pad retainer from the torque member, lift it in the direction indicated by the arrow in the illustration.

To install:

➡Be sure to use new fasteners, as required.

8. Push pistons in so that the pad is firmly installed, using a suitable tool.

9. Mount the brake caliper to torque member.

10. Attach pad retainer to torque member.

➡Securely assemble the pad retainers so that they are not being lifted up from the torque member. Both inner and outer pads have a pad return system on the pad retainer. Install the pad return lever securely to the pad wear sensor.

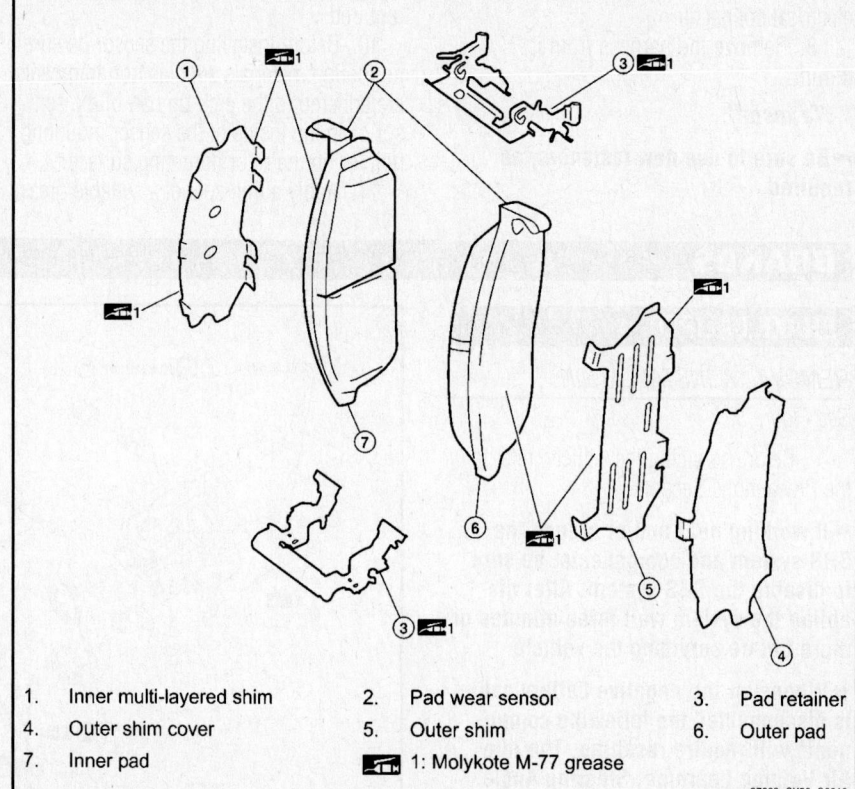

1.	Inner multi-layered shim	2.	Pad wear sensor	3.	Pad retainer
4.	Outer shim cover	5.	Outer shim	6.	Outer pad
7.	Inner pad		1: Molykote M-77 grease		

37663_QX56_G0019

Fig. 5 Front brake pads and related components

11. Lubricate lower sliding pin bolt with a thin layer of silicone grease and install. Torque to specification.

12. Install the wheel and tire assembly.

13. Be sure to perform the reconnect/relearn procedures.

BRAKES

BRAKE CALIPER

REMOVAL & INSTALLATION

See Figure 6.

1. Before servicing the vehicle, refer to the Precautions Section.

➡If working near and/or around the SRS system and components, be sure to disable the SRS system. After disabling the system wait three minutes or more before servicing the vehicle.

➡Whenever the negative battery cable is disconnected the following components will require resetting. The Idle Air Volume Learning, Steering Angle Sensor Neutral Position, Sunroof Memory Reset/Initialization, Automatic Drive Positioner System, Audio presets and Navigation. Use the CONSULT-III diagnostic tool, or equivalent to perform the required resets.

2. Disconnect the negative battery cable.

3. Drain brake fluid as necessary.

4. Raise and safely support the vehicle.

5. Drain brake fluid as necessary.

6. Remove or disconnect the following:

- Wheel and tire assembly
- Union bolt
- Mounting bolts
- Brake caliper assembly. Discard the copper washers.

To install:

➡Be sure to use new fasteners, as required.

7. Install or connect the following:
- Brake caliper assembly and tighten mounting bolts to specification
- Union bolt and tighten to 13 ft. lbs. (18 Nm)

8. Fill the master cylinder and bleed the brake system.

9. Install the wheel and tire assemblies.

10. Be sure to perform the reconnect/relearn procedures.

REAR DISC BRAKES

DISC BRAKE PADS

REMOVAL & INSTALLATION

See Figure 7.

1. Before servicing the vehicle, refer to the Precautions Section.

➡If working near and/or around the SRS system and components, be sure to disable the SRS system. After disabling the system wait three minutes or more before servicing the vehicle.

➡Whenever the negative battery cable is disconnected the following components will require resetting. The Idle Air Volume Learning, Steering Angle Sensor Neutral Position, Sunroof Memory Reset/Initialization, Automatic Drive Positioner System, Audio presets

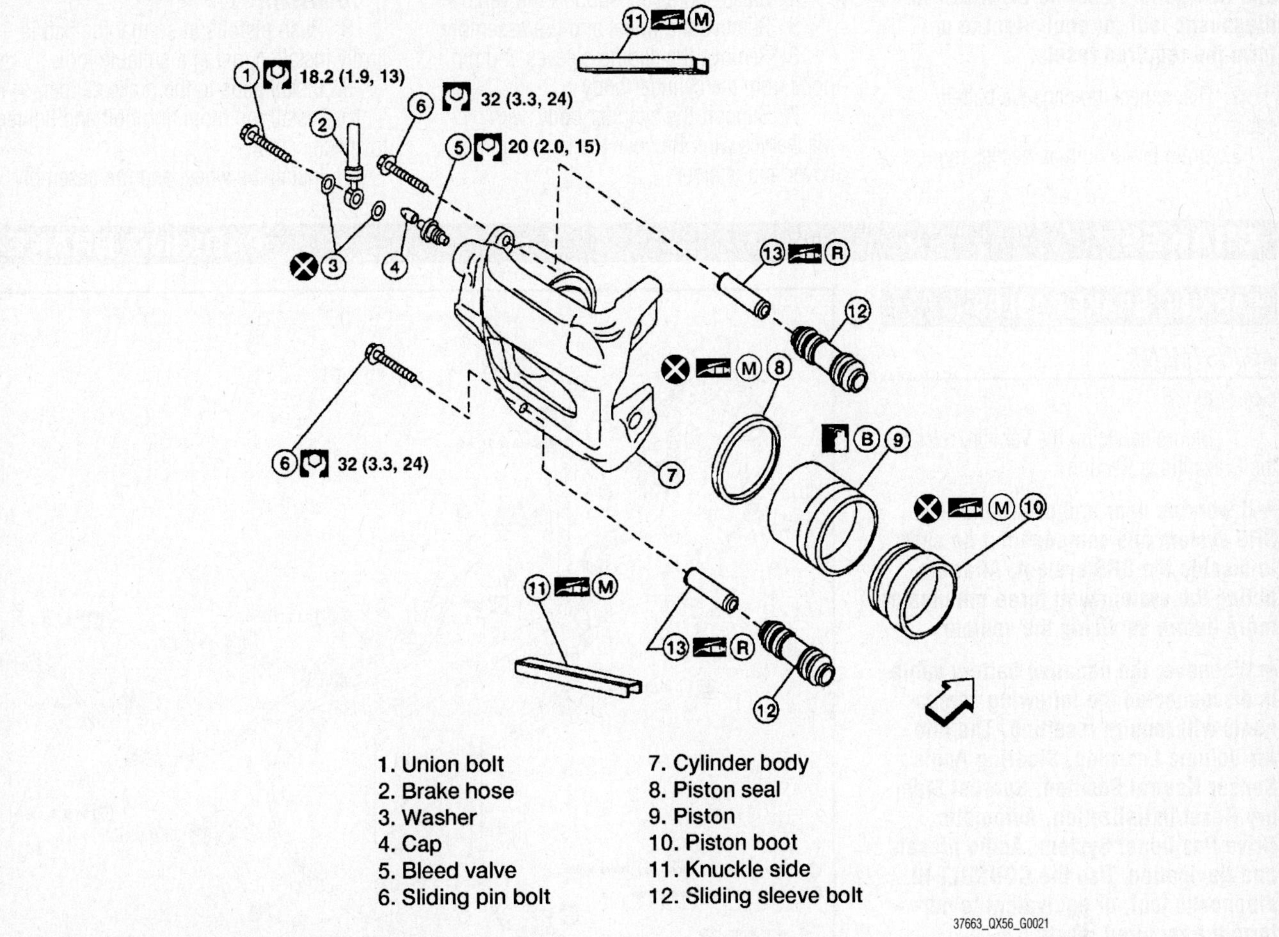

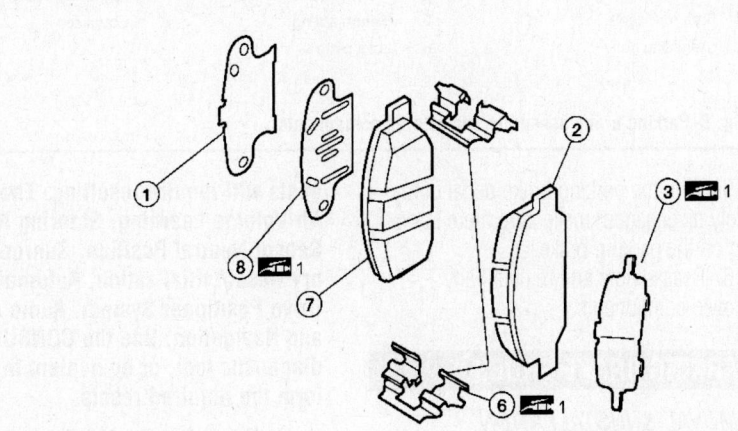

18.2 (1.9, 13)
32 (3.3, 24)
20 (2.0, 15)
32 (3.3, 24)
1. Union bolt
2. Brake hose
3. Washer
4. Cap
5. Bleed valve
6. Sliding pin bolt
7. Cylinder body
8. Piston seal
9. Piston
10. Piston boot
11. Knuckle side
12. Sliding sleeve bolt
37663_QX56_G0021

Fig. 6 Rear brake caliper and related components

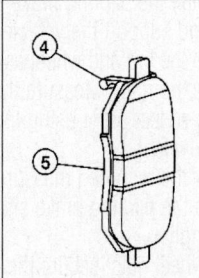

1. Inner shim cover
2. Outer pad
3. Outer multi-layered shim
4. Pad wear sensor
5. Inner pad (RH)
6. Pad retainer
7. Inner pad
8. Inner shim
1: Molykote M-77 grease
37663_QX56_G0022

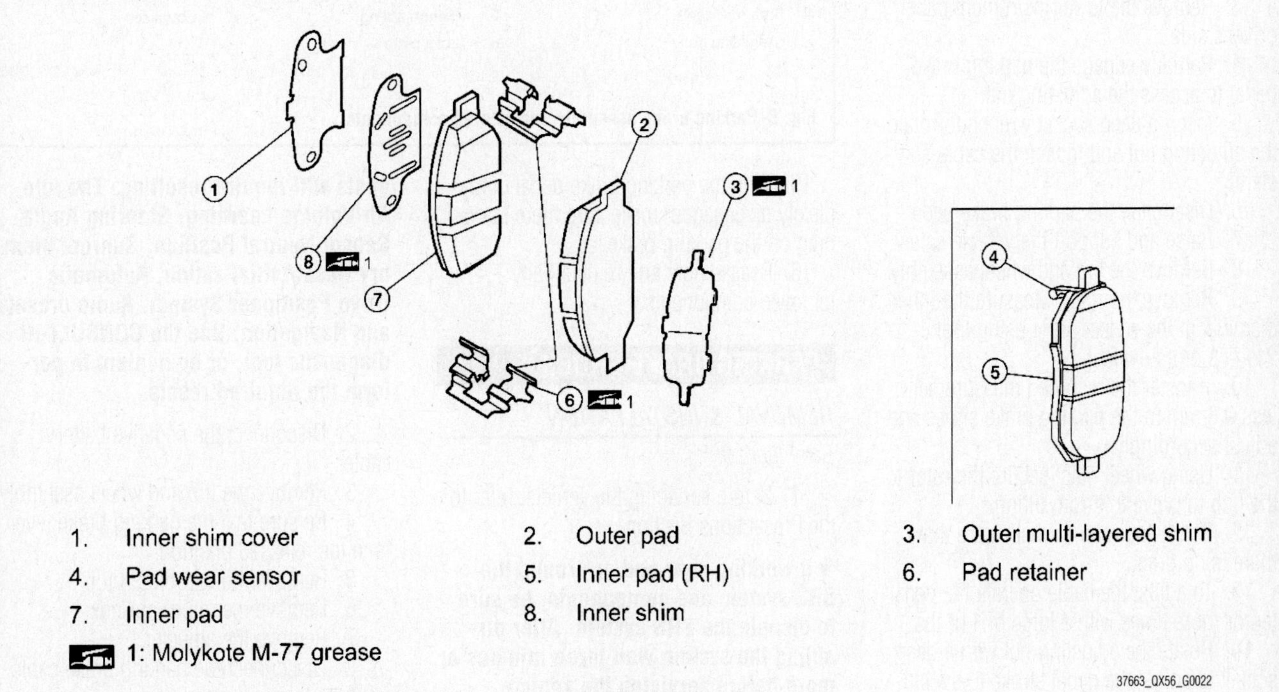

Fig. 7 Rear brake pads and related components

and Navigation. Use the CONSULT-III diagnostic tool, or equivalent to perform the required resets.

2. Disconnect the negative battery cable.
3. Drain brake fluid as necessary.
4. Raise and safely support the vehicle.
5. Remove the wheel and tire assembly.
6. Remove the sliding sleeves and pin bolts from the cylinder body
7. Support the cylinder body, with mechanics wire. Remove the pads, shims, cover and retainer.

To install:

8. Push pistons in so that the pad is firmly installed, using a suitable tool.
9. Install pads to the brake caliper.
10. Install top mounting bolt and tighten to specification.
11. Install the wheel and tire assembly.

BRAKES — PARKING BRAKE

PARKING BRAKE CABLES

ADJUSTMENT

See Figure 8.

1. Before servicing the vehicle, refer to the Precautions Section.

➡**If working near and/or around the SRS system and components, be sure to disable the SRS system. After disabling the system wait three minutes or more before servicing the vehicle.**

➡**Whenever the negative battery cable is disconnected the following components will require resetting. The Idle Air Volume Learning, Steering Angle Sensor Neutral Position, Sunroof Memory Reset/Initialization, Automatic Drive Positioner System, Audio presets and Navigation. Use the CONSULT-III diagnostic tool, or equivalent to perform the required resets.**

2. Disconnect the negative battery cable.
3. Remove the lower instrument panel, driver's side.
4. Partially engage the parking brake pedal to access the adjusting nut.
5. Insert a deep socket wrench to rotate the adjusting nut and loosen the cable sufficiently.
6. Disengage the parking brake pedal.
7. Raise and support the vehicle safely.
8. Remove the tire and wheel assembly.
9. Remove the rotor. Measure the inner diameter at the widest point using tool J-21177A or equivalent.
10. Transfer the recorded measurement less 0.6 mm to the parking brake shoes and adjust accordingly.
11. Using wheel nuts, secure the rotor to the hub to prevent it from tilting.
12. Rotate the rotor to make sure that there is no drag.
13. To adjust the cable operate the pedal ten or more times with a force of 110 lbs.
14. Rotate the adjusting nut with a deep socket to adjust the pedal stroke to specification. Specification is 3–4 notches with a force of 44.1 lbs.

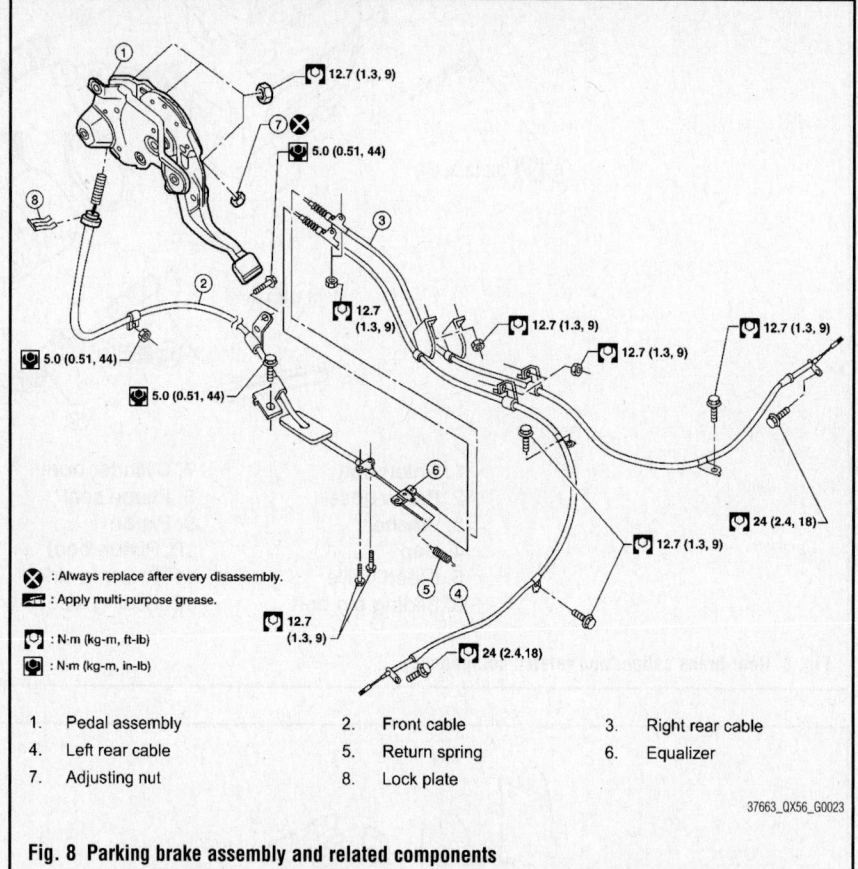

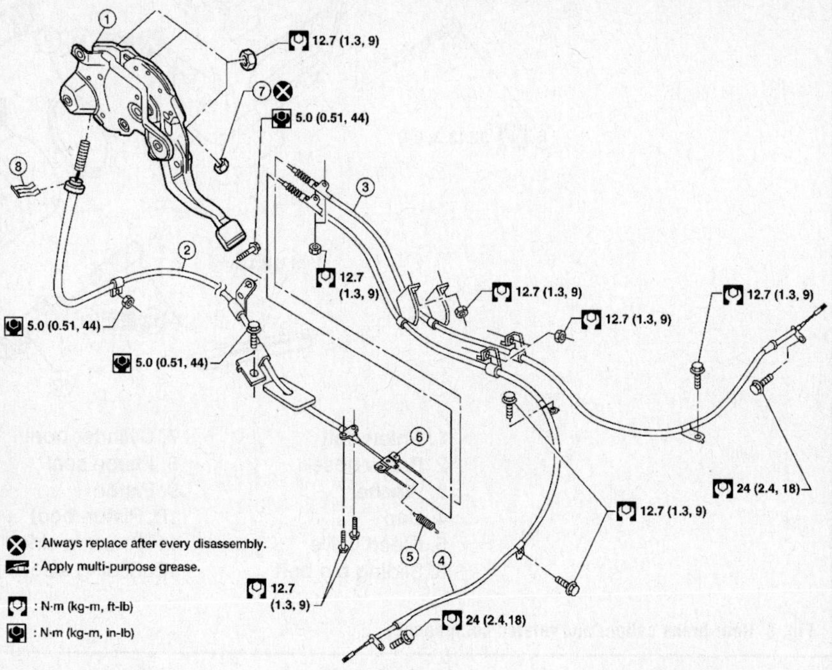
12.7 (1.3, 9)
5.0 (0.51, 44)
12.7 (1.3, 9)
5.0 (0.51, 44)
5.0 (0.51, 44)
12.7 (1.3, 9)
12.7 (1.3, 9)
12.7 (1.3, 9)
24 (2.4, 18)
12.7 (1.3, 9)
12.7 (1.3, 9)
24 (2.4,18)
: Always replace after every disassembly.
: Apply multi-purpose grease.
: N·m (kg-m, ft-lb)
: N·m (kg-m, in-lb)

1.	Pedal assembly	2.	Front cable	3.	Right rear cable
4.	Left rear cable	5.	Return spring	6.	Equalizer
7.	Adjusting nut	8.	Lock plate		

37663_QX56_G0023

Fig. 8 Parking brake assembly and related components

15. With the parking brake pedal completely disengaged, make sure there is no drag on the parking brake.
16. Reassemble and reinstall any removed components.

PARKING BRAKE SHOES

REMOVAL & INSTALLATION

See Figure 9.

1. Before servicing the vehicle, refer to the Precautions Section.

➡**If working near and/or around the SRS system and components, be sure to disable the SRS system. After disabling the system wait three minutes or more before servicing the vehicle.**

➡**Whenever the negative battery cable is disconnected the following components will require resetting. The Idle Air Volume Learning, Steering Angle Sensor Neutral Position, Sunroof Memory Reset/Initialization, Automatic Drive Positioner System, Audio presets and Navigation. Use the CONSULT-III diagnostic tool, or equivalent to perform the required resets.**

2. Disconnect the negative battery cable.
3. Remove the tire and wheel assembly.
4. Be sure that the parking brake lever is in the released position.
5. Remove the rear disc rotor.
6. Remove the return springs.
7. Remove the adjuster.
8. Disconnect the parking brake cable from the toggle lever.
9. Remove the retainers.
10. Remove the anti rattle pins and shoes.

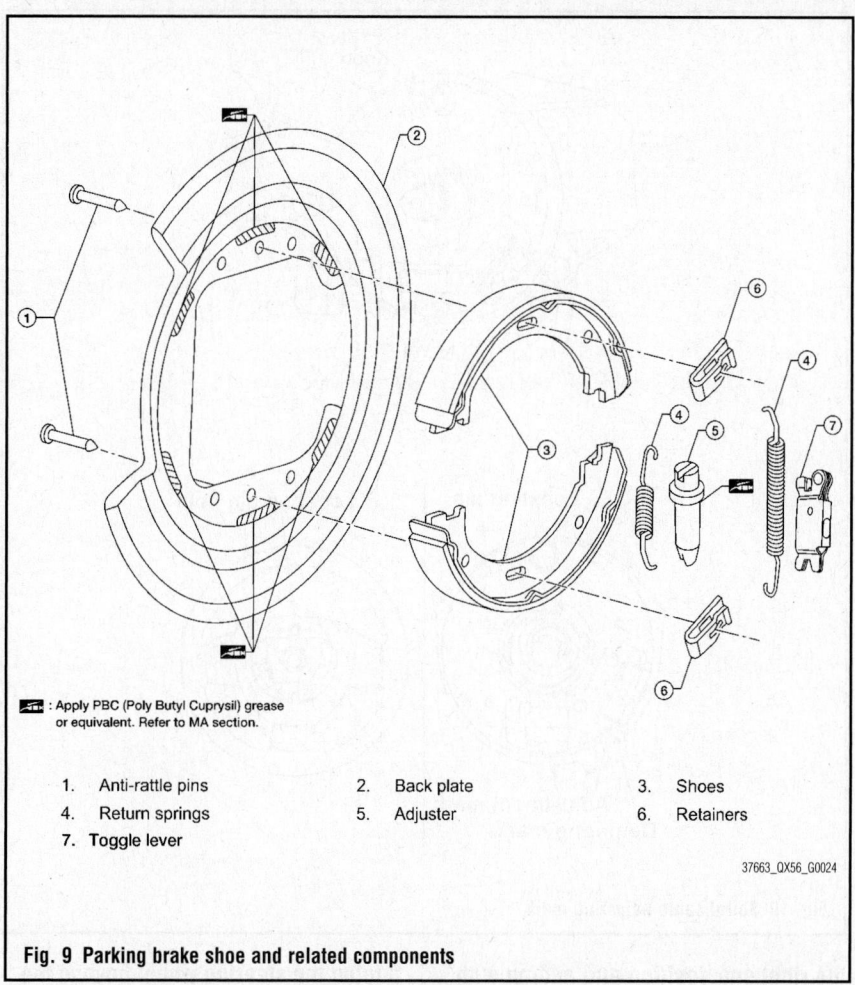

: Apply PBC (Poly Butyl Cuprysil) grease or equivalent. Refer to MA section.

1. Anti-rattle pins
2. Back plate
3. Shoes
4. Return springs
5. Adjuster
6. Retainers
7. Toggle lever

37663_QX56_G0024

Fig. 9 Parking brake shoe and related components

To install:

→Be sure to use new fasteners, as required.

11. Apply brake grease to the specified points during reassembly, see illustration for locating points.

12. Install the adjuster so that the threaded part expands when rotating it in the proper direction.

13. Continue the installation in the reverse order of the removal procedure.

14. Adjust the parking brake.

15. Perform the parking brake burnishing operation.

16. Be sure to perform the reconnect/relearn procedures.

ADJUSTMENT

Perform the parking brake burnishing operation by driving the vehicle forward under the following conditions: vehicle speed 25 mph forward direction, parking brake operating force 44.1 lbs set and apply time of 30 seconds. After parking brake burnishing operation, recheck parking brake adjustment, correct as required.

→To prevent the brake lining from getting to hot, allow a cool off period of five minutes between operations. Do not perform excessive break-in operations, because it may cause uneven or early wear of the lining.

CHASSIS ELECTRICAL · AIR BAG (SUPPLEMENTAL RESTRAINT SYSTEM)

GENERAL INFORMATION

❋❋ CAUTION

These vehicles are equipped with an air bag system. The system must be disarmed before performing service on, or around, system components, the steering column, instrument panel components, wiring and sensors. Failure to follow the safety precautions and the disarming procedure could result in accidental air bag deployment, possible injury and unnecessary system repairs.

PRECAUTIONS

Disconnect and isolate the battery negative cable before beginning any airbag system component diagnosis, testing, removal, or installation procedures. Allow system capacitor to discharge for two minutes before beginning any component service. This will disable the airbag system. Failure to disable the airbag system may result in accidental airbag deployment, personal injury, or death.

DISARMING THE SYSTEM

1. Before servicing the vehicle, refer to the Precautions Section.

2. Disconnect both battery cables.

3. Wait at least 3 minutes before working on the vehicle. The air bag system is designed to retain enough power to deploy the air bag for a short time after the battery has been disconnected.

ARMING THE SYSTEM

→Once repair work has been completed, return the ignition switch to the LOCK position, before connecting the battery cables. At this time the steering lock mechanism will engage. Install the CONSULT-III diagnostic tool, or equivalent, and follow the directions on the screen of the tool and perform self diagnosis check.

CLOCKSPRING CENTERING

See Figure 10.

→Before servicing, or working around, the SRS system, turn the ignition switch OFF, disconnect both battery cables and wait at least three minutes. When servicing, or working around, the SRS system do not work directly in front of the air bag module.

1. Before servicing the vehicle, refer to the Precautions Section.

→If working near and/or around the SRS system and components, be sure to disable the SRS system. After disabling the system wait three minutes or more before servicing the vehicle.

→Whenever the negative battery cable is disconnected the following components will require resetting. The Idle Air Volume Learning, Steering Angle Sensor Neutral Position, Sunroof Memory Reset/Initialization, Automatic Drive Positioner System,

Audio presets and Navigation. Use the CONSULT-III diagnostic tool, or equivalent to perform the required resets.

2. Disconnect the negative battery cable. Disconnect the positive battery cable.

3. Position the front wheels in the straight ahead position.

4. Remove the air bag module.

5. Remove the steering wheel.

6. Remove the upper and lower steering column covers.

7. Remove the wiper washer switch connector. Pinch the tabs at the wiper and washer switch base and slide the switch away from the steering column to remove it.

8. While pressing the tabs, pull the headlight and turn signal switch toward the driver's door and disconnect it from the base.

9. Remove the spiral cable retaining screws, release the clip and remove the spiral cable.

➡ **Do not disassemble the spiral cable. Do not apply lubricant to the spiral cable.**

10. Remove the spiral cable connectors.

➡ **With the steering linkage disconnected, the spiral cable may snap by turning the steering wheel beyond the limited number of turns. The spiral cable can be turned counterclockwise about 2.5 turns from the right end position.**

To install:

➡ **Be sure to use new fasteners, as required.**

11. Installation is the reverse of the removal procedure.

12. Be sure to align the spiral cable correctly when installing the steering wheel. Make sure that the spiral cable is in the neutral position.

❊❊ WARNING

The neutral position is detected by turning to the left 2.5 revolutions from

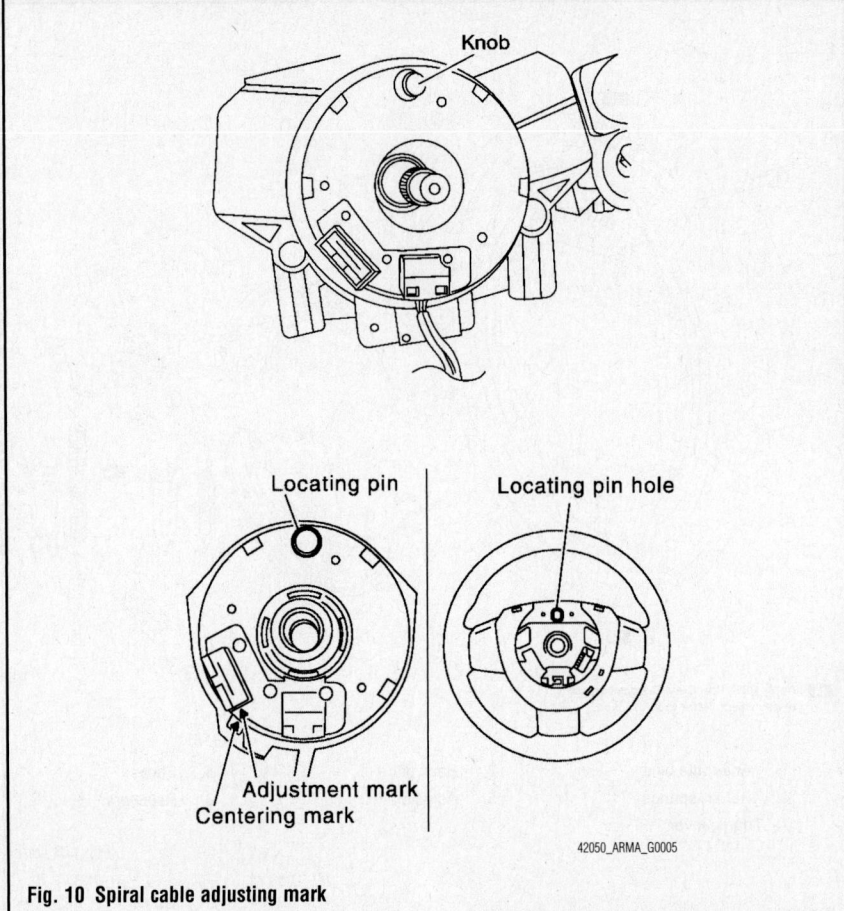

Fig. 10 Spiral cable adjusting mark

the right end position and ending with the knob at the top. The spiral cable may snap due to steering operation if the cable is installed incorrectly. Also, with the steering linkage disconnected the cable may snap by turning the steering wheel beyond the limited number of turns (2.5 from the neutral position to both the left and right).

13. If equipped with VDC adjust the steering angle sensor.

14. Use the CONSULT-III® tool and perform self diagnosis to ensure no malfunction is detected.

➡ **With the steering linkage disconnected, the spiral cable may snap by**

turning the steering wheel beyond the limited number of turns. The spiral cable can be turned counterclockwise about 2.5 turns from the right end position.

15. Tighten the steering wheel retaining nut to 25 ft. lbs. (33.9 Nm).

16. When reinstalling the air bag module, be sure to use new bolts. Tighten the bolts to 8 ft. lbs. (10.8 Nm).

17. Adjust the steering angle sensor neutral position, using the CONSULT-III diagnostic tool, or equivalent.

18. Be sure to perform the reconnect/relearn procedures.

DRIVE TRAIN

FRONT HALFSHAFT

REMOVAL & INSTALLATION

See Figure 11.

1. Before servicing the vehicle, refer to the Precautions Section.

➡**If working near and/or around the SRS system and components, be sure to disable the SRS system. After disabling the system wait three minutes or more before servicing the vehicle.**

➡**Whenever the negative battery cable is disconnected the following components will require resetting. The Idle Air Volume Learning, Steering Angle Sensor Neutral Position, Sunroof Memory Reset/Initialization, Automatic Drive Positioner System, Audio presets and Navigation. Use the CONSULT-III diagnostic tool, or equivalent to perform the required resets.**

2. Disconnect the negative battery cable.

3. Remove or disconnect the following:
   • Wheel and tire assembly
   • Engine splash guard
   • Wheel speed sensor harness from mount on knuckle, and reposition
   • Brake caliper, do not disconnect brake hose or allow caliper to hang by hose
   • Coil spring and shock absorber
   • Cotter pin and halfshaft nut
   • Halfshaft from front differential
   • Halfshaft from hub and bearing assembly

To install:

➡**Be sure to use new fasteners, as required.**

4. Install or connect the following:
   • Halfshaft into hub
   • Halfshaft into front differential
   • Halfshaft nut and tighten to 101 ft. lbs. (137 Nm) and replace cotter pin
   • Wheel speed sensor harness
   • Brake caliper
   • Coil spring and shock absorber
   • Engine splash guard
   • Wheel and tire assembly

5. Be sure to perform the reconnect/relearn procedures.

PINION SEAL

REMOVAL & INSTALLATION

1. Before servicing the vehicle, refer to the Precautions Section.

➡**If working near and/or around the SRS system and components, be sure to disable the SRS system. After disabling the system wait three minutes or more before servicing the vehicle.**

➡**Whenever the negative battery cable is disconnected the following components will require resetting. The Idle Air Volume Learning, Steering Angle Sensor Neutral Position, Sunroof Memory Reset/Initialization, Automatic Drive Positioner System, Audio presets and Navigation. Use the CONSULT-III diagnostic tool, or equivalent to perform the required resets.**

2. Disconnect the negative battery cable.

3. Remove or disconnect the following:
   • Front driveshaft
   • Halfshafts

4. Measure and record the pinion bearing preload using special tool J-25765-A.

5. Loosen the pinion nut while holding the companion flange using special tool J-44195.

6. Remove the companion flange using a suitable tool.

7. Using a punch or drill, place a small hole in the case.

8. Remove the seal using special tool SP8P or equivalent.

To install:

➡**Be sure to use new fasteners, as required.**

9. Press front seal into carrier using a suitable tool.

10. Install companion flange and new pinion nut. Tighten pinion nut until there is no end play and until recorded pinion bearing preload is met plus an additional 5 inch lbs. (0.5 Nm).

11. Install or connect the following:
    • Halfshafts
    • Front driveshaft

12. Be sure to perform the reconnect/relearn procedures.

REAR AXLE HOUSING

REMOVAL & INSTALLATION

See Figure 12.

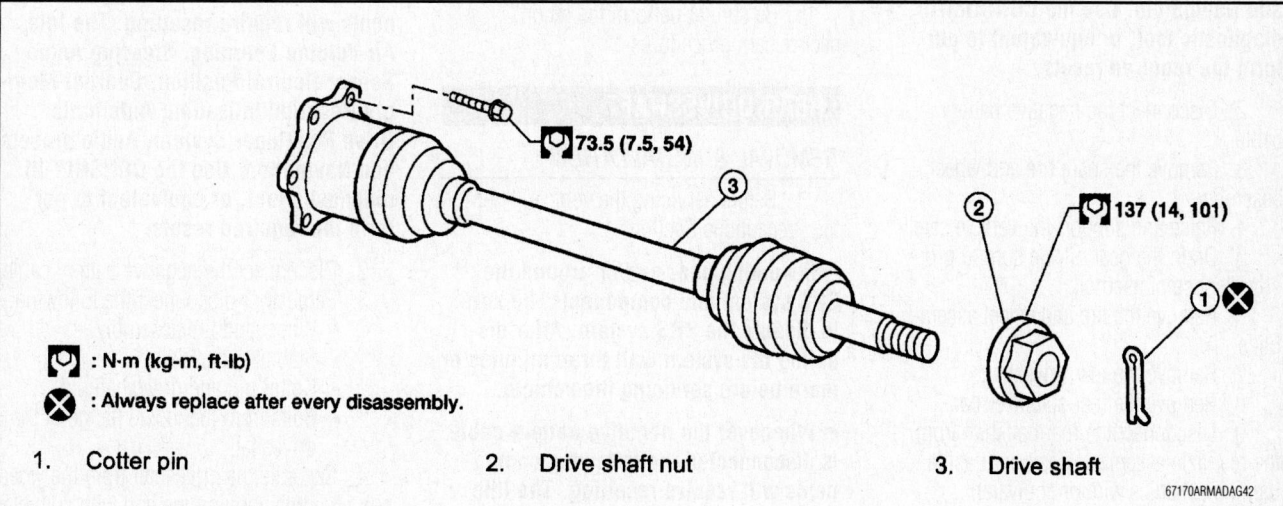

🔧 73.5 (7.5, 54)

🔧 137 (14, 101)

🔧 : N·m (kg-m, ft-lb)

❎ : Always replace after every disassembly.

| 1. | Cotter pin | 2. | Drive shaft nut | 3. | Drive shaft |

67170ARMADAG42

Fig. 11 Front halfshaft and related components

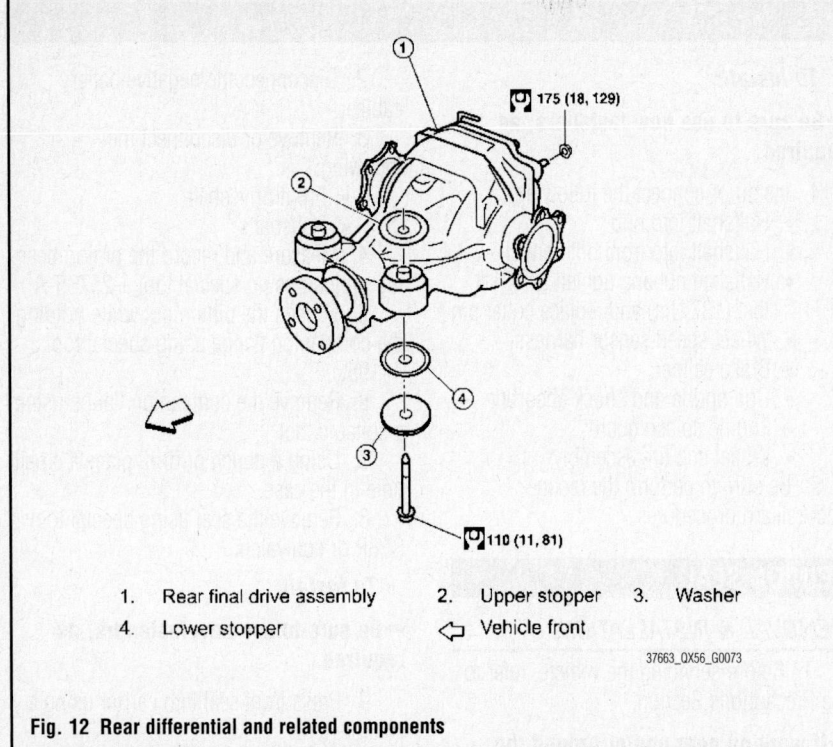

1. Rear final drive assembly   2. Upper stopper   3. Washer
4. Lower stopper   ⇦ Vehicle front

37663_QX56_G0073

Fig. 12 Rear differential and related components

1. Before servicing the vehicle, refer to the Precautions Section.

➡ If working near and/or around the SRS system and components, be sure to disable the SRS system. After disabling the system wait three minutes or more before servicing the vehicle.

➡ Whenever the negative battery cable is disconnected the following components will require resetting. The Idle Air Volume Learning, Steering Angle Sensor Neutral Position, Sunroof Memory Reset/Initialization, Automatic Drive Positioner System, Audio presets and Navigation. Use the CONSULT-III diagnostic tool, or equivalent to perform the required resets.

2. Disconnect the negative battery cable.
3. Remove the spare tire and wheel assembly.
4. Raise and support the vehicle safely.
5. Drain the gear oil. Be sure to properly dispose of used oil.
6. Remove the tire and wheel assemblies.
7. Remove the rear driveshaft.
8. Remove the rear stabilizer bar.
9. Disconnect the rear halfshaft from the rear axle assembly. Position it aside using mechanics wire or equivalent.
10. Disconnect the breather hose from the axle cover.

11. Position a suitable jack under the assembly.

➡ Do not position the jack under the aluminum cover.

12. Remove the rear axle assembly retaining bolts and nuts.
13. Carefully remove the assembly from the vehicle.

To install:

➡ Be sure to use new fasteners, as required.

14. Installation is the reverse of the removal procedure.
15. Be sure to perform the reconnect/relearn procedures.

REAR DRIVESHAFT

REMOVAL & INSTALLATION

1. Before servicing the vehicle, refer to the Precautions Section.

➡ If working near and/or around the SRS system and components, be sure to disable the SRS system. After disabling the system wait three minutes or more before servicing the vehicle.

➡ Whenever the negative battery cable is disconnected the following components will require resetting. The Idle Air Volume Learning, Steering Angle Sensor Neutral Position, Sunroof Mem-

ory Reset/Initialization, Automatic Drive Positioner System, Audio presets and Navigation. Use the CONSULT-III diagnostic tool, or equivalent to perform the required resets.

2. Disconnect the negative battery cable.
3. Position the selector lever in the N position.
4. Release the parking brake.
5. Raise and support the vehicle safely.
6. Matchmark the driveshaft and companion flange.
7. Remove the center support bearing bracket nuts, if equipped.
8. Remove the driveshaft. Discard the nuts.

To install:

➡ Be sure to use new fasteners, as required.

9. Installation is the reverse of the removal procedure.
10. Check for vehicle vibration, correct as required.
11. Be sure to perform the reconnect/relearn procedures.

HALFSHAFTS

REMOVAL & INSTALLATION

See Figure 13.

1. Before servicing the vehicle, refer to the Precautions Section.

➡ If working near and/or around the SRS system and components, be sure to disable the SRS system. After disabling the system wait three minutes or more before servicing the vehicle.

➡ Whenever the negative battery cable is disconnected the following components will require resetting. The Idle Air Volume Learning, Steering Angle Sensor Neutral Position, Sunroof Memory Reset/Initialization, Automatic Drive Positioner System, Audio presets and Navigation. Use the CONSULT-III diagnostic tool, or equivalent to perform the required resets.

2. Disconnect the negative battery cable.
3. Remove or disconnect the following:
   • Wheel and tire assembly
   • Stabilizer bar clamp
   • Cotter pin and driveshaft nut
   • Bolts from the inside flange of the driveshaft
4. Separate the driveshaft from the wheel hub by lightly tapping the end with suitable hammer and wood block.
5. Remove the halfshaft.

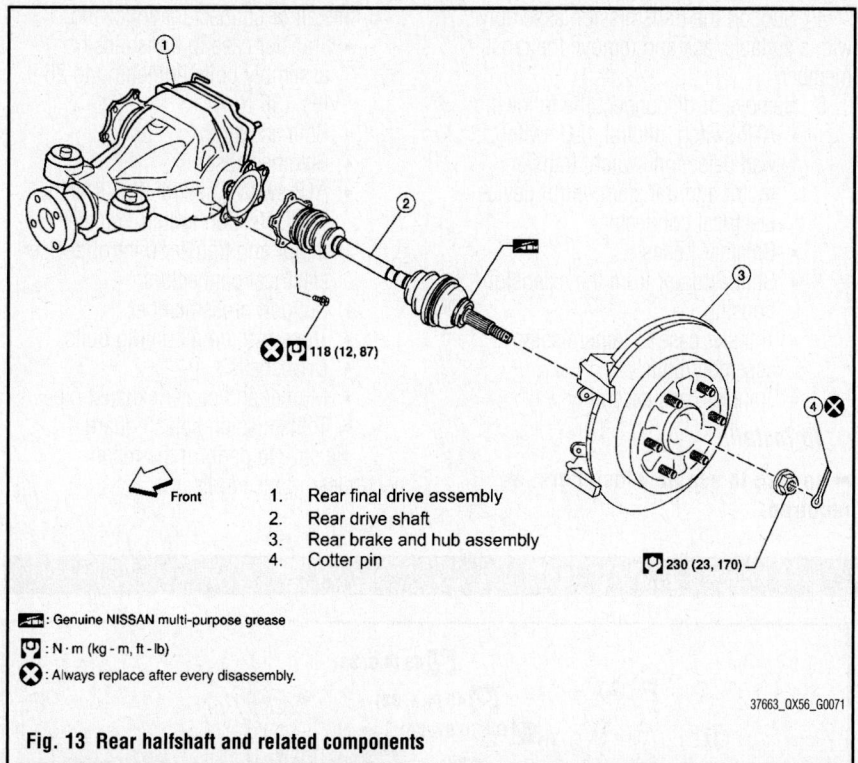

1. Rear final drive assembly
2. Rear drive shaft
3. Rear brake and hub assembly
4. Cotter pin

Front

⊗[118 (12, 87)]

230 (23, 170)

: Genuine NISSAN multi-purpose grease

: N · m (kg - m, ft - lb)

⊗ : Always replace after every disassembly.

37663_QX56_G0071

Fig. 13 Rear halfshaft and related components

✳✳ WARNING

Do not excessively extend the slide joint.

To install:

➡**Be sure to use new fasteners, as required.**

➡**Do not reuse the halfshaft inside flange bolts and cotter pin.**

6. Install or connect the following:
   • Halfshaft
   • Bolts for the inside flange and tighten to 87 ft. lbs. (118 Nm)
   • Driveshaft nut and tighten nut to 101 ft. lbs. (137 Nm) and replace cotter pin
   • Stabilizer bar clamp
   • Wheel and tire assembly
7. Be sure to perform the reconnect/relearn procedures.

REAR PINION SEAL

REMOVAL & INSTALLATION

See Figure 14.

1. Before servicing the vehicle, refer to the Precautions Section.

➡**If working near and/or around the SRS system and components, be sure to disable the SRS system. After disabling the system wait three minutes or more before servicing the vehicle.**

➡**Whenever the negative battery cable is disconnected the following components will require resetting. The Idle Air Volume Learning, Steering Angle Sensor Neutral Position, Sunroof Memory Reset/Initialization, Automatic Drive Positioner System, Audio presets and Navigation. Use the CONSULT-III diagnostic tool, or equivalent to perform the required resets.**

2. Disconnect the negative battery cable.
3. Raise and safely support the vehicle.
4. Remove the rear driveshaft.
5. Measure and record the total preload.
6. Matchmark the drive pinion to position 'B' on the companion flange.

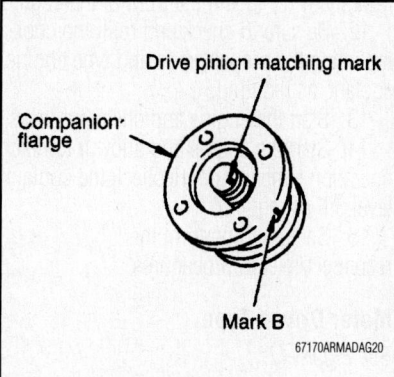

Drive pinion matching mark

Companion flange

Mark B

67170ARMADAG20

Fig. 14 Companion flange marking

---

7. Remove the drive pinion nut using suitable tool.
8. Remove the companion flange using suitable tool.
9. Remove the rear pinion seal using special tool J-34286.

To install:

➡**Be sure to use new fasteners, as required.**

10. Press the rear pinion seal into the carrier using suitable tool.
11. Align the matchmark on the companion flange to the drive pinion and install the companion flange.
12. Lubricate the drive pinion threads and seating surfaces of the drive pinion nut with grease.
13. Using a new drive pinion nut, tighten to 124–274 ft. lbs. (167–372 Nm).

➡**Final torque is determined when adjusting total preload using special tool J-25765-A.**

14. Install rear driveshaft.
15. Be sure to perform the reconnect/relearn procedures.

TRANSFER CASE ASSEMBLY

REMOVAL & INSTALLATION

See Figure 15.

1. Before servicing the vehicle, refer to the Precautions Section.

➡**If working near and/or around the SRS system and components, be sure to disable the SRS system. After disabling the system wait three minutes or more before servicing the vehicle.**

➡**Whenever the negative battery cable is disconnected the following components will require resetting. The Idle Air Volume Learning, Steering Angle Sensor Neutral Position, Sunroof Memory Reset/Initialization, Automatic Drive Positioner System, Audio presets and Navigation. Use the CONSULT-III diagnostic tool, or equivalent to perform the required resets.**

2. Disconnect the negative battery cable.
3. Remove or disconnect the following:
   • Transmission splash guard
   • Center exhaust pipe and muffler
   • Front and rear driveshafts

➡**Plug rear oil seal after removing rear driveshaft.**

   • Transmission assembly mounting bolts

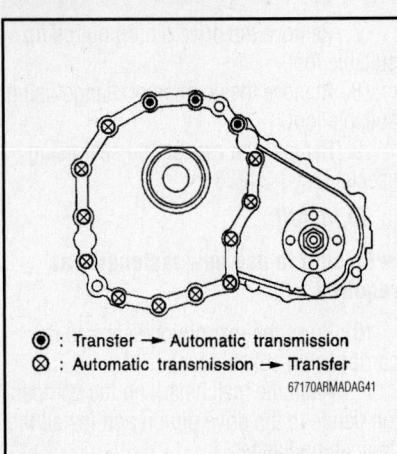

● : Transfer → Automatic transmission
⊗ : Automatic transmission → Transfer

67170ARMADAG41

Fig. 15 Transfer case mounting bolt locations

4. Support the transmission assembly with a suitable jack and remove the cross-member.

5. Remove or disconnect the following:
- ATP switch, neutral 4LO switch, wait detection switch, transfer motor and transfer control device electrical connectors
- Breather hoses
- Shift actuator from the extension housing
- Transfer case to transmission assembly bolts
- Transfer case assembly

To install:

➡**Be sure to use new fasteners, as required.**

6. Install or connect the following:
- Transfer case to transmission assembly bolts tightening to 26 ft. lbs. (36 Nm)
- Shift actuator
- Breather hoses
- ATP switch, neutral 4LO switch, wait detection switch, transfer motor and transfer control device electrical connectors
- Support crossmember
- Transmission mounting bolts
- Driveshafts
- Muffler and center exhaust pipe
- Transmission splash guard

7. Be sure to perform the reconnect/relearn procedures.

ENGINE COOLING

ENGINE FAN

REMOVAL & INSTALLATION

Crankshaft Driven Type

See Figure 16.

➡**Never remove the radiator cap when the engine is hot. Serious burns could occur from high-pressure engine coolant escaping from the radiator.**

1. Before servicing the vehicle, refer to the Precautions Section.

➡**If working near and/or around the SRS system and components, be sure to disable the SRS system. After disabling the system wait three minutes or more before servicing the vehicle.**

➡**Whenever the negative battery cable is disconnected the following components will require resetting. The Idle Air Volume Learning, Steering Angle Sensor Neutral Position, Sunroof Memory Reset/Initialization, Automatic Drive Positioner System, Audio presets and Navigation. Use the CONSULT-III diagnostic tool, or equivalent to perform the required resets.**

2. Disconnect the negative battery cable.
3. Make sure the engine is cold.
4. Remove the air duct and resonator assembly.
5. Remove the engine front undercover.
6. Remove the lower radiator shroud.
7. Remove the drive belt.
8. Remove the cooling fan retaining bolts.
9. Remove the cooling fan from its mounting.

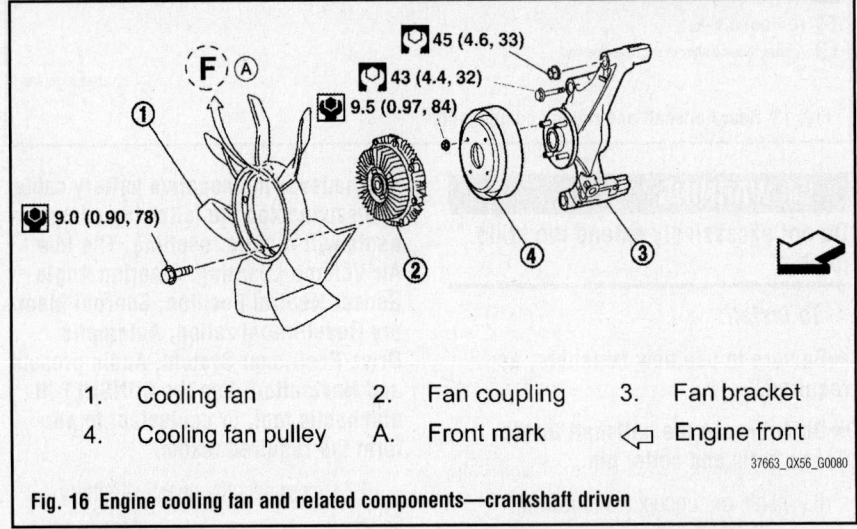

45 (4.6, 33)
43 (4.4, 32)
9.5 (0.97, 84)
9.0 (0.90, 78)

| 1. | Cooling fan | 2. | Fan coupling | 3. | Fan bracket |
| 4. | Cooling fan pulley | A. | Front mark | ⇦ | Engine front |

37663_QX56_G0080

Fig. 16 Engine cooling fan and related components—crankshaft driven

To install:

➡**Be sure to use new fasteners, as required.**

10. Installation is the reverse of the removal procedure.

11. Be sure to install the fan with its front mark "F" facing the front of the engine.

12. Be sure to check and refill the cooling using the proper grade and type engine coolant, as required.

13. Start the engine and check for leaks.

14. Start the engine and allow it to reach operation temperature. Recheck the coolant level, fill as required.

15. Be sure to perform the reconnect/relearn procedures.

Motor Driven Type

See Figure 17.

➡**Never remove the radiator cap when the engine is hot. Serious**

burns could occur from high-pressure engine coolant escaping from the radiator.

1. Before servicing the vehicle, refer to the Precautions Section.

➡**If working near and/or around the SRS system and components, be sure to disable the SRS system. After disabling the system wait three minutes or more before servicing the vehicle.**

➡**Whenever the negative battery cable is disconnected the following components will require resetting. The Idle Air Volume Learning, Steering Angle Sensor Neutral Position, Sunroof Memory Reset/Initialization, Automatic Drive Positioner System, Audio presets and Navigation. Use the CONSULT-III diagnostic tool, or equivalent to perform the required resets.**

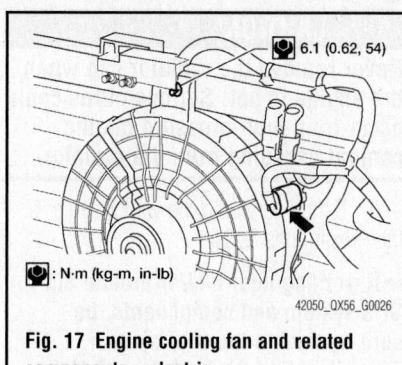

6.1 (0.62, 54)

N·m (kg-m, in-lb)

42050_QX56_G0026

Fig. 17 Engine cooling fan and related components—electric

2. Disconnect the negative battery cable.

3. Remove the front bumper fascia.

4. Disconnect the harness connector from the fan motor.

5. Remove the retaining bolt.

6. Remove the fan grille and motor assembly.

To install:

➡**Be sure to use new fasteners, as required.**

7. Installation is the reverse of the removal procedure.

8. Be sure to check and refill the cooling using the proper grade and type engine coolant, as required.

9. Start the engine and check for leaks.

10. Start the engine and allow it to reach operation temperature. Recheck the coolant level, fill as required.

11. Be sure to perform the reconnect/relearn procedures.

RADIATOR

REMOVAL & INSTALLATION

※※ CAUTION

Never remove the radiator cap when the engine is hot. Serious burns could occur from high-pressure engine coolant escaping from the radiator.

1. Before servicing the vehicle, refer to the Precautions Section.

➡**If working near and/or around the SRS system and components, be sure to disable the SRS system. After disabling the system wait three minutes or more before servicing the vehicle.**

➡**Whenever the negative battery cable is disconnected the following components will require resetting. The Idle Air Volume Learning, Steering Angle Sensor Neutral Position, Sunroof Memory Reset/Initialization, Automatic Drive Positioner System, Audio presets**

and Navigation. Use the CONSULT-III diagnostic tool, or equivalent to perform the required resets.

2. Disconnect the negative battery cable.

3. Make sure the engine is cold before removing the radiator.

4. Remove the engine room cover.

5. Remove the air cleaner and air duct assembly.

6. Drain the engine coolant. Be sure to properly dispose of the used coolant.

7. Disconnect and plug the automatic transmission fluid lines.

8. Disconnect the upper radiator hose. Do not allow coolant to contact the drive belts.

9. Disconnect the lower radiator hose. Do not allow coolant to contact the drive belts.

10. To remove the lower radiator shroud, release the tabs and pull the lower radiator shroud rearwards and down.

11. Remove the radiator shroud upper bolts and remove the upper radiator shroud.

12. Remove the air conditioning condenser bolts and brackets.

➡**Lift the condenser up and forward to remove it from the radiator.**

13. Remove the transmission fluid cooler bolts and the fluid cooler from the radiator. Position it to the side.

14. Lift up and remove the radiator. Be careful not to damage or scratch the air conditioning condenser and radiator core when removing the radiator.

To install:

➡**Be sure to use new fasteners, as required.**

15. Installation is the reverse of the removal procedure.

16. Be sure to refill the cooling using the proper grade and type engine coolant.

17. Start the engine and check for leaks.

18. Start the engine and allow it to reach operation temperature. Recheck the coolant level, fill as required.

19. Be sure to perform the reconnect/relearn procedures.

THERMOSTAT

REMOVAL & INSTALLATION

See Figures 18 and 19.

※※ CAUTION

Never remove the radiator cap when the engine is hot. Serious burns could occur from high-pressure engine coolant escaping from the radiator.

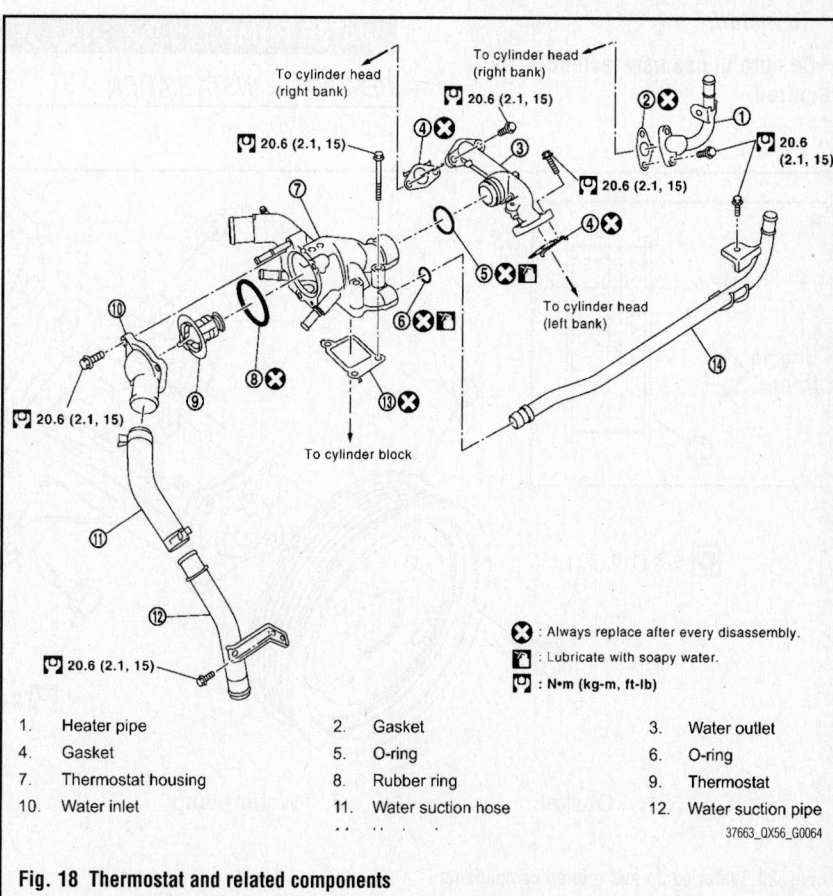

To cylinder head (right bank)
To cylinder head (right bank)
20.6 (2.1, 15)
20.6 (2.1, 15)
20.6 (2.1, 15)
20.6 (2.1, 15)
To cylinder head (left bank)
To cylinder block
20.6 (2.1, 15)
20.6 (2.1, 15)

: Always replace after every disassembly.

: Lubricate with soapy water.

: N·m (kg-m, ft-lb)

1.	Heater pipe	2.	Gasket	3.	Water outlet
4.	Gasket	5.	O-ring	6.	O-ring
7.	Thermostat housing	8.	Rubber ring	9.	Thermostat
10.	Water inlet	11.	Water suction hose	12.	Water suction pipe

37663_QX56_G0064

Fig. 18 Thermostat and related components

1. Before servicing the vehicle, refer to the Precautions Section.

➡ **If working near and/or around the SRS system and components, be sure to disable the SRS system. After disabling the system wait three minutes or more before servicing the vehicle.**

➡ **Whenever the negative battery cable is disconnected the following components will require resetting. The Idle Air Volume Learning, Steering Angle Sensor Neutral Position, Sunroof Memory Reset/Initialization, Automatic Drive Positioner System, Audio presets and Navigation. Use the CONSULT-III diagnostic tool, or equivalent to perform the required resets.**

2. Disconnect the negative battery cable.

3. Make sure the engine is cold.

4. Remove the engine room cover.

5. Remove the air duct and resonator assembly.

6. Disconnect the water suction hose from the water inlet.

7. Remove the water inlet and thermostat.

➡ **To remove the thermostat housing, water outlet and heater pipe, you will first have to remove the intake manifold.**

To install:

➡ **Be sure to use new fasteners, as required.**

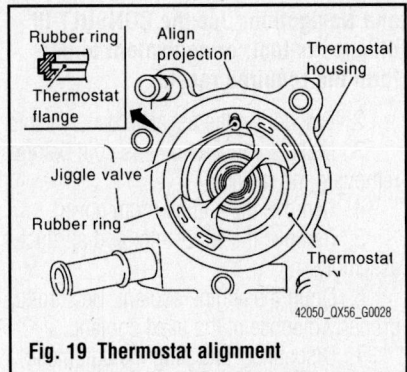

Fig. 19 Thermostat alignment

8. Installation is the reverse of the removal procedure.

9. Be sure to use a new gasket.

10. Install the thermostat with the whole circumference of each flange part fitting securely inside the rubber ring, as shown in the illustration.

11. Be sure to perform the reconnect/relearn procedures.

12. Install the thermostat with the jiggle valve facing upward.

13. Be sure to refill the cooling using the proper grade and type engine coolant.

14. Start the engine and check for leaks.

15. Start the engine and allow it to reach operation temperature. Recheck the coolant level, fill as required.

WATER PUMP

REMOVAL & INSTALLATION

See Figure 20.

✲✲✲ CAUTION

Never remove the radiator cap when the engine is hot. Serious burns could occur from high-pressure engine coolant escaping from the radiator.

1. Before servicing the vehicle, refer to the Precautions Section.

➡ **If working near and/or around the SRS system and components, be sure to disable the SRS system. After disabling the system wait three minutes or more before servicing the vehicle.**

➡ **Whenever the negative battery cable is disconnected the following components will require resetting. The Idle Air Volume Learning, Steering Angle Sensor Neutral Position, Sunroof Memory Reset/Initialization, Automatic Drive Positioner System, Audio presets and Navigation. Use the CONSULT-III diagnostic tool, or equivalent to perform the required resets.**

2. Disconnect the negative battery cable.

3. Make sure the engine is cold.

4. Drain the cooling system.

5. Remove or disconnect the following:
   - Engine room cover
   - Air intake assembly
   - Accessory drive belt

➡ **Leave tensioner pulley in its fixed position.**

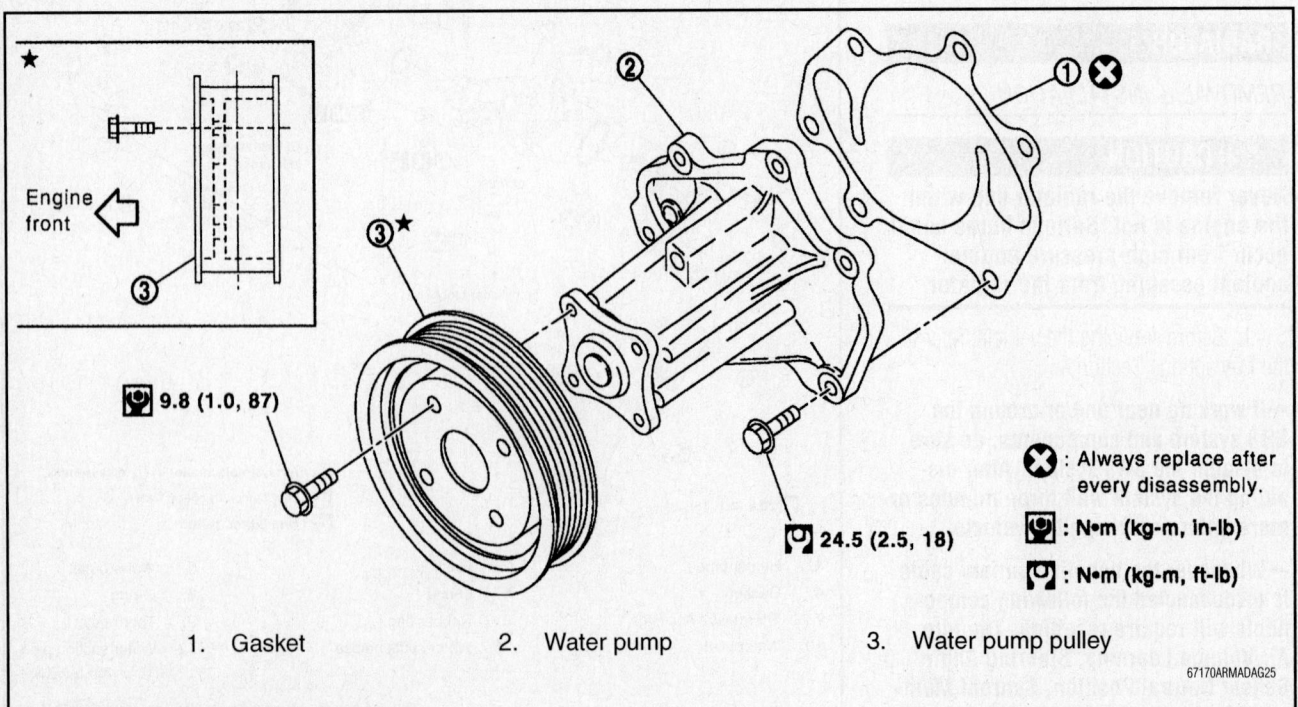

9.8 (1.0, 87)

24.5 (2.5, 18)

❌ : Always replace after every disassembly.

🔧 : N•m (kg-m, in-lb)

🔧 : N•m (kg-m, ft-lb)

1. Gasket   2. Water pump   3. Water pump pulley

Engine front

67170ARMADAG25

Fig. 20 Water pump and related components

- Cooling fan
- Water pump pulley
- Water pump

To install:

➡Be sure to use new fasteners, as required.

6. Install or connect the following:
- Water pump with a new gasket. Tighten bolts to 18 ft. lbs. (25 Nm)
- Water pump pulley and tighten bolts to 87 inch lbs. (10 Nm)
- Accessory drive belt

- Air intake assembly
- Engine splash guard
7. Refill the cooling system.
8. Start the engine and check for leaks.
9. Be sure to perform the reconnect/relearn procedures.

ENGINE ELECTRICAL   CHARGING SYSTEM

ALTERNATOR

REMOVAL & INSTALLATION
See Figure 21.

1. Before servicing the vehicle, refer to the Precautions Section.

➡If working near and/or around the SRS system and components, be sure to disable the SRS system. After disabling the system wait three minutes or more before servicing the vehicle.

➡Whenever the negative battery cable is disconnected the following components will require resetting. The Idle Air Volume Learning, Steering Angle Sensor Neutral Position, Sunroof Memory Reset/Initialization, Automatic Drive Positioner System, Audio presets and Navigation. Use the CONSULT-III

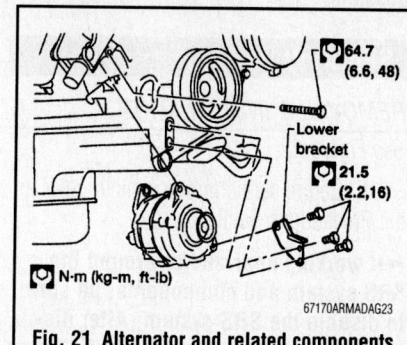

Fig. 21 Alternator and related components

diagnostic tool, or equivalent to perform the required resets.

2. Disconnect the negative battery cable.
3. Remove or disconnect the following:
- Fan shroud
- Drive belt

- Lower alternator bracket
- Alternator upper bolt
- Alternator harness connectors
- Alternator

To install:

➡Be sure to use new fasteners, as required.

4. Install or connect the following:
- Alternator
- Alternator harness connectors
- Upper bolt, tighten to 48 ft. lbs. (65 Nm)
- Lower bracket, tighten to 16 ft. lbs (22 Nm)
- Drive belt
- Fan shroud
- Negative battery cable
5. Be sure to perform the reconnect/relearn procedures.

ENGINE ELECTRICAL   IGNITION SYSTEM

FIRING ORDER

1–8–7–3–6–5–4–2

IGNITION COIL

REMOVAL & INSTALLATION
See Figure 22.

1. Before servicing the vehicle, refer to the Precautions Section.

➡If working near and/or around the SRS system and components, be sure

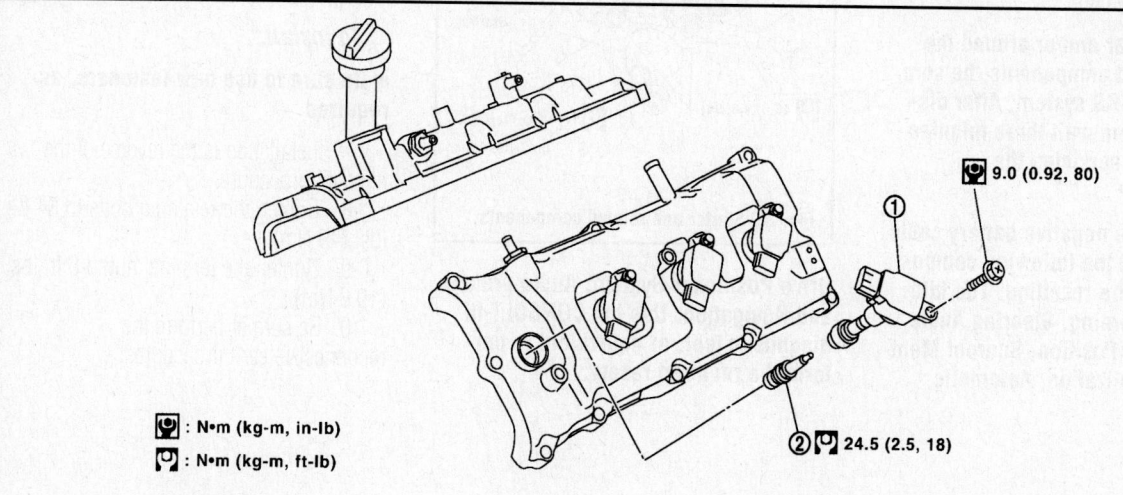

: N•m (kg-m, in-lb)

: N•m (kg-m, ft-lb)

1. Ignition coil
2. Spark plug

Fig. 22 Ignition coil and spark plugs

to disable the SRS system. After disabling the system wait three minutes or more before servicing the vehicle.

➡Whenever the negative battery cable is disconnected the following components will require resetting. The Idle Air Volume Learning, Steering Angle Sensor Neutral Position, Sunroof Memory Reset/Initialization, Automatic Drive Positioner System, Audio presets and Navigation. Use the CONSULT-III diagnostic tool, or equivalent to perform the required resets.

2. Disconnect the negative battery cable.
3. Remove the engine room cover. Remove the air cleaner assembly, as required.
4. Disconnect the harness connector from the ignition coil.
5. Remove the ignition coil retaining bolt.
6. Remove the ignition coil.

To install:

➡Be sure to use new fasteners, as required.

7. Install the ignition coil, torque the retaining bolt to 80 inch lbs. (9 Nm).
8. Connect the harness coil.

9. Connect the negative battery cable.
10. Be sure to perform the reconnect/relearn procedures.

IGNITION TIMING

ADJUSTMENT

The ignition timing is controlled by the Powertrain Control Module (PCM). No adjustment is necessary or possible.

SPARK PLUGS

REMOVAL & INSTALLATION

See Figure 22.

1. Before servicing the vehicle, refer to the Precautions Section.

➡If working near and/or around the SRS system and components, be sure to disable the SRS system. After disabling the system wait three minutes or more before servicing the vehicle.

➡Whenever the negative battery cable is disconnected the following components will require resetting. The Idle Air Volume Learning, Steering Angle Sensor Neutral Position, Sunroof Memory Reset/Initialization, Automatic

Drive Positioner System, Audio presets and Navigation. Use the CONSULT-III diagnostic tool, or equivalent to perform the required resets.

2. Disconnect the negative battery cable.
3. Disconnect the harness connector from the ignition coil.
4. Remove the ignition coil retaining bolt.
5. Remove the ignition coil.
6. Remove the spark plug using a spark plug socket and wrench.

To install:

➡Be sure to use new fasteners, as required.

7. Be sure the spark plug gap is to specification (0.043 in.).
8. Carefully install the spark plug and torque to specification, 18 ft. lbs.
9. Install the ignition coil, torque the retaining bolt to 80 inch lbs. (9 Nm).
10. Connect the harness coil.
11. Connect the negative battery cable.
12. Be sure to perform the reconnect/relearn procedures.

BRAKES

STARTER

REMOVAL & INSTALLATION

See Figure 23.

1. Before servicing the vehicle, refer to the Precautions Section.

➡If working near and/or around the SRS system and components, be sure to disable the SRS system. After disabling the system wait three minutes or more before servicing the vehicle.

➡Whenever the negative battery cable is disconnected the following components will require resetting. The Idle Air Volume Learning, Steering Angle Sensor Neutral Position, Sunroof Memory Reset/Initialization, Automatic

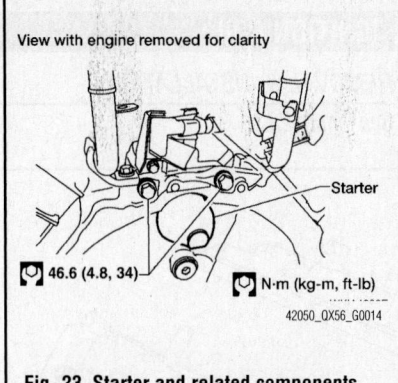

View with engine removed for clarity

Starter

46.6 (4.8, 34)   N·m (kg-m, ft-lb)

42050_QX56_G0014

Fig. 23 Starter and related components

Drive Positioner System, Audio presets and Navigation. Use the CONSULT-III diagnostic tool, or equivalent to perform the required resets.

STARTING SYSTEM

2. Disconnect the negative battery cable.
3. Remove the intake manifold.
4. Remove the starter harness connectors.
5. Remove the starter retaining bolts.
6. Remove the starter from its mounting.

To install:

➡Be sure to use new fasteners, as required.

7. Installation is the reverse of the removal procedure.
8. Tighten the retaining bolts to 34 ft. lbs. (46 Nm).
9. Tighten the terminal nut to 8 ft. lbs. (10.8 Nm).
10. Be sure to perform the reconnect/relearn procedures.

ENGINE MECHANICAL

ACCESSORY DRIVE BELTS

ACCESSORY BELT ROUTING

See Figure 24.

Refer to the accompanying illustration.

INSPECTION

Inspect the drive belt for signs of glazing or cracking. A glazed belt will be perfectly smooth from slippage, while a good belt will have a slight texture of fabric visible. Cracks will usually start at the inner edge of the belt and run outward. All worn or damaged drive belts should be replaced immediately.

ADJUSTMENT

Drive belt tension is not necessary, as it is automatically adjusted by the auto tensioner.

REMOVAL & INSTALLATION

See Figure 24.

1. Before servicing the vehicle, refer to the Precautions Section.

➡ If working near and/or around the SRS system and components, be sure to disable the SRS system. After disabling the system wait three minutes or more before servicing the vehicle.

➡ Whenever the negative battery cable is disconnected the following components will require resetting. The Idle Air Volume Learning, Steering Angle Sensor Neutral Position, Sunroof Memory Reset/Initialization, Automatic Drive Positioner System, Audio presets and Navigation. Use the CONSULT-III diagnostic tool, or equivalent to perform the required resets.

2. Disconnect the negative battery cable.
3. Remove the engine room cover.
4. Remove the air duct and resonator assembly.
5. Install special tool J-46535, or equivalent on the auto tensioner pulley bolt and move it upward.

➡ Avoid placing your hand in a location where pinching may occur if the holding tool accidentally comes off.

6. Remove the drive belt from the vehicle.

To install:

➡ Be sure to use new fasteners, as required.

7. Installation is the reverse of the removal procedure.
8. Be sure that the belt is securely installed around all pulleys.
9. Rotate the crankshaft several times clockwise to equalize belt tension between the pulleys.
10. Make sure that the belt tension is within the allowable working range, using the indicator notch on the auto tensioner.
11. Be sure to perform the reconnect/relearn procedures.

CAMSHAFT AND VALVE LIFTERS

REMOVAL & INSTALLATION

See Figures 25 through 40.

1. Before servicing the vehicle, refer to the Precautions Section.

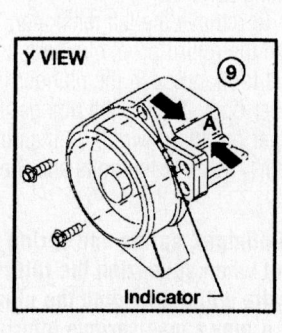
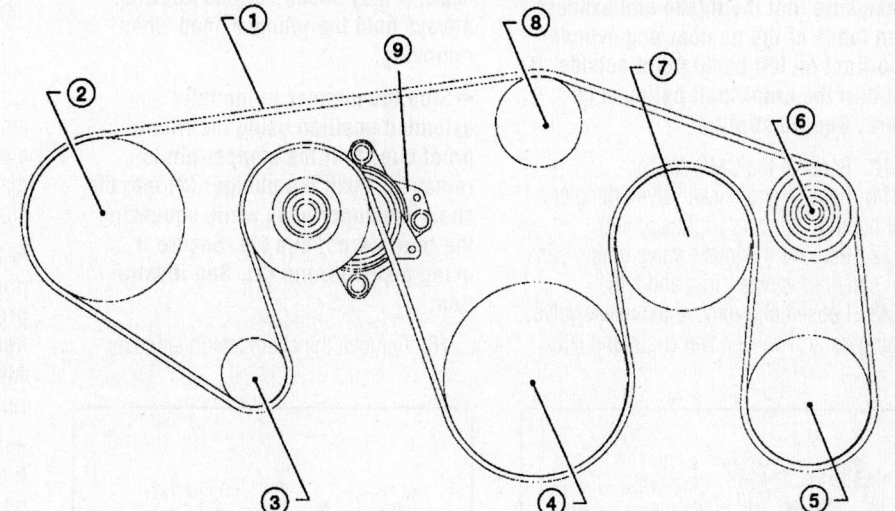

1.	Drive Belt	2.	Power Steering Pump Pulley	3.	Generator pulley
4.	Crankshaft Pulley	5.	A/C Compressor	6.	Idler Pulley
7.	Cooling Fan Pulley	8.	Water Pump Pulley	9.	Drive Belt Tensioner

67162-QX56-G47

Fig. 24 Accessory drive belt routing

➡️If working near and/or around the SRS system and components, be sure to disable the SRS system. After disabling the system wait three minutes or more before servicing the vehicle.

➡️Whenever the negative battery cable is disconnected the following components will require resetting. The Idle Air Volume Learning, Steering Angle Sensor Neutral Position, Sunroof Memory Reset/Initialization, Automatic Drive Positioner System, Audio presets and Navigation. Use the CONSULT-III diagnostic tool, or equivalent to perform the required resets.

2. Disconnect the negative battery cable.

3. Remove the engine cover.

4. Remove the air cleaner assembly.

5. Remove the power steering reservoir tank bolts. Position the unit to the side.

6. Remove the valve covers.

7. Remove the spark plugs.

8. Remove the drive belt.

9. Be sure that the number one cylinder is at TDC on the compression stroke.

➡️Turn the crankshaft pulley clockwise to align the TDC identification notch (without paint mark) with the timing indicator on the front cover. At this time make sure that the intake and exhaust cam lobes of the number one cylinder (top front on left bank) point outside. If not turn the crankshaft pulley once more. See illustration

10. Remove the CMP sensor.

11. Remove the intake valve timing control position sensors (right and left).

12. Remove the intake valve timing control solenoid valves (right and left).

13. Loosen and remove the intake valve timing control valve cover (right and left)

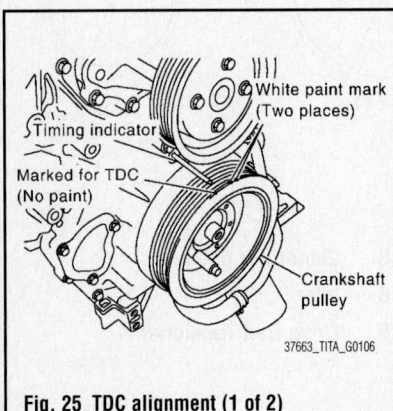

Fig. 25 TDC alignment (1 of 2)

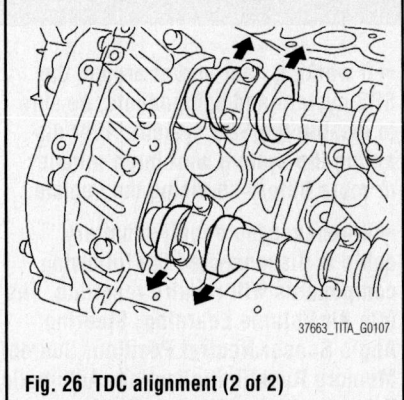

Fig. 26 TDC alignment (2 of 2)

bolts in the reverse order of the tightening sequence.

14. Paint alignment marks on the right bank (A) timing chain links (C) and left bank (B) timing chain links (D) and align with the camshaft sprocket alignment marks (E) and (F). See illustration.

15. To remove the left tensioner, squeeze the return proof clip ends using a suitable tool and push the plunger into the tensioner body. Secure the plunger using a stopper pin (hard wire 0.04 inch in diameter). Remove the bolts and the tensioner.

➡️The plunger, spring and spring seat pop out when squeezing the return proof clip without holding the plunger head. It may cause serious injuries. Always hold the plunger head when removing.

➡️Stop the plunger in the fully extended position using the return proof clip (1) if the stopper pin is removed. Push the plunger (2) into the chain tensioner body while squeezing the return proof clip (1). Secure it using a stopper pin (3). See illustration.

16. Remove the chain tensioner cover

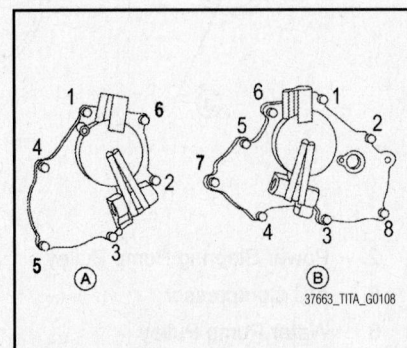

Fig. 27 Intake valve timing control solenoid cover tightening sequence

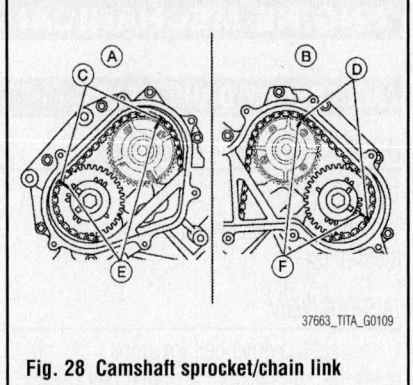

Fig. 28 Camshaft sprocket/chain link alignment

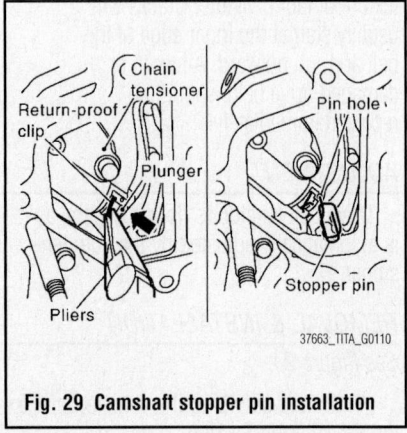

Fig. 29 Camshaft stopper pin installation

from the front cover, using tool KV10111100, or equivalent. Do not damage the mating surfaces.

17. To remove the left tensioner, squeeze the return proof clip ends using a suitable tool and push the plunger into the tensioner body. Secure the plunger using a stopper pin (hard wire 0.04 inch in diameter). Remove the bolts and the tensioner.

➡️The plunger, spring and spring seat pop out when squeezing the return proof clip without holding the plunger head. It may cause serious injuries. Always hold the plunger head when removing.

➡️If it is difficult to push the plunger on the tensioner, remove the plunger under the extended condition.

18. Loosen the camshaft sprocket bolts and remove the sprockets.

➡️To avoid interference between the valves and pistons, do not turn the crankshaft or camshaft with the timing chain disconnected.

19. Remove the front cover bolts. See illustration for location (arrow).

20. Remove the camshaft bracket bolts

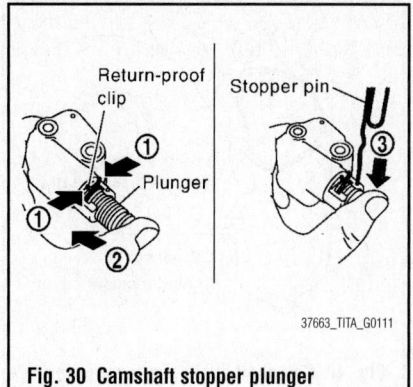

Fig. 30 Camshaft stopper plunger retention

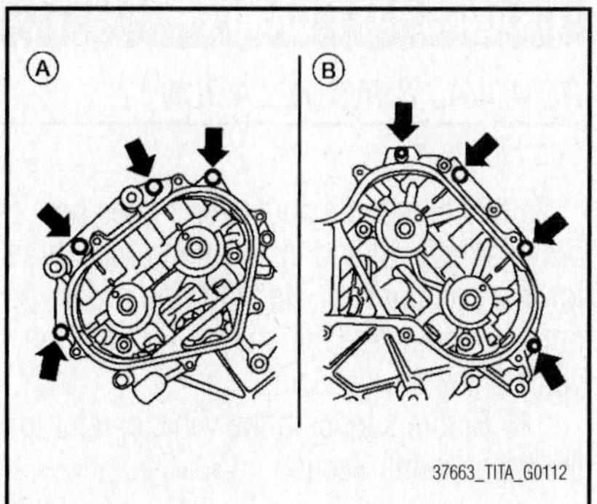

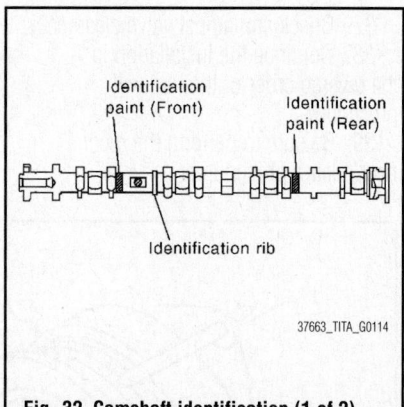

Fig. 31 Camshaft front cover bolt location (arrow)

in the reverse order of the tightening sequence. Remove the number one camshaft bracket. The bottom of the front surface of the bracket will be stuck because of liquid gasket.

21. Remove the camshaft. Remove the lifters, as necessary.

To install:

➡**Be sure to use new fasteners, as required.**

22. Install the camshafts. Be sure that the camshafts are properly identified. See illustrations.

23. Install the dowel pins at the front of the camshaft. See illustration for proper direction.

24. Install the camshaft brackets.

➡**Install by referring to the illustration location mark on the upper surface. Install so that the installation mark can be correctly read when viewed from the intake manifold side.**

25. To install the number one camshaft bracket, apply liquid gasket as shown in the illustration. Be sure to wipe off any excessive gasket after installation.

26. Apply liquid gasket to the back side of the left front cover and the right front

cover. Bead diameter should be 0.102–0.142 inch. Position the number one camshaft bracket close to the mounting position and then install it to prevent from touching gasket applied to each surface.

27. Temporarily tighten the right and left front cover bolts.

28. Tighten the camshaft bracket bolts to specification and in the proper sequence.

29. Tighten the right and left front cover bolts to 8 ft. lbs.

30. Install the camshaft sprockets aligning them with the matching marks painted on the timing chain and the camshaft sprockets, before removal. Align the sprocket key groove with the dowel pin on the camshaft front edge at the same time. Temporarily tighten the sprocket bolts.

31. Install the intake VTC and the exhaust side camshaft sprockets by selectively using the groove of the dowel pin according to the bank for the exhaust side camshaft sprockets, (common part used for both exhaust banks).

➡**Use the groove marked "R" for right bank and "L" for left bank.**

32. Lock the hex part of the camshaft in the same way as for removal. Tighten the sprocket bolts.

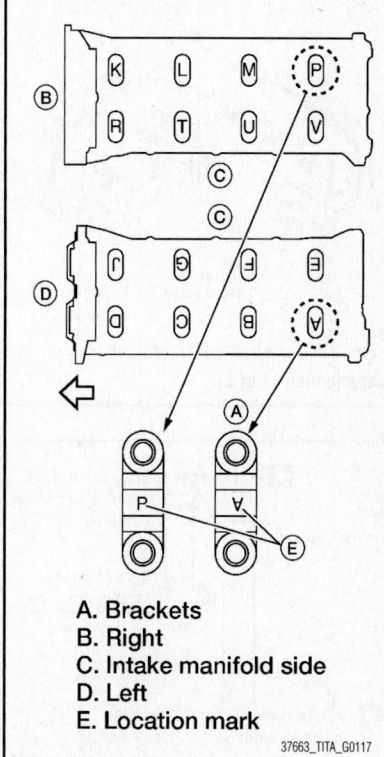

A. Brackets
B. Right
C. Intake manifold side
D. Left
E. Location mark

Fig. 35 Camshaft bracket installation and identification

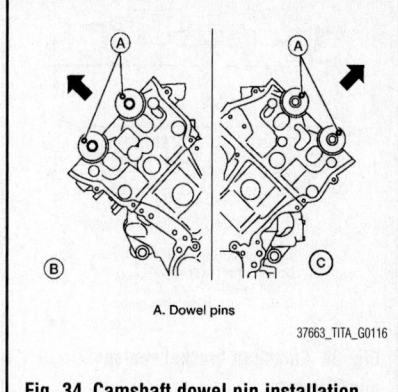

A. Dowel pins

Fig. 34 Camshaft dowel pin installation

Fig. 32 Camshaft identification (1 of 2)

Identification paint (Front)
Identification paint (Rear)
Identification rib

Bank	INT EXH	Identification paint (front)	Identification paint (rear)	Identification rib
RH	INT	Pink	—	Yes
	EXH	—	Orange	Yes
LH	INT	Pink	—	No
	EXH	—	Orange	No

Fig. 33 Camshaft identification (2 of 2)

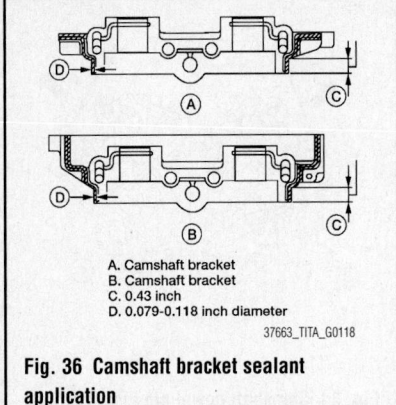

Fig. 36 Camshaft bracket sealant application

A. Camshaft bracket
B. Camshaft bracket
C. 0.43 inch
D. 0.079-0.118 inch diameter

37663_TITA_G0118

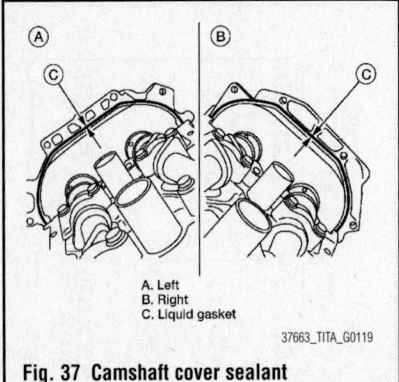

A. Left
B. Right
C. Liquid gasket

37663_TITA_G0119

Fig. 37 Camshaft cover sealant application (1 of 2)

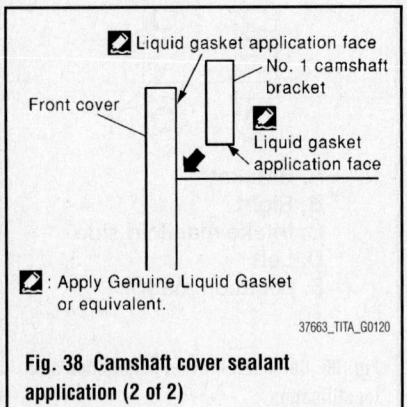

Liquid gasket application face

No. 1 camshaft bracket

Front cover

Liquid gasket application face

: Apply Genuine Liquid Gasket or equivalent.

37663_TITA_G0120

Fig. 38 Camshaft cover sealant application (2 of 2)

33. Check that the timing marks are properly aligned.

34. To install the chain tensioner, compress the plunger and hold it using a stopper pin. Loosen the slack guide timing chain by rotating the camshaft hex part if mounting space is small. Tighten the tensioner bolts to 61 inch lbs.

35. Remove the stopper pin and release the plunger, then apply tension to the chain.

36. Install the chain tensioner cover onto the front cover. Apply liquid gasket. See illustration. Tighten the bolts to 80 inch lbs.

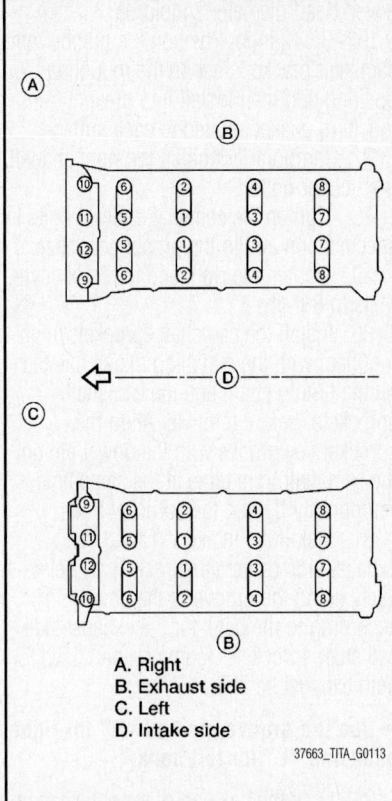

A. Right
B. Exhaust side
C. Left
D. Intake side

37663_TITA_G0113

Fig. 39 Camshaft bracket bolt tightening sequence

37. Check and adjust valve clearances.

38. Continue the installation in the reverse order of the removal procedure.

39. Be sure to perform the reconnect/relearn procedures.

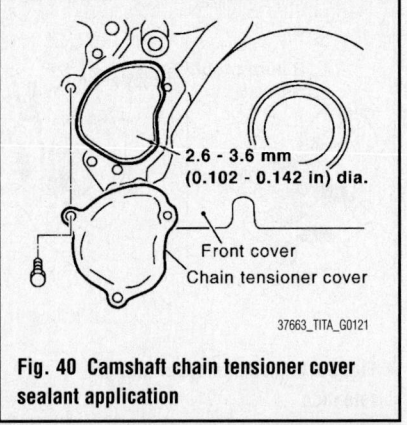

2.6 - 3.6 mm (0.102 - 0.142 in) dia.

Front cover
Chain tensioner cover

37663_TITA_G0121

Fig. 40 Camshaft chain tensioner cover sealant application

CRANKSHAFT DAMPER

REMOVAL & INSTALLATION

See Figure 41.

At this time the manufacturer does not provide removal and installation procedures for this component. The following procedure is a guideline and may differ from the vehicle you are servicing.

1. Before servicing the vehicle, refer to the Precautions Section.

➡**If working near and/or around the SRS system and components, be sure to disable the SRS system. After disabling the system wait three minutes or more before servicing the vehicle.**

➡**Whenever the negative battery cable is disconnected the following components will require resetting. The Idle Air Volume Learning, Steering Angle**

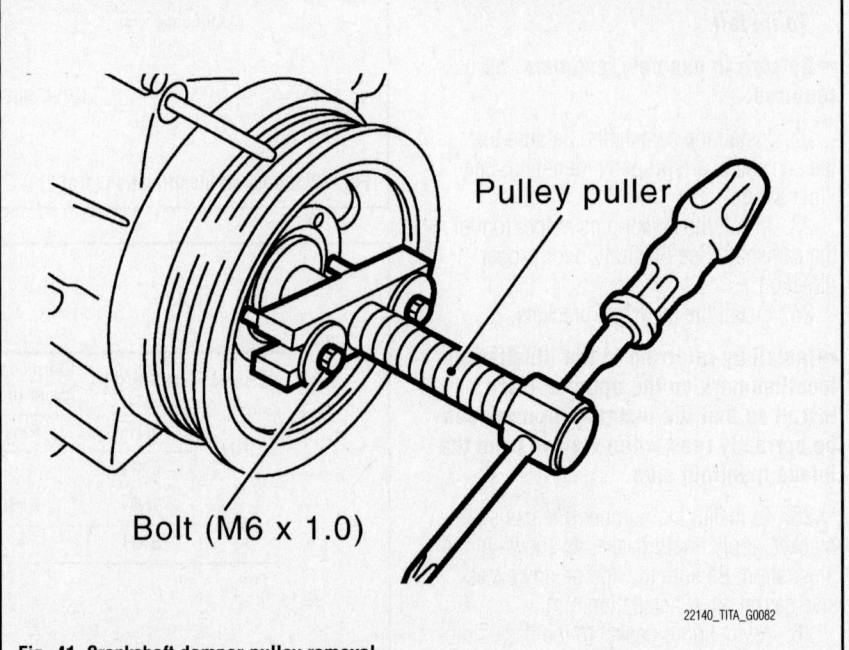

Pulley puller

Bolt (M6 x 1.0)

22140_TITA_G0082

Fig. 41 Crankshaft damper pulley removal

Sensor Neutral Position, Sunroof Memory Reset/Initialization, Automatic Drive Positioner System, Audio presets and Navigation. Use the CONSULT-III diagnostic tool, or equivalent to perform the required resets.

2. Disconnect the negative battery cable.

3. Remove the engine cover, engine undercover and air cleaner assembly, as required for access.

4. Remove the drive belt.

5. Remove the necessary components to gain access to the crankshaft damper.

6. Remove the crankshaft pulley using suitable tool.

7. Set the bolts in the two bolt holes 0.04 inch (M6 x 1.0 mm) on the front surface.

8. Remove the crankshaft pulley from the crankshaft using tool.

To install:

9. Install the crankshaft damper pulley.

10. Tighten the crankshaft pulley bolt as follows:
- Step 1: 69 ft. lbs. (93 Nm)
- Step 2: Additional 90° (angle tightening)

11. Be sure to perform the reconnect/relearn procedures.

CRANKSHAFT FRONT SEAL

REMOVAL & INSTALLATION

See Figure 42.

1. Before servicing the vehicle, refer to the Precautions Section.

➡️**If working near and/or around the SRS system and components, be sure to disable the SRS system. After disabling the system wait three minutes or more before servicing the vehicle.**

➡️**Whenever the negative battery cable is disconnected the following components will require resetting. The Idle Air Volume Learning, Steering Angle Sensor Neutral Position, Sunroof Memory Reset/Initialization, Automatic Drive Positioner System, Audio presets and Navigation. Use the CONSULT-III diagnostic tool, or equivalent to perform the required resets.**

2. Disconnect the negative battery cable.

3. Remove the engine cover, engine undercover and air cleaner assembly, as required for access.

4. Remove the drive belt.

5. Remove the radiator.

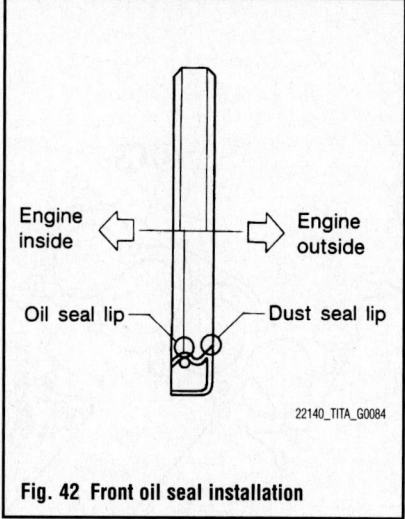

Fig. 42 Front oil seal installation

6. Remove the necessary components to gain access to the crankshaft damper.

7. Remove the crankshaft pulley.

8. Remove the oil seal using a suitable tool.

To install:

9. Apply new engine oil to both the oil seal lip and dust seal lip of the new front oil seal.

10. Install the front oil seal so that each seal lip is oriented as shown.

11. Install the crankshaft damper pulley.

12. Tighten the crankshaft pulley bolt as follows:
- Step 1: 69 ft. lbs. (93 Nm)
- Step 2: Additional 90° (angle tightening)

13. Be sure to fill the cooling system with the proper grade and type engine coolant.

14. Be sure to perform the reconnect/relearn procedures.

CYLINDER HEAD

REMOVAL & INSTALLATION

See Figures 43 and 44.

➡️**The engine must be removed from the vehicle to perform this procedure. Be sure that the engine is secured in a suitable holding fixture before performing this procedure.**

1. Before servicing the vehicle, refer to the Precautions Section.

➡️**If working near and/or around the SRS system and components, be sure to disable the SRS system. After disabling the system wait three minutes or more before servicing the vehicle.**

➡️**Whenever the negative battery cable is disconnected the following components will require resetting. The Idle Air Volume Learning, Steering Angle Sensor Neutral Position, Sunroof Memory Reset/Initialization, Automatic Drive Positioner System, Audio presets and Navigation. Use the CONSULT-III diagnostic tool, or equivalent to perform the required resets.**

2. Disconnect the negative battery cable.

3. Remove the engine room cover and air cleaner assembly.

4. Remove or disconnect the following:
- Engine assembly
- Belt tensioner
- Idler pulley
- Thermostat housing and hose
- Oil pan and strainer
- Fuel tube and injector assembly
- Intake manifold
- Ignition coil
- Rocker cover
- Crankshaft pulley
- Front engine cover
- Oil pump
- Timing chain
- Camshaft sprockets
- Camshafts
- Cylinder head, removing bolts in reverse order of installation sequence

To install:

➡️**Be sure to use new fasteners, as required.**

5. Install the cylinder head with a new gasket.

6. Tighten the bolts in sequence to specification.

7. Install or connect the following:
- Camshaft
- Camshaft sprockets
- Timing chain
- Oil pump
- Front engine cover
- Crankshaft pulley
- Rocker cover
- Ignition coil
- Intake manifold
- Fuel tube and injector assembly
- Oil pain and strainer
- Thermostat housing and hose
- Idler pulley
- Belt tensioner
- Engine assembly

8. Start the engine and check for leaks.

9. Be sure to perform the reconnect/relearn procedures.

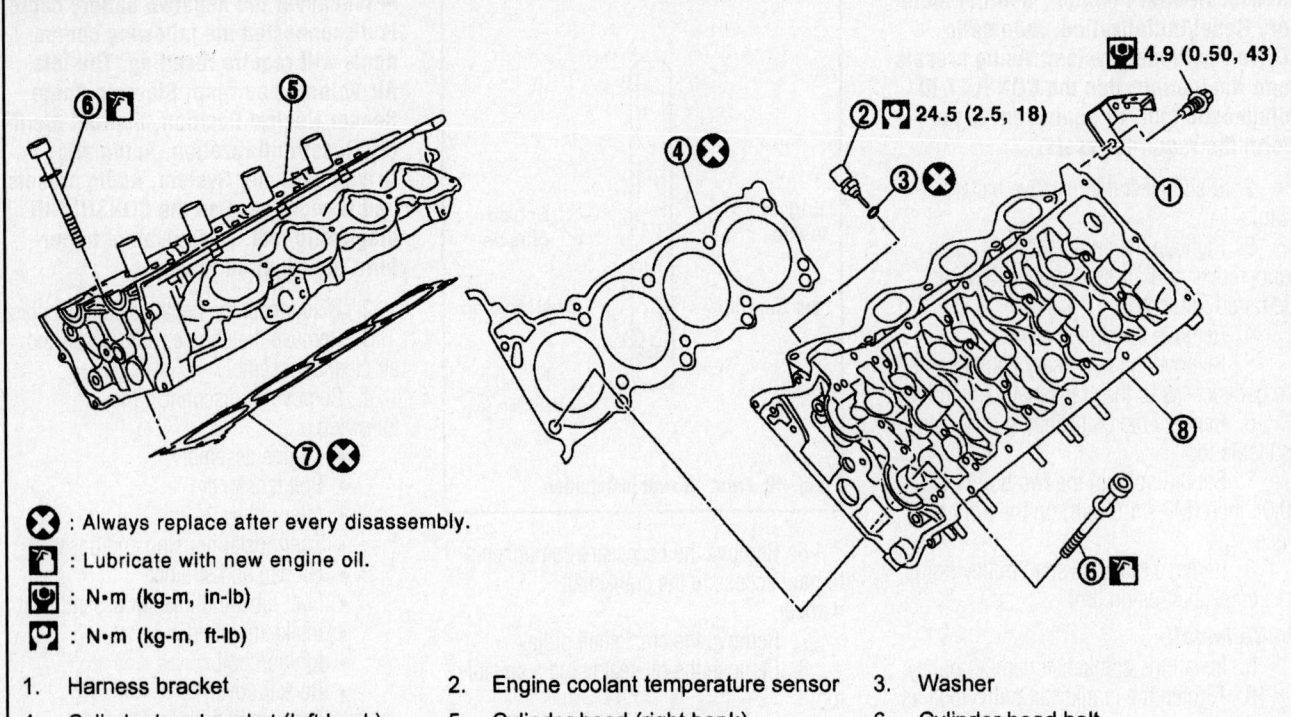

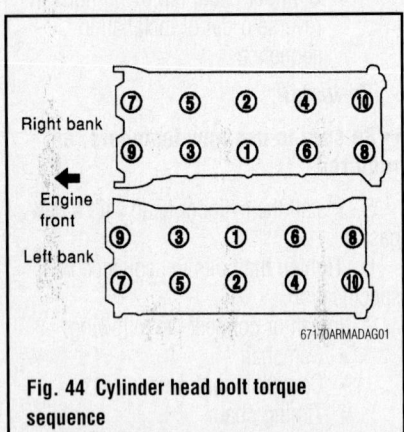

⊗ : Always replace after every disassembly.

⬚ : Lubricate with new engine oil.

⬚ : N·m (kg-m, in-lb)

⬚ : N·m (kg-m, ft-lb)

1. Harness bracket
2. Engine coolant temperature sensor
3. Washer
4. Cylinder head gasket (left bank)
5. Cylinder head (right bank)
6. Cylinder head bolt
7. Cylinder head gasket (right bank)
8. Cylinder head (left bank)

67170ARMADAG28

Fig. 43 Cylinder heads and related components

Fig. 44 Cylinder head bolt torque sequence

67170ARMADAG01

EXHAUST MANIFOLD

REMOVAL & INSTALLATION

See Figure 45.

1. Before servicing the vehicle, refer to the Precautions Section.

➡ If working near and/or around the SRS system and components, be sure to disable the SRS system. After disabling the system wait three minutes or more before servicing the vehicle.

➡ Whenever the negative battery cable is disconnected the following compo-

nents will require resetting. The Idle Air Volume Learning, Steering Angle Sensor Neutral Position, Sunroof Memory Reset/Initialization, Automatic Drive Positioner System, Audio presets and Navigation. Use the CONSULT-III diagnostic tool, or equivalent to perform the required resets.

2. Disconnect the negative battery cable.

3. Raise and support the vehicle safely.

4. Remove the engine under cover, if equipped.

5. Remove the front final drive assembly, if equipped.

6. Remove the main muffler and center exhaust tube.

7. Remove the front exhaust tubes.

8. Remove the tire and wheel assemblies.

9. Remove the fender protectors.

10. Remove the A/F sensors. Do not drop the sensors. If the sensor is dropped it must be replaced.

11. Properly support the engine.

12. Remove the engine mounting insulator.

13. Remove the exhaust manifold cover.

14. Remove the engine mounting bracket.

15. On right side remove the oil level dipstick.

16. Remove the left exhaust manifold nuts/bolts in the reverse order of the installation sequence.

17. Remove the exhaust manifold. Discard the gaskets.

To install:

➡ Be sure to use new fasteners, as required.

18. Installation is the reverse of the removal procedure.

19. Install new gaskets with the top of

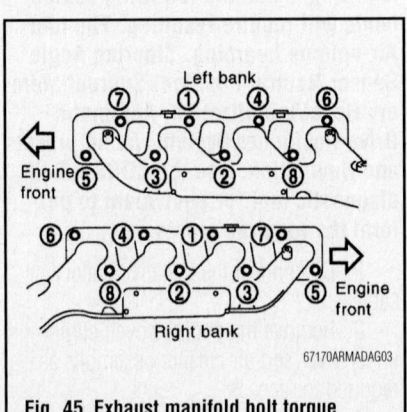

67170ARMADAG03

Fig. 45 Exhaust manifold bolt torque sequence

the triangular UP mark on it facing up and its coated face (gray side) toward the exhaust manifold side.

20. Tighten the retaining nuts/bolts to specification and in the proper sequence.

21. Be sure to perform the reconnect/relearn procedures.

FLEXPLATE

REMOVAL & INSTALLATION

See Figure 46.

1. Before servicing the vehicle, refer to the Precautions Section.

➡ If working near and/or around the SRS system and components, be sure to disable the SRS system. After disabling the system wait three minutes or more before servicing the vehicle.

➡ Whenever the negative battery cable is disconnected the following components will require resetting. The Idle Air Volume Learning, Steering Angle Sensor Neutral Position, Sunroof Memory Reset/Initialization, Automatic

Drive Positioner System, Audio presets and Navigation. Use the CONSULT-III diagnostic tool, or equivalent to perform the required resets.

2. Disconnect the negative battery cable.

3. Remove the transmission.

4. Remove the flexplate retaining bolts.

5. Remove the flexplate from the engine.

To install:

➡ Be sure to use new fasteners, as required.

6. Align the dowel pin of the crankshaft rear end with the pin holes of each part.

7. Install the flexplate, reinforcement plate and pilot converter.

8. Face the chamfered or rounded edge side to the crankshaft.

9. Continue the installation in the reverse order of the removal procedure.

10. Be sure to perform the reconnect/relearn procedures.

INTAKE MANIFOLD

REMOVAL & INSTALLATION

See Figures 47 and 48.

1. Before servicing the vehicle, refer to the Precautions Section.

➡ If working near and/or around the SRS system and components, be sure to disable the SRS system. After disabling the system wait three minutes or more before servicing the vehicle.

➡ Whenever the negative battery cable is disconnected the following components will require resetting. The Idle Air Volume Learning, Steering Angle Sensor Neutral Position, Sunroof Memory Reset/Initialization, Automatic Drive Positioner System, Audio presets and Navigation. Use the CONSULT-III diagnostic tool, or equivalent to perform the required resets.

2. Disconnect the negative battery cable.

3. Drain the cooling system.

4. Relieve the fuel system pressure.

5. Remove or disconnect the following:
- Engine room cover
- Air intake assembly
- Fuel tube quick connector using special tool J-45488
- Wiring harnesses and brackets from manifold
- Vacuum hoses
- PCV hose and tube
- Electric throttle control actuator, loosening bolts diagonally
- Fuel injectors
- Fuel tube assembly
- Intake manifold, removing bolts in reverse order of installation

To install:

➡ Be sure to use new fasteners, as required.

6. Install the intake manifold with new gaskets. Tighten the bolts in order as shown.

7. Install or connect the following:
- Fuel tube assembly
- Fuel injectors
- Electronic throttle control actuator, tightening the bolts in several steps
- PCV hose
- Vacuum hoses
- Wiring harnesses

8. Connect the fuel tube as follows:
   a. Apply a thin layer of engine oil

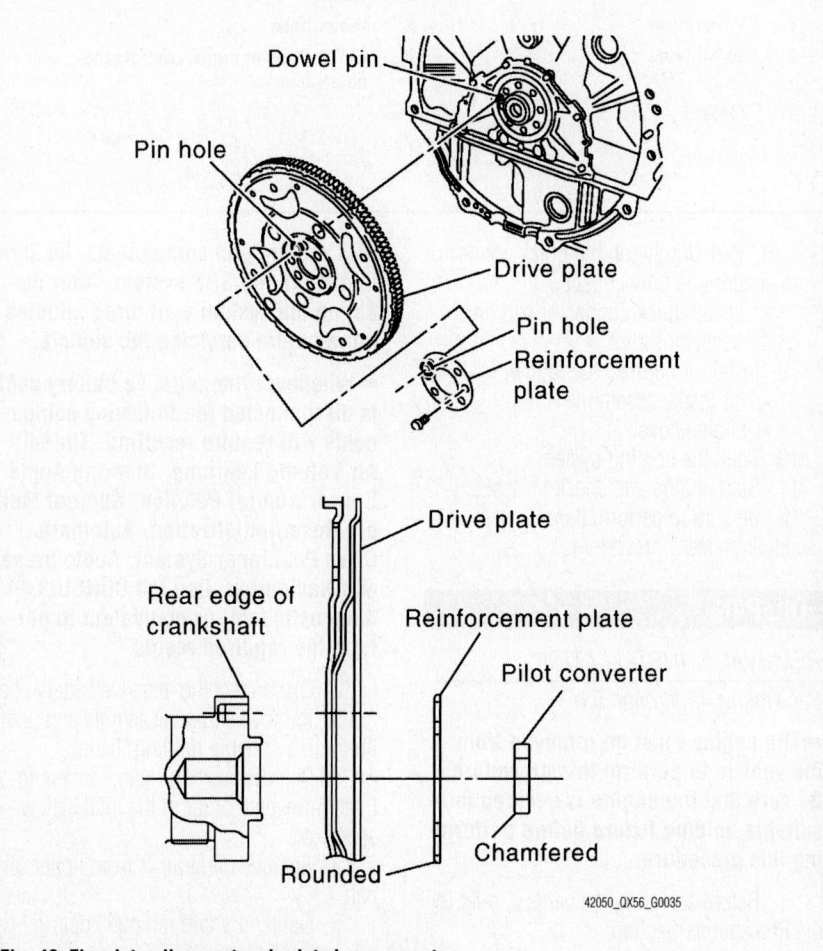

Dowel pin

Pin hole

Drive plate

Pin hole
Reinforcement plate

Drive plate

Rear edge of crankshaft

Reinforcement plate

Pilot converter

Rounded

Chamfered

42050_QX56_G0035

Fig. 46 Flexplate alignment and related components

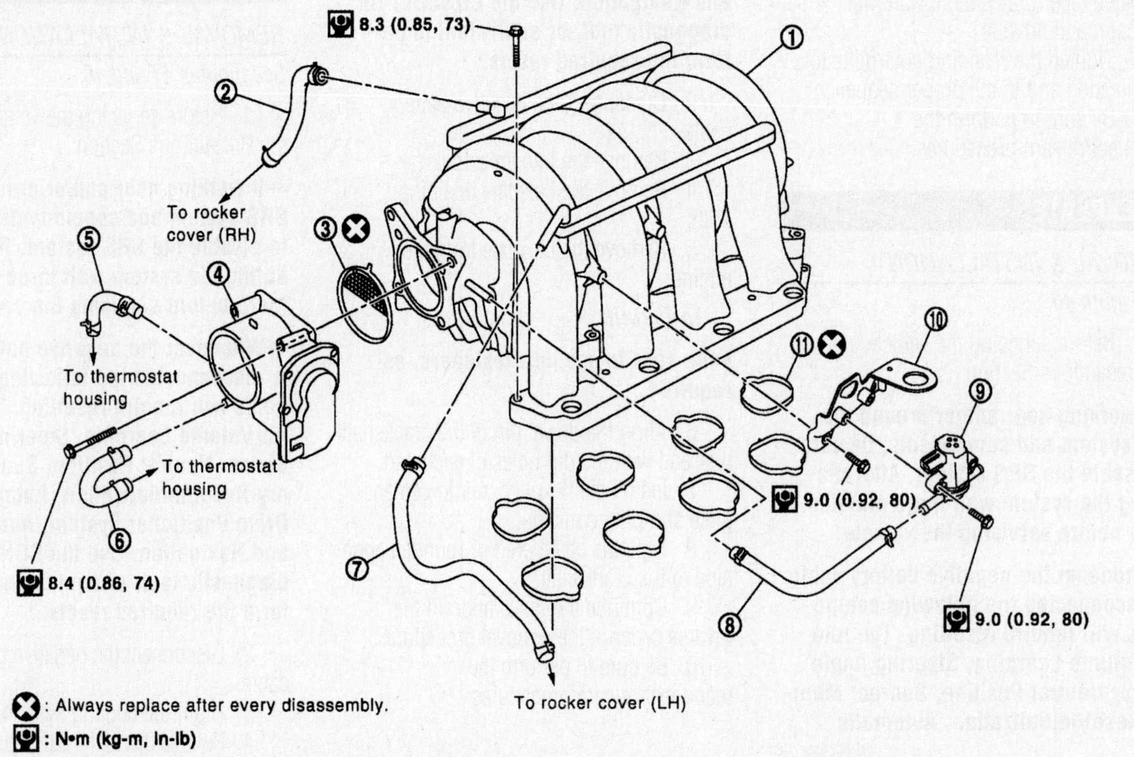

8.3 (0.85, 73)

2 To rocker cover (RH)

3 ✕

5

4

To thermostat housing

To thermostat housing

8.4 (0.86, 74)

6

7

1

10

11 ✕

9

9.0 (0.92, 80)

8

9.0 (0.92, 80)

To rocker cover (LH)

✕ : Always replace after every disassembly.

🔧 : N•m (kg-m, in-lb)

1.	Intake manifold	2.	PCV hose	3.	Gasket
4.	Electric throttle control actuator	5.	Water hose	6.	Water hose
7.	PCV hose	8.	EVAP hose	9.	EVAP canister purge control solenoid valve
10.	Bracket	11.	Gasket		

67170ARMADAG29

Fig. 47 Intake manifold and related components

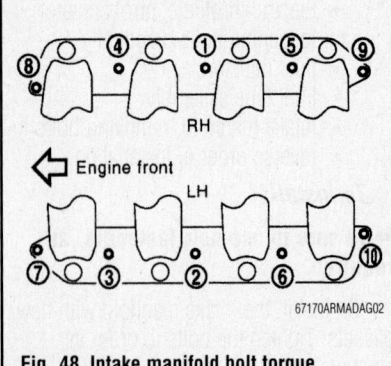

RH

⬅ Engine front

LH

67170ARMADAG02

Fig. 48 Intake manifold bolt torque sequence

on the tube from tip end to spool end.

b. Insert tube into quick connector past the white identification mark.

c. Insert tube into quick connector until top spool is completely inside the connector and 2nd level spool is exposed right below the connector.

d. Pull slightly on the quick connector to ensure it is fully engaged.

e. Install quick connector cap on quick connector joint.

9. Install or connect the following:
- Air intake assembly
- Engine cover

10. Refill the cooling system.

11. Start engine and check for leaks.

12. Be sure to perform the reconnect/relearn procedures.

OIL PAN

REMOVAL & INSTALLATION
See Figures 49 through 53.

➡ **The engine must be removed from the vehicle to perform this procedure. Be sure that the engine is secured in a suitable holding fixture before performing this procedure.**

1. Before servicing the vehicle, refer to the Precautions Section.

➡ **If working near and/or around the**

SRS system and components, be sure to disable the SRS system. After disabling the system wait three minutes or more before servicing the vehicle.

➡ **Whenever the negative battery cable is disconnected the following components will require resetting. The Idle Air Volume Learning, Steering Angle Sensor Neutral Position, Sunroof Memory Reset/Initialization, Automatic Drive Positioner System, Audio presets and Navigation. Use the CONSULT-III diagnostic tool, or equivalent to perform the required resets.**

2. Disconnect the negative battery cable.

3. Remove engine assembly and position it in a suitable holding fixture.

4. Remove lower oil pan, loosening bolts in reverse order of the installation sequence.

5. Remove oil strainer from upper oil pan.

6. Gently pry and remove upper oil pan from engine block.

To install:

➡ **Be sure to use new fasteners, as required.**

7. Apply liquid gasket to upper oil pan mating surfaces.

8. Install new O-rings to oil pump and front cover side.

9. Tighten upper oil pan bolts to specification and in the proper sequence.

10. Install or connect the following:
- Rear plate cover
- Oil strainer to upper oil pan

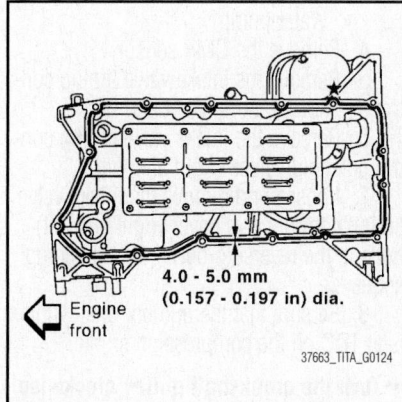

4.0 - 5.0 mm
(0.157 - 0.197 in) dia.

Engine
front

37663_TITA_G0124

Fig. 49 Upper oil pan sealant application

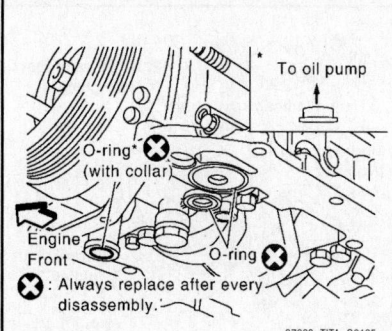

To oil pump

O-ring*
(with collar)

Engine
Front

O-ring*

⊗ : Always replace after every disassembly.

37663_TITA_G0125

Fig. 50 Upper oil pan O-ring installation

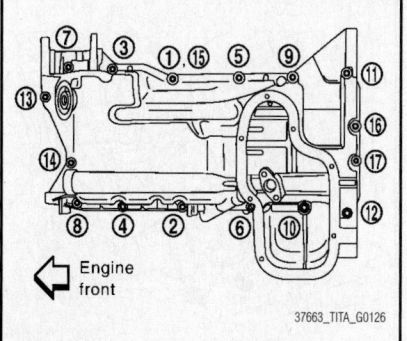

Engine
front

37663_TITA_G0126

Fig. 51 Upper oil pan bolt tightening sequence

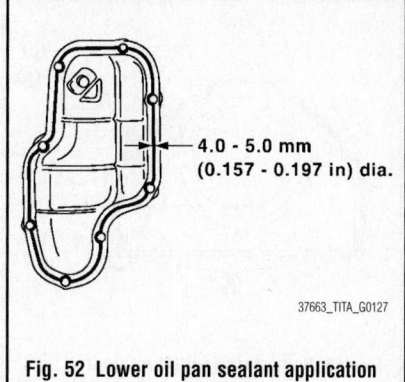

4.0 - 5.0 mm
(0.157 - 0.197 in) dia.

37663_TITA_G0127

Fig. 52 Lower oil pan sealant application

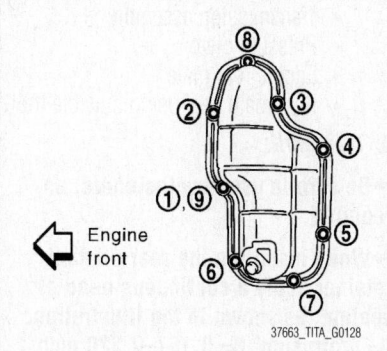

Engine
front

37663_TITA_G0128

Fig. 53 Lower oil pan bolt tightening sequence

- Lower oil pan, tightening bolts in proper sequence and to specification

11. Continue the installation in the reverse order of the removal procedure.

12. Be sure to perform the reconnect/relearn procedures.

OIL PUMP

REMOVAL & INSTALLATION

See Figure 54.

1. Before servicing the vehicle, refer to the Precautions Section.

➡ **If working near and/or around the SRS system and components, be sure to disable the SRS system. After disabling the system wait three minutes or more before servicing the vehicle.**

➡ **Whenever the negative battery cable is disconnected the following components will require resetting. The Idle Air Volume Learning, Steering Angle Sensor Neutral Position, Sunroof Memory Reset/Initialization, Automatic Drive Positioner System, Audio presets and Navigation. Use the CONSULT-III diagnostic tool, or equivalent to perform the required resets.**

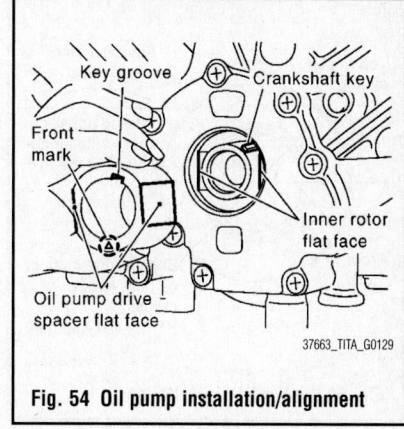

Key groove   Crankshaft key

Front
mark

Inner rotor
flat face

Oil pump drive
spacer flat face

37663_TITA_G0129

Fig. 54 Oil pump installation/alignment

2. Disconnect the negative battery cable.

3. Remove or disconnect the following:
- Timing chain cover
- Oil pump drive spacer
- Oil pump

To install:

➡ **Be sure to use new fasteners, as required.**

4. Install or connect the following:
- Oil pump
- Oil pump drive spacer
- Timing chain cover

➡ **When inserting the oil pump drive spacer, align the crankshaft key and the flat face of the inner rotor. If they are not aligned rotate the pump inner rotor by hand. Make sure that each part is aligned and tap lightly until it reaches the end.**

5. Be sure to perform the reconnect/relearn procedures.

PISTON AND RING

POSITIONING

See Figures 55 and 56.

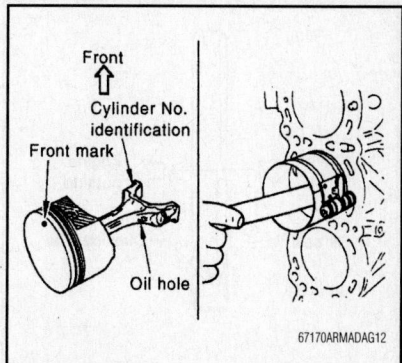

Front

Cylinder No.
identification

Front mark

Oil hole

67170ARMADAG12

Fig. 55 Piston and rod positioning and identification

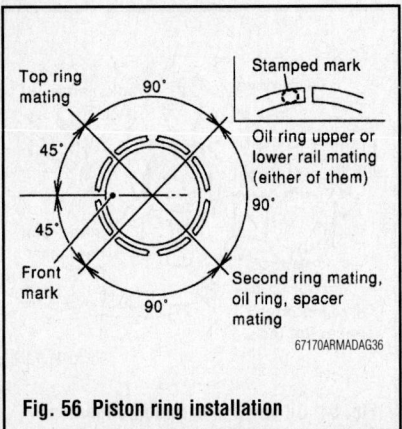

Fig. 56 Piston ring installation

REAR MAIN SEAL

REMOVAL & INSTALLATION

See Figures 57 and 58.

At this time the manufacturer does not provide removal and installation procedures for this component. The following procedure is a guideline and may differ from the vehicle you are servicing.

1. Before servicing the vehicle, refer to the Precautions Section.

➡️If working near and/or around the SRS system and components, be sure to disable the SRS system. After disabling the system wait three minutes or more before servicing the vehicle.

➡️Whenever the negative battery cable is disconnected the following components will require resetting. The Idle Air Volume Learning, Steering Angle Sensor Neutral Position, Sunroof Memory Reset/Initialization, Automatic Drive Positioner System, Audio presets and Navigation. Use the CONSULT-III diagnostic tool, or equivalent to perform the required resets.

2. Disconnect the negative battery cable.
3. Remove or disconnect the following:

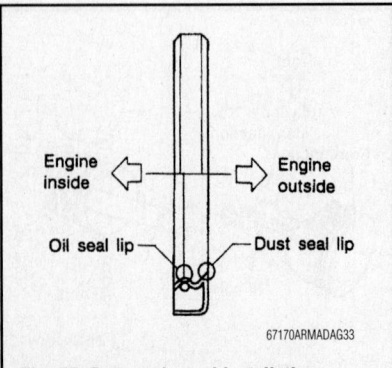

Fig. 57 Rear main seal installation positioning

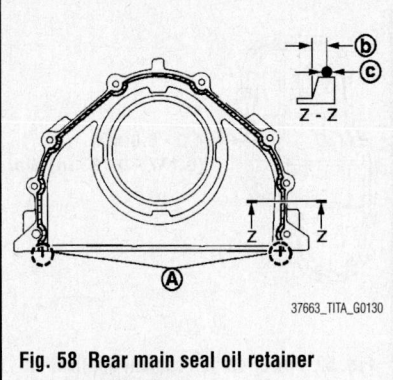

Fig. 58 Rear main seal oil retainer sealant application

• Transmission assembly
• Pressure plate
• Engine rear plate
• Rear main seal using suitable tool

To install:

➡️Be sure to use new fasteners, as required.

➡️When installing the rear oil seal retainer apply a continuous bead of sealant, as shown in the illustration. A= protrusion. B= 0.157–0.220 inch sealant. C= 0.134–0.173 inch sealant.

4. Install or connect the following:
• Rear main seal using suitable tool
• Engine rear plate
• Pressure plate
• Transmission assembly
5. Be sure to perform the reconnect/relearn procedures.

TIMING CHAIN FRONT COVER

REMOVAL & INSTALLATION

See Figures 59 through 64.

➡️The engine must be removed from the vehicle to perform this procedure. Be sure that the engine is secured in a suitable holding fixture before performing this procedure.

1. Before servicing the vehicle, refer to the Precautions Section.

➡️If working near and/or around the SRS system and components, be sure to disable the SRS system. After disabling the system wait three minutes or more before servicing the vehicle.

➡️Whenever the negative battery cable is disconnected the following components will require resetting. The Idle Air Volume Learning, Steering Angle Sensor Neutral Position, Sunroof Memory Reset/Initialization, Automatic

Drive Positioner System, Audio presets and Navigation. Use the CONSULT-III diagnostic tool, or equivalent to perform the required resets.

2. Disconnect the negative battery cable.
3. Remove or disconnect the following:
• Engine assembly
• Drive belt auto tensioner
• Idler pulley
• Thermostat housing and water hose
• Power steering pump bracket
• Oil pan (upper and lower)
• Oil strainer
• Alternator and bracket
• Rocker cover
• Water pump
4. Remove the CMP sensor.
5. Remove the intake valve timing control position sensors (right and left).
6. Remove the intake valve timing control solenoid valves (right and left).
7. Loosen and remove the intake valve timing control valve cover (right and left) bolts in the reverse order of the tightening sequence.
8. Be sure that the number one cylinder is at TDC on the compression stroke.

➡️Turn the crankshaft pulley clockwise to align the TDC identification notch (without paint mark) with the timing

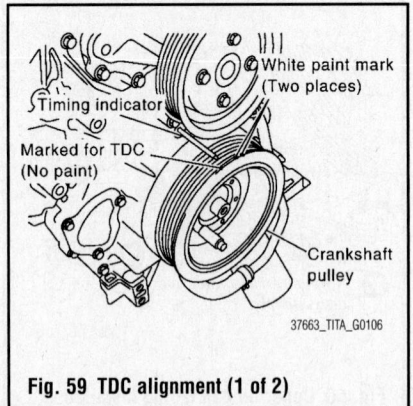

Fig. 59 TDC alignment (1 of 2)

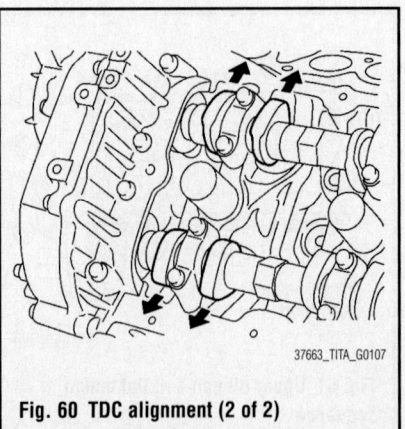

Fig. 60 TDC alignment (2 of 2)

indicator on the front cover. At this time make sure that the intake and exhaust cam lobes of the number one cylinder (top front on left bank) point outside. If not turn the crankshaft pulley once more. See illustration

9. Remove the crankshaft pulley.
10. Remove the front cover retaining bolts. Remove the front cover.
11. Discard the gasket.

To install:

→Be sure to use new fasteners, as required.

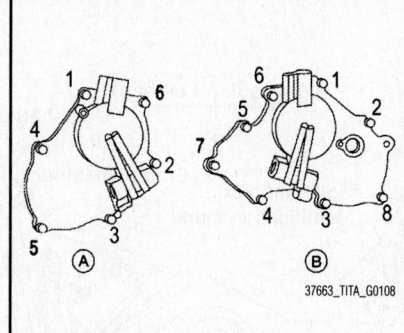

Fig. 64 Intake valve timing control solenoid cover tightening sequence

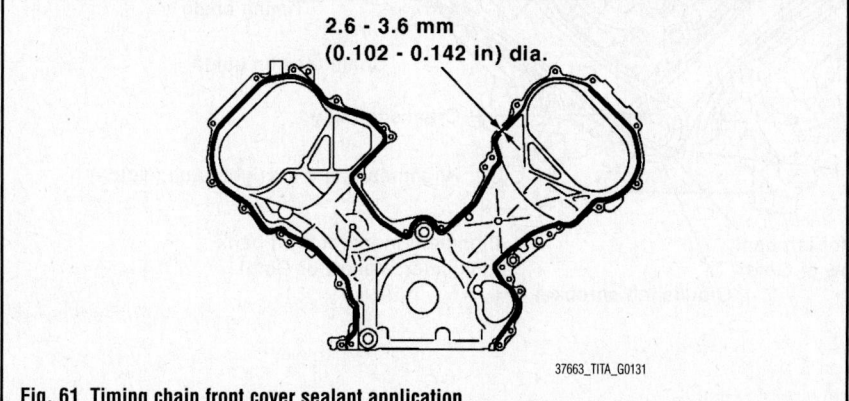

Fig. 61 Timing chain front cover sealant application

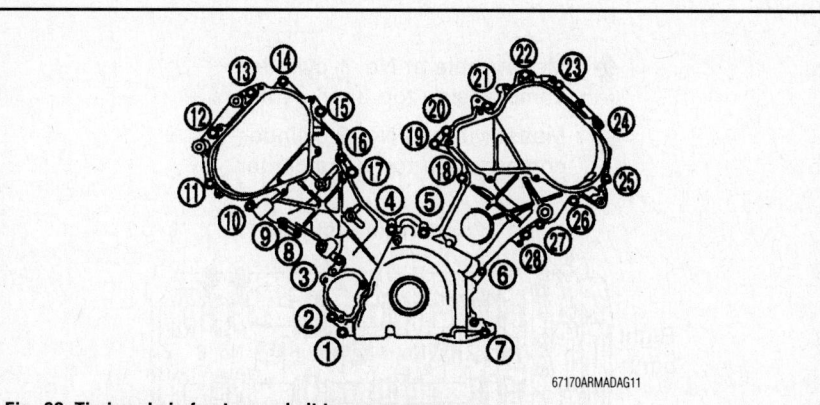

Fig. 62 Timing chain front cover bolt torque sequence

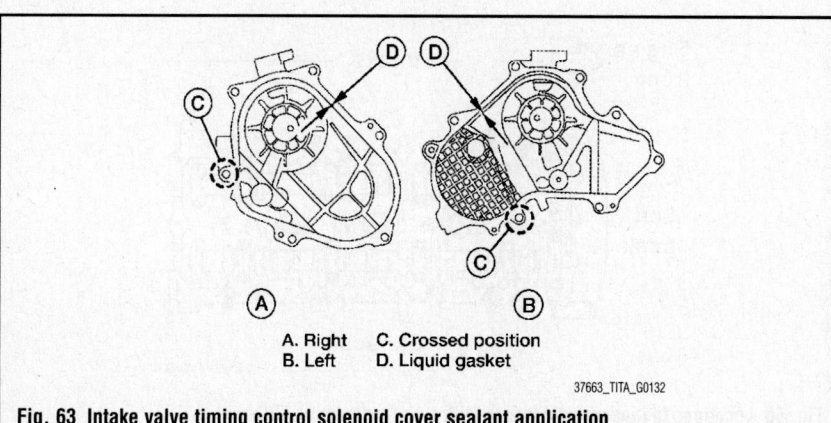

A. Right   C. Crossed position
B. Left   D. Liquid gasket

37663_TITA_G0132

Fig. 63 Intake valve timing control solenoid cover sealant application

12. Installation is the reverse of the removal procedure.
13. Apply sealant to the front cover. See illustration.
14. Install retaining bolts and tighten to specification and in the proper sequence.
15. Apply liquid gasket to the valve timing control solenoid covers.

→The start and end of the sealant application should be crossed at position "C", see illustration, that cannot be seen after attaching the intake valve timing control solenoid valve cover.

16. Install the intake valve timing control solenoid covers. Tighten bolts in proper sequence.
17. Continue the installation in the reverse order of the removal procedure.
18. Be sure to perform the reconnect/relearn procedures.

TIMING CHAIN & SPROCKETS

REMOVAL & INSTALLATION

See Figure 65.

→The engine must be removed from the vehicle to perform this procedure. Be sure that the engine is secured in a suitable holding fixture before performing this procedure.

1. Before servicing the vehicle, refer to the Precautions Section.

→If working near and/or around the SRS system and components, be sure to disable the SRS system. After disabling the system wait three minutes or more before servicing the vehicle.

→Whenever the negative battery cable is disconnected the following components will require resetting. The Idle Air Volume Learning, Steering Angle Sensor Neutral Position, Sunroof Memory Reset/Initialization, Automatic Drive Positioner System, Audio presets and Navigation. Use the CONSULT-III diagnostic tool, or equivalent to perform the required resets.

2. Disconnect the negative battery cable.
3. Remove the front cover.
4. Remove the oil pump drive spacer.
5. Remove the oil pump.
   - Timing chain tensioner
   - Chain tension guide and slack guide
   - Timing chain
   - Camshaft sprocket

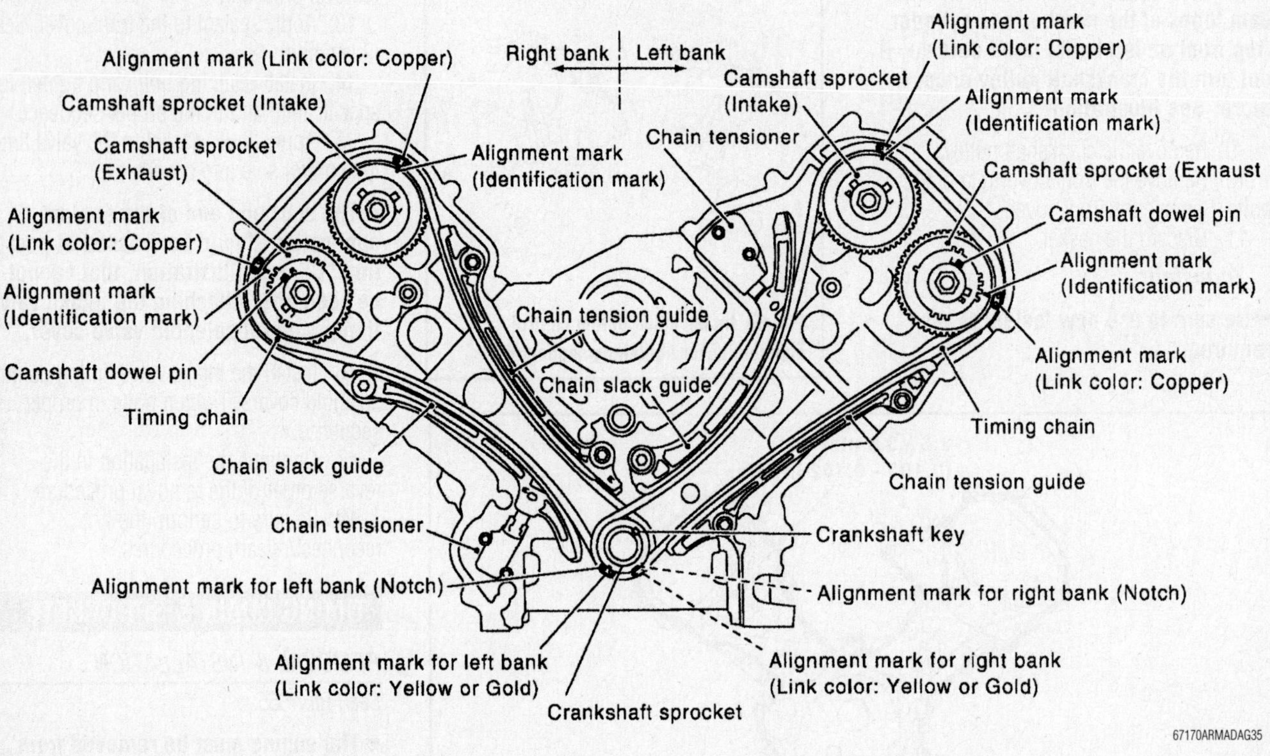

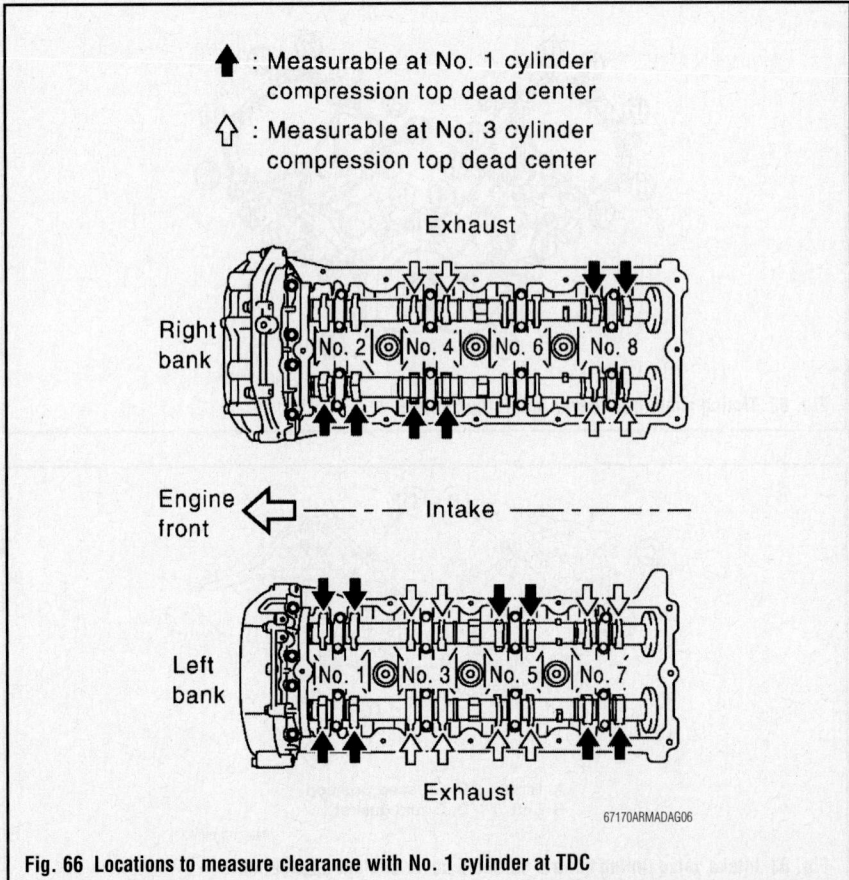

Fig. 65 Timing chains and related components

To install:

➡ **Be sure to use new fasteners, as required.**

6. Ensure that the crankshaft key and dowel pin of each camshaft are facing the same direction.

7. Install or connect the following:
- Camshaft sprockets
- Timing chain
- Chain tension guide and slack guide
- Oil pump
- Oil pump drive spacer
- Front oil seal, using suitable tool

8. Continue the installation in the reverse order of the removal procedure.

9. Be sure to perform the reconnect/relearn procedures.

VALVE LASH

ADJUSTMENT

See Figures 66 and 67.

1. Before servicing the vehicle, refer to the Precautions Section.

➡ **Perform the following inspection after removal, installation or replacement of camshaft or valve-related**

Fig. 66 Locations to measure clearance with No. 1 cylinder at TDC

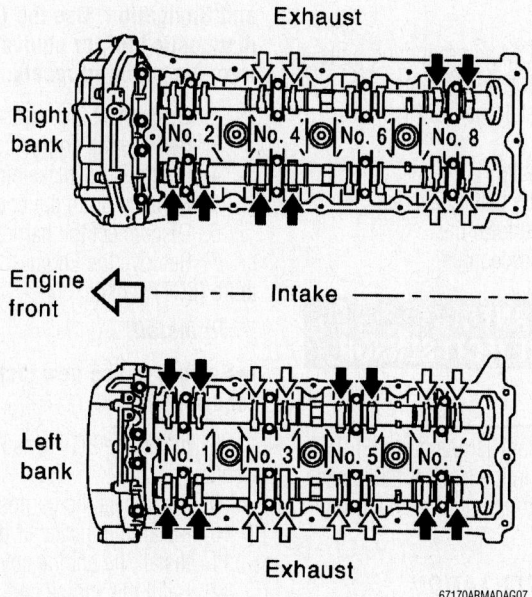

↑ (filled): Measurable at No. 1 cylinder compression top dead center

↑ (open): Measurable at No. 3 cylinder compression top dead center

Exhaust

Right bank   No. 2 No. 4 No. 6 No. 8

Engine front ⟵ - - - Intake

Left bank   No. 1 No. 3 No. 5 No. 7

Exhaust

67170ARMADAG07

Fig. 67 Locations to measure clearance with No. 3 cylinder at TDC

parts, or if there are unusual engine conditions due to changes in valve clearance over time (starting, idling, and/or noise).

2. Run engine to operating temperature.
3. Remove or disconnect the following:

- Battery cover, if equipped
- Engine room cover
- Air intake assembly
- Left and right rocker covers

4. Turn the crankshaft pulley clockwise to Top Dead Center (TDC) identification notch with timing indicator.

5. Ensure that both the intake and exhaust cam noses of the No. 1 cylinder face outside.

6. Measure the valve clearances at locations shown in figure.

7. Turn the crankshaft pulley clockwise 270 degrees from the position of No. 1 cylinder compression to obtain No. 3 cylinder compression TDC.

8. Measure the valve clearances at locations shown in the figure.

9. Turn crankshaft pulley clockwise 90 degrees and measure the intake and exhaust valve clearance of No. 6 cylinder and exhaust valve clearance of No. 2 cylinder.

10. To adjust the valves, remove camshaft and valve lifter(s) out of specification.

11. Install replacement valve lifter(s).

12. Install the camshaft.

13. Manually turn the crankshaft pulley several turns.

14. Recheck valve clearances with engine at operating temperature.

INSPECTION

1. Before servicing the vehicle, refer to the Precautions Section.

2. Remove camshaft and valve lifter(s) out of specification.

3. Install replacement valve lifter(s).

4. Install the camshaft.

5. Manually turn the crankshaft pulley several turns.

6. Recheck valve clearances with engine at operating temperature.

ENGINE PERFORMANCE & EMISSION CONTROLS

CAMSHAFT POSITION (CMP) SENSOR

LOCATION

The Camshaft Position (CMP) sensor is located on the right front of the timing cover, facing the engine.

REMOVAL & INSTALLATION

1. Before servicing the vehicle, refer to the Precautions Section.

➡**If working near and/or around the SRS system and components, be sure to disable the SRS system. After disabling the system wait three minutes or more before servicing the vehicle.**

➡**Whenever the negative battery cable is disconnected the following compo-**

nents will require resetting. The Idle Air Volume Learning, Steering Angle Sensor Neutral Position, Sunroof Memory Reset/Initialization, Automatic Drive Positioner System, Audio presets and Navigation. Use the CONSULT-III diagnostic tool, or equivalent to perform the required resets.

2. Disconnect the negative battery cable.

3. Remove the engine cover.

4. Remove air intake duct.

5. Disconnect the camshaft position sensor.

6. Remove the bolt and the Camshaft Position (CMP) sensor.

To install:

7. Install the CMP sensor and tighten the bolt.

8. Reconnect the camshaft electrical sensor.

9. Install the air intake duct.

10. Install the engine cover.

11. Be sure to perform the reconnect/relearn procedures.

CRANKSHAFT POSITION (CKP) SENSOR

LOCATION

The Crankshaft Position (CKP) sensor is located on the transmission assembly facing the gear teeth (cogs) of the signal plate.

REMOVAL & INSTALLATION

1. Before servicing the vehicle, refer to the Precautions Section.

➡**If working near and/or around the**

SRS system and components, be sure to disable the SRS system. After disabling the system wait three minutes or more before servicing the vehicle.

➡Whenever the negative battery cable is disconnected the following components will require resetting. The Idle Air Volume Learning, Steering Angle Sensor Neutral Position, Sunroof Memory Reset/Initialization, Automatic Drive Positioner System, Audio presets and Navigation. Use the CONSULT-III diagnostic tool, or equivalent to perform the required resets.

2. Disconnect the negative battery cable.
3. Raise and support the vehicle safely.
4. Disconnect the Crankshaft Position (CKP) sensor connector.
5. Remove the mounting bolt and CKP sensor.

To install:
6. Install the CKP sensor and tighten the mounting bolt.
7. Reconnect the CKP sensor connector.
8. Lower the vehicle.
9. Be sure to perform the reconnect/relearn procedures.

ELECTRONIC CONTROL MODULE (ECM)

LOCATION

The Electronic Control Module (ECM) is located in the engine room passenger side behind battery.

REMOVAL & INSTALLATION

At this time the manufacturer does not provide removal and installation procedures for this component. The following procedure is a guideline and may differ from the vehicle you are servicing.
1. Before servicing the vehicle, refer to the Precautions Section.

➡If working near and/or around the SRS system and components, be sure to disable the SRS system. After disabling the system wait three minutes or more before servicing the vehicle.

➡Whenever the negative battery cable is disconnected the following components will require resetting. The Idle Air Volume Learning, Steering Angle Sensor Neutral Position, Sunroof Memory Reset/Initialization, Automatic Drive Positioner System, Audio presets and Navigation. Use the CONSULT-III diagnostic tool, or equivalent to perform the required resets.

2. Disconnect the negative battery cable.
3. Disconnect the positive battery cable and remove the battery, as required..
4. Carefully remove the Electronic Control Module (ECM) harness connectors.
5. Remove the ECM mounting bolts and the ECM.

To install:
6. Install the ECM and mounting bolts and tighten to 62 inch lbs. (7 Nm).
7. Carefully install the ECM harness connectors.
8. Install the battery.
9. Reconnect the battery cables.
10. Be sure to perform the reconnect/relearn procedures.

ENGINE COOLANT TEMPERATURE (ECT) SENSOR

LOCATION

The Engine Coolant Temperature (ECT) sensor is mounted in the front of the intake manifold. It is just to the right of the throttle body.

REMOVAL & INSTALLATION

1. Before servicing the vehicle, refer to the Precautions Section.

➡If working near and/or around the SRS system and components, be sure to disable the SRS system. After disabling the system wait three minutes or more before servicing the vehicle.

➡Whenever the negative battery cable is disconnected the following components will require resetting. The Idle Air Volume Learning, Steering Angle Sensor Neutral Position, Sunroof Memory Reset/Initialization, Automatic Drive Positioner System, Audio presets and Navigation. Use the CONSULT-III diagnostic tool, or equivalent to perform the required resets.

2. Disconnect the negative battery cable.
3. Remove the engine cover.
4. Remove the intake air duct.
5. Partially drain the cooling system.
6. Disconnect the harness connector.
7. Remove the Engine Coolant Temperature (ECT) sensor.

To install:

➡Be sure to use new fasteners, as required.

8. Install the ECT sensor and carefully tighten.
9. Reconnect the harness connector.
10. Install the intake air duct.
11. Install the engine cover.
12. Refill the engine coolant.
13. Be sure to perform the reconnect/relearn procedures.

HEATED OXYGEN SENSOR (HO2S)

LOCATION
See Figure 68.

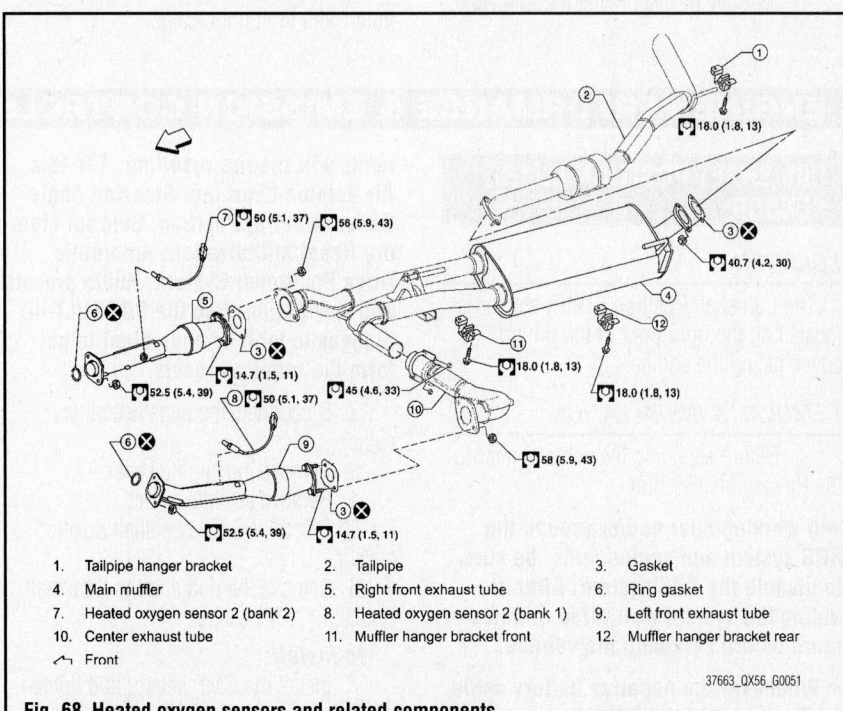

1. Tailpipe hanger bracket
2. Tailpipe
3. Gasket
4. Main muffler
5. Right front exhaust tube
6. Ring gasket
7. Heated oxygen sensor 2 (bank 2)
8. Heated oxygen sensor 2 (bank 1)
9. Left front exhaust tube
10. Center exhaust tube
11. Muffler hanger bracket front
12. Muffler hanger bracket rear
↰ Front

37663_QX56_G0051

Fig. 68 Heated oxygen sensors and related components

The Heated Oxygen (HO2S) sensors are located after the exhaust manifold converter assembly, in the lower part of the exhaust system.

REMOVAL & INSTALLATION

See Figure 68.

1. Before servicing the vehicle, refer to the Precautions Section.

➡**If working near and/or around the SRS system and components, be sure to disable the SRS system. After disabling the system wait three minutes or more before servicing the vehicle.**

➡**Whenever the negative battery cable is disconnected the following components will require resetting. The Idle Air Volume Learning, Steering Angle Sensor Neutral Position, Sunroof Memory Reset/Initialization, Automatic Drive Positioner System, Audio presets and Navigation. Use the CONSULT-III diagnostic tool, or equivalent to perform the required resets.**

2. Disconnect the negative battery cable.
3. Raise and safely support the vehicle.
4. Remove the engine undercover, as needed.
5. Unplug the Heated Oxygen (HO2S) sensor harness.
6. Using an O2 wrench remove the HO2S sensor.

➡**Lower the exhaust in needed.**

To install:

➡**Be sure to use new fasteners, as required.**

7. Install the HO2S sensor and tighten to 37 ft. lbs. (50 Nm).
8. Install the harness connector.
9. Keep the harness connector and wiring away from exhaust system.
10. Be sure to perform the reconnect/relearn procedures.

INTAKE AIR TEMPERATURE (IAT) SENSOR

LOCATION

The Intake Air Temperature (IAT) sensor is integral to the mass air flow sensor, and is mounted on the air filter housing lid.

REMOVAL & INSTALLATION

See Figure 69.

1. Before servicing the vehicle, refer to the Precautions Section.

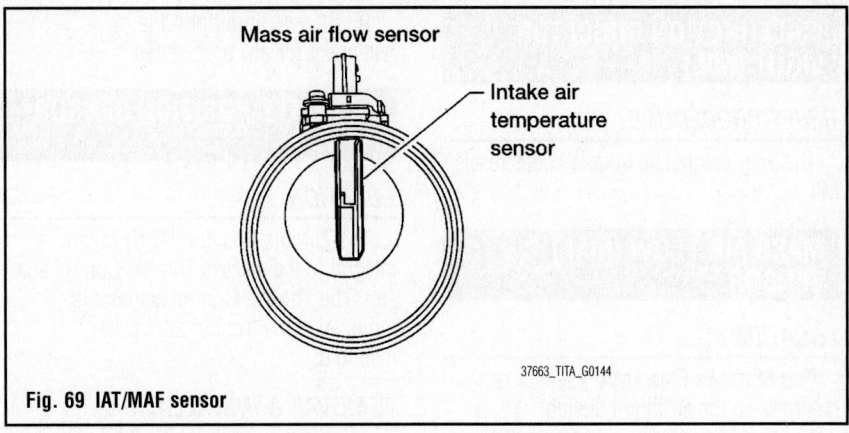

Mass air flow sensor

Intake air temperature sensor

37663_TITA_G0144

Fig. 69 IAT/MAF sensor

➡**If working near and/or around the SRS system and components, be sure to disable the SRS system. After disabling the system wait three minutes or more before servicing the vehicle.**

➡**Whenever the negative battery cable is disconnected the following components will require resetting. The Idle Air Volume Learning, Steering Angle Sensor Neutral Position, Sunroof Memory Reset/Initialization, Automatic Drive Positioner System, Audio presets and Navigation. Use the CONSULT-III diagnostic tool, or equivalent to perform the required resets.**

2. Disconnect the negative battery cable.
3. Remove the engine room cover.
4. Remove the Intake Air Temperature (IAT/MAF) sensor harness.
5. Remove the mounting screws and the IAT/MAF sensor.

To install:

6. Install the IAT/MAF sensor.
7. Install the harness connector.
8. Install the engine room cover.
9. Be sure to perform the reconnect/relearn procedures.

KNOCK SENSOR (KS)

LOCATION

See Figure 70.

The Knock (KS) sensors are mounted under the intake manifold on the cylinder block.

REMOVAL & INSTALLATION

See Figure 70.

At this time the manufacturer does not provide removal and installation procedures for this component. The intake manifold will have to be removed to service this component.

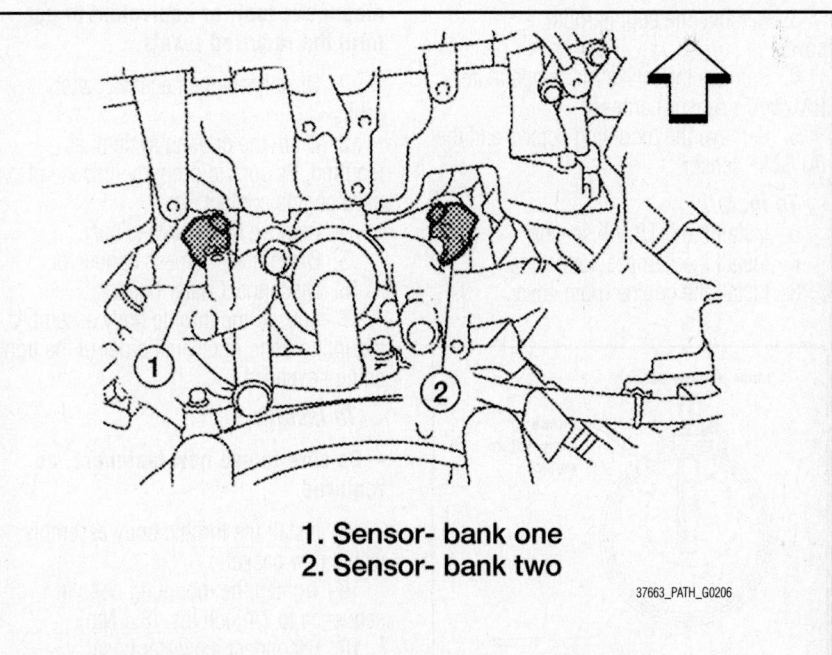

1. Sensor- bank one
2. Sensor- bank two

37663_PATH_G0206

Fig. 70 Knock Sensor (KS) location (view with engine removed from vehicle)

MALFUNCTION INDICATOR LIGHT (MIL)

RESET PROCEDURE

Clearing diagnostic trouble codes resets MIL.

MASS AIR FLOW (MAF) SENSOR

LOCATION

The Mass Air Flow (MAF) sensor is mounted on the air filter housing.

REMOVAL & INSTALLATION

See Figure 71.

1. Before servicing the vehicle, refer to the Precautions Section.

➡**If working near and/or around the SRS system and components, be sure to disable the SRS system. After disabling the system wait three minutes or more before servicing the vehicle.**

➡**Whenever the negative battery cable is disconnected the following components will require resetting. The Idle Air Volume Learning, Steering Angle Sensor Neutral Position, Sunroof Memory Reset/Initialization, Automatic Drive Positioner System, Audio presets and Navigation. Use the CONSULT-III diagnostic tool, or equivalent to perform the required resets.**

2. Disconnect the negative battery cable.
3. Remove the engine room cover.
4. Remove the Intake Air Temperature (IAT/MAF) sensor harness.
5. Remove the mounting screws and the IAT/MAF sensor.

To install:

6. Install the IAT/MAF sensor.
7. Install the harness connector.
8. Install the engine room cover.

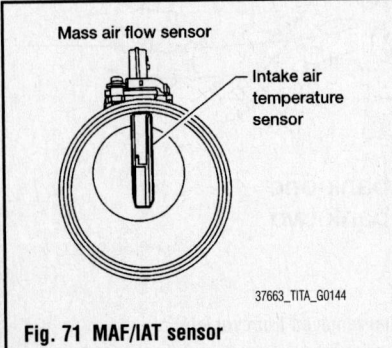

Mass air flow sensor

Intake air temperature sensor

37663_TITA_G0144

Fig. 71 MAF/IAT sensor

9. Be sure to perform the reconnect/relearn procedures.

THROTTLE POSITION SENSOR (TPS)

LOCATION

The Throttle Position (TPS) sensor is integral to the electric Throttle Control actuator. The Throttle Control actuator is mounted at the front of the intake manifold.

REMOVAL & INSTALLATION

See Figure 72.

At this time the manufacturer does not provide removal and installation procedures for this component. The following procedure is a guideline and may differ from the vehicle you are servicing.

1. Before servicing the vehicle, refer to the Precautions Section.

➡**If working near and/or around the SRS system and components, be sure to disable the SRS system. After disabling the system wait three minutes or more before servicing the vehicle.**

➡**Whenever the negative battery cable is disconnected the following components will require resetting. The Idle Air Volume Learning, Steering Angle Sensor Neutral Position, Sunroof Memory Reset/Initialization, Automatic Drive Positioner System, Audio presets and Navigation. Use the CONSULT-III diagnostic tool, or equivalent to perform the required resets.**

2. Disconnect the negative battery cable.
3. Drain the cooling system, as required. Be sure to properly dispose of used engine coolant.
4. Remove the air intake duct.
5. Disconnect harness connector.
6. Disconnect water hoses.
7. Loosen the throttle body assembly mounting bolts in reverse order of the tightening sequence.

To install:

➡**Be sure to use new fasteners, as required.**

8. Install the throttle body assembly with a new gasket.
9. Tighten the mounting bolts in sequence to 74 inch lbs. (8.4 Nm).
10. Reconnect the water hose.
11. Reconnect the harness connector.

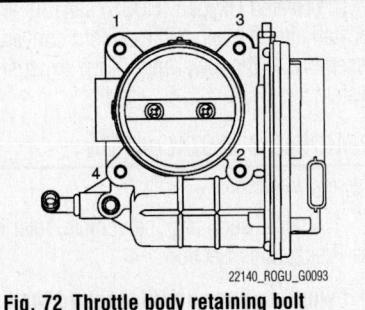

22140_ROGU_G0093

Fig. 72 Throttle body retaining bolt tightening sequence

12. Reconnect the air intake duct.
13. Fill the cooling system with the proper grade and type engine coolant.
14. Be sure to perform the reconnect/relearn procedures.

VARIABLE CAMSHAFT TIMING OIL CONTROL SOLENOID

LOCATION

The Intake Valve Timing Control solenoid is located in the front timing chain cover.

REMOVAL & INSTALLATION

At this time the manufacturer does not provide removal and installation procedures for this component.

VEHICLE SPEED SENSOR (VSS)

LOCATION

The VSS sensor is located at the rear of the transmission case, under the tail shaft. On 4WD vehicles this component is located under the transfer case.

REMOVAL & INSTALLATION

1. Before servicing the vehicle, refer to the Precautions Section.

➡**If working near and/or around the SRS system and components, be sure to disable the SRS system. After disabling the system wait three minutes or more before servicing the vehicle.**

➡**Whenever the negative battery cable is disconnected the following components will require resetting. The Idle Air Volume Learning, Steering Angle Sensor Neutral Position, Sunroof Memory Reset/Initialization, Automatic Drive Positioner System, Audio presets and Navigation. Use the CONSULT-III diagnostic tool, or equivalent to perform the required resets.**

2. Disconnect the negative battery cable.

3. Raise and safely support the vehicle.
4. Disconnect the sensor harness.
5. Remove the mounting bolt and the sensor.

To install:

➡**Be sure to use new fasteners, as required.**

6. Apply a small amount of transmission fluid to the sensor O-ring.

7. Install the Speed sensor and tighten the mounting bolt to 51 inch lbs. (5.8 Nm).
8. Be sure to perform the reconnect/relearn procedures.

FUEL

GASOLINE FUEL INJECTION SYSTEM

FUEL SYSTEM SERVICE PRECAUTIONS

Safety is the most important factor when performing not only fuel system maintenance, but any type of maintenance. Failure to conduct maintenance and repairs in a safe manner may result in serious personal injury or death. Work on a vehicle's fuel system components can be accomplished safely and effectively by adhering to the following rules and guidelines.

• To avoid the possibility of fire and personal injury, always disconnect the negative battery cable unless the repair or test procedure requires that battery voltage be applied.

• Always relieve the fuel system pressure prior to disconnecting any fuel system component (injector, fuel rail, pressure regulator, etc.) fitting or fuel line connection. Exercise extreme caution whenever relieving fuel system pressure to avoid exposing skin, face and eyes to fuel spray. Please be advised that fuel under pressure may penetrate the skin or any part of the body that it contacts.

• Always place a shop towel or cloth around the fitting or connection prior to loosening to absorb any excess fuel due to spillage. Ensure that all fuel spillage is quickly removed from engine surfaces. Ensure that all fuel-soaked cloths or towels are deposited into a flame-proof waste container with a lid.

• Always keep a dry chemical (Class B) fire extinguisher near the work area.

• Do not allow fuel spray or fuel vapors to come into contact with a spark or open flame.

• Always use a second wrench when loosening or tightening fuel line connection fittings. This will prevent unnecessary stress and torsion on fuel piping. Always follow the proper torque specifications.

• Always replace worn fuel fitting O-rings with new ones. Do not substitute fuel hose where rigid pipe is installed.

FUEL SYSTEM PRESSURE

RELIEVING

With CONSULT-III®

1. Turn ignition switch **ON**.
2. Perform "FUEL PRESSURE RELEASE" in "WORK SUPPORT" mode with CONSULT-III®.
3. Start engine.
4. After engine stalls, turn over the engine two or three times to release all fuel pressure.
5. Turn ignition switch **OFF**.

Without CONSULT-III®

See Figure 73.

1. Before servicing the vehicle, refer to the Precautions Section.

➡**If working near and/or around the SRS system and components, be sure to disable the SRS system. After disabling the system wait three minutes or more before servicing the vehicle.**

➡**Whenever the negative battery cable is disconnected the following components will require resetting. The Idle Air Volume Learning, Steering Angle Sensor Neutral Position, Sunroof Memory Reset/Initialization, Automatic Drive Positioner System, Audio presets and Navigation. Use the CONSULT-III diagnostic tool, or equivalent to perform the required resets.**

2. Remove fuel pump fuse located in IPDM E/R.
3. Start engine.
4. After engine stalls, turn over engine two or three times to release all fuel pressure.
5. Turn ignition switch **OFF**.
6. Disconnect the negative battery cable.
7. Reinstall fuel pump fuse after servicing fuel system.
8. Be sure to perform the reconnect/relearn procedures.

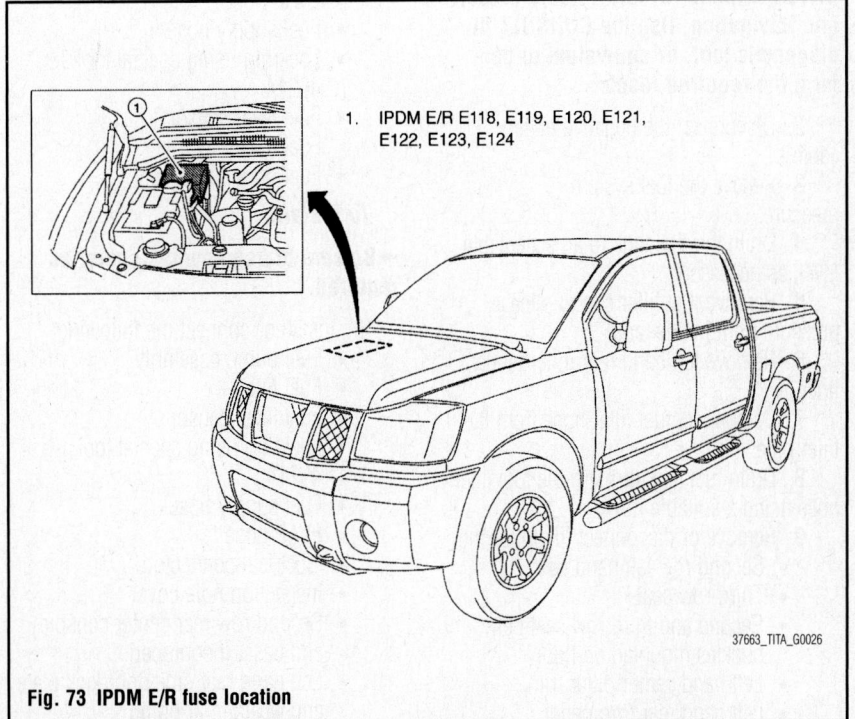

1. IPDM E/R E118, E119, E120, E121, E122, E123, E124

37663_TITA_G0026

Fig. 73 IPDM E/R fuse location

FUEL FILTER

REMOVAL & INSTALLATION

➡The fuel filter is part of the fuel pump assembly.

FUEL LEVEL SENDING UNIT

LOCATION

This component is located on the fuel pump module.

REMOVAL & INSTALLATION

This component is located on the fuel pump module. This component is removed along with the fuel pump module.

FUEL PUMP MODULE

REMOVAL & INSTALLATION

See Figure 74.

1. Before servicing the vehicle, refer to the Precautions Section.

➡If working near and/or around the SRS system and components, be sure to disable the SRS system. After disabling the system wait three minutes or more before servicing the vehicle.

➡Whenever the negative battery cable is disconnected the following components will require resetting. The Idle Air Volume Learning, Steering Angle Sensor Neutral Position, Sunroof Memory Reset/Initialization, Automatic Drive Positioner System, Audio presets and Navigation. Use the CONSULT-III diagnostic tool, or equivalent to perform the required resets.

2. Disconnect the negative battery cable.
3. Relieve the fuel system pressure.
4. Drain the fuel tank to an acceptable level, as necessary.
5. Remove fuel filler cap to release pressure from inside tank.
6. Remove left hand rear inner fender liner.
7. Disconnect fuel filler hose from fuel filler pipe.
8. Drain fuel tank through the fuel filler hose using a suitable hose.
9. Remove or disconnect the following:
   - Second row left hand seat
   - Third row seat
   - Second and third row seat belt buckles mounted on floor
   - Left hand center pillar trim
   - Left hand rear trim panel

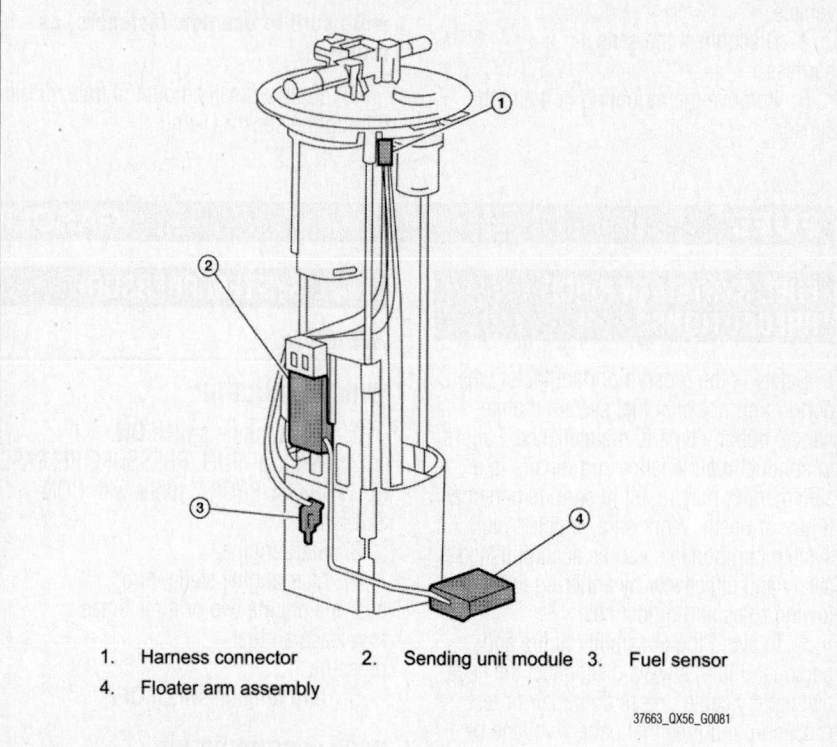

1. Harness connector 2. Sending unit module 3. Fuel sensor
4. Floater arm assembly

37663_QX56_G0081

Fig. 74 Fuel pump module and related components

   - Left hand rear side door kick plate and weather stripping
   - Second row rear center console and base, if equipped
   - Inspection hole cover under carpet by turning retainers 90 degrees
   - Electrical connectors
   - EVAP hose
   - Fuel supply hose
   - Lockring using special tool J-46214
   - Fuel level sensor
   - Fuel filter
   - Fuel pump assembly

To install:

➡Be sure to use new fasteners, as required.

10. Install or connect the following:
   - Fuel pump assembly
   - Fuel filter
   - Fuel level sensor
   - Lockring using special tool J-46214
   - Fuel supply hose
   - EVAP hose
   - Electrical connectors
   - Inspection hole cover
   - Second row rear center console and base, if equipped
   - Left hand rear side door kick plate and weather stripping

   - Left hand rear trim panel
   - Left hand center pillar trim
   - Second and third row seat belt buckles
   - Third row seat
   - Second row left hand seat
   - Fuel filler hose to fuel filler pipe
   - Left hand rear inner fender liner

11. Start the engine and check for leaks.

12. Be sure to perform the reconnect/relearn procedures.

FUEL RAIL AND INJECTOR

REMOVAL & INSTALLATION

See Figure 75.

1. Before servicing the vehicle, refer to the Precautions Section.

➡If working near and/or around the SRS system and components, be sure to disable the SRS system. After disabling the system wait three minutes or more before servicing the vehicle.

➡Whenever the negative battery cable is disconnected the following components will require resetting. The Idle Air Volume Learning, Steering Angle Sensor Neutral Position, Sunroof Memory Reset/Initialization, Automatic Drive Positioner System, Audio presets and **Navigation. Use the CONSULT-III diagnostic tool, or equivalent to perform the required resets.**

2. Disconnect the negative battery cable.

3. Remove engine cover. Remove the air cleaner assembly.

4. Relieve fuel system pressure.

5. Remove or disconnect the following:

- Fuel injector harness connectors
- Fuel hose assembly from right and left fuel rails

- Fuel injectors with fuel rail as an assembly
- Fuel injector from fuel rail

To install:

➡**Be sure to use new fasteners, as required.**

6. Install or connect the following:

➡**Always use a new O-ring when reinstalling the fuel injector to the fuel rail.**

- New clip onto the fuel injector
- Fuel injector to fuel rail

- Fuel injectors and fuel rail as an assembly to the intake manifold. Tighten the bolts to 8 ft. lbs. (11 Nm).
- Fuel hose assembly
- Fuel injector harness connectors
- Negative battery cable
- Engine cover

7. Start engine and check for leaks.

8. Be sure to perform the reconnect/relearn procedures.

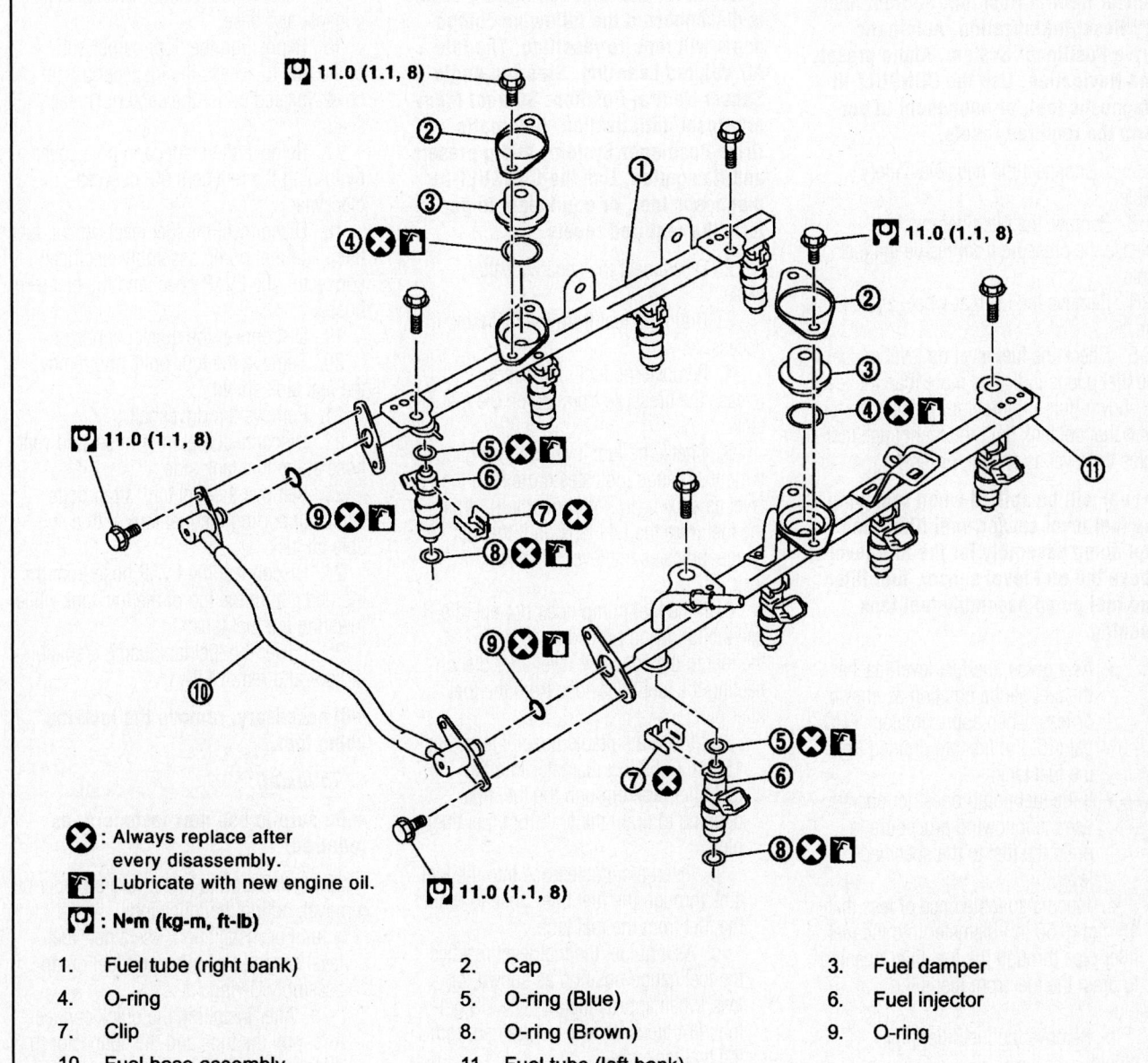

🅧 : Always replace after every disassembly.

🛢 : Lubricate with new engine oil.

🔧 : N•m (kg-m, ft-lb)

1.	Fuel tube (right bank)	2.	Cap	3.	Fuel damper
4.	O-ring	5.	O-ring (Blue)	6.	Fuel injector
7.	Clip	8.	O-ring (Brown)	9.	O-ring
10.	Fuel hose assembly	11.	Fuel tube (left bank)		

67170ARMADAG38

Fig. 75 Fuel injectors/rail and related components

FUEL TANK

DRAINING

1. Before servicing the vehicle, refer to the Precautions Section.

➡ **If working near and/or around the SRS system and components, be sure to disable the SRS system. After disabling the system wait three minutes or more before servicing the vehicle.**

➡ **Whenever the negative battery cable is disconnected the following components will require resetting. The Idle Air Volume Learning, Steering Angle Sensor Neutral Position, Sunroof Memory Reset/Initialization, Automatic Drive Positioner System, Audio presets and Navigation. Use the CONSULT-III diagnostic tool, or equivalent to perform the required resets.**

2. Disconnect the negative battery cable.

3. Remove the fuel filler cap to release the pressure from inside the fuel tank.

4. Remove the LH rear wheel and tire.

5. Check the fuel level on level gauge. If the fuel gauge indicates more than the level as shown (full or almost full), drain the fuel from the fuel tank until the fuel gauge indicates the level as shown, or less.

➡ **Fuel will be spilled when removing the fuel level sensor, fuel filter, and fuel pump assembly for the fuel level is above the fuel level sensor, fuel filter, and fuel pump assembly fuel tank opening.**

- As a guide, the fuel level reaches the fuel gauge position as shown, or less, when approximately 4 US gal (15L) of fuel are drained from the fuel tank.
- If the fuel pump does not operate, use the following procedure to drain the fuel to the specified level.

  a. Insert a suitable hose of less than 15 mm (0.59 in.) diameter into the fuel filler pipe through the fuel filler opening to drain the fuel from fuel filler pipe.

  b. Remove the fuel filler pipe shield.

  c. Disconnect the fuel filler hose from the fuel filler pipe.

  d. Insert a suitable hose into the fuel tank through the fuel filler hose to drain the fuel from the fuel tank.

6. Release the fuel pressure from the fuel lines.

7. Be sure to perform the reconnect/relearn procedures.

REMOVAL & INSTALLATION

1. Before servicing the vehicle, refer to the Precautions Section.

➡ **If working near and/or around the SRS system and components, be sure to disable the SRS system. After disabling the system wait three minutes or more before servicing the vehicle.**

➡ **Whenever the negative battery cable is disconnected the following components will require resetting. The Idle Air Volume Learning, Steering Angle Sensor Neutral Position, Sunroof Memory Reset/Initialization, Automatic Drive Positioner System, Audio presets and Navigation. Use the CONSULT-III diagnostic tool, or equivalent to perform the required resets.**

2. Disconnect the negative battery cable.

3. Drain the fuel from the fuel tank, if necessary.

4. Remove the fuel filler cap to release the pressure from inside the fuel tank.

5. Check the fuel level on level gauge. If the fuel gauge indicates more than the level as shown (full or almost full), drain the fuel from the fuel tank until the fuel gauge indicates the level as shown, or less.

6. If the fuel pump does not operate, use the following procedure to drain the fuel to the specified level after disconnecting the fuel filler hose from the fuel filler pipe:

  a. Insert a suitable hose of less than 15 mm (0.59 in.) diameter into the fuel filler pipe through the fuel filler opening to drain the fuel from fuel filler pipe.

  b. Insert a suitable hose into the fuel tank through the fuel filler hose to drain the fuel from the fuel tank.

  c. As a guide, the fuel level reaches the fuel gauge position as shown, or less, when approximately 3 ¾ US gallons (14 liters) of fuel are drained from the fuel tank.

7. Remove the LH rear wheel and tire.

8. Remove the four clips and remove the rear fender protector, front.

9. Disconnect the fuel filler hose from the fuel filler pipe and disconnect the vent hose quick connector.

10. Release the fuel pressure from the fuel lines.

11. Disconnect the battery negative terminal.

12. Remove the second row seat and the third row LH seat.

13. Remove the second and third row rear seat belt buckles mounted on the floor.

14. Remove the LH center pillar trim, the LH rear trim panel, and the LH rear side door kick plate and weather stripping.

15. Remove the second row rear center console and base.

16. Reposition the floor carpet out of the way to access the inspection hole cover, located under the center LH rear seat.

17. Remove the inspection hole cover by turning the retainers 90_degrees clockwise.

18. Disconnect the fuel level sensor, fuel filter, and fuel pump assembly electrical connector, the EVAP hose, and the fuel feed hose.

19. Disconnect the quick connector

20. Remove the four bolts and remove the fuel tank shield.

21. Remove the driveshaft.

22. Disconnect fuel filler hose, and vent hose at the fuel tank side.

23. Remove the fuel tank strap bolts while supporting the fuel tank with a suitable lift jack.

24. Disconnect the EVAP hose from the molded clip in the top of the fuel tank while lowering the fuel tank.

25. Lower the fuel tank using a suitable lift jack and remove it.

➡ **If necessary, remove the lockring using tool.**

To install:

➡ **Be sure to use new fasteners, as required.**

26. Installation is in the reverse order of removal, noting the following:

  a. For installation, use a new fuel level sensor, fuel filter, and fuel pump assembly O-ring.

  b. After installing the quick connectors, pull the tube and the connector to make sure they are securely connected. Visually inspect the connector to make sure the two retainer tabs are securely connected.

27. Be sure to perform the reconnect/relearn procedures.

IDLE SPEED

ADJUSTMENT

There is no idle adjustment available or necessary.

THROTTLE BODY

REMOVAL & INSTALLATION

1. Before servicing the vehicle, refer to the Precautions Section.

➡If working near and/or around the SRS system and components, be sure to disable the SRS system. After disabling the system wait three minutes or more before servicing the vehicle.

➡Whenever the negative battery cable is disconnected the following components will require resetting. The Idle Air Volume Learning, Steering Angle Sensor Neutral Position, Sunroof Memory Reset/Initialization, Automatic Drive Positioner System, Audio presets and Navigation. Use the CONSULT-III diagnostic tool, or equivalent to perform the required resets.

2. Disconnect the negative battery cable.
3. Partially drain the engine coolant.
4. Remove the engine room cover.
5. Remove the air duct and resonator assembly.
6. Drain the engine coolant. Be sure to properly dispose of used coolant.
7. Disconnect the hoses from the unit.
8. Remove the 4 mounting bolts.
9. Remove electric throttle control actuator by loosening bolts diagonally.
10. Remove the old gasket and discard it.

To install:

➡Be sure to use new fasteners, as required.

11. Install a new gasket and the throttle body.
12. Install the 4 mounting bolts in alternate sequence and tighten to 74 inch lbs. (8.4 Nm).
13. Reconnect the hoses to the throttle body.
14. Reconnect the air duct and resonator assembly.
15. As required, fill the cooling system.
16. Install the engine cover.
17. Be sure to perform the reconnect/relearn procedures.

HEATING & AIR CONDITIONING SYSTEM

BLOWER MOTOR

REMOVAL & INSTALLATION
See Figure 76.

1. Before servicing the vehicle, refer to the Precautions Section.

➡If working near and/or around the SRS system and components, be sure to disable the SRS system. After disabling the system wait three minutes or more before servicing the vehicle.

➡Whenever the negative battery cable is disconnected the following components will require resetting. The Idle Air Volume Learning, Steering Angle Sensor Neutral Position, Sunroof Memory Reset/Initialization, Automatic Drive Positioner System, Audio presets and Navigation. Use the CONSULT-III diagnostic tool, or equivalent to perform the required resets.

2. Remove the glove box assembly.
3. Disconnect the front blower motor electrical connector.
4. Remove the blower retaining screws.
5. Remove the blower motor from its mounting.

To install:

➡Be sure to use new fasteners, as required.

6. Installation is the reverse of the removal procedure.
7. Be sure to perform the reconnect/relearn procedures.

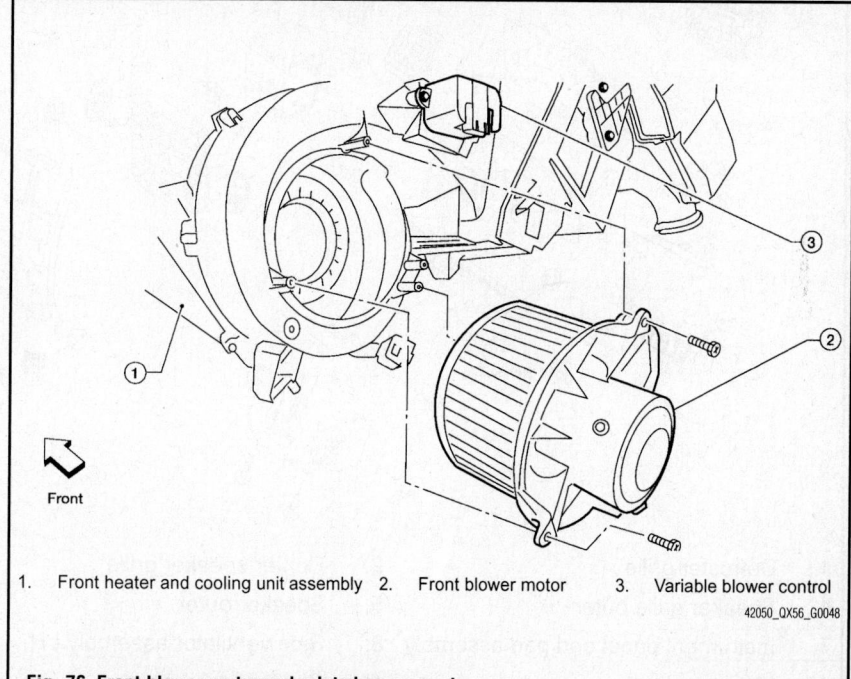

Front

| 1. | Front heater and cooling unit assembly | 2. | Front blower motor | 3. | Variable blower control |

42050_QX56_G0048

Fig. 76 Front blower motor and related components

HEATER/COOLING UNIT

REMOVAL & INSTALLATION
See Figures 77 through 79.

1. Before servicing the vehicle, refer to the Precautions Section.

➡If working near and/or around the SRS system and components, be sure to disable the SRS system. After disabling the system wait three minutes or more before servicing the vehicle.

➡Whenever the negative battery cable is disconnected the following components will require resetting. The Idle Air Volume Learning, Steering Angle Sensor Neutral Position, Sunroof Memory Reset/Initialization, Automatic Drive Positioner System, Audio presets and Navigation. Use the CONSULT-III diagnostic tool, or equivalent to perform the required resets.

2. Position the front seats in the rearmost position.

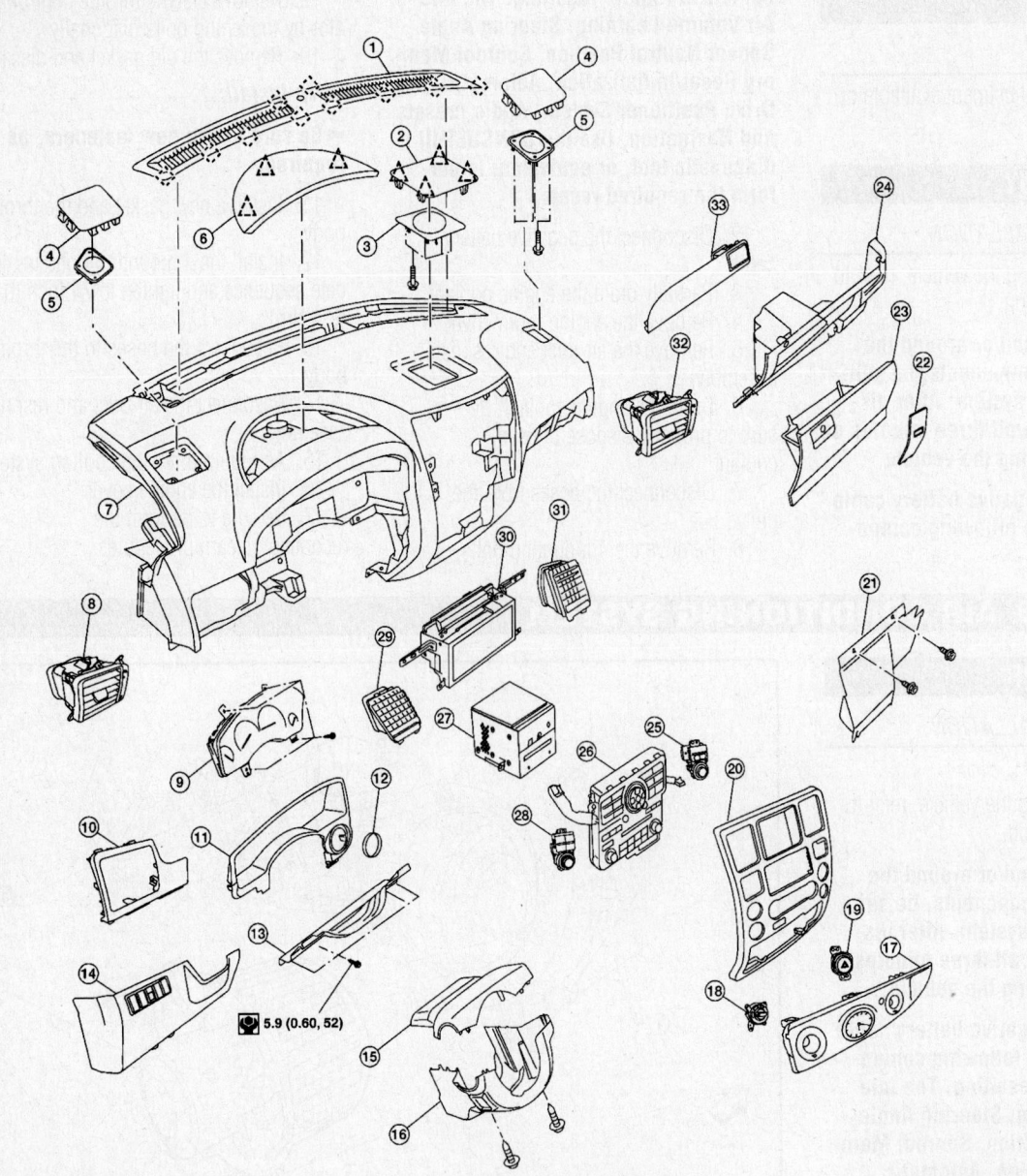

Fig. 77 Instrument panel and related components

5.9 (0.60, 52)

1. Defroster grille	2. Center speaker grille	3. Speaker center
4. Speaker grille outer	5. Speaker outer	6. Instrument panel upper cover
7. Instrument panel and pad assembly	8. Side ventilator assembly LH	9. Combination meter
10. Instrument upper panel LH	11. Cluster lid A	12. Key cylinder escutcheon
13. Lower knee protector	14. Instrument lower panel LH	15. Steering column cover upper
16. Steering column cover lower	17. Cluster lid C lower	18. 4WD switch (if equipped)
19. Hazard switch	20. Cluster lid C	21. Instrument lower cover RH
22. Fuse access cover	23. Glove box	24. Instrument lower panel RH
25. Audio switch RH	26. A/C and AV switch assembly	27. Audio unit
28. Audio switch LH	29. Center ventilator assembly LH	30. Display assembly
31. Center ventilator assembly RH	32. Side ventilator assembly RH	33. Instrument upper panel RH

☐ Metal clip

△ Clip

37663_QX56_G0005

3. Disconnect the negative battery cable.

4. Properly discharge the A/C system.

5. Drain the cooling system. Be sure to properly dispose of used coolant.

6. Disconnect the heater hoses at the heater core.

7. Disconnect and plug the refrigerant lines at the evaporator core.

8. Remove the instrument panel.

9. Remove the center console.

10. Disconnect the instrument panel wire harness at the right and left in-line connector brackets, and the fuse block (JB) electrical connectors.

11. Disconnect the steering member from each side of the vehicle body.

12. Remove the heater/cooling unit with it attached to the steering member from the vehicle.

➡**Use care not to damage the seats or interior trim panels.**

13. Remove the heater/cooling unit from the steering member.

To install:

➡**Be sure to use new fasteners, as required.**

14. Installation is the reverse of the removal procedure.

15. Be sure to use new O-rings coated with clean refrigerant oil, as required.

16. Fill the cooling system with the proper grade and the coolant.

17. Properly recharge the A/C system.

18. Start the engine and check for leaks, correct as required.

19. Be sure to perform the reconnect/relearn procedures.

HEATER CORE

REMOVAL & INSTALLATION
See Figure 79.

1. Before servicing the vehicle, refer to the Precautions Section.

➡**If working near and/or around the SRS system and components, be sure to disable the SRS system. After disabling the system wait three minutes or more before servicing the vehicle.**

➡**Whenever the negative battery cable is disconnected the following components will require resetting. The Idle Air Volume Learning, Steering Angle Sensor Neutral Position, Sunroof Mem-ory Reset/Initialization, Automatic Drive Positioner System, Audio presets and Navigation. Use the CONSULT-III diagnostic tool, or equivalent to perform the required resets.**

2. Position the front seats in the rear-most position.

3. Disconnect the negative battery cable.

4. Properly discharge the A/C system.

5. Drain the cooling system. Be sure to properly dispose of used coolant.

6. Remove the heater/cooling unit, as outlined in this section.

7. Remove the heater core from the heater/cooling unit.

To install:

➡**Be sure to use new fasteners, as required.**

8. Installation is the reverse of the removal procedure.

9. Be sure to use new O-rings coated with clean refrigerant oil, as required.

10. Fill the cooling system with the proper grade and the coolant.

11. Properly recharge the A/C system.

12. Start the engine and check for leaks, correct as required.

13. Be sure to perform the reconnect/relearn procedures.

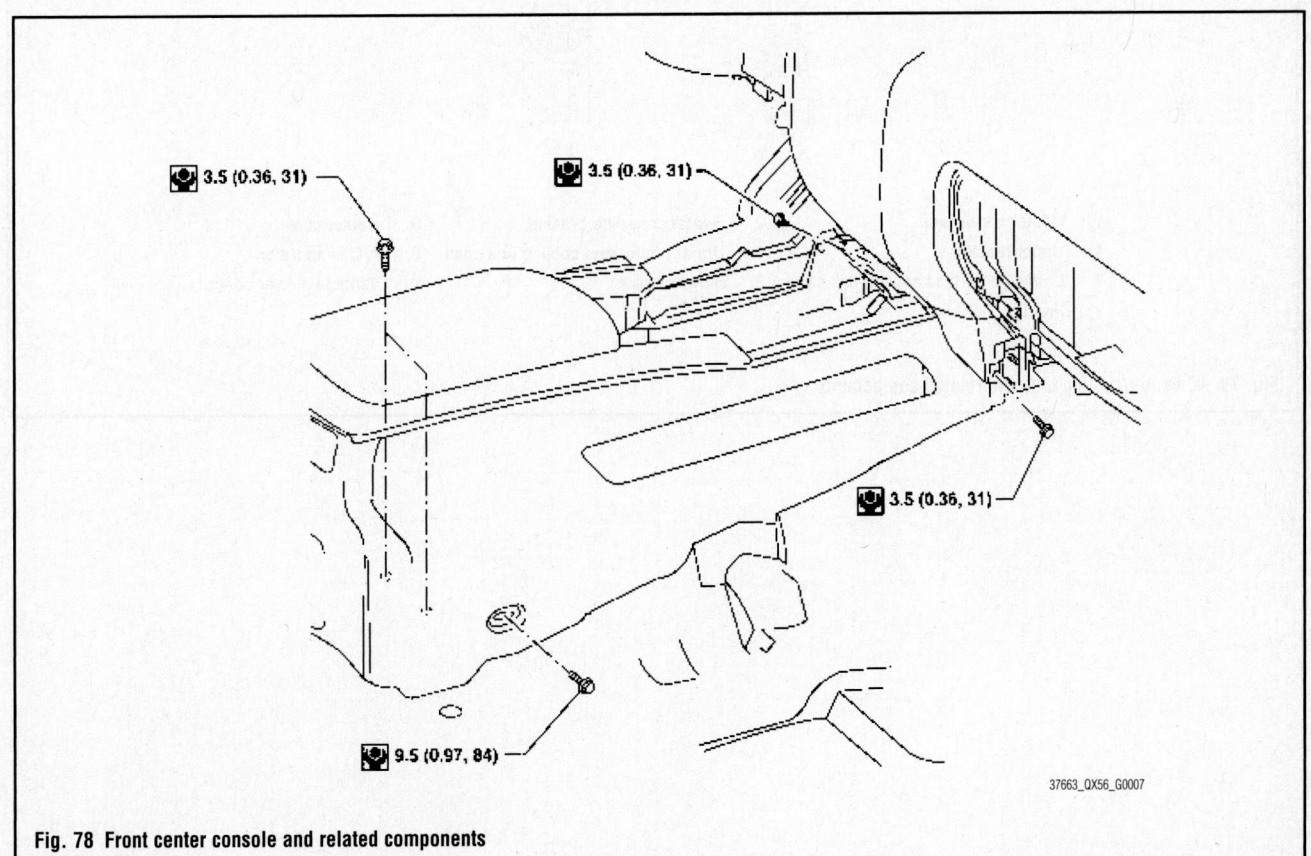

3.5 (0.36, 31)

3.5 (0.36, 31)

3.5 (0.36, 31)

9.5 (0.97, 84)

37663_QX56_G0007

Fig. 78 Front center console and related components

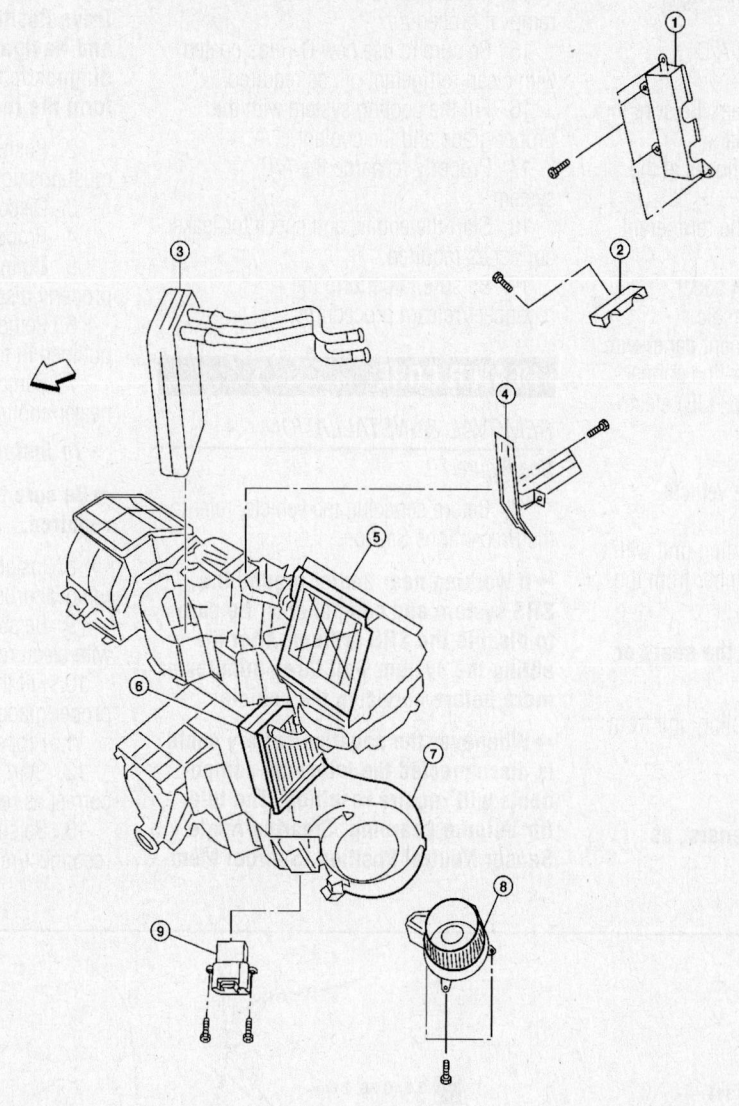

1. Heater core cover
2. Heater core pipe bracket
3. Heater core
4. Upper bracket
5. Upper heater and cooling unit case
6. A/C evaporator
7. Lower heater and cooling unit case
8. Blower motor
9. Variable blower control
⟵ Front

37663_QX56_G0089

Fig. 79 AC heater/cooling unit and related components

STEERING

POWER RACK & PINION STEERING GEAR

REMOVAL & INSTALLATION

See Figures 80 and 81.

1. Before servicing the vehicle, refer to the Precautions Section.
2. Ensure the wheels are in the straight-ahead position.
3. Remove or disconnect the following:
   - Wheels and tires
   - Engine splash guard
4. On 4WD, remove front final drive and support the halfshafts.
5. Remove cotter pin at steering outer socket and loosen mounting nut.
6. With the steering wheel in the straight ahead position, make sure that the slit of the lower joint (A) fits with the projection on the rear cover cap (B), while checking that the mark on the steering gear assembly aligns with the mark on the rear cover cap. See illustration.
7. Remove steering outer socket from steering knuckle using special tool J-25730-A.
8. Remove or disconnect the following:
   - Oil pipes from steering gear assembly
   - Lower joint mounting bolt from lower shaft
   - Mounting bolts and nuts from steering gear assembly
   - Steering gear assembly

To install:

➡**Be sure to use new fasteners, as required.**

9. Installation is the reverse of the removal procedure.
10. With the steering wheel in the straight ahead position, make sure that the

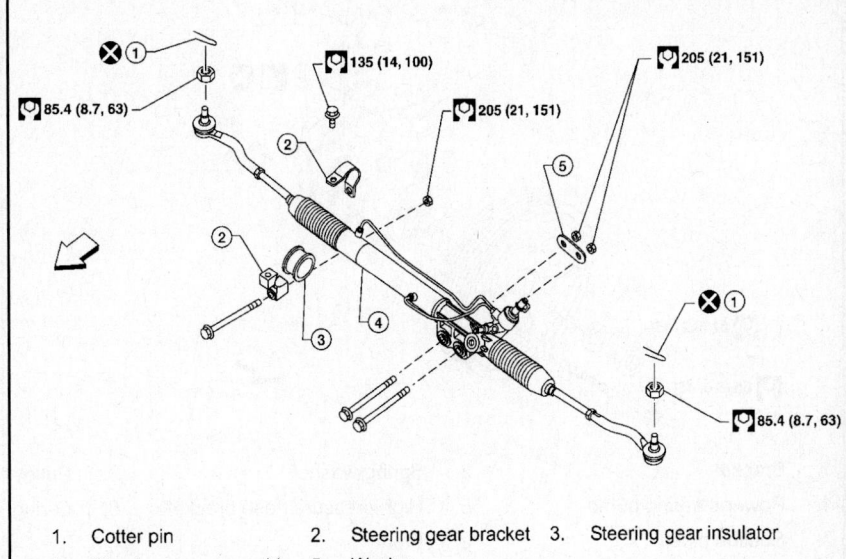

1.	Cotter pin	2.	Steering gear bracket	3.	Steering gear insulator
4.	Steering gear assembly	5.	Washer		

37663_QX56_G0098

Fig. 81 Power steering gear and related components

slit of the lower joint (A) fits with the projection on the rear cover cap (B), while checking that the mark on the steering gear assembly aligns with the mark on the rear cover cap. See illustration.

11. Check the wheel alignment and adjust as necessary.
12. Adjust the steering angle sensor neutral position, using the CONSULT-III diagnostic tool, or equivalent.
13. Be sure to perform the reconnect/relearn procedures.

POWER STEERING PUMP

REMOVAL & INSTALLATION

See Figure 82.

1. Before servicing the vehicle, refer to the Precautions Section.

➡**If working near and/or around the SRS system and components, be sure to disable the SRS system. After disabling the system wait three minutes or more before servicing the vehicle.**

➡**Whenever the negative battery cable is disconnected the following components will require resetting. The Idle Air Volume Learning, Steering Angle Sensor Neutral Position, Sunroof Memory Reset/Initialization, Automatic Drive Positioner System, Audio presets and Navigation. Use the CONSULT-III**

diagnostic tool, or equivalent to perform the required resets.

2. Disconnect the negative battery cable.
3. Drain the power steering fluid into a suitable container. Properly discard the used fluid.
4. Remove the engine room cover.
5. Remove the air duct assembly.
6. Remove the power steering reservoir tank.
7. Remove the drive belt.
8. Disconnect the pressure sensor electrical connector.
9. Remove the high pressure and the low pressure lines from the power steering fluid pump.
10. Remove the pump mounting bolts.
11. Remove the pump from the vehicle.

To install:

➡**Be sure to use new fasteners, as required.**

12. Installation is the reverse of the removal procedure.
13. Bleed the power steering system.

➡**The drive belt tension is automatic and requires no adjustment.**

14. Be sure to perform the reconnect/relearn procedures.

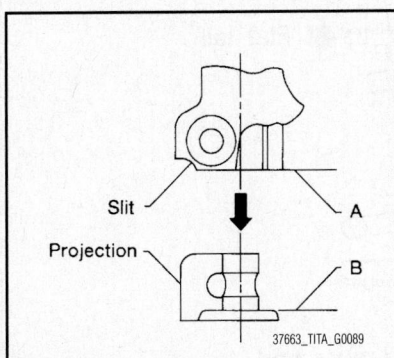

37663_TITA_G0089

Fig. 80 Steering gear lower joint alignment

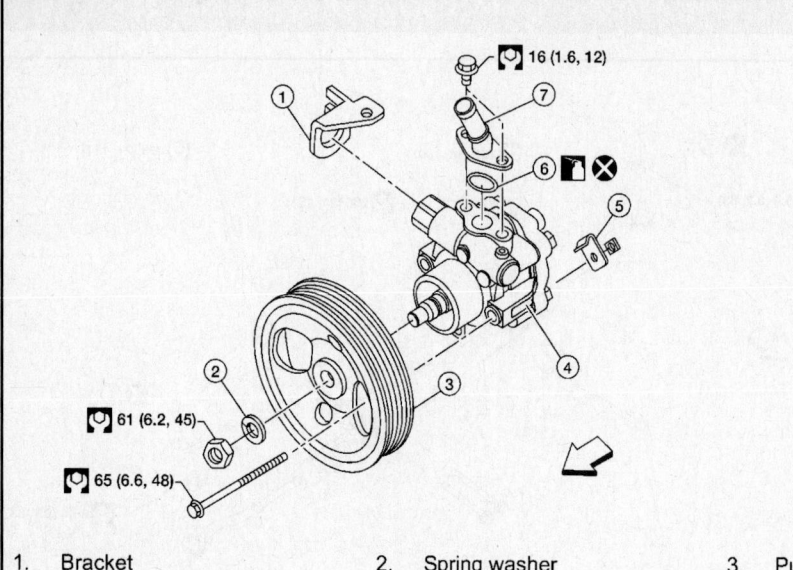

1. Bracket
2. Spring washer
3. Pulley
4. Power steering pump
5. High pressure hose bracket
6. O-ring
7. Suction pipe
⇦ Front

37663_QX56_G0100

Fig. 82 Power steering pump and related components

BLEEDING

1. Before servicing the vehicle, refer to the Precautions Section.
2. Stop the engine.
3. Turn the steering wheel fully to the right and left several times.

➡**Do not allow the fluid level in the reservoir tank to go below the MIN level line. Check and add fluid as needed.**

4. Run the engine at idle speed. Turn the steering wheel fully to the right and then fully to the left. Hold for about three seconds. Check for fluid leakage.
5. Repeat the above step several times at three second intervals.

➡**Do not hold the steering wheel in the locked position for more than ten seconds.**

6. Check for air bubbles or cloudy fluid. If found, repeat the bleeding procedure.
7. Stop the engine and check the fluid level. Correct as required.

SUSPENSION   FRONT SUSPENSION

COIL SPRING

REMOVAL & INSTALLATION

See Figure 83.

1. Before servicing the vehicle, refer to the Precautions Section.

➡**If working near and/or around the SRS system and components, be sure to disable the SRS system. After disabling the system wait three minutes or more before servicing the vehicle.**

➡**Whenever the negative battery cable is disconnected the following components will require resetting. The Idle Air Volume Learning, Steering Angle Sensor Neutral Position, Sunroof Memory Reset/Initialization, Automatic Drive Positioner System, Audio presets and Navigation. Use the CONSULT-III diagnostic tool, or equivalent to perform the required resets.**

2. Disconnect the negative battery cable.
3. Raise and safely support the vehicle.
4. Remove or disconnect the following:
   - Wheel and tire assembly
   - Lower shock absorber bolt
   - Upper shock absorber bolts
   - Coil spring and shock absorber assembly

5. Secure the shock absorber in a vice and loosen (without removing) the piston rod locknut.
6. Install a spring compressor and tighten until the shock absorber mounting insulator can be turned by hand.
7. Remove piston rod locknut and remove shock absorber from the coil spring.

To install:

➡**Be sure to use new fasteners, as required.**

8. Install upper mounting insulator in line with the lower shock absorber mount and step in shock absorber lower seat as shown in figure.
9. Tighten the new piston rod locknut to 40 ft. lbs. (54 Nm).
10. Install or connect the following:
   - Coil spring and shock absorber assembly
   - Upper shock absorber bolts and tighten to 22 ft. lbs (30 Nm)

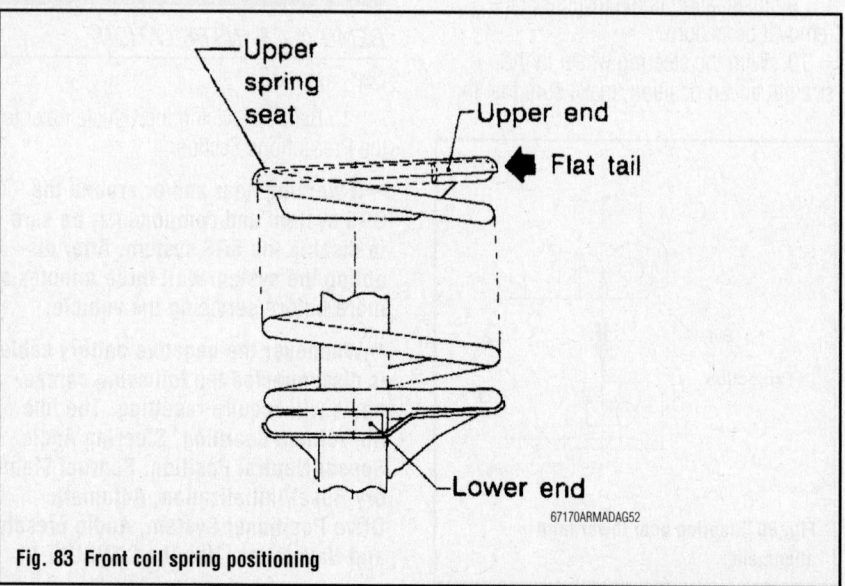

67170ARMADAG52

Fig. 83 Front coil spring positioning

- Lower shock absorber bolt and tighten to 99 ft. lbs. (134 Nm)
- Wheel and tire assembly

11. Check wheel alignment and adjust as necessary.

12. Be sure to perform the reconnect/relearn procedures.

LOWER BALL JOINT

REMOVAL & INSTALLATION

At this time the manufacturer does not provide removal and installation procedures for this component. The upper ball joint is part of the upper control arm assembly.

LOWER CONTROL ARM

REMOVAL & INSTALLATION

See Figure 84.

➡Nissan/Infiniti refers to the lower control arm as a lower link.

1. Before servicing the vehicle, refer to the Precautions Section.

➡If working near and/or around the SRS system and components, be sure to disable the SRS system. After disabling the system wait three minutes or more before servicing the vehicle.

➡Whenever the negative battery cable is disconnected the following components will require resetting. The Idle Air Volume Learning, Steering Angle Sensor Neutral Position, Sunroof Memory Reset/Initialization, Automatic Drive Positioner System, Audio presets and Navigation. Use the CONSULT-III diagnostic tool, or equivalent to perform the required resets.

2. Disconnect the negative battery cable.

3. Raise and safely support the vehicle.

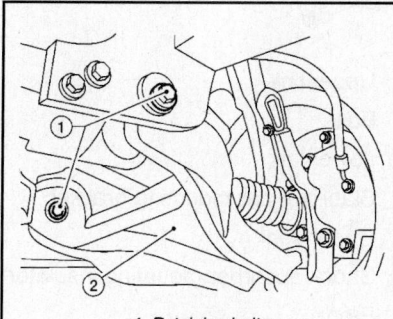

1. Retaining bolts
2. Control arm

37663_QX56_G0107

Fig. 84 Front lower control arm (lower link) and related components

4. Remove the tire and wheel assembly.

5. Remove the lower shock absorber retaining bolt.

6. Remove the stabilizer bar connecting rod lower nut.

7. Remove the pinch bolt from the steering knuckle, than separate the lower link ball joint from the steering knuckle.

8. Remove the lower link bolts and nuts.

9. Remove the lower link.

To install:

➡Be sure to use new fasteners, as required.

10. Installation is the reverse of the removal procedure.

11. Check and adjust alignment, as required.

12. Be sure to perform the reconnect/relearn procedures.

SHOCK ABSORBERS

REMOVAL & INSTALLATION

See Figures 85 and 86.

1. Before servicing the vehicle, refer to the Precautions Section.

➡If working near and/or around the SRS system and components, be sure to disable the SRS system. After disabling the system wait three minutes or more before servicing the vehicle.

➡Whenever the negative battery cable is disconnected the following components will require resetting. The Idle Air Volume Learning, Steering Angle Sensor Neutral Position, Sunroof Memory Reset/Initialization, Automatic Drive Positioner System, Audio presets and Navigation. Use the CONSULT-III diagnostic tool, or equivalent to perform the required resets.

2. Disconnect the negative battery cable.

3. Raise and safely support the vehicle.

4. Remove or disconnect the following:
- Wheel and tire assembly
- Lower shock absorber bolt
- Upper shock absorber bolts
- Coil spring and shock absorber assembly

To install:

➡Be sure to use new fasteners, as required.

5. Installation is the reverse of removal procedure.

6. Install upper mounting insulator in line with the lower shock absorber mount

and step in shock absorber lower seat as shown in figure.

7. Check wheel alignment and adjust as necessary.

8. Be sure to perform the reconnect/relearn procedures.

STEERING KNUCKLE

REMOVAL & INSTALLATION

See Figure 87.

1. Before servicing the vehicle, refer to the Precautions Section.

➡If working near and/or around the SRS system and components, be sure to disable the SRS system. After disabling the system wait three minutes or more before servicing the vehicle.

➡Whenever the negative battery cable is disconnected the following components will require resetting. The Idle Air Volume Learning, Steering Angle Sensor Neutral Position, Sunroof Memory Reset/Initialization, Automatic Drive Positioner System, Audio presets and Navigation. Use the CONSULT-III diagnostic tool, or equivalent to perform the required resets.

2. Disconnect the negative battery cable.

3. Raise and support the vehicle safely.

4. Remove the tire and wheel assembly.

5. Remove the brake caliper from its mounting and position it to the side.

➡Do not disconnect the hydraulic lines. It is not necessary to remove the bolts on the torque member and brake hose except for disassembly or replacement of the caliper. In this case hang the caliper to the side with mechanics wire so that the brake hose is not under tension. Avoid depressing the brake pedal with the caliper removed.

6. Put alignment marks on the rotor and wheel hub and bearing assembly. Remove the rotor.

7. Remove the ABS sensor from the steering knuckle. Do not pull on the ABS sensor harness.

8. Remove the cotter pin. Remove the locknut from the halfshaft.

9. Remove the steering outer shaft socket cotter pin at the steering knuckle. Loosen the mounting nut.

10. Disconnect the steering outer socket from the steering knuckle.

➡To prevent damage to the threads

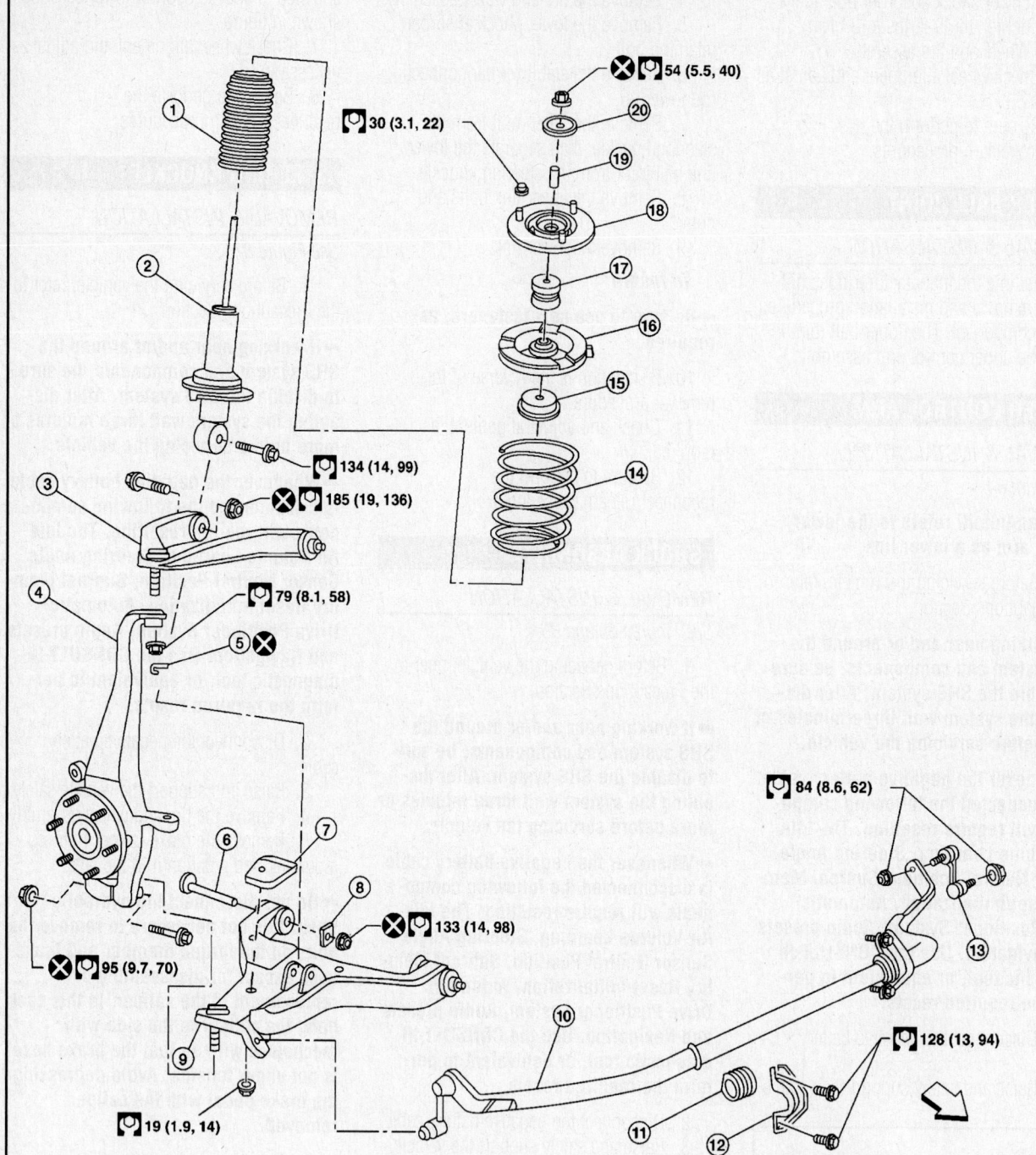

54 (5.5, 40)

30 (3.1, 22)

20

19

18

17

16

15

14

134 (14, 99)

185 (19, 136)

79 (8.1, 58)

84 (8.6, 62)

95 (9.7, 70)

133 (14, 98)

13

128 (13, 94)

19 (1.9, 14)

10

11

12

1.	Dust cover	2.	Shock absorber	3.	Upper link
4.	Steering knuckle	5.	Cotter pin	6.	Bolt
7.	Jounce bumper	8.	Washer	9.	Lower link
10.	Stabilizer bar	11.	Stabilizer bar bushing	12.	Stabilizer bar mounting bracket
13.	Connecting rod	14.	Coil spring	15.	Upper seat
16.	Upper spring seat	17.	Shock absorber bushing	18.	Shock absorber mounting insulator
19.	Spacer	20.	Washer	⇦	Front

37663_QX56_G0104

Fig. 85 Front suspension and related components

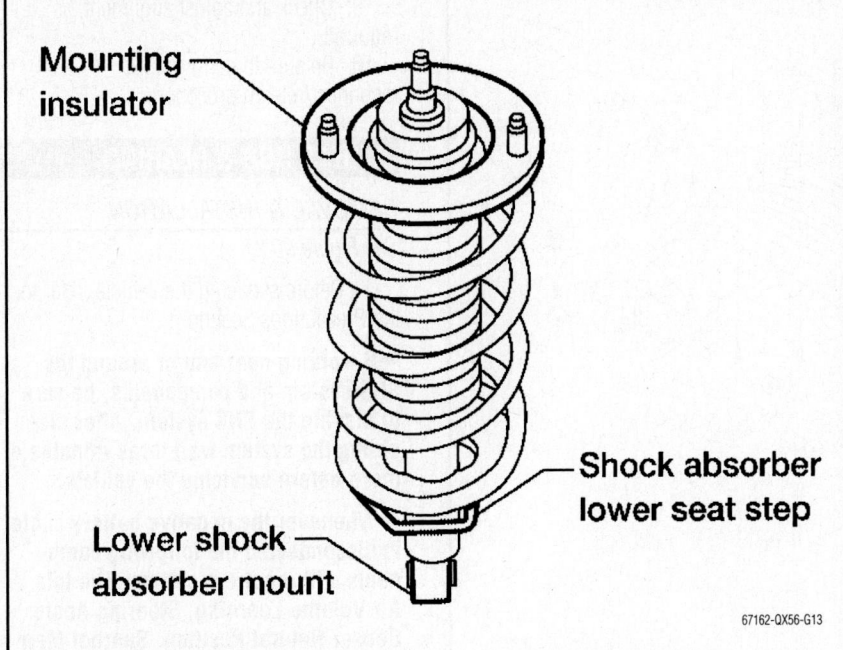

Mounting insulator

Shock absorber lower seat step

Lower shock absorber mount

67162-QX56-G13

Fig. 86 Shock absorber alignment

and to prevent the tool from coming off suddenly, temporarily loosely install the mounting nut.

11. Remove the halfshaft.
12. Remove the wheel hub and bearing assembly bolts.
13. Remove the splash guard and wheel hub and bearing assembly from the steering knuckle.
14. Support the lower control arm assembly, using a suitable jack.
15. Remove the cotter pin and nut from the upper ball joint.
16. Separate the upper link ball joint from the steering knuckle using tool J-24319-01 or equivalent.
17. Remove the pinch bolt from the steering knuckle. Remove the steering knuckle from the lower control arm ball joint.
18. Remove the steering knuckle from the vehicle.

To install:

➡ **Be sure to use new fasteners, as required.**

19. Installation is the reverse of the removal procedure.

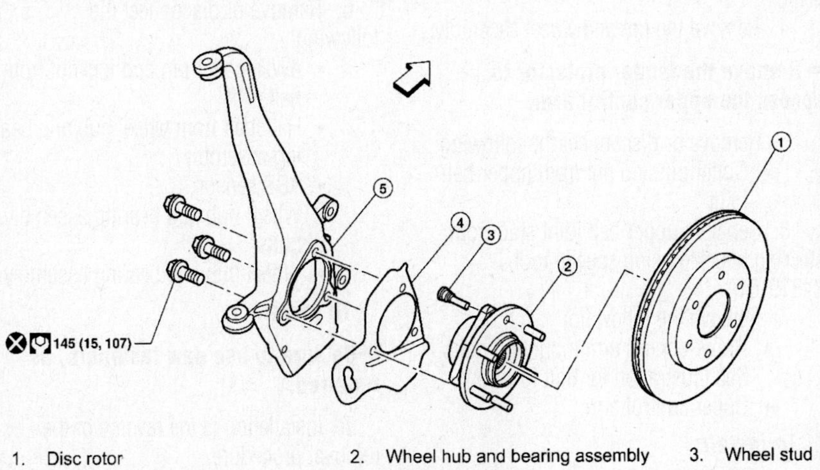

20. Be sure to use the alignment marks made during the removal procedure when reinstalling removed components.
21. Check and adjust the front end alignment, as required.
22. Be sure to perform the reconnect/relearn procedures.

STABILIZER BAR

REMOVAL & INSTALLATION

1. Before servicing the vehicle, refer to the Precautions Section.

➡ **If working near and/or around the SRS system and components, be sure to disable the SRS system. After disabling the system wait three minutes or more before servicing the vehicle.**

➡ **Whenever the negative battery cable is disconnected the following components will require resetting. The Idle Air Volume Learning, Steering Angle Sensor Neutral Position, Sunroof Memory Reset/Initialization, Automatic Drive Positioner System, Audio presets and Navigation. Use the CONSULT-III diagnostic tool, or equivalent to perform the required resets.**

2. Disconnect the negative battery cable.
3. Raise and safely support the vehicle.
4. Remove the tire and wheel assembly.
5. Remove the engine under cover.
6. Remove the stabilizer bar mounting bracket retaining bolts and rubber bushings.
7. Remove the connecting rod nuts.
8. Remove the stabilizer bar from the vehicle.

To install:

➡ **Be sure to use new fasteners, as required.**

9. Installation is the reverse of removal procedure.
10. Be sure to perform the reconnect/relearn procedures.

UPPER BALL JOINT

REMOVAL & INSTALLATION

At this time the manufacturer does not provide removal and installation procedures for this component. The upper ball joint is part of the upper control arm assembly.

UPPER CONTROL ARM

REMOVAL & INSTALLATION

See Figures 88 and 89.

⊗ ⟨⟩ 145 (15, 107)

1. Disc rotor
4. Splash guard
2. Wheel hub and bearing assembly
5. Steering knuckle
3. Wheel stud
◁ Front

37663_QX56_G0108

Fig. 87 Front steering knuckle and related components

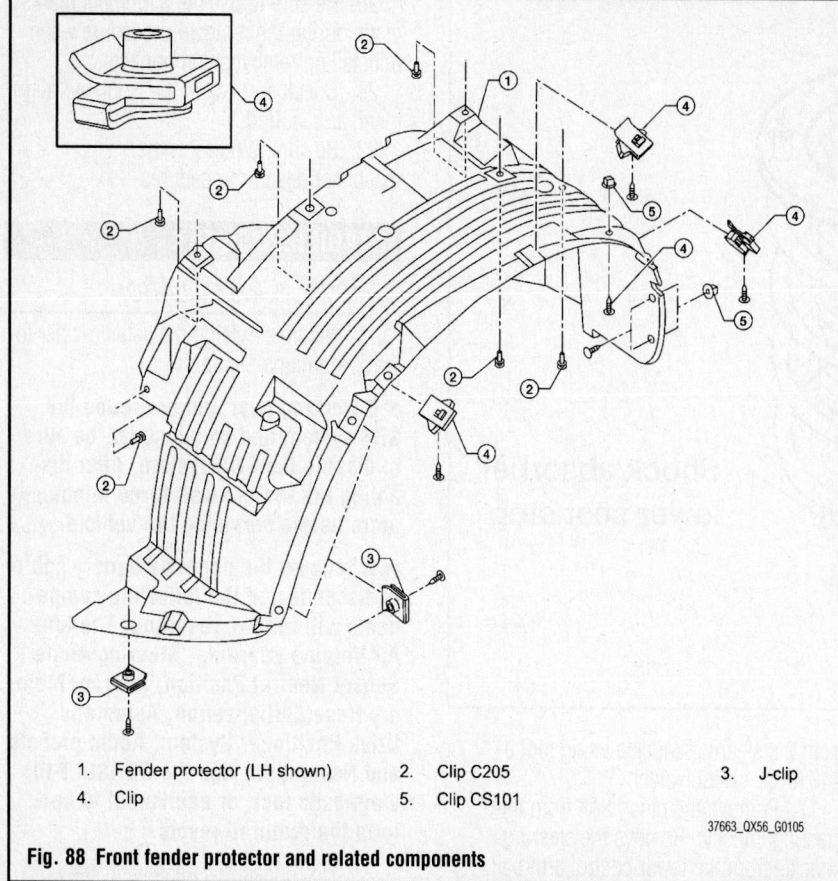

| 1. | Fender protector (LH shown) | 2. | Clip C205 | 3. | J-clip |
| 4. | Clip | 5. | Clip CS101 | | |

37663_QX56_G0105

Fig. 88 Front fender protector and related components

➡️Nissan/Infiniti refers to the upper control arm as a upper link.

1. Before servicing the vehicle, refer to the Precautions Section.

➡️If working near and/or around the SRS system and components, be sure to disable the SRS system. After disabling the system wait three minutes or more before servicing the vehicle.

➡️Whenever the negative battery cable is disconnected the following components will require resetting. The Idle Air Volume Learning, Steering Angle Sensor Neutral Position, Sunroof Memory Reset/Initialization, Automatic Drive Positioner System, Audio presets and Navigation. Use the CONSULT-III diagnostic tool, or equivalent to perform the required resets.

2. Disconnect the negative battery cable.

3. Raise and safely support the vehicle.

4. Remove the tire and wheel assembly.

➡️Remove the fender protector to access the upper control arm.

5. Remove or disconnect the following:
- Cotter pin and nut from upper ball joint

6. Separate upper ball joint stud from steering knuckle using special tool J-24319-01.

7. Remove the following:
- Upper control arm mounting bolts. See illustration for bolt locations.
- Upper control arm

To install:

➡️Be sure to use new fasteners, as required.

8. Installation is the reverse of the removal procedure.

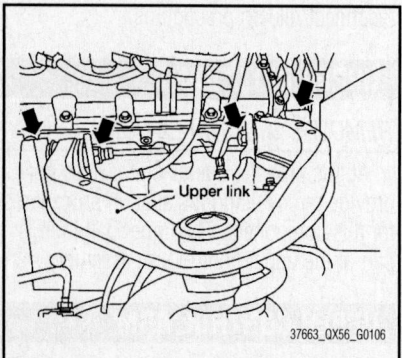

37663_QX56_G0106

Fig. 89 Front upper control arm (upper link) and related components

9. Check and adjust alignment, as required.

10. Be sure to perform the reconnect/relearn procedures.

WHEEL HUB & BEARING

REMOVAL & INSTALLATION

See Figure 90.

1. Before servicing the vehicle, refer to the Precautions Section.

➡️If working near and/or around the SRS system and components, be sure to disable the SRS system. After disabling the system wait three minutes or more before servicing the vehicle.

➡️Whenever the negative battery cable is disconnected the following components will require resetting. The Idle Air Volume Learning, Steering Angle Sensor Neutral Position, Sunroof Memory Reset/Initialization, Automatic Drive Positioner System, Audio presets and Navigation. Use the CONSULT-III diagnostic tool, or equivalent to perform the required resets.

2. Disconnect the negative battery cable.

3. raise and safely support the vehicle.

4. Remove or disconnect the following:
- Wheel and tire assembly
- Engine splash guard
- Brake caliper without disconnecting the hydraulic lines, and reposition aside with wire

5. Matchmark the brake rotor to the wheel hub and remove the brake rotor.

6. Remove or disconnect the following:
- 4WD, cotter pin and locknut from halfshaft
- Halfshaft from wheel hub and bearing assembly
- ABS sensor
- Wheel hub and bearing assembly bolts
- Wheel hub and bearing assembly

To install:

➡️Be sure to use new fasteners, as required.

7. Installation is the reverse of the removal procedure.

8. Check and adjust alignment, as required.

9. Be sure to perform the reconnect/relearn procedures.

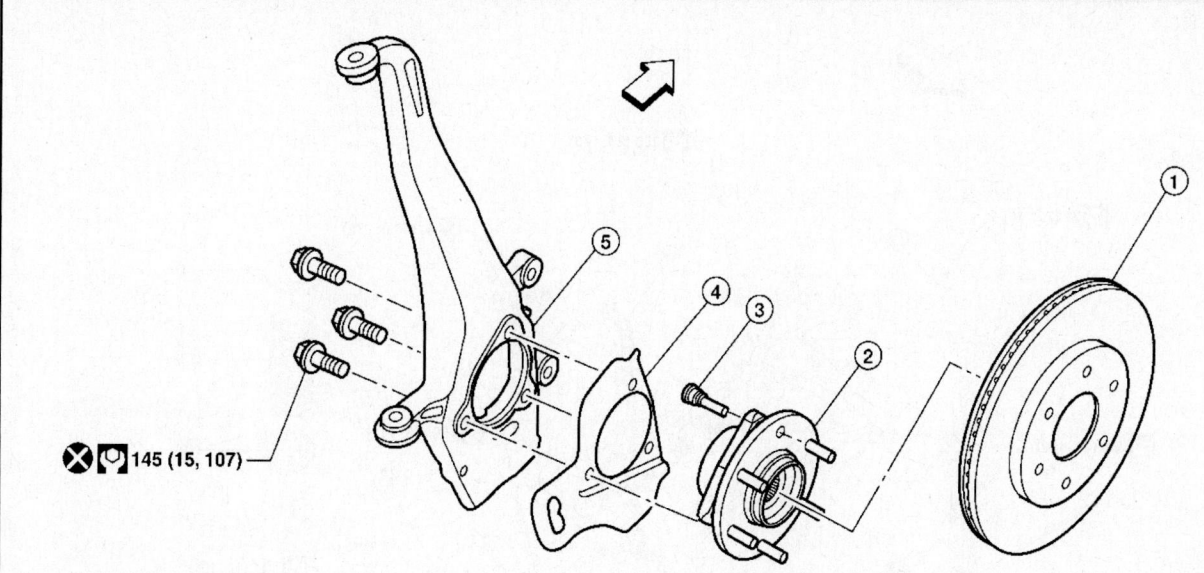

1. Disc rotor
4. Splash guard
2. Wheel hub and bearing assembly
5. Steering knuckle
3. Wheel stud
⇦ Front

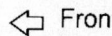 145 (15, 107)

37663_QX56_G0108

Fig. 90 Front hub/bearing assembly and related components

SUSPENSION

COIL SPRING

REMOVAL & INSTALLATION

See Figures 91 and 92.

1. Before servicing the vehicle, refer to the Precautions Section.

➡**If working near and/or around the SRS system and components, be sure to disable the SRS system. After disabling the system wait three minutes or more before servicing the vehicle.**

➡**Whenever the negative battery cable is disconnected the following components will require resetting. The Idle Air Volume Learning, Steering Angle Sensor Neutral Position, Sunroof Memory Reset/Initialization, Automatic Drive Positioner System, Audio presets and Navigation. Use the CONSULT-III diagnostic tool, or equivalent to perform the required resets.**

2. Disconnect the negative battery cable.
3. Raise and safely support the vehicle.
4. Remove the tire and wheel assembly.
5. Release the air pressure from the rear load leveling air suspension system using the CONSULT-III® "EXHAUST SOLENOID" active test.

6. Remove the height sensor arm bracket bolt from the left-hand rear lower link.
7. Place a suitable jack under the rear lower link and relieve the coil spring tension.
8. Loosen the rear lower link adjusting bolt and nut connected to the rear suspension member.
9. Remove the rear lower link bolt and nut from the knuckle.
10. Slowly lower the jack to relieve the coil spring tension.
11. Remove the coil spring.

To install:

➡**Be sure to use new fasteners, as required.**

12. Installation is the reverse of the removal procedure.

➡**When installing the rubber seats for the coil spring, ensure the embossed arrow points outward toward the wheel.**

13. Be sure to perform the reconnect/relearn procedures.

CONTROL ARMS/LINKS

REMOVAL & INSTALLATION

Suspension Arm
See Figure 91.

REAR SUSPENSION

1. Before servicing the vehicle, refer to the Precautions Section.

➡**If working near and/or around the SRS system and components, be sure to disable the SRS system. After disabling the system wait three minutes or more before servicing the vehicle.**

➡**Whenever the negative battery cable is disconnected the following components will require resetting. The Idle Air Volume Learning, Steering Angle Sensor Neutral Position, Sunroof Memory Reset/Initialization, Automatic Drive Positioner System, Audio presets and Navigation. Use the CONSULT-III diagnostic tool, or equivalent to perform the required resets.**

2. Disconnect the negative battery cable.
3. Raise and support the vehicle safely.
4. Remove the tire and wheel assemblies.
5. Remove the rear suspension member.

➡**It is necessary to remove the rear suspension member in order to remove the front upper bolt from the suspension arm.**

6. Remove the shock absorber upper end bolt.

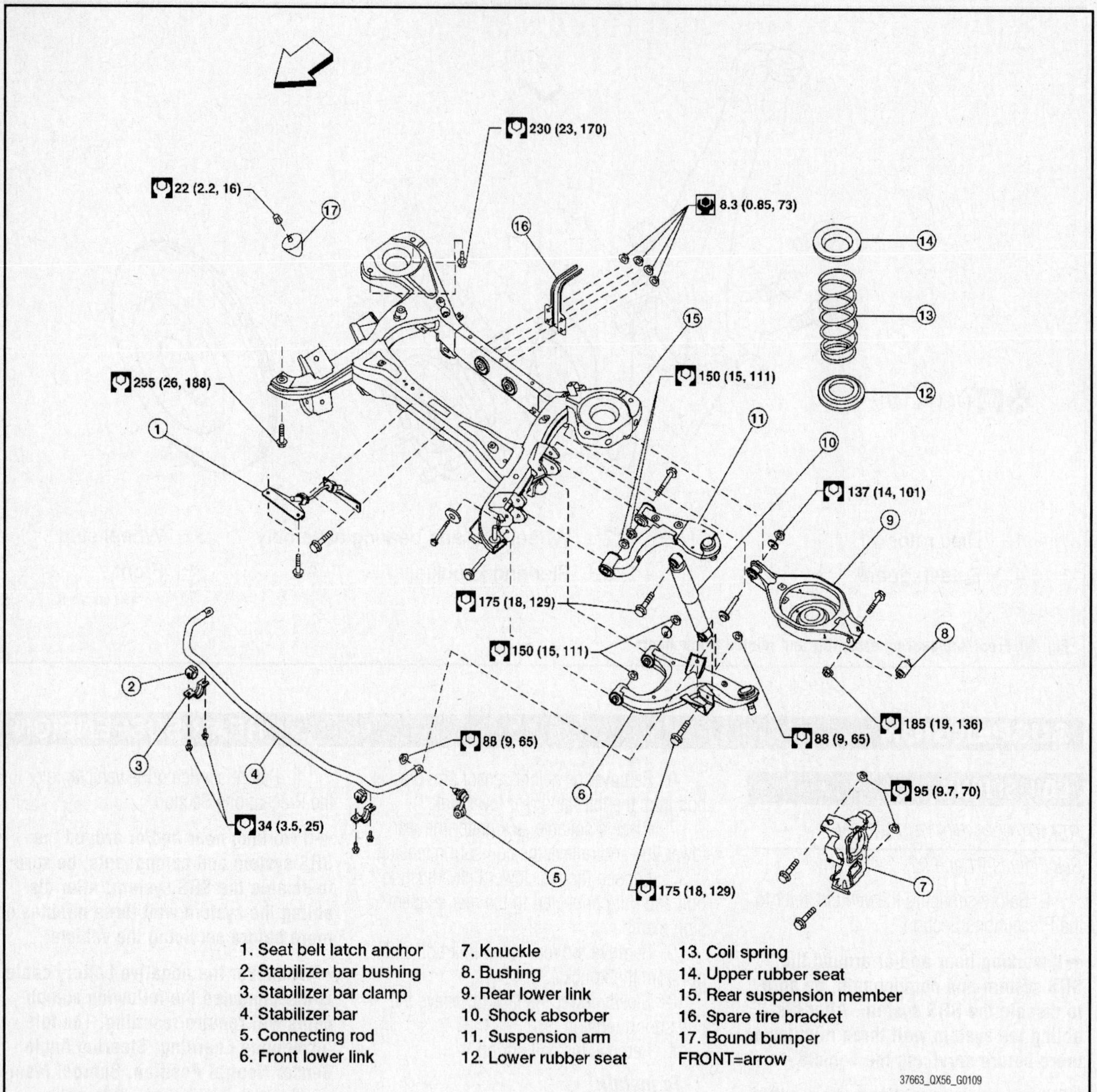

230 (23, 170)

22 (2.2, 16)   17

16   8.3 (0.85, 73)

15   14

255 (26, 188)   13

1   150 (15, 111)   12

11

10   137 (14, 101)

9

175 (18, 129)

150 (15, 111)   8

2

88 (9, 65)   88 (9, 65)   185 (19, 136)

3

4   6   95 (9.7, 70)

5

34 (3.5, 25)   175 (18, 129)   7

1. Seat belt latch anchor   7. Knuckle   13. Coil spring
2. Stabilizer bar bushing   8. Bushing   14. Upper rubber seat
3. Stabilizer bar clamp   9. Rear lower link   15. Rear suspension member
4. Stabilizer bar   10. Shock absorber   16. Spare tire bracket
5. Connecting rod   11. Suspension arm   17. Bound bumper
6. Front lower link   12. Lower rubber seat   FRONT=arrow

37663_QX56_G0109

Fig. 91 Rear suspension and related components

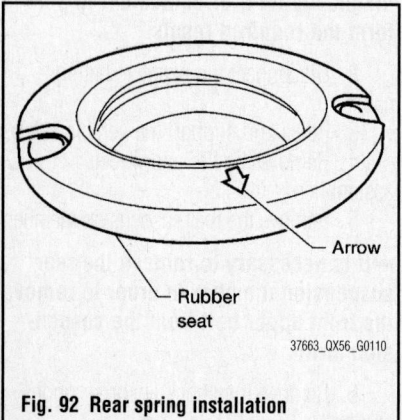

— Arrow

— Rubber
seat

37663_QX56_G0110

Fig. 92 Rear spring installation

7. Remove the suspension arm upper nuts and bolts on the suspension member side.

8. Remove the suspension arm pinch bolt and nut on the knuckle side.

9. Disconnect the suspension arm from the knuckle.

✻✻ WARNING

If necessary, use a soft hammer. Do not damage the ball joint with the soft hammer.

10. Remove the suspension arm.

To install:

11. Installation is the reverse of the removal procedure.

12. Perform the final tightening of the nuts and bolts for the links (rubber bushing) with the vehicle in the unladen condition with the tires on level ground.

➡**Unladen condition means that the fuel tank, engine coolant and lubricants are at the full specification and the spare tire, jack, hand tools and mats are in their designated positions.**

13. Check and adjust the alignment, as required.

14. Be sure to perform the reconnect/relearn procedures.

Front Lower Link

See Figure 91.

1. Before servicing the vehicle, refer to the Precautions Section.

➡ **If working near and/or around the SRS system and components, be sure to disable the SRS system. After disabling the system wait three minutes or more before servicing the vehicle.**

➡ **Whenever the negative battery cable is disconnected the following components will require resetting. The Idle Air Volume Learning, Steering Angle Sensor Neutral Position, Sunroof Memory Reset/Initialization, Automatic Drive Positioner System, Audio presets and Navigation. Use the CONSULT-III diagnostic tool, or equivalent to perform the required resets.**

2. Disconnect the negative battery cable.
3. Raise and support the vehicle safely.
4. Remove the tire and wheel assemblies.
5. Release the air pressure from the rear load leveling air suspension system using the CONSULT-III® "EXHAUST SOLENOID" active test.
6. Remove the shock absorber lower end bolt.
7. Remove the adjusting bolt and nut, and the bolt and nut from the front lower link and rear suspension member.
8. Remove the front lower link pinch bolt and nut on the knuckle side.
9. Disconnect the front lower link from the knuckle.

❋❋ WARNING

If necessary, use a soft hammer. Do not damage the ball joint with the soft hammer.

10. Remove the front lower link.

To install:

11. Installation is the reverse of the removal procedure.
12. Perform the final tightening of the nuts and bolts for the links (rubber bushing) with the vehicle in the unladen condition with the tires on level ground.

➡ **Unladen condition means that the fuel tank, engine coolant and lubricants are at the full specification and the spare tire, jack, hand tools and mats are in their designated positions.**

13. Check and adjust the alignment, as required.

14. Be sure to perform the reconnect/relearn procedures.

Rear Lower Link and Spring

See Figure 91.

1. Before servicing the vehicle, refer to the Precautions Section.

➡ **If working near and/or around the SRS system and components, be sure to disable the SRS system. After disabling the system wait three minutes or more before servicing the vehicle.**

➡ **Whenever the negative battery cable is disconnected the following components will require resetting. The Idle Air Volume Learning, Steering Angle Sensor Neutral Position, Sunroof Memory Reset/Initialization, Automatic Drive Positioner System, Audio presets and Navigation. Use the CONSULT-III diagnostic tool, or equivalent to perform the required resets.**

2. Disconnect the negative battery cable.
3. Raise and support the vehicle safely.
4. Remove the tire and wheel assemblies.
5. Release the air pressure from the rear load leveling air suspension system using the CONSULT-III® "EXHAUST SOLENOID" active test.
6. Remove the height sensor arm bracket bolt from the left-hand rear lower link.
7. Place a suitable jack under the rear lower link and relieve the coil spring tension.
8. Loosen the rear lower link adjusting bolt and nut connected to the rear suspension member.
9. Remove the rear lower link bolt and nut from the knuckle.
10. Slowly lower the jack to relieve the coil spring tension.
11. Remove the coil spring.
12. Remove the upper rubber seat, coil spring and lower rubber seat from the rear lower link.
13. Remove the rear lower link adjusting bolt and nut from the rear suspension member.
14. Remove the rear lower link from its mounting.

To install:

15. Installation is the reverse of the removal procedure.
16. When installing the upper and lower rubber seats for the rear coil springs, the arrow embossed on the rubber seats must point out toward the wheel and tire assembly.

17. Tighten the rear lower link bolt to knuckle to 70 ft. lbs. (95 Nm).
18. Tighten the rear lower link adjusting bolt to rear suspension member to 101 ft. lbs. (137 Nm).
19. Tighten the height sensor arm bracket bolt to left-head rear lower link to 9 ft. lbs. (12 Nm).
20. Perform the final tightening of the nuts and bolts for the links (rubber bushing) with the vehicle in the unladen condition with the tires on level ground.

➡ **Unladen condition means that the fuel tank, engine coolant and lubricants are at the full specification and the spare tire, jack, hand tools and mats are in their designated positions.**

21. Check and adjust the alignment, as required.
22. Be sure to perform the reconnect/relearn procedures.

SHOCK ABSORBER

REMOVAL & INSTALLATION

See Figure 93.

1. Before servicing the vehicle, refer to the Precautions Section.

➡ **If working near and/or around the SRS system and components, be sure to disable the SRS system. After disabling the system wait three minutes or more before servicing the vehicle.**

➡ **Whenever the negative battery cable is disconnected the following components will require resetting. The Idle Air Volume Learning, Steering Angle Sensor Neutral Position, Sunroof Memory Reset/Initialization, Automatic**

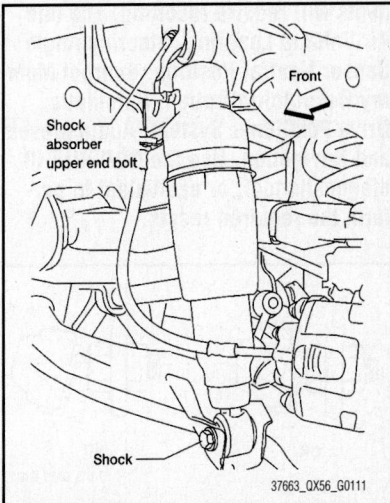

37663_QX56_G0111

Fig. 93 Rear shock absorber and related components

Drive Positioner System, Audio presets and Navigation. Use the CONSULT-III diagnostic tool, or equivalent to perform the required resets.

2. Disconnect the negative battery cable.
3. Raise and safely support the vehicle.
4. Remove the tire and wheel assemblies.
5. Release the air pressure from the rear load leveling air suspension system using the CONSULT-III® "EXHAUST SOLENOID" active test.
6. Remove or disconnect the following:
- Rear fender protector
- Rear load leveling air suspension hose from the shock absorber
- Shock absorber upper and lower end bolts
- Shock absorber

To install:

➡**Be sure to use new fasteners, as required.**

7. Installation is the reverse of the removal procedure.
8. Be sure to perform the reconnect/relearn procedures.

STABILIZER BAR

REMOVAL & INSTALLATION

See Figures 91 and 94.

1. Before servicing the vehicle, refer to the Precautions Section.

➡**If working near and/or around the SRS system and components, be sure to disable the SRS system. After disabling the system wait three minutes or more before servicing the vehicle.**

➡**Whenever the negative battery cable is disconnected the following components will require resetting. The Idle Air Volume Learning, Steering Angle Sensor Neutral Position, Sunroof Memory Reset/Initialization, Automatic Drive Positioner System, Audio presets and Navigation. Use the CONSULT-III diagnostic tool, or equivalent to perform the required resets.**

2. Disconnect the negative battery cable.
3. Raise and safely support the vehicle.
4. Disconnect the bar ends from the connecting rods.
5. Remove the bar clamps. Remove the bar bushings.
6. Remove the stabilizer bar.

To install:

➡**Be sure to use new fasteners, as required.**

7. Installation is the reverse of the removal procedure.

➡**Install the stabilizer bar with the ball joint sockets properly aligned. Install the bar bushing and clamp so that they are positioned inside of the sideslip prevention clamp on the bar. See illustration**

8. Be sure to perform the reconnect/relearn procedures.

WHEEL HUB & BEARING (SEALED UNIT)

REMOVAL & INSTALLATION

See Figure 95.

1. Before servicing the vehicle, refer to the Precautions Section.

➡**If working near and/or around the SRS system and components, be sure to disable the SRS system. After disabling the system wait three minutes or more before servicing the vehicle.**

➡**Whenever the negative battery cable is disconnected the following compo-**

nents will require resetting. The Idle Air Volume Learning, Steering Angle Sensor Neutral Position, Sunroof Memory Reset/Initialization, Automatic Drive Positioner System, Audio presets and Navigation. Use the CONSULT-III diagnostic tool, or equivalent to perform the required resets.

2. Disconnect the negative battery cable.
3. Raise and support the vehicle safely.
4. Remove or disconnect the following:
- Wheel and tire assembly
- Brake caliper without disconnecting the hydraulic lines, and reposition aside with wire
- Brake rotor
- Cotter pin and nut from driveshaft
- Driveshaft
- Wheel hub and bearing assembly bolts
5. Pulling out the wheel hub and bearing assembly slightly, remove the ABS sensor.
6. Remove the wheel hub and bearing assembly.

To install:

➡**Be sure to use new fasteners, as required.**

7. Install or connect the following:
- ABS sensor
- Wheel hub and bearing assembly, using new bolts
- Driveshaft
- Lock nut and new cotter pin
- Brake rotor
- Brake caliper
- Wheel and tire assembly
8. Be sure to perform the reconnect/relearn procedures.

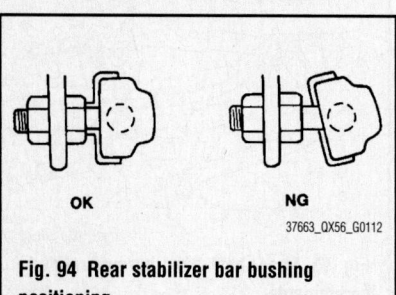

Fig. 94 Rear stabilizer bar bushing positioning

OK   NG

37663_QX56_G0112

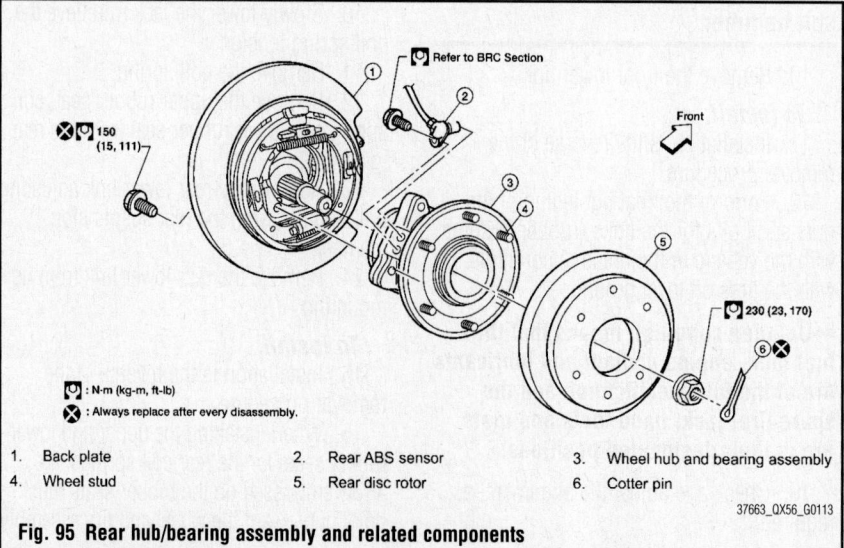

: N·m (kg-m, ft-lb)
: Always replace after every disassembly.

| 1. | Back plate | 2. | Rear ABS sensor | 3. | Wheel hub and bearing assembly |
| 4. | Wheel stud | 5. | Rear disc rotor | 6. | Cotter pin |

37663_QX56_G0113

Fig. 95 Rear hub/bearing assembly and related components

INFINITI

Diagnostic Trouble Codes

6

DIAGNOSTIC TROUBLE CODES

OBD II VEHICLE APPLICATIONS

INFINITI

EX35
2009–2010
• 3.5L V6 .... ID VQ35HR

FX35, FX50
2009–2010
• 3.5L V6 .... ID VQ354HR
• 4.5L V8 .... ID VK50VE

G37
2009–2010
• 3.7L V6 .... ID VQ37VHR

M35 & M45
2009–2010
• 3.5L V6 .... ID VQ35HR
• 4.5L V8 .... ID VK45DE

QX56
2009–2010
• 5.6L V8 .... ID VK56DE

OBD II Trouble Code List (P0XXX Codes)

DTC	Trouble Code Title, Conditions & Possible Causes
DTC: P0011 **1T ECM, MIL: Yes** **Year:** 2009, 2010 **Model:** M45, QX56 **Engine:** 4.5L V8, 5.6L V8 **Transmission:** All	**Intake Valve Timing Control Performance (Bank 1):** **NOTE: If DTC P0011 is displayed with DTC P0075, P0081, P1140 or P1145, first perform the trouble diagnosis for any other DTC before proceeding with P0011.** Condition A: The alignment of the intake valve timing control has been misresisted. Condition B: There is a gap between angle of target and phase-control angle degree. **Possible Causes:** • Harness or connectors • Intake valve timing control solenoid valve circuit is open or shorted. • Intake valve timing control position sensor circuit is open or shorted. • Intake valve timing control solenoid valve • Intake valve timing control position sensor • Crankshaft position sensor (POS) • Camshaft position sensor (PHASE) • Accumulation of debris to the signal pick-up portion of the camshaft sprocket • Timing chain installation • Foreign matter caught in the oil groove for intake valve timing control
DTC: P0011 **1T ECM, MIL: Yes** **Year:** 2009, 2010 **Model:** EX35, FX35, FX50, G37, M35 **Engine:** 3.5L V6, 3.7L V6, 5.0L V8 **Transmission:** All	**Intake Valve Timing Control Performance (Includes Hybrid Models):** HYBRID MODELS CAUTION: Hybrid systems use very high-voltage battery systems. Before starting any service work involving the battery system, turn the ignition switch OFF and then remove the service plug from pocket in the trunk. After removing the service plug, wait 10 minutes before touching any of the high-voltage connectors and terminals. **NOTE: On V6 and V8 engines, this applies to Bank 1.** **NOTE: When the malfunction is detected, the ECM enters fail-safe mode.** **NOTE: If DTC P0011 is displayed with DTC P0075, P0081, or P0524, perform the appropriate trouble diagnosis first.** • There is a gap between angle of target and phase-control angle degree. **Possible Causes:** • Crankshaft position sensor (POS) • Camshaft position sensor (PHASE) • Intake valve control solenoid valve • Accumulation of debris to the signal pick-up portion of the camshaft • Timing chain installation • Foreign matter caught in the oil groove for intake valve timing control
DTC: P0014 **1T ECM, MIL: Yes** **Year:** 2009, 2010 **Model:** EX35, FX35, FX50, M35 **Engine:** 3.5L V6, 5.0L V8 **Transmission:** All	**Exhaust Valve Timing (EVT) Control Performance (Bank 1):** **NOTE: If DTC P0014 or P0024 is displayed with DTC P0078, P0084, P1078 or P1084 first perform trouble diagnosis for respective DTC before proceeding with P0014.** • There is a gap between angle of target and phase-control angle degree. **Possible Causes:** • Crankshaft position sensor • Camshaft position sensor • EVT control position sensor • EVT control magnet retarder • Accumulation of debris to the signal pick-up portion of the camshaft • Timing chain installation • EVT control pulley assembly
DTC: P0021 **1T ECM, MIL: Yes** **Year:** 2009, 2010 **Model:** G37 **Engine:** 3.7L V6 **Transmission:** All	**Intake Valve Timing Control Performance (Bank 2):** **NOTE: If DTC P0011 or P0021 is displayed with DTC P0075, P0081 or P0524, first perform the trouble diagnosis for the applicable DTC before proceeding with P0011.** • There is a gap between angle of target and phase-control angle degree. **Possible Causes:** • Crankshaft position sensor (POS) • Camshaft position sensor (PHASE) • Intake valve control solenoid valve • Accumulation of debris to the signal pick-up portion of the camshaft • Timing chain installation • Foreign matter caught in the oil groove for intake valve timing control

DTC	Trouble Code Title, Conditions & Possible Causes
DTC: P0021 **1T ECM, MIL: Yes** **Year:** 2009, 2010 **Model:** EX35, FX35, FX50, M35 **Engine:** 3.5L V6, 5.0L V8 **Transmission:** All	**Intake Valve Timing Control Performance (Bank 2):** • If DTC P0011 or P0021 is displayed with DTC P0075 or P0081, first perform trouble diagnosis for DTC P0075 or P0081. • There is a gap between angle of target and phase-control angle degree. • When the malfunction is detected, the ECM enters fail-safe mode. **Possible Causes:** • Crankshaft position sensor (POS) • Camshaft position sensor (PHASE) • Intake valve timing control solenoid valve • Accumulation of debris to the signal pick-up portion of the camshaft • Timing chain installation • Foreign matter caught in the oil groove for intake valve timing control
DTC: P0021 **1T ECM, MIL: Yes** **Year:** 2009, 2010 **Model:** M45, QX56 **Engine:** 4.5L V8, 5.6L V8 **Transmission:** All	**Intake Valve Timing Control Performance (Bank 2):** NOTE: If DTC P0021 is displayed with DTC P0075, P0081, P1140 or P1145, first perform the trouble diagnosis for any other DTC before proceeding with P0021. Condition A: The alignment of the intake valve timing control has been misresistered. Condition B: There is a gap between angle of target and phase-control angle degree. **Possible Causes:** • Harness or connectors • Intake valve timing control solenoid valve circuit is open or shorted. • Intake valve timing control position sensor circuit is open or shorted. • Intake valve timing control solenoid valve • Intake valve timing control position sensor • Crankshaft position sensor (POS) • Camshaft position sensor (PHASE) • Accumulation of debris to the signal pick-up portion of the camshaft sprocket • Timing chain installation • Foreign matter caught in the oil groove for intake valve timing control
DTC: P0024 **1T ECM, MIL: Yes** **Year:** 2009, 2010 **Model:** EX35, FX35, FX50, M35 **Engine:** 3.5L V6, 5.0L V8 **Transmission:** All	**Exhaust Valve Timing (EVT) Control Performance (Bank 2):** NOTE: If DTC P0014 or P0024 is displayed with DTC P0078, P0084, P1078 or P1084, first perform trouble diagnosis for the respective DTC before proceeding with P0024. • There is a gap between angle of target and phase-control angle degree. **Possible Causes:** • Crankshaft position sensor • Camshaft position sensor • EVT control position sensor • EVT control magnet retarder • Accumulation of debris to the signal pick-up portion of the camshaft • Timing chain installation • EVT control pulley assembly
DTC: P0031 **1T ECM, MIL: Yes** **Year:** 2009, 2010 **Model:** EX35, FX35, FX50, G37, M35, M45, QX56 **Engine:** 3.5L V6, 3.7L V6, 4.5L V8, 5.0L V8, 5.6L V8 **Transmission:** All	**Air Fuel Ratio (A/F) Sensor 1 Heater Control Circuit Low (Includes Hybrid Models):** HYBRID MODELS CAUTION: Hybrid systems use very high-voltage battery systems. Before starting any service work involving the battery system, turn the ignition switch OFF and then remove the service plug from pocket in the trunk. After removing the service plug, wait 10 minutes before touching any of the high-voltage connectors and terminals. NOTE: On V6 and V8 engines, this applies to Bank 1. • The ECM performs ON/OFF duty control of the A/F sensor 1 heater corresponding to the engine operating condition to keep the temperature of A/F sensor 1 element at the specified range. • The current amperage in the A/F sensor 1 heater circuit is out of the normal range. (An excessively low voltage signal is sent to ECM through the A/F sensor heater.) **Possible Causes:** • Harness or connectors • The A/F sensor 1 heater circuit is open or shorted. • A/F sensor 1 heater

DTC	Trouble Code Title, Conditions & Possible Causes
DTC: P0032 **1T ECM, MIL:** Yes **Year:** 2009, 2010 **Model:** EX35, FX35, FX50, G37, M35, M45, QX56 **Engine:** 3.5L V6, 3.7L V6, 4.5L V8, 5.0L V8, 5.6L V8 **Transmission:** All	**Air Fuel Ratio (A/F) Sensor 1 Heater (Bank 1) Control Circuit High (Includes Hybrid Models):** HYBRID MODELS CAUTION: Hybrid systems use very high-voltage battery systems. Before starting any service work involving the battery system, turn the ignition switch OFF and then remove the service plug from pocket in the trunk. After removing the service plug, wait 10 minutes before touching any of the high-voltage connectors and terminals. **NOTE: On V6 and V8 engines, this applies to Bank 1.** • The ECM performs ON/OFF duty control of the A/F sensor 1 heater corresponding to the engine operating condition to keep the temperature of A/F sensor 1 element at the specified range. • The current amperage in the A/F sensor 1 heater circuit is out of the normal range. (An excessively high voltage signal is sent to ECM through the A/F sensor heater.) **Possible Causes:** • Harness or connectors • A/F sensor 1 heater circuit is shorted. • A/F sensor 1 heater
DTC: P0037 **1T ECM, MIL:** Yes **Year:** 2009, 2010 **Model:** EX35, FX35, FX50, G37, M35, M45, QX56 **Engine:** 3.5L V6, 3.7L V6, 4.5L V8, 5.0L V8, 5.6L V8 **Transmission:** All	**Heated Oxygen Sensor 2 Heater Control Circuit Low (Includes Hybrid Models):** HYBRID MODEL CAUTION: Hybrid systems use very high-voltage battery systems. Before starting any service work involving the battery system, turn the ignition switch OFF and then remove the service plug from pocket in the trunk. After removing the service plug, wait 10 minutes before touching any of the high-voltage connectors and terminals. **NOTE: On V6 and V8 engines, this applies to Bank 1.** • The current amperage in the heated oxygen sensor 2 heater circuit is out of the normal range. (An excessively low voltage signal is sent to ECM via the heated oxygen sensor 2 heater.) **Possible Causes:** • Harness or connectors • The heated oxygen sensor 2 heater circuit is open or shorted. • Heated oxygen sensor 2 heater
DTC: P0038 **1T ECM, MIL:** Yes **Year:** 2009, 2010 **Model:** EX35, FX35, FX50, G37, M35, M45, QX56 **Engine:** 3.5L V6, 3.7L V6, 4.5L V8, 5.0L V8, 5.6L V8 **Transmission:** All	**Heated Oxygen Sensor 2 Heater Control Circuit High (Includes Hybrid Models):** HYBRID MODELS CAUTION: Hybrid systems use very high-voltage battery systems. Before starting any service work involving the battery system, turn the ignition switch OFF and then remove the service plug from pocket in the trunk. After removing the service plug, wait 10 minutes before touching any of the high-voltage connectors and terminals. **NOTE: On V6 and V8 engines, this applies to Bank 1.** • The current amperage in the heated oxygen sensor 2 heater circuit is out of the normal range. (An excessively high voltage signal is sent to ECM via the heated oxygen sensor 2 heater.) **Possible Causes:** • Harness or connectors • The heated oxygen sensor 2 heater circuit is shorted. • Heated oxygen sensor 2 heater
DTC: P0051 **1T ECM, MIL:** Yes **Year:** 2009, 2010 **Model:** EX35, FX35, FX50, G37, M35, M45, QX56 **Engine:** 3.5L V6, 3.7L V6, 4.5L V8, 5.0L V8, 5.6L V8 **Transmission:** All	**Air Fuel Ratio (A/F) Sensor 1 (Bank 2) Heater Control Circuit Low:** **NOTE: On V6 and V8 engines, this applies to Bank 2.** • The ECM performs ON/OFF duty control of the A/F sensor 1 heater corresponding to the engine operating condition to keep the temperature of A/F sensor 1 element at the specified range. • The current amperage in the A/F sensor 1 heater circuit is out of the normal range. (An excessively low voltage signal is sent to ECM through the A/F sensor heater.) **Possible Causes:** • Harness or connectors • The A/F sensor 1 heater circuit is open or shorted. • A/F sensor 1 heater
DTC: P0052 **1T ECM, MIL:** Yes **Year:** 2009, 2010 **Model:** EX35, FX35, FX50, G37, M35, M45, QX56 **Engine:** 3.5L V6, 3.7L V6, 4.5L V8, 5.0L V8, 5.6L V8 **Transmission:** All	**Air Fuel Ratio (A/F) Sensor 1 (Bank 2) Heater Control Circuit High:** **NOTE: On V6 and V8 engines, this applies to Bank 2.** • The ECM performs ON/OFF duty control of the A/F sensor 1 heater corresponding to the engine operating condition to keep the temperature of A/F sensor 1 element at the specified range. • The current amperage in the A/F sensor 1 heater circuit is out of the normal range. (An excessively high voltage signal is sent to ECM through the A/F sensor heater.) **Possible Causes:** • Harness or connectors (The A/F sensor 1 heater circuit is shorted.) • A/F sensor 1 heater

DTC	Trouble Code Title, Conditions & Possible Causes

DTC: P0057
1T ECM, MIL: Yes
Year: 2009, 2010
Model: EX35, FX35, FX50, G37, M35, M45, QX56
Engine: 3.5L V6, 3.7L V6, 4.5L V8, 5.0L V8, 5.6L V8
Transmission: All
	Heated Oxygen Sensor 2 (Bank 2) Heater Control Circuit Low:
	NOTE: On V6 and V8 engines, this applies to Bank 2.
	    • The current amperage in the heated oxygen sensor 2 heater circuit is out of the normal range. (An excessively low voltage signal is sent to ECM via the heated oxygen sensor 2 heater.)
	Possible Causes:
	    • Harness or connectors
	    • The heated oxygen sensor 2 heater circuit is open or shorted.
	    • Heated oxygen sensor 2 heater

DTC: P0058
1T ECM, MIL: Yes
Year: 2009, 2010
Model: EX35, FX35, FX50, G37, M35, M45, QX56
Engine: 3.5L V6, 3.7L V6, 4.5L V8, 5.0L V8, 5.6L V8
Transmission: All
	Heated Oxygen Sensor 2 (Bank 2) Heater Control Circuit High:
	NOTE: On V6 and V8 engines, this applies to Bank 2.
	    • The current amperage in the heated oxygen sensor 2 heater circuit is out of the normal range. (An excessively high voltage signal is sent to ECM via the heated oxygen sensor 2 heater.)
	Possible Causes:
	    • Harness or connectors
	    • The heated oxygen sensor 2 heater circuit is shorted.
	    • Heated oxygen sensor 2 heater

DTC: P0075
1T ECM, MIL: Yes
Year: 2009, 2010
Model: EX35, FX35, FX50, G37, M35, M45, QX56
Engine: 3.5L V6, 3.7L V6, 4.5L V8, 5.0L V8, 5.6L V8
Transmission: All
	Intake Valve Timing Control Solenoid Valve Circuit (Includes Hybrid Models):
	HYBRID MODELS CAUTION: Hybrid systems use very high-voltage battery systems. Before starting any service work involving the battery system, turn the ignition switch OFF and then remove the service plug from pocket in the trunk. After removing the service plug, wait 10 minutes before touching any of the high-voltage connectors and terminals.
	NOTE: On V6 and V8, this IVT is on bank 1.
	    • An improper voltage is sent to the ECM via the intake valve timing control solenoid valve.
	Possible Causes:
	    • Harness or connectors.
	    • Intake valve timing control solenoid valve circuit is open or shorted.
	    • Intake valve timing control solenoid valve has failed.

DTC: P0078
1T ECM, MIL: Yes
Year: 2009, 2010
Model: EX35, FX35, FX50, M35
Engine: 3.5L V6, 5.0L V8
Transmission: All
	Exhaust Valve Timing (EVT) Control Magnet Retarder (Bank 1) Circuit:
	    • An improper voltage is sent to the ECM via the exhaust valve timing control magnet retarder.
	Possible Causes:
	    • Harness or connectors
	    • Exhaust valve timing control magnet retarder circuit is open or shorted.
	    • Exhaust valve timing control magnet retarder

DTC: P0081
1T ECM, MIL: Yes
Year: 2009, 2010
Model: EX35, FX35, FX50, G37, M35, M45, QX56
Engine: 3.5L V6, 3.7L V6, 4.5L V8, 5.0L V8, 5.6L V8
Transmission: All
	Intake Valve Timing Control Solenoid Valve Circuit (Bank 2):
	NOTE: Applies to V6 and V8 engines only.
	    • An improper voltage is sent to the ECM via the intake valve timing control solenoid valve.
	Possible Causes:
	    • Harness or connectors.
	    • Intake valve timing control solenoid valve circuit is open or shorted.
	    • Intake valve timing control solenoid valve has failed.

DTC: P0084
1T ECM, MIL: Yes
Year: 2009, 2010
Model: EX35, FX35, FX50, M35
Engine: 3.5L V6, 5.0L V8
Transmission: All
	Exhaust Valve Timing (EVT) Timing Control (Magnet Retarder) Solenoid (Bank 2) Circuit:
	    • An improper voltage is sent to the ECM via the exhaust valve timing control (magnet retarder).
	Possible Causes:
	    • Harness or connectors
	    • Exhaust valve timing control magnet retarder circuit is open or shorted.
	    • Exhaust valve timing control magnet retarder

DTC	Trouble Code Title, Conditions & Possible Causes

DTC: P0101
1T ECM, MIL: Yes
Year: 2009, 2010
Model: EX35, FX35, FX50, G37, M35, M45, QX56
Engine: 3.5L V6, 3.7L V6, 4.5L V8, 5.0L V8, 5.6L V8
Transmission: All
	Mass Air Flow Sensor Circuit Range/Performance (Includes Hybrid Models):
	HYBRID MODELS CAUTION: Hybrid systems use very high-voltage battery systems. Before starting any service work involving the battery system, turn the ignition switch OFF and then remove the service plug from pocket in the trunk. After removing the service plug, wait 10 minutes before touching any of the high-voltage connectors and terminals.
	    • Condition A: A high voltage from the sensor is sent to ECM under light load driving condition.
	    • Condition B: A low voltage from the sensor is sent to ECM under heavy load driving condition.
	Possible Causes:
	    • Condition A:
	    • Harness or connectors
	    • The sensor circuit is open or shorted.
	    • Mass air flow sensor
	    • EVAP control system pressure sensor
	    • Intake air temperature sensor
	    • Condition B:
	    • Harness or connectors
	    • The sensor circuit is open or shorted.
	    • Intake air leakage
	    • Mass air flow sensor
	    • EVAP control system pressure sensor
	    • Intake air temperature sensor

DTC: P0102
1T ECM, MIL: Yes
Year: 2009, 2010
Model: EX35, FX35, FX50, G37, M35, M45, QX56
Engine: 3.5L V6, 3.7L V6, 4.5L V8, 5.0L V8, 5.6L V8
Transmission: All
	Mass Air Flow Circuit Low Input (Includes Hybrid Models):
	HYBRID MODELS CAUTION: Hybrid systems use very high-voltage battery systems. Before starting any service work involving the battery system, turn the ignition switch OFF and then remove the service plug from pocket in the trunk. After removing the service plug, wait 10 minutes before touching any of the high-voltage connectors and terminals.
	NOTE: On V6 and V8 engines, this DTC is for Bank 1.
	    • An excessively low voltage from the sensor is sent to ECM.
	Possible Causes:
	    • Harness or connectors
	    • The sensor circuit is open or shorted.
	    • Intake air leakage
	    • Mass air flow sensor

DTC: P0103
1T ECM, MIL: Yes
Year: 2009, 2010
Model: EX35, FX35, FX50, G37, M35, M45, QX56
Engine: 3.5L V6, 3.7L V6, 4.5L V8, 5.0L V8, 5.6L V8
Transmission: All
	Mass Air Flow Sensor Circuit High Input (Includes Hybrid Models):
	HYBRID MODELS CAUTION: Hybrid systems use very high-voltage battery systems. Before starting any service work involving the battery system, turn the ignition switch OFF and then remove the service plug from pocket in the trunk. After removing the service plug, wait 10 minutes before touching any of the high-voltage connectors and terminals.
	NOTE: On V6 and V8 engines, this DTC is for Bank 1.
	    • An excessively high voltage from the sensor is sent to ECM.
	Possible Causes:
	    • Harness or connectors
	    • The sensor circuit is open or shorted.
	    • Mass air flow sensor

DTC: P010A
1T ECM, MIL: Yes
Year: 2009, 2010
Model: G37
Engine: 3.7L V6
Transmission: All
	Manifold Absolute Pressure Sensor Circuit:
	NOTE: If DTC P010A is displayed with DTC P0643, first perform the trouble diagnosis for DTC P0643.
	    • An excessively low voltage from the sensor is sent to ECM.
	    • An excessively high voltage from the sensor is sent to ECM.
	Possible Causes:
	    • Harness or connectors
	    • The sensor circuit is open or shorted.
	    • Manifold absolute pressure (MAP) sensor

DTC	Trouble Code Title, Conditions & Possible Causes

DTC: P010B
1T ECM, MIL: Yes
Year: 2009, 2010
Model: EX35, FX35, FX50, G37, M35
Engine: 3.5L V6, 3.7L V6, 5.0L V8
Transmission: All
	Mass Air Flow Sensor (Bank 2) Circuit/Range Performance:
	    • Condition A: A high voltage from the sensor is sent to ECM under light load driving condition.
	    • Condition B: A low voltage from the sensor is sent to ECM under heavy load driving condition.
	Possible Causes:
	    • Condition A:
	    • Harness or connectors
	    • The sensor circuit is open or shorted.
	    • Mass air flow sensor
	    • EVAP control system pressure sensor
	    • Condition B:
	    • Harness or connectors
	    • The sensor circuit is open or shorted.
	    • Intake air leaks
	    • Mass air flow sensor
	    • EVAP control system pressure sensor
	    • Intake air temperature sensor

DTC: P010C
1T ECM, MIL: Yes
Year: 2009, 2010
Model: EX35, FX35, FX50, G37, M35
Engine: 3.5L V6, 3.7L V6, 5.0L V8
Transmission: All
	Mass Air Flow Sensor (Bank 2) Circuit Low Input:
	    • An excessively low voltage from the sensor is sent to ECM.
	Possible Causes:
	    • Harness or connectors
	    • The sensor circuit is open or shorted.
	    • Intake air leaks
	    • Mass air flow sensor

DTC: P010D
1T ECM, MIL: Yes
Year: 2009, 2010
Model: EX35, FX35, FX50, G37, M35
Engine: 3.5L V6, 3.7L V6, 5.0L V8
Transmission: All
	Mass Air Flow Sensor (Bank 2) Circuit High Input:
	An excessively high voltage from the sensor is sent to ECM.
	Possible Causes:
	    • Harness or connectors
	    • The sensor circuit is open or shorted.
	    • Mass air flow sensor

DTC: P0112
1T ECM, MIL: Yes
Year: 2009, 2010
Model: EX35, FX35, FX50, G37, M35, M45, QX56
Engine: 3.5L V6, 3.7L V6, 4.5L V8, 5.0L V8, 5.6L V8
Transmission: All
	Intake Air Temperature Sensor Circuit Low Input (Includes Hybrid Models):
	HYBRID MODELS CAUTION: Hybrid systems use very high-voltage battery systems. Before starting any service work involving the battery system, turn the ignition switch OFF and then remove the service plug from pocket in the trunk. After removing the service plug, wait 10 minutes before touching any of the high-voltage connectors and terminals.
	NOTE: On V6 and V8 engines, this DTC is for Bank 1.
	    • An excessively low voltage from the sensor is sent to ECM.
	Possible Causes:
	    • Harness or connectors
	    • Intake air temperature sensor circuit is open or shorted.
	    • Intake air temperature sensor

DTC: P0113
1T ECM, MIL: Yes
Year: 2009, 2010
Model: EX35, FX35, FX50, G37, M35, M45, QX56
Engine: 3.5L V6, 3.7L V6, 4.5L V8, 5.0L V8, 5.6L V8
Transmission: All
	Intake Air Temperature Sensor Circuit High Input (Includes Hybrid Models):
	HYBRID MODELS CAUTION: Hybrid systems use very high-voltage battery systems. Before starting any service work involving the battery system, turn the ignition switch OFF and then remove the service plug from pocket in the trunk. After removing the service plug, wait 10 minutes before touching any of the high-voltage connectors and terminals.
	NOTE: On V6 and V8 engines, this DTC is for Bank 1.
	    • An excessively high voltage from the sensor is sent to ECM.
	Possible Causes:
	    • Harness or connectors
	    • Intake air temperature sensor circuit is open or shorted.
	    • Intake air temperature sensor has failed

DTC: P0116
1T ECM, MIL: Yes
Year: 2009, 2010
Model: EX35, FX35, FX50, G37, M35, M45, QX56
Engine: 3.5L V6, 3.7L V6, 4.5L V8, 5.0L V8, 5.6L V8
Transmission: All
	Engine Coolant Temperature Sensor Circuit Range/Performance (Includes Hybrid Models):
	HYBRID MODELS CAUTION: Hybrid systems use very high-voltage battery systems. Before starting any service work involving the battery system, turn the ignition switch OFF and then remove the service plug from pocket in the trunk. After removing the service plug, wait 10 minutes before touching any of the high-voltage connectors and terminals.
	NOTE: If DTC P0116 is displayed with P0117 or P0118, first perform the trouble diagnosis for DTC P0117, P0118.
	    • Engine coolant temperature signal from engine coolant temperature sensor does not fluctuate, even when some time has passed after starting the engine with pre-warming up condition.
	Possible Causes:
	    • Harness or connectors
	    • High or low resistance in the circuit.
	    • Engine coolant temperature sensor

DTC	Trouble Code Title, Conditions & Possible Causes
DTC: P0117 **1T ECM, MIL: Yes** **Year:** 2009, 2010 **Model:** EX35, FX35, FX50, G37, M35, M45, QX56 **Engine:** 3.5L V6, 3.7L V6, 4.5L V8, 5.0L V8, 5.6L V8 **Transmission:** All	**Engine Coolant Temperature Circuit Low Input (Includes Hybrid Models):** HYBRID MODELS CAUTION: Hybrid systems use very high-voltage battery systems. Before starting any service work involving the battery system, turn the ignition switch OFF and then remove the service plug from pocket in the trunk. After removing the service plug, wait 10 minutes before touching any of the high-voltage connectors and terminals. • An excessively low voltage from the sensor is sent to ECM. **Possible Causes:** • Harness or connectors • Engine coolant temperature sensor circuit is open or shorted. • Engine coolant temperature sensor has failed
DTC: P0118 **1T ECM, MIL: Yes** **Year:** 2009, 2010 **Model:** EX35, FX35, FX50, G37, M35, M45, QX56 **Engine:** 3.5L V6, 3.7L V6, 4.5L V8, 5.0L V8, 5.6L V8 **Transmission:** All	**Engine Coolant Temperature Sensor Circuit High Input (Includes Hybrid Models):** HYBRID MODELS CAUTION: Hybrid systems use very high-voltage battery systems. Before starting any service work involving the battery system, turn the ignition switch OFF and then remove the service plug from pocket in the trunk. After removing the service plug, wait 10 minutes before touching any of the high-voltage connectors and terminals. • An excessively high voltage from the sensor is sent to ECM. **Possible Causes:** • Harness or connectors • Engine coolant temperature sensor circuit is open or shorted. • Engine coolant temperature sensor has failed
DTC: P0122 **1T ECM, MIL: Yes** **Year:** 2009, 2010 **Model:** EX35, FX35, FX50, G37, M35, M45, QX56 **Engine:** 3.5L V6, 3.7L V6, 4.5L V8, 5.0L V8, 5.6L V8 **Transmission:** All	**Throttle Position (TP) Sensor 2 Circuit Low Input (Includes Hybrid Models):** HYBRID MODELS CAUTION: Hybrid systems use very high-voltage battery systems. Before starting any service work involving the battery system, turn the ignition switch OFF and then remove the service plug from pocket in the trunk. After removing the service plug, wait 10 minutes before touching any of the high-voltage connectors and terminals. **NOTE: If this DTC is displayed with DTC P0643, first perform the trouble diagnosis for DTC P0643.** **NOTE: On V6 and V8 engines, this DTC is for Bank 1.** • An excessively low voltage from the TP sensor 2 is sent to ECM. **Possible Causes:** • Harness or connectors • TP sensor 2 circuit is open or shorted. • Electric throttle control actuator (TP sensor 2) • Accelerator pedal position sensor (APP sensor 2)
DTC: P0123 **1T ECM, MIL: Yes** **Year:** 2009, 2010 **Model:** EX35, FX35, FX50, G37, M35, M45, QX56 **Engine:** 3.5L V6, 3.7L V6, 4.5L V8, 5.0L V8, 5.6L V8 **Transmission:** All	**Throttle Position Sensor 2 Circuit High Input (Includes Hybrid Models):** HYBRID MODELS CAUTION: Hybrid systems use very high-voltage battery systems. Before starting any service work involving the battery system, turn the ignition switch OFF and then remove the service plug from pocket in the trunk. After removing the service plug, wait 10 minutes before touching any of the high-voltage connectors and terminals. **NOTE: If DTC P0122 or P0123 is displayed with DTC P0643, first perform the trouble diagnosis for DTC P0643.** **NOTE: On 3.7L engine, this DTC is for Bank 1.** • An excessively high voltage from the TP sensor 2 is sent to ECM. • When the malfunction is detected, ECM enters fail-safe mode and the MIL lights up. **Possible Causes:** • Harness or connectors • TP sensor 2 circuit is open or shorted. • Electric throttle control actuator (TP sensor 2) • Accelerator pedal position sensor (APP sensor 2)
DTC: P0125 **1T ECM, MIL: Yes** **Year:** 2009, 2010 **Model:** EX35, FX35, FX50, G37, M35, M45, QX56 **Engine:** 3.5L V6, 3.7L V6, 4.5L V8, 5.0L V8, 5.6L V8 **Transmission:** All	**Insufficient Engine Coolant Temperature for Closed Loop Fuel Control (Includes Hybrid Models):** HYBRID MODELS CAUTION: Hybrid systems use very high-voltage battery systems. Before starting any service work involving the battery system, turn the ignition switch OFF and then remove the service plug from pocket in the trunk. After removing the service plug, wait 10 minutes before touching any of the high-voltage connectors and terminals. • If DTC P0125 is displayed with P0116, P0117 or P0118, first perform the trouble diagnosis for the appropriate DTC, then proceed with P0125. • Voltage sent to ECM from the sensor is not practical, even when some time has passed after starting the engine. • Engine coolant temperature is insufficient for closed loop fuel control. **Possible Causes:** • Harness or connectors • High resistance in the circuit • Engine coolant temperature sensor • Thermostat

DTC	Trouble Code Title, Conditions & Possible Causes
DTC: P0127 **1T ECM, MIL: Yes** **Year:** 2009, 2010 **Model:** EX35, FX35, FX50, G37, M35, M45, QX56 **Engine:** 3.5L V6, 3.7L V6, 4.5L V8, 5.0L V8, 5.6L V8 **Transmission:** All	**Intake Air Temperature Too High (Includes Hybrid Models):** HYBRID MODELS CAUTION: Hybrid systems use very high-voltage battery systems. Before starting any service work involving the battery system, turn the ignition switch OFF and then remove the service plug from pocket in the trunk. After removing the service plug, wait 10 minutes before touching any of the high-voltage connectors and terminals. • Rationally incorrect voltage from the sensor is sent to ECM, compared with the voltage signal from engine coolant temperature sensor. **Possible Causes:** • Harness or connectors • Intake temperature sensor circuit is open or shorted • Intake air temperature sensor
DTC: P0128 **1T ECM, MIL: Yes** **Year:** 2009, 2010 **Model:** EX35, FX35, FX50, G37, M35, M45, QX56 **Engine:** 3.5L V6, 3.7L V6, 4.5L V8, 5.0L V8, 5.6L V8 **Transmission:** All	**Thermostat Function :** **NOTE: If DTC P0128 is displayed with DTC P0300, P0301, P0302, P0303, P0304, P0305, P0306, P0307 or P0308 first perform the trouble diagnosis for this DTC before continuing with P0128 diagnosis.** • The engine coolant temperature does not reach to specified temperature even though the engine has run long enough. **Possible Causes:** • Thermostat • Leakage from sealing portion of thermostat • Engine coolant temperature sensor
DTC: P0130 **1T ECM, MIL: Yes** **Year:** 2009, 2010 **Model:** EX35, FX35, FX50, G37, M35, QX56 **Engine:** 3.5L V6, 3.7L V6, 5.0L V8, 5.6L V8 **Transmission:** All	**Air Fuel Ratio (A/F) Sensor 1 Circuit (Includes Hybrid Models):** HYBRID MODELS CAUTION: Hybrid systems use very high-voltage battery systems. Before starting any service work involving the battery system, turn the ignition switch OFF and then remove the service plug from pocket in the trunk. After removing the service plug, wait 10 minutes before touching any of the high-voltage connectors and terminals. **NOTE: On V6 and V8 engines, this applies to Bank 1.** • Condition A: The A/F signal computed by ECM from the A/F sensor 1 signal is constantly in the range other than approx. 2.2V. • Condition B: The A/F signal computed by ECM from the A/F sensor 1 signal is constantly approx. 2.2V. **Possible Causes:** • Harness or connectors • Air fuel ratio (A/F) sensor 1 circuit is open or shorted. • Air fuel ratio (A/F) sensor 1
DTC: P0131 **1T ECM, MIL: Yes** **Year:** 2009, 2010 **Model:** EX35, FX35, FX50, G37, M35, M45, QX56 **Engine:** 3.5L V6, 3.7L V6, 4.5L V8, 5.0L V8, 5.6L V8 **Transmission:** All	**Air Fuel Ratio (A/F) Sensor 1 Circuit Low Voltage (Includes Hybrid Models):** HYBRID MODELS CAUTION: Hybrid systems use very high-voltage battery systems. Before starting any service work involving the battery system, turn the ignition switch OFF and then remove the service plug from pocket in the trunk. After removing the service plug, wait 10 minutes before touching any of the high-voltage connectors and terminals. **NOTE: On V6 and V8 engines, this applies to Bank 1.** • To judge the malfunction, the diagnosis checks that the A/F signal computed by ECM from the A/F sensor 1 signal is not inordinately low. • The A/F signal computed by ECM from the A/F sensor 1 signal is constantly approx. 0V. **Possible Causes:** • Harness or connectors • Air fuel ratio (A/F) sensor circuit is open or shorted. • Air fuel ratio (A/F) sensor 1
DTC: P0132 **1T ECM, MIL: Yes** **Year:** 2009, 2010 **Model:** EX35, FX35, FX50, G37, M35, M45, QX56 **Engine:** 3.5L V6, 3.7L V6, 4.5L V8, 5.0L V8, 5.6L V8 **Transmission:** All	**Air Fuel Ratio (A/F) Sensor 1 Circuit High Voltage (Includes Hybrid Models):** HYBRID MODELS CAUTION: Hybrid systems use very high-voltage battery systems. Before starting any service work involving the battery system, turn the ignition switch OFF and then remove the service plug from pocket in the trunk. After removing the service plug, wait 10 minutes before touching any of the high-voltage connectors and terminals. **NOTE: On V6 and V8 engines, this applies to Bank 1.** • To judge the malfunction, the diagnosis checks that the A/F signal computed by ECM from the A/F sensor 1 signal is not inordinately low. • The A/F signal computed by ECM from the A/F sensor 1 signal is constantly approx. 5V. **Possible Causes:** • Harness or connectors • Air fuel ratio (A/F) sensor circuit is open or shorted. • Air fuel ratio (A/F) sensor 1

DTC	Trouble Code Title, Conditions & Possible Causes

DTC: P0133
1T ECM, MIL: Yes
Year: 2009, 2010
Model: EX35, FX35, FX50, G37, M35, M45, QX56
Engine: 3.5L V6, 3.7L V6, 4.5L V8, 5.0L V8, 5.6L V8
Transmission: All
	Air Fuel Ratio (A/F) Sensor 1 Circuit Slow Response (Includes Hybrid Models):
	HYBRID MODELS CAUTION: Hybrid systems use very high-voltage battery systems. Before starting any service work involving the battery system, turn the ignition switch OFF and then remove the service plug from pocket in the trunk. After removing the service plug, wait 10 minutes before touching any of the high-voltage connectors and terminals.
	NOTE: On V6 and V8 engines, this applies to Bank 1.
	   • To judge the malfunction of A/F sensor 1, this diagnosis measures response time of the A/F signal computed by ECM from the A/F sensor 1 signal. The time is compensated by engine operating (speed and load), fuel feedback control constant, and the A/F sensor 1 temperature index. Judgment is based on whether the compensated time (the A/F signal cycling time index) is inordinately long or not.
	   • The response of the A/F signal computed by ECM from A/F sensor 1 signal takes more than the specified time.
	Possible Causes:
	   • Harness or connectors
	   • Air fuel ratio (A/F) sensor circuit is open or shorted.
	   • Air fuel ratio (A/F) sensor 1
	   • Air fuel ratio (A/F) sensor heater 1
	   • Fuel pressure
	   • Fuel injector
	   • Intake air leaks
	   • Exhaust gas leaks
	   • PCV valve
	   • Mass air flow sensor

DTC: P0137
1T ECM, MIL: Yes
Year: 2009, 2010
Model: EX35, FX35, FX50, G37, M35, M45, QX56
Engine: 3.5L V6, 3.7L V6, 4.5L V8, 5.0L V8, 5.6L V8
Transmission: All
	Heated Oxygen Sensor 2 Circuit Low Voltage (Includes Hybrid Models):
	HYBRID MODELS CAUTION: Hybrid systems use very high-voltage battery systems. Before starting any service work involving the battery system, turn the ignition switch OFF and then remove the service plug from pocket in the trunk. After removing the service plug, wait 10 minutes before touching any of the high-voltage connectors and terminals.
	NOTE: On V6 and V8 engines, this applies to Bank 1.
	   • The maximum voltage from the sensor is not reached to the specified voltage
	Possible Causes:
	   • Harness or connectors
	   • Heated oxygen sensor 2 circuit open or shorted.
	   • Heated oxygen sensor 2
	   • Fuel pressure
	   • Fuel injector
	   • Intake air leaks

DTC: P0138
1T ECM, MIL: Yes
Year: 2009, 2010
Model: EX35, FX35, FX50, G37, M35, M45, QX56
Engine: 3.5L V6, 3.7L V6, 4.5L V8, 5.0L V8, 5.6L V8
Transmission: All
	Heated Oxygen Sensor 2 Circuit High Voltage (Includes Hybrid Models):
	HYBRID MODELS CAUTION: Hybrid systems use very high-voltage battery systems. Before starting any service work involving the battery system, turn the ignition switch OFF and then remove the service plug from pocket in the trunk. After removing the service plug, wait 10 minutes before touching any of the high-voltage connectors and terminals.
	NOTE: On V6 and V8 engines, this applies to Bank 1.
	   • Condition A: An excessively high voltage from the sensor is sent to ECM, or,
	   • Condition B: The minimum voltage from the sensor is not reached to the specified voltage.
	Possible Causes:
	   • Condition A:
	   • Harness or connectors
	   • Heated oxygen sensor 2 circuit is open or shorted.
	   • Heated oxygen sensor 2
	   • Condition B:
	   • Harness or connectors
	   • Heated oxygen sensor circuit is open or shorted.
	   • Heated oxygen sensor 2
	   • Fuel pressure
	   • Fuel injector

DTC	Trouble Code Title, Conditions & Possible Causes

DTC: P0139
1T ECM, MIL: Yes
Year: 2009, 2010
Model: EX35, FX35, FX50, G37, M35, M45, QX56
Engine: 3.5L V6, 3.7L V6, 4.5L V8, 5.0L V8, 5.6L V8
Transmission: All
	Heated Oxygen Sensor 2 Circuit Slow Response (Includes Hybrid Models):
	HYBRID MODELS CAUTION: Hybrid systems use very high-voltage battery systems. Before starting any service work involving the battery system, turn the ignition switch OFF and then remove the service plug from pocket in the trunk. After removing the service plug, wait 10 minutes before touching any of the high-voltage connectors and terminals.
	NOTE: On 2.0L & 2.5L, Sensor 2; on 3.5L, Sensor 2 Bank 1
	    • It takes more time for the sensor to respond between rich and lean than the specified time.
	Possible Causes:
	    • Harness or connectors
	    • Heated oxygen sensor circuit is open or shorted.
	    • Heated oxygen sensor 2
	    • Fuel pressure
	    • Fuel injector
	    • Intake air leaks

DTC: P0150
1T ECM, MIL: Yes
Year: 2009, 2010
Model: EX35, FX35, FX50, G37, M35, M45, QX56
Engine: 3.5L V6, 3.7L V6, 4.5L V8, 5.0L V8, 5.6L V8
Transmission: All
	Air Fuel Ratio (A/F) Sensor 1 Bank 2 Circuit:
	    • The A/F signal computed by ECM from the A/F sensor 1 bank 2 signal is constantly in the range other than approx. 2.2V.
	    • The A/F signal computed by ECM from the A/F sensor 1 bank 2 signal is constantly approx. 2.2V.
	Possible Causes:
	    • Harness or connectors
	    • The A/F sensor 1 circuit is open or shorted.
	    • A/F sensor 1 bank 2

DTC: P0151
1T ECM, MIL: Yes
Year: 2009, 2010
Model: EX35, FX35, FX50, G37, M35, M45, QX56
Engine: 3.5L V6, 3.7L V6, 4.5L V8, 5.0L V8, 5.6L V8
Transmission: All
	Air Fuel Ratio (A/F) Sensor 1 Bank 2 Circuit Low Voltage:
	    • To judge the malfunction, the diagnosis checks that the A/F signal computed by ECM from the A/F sensor 1 signal is not inordinately low.
	    • The A/F signal computed by ECM from the A/F sensor 1 signal is constantly approx. 0V.
	Possible Causes:
	    • Harness or connectors
	    • Air fuel ratio (A/F) sensor circuit is open or shorted.
	    • Air fuel ratio (A/F) sensor 1

DTC: P0152
1T ECM, MIL: Yes
Year: 2009, 2010
Model: EX35, FX35, FX50, G37, M35, M45, QX56
Engine: 3.5L V6, 3.7L V6, 4.5L V8, 5.0L V8, 5.6L V8
Transmission: All
	Air Fuel Ratio (A/F) Sensor 1 (Bank 2) Circuit High Voltage:
	NOTE: On V6 and V8 engines, this applies to Bank 2.
	    • Engine: After warming up
	    • Maintaining engine speed at 2,000 rpm
	    • The A/F signal computed by ECM from the A/F sensor 1 signal is constantly approx. 5V.
	Possible Causes:
	    • Harness or connectors
	    • Air fuel ratio (A/F) sensor circuit is open or shorted.
	    • Air fuel ratio (A/F) sensor 1

DTC: P0153
1T ECM, MIL: Yes
Year: 2009, 2010
Model: EX35, FX35, FX50, G37, M35, M45, QX56
Engine: 3.5L V6, 3.7L V6, 4.5L V8, 5.0L V8, 5.6L V8
Transmission: All
	Air Fuel Ratio (A/F) Sensor 1 Bank 2 Circuit Slow Response:
	    • The response of the A/F signal computed by ECM from A/F sensor 1 signal takes more than the specified time.
	Possible Causes:
	    • Harness or connectors
	    • Air fuel ratio (A/F) sensor circuit is open or shorted.
	    • Air fuel ratio (A/F) sensor 1
	    • Air fuel ratio (A/F) sensor heater 1
	    • Fuel pressure
	    • Fuel injector
	    • Intake air leaks
	    • Exhaust gas leaks
	    • PCV valve
	    • Mass air flow sensor

DTC: P0157
1T ECM, MIL: Yes
Year: 2009, 2010
Model: EX35, FX35, FX50, G37, M35, M45, QX56
Engine: 3.5L V6, 3.7L V6, 4.5L V8, 5.0L V8, 5.6L V8
Transmission: All
	Heated Oxygen Sensor 2 Bank 2 Circuit Low Voltage:
	    • The maximum voltage from the sensor does not reach the specified voltage.
	Possible Causes:
	    • Harness or connectors
	    • Sensor circuit is open or shorted
	    • Heated oxygen sensor 2
	    • Fuel pressure
	    • Fuel injector
	    • Intake air leakage

DTC	Trouble Code Title, Conditions & Possible Causes
DTC: P0158 **1T ECM, MIL: Yes** **Year:** 2009, 2010 **Model:** EX35, FX35, FX50, G37, M35, M45, QX56 **Engine:** 3.5L V6, 3.7L V6, 4.5L V8, 5.0L V8, 5.6L V8 **Transmission:** All	**Heated Oxygen Sensor 2 Bank 2 Circuit High Voltage:** • Condition A: An excessively high voltage from the sensor is sent to ECM, or, • Condition B: The minimum voltage from the sensor is not reached to the specified voltage. **Possible Causes:** • Condition A: • Harness or connectors (Heated oxygen sensor 2 circuit is open or shorted.) • Heated oxygen sensor 2 • Condition B: • Harness or connectors (Heated oxygen sensor circuit is open or shorted.) • Heated oxygen sensor 2 • Fuel pressure • Fuel injector
DTC: P0159 **1T ECM, MIL: Yes** **Year:** 2009, 2010 **Model:** EX35, FX35, FX50, G37, M35, M45, QX56 **Engine:** 3.5L V6, 3.7L V6, 4.5L V8, 5.0L V8, 5.6L V8 **Transmission:** All	**Heated Oxygen Sensor 2 Bank 2 Circuit Slow Response:** • It takes more time for the sensor to respond between rich and lean than the specified time. **Possible Causes:** • Harness or connectors • Heated oxygen sensor circuit is open or shorted. • Heated oxygen sensor 2 • Fuel pressure • Fuel injector • Intake air leaks
DTC: P0171 **1T ECM, MIL: Yes** **Year:** 2009, 2010 **Model:** EX35, FX35, FX50, G37, M35, M45, QX56 **Engine:** 3.5L V6, 3.7L V6, 4.5L V8, 5.0L V8, 5.6L V8 **Transmission:** All	**Fuel Injection System Too Lean (Includes Hybrid Models):** HYBRID MODELS CAUTION: Hybrid systems use very high-voltage battery systems. Before starting any service work involving the battery system, turn the ignition switch OFF and then remove the service plug from pocket in the trunk. After removing the service plug, wait 10 minutes before touching any of the high-voltage connectors and terminals. **NOTE: On V6 and V8 engines, this applies to Bank 1.** • Fuel injection system does not operate properly. • The amount of mixture ratio compensation is too large. (The mixture ratio is too lean.) **Possible Causes:** • Intake air leaks • Air fuel ratio (A/F) sensor 1 • Fuel injector • Exhaust gas leaks • Incorrect fuel pressure • Lack of fuel • Mass air flow sensor • Incorrect PCV hose connection
DTC: P0172 **1T ECM, MIL: Yes** **Year:** 2009, 2010 **Model:** EX35, FX35, FX50, G37, M35, M45, QX56 **Engine:** 3.5L V6, 3.7L V6, 4.5L V8, 5.0L V8, 5.6L V8 **Transmission:** All	**Fuel Injection System Too Rich (Includes Hybrid Models):** HYBRID MODELS CAUTION: Hybrid systems use very high-voltage battery systems. Before starting any service work involving the battery system, turn the ignition switch OFF and then remove the service plug from pocket in the trunk. After removing the service plug, wait 10 minutes before touching any of the high-voltage connectors and terminals. **NOTE: On V6 and V8 engines, this applies to Bank 1.** • Fuel injection system does not operate properly. • The amount of mixture ratio compensation is too large. (The mixture ratio is too rich.) **Possible Causes:** • Air fuel ratio (A/F) sensor 1 • Fuel injector • Exhaust gas leaks • Incorrect fuel pressure • Mass air flow sensor
DTC: P0174 **1T ECM, MIL: Yes** **Year:** 2009, 2010 **Model:** EX35, FX35, FX50, G37, M35, M45, QX56 **Engine:** 3.5L V6, 3.7L V6, 4.5L V8, 5.0L V8, 5.6L V8 **Transmission:** All	**Fuel Injection System Too Lean (Bank 2):** • Fuel injection system does not operate properly. • The amount of mixture ratio compensation is too large. (The mixture ratio is too lean.) **Possible Causes:** • Intake air leaks • Air fuel ratio (A/F) sensor 1 • Fuel injector • Exhaust gas leaks • Incorrect fuel pressure • Lack of fuel • Mass air flow sensor • Incorrect PCV hose connection

DTC	Trouble Code Title, Conditions & Possible Causes
DTC: P0175 **1T ECM, MIL: Yes** **Year:** 2009, 2010 **Model:** EX35, FX35, FX50, G37, M35, M45, QX56 **Engine:** 3.5L V6, 3.7L V6, 4.5L V8, 5.0L V8, 5.6L V8 **Transmission:** All	**Fuel Injection System Too Rich (Bank 2):** • Fuel injection system does not operate properly. • The amount of mixture ratio compensation is too large. (The mixture ratio is too rich.) **Possible Causes:** • Air fuel ratio (A/F) sensor 1 • Fuel injector • Exhaust gas leaks • Incorrect fuel pressure • Mass air flow sensor
DTC: P0181 **1T ECM, MIL: Yes** **Year:** 2009, 2010 **Model:** EX35, FX35, FX50, G37, M35, M45, QX56 **Engine:** 3.5L V6, 3.7L V6, 4.5L V8, 5.0L V8, 5.6L V8 **Transmission:** All	**Fuel Tank Temperature Sensor Circuit Range/Performance (Includes Hybrid Models):** HYBRID MODELS CAUTION: Hybrid systems use very high-voltage battery systems. Before starting any service work involving the battery system, turn the ignition switch OFF and then remove the service plug from pocket in the trunk. After removing the service plug, wait 10 minutes before touching any of the high-voltage connectors and terminals. • Rationally incorrect voltage from the sensor is sent to ECM, compared with the voltage signals from engine coolant temperature sensor and intake air temperature sensor. **Possible Causes:** • Harness or connectors • Fuel tank temperature sensor circuit is open or shorted • Fuel tank temperature sensor
DTC: P0182 **1T ECM, MIL: Yes** **Year:** 2009, 2010 **Model:** EX35, FX35, FX50, G37, M35, M45, QX56 **Engine:** 3.5L V6, 3.7L V6, 4.5L V8, 5.0L V8, 5.6L V8 **Transmission:** All	**Fuel Tank Temperature Sensor Circuit Low Input (Includes Hybrid Models):** HYBRID MODELS CAUTION: Hybrid systems use very high-voltage battery systems. Before starting any service work involving the battery system, turn the ignition switch OFF and then remove the service plug from pocket in the trunk. After removing the service plug, wait 10 minutes before touching any of the high-voltage connectors and terminals. • An excessively low voltage from the sensor is sent to ECM. **Possible Causes:** • Harness or connectors • Fuel tank temperature sensor circuit is open or shorted. • Fuel tank temperature sensor
DTC: P0183 **1T ECM, MIL: Yes** **Year:** 2009, 2010 **Model:** EX35, FX35, FX50, G37, M35, M45, QX56 **Engine:** 3.5L V6, 3.7L V6, 4.5L V8, 5.0L V8, 5.6L V8 **Transmission:** All	**Fuel Tank Temperature Sensor Circuit High Input (Includes Hybrid Models):** HYBRID MODELS CAUTION: Hybrid systems use very high-voltage battery systems. Before starting any service work involving the battery system, turn the ignition switch OFF and then remove the service plug from pocket in the trunk. After removing the service plug, wait 10 minutes before touching any of the high-voltage connectors and terminals. • An excessively high voltage from the sensor is sent to ECM. **Possible Causes:** • Harness or connectors • Fuel tank temperature sensor circuit is open or shorted. • Fuel tank temperature sensor
DTC: P0196 **1T ECM, MIL: Yes** **Year:** 2009, 2010 **Model:** EX35, FX35, FX50, G37, M35 **Engine:** 3.5L V6, 3.7L V6, 5.0L V8 **Transmission:** All	**Engine Oil Temperature (EOT) Sensor Range/Performance:** NOTE: If DTC P0196 is displayed with P0197 or P0198, first perform the trouble diagnosis for DTC P0197, P0198. • Rationally incorrect voltage from the sensor is sent to ECM, compared with the voltage signals from engine coolant temperature sensor and intake air temperature sensor. **Possible Causes:** • Harness or connectors (The sensor circuit is open or shorted) • Engine oil temperature sensor
DTC: P0197 **1T ECM, MIL: Yes** **Year:** 2009, 2010 **Model:** EX35, FX35, FX50, G37, M35 **Engine:** 3.5L V6, 3.7L V6, 5.0L V8 **Transmission:** All	**Engine Oil Temperature (EOT) Sensor Circuit Low Input:** • An excessively low voltage from the sensor is sent to ECM. **Possible Causes:** • Harness or connectors (Engine oil temperature sensor circuit is open or shorted.) • Engine oil temperature sensor.
DTC: P0198 **1T ECM, MIL: Yes** **Year:** 2009, 2010 **Model:** EX35, FX35, FX50, G37, M35 **Engine:** 3.5L V6, 3.7L V6, 5.0L V8 **Transmission:** All	**Engine Oil Temperature (EOT) Sensor Circuit High Input:** • An excessively high voltage from the sensor is sent to ECM. **Possible Causes:** • Harness or connectors (Engine oil temperature sensor circuit is open or shorted.) • Engine oil temperature sensor

DTC	Trouble Code Title, Conditions & Possible Causes
DTC: P0222 **1T ECM, MIL: Yes** **Year:** 2009, 2010 **Model:** EX35, FX35, FX50, G37, M35, M45, QX56 **Engine:** 3.5L V6, 3.7L V6, 4.5L V8, 5.0L V8, 5.6L V8 **Transmission:** All	**Throttle Position (TP) Sensor 1 Circuit Low Input (Includes Hybrid Models):** HYBRID MODELS CAUTION: Hybrid systems use very high-voltage battery systems. Before starting any service work involving the battery system, turn the ignition switch OFF and then remove the service plug from pocket in the trunk. After removing the service plug, wait 10 minutes before touching any of the high-voltage connectors and terminals. **NOTE: If DTC P0222 or P0223 is displayed with DTC P0643, first perform the trouble diagnosis for DTC P0643.** **NOTE: On V6 and V8 engines, this DTC is for Bank 1.** • An excessively low voltage from the TP sensor 1 is sent to ECM. **Possible Causes:** • Harness or connectors • TP sensor 1 circuit is open or shorted. • APP sensor circuit is shorted (if equipped) • Electric throttle control actuator (TP sensor 1) • Accelerator pedal position sensor (APP sensor 2)
DTC: P0223 **1T ECM, MIL: Yes** **Year:** 2009, 2010 **Model:** EX35, FX35, FX50, G37, M35, M45, QX56 **Engine:** 3.5L V6, 3.7L V6, 4.5L V8, 5.0L V8, 5.6L V8 **Transmission:** All	**Throttle Position (TP) Sensor 1 Circuit High Input (Includes Hybrid Models):** HYBRID MODELS CAUTION: Hybrid systems use very high-voltage battery systems. Before starting any service work involving the battery system, turn the ignition switch OFF and then remove the service plug from pocket in the trunk. After removing the service plug, wait 10 minutes before touching any of the high-voltage connectors and terminals. **NOTE: If DTC P0222 or P0223 is displayed with DTC P0643, first perform the trouble diagnosis for DTC P0643.** **NOTE: On V6 and V8 engines, this DTC is for Bank 1.** • An excessively high voltage from the TP sensor 1 is sent to ECM. **Possible Causes:** • Harness or connectors • TP sensor 1 circuit is open or shorted. • APP sensor 2 circuit is shorted. • Electric throttle control actuator (TP sensor 1) • Accelerator pedal position sensor (APP sensor 2)
DTC: P0225 **T ECM, MIL: Yes** **Year:** 2009, 2010 **Model:** FX50 **Engine:** 5.0L V8 **Transmission:** All	**Closed Throttle Position Learning Performance (Bank 2):** **NOTE: DTC P0225 is displayed with another DTC for electric throttle control actuator. Perform the trouble diagnosis for the corresponding DTC.** • Closed throttle position learning value is excessively low. • Closed throttle position learning is not performed successfully, repeatedly. **Possible Causes:** • Electric throttle control actuator (TP sensor 1 and 2)
DTC: P0227 **1T ECM, MIL: Yes** **Year:** 2009, 2010 **Model:** EX35, FX35, FX50, G37, M35 **Engine:** 3.5L V6, 3.7L V6, 5.0L V8 **Transmission:** All	**Throttle Position Sensor 2 (Bank 2) Circuit Low Input:** **NOTE: If DTC P0122, P0123, P0227 or P0228 is displayed with DTC P0643, first perform the trouble diagnosis for DTC P0643.** • An excessively low voltage from the TP sensor 2 is sent to ECM. **Possible Causes:** • Harness or connectors • TP sensor 2 circuit is open or shorted. • Electric throttle control actuator (TP sensor 2)
DTC: P0228 **1T ECM, MIL: Yes** **Year:** 2009, 2010 **Model:** EX35, FX35, FX50, G37, M35 **Engine:** 3.5L V6, 3.7L V6, 5.0L V8 **Transmission:** All	**Throttle Position Sensor 2 (Bank 2) Circuit High Input:** **NOTE: If DTC P0122, P0123, P0227 or P0228 is displayed with DTC P0643, first perform the trouble diagnosis for DTC P0643.** • An excessively high voltage from the TP sensor 2 is sent to ECM. **Possible Causes:** • Harness or connectors • TP sensor 2 circuit is open or shorted. • Electric throttle control actuator (TP sensor 2)

DTC	Trouble Code Title, Conditions & Possible Causes

DTC: P0300
1T ECM, MIL: Yes
Year: 2009, 2010
Model: EX35, FX35, FX50, G37, M35, M45, QX56
Engine: 3.5L V6, 3.7L V6, 4.5L V8, 5.0L V8, 5.6L V8
Transmission: All
	Multiple Cylinder Misfire Detected (Includes Hybrid Models):
	HYBRID MODELS CAUTION: Hybrid systems use very high-voltage battery systems. Before starting any service work involving the battery system, turn the ignition switch OFF and then remove the service plug from pocket in the trunk. After removing the service plug, wait 10 minutes before touching any of the high-voltage connectors and terminals.
	  • Multiple cylinder misfire.
	  • One Trip Detection Logic (Three Way Catalyst Damage)
	NOTE: On the 1st trip, when a misfire condition occurs that can damage the three way catalyst (TWC) due to overheating, the MIL will blink.
	  • When a misfire condition occurs, the ECM monitors the CKP sensor (POS) signal every 200 engine revolutions for a change.
	  • When the misfire condition decreases to a level that will not damage the TWC, the MIL will turn off.
	  • If another misfire condition occurs that can damage the TWC on a second trip, the MIL will blink.
	  • Two Trip Detection Logic (Exhaust quality deterioration)
	NOTE: For misfire conditions that will not damage the TWC (but will affect vehicle emissions), the MIL will only light when the misfire is detected on a second trip. During this condition, the ECM monitors the CKP sensor signal every 1,000 engine revolutions.
	  • A misfire malfunction can be detected in any one cylinder or in multiple cylinders.
	Possible Causes:
	  • Improper spark plug
	  • Insufficient compression
	  • Incorrect fuel pressure
	  • Fuel injector circuit is open or shorted
	  • Fuel injector
	  • Intake air leak
	  • The ignition signal circuit is open or shorted
	  • Lack of fuel
	  • Drive plate or flywheel
	  • Air fuel ratio (A/F) sensor 1
	  • Incorrect PCV hose connection

DTC: P0301
1T ECM, MIL: Yes
Year: 2009, 2010
Model: EX35, FX35, FX50, G37, M35, M45, QX56
Engine: 3.5L V6, 3.7L V6, 4.5L V8, 5.0L V8, 5.6L V8
Transmission: All
	No.1 Cylinder Misfire Detected (Includes Hybrid Models):
	HYBRID MODELS CAUTION: Hybrid systems use very high-voltage battery systems. Before starting any service work involving the battery system, turn the ignition switch OFF and then remove the service plug from pocket in the trunk. After removing the service plug, wait 10 minutes before touching any of the high-voltage connectors and terminals.
	  • No. 1 cylinder misfires.
	1. One Trip Detection Logic (Three Way Catalyst Damage)
	  • On the 1st trip, when a misfire condition occurs that can damage the three way catalyst (TWC) due to overheating, the MIL will blink.
	  • When a misfire condition occurs, the ECM monitors the CKP sensor (POS) signal every 200 engine revolutions for a change.
	  • When the misfire condition decreases to a level that will not damage the TWC, the MIL will turn off.
	2. Two Trip Detection Logic (Exhaust quality deterioration)
	  • For misfire conditions that will not damage the TWC (but will affect vehicle emissions), the MIL will only light when the misfire is detected on a second trip.
	  • During this condition, the ECM monitors the CKP sensor signal every 1,000 engine revolutions.
	  • A misfire malfunction can be detected on any one cylinder or on multiple cylinders.
	  • If another misfire condition occurs that can damage the TWC on a second trip, the MIL will blink.
	Possible Causes:
	  • Improper spark plug
	  • Insufficient compression
	  • Incorrect fuel pressure
	  • Fuel injector circuit is open or shorted
	  • Fuel injector
	  • Intake air leak
	  • The ignition signal circuit is open or shorted
	  • Lack of fuel
	  • Drive plate or flywheel
	  • Air fuel ratio (A/F) sensor 1
	  • Incorrect PCV hose connection

DTC	Trouble Code Title, Conditions & Possible Causes

DTC: P0302
1T ECM, MIL: Yes
Year: 2009, 2010
Model: EX35, FX35, FX50, G37, M35, M45, QX56
Engine: 3.5L V6, 3.7L V6, 4.5L V8, 5.0L V8, 5.6L V8
Transmission: All
	No. 2 Cylinder Misfire Detected (Includes Hybrid Models):
	HYBRID MODELS CAUTION: Hybrid systems use very high-voltage battery systems. Before starting any service work involving the battery system, turn the ignition switch OFF and then remove the service plug from pocket in the trunk. After removing the service plug, wait 10 minutes before touching any of the high-voltage connectors and terminals.
	    • No. 2 cylinder misfires.
	1. One Trip Detection Logic (Three Way Catalyst Damage)
	    • On the 1st trip, when a misfire condition occurs that can damage the three way catalyst (TWC) due to overheating, the MIL will blink.
	    • When a misfire condition occurs, the ECM monitors the CKP sensor (POS) signal every 200 engine revolutions for a change.
	    • When the misfire condition decreases to a level that will not damage the TWC, the MIL will turn off.
	2. Two Trip Detection Logic (Exhaust quality deterioration)
	    • For misfire conditions that will not damage the TWC (but will affect vehicle emissions), the MIL will only light when the misfire is detected on a second trip.
	    • During this condition, the ECM monitors the CKP sensor signal every 1,000 engine revolutions.
	    • A misfire malfunction can be detected on any one cylinder or on multiple cylinders.
	    • If another misfire condition occurs that can damage the TWC on a second trip, the MIL will blink.
	Possible Causes:
	    • Improper spark plug
	    • Insufficient compression
	    • Incorrect fuel pressure
	    • Fuel injector circuit is open or shorted
	    • Fuel injector
	    • Intake air leak
	    • The ignition signal circuit is open or shorted
	    • Lack of fuel
	    • Drive plate or flywheel
	    • Air fuel ratio (A/F) sensor 1
	    • Incorrect PCV hose connection

DTC: P0303
1T ECM, MIL: Yes
Year: 2009, 2010
Model: EX35, FX35, FX50, G37, M35, M45, QX56
Engine: 3.5L V6, 3.7L V6, 4.5L V8, 5.0L V8, 5.6L V8
Transmission: All
	No. 3 Cylinder Misfire Detected (Includes Hybrid Models):
	HYBRID MODELS CAUTION: Hybrid systems use very high-voltage battery systems. Before starting any service work involving the battery system, turn the ignition switch OFF and then remove the service plug from pocket in the trunk. After removing the service plug, wait 10 minutes before touching any of the high-voltage connectors and terminals.
	    • No. 3 cylinder misfires.
	1. One Trip Detection Logic (Three Way Catalyst Damage)
	    • On the 1st trip, when a misfire condition occurs that can damage the three way catalyst (TWC) due to overheating, the MIL will blink.
	    • When a misfire condition occurs, the ECM monitors the CKP sensor (POS) signal every 200 engine revolutions for a change.
	    • When the misfire condition decreases to a level that will not damage the TWC, the MIL will turn off.
	2. Two Trip Detection Logic (Exhaust quality deterioration)
	    • For misfire conditions that will not damage the TWC (but will affect vehicle emissions), the MIL will only light when the misfire is detected on a second trip.
	    • During this condition, the ECM monitors the CKP sensor signal every 1,000 engine revolutions.
	    • A misfire malfunction can be detected on any one cylinder or on multiple cylinders.
	    • If another misfire condition occurs that can damage the TWC on a second trip, the MIL will blink.
	Possible Causes:
	    • Improper spark plug
	    • Insufficient compression
	    • Incorrect fuel pressure
	    • Fuel injector circuit is open or shorted
	    • Fuel injector
	    • Intake air leak
	    • The ignition signal circuit is open or shorted
	    • Lack of fuel
	    • Drive plate or flywheel
	    • Air fuel ratio (A/F) sensor 1
	    • Incorrect PCV hose connection

DTC	Trouble Code Title, Conditions & Possible Causes

DTC: P0304
1T ECM, MIL: Yes
Year: 2009, 2010
Model: EX35, FX35, FX50, G37, M35, M45, QX56
Engine: 3.5L V6, 3.7L V6, 4.5L V8, 5.0L V8, 5.6L V8
Transmission: All
	No. 4 Cylinder Misfire Detected (Includes Hybrid Models):
	HYBRID MODELS CAUTION: Hybrid systems use very high-voltage battery systems. Before starting any service work involving the battery system, turn the ignition switch OFF and then remove the service plug from pocket in the trunk. After removing the service plug, wait 10 minutes before touching any of the high-voltage connectors and terminals.
	  • No. 4 cylinder misfires.
	  • The misfire detection logic consists of the following two conditions.
	1. One Trip Detection Logic (Three Way Catalyst Damage)
	  • On the 1st trip, when a misfire condition occurs that can damage the three way catalyst (TWC) due to overheating, the MIL will blink.
	  • When a misfire condition occurs, the ECM monitors the CKP sensor (POS) signal every 200 engine revolutions for a change.
	  • When the misfire condition decreases to a level that will not damage the TWC, the MIL will turn off.
	  • If another misfire condition occurs that can damage the TWC on a second trip, the MIL will blink.
	  When the misfire condition decreases to a level that will not damage the TWC, the MIL will remain on.
	  If another misfire condition occurs that can damage the TWC, the MIL will begin to blink again.
	2. Two Trip Detection Logic (Exhaust quality deterioration)
	  • For misfire conditions that will not damage the TWC (but will affect vehicle emissions), the MIL will only light when the misfire is detected on a second trip.
	  • During this condition, the ECM monitors the CKP sensor signal every 1,000 engine revolutions.
	  • A misfire malfunction can be detected on any one cylinder or on multiple cylinders.
	Possible Causes:
	  • Improper spark plug
	  • Insufficient compression
	  • Incorrect fuel pressure
	  • Fuel injector circuit is open or shorted
	  • Fuel injector
	  • Intake air leak
	  • The ignition signal circuit is open or shorted
	  • Lack of fuel
	  • Drive plate or flywheel
	  • Air fuel ratio (A/F) sensor 1
	  • Incorrect PCV hose connection

DTC: P0305
1T ECM, MIL: Yes
Year: 2009, 2010
Model: EX35, FX35, FX50, G37, M35, M45, QX56
Engine: 3.5L V6, 3.7L V6, 4.5L V8, 5.0L V8, 5.6L V8
Transmission: All
	No. 5 Cylinder Misfire Detected:
	  • No. 5 cylinder misfires.
	  • The misfire detection logic consists of the following two conditions.
	1. One Trip Detection Logic (Three Way Catalyst Damage):
	  - On the first trip, when a misfire condition occurs that can damage the three way catalyst (TWC) due to overheating, the MIL will blink.
	  - When a misfire condition occurs, the ECM monitors the CKP sensor signal every 200 engine revolutions for a change.
	  - When the misfire condition decreases to a level that will not damage the TWC, the MIL will turn off.
	  - If another misfire condition occurs that can damage the TWC on a second trip, the MIL will blink.
	  - When the misfire condition decreases to a level that will not damage the TWC, the MIL will remain on.
	  - If another misfire condition occurs that can damage the TWC, the MIL will begin to blink again.
	2. Two Trip Detection Logic (Exhaust quality deterioration):
	  - For misfire conditions that will not damage the TWC (but will affect vehicle emissions), the MIL will only light when the misfire is detected on a second trip.
	  - During this condition, the ECM monitors the CKP sensor signal every 1,000 engine revolutions.
	  - A misfire malfunction can be detected in any one cylinder or in multiple cylinders.
	Possible Causes:
	  • Improper spark plug
	  • Insufficient compression
	  • Incorrect fuel pressure
	  • The fuel injector circuit is open or shorted
	  • Fuel injector
	  • Intake air leakage
	  • The ignition signal circuit is open or shorted
	  • Lack of fuel
	  • Signal plate
	  • A/F sensor 1
	  • Incorrect PCV hose connection

DTC	Trouble Code Title, Conditions & Possible Causes

DTC: P0306
1T ECM, MIL: Yes
Year: 2009, 2010
Model: EX35, FX35, FX50, G37, M35, M45, QX56
Engine: 3.5L V6, 3.7L V6, 4.5L V8, 5.0L V8, 5.6L V8
Transmission: All
	No. 6 Cylinder Misfire Detected:
	• No. 6 cylinder misfires.
	• The misfire detection logic consists of the following two conditions.
	1. One Trip Detection Logic (Three Way Catalyst Damage):
	  - On the first trip, when a misfire condition occurs that can damage the three way catalyst (TWC) due to overheating, the MIL will blink.
	  - When a misfire condition occurs, the ECM monitors the CKP sensor signal every 200 engine revolutions for a change.
	  - When the misfire condition decreases to a level that will not damage the TWC, the MIL will turn off.
	  - If another misfire condition occurs that can damage the TWC on a second trip, the MIL will blink.
	  - When the misfire condition decreases to a level that will not damage the TWC, the MIL will remain on.
	  - If another misfire condition occurs that can damage the TWC, the MIL will begin to blink again.
	2. Two Trip Detection Logic (Exhaust quality deterioration):
	  - For misfire conditions that will not damage the TWC (but will affect vehicle emissions), the MIL will only light when the misfire is detected on a second trip.
	  - During this condition, the ECM monitors the CKP sensor signal every 1,000 engine revolutions.
	  - A misfire malfunction can be detected in any one cylinder or in multiple cylinders.
	Possible Causes:
	• Improper spark plug
	• Insufficient compression
	• Incorrect fuel pressure
	• The fuel injector circuit is open or shorted
	• Fuel injector
	• Intake air leakage
	• The ignition signal circuit is open or shorted
	• Lack of fuel
	• Signal plate
	• A/F sensor 1
	• Incorrect PCV hose connection

DTC: P0307
1T ECM, MIL: Yes
Year: 2009, 2010
Model: FX50, M45, QX56
Engine: 4.5L V8, 5.0L V8, 5.6L V8
Transmission: All
	No. 7 Cylinder Misfire Detected:
	The misfire detection logic consists of the following two conditions:
	• One Trip Detection Logic (Three Way Catalyst Damage)
	  - On the 1st trip that a misfire condition occurs that can damage the three way catalyst (TWC) due to overheating, the MIL will blink.
	  - When a misfire condition occurs, the ECM monitors the CKP sensor signal every 200 engine revolutions for a change.
	  - When the misfire condition decreases to a level that will not damage the TWC, the MIL will turn off.
	  - If another misfire condition occurs that can damage the TWC on a second trip, the MIL will blink.
	  - When the misfire condition decreases to a level that will not damage the TWC, the MIL will remain on.
	  - If another misfire condition occurs that can damage the TWC, the MIL will begin to blink again.
	• Two Trip Detection Logic (Exhaust quality deterioration)
	  - For misfire conditions that will not damage the TWC (but will affect vehicle emissions), the MIL will only light when the misfire is detected on a second trip.
	  - During this condition, the ECM monitors the CKP sensor signal every 1,000 engine revolutions.
	  - A misfire malfunction can be detected on any one cylinder or on multiple cylinders.
	• No. 7 cylinder misfires.
	Possible Causes:
	• Improper spark plug
	• Insufficient compression
	• Incorrect fuel pressure
	• The fuel injector circuit is open or shorted
	• Fuel injector
	• Intake air leak
	• The ignition signal circuit is open or shorted
	• Lack of fuel
	• Signal plate
	• Air fuel ratio (A/F) sensor 1
	• Incorrect PCV hose connection

DTC	Trouble Code Title, Conditions & Possible Causes
DTC: P0308 **1T ECM, MIL: Yes** **Year:** 2009, 2010 **Model:** FX50, M45, QX56 **Engine:** 4.5L V8, 5.0L V8, 5.6L V8 **Transmission:** All	**No. 8 Cylinder Misfire Detected:** The misfire detection logic consists of the following two conditions: • One Trip Detection Logic (Three Way Catalyst Damage) - On the 1st trip that a misfire condition occurs that can damage the three way catalyst (TWC) due to overheating, the MIL will blink. - When a misfire condition occurs, the ECM monitors the CKP sensor signal every 200 engine revolutions for a change. - When the misfire condition decreases to a level that will not damage the TWC, the MIL will turn off. - If another misfire condition occurs that can damage the TWC on a second trip, the MIL will blink. - When the misfire condition decreases to a level that will not damage the TWC, the MIL will remain on. - If another misfire condition occurs that can damage the TWC, the MIL will begin to blink again. • Two Trip Detection Logic (Exhaust quality deterioration) - For misfire conditions that will not damage the TWC (but will affect vehicle emissions), the MIL will only light when the misfire is detected on a second trip. - During this condition, the ECM monitors the CKP sensor signal every 1,000 engine revolutions. - A misfire malfunction can be detected on any one cylinder or on multiple cylinders. • No. 8 cylinder misfires. **Possible Causes:** • Improper spark plug • Insufficient compression • Incorrect fuel pressure • The fuel injector circuit is open or shorted • Fuel injector • Intake air leak • The ignition signal circuit is open or shorted • Lack of fuel • Signal plate • Air fuel ratio (A/F) sensor 1 • Incorrect PCV hose connection
DTC: P0327 **1T ECM, MIL: Yes** **Year:** 2009, 2010 **Model:** EX35, FX35, FX50, G37, M35, M45, QX56 **Engine:** 3.5L V6, 3.7L V6, 4.5L V8, 5.0L V8, 5.6L V8 **Transmission:** All	**Knock Sensor Circuit Low Input (Includes Hybrid Models):** HYBRID MODELS CAUTION: Hybrid systems use very high-voltage battery systems. Before starting any service work involving the battery system, turn the ignition switch OFF and then remove the service plug from pocket in the trunk. After removing the service plug, wait 10 minutes before touching any of the high-voltage connectors and terminals. **NOTE: On V6 and V8 engines, this applies to Bank 1.** • An excessively low voltage from the sensor is sent to ECM. **Possible Causes:** • Harness or connectors • Knock sensor circuit is open or shorted. • Knock sensor has failed
DTC: P0328 **1T ECM, MIL: Yes** **Year:** 2009, 2010 **Model:** EX35, FX35, FX50, G37, M35, M45, QX56 **Engine:** 3.5L V6, 3.7L V6, 4.5L V8, 5.0L V8, 5.6L V8 **Transmission:** All	**Knock Sensor Circuit High Input (Includes Hybrid Models):** HYBRID MODELS CAUTION: Hybrid systems use very high-voltage battery systems. Before starting any service work involving the battery system, turn the ignition switch OFF and then remove the service plug from pocket in the trunk. After removing the service plug, wait 10 minutes before touching any of the high-voltage connectors and terminals. **NOTE: On V6 and V8 engines, this applies to Bank 1.** • An excessively high voltage from the sensor is sent to ECM. **Possible Causes:** • Harness or connectors • Knock sensor circuit is open or shorted. • Knock sensor
DTC: P0332 **1T ECM, MIL: Yes** **Year:** 2009, 2010 **Model:** EX35, FX35, FX50, G37, M35, M45, QX56 **Engine:** 3.5L V6, 3.7L V6, 4.5L V8, 5.0L V8, 5.6L V8 **Transmission:** All	**Knock Sensor (KS) Bank 2 Sensor Circuit Low Input:** • An excessively low voltage from the sensor is sent to ECM. **Possible Causes:** • Harness or connectors • The sensor circuit is open or shorted. • Knock sensor

DTC	Trouble Code Title, Conditions & Possible Causes

DTC: P0333
1T ECM, MIL: Yes
Year: 2009, 2010
Model: EX35, FX35, FX50, G37, M35, M45, QX56
Engine: 3.5L V6, 3.7L V6, 4.5L V8, 5.0L V8, 5.6L V8
Transmission: All
	Knock Sensor (Bank 2) Circuit High Input:
	  • An excessively high voltage from the sensor is sent to ECM.
	Possible Causes:
	  • Harness or connectors
	  • The sensor circuit is open or shorted.
	  • Knock sensor

DTC: P0335
1T ECM, MIL: Yes
Year: 2009, 2010
Model: EX35, FX35, FX50, G37, M35, M45, QX56
Engine: 3.5L V6, 3.7L V6, 4.5L V8, 5.0L V8, 5.6L V8
Transmission: All
	Crankshaft Position Sensor (POS) Circuit (Includes Hybrid Models):
	HYBRID MODELS CAUTION: Hybrid systems use very high-voltage battery systems. Before starting any service work involving the battery system, turn the ignition switch OFF and then remove the service plug from pocket in the trunk. After removing the service plug, wait 10 minutes before touching any of the high-voltage connectors and terminals.
	  • The crankshaft position sensor (POS) signal is not detected by the ECM during the first few seconds of engine cranking.
	  • The proper pulse signal from the crankshaft position sensor (POS) is not sent to ECM while the engine is running.
	  • The crankshaft position sensor (POS) signal is not in the normal pattern during engine running.
	Possible Causes:
	  • Harness or connectors
	  • CKP sensor (POS) circuit is open or shorted.
	  • CMP sensor (PHASE) (bank 2) circuit is shorted.
	  • EVT control position sensor (bank 2) circuit is shorted.
	  • Battery current sensor circuit is shorted.
	  • APP sensor 2 circuit is shorted.
	  • Crankshaft position sensor (POS) circuit is open or shorted.
	  • Accelerator pedal position sensor circuit is shorted.
	  • Refrigerant pressure sensor circuit is shorted.
	  • EVAP control system pressure sensor circuit is sorted.
	  • Tumble control valve position sensor circuit is shorted.
	  • Crankshaft position sensor (POS)
	  • Accelerator pedal position sensor
	  • Camshaft position sensor (PHASE) (bank 2)
	  • Exhaust valve timing control position sensor (bank 2)
	  • Battery current sensor
	  • Refrigerant pressure sensor
	  • EVAP control system pressure sensor
	  • Tumble control valve position sensor
	  • Signal plate
	  • Battery current sensor

DTC: P0340
1T ECM, MIL: Yes
Year: 2009, 2010
Model: EX35, FX35, FX50, G37, M35, M45, QX56
Engine: 3.5L V6, 3.7L V6, 4.5L V8, 5.0L V8, 5.6L V8
Transmission: All
	Camshaft Position Sensor Circuit (Includes Hybrid Models):
	HYBRID MODELS CAUTION: Hybrid systems use very high-voltage battery systems. Before starting any service work involving the battery system, turn the ignition switch OFF and then remove the service plug from pocket in the trunk. After removing the service plug, wait 10 minutes before touching any of the high-voltage connectors and terminals.
	NOTE: On V6 and V8 engines, this applies to Bank 1.
	NOTE: If DTC P0340 is displayed with DTC P0643, first perform the trouble diagnosis for DTC P0643.
	  • The cylinder No. signal is not sent to ECM for the first few seconds during engine cranking.
	  • The cylinder No. signal is not set to ECM during engine running.
	  • The cylinder No. signal is not in the normal pattern during engine running.
	Possible Causes:
	  • Harness or connectors
	  • CMP sensor (PHASE) (bank 1) circuit is open or shorted.
	  • Camshaft position sensor (PHASE) (bank 1)
	  • Camshaft (INT)
	  • Starter motor
	  • Starting system circuit
	  • Dead (Weak) battery

DTC	Trouble Code Title, Conditions & Possible Causes

DTC: P0345
1T ECM, MIL: Yes
Year: 2009, 2010
Model: EX35, FX35, FX50, G37, M35
Engine: 3.5L V6, 3.7L V6, 5.0L V8
Transmission: All
	Camshaft Position (CMP) Sensor Bank 2 Circuit:
	NOTE: If DTC P0340 or P0345 is displayed with DTC P0643, first perform the trouble diagnosis for DTC P0643.
	    • The cylinder No. signal is not sent to ECM for the first few seconds during engine cranking.
	    • The cylinder No. signal is not sent to ECM during engine running.
	    • The cylinder No. signal is not in the normal pattern during engine running.
	Possible Causes:
	    • Harness or connectors
	    • CMP sensor (PHASE) (bank 2) circuit is open or shorted.
	    • CKP sensor (POS) circuit is shorted.
	    • EVT control position sensor (bank 2) circuit is shorted.
	    • Battery current sensor circuit is shorted.
	    • APP sensor 2 circuit is shorted.
	    • EVAP control system pressure sensor circuit is shorted.
	    • Refrigerant pressure sensor circuit is shorted.
	    • Camshaft position sensor (PHASE)
	    • (bank 2)
	    • Crankshaft position sensor (POS)
	    • Exhaust valve timing control position sensor (bank 2)
	    • Battery current sensor
	    • Accelerator pedal position sensor
	    • EVAP control system pressure sensor
	    • Refrigerant pressure sensor
	    • Camshaft (INT)
	    • Starter motor
	    • Starting system circuit
	    • Dead (Weak) battery

DTC: P0420
1T ECM, MIL: Yes
Year: 2009, 2010
Model: EX35, FX35, FX50, G37, M35, M45, QX56
Engine: 3.5L V6, 3.7L V6, 4.5L V8, 5.0L V8, 5.6L V8
Transmission: All
	Catalyst System Efficiency Below Threshold (Includes Hybrid Models):
	HYBRID MODELS CAUTION: Hybrid systems use very high-voltage battery systems. Before starting any service work involving the battery system, turn the ignition switch OFF and then remove the service plug from pocket in the trunk. After removing the service plug, wait 10 minutes before touching any of the high-voltage connectors and terminals.
	NOTE: On models with dual exhaust, this DTC refers to Bank 1.
	    • Three way catalyst (manifold) does not operate properly.
	    • Three way catalyst (manifold) does not have enough oxygen storage capacity.
	Possible Causes:
	    • Three way catalyst (manifold)
	    • Exhaust tube
	    • Intake air leaks
	    • Fuel injector
	    • Fuel injector leaks
	    • Spark plug
	    • Improper ignition timing

DTC: P0430
1T ECM, MIL: Yes
Year: 2009, 2010
Model: EX35, FX35, FX50, G37, M35, M45, QX56
Engine: 3.5L V6, 3.7L V6, 4.5L V8, 5.0L V8, 5.6L V8
Transmission: All
	Catalyst System Efficiency Below Threshold (Bank 2):
	    • Three way catalyst (manifold) does not operate properly.
	    • Three way catalyst (manifold) does not have enough oxygen storage capacity.
	Possible Causes:
	    • Three way catalyst (manifold)
	    • Exhaust tube
	    • Intake air leaks
	    • Fuel injector
	    • Fuel injector leaks
	    • Spark plug
	    • Improper ignition timing

DTC	Trouble Code Title, Conditions & Possible Causes
DTC: P0441 **1T ECM, MIL: Yes** **Year:** 2009, 2010 **Model:** EX35, FX35, FX50, G37, M35, M45, QX56 **Engine:** 3.5L V6, 3.7L V6, 4.5L V8, 5.0L V8, 5.6L V8 **Transmission:** All	**EVAP Control System Incorrect Purge Flow (Includes Hybrid Models):** HYBRID MODELS CAUTION: Hybrid systems use very high-voltage battery systems. Before starting any service work involving the battery system, turn the ignition switch OFF and then remove the service plug from pocket in the trunk. After removing the service plug, wait 10 minutes before touching any of the high-voltage connectors and terminals. **NOTE: If DTC P0441 is displayed with other DTC such as P2122, P2123 P2127, P2128, P2138, first perform trouble diagnosis for other DTC.** • Under normal conditions (non-closed throttle), sensor output voltage indicates if pressure drop and purge flow are adequate. If not, a malfunction is determined. • EVAP control system does not operate properly – EVAP control system has a leak between intake manifold and EVAP control system pressure sensor. **Possible Causes:** • EVAP canister purge volume control solenoid valve stuck closed • EVAP control system pressure sensor and the circuit • Loose, disconnected or improper connection of rubber tube • Blocked rubber tube • Cracked EVAP canister • EVAP canister purge volume control solenoid valve circuit • Accelerator pedal position sensor • Blocked purge port • EVAP canister vent control valve • Drain filter
DTC: P0442 **1T ECM, MIL: Yes** **Year:** 2009, 2010 **Model:** EX35, FX35, FX50, G37, M35, M45, QX56 **Engine:** 3.5L V6, 3.7L V6, 4.5L V8, 5.0L V8, 5.6L V8 **Transmission:** All	**EVAP Control System Small Leak Detected (Negative Pressure) (Includes Hybrid Models):** **NOTE: If DTC P0442 is displayed with DTC P0456, first perform the trouble diagnosis for DTC P0456.** • EVAP control system has a leak, EVAP control system does not operate properly. **Possible Causes:** • Incorrect fuel tank vacuum relief valve • Incorrect fuel filler cap used • Fuel filler cap remains open or fails to close. • Foreign matter caught in fuel filler cap. • Leak is in line between intake manifold and EVAP canister purge volume control solenoid valve. • Foreign matter caught in EVAP canister vent control valve. • EVAP canister or fuel tank leaks • EVAP purge line (pipe and rubber tube) leaks • EVAP purge line rubber tube bent • Loose or disconnected rubber tube • EVAP canister vent control valve and the circuit • EVAP canister purge volume control solenoid valve and the circuit • Fuel tank temperature sensor • O-ring of EVAP canister vent control valve is missing or damaged • Drain filter • EVAP canister is saturated with water • EVAP control system pressure sensor • Fuel level sensor and the circuit • Refueling EVAP vapor cut valve • ORVR system leaks
DTC: P0443 **1T ECM, MIL: Yes** **Year:** 2009, 2010 **Model:** EX35, FX35, G37, M35, M45, QX56 **Engine:** 3.5L V6, 3.7L V6, 4.5L V8, 5.6L V8 **Transmission:** All	**EVAP Canister Purge Volume Control Solenoid Valve (Includes Hybrid Models):** HYBRID MODELS CAUTION: Hybrid systems use very high-voltage battery systems. Before starting any service work involving the battery system, turn the ignition switch OFF and then remove the service plug from pocket in the trunk. After removing the service plug, wait 10 minutes before touching any of the high-voltage connectors and terminals. • Condition A: The canister purge flow is detected during the vehicle is stopped while the engine is running, even when EVAP canister purge volume control solenoid valve is completely closed. • Condition B: The canister purge flow is detected during the specified driving conditions, even when EVAP canister purge volume control solenoid valve is completely closed. **Possible Causes:** • EVAP control system pressure sensor • EVAP canister purge volume control solenoid valve (EVAP canister purge volume control solenoid valve is stuck open.) • EVAP canister vent control valve • Drain filter • EVAP canister • Hoses (Hoses are connected incorrectly or clogged.)

DTC	Trouble Code Title, Conditions & Possible Causes
DTC: P0443 **1T ECM, MIL: Yes** **Year:** 2009, 2010 **Model:** FX50 **Engine:** 5.0L V8 **Transmission:** All	**EVAP Canister Purge Volume Control Solenoid Valve:** • The canister purge flow is detected during the specified driving conditions, even when EVAP canister purge volume control solenoid valve is completely closed. **Possible Causes:** • EVAP control system pressure sensor • EVAP canister purge volume control solenoid valve (valve is stuck open) • EVAP canister vent control valve • EVAP canister • Hoses are connected incorrectly or clogged
DTC: P0444 **1T ECM, MIL: Yes** **Year:** 2009, 2010 **Model:** EX35, FX35, FX50, G37, M35, M45, QX56 **Engine:** 3.5L V6, 3.7L V6, 4.5L V8, 5.0L V8, 5.6L V8 **Transmission:** All	**EVAP Canister Purge Volume Control Solenoid Valve Circuit Open (Includes Hybrid Models):** HYBRID MODELS CAUTION: Hybrid systems use very high-voltage battery systems. Before starting any service work involving the battery system, turn the ignition switch OFF and then remove the service plug from pocket in the trunk. After removing the service plug, wait 10 minutes before touching any of the high-voltage connectors and terminals. • An excessively low voltage signal is sent to ECM through the valve **Possible Causes:** • Harness or connectors • EVAP canister purge volume control solenoid valve circuit is open or shorted. • EVAP canister purge volume control solenoid valve
DTC: P0445 **1T ECM, MIL: Yes** **Year:** 2009, 2010 **Model:** EX35, FX35, FX50, G37, M35, M45, QX56 **Engine:** 3.5L V6, 3.7L V6, 4.5L V8, 5.0L V8, 5.6L V8 **Transmission:** All	**EVAP Canister Purge Volume Control Solenoid Valve Circuit Shorted (Includes Hybrid Models):** HYBRID MODELS CAUTION: Hybrid systems use very high-voltage battery systems. Before starting any service work involving the battery system, turn the ignition switch OFF and then remove the service plug from pocket in the trunk. After removing the service plug, wait 10 minutes before touching any of the high-voltage connectors and terminals. • An excessively high voltage signal is sent to ECM through the valve **Possible Causes:** • Harness or connectors • EVAP canister purge volume control solenoid valve circuit is shorted. • EVAP canister purge volume control solenoid valve
DTC: P0447 **1T ECM, MIL: Yes** **Year:** 2009, 2010 **Model:** EX35, FX35, FX50, G37, M35, M45, QX56 **Engine:** 3.5L V6, 3.7L V6, 4.5L V8, 5.0L V8, 5.6L V8 **Transmission:** All	**EVAP Canister Vent Control Valve Circuit Open (Includes Hybrid Models):** HYBRID MODELS CAUTION: Hybrid systems use very high-voltage battery systems. Before starting any service work involving the battery system, turn the ignition switch OFF and then remove the service plug from pocket in the trunk. After removing the service plug, wait 10 minutes before touching any of the high-voltage connectors and terminals. • An improper voltage signal is sent to ECM through EVAP canister vent control valve. **Possible Causes:** • Harness or connectors • EVAP canister vent control valve circuit is open or shorted. • EVAP canister vent control valve • Drain filter (if equipped)
DTC: P0448 **1T ECM, MIL: Yes** **Year:** 2009, 2010 **Model:** EX35, FX35, FX50, G37, M35, M45, QX56 **Engine:** 3.5L V6, 3.7L V6, 4.5L V8, 5.0L V8, 5.6L V8 **Transmission:** All	**EVAP Canister Vent Control Valve Closed (Includes Hybrid Models):** HYBRID MODELS CAUTION: Hybrid systems use very high-voltage battery systems. Before starting any service work involving the battery system, turn the ignition switch OFF and then remove the service plug from pocket in the trunk. After removing the service plug, wait 10 minutes before touching any of the high-voltage connectors and terminals. • EVAP canister vent control valve remains closed under specified driving conditions. **Possible Causes:** • EVAP canister vent control valve • EVAP control system pressure sensor and the circuit • Blocked rubber tube to EVAP canister vent control valve • EVAP canister is saturated with water • Drain filter (if equipped)

DTC	Trouble Code Title, Conditions & Possible Causes

DTC: P0451
1T ECM, MIL: Yes
Year: 2009, 2010
Model: EX35, FX35, FX50, G37, M35, M45, QX56
Engine: 3.5L V6, 3.7L V6, 4.5L V8, 5.0L V8, 5.6L V8
Transmission: All
	EVAP Control System Pressure Sensor Performance (Includes Hybrid Models):
	HYBRID MODELS CAUTION: Hybrid systems use very high-voltage battery systems. Before starting any service work involving the battery system, turn the ignition switch OFF and then remove the service plug from pocket in the trunk. After removing the service plug, wait 10 minutes before touching any of the high-voltage connectors and terminals.
	    • ECM detects a sloshing signal from the EVAP control system pressure sensor
	Possible Causes:
	    • Harness or connectors
	    • EVAP control system pressure sensor circuit is shorted.
	    • CKP sensor (POS) circuit is shorted.
	    • CMP sensor (PHASE) (bank 2) circuit is shorted.
	    • EVT control position sensor (bank 2) circuit is shorted.
	    • Battery current sensor circuit is shorted.
	    • APP sensor 2 circuit is shorted.
	    • Refrigerant pressure sensor circuit is shorted.
	    • EVAP control system pressure sensor
	    • Crankshaft position sensor (POS)
	    • Camshaft position sensor (PHASE) (bank 2)
	    • Exhaust valve timing control position sensor (bank 2)
	    • Battery current sensor
	    • Accelerator pedal position sensor
	    • Refrigerant pressure sensor

DTC: P0452
1T ECM, MIL: Yes
Year: 2009, 2010
Model: EX35, FX35, FX50, G37, M35, M45, QX56
Engine: 3.5L V6, 3.7L V6, 4.5L V8, 5.0L V8, 5.6L V8
Transmission: All
	EVAP Control System Pressure Sensor Low Input (Includes Hybrid Models):
	HYBRID MODELS CAUTION: Hybrid systems use very high-voltage battery systems. Before starting any service work involving the battery system, turn the ignition switch OFF and then remove the service plug from pocket in the trunk. After removing the service plug, wait 10 minutes before touching any of the high-voltage connectors and terminals.
	    • An excessively low voltage from the sensor is sent to ECM.
	Possible Causes:
	    • Harness or connectors
	    • EVAP control system pressure sensor circuit is shorted.
	    • CKP sensor (POS) circuit is shorted.
	    • CMP sensor (PHASE) (bank 2) circuit is shorted.
	    • EVT control position sensor (bank 2) circuit is shorted.
	    • Battery current sensor circuit is shorted.
	    • APP sensor 2 circuit is shorted.
	    • Refrigerant pressure sensor circuit is shorted.
	    • EVAP control system pressure sensor
	    • Crankshaft position sensor (POS)
	    • Camshaft position sensor (PHASE) (bank 2)
	    • Exhaust valve timing control position sensor (bank 2)
	    • Battery current sensor
	    • Accelerator pedal position sensor
	    • Refrigerant pressure sensor

DTC	Trouble Code Title, Conditions & Possible Causes
DTC: P0453 **1T ECM, MIL: Yes** **Year:** 2009, 2010 **Model:** EX35, FX35, FX50, G37, M35, M45, QX56 **Engine:** 3.5L V6, 3.7L V6, 4.5L V8, 5.0L V8, 5.6L V8 **Transmission:** All	**EVAP Control System Pressure Sensor High Input (Includes Hybrid Models):** HYBRID MODELS CAUTION: Hybrid systems use very high-voltage battery systems. Before starting any service work involving the battery system, turn the ignition switch OFF and then remove the service plug from pocket in the trunk. After removing the service plug, wait 10 minutes before touching any of the high-voltage connectors and terminals. <ul><li>An excessively high voltage from the sensor is sent to ECM.</li></ul>**Possible Causes:** <ul><li>Harness or connectors</li><li>EVAP control system pressure sensor circuit is open or shorted.</li><li>CKP sensor (POS) circuit is shorted.</li><li>CMP sensor (PHASE) (bank 2) circuit is shorted.</li><li>EVT control position sensor (bank 2) circuit is shorted.</li><li>Battery current sensor circuit is shorted.</li><li>APP sensor 2 circuit is shorted.</li><li>Refrigerant pressure sensor circuit is shorted.</li><li>EVAP control system pressure sensor</li><li>Crankshaft position sensor (POS)</li><li>Camshaft position sensor (PHASE)</li><li>(bank 2)</li><li>Exhaust valve timing control position sensor (bank 2)</li><li>Battery current sensor</li><li>Accelerator pedal position sensor</li><li>Refrigerant pressure sensor</li><li>EVAP canister vent control valve</li><li>EVAP canister</li><li>Rubber hose from EVAP canister vent control valve to vehicle frame</li></ul>
DTC: P0455 **1T ECM, MIL: Yes** **Year:** 2009, 2010 **Model:** EX35, FX35, FX50, G37, M35, M45, QX56 **Engine:** 3.5L V6, 3.7L V6, 4.5L V8, 5.0L V8, 5.6L V8 **Transmission:** All	**EVAP Control System Gross Leak Detected (Includes Hybrid Models):** <ul><li>EVAP control system has a very large leak, such as fuel filler cap fell off.</li><li>EVAP control system does not operate properly.</li></ul>CAUTION: Never remove fuel filler cap during the DTC Confirmation Procedure. **Possible Causes:** <ul><li>Fuel filler cap remains open or fails to close.</li><li>Incorrect fuel tank vacuum relief valve</li><li>Incorrect fuel filler cap used</li><li>Foreign matter caught in fuel filler cap</li><li>Leak is in line between intake manifold and EVAP canister purge volume control solenoid valve</li><li>Foreign matter caught in EVAP canister vent control valve.</li><li>EVAP canister or fuel tank leaks</li><li>EVAP purge line (pipe and rubber tube) leaks</li><li>EVAP purge line rubber tube bent.</li><li>Loose or disconnected rubber tube</li><li>EVAP canister vent control valve and the circuit</li><li>Drain filter</li><li>EVAP canister purge volume control solenoid valve and the circuit</li><li>Fuel tank temperature sensor</li><li>O-ring of EVAP canister vent control valve is missing or damaged.</li><li>EVAP control system pressure sensor</li><li>Refueling EVAP vapor cut valve</li><li>ORVR system leaks</li></ul>

DTC	Trouble Code Title, Conditions & Possible Causes

DTC: P0456
1T ECM, MIL: Yes
Year: 2009, 2010
Model: EX35, FX35, FX50, G37, M35, M45, QX56
Engine: 3.5L V6, 3.7L V6, 4.5L V8, 5.0L V8, 5.6L V8
Transmission: All
	Evaporative Emission Control System Very Small Leak (Negative Pressure Check) (Includes Hybrid Models):
	HYBRID MODELS CAUTION: Hybrid systems use very high-voltage battery systems. Before starting any service work involving the battery system, turn the ignition switch OFF and then remove the service plug from pocket in the trunk. After removing the service plug, wait 10 minutes before touching any of the high-voltage connectors and terminals.
	NOTE: If ECM judges a leak which corresponds to a very small leak, the very small leak P0456 will be detected.
	NOTE: If ECM judges a leak equivalent to a small leak, EVAP small leak P0442 will be detected.
	NOTE: If ECM judges there are no leaks, the diagnosis will be OK.
	    • If DTC P0456 is displayed with DTC P0442, first perform the trouble diagnosis for DTC P0456.
	    • This diagnosis detects very small leakage in the EVAP line between fuel tank and EVAP canister purge volume control solenoid valve, using the intake manifold vacuum in the same way as conventional EVAP small leakage diagnosis.
	NOTE: If ECM judges a leakage which corresponds to a very small leakage, the very small leakage P0456 will be detected.
	    • If ECM judges a leakage equivalent to a small leakage, EVAP small leakage P0442 will be detected.
	    • If ECM judges that there are no leakage, the diagnosis will be OK.
	    • EVAP system has a very small leak.
	    • EVAP system does not operate properly.
	Possible Causes:
	    • Incorrect fuel tank vacuum relief valve
	    • Incorrect fuel filler cap used
	    • Fuel filler cap remains open or fails to close.
	    • Foreign matter caught in fuel filler cap
	    • Leak is in line between intake manifold and EVAP canister purge volume control solenoid valve.
	    • Foreign matter caught in EVAP canister vent control valve
	    • EVAP canister or fuel tank leaks
	    • EVAP purge line (pipe and rubber tube) leaks
	    • EVAP purge line rubber tube bent
	    • Loose or disconnected rubber tube
	    • EVAP canister vent control valve and the circuit
	    • EVAP canister purge volume control solenoid valve and the circuit
	    • Fuel tank temperature sensor
	    • Drain filter
	    • O-ring of EVAP canister vent control valve is missing or damaged
	    • EVAP canister is saturated with water
	    • EVAP control system pressure sensor
	    • Refueling EVAP vapor cut valve
	    • ORVR system leaks
	    • Fuel level sensor and the circuit
	    • Foreign matter caught in EVAP canister purge volume control solenoid valve

DTC: P0460
1T ECM, MIL: Yes
Year: 2009, 2010
Model: EX35, FX35, FX50, G37, M35, M45, QX56
Engine: 3.5L V6, 3.7L V6, 4.5L V8, 5.0L V8, 5.6L V8
Transmission: All
	Fuel Level Sensor Circuit Noise (Includes Hybrid Models):
	HYBRID MODELS CAUTION: Hybrid systems use very high-voltage battery systems. Before starting any service work involving the battery system, turn the ignition switch OFF and then remove the service plug from pocket in the trunk. After removing the service plug, wait 10 minutes before touching any of the high-voltage connectors and terminals.
	NOTE: If DTC P0461 is displayed with DTC UXXXX, first perform the trouble diagnosis for DTC UXXXX.
	NOTE: If DTC P0460 is displayed with DTC P0607, first perform the trouble diagnosis for DTC P0607.
	    • When the vehicle is parked, naturally the fuel level in the fuel tank is stable. It means that output signal of the fuel level sensor does not change. If ECM senses sloshing signal from the sensor, fuel level sensor malfunction is detected.
	    • Even though the vehicle is parked, a signal being varied is sent from the fuel level sensor to ECM.
	Possible Causes:
	    • Harness or connectors
	    • CAN communication line is open or shorted
	    • Fuel level sensor circuit is open or shorted
	    • Combination meter (unified meter and A/C amp.)
	    • Fuel level sensor

DTC	Trouble Code Title, Conditions & Possible Causes
DTC: P0461 **1T ECM, MIL: Yes** **Year:** 2009, 2010 **Model:** EX35, FX35, FX50, G37, M35, M45, QX56 **Engine:** 3.5L V6, 3.7L V6, 4.5L V8, 5.0L V8, 5.6L V8 **Transmission:** All	**Fuel Level Sensor Circuit Range/Performance (Includes Hybrid Models):** HYBRID MODELS CAUTION: Hybrid systems use very high-voltage battery systems. Before starting any service work involving the battery system, turn the ignition switch OFF and then remove the service plug from pocket in the trunk. After removing the service plug, wait 10 minutes before touching any of the high-voltage connectors and terminals. **NOTE: If DTC P0461 is displayed with DTC U1000 or U1001, first perform the trouble diagnosis for appropriate "U" code.** **NOTE: If DTC P0461 is displayed with DTC P0607, first perform the trouble diagnosis for DTC P0607.** • This diagnosis detects the fuel gauge malfunction of the gauge not moving even after a long distance has been driven. Driving long distances naturally affect fuel gauge level. • The output signal of the fuel level sensor does not change within the specified range even though the vehicle has been driven a long distance. **Possible Causes:** • Harness or connectors • CAN communication line is open or shorted • Fuel level sensor circuit is open or shorted • Combination meter (unified meter and A/C amp.) • Fuel level sensor
DTC: P0462 **1T ECM, MIL: Yes** **Year:** 2009, 2010 **Model:** EX35, FX35, FX50, G37, M35, M45, QX56 **Engine:** 3.5L V6, 3.7L V6, 4.5L V8, 5.0L V8, 5.6L V8 **Transmission:** All	**Fuel Level Sensor Circuit Low Input (Includes Hybrid Models):** HYBRID MODELS CAUTION: Hybrid systems use very high-voltage battery systems. Before starting any service work involving the battery system, turn the ignition switch OFF and then remove the service plug from pocket in the trunk. After removing the service plug, wait 10 minutes before touching any of the high-voltage connectors and terminals. **NOTE: If DTC P0462 or P0463 is displayed with DTC UXXXX, first perform the trouble diagnosis for DTC UXXXX.** **NOTE: If DTC P0462 or P0463 is displayed with DTC P0607, first perform the trouble diagnosis for DTC P0607.** • An excessively low voltage from the sensor is sent to ECM. **Possible Causes:** • Harness or connectors • CAN communication line is open or shorted • Fuel level sensor circuit is open or shorted • Combination meter (unified meter and A/C amp.) • Fuel level sensor
DTC: P0463 **1T ECM, MIL: Yes** **Year:** 2009, 2010 **Model:** EX35, FX35, FX50, G37, M35, M45, QX56 **Engine:** 3.5L V6, 3.7L V6, 4.5L V8, 5.0L V8, 5.6L V8 **Transmission:** All	**Fuel Level Sensor Circuit High Input (Includes Hybrid Models):** HYBRID MODELS CAUTION: Hybrid systems use very high-voltage battery systems. Before starting any service work involving the battery system, turn the ignition switch OFF and then remove the service plug from pocket in the trunk. After removing the service plug, wait 10 minutes before touching any of the high-voltage connectors and terminals. **NOTE: If DTC P0462 or P0463 is displayed with DTC UXXXX, first perform the trouble diagnosis for DTC UXXXX.** **NOTE: If DTC P0462 or P0463 is displayed with DTC P0607, first perform the trouble diagnosis for DTC P0607.** • An excessively high voltage from the sensor is sent to ECM. **Possible Causes:** • Harness or connectors • CAN communication line is open or shorted • Fuel level sensor circuit is open or shorted • Combination meter (unified meter and A/C amp.) • Fuel level sensor
DTC: P0500 **1T ECM, MIL: Yes** **Year:** 2009, 2010 **Model:** EX35, FX35, FX50, G37, M35, M45, QX56 **Engine:** 3.5L V6, 3.7L V6, 4.5L V8, 5.0L V8, 5.6L V8 **Transmission:** All	**Vehicle Speed Sensor Input (Includes Hybrid Models):** HYBRID MODELS CAUTION: Hybrid systems use very high-voltage battery systems. Before starting any service work involving the battery system, turn the ignition switch OFF and then remove the service plug from pocket in the trunk. After removing the service plug, wait 10 minutes before touching any of the high-voltage connectors and terminals. **NOTE: If DTC P0500 is displayed with DTC UXXXX, first perform the trouble diagnosis for DTC UXXXX.** **NOTE: If DTC P0500 is displayed with DTC P0607, first perform the trouble diagnosis for DTC P0607.** • When in fail-safe mode, the cooling fan operates (High) while engine is running. • The vehicle speed signal sent to ECM is almost 0 km/h (0 MPH) even when vehicle is being driven. **Possible Causes:** • Harness or connectors • The CAN communication line is open or shorted • The vehicle speed signal circuit is open or shorted • Wheel sensor • Unified meter and A/C amp. (combination meter) • ABS actuator and electric unit (control unit)

DTC	Trouble Code Title, Conditions & Possible Causes
DTC: P0506 **1T ECM, MIL: Yes** **Year:** 2009, 2010 **Model:** EX35, FX35, FX50, G37, M35, M45, QX56 **Engine:** 3.5L V6, 3.7L V6, 4.5L V8, 5.0L V8, 5.6L V8 **Transmission:** All	**Idle Speed Control System RPM Lower Than Expected (Includes Hybrid Models):** HYBRID MODELS CAUTION: Hybrid systems use very high-voltage battery systems. Before starting any service work involving the battery system, turn the ignition switch OFF and then remove the service plug from pocket in the trunk. After removing the service plug, wait 10 minutes before touching any of the high-voltage connectors and terminals. **NOTE: If DTC P0506 is displayed with other DTC, first perform the trouble diagnosis for the other DTC.** • The idle speed is less than the target idle speed by 100 rpm or more. **Possible Causes:** • Electric throttle control actuator • Intake air leak
DTC: P0507 **1T ECM, MIL: Yes** **Year:** 2009, 2010 **Model:** EX35, FX35, FX50, G37, M35, M45, QX56 **Engine:** 3.5L V6, 3.7L V6, 4.5L V8, 5.0L V8, 5.6L V8 **Transmission:** All	**Idle Speed Control System RPM Higher Than Expected (Includes Hybrid Models):** HYBRID MODELS CAUTION: Hybrid systems use very high-voltage battery systems. Before starting any service work involving the battery system, turn the ignition switch OFF and then remove the service plug from pocket in the trunk. After removing the service plug, wait 10 minutes before touching any of the high-voltage connectors and terminals. **NOTE: If DTC P0507 is displayed with other DTC, first perform the trouble diagnosis for the other DTC.** • The idle speed is more than the target idle speed by 200 rpm or more. **Possible Causes:** • Electric throttle control actuator • Intake air leak • PCV system
DTC: P0524 **1T ECM, MIL: Yes** **Year:** 2009, 2010 **Model:** FX50, G37 **Engine:** 3.7L V6, 5.0L V8 **Transmission:** All	**Engine Oil Pressure Too Low:** Engine oil pressure is low because there is a gap between angle of target and phase-control angle. **Possible Causes:** • Engine oil pressure or level too low • Crankshaft position sensor (POS) • Camshaft position sensor (PHASE) • Intake valve control solenoid valve • Accumulation of debris to the signal pick-up portion of the camshaft • Timing chain installation • Foreign matter caught in the oil groove for intake valve timing control
DTC: P0550 **1T ECM** **Year:** 2009, 2010 **Model:** EX35, FX35, FX50, G37, M35, M45, QX56 **Engine:** 3.5L V6, 3.7L V6, 4.5L V8, 5.0L V8, 5.6L V8 **Transmission:** All	**Power Steering Pressure Sensor Circuit:** • The MIL will not illuminate for this diagnosis. **NOTE: If DTC P0550 is displayed with DTC P0643, first perform the trouble diagnosis for DTC P0643.** • An excessively low or high voltage from the sensor is sent to ECM. **Possible Causes:** • Harness or connectors • The sensor circuit is open or shorted • Power steering pressure sensor
DTC: P0555 **T ECM, MIL: Yes** **Year:** 2009, 2010 **Model:** G37 **Engine:** 3.7L V6 **Transmission:** All	**Brake Booster Pressure Sensor Circuit:** • An excessively low voltage from the sensor is sent to ECM. • An excessively high voltage from the sensor is sent to ECM. **Possible Causes:** • Harness or connectors • The sensor circuit is open or shorted. • CKP sensor (POS) circuit is shorted. • APP sensor 2 circuit is shorted. • EVAP control system pressure sensor circuit is shorted. • Refrigerant pressure sensor circuit is shorted. • Gear lever position sensor circuit is shorted. • Brake booster pressure sensor • Crankshaft position sensor (POS) • Accelerator pedal position sensor • EVAP control system pressure sensor • Refrigerant pressure sensor • Gear lever position sensor

DTC	Trouble Code Title, Conditions & Possible Causes
DTC: P0603 **1T ECM, MIL: Yes** **Year:** 2009, 2010 **Model:** EX35, FX35, FX50, G37, M35, M45, QX56 **Engine:** 3.5L V6, 3.7L V6, 4.5L V8, 5.0L V8, 5.6L V8 **Transmission:** All	**ECM Power Supply Circuit (Includes Hybrid Models):** HYBRID MODELS CAUTION: Hybrid systems use very high-voltage battery systems. Before starting any service work involving the battery system, turn the ignition switch OFF and then remove the service plug from pocket in the trunk. After removing the service plug, wait 10 minutes before touching any of the high-voltage connectors and terminals. • ECM back-up RAM system does not function properly. **Possible Causes:** • Harness or connectors • The ECM power supply (back-up) circuit is open or shorted. • ECM
DTC: P0605 **1T ECM, MIL: Yes** **Year:** 2009, 2010 **Model:** EX35, FX35, FX50, G37, M35, M45, QX56 **Engine:** 3.5L V6, 3.7L V6, 4.5L V8, 5.0L V8, 5.6L V8 **Transmission:** All	**Engine Control Module (ECM) (Includes Hybrid Models):** HYBRID MODELS CAUTION: Hybrid systems use very high-voltage battery systems. Before starting any service work involving the battery system, turn the ignition switch OFF and then remove the service plug from pocket in the trunk. After removing the service plug, wait 10 minutes before touching any of the high-voltage connectors and terminals. A. ECM calculation function is malfunctioning. B. ECM EEP-ROM system is malfunctioning. C. ECM self shut-off function is malfunctioning. **Possible Causes:** • ECM
DTC: P0607 **1T ECM, MIL: Yes** **Year:** 2009, 2010 **Model:** EX35, FX50, G37, M35, M45, QX56 **Engine:** 3.5L V6, 3.7L V6, 4.5L V8, 5.0L V8, 5.6L V8 **Transmission:** All	**CAN Communication Bus (Includes Hybrid Models):** HYBRID MODELS CAUTION: Hybrid systems use very high-voltage battery systems. Before starting any service work involving the battery system, turn the ignition switch OFF and then remove the service plug from pocket in the trunk. After removing the service plug, wait 10 minutes before touching any of the high-voltage connectors and terminals. • When detecting error during the initial diagnosis of CAN controller of ECM. **Possible Causes:** • ECM has failed
DTC: P0615 **T TCM, TCIL: Yes** **Year:** 2009, 2010 **Model:** EX35, FX35, FX50, G37, M35, M45, QX56 **Engine:** 3.5L V6, 3.7L V6, 4.5L V8, 5.0L V8, 5.6L V8 **Transmission:** All	**Starter Relay Circuit:** • This is not an OBD-II self-diagnostic item • This DTC will set if the starter monitor value is OFF when the ignition switch is ON at the "P" and "N" positions. **Possible Causes:** • Harness or connectors • Starter relay and TCM circuit is open or shorted. • Starter relay circuit
DTC: P0643 **1T ECM, MIL: Yes** **Year:** 2009, 2010 **Model:** EX35, FX35, FX50, G37, M35, M45, QX56 **Engine:** 3.5L V6, 3.7L V6, 4.5L V8, 5.0L V8, 5.6L V8 **Transmission:** All	**Sensor Power Supply Circuit Short (Includes Hybrid Models):** HYBRID MODELS CAUTION: Hybrid systems use very high-voltage battery systems. Before starting any service work involving the battery system, turn the ignition switch OFF and then remove the service plug from pocket in the trunk. After removing the service plug, wait 10 minutes before touching any of the high-voltage connectors and terminals. • ECM detects a voltage of power source for sensor is excessively low or high. **NOTE: When the malfunction is detected, ECM enters fail-safe mode and the MIL illuminates.** • ECM stops the electric throttle control actuator control, throttle valve is maintained at a fixed opening (approx. 5 degrees) by the return spring. **Possible Causes:** • Harness or connectors • APP sensor 1 circuit is shorted. • TP sensor circuit is shorted. • CMP sensor (PHASE) (bank 1) circuit is shorted. • EVT control position sensor (bank 1) circuit is shorted. • PSP sensor circuit is shorted. • Accelerator pedal position sensor • Throttle position sensor • Camshaft position sensor (PHASE) (bank 1) • Exhaust valve timing control position sensor (bank 1) • Power steering pressure sensor
DTC: P0700 **T TCM, MIL: Yes, TCIL: Yes** **Year:** 2009, 2010 **Model:** EX35, M35, M45, QX56 **Engine:** 3.5L V6, 4.5L V8, 5.6L V8 **Transmission:** All	**TCM:** • This is an OBD-II self-diagnostic item. • Diagnostic trouble code P0700 is detected when the TCM is malfunctioning. **Possible Causes:** • TCM

DTC	Trouble Code Title, Conditions & Possible Causes
DTC: P0705 **T TCM, TCIL: Yes** **Year:** 2009, 2010 **Model:** EX35, FX35, FX50, G37, M35, M45, QX56 **Engine:** 3.5L V6, 3.7L V6, 4.5L V8, 5.0L V8, 5.6L V8 **Transmission:** All	**Transmission Range Switch A:** • Transmission range switch 1 – 4 signals input with impossible pattern. • "P" position is detected from "N" position without any other position being detected in between. **Possible Causes:** • Harness or connectors • Transmission range switches 1, 2, 3, 4 and TCM circuit is open or shorted. • Transmission range switches 1, 2, 3 and 4
DTC: P0710 **T TCM, TCIL: Yes** **Year:** 2009, 2010 **Model:** FX35, FX50, G37 **Engine:** 3.5L V6, 3.7L V6, 5.0L V8 **Transmission:** All	**Transmission Fluid Temperature Sensor A Circuit:** • Set DTC when the A/T fluid temperature sensor is -40 °C (-40 °F) or less for 5 seconds while driving the vehicle at the vehicle speed 10 km/h (7 MPH) or more. • Set DTC when the A/T fluid temperature sensor is 180 °C (356 °F) or more for 5 seconds. **Possible Causes:** • Harness or connectors • Sensor circuit is open • A/T fluid temperature sensor
DTC: P0717 **T TCM, TCIL: Yes** **Year:** 2009, 2010 **Model:** FX35, FX50, G37 **Engine:** 3.5L V6, 3.7L V6, 5.0L V8 **Transmission:** All	**Input/Turbine Speed Sensor A Circuit No Signal:** The revolution of input speed sensor 1 and/or 2 is 270 rpm or less. **Possible Causes:** • Harness or connectors • Sensor circuit is open. • Input speed sensor 1 and/or 2
DTC: P0717 **1T TCM, TCIL: Yes** **Year:** 2009, 2010 **Model:** EX35, M35, M45, QX56 **Engine:** 3.5L V6, 4.5L V8, 5.6L V8 **Transmission:** All	**Input Speed Sensor A (Turbine Revolution Sensor):** • The input speed sensor detects input shaft rpm (revolutions per minute). It is located on the input side of the automatic transmission. Monitors revolution of sensor 1 and sensor 2 for non-standard conditions. • This is an OBD-II self-diagnostic item. • Diagnostic trouble code P0717 is detected under the following conditions: - When TCM does not receive the proper voltage signal from the sensor. - When TCM detects an irregularity only at position of 4th gear for input speed sensor 2. **Possible Causes:** • Harness or connectors • The sensor circuit is open or shorted. • Input speed sensor 1, 2
DTC: P0720 **T TCM, TCIL: Yes** **Year:** 2009, 2010 **Model:** EX35, M35 **Engine:** 3.5L V6 **Transmission:** All	**Output Speed Sensor Circuit:** • Signal from vehicle speed sensor CVT [output speed sensor (secondary speed sensor)] is not input due to open or short circuit. • An unexpected signal is input during running. • After ignition switch is turned ON, unexpected signal input from vehicle speed signal before the vehicle starts moving. **Possible Causes:** • Harness or connectors • Sensor circuit is open or shorted. • Secondary (output) speed sensor • Vehicle speed signal
DTC: P0720 **T TCM, TCIL: Yes** **Year:** 2009, 2010 **Model:** FX35, FX50, G37 **Engine:** 3.5L V6, 3.7L V6, 5.0L V8 **Transmission:** All	**Output Speed Sensor Circuit:** • The output speed sensor recognizes that the vehicle speed is 5 km/h (3 MPH) or less even if the vehicle speed signal recognizes that the vehicle speed is 20 km/h (12 MPH) or more. (Only when starts after the ignition switch is turned ON.) • The vehicle speed recognized by the output speed sensor decelerates 36 km/h (23 MPH) or more during 60 msec when the output speed sensor recognizes that the vehicle speed is 36 km/h (23 MPH) or more and the vehicle speed signal recognizes that the vehicle speed is 24 km/h (15 MPH) or less. • The vehicle speed of output speed sensor decelerates 36 km/h (23 MPH) or more even if the vehicle speed of vehicle speed signal accelerates or decelerates 24 km/h (15 MPH) or less during 60 msec when the output speed sensor recognizes that the vehicle speed is 36 km/h (23 MPH) or more. **Possible Causes:** • Harness or connectors • Sensor circuit is open. • Output speed sensor

DTC	Trouble Code Title, Conditions & Possible Causes
DTC: P0720 **1T TCM, TCIL: Yes** **Year:** 2009, 2010 **Model:** M45, QX56 **Engine:** 4.5L V8, 5.6L V8 **Transmission:** All	**Vehicle Speed Sensor A/T (Revolution Sensor/Output Speed Sensor):** • This is an OBD-II self-diagnostic item. • Diagnostic trouble code P0720 is detected under the following conditions: - When TCM does not receive the proper voltage signal from the sensor. - After ignition switch is turned "ON", irregular signal input from vehicle speed sensor MTR before the vehicle starts moving. **Possible Causes:** • Harness or connectors • The sensor circuit is open or shorted. • Revolution sensor • Vehicle speed sensor MTR
DTC: P0725 **T TCM, TCIL: Yes** **Year:** 2009, 2010 **Model:** EX35, FX35, FX50, G37, M35, M45 **Engine:** 3.5L V6, 3.7L V6, 4.5L V8, 5.0L V8 **Transmission:** All	**Engine Speed Input Circuit:** • TCM does not receive the CAN communication (ignition) signal from the ECM. • The engine speed is more less 150 rpm even if the vehicle speed is more than 10 km/h (7 MPH). **Possible Causes:** • Harness or connectors • ECM to TCM circuit is open or shorted.
DTC: P0725 **1T TCM, TCIL: Yes** **Year:** 2009, 2010 **Model:** QX56 **Engine:** 5.6L V8 **Transmission:** All	**Engine Speed Signal:** • The engine speed signal is sent from the ECM to the TCM. • This is not an OBD-II self-diagnostic item. • Diagnostic trouble code P0725 is detected when TCM does not receive the engine speed signal or ignition signal (input by CAN communication) from ECM. • The engine speed signal is sent with the engine running and should closely match the tachometer reading. **Possible Causes:** • Harness or connectors • The ECM to the TCM circuit is open or shorted.
DTC: P0729 **T TCM, TCIL: Yes** **Year:** 2009, 2010 **Model:** FX35, FX50, G37 **Engine:** 3.5L V6, 3.7L V6, 5.0L V8 **Transmission:** All	**Gear 6 Incorrect Ratio:** The gear ratio is: • 0.914 or more • 0.813 or less **Possible Causes:** • Input clutch solenoid valve • Direct clutch solenoid valve • High and low reverse clutch solenoid valve • Front brake solenoid valve • Low brake solenoid valve • 2346 brake solenoid valve • Anti-interlock solenoid valve • Each clutch and brake • Output speed sensor • Input speed sensor 1, 2 • Hydraulic control circuit
DTC: P0730 **T TCM, TCIL: Yes** **Year:** 2009, 2010 **Model:** FX35, FX50, G37 **Engine:** 3.5L V6, 3.7L V6, 5.0L V8 **Transmission:** All	**Incorrect Gear Ratio:** • The revolution of under drive sun gear is 8,000 rpm or more. **NOTE: Not detected when in "P" or "N" position and during a shift to "P" or "N" position.** **Possible Causes:** • 2346 brake solenoid valve • Front brake solenoid valve • Input speed sensor 1, 2

DTC	Trouble Code Title, Conditions & Possible Causes
DTC: P0731 **T TCM, TCIL: Yes** **Year:** 2009, 2010 **Model:** EX35, FX35, FX50, G37, M35, M45, QX56 **Engine:** 3.5L V6, 3.7L V6, 4.5L V8, 5.0L V8, 5.6L V8 **Transmission:** All	**A/T 1st Gear Function:** • This is an OBD-II self-diagnostic item. • Diagnostic trouble code P0731 is detected when TCM detects any inconsistency in the actual gear ratio. **Possible Causes:** • Harness or connectors • Solenoid circuits are open or shorted. • Input clutch solenoid valve • Direct clutch solenoid valve • High and low reverse clutch solenoid valve • Front brake solenoid valve • Low brake solenoid valve • 2346 brake solenoid valve • Anti-interlock solenoid valve • Each clutch and brake • Output speed sensor • Input speed sensor 1, 2 • Hydraulic control circuit
DTC: P0732 **T TCM, TCIL: Yes** **Year:** 2009, 2010 **Model:** EX35, FX35, FX50, G37, M35, M45, QX56 **Engine:** 3.5L V6, 3.7L V6, 4.5L V8, 5.0L V8, 5.6L V8 **Transmission:** All	**A/T 2nd Gear Function:** • This malfunction is detected when the A/T does not shift into 2GR position as instructed by TCM. This is not only caused by electrical malfunction (circuits open or shorted) but mechanical malfunction such as control valve sticking, improper solenoid valve operation. • This is an OBD-II self-diagnostic item. • Diagnostic trouble code P0732 is detected when TCM detects any inconsistency in the actual gear ratio. **Possible Causes:** • Harness or connectors • Solenoid circuits are open or shorted. • Input clutch solenoid valve • Direct clutch solenoid valve • High and low reverse clutch solenoid valve • Front brake solenoid valve • Low brake solenoid valve • 2346 brake solenoid valve • Anti-interlock solenoid valve • Each clutch and brake • Output speed sensor • Input speed sensor 1, 2 • Hydraulic control circuit
DTC: P0733 **T TCM, TCIL: Yes** **Year:** 2009, 2010 **Model:** EX35, FX35, FX50, G37, M35, M45, QX56 **Engine:** 3.5L V6, 3.7L V6, 4.5L V8, 5.0L V8, 5.6L V8 **Transmission:** All	**A/T 3rd Gear Function:** • This malfunction is detected when the A/T does not shift into 3GR position as instructed by TCM. This is not only caused by electrical malfunction (circuits open or shorted) but mechanical malfunction such as control valve sticking, improper solenoid valve operation. • This is an OBD-II self-diagnostic item. • Diagnostic trouble code P0733 is detected when TCM detects any inconsistency in the actual gear ratio. **Possible Causes:** • Harness or connectors • Solenoid circuits are open or shorted. • Input clutch solenoid valve • Direct clutch solenoid valve • High and low reverse clutch solenoid valve • Front brake solenoid valve • Low brake solenoid valve • 2346 brake solenoid valve • Anti-interlock solenoid valve • Each clutch and brake • Output speed sensor • Input speed sensor 1, 2 • Hydraulic control circuit

DTC	Trouble Code Title, Conditions & Possible Causes
DTC: P0734 **T TCM, TCIL: Yes** **Year:** 2009, 2010 **Model:** EX35, FX35, FX50, G37, M35, M45, QX56 **Engine:** 3.5L V6, 3.7L V6, 4.5L V8, 5.0L V8, 5.6L V8 **Transmission:** All	**4th Gear Incorrect Ratio:** • This malfunction is detected when the A/T does not shift into 4GR position as instructed by TCM. This is not only caused by electrical malfunction (circuits open or shorted) but mechanical malfunction such as control valve sticking, improper solenoid valve operation. • This is an OBD-II self-diagnostic item. • P0734 is detected when TCM detects any inconsistency in the actual gear ratio. **Possible Causes:** • Input clutch solenoid valve • Direct clutch solenoid valve • High and low reverse clutch solenoid valve • Front brake solenoid valve • Low brake solenoid valve • 2346 brake solenoid valve • Anti-interlock solenoid valve • Each clutch and brake • Output speed sensor • Input speed sensor 1, 2 • Hydraulic control circuit
DTC: P0735 **T TCM, TCIL: Yes** **Year:** 2009, 2010 **Model:** EX35, FX35, FX50, G37, M35, M45, QX56 **Engine:** 3.5L V6, 3.7L V6, 4.5L V8, 5.0L V8, 5.6L V8 **Transmission:** All	**5th Gear Incorrect Ratio:** This malfunction is detected when the A/T does not shift into 5GR position as instructed by TCM. This is not only caused by electrical malfunction (circuits open or shorted) but mechanical malfunction such as control valve sticking, improper solenoid valve operation. • This is an OBD-II self-diagnostic item. • Diagnostic trouble code P0735 is detected when TCM detects any inconsistency in the actual gear ratio. **Possible Causes:** • Input clutch solenoid valve • Direct clutch solenoid valve • High and low reverse clutch solenoid valve • Front brake solenoid valve • Low brake solenoid valve • 2346 brake solenoid valve • Anti-interlock solenoid valve • Each clutch and brake • Output speed sensor • Input speed sensor 1, 2 • Hydraulic control circuit
DTC: P0740 **T TCM, TCIL: Yes** **Year:** 2009, 2010 **Model:** EX35, M35, M45 **Engine:** 3.5L V6, 4.5L V8 **Transmission:** All	**Torque Converter Clutch (TCC) Circuit Open/Short:** • Normal voltage is not applied to solenoid due to open or short circuit. **Possible Causes:** • Torque converter clutch solenoid valve • Harness or connectors • Solenoid circuit is open or shorted.
DTC: P0740 **T TCM, TCIL: Yes** **Year:** 2009, 2010 **Model:** FX35, FX50, G37 **Engine:** 3.5L V6, 3.7L V6, 5.0L V8 **Transmission:** All	**Torque Converter Clutch Circuit - Open:** • The torque converter clutch solenoid valve monitor value is 0.4 A or less when the torque converter clutch solenoid valve command value is more than 0.75 A. **Possible Causes:** • Harness or connectors • Solenoid valve circuit is open or shorted. • Torque converter clutch solenoid valve
DTC: P0740 **1T TCM, TCIL: Yes** **Year:** 2009, 2010 **Model:** QX56 **Engine:** 5.6L V8 **Transmission:** All	**Torque Converter Clutch Solenoid Valve:** • Diagnostic trouble code P0740 is detected under the following conditions: - TCM detects an improper voltage drop when it tries to operate the solenoid valve. - When TCM detects as irregular by comparing target value with monitor value. **Possible Causes:** • Torque converter clutch solenoid valve • Harness or connectors • Solenoid circuit is open or shorted.

DTC	Trouble Code Title, Conditions & Possible Causes
DTC: P0744 **T TCM, TCIL: Yes** **Year:** 2009, 2010 **Model:** FX35, FX50, G37 **Engine:** 3.5L V6, 3.7L V6, 5.0L V8 **Transmission:** All	**Torque Converter Clutch Circuit Intermittent:** • The lock-up is not performed properly within the lock-up area. • When A/T cannot perform lock-up even if electrical circuit is good. • When TCM detects as irregular by comparing difference value with slip rotation. **Possible Causes:** • Harness or connectors • Torque converter clutch solenoid valve • Torque converter • Input speed sensor 1, 2 • Hydraulic control circuit
DTC: P0744 **T TCM, TCIL: Yes** **Year:** 2009, 2010 **Model:** EX35, M35 **Engine:** 3.5L V6 **Transmission:** All	**Torque Converter Clutch Circuit Intermittent:** • Transmission cannot perform lock-up even if electrical circuit is good. • TCM detects as irregular by comparing difference value with slip rotation. • There is a big difference between engine speed and primary speed sensor when TCM lock-up signal is on. **Possible Causes:** • Torque converter clutch solenoid valve • Hydraulic control circuit
DTC: P0744 **T TCM, TCIL: Yes** **Year:** 2009, 2010 **Model:** QX56 **Engine:** 5.6L V8 **Transmission:** All	**A/T TCC S/V Function (Lock-Up):** This malfunction is detected when the A/T does not lock-up or does not shift to 5th gear. This is not only caused by electrical malfunction (circuits open or shorted) but also by mechanical malfunction such as control valve sticking, improper solenoid valve operation, etc. • This is an OBD-II self-diagnostic item. • Diagnostic trouble code P0744 is detected under the following conditions: - When A/T cannot perform lock-up even if electrical circuit is good. - When TCM detects as irregular by comparing difference value with slip rotation. **Possible Causes:** • Harness or connectors • The solenoid circuit is open or shorted. • Torque converter clutch solenoid valve • Hydraulic control circuit
DTC: P0745 **T TCM, TCIL: Yes** **Year:** 2009, 2010 **Model:** M45, QX56 **Engine:** 4.5L V8, 5.6L V8 **Transmission:** All	**Pressure Control Solenoid A:** The line pressure solenoid valve regulates the oil pump discharge pressure to suit the driving condition in response to a signal sent from the TCM. • This is an OBD-II self-diagnostic item. • Diagnostic trouble code P0745 is detected under the following conditions: - When TCM detects an improper voltage drop when it tries to operate the solenoid valve. - When TCM detects as irregular by comparing target value with monitor value. **Possible Causes:** • Harness or connectors • The solenoid circuit is open or shorted. • Line pressure solenoid valve
DTC: P0745 **1T TCM, TCIL: Yes** **Year:** 2009, 2010 **Model:** EX35, M35 **Engine:** 3.5L V6 **Transmission:** All	**Pressure Control Solenoid Valve A:** • The pressure control solenoid valve A (line pressure solenoid valve) regulates the oil pump discharge pressure to suit the driving condition in response to a signal sent from the TCM. • This is an OBD-II self-diagnostic item. • Diagnostic trouble code P0745 is detected under the following conditions: - TCM detects an improper voltage drop when it tries to operate the solenoid valve. - When TCM compares target value with monitor value and detects an irregularity. **Possible Causes:** • Harness or connectors (Solenoid circuit is open or shorted.) • Pressure control solenoid valve A (Line pressure solenoid valve)
DTC: P0745 **T TCM, TCIL: Yes** **Year:** 2009, 2010 **Model:** FX35, G37 **Engine:** 3.5L V6, 3.7L V6 **Transmission:** All	**Pressure Control Solenoid A:** The line pressure solenoid valve monitor value is 0.4 A or less when the line pressure solenoid valve command value is more than 0.75 A. **Possible Causes:** • Harness or connectors • Solenoid valve circuit is open or shorted. • Line pressure solenoid valve

DTC	Trouble Code Title, Conditions & Possible Causes
DTC: P0750 **T TCM, TCIL: Yes** **Year:** 2009, 2010 **Model:** FX35, FX50, G37 **Engine:** 3.5L V6, 3.7L V6, 5.0L V8 **Transmission:** All	**Shift Solenoid A:** • The anti-interlock solenoid valve monitor value is ON when the anti-interlock solenoid valve command value is OFF. • The anti-interlock solenoid valve monitor value is OFF when the anti-interlock solenoid valve command value is ON. **Possible Causes:** • Harness or connectors • Solenoid valve circuit is open or shorted. • Anti-interlock solenoid valve
DTC: P0775 **T TCM, TCIL: Yes** **Year:** 2009, 2010 **Model:** FX35, FX50, G37 **Engine:** 3.5L V6, 3.7L V6, 5.0L V8 **Transmission:** All	**Pressure Control Solenoid B:** The input clutch solenoid valve monitor value is 0.4 A or less when the input clutch solenoid valve command value is more than 0.75 A. **Possible Causes:** • Harness or connectors • Solenoid valve circuit is open or shorted. • Input clutch solenoid valve
DTC: P0780 **T TCM, TCIL: Yes** **Year:** 2009, 2010 **Model:** FX35, FX50, G37 **Engine:** 3.5L V6, 3.7L V6, 5.0L V8 **Transmission:** All	**Shift Error:** • When shifting from 3GR to 4GR with the selector lever in "D" position, the gear ratio does not shift to 1.412 (gear ratio of 4GR). • When shifting from 5GR to 6GR or 6GR to 7GR, the engine speed exceeds the prescribed speed. • The shift change time from 4GR to 3GR is 0.2 second or less. **Possible Causes:** • Anti-interlock solenoid valve • Low brake solenoid valve • Hydraulic control circuit
DTC: P0795 **T TCM, TCIL: Yes** **Year:** 2009, 2010 **Model:** FX35, FX50, G37 **Engine:** 3.5L V6, 3.7L V6, 5.0L V8 **Transmission:** All	**Pressure Control Solenoid C:** The front brake solenoid valve monitor value is 0.4 A or less when the front brake solenoid valve command value is more than 0.75 A. **Possible Causes:** • Harness or connectors • Solenoid valve circuit is open or shorted. • Front brake solenoid valve
DTC: P0850 **1T TCM, TCIL: Yes** **Year:** 2009, 2010 **Model:** G37 **Engine:** 3.7L V6 **Transmission:** All	**Park/Neutral Position (PNP) Switch:** • The signal of the park/neutral position (PNP) does not change during driving after the engine is started. **Possible Causes:** • Harness or connectors • The park/neutral position (PNP) signal circuit is open or shorted. • Park/neutral position (PNP) switch (M/T) • Transmission range switch (A/T) • Combination meter • TCM (A/T)
DTC: P0850 **1T ECM, MIL: Yes** **Year:** 2009, 2010 **Model:** EX35, FX35, FX50, M35, M45, QX56 **Engine:** 3.5L V6, 4.5L V8, 5.0L V8, 5.6L V8 **Transmission:** All	**Park/Neutral Position Switch:** • CVT & M/T: When the shift lever position is P or N (CVT), Neutral (M/T), park/neutral position (PNP) switch is ON. ECM detects the position because the continuity of the line (the ON signal) exists. • A/T: The signal of the park/neutral position (PNP) switch does not change in the process of engine starting and driving. **Possible Causes:** • Harness or connectors • Park/neutral position (PNP) switch circuit is open or shorted. • Park/neutral position (PNP) switch

OBD II Trouble Code List (P1XXX Codes)

DTC	Trouble Code Title, Conditions & Possible Causes
DTC: P100A **1T ECM, MIL: Yes** **Year:** 2009, 2010 **Model:** FX50, G37 **Engine:** 3.7L V6, 5.0L V8 **Transmission:** All	**Variable Valve Event & Lift (VVEL) Response Malfunction (Bank 1):** NOTE: If DTC P100A or P100B is displayed with DTC P1090 or P1093, first perform the trouble diagnosis for DTC P1090 or P1093. • Actual event response to target is poor. **Possible Causes:** • Harness or connectors • VVEL actuator motor circuit is open or shorted. • VVEL actuator motor • VVEL actuator sub assembly • VVEL ladder assembly • VVEL control module
DTC: P100B **1T ECM, MIL: Yes** **Year:** 2009, 2010 **Model:** FX50, G37 **Engine:** 3.7L V6, 5.0L V8 **Transmission:** All	**Variable Valve Event & Lift (VVEL) Response Malfunction (Bank 2):** NOTE: If DTC P100A or P100B is displayed with DTC P1090 or P1093, first perform the trouble diagnosis for DTC P1090 or P1093. • Actual event response to target is poor. **Possible Causes:** • Harness or connectors • VVEL actuator motor circuit is open or shorted. • VVEL actuator motor • VVEL actuator sub assembly • VVEL ladder assembly • VVEL control module
DTC: P1078 **1T ECM, MIL: Yes** **Year:** 2009, 2010 **Model:** EX35, FX35, FX50, M35 **Engine:** 3.5L V6, 5.0L V8 **Transmission:** All	**Exhaust Valve Timing Control Position Sensor (Bank 1) Circuit:** NOTE: If this DTC is displayed with DTC P0643, first perform the trouble diagnosis for DTC P0643. • An excessively high or low voltage from the sensor is sent to ECM. **Possible Causes:** • Harness or connectors • The sensor circuit is open or shorted. • Exhaust valve timing (EVT) control position sensor • Crankshaft position sensor (POS) • Camshaft position sensor (PHASE) (bank 1) • Accumulation of debris to the signal pick-up portion of the camshaft
DTC: P1084 **1T ECM, MIL: Yes** **Year:** 2009, 2010 **Model:** EX35, FX35, M35 **Engine:** 3.5L V6 **Transmission:** All	**Exhaust Valve Timing Control Position Sensor (Bank 2) Circuit:** NOTE: If this DTC is displayed with DTC P0643, first perform the trouble diagnosis for DTC P0643. • An excessively high or low voltage from the sensor is sent to ECM. **Possible Causes:** • Harness or connectors • EVT control position sensor (bank 2) circuit is open or shorted • CKP sensor (POS) circuit is shorted. • CMP sensor (PHASE) (bank 2) circuit is shorted. • Battery current sensor circuit is shorted. • APP sensor 2 circuit is shorted. • EVAP control system pressure sensor circuit is shorted. • Refrigerant pressure sensor circuit is shorted. • Exhaust valve timing control position sensor (bank 2) • Crankshaft position sensor (POS) • Camshaft position sensor (PHASE) (bank 2) • Battery current sensor • Accelerator pedal position sensor • EVAP control system pressure sensor • Refrigerant pressure sensor • Accumulation of debris to the signal pick-up portion of the camshaft

DTC	Trouble Code Title, Conditions & Possible Causes

DTC: P1087
T ECM, MIL: Yes
Year: 2009, 2010
Model: FX50, G37
Engine: 3.7L V6, 5.0L V8
Transmission: All
	Variable Valve Event & Lift (VVEL) Small Event Angle Malfunction (Bank 1):
	NOTE: If DTC P1087 or P1088 is displayed with DTC P1090 or P1093, perform the diagnosis for P1090 or P1093 first.
	  • The event angle of VVEL control shaft is always small.
	Possible Causes:
	  • Harness or connectors
	  • VVEL actuator motor circuit is open or shorted.
	  • VVEL actuator motor
	  • VVEL actuator sub assembly
	  • VVEL ladder assembly
	  • VVEL control module

DTC: P1088
T ECM, MIL: Yes
Year: 2009, 2010
Model: FX50, G37
Engine: 3.7L V6, 5.0L V8
Transmission: All
	Variable Valve Event & Lift (VVEL) Small Event Angle Malfunction (Bank 2):
	NOTE: If DTC P1087 or P1088 is displayed with DTC P1090 or P1093, perform the diagnosis for P1090 or P1093 first.
	  • The event angle of VVEL control shaft is always small.
	Possible Causes:
	  • Harness or connectors
	  • VVEL actuator motor circuit is open or shorted.
	  • VVEL actuator motor
	  • VVEL actuator sub assembly
	  • VVEL ladder assembly
	  • VVEL control module

DTC: P1089
T , MIL: Yes
Year: 2009, 2010
Model: FX50, G37
Engine: 3.7L V6, 5.0L V8
Transmission: All
	Variable Valve Event & Lift (VVEL) Control Shaft Position Sensor (Bank 1) Circuit:
	NOTE: If DTC P1089 or P1092 is displayed with DTC P1608, first perform the trouble diagnosis for DTC P1608.
	  • An excessively low voltage from the sensor is sent to VVEL control module.
	  • An excessively high voltage from the sensor is sent to VVEL control module.
	  • Rationally incorrect voltage is sent to VVEL control module compared with the signals from VVEL control shaft position sensor 1 and VVEL control shaft position sensor 2.
	Possible Causes:
	  • Harness or connectors
	  • VVEL control shaft position sensor circuit is open or shorted.
	  • VVEL control shaft position sensor
	  • VVEL control module

DTC: P1090
T , MIL: Yes
Year: 2009, 2010
Model: FX50, G37
Engine: 3.7L V6, 5.0L V8
Transmission: All
	Variable Valve Event & Lift (VVEL) System Performance (Bank 1) :
	NOTE: If DTC P1090 or P1093 is displayed with DTC P1091, first perform the trouble diagnosis for DTC P1091.
	  • Event angle difference between the actual and the target is detected.
	  • Abnormal current is sent to VVEL actuator motor.
	Possible Causes:
	  • Harness or connectors
	  • VVEL actuator motor circuit is open or shorted.
	  • VVEL actuator motor
	  • VVEL actuator sub assembly
	  • VVEL ladder assembly
	  • VVEL control module

DTC: P1091
T ECM, MIL: Yes
Year: 2009, 2010
Model: FX50, G37
Engine: 3.7L V6, 5.0L V8
Transmission: All
	Variable Valve Event & Lift (VVEL) Actuator Motor Relay Circuit:
	  • VVEL control module detects the VVEL actuator motor relay is stuck OFF.
	  • VVEL control module detects the VVEL actuator motor relay is stuck ON.
	Possible Causes:
	  • Harness or connectors
	  • VVEL actuator motor relay circuit is open or shorted.
	  • Abort circuit is open or shorted.
	  • VVEL actuator motor relay
	  • VVEL control module
	  • ECM

DTC	Trouble Code Title, Conditions & Possible Causes
DTC: P1092 **T ECM, MIL: Yes** **Year:** 2009, 2010 **Model:** FX50, G37 **Engine:** 3.7L V6, 5.0L V8 **Transmission:** All	**Variable Valve Event & Lift (VVEL) Control Shaft Position Sensor (Bank 2) Circuit:** **NOTE: If DTC P1089 or P1092 is displayed with DTC P1608, first perform the trouble diagnosis for DTC P1608.** • An excessively low voltage from the sensor is sent to VVEL control module. • An excessively high voltage from the sensor is sent to VVEL control module. • Rationally incorrect voltage is sent to VVEL control module compared with the signals from VVEL control shaft position sensor 1 and VVEL control shaft position sensor 2. **Possible Causes:** • Harness or connectors • VVEL control shaft position sensor circuit is open or shorted. • VVEL control shaft position sensor • VVEL control module
DTC: P1093 **T ECM, MIL: Yes** **Year:** 2009, 2010 **Model:** FX50, G37 **Engine:** 3.7L V6, 5.0L V8 **Transmission:** All	**Variable Valve Event & Lift (VVEL) System Performance (Bank 2) :** **NOTE: If DTC P1090 or P1093 is displayed with DTC P1091, first perform the trouble diagnosis for DTC P1091** • Event angle difference between the actual and the target is detected. • Abnormal current is sent to VVEL actuator motor. **Possible Causes:** • Harness or connectors • VVEL actuator motor circuit is open or shorted. • VVEL actuator motor • VVEL actuator sub assembly • VVEL ladder assembly • VVEL control module
DTC: P1140 **1T ECM, MIL: Yes** **Year:** 2009, 2010 **Model:** M45, QX56 **Engine:** 4.5L V8, 5.6L V8 **Transmission:** All	**Intake Valve Timing Control Position Sensor Circuit (Bank 1):** • An excessively high or low voltage from the sensor is sent to ECM. **Possible Causes:** • Harness or connectors • Intake valve timing control position sensor circuit is open or shorted • Intake valve timing control position sensor • Crankshaft position sensor (POS) • Camshaft position sensor (PHASE) • Accumulation of debris to the signal pick-up portion of the camshaft sprocket
DTC: P1145 **1T ECM, MIL: Yes** **Year:** 2009, 2010 **Model:** M45, QX56 **Engine:** 4.5L V8, 5.6L V8 **Transmission:** All	**Intake Valve Timing Control Position Sensor Circuit (Bank 2):** An excessively high or low voltage from the sensor is sent to ECM. **Possible Causes:** • Harness or connectors • Intake valve timing control position sensor circuit is open or shorted • Intake valve timing control position sensor • Crankshaft position sensor (POS) • Camshaft position sensor (PHASE) • Accumulation of debris to the signal pick-up portion of the camshaft sprocket
DTC: P1148 **1T ECM, MIL: Yes** **Year:** 2009, 2010 **Model:** EX35, FX35, FX50, G37, M35, M45, QX56 **Engine:** 3.5L V6, 3.7L V6, 4.5L V8, 5.0L V8, 5.6L V8 **Transmission:** All	**Closed Loop Control Function (Includes Hybrid Models):** HYBRID MODELS CAUTION: Hybrid systems use very high-voltage battery systems. Before starting any service work involving the battery system, turn the ignition switch OFF and then remove the service plug from pocket in the trunk. After removing the service plug, wait 10 minutes before touching any of the high-voltage connectors and **NOTE: On V6 and V8 engines, this applies to Bank 1, except 3.7L which is Bank 2.** **NOTE: DTC P1148 or P1168 is displayed with another DTC for A/F sensor 1. Perform the trouble diagnosis for the corresponding DTC.** • The closed loop control function for bank 1 does not operate even when vehicle is being driven in the specified condition. **Possible Causes:** • Harness or connectors • Air fuel ratio (A/F) sensor (sensor 1 on V6 and V8) circuit is open or shorted. • Air fuel ratio (A/F) sensor (sensor 1 on V6 and V8) • Air fuel ratio (A/F) sensor (sensor 1 on V6 and V8) heater

DTC	Trouble Code Title, Conditions & Possible Causes
DTC: P1168 **1T ECM, MIL: Yes** **Year:** 2009, 2010 **Model:** EX35, FX35, FX50, G37, M35, M45, QX56 **Engine:** 3.5L V6, 3.7L V6, 4.5L V8, 5.0L V8, 5.6L V8 **Transmission:** All	**Closed Loop Control Function (Bank 2):** NOTE: If DTC P1148 or P1168 is displayed with another DTC for air fuel ratio (A/F) sensor 2. Perform the trouble diagnosis for the corresponding DTC. • The closed loop control function for bank 2 does not operate even when vehicle is being driven in the specified condition. **Possible Causes:** • Harness or connectors • Air fuel ratio (A/F) sensor 1 or 2 circuit is open or shorted. • Air fuel ratio (A/F) sensor 1 or 2 • Air fuel ratio (A/F) sensor 1 or 2 heater
DTC: P1211 **1T ECM** **Year:** 2009, 2010 **Model:** EX35, FX35, FX50, G37, M35, M45, QX56 **Engine:** 3.5L V6, 3.7L V6, 4.5L V8, 5.0L V8, 5.6L V8 **Transmission:** All	**TCS Control Unit:** • Freeze frame data is not stored in the ECM for this self-diagnosis. • The MIL will not illuminate for this self-diagnosis. • ECM receives malfunction information from "ABS actuator and electric unit (control unit)". **Possible Causes:** • ABS actuator and electric unit (control unit) • TCS related parts
DTC: P1212 **1T ECM, MIL: Yes** **Year:** 2009, 2010 **Model:** EX35, FX35, FX50, G37, M35, M45, QX56 **Engine:** 3.5L V6, 3.7L V6, 4.5L V8, 5.0L V8, 5.6L V8 **Transmission:** All	**TCS Communication Line:** NOTE: If DTC P1212 is displayed with DTC UXXXX, first perform the trouble diagnosis for DTC UXXXX. NOTE: If DTC P1212 is displayed with DTC P0607, first perform the trouble diagnosis for DTC P0607. NOTE: Be sure to erase the malfunction information such as DTC not only for "ABS actuator and electric unit (control unit)" but also for ECM after TCS related repair. • Freeze frame data is not stored in the ECM for this self-diagnosis. • The MIL will not illuminate for this self-diagnosis. • ECM cannot receive the information from "ABS actuator and electric unit (control unit)". **Possible Causes:** • Harness or connectors • The CAN communication line is open or shorted. • ABS actuator and electric unit (control unit) • Dead (Weak) battery
DTC: P1217 **1T ECM, MIL: Yes** **Year:** 2009, 2010 **Model:** EX35, FX35, FX50, G37, M35, M45, QX56 **Engine:** 3.5L V6, 3.7L V6, 4.5L V8, 5.0L V8, 5.6L V8 **Transmission:** All	**Engine Over Temperature (Overheat) (Includes Hybrid Models):** HYBRID MODELS CAUTION: Hybrid systems use very high-voltage battery systems. Before starting any service work involving the battery system, turn the ignition switch OFF and then remove the service plug from pocket in the trunk. After removing the service plug, wait 10 minutes before touching any of the high-voltage connectors and terminals. NOTE: If DTC P1217 is displayed with DTC UXXXX, first perform the trouble diagnosis for DTC UXXXX. NOTE: If DTC P1217 is displayed with DTC P0607, first perform the trouble diagnosis for DTC P0607. • The ECM controls cooling fan relays through CAN communication line. • Cooling fan does not operate properly (overheat). • Cooling fan system does not operate properly (overheat). • Engine coolant was not added to the system using the proper filling method. • Engine coolant is not within the specified range. **Possible Causes:** • Harness or connectors • Cooling fan circuit is open or shorted. • Cooling fan motor • IPDM E/R (cooling fan relays-1, -2 and -3) • Cooling fan relays-4 and -5 • Radiator hose • Radiator • Reservoir tank • Radiator cap • Water pump • Thermostat
DTC: P1220 **1T ECM, MIL: Yes** **Year:** 2009, 2010 **Model:** FX50, M45 **Engine:** 4.5L V8, 5.0L V8 **Transmission:** All	**Fuel Pump Control Module (FPCM):** • An improper voltage signal from the FPCM, which is supplied to a point between the fuel pump and the dropping resistor, is detected by ECM. • During engine cranking, the signal voltage of the FPCM to the ECM is too low. **Possible Causes:** • Harness or connectors • FPCM circuit is shorted • Dropping resistor • FPCM

DTC	Trouble Code Title, Conditions & Possible Causes
DTC: P1225 **1T ECM, MIL: Yes** **Year:** 2009, 2010 **Model:** EX35, FX35, FX50, G37, M35, M45, QX56 **Engine:** 3.5L V6, 3.7L V6, 4.5L V8, 5.0L V8, 5.6L V8 **Transmission:** All	**Closed Throttle Position Learning Performance (Includes Hybrid Models):** HYBRID MODELS CAUTION: Hybrid systems use very high-voltage battery systems. Before starting any service work involving the battery system, turn the ignition switch OFF and then remove the service plug from pocket in the trunk. After removing the service plug, wait 10 minutes before touching any of the high-voltage connectors and terminals. **NOTE: For V6 and V8, this DTC is for Bank 1.** • Closed throttle position learning value is excessively low. **Possible Causes:** • Electric throttle control actuator (TP sensor 1 and 2)
DTC: P1226 **1T ECM, MIL: Yes** **Year:** 2009, 2010 **Model:** EX35, FX35, FX50, G37, M35, M45, QX56 **Engine:** 3.5L V6, 3.7L V6, 4.5L V8, 5.0L V8, 5.6L V8 **Transmission:** All	**Closed Throttle Position Learning Performance (Includes Hybrid Models):** HYBRID MODELS CAUTION: Hybrid systems use very high-voltage battery systems. Before starting any service work involving the battery system, turn the ignition switch OFF and then remove the service plug from pocket in the trunk. After removing the service plug, wait 10 minutes before touching any of the high-voltage connectors and terminals. **NOTE: On V6 and V8, this DTC is for Bank 1.** • Closed throttle position learning is not performed successfully, repeatedly. **Possible Causes:** • Electric throttle control actuator (TP sensor 1 and 2)
DTC: P1233 **T ECM, MIL: Yes** **Year:** 2009, 2010 **Model:** EX35, FX35, FX50, G37, M35 **Engine:** 3.5L V6, 3.7L V6, 5.0L V8 **Transmission:** All	**Electric Throttle Control Performance (Bank 2):** **NOTE: If DTC P1233 or P2101 is displayed with DTC P1238, P1290, P2100 or 2119, first perform the trouble diagnosis for DTC P1238, P2119 or P1290, P2100.** • Electric throttle control function does not operate properly **Possible Causes:** • Harness or connectors • Throttle control motor circuit is open or shorted • Electric throttle control actuator
DTC: P1234 **1T ECM, MIL: Yes** **Year:** 2009, 2010 **Model:** EX35, FX35, FX50, G37, M35 **Engine:** 3.5L V6, 3.7L V6, 5.0L V8 **Transmission:** All	**Closed Throttle Position Learning Performance (Bank 2):** • Closed throttle position learning value is excessively low. **Possible Causes:** • Electric throttle control actuator (TP sensor 1 and 2)
DTC: P1235 **1T ECM, MIL: Yes** **Year:** 2009, 2010 **Model:** EX35, FX35, FX50, G37, M35 **Engine:** 3.5L V6, 3.7L V6, 5.0L V8 **Transmission:** All	**Closed Throttle Position Learning Performance (Bank 2):** • Closed throttle position learning is not performed successfully, repeatedly. **Possible Causes:** • Electric throttle control actuator (TP sensor 1 and 2)
DTC: P1236 **T ECM, MIL: Yes** **Year:** 2009, 2010 **Model:** EX35, FX35, FX50, G37, M35 **Engine:** 3.5L V6, 3.7L V6, 5.0L V8 **Transmission:** All	**Throttle Control Motor (Bank 2) Circuit Short:** ECM detects short in both circuits between ECM and throttle control motor. **Possible Causes:** • Harness or connectors • (Throttle control motor circuit is shorted. • Electric throttle control actuator (Throttle control motor)
DTC: P1238 **T ECM, MIL: Yes** **Year:** 2009, 2010 **Model:** EX35, FX35, FX50, G37, M35 **Engine:** 3.5L V6, 3.7L V6, 5.0L V8 **Transmission:** All	**Electrical Throttle Control Actuator (Bank 2):** Condition A: Electric throttle control actuator does not function properly due to the return spring malfunction. Condition B: Throttle valve opening angle in fail-safe mode is not in specified range. Condition CL ECM detect the throttle valve is stuck open. **Possible Causes:** • Electric throttle control actuator
DTC: P1239 **T ECM, MIL: Yes** **Year:** 2009, 2010 **Model:** EX35, FX35, FX50, G37, M35 **Engine:** 3.5L V6, 3.7L V6, 5.0L V8 **Transmission:** All	**Throttle Position Sensor (Bank 2) Circuit Range/Performance:** • Rationally incorrect voltage is sent to ECM compared with the signals from TP sensor 1 and TP sensor 2. **Possible Causes:** • Harness or connector • TP sensor 1 and 2 circuit is open or shorted. • Electric throttle control actuator (TP sensor 1 and 2)

DTC	Trouble Code Title, Conditions & Possible Causes
DTC: P1290 **T ECM, MIL: Yes** **Year:** 2009, 2010 **Model:** EX35, FX35, FX50, G37, M35 **Engine:** 3.5L V6, 3.7L V6, 5.0L V8 **Transmission:** All	**Throttle Control Motor Relay Circuit Open (Bank 2):** • ECM detects a voltage of power source for throttle control motor is excessively low. **Possible Causes:** • Harness or connectors • Throttle control motor relay circuit is open • Throttle control motor relay
DTC: P1421 **1T ECM, MIL: Yes** **Year:** 2009, 2010 **Model:** EX35, FX35, FX50, G37, M35, M45, QX56 **Engine:** 3.5L V6, 3.7L V6, 4.5L V8, 5.0L V8, 5.6L V8 **Transmission:** All	**Cold Start Emission Reduction Strategy Monitoring (Includes Hybrid Models):** HYBRID MODELS CAUTION: Hybrid systems use very high-voltage battery systems. Before starting any service work involving the battery system, turn the ignition switch OFF and then remove the service plug from pocket in the trunk. After removing the service plug, wait 10 minutes before touching any of the high-voltage connectors and terminals. **NOTE: If DTC P1421 is displayed with other DTC, first perform the trouble diagnosis for other DTC.** • ECM does not control ignition timing and engine idle speed properly when engine is started with pre-warming up condition. **Possible Causes:** • Lack of intake air volume • Fuel injection system • ECM
DTC: P1550 **1T ECM, MIL: Yes** **Year:** 2009, 2010 **Model:** EX35, FX35, FX50, G37, M35, M45, QX56 **Engine:** 3.5L V6, 3.7L V6, 4.5L V8, 5.0L V8, 5.6L V8 **Transmission:** All	**Battery Current Sensor Circuit Range/Performance:** • The MIL will not illuminate for this diagnosis. **NOTE: If DTC P1550 is displayed with DTC P0643, first perform the trouble diagnosis for DTC P0643.** • The output voltage of the battery current sensor remains within the specified range while engine is running. **Possible Causes:** • Harness or connectors • Battery current sensor circuit is open or shorted. • CKP sensor (POS) circuit is shorted. • CMP sensor (PHASE) (bank 2) circuit is shorted. • EVT control position sensor (bank 2) circuit is shorted. • APP sensor 2 circuit is shorted. • EVAP control system pressure sensor circuit is shorted. • Refrigerant pressure sensor circuit is shorted. • Battery current sensor • Crankshaft position sensor (POS) • Camshaft position sensor (PHASE) (bank 2) • Exhaust valve timing control position sensor (bank 2) • Accelerator pedal position sensor • EVAP control system pressure sensor • Refrigerant pressure sensor
DTC: P1551 **1T ECM, MIL: Yes** **Year:** 2009, 2010 **Model:** EX35, FX35, FX50, G37, M35, M45, QX56 **Engine:** 3.5L V6, 3.7L V6, 4.5L V8, 5.0L V8, 5.6L V8 **Transmission:** All	**Battery Current Sensor Circuit Low Input:** • The MIL will not illuminate for this diagnosis. **NOTE: If DTC P1551 or P1552 is displayed with DTC P0643, first perform the trouble diagnosis for DTC P0643.** • An excessively low voltage from the sensor is sent to ECM. **Possible Causes:** • Harness or connectors • Battery current sensor circuit is open or shorted. • CKP sensor (POS) circuit is shorted. • CMP sensor (PHASE) (bank 2) circuit is shorted. • EVT control position sensor (bank 2) circuit is shorted. • APP sensor 2 circuit is shorted. • EVAP control system pressure sensor circuit is shorted. • Refrigerant pressure sensor circuit is shorted. • Battery current sensor • Crankshaft position sensor (POS) • Camshaft position sensor (PHASE) • (bank 2) • Exhaust valve timing control position • sensor (bank 2) • Accelerator pedal position sensor • EVAP control system pressure sensor • Refrigerant pressure sensor

DTC	Trouble Code Title, Conditions & Possible Causes
DTC: P1552 **1T ECM, MIL: Yes** **Year:** 2009, 2010 **Model:** EX35, FX35, FX50, G37, M35, M45, QX56 **Engine:** 3.5L V6, 3.7L V6, 4.5L V8, 5.0L V8, 5.6L V8 **Transmission:** All	**Battery Current Sensor Circuit High Input:** • The MIL will not illuminate for this diagnosis. **NOTE: If DTC P1551 or P1552 is displayed with DTC P0643, first perform the trouble diagnosis for DTC P0643.** • An excessively high voltage from the sensor is sent to ECM. **Possible Causes:** • Harness or connectors • Battery current sensor circuit is open or shorted. • CKP sensor (POS) circuit is shorted. • CMP sensor (PHASE) (bank 2) circuit is shorted. • EVT control position sensor (bank 2) circuit is shorted. • APP sensor 2 circuit is shorted. • EVAP control system pressure sensor circuit is shorted. • Refrigerant pressure sensor circuit is shorted. • Battery current sensor • Crankshaft position sensor (POS) • Camshaft position sensor (PHASE) • (bank 2) • Exhaust valve timing control position • sensor (bank 2) • Accelerator pedal position sensor • EVAP control system pressure sensor • Refrigerant pressure sensor
DTC: P1553 **1T ECM, MIL: Yes** **Year:** 2009, 2010 **Model:** EX35, FX35, FX50, G37, M35, M45, QX56 **Engine:** 3.5L V6, 3.7L V6, 4.5L V8, 5.0L V8, 5.6L V8 **Transmission:** All	**Battery Current Sensor Performance:** • The MIL will not illuminate for this diagnosis. **NOTE: If DTC P1553 is displayed with DTC P0643, first perform the trouble diagnosis for DTC P0643.** • The signal voltage transmitted from the sensor to ECM is higher than the amount of the maximum power generation. **Possible Causes:** • Harness or connectors • Battery current sensor circuit is open or shorted. • CKP sensor (POS) circuit is shorted. • CMP sensor (PHASE) (bank 2) circuit is shorted. • EVT control position sensor (bank 2) circuit is shorted. • APP sensor 2 circuit is shorted. • EVAP control system pressure sensor circuit is shorted. • Refrigerant pressure sensor circuit is shorted. • Battery current sensor • Crankshaft position sensor (POS) • Camshaft position sensor (PHASE) (bank 2) • Exhaust valve timing control position sensor (bank 2) • Accelerator pedal position sensor • EVAP control system pressure sensor • Refrigerant pressure sensor
DTC: P1554 **1T ECM, MIL: Yes** **Year:** 2009, 2010 **Model:** EX35, FX35, FX50, G37, M35, M45, QX56 **Engine:** 3.5L V6, 3.7L V6, 4.5L V8, 5.0L V8, 5.6L V8 **Transmission:** All	**Battery Current Sensor Performance:** • The MIL will not illuminate for this diagnosis. **NOTE: If DTC P1554 is displayed with DTC P0643, first perform the trouble diagnosis for DTC P0643.** • The output voltage of the battery current sensor is lower than the specified value while the battery voltage is high enough. **Possible Causes:** • Harness or connectors • Battery current sensor circuit is open or shorted. • CKP sensor (POS) circuit is shorted. • CMP sensor (PHASE) (bank 2) circuit is shorted. • EVT control position sensor (bank 2) circuit is shorted. • APP sensor 2 circuit is shorted. • EVAP control system pressure sensor circuit is shorted. • Refrigerant pressure sensor circuit is shorted. • Battery current sensor • Crankshaft position sensor (POS) • Camshaft position sensor (PHASE) (bank 2) • Exhaust valve timing control position sensor (bank 2) • Accelerator pedal position sensor • EVAP control system pressure sensor • Refrigerant pressure sensor

DTC	Trouble Code Title, Conditions & Possible Causes

DTC: P1564
1T ECM
Year: 2009, 2010
Model: EX35, FX35, FX50, G37, M35, QX56
Engine: 3.5L V6, 3.7L V6, 5.0L V8, 5.6L V8
Transmission: All
	ASCD Steering Switch Malfunction (Includes Hybrid Models):
	HYBRID MODELS CAUTION: Hybrid systems use very high-voltage battery systems. Before starting any service work involving the battery system, turn the ignition switch OFF and then remove the service plug from pocket in the trunk. After removing the service plug, wait 10 minutes before touching any of the high-voltage connectors and terminals.
	    • This self-diagnosis has the one trip detection logic.
	    • The MIL will not illuminate for this self-diagnosis.
	NOTE: If DTC P1564 is displayed with DTC P0605, first perform the trouble diagnosis for DTC P0605.
	    • An excessively high voltage signal from the ASCD steering switch is sent to ECM.
	    • ECM detects that input signal from the ASCD steering switch is out of the specified range.
	    • ECM detects that the ASCD steering switch is stuck ON.
	Possible Causes:
	    • Harness or connectors
	    • ASCD switch circuit is open or shorted.
	    • ASCD steering switch
	    • ECM

DTC: P1564
T ECM, MIL: Yes
Year: 2009, 2010
Model: EX35, FX35, FX50, G37, M35, QX56
Engine: 3.5L V6, 3.7L V6, 5.0L V8, 5.6L V8
Transmission: All
	ICC Steering Switch:
	NOTE: If DTC P1564 is displayed with DTC P0605, first perform the trouble diagnosis for DTC P0605.
	    • An excessively high voltage signal from the ICC steering switch is sent to ECM.
	    • ECM detects that input signal from the ICC steering switch is out of the specified range.
	    • ECM detects that the ICC steering switch is stuck ON.
	Possible Causes:
	    • Harness or connectors
	    • The switch circuit is open or shorted.
	    • ICC steering switch
	    • ECM

DTC: P1568
T ECM, MIL: Yes
Year: 2009, 2010
Model: EX35, FX35, FX50, G37, M35, QX56
Engine: 3.5L V6, 3.7L V6, 5.0L V8, 5.6L V8
Transmission: All
	ICC Function:
	NOTE:
	    • If DTC P1568 is displayed with DTC UXXXX, first perform the trouble diagnosis for DTC UXXXX.
	    • If this DTC is displayed with DTC P0605 or P0607, first perform the trouble diagnosis for DTC P0605 or P0607.
	    • ECM detects a difference between signals from ICC sensor integrated unit is out of specified range.
	Possible Causes:
	    • Harness or connectors
	    • The CAN communication line is open or shorted.
	    • ICC sensor integrated unit
	    • ECM

DTC: P1568
1T CCM
Year: 2009, 2010
Model: FX35
Engine: 3.5L V6
Transmission: All
	ASCD Command Valve Circuit Malfunction:
	If DTC P1568 is displayed with DTC U1000 and/or U1001, diagnose the cause of the DTC U1000 and/or U1001 first. If DTC P1568 is displayed with DTC P0605, diagnose the cause of DTC P0605 first.
	Engine started, and PCM detected that the signals from the command valve unit (ICC unit) were out of range.
	Possible Causes:
	    • CAN communication line is open, shorted to ground or power
	    • ICC unit is damaged or has failed
	    • PCM has failed

DTC	Trouble Code Title, Conditions & Possible Causes
DTC: P1572 **1T ECM, MIL: Yes** **Year:** 2009, 2010 **Model:** EX35, FX35, FX50, G37, M35 **Engine:** 3.5L V6, 3.7L V6, 5.0L V8 **Transmission:** All	**ACSD Brake Switch Malfunction (Includes Hybrid Models):** HYBRID MODELS CAUTION: Hybrid systems use very high-voltage battery systems. Before starting any service work involving the battery system, turn the ignition switch OFF and then remove the service plug from pocket in the trunk. After removing the service plug, wait 10 minutes before touching any of the high-voltage connectors and terminals. • This self-diagnosis has the one trip detection logic. • The MIL will not illuminate for this self-diagnosis. **NOTE: If DTC P1572 is displayed with DTC P0605, first perform the trouble diagnosis for DTC P0605.** • This self-diagnosis has the one trip detection logic. **NOTE: When malfunction A is detected, the DTC is not stored in ECM memory. And in that case, 1st trip DTC and 1st trip freeze frame data are displayed. 1st trip DTC is erased when ignition switch is turned OFF. And even when Malfunction A is detected in two consecutive trips, DTC is not stored in ECM memory.** • Malfunction A: When the vehicle speed is above 19 MPH, ON signals from the stop lamp switch and the ASCD brake switch are sent to ECM at the same time. • Malfunction B: ASCD brake switch signal is not sent to ECM for extremely long time while the vehicle is being driven. **Possible Causes:** • Stop lamp switch circuit is shorted. • ASCD brake switch circuit is shorted. • ASCD clutch switch circuit is shorted (M/T). • Stop lamp switch • ASCD brake switch • ASCD clutch switch (M/T) • Incorrect stop lamp switch installation • Incorrect ASCD brake switch installation • Incorrect ASCD clutch switch installation (M/T) • ECM
DTC: P1572 **1T ECM, MIL: Yes** **Year:** 2009, 2010 **Model:** EX35, FX35, FX50, G37, M35, QX56 **Engine:** 3.5L V6, 3.7L V6, 5.0L V8, 5.6L V8 **Transmission:** All	**ICC Brake Switch:** **NOTE: If DTC P1572 is displayed with DTC P0605, first perform the trouble diagnosis for DTC P0605.** • This self-diagnosis has the one trip detection logic. When malfunction A is detected, DTC is not stored in ECM memory. And in that case, 1st trip DTC and 1st trip freeze frame data are displayed. 1st trip DTC is erased when ignition switch OFF. And even when malfunction A is detected in two consecutive trips, DTC is not stored in ECM memory. • Condition A: ON signals from the stop lamp switch and the ICC brake switch are sent to ECM at the same time. • Condition B: ICC brake switch signal is not sent to ECM for extremely long time while the vehicle is driving. **Possible Causes:** • Harness or connectors • The stop lamp switch circuit is shorted. • The ICC brake switch circuit is shorted. • Stop lamp switch • ICC brake switch • ICC brake hold relay • Incorrect stop lamp switch installation • Incorrect ICC brake switch installation • ECM
DTC: P1574 **T ECM, MIL: Yes** **Year:** 2009, 2010 **Model:** EX35, FX35, FX50, G37, M35, QX56 **Engine:** 3.5L V6, 3.7L V6, 5.0L V8, 5.6L V8 **Transmission:** All	**ICC Vehicle Speed Sensor:** **NOTE: If DTC P1574 is displayed with DTC UXXXX, first perform the trouble diagnosis for DTC UXXXX.** **NOTE: If this DTC is displayed with DTC P0500, P0605 or P0607, first perform the trouble diagnosis for these other DTC(s) first.** • ECM detects a difference between two vehicle speed signals is out of the specified range. **Possible Causes:** • Harness or connectors • The CAN communication line is open or shorted. • Unified meter and A/C amp. • ABS actuator and electric unit (control unit) • Wheel sensor • TCM • ECM

DTC	Trouble Code Title, Conditions & Possible Causes

DTC: P1574
1T ECM, MIL: Yes
Year: 2009, 2010
Model: EX35, FX35, FX50, G37, M35, QX56
Engine: 3.5L V6, 3.7L V6, 5.0L V8, 5.6L V8
Transmission: All
	ASCD Vehicle Speed Sensor Malfunction (Includes Hybrid Models):
	HYBRID MODELS CAUTION: Hybrid systems use very high-voltage battery systems. Before starting any service work involving the battery system, turn the ignition switch OFF and then remove the service plug from pocket in the trunk. After removing the service plug, wait 10 minutes before touching any of the high-voltage connectors and terminals.
	    • The MIL will not illuminate for this self-diagnosis.
	NOTE: If DTC P1574 is displayed with DTC UXXXX, first perform the trouble diagnosis for DTC UXXXX.
	NOTE: If DTC P1574 is displayed with DTC P0500, P0605 and/or P0607, first perform the trouble diagnosis for these DTCs before continuing with DTC P1574.
	    • ECM detects a difference between two vehicle speed signals is out of the specified range.
	Possible Causes:
	    • Harness or connectors
	    • CAN communication line is open or shorted.
	    • Combination meter circuit is open or shorted.
	    • TCM (CVT models)
	    • Unified meter and A/C amp.
	    • Combination meter
	    • ABS actuator and electric unit (control unit)
	    • ECM

DTC: P1606
T ECM, MIL: Yes
Year: 2009, 2010
Model: FX50, G37
Engine: 3.7L V6, 5.0L V8
Transmission: All
	Variable Valve Event & Lift (VVEL) Control Module:
	    • VVEL control module calculation function is malfunctioning.
	    • VVEL EEPROM system is malfunctioning.
	Possible Causes:
	    • VVEL control module

DTC: P1607
T ECM, MIL: Yes
Year: 2009, 2010
Model: FX50, G37
Engine: 3.7L V6, 5.0L V8
Transmission: All
	Variable Valve Event & Lift (VVEL) Control Module Circuit:
	    • The internal circuit of the VVEL control module is malfunctioning.
	Possible Causes:
	    • VVEL control module

DTC: P1608
T ECM, MIL: Yes
Year: 2009, 2010
Model: FX50, G37
Engine: 3.7L V6, 5.0L V8
Transmission: All
	Variable Valve Event & Lift (VVEL) Sensor Power Supply Circuit:
	VVEL control module detects a voltage of power source for sensor is excessively low or high.
	Possible Causes:
	    • Harness or connectors
	    • VVEL control shaft position sensor power supply circuit is open or shorted.
	    • VVEL control shaft position sensor
	    • VVEL control module

DTC: P1705
1T TCM, TCIL: Yes
Year: 2009, 2010
Model: EX35, M35
Engine: 3.5L V6
Transmission: All
	Throttle Position/Accelerator Pedal Position Sensor Circuit:
	Electric throttle control actuator consists of throttle control motor, accelerator pedal position sensor, throttle position sensor etc. The actuator sends a signal to the ECM, and ECM sends the signal to TCM with CAN communication.
	    • This is not an OBD-II self-diagnostic item.
	    • Diagnostic trouble code P1705 is detected when TCM does not receive the proper accelerator pedal position signals (input by CAN communication) from ECM.
	Possible Causes:
	    • ECM
	    • Harness or connectors
	    • CAN communication line is open or shorted.

DTC: P1705
T TCM, TCIL: Yes
Year: 2009, 2010
Model: FX35, FX50, G37, QX56
Engine: 3.5L V6, 3.7L V6, 5.0L V8, 5.6L V8
Transmission: All
	Accelerator Pedal Position (APP) Sensor Signal Circuit:
	TCM detects improper accelerator pedal position signals received from ECM via CAN communication.
	Possible Causes:
	    • Harness or connectors
	    • Sensor circuit is open or shorted.

DTC	Trouble Code Title, Conditions & Possible Causes
DTC: P1705 **T TCM, TCIL:** Yes **Year:** 2009, 2010 **Model:** M45 **Engine:** 4.5L V8 **Transmission:** All	**Throttle Position (TP) Sensor/Accelerator Pedal Position (APP) Sensor:** Electric throttle control actuator consists of throttle control motor, accelerator pedal position sensor, throttle position sensor, etc. The actuator sends a signal to the ECM, and ECM sends signals to TCM with CAN communication. • This is not an OBD-II self-diagnostic item. • Diagnostic trouble code P1705 is detected when TCM does not receive the proper accelerator pedal position signals (input by CAN communication) from ECM. **Possible Causes:** • Harness or connectors • The sensor circuit is open or shorted.
DTC: P1710 **T TCM, TCIL:** Yes **Year:** 2009, 2010 **Model:** EX35, M35, QX56 **Engine:** 3.5L V6, 5.6L V8 **Transmission:** All	**A/T Fluid Temperature Sensor Circuit:** • This is an OBD-II self-diagnostic item. • Diagnostic trouble code P1710 will be detected when TCM receives an excessively low or high voltage from the sensor. • A/T fluid temperature does not rise to the specified temperature while driving. **Possible Causes:** • Harness or connectors • The sensor circuit is open or shorted. • A/T fluid temperature sensor 1
DTC: P1715 **1T TCM** **Year:** 2009, 2010 **Model:** EX35, FX50, M35 **Engine:** 3.5L V6, 5.0L V8 **Transmission:** All	**Input Speed Sensor (Primary Speed Sensor/TCM Output):** • The MIL will not illuminate for this self-diagnosis. **NOTE: If DTC P1715 is displayed with DTC UXXXX, first perform the trouble diagnosis for DTC UXXXX.** **NOTE: If DTC P1715 is displayed with DTC P0335, P0340, P0605 and/or P0607, first perform the trouble diagnosis for the appropriate DTC before proceeding with P1715 diagnosis.** • Sensor signal is different from the theoretical value calculated by ECM from secondary speed sensor signal and engine rpm signal. **Possible Causes:** • Harness or connectors • CAN communication line is open or shorted • Primary speed sensor circuit is open or shorted • TCM
DTC: P1721 **T TCM, TCIL:** Yes **Year:** 2009, 2010 **Model:** FX35, FX50, G37 **Engine:** 3.5L V6, 3.7L V6, 5.0L V8 **Transmission:** All	**Vehicle Speed Signal Circuit:** • The vehicle speed signal recognizes that the vehicle speed is 5 km/h (3 MPH) or less even if the output speed sensor recognizes that the vehicle speed is 20 km/h (12 MPH) or more. (Only when starts after the ignition switch is turned ON.) • The vehicle speed recognized by the vehicle speed signal decelerates 36 km/h (23 MPH) or more during 60 msec when the vehicle speed signal recognizes that the vehicle speed is 36 km/h (23 MPH) or more and the output speed sensor recognizes that the vehicle speed is 24 km/h (15 MPH) or less. • The vehicle speed of vehicle speed signal decelerates 36 km/h (23 MPH) or more even if the vehicle speed of output speed sensor accelerates or decelerates 24 km/h (15 MPH) or less during 60 msec when the vehicle speed sensor recognizes that the vehicle speed is 36 km/h (23 MPH) or more. **Possible Causes:** • Harness or connectors • Sensor circuit is open or shorted.
DTC: P1721 **T TCM, TCIL:** Yes **Year:** 2009, 2010 **Model:** M45, QX56 **Engine:** 4.5L V8, 5.6L V8 **Transmission:** All	**Vehicle Speed Sensor MTR:** • This is not an OBD-II self-diagnostic item. • Diagnostic trouble code P1721 is detected when TCM does not receive the proper vehicle speed sensor MTR signal (input by CAN communication) from combination meter (unified meter and A/C amp). **Possible Causes:** • Harness or connectors • The sensor circuit is open or shorted.
DTC: P1721 **T TCM, TCIL:** Yes **Year:** 2009, 2010 **Model:** EX35, M35 **Engine:** 3.5L V6 **Transmission:** All	**Vehicle Speed Signal:** • Signal (CAN communication) from vehicle speed signal not input due to cut line or the like. • Unexpected signal input during running. **Possible Causes:** • Harness or connectors • Sensor circuit is open or shorted.

DTC	Trouble Code Title, Conditions & Possible Causes
DTC: P1730 **T TCM, TCIL: Yes** **Year:** 2009, 2010 **Model:** EX35, M35, M45 **Engine:** 3.5L V6, 4.5L V8 **Transmission:** All	**A/T Interlock:** • This is an OBD-II self-diagnostic item. • Diagnostic trouble code P1730 is detected when TCM does not receive the proper voltage signal from the sensor and switch. • TCM monitors and compares gear position and conditions of each ATF pressure switch when gear is steady. **Possible Causes:** • Harness or connectors • The solenoid and switch circuit is open or shorted. • Low coast brake solenoid valve • ATF pressure switch 2
DTC: P1730 **2T PCM, MIL: Yes, TCIL: Yes** **Year:** 2009, 2010 **Model:** QX56 **Engine:** 5.6L V8 **Transmission:** All	**Problem in Shift Control System:** With the engine running and in Drive position allow transmission to shift to 5th gear. Shift solenoid A or D stuck OFF, Shift solenoid B stuck ON, Shift Valves A, B or D stuck. **Possible Causes:** • Low or dirty transmission fluid. • Repair the hydraulic system related to shift valves A, B, and D, or replace the transmission
DTC: P1730 **T TCM, TCIL: Yes** **Year:** 2009, 2010 **Model:** FX35, FX50, G37 **Engine:** 3.5L V6, 3.7L V6, 5.0L V8 **Transmission:** All	**Interlock:** • The output sensor detects the deceleration of 12 km/h (7 MPH) or more for 1 second. **Possible Causes:** • Harness or connectors • Solenoid valve circuit is open or shorted. • Input clutch solenoid valve • Direct clutch solenoid valve • High and low reverse clutch solenoid valve • Front brake solenoid valve • Low brake solenoid valve • 2346 brake solenoid valve • Anti-interlock solenoid valve • Each clutch and brake • Hydraulic control circuit
DTC: P1731 **T TCM, TCIL: Yes** **Year:** 2009, 2010 **Model:** EX35, M35 **Engine:** 3.5L V6 **Transmission:** All	**1st Engine Braking:** • ATF pressure switch 2 and solenoid current is monitor and if a pattern is detected having engine braking 1GR other than in the M1 position, a malfunction is detected. **Possible Causes:** • Harness or connectors • Sensor circuit is open or shorted. • Low coast brake solenoid valve • ATF pressure switch 2
DTC: P1731 **T TCM, TCIL: Yes** **Year:** 2009, 2010 **Model:** M45, QX56 **Engine:** 4.5L V8, 5.6L V8 **Transmission:** All	**A/T 1st Engine Braking:** • This is not an OBD-II self-diagnostic item. • Diagnostic trouble code P1731 is detected under the following conditions. - When TCM does not receive the proper voltage signal from the sensor. - When TCM monitors each ATF pressure switch and solenoid monitor value, and detects as irregular when engine brake of 1st gear acts other than at "1" position. **Possible Causes:** • Harness or connectors • The sensor circuit is open or shorted. • Low coast brake solenoid valve • ATF pressure switch 2

DTC	Trouble Code Title, Conditions & Possible Causes
DTC: P1734 **T TCM, TCIL: Yes** **Year:** 2009, 2010 **Model:** FX35, FX50, G37 **Engine:** 3.5L V6, 3.7L V6, 5.0L V8 **Transmission:** All	**7th Gear Incorrect Ratio:** • DTC is set when any inconsistency is recognized in gear ratio. **Possible Causes:** • Input clutch solenoid valve • Direct clutch solenoid valve • High and low reverse clutch solenoid valve • Front brake solenoid valve • Low brake solenoid valve • 2346 brake solenoid valve • Anti-interlock solenoid valve • Each clutch and brake • Output speed sensor • Input speed sensor 1, 2 • Hydraulic control circuit
DTC: P1752 **T TCM, TCIL: Yes** **Year:** 2009, 2010 **Model:** M45, QX56 **Engine:** 4.5L V8, 5.6L V8 **Transmission:** All	**Input Clutch Solenoid Valve:** • This is an OBD-II self-diagnostic item. • Diagnostic trouble code P1752 is detected under the following conditions: - When TCM detects an improper voltage drop when it tries to operate the solenoid valve. - When TCM detects as irregular by comparing target value with monitor value. **Possible Causes:** • Harness or connectors • The solenoid circuit is open or shorted. • Input clutch solenoid valve
DTC: P1752 **T TCM, TCIL: Yes** **Year:** 2009, 2010 **Model:** EX35, M35 **Engine:** 3.5L V6 **Transmission:** All	**Input Clutch Solenoid:** • Normal voltage not applied to solenoid due to cut line, short, or the like. • TCM detects as irregular by comparing target value with monitor value. **Possible Causes:** • Harness or connectors • Solenoid circuit is open or shorted. • Input clutch solenoid valve
DTC: P1757 **T TCM, TCIL: Yes** **Year:** 2009, 2010 **Model:** EX35, M35, M45, QX56 **Engine:** 3.5L V6, 4.5L V8, 5.6L V8 **Transmission:** All	**Front Brake Solenoid Valve:** • This is an OBD-II self-diagnostic item. • Diagnostic trouble code P1757 is detected under the following conditions: - When TCM detects an improper voltage drop when it tries to operate the solenoid valve. - When TCM detects as irregular by comparing target value with monitor value. **Possible Causes:** • Harness or connectors • The solenoid circuit is open or shorted. • Front brake solenoid valve
DTC: P1762 **T TCM, TCIL: Yes** **Year:** 2009, 2010 **Model:** EX35, M35, M45, QX56 **Engine:** 3.5L V6, 4.5L V8, 5.6L V8 **Transmission:** All	**Direct Clutch Solenoid Valve:** • This is an OBD-II self-diagnostic item. • Diagnostic trouble code P1762 will be detected under the following conditions: - When TCM detects an improper voltage drop when it tries to operate the solenoid valve. - When TCM detects as irregular by comparing target value with monitor value. **Possible Causes:** • Harness or connectors • The solenoid circuit is open or shorted. • Direct clutch solenoid valve
DTC: P1767 **T TCM, TCIL: Yes** **Year:** 2009, 2010 **Model:** EX35, M35, M45, QX56 **Engine:** 3.5L V6, 4.5L V8, 5.6L V8 **Transmission:** All	**High & Low Reverse Clutch Solenoid Valve:** • This is an OBD-II self-diagnostic item. • Diagnostic trouble code P1767 will be detected under the following conditions: - When TCM detects an improper voltage drop when it tries to operate the solenoid valve. - When TCM detects as irregular by comparing target value with monitor value. **Possible Causes:** • Harness or connectors • The solenoid circuit is open or shorted. • High and low reverse clutch solenoid valve

DTC	Trouble Code Title, Conditions & Possible Causes
DTC: P1772 **T TCM, TCIL: Yes** **Year:** 2009, 2010 **Model:** EX35, M35, M45, QX56 **Engine:** 3.5L V6, 4.5L V8, 5.6L V8 **Transmission:** All	**Low Coast Brake Solenoid Valve:** • This is an OBD-II self-diagnostic item. • Diagnostic trouble code P1772 will be set when the TCM detects an improper voltage drop when it tries to operate the solenoid valve. **Possible Causes:** • Harness or connectors • The solenoid circuit is open or shorted. • Low coast brake solenoid valve
DTC: P1774 **T TCM, TCIL: Yes** **Year:** 2009, 2010 **Model:** EX35, M35, M45, QX56 **Engine:** 3.5L V6, 4.5L V8, 5.6L V8 **Transmission:** All	**Low Coast Brake Solenoid Valve Function:** • This is an OBD-II self-diagnostic item. • Diagnostic trouble code P1774 will be detected under the following conditions: - TCM detects an improper voltage drop when it tries to operate the solenoid valve. - When TCM detects that actual gear ratio is irregular, and relation between gear position and condition of ATF pressure switch 2 is irregular during depressing accelerator pedal (other than during shift change). - When TCM detects that relation between gear position and condition of ATF pressure switch 2 is irregular during releasing accelerator pedal. (Other than during shift change) **Possible Causes:** • Harness or connectors • The solenoid and switch circuits are open or shorted. • Low coast brake solenoid valve • ATF pressure switch 2
DTC: P1805 **1T ECM** **Year:** 2009, 2010 **Model:** EX35, FX35, FX50, G37, M35, QX56 **Engine:** 3.5L V6, 3.7L V6, 5.0L V8, 5.6L V8 **Transmission:** All	**Brake Switch Signal Malfunction (Includes Hybrid Models):** HYBRID MODELS CAUTION: Hybrid systems use very high-voltage battery systems. Before starting any service work involving the battery system, turn the ignition switch OFF and then remove the service plug from pocket in the trunk. After removing the service plug, wait 10 minutes before touching any of the high-voltage connectors and terminals. • The MIL may not illuminate for this self-diagnosis. • A brake switch signal is not sent to ECM for extremely long time while the vehicle is being driven. **Possible Causes:** • Harness or connectors • Stop lamp switch circuit is open or shorted. • Stop lamp switch
DTC: P1815 **T TCM, TCIL: Yes** **Year:** 2009, 2010 **Model:** EX35, FX35, FX50, G37, M35, M45 **Engine:** 3.5L V6, 3.7L V6, 4.5L V8, 5.0L V8 **Transmission:** All	**Manual Mode Switch Circuit:** • TCM monitors manual mode, non manual mode, up or down switch signal, and detects as irregular when impossible input pattern occurs 2 seconds or more. • Shift up/down signal of paddle shifter continuously remains ON for 60 seconds. **Possible Causes:** • Harness or connectors • These switches circuit is open or shorted. • Manual mode switch (into A/T shift selector) • Manual mode shift-up switch (into A/T shift selector) • Manual mode shift-down switch (into A/T shift selector) • Paddle shifter

OBD II Trouble Code List (P2XXX Codes)

DTC	Trouble Code Title, Conditions & Possible Causes
DTC: P2100 **1T ECM, MIL: Yes** **Year:** 2009, 2010 **Model:** EX35, FX35, FX50, G37, M35, QX56 **Engine:** 3.5L V6, 3.7L V6, 5.0L V8, 5.6L V8 **Transmission:** All	**Throttle Control Motor Relay Circuit is Open (Includes Hybrid Models):** HYBRID MODELS CAUTION: Hybrid systems use very high-voltage battery systems. Before starting any service work involving the battery system, turn the ignition switch OFF and then remove the service plug from pocket in the trunk. After removing the service plug, wait 10 minutes before touching any of the high-voltage connectors and terminals. **NOTE: On V6 and V8, this DTC is for Bank 1.** • These self-diagnoses have the one trip detection logic. • ECM detects that the voltage of power source for throttle control motor is excessively low. **Possible Causes:** • Harness or connectors • Throttle control motor relay circuit is open. • Throttle control motor relay

DTC	Trouble Code Title, Conditions & Possible Causes
DTC: P2101 **1T ECM, MIL: Yes** **Year:** 2009, 2010 **Model:** EX35, FX35, FX50, G37, M35, QX56 **Engine:** 3.5L V6, 3.7L V6, 5.0L V8, 5.6L V8 **Transmission:** All	**Electric Throttle Control Performance (Includes Hybrid Models):** HYBRID MODELS CAUTION: Hybrid systems use very high-voltage battery systems. Before starting any service work involving the battery system, turn the ignition switch OFF and then remove the service plug from pocket in the trunk. After removing the service plug, wait 10 minutes before touching any of the high-voltage connectors and terminals. **NOTE: On V6 and V8, this DTC refers to Bank 1.** **NOTE: If DTC P1233 or P2101 is displayed with DTC P1238, P1290, P2100 or 2119, first perform the trouble diagnosis for DTC P1238, P2119 or P1290, P2100.** • Electric throttle control function does not operate properly. **Possible Causes:** • Harness or connectors • Throttle control motor circuit is open or shorted • Electric throttle control actuator
DTC: P2103 **1T ECM, MIL: Yes** **Year:** 2009, 2010 **Model:** EX35, FX35, FX50, G37, M35, QX56 **Engine:** 3.5L V6, 3.7L V6, 5.0L V8, 5.6L V8 **Transmission:** All	**Throttle Control Motor Relay Circuit is Short (Includes Hybrid Models):** HYBRID MODELS CAUTION: Hybrid systems use very high-voltage battery systems. Before starting any service work involving the battery system, turn the ignition switch OFF and then remove the service plug from pocket in the trunk. After removing the service plug, wait 10 minutes before touching any of the high-voltage connectors and terminals. • ECM detects the throttle control motor relay is stuck ON. **Possible Causes:** • Harness or connectors • Throttle control motor relay circuit is shorted • Throttle control motor relay
DTC: P2118 **1T ECM, MIL: Yes** **Year:** 2009, 2010 **Model:** EX35, FX35, FX50, G37, M35, QX56 **Engine:** 3.5L V6, 3.7L V6, 5.0L V8, 5.6L V8 **Transmission:** All	**Throttle Control Motor Circuit Short (Includes Hybrid Models):** HYBRID MODELS CAUTION: Hybrid systems use very high-voltage battery systems. Before starting any service work involving the battery system, turn the ignition switch OFF and then remove the service plug from pocket in the trunk. After removing the service plug, wait 10 minutes before touching any of the high-voltage connectors and terminals. **NOTE: On V6 and V8, this DTC is for Bank 1.** • ECM detects short in both circuits between ECM and throttle control motor. **Possible Causes:** • Harness or connectors • Throttle control motor circuit is shorted. • Electric throttle control actuator (Throttle control motor)
DTC: P2119 **1T ECM, MIL: Yes** **Year:** 2009, 2010 **Model:** EX35, FX35, FX50, G37, M35, QX56 **Engine:** 3.5L V6, 3.7L V6, 5.0L V8, 5.6L V8 **Transmission:** All	**Electric Throttle Control Actuator (Includes Hybrid Models):** HYBRID MODELS CAUTION: Hybrid systems use very high-voltage battery systems. Before starting any service work involving the battery system, turn the ignition switch OFF and then remove the service plug from pocket in the trunk. After removing the service plug, wait 10 minutes before touching any of the high-voltage connectors and terminals. **NOTE: When the malfunction is detected, ECM enters fail-safe mode and the MIL illuminates.** **NOTE: On V6 and V8, this DTC is for Bank 1.** • Malfunction A: Electric throttle control actuator does not function properly due to the return spring malfunction. ECM controls the electric throttle actuator by regulating the throttle opening around the idle position. The engine speed will not rise more than 2,000 rpm. • Malfunction B: Throttle valve opening angle in fail-safe mode is not in specified range. ECM controls the electric throttle control actuator by regulating the throttle opening to 20 degrees or less. • Malfunction C: ECM detects the throttle valve is stuck open. While the vehicle is driving, it slows down gradually by fuel cut. After the vehicle stops, the engine stalls. • The engine can restart in N or P position (CVT), neutral (M/T), and engine speed will not exceed 1,000 rpm or more. **Possible Causes:** • Electric throttle control actuator
DTC: P2122 **1T ECM, MIL: Yes** **Year:** 2009, 2010 **Model:** EX35, FX35, FX50, G37, M35, QX56 **Engine:** 3.5L V6, 3.7L V6, 5.0L V8, 5.6L V8 **Transmission:** All	**Accelerator Pedal Position Sensor 1 Circuit Low Input:** **NOTE: If DTC P2122 or P2123 is displayed with DTC P0643, first perform the trouble diagnosis for DTC P0643.** • An excessively low voltage from the APP sensor 1 is sent to ECM. **Possible Causes:** • Harness or connectors • APP sensor 1 circuit is open or shorted. • Accelerator pedal position sensor 1 (APP sensor 1)

DTC	Trouble Code Title, Conditions & Possible Causes

DTC: P2123
1T ECM, MIL: Yes
Year: 2009, 2010
Model: EX35, FX35, FX50, G37, M35, QX56
Engine: 3.5L V6, 3.7L V6, 5.0L V8, 5.6L V8
Transmission: All
	Accelerator Pedal Position Sensor 1 Circuit High Input:
	NOTE: If DTC P2122 or P2123 is displayed with DTC P0643, first perform the trouble diagnosis for DTC P0643.
	  • An excessively high voltage from the APP sensor 1 is sent to ECM.
	  • When the malfunction is detected, ECM enters fail-safe mode and the MIL illuminates.
	Possible Causes:
	  • Harness or connectors
	  • APP sensor 1 circuit is open or shorted.
	  • Accelerator pedal position sensor 1 (APP sensor 1)

DTC: P2127
1T ECM, MIL: Yes
Year: 2009, 2010
Model: EX35, FX35, FX50, G37, M35, QX56
Engine: 3.5L V6, 3.7L V6, 5.0L V8, 5.6L V8
Transmission: All
	Accelerator Pedal Position Sensor 2 Circuit Low Input:
	  • An excessively low voltage from the APP sensor 2 is sent to ECM.
	  • When the malfunction is detected, ECM enters fail-safe mode and the MIL illuminates.
	Possible Causes:
	  • Harness or connectors
	  • APP sensor 2 circuit is open or shorted.
	  • Crankshaft position sensor (POS) circuit is shorted.
	  • Refrigerant pressure sensor circuit is shorted.
	  • EVAP control system pressure sensor circuit is shorted.
	  • Battery current sensor circuit is shorted.
	  • Accelerator pedal position sensor (APP sensor 2)
	  • Electric throttle control actuator (TP sensor)
	  • Crankshaft position sensor (POS)
	  • Refrigerant pressure sensor
	  • EVAP control system pressure sensor
	  • Battery current sensor

DTC: P2128
1T ECM, MIL: Yes
Year: 2009, 2010
Model: EX35, FX35, FX50, G37, M35, QX56
Engine: 3.5L V6, 3.7L V6, 5.0L V8, 5.6L V8
Transmission: All
	Accelerator Pedal Position Sensor 2 Circuit High Input:
	  • An excessively high voltage from the APP sensor 2 is sent to ECM.
	  • When the malfunction is detected, ECM enters fail-safe mode and the MIL illuminates.
	Possible Causes:
	  • Harness or connectors
	  • APP sensor 2 circuit is open or shorted.
	  • Crankshaft position sensor (POS) circuit is shorted.
	  • Refrigerant pressure sensor circuit is shorted.
	  • EVAP control system pressure sensor circuit is shorted.
	  • Battery current sensor circuit is shorted.
	  • Accelerator pedal position sensor (APP sensor 2)
	  • Electric throttle control actuator (TP sensor)
	  • Crankshaft position sensor (POS)
	  • Refrigerant pressure sensor
	  • EVAP control system pressure sensor
	  • Battery current sensor

DTC: P2132
1T ECM, MIL: Yes
Year: 2009, 2010
Model: EX35, FX35, FX50, G37, M35
Engine: 3.5L V6, 3.7L V6, 5.0L V8
Transmission: All
	Throttle Position Sensor 1 (Bank 2) Circuit Low Input:
	An excessively low voltage from the TP sensor 1 is sent to ECM.
	Possible Causes:
	  • Harness or connectors
	  • TP sensor 1 circuit is open or shorted.
	  • Electric throttle control actuator (TP sensor 1)

DTC: P2133
1T ECM, MIL: Yes
Year: 2009, 2010
Model: EX35, FX35, FX50, G37, M35
Engine: 3.5L V6, 3.7L V6, 5.0L V8
Transmission: All
	Throttle Position Sensor 1 (Bank 2) Circuit High Input:
	An excessively high voltage from the TP sensor 1 is sent to ECM.
	Possible Causes:
	  • Harness or connectors
	  • TP sensor 1 circuit is open or shorted.
	  • Electric throttle control actuator (TP sensor 1)

DTC	Trouble Code Title, Conditions & Possible Causes
DTC: P2135 **1T ECM, MIL: Yes** **Year:** 2009, 2010 **Model:** EX35, FX35, FX50, G37, M35, QX56 **Engine:** 3.5L V6, 3.7L V6, 5.0L V8, 5.6L V8 **Transmission:** All	**Throttle Position Sensor Circuit Range/Performance (Includes Hybrid Models):** HYBRID MODELS CAUTION: Hybrid systems use very high-voltage battery systems. Before starting any service work involving the battery system, turn the ignition switch OFF and then remove the service plug from pocket in the trunk. After removing the service plug, wait 10 minutes before touching any of the high-voltage connectors and terminals. **NOTE: If DTC P2135 is displayed with DTC P0643, first perform the trouble diagnosis for DTC P0643.** **NOTE: On V6 and V8, this DTC refers to Bank 1.** • Rationally incorrect voltage is sent to ECM compared with the signals from TP sensor 1 and TP sensor 2. • When the malfunction is detected, the ECM enters fail-safe mode and the MIL illuminates. **Possible Causes:** • Harness or connectors • TP sensor 1 or 2 circuit is open or shorted. • Electric throttle control actuator (TP sensor 1 or 2) • Accelerator pedal position sensor (APP sensor 2)
DTC: P2138 **1T ECM, MIL: Yes** **Year:** 2009, 2010 **Model:** EX35, FX35, FX50, G37, M35, QX56 **Engine:** 3.5L V6, 3.7L V6, 5.0L V8, 5.6L V8 **Transmission:** All	**Accelerator Pedal Position Sensor Circuit Range/Performance:** **NOTE: If DTC P2138 is displayed with DTC P0643, first perform the trouble diagnosis for DTC P0643.** • Rationally incorrect voltage is sent to ECM compared with the signals from APP sensor 1 and APP sensor 2. • When the malfunction is detected, ECM enters fail-safe mode and the MIL illuminates. **Possible Causes:** • Harness or connector • APP sensor 1 or 2 circuit is open or shorted. • Crankshaft position sensor (POS) circuit is shorted. • Refrigerant pressure sensor circuit is shorted. • EVAP control system sensor circuit is shorted. • Battery current sensor circuit is shorted. • Accelerator pedal position sensor (APP sensor 1 or 2) • Electric throttle control actuator (TP sensor) • Crankshaft position sensor (POS) • Camshaft position sensor (PHASE) (bank 2) • Exhaust valve timing control position sensor (bank 2) • Refrigerant pressure sensor • EVAP control system pressure sensor • Battery current sensor
DTC: P2713 **T TCM, TCIL: Yes** **Year:** 2009, 2010 **Model:** FX35, FX50, G37 **Engine:** 3.5L V6, 3.7L V6, 5.0L V8 **Transmission:** All	**Pressure Control Solenoid D:** The high and low reverse clutch solenoid valve monitor value is 0.4 A or less when the high and low reverse clutch solenoid valve command value is more than 0.75 A. **Possible Causes:** • Harness or connectors • Solenoid valve circuit is open or shorted. • High and low reverse clutch solenoid valve
DTC: P2722 **T TCM, TCIL: Yes** **Year:** 2009, 2010 **Model:** FX35, FX50, G37 **Engine:** 3.5L V6, 3.7L V6, 5.0L V8 **Transmission:** All	**Pressure Control Solenoid E:** The low brake solenoid valve monitor value is 0.4 A or less when the low brake solenoid valve command value is more than 0.75 A. **Possible Causes:** • Harness or connectors • Solenoid valve circuit is open or shorted. • Low brake solenoid valve
DTC: P2731 **T TCM, TCIL: Yes** **Year:** 2009, 2010 **Model:** FX35, FX50, G37 **Engine:** 3.5L V6, 3.7L V6, 5.0L V8 **Transmission:** All	**Pressure Control Solenoid F:** The 2346 brake solenoid valve monitor value is 0.4 A or less when the 2346 brake solenoid valve command value is more than 0.75 A. **Possible Causes:** • Harness or connectors • Solenoid valve circuit is open or shorted. • 2346 brake solenoid valve
DTC: P2807 **T TCM, TCIL: Yes** **Year:** 2009, 2010 **Model:** FX35, FX50, G37 **Engine:** 3.5L V6, 3.7L V6, 5.0L V8 **Transmission:** All	**Pressure Control Solenoid G:** The direct clutch solenoid valve monitor value is 0.4 A or less when the direct clutch solenoid valve command value is more than 0.75 A. **Possible Causes:** • Harness or connectors • Solenoid valve circuit is open or shorted. • Direct clutch solenoid valve

DTC	Trouble Code Title, Conditions & Possible Causes
DTC: P2A00 **1T ECM, MIL: Yes** **Year:** 2009, 2010 **Model:** EX35, FX35, FX50, G37, M35, QX56 **Engine:** 3.5L V6, 3.7L V6, 5.0L V8, 5.6L V8 **Transmission:** All	**Air Fuel (A/F) Sensor 1 Circuit Range/Performance (Includes Hybrid Models):** HYBRID MODELS CAUTION: Hybrid systems use very high-voltage battery systems. Before starting any service work involving the battery system, turn the ignition switch OFF and then remove the service plug from pocket in the trunk. After removing the service plug, wait 10 minutes before touching any of the high-voltage connectors and terminals. **NOTE: On V6 and V8 engines, this applies to Bank 1.** • To judge the malfunction, the A/F signal computed by ECM from the A/F sensor 1 signal is monitored not to be shifted to LEAN side or RICH side. • The output voltage computed by ECM from the A/F sensor 1 signal is shifted to the lean side for a specified period. • The A/F signal computed by ECM from the A/F sensor 1 signal is shifted to the rich side for a specified period. **Possible Causes:** • Air fuel ratio (A/F) sensor 1 • Air fuel ratio (A/F) sensor 1 heater • Fuel pressure • Fuel injector • Intake air leaks
DTC: P2A03 **1T ECM, MIL: Yes** **Year:** 2009, 2010 **Model:** EX35, FX35, FX50, G37, M35, QX56 **Engine:** 3.5L V6, 3.7L V6, 5.0L V8, 5.6L V8 **Transmission:** All	**Air Fuel (A/F) Ratio Sensor 1 Circuit Range/Performance:** **NOTE: On V6 and V8 engines, this applies to Bank 2.** • The output voltage computed by ECM from the A/F sensor 1 signal is shifted to the lean side for a specified period. • The A/F signal computed by ECM from the A/F sensor 1 signal is shifted to the rich side for a specified period. • To judge the malfunction, the A/F signal computed by ECM from the A/F sensor 1 signal is monitored so it will not shift to LEAN side or RICH side. **Possible Causes:** • Air fuel ratio (A/F) sensor 1 • Air fuel ratio (A/F) sensor 1 heater • Fuel pressure • Fuel injector • Intake air leaks

OBD II Trouble Code List (U0XXX Codes)

DTC	Trouble Code Title, Conditions & Possible Causes
DTC: U0101 **1T ECM, MIL: Yes** **Year:** 2009, 2010 **Model:** EX35, FX35, FX50, G37, M35, M45 **Engine:** 3.5L V6, 3.7L V6, 4.5L V8, 5.0L V8 **Transmission:** All	**Lost communication with TCM:** When ECM is not transmitting or receiving CAN communication signal of OBD (emission-related diagnosis) with TCM for 2 seconds or more. **Possible Causes:** • CAN communication line between TCM and ECM • CAN communication line is open or shorted
DTC: U0113 **T ECM, MIL: Yes** **Year:** 2009, 2010 **Model:** FX50, G37 **Engine:** 3.7L V6, 5.0L V8 **Transmission:** All	**Lost Communication with Variable Valve Event & Lift (VVEL) Control Module:** **NOTE: If DTC U0113 or U1003 is displayed with DTC P0607, first perform the trouble diagnosis for DTC P0607.** • CAN communication signal of OBD (emission related diagnosis) is not received VVEL control module and ECM for 2 seconds or more. **Possible Causes:** • Harness or connectors • VVEL CAN communication line is open or shorted • ECM • VVEL control module
DTC: U0164 **T ECM, MIL: Yes** **Year:** 2009, 2010 **Model:** EX35, FX35, G37, M35 **Engine:** 3.5L V6, 3.7L V6 **Transmission:** All	**Lost Communication with Unified Meter & A/C Amp or Combination Meter:** • When ECM is not transmitting or receiving CAN communication signal of OBD (emission related diagnosis) with unified meter and A/C amp. or the combination meter for 2 seconds or more. **Possible Causes:** • CAN communication line between Unified meter and A/C amp. and ECM • CAN communication line open or shorted
DTC: U0300 **T TCM, TCIL: Yes** **Year:** 2009, 2010 **Model:** G37 **Engine:** 3.7L V6 **Transmission:** All	**Internal Control Module Software Incompatibility:** When the amount of data transmitted from each control unit is smaller than the specified amount. **Possible Causes:** • Control units other than TCM.

OBD II Trouble Code List (U1XXX Codes)

DTC	Trouble Code Title, Conditions & Possible Causes
DTC: U1000 **1T TCM, TCIL: Yes** **Year:** 2009, 2010 **Model:** EX35, FX35, FX50, G37, M35, M45, QX56 **Engine:** 3.5L V6, 3.7L V6, 4.5L V8, 5.0L V8, 5.6L V8 **Transmission:** All	**CAN Communication Line:** • This is an OBD-II self-diagnostic item. • Diagnostic trouble code U1000 is detected when ECM or TCM cannot communicate to other control units for 2 seconds or more. **Possible Causes:** • Harness or connectors • CAN communication line is open or shorted. • TCM
DTC: U1001 **1T ECM, MIL: Yes, TCIL: Yes** **Year:** 2009, 2010 **Model:** EX35, FX35, FX50, G37, M35, M45, QX56 **Engine:** 3.5L V6, 3.7L V6, 4.5L V8, 5.0L V8, 5.6L V8 **Transmission:** All	**CAN Communication Line (Includes Hybrid models):** HYBRID CAUTION: Hybrid systems use very high-voltage battery systems. Before starting any service work involving the battery system, turn the ignition switch OFF and then remove the service plug from pocket in the trunk. After removing the service plug, wait 10 minutes before touching any of the high-voltage connectors and terminals. • When ECM is not transmitting or receiving CAN communication signal other than OBD (emission related diagnosis) for 2 seconds or more. **Possible Causes:** • Harness or connectors • CAN communication line is open or shorted
DTC: U1003 **T ECM, MIL: Yes** **Year:** 2009, 2010 **Model:** FX50, G37 **Engine:** 3.7L V6, 5.0L V8 **Transmission:** All	**Lost Communication with Variable Valve Event & Lift (VVEL) Control Module:** **NOTE: If DTC U0113 or U1003 is displayed with DTC P0607, first perform the trouble diagnosis for DTC P0607.** • CAN communication signal other than OBD (emission related diagnosis) is not received between VVEL control module and ECM for 2 seconds or more. **Possible Causes:** • Harness or connectors • VVEL CAN communication line is open or shorted • ECM • VVEL control module
DTC: U1024 **T ECM, MIL: Yes** **Year:** 2009, 2010 **Model:** FX50, G37 **Engine:** 3.7L V6, 5.0L V8 **Transmission:** All	**Variable Valve Event & Lift (VVEL) CAN Communication:** • When VVEL control module cannot transmit/receive can communication signal from ECM. • When detecting error during the initial diagnosis of CAN controller of VVEL control module. **Possible Causes:** • Harness or connectors • CAN communication line is open or shorted • ECM • VVEL control module

SPECIFICATIONS AND MAINTENANCE CHARTS

ENGINE AND VEHICLE IDENTIFICATION

	Engine							Model Year	
Code ①	Liters (cc)	Cu. In.	Cyl.	Fuel Sys.	Engine Type	Eng. Mfg.		Code ②	Year
VQ35HR	3.5 (3498)	213	6	MFI	DOHC	Nissan		9	2009

MFI: Multi-port Fuel Injection

DOHC: Double Overhead Camshaft

① The Engine Code is stamped on the engine block near the starter.

② 10th position of the Vehicle Identification Number (VIN)

37663_350Z_C0001

GENERAL ENGINE SPECIFICATIONS

Year	Model	Engine Displacement Liters	Engine Series (ID/VIN)	Net Horsepower @ rpm	Net Torque @ rpm (ft. lbs.)	Bore x Stroke (in.)	Compression Ratio	Oil Pressure @ rpm
2009	350Z	3.5	VQ35HR	306@6000	268@4800	3.76X3.20	10.0:1	43@2000

37663_350Z_C0002

ENGINE TUNE-UP SPECIFICATIONS

Year	Model	Engine Displacement Liters	Engine ID/VIN	Spark Plug Gap (in.)	Ignition Timing (deg.) MT	AT	Fuel Pump (psi) ①	Idle Speed (rpm) MT	AT ②	Valve Clearance (in.) Intake ③	Exhaust ③
2009	350Z	3.5	VQ35HR	0.043	16B	16B	51	600-700	600-700	0.010-0.013	0.011-0.015

NOTE: The Vehicle Emission Control Information label often reflects specification changes made during production.

The label figures must be used if they differ from those in this chart.

NA: Not Available

B: Before top dead center

① System pressure at idle with vacuum hose connected; should increase to 43 psi when disconnected

② Automatic transmission in neutral

③ Engine cold

37663_350Z_C0003

CAPACITIES

Year	Model	Engine ID/VIN	Engine Displacement Liters	Engine Oil with Filter (qts.)	Transmission (pts.)		Drive Axle Rear (pts.)	Fuel Tank (gal.)	Cooling System (qts.)
					Man	Auto.			
2009	350Z	VQ35HR	3.5	5.0	6.25	21.3	3.0	20.0	9.25

NOTE: All capacities are approximate. Add fluid gradually and check to be sure a proper fluid level is obtained.

37663_350Z_C0004

FLUID SPECIFICATIONS

Year	Model	Engine Displacement Liters	Engine ID/VIN	Engine Oil	Auto. Trans.	Manual Trans.	Power Steering Fluid	Brake Master Cylinder
2009	350Z	3.5	VQ35HR	5W-30	①	②	③	DOT 3

DOT: Department Of Transportation

① Nissan Matic J ATF

② Nissan (MTF) HQ MULTI 75W-85 or API GL-4, Viscosity 75W-85 or 75W-90

③ Nissan PSF or equivalent. Canada Nissan transmission fluid or DEXRON VI type ATF may be used

37663_350Z_C0013

VALVE SPECIFICATIONS

Year	Engine ID/VIN	Engine Displacement Liters	Seat Angle (deg.)	Face Angle (deg.)	Spring Test Pressure (lbs. @ in.)	Spring Installed Height (in.)	Stem-to-Guide Clearance (in.)		Stem Diameter (in.)	
							Intake	Exhaust	Intake	Exhaust
2009	VQ35HR	3.5	45.15-45.45	45	37-42@ 1.4567	1.466	0.0008-0.0021	0.0012-0.0022	0.2348-0.2354	0.2344-0.2350

37663_350Z_C0005

CAMSHAFT AND BEARING SPECIFICATIONS CHART

All measurements are given in inches.

Year	Engine Displ. Liters	Engine ID/VIN	Journal Dia.	Brg. Oil Clearance	Shaft End-play	Runout	Journal-to-Bore Clearance	Lobe Lift	
								Intake	Exhaust
2009	3.5	VQ35HR	①	②	0.0045-0.0074	0.0008	NA	1.8061-1.8132	1.8061-1.8136

NA: Not Available

① Front No. 1: 1.0221- 1.0218

   No. 2, 3, 4: 0.9230- 0.9238

② Front No. 1: 0.0018- 0.0034

   No. 2, 3, 4: 0.0014- 0.0030

③ Non active fuel management cylinders: 0.2900

   Active fuel management cylinders: 0.2950

37663_350Z_C0014

CRANKSHAFT AND CONNECTING ROD SPECIFICATIONS

All measurements are given in inches.

Year	Engine Displacement Liters	Engine ID/VIN	Crankshaft				Connecting Rod		
			Main Brg. Journal Dia.	Main Brg. Oil Clearance	Shaft End-play	Thrust on No.	Journal Diameter	Oil Clearance	Side Clearance
2009	3.5	VQ35HR	2.5571-2.5581*	0.0014-0.0018	0.0039-0.0098	3	2.1242-2.1250*	0.0016-0.0021	0.0079-0.0138

* Based upon grade

37663_350Z_C0008

PISTON AND RING SPECIFICATIONS

All measurements are given in inches.

Year	Engine Displ. Liters	Engine ID/VIN	Piston Clearance	Ring Gap			Ring Side Clearance		
				Top Compression	Bottom Compression	Oil Control	Top Compression	Bottom Compression	Oil Control
2009	3.5	VQ35HR	0.0004-0.0012	0.0091-0.0130	0.0130-0.0189	0.0067-0.0185	0.0016-0.0031	0.0012-0.0028	0.0022-0.0061

37663_350Z_C0007

TORQUE SPECIFICATIONS

All readings in ft. lbs.

Year	Engine Displacement Liters	Engine ID/VIN	Cylinder Head Bolts	Main Bearing Bolts	Rod Bearing Bolts	Crankshaft Damper Bolts	Flywheel Bolts	Manifold Intake	Manifold Exhaust	Spark Plugs	Oil Drain Plug
2009	3.5	VQ35HR	①	②	③	④	65	⑤	⑥	18	25

① Step 1: 77 ft. lbs.

   Step 2: Loosen bolts completely

   Step 3: 30 ft. lbs.

   Step 4: Tighten an additional 95 degrees

   Step 5: Tighten an additional 95 degrees

② Step 1: 18 ft. lbs.

   Step 2: 26 ft.lbs.

   Step 3: Tighten an additional 90 degrees

③ Step 1: 21 ft. lbs.

   Step 2: Loosen bolts completely

   Step 3: 15 ft.lbs.

   Step 4: Tighten an additional 90 degrees

④ Step 1: 33 ft. lbs.

   Step 2: Tighten an additional 90 degrees

⑤ Step 1: 5 ft. lbs.

   Step 2: 20 ft. lbs.

⑥ Step 1: 11 ft. lbs. in 2 steps

37663_350Z_C0006

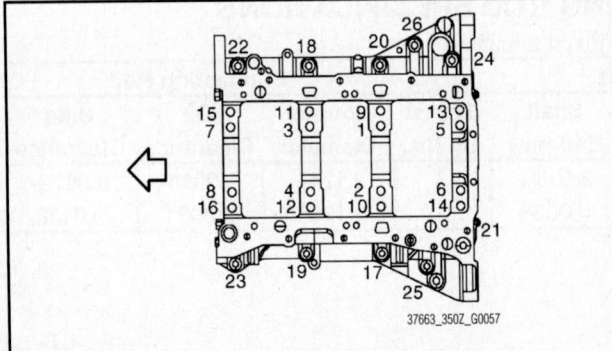

37663_350Z_G0057

Fig. 1 Main bearing (lower cylinder block bolts) tightening sequence

WHEEL ALIGNMENT

Year	Model		Caster Range (+/-Deg.)	Caster Preferred Setting (Deg.)	Camber Range (+/-Deg.)	Camber Preferred Setting (Deg.)	Toe-in (in.)
2009	350Z	F	0.75	+8.17	0.75 / 0.40	-0.58	0.04 +/- 0.04
		R	—	—	0.50	-1.58	0.075 +/- 0.03

37663_350Z_C0009

TIRE, WHEEL AND BALL JOINT SPECIFICATIONS

| Year | Model | OEM Tires | | Tire Pressures (psi) | | Wheel Size | Lug Nut Torque (ft. lbs.) |
		Standard	Optional	Front	Rear		
2009	Front	225/45R18	245/45R18	35	35	8JJ/8.5JJ	80
	Rear	245/40R18	265/35R19	35	35	9JJ/10JJ	80

OEM: Original Equipment Manufacturer

PSI: Pounds Per Square Inch

37663_350Z_C0010

BRAKE SPECIFICATIONS
All measurements in inches unless noted

| Year | Model | | Brake Disc | | | Minimum Lining Thickness | | Brake Caliper | |
			Original Thickness	Minimum Thickness	Maximum Run-out	Front	Rear	Bracket Bolts (ft. lbs.)	Mounting Bolts (ft. lbs.)
2009	350Z ①	F	1.102	1.024	0.001	0.079	0.079	113-114	20
		R	0.630	0.551	0.008	0.079	0.079	62	32
	350Z ②	F	1.181	1.118	0.002	0.079	0.079	—	113
		R	0.866	0.795	0.002	0.079	0.079	53-71	—

NA: Not Available

① Except Brembo

② Brembo

37663_350Z_C0011

SCHEDULED MAINTENANCE INTERVALS
NISSAN—350Z

TO BE SERVICED	TYPE OF SERVICE	VEHICLE MILEAGE INTERVAL (x1000)												
		7.5	15	22.5	30	37.5	45	52.5	60	67.5	75	82.5	90	97.5
Engine oil & filter	R	✓	✓	✓	✓	✓	✓	✓	✓	✓	✓	✓	✓	✓
Brake lines & cables	S/I		✓		✓		✓		✓		✓		✓	
Brake pads & discs	S/I		✓		✓		✓		✓		✓		✓	
Driveshaft boots	S/I		✓		✓		✓		✓		✓		✓	
Exhaust system	S/I				✓				✓				✓	
Transmission or transaxle fluid	S/I		✓		✓		✓		✓		✓		✓	
Air cleaner filter	R				✓				✓				✓	
Spark plugs (except platinum)	R				✓				✓				✓	
Spark plugs (platinum tip)	R								✓					
Steering gear & linkage, axle & suspension parts	S/I				✓				✓				✓	
Engine coolant	R								✓					
Drive belts	S/I								✓					
Fuel lines	S/I								✓					
Vapor lines	S/I								✓					

R: Replace   S/I: Service or Inspect

FREQUENT OPERATION MAINTENANCE (SEVERE SERVICE)

If a vehicle is operated under any of the following conditions it is considered severe service:

- Extremely dusty areas.
- 50% or more of the vehicle operation is in 32°C (90°F) or higher temperatures, or constant operation in temperatures below 0°C (32°F).
- Prolonged idling (vehicle operation in stop and go traffic).
- Frequent short running periods (engine does not warm to normal operating temperatures).
- Police, taxi, delivery usage or trailer towing usage.

Oil & oil filter: change every 3750 miles.

Brake pads & discs: service or inspect every 7500 miles.

Driveshaft boots: service or inspect every 7500 miles.

Exhaust system: service or inspect every 7500 miles.

Steering gear & linkage, axle & suspension parts: service or inspect every 7500 miles.

Steering linkage ball joints & front suspension ball joints: service or inspect every 7500 miles.

Air cleaner filter: service or inspect every 15,000 miles.

37663_350Z_C0012

BRAKES — INFORMATION AND PRECAUTIONS

ANTI-LOCK SYSTEMS

- Certain components within the ABS system are not intended to be serviced or repaired individually.

- Do not use rubber hoses or other parts not specifically specified for and ABS system. When using repair kits, replace all parts included in the kit. Partial or incorrect repair may lead to functional problems and require the replacement of components.

- Lubricate rubber parts with clean, fresh brake fluid to ease assembly. Do not use shop air to clean parts; damage to rubber components may result.

- Use only DOT 3 brake fluid from an unopened container.

- If any hydraulic component or line is removed or replaced, it may be necessary to bleed the entire system.

- A clean repair area is essential. Always clean the reservoir and cap thoroughly before removing the cap. The slightest amount of dirt in the fluid may plug an orifice and impair the system function. Perform repairs after components have been thoroughly cleaned; use only denatured alcohol to clean components. Do not allow ABS components to come into contact with any substance containing mineral oil; this includes used shop rags.

- The Anti-Lock control unit is a microprocessor similar to other computer units in the vehicle. Ensure that the ignition switch is **OFF** before removing or installing controller harnesses. Avoid static electricity discharge at or near the controller.

- If any arc welding is to be done on the vehicle, the control unit should be unplugged before welding operations begin.

DISC AND DRUM SYSTEMS

> **❊❊ CAUTION**
>
> **Dust and dirt accumulating on brake parts during normal use may contain asbestos fibers from production or aftermarket brake linings. Breathing excessive concentrations of asbestos fibers can cause serious bodily harm. Exercise care when servicing brake parts. Do not sand or grind brake lining unless equipment used is designed to contain the dust residue. Do not clean brake parts with compressed air or by dry brushing. Cleaning should be done by dampening the brake components with a fine mist of water, then wiping the brake components clean with a dampened cloth. Dispose of cloth and all residue containing asbestos fibers in an impermeable container with the appropriate label. Follow practices prescribed by the Occupational Safety and Health Administration (OSHA) and the Environmental Protection Agency (EPA) for the handling, processing, and disposing of dust or debris that may contain asbestos fibers.**

BRAKES — BLEEDING THE BRAKE SYSTEM

BLEEDING PROCEDURE

BLEEDING PROCEDURE

> **❊❊ WARNING**
>
> **Be careful not to splash brake fluid on painted areas; it may cause paint damage. If brake fluid is splashed on painted areas, wash it away with water immediately. All hoses must be free from excessive bending, twisting and pulling.**

> **❊❊ WARNING**
>
> **Pay attention to the following:**

- Carefully monitor brake fluid level at master cylinder during bleeding operation.
- If master cylinder is suspected to have air inside, bleed air from master cylinder first.
- Fill reservoir with new brake fluid DOT 3. Make sure it is full at all times while bleeding air out of system.
- Place a container under master cylinder to avoid spillage of brake fluid.
- For models with ABS, turn ignition switch OFF and disconnect ABS actuator connector or battery cable.

1. Bleed the air from the brake system in the following order:
   - Right rear brake
   - Left front brake
   - Left rear brake
   - Right front brake
2. Connect a transparent vinyl tube to air bleeder valve.
3. Fully depress brake pedal several times.
4. With brake pedal depressed, open air bleeder valve to release air.
5. Close air bleeder valve.
6. Release brake pedal slowly.
7. Repeat steps 2 through 5 until clear brake fluid comes out of air bleeder valve.
8. Tighten air bleeder valve to 61–78 inch lbs. (7–9 Nm).

BLEEDING THE ABS SYSTEM

➡**Carefully monitor brake fluid level at master cylinder during bleeding operation. Fill reservoir with new brake fluid. Make sure it is full at all times while bleeding air out of system. Place a container under master cylinder to avoid spillage of brake fluid. Do not loosen the connecting portion of the actuator during air bleeding.**

1. Disconnect battery negative terminal.
2. Connect a transparent vinyl tube and container to air bleeder valve.
3. Fully depress brake pedal several times.
4. With brake pedal depressed, open air bleeder valve to release air.
5. Close air bleeder valve.
6. Release brake pedal slowly.
7. Tighten air bleeder valve to 69 inch lbs. (8 Nm).
8. Repeat steps 3 through 6 until no more air bubbles come out of air bleeder valve.
9. Bleed the brake hydraulic system air bleeder valves in the following order:
   - Right rear brake
   - Left front brake
   - Left rear brake
   - Right front brake

WHEEL SPEED SENSORS

REMOVAL & INSTALLATION

See Figure 2.

❋❋ WARNING

Be careful not to damage sensor edge and sensor rotor teeth. When removing the front or rear wheel hub assembly, first remove the ABS wheel sensor from the assembly. Failure to do so may result in damage to the sensor wires making the sensor inoperative.

❋❋ WARNING

Pull out the wheel sensor, being careful to turn it as little as possible. Do not pull on the wheel sensor harness. Installation should be performed while paying attention to the following and then tighten mounting bolts and nuts to the specified torque.

❋❋ WARNING

Check if foreign objects such as iron fragments are adhered to the pick-up part of the sensor or to the inside of the hole for the wheel sensor, or if a foreign object is caught in the mating surface of the sensor rotor. If something wrong is found, fix it and then install the wheel sensor.

1. Before servicing the vehicle, refer to the Precautions Section.
2. Remove appropriate wheel.
3. Disconnect wheel sensor harness connector and remove harness wire from attachment points.
4. Remove wheel sensor bolt and wheel sensor.

To install:

5. To install, reverse removal procedure.
6. When installing front sensor, be sure to press rubber grommets in until they lock at the three locations (2 at shock absorbers and 1 at body panel). When installed, harness must not be twisted. White line on harness (shaded part) must be visible from front.

WHEEL SPEED SENSOR RINGS (TOOTHED RINGS)

REMOVAL & INSTALLATION

The sensor rotor cannot be disassembled. To replace the sensor rotor, replace the hub bearing assembly.

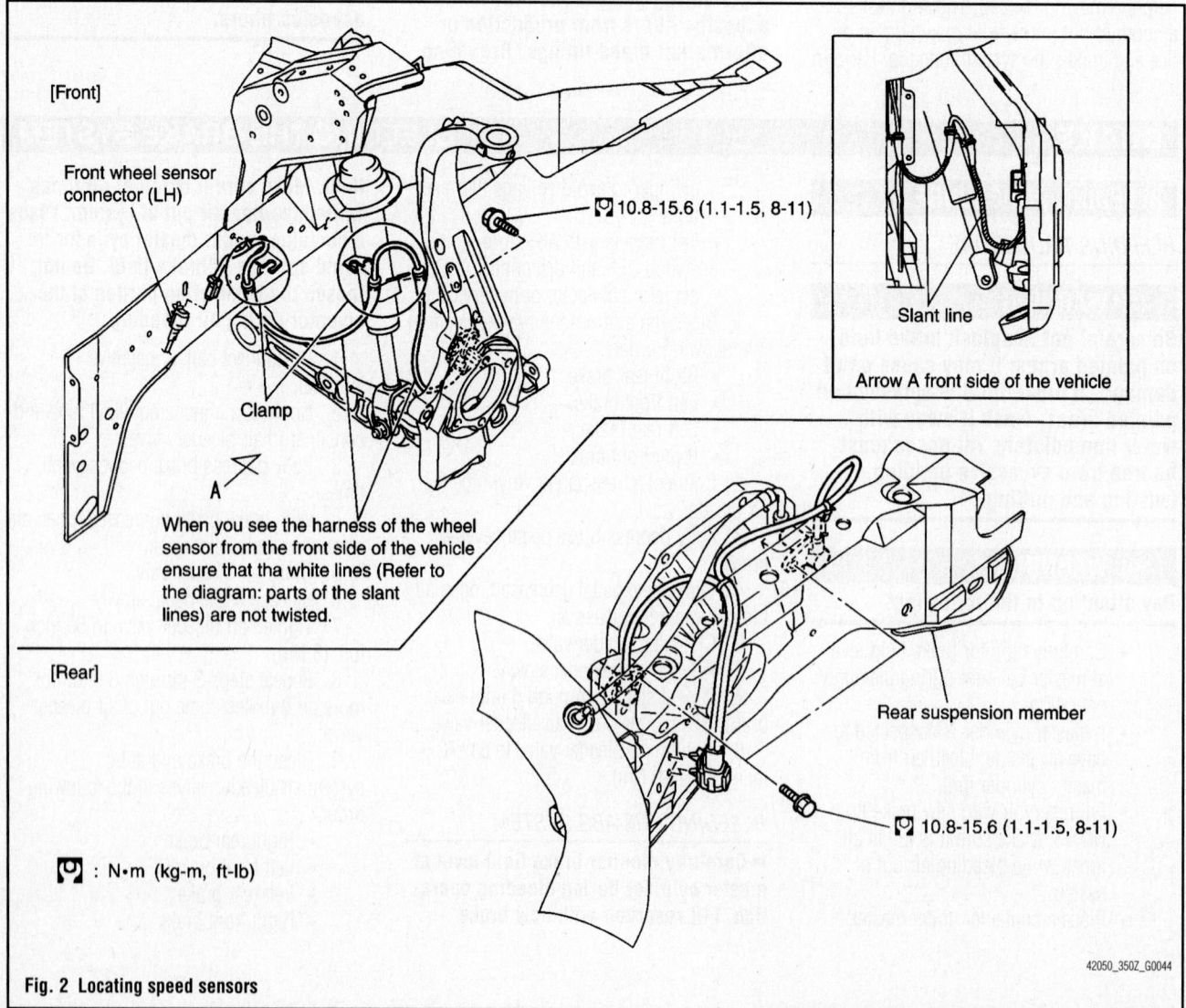

[Front]

Front wheel sensor connector (LH)

10.8-15.6 (1.1-1.5, 8-11)

Clamp

A

When you see the harness of the wheel sensor from the front side of the vehicle ensure that tha white lines (Refer to the diagram: parts of the slant lines) are not twisted.

Slant line

Arrow A front side of the vehicle

Rear suspension member

10.8-15.6 (1.1-1.5, 8-11)

[Rear]

: N•m (kg-m, ft-lb)

Fig. 2 Locating speed sensors

42050_350Z_G0044

BRAKES

BRAKE CALIPER

REMOVAL & INSTALLATION

With Brembo® Caliper

1. Before servicing the vehicle, refer to the Precautions Section.
2. Remove tires from vehicle.
3. Fasten disc rotor using wheel nut.
4. Drain brake fluid gradually (from bleed valve while depressing brake pedal).
5. Remove union bolt, and then remove brake hose from caliper assembly.
6. Remove torque member mounting bolts (from torque member), and remove caliper assembly (from vehicle with a power tool).

⁕⁕ WARNING

Do not drop brake pads.

7. Remove disc rotor.

➡**Put matching marks on both disc rotor and wheel hub when removing disc rotor.**

To install:

8. Install disc rotor.

➡**Align the matching marks of disc rotor and wheel hub, which were marked at the time of removal when reusing disc rotor.**

9. Install caliper assembly to vehicle, and tighten torque member mounting bolts to 113 ft. lbs. (154 Nm).

➡**Before installing torque member to vehicle, wipe oil and grease on washer seats on steering knuckle and mounting surface of torque member.**

10. Install a projection of brake hose metal fitting by aligning with protrusions on cylinder body, and tighten union bolt to 13 ft. lbs. (18 Nm).

⁕⁕ WARNING

Refill with new brake fluid "DOT 3". Never reuse drained brake fluid.

11. Bleed the brake system and top off the master cylinder as necessary.
12. Install tires to vehicle.

Without Brembo® Caliper

See Figure 3.

1. Before servicing the vehicle, refer to the Precautions Section.
2. Remove or disconnect the following:
   - Front wheels
   - Brake fluid hose
   - Sliding pin bolts
   - Caliper assembly from the vehicle

To install:

3. Use a large C-clamp to press the caliper piston back into the caliper.

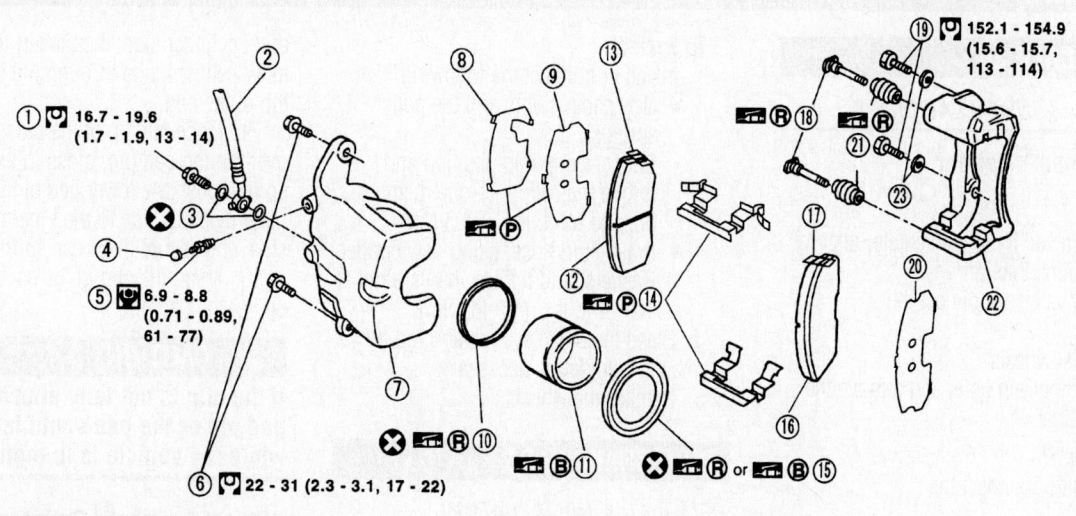

① ⊡ 16.7 - 19.6 (1.7 - 1.9, 13 - 14)

⑤ ⊡ 6.9 - 8.8 (0.71 - 0.89, 61 - 77)

⑥ ⊡ 22 - 31 (2.3 - 3.1, 17 - 22)

⑲ ⊡ 152.1 - 154.9 (15.6 - 15.7, 113 - 114)

⊡Ⓟ : PBC (Poly Butyl Cuprysil) gress or silicone-based grease
⊡Ⓡ : Rubber Gress
⊡Ⓑ : Brake fluid
⊡ : N•m (kg-m, ft-lb)
⊡ : N•m (kg-m, in-lb)
❌ : Always replace after every disassembly.

1.	Union bolt	2.	Brake hose	3.	Copper washer
4.	Cap	5.	Bleed valve	6.	Sliding pin bolt
7.	Cylinder body	8.	Inner shim cover	9.	Inner shim
10.	Piston seal	11.	Piston	12.	Inner pad
13.	Pad wear sensor	14.	Pad retainer	15.	Piston boot
16.	Outer pad	17.	Pad wear sensor	18.	Sliding pin bolt
19.	Torque member bolts	20.	Outer shim	21.	Slide pin boot
22.	Torque member	23.	Washer		

Fig. 3 Front brake caliper assembly—with CLZ25VD caliper

09482_ZMAX_G0040

4. Install or connect the following:
- New pads, new shims and pad retainers
- Brake caliper and torque the sliding pin bolts to 17–22 ft. lbs. (22–31 Nm)
- Brake line to the caliper, using new copper washers, and torque the connecting bolt to 12–14 ft. lbs. (17–20 Nm)
- Wheels

5. Bleed the brake system and top off the master cylinder as necessary.

DISC BRAKE PADS

REMOVAL & INSTALLATION

With Brembo® Caliper

1. Remove tires from vehicle with a power tool.
2. Remove the clip from the pad pin.
3. Remove the pad pin while holding down the cross spring, then remove the cross spring from the caliper.
4. Using pliers, remove the pad from the caliper.

To install:
5. Insert the piston to the position where the pad is attached.
6. Attach pad.

➡**The side of the shim with the larger cutouts should be on the entry side of the disc rotor spin.**

7. Insert the upper pad pin from the inner cylinder side, then insert firmly to the outer cylinder side through the hole in the top of the pad.
8. Place the top of the cross spring over the top pad pin, press in the cross spring, push the lower pad pin from the inner cylinder side to the outer cylinder side, and secure the cross spring.
9. Insert the clip in the small hole at the end of the pad pin.

※ CAUTION

If the clip is not fully attached, the pad pin or the pad could fall out while the vehicle is in motion.

10. Install tires to vehicle.

Without Brembo® Caliper

1. Before servicing the vehicle, refer to the Precautions Section.
2. Remove or disconnect the following:
- Front wheels
- Sliding pin bolts
- Rotate the caliper up and remove the brake pads

To install:
3. Install or connect the following:
- New pads, new shims and pad retainers
- Brake caliper
- Wheels

4. Bleed the brake system and top off the master cylinder as necessary.

BRAKES

BRAKE CALIPER

REMOVAL & INSTALLATION

With Brembo® Caliper

See Figure 4.

1. Before servicing the vehicle, refer to the Precautions Section.
2. Remove or disconnect the following:
- Rear wheels
- Caliper pin bolts and remove the caliper

To install:
3. Install or connect the following:
- New pads, shims and the pad springs
- Caliper body into position and torque the caliper mounting bolts to 62 ft. lbs. (84.3 Nm)

Without Brembo® Caliper

1. Before servicing the vehicle, refer to the Precautions Section.
2. Remove or disconnect the following:
- Rear wheels
- Brake fluid hose from the caliper
- Caliper torque member bolts and remove the caliper

To install:
3. Install or connect the following:
- New pads, shims and the pad springs
- Caliper body into position and torque the caliper torque member bolts to 62 ft. lbs. (84.3 Nm)
- Brake fluid hose, using new copper washers, and tighten the flare nut to 12–14 ft. lbs. (17–20 Nm)

4. Bleed the brake system and top off the master cylinder as necessary.
5. Replace the wheels.

DISC BRAKE PADS

REMOVAL & INSTALLATION

With Brembo® Caliper

1. Remove tires from vehicle with a power tool.
2. Remove the clip from the pad pin.
3. Remove the pad pin while holding down the cross spring, then remove the cross spring from the caliper.
4. Using pliers, remove the pad from the caliper.

To install:
5. Insert the piston to the position where the pad is attached.
6. Attach pad and shim cover.
7. Attach the pad with wear sensor to the outer side.
8. Insert the upper pad pin from the outer cylinder side, then insert firmly to the inner cylinder side through the hole in the top of the pad.
9. Place the top of the cross spring over the top pad pin, press in the cross spring, push the lower pad pin from the outer cylinder side to the inner cylinder side, and secure the cross spring.
10. Insert the clip in the small hole at the end of the pad pin.

※ CAUTION

If the clip is not fully attached, the pad pin or the pad could fall out while the vehicle is in motion.

Without Brembo® Calipers

1. Before servicing the vehicle, refer to the Precautions Section.
2. Remove or disconnect the following:
- Rear wheels
- Top sliding pin bolt
- Rotate the caliper up and remove the brake pads

To install:
3. Install or connect the following:
- New pads, new shims and pad retainers
- Brake caliper Tighten sliding bolt to 32 ft. lbs. (43 Nm)
- Wheels

4. Bleed the brake system and top off the master cylinder as necessary.

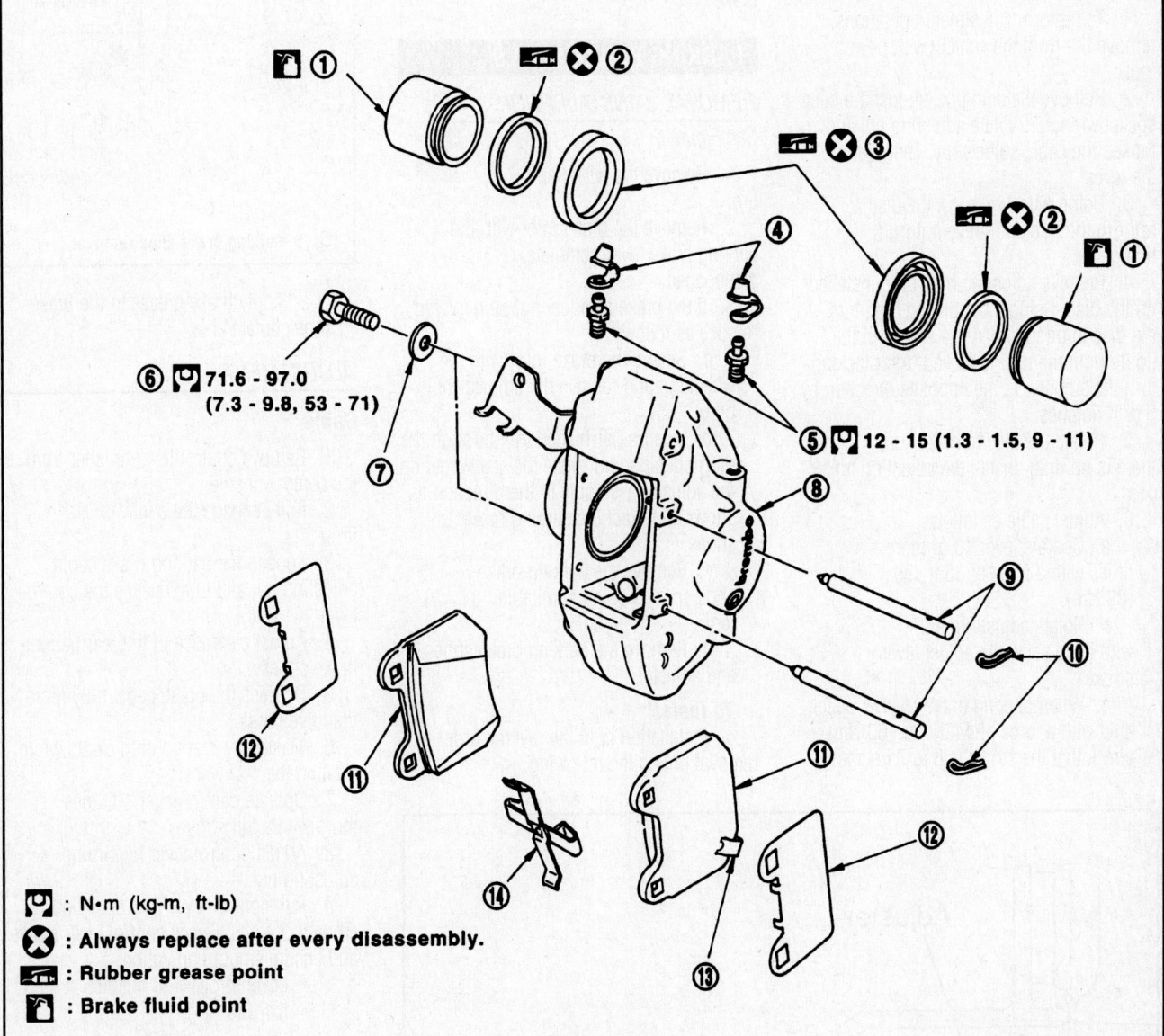

6 🔧 71.6 - 97.0
(7.3 - 9.8, 53 - 71)

5 🔧 12 - 15 (1.3 - 1.5, 9 - 11)

🔧 : N·m (kg-m, ft-lb)

❌ : Always replace after every disassembly.

▥ : Rubber grease point

🛢 : Brake fluid point

1. Piston	2. Piston seal	3. Piston boot
4. Cap	5. Bleed valve	6. Bolt
7. Washer	8. Caliper	9. Pad pins
10. Clips	11. Brake pad	12. Shim cover
13. Pad wear sensor	14. Cross spring	

09482_ZMAX_G0041

Fig. 4 Rear brake caliper assembly—with OPB13VB/Brembo® caliper

PARKING BRAKE CABLES

ADJUSTMENT

See Figure 5.

1. To perform adjustment operations, remove tire from the vehicle with power tool.

2. Remove the coin pocket. Insert a deep socket wrench to rotate adjusting nut and loosen the cable sufficiently. Then, return the lever.

3. Using wheel nuts, fix the disc rotor to the hub and prevent it from tilting.

4. Remove adjusting hole plug installed on the disc. Using a flat bladed tool, turn the disc in direction "A" as shown in the figure until the disc is locked. After locking, turn the adjuster in the opposite direction by 5 or 6 notches.

5. Rotate the disc to make sure there is no drag. Install the adjusting hole plug.

6. Adjust cable as follows:

   a. Operate lever 10 or more times with a force of 66 ft. lbs. (89 Nm).

   b. Rotate adjusting nut with deep socket to adjust lever stroke.

   c. When parking brake lever is operated with a force of 44 ft. lbs. (60 Nm), check that the stroke is 6 to 7 notches

(Check it by listening and counting the ratchet clicks).

7. With the lever completely returned, make sure there is no drag on the rear brake.

PARKING BRAKE SHOES

REMOVAL & INSTALLATION

See Figure 6.

1. Remove the wheel and tire.

2. Remove the brake rotor with the parking brake lever completely disengaged.

3. If the brake rotor cannot be removed, remove as follows:

   a. Secure the brake rotor with the wheel nut and remove the adjuster hole plug.

   b. Insert a flat-bladed tool through the plug opening and rotate the star wheel on the adjuster assembly in the direction as shown to retract the parking brake shoes.

   c. Remove the parking brake shoe springs using a suitable tool.

   d. Remove the parking brake shoes and adjuster.

To install:

4. Installation is in the reverse order of removal noting the following:

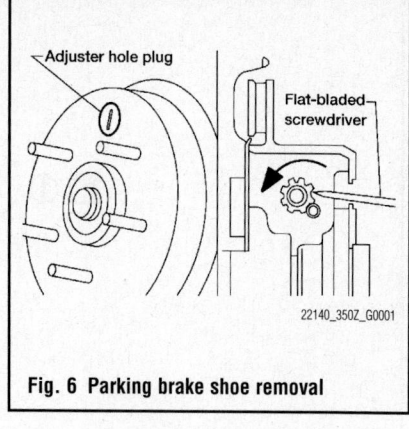

Fig. 6 Parking brake shoe removal

   a. Apply brake grease to the brake shoe contact area.

ADJUSTMENT

Cable

1. Remove control lever finisher from the center console.

2. Fully engage the control lever.

3. Loosen the parking brake cable adjusting nut and fully release the control lever.

4. Adjust clearance of the rear parking brake shoes.

5. Depress the brake pedal fully more than five times.

6. Make sure that no drag exists while rotating the rear wheels.

7. Operate control lever 10 times or more with a full stroke.

8. Adjust control lever by turning adjusting nut.

9. Pull control lever with specified amount of force. Check control lever stroke and ensure smooth operation:

   • Lever stroke 7–8 notches at 44 ft. lbs. (60 Nm)

Shoes

1. Make sure the parking brake control lever is fully released and parking brake cable adjusting nut is loosened.

2. Remove the adjuster hole plug installed on the rotor. Using a suitable tool, turn the adjuster as shown until the rotor is locked. After locking, turn the adjuster in the opposite direction 5 or 6 notches.

3. Rotate the rotor to make sure that there is no drag. Install the adjuster hole plug.

4. After adjusting the clearance of the rear shoes, adjust the parking brake cable.

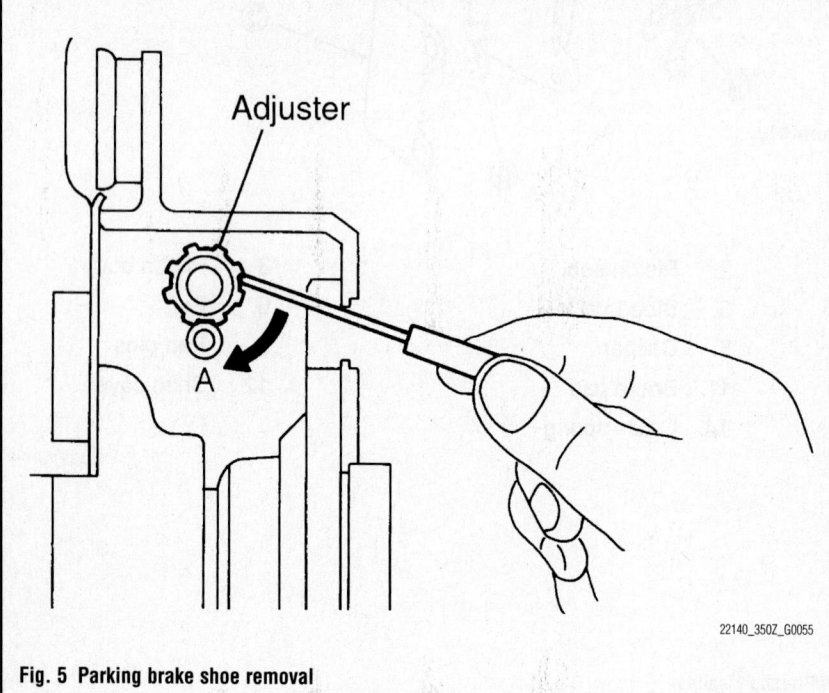

Fig. 5 Parking brake shoe removal

CHASSIS ELECTRICAL — AIR BAG (SUPPLEMENTAL RESTRAINT SYSTEM)

GENERAL INFORMATION

❊❊ CAUTION

These vehicles are equipped with an air bag system. The system must be disarmed before performing service on, or around, system components, the steering column, instrument panel components, wiring and sensors. Failure to follow the safety precautions and the disarming procedure could result in accidental air bag deployment, possible injury or death, or unnecessary system repairs.

PRECAUTIONS

Disconnect and isolate the battery negative cable before beginning any airbag system component diagnosis, testing, removal, or installation procedures. Allow system capacitor to discharge for two minutes before beginning any component service. This will disable the airbag system. Failure to disable the airbag system may result in accidental airbag deployment, personal injury, or death.

DISARMING THE SYSTEM

All SRS electrical wiring harnesses and connectors are covered with YELLOW outer insulation. Do not use electrical test equipment on any circuit related to the SRS (air bag) sensors. When installing SRS components, always install with the arrow marks facing the front of the vehicle.

To disarm the SRS system turn the ignition switch to **OFF** position. Then, disconnect the both battery cables starting with the negative cable first and wait at least 10 minutes after the cables are disconnected. Be sure to insulate the battery terminal ends.

ARMING THE SYSTEM

To arm the SRS system turn the ignition switch to **OFF** position. Connect the both battery cables starting with the positive cable first.

The SRS or air bag system is equipped with a self-diagnostic operation. After turning the ignition key to the ON or START position, the AIR BAG warning lamp will illuminate for 7 seconds. After 7 seconds, the AIR BAG lamp will extinguish if no malfunction is detected. If the AIR BAG lamp does not extinguish after 7 seconds, check the SRS self-diagnostic system for a malfunction.

CLOCKSPRING CENTERING

Align spiral cable correctly when installing steering wheel. Make sure that the spiral cable is in the neutral position. The neutral position is detected by turning left 2.5 revolutions from the right end position and ending with the knob at the top.

DRIVE TRAIN

DRIVEN DISC & PRESSURE PLATE

REMOVAL & INSTALLATION

See Figure 7.

1. Before servicing the vehicle, refer to the Precautions Section.
2. Remove the transmission assembly.
3. Insert a clutch disc centering tool into the clutch disc hub for support.

4. Remove or disconnect the following:
   - Pressure plate bolts evenly in reverse order of the tightening sequence, a little at a time to prevent distortion

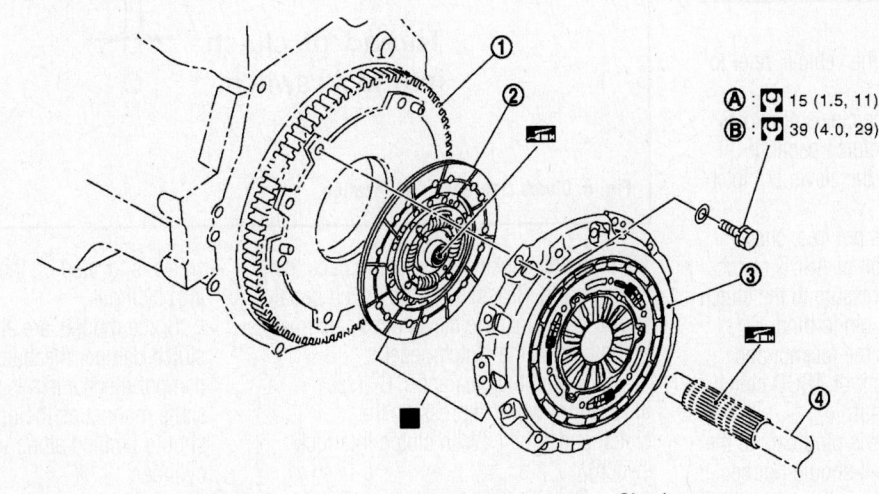

1. Flywheel
2. Clutch disc
3. Clutch cover
4. Main drive gear
A. First step
B. Final step

■ : Replace the parts as a set.

: Apply lithium-based grease including molybdenum disulphide

A : 15 (1.5, 11)
B : 39 (4.0, 29)

37663_350Z_G0037

Fig. 7 Clutch assembly and related components

- Clutch assembly
- Throw-out bearing from the clutch lever

To install:

5. Apply a light coating of chassis lube to the clutch disc spleens, input shaft and pilot bearing. Use a disc centering tool to aid installation.

6. Install the disc and pressure plate.

7. Torque the pressure plate bolts in a crisscross pattern in the following 2 steps:
- Step 1: 11 ft. lbs.
- Step 2: 29 ft. lbs.

8. Install new throw-out bearing in the clutch release lever. Remove the clutch disc centering tool.

9. Install transaxle into the vehicle. If the mating surfaces will not come together, do not force the units together. Remove the transmission and recheck that the disc is centered.

✳✳ WARNING

DO NOT draw the transmission to the engine with the bolts. This may damage the clutch and/or transaxle. Also, be careful not to move the throw-out bearing when installing the transaxle.

10. After the transmission is installed, connect the clutch cable and check operation before complete reassembly.

11. Adjust the clutch pedal as necessary.

ADJUSTMENTS

See Figure 8.

1. Before servicing the vehicle, refer to the Precautions Section.

2. Check to see if the clevis pin floats freely in the bore of the clutch pedal. It should not be bound by the clevis or clutch pedal.

3. If the clevis pin is not free, check that the pedal stopper bolt or ASCD clutch switch is not applying pressure to the clutch pedal causing the clevis pin to bind.

4. To adjust, loosen the locknut and turn the pedal stopper bolt or ASCD clutch switch. Tighten the locknut.

5. Verify that the clevis pin floats in the bore of the clutch pedal. It should not be bound by the clutch pedal.

6. If the clevis pin is still not free, remove the pin and check for deformation or damage. Replace the pin, if required. Leave the pin removed for the next step.

7. Check the clutch pedal stroke for free range of movement. With the pin removed, manually move the clutch pedal up and down to determine if it moves freely.

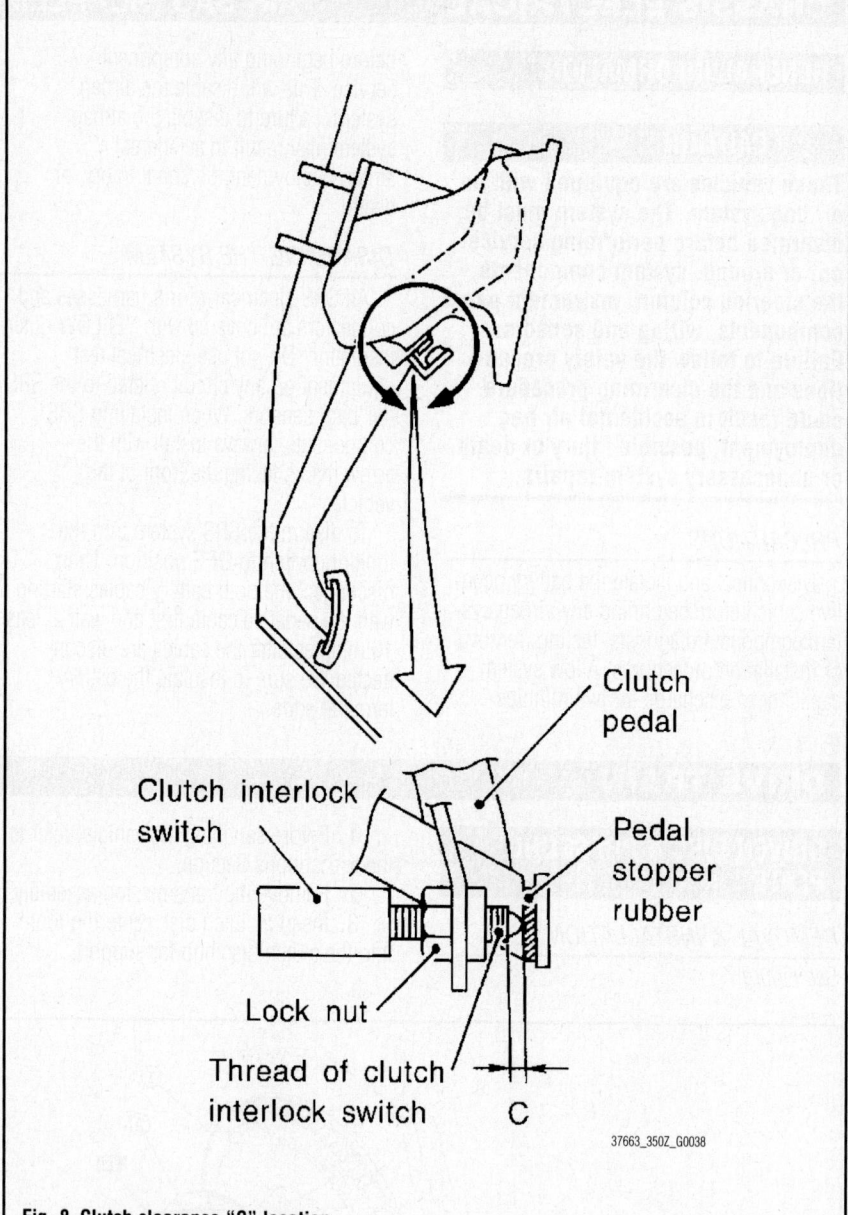

Fig. 8 Clutch clearance "C" location

37663_350Z_G0038

8. If any sticking is noted, replace the related parts. Reassemble the clutch pedal and re-verify that the clevis pin floats freely in the bore of the clutch pedal.

9. Adjust the clearance "C" (see illustration) while depressing the clutch pedal fully, (With clutch interlock switch).

10. Specification should be 0.004—0.039 inch.

HYDRAULIC SYSTEM BLEEDING

BLEEDING PROCEDURE

Bleeding is required to remove air trapped in the hydraulic system. The bleed screw is located on the clutch slave (operating) cylinder.

Some models are also equipped with a clutch damper mechanism. The clutch damper mechanism is bled in exactly the same manner as the operating cylinder. It should be bled along with the operating cylinder.

1. Before servicing the vehicle, refer to the Precautions Section.

2. Remove the bleed screw dust cap.

3. Attach a transparent vinyl tube to the bleed screw, immersing the free end in a clean container of clean brake fluid.

4. Fill the master cylinder with the proper fluid.

5. Open the bleed screw about ¾ turn.

6. Depress the clutch pedal quickly. Hold it down. Have an assistant tighten the bleed screw. Allow the pedal to return slowly.

7. Repeat the above procedure until no more air bubbles are seen in the fluid container.

8. Remove the bleed tube.

9. Replace the dust cap and refill the master cylinder.

10. Bleed the clutch damper, if equipped.

REAR AXLE HOUSING

REMOVAL & INSTALLATION

See Figures 9 through 10.

1. Before servicing the vehicle, refer to the Precautions Section.

2. Raise and safely support the vehicle.

3. Matchmark and remove the driveshaft. Never impact or damage the driveshaft tube.

4. Remove the rear stabilizer bar.

5. Remove the halfshaft from the final drive. Suspend the component with mechanics wire, do not allow it to just hang unsupported.

6. Remove the breather hose from the final drive.

7. Remove the rear wheel sensor.

8. Proper support the rear final drive assembly.

➡**Never position the support tool on the rear cover (aluminum case).**

9. Remove the mounting bolts and nuts connecting the suspension member.

10. Remove the rear final drive assembly.

➡**Be sure to secure the assembly to the support tool before removal.**

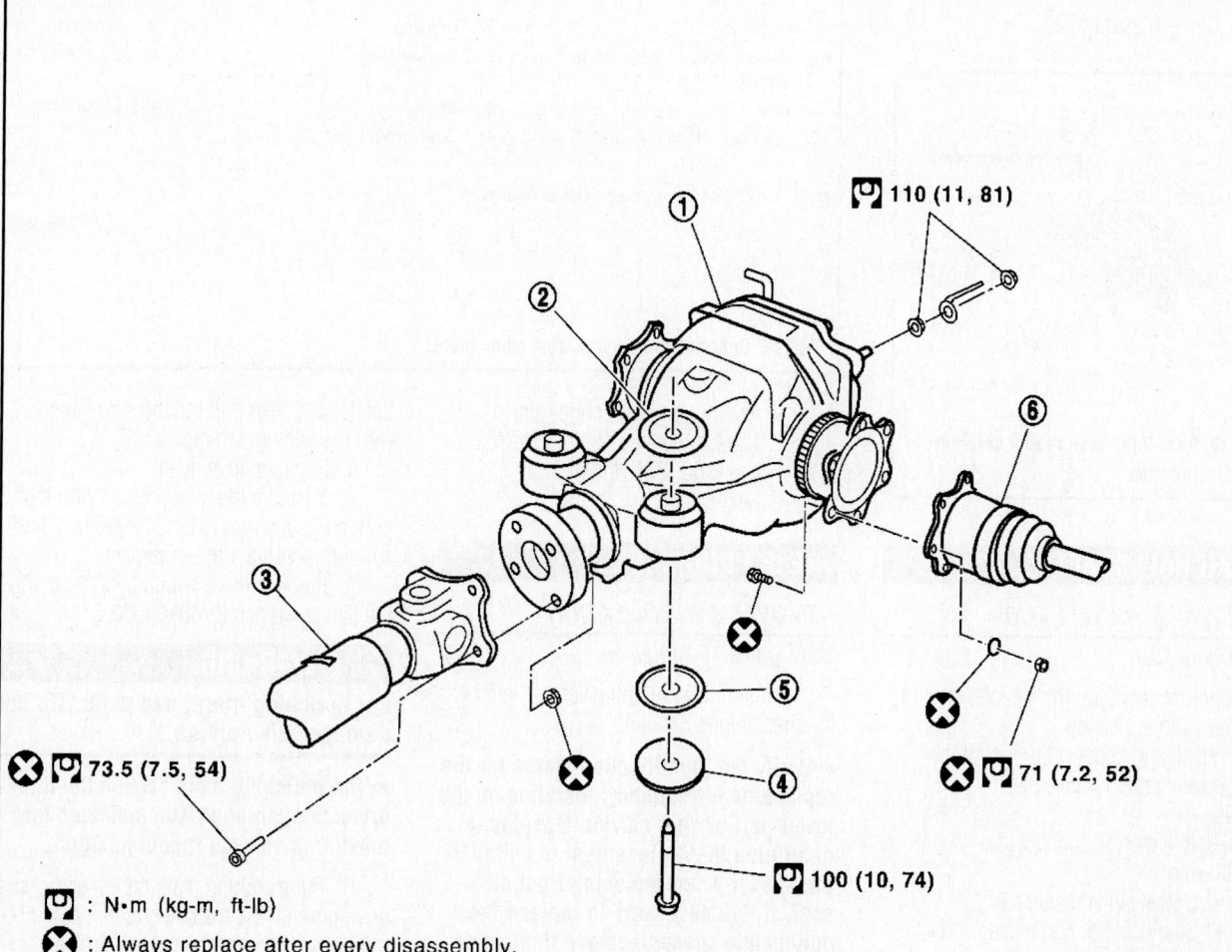

110 (11, 81)

73.5 (7.5, 54)

71 (7.2, 52)

100 (10, 74)

: N•m (kg-m, ft-lb)

: Always replace after every disassembly.

| 1. | Rear final drive assembly | 2. | Upper stopper | 3. | Propeller shaft |
| 4. | Washer | 5. | Lower stopper | 6. | Drive shaft |

37663_350Z_G0041

Fig. 9 Rear axle housing and related components

To install:

➡ **Be sure to use new fasteners, as required.**

11. Installation is the reverse of the removal procedure.

12. When installing the breather hoses, make sure that there are no pinched or restricted area on the hose caused by bending or winding when installing it.

13. For installation, the vehicle side end shall be inserted to suspension member. Install metal connector side of this hose to rear cover by inserting it with aiming painted marking to the front of the vehicle.

14. If oil leaks while removing the final drive assembly, check oil level after installation.

15. Correct, as required.

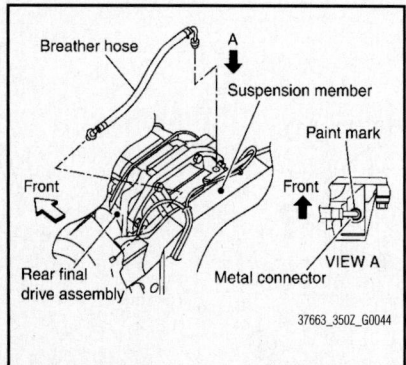

Fig. 10 Rear final drive proper breather hose installation

REAR HALFSHAFTS

REMOVAL & INSTALLATION

See Figure 13.

1. Before servicing the vehicle, refer to the Precautions Section.

2. Raise and support the rear of the vehicle safely and remove the wheels.

3. Remove or disconnect the following:
   - Cotter pin and axle nut
   - Stabilizer bar connecting rod

4. Remove the nuts and bolts between the side flange and the drive shaft.

5. Use a suitable puller and remove the drive shaft from the axle.

To install:

6. Install or connect the following:
   - New seal into the transaxle and install halfshaft
   - Tighten the side flange bolts to 47–58 ft. lbs. (63–79 Nm)
   - Stabilizer bar connecting rod

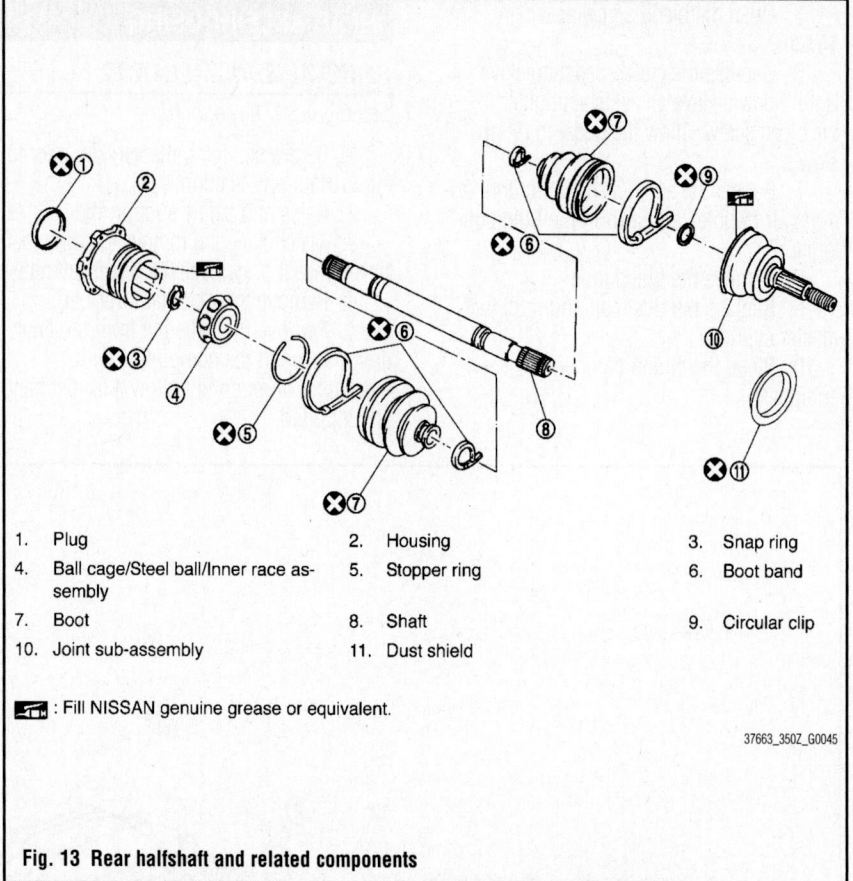

1. Plug
2. Housing
3. Snap ring
4. Ball cage/Steel ball/Inner race assembly
5. Stopper ring
6. Boot band
7. Boot
8. Shaft
9. Circular clip
10. Joint sub-assembly
11. Dust shield

▥ : Fill NISSAN genuine grease or equivalent.

37663_350Z_G0045

Fig. 13 Rear halfshaft and related components

- Axle nut and new cotter pin. Tighten the axle nut to 152–202 ft. lbs. (206–274 Nm)
- Rear wheels

REAR PINION SEAL

REMOVAL & INSTALLATION

See Figures 14 through 18.

1. Before servicing the vehicle, refer to the Precautions Section.

➡ **Verify the identification stamp for the replacement frequency, installed on the lower part of gear carrier. Use this to determine the replacement of collapsible spacer when replacing front oil seal. If it is necessary to replace the collapsible spacer, remove final drive assembly and disassemble to replace front oil seal and collapsible spacer.**

2. Drain gear oil.

3. Raise and support the rear of the vehicle safely and remove the wheels.

4. Remove or disconnect the following:
   - ABS rear wheel sensor
   - Rear halfshaft, and suspend with mechanics wire

5. Install attachments [A: KV40104100 and B: ST36230000 (J-25840-A)] to side

flange, and then pull out the side flange with the sliding hammer.

6. Remove drive shaft.

7. Measure the total preload with the preload gauge tool [ST3127S000 (J-25765-A)] and record the measurement.

8. Remove drive pinion lock nut using the flange wrench KV40104000.

❋❋ WARNING

For matching mark, use paint. Do not damage drive pinion.

➡ **The matching mark "A" on the final drive companion flange indicates the maximum vertical runout position.**

9. Put matching mark on the end of the drive pinion. The matching mark should be in line with the matching mark on the companion flange.

10. Remove companion flange using a puller.

11. Remove front oil seal using the puller [A: KV381054S0 (J-34286)].

To install:

➡ **Do not reuse oil seal. Do not incline oil seal when installing.**

12. Apply multi-purpose grease to front oil seal lips.

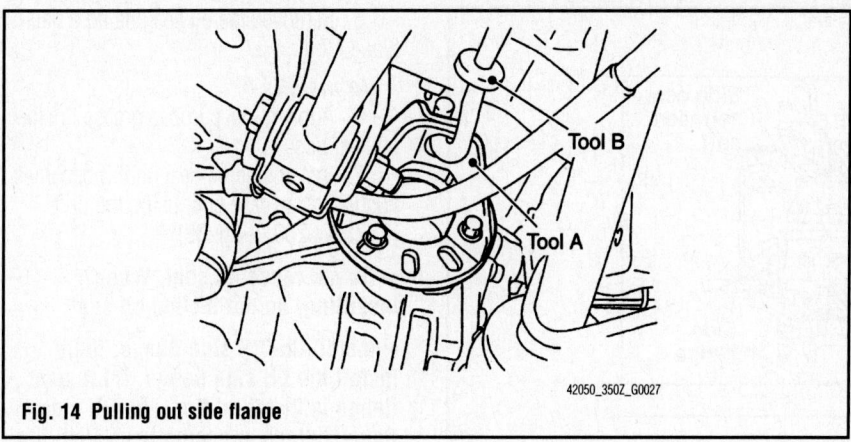

Fig. 14 Pulling out side flange

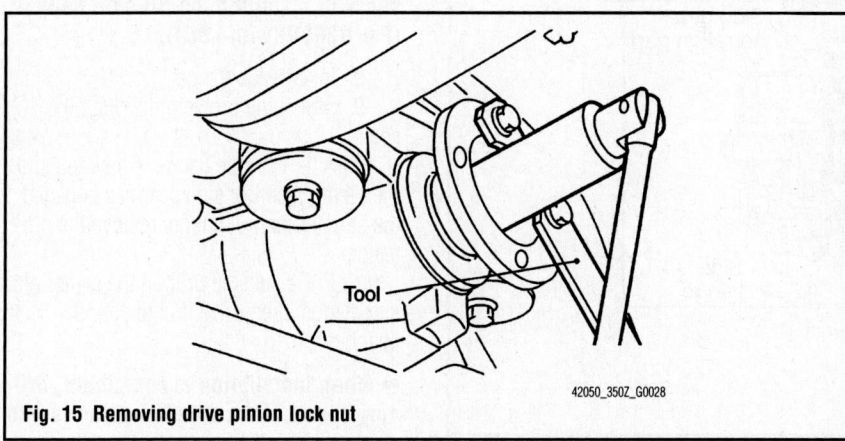

Fig. 15 Removing drive pinion lock nut

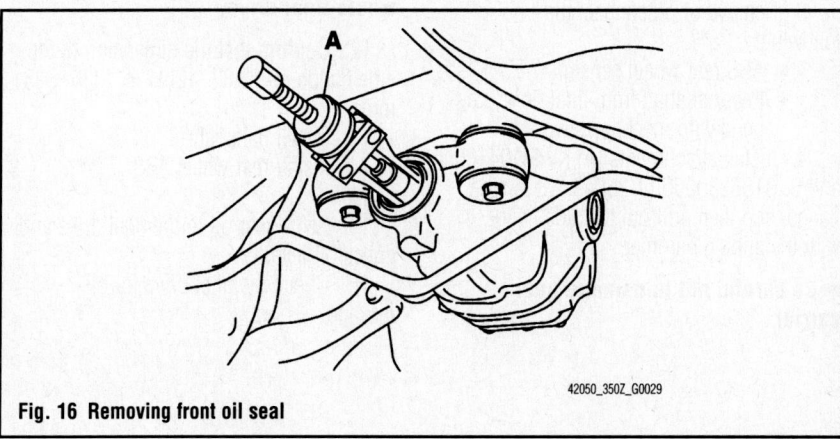

Fig. 16 Removing front oil seal

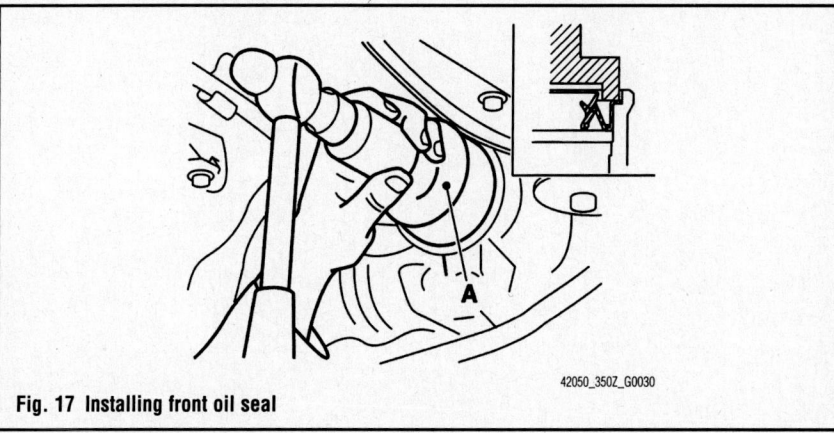

Fig. 17 Installing front oil seal

13. Install front oil seal using the drift [A:ST30720000 (J-25405)] as shown in the illustration.

14. Align the matching mark of drive pinion with the matching mark of companion flange, and then install the companion flange.

➡**Do not reuse drive pinion lock nut.**

15. Apply anti-corrosion oil to the thread and seat of new drive pinion lock nut, and temporarily tighten drive pinion lock nut to drive pinion.

➡**Total preload torque should equal the measurement taken during removal plus an additional 3 inch lbs. (0.1 to 0.4 Nm).**

16. Tighten to drive pinion lock nut to 09–238 ft. lbs. (147–323 Nm), while adjusting total reload torque.

➡**Adjust to the lower limit of the drive pinion lock nut tightening torque first. If the preload torque exceeds the specified value, replace collapsible spacer and tighten it again to adjust. Do not loosen drive pinion lock nut to adjust the preload torque.**

17. Make a stamping for identification of front oil seal replacement frequency. Be sure to make a stamping after replacing front oil seal.

18. Install drive shaft.

➡**Install the RH side flange, then install the LH side flange. If LH side flange is installed first, the RH side flange comes out sometimes from the shock of installing the RH side flange [For R200V (with LSD)].**

19. Install side flange.

20. Attach the protector to side oil seal.

21. After the side flange is inserted and the serrated part of side gear has engaged the serrated part of flange, remove the protector.

22. Put a suitable drift on the center of side flange, then drive it until sound changes.

➡**When installation is completed, driving sound of the side flange turns into a sound which seems to affect the whole final drive.**

23. Confirm that the dimension of the side flange is 12.83–12.91 in. (326–328 mm) as shown in the illustration.

24. Install halfshaft.

25. Install rear wheel ABS sensor.

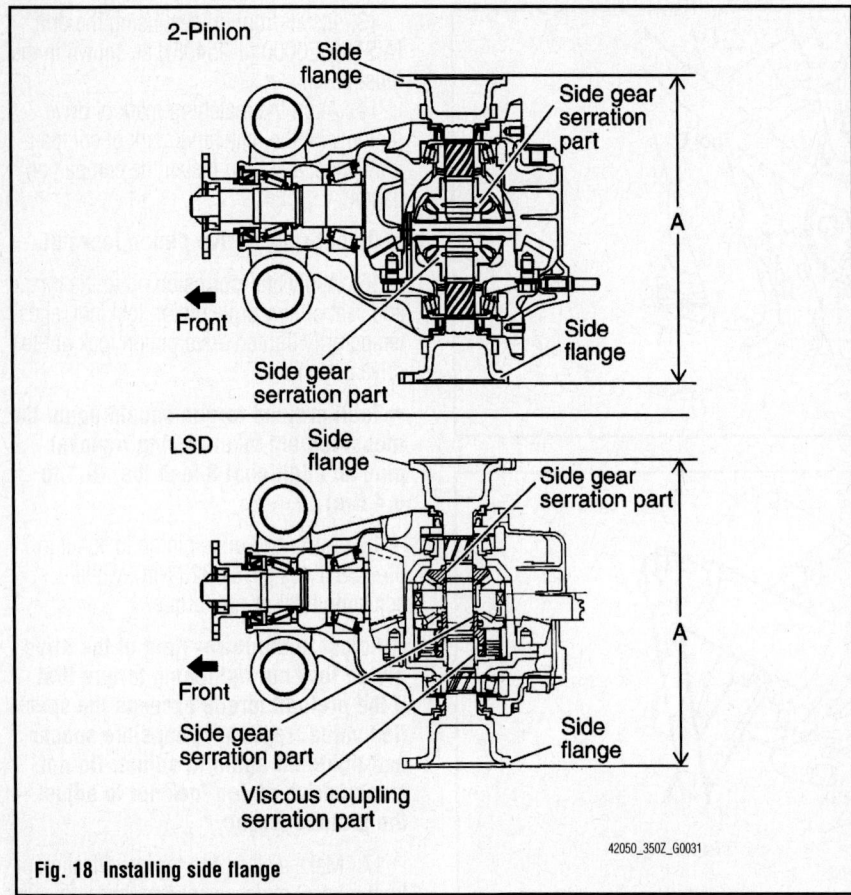

2-Pinion

Side flange

Side gear serration part

Side gear serration part

Side flange

Front

A

LSD

Side flange

Side gear serration part

Side gear serration part

Side flange

Front

Viscous coupling serration part

A

42050_350Z_G0031

Fig. 18 Installing side flange

26. Refill gear oil to the final drive and check oil level.

27. Check the final drive for oil leakage.

Side

1. Before servicing the vehicle, refer to the Precautions Section.

2. Raise and support the rear of the vehicle safely and remove the wheels.

3. Remove or disconnect the following:
  • ABS rear wheel sensor
  • Rear halfshaft from final drive, suspend with mechanics wire

4. Install attachments [A: KV40104100 and B: ST36230000 (J-25840-A)] to side flange, and then pull out the side flange with the sliding hammer.

➡**Be careful not to damage gear carrier.**

5. Remove side oil seal, using a flat-bladed tool.

To install:

6. Apply multi-purpose grease to side oil seal lips.

7. Install side oil seal until it becomes flush with the case end, using the drift [KV38100200 (J-26233)].

➡**Do not reuse oil seal. When installing, do not incline oil seal.**

➡**Install the RH side flange, then install the LH side flange. If LH side flange is installed first, the RH side flange comes out sometimes from the shock of installing the RH side flange (For R200V (with LSD)).**

8. Install side flange.

9. Attach the protector to side oil seal.

10. After the side flange is inserted and the serrated part of side gear has engaged the serrated part of flange, remove the protector.

11. Put a suitable drift on the center of side flange, then drive it until sound changes.

➡**When installation is completed, driving sound of the side flange turns into a sound which seems to affect the whole final drive.**

12. Confirm that the dimension of the side flange is: 12.83–12.91 in. (326–328 mm).

13. Install halfshaft

14. Install rear wheel ABS sensor.

15. Refill gear oil to the final drive and check oil level.

16. Check the final drive for oil leakage.

ENGINE COOLING

ENGINE FAN

REMOVAL & INSTALLATION
See Figures 19 and 20.

1. Before servicing the vehicle, refer to the Precautions Section.

➡If working near and/or around the SRS system and components, be sure to disable the SRS system. After disabling the system wait three minutes or more before servicing the vehicle.

➡Before disconnecting the battery, lower both the driver's and passenger's windows. This will prevent any interference between the window edge and the vehicle when the door is opened or closed. During normal operation the window slightly raises and lowers automatically to prevent any window to vehicle interference. The automatic window function will not work with the battery disconnected.

➡Never drain the engine coolant when the engine is hot. Wrap a thick cloth around the cap and carefully remove it. First turn the cap a quarter of a turn to release any pressure, then turn it all the way. Do not allow coolant to come in contact with the drive belts.

2. Remove the engine cover.
3. Drain the engine coolant. Be sure to properly dispose of used coolant.
4. Remove the air cleaner assembly.
5. Remove the reservoir tank.
6. Disconnect the upper radiator hose at the radiator.
7. Disconnect the fan motor harness connectors at the right lower portion of the fan shroud.

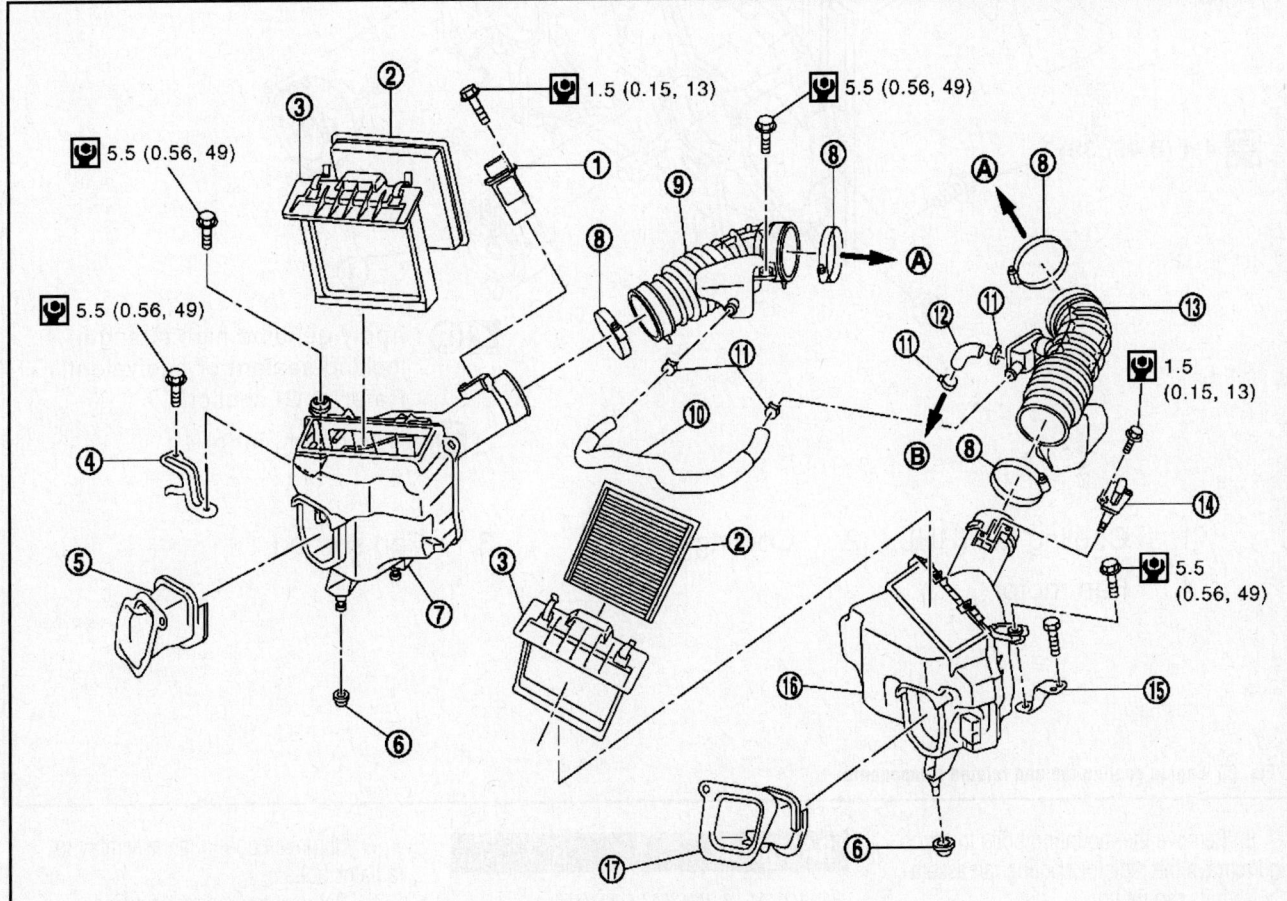

1. Mass air flow sensor (RH)	2. Air cleaner filter	3. Holder
4. Bracket	5. Air duct (inlet)	6. Grommet
7. Air cleaner case (RH)	8. Clamp	9. Air duct (RH)
10. Air hose	11. Clamp	12. PCV hose
13. Air duct (LH)	14. Mass air flow sensor (LH)	15. Bracket
16. Air cleaner case (LH)	17. Air duct (inlet)	

37663_350Z_G0051

Fig. 19 Air cleaner assembly and related components

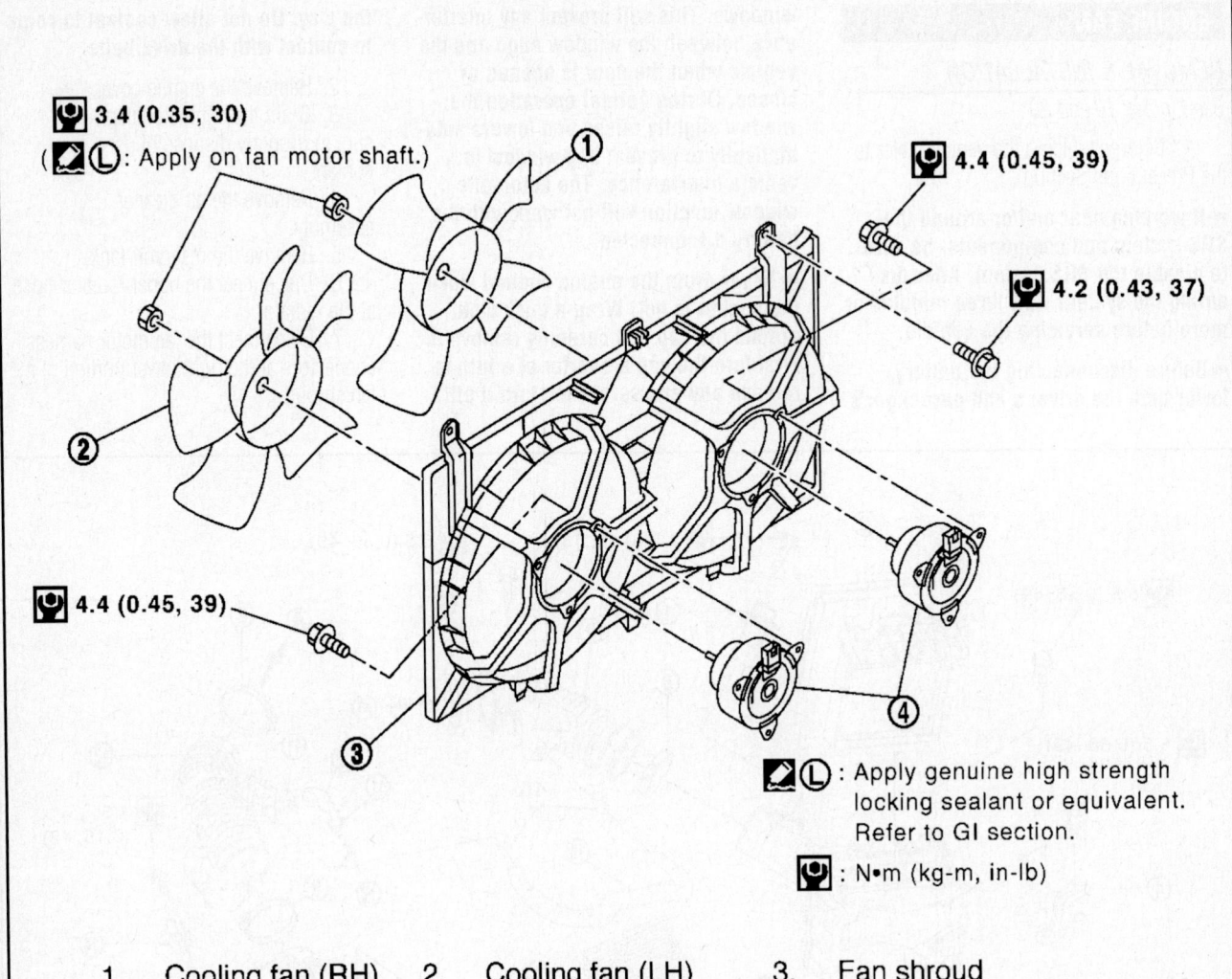

3.4 (0.35, 30)
(: Apply on fan motor shaft.)

4.4 (0.45, 39)

4.2 (0.43, 37)

4.4 (0.45, 39)

 : Apply genuine high strength locking sealant or equivalent. Refer to GI section.

 : N•m (kg-m, in-lb)

1. Cooling fan (RH)   2. Cooling fan (LH)   3. Fan shroud
4. Fan motor

37663_350Z_G0050

Fig. 20 Engine cooling fan and related components

8. Remove the mounting bolts to lift up and remove the radiator cooling fan assembly from its mounting.

To install:

➡**Be sure to use new fasteners, as required.**

9. Installation is the reverse of the removal procedure.
10. Be sure to fill the engine with the proper grade and type coolant.
11. Start the engine and check for leaks.
12. Correct, as required.

RADIATOR

REMOVAL & INSTALLATION

1. Drain the coolant from the radiator.
2. Remove fresh air duct.
3. Disconnect radiator upper and lower hoses.
4. Remove the CVT fluid cooler hoses, if equipped. Plug hoses to avoid leakage of CVT fluid.
5. Disconnect the reservoir tank hose.
6. Remove the radiator upper clips by pulling the tabs outside to release the lock, do not pull locks excessively.

7. Remove cooling fan assembly to radiator bolts.
8. Remove the radiator assembly.

➡**Do not damage or scratch air conditioner condenser and radiator core when removing.**

To install:

9. Installation is in the reverse order of removal, paying attention to the following:
- Install the rubber on mounting pin of radiator core
- Align the radiator upper clip with the radiator core connector, then insert the radiator upper clip

straight into the radiator core connections until a click is heard

- After connecting the radiator upper clip, visually confirm that the two radiator upper clips are connected to the radiator core connections
- Move the radiator upper clip and the radiator core forward and backward to make sure they are securely connected

THERMOSTAT

REMOVAL & INSTALLATION

See Figure 21.

1. Before servicing the vehicle, refer to the Precautions Section.
2. Drain the cooling system.
3. Remove or disconnect the following:
   - Negative battery cable
   - Engine cover, if equipped
   - Water drain plug on water pump side of cylinder block
   - Lower radiator hose
   - Water inlet and thermostat assembly

To install:

4. To install, reverse removal procedure.
5. Install thermostat with jiggle valve facing upward and tighten bolts to 75–99 inch lbs. (8.4–11.2 Nm).
6. Run engine and check for leaks.

WATER PUMP

REMOVAL & INSTALLATION

See Figure 22.

1. Before servicing the vehicle, refer to the Precautions Section.
2. Drain the cooling system.
3. Remove or disconnect the following:
   - Negative battery cable
   - Accessory drive belts
   - Radiator hoses
   - Cooling fan
   - Water drain plug on water pump side of block
   - Timing chain tensioner cover
   - Water pump cover
   - Primary timing chain tensioner
   - Water pump mounting bolts
4. Turn the crankshaft pulley counterclockwise until the timing chain slack on the water pump pulley is at maximum.
5. Place M8 bolts in the upper and lower M8 threaded holes of the water pump.
6. Tighten each bolt by turning alternately ½ turn until they reach the timing chain rear case. Be sure to turn each bolt ½ turn at a time to prevent damage.
7. Lift up the water pump and remove it.
8. When removing the water pump, do not allow the water pump gear to hit the timing chain.

9. Remove and discard the O-rings from the water pump.
10. Clean all traces of liquid gasket from the water pump and covers.

To install:

11. Install the water pump using new O-rings to the engine block. Lubricate the inner O-ring with clean engine oil and the outer O-ring with engine coolant. Ensure the water pump sprocket and timing chain are engaged. Torque the 3 water pump mounting bolts evenly to 85 inch lbs. (10 Nm).
12. Rotate the crankshaft pulley clockwise so the timing chain on the tensioner side is loose.
13. Install the primary timing chain tensioner.
14. Apply a continuous [0.091–0.130 in. (2.3–3.3mm)] bead of liquid sealant to the mating surfaces of the timing chain tensioner and water pump covers.
15. Install the timing chain tensioner and water pump covers to the engine block. Torque the bolts to 97 inch lbs. (11 Nm).
16. Install or connect the following:
    - Water drain plug
    - Cooling fan
    - Radiator hoses
    - Accessory drive belts
    - Negative battery cable
17. Fill the cooling system.
18. Start the engine, check for leaks, and repair if necessary.

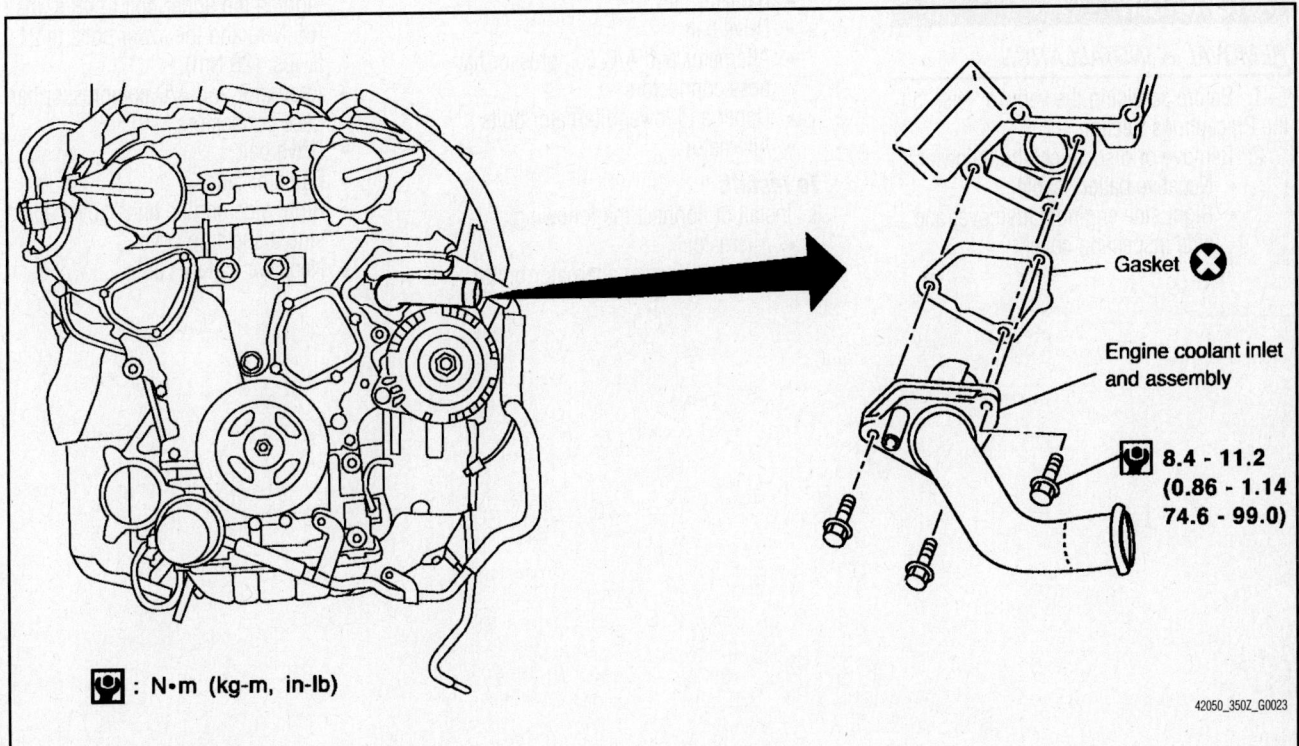

Gasket ⊗

Engine coolant inlet and assembly

8.4 - 11.2
(0.86 - 1.14
74.6 - 99.0)

: N•m (kg-m, in-lb)

42050_350Z_G0023

Fig. 21 Thermostat location and components

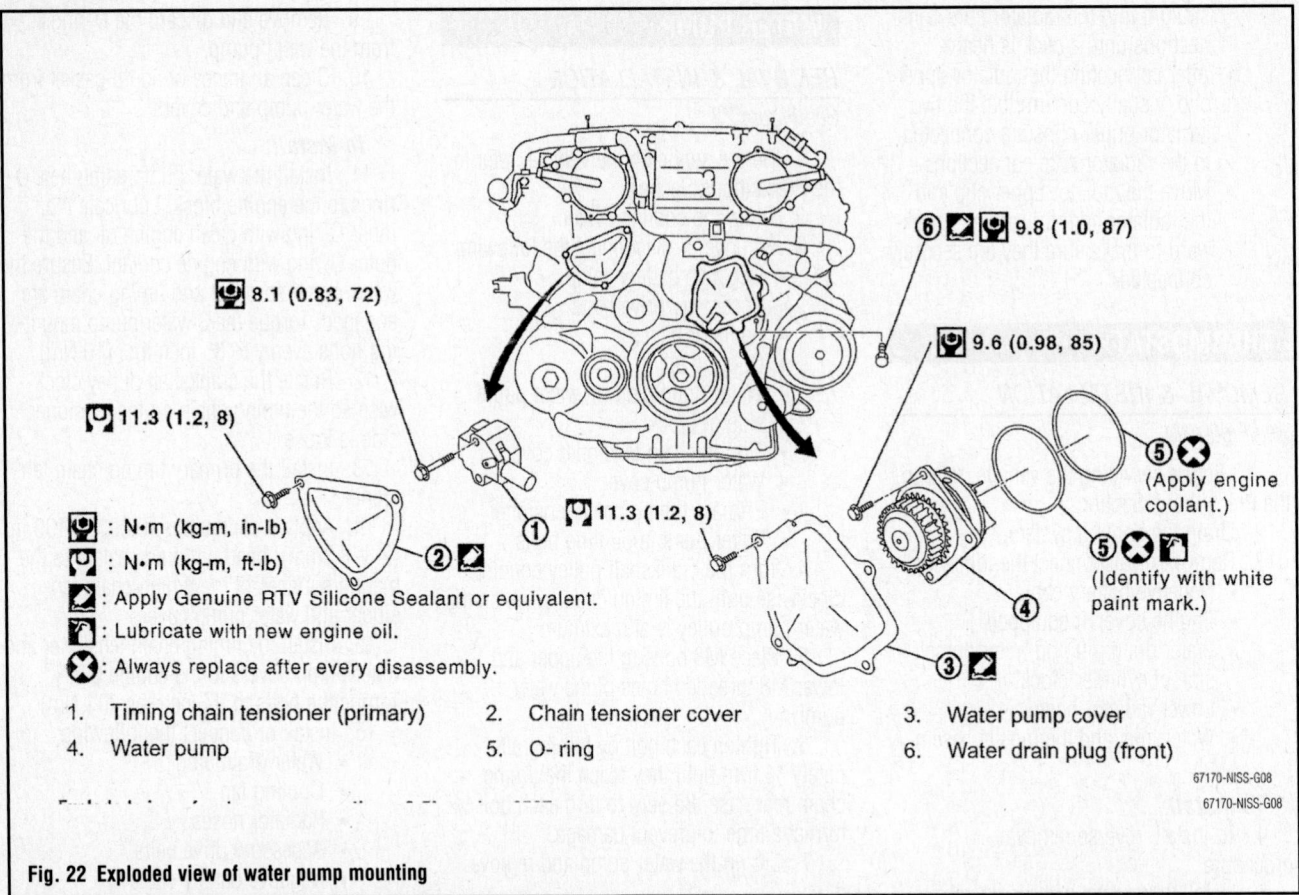

6 ⬛✏️🔧 9.8 (1.0, 87)

8.1 (0.83, 72)

9.6 (0.98, 85)

11.3 (1.2, 8)

11.3 (1.2, 8)

5 ✖️ (Apply engine coolant.)

5 ✖️🔧 (Identify with white paint mark.)

2 ✏️

3 ✏️

1

4

🔧 : N•m (kg-m, in-lb)

🔧 : N•m (kg-m, ft-lb)

✏️ : Apply Genuine RTV Silicone Sealant or equivalent.

🔧 : Lubricate with new engine oil.

✖️ : Always replace after every disassembly.

1. Timing chain tensioner (primary)
4. Water pump
2. Chain tensioner cover
5. O-ring
3. Water pump cover
6. Water drain plug (front)

67170-NISS-G08
67170-NISS-G08

Fig. 22 Exploded view of water pump mounting

ENGINE ELECTRICAL

CHARGING SYSTEM

ALTERNATOR

REMOVAL & INSTALLATION

1. Before servicing the vehicle, refer to the Precautions Section.
2. Remove or disconnect the following:
   • Negative battery cable
   • Right side engine undercover and side inspection cover
   • Radiator fan
   • Drive belt
   • Alternator and A/C compressor harness connectors
   • Upper and lower alternator bolts
   • Alternator

To install:
3. Install or connect the following:
   • Alternator
   • Upper and lower alternator bolts.

Tighten the upper bolt to 48 ft. lbs. (65 Nm) and the lower bolts to 21 ft. lbs. (28 Nm).
   • Alternator and A/C compressor harness connectors
   • Drive belt
   • Radiator fan
   • Right side engine undercover and side inspection cover
   • Negative battery cable

ENGINE ELECTRICAL   **IGNITION SYSTEM**

FIRING ORDER

See Figure 23.

Firing order: 1–2–3–4–5–6

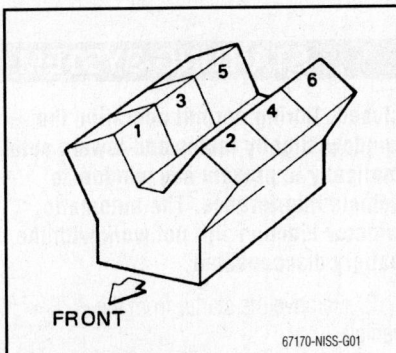

Fig. 23 Distributorless ignition system (one coil on each cylinder)

IGNITION COIL

REMOVAL & INSTALLATION

See Figure 24.

1. Disconnect the negative battery cable.
2. Remove the engine cover (if equipped) with power tool.
3. Remove the air cleaner assembly and air intake tubes.
4. Move aside harness, harness bracket, and hoses located above ignition coil.

5. Disconnect harness connector from ignition coil.
6. Remove the six ignition coils.

To install:

7. To install, reverse removal procedure.

IGNITION TIMING

INSPECTION

➡**The ignition timing is not adjustable. If not within specifications, further diagnostic inspection is required. The following procedure is for viewing the ignition timing setting.**

Visually check the air cleaner, intake hoses, ducts, Exhaust Gas Recirculation (EGR) valve operation and electrical connections prior to the adjustment of the ignition timing. Correct or repair any problem as required. Be sure to inspect the throttle valve and Throttle Position (TP) sensor for proper operation.

1. Before servicing the vehicle, refer to the Precautions Section.
2. Locate the timing marks on the crankshaft pulley and the front of the engine.
3. Clean the timing marks.

➡**The ignition timing specification is 15°, plus or minus 5° Before Top Dead Center (BTDC).**

4. Using chalk or white paint, color the mark on the crankshaft pulley and the mark on the scale that will indicate the correct timing when aligned with the notch on the crankshaft pulley.
5. Attach a tachometer to the engine.
6. Attach a timing light to the engine to number 1 cylinder ignition wire.
7. Turn all electrical equipment and accessories **OFF**.
8. Check to be sure all of the wires clear the fan, then, start the engine and allow it to reach normal operating temperatures.
9. Block the front wheels and set the parking brake. Shift the transmission into **NEUTRAL** for manual transmission and automatic transmissions. Do not stand in front of the vehicle when making adjustments.
10. Perform the following procedures:
    a. Race the engine at 2000 rpm for about 2 minutes under a no-load condition; be sure all of the accessories are turned **OFF**.
    b. Perform on board engine diagnostics and repair any fault code.
    c. Race the engine at 2000 rpm for about 2 minutes under a no-load condition.
    d. Turn the engine **OFF** and disconnect the TP sensor.
    e. Start and race the engine 2 to 3 times under no-load, then run the engine at idle speed.

➡**The ignition timing specification is 15°, plus or minus 5° BTDC.**

11. Aim the timing light at the timing marks. If the marks on the pulley and the engine are aligned when the light flashes, the timing is correct. Turn the engine **OFF** and remove the tachometer and the timing light. If the marks are not in alignment, proceed with the following steps:
    a. Turn the engine **OFF**.
    b. Check the Camshaft Position (CMP) sensor (PHASE), Crankshaft Position (CKP) sensor (REF) and CKP sensor (POS). Replace if necessary.
    c. If the ignition timing is still not correct, substitute a known good Electronic Control Module (ECM).

➡**The ECM may be the cause of the problem but this is rarely the case.**

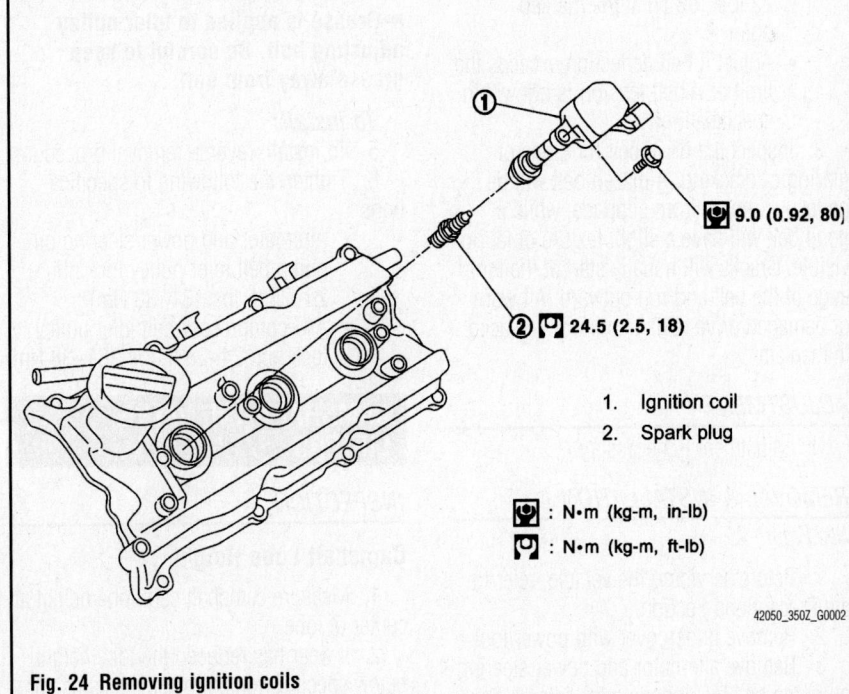

9.0 (0.92, 80)

24.5 (2.5, 18)

1. Ignition coil
2. Spark plug

: N•m (kg-m, in-lb)

: N•m (kg-m, ft-lb)

42050_350Z_G0002

Fig. 24 Removing ignition coils

12. Turn the engine **OFF** and remove the tachometer and the timing light.

ADJUSTMENT

No timing adjustment is necessary.

ENGINE ELECTRICAL

STARTER

REMOVAL & INSTALLATION

1. Remove or disconnect the following:
   - Negative battery cable
   - Air duct
   - Harness protector from the harness
   - Starter wiring at the starter
   - Starter-to-engine bolts
   - Starter from the vehicle

To install:

2. Install or connect the following:
   - Starter
   - Starter wiring
   - Harness protector

SPARK PLUGS

REMOVAL & INSTALLATION

1. Disconnect the negative battery cable.
2. Disconnect ignition wires from coil packs.
   - Air duct
   - Negative battery cable

SOLENOID OR RELAY REPLACEMENT

1. Before servicing the vehicle, refer to the Precautions Section.

➡**If working near and/or around the SRS system and components, be sure to disable the SRS system. After disabling the system wait three minutes or more before servicing the vehicle.**

➡**Before disconnecting the battery, lower both the driver's and passenger's windows. This will prevent any interference between the window edge and the vehicle when the door is opened or**

3. Remove spark plugs with spark plug socket.
4. Install spark plugs and torque to 14–22 ft. lbs. (19–30 Nm).
5. Reconnect ignition wires according to numbers indicated on them.

STARTING SYSTEM

closed. During normal operation the window slightly raises and lowers automatically to prevent any window to vehicle interference. The automatic window function will not work with the battery disconnected.

2. Remove the starter from the vehicle.
3. Separate the solenoid from the starter.

To install:

➡**Be sure to use new fasteners, as required.**

4. Installation is the reverse of the removal procedure.

ENGINE MECHANICAL

ACCESSORY DRIVE BELTS

ACCESSORY BELT ROUTING

See Figure 25.

Refer to the accompanying illustration.

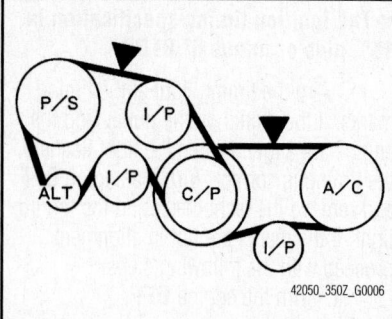

Fig. 25 Accessory drive belt routing—with A/C

42050_350Z_G0006

INSPECTION

1. Inspect belts for cracks, fraying, wear and oil. If necessary, replace.
2. Inspect drive belt deflection at a point on the belt midway between pulleys.
   - Inspection should be done only when engine is cold, or over 30

minutes after engine is stopped
   - Measure belt tension with tension gauge (BT3373-F or equivalent) at points marked
   - When measuring deflection, apply 22 lbs. (98 N) at the marked point
   - Adjust if belt deflection exceeds the limit or if belt tension is not within specifications

3. Inspect the drive belt for signs of glazing or cracking. A glazed belt will be perfectly smooth from slippage, while a good belt will have a slight texture of fabric visible. Cracks will usually start at the inner edge of the belt and run outward. All worn or damaged drive belts should be replaced immediately.

ADJUSTMENT

No adjustment is necessary.

REMOVAL & INSTALLATION

See Figure 25.

1. Before servicing the vehicle, refer to the Precautions Section.
2. Remove undercover with power tool.
3. Remove alternator and power steering oil pump belt by loosening the idler pulley

lock nut located between the alternator and crankshaft pulley.

4. Remove A/C compressor belt by loosening the idler pulley lock nut located between the crankshaft pulley and A/C pulley.

➡**Grease is applied to idler pulley adjusting bolt. Be careful to keep grease away from belt.**

To install:

5. To install, reverse removal procedure.
6. Tighten the following to specifications:
   - Alternator and power steering oil pump belt idler pulley lock nut: 24–28 ft. lbs. (31–38 Nm)
   - A/C compressor belt idler pulley lock nut: 24–28 ft. lbs. (31–38 Nm)

CAMSHAFT AND VALVE LIFTERS

INSPECTION

Camshaft Lobe Height

1. Measure camshaft cam lobe height at center of lobe.
2. If wear has reduced the lobe height below specifications, replace the camshaft.

Journal Oil Clearance

1. Measure outer diameter of camshaft journal. Standard diameter, journal number 1: 1.0211 to 1.0218. Standard diameter, journal numbers 2,3,4: 0.9230 to 0.09238.

2. If wear has reduced the diameter below specifications, replace the camshaft.

Runout

See Figure 26.

1. Put V-block on precise flat bed and support No. 2 and No. 4 journal of camshaft as shown.

2. Set dial gauges vertically to No. 3 journal as shown.

3. Turn camshaft in one direction slowly by hand, measure the camshaft runout on the dial gauges. Standard limit: less than 0.0008 in. (0.02 mm).

4. If actual runout exceeds the limit, replace the camshaft.

Valve Lifter

1. Check if the surface of the valve lifter has any excessive wear or cracks, and replace as necessary.

2. Measure the outer diameter of the valve lifter. Standard diameter: 1.3377 to 1.3381 in.

REMOVAL & INSTALLATION

See Figures 27 through 30.

1. Before servicing the vehicle, refer to the Precautions Section.
2. Relieve the fuel system pressure.
3. Drain the engine oil.
4. Drain the cooling system.
5. Disconnect the negative battery cable.
6. Remove the timing chain case, camshaft sprockets, timing chain and rear timing chain case.
7. Remove or disconnect the following:

- Camshaft Position sensors (PHASE) from the back of the cylinder heads
- Intake valve timing control solenoid valves and discard the gaskets
- Camshaft bearing caps in the reverse of the tightening sequence
- Camshafts
- Valve lifters
- Secondary timing chain tensioners

To install:

➡**Before installing the camshaft brackets, apply RTV sealant to the mating surface of the No. 1 journal head.**

8. Lubricate the valve lifters with clean engine oil and install the lifters into the bore from which they were removed.

9. Ensure the crankshaft is set to TDC for the No. 1 cylinder.

10. Install the camshaft tensioners using new O-rings on both sides of the cylinder heads. The sliding part faces downward on the right head and upward on the left head. Torque the bolts to 75 inch lbs. (8.4 Nm).

➡**The camshafts can be identified by the paint marks on the camshaft. The intake camshafts have a PINK paint mark and the exhaust camshafts have a ORANGE paint mark.**

➡**Install the camshafts so the large and small pin holes are located on the front face of the camshafts at 180° intervals.**

11. Install the bearing caps aligning the stamp marks on the caps as shown.

12. Torque the camshaft bearing caps as follows:

   a. Bolts No. 7 to 10: 17 inch lbs. (2 Nm).

   b. Bolts No. 1 to 6: 17 inch lbs. (2 Nm).

   c. Bolts No. 1 to 10: 52 inch lbs. (6 Nm).

   d. Bolts No. 1 to 10: 92 inch lbs. (10.4 Nm).

13. Check and adjust the valve clearance.

14. Install or connect the following:

- Intake valve timing control solenoid valves, using new gaskets, and tightening the bolts to 8 ft. lbs. (11.3 Nm)
- Camshaft Position sensors (PHASE) and tighten the bolts to 85 inch lbs. (9.6 Nm)
- Timing chain case, camshaft sprockets, timing chain and rear timing chain case
- Negative battery cable

15. Fill the cooling system.
16. Fill the engine with clean oil.

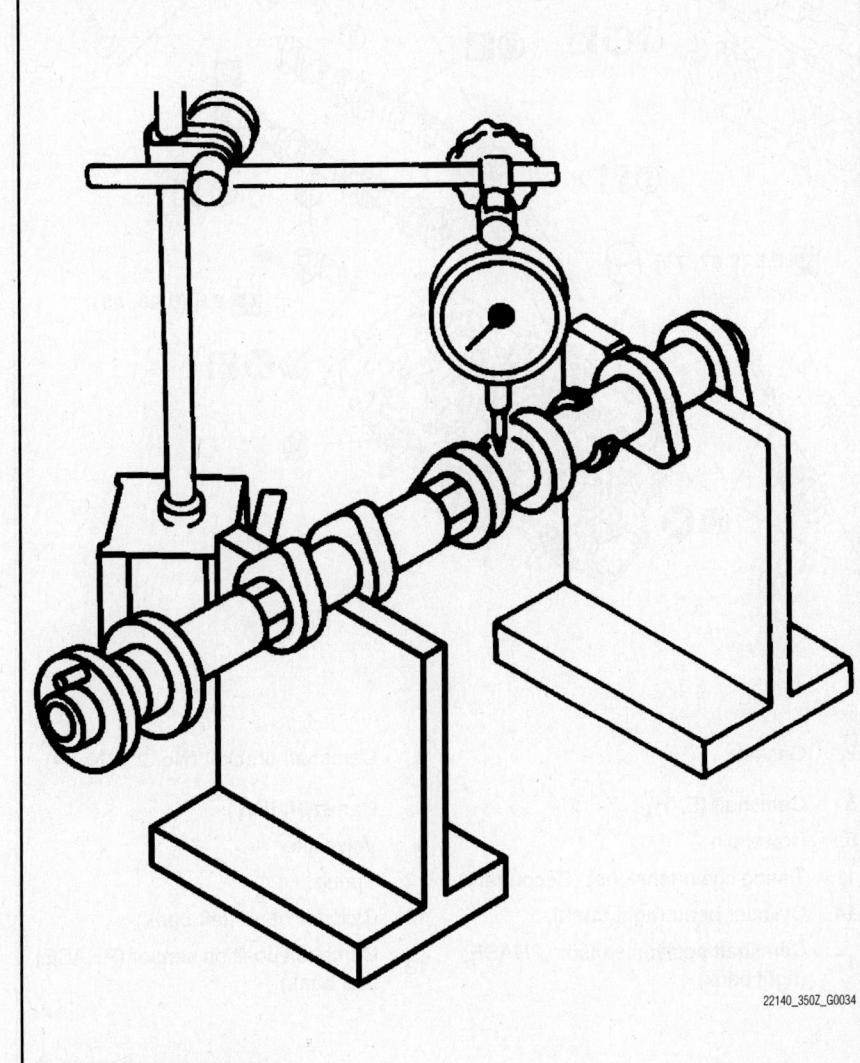

22140_350Z_G0034

Fig. 26 Camshaft inspection

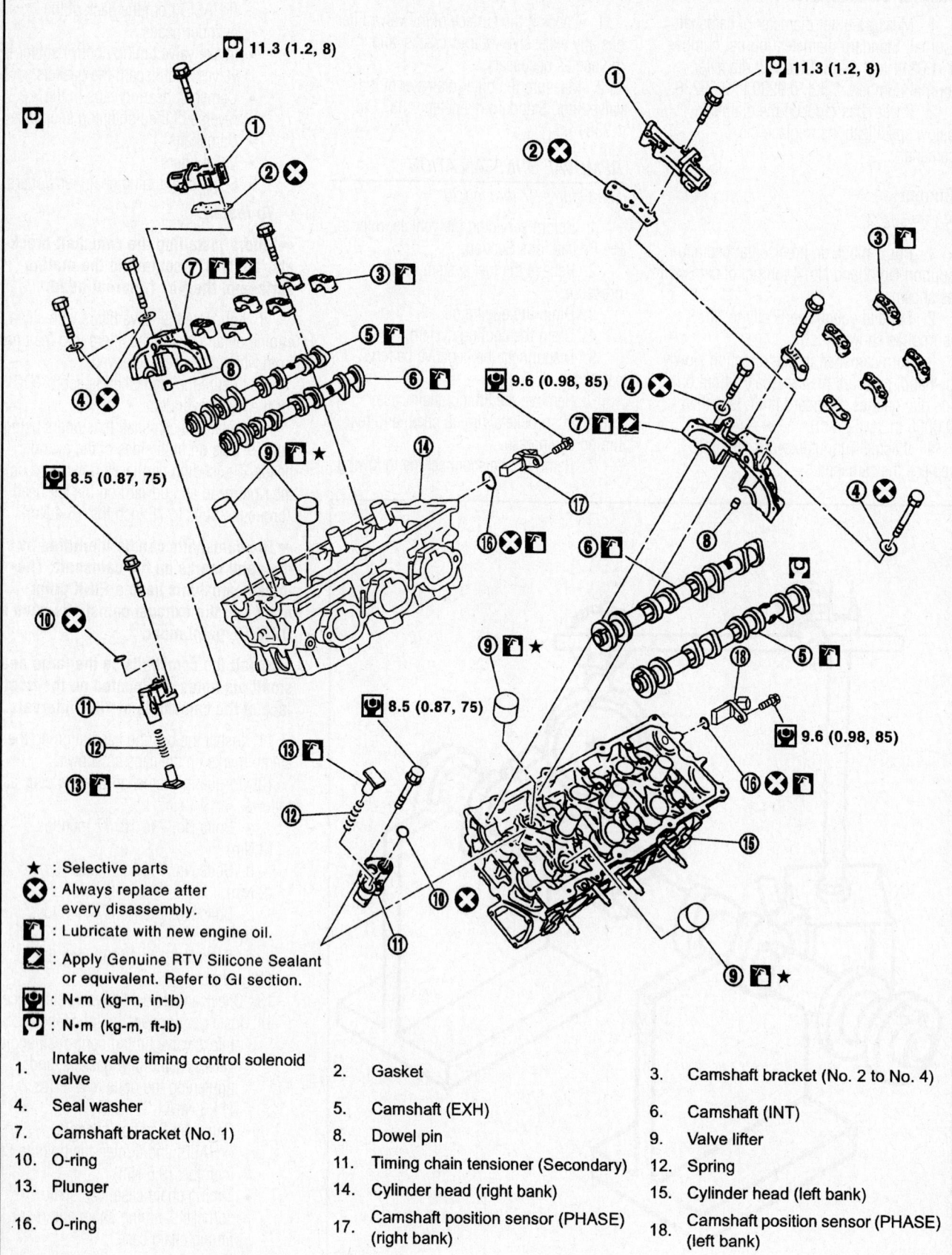

11.3 (1.2, 8)

11.3 (1.2, 8)

9.6 (0.98, 85)

8.5 (0.87, 75)

8.5 (0.87, 75)

9.6 (0.98, 85)

★ : Selective parts

⊗ : Always replace after every disassembly.

: Lubricate with new engine oil.

: Apply Genuine RTV Silicone Sealant or equivalent. Refer to GI section.

: N•m (kg-m, in-lb)

: N•m (kg-m, ft-lb)

1. Intake valve timing control solenoid valve	2. Gasket	3. Camshaft bracket (No. 2 to No. 4)
4. Seal washer	5. Camshaft (EXH)	6. Camshaft (INT)
7. Camshaft bracket (No. 1)	8. Dowel pin	9. Valve lifter
10. O-ring	11. Timing chain tensioner (Secondary)	12. Spring
13. Plunger	14. Cylinder head (right bank)	15. Cylinder head (left bank)
16. O-ring	17. Camshaft position sensor (PHASE) (right bank)	18. Camshaft position sensor (PHASE) (left bank)

67170-NISS-G29

Fig. 27 Exploded view of camshaft assemblies

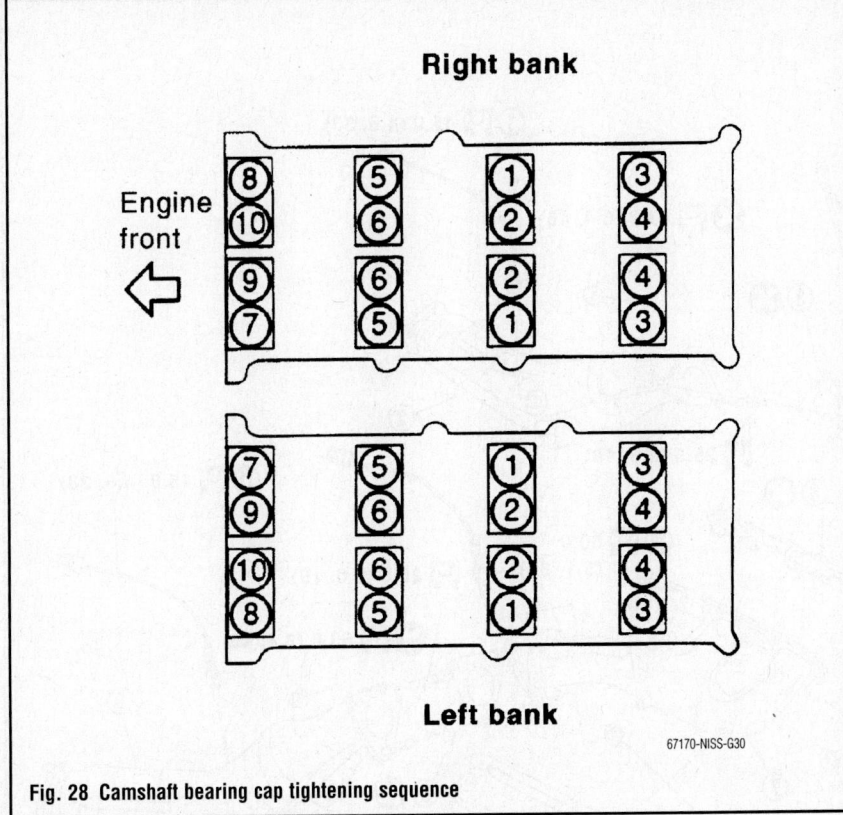

Fig. 28 Camshaft bearing cap tightening sequence

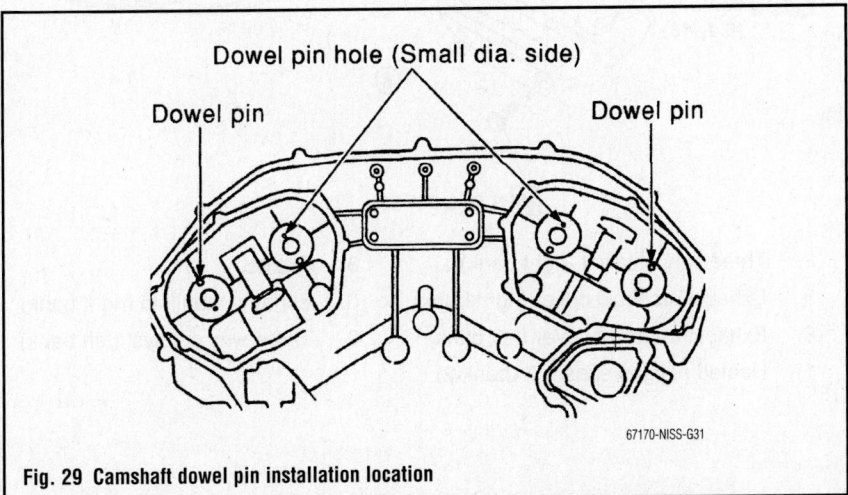

Fig. 29 Camshaft dowel pin installation location

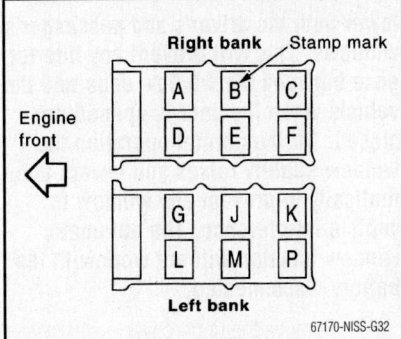

Fig. 30 Camshaft bearing cap stamp positioning

➥If working near and/or around the SRS system and components, be sure to disable the SRS system. After disabling the system wait three minutes or more before servicing the vehicle.

➥Before disconnecting the battery, lower both the driver's and passenger's windows. This will prevent any interference between the window edge and the vehicle when the door is opened or closed. During normal operation the window slightly raises and lowers automatically to prevent any window to vehicle interference. The automatic window function will not work with the battery disconnected.

2. Raise and safely support the vehicle.
3. Properly support the exhaust system.
4. Disconnect the oxygen sensor electrical wires. As required, remove the sensor from its mounting.

➥Be careful not to drop or damage the sensor. If a sensor has been dropped, it must be replaced.

5. Remove the converter retaining bolts and nuts.
6. Remove the component from the vehicle.
7. Discard the gaskets.

To install:

➥Be sure to use new fasteners, as required.

8. Installation is the reverse of the removal procedure.
9. Be sure to use new gaskets, as required.

CRANKSHAFT DAMPER

REMOVAL & INSTALLATION

1. Before servicing the vehicle, refer to the Precautions Section.

➥If working near and/or around the SRS system and components, be sure to disable the SRS system. After disabling the system wait three minutes or more before servicing the vehicle.

➥Before disconnecting the battery, lower both the driver's and passenger's windows. This will prevent any interference between the window edge and the vehicle when the door is opened or closed. During normal operation the window slightly raises and lowers automatically to prevent any window to vehicle interference. The automatic window function will not work with the battery disconnected.

17. Start the vehicle, check for leaks and repair if necessary.

CATALYTIC CONVERTER

REMOVAL & INSTALLATION

See Figure 31.

At this time the manufacturer does not provide removal and installation procedures for this component. The following procedure is a guideline and may differ from the vehicle you are servicing.

1. Before servicing the vehicle, refer to the Precautions Section.

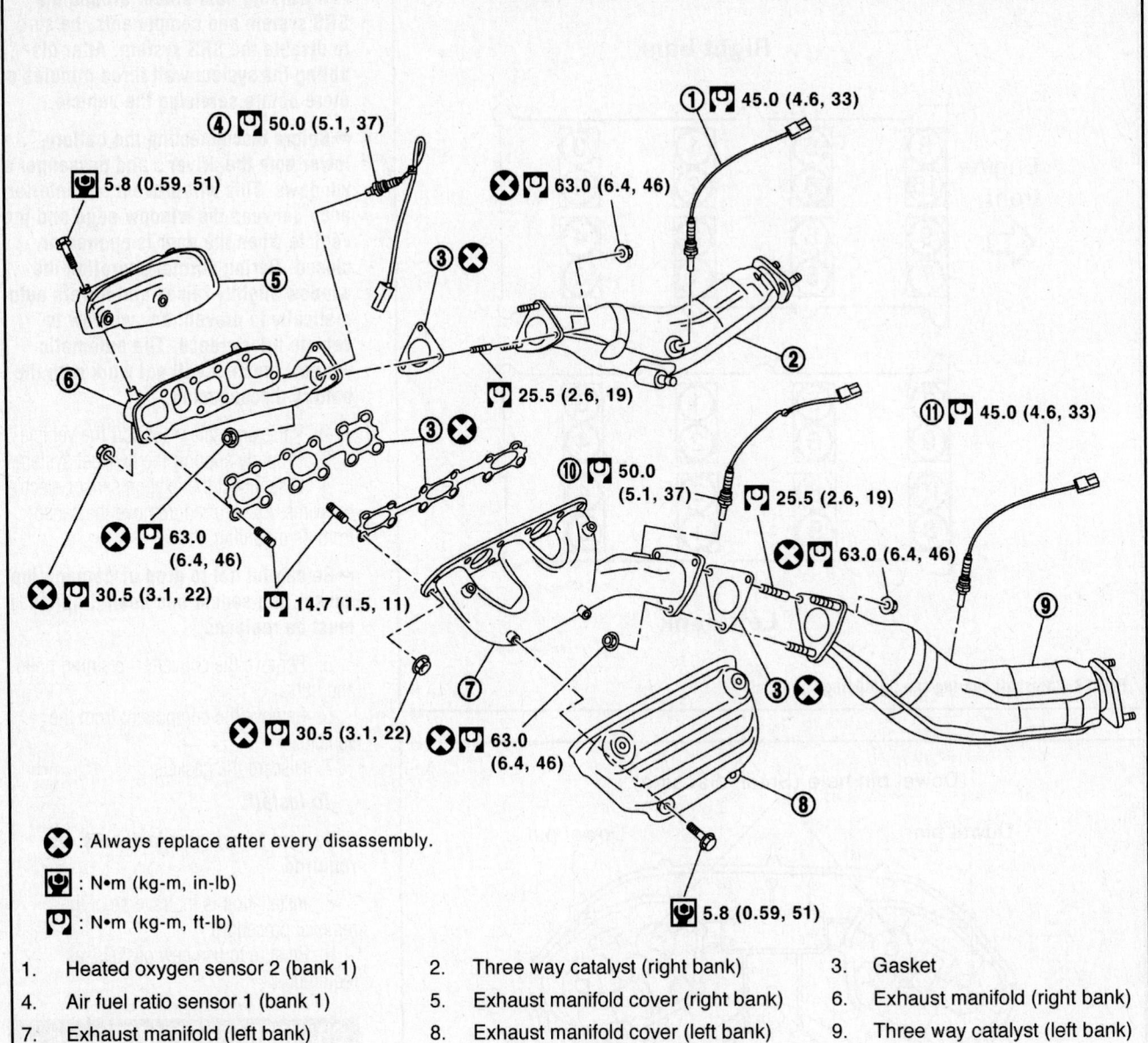

① 🔩 45.0 (4.6, 33)

④ 🔩 50.0 (5.1, 37)

❌ 🔩 63.0 (6.4, 46)

🔧 5.8 (0.59, 51)

⑤

③ ❌

⑥

②

🔩 25.5 (2.6, 19)

❌ 🔩 63.0
(6.4, 46)

❌ 🔩 30.5 (3.1, 22)

🔩 14.7 (1.5, 11)

③ ❌

⑦

❌ 🔩 30.5 (3.1, 22)

❌ 🔩 63.0
(6.4, 46)

⑩ 🔩 50.0
(5.1, 37)

🔩 25.5 (2.6, 19)

❌ 🔩 63.0 (6.4, 46)

③ ❌

⑧

⑪ 🔩 45.0 (4.6, 33)

⑨

🔧 5.8 (0.59, 51)

❌ : Always replace after every disassembly.

🔧 : N•m (kg-m, in-lb)

🔩 : N•m (kg-m, ft-lb)

1. Heated oxygen sensor 2 (bank 1)	2. Three way catalyst (right bank)	3. Gasket
4. Air fuel ratio sensor 1 (bank 1)	5. Exhaust manifold cover (right bank)	6. Exhaust manifold (right bank)
7. Exhaust manifold (left bank)	8. Exhaust manifold cover (left bank)	9. Three way catalyst (left bank)
10. Air fuel ratio sensor 1 (bank 2)	11. Heated oxygen sensor 2 (bank 2)	

41311_350Z_G0033

Fig. 31 Catalytic converter and related components

2. Remove the undercover.
3. Remove the drive belts.
4. Remove the radiator cooling fan assembly.
5. Remove the crankshaft damper.

To install:

➡**Be sure to use new fasteners, as required**

6. Installation is the reverse of the removal procedure.
7. Be sure to tighten the bolt to specification.

CRANKSHAFT FRONT SEAL

REMOVAL & INSTALLATION

See Figure 32.

1. Before servicing the vehicle, refer to the Precautions Section.

➡**If working near and/or around the SRS system and components, be sure to disable the SRS system. After disabling the system wait three minutes or more before servicing the vehicle.**

➡**Before disconnecting the battery,**

lower both the driver's and passenger's windows. This will prevent any interference between the window edge and the vehicle when the door is opened or closed. During normal operation the window slightly raises and lowers automatically to prevent any window to vehicle interference. The automatic window function will not work with the battery disconnected.

2. Remove the undercover.
3. Remove the drive belts.
4. Remove the radiator cooling fan assembly.

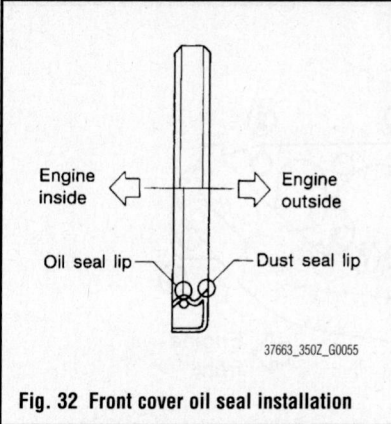

Fig. 32 Front cover oil seal installation

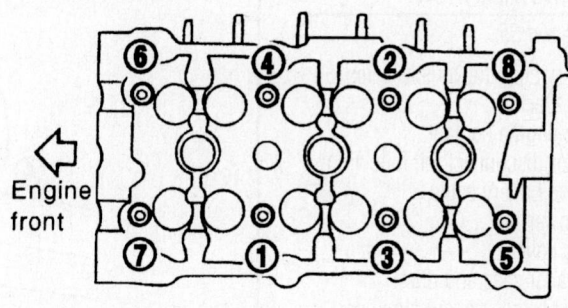

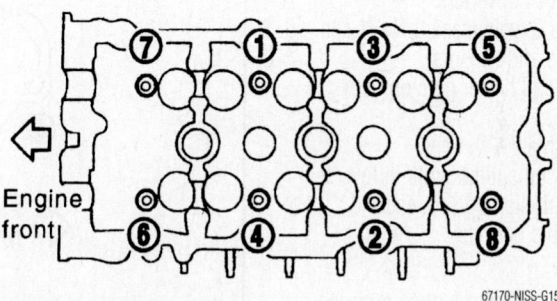

Fig. 33 Cylinder head bolt torque sequence

5. Remove the crankshaft damper.

6. Remove the oil seal using a suitable tool. Discard the seal.

➡**Be sure not to damage the timing chain case and crankshaft.**

To install:

➡**Be sure to use new fasteners, as required**

7. Apply clean engine oil to both the seal lip and the dust seal lip.

8. Install the seal so that each seal lip is positioned as shown in the illustration.

9. Using a drift (outer diameter 2.36 inch, inner diameter 1.97 inch), press fit the seal in place until the height of the seal is level with the mounting surface.

➡ **Be sure not to damage the timing chain case and crankshaft. Press fit straight and avoid causing burrs or tilting the seal.**

10. Continue the installation is the reverse of the removal procedure.

CYLINDER HEAD

REMOVAL & INSTALLATION

See Figure 33.

➡**You must remove the engine from the vehicle in order to remove the cylinder head for this procedure.**

1. Before servicing the vehicle, refer to the Precautions Section.

2. Relieve the fuel system pressure.

3. Drain the engine oil.

4. Drain the cooling system.

➡**Before detaching any hoses or connectors, note the locations for reassembly.**

5. Remove or disconnect the following:

- Negative battery cable
- Engine assembly
- Timing chain
- Camshafts
- Fuel injector assembly
- Intake manifold
- Exhaust manifold
- Thermostat housing
- Cylinder head bolts in the reverse of the tightening sequence
- Cylinder head gaskets

To install:

6. Turn the crankshaft until the No. 1 piston is a Top Dead Center (TDC) on compression stroke. The crankshaft key should face toward the right bank.

7. Using new head gaskets, install the cylinder heads.

➡**If possible, replacement of the head bolts is suggested.**

8. If replacement of the head bolts is not possible, perform the following bolt measurement:

a. Measure the diameter of the head bolt 0.43 in. (11 mm) from the bottom of the bolt.

b. Measure the diameter of the head bolt 1.89 in. (48 mm) from the bottom of the bolt.

c. Whenever the size difference between the 2 measurements exceeds 0.0043 in. (0.11 mm), the head bolts must be replaced.

9. Install the cylinder head bolts and torque in sequence as follows:

a. Step 1: 72 ft. lbs. (98 Nm).

b. Step 2: Completely loosen all bolts.

c. Step 3: 26–32 ft. lbs. (34–44 Nm).

d. Step 4: plus 90–95 degrees clockwise.

e. Step 5: plus 90–95 degrees clockwise.

10. Install or connect the following:

- Thermostat housing
- Exhaust manifold
- Intake manifold
- Fuel injector assembly
- Camshafts
- Timing chain
- Engine assembly
- Negative battery cable

11. Fill the cooling system.

12. Fill the engine with clean oil.

13. Start the vehicle, check for leaks and repair if necessary.

EXHAUST MANIFOLD

REMOVAL & INSTALLATION

See Figures 34 and 35.

1. Before servicing the vehicle, refer to the Precautions Section.
2. Drain the engine coolant.
3. Remove or disconnect the following:
   - Negative battery cable
   - Strut tower bar
   - Engine cover
   - Air cleaner case and duct
   - Heated Oxygen sensor No. 2 connectors and sensors
   - Exhaust mounting bracket between transmission and catalytic converters
   - Catalytic converters
   - Heated Oxygen sensor No. 1 connectors and sensors
   - Water and heater pipe on both sides
   - Heat shield
   - Exhaust manifold bolts in reverse of the tightening sequence
   - Manifold gaskets

To install:

4. Clean all gasket mounting surfaces. Install new gaskets noting the correct placement.
5. Installation is the reverse of the removal procedure noting the following:
   - Exhaust manifold and torque the nuts in sequence to 22 ft. lbs. (30 Nm)
   - Heat shields and torque the bolts in steps to 51 inch lbs. (6 Nm)
   - Exhaust manifolds to the exhaust pipes and torque the nuts to 46 ft. lbs. (63 Nm)

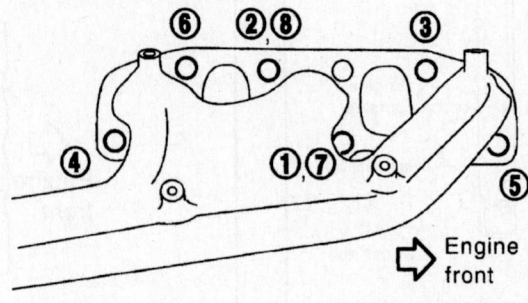

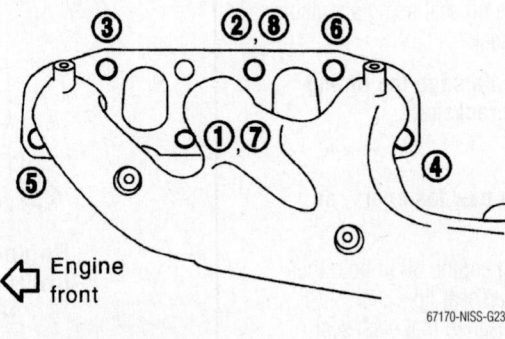

Fig. 35 Exhaust manifold bolt tightening sequence

- Oxygen sensors and torque the fastener to 33 ft. lbs. (45 Nm)

FLEXPLATE/FLYWHEEL

REMOVAL & INSTALLATION

1. Remove the transmission.
2. Remove drive plate.

To install:

3. Installation is reverse of removal.
   a. Tighten bolts to 65 ft. lbs. (88 Nm).

INTAKE MANIFOLD

REMOVAL & INSTALLATION

See Figures 36 through 40.

1. Before servicing the vehicle, refer to the Precautions Section.
2. Drain the cooling system.
3. Release the fuel system pressure.
4. Disconnect the negative battery cable.
5. Remove or disconnect the following:
   - Strut tower bar
   - Engine cover
   - Air cleaner case and duct
   - Electronic throttle control actuator bolt in the sequence shown

- Fuel injector and fuel tube assembly
- Vacuum and water hoses from intake manifold collector
- EVAP solenoid valve bracket
- Upper intake manifold collector bolts in reverse of the tightening sequence
- PCV hose

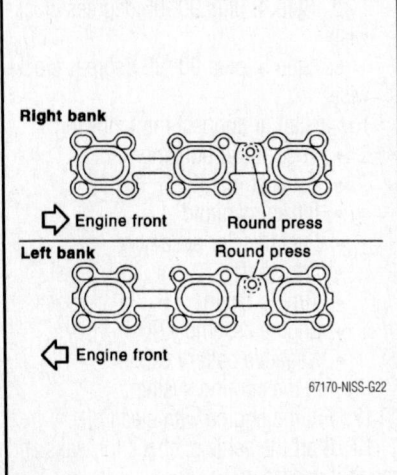

Fig. 34 Exhaust manifold gasket identification

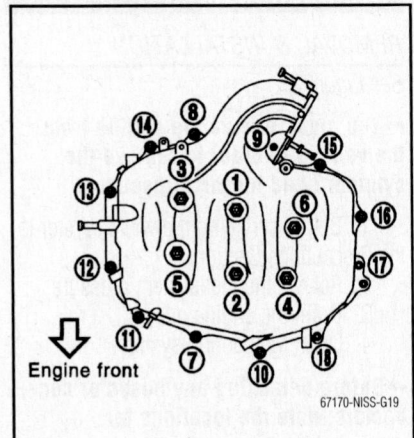

Fig. 36 Upper intake manifold collector tightening sequence

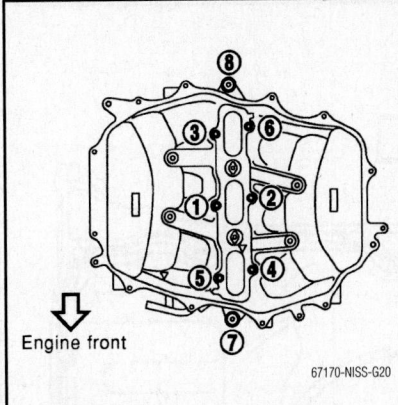

Fig. 37 Lower intake manifold collector tightening sequence

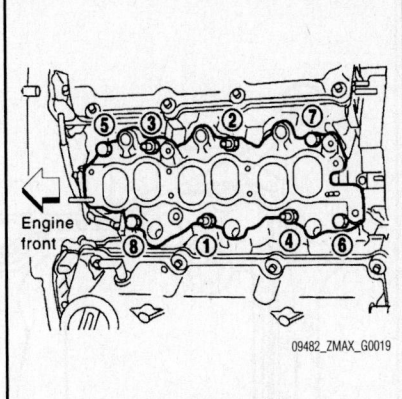

Fig. 39 Intake manifold torque sequence

- Lower intake manifold collector bolts in reverse of the tightening sequence
6. Remove the fuel injector assembly by performing the following procedures:
   a. Detach the electrical connectors from the fuel injectors.
   b. Disconnect the fuel lines from the fuel injector assembly.
   c. Remove the fuel rail-to-cylinder head bolts.
   d. Remove the fuel rail assembly from the engine.
7. Remove or disconnect the following:
   - Intake manifold bolts/nuts in the reverse of the installation sequence
   - Intake manifold from the engine and discard the gaskets
8. Clean all gasket mounting surfaces.

To install:

9. Using new gaskets, install the intake manifold to the engine.
10. If necessary, tighten the intake manifold stud bolts to 87–104 inch lbs. (10–12 Nm).
11. Torque the intake manifold bolts in sequence as follows:

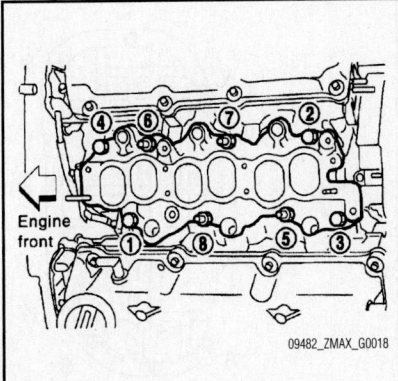

Fig. 38 Intake manifold bolt loosening sequence

a. Step 1: 4–7 ft. lbs. (5–10 Nm).
b. Step 2: 20–23 ft. lbs. (26–31 Nm).
c. Step 3: Tighten again to 20–23 ft. lbs. (26–31 Nm).
12. Install or connect the following:
   a. Install the lower intake manifold collector bolts and tighten to 10 ft. lbs. (14 Nm)
   b. Install PCV hose.
   c. Install upper intake manifold collector bolts and tighten to 10 ft. lbs. (14 Nm)
   d. Install EVAP solenoid valve bracket.
   e. Install vacuum and water hoses to intake manifold collector.
   f. Install fuel injector and fuel tube assembly.
   g. Install electronic throttle control actuator bolts in the sequence shown and tighten to 64–85 inch lbs. (7–10 Nm).
   h. Install air cleaner case and duct.
   i. Install engine cover.
   j. Install strut tower bar and tighten to 24 ft. lbs. (32 Nm).
13. Connect the negative battery cable.
14. Fill the cooling system.

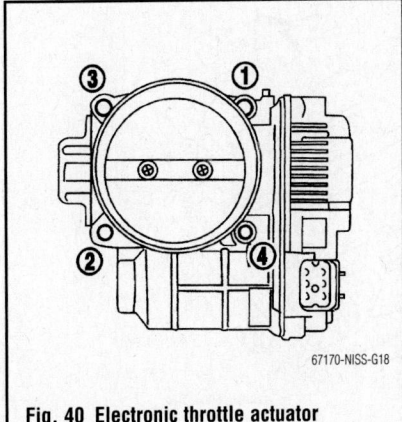

Fig. 40 Electronic throttle actuator removal and installation sequence

15. Start the vehicle, check for leaks and repair if necessary.

OIL PAN

REMOVAL & INSTALLATION

See Figures 41 through 45.

➡ **When removing oil pan (lower) only, remove engine assembly is not necessary. Perform step 1, 2 and 10.**

1. Drain engine oil.
2. Remove undercover with power tool.
3. Remove engine assembly from the vehicle, and separate front suspension member and transmission from engine.
4. Remove alternator.
5. Remove starter motor.
6. Remove idler pulley and bracket assembly.
7. Remove oil filter as necessary.
8. Remove oil temperature sensor as necessary.
9. Remove oil pan (lower) as follows:
   a. Loosen mounting bolts with power tool in reverse order as shown in the figure to remove.
   b. Insert seal cutter (SST) between oil pan (upper) and oil pan (lower).

➡ **Be careful not to damage the mating surfaces. Never insert screwdriver, this will damage the mating surface.**

10. Remove oil strainer.
11. Remove rear cover plate.
12. Loosen mounting bolts with power tool in reverse order as shown in the figure to remove oil pan (upper).

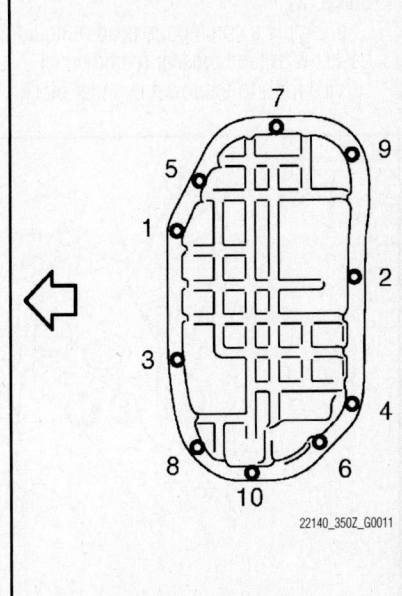

Fig. 41 Lower oil pan bolt sequence

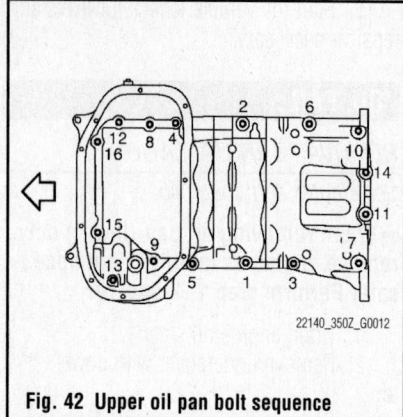

Fig. 42 Upper oil pan bolt sequence

13. Insert seal cutter [SST: KV10111100 (J37228)] between oil pan(upper) and cylinder block. Slide seal cutter by tapping on the side of tool with hammer. Remove oil pan (upper).

※※ WARNING

Be careful not to damage the mating surfaces. Never insert screwdriver, this will damage the mating surface.

14. Remove O-rings (2) from bottom of lower cylinder block (1) and oil pump (3).

To install:

15. Install oil pan (upper) as follows:

a. Use scraper to remove old liquid gasket from mating surfaces.

b. Also remove the old liquid gasket from mating surface of lower cylinder block.

c. Remove old liquid gasket from the bolt holes and threads.

d. Install new O-rings (2) on the bottom of lower cylinder block (1) and oil pump (3).

e. Apply a continuous bead of liquid gasket with tube presser (commercial service tool) to the lower cylinder block

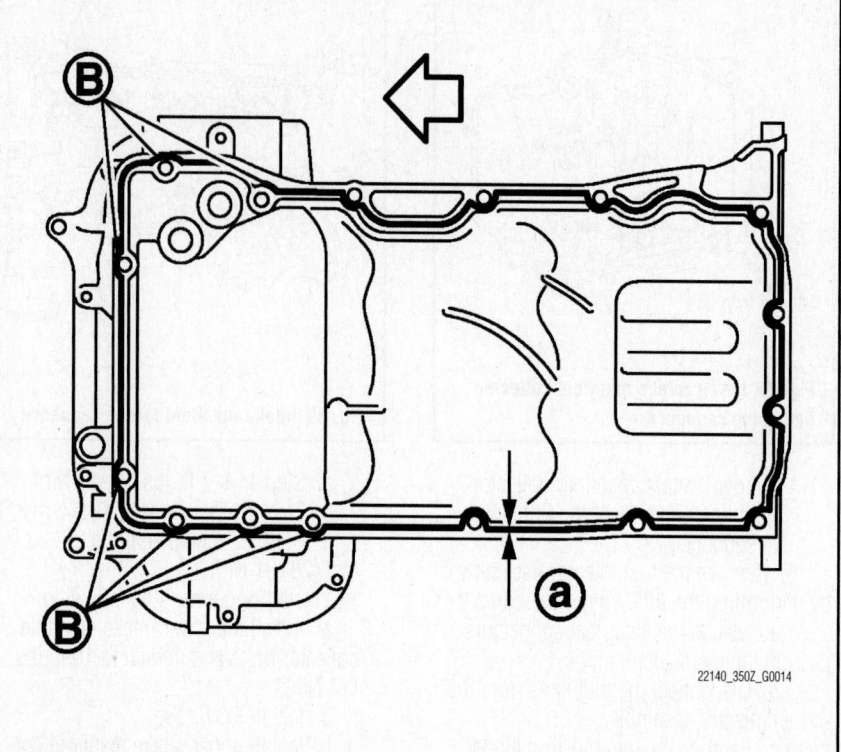

Fig. 44 Sealant location

mating surface of oil pan (upper) to a limited portion as shown in the figure.

➡**For bolt holes (B) (7 locations), apply liquid gasket outside the holes. Attaching should be done within 5 minutes after coating.**

f. Install oil pan (upper).

g. Tighten mounting bolts in numerical order as shown in the figure.

➡**There are two types of mounting bolts. Install the bolts to where they were removed from.**

h. Tighten transmission joint bolts.

16. Install oil strainer to oil pump. Apply locking sealant to the thread of mounting bolts.

17. Install oil pan (lower) as follows:

a. Use scraper to remove old liquid gasket from mating surfaces.

b. Also remove the old liquid gasket from mating surface of upper oil pan.

c. Remove old liquid gasket from the bolt holes and threads.

d. Apply a continuous bead of liquid gasket with tube presser (commercial service tool) to the oil pan (lower).

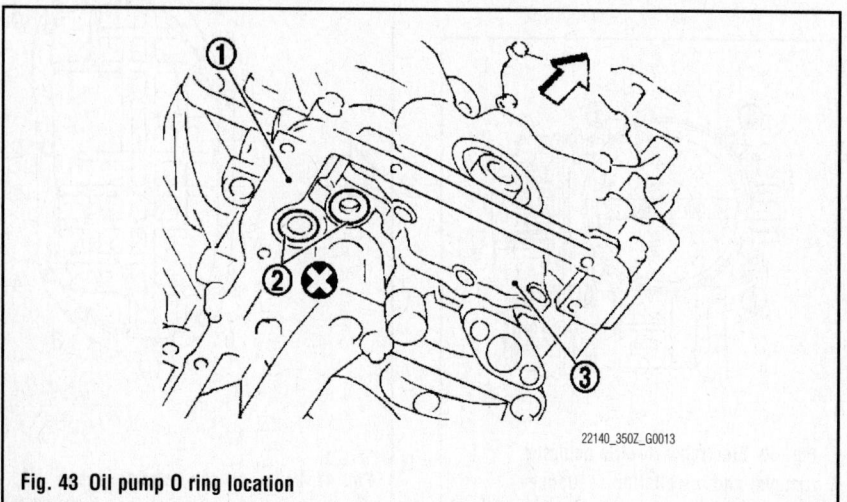

Fig. 43 Oil pump O ring location

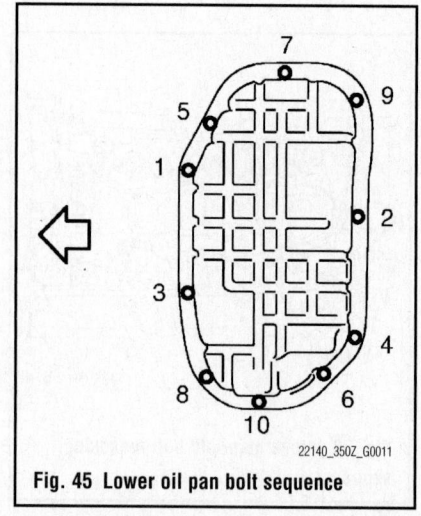

Fig. 45 Lower oil pan bolt sequence

e. Install oil pan (lower). Tighten mounting bolts in numerical order as shown in the figure.

18. Install oil pan drain plug.

19. Install in the reverse order of removal after this step.

OIL PUMP

REMOVAL & INSTALLATION

See Figure 46.

1. Before servicing the vehicle, refer to the Precautions Section.

2. Drain the engine oil

3. Remove or disconnect the following:
- Negative battery cable
- Drive belts
- Camshaft Position (CMP) sensor (PHASE) and the Crankshaft Position (CKP) sensor (REF)/(POS)
- Engine lower covers
- Crankshaft pulley
- Front exhaust tube and support
- Right side mounting insulator and bracket
- Center member
- A/C compressor and move it aside
- Oil pans
- Water pump cover
- Front cover
- Timing chain
- Oil pump assembly

4. Clean all mating surfaces.

To install:

5. Install or connect the following:
- Oil pump
- Timing chain
- Front cover and torque the long bolt to 62 inch lbs. (7 Nm) and the short bolt to 61 inch lbs. (6.5 Nm)
- Water pump cover
- Oil pans
- A/C compressor
- Center member
- Right side mounting insulator and bracket
- Front exhaust tube and support
- Crankshaft pulley
- CMP and CKP sensors
- Engine lower covers
- Drive belts
- Negative battery cable

6. Fill the engine with clean oil.

7. Start the vehicle, check for leaks and repair if necessary.

INSPECTION

1. Clearance between outer rotor and oil pump body: 0.0045–0.0079 in. (0.114–0.200 mm).

2. Tip clearance between inner rotor and outer rotor: Below 0.0071 in. (0.180 mm).

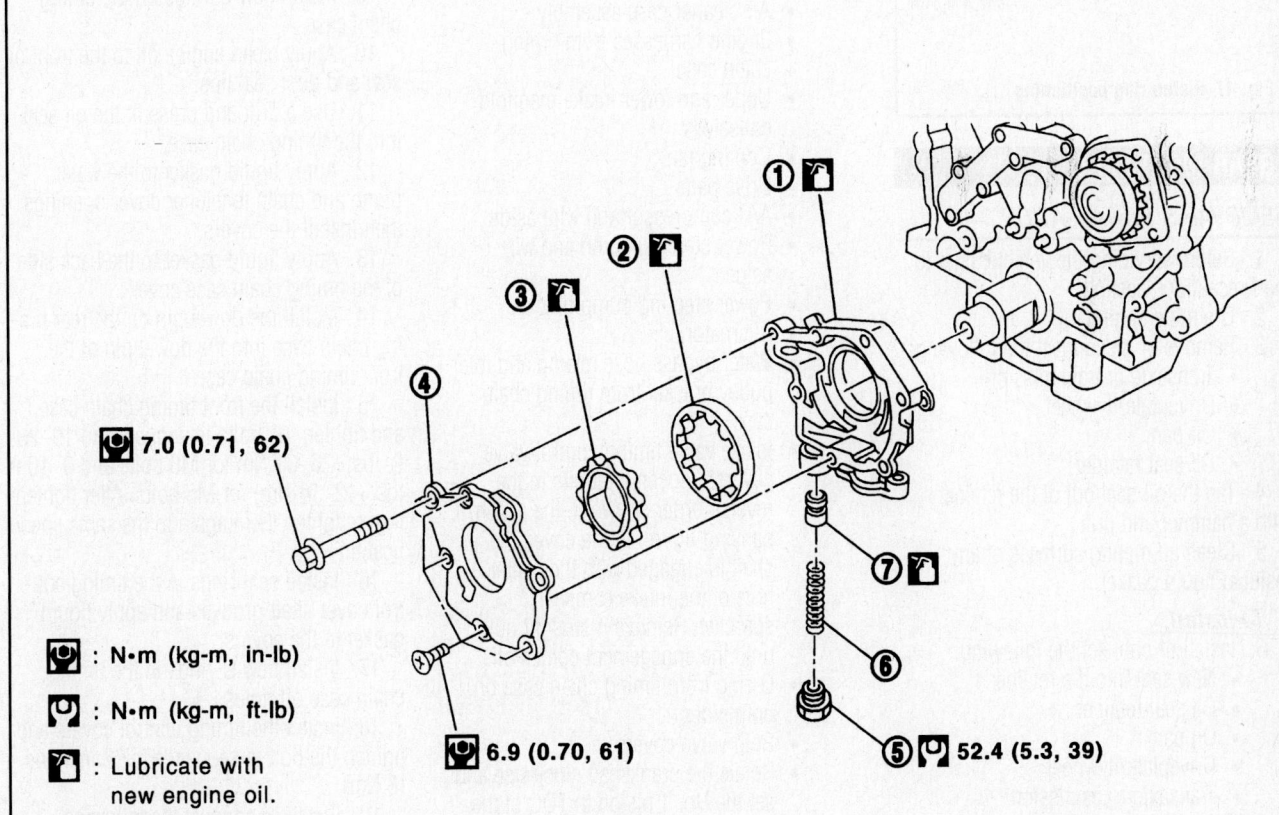

7.0 (0.71, 62)

🔲 : N•m (kg-m, in-lb)

🔲 : N•m (kg-m, ft-lb)

🔲 : Lubricate with new engine oil.

6.9 (0.70, 61)

52.4 (5.3, 39)

1. Oil pump body
2. Oil pump outer rotor
3. Oil pump inner rotor
4. Oil pump cover
5. Regulator valve plug
6. Regulator valve spring
7. Regulator valve

09482_ZMAX_G0029

Fig. 46 Exploded view of the oil pump assembly

3. Side clearance with a straight-edge between inner rotor and oil pump body: 0.0012–0.0028 in. (0.030–0.070 mm).

4. Side clearance with a straight-edge between outer rotor and oil pump body: 0.0020–0.0043 in. (0.050–0.110 mm).

PISTON AND RING

POSITIONING

See Figure 47.

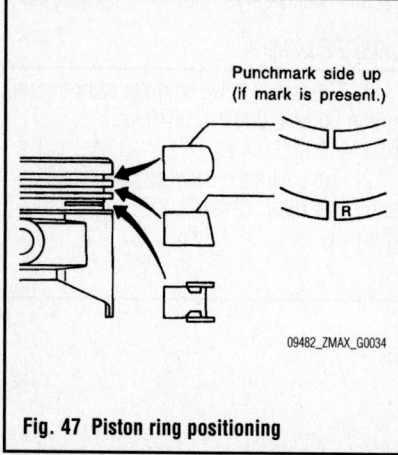

Fig. 47 Piston ring positioning

REAR MAIN SEAL

REMOVAL & INSTALLATION

1. Before servicing the vehicle, refer to the Precautions section.
2. Drain the engine oil.
3. Remove or disconnect the following:
   • Transaxle or transmission
   • Driveplate/flywheel
   • Oil pan
   • Oil seal retainer
4. Tap the oil seal out of the retainer with a hammer and drift.
5. Clean all mating surfaces of any residual liquid gasket.

To install:
6. Install or connect the following:
   • New seal into the retainer
   • Oil seal retainer
   • Oil pan
   • Driveplate/flywheel
   • Transaxle/transmission
7. Fill the engine with clean oil.
8. Start the vehicle, check for leaks and repair if necessary.

TIMING CHAIN FRONT COVER

REMOVAL & INSTALLATION

See Figures 48 and 49.

1. Before servicing the vehicle, refer to the Precautions Section.

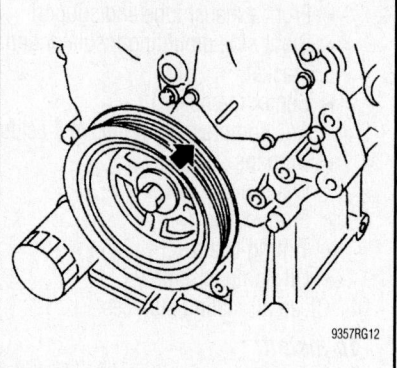

Fig. 48 Set the No. 1 piston to Top Dead Center (TDC)

2. Drain the engine oil.
3. Drain the cooling system.
4. Relieve the fuel system pressure.
5. Remove or disconnect the following:
   • Negative battery cable
   • Engine cover
   • Air cleaner case assembly
   • Engine harnesses from timing chain case
   • Upper and lower intake manifold collectors
   • Cooling fan
   • Drive belts
   • A/C compressor and wire aside
   • Power steering pump and wire aside
   • Power steering pump bracket
   • Alternator
   • Water bypass hose, clamp and idler pulley bracket from timing chain case
   • Intake valve timing control valve covers. Loosen the bolts in the reverse order shown in the accompanying figure. In the cover, the shaft is engaged with the center hole of the intake camshaft sprocket. Remove it straight out until the engagement comes off.
   • O-ring from timing chain case on both sides
   • Both valve covers
   • Rotate the crankshaft clockwise and set the No. 1 piston to TDC of the compression stroke
6. Make sure the intake and exhaust camshaft lobes are facing toward the inside of the cylinder head.
7. Remove the starter and lock the flywheel through the starter mounting hole.
8. Loosen the crankshaft pulley bolt.
   • Crankshaft pulley using a suitable puller
   • Upper and lower oil pans

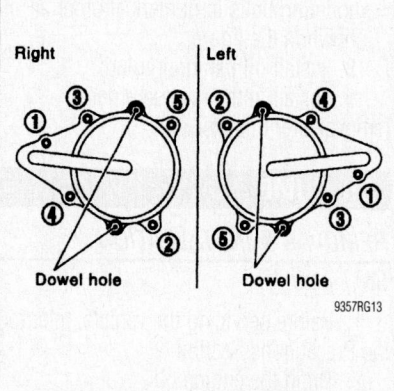

Fig. 49 Tighten the intake valve timing control valve cover bolts in sequence

   • Front timing chain cover bolts in reverse order of the tightening sequence
   • Front timing chain case

To install:
9. Install new O-rings on the timing chain case.
10. Apply clean engine oil to the front oil seal and dust seal lips.
11. Use a drift and press fit the oil seal into the timing chain case.
12. Apply liquid gasket to the water pump and chain tensioner cover openings, then install the covers.
13. Apply liquid gasket to the back side of the timing chain case cover.
14. Install the dowel pin on the rear timing chain case into the dowel pin of the front timing chain case.
15. Install the front timing chain case and tighten the bolts in sequence to 19–23 ft. lbs. (26–31 Nm for M8 bolts and 9–10 ft. lbs. (12–14 Nm) for M6 bolts. After tightening, retighten them again to the same specification.
16. Install seal rings in the timing control cover shaft grooves and apply liquid gasket to the covers.
17. Install new O-rings in the timing chain case oil holes.
18. Install the timing control covers and tighten the bolt in sequence to 72 inch lbs. (8 Nm).
19. Install or connect the following:
   • Upper and lower oil pans
   • Valve covers
   • Crankshaft pulley and tighten the bolt to 29–36 ft. lbs. (40–49 Nm), plus an additional 60°
20. The remainder of the installation is the reverse of the removal procedure.
21. Fill the cooling system.
22. Fill the engine with clean oil.

23. Start the vehicle, check for leaks, and repair if necessary.

TIMING CHAIN & SPROCKETS

REMOVAL & INSTALLATION

See Figures 48, 50 through 55.

1. Before servicing the vehicle, refer to the Precautions Section.
2. Drain the engine oil.
3. Drain the cooling system.
4. Relieve the fuel system pressure.
5. Remove or disconnect the following:

- Negative battery cable
- Engine cover
- Air cleaner case assembly
- Engine harnesses from timing chain case
- Upper and lower intake manifold collectors
- Cooling fan
- Drive belts
- A/C compressor and wire aside
- Power steering pump and wire aside

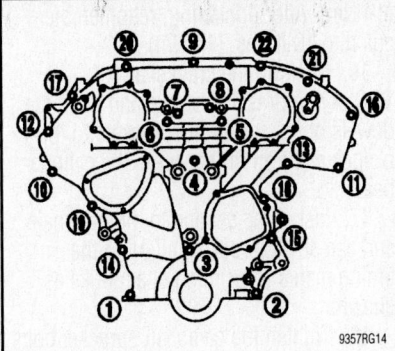

Fig. 50 Remove the front timing chain case mounting bolts in the reverse of the sequence shown

- Power steering pump bracket
- Alternator
- Water bypass hose, clamp and idler pulley bracket from timing chain case
- Intake valve timing control valve covers. Loosen the bolts in the reverse order of the tightening

sequence. In the cover, the shaft is engaged with the center hole of the intake camshaft sprocket. Remove it straight out until the engagement comes off.

- O-ring from timing chain case on both sides
- Both valve covers
- Rotate the crankshaft clockwise and set the No. 1 piston to TDC of the compression stroke

6. Make sure the intake and exhaust camshaft lobes are facing toward the inside of the cylinder head.
7. Remove the starter and lock the flywheel through the starter mounting hole.
8. Loosen the crankshaft pulley bolt.

- Crankshaft pulley using a suitable puller
- Upper and lower oil pans
- Front timing chain cover bolts in reverse order of the tightening sequence
- Front timing chain case

Camshaft sprocket (INT) — Internal chain guide — Camshaft sprocket (INT)

Mating mark (yellow link) — Mating mark (yellow link)

Mating mark (punched) — Mating mark (punched)

Mating mark (back side)

Timing chain tensioner (secondary)

Mating mark (punched) — Mating mark (gold link) — Mating mark (punched)

Mating mark (gold link) — Timing chain tensioner (secondary)

Mating mark (gold link)

Timing chain (secondary) — Timing chain (secondary)

Camshaft sprocket (EXH) — Crankshaft key — Camshaft sprocket (EXH)

Timing chain tensioner (primary)

Tension guide

Slack guide

Water pump

Timing chain (primary) — Crankshaft sprocket

Mating mark (notched) — Mating mark (orange link)

67170-NISS-G43

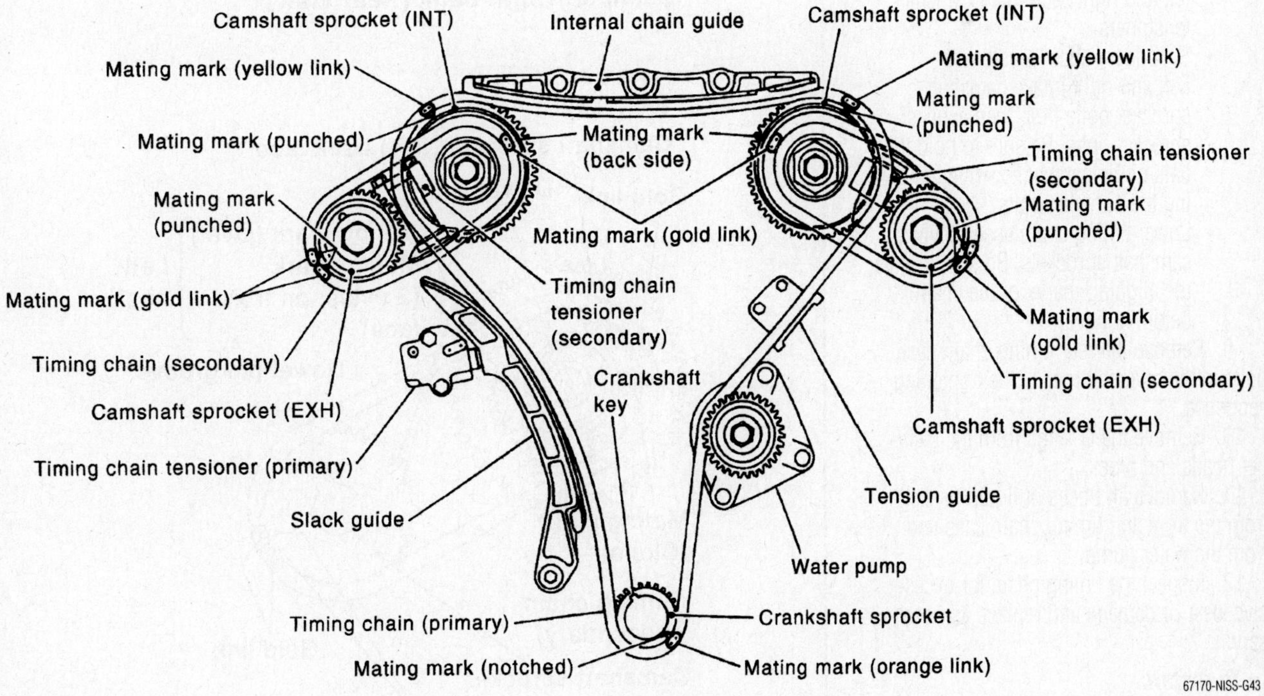

Fig. 51 Timing chain alignment marks

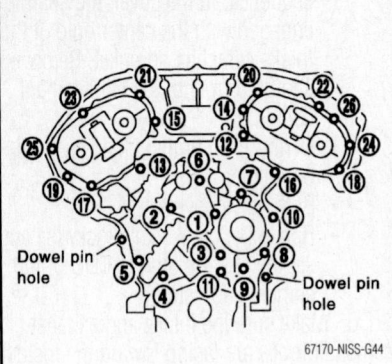

Fig. 52 Rear timing chain case tightening sequence

- O-rings from rear timing chain cover
- Water pump and chain tensioner cover from rear cover
- Pry out front oil seal
- Insert a suitable stopper pin for the left and right primary camshaft tensioners
- Primary chain tensioners
- Timing chain guide, slack guide and tension guide
- Primary timing chain and crankshaft sprocket
- Insert a suitable stopper pin for the left and right secondary camshaft tensioners
- Secondary chain tensioners
- Left and right intake camshaft sprocket bolts first, then exhaust sprocket bolts. Be sure to hold the flats of the camshafts while removing the sprocket bolts
- Lower timing chain assembly with camshaft sprockets. Be sure to note the aligning marks of the chain before removal

9. Remove the rear timing chain case bolts in the reverse order of the tightening sequence.

10. Remove the O-rings from the cylinder heads and block.

11. Remove all traces of liquid gasket from the front and timing chain case and from the water pump.

12. Inspect the timing chain for excessive wear or damage and replace as necessary.

To install:

13. Install new O-rings to the cylinder heads and block.

14. Apply a bead of sealant to the back side of the rear timing chain case.

15. Install the rear timing chain case and tighten the bolts in sequence to 10 ft. lbs.

(14 Nm). After tightening, retighten them again to 10 ft. lbs. (14 Nm).

16. Position the crankshaft to TDC of compression stroke and align the dowels of the camshaft sprockets to the 12 o'clock position in respect to the cylinder head.

17. Install the secondary timing chain and camshaft sprockets aligning the timing marks and timing chain links as shown.

18. Tighten the camshaft sprocket bolts to 76 ft. lbs. (103 Nm).

19. Remove the pins from the secondary timing chain tensioners.

20. Install the primary timing chain tension guide.

21. Install the crankshaft sprocket with the mating marks on the front side.

22. Install the primary timing chain and so the mating marks are aligned as shown.

23. Install the internal chain guide and slack guide.

24. Install the primary timing chain tensioner and tighten the bolt to 72 inch lbs. (8 Nm). Remove the stopper pin.

25. Double check that the mating marks on the timing chain and sprockets are in the correct locations.

26. Install new O-rings on the timing chain case.

27. Apply clean engine oil to the front oil seal and dust seal lips.

28. Use a drift and press fit the oil seal into the timing chain case.

29. Apply liquid gasket to the water pump and chain tensioner cover openings, then install the covers.

30. Apply liquid gasket to the back side of the timing chain case cover.

31. Install the dowel pin on the rear timing chain case into the dowel pin of the front timing chain case.

32. Install the front timing chain case and tighten the bolts in sequence to 19–23 ft. lbs. (26–31 Nm for M8 bolts and 9–10 ft. lbs. (12–14 Nm) for M6 bolts. After tightening, retighten them again to the same specification.

33. Install seal rings in the timing control cover shaft grooves and apply liquid gasket to the covers.

34. Install new O-rings in the timing chain case oil holes.

35. Install the timing control covers and tighten the bolt in sequence to 72 inch lbs. (8 Nm).

36. Install or connect the following:

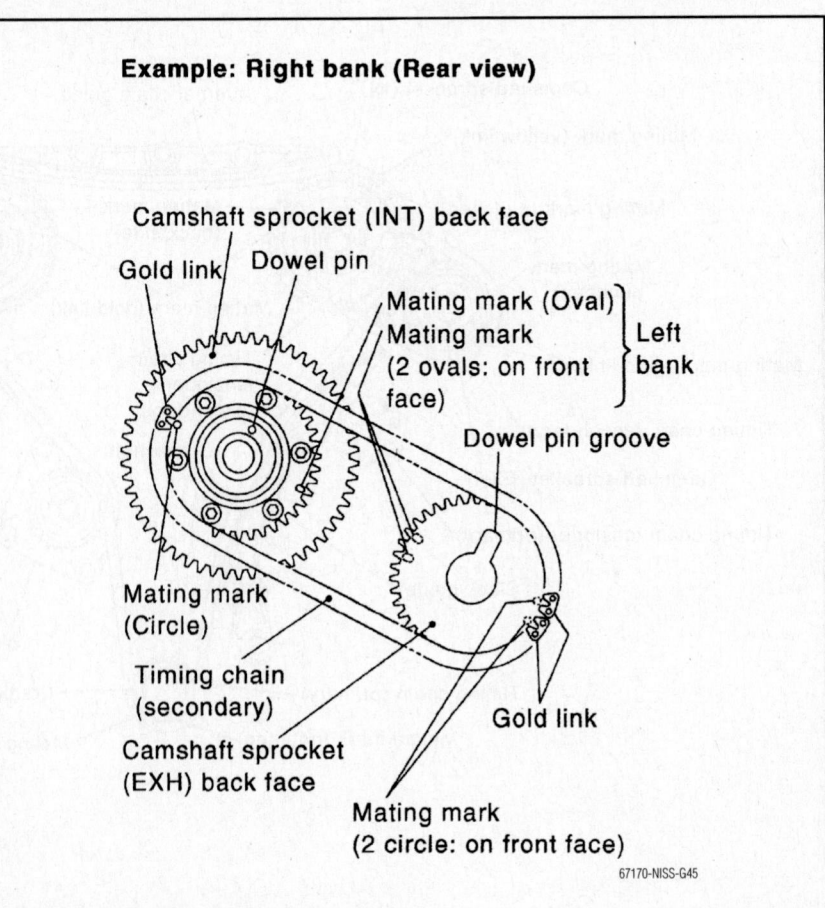

Example: Right bank (Rear view)

Camshaft sprocket (INT) back face

Gold link Dowel pin

Mating mark (Oval)
Mating mark (2 ovals: on front face) } Left bank

Dowel pin groove

Mating mark (Circle)

Timing chain (secondary)

Camshaft sprocket (EXH) back face

Gold link

Mating mark (2 circle: on front face)

67170-NISS-G45

Fig. 53 Secondary timing chain alignment marks

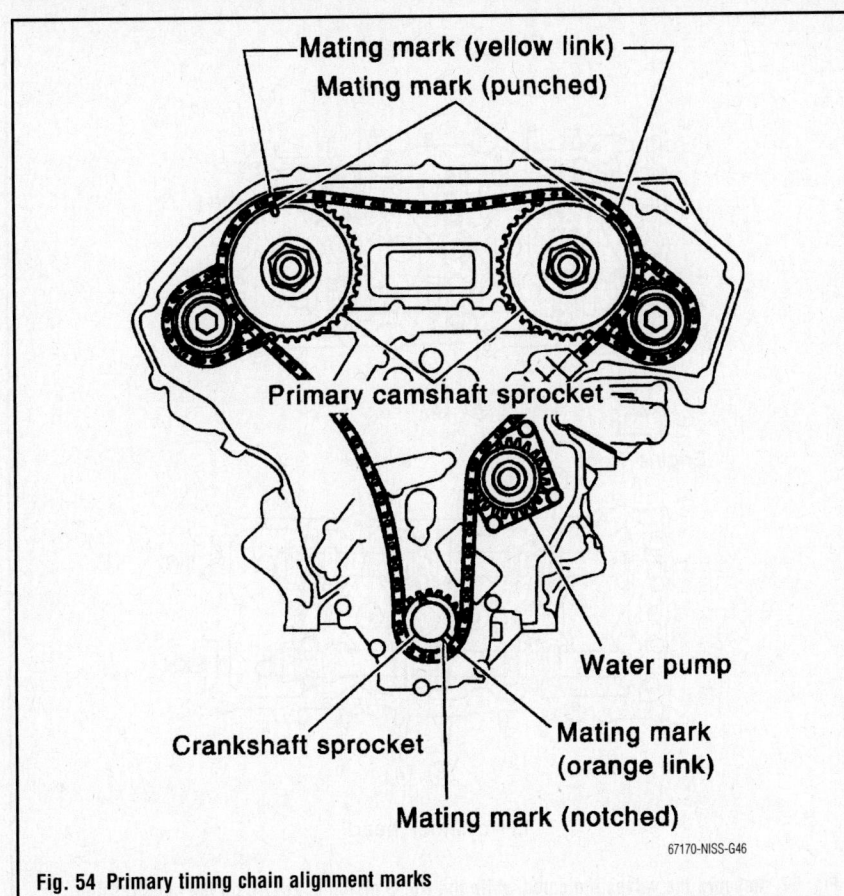

Fig. 54 Primary timing chain alignment marks

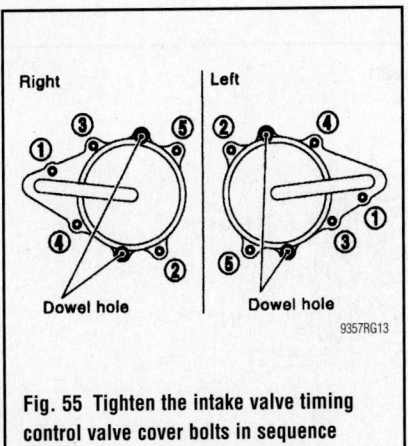

Fig. 55 Tighten the intake valve timing control valve cover bolts in sequence

- Upper and lower oil pans
- Valve covers
- Crankshaft pulley and tighten the bolt to 29–36 ft. lbs. (40–49 Nm), plus an additional 60°

37. The remainder of the installation is the reverse of the removal procedure.

38. Fill the cooling system.

39. Fill the engine with clean oil.

40. Start the vehicle, check for leaks, and repair if necessary.

VALVE LASH

ADJUSTMENT

See Figures 56 through 58.

1. Before servicing the vehicle, refer to the Precautions Section.

2. Remove or disconnect the following:
- Intake manifold collector
- Left and right rocker covers
- Spark plugs

3. Set the No. 1 cylinder at Top Dead Center (TDC) on its compression stroke. Align the pointer with the TDC mark on the crankshaft pulley. Check that the valve lifters on the No. 1 cylinder are loose and valve lifters on the No. 4 cylinder are tight. If not, turn the crankshaft 1 revolution (360 degrees) and align the pointer with the TDC mark on the crankshaft pulley.

4. Check the following valves:
- Both No. 1 intake valves
- Both No. 2 exhaust valves
- Both No. 3 exhaust valves
- Both No. 6 intake valves

5. Using a feeler gauge, measure the clearance between the valve lifter and the camshaft. Record any valve clearance measurements that are out of specification. Intake valve clearance (cold) is 0.010–0.013 in.

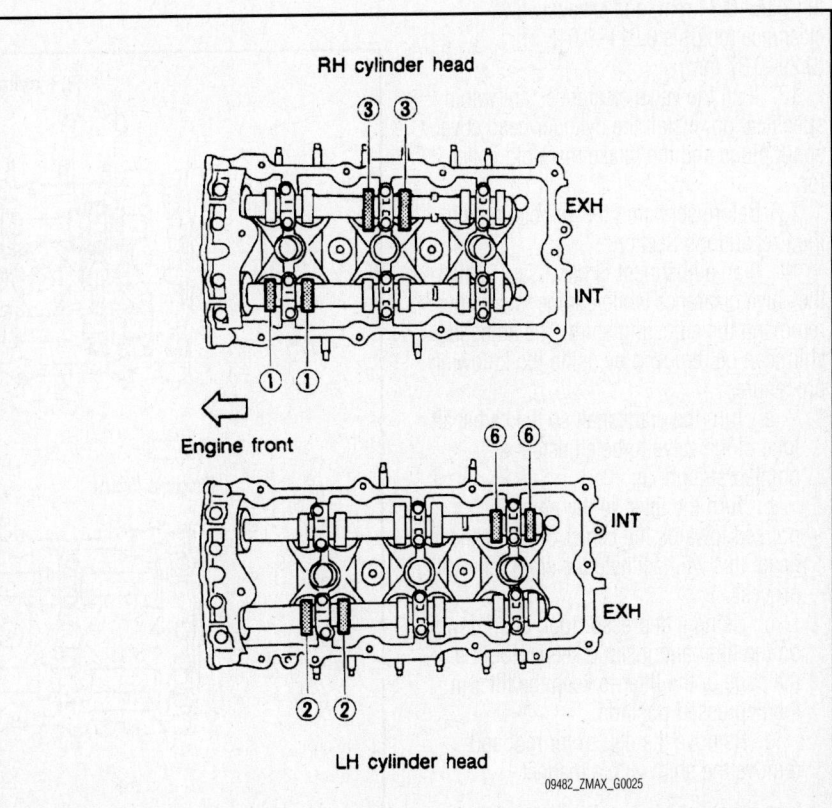

Fig. 56 Measure the valves indicated while the No. 1 piston is at TDC on the compression stroke

(0.26–0.34 mm) and exhaust valve clearance (cold) is 0.011–0.015 in. (0.29–0.37 mm).

6. Turn the crankshaft 240 degrees and set the No. 3 cylinder to TDC of its compression stroke.

7. Check the following valves:
- Both No. 2 intake valves
- Both No. 3 intake valves
- Both No. 4 exhaust valves
- Both No. 5 exhaust valves

8. Using a feeler gauge, measure the clearance between the valve lifter and the camshaft. Record any valve clearance measurements that are out of specification. Intake valve clearance (cold) is 0.010–0.013 in. (0.26–0.34 mm) and exhaust valve clearance (cold) is 0.011–0.015 in. (0.29–0.37 mm).

9. Turn the crankshaft 240 degrees and set the No. 5 cylinder to TDC of its compression stroke.

10. Check the following valves:
- Both No. 1 exhaust valves
- Both No. 4 intake valves
- Both No. 5 intake valves
- Both No. 6 exhaust valves

11. Using a feeler gauge, measure the clearance between the valve lifter and the camshaft. Record any valve clearance measurements that are out of specification. Intake valve clearance (cold) is 0.010–0.013 in. (0.26–0.34 mm) and exhaust valve clearance (cold) is 0.011–0.015 in. (0.29–0.37 mm).

12. If all the valve clearances are within specification, install the cylinder head cover, spark plugs and the intake manifold collector.

13. Before servicing the vehicle, refer to the Precautions Section.

14. If an adjustment is necessary, adjust the valve clearance while engine is cold by removing the adjusting shim. The adjusting shim can be removed by using the following procedures:

a. Turn the crankshaft so the camshaft lobe of the valve to be adjusted is pointed straight up.

b. Turn the lifter so the notch is pointed towards the center of the cylinder head; this will facilitate the shim removal process.

c. Using a depressor tool, push down on the lifter and insert a keeper tool on the edge of the lifter to keep the lifter in the depressed position.

d. Remove the depressor tool and remove the shim with a magnet.

➡Compressed air can be blown into the hole of the lifter to separate the adjusting shim from the lifter.

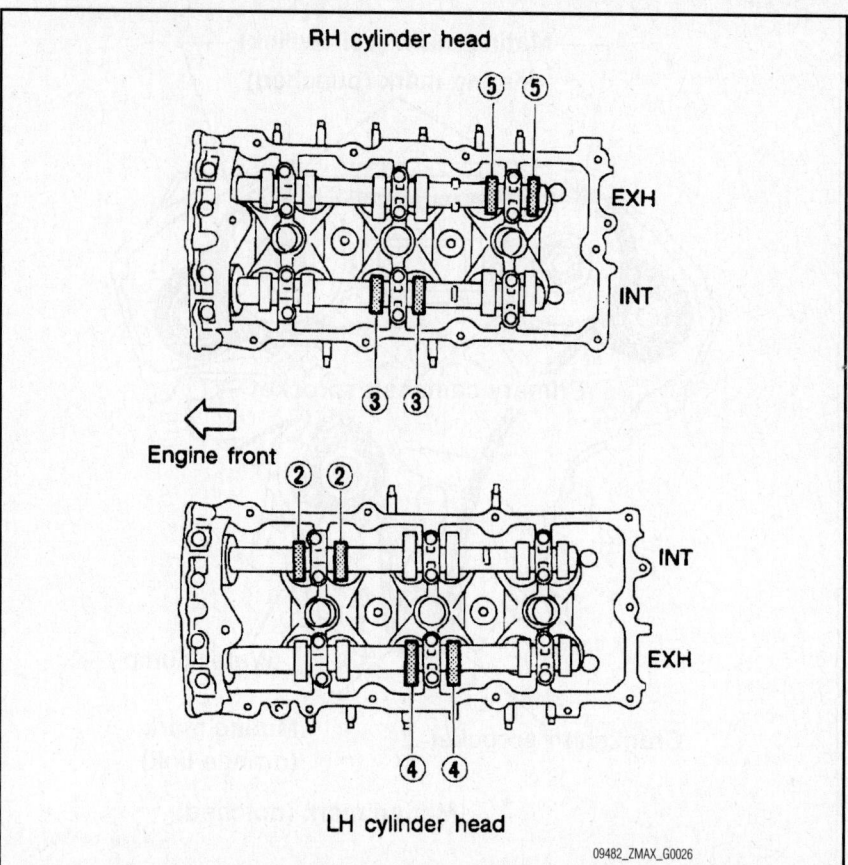

Fig. 57 Measure the valves indicated while the No. 3 piston is at TDC on the compression stroke

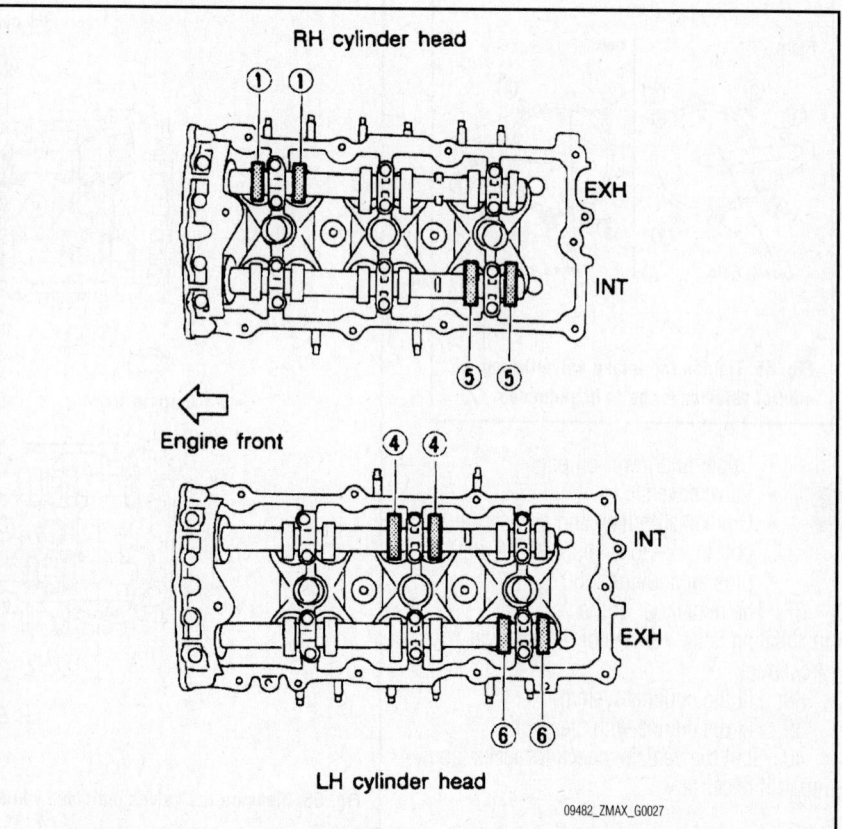

Fig. 58 Measure the valves indicated while the No. 5 piston is at TDC on compression

15. Determine the replacement adjusting shim size by using the following procedures and formula:

a. Using a micrometer determine thickness of the removed shim.

b. Calculate the thickness of a new adjusting shim so valve clearance is within the specified values.

c. R = thickness of the removed shim.

d. N = thickness of the new shim.

e. M = measured valve clearance.

- Intake shim determination formula:

$N = R + (M$ minus 0.0118 in. or 0.30 mm)

- Exhaust shim determination formula: $N = R + (M$ minus 0.0130 in. or 0.33 mm)

16. Shims are available in 64 sizes from 0.0913–0.1161 in. (2.32–2.95 mm) in steps of 0.004 in. (0.01 mm). The thickness is stamped on the shim; this side is always installed facing down. Select new shims with thickness as close as possible to calculated valve and install it in the lifter.

17. Install the new shim onto the lifter.

18. Depress the lifter and remove the keeper tool. Remove the depressor tool and recheck the valve clearance. Repeat this procedure for any other valves requiring adjustment.

19. When all valve adjustments are finished, install the cylinder head cover, spark plugs and the intake manifold collector.

➡ **Check and adjust the valve clearances while the engine is cold and not running.**

ENGINE PERFORMANCE & EMISSION CONTROLS

ACCELERATOR PEDAL POSITION (APP) SENSOR

REMOVAL & INSTALLATION

1. Unplug Accelerator Pedal Position (APP) sensor connector.

2. Remove accelerator pedal retaining bolts.

To install:

3. Installation is reverse of removal.

CAMSHAFT POSITION (CMP) SENSOR

LOCATION

See Figure 59.

Refer to the accompanying illustration.

REMOVAL & INSTALLATION

1. Loosen the fixing bolt of the sensor.

2. Disconnect Camshaft Position (CMP) sensor harness connector.

3. Remove the sensor.

To install:

4. Installation is reverse of removal.

CRANKSHAFT POSITION (CKP) SENSOR

LOCATION

See Figure 60.

This sensor is located on the engine oil pan.

REMOVAL & INSTALLATION

1. Loosen the fixing bolt of the sensor.

2. Disconnect Crankshaft Position (CKP) sensor (POS) harness connector.

3. Remove the sensor.

To install:

4. Installation is reverse of removal.

ELECTRONIC CONTROL MODULE (ECM)

LOCATION

The ECM is located on the passenger's side of the vehicle at the kick panel.

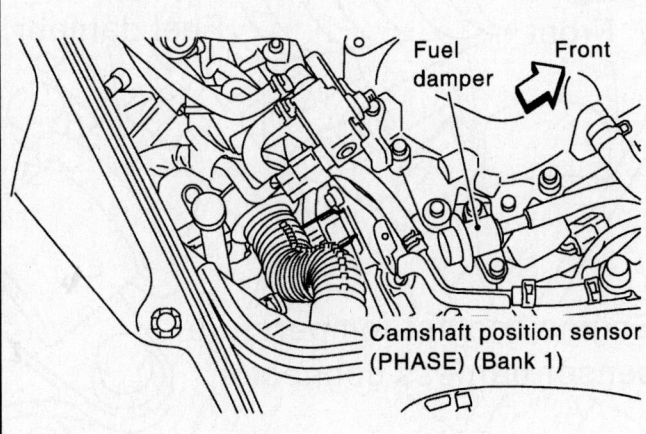
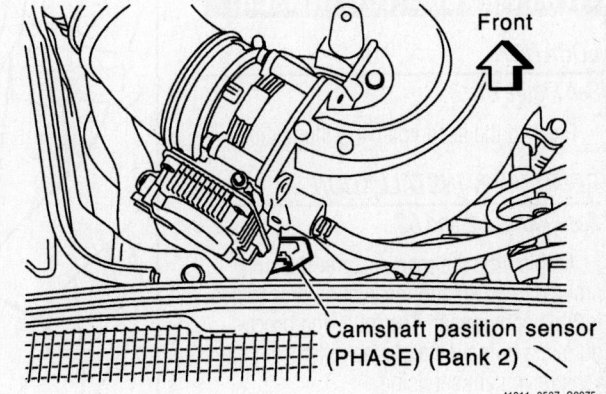

41311_350Z_G0075

Fig. 59 Camshaft Position (CMP) sensor location

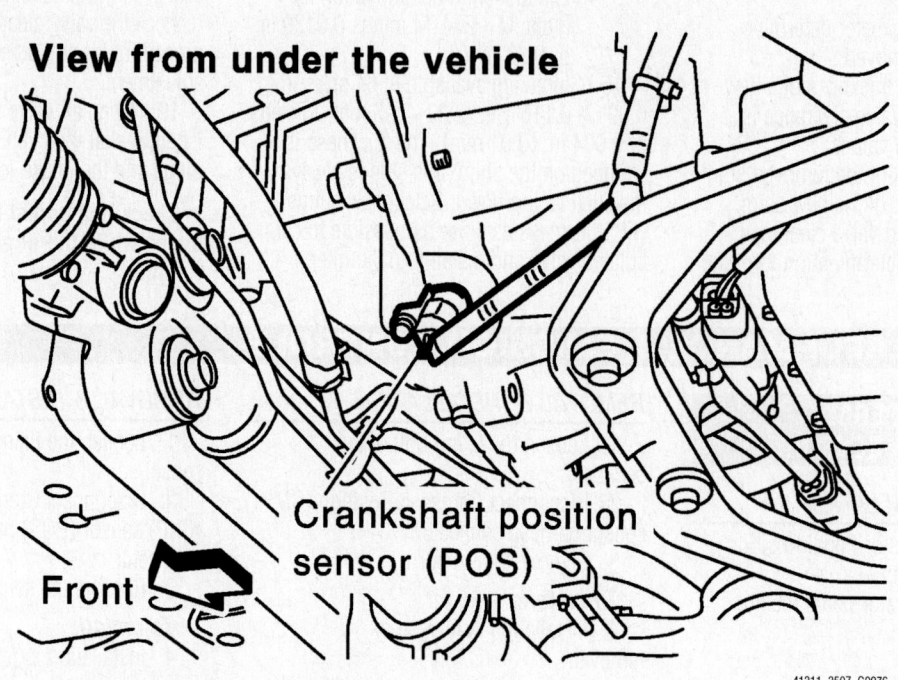

View from under the vehicle

Crankshaft position
sensor (POS)

Front

41311_350Z_G0076

Fig. 60 Crankshaft Position (CKP) sensor location

REMOVAL & INSTALLATION

1. Remove glove box.
2. Disconnect electrical connectors.
3. Remove attachment.

To install:

4. Installation is reverse of removal.

ENGINE COOLANT TEMPERATURE (ECT) SENSOR

LOCATION

See Figure 61.

Refer to the accompanying illustration.

REMOVAL & INSTALLATION

See Figures 62 and 63.

At this time the manufacturer does not provide removal and installation procedures for this component. The following procedure is a guideline and may differ from the vehicle you are servicing.

1. Before servicing the vehicle, refer to the Precautions Section.

➡**If working near and/or around the SRS system and components, be sure to disable the SRS system. After disabling the system wait three minutes or more before servicing the vehicle.**

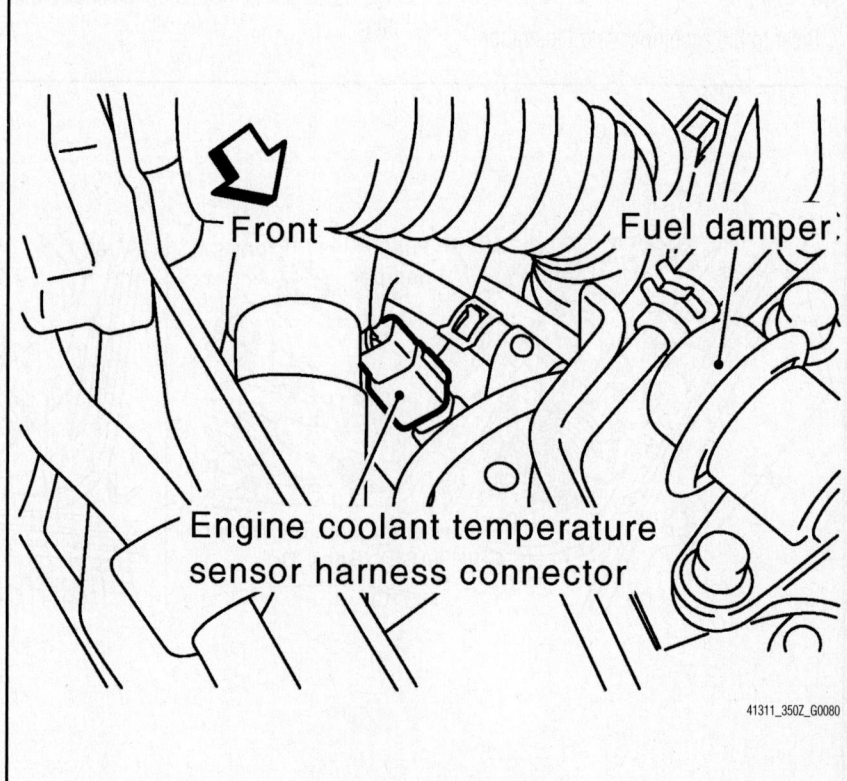

Front   Fuel damper

Engine coolant temperature
sensor harness connector

41311_350Z_G0080

Fig. 61 Engine Coolant Temperature (ECT) sensor location

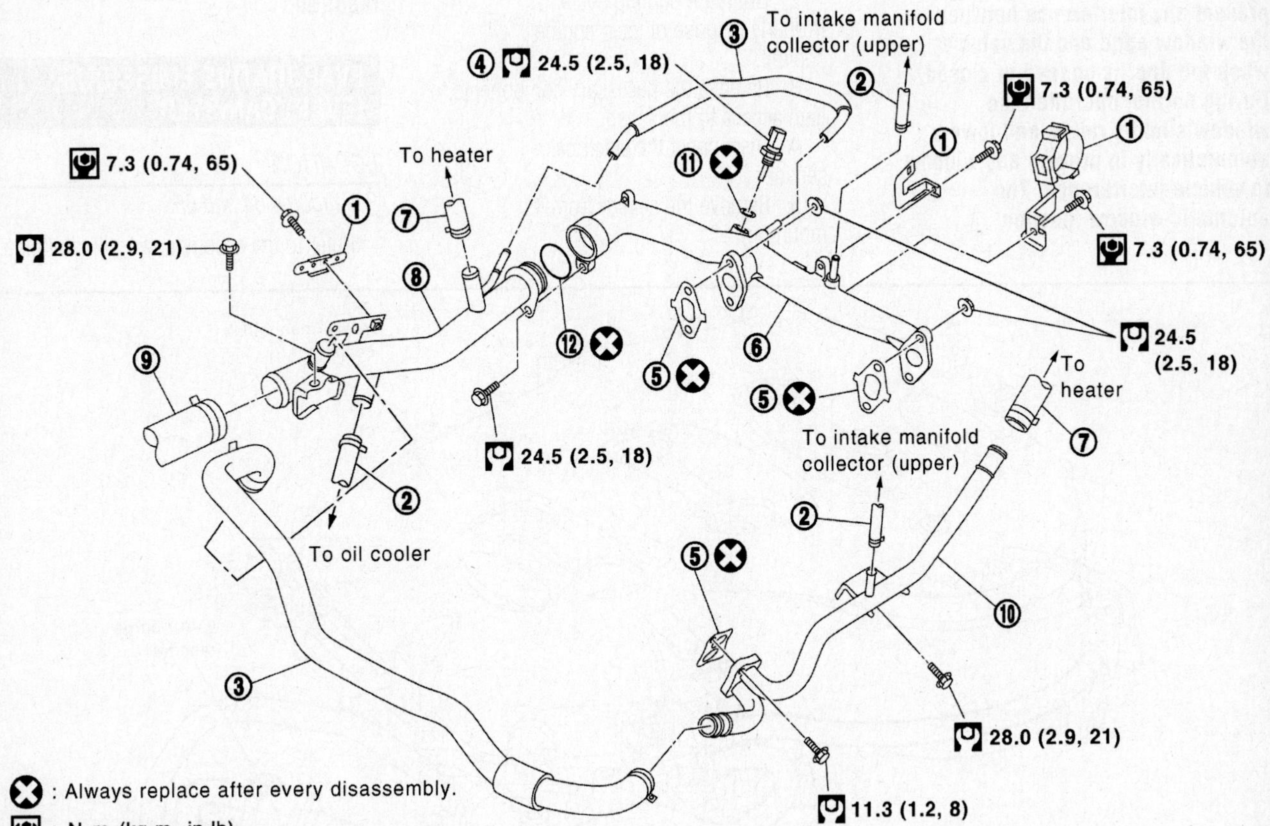

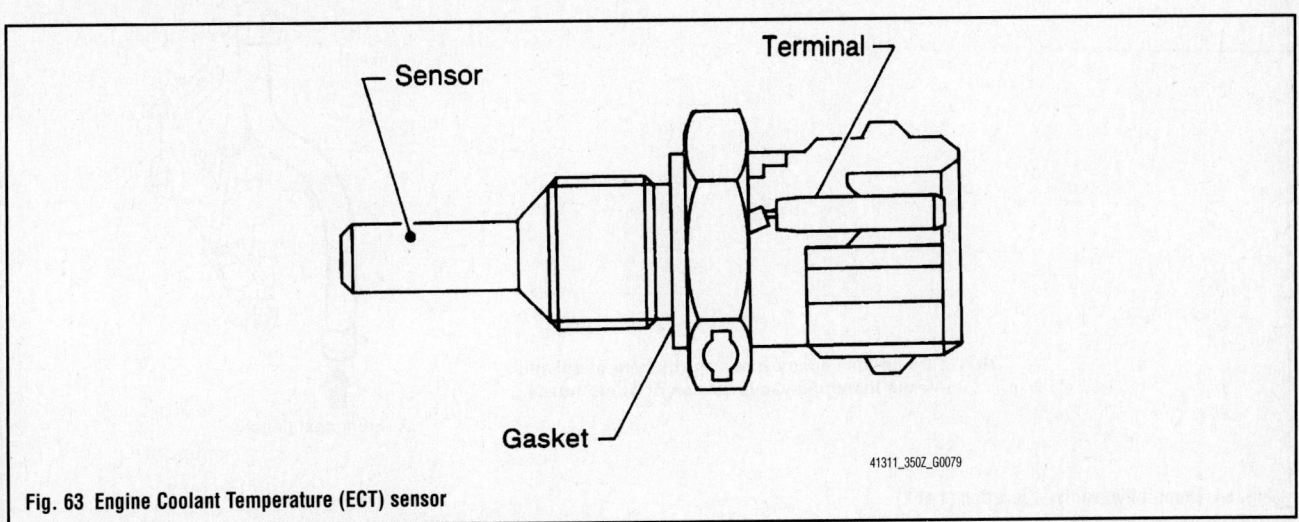

Fig. 62 Engine Coolant Temperature (ECT) sensor and related components

⊗ : Always replace after every disassembly.

⊡ : N·m (kg-m, in-lb)

⊡ : N·m (kg-m, ft-lb)

1. Harness bracket
2. Water hose
3. Water bypass hose
4. Engine coolant temperature sensor
5. Gasket
6. Water outlet
7. Heater hose
8. Water pipe
9. Radiator hose (upper)
10. Heater pipe
11. Washer
12. O-ring

41311_350Z_G0081

Fig. 63 Engine Coolant Temperature (ECT) sensor

41311_350Z_G0079

➡️Before disconnecting the battery, lower both the driver's and passenger's windows. This will prevent any interference between the window edge and the vehicle when the door is opened or closed. During normal operation the window slightly raises and lowers automatically to prevent any window to vehicle interference. The automatic window function will not work with the battery disconnected.

2. Drain the cooling system. Properly dispose of used engine coolant.

3. Remove the necessary components to gain access to the sensor.

4. Disconnect the electrical sensor.

5. Remove the sensor from its mounting.

To install:

➡️Be sure to use new fasteners, as required.

EVAPORATIVE EMISSIONS (EVAP) CANISTER

LOCATION

See Figures 64 and 65.

Refer to the accompanying illustrations.

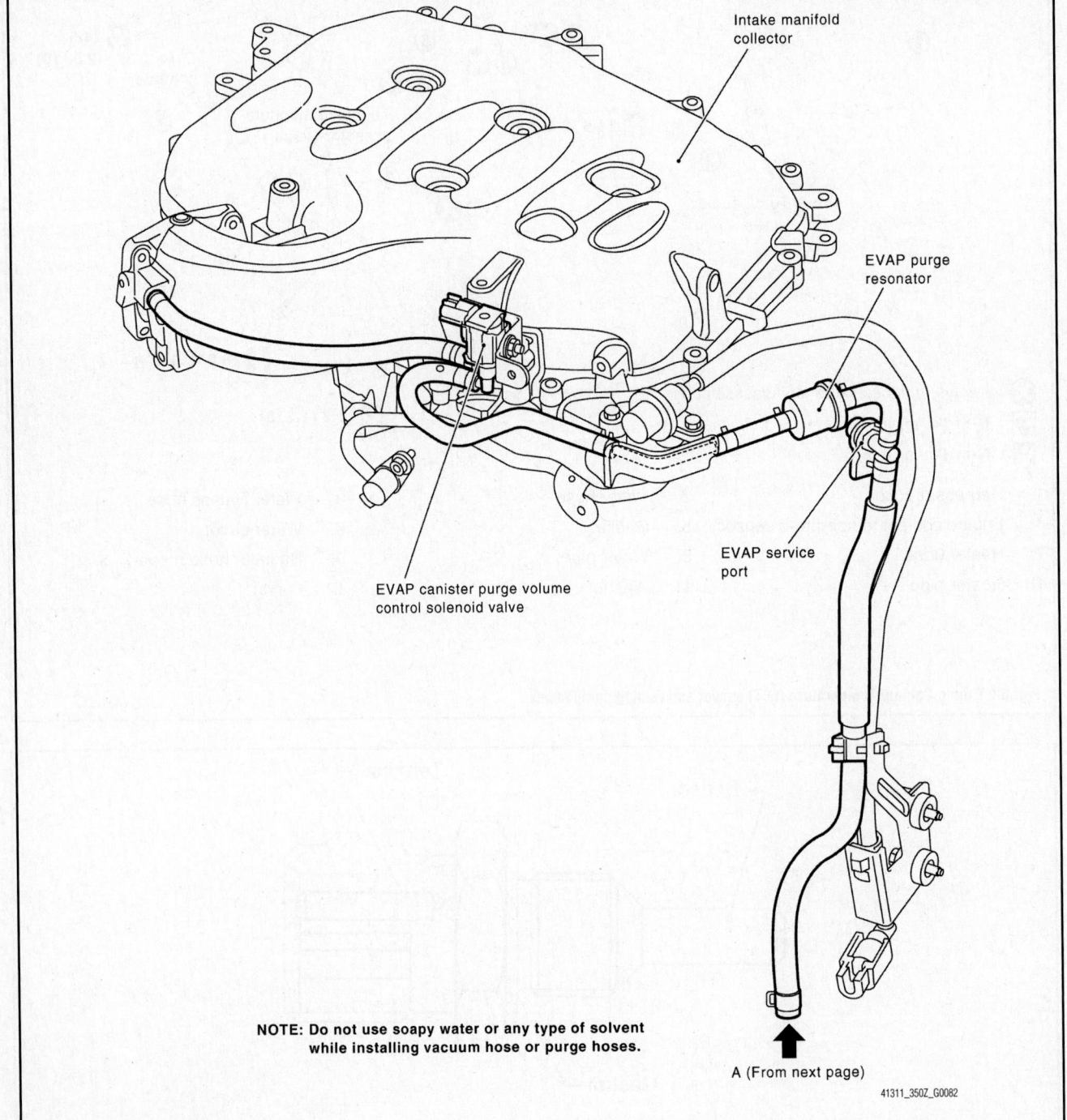

Intake manifold collector

EVAP purge resonator

EVAP service port

EVAP canister purge volume control solenoid valve

NOTE: Do not use soapy water or any type of solvent while installing vacuum hose or purge hoses.

A (From next page)

41311_350Z_G0082

Fig. 64 Evaporative canister location (1 of 2)

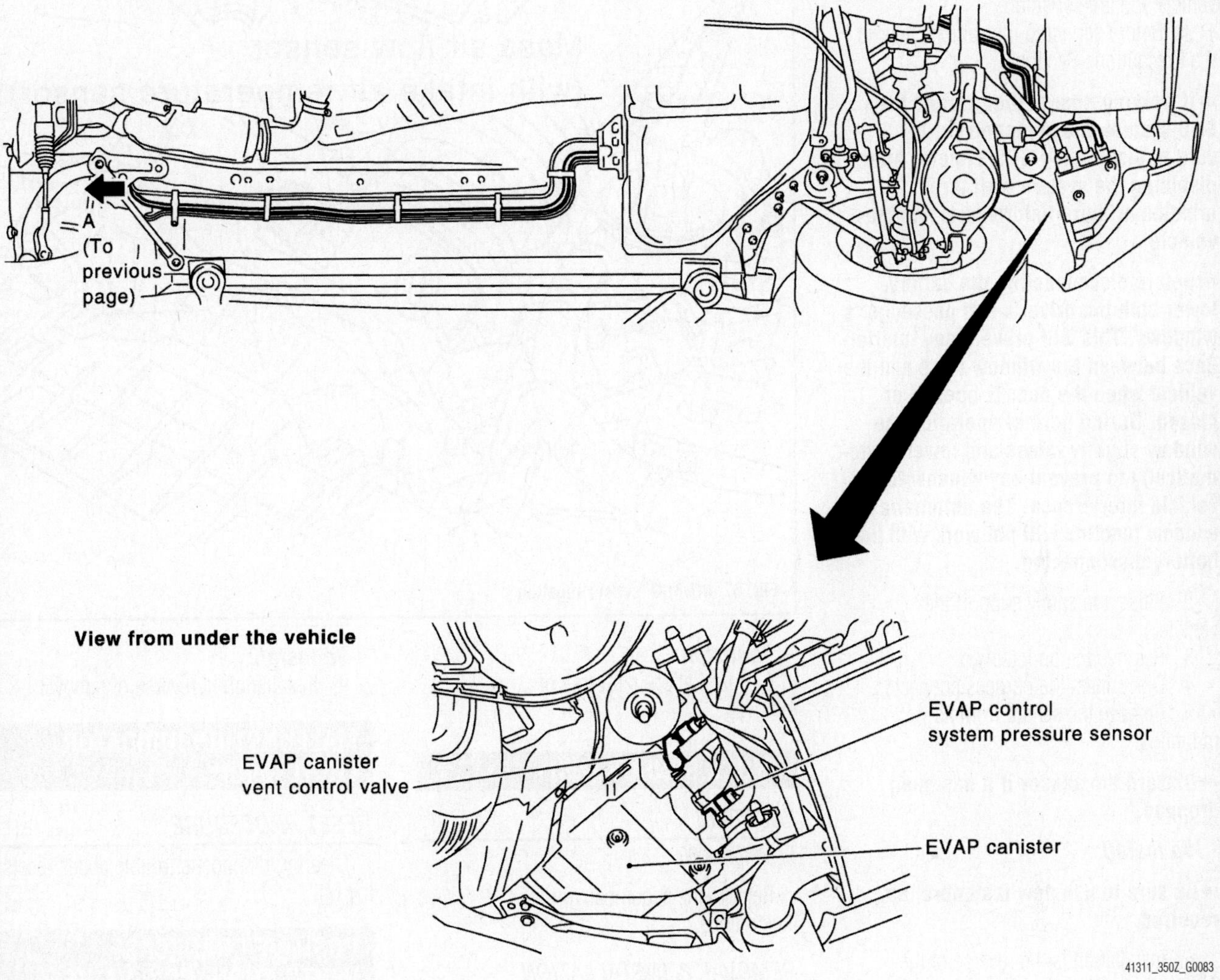

View from under the vehicle

EVAP control
system pressure sensor

EVAP canister
vent control valve

EVAP canister

41311_350Z_G0083

Fig. 65 Evaporative canister location (2 of 2)

REMOVAL & INSTALLATION

See Figure 66.

At this time the manufacturer does not provide removal and installation procedures for this component.

HEATED OXYGEN (HO2S) SENSOR

LOCATION

This sensor is located in the exhaust stream, after the catalytic converter. Two sensors are used, one for each cylinder bank.

REMOVAL & INSTALLATION

At this time the manufacturer does not provide removal and installation procedures

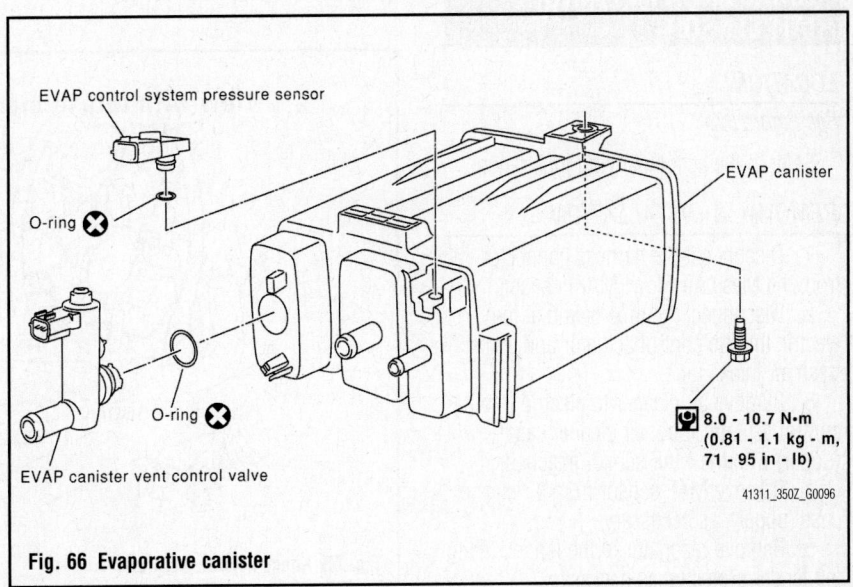

EVAP control system pressure sensor

O-ring

EVAP canister

EVAP canister vent control valve

O-ring

8.0 - 10.7 N•m
(0.81 - 1.1 kg - m,
71 - 95 in - lb)

41311_350Z_G0096

Fig. 66 Evaporative canister

for this component. The following procedure is a guideline and may differ from the vehicle you are servicing.

1. Before servicing the vehicle, refer to the Precautions Section.

➡If working near and/or around the SRS system and components, be sure to disable the SRS system. After disabling the system wait three minutes or more before servicing the vehicle.

➡Before disconnecting the battery, lower both the driver's and passenger's windows. This will prevent any interference between the window edge and the vehicle when the door is opened or closed. During normal operation the window slightly raises and lowers automatically to prevent any window to vehicle interference. The automatic window function will not work with the battery disconnected.

2. Raise and safely support the vehicle.
3. Remove the undercover.
4. Disconnect the harness connector.
5. Remove the sensor from its mounting.

➡Discard the sensor if it has been dropped.

To install:

➡Be sure to use new fasteners, as required.

6. Installation is the reverse of the removal procedure.
7. Before installing a new sensor coat the threads with an approved anti-seize lubricant.

INTAKE AIR TEMPERATURE (IAT) SENSOR

LOCATION
See Figure 67.

Refer to the accompanying illustration.

REMOVAL & INSTALLATION

1. Disconnect the harness connector from the Mass Air Flow (MAF) sensor.
2. Disconnect the tube clamp at the electric throttle control actuator and at the fresh air intake tube.
3. Remove air cleaner to electric throttle control actuator tube, air cleaner case (upper) with the MAF sensor attached.
4. Remove MAF sensor from air cleaner case (upper), as necessary.
5. Remove resonator in the fender, lifting left fender protector, as necessary.

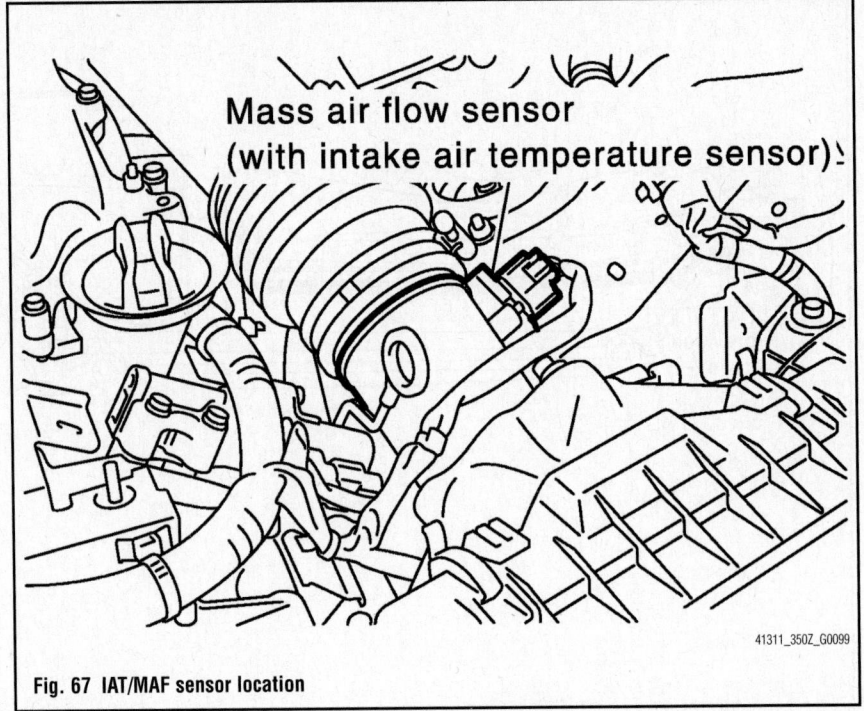

Mass air flow sensor
(with intake air temperature sensor)

41311_350Z_G0099

Fig. 67 IAT/MAF sensor location

To install:
6. Installation is reverse of removal.

KNOCK SENSOR (KS)

LOCATION
See Figure 68.

Refer to the accompanying illustration.

REMOVAL & INSTALLATION

1. Remove the engine assembly.
2. Install the engine on the engine stand.
3. Remove the knock sensor.

To install:
4. Installation is reverse of removal.

MALFUNCTION INDICATOR LIGHT (MIL)

RESET PROCEDURE

Clearing diagnostic trouble codes resets the MIL.

POSITIVE CRANKCASE VENTILATION (PCV) VALVE

LOCATION
See Figure 69.

Refer to the accompanying illustration.

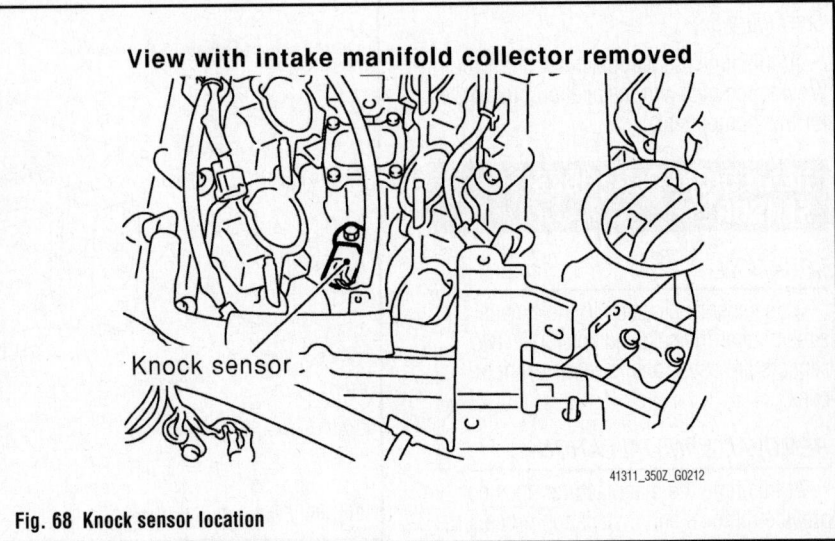

View with intake manifold collector removed

Knock sensor

41311_350Z_G0212

Fig. 68 Knock sensor location

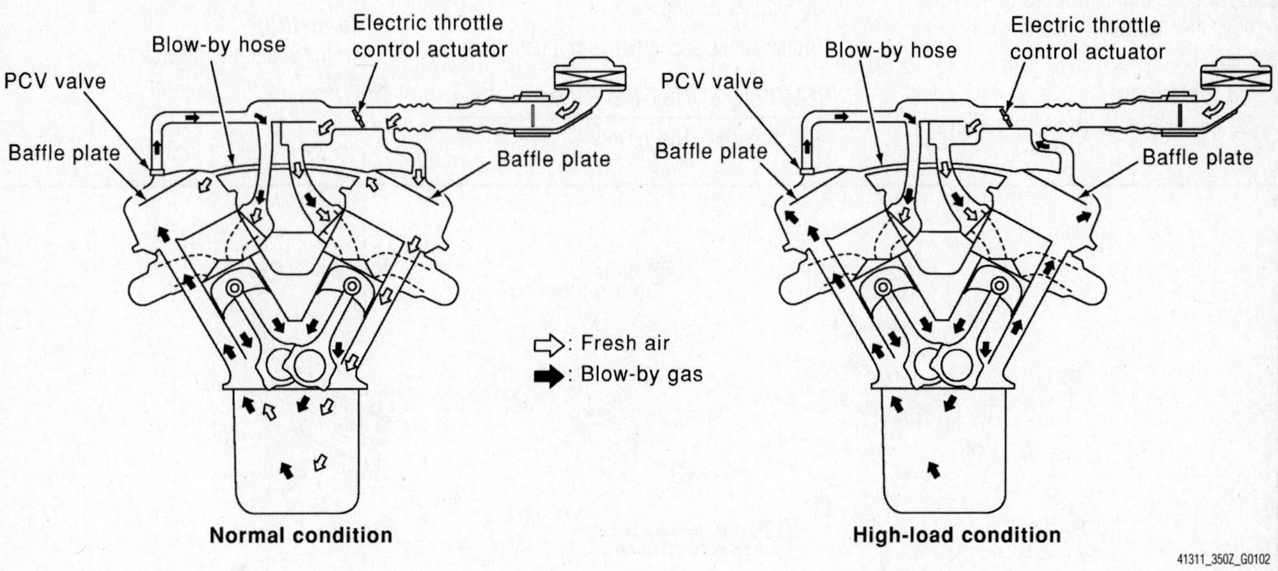

Electric throttle
control actuator

Blow-by hose

PCV valve

Baffle plate

Baffle plate

⇨ : Fresh air
➡ : Blow-by gas

Normal condition

Electric throttle
control actuator

Blow-by hose

PCV valve

Baffle plate

Baffle plate

High-load condition

41311_350Z_G0102

Fig. 69 PCV valve location

REMOVAL & INSTALLATION

At this time the manufacturer does not provide removal and installation procedures for this component. The following procedure is a guideline and may differ from the vehicle you are servicing.

1. Before servicing the vehicle, refer to the Precautions Section.

➡**If working near and/or around the SRS system and components, be sure to disable the SRS system. After disabling the system wait three minutes or more before servicing the vehicle.**

➡**Before disconnecting the battery, lower both the driver's and passenger's windows. This will prevent any interference between the window edge and the vehicle when the door is opened or closed. During normal operation the window slightly raises and lowers automatically to prevent any window to vehicle interference. The automatic window function will not work with the battery disconnected.**

2. Remove the necessary components in order to gain access to the component.
3. Disconnect the PCV hose.

4. Remove the valve from its mounting.
To install:

➡**Be sure to use new fasteners, as required.**

5. Installation is the reverse of the removal procedure.

THROTTLE POSITION SENSOR (TPS)

LOCATION

See Figure 70.

Refer to the accompanying illustration.

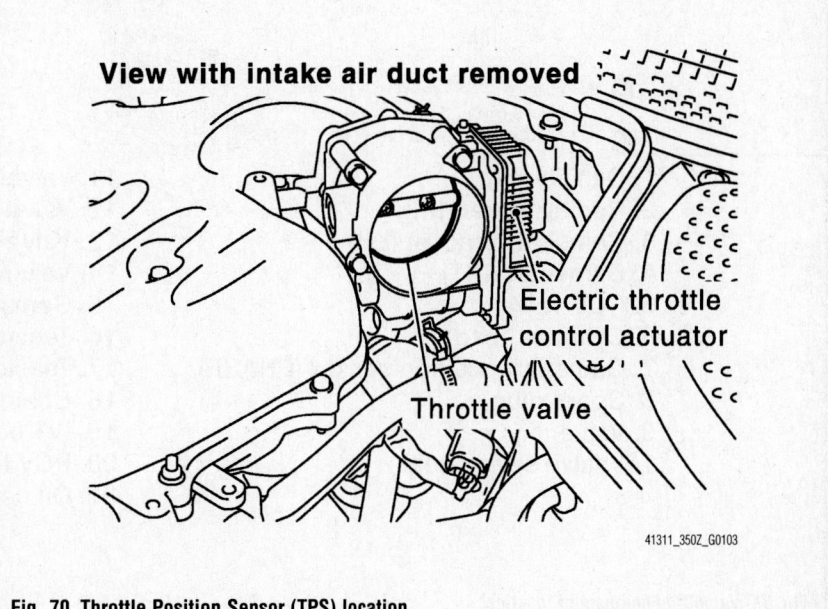

View with intake air duct removed

Electric throttle
control actuator

Throttle valve

41311_350Z_G0103

Fig. 70 Throttle Position Sensor (TPS) location

REMOVAL & INSTALLATION

1. Remove the electrical connector.

2. Remove the electric throttle control actuator bolts in the order as shown and remove the electric throttle control actuator.

To install:

3. Installation is reverse of removal.

VARIABLE CAMSHAFT TIMING OIL CONTROL SOLENOID

LOCATION

See Figures 71 and 72.

Refer to the accompanying illustrations.

REMOVAL & INSTALLATION

1. Remove valve covers

2. Remove the IVT control solenoid valves.

3. Discard the IVT control solenoid valve gaskets and use new gaskets for installation.

To install:

4. Installation is reverse of removal.

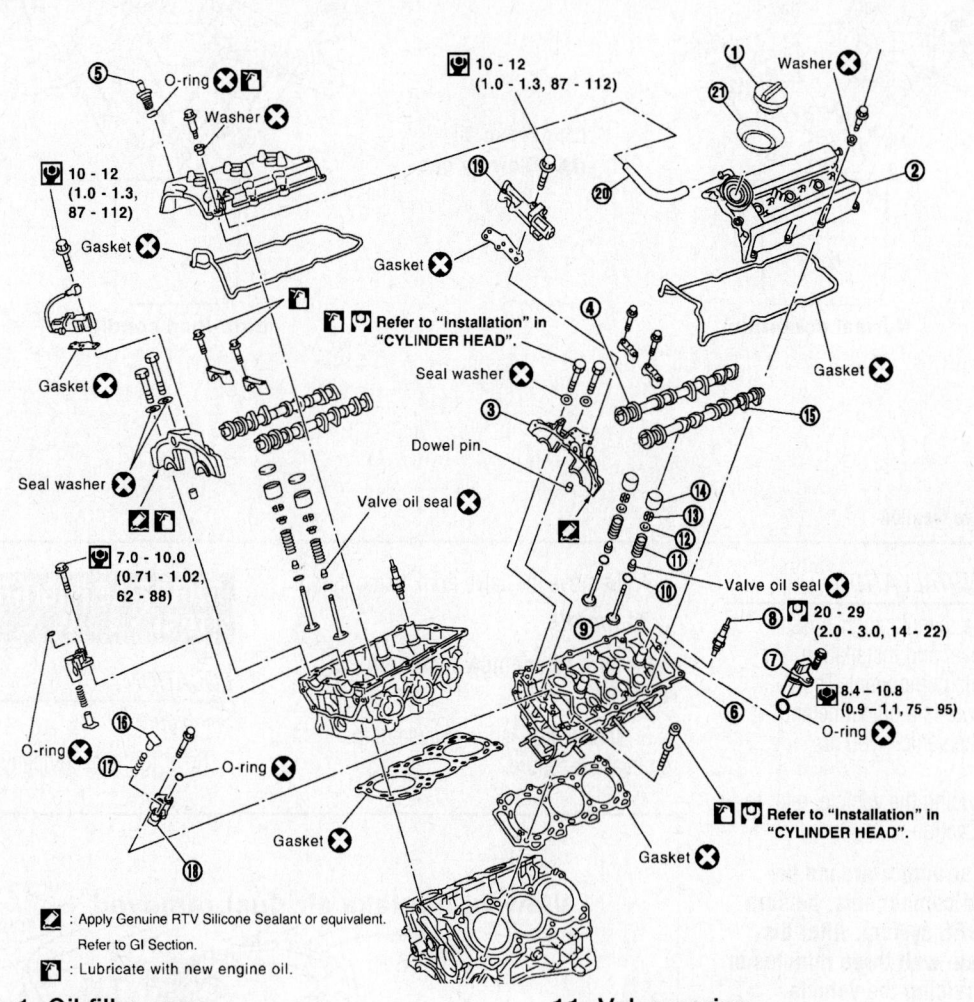

: Apply Genuine RTV Silicone Sealant or equivalent. Refer to GI Section.

: Lubricate with new engine oil.

1. Oil filler cap
2. Rocker cover (LH)
3. Camshaft bracket (LH)
4. Camshaft (INT)
5. PCV valve
6. Cylinder head
7. Camshaft position sensor (PHASE)
8. Spark plug
9. Valve
10. Valve spring seat
11. Valve spring
12. Valve spring retainer
13. Valve collet
14. Valve lifter
15. Camshaft (EXH)
16. Tensioner sleeve
17. Tensioner spring
18. Chain tensioner
19. IVT control solenoid valve
20. PCV hose
21. Oil catcher

22140_350Z_G0037

Fig. 71 Valvetrain component location

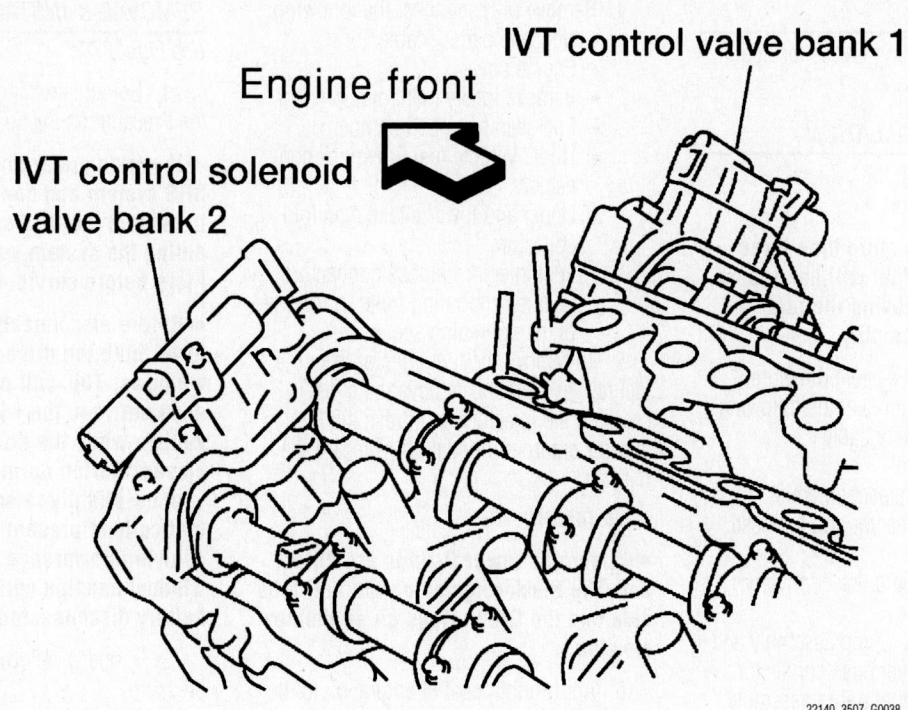

IVT control valve bank 1

Engine front

IVT control solenoid
valve bank 2

22140_350Z_G0038

Fig. 72 IVT solenoid location

FUEL   **GASOLINE FUEL INJECTION SYSTEM**

FUEL SYSTEM SERVICE PRECAUTIONS

Safety is the most important factor when performing not only fuel system maintenance, but any type of maintenance. Failure to conduct maintenance and repairs in a safe manner may result in serious personal injury or death. Work on a vehicle's fuel system components can be accomplished safely and effectively by adhering to the following rules and guidelines.

• To avoid the possibility of fire and personal injury, always disconnect the negative battery cable unless the repair or test procedure requires that battery voltage be applied.

• Always relieve the fuel system pressure prior to disconnecting any fuel system component (injector, fuel rail, pressure regulator, etc.) fitting or fuel line connection. Exercise extreme caution whenever relieving fuel system pressure to avoid exposing skin, face and eyes to fuel spray. Please be advised that fuel under pressure may penetrate the skin or any part of the body that it contacts.

• Always place a shop towel or cloth around the fitting or connection prior to loosening to absorb any excess fuel due to spillage. Ensure that all fuel spillage is quickly removed from engine surfaces. Ensure that all fuel-soaked cloths or towels are deposited into a flame-proof waste container with a lid.

• Always keep a dry chemical (Class B) fire extinguisher near the work area.

• Do not allow fuel spray or fuel vapors to come into contact with a spark or open flame.

• Always use a second wrench when loosening or tightening fuel line connection fittings. This will prevent unnecessary stress and torsion on fuel piping. Always follow the proper torque specifications.

• Always replace worn fuel fitting O-rings with new ones. Do not substitute fuel hose where rigid pipe is installed.

FUEL SYSTEM PRESSURE

RELIEVING

The fuel pump fuse is located in the dash fuse box or in the engine compartment fuse box. Check the lid of the fuse box for exact location.

1. Before servicing the vehicle, refer to the Precautions Section.
2. Remove the fuel pump fuse.
3. Start the engine.
4. Start the engine and run until the engine stalls.
5. After the engine stalls, try to restart the engine. If the engine will not start, the fuel pressure has been released.
6. Turn the ignition switch **OFF**. Reinstall the fuel pump fuse into the fuse block.

➡**Do not crank the engine or turn the ignition switch ON after the fuel pump fuse has been reinstalled, or the fuel pressure will be re-established.**

FUEL FILTER

REMOVAL & INSTALLATION

The fuel filter is attached to the fuel pump assembly. The fuel pump must be removed before the filter can be serviced.

FUEL LEVEL SENDING UNIT

LOCATION

The fuel level sending unit is located in the fuel tank.

REMOVAL & INSTALLATION

1. Before servicing the vehicle, refer to the Precautions Section.

➡ **If the fuel tank is more than three quarter full, some fuel will have to be drained before removing the fuel pump/fuel filter assembly.**

2. Relieve the fuel system pressure.
3. Remove or disconnect the following:
   - Negative battery cable
   - Rear floor box
   - Fuel pump inspection cover
   - Harness connector and fuel feed tube
   - Fuel feed tube quick connector
   - Retainer
   - Raise the fuel pump assembly and remove the fuel hose connector
   - Remove the fuel pump assembly
4. Reverse the removal procedure to install.

FUEL PUMP

REMOVAL & INSTALLATION

1. Before servicing the vehicle, refer to the Precautions Section.

➡ **If the fuel tank is more than three quarter full, some fuel will have to be drained before removing the fuel pump/fuel filter assembly.**

2. Relieve the fuel system pressure.
3. Remove or disconnect the following:
   - Negative battery cable
   - Rear floor box
   - Fuel pump inspection cover
   - Harness connector and fuel feed tube
   - Fuel feed tube quick connector
   - Retainer
   - Raise the fuel pump assembly and remove the fuel hose connector
   - Remove the fuel pump assembly
4. Reverse the removal procedure to install.

FUEL RAIL AND INJECTOR

REMOVAL & INSTALLATION

See Figure 73.

1. Before servicing the vehicle, refer to the Precautions Section.

2. Relieve the fuel system pressure.
3. Drain coolant.
4. Remove or disconnect the following:
   - Negative battery cable
   - Engine cover
   - Remove intake manifold collector
   - Fuel feed hose and damper
   - Under vehicle fuel line quick connectors
   - Upper and lower intake manifold collectors
   - Fuel injector harness connectors
   - Fuel rail mounting bolts in reverse of the tightening sequence
5. To remove the fuel injector from the fuel rail, expand and remove the clips securing the injectors and press the fuel injector out from the fuel rail. Discard the O-rings.

To install:

➡ **Upper and lower O-rings are different. The Black rings go on the fuel tube side and the Green rings go on the nozzle side.**

6. Apply a thin coat of engine oil to the new O-rings, install them on the injectors, and then press the injector into the fuel rail.
7. Install or connect the following:
   - New injector retaining clips
   - Fuel rail mounting bolts in sequence and tighten to 8 ft. lbs. (11 Nm), then to 16–19 ft. lbs. (21–27 Nm)
   - Fuel injector harness connectors
   - Upper and lower intake manifold collectors
   - Under vehicle fuel line quick connectors
   - Fuel feed hose and damper
   - Engine cover
   - Negative battery cable
8. Start the engine and check for leaks.

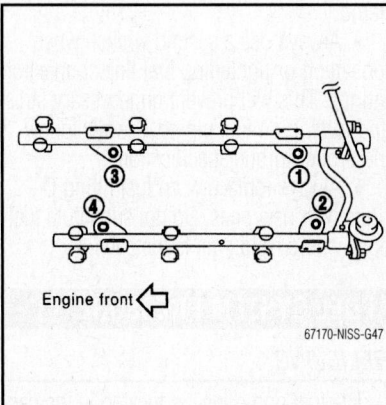

Engine front

67170-NISS-G47

Fig. 73 Fuel rail tightening sequence

FUEL TANK

REMOVAL & INSTALLATION

See Figure 74.

1. Before servicing the vehicle, refer to the Precautions Section.

➡ **If working near and/or around the SRS system and components, be sure to disable the SRS system. After disabling the system wait three minutes or more before servicing the vehicle.**

➡ **Before disconnecting the battery, lower both the driver's and passenger's windows. This will prevent any interference between the window edge and the vehicle when the door is opened or closed. During normal operation the window slightly raises and lowers automatically to prevent any window to vehicle interference. The automatic window function will not work with the battery disconnected.**

2. Properly relieve the fuel system pressure.
3. Raise and support the vehicle safely.
4. Drain the fuel tank. Be sure to properly dispose of gasoline stored in the fuel tank.
5. Remove the woofer or rear floor box mat and the rear floor box.
6. Remove the inspection hole cover for the main and sub fuel level sensor units.
7. Disconnect the harness connector and the quick connector.
8. Remove the stay tunnel.
9. Remove the exhaust front tube, center muffler and main muffler.
10. Matchmark and remove the driveshaft. Do not drop the driveshaft.
11. Remove the parking brake rear brake cables.
12. Remove the rear suspension assembly.
13. Remove the fuel tank protector.
14. Disconnect the fuel filler hose, vent hose and EVAP hoses at the fuel tank side.
15. Properly support the fuel tank, using a suitable jack.
16. Remove the mounting band bolts.
17. Lower the jack, check that all required lines, hoses and wires are disconnected.
18. Remove the fuel tank from the vehicle.

To install:

➡ **Be sure to use new fasteners, as required.**

19. Position the fuel tank to its mounting.

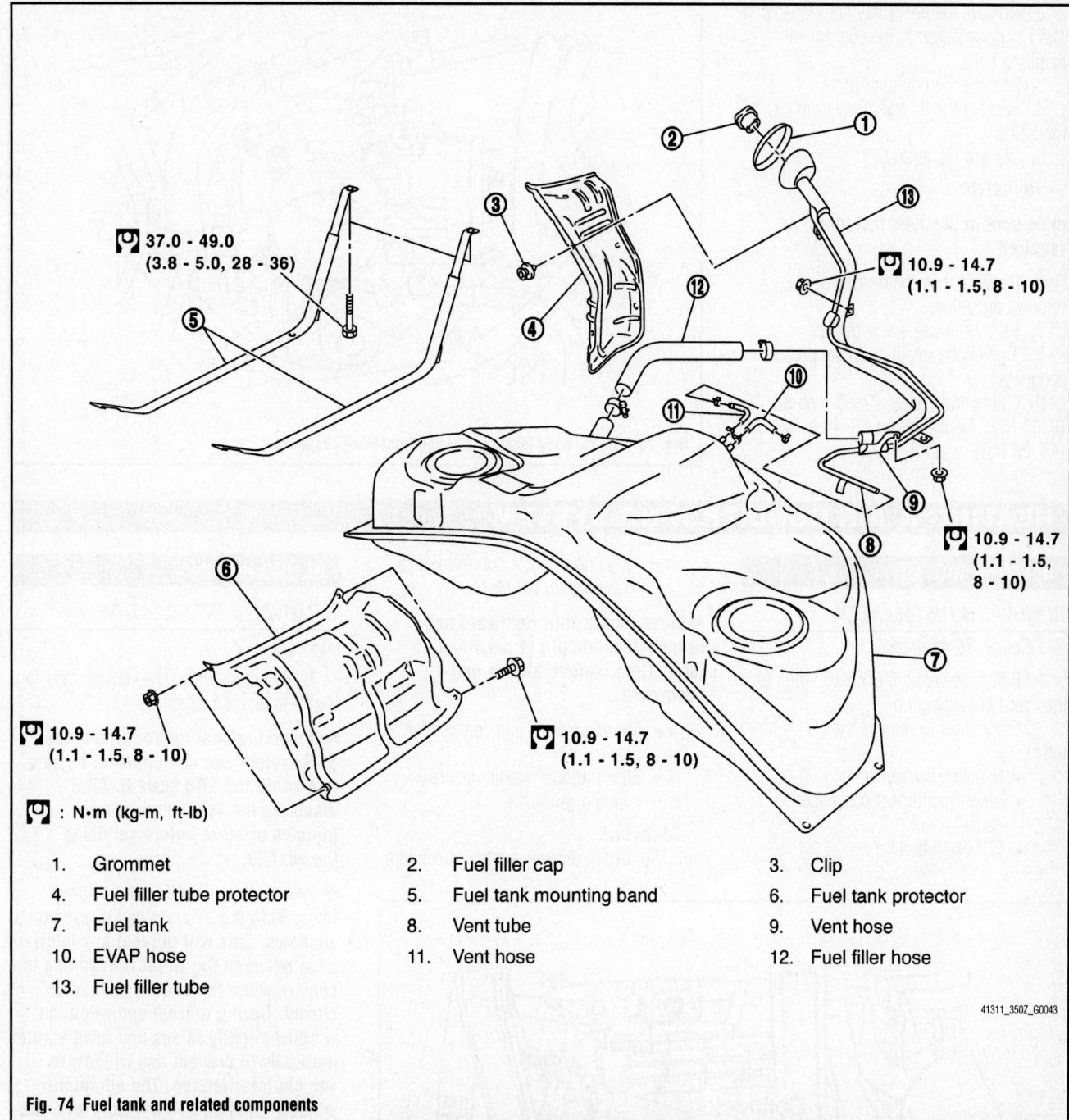

🔧 37.0 - 49.0
(3.8 - 5.0, 28 - 36)

🔧 10.9 - 14.7
(1.1 - 1.5, 8 - 10)

🔧 10.9 - 14.7
(1.1 - 1.5,
8 - 10)

🔧 10.9 - 14.7
(1.1 - 1.5, 8 - 10)

🔧 10.9 - 14.7
(1.1 - 1.5, 8 - 10)

🔧 : N•m (kg-m, ft-lb)

1. Grommet	2. Fuel filler cap	3. Clip
4. Fuel filler tube protector	5. Fuel tank mounting band	6. Fuel tank protector
7. Fuel tank	8. Vent tube	9. Vent hose
10. EVAP hose	11. Vent hose	12. Fuel filler hose
13. Fuel filler tube		

41311_350Z_G0043

Fig. 74 Fuel tank and related components

20. Install the retaining straps and bolts.

21. Continue the installation in the reverse order of the removal procedure.

22. Turn the ignition switch ON (with the engine stopped). Check all connections for leakage by applying fuel pressure to the lines.

23. Start the engine and check for leaks. Correct, as required.

IDLE SPEED

ADJUSTMENT

Idle speed is not adjustable.

THROTTLE BODY

REMOVAL & INSTALLATION

See Figure 75.

At this time the manufacturer does not provide removal and installation procedures for this component. The following procedure is a guideline and may differ from the vehicle you are servicing.

1. Before servicing the vehicle, refer to the Precautions Section.

➡**If working near and/or around the SRS system and components, be sure** to disable the SRS system. After disabling the system wait three minutes or more before servicing the vehicle.

➡**Before disconnecting the battery, lower both the driver's and passenger's windows. This will prevent any interference between the window edge and the vehicle when the door is opened or closed. During normal operation the window slightly raises and lowers automatically to prevent any window to vehicle interference. The automatic window function will not work with the battery disconnected.**

2. Remove the necessary components in order to gain access to the component retaining bolts.

3. Remove the retaining bolts.

4. Remove the throttle body from its mounting.

5. Discard the gasket.

To install:

→**Be sure to use new fasteners, as required.**

6. Installation is the reverse of the removal procedure.

7. Be sure to use a new gasket.

8. Tighten the retaining in the proper sequence.

9. Tighten first pass: 79–95 inch lbs. (9–11 Nm). Second pass: 13–16 ft. lbs. (18–22 Nm)

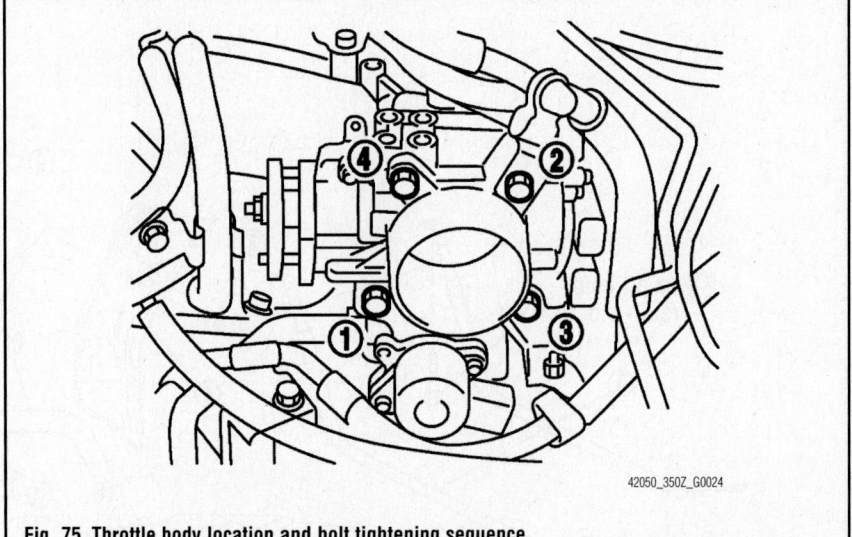

42050_350Z_G0024

Fig. 75 Throttle body location and bolt tightening sequence

HEATING & AIR CONDITIONING SYSTEM

BLOWER MOTOR

REMOVAL & INSTALLATION

See Figures 76 through 78.

1. Before servicing the vehicle, refer to the Precautions Section.

2. Disconnect or remove the following:
   - Negative battery cable
   - Lower instrument passenger panel
   - ECM with bracket
   - Intake door motor connector and blower motor connector

→**Move blower unit rightward and remove location pin (1 part) and joint. Then remove blower unit downward.**

   - Mounting bolts and screws then blower unit
   - Blower motor mounting screws then blower motor

To install:

3. To install, reverse removal procedure.

BLOWER UNIT

REMOVAL & INSTALLATION

See Figure 79.

1. Before servicing the vehicle, refer to the Precautions Section.

→**If working near and/or around the SRS system and components, be sure to disable the SRS system. After disabling the system wait three minutes or more before servicing the vehicle.**

→**Before disconnecting the battery, lower both the driver's and passenger's windows. This will prevent any interference between the window edge and the vehicle when the door is opened or closed. During normal operation the window slightly raises and lowers automatically to prevent any window to vehicle interference. The automatic window function will not work with the battery disconnected.**

2. Remove the remote keyless entry receiver.

3. Remove the ECM with the bracket attached.

4. Disconnect the intake door motor and blower motor connector.

5. Remove the blower motor mounting screws and remove the component from the vehicle.

→**Move the unit rightward, and remove the location pin and joint. Then remove the unit downward.**

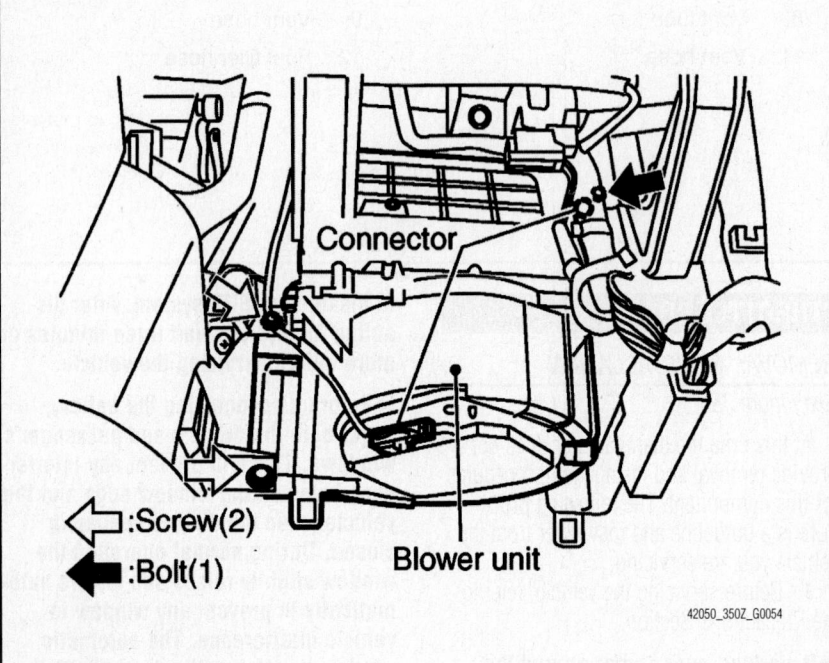

Connector

⇦ :Screw(2)

◀ :Bolt(1)

Blower unit

42050_350Z_G0054

Fig. 76 Removing blower unit

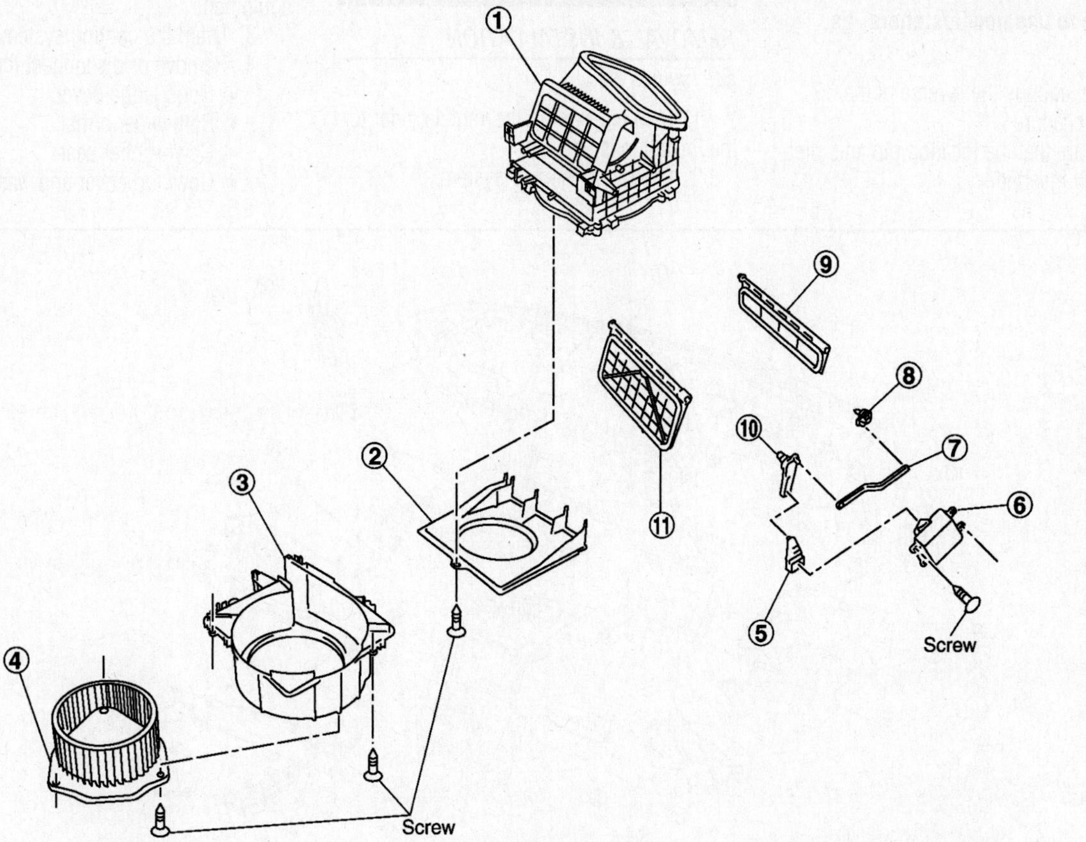

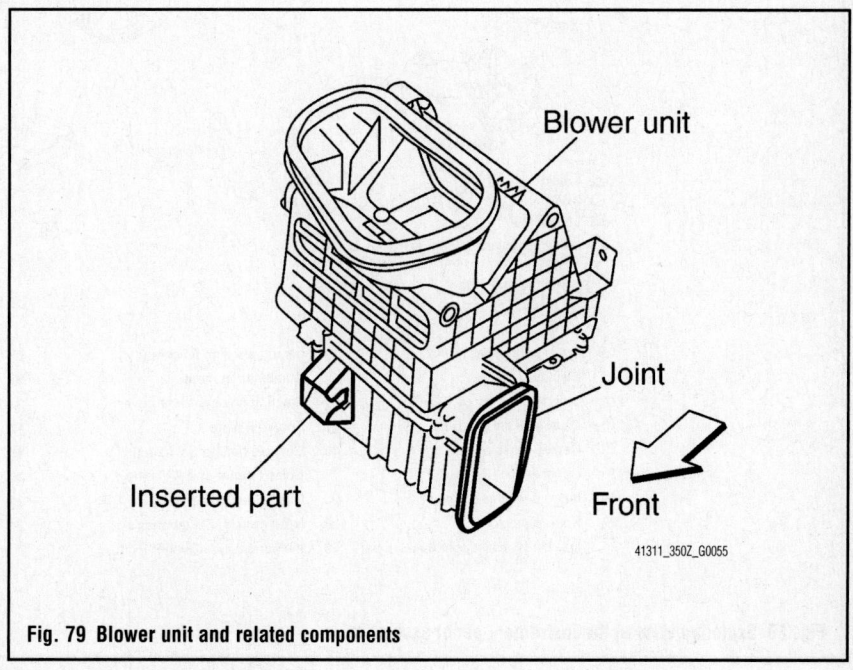

1. Intake upper case
2. Intake bell mouth
3. Intake lower case
4. Blower motor assembly
5. Intake door lever 2
6. Intake door motor
7. Intake door link
8. Intake door lever 3
9. Intake door 2
10. Intake door lever 1
11. Intake door 1

42050_350Z_G0055

Fig. 77 Exploded view of blower motor and components

42050_350Z_G0036

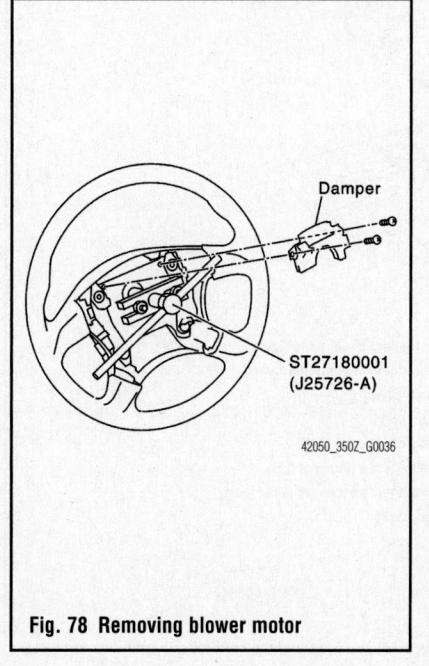

Fig. 78 Removing blower motor

41311_350Z_G0055

Fig. 79 Blower unit and related components

To install:

➡**Be sure to use new fasteners, as required.**

6. Installation is the reverse of the removal procedure.

7. Be sure that the location pin and joint are securely inserted.

HEATER CORE

REMOVAL & INSTALLATION

See Figures 80.

1. Before servicing the vehicle, refer to the Precautions Section.

2. Discharge the A/C system

using approved recycling equipment.

3. Drain the cooling system.

4. Remove or disconnect the following:
- Hood ledge cover
- Both wiper arms
- Cowl rubber seal
- Cowl top cover and washer hose

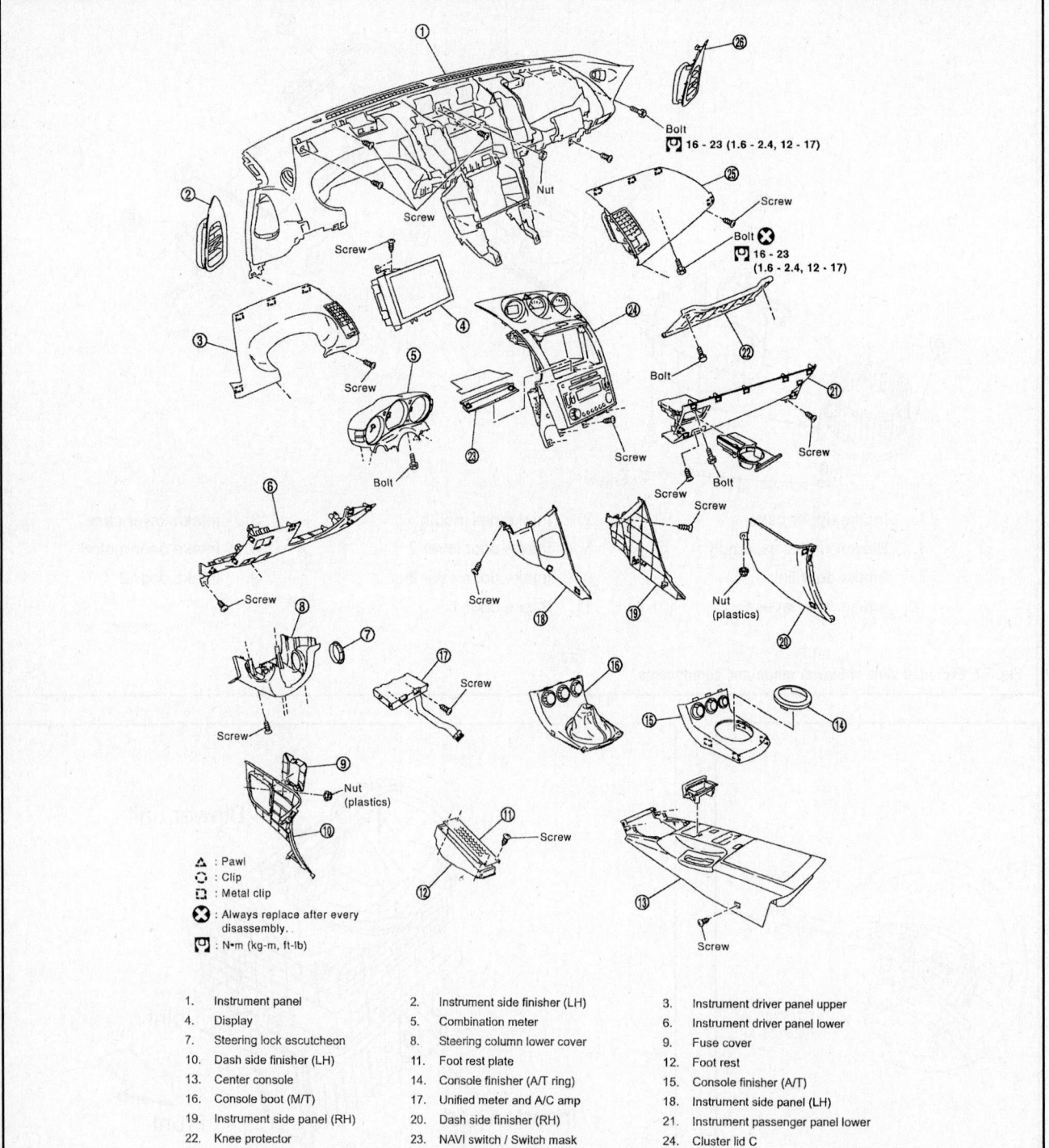

△ : Pawl
◇ : Clip
▱ : Metal clip
⊗ : Always replace after every disassembly.
⧉ : N•m (kg-m, ft-lb)

1. Instrument panel	2. Instrument side finisher (LH)	3. Instrument driver panel upper
4. Display	5. Combination meter	6. Instrument driver panel lower
7. Steering lock escutcheon	8. Steering column lower cover	9. Fuse cover
10. Dash side finisher (LH)	11. Foot rest plate	12. Foot rest
13. Center console	14. Console finisher (A/T ring)	15. Console finisher (A/T)
16. Console boot (M/T)	17. Unified meter and A/C amp	18. Instrument side panel (LH)
19. Instrument side panel (RH)	20. Dash side finisher (RH)	21. Instrument passenger panel lower
22. Knee protector	23. NAVI switch / Switch mask	24. Cluster lid C
25. Instrument passenger panel upper	26. Instrument side finisher (RH)	

67170-NISS-G010
67170-NISS-G10

Fig. 80 Exploded view of the instrument panel assembly

- Evaporator lines from the firewall and cap openings
- Electronic throttle control assembly
- Heater hoses
- Kick panels on both sides
- Foot rests
- Passenger side lower instrument panel cover
- Instrument panel side finish panels on both sides
- Cluster Lid C
- Data link connector
- Hood lock cable
- Steering column lower cover
- 4 bolts and combination meter

5. Position the steering wheel in the straight-ahead position.

6. Turn the ignition switch OFF.

7. Disconnect the negative (() battery cable; then, the positive (+) battery cable.

8. Remove the driver's side SRS and steering wheel by performing the following procedure:

   a. Remove both the left and right side lids from the steering wheel.

   b. Using a tamper resistant Torx® wrench (T50), remove the special bolts from both sides of the steering wheel.

   c. Remove steering wheel switch sub-harness connector.

   d. Remove air bag harness connector.

   e. Carefully remove the air bag module and store it face up.

9. Remove or disconnect the following:

- Steering wheel
- Steering column upper cover
- Spiral cable connector
- Combination switch
- Automatic transmission console finisher panel, if equipped
- Manual transmission shift knob, if equipped
- Center console
- Cup holder
- Passenger side lower instrument panel cover
- Instrument panel side cover
- Navigation switch cover panel and switch connector
- Audio cluster lid
- Audio unit and meter assembly
- Display unit
- Garnish panels and side finishers

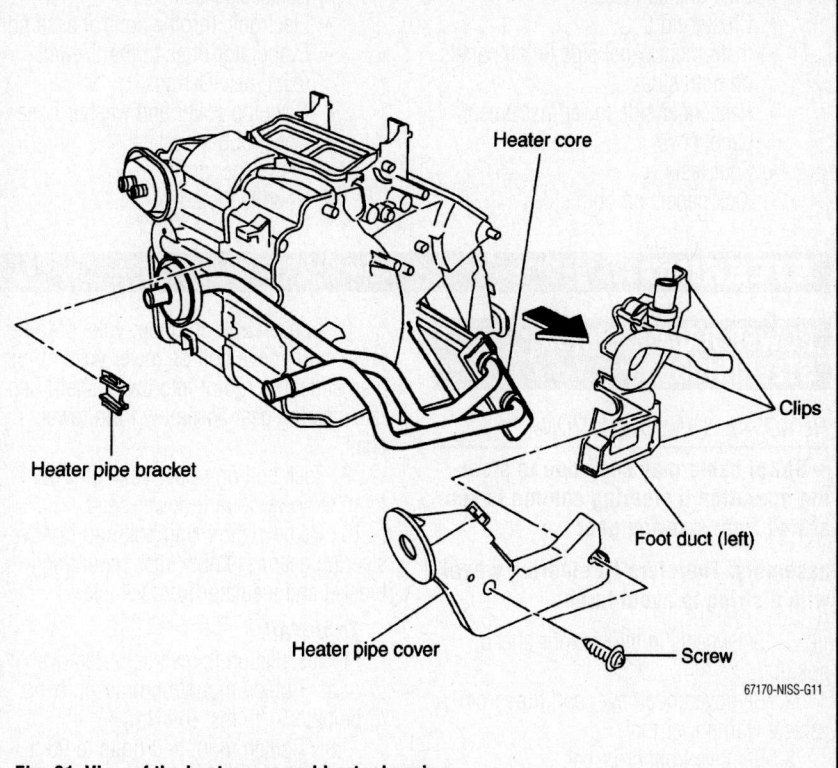

Fig. 81 View of the heater core and heater housing

Labels: Heater core; Clips; Foot duct (left); Screw; Heater pipe cover; Heater pipe bracket; 67170-NISS-G11

- Passenger air bag connector
- Passenger air bag bolt and passenger air bag
- ECM and bracket
- Intake door motor and blower motor connectors
- 2 screws, 1 bolt and the blower unit
- Left and right instrument panel stays
- Defroster and ventilation ducts
- Heating and A/C unit
- Heater pipe cover, support, and grommet
- Slide the heater core out of the heating and A/C unit

To install:

10. Install or connect the following:

- Heater core to the heater and A/C unit
- Heater pipe cover, support, and grommet
- Heater and A/C unit and tighten the bolts to 61 inch lbs. (7 Nm)
- Defroster and ventilation ducts
- Left and right instrument panel stays
- 2 screws, 1 bolt and the blower unit
- Intake door motor and blower motor connectors
- ECM and bracket

- Passenger air bag and tighten the bolt to 15–21 ft. lbs. (20–29 Nm)
- Passenger air bag connector
- Garnish panels and side finishers
- Display unit
- Audio unit and meter assembly
- Audio cluster lid
- Navigation switch cover panel and switch connector
- Instrument panel side cover
- Passenger side lower instrument panel cover
- Cup holder
- Center console
- Manual transmission shift knob, if equipped
- Automatic transmission console finisher panel, if equipped
- Combination switch
- Spiral cable connector
- Steering column upper cover
- Steering wheel and tighten the bolt to 22–28 ft. lbs. (30–39 Nm)
- Driver air bag module and tighten the bolts to 83 inch lbs. (9.4 Nm)
- Air bag harness connector
- Steering wheel switch sub-harness connector
- Both the left and right side lids to steering wheel
- 4 bolts and combination meter
- Steering column lower cover
- Hood lock cable

- Data link connector
- Cluster lid C
- Instrument panel side finish panels on both sides
- Passenger side lower instrument panel cover
- Foot rests
- Kick panels on both sides

- Heater hoses
- Electronic throttle control assembly
- Evaporator lines to the firewall using new O-rings
- Cowl top cover and washer hose
- Cowl rubber seal
- Both wiper arms
- Hood ledge cover

11. Refill the cooling system.
12. Connect the positive (+) battery cable, then the negative (()) battery cable.
13. Evacuate, charge, and leak test the air conditioning system refrigerant.
14. Operate the engine to normal operating temperatures, then check the climate control operation and check for leaks.

STEERING

POWER RACK & PINION STEERING GEAR

REMOVAL & INSTALLATION

➡**Spiral cable may snap due to steering operation if steering column is separated from steering gear**

assembly. Therefore fix steering wheel with a string to avoid turns.

1. Set wheels in the straight-ahead position.
2. Remove undercover and tires from vehicle with power tool.
3. Remove front crossbar.
4. Confirm slit of lower joints fits with the projection on rear cover cap, furthermore marking position on steering gear assembly nearly fits with the projection on rear cover cap.
5. Remove cotter pin at steering outer socket, then loosen mounting nut.
6. Use a ball joint remover (SST) to remove steering outer socket from steering knuckle.
7. Remove oil pipes (high pressure side and low pressure side) from steering gear assembly, then drain fluid from pipes.

8. Loosen bolt on upper yoke of lower joint and remove bolt on lower yoke of joint, then slide lower joint into lower shaft. Separate steering gear assembly from lower shaft.
9. Tack bolt on upper yoke of lower joint, fix lower joint to lower shaft.
10. Remove the fixing bolt and remove steering gear assembly, rack mounting bracket and insulator from vehicle.

To install:
11. Installation is reverse of removal.
   a. Tighten mounting bracket fixing bolts to 51 ft. lbs. (69 Nm).
   b. Tighten mounting bolts to 96 ft. lbs. (130 Nm).
   c. Tighten yolk shaft bolts to 18–21 ft. lbs. (23.5–29.4 Nm).

POWER STEERING PUMP

REMOVAL & INSTALLATION

1. Remove engine cover.
2. Remove air cleaner box.
3. Drain water from radiator upper tank, then remove radiator upper hose.
4. Remove radiator fan shroud.
5. Loosen idler pulley, then remove belt.

6. Drain power steering fluid from reservoir tank.
7. Remove piping of high pressure and low pressure (drain fluid).
8. Remove bolt common to water pump and power steering pump.
9. Remove bolt then remove power steering pump.

To install:
10. Installation is reverse of removal.

BLEEDING

1. Stop engine, and then turn steering wheel fully to right and left several times.
2. Do not allow steering fluid reservoir tank to go below the low-level line. Check tank frequenter and add fluid as needed.
3. Run engine at idle speed. Turn steering wheel fully to the right and then fully to the left, and keep for about 3 seconds. Then check whether a fluid leakage has occurred.
4. Repeat the 2nd procedure several times at about three seconds interval Check generation of air bubbles and cloud in fluid.
5. If air bubbles and the cloud don't fade, stop engine, hold air bleeding until air bubbles and the cloud fade. Perform the 2nd and the 3rd procedures again.
6. Stop engine, check fluid level.

SUSPENSION

FRONT SUSPENSION

CONTROL LINKS

REMOVAL & INSTALLATION

1. Remove the wheel and tire.
2. Remove undercover.
3. Disconnect the connecting rod end at the stabilizer bar using power tool.

➡**Prevent the stabilizer connecting rod from turning by inserting a hex wrench into the end of the ball stud, then remove nut.**

To install:
4. Installation is in the reverse order of removal.
   a. Tighten nuts to 66 ft. lbs. (89 Nm).

LOWER BALL JOINT

REMOVAL & INSTALLATION

The ball joint is an integral part of the lower control arm (transverse link). If the ball joint is defective the control arm (transverse link) must be replaced.

LOWER CONTROL ARM

REMOVAL & INSTALLATION

1. Before servicing the vehicle, refer to the Precautions Section.
2. Raise and safely support the vehicle.
3. Remove front wheels.
4. Remove mounting nut and washer

on lower portion of stabilizer connecting rod with power tool.
5. Remove mounting nut between transverse link and shock absorber on lower position.
6. Remove mounting nut between transverse link and front suspension member with power tool.
7. Remove transverse link from steering knuckle.
8. Remove transverse link from vehicle.
9. Check transverse link and bushing for deformation, cracks, or damage. If any non-standard condition is found, replace it.

To install:
10. To install, reverse removal procedure.

11. Perform final tightening of front suspension member installation position and shock absorber lower side (rubber bushing) under unladen condition with tires on ground. Check wheel alignment.

STEERING KNUCKLE

REMOVAL & INSTALLATION

See Figure 82.

➡ **The steering knuckle and wheel bearing must be replaced as an assembly.**

1. Before servicing the vehicle, refer to the Precautions Section.

2. Remove or disconnect the following:
   - Front wheel
   - Engine undercover
   - Brake caliper and wire aside
   - Brake rotor
   - ABS sensor
   - Brake hose bracket
   - Loosen steering outer socket nut
   - Separate outer socket from steering knuckle
   - Upper link from knuckle
   - Transverse link from knuckle
   - Compression rod from knuckle
   - Loosen steering knuckle nut
   - Knuckle and hub assembly from the vehicle

To install:

3. Install the wheel hub to the knuckle.
4. Install or connect the following:
   - Knuckle and hub assembly
   - Tighten the steering knuckle/hub nut to 58–72 ft. lbs. (79–98 Nm)
   - Compression rod to knuckle and tighten the nut to 56–69 ft. lbs. (75–94 Nm)
   - Transverse link to knuckle and tighten the nut to 56–69 ft. lbs. (75–94 Nm)
   - Upper link to knuckle and tighten the nut to 40–46 ft. lbs. (54–64 Nm)
   - Install outer socket to steering knuckle

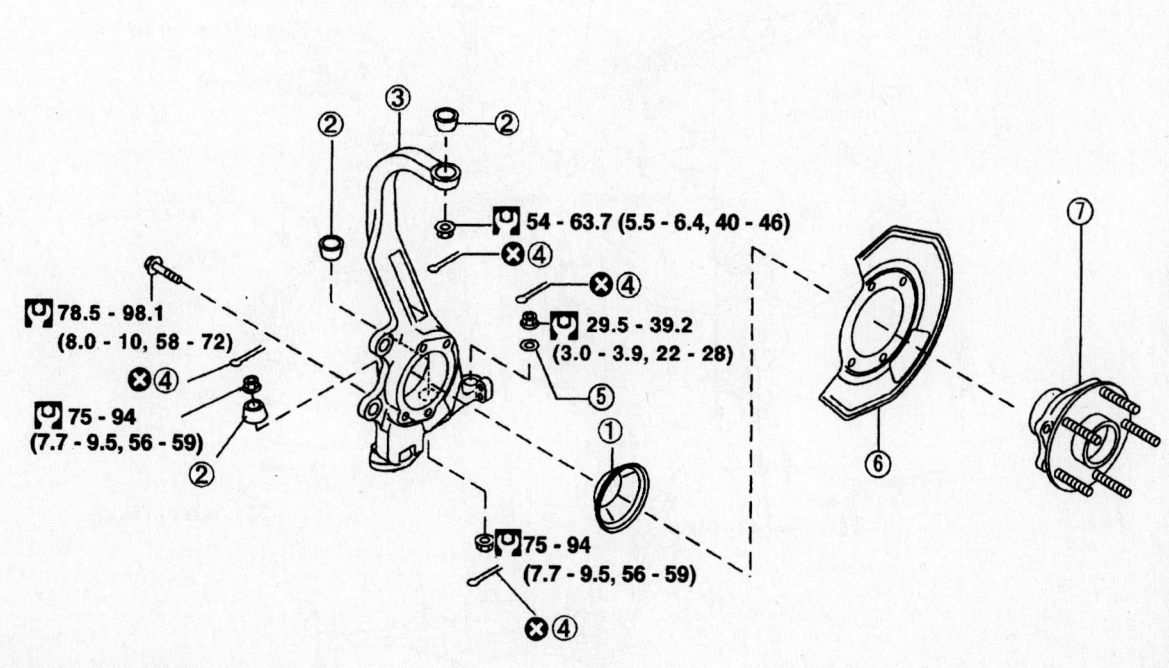

78.5 - 98.1 (8.0 - 10, 58 - 72)

75 - 94 (7.7 - 9.5, 56 - 59)

54 - 63.7 (5.5 - 6.4, 40 - 46)

29.5 - 39.2 (3.0 - 3.9, 22 - 28)

75 - 94 (7.7 - 9.5, 56 - 59)

: N•m (kg-m, ft-lb)

: Always replace after disassembly

1. Hub cap
2. Ball seat
3. Steering knuckle
4. Cotter pin
5. Washer
6. Splash guard
7. Wheel hub and bearing assembly

67170-NISS-G55

Fig. 82 Exploded view of the front steering knuckle and wheel bearing assembly

- Steering outer socket nut
- Brake hose bracket
- ABS sensor
- Brake rotor
- Brake caliper
- Engine undercover
- Front wheel

STRUT

REMOVAL & INSTALLATION

See Figure 83.

Use this head if OEM does not specify MacPherson strut

1. Before servicing the vehicle, refer to the Precautions Section.
2. Raise and safely support the vehicle.
3. Remove or disconnect the following:
   - Wheel
   - Engine undercover

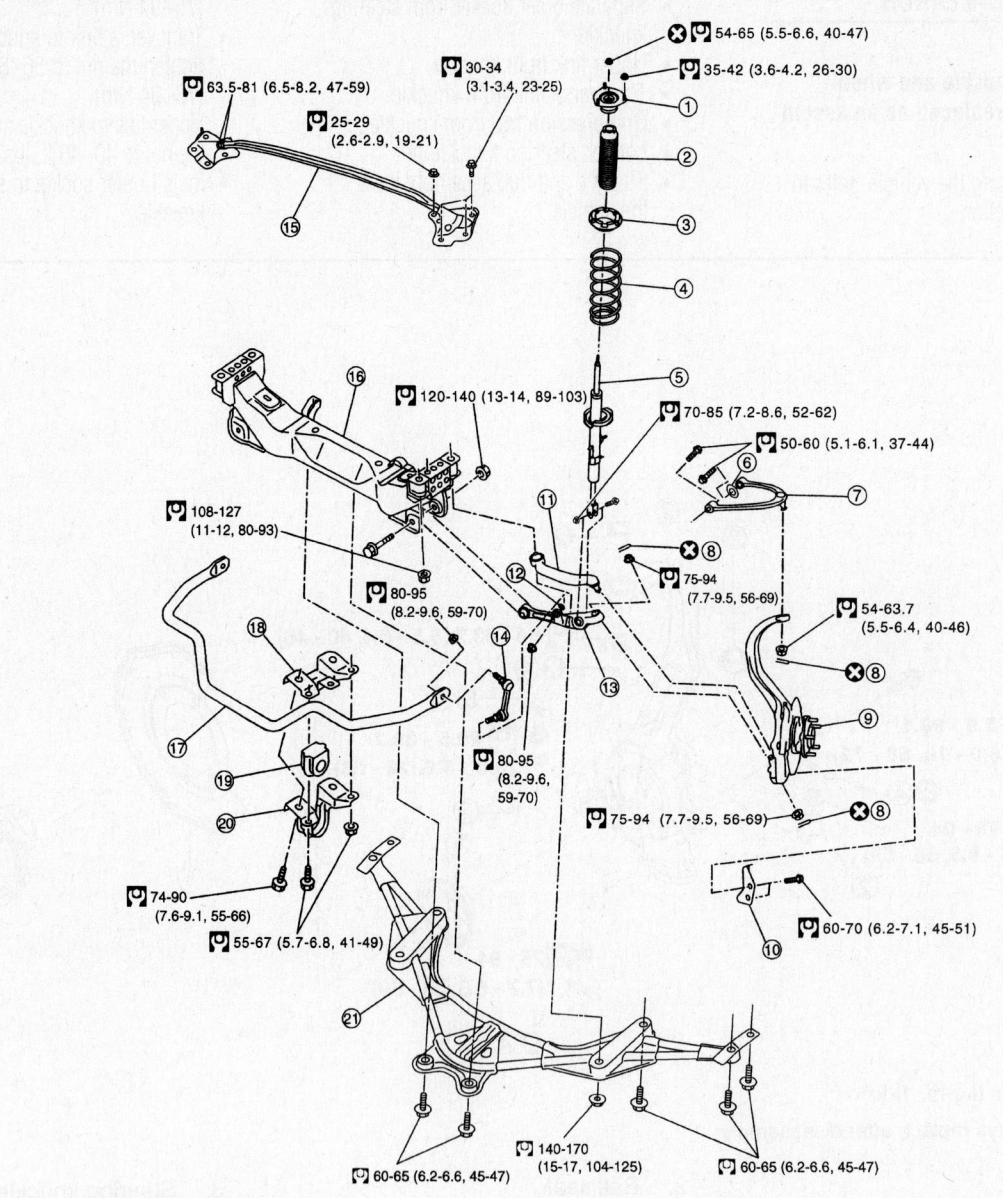

📏 : N•m (kg-m, ft-lb)

✖ : Always replace after every disassembly

1. Mounting insulator	2. Bound bumper	3. Spring upper seat
4. Coil spring	5. Shock absorber	6. Stopper rubber
7. Upper link	8. Cotter pin	9. Front axle
10. Steering stopper bracket	11. Compression rod	12. Washer
13. Transverse link	14. Stabilizer connecting rod	15. Tower bar
16. Front suspension member	17. Stabilizer bar	18. Stabilizer clamp bracket
19. Stabilizer bushing	20. Stabilizer clamp	21. Front cross bar

67170-NISS-G53

Fig. 83 Exploded view of the front suspension

- Anti-Lock Brake (ABS) wheel sensor and move it out of the way
- Brake hose from the strut
- Strut-to-transverse link nut and bolt

4. Open the hood and remove the strut attaching nuts while holding the strut.

✳✳ CAUTION

Do not remove the center locknut from the strut assembly until the strut is safely compressed.

5. Remove strut from the vehicle.
6. Place the strut assembly in a vise with a holding tool or in a spring compressor.
7. Loosen the piston rod locknut.

✳✳ CAUTION

Do not remove the piston rod locknut, the spring is under tension and can cause serious personal injury.

8. Compress the spring with the spring compressor, then remove the piston rod locknut

➡**Before removing the strut from the coil spring, note the positioning of the strut in relationship to the coil spring for reassembly.**

9. Remove or disconnect the following:
- Strut mounting insulator bracket, strut mounting bearing, upper spring seat and the upper spring rubber seat
- Strut, leaving the coil spring compressed
- Piston boot and rebound bumper from the strut

To install:
10. Install or connect the following:
- Rebound bumper and the boot to the strut piston
- Strut into the coil spring, be sure the strut and spring are properly positioned
- Upper spring rubber seat, upper spring seat, strut mounting bearing and the strut mounting insulator bracket. Be sure that the cutout on the upper spring seat is facing the outside of the vehicle
- Piston rod locknut. Remove the tool and torque the piston rod locknut to 40–47 ft. lbs. (54–65 Nm)
- Strut into the strut tower
- New attaching nuts and torque to 26–30 ft. lbs. (35–42 Nm)
- Bolts attaching the strut to the transverse link and torque to 52–62 ft. lbs. (70–80 Nm)

- ABS wheel sensor
- Brake hose to the strut
- Engine undercover
- Front wheels

11. Lower the vehicle.
12. Check and/or adjust the wheel alignment as necessary.

STABILIZER BAR

REMOVAL & INSTALLATION
See Figure 84.

1. Before servicing the vehicle, refer to the Precautions Section.
2. Raise and safely support the vehicle.
3. Remove or disconnect the following:
- Undercover
- Mounting nut on upper portion of stabilizer connecting rod with power tool
- Fixing bolts and nuts, then remove stabilizer clamp, stabilizer bushing, and stabilizer clamp bracket
- Stabilizer bar from vehicle

To install:
4. To install, reverse removal procedure.
5. When installing stabilizer bracket and stabilizer clamp, refer to illustration and perform the following:
- (1) fully tighten
- (2) temporarily tighten
- (3) temporarily tighten
- (2) fully tighten
- (3) fully tighten
- (4), (5) temporarily tighten then fully tighten

TOWER BAR

REMOVAL & INSTALLATION
See Figure 85.

1. Fix center bolt, and then loosen nut in the right and left side.
2. Loosen center bolt 660 degrees (or turn bolt 1.7 times) to place the black mark of center bolt above.
3. Remove tower bar fixing bolts and nuts, and remove tower bar from vehicle with power tool.

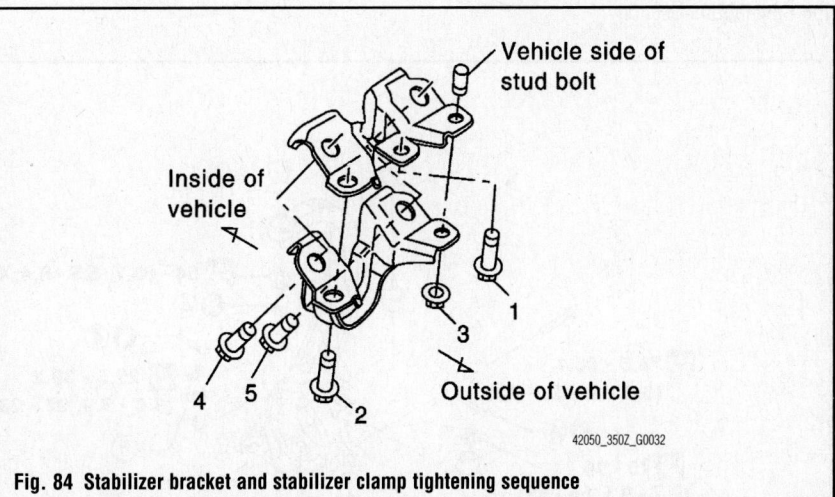

Fig. 84 Stabilizer bracket and stabilizer clamp tightening sequence

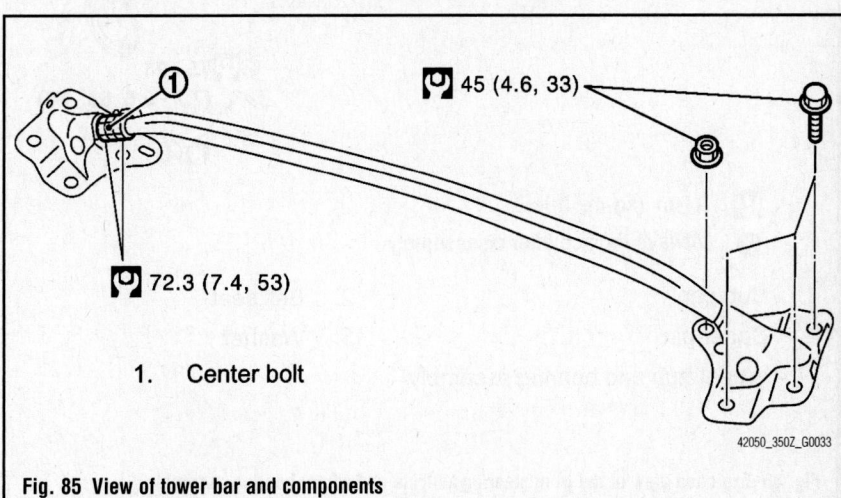

Fig. 85 View of tower bar and components

To install:

4. Install tower bar.

➡ **If it is hard to install tower bar, install it turning center bolt.**

5. Tighten center bolt 660 degrees, (or turn bolt 1.7 times) to place the black mark of center bolt above.

➡ **The space between tower bar and engine collector should be between 0.394 in. (10.0 mm) and 0.669 in. (17.0 mm).**

6. Tighten both nut of the right and left side of center bolt.

UPPER BALL JOINT

REMOVAL & INSTALLATION

The upper ball joint is removed with the upper control arm.

UPPER CONTROL ARM

REMOVAL & INSTALLATION

1. Before servicing the vehicle, refer to the Precautions Section.

2. Raise and safely support the vehicle.
3. Remove or disconnect the following:
   - Wheel
   - Engine undercover
   - Shock absorber
   - Cotter pin of upper link ball joint, loosen mounting nut
   - Using a ball joint remover, upper steering link from steering knuckle

❋❋ WARNING

Tighten temporarily mounting nut to prevent damage to threads and to prevent ball joint remover from coming off.

   - Bolts holding upper link to body
   - Upper link

To install:

4. To install, reverse removal procedure.

WHEEL HUB & BEARING

REMOVAL & INSTALLATION
See Figure 86.

➡ **If the wheel bearing is damaged, the steering knuckle and bearing must be replaced as an assembly.**

1. Before servicing the vehicle, refer to the Precautions Section.
2. Remove or disconnect the following:
   - Front wheel
   - Engine undercover
   - Brake caliper and wire aside
   - Brake rotor
   - ABS sensor
   - Brake hose bracket
   - Loosen steering outer socket nut
   - Separate outer socket from steering knuckle
   - Upper link from knuckle
   - Transverse link from knuckle
   - Compression rod from knuckle
   - Loosen steering knuckle nut
   - Knuckle and hub assembly from the vehicle
   - Separate the wheel hub from the knuckle

To install:

3. Install the wheel hub to the knuckle.
4. Install or connect the following:

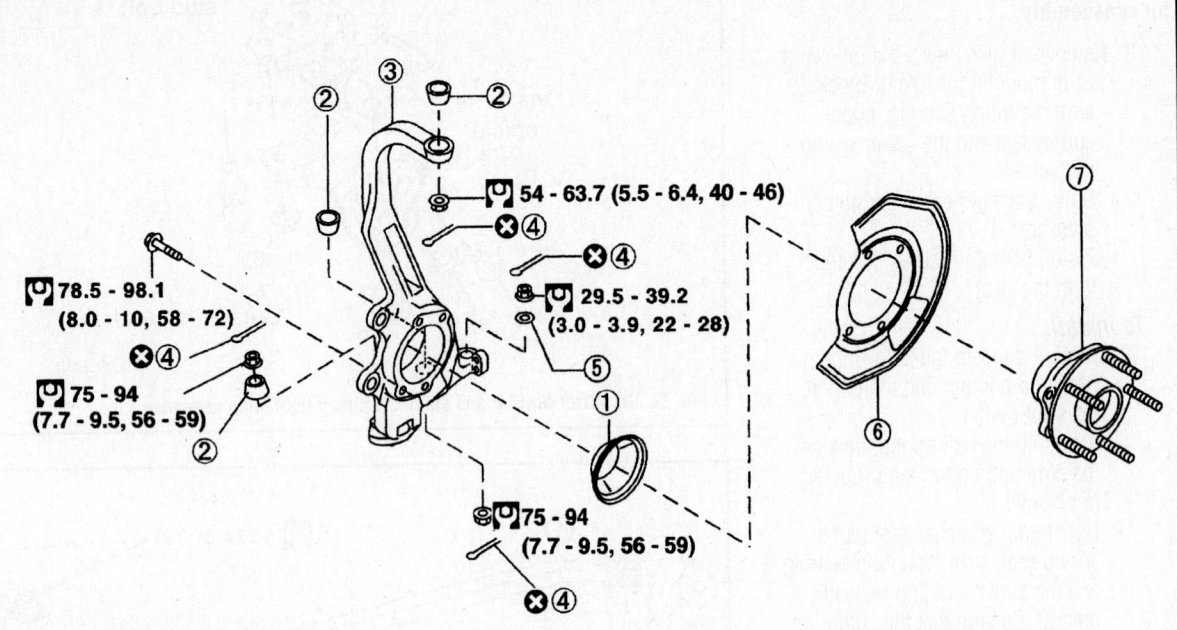

78.5 - 98.1 (8.0 - 10, 58 - 72)

75 - 94 (7.7 - 9.5, 56 - 59)

54 - 63.7 (5.5 - 6.4, 40 - 46)

29.5 - 39.2 (3.0 - 3.9, 22 - 28)

75 - 94 (7.7 - 9.5, 56 - 59)

[⊡] : N•m (kg-m, ft-lb)

[✕] : Always replace after disassembly

1. Hub cap
4. Cotter pin
7. Wheel hub and bearing assembly

2. Ball seat
5. Washer

3. Steering knuckle
6. Splash guard

67170-NISS-G55

Fig. 86 Exploded view of the front steering knuckle and wheel bearing assembly

- Knuckle and hub assembly
- Tighten the steering knuckle/hub nut to 58–72 ft. lbs. (79–98 Nm)
- Compression rod to knuckle and tighten the nut to 56–69 ft. lbs. (75–94 Nm)
- Transverse link to knuckle and tighten the nut to 56–69 ft. lbs. (75–94 Nm)
- Upper link to knuckle and tighten

the nut to 40–46 ft. lbs. (54–64 Nm)
- Install outer socket to steering knuckle
- Steering outer socket nut
- Brake hose bracket
- ABS sensor
- Brake rotor
- Brake caliper
- Engine undercover
- Front wheel

ADJUSTMENT

➡**Whenever the hub or bearing assemblies are removed, the wheel bearing must be replaced.**

Never reuse the old bearing assembly.

The wheel bearings are sealed and are not adjustable. If defective, replacement is the only option.

SUSPENSION

COIL SPRING

REMOVAL & INSTALLATION

See Figure 87.

1. Before servicing the vehicle, refer to the Precautions Section.
2. Place a jack under the rear lower link.
3. Remove the rear wheel.
4. Loosen the lower link nut and bolt on the suspension member side, then remove the bolt and nut on the axle side.
5. Lower the jack slowly and remove the upper seat, coil spring and rubber sheet from the lower link.
6. Remove the lower link nut and bolt on the axle side to remove the lower link.

To install:

7. Reverse the removal procedure and tighten the lower link nut and bolts to 48–59 ft. lbs. (65–80 Nm).

CONTROL ARMS/LINKS

REMOVAL & INSTALLATION

1. Remove tire with power tool.
2. Remove drive shaft.
3. Remove fixing nuts and bolts between suspension arm and rear suspension member.
4. Remove cotter pin of suspension arm ball joint, and loosen nut.
5. Use a ball joint remover (suitable tool) to remove suspension arm from axle. Be careful not to damage ball joint boot.

➡**Tighten temporarily mounting nut to prevent damage to threads and to prevent ball joint remover (suitable tool) from coming off.**

6. Remove suspension arm and stopper rubber from vehicle.

To install:

7. Installation is in the reverse order of removal.
   a. Tighten fixing nuts and bolts to 53 ft. lbs. (72.5 Nm).

b. Tighten ball joint nut to 62 ft. lbs. (84.5 Nm).

SHOCK ABSORBER

REMOVAL & INSTALLATION

1. Before servicing the vehicle, refer to the Precautions Section.
2. Set a floor jack on the rear lower link to remove the lower shock absorber nut and bolt.
3. Remove the rear wheel.
4. Remove fixing bolt in lower side of shock absorber assembly with power tool.
5. Remove mounting seal bracket fixing nuts of shock absorber upper side with power tool and remove shock absorber from vehicle.

To install:

6. Reverse the removal procedure and tighten the upper shock nuts to 21 ft. lbs. (28 Nm, and the lower bolt to 74–88 ft. lbs. (100–120 Nm).

TESTING

1. Check the shock for deformation, cracks, damage and replace as required.
2. Check the piston rod for damage, uneven wear or distortion and replace as required.
3. Check the welded and sealed areas for oil leakage and replace as required.

STABILIZER BAR

REMOVAL & INSTALLATION

1. Before servicing the vehicle, refer to the Precautions Section.
2. Raise and safely support the vehicle.
3. Remove fixing bolts and remove stabilizer connecting rod mount bracket from suspension arm.
4. Remove lower side fixing nut on stabilizer connecting rod and remove stabilizer connecting rod from stabilizer bar with power tool.

REAR SUSPENSION

5. Remove fixing nut on stabilizer clamp and remove stabilizer from vehicle with power tool.

To install:

6. To install, reverse removal procedure.

➡**Stabilizer bar uses pillow ball type connecting rod, position ball joint with case on pillow ball head parallel to stabilizer bar.**

WHEEL HUB & BEARING (SEALED UNIT)

REMOVAL & INSTALLATION

1. Remove tires from vehicle with power tool.
2. Remove cotter pin. Then remove lock nut from drive shaft.
3. Remove brake caliper with power tool. Hang it in a place where it will not interfere with work.
4. Remove disc rotor and remove parking cable and parking brake shoe from back plate.
5. Remove fixing bolts and nuts in axle side of radius rod, front lower link with power tool.
6. Remove fixing bolt and nut in axle side of rear lower link with power tool. Then remove coil spring.
7. Remove fixing bolt and nut in axle side of shock absorber with power tool.
8. Using a puller (suitable tool), remove axle from drive shaft.
9. Remove suspension arm and cotter pin at axle, then loosen mounting nut.
10. Use a ball joint remover (suitable tool) to remove suspension arm from axle. Be careful not to damage ball joint boot.
11. Remove knuckle assembly
12. Remove wheel bearing fixing bolts and anchor block fixing nuts, and remove wheel hub and bearing assembly, back plate and anchor block from axle.
13. Using a drift (SST) and a puller (suitable tool), press wheel hub out to remove from wheel bearing.

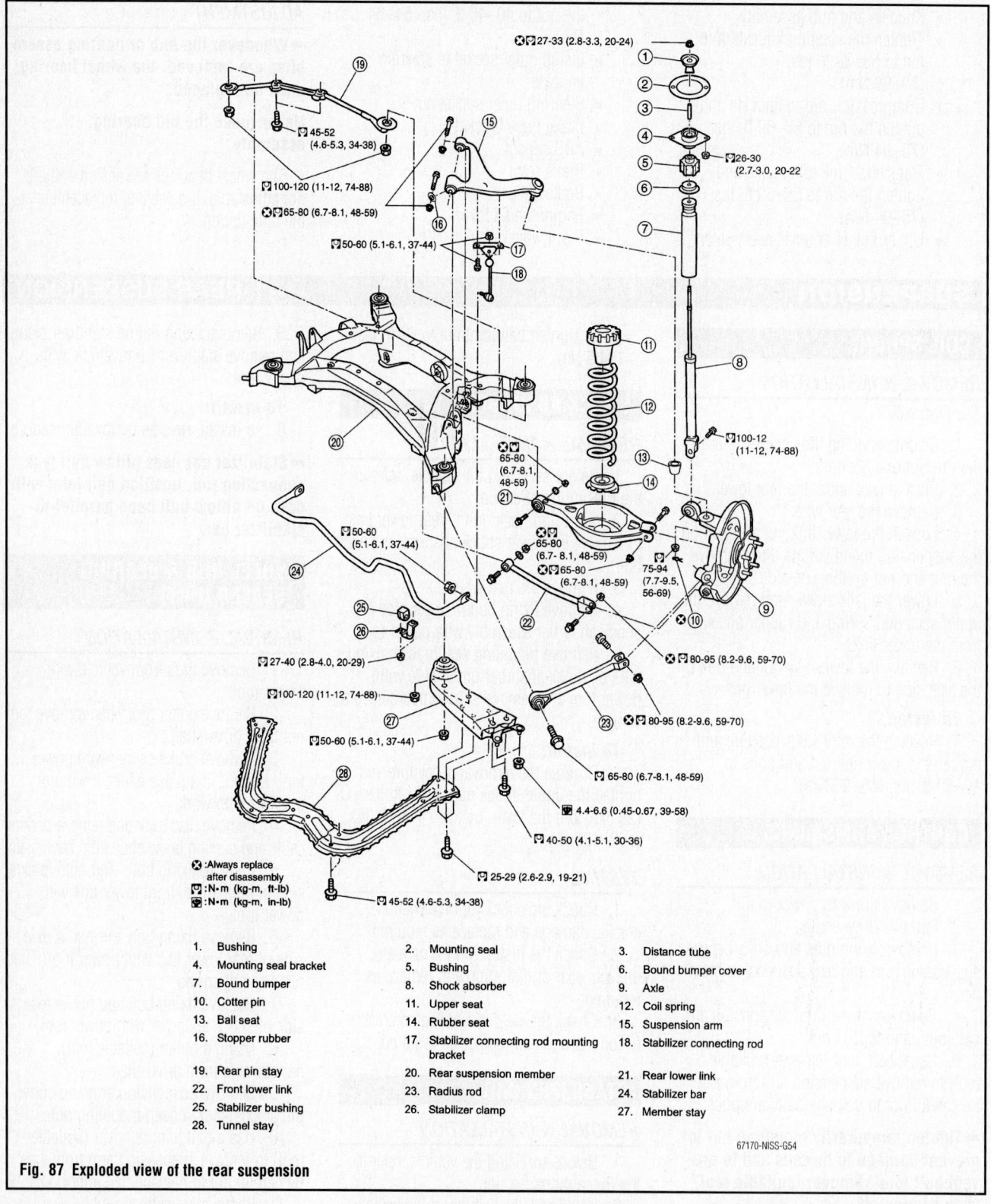

27-33 (2.8-3.3, 20-24)

45-52
(4.6-5.3, 34-38)

100-120 (11-12, 74-88)

65-80 (6.7-8.1, 48-59)

50-60 (5.1-6.1, 37-44)

26-30
(2.7-3.0, 20-22)

100-12
(11-12, 74-88)

65-80
(6.7-8.1,
48-59)

50-60
(5.1-6.1, 37-44)

65-80
(6.7-8.1, 48-59)

65-80
(6.7-8.1, 48-59)

75-94
(7.7-9.5,
56-69)

80-95 (8.2-9.6, 59-70)

27-40 (2.8-4.0, 20-29)

100-120 (11-12, 74-88)

50-60 (5.1-6.1, 37-44)

80-95 (8.2-9.6, 59-70)

65-80 (6.7-8.1, 48-59)

4.4-6.6 (0.45-0.67, 39-58)

40-50 (4.1-5.1, 30-36)

25-29 (2.6-2.9, 19-21)

45-52 (4.6-5.3, 34-38)

:Always replace
after disassembly
N·m (kg-m, ft-lb)
N·m (kg-m, in-lb)

1. Bushing	2. Mounting seal	3. Distance tube
4. Mounting seal bracket	5. Bushing	6. Bound bumper cover
7. Bound bumper	8. Shock absorber	9. Axle
10. Cotter pin	11. Upper seat	12. Coil spring
13. Ball seat	14. Rubber seat	15. Suspension arm
16. Stopper rubber	17. Stabilizer connecting rod mounting bracket	18. Stabilizer connecting rod
19. Rear pin stay	20. Rear suspension member	21. Rear lower link
22. Front lower link	23. Radius rod	24. Stabilizer bar
25. Stabilizer bushing	26. Stabilizer clamp	27. Member stay
28. Tunnel stay		

67170-NISS-G54

Fig. 87 Exploded view of the rear suspension

14. Using a drift (SST) and a puller (suitable tool), press wheel bearing outer side inner race out to remove from wheel hub.

To install:

15. Press fit a wheel hub into wheel bearing with a drift (SST).

➡**Press fit a drift (SST) while holding it against wheel bearing inner side inner race. Wheel bearing cannot be reused. Do not attempt to reuse it.**

16. Install back plate and wheel hub and bearing assembly.

17. Install anchor block onto axle.

18. The remaining installation is in the reverse order of removal.

ADJUSTMENT

The wheel bearings are sealed and are not adjustable. If defective, replacement is the only option.

SPECIFICATIONS AND MAINTENANCE CHARTS

ENGINE AND VEHICLE IDENTIFICATION

Engine							Model Year	
Code ①	Liters (cc)	Cu. In.	Cyl.	Fuel Sys.	Engine Type	Eng. Mfg.	Code ②	Year
VQ37VHR	3696	225	6	MFI	DOHC	Nissan	9	2009
							A	2010

MFI: Multi-port Fuel Injection

DOHC: Double Overhead Camshaft

① The Engine Code is stamped on the engine block near the starter.

② 10th position of the Vehicle Identification Number (VIN)

37663_370Z_C0001

GENERAL ENGINE SPECIFICATIONS

Year	Model	Engine Displacement Liters	Engine Series (ID/VIN)	Net Horsepower @ rpm	Net Torque @ rpm (ft. lbs.)	Bore x Stroke (in.)	Compression Ratio	Oil Pressure @ rpm
2009	370Z	3.7	VQ37VHR	332@7000	270@5200	3.760X3386	11.0:1	43@2000
2010	370Z	3.7	VQ37VHR	332@7000	270@5200	3.760X3386	11.0:1	43@2000

37663_370Z_C0002

ENGINE TUNE-UP SPECIFICATIONS

Year	Model	Engine Displacement Liters	Engine ID/VIN	Spark Plug Gap (in.)	Ignition Timing (deg.) MT	AT	Fuel Pump (psi) ①	Idle Speed (rpm) MT	AT ②	Valve Clearance (in.) Intake ③	Exhaust ③
2009	370Z	3.7	VQ37VHR	0.043	④	④	51	600-700	600-700	0.010-0.013	0.011-0.015
2010	370Z	3.7	VQ37VHR	0.043	④	④	51	600-700	600-700	0.010-0.013	0.011-0.015

NOTE: The Vehicle Emission Control Information label often reflects specification changes made during production.

The label figures must be used if they differ from those in this chart.

NA: Not Available

B: Before top dead center

① System pressure at idle with vacuum hose connected; should increase to 43 psi when disconnected

② Automatic transmission in park or neutral

③ Engine cold

④ 5-15 degrees BTDC

37663_370Z_C0003

CAPACITIES

Year	Model	Engine ID/VIN	Engine Displacement Liters	Engine Oil with Filter (qts.)	Transmission (pts.) Man	Transmission (pts.) Auto.	Drive Axle Rear (pts.)	Fuel Tank (gal.)	Cooling System (qts.)
2009	370Z	VQ37VHR	3.7	5.0	6.0	19.5	3.0	19.0	9.0
2010	370Z	VQ37VHR	3.7	5.0	6.0	19.5	3.0	19.0	9.0

NOTE: All capacities are approximate. Add fluid gradually and check to be sure a proper fluid level is obtained.

37663_370Z_C0004

FLUID SPECIFICATIONS

Year	Model	Engine Displacement Liters	Engine ID/VIN	Engine Oil	Auto. Trans.	Manual Trans.	Power Steering Fluid	Brake Master Cylinder
2009	370Z	3.7	VQ37VHR	5W-30	①	②	③	DOT 3
2010	370Z	3.7	VQ37VHR	5W-30	①	②	③	DOT 3

DOT: Department Of Transportation

① Nissan Matic S ATF

② Nissan (MTF) HQ MULTI 75W-85 or API GL-4, Viscosity 75W-85

③ Nissan PSF or equivalent. DEXRON VI type ATF may be used

37663_370Z_C0013

VALVE SPECIFICATIONS

Year	Engine ID/VIN	Engine Displacement Liters	Seat Angle (deg.)	Face Angle (deg.)	Spring Test Pressure (lbs. @ in.)	Spring Installed Height (in.)	Stem-to-Guide Clearance (in.) Intake	Stem-to-Guide Clearance (in.) Exhaust	Stem Diameter (in.) Intake	Stem Diameter (in.) Exhaust
2009	VQ37VHR	3.7	45.15-45.45	45	①	②	0.0008-0.0021	0.0012-0.0022	0.2348-0.2354	0.2347-0.2350
2010	VQ37VHR	3.7	45.15-45.45	45	①	②	0.0008-0.0021	0.0012-0.0022	0.2348-0.2354	0.2347-0.2350

① Intake: 43-48 lb at 1.6102 inch. Exhaust: 37-42 lb at 1.4567 inch.

② Intake: 1.7976. Exhaust: 1.7264.

37663_370Z_C0005

CAMSHAFT AND BEARING SPECIFICATIONS CHART

All measurements are given in inches.

Year	Engine Displ. Liters	Engine ID/VIN	Journal Dia.	Brg. Oil Clearance	Shaft End-play	Runout	Journal-to-Bore Clearance	Lobe Lift Intake	Lobe Lift Exhaust
2009	3.7	VQ37VHR	① ③	② ③	0.0045-0.0074	0.0008	NA	NA	④ ③
2010	3.7	VQ37VHR	① ③	② ③	0.0045-0.0074	0.0008	NA	NA	④ ③

NA: Not Available

① Front No. 1: 1.0211- 1.0218

  No. 2, 3, 4: 0.9230- 0.9238

② Front No. 1: 0.0018- 0.0034

  No. 2, 3, 4: 0.0014- 0.0030

③ Specification is for exhaust camshaft.

④ 1.7722- 1.7797 bank one. 1.8400- 1.8474 bank two.

37663_370Z_C0014

CRANKSHAFT AND CONNECTING ROD SPECIFICATIONS

All measurements are given in inches.

Year	Engine Displacement Liters	Engine ID/VIN	Crankshaft Main Brg. Journal Dia.	Main Brg. Oil Clearance	Shaft End-play	Thrust on No.	Connecting Rod Journal Diameter	Oil Clearance	Side Clearance
2009	3.7	VQ37VHR	2.5581-2.5580*	0.0014-0.0018	0.0039-0.0098	3	2.1241-2.1241*	0.0016-0.0021	0.0079-0.0138
2010	3.7	VQ37VHR	2.5581-2.5580*	0.0014-0.0018	0.0039-0.0098	3	2.1241-2.1241*	0.0016-0.0021	0.0079-0.0138

* Based upon grade. Grade "A" specification. Specification for other grades will differ.

37663_370Z_C0008

PISTON AND RING SPECIFICATIONS

All measurements are given in inches.

Year	Engine Displ. Liters	Engine ID/VIN	Piston Clearance	Ring Gap Top Compression	Ring Gap Bottom Compression	Ring Gap Oil Control	Ring Side Clearance Top Compression	Ring Side Clearance Bottom Compression	Ring Side Clearance Oil Control
2009	3.7	VQ37VHR	0.0004-0.0012	0.0091-0.0130	0.0091-0.0130	0.0067-0.0185	0.0016-0.0031	0.0012-0.0028	0.0022-0.0061
2010	3.7	VQ37VHR	0.0004-0.0012	0.0091-0.0130	0.0091-0.0130	0.0067-0.0185	0.0016-0.0031	0.0012-0.0028	0.0022-0.0061

37663_370Z_C0007

TORQUE SPECIFICATIONS

All readings in ft. lbs.

Year	Engine Displacement Liters	Engine ID/VIN	Cylinder Head Bolts	Main Bearing Bolts	Rod Bearing Bolts	Crankshaft Damper Bolts	Flywheel Bolts	Manifold		Spark Plugs	Oil Drain Plug
								Intake	Exhaust		
2009	3.7	VQ37VHR	①	②	③	④	65	⑤	⑥	18	25
2010	3.7	VQ37VHR	①	②	③	④	65	⑤	⑥	18	25

① Step 1: 77 ft. lbs.

  Step 2: Loosen bolts completely

  Step 3: 30 ft. lbs.

  Step 4: Tighten an additional 95 degrees

  Step 5: Tighten an additional 95 degrees

② Step 1: 18 ft. lbs.

  Step 2: 26 ft.lbs.

  Step 3: Tighten an additional 90 degrees

③ Step 1: 21 ft. lbs.

  Step 2: Loosen bolts completely

  Step 3: 18 ft.lbs.

  Step 4: Tighten an additional 90 degrees

④ Step 1: 33 ft. lbs.

  Step 2: Tighten an additional 90 degrees

⑤ Step 1: 5 ft. lbs.

  Step 2: 19 ft. lbs.

⑥ Step 1: 11 ft. lbs. in 2 steps

37663_370Z_C0006

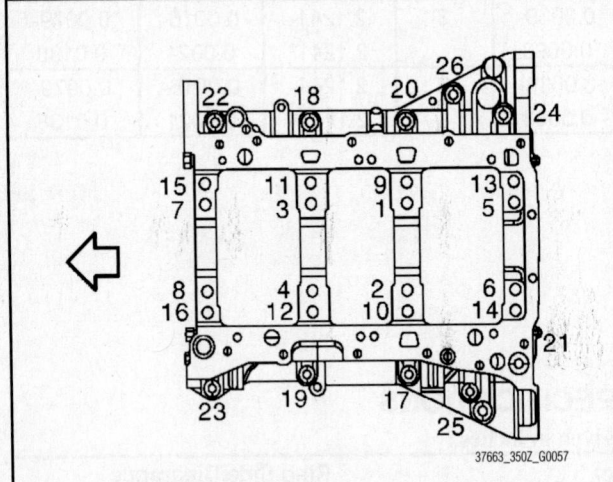

Fig. 1 Main bearing (lower cylinder block bolts) tightening sequence

37663_350Z_G0057

WHEEL ALIGNMENT

Year	Model		Caster Range (+/-Deg.)	Caster Preferred Setting (Deg.)	Camber Range (+/-Deg.)	Camber Preferred Setting (Deg.)	Toe-in (in.)
2009	370Z	F	①	①	②	②	③
		R	—	—	④	④	⑤ ⑥
2010	370Z	F	①	①	②	②	③
		R	—	—	④	④	⑤ ⑥

① Min 4 degrees 25' (4.42 degrees)
   Max 5 degrees 55' (5.91 degrees)
② Min -1 degree 25' (-1.41 degrees)
   Max 0 degree 40' (0.08 degrees)
③ Min 0.04 inch
   Nominal 0.08 inch
   Max 0.11 inch

④ Min -2 degrees 10' (-2.16 degrees)
   Max -1 degrees 10' (-1.17 degrees)
⑤ Min 0.079 inch 18 inch wheel
   Nominal 0.150 inch 18 inch wheel
   Max 0.221 inch 18 inch wheel
⑥ Min 0.079 inch 19 inch wheel
   Nominal 0.146 inch 19 inch wheel
   Max 0.213 inch 19 inch wheel

37663_370Z_C0009

TIRE, WHEEL AND BALL JOINT SPECIFICATIONS

Year	Model	OEM Tires Standard	OEM Tires Optional	Tire Pressures (psi) Front	Tire Pressures (psi) Rear	Wheel Size	Lug Nut Torque (ft. lbs.)
2009	Front	225/50W5R18	245/40WR19	①	②	NA	80
	Rear	245/45WR18	275/35WR19	①	②	NA	80
2010	Front	225/50W5R18	245/40WR19	①	②	NA	80
	Rear	245/45WR18	275/35WR19	①	②	NA	80

Note: If specification differes from vehicle placard, use specification given on vehicle placard.
NA: Not Available
OEM: Original Equipment Manufacturer
PSI: Pounds Per Square Inch
① See vehicle placard for specification

37663_370Z_C0010

BRAKE SPECIFICATIONS

All measurements in inches unless noted

Year	Model		Brake Disc Original Thickness	Brake Disc Minimum Thickness	Brake Disc Maximum Run-out	Minimum Lining Thickness Front	Minimum Lining Thickness Rear	Brake Caliper Bracket Bolts (ft. lbs.)	Brake Caliper Mounting Bolts (ft. lbs.)
2009	370Z ①	F	1.1024	NA	0.0014	0.079	0.079	NA	NA
		R	0.5510	NA	0.0022	0.079	0.079	NA	NA
	370Z ②	F	1.1810	NA	0.0014	0.079	0.079	NA	NA
		R	0.7090	NA	0.0022	0.079	0.079	NA	NA
2010	370Z ①	F	1.1024	NA	0.0014	0.079	0.079	NA	NA
		R	0.5510	NA	0.0022	0.079	0.079	NA	NA
	370Z ②	F	1.1810	NA	0.0014	0.079	0.079	NA	NA
		R	0.7090	NA	0.0022	0.079	0.079	NA	NA

NA: Not Available
① One piston type rear. Two piston type front.
② Two piston type rear. Four piston type front.

37663_370Z_C0011

SCHEDULED MAINTENANCE INTERVALS
NISSAN—370Z

TO BE SERVICED	TYPE OF SERVICE	VEHICLE MILEAGE INTERVAL (x1000)												
		7.5	15	22.5	30	37.5	45	52.5	60	67.5	75	82.5	90	97.5
Engine oil & filter	R	✓	✓	✓	✓	✓	✓	✓	✓	✓	✓	✓	✓	✓
Brake lines & cables	S/I		✓		✓		✓		✓		✓		✓	
Brake pads & discs	S/I		✓		✓		✓		✓		✓		✓	
Cabin Filter	R		✓		✓		✓		✓		✓		✓	
Driveshaft boots	S/I		✓		✓		✓		✓		✓		✓	
Exhaust system	S/I				✓				✓				✓	
Transmission fluid	S/I		✓		✓		✓		✓		✓		✓	
Air cleaner filter	R				✓				✓				✓	
Spark plugs (except platinum)	R				✓				✓				✓	
Spark plugs (platinum tip) exc. Nismo 370Z	R	replace every 105,000 miles												
Spark plugs (platinum tip) Nismo 370Z	R								✓					
Steering gear & linkage, axle & suspension parts	S/I				✓				✓				✓	
Engine coolant	R								✓				✓	
Drive belts	S/I								✓		✓		✓	
Fuel lines	S/I								✓					
Vapor lines	S/I								✓					

R: Replace   S/I: Service or Inspect

FREQUENT OPERATION MAINTENANCE (SEVERE SERVICE)

If a vehicle is operated under any of the following conditions it is considered severe service:

- Extremely dusty areas.

- 50% or more of the vehicle operation is in 32°C (90°F) or higher temperatures, or constant operation in temperatures below 0°C (32°F).

- Prolonged idling (vehicle operation in stop and go traffic).

- Frequent short running periods (engine does not warm to normal operating temperatures).

- Police, taxi, delivery usage or trailer towing usage.

Oil & oil filter: change every 3750 miles.

Brake pads & discs: service or inspect every 7500 miles.

Driveshaft boots: service or inspect every 7500 miles.

Exhaust system: service or inspect every 7500 miles.

Steering gear & linkage, axle & suspension parts: service or inspect every 7500 miles.

Steering linkage ball joints & front suspension ball joints: service or inspect every 7500 miles.

Air cleaner filter: service or inspect every 15,000 miles.

37663_370Z_C0012

BRACES — INFORMATION AND PRECAUTIONS

ANTI-LOCK SYSTEMS

• Certain components within the ABS system are not intended to be serviced or repaired individually.

• Do not use rubber hoses or other parts not specifically specified for and ABS system. When using repair kits, replace all parts included in the kit. Partial or incorrect repair may lead to functional problems and require the replacement of components.

• Lubricate rubber parts with clean, fresh brake fluid to ease assembly. Do not use shop air to clean parts; damage to rubber components may result.

• Use only DOT 3 brake fluid from an unopened container.

• If any hydraulic component or line is removed or replaced, it may be necessary to bleed the entire system.

• A clean repair area is essential. Always clean the reservoir and cap thoroughly before removing the cap. The slightest amount of dirt in the fluid may plug an orifice and impair the system function. Perform repairs after components have been thoroughly cleaned; use only denatured alcohol to clean components. Do not allow ABS components to come into contact with any substance containing mineral oil; this includes used shop rags.

• The Anti-Lock control unit is a microprocessor similar to other computer units in the vehicle. Ensure that the ignition switch is **OFF** before removing or installing controller harnesses. Avoid static electricity discharge at or near the controller.

• If any arc welding is to be done on the vehicle, the control unit should be unplugged before welding operations begin.

DISC AND DRUM SYSTEMS

✳✳ CAUTION

Dust and dirt accumulating on brake parts during normal use may contain asbestos fibers from production or aftermarket brake linings. Breathing excessive concentrations of asbestos fibers can cause serious bodily harm. Exercise care when servicing brake parts. Do not sand or grind brake lining unless equipment used is designed to contain the dust residue. Do not clean brake parts with compressed air or by dry brushing. Cleaning should be done by dampening the brake components with a fine mist of water, then wiping the brake components clean with a dampened cloth. Dispose of cloth and all residue containing asbestos fibers in an impermeable container with the appropriate label. Follow practices prescribed by the Occupational Safety and Health Administration (OSHA) and the Environmental Protection Agency (EPA) for the handling, processing, and disposing of dust or debris that may contain asbestos fibers.

BRAKES — BLEEDING THE BRAKE SYSTEM

BLEEDING PROCEDURE

BLEEDING PROCEDURE

1. Before servicing the vehicle, refer to the Precautions Section.

➡ If working near and/or around the SRS system and components, be sure to disable the SRS system. After disabling the system wait three minutes or more before servicing the vehicle.

➡ Before disconnecting the battery, lower both the driver's and passenger's windows. This will prevent any interference between the window edge and the vehicle when the door is opened or closed. During normal operation the window slightly raises and lowers automatically to prevent any window to vehicle interference. The automatic window function will not work with the battery disconnected.

✳✳ WARNING

Be careful not to splash brake fluid on painted areas; it may cause paint damage. If brake fluid is splashed on painted areas, wash it away with water immediately. All hoses must be free from excessive bending, twisting and pulling.

➡ Turn the ignition switch off and disconnect the ABS actuator and electric control unit connector, or the negative battery cable before performing the work.

➡ Monitor the fluid level in the reservoir while performing the work. Always use new brake fluid. Be sure to use the proper grade and type fluid.

➡ Cover the crowfoot and flare nut wrench with a shop towel to prevent damage to the front four piston type caliper and rear two piston type caliper.

2. Connect a vinyl tube to the bleeder valve of the right rear brake.

3. Fully depress the brake pedal four or five times.

4. Loosen the bleeder valve and bleed the air with the brake pedal depressed, quickly tighten the bleeder valve.

5. Repeat the above step until all air is expelled out of the brake line.

6. Tighten the bleeder valve.

7. Perform the above procedure to the brakes in the following sequence.

8. Right rear brake, left front brake, left rear brake and right front brake.

9. Check and refill the master cylinder, as required.

BLEEDING THE ABS SYSTEM

Refer to "Bleeding the Brake System".

BRAKES — ANTI-LOCK BRAKE SYSTEM (ABS)

WHEEL SPEED SENSORS

REMOVAL & INSTALLATION

See Figures 2 and 3.

At this time the manufacturer does not provide removal and installation procedures for this component. The following procedure is a guideline and may differ from the vehicle you are servicing.

1. Before servicing the vehicle, refer to the Precautions Section.

➡ If working near and/or around the SRS system and components, be sure to disable the SRS system. After disabling the system wait three minutes or more before servicing the vehicle.

➡ Before disconnecting the battery, lower both the driver's and passenger's

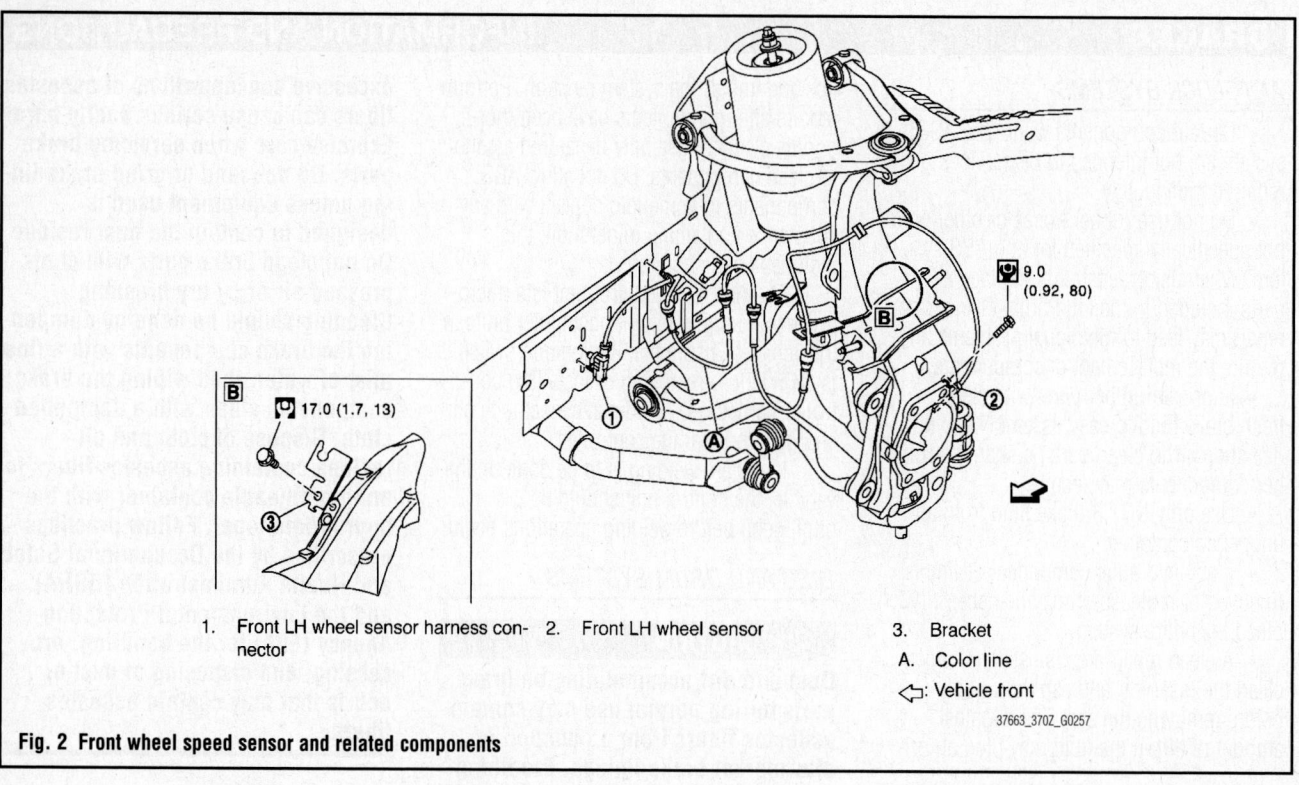

1. Front LH wheel sensor harness connector
2. Front LH wheel sensor
3. Bracket
A. Color line
⇦: Vehicle front

37663_370Z_G0257

Fig. 2 Front wheel speed sensor and related components

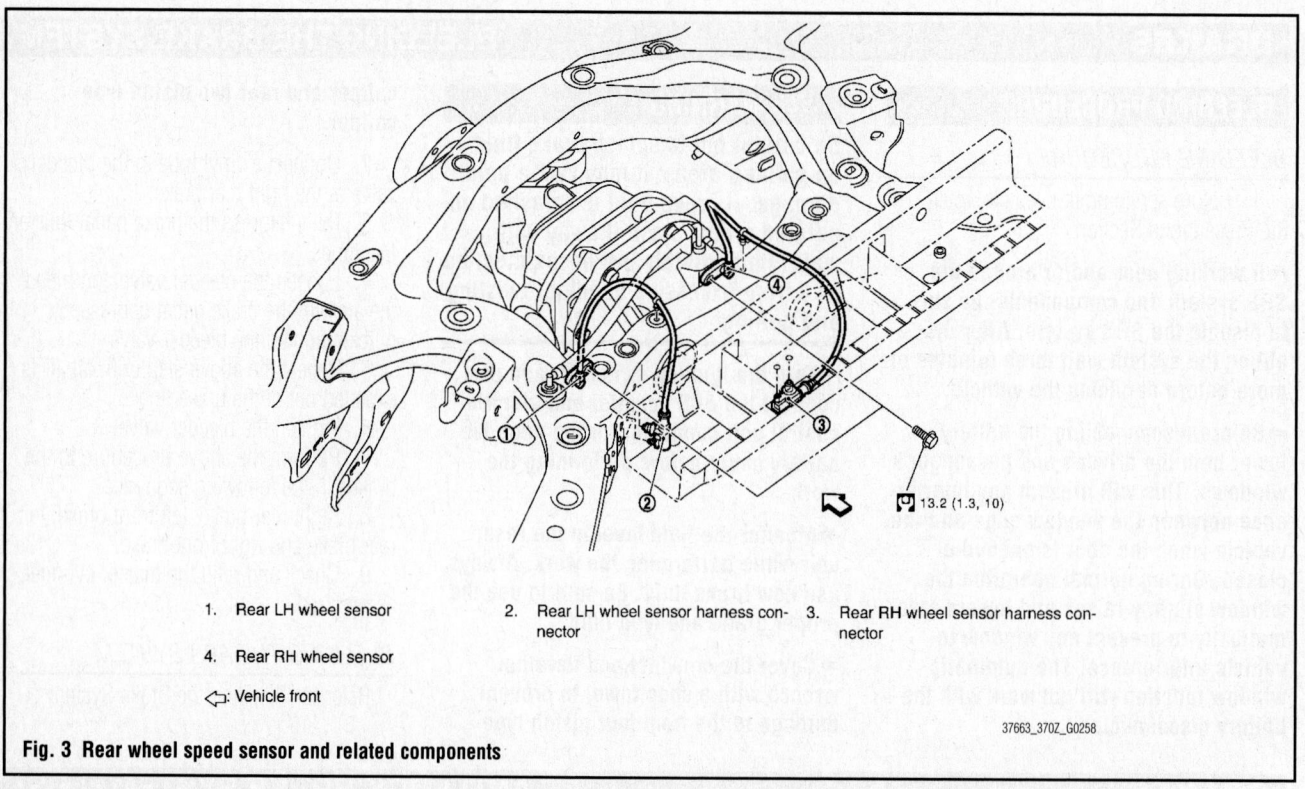

1. Rear LH wheel sensor
2. Rear LH wheel sensor harness connector
3. Rear RH wheel sensor harness connector
4. Rear RH wheel sensor
⇦: Vehicle front

37663_370Z_G0258

Fig. 3 Rear wheel speed sensor and related components

windows. This will prevent any interference between the window edge and the vehicle when the door is opened or closed. During normal operation the window slightly raises and lowers automatically to prevent any window to vehicle interference. The automatic

window function will not work with the battery disconnected.

2. Raise and support the vehicle safely.
3. Remove the tire and wheel assembly, as required.
4. Never twist or bend sensor harness when removing it.

5. Pull the wheel sensor out without pulling the harness.
6. Be careful not to damage the sensor edges or rotor teeth.
7. Remove the sensor first, before removing the wheel hub and bearing assembly.

To install:

➡ **Be sure to use new fasteners, as required.**

8. Installation is the reverse of the removal procedure.

WHEEL SPEED SENSOR RINGS (TOOTHED RINGS)

REMOVAL & INSTALLATION

See Figure 4.

The front sensor rotor cannot be disassembled. To replace the sensor rotor, replace the hub bearing assembly.

At this time the manufacturer does not provide removal and installation procedures for this component, refer to the illustration as required.

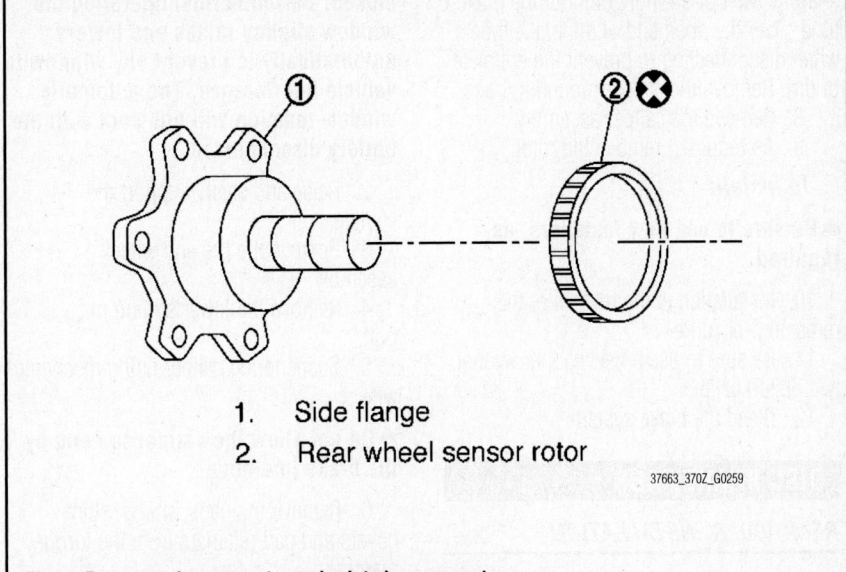

1. Side flange
2. Rear wheel sensor rotor

37663_370Z_G0259

Fig. 4 Rear speed sensor rotor and related components

BRAKES

FRONT DISC BRAKES

BRAKE CALIPER

REMOVAL & INSTALLATION

See Figures 5 and 6.

1. Before servicing the vehicle, refer to the Precautions Section.

➡ **If working near and/or around the SRS system and components, be sure to disable the SRS system. After disabling the system wait three minutes or more before servicing the vehicle.**

➡ **Before disconnecting the battery, lower both the driver's and passenger's windows. This will prevent any interference between the window edge and the vehicle when the door is opened or closed. During normal operation the**

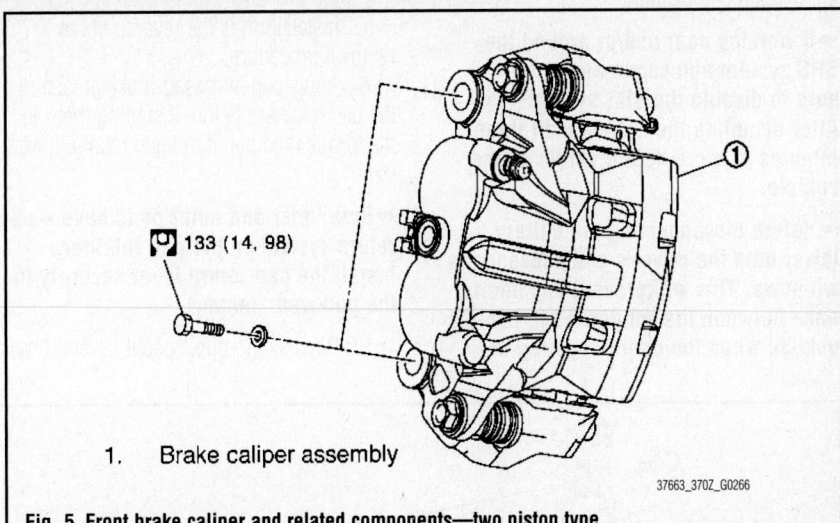

133 (14, 98)

1. Brake caliper assembly

37663_370Z_G0266

Fig. 5 Front brake caliper and related components—two piston type

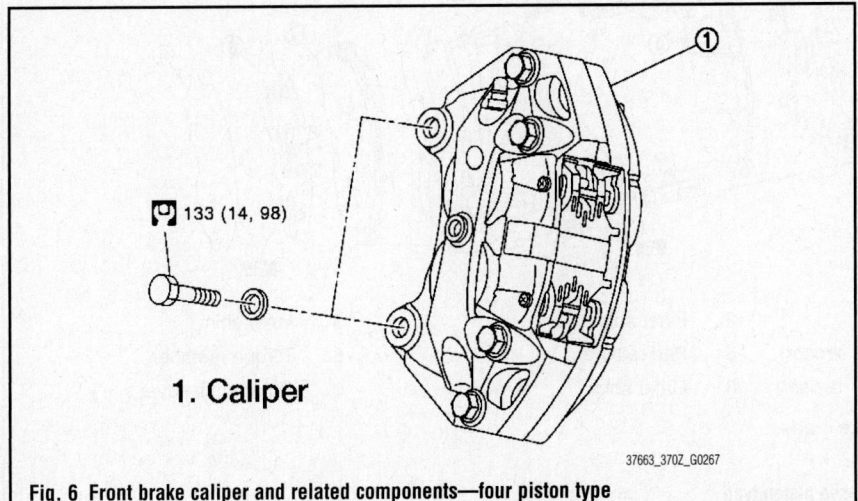

133 (14, 98)

1. Caliper

37663_370Z_G0267

Fig. 6 Front brake caliper and related components—four piston type

window slightly raises and lowers automatically to prevent any window to vehicle interference. The automatic window function will not work with the battery disconnected.

2. Raise and safely support the vehicle.

3. Remove the tire and wheel assembly.

4. Hold the rotor, using the wheel nuts.

5. Drain the brake fluid.

6. On two piston caliper, remove the union bolt and copper washer. Discard the washer. Disconnect the brake hose from the caliper. Remove the torque member mounting bolts.

7. On four piston caliper, cover the flare nut wrench with a shop towel. Never scratch the flare nut and the brake tube. Never bend

sharply, twist or strongly pull out the brake tube. Over the open end of the brake tube when disconnecting to prevent the entrance of dirt. Remove the caliper mounting bolts.

8. Remove the caliper assembly.
9. As required, remove the rotor.

To install:

➡**Be sure to use new fasteners, as required.**

10. Installation is the reverse of the removal procedure.
11. Be sure to use a new cooper washer, two piston caliper.
12. Bleed the brake system.

DISC BRAKE PADS

REMOVAL & INSTALLATION

Two Piston Caliper

See Figure 7.

1. Before servicing the vehicle, refer to the Precautions Section.

➡**If working near and/or around the SRS system and components, be sure to disable the SRS system. After disabling the system wait three minutes or more before servicing the vehicle.**

➡**Before disconnecting the battery, lower both the driver's and passenger's windows. This will prevent any interference between the window edge and the vehicle when the door is opened or closed. During normal operation the window slightly raises and lowers automatically to prevent any window to vehicle interference. The automatic window function will not work with the battery disconnected.**

2. Raise and safely support the vehicle.
3. Remove the tire and wheel assembly.
4. Remove the lower sliding pin bolt.
5. Suspend the caliper using mechanics wire.

➡**Do not allow the caliper to hang by the brake line hose.**

6. Remove the pads, shims, shim covers and pad retainers from the torque member.

To install:

➡**Be sure to use new fasteners, as required.**

7. Installation is the reverse of the removal procedure.
8. Apply copper based brake grease to the pad retainers before installing them to the torque member, if the pad retainers were removed.

➡**Both inner and outer pads have a pad return system on the pad retainer. Install the pad return lever securely to the pad wear sensor.**

9. Depress the brake pedal several times to seat the pads and check that no drag feel is present for the disc rotor.

Four Piston Caliper

See Figure 8.

1. Before servicing the vehicle, refer to the Precautions Section.

➡**If working near and/or around the SRS system and components, be sure to disable the SRS system. After disabling the system wait three minutes or more before servicing the vehicle.**

➡**Before disconnecting the battery, lower both the driver's and passenger's windows. This will prevent any interference between the window edge and the vehicle when the door is opened or closed. During normal operation the window slightly raises and lowers automatically to prevent any window to vehicle interference. The automatic window function will not work with the battery disconnected.**

2. Raise and safely support the vehicle.
3. Remove the tire and wheel assembly.
4. Remove the clips from the pad shims.
5. Remove the pad pins while holding down the cross spring. Remove the cross spring from the caliper.
6. Using a pliers, remove the brake pads and shims from the caliper.

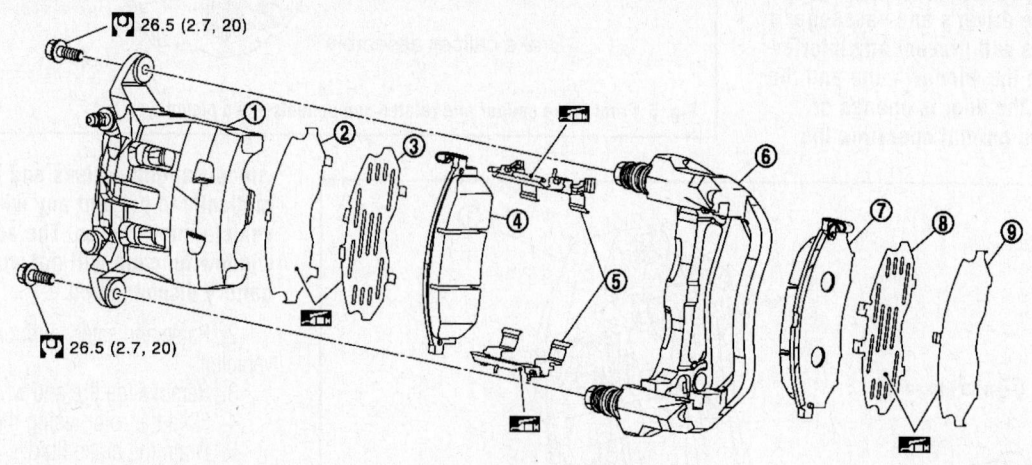

26.5 (2.7, 20)

26.5 (2.7, 20)

1.	Cylinder body	2.	Inner shim cover	3.	Inner shim
4.	Inner pad (with pad wear sensor)	5.	Pad retainer	6.	Torque member
7.	Outer pad (with pad wear sensor)	8.	Outer shim	9.	Outer shim cover

🔧: Apply copper based brake grease.

37663_370Z_G0264

Fig. 7 Front brake pads and related components—two piston type

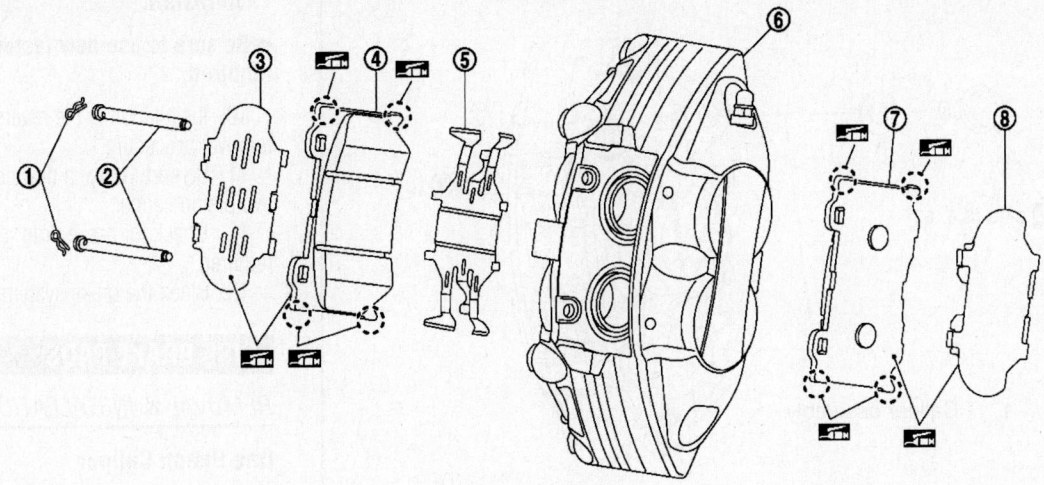

1. Clip
2. Pad pin
3. Inner shim
4. Inner pad (with pad wear sensor)*
5. Cross spring
6. Caliper
7. Outer pad
8. Outer shim

*: Some vehicles has pad wear sensor only for one side.

🔧: Apply copper based brake grease.

37663_370Z_G0265

Fig. 8 Front brake pads and related components—four piston type

To install:

➡**Be sure to use new fasteners, as required.**

7. Installation is the reverse of the removal procedure.

8. Apply copper based brake grease to the mating surfaces between the pads and shims. Install the shims to the brake pads.

9. Install the pads to the caliper.

10. Install the upper pad pin from the inner side, then install firmly to the outer side through the hole in the top of the brake pad.

11. Place the top of the cross spring over the upper pad pin, press the cross spring, install the lower pad pin from the inner side to the outer side and secure the cross spring.

12. Install the clips to the pad pins.

➡**If the clip is not fully attached, pad pin or brake pad could fall out while the vehicle is in motion.**

13. Depress the brake pedal several times to seat the pads and check that no drag feel is present for the disc rotor.

BRAKES   **REAR DISC BRAKES**

BRAKE CALIPER

REMOVAL & INSTALLATION

See Figures 9 and 10.

1. Before servicing the vehicle, refer to the Precautions Section.

➡**If working near and/or around the SRS system and components, be sure to disable the SRS system. After disabling the system wait three minutes or more before servicing the vehicle.**

➡**Before disconnecting the battery, lower both the driver's and passenger's windows. This will prevent any interference between the window edge and the vehicle when the door is opened or closed. During normal operation the window slightly raises and lowers auto-**

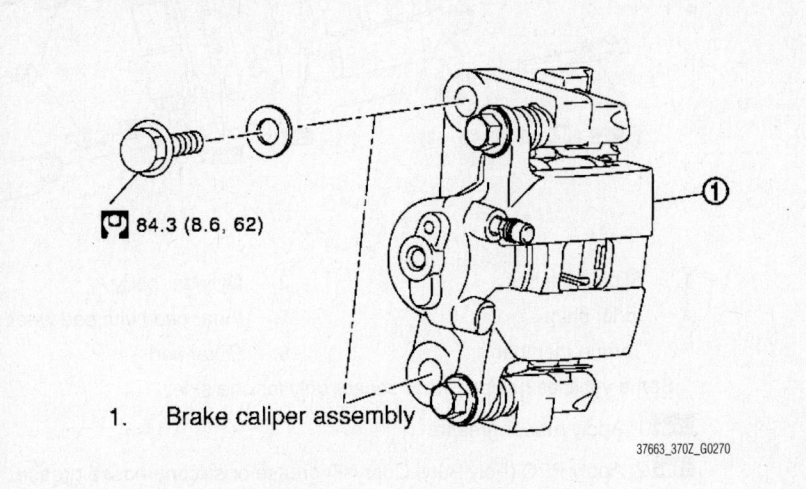

🔧 84.3 (8.6, 62)

1. Brake caliper assembly

37663_370Z_G0270

Fig. 9 Rear brake caliper and related components—one piston type

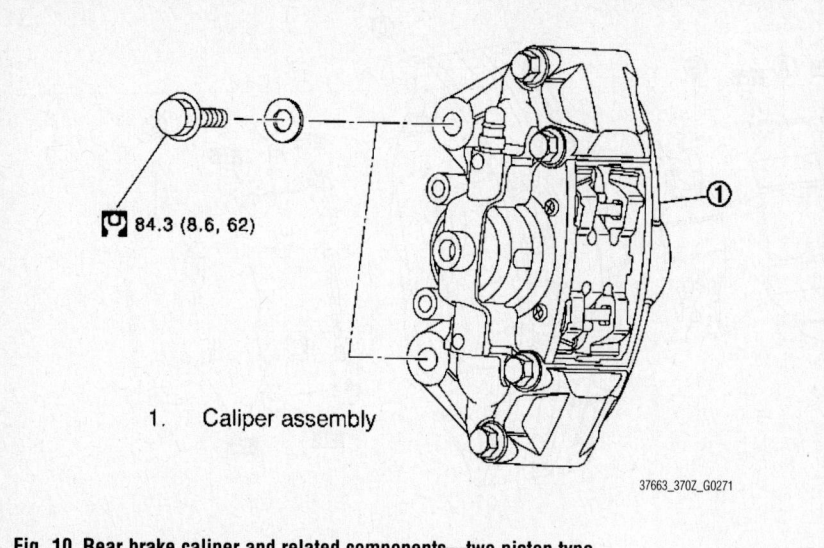

84.3 (8.6, 62)

1. Caliper assembly

37663_370Z_G0271

Fig. 10 Rear brake caliper and related components—two piston type

matically to prevent any window to vehicle interference. The automatic window function will not work with the battery disconnected.

2. Raise and safely support the vehicle.

3. Remove the tire and wheel assembly.

4. Hold the rotor, using the wheel nuts.

5. Drain the brake fluid.

6. On two piston caliper, remove the union bolt and copper washer. Discard the

washer. Disconnect the brake hose from the caliper. Remove the torque member mounting bolts.

7. On four piston caliper, cover the flare nut wrench with a shop towel. Never scratch the flare nut and the brake tube. Never bend sharply, twist or strongly pull out the brake tube. Over the open end of the brake tube when disconnecting to prevent the entrance of dirt. Remove the caliper mounting bolts.

8. Remove the caliper assembly.

9. As required, remove the rotor.

To install:

➡ Be sure to use new fasteners, as required.

10. Installation is the reverse of the removal procedure.

11. Be sure to use a new cooper washer, two piston caliper.

12. Check for brake drag, correct as required.

13. Bleed the brake system.

DISC BRAKE PADS

REMOVAL & INSTALLATION

One Piston Caliper

See Figure 11.

1. Before servicing the vehicle, refer to the Precautions Section.

➡ If working near and/or around the SRS system and components, be sure to disable the SRS system. After disabling the system wait three minutes or more before servicing the vehicle.

➡ Before disconnecting the battery, lower both the driver's and passenger's windows. This will prevent any interference between the window edge and the vehicle when the door is opened or closed. During normal operation the window slightly raises and lowers auto-

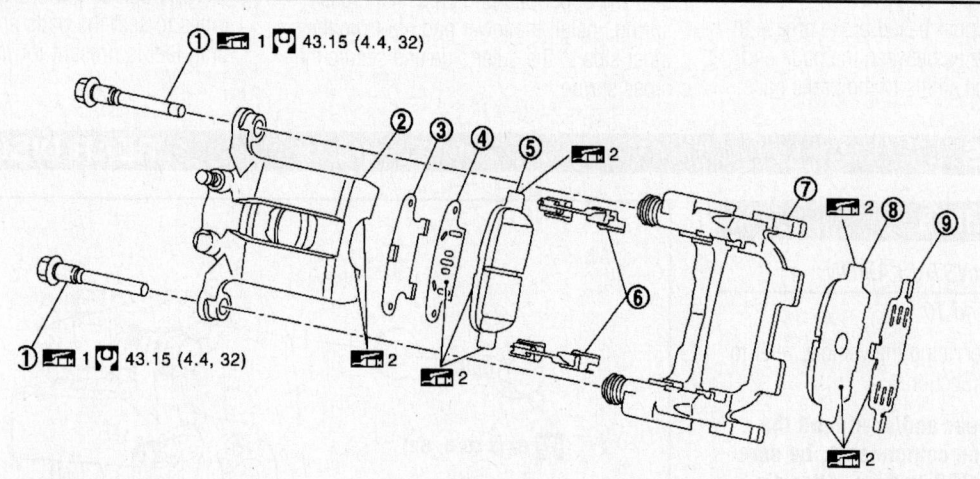

1: 43.15 (4.4, 32)

1. Sliding pin bolt
4. Inner shim
7. Torque member

2. Cylinder body
5. Inner pad (with pad wear sensor)*
8. Outer pad

3. Inner shim cover
6. Pad retainer
9. Outer shim

*: Some vehicles has pad wear sensor only for one side.

1: Apply rubber grease.

2: Apply PBC (Poly Butyl Cuprysil) grease or silicone-based grease.

37663_370Z_G0268

Fig. 11 Rear brake pads and related components—one piston type

matically to prevent any window to vehicle interference. The automatic window function will not work with the battery disconnected.

2. Raise and safely support the vehicle.
3. Remove the tire and wheel assembly.
4. Remove the lower sliding pin bolt.
5. Suspend the caliper using mechanics wire.

➡ **Do not allow the caliper to hang by the brake line hose.**

6. Remove the pads, shims, shim covers and pad retainers from the torque member.

To install:

➡ **Be sure to use new fasteners, as required.**

7. Installation is the reverse of the removal procedure.
8. Apply PBC grease to the mating surfaces between the pads and shims.
9. Apply PBC grease to the mating surfaces between the pad retainers and the pads before installing them to the brake pads.
10. Depress the brake pedal several times to seat the pads and check that no drag feel is present for the disc rotor.

Two Piston Caliper

See Figure 12.

1. Before servicing the vehicle, refer to the Precautions Section.

➡ **If working near and/or around the SRS system and components, be sure to disable the SRS system. After disabling the system wait three minutes or more before servicing the vehicle.**

➡ **Before disconnecting the battery, lower both the driver's and passenger's windows. This will prevent any interference between the window edge and the vehicle when the door is opened or closed. During normal operation the window slightly raises and lowers automatically to prevent any window to vehicle interference. The automatic window function will not work with the battery disconnected.**

2. Raise and safely support the vehicle.
3. Remove the tire and wheel assembly.
4. Remove the clips from the pad shims.
5. Remove the pad pins while holding down the cross spring. Remove the cross spring from the caliper.

6. Using a pliers, remove the brake pads and shims from the caliper.

To install:

➡ **Be sure to use new fasteners, as required.**

7. Installation is the reverse of the removal procedure.
8. Apply copper based brake grease to the mating surfaces between the pads and caliper, between the pads and pad pins and between the pads and cross spring.
9. Install the pads to the caliper.
10. Install the upper pad pin from the inner side, then install firmly to the outer side through the hole in the top of the brake pad.
11. Place the top of the cross spring over the upper pad pin, press the cross spring, install the lower pad pin from the inner side to the outer side and secure the cross spring.
12. Install the clips to the pad pins.

➡ **If the clip is not fully attached, pad pin or brake pad could fall out while the vehicle is in motion.**

13. Depress the brake pedal several times to seat the pads and check that no drag feel is present for the disc rotor.

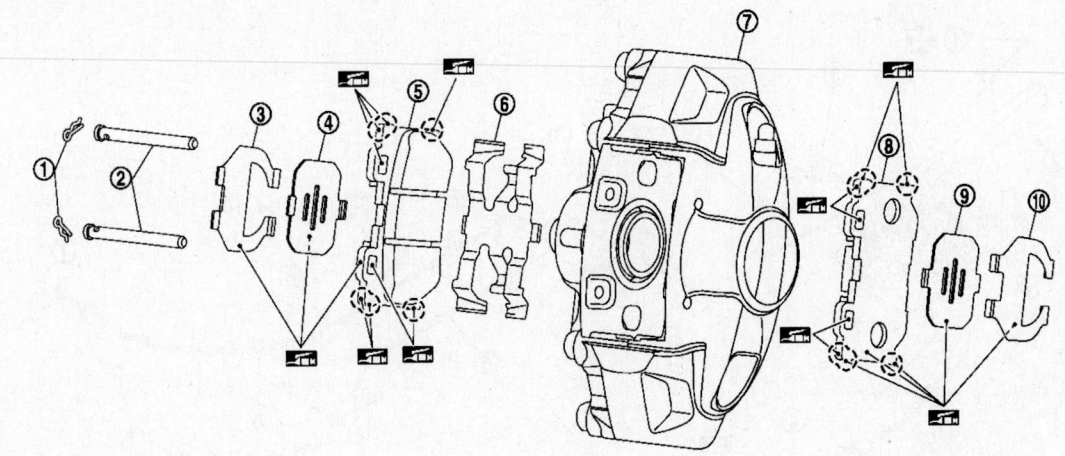

1.	Clip	2.	Pad pin	3.	Inner shim cover
4.	Inner shim	5.	Inner pad (with pad wear sensor)*	6.	Cross spring
7.	Caliper	8.	Outer pad	9.	Outer shim
10.	Outer shim cover				

*: Some vehicles has pad wear sensor only for one side.

🔲: Apply copper based brake grease.

Fig. 12 Rear brake pads and related components—two piston type

37663_370Z_G0269

PARKING BRAKE CABLES

ADJUSTMENT

See Figures 13 and 14.

1. To perform adjustment operations, remove tire from the vehicle with power tool.

2. Remove the coin pocket. Insert a deep socket wrench to rotate adjusting nut (1) and loosen the cable sufficiently. Then, return the lever.

3. Using wheel nuts, fix the disc rotor to the hub and prevent it from tilting.

4. Remove adjusting hole plug installed on the disc. Using a flat bladed tool, turn the disc in direction "A" as shown in the figure until the disc is locked. After locking, turn the adjuster in the opposite direction by 5 or 6 notches.

5. Rotate the disc to make sure there is no drag. Install the adjusting hole plug.

6. Adjust cable as follows:

   a. Operate lever 10 or more times with a force of 66 ft. lbs. (89 Nm).

   b. Rotate adjusting nut with deep socket to adjust lever stroke.

   c. When parking brake lever is oper-

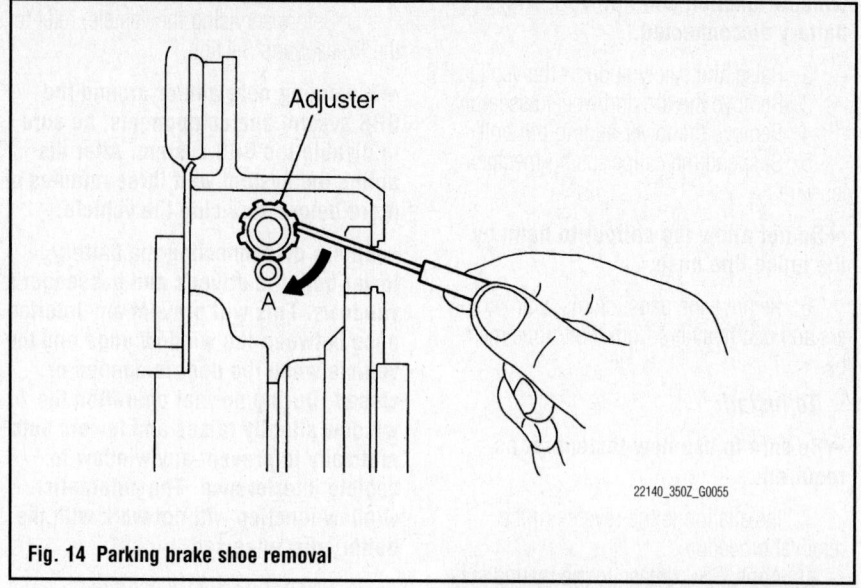

Fig. 14 Parking brake shoe removal

ated with a force of 44 ft. lbs. (60 Nm), check that the stroke is 6 to 7 notches (Check it by listening and counting the ratchet clicks).

7. With the lever completely returned, make sure there is no drag on the rear brake.

PARKING BRAKE SHOES

REMOVAL & INSTALLATION

See Figure 15.

1. Before servicing the vehicle, refer to the Precautions Section.

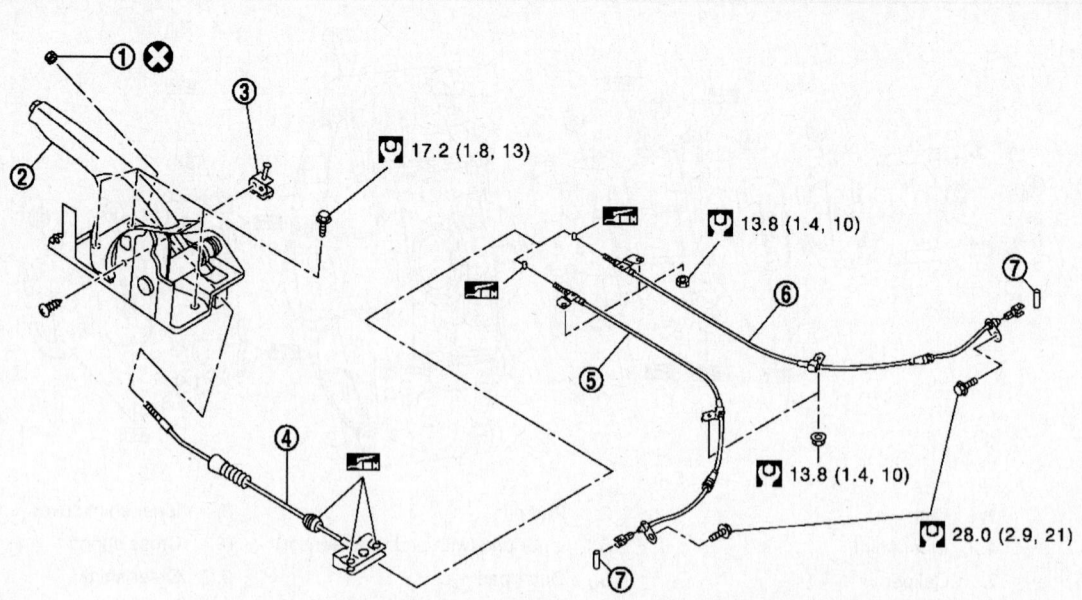

1. Adjusting nut   2. Device assembly   3. Parking brake switch
4. Front cable   5. Rear cable (LH)   6. Rear cable (RH)
7. Pin

: Apply multi-purpose grease.

Fig. 13 Parking brake cables and related components

➡ If working near and/or around the SRS system and components, be sure to disable the SRS system. After disabling the system wait three minutes or more before servicing the vehicle.

➡ Before disconnecting the battery, lower both the driver's and passenger's windows. This will prevent any interference between the window edge and the vehicle when the door is opened or closed. During normal operation the window slightly raises and lowers automatically to prevent any window to vehicle interference. The automatic window function will not work with the battery disconnected.

➡ Be sure to use new fasteners, as required.

2. Raise and safely support the vehicle.

3. Remove the wheel and tire.

4. Remove the brake rotor with the parking brake lever completely disengaged.

5. If the brake rotor cannot be removed, remove as follows:

    a. Secure the brake rotor with the wheel nut and remove the adjuster hole plug.

    b. Insert a flat-bladed tool through the plug opening and rotate the star wheel on the adjuster assembly in the direction as shown to retract the parking brake shoes.

    c. Remove the parking brake shoe springs using a suitable tool.

    d. Remove the parking brake shoes and adjuster.

To install:

6. Installation is in the reverse order of removal noting the following:

    a. Apply brake grease to the brake shoe contact area.

ADJUSTMENT

Cable

See Figure 16.

1. Remove control lever finisher from the center console.

2. Fully engage the control lever.

3. Loosen the parking brake cable adjusting nut and fully release the control lever.

4. Adjust clearance of the rear parking brake shoes.

5. Depress the brake pedal fully more than five times.

6. Make sure that no drag exists while rotating the rear wheels.

7. Operate control lever 10 times or more with a full stroke.

8. Adjust control lever by turning adjusting nut.

9. Pull control lever with specified amount of force. Check control lever stroke and ensure smooth operation:

- Lever stroke 7–8 notches at 44 ft. lbs. (60 Nm)

Shoes

1. Make sure the parking brake control lever is fully released and parking brake cable adjusting nut is loosened.

2. Remove the adjuster hole plug installed on the rotor. Using a suitable tool, turn the adjuster as shown until the rotor is locked. After locking, turn the adjuster in the opposite direction 5 or 6 notches.

3. Rotate the rotor to make sure that there is no drag. Install the adjuster hole plug.

4. After adjusting the clearance of the rear shoes, adjust the parking brake cable.

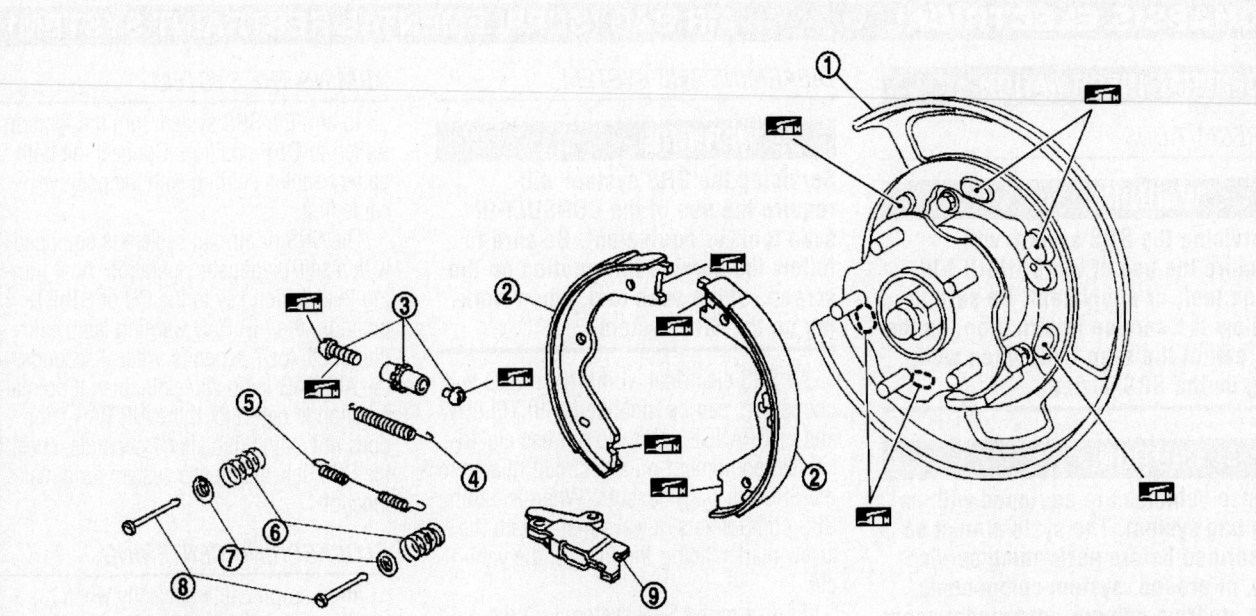

1. Back plate	2. Brake shoe	3. Adjuster
4. Adjuster spring	5. Return spring	6. Anti-rattle spring
7. Retainer	8. Anti-rattle pin	9. Toggle lever

🔧: Apply PBC (Poly Butyl Cuprysil) grease or silicone-based grease.

37663_370Z_G0263

Fig. 15 Parking brake shoes and related components

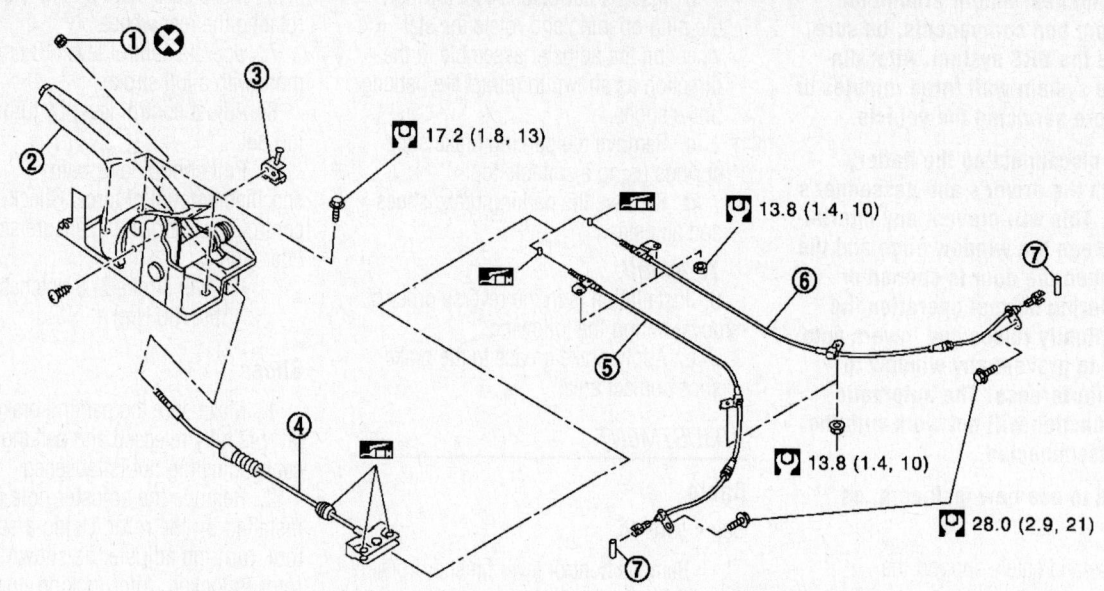

1. Adjusting nut
2. Device assembly
3. Parking brake switch
4. Front cable
5. Rear cable (LH)
6. Rear cable (RH)
7. Pin

🔧: Apply multi-purpose grease.

37663_370Z_G0108

Fig. 16 Parking brake cables and related components

CHASSIS ELECTRICAL

AIR BAG (SUPPLEMENTAL RESTRAINT SYSTEM)

GENERAL INFORMATION

PRECAUTIONS

❋❋ WARNING

Servicing the SRS system will require the use of the CONSULT-III scan tool, or equivalent. Be sure to follow the service information, on the screen of the scan tool, when working on the SRS system.

❋❋ CAUTION

These vehicles are equipped with an air bag system. The system must be disarmed before performing service on, or around, system components, the steering column, instrument panel components, wiring and sensors. Failure to follow the safety precautions and the disarming procedure could result in accidental air bag deployment, possible injury or death, or unnecessary system repairs.

DISARMING THE SYSTEM

❋❋ WARNING

Servicing the SRS system will require the use of the CONSULT-III scan tool, or equivalent. Be sure to follow the service information on the screen, of the scan tool, when working on the SRS system.

All SRS electrical wiring harnesses and connectors can be identified with YELLOW and or ORANGE color. Do not use electrical test equipment on any circuit related to the SRS (air bag) sensors. When installing SRS components, always install with the arrow marks facing the front of the vehicle.

To disarm the SRS system turn the ignition switch to **OFF** position. Then, disconnect the both battery cables starting with the negative cable first and wait at least 10 minutes after the cables are disconnected. Be sure to insulate the battery terminal ends.

ARMING THE SYSTEM

To arm the SRS system turn the ignition switch to **OFF** position. Connect the both battery cables starting with the positive cable first.

The SRS or air bag system is equipped with a self-diagnostic operation. After turning the ignition key to the ON or START position, the AIR BAG warning lamp will illuminate for 7 seconds. After 7 seconds, the AIR BAG lamp will extinguish if no malfunction is detected. If the AIR BAG lamp does not extinguish after 7 seconds, check the SRS self-diagnostic system for a malfunction.

CLOCKSPRING CENTERING

Align spiral cable correctly when installing steering wheel. Make sure that the spiral cable is in the neutral position. The neutral position is detected by turning left 2.5 revolutions from the right end position and ending with the knob at the top.

DRIVE TRAIN

CLUTCH

REMOVAL & INSTALLATION

See Figures 17 and 18.

1. Before servicing the vehicle, refer to the Precautions Section.

➡ **If working near and/or around the SRS system and components, be sure to disable the SRS system. After disabling the system wait three minutes or more before servicing the vehicle.**

➡ **Before disconnecting the battery, lower both the driver's and passenger's windows. This will prevent any interference between the window edge and the vehicle when the door is opened or closed. During normal operation the window slightly raises and lowers automatically to prevent any window to vehicle interference. The automatic window function will not work with the battery disconnected.**

➡ **Never reuse the concentric slave cylinder body and tube. Never put grease on the disc or pressure plate. Never clean the disc with solvent. Never drop the disc. Replace the clutch and clutch pressure plate as a set.**

2. Raise and support the vehicle safely.

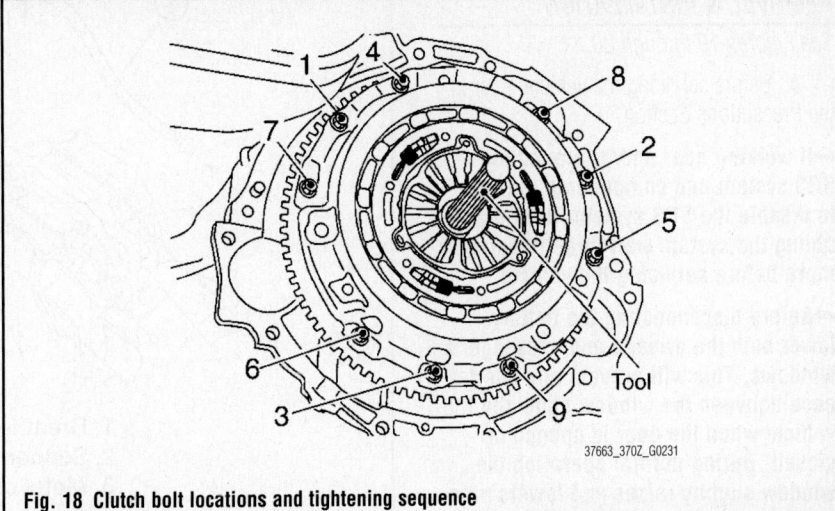
Fig. 18 Clutch bolt locations and tightening sequence

3. Remove the transmission assembly.
4. Remove the clutch cover and disc.

To install:

➡ **Be sure to use new fasteners, as required.**

5. Installation is the reverse of the removal procedure.
6. Be sure to apply grease to the specified points, see illustration.
7. Be sure to tighten the clutch pressure plate retaining bolts to specification and in the proper sequence. Tighten in two steps.

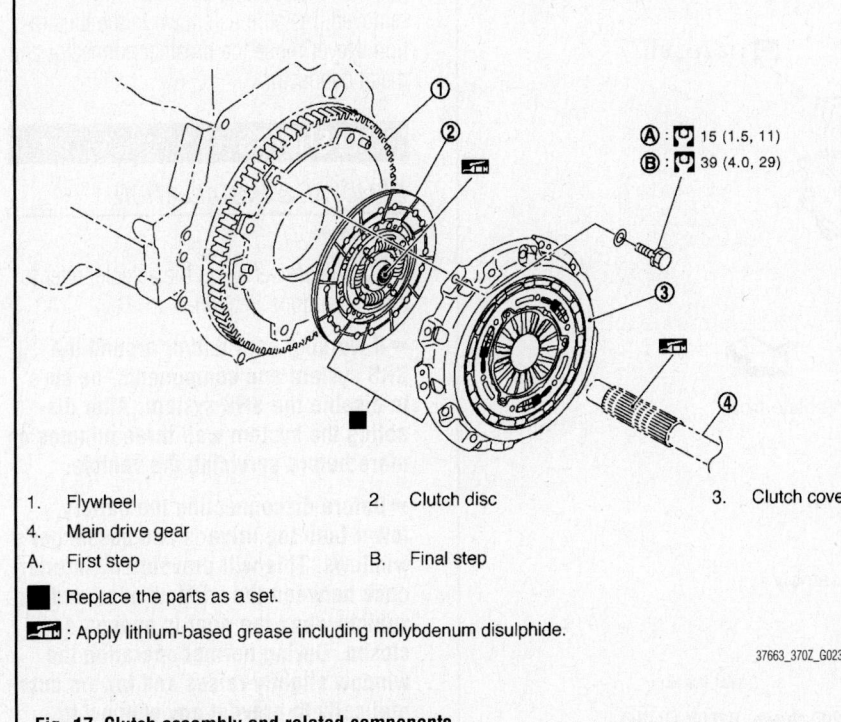

1. Flywheel
2. Clutch disc
3. Clutch cover
4. Main drive gear
A. First step
B. Final step
■ : Replace the parts as a set.
▆ : Apply lithium-based grease including molybdenum disulphide.

Ⓐ : 15 (1.5, 11)
Ⓑ : 39 (4.0, 29)

Fig. 17 Clutch assembly and related components

HYDRAULIC SYSTEM BLEEDING

BLEEDING PROCEDURE

At this time the manufacturer does not provide service information for bleeding the system. The following procedure is a guideline and may differ from the vehicle you are servicing.

Bleeding is required to remove air trapped in the hydraulic system. The bleed screw is located on the clutch slave (operating) cylinder.

Some models are also equipped with a clutch damper mechanism. The clutch damper mechanism is bled in exactly the same manner as the operating cylinder. It should be bled along with the operating cylinder.

1. Before servicing the vehicle, refer to the Precautions Section.
2. Remove the bleed screw dust cap.
3. Attach a transparent vinyl tube to the bleed screw, immersing the free end in a clean container of clean brake fluid.
4. Fill the master cylinder with the proper fluid.
5. Open the bleed screw about ¾ turn.
6. Depress the clutch pedal quickly. Hold it down. Have an assistant tighten the bleed screw. Allow the pedal to return slowly.
7. Repeat the above procedure until no more air bubbles are seen in the fluid container.
8. Remove the bleed tube.
9. Replace the dust cap and refill the master cylinder.
10. Bleed the clutch damper, if equipped.

REAR AXLE HOUSING

REMOVAL & INSTALLATION

See Figures 19 through 20.

1. Before servicing the vehicle, refer to the Precautions Section.

➡️**If working near and/or around the SRS system and components, be sure to disable the SRS system. After disabling the system wait three minutes or more before servicing the vehicle.**

➡️**Before disconnecting the battery, lower both the driver's and passenger's windows. This will prevent any interference between the window edge and the vehicle when the door is opened or closed. During normal operation the window slightly raises and lowers automatically to prevent any window to vehicle interference. The automatic window function will not work with the battery disconnected.**

2. Raise and support the vehicle safely.
3. Remove the center muffler.
4. Remove the diag brace.
5. Remove the stabilizer bar.
6. Remove the driveshaft.
7. Remove the rear halfshafts from the final drive.
8. Remove the breather hose from the final drive. Discard the hose clamp.

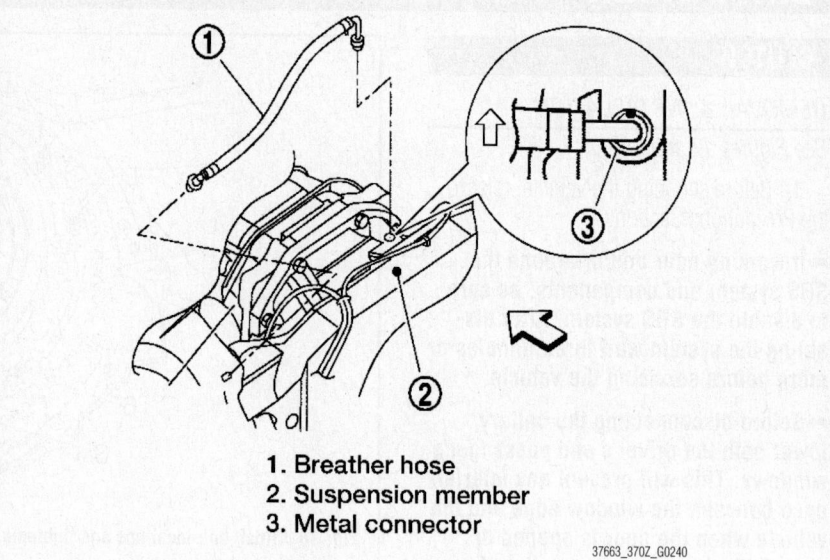

1. Breather hose
2. Suspension member
3. Metal connector

37663_370Z_G0240

Fig. 20 Breather connector installation

9. Remove the rear wheel speed sensors.
10. Position a suitable jack under the assembly.

➡️**Never place the jack on the rear cover.**

11. Remove the mounting bolts and nuts connecting the suspension member and remove the final drive assembly. Be sure the assembly is secured in the jack before removing it.

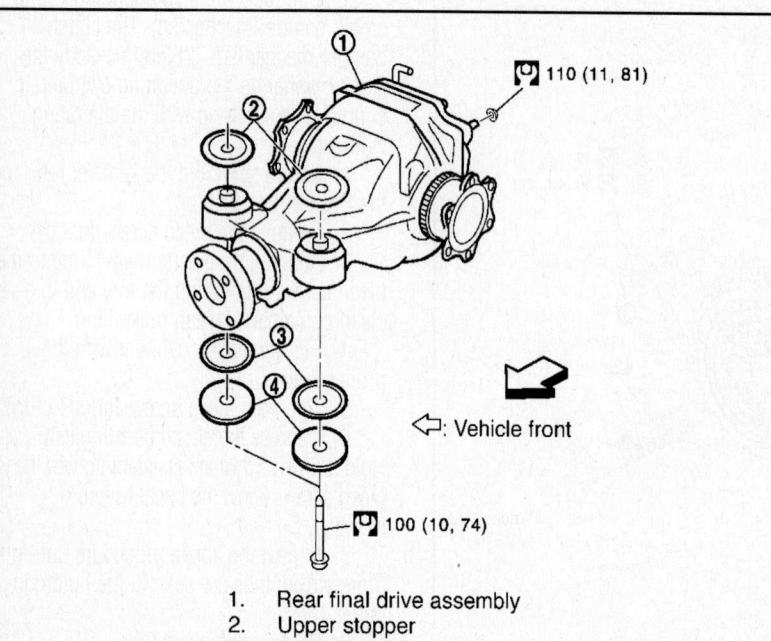

110 (11, 81)

⬅️: Vehicle front

100 (10, 74)

1. Rear final drive assembly
2. Upper stopper
3. Lower stopper
4. Washer

37663_370Z_G0241

Fig. 19 Rear drive axle and related components—R200 shown, R200V similar

To install:

➡️**Be sure to use new fasteners, as required.**

12. Installation is the reverse of the removal procedure.
13. Be sure that there are no pinched or restricted areas on the breather hose caused by bending or winding, when installing it.
14. When installing the new hose clamp, install it at the final drive side with the tab facing downward.
15. If the breather connector was removed, install it as shown in the illustration. Never reuse the breather connector and metal connector.

HALFSHAFTS

REMOVAL & INSTALLATION

See Figure 21.

1. Before servicing the vehicle, refer to the Precautions Section.

➡️**If working near and/or around the SRS system and components, be sure to disable the SRS system. After disabling the system wait three minutes or more before servicing the vehicle.**

➡️**Before disconnecting the battery, lower both the driver's and passenger's windows. This will prevent any interference between the window edge and the vehicle when the door is opened or closed. During normal operation the window slightly raises and lowers automatically to prevent any window to vehicle interference. The automatic**

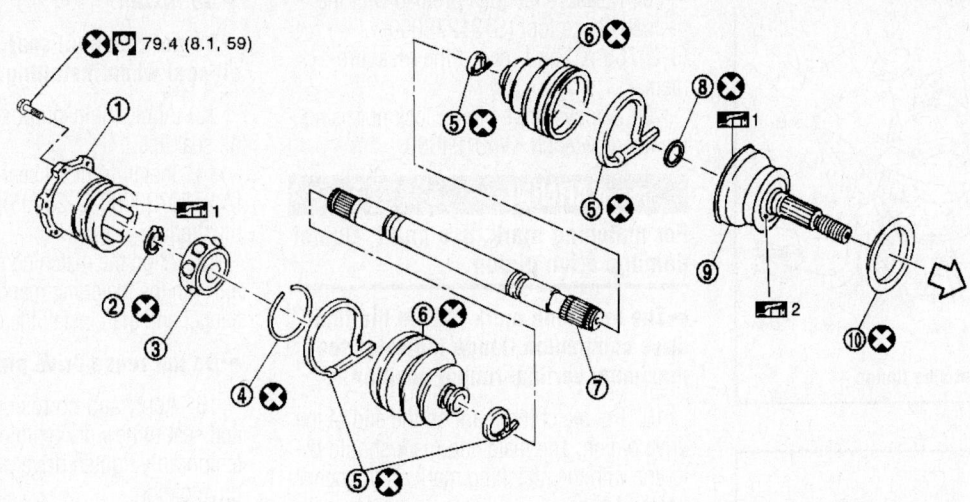

79.4 (8.1, 59)

1. Housing
2. Snap ring
3. Ball cage/steel ball/inner race assembly
4. Stopper ring
5. Boot band
6. Boot
7. Shaft
8. Circular clip
9. Joint sub-assembly
10. Dust shield
◁: Wheel side

1: NISSAN genuine grease or an equivalent.

2: Apply paste [service parts (440037S000)].

37663_370Z_G0245

Fig. 21 Rear halfshaft exploded view

window function will not work with the battery disconnected.

2. raise and support the vehicle safely.

3. Remove the tire and wheel assemblies.

4. Remove and discard the cotter pin.

5. Loosen the wheel hub locknut.

6. Matchmark the halfshaft and wheel hub and bearing.

7. Remove the diag brace.

8. Remove the center muffler.

9. Patch wheel hub locknut with a piece of wood. Hammer the wood to disengage the wheel hub and bearing assembly from the halfshaft.

→**Never position the halfshaft at an extreme angle. Do not overextend the slide joint. Properly support the halfshaft, do not allow it to hang unsupported.**

10. Use a suitable puller if the wheel hub and bearing assembly and halfshaft cannot be separated even after performing the above procedure.

11. Remove the wheel locknut.

12. Remove the mounting bolts between the side flange and the halfshaft.

To install:

→**Be sure to use new fasteners, as required.**

13. Installation is the reverse of the removal procedure.

14. Clean the matching surface of the halfshaft and wheel hub and bearing assembly.

15. Apply paste part number 440037S000 or equivalent to the surface of the sub joint assembly of the halfshaft.

→**Apply the paste, about 0.04–0.10 ounce) to cover the entire flat surface of the sub joint assembly of the halfshaft.**

16. The wheel hub locknut tightening specification is 136 ft. lbs. (185 Nm).

17. Perform a final tightening of nuts and bolts of each removed component with the vehicle in an unladen position.

PINION SEAL

REMOVAL & INSTALLATION

See Figures 22 through 26.

At this time the manufacturer does not provide removal and installation procedures

for this component. The following procedure is a guideline and may differ from the vehicle you are servicing.

1. Before servicing the vehicle, refer to the Precautions Section.

→**Verify the identification stamp for the replacement frequency, installed on the lower part of gear carrier. Use this to determine the replacement of collapsible spacer when replacing front oil seal. If it is necessary to replace the collapsible spacer, remove final drive assembly and disassemble to replace front oil seal and collapsible spacer.**

2. Drain gear oil.

3. Raise and support the rear of vehicle safely and remove the wheels.

4. Remove the center muffler.

5. Remove or disconnect the following:

- ABS rear wheel sensor
- Rear halfshaft, and suspend with mechanics wire

6. Install attachments [A: KV40104100 and B: ST36230000 (J-25840-A)] to side flange, and then pull out the side flange with the sliding hammer.

7. Remove drive shaft.

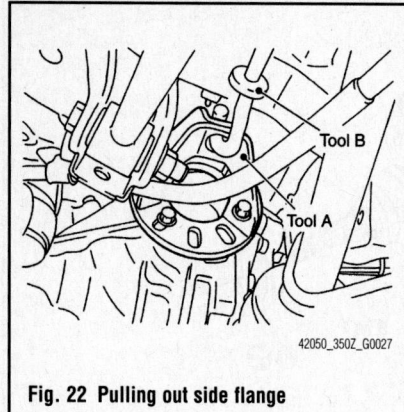

Fig. 22 Pulling out side flange

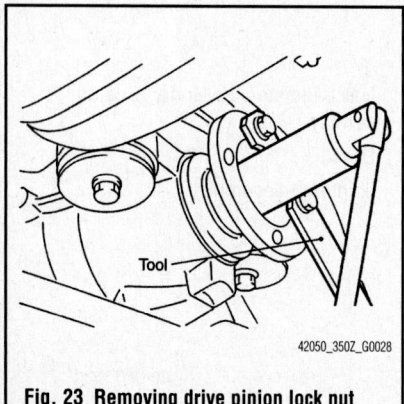

Fig. 23 Removing drive pinion lock nut

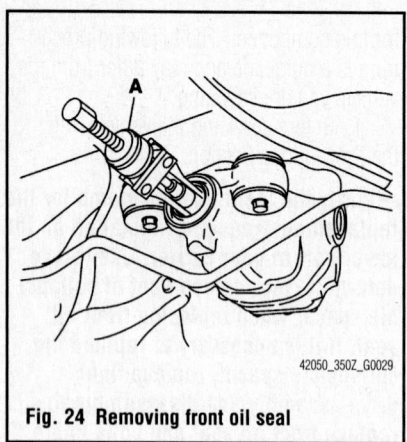

Fig. 24 Removing front oil seal

Fig. 25 Installing front oil seal

8. Measure the total preload with the preload gauge tool [ST3127S000 (J-25765-A)] and record the measurement.

9. Remove drive pinion lock nut using the flange wrench KV40104000.

⁂⁂ WARNING

For matching mark, use paint. Do not damage drive pinion.

➡**The matching mark "A" on the final drive companion flange indicates the maximum vertical runout position.**

10. Put matching mark on the end of the drive pinion. The matching mark should be in line with the matching mark on the companion flange.

11. Remove companion flange using a puller.

12. Remove front oil seal using the puller [A: KV381054S0 (J-34286)].

To install:

➡**Do not reuse oil seal. Do not incline oil seal when installing.**

13. Apply multi-purpose grease to front oil seal lips.

14. Install front oil seal using the drift [A:ST30720000 (J-25405)] as shown in the illustration.

15. Align the matching mark of drive pinion with the matching mark of companion flange, and then install the companion flange.

➡**Do not reuse drive pinion lock nut.**

16. Apply anti-corrosion oil to the thread and seat of new drive pinion lock nut, and temporarily tighten drive pinion lock nut to drive pinion.

➡**Total preload torque should equal the measurement taken during removal plus an additional 3 inch lbs. (0.1 to 0.4 Nm).**

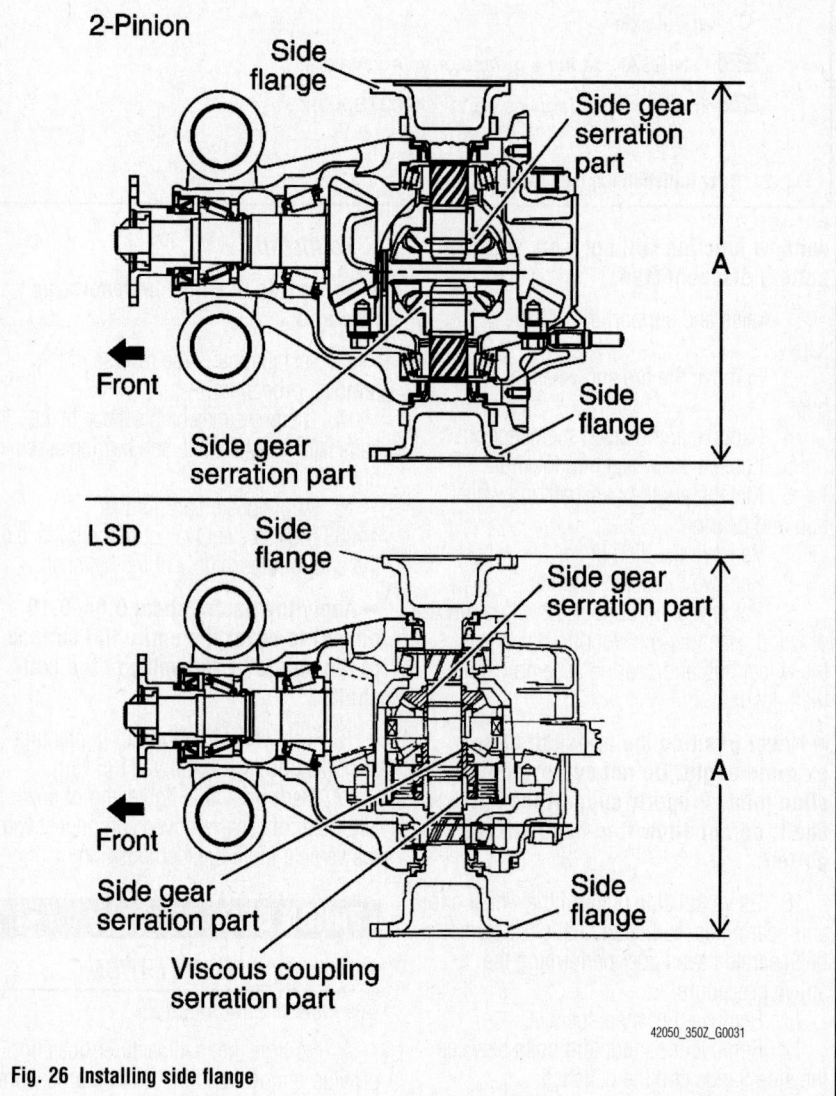

Fig. 26 Installing side flange

17. Tighten to drive pinion lock nut to 09–238 ft. lbs. (147–323 Nm), while adjusting total reload torque.

➡ **Adjust to the lower limit of the drive pinion lock nut tightening torque first. If the preload torque exceeds the specified value, replace collapsible spacer and tighten it again to adjust. Do not loosen drive pinion lock nut to adjust the preload torque.**

18. Make a stamping for identification of front oil seal replacement frequency. Be sure to make a stamping after replacing front oil seal.

19. Install drive shaft.

➡ **Install the RH side flange, then install the LH side flange. If LH side flange is installed first, the RH side flange comes out sometimes from the shock of installing the RH side flange [For R200V (with LSD)].**

20. Install side flange.
21. Attach the protector to side oil seal.
22. After the side flange is inserted and the serrated part of side gear has engaged the serrated part of flange, remove the protector.
23. Put a suitable drift on the center of side flange, then drive it until sound changes.

➡ **When installation is completed, driving sound of the side flange turns into a sound which seems to affect the whole final drive.**

24. Confirm that the dimension of the side flange is 12.83–12.91 in. (326–328 mm) as shown in the illustration.
25. Install halfshaft.
26. Install rear wheel ABS sensor.
27. Refill gear oil to the final drive and check oil level.
28. Check the final drive for oil leakage.

ENGINE COOLING

ENGINE FAN

REMOVAL & INSTALLATION

1. Before servicing the vehicle, refer to the Precautions Section.

➡ **If working near and/or around the SRS system and components, be sure to disable the SRS system. After disabling the system wait three minutes or more before servicing the vehicle.**

➡ **Before disconnecting the battery, lower both the driver's and passenger's windows. This will prevent any interference between the window edge and the vehicle when the door is opened or closed. During normal operation the window slightly raises and lowers automatically to prevent any window to vehicle interference. The automatic window function will not work with the battery disconnected.**

2. Remove the reservoir tank.
3. Disconnect the crash zone sensor harness clips from the fan shroud, move the harness to the side.
4. Disconnect the harness connector from the cooling fan control module, move the harness to the side.
5. Remove the undercover.
6. Disconnect and plug the automatic transmission fluid cooler hose from the fan shroud, if equipped.
7. Remove the cooling fan from under the vehicle.

To install:

➡ **Be sure to use new fasteners, as required.**

8. Installation is the reverse of the removal procedure.

RADIATOR

REMOVAL & INSTALLATION
See Figure 27.

1. Before servicing the vehicle, refer to the Precautions Section.

➡ **If working near and/or around the SRS system and components, be sure to disable the SRS system. After disabling the system wait three minutes or more before servicing the vehicle.**

➡ **Before disconnecting the battery, lower both the driver's and passenger's windows. This will prevent any interference between the window edge and the vehicle when the door is opened or closed. During normal operation the window slightly raises and lowers automatically to prevent any window to vehicle interference. The automatic window function will not work with the battery disconnected.**

➡ **Never change the engine coolant when the engine is hot. Wrap a thick cloth around the cap and carefully remove it. First turn the cap a quarter of a turn to release any pressure, then turn it all the way. Do not allow coolant to come in contact with the drive belts.**

2. Remove the engine under cover.
3. Drain the engine coolant. Be sure to properly dispose of used coolant.
4. As required, properly discharge the air conditioning system.
5. Remove the air cleaner assembly.
6. Remove the reservoir tank.
7. Remove the bumper center upper finisher and bumper fascia assembly.
8. Properly support the hood, with a proper support tool.

9. Disconnect the harness clips and hood lock control cable clips from the bumper retainer.
10. Remove the front bumper retainer.
11. Remove the horn. Remove the hood locks.
12. Remove the front combination lamp, left side.
13. Remove the hood lock, right and left brackets. Remove the hood lock stay assembly.
14. Remove the condenser pipe assembly.
15. Remove the hood lock stay mounting bolt.
16. Disconnect and plug the automatic transmission cooler lines at the radiator, if equipped.
17. Remove the upper and lower radiator hoses.
18. Remove the radiator water inlet pipe.
19. Rotate the two upper radiator mount brackets ninety degrees, in the direction shown in the illustration.
20. To remove the radiator and condenser assembly, lift up and pull the assembly forward and then remove the

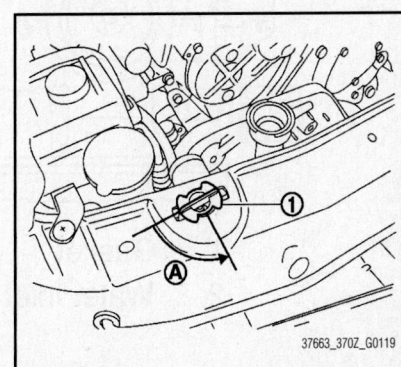

37663_370Z_G0119

Fig. 27 Radiator upper mount bracket removal

lower mounting rubber from the radiator support.

21. Remove the assembly from the front of the radiator core support.

To install:

→Be sure to use new fasteners, as required.

22. Installation is the reverse of the removal procedure.

23. Be sure to fill the cooling system with the proper grade and type engine coolant.

24. Check that there is no fluid leakage at the automatic transmission lines, if equipped.

THERMOSTAT

REMOVAL & INSTALLATION
See Figure 28.

1. Before servicing the vehicle, refer to the Precautions Section.

→If working near and/or around the SRS system and components, be sure to disable the SRS system. After dis-

abling the system wait three minutes or more before servicing the vehicle.

→Before disconnecting the battery, lower both the driver's and passenger's windows. This will prevent any interference between the window edge and the vehicle when the door is opened or closed. During normal operation the window slightly raises and lowers automatically to prevent any window to vehicle interference. The automatic window function will not work with the battery disconnected.

→Never change the engine coolant when the engine is hot. Wrap a thick cloth around the cap and carefully remove it. First turn the cap a quarter of a turn to release any pressure, then turn it all the way. Do not allow coolant to come in contact with the drive belts.

2. Remove the engine undercover.
3. Drain the coolant. Be sure to properly dispose of used engine coolant.
4. Remove the air cleaner assembly.
5. Remove the reservoir assembly.

6. Disconnect the radiator hose from the water inlet and thermostat assembly.

7. Disconnect the intake valve timing control solenoid harness connector (bank 2) and remove the component.

8. Remove the water inlet and thermostat assembly retaining bolts.

9. Remove the component from its mounting. Discard the gasket.

→Never disassemble the water inlet and thermostat assembly, replace them as a complete unit.

To install:

→Be sure to use new fasteners, as required.

10. Installation is the reverse of the removal procedure.

11. Be sure to fill the cooling system with the proper grade and type engine coolant.

WATER PUMP

REMOVAL & INSTALLATION
See Figures 29 through 33.

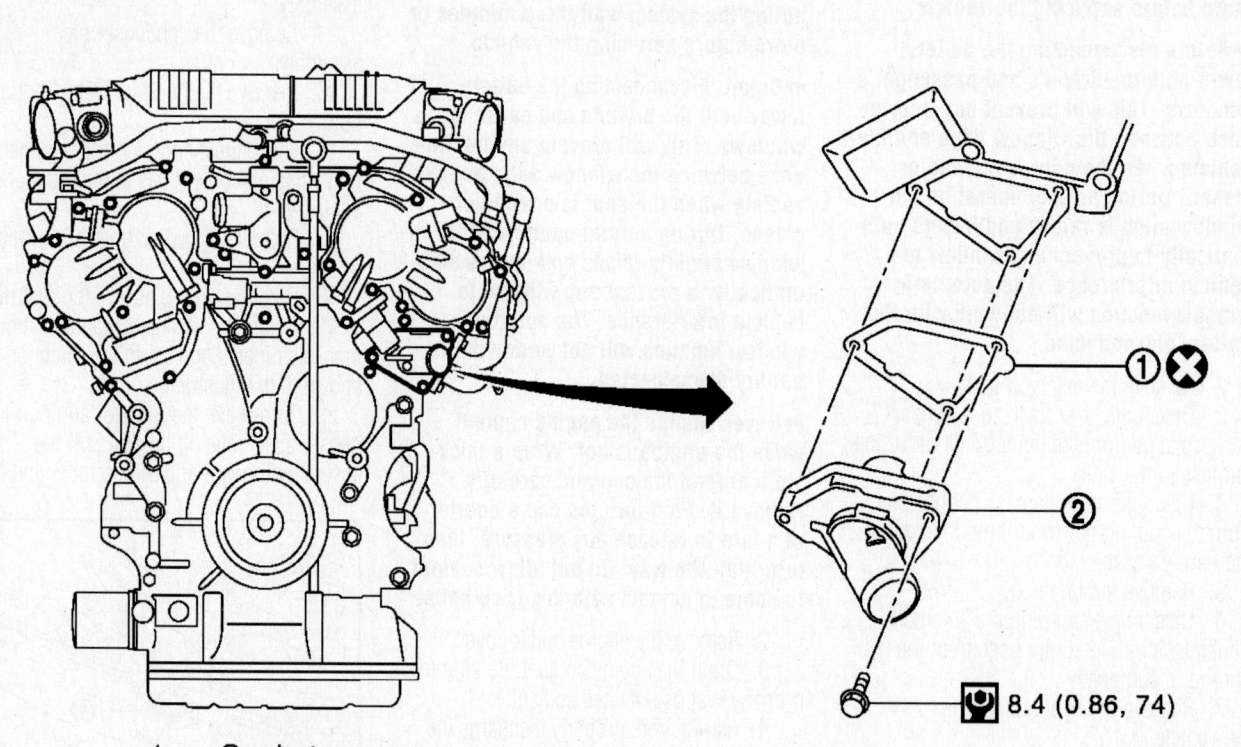

1. Gasket
2. Water inlet and thermostat assembly

8.4 (0.86, 74)

Fig. 28 Thermostat and related components

37663_370Z_G0031

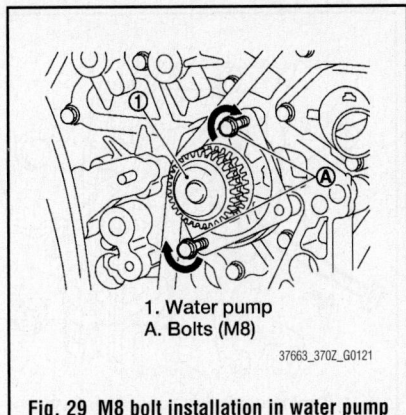

1. Water pump
A. Bolts (M8)

37663_370Z_G0121

Fig. 29 M8 bolt installation in water pump

1. Before servicing the vehicle, refer to the Precautions Section.

➡️**If working near and/or around the SRS system and components, be sure to disable the SRS system. After disabling the system wait three minutes or more before servicing the vehicle.**

➡️**Before disconnecting the battery, lower both the driver's and passenger's windows. This will prevent any interference between the window edge and the** vehicle when the door is opened or closed. During normal operation the window slightly raises and lowers automatically to prevent any window to vehicle interference. The automatic window function will not work with the battery disconnected.

➡️**Never change the engine coolant when the engine is hot. Wrap a thick cloth around the cap and carefully remove it. First turn the cap a quarter of a turn to release any pressure, then turn it all the way. Do not allow coolant to come in contact with the drive belts.**

2. Remove the engine cover.
3. Properly relieve the fuel system pressure.
4. Disconnect the negative battery cable.
5. Remove the air cleaner assembly.
6. Remove the reservoir tank.
7. Separate the engine harness by removing their brackets from the front timing chain case.
8. Remove the engine undercover.
9. Drain the engine oil. Be sure to properly dispose of used oil.
10. Drain the engine coolant. Be sure to properly dispose of used coolant.

11. Remove the radiator hoses.
12. Remove the cooling fan assembly.
13. Remove the front timing chain case cover.
14. To remove the timing chain tensioner (primary), remove the lower mounting bolt. Loosen the upper mounting bolt slowly and then turn the timing chain tensioner (primary) on the upper mounting bolt so that the plunger is fully expanded. Remove the upper mounting bolt and then remove the timing chain tensioner (primary).

➡️**Even if the plunger is fully expanded, it does not drop from the body of the timing chain tensioner (primary).**

15. To remove the water pump, remove the three retaining bolts. Secure a gap between the water pump gear and the timing chain, by turning the crankshaft counterclockwise until the timing chain looseness on the water pump sprocket becomes maximum.
16. Screw M8 bolts (pitch 0.0492 inch) in length approximately 1.97 inch into the water pump upper and lower mounting bolt holes until they reach the timing chain case. Then alternately tighten each bolt for a half a turn, and pull out the water pump.

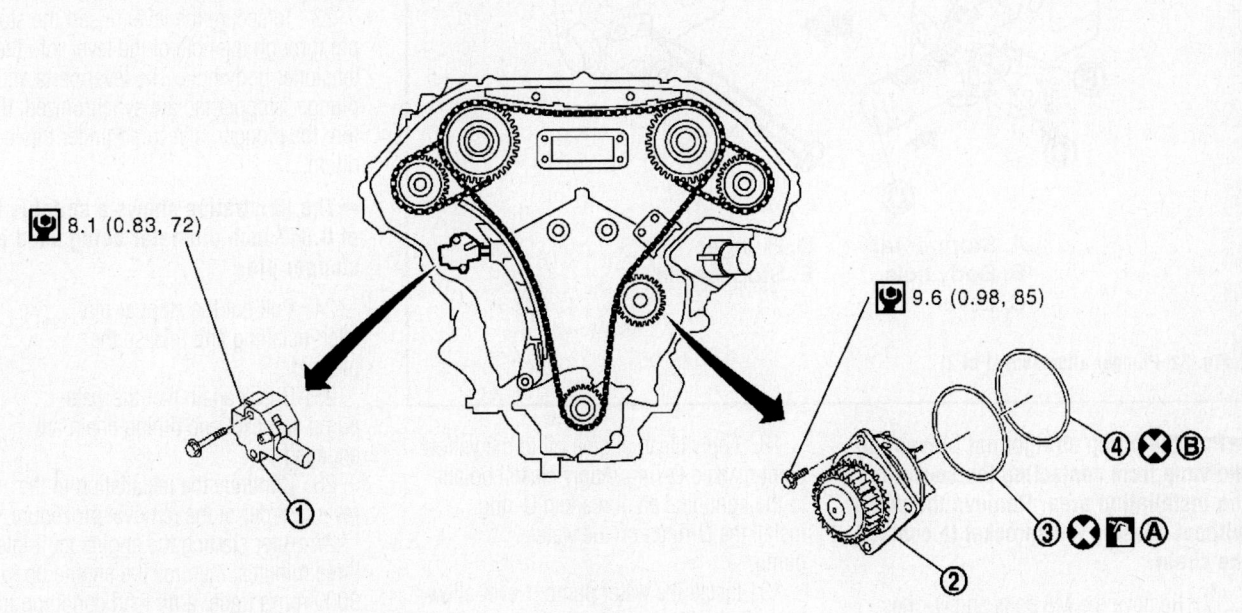

8.1 (0.83, 72)

9.6 (0.98, 85)

1. Timing chain tensioner (primary)
4. O-ring

2. Water pump

3. O-ring

A. Identify with yellow paint mark

B. Identify with light blue paint mark
Apply engine coolant

37663_370Z_G0120

Fig. 30 Water pump and related components

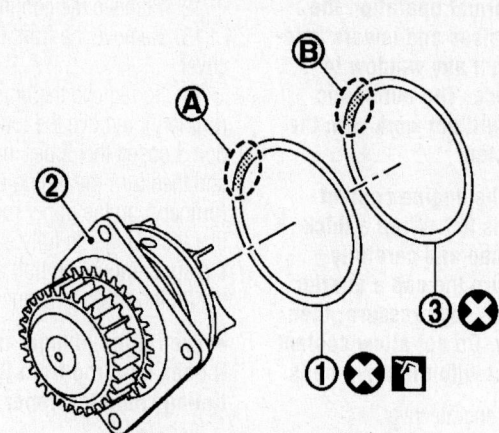

1. O-ring (yellow paint mark)
2. Water pump
3. O-ring (light blue paint mark)
A. Yellow paint mark
B. Light blue paint mark

37663_370Z_G0122

Fig. 31 Water pump and O-ring positioning and color identification

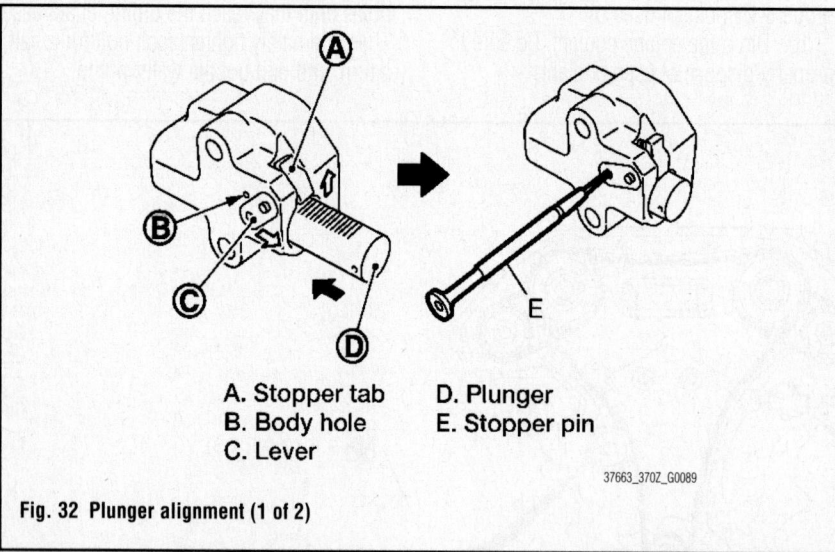

A. Stopper tab   D. Plunger
B. Body hole   E. Stopper pin
C. Lever

37663_370Z_G0089

Fig. 32 Plunger alignment (1 of 2)

➡Pull the pump straight out preventing the vane from contacting the socket in the installation area. Remove the pump without causing the sprocket to contact the chain.

17. Remove the M8 bolts and O-rings from the water pump.

To install:

➡Be sure to use new fasteners, as required.

18. Apply clean engine oil to the yellow paint marked O-ring. Apply clean coolant to the light blue paint marked O-ring. Install the O-rings on the water pump.

19. Install the water pump. Never allow the cylinder block to nip the O-rings. Check that the chain and pump sprocket are engaged.

20. Tighten the mounting bolts alternately and securely.

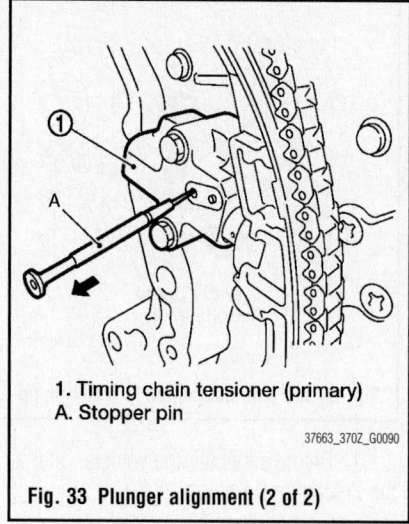

1. Timing chain tensioner (primary)
A. Stopper pin

37663_370Z_G0090

Fig. 33 Plunger alignment (2 of 2)

21. To install the timing chain tensioner (primary) pull the plunger stopper tab up (or turn lever downward) so as to remove the plunger stopper tab from the ratchet of the plunger. Note that the plunger stopper tab and lever are synchronized.

22. Push the plunger into the inside of the tensioner body. Hold the plunger in the fully compressed position by engaging the plunger stopper tab with the tip of the ratchet.

23. To secure the lever, insert the stopper pin through the hole of the lever into the tensioner body hole. The lever parts and the plunger stopper tab are synchronized, therefore the plunger is secured under this condition.

➡The illustration shows a suitable tool of 0.047 inch diameter being used as a stopper pin.

24. Pull out the stopper pin after installing and release the plunger.

25. Check again that the water pump sprocket and timing chain are engaged.

26. Continue the installation in the reverse order of the removal procedure.

27. After starting the engine let it idle for three minutes, then rev the engine up to 3000 rpm's under a no load condition to purge air from the high pressure chamber of the chain tensioner. The engine may produce a rattling noise. This indicates that there is still air in the chamber and not a matter of concern.

ENGINE ELECTRICAL

CHARGING SYSTEM

ALTERNATOR

REMOVAL & INSTALLATION

1. Before servicing the vehicle, refer to the Precautions Section.

➡If working near and/or around the SRS system and components, be sure to disable the SRS system. After disabling the system wait three minutes or more before servicing the vehicle.

➡Before disconnecting the battery, lower both the driver's and passenger's windows. This will prevent any interference between the window edge and the vehicle when the door is opened or closed. During normal operation the window slightly raises and lowers automatically to prevent any window to vehicle interference. The automatic window function will not work with the battery disconnected.

2. Disconnect the negative battery cable.
3. Remove the engine front undercover.
4. Remove the cooling fan assembly.
5. Remove the drive belt.
6. Disconnect the alternator electrical connectors.
7. Remove the harness bracket bolts.
8. Remove the oil pressure switch har-

ness clip. Disconnect the electrical connectors from the oil pressure switch and the oil temperature sensor.

9. Remove the alternator mounting bolts.
10. Remove the alternator from the vehicle, in the downward direction.

To install:

➡Be sure to use new fasteners, as required.

11. Installation is the reverse of the removal procedure.
12. Check belt tension, as required.

ENGINE ELECTRICAL

IGNITION SYSTEM

FIRING ORDER

See Figure 34.

Firing order: 1–2–3–4–5–6

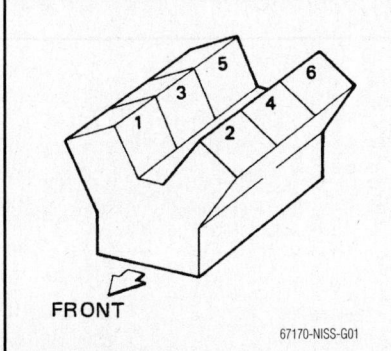

Fig. 34 Distributorless ignition system (one coil on each cylinder)

FRONT

67170-NISS-G01

IGNITION COIL

REMOVAL & INSTALLATION

At this time the manufacturer does not provide removal and installation procedures for this component. The following procedure is a guideline and may differ from the vehicle you are servicing.

1. Before servicing the vehicle, refer to the Precautions Section.

➡If working near and/or around the SRS system and components, be sure to disable the SRS system. After disabling the system wait three minutes or more before servicing the vehicle.

➡Before disconnecting the battery, lower both the driver's and passenger's

windows. This will prevent any interference between the window edge and the vehicle when the door is opened or closed. During normal operation the window slightly raises and lowers automatically to prevent any window to vehicle interference. The automatic window function will not work with the battery disconnected.

2. Remove the engine undercover.
3. Remove the air cleaner assembly.
4. Remove the intake manifold collector.
5. Remove the necessary components in order to gain access to the ignition coils.
6. Disconnect the electrical connectors.
7. Remove the retainer bolt.
8. Remove the component from its mounting.

To install:

➡Be sure to use new fasteners, as required.

9. Installation is the reverse of the removal procedure.

IGNITION TIMING

INSPECTION

➡The ignition timing is not adjustable. If not within specifications, further diagnostic inspection is required. You will need the CONSULT-III diagnostic tool, or equivalent. Follow the directions on the screen of the tool, as needed.

ADJUSTMENT

No timing adjustment is necessary.

SPARK PLUGS

REMOVAL & INSTALLATION

1. Before servicing the vehicle, refer to the Precautions Section.

➡If working near and/or around the SRS system and components, be sure to disable the SRS system. After disabling the system wait three minutes or more before servicing the vehicle.

➡Before disconnecting the battery, lower both the driver's and passenger's windows. This will prevent any interference between the window edge and the vehicle when the door is opened or closed. During normal operation the window slightly raises and lowers automatically to prevent any window to vehicle interference. The automatic window function will not work with the battery disconnected.

2. Remove the engine undercover.
3. Remove the air cleaner assembly.
4. Remove the intake manifold collector.
5. Remove the necessary components in order to gain access to the ignition coils.
6. Disconnect the electrical connectors.
7. Remove the retainer bolt.
8. Remove the component from its mounting.
9. Remove the spark plug from its mounting.

To install:

➡Be sure to use new fasteners, as required.

10. Installation is the reverse of the removal procedure.

STARTER

REMOVAL & INSTALLATION

See Figure 35.

1. Before servicing the vehicle, refer to the Precautions Section.

➡**If working near and/or around the SRS system and components, be sure to disable the SRS system. After disabling the system wait three minutes or more before servicing the vehicle.**

➡**Before disconnecting the battery, lower both the driver's and passenger's windows. This will prevent any interference between the window edge and the vehicle when the door is opened or closed. During normal operation the window slightly raises and lowers automatically to prevent any window to vehicle interference. The automatic window function will not work with the battery disconnected.**

2. Disconnect the negative battery cable.

3. Remove the engine undercover.

4. Disconnect the starter electrical connections.

5. Remove the starter retaining bolts and harness bracket.

6. If equipped with automatic transmission, remove the fluid cooler tube clips and bracket. Move the fluid cooler tube downward.

7. Remove the component from the vehicle in the forward direction.

To install:

➡**Be sure to use new fasteners, as required.**

8. Installation is the reverse of the removal procedure.

SOLENOID OR RELAY REPLACEMENT

1. Before servicing the vehicle, refer to the Precautions Section.

➡**If working near and/or around the SRS system and components, be sure to disable the SRS system. After disabling the system wait three minutes or more before servicing the vehicle.**

➡**Before disconnecting the battery, lower both the driver's and passenger's windows. This will prevent any interference between the window edge and the vehicle when the door is opened or closed. During normal operation the window slightly raises and lowers automatically to prevent any window to vehicle interference. The automatic window function will not work with the battery disconnected.**

2. Remove the starter from the vehicle.

3. Separate the solenoid from the starter.

To install:

➡**Be sure to use new fasteners, as required.**

4. Installation is the reverse of the removal procedure.

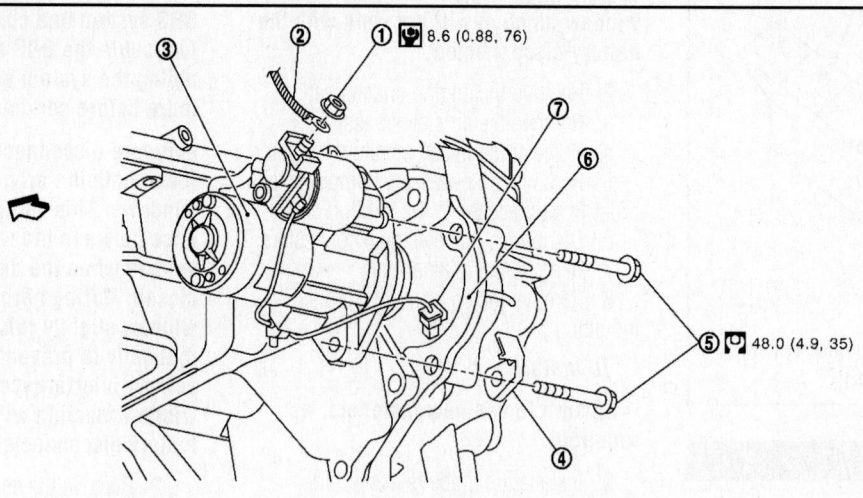

1. "B" terminal nut
2. "B" terminal harness
3. Starter motor
4. Harness clip bracket
5. Starter motor mounting bolt
6. Converter housing (A/T models) Transmission case (M/T models)
7. "S" connector
⬅: Engine front

37663_370Z_G0128

Fig. 35 Starter and related components

ACCESSORY DRIVE BELTS

ACCESSORY BELT ROUTING

See Figure 36.

Refer to the accompanying illustration.

INSPECTION

1. Inspect belts for cracks, fraying, wear and oil. If necessary, replace.

ADJUSTMENT

No adjustment is necessary.

REMOVAL & INSTALLATION

See Figure 37.

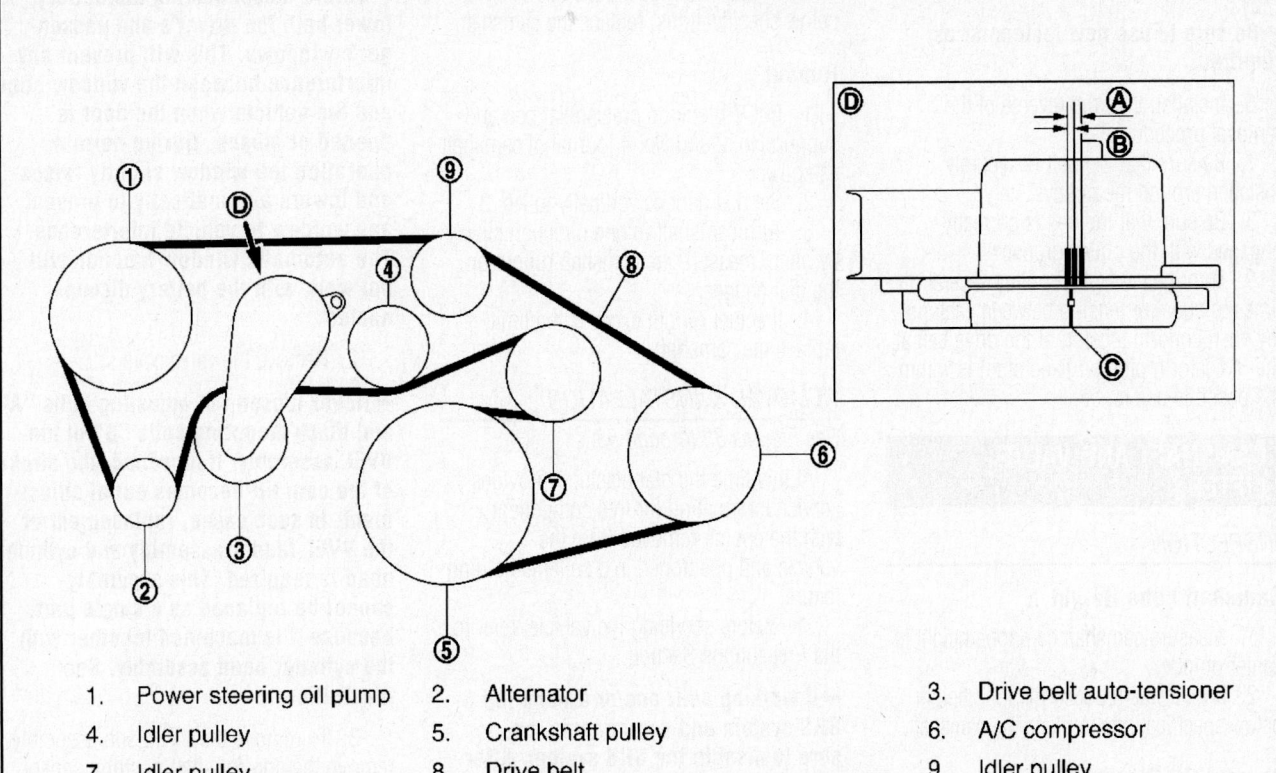

1. Power steering oil pump
4. Idler pulley
7. Idler pulley
A. Possible use range
D. View D

2. Alternator
5. Crankshaft pulley
8. Drive belt
B. Range when new drive belt is installed

3. Drive belt auto-tensioner
6. A/C compressor
9. Idler pulley
C. Indicator

37663_370Z_G0001

Fig. 36 Drive belt routing

1. Before servicing the vehicle, refer to the Precautions Section.

➡️**If working near and/or around the SRS system and components, be sure** to disable the SRS system. After disabling the system wait three minutes or more before servicing the vehicle.

➡️**Before disconnecting the battery,** lower both the driver's and passenger's windows. This will prevent any interference between the window edge and the vehicle when the door is opened or closed. During normal operation the window slightly raises and lowers automatically to prevent any window to vehicle interference. The automatic window function will not work with the battery disconnected.

2. Remove the engine undercover.
3. While securely holding the square hole in the pulley center of the auto tensioner with a spinner handle, move the spinner handle in the direction of the arrow (loosening the drive belt). See illustration.

➡️**Never place your hand in a location where pinching may occur if the holding tool accidentally comes off.**

4. Under the above condition insert a metallic bar about 0.24 inch in diameter through the holding boss to lock the auto tensioner pulley arm.
5. Remove the drive belt.

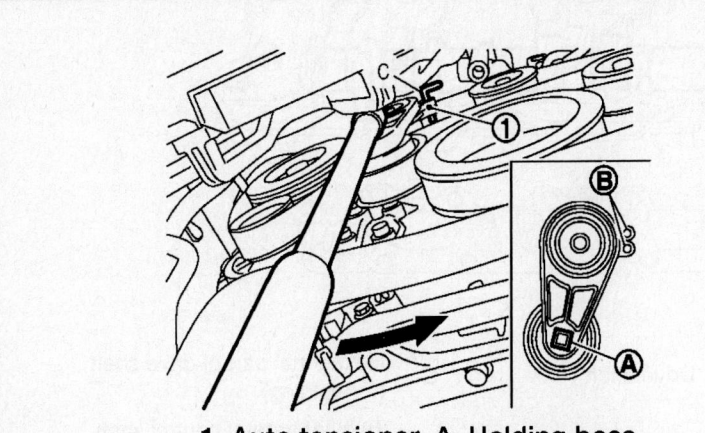

1. Auto tensioner
A. Holding boss
B. Square hole
C Hex wrench

37663_370Z_G0002

Fig. 37 Drive belt removal

To install:

➡ **Be sure to use new fasteners, as required.**

6. Installation is the reverse of the removal procedure.

7. Be sure that the belt is securely installed around the pulleys.

8. Be sure that the belt is correctly engaged with the pulley groove.

9. Turn the crankshaft pulley several times to equalize tension between each pulley, then confirm tension of the drive belt at the indicator (notch on fixed side) is within the possible use range.

CAMSHAFT AND VALVE LIFTERS

INSPECTION

Camshaft Lobe Height

1. Measure camshaft cam lobe height at center of lobe.

2. If wear has reduced the lobe height below specifications, replace the camshaft.

Journal Oil Clearance

1. Measure outer diameter of camshaft journal.

2. If wear has reduced the diameter below specifications, replace the camshaft.

Runout

1. Put V-block on precise flat bed and support No. 2 and No. 4 journal of camshaft as shown.

2. Set dial gauges vertically to No. 3.

3. Turn camshaft in one direction slowly by hand, measure the camshaft runout on the dial gauges.

4. If actual runout exceeds the limit, replace the camshaft.

REMOVAL & INSTALLATION

See Figures 38 through 53.

At this time the manufacturer provides service information for this component with the engine removed from the vehicle and positioned in a suitable holding fixture.

1. Before servicing the vehicle, refer to the Precautions Section.

➡ **If working near and/or around the SRS system and components, be sure to disable the SRS system. After disabling the system wait three minutes or more before servicing the vehicle.**

➡ **Before disconnecting the battery, lower both the driver's and passenger's windows. This will prevent any interference between the window edge and the vehicle when the door is opened or closed. During normal operation the window slightly raises and lowers automatically to prevent any window to vehicle interference. The automatic window function will not work with the battery disconnected.**

2. Remove the valve covers.

➡ **Never loosen the adjusting bolts "A" and black mounting bolts "B" of the VVEL assembly. If loosened, the stroke of the cam lift becomes out of adjustment. In such cases, replacement of the VVEL ladder assembly and cylinder head is required. This assembly cannot be replaced as a single part, because it is machined together with the cylinder head assembly. See illustration.**

3. To remove the VVEL sub assembly, remove the control shaft position sensor. Fix two flat areas of the shaft with a wrench to remove the mounting bolts of the shaft.

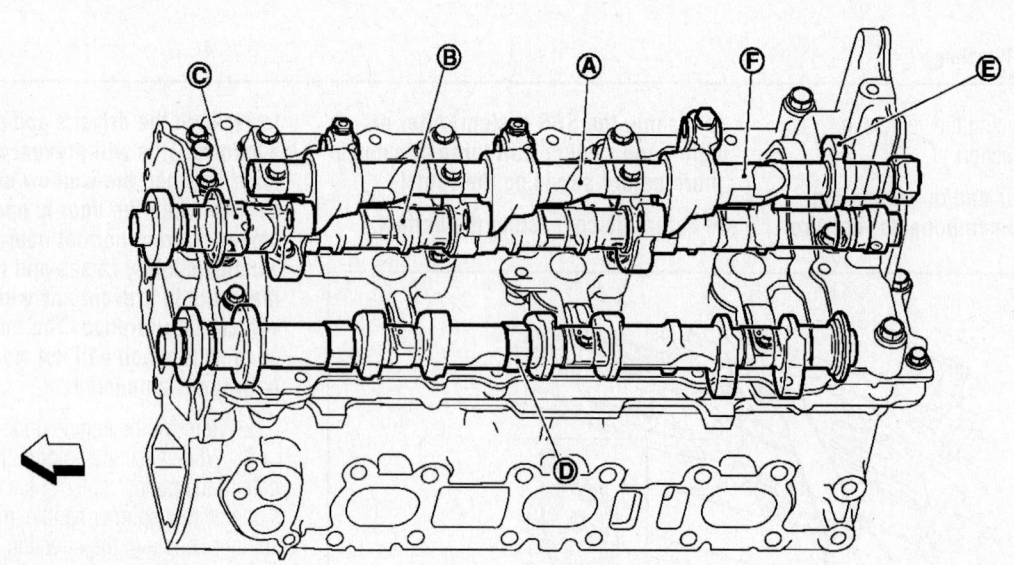

| A. | Control shaft | B. | Drive shaft | C. | Hexagonal part of drive shaft (for holding) |
| D. | Hexagonal part of camshaft (EXH) (for holding) | E. | Stopper of control shaft | F. | Two flat area of control shaft (for holding) |

⟸ : Engine front

37663_370Z_G0005

Fig. 38 VVEL ladder assembly and related components (1 of 2)

➡ During this operation never allow the wrench to interfere with other parts. Fix the control shaft to prevent interference of the stopper surface.

➡ The VVEL actuator sub assembly and control shaft position sensor are non-reusable. Never remove them unless required.

4. Remove the VVEL sub actuator assembly by loosening the mounting bolts in the reverse order of the tightening sequence.

5. Remove the actuator bracket by loosening the mounting bolts in the reverse order of the tightening sequence.

➡ When removing, be sure to properly dispose of any oil, as required. When installing be careful with the VVEL actuator subassembly (bank 2) mounting bolt No. 1, because the length is different from the other.

6. Remove the timing chain case, camshaft sprockets and timing chain.

7. Remove the rear timing chain case.

8. Remove the VVEL ladder assembly.

9. Loosen the mounting bolts (gold color) in the reverse order of the tightening sequence.

➡ Never loosen the adjusting bolts and mounting bolts (black color). When removing the VVEL ladder assembly, hold the driveshaft from below as not to drop it.

10. Remove the camshaft (EXH).

11. Remove the valve lifter. Be sure to identify and store properly for reinstallation.

12. Remove the chain tensioners (secondary) from the cylinder head. Remove the component with its stopper pin attached. The stopper pin should be attached when removed.

13. If necessary, remove the oil filter from the cylinder head.

To install:

➡ Be sure to use new fasteners, as required.

14. Install the secondary timing chain tensioners.

15. Install the oil filter, if removed.

16. Install the valve lifters, in their original position.

17. Install the camshaft (EXH).

➡ Note identification mark distinction between camshaft (EXH).

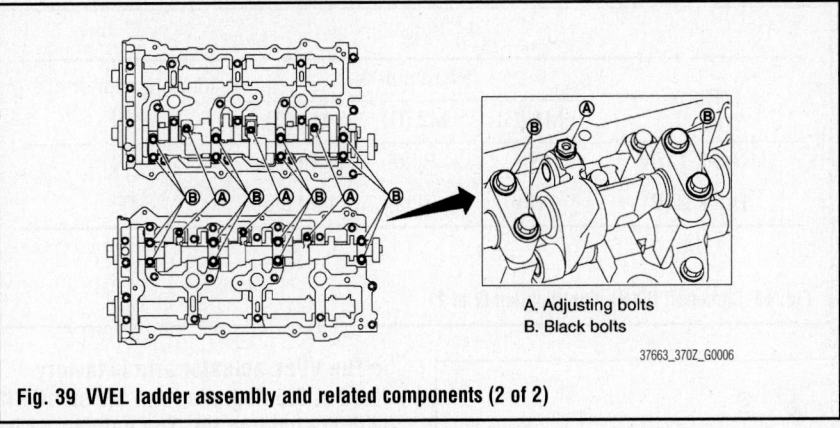

A. Adjusting bolts
B. Black bolts

37663_370Z_G0006

Fig. 39 VVEL ladder assembly and related components (2 of 2)

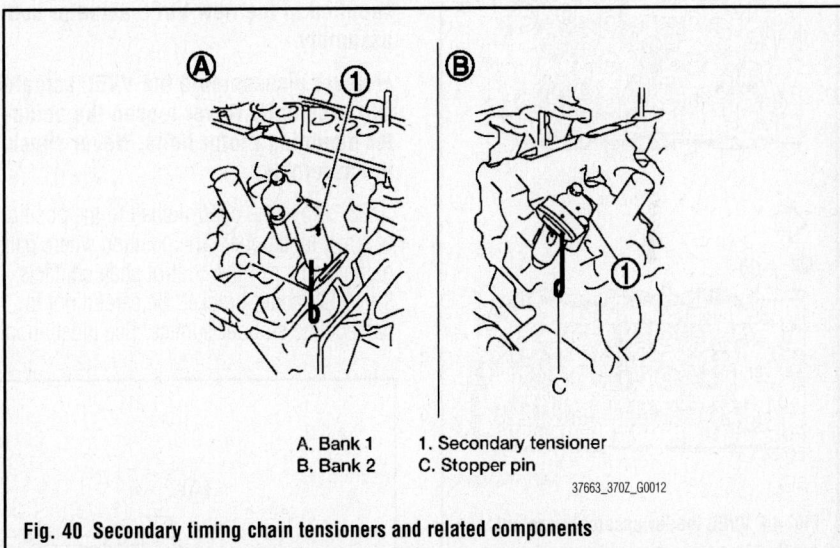

A. Bank 1
B. Bank 2
1. Secondary tensioner
C. Stopper pin

37663_370Z_G0012

Fig. 40 Secondary timing chain tensioners and related components

18. To install the VVEL ladder assembly, apply a continuous bead of liquid gasket (0.134–0.173 inch) to the cylinder head, as shown in the illustration.

19. Tighten the mounting bolts in the proper sequence to specification.

20. Specification is first pass 1 ft. lb. (1.96 Nm). Second pass 4 ft. lbs. (5.88 Nm). Third pass 8 ft. lbs. (10.4 Nm).

21. Measure the difference in levels between the front faces of the VVEL ladder assembly and the cylinder head. Specification should be -0.0055–0.0055 inch (-0.14 –0.14 mm).

➡ Measure two positions (both intake and exhaust side) for a single bank. If

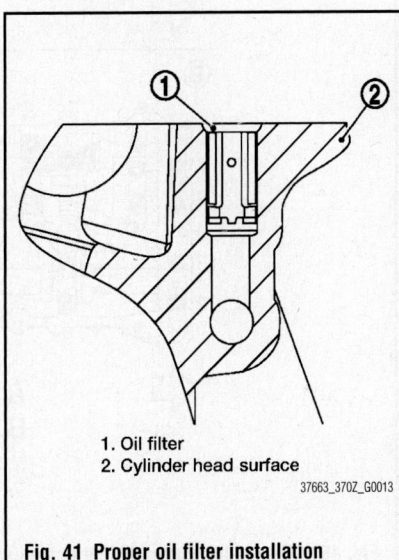

1. Oil filter
2. Cylinder head surface

37663_370Z_G0013

Fig. 41 Proper oil filter installation

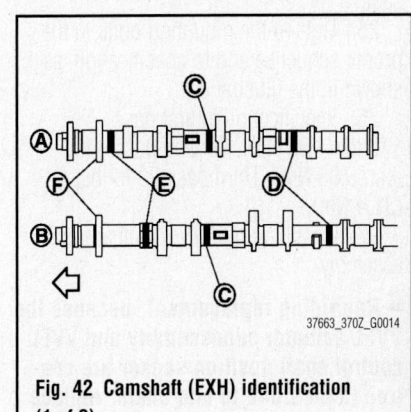

37663_370Z_G0014

Fig. 42 Camshaft (EXH) identification (1 of 2)

| Bank | Paint marks | | | Identification mark |
	M1 (C)	M2 (D)	M3 (E)	(F)
Bank 1 (A)	No	Blue	Light blue	1N
Bank 2 (B)	No	Blue	Light blue	1P

37663_370Z_G0015

Fig. 43 Camshaft (EXH) identification (2 of 2)

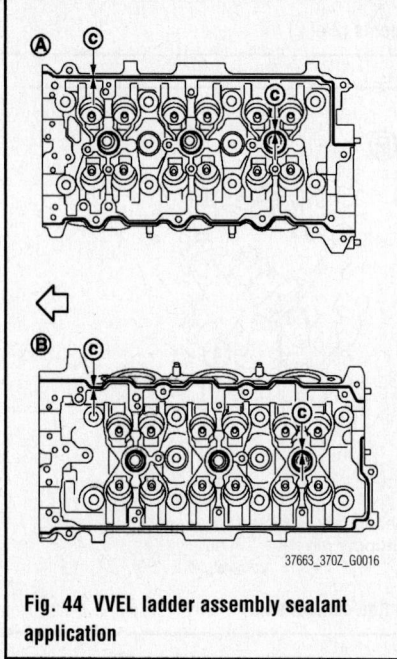

37663_370Z_G0016

Fig. 44 VVEL ladder assembly sealant application

the measured value is out of standard, reinstall the VVEL ladder assembly.

22. Install the rear timing case.

23. Install the timing chain sprockets, timing chain and front case.

24. To install the rear actuator bracket, apply a continuous bead of liquid gasket (0.134–0.173 inch) to the rear actuator bracket as shown in the illustration. Never apply gasket material to the oil passage.

25. Tighten the mounting bolts in the proper sequence and to specification, as shown in the illustration.

26. Specification is first pass 1 ft. lb. (1.96 Nm). Second pass 4 ft. lbs. (5.88 Nm). Third pass 23 ft. lbs. (31.4 Nm).

27. Install the new VVEL actuator sub assembly.

➡**Regarding replacement, because the VVEL actuator subassembly and VVEL control shaft position sensor are controlled on a one to one basis, replace them as a set.**

➡**The VVEL actuator arm is factory fixed at 5.5 degrees from the small lift with the holding jig. The holding jig is supplied in the new VVEL actuator subassembly.**

➡**Never disassemble the VVEL actuator sub assembly. Never loosen the actuator mounting motor bolts. Never shock the assembly.**

28. Move the control shaft to the position of small lift stopper. The position where part of the stopper of the control shaft contacts the VVEL ladder bracket. Be careful not to damage the stopper surface. See illustration.

29. If the control shaft cannot be moved, set the crankshaft in the following position. Bank 1: Turn 120 degrees from No. 1 cylinder at TDC. Bank 2: No. 1 cylinder at TDC.

30. Hold two flat areas of the control shaft with a wrench, and rotate the control shaft (5.5 degrees from the stopper) to the large lift side. This is for aligning the bolt hole of the control shaft and the hole of the VVEL actuator arm.

31. Apply a continuous bead of sealant, as shown in the illustration to the VVEL actuator subassembly.

32. Install the new VVEL actuator subassembly. Tighten the mounting bolts to specification and in the proper sequence.

➡**When installing be careful with the VVEL actuator subassembly (bank 2) mounting No.1 mounting bolt. The length is different. Be sure to check that the VVEL actuator subassembly is in contact with the cylinder head before tightening the mounting bolts.**

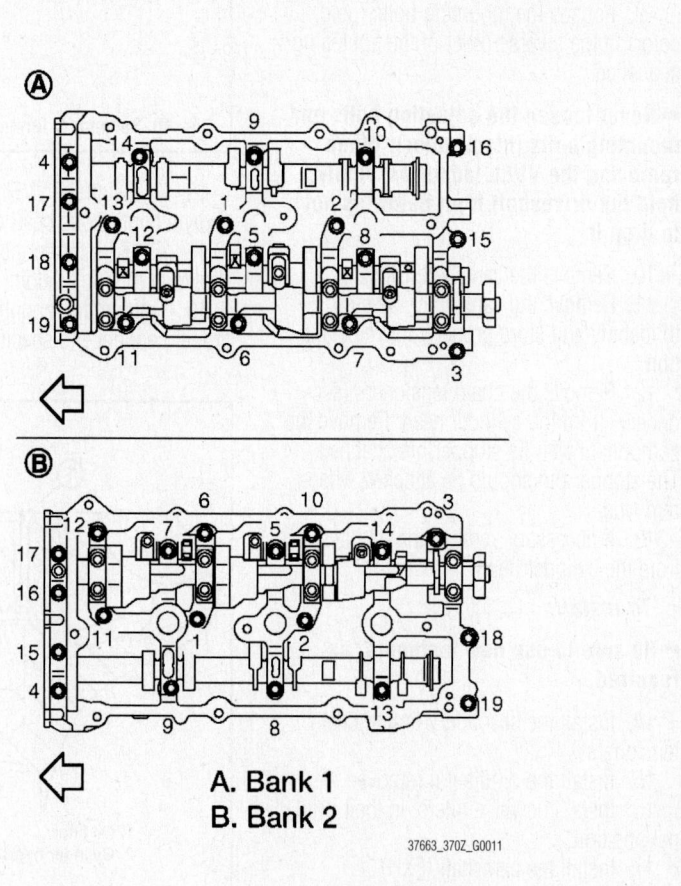

A. Bank 1
B. Bank 2

37663_370Z_G0011

Fig. 45 VVEL ladder assembly bolt tightening locations and tightening sequence

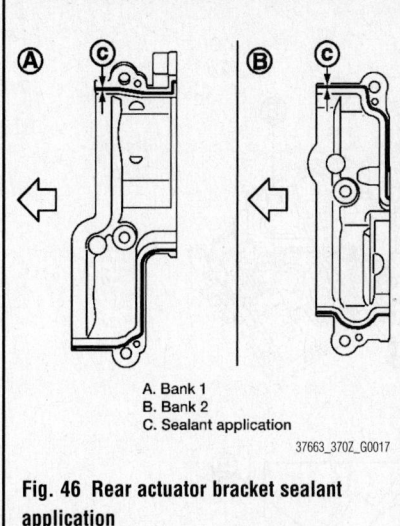

A. Bank 1
B. Bank 2
C. Sealant application

37663_370Z_G0017

Fig. 46 Rear actuator bracket sealant application

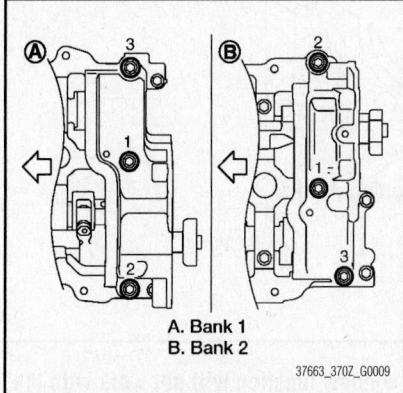

A. Bank 1
B. Bank 2

37663_370Z_G0009

Fig. 47 VVEL actuator bracket mounting bolt tightening locations and tightening sequence

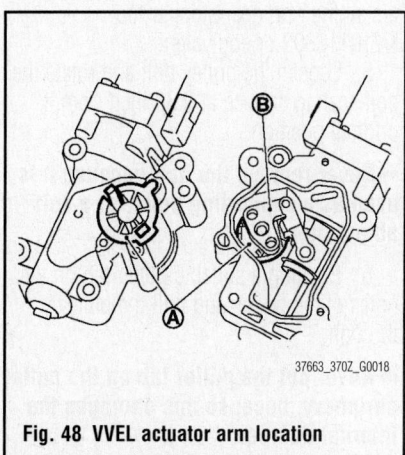

37663_370Z_G0018

Fig. 48 VVEL actuator arm location

33. Remove the jig. Check that the VVEL actuator arm bolt hole is aligned with the control shaft tapped hole. If not turn the control shaft for alignment.

34. Fix the two flat areas of the control shaft with a wrench to install the mounting bolts of the control shaft.

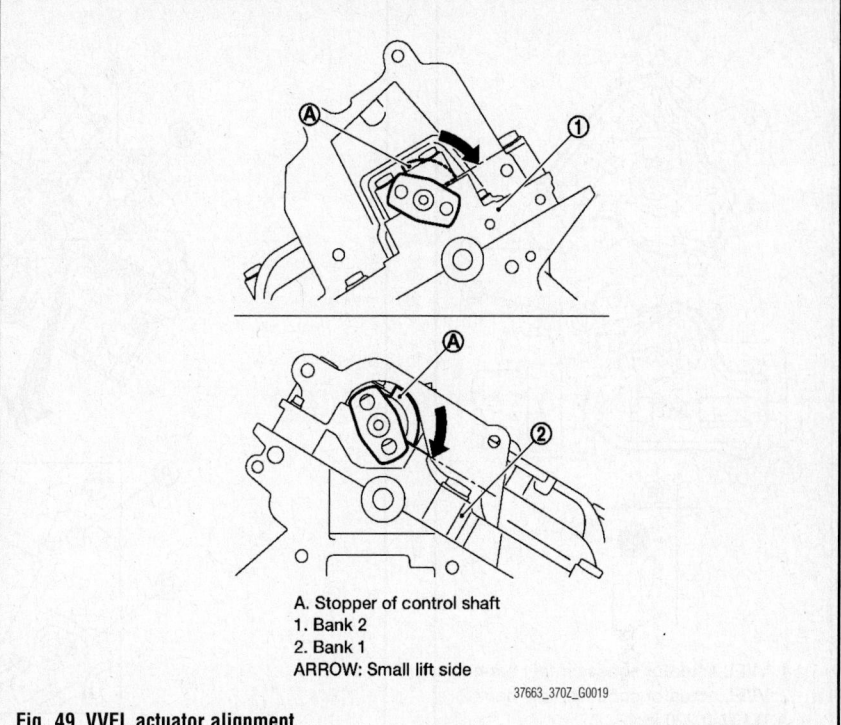

A. Stopper of control shaft
1. Bank 2
2. Bank 1
ARROW: Small lift side

37663_370Z_G0019

Fig. 49 VVEL actuator alignment

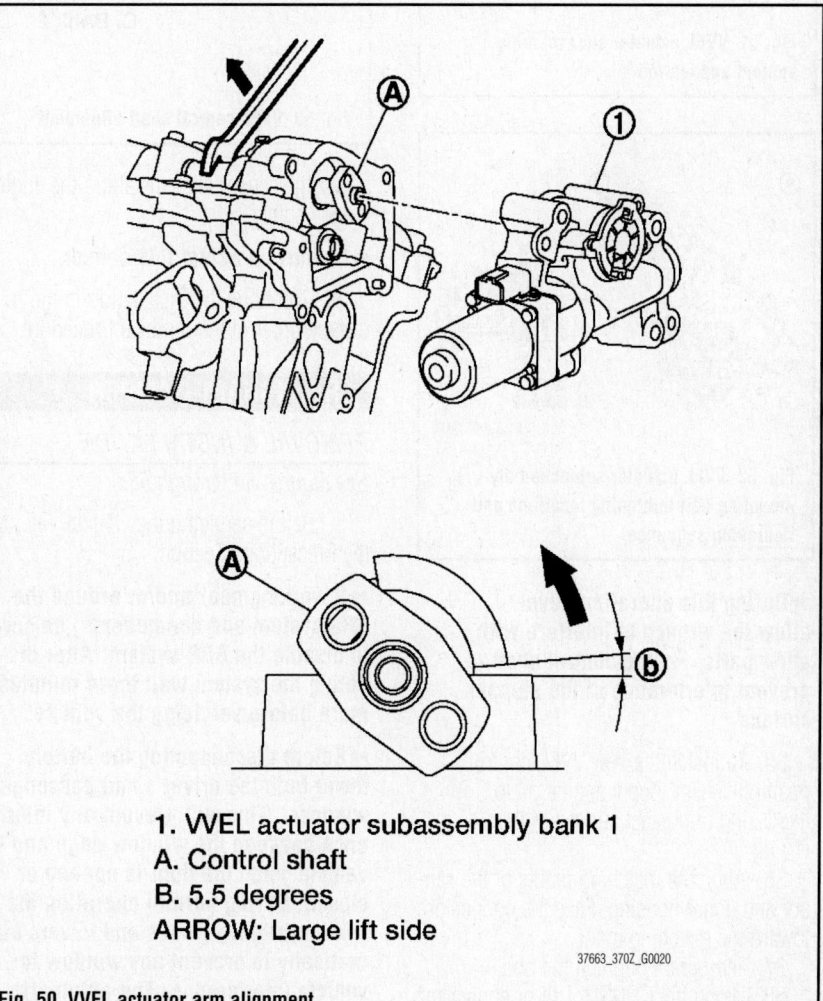

1. VVEL actuator subassembly bank 1
A. Control shaft
B. 5.5 degrees
ARROW: Large lift side

37663_370Z_G0020

Fig. 50 VVEL actuator arm alignment

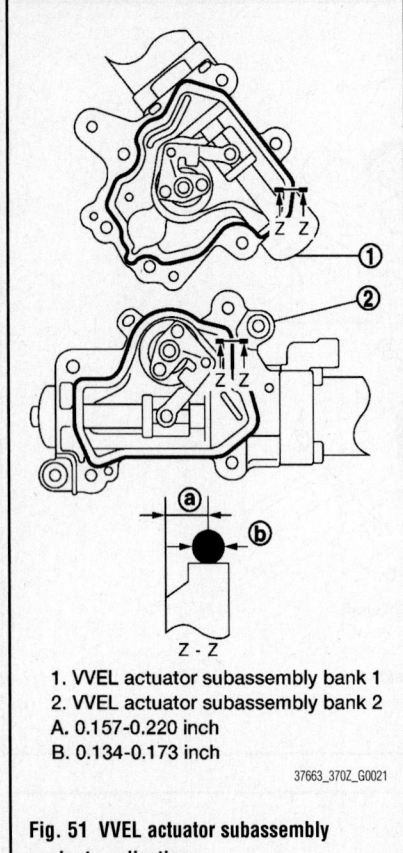

1. VVEL actuator subassembly bank 1
2. VVEL actuator subassembly bank 2
A. 0.157-0.220 inch
B. 0.134-0.173 inch

37663_370Z_G0021

Fig. 51 VVEL actuator subassembly sealant application

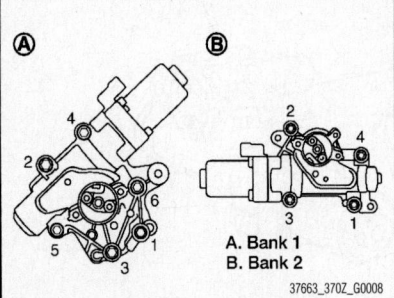

A. Bank 1
B. Bank 2

37663_370Z_G0008

Fig. 52 VVEL actuator sub assembly mounting bolt tightening locations and tightening sequence

➡**During this operation never allow the wrench to interfere with other parts. Fix the control shaft to prevent interference of the stopper surface.**

35. To install the new VVEL control position sensor. Apply engine oil to the O-ring or contact surface of the O-ring.

36. Align the matching marks of the sensor and upper housing. Face the connector toward the matching mark.

37. Temporarily tighten the bolt.

38. Using the CONSULT III or equivalent,

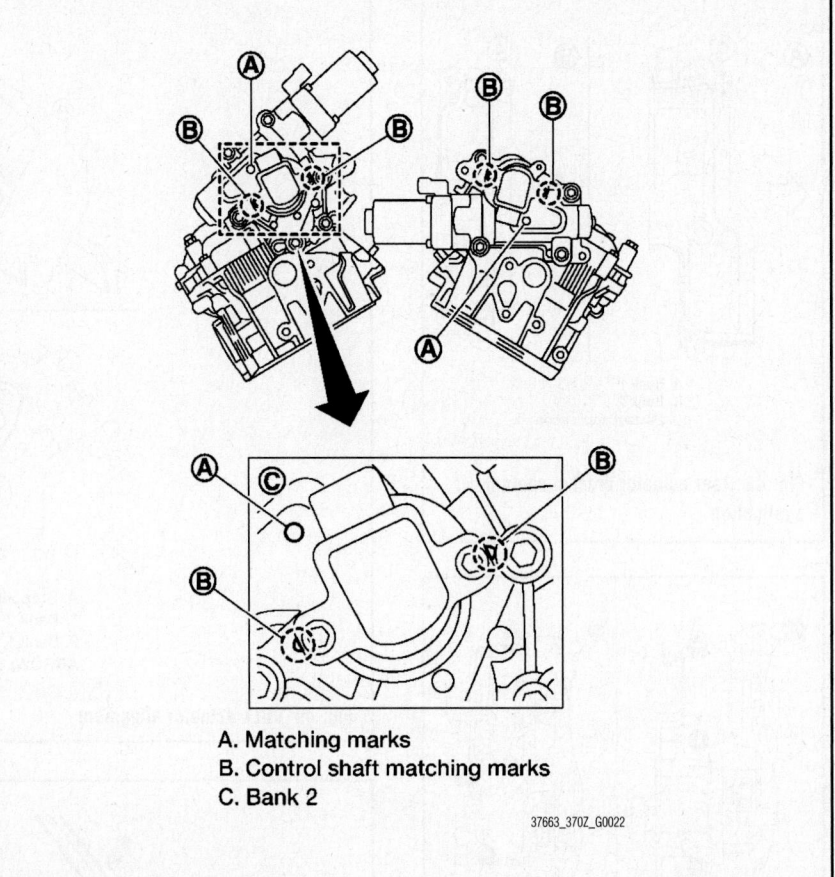

A. Matching marks
B. Control shaft matching marks
C. Bank 2

37663_370Z_G0022

Fig. 53 VVEL control shaft alignment

adjust the sensor after installing the engine in the vehicle.

➡**Be sure to adjust this sensor.**

39. Continue the installation in the reverse order of the removal procedure.

CRANKSHAFT DAMPER

REMOVAL & INSTALLATION
See Figures 54 through 56.

1. Before servicing the vehicle, refer to the Precautions Section.

➡**If working near and/or around the SRS system and components, be sure to disable the SRS system. After disabling the system wait three minutes or more before servicing the vehicle.**

➡**Before disconnecting the battery, lower both the driver's and passenger's windows. This will prevent any interference between the window edge and the vehicle when the door is opened or closed. During normal operation the window slightly raises and lowers automatically to prevent any window to vehicle interference. The automatic**

window function will not work with the battery disconnected.

2. Remove the undercover.

3. Remove the drive belt.

4. Remove the rear cover plate and install the ring gear stopper tool KV10118600 or equivalent.

5. Loosen the pulley bolt and rotate the bolt seating surface at 0.39 inch from its original position.

➡**Never remove the bolt because it is used as a supporting point for a suitable puller.**

6. Position a suitable puller tab on the holes of the pulley and pull the pulley through.

➡**Never put the puller tab on the pulley periphery, because this damages the internal damper.**

7. Remove the crankshaft damper.

To install:

➡**Be sure to use new fasteners, as required**

8. Installation is the reverse of the removal procedure.

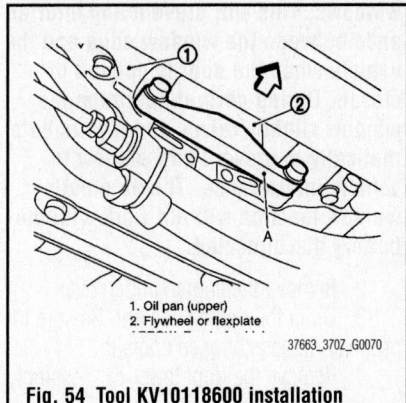

1. Oil pan (upper)
2. Flywheel or flexplate

37663_370Z_G0070

Fig. 54 Tool KV10118600 installation

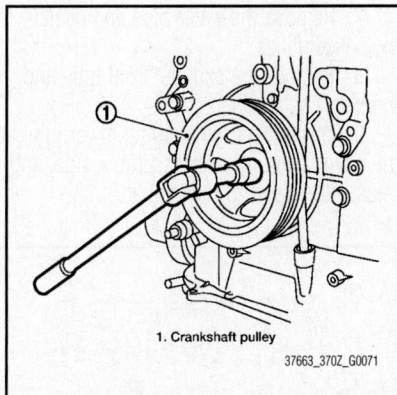

1. Crankshaft pulley

37663_370Z_G0071

Fig. 55 Crankshaft pulley removal (1 of 2)

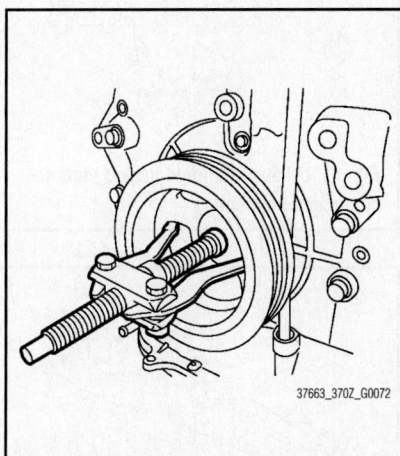

37663_370Z_G0072

Fig. 56 Crankshaft pulley removal (2 of 2)

9. Be sure to tighten the bolt to specification.

CRANKSHAFT FRONT SEAL

REMOVAL & INSTALLATION

See Figure 57.

1. Before servicing the vehicle, refer to the Precautions Section.

➡**If working near and/or around the SRS system and components, be sure to disable the SRS system. After dis-**

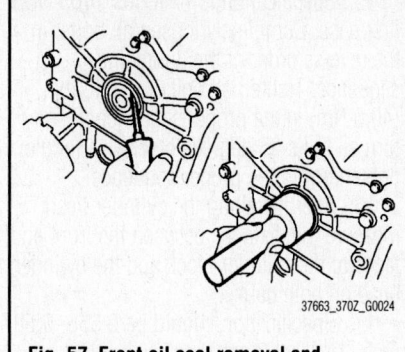

37663_370Z_G0024

Fig. 57 Front oil seal removal and installation

abling the system wait three minutes or more before servicing the vehicle.

➡**Before disconnecting the battery, lower both the driver's and passenger's windows. This will prevent any interference between the window edge and the vehicle when the door is opened or closed. During normal operation the window slightly raises and lowers automatically to prevent any window to vehicle interference. The automatic window function will not work with the battery disconnected.**

2. Remove the undercover.
3. Remove the drive belt.
4. Remove the crankshaft damper.
5. Remove the front seal, using a suitable tool. Be careful not to damage the front timing chain case and crankshaft.
6. Discard the seal.

To install:

➡**Be sure to use new fasteners, as required**

7. Apply clean engine oil to both the seal lip and the dust seal lip.
8. Using a suitable drift, press fit the new seal into position until the height of the seal is level with the mounting surface. Check that the garter spring is in position and the seal lips are not inverted.

➡ **Be careful not to damage the front timing chain case and crankshaft. Press fit straight ahead and avoid causing burrs or tilting the seal.**

9. Continue the installation is the reverse of the removal procedure.

CYLINDER HEAD

REMOVAL & INSTALLATION

See Figures 58 through 60.

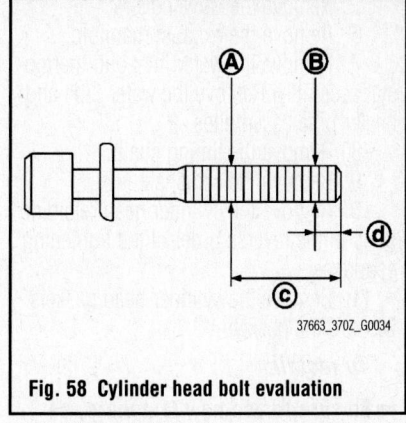

37663_370Z_G0034

Fig. 58 Cylinder head bolt evaluation

At this time the manufacturer provides service information for this component with the engine removed from the vehicle and positioned in a suitable holding fixture.

1. Before servicing the vehicle, refer to the Precautions Section.

➡**If working near and/or around the SRS system and components, be sure to disable the SRS system. After disabling the system wait three minutes or more before servicing the vehicle.**

➡**Before disconnecting the battery, lower both the driver's and passenger's windows. This will prevent any interference between the window edge and the vehicle when the door is opened or closed. During normal operation the window slightly raises and lowers automatically to prevent any window to vehicle interference. The automatic window function will not work with the battery disconnected.**

2. Remove the intake manifold collector.
3. Remove the fuel tube and fuel injector assembly.
4. Remove the intake manifold.

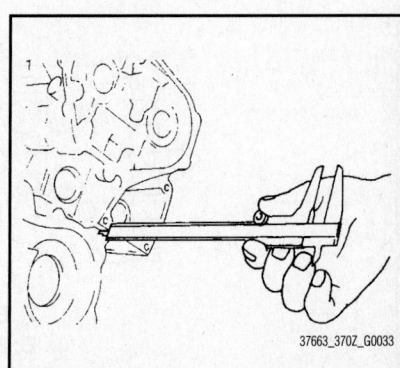

37663_370Z_G0033

Fig. 59 Cylinder head to cylinder block measurement check

5. Remove the valve covers.

6. Remove the exhaust manifold.

7. Remove the water inlet and thermostat assembly. Remove the water pipe and heater pipe assemblies.

8. Remove the timing chain.

9. Remove the camshaft.

10. Remove the cylinder head retaining bolts in the reverse order of the tightening sequence.

11. Remove the cylinder head gaskets. Discard the gaskets.

To install:

➡**Be sure to use new fasteners, as required.**

12. Installation is the reverse of the removal procedure.

13. Be sure to use new gaskets.

➡**If the old bolts are being reused check their outer diameter before installation. Out of spec bolts must be replaced. "B" minus "A" should be 0.0071 inch. "C" should be 1.89 inch. "D" should be 0.43 inch.**

14. Tighten the cylinder head bolts in the proper sequence and to specification. Coat the bolt threads with clean engine oil before installation.

15. Specification is 77 ft. lbs. (105 NM) first pass. Completely loosen all bolts, in the reverse order of the tightening sequence. Tighten all bolts to 30 ft. lbs. (40.0 Nm) in the proper sequence. Finally turn all bolts 95 degrees clockwise (angle tightening) in the proper sequence.

16. After installing the cylinder head measure the distance between the front end faces of the cylinder block and the cylinder head on both banks.

17. Specification should be 0.555–0.587 inch (14.1–14.9 mm).

18. Continue the installation in the reverse order of the removal procedure.

EXHAUST MANIFOLD

REMOVAL & INSTALLATION

See Figures 61 and 62.

1. Before servicing the vehicle, refer to the Precautions Section.

➡**If working near and/or around the SRS system and components, be sure to disable the SRS system. After disabling the system wait three minutes or more before servicing the vehicle.**

➡**Before disconnecting the battery, lower both the driver's and passenger's windows. This will prevent any interference between the window edge and the vehicle when the door is opened or closed. During normal operation the window slightly raises and lowers automatically to prevent any window to vehicle interference. The automatic window function will not work with the battery disconnected.**

2. Remove the engine undercover.

3. Drain the engine coolant. Be sure to properly dispose of used coolant.

4. Remove the front tower bar assembly.

5. Remove the engine cover.

6. Remove the air cleaner assembly.

7. Remove the water pipe and heater pipe assemblies.

8. Remove the exhaust front tube and three way catalysts.

9. Disconnect the steering lower joint at the power steering gear assembly side, and release the steering lower shaft.

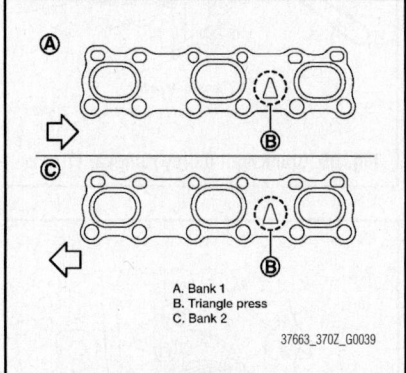

A. Bank 1
B. Triangle press
C. Bank 2

37663_370Z_G0039

Fig. 61 Exhaust manifold gasket identification and positioning

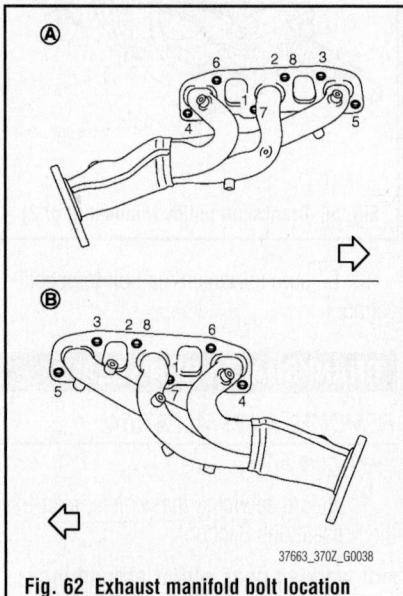

37663_370Z_G0038

Fig. 62 Exhaust manifold bolt location and tightening sequence

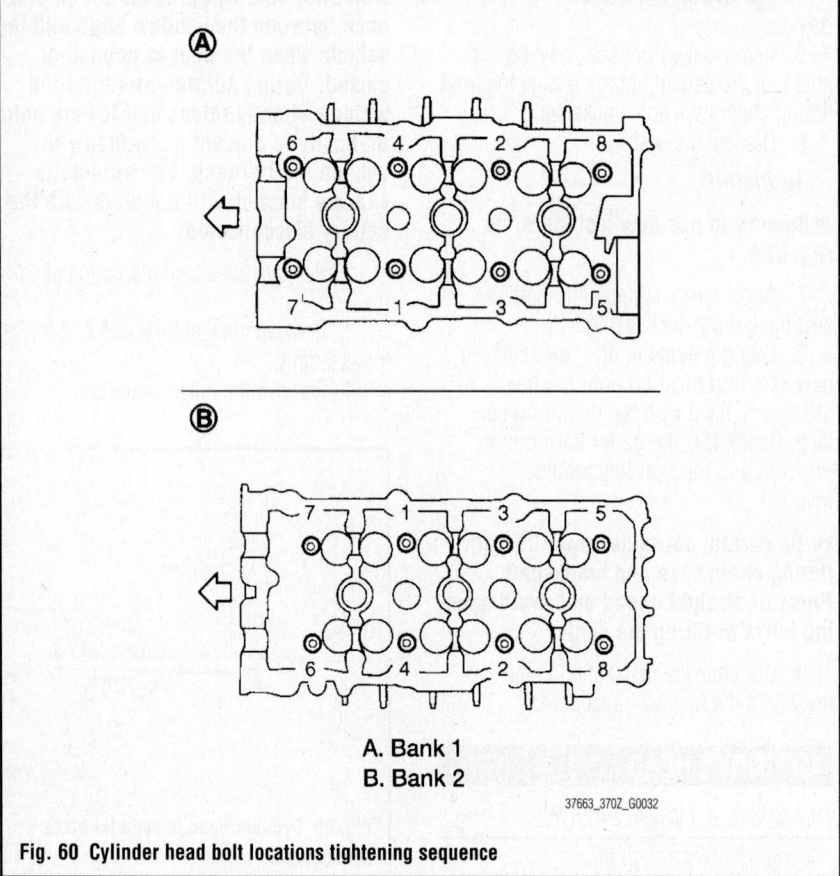

A. Bank 1
B. Bank 2

37663_370Z_G0032

Fig. 60 Cylinder head bolt locations tightening sequence

10. Disconnect the air fuel sensor electrical connectors. Using the proper tool remove the sensors. Be careful not to drop or damage the sensors, or they will have to be replaced.

11. Remove the exhaust manifold cover.

12. Remove the retaining nuts in the reverse order of the installation sequence.

13. Remove the exhaust manifold. Discard the gaskets.

To install:

➡**Be sure to use new fasteners, as required.**

14. Installation is the reverse of the removal procedure.

15. Be sure to use new gaskets. Install the gasket as shown in the illustration.

16. Tighten the nuts to specification and in the sequence shown in the illustration.

17. When installing the oxygen sensors, be sure to coat them with anti-seize compound.

18. Continue the installation in the reverse order of the removal procedure.

FLEXPLATE

REMOVAL & INSTALLATION

See Figure 63.

1. Before servicing the vehicle, refer to the Precautions Section.

➡**If working near and/or around the SRS system and components, be sure to disable the SRS system. After disabling the system wait three minutes or more before servicing the vehicle.**

➡**Before disconnecting the battery, lower both the driver's and passenger's windows. This will prevent any interference between the window edge and the vehicle when the door is opened or closed. During normal operation the window slightly raises and lowers automatically to prevent any window to vehicle interference. The automatic window function will not work with the battery disconnected.**

2. Raise and safely support the vehicle.
3. Remove the transmission assembly.
4. Remove the retaining bolts.
5. Remove the component from its mounting.

To install:

➡**Be sure to use new fasteners, as required.**

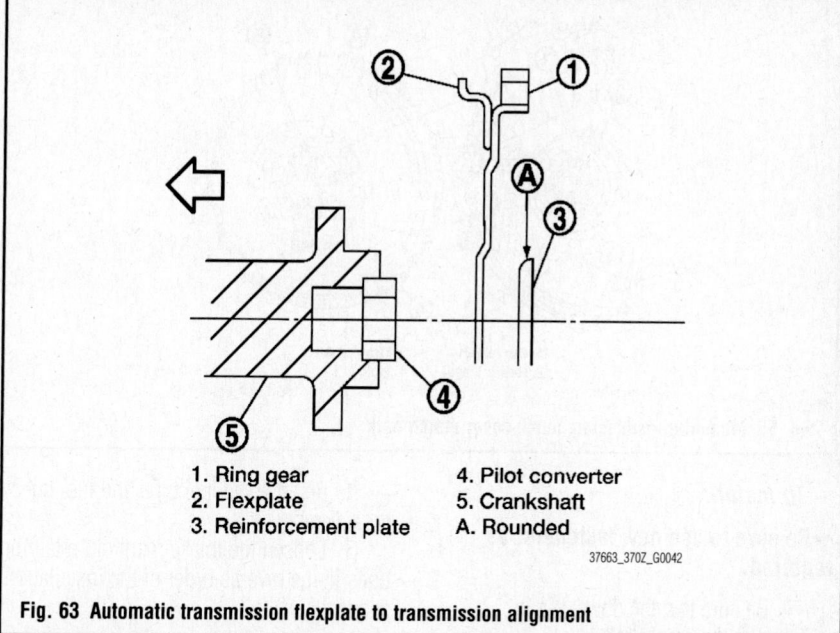

1. Ring gear
2. Flexplate
3. Reinforcement plate
4. Pilot converter
5. Crankshaft
A. Rounded

37663_370Z_G0042

Fig. 63 Automatic transmission flexplate to transmission alignment

6. Be sure to correctly align the crankshaft side dowel pin and the flexplate side dowel pin hole.

➡**If not aligned correctly the MIL light will illuminate.**

7. Install the flexplate and reinforcement plate. Be sure to hold the ring gear with the ring gear stopper tool KV10118600 or equivalent.

8. Tighten the retaining bolts to specification in a crisscross pattern.

9. Continue the installation in the reverse order of the removal procedure.

FLYWHEEL

REMOVAL & INSTALLATION

See Figures 64 and 65.

1. Before servicing the vehicle, refer to the Precautions Section.

➡**If working near and/or around the SRS system and components, be sure to disable the SRS system. After disabling the system wait three minutes or more before servicing the vehicle.**

➡**Before disconnecting the battery, lower both the driver's and passenger's windows. This will prevent any interference between the window edge and the vehicle when the door is opened or closed. During normal operation the window slightly raises and lowers automatically to prevent any window to vehicle interference. The automatic window function will not work with the battery disconnected.**

2. Raise and safely support the vehicle.
3. Remove the transmission assembly.
4. Remove the retaining bolts.
5. Remove the component from its mounting.

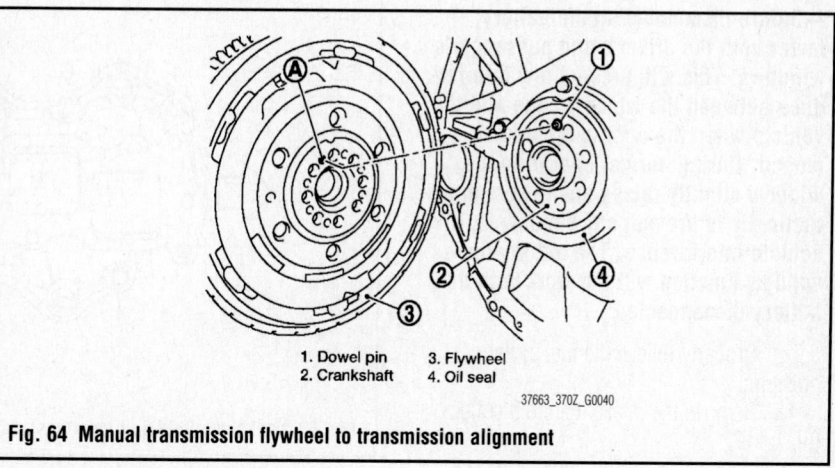

1. Dowel pin
2. Crankshaft
3. Flywheel
4. Oil seal

37663_370Z_G0040

Fig. 64 Manual transmission flywheel to transmission alignment

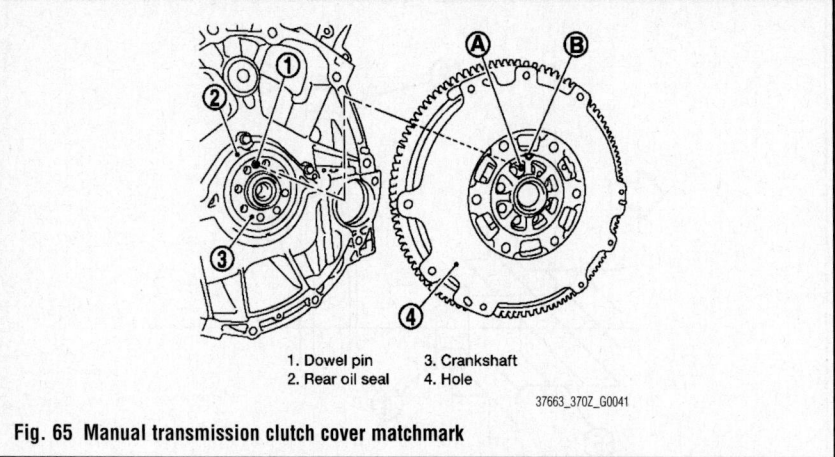

1. Dowel pin   3. Crankshaft
2. Rear oil seal   4. Hole

37663_370Z_G0041

Fig. 65 Manual transmission clutch cover matchmark

To install:

➡️**Be sure to use new fasteners, as required.**

6. Be sure that the dowel pin is installed in the crankshaft.

7. Be sure to correctly align the crankshaft side dowel pin and the flywheel side dowel pin hole.

8. There is a matchmark on the clutch cover side of the flywheel.

9. Tighten the retaining bolts to specification in a crisscross pattern.

10. Continue the installation in the reverse order of the removal procedure.

INTAKE MANIFOLD

REMOVAL & INSTALLATION

See Figures 66 and 67.

1. Before servicing the vehicle, refer to the Precautions Section.

➡️**If working near and/or around the SRS system and components, be sure to disable the SRS system. After disabling the system wait three minutes or more before servicing the vehicle.**

➡️**Before disconnecting the battery, lower both the driver's and passenger's windows. This will prevent any interference between the window edge and the vehicle when the door is opened or closed. During normal operation the window slightly raises and lowers automatically to prevent any window to vehicle interference. The automatic window function will not work with the battery disconnected.**

2. Properly relieve the fuel system pressure.

3. Remove the intake manifold collector.

4. Remove the fuel tube and fuel injector assembly.

5. Loosen the intake manifold retaining bolts in the reverse order of the installation sequence.

6. Remove the component from its mounting. Discard the gasket.

➡️**Matchmark the manifold and the cylinder head for proper installation, as these components need to be installed in a specified direction.**

To install:

➡️**Be sure to use new fasteners, as required.**

7. Installation is the reverse of the removal procedure.

➡️**Be sure to use the marks made in the removal for proper alignment and installation.**

8. If the stud bolts were removed, install and tighten to 8 ft. lbs. (10.8 Nm).

9. Tighten all mounting bolts to specification and in the proper sequence.

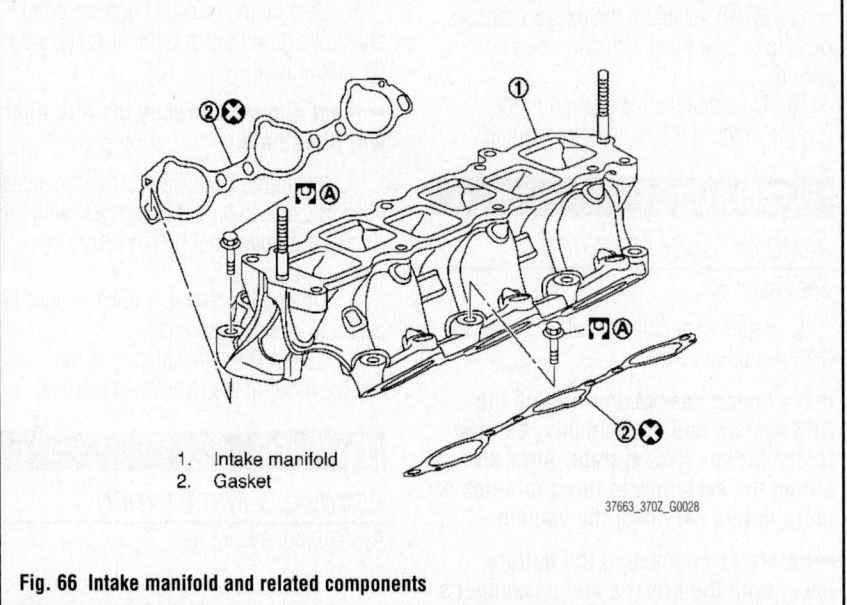

1. Intake manifold
2. Gasket

37663_370Z_G0028

Fig. 66 Intake manifold and related components

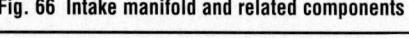
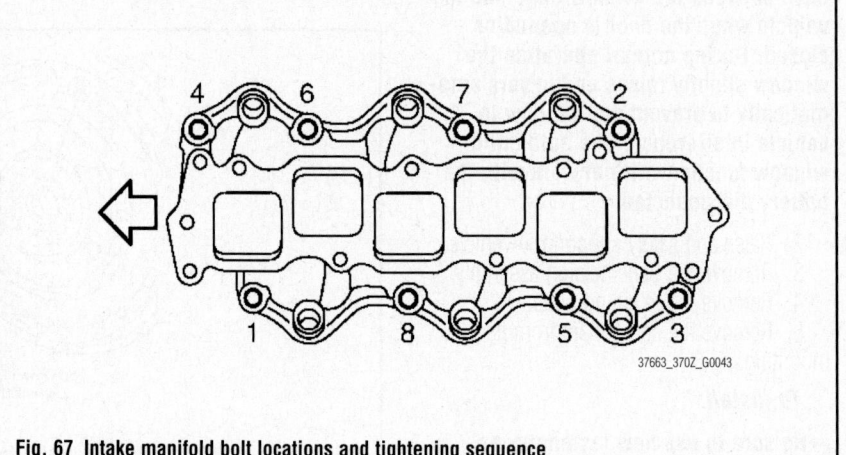

37663_370Z_G0043

Fig. 67 Intake manifold bolt locations and tightening sequence

10. Specification is 5 ft. lbs. (7.4 Nm) step one. 19 ft. lbs. (25.5 Nm) step two.

INTAKE MANIFOLD COLLECTOR

REMOVAL & INSTALLATION

See Figures 68 and 69.

1. Before servicing the vehicle, refer to the Precautions Section.

➡ **If working near and/or around the SRS system and components, be sure to disable the SRS system. After disabling the system wait three minutes or more before servicing the vehicle.**

➡ **Before disconnecting the battery, lower both the driver's and passenger's windows. This will prevent any interference between the window edge and the vehicle when the door is opened or closed. During normal operation the window slightly raises and lowers automatically to prevent any window to vehicle interference. The automatic window function will not work with the battery disconnected.**

2. Remove the engine undercover.
3. Drain the engine coolant. Be sure to properly dispose of used coolant.
4. Remove the front tower bar assembly.
5. Remove the engine cover.
6. Remove the air cleaner assembly.
7. To remove the electric throttle control actuator, Disconnect the water hoses from the component. Disconnect the harness connector.

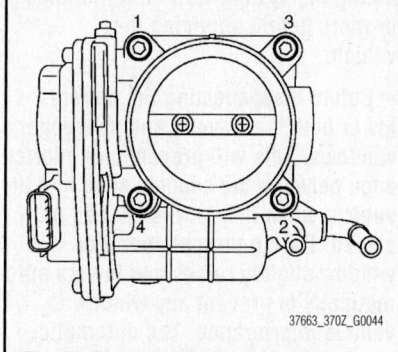

Fig. 69 Throttle control actuator bolt locations and tightening sequence

8. Loosen the mounting bolts in the reverse order of the tightening sequence.

➡ **When removing only the intake manifold collector, move the electric throttle control actuator without disconnecting the water hose. Handel the component carefully to avoid shock damage.**

9. Disconnect the vacuum hose, PCV hose, and EVAP hose from the intake manifold collector.
10. Remove the EVAP canister purge volume control solenoid valve and EVAP tube assembly from the manifold collector.
11. Loosen the retaining bolts and nuts in the reverse order of the tightening sequence and remove the component from its mounting.

To install:

➡ **Be sure to use new fasteners, as required.**

12. Installation is the reverse of the removal procedure.
13. Tighten the retaining bolts to specification and in the proper sequence.
14. Specification is 8 ft. lbs. (10.8 Nm).
15. Tighten the electric throttle actuator retaining bolts in the proper sequence.
16. Using the CONSULT-III diagnostic tool or equivalent, perform the throttle valve closed position learning and the idle air volume learning procedures.

OIL PAN

REMOVAL & INSTALLATION

Lower

See Figures 70 through 73.

1. Before servicing the vehicle, refer to the Precautions Section.

➡ **If working near and/or around the SRS system and components, be sure to disable the SRS system. After disabling the system wait three minutes or more before servicing the vehicle.**

➡ **Before disconnecting the battery, lower both the driver's and passenger's windows. This will prevent any interference between the window edge and the vehicle when the door is opened or closed. During normal operation the window slightly raises and lowers automatically to prevent any window to vehicle interference. The automatic window function will not work with the battery disconnected.**

2. Raise and support the vehicle safely.
3. Remove the engine undercover.
4. Drain the engine oil. Be sure to properly dispose of used oil.
5. Loosen the oil pan retaining bolts in the reverse order of the installation.
6. Insert a seal cutter tool between the upper and lower oil pan sealing surfaces. Never use a screwdriver. Slide the tool, by tapping the side of the tool with a hammer. Remove the lower oil pan.
7. Discard the gasket.

To install:

➡ **Be sure to use new fasteners, as required.**

8. Be sure to use a new gasket.
9. Apply a continuous bead of RTV sealant as shown in the illustration. Install within five minutes.
10. Tighten retaining bolts in the proper sequence and to the proper specification.

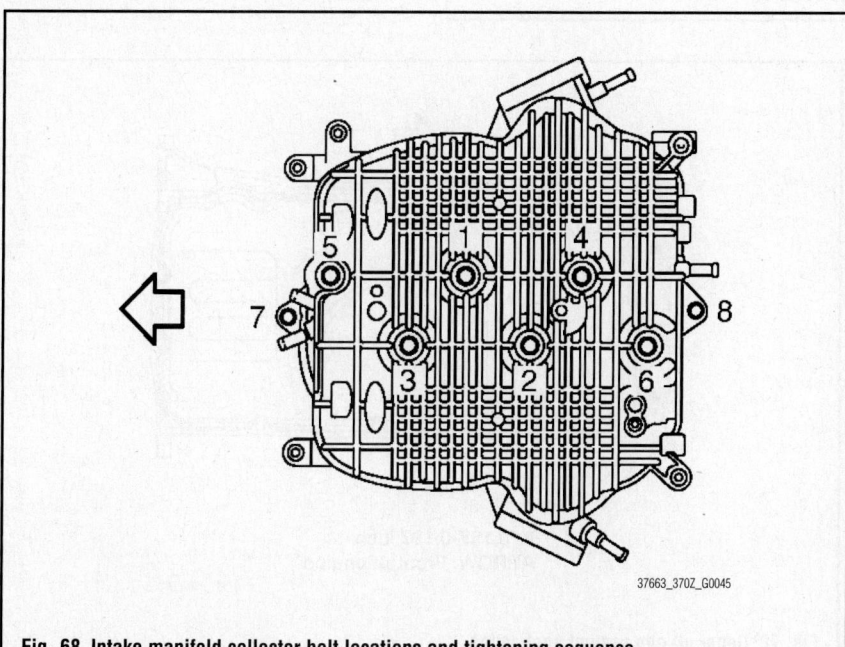

Fig. 68 Intake manifold collector bolt locations and tightening sequence

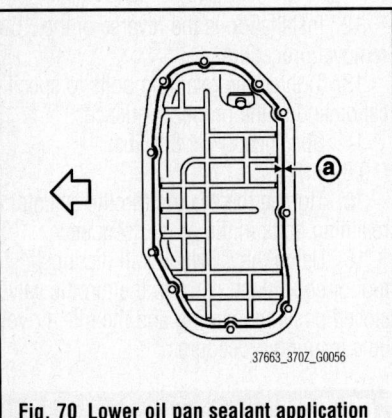

37663_370Z_G0056

Fig. 70 Lower oil pan sealant application

abling the system wait three minutes or more before servicing the vehicle.

→Before disconnecting the battery, lower both the driver's and passenger's windows. This will prevent any interference between the window edge and the vehicle when the door is opened or closed. During normal operation the window slightly raises and lowers automatically to prevent any window to vehicle interference. The automatic window function will not work with the battery disconnected.

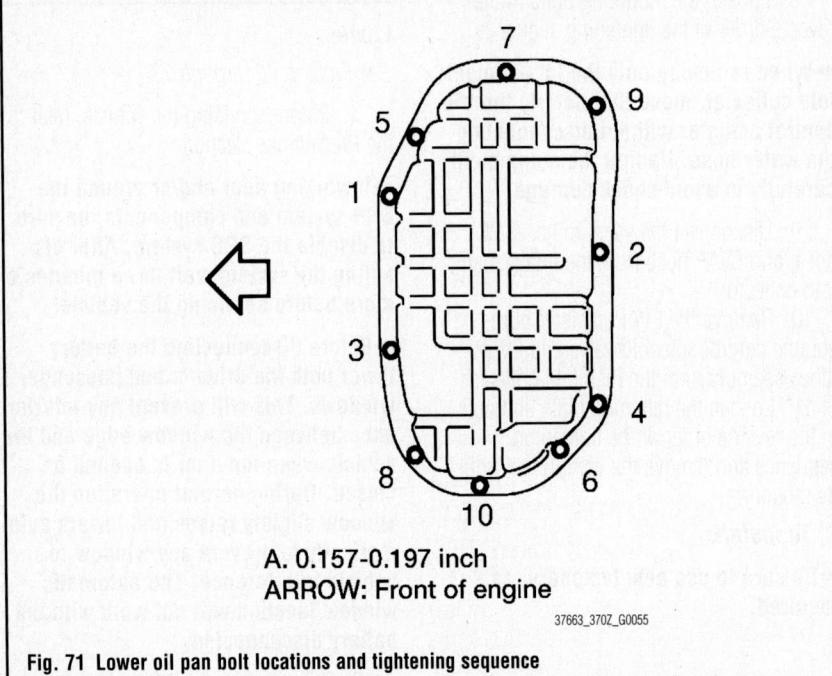

A. 0.157-0.197 inch
ARROW: Front of engine

37663_370Z_G0055

Fig. 71 Lower oil pan bolt locations and tightening sequence

11. Continue the installation in the reverse order of the removal procedure.

12. Wait at least thirty minutes after installation before fill the crankcase with clean engine oil.

13. Start the engine and check for leaks. Correct as required.

Upper

See Figures 85 and 86.

At this time the manufacturer provides service information for this component with the engine removed from the vehicle and positioned in a suitable holding fixture.

1. Before servicing the vehicle, refer to the Precautions Section.

→If working near and/or around the SRS system and components, be sure to disable the SRS system. After dis-

2. Drain the engine oil. Be sure to properly dispose of used oil.

3. Remove the oil level gauge, oil pressure switch and oil temperature sensor.

4. Remove the lower oil pan.

5. Remove the oil strainer.

6. Loosen the retaining bolts in the reverse order of the installation sequence.

7. Insert a seal cutter tool between the upper and lower oil pan sealing surfaces. Never use a screwdriver. Slide the tool, by tapping the side of the tool with a hammer. Remove the upper oil pan.

8. Discard the gasket.

9. Remove the O-rings from the bottom of the lower cylinder block and oil pump. Discard the O-rings.

To install:

→Be sure to use new fasteners, as required.

10. Be sure to use a new gasket and O-rings.

11. Apply a continuous bead of RTV sealant as shown in the illustration. Install within five minutes.

→For bolt holes with the triangle (see illustration) apply liquid gasket outside the holes.

12. Tighten retaining bolts in the proper sequence and to the proper specification.

→There are two size bolts. Be careful to install the wrong bolt, in the wrong hole.

13. Continue the installation in the reverse order of the removal procedure.

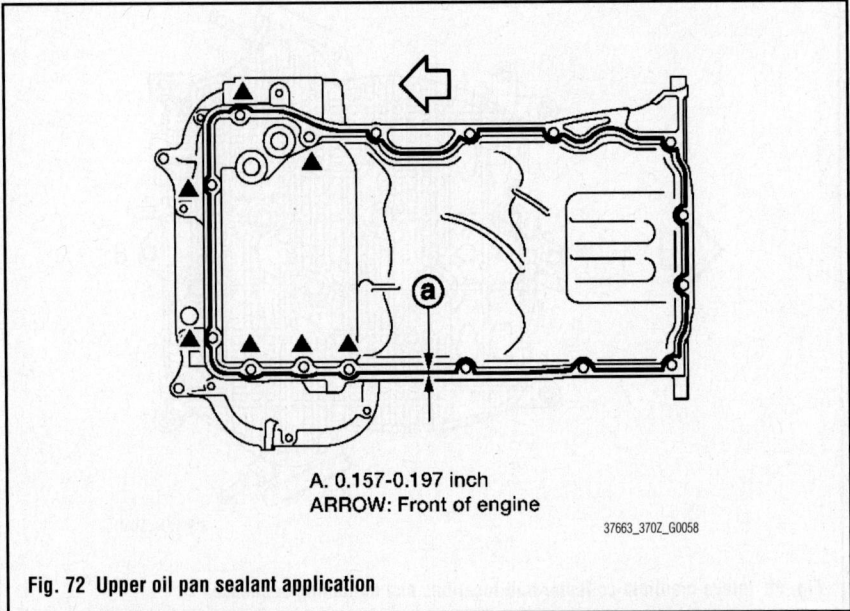

A. 0.157-0.197 inch
ARROW: Front of engine

37663_370Z_G0058

Fig. 72 Upper oil pan sealant application

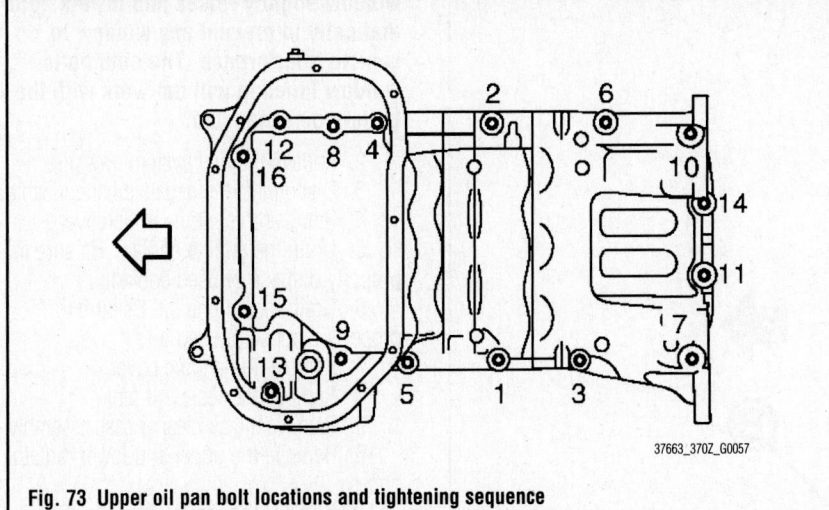

Fig. 73 Upper oil pan bolt locations and tightening sequence

14. Wait at least thirty minutes after installation before fill the crankcase with clean engine oil.

15. Start the engine and check for leaks. Correct as required.

OIL PUMP

REMOVAL & INSTALLATION

1. Before servicing the vehicle, refer to the Precautions Section.

➡If working near and/or around the SRS system and components, be sure to disable the SRS system. After disabling the system wait three minutes or more before servicing the vehicle.

➡Before disconnecting the battery, lower both the driver's and passenger's windows. This will prevent any interference between the window edge and the vehicle when the door is opened or closed. During normal operation the window slightly raises and lowers automatically to prevent any window to vehicle interference. The automatic window function will not work with the battery disconnected.

2. Remove the upper and lower oil pan assemblies.

3. Remove the timing chain cover and the primary timing chain.

4. Remove the oil pump assembly from its mounting.

To install:

➡Be sure to use new fasteners, as required.

5. Installation is the reverse of the removal procedure.

➡When installing, align the crankshaft flat surfaces with the oil pump inner rotor flat surfaces.

INSPECTION

1. Clearance between outer rotor and oil pump body: 0.0045–0.0079 in. (0.114–0.200 mm).

2. Tip clearance between inner rotor and outer rotor: Below 0.0071 in. (0.180 mm).

3. Side clearance with a straight-edge between inner rotor and oil pump body: 0.0012–0.0028 in. (0.030–0.070 mm).

4. Side clearance with a straight-edge between outer rotor and oil pump body: 0.0020–0.0043 in. (0.050–0.110 mm).

PISTON AND RING

POSITIONING

See Figure 74.

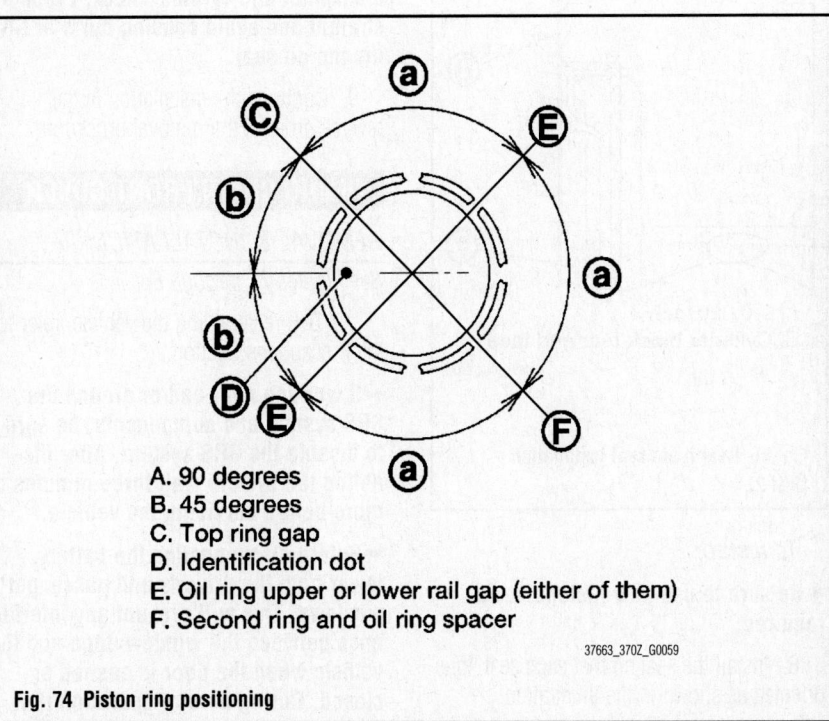

A. 90 degrees
B. 45 degrees
C. Top ring gap
D. Identification dot
E. Oil ring upper or lower rail gap (either of them)
F. Second ring and oil ring spacer

Fig. 74 Piston ring positioning

REAR MAIN SEAL

REMOVAL & INSTALLATION

See Figures 75 and 76.

1. Before servicing the vehicle, refer to the Precautions Section.

➡If working near and/or around the SRS system and components, be sure to disable the SRS system. After disabling the system wait three minutes or more before servicing the vehicle.

➡Before disconnecting the battery, lower both the driver's and passenger's windows. This will prevent any interference between the window edge and the vehicle when the door is opened or closed. During normal operation the window slightly raises and lowers automatically to prevent any window to vehicle interference. The automatic window function will not work with the battery disconnected.

2. Raise and safely support the vehicle.

3. Remove the transmission from the vehicle.

4. Remove the flexplate or flywheel.

5. Using a suitable tool, carefully remove the seal from its mounting.

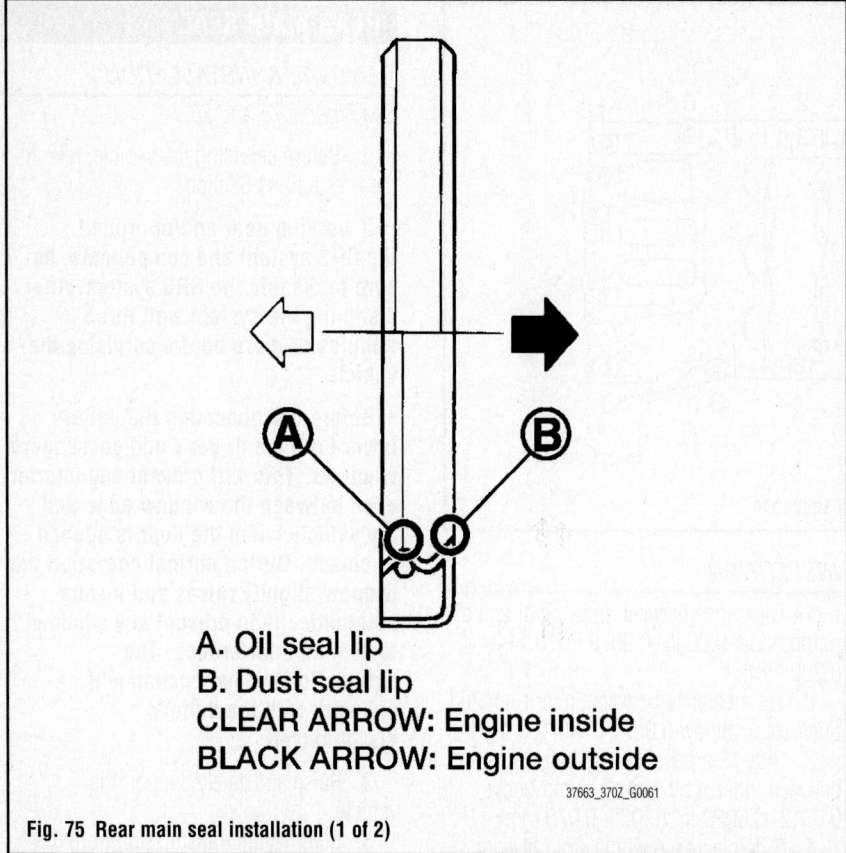

A. Oil seal lip
B. Dust seal lip
CLEAR ARROW: Engine inside
BLACK ARROW: Engine outside

37663_370Z_G0061

Fig. 75 Rear main seal installation (1 of 2)

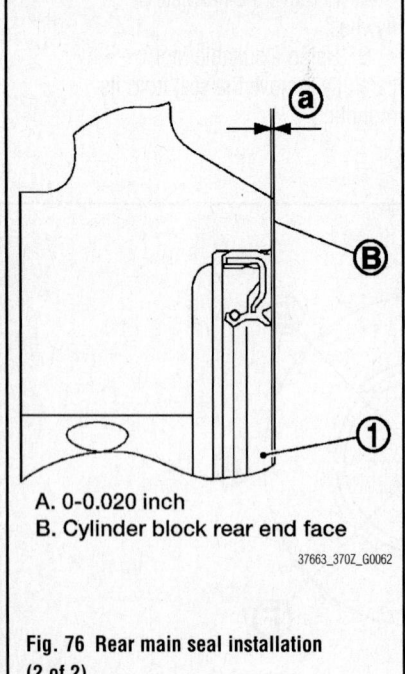

A. 0-0.020 inch
B. Cylinder block rear end face

37663_370Z_G0062

Fig. 76 Rear main seal installation (2 of 2)

To install:

➡Be sure to use new fasteners, as required.

6. Install the seal so that each seal lip is oriented as shown in the illustration.

7. Press the seal into position as shown in the illustration.

8. Using a suitable drift, press fit the seal until the height of the seal is level with the mounting surface.

➡Be careful to avoid damage to the crankshaft and cylinder block. Press fit straight and avoid causing burrs or tilting the oil seal.

9. Continue the installation in the reverse order of the removal procedure.

TIMING CHAIN FRONT COVER

REMOVAL & INSTALLATION

See Figures 77 through 84.

1. Before servicing the vehicle, refer to the Precautions Section.

➡If working near and/or around the SRS system and components, be sure to disable the SRS system. After disabling the system wait three minutes or more before servicing the vehicle.

➡Before disconnecting the battery, lower both the driver's and passenger's windows. This will prevent any interference between the window edge and the vehicle when the door is opened or closed. During normal operation the window slightly raises and lowers automatically to prevent any window to vehicle interference. The automatic window function will not work with the battery disconnected.

2. Relieve the fuel system pressure.
3. Disconnect the negative battery cable.
4. Remove the engine undercover.
5. Drain the engine coolant. Be sure to properly dispose of used coolant.
6. Drain the engine oil. Be sure to properly dispose of used oil.
7. Remove the engine cover.
8. Remove the reservoir tank.
9. Remove the air cleaner case assembly.
10. Remove the upper and lower radiator hoses.
11. Remove the radiator cooling fan assembly.
12. Remove the drive belt.
13. Separate the engine harnesses by removing their brackets from the timing chain cover.
14. Remove the intake manifold collector.
15. Remove the fuel sub mounting bolt.
16. Remove the oil level gauge and guide.
17. Remove the air conditioning com-

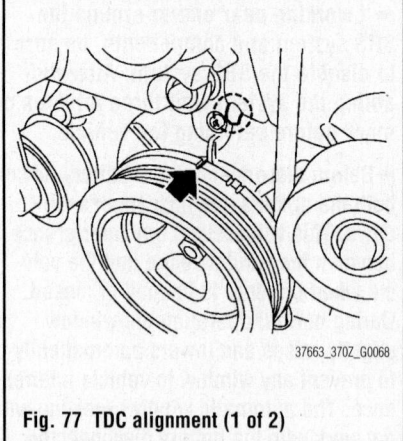

37663_370Z_G0068

Fig. 77 TDC alignment (1 of 2)

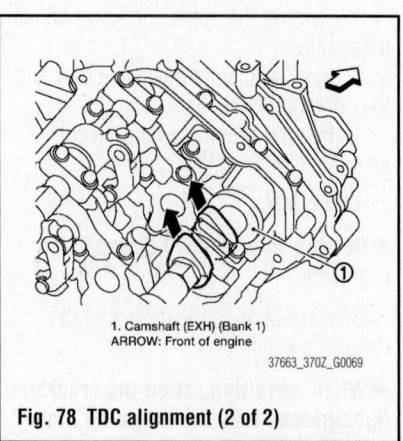

1. Camshaft (EXH) (Bank 1)
ARROW: Front of engine

37663_370Z_G0069

Fig. 78 TDC alignment (2 of 2)

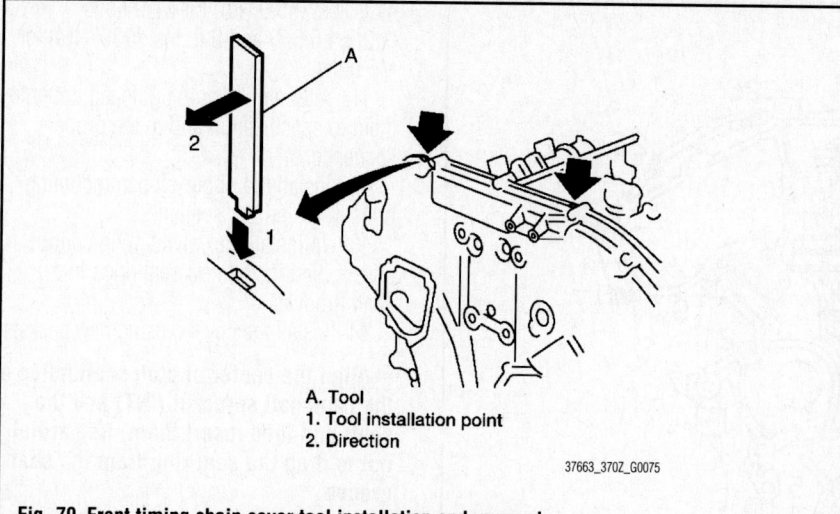

A. Tool
1. Tool installation point
2. Direction

37663_370Z_G0075

Fig. 79 Front timing chain cover tool installation and removal

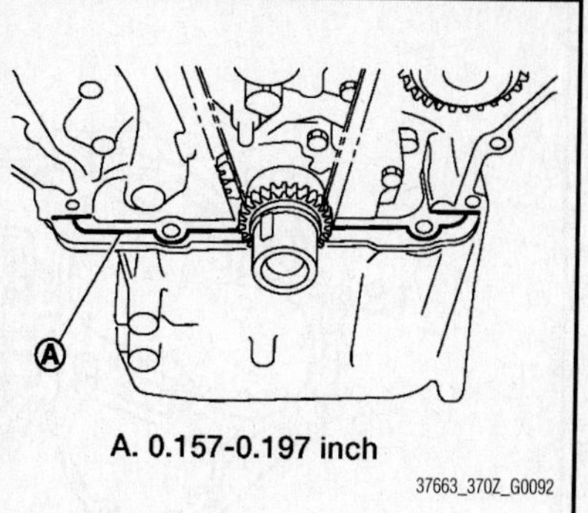

pressor from the bracket. Secure it to the side. Do not discharge the refrigerant or disconnect the refrigerant hoses.

18. Remove the power steering fluid pump from the bracket with the hoses connected, secure it to the side. Remove the power steering oil pump bracket.

19. Remove the idler pulley, drive belt auto tensioner and bracket.

20. Remove the alternator and alternator bracket.

21. Remove the front water outlet.

22. Remove the camshaft position sensor.

➡**Do not drop the sensor. Never disassemble it. Never allow metal powder to adhere to the magnetic portion of the sensor. Never store the sensor where it is exposed to magnetism.**

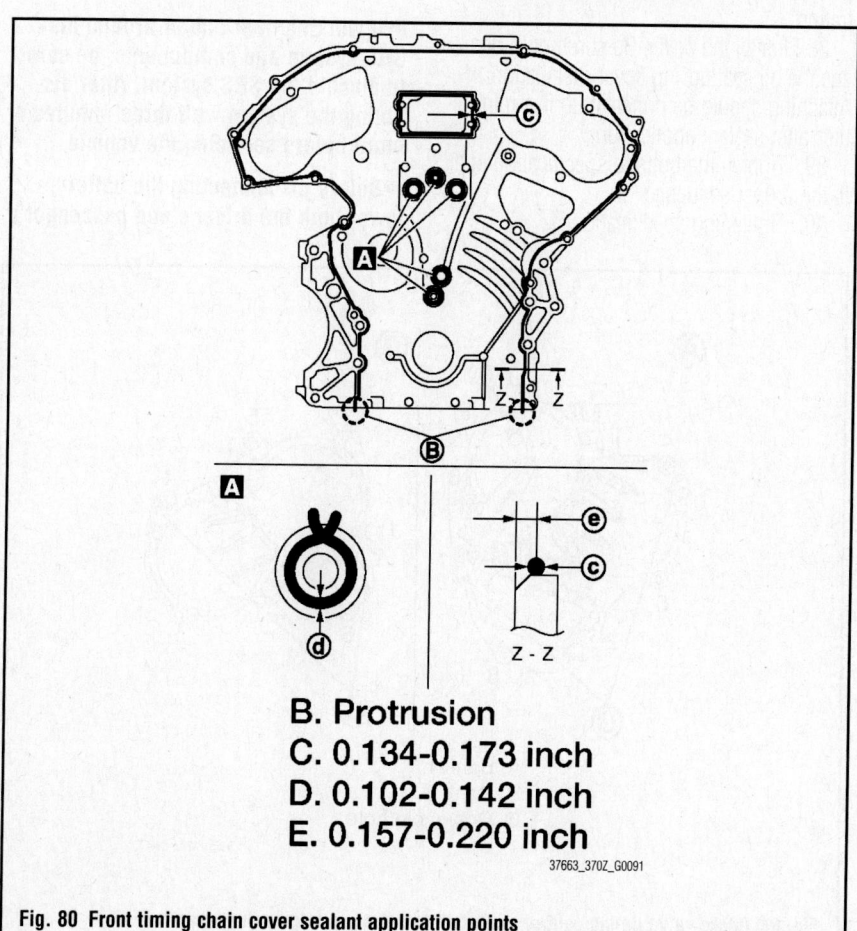

B. Protrusion
C. 0.134-0.173 inch
D. 0.102-0.142 inch
E. 0.157-0.220 inch

37663_370Z_G0091

Fig. 80 Front timing chain cover sealant application points

A. 0.157-0.197 inch

37663_370Z_G0092

Fig. 81 Oil pan (upper) sealant application points

23. To remove the intake valve timing control covers and gasket, disconnect the intake valve timing control solenoid valve harness connector. Loosen the mounting bolts in the reverse order of the tightening sequence.

➡**The shaft is internally jointed with the camshaft sprocket (INT) center hole. When removing, keep it horizontal until it is completely disconnected.**

24. The shaft is engaged with the camshaft sprocket (INT) center hole on the inside. Pull straight out so that it does not tilt until the joint is disengaged.

25. Remove the intake valve timing control solenoid valve, if necessary.

➡**This valve is not reusable. Never remove it unless required.**

26. Remove the valve covers.

27. To position the engine to TDC on the compression stroke, rotate the crankshaft pulley clockwise to align the timing mark (grooved line without color) with the timing indicator. Check that the exhaust cam noses on No. 1 cylinder (engine front side of bank 1) is located as shown in the illustration. If not turn the crankshaft on complete revolution and align.

28. Remove the crankshaft pulley.

29. Remove the lower oil pan.

30. Loosen the mounting bolts in the front of the upper oil pan in the reverse order of the installation sequence.

31. To remove the case cover, loosen the mounting bolts in the reverse order of the installation sequence.

32. Insert a suitable tool (KV10111100) into the notch at the top of the front timing chain cover, as shown in the illustration. Pry off the case by moving the tool, as shown in the illustration.

➡**Never use a screwdriver or similar item. After removal handle the cover carefully so it does not tilt, cant or warp under load.**

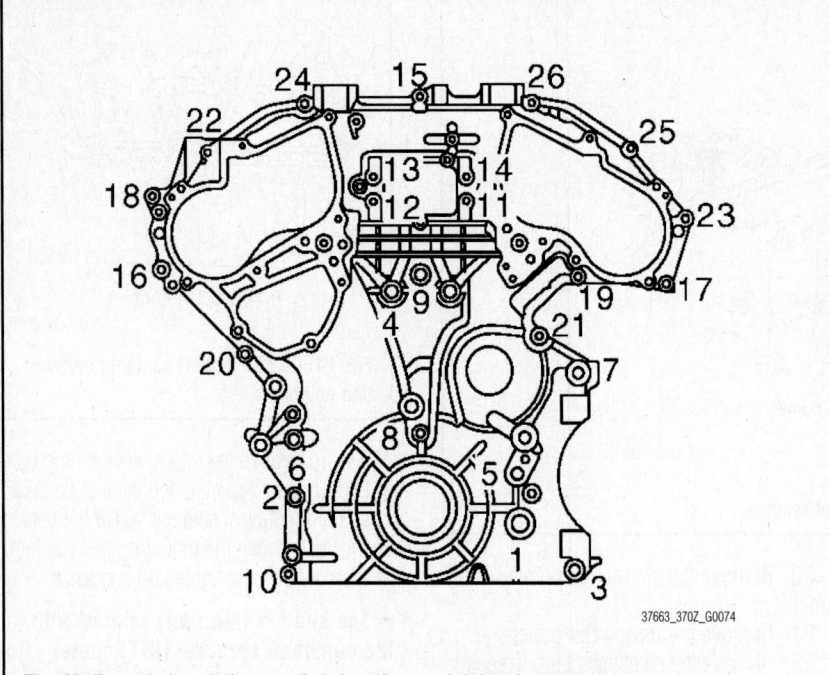

Fig. 82 Front timing chain cover bolt locations and tightening sequence

37663_370Z_G0074

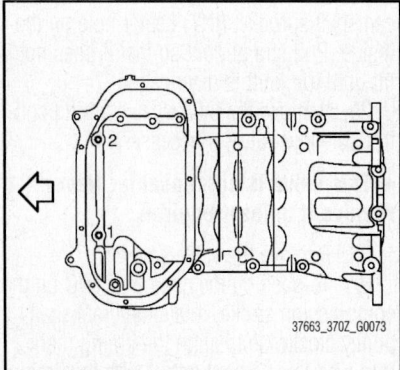

37663_370Z_G0073

Fig. 83 Upper oil pan front bolt locations and tightening sequence

33. Using the proper tool, remove the front case seal, as required.

To install:

➡**Be sure to use new fasteners, as required.**

34. Install new O-rings on the rear timing chain case cover.

➡**Be sure that the O-rings remain in place during installation to the rear case cover.**

35. Install a new oil seal in the front timing chain cover. Coat the seal with clean engine oil before installation. Press fit the seal into position until it becomes flush with the front timing chain case end face. Check that the garter spring is in position and that the seal lip is not inverted.

36. Apply a continuous bead of sealant as indicated in the illustration.

37. Apply sealant to the top surface of the upper oil pan, as indicated in the illustration.

38. Install the cover. Be sure not to damage the oil seal during cover installation. Attaching should be done within five minutes after sealant application.

39. Tighten the bolts to specification and in the proper sequence.

40. Tightening specification is

41 ft. lbs. (55.0 Nm) for M10 bolts (1,2,3,4,5,6,7) and 9 ft. lbs. (12.7 NM) for M6 bolts.

41. After all bolts are tightened, retighten them to specification and in the proper sequence.

42. Install the upper oil pan mounting bolts in the proper sequence.

43. To install the valve timing control covers, first install new seal rings in the shaft grooves.

44. Install the covers, using new gaskets.

➡**Align the center of both shaft holes of the camshaft sprocket (INT) and the shaft and then insert them. Be careful not to drop the seal ring from the shaft groove.**

45. Tighten the mounting bolts in the sequence shown in the illustration.

46. Continue the installation in the reverse order of the removal procedure.

TIMING CHAIN & SPROCKETS

REMOVAL & INSTALLATION

See Figures 85 through 100.

1. Before servicing the vehicle, refer to the Precautions Section.

➡**If working near and/or around the SRS system and components, be sure to disable the SRS system. After disabling the system wait three minutes or more before servicing the vehicle.**

➡**Before disconnecting the battery, lower both the driver's and passenger's**

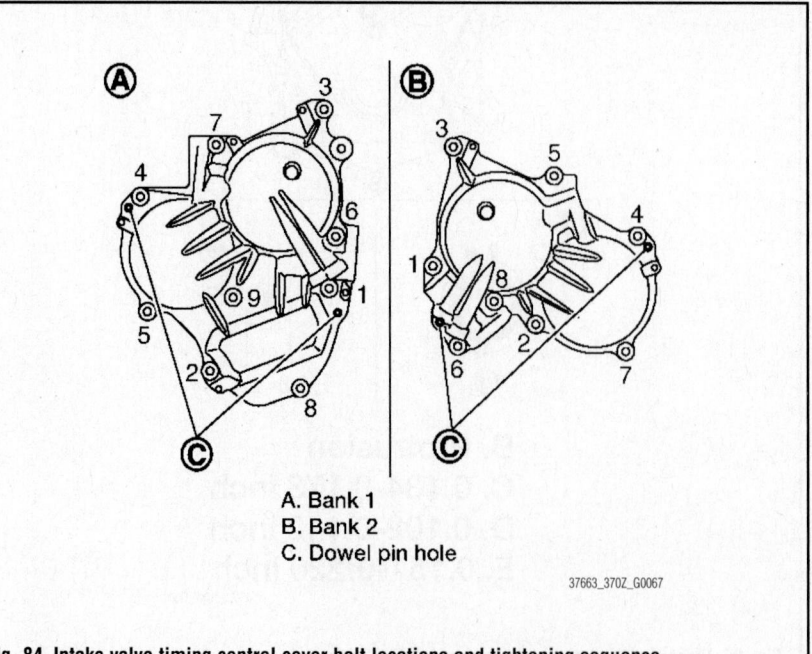

A. Bank 1
B. Bank 2
C. Dowel pin hole

37663_370Z_G0067

Fig. 84 Intake valve timing control cover bolt locations and tightening sequence

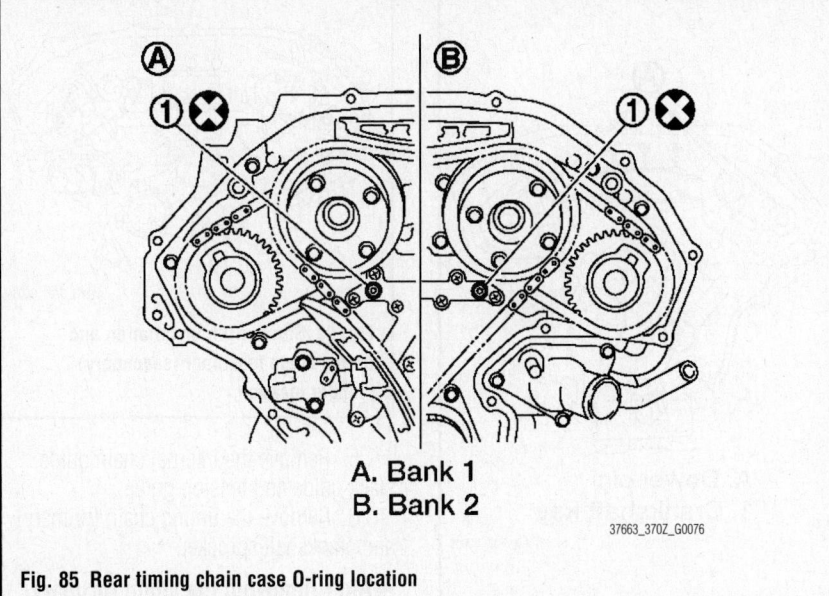

A. Bank 1
B. Bank 2

37663_370Z_G0076

Fig. 85 Rear timing chain case O-ring location

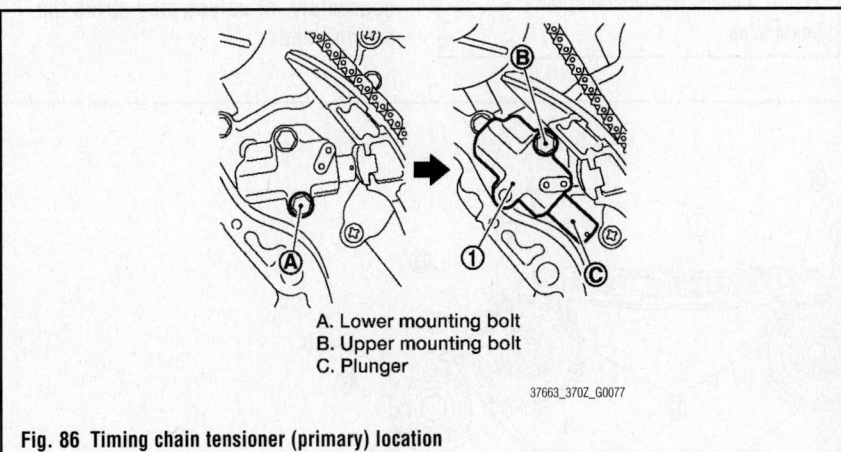

A. Lower mounting bolt
B. Upper mounting bolt
C. Plunger

37663_370Z_G0077

Fig. 86 Timing chain tensioner (primary) location

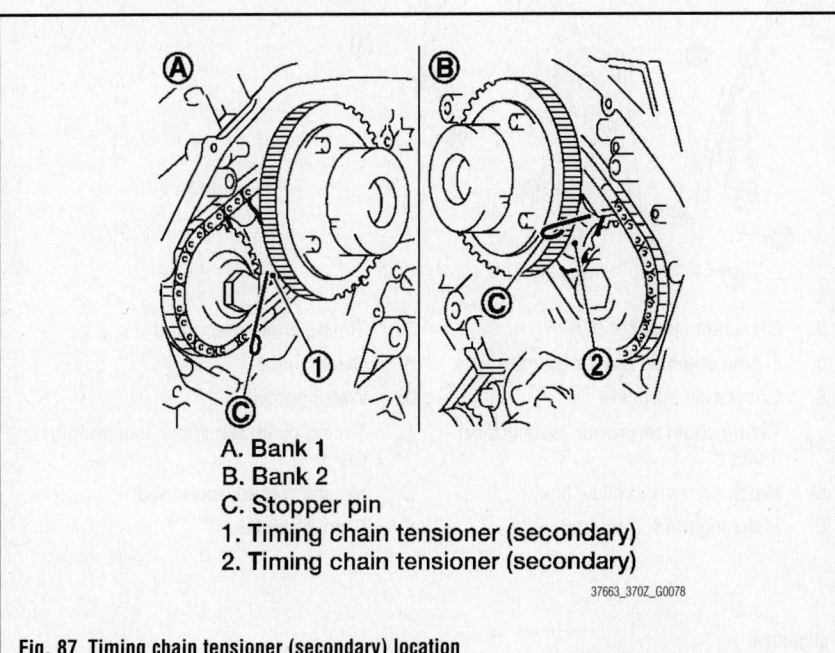

A. Bank 1
B. Bank 2
C. Stopper pin
1. Timing chain tensioner (secondary)
2. Timing chain tensioner (secondary)

37663_370Z_G0078

Fig. 87 Timing chain tensioner (secondary) location

windows. This will prevent any interference between the window edge and the vehicle when the door is opened or closed. During normal operation the window slightly raises and lowers automatically to prevent any window to vehicle interference. The automatic window function will not work with the battery disconnected.

2. Remove the timing chain cover.

3. Remove the O-ring from the rear timing chain case.

4. To remove the timing chain tensioner (primary), remove the lower mounting bolt. Loosen the upper mounting bolt slowly and then turn the timing chain tensioner (primary) on the upper mounting bolt so that the plunger is fully expanded. Remove the upper mounting bolt and then remove the timing chain tensioner (primary).

➡Even if the plunger is fully expanded, it does not drop from the body of the timing chain tensioner (primary).

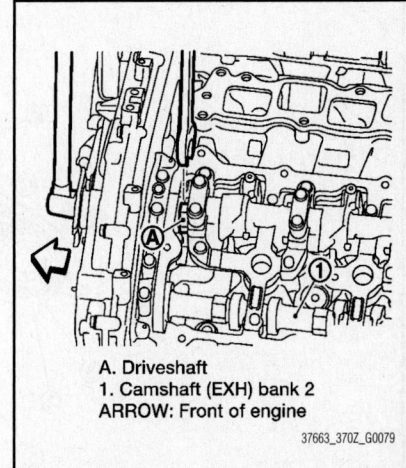

A. Driveshaft
1. Camshaft (EXH) bank 2
ARROW: Front of engine

37663_370Z_G0079

Fig. 88 Camshaft sprocket (INT) removal

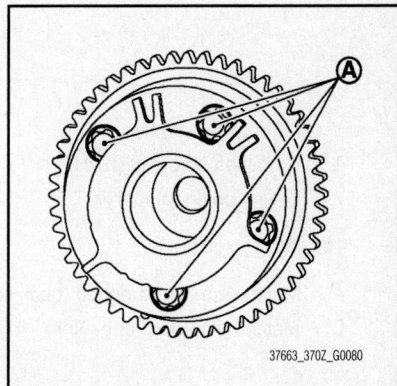

37663_370Z_G0080

Fig. 89 Camshaft sprocket (INT) bolt "A" location

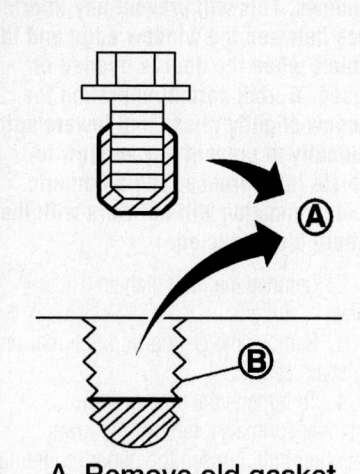

A. Remove old gasket
   that is stuck
B. Bolt hole

37663_370Z_G0081

**Fig. 90 Removing gasket material from
bolt hole**

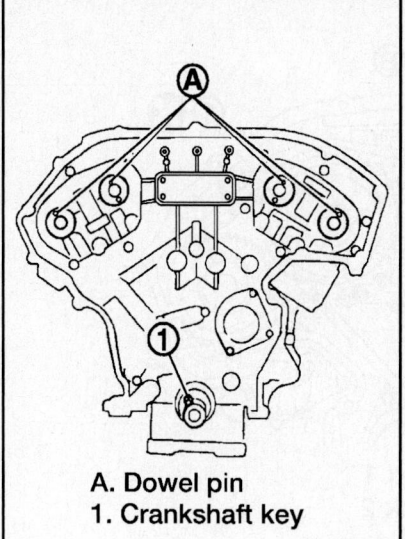

A. Dowel pin
1. Crankshaft key

37663_370Z_G0083

**Fig. 92 Camshafts and crankshaft
positioning**

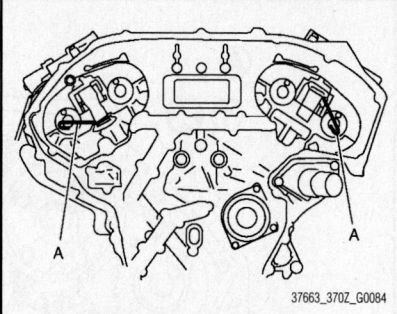

37663_370Z_G0084

**Fig. 93 Stopper pin installation and
timing chain tensioner (secondary)
plunger location**

5. Remove the internal chain guide,
slack guide and tension guide.
6. Remove the timing chain (primary)
and crankshaft sprocket.

➡**After removing the chain (primary),
never turn the crankshaft and camshaft
separately, or valves may strike the
piston heads.**

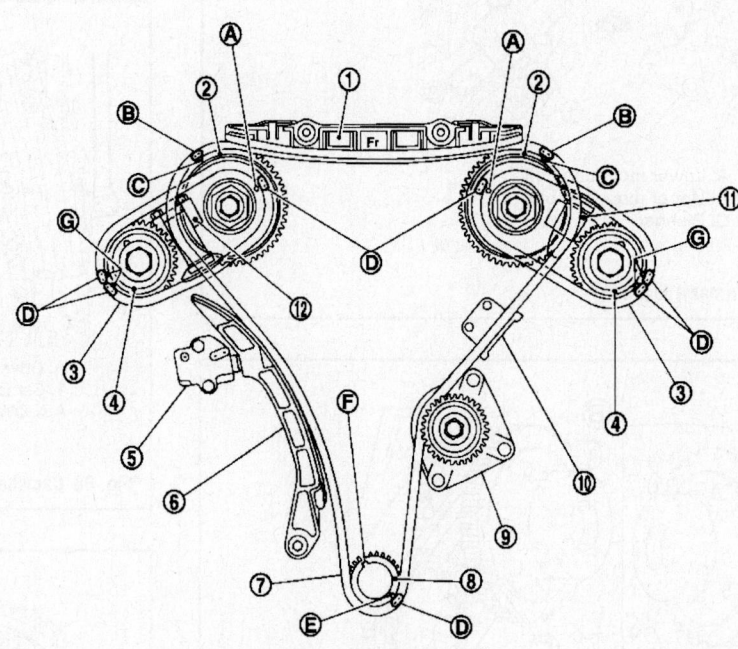

1.	Internal chain guide	2.	Camshaft sprocket (INT)	3.	Timing chain (secondary)
4.	Camshaft sprocket (EXH)	5.	Timing chain tensioner (primary)	6.	Slack guide
7.	Timing chain (primary)	8.	Crankshaft sprocket	9.	Water pump
10.	Tension guide	11.	Timing chain tensioner (secondary) (bank 2)	12.	Timing chain tensioner (secondary) (bank 1)
A.	Matching mark [punched (back side)]	B.	Matching mark (yellow link)	C.	Matching mark (punched)
D.	Matching mark (orange link)	E.	Matching mark (notched)	F.	Crankshaft key

37663_370Z_G0082

Fig. 91 Timing chains and sprockets (relationship) alignment

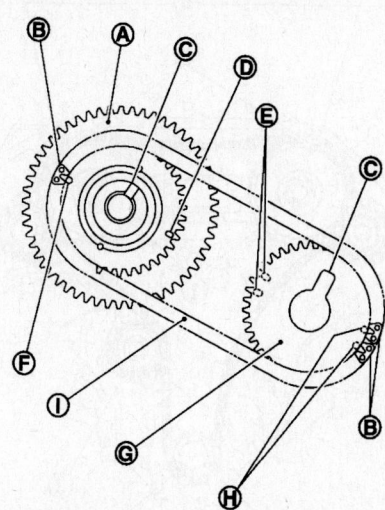

A. Camshaft sprocket (INT) back face
B. Orange link
C. Dowel groove
D. Matching mark (oval)
E. Matching mark (2 oval: on front face)
F. Matching mark (circle)
G. Camshaft sprocket (EXH) back face
H. Matching mark (2 oval: on front face)
I. Timing chain (secondary)

37663_370Z_G0085

Fig. 94 Timing chain tensioner (secondary) and camshaft sprockets—bank 1 rear view

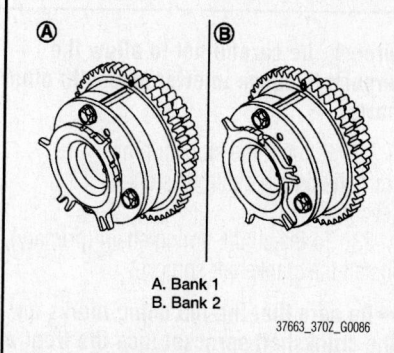

A. Bank 1
B. Bank 2

37663_370Z_G0086

Fig. 95 Camshaft sprocket (INT) signal plate orientation

7. To remove the timing chain (secondary) and camshaft sprockets attach a suitable stopper pin (0.020 inch hard metal) to the chain tensioners (secondary).

8. For removal of the chain tensioners (secondary) refer to the illustration.

➡Removing VVEL ladder assembly is required.

9. Remove the camshaft sprocket (EXH) mounting bolt.

➡Secure the hexagonal portion of the camshaft (EXH) using a wrench to loosen mounting bolt. Never loosen the mounting bolt by securing anything other than camshaft (EXH) hexagonal portion or with tensioning the timing chain.

10. Remove the camshaft sprocket (INT) mounting bolt.

➡Secure the hexagonal portion (located between journal No. 1 and journal No. 2) of the driveshaft using a wrench to loosen the mounting bolt. Never loosen the mounting bolt by securing anything other than the driveshaft hexagonal portion or with tensioning the timing chain. When holding the hexagonal part of the driveshaft on the intake side with a wrench, be careful not to allow the wrench to cause interference with other parts.

➡Never disassemble the camshaft sprocket (INT). Never loosen bolts (A) as shown in the illustration.

11. Remove the timing chain (secondary) along with the camshaft sprockets.

12. Remove all traces of gasket material from the front and rear timing chain covers. Be sure to remove all material from the bolt holes and threads. See illustration

To install:

➡Be sure to use new fasteners, as required.

➡See illustration that shows the relationship between the matching mark on each timing chain and that on the corresponding sprocket with components installed.

13. Check that the dowel pin and crankshaft key are located as shown in the illustration (engine at TDC on the compression stroke).

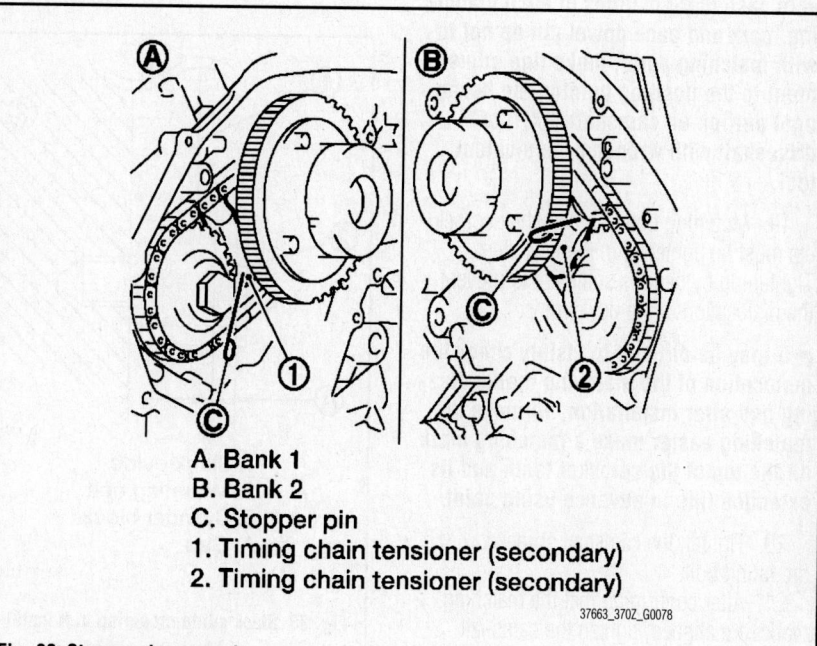

A. Bank 1
B. Bank 2
C. Stopper pin
1. Timing chain tensioner (secondary)
2. Timing chain tensioner (secondary)

37663_370Z_G0078

Fig. 96 Stopper pin removal

➡Though the camshaft does not stop at the position as shown, for placement of cam noses, it is generally accepted that the camshaft is placed in the same direction as that of the illustration.

➡Matching marks between the chain and sprockets slip easily. Confirm all matching mark positions repeatedly during the installation process.

14. To install the timing chains (secondary) and camshaft sprockets, push the plunger of the timing chain tensioner (secondary) and keep it pressed in with a stopper pin.

15. Install the timing chains (secondary) and camshaft sprockets. See illustration.

➡Illustration shows bank 1 (rear view)

16. Align the matching marks on the chain (secondary) (orange link) with the ones on the intake and exhaust camshaft sprockets (punched), and install them.

➡Matching marks for camshaft sprockets (INT) are on the back side of the camshaft sprockets (secondary). There are two types of matching marks, the circle and the oval. They should be used for bank 1 (circle) and bank 2 (oval) respectively.

17. Shape (orientation of signal plate) of camshaft sprocket (INT) varies depending on the bank position. See illustration.

18. Align dowel pin camshafts with the pin groove on sprockets and install them.

➡In case that positions of each matching mark and each dowel pin do not fit with matching parts, make fine adjustment to the position holding the hexagonal portion on camshaft (EXH) or driveshaft with wrench or equivalent tool.

19. Mounting bolts for camshaft sprockets must be tightened in the next step. Tightening by hand is sufficient to prevent the dislocation of the dowel pins.

➡It may be difficult to visibly check the dislocation of the matching marks during and after installation. To make the matching easier make a matching mark on the top of the sprocket teeth and its extended line in advance using paint.

20. Tighten the camshaft sprocket (EXH) mounting bolt.

21. After confirming that the matching marks are aligned, tighten the camshaft sprocket (INT) mounting bolt. Secure the

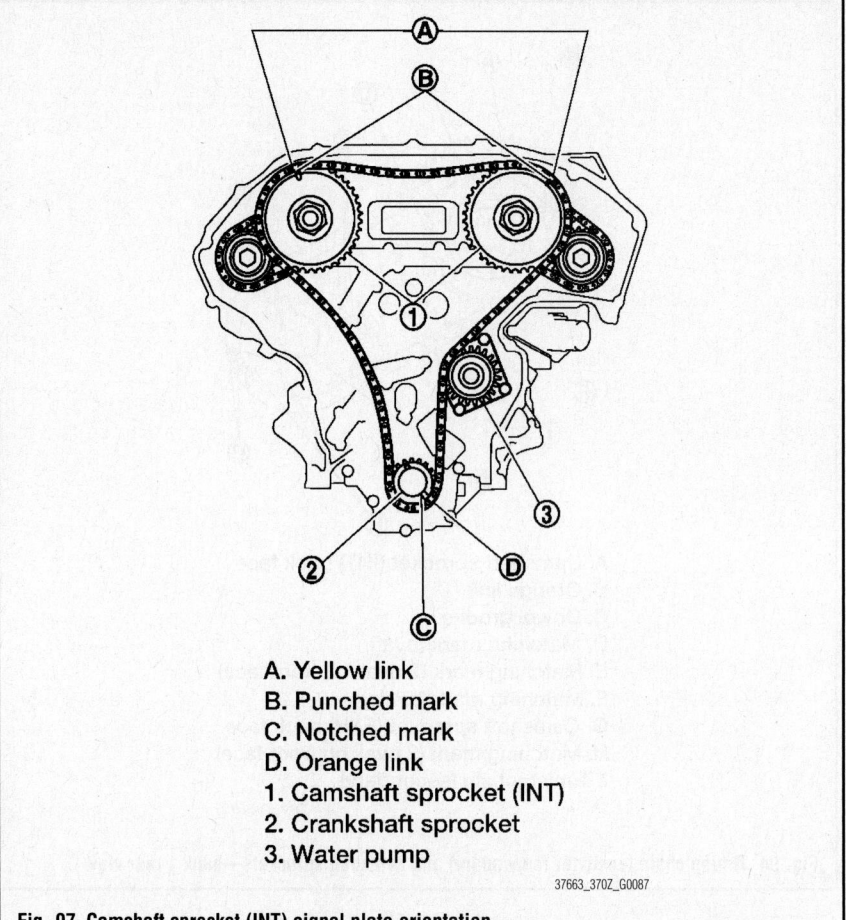

A. Yellow link
B. Punched mark
C. Notched mark
D. Orange link
1. Camshaft sprocket (INT)
2. Crankshaft sprocket
3. Water pump

37663_370Z_G0087

Fig. 97 Camshaft sprocket (INT) signal plate orientation

hexagonal portion (located between journal No. 1 and Journal No. 2) of the driveshaft using a wrench to tighten the mounting bolt.

➡ When holding the hexagonal part of the driveshaft on the intake side with a

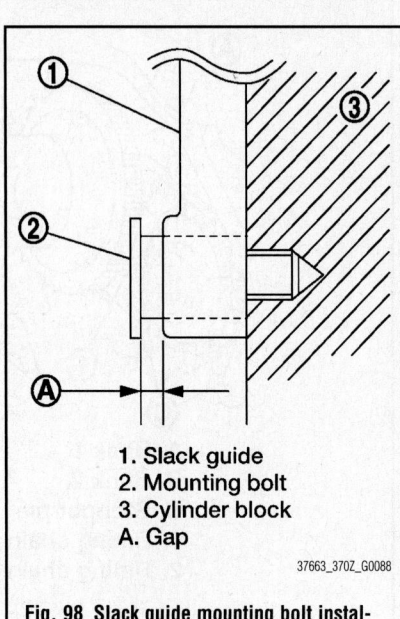

1. Slack guide
2. Mounting bolt
3. Cylinder block
A. Gap

37663_370Z_G0088

Fig. 98 Slack guide mounting bolt installation

wrench, be careful not to allow the wrench to cause interference with other parts.

22. Pull out the stopper pins from the timing chain tensioners (secondary).

23. To install the timing chain (primary), install the crankshaft sprocket.

➡Be sure that the matching marks on the crankshaft sprocket face the front of the engine.

24. Install the timing chain (primary) so that the matching mark (punched) on the camshaft sprocket (INT) is aligned with the yellow link on the timing chain while the matching mark (notched) on the crankshaft sprocket is aligned with the orange link on the timing chain. See illustration.

➡When it is difficult to align the matching marks of the timing chain (primary) with each sprocket, gradually turn the driveshaft sing a wrench on the hexagonal portion to align it with the matching marks.

25. Install the internal chain guide, slack guide and tension guide.

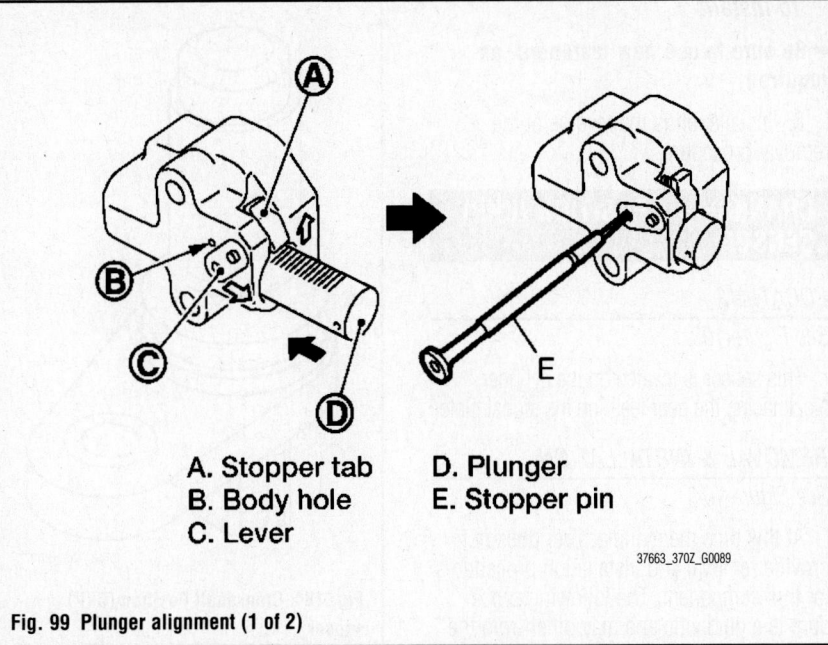

A. Stopper tab D. Plunger
B. Body hole E. Stopper pin
C. Lever

37663_370Z_G0089

Fig. 99 Plunger alignment (1 of 2)

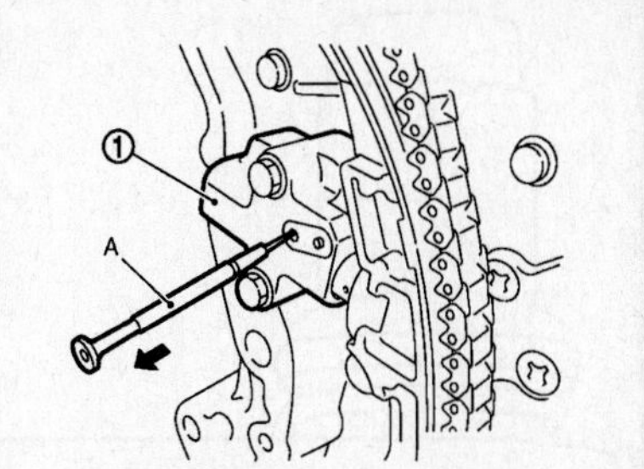

1. Timing chain tensioner (primary)
A. Stopper pin

37663_370Z_G0090

Fig. 100 Plunger alignment (2 of 2)

➡ Never overtighten the slack guide mounting bolt. It is normal for a gap to exist under the bolt seats when mounting bolts are tightened to specification.

26. To install the timing chain tensioner (primary) pull the plunger stopper tab up (or turn lever downward) so as to remove the plunger stopper tab from the ratchet of the plunger. Note that the plunger stopper tab and lever are synchronized.

27. Push the plunger into the inside of the tensioner body. Hold the plunger in the fully compressed position by engaging plunger stopper tab with the tip of the ratchet.

28. To secure the lever, insert the stopper pin through the hole of the lever into the tensioner body hole. The lever parts and the plunger stopper tab are synchronized, therefore the plunger is secured under this condition.

➡ The illustration shows a suitable tool of 0.047 inch diameter being used as a stopper pin.

29. Pull out the stopper pin after installing and release the plunger.

30. Check again that the matching marks on the sprockets and the timing chain have not slipped out of alignment.

31. Install the timing chain cover.

ENGINE PERFORMANCE & EMISSION CONTROLS

CAMSHAFT POSITION (CMP) SENSOR

LOCATION

See Figure 101.

Refer to the accompanying illustration.

REMOVAL & INSTALLATION

See Figure 102.

At this time the manufacturer does not provide removal and installation procedures for this component. The following procedure is a guideline and may differ from the vehicle you are servicing.

1. Before servicing the vehicle, refer to the Precautions Section.

➡ If working near and/or around the SRS system and components, be sure to disable the SRS system. After disabling the system wait three minutes or more before servicing the vehicle.

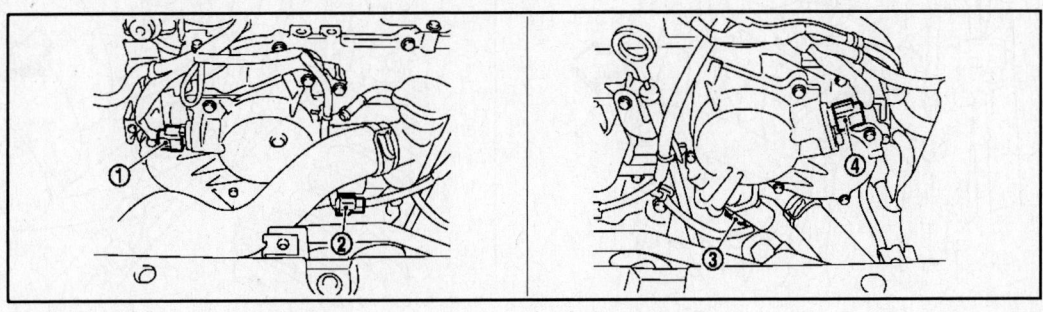

1. Camshaft position sensor (PHASE) (bank 1)
2. Intake valve timing control solenoid valve (bank 1) harness connector
3. Intake valve timing control solenoid valve (bank 2) harness connector
4. Camshaft position sensor (PHASE) (bank 2)

37663_370Z_G0135

Fig. 101 Camshaft Position (CMP) sensor location

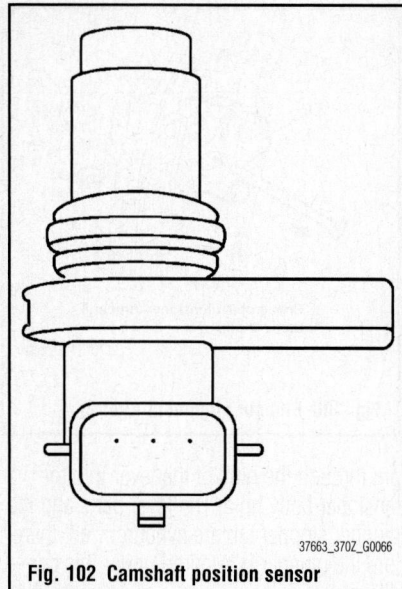

Fig. 102 Camshaft position sensor

➡Before disconnecting the battery, lower both the driver's and passenger's windows. This will prevent any interference between the window edge and the vehicle when the door is opened or closed. During normal operation the window slightly raises and lowers automatically to prevent any window to vehicle interference. The automatic window function will not work with the battery disconnected.

2. Remove the necessary components in order to gain access to the sensor.
3. Disconnect the electrical connector.
4. Remove the retaining bolt.
5. Remove the component from the vehicle.

To install:

➡Be sure to use new fasteners, as required.

6. Installation is the reverse of the removal procedure.

CRANKSHAFT POSITION (CKP) SENSOR

LOCATION

See Figure 103.

This sensor is located on the cylinder block facing the gear teeth on the signal plate.

REMOVAL & INSTALLATION

See Figure 104.

At this time the manufacturer does not provide removal and installation procedures for this component. The following procedure is a guideline and may differ from the vehicle you are servicing.

1. Before servicing the vehicle, refer to the Precautions Section.

➡If working near and/or around the SRS system and components, be sure to disable the SRS system. After disabling the system wait three minutes or more before servicing the vehicle.

➡Before disconnecting the battery, lower both the driver's and passenger's windows. This will prevent any interference between the window edge and the vehicle when the door is opened or closed. During normal operation the window slightly raises and lowers automatically to prevent any window to

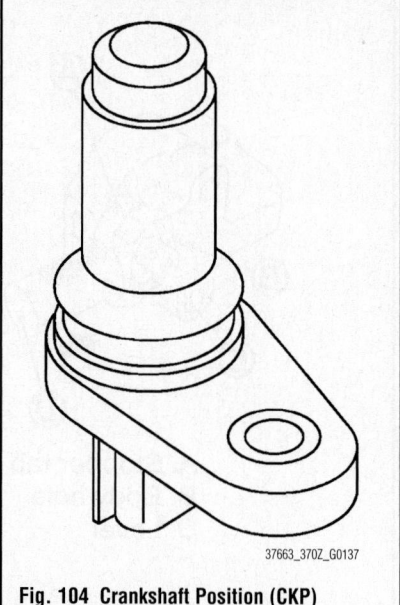

Fig. 104 Crankshaft Position (CKP) sensor

vehicle interference. The automatic window function will not work with the battery disconnected.

2. Remove the necessary components in order to gain access to the sensor.
3. Disconnect the electrical connector.
4. Remove the retaining bolt.
5. Remove the component from the vehicle.

To install:

➡Be sure to use new fasteners, as required.

6. Installation is the reverse of the removal procedure.

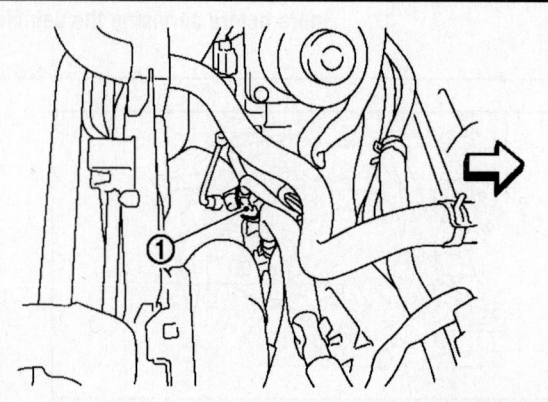

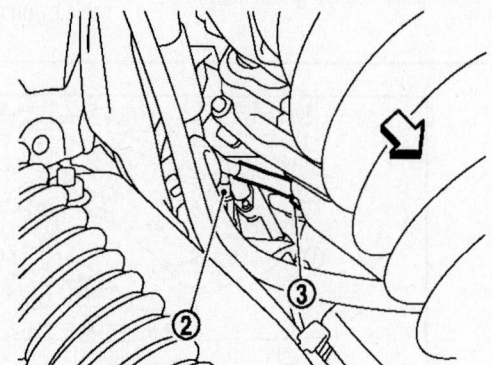

⇦ : Vehicle front

1. Engine coolant temperature sensor
2. A/F sensor 1 (bank 1)
3. Crankshaft position sensor (POS)

Fig. 103 Crankshaft Position (CKP) sensor location

ELECTRONIC CONTROL MODULE (ECM)

LOCATION

The ECM is located on the passenger's side of the vehicle at the kick panel.

REMOVAL & INSTALLATION

1. Before servicing the vehicle, refer to the Precautions Section.

➡**If working near and/or around the SRS system and components, be sure to disable the SRS system. After disabling the system wait three minutes or more before servicing the vehicle.**

➡**Before disconnecting the battery, lower both the driver's and passenger's windows. This will prevent any interference between the window edge and the vehicle when the door is opened or closed. During normal operation the window slightly raises and lowers automatically to prevent any window to vehicle interference. The automatic window function will not work with the battery disconnected.**

2. Remove glove box.
3. Disconnect electrical connectors.
4. Remove the component.

To install:

➡**Be sure to use new fasteners, as required.**

5. Installation is the reverse of the removal procedure.

ENGINE COOLANT TEMPERATURE (ECT) SENSOR

LOCATION

See Figure 105.

Refer to the accompanying illustration.

REMOVAL & INSTALLATION

See Figure 106.

At this time the manufacturer does not provide removal and installation procedures for this component. The following procedure is a guideline and may differ from the vehicle you are servicing.

1. Before servicing the vehicle, refer to the Precautions Section.

➡**If working near and/or around the SRS system and components, be sure to disable the SRS system. After disabling the system wait three minutes or more before servicing the vehicle.**

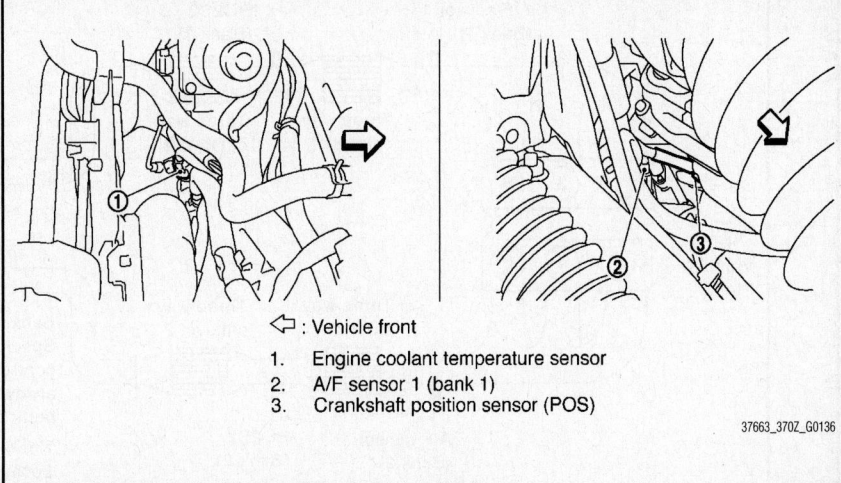

◁ : Vehicle front
1. Engine coolant temperature sensor
2. A/F sensor 1 (bank 1)
3. Crankshaft position sensor (POS)

37663_370Z_G0136

Fig. 105 Engine coolant temperature sensor location

➡**Before disconnecting the battery, lower both the driver's and passenger's windows. This will prevent any interference between the window edge and the vehicle when the door is opened or closed. During normal operation the window slightly raises and lowers automatically to prevent any window to vehicle interference. The automatic window function will not work with the battery disconnected.**

2. Drain the cooling system. Properly dispose of used engine coolant.
3. Remove the necessary components to gain access to the sensor.
4. Disconnect the electrical sensor.
5. Remove the sensor from its mounting.

To install:

➡**Be sure to use new fasteners, as required.**

6. Installation is the reverse of the removal procedure.

HEATED OXYGEN SENSOR (HO2S)

LOCATION

See Figure 107.

This sensor is located in the exhaust stream, after the catalytic converter. Two sensors are used, one for each cylinder bank.

REMOVAL & INSTALLATION

See Figure 108.

At this time the manufacturer does not provide removal and installation procedures for this component. The following procedure is a guideline and may differ from the vehicle you are servicing.

1. Before servicing the vehicle, refer to the Precautions Section.

➡**If working near and/or around the SRS system and components, be sure to disable the SRS system. After disabling the system wait three minutes or more before servicing the vehicle.**

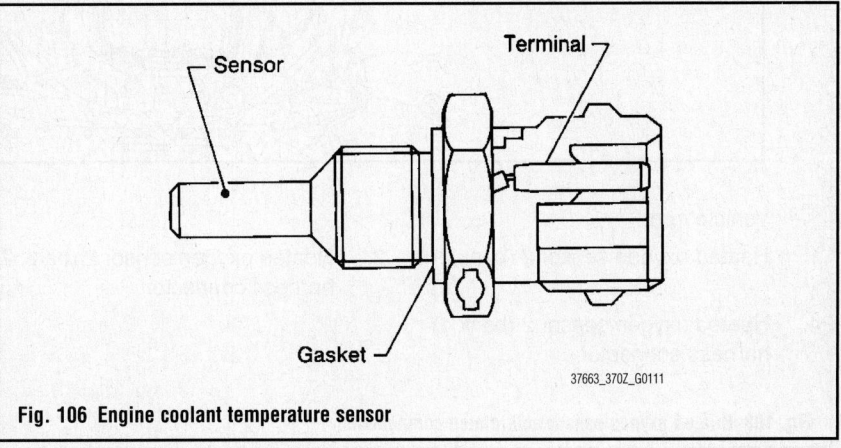

37663_370Z_G0111

Fig. 106 Engine coolant temperature sensor

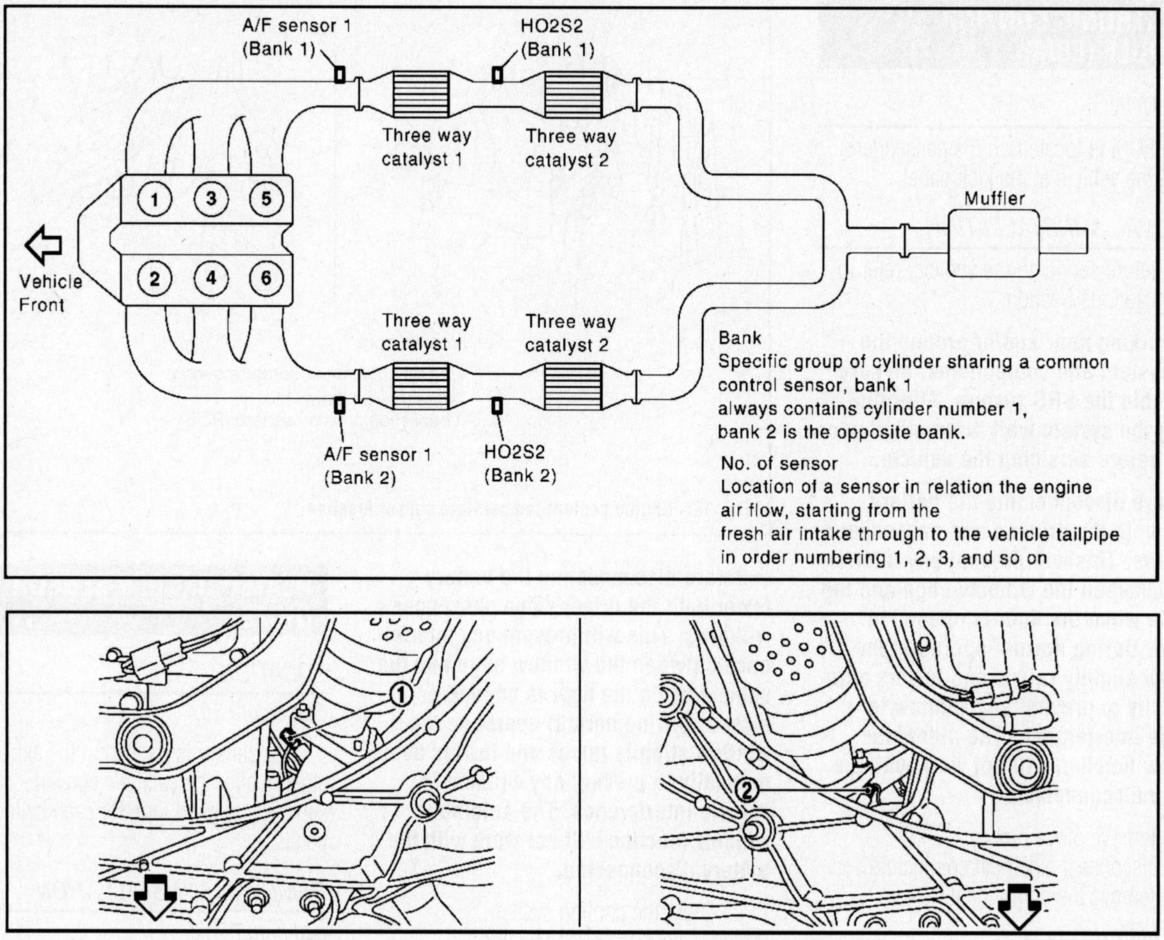

A/F sensor 1
(Bank 1)

HO2S2
(Bank 1)

Three way
catalyst 1

Three way
catalyst 2

Muffler

1 3 5

2 4 6

Vehicle
Front

Three way
catalyst 1

Three way
catalyst 2

A/F sensor 1
(Bank 2)

HO2S2
(Bank 2)

Bank
Specific group of cylinder sharing a common
control sensor, bank 1
always contains cylinder number 1,
bank 2 is the opposite bank.

No. of sensor
Location of a sensor in relation the engine
air flow, starting from the
fresh air intake through to the vehicle tailpipe
in order numbering 1, 2, 3, and so on

⬅ : Vehicle front

1.   A/F sensor 1 (bank 1)   2.   A/F sensor 1 (bank 2)

37663_370Z_G0132

Fig. 107 Oxygen sensors and related components

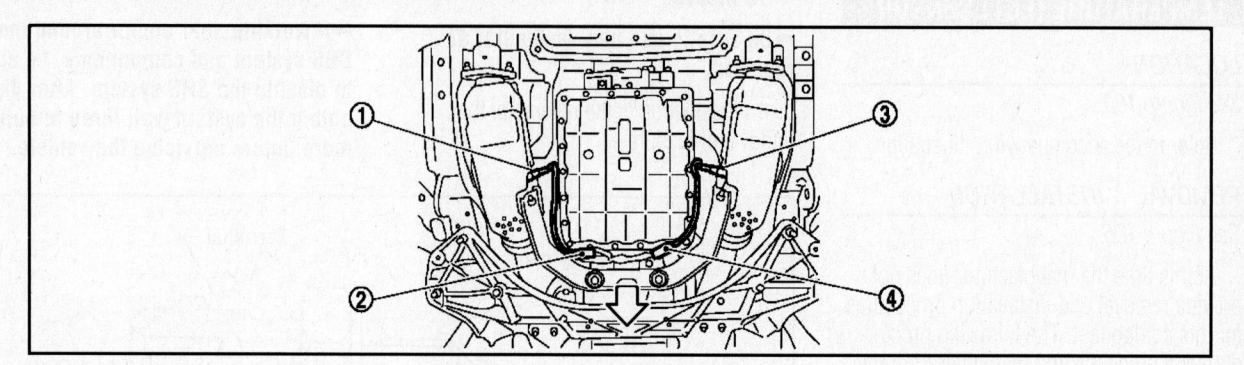

⬅ : Vehicle front

1.   Heated oxygen sensor 2 (bank 2)   2.   Heated oxygen sensor 2 (bank 2)   3.   Heated oxygen sensor 2 (bank 1)
                                              harness connector

4.   Heated oxygen sensor 2 (bank 1)
     harness connector

37663_370Z_G0134

Fig. 108 Heated oxygen sensor and related components

→Before disconnecting the battery, lower both the driver's and passenger's windows. This will prevent any interference between the window edge and the vehicle when the door is opened or closed. During normal operation the window slightly raises and lowers automatically to prevent any window to vehicle interference. The automatic window function will not work with the battery disconnected.

2. Raise and safely support the vehicle.
3. Remove the undercover.
4. Disconnect the harness connector.
5. Remove the sensor from its mounting.

→Discard the sensor if it has been dropped.

To install:

→Be sure to use new fasteners, as required.

6. Installation is the reverse of the removal procedure.
7. Before installing a new sensor coat the threads with an approved anti-seize lubricant.

INTAKE AIR TEMPERATURE (IAT) SENSOR

LOCATION

This sensor is built into the MAF sensor and is services with that component.

REMOVAL & INSTALLATION
See Figure 109.

At this time the manufacturer does not provide removal and installation procedures for this component. The following procedure is a guideline and may differ from the vehicle you are servicing.

1. Before servicing the vehicle, refer to the Precautions Section.

→If working near and/or around the SRS system and components, be sure to disable the SRS system. After disabling the system wait three minutes or more before servicing the vehicle.

→Before disconnecting the battery, lower both the driver's and passenger's windows. This will prevent any interference between the window edge and the

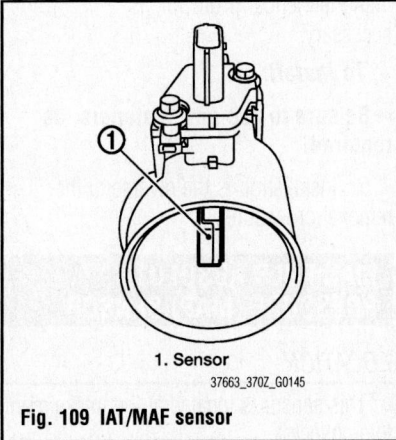

1. Sensor

37663_370Z_G0145

Fig. 109 IAT/MAF sensor

vehicle when the door is opened or closed. During normal operation the window slightly raises and lowers automatically to prevent any window to vehicle interference. The automatic window function will not work with the battery disconnected.

2. Disconnect the harness connector from the IAT/MAF sensor.
3. Disconnect the tube clamp at the electric throttle control actuator and at the fresh air intake tube.
4. Remove air cleaner to electric throttle control actuator tube, air cleaner case (upper) with the IAT/MAF sensor attached.
5. Remove IAT/MAF sensor from air cleaner case (upper), as necessary.
6. Remove resonator in the fender, lifting left fender protector, as necessary.

To install:

→Be sure to use new fasteners, as required.

7. Installation is the reverse of the removal procedure.

KNOCK SENSOR (KS)

LOCATION

The knock sensors are located under the intake manifold. There are two of them.

REMOVAL & INSTALLATION
See Figure 110.

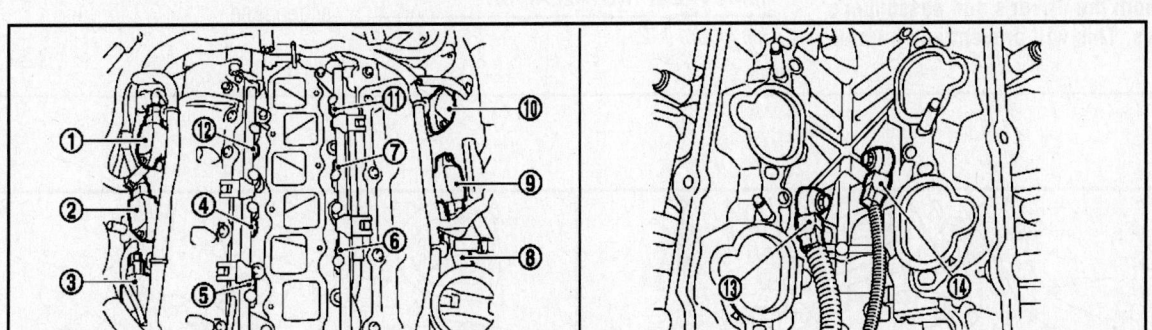

1. Ignition coil No.5 (with power transistor)	2. Ignition coil No.3 (with power transistor)	3. Ignition coil No.1 (with power transistor)
4. Fuel injector No.3	5. Fuel injector No.1	6. Fuel injector No.2
7. Fuel injector No.4	8. Ignition coil No.2 (with power transistor)	9. Ignition coil No.4 (with power transistor)
10. Ignition coil No.6 (with power transistor)	11. Fuel injector No.6	12. Fuel injector No.5
13. Knock sensor (bank 1)	14. Knock sensor (bank 2)	

37663_370Z_G0144

Fig. 110 Knock sensors and related components

At this time the manufacturer does not provide removal and installation procedures for this component, refer to the illustration as required.

MALFUNCTION INDICATOR LIGHT (MIL)

RESET PROCEDURE

Clearing diagnostic trouble codes resets the MIL.

MASS AIR FLOW (MAF) SENSOR

LOCATION

This sensor is placed in the stream of the air intake. It is incorporated along with the IAT sensor.

REMOVAL & INSTALLATION

See Figure 111.

At this time the manufacturer does not provide removal and installation procedures for this component. The following procedure is a guideline and may differ from the vehicle you are servicing.

1. Before servicing the vehicle, refer to the Precautions Section.

➡ If working near and/or around the SRS system and components, be sure to disable the SRS system. After disabling the system wait three minutes or more before servicing the vehicle.

➡ Before disconnecting the battery, lower both the driver's and passenger's windows. This will prevent any interference between the window edge and the vehicle when the door is opened or closed. During normal operation the window slightly raises and lowers automatically to prevent any window to vehicle interference. The automatic window function will not work with the battery disconnected.

2. Disconnect the harness connector from the MAF/IAT sensor.
3. Disconnect the tube clamp at the electric throttle control actuator and at the fresh air intake tube.
4. Remove air cleaner to electric throttle control actuator tube, air cleaner case (upper) with the MAF/IAT sensor attached.
5. Remove MAF/IAT sensor from air cleaner case (upper), as necessary.
6. Remove resonator in the fender, lifting left fender protector, as necessary.

To install:

➡ Be sure to use new fasteners, as required.

7. Installation is the reverse of the removal procedure.

MANIFOLD ABSOLUTE PRESSURE (MAP) SENSOR

LOCATION

This sensor is located at the intake manifold collector.

REMOVAL & INSTALLATION

See Figure 112.

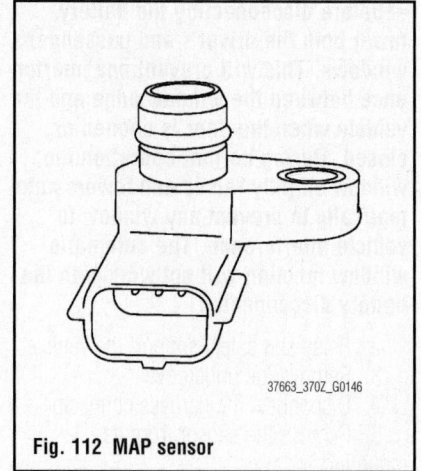

37663_370Z_G0146

Fig. 112 MAP sensor

At this time the manufacturer does not provide removal and installation procedures for this component, refer to the illustration as required.

THROTTLE POSITION SENSOR (TPS)

LOCATION

See Figure 113.

This sensor is located in the main hose connecting the intake manifold collector . There are two of these sensors.

REMOVAL & INSTALLATION

See Figures 35 and 113.

At this time the manufacturer does not provide removal and installation procedures for this component, refer to the illustration as required.

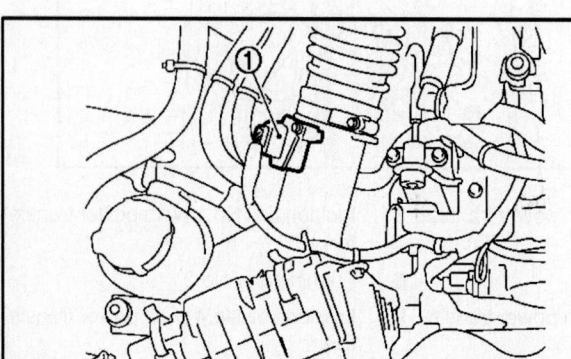
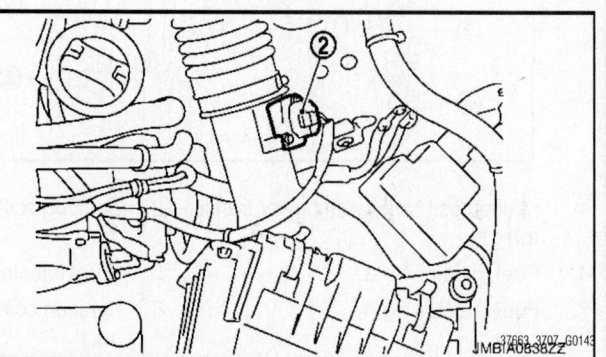

JMBIA0838ZZ

1. Mass air flow sensor (with intake air temperature sensor) (bank 1)
2. Mass air flow sensor (with intake air temperature sensor) (bank 2)

Fig. 114 MAF sensor location

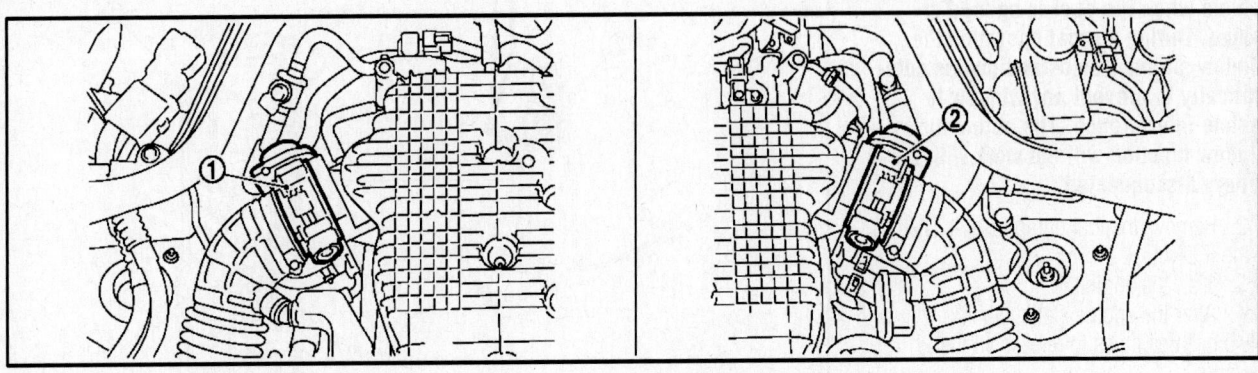

1. Electric throttle control actuator (bank 1)

2. Electric throttle control actuator (bank 2)

37663_370Z_G0148

Fig. 113 TPS sensor (electronic throttle control) location

FUEL

GASOLINE FUEL INJECTION SYSTEM

FUEL SYSTEM SERVICE PRECAUTIONS

Safety is the most important factor when performing not only fuel system maintenance, but any type of maintenance. Failure to conduct maintenance and repairs in a safe manner may result in serious personal injury or death. Work on a vehicle's fuel system components can be accomplished safely and effectively by adhering to the following rules and guidelines.

• To avoid the possibility of fire and personal injury, always disconnect the negative battery cable unless the repair or test procedure requires that battery voltage be applied.

• Always relieve the fuel system pressure prior to disconnecting any fuel system component (injector, fuel rail, pressure regulator, etc.) fitting or fuel line connection. Exercise extreme caution whenever relieving fuel system pressure to avoid exposing skin, face and eyes to fuel spray. Please be advised that fuel under pressure may penetrate the skin or any part of the body that it contacts.

• Always place a shop towel or cloth around the fitting or connection prior to loosening to absorb any excess fuel due to spillage. Ensure that all fuel spillage is

quickly removed from engine surfaces. Ensure that all fuel-soaked cloths or towels are deposited into a flame-proof waste container with a lid.

• Always keep a dry chemical (Class B) fire extinguisher near the work area.

• Do not allow fuel spray or fuel vapors to come into contact with a spark or open flame.

• Always use a second wrench when loosening or tightening fuel line connection fittings. This will prevent unnecessary stress and torsion on fuel piping. Always follow the proper torque specifications.

• Always replace worn fuel fitting O-

rings with new ones. Do not substitute fuel hose where rigid pipe is installed.

FUEL SYSTEM PRESSURE

RELIEVING

See Figure 114.

1. Before servicing the vehicle, refer to the Precautions Section.

→**If working near and/or around the SRS system and components, be sure to disable the SRS system. After disabling the system wait three minutes or more before servicing the vehicle.**

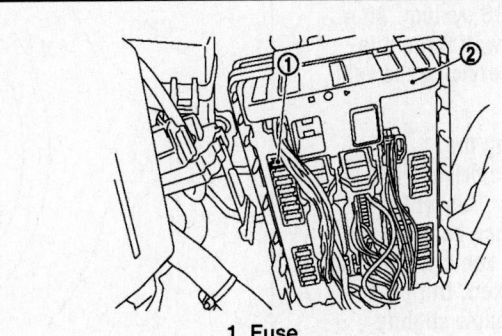

1. Fuse
2. IPDM E/R

37663_370Z_G0093

Fig. 114 Fuel pump fuse location

→Before disconnecting the battery, lower both the driver's and passenger's windows. This will prevent any interference between the window edge and the vehicle when the door is opened or closed. During normal operation the window slightly raises and lowers automatically to prevent any window to vehicle interference. The automatic window function will not work with the battery disconnected.

2. Remove the fuel pump fuse, located in the IPDM E/R.
3. Start the engine.
4. After the engine stalls, crank it two or three times to release all fuel pressure.
5. Turn the ignition switch off.
6. Reinstall the fuel pump fuse after servicing the fuel system.

FUEL FILTER

REMOVAL & INSTALLATION

The fuel filter is attached to the fuel pump assembly. The fuel pump must be removed before the filter can be serviced.

FUEL LEVEL SENDING UNIT

LOCATION

The fuel level sending unit is attached to the fuel pump assembly. The fuel level sending unit is located in the fuel tank.

REMOVAL & INSTALLATION

See Figure 115.

1. Before servicing the vehicle, refer to the Precautions Section.

→If working near and/or around the SRS system and components, be sure to disable the SRS system. After disabling the system wait three minutes or more before servicing the vehicle.

→Before disconnecting the battery, lower both the driver's and passenger's windows. This will prevent any interference between the window edge and the vehicle when the door is opened or closed. During normal operation the window slightly raises and lowers automatically to prevent any window to vehicle interference. The automatic window function

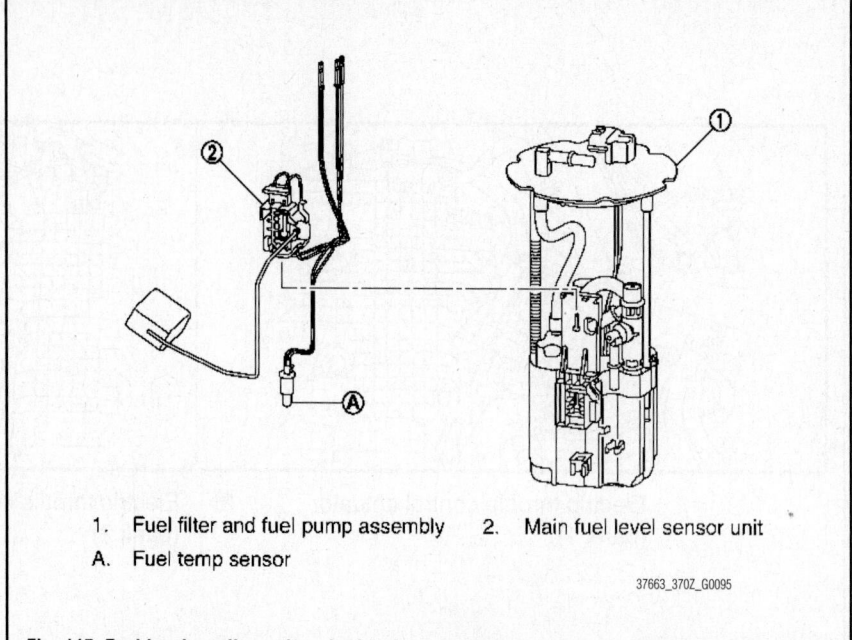

1. Fuel filter and fuel pump assembly
A. Fuel temp sensor
2. Main fuel level sensor unit

37663_370Z_G0095

Fig. 115 Fuel level sending unit and related components

will not work with the battery disconnected.

2. Properly relieve the fuel system pressure.
3. Remove the fuel pump module.
4. Discard the gasket.
5. Service the fuel level sending unit, as required.

→This component cannot be disassembled and should be replaced as a unit.

To install:

→Be sure to use new fasteners, as required.

6. Installation is the reverse of the removal procedure.

FUEL PUMP MODULE

REMOVAL & INSTALLATION

See Figures 116 through 122.

1. Before servicing the vehicle, refer to the Precautions Section.

→If working near and/or around the SRS system and components, be sure to disable the SRS system. After disabling the system wait three minutes or more before servicing the vehicle.

→Before disconnecting the battery, lower both the driver's and passenger's windows. This will prevent any interference between the window edge and the

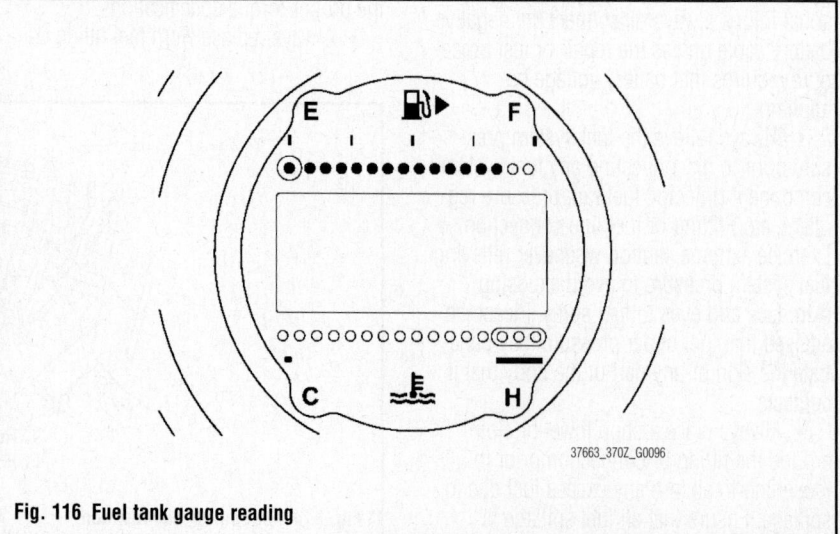

37663_370Z_G0096

Fig. 116 Fuel tank gauge reading

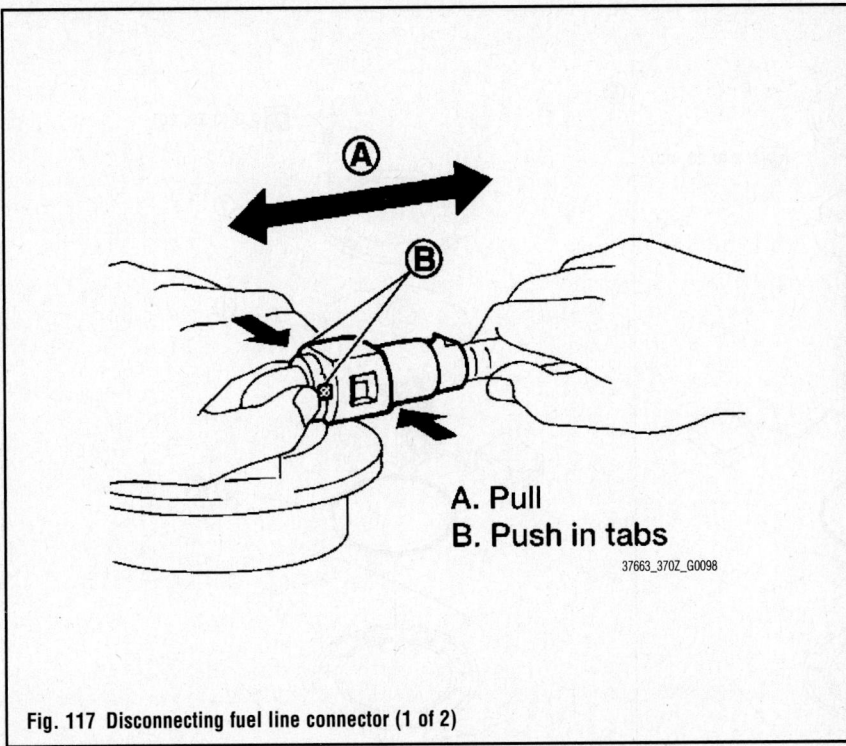

A. Pull
B. Push in tabs

37663_370Z_G0098

Fig. 117 Disconnecting fuel line connector (1 of 2)

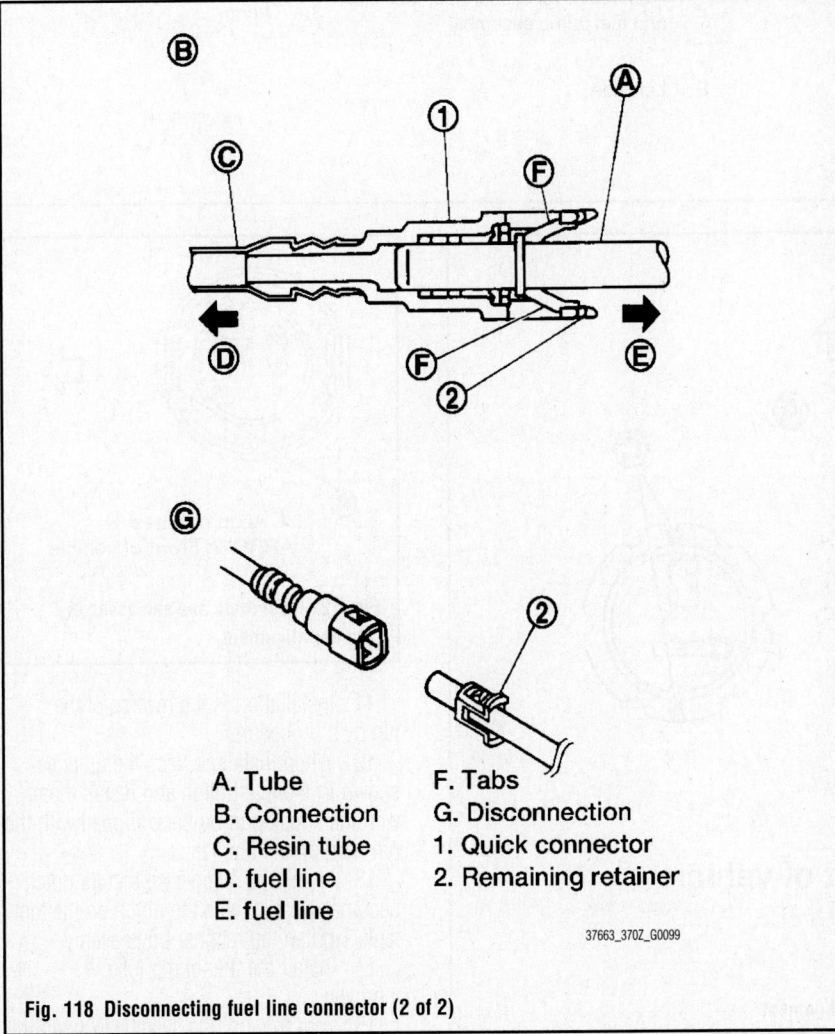

A. Tube
B. Connection
C. Resin tube
D. fuel line
E. fuel line
F. Tabs
G. Disconnection
1. Quick connector
2. Remaining retainer

37663_370Z_G0099

Fig. 118 Disconnecting fuel line connector (2 of 2)

vehicle when the door is opened or closed. During normal operation the window slightly raises and lowers automatically to prevent any window to vehicle interference. The automatic window function will not work with the battery disconnected.

2. Drain the fuel tank to an acceptable level. If the fuel level indicates more than the level shown in the illustration, full or almost full, drain the fuel from the tank until the fuel gauge indicates a level as shown in the illustration.

➡Because fuel will be spilled when removing the main and sub fuel level sensor units for the top of the fuel is above the main and sub fuel level sensor units installed surface. As a guide, fuel level becomes the position as shown in the illustration when approximately 3 3/8 gallons of fuel are removed from the tank.

3. Properly relieve the fuel system pressure.
4. Remove the fuel tank cap.
5. Remove the rear parcel shelf covers, both left and right.
6. Remove the inspection hole cover.

➡Right side for main fuel level sensor, fuel filter and fuel pump assembly. Left side for sub fuel level sensor unit.

7. Disconnect and fuel feed tube. Disconnect the harness connector.
8. Disconnect the quick connector. Hold the sides of the connector, push in the tabs and pull out the fuel feed tube.

➡The quick connector can be disconnected when the tabs are completely depressed. Never twist it more than necessary. Never use tools to disconnect the quick connector. Cover the fuel line openings to prevent dirt from entering the fuel system.

9. To remove the main fuel level sensor unit, fuel filter and fuel pump assembly, remove the retainer. Raise the unit and disconnect the quick connector.
10. To remove the sub fuel level sensor, remove the retainer. Raise the component and remove it.

To install:

➡Be sure to use new fasteners, as required.

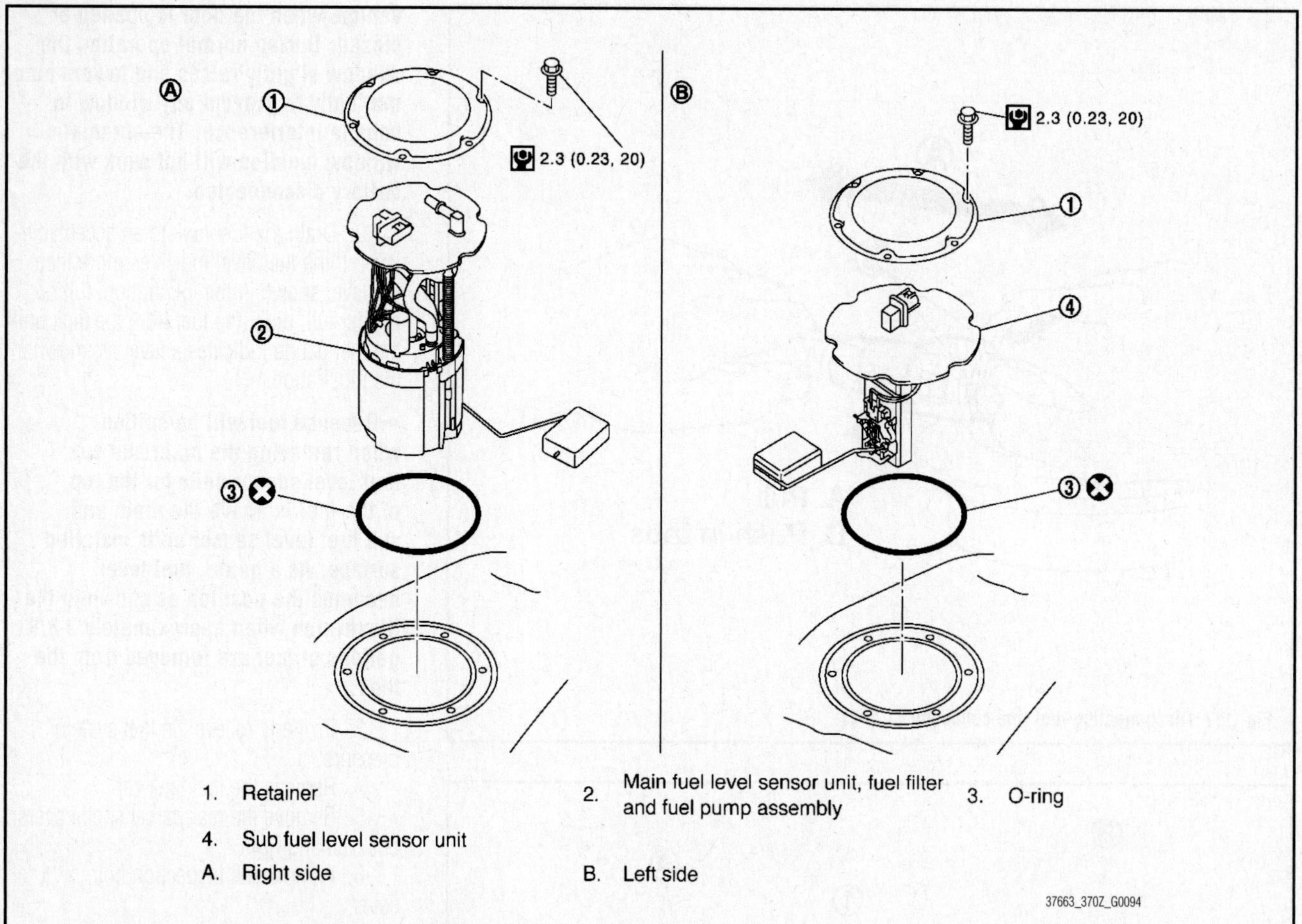

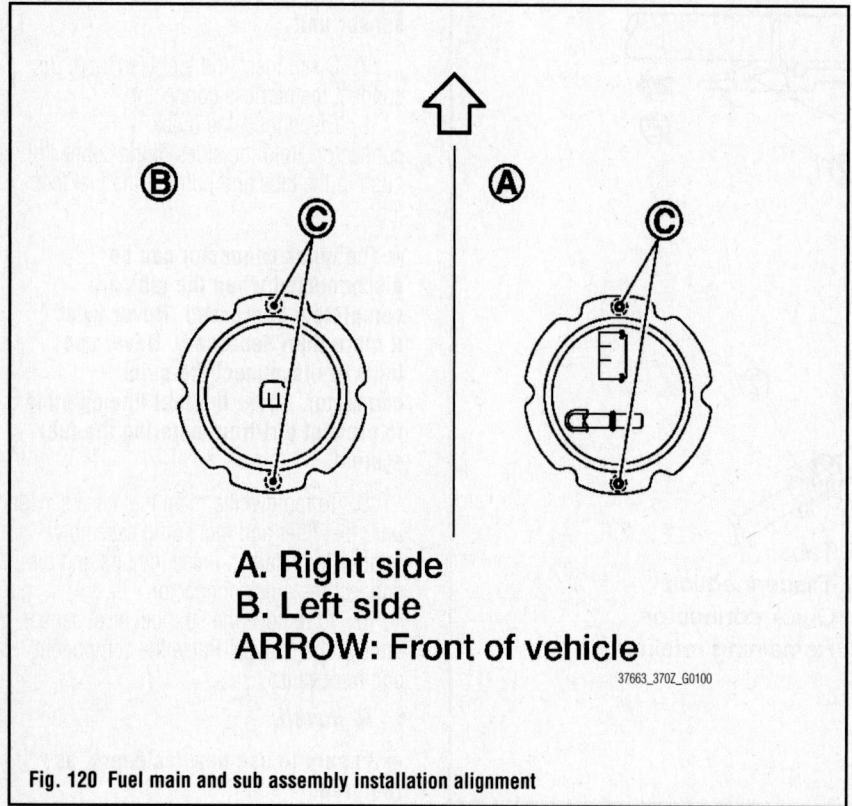

Ⓐ
①
2.3 (0.23, 20)
②
③
Ⓑ
2.3 (0.23, 20)
①
④
③
1. Retainer
2. Main fuel level sensor unit, fuel filter and fuel pump assembly
3. O-ring
4. Sub fuel level sensor unit
A. Right side
B. Left side
37663_370Z_G0094

Fig. 119 Fuel pump module and related components

Ⓑ
Ⓐ
Ⓒ
Ⓒ
A. Right side
B. Left side
ARROW: Front of vehicle
37663_370Z_G0100

Fig. 120 Fuel main and sub assembly installation alignment

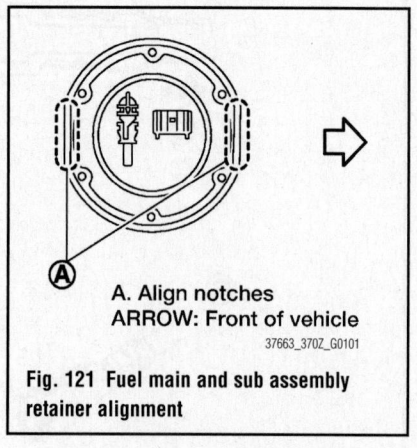

Ⓐ
A. Align notches
ARROW: Front of vehicle
37663_370Z_G0101

Fig. 121 Fuel main and sub assembly retainer alignment

11. Installation is the reverse of the removal procedure.
12. When installing, face the units as shown in the illustration and install them with the knock pun on back aligned with the pin hole on the fuel tank.
13. Install the retainer so that its notch becomes parallel with the notch on the fuel tank. Tighten the retainer bolts evenly.
14. Install the fuel pump fuse, if removed.
15. Turn the ignition switch ON (with the

engine stopped), check all connections for fuel leakage, correct as required.

16. Start the engine and let it idle, check for fuel leaks. Correct as required.

FUEL RAIL AND INJECTOR

REMOVAL & INSTALLATION

See Figures 122 through 126.

1. Before servicing the vehicle, refer to the Precautions Section.

➡**If working near and/or around the SRS system and components, be sure to disable the SRS system. After disabling the system wait three minutes or more before servicing the vehicle.**

➡**Before disconnecting the battery, lower both the driver's and passenger's windows. This will prevent any interference between the window edge and the vehicle when the door is opened or closed. During normal operation the window slightly raises and lowers automatically to prevent any window to vehicle interference. The automatic window function will not work with the battery disconnected.**

2. Properly relieve the fuel system pressure.

3. Disconnect the negative battery cable.

4. Remove the engine cover.

5. Remove the air cleaner assembly.

6. Remove the intake manifold collector.

7. Remove the fuel feed hose (with damper) from the fuel sub tube. Remove the harness bracket.

➡**There is no fuel return route. Plug the lines to prevent fuel leakage. Never separate the damper and the hose.**

➡**When separating the fuel feed hose (with damper) and centralized under floor piping connection, disconnect the quick connector, as shown in the illustrations. Disconnect the quick connector by using a quick connector tool, J-45488 or equivalent.**

8. To disconnect the quick connector from the centralized under floor piping, with the sleeve side of the quick connector release facing the quick connector, install the quick connector release onto the centralized floor piping. Insert the quick connector release into the quick connector, until the sleeve contacts and stops. Hold the quick connector and release on that position. Draw and pull out the quick connector straight from the centralized under floor piping. Never bend or twist the connection between the quick connector and fuel feed hose (with damper) during removal and/or installation. Be sure to have a catch pan available to catch spilled fuel. Cover the fuel line openings to prevent dirt from entering the fuel system.

➡**Inserting the quick connector release hard will not disconnect the quick connector. Hold the quick connector release where it contacts and goes no further.**

➡**Pull the quick connector holding as shown in the removal illustration. Never pull with lateral force. O-ring inside quick connector may be damaged.**

9. Remove the sub tube mounting bolt.

10. Disconnect the harness connector from the fuel injector.

11. Loosen the mounting bolts in the reverse order of the tightening sequence.

12. Remove the assembly from its mounting.

13. Remove the fuel injector from the rail, as required. Discard the O-rings.

14. Do not remove the fuel sub tube and fuel damper.

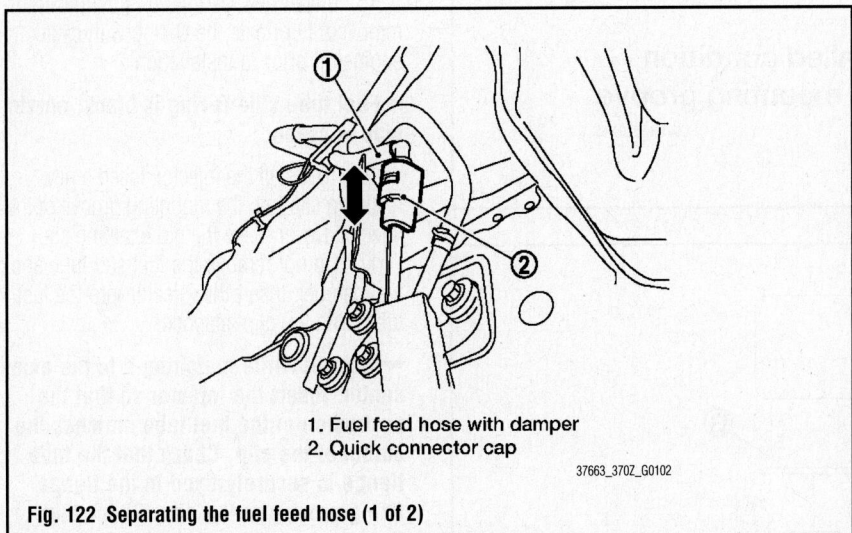

1. Fuel feed hose with damper
2. Quick connector cap

37663_370Z_G0102

Fig. 122 Separating the fuel feed hose (1 of 2)

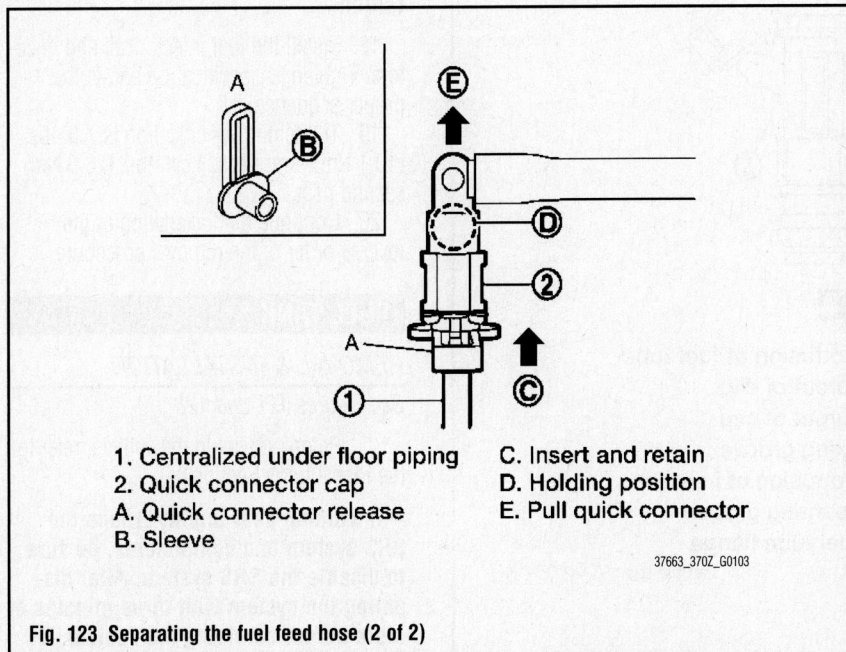

1. Centralized under floor piping
2. Quick connector cap
A. Quick connector release
B. Sleeve

C. Insert and retain
D. Holding position
E. Pull quick connector

37663_370Z_G0103

Fig. 123 Separating the fuel feed hose (2 of 2)

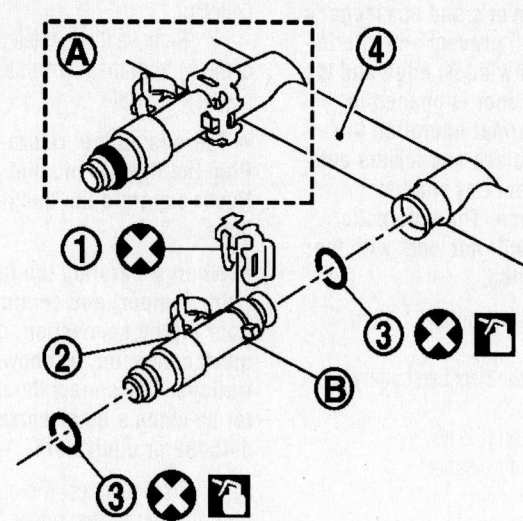

1. Retainer   4. Rail
2. Injector   A. Installed condition
3. O-ring   B. Clip mounting groove

37663_370Z_G0105

Fig. 124 Fuel injector and related components

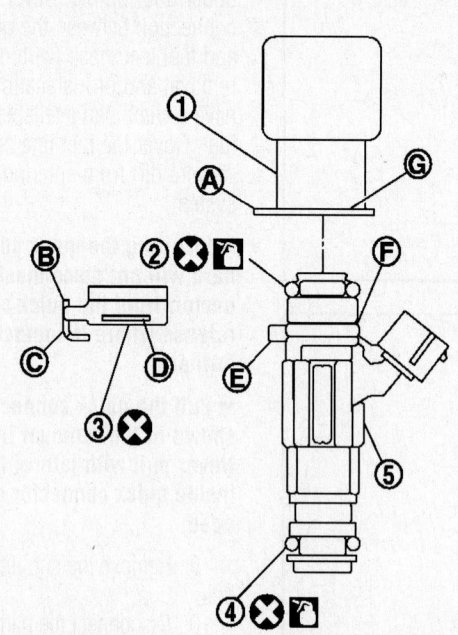

1. Fuel tube   A. Protrusion of fuel tube
2. Black O-ring   B. Cutout of clip
3. Clip   C. Cutout of clip
4. Green O-ring   D. Fixing groove
5. Injector   E. Protrusion of injector
                F. Mounting groove
                G. Fuel tube flange

37663_370Z_G0106

Fig. 125 Fuel injector installation

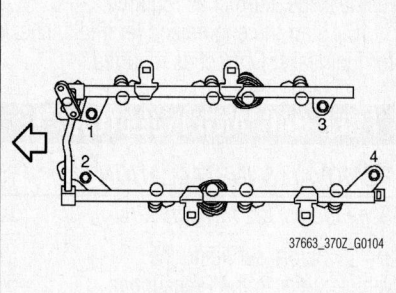

37663_370Z_G0104

Fig. 126 Fuel injector rail bolt locations and tightening sequence

To install:

➡ Be sure to use new fasteners, as required.

15. Install new O-rings on the injector, if removed. Lubricate the O-ring with clean engine oil prior to installation.

➡ Fuel tube side O-ring is black, nozzle side is green.

16. To install the injector, insert a new retaining clip into the mounting groove of the injector. Never reuse the old retaining clip.

17. Do not remove the fuel sub tube and fuel damper. Insert the injector into the fuel tube, with the clip attached.

➡ Insert it while matching it to the axial center. Insert the injector so that the protrusion of the fuel tube matches the cutout of the clip. Check that the tube flange is securely fixed in the flange fixing groove on the clip. Check that the injector does not rotate. See illustration.

18. Install the fuel injector rail and injectors. Tighten to specification and in the proper sequence.

19. Tightening specification is 7 ft. lbs. (10.1 Nm), first pass. 17 ft. lbs. (23.6 Nm), second pass.

20. Continue the installation in the reverse order of the removal procedure.

FUEL TANK

REMOVAL & INSTALLATION

See Figures 127 and 128.

1. Before servicing the vehicle, refer to the Precautions Section.

➡ If working near and/or around the SRS system and components, be sure to disable the SRS system. After disabling the system wait three minutes or more before servicing the vehicle.

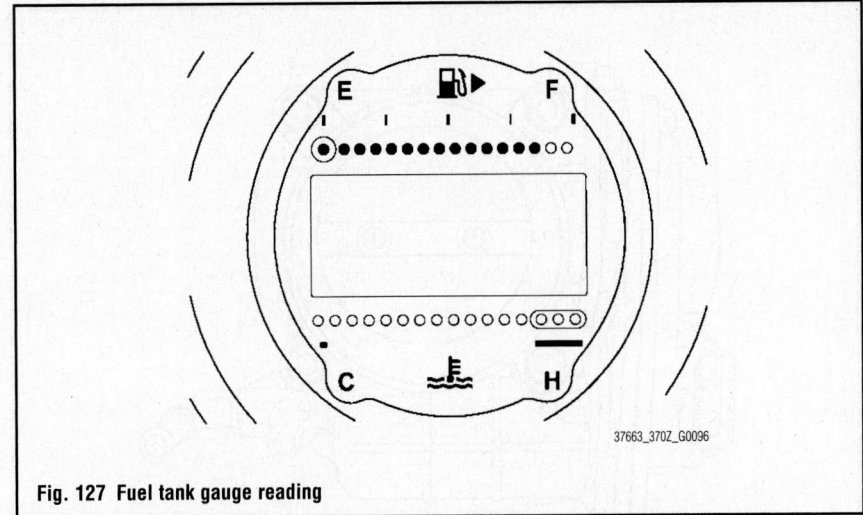

Fig. 127 Fuel tank gauge reading

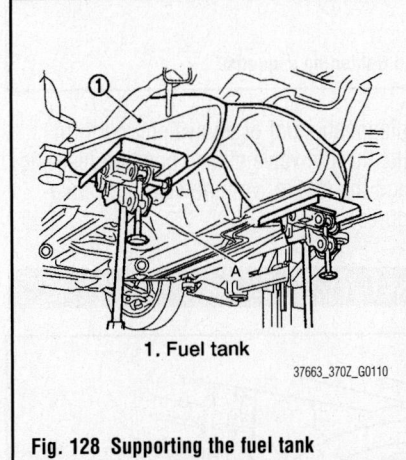

1. Fuel tank

Fig. 128 Supporting the fuel tank

➡**Before disconnecting the battery, lower both the driver's and passenger's windows. This will prevent any interference between the window edge and the vehicle when the door is opened or closed. During normal operation the window slightly raises and lowers automatically to prevent any window to vehicle interference. The automatic window function will not work with the battery disconnected.**

2. Drain the fuel tank to an acceptable level. If the fuel level indicates more than the level shown in the illustration, full or almost full, drain the fuel from the tank until the fuel gauge indicates a level as shown in the illustration.

➡**Because fuel will be spilled when removing the main and sub fuel level sensor units for the top of the fuel is above the main and sub fuel level sensor units installed surface. As a guide, fuel level becomes the position as shown in the illustration when approxi-**

mately 3 3/8 gallons of fuel are removed from the tank.

3. Properly relieve the fuel system pressure.
4. Remove the fuel tank cap.
5. Remove the rear parcel shelf covers, both left and right.
6. Remove the inspection hole cover.

➡**Right side for main fuel level sensor, fuel filter and fuel pump assembly. Left side for sub fuel level sensor unit.**

7. Disconnect and fuel feed tube. Disconnect the harness connector.
8. Disconnect the quick connector. Hold the sides of the connector, push in the tabs and pull out the fuel feed tube.

➡**The quick connector can be disconnected when the tabs are completely depressed. Never twist it more than necessary. Never use tools to disconnect the quick connector. Cover the fuel line openings to prevent dirt from entering the fuel system.**

9. Remove the exhaust front tube, center muffler and main muffler.
10. Remove the driveshaft.
11. Remove the parking brake cables.
12. Remove the rear suspension member assembly.

➡**For this service, halfshaft, final drive and rear suspension member are not required to be separated from one another during removal.**

13. Disconnect the fuel filler hose, vent hose and EVAP hose at the tube side.
14. Properly support the lower part of the fuel tank, with a suitable jack.

➡**Support the position that the fuel tank retaining straps do not engage.**

15. Remove the fuel tank mounting bands.
16. Carefully lower the tank assembly from its mounting. Check that all required hoses and electrical connectors are disconnected before fully lowering the assembly to the ground.

To install:

➡**Be sure to use new fasteners, as required.**

17. Installation is the reverse of the removal procedure.
18. Install the fuel pump fuse, if removed.
19. Turn the ignition switch ON (with the engine stopped), check all connections for fuel leakage, correct as required.
20. Start the engine and let it idle, check for fuel leaks. Correct as required.
21. Check and adjust the rear wheel alignment, as required.

IDLE SPEED

ADJUSTMENT

Idle speed is not adjustable.

THROTTLE BODY

REMOVAL & INSTALLATION
See Figures 68 and 129.

At this time the manufacturer does not provide removal and installation procedures for this component. The following procedure is a guideline and may differ from the vehicle you are servicing.

1. Before servicing the vehicle, refer to the Precautions Section.

➡**If working near and/or around the SRS system and components, be sure to disable the SRS system. After disabling the system wait three minutes or more before servicing the vehicle.**

➡**Before disconnecting the battery, lower both the driver's and passenger's windows. This will prevent any interference between the window edge and the vehicle when the door is opened or closed. During normal operation the window slightly raises and lowers automatically to prevent any window to vehicle interference. The automatic window function will not work with the battery disconnected.**

2. Remove the engine undercover.
3. Drain the engine coolant. Be sure to properly dispose of used coolant.

4. Remove the front tower bar assembly.

5. Remove the engine cover.

6. Remove the air cleaner assembly.

7. To remove the electric throttle control actuator, Disconnect the water hoses from the component. Disconnect the harness connector.

8. Loosen the mounting bolts in the reverse order of the tightening sequence.

9. Remove the component from its mounting.

➡ When removing only the intake manifold collector, move the electric throttle control actuator without disconnecting the water hose. Handel the component carefully to avoid shock damage.

To install:

➡ Be sure to use new fasteners, as required.

10. Installation is the reverse of the removal procedure.

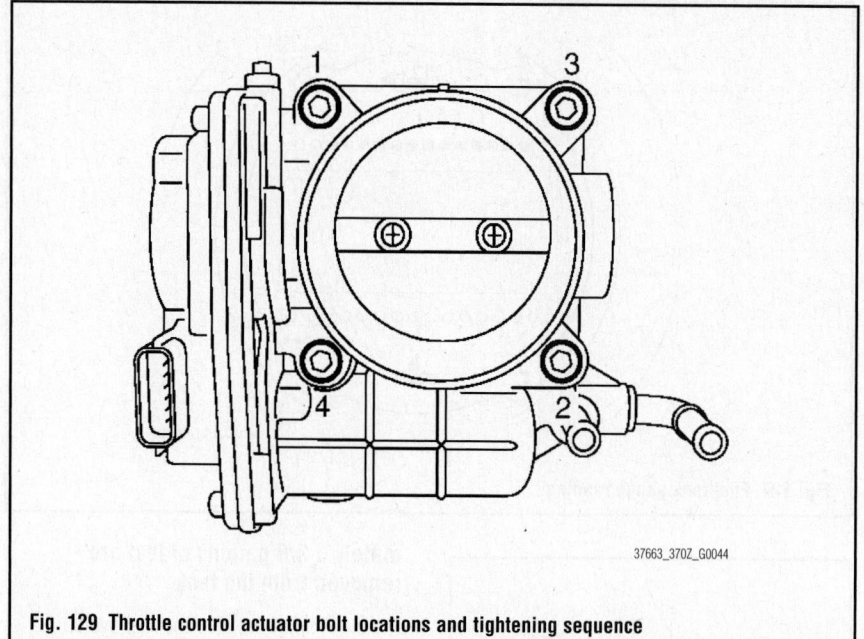

Fig. 129 Throttle control actuator bolt locations and tightening sequence

11. Tighten the electric throttle actuator retaining bolts in the proper sequence.

12. Using the CONSULT-III diagnostic tool or equivalent, perform the throttle valve closed position learning and the idle air volume learning procedures.

HEATING & AIR CONDITIONING SYSTEM

BLOWER MOTOR

REMOVAL & INSTALLATION

See Figure 130.

1. Before servicing the vehicle, refer to the Precautions Section.

➡ If working near and/or around the SRS system and components, be sure to disable the SRS system. After disabling the system wait three minutes or more before servicing the vehicle.

➡ Before disconnecting the battery, lower both the driver's and passenger's windows. This will prevent any interference between the window edge and the vehicle when the door is opened or closed. During normal operation the window slightly raises and lowers automatically to prevent any window to vehicle interference. The automatic window function will not work with the battery disconnected.

2. Disconnect the blower motor electrical connector.

3. Remove the blower motor retaining screws.

4. Remove the component from its mounting.

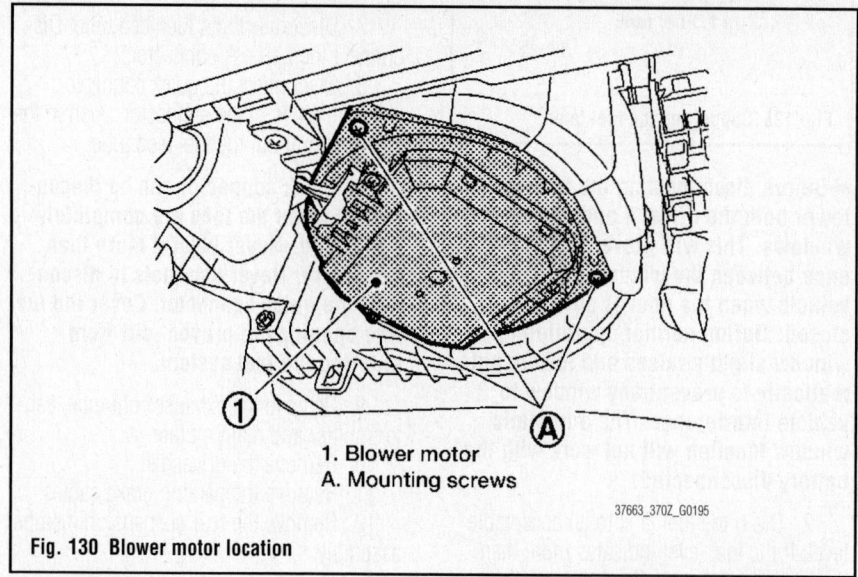

1. Blower motor
A. Mounting screws

Fig. 130 Blower motor location

To install:

➡ Be sure to use new fasteners, as required.

5. Installation is the reverse of the removal procedure.

HEATER CORE

REMOVAL & INSTALLATION

See Figure 131.

1. Before servicing the vehicle, refer to the Precautions Section.

➡ If working near and/or around the SRS system and components, be sure to disable the SRS system. After disabling the system wait three minutes or more before servicing the vehicle.

➡ Before disconnecting the battery, lower both the driver's and passenger's windows. This will prevent any interfer-

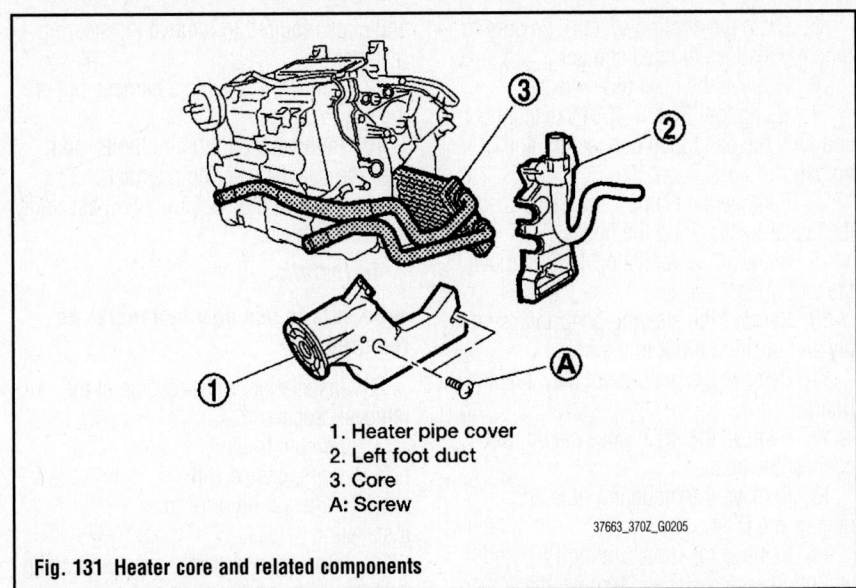

1. Heater pipe cover
2. Left foot duct
3. Core
A: Screw

37663_370Z_G0205

Fig. 131 Heater core and related components

ence between the window edge and the vehicle when the door is opened or closed. During normal operation the window slightly raises and lowers automatically to prevent any window to vehicle interference. The automatic

window function will not work with the battery disconnected.

2. Drain the radiator. Properly dispose of used coolant.

3. Properly discharge the air conditioning system.

4. Remove the heater/cooling unit assembly.

5. Remove the heater piping grommet and heater pipe bracket.

6. Remove the mounting screws and heater pipe cover.

7. Remove the left foot duct.

8. Slide the core leftward and remove.

To install:

➡ **Be sure to use new fasteners, as required.**

9. Installation is the reverse of the removal procedure.

10. Be sure to use new O-rings, coated with clean refrigerant oil prior to installation.

11. Properly refill the cooling system.

12. Properly charge the air conditioning system.

13. Start the engine and check the system for proper operation and refrigerant leakage.

HEATER AND COOLING UNIT

REMOVAL & INSTALLATION

See Figures 132 and 133.

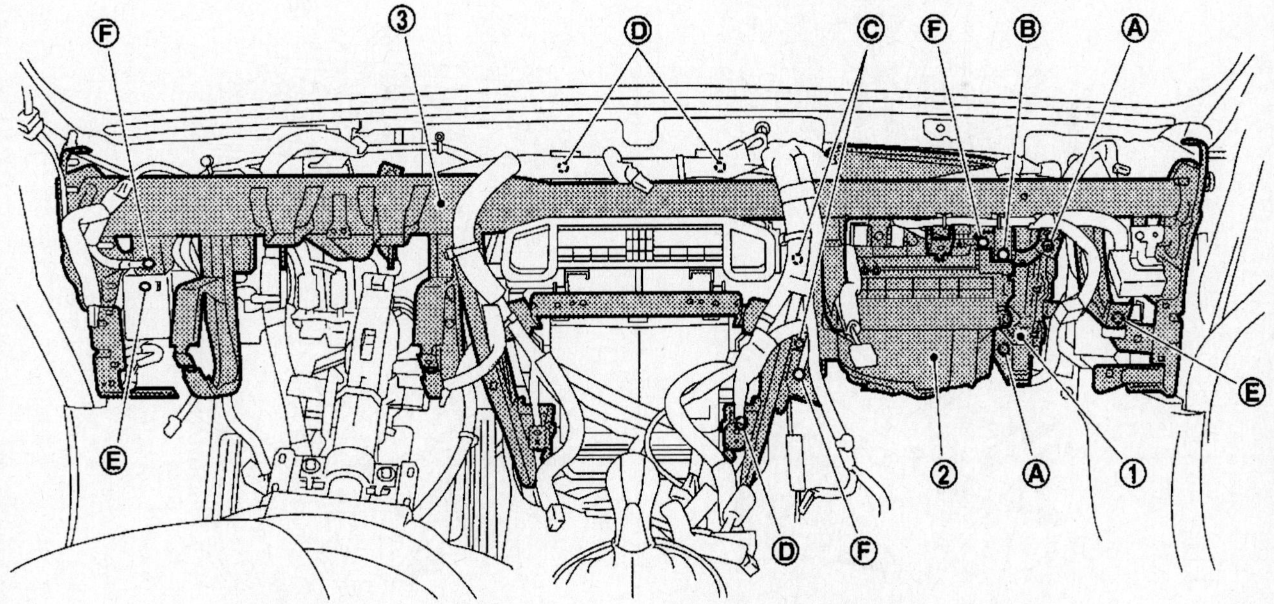

1. ECM
2. Heater/cooling unit
3. Bar

A. Mounting bolts
B. Mounting bolt
C. Mounting screws
D. Mounting bolts.
E. Mounting screws
F. Ground bolts

37663_370Z_G0204

Fig. 132 Heater/cooling unit component disconnection points

1. Before servicing the vehicle, refer to the Precautions Section.

➡ If working near and/or around the SRS system and components, be sure to disable the SRS system. After disabling the system wait three minutes or more before servicing the vehicle.

➡ Before disconnecting the battery, lower both the driver's and passenger's windows. This will prevent any interference between the window edge and the vehicle when the door is opened or closed. During normal operation the window slightly raises and lowers automatically to prevent any window to vehicle interference. The automatic window function will not work with the battery disconnected.

2. Set the temperature to 60 degrees F.

3. Disconnect the negative battery cable.

4. Properly discharge the air conditioning system

5. Drain the cooling system. Be sure to properly dispose of used coolant.

6. Remove the cowl top cover.

7. Using tool SST: J-45815 disconnect and plug the one touch connectors at the housing.

8. Remove the clamps and disconnect the heater hoses. Plug the hoses.

9. Remove the ventilator duct. Remove the foot grille.

10. Remove the steering column assembly and position it out of the way.

11. Remove the instrument stay, left and right.

12. Remove the drain hose clamp, disconnect the hose.

13. Remove the mounting nuts and remove the ECM.

14. Remove the mounting bolt (B) and mounting screws (C) and remove the heater/cooling unit.

15. Remove the mounting bolts (D), mounting screws (E) and ground bolts (F).

16. Disconnect the harness connectors

and clips required to remove the steering member.

17. Remove the vehicle harness out of the way.

18. Remove the mounting bolts, and then remove the steering member.

19. Remove the heater/cooling assembly from the vehicle.

To install:

➡ Be sure to use new fasteners, as required.

20. Installation is the reverse of the removal procedure.

21. Be sure to use new O-rings, coated with clean refrigerant oil prior to installation.

22. Properly charge the air conditioning system.

23. Properly refill the cooling system.

24. Start the engine and check the system for proper operation and refrigerant leakage.

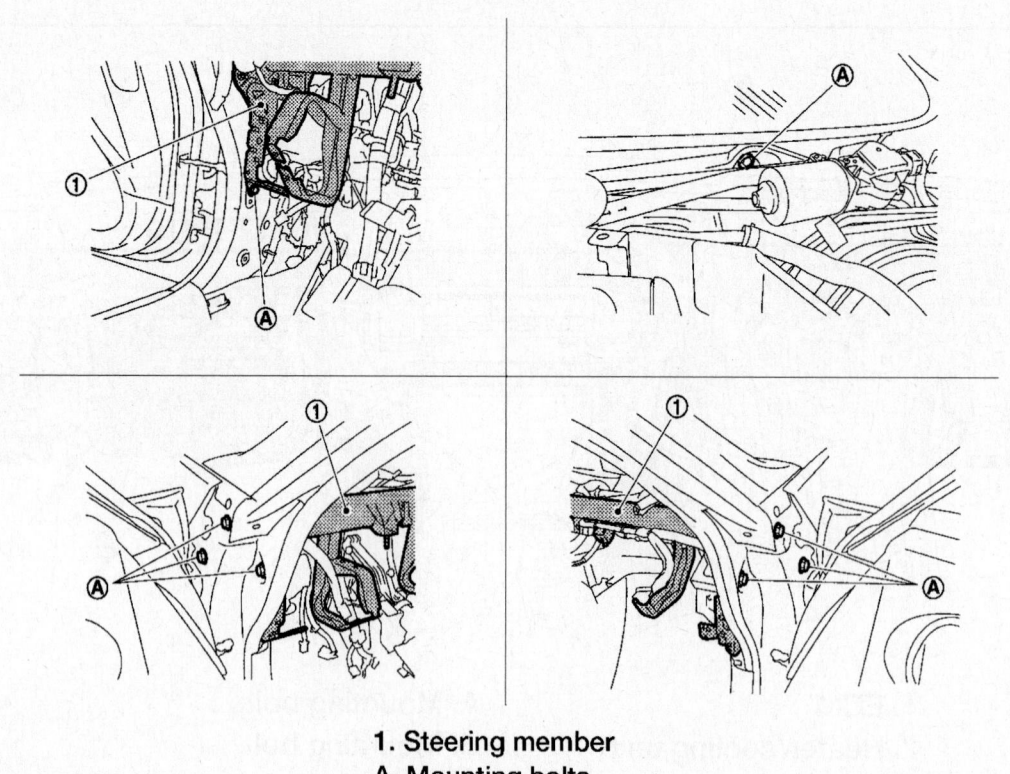

1. Steering member
A. Mounting bolts

37663_370Z_G0202

Fig. 133 Steering member and related components

STEERING

POWER RACK & PINION STEERING GEAR

REMOVAL & INSTALLATION

See Figures 134 through 138.

1. Before servicing the vehicle, refer to the Precautions Section.

➡If working near and/or around the SRS system and components, be sure to disable the SRS system. After disabling the system wait three minutes or more before servicing the vehicle.

➡Before disconnecting the battery, lower both the driver's and passenger's windows. This will prevent any interference between the window edge and the vehicle when the door is opened or closed. During normal operation the window slightly raises and lowers automatically to prevent any window to vehicle interference. The automatic window function will not work with the battery disconnected.

2. Position the front tires in the straight ahead position.
3. Raise and support the vehicle safely.
4. Remove the tire and wheel assemblies.
5. Remove the front suspension member.
6. Remove the cotter pin and loosen the locknut.
7. Remove the outer steering socket from the steering knuckle, so as not to damage the ball joint boot, using a suitable removal tool.

➡Temporarily tighten the nut to prevent damage to the threads and prevent the ball joint remover from sudden drop.

8. Disconnect and plug the power steering fluid lines. Drain the power steering fluid. Be sure to correctly dispose of used fluid.
9. Remove the power steering solenoid valve harness connector.
10. Remove the rack stay.
11. Remove the lower joint fixing bolts (steering gear side).
12. Separate the lower shaft from the steering gear assembly by sliding the slide shaft. See illustration.

➡The spiral cable may be cut if the steering wheel turns while separating the steering column assembly and steering gear assembly. Always lock

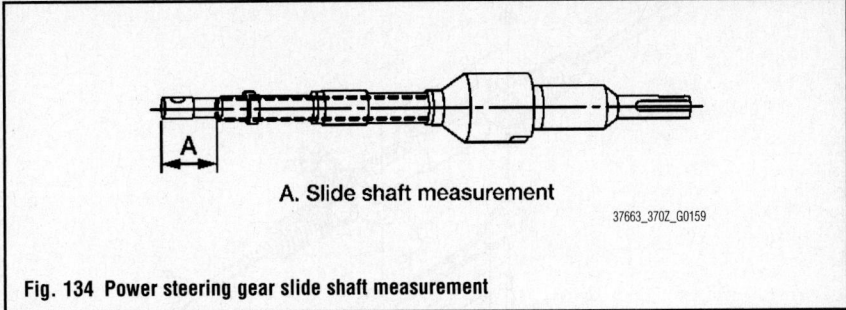

A. Slide shaft measurement

37663_370Z_G0159

Fig. 134 Power steering gear slide shaft measurement

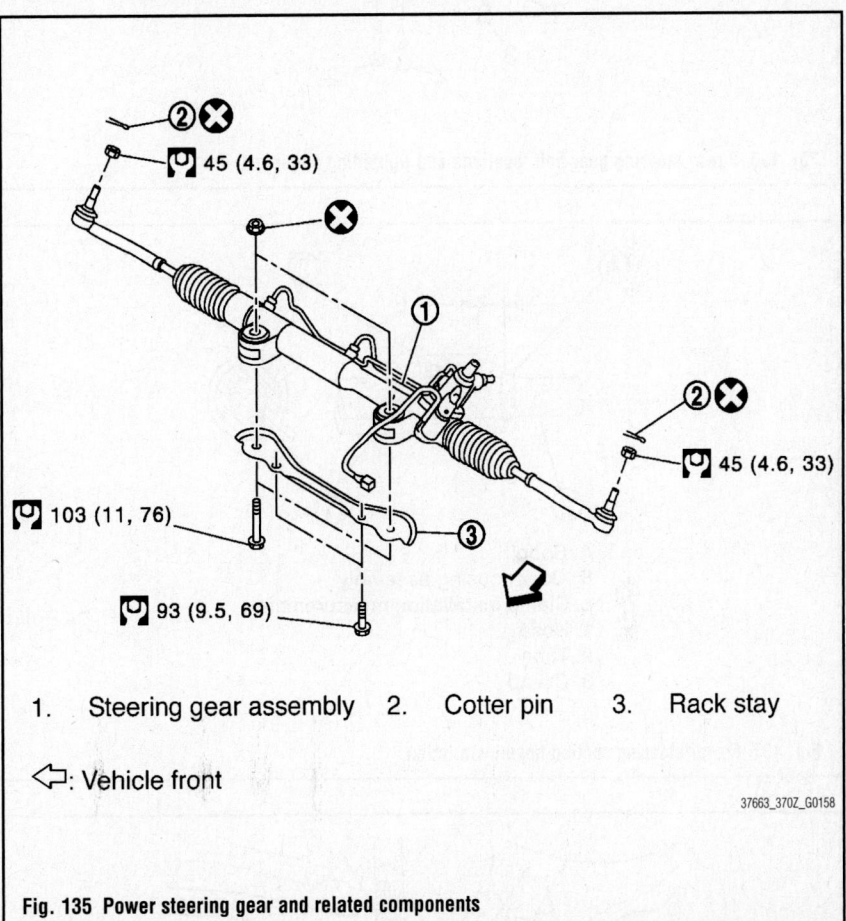

1. Steering gear assembly   2. Cotter pin   3. Rack stay

⇦: Vehicle front

37663_370Z_G0158

Fig. 135 Power steering gear and related components

the steering wheel using string to avoid turning.

13. Remove the steering gear retaining bolts.
14. Remove the gear assembly from its mounting.

To install:

➡Be sure to use new fasteners, as required.

15. Installation is the reverse of the removal procedure.

➡The spiral cable may be cut if the steering wheel turns while separating

the steering column assembly and steering gear assembly. Always lock the steering wheel using string to avoid turning.

16. Tighten the mounting bolts in the order shown in the illustration.
17. Tighten step one, temporary and in step two final.
18. When installing the suction hoses, refer to the illustration.

➡Never apply fluid to the hose and the tube. Insert the hose securely until it contacts the spool of the tube. Install the clamp at the hose (0.12–

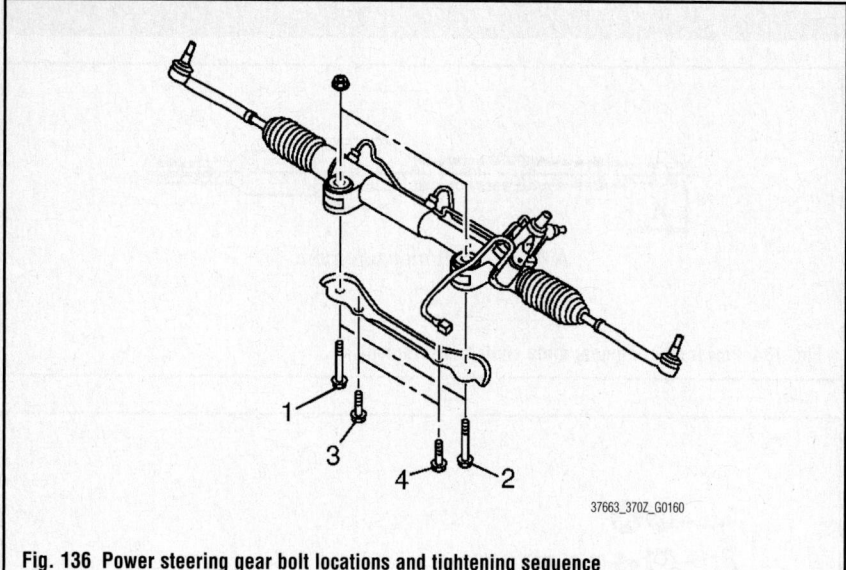

Fig. 136 Power steering gear bolt locations and tightening sequence

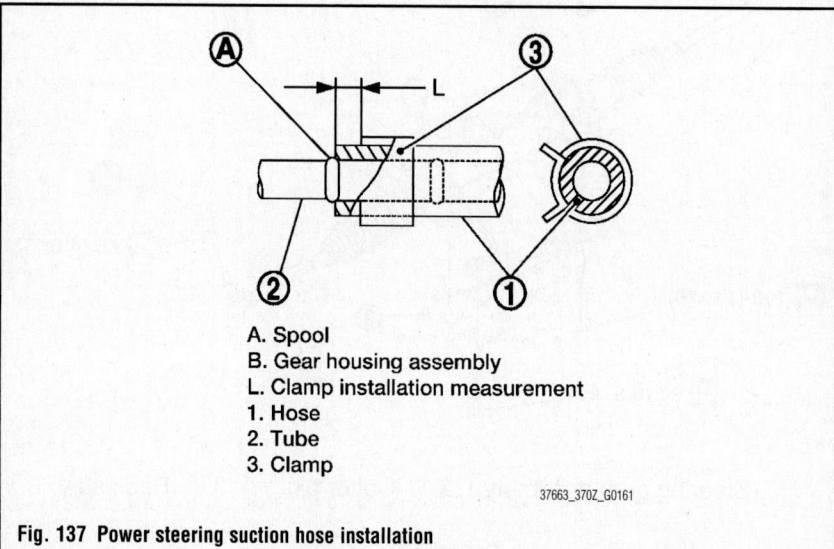

A. Spool
B. Gear housing assembly
L. Clamp installation measurement
1. Hose
2. Tube
3. Clamp

Fig. 137 Power steering suction hose installation

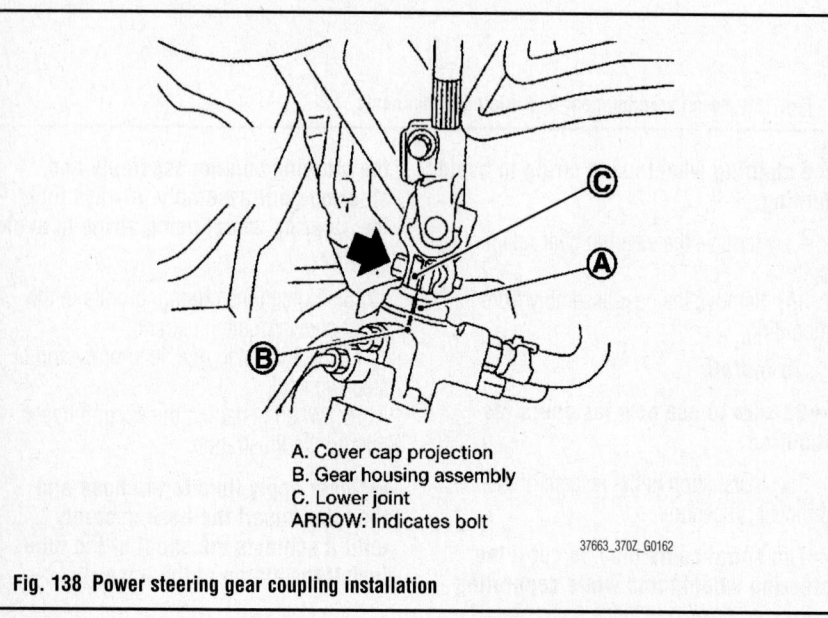

A. Cover cap projection
B. Gear housing assembly
C. Lower joint
ARROW: Indicates bolt

Fig. 138 Power steering gear coupling installation

0.31 inch from the edge of the hose.

19. To install the lower joint to the steering gear, see the illustration. Position the rack of the steering gear in the neutral position.

➡To get to the neutral position turn the sub gear assembly and measure the distance of the inner socket, then measure the intermediate position and distance. Align the rear cover cap projection with the marking position of the gear housing assembly. Install the slip part of the lower joint aligning with the rear cover cap projection. Make sure that the slit part of the lower joint is aligned with the rear cover cap protection and the marking position of the gear housing assembly. Make sure that there is no clearance between the lower joint, gear housing assembly and mounting bolt.

20. Fill the system. Bleed the system.
21. Perform a final tightening of nuts and bolts of each removed component with the vehicle in an unladen position.
22. Check and adjust the wheel alignment, as required.
23. Adjust the neutral position of the steering angle sensor using the CONSULT-III diagnostic tool, or equivalent, after checking the wheel alignment.

POWER STEERING PUMP

REMOVAL & INSTALLATION
See Figures 139 through 141.

1. Before servicing the vehicle, refer to the Precautions Section.

➡If working near and/or around the SRS system and components, be sure to disable the SRS system. After disabling the system wait three minutes or more before servicing the vehicle.

➡Before disconnecting the battery, lower both the driver's and passenger's windows. This will prevent any interference between the window edge and the vehicle when the door is opened or closed. During normal operation the window slightly raises and lowers automatically to prevent any window to vehicle interference. The automatic window function will not work with the battery disconnected.

2. Drain the power steering fluid from the reservoir. Be sure to properly dispose of used fluid.
3. Remove the air cleaner assembly.

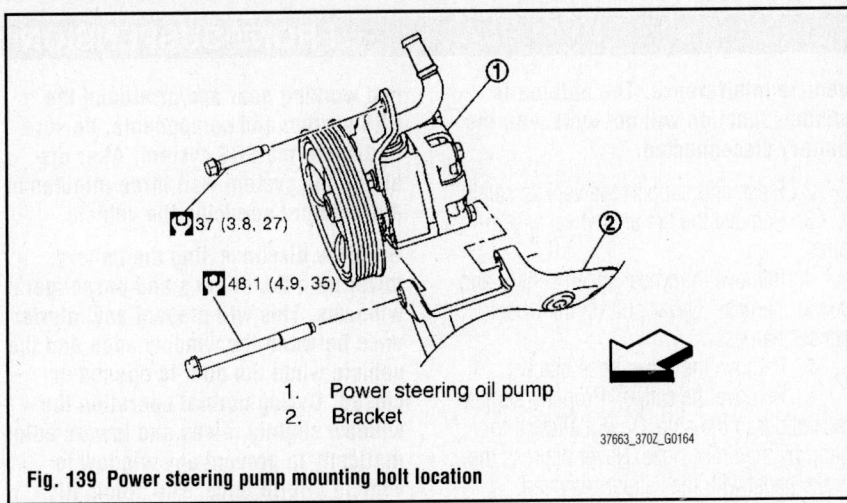

1. Power steering oil pump
2. Bracket

37663_370Z_G0164

Fig. 139 Power steering pump mounting bolt location

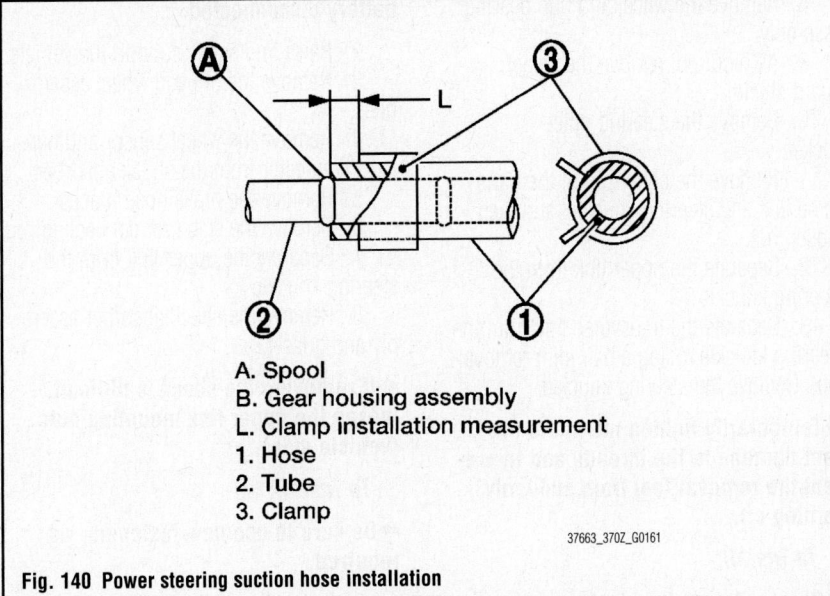

A. Spool
B. Gear housing assembly
L. Clamp installation measurement
1. Hose
2. Tube
3. Clamp

37663_370Z_G0161

Fig. 140 Power steering suction hose installation

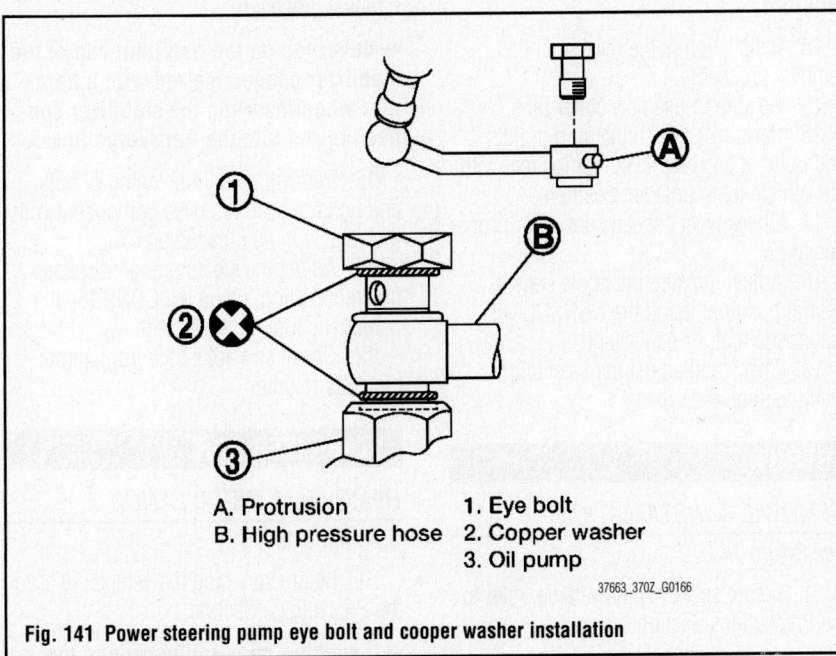

A. Protrusion   1. Eye bolt
B. High pressure hose 2. Copper washer
                       3. Oil pump

37663_370Z_G0166

Fig. 141 Power steering pump eye bolt and cooper washer installation

4. Loosen the drive belt.
5. Remove the belt from the steering pump.
6. Remove the pressure sensor connector.
7. Remove the copper washers and eye bolt. Drain fluid from their pipings.
8. Remove the suction hose. Drain fluid from their pipings.
9. Remove the cooling fan assembly.
10. Remove the pump retaining bolts.
11. Remove the pump from the vehicle.

To install:

➡ **Be sure to use new fasteners, as required.**

12. Installation is the reverse of the removal procedure.
13. When installing the suction hoses see note below.

➡ **Never apply fluid to the hose and the tube. Insert the hose securely until it contacts the spool of the tube. Install the clamp at the hose (0.12–0.31 inch from the edge of the hose.**

14. When installing the eye bolt and copper washer to the pump, see illustration.

➡ **Never reuse the cooper washer. Apply clean power steering fluid around the washers, then install the eye bolt. Install the eye bolt with the eye joint (assembled to the high pressure hose) protrusion facing with pump side cutout, and then tighten it to specification.**

15. Adjust the belt tension.
16. Fill the system with the proper grade and type fluid. Bleed the system.
17. Check for fluid leakage, correct as required.

BLEEDING

1. Stop engine, and then turn steering wheel fully to right and left several times.
2. Do not allow steering fluid reservoir tank to go below the low-level line. Check tank frequenter and add fluid as needed.
3. Run engine at idle speed. Turn steering wheel fully to the right and then fully to the left, and keep for about 3 seconds. Then check whether a fluid leakage has occurred.
4. Repeat the 2nd procedure several times at about three seconds interval Check generation of air bubbles and cloud in fluid.
5. If air bubbles and the cloud don't fade, stop engine, hold air bleeding until air bubbles and the cloud fade. Perform the 2nd and the 3rd procedures again.
6. Stop engine, check fluid level.

FRONT PERFORMANCE DAMPER

REMOVAL & INSTALLATION

See Figure 142.

1. Before servicing the vehicle, refer to the Precautions Section.

➡If working near and/or around the SRS system and components, be sure to disable the SRS system. After disabling the system wait three minutes or more before servicing the vehicle.

➡Before disconnecting the battery, lower both the driver's and passenger's windows. This will prevent any interference between the window edge and the vehicle when the door is opened or closed. During normal operation the window slightly raises and lowers automatically to prevent any window to vehicle interference. The automatic window function will not work with the battery disconnected.

2. Remove the front bumper fascia.
3. Remove the bolts and nuts from the performance damper.

To install:

➡Be sure to use new fasteners, as required.

4. Installation is the reverse of the removal procedure.
5. When installing, check all clearances to be sure there are no areas of interference.

STEERING KNUCKLE

REMOVAL & INSTALLATION

See Figure 147.

1. Before servicing the vehicle, refer to the Precautions Section.

➡If working near and/or around the SRS system and components, be sure to disable the SRS system. After disabling the system wait three minutes or more before servicing the vehicle.

➡Before disconnecting the battery, lower both the driver's and passenger's windows. This will prevent any interference between the window edge and the vehicle when the door is opened or closed. During normal operation the window slightly raises and lowers automatically to prevent any window to

vehicle interference. The automatic window function will not work with the battery disconnected.

2. Raise and support the vehicle safely.
3. Remove the tire and wheel assemblies.
4. Remove the wheel speed sensor and sensor harness. Never pull on the wheel sensor harness.
5. Remove the brake hose bracket.
6. Remove the caliper. Properly support the caliper to the side. Do not allow it to hang by the brake hose. Never depress the brake pedal with the caliper removed.
7. Remove the brake rotor.
8. Remove the wheel and hub bearing assembly.
9. As required, remove the splash guard shield.
10. Remove the steering outer socket.
11. Remove the cotter pin of the transverse link and steering knuckle, and then loosen nut.
12. Separate the upper link from the steering knuckle.
13. Separate the transverse link from the steering knuckle, using a ball joint removal tool. Remove the steering knuckle.

➡Temporarily tighten the nut to prevent damage to the threads and to prevent the removal tool from suddenly coming off.

To install:

➡Be sure to use new fasteners, as required.

14. Installation is the reverse of the removal procedure.
15. Be sure to use new cotter pins.
16. Perform a final tightening of nuts and bolts of each removed component with the vehicle in an unladen position.
17. Check wheel speed sensor for proper operation.
18. Adjust the steering angle sensor neutral position, using the CONSULT-III diagnostic tool, or equivalent.
19. Check and adjust the front alignment, as required.

STRUT

REMOVAL & INSTALLATION

See Figure 143.

1. Before servicing the vehicle, refer to the Precautions Section.

➡If working near and/or around the SRS system and components, be sure to disable the SRS system. After disabling the system wait three minutes or more before servicing the vehicle.

➡Before disconnecting the battery, lower both the driver's and passenger's windows. This will prevent any interference between the window edge and the vehicle when the door is opened or closed. During normal operation the window slightly raises and lowers automatically to prevent any window to vehicle interference. The automatic window function will not work with the battery disconnected.

2. Raise and safely support the vehicle.
3. Remove the tire and wheel assemblies.
4. Remove the wheel sensor and harness connector from the shock absorber.
5. Remove the brake hose bracket.
6. Remove the stabilizer connecting rod.
7. Separate the upper link from the steering knuckle.
8. Remove the shock absorber assembly and gusset.

➡If removing the shock is difficult, loosen the upper link mounting nuts (vehicle side).

To install:

➡Be sure to use new fasteners, as required.

9. Installation is the reverse of the removal procedure.

➡Never tap on the ball joint cap of the stabilizer connecting rod with a hammer when inserting the stabilizer connecting rod into the transverse link.

10. Perform a final tightening of nuts and bolts of each removed component with the vehicle in an unladen position.
11. Adjust the steering angle sensor neutral position, using the CONSULT-III diagnostic tool, or equivalent.
12. Check and adjust the front alignment, as required.

STABILIZER BAR

REMOVAL & INSTALLATION

See Figure 144.

1. Before servicing the vehicle, refer to the Precautions Section.

➡If working near and/or around the

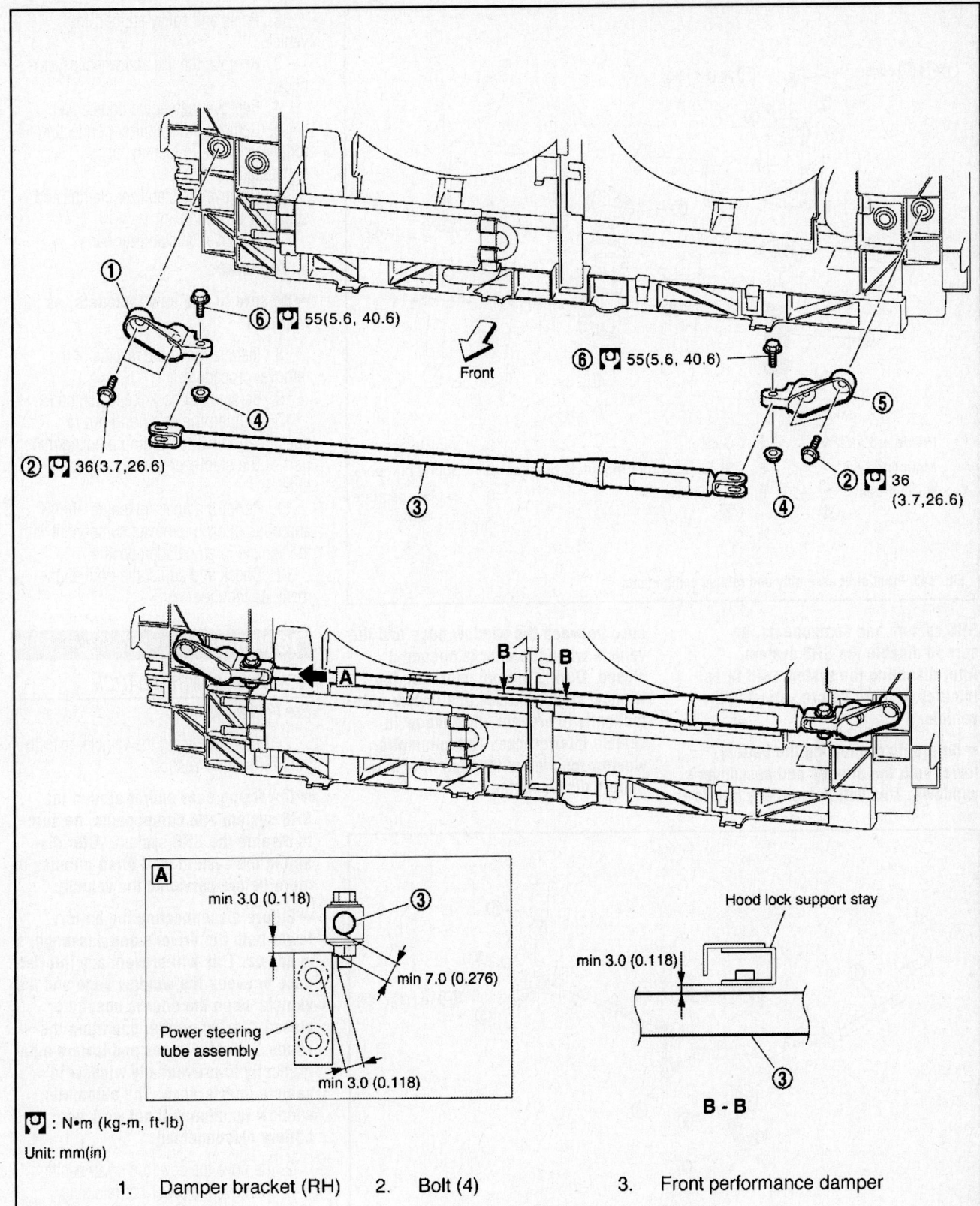

min 3.0 (0.118)

min 7.0 (0.276)

Power steering
tube assembly

min 3.0 (0.118)

Hood lock support stay

min 3.0 (0.118)

B - B

⬤ : N•m (kg-m, ft-lb)

Unit: mm(in)

1.	Damper bracket (RH)	2.	Bolt (4)	3.	Front performance damper
4.	Nut (2)	5.	Damper bracket (LH)	6.	Bolt (2)

37663_370Z_G0185

Fig. 142 Front performance damper and related components

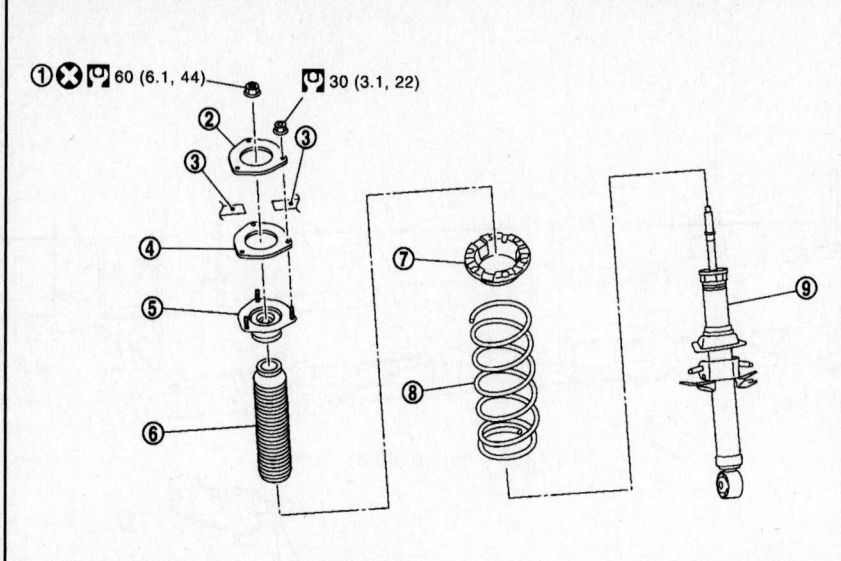

1.	Piston rod lock nut	2.	Gusset	3.	Vehicle body
4.	Mounting seal	5.	Shock absorber mounting bracket	6.	Bound bumper
7.	Rubber seat	8.	Coil spring	9.	Shock absorber

37663_370Z_G0176

Fig. 143 Front strut assembly and related components

SRS system and components, be sure to disable the SRS system. After disabling the system wait three minutes or more before servicing the vehicle.

➡Before disconnecting the battery, lower both the driver's and passenger's windows. This will prevent any interfer-ence between the window edge and the vehicle when the door is opened or closed. During normal operation the window slightly raises and lowers auto-matically to prevent any window to vehicle interference. The automatic window function will not work with the battery disconnected.

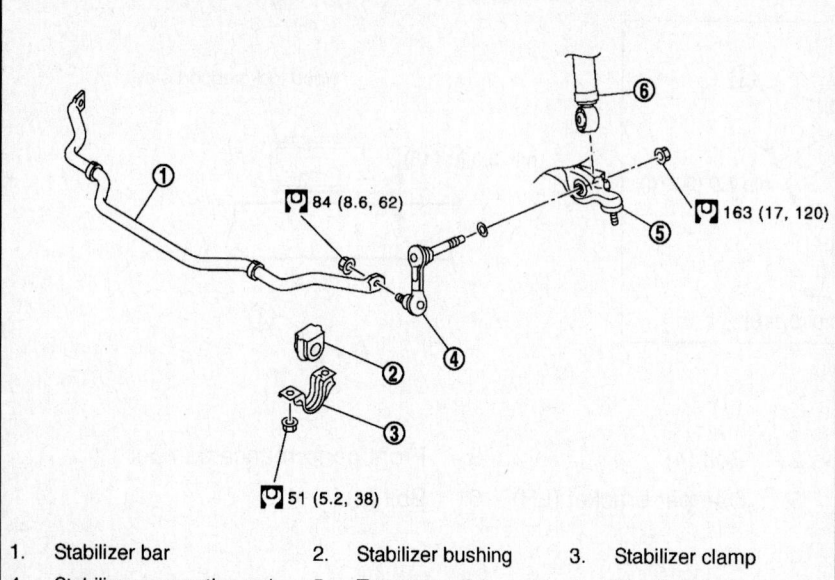

1.	Stabilizer bar	2.	Stabilizer bushing	3.	Stabilizer clamp
4.	Stabilizer connecting rod	5.	Transverse link	6.	Shock absorber

37663_370Z_G0178

Fig. 144 Front stabilizer bar and related components

2. Raise and safely support the vehicle.
3. Remove the tire and wheel assem-blies.
4. Remove the engine undercover.
5. Remove the stabilizer connecting rod. Matchmark to identify for reinstallation.
6. Remove the stabilizer clamps and stabilizer bushings.
7. Remove the stabilizer bar.

To install:

➡Be sure to use new fasteners, as required.

8. Installation is the reverse of the removal procedure.
9. Be sure to check the matchmarks.
10. Tighten the mounting nut to specification while holding a hexagonal part of the stabilizer connecting rod side.
11. Perform a final tightening of nuts and bolts of each removed component with the vehicle in an unladen position.
12. Check and adjust the front align-ment, as required.

TOWER BAR

REMOVAL & INSTALLATION
See Figure 145.

1. Before servicing the vehicle, refer to the Precautions Section.

➡If working near and/or around the SRS system and components, be sure to disable the SRS system. After dis-abling the system wait three minutes or more before servicing the vehicle.

➡Before disconnecting the battery, lower both the driver's and passenger's windows. This will prevent any interfer-ence between the window edge and the vehicle when the door is opened or closed. During normal operation the window slightly raises and lowers auto-matically to prevent any window to vehicle interference. The automatic window function will not work with the battery disconnected.

2. Remove the cowl top cover center.
3. Remove the tower bar retaining nuts.
4. Remove the component from the vehicle.

To install:

➡Be sure to use new fasteners, as required.

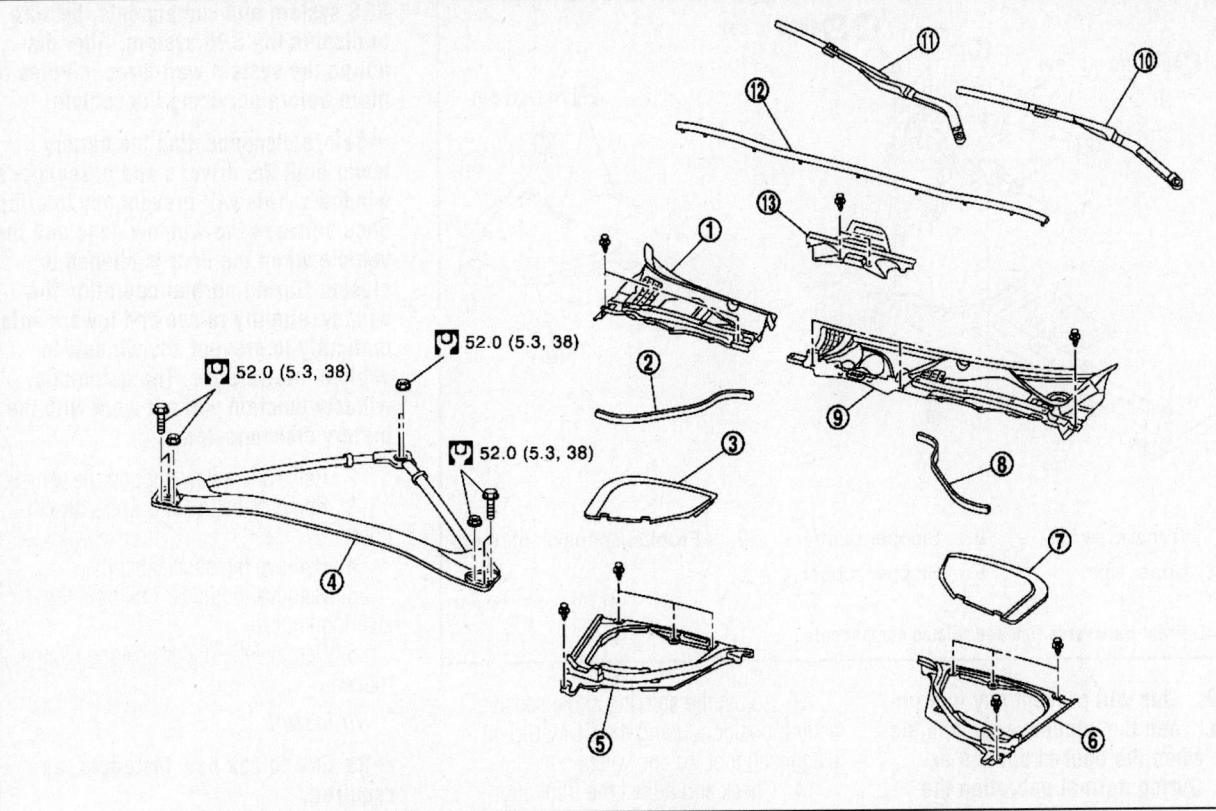

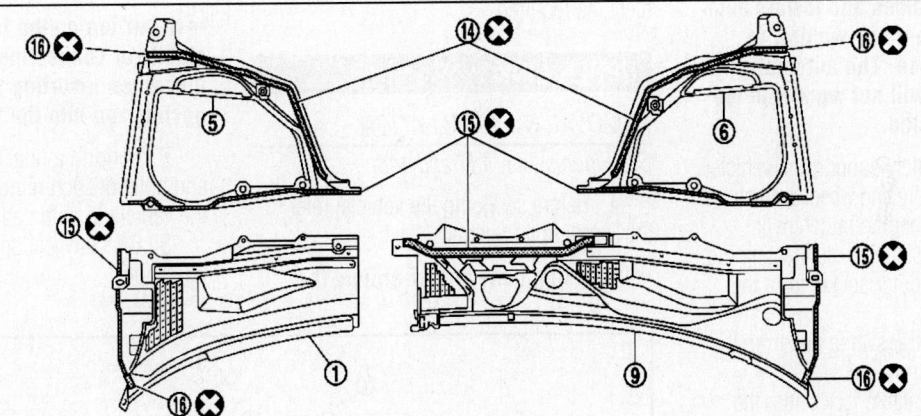

1. Cowl top cover RH	2. Cowl top cover seal RH	3. Battery cover
4. Front tower bar assembly	5. Hoodledge cover RH	6. Hoodledge cover LH
7. Brake master cylinder cover	8. Cowl top cover seal LH	9. Cowl top cover LH
10. Wiper arm and blade LH	11. Wiper arm and blade RH	12. Cowl top cover seal
13. Cowl top cover center	14. EPT sealer [t: 5.0 mm (0.197 in)]	15. EPT sealer [t:3.0 mm (0.118 in)]
16. EPT sealer [t: 10.0 mm (0.394 in)]		

37663_370Z_G0035

Fig. 145 Front tower bar and related components

5. Installation is the reverse of the removal procedure.

6. Perform a final tightening of nuts and bolts of each removed component with the vehicle in an unladen position.

7. Check and adjust the front alignment, as required.

TRANSVERSE LINK

REMOVAL & INSTALLATION

See Figure 146.

1. Before servicing the vehicle, refer to the Precautions Section.

➡If working near and/or around the SRS system and components, be sure to disable the SRS system. After disabling the system wait three minutes or more before servicing the vehicle.

➡Before disconnecting the battery, lower both the driver's and passenger's

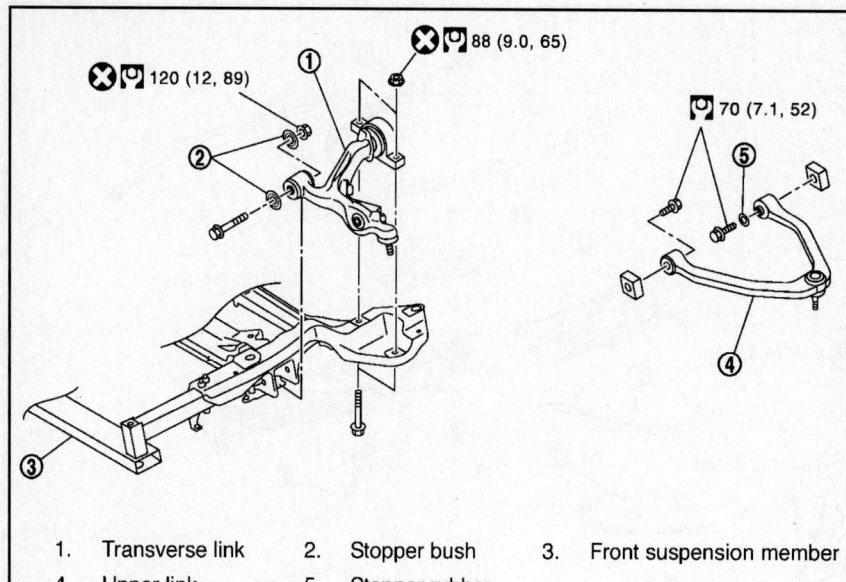

1. Transverse link
2. Stopper bush
3. Front suspension member
4. Upper link
5. Stopper rubber

37663_370Z_G0182

Fig. 146 Front transverse link and related components

windows. This will prevent any interference between the window edge and the vehicle when the door is opened or closed. During normal operation the window slightly raises and lowers automatically to prevent any window to vehicle interference. The automatic window function will not work with the battery disconnected.

2. Raise and safely support the vehicle.
3. Remove the tire and wheel assemblies.
4. Remove the engine undercover.
5. Remove the stabilizer connecting rod.
6. Remove the outer socket from the steering knuckle.
7. Remove the transverse link from the steering knuckle.
8. Position a suitable jack under the transverse link.
9. Remove the transverse link from its mounting.

To install:

➡Be sure to use new fasteners, as required.

10. Installation is the reverse of the removal procedure.

➡Never tap on the ball joint cap of the stabilizer connecting rod with a hammer when inserting the stabilizer connecting rod into the transverse link.

11. Perform a final tightening of nuts and bolts of each removed component with the vehicle in an unladen position.
12. Check wheel speed sensor for proper operation.

13. Adjust the steering angle sensor neutral position, using the CONSULT-III diagnostic tool, or equivalent.
14. Check and adjust the front alignment, as required.

UPPER LINK

REMOVAL & INSTALLATION
See Figures 141, 147 and 148.

1. Before servicing the vehicle, refer to the Precautions Section.

➡If working near and/or around the

SRS system and components, be sure to disable the SRS system. After disabling the system wait three minutes or more before servicing the vehicle.

➡Before disconnecting the battery, lower both the driver's and passenger's windows. This will prevent any interference between the window edge and the vehicle when the door is opened or closed. During normal operation the window slightly raises and lowers automatically to prevent any window to vehicle interference. The automatic window function will not work with the battery disconnected.

2. Raise and safely support the vehicle.
3. Remove the tire and wheel assemblies.
4. Remove the shock absorber.
5. Remove the upper link from the steering knuckle.
6. Remove the upper link and stopper rubber.

To install:

➡Be sure to use new fasteners, as required.

➡Never tap on the ball joint cap of the stabilizer connecting rod with a hammer when inserting the stabilizer connecting rod into the transverse link.

7. Perform a final tightening of nuts and bolts of each removed component with the vehicle in an unladen position.
8. Check wheel speed sensor for proper operation.

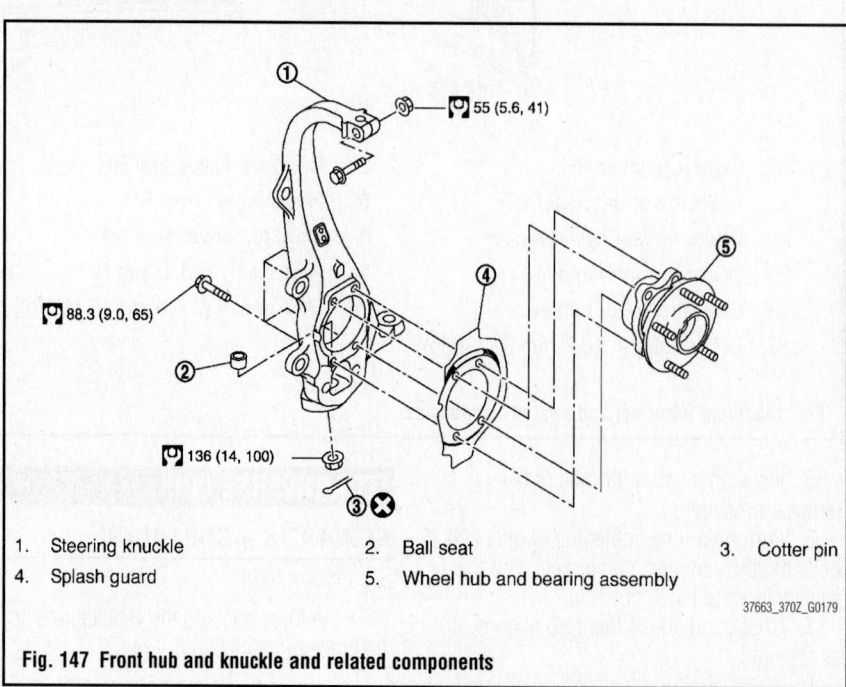

1. Steering knuckle
2. Ball seat
3. Cotter pin
4. Splash guard
5. Wheel hub and bearing assembly

37663_370Z_G0179

Fig. 147 Front hub and knuckle and related components

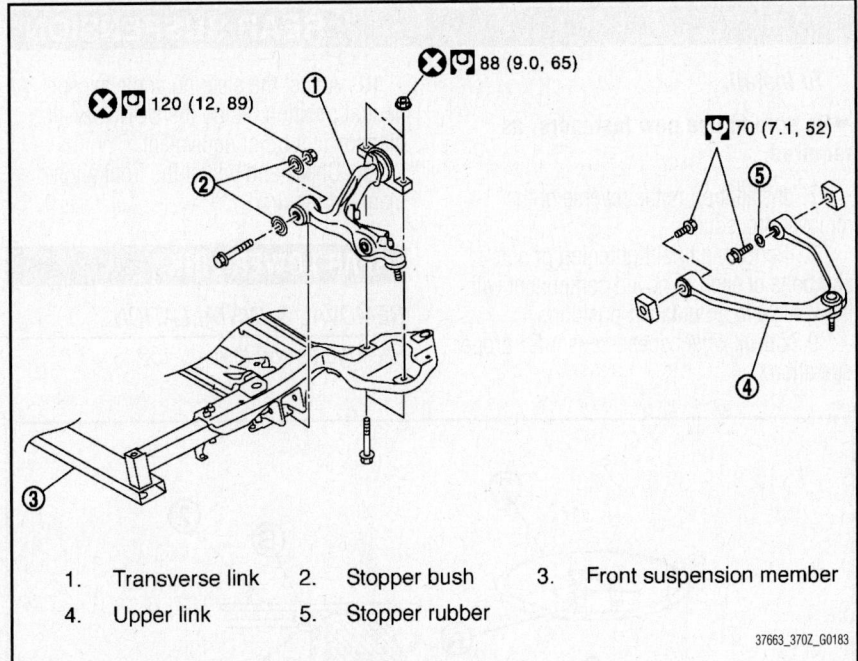

1. Transverse link 2. Stopper bush 3. Front suspension member
4. Upper link 5. Stopper rubber

37663_370Z_G0183

Fig. 148 Front upper link and related components

9. Adjust the steering angle sensor neutral position, using the CONSULT-III diagnostic tool, or equivalent.

10. Check and adjust the front alignment, as required.

WHEEL HUB & BEARING

REMOVAL & INSTALLATION

See Figure 149.

1. Before servicing the vehicle, refer to the Precautions Section.

➡**If working near and/or around the SRS system and components, be sure to disable the SRS system. After disabling the system wait three minutes or more before servicing the vehicle.**

➡**Before disconnecting the battery, lower both the driver's and passenger's windows. This will prevent any interference between the window edge and the vehicle when the door is opened or closed. During normal operation the window slightly raises and lowers automatically to prevent any window to vehicle interference. The automatic window function will not work with the battery disconnected.**

2. Raise and support the vehicle safely.

3. Remove the tire and wheel assemblies.

4. Remove the wheel speed sensor and sensor harness. Never pull on the wheel sensor harness.

5. Remove the brake hose bracket.

6. Remove the caliper. Properly support the caliper to the side. Do not allow it to hang by the brake hose. Never depress the brake pedal with the caliper removed.

7. Remove the brake rotor.

8. Remove the wheel and hub bearing assembly.

9. As required, remove the splash guard shield.

To install:

➡**Be sure to use new fasteners, as required.**

10. Installation is the reverse of the removal procedure.

11. Perform a final tightening of nuts and bolts of each removed component with the vehicle in an unladen position.

12. Check wheel speed sensor for proper operation.

13. Adjust the steering angle sensor neutral position, using the CONSULT-III diagnostic tool, or equivalent.

14. Check and adjust the front alignment, as required.

ADJUSTMENT

These bearings are not adjustable. If defective, they must be replaced.

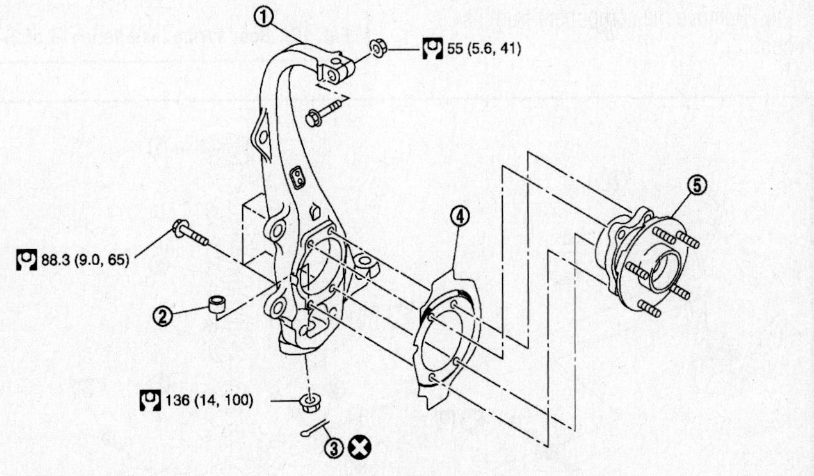

1. Steering knuckle 2. Ball seat 3. Cotter pin
4. Splash guard 5. Wheel hub and bearing assembly

37663_370Z_G0179

Fig. 149 Front hub and knuckle and related components

FRONT LOWER LINK

REMOVAL & INSTALLATION

See Figure 150.

1. Before servicing the vehicle, refer to the Precautions Section.

➡ **If working near and/or around the SRS system and components, be sure to disable the SRS system. After disabling the system wait three minutes or more before servicing the vehicle.**

➡ **Before disconnecting the battery, lower both the driver's and passenger's windows. This will prevent any interference between the window edge and the vehicle when the door is opened or closed. During normal operation the window slightly raises and lowers automatically to prevent any window to vehicle interference. The automatic window function will not work with the battery disconnected.**

2. Raise and support the vehicle safely.
3. Remove the tire and wheel assemblies.
4. Position a suitable jack under the rear axle assembly to relieve tension on the coil spring.
5. Remove the lower link retaining bolts.
6. Remove the component from its mounting.

To install:

➡ **Be sure to use new fasteners, as required.**

7. Installation is the reverse of the removal procedure.
8. Perform a final tightening of nuts and bolts of each removed component with the vehicle in an unladen position.
9. Check wheel speed sensor for proper operation.

10. Adjust the steering angle sensor neutral position, using the CONSULT-III diagnostic tool, or equivalent.
11. Check and adjust the front alignment, as required.

REAR LOWER LINK

REMOVAL & INSTALLATION

See Figures 150 through 153.

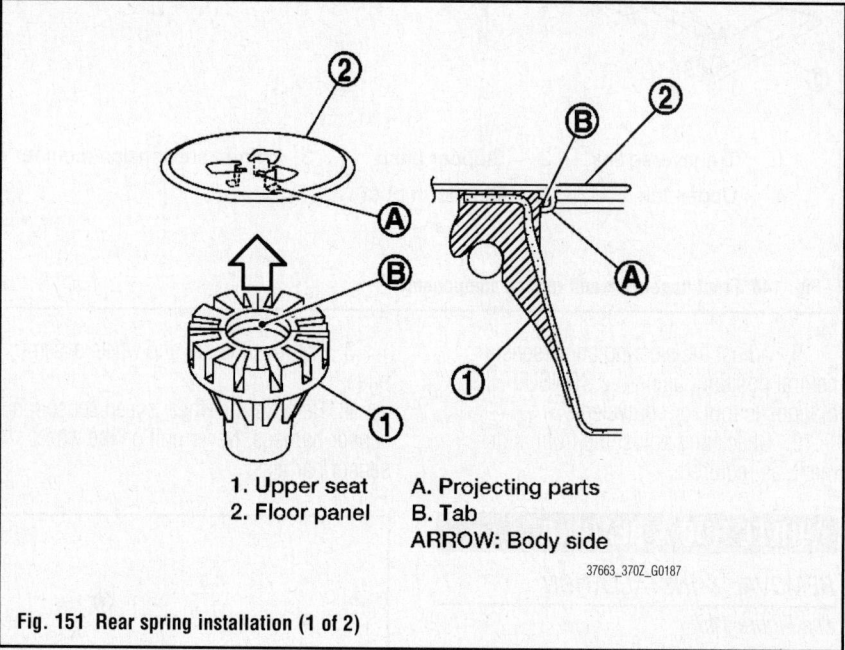

1. Upper seat
2. Floor panel

A. Projecting parts
B. Tab
ARROW: Body side

37663_370Z_G0187

Fig. 151 Rear spring installation (1 of 2)

1. Before servicing the vehicle, refer to the Precautions Section.

➡ **If working near and/or around the SRS system and components, be sure to disable the SRS system. After disabling the system wait three minutes or more before servicing the vehicle.**

➡ **Before disconnecting the battery, lower both the driver's and passenger's windows. This will prevent any interference between the window edge and the vehicle when the door is opened or closed. During normal operation the window slightly raises and lowers automatically to prevent any window to vehicle interference. The automatic window function will not work with the battery disconnected.**

2. Raise and support the vehicle safely.
3. Remove the tire and wheel assemblies.

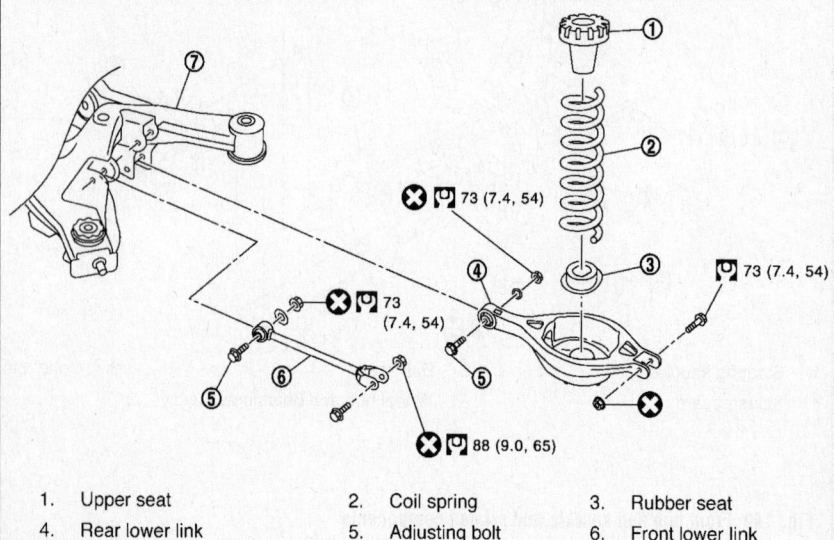

1.	Upper seat	2.	Coil spring	3.	Rubber seat
4.	Rear lower link	5.	Adjusting bolt	6.	Front lower link
7.	Rear suspension member				

37663_370Z_G0186

Fig. 150 Front and rear lower link and related components

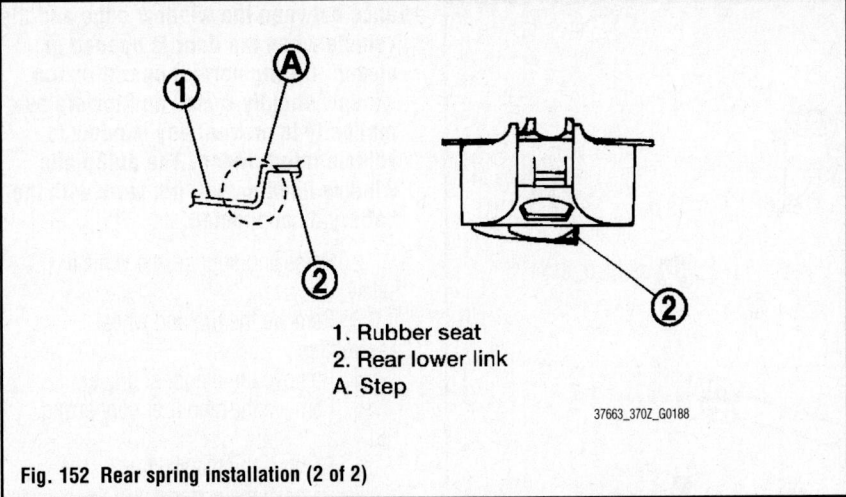

1. Rubber seat
2. Rear lower link
A. Step

37663_370Z_G0188

Fig. 152 Rear spring installation (2 of 2)

4. Position a suitable jack under the rear lower link to relieve the coil spring tension.

5. Loosen the rear lower link mounting nuts (rear suspension member side).

6. Remove the rear lower link (axle housing side).

7. Slowly lower the jack and remove the upper seat, coil spring and rubber sheet from the rear lower link.

8. Remove the rear lower link.

To install:

➡**Be sure to use new fasteners, as required.**

9. Installation is the reverse of the removal procedure.

10. Be sure that the upper seat is attached as indicated in the illustration.

➡**Make sure that the projecting parts of the floor panel is securely fitted with the upper seat tab.**

11. Match up the rubber seat indentations and the rear lower link grooves and attach.

12. Install the coil spring by aligning the lower end of the large diameter side to the step between the rubber seat and the rear lower link.

➡**Make sure that the spring is not upside down. The top and bottom are indicated by paint color.**

13. Perform a final tightening of nuts

and bolts of each removed component with the vehicle in an unladen position.

14. Check wheel speed sensor for proper operation.

15. Adjust the steering angle sensor neutral position, using the CONSULT-III diagnostic tool, or equivalent.

16. Check and adjust the front alignment, as required.

SHOCK ABSORBER

REMOVAL & INSTALLATION

See Figure 154.

1. Before servicing the vehicle, refer to the Precautions Section.

➡**If working near and/or around the SRS system and components, be sure to disable the SRS system. After disabling the system wait three minutes or more before servicing the vehicle.**

➡**Before disconnecting the battery, lower both the driver's and passenger's windows. This will prevent any interference between the window edge and the vehicle when the door is opened or closed. During normal operation the window slightly raises and lowers automatically to prevent any window to vehicle interference. The automatic window function will not work with the battery disconnected.**

2. Raise and support the vehicle safely.

3. Remove the tire and wheel assemblies.

4. Position a suitable jack under the rear axle assembly to relieve the coil spring tension.

5. Gradually lower the jack and separate the shock absorber (lower side) from the axle housing.

6. Remove the shock absorber mounting nuts (upper side), and then remove the shock absorber.

To install:

➡**Be sure to use new fasteners, as required.**

7. Installation is the reverse of the removal procedure.

8. Perform a final tightening of nuts and bolts of each removed component with the vehicle in an unladen position.

9. Check wheel speed sensor for proper operation.

10. Adjust the steering angle sensor neutral position, using the CONSULT-III diagnostic tool, or equivalent.

11. Check and adjust the front alignment, as required.

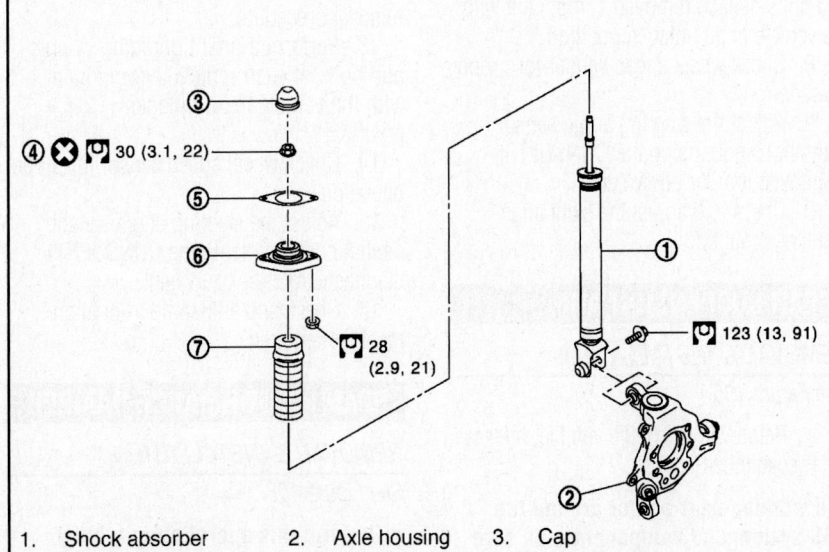

④ ✕ 🔧 30 (3.1, 22)

🔧 28 (2.9, 21)

🔧 123 (13, 91)

1. Shock absorber
2. Axle housing
3. Cap
4. Piston rod lock nut
5. Mounting seal
6. Shock absorber mounting bracket
7. Bound bumper cover

37663_370Z_G0189

Fig. 153 Rear shock absorber and related components

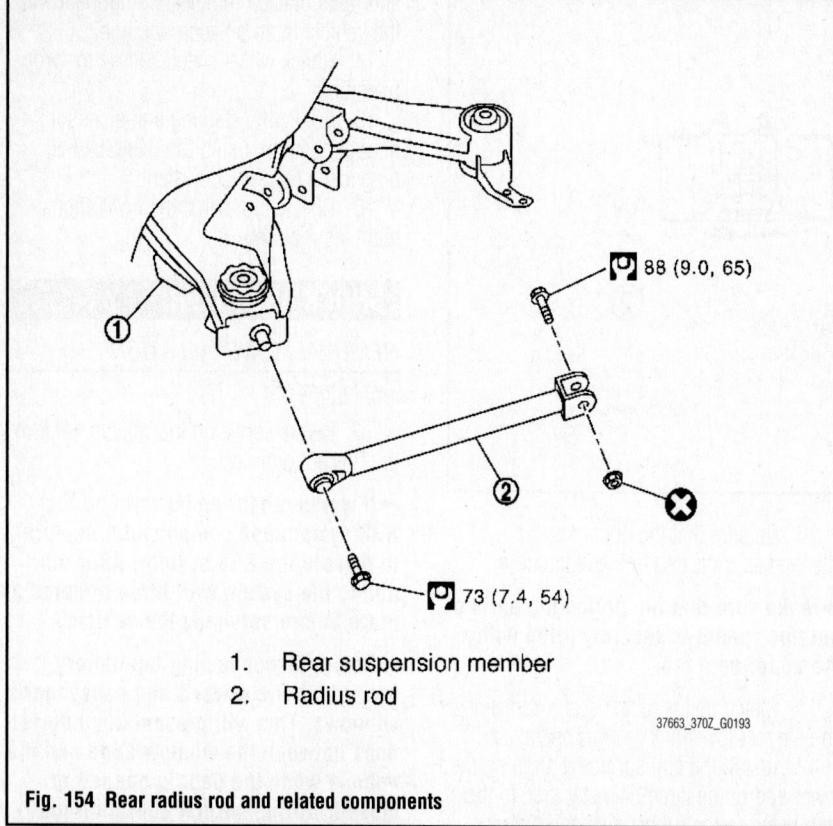

88 (9.0, 65)

73 (7.4, 54)

1. Rear suspension member
2. Radius rod

37663_370Z_G0193

Fig. 154 Rear radius rod and related components

RADIUS ARM

REMOVAL & INSTALLATION

See Figure 154.

1. Before servicing the vehicle, refer to the Precautions Section.

➡ **If working near and/or around the SRS system and components, be sure to disable the SRS system. After disabling the system wait three minutes or more before servicing the vehicle.**

➡ **Before disconnecting the battery, lower both the driver's and passenger's windows. This will prevent any interference between the window edge and the vehicle when the door is opened or closed. During normal operation the window slightly raises and lowers automatically to prevent any window to vehicle interference. The automatic window function will not work with the battery disconnected.**

2. Raise and support the vehicle safely.
3. Remove the tire and wheel assemblies.
4. Remove the radius rod retaining nuts.
5. Remove the component from the vehicle.

To install:

➡ **Be sure to use new fasteners, as required.**

6. Installation is the reverse of the removal procedure.
7. Perform a final tightening of nuts and bolts of each removed component with the vehicle in an unladen position.
8. Check wheel speed sensor for proper operation.
9. Adjust the steering angle sensor neutral position, using the CONSULT-III diagnostic tool, or equivalent.
10. Check and adjust the front alignment, as required.

REAR SUSPENSION ARM

REMOVAL & INSTALLATION

See Figure 155.

1. Before servicing the vehicle, refer to the Precautions Section.

➡ **If working near and/or around the SRS system and components, be sure to disable the SRS system. After disabling the system wait three minutes or more before servicing the vehicle.**

➡ **Before disconnecting the battery, lower both the driver's and passenger's windows. This will prevent any interfer-**

ence between the window edge and the vehicle when the door is opened or closed. During normal operation the window slightly raises and lowers automatically to prevent any window to vehicle interference. The automatic window function will not work with the battery disconnected.

2. Raise and support the vehicle safely.
3. Remove the tire and wheel assemblies.
4. Remove the diagonal brace.
5. Remove the stabilizer connecting rod.
6. Remove the halfshaft.
7. Remove the cotter pin of the suspension arm ball joint, and loosen the nut.
8. Remove the suspension arm (rear suspension member side).
9. Use a ball joint removal tool to remove the suspension arm from the axle housing.

➡ **Be careful not to damage the ball joint boot. Temporarily tighten the nut to prevent damage to the threads and to prevent the tool from coming off.**

10. Remove the suspension arm.
11. Remove the stabilizer connecting rod mounting bracket.

To install:

➡ **Be sure to use new fasteners, as required.**

12. Installation is the reverse of the removal procedure.
13. Perform a final tightening of nuts and bolts of each removed component with the vehicle in an unladen position.
14. Check wheel speed sensor for proper operation.
15. Adjust the steering angle sensor neutral position, using the CONSULT-III diagnostic tool, or equivalent.
16. Check and adjust the front alignment, as required.

REAR PERFORMANCE DAMPER

REMOVAL & INSTALLATION

See Figure 156.

➡ **Perform this operation in a level place while the vehicle is in the unladen position, in running order. Never tighten bolts while the vehicle is raised or jacked up.**

1. Before servicing the vehicle, refer to the Precautions Section.

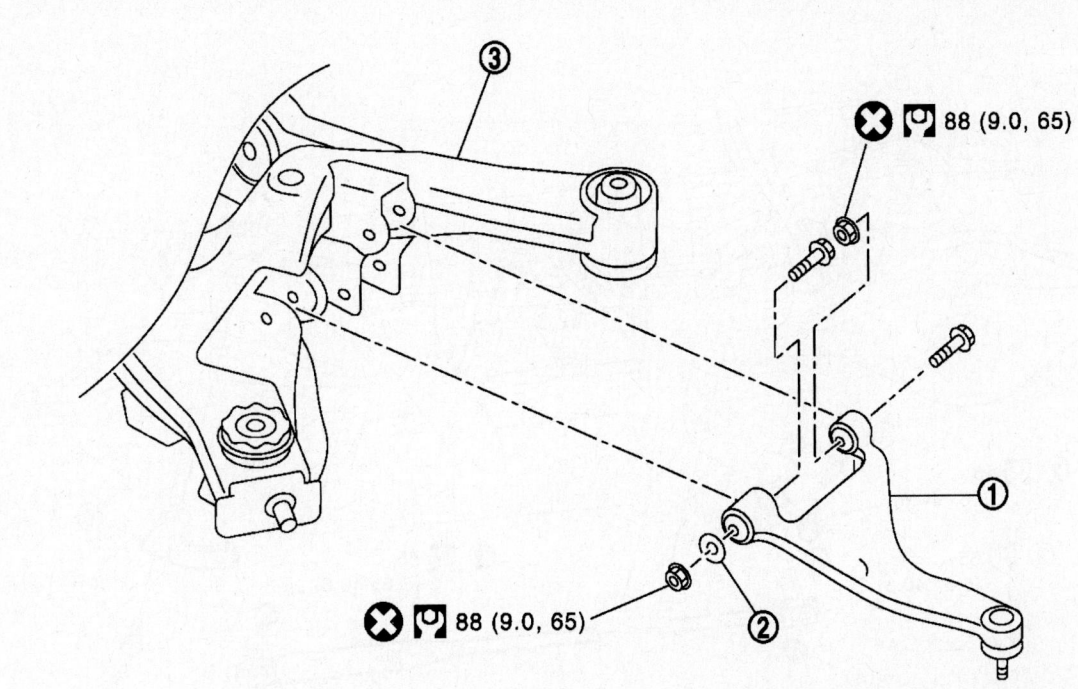

❌ 🔧 88 (9.0, 65)

❌ 🔧 88 (9.0, 65)

1.   Suspension arm   2.   Stopper rubber   3.   Rear suspension member

37663_370Z_G0190

Fig. 155 Rear suspension arm (upper) and related components

➡If working near and/or around the SRS system and components, be sure to disable the SRS system. After disabling the system wait three minutes or more before servicing the vehicle.

➡Before disconnecting the battery, lower both the driver's and passenger's windows. This will prevent any interference between the window edge and the vehicle when the door is opened or closed. During normal operation the window slightly raises and lowers automatically to prevent any window to vehicle interference. The automatic window function will not work with the battery disconnected.

2.   Raise and support the vehicle safely.

3.   Remove the bolts and then remove the right and left damper brackets.

To install:

➡Be sure to use new fasteners, as required.

4.   Installation is the reverse of the removal procedure.

STABILIZER BAR

REMOVAL & INSTALLATION

See Figure 157.

1.   Before servicing the vehicle, refer to the Precautions Section.

➡If working near and/or around the SRS system and components, be sure to disable the SRS system. After disabling the system wait three minutes or more before servicing the vehicle.

➡Before disconnecting the battery, lower both the driver's and passenger's windows. This will prevent any interference between the window edge and the vehicle when the door is opened or closed. During normal operation the window slightly raises and lowers automatically to prevent any window to vehicle interference. The automatic window function will not work with the battery disconnected.

2.   Raise and support the vehicle safely.

3.   Remove the tire and wheel assemblies.

4.   Remove the diagonal brace.

5.   Remove the stabilizer connecting rods.

6.   Remove the stabilizer clamps. Remove the stabilizer bushings.

7.   Remove the stabilizer bar.

8.   Remove the stabilizer connecting rod mounting brackets from the suspension arm.

To install:

➡Be sure to use new fasteners, as required.

9.   Installation is the reverse of the removal procedure.

10.   Perform a final tightening of nuts and bolts of each removed component with the vehicle in an unladen position.

11.   Check wheel speed sensor for proper operation.

12.   Adjust the steering angle sensor neutral position, using the CONSULT-III diagnostic tool, or equivalent.

13.   Check and adjust the front alignment, as required.

WHEEL HUB & BEARING

REMOVAL & INSTALLATION

See Figure 158.

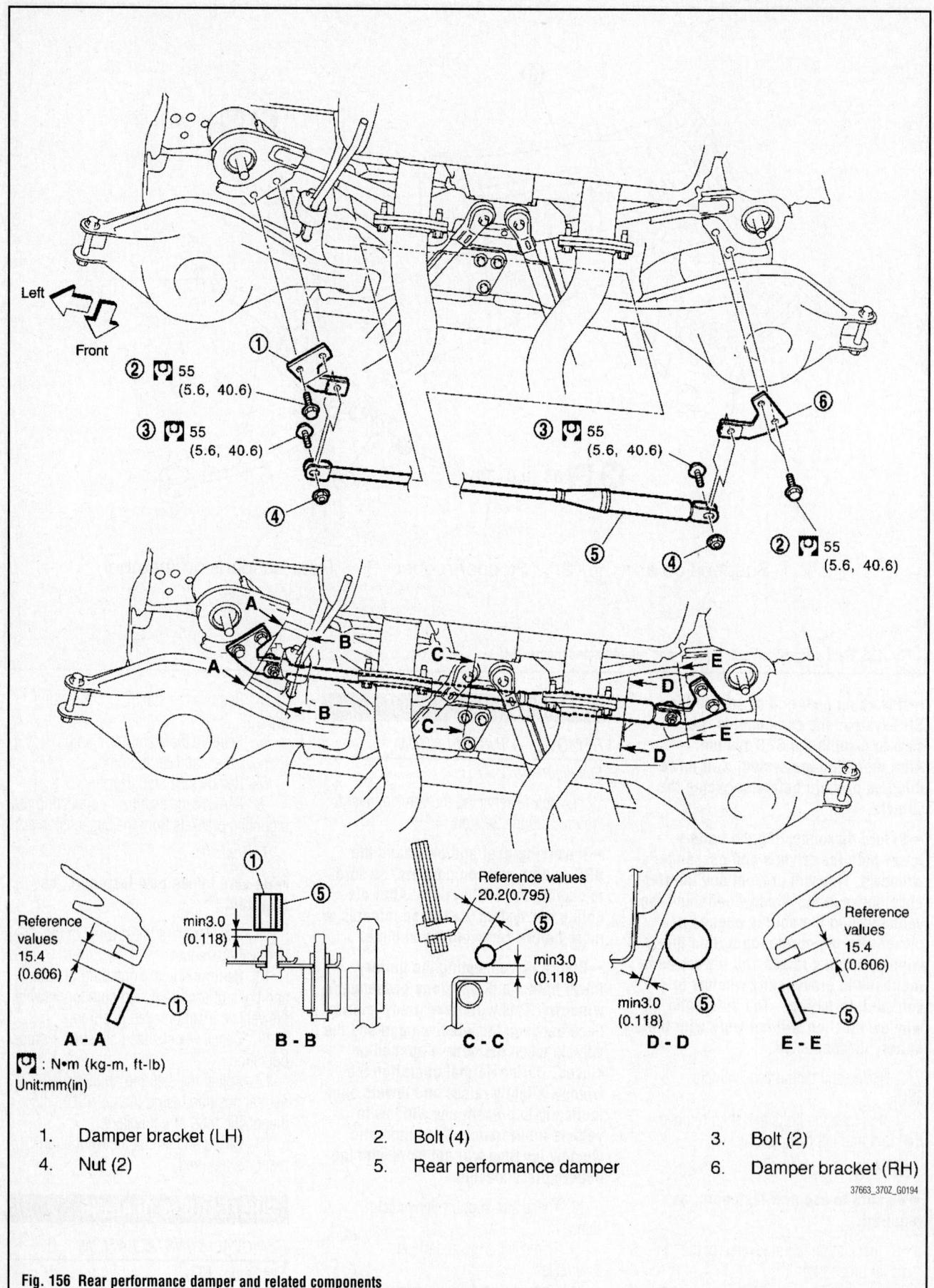

Reference values
20.2(0.795)

min3.0
(0.118)

Reference
values
15.4
(0.606)

min3.0
(0.118)

min3.0
(0.118)

Reference
values
15.4
(0.606)

A - A   B - B   C - C   D - D   E - E

⬚ : N•m (kg-m, ft-lb)
Unit:mm(in)

1.	Damper bracket (LH)	2.	Bolt (4)	3.	Bolt (2)
4.	Nut (2)	5.	Rear performance damper	6.	Damper bracket (RH)

37663_370Z_G0194

Fig. 156 Rear performance damper and related components

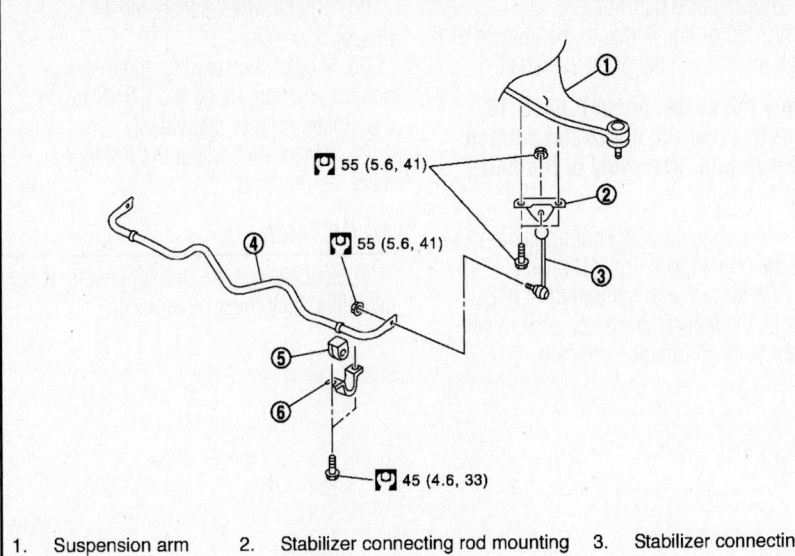

Fig. 157 Rear stabilizer bar and related components

1. Suspension arm
2. Stabilizer connecting rod mounting bracket
3. Stabilizer connecting rod
4. Stabilizer bar
5. Stabilizer bushing
6. Stabilizer clamp

37663_370Z_G0191

1. Before servicing the vehicle, refer to the Precautions Section.

➡️If working near and/or around the SRS system and components, be sure to disable the SRS system. After disabling the system wait three minutes or more before servicing the vehicle.

➡️Before disconnecting the battery, lower both the driver's and passenger's windows. This will prevent any interference between the window edge and the vehicle when the door is opened or closed. During normal operation the window slightly raises and lowers automatically to prevent any window to vehicle interference. The automatic window function will not work with the battery disconnected.

2. Raise and support the vehicle safely.

3. Remove the tire and wheel assembly.

4. Remove the caliper. Properly support the caliper to the side. Do not allow it to hang by the brake hose. Never depress the brake pedal with the caliper removed.

5. Remove the brake rotor.

6. Remove the cotter pin. Loosen the wheel hub locknut.

7. matchmark the halfshaft and the wheel hub and bearing assembly.

8. Remove the locknut.

9. Remove the parking brake shoe and parking brake cable.

10. Remove the stabilizer connecting rod (upper side).

11. Remove the coil spring.

12. Properly position a suitable jack under the axle housing.

13. Remove the radius rod.

14. Remove the shock absorber (lower side).

15. Remove the front lower link (axle housing side). Remove the rear lower link (axle housing side)

16. Separate the axle housing from the suspension arm, using the proper tool and remove the axle housing.

➡️Be careful not to damage the ball joint boot. Temporarily tighten the nut to prevent damage to the threads and to prevent the tool from coming off.

➡️Never place the halfshaft at an extreme angle. Be careful not to overextend the slide joint. Never allow the halfshaft to hang down with proper support.

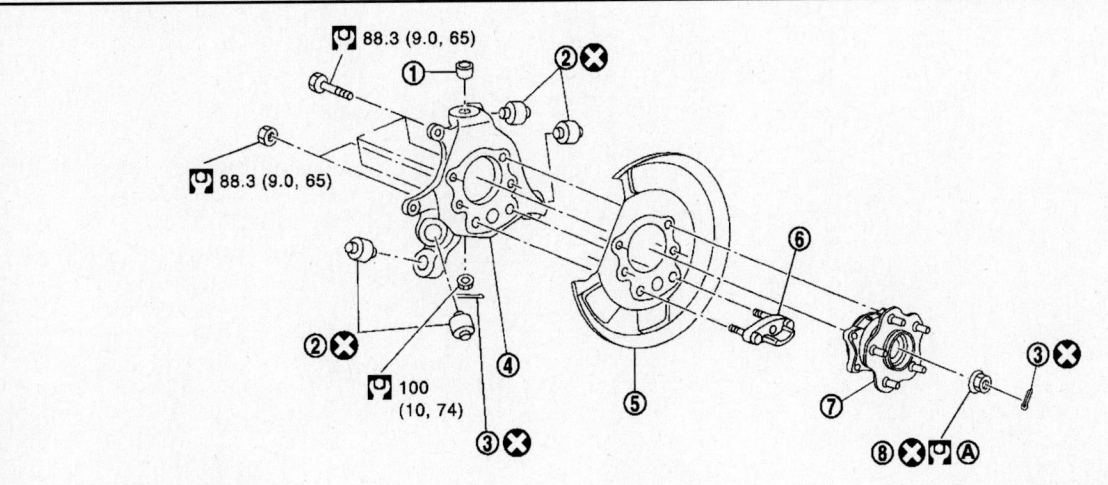

1. Ball seat
2. Bushing
3. Cotter pin
4. Axle housing
5. Back plate
6. Anchor block
7. Wheel hub and bearing assembly
8. Wheel hub lock nut

37663_370Z_G0192

Fig. 158 Rear wheel bearing and hub assembly and related components

17. Remove the wheel hub and bearing assembly.

To install:

→**Be sure to use new fasteners, as required.**

18. Installation is the reverse of the removal procedure.

19. Use the matchmarks to align removed components.

20. Clean the matching surface of the halfshaft and wheel hub and bearing assembly.

21. Apply paste part number 440037S000 or equivalent to the surface of the sub joint assembly of the halfshaft.

→**Apply the paste, about 0.04–0.10 ounce) to cover the entire flat surface of the sub joint assembly of the half-shaft.**

22. The wheel hub locknut tightening specification is 136 ft. lbs. (185 Nm).

23. Perform a final tightening of nuts and bolts of each removed component with the vehicle in an unladen position.

24. Check wheel speed sensor for proper operation.

25. Adjust the steering angle sensor neutral position, using the CONSULT-III diagnostic tool, or equivalent.

26. Check and adjust the front alignment, as required.

ADJUSTMENT

These bearings are not adjustable. If defective, they must be replaced.

NISSAN

Altima

9

SPECIFICATIONS AND MAINTENANCE CHARTS

ENGINE AND VEHICLE IDENTIFICATION

	Engine						Model Year	
Code ①	Liters (cc)	Cu. In.	Cyl.	Fuel Sys.	Engine Type	Eng. Mfg.	Code ②	Year
QR25DE	2.5 (2488)	152	4	MFI	DOHC	Nissan	9	2009
VQ35DE	3.5 (3498)	213	6	MFI	DOHC	Nissan	A	2010

MFI: Multi-port Fuel Injection

DOHC: Double Overhead Camshaft

① The Engine Code is stamped on the engine block near the starter.

② 10th position of the Vehicle Identification Number (VIN)

37663_ALTI_C0001

GENERAL ENGINE SPECIFICATIONS

Year	Model	Engine Displacement Liters	Engine Series (ID/VIN)	Net Horsepower @ rpm	Net Torque @ rpm (ft. lbs.)	Bore x Stroke (in.)	Com-pression Ratio	Oil Pressure @ rpm
2009	Altima	2.5	QR25DE	175@5600	180@3900	3.50X3.94	9.5:1	43@2000
	Altima	3.5	VQ35DE	270@6000	258@4400	3.76X3.20	10.3:1	43@2000
2010	Altima	2.5	QR25DE	175@5600	180@3900	3.50X3.94	9.5:1	43@2000
	Altima	3.5	VQ35DE	270@6000	258@4400	3.76X3.20	10.3:1	43@2000

37663_ALTI_C0002

ENGINE TUNE-UP SPECIFICATIONS

Year	Engine Displacement Liters	Engine ID/VIN	Spark Plug Gap (in.)	Ignition Timing (deg.) MT	AT	Fuel Pump (psi) ①	Idle Speed (rpm) MT	AT ②	Valve Clearance Intake ③	Exhaust ③
2009	2.5	QR25DE	0.043	15B	15B	51	650-750	650-750	0.009-0.013	0.010-0.013
	3.5	VQ35DE	0.043	12B	12B	51	550-650	550-650	0.010-0.013	0.011-0.015
2010	2.5	QR25DE	0.043	15B	15B	51	650-750	650-750	0.009-0.013	0.010-0.013
	3.5	VQ35DE	0.043	12B	12B	51	550-650	550-650	0.010-0.013	0.011-0.015

B: Before top dead center

① At idle

② Automatic transmission in neutral

③ Engine cold

37663_ALTI_C0003

CAPACITIES

Year	Model	Engine ID/VIN	Engine Displacement Liters	Engine Oil with Filter (qts.)	Transmission (pts.) 5-Spd	Auto.	Drive Axle Rear (pts.)	Fuel Tank (gal.)	Cooling System (qts.)
2009	Altima	QR25DE	2.5	4.5	7.2	17.5	—	20.0	8.0
	Altima	VQ35DE	3.5	4.5	3.6	10.8	—	20.0	8.5
2010	Altima	QR25DE	2.5	4.5	7.2	17.5	—	20.0	8.0
	Altima	VQ35DE	3.5	4.5	3.6	10.8	—	20.0	8.5

37663_ALTI_C0004

FLUID SPECIFICATIONS

Year	Model	Engine Displacement Liters	Engine Oil	Man. Trans.	Auto. Trans.	Power Steering Fluid	Brake Master Cylinder	Cooling System
2009	Altima	2.5	5W-30	75W-80	Nissan NS-2	Dexron IV	DOT 3	N-LL
	Altima	3.5	5W-30	75W-80	Nissan NS-2	Dexron IV	DOT 3	N-LL
2010	Altima	2.5	5W-30	75W-80	Nissan NS-2	Dexron IV	DOT 3	N-LL
	Altima	3.5	5W-30	75W-80	Nissan NS-2	Dexron IV	DOT 3	N-LL

N-LL: Nissan Long Life coolant

37663_ALTI_C0005

VALVE SPECIFICATIONS

Year	Engine ID/VIN	Engine Displacement Liters	Seat Angle (deg.)	Face Angle (deg.)	Spring Test Pressure (lbs. @ in.)	Spring Installed Height (in.)	Stem-to-Guide Clearance (in.) Intake	Exhaust	Stem Diameter (in.) Intake	Exhaust
2009	QR25DE	2.5	45.15-45.45	—	34-39@1.39	1.390	0.0008-0.0021	0.0012-0.0025	0.2348-0.2354	0.2344-0.2350
	VQ35DE	3.5	45.15-45.45	—	37-42@1.457	1.457	0.0008-0.0021	0.0016-0.0028	0.2348-0.2354	0.2344-0.2350
2010	QR25DE	2.5	45.15-45.45	—	34-39@1.39	1.390	0.0008-0.0021	0.0012-0.0025	0.2348-0.2354	0.2344-0.2350
	VQ35DE	3.5	45.15-45.45	—	37-42@1.457	1.457	0.0008-0.0021	0.0016-0.0028	0.2348-0.2354	0.2344-0.2350

37663_ALTI_C0006

CAMSHAFT SPECIFICATIONS
All measurements in inches unless noted

Year	Engine Displacement Liters	Engine Code/ID	Journal Dia.	Brg. Oil Clearance	Shaft End-play	Circle Runout	Lobe Height Intake	Lobe Height Exhaus
2009	2.5	QR25DE	①	0.0018-0.0034	0.0045-0.0074	0.0016	1.7644-1.7718	1.7313-1.7388
	3.5	VQ35DE	②	③	0.0045-0.0074	0.0008	1.7904-1.7978	1.7904-1.7978
2010	2.5	QR25DE	①	0.0018-0.0034	0.0045-0.0074	0.0016	1.7644-1.7718	1.7313-1.7388
	3.5	VQ35DE	②	③	0.0045-0.0074	0.0008	1.7904-1.7978	1.7904-1.7978

① No. 1: 1.0998-1.1006
All others: 0.9926-0.9234

② No. 1: 1.0211-1.0218
All others: 0.9230-0.9238

③ No. 1: 0.0018-0.0034
All others: 0.0014-0.0030

37663_ALTI_C0007

CRANKSHAFT AND CONNECTING ROD SPECIFICATIONS
All measurements are given in inches.

Year	Engine Displacement Liters	Engine ID/VIN	Crankshaft Main Brg. Journal Dia.	Crankshaft Main Brg. Oil Clearance	Crankshaft Shaft End-play	Crankshaft Thrust on No.	Connecting Rod Journal Diameter	Connecting Rod Oil Clearance	Connecting Rod Side Clearance
2009	2.5	QR25DE	2.3206-2.3216	①	0.0039-0.0102	3	1.8898-1.8903	0.0002-0.0007	0.0079-0.0138
	3.5	VQ35DE	2.3603-2.3612	0.0014-0.0018	0.0039-0.0098	3	2.1654-2.1659	0.0002-0.0007	0.0079-0.0138
2010	2.5	QR25DE	2.3206-2.3216	①	0.0039-0.0102	3	1.8898-1.8903	0.0002-0.0007	0.0079-0.0138
	3.5	VQ35DE	2.3603-2.3612	0.0014-0.0018	0.0039-0.0098	3	2.1654-2.1659	0.0002-0.0007	0.0079-0.0138

① Nos. 1, 3, 5 : 0.0005-0.0009
Nos. 2, 4 : 0.0007-0.0011

37663_ALTI_C0008

PISTON AND RING SPECIFICATIONS
All measurements are given in inches.

Year	Engine Displacement Liters	Engine ID/VIN	Piston Clearance	Ring Gap Top Compression	Ring Gap Bottom Compression	Ring Gap Oil Control	Ring Side Clearance Top Compression	Ring Side Clearance Bottom Compression	Ring Side Clearance Oil Control
2009	2.5	QR25DE	0.0004-0.0012	0.0083-0.0122	0.0146-0.0205	0.0079-0.0177	0.0016-0.0031	0.0012-0.0028	0.0018-0.0049
	3.5	VQ35DE	0.0004-0.0012	0.0091-0.0130	0.0091-0.0130	0.0079-0.0177	0.0018-0.0031	0.0012-0.0028	0.0026-0.0049
2010	2.5	QR25DE	0.0004-0.0012	0.0083-0.0122	0.0146-0.0205	0.0079-0.0177	0.0016-0.0031	0.0012-0.0028	0.0018-0.0049
	3.5	VQ35DE	0.0004-0.0012	0.0091-0.0130	0.0091-0.0130	0.0079-0.0177	0.0018-0.0031	0.0012-0.0028	0.0026-0.0049

37663_ALTI_C0009

TORQUE SPECIFICATIONS

All readings in ft. lbs.

Year	Engine Displacement Liters	Engine ID/VIN	Cylinder Head Bolts	Main Bearing Bolts	Rod Bearing Bolts	Crankshaft Damper Bolts	Flywheel Bolts	Manifold Intake	Manifold Exhaust	Spark Plugs	Oil Drain Plug
2009	2.5	QR25DE	①	②	③	NA	76-83	13-15	29-32	18	25
	3.5	VQ35DE	④	⑤	⑥	⑦	61-69	⑧	21-24	18	25
2010	2.5	QR25DE	①	②	③	NA	76-83	13-15	29-32	18	25
	3.5	VQ35DE	④	⑤	⑥	⑦	61-69	⑧	21-24	18	25

NA: Not available

① Step 1: 72 ft. lbs.

Step 2: Loosen completely, then retorque to 29 ft. lbs.

Step 3: Turn each bolt, in sequence, an additional 75 degrees

Step 4: Turn each bolt, in sequence, an additional 75 degrees

② Tighten bolts 11-22 to 19 ft. lbs.

Step 2: Tighten bolts 1-10 to 29 ft. lbs.

Step 3: Tighten bolts 1-10 an additional 60 degrees

③ Step 1: Tighten bolts to 22 ft. lbs.

Step 2: Loosen all bolts

Step 3: tighten bolts to 14 ft. lbs.

Step 4: Tighten bolts an additional 90 degrees

④ Step 1: Tighten to 72 ft. lbs.

Step 2: Loosen bolts completely in reverse order

Step 3: Tighten to 29 ft. lbs.

Step 4: Tighten an additional 103 degrees

Step 5: Repeat Step 4

⑤ Step 1: Shift crankshaft to align bearing beam

Step 2: Tighten all bolts to 24-28 ft. lbs.

Step 3: Tighten an additional 90-95 degrees

⑥ Step 1: Tighten to 14-15 ft. lbs.

Step 2: Tighten an additional 90-95 degrees

⑦ Step 1: Tighten to 32 ft. lbs. an additional 90 degrees

⑧ Step 1: Tighten to 65 inch lbs.

Step 2: tighten to 19 ft. lbs.

37663_ALTI_C0010

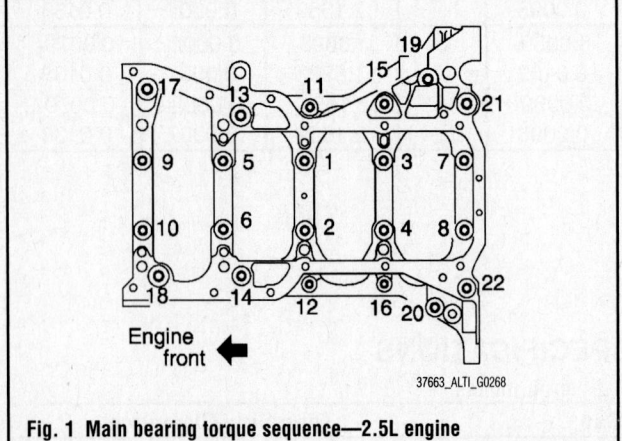

Fig. 1 Main bearing torque sequence—2.5L engine

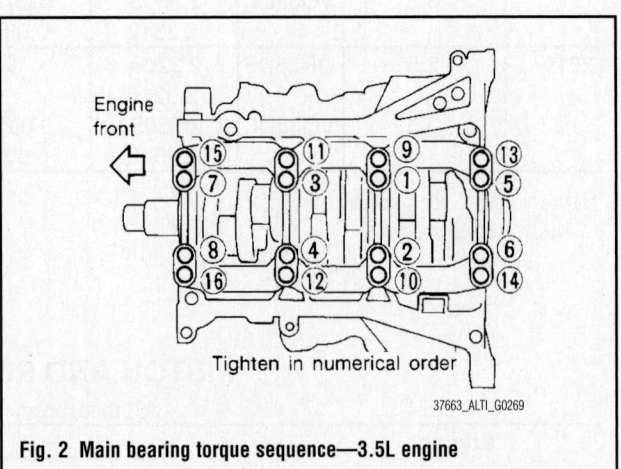

Fig. 2 Main bearing torque sequence—3.5L engine

WHEEL ALIGNMENT

			Caster		Camber			
			Range (+/-Deg.)	Preferred Setting (Deg.)	Range (+/-Deg.)	Preferred Setting (Deg.)	Toe-in (in.)
Year	Model						
2009	Altima	F	0.75	4.90	①	②	0.04 +/- 0.04
		R	—	—	0.30	-1.25	0.09 +/- 0.06
2010	Altima	F	0.75	4.90	①	②	0.04 +/- 0.04
		R	—	—	0.50	-0.62	0.09 +/- 0.06

① Minus 0.25 degrees, plus 0.75 degrees
② Left, -.050 degrees; Right -0.75 degrees

37663_ALTI_C0011

TIRE, WHEEL AND BALL JOINT SPECIFICATIONS

		OEM Tires		Tire Pressures (psi)		Wheel	Lug Nut Torque
Year	Model	Standard	Optional	Front	Rear	Size	(ft. lbs.)
2009	Altima 2.5	P215/60R16	None	32	32	6.5-JJ	80
	Altima 3.5 SL	P215/55R17	None	33	33	7-JJ	80
	Altima 3.5 SE	P235/45R18	None	33	33	7.5-JJ	80
2010	Altima 2.5	P215/60R16	None	32	32	6.5-JJ	80
	Altima 3.5 SL	P215/55R17	None	33	33	7-JJ	80
	Altima 3.5 SE	P235/45R18	None	33	33	7.5-JJ	80

OEM: Original Equipment Manufacturer
PSI: Pounds Per Square Inch

37663_ALTI_C0012

BRAKE SPECIFICATIONS
All measurements in inches unless noted

			Brake Disc			Brake Drum Diameter			Minimum Lining Thickness		Brake Caliper	
			Original Thickness	Minimum Thickness	Maximum Run-out	Original Inside Diameter	Max. Wear Limit	Maximum Machine Diameter	Front	Rear	Bracket Bolts (ft. lbs.)	Mounting Bolts (ft. lbs.)
Year	Model											
2009	Altima	F	1.024	0.945	0.002	—	—	—	0.079	—	98	20
		R	0.354	0.315	0.002	—	—	—	—	0.039	62	32
2010	Altima	F	1.020	0.945	0.002	—	—	—	0.079	—	98	20
		R	0.354	0.315	0.002	—	—	—	—	0.039	62	32

37663_ALTI_C0013

SCHEDULED MAINTENANCE INTERVALS
Nissan—Altima

TO BE SERVICED	TYPE OF SERVICE	VEHICLE MILEAGE INTERVAL (x1000)												
		7.5	15	22.5	30	37.5	45	52.5	60	67.5	75	82.5	90	97.5
Engine oil & filter	R	✓	✓	✓	✓	✓	✓	✓	✓	✓	✓	✓	✓	✓
Brake lines & cables	S/I		✓		✓		✓		✓		✓		✓	
Brake pads, discs, drums & linings	S/I		✓		✓		✓		✓		✓		✓	
Driveshaft boots	S/I		✓		✓		✓		✓		✓		✓	
Exhaust system	S/I				✓				✓				✓	
Transaxle fluid	S/I		✓		✓		✓		✓		✓		✓	
Air cleaner filter	R				✓				✓				✓	
Spark plugs (except platinum)	R				✓				✓				✓	
Spark plugs (iridium and platinum)	R	Replace every 105,000 miles												
Steering gear & linkage, axle & suspension parts	S/I				✓				✓				✓	
Engine coolant	R	Replace every 60,000 miles, then every 30,000 miles												
Inverter coolant	R	Replace every 60,000 miles, then every 30,000 miles												
Drive belts	S/I								✓					
Fuel lines	S/I								✓					
Vapor lines	S/I								✓					
Cabin microfilter	R		✓		✓		✓		✓		✓		✓	
Valve adjustment	S/I	As needed												

R: Replace   S/I: Service or Inspect

FREQUENT OPERATION MAINTENANCE (SEVERE SERVICE)

If a vehicle is operated under any of the following conditions it is considered severe service:

- Extremely dusty areas.

- 50% or more of the vehicle operation is in 32°C (90°F) or higher temperatures, or constant operation in temperatures below 0°C (32°F).

- Prolonged idling (vehicle operation in stop and go traffic).

- Frequent short running periods (engine does not warm to normal operating temperatures).

- Police, taxi, delivery usage or trailer towing usage.

Oil & oil filter: change every 3750 miles.

Brake pads & discs: service or inspect every 7500 miles.

Driveshaft boots: service or inspect every 7500 miles.

Exhaust system: service or inspect every 7500 miles.

Steering gear & linkage, axle & suspension parts: service or inspect every 7500 miles.

Steering linkage ball joints & front suspension ball joints: service or inspect every 7500 miles.

Air cleaner filter: service or inspect every 15,000 miles.

37663_ALTI_C0014

BRAKES INFORMATION AND PRECAUTIONS

ANTI-LOCK SYSTEMS

- Certain components within the ABS system are not intended to be serviced or repaired individually.
- Do not use rubber hoses or other parts not specifically specified for and ABS system. When using repair kits, replace all parts included in the kit. Partial or incorrect repair may lead to functional problems and require the replacement of components.
- Lubricate rubber parts with clean, fresh brake fluid to ease assembly. Do not use shop air to clean parts; damage to rubber components may result.
- Use only DOT 3 brake fluid from an unopened container.
- If any hydraulic component or line is removed or replaced, it may be necessary to bleed the entire system.
- A clean repair area is essential. Always clean the reservoir and cap thoroughly before removing the cap. The slightest amount of dirt in the fluid may plug an orifice and impair the

system function. Perform repairs after components have been thoroughly cleaned; use only denatured alcohol to clean components. Do not allow ABS components to come into contact with any substance containing mineral oil; this includes used shop rags.
- The Anti-Lock control unit is a microprocessor similar to other computer units in the vehicle. Ensure that the ignition switch is **OFF** before removing or installing controller harnesses. Avoid static electricity discharge at or near the controller.
- If any arc welding is to be done on the vehicle, the control unit should be unplugged before welding operations begin.

DISC AND DRUM SYSTEMS

> ※※ **CAUTION**
> Dust and dirt accumulating on brake parts during normal use may contain asbestos fibers from production or aftermarket brake linings. Breathing

excessive concentrations of asbestos fibers can cause serious bodily harm. Exercise care when servicing brake parts. Do not sand or grind brake lining unless equipment used is designed to contain the dust residue. Do not clean brake parts with compressed air or by dry brushing. Cleaning should be done by dampening the brake components with a fine mist of water, then wiping the brake components clean with a dampened cloth. Dispose of cloth and all residue containing asbestos fibers in an impermeable container with the appropriate label. Follow practices prescribed by the Occupational Safety and Health Administration (OSHA) and the Environmental Protection Agency (EPA) for the handling, processing, and disposing of dust or debris that may contain asbestos fibers.

BRAKES BLEEDING THE BRAKE SYSTEM

BLEEDING PROCEDURE

BLEEDING PROCEDURE

➡ **While bleeding, pay attention to master cylinder fluid level.**

> ※※ **CAUTION**
> Before working, disconnect connectors of ABS actuator and electric unit (control unit) or battery cable from the negative terminal.

1. Connect a vinyl tube to rear right brake caliper bleed valve.
2. Fully depress brake pedal 4 or 5 times.
3. With brake pedal depressed, loosen bleed valve to bleed air in brake line, and then tighten it immediately.
4. Repeat steps 2 and 3 until all of the air is out of the brake line.
5. Tighten the bleed valve.
6. From step 1 to 5, with master

cylinder reservoir tank filled at least half way, bleed air from brake hydraulic line bleed valves in the following order:

- Rear right brake
- Front left brake
- Rear left brake
- Front right brake

BLEEDING THE ABS SYSTEM

Refer to "Bleeding Procedure".

BRAKES ANTI-LOCK BRAKE SYSTEM (ABS)

WHEEL SPEED SENSORS

REMOVAL & INSTALLATION
See Figure 3.

> ※※ **CAUTION**
> Note the following:

- Be careful not to damage wheel sensor edge and sensor rotor teeth.
- When removing the front or rear wheel hub, first remove the wheel sensor from the wheel hub. Failure to do so may result in damage to the wheel sensor wires making the sensor inoperative.

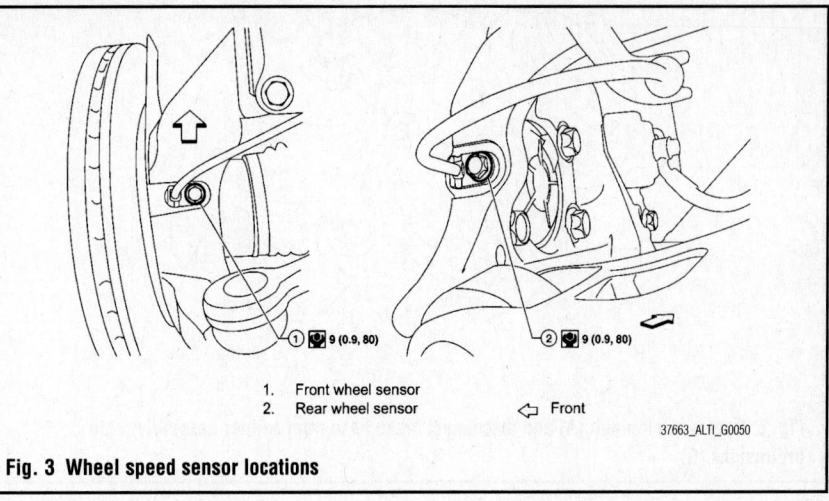

1. Front wheel sensor
2. Rear wheel sensor

⬅ Front

37663_ALTI_G0050

Fig. 3 Wheel speed sensor locations

- Pull out the wheel sensor, being careful to turn it as little as possible. Do not pull on the wheel sensor harness.
- Before installation, check if foreign objects such as iron fragments are adhered to the pick-up part of the sensor or to the inside of the hole in the wheel hub for the wheel sensor, or if a foreign object is caught in the surface of the mating surface for the sensor rotor. Fix as necessary and then install the wheel sensor.

Front Wheel Sensor

1. Remove front wheel and tire.
2. Partially front wheel fender protector.
3. Remove wheel sensor bolt and wheel sensor.
4. Remove harness wire from mounts and disconnect wheel sensor harness connector.

To install:

5. Installation is in the reverse order of removal.

Rear Wheel Sensor

See Figure 4.

➡**Both rear wheel sensors share one harness and must be replaced as an assembly.**

1. Remove rear wheel and tire.
2. Remove wheel sensor bolts and wheel sensors from both rear wheel hub and bearing assemblies.
3. Remove harness wire from mounts and harness wire clips from rear suspension member.
4. Disconnect wheel sensor harness connector.

To install:

5. Installation is in the reverse order of removal.

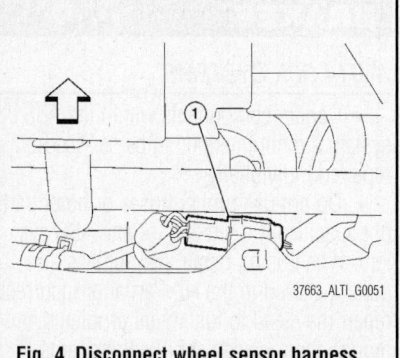

37663_ALTI_G0051

Fig. 4 Disconnect wheel sensor harness connector (1)

WHEEL SPEED SENSOR ROTORS

REMOVAL & INSTALLATION

The front and rear wheel sensor rotors are an integral part of the wheel hubs and cannot be disassembled. When replacing the sensor rotor, replace the wheel hub.

BRAKES

FRONT DISC BRAKES

BRAKE CALIPER

REMOVAL & INSTALLATION

See Figures 5 through 8.

1. Remove front wheel and tires.
2. Drain brake fluid.
3. Remove union bolt and disconnect brake hose from caliper assembly. Discard the copper washers.

❊❊ CAUTION

Do not reuse copper washers.

4. Remove torque member bolts, and remove brake caliper assembly.

❊❊ CAUTION

Do not drop brake pad.

5. Remove disc rotor. If reusing

the disc rotor apply matching marks as shown.

❊❊ CAUTION

Put matching marks on wheel hub assembly and disc rotor, if it is necessary to remove disc rotor.

To install:

6. Install disc rotor, align the matching marks if installing the original disc rotor as shown.

❊❊ CAUTION

Align the marks on disc rotor and wheel hub at the time of installation when reusing disc rotor.

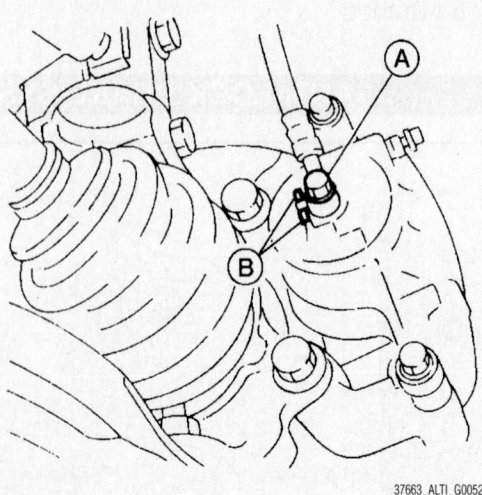

37663_ALTI_G0052

Fig. 5 Remove union bolt (A) and disconnect brake hose from caliper assembly; note protrusions (B)

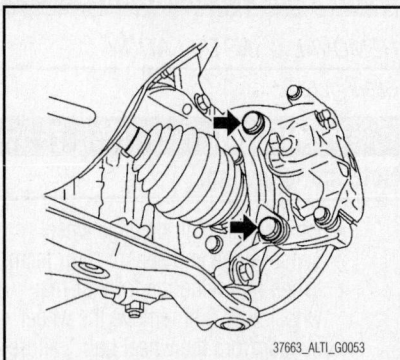

37663_ALTI_G0053

Fig. 6 Remove torque member bolts, and remove brake caliper assembly

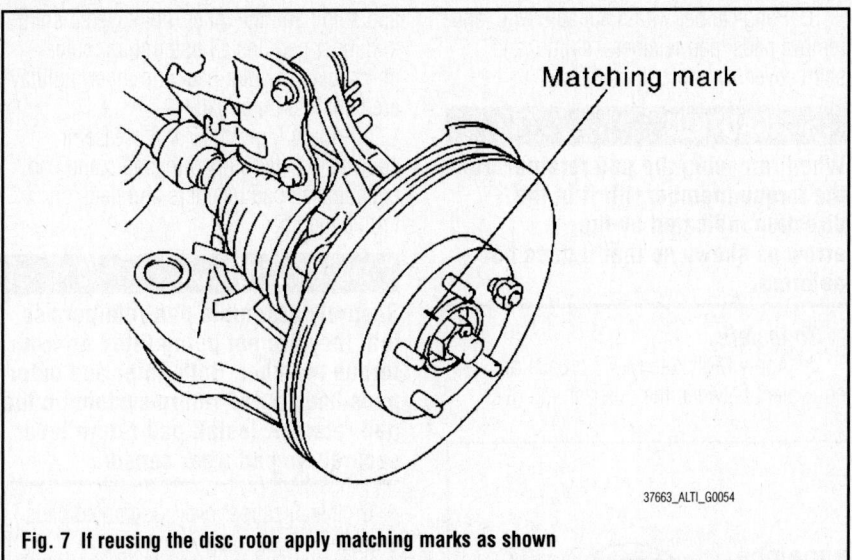

Fig. 7 If reusing the disc rotor apply matching marks as shown

Matching mark

37663_ALTI_G0054

7. Install brake caliper assembly to vehicle, and tighten torque member bolts to the specified torque.

❊❊ **CAUTION**

Do not allow oil or any moisture on all contact surfaces between steering knuckle and caliper assembly, bolts, and washer.

8. Install brake hose to brake caliper assembly with new copper washers. Align the brake hose tab between the protrusions on the caliper assembly as shown. Tighten union bolt to the specified torque.

❊❊ **CAUTION**

Do not reuse copper washers.

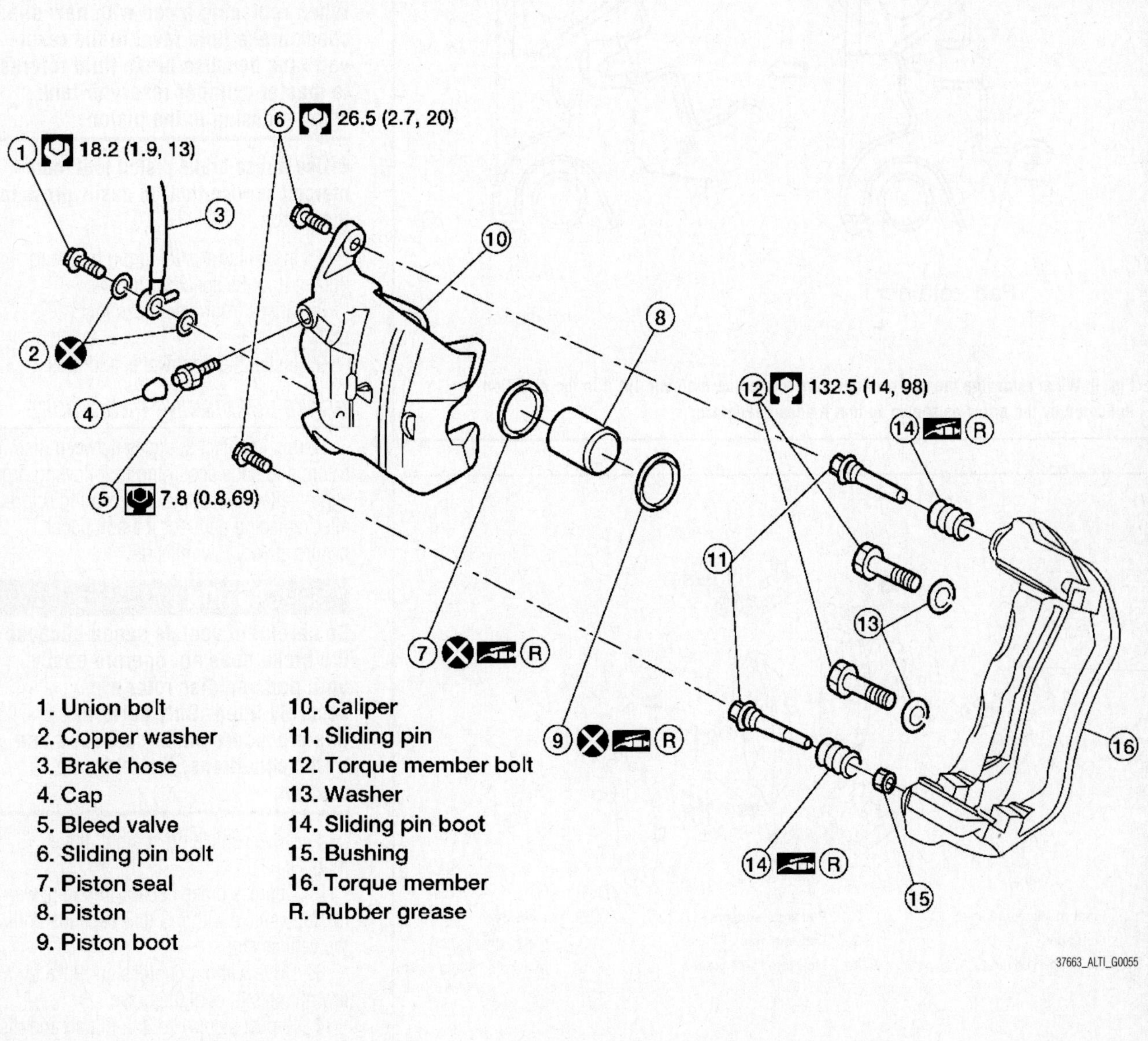

① 18.2 (1.9, 13)
⑥ 26.5 (2.7, 20)
⑤ 7.8 (0.8, 69)
⑫ 132.5 (14, 98)

1. Union bolt
2. Copper washer
3. Brake hose
4. Cap
5. Bleed valve
6. Sliding pin bolt
7. Piston seal
8. Piston
9. Piston boot
10. Caliper
11. Sliding pin
12. Torque member bolt
13. Washer
14. Sliding pin boot
15. Bushing
16. Torque member
R. Rubber grease

37663_ALTI_G0055

Fig. 8 Exploded view of front brake caliper assembly

9. Refill with new brake fluid and bleed air from the brake hydraulic system.
10. Check front disc brakes for drag.
11. Install front wheel and tires.

DISC BRAKE PADS

REMOVAL & INSTALLATION

See Figures 9 and 10.

1. Remove the front wheel and tires.
2. Remove lower sliding pin bolt.

3. Hang caliper with a suitable wire, and remove pads, pad retainers, shims, and shim cover from torque member.

✳✳ CAUTION

When removing the pad retainer from the torque member, lift it in the direction indicated by the arrow as shown so that it does not deform.

To install:

4. Apply Molykote M-77 grease or equivalent between the outer shim cover

and shim; and the inner multilayered shim and inner pad. Install outer shim, outer shim cover to outer pad, and inner multilayered shim to inner pad.

5. Apply Molykote 7439 grease or equivalent between pad retainers and pad ends. Install pad retainers and pads on torque member.

✳✳ CAUTION

Securely assemble pad retainers so that they are not being lifted up from torque member. Both inner and outer pads have a pad return system on the pad retainer. Install pad return lever securely to pad wear sensor.

6. Install caliper over assembled pads on to the torque member.

✳✳ CAUTION

When replacing a pad with new one, check brake fluid level in the reservoir tank because brake fluid returns to master cylinder reservoir tank when pressing in the piston.

➡ **Use a disc brake piston tool (commercial service tool) to easily press in the piston.**

7. Install lower sliding pin bolt, and tighten it to the specified torque.
8. Check front disc brake for drag.
9. Install the front wheel and tires.

BRAKE BURNISHING PROCEDURE

Burnish contact surfaces between disc rotors and pads according to following procedure after refinishing or replacing rotors, after replacing pads, or if a soft pedal occurs at very low mileage.

✳✳ CAUTION

Be careful of vehicle speed because the brake does not operate easily until pad and disc rotor are securely fitted. Only perform this procedure under safe road and traffic conditions. Use extreme caution.

1. Drive vehicle on straight, flat road.
2. Depress brake pedal with the power to stop vehicle within 3 to 5 seconds until the vehicle stops.
3. Drive without depressing brake for a few minutes to cool the brake.
4. Repeat steps 1 to 3 until pad and disc rotor are securely fitted.

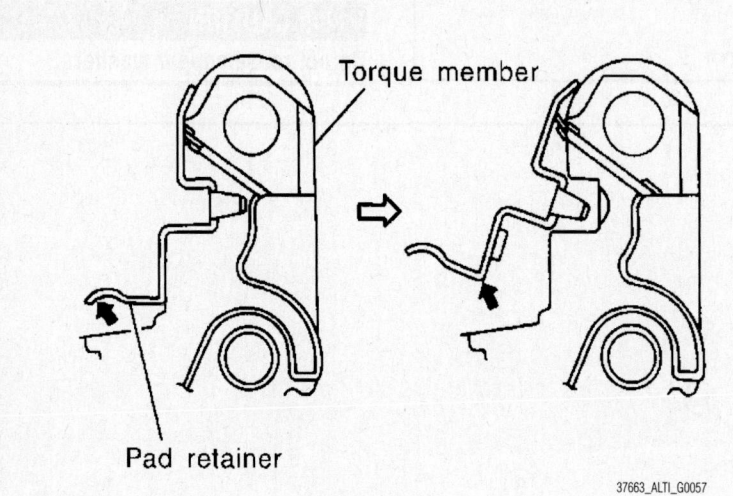

Torque member

Pad retainer

37663_ALTI_G0057

Fig. 9 When removing the pad retainer from the torque member, lift it in the direction indicated by the arrow as shown so that it does not deform

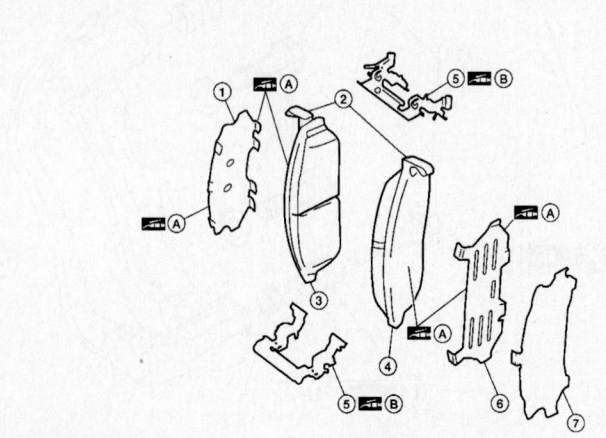

1.	Inner multilayered shim	2.	Pad wear sensors	3.	Inner pad
4.	Outer pad	5.	Pad retainers	6.	Outer shim
7.	Outer shim cover	A.	Molykote M-77 grease	B.	Molykote 7439 grease

37663_ALTI_G0058

Fig. 10 Exploded view of front brake pad components

BRAKES

BRAKE CALIPER

REMOVAL & INSTALLATION

See Figures 11 through 13.

1. Remove rear wheel and tires.
2. Fasten disc rotor using a wheel nut.
3. Drain brake fluid.

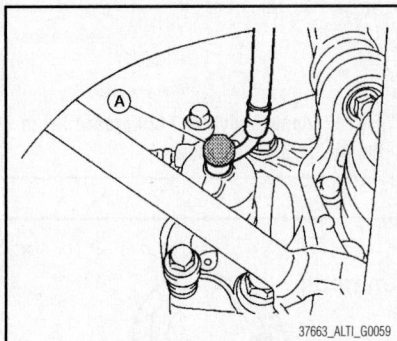

Fig. 11 Remove union bolt (A) and then disconnect brake hose from caliper

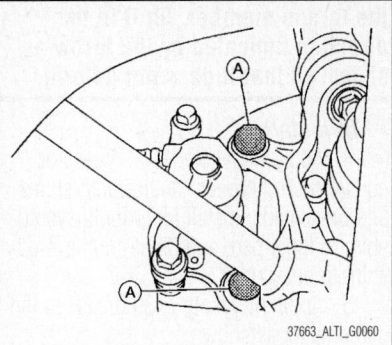

Fig. 12 Remove the two torque member bolts (A), and then remove the torque member, caliper and pads as an assembly

4. Remove union bolt and then disconnect brake hose from caliper. Discard the copper washers.
5. Remove the two torque member bolts, and then remove the torque member, caliper and pads as an assembly.

✳✳ CAUTION

Do not drop the brake pad and multi-layered shim assemblies.

6. Remove the two sliding pin bolts and separate the caliper from the torque member.
7. Remove the brake pad and multilayered shim assemblies from the caliper.
8. Remove the disc rotor.

✳✳ CAUTION

Put matching marks on wheel hub assembly and disc rotor, if it necessary to reuse the disc rotor.

To install:
9. Install the disc rotor.

✳✳ CAUTION

Alignment marks of disc rotor and wheel hub put at the time of removal when reusing disc rotor.

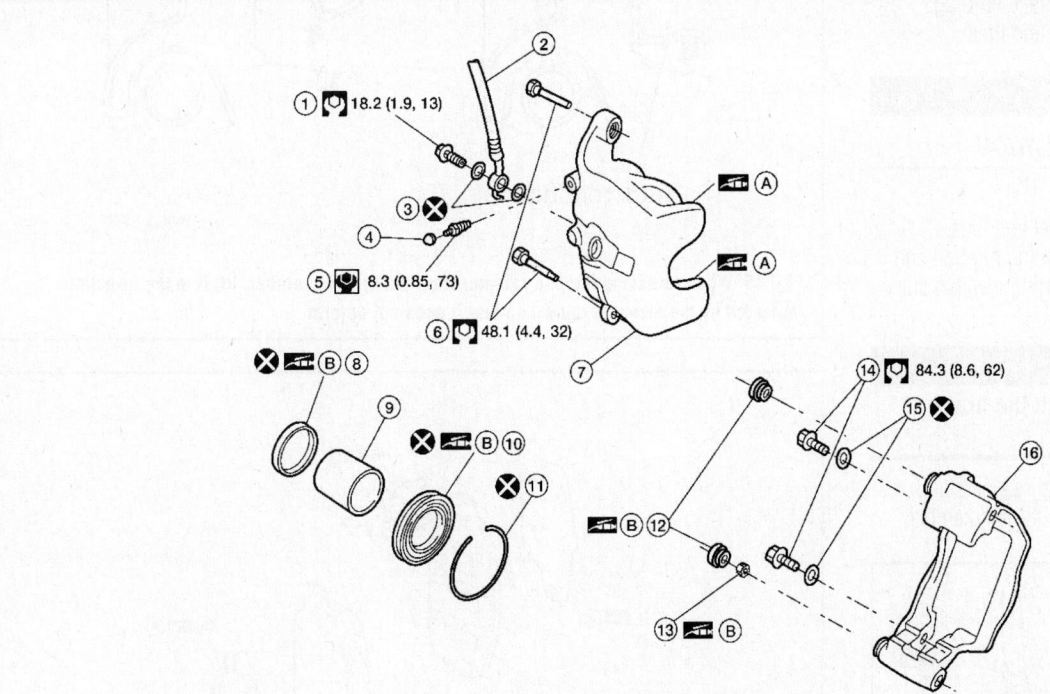

1. Union bolt	2. Brake hose	3. Copper washer
4. Cap	5. Bleed valve	6. Sliding pin bolt
7. Caliper	8. Piston seal	9. Piston
10. Piston boot	11. Retaining ring	12. Sliding pin boot
13. Bushing	14. Torque member bolt	15. Washer
16. Torque member	A. PBC (Poly Butyl Cuprysil) grease or silicone-based grease	B. Rubber grease

37663_ALTI_G0061

Fig. 13 Exploded view of rear disc brake assembly

10. Install the brake pad and multilayered shim assemblies on the caliper.

11. Install the caliper and pad assembly on the torque member, then tighten the two sliding pin bolts to the specified torque.

12. Install the torque member, pads and brake caliper assembly, and tighten the torque member bolts (A) to the specified torque.

※※ CAUTION

Before installing wipe off all oil and moisture on all mating surfaces of rear axle and torque member, threads, bolts and washers.

13. Align the L-shaped pin on the brake hose in the hole in the caliper, then install the brake hose with new copper washers and tighten the union bolt to the specified torque.

※※ CAUTION

Do not reuse copper washers.

14. Refill with new brake fluid and bleed air.

15. Check rear disc brake for drag.

16. Install rear wheel and tires.

DISC BRAKE PADS

REMOVAL & INSTALLATION

See Figures 14 through 18.

1. Remove rear wheel and tires.

2. Remove upper sliding pin bolt and swing caliper out supporting it with a suitable wire.

※※ CAUTION

Do not twist or stretch the brake hose.

3. Remove pads, pad retainers and multilayered shims from torque member.

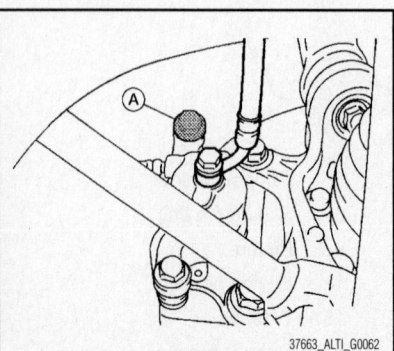

Fig. 14 Remove upper sliding pin bolt (A) and swing caliper out supporting it with a suitable wire

※※ CAUTION

When removing the pad retainer from the torque member, lift it in the direction indicated by the arrow as shown so that it does not deform.

To install:

4. Apply Molykote M-77 grease or equivalent to between multilayered shims and brake pads. Install inner multilayered shim to inner pad, and outer multilayered shim to outer pad.

5. Apply Molykote 7439 grease to the pad retainer as shown.

6. Attach pad retainers to torque member, then install brake pads and multilayered shim assemblies.

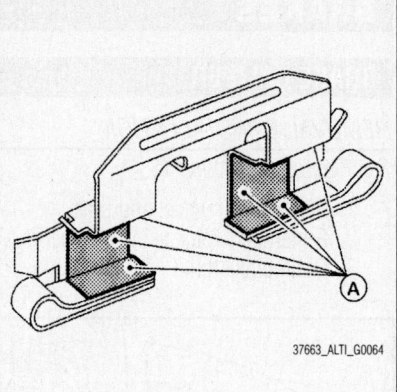

Fig. 17 Apply Molykote 7439 grease (A) to the pad retainer as shown

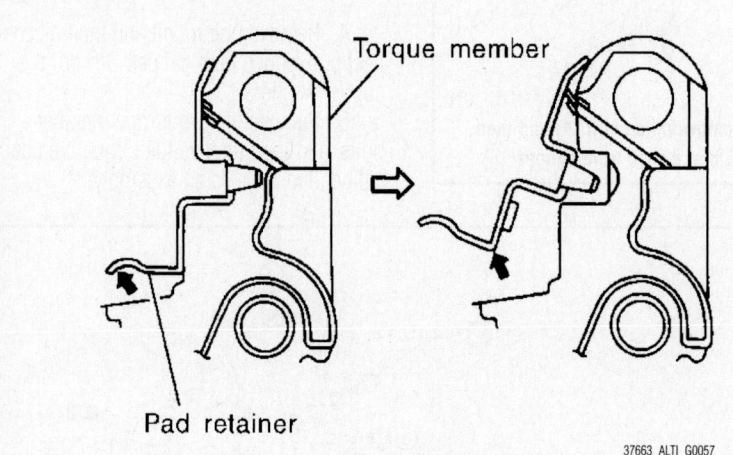

Fig. 15 When removing the pad retainer from the torque member, lift it in the direction indicated by the arrow as shown so that it does not deform

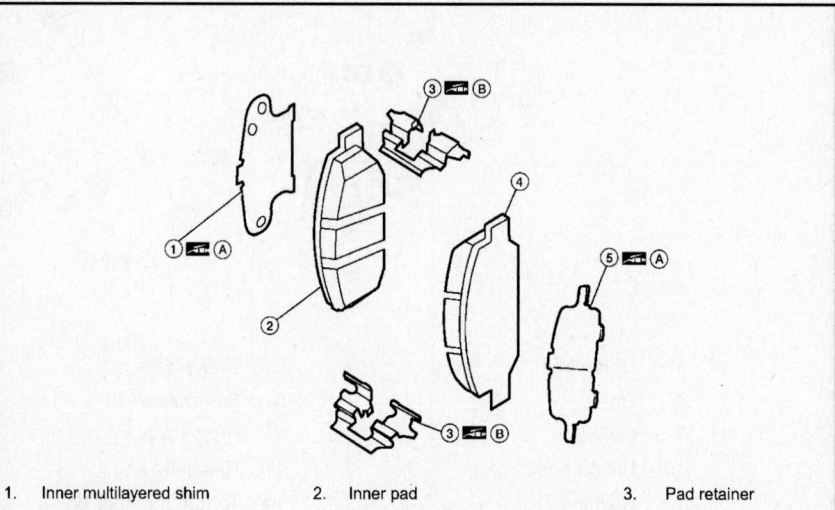

1.	Inner multilayered shim	2.	Inner pad	3.	Pad retainer
4.	Outer pad	5.	Outer multilayered shim	A.	Molykote M-77 grease
B.	Molykote 7439 grease				

Fig. 16 Exploded view of rear brake pad components

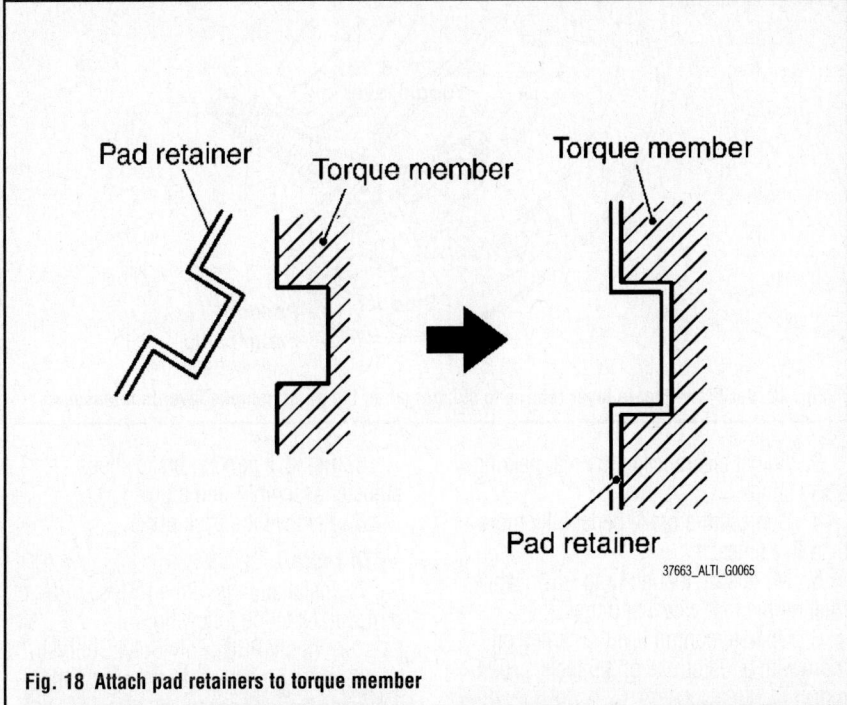

Fig. 18 Attach pad retainers to torque member

※※ **CAUTION**

When attaching pad retainer, attach it firmly so that it is flush with torque member as shown.

7. Press in piston until pads can be installed, and then install caliper to torque member.

※※ **CAUTION**

In the case of replacing a pad with new one, check a brake fluid level in the reservoir tank because brake fluid returns to master cylinder reservoir tank when pressing piston in.

➡ **Use a disc brake piston tool (commercial service tool) to easily press piston.**

8. Install upper sliding pin bolt and tighten to the specified torque.
9. Check rear disc brake for drag.
10. Install rear wheel and tires.

BRAKE BURNISHING PROCEDURE

Burnish contact surfaces between disc rotors and pads according to following procedure after refinishing or replacing rotors, after replacing pads, or if a soft pedal occurs at very low mileage.

※※ **CAUTION**

Be careful of vehicle speed because the brake does not operate easily until pad and disc rotor are securely fitted. Only perform this procedure under safe road and traffic conditions. Use extreme caution.

1. Drive vehicle on straight, flat road.
2. Depress brake pedal with the power to stop vehicle within 3 to 5 seconds until the vehicle stops.
3. Drive without depressing brake for a few minutes to cool the brake.
4. Repeat steps 1 to 3 until pad and disc rotor are securely fitted.

BRAKES   **PARKING BRAKE**

PARKING BRAKE CABLES

ADJUSTMENT

Pedal Type

See Figures 19 and 20.

1. Remove rear wheel and tires.
2. Insert a deep socket wrench onto adjusting nut. Rotate adjusting nut to fully loosen cable, and then release parking brake pedal.
3. Secure disc rotor to hub using wheel nut so as not to tilt disc rotor.
4. Remove adjuster hole plug installed on the disc rotor. Turn the adjuster in direction using a suitable tool or a flat-bladed screwdriver as shown, until disc rotor is locked. Turn the adjuster in the opposite direction by 5 or 6 notches after locking.
5. Rotate disc rotor to make sure that there is no drag. Install the adjuster hole plug.

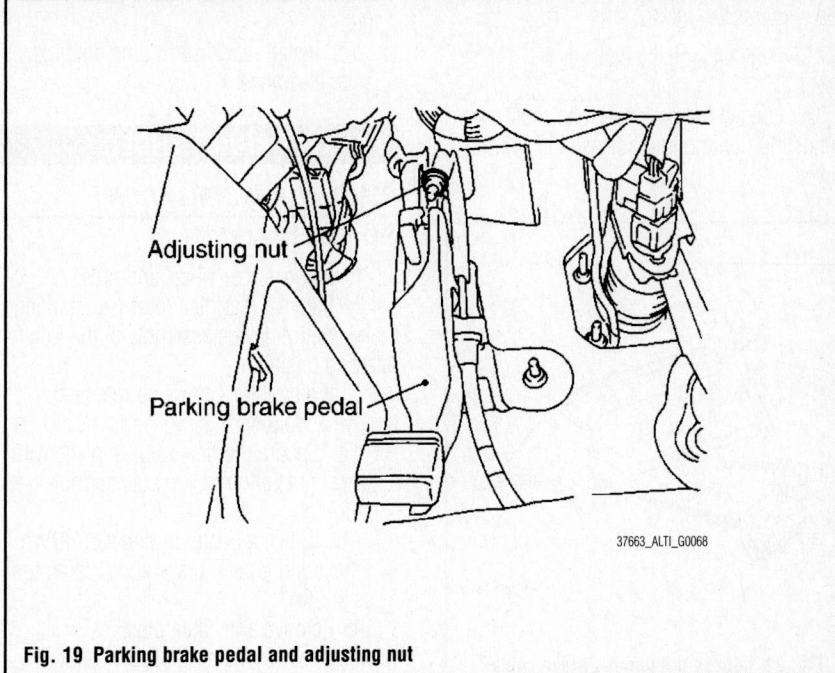

Fig. 19 Parking brake pedal and adjusting nut

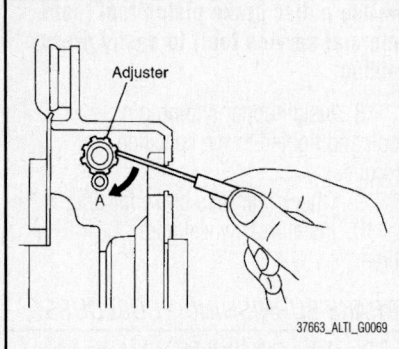

Fig. 20 Turn the adjuster in direction (A) using a suitable tool or a flat-bladed screwdriver as shown, until disc rotor is locked

6. Adjust parking brake cable with the following procedure.

a. Operate parking brake pedal 10 or more times with a full stroke of 7.6 inches (194.3 mm).

b. Rotate adjusting nut to adjust parking brake pedal stroke using a deep socket wrench.

c. Operate parking brake pedal with a force of 66 lbs. (294 N), make sure the pedal stroke is within the specified number (4 to 5) of notches. Check it by listening and counting the ratchet clicks.

d. Make sure that there is no drag on the parking brake with the parking brake pedal completely released.

Lever Type

See Figures 21 and 22.

1. Fully engage the control lever.

2. Loosen the parking brake cable adjusting nut and fully release the control lever.

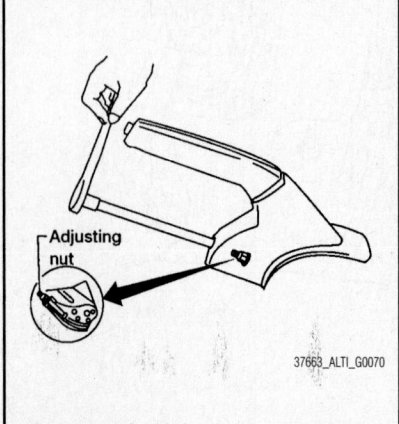

Fig. 21 Loosen the parking brake cable adjusting nut

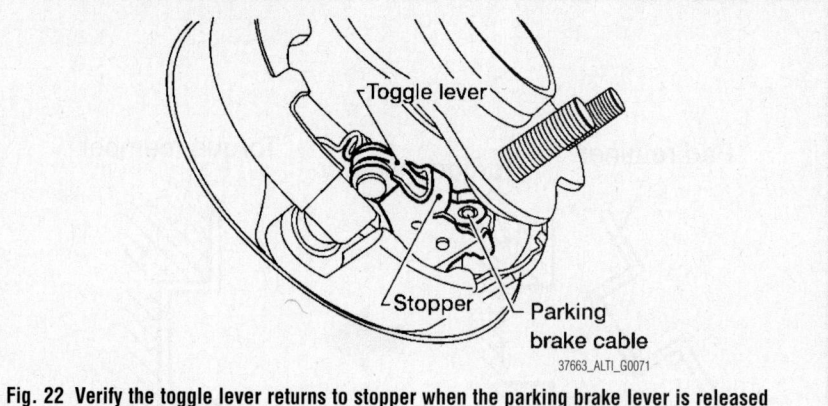

Fig. 22 Verify the toggle lever returns to stopper when the parking brake lever is released

3. Adjust clearance of the rear parking brake shoes.

4. Depress the brake pedal fully more than five times.

5. Make sure that no drag exists while rotating the rear wheel and tires.

6. Operate control lever 10 times or more with a full stroke of 3.9 inches (99.5 mm).

7. Adjust control lever by turning adjusting nut.

8. Pull control lever with a force of 66 lbs. (294 N). Check control lever stroke and ensure smooth operation.

9. After adjustment, check that there is no drag while the control lever is being released. If drag exists, perform the following:

a. Remove the rear disc rotor. Verify the toggle lever returns to stopper when the parking brake lever is released.

b. If toggle lever does not return to stopper, loosen adjusting nut.

c. Install rear disc rotor and adjust shoe clearance.

PARKING BRAKE SHOES

REMOVAL & INSTALLATION

See Figures 23 and 24.

1. Remove rear wheel and tires.

2. Remove rear disc rotor with parking brake control device assembly in the fully released position.

3. If disc rotor cannot be removed, remove as follows:

a. Secure the disc rotor in place with wheel nuts and remove adjuster hole plug.

b. Rotate adjuster in direction (B) to retract and loosen brake shoe, using tool as shown.

4. Remove anti-rattle pins, retainers, anti-rattle springs, and return springs.

5. Remove parking brake shoes, adjuster assembly, and toggle lever.

6. Remove the back plate.

To install:

7. Installation is in the reverse order of removal. Note the following:

a. Apply PBC (Poly Butyl Cuprysil) grease or equivalent to the specified points during assembly.

b. Assemble adjusters so that threaded part is expanded when rotating it in the direction shown.

c. Shorten adjuster by rotating it as shown.

d. Check shoe sliding surface and drum inner surface for grease. Wipe it off if it adhere on the surfaces.

e. Perform break-in operation as follows after replacing brake shoes or disc rotors, or if brakes do not function well.

8. Adjust parking brake control device assembly stroke to the specified amount.

9. Perform parking brake break-in (drag run) operation by driving vehicle under the following conditions:

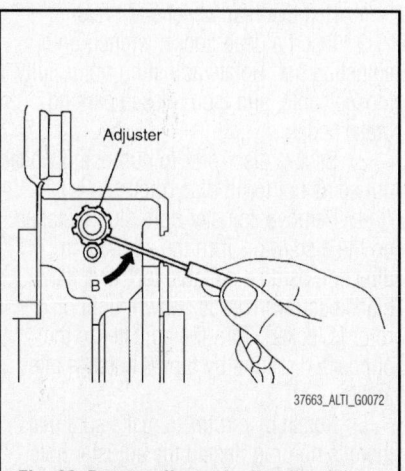

Fig. 23 Rotate adjuster in direction (B) to retract and loosen brake shoe

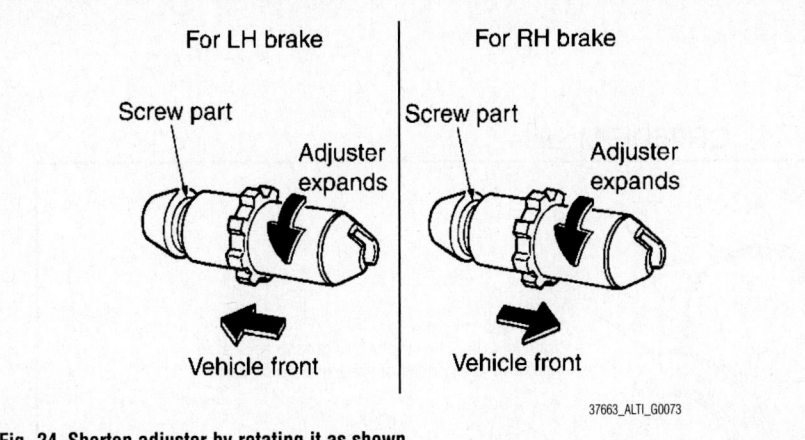

Fig. 24 Shorten adjuster by rotating it as shown

a. Drive the vehicle forward.

b. Maintain vehicle speed at approximately 40 km/h (25 MPH) keeping it constant in forward direction.

c. Apply the parking brake at an operating force of approximately 88 lbs. (400 N) with constant force.

d. Release the parking brake after approximately 10 seconds.

10. Check parking brake control device assembly stroke of parking brake. Readjust as necessary if it is outside the standard specifications.

CHASSIS ELECTRICAL

GENERAL INFORMATION

✳✳ CAUTION

These vehicles are equipped with an air bag system. The system must be disarmed before performing service on, or around, system components, the steering column, instrument panel components, wiring and sensors. Failure to follow the safety precautions and the disarming procedure could result in accidental air bag deployment, possible injury and unnecessary system repairs.

SERVICE PRECAUTIONS

✳✳ CAUTION

Disconnect and isolate the battery negative cable before beginning any

AIR BAG (SUPPLEMENTAL RESTRAINT SYSTEM)

airbag system component diagnosis, testing, removal, or installation procedures. Wait at least 90 seconds after the ignition switch is turned off and the negative (-) terminal cable is disconnected from the battery before starting the operation. The SRS is equipped with a backup power source, so if work is started within 90 seconds after disconnecting the negative (-) terminal cable from the battery, the SRS may be deployed. Failure to disable the airbag system may result in accidental airbag deployment, personal injury, or death.

DISARMING THE SYSTEM

✳✳ CAUTION

Before servicing the SRS, turn ignition switch OFF, disconnect both

battery cables and wait at least 3 minutes. For approximately 3 minutes after the cables are removed, it is still possible for the air bag and seat belt pretensioner to deploy. Therefore, do not work on any SRS connectors or wires until at least 3 minutes have passed.

ARMING THE SYSTEM

When the repair work is completed, reconnect both battery cables. With the brake pedal released, turn the push-button ignition switch from ACC position to ON position, then to LOCK position. (The steering wheel will lock when the push-button ignition switch is turned to LOCK position.)

DRIVE TRAIN

CLUTCH DRIVEN DISC & PRESSURE PLATE

REMOVAL & INSTALLATION

See Figures 25 and 26.

✳✳ CAUTION

Note the following:

- If transaxle assembly is removed from the vehicle, always replace CSC (Concentric Slave Cylinder). Return CSC insert to original position to remove transaxle assembly.

Dust on clutch disc sliding parts may damage CSC seal and may cause clutch fluid leakage.

- Be careful not to apply any grease to the clutch disc facing, pressure plate surface and flywheel surface.
- If flywheel is removed, align dowel pin with the smallest hole of flywheel (for VQ35DE engine models).
- Replace clutch disc and clutch

cover as a set (for VQ35DE engine models).

1. Remove transaxle assembly from the vehicle.

2. Loosen clutch cover bolts evenly. Then remove clutch cover and clutch disc.

To install:

3. Clean clutch disc and input shaft splines to remove grease and dust caused by abrasion.

4. Apply recommended grease to clutch disc and input shaft splines.

QR25DE Models

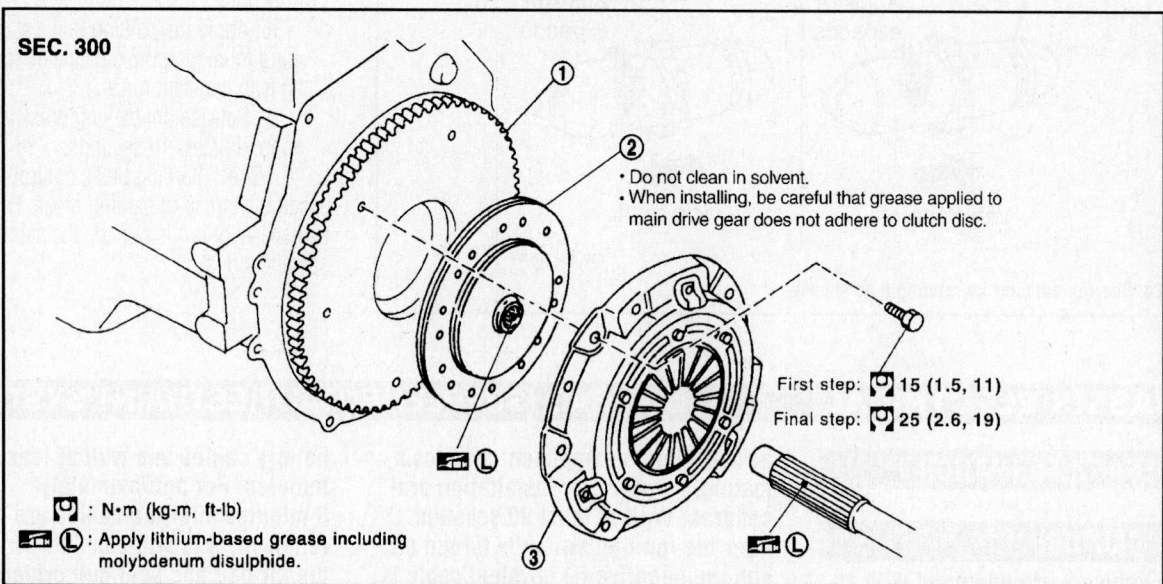

SEC. 300

• Do not clean in solvent.
• When installing, be careful that grease applied to main drive gear does not adhere to clutch disc.

First step: 🔧 15 (1.5, 11)
Final step: 🔧 25 (2.6, 19)

🔧 : N·m (kg-m, ft-lb)

🛠️ Ⓛ : Apply lithium-based grease including molybdenum disulphide.

1. Flywheel
2. Clutch disc
3. Clutch cover

VQ35DE Models

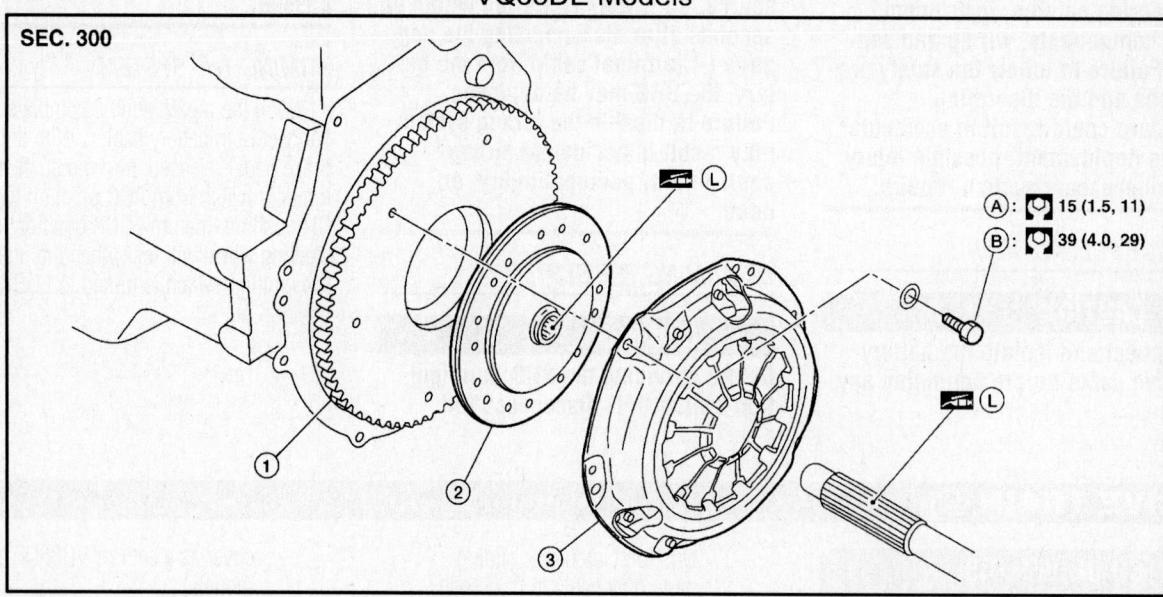

SEC. 300

A: 🔧 15 (1.5, 11)
B: 🔧 39 (4.0, 29)

1. Flywheel
A. First step

2. Clutch disc *1, *2
B. Final Step

3. Clutch cover

🛠️ Ⓛ: Apply lithium-based grease including molybdenum disulphide

*1. Do not clean in solvent.

*2. When installing, be careful that grease applied to input shaft does not adhere to clutch disc.

37663_ALTI_G0102

Fig. 25 Exploded views of clutch assemblies

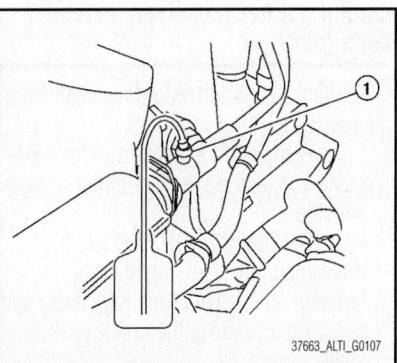

Fig. 26 Tighten clutch cover bolts evenly in two steps in the order shown

※ **CAUTION**

Be sure to apply grease to the points specified. Otherwise, noise, poor disengagement, or damage to the clutch may result. Excessive grease may cause slip or shudder. If it adheres to CSC seal, it will cause clutch fluid leakage. Wipe off excess grease.

5. Install clutch disc using Tool KV30101000.
6. Install clutch cover. Pre-tighten clutch cover bolts.
7. Tighten clutch cover bolts evenly in two steps in the order shown.
8. Install transaxle assembly.

HYDRAULIC SYSTEM BLEEDING

BLEEDING PROCEDURE
See Figures 27 through 30.

※ **CAUTION**

Do not spill clutch fluid onto painted surfaces. If it spills, wipe up immediately and wash the affected area with water.

➡ **Note the following:**

- Do not use a vacuum assist or any other type of power bleeder on this system. Use of vacuum assist or power bleeder will not purge all the air from the system.
- Carefully monitor clutch fluid level in reservoir tank during bleeding operation.
- First bleed the air from the bleeding connector on the CSC and then from the air bleed connector valve.

1. Fill master cylinder reservoir tank with new clutch fluid.
2. Connect a transparent vinyl tube and container to the bleeding connector on the CSC.
3. Depress and release the clutch pedal slowly and fully 15 times at an interval of two to three seconds and release the clutch pedal.

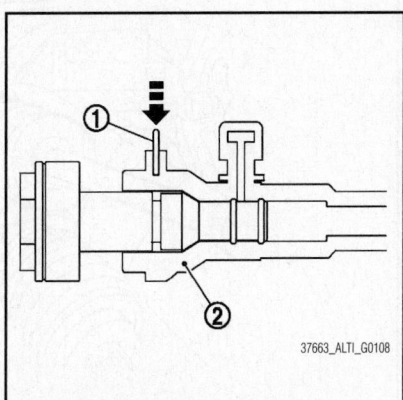

Fig. 27 Connect a transparent vinyl tube and container to the bleeding connector (1) on the CSC

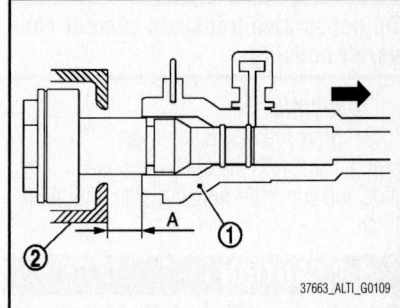

Fig. 28 Push in the lock pin (1) of the bleeding connector (2) and hold it in

4. Bleed the air from the clutch system according to the following:
   a. Push in the lock pin of the bleeding connector, and hold it in.

※ **CAUTION**

Hold the lock pin in to prevent the bleeding connector from separating when fluid pressure is applied.

   b. Slide the bleeding connector away from the transaxle housing to a distance of 0.39 inches (10 mm) to allow air to bleed from the clutch system.
   c. Depress the clutch pedal and hold it down.

※ **CAUTION**

Hold the clutch pedal down to prevent air from getting back into the clutch system.

5. Return the bleeding connector and lock pin to their original positions.
6. Release the clutch pedal and wait for five seconds.

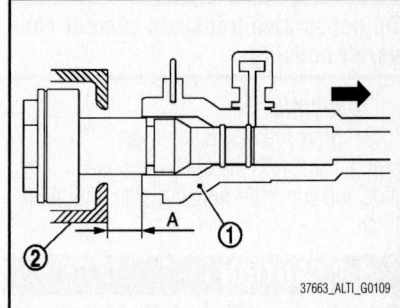

Fig. 29 Slide the bleeding connector (1) away from the transaxle housing (2) to the specified distance (A) of 0.39 inches (10 mm)

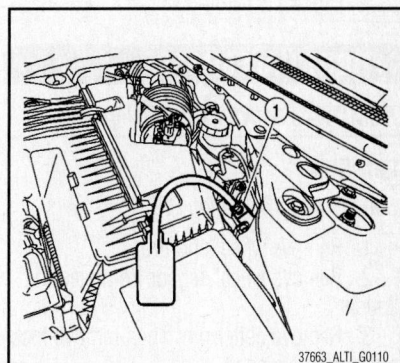

Fig. 30 Connect a transparent vinyl tube and container to the air bleed connector valve (1)

7. Repeat steps 3 through 6 until no air bubbles can be observed in the clutch fluid.

8. Connect a transparent vinyl tube and container to the air bleed connector valve.

9. Fully depress the clutch pedal several times.

10. With clutch pedal depressed, open the air bleed connector valve.

11. Close the air bleed connector valve.

12. Release the clutch pedal and wait for five seconds.

13. Repeat steps 9 through 12 until no air bubble can be observed in the clutch fluid.

14. Check clutch fluid level in reservoir tank.

FRONT DIFFERENTIAL SIDE OIL SEAL

REMOVAL & INSTALLATION

1. Remove drive shaft assembly.
2. Remove the differential side oil seal using suitable tool.

❊❊ CAUTION

Do not scratch transaxle case or converter housing.

To install:

3. Drive the new differential side oil seal into the transaxle case side and converter housing side until it is flush.

❊❊ CAUTION

Do not reuse differential side oil seals. Apply specified NISSAN CVT fluid to side oil seals.

4. Install drive shaft assembly.
5. Check CVT fluid level.

FRONT HALFSHAFT

REMOVAL & INSTALLATION

Left Side

See Figures 31 through 33.

1. Remove wheel and tire.
2. Remove wheel sensor from steering knuckle.
3. Remove cotter pin. Then remove lock nut from drive shaft.
4. Remove brake hose lock plate. Then remove brake hose from strut.

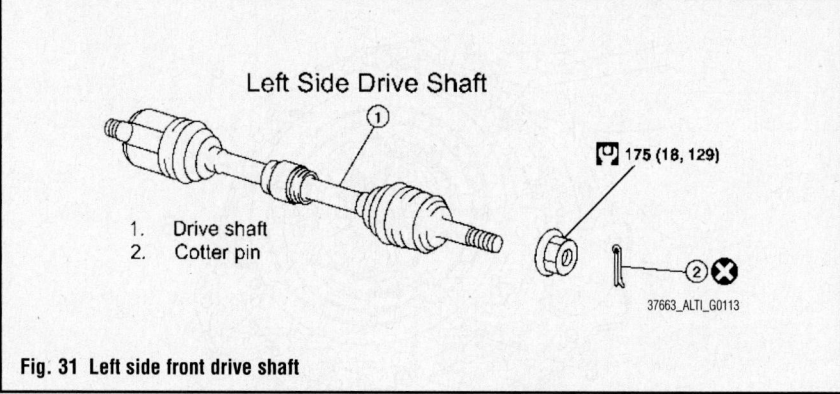

1. Drive shaft
2. Cotter pin

37663_ALTI_G0113

Fig. 31 Left side front drive shaft

5. Remove brake caliper, leaving hydraulic brake line attached. Hang caliper aside using wire.

6. Remove front strut to steering knuckle bolts and nuts, then separate steering knuckle front strut.

7. Remove drive shaft from wheel hub and bearing assembly, using a puller or suitable tool.

❊❊ CAUTION

When removing drive shaft, do not apply an excessive angle to drive shaft joint. Also be careful not to excessively extend slide joint.

8. Remove the left side drive shaft from the transaxle.

   a. Remove drive shaft from transaxle using Tool and drive shaft puller or suitable tool.

   b. Set Tool KV40107500 and a drive shaft puller or suitable tool between drive shaft (slide joint side) and transaxle as shown, then remove drive shaft.

To install:

9. Installation is in the reverse order of removal. Note the following:

❊❊ CAUTION

Do not reuse non-reusable parts.

   a. In order to prevent damage to differential side oil seal, place Tool KV38106700 (J-34296) (A) onto oil seal before inserting drive shaft as shown.

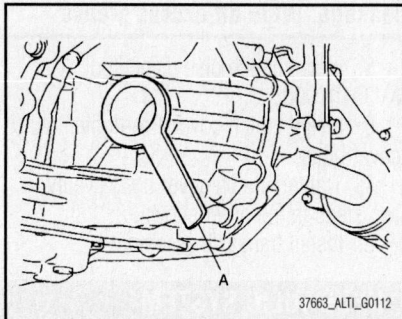

37663_ALTI_G0112

Fig. 33 Place Tool KV38106700 (J-34296) (A) onto oil seal before inserting drive shaft as shown

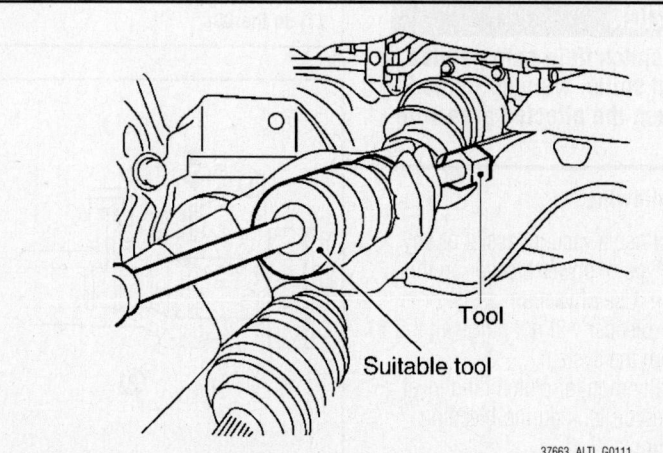

Tool

Suitable tool

37663_ALTI_G0111

Fig. 32 Set Tool and a drive shaft puller or suitable tool between drive shaft (slide joint side) and transaxle as shown, then remove drive shaft

Slide drive shaft into slide joint and tap with a hammer to install securely.

b. Install new circlip on drive shaft in the circular clip groove on transaxle side.

> ※※ **CAUTION**
>
> **Make sure the new circlip on the drive shaft is securely fastened.**

c. After its insertion, try to pull the flange out of the slide joint by hand.

> ※※ **CAUTION**
>
> **If it pulls out, the circlip is not properly meshed with the transaxle side gear.**

Right Side

See Figures 34 and 35.

1. Remove wheel and tire.
2. Remove wheel sensor from steering knuckle.
3. Remove cotter pin. Then remove lock nut from drive shaft.
4. Remove brake hose lock plate. Then remove brake hose from strut.
5. Remove brake caliper, leaving hydraulic brake line attached. Hang caliper aside using wire.
6. Remove front strut to steering knuckle bolts and nuts, then separate steering knuckle front strut.

7. Remove drive shaft from wheel hub and bearing assembly, using a puller or suitable tool.

> ※※ **CAUTION**
>
> **When removing drive shaft, do not apply an excessive angle to drive shaft joint. Also be careful not to excessively extend slide joint.**

8. Remove the retaining bracket bolts, and separate drive shaft from transaxle.

To install:

9. Installation is in the reverse order of removal. Note the following:

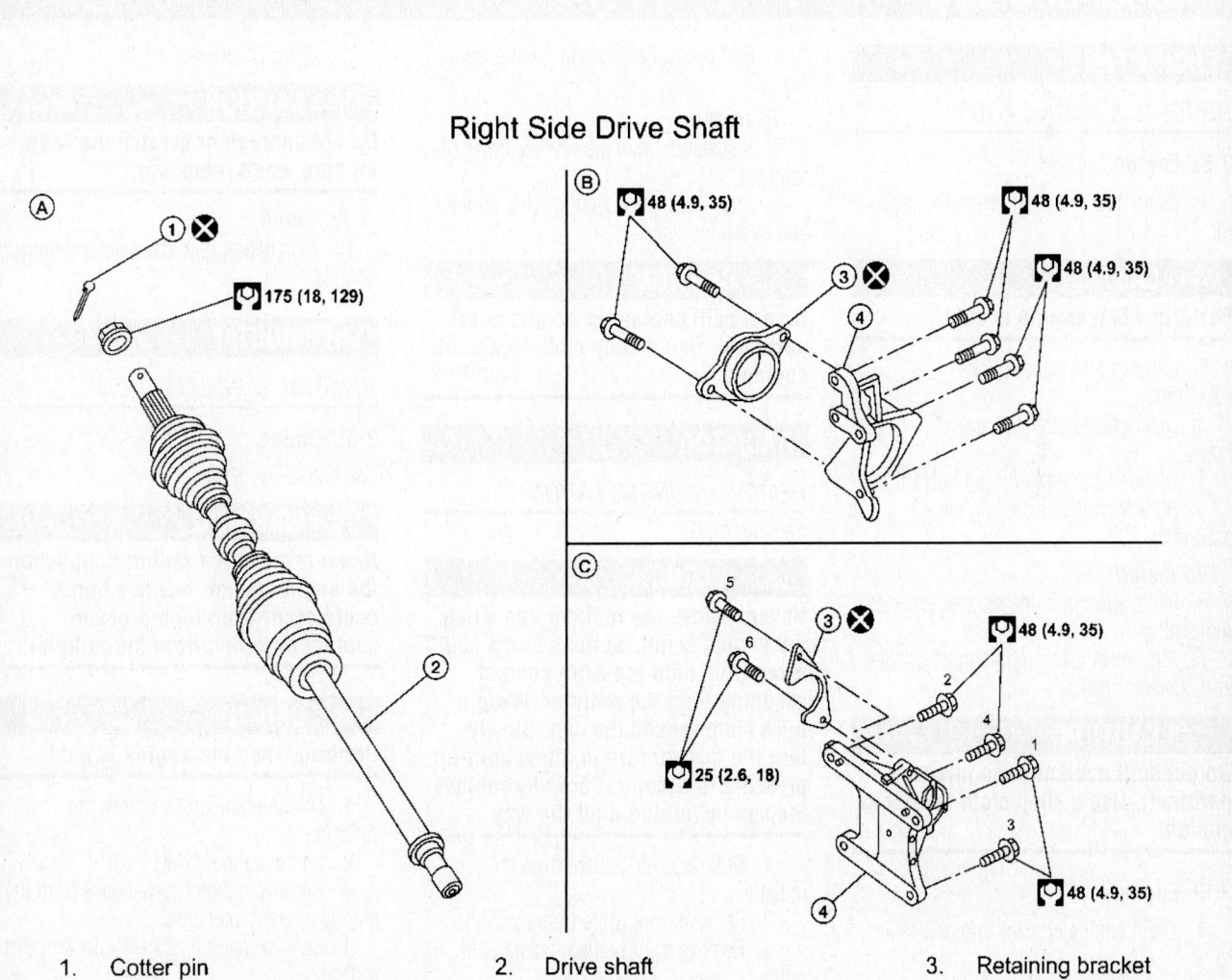

Right Side Drive Shaft

1. Cotter pin	2. Drive shaft
4. Support bearing bracket	A. Front RH drive shaft
C. QR25DE engine	3. Retaining bracket
	B. VQ35DE engine

37663_ALTI_G0115

Fig. 34 Exploded view of right side drive shaft assembly

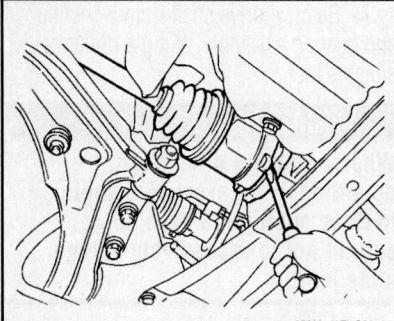

Fig. 35 Remove the retaining bracket bolts, and separate drive shaft from transaxle

37663_ALTI_G0114

✳✳ CAUTION

Do not reuse non-reusable parts.

a. Tighten retaining bracket bolts and support bearing bracket bolts to specifications.

b. For QR25DE models, install the retaining bracket with the notch facing up and follow the bolt tightening order.

c. In order to prevent damage to differential side oil seal, place Tool (A) onto oil seal before inserting drive shaft as shown. Slide drive shaft into slide joint and tap with a hammer to install securely.

d. Install new circlip on drive shaft in the circular clip groove on transaxle side.

✳✳ CAUTION

Make sure the new circlip on the drive shaft is securely fastened.

e. After its insertion, try to pull the flange out of the slide joint by hand.

✳✳ CAUTION

If it pulls out, the circlip is not properly meshed with the transaxle side gear.

ENGINE COOLING

ENGINE FAN

REMOVAL & INSTALLATION

2.5L Engine

1. Drain engine coolant from the radiator.

✳✳ CAUTION

Perform when engine is cold.

2. Remove air cleaner duct assembly.
3. Disconnect upper radiator hose.
4. Disconnect fan motor connectors.
5. Remove radiator cooling fan assembly.

To install:

6. Installation is in the reverse order of removal.
7. After installation refill engine coolant and check for leaks.

✳✳ CAUTION

Do not spill coolant in engine compartment. Use a shop cloth to absorb coolant.

3.5L Engine

1. Drain engine coolant from the radiator.

✳✳ CAUTION

Perform when engine is cold.

2. Remove CVT control module (if equipped).
3. Remove battery tray.
4. Disconnect upper radiator hose.
5. Disconnect fan motor connectors.

6. Remove radiator cooling fan assembly.

To install:

7. Installation is in the reverse order of removal.
8. After installation refill engine coolant and check for leaks.

✳✳ CAUTION

Do not spill coolant in engine compartment. Use a shop cloth to absorb coolant.

RADIATOR

REMOVAL & INSTALLATION
See Figure 42.

✳✳ WARNING

Never remove the radiator cap when the engine is hot. Serious burns could occur from high pressure coolant escaping from the radiator. Wrap a thick cloth around the cap. Slowly turn it a quarter turn to allow built-up pressure to escape. Carefully remove the cap by turning it all the way.

1. Drain engine coolant from the radiator.
2. Remove front grille (Sedan only).
3. Remove front bumper fascia (Coupe only).
4. Remove engine undercover.
5. Remove front air duct.
6. Remove A/C condenser.
7. Disconnect upper and lower radiator hoses.
8. Disconnect the CVT oil cooler hoses, if equipped. Plug the hoses to prevent CVT oil loss.

9. Remove radiator.

✳✳ CAUTION

Do not damage or scratch the radiator core when removing.

To install:

10. Installation is in the reverse order of removal.

THERMOSTAT

REMOVAL & INSTALLATION

2.5L Engine
See Figure 36.

✳✳ WARNING

Never remove the radiator cap when the engine is hot. Serious burns could occur from high pressure coolant escaping from the radiator.

✳✳ CAUTION

Perform when the engine is cold.

1. Drain engine coolant from the radiator.
2. Remove the air duct.
3. Remove radiator lower hose from the engine coolant inlet side.
4. Remove engine coolant inlet and thermostat.

To install:

5. Installation is in the reverse order of removal.
6. Install the engine coolant temperature sensor.

➡ **Use Genuine RTV Silicone Sealant or equivalent.**

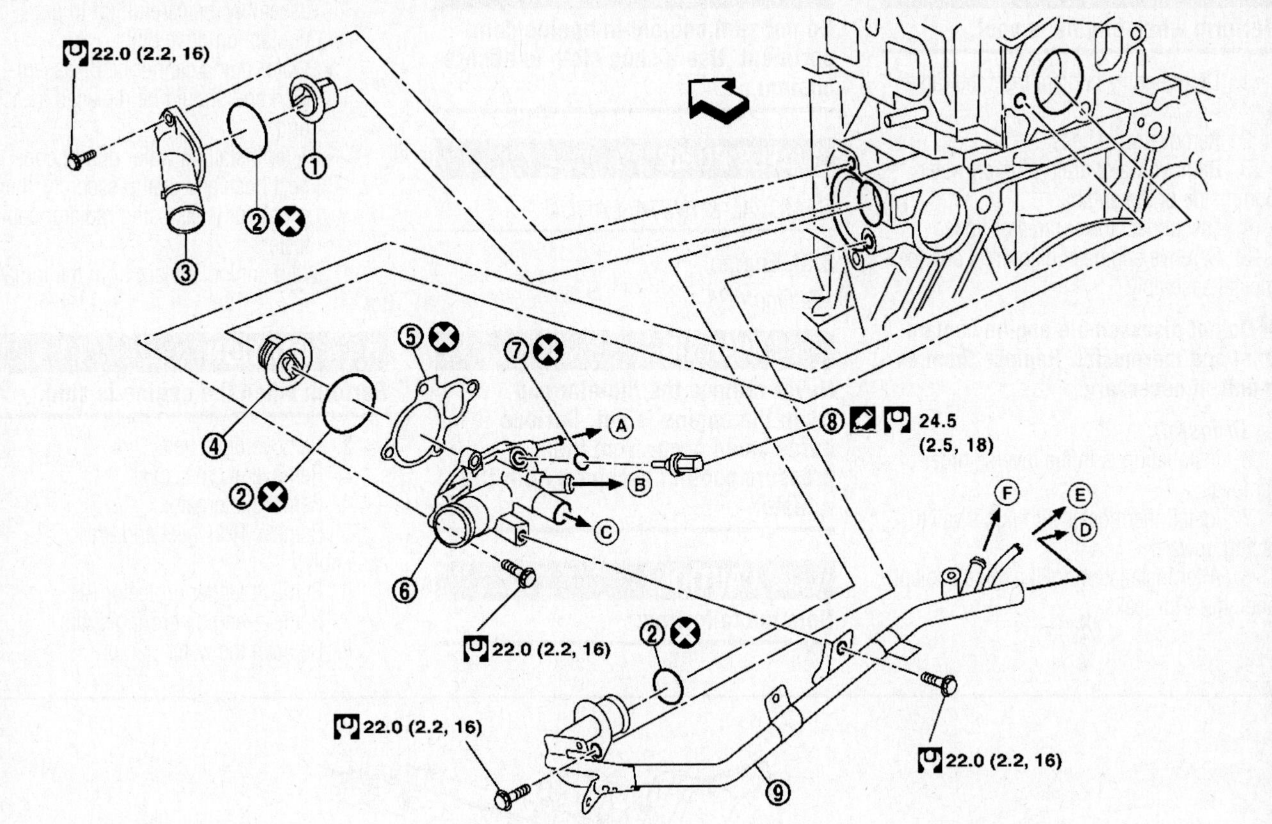

Fig. 36 showing exploded view of thermostat housing assembly

22.0 (2.2, 16)

1

2

3

4

5

7

8 24.5 (2.5, 18)

A

B

C

2

6

F

E

D

22.0 (2.2, 16)

2

22.0 (2.2, 16)

22.0 (2.2, 16)

9

37663_ALTI_G0128

1.	Thermostat	2.	O-ring	3.	Engine coolant inlet
4.	Water control valve	5.	Gasket	6.	Engine coolant outlet
7.	Copper washer	8.	Engine coolant temperature sensor	9.	Heater pipe
A.	To electric throttle control	B.	To oil cooler	C.	To heater
D.	To heater	E.	To electric throttle control	F.	To oil cooler

Engine front

Fig. 36 Exploded view of thermostat housing assembly

7. Install the thermostat with the whole circumference of the flange part fitting securely inside the rubber ring.

8. Install the thermostat with the jiggle valve facing upwards. The position deviation may be within the range of ±10°.

9. If necessary, to install the heater pipe, first apply a mild detergent to the O-ring and then quickly insert the pipe into the housing.

3.5L Engine

See Figure 37.

✴✴ WARNING

Never remove the radiator cap when the engine is hot. Serious burns could occur from high pressure coolant escaping from the radiator.

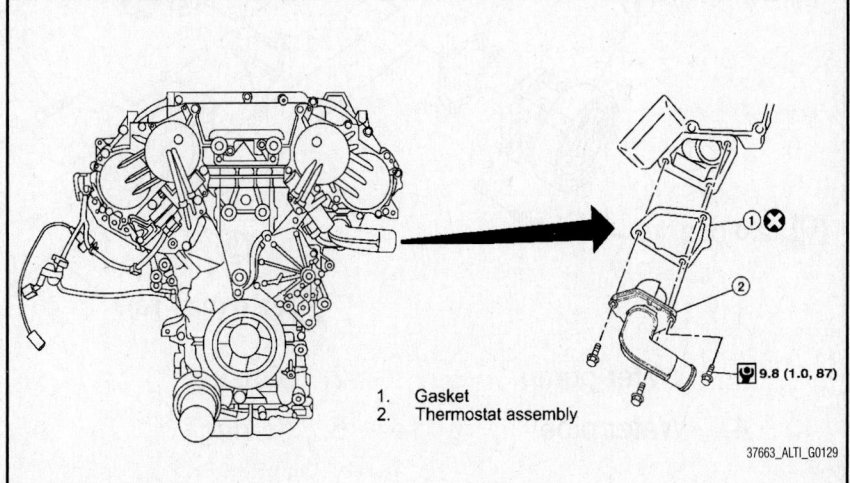

1. Gasket
2. Thermostat assembly

9.8 (1.0, 87)

37663_ALTI_G0129

Fig. 37 Exploded view of thermostat housing assembly

❋❋ **CAUTION**

Perform when engine is cool.

1. Drain engine coolant from the radiator.
2. Remove drive belts.
3. Remove water drain plug on water pump side of the engine.
4. Disconnect lower radiator hose.
5. Remove engine coolant inlet and thermostat assembly.

➡**Do not disassemble engine coolant inlet and thermostat. Replace them as a unit, if necessary.**

To install:

6. Installation is in the reverse order of removal.
7. Install thermostat with jiggle valve facing upward.
8. After installation refill engine coolant and check for leaks.

❋❋ **CAUTION**

Do not spill coolant in engine compartment. Use a shop cloth to absorb coolant.

WATER PUMP

REMOVAL & INSTALLATION

2.5L Engine
See Figure 38.

❋❋ **WARNING**

Never remove the radiator cap when the engine is hot. Serious burns could occur from high pressure coolant escaping from the radiator.

❋❋ **CAUTION**

Note the following:

- When removing water pump assembly, be careful not to get coolant on drive belt.
- Water pump cannot be disassembled and should be replaced as a unit.
- After installing water pump, connect hose and clamp securely, then check for leaks using radiator cap tester.

1. Drain engine coolant from the radiator.

❋❋ **CAUTION**

Perform when the engine is cold.

2. Remove drive belt.
3. Remove engine cover.
4. Remove alternator.
5. Remove RH wheel and tire assembly.
6. Remove fender protector RH.
7. Remove engine ground strap.
8. Remove the water pump.

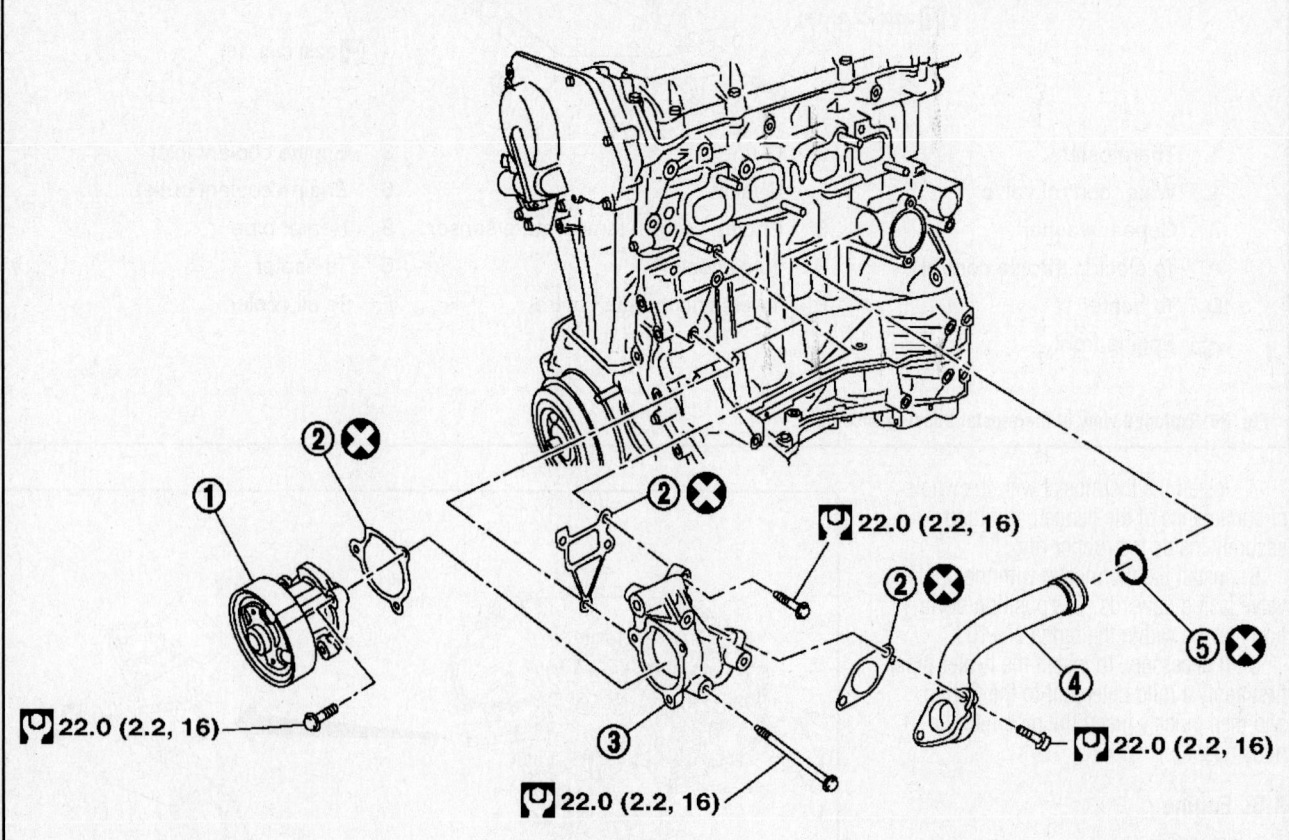

1. Water pump
2. Gaskets
3. Water pump housing
4. Water pipe
5. O-ring

37663_ALTI_G0130

Fig. 38 Exploded view of water pump assembly

Handle the water pump vane so that it does not contact any other parts. Water pump cannot be disassembled and should be replaced as an assembly.

➡If it is necessary to remove the water pipe, the exhaust manifold and three way catalyst assembly must be removed.

To install:

9. Installation is in the reverse order of removal.

10. When inserting water pipe end to cylinder block, apply a neutral detergent to O-ring. Then insert it immediately.

11. After installation refill engine coolant and check for leaks.

3.5L Engine

See Figures 39 through 44.

Never remove the radiator cap when the engine is hot. Serious burns could occur from high pressure coolant escaping from the radiator.

Note the following:

- When removing water pump assembly, be careful not to get coolant on drive belt.
- Water pump cannot be disassembled and should be replaced as a unit.
- After installing water pump, connect hose and clamp securely, then check for leaks using radiator cap tester.

1. Drain engine coolant from the radiator.

Perform when the engine is cold.

2. Remove engine coolant reservoir tank.

3. Remove RH wheel and tire.

4. Remove the fender protector (RH).

5. Remove drive belts.

6. Remove the drive belt auto tensioner and the idler pulley.

7. Support engine and remove the front engine insulator and bracket.

8. Remove water drain plug on water pump side of cylinder block.

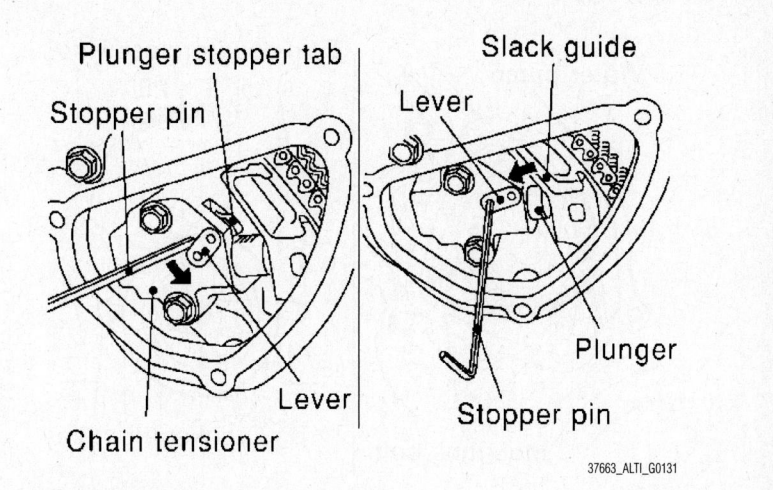

Fig. 39 Insert the stopper pin into the tensioner body hole to hold the lever and keep the plunger stopper tab released

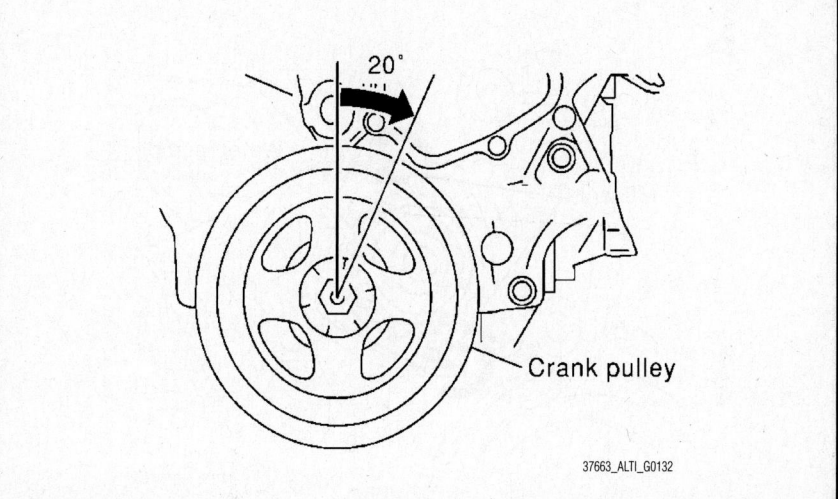

Fig. 40 Make a gap between water pump gear and timing chain, by turning the crankshaft pulley approximately 20° clockwise

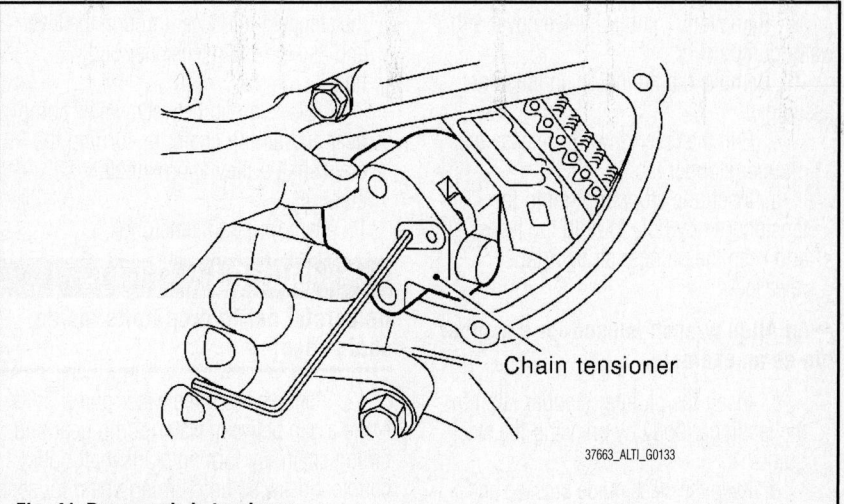

Fig. 41 Remove chain tensioner

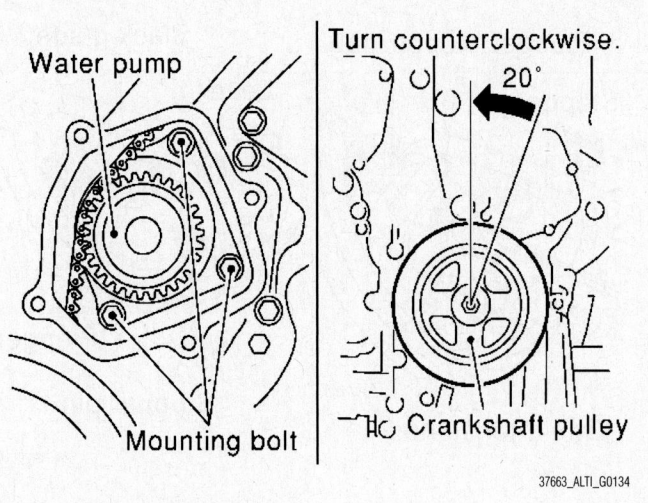

Fig. 42 Remove the three water pump bolts

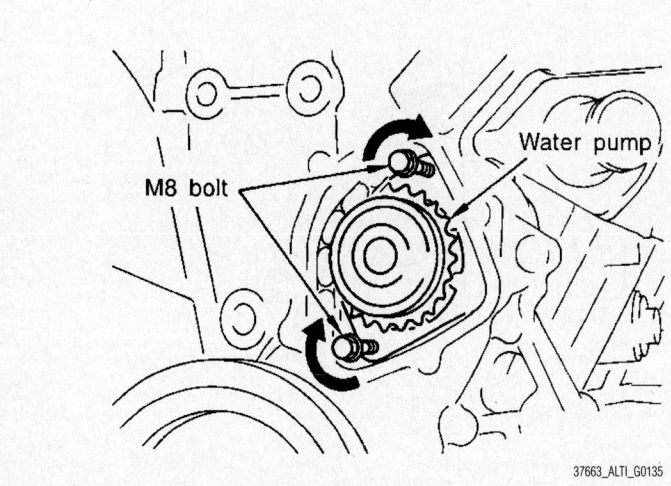

Fig. 43 Screw bolts into water pump upper and lower bolt holes until they reach the timing chain case

9. Remove IVT control valve cover and water pump cover.

10. Remove the timing chain tensioner assembly.

  a. Pull the lever down to release the plunger stopper tab.

  b. Insert the stopper pin into the tensioner body hole to hold the lever and keep the plunger stopper tab released.

➡**An Allen wrench is used for a stopper pin as an example.**

  c. Insert the plunger stopper tab into the tensioner body by pressing the slack guide.

  d. Keep the slack guide pressed and hold the plunger stopper tab in by pushing the stopper pin deeper through the lever and into the chain tensioner body hole.

  e. Make a gap between water pump gear and timing chain, by turning the crankshaft pulley approximately 20° clockwise.

11. Remove chain tensioner.

※※ CAUTION

Be careful not to drop bolts inside chain case.

12. Remove the three water pump bolts. Make a gap between water pump gear and timing chain, by turning crankshaft pulley counterclockwise until timing chain loosens on water pump sprocket.

13. Screw bolts into water pump upper and lower bolt holes until they reach the timing chain case. Then, alternately tighten each bolt for a half turn, and pull out the water pump.

※※ CAUTION

Pull straight out while preventing vane from contacting socket in installation area. Remove water pump without causing sprocket to contact timing chain.

14. Remove bolts and O-rings from water pump.

  To install:
15. Install new O-rings to water pump.

16. Apply engine oil and coolant to the O-rings as shown. Locate the O-ring with white paint mark to engine front side.

17. Install the water pump.

※※ CAUTION

Do not allow cylinder block to interfere with the O-rings when installing the water pump.

18. Check that timing chain and water pump sprocket are engaged.

19. Insert water pump by tightening bolts alternately and evenly.

20. Remove dust and foreign material completely from backside of chain tensioner and from installation area of rear timing chain case.

21. Turn the crankshaft pulley approximately 20° clockwise so that the timing chain on the timing chain tensioner side is loose.

➡**When installing the timing chain tensioner, engine oil should be applied to the oil hole and tensioner.**

22. Install the timing chain tensioner.

23. Remove the stopper pin.

24. Install IVT control valve cover and water pump cover.

  a. Before installing, remove all traces of sealant from mating surface of water pump cover and IVT control valve cover using a scraper. Also remove traces of sealant from the mating surface of the front cover.

  b. Apply a continuous bead of RTV Silicone Sealant or equivalent, to mating surface of IVT control valve cover and water pump cover.

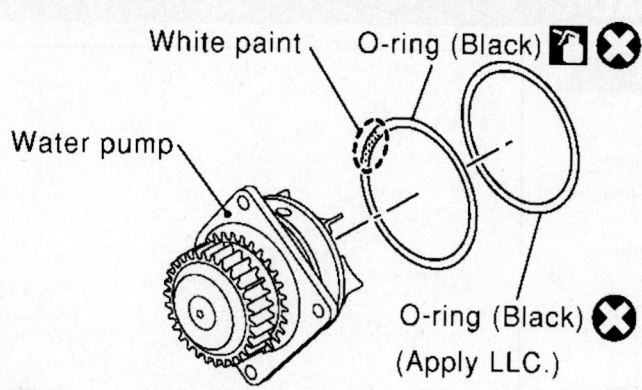

White paint — O-ring (Black)

Water pump

O-ring (Black)
(Apply LLC.)

❌ : Always replace after every disassembly.

🛢 : Lubricate with new engine oil.

37663_ALTI_G0136

Fig. 44 Locate the O-ring with white paint mark to engine front side

25. Install water drain plug on water pump side of cylinder block.

26. Install idler pulley.

27. Installation of remaining components is in the reverse order of removal.

28. After installation refill engine coolant and check for leaks.

> ✳✳ **CAUTION**
>
> **Do not spill coolant in engine compartment. Use a shop cloth to absorb coolant.**

29. After starting engine, let idle for three minutes, then rev engine up to 3,000 rpm under no load to purge air from the high-pressure chamber of the chain tensioner. The engine may produce a rattling noise. This indicates that air still remains in the chamber and is not a matter of concern.

ENGINE ELECTRICAL

ALTERNATOR

REMOVAL & INSTALLATION

2.5L Engine

1. Disconnect the battery negative terminal.

2. Remove engine side undercover.

3. Remove drive belt.

4. Remove "B" terminal nut.

5. Remove air intake duct.

6. Disconnect alternator harness connectors.

7. Remove alternator ground harness bolt.

8. Remove alternator bolts.

9. Remove alternator assembly upward.

To install:

10. Installation is in the reverse order of removal.

> ✳✳ **CAUTION**
>
> **Be sure to tighten "B" terminal nut carefully.**

11. Install alternator and check tension of belt.

➡ **For this model, the power generation voltage variable control system that** controls the power generation voltage of the alternator has been adopted. Therefore, the power generation voltage variable control system operation inspection should be performed after replacing the alternator, and then make sure that the system operates normally.

3.5L Engine

1. Disconnect the battery negative terminal.

2. Partially drain engine coolant.

3. Remove engine room cover.

4. Remove RH front wheel and tire assembly.

5. Remove engine side undercover.

6. Remove air cleaner and duct assembly.

7. Remove battery tray.

8. Remove cooling fan assembly.

9. Evacuate A/C system.

10. Remove drive belt.

11. Remove the A/C compressor.

12. Remove idler pulley.

13. Remove A/C idler pulley.

14. Disconnect oil pressure switch.

15. Disconnect the alternator harness connectors.

16. Remove the alternator bolt and nuts.

CHARGING SYSTEM

17. Slide the alternator out and remove.

To install:

18. Installation is in the reverse order of removal.

> ✳✳ **CAUTION**
>
> **Be sure to tighten "B" terminal nut carefully.**

19. Install alternator and check tension of belt.

20. For this model, the power generation voltage variable control system that controls the power generation voltage of the alternator has been adopted. Therefore, the power generation voltage variable control system operation inspection should be performed after replacing the alternator, and then make sure that the system operates normally.

21. Make sure that alternator pulley does not rattle.

22. Make sure that alternator pulley nut is tight. Tighten to 87 ft. lbs. (118 Nm).

VOLTAGE REGULATOR

The voltage regulator is an integral part of the alternator. Refer to the alternator when servicing this component.

FIRING ORDER

See Figure 45.

The firing order for the 2.5L engine is 1–3–4–2. The 2.5L engine is an in-line 4 cylinder engine with cylinders numbered in order front to back of the engine.

The firing order for the 3.5L engine is 1–2–3–4–5–6.

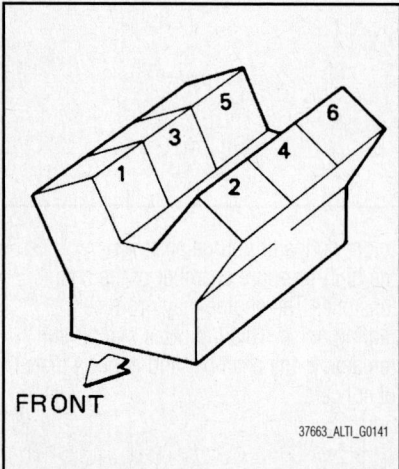

Fig. 45 Cylinder number locations—3.5L engine

IGNITION COIL

REMOVAL & INSTALLATION

2.5L Engine

See Figure 46.

1. Remove the engine cover.
2. Disconnect the harness connector from the ignition coil.
3. Remove the ignition coil.

✳✳ CAUTION

Do not drop or shock it.

To install:

4. Installation is in the reverse order of removal.

3.5L Engine

Left Side

See Figure 47.

1. Remove the engine cover.
2. Disconnect ignition coil connector.
3. Remove the ignition coil.

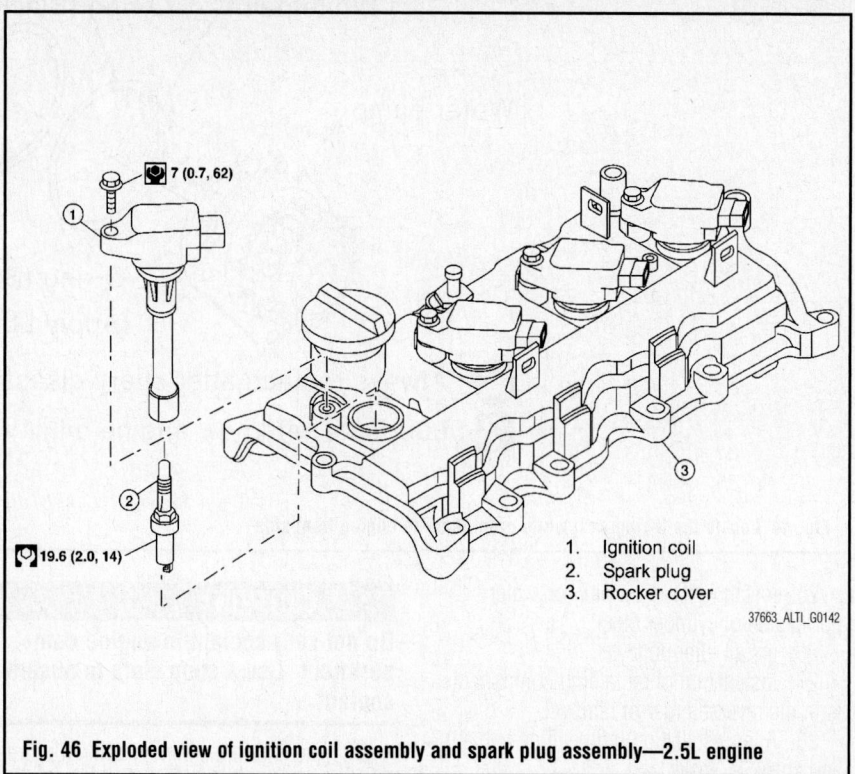

Fig. 46 Exploded view of ignition coil assembly and spark plug assembly—2.5L engine

1. Ignition coil
2. Spark plug
3. Rocker cover

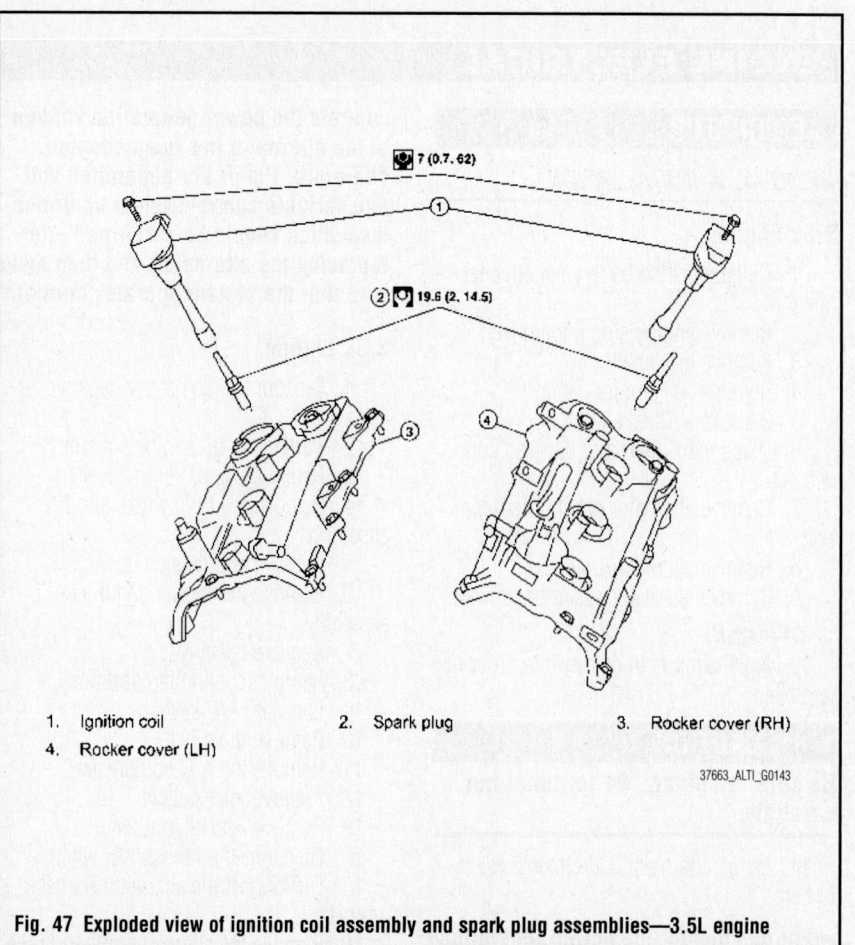

Fig. 47 Exploded view of ignition coil assembly and spark plug assemblies—3.5L engine

1. Ignition coil
2. Spark plug
3. Rocker cover (RH)
4. Rocker cover (LH)

❊❊ CAUTION

Do not drop or shock it.

To install:
4. Installation is in the reverse order of removal.

Right Side
1. Remove the intake manifold collector.
2. Disconnect ignition coil connector.
3. Remove the ignition coil.

❊❊ CAUTION

Never shock ignition coil.

To install:
4. Installation is in the reverse order of removal.

IGNITION TIMING

INSPECTION

The ignition timing is controlled by the ECM. No adjustment is necessary or possible.

SPARK PLUGS

REMOVAL & INSTALLATION
See Figures 46 and 47.

1. Remove the ignition coil.
2. Remove the spark plug with a suitable spark plug wrench.
3. Installation is in the reverse order of removal.

ENGINE ELECTRICAL

STARTING SYSTEM

STARTER

REMOVAL & INSTALLATION

2.5L Engine

With Manual Transaxle
See Figure 48.

1. Disconnect the negative battery terminal.
2. Disconnect the starter motor harness connectors.
3. Remove the two starter motor bolts.
4. Remove the starter motor.

To install:
5. Installation is in the reverse order of removal.

With CVT Transaxle
See Figure 49.

1. Remove the battery and battery tray bracket.
2. Remove the air cleaner assembly ducts.
3. Disconnect the following:
   - ECM
   - TCM
4. Disconnect the starter motor harness connectors.
5. Remove the two starter motor bolts.
6. Remove the starter motor.

To install:
7. Installation is in the reverse order of removal.

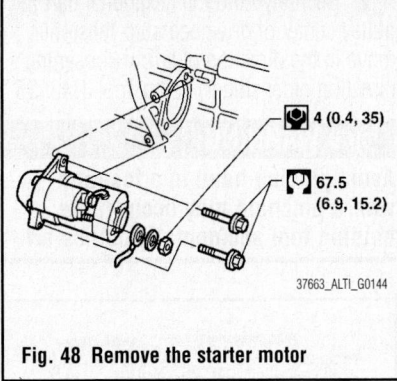

Fig. 48 Remove the starter motor

3.5L Engine

With Manual Transaxle
See Figure 48.

1. Disconnect the negative battery terminal.
2. Disconnect the starter motor harness connectors.
3. Remove the two starter motor bolts.
4. Remove the starter motor.

To install:
5. Installation is in the reverse order of removal.

With CVT Transaxle
See Figure 50.

1. Disconnect the negative and positive battery terminal.
2. Remove the air cleaner assembly and air ducts.
3. Disconnect the following:
   - ECM
   - TCM

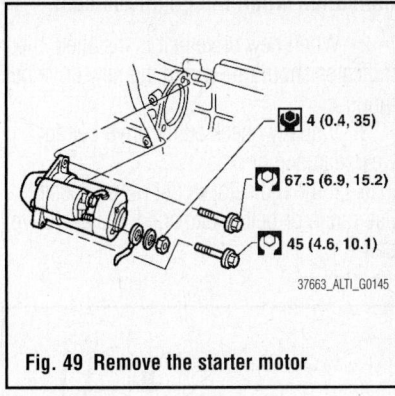

Fig. 49 Remove the starter motor

4. Remove the battery tray.
5. Disconnect the starter motor harness connectors.
6. Remove the two starter motor bolts.
7. Remove the starter motor.

To install:
8. Installation is in the reverse order of removal.

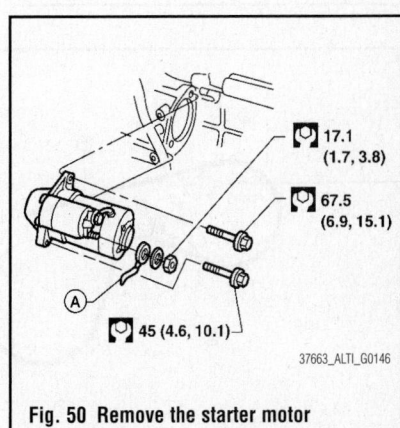

Fig. 50 Remove the starter motor

ENGINE MECHANICAL

ACCESSORY DRIVE BELTS

ACCESSORY BELT ROUTING

See Figures 51 and 52.

Refer to the accompanying illustrations.

INSPECTION

1. Check that the indicator of drive belt auto-tensioner is within the possible use range.

➡**Check the drive belt auto-tensioner indication when the engine is cold.**

2. When new drive belt is installed, the indicator should be within the new drive belt range.

3. Visually check entire drive belt for wear, damage or cracks.

4. If the indicator is out of the possible use range or belt is damaged, replace drive belt.

ADJUSTMENT

Belt tension is not manually adjustable, it is automatically adjusted by the drive belt auto-tensioner. No adjustment is necessary. If the belt is out of adjustment, replace the accessory drive belt.

REMOVAL & INSTALLATION

2.5L Engine

See Figure 53.

1. Remove the fender protector side cover RH.

2. Securely hold the hexagonal part in pulley center of drive belt auto-tensioner, move in the direction of arrow (loosening direction of tensioner) using Tool J-46535.

✳✳ WARNING

Avoid placing hand in a location where pinching may occur if the holding tool accidentally comes off.

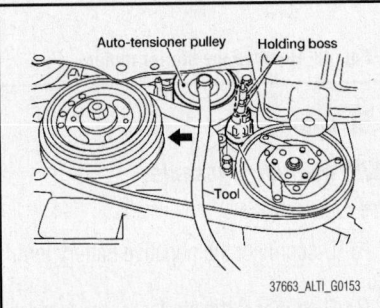

37663_ALTI_G0153

Fig. 53 Securely hold the hexagonal part in pulley center of drive belt auto-tensioner, move in the direction of arrow

✳✳ CAUTION

Do not loosen the auto-tensioner pulley bolt. (Do not turn it counterclockwise.) If turned counterclockwise, the complete auto-tensioner must be replaced as a unit, including pulley.

3. Insert a rod approximately 0.24 inches (6 mm) in diameter through the rear of tensioner into retaining boss to lock tensioner pulley.

4. Leave tensioner pulley arm locked until drive belt is installed again.

5. Loosen drive belt from water pump pulley and then remove it from the other pulleys.

To install:

6. Install the drive belt onto all of the pulleys except for the water pump pulley. Then install the drive belt onto water pump pulley last.

✳✳ CAUTION

Confirm belts are completely set on the pulleys.

7. Release tensioner, and apply tension to drive belt.

✳✳ WARNING

Avoid placing hand in a location where pinching may occur if the holding tool accidentally comes off.

✳✳ CAUTION

Do not loosen the auto-tensioner pulley bolt. Don't turn it counterclockwise. If turned counterclockwise, the complete auto-tensioner must be replaced as a unit, including pulley.

8. Turn crankshaft pulley clockwise several times to equalize tension between each pulley.

9. Confirm tension of drive belt at indicator is within the allowable use range.

3.5L Engine

See Figure 54.

1. Remove the front RH wheel and tire.

2. Remove the front RH side cover.

3. While securely holding the hexagonal part in pulley center of drive belt auto-tensioner, move in the direction of arrow (loosening direction of tensioner) using suitable tool.

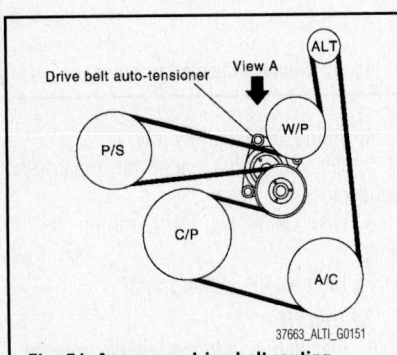

37663_ALTI_G0151

Fig. 51 Accessory drive belt routing—2.5L engine

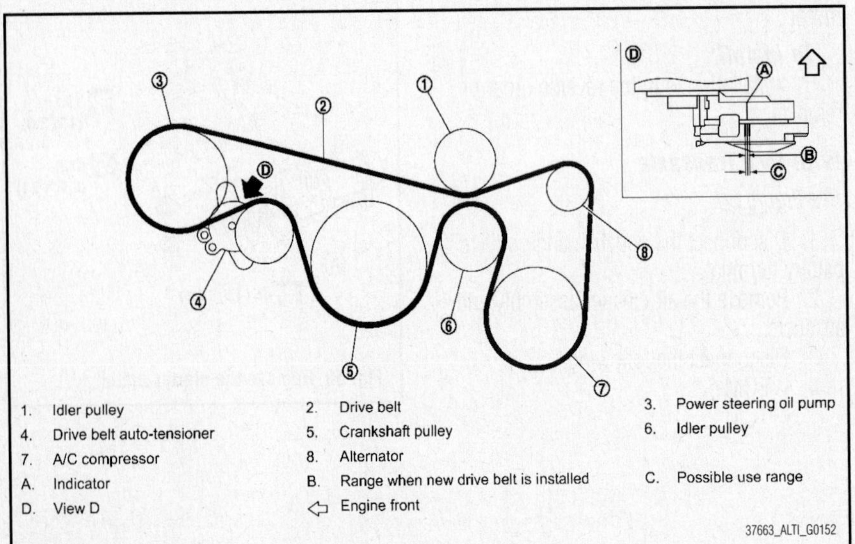

1.	Idler pulley	2.	Drive belt	3.	Power steering oil pump
4.	Drive belt auto-tensioner	5.	Crankshaft pulley	6.	Idler pulley
7.	A/C compressor	8.	Alternator		
A.	Indicator	B.	Range when new drive belt is installed	C.	Possible use range
D.	View D	⇦	Engine front		

37663_ALTI_G0152

Fig. 52 Accessory drive belt routing—3.5L engine

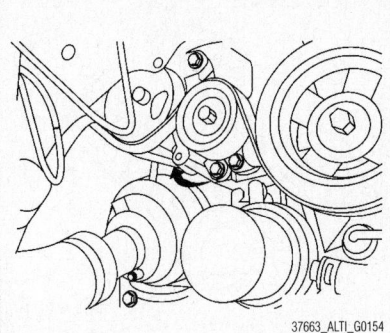

Fig. 54 While securely holding the hexag-onal part in pulley center of drive belt auto-tensioner, move in the direction of arrow

> ※ **WARNING**
>
> Avoid placing hand in a location where pinching may occur if the holding tool accidentally comes off.

> ※ **CAUTION**
>
> Do not loosen the auto-tensioner pulley bolt. (Do not turn it counter-clockwise.) If turned counterclock-wise, the complete auto-tensioner must be replaced as a unit, including pulley.

4. Insert a rod approximately 0.24 inches (6 mm) in diameter through the rear of tensioner into retaining boss to lock tensioner pulley.
5. Leave tensioner pulley arm locked until belt is installed again.
6. Loosen drive belt from water pump pulley and then remove it from the other pulleys.

To install:
7. Install the drive belt onto all of the pulleys.

> ※ **CAUTION**
>
> Confirm belts are completely set on the pulleys.

8. Release tensioner, and apply tension to belt.

> ※ **WARNING**
>
> Avoid placing hand in a location where pinching may occur if the holding tool accidentally comes off.

> ※ **CAUTION**
>
> Do not loosen the auto-tensioner pul-ley bolt. (Don't turn it counterclock-wise. If turned counterclockwise, the complete auto-tensioner must be replaced as a unit, including pulley.

9. Turn crankshaft pulley clockwise several times to equalize tension between each pulley.
10. Confirm tension of belt at indicator is within the possible use range.

CAMSHAFT AND VALVE LIFTERS

REMOVAL & INSTALLATION

2.5L Engine

See Figures 55 through 71.

1. Remove the rocker cover.
2. Remove the front right side tire and wheel.
3. Remove the RH splash shield.
4. Remove the drive belt.
5. Remove the power steering reservoir.
6. Remove the coolant overflow reser-voir tank.
7. Disconnect variable timing control solenoid and camshaft sensor harness connectors.
8. Remove camshaft sensor.
9. Remove camshaft sensor bracket.
10. Loosen the IVT control cover bolts in the order as shown.
11. Remove the IVT control cover by cutting the sealant using Tool KV10111100 (J-37228).
12. Set the No.1 cylinder at TDC on its compression stroke with the following procedure:
    a. Open the splash cover on RH under cover.
    b. Rotate crankshaft pulley clockwise, and align mating marks for TDC with timing indicator on front cover, as shown.
    c. At the same time, make sure that the mating marks on camshaft sprockets are lined up with the yellow links in the timing chain, as shown.
    d. If not, rotate crankshaft pulley one more turn to line up the mating marks to the yellow links, as shown.
13. Pull the timing chain guide out between the camshaft sprockets through front cover.

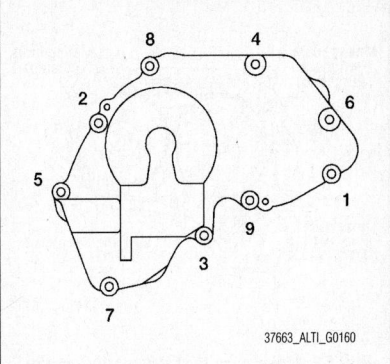

Fig. 55 Loosen the IVT control cover bolts in the order as shown

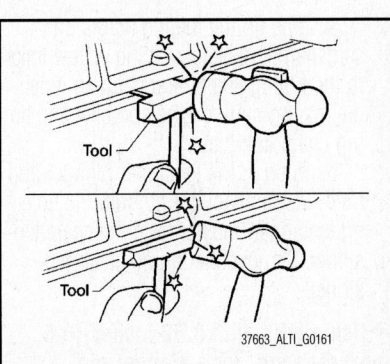

Fig. 56 Remove the IVT control cover by cutting the sealant

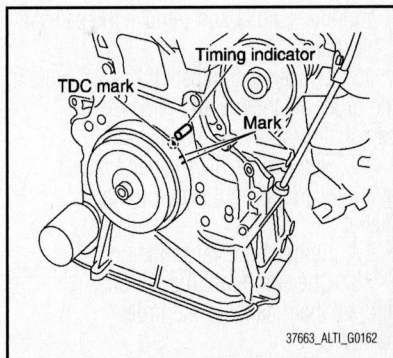

Fig. 57 Set the No.1 cylinder at TDC on its compression stroke

14. Remove camshaft sprockets with the following procedure.

> ※ **CAUTION**
>
> Do not rotate the crankshaft or camshaft while the timing chain is removed. It causes interference between valve and piston.

➡ Chain tension holding work is not necessary. Crankshaft sprocket and timing chain do not disconnect structurally while front cover is attached.

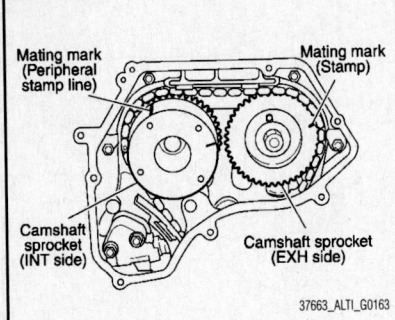

Fig. 58 Make sure that the mating marks on camshaft sprockets are lined up with the yellow links in the timing chain

a. Line up the mating marks on camshaft sprockets with the yellow links in the timing chain, and paint an indelible mating mark on the sprocket and timing chain link plate.

b. Push in the tensioner plunger and hold. Insert a stopper pin into the hole on tensioner body to hold the chain tensioner. Remove the timing chain tensioner.

➡**Use a wire with 0.02 inches (0.5 mm) diameter for a stopper pin.**

c. Secure the hexagonal part of camshaft with a suitable tool.

d. Loosen the camshaft sprocket mounting bolts and remove the camshaft sprockets.

15. Loosen the camshaft bracket bolts in the order as shown, and remove the camshaft brackets and camshafts.

16. Remove No.1 camshaft bracket by slightly tapping it with a rubber mallet.

17. Remove the valve lifters.

18. Check mounting positions, and set them aside in the order removed.

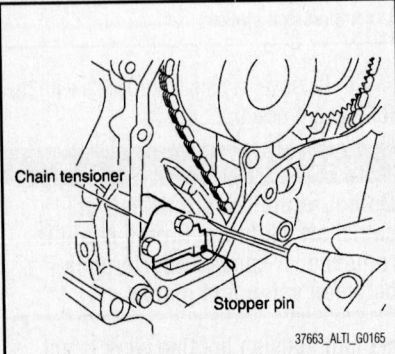

Fig. 59 Push in the tensioner plunger and hold; insert a stopper pin into the hole on tensioner body to hold the chain tensioner

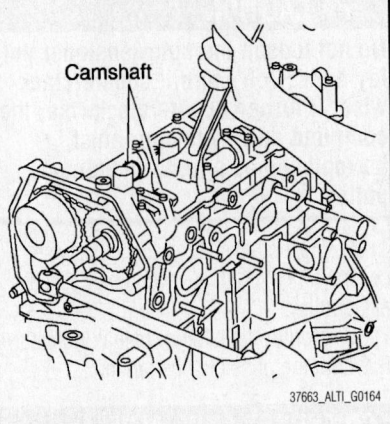

Fig. 60 Loosen the camshaft sprocket mounting bolts and remove the camshaft sprockets

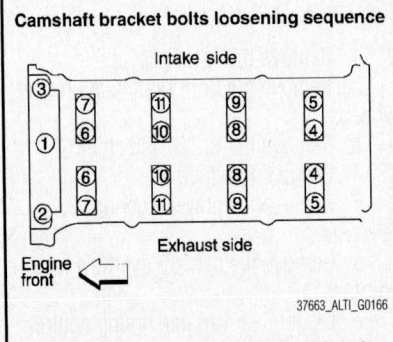

Fig. 61 Loosen the camshaft bracket bolts in the order as shown

To install:

19. Install the valve lifter.

➡**Install them in the same position from which they were removed.**

20. Install the camshafts.

a. The distinction between the intake and exhaust camshafts is in a difference of shapes of the back end:
- A: Exhaust
- B: Intake Signal plate for the camshaft position sensor (PHASE)

b. Install camshafts so that the dowel pins on the front side are positioned as shown.

21. Install camshaft brackets.

a. Install by referring to identification mark on upper surface mark.

b. Install so that identification mark can be correctly read when viewed from the exhaust side.

c. Install No. 1 camshaft bracket as follows.

d. Apply sealant to No.1 camshaft bracket as shown.

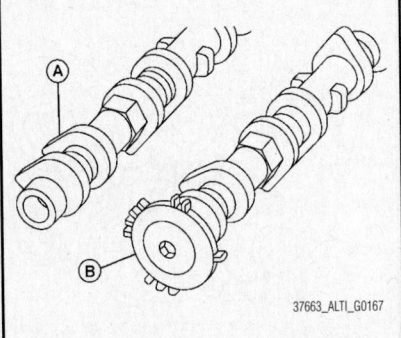

Fig. 62 A: Exhaust; B: Intake Signal plate for the camshaft position sensor (PHASE)

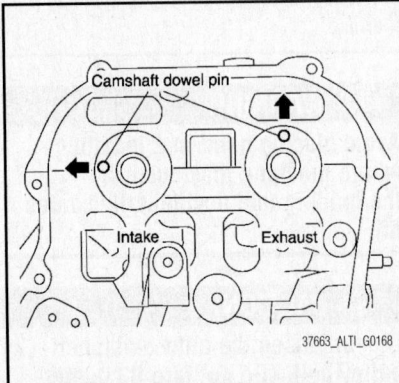

Fig. 63 Install camshafts so that the dowel pins on the front side are positioned as shown

➡**Use Genuine Silicone RTV Sealant, or equivalent.**

✳✳ CAUTION

After installation, be sure to wipe off any excessive sealant leaking from part (A) (both on right and left sides).

e. Apply sealant to camshaft bracket contact surface on the front cover backside.

f. Apply sealant to the outside of bolt hole on front cover.

g. Position the No.1 camshaft bracket near the mounting position, and install it without disturbing the sealant applied to the surfaces.

22. Tighten camshaft bracket bolts in four steps in the order as shown.

a. Step 1: Bolts 9–11: 17 inch lbs. (2. Nm)

b. Step 2: Bolts 1–8: 17 inch lbs. (2. Nm)

c. Step 3: Bolts 1–11: 52 inch lbs. (6 Nm)

d. Step 4: Bolts 1–11: 92 inch lbs. (10 Nm)

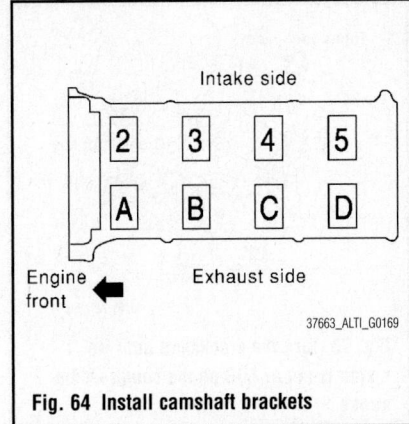

Fig. 64 Install camshaft brackets

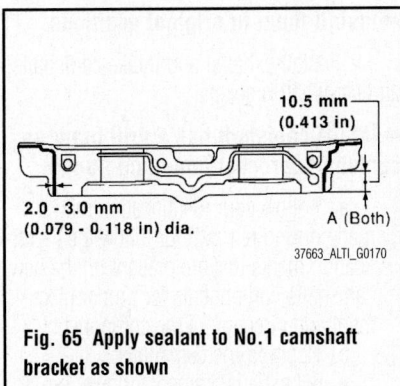

Fig. 65 Apply sealant to No.1 camshaft bracket as shown

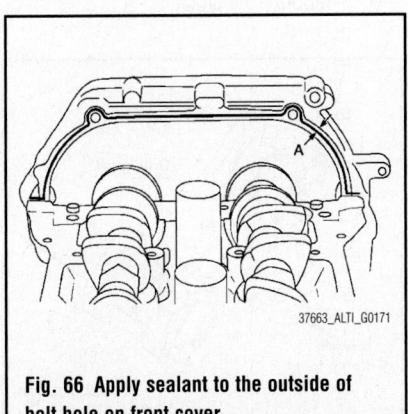

Fig. 66 Apply sealant to the outside of bolt hole on front cover

⚠ **CAUTION**

After tightening camshaft bracket bolts, be sure to wipe off excessive sealant from the mating surface of rocker cover and the mating surface of front cover, when installed without the front cover.

23. Install camshaft sprockets.
  a. Install them by lining up the mating marks on each camshaft sprocket with the ones painted on the timing chain during removal.
  b. Before installation of chain tensioner, it is possible to re-match the marks on timing chain with the ones on each sprocket.

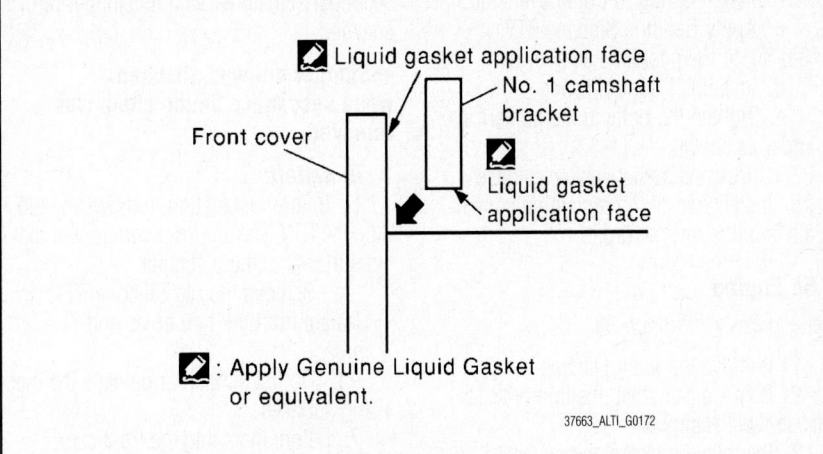

Fig. 67 Position the No.1 camshaft bracket near the mounting position, and install it without disturbing the sealant applied to the surfaces

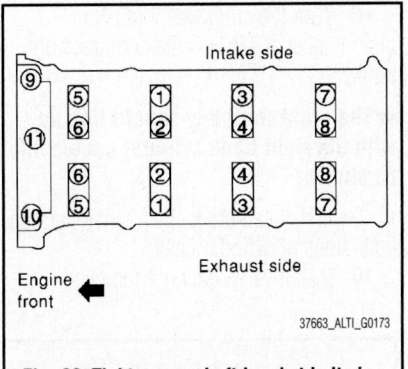

Fig. 68 Tighten camshaft bracket bolts in four steps in the order as shown

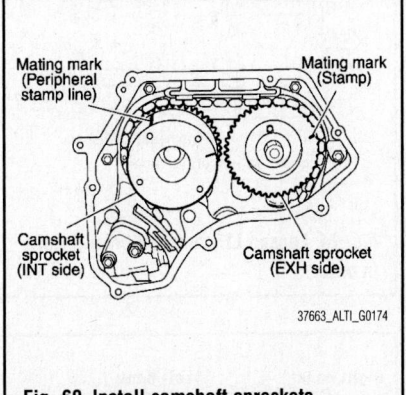

Fig. 69 Install camshaft sprockets

⚠ **CAUTION**

Aligned mating marks could slip. Therefore, after matching them, hold the timing chain in place by hand. Before and after installing chain tensioner, check again to make sure that mating marks have not slipped.

24. Install chain tensioner.

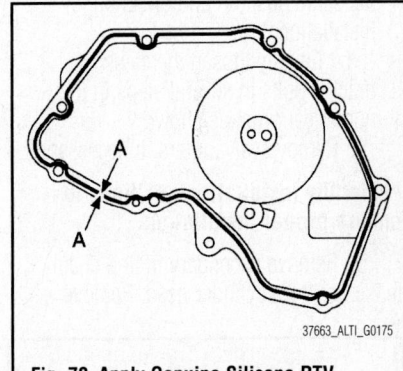

Fig. 70 Apply Genuine Silicone RTV Sealant to the positions as shown

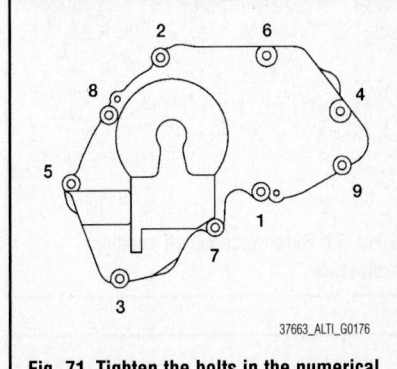

Fig. 71 Tighten the bolts in the numerical order as shown

⚠ **CAUTION**

After installation, pull the stopper pin off completely, and make sure that the tensioner is fully released.

25. Install chain guide.
26. Install IVT control cover with the following procedure.
  a. Install IVT control solenoid valve to intake valve timing control cover.

b. Install O-ring to front cover side.

c. Apply Genuine Silicone RTV Sealant to the positions as shown.

d. Install IVT control cover.

e. Tighten the bolts in the numerical order as shown.

27. Check and adjust valve clearances.

28. Installation of the remaining components is in the reverse order.

3.5L Engine

See Figures 72 through 83.

1. Remove the timing chains.

2. Remove camshaft position brackets (RH shown LH similar).

3. Remove the intake and exhaust camshaft brackets and the camshafts.

a. Mark the camshafts, camshaft brackets, and bolts so they are placed in the same position and direction for installation.

b. Equally loosen the camshaft bracket bolts in several steps in the numerical order as shown.

4. Remove valve lifters, if necessary.

➡**Identify installation positions to ensure proper installation.**

5. Remove secondary timing chain tensioner from cylinder head. Remove

secondary tensioner with its stopper pin attached.

➡**Stopper pin was attached when secondary timing chain was removed.**

To install:

6. Before installation, remove any old Silicone RTV Sealant from component mating surfaces using a scraper.

a. Remove the old Silicone RTV Sealant from the bolt holes and threads.

b. Do not scratch or damage the mating surfaces.

7. Before installing the front cam bracket, remove the old Silicone RTV Sealant from the mating surface using a scraper. Do not scratch or damage the mating surface.

8. Turn the crankshaft until No. 1 piston is set at TDC on the compression stroke.

➡**The crankshaft key should line up with the right bank cylinder center line as shown.**

9. Install camshaft chain tensioners on both sides of cylinder head.

10. Install valve lifters, if removed.

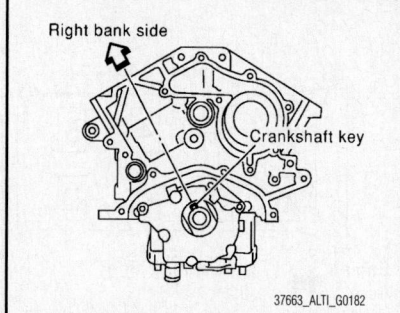

Fig. 76 Turn the crankshaft until No. 1 piston is set at TDC on the compression stroke

➡**Install them in original positions.**

11. Install exhaust and intake camshafts and camshaft brackets.

➡**Intake camshaft has a drill mark on camshaft sprocket mounting flange.**

a. Follow your identification marks made during removal, or follow the identification marks that are present on the new camshafts components for proper placement and direction of the components.

b. Position the camshafts:

• RH exhaust camshaft dowel pin at about 10 o'clock.

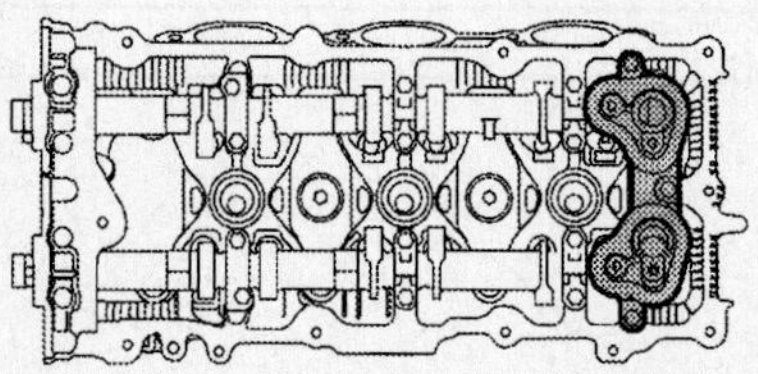

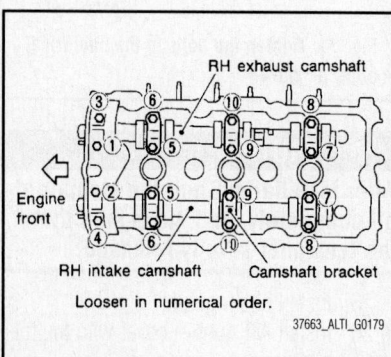

Fig. 72 Remove camshaft position brackets

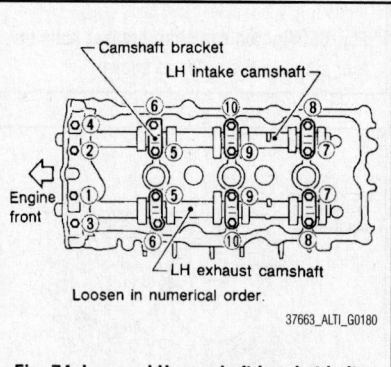

Fig. 74 Loosen LH camshaft bracket bolts in order

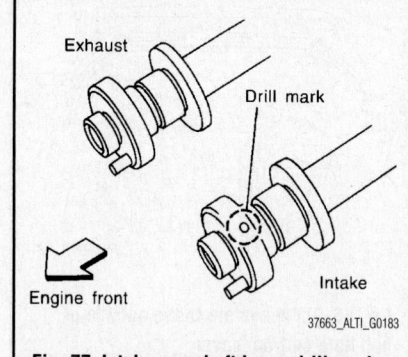

Fig. 77 Intake camshaft has a drill mark on camshaft sprocket mounting flange

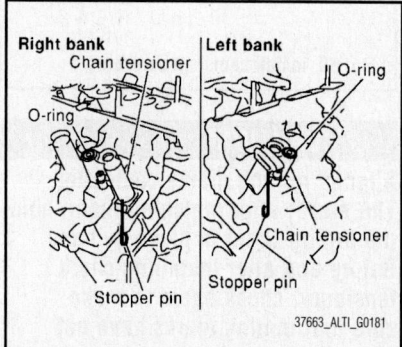

Fig. 73 Loosen RH camshaft bracket bolts in order

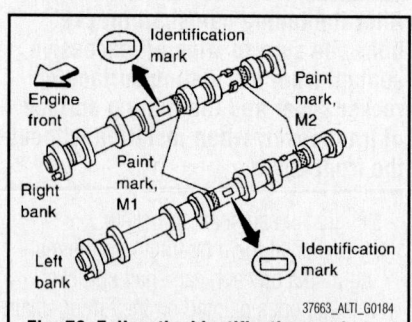

Fig. 75 Remove secondary timing chain tensioner from cylinder head

Fig. 78 Follow the identification marks that are present on the new camshafts components for proper placement and direction of the components

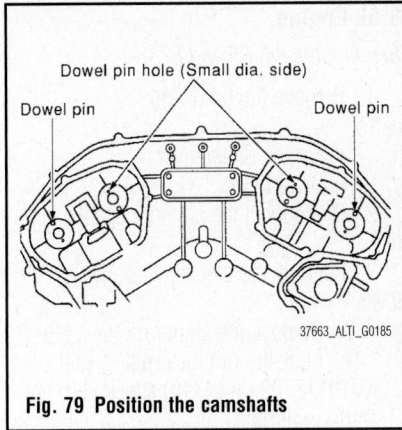

Fig. 79 Position the camshafts

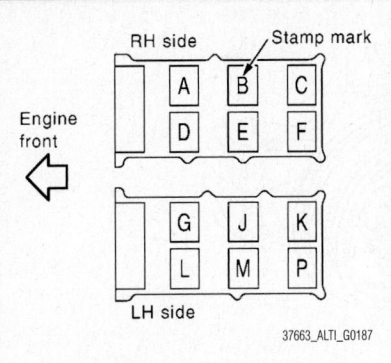

Fig. 81 Align the stamp marks as shown

REMOVAL & INSTALLATION

2.5L Engine

1. Remove crankshaft pulley with the following procedure:

    a. Hold the crankshaft pulley using suitable tool, then loosen the crankshaft pulley bolt, and pull the pulley out about 0.39 inches (10 mm).

    b. Attach suitable pulley puller in the M 6 (0.24 inch diameter) thread hole on crankshaft pulley, and remove crankshaft pulley using a suitable puller.

2. Installation is the reverse of removal.

3.5L Engine

See Figures 84 and 85.

1. Remove the following parts:
   - Engine undercover.
   - Drive belts.

2. Disconnect the battery negative terminal.

3. Remove the crankshaft pulley as follows:

    a. Remove the starter motor.

    b. Lock the ring gear using Tool KV10117700 (J-44716) attached to the starter bolt hole.

> **✳✳ CAUTION**
>
> **Do not damage the ring gear teeth, or the signal plate teeth behind the ring gear when setting the stopper.**

    c. Loosen crankshaft pulley bolt using Tool KV10109300 and locate bolt seating surface at 10 mm (0.39 in) from its original position.

    d. Position a pulley puller at recess hole of crankshaft pulley to remove crankshaft pulley.

- LH exhaust camshaft dowel pin at about 2 o'clock.

    c. Before installing camshaft brackets, apply sealant to mating surface of No. 1 camshaft bracket.

    d. Before installation, wipe off any protruding sealant.

    e. Install camshaft brackets in their original positions and direction.

    f. Align the stamp marks as shown.

    g. If checking and adjusting any part of valve assembly or camshaft, check valve clearance according to the reference data.

    h. Tighten the camshaft brackets in the three steps:

- Step 1: Tighten Bolts 7–10, then tighten 1–6 in numerical order to 17 inch lbs. (2 Nm).
- Step 2: Tighten all bolts in numerical order to 52 inch lbs. (6 Nm).
- Step 3: Tighten all bolts in numerical order to 8 ft. lbs. (10 Nm).

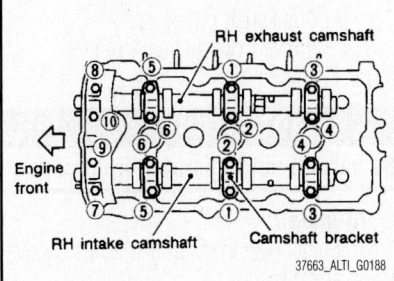

Fig. 82 Tighten the RH camshaft brackets in the three steps, in numerical order

12. Measure difference in levels between front end faces of No. 1 camshaft bracket and cylinder head.

> **✳✳ CAUTION**
>
> **If measurement is outside the specified range of -0.0055 inches (-0.14 mm), re-install camshaft and camshaft bracket.**

13. Install camshaft position sensor (PHASE) (RH and LH bank.)

14. Install the fuel rail and injectors.

15. Install the timing chains.

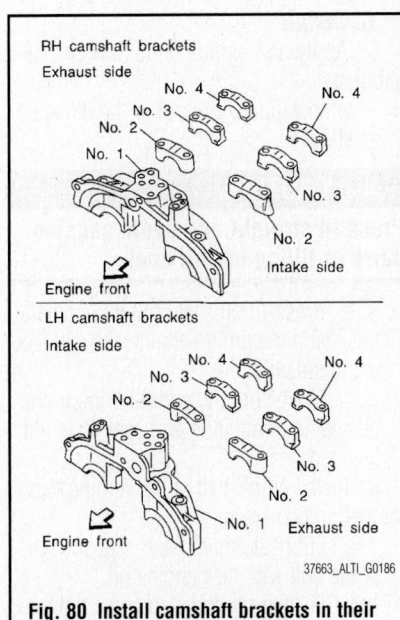

Fig. 80 Install camshaft brackets in their original positions and direction

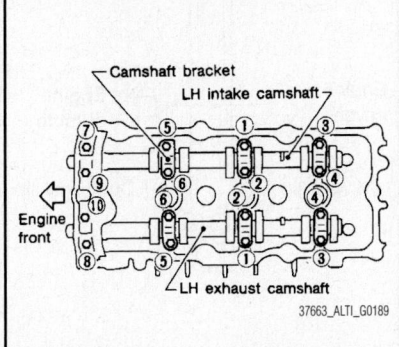

Fig. 83 Tighten the LH camshaft brackets in the three steps, in numerical order

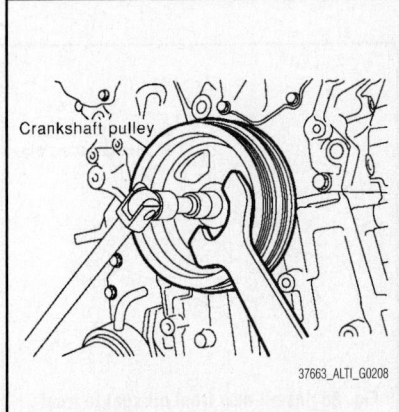

Fig. 84 Loosen crankshaft pulley bolt

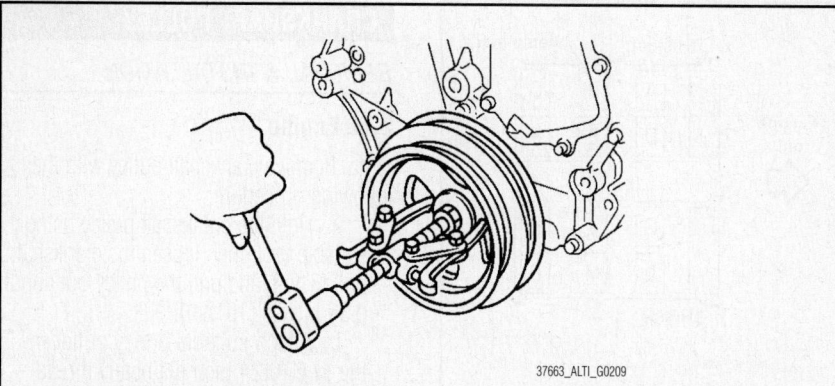

Fig. 85 Position a pulley puller at recess hole of crankshaft pulley to remove crankshaft pulley

✳✳ CAUTION

Do not use a puller claw on crankshaft pulley periphery.

To install:

4. Install crankshaft pulley and tighten the bolt in two steps.

a. Lubricate thread and seat surface of the bolt with new engine oil.

b. For the second step angle tighten using Tool KV10112100 (BT-8653-A).

c. Step 1: 32 ft. lbs. (44 Nm)

d. Step 2: Tighten an additional 84–90 degrees

5. Remove tool attached to the starter bolt hole.

6. Installation of the remaining components is in reverse order of removal.

CRANKSHAFT FRONT SEAL

REMOVAL & INSTALLATION

2.5L Engine

See Figures 86 and 87.

1. Remove the following parts:
   - RH front wheel
   - Engine under cover

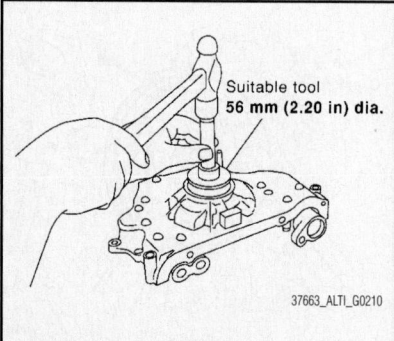

Fig. 86 Install new front oil seal to front cover using suitable tool

- Drive belts
- Crankshaft pulley

2. Remove front oil seal from front cover.

✳✳ CAUTION

Be careful not to scratch front cover.

To install:

3. Install new front oil seal to front cover using suitable tool.

a. Install new oil seal in until it is flush with front end surface of front cover.

✳✳ CAUTION

Do not reuse oil seal. Be careful not to cause damage to circumference of oil seal.

b. Install new oil seal in the direction shown.

4. Installation of the remaining components is in the reverse order of removal.

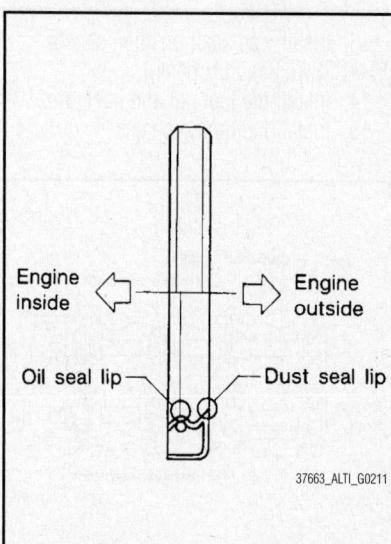

Fig. 87 Install new oil seal in the direction shown

3.5L Engine

See Figures 84, 85 and 87.

1. Remove the following parts:
   - Engine undercover.
   - Drive belts.

2. Disconnect the battery negative terminal.

3. Remove the crankshaft pulley as follows:

a. Remove the starter motor.

b. Lock the ring gear using Tool KV10117700 (J-44716) attached to the starter bolt hole.

✳✳ CAUTION

Do not damage the ring gear teeth, or the signal plate teeth behind the ring gear when setting the stopper.

c. Loosen crankshaft pulley bolt using Tool KV10109300 and locate bolt seating surface at 10 mm (0.39 in) from its original position.

d. Position a pulley puller at recess hole of crankshaft pulley to remove crankshaft pulley.

✳✳ CAUTION

Do not use a puller claw on crankshaft pulley periphery.

4. Remove front oil seal from front cover.

✳✳ CAUTION

Be careful not to damage front cover or crankshaft.

To install:

5. Apply new engine oil to new oil seal and install.

a. Install new oil seal in the direction as shown.

✳✳ CAUTION

Press fit straight and avoid causing burrs or tilting the oil seal.

b. Press-fit oil seal until it becomes flush with the timing chain case end face, using suitable tool.

c. Make sure the garter spring in the oil seal is in position and seal lip is not inverted.

6. Install crankshaft pulley and tighten the bolt in two steps.

a. Lubricate thread and seat surface of the bolt with new engine oil.

b. For the second step angle tighten using Tool KV10112100 (BT-8653-A).

c. Step 1: 32 ft. lbs. (44 Nm)
d. Step 2: Tighten an additional 84–90 degrees

7. Remove tool attached to the starter bolt hole.

8. Installation of the remaining components is in reverse order of removal.

CYLINDER HEAD

REMOVAL & INSTALLATION

2.5L Engine

See Figures 88 through 90.

1. Remove the engine and transaxle assembly.
2. Remove the timing chain.
3. Remove the camshafts.
4. Remove the ignition coils and spark plugs.
5. Remove the exhaust manifold and three way catalyst.
6. Remove cylinder head by loosening bolts in the order as shown.
7. If necessary to transfer to new cylinder head or remove for reconditioning, remove the intake manifold collector, intake manifold, and fuel tube assembly.

To install:
8. Install a new cylinder head gasket.
9. Install the cylinder head.
10. Follow the steps below to tighten the cylinder head bolts in the numerical order as shown.
   a. Apply new engine oil to the threads and the seating surfaces of bolts.
   b. Step 1: 72 ft. lbs. (98 Nm)
   c. Step 2: Loosen all bolts in reverse order of tightening.

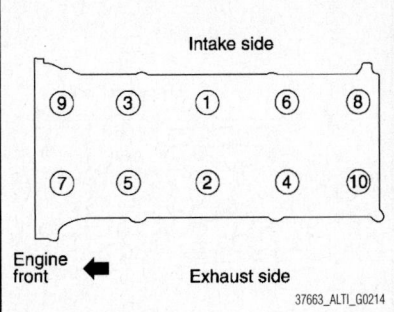

Fig. 89 Tighten the cylinder head bolts in the numerical order

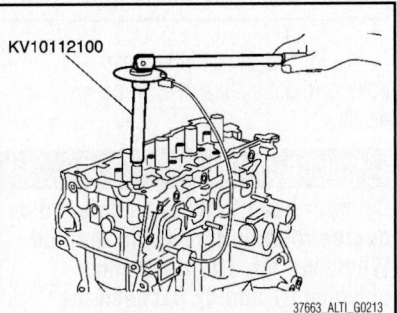

Fig. 90 Check and confirm the tightening angle by using angle wrench or protractor

d. Step 3: 29 ft. lbs. (39 Nm)
e. Step 4 75 degrees clockwise
f. Step 5 75 degrees clockwise

✳✳ CAUTION

If cylinder head bolts are re-used, check their outer diameters before installation. Follow the Outer Diameter of Cylinder Head Bolts measure-

ment procedure. Check and confirm the tightening angle by using angle wrench or protractor. Avoid judgment by visual inspection without the tool.

11. Installation of the remaining components is in the reverse order of removal.

3.5L Engine

See Figures 91 through 96.

1. Remove the engine from the vehicle.
2. Remove the RH and LH exhaust manifold and three way catalysts.
3. Remove the rear timing chain case.
4. Remove the intake manifold.
5. Remove the intake and exhaust camshafts.
6. Remove the coolant outlet housing.
7. Remove the RH and LH cylinder head bolts.
   a. The bolts should be loosened gradually in three stages.
   b. Loosen the bolts in the numerical order as shown.
8. Remove cylinder heads and gaskets.

➡**Discard the cylinder head gaskets and use new gaskets for installation.**

To install:
9. Turn the crankshaft until No. 1 piston is set at TDC on the compression

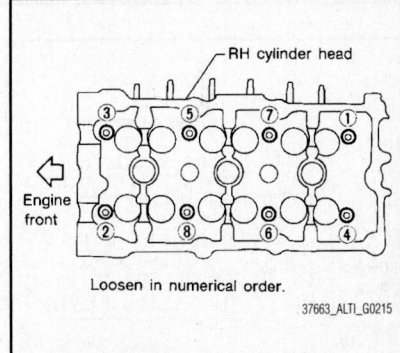

Fig. 91 Loosen the bolts in the numerical order as shown—RH

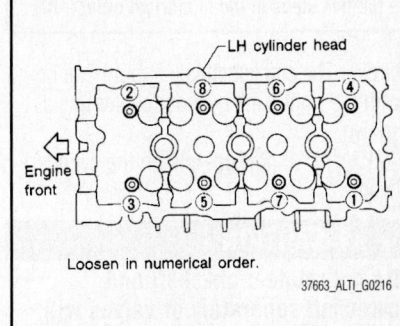

Fig. 92 Loosen the bolts in the numerical order as shown—LH

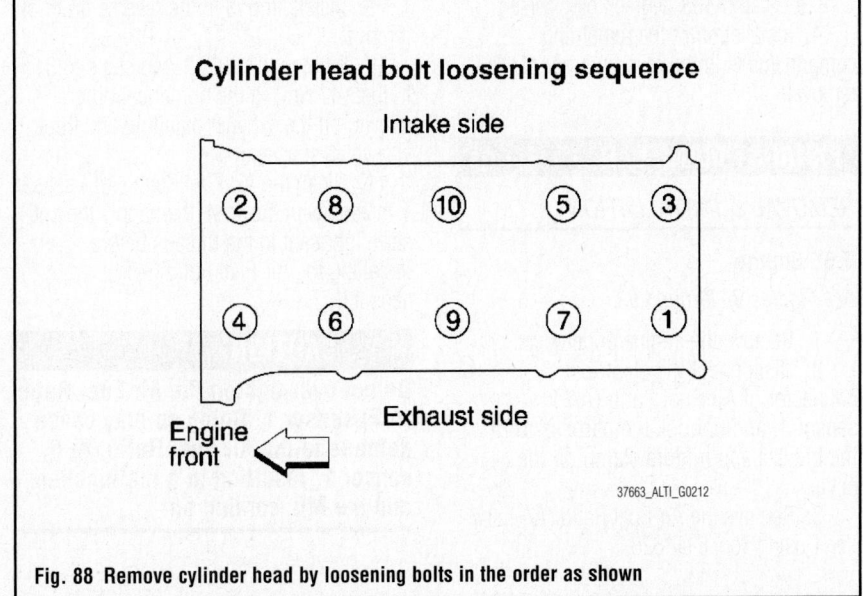

Fig. 88 Remove cylinder head by loosening bolts in the order as shown

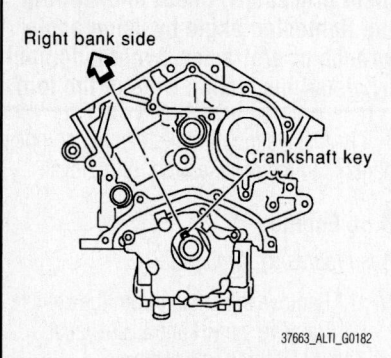

Fig. 93 Turn the crankshaft until No. 1 piston is set at TDC on the compression stroke

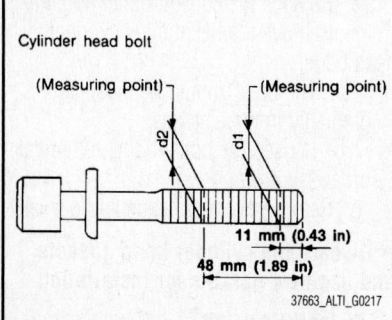

Fig. 94 Whenever the size difference between d1 and d2 exceeds the limit, replace the bolts with new ones

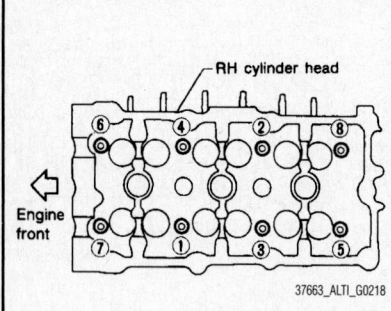

Fig. 95 Tighten the cylinder head bolts in the five steps in the numerical order—RH

stroke. The crankshaft key should line up with the right bank cylinder center line as shown.

10. Install new gaskets on the cylinder heads.

✳✳ CAUTION

Do not rotate crankshaft and camshaft separately or valves will strike piston heads.

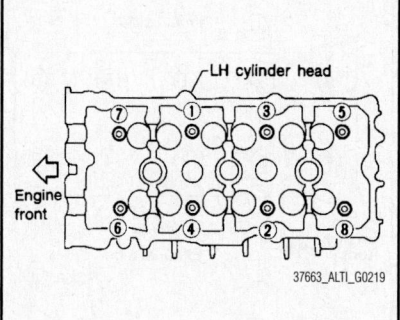

Fig. 96 Tighten the cylinder head bolts in the five steps in the numerical order—LH

11. Inspect the cylinder head bolts before installing the cylinder heads.

✳✳ CAUTION

Cylinder head bolts are tightened by degree rotation tightening method. Whenever the size difference between d1 and d2 exceeds the limit, replace the bolts with new ones.

12. Lubricate threads and seat surfaces of the bolts with new engine oil.

13. Install the cylinder heads on the cylinder block. Tighten the cylinder head bolts in the five steps in the numerical order as shown using Tool KV10112100 (BT-8653-A).

    a. Step 1: 72 ft. lbs. (98 Nm)

    b. Step 2: Loosen all bolts in reverse order of tightening.

    c. Step 3: 29 ft. lbs. (39 Nm)

    d. Step 4 103 degrees clockwise

    e. Step 5 103 degrees clockwise

14. Installation of the remaining components is in the reverse order of removal.

EXHAUST MANIFOLD

REMOVAL & INSTALLATION

2.5L Engine

See Figures 97 through 99.

1. Remove the engine undercover.

2. Disconnect the electrical connector of Air Fuel Ratio (A/F) sensor 1, and unhook the harness from the bracket and middle clamp on the cover.

3. Remove the Air Fuel Ratio (A/F) sensor 1 using Tool J-44626.

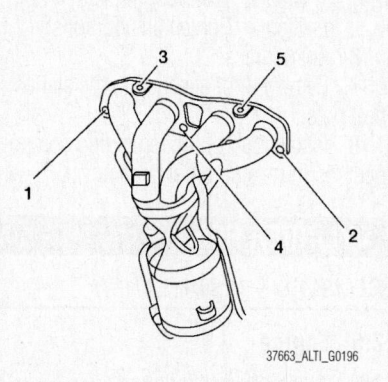

Fig. 97 Loosen the nuts in the reverse order as shown

✳✳ CAUTION

Be careful not to damage Air Fuel Ratio (A/F) sensor. Discard any Air Fuel Ratio (A/F) sensor which has been dropped from a height of more than 20 inches (0.5 m) onto a hard surface such as a concrete floor; replace with a new one.

4. Remove the exhaust manifold cover (lower).

5. Remove the exhaust front tube.

6. Remove the exhaust manifold cover (upper).

7. Loosen the nuts in the reverse order as shown, on the exhaust manifold and three way catalyst assembly.

8. Remove the exhaust manifold and three way catalyst assembly and gasket. Discard the gasket.

To install:

9. Installation is in the reverse order of removal.

10. Tighten the nuts in two steps to 31 ft. lbs. (42 Nm) in the numerical order shown, on the exhaust manifold and three way catalyst assembly.

11. Clean the Air Fuel Ratio (A/F) sensor 1 threads with the Tool, then apply the anti-seize lubricant to the threads before installing the Air Fuel Ratio (A/F) sensor 1.

✳✳ CAUTION

Do not over-tighten the Air Fuel Ratio (A/F) sensor 1. Doing so may cause damage to the Air Fuel Ratio (A/F) sensor 1, resulting in a malfunction and the MIL coming on.

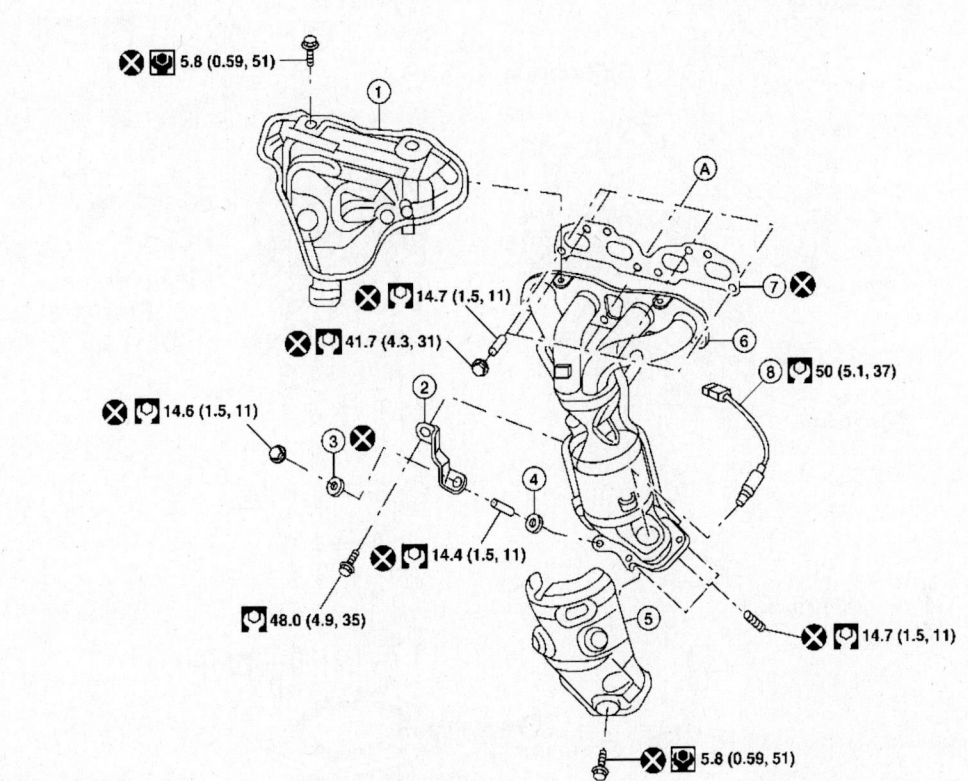

1. Exhaust manifold cover (upper)
2. Exhaust manifold stay
3. Manifold yoke (type B)
4. Manifold yoke (type A)
5. Exhaust manifold cover (lower)
6. Exhaust manifold and three way catalyst assembly
7. Exhaust manifold gasket
8. Air fuel ratio (A/F) sensor 1
A. To cylinder head

37663_ALTI_G0197

Fig. 98 Exploded view of exhaust manifold and three way catalyst

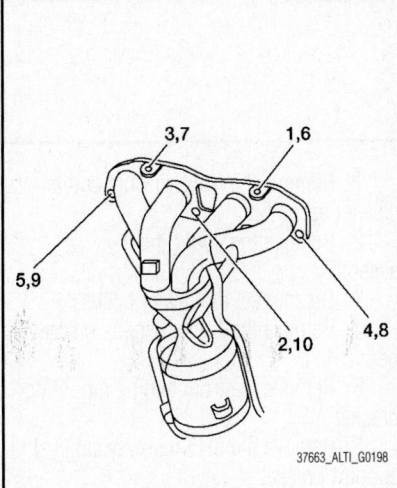

37663_ALTI_G0198

Fig. 99 Tighten the nuts to specification in the numerical order shown

3.5L Engine

Right Side

See Figures 100 and 101.

> ☀ **WARNING**
>
> **Perform the work when the exhaust and cooling system have completely cooled down.**

1. Remove the front air duct and air cleaner assembly.
2. Disconnect the EVAP vacuum hose and brake booster vacuum hose.
3. Remove cowl top.
4. Remove the front suspension member.
5. Remove the rear engine mounting bracket.
6. Remove the RH three way catalyst support bracket.
7. Remove heated oxygen sensor 2 (bank 1) and Air Fuel Ratio (A/F) sensor 1 (bank 1).

   a. Remove harness connector of each sensor, and disconnect the harness from the bracket and middle clamp.

   b. Remove both heated oxygen sensor and Air Fuel Ratio (A/F) sensor using Tool.

> ☀ **CAUTION**
>
> **Be careful not to damage heated oxygen sensor or Air Fuel Ratio (A/F) sensor. Discard any heated oxygen sensor which has been dropped from a height of more than 20 inches (0.5 m) onto a hard surface such as a concrete floor; replace with a new sensor.**

8. Remove exhaust manifold and three way catalyst heat shields.

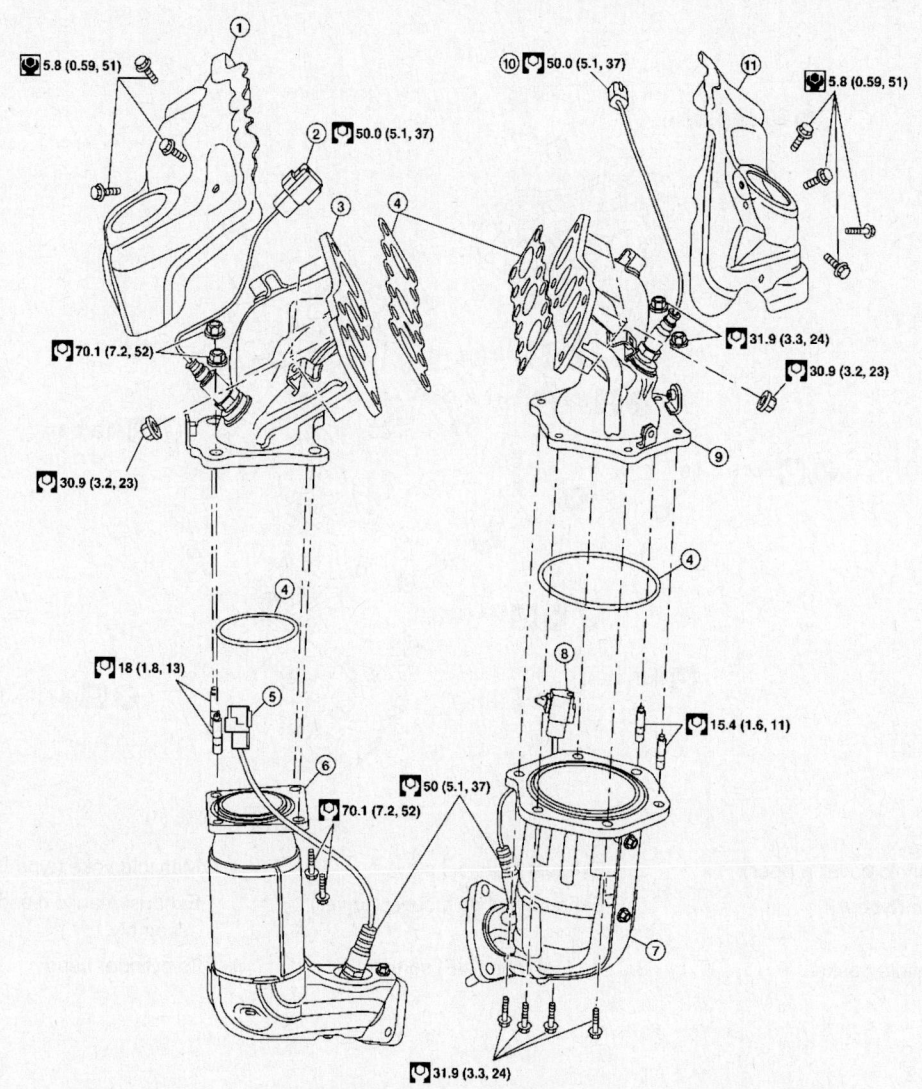

Fig. 100 Exploded view of exhaust manifold and three way catalyst

1. Exhaust manifold heat shield (RH)
2. Air fuel ratio (A/F) sensor 1 (bank 1)
3. Exhaust manifold (RH)
4. Gaskets
5. Heated oxygen sensor 2 (bank 1)
6. Three way catalyst (manifold) (bank 1)
7. Three way catalyst (manifold) (bank 2)
8. Heated oxygen sensor 2 (bank 2)
9. Exhaust manifold (LH)
10. Air fuel ratio (A/F) sensor 1 (bank 2)
11. Exhaust manifold heat shield (LH)

37663_ALTI_G0199

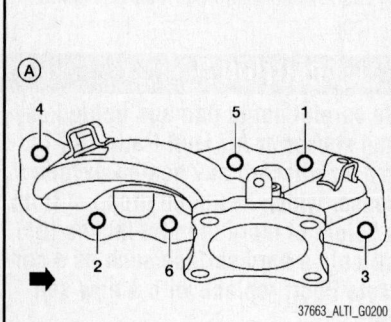

Fig. 101 Loosen the exhaust manifold nuts in the order as shown

37663_ALTI_G0200

9. Remove the three way catalyst (manifold) (bank 1) by loosening the bolts first and then removing the nuts and through bolts.

10. Remove the exhaust manifold RH (A). Loosen the exhaust manifold nuts in the order as shown.

Left Side

See Figures 102 through 105.

❋❋❋ WARNING

Perform the work when the exhaust and cooling system have completely cooled down.

1. Remove the front air duct and air cleaner assembly.

2. Remove the cooling fan assembly.

3. Disconnect the heater hoses.

4. Remove the front suspension member.

5. Remove the front engine mounting bracket.

6. Remove the LH three way catalyst support bracket.

7. Remove heated oxygen sensor 2 (bank 2) and Air Fuel Ratio (A/F) sensor 1 (bank 2).

a. Remove harness connector of each

sensor, and disconnect the harness from the bracket and middle clamp.

b. Remove both heated oxygen sensor and Air Fuel Ratio (A/F) sensor using Tool.

✳✳ CAUTION

Be careful not to damage heated oxygen sensor or Air Fuel Ratio (A/F) sensor. Discard any heated oxygen sensor which has been dropped from a height of more than 20 inches (0.5 m) onto a hard surface such as a concrete floor; replace with a new sensor.

8. Remove exhaust manifold and three way catalyst heat shields.

9. Remove the three way catalyst (manifold) (bank 2) by loosening the bolts first and then removing the nuts and through bolts.

10. Remove the exhaust manifold LH (B). Loosen the exhaust manifold nuts in the order as shown.

To install:

11. Installation is in the reverse order of removal.

12. Install the exhaust manifold nuts in the order as shown RH (A) and LH (B).

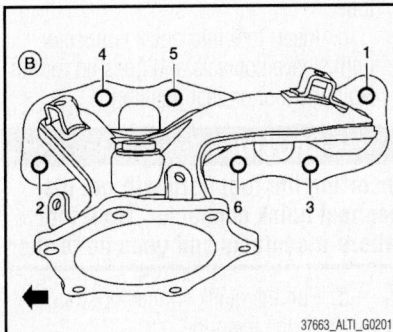

Fig. 102 Loosen the exhaust manifold nuts in the order as shown

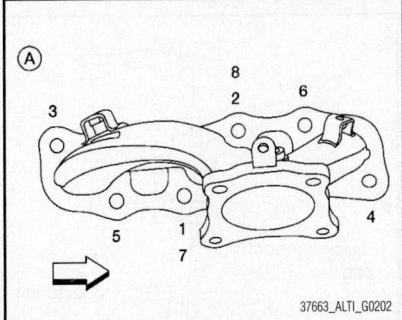

37663_ALTI_G0202

Fig. 103 Install the exhaust manifold nuts in the order as shown RH

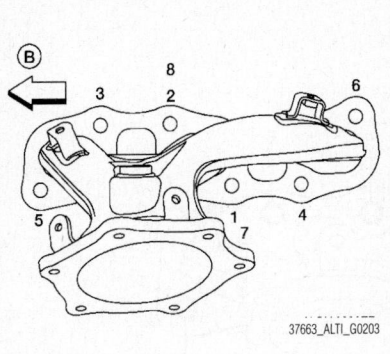

37663_ALTI_G0203

Fig. 104 Install the exhaust manifold nuts in the order as shown LH

13. Install RH and LH three way catalyst support brackets.

14. Hand tighten the three way catalyst support bracket bolts to seat the support brackets.

15. Tighten the bolts to 16 ft. lbs. (22 Nm) in the numerical order as shown.

✳✳ CAUTION

Do not tighten if support brackets do not fit tightly against oil pan and three way catalysts.

✳✳ CAUTION

Before installing a heated oxygen sensor or Air Fuel Ratio (A/F) sensor, clean the exhaust manifold threads using the oxygen sensor thread

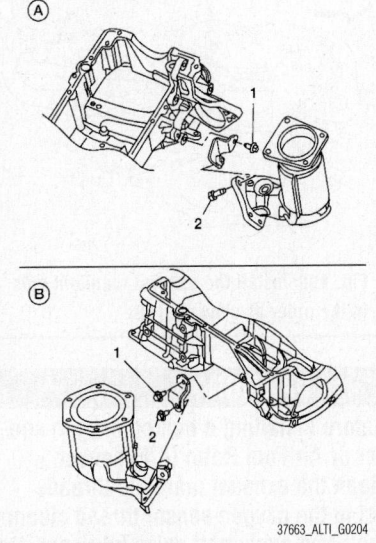

37663_ALTI_G0204

Fig. 105 Install RH (A) and LH (B) three way catalyst support brackets

cleaner tool, and apply anti-seize lubricant. Do not over-tighten the Air Fuel Ratio (A/F) sensor or heated oxygen sensors. Doing so may cause damage.

CVT Models

See Figures 106 through 108.

✳✳ WARNING

Note the following:

- Perform the work when the exhaust and cooling system have completely cooled down.
- When removing the front and rear engine mounting through bolts and nuts, lift the engine up slightly for safety.

1. Remove the engine and transaxle assembly.

2. Remove the RH and LH three way catalyst support brackets.

3. Remove heated oxygen sensor 2 (bank 1), heated oxygen sensor 2 (bank 2), Air Fuel Ratio (A/F) sensor 1 (bank 1) and Air Fuel Ratio (A/F) sensor 1 (bank 2).

a. Remove harness connector of each sensor, and disconnect the harness from the bracket and middle clamp.

b. Remove both heated oxygen sensors and air fuel ratio (A/F) sensors using Tool.

✳✳ CAUTION

Be careful not to damage heated oxygen sensors or Air Fuel Ratio (A/F) sensors. Discard any heated oxygen sensor which has been dropped from a height of more than 20 inches (0.5 m) onto a hard surface such as a concrete floor; replace with a new sensor.

4. Remove exhaust manifold and three way catalyst heat shields.

5. Remove the three way catalyst (manifold) (bank 1) and three way catalyst (manifold) (bank 2) by loosening the bolts first and then removing the nuts and through bolts.

6. Remove the exhaust manifolds RH and LH. Loosen the exhaust manifold nuts in the order as shown.

To install:

7. Installation is in the reverse order of removal.

8. Install the exhaust manifold nuts in the order as shown RH and LH.

9. Install RH (A) and LH (B) three way catalyst support brackets.

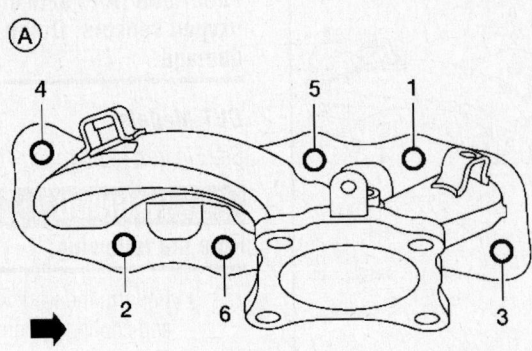

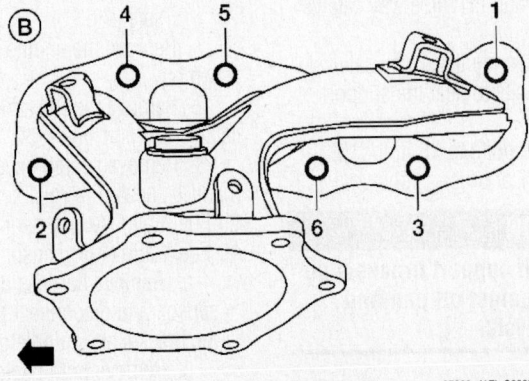

Fig. 106 Remove the exhaust manifolds RH (A) and LH (B); loosen the exhaust manifold nuts in the order as shown

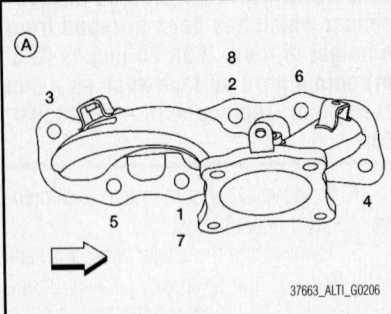

Fig. 107 Install the exhaust manifold nuts in the order as shown RH (A)

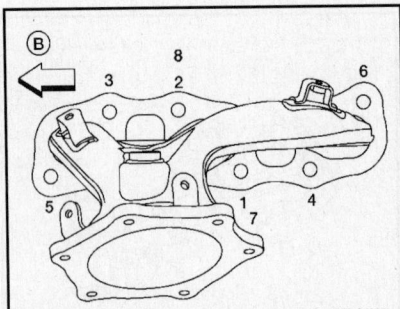

Fig. 108 Install the exhaust manifold nuts in the order as shown LH (B)

10. Hand tighten the three way catalyst support bracket bolts to seat the support brackets.

11. Tighten the bolts to 16 ft. lbs. (22 Nm) in the numerical order as shown.

> ❊❊ **CAUTION**
>
> Do not tighten if support brackets do not fit tightly against oil pan and three way catalysts.

> ❊❊ **CAUTION**
>
> Before installing a heated oxygen sensor or Air Fuel Ratio (A/F) sensor, clean the exhaust manifold threads using the oxygen sensor thread cleaner tool, and apply anti-seize lubricant. Do not over-tighten the Air Fuel Ratio (A/F) sensor or heated oxygen sensors. Doing so may cause damage.

INTAKE MANIFOLD

REMOVAL & INSTALLATION

2.5L Engine

See Figures 109 through 112.

> ❊❊ **WARNING**
>
> To avoid the danger of being scalded, never drain the coolant when the engine is hot.

1. Release the fuel pressure.
2. Drain coolant when engine is cooled.
3. Disconnect the MAF sensor electrical connector.
4. Remove air cleaner and air duct assembly.
5. Remove cowl top finisher.
6. Disconnect the following components at the intake side:
   - PCV hose
   - EVAP hose and EVAP canister purge volume control solenoid
   - Electric throttle control actuator
   - Brake booster vacuum hose
7. Disconnect the fuel quick connector on the engine side.
   a. Remove quick connector cap.
   b. With the sleeve side of tool facing quick connector, install tool onto fuel tube.
   c. Insert tool into quick connector until sleeve contacts and goes no further. Hold the tool on that position.

> ❊❊ **CAUTION**
>
> Inserting the tool hard will not disconnect quick connector. Hold tool where it contacts and goes no further.

   d. Pull the quick connector straight out from the fuel tube.

> ❊❊ **CAUTION**
>
> Note the following:

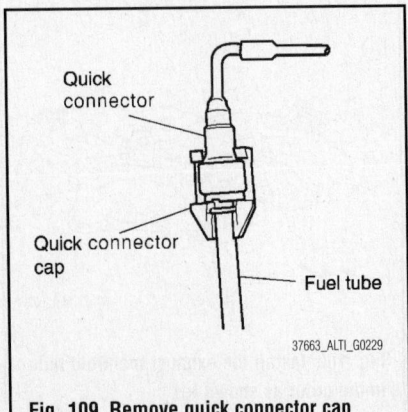

Fig. 109 Remove quick connector cap

- Pull quick connector holding it at the A position, as shown.
- Do not pull with lateral force applied. O-ring inside quick connector may be damaged.
- Prepare container and cloth beforehand as fuel will leak out.
- Avoid fire and sparks.
- Be sure to cover openings of disconnected pipes with plug or plastic bag to avoid fuel leakage and entry of foreign materials.

8. When removing fuel hose quick connector at vehicle piping side, perform as follows.

a. Remove quick connector cap.

b. Hold the sides of the connector, push in tabs and pull out the tube. (The figure is shown for reference only.)

c. If the connector and the tube are stuck together, push and pull several times until they start to move. Then disconnect them by pulling.

✻✻ CAUTION

Note the following:

- The tube can be removed when the tabs are completely depressed. Do not twist it more than necessary.
- Do not use any tools to remove the quick connector.
- Keep the resin tube away from heat. Be especially careful when welding near the tube.
- Prevent acid liquid such as battery electrolyte etc. from getting on the resin tube.
- Do not bend or twist the tube during installation and removal.
- Do not remove the remaining retainer on tube.
- When the tube is replaced, also replace the retainer with a new one. Retainer color: Green.

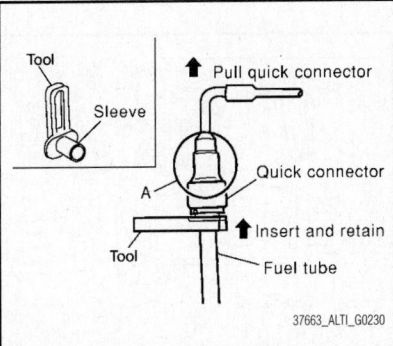

Fig. 110 With the sleeve side of tool facing quick connector, install tool onto fuel tube

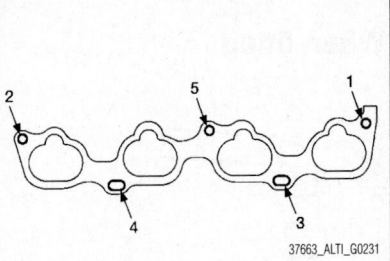

Fig. 111 Remove the bolts and nuts in the order shown

- To keep clean the connecting portion and to avoid damage and foreign materials, cover them completely with plastic bags or something similar.

9. Disconnect electric throttle control actuator coolant hoses.

10. Loosen bolts diagonally, and remove the electric throttle control actuator.

✻✻ CAUTION

Handle carefully to avoid any damage.

11. Remove the bolts and nuts in the order shown and remove the intake manifold assembly.

✻✻ CAUTION

Cover engine openings to avoid entry of foreign materials.

To install:

12. Installation is in the reverse order of removal. Follow the tightening sequences below.

13. Tighten intake manifold bolts and nuts.

a. Tighten in numerical order as shown.

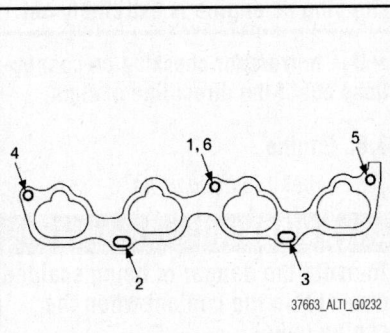

Fig. 112 Tighten in numerical order as shown

✻✻ CAUTION

After tightening the five bolts in the order shown, the 1, 6 position designates that the first bolt tightened is to be retightened to specification.

14. Install the Electric Throttle Control Actuator:

a. Tighten the bolts of electric throttle control actuator equally and diagonally in several steps.

b. After installation perform procedure in "INSPECTION AFTER INSTALLATION".

Connecting Quick Connector on the Fuel Hose (Engine Side)

See Figures 113 and 114.

1. Make sure no foreign substances are deposited in and around the fuel tube and quick connector, and there is no damage to them.

2. Thinly apply new engine oil around the fuel tube tip end.

3. Align center to insert quick connector straight into fuel tube.

4. Insert fuel tube into quick connector until the top spool on fuel tubes is inserted completely and the second level spool is positioned slightly below the quick connector bottom end.

✻✻ CAUTION

Note the following:

- Hold at position A as shown, when inserting the fuel tube into the quick connector.
- Carefully align to center to avoid inclined insertion to prevent damage to the O-ring inside the quick connector.
- Insert the fuel tube until you hear a "click" sound and actually feel the engagement.
- To avoid misidentification of engagement with a similar sound, be sure to perform the next step.

5. Before clamping the fuel hose with the hose clamp, pull the quick connector hard by hand, holding at the A position, as shown. Make sure it is completely engaged (connected) so that it does not come off of the fuel tube.

➡Recommended pulling force is 11 lbs. (50 N).

6. Install quick connector cap on quick connector joint.

7. Direct arrow mark on quick connector cap to upper side (fuel hose side).

8. Install fuel hose to hose clamp.

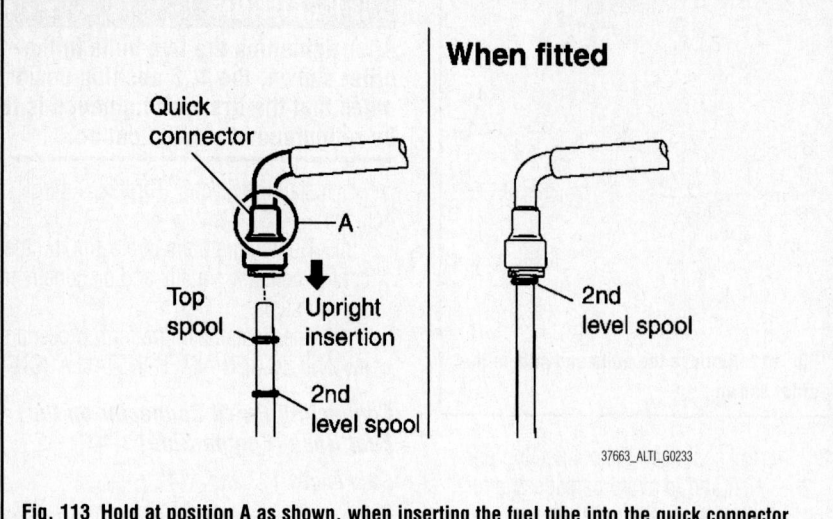

When fitted

Fig. 113 Hold at position A as shown, when inserting the fuel tube into the quick connector

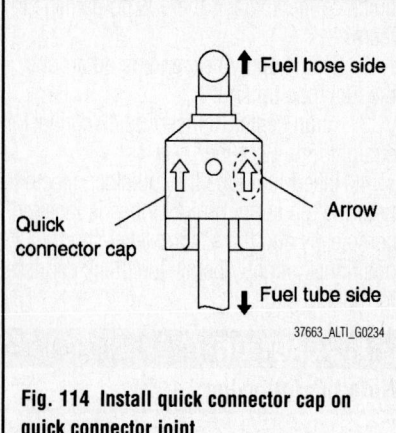

Fig. 114 Install quick connector cap on quick connector joint

Connecting Quick Connector on the Fuel Hose (Vehicle Piping Side)

See Figure 115.

1. Make sure no foreign substances are deposited in and around the fuel tube and quick connector, and there is no damage to them.

2. Align center to insert quick connector straight into fuel tube.

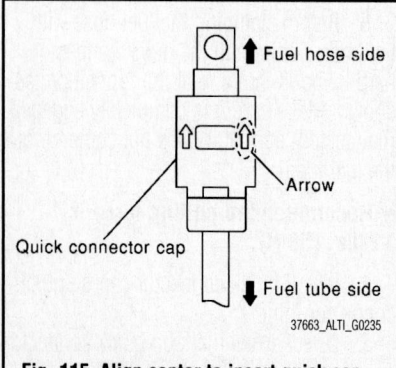

Fig. 115 Align center to insert quick connector straight into fuel tube

a. Insert fuel tube until a click is heard.

b. Install quick connector cap on quick connector joint. Direct arrow mark on quick connector cap upper side.

c. Install fuel hose to hose clamp.

Inspection After Installation

1. Make sure there is no fuel leakage at connections as follows:

a. Apply fuel pressure to fuel lines by turning ignition switch ON (with engine stopped). Then check for fuel leaks at connections.

b. Start the engine and rev it up and check for fuel leaks at connections.

c. Perform procedures for "Throttle Valve Closed Position Learning" after finishing repairs.

d. If electric throttle control actuator is replaced, perform procedures for "Idle Air Volume Learning" after finishing repairs.

> ✳✳ **WARNING**
>
> **Do not touch engine immediately after stopping as engine is extremely hot.**

➡**Use mirrors for checking on connections out of the direct line of sight.**

3.5L Engine

See Figures 116 through 120.

> ✳✳ **WARNING**
>
> **To avoid the danger of being scalded, never drain the coolant when the engine is hot.**

1. Release the fuel pressure.
2. Disconnect the battery negative terminal.

3. Remove the cowl top.
4. Remove the engine cover.
5. Remove front air duct and air duct hose.
6. If necessary, remove the electric throttle control actuator bolts in the reverse order as shown and remove the electric throttle control actuator and position aside.

> ✳✳ **CAUTION**
>
> **Handle carefully to avoid any shock to the electric throttle control actuator. Do not disassemble.**

7. If necessary, remove power valve bolts in the reverse order as shown and remove the power valves.
8. Disconnect the following:
   - Power brake booster vacuum hose
   - Fuel injector electrical connectors
   - PCV hose
   - Electric throttle control actuator electrical connector
   - EVAP canister purge hose

> ✳✳ **CAUTION**
>
> **Cover any engine openings to avoid the entry of any foreign material.**

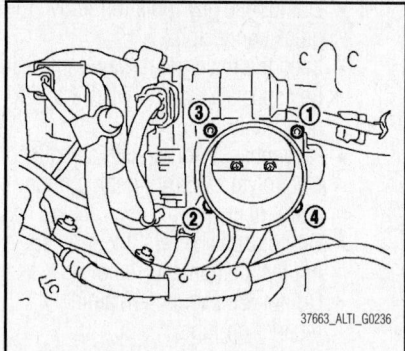

Fig. 116 Remove the electric throttle control actuator bolts in the reverse order as shown

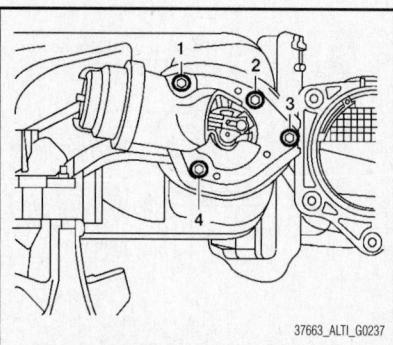

Fig. 117 Remove power valve bolts in the reverse order as shown

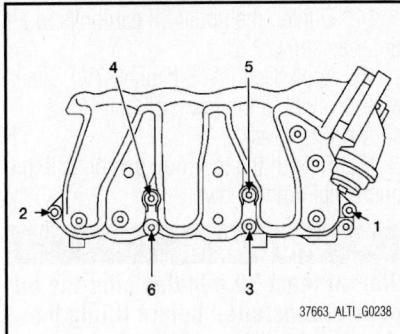

Fig. 118 Loosen the intake manifold collector bolts in the order as shown

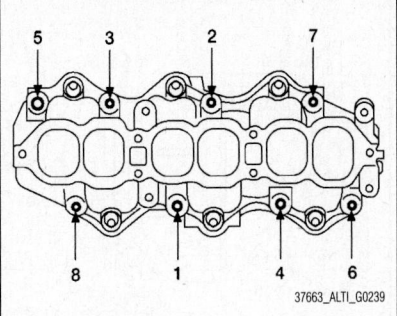

Fig. 119 Loosen the bolts in the order as shown

9. Remove the EVAP canister purge volume solenoid valve bracket bolt. Position the valve aside.

10. Loosen the intake manifold collector bolts in the order as shown, and remove the intake manifold collector and gasket.

11. If necessary remove the following components:

- VIAS control solenoid valve
- EVAP canister purge volume control solenoid valve

12. Disconnect fuel tube quick connector at vehicle piping side.

13. To remove the quick connector cap, hold the sides of the connector, push in the tabs and pull out the tube.

❈❈ CAUTION

Note the following:

- The tube can be removed when the tabs are completely depressed. Do not twist it more than necessary.
- Do not use any tools to remove the quick connector.
- Keep the resin tube away from heat. Be especially careful when welding near the tube.
- Prevent acid liquids such as battery electrolyte, etc. from getting on the resin tube.
- Do not bend or twist the tube during removal or installation.
- Do not remove the remaining retainer on the tube
- When the tube is replaced, also replace the retainer with a new one.
- To keep the connecting portion clean and to avoid damage and foreign materials entering, cover the ends of the fuel tubes with plastic bags or something similar.

➡**If the connector and the tube are stuck together, push and pull several times until they start to move. Then disconnect them by pulling.**

14. Remove the fuel rail with the fuel injectors attached, from the intake manifold. Remove the fuel injector O-rings and use new O-rings for installation.

15. Loosen the bolts in the order as shown, and remove the intake manifold.

To install:

16. Installation is in the reverse order of removal. Follow the procedure below for specific tightening sequences and procedures.

17. Install intake manifold bolts in two steps in the numerical order as shown.

    a. Step 1: 65 inch lbs. (7 Nm)
    b. Step 2: 19 ft. lbs. (25 Nm)

➡**After installation, it is necessary to re-calibrate the electric throttle control actuator.**

18. Perform the "Throttle Valve Closed Position Learning" when harness connector of the electric throttle control actuator is disconnected.

19. Perform the "Idle Air Volume Learning" when the electric throttle control actuator is replaced.

20. Install the quick connector as follows:

    a. Make sure no foreign substances are deposited in and around the fuel tube

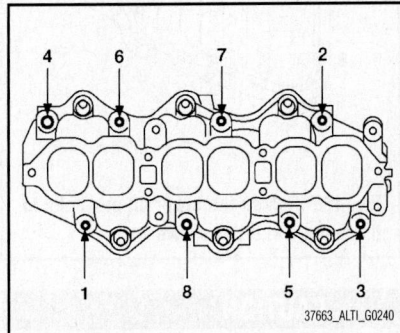

Fig. 120 Install intake manifold bolts in two steps in the numerical order as shown

and quick connector and that there is no damage.

    b. Align the center to insert the quick connector straight onto the fuel tube.

    c. Insert the fuel tube until a click is heard.

    d. Install the quick connector cap on the quick connector joint. Align the arrow mark on the quick connector cap to the upper side.

    e. Install the fuel hose into the hose clamp.

Inspection After Installation

1. Apply fuel pressure to fuel lines by turning ignition switch ON (with engine stopped). Then check for fuel leaks at connections.

2. Start the engine and rev it up and check for fuel leaks at connections.

❈❈ CAUTION

Do not touch engine immediately after stopping as engine is extremely hot.

➡**Use mirrors for checking on connections out of the direct line of sight.**

OIL PAN

REMOVAL & INSTALLATION

2.5L Engine

See Figures 121 through 126.

❈❈ WARNING

To avoid the danger of being scalded, never drain the engine oil when the engine is hot.

1. Drain engine oil.
2. Remove the front exhaust tube.
3. Remove power steering cooler hose bracket from suspension member.
4. Remove the front suspension member for clearance to remove the oil pan.
5. Remove the lower oil pan bolts in the order as shown.
6. Remove the lower oil pan using Tool KV10111100 (J-37228).

➡**Tap gently to cut sealant around the pan; do not damage the mating surface using Tool.**

7. Remove the oil strainer.
8. Remove rear plate cover, and four engine-to transaxle bolts.
9. Loosen the upper oil pan bolts in the order shown to remove upper oil pan.
10. Remove upper oil pan using Tool KV10111100 (J-37228).

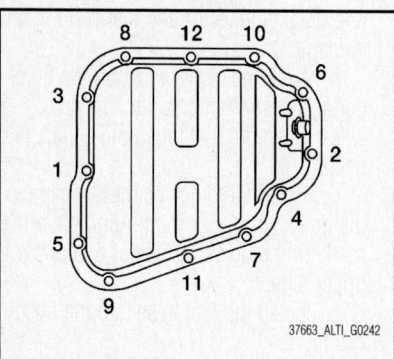

Fig. 121 Remove the lower oil pan bolts in the order

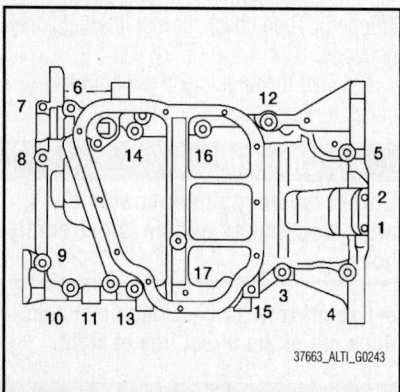

Fig. 122 Loosen the upper oil pan bolts in the order

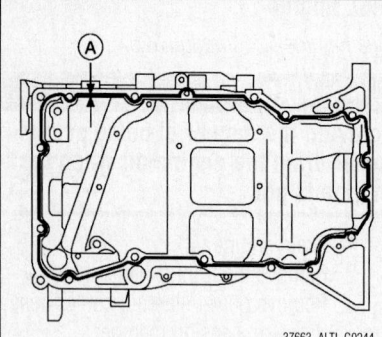

Fig. 123 Apply Genuine Silicone RTV Sealant or equivalent to the upper oil pan (A)

➡Tap gently to cut sealant around the pan; do not damage the mating surface using Tool.

11. Clean the oil strainer screen to remove any foreign material.

To install:

12. Installation is in the reverse order of removal.

13. Apply Genuine Silicone RTV Sealant or equivalent to the upper oil pan as shown.

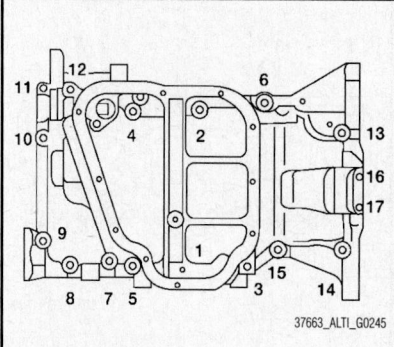

Fig. 124 Tighten the upper oil pan bolts in the order as shown

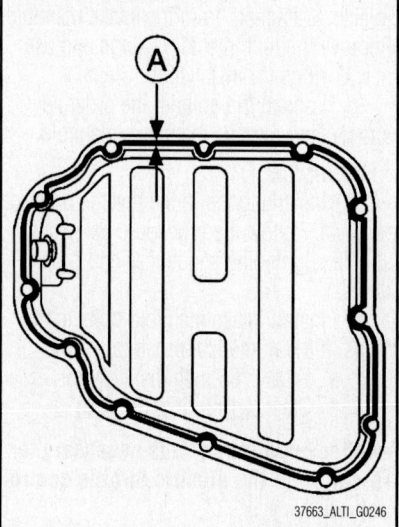

Fig. 125 Apply Genuine Silicone RTV Sealant or equivalent to the lower oil pan (A)

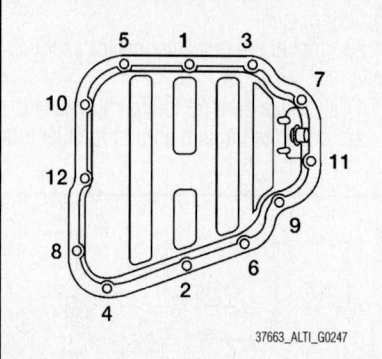

Fig. 126 Tighten the lower oil pan bolts in the numerical order shown

✳✳ CAUTION

Install two new O-rings in the upper pan.

14. Tighten the upper oil pan bolts in the order as shown.

15. Apply Genuine Silicone RTV Sealant or equivalent to the lower oil pan (A) as shown.

16. Tighten the lower oil pan bolts in the numerical order shown.

✳✳ CAUTION

Wait at least 30 minutes after the oil pans are installed before filling the engine with oil.

17. Check for any engine oil leaks with the engine at operating temperature and running at idle.

3.5L Engine

Lower Oil Pan

See Figures 127 and 128.

✳✳ WARNING

You should not remove the oil pan until the exhaust system and cooling system have completely cooled off.

1. Drain the engine oil.
2. Loosen the lower oil pan bolts in order as shown.
3. Remove the lower oil pan.
   a. Insert Tool KV10111100 (J-37228) between the lower oil pan and the upper oil pan.

✳✳ CAUTION

Be careful not to damage the mating surface. Do not insert a screwdriver, this will damage the mating surfaces.

   b. Slide the Tool by tapping its side with a hammer to remove the lower oil pan from the upper oil pan.
4. If re-installing the original lower oil pan, remove the old sealant from the mating surfaces using a scraper.

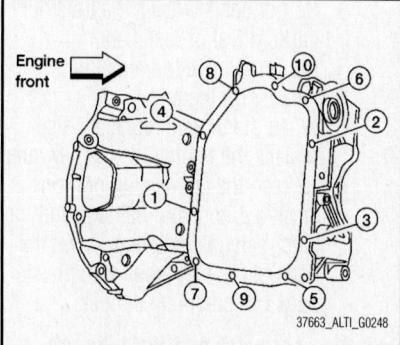

Fig. 127 Loosen the lower oil pan bolts in order as shown

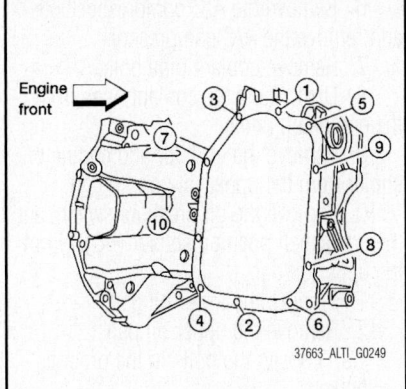

Fig. 128 Tighten the lower oil pan bolts in order as shown

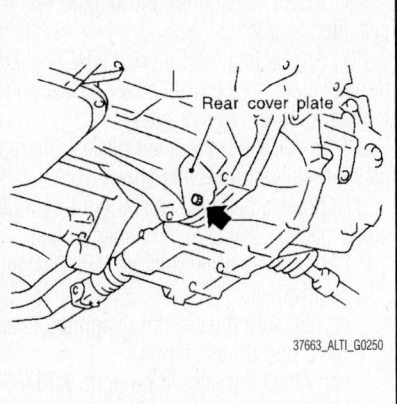

Fig. 129 Remove the rear plate cover from the upper oil pan

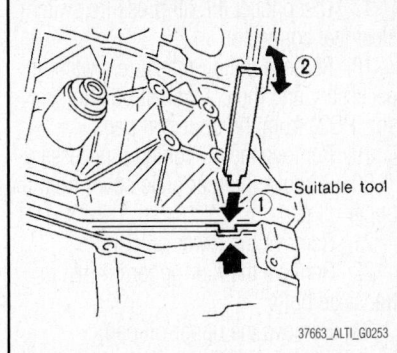

Fig. 132 Insert an appropriate size tool into the notch (1) of the upper oil pan; pry off the upper oil pan by moving the tool up and down (2)

a. Also remove the old sealant from mating surface of the upper oil pan.

b. Remove the old sealant from the bolt holes and threads.

✳✳ CAUTION

Do not scratch or damage the mating surfaces when cleaning off the old sealant.

5. Clean oil strainer if any object is attached.

To install:

6. Apply a continuous bead of sealant to the lower oil pan.

a. Use Genuine Silicone RTV Sealant, or equivalent.

b. Installation must be done within 5 minutes after applying sealant.

7. Install the lower oil pan. Tighten the lower oil pan bolts in order as shown.

a. Wait at least 30 minutes before refilling the engine with oil.

b. Start the engine and check for leaks.

c. Inspect the engine oil level.

Upper Oil Pan M/T Models

See Figures 129 through 134.

✳✳ WARNING

Note the following:

- You should not remove the oil pan until the exhaust system and cooling system have completely cooled off.
- When removing the front and rear engine through bolts and nuts, lift the engine up slightly for safety.

✳✳ CAUTION

When removing the upper oil pan from the engine, first remove the

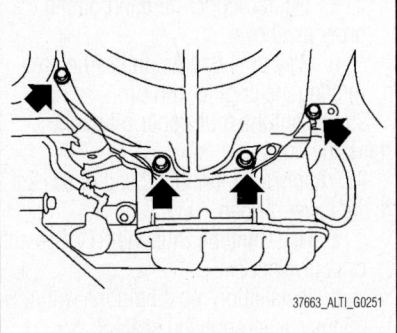

Fig. 130 Remove the four upper oil pan to transaxle bolts

crankshaft position sensor (POS). Be careful not to damage sensor edges or signal plate teeth.

1. Drain the engine coolant.
2. Drain engine oil.
3. Disconnect the battery negative terminal.
4. Remove the engine room cover.
5. Remove the front air duct, air duct hose and air cleaner assembly.
6. Remove the cowl top.

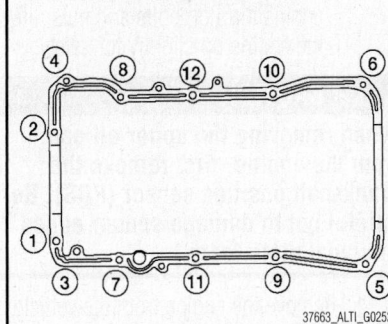

Fig. 131 Loosen the bolts in the order as shown

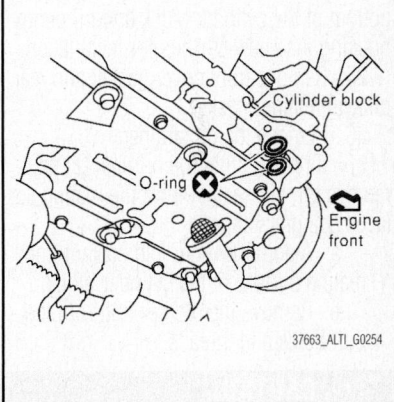

Fig. 133 Remove the O-ring seals from the bottom of the cylinder block and oil pump housing

7. Remove the intake manifold collector.
8. Remove the cooling fan assembly.
9. Remove the front suspension member.
10. Disconnect the heated oxygen sensors and air flow ratio (A/F) sensors and remove the two catalytic convertors from the exhaust manifolds.
11. Remove the oil gauge and oil gauge guide from the oil pan.
12. Remove the drive belt.
13. Remove the A/C compressor with piping attached, and position it out of the way securely with wire.

✳✳ CAUTION

Do not pull on or crimp the A/C lines and hoses.

14. Remove coolant pipe bolts.
15. Disconnect the coolant lines from the engine oil cooler and plug them to prevent coolant loss.
16. Remove the oil filter.

17. Disconnect the oil pressure switch electrical connector.
18. Remove the oil pressure switch, if necessary and the crankshaft position sensor (POS) from the upper oil pan.
19. Remove the oil cooler, if necessary.
20. Remove the rear plate cover from the upper oil pan.
21. Remove the lower oil pan.
22. Remove the four upper oil pan to transaxle bolts.
23. Remove the upper oil pan.
 a. Loosen the bolts in the order as shown.
 b. Insert an appropriate size tool into the notch of the upper oil pan as shown.
 c. Pry off the upper oil pan by moving the tool up and down as shown.
24. Remove the O-ring seals from the bottom of the cylinder block and oil pump housing, use new O-rings for installation.
25. Remove front cover gasket and rear oil seal retainer gasket.
26. Remove the oil strainer.
27. If re-installing the original oil pan, remove the old sealant from the mating surfaces using a scraper.
 a. Also remove the old sealant from mating surface of the cylinder block.
 b. Remove the old sealant from the bolt holes and threads.

⁂ CAUTION

Do not scratch or damage the mating surfaces when cleaning off the old sealant.

28. Clean oil strainer if any object is attached.

To install:

⁂ CAUTION

Wait at least 30 minutes before refilling the engine with oil.

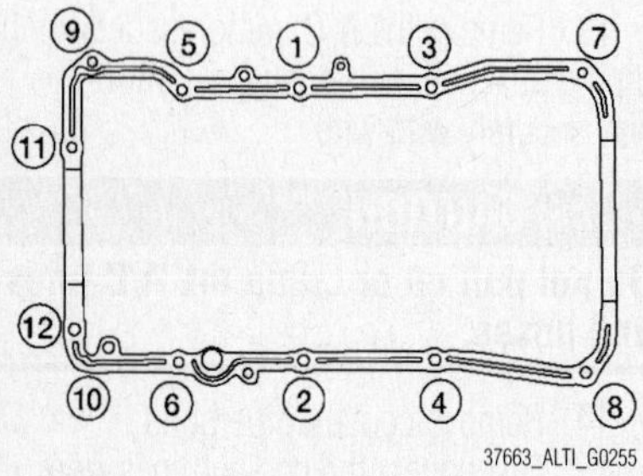

Fig. 134 Tighten upper oil pan bolts in the order as shown

29. Install oil strainer and tighten bolt to specified torque.
30. Apply Genuine Silicone RTV Sealant or equivalent, to the front cover gasket and the rear oil seal retainer gasket.
31. Install the front cover gasket and rear oil seal retainer gasket as shown.
32. Apply a bead of sealant to the cylinder block mating surface of the upper oil pan.
 a. Use Genuine Silicone RTV Sealant, or equivalent.
 b. Be sure the sealant is applied to a limited portion as shown.
 c. Attaching should be done within 5 minutes after coating.
33. Install new O-rings on the cylinder block and oil pump body.
34. Install the upper oil pan.
 a. Tighten upper oil pan bolts in the order as shown.
 b. Wait at least 30 minutes before refilling the engine with oil.
35. Install the four upper oil pan to transaxle bolts.
36. Apply a continuous bead of sealant to the lower oil pan.
 a. Use Genuine Silicone RTV Sealant, or equivalent.
 b. Installation must be done within 5 minutes after applying sealant.
37. Install the lower oil pan.
38. Install rear plate cover.
39. Installation of the remaining components is in the reverse order of removal.
 a. Start the engine and check for leaks.
 b. Inspect the engine oil level.

Upper Oil Pan CVT Models

See Figures 134 through 136.

⁂ WARNING

Note the following:

- You should not remove the oil pan until the exhaust system and cooling system have completely cooled off.
- When removing the front and rear engine through bolts and nuts, lift the engine up slightly for safety.

⁂ CAUTION

When removing the upper oil pan from the engine, first remove the crankshaft position sensor (POS). Be careful not to damage sensor edges or signal plate teeth.

1. Remove the engine from the vehicle.
2. Drain the engine oil.
3. Remove the oil dipstick.
4. Remove the drive belt.
5. Disconnect the A/C compressor harness connector.
6. Remove the A/C compressor bolts and remove the A/C compressor.
7. Remove coolant pipe bolts.
8. Disconnect the coolant lines from the engine oil cooler.
9. Remove the oil filter and engine oil cooler from the upper oil pan.
10. Remove the oil pressure switch, and the crankshaft position sensor (POS) from the upper oil pan.
11. Remove the lower oil pan.
12. Remove the upper oil pan.
 a. Loosen the bolts in the order as shown.
 b. Insert an appropriate size tool into the notch of the upper oil pan.
 c. Pry off the upper oil pan by moving the tool up and down.
13. Remove the O-ring seals from the bottom of the cylinder block and oil pump housing. Use new O-rings for installation.
14. Remove front cover gasket and rear oil seal retainer gasket.
15. Remove the oil strainer.
16. If re-installing the original oil pan, remove the old sealant from the mating surfaces using a scraper.
 a. Also remove the old sealant from mating surface of the cylinder block.

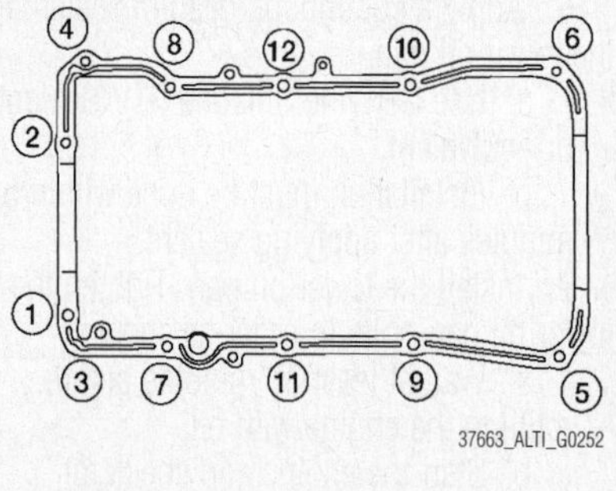

Fig. 135 Loosen the bolts in the order as shown

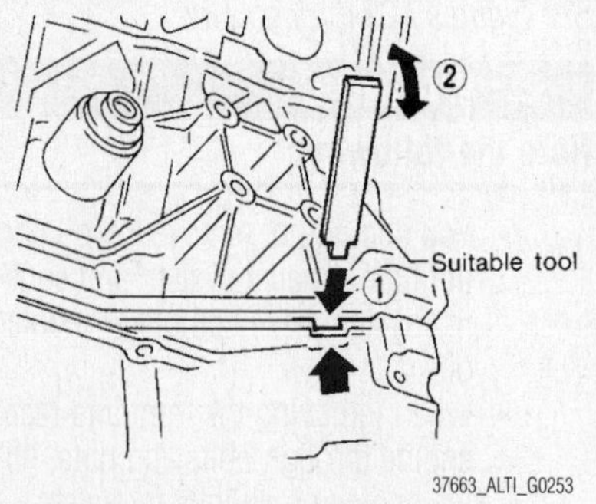

Fig. 136 Insert an appropriate size tool into the notch (1) of the upper oil pan; pry off the upper oil pan by moving the tool up and down (2)

b. Remove the old sealant from the bolt holes and threads.

✳✳ CAUTION

Do not scratch or damage the mating surfaces when cleaning off the old sealant.

17. Clean oil strainer if any object is attached.

To install:

✳✳ CAUTION

Wait at least 30 minutes before refilling the engine with oil.

18. Install oil strainer and tighten bolt to specified torque.
19. Apply Genuine Silicone RTV Sealant or equivalent, to the front cover gasket and the rear oil seal retainer gasket as shown.
20. Install the front cover gasket and rear oil seal retainer gasket as shown.
21. Apply a bead of sealant to the cylinder block mating surface of the upper oil pan to a limited portion as shown.
    a. Use Genuine Silicone RTV Sealant, or equivalent.
    b. Attaching should be done within 5 minutes after coating.
22. Install new O-rings on the cylinder block and oil pump body.
23. Install the upper oil pan.
    a. Tighten upper oil pan bolts in the order as shown.
    b. Wait at least 30 minutes before refilling the engine with oil.
24. Install the lower oil pan.
25. Installation of the remaining components is in the reverse order of removal.
    a. Start the engine and check for leaks.
    b. Inspect the engine oil level.

OIL PUMP

REMOVAL & INSTALLATION

2.5L Engine

The oil pump is part of the front cover. For removal and installation of the oil pump, it is necessary to remove and install the front cover. Refer to Front Cover section for Removal and Installation.

3.5L Engine

1. Remove the engine from the vehicle.
2. Remove the upper oil pan.
3. Remove the timing chain.
4. Remove oil pump assembly.

To install:

5. Installation is in the reverse order of removal.

PISTON AND RING

POSITIONING

See Figures 137 through 139.

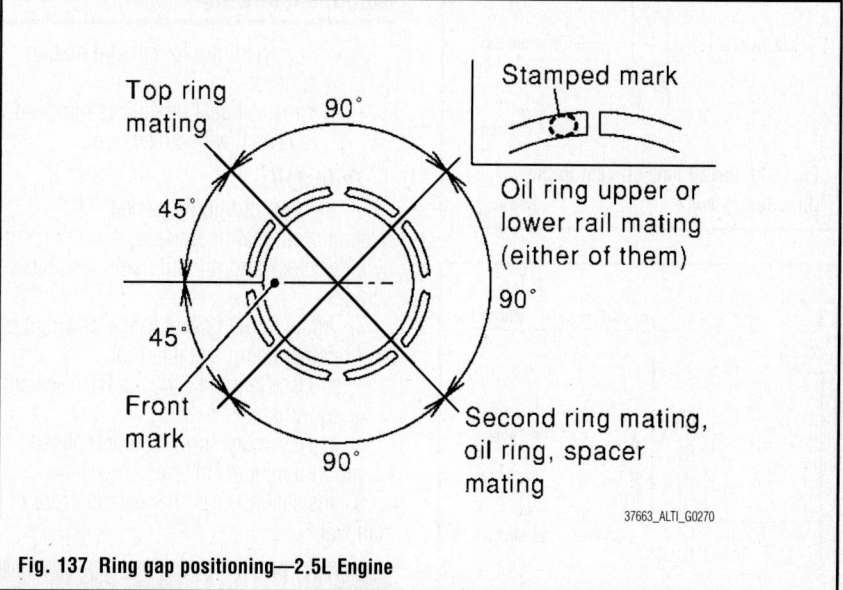

Fig. 137 Ring gap positioning—2.5L Engine

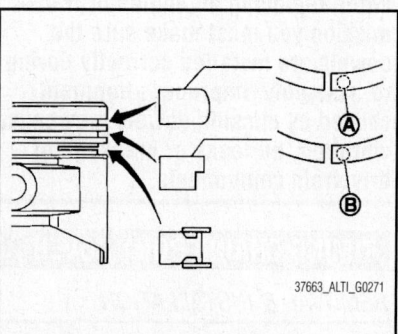

Fig. 138 Piston ring installation orientation—3.5L Engine

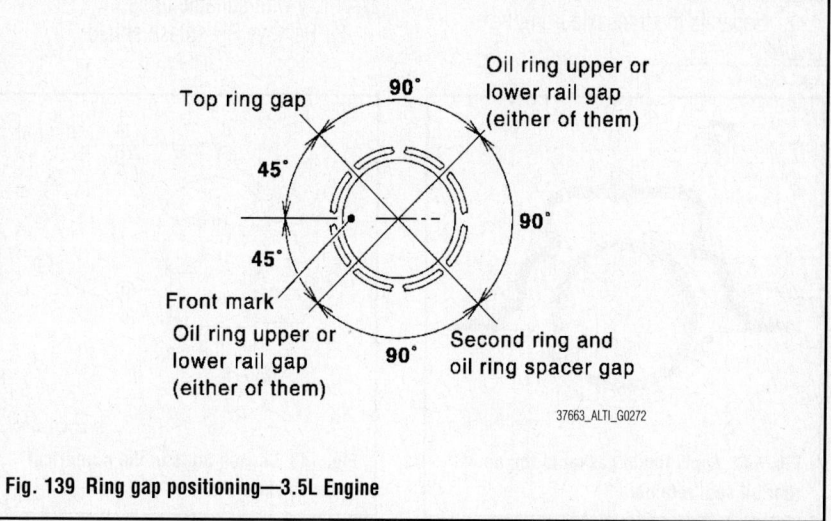

Fig. 139 Ring gap positioning—3.5L Engine

REAR MAIN SEAL

REMOVAL & INSTALLATION

2.5L Engine

See Figures 140 and 141.

1. Remove the transaxle.
2. Remove flywheel (MT) or drive plate (CVT).
3. Remove rear oil seal using suitable tool.

✳✳ CAUTION

Be careful not to scratch rear oil seal retainer.

To install:

4. Apply new engine oil to new oil seal and install it using a suitable tool.
    a. Install new oil seal in the direction shown.

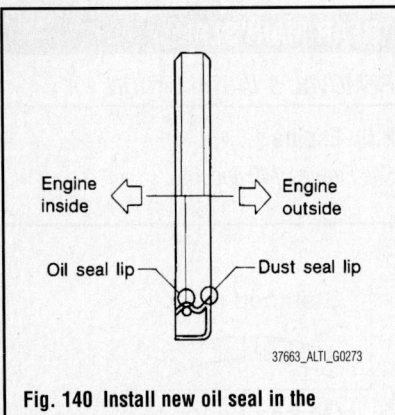

Fig. 140 Install new oil seal in the direction shown

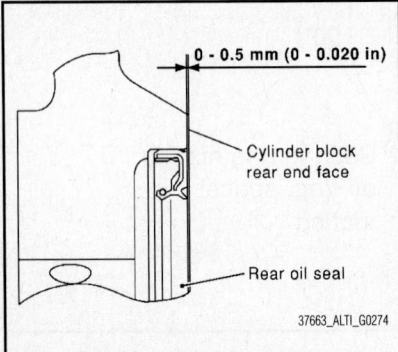

Fig. 141 Press in the new oil seal to the specified depth as shown

b. Press fit new oil seal straight using a suitable drift, to avoid causing burrs or tilting.

c. Press in the new oil seal to the specified depth as shown.

5. Installation of the remaining components is in the reverse order of removal.

3.5L Engine

See Figure 142.

1. Remove the engine and transaxle assembly.

2. Separate the transaxle from the engine.

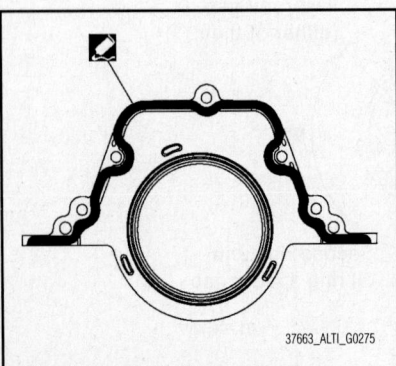

Fig. 142 Apply liquid gasket to the new rear oil seal retainer

3. Remove flywheel (M/T) or drive plate (CVT).

4. Remove the upper oil pan.

5. Remove rear oil seal retainer using Tool KV10111100 (J-37228).

✳✳ CAUTION

Note the following:

- Be careful not to damage mating surface.
- If rear oil seal retainer is removed, replace it with a new one.

To install:

6. Remove old liquid gasket material from mating surface of cylinder block and oil pan using a suitable scraper.

7. Apply liquid gasket to the new rear oil seal retainer using suitable tool.

   a. Use Genuine Silicone RTV Sealant or equivalent.

   b. Assembly should be completed within 5 minutes after coating.

8. Installation is in the reverse order of removal.

✳✳ CAUTION

When replacing an engine or transmission you must make sure the dowels are installed correctly during re-assembly. Improper alignment caused by missing dowels may cause vibration, oil leaks or breakage of drivetrain components.

TIMING CHAIN FRONT COVER

REMOVAL & INSTALLATION

2.5L Engine

See Figures 143 through 151.

1. Support the engine and transaxle assembly with suitable tools.

2. Remove RH splash shield.

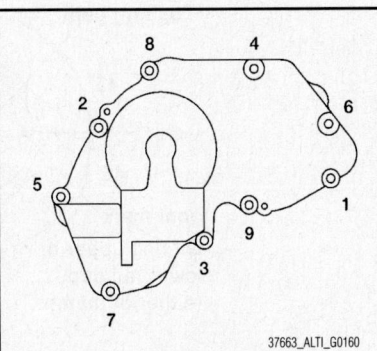

Fig. 143 Loosen bolts in the numerical order as shown

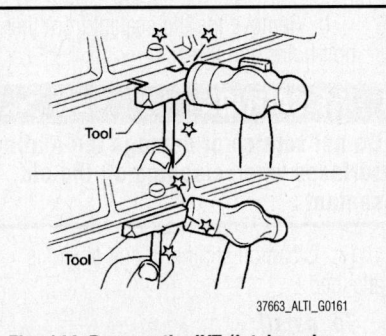

Fig. 144 Remove the IVT (intake valve timing) control cover using Tool KV10111100 (J-37228)

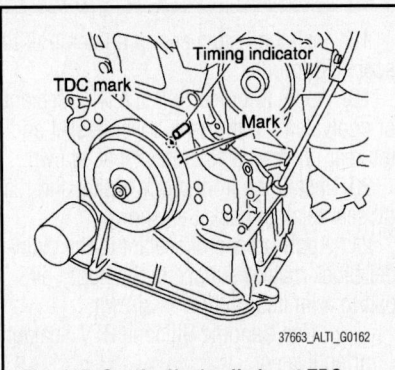

Fig. 145 Set the No.1 cylinder at TDC on the compression stroke

3. Remove the upper and lower oil pan, and oil strainer.

4. Remove the alternator.

5. Remove the engine cover.

6. Disconnect variable timing control solenoid harness connector.

7. Remove the engine ground.

8. Remove the coolant overflow reservoir tank.

9. Position the RH engine compartment fuse and relay box aside.

10. Remove the RH engine mount and bracket.

11. Loosen bolts in the numerical order as shown.

12. Remove the IVT (intake valve timing) control cover using Tool KV10111100 (J-37228).

13. Pull chain guide between camshaft sprockets out through front cover.

14. Set the No.1 cylinder at TDC on the compression stroke with the following procedure:

   a. Rotate the crankshaft pulley clockwise and align the mating marks to the timing indicator on the front cover.

   b. At the same time, make sure that the mating marks on the camshaft sprockets are lined up as shown.

   c. If not lined up, rotate the crankshaft pulley one more turn to line

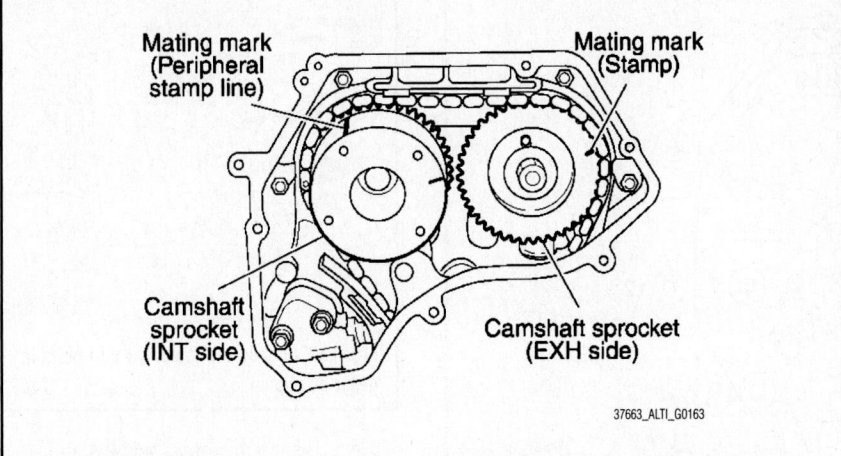

Fig. 146 Make sure that the mating marks on the camshaft sprockets are lined up as shown

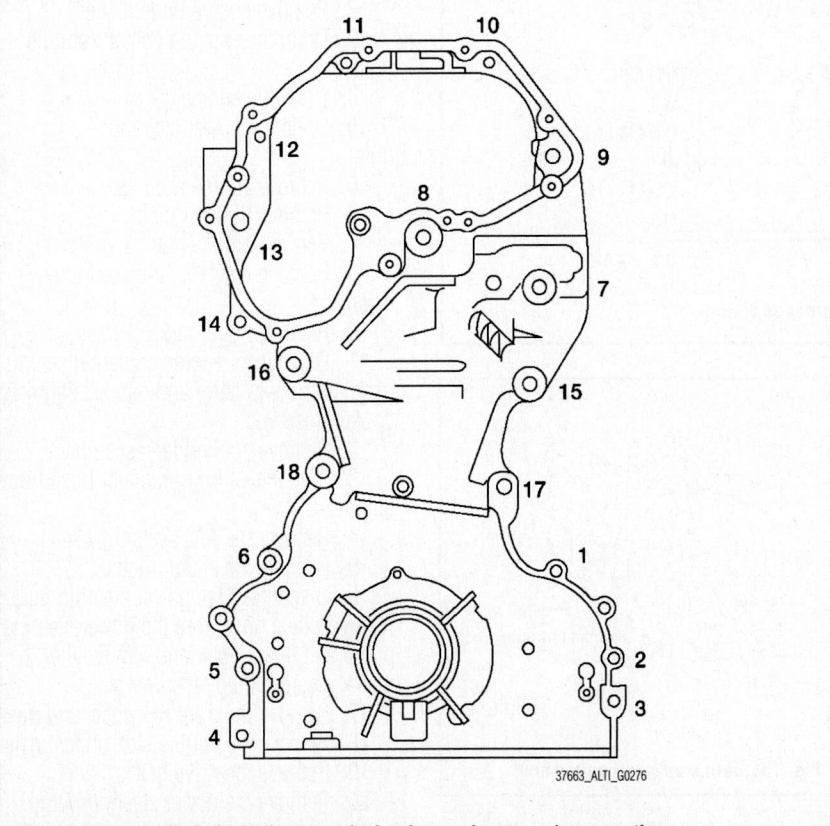

Fig. 147 Loosen the bolts in the numerical order as shown, and remove them

up the mating marks to the positions as shown.

15. Remove crankshaft pulley with the following procedure:

 a. Hold the crankshaft pulley using suitable tool, then loosen the crankshaft pulley bolt, and pull the pulley out about 0.39 inches (10 mm).

 b. Attach suitable pulley puller in the thread hole on crankshaft pulley, and remove crankshaft pulley using a suitable puller.

16. Remove the front cover with the following procedure:

 a. Loosen the bolts in the numerical order as shown, and remove them.

 b. Remove the front cover.

※ CAUTION

Be careful not to damage the mounting surface.

To install:

17. Install front cover with the following procedure:

 a. Install O-rings to cylinder head and cylinder block.

 b. Apply Genuine Silicone RTV Sealant or equivalent, to front cover.

 c. Make sure the mating marks on the timing chain and each sprocket are still aligned. Then install the front cover.

※ CAUTION

Be careful not to damage the front oil seal during installation with the front end of the crankshaft.

 d. Tighten front cover bolts in the numerical order as shown.

 e. After all bolts are tightened, retighten them to the specified torque:

- Bolts A: 36 ft. lbs. (49 Nm)
- Bolts B: 9 ft. lbs. (13 Nm)
- Bolts C: 9 ft. lbs. (13 Nm)
- Bolts D: 36 ft. lbs. (49 Nm)
- Dowel pins

※ CAUTION

Wipe off any excess sealant leaking at the surface for installing the oil pan.

18. Install the chain guide between the camshaft sprockets.

19. Install IVT cover with the following procedure:

 a. Install IVT solenoid valve to IVT cover.

 b. Install new O-ring to front cover.

 c. Apply Silicone RTV Sealant to the IVT cover as shown.

 d. Tighten the IVT cover bolts in the numerical order as shown.

20. Insert crankshaft pulley by aligning with crankshaft key.

 a. Tap its center with a plastic hammer to insert.

 b. Do not tap the belt hook.

21. Tighten crankshaft pulley bolts.

 a. Secure crankshaft pulley with tool to tighten the bolt.

 b. Perform angle tightening with the following procedure:

 c. Apply new engine oil to threads and seat surfaces of bolts.

 d. Tighten to initial specifications: 31 ft. lbs. (42 Nm).

 e. Apply a paint mark on the front cover, mating with any one of six easy to recognize stamp marks on bolt flange.

 f. Turn crankshaft pulley bolt another 60° to 66° (Target: 60°).

 g. Check vertical mounting angle with movement of one stamp mark.

22. Installation of the remaining components is in the reverse order of removal.

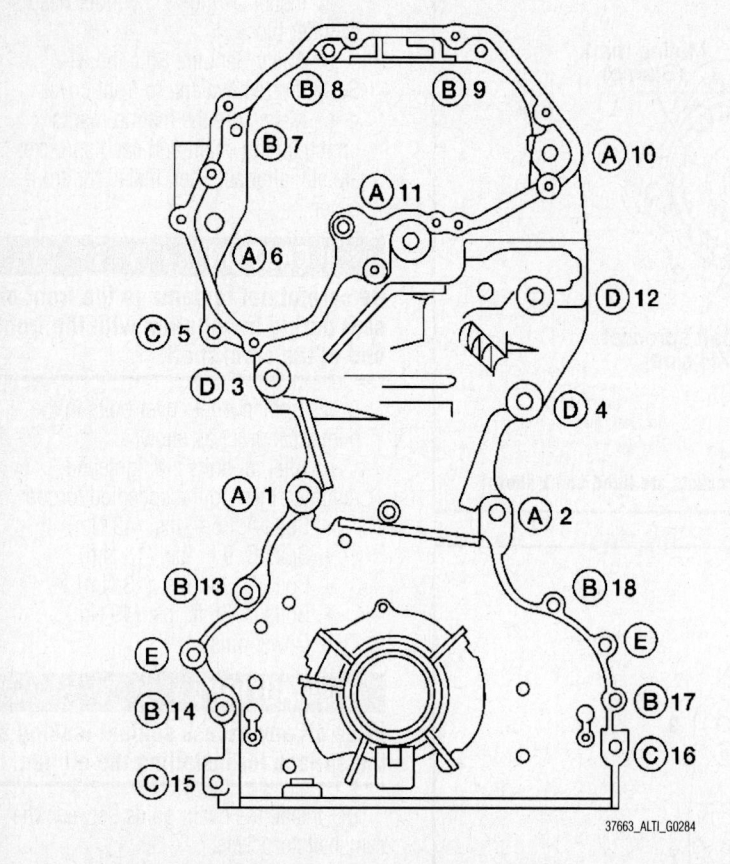

Fig. 148 Tighten front cover bolts in the numerical order as shown

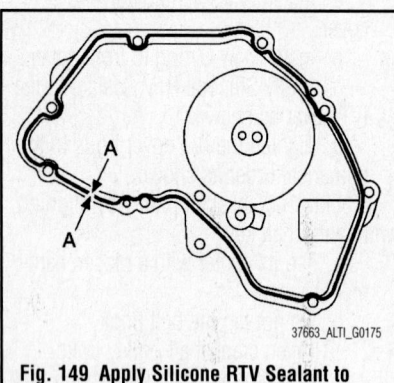

Fig. 149 Apply Silicone RTV Sealant to the IVT cover as shown

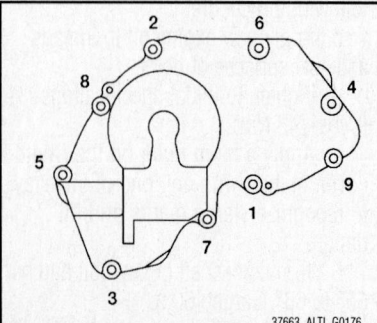

Fig. 150 Tighten the IVT cover bolts in the numerical order as shown

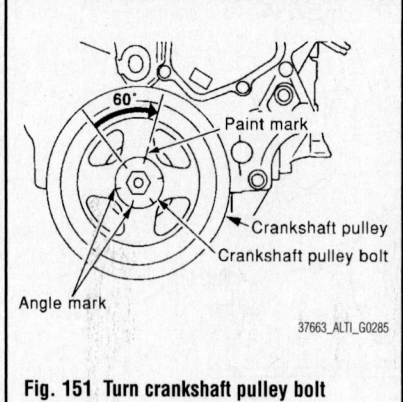

Fig. 151 Turn crankshaft pulley bolt

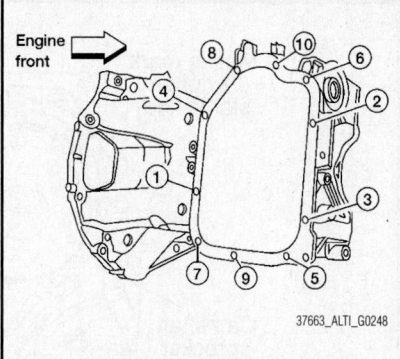

Fig. 152 Loosen the lower oil pan bolts in order as shown

3.5L Engine

See Figures 152 through 162.

➡**Note the following:**

- This section describes the proce- dure for removal/installation of the front timing chain case without removing the oil pan (upper) from the vehicle.
- When rear timing chain case must be removed, remove the engine from the vehicle. Then remove front timing chain case, timing chain related parts, and rear timing chain

case in this order, and install in reverse order of removal.

1. Disconnect the battery negative terminal.
2. Remove engine under cover.
3. Drain the engine coolant from the radiator.
4. Drain the engine oil.
5. Drain the power steering fluid.
6. Remove engine room cover.
7. Remove front air duct.
8. Remove battery tray.
9. Remove cowl top and cowl top extension.
10. Remove upper radiator hose.
11. Disconnect engine coolant reservoir hose from the radiator and remove engine coolant reservoir.
12. Remove cooling fan assembly.
13. Disconnect lower radiator hose from engine.
14. Recover the A/C system R134a.
15. Remove the starter motor.
16. Disconnect the power steering fluid reservoir tank hose from the power steering pump and fluid cooler and remove the power steering fluid reservoir tank.
17. Remove the front RH wheel and tire.
18. Remove the engine side under cover.
19. Remove the drive belt.
20. Remove the power steering pump.
21. Remove the lower oil pan. Loosen the lower oil pan bolts in order as shown.
   a. Insert Tool between the lower oil pan and the upper oil pan.
   b. Be careful not to damage the mat- ing surface.

✷✷ CAUTION

Do not insert a screwdriver, this will damage the mating surfaces.

   c. Slide the Tool by tapping its side with a hammer to remove the lower oil pan from the upper oil pan.

22. Remove upper oil pan bolts in reverse order as shown.

23. Remove the alternator.

24. Disconnect the A/C tubes from the A/C compressor and position aside.

25. Remove the A/C compressor bolts and remove the A/C compressor.

26. Remove the alternator bracket.

27. Support the engine with suitable jack and remove the RH engine insulator, mount and bracket.

28. Remove the rocker covers, if necessary.

➡**Necessary only when removing timing chains.**

29. If removing the timing chains, obtain compression TDC of No. 1 cylinder as follows:

    a. Rotate crankshaft pulley clockwise to align timing mark (grooved line without color) with timing indicator.

    b. Check that intake and exhaust camshaft lobes on No. 1 cylinder (right bank of engine) are located as shown.

    c. If not, turn the crankshaft one revolution (360°) and align as shown.

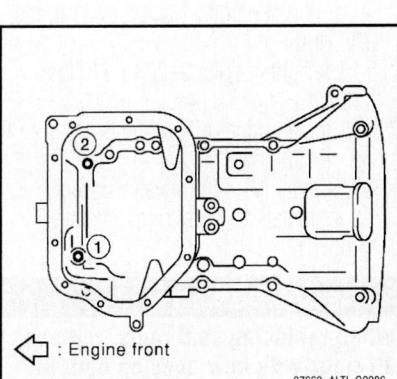

: Engine front

37663_ALTI_G0286

Fig. 153 Remove upper oil pan bolts in reverse order as shown

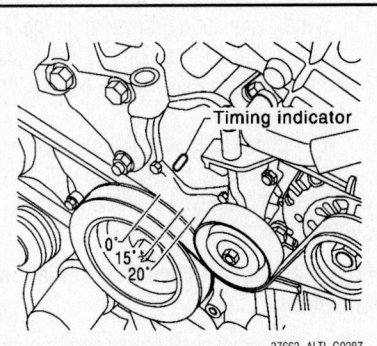

37663_ALTI_G0287

Fig. 154 Rotate crankshaft pulley clockwise to align timing mark (grooved line without color) with timing indicator

30. Lock the drive plate using Tool KV10117700 (J-44716) attached to the starter bolt hole.

⚹⚹ **CAUTION**

Do not damage the ring gear teeth, or the signal plate teeth behind the ring gear, when setting the Tool.

31. Remove the crankshaft pulley as follows:

    a. Loosen crankshaft pulley bolt using suitable tool and locate bolt seating surface at 0.39 inches (10 mm) from its original position.

    b. Position a pulley puller at recess hole of crankshaft pulley to remove crankshaft pulley.

⚹⚹ **CAUTION**

Do not use a puller claw on crankshaft pulley periphery.

32. Remove engine oil cooler tube bolts and bracket.

33. Disconnect the oil pressure switch harness connector.

34. Disconnect valve timing control harness connector.

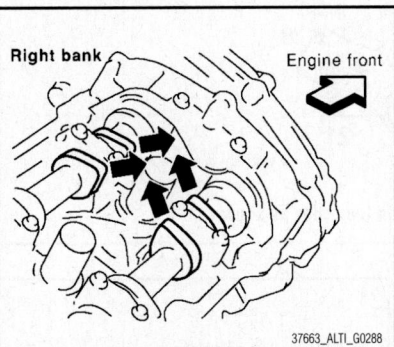

Right bank   Engine front

37663_ALTI_G0288

Fig. 155 Check that intake and exhaust camshaft lobes on No. 1 cylinder (right bank of engine) are located as shown

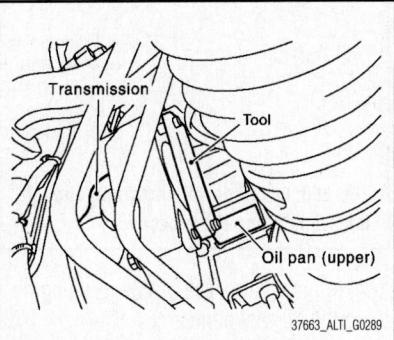

Transmission   Tool

Oil pan (upper)

37663_ALTI_G0289

Fig. 156 Lock the drive plate using Tool KV10117700 (J-44716) attached to the starter bolt hole

35. Remove the Bank 1 (RH) and Bank 2 (LH) IVT covers.

36. Loosen the IVT cover bolts in the reverse order as shown.

⚹⚹ **CAUTION**

The shaft in the IVT cover is inserted into the center hole of the intake camshaft sprocket. Remove the IVT cover by pulling straight out until the IVT cover disengages from the camshaft sprocket.

37. Remove the A/C idler pulley and bracket and the drive belt auto-tensioner.

38. If necessary, remove the idler pulley and water pump cover.

39. Remove the front timing chain case.

    a. Loosen the front timing chain case bolts in the order as shown.

    b. Insert the appropriate size tool into the notch at the top of the front timing chain case.

    c. Pry off the case by moving the suitable tool back and forth.

⚹⚹ **CAUTION**

Do not use a screwdriver or similar tool. After removal, handle carefully so it does not bend, or warp under a load.

40. Remove O-rings from rear timing chain case.
- A: Bank 1
- B: Bank 2

⚹⚹ **CAUTION**

Use new O-rings for installation.

41. Remove the front oil seal from the front timing chain case using a suitable tool.

⚹⚹ **CAUTION**

Do not damage the front cover.

42. Remove all old Silicone RTV Sealant from all the bolt holes and bolts.

⚹⚹ **CAUTION**

Do not damage the threads or mating surfaces.

43. Use a scraper to remove all of the old Silicone RTV Sealant from the front timing chain case and opposite mating surfaces.

⚹⚹ **CAUTION**

Do not damage the mating surfaces.

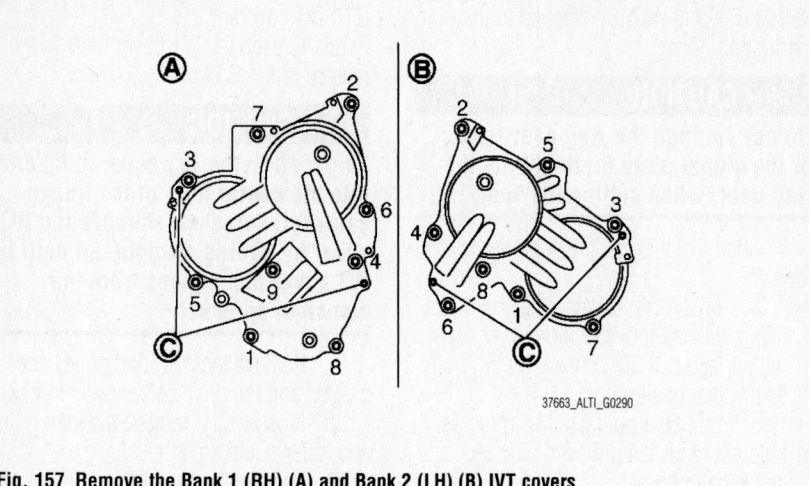

Fig. 157 Remove the Bank 1 (RH) (A) and Bank 2 (LH) (B) IVT covers

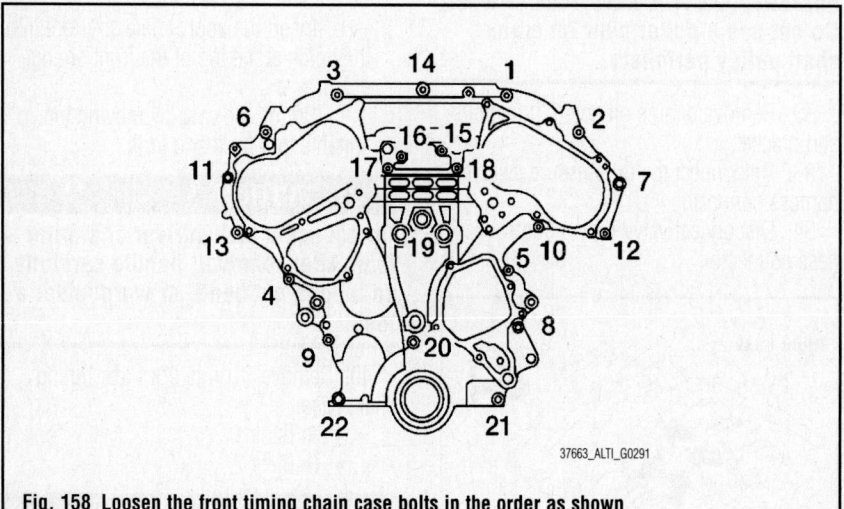

Fig. 158 Loosen the front timing chain case bolts in the order as shown

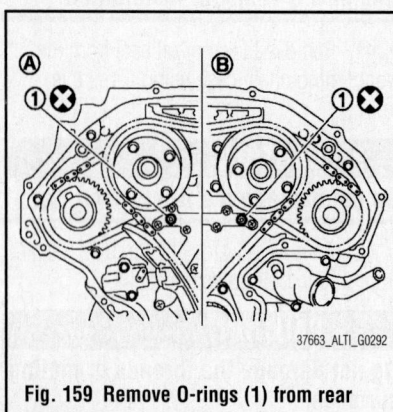

Fig. 159 Remove O-rings (1) from rear timing chain case of Bank 1 (A) and Bank 2 (B)

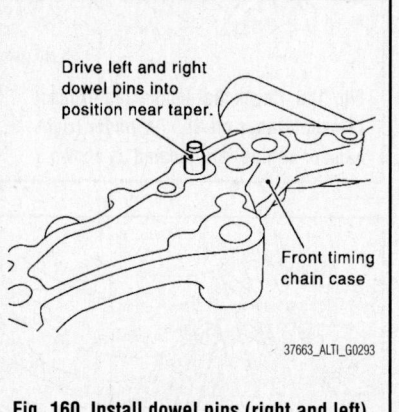

Fig. 160 Install dowel pins (right and left) into front timing chain case

To install:

44. Install dowel pins (right and left) into front timing chain case up to a point close to taper in order to shorten protrusion length.

45. Install the new front oil seal on the front timing chain case. Apply new engine oil to the oil seal edges.

a. Install the new front oil seal so that it becomes flush with the face with front timing chain case using suitable drift.

✳✳ CAUTION

Press fit straight and avoid causing burrs or tilting the oil seal.

➡**Make sure the garter spring is in position and seal lip is not inverted.**

46. Install new O-rings on rear timing chain case.

✳✳ CAUTION

Use new O-rings for installation.

47. Apply Silicone RTV Sealant to front timing chain case.

➡**Use Genuine Silicone RTV Sealant, or equivalent. Before installation, wipe off the protruding sealant.**

48. Install dowel pin on the rear timing chain case into dowel pin hole in front timing chain case.

49. Loosely install the front timing chain case bolts.

50. Tighten the front timing chain case bolts in the order as shown.

51. Retighten the front timing chain case bolts in the order as shown.

a. Tighten bolts 1 and 2 to 21 ft. lbs. (28 Nm).

b. Tighten bolts 3–22 to 9 ft. lbs. (13 Nm).

52. Install upper oil pan bolts.

53. Install lower oil pan.

54. Install IVT control valve covers.

a. Install new seal rings in shaft grooves.

✳✳ CAUTION

When replacing seal rings, replace all rings with new ones on both RH and LH IVT control valve covers.

b. Install IVT covers with a new gasket to front timing chain case.

c. Being careful not to move seal ring from the installation groove, align the dowel pins on the front timing chain case with the holes to install valve timing control covers.

d. Tighten bolts in the numerical order as shown to 8 ft. lbs. (11 Nm).

55. Apply liquid gasket and install the water pump cover, if removed.

➡**Use Genuine Silicone RTV Sealant or equivalent.**

56. Install crankshaft pulley and tighten the bolt in two steps.

a. Lubricate thread and seat surface of the bolt with new engine oil.

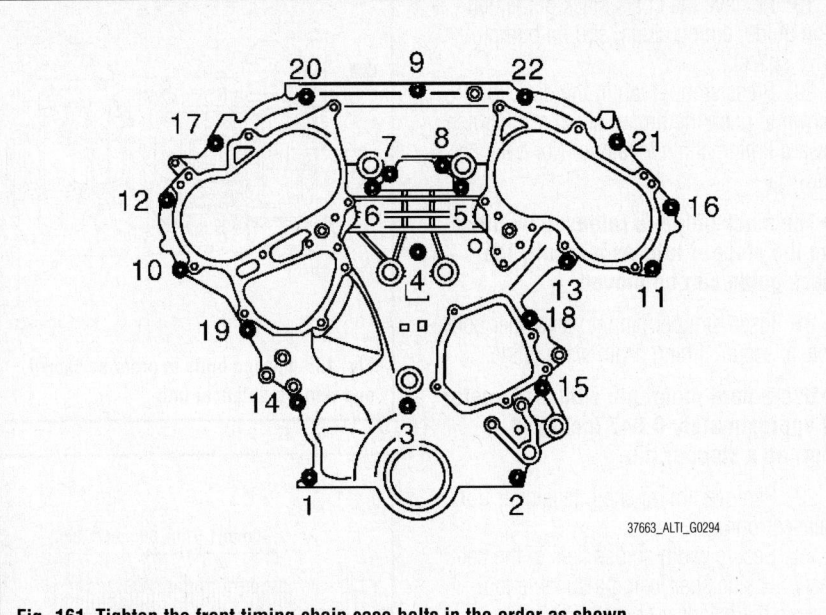

Fig. 161 Tighten the front timing chain case bolts in the order as shown

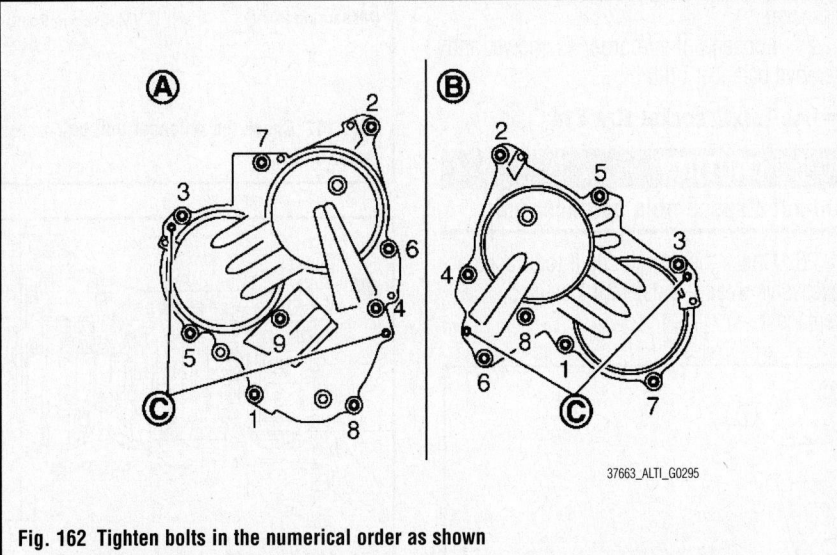

Fig. 162 Tighten bolts in the numerical order as shown

b. Apply a paint mark for the second step of angle tightening.

c. Step 1: 32 ft. lbs. (44 Nm)

d. Step 2: Additional 84–90 degrees clockwise

57. Rotate crankshaft pulley in normal direction (clockwise when viewed from front) to confirm it turns smoothly.

58. Installation of the remaining components is in reverse order of removal.

TIMING CHAIN & SPROCKETS

REMOVAL & INSTALLATION

2.5L Engine

See Figures 143 through 151 and 163 through 171.

1. Support the engine and transaxle assembly with suitable tools.
2. Remove RH splash shield.
3. Remove the upper and lower oil pan, and oil strainer.
4. Remove the alternator.
5. Remove the engine cover.
6. Disconnect variable timing control solenoid harness connector.
7. Remove the engine ground.
8. Remove the coolant overflow reservoir tank.
9. Position the RH engine compartment fuse and relay box aside.
10. Remove the RH engine mount and bracket.
11. Loosen bolts in the numerical order as shown.

12. Remove the IVT (intake valve timing) control cover using Tool KV10111100 (J-37228).

13. Pull chain guide between camshaft sprockets out through front cover.

14. Set the No.1 cylinder at TDC on the compression stroke with the following procedure:

a. Rotate the crankshaft pulley clockwise and align the mating marks to the timing indicator on the front cover.

b. At the same time, make sure that the mating marks on the camshaft sprockets are lined up as shown.

c. If not lined up, rotate the crankshaft pulley one more turn to line up the mating marks to the positions as shown.

15. Remove crankshaft pulley with the following procedure:

a. Hold the crankshaft pulley using suitable tool, then loosen the crankshaft pulley bolt, and pull the pulley out about 0.39 inches (10 mm).

b. Attach suitable pulley puller in the thread hole on crankshaft pulley, and remove crankshaft pulley using a suitable puller.

16. Remove the front cover with the following procedure:

a. Loosen the bolts in the numerical order as shown, and remove them.

b. Remove the front cover.

✳✳ CAUTION

Be careful not to damage the mounting surface.

17. Remove front oil seal using suitable tool, if necessary.

18. Remove timing chain with the following procedure:

a. Push in the tensioner plunger. Insert a stopper pin into the hole on the

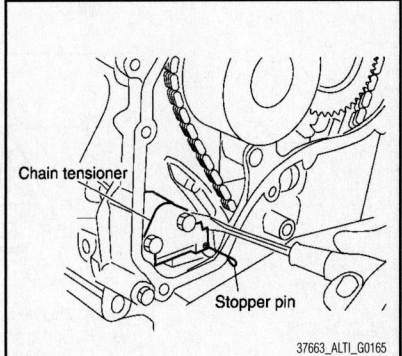

Fig. 163 Insert a stopper pin into the hole on the tensioner body to secure the chain tensioner plunger and remove chain tensioner

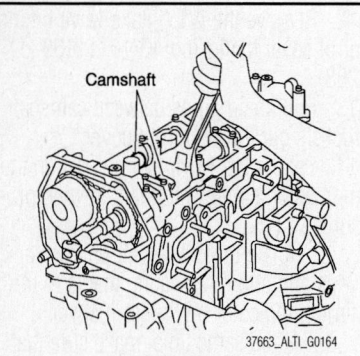

Fig. 164 Secure hexagonal part of the camshaft with a wrench and loosen the camshaft sprocket bolt

tensioner body to secure the chain tensioner plunger and remove chain tensioner.

b. Use a wire of 0.02 inches (0.5 mm) diameter as a stopper pin.

c. Remove the timing chain.

d. Secure hexagonal part of the camshaft with a wrench and loosen the camshaft sprocket bolt and remove the camshaft sprocket for both camshafts.

❋❋ CAUTION

Do not rotate the crankshaft or camshafts while the timing chain is removed. It can cause damage to the valve and piston.

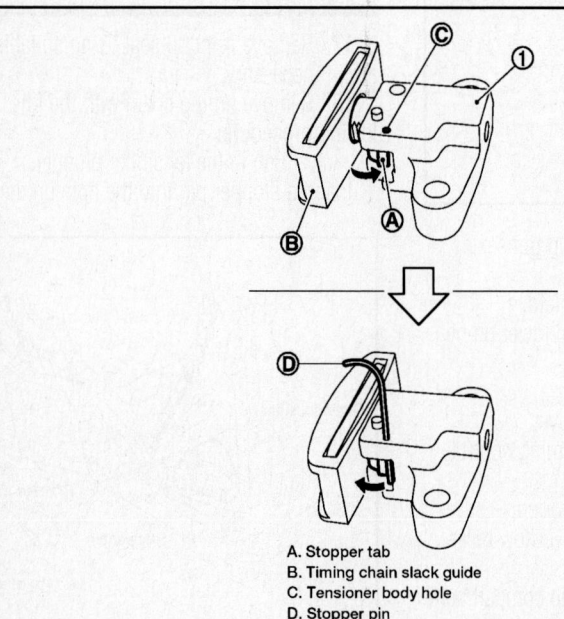

A. Stopper tab
B. Timing chain slack guide
C. Tensioner body hole
D. Stopper pin
1. Balancer unit

Fig. 165 Press stopper tab in the direction shown to push the timing chain slack guide toward timing chain tensioner (for balancer unit)

19. Remove the chain slack guide, tension guide, timing chain, and oil pump drive spacer.

20. Press stopper tab in the direction shown to push the timing chain slack guide toward timing chain tensioner (for balancer unit).

➡**The slack guide is released by pressing the stopper tab. As a result, the slack guide can be moved.**

21. Insert stopper pin into tensioner body hole to secure timing chain slack guide.

➡**Use a hard metal pin with a diameter of approximately 0.047 inches (1.2 mm) as a stopper pin.**

22. Remove timing chain tensioner (for balancer unit).

23. Secure width across flats of the balancer LH side shaft using a suitable tool. Loosen the balancer sprocket bolt.

24. Remove balancer unit timing chain, balancer unit sprocket and crankshaft sprocket.

25. Loosen bolts in order as shown, and remove balancer unit.

➡**Use Torx® socket size E14.**

❋❋ CAUTION

Do not disassemble balancer unit.

26. Check the timing chain for cracks or excessive wear. If a defect is detected, replace it.

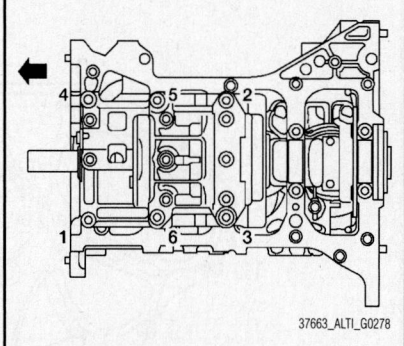

Fig. 166 Loosen bolts in order as shown, and remove balancer unit

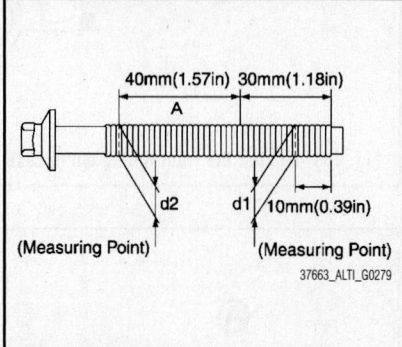

Fig. 167 Check the balancer unit bolt outer diameter

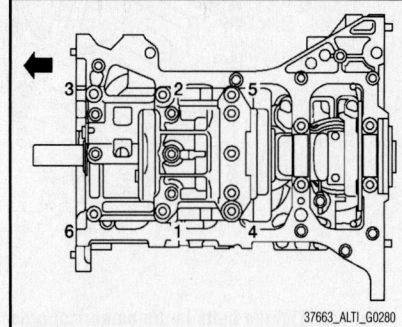

Fig. 168 Install the balancer unit and tighten the bolts in the numerical order as shown

27. Check the balancer unit bolt outer diameter.

a. Measure outer diameters (d1, d2) at the two positions as shown.

b. Measure d2 within the range A.

c. If the value difference (d1 - d2) exceeds the limit, replace it with a new one.

To install:

➡**There may be two color variations of the link marks (link colors) on the timing chain. There are 26 links between the gold/yellow mating marks on the timing chain; and 64 links between the camshaft sprocket gold/yellow link and** the crankshaft sprocket orange/blue link, on the timing chain side without the tensioner.

28. Make sure the crankshaft key points straight up.

29. Install the balancer unit and tighten the bolts in the numerical order as shown.

    a. Step 1: Tighten bolts 1–5 to 31 ft. lbs. (42 Nm). Tighten bolt 6 to 27 ft. lbs. (36 Nm).

    b. Step 2: Tighten bolts 1–5 an additional 120°. Tighten bolt 6 an additional 90°.

    c. Step 3: Loosen all bolts in reverse order of tightening.

    d. Step 4: Tighten bolts 1–5 to 31 ft. lbs. (42 Nm). Tighten bolt 6 to 27 ft. lbs. (36 Nm).

    e. Step 5: Tighten bolts 1–5 an additional 120°. Tighten bolt 6 an additional 90°.

✳✳ CAUTION

Note the following:

- When reusing a bolt, check its outer diameter before installation. Follow the Balancer Unit Bolt Outer Diameter procedure.
- Apply new engine oil to threads and seating surfaces of bolts.
- Check tightening angle with an angle wrench or a protractor.
- Do not make judgment by visual check alone.

30. Install the crankshaft sprocket and timing chain for the balancer unit.

    a. Make sure that the crankshaft sprocket is positioned with mating marks

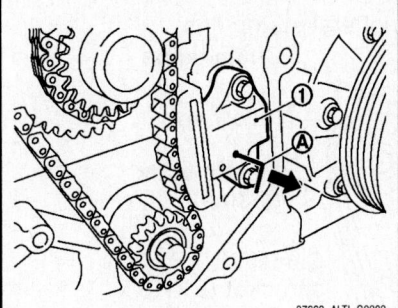

Fig. 170 Install timing chain tensioner (for balancer unit)

on the block and sprocket meeting at the top.

    b. Install it by lining up mating marks on each sprocket and timing chain.

31. Install timing chain tensioner (for balancer unit).

    a. Fix the plunger at the most compressed position using a stopper pin, and then install it.

    b. Securely pull out the stopper pin after installing the timing chain tensioner (for balancer unit).

    c. Check matching mark position of balancer unit drive chain and each sprocket again.

32. Install timing chain and related parts.

    a. Install by lining up mating marks on each sprocket and timing chain as shown.

    b. Before and after installing timing chain tensioner, check again to make sure the mating marks have not slipped.

    c. After installing timing chain tensioner, remove the stopper pin, and make sure that the tensioner moves freely.

✳✳ CAUTION

Note the following:

- For the following note, after the mating marks are aligned, keep them aligned by holding them by hand.
- To avoid skipped teeth, do not move crankshaft and camshaft until front cover is installed.

➡ Before installing chain tensioner, it is possible to change the position of mating mark on timing chain for that of each sprocket for alignment.

33. Install new front oil seal to front cover, using suitable tool.

    a. Install new oil seal in until it is flush with front end surface of front cover.

✳✳ CAUTION

Do not reuse oil seal. Be careful not to cause damage to circumference of oil seal.

34. Install front cover with the following procedure:

    a. Install O-rings to cylinder head and cylinder block.

    b. Apply Genuine Silicone RTV Sealant or equivalent, to front cover.

    c. Make sure the mating marks on the timing chain and each sprocket are still aligned. Then install the front cover.

✳✳ CAUTION

Be careful not to damage the front oil seal during installation with the front end of the crankshaft.

    d. Tighten front cover bolts in the numerical order as shown.

    e. After all bolts are tightened, retighten them to the specified torque:

- Bolts A: 36 ft. lbs. (49 Nm)
- Bolts B: 9 ft. lbs. (13 Nm)
- Bolts C: 9 ft. lbs. (13 Nm)
- Bolts D: 36 ft. lbs. (49 Nm)
- Dowel pins

✳✳ CAUTION

Wipe off any excess sealant leaking at the surface for installing the oil pan.

35. Install the chain guide between the camshaft sprockets.

36. Install IVT cover with the following procedure:

    a. Install IVT solenoid valve to IVT cover.

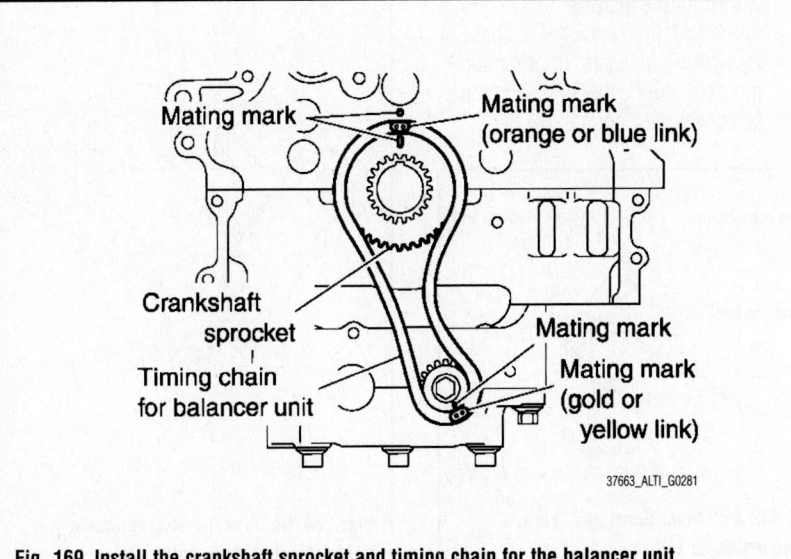

Fig. 169 Install the crankshaft sprocket and timing chain for the balancer unit

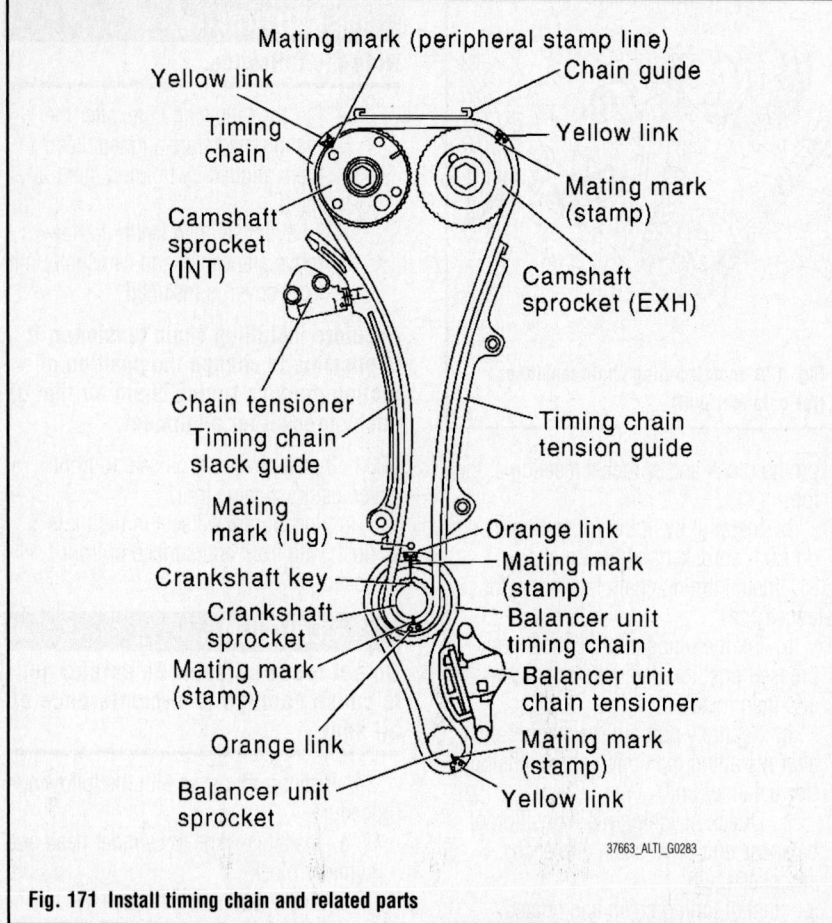

Fig. 171 Install timing chain and related parts

b. Install new O-ring to front cover.

c. Apply Silicone RTV Sealant to the IVT cover as shown.

d. Tighten the IVT cover bolts in the numerical order as shown.

37. Insert crankshaft pulley by aligning with crankshaft key.

a. Tap its center with a plastic hammer to insert.

b. Do not tap the belt hook.

38. Tighten crankshaft pulley bolts.

a. Secure crankshaft pulley with tool to tighten the bolt.

b. Perform angle tightening with the following procedure:

c. Apply new engine oil to threads and seat surfaces of bolts.

d. Tighten to initial specifications: ·31 ft. lbs. (42 Nm).

e. Apply a paint mark on the front cover, mating with any one of six easy to recognize stamp marks on bolt flange.

f. Turn crankshaft pulley bolt another 60° to 66° (Target: 60°).

g. Check vertical mounting angle with movement of one stamp mark.

39. Installation of the remaining components is in the reverse order of removal.

3.5L Engine

See Figures 172 through 183.

✳✳ CAUTION
Note the following:

- After removing timing chains, do not turn the crankshaft and camshaft separately, or the valves will strike the pistons.
- When installing camshafts, chain tensioners, oil seals, or other sliding parts, lubricate contacting surfaces with new engine oil.

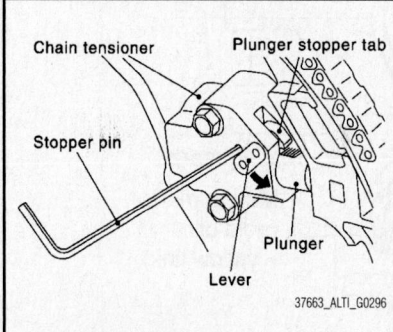

Fig. 172 Pull lever down and release plunger stopper tab

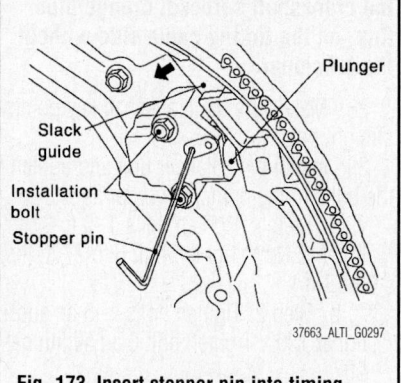

Fig. 173 Insert stopper pin into timing chain tensioner (primary) body hole to hold lever, and keep the tab released

- Apply new engine oil to bolt threads and seat surfaces when installing camshaft sprockets, camshaft brackets, and crankshaft pulley.

1. Remove front timing chain case.

2. Remove the intake manifold collector.

3. Remove the engine oil dipstick.

4. Place paint marks on the timing chain and sprockets to indicate the correct position of the components for installation.

5. Remove the timing chain tensioner (primary).

a. Pull lever down and release plunger stopper tab. Plunger stopper tab can be pushed up to release (coaxial structure with lever).

b. Insert stopper pin into timing chain tensioner (primary) body hole to hold lever, and keep the tab released. An Allen

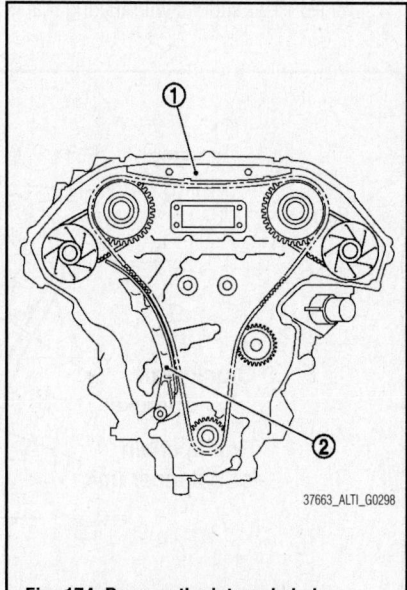

Fig. 174 Remove the internal chain guide (1), and slack guide (2)

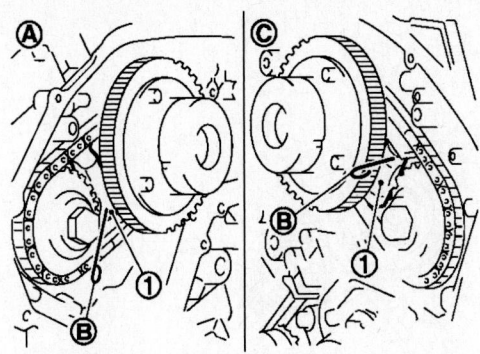

A. Right bank   1. Timing chain tensioners (secondary)
B. Stopper pin
C. Left bank

37663_ALTI_G0299

Fig. 175 Attach a suitable stopper pin to the right and left timing chain tensioners (secondary)

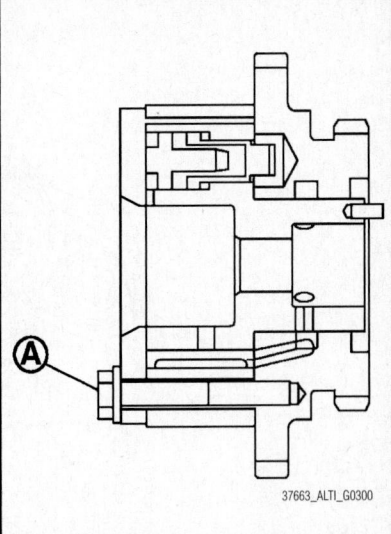

37663_ALTI_G0300

Fig. 176 Do not disassemble the camshaft sprockets (never loosen bolts (A) as shown)

wrench is used for a stopper pin as an example.

c. Insert plunger into tensioner body by pressing the slack guide.

d. Keep the slack guide pressed and hold it by pushing the stopper pin through the lever hole and body hole.

e. Remove the bolts and remove the timing chain tensioner (primary).

6. Remove the internal chain guide and slack guide.

7. Remove timing chain (primary) and crankshaft sprocket.

> ❈❈ **CAUTION**
>
> **After removing timing chains, do not turn the crankshaft and camshaft separately, or the valves will strike the pistons.**

8. Attach a suitable stopper pin to the right and left timing chain tensioners (secondary).

9. Remove the timing chains (secondary) with camshaft sprockets (INT) and (EXH).

a. Insert metal or resin plate [0.5 mm (0.020 in)] into guide between timing chain (secondary) and timing chain tensioner (secondary) plunger. Remove camshaft sprocket and timing chain (secondary) with timing chain removed from guide groove.

> ❈❈ **CAUTION**
>
> **Timing chain tensioner plunger can move while stopper pin is inserted in timing chain tensioner. Plunger can come out of tensioner when timing chain is removed. Use caution during removal.**

b. Apply paint to the timing chain and

camshaft sprockets for alignment during installation.

c. Remove the camshaft sprocket (INT) and (EXH) bolts.

d. Hold the hexagonal portion of the camshaft using a wrench to loosen the bolts.

e. Handle the sprockets as an assembly.

f. Remove timing chains (secondary).

> ❈❈ **CAUTION**
>
> **Avoid impact or dropping the camshaft sprockets. Do not disassemble the camshaft sprockets (never loosen bolts as shown).**

10. Remove the tension guide.

11. Check for cracks and any excessive wear of the timing chain. Replace the timing chain as necessary.

To install:

➡**The figure shows the relationship between the mating mark on each timing chain and that on the corresponding sprocket, the components installed.**

12. Install the tension guide.

13. Position the crankshaft so No. 1 piston is set at TDC on the compression stroke.

➡**Make sure that the dowel pin hole, dowel pin and crankshaft key are located as shown.**

- Camshaft dowel pin hole (intake side): at cylinder head upper face side in each bank.
- Camshaft dowel pin (exhaust side): at cylinder head upper face side in each bank.

- Crankshaft key: at cylinder head side of RH bank.

> ❈❈ **CAUTION**
>
> **Hole on small diameter side must be used for intake camshaft sprocket dowel pin. Do not misidentify (ignore big diameter side).**

14. Install the timing chains (secondary) and camshaft sprockets.

> ❈❈ **CAUTION**
>
> **Matching marks between the timing chain and sprockets slip easily. Confirm all matching mark positions repeatedly during the installation process.**

a. Push the sleeve of the chain tensioner (secondary) and keep it pressed in with a stopper pin.

b. Align the matching marks on the timing chain (secondary), with the ones on the camshaft sprockets (INT) and (EXH) (stamped), and install them.

c. Matching marks for the camshaft sprocket (INT) are on the back side of the secondary sprocket.

d. There are two types of matching marks, round and oval types. They should be used for the RH and LH banks, respectively.

- RH bank: use round type.
- LH bank: use oval type.

e. Align the dowel pin with and pin hole on the camshaft sprocket (INT) side,

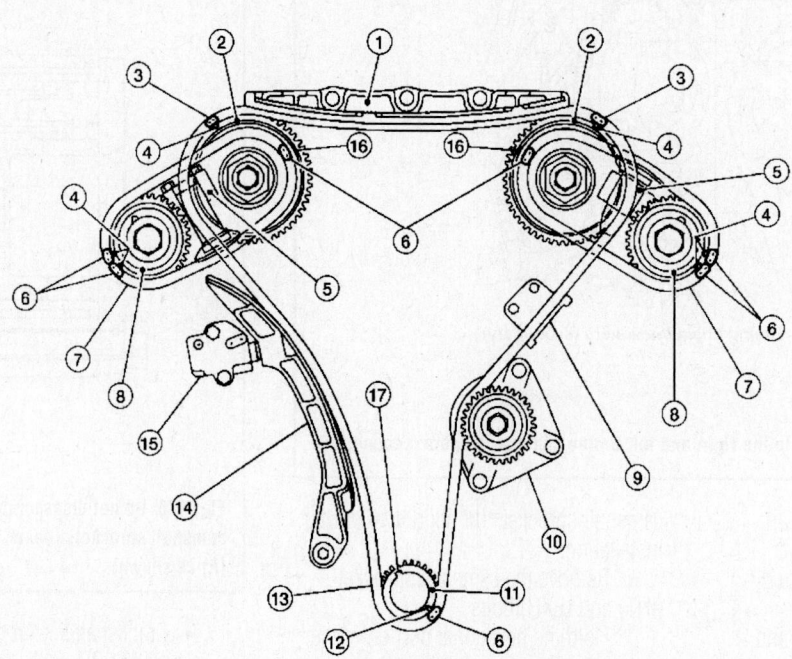

1. Internal chain guide
2. Camshaft sprocket (INT)
3. Mating mark (pink link)
4. Mating mark (punched)
5. Timing chain tensioner (secondary)
6. Mating mark (orange link)
7. Timing chain (secondary)
8. Camshaft sprocket (EXH)
9. Tension guide
10. Water pump
11. Crankshaft sprocket
12. Mating mark (notched)
13. Timing chain (primary)
14. Slack guide
15. Timing chain tensioner (primary)
16. Mating mark (back side)
17. Crankshaft key

37663_ALTI_G0302

Fig. 177 Mating mark locations and relationships to timing chain assembly

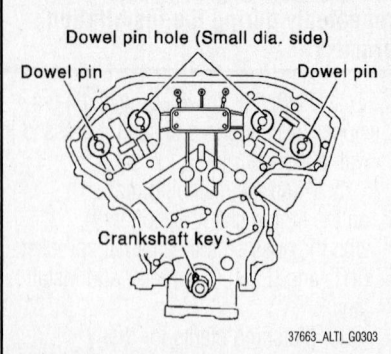

37663_ALTI_G0303

Fig. 178 Position the crankshaft so No. 1 piston is set at TDC on the compression stroke

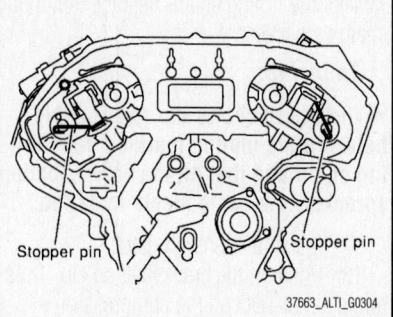

37663_ALTI_G0304

Fig. 179 Push the sleeve of the chain tensioner (secondary) and keep it pressed in with a stopper pin

and dowel pin groove with the dowel pin on the camshaft sprocket (EXH) side, and install them.

f. On the intake side, align the pin hole on the small diameter side of the camshaft front end with the dowel pin on

the back side of the camshaft sprocket, and install them.

g. On the exhaust side, align the dowel pin on the camshaft front end with the dowel pin groove on the camshaft sprocket, and install them.

h. Camshaft sprocket bolts must be tightened in the next step. Tightening them by hand is enough to prevent the dislocation of the dowel pins and dowel pin grooves.

➡It may be difficult to visually check the dislocation of mating marks during and after installation. To make the matching easier, make a mating mark on the sprocket teeth in advance with paint.

15. After confirming the mating marks are aligned, tighten the camshaft sprocket bolts.

➡Hold the camshaft using a wrench at the hexagonal portion to tighten the bolts.

16. Pull the stopper pins out from the secondary timing chain tensioners.

17. Install the crankshaft sprocket on the crankshaft.

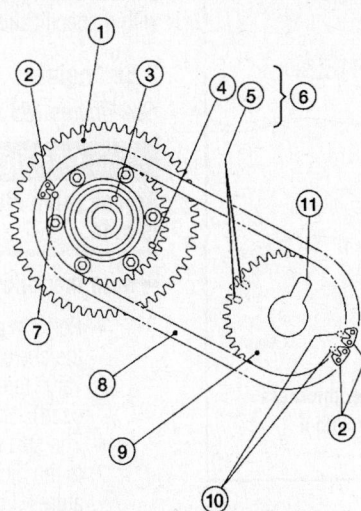

1. Camshaft sprocket (INT) side
2. Secondary timing chain orange link
3. Dowel pin
4. Matching marks
5. Matching marks
6. LH bank: oval type (4 and 5)
7. Matching marks
8. Timing chain (secondary)
9. Camshaft sprocket
10. Matching marks
11. Dowel pin grooves

37663_ALTI_G0305

Fig. 180 Align the matching marks on the timing chain (secondary), with the ones on the camshaft sprockets (INT) and (EXH) (stamped), and install them

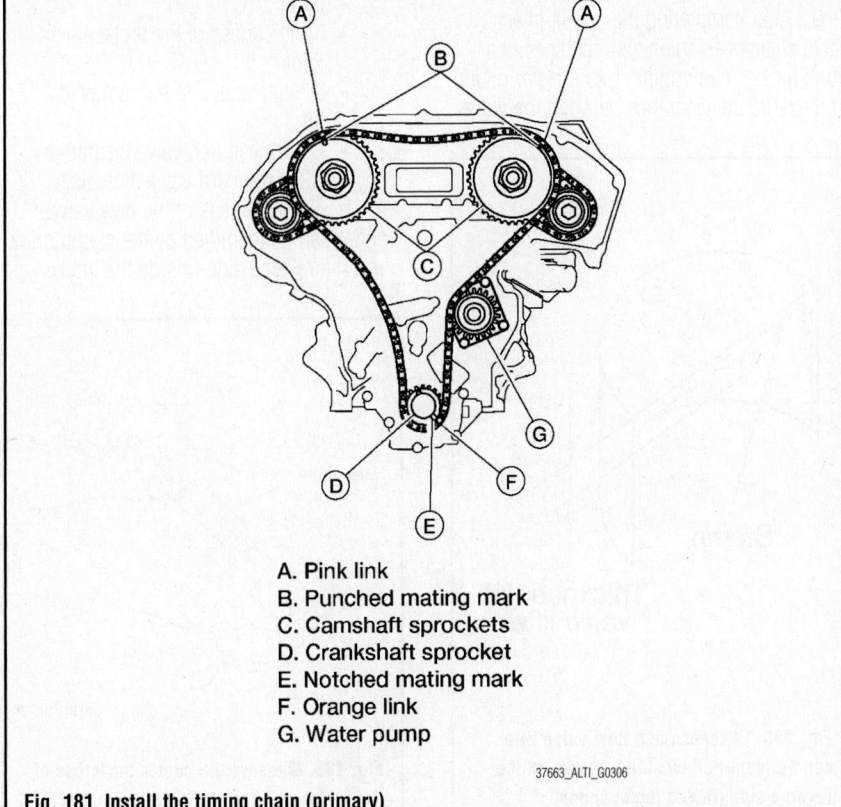

A. Pink link
B. Punched mating mark
C. Camshaft sprockets
D. Crankshaft sprocket
E. Notched mating mark
F. Orange link
G. Water pump

37663_ALTI_G0306

Fig. 181 Install the timing chain (primary)

➡**Make sure the mating marks on the crankshaft sprocket face the front of the engine.**

18. Install the timing chain (primary).

a. Install timing chain (primary) so the mating mark (punched) on camshaft sprocket is aligned with the pink link on the timing chain, while the mating mark (notched) on the crankshaft sprocket is aligned with the orange one on the timing chain, as shown.

b. When it is difficult to align mating marks of the timing chain (primary) with each sprocket, gradually turn the camshaft using a wrench on the hexagonal portion to align it with the mating marks.

c. During alignment, be careful to prevent dislocation of mating mark alignments of the secondary timing chains.

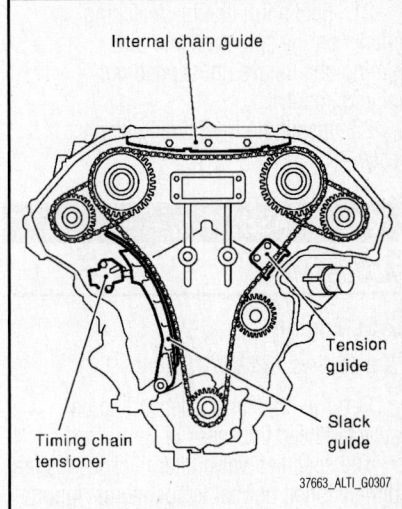

37663_ALTI_G0307

Fig. 182 Install the internal chain guide and slack guide

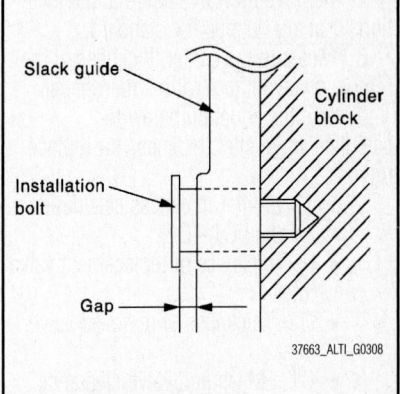

37663_ALTI_G0308

Fig. 183 It is normal for a gap to exist under the bolt seats when the bolts are tightened to specification

19. Install the internal chain guide and slack guide.

> ### ❊❊ CAUTION
>
> **Do not overtighten the slack guide bolts. It is normal for a gap to exist under the bolt seats when the bolts are tightened to specification.**

20. Install the timing chain tensioner (primary) for the slack guide.

   a. When installing the timing chain tensioner (primary), push in the sleeve and keep it pressed in with the stopper pin.

   b. Remove any dirt and foreign materials completely from the back and the mounting surfaces of the timing chain tensioner (primary).

   c. After installation, pull out the stopper pin while pressing the slack guide.

21. Reconfirm that the matching marks on the sprockets and the timing chain have not slipped out of alignment.

22. Install the front timing chain case.

VALVE LASH

ADJUSTMENT

2.5L Engine

See Figures 184 and 185.

Perform adjustment depending on selected head thickness of valve lifter.

The specified valve lifter thickness is the dimension at normal temperatures. Ignore dimensional differences caused by temperature. Use the specifications for hot engine condition to adjust.

1. Remove camshaft.
2. Remove the valve lifters at the locations that are outside the standard.
3. Measure the center thickness of the removed valve lifters with a micrometer.
4. Use the equation below to calculate valve lifter thickness for replacement.

   a. Valve lifter thickness calculation:
   - $t = t1 + (C1 - C2)$
   - t = Thickness of replacement valve lifter.
   - t1 = Thickness of removed valve lifter.
   - C1 = Measured valve clearance.
   - C2 = Standard valve clearance.

   b. Thickness of a new valve lifter can be identified by stamp marks on the

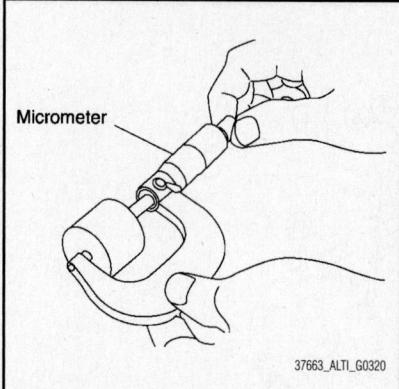

Fig. 184 Measure the center thickness of the removed valve lifters with a micrometer

reverse side (inside the cylinder). Stamp mark 696 indicates a thickness of 0.2740 inches (6.96 mm).

➡**Available thickness of valve lifter: 26 sizes with a range of 0.3102 to 0.3299 inches (7.88 to 8.38 mm), in steps of 0.0008 inches (0.02 mm), when assembled at the factory.**

5. Install the selected valve lifter.
6. Install camshaft.
7. Manually turn crankshaft pulley a few turns.
8. Check that valve clearances for cold engine are within specifications, by referring to the specified values.
9. After completing the repair, check valve clearances again with the specifications for warmed engine. Use a feeler gauge to measure the clearance between the valve

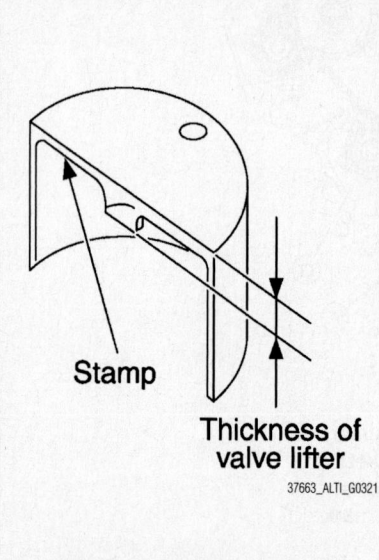

Fig. 185 Thickness of a new valve lifter can be identified by stamp marks on the reverse side (inside the cylinder)

and camshaft. Make sure the values are within specifications.

3.5L Engine

See Figures 185 and 186.

> ### ❊❊ CAUTION
>
> **Adjust valve clearance while engine is cold.**

➡**Note the following:**

- Perform adjustment by selecting the correct head thickness of the valve lifter (adjusting shims are not used).
- The specified valve lifter thickness is the dimension at normal temperatures. Ignore dimensional differences caused by temperature. Use specifications for hot engine condition to confirm valve clearances.

1. Remove the camshaft.
2. Remove the valve lifter that was measured as being outside the standard specifications.
3. Measure the center thickness of the removed lifter with a micrometer, as shown.
4. Use the equation below to calculate the replacement valve lifter thickness.

   a. Valve lifter thickness calculation equation:
   - $t = t1 + (C1 - C2)$
   - t = thickness of the replacement lifter
   - t1 = thickness of the removed lifter
   - C1 = measured valve clearance
   - C2 = standard valve clearance

   b. The thickness of the new valve lifter can be identified by the stamp mark on the reverse side (inside the lifter).

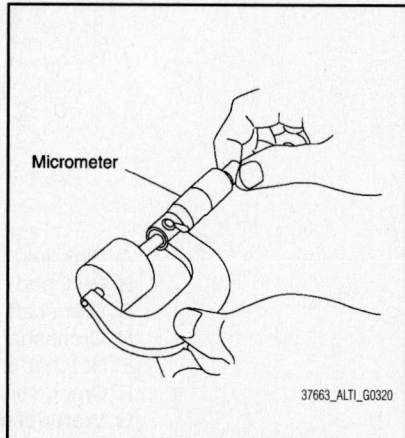

Fig. 186 Measure the center thickness of the removed lifters with a micrometer

➡**Available thickness of the valve lifter (factory setting): 0.3102 –0.3307 inches (7.88–8.40 mm), in 0.0008inches (0.02 mm) increments, in 27 sizes (intake / exhaust).**

5. Install the selected replacement valve lifter.

6. Install the camshaft.

7. Rotate the crankshaft a few turns by hand.

8. Confirm that the valve clearances are within specification.

9. After the engine has been run to full operating temperature, confirm that the valve clearances are within specification.

a. Intake: Cold: 0.010–0.013 inches (0.26–0.34 mm)

b. Intake: Hot: 0.012–0.016 inches (0.304–0.416 mm)

c. Exhaust Cold: 0.011–0.015 inches (0.29–0.37 mm)

d. Exhaust: Hot: 0.012–0.017 inches (0.308–0.432 mm)

INSPECTION

2.5L Engine

See Figures 187 through 189.

Perform this inspection as follows after removal, installation, or replacement of the camshaft or any valve related parts, or if there are any unusual engine conditions due to changes in valve clearance over time (starting, idling, and/or noise).

1. Warm up the engine, then stop it.

2. Remove the fender protector side cover RH.

3. Remove the rocker cover.

4. Turn crankshaft pulley in normal direction (clockwise when viewed from

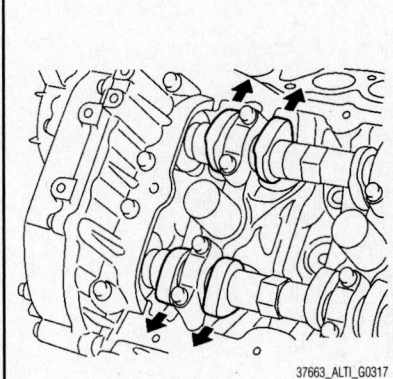

Fig. 187 Check that the both intake and exhaust cam lobes of No. 1 cylinder face outside

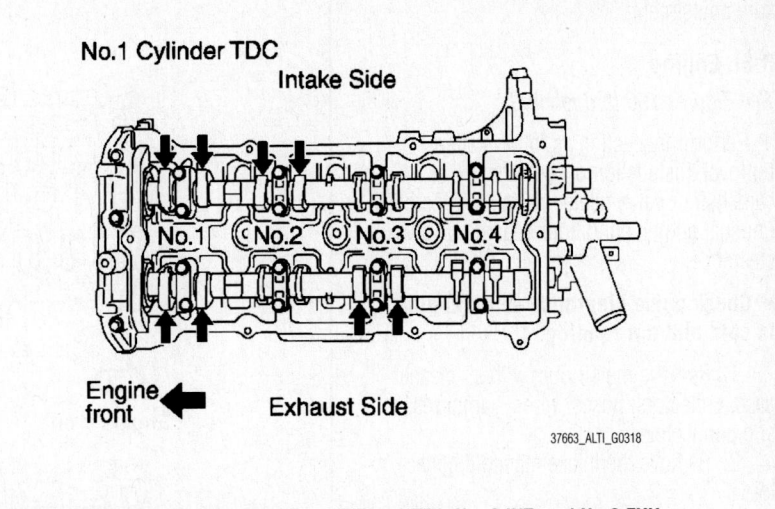

Fig. 188 Measure valve clearances at No. 1 INT and EXH, No. 2 INT, and No.3 EXH

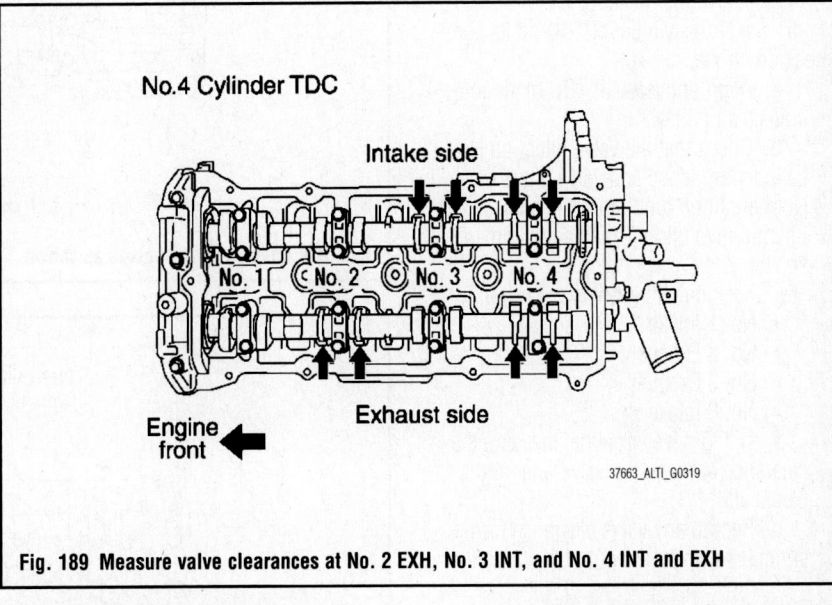

Fig. 189 Measure valve clearances at No. 2 EXH, No. 3 INT, and No. 4 INT and EXH

front) to align TDC identification mark (without paint mark) with timing indicator.

5. At this time, check that the both intake and exhaust cam lobes of No. 1 cylinder face outside.

➡**If they do not face outside, turn crankshaft pulley once more.**

6. Measure valve clearances with a feeler gauge at No. 1 INT and EXH, No. 2 INT, and No.3 EXH with No.1 cylinder compression TDC.

a. Use a feeler gauge to measure the clearance between valve and camshaft.

b. Valve clearance standards:
- Cold: Intake: 0.009–0.013 inches (0.24–0.32 mm)
- Cold: Exhaust: 0.010–0.013 inches (0.26–0.34 mm)
- Hot: Intake: 0.012–0.016 inches (0.304–0.416 mm)
- Hot Exhaust: 0.012–0.017 inches (0.308–0.432 mm)

❋❋ **CAUTION**

If inspection was carried out with cold engine, check that values with fully warmed up engine are still within specifications.

7. Turn crankshaft one complete revolution (360°) and align mark on crankshaft pulley with pointer.

8. Measure valve clearances with a feeler gauge at No. 2 EXH, No. 3 INT, and No. 4 INT and EXH with No.4 cylinder compression TDC.

9. If out of specifications, make necessary adjustment.

3.5L Engine

See Figures 190 through 193.

Perform inspection as follows after removal, installation or replacement of camshaft or valve related parts, or if there is unusual engine conditions regarding valve clearance.

➡**Check valve clearance while engine is cold and not running.**

1. Remove the air duct with air cleaner case, collectors, hoses, wires, harnesses, and connectors.

2. Remove the intake manifold collectors.

3. Remove the ignition coils and spark plugs.

4. Remove the rocker covers.

5. Set No.1 cylinder at TDC on its compression stroke.

  a. Align pointer with TDC mark on crankshaft pulley.

  b. Check that the valve lifters on No.1 cylinder are loose and valve lifters on No.4 are tight. If not, turn the crankshaft one full revolution (360°) and align as shown.

6. Check only the valves as shown.

- No. 1 Intake
- No. 2 Exhaust
- No. 3 Exhaust
- No. 6 Intake

  a. Using a feeler gauge, measure the clearance between the valve lifter and camshaft.

  b. Record any valve clearance measurements which are out of specification.

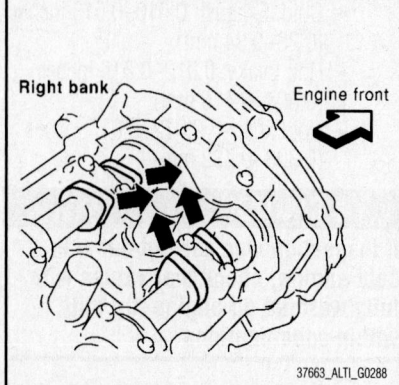

Fig. 190 Check that the valve lifters on No.1 cylinder are loose and valve lifters on No.4 are tight, if not, turn the crankshaft one full revolution and align as shown

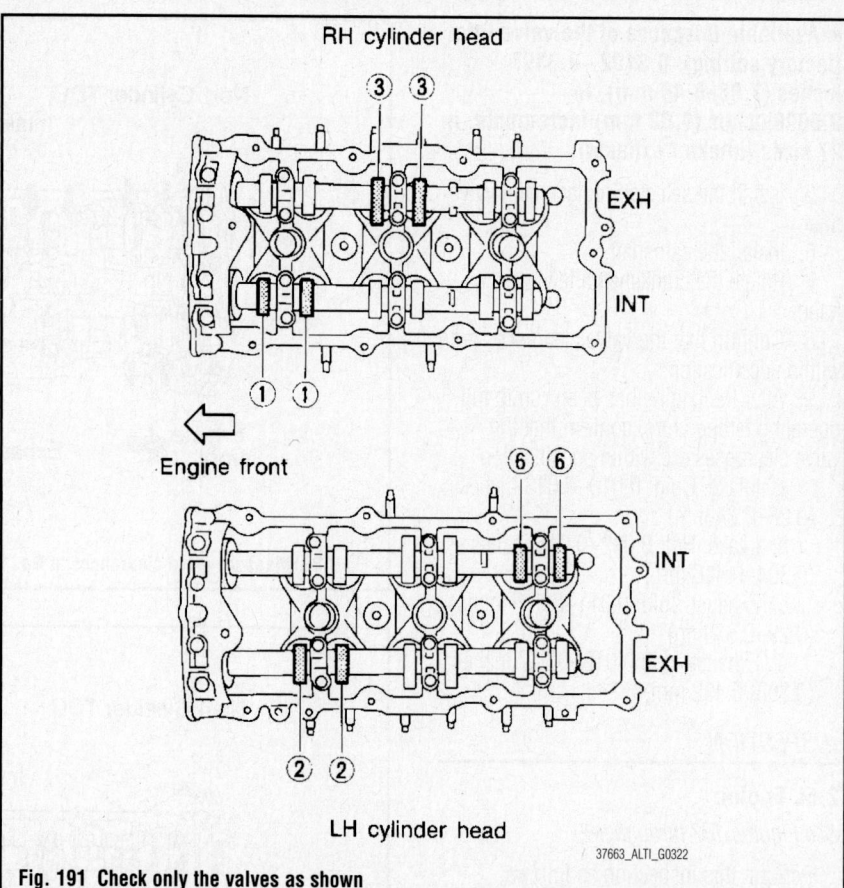

Fig. 191 Check only the valves as shown

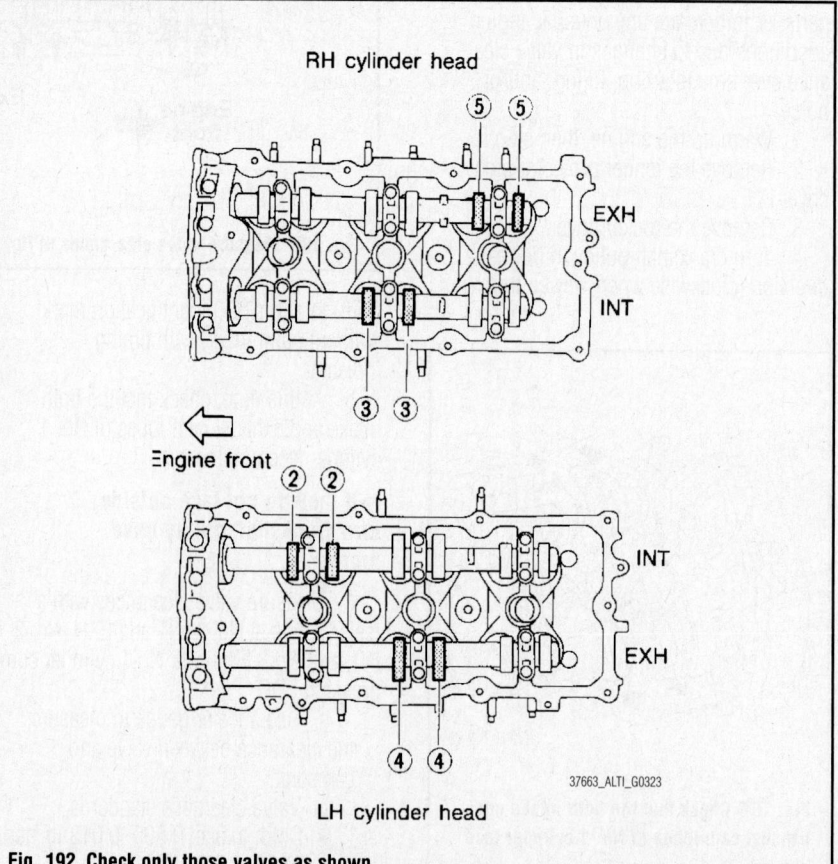

Fig. 192 Check only those valves as shown

They will be used later to determine the required replacement lifter size.

7. Turn crankshaft 240°.

8. Set No.3 cylinder at TDC on its compression stroke.

9. Check only those valves as shown.
- No. 2 Intake
- No. 3 Intake
- No. 4 Exhaust
- No. 5 exhaust

10. Turn the crankshaft 240° and align as above.

11. Set No.5 cylinder at TDC on its compression stroke.

12. Check only those valves as shown.
- No. 1 Exhaust
- No. 4 Intake
- No. 5 Intake
- No. 6 Exhaust

13. If all valve clearances are within specification, install the following components:
- Intake manifold collectors
- Rocker covers
- All spark plugs
- All ignition coils

14. If the valve clearances are out of specification, adjust the valve clearances.

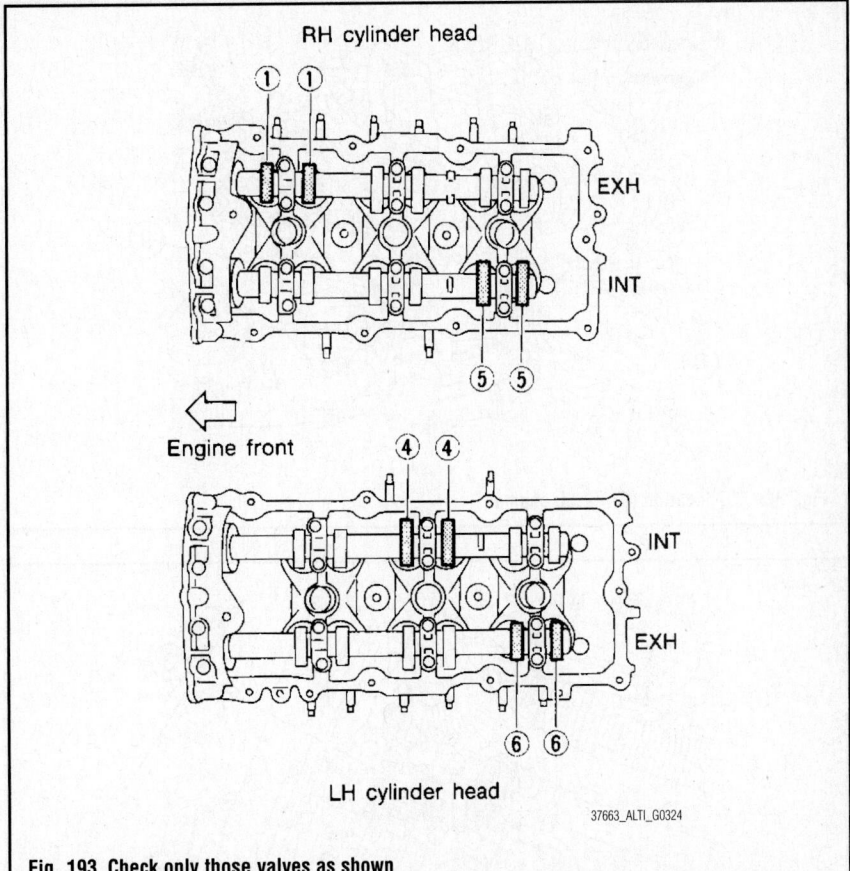

Fig. 193 Check only those valves as shown

ENGINE PERFORMANCE & EMISSION CONTROLS

CAMSHAFT POSITION (CMP) SENSOR

REMOVAL & INSTALLATION

1. Turn ignition switch OFF.
2. Loosen the fixing bolt of the sensor.
3. Disconnect camshaft position sensor (PHASE) harness connector.
4. Remove the sensor.
5. Installation is the reverse of removal.

CRANKSHAFT POSITION (CKP) SENSOR

REMOVAL & INSTALLATION

1. Turn ignition switch OFF.
2. Loosen the fixing bolt of the sensor.
3. Disconnect crankshaft position sensor (POS) harness connector.
4. Remove the sensor.
5. Installation is the reverse of removal.

ELECTRONIC CONTROL MODULE (ECM)

LOCATION
See Figure 194.

The ECM is located in the engine compartment, adjacent to the battery.

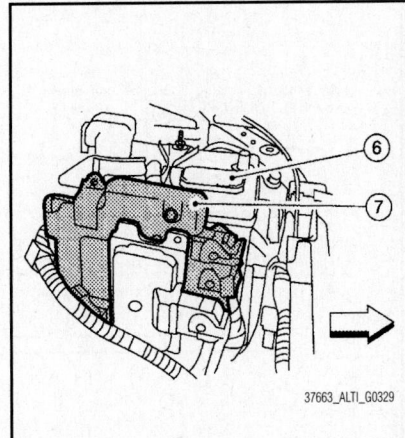

Fig. 194 Location of battery (6) and ECM (7)

REMOVAL & INSTALLATION

➡The manufacturer does not provide a specific Removal and Installation procedure for this component. Refer to the graphic(s) when servicing this component.

ENGINE COOLANT TEMPERATURE (ECT) SENSOR

LOCATION

See Figures 195 and 196.

Refer to the accompanying illustrations.

REMOVAL & INSTALLATION

1. Turn ignition switch OFF.
2. Disconnect engine coolant temperature sensor harness connector.
3. Remove engine coolant temperature sensor.
4. Installation is the reverse of removal.

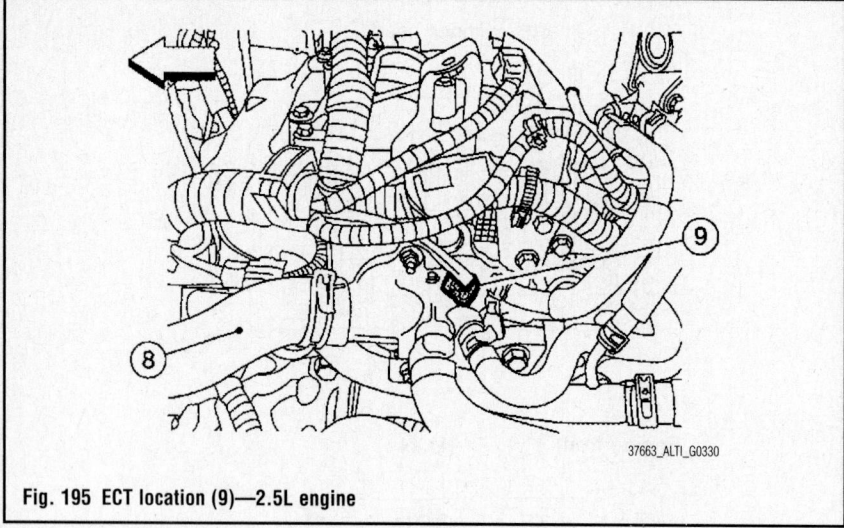

Fig. 195 ECT location (9)—2.5L engine

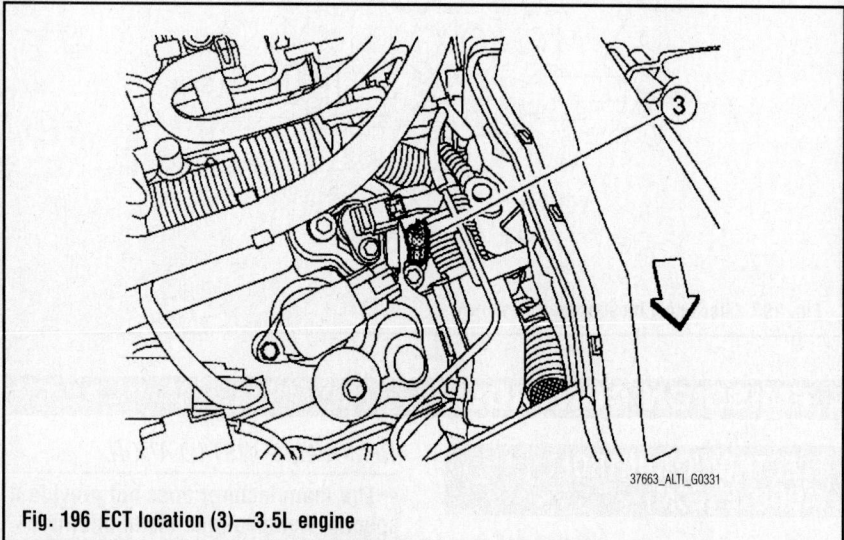

Fig. 196 ECT location (3)—3.5L engine

HEATED OXYGEN (HO2S) SENSOR

LOCATION

2.5L Engine

See Figure 197.

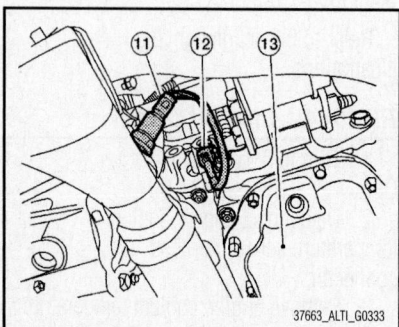

Fig. 197 Heated oxygen sensor 2 (11) showing the connector (12) and oil pan (13)—2.5L engine

3.5L Engine

See Figure 198.

Heated oxygen sensor 2 is located downstream of the three way catalyst (manifold).

REMOVAL & INSTALLATION

➥**The manufacturer does not provide a specific Removal and Installation procedure for this component. Refer to the graphic(s) when servicing this component.**

INTAKE AIR TEMPERATURE (IAT) SENSOR

LOCATION

The Intake Air Temperature (IAT) sensor is an integral part of the Mass Air Flow (MAF) sensor/Intake Air Temperature (IAT) sensor assembly which is mounted on the air intake duct. Refer to the Mass Air Flow section for information regarding servicing this component.

REMOVAL & INSTALLATION

The Intake Air Temperature (IAT) sensor is an integral part of the Mass Air Flow (MAF) sensor/Intake Air Temperature (IAT) sensor assembly which is mounted on the air intake duct. Refer to the Mass Air Flow section for information regarding servicing this component.

KNOCK SENSOR (KS)

LOCATION

2.5L Engine

See Figure 199.

The knock sensor is attached to the cylinder block.

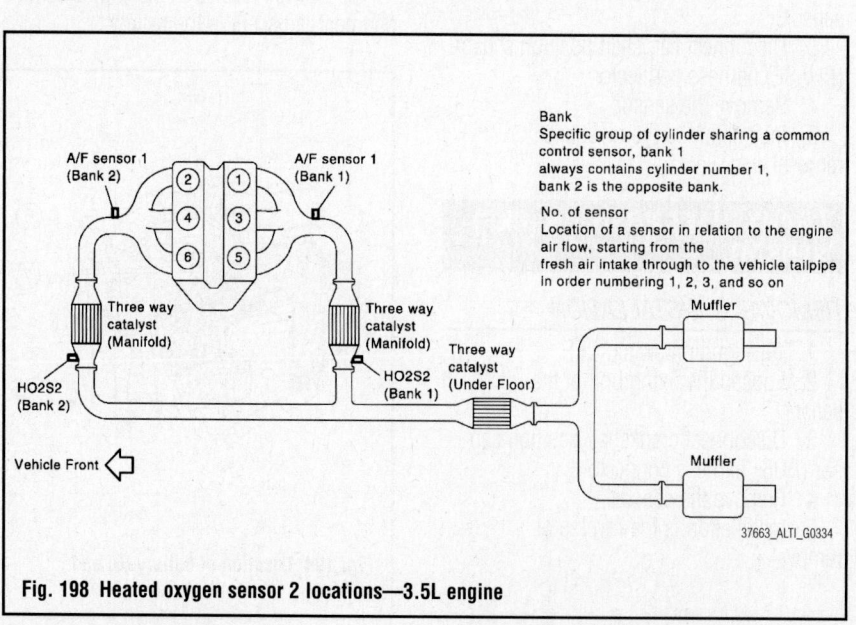

Fig. 198 Heated oxygen sensor 2 locations—3.5L engine

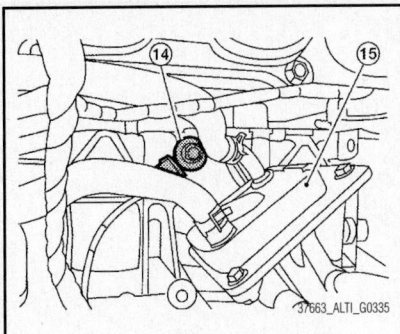

Fig. 199 Knock sensor (14) location near engine oil cooler (15)—2.5L engine

3.5L Engine

See Figure 200.

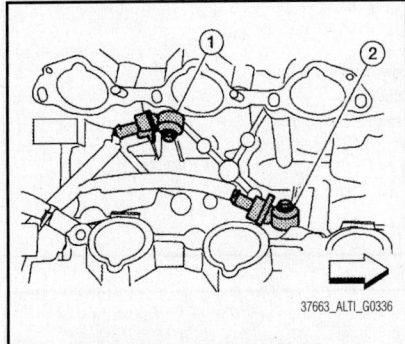

Fig. 200 Knock sensor (bank 2) (1) and knock sensor (bank 1) (2)—3.5L engine

REMOVAL & INSTALLATION

➡The manufacturer does not provide a specific Removal and Installation procedure for this component. Refer to the graphic(s) when servicing this component.

MASS AIR FLOW (MAF) SENSOR

LOCATION

2.5L Engine

See Figure 201.

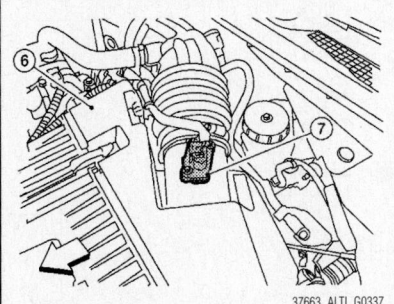

Fig. 201 MAF sensor/IAT sensor location—2.5L engine

3.5L Engine

See Figure 202.

REMOVAL & INSTALLATION

➡The manufacturer does not provide a specific Removal and Installation procedure for this component. Refer to the graphic(s) when servicing this component.

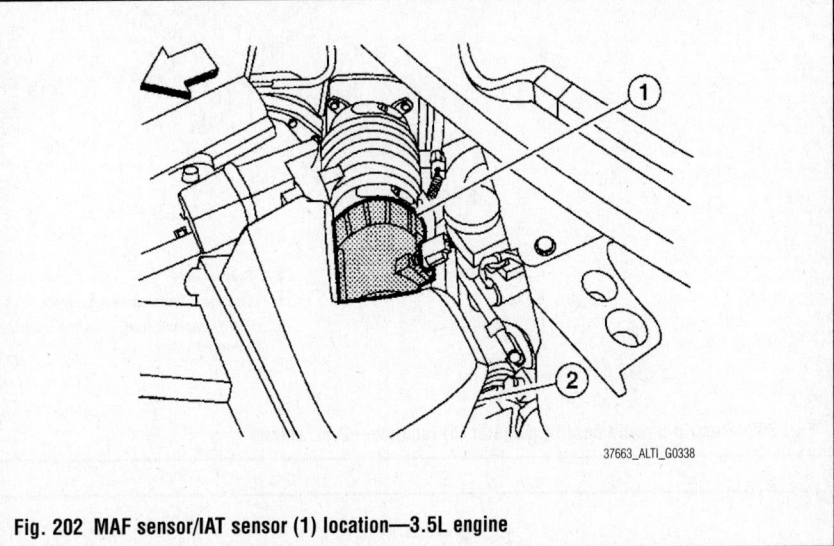

Fig. 202 MAF sensor/IAT sensor (1) location—3.5L engine

THROTTLE CONTROL ACTUATOR (TAC)

LOCATION

See Figures 203 and 204.

The electrical throttle control actuator is located between the engine air intake duct and the intake manifold.

REMOVAL & INSTALLATION

1. Disconnect the engine air intake duct from the electric throttle control actuator.
2. Disconnect the electric throttle control actuator electrical connector.
3. Disconnect electric throttle control actuator coolant hoses.
4. Loosen bolts diagonally, and remove the electric throttle control actuator.

✳✳ CAUTION

Handle carefully to avoid any damage.

5. Installation is the reverse of removal.
6. Tighten the bolts of electric throttle control actuator equally and diagonally in several steps.

7. After installation perform procedure in "INSPECTION AFTER INSTALLATION".

Inspection After Installation

1. Make sure there is no fuel leakage at connections as follows:
   a. Apply fuel pressure to fuel lines by turning ignition switch ON (with engine stopped). Then check for fuel leaks at connections.
   b. Start the engine and rev it up and check for fuel leaks at connections.

✳✳ WARNING

Do not touch engine immediately after stopping as engine is extremely hot.

➡Use mirrors for checking on connections out of the direct line of sight.

THROTTLE POSITION SENSOR (TPS)

LOCATION

Electric throttle control actuator consists of throttle control motor, accelerator pedal position sensor, throttle position sensor etc. The actuator sends a signal to the ECM, and ECM sends the signal to TCM with CAN communication.

REMOVAL & INSTALLATION

Refer to Throttle Control Actuator section when servicing this component.

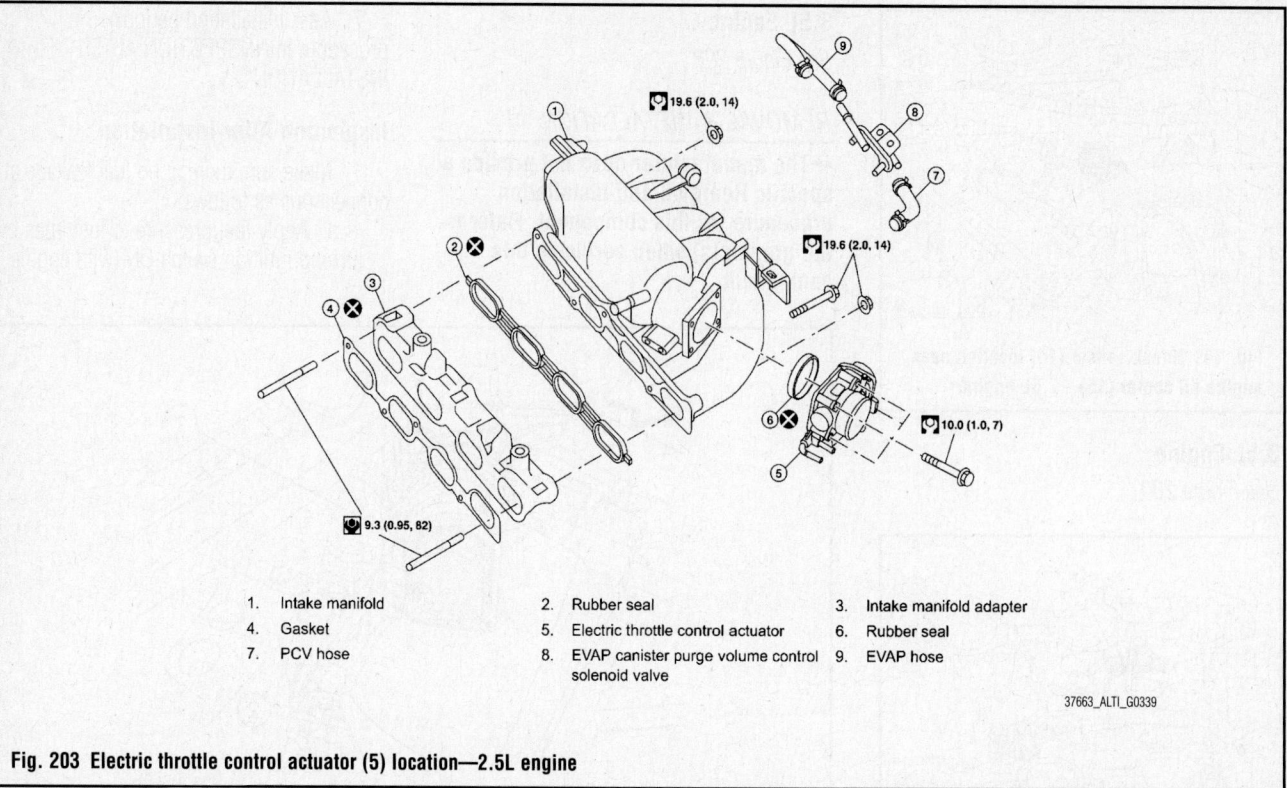

1. Intake manifold
2. Rubber seal
3. Intake manifold adapter
4. Gasket
5. Electric throttle control actuator
6. Rubber seal
7. PCV hose
8. EVAP canister purge volume control solenoid valve
9. EVAP hose

37663_ALTI_G0339

Fig. 203 Electric throttle control actuator (5) location—2.5L engine

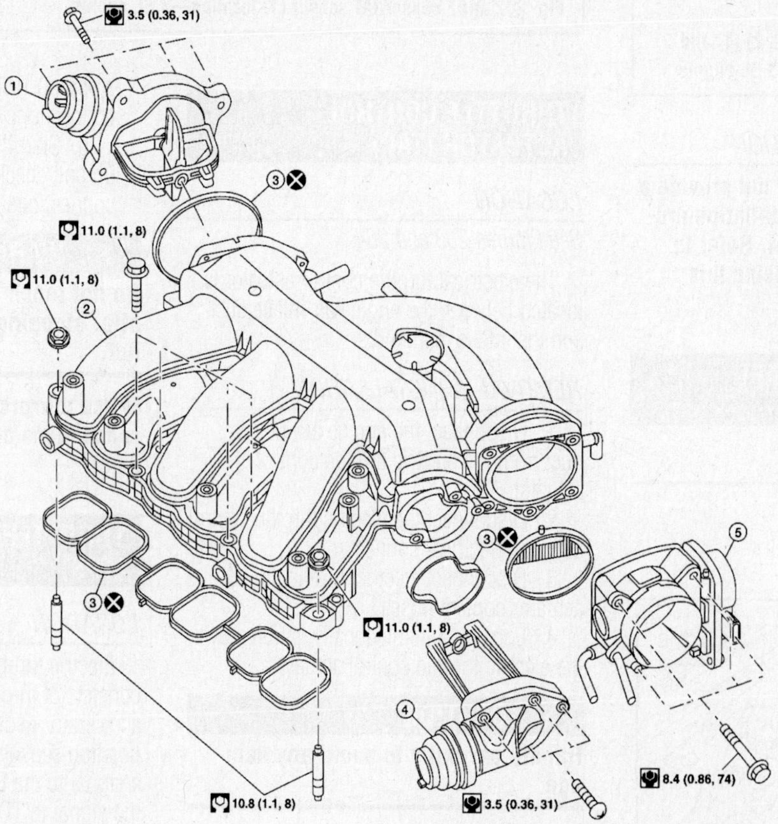

1. Power valve (RH)
2. Intake manifold collector
3. Gasket
4. Power valve (LH)
5. Electric throttle control actuator

37663_ALTI_G0340

Fig. 204 Electric throttle control actuator (5) location—3.5L engine

FUEL

FUEL SYSTEM SERVICE PRECAUTIONS

Safety is the most important factor when performing not only fuel system maintenance, but any type of maintenance. Failure to conduct maintenance and repairs in a safe manner may result in serious personal injury or death. Work on a vehicle's fuel system components can be accomplished safely and effectively by adhering to the following rules and guidelines.

• To avoid the possibility of fire and personal injury, always disconnect the negative battery cable unless the repair or test procedure requires that battery voltage be applied.

• Always relieve the fuel system pressure prior to disconnecting any fuel system component (injector, fuel rail, pressure regulator, etc.) fitting or fuel line connection. Exercise extreme caution whenever relieving fuel system pressure to avoid exposing skin, face and eyes to fuel spray. Please be advised that fuel under pressure may penetrate the skin or any part of the body that it contacts.

• Always place a shop towel or cloth around the fitting or connection prior to loosening to absorb any excess fuel due to spillage. Ensure that all fuel spillage is quickly removed from engine surfaces. Ensure that all fuel-soaked cloths or towels are deposited into a flame-proof waste container with a lid.

• Always keep a dry chemical (Class B) fire extinguisher near the work area.

• Do not allow fuel spray or fuel vapors to come into contact with a spark or open flame.

• Always use a second wrench when loosening or tightening fuel line connection fittings. This will prevent unnecessary stress and torsion on fuel piping. Always follow the proper torque specifications.

• Always replace worn fuel fitting O-rings with new ones. Do not substitute fuel hose where rigid pipe is installed.

FUEL SYSTEM PRESSURE

RELIEVING

With CONSULT-III

1. Turn ignition switch ON.
2. Perform "FUEL PRESSURE RELEASE" in "WORK SUPPORT" mode with CONSULT-III.
3. Start engine.

4. After engine stalls, crank it two or three times to release all fuel pressure.
5. Turn ignition switch OFF.

With CONSULT-III

1. Remove fuel pump fuse located in IPDM E/R.
2. Start engine.
3. After engine stalls, crank it two or three times to release all fuel pressure.
4. Turn ignition switch OFF.
5. Reinstall fuel pump fuse after servicing fuel system.

FUEL FILTER

REMOVAL & INSTALLATION
See Figures 205 through 208.

1. Unscrew the fuel filler cap to release the pressure inside the fuel tank.
2. Release the fuel pressure from the fuel lines.

3. Disconnect the battery negative terminal.
4. Remove the rear seat bottom.
5. Turn the four retainers 90° in a clockwise direction and remove the fuel pump inspection hole cover.
6. Disconnect the fuel level sensor, fuel filter, and fuel pump assembly electrical connector, EVAP hose quick connector, and the fuel feed hose quick connector from the fuel level sensor unit, fuel filter, and fuel pump assembly.
7. Remove the quick connector as follows:

   a. Hold the sides of the connector, push in tabs and pull out the tube.

   b. If the connector and the tube are stuck together, push and pull several times until they start to move. Then disconnect them by pulling.

✷✷ CAUTION
Note the following:

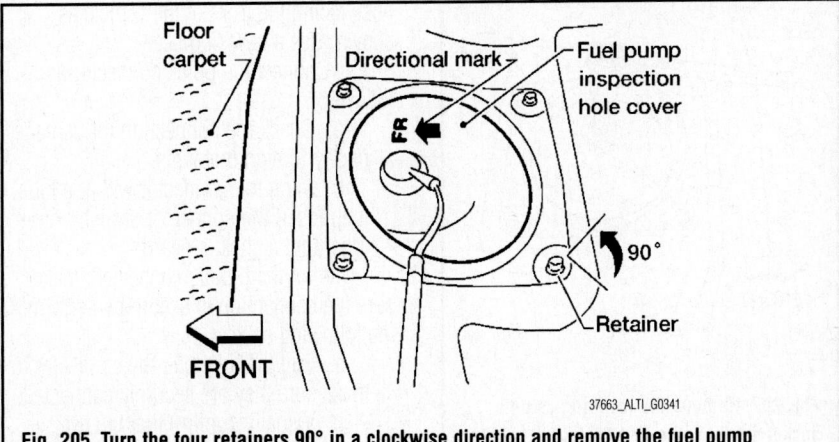

Fig. 205 Turn the four retainers 90° in a clockwise direction and remove the fuel pump inspection hole cover

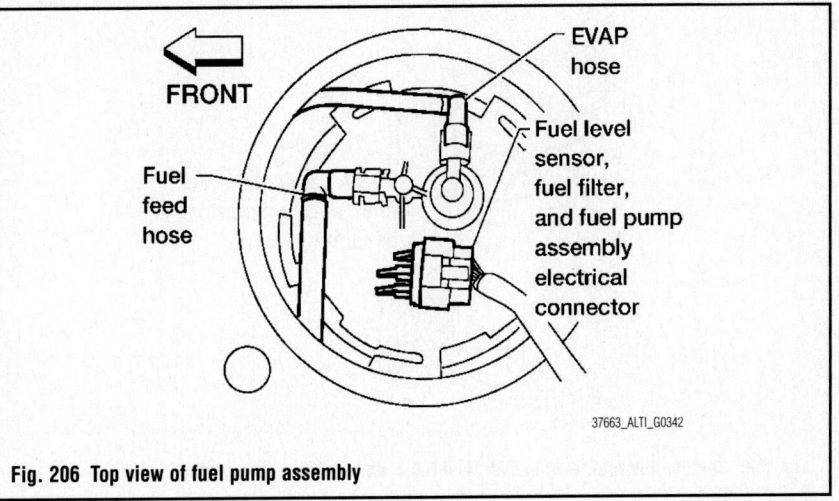

Fig. 206 Top view of fuel pump assembly

- The tube can be removed when the tabs are completely depressed. Do not twist it more than necessary.
- Do not use any tools to remove the quick connector.
- Keep the resin tube away from heat. Be especially careful when welding near the tube.
- Prevent acid liquid such as battery electrolyte, etc. from getting on the resin tube.
- Do not bend or twist the tube during installation and removal.
- Only when the tube is replaced, remove the remaining retainer on the tube or fuel level sensor, fuel filter, and fuel pump assembly.
- When the tube or fuel level sensor, fuel filter, and fuel pump assembly is replaced, also replace the retainer with a new one (green colored retainer).
- To keep the connecting portion clean and to avoid damage and foreign materials, cover them completely with plastic bags or something similar.

8. Remove the lock ring using a socket drive handle and Tool KV991J0090 (J-46214) as shown.

※※ CAUTION

Discard the lock ring, do not reuse the lock ring. Discard the ring seal, do not reuse the ring seal.

9. Remove the fuel level sensor, fuel filter, and fuel pump assembly.

※※ CAUTION

Do not bend the float arm during removal. Discard the ring seal, do not reuse the ring seal.

10. Inspect the fuel level sensor, fuel filter, and fuel pump for any defects and foreign materials. Replace as necessary.

To install:

11. Installation is in the reverse order of removal.

12. Install the fuel level sensor, fuel filter, and fuel pump assembly with the fuel feed hose facing the front of the vehicle as shown. Use a new ring seal.

13. Connect the quick connector as follows:

   a. Check the connection for damage or any foreign materials.

   b. Align the connector with the tube, then insert the connector straight into the tube until a click is heard.

14. After the tube is connected, make sure the connection is secure by performing the following checks:

   a. Pull the tube and the connector to make sure they are securely connected.

   b. Visually confirm that the two retainer tabs are connected to the quick connector.

15. Turn the ignition switch to ON (without starting the engine) to apply fuel pressure to the fuel system, then check the connections for fuel leaks.

16. Start the engine and let it idle and check for fuel leaks at the fuel system connections.

FUEL LEVEL SENSOR UNIT

LOCATION

The fuel level sensor unit is an integral part of the fuel pump assembly.

REMOVAL & INSTALLATION

See Figures 205 through 208.

1. Unscrew the fuel filler cap to release the pressure inside the fuel tank.

2. Release the fuel pressure from the fuel lines.

3. Disconnect the battery negative terminal.

4. Remove the rear seat bottom.

5. Turn the four retainers 90° in a clockwise direction and remove the fuel pump inspection hole cover.

6. Disconnect the fuel level sensor, fuel filter, and fuel pump assembly electrical connector, EVAP hose quick connector, and the fuel feed hose quick connector from the fuel level sensor unit, fuel filter, and fuel pump assembly.

7. Remove the quick connector as follows:

   a. Hold the sides of the connector, push in tabs and pull out the tube.

   b. If the connector and the tube are stuck together, push and pull several times until they start to move. Then disconnect them by pulling.

※※ CAUTION

Note the following:

- The tube can be removed when the tabs are completely depressed. Do not twist it more than necessary.
- Do not use any tools to remove the quick connector.
- Keep the resin tube away from heat. Be especially careful when welding near the tube.
- Prevent acid liquid such as battery electrolyte, etc. from getting on the resin tube.
- Do not bend or twist the tube during installation and removal.
- Only when the tube is replaced, remove the remaining retainer on the tube or fuel level sensor, fuel filter, and fuel pump assembly.

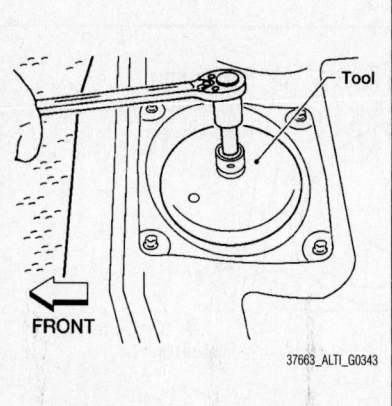

37663_ALTI_G0343

Fig. 207 Remove the lock ring using a socket drive handle and Tool KV991J0090 (J-46214)

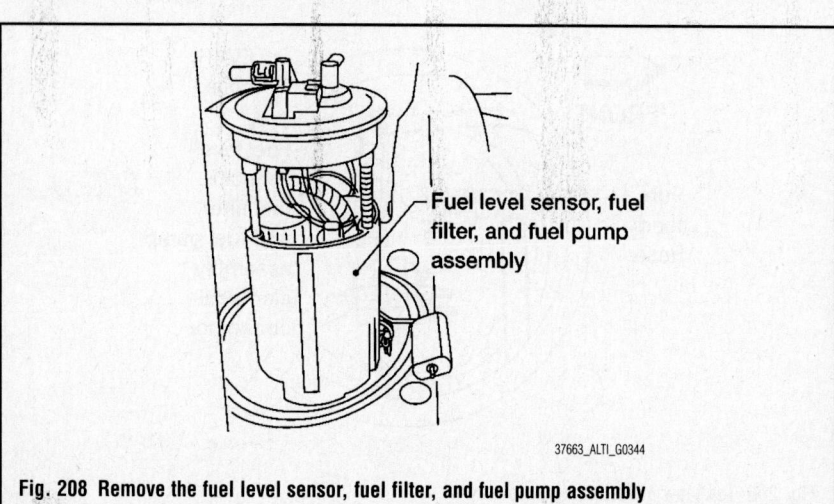

Fuel level sensor, fuel filter, and fuel pump assembly

37663_ALTI_G0344

Fig. 208 Remove the fuel level sensor, fuel filter, and fuel pump assembly

- When the tube or fuel level sensor, fuel filter, and fuel pump assembly is replaced, also replace the retainer with a new one (green colored retainer).
- To keep the connecting portion clean and to avoid damage and foreign materials, cover them completely with plastic bags or something similar.

8. Remove the lock ring using a socket drive handle and Tool KV991J0090 (J-46214) as shown.

❊❊ CAUTION

Discard the lock ring, do not reuse the lock ring. Discard the ring seal, do not reuse the ring seal.

9. Remove the fuel level sensor, fuel filter, and fuel pump assembly.

❊❊ CAUTION

Do not bend the float arm during removal. Discard the ring seal, do not reuse the ring seal.

10. Inspect the fuel level sensor, fuel filter, and fuel pump for any defects and foreign materials. Replace as necessary.

To install:

11. Installation is in the reverse order of removal.

12. Install the fuel level sensor, fuel filter, and fuel pump assembly with the fuel feed hose facing the front of the vehicle as shown. Use a new ring seal.

13. Connect the quick connector as follows:

a. Check the connection for damage or any foreign materials.

b. Align the connector with the tube, then insert the connector straight into the tube until a click is heard.

14. After the tube is connected, make sure the connection is secure by performing the following checks:

a. Pull the tube and the connector to make sure they are securely connected.

b. Visually confirm that the two retainer tabs are connected to the quick connector.

15. Turn the ignition switch to ON (without starting the engine) to apply fuel pressure to the fuel system, then check the connections for fuel leaks.

16. Start the engine and let it idle and check for fuel leaks at the fuel system connections.

FUEL PUMP ASSEMBLY

REMOVAL & INSTALLATION

See Figures 205 through 208.

1. Unscrew the fuel filler cap to release the pressure inside the fuel tank.

2. Release the fuel pressure from the fuel lines.

3. Disconnect the battery negative terminal.

4. Remove the rear seat bottom.

5. Turn the four retainers 90° in a clockwise direction and remove the fuel pump inspection hole cover.

6. Disconnect the fuel level sensor, fuel filter, and fuel pump assembly electrical connector, EVAP hose quick connector, and the fuel feed hose quick connector from the fuel level sensor unit, fuel filter, and fuel pump assembly.

7. Remove the quick connector as follows:

a. Hold the sides of the connector, push in tabs and pull out the tube.

b. If the connector and the tube are stuck together, push and pull several times until they start to move. Then disconnect them by pulling.

❊❊ CAUTION

Note the following:

- The tube can be removed when the tabs are completely depressed. Do not twist it more than necessary.
- Do not use any tools to remove the quick connector.
- Keep the resin tube away from heat. Be especially careful when welding near the tube.
- Prevent acid liquid such as battery electrolyte, etc. from getting on the resin tube.
- Do not bend or twist the tube during installation and removal.
- Only when the tube is replaced, remove the remaining retainer on the tube or fuel level sensor, fuel filter, and fuel pump assembly.
- When the tube or fuel level sensor, fuel filter, and fuel pump assembly is replaced, also replace the retainer with a new one (green colored retainer).
- To keep the connecting portion clean and to avoid damage and foreign materials, cover them completely with plastic bags or something similar.

8. Remove the lock ring using a socket drive handle and Tool KV991J0090 (J-46214) as shown.

❊❊ CAUTION

Discard the lock ring, do not reuse the lock ring. Discard the ring seal, do not reuse the ring seal.

9. Remove the fuel level sensor, fuel filter, and fuel pump assembly.

❊❊ CAUTION

Do not bend the float arm during removal. Discard the ring seal, do not reuse the ring seal.

10. Inspect the fuel level sensor, fuel filter, and fuel pump for any defects and foreign materials. Replace as necessary.

To install:

11. Installation is in the reverse order of removal.

12. Install the fuel level sensor, fuel filter, and fuel pump assembly with the fuel feed hose facing the front of the vehicle as shown. Use a new ring seal.

13. Connect the quick connector as follows:

a. Check the connection for damage or any foreign materials.

b. Align the connector with the tube, then insert the connector straight into the tube until a click is heard.

14. After the tube is connected, make sure the connection is secure by performing the following checks:

a. Pull the tube and the connector to make sure they are securely connected.

b. Visually confirm that the two retainer tabs are connected to the quick connector.

15. Turn the ignition switch to ON (without starting the engine) to apply fuel pressure to the fuel system, then check the connections for fuel leaks.

16. Start the engine and let it idle and check for fuel leaks at the fuel system connections.

FUEL RAIL AND INJECTOR

REMOVAL & INSTALLATION

2.5L Engine

See Figures 209 through 211.

1. Remove engine room cover.
2. Release the fuel pressure.
3. Remove the front air duct.
4. Disconnect the fuel hose quick connector at the fuel tube side.

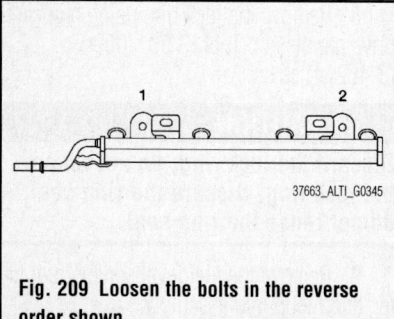

Fig. 209 Loosen the bolts in the reverse order shown

✳✳ CAUTION

Prepare a container and cloth for catching any spilled fuel. This operation should be performed in a place that is free from any open flames. While hoses are disconnected seal their openings with vinyl bag or similar material to prevent foreign material from entering them.

5. Remove the intake manifold.

6. Disconnect sub-harness for injector at engine front side, and remove it from bracket.

7. Loosen the bolts in the reverse order shown, then remove fuel tube and fuel injectors as an assembly.

8. Remove the fuel injectors from the fuel tube.

   a. Release the clip and remove the fuel injector.

   b. Pull fuel injector straight out of the fuel tube.

✳✳ CAUTION

Be careful not to damage the nozzle. Avoid any impact, such as dropping the fuel injector. Do not disassemble or adjust the fuel injector.

To install:

9. Install new O-rings on the fuel injector, the fuel side black O-ring and the nozzle side green O-ring.

✳✳ CAUTION

Upper and lower O-rings are different. Be careful not to confuse them. Fuel tube side: black O-ring; Nozzle side: green O-ring.

✳✳ CAUTION

Note the following:

- Lubricate the O-rings lightly with new engine oil.
- Handle O-rings with bare hands only. Do not wear gloves.

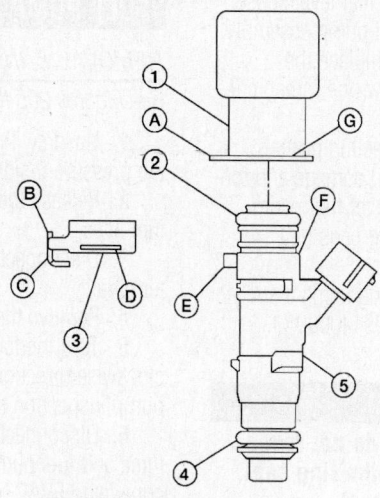

A. Projection of fuel tube
B. Notch of clip
C. Notch of clip
D. Flange fixing groove
E. Projection of fuel injector
F. Clip mounting groove
G. fuel tube flange
1. fuel tube
2. Black O-ring
3. Clip
4. Green O-ring
5. Fuel injector

Fig. 210 Install the fuel injector into the fuel tube

- Do not clean O-rings with solvent.
- Make sure that O-ring and its mating part are free of foreign material.
- Be careful not to scratch O-rings during installation.
- Do not twist or stretch the O-ring. If the O-ring was stretched while it is attached, do not insert it into the fuel tube immediately.

10. Install the fuel injector into the fuel tube with the following procedure:

   a. Do not reuse the clip, replace it with a new one.

   b. Insert the new clip into the clip mounting groove on fuel injector.

   c. Insert the clip so that projection of fuel injector matches notch of the clip.

   d. Fuel tube side: black O-ring

   e. Nozzle side: green O-ring

11. Insert fuel injector into fuel tube with clip attached.

   a. Insert it while matching it to the axial center.

   b. Insert fuel injector so that projection of fuel tube matches notch of the clip.

   c. Make sure that fuel tube flange is securely fixed in flange fixing groove on the clip.

   d. Make sure that installation is complete by checking that fuel injector does not rotate or come off.

12. Install fuel tube assembly.

   a. Insert the tip of each fuel injector into intake manifold.

   b. Tighten the bolts in two steps in the numerical order as shown.

   c. Step 1: 7 ft. lbs. (10 Nm)

   d. Step 2: 16 ft. lbs. (22 Nm)

✳✳ CAUTION

After properly connecting fuel tube assembly to injector and fuel hose, check connection for fuel leakage.

13. Install the intake manifold.

14. Connect the fuel hose quick connector.

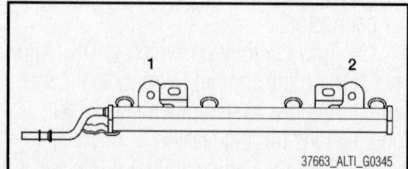

Fig. 211 Tighten the bolts in the order shown

15. Installation of the remaining components is in the reverse order of removal.

16. Make sure there is no fuel leakage at connections as follows:

　a. Apply fuel pressure to fuel lines by turning ignition switch ON (with engine stopped). Then check for fuel leaks at connections.

　b. Start the engine and rev it up and check for fuel leaks at connections.

17. Perform procedures for "Throttle Valve Closed Position Learning" after finishing repairs.

18. If electric throttle control actuator is replaced, perform procedures for "Idle Air Volume Learning" after finishing repairs.

✷✷ WARNING

Do not touch engine immediately after stopping as engine is extremely hot.

➡**Use mirrors for checking on connections out of the direct line of sight.**

3.5L Engine

See Figures 212 through 217.

✷✷ WARNING

Note the following:

- Put a "CAUTION: FLAMMABLE" sign in the workshop.
- Be sure to work in a well ventilated area and furnish workshop with a CO_2 fire extinguisher.
- Never smoke while servicing fuel system. Keep open flames and sparks away from the work area.
- To avoid the danger of being scalded, never drain engine coolant when engine is hot.

1. Remove engine cover.

2. Release the fuel pressure.

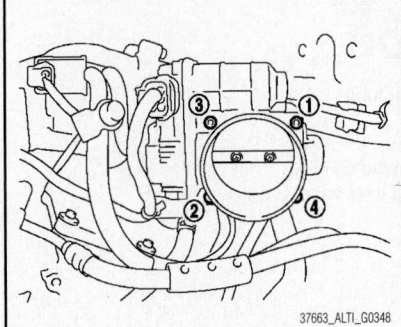

Fig. 212 Remove the electric throttle control actuator bolts in the reverse order as shown

3. Disconnect the battery negative terminal.

4. Remove front wiper arm and extension cowl top.

5. Remove the electric throttle control actuator bolts in the reverse order as shown and remove the electric throttle control actuator.

✷✷ CAUTION

Handle carefully to avoid any shock to the electric throttle control actuator. Do not disassemble.

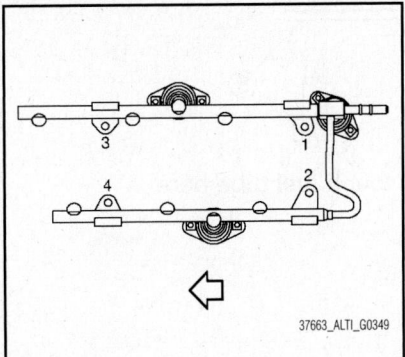

Fig. 213 Loosen bolts in reverse order as shown

6. Remove intake manifold collector.

7. When separating fuel feed hose and fuel tube connection, disconnect quick connector as follows:

　a. Remove quick connector cap from quick connector.

　b. Disconnect quick connector from fuel tube as follows:

✷✷ CAUTION

Disconnect quick connector by using the quick connector release (commercial service tool: J-45488, not by picking out retainer tabs.

8. With the sleeve side of quick connector release facing to quick connector, install the quick connector release onto fuel tube.

9. Insert the quick connector release into quick connector until sleeve contacts and goes no further. Hold quick connector release on that position.

✷✷ CAUTION

Inserting quick connector release hard will not disconnect quick connector. Hold quick connector release where it contacts and goes no further.

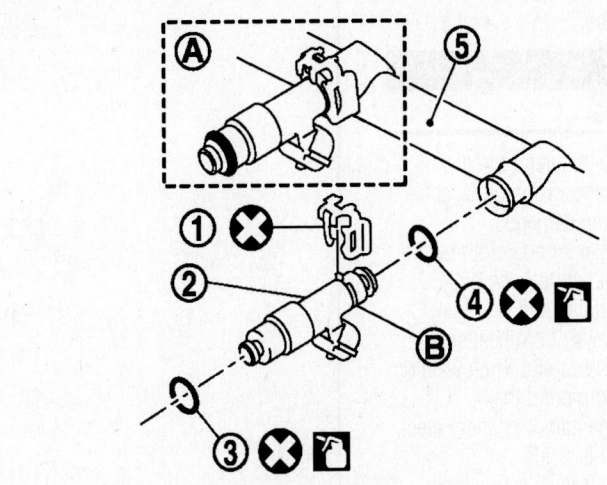

1. Clip
2. Fuel injector
3. O-ring (green)
4. O-ring (black)
5. Fuel tube
A. Installed condition
B. Clip groove

Fig. 214 Remove fuel injector from fuel tube

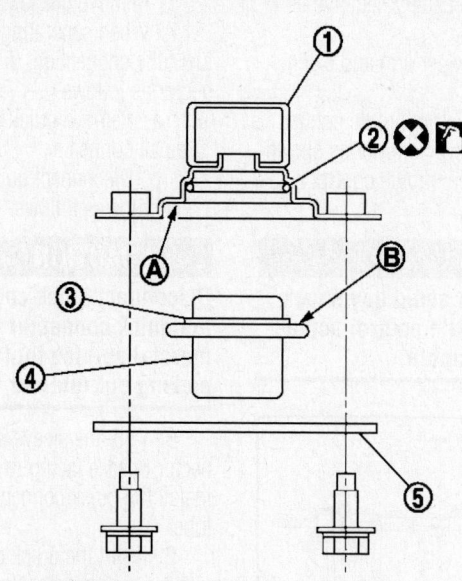

A. Fuel damper rim must touch fuel tube here
B. Fuel damper rim
1. Fuel tube
2. O-ring
3. Spacer
4. Fuel damper
5. Fuel damper cap

37663_ALTI_G0351

Fig. 215 Install fuel damper

10. Draw and pull out quick connector straight from fuel tube.

※※ CAUTION
Note the following:

- Never pull with lateral force applied. O-ring inside quick connector may be damaged.
- Prepare container and cloth beforehand as fuel will leakage out.
- Avoid fire and sparks.
- Keep parts away from heat source. Especially, be careful when welding is performed around them.
- Never expose parts to battery electrolyte or other acids.
- Never bend or twist connection between quick connector and fuel feed hose (with damper) during installation/removal.
- To keep clean the connecting portion and to avoid damage and foreign materials, cover them completely with plastic bags or something similar.

11. Disconnect harness connector from fuel injector.

12. Loosen bolts in reverse order as shown, and remove fuel tube and fuel injector assembly.

※※ CAUTION
Never tilt fuel tube, or remaining fuel in pipes may flow out from pipes.

13. Remove fuel injector from fuel tube as follows:
   a. Open and remove clip.
   b. Remove fuel injector from fuel tube by pulling straight.

※※ CAUTION
Note the following:

- Be careful with remaining fuel that may go out from fuel tube.
- Be careful not to damage injector nozzle during removal.
- Never bump or drop fuel injector.
- Never disassemble fuel injector.

14. Remove fuel damper from fuel tube.

To install:

15. Install fuel damper as follows:
   a. Install new O-ring to fuel tube as shown. When handling new O-ring, be careful of the following caution:

※※ CAUTION
Note the following:

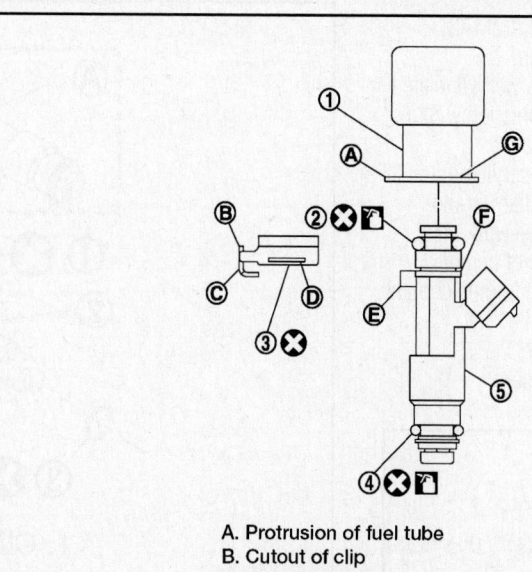

A. Protrusion of fuel tube
B. Cutout of clip
C. Cutout of clip
D. Flange fixing groove on clip
E. Protrusion of fuel injector
F. Clip groove
G. Fuel tube flange
1. Fuel tube
2. O-ring: Black
3. Clip
4. O-ring: Green
5. Fuel injector

37663_ALTI_G0352

Fig. 216 Install fuel injector to fuel tube

- Handle O-ring with bare hands. Never wear gloves.
- Lubricate O-ring with new engine oil.
- Never clean O-ring with solvent.
- Check that O-ring and its mating part are free of foreign material.
- When installing O-ring, be careful not to scratch it with tool or fingernails. Also be careful not to twist or stretch O-ring. If O-ring was stretched while it was being attached, never insert it quickly into fuel tube.
- Insert new O-ring straight into fuel tube. Never twist it.

b. Install spacer to fuel damper.

c. Insert fuel damper straight into fuel tube.

✳ CAUTION

Note the following:

- Insert straight, checking that the axis is lined up.
- Never pressure-fit with excessive force.
- Insert fuel damper until it is touching the fuel tube.

d. Tighten bolts evenly in turn.

e. After tightening bolts, check that there is no gap between fuel damper cap and fuel tube.

16. Install new O-rings to fuel injector paying attention to the following.

✳ CAUTION

Note the following:

- Upper and lower O-ring are different. Be careful not to confuse them.
- Fuel tube side: Black
- Nozzle side: Green
- Lubricate O-ring with new engine oil.
- Never clean O-ring with solvent.
- Check that O-ring and its mating part are free of foreign material.
- When installing O-ring, be careful not to scratch it with tool or fingernails. Also be careful not to twist or stretch O-ring. If O-ring was stretched while it was being attached, never insert it quickly into fuel tube.
- Insert O-ring straight into fuel injector. Never decenter or twist it.

17. Install fuel injector to fuel tube as follows:

a. Insert clip into clip groove on fuel injector.

b. Insert clip so that protrusion of fuel injector matches cutout of clip.

✳ CAUTION

Never reuse clip. Replace it with new one. Be careful to keep clip from interfering with O-ring. If interference occurs, replace O-ring.

c. Insert fuel injector into fuel tube with clip attached.

d. Insert it while matching it to the axial center.

e. Insert fuel injector so that protrusion of fuel tube matches cutout of clip.

f. Check that fuel tube flange is securely fixed in flange fixing groove on clip.

g. Check that installation is complete by checking that fuel injector does not rotate or come off.

h. Check that protrusions of fuel injectors and fuel tubes are aligned with cutouts of clips after installation.

18. Install fuel tube and fuel injector assembly to intake manifold.

✳ CAUTION

Be careful not to let tip of injector nozzle come in contact with other parts.

19. Tighten bolts in two steps in numerical order as shown.

a. Step 1: 7 ft. lbs. (10 Nm)

b. Step 2: 16 ft. lbs. (22 Nm)

20. Connect fuel injector harness.

21. Install intake manifold collector.

22. Connect quick connector between fuel feed hose and fuel tube connection with the following procedure:

a. Check no foreign substances are deposited in and around fuel tube and quick connector, and no damage on them.

b. Thinly apply new engine oil around fuel tube from tip end to spool end.

c. Align center to insert quick connector straightly into fuel tube.

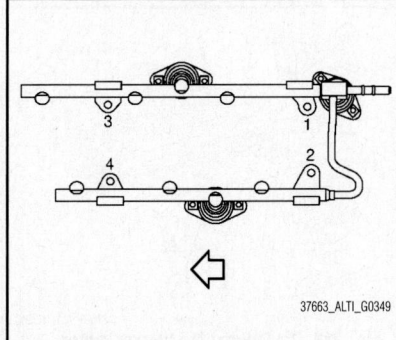

37663_ALTI_G0349

Fig. 217 Tighten bolts in two steps in numerical order

d. Insert quick connector to fuel tube until top spool is completely inside quick connector, and 2nd level spool exposes right below quick connector.

✳ CAUTION

Note the following:

- Hold (A) position as shown in the figure when inserting fuel tube into quick connector.
- Carefully align center to avoid inclined insertion to prevent damage to O-ring inside quick connector.
- Insert until you hear a "click" sound and actually feel the engagement.
- To avoid misidentification of engagement with a similar sound, be sure to perform the next step.

e. Pull quick connector by hand holding position. Check it is completely engaged (connected) so that it does not come out from fuel tube.

f. Install quick connector cap to quick connector.

g. Install quick connector cap with arrow on surface facing in direction of quick connector (fuel feed hose side).

✳ CAUTION

If quick connector cap cannot be installed smoothly, quick connector may have not been installed correctly. Check connection again.

h. Secure fuel feed hose to clamp of quick connector cap.

23. Installation is in the reverse order of removal.

24. Make sure there is no fuel leakage at connections as follows:

a. Apply fuel pressure to fuel lines by turning ignition switch ON (with engine stopped). Then check for fuel leaks at connections.

b. Start the engine and rev it up and check for fuel leaks at connections.

✳ CAUTION

Do not touch engine immediately after stopping as engine is extremely hot.

➡Use mirrors for checking on connections out of the direct line of sight.

25. Perform procedures for "Throttle Valve Closed Position Learning" after finishing repairs.

26. If electric throttle control actuator is replaced, perform procedures for "Idle Air Volume Learning" after finishing repairs.

FUEL TANK

REMOVAL & INSTALLATION

See Figures 205 and 206.

1. Disconnect the battery negative terminal.

2. Check the fuel level with the vehicle on a level surface. If the fuel gauge indicates more than the level (7/8 full), drain the fuel from the fuel tank until the fuel gauge indicates a level at or below (7/8 full).

3. In case the fuel pump does not operate, use the following procedure.

   a. Insert fuel tubing of less than 0.98 inches (25 mm) diameter into the fuel filler tube through the fuel filler opening to drain fuel from the fuel filler tube.

   b. Disconnect the fuel filler hose from the fuel filler tube.

   c. Insert fuel tubing into the fuel tank through the fuel filler hose to drain fuel from the fuel tank.

   d. As a guide, the fuel level reaches or is less than the level on the fuel gauge as shown, when approximately 2 5/8 US gal. (10L) of fuel is drained from a full fuel tank.

4. Open the fuel filler cap to release the pressure inside the fuel tank.

5. Release fuel pressure from fuel line.

6. Remove rear seat bottom.

7. Turn the four retainers 90° in a clockwise direction and remove the fuel pump inspection hole cover.

8. Disconnect the fuel level sensor, fuel filter, and fuel pump assembly electrical connector, EVAP hose quick connector, and the fuel feed hose quick connector from the fuel level sensor unit, fuel filter, and fuel pump assembly.

9. Disconnect the quick connectors as follows:

   a. Hold the sides of the connector, push in tabs and pull out the tube.

   b. If the connector and the tube are stuck together, push and pull several times until they start to move. Then disconnect them by pulling.

✳✳ CAUTION
Note the following:

- The tube can be removed when the tabs are completely depressed. Do not twist it more than necessary.
- Do not use any tools to remove the quick connector.
- Keep the resin tube away from heat. Be especially careful when welding near the tube.
- Prevent acid liquid such as battery electrolyte, from getting on the resin tube.
- Do not bend or twist the tube during installation and removal.
- Only when the tube is replaced, remove the remaining retainer on the tube or fuel level sensor, fuel filter, and fuel pump assembly.
- When the tube or fuel level sensor, fuel filter, and fuel pump assembly is replaced, also replace the retainer with a new one (green colored retainer).
- To keep the connecting portion clean and to avoid damage and foreign materials, cover them completely with plastic bags or something similar.

10. Remove the center exhaust tube, with muffler(s).

11. Disconnect the fuel filler hose and the recirculation hose at the fuel tank side.

12. Disconnect the three parking brake cable mounting brackets on each cable and position the cables out of the way.

13. Remove the fuel tank protector.

14. Disconnect the fuel tank mounting straps while supporting the fuel tank.

15. Remove the fuel tank.

16. If replacing the fuel tank, remove the fuel level sensor, fuel filter and fuel pump assembly to transfer to the new fuel tank.

To install:

17. Install in the reverse order of removal paying attention to the following.

18. Before tightening the fuel tank mounting straps, temporarily install the filler hose and the recirculation hose.

19. Tighten all fuel tank mounting strap bolts to specification, then tighten the hose clamps.

20. Connect the quick connector as follows:

   a. Check the connection for damage or any foreign materials.

   b. Align the connector with the tube, then insert the connector straight into the tube until a click is heard.

21. After the tube is connected, make sure the connection is secure by performing the following checks:

   a. Pull on the tube and the connector to make sure they are securely connected.

   b. Visually confirm that the two retainer tabs are connected to the quick connector.

22. Use the following procedure to check for fuel leaks.

   a. Turn the ignition switch ON (without starting the engine). Then check the connections for fuel leaks by applying fuel pressure to the fuel piping.

   b. Run the engine and check for fuel leaks at the fuel system tube and hose connections.

IDLE SPEED

ADJUSTMENT

The idle speed is controlled by the ECM. No adjustment is necessary or possible.

HEATING & AIR CONDITIONING SYSTEM

BLOWER MOTOR

REMOVAL & INSTALLATION

See Figure 218.

1. Remove the glove box.

2. Disconnect the blower motor connector.

3. Remove the three blower motor screws and remove the blower motor from the blower unit.

To install:

4. Installation is in the reverse order of removal.

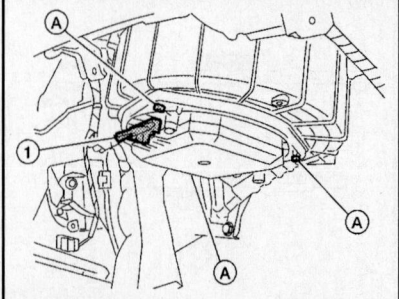

37663_ALTI_G0362

Fig. 218 Disconnect the blower motor connector (1); remove the three blower motor screws (A)

HEATER CORE

REMOVAL & INSTALLATION

1. Remove the heater and cooling unit assembly.

2. Remove the heater grommet, heater pipe support and heater pipe cover.

3. Remove the heater and cooling unit foot duct LH.

4. Remove the heater core.

To install:

5. Installation is in the reverse order of removal.

→Make sure that the aspirator hose is securely attached to the aspirator on the heater and cooling unit foot duct LH.

HEATER & COOLING UNIT ASSEMBLY

REMOVAL & INSTALLATION

1. Discharge the refrigerant from the A/C system.
2. Drain the engine coolant from the cooling system.
3. Disconnect the negative battery terminal.
4. Remove the wiper motor and linkage.
5. Remove the upper cowl (for VQ35DE only).
6. Remove the strut tower bar (for VQ35DE only).

7. Remove the lower RH cowl (for VQ35DE only).
8. Disconnect the heater hoses from the heater core pipes.

※※ CAUTION
Cap or wrap the pipe joint with a suitable material such as vinyl tape to avoid the entry of contaminants into the system.

9. Disconnect the refrigerant lines from the expansion valve.

※※ CAUTION
Cap or wrap the line joint with a suitable material such as vinyl tape to avoid the entry of contaminants into the system.

10. Remove the instrument panel assembly.
11. Remove the steering column assembly.
12. Disconnect the drain hose.
13. Remove the heater and cooling unit assembly attached to the steering member as one assembly from the vehicle.
14. Remove the blower unit from the heater and cooling unit and steering member assembly.
15. Remove the heater and cooling unit from the steering member.

To install:
16. Installation is in the reverse order of removal.
17. Fill the radiator with the specified water and coolant mixture.
18. Recharge the A/C system.

STEERING

POWER RACK & PINION STEERING GEAR

REMOVAL & INSTALLATION
See Figures 219 through 222.

1. Remove the front tires.
2. Remove undercover.
3. Remove lower side bolt of lower joint.
4. Remove cotter pin and loosen the nut.
5. Remove steering outer socket from steering knuckle so as not to damage ball

joint boot using the Tool HT72520000 (J-25730-A).

※※ CAUTION
Temporarily tighten the nut to prevent damage to threads and to prevent the Tool from suddenly coming off.

6. Remove high and low pressure piping of hydraulic piping, and then drain power steering fluid.
7. Remove steering hydraulic piping bracket from front suspension member.

8. Remove SSPS valve harness connector.
9. Remove bolts and nuts of steering gear assembly, and then remove steering gear assembly from vehicle.
10. Check for fluid leaks or damage to steering gear. If any exist, replace steering gear as an assembly.

To install:
11. Installation is in the reverse order of removal.
12. When installing lower joint to steering gear assembly, follow the procedure listed below.

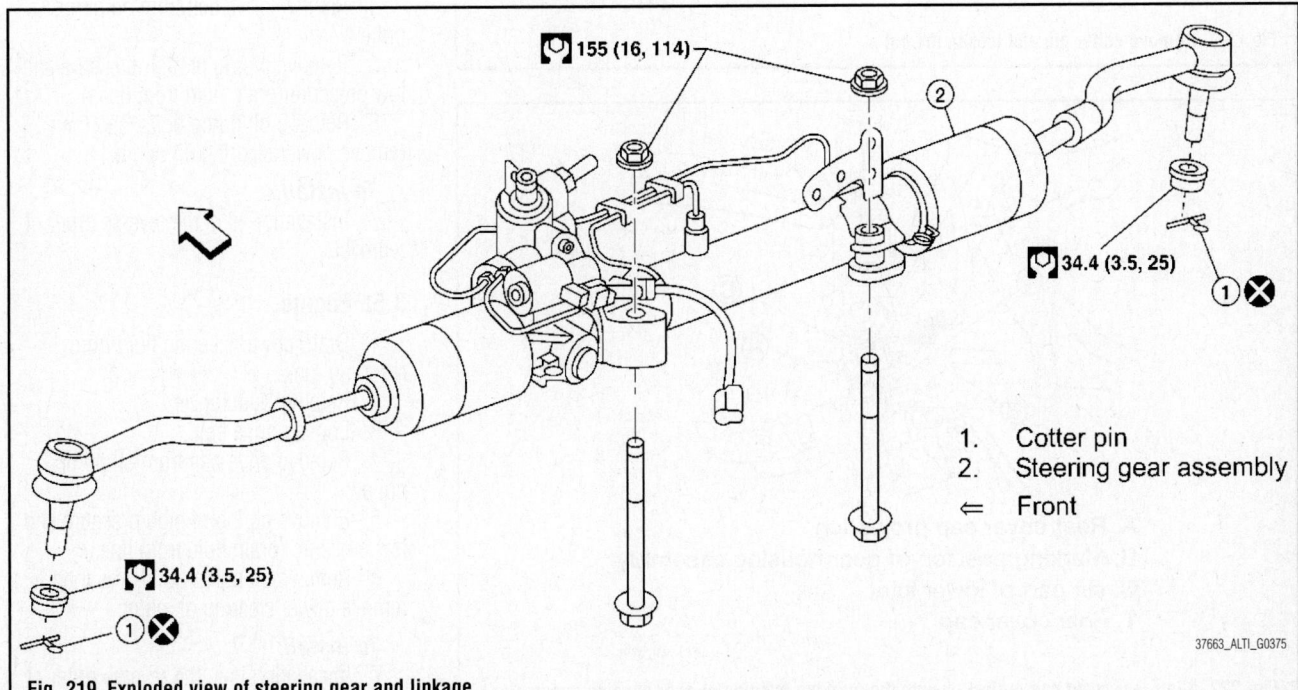

1. Cotter pin
2. Steering gear assembly
⇐ Front

Fig. 219 Exploded view of steering gear and linkage

37663_ALTI_G0375

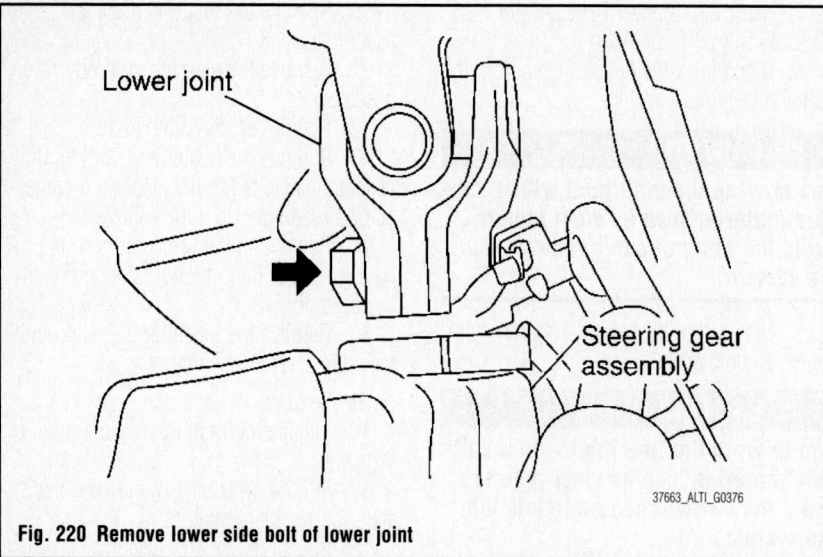

Fig. 220 Remove lower side bolt of lower joint

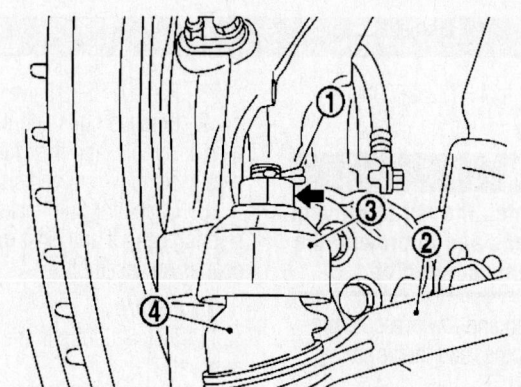

1. Cotter pin   3. Steering knuckle
2. Steering outer socket 4. Ball joint boot

Fig. 221 Remove cotter pin and loosen the nut

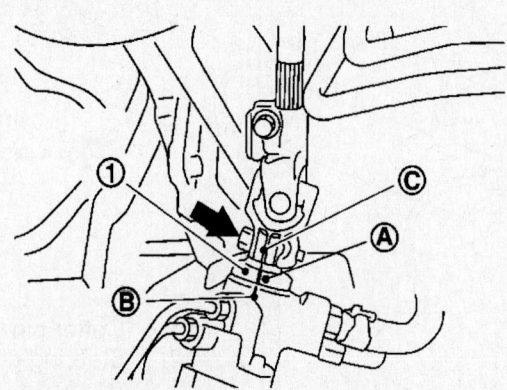

A. Rear cover cap projection
B. Marking position of gear housing assembly
C. slit part of lower joint
1. Rear cover cap

Fig. 222 Align rear cover cap projection with the marking position of gear housing assembly

a. Set rack of steering gear in the neutral position.

➡To get the neutral position of rack, turn gear-sub assembly and measure the distance of inner socket, and then measure the intermediate position of the distance.

b. Align rear cover cap projection with the marking position of gear housing assembly.

c. Install slit part of lower joint aligning with the projection of rear cover cap. Make sure that the slit part of lower joint is aligned with both the projection of rear cover cap and the marking position of gear housing assembly.

13. After installation, bleed air from the steering hydraulic system.

14. Perform final tightening of nuts and bolts on each part under unladen conditions with tires on level ground when removing steering gear assembly. Check wheel alignment.

15. Make sure that steering wheel operates smoothly by turning several times from full left stop to full right stop.

POWER STEERING PUMP

REMOVAL & INSTALLATION

2.5L Engine

1. Drain power steering fluid from reservoir tank.
2. Remove undercover.
3. Loosen drive belt.
4. Remove drive belt from oil pump pulley.
5. Remove piping of high pressure and low pressure (drain fluid from lines).
6. Remove oil pump bolts, and then remove power steering oil pump.

To install:

7. Installation is in the reverse order of removal.

3.5L Engine

1. Drain power steering fluid from reservoir tank.
2. Remove undercover.
3. Loosen drive belt.
4. Remove drive belt from oil pump pulley.
5. Remove piping of high pressure and low pressure (drain fluid from lines).
6. Remove oil pump bolts, and then remove power steering oil pump.

To install:

7. Installation is in the reverse order of removal.

LOWER CONTROL ARM (TRANSVERSE LINK)

REMOVAL & INSTALLATION

➡Nissan refers to the Lower Control Arm as Transverse Link. Refer to the exploded view of front suspension while servicing this component.

1. Remove wheel and tire.
2. Remove steering knuckle from transverse link.
3. Remove mounting nuts and washers on lower portion of stabilizer connecting rod.
4. Slightly loosen transverse link mounting bolts.
5. Remove transverse link bolts and nuts, and remove transverse link from suspension member.

To install:
6. Installation is in the reverse order of removal.
7. Refer to "Exploded View" for tightening torque.
8. Tighten transverse link bolts with vehicle unladen and all four tires on flat, level ground.
9. After installation, check wheel alignment.

STEERING KNUCKLE

REMOVAL & INSTALLATION

Refer to Wheel Hub, Steering Knuckle & Bearing (sealed unit) Section for Removal and Installation of the Steering Knuckle.

STRUT

REMOVAL & INSTALLATION

See Figure 223.

➡Refer to the exploded view of front suspension while servicing this component.

1. Remove wheel and tire.
2. Remove brake caliper and reposition aside using wire.

❊❊ CAUTION

Avoid depressing brake pedal with brake caliper removed.

3. Remove wheel sensor electrical harness from strut.
4. Remove brake hose lock plate.
5. Remove steering knuckle to strut bolts and nuts.

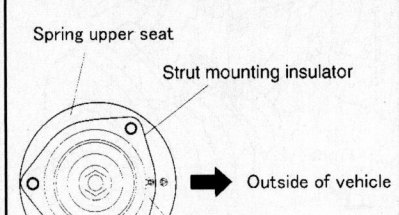

Fig. 223 Assemble upper mounting plate with its notch facing toward the outside

6. Remove bolt on strut tower bar then bolts on strut tower and remove strut from vehicle.
7. Check the strut for any oil leakage or other damage and replace as necessary.

To install:
8. Installation is in the reverse order of removal.
9. Refer to "Exploded View" for tightening torque.
10. Be sure arrows on strut mount insulator and spring upper seat are positioned as shown. Also be sure notch in strut spacer is positioned as shown. Then install strut.
11. Assemble upper mounting plate with its notch facing toward the outside.

STABILIZER BAR

REMOVAL & INSTALLATION

See Figures 224 through 226.

1. Remove steering gear.
2. Remove mounting nuts on upper portion of stabilizer connecting rod.
3. Remove stabilizer clamp bolts.
4. Remove stabilizer from the vehicle.
5. Check stabilizer, connecting rod, bushing and clamp for deformation, cracks and damage, and replace if necessary.

To install:
6. Installation is in the reverse order of removal.
7. Refer to "Exploded View" for tightening torque.
8. When installing stabilizer, make sure that the clamps are facing in the direction shown.
9. Make sure the cut surface of the bushing faces the rear.
10. Stabilizer uses pillow ball type connecting rod. Position ball joint with case on pillow ball head parallel to stabilizer.

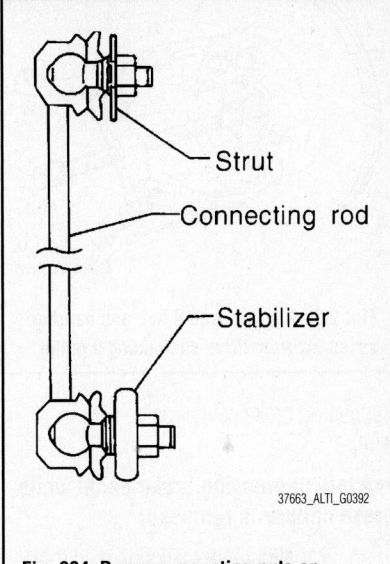

Fig. 224 Remove mounting nuts on upper portion of stabilizer connecting rod

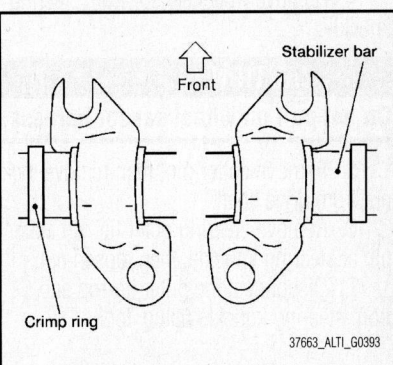

Fig. 225 When installing stabilizer, make sure that the clamps are facing in the direction shown

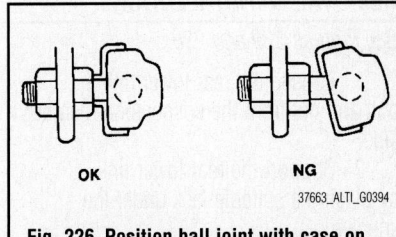

Fig. 226 Position ball joint with case on pillow ball head parallel to stabilizer

WHEEL HUB, STEERING KNUCKLE & BEARING

REMOVAL & INSTALLATION

See Figures 227 and 228.

1. Remove wheel and tire from vehicle.
2. Remove brake caliper, leaving brake caliper hydraulic lines connected.

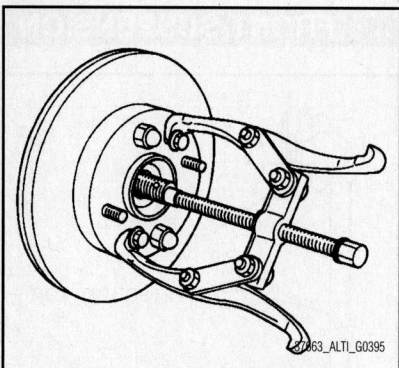

Fig. 227 Remove wheel hub and bearing assembly from drive shaft using a puller

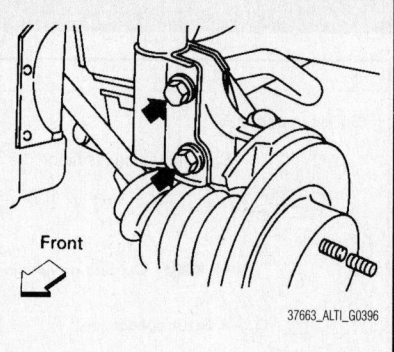

Front

Fig. 228 Remove the lower strut bolts and nuts

Reposition brake caliper aside with wire.

➡**Avoid depressing brake pedal while brake caliper is removed.**

3. Put alignment marks on disc rotor and wheel hub and bearing assembly, then remove disc rotor.

4. Remove wheel sensor from steering knuckle.

✳✳ CAUTION

Do not pull on wheel sensor harness.

5. Remove cotter pin, then remove lock nut from drive shaft.

6. Remove steering outer tie-rod cotter pin at steering knuckle, then loosen nut.

7. Disconnect the outer tie-rod end from steering knuckle using Tool

HT2520000 (J-25730-A). Be careful not to damage ball joint boot.

✳✳ CAUTION

To prevent damage to threads and to prevent tool from coming off suddenly, temporarily tighten mounting nut.

8. Remove transverse link and steering knuckle pinch bolt and nut.

9. Remove wheel hub and bearing assembly from drive shaft using a puller or suitable tool.

✳✳ CAUTION

When removing wheel hub and bearing assembly, do not apply an excessive angle to drive shaft joint. Also be careful not to excessively extend

slide joint. Support drive shaft when removing.

10. Remove wheel hub and bearing assembly bolts.

11. Remove splash guard and wheel hub and bearing assembly from steering knuckle.

12. Remove the lower strut bolts and nuts.

13. Remove steering knuckle from vehicle.

14. Check for deformity, cracks and damage on each part, replace if necessary.

15. Check for boot breakage, axial looseness, and torque of transverse link ball joint and repair as necessary.

To install:

16. Installation is in the reverse order of removal.

✳✳ CAUTION

Do not reuse non-reusable parts.

17. When installing wheel hub and bearing assembly to steering knuckle, align cutout in toner ring cover with wheel sensor mounting hole in steering knuckle.

18. When installing disc rotor on wheel hub and bearing assembly, align the marks.

ADJUSTMENT

No adjustment is possible. If there is too much play, replace the bearing and/or hub.

SUSPENSION   REAR SUSPENSION

COIL SPRING

REMOVAL & INSTALLATION

See Figures 229 and 230.

1. Loosen the rear lower link bolt and nut from the suspension member side.

2. Support the rear lower link by placing a suitable jack under the knuckle.

3. Remove the rear lower link adjusting bolt and nut from the suspension member side.

➡**Do not reuse the adjusting nut, use a new adjusting nut for installation.**

4. Slowly lower the jack to lower the rear lower link and coil spring.

5. Remove the upper rubber seat, coil spring, and lower rubber seat from the rear lower link.

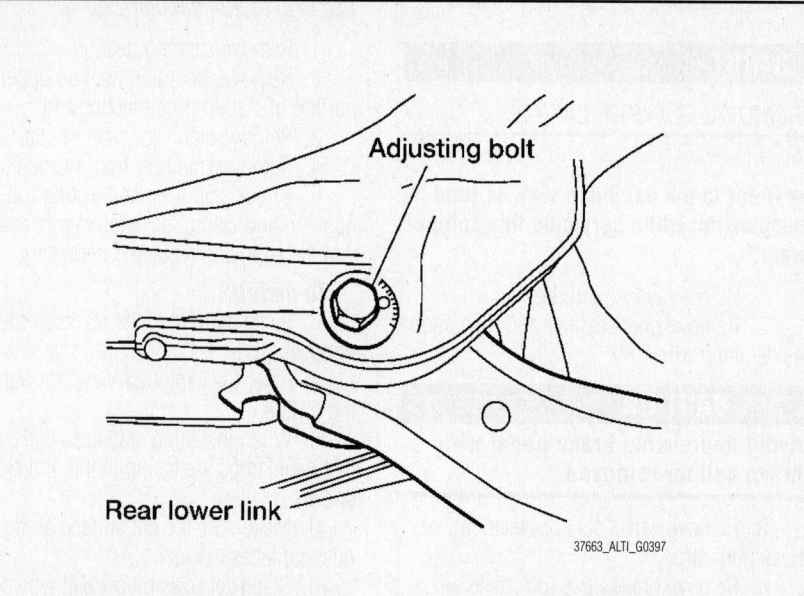

Adjusting bolt

Rear lower link

Fig. 229 Remove the rear lower link adjusting bolt and nut from the suspension member side

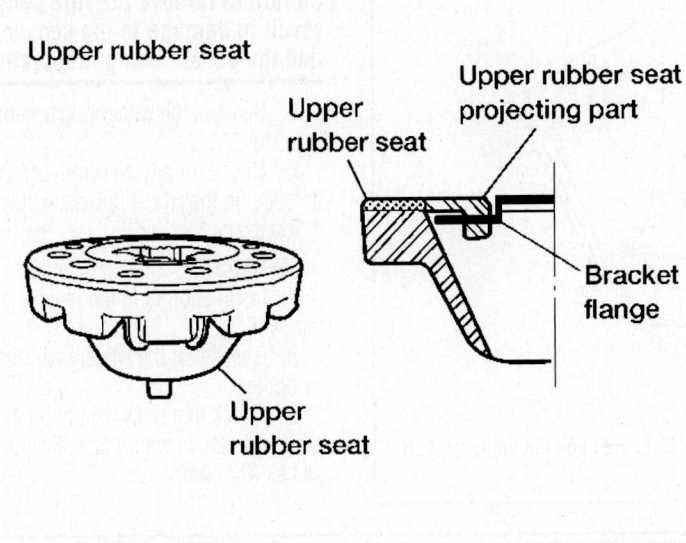

Upper rubber seat

Upper rubber seat

Upper rubber seat projecting part

Upper rubber seat

Bracket flange

Lower rubber seat

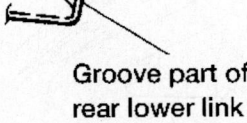

Position the hollow of lower rubber seat

Groove part of rear lower link

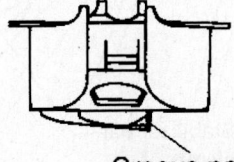

Groove part of rear lower link

37663_ALTI_G0398

Fig. 230 Proper orientation of coil spring mounting components

6. Remove rear lower link bolt and nut from the suspension member side.

7. Remove the rear lower link.

To install:

8. Installation is in the reverse order of removal.

➡ **Do not reuse the adjusting nut, use a new adjusting nut for installation.**

9. Check that the projecting part inside the upper rubber seat and the bracket flange are attached as shown.

10. Check that the projection part outside the upper rubber seat is directed toward the front of the vehicle.

11. Position the hollow of the lower rubber seat with the groove part of the rear lower link.

12. Install the coil spring so that the side with the two paint markers is directed toward the lower side.

13. Check the rear wheel alignment and adjust if necessary.

CONTROL ARMS/LINKS

REMOVAL & INSTALLATION

Front Lower Link

See Figure 231.

1. Remove the front lower link nut and bolt from the knuckle side and the adjusting bolt and nut from the suspension member side.

➡ **Do not reuse the adjusting nut, use a new adjusting nut for installation.**

2. Remove the front lower link.

To install:

3. Installation is in the reverse order of removal.

➡ **Do not reuse the adjusting nut, use a new adjusting nut for installation.**

4. Check the rear wheel alignment and adjust if necessary.

Radius Rod

➡ **Refer to Exploded View of Rear Suspension while servicing this component.**

1. Remove the rear suspension assembly.
2. Remove the radius rod.

To install:

3. Installation is in the reverse order of removal.

4. Check the rear wheel alignment and adjust if necessary.

Rear Lower Link

Refer to Coil Spring Removal and Installation when servicing this component.

Suspension Arm

➡ **Refer to Exploded View of Rear Suspension while servicing this component.**

1. Remove the rear suspension assembly.

2. Remove the connecting rod bracket from the suspension arm.

3. Remove the two suspension arm bolts and nuts from the suspension member side of the suspension arm.

4. Remove the ball joint cotter pin and lock nut.

➡ **Discard the cotter pin, use a new cotter pin for installation.**

5. Remove the suspension arm from the knuckle using Tool HT2520000 (J-25730-A).

✳✳ CAUTION
Do not damage ball joint when removing.

To install:

6. Installation is in the reverse order of removal.

➡ **Discard the cotter pin, use a new cotter pin for installation.**

7. Check the rear wheel alignment and adjust if necessary.

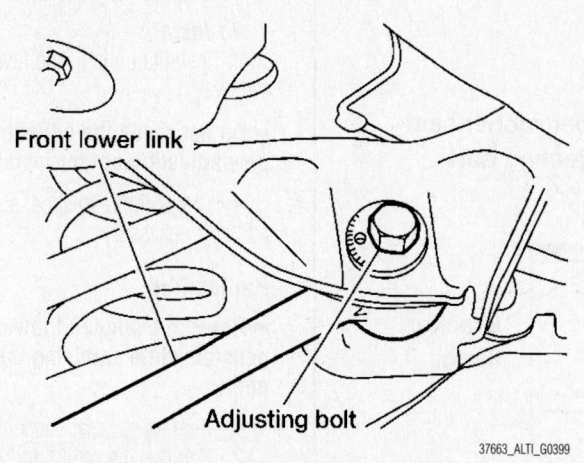

Fig. 231 Remove the front lower link nut and bolt from the knuckle side and the adjusting bolt and nut from the suspension member side

STABILIZER BAR

REMOVAL & INSTALLATION

See Figure 232.

1. Disconnect the stabilizer bar from connecting rod.
2. Remove the stabilizer bar clamps and bushings.
3. Remove the stabilizer bar.

To install:

4. Installation is in the reverse order of removal.

WHEEL HUB & BEARING (SEALED UNIT)

REMOVAL & INSTALLATION

See Figure 233.

❊❊ CAUTION

Wheel hub assembly does not require maintenance. If any of the following symptoms are noted, replace the wheel hub assembly:

- A growling noise is emitted from the wheel hub assembly while driving.
- The wheel hub assembly drags or turns roughly.

1. Remove the rear wheel and tire.
2. Remove the brake caliper assembly.
   a. The brake hose does not need to be disconnected from the brake caliper.
   b. Suspend the brake caliper assembly using wire, do not stretch the brake hose.
   c. Do not depress the brake pedal, or the caliper piston will pop out.
   d. Do not twist the brake hose.

3. Remove the brake rotor.
4. Remove the rear ABS sensor, then move it away from the wheel hub assembly.

❊❊ CAUTION

Failure to remove the ABS sensor may result in damage to the sensor wires and the sensor being inoperative.

5. Remove the wheel hub assembly from knuckle.
6. Check for any deformity, cracks, or damage on the wheel hub assembly, replace if necessary.

To install:

7. Installation is in the reverse order of removal.
8. Check that the wheel bearings operate smoothly.
9. Check that the wheel hub bearing axial end play is within specification: 0.004 inches (0.1 mm).

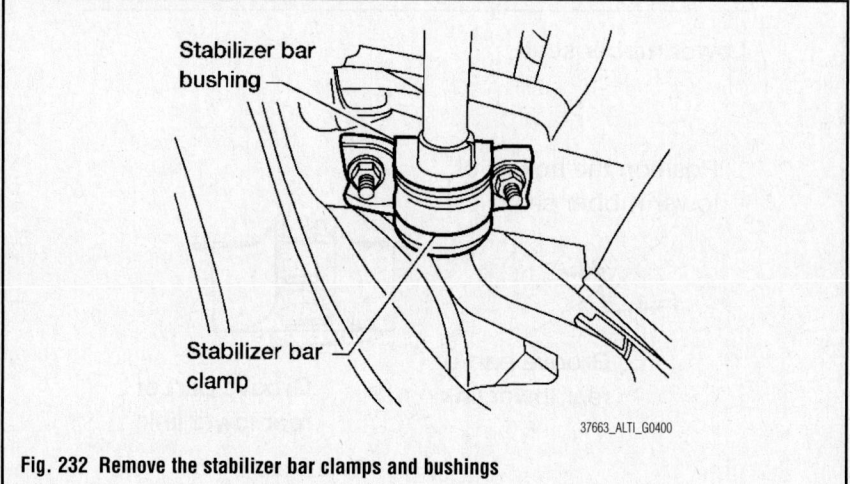

Fig. 232 Remove the stabilizer bar clamps and bushings

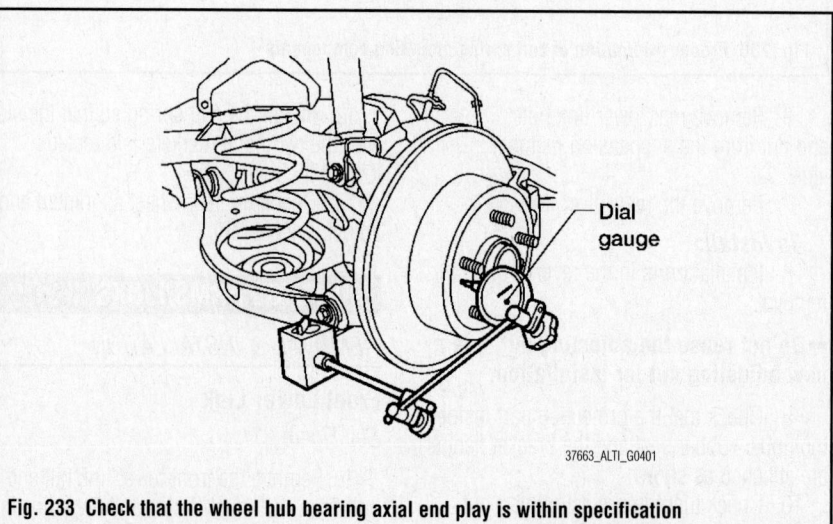

Fig. 233 Check that the wheel hub bearing axial end play is within specification

ADJUSTMENT

No adjustment is possible. If there is too much play, replace the bearing and/or hub.

NISSAN

Altima Hybrid

10

SPECIFICATIONS AND MAINTENANCE CHARTS

ENGINE AND VEHICLE IDENTIFICATION

		Engine							Model Year	
Code ①	Liters (cc)	Cu. In.	Cyl.	Fuel Sys.	Engine Type	Eng. Mfg.		Code ②	Year	
QR25DE	2.5 (2488)	152	4	MFI	DOHC	Nissan		7	2007	
								8	2008	
								9	2009	
								A	2010	

MFI: Multi-port Fuel Injection

DOHC: Double Overhead Camshaft

① The Engine Code is stamped on the engine block near the starter.

② 10th position of the Vehicle Identification Number (VIN)

37663_ALTH_C0001

GENERAL ENGINE SPECIFICATIONS

Year	Model	Engine Displacement Liters	Engine Series (ID/VIN)	Net Horsepower @ rpm	Net Torque @ rpm (ft. lbs.)	Bore x Stroke (in.)	Com-pression Ratio	Oil Pressure @ rpm
2007	Altima Hybrid	2.5	QR25DE	158@6200	162@2800	3.50X3.94	9.5:1	43@2000
2008	Altima Hybrid	2.5	QR25DE	158@6200	162@2800	3.50X3.94	9.5:1	43@2000
2009	Altima Hybrid	2.5	QR25DE	158@6200	162@2800	3.50X3.94	9.5:1	43@2000
2010	Altima Hybrid	2.5	QR25DE	158@6200	162@2800	3.50X3.94	9.5:1	43@2000

37663_ALTH_C0002

ENGINE TUNE-UP SPECIFICATIONS

Year	Engine Displacement Liters	Engine ID/VIN	Spark Plug Gap (in.)	Ignition Timing (deg.) MT	AT	Fuel Pump (psi) ①	Idle Speed (rpm) MT	AT ②	Valve Clearance Intake ③	Exhaust ③
2007	2.5	QR25DE	0.043	NA	15B	51	NA	650-750	0.009-0.013	0.010-0.013
2008	2.5	QR25DE	0.043	NA	15B	51	NA	650-750	0.009-0.013	0.010-0.013
2009	2.5	QR25DE	0.043	NA	15B	51	NA	650-750	0.009-0.013	0.010-0.013
2010	2.5	QR25DE	0.043	NA	15B	51	NA	650-750	0.009-0.013	0.010-0.013

B: Before top dead center

① At idle

② Automatic transmission in neutral

③ Engine cold

37663_ALTH_C0003

CAPACITIES

Year	Model	Engine ID/VIN	Engine Displacement Liters	Engine Oil with Filter (qts.)	Transmission (pts.) 5-Spd	Transmission (pts.) Auto.	Drive Axle Rear (pts.)	Fuel Tank (gal.)	Cooling System (qts.)
2007	Altima Hybrid	QR25DE	2.5	4.8	—	8.66	—	20.0	①
2008	Altima Hybrid	QR25DE	2.5	4.8	—	8.66	—	20.0	①
2009	Altima Hybrid	QR25DE	2.5	4.8	—	8.66	—	20.0	②
2010	Altima Hybrid	QR25DE	2.5	4.8	—	8.66	—	20.0	②

① 2007-2008: Engine coolant, 8 qts.: Inverter coolant, 3 qts.

② 2009-2010: Engine coolant, 8-1/8 qts.: Inverter coolant, 3-3/8 qts.

37663_ALTH_C0004

FLUID SPECIFICATIONS

Year	Model	Engine Displ. Liters	Engine Oil ①	Man. Trans.	Auto. Trans.	Drive Axle Front	Drive Axle Rear	Transfer Case	Power Steering Fluid	Brake Master Cylinder	Cooling System
2007	Altima	2.5	0W-20	NA	②	NA	NA	NA	NA	DOT 3	N-LL
2008	Altima	2.5	0W-20	NA	②	NA	NA	NA	NA	DOT 3	N-LL
2009	Altima	2.5	0W-20	NA	②	NA	NA	NA	NA	DOT 3	N-LL
2010	Altima	2.5	0W-20	NA	②	NA	NA	NA	NA	DOT 3	N-LL

NA: Not Applicable

N-LL: Nissan Long Life coolant

① For warm and hot climates, if 0W-20 is not available, 5W-20 or 5W-30 is applicable.

② Genuine NISSAN Matic W ATF; Using transaxle fluid other than Genuine NISSAN Matic W ATF will damage CVT.

37663_ALTH_C0005

VALVE SPECIFICATIONS

Year	Engine ID/VIN	Engine Displacement Liters	Seat Angle (deg.)	Face Angle (deg.)	Spring Test Pressure (lbs. @ in.)	Spring Installed Height (in.)	Stem-to-Guide Clearance (in.) Intake	Stem-to-Guide Clearance (in.) Exhaust	Stem Diameter (in.) Intake	Stem Diameter (in.) Exhaust
2007	QR25DE	2.5	45.15-45.45	NA	34-39@1.39	1.390	0.0008-0.0021	0.0012-0.0025	0.2348-0.2354	0.2344-0.2350
2008	QR25DE	2.5	45.15-45.45	NA	34-39@1.39	1.390	0.0008-0.0021	0.0012-0.0025	0.2348-0.2354	0.2344-0.2350
2009	QR25DE	2.5	45.15-45.45	NA	34-39@1.39	1.390	0.0008-0.0021	0.0012-0.0025	0.2348-0.2354	0.2344-0.2350
2010	QR25DE	2.5	45.15-45.45	NA	34-39@1.39	1.390	0.0008-0.0021	0.0012-0.0025	0.2348-0.2354	0.2344-0.2350

NA: Not Available

37663_ALTH_C0006

CAMSHAFT AND BEARING SPECIFICATIONS

All measurements are given in inches.

Year	Engine Displacement Liters	Engine ID/VIN	Journal Diameter	Brg. Oil Clearance	Shaft End-play	Runout	Journal Bore	Lobe Lift Intake	Lobe Lift Exhaust
2007	2.5	QR25DE	①	0.0018-0.0034	0.045-0.0074	0.0059	②	1.7644-1.7718	1.7313-1.7388
2008	2.5	QR25DE	①	0.0018-0.0034	0.045-0.0074	0.0059	②	1.7644-1.7718	1.7313-1.7388
2009	2.5	QR25DE	①	0.0018-0.0034	0.045-0.0074	0.0059	②	1.7644-1.7718	1.7313-1.7388
2010	2.5	QR25DE	①	0.0018-0.0034	0.045-0.0074	0.0059	②	1.7644-1.7718	1.7313-1.7388

① No. 1: 1.0998-1.1006 inches
  Nos. 2, 3, 4, 5: 0.9226-0.9234 inches

② No. 1: 1.1024-1.1032 inches
  Nos. 2, 3, 4, 5: 0.9252-0.9260 inches

37663_ALTH_C0007

CRANKSHAFT AND CONNECTING ROD SPECIFICATIONS

All measurements are given in inches.

Year	Engine Displacement Liters	Engine ID/VIN	Crankshaft Main Brg. Journal Dia.	Crankshaft Main Brg. Oil Clearance	Crankshaft Shaft End-play	Thrust on No.	Connecting Rod Journal Diameter	Connecting Rod Oil Clearance	Connecting Rod Side Clearance
2007	2.5	QR25DE	2.1636-2.1645	①	0.0039-0.0102	3	1.8898-1.8903	0.0014-0.0018	0.0079-0.0138
2008	2.5	QR25DE	2.1636-2.1645	①	0.0039-0.0102	3	1.8898-1.8903	0.0014-0.0018	0.0079-0.0138
2009	2.5	QR25DE	2.1636-2.1645	①	0.0039-0.0102	3	1.8898-1.8903	0.0014-0.0018	0.0079-0.0138
2010	2.5	QR25DE	2.1636-2.1645	①	0.0039-0.0102	3	1.8898-1.8903	0.0014-0.0018	0.0079-0.0138

① Nos. 1, 3, 5 : 0.0005-0.0009
  Nos. 2, 4 : 0.0007-0.0011

37663_ALTH_C0008

PISTON AND RING SPECIFICATIONS

All measurements are given in inches.

Year	Engine Displacement Liters	Engine ID/VIN	Piston Clearance	Ring Gap Top Compression	Ring Gap Bottom Compression	Ring Gap Oil Control	Ring Side Clearance Top Compression	Ring Side Clearance Bottom Compression	Ring Side Clearance Oil Control
2007	2.5	QR25DE	0.0004-0.0012	0.0083-0.0122	0.0146-0.0205	0.0079-0.0177	0.0016-0.0031	0.0012-0.0028	0.0018-0.0049
2008	2.5	QR25DE	0.0004-0.0012	0.0083-0.0122	0.0146-0.0205	0.0079-0.0177	0.0016-0.0031	0.0012-0.0028	0.0018-0.0049
2009	2.5	QR25DE	0.0004-0.0012	0.0083-0.0122	0.0146-0.0205	0.0079-0.0177	0.0016-0.0031	0.0012-0.0028	0.0018-0.0049
2010	2.5	QR25DE	0.0004-0.0012	0.0083-0.0122	0.0146-0.0205	0.0079-0.0177	0.0016-0.0031	0.0012-0.0028	0.0018-0.0049

37663_ALTH_C0009

TORQUE SPECIFICATIONS
All readings in ft. lbs.

Year	Engine Displacement Liters	Engine ID/VIN	Cylinder Head Bolts	Main Bearing Bolts	Rod Bearing Bolts	Crankshaft Damper Bolts	Flywheel Bolts	Manifold		Spark Plugs	Oil Drain Plug
								Intake	Exhaust		
2007	2.5	QR25DE	①	②	③	④	80	14	31	18	25
2008	2.5	QR25DE	①	②	③	④	80	14	31	18	25
2009	2.5	QR25DE	①	②	③	④	80	14	31	18	25
2010	2.5	QR25DE	①	②	③	④	80	14	31	18	25

① Step 1: 72 ft. lbs.

  Step 2: Loosen completely, then retorque to 29 ft. lbs.

  Step 3: Turn each bolt, in sequence, an additional 75 degrees

  Step 4: Turn each bolt, in sequence, an additional 75 degrees

② Tighten bolts 11-22 to 19 ft. lbs.

  Step 2: Tighten bolts 1-10 to 29 ft. lbs.

  Step 3: Tighten bolts 1-10 an additional 60 degrees

③ Step 1: Tighten bolts to 20 ft. lbs.

  Step 2: Loosen all bolts

  Step 3: tighten bolts to 14 ft. lbs.

  Step 4: Tighten bolts an additional 90 degrees

④ Step 1: 31 ft. lbs.

  Step 2: 60 degrees

37663_ALTH_C0010

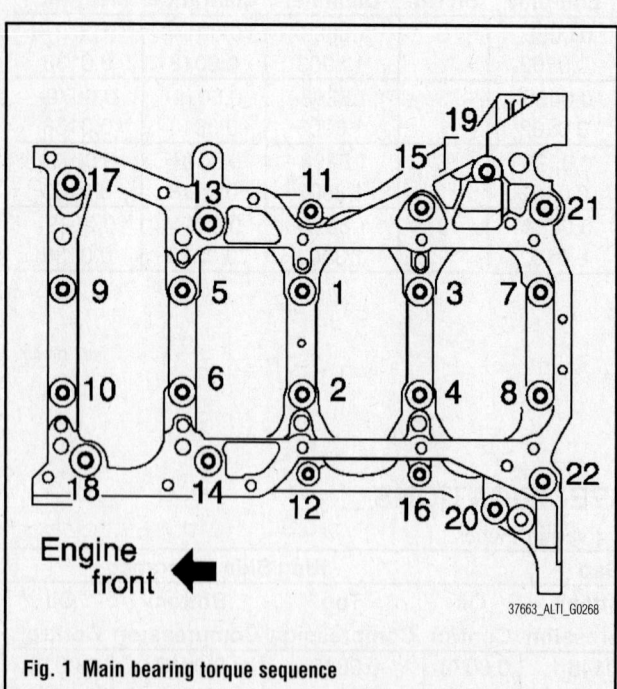

Fig. 1 Main bearing torque sequence

WHEEL ALIGNMENT

Year	Model		Caster Range (+/-Deg.)	Caster Preferred Setting (Deg.)	Camber Range (+/-Deg.)	Camber Preferred Setting (Deg.)	Toe-in (in.)
2007	Altima Hybrid	F	0.75	+5.00	0.75	①	0.04 +/- 0.04
		R			NA	NA	NA
2008	Altima Hybrid	F	0.75	+5.00	0.75	①	0.04 +/- 0.04
		R			NA	NA	NA
2009	Altima Hybrid	F	0.55	+5.00	0.75	①	0.04 +/- 0.04
		R			NA	NA	NA
2010	Altima Hybrid	F	0.55	+5.00	0.75	①	0.04 +/- 0.04
		R			NA	NA	NA

NA Not Available

① Left, -0.40; Right, -0.65

37663_ALTH_C0011

TIRE, WHEEL AND BALL JOINT SPECIFICATIONS

Year	Model	OEM Tires Standard	OEM Tires Optional	Tire Pressures (psi) Front	Tire Pressures (psi) Rear	Wheel Size	Lug Nut Torque (ft. lbs.)
2007	Altima Hybrid	P215/60TR16	None	35	35	6.5-JJ	83
2008	Altima Hybrid	P215/60TR16	None	35	35	6.5-JJ	83
2009	Altima Hybrid	P215/60TR16	None	35	35	6.5-JJ	83
2010	Altima Hybrid	P215/60TR16	None	35	35	6.5-JJ	83

OEM: Original Equipment Manufacturer

PSI: Pounds Per Square Inch

37663_ALTH_C0012

BRAKE SPECIFICATIONS
All measurements in inches unless noted

Year	Model		Brake Disc Original Thickness	Brake Disc Minimum Thickness	Brake Disc Maximum Run-out	Brake Drum Diameter Original Inside Diameter	Brake Drum Diameter Max. Wear Limit	Brake Drum Diameter Maximum Machine Diameter	Minimum Lining Thickness Front	Minimum Lining Thickness Rear	Brake Caliper Bracket Bolts (ft. lbs.)	Brake Caliper Mounting Bolts (ft. lbs.)
2007	Altima Hybrid	F	1.024	0.945	0.001	—	—	—	0.079	—	98	20
		R	0.354	0.315	0.001	—	—	—	—	0.039	62	32
2008	Altima Hybrid	F	1.024	0.945	0.001	—	—	—	0.079	—	98	20
		R	0.354	0.315	0.001	—	—	—	—	0.039	62	32
2009	Altima Hybrid	F	1.024	0.945	0.001	—	—	—	0.079	—	98	20
		R	0.354	0.315	0.001	—	—	—	—	0.039	62	32
2010	Altima Hybrid	F	1.024	0.945	0.001	—	—	—	0.079	—	98	20
		R	0.354	0.315	0.001	—	—	—	—	0.039	62	32

37663_ALTH_C0013

SCHEDULED MAINTENANCE INTERVALS
Nissan—Altima Hybrid

TO BE SERVICED	TYPE OF SERVICE	VEHICLE MILEAGE INTERVAL (x1000)												
		7.5	15	22.5	30	37.5	45	52.5	60	67.5	75	82.5	90	97.5
Engine oil & filter	R	✓	✓	✓	✓	✓	✓	✓	✓	✓	✓	✓	✓	✓
Brake lines & cables	S/I		✓		✓		✓		✓		✓		✓	
Brake pads, discs, drums & linings	S/I		✓		✓		✓		✓		✓		✓	
Driveshaft boots	S/I		✓		✓		✓		✓		✓		✓	
Exhaust system	S/I				✓				✓				✓	
Transaxle fluid	S/I		✓		✓		✓		✓		✓		✓	
Air cleaner filter	R				✓				✓				✓	
Spark plugs (except platinum)	R				✓				✓				✓	
Spark plugs (iridium and platinum)	R	Replace every 105,000 miles												
Steering gear & linkage, axle & suspension parts	S/I			✓					✓				✓	
Engine coolant	R	Replace every 60,000 miles, then every 30,000 miles												
Inverter coolant	R	Replace every 60,000 miles, then every 30,000 miles												
Drive belts	S/I								✓					
Fuel lines	S/I								✓					
Vapor lines	S/I								✓					
Cabin microfilter	R		✓		✓		✓		✓		✓		✓	
Valve adjustment	S/I	As needed												

R: Replace   S/I: Service or Inspect

FREQUENT OPERATION MAINTENANCE (SEVERE SERVICE)

If a vehicle is operated under any of the following conditions it is considered severe service:

- Extremely dusty areas.

- 50% or more of the vehicle operation is in 32°C (90°F) or higher temperatures, or constant operation in temperatures below 0°C (32°F).

- Prolonged idling (vehicle operation in stop and go traffic).

- Frequent short running periods (engine does not warm to normal operating temperatures).

- Police, taxi, delivery usage or trailer towing usage.

Oil & oil filter: change every 3750 miles.

Brake pads & discs: service or inspect every 7500 miles.

Driveshaft boots: service or inspect every 7500 miles.

Exhaust system: service or inspect every 7500 miles.

Steering gear & linkage, axle & suspension parts: service or inspect every 7500 miles.

Steering linkage ball joints & front suspension ball joints: service or inspect every 7500 miles.

Air cleaner filter: service or inspect every 15,000 miles.

37663_ALTH_C0014

BRAKES — **INFORMATION AND PRECAUTIONS**

ANTI-LOCK SYSTEMS

• Certain components within the ABS system are not intended to be serviced or repaired individually.

• Do not use rubber hoses or other parts not specifically specified for and ABS system. When using repair kits, replace all parts included in the kit. Partial or incorrect repair may lead to functional problems and require the replacement of components.

• Lubricate rubber parts with clean, fresh brake fluid to ease assembly. Do not use shop air to clean parts; damage to rubber components may result.

• Use only DOT 3 brake fluid from an unopened container.

• If any hydraulic component or line is removed or replaced, it may be necessary to bleed the entire system.

• A clean repair area is essential. Always clean the reservoir and cap thoroughly before removing the cap. The slightest amount of dirt in the fluid may plug an ori-

fice and impair the system function. Perform repairs after components have been thoroughly cleaned; use only denatured alcohol to clean components. Do not allow ABS components to come into contact with any substance containing mineral oil; this includes used shop rags.

• The Anti-Lock control unit is a microprocessor similar to other computer units in the vehicle. Ensure tha the ignition switch is **OFF** before removing or installing controller harnesses. Avoid static electricity discharge at or near the controller.

• If any arc welding is to be done on the vehicle, the control unit should be unplugged before welding operations begin.

DISC AND DRUM SYSTEMS

✳✳ CAUTION

Dust and dirt accumulating on brake parts during normal use may contain asbestos fibers from production or

aftermarket brake linings. Breathing excessive concentrations of asbestos fibers can cause serious bodily harm. Exercise care when servicing brake parts. Do not sand or grind brake lining unless equipment used is designed to contain the dust residue. Do not clean brake parts with compressed air or by dry brushing. Cleaning should be done by dampening the brake components with a fine mist of water, then wiping the brake components clean with a dampened cloth. Dispose of cloth and all residue containing asbestos fibers in an impermeable container with the appropriate label. Follow practices prescribed by the Occupational Safety and Health Administration (OSHA) and the Environmental Protection Agency (EPA) for the handling, processing, and disposing of dust or debris that may contain asbestos fibers.

BRAKES — **BLEEDING THE BRAKE SYSTEM**

BLEEDING PROCEDURE

BLEEDING PROCEDURE

✳✳ CAUTION

If any DTC is indicated, erase the indicated DTC. After the procedure of air bleed, perform initialization of linear solenoid valve.

➡**The brake warning buzzer may be activated during the air bleed procedure. The work can be continued, as it is normal.**

Air Release Of Static Pressure System (Front Wheel)

✳✳ CAUTION

Monitor the fluid level in the reservoir tank during the air bleeding. Always use new brake fluid for refilling. Never reuse the drained brake fluid.

1. Turn ignition switch OFF.
2. Connect CONSULT-III.
3. Turn ignition switch (READY).
4. When performing air bleed of the static pressure system and suction drain system, remove 2 relays for brake actuator motor beforehand.
5. Connect a vinyl tube to the bleeder valve of the front brake.

6. When performing air bleed, following conditions are required:
   • ABS relay No.1 and No.2: ON
   • Parking brake: ON
   • Shift position: P range
   • Vehicle speed: 0 km/h (0 MPH)
   • Normal power supply voltage
   • Normal communication with HV
   • No failure of brake system (except following items):
   • Motor relay
   • Accumulator
   • Fluid level switch
   • Calibration for each sensors and linear solenoid
   • Test mode diagnostic code
7. Select "AIR REL INHIBIT" in "ACTIVE TEST".
8. Loosen the bleeder valve and bleed air with the brake pedal depressed.

➡**Air bleeding is allowed to start from either right or left.**

9. After a complete air bleeding, tighten bleeder valve to the specified torque.
10. Check that the fluid level is the reservoir tank is within the specified range after air bleeding.

Air Release Of Suction Drain System

✳✳ CAUTION

Monitor the fluid level in the reservoir tank during the air bleeding. Perform the air bleed procedure within 30 seconds after the transmission of the signal from CONSULT-III. When the air bleed is performed afterward, the re-transmission of the signal from CONSULT-III is needed.

➡**Air bleed from the bleeder valve is not necessary since this operation is to return brake fluid (air).**

1. Turn ignition switch OFF.
2. Connect CONSULT-III.
3. Turn ignition switch (READY).
4. When performing air bleed, following conditions are required:
   • ABS relay No.1 and No.2: ON
   • Parking brake: ON
   • Shift position: P range
   • Vehicle speed: 0 km/h (0 MPH)
   • Normal power supply voltage
   • Normal communication with HV
   • No failure of brake system (except following items):
   • Motor relay
   • Accumulator
   • Fluid level switch
   • Calibration for each sensors and linear solenoid
   • Test mode diagnostic code

5. Select "AIR REL DRAIN" in "ACTIVE TEST".

6. Step on the brake pedal and return brake fluid to reservoir tank.

7. Ensure that no air (bubble) is contained in the brake fluid circulated from reservoir tank.

Air Release Of Rear Wheel System

※※ CAUTION

Monitor the fluid level in the reservoir tank during the air bleeding. Always use new brake fluid for refilling. Never reuse the drained brake fluid.

1. Turn ignition switch OFF.
2. Connect 2 motor relays.
3. Connect CONSULT-III.
4. Turn ignition switch (READY).

➡**If CONSULT-III is frozen, erase the DTC.**

5. Confirm accumulator pressure level by using "DATA MONITOR" in CONSULT-III.
   a. Select "ACC PRESS SEN" in "DATA MONITOR".
   b. Ensure that this voltage is over 3.42 V.
   c. If voltage is under 3.42 V, then step on the brake pedal several time.

6. When performing air bleed, following conditions must be met:
   - ABS relay No.1 and No.2: ON
   - Parking brake: ON
   - Shift position: P range
   - Vehicle speed: 0 km/h (0 MPH)
   - Normal power supply voltage
   - Normal communication with HV
   - ABS motor relay No.1 and No.2 are set
   - No failure of brake system (except following items):
   - Motor relay
   - Accumulator
   - Fluid level switch
   - Calibration for each sensors and linear solenoid
   - Test mode diagnostic code

7. Connect a vinyl tube to the bleeder valve of the rear brake.

8. Select "AIR REL INHIBIT" in "ACTIVE TEST".

9. Loosen the bleeder valve and bleed air with the brake pedal depressed.

10. Ensure that there is no air leakage from the bleeder.

11. After a complete air bleeding, tighten bleeder valve to the specified torque.

12. Check that the fluid level is the reservoir tank is within the specified range after air bleeding.

Air Release Of Power Supply System

※※ CAUTION
Note the following:

- Monitor the fluid level in the reservoir tank during the air bleeding.
- Always use new brake fluid for refilling. Never reuse the drained brake fluid.
- Perform the air bleed procedure within 10 seconds after the transmission of the signal from CONSULT-III. When the air bleed is performed afterward, the re-transmission of the signal from CONSULT-III is needed.

➡**No need to step on the brake pedal. Air bleeding is necessary for the front left brake only.**

1. Turn ignition switch OFF.
2. Connect CONSULT-III.
3. Turn ignition switch (READY).
4. Connect a vinyl tube to the bleeder valve of the front left brake.
5. When performing air bleed, following conditions must be met:
   - ABS relay No.1 and No.2: ON
   - Parking brake: ON
   - Shift position: P range
   - Vehicle speed: 0 km/h (0 MPH)
   - Normal power supply voltage
   - Normal communication with HV
   - ABS motor relay No.1 and No.2 are set
   - No failure of brake system (except following items):
   - Motor relay
   - Accumulator
   - Fluid level switch
   - Calibration for each sensors and linear solenoid
   - Test mode diagnostic code

6. Select "AIR REL PWR SPLY 2" in "ACTIVE TEST".

7. Loosen the bleeder valve.

8. Ensure that there is no air leakage from the bleeder.

9. After a complete air bleeding, tighten bleeder valve to the specified torque.

Air Release Of Stroke Simulator System

Air Bleed of Stroke Simulator System 1

※※ CAUTION

Perform the air bleed procedure within 30 seconds after the transmission of the signal from CONSULTIII. When the air bleed is performed

afterward, the re-transmission of the signal from CONSULT-III is needed.

➡**Air bleed from the bleeder is not necessary in this stage. This process is performed to send air contained in the stroke simulator to piping. Pedal operation only and no need of air bleed from the bleeder.**

1. Turn ignition switch OFF.
2. Connect CONSULT-III.
3. Turn ignition switch (READY).
4. When performing air bleed, following conditions must be met:
   - ABS relay No.1 and No.2: ON
   - Parking brake: ON
   - Shift position: P range
   - Vehicle speed: 0 km/h (0 MPH)
   - Normal power supply voltage
   - Normal communication with HV
   - ABS motor relay No.1 and No.2 are set
   - No failure of brake system (except following items):
   - Motor relay
   - Accumulator
   - Fluid level switch
   - Calibration for each sensors and linear solenoid
   - Test mode diagnostic code

5. Select "AIR REL STROKE SIM" in "ACTIVE TEST".

6. Step on the brake pedal 20 times with its stroke fully within continuously 20 to 30 seconds.

Air Release of Stroke Simulator System 2

➡**Air bleeding is necessary for the front left brake only.**

1. Connect a vinyl tube to the bleeder valve of the front left brake.
2. When performing air bleed, following conditions must be met:
   - ABS relay No.1 and No.2: ON
   - Parking brake: ON
   - Shift position: P range
   - Vehicle speed: 0 km/h (0 MPH)
   - Normal power supply voltage
   - Normal communication with HV
   - ABS motor relay No.1 and No.2 are set
   - No failure of brake system (except following items):
   - Motor relay
   - Accumulator
   - Fluid level switch
   - Calibration for each sensors and linear solenoid
   - Test mode diagnostic code

3. Select "AIR REL INHIBIT" in "ACTIVE TEST".

4. Loosen the bleeder valve and bleed air with the brake pedal depressed.

5. Ensure that there is no air from the bleeder.

6. Tighten the bleeder valve to the specified torque.

7. Return to previous step "Air Release of Stroke Simulator System 1". Repeat "Air Release of Stroke Simulator System 1" and "Air Release of Stroke Simulator System 2" at least 3 times.

Air Release Of High-Pressure Line

> ❊❊ **CAUTION**
>
> **Be careful with fluid level in the reservoir tank because a large**

amount of brake fluid flows back to the reservoir tank.

➡**Air bleed from the bleeder is not necessary in this stage.**

1. Turn ignition switch OFF.
2. Connect CONSULT-III.
3. Turn ignition switch (READY).
4. When performing air bleed, following conditions must be met:

- ABS relay No.1 and No.2: ON
- Parking brake: ON
- Shift position: P range
- Vehicle speed: 0 km/h (0 MPH)
- Normal power supply voltage
- Normal communication with HV
- ABS motor relay No.1 and No.2 are set

- No failure of brake system (except following items):
- Motor relay
- Accumulator
- Fluid level switch
- Calibration for each sensors and linear solenoid
- Test mode diagnostic code

5. Select "ACC 0 DOWN" in "ACTIVE TEST".

➡**Return air remaining in the high-pressure line to reservoir tank and open atmosphere.**

6. Repeat 5 times to ensure the circulation of brake fluid since visual judgment of completion is difficult.

7. Fill the brake fluid to the MAX line after completing this operation, with "ACC 0 DOWN" condition.

BRAKES   **ANTI-LOCK BRAKE SYSTEM (ABS)**

WHEEL SPEED SENSORS

REMOVAL & INSTALLATION

See Figure 2.

> ❊❊ **CAUTION**
>
> **Note the following:**

- Be careful not to damage wheel sensor edge and sensor rotor teeth.
- When removing the front or rear wheel hub, first remove the wheel sensor from the wheel hub. Failure to do so may result in damage to the wheel sensor wires making the sensor inoperative.

- Pull out the wheel sensor, being careful to turn it as little as possible. Do not pull on the wheel sensor harness.
- Before installation, check if foreign objects such as iron fragments are adhered to the pick-up part of the sensor or to the inside of the hole

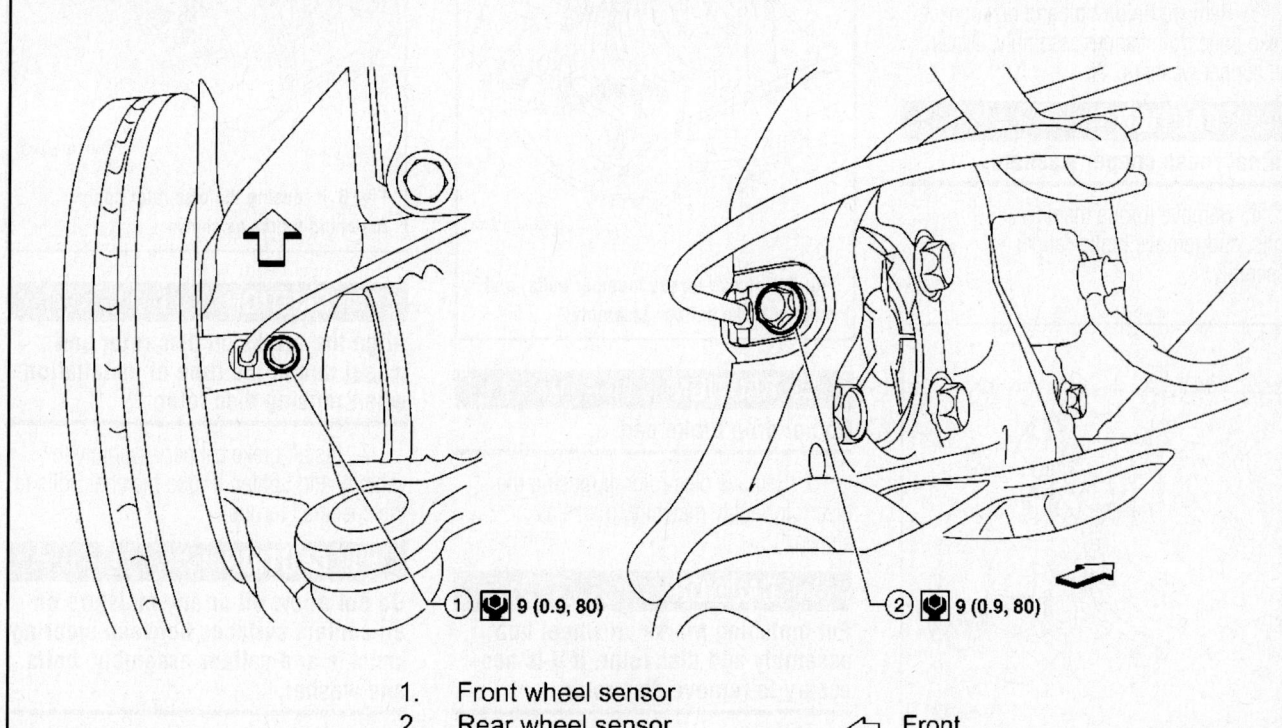

1. Front wheel sensor
2. Rear wheel sensor   ⇐ Front

37663_ALTI_G0050

Fig. 2 Wheel speed sensor locations

in the wheel hub for the wheel sensor, or if a foreign object is caught in the surface of the mating surface for the sensor rotor. Fix as necessary and then install the wheel sensor.

Front Wheel Sensor

1. Remove front wheel and tire.
2. Partially front wheel fender protector.
3. Remove wheel sensor bolt and wheel sensor.
4. Remove harness wire from mounts and disconnect wheel sensor harness connector.

To install:
5. Installation is in the reverse order of removal.

Rear Wheel Sensor

See Figure 3.

➡ **Both rear wheel sensors share one harness and must be replaced as an assembly.**

1. Remove rear wheel and tire.

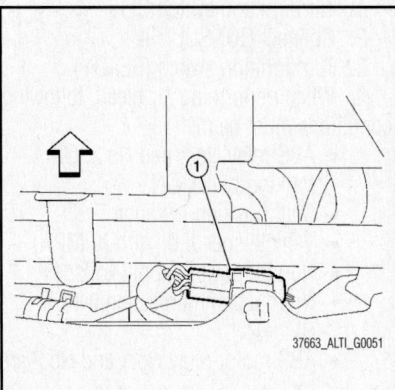

Fig. 3 Disconnect wheel sensor harness connector (1)

2. Remove wheel sensor bolts and wheel sensors from both rear wheel hub and bearing assemblies.
3. Remove harness wire from mounts and harness wire clips from rear suspension member.
4. Disconnect wheel sensor harness connector.

To install:
5. Installation is in the reverse order of removal.

WHEEL SPEED SENSOR RINGS (TOOTHED RINGS)

REMOVAL & INSTALLATION

The front and rear wheel sensor rotors are an integral part of the wheel hubs and cannot be disassembled. When replacing the sensor rotor, replace the wheel hub.

BRAKES   FRONT DISC BRAKES

BRAKE CALIPER

REMOVAL & INSTALLATION

See Figures 4 through 7.

1. Remove front wheel and tires.
2. Drain brake fluid.
3. Remove union bolt and disconnect brake hose from caliper assembly. Discard the copper washers.

✳✳ CAUTION
Do not reuse copper washers.

4. Remove torque member bolts, and remove brake caliper assembly.

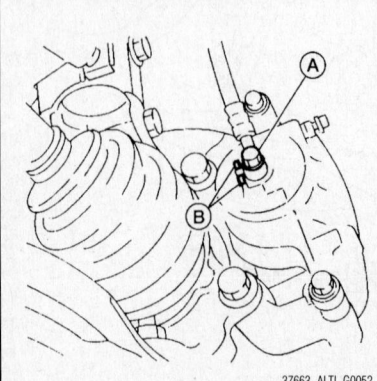

Fig. 4 Remove union bolt (A) and disconnect brake hose from caliper assembly; note protrusions (B)

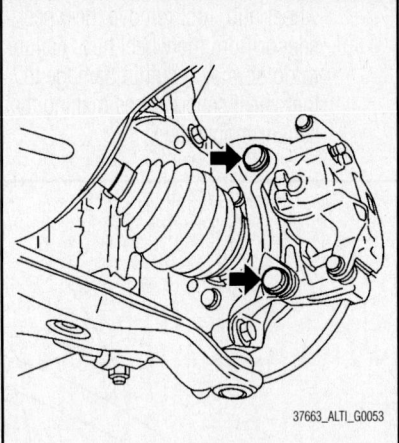

Fig. 5 Remove torque member bolts, and remove brake caliper assembly

✳✳ CAUTION
Do not drop brake pad.

5. Remove disc rotor. If reusing the disc rotor apply matching marks as shown.

✳✳ CAUTION
Put matching marks on wheel hub assembly and disc rotor, if it is necessary to remove disc rotor.

To install:
6. Install disc rotor, align the matching marks if installing the original disc rotor as shown.

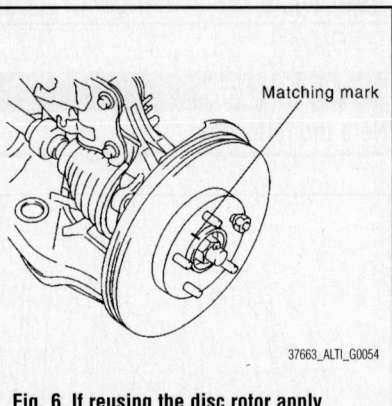

Fig. 6 If reusing the disc rotor apply matching marks as shown

✳✳ CAUTION
Align the marks on disc rotor and wheel hub at the time of installation when reusing disc rotor.

7. Install brake caliper assembly to vehicle, and tighten torque member bolts to the specified torque.

✳✳ CAUTION
Do not allow oil or any moisture on all contact surfaces between steering knuckle and caliper assembly, bolts, and washer.

8. Install brake hose to brake caliper assembly with new copper washers. Align the brake hose tab between the protrusions on the caliper assembly as

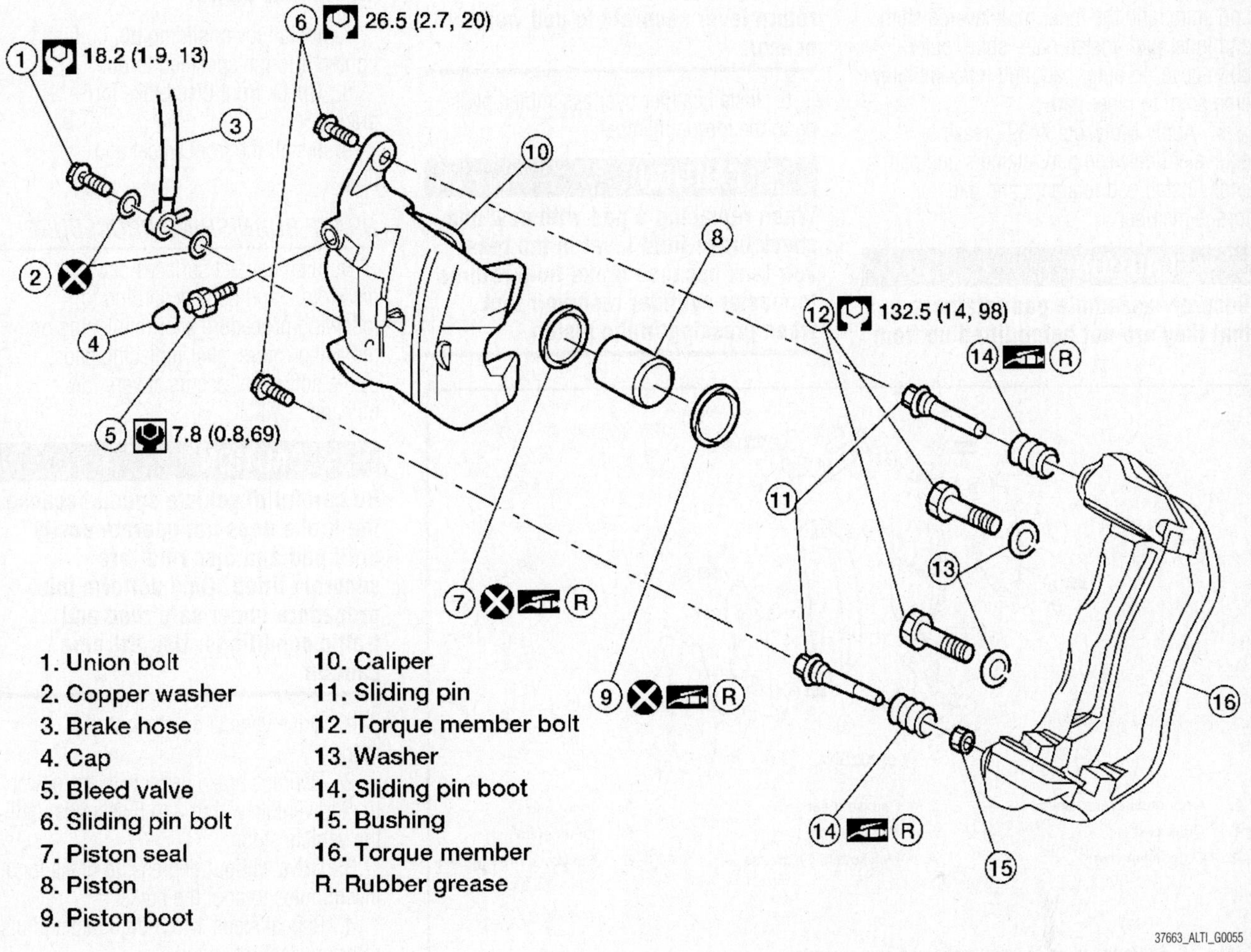

1. Union bolt 🔧 18.2 (1.9, 13)
6. Sliding pin bolt 🔧 26.5 (2.7, 20)
2. Copper washer ✕
4. Cap
5. Bleed valve 🔧 7.8 (0.8, 69)
10. Caliper
8. Piston
12. Torque member bolt 🔧 132.5 (14, 98)
14. Sliding pin boot ⬅ R
13. Washer
11. Sliding pin
7. Piston seal ✕ ⬅ R
9. Piston boot ✕ ⬅ R
14. ⬅ R
16. Torque member
15. Bushing

1. Union bolt
2. Copper washer
3. Brake hose
4. Cap
5. Bleed valve
6. Sliding pin bolt
7. Piston seal
8. Piston
9. Piston boot
10. Caliper
11. Sliding pin
12. Torque member bolt
13. Washer
14. Sliding pin boot
15. Bushing
16. Torque member
R. Rubber grease

37663_ALTI_G0055

Fig. 7 Exploded view of front brake caliper assembly

shown. Tighten union bolt to the specified torque.

❉❉ CAUTION

Do not reuse copper washers.

9. Refill with new brake fluid and bleed air from the brake hydraulic system.
10. Check front disc brakes for drag.
11. Install front wheel and tires.

DISC BRAKE PADS

REMOVAL AND INSTALLATION

See Figures 8 through 10.

1. Remove the front wheel and tires.
2. Remove lower sliding pin bolt.
3. Hang caliper with a suitable wire, and remove pads, pad retainers, shims, and shim cover from torque member.

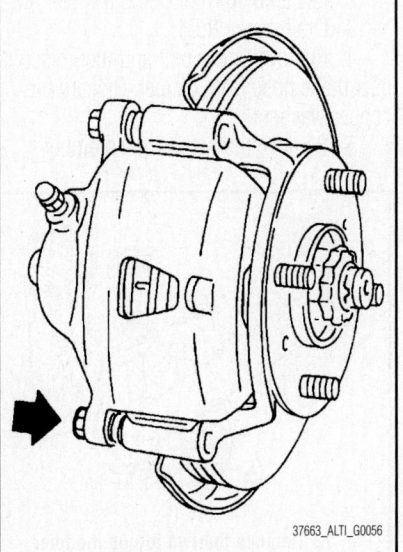

37663_ALTI_G0056

Fig. 8 Remove lower sliding pin bolt

❉❉ CAUTION

When removing the pad retainer from the torque member, lift it in the direction indicated by the arrow as shown so that it does not deform.

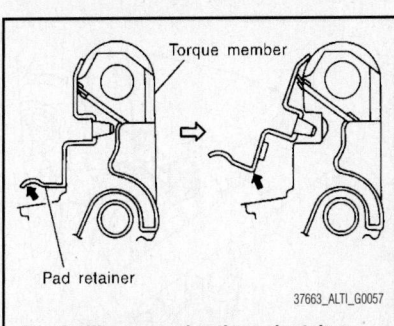

Torque member

Pad retainer

37663_ALTI_G0057

Fig. 9 When removing the pad retainer from the torque member, lift it in the direction indicated by the arrow as shown so that it does not deform

To install:

4. Apply Molykote M-77 grease or equivalent between the outer shim cover and shim; and the inner multilayered shim and inner pad. Install outer shim, outer shim cover to outer pad, and inner multilayered shim to inner pad.

5. Apply Molykote 7439 grease or equivalent between pad retainers and pad ends. Install pad retainers and pads on torque member.

❋❋ CAUTION
Securely assemble pad retainers so that they are not being lifted up from

torque member. Both inner and outer pads have a pad return system on the pad retainer. Install pad return lever securely to pad wear sensor.

6. Install caliper over assembled pads on to the torque member.

❋❋ CAUTION
When replacing a pad with new one, check brake fluid level in the reservoir tank because brake fluid returns to master cylinder reservoir tank when pressing in the piston.

➡ Use a disc brake piston tool (commercial service tool) to easily press in the piston.

7. Install lower sliding pin bolt, and tighten it to the specified torque.

8. Check front disc brake for drag.

9. Install the front wheel and tires.

BRAKE BURNISHING PROCEDURE

Burnish contact surfaces between disc rotors and pads according to following procedure after refinishing or replacing rotors, after replacing pads, or if a soft pedal occurs at very low mileage.

❋❋ CAUTION
Be careful of vehicle speed because the brake does not operate easily until pad and disc rotor are securely fitted. Only perform this procedure under safe road and traffic conditions. Use extreme caution.

1. Drive vehicle on straight, flat road.

2. Depress brake pedal with the power to stop vehicle within 3 to 5 seconds until the vehicle stops.

3. Drive without depressing brake for a few minutes to cool the brake.

4. Repeat steps 1 to 3 until pad and disc rotor are securely fitted.

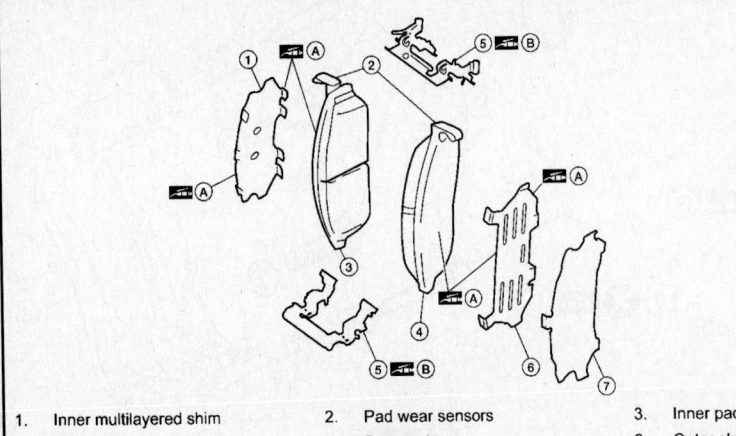

1.	Inner multilayered shim	2.	Pad wear sensors	3.	Inner pad
4.	Outer pad	5.	Pad retainers	6.	Outer shim
7.	Outer shim cover	A.	Molykote M-77 grease	B.	Molykote 7439 grease

37663_ALTI_G0058

Fig. 10 Exploded view of front brake pad components

BRAKES

BRAKE CALIPER

REMOVAL & INSTALLATION

See Figures 11 through 13.

1. Remove rear wheel and tires.

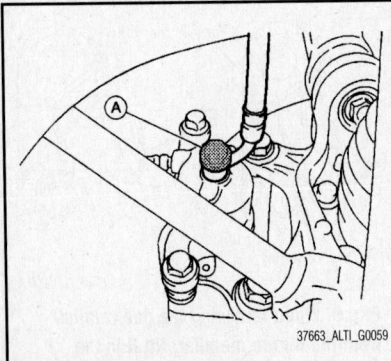

37663_ALTI_G0059

Fig. 11 Remove union bolt (A) and then disconnect brake hose from caliper

2. Fasten disc rotor using a wheel nut.

3. Drain brake fluid.

4. Remove union bolt and then disconnect brake hose from caliper. Discard the copper washers.

5. Remove the two torque member

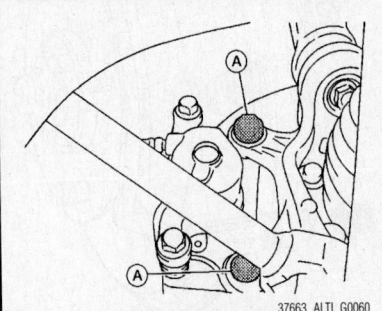

37663_ALTI_G0060

Fig. 12 Remove the two torque member bolts (A), and then remove the torque member, caliper and pads as an assembly

REAR DISC BRAKES

bolts, and then remove the torque member, caliper and pads as an assembly.

❋❋ CAUTION
Do not drop the brake pad and multilayered shim assemblies.

6. Remove the two sliding pin bolts and separate the caliper from the torque member.

7. Remove the brake pad and multilayered shim assemblies from the caliper.

8. Remove the disc rotor.

❋❋ CAUTION
Put matching marks on wheel hub assembly and disc rotor, if it necessary to reuse the disc rotor.

To install:

9. Install the disc rotor.

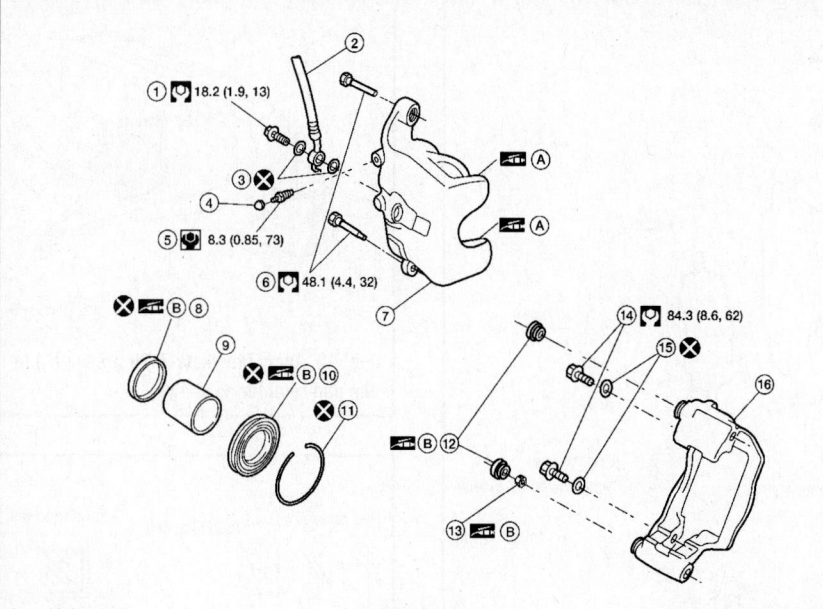

1. Union bolt
2. Brake hose
3. Copper washer
4. Cap
5. Bleed valve
6. Sliding pin bolt
7. Caliper
8. Piston seal
9. Piston
10. Piston boot
11. Retaining ring
12. Sliding pin boot
13. Bushing
14. Torque member bolt
15. Washer
16. Torque member
A. PBC (Poly Butyl Cuprysil) grease or silicone-based grease
B. Rubber grease

37663_ALTI_G0061

Fig. 13 Exploded view of rear disc brake assembly

> ※※ **CAUTION**
> **Alignment marks of disc rotor and wheel hub put at the time of removal when reusing disc rotor.**

10. Install the brake pad and multilayered shim assemblies on the caliper.

11. Install the caliper and pad assembly on the torque member, then tighten the two sliding pin bolts to the specified torque.

12. Install the torque member, pads and brake caliper assembly, and tighten the torque member bolts (A) to the specified torque.

> ※※ **CAUTION**
> **Before installing wipe off all oil and moisture on all mating surfaces of rear axle and torque member, threads, bolts and washers.**

13. Align the L-shaped pin on the brake hose in the hole in the caliper, then install the brake hose with new copper washers and tighten the union bolt to the specified torque.

> ※※ **CAUTION**
> **Do not reuse copper washers.**

14. Refill with new brake fluid and bleed air.

15. Check rear disc brake for drag.

16. Install rear wheel and tires.

DISC BRAKE PADS

REMOVAL AND INSTALLATION

See Figures 14 through 18.

1. Remove rear wheel and tires.

2. Remove upper sliding pin bolt and swing caliper out supporting it with a suitable wire.

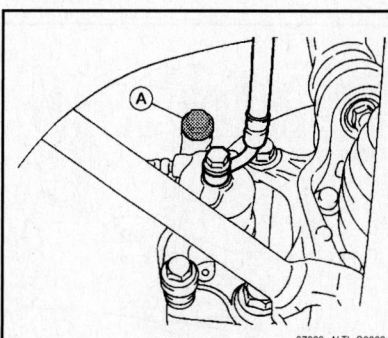

37663_ALTI_G0062

Fig. 14 Remove upper sliding pin bolt (A) and swing caliper out supporting it with a suitable wire

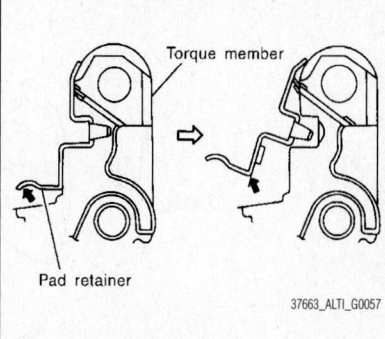

37663_ALTI_G0057

Fig. 15 When removing the pad retainer from the torque member, lift it in the direction indicated by the arrow as shown so that it does not deform

> ※※ **CAUTION**
> **Do not twist or stretch the brake hose.**

3. Remove pads, pad retainers and multilayered shims from torque member.

> ※※ **CAUTION**
> **When removing the pad retainer from the torque member, lift it in the direction indicated by the arrow as shown so that it does not deform.**

To install:

4. Apply Molykote M-77 grease or equivalent to between multilayered shims and brake pads. Install inner multilayered shim to inner pad, and outer multilayered shim to outer pad.

5. Apply Molykote 7439 grease to the pad retainer as shown.

6. Attach pad retainers to torque member, then install brake pads and multilayered shim assemblies.

> ※※ **CAUTION**
> **When attaching pad retainer, attach it firmly so that it is flush with torque member as shown.**

7. Press in piston until pads can be installed, and then install caliper to torque member.

> ※※ **CAUTION**
> **In the case of replacing a pad with new one, check a brake fluid level in the reservoir tank because brake fluid returns to master cylinder reservoir tank when pressing piston in.**

➡Use a disc brake piston tool (commercial service tool) to easily press piston.

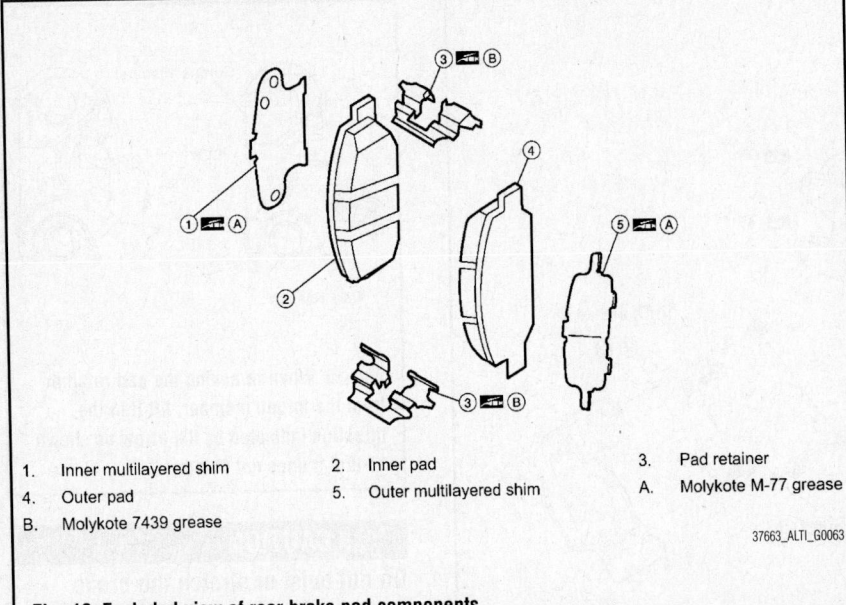

1. Inner multilayered shim
2. Inner pad
3. Pad retainer
4. Outer pad
5. Outer multilayered shim
A. Molykote M-77 grease
B. Molykote 7439 grease

37663_ALTI_G0063

Fig. 16 Exploded view of rear brake pad components

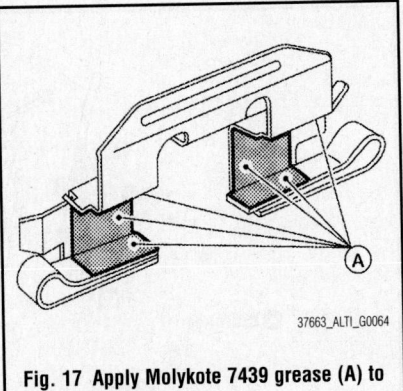

37663_ALTI_G0064

Fig. 17 Apply Molykote 7439 grease (A) to the pad retainer as shown

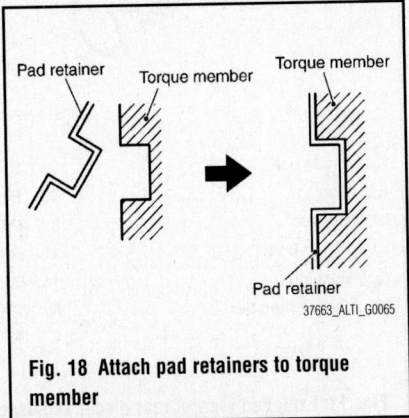

37663_ALTI_G0065

Fig. 18 Attach pad retainers to torque member

8. Install upper sliding pin bolt and tighten to the specified torque.
9. Check rear disc brake for drag.
10. Install rear wheel and tires.

BRAKE BURNISHING PROCEDURE

Burnish contact surfaces between disc rotors and pads according to following procedure after refinishing or replacing rotors, after replacing pads, or if a soft pedal occurs at very low mileage.

✳✳ CAUTION

Be careful of vehicle speed because the brake does not operate easily until pad and disc rotor are securely fitted. Only perform this procedure under safe road and traffic conditions. Use extreme caution.

1. Drive vehicle on straight, flat road.
2. Depress brake pedal with the power to stop vehicle within 3 to 5 seconds until the vehicle stops.

3. Drive without depressing brake for a few minutes to cool the brake.
4. Repeat steps 1 to 3 until pad and disc rotor are securely fitted.

BRAKES

PARKING BRAKE

PARKING BRAKE CABLES

ADJUSTMENT

See Figures 19 and 20.

1. Remove rear wheel and tires.
2. Insert a deep socket wrench onto adjusting nut. Rotate adjusting nut to fully loosen cable, and then release parking brake pedal.
3. Secure disc rotor to hub using wheel nut so as not to tilt disc rotor.
4. Remove adjuster hole plug installed on the disc rotor. Turn the adjuster in direction using a suitable tool or a flat-bladed screwdriver as shown, until disc rotor is locked. Turn the adjuster in the opposite direction by 5 or 6 notches after locking.
5. Rotate disc rotor to make sure that there is no drag. Install the adjuster hole plug.
6. Adjust parking brake cable with the following procedure.

a. Operate parking brake pedal 10 or more times with a full stroke of 7.6 inches (194.3 mm).
b. Rotate adjusting nut to adjust parking brake pedal stroke using a deep socket wrench.
c. Operate parking brake pedal with a

force of 66 lbs. (294 N), make sure the pedal stroke is within the specified number (4 to 5) of notches. Check it by listening and counting the ratchet clicks.
d. Make sure that there is no drag on the parking brake with the parking brake pedal completely released.

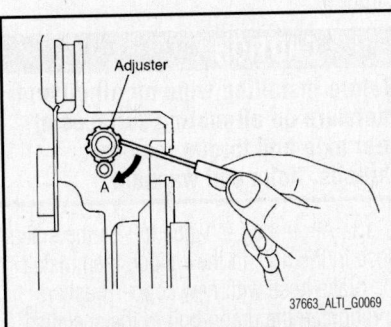

37663_ALTI_G0068

Fig. 19 Parking brake pedal and adjusting nut

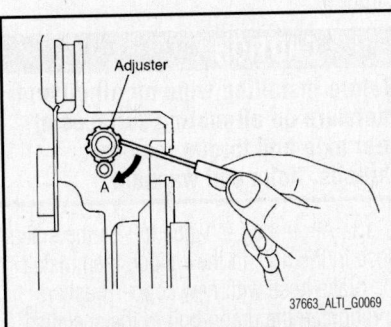

37663_ALTI_G0069

Fig. 20 Turn the adjuster in direction (A) using a suitable tool or a flat-bladed screwdriver as shown, until disc rotor is locked

PARKING BRAKE SHOES

REMOVAL & INSTALLATION

See Figures 21 and 22.

1. Remove rear wheel and tires.
2. Remove rear disc rotor with parking brake control device assembly in the fully released position.
3. If disc rotor cannot be removed, remove as follows:
    a. Secure the disc rotor in place with wheel nuts and remove adjuster hole plug.
    b. Rotate adjuster in direction (B) to

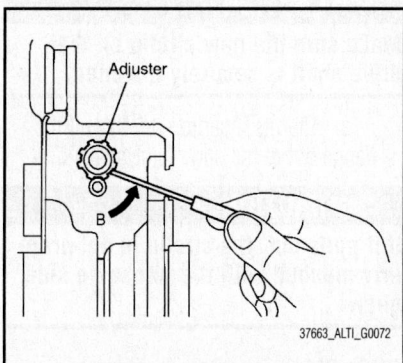

Fig. 21 Rotate adjuster in direction (B) to retract and loosen brake shoe

retract and loosen brake shoe, using tool as shown.

4. Remove anti-rattle pins, retainers, anti-rattle springs, and return springs.
5. Remove parking brake shoes, adjuster assembly, and toggle lever.
6. Remove the back plate.

To install:

7. Installation is in the reverse order of removal. Note the following:
    a. Apply PBC (Poly Butyl Cuprysil) grease or equivalent to the specified points during assembly.

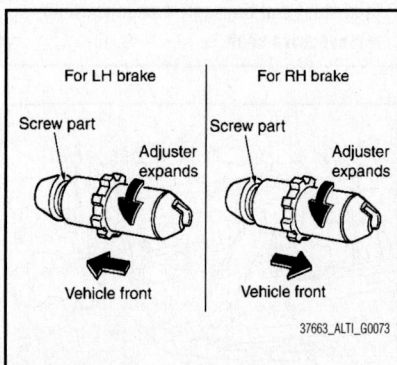

Fig. 22 Shorten adjuster by rotating it as shown

b. Assemble adjusters so that threaded part is expanded when rotating it in the direction shown.
    c. Shorten adjuster by rotating it as shown.
    d. Check shoe sliding surface and drum inner surface for grease. Wipe it off if it adhere on the surfaces.
    e. Perform break-in operation as follows after replacing brake shoes or disc rotors, or if brakes do not function well.
8. Adjust parking brake control device assembly stroke to the specified amount.
9. Perform parking brake break-in (drag run) operation by driving vehicle under the following conditions:
    a. Drive the vehicle forward.
    b. Maintain vehicle speed at approximately 40 km/h (25 MPH) keeping it constant in forward direction.
    c. Apply the parking brake at an operating force of approximately 88 lbs. (400 N) with constant force.
    d. Release the parking brake after approximately 10 seconds.
10. Check parking brake control device assembly stroke of parking brake. Readjust as necessary if it is outside the standard specifications.

CHASSIS ELECTRICAL

GENERAL INFORMATION

PRECAUTIONS

The Supplemental Restraint System such as "AIR BAG" and "SEAT BELT PRE-TEN-SIONER", used along with a front seat belt, helps to reduce the risk or severity of injury to the driver and front passenger for certain types of collision. This system includes seat belt switch inputs and dual stage front air bag modules. The SRS system uses the seat belt switches to determine the front air bag deployment, and may only deploy one front air bag, depending on the severity of a collision and whether the front occupants are belted or unbelted.

❊❊ WARNING

Note the following:

- To avoid rendering the SRS inoperative, which could increase the risk

AIR BAG (SUPPLEMENTAL RESTRAINT SYSTEM)

of personal injury or death in the event of a collision which would result in air bag inflation, all maintenance must be performed by an authorized NISSAN/INFINITI dealer.

- Improper maintenance, including incorrect removal and installation of the SRS, can lead to personal injury caused by unintentional activation of the system. For removal of Spiral Cable and Air Bag Module, see the SR section.

- Do not use electrical test equipment on any circuit related to the SRS unless instructed to in this Service Manual. SRS wiring harnesses can be identified by yellow and/or orange harnesses or harness connectors.

DISARMING THE SYSTEM

❊❊ CAUTION

Before servicing the SRS, turn ignition switch OFF, disconnect both battery cables and wait at least 3 minutes. For approximately 3 minutes after the cables are removed, it is still possible for the air bag and seat belt pretensioner to deploy. Therefore, do not work on any SRS connectors or wires until at least 3 minutes have passed.

ARMING THE SYSTEM

When the repair work is completed, re-connect both battery cables. With the brake pedal released, turn the push-button ignition switch from ACC position to ON position, then to LOCK position. (The steering wheel will lock when the push-button ignition switch is turned to LOCK position.)

DRIVE TRAIN

HALFSHAFTS

REMOVAL & INSTALLATION

Left Side

See Figures 23 through 25.

1. Remove wheel and tire.
2. Remove wheel sensor from steering knuckle.
3. Remove cotter pin. Then remove lock nut from drive shaft.
4. Remove brake hose lock plate. Then remove brake hose from strut.
5. Remove brake caliper, leaving hydraulic brake line attached. Hang caliper aside using wire.
6. Remove front strut to steering knuckle bolts and nuts, then separate steering knuckle front strut.
7. Remove drive shaft from wheel hub and bearing assembly, using a puller or suitable tool.

✳✳ CAUTION

When removing drive shaft, do not apply an excessive angle to drive shaft joint. Also be careful not to excessively extend slide joint.

8. Remove the left side drive shaft from the transaxle.

   a. Remove drive shaft from transaxle using Tool and drive shaft puller or suitable tool.

   b. Set Tool KV40107500 and a drive shaft puller or suitable tool between drive shaft (slide joint side) and transaxle as shown, then remove drive shaft.

To install:

9. Installation is in the reverse order of removal. Note the following:

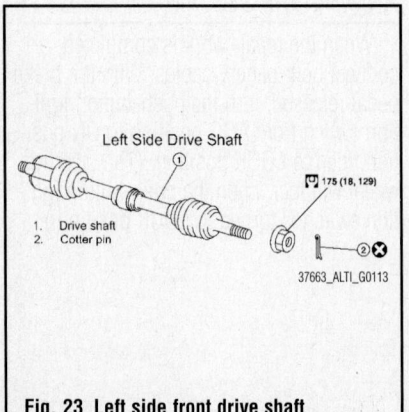

1. Drive shaft
2. Cotter pin

37663_ALTI_G0113

Fig. 23 Left side front drive shaft

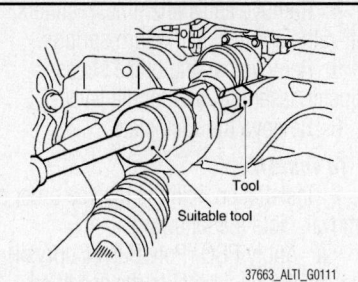

37663_ALTI_G0111

Fig. 24 Set Tool and a drive shaft puller or suitable tool between drive shaft (slide joint side) and transaxle as shown, then remove drive shaft

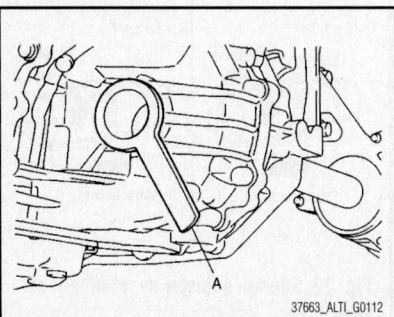

37663_ALTI_G0112

Fig. 25 Place Tool KV38106700 (J-34296) (A) onto oil seal before inserting drive shaft as shown

✳✳ CAUTION

Do not reuse non-reusable parts.

   a. In order to prevent damage to differential side oil seal, place Tool KV38106700 (J-34296) (A) onto oil seal before inserting drive shaft as shown. Slide drive shaft into slide joint and tap with a hammer to install securely.

   b. Install new circlip on drive shaft in the circular clip groove on transaxle side.

✳✳ CAUTION

Make sure the new circlip on the drive shaft is securely fastened.

   c. After its insertion, try to pull the flange out of the slide joint by hand.

✳✳ CAUTION

If it pulls out, the circlip is not properly meshed with the transaxle side gear.

Right Side

See Figures 26 and 27.

1. Remove wheel and tire.
2. Remove wheel sensor from steering knuckle.

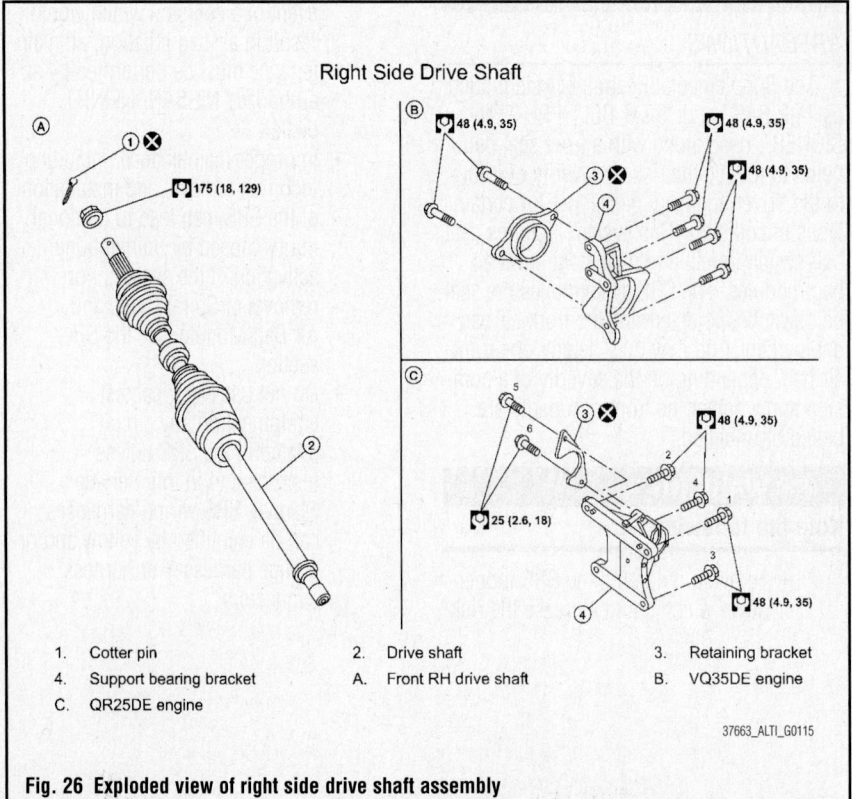

Right Side Drive Shaft

1. Cotter pin
2. Drive shaft
3. Retaining bracket
4. Support bearing bracket
A. Front RH drive shaft
B. VQ35DE engine
C. QR25DE engine

37663_ALTI_G0115

Fig. 26 Exploded view of right side drive shaft assembly

3. Remove cotter pin. Then remove lock nut from drive shaft.

4. Remove brake hose lock plate. Then remove brake hose from strut.

5. Remove brake caliper, leaving hydraulic brake line attached. Hang caliper aside using wire.

6. Remove front strut to steering knuckle bolts and nuts, then separate steering knuckle front strut.

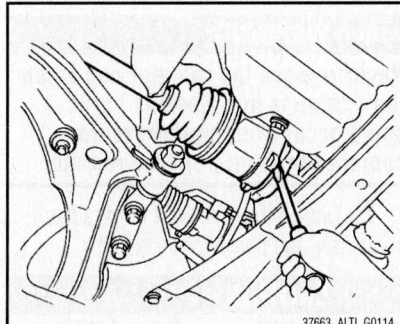

Fig. 27 Remove the retaining bracket bolts, and separate drive shaft from transaxle

7. Remove drive shaft from wheel hub and bearing assembly, using a puller or suitable tool.

> ※※ **CAUTION**
> **When removing drive shaft, do not apply an excessive angle to drive shaft joint. Also be careful not to excessively extend slide joint.**

8. Remove the retaining bracket bolts, and separate drive shaft from transaxle.

To install:

9. Installation is in the reverse order of removal. Note the following:

> ※※ **CAUTION**
> **Do not reuse non-reusable parts.**

a. Tighten retaining bracket bolts and support bearing bracket bolts to specifications.

b. For QR25DE models, install the retaining bracket with the notch facing up and follow the bolt tightening order.

c. In order to prevent damage to differ-

ential side oil seal, place Tool (A) onto oil seal before inserting drive shaft as shown. Slide drive shaft into slide joint and tap with a hammer to install securely.

d. Install new circlip on drive shaft in the circular clip groove on transaxle side.

> ※※ **CAUTION**
> **Make sure the new circlip on the drive shaft is securely fastened.**

e. After its insertion, try to pull the flange out of the slide joint by hand.

> ※※ **CAUTION**
> **If it pulls out, the circlip is not properly meshed with the transaxle side gear.**

CV-BOOTS INSPECTION

1. Check the rubber boot for cracks and/or tears.

2. Check the metal boot bands for breaks, looseness or other damage.

3. Replace any defective parts.

ENGINE COOLING

ENGINE FAN

REMOVAL & INSTALLATION

1. Disconnect the 12-volt battery negative terminal.

2. Drain engine coolant from radiator, condenser and liquid tank assembly.

> ※※ **CAUTION**
> **Perform when engine and inverter are cold.**

3. Remove inverter upper bracket.

4. Drain inverter coolant from sub-radiator.

5. Remove air cleaner duct (front).

6. Disconnect radiator upper hose.

7. Remove sub radiator coolant reservoir tank.

8. Disconnect ECM.

9. Remove ECM and bracket assembly.

10. Disconnect radiator cooling fan controller.

11. Remove radiator cooling fan assembly.

To install:

12. Installation is in the reverse order of removal.

➡Radiator cooling fan is controlled by ECM.

RADIATOR

REMOVAL & INSTALLATION

> ※※ **WARNING**
> **Never remove the radiator cap when the engine and inverter are hot. Serious burns could occur from high pressure coolant escaping from the radiator, condenser and liquid tank assembly. Wrap a thick cloth around the cap. Slowly turn it a quarter turn to allow built-up pressure to escape. Carefully remove the cap by turning it all the way.**

1. Remove engine under cover.

2. Drain engine coolant from radiator, condenser and liquid tank assembly.

3. Remove air cleaner duct (front).

4. Remove radiator, condenser and liquid tank assembly upper hose and lower hose.

5. Remove coolant reservoir hose.

6. Drain inverter coolant from sub-radiator.

7. Remove front bumper reinforcement.

8. Remove sub radiator.

9. Discharge A/C system.

10. Remove high side junction pipe assembly.

11. Remove the refrigerant pressure sensor for installation on new radiator, condenser and liquid tank assembly.

12. Remove both radiator, condenser and liquid tank assembly clips.

13. Remove radiator, condenser and liquid tank assembly.

> ※※ **CAUTION**
> **Note the following:**

- Do not damage or scratch the radiator, condenser and liquid tank assembly and sub radiator core when removing.
- When removing refrigerant components from a vehicle, immediately cap (seal) the component to minimize the entry of moisture from the atmosphere.
- When installing refrigerant components to vehicle, never remove the caps (unseal) until just before connecting the components. Connect all refrigerant loop components as quickly as possible to minimize the entry of moisture into system

To install:

14. Installation is in the reverse order of removal.

THERMOSTAT

REMOVAL & INSTALLATION

See Figure 28.

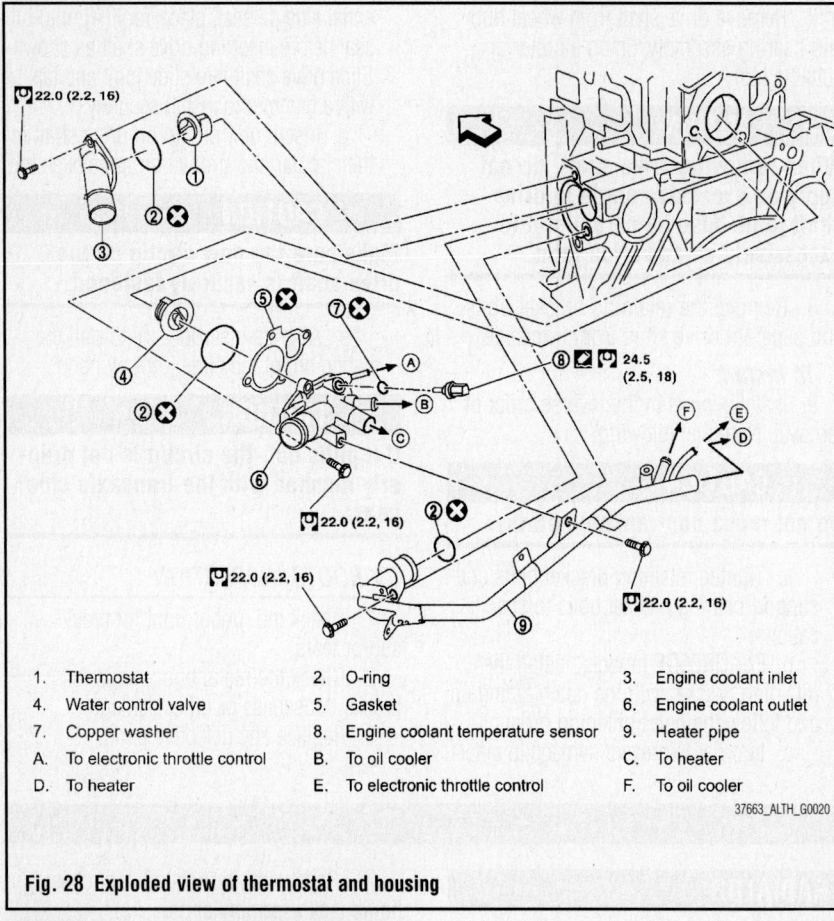

1.	Thermostat	2.	O-ring	3.	Engine coolant inlet
4.	Water control valve	5.	Gasket	6.	Engine coolant outlet
7.	Copper washer	8.	Engine coolant temperature sensor	9.	Heater pipe
A.	To electronic throttle control	B.	To oil cooler	C.	To heater
D.	To heater	E.	To electronic throttle control	F.	To oil cooler

37663_ALTH_G0020

Fig. 28 Exploded view of thermostat and housing

✳✳ WARNING

Never remove the radiator cap when the engine is hot. Serious burns could occur from high pressure coolant escaping from the radiator.

✳✳ CAUTION

Perform when the engine is cold.

1. Drain engine coolant from radiator, condenser and liquid tank assembly.

2. Remove air cleaner duct (front).

3. Remove radiator lower hose from the engine coolant inlet side.

4. Remove engine coolant inlet and thermostat.

To install:

5. Installation is in the reverse order of removal.

6. Install the engine coolant temperature sensor.

➡**Use Genuine RTV Silicone Sealant or equivalent.**

7. Install the thermostat with the whole circumference of the flange part fitting securely inside the rubber ring.

8. Install the thermostat with the jiggle valve facing upwards. The position deviation may be within the range of ±10°.

9. If necessary, to install the heater pipe, first apply a mild detergent to the O-ring and then quickly insert the pipe into the housing.

10. Use a new gasket and O-ring for installation.

WATER PUMP

REMOVAL & INSTALLATION

See Figure 29.

✳✳ WARNING

Never remove the radiator cap when the engine is hot. Serious burns could occur from high pressure coolant escaping from the radiator.

1. Drain engine coolant from radiator, condenser and liquid tank assembly.

✳✳ CAUTION

Perform when the engine is cold.

2. Remove drive belt.
3. Remove engine cover.
4. Remove air cleaner duct (front).
5. Remove engine coolant reservoir.
6. Remove RH wheel and tire assembly.
7. Remove fender protector.
8. Remove engine ground strap.
9. Remove idler pulley bracket.
10. Remove the water pump.

✳✳ CAUTION

Handle the water pump vane so that it does not contact any other parts.

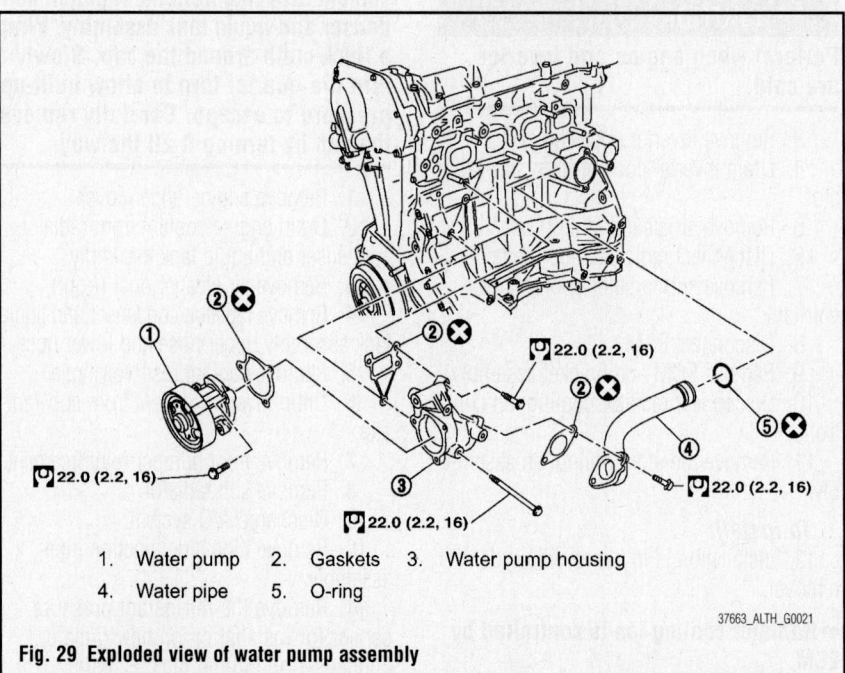

1.	Water pump	2.	Gaskets	3.	Water pump housing
4.	Water pipe	5.	O-ring		

37663_ALTH_G0021

Fig. 29 Exploded view of water pump assembly

※※ **CAUTION**

Water pump cannot be disassembled and should be replaced as an assembly.

→**If necessary, the exhaust manifold**

and three way catalyst assembly must be removed to remove the water pipe.

To install:

11. Installation is in the reverse order of removal.

12. When inserting water pipe end to cylinder block, apply a neutral detergent to O-ring. Then insert it immediately.

13. After installing the water pump, check for leaks using the radiator cap tester.

ENGINE ELECTRICAL

Refer to Engine Hybrid System section for information regarding the charging system of this vehicle.

ALTERNATOR

REMOVAL & INSTALLATION

Refer to Engine Hybrid System section for information regarding

CHARGING SYSTEM

the charging system of this vehicle.

ENGINE ELECTRICAL

PRECAUTIONS

※※ **CAUTION**

Observe the following precautions to ensure safe and proper servicing. These precautions are not described in each individual section.

PRECAUTION FOR HIGH-VOLTAGE CIRCUIT

→**The hybrid system contains a 244.8 V high-voltage system with a strong alkali solution of potassium hydroxide.**

※※ **WARNING**

Be sure to follow the instructions in this manual to handle the system correctly. Failure to do so may result in serious injury or electrocution.

1. Engineer must undergo special training to be able to perform high-voltage system inspection and servicing.

2. High-voltage cables are colored orange. The HV battery and other high-voltage components have "high voltage" caution labels. Do not carelessly touch these wires and components.

3. Before inspecting or servicing the high-voltage system, be sure to follow safety measures, such as wearing insulated gloves and removing the service plug to prevent electrocution. Carry the removed service plug in your pocket to prevent other technicians from reinstalling it while you are servicing the vehicle.

   a. Before removing the service plug, confirm ignition switch off.

   b. Do not put the vehicle into the ON (READY) state after removing the service plug grip as the ECU may be damaged.

ON (READY): The condition which the ready indicator lamp illuminates and vehicle is ready to be driven.

   c. Turn the ignition switch off, wear insulated gloves, and disconnect the negative terminal of the auxiliary battery before touching any of the orange-colored wires of the high-voltage system.

   d. Turn the ignition switch off before performing any resistance checks.

   e. Turn the ignition switch off before disconnecting or reconnecting any connectors.

4. After removing the service plug, wait 10 minutes before touching any of the high-voltage connectors and terminals.

→**10 minutes are required to discharge the high-voltage condenser inside the inverter.**

5. Before wearing insulated gloves, make sure that they are not cracked, ruptured, torn, or damaged in any way. Do not wear wet insulated gloves.

6. When servicing the vehicle, do not carry metal objects like mechanical pencils or scales that can be dropped accidentally and cause a short circuit.

7. Before touching a bare high-voltage terminal, wear insulated gloves and use a tester to make sure that the terminal voltage is 0 V.

8. After disconnecting or exposing a high-voltage connector or terminal, insulate it immediately using insulation tape.

9. The screw of a high-voltage terminal should be tightened firmly to the specified torque. Both insufficient and excessive torque can cause failure.

10. Use the "CAUTION: high-voltage. DO NOT TOUCH DURING OPERATION" sign to notify other engineers that a high-voltage system is being inspected and/or repaired.

ENGINE HYBRID SYSTEM

11. After servicing the high-voltage system and before reinstalling the service plug, check again that you have not left a part or tool inside, that the high-voltage terminal screws are firmly tightened, and that the connectors are correctly connected.

12. Do not place the battery upside down while removing and installing it.

※※ **WARNING**

When engaged in operations such as removal and installation related to high-voltage equipment, use personal protective equipment to avoid death or serious personal injury from electric shock.

AUXILIARY BATTERY

REMOVAL & INSTALLATION
See Figure 30.

1. Remove trunk side finisher (RH).

2. Loosen 12-volt battery terminal nuts, and disconnect both 12-volt battery terminals.

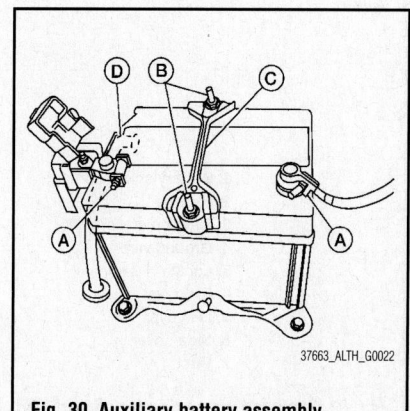

37663_ALTH_G0022

Fig. 30 Auxiliary battery assembly

❈❈ CAUTION

When disconnecting, disconnect the 12-volt battery negative terminal first.

3. Remove the 12-volt battery ventilation tube.

4. Remove 12-volt battery frame nuts and 12-volt battery frame.

5. Remove 12-volt battery.

To install:

6. Installation is the reverse order of removal.

❈❈ CAUTION

When connecting, connect the 12-volt battery positive terminal first.

BATTERY SMART UNIT

REMOVAL & INSTALLATION

See Figure 31.

1. Remove the HV relay assembly from the HV battery assembly.

2. Remove the bolt from the battery smart unit.

3. Disconnect the connectors from the battery smart unit and remove it from the HV battery assembly.

To install:

4. Installation is in the reverse order of removal.

FRAME WIRE

REMOVAL & INSTALLATION

Frame Wire (Main)

See Figures 32 and 33.

1. Disconnect the positive 12 volt terminal from the 12 volt battery.

2. Remove the rear seat.

3. Remove the fuel tank.

4. Remove the 12 volt positive battery cable retaining clips from the trunk compartment.

5. Disconnect the DC/DC converter connectors.

6. Remove the DC/DC converter harness retaining clip from the HV battery assembly.

7. Remove the frame wire from the HV battery assembly.

8. Disconnect the 12 volt terminal and cable retaining clip from the HV battery assembly.

9. Remove the frame wire harness retaining clips from the vehicle interior.

10. Remove the air cleaner and air duct.

11. Remove the inverter cover and terminal cover from the inverter.

12. Remove the frame wire inverter connector bolt and disconnect the frame wire inverter connector from the inverter.

13. Remove the HV fuse box cover from the HV fuse box.

14. Remove the HV fuse box terminal cap and nuts from the HV fuse box.

15. Open the HV fuse box side cover and remove the harness retaining clip and HV fuse box terminals from the HV fuse box.

16. Disconnect the EPS ECU connectors.

17. Remove the EPS ECU harness retaining clips from the engine room.

18. Remove the EPS ECU bonding wire bolt.

19. Remove the frame wire harness retaining clips from the engine room.

20. Remove the RH member pin stay.

21. Remove the frame wire retainer nuts and bolts from the underside of vehicle.

22. Remove the frame wire harness assembly with grommet from floor pass through and underside of vehicle.

23. Remove the frame wire harness from the engine room clip and remove the frame wire harness from the engine room.

To install:

24. Installation is in the reverse order of removal.

Frame Wire (Electric Compressor)

1. Remove the air cleaner and air duct.

2. Remove the front terminal cover bolt from the inverter cover and disconnect the electric compressor inverter connector from the inverter.

3. Disconnect the electric compressor connector from the electric compressor.

4. Remove the front engine mounting insulator and bracket bolts.

5. Remove the electric compressor frame wire harness clips and electric compressor frame wire harness from the vehicle.

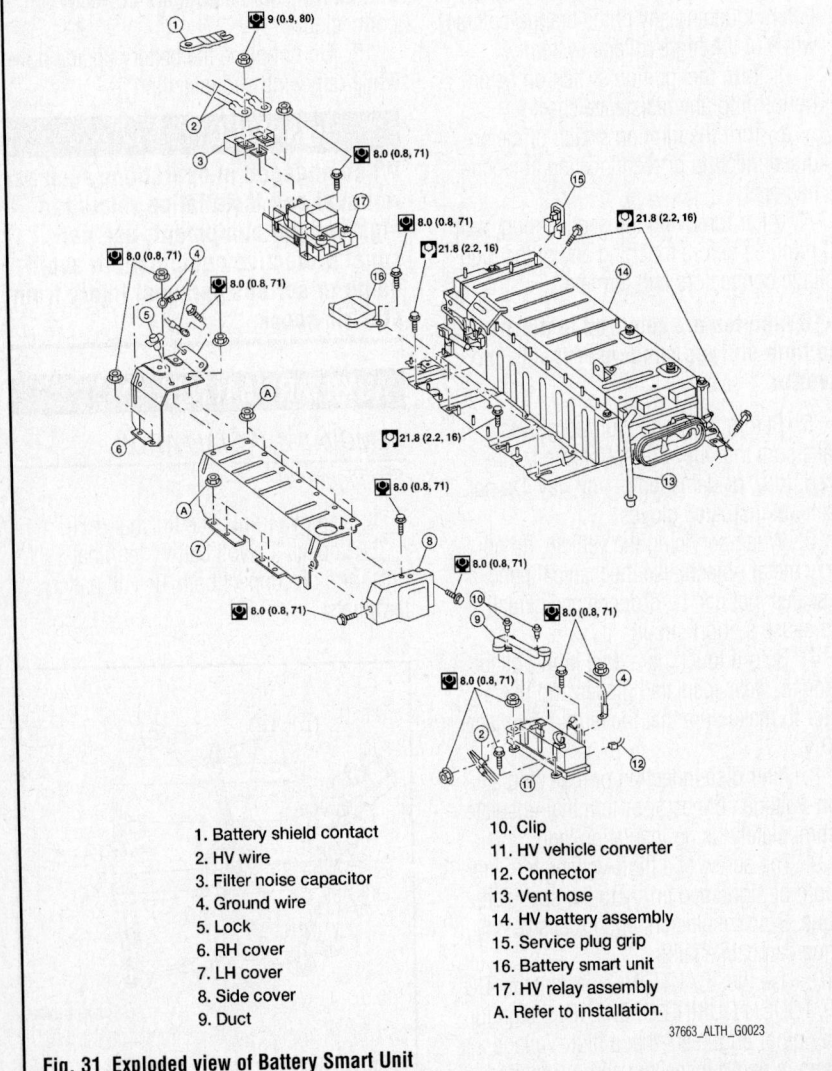

9 (0.9, 80)
8.0 (0.8, 71)
8.0 (0.8, 71)
8.0 (0.8, 71)
21.8 (2.2, 16)
8.0 (0.8, 71)
21.8 (2.2, 16)
21.8 (2.2, 16)
8.0 (0.8, 71)
8.0 (0.8, 71)
8.0 (0.8, 71)
8.0 (0.8, 71)

1. Battery shield contact
2. HV wire
3. Filter noise capacitor
4. Ground wire
5. Lock
6. RH cover
7. LH cover
8. Side cover
9. Duct
10. Clip
11. HV vehicle converter
12. Connector
13. Vent hose
14. HV battery assembly
15. Service plug grip
16. Battery smart unit
17. HV relay assembly
A. Refer to installation.

37663_ALTH_G0023

Fig. 31 Exploded view of Battery Smart Unit

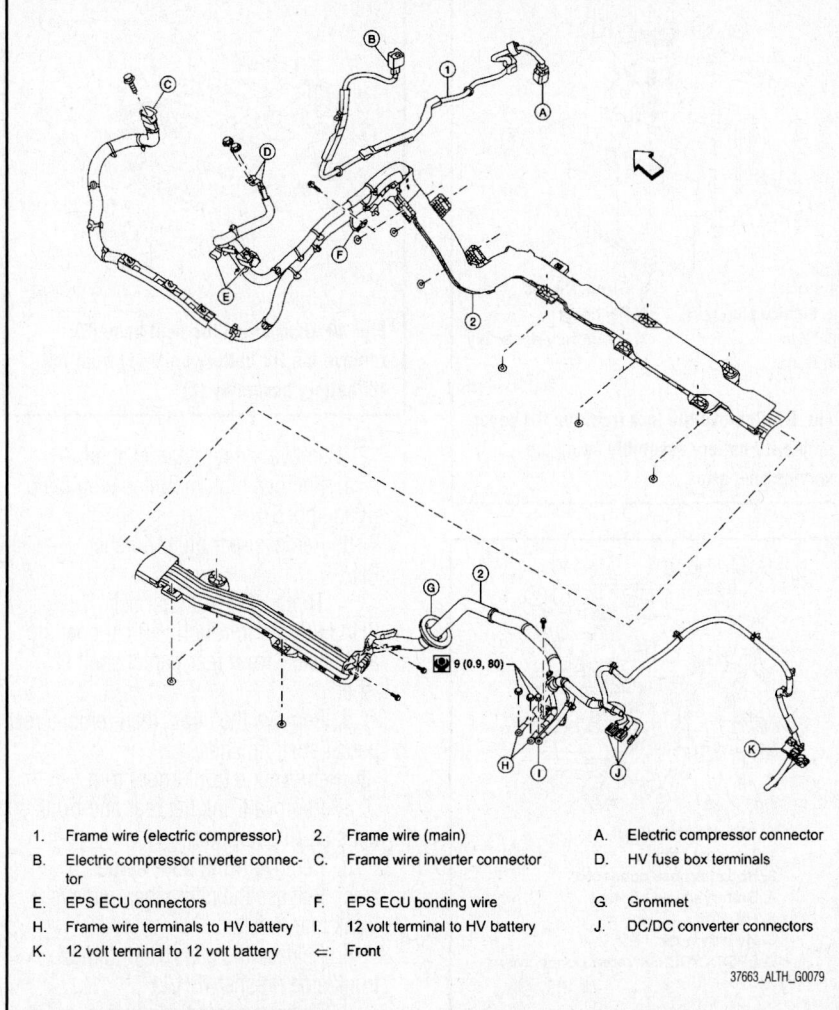

1. Frame wire (electric compressor)
2. Frame wire (main)
A. Electric compressor connector
B. Electric compressor inverter connector
C. Frame wire inverter connector
D. HV fuse box terminals
E. EPS ECU connectors
F. EPS ECU bonding wire
G. Grommet
H. Frame wire terminals to HV battery
I. 12 volt terminal to HV battery
J. DC/DC converter connectors
K. 12 volt terminal to 12 volt battery
⇐: Front

37663_ALTH_G0079

Fig. 32 Exploded view of frame wire

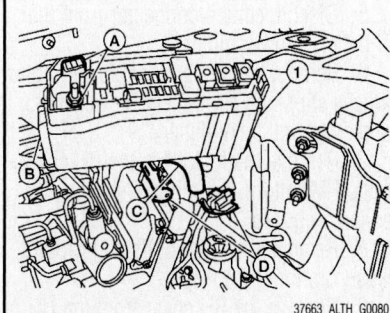

37663_ALTH_G0080

Fig. 33 Remove the HV fuse box cover from the HV fuse box

To install:
6. Installation is in the reverse order of removal.

HV BATTERY ASSEMBLY

REMOVAL & INSTALLATION

2007–08 Models

See Figure 34.

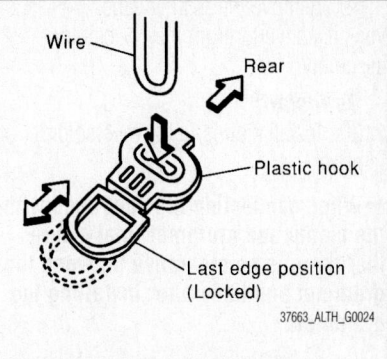

37663_ALTH_G0024

Fig. 34 Pull the lock at the front bottom of the seat cushion forward

❋❋ **CAUTION**

Do not tilt the HV battery more than 30° for extremely long time. Do not tilt the HV battery more than 60°.

1. Remove the rear seat.
   a. Remove the rear seat cushion trim and pad.
   b. Pull the lock at the front bottom of the seat cushion forward (one for each side), and pull the seat cushion upward to release the wire from the plastic hook, then pull the seat cushion forward to remove.
   c. Remove the seatback anchor bolts.
   d. Lift the seatback off the rear parcel panel front hangers and remove the seatback assembly.
2. Remove the trunk room trim.
   a. Release the latch, then position rear seatback (RH/LH) to the folded down position.
   b. Release the clips, then remove trunk floor carpet.
   c. Remove trunk net rear and trunk net side (if installed).
   d. Release the clips, then remove trunk rear finisher.
   e. Release the clips, then remove trunk side finisher (RH/LH).
   f. Remove spare tire cover and trunk floor board (LH).
   g. Remove the trunk lid rubber bumper (RH/LH), then release the clips and remove trunk lid finisher.
3. Remove the inlet and outlet cooling ducts.
4. Remove the lock using the service plug grip.
5. Disconnect the ground wire.
6. Remove the RH cover from the HV battery assembly.
7. Remove the LH cover from the HV battery assembly.
8. Disconnect the connectors from the high voltage vehicle converter
9. Remove the harness clips and harness from the HV battery assembly.
10. Disconnect the body harness connector from the HV battery assembly.
11. Disconnect the connector from the HV battery blower motor.
12. Remove the HV battery blower motor harness clips and HV battery blower motor harness from the HV battery.
13. Remove the frame wire and 12V battery harness from the HV battery.
14. Disconnect the vent hose from the vehicle.
15. Remove the HV battery bolts from the HV battery.
16. Remove the HV battery from the vehicle.
17. Remove the HV battery cooling ducts and HV battery blower motor from the battery assembly.
18. Disconnect the high voltage vehicle converter harness connector from the HV battery assembly.

19. Remove the high voltage vehicle converter from the HV battery assembly.

To install:

20. Installation is in the reverse order of removal.

➡ **When connecting the vent hose with the clamp and grommet, make sure that there is no clearance between the grommet and body after installing the grommet.**

2009–10 Models

See Figures 35 through 43.

⁂ **CAUTION**

Do not tilt the HV battery more than 30° for extremely long time. Do not tilt the HV battery more than 60°.

1. Remove the rear seat.
   a. Remove the rear seat cushion trim and pad.
   b. Pull the lock at the front bottom of the seat cushion forward (one for each side), and pull the seat cushion upward to release the wire from the plastic hook,

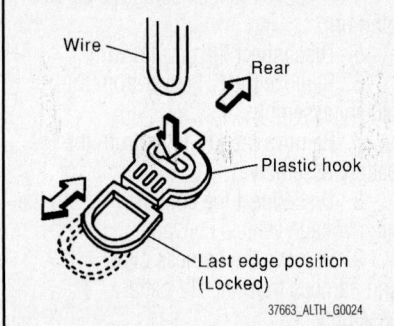

Fig. 35 Pull the lock at the front bottom of the seat cushion forward

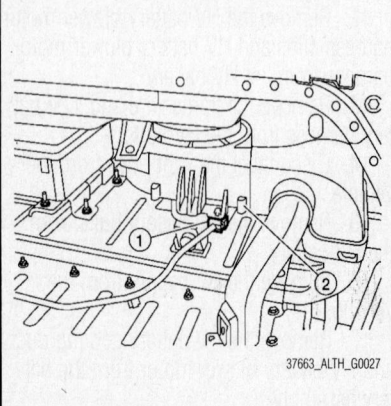

Fig. 36 Disconnect the connector (1) from the HV battery blower motor (2)

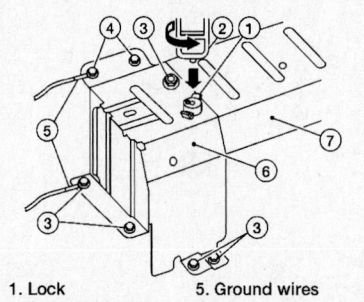

1. Lock
2. Service plug grip
3. Nuts
4. Bolts
5. Ground wires
6. RH cover
7. HV battery assembly

37663_ALTH_G0028

Fig. 37 Remove the lock from the RH cover on the HV battery assembly using the service plug grip

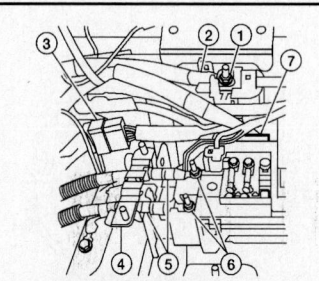

1. 12 volt terminal nut
2. Terminal cable
3. Body harness connector
4. Battery shield contact
5. HV wires
6. HV wire nuts
7. EPS DC/DC converter connector

37663_ALTH_G0029

Fig. 38 Remove the terminal cover and 12 volt terminal nut, then remove the terminal cable and 12 volt harness from the HV battery assembly

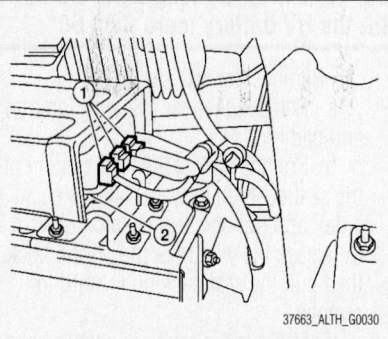

37663_ALTH_G0030

Fig. 39 Disconnect the electrical connectors (1) from the EPS DC/DC converter (2)

then pull the seat cushion forward to remove.
   c. Remove the seatback anchor bolts.
   d. Lift the seatback off the rear parcel panel front hangers and remove the seatback assembly.

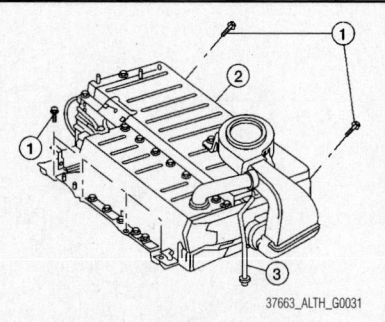

37663_ALTH_G0031

Fig. 40 Disconnect the vent hose (3); remove the HV battery bolts (1) from the HV battery assembly (2)

2. Remove the rear parcel shelf.
   a. Remove high mounted stop lamp (if equipped).
   b. Remove rear pillar finisher RH/LH.
   c. Thread the rear seat belt RH/LH/Center through vertical opening and release from rear parcel shelf finisher.
   d. Remove the clips, then remove rear parcel shelf finisher.
3. Remove the trunk room trim.
   a. Remove trunk net rear and trunk net side (if equipped).
   b. Remove trunk floor carpet.
   c. Release the clips, then remove trunk rear finisher.
   d. Release the clips, then remove trunk side finisher (RH/LH).
   e. Remove spare tire cover.
4. Remove the inlet and outlet cooling ducts.
5. Disconnect the connector from the HV battery blower motor.
6. Remove the HV battery blower motor harness clips and HV battery blower motor harness from the HV battery.
7. Remove the lock from the RH cover on the HV battery assembly using the service plug grip.
8. Remove the nuts, bolts and ground wires from the RH cover.
9. Remove the RH cover from the HV battery assembly.
10. Remove the terminal cover and 12 volt terminal nut, then remove the terminal cable and 12 volt harness from the HV battery assembly.
11. Remove the battery shield contact, HV wire nuts and HV wires from the HV battery assembly.
12. Disconnect the body harness connector from the HV battery assembly.
13. Disconnect the EPS DC/DC converter connector from the HV battery assembly.

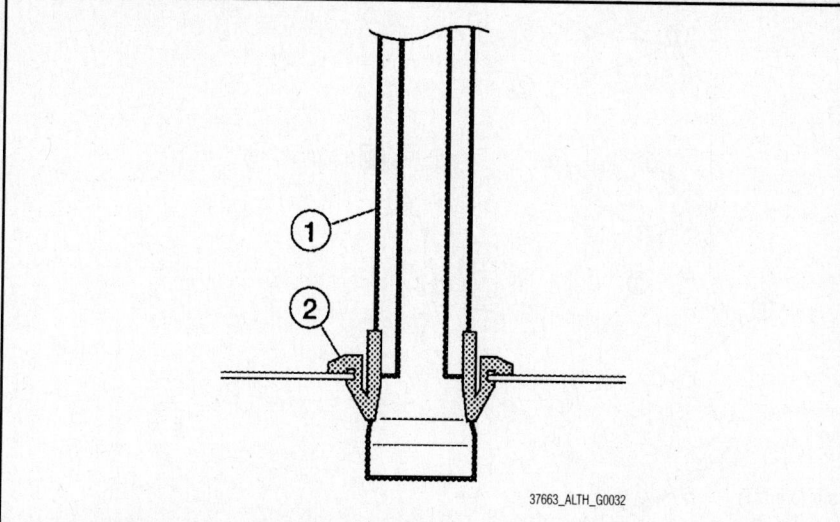

Fig. 41 When connecting the vent hose (1), make sure that there is no clearance between the grommet (2) and body after installing the grommet

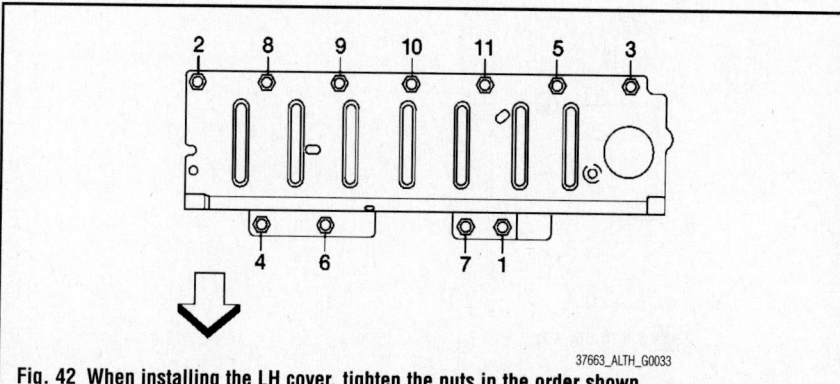

Fig. 42 When installing the LH cover, tighten the nuts in the order shown

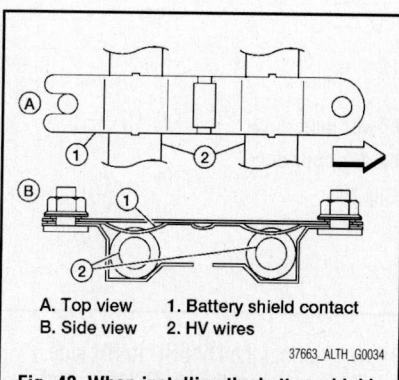

A. Top view 1. Battery shield contact
B. Side view 2. HV wires

Fig. 43 When installing the battery shield contact, position as shown

14. Remove the harnesses from the HV battery assembly.
15. Disconnect the electrical connectors from the EPS DC/DC converter.
16. Remove the harness clips and harness from the HV battery assembly.
17. Disconnect the vent hose from the vehicle.
18. Remove the HV battery bolts from the HV battery assembly.

19. Remove the HV battery assembly from the vehicle.
20. If necessary, remove the following components from the HV battery assembly:
- The HV battery blower motor and cooling ducts.
- The EPS DC/DC converter.
- The HV relay assembly.
- The battery smart unit.
- The HV vehicle converter.

To install:
21. Installation is in the reverse order of removal.

➡Note the following:

- When connecting the vent hose, make sure that there is no clearance between the grommet and body after installing the grommet.
- When installing the LH cover, tighten the nuts in the order shown.
- When installing the battery shield contact, position as shown.

- When installing the lock to the RH cover, push the lock into the hole and ensure it is locked.

HV BATTERY BLOWER MOTOR

REMOVAL & INSTALLATION

See Figures 35 and 44.

1. Remove the rear seat.
a. Remove the rear seat cushion trim and pad.
b. Pull the lock at the front bottom of the seat cushion forward (one for each side), and pull the seat cushion upward to release the wire from the plastic hook, then pull the seat cushion forward to remove.
c. Remove the seatback anchor bolts.
d. Lift the seatback off the rear parcel panel front hangers and remove the seatback assembly.
2. Remove the rear parcel shelf finisher.
a. Remove high mounted stop lamp (if equipped).
b. Remove rear pillar finisher RH/LH.
c. Thread the rear seat belt RH/LH/Center through vertical opening and release from rear parcel shelf finisher.
d. Remove the clips, then remove rear parcel shelf finisher.
3. Remove the trunk room trim.
a. Remove trunk net rear and trunk net side (if equipped).
b. Remove trunk floor carpet.
c. Release the clips, then remove trunk rear finisher.
d. Release the clips, then remove trunk side finisher (RH/LH).
e. Remove spare tire cover.
4. Remove the upper and lower inlet duct clips and bolts.
5. Remove the upper and lower inlet duct from the package shelf and HV battery blower motor.
6. Remove the front duct clip and remove the front duct from the rear upper duct and HV battery assembly.
7. Separate the rear upper duct from the rear lower duct and remove the rear upper duct from the HV battery blower motor.
8. Disconnect the HV battery blower motor harness connector from the HV battery blower motor.
9. Remove the HV battery blower motor from the HV battery assembly.

To install:
10. Installation is in the reverse order of removal.

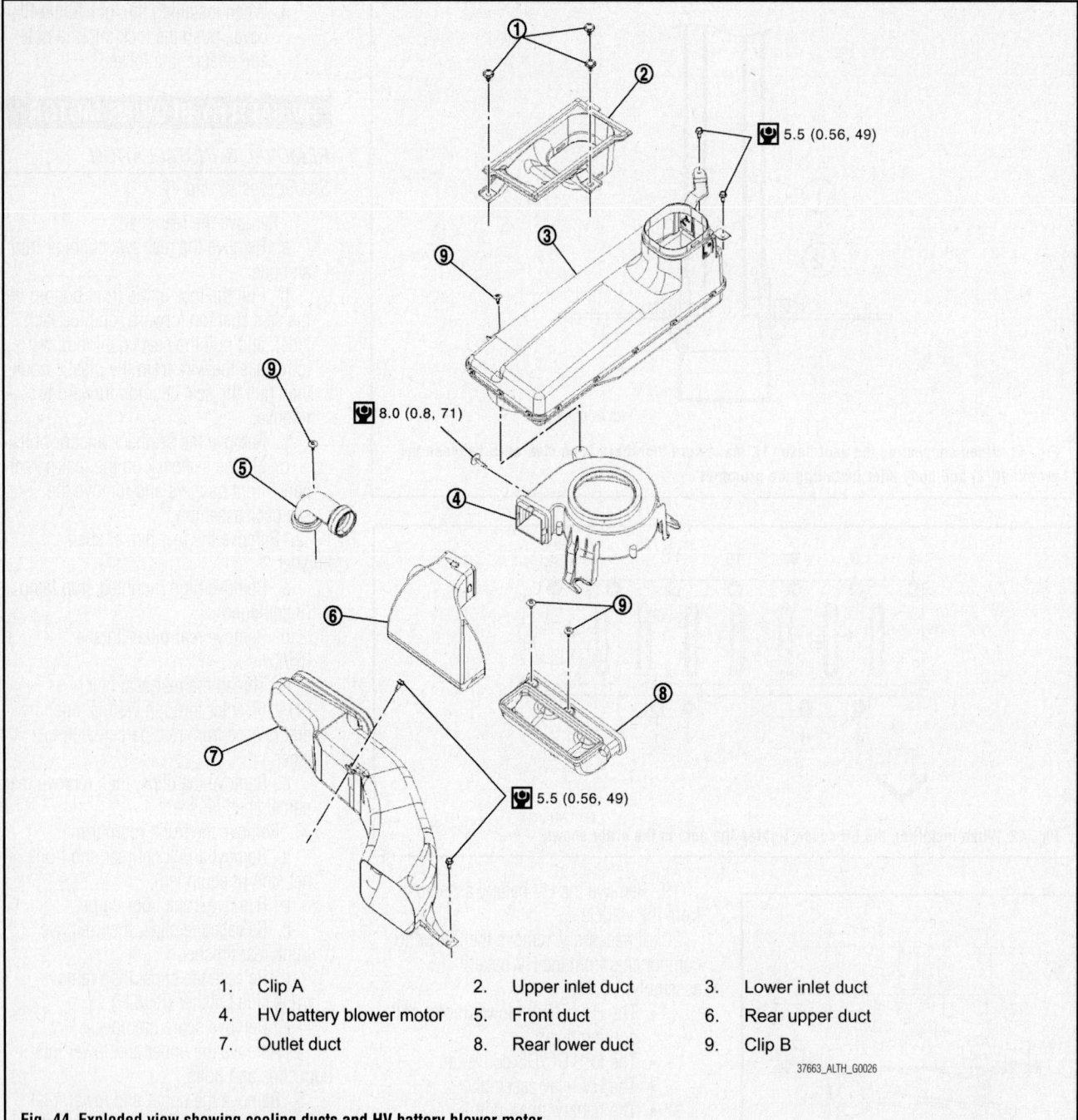

1. Clip A	2. Upper inlet duct	3. Lower inlet duct
4. HV battery blower motor	5. Front duct	6. Rear upper duct
7. Outlet duct	8. Rear lower duct	9. Clip B

37663_ALTH_G0026

Fig. 44 Exploded view showing cooling ducts and HV battery blower motor

HV ECU

REMOVAL & INSTALLATION

Precaution For Replacing Hybrid Vehicle Control ECU

See Figures 45 through 48.

When replacing the hybrid vehicle control ECU, never remove the waterproof sheet.

➡**The hybrid vehicle control ECU is covered with a waterproof sheet. If the waterproof sheet is peeled off, the labels on the hybrid vehicle control ECU will be removed together with the waterproof sheet. Consequently important data printed on the label for warranty procedure will be lost.**

1. Remove the console side finisher LH.

2. Remove the LH bolts from the HV ECU.

3. Remove the instrument side panel RH.

4. Remove the RH bolts from the HV ECU.

5. Disconnect the drain hose from the heater and cooling unit assembly.

6. Pull out the HV ECU to RH side.

7. Disconnect the HV ECU harness connector from the HV ECU, and remove the HV ECU from the vehicle.

8. If necessary, remove the screws and HV ECU brackets from the HV ECU.

To install:

9. Installation is in the reverse order of removal.

➡**When tightening the bolts, perform the following procedure and refer to "Exploded View".**

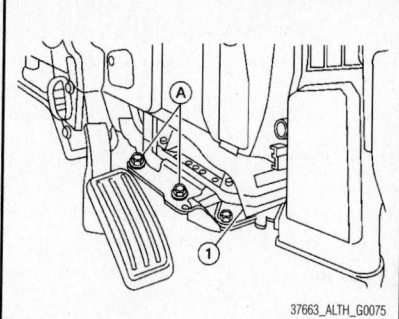

Fig. 45 Remove the LH bolts (A) from the HV ECU (1)

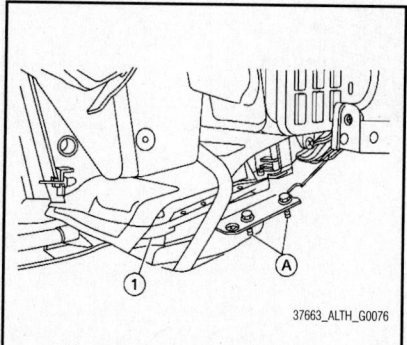

Fig. 46 Remove the RH bolts (A) from the HV ECU (1)

a. Temporarily tighten bolt A first.

b. Tighten the other bolts in numerical order to the specified torque.

c. Tighten bolt A to the specified torque.

10. If installing a new HV ECU, apply the waterproof sheet to the HV ECU as shown. Center the waterproof sheet on the HV ECU and press down on the adhesive area to secure the waterproof sheet to the HV ECU.

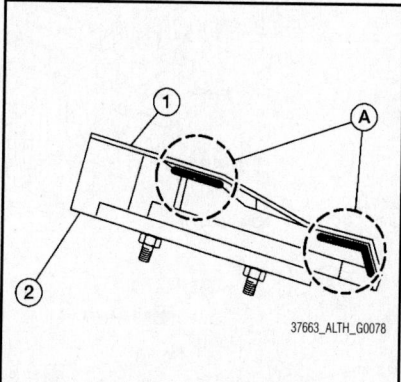

Fig. 48 Apply the waterproof sheet (1) to the HV ECU (2) and press down on the adhesive area (A)

HV RELAY ASSEMBLY

REMOVAL & INSTALLATION

2007–08 Models

1. Remove the frame wires from the HV battery assembly.

2. Remove the lock using the service plug grip.

3. Disconnect the ground wire.

4. Remove the RH cover from the HV battery assembly.

5. Remove the LH cover from the HV battery assembly.

6. Remove the filter noise capacitor.

7. Disconnect the connectors from the HV relay assembly.

8. Remove the bolts and the HV relay assembly from the HV battery assembly.

To install:

9. Installation is in the reverse order of removal.

2009–10 Models

See Figures 49 through 51.

1. Remove the HV wires from the HV battery assembly.

2. Remove the side cover and LH cover from the HV battery assembly.

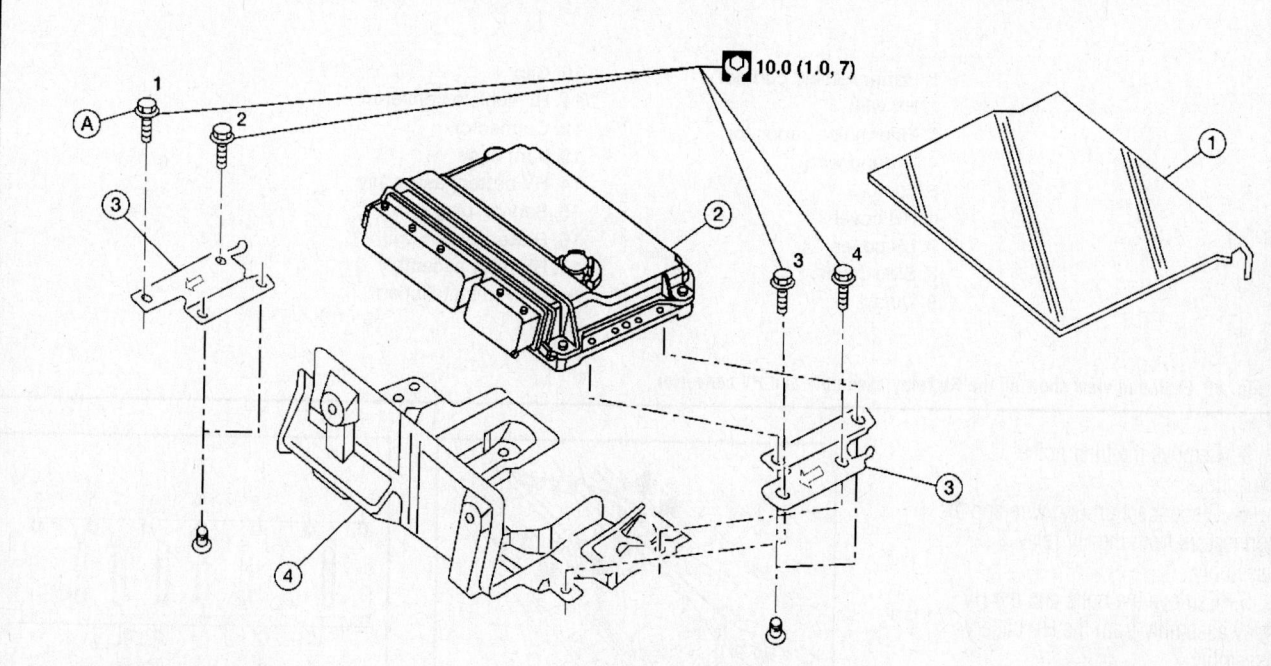

1. Waterproof sheet
4. Mounting bracket
2. HV ECU
A. Bolt
3. HV ECU bracket
⇐: Front

Fig. 47 Exploded view of HV ECU assembly

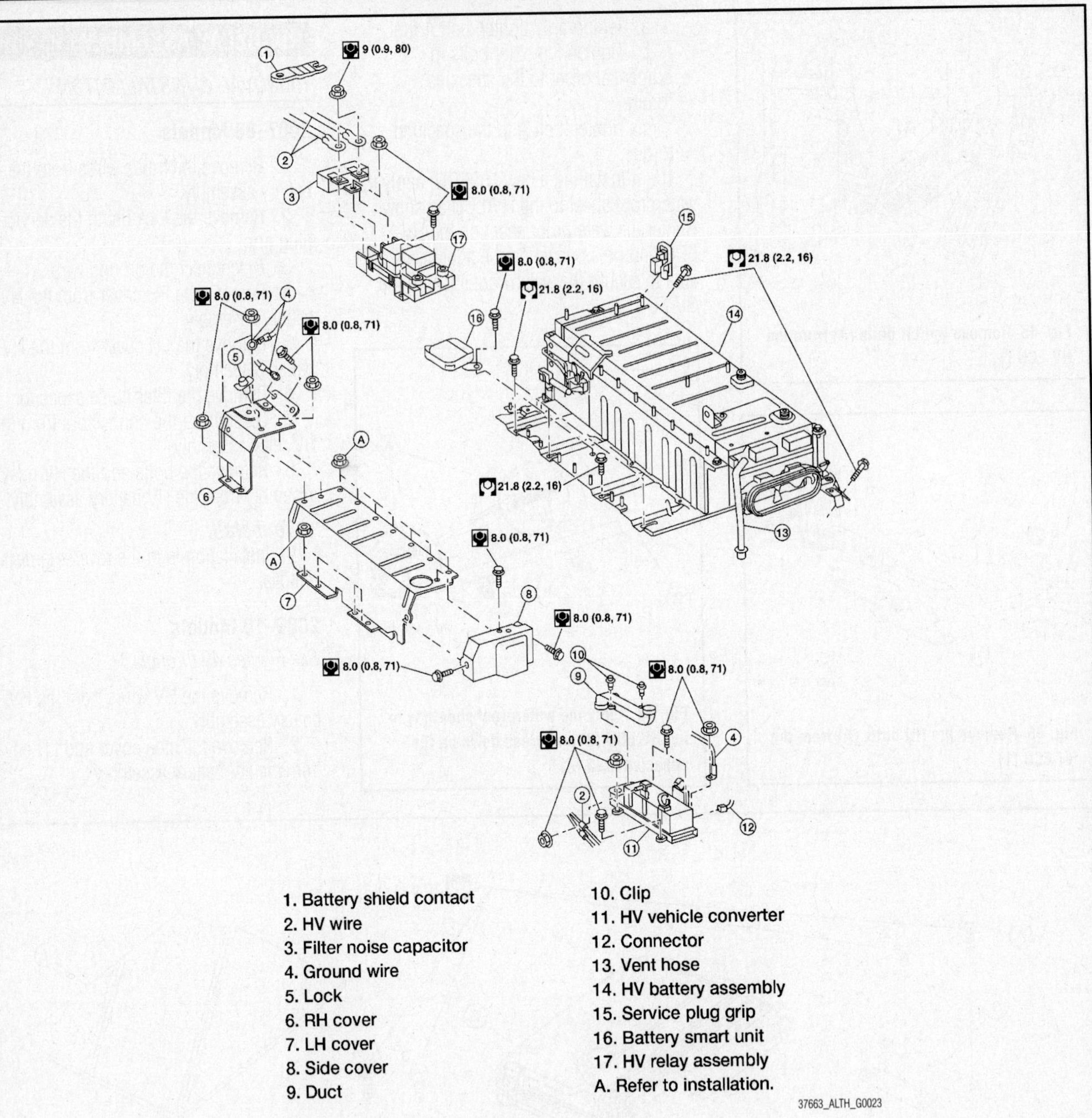

9 (0.9, 80)

8.0 (0.8, 71)

8.0 (0.8, 71)

21.8 (2.2, 16)

8.0 (0.8, 71)

8.0 (0.8, 71)

21.8 (2.2, 16)

21.8 (2.2, 16)

21.8 (2.2, 16)

8.0 (0.8, 71)

8.0 (0.8, 71)

8.0 (0.8, 71)

8.0 (0.8, 71)

8.0 (0.8, 71)

1. Battery shield contact	10. Clip
2. HV wire	11. HV vehicle converter
3. Filter noise capacitor	12. Connector
4. Ground wire	13. Vent hose
5. Lock	14. HV battery assembly
6. RH cover	15. Service plug grip
7. LH cover	16. Battery smart unit
8. Side cover	17. HV relay assembly
9. Duct	A. Refer to installation.

37663_ALTH_G0023

Fig. 49 Exploded view showing the HV relay assembly and HV converter

3. Remove the filter noise capacitor.

4. Disconnect ground wire and the connectors from the HV relay assembly.

5. Remove the bolts and the HV relay assembly from the HV battery assembly.

To install:

6. Installation is in the reverse order of removal.

7. When installing the LH cover, tighten the nuts in the order shown.

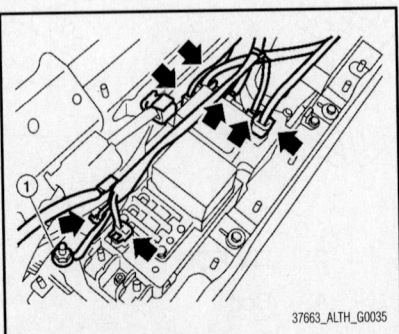

37663_ALTH_G0035

Fig. 50 Disconnect ground wire (1), and the connectors from the HV relay assembly

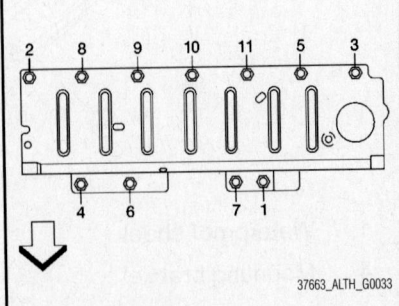

37663_ALTH_G0033

Fig. 51 When installing the LH cover, tighten the nuts in the order shown

HV VEHICLE CONVERTER

REMOVAL & INSTALLATION

2007–08 Models

See Figure 52.

1. Pull the service plug to disconnect high voltage battery.
2. Remove rear seat.
3. Remove EPS DC/DC converter cover nuts and remove the cover.
4. Remove nut of shield earth.
5. Disconnect EPS motor power line (245 V) connector and clip.

6. Disconnect EPS motor power line ground.
7. Disconnect EPS motor power line (42 V) connector and clip.
8. Remove EPS DC/DC converter nuts and remove the converter assembly.

To install:

9. Installation is in the reverse order of removal.

2009–10 Models

See Figure 49.

1. Disconnect the connectors from the HV relay assembly.

2. Remove the HV wire nut and HV wire from the HV vehicle converter.
3. Remove the ground wire nut and ground wire from the HV vehicle converter.
4. Remove the HV vehicle converter.
   a. Remove the HV vehicle converter nut and bolts from the HV vehicle converter.
   b. Disconnect the connector from the back of the HV vehicle converter.
   c. Remove the HV vehicle converter.
5. Remove the clips and duct from the HV vehicle converter.

To install:

6. Installation is in the reverse order of removal.

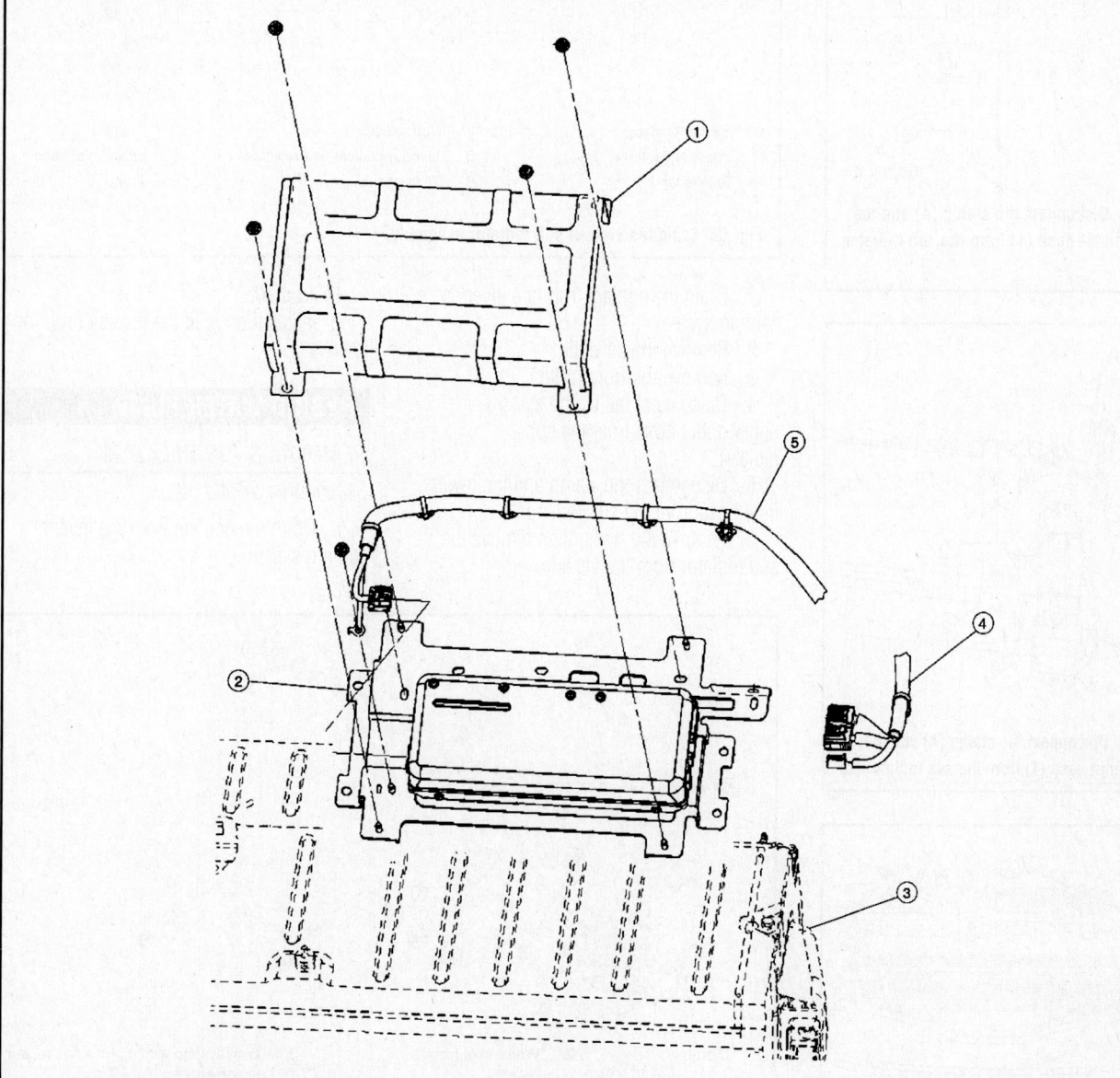

1.	EPS DC/DC converter cover	2.	EPS DC/DC converter	3.	High voltage battery assembly
4.	EPS motor power line (42 V)	5.	EPS motor power line (245 V)		

37663_ALTH_G0104

Fig. 52 Exploded view of HV vehicle converter

HYBRID SUB RADIATOR

REMOVAL & INSTALLATION

See Figures 53 through 56.

✳✳ CAUTION

Do not damage or scratch the radiator and condenser assembly and sub radiator core when removing.

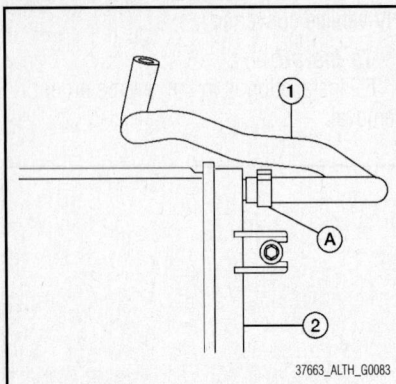

Fig. 53 Disconnect the clamp (A) and the upper outlet hose (1) from the sub radiator (2)

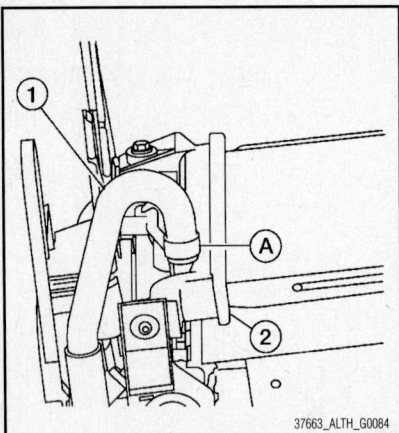

Fig. 54 Disconnect the clamp (A) and the lower inlet hose (1) from the sub radiator (2)

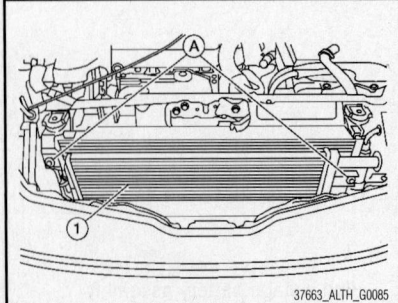

Fig. 55 Remove the bolts (A), then remove the sub radiator (1) from the vehicle

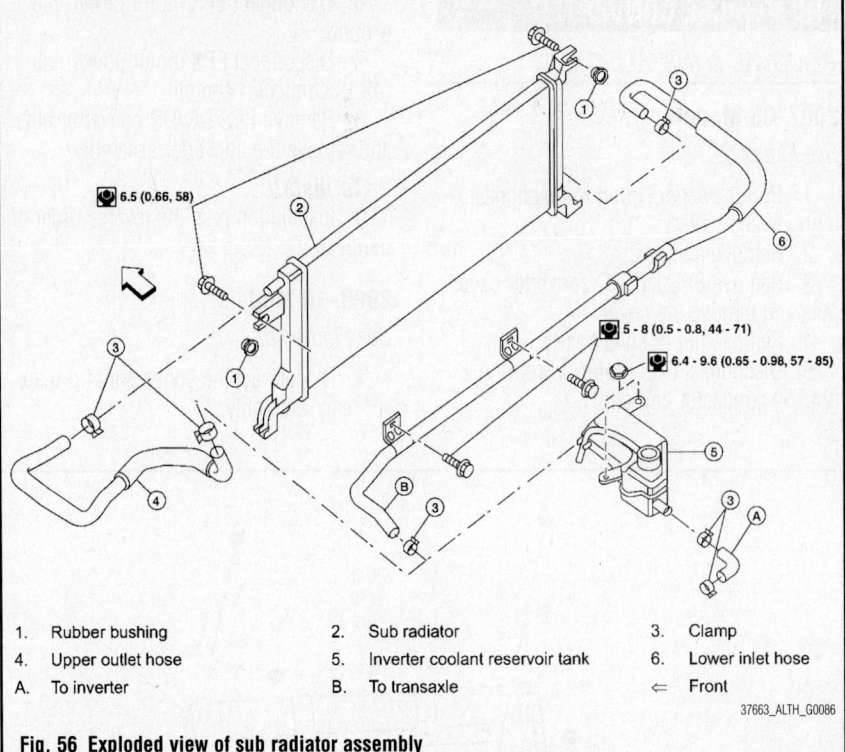

1. Rubber bushing	2. Sub radiator	3. Clamp
4. Upper outlet hose	5. Inverter coolant reservoir tank	6. Lower inlet hose
A. To inverter	B. To transaxle	⇐ Front

Fig. 56 Exploded view of sub radiator assembly

1. Drain the coolant from the inverter cooling system.
2. Remove the air duct.
3. Remove the front grille.
4. Disconnect the clamp and the upper outlet hose from the sub radiator.
5. Disconnect the clamp and the lower inlet hose from the sub radiator.
6. Remove the bolts, then remove the sub radiator from the vehicle.

To install:

7. Installation is in the reverse order of removal.

HYBRID WATER PUMP

REMOVAL & INSTALLATION

See Figures 57 and 58.

1. Drain the coolant from the inverter cooling system.

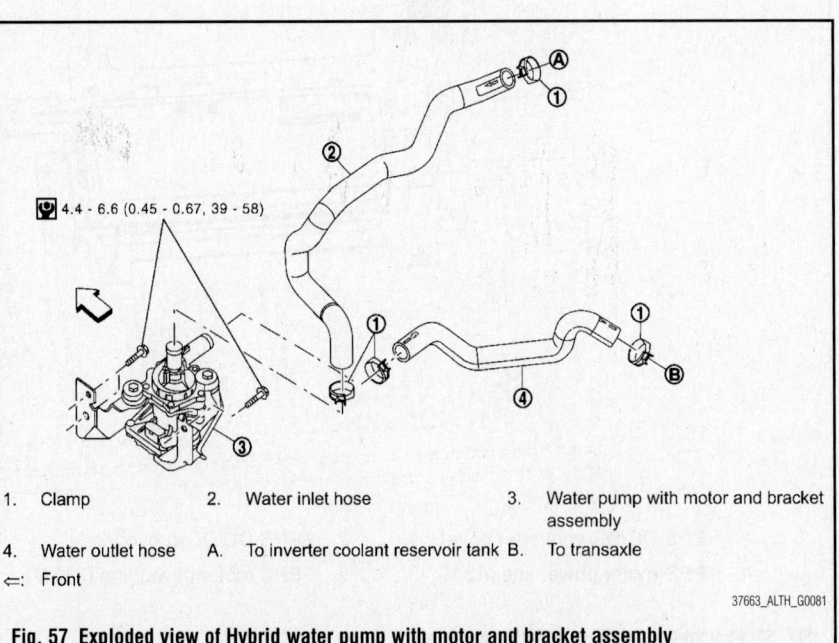

1. Clamp	2. Water inlet hose	3. Water pump with motor and bracket assembly
4. Water outlet hose	A. To inverter coolant reservoir tank	B. To transaxle
⇐: Front		

Fig. 57 Exploded view of Hybrid water pump with motor and bracket assembly

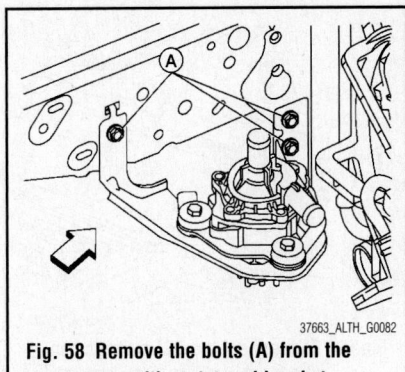

Fig. 58 Remove the bolts (A) from the water pump with motor and bracket assembly and remove from the vehicle

2. Disconnect the water inlet hose and water outlet hose from the water pump with motor and bracket assembly.

3. Remove the bolts from the water pump with motor and bracket assembly and remove from the vehicle.

To install:

4. Installation is in the reverse order of removal.

➡**Do not use the power tool.**

INVERTER WITH CONVERTER ASSEMBLY

REMOVAL & INSTALLATION

See Figures 59 through 64.

1. Drain the coolant from the inverter cooling system.
2. Remove the inverter cover.
3. Remove the air cleaner and air duct.
4. Remove the nuts and bolts from the upper center bracket.
5. Remove the inverter upper center bracket.
6. Remove the hoses and bolts from the inverter cooling reservoir tank.
7. Remove the inverter cooling reservoir tank from the vehicle.
8. Disconnect the MG1 and MG2 connectors from the inverter as follows:
    a. Remove bolts G, I, J and L as shown.
    b. Remove bolts H and K as shown.
    c. Disconnect the MG1 and MG2 connectors from the inverter.
9. Remove the MG1 and MG2 harness clips from the bracket and set the MG1 and MG2 harness aside.
10. Remove the coolant hoses from the inverter.
11. Remove the upper RH bracket bolt and bracket from the inverter.

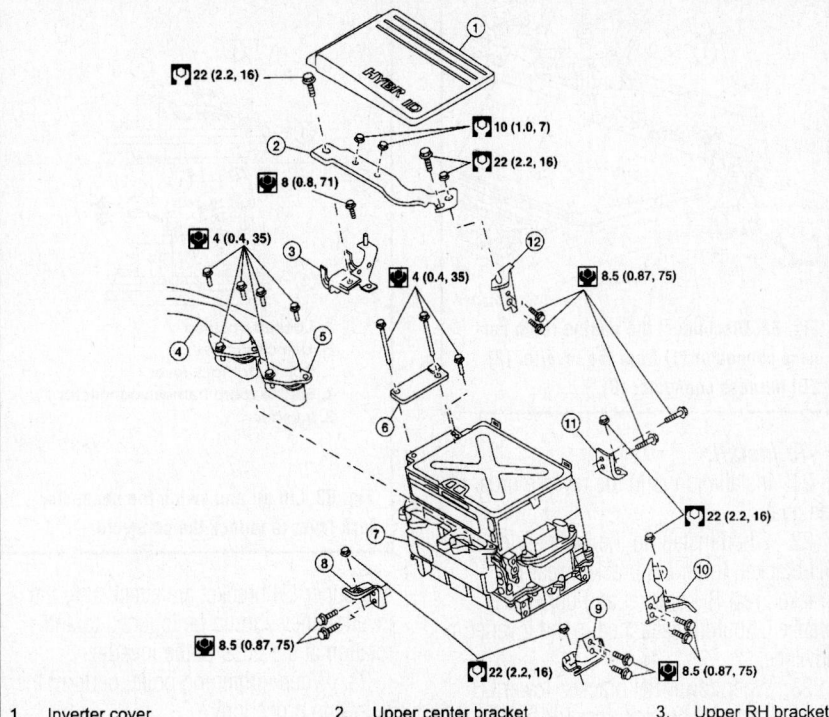

1.	Inverter cover	2.	Upper center bracket	3.	Upper RH bracket
4.	MG1 harness connector	5.	MG2 harness connector	6.	Terminal cover
7.	Inverter	8.	Lower RH bracket	9.	Lower LH bracket
10.	Rear LH bracket	11.	Rear RH bracket	12.	Upper LH bracket

Fig. 59 Exploded view of inverter with converter assembly

12. Remove the terminal cover bolt and terminal cover from the inverter.
13. Disconnect the electric compressor inverter connector from the inverter.
14. Remove the frame wire inverter connector bolt and disconnect the frame wire inverter connector from the inverter.
15. Disconnect the engine room harness connector from the inverter; EGI harness connector
    a. Lift up and swing the connector lock lever to unlock the connector.
    b. Pull up on the engine room harness connector to disconnect it from the inverter.

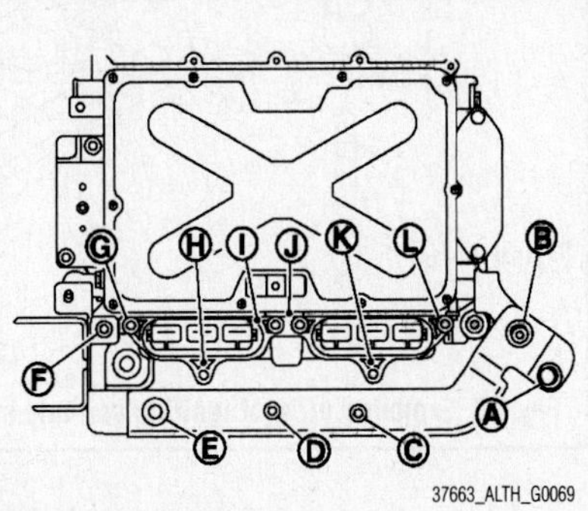

16. Remove the engine room harness clip from the bracket and set the engine room harness aside.
17. Disconnect the EGI harness connector from the inverter.
18. Remove the inverter nuts.
19. Remove the inverter from the vehicle.
20. Remove any necessary brackets from the inverter.

Fig. 60 Disconnect the MG1 and MG2 connectors from the inverter

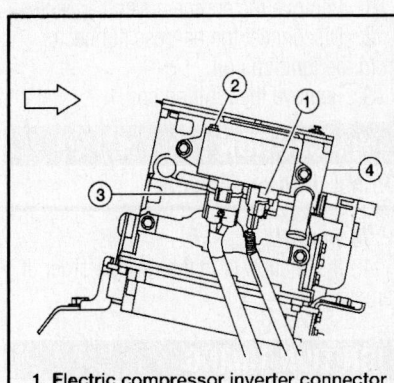

1. Electric compressor inverter connector
2. Frame wire inverter connector bolt
3. Frame wire inverter connector
4. Inverter

Fig. 61 Disconnect the electric compressor inverter connector from the inverter

Fig. 62 Disconnect the engine room harness connector (1) from the inverter (2); EGI harness connector (3)

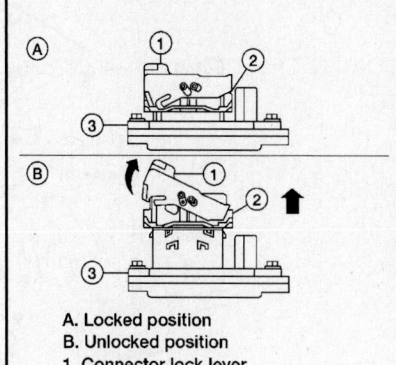

A. Locked position
B. Unlocked position
1. Connector lock lever
2. Engine room harness connector
3. Inverter

37663_ALTH_G0073

Fig. 63 Lift up and swing the connector lock lever to unlock the connector

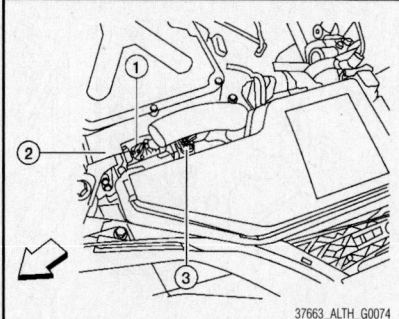

Fig. 64 Disconnect the EGI harness connector (3) from the inverter (2); EGI engine room connector (1)

To install:

21. Installation is in the reverse order of removal.

22. When installing the inverter, lower RH bracket, lower LH bracket, rear LH bracket, rear RH bracket and upper LH bracket should be attached to the inverter in advance.

23. When lower RH bracket, lower LH bracket, rear LH bracket, rear RH bracket

and upper LH bracket are attached to the inverter, they should be touched to anti-rotation at the boss of the inverter.

24. When tightening bolts, perform the following procedure:

a. Temporarily tighten the bolts A, B, E, F.

b. Connect MG1 harness connector and MG2 harness connector.

c. Fully tighten the bolts H, K.

d. Fully tighten the bolts G, I, J and L.

e. Fully tighten the bolt F.

f. Fully tighten the bolts A, B, C, D, E.

ENGINE ELECTRICAL

FIRING ORDER

The firing order for the 2.5L engine is 1–3–4–2. The 2.5L engine is an in-line 4 cylinder engine with cylinders numbered in order front to back of the engine.

IGNITION COIL

REMOVAL & INSTALLATION

See Figure 65.

1. Remove the engine cover.
2. Disconnect the harness connector from the ignition coil.
3. Remove the ignition coil.

❋❋ CAUTION

Do not drop or shock it.

To install:

4. Installation is in the reverse order of removal.

IGNITION TIMING

ADJUSTMENT

The ignition timing is controlled by the

ECM. No adjustment is necessary or possible.

SPARK PLUGS

REMOVAL & INSTALLATION

See Figure 65.

IGNITION SYSTEM

1. Remove the ignition coil.
2. Remove the spark plug with a suitable spark plug wrench.
3. Installation is in the reverse order of removal.

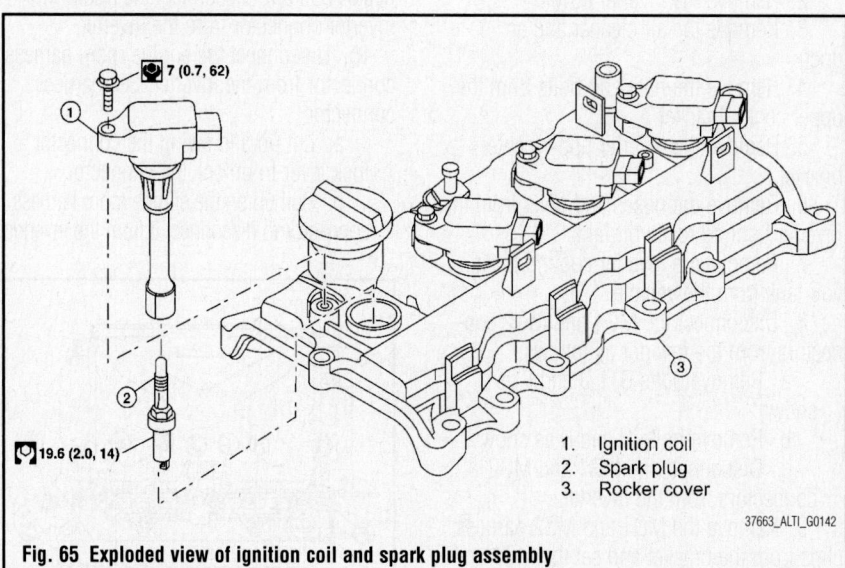

1. Ignition coil
2. Spark plug
3. Rocker cover

37663_ALTI_G0142

Fig. 65 Exploded view of ignition coil and spark plug assembly

ENGINE ELECTRICAL

STARTING SYSTEM

STARTER

REMOVAL & INSTALLATION

Refer to Engine Hybrid
System section for information

regarding the starting system of this
vehicle.

ENGINE MECHANICAL

ACCESSORY DRIVE BELTS

ACCESSORY BELT ROUTING

See Figure 66.

Refer to the accompanying illustration.

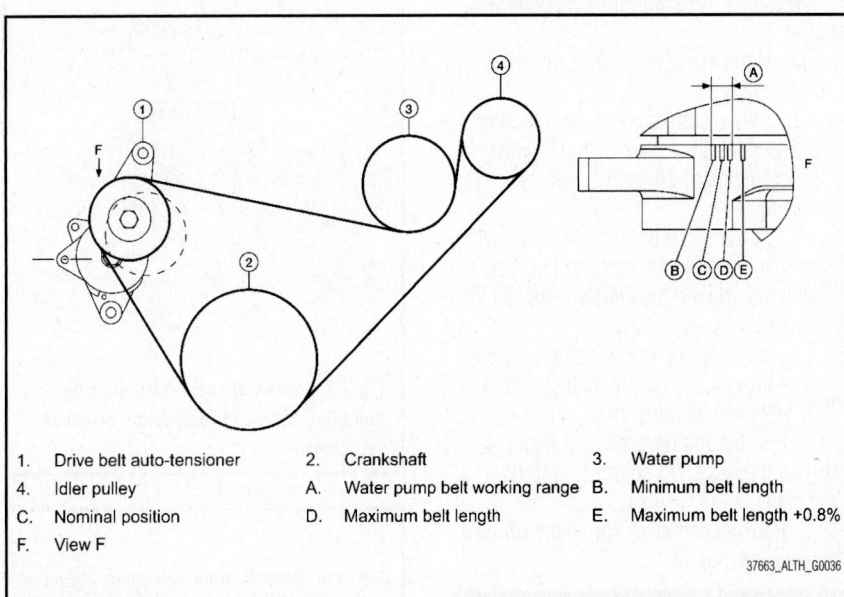

1.	Drive belt auto-tensioner	2.	Crankshaft	3.	Water pump
4.	Idler pulley	A.	Water pump belt working range	B.	Minimum belt length
C.	Nominal position	D.	Maximum belt length	E.	Maximum belt length +0.8%
F.	View F				

37663_ALTH_G0036

Fig. 66 Accessory drive belt routing

INSPECTION

> ※※ **WARNING**
>
> **Inspect the drive belt only when the Hybrid System is off.**

1. Make sure that the stamp mark of drive belt auto-tensioner is within the usable range.
2. Check the drive belt auto-tensioner indicator (notch) when the engine is cold.
3. When the new drive belt is installed, check that the belt is in range.
4. Visually check entire belt for wear, damage or cracks.
5. If the indicator is out of allowable use range or belt is damaged, replace the belt.

ADJUSTMENT

Belt tension is not manually adjustable, it is automatically adjusted by the drive belt auto-tensioner.

REMOVAL & INSTALLATION

1. Remove the fender protector side cover (RH).
2. Securely hold the hexagonal part in pulley center of drive belt auto-tensioner, using suitable tool.

> ※※ **WARNING**
>
> **Avoid placing hand in a location where pinching may occur if the holding tool accidentally comes off.**

> ※※ **CAUTION**
>
> **Do not loosen the auto-tensioner pulley bolt. (Do not turn it counterclockwise) If turned counterclockwise, the complete auto-tensioner must be replaced as a unit, including pulley.**

3. Insert a rod approximately 0.24 inches (6 mm) in diameter through the rear of tensioner into retaining boss to lock tensioner pulley.
4. Leave tensioner pulley arm locked until belt is installed again.

5. Loosen auxiliary drive belt from water pump pulley in sequence, and remove it.

   To install:
6. Hook the auxiliary drive belt onto all of the pulleys except for the water pump pulley. Hook the drive belt onto water pump pulley last.

> ※※ **CAUTION**
>
> **Confirm belts are completely set on the pulleys.**

7. Release tensioner, and apply tension to belt.

> ※※ **WARNING**
>
> **Avoid placing hand in a location where pinching may occur if the holding tool accidentally comes off.**

> ※※ **CAUTION**
>
> **Do not loosen the auto-tensioner pulley bolt. (Don't turn it counterclockwise). If turned counterclockwise, the complete auto-tensioner must be replaced as a unit, including pulley.**

8. Turn crankshaft pulley clockwise several times to equalize tension between each pulley.
9. Confirm tensions of belt at indicator is within the allowable use range.

CAMSHAFT AND VALVE LIFTERS

REMOVAL & INSTALLATION

See Figures 67 through 83.

1. Remove the rocker cover.
2. Remove the front right side tire and wheel.
3. Remove the RH splash shield.
4. Remove the drive belt.
5. Remove the power steering reservoir.
6. Remove the coolant overflow reservoir tank.

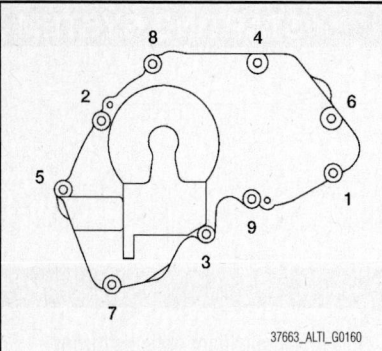

Fig. 67 Loosen the IVT control cover bolts in the order as shown

7. Disconnect variable timing control solenoid and camshaft sensor harness connectors.

8. Remove camshaft sensor.

9. Remove camshaft sensor bracket.

10. Loosen the IVT control cover bolts in the order as shown.

11. Remove the IVT control cover by cutting the sealant using Tool KV10111100 (J-37228).

12. Set the No.1 cylinder at TDC on its

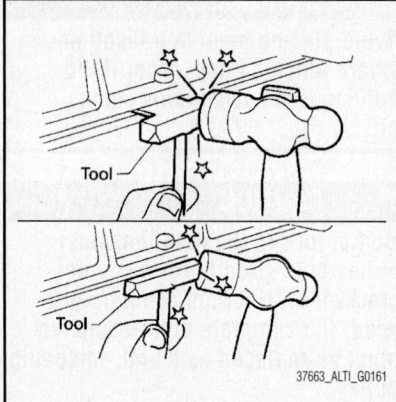

Fig. 68 Remove the IVT control cover by cutting the sealant

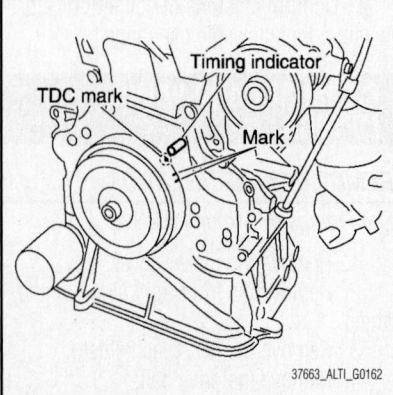

Fig. 69 Set the No.1 cylinder at TDC on its compression stroke

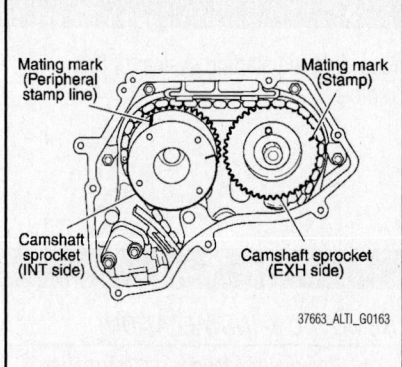

Fig. 70 Make sure that the mating marks on camshaft sprockets are lined up with the yellow links in the timing chain

compression stroke with the following procedure:

a. Open the splash cover on RH under cover.

b. Rotate crankshaft pulley clockwise, and align mating marks for TDC with timing indicator on front cover, as shown.

c. At the same time, make sure that the mating marks on camshaft sprockets are lined up with the yellow links in the timing chain, as shown.

d. If not, rotate crankshaft pulley one more turn to line up the mating marks to the yellow links, as shown.

13. Pull the timing chain guide out between the camshaft sprockets through front cover.

14. Remove camshaft sprockets with the following procedure:

✳✳ CAUTION

Do not rotate the crankshaft or camshaft while the timing chain is removed. It causes interference between valve and piston.

➡**Chain tension holding work is not necessary. Crankshaft sprocket and timing chain do not disconnect structurally while front cover is attached.**

a. Line up the mating marks on camshaft sprockets with the yellow links in the timing chain, and paint an indelible mating mark on the sprocket and timing chain link plate.

b. Push in the tensioner plunger and hold. Insert a stopper pin into the hole on tensioner body to hold the chain tensioner. Remove the timing chain tensioner.

➡**Use a wire with 0.02 inches (0.5 mm) diameter for a stopper pin.**

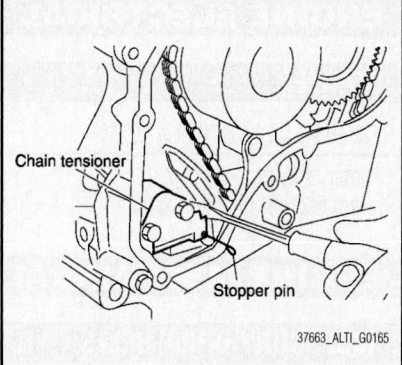

Fig. 71 Push in the tensioner plunger and hold; insert a stopper pin into the hole on tensioner body to hold the chain tensioner

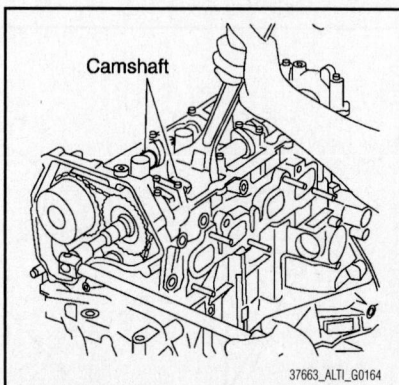

Fig. 72 Loosen the camshaft sprocket mounting bolts and remove the camshaft sprockets

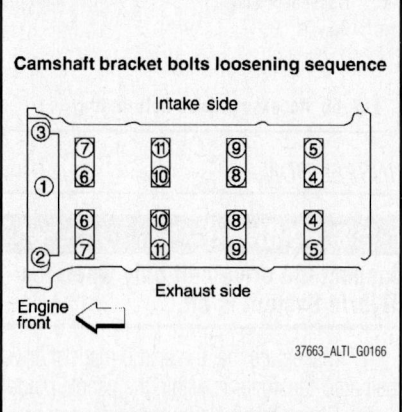

Fig. 73 Loosen the camshaft bracket bolts in the order as shown

c. Secure the hexagonal part of camshaft with a suitable tool.

d. Loosen the camshaft sprocket mounting bolts and remove the camshaft sprockets.

15. Loosen the camshaft bracket bolts in the order as shown, and remove the camshaft brackets and camshafts.

16. Remove No.1 camshaft bracket

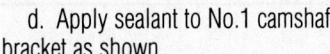
by slightly tapping it with a rubber mallet.

17. Remove the valve lifters.

18. Check mounting positions, and set them aside in the order removed.

To install:

19. Install the valve lifter.

➡**Install them in the same position from which they were removed.**

20. Install the camshafts.

a. The distinction between the intake and exhaust camshafts is in a difference of shapes of the back end:

- A: Exhaust
- B: Intake Signal plate for the camshaft position sensor (PHASE)

b. Install camshafts so that the dowel pins on the front side are positioned as shown.

21. Install camshaft brackets.

a. Install by referring to identification mark on upper surface mark.

b. Install so that identification mark can be correctly read when viewed from the exhaust side.

c. Install No. 1 camshaft bracket as follows.

d. Apply sealant to No.1 camshaft bracket as shown.

➡**Use Genuine Silicone RTV Sealant, or equivalent.**

✳✳ CAUTION

After installation, be sure to wipe off any excessive sealant leaking from part (A) (both on right and left sides).

e. Apply sealant to camshaft bracket contact surface on the front cover backside.

f. Apply sealant to the outside of bolt hole on front cover.

g. Position the No.1 camshaft bracket near the mounting position, and install it without disturbing the sealant applied to the surfaces.

22. Tighten camshaft bracket bolts in four steps in the order as shown.

a. Step 1: Bolts 9–11: 17 inch lbs. (2. Nm)

b. Step 2: Bolts 1–8: 17 inch lbs. (2. Nm)

c. Step 3: Bolts 1–11: 52 inch lbs. (6 Nm)

d. Step 4: Bolts 1–11: 92 inch lbs. (10 Nm)

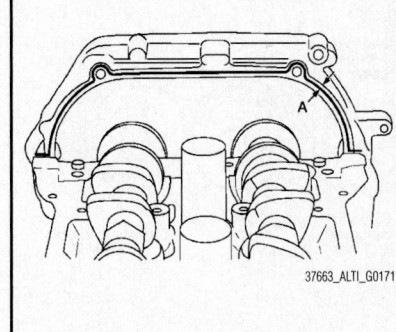

Fig. 78 Apply sealant to the outside of bolt hole on front cover

✳✳ CAUTION

After tightening camshaft bracket bolts, be sure to wipe off excessive sealant from the mating surface of rocker cover and the mating surface of front cover, when installed without the front cover.

23. Install camshaft sprockets.

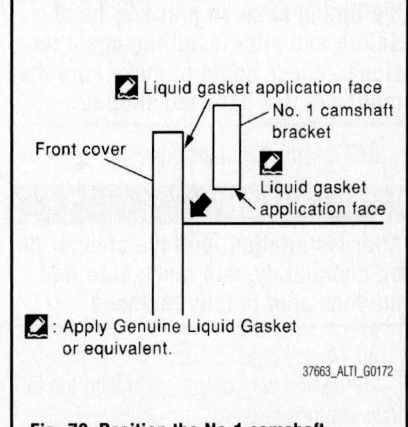

Fig. 79 Position the No.1 camshaft bracket near the mounting position, and install it without disturbing the sealant applied to the surfaces

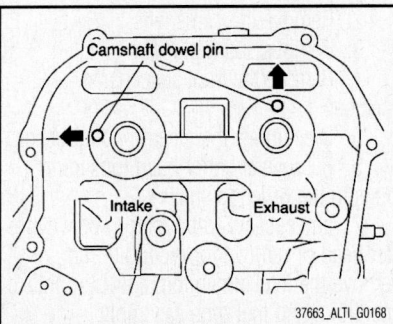

Fig. 74 A: Exhaust; B: Intake Signal plate for the camshaft position sensor (PHASE)

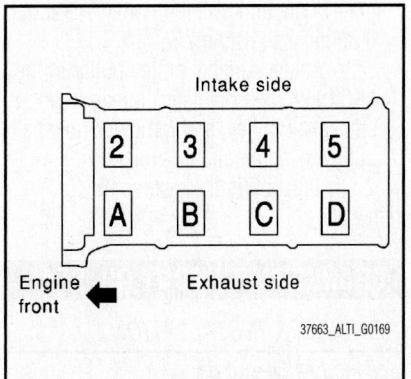

Fig. 76 Install camshaft brackets

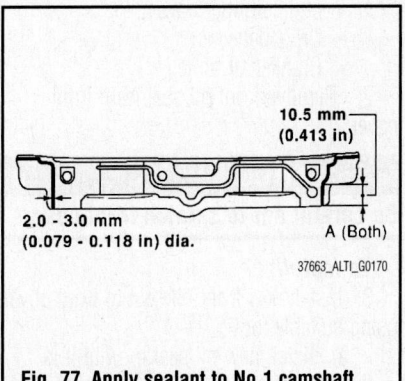

Fig. 75 Install camshafts so that the dowel pins on the front side are positioned as shown

Fig. 77 Apply sealant to No.1 camshaft bracket as shown

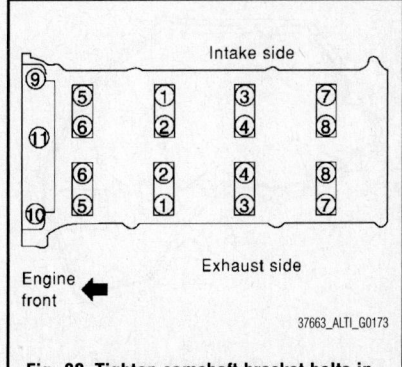

Fig. 80 Tighten camshaft bracket bolts in four steps in the order as shown

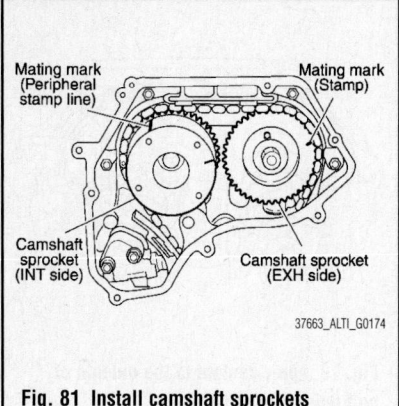

Fig. 81 Install camshaft sprockets

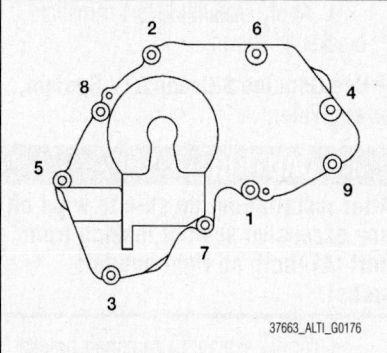

Fig. 83 Tighten the bolts in the numerical order as shown

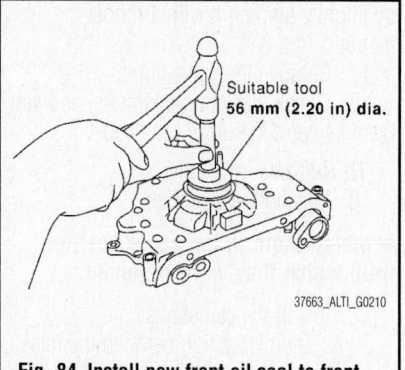

Fig. 84 Install new front oil seal to front cover using suitable tool

a. Install them by lining up the mating marks on each camshaft sprocket with the ones painted on the timing chain during removal.

b. Before installation of chain tensioner, it is possible to re-match the marks on timing chain with the ones on each sprocket.

✸✸ CAUTION

Aligned mating marks could slip. Therefore, after matching them, hold the timing chain in place by hand. Before and after installing chain tensioner, check again to make sure that mating marks have not slipped.

24. Install chain tensioner.

✸✸ CAUTION

After installation, pull the stopper pin off completely, and make sure that the tensioner is fully released.

25. Install chain guide.
26. Install IVT control cover with the following procedure.

 a. Install IVT control solenoid valve to intake valve timing control cover.

 b. Install O-ring to front cover side.

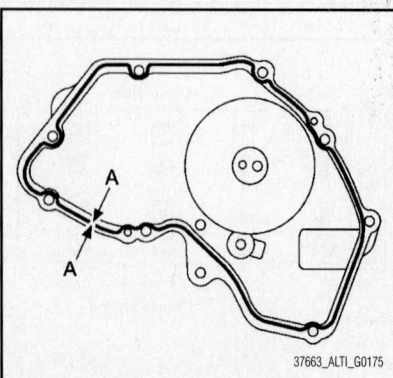

Fig. 82 Apply Genuine Silicone RTV Sealant to the positions as shown

c. Apply Genuine Silicone RTV Sealant to the positions as shown.

d. Install IVT control cover.

e. Tighten the bolts in the numerical order as shown.

27. Check and adjust valve clearances.
28. Installation of the remaining components is in the reverse order.

CRANKSHAFT DAMPER

REMOVAL & INSTALLATION

1. Remove crankshaft pulley with the following procedure:

 a. Hold the crankshaft pulley using suitable tool, then loosen the crankshaft pulley bolt, and pull the pulley out about 0.39 inches (10 mm).

 b. Attach suitable pulley puller in the M 6 (0.24 inch diameter) thread hole on crankshaft pulley, and remove crankshaft pulley using a suitable puller.

2. Installation is the reverse of removal.

CRANKSHAFT FRONT OIL SEAL

REMOVAL & INSTALLATION

See Figures 84 and 85.

1. Remove the following parts:
   • RH front wheel
   • Engine under cover
   • Drive belts
   • Crankshaft pulley

2. Remove front oil seal from front cover.

✸✸ CAUTION

Be careful not to scratch front cover.

To install:

3. Install new front oil seal to front cover using suitable tool.

 a. Install new oil seal in until it is flush with front end surface of front cover.

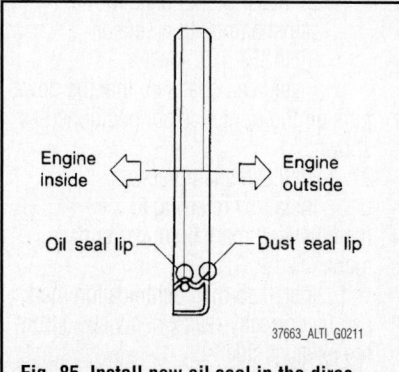

Fig. 85 Install new oil seal in the direction shown

✸✸ CAUTION

Do not reuse oil seal. Be careful not to cause damage to circumference of oil seal.

 b. Install new oil seal in the direction shown.

4. Installation of the remaining components is in the reverse order of removal.

CYLINDER HEAD

REMOVAL & INSTALLATION

See Figures 86 and 87.

1. Remove the timing chain.
2. Remove the camshafts.
3. Remove spark plugs.
4. Remove the front suspension member.
5. Disconnect the electric compressor.
6. Remove cylinder head loosening bolts in the order as shown.
7. If necessary to transfer to new cylinder head or remove for reconditioning, remove the intake manifold collector, intake manifold, and fuel tube assembly.

To install:

8. Install a new cylinder head gasket.
9. Follow the steps below to tighten the

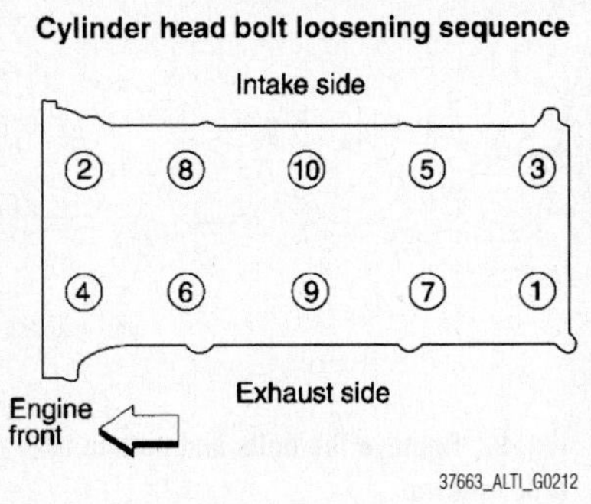

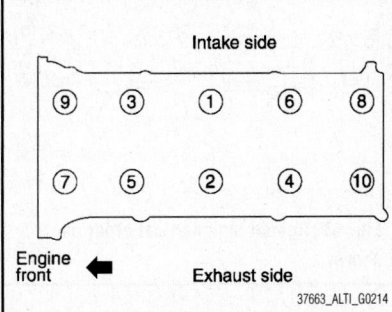

Fig. 86 Remove cylinder head loosening bolts in the order as shown

Fig. 87 Tighten the cylinder head bolts in the numerical order

cylinder head bolts in the numerical order as shown.

    a. Apply new engine oil to the threads and the seating surfaces of bolts.

    b. Step 1: 72 ft. lbs. (98 Nm)

    c. Step 2: Loosen all bolts in reverse order of tightening.

    d. Step 3: 29 ft. lbs. (39 Nm)

    e. Step 4 75 degrees clockwise

    f. Step 5 75 degrees clockwise

❊❊ CAUTION
Note the following:

- If cylinder head bolts are re-used, check their outer diameters before installation. Follow the Outer Diameter of Cylinder Head Bolts procedure.
- Check and confirm the tightening angle by using angle wrench or protractor. Avoid judgment by visual inspection without the tool.

10. Installation of the remaining components is in the reverse order of removal.

EXHAUST MANIFOLD

REMOVAL & INSTALLATION
See Figures 88 and 89.

1. Remove the engine undercover.
2. Disconnect the electrical connector of Air Fuel Ratio (A/F) sensor 1, and unhook the harness from the bracket and middle clamp on the cover.
3. Remove the Air Fuel Ratio (A/F) sensor 1 using Tool J-44626.

❊❊ CAUTION

Be careful not to damage Air Fuel Ratio (A/F) sensor. Discard any Air Fuel Ratio (A/F) sensor which has been dropped from a height of more than 20 inches (0.5 m) onto a hard surface such as a concrete floor; replace with a new one.

4. Remove the exhaust manifold cover (lower).
5. Remove the exhaust front tube.
6. Remove the exhaust manifold cover (upper).
7. Loosen the nuts in the reverse order as shown, on the exhaust manifold and three way catalyst assembly.
8. Remove the exhaust manifold and

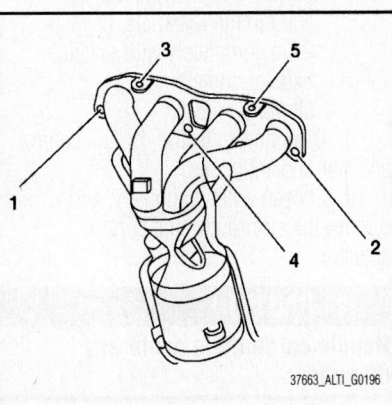

Fig. 88 Loosen the nuts in the reverse order as shown

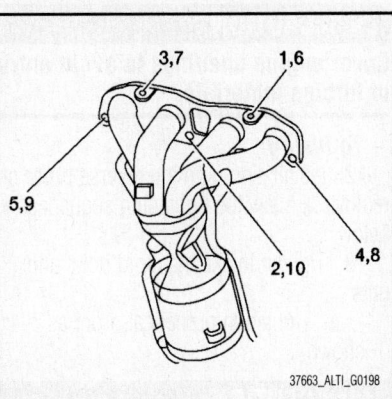

Fig. 89 Tighten the nuts to specification in the numerical order shown

three way catalyst assembly and gasket. Discard the gasket.

To install:

9. Installation is in the reverse order of removal.

10. Tighten the nuts in two steps to 31 ft. lbs. (42 Nm) in the numerical order shown, on the exhaust manifold and three way catalyst assembly.

11. Clean the Air Fuel Ratio (A/F) sensor 1 threads with the Tool, then apply the anti-seize lubricant to the threads before installing the Air Fuel Ratio (A/F) sensor 1.

❊❊ CAUTION

Do not over-tighten the Air Fuel Ratio (A/F) sensor 1. Doing so may cause damage to the Air Fuel Ratio (A/ F) sensor 1, resulting in a malfunction and the MIL coming on.

INTAKE MANIFOLD

REMOVAL & INSTALLATION
See Figures 90 through 93.

❊❊ WARNING

To avoid the danger of being scalded, never drain the coolant when the engine is hot.

1. Release the fuel pressure.
2. Drain coolant when engine is cooled.
3. Disconnect the MAF sensor electrical connector.
4. Remove air cleaner and air duct assembly.
5. Remove cowl top finisher.
6. Disconnect the following components at the intake side:
   - PCV hose
   - EVAP hose and EVAP canister purge volume control solenoid
   - Electric throttle control actuator
7. Disconnect the fuel quick connector on the engine side.

    a. Remove quick connector cap.

    b. With the sleeve side of tool facing quick connector, install tool onto fuel tube.

    c. Insert tool into quick connector until sleeve contacts and goes no further. Hold the tool on that position.

❊❊ CAUTION

Inserting the tool hard will not disconnect quick connector. Hold tool where it contacts and goes no further.

d. Pull the quick connector straight out from the fuel tube.

※※ CAUTION

Note the following:

- Pull quick connector holding it at the A position, as shown.
- Do not pull with lateral force applied. O-ring inside quick connector may be damaged.
- Prepare container and cloth beforehand as fuel will leak out.
- Avoid fire and sparks.
- Be sure to cover openings of disconnected pipes with plug or plastic bag to avoid fuel leakage and entry of foreign materials.

8. When removing fuel hose quick connector at vehicle piping side, perform as follows.

a. Remove quick connector cap.

b. Hold the sides of the connector, push in tabs and pull out the tube. (The figure is shown for reference only.)

c. If the connector and the tube are

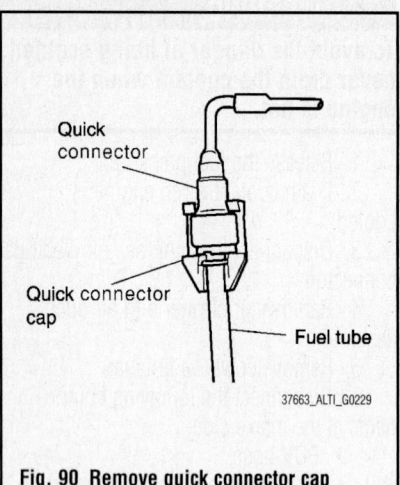

Fig. 90 Remove quick connector cap

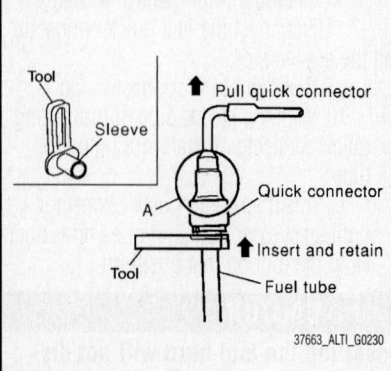

Fig. 91 With the sleeve side of tool facing quick connector, install tool onto fuel tube

stuck together, push and pull several times until they start to move. Then disconnect them by pulling.

※※ CAUTION

Note the following:

- The tube can be removed when the tabs are completely depressed. Do not twist it more than necessary.
- Do not use any tools to remove the quick connector.
- Keep the resin tube away from heat. Be especially careful when welding near the tube.
- Prevent acid liquid such as battery electrolyte etc. from getting on the resin tube.
- Do not bend or twist the tube during installation and removal.
- Do not remove the remaining retainer on tube.
- When the tube is replaced, also replace the retainer with a new one. Retainer color: Green.
- To keep clean the connecting portion and to avoid damage and foreign materials, cover them completely with plastic bags or something similar.

9. Disconnect electric throttle control actuator coolant hoses.

10. Loosen bolts diagonally, and remove the electric throttle control actuator.

※※ CAUTION

Handle carefully to avoid any damage.

11. Remove the bolts and nuts in the order shown and remove the intake manifold assembly.

※※ CAUTION

Cover engine openings to avoid entry of foreign materials.

To install:

12. Installation is in the reverse order of removal. Follow the tightening sequences below.

13. Tighten intake manifold bolts and nuts.

a. Tighten in numerical order as shown.

※※ CAUTION

After tightening the five bolts in the order shown, the 1, 6 position desig-

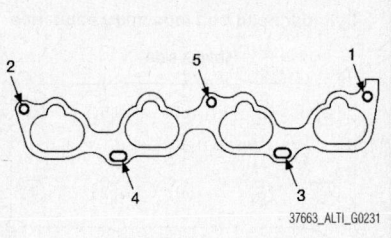

Fig. 92 Remove the bolts and nuts in the order shown

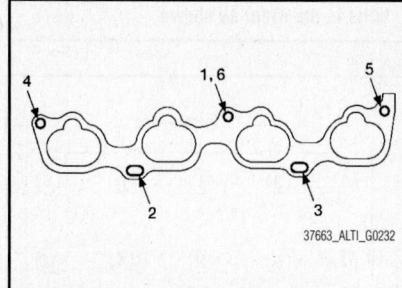

Fig. 93 Tighten in numerical order as shown

nates that the first bolt tightened is to be retightened to specification.

14. Install the Electric Throttle Control Actuator:

a. Tighten the bolts of electric throttle control actuator equally and diagonally in several steps.

b. After installation perform procedure in "INSPECTION AFTER INSTALLATION".

Connecting quick connector on the fuel house (Engine Side)

See Figures 94 and 95.

1. Make sure no foreign substances are deposited in and around the fuel tube and quick connector, and there is no damage to them.

2. Thinly apply new engine oil around the fuel tube tip end.

3. Align center to insert quick connector straight into fuel tube.

4. Insert fuel tube into quick connector until the top spool on fuel tubes is inserted completely and the second level spool is positioned slightly below the quick connector bottom end.

※※ CAUTION

Note the following:

- Hold at position A as shown, when inserting the fuel tube into the quick connector.

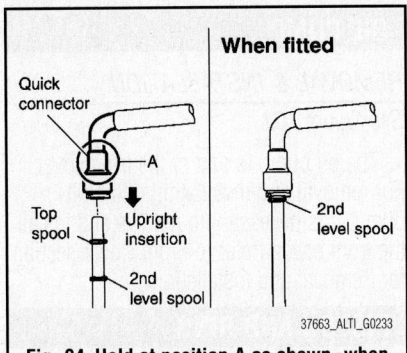

Fig. 94 Hold at position A as shown, when inserting the fuel tube into the quick connector

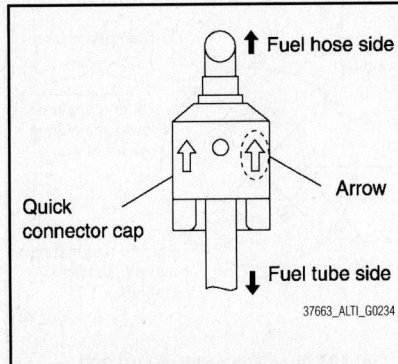

Fig. 95 Install quick connector cap on quick connector joint

- Carefully align to center to avoid inclined insertion to prevent damage to the O-ring inside the quick connector.
- Insert the fuel tube until you hear a "click" sound and actually feel the engagement.
- To avoid misidentification of engagement with a similar sound, be sure to perform the next step.

5. Before clamping the fuel hose with the hose clamp, pull the quick connector hard by hand, holding at the A position, as shown. Make sure it is completely engaged (connected) so that it does not come off of the fuel tube.

➡ **Recommended pulling force is 11 lbs. (50 N).**

6. Install quick connector cap on quick connector joint.
7. Direct arrow mark on quick connector cap to upper side (fuel hose side).
8. Install fuel hose to hose clamp.

Connecting quick connector on the fuel house (Vehicle Piping Side)

See Figure 96.

1. Make sure no foreign substances are deposited in and around the fuel tube and

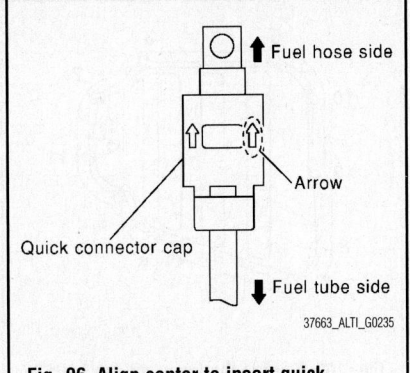

Fig. 96 Align center to insert quick connector straight into fuel tube

quick connector, and there is no damage to them.

2. Align center to insert quick connector straight into fuel tube.

   a. Insert fuel tube until a click is heard.

   b. Install quick connector cap on quick connector joint. Direct arrow mark on quick connector cap upper side.

   c. Install fuel hose to hose clamp.

Inspection after installation

1. Make sure there is no fuel leakage at connections as follows:

   a. Apply fuel pressure to fuel lines by turning ignition switch ON (with engine stopped). Then check for fuel leaks at connections.

   b. Start the engine and rev it up and check for fuel leaks at connections.

   c. Perform procedures for "Throttle Valve Closed Position Learning" after finishing repairs.

   d. If electric throttle control actuator is replaced, perform procedures for "Idle Air Volume Learning" after finishing repairs.

✼✼ WARNING

Do not touch engine immediately after stopping as engine is extremely hot.

➡ **Use mirrors for checking on connections out of the direct line of sight.**

OIL PAN

REMOVAL & INSTALLATION
See Figures 97 through 102.

✼✼ WARNING

To avoid the danger of being scalded, never drain the engine oil when the engine is hot.

1. Drain engine oil.
2. Remove the front exhaust tube.
3. Support the engine from above and underneath with suitable hoist and or jack.
4. Remove the front suspension member for clearance to remove the oil pan.
5. Remove the lower oil pan bolts in the order as shown.
6. Remove the lower oil pan using Tool KV10111100 (J-37228).

➡ **Tap gently to cut sealant around the pan; do not damage the mating surface using Tool.**

7. Remove the oil strainer.
8. Remove rear plate cover, and four engine-to transaxle bolts.
9. Loosen the upper oil pan bolts in the order shown to remove upper oil pan.
10. Remove upper oil pan using Tool KV10111100 (J-37228).

➡ **Tap gently to cut sealant around the pan; do not damage the mating surface using Tool.**

11. Clean the oil strainer screen to remove any foreign material.

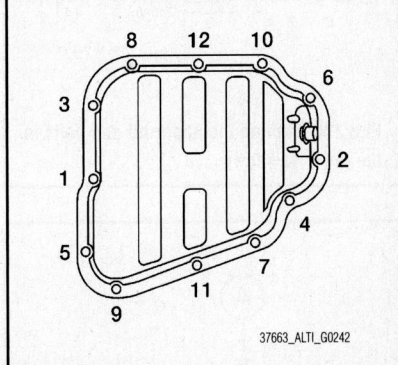

Fig. 97 Remove the lower oil pan bolts in the order

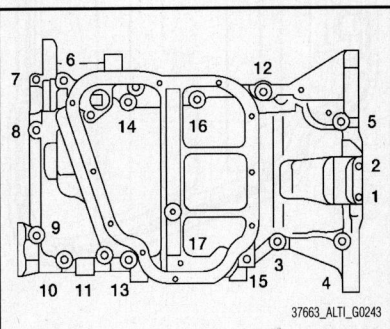

Fig. 98 Loosen the upper oil pan bolts in the order

To install:

12. Installation is in the reverse order of removal.

13. Apply Genuine Silicone RTV Sealant or equivalent to the upper oil pan as shown.

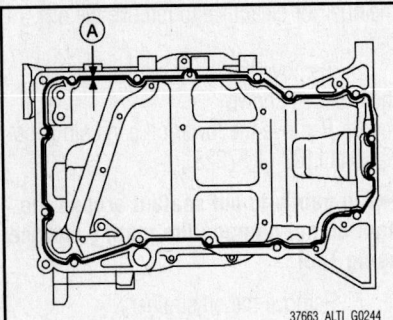

Fig. 99 Apply Genuine Silicone RTV Sealant or equivalent to the upper oil pan (A)

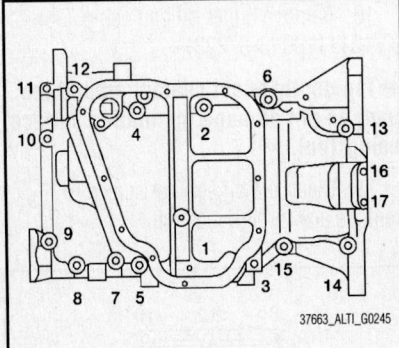

Fig. 100 Tighten the upper oil pan bolts in the order as shown

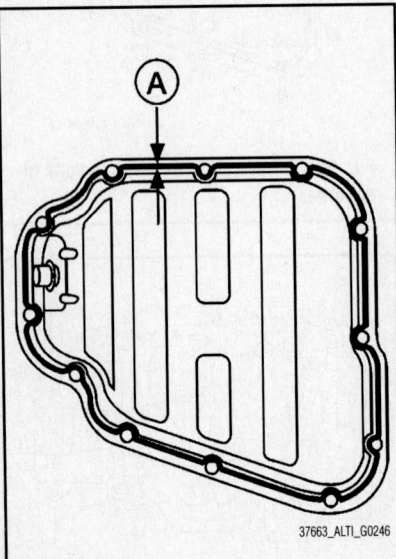

Fig. 101 Apply Genuine Silicone RTV Sealant or equivalent to the lower oil pan (A)

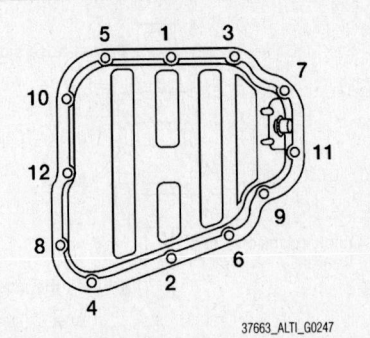

Fig. 102 Tighten the lower oil pan bolts in the numerical order shown

❄ CAUTION

Install two new O-rings in the upper pan.

14. Tighten the upper oil pan bolts in the order as shown.

15. Apply Genuine Silicone RTV Sealant or equivalent to the lower oil pan (A) as shown.

16. Tighten the lower oil pan bolts in the numerical order shown.

❄ CAUTION

Wait at least 30 minutes after the oil pans are installed before filling the engine with oil.

17. Check for any engine oil leaks with the engine at operating temperature and running at idle.

OIL PUMP

REMOVAL & INSTALLATION

See Figure 103.

The oil pump is part of the front cover. For removal and installation of the oil pump, it is necessary to remove and install the front cover. Refer to Front Cover section for Removal and Installation.

PISTON AND RING

POSITIONING

See Figure 104.

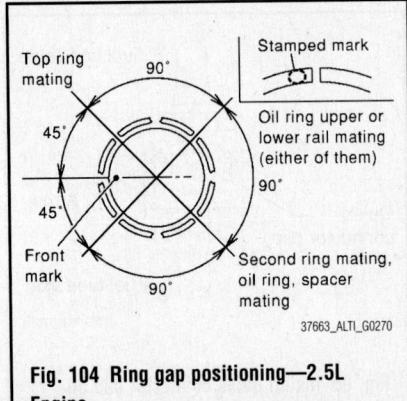

Fig. 104 Ring gap positioning—2.5L Engine

REAR MAIN SEAL

REMOVAL & INSTALLATION

See Figures 105 and 106.

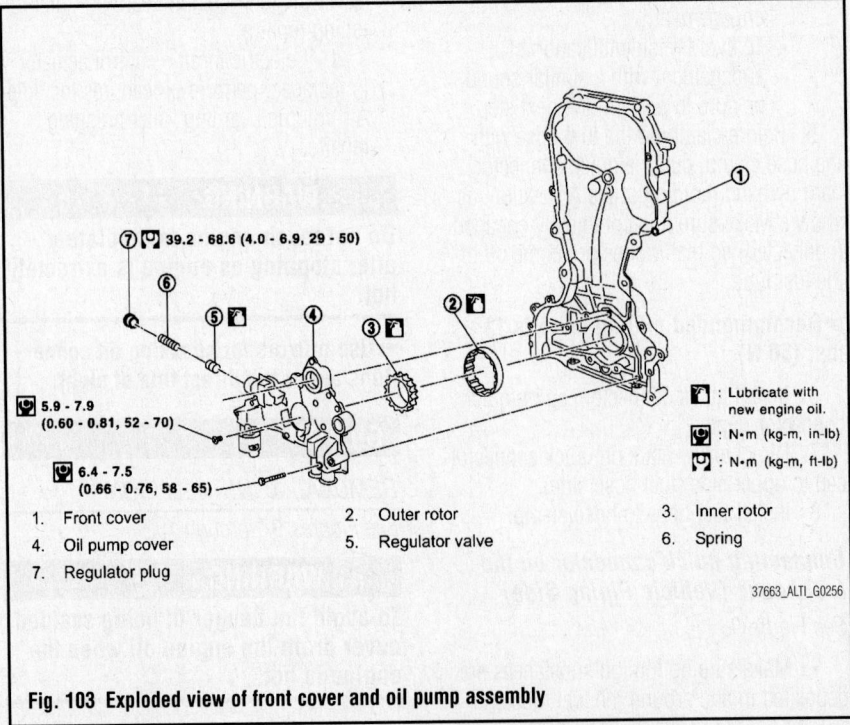

⊡ 39.2 - 68.6 (4.0 - 6.9, 29 - 50)	
⊡ 5.9 - 7.9 (0.60 - 0.81, 52 - 70)	
⊡ 6.4 - 7.5 (0.66 - 0.76, 58 - 65)	

1. Front cover
2. Outer rotor
3. Inner rotor
4. Oil pump cover
5. Regulator valve
6. Spring
7. Regulator plug

▣ : Lubricate with new engine oil.
⊡ : N·m (kg-m, in-lb)
⊡ : N·m (kg-m, ft-lb)

Fig. 103 Exploded view of front cover and oil pump assembly

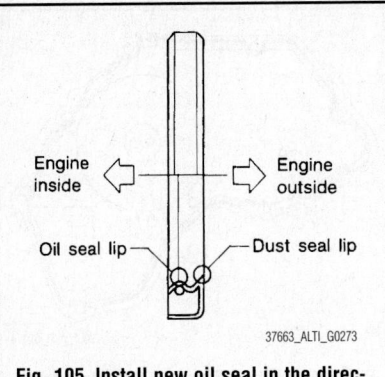

Fig. 105 Install new oil seal in the direction shown

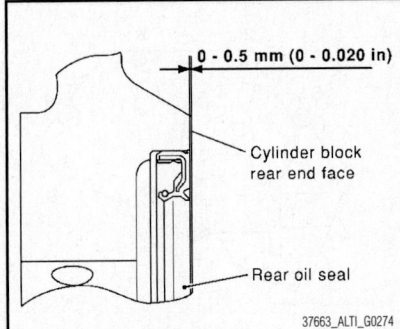

Fig. 106 Press in the new oil seal to the specified depth as shown

1. Remove the transaxle.
2. Remove drive plate (CVT).
3. Remove rear oil seal using suitable tool.

❋❋ CAUTION

Be careful not to scratch rear oil seal retainer.

To install:

4. Apply new engine oil to new oil seal and install it using a suitable tool.

   a. Install new oil seal in the direction shown.

   b. Press fit new oil seal straight using a suitable drift, to avoid causing burrs or tilting.

   c. Press in the new oil seal to the specified depth as shown.

5. Installation of the remaining components is in the reverse order of removal.

TIMING CHAIN FRONT COVER

REMOVAL & INSTALLATION

See Figures 107 through 115.

1. Support the engine and transaxle assembly with suitable tools.
2. Remove RH splash shield.
3. Remove the upper and lower oil pan, and oil strainer.

4. Remove the alternator.
5. Remove the engine cover.
6. Disconnect variable timing control solenoid harness connector.
7. Remove the engine ground.
8. Remove the coolant overflow reservoir tank.
9. Position the RH engine compartment fuse and relay box aside.
10. Remove the RH engine mount and bracket.
11. Loosen bolts in the numerical order as shown.

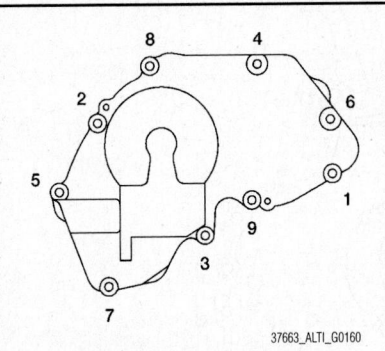

Fig. 107 Loosen bolts in the numerical order as shown

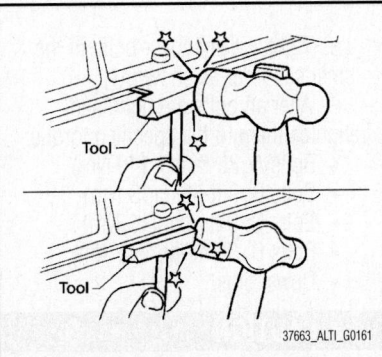

Fig. 108 Remove the IVT (intake valve timing) control cover using Tool KV10111100 (J-37228)

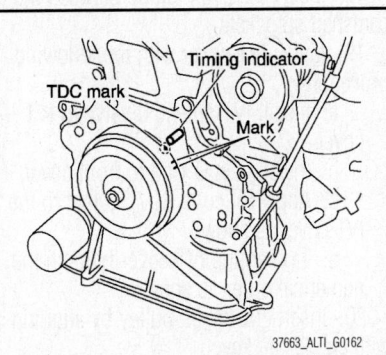

Fig. 109 Set the No.1 cylinder at TDC on the compression stroke

12. Remove the IVT (intake valve timing) control cover using Tool KV10111100 (J-37228).
13. Pull chain guide between camshaft sprockets out through front cover.
14. Set the No.1 cylinder at TDC on the compression stroke with the following procedure:

   a. Rotate the crankshaft pulley clockwise and align the mating marks to the timing indicator on the front cover.

   b. At the same time, make sure that the mating marks on the camshaft sprockets are lined up as shown.

   c. If not lined up, rotate the crankshaft pulley one more turn to line up the mating marks to the positions as shown.

15. Remove crankshaft pulley with the following procedure:

   a. Hold the crankshaft pulley using suitable tool, then loosen the crankshaft pulley bolt, and pull the pulley out about 0.39 inches (10 mm).

   b. Attach suitable pulley puller in the thread hole on crankshaft pulley, and remove crankshaft pulley using a suitable puller.

16. Remove the front cover with the following procedure:

   a. Loosen the bolts in the numerical order as shown, and remove them.

   b. Remove the front cover.

❋❋ CAUTION

Be careful not to damage the mounting surface.

To install:

17. Install front cover with the following procedure:

   a. Install O-rings to cylinder head and cylinder block.

   b. Apply Genuine Silicone RTV Sealant or equivalent, to front cover.

   c. Make sure the mating marks on the timing chain and each sprocket

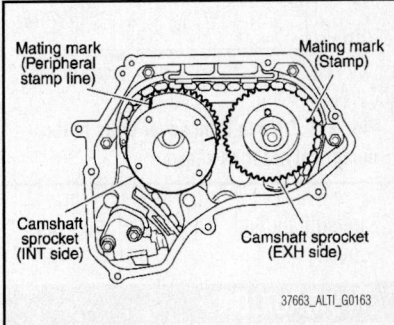

Fig. 110 Make sure that the mating marks on the camshaft sprockets are lined up as shown

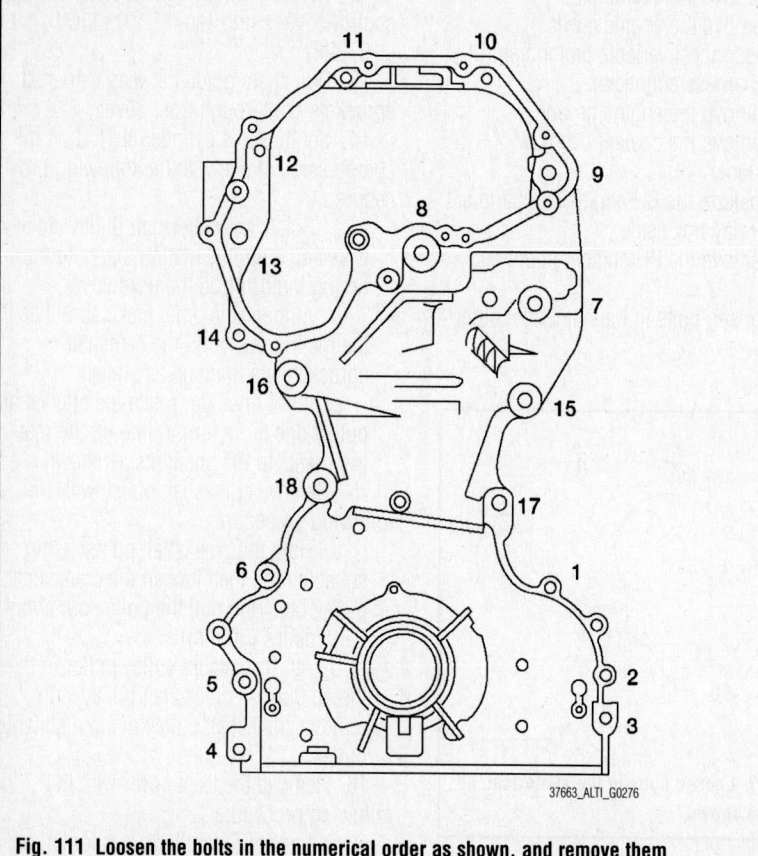

Fig. 111 Loosen the bolts in the numerical order as shown, and remove them

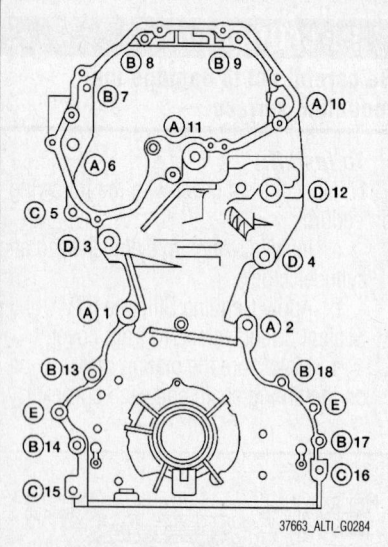

Fig. 112 Tighten front cover bolts in the numerical order as shown

are still aligned. Then install the front cover.

※※ CAUTION

Be careful not to damage the front oil seal during installation with the front end of the crankshaft.

d. Tighten front cover bolts in the numerical order as shown.

e. After all bolts are tightened, retighten them to the specified torque:
- Bolts A: 36 ft. lbs. (49 Nm)
- Bolts B: 9 ft. lbs. (13 Nm)
- Bolts C: 9 ft. lbs. (13 Nm)
- Bolts D: 36 ft. lbs. (49 Nm)
- Dowel pins

※※ CAUTION

Wipe off any excess sealant leaking at the surface for installing the oil pan.

18. Install the chain guide between the camshaft sprockets.

19. Install IVT cover with the following procedure:

a. Install IVT solenoid valve to IVT cover.

b. Install new O-ring to front cover.

c. Apply Silicone RTV Sealant to the IVT cover as shown.

d. Tighten the IVT cover bolts in the numerical order as shown.

20. Insert crankshaft pulley by aligning with crankshaft key.

a. Tap its center with a plastic hammer to insert.

b. Do not tap the belt hook.

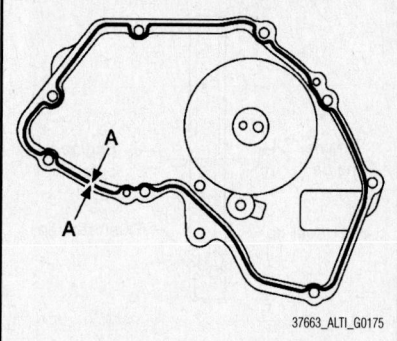

Fig. 113 Apply Silicone RTV Sealant to the IVT cover as shown

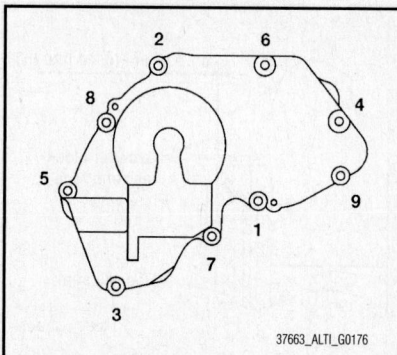

Fig. 114 Tighten the IVT cover bolts in the numerical order as shown

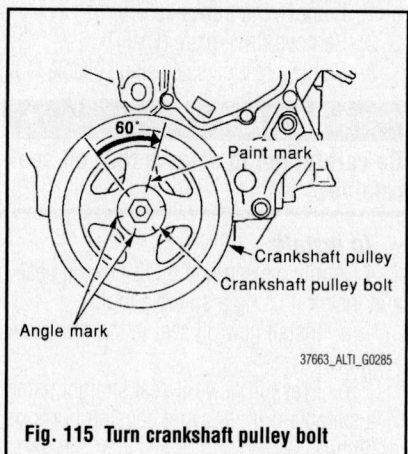

Fig. 115 Turn crankshaft pulley bolt

21. Tighten crankshaft pulley bolts.

a. Secure crankshaft pulley with tool to tighten the bolt.

b. Perform angle tightening with the following procedure:

c. Apply new engine oil to threads and seat surfaces of bolts.

d. Tighten to initial specifications: 31 ft. lbs. (42 Nm).

e. Apply a paint mark on the front cover, mating with any one of six easy to recognize stamp marks on bolt flange.

f. Turn crankshaft pulley bolt another 60° to 66° (Target: 60°).

g. Check vertical mounting angle with movement of one stamp mark.

22. Installation of the remaining components is in the reverse order of removal.

TIMING CHAIN & SPROCKETS

REMOVAL & INSTALLATION

See Figures 107 through 124.

1. Support the engine and transaxle assembly with suitable tools.

2. Remove RH splash shield.

3. Remove the upper and lower oil pan, and oil strainer.

4. Remove the alternator.

5. Remove the engine cover.

6. Disconnect variable timing control solenoid harness connector.

7. Remove the engine ground.

8. Remove the coolant overflow reservoir tank.

9. Position the RH engine compartment fuse and relay box aside.

10. Remove the RH engine mount and bracket.

11. Loosen bolts in the numerical order as shown.

12. Remove the IVT (intake valve timing) control cover using Tool KV10111100 (J-37228).

13. Pull chain guide between camshaft sprockets out through front cover.

14. Set the No.1 cylinder at TDC on the compression stroke with the following procedure:

a. Rotate the crankshaft pulley clockwise and align the mating marks to the timing indicator on the front cover.

b. At the same time, make sure that the mating marks on the camshaft sprockets are lined up as shown.

c. If not lined up, rotate the crankshaft pulley one more turn to line up the mating marks to the positions as shown.

15. Remove crankshaft pulley with the following procedure:

a. Hold the crankshaft pulley using suitable tool, then loosen the crankshaft pulley bolt, and pull the pulley out about 0.39 inches (10 mm).

b. Attach suitable pulley puller in the thread hole on crankshaft pulley, and remove crankshaft pulley using a suitable puller.

16. Remove the front cover with the following procedure:

a. Loosen the bolts in the numerical order as shown, and remove them.

b. Remove the front cover.

✳✳ CAUTION

Be careful not to damage the mounting surface.

17. Remove front oil seal using suitable tool, if necessary.

18. Remove timing chain with the following procedure:

a. Push in the tensioner plunger. Insert a stopper pin into the hole on the tensioner body to secure the chain tensioner plunger and remove chain tensioner.

b. Use a wire of 0.02 inches (0.5 mm) diameter as a stopper pin.

c. Remove the timing chain.

d. Secure hexagonal part of the camshaft with a wrench and loosen the camshaft sprocket bolt and remove the camshaft sprocket for both camshafts.

✳✳ CAUTION

Do not rotate the crankshaft or camshafts while the timing chain is

removed. **It can cause damage to the valve and piston.**

19. Remove the chain slack guide, tension guide, timing chain, and oil pump drive spacer.

20. Press stopper tab in the direction shown to push the timing chain slack guide toward timing chain tensioner (for balancer unit).

➡**The slack guide is released by pressing the stopper tab. As a result, the slack guide can be moved.**

21. Insert stopper pin into tensioner body hole to secure timing chain slack guide.

➡**Use a hard metal pin with a diameter of approximately 0.047 inches (1.2 mm) as a stopper pin.**

22. Remove timing chain tensioner (for balancer unit).

23. Secure width across flats of the balancer LH side shaft using a suitable tool. Loosen the balancer sprocket bolt.

24. Remove balancer unit timing chain, balancer unit sprocket and crankshaft sprocket.

25. Loosen bolts in order as shown, and remove balancer unit.

➡**Use TORX® socket size E14.**

✳✳ CAUTION

Do not disassemble balancer unit.

26. Check the timing chain for cracks or excessive wear. If a defect is detected, replace it.

27. Check the balancer unit bolt outer diameter.

a. Measure outer diameters (d1, d2) at the two positions as shown.

b. Measure d2 within the range A.

c. If the value difference (d1 - d2) exceeds the limit, replace it with a new one.

To install:

➡**There may be two color variations of the link marks (link colors) on the timing chain. There are 26 links between the gold/yellow mating marks on the timing chain; and 64 links between the camshaft sprocket gold/yellow link and the crankshaft sprocket orange/blue link, on the timing chain side without the tensioner.**

28. Make sure the crankshaft key points straight up.

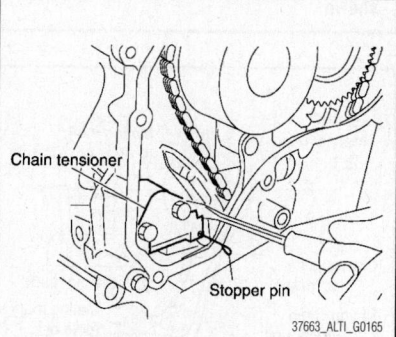

37663_ALTI_G0165

Fig. 116 Insert a stopper pin into the hole on the tensioner body to secure the chain tensioner plunger and remove chain tensioner

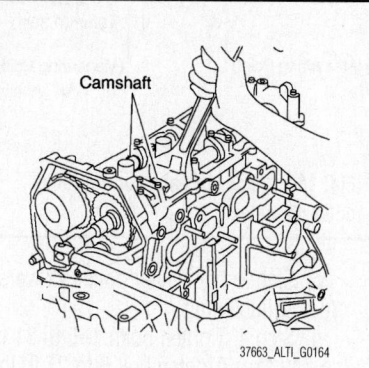

Camshaft

37663_ALTI_G0164

Fig. 117 Secure hexagonal part of the camshaft with a wrench and loosen the camshaft sprocket bolt

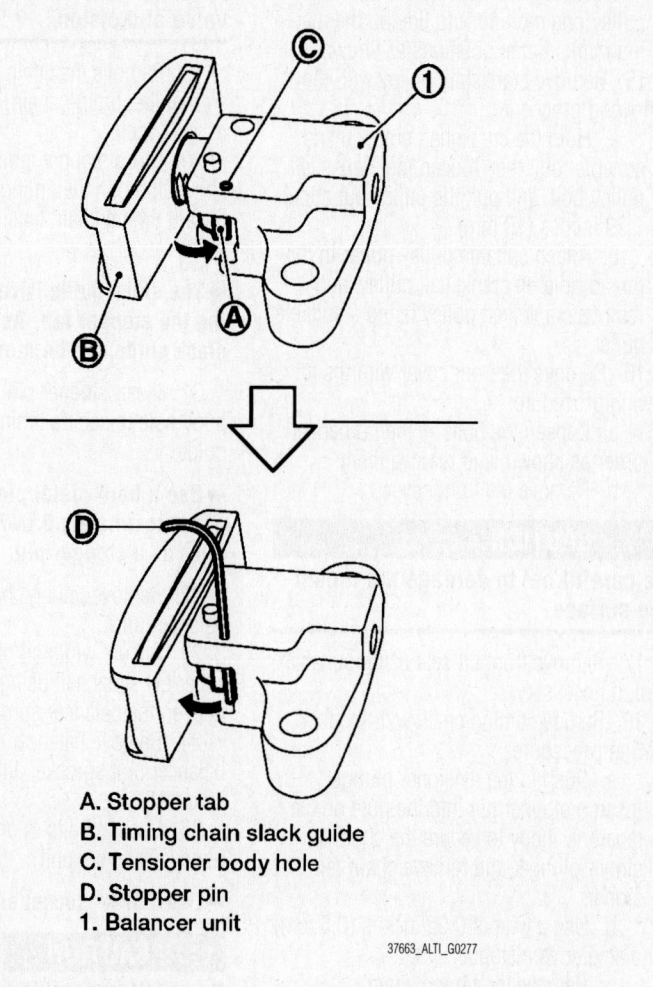

A. Stopper tab
B. Timing chain slack guide
C. Tensioner body hole
D. Stopper pin
1. Balancer unit

37663_ALTI_G0277

Fig. 118 Press stopper tab in the direction shown to push the timing chain slack guide toward timing chain tensioner (for balancer unit)

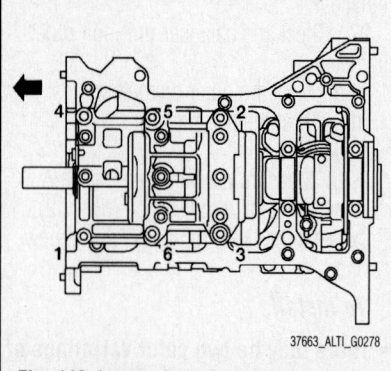

37663_ALTI_G0278

Fig. 119 Loosen bolts in order as shown, and remove balancer unit

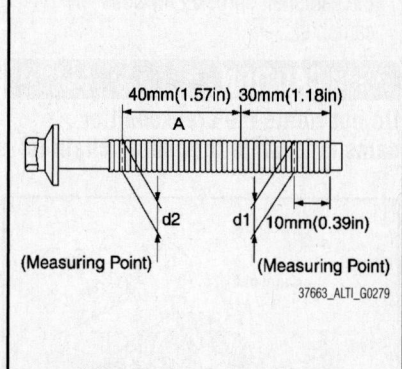

37663_ALTI_G0279

Fig. 120 Check the balancer unit bolt outer diameter

29. Install the balancer unit and tighten the bolts in the numerical order as shown.
   a. Step 1: Tighten bolts 1–5 to 31 ft. lbs. (42 Nm). Tighten bolt 6 to 27 ft. lbs. (36 Nm).
   b. Step 2: Tighten bolts 1–5 an additional 120°. Tighten bolt 6 an additional 90°.

   c. Step 3: Loosen all bolts in reverse order of tightening.
   d. Step 4: Tighten bolts 1–5 to 31 ft. lbs. (42 Nm). Tighten bolt 6 to 27 ft. lbs. (36 Nm).
   e. Step 5: Tighten bolts 1–5 an additional 120°. Tighten bolt 6 an additional 90°.

✳✳ CAUTION

Note the following:

- When reusing a bolt, check its outer diameter before installation. Follow the Balancer Unit Bolt Outer Diameter procedure.
- Apply new engine oil to threads and seating surfaces of bolts.
- Check tightening angle with an angle wrench or a protractor.
- Do not make judgment by visual check alone.

30. Install the crankshaft sprocket and timing chain for the balancer unit.
   a. Make sure that the crankshaft sprocket is positioned with mating marks on the block and sprocket meeting at the top.
   b. Install it by lining up mating marks on each sprocket and timing chain.

31. Install timing chain tensioner (for balancer unit).
   a. Fix the plunger at the most compressed position using a stopper pin, and then install it.
   b. Securely pull out the stopper pin after installing the timing chain tensioner (for balancer unit).

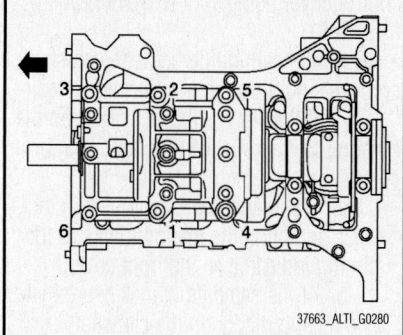

37663_ALTI_G0280

Fig. 121 Install the balancer unit and tighten the bolts in the numerical order as shown

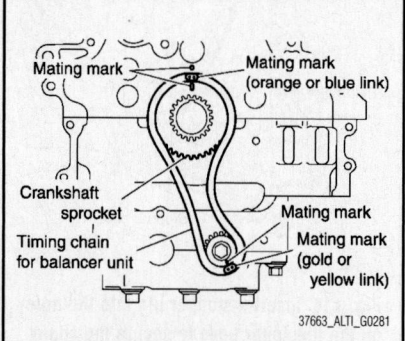

37663_ALTI_G0281

Fig. 122 Install the crankshaft sprocket and timing chain for the balancer unit

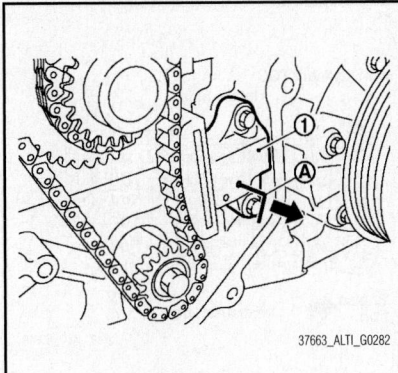

Fig. 123 Install timing chain tensioner (for balancer unit)

c. Check matching mark position of balancer unit drive chain and each sprocket again.

32. Install timing chain and related parts.

a. Install by lining up mating marks on each sprocket and timing chain as shown.

b. Before and after installing timing chain tensioner, check again to make sure the mating marks have not slipped.

c. After installing timing chain tensioner, remove the stopper pin, and make sure that the tensioner moves freely.

✽✽ CAUTION

Note the following:

- For the following note, after the mating marks are aligned, keep them aligned by holding them by hand.
- To avoid skipped teeth, do not move crankshaft and camshaft until front cover is installed.

➡Before installing chain tensioner, it is possible to change the position of mating mark on timing chain for that of each sprocket for alignment.

33. Install new front oil seal to front cover, using suitable tool.

a. Install new oil seal in until it is flush with front end surface of front cover.

✽✽ CAUTION

Do not reuse oil seal. Be careful not to cause damage to circumference of oil seal.

34. Install front cover with the following procedure:

a. Install O-rings to cylinder head and cylinder block.

b. Apply Genuine Silicone RTV Sealant or equivalent, to front cover.

c. Make sure the mating marks on the timing chain and each sprocket are still aligned. Then install the front cover.

✽✽ CAUTION

Be careful not to damage the front oil seal during installation with the front end of the crankshaft.

d. Tighten front cover bolts in the numerical order as shown.

e. After all bolts are tightened, retighten them to the specified torque:

- Bolts A: 36 ft. lbs. (49 Nm)
- Bolts B: 9 ft. lbs. (13 Nm)
- Bolts C: 9 ft. lbs. (13 Nm)
- Bolts D: 36 ft. lbs. (49 Nm)
- Dowel pins

✽✽ CAUTION

Wipe off any excess sealant leaking at the surface for installing the oil pan.

35. Install the chain guide between the camshaft sprockets.

36. Install IVT cover with the following procedure:

a. Install IVT solenoid valve to IVT cover.

b. Install new O-ring to front cover.

c. Apply Silicone RTV Sealant to the IVT cover as shown.

d. Tighten the IVT cover bolts in the numerical order as shown.

37. Insert crankshaft pulley by aligning with crankshaft key.

a. Tap its center with a plastic hammer to insert.

b. Do not tap the belt hook.

38. Tighten crankshaft pulley bolts.

a. Secure crankshaft pulley with tool to tighten the bolt.

b. Perform angle tightening with the following procedure:

c. Apply new engine oil to threads and seat surfaces of bolts.

d. Tighten to initial specifications: 31 ft. lbs. (42 Nm).

e. Apply a paint mark on the front cover, mating with any one of six easy to recognize stamp marks on bolt flange.

f. Turn crankshaft pulley bolt another 60° to 66° (Target: 60°).

g. Check vertical mounting angle with movement of one stamp mark.

39. Installation of the remaining components is in the reverse order of removal.

VALVE LASH

ADJUSTMENT

See Figures 125 and 126.

Perform adjustment depending on selected head thickness of valve lifter.

The specified valve lifter thickness is the dimension at normal temperatures. Ignore dimensional differences caused by temperature. Use the specifications for hot engine condition to adjust.

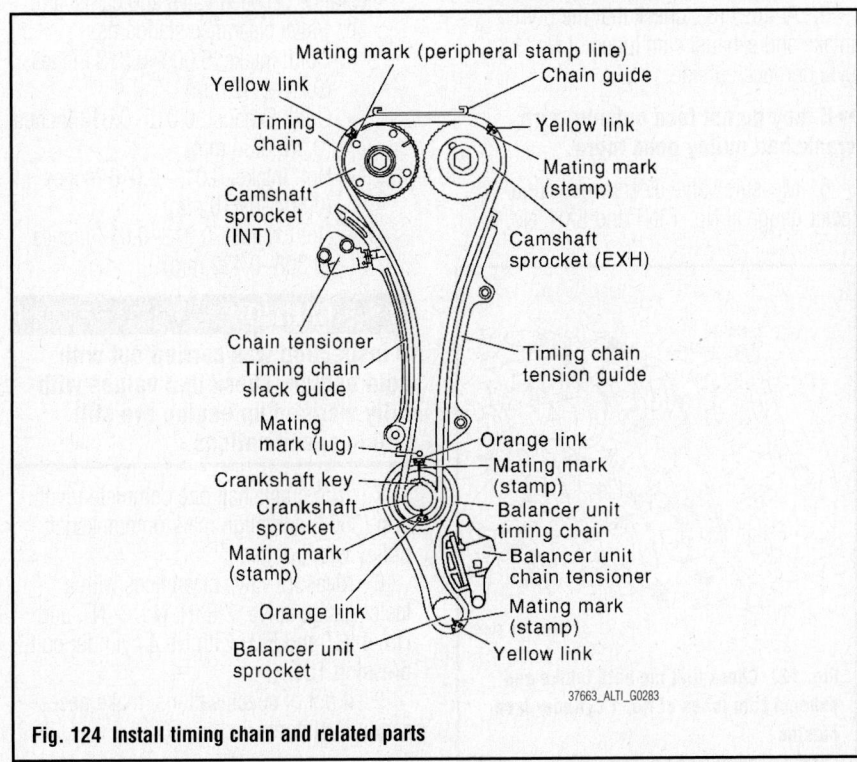

Fig. 124 Install timing chain and related parts

1. Remove camshaft.
2. Remove the valve lifters at the locations that are outside the standard.
3. Measure the center thickness of the removed valve lifters with a micrometer.
4. Use the equation below to calculate valve lifter thickness for replacement.

   a. Valve lifter thickness calculation:
- $t = t_1 + (C_1 - C_2)$
- t = Thickness of replacement valve lifter.
- t_1 = Thickness of removed valve lifter.
- C_1 = Measured valve clearance.
- C_2 = Standard valve clearance.

   b. Thickness of a new valve lifter can be identified by stamp marks on the reverse side (inside the cylinder). Stamp mark 696 indicates a thickness of 0.2740 inches (6.96 mm).

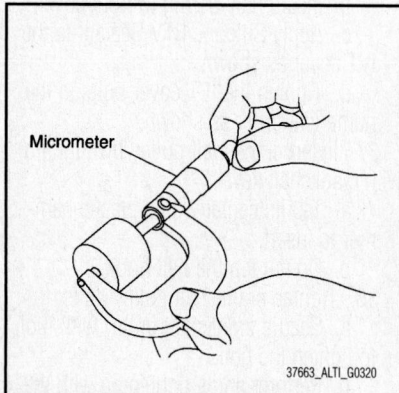

Fig. 125 Measure the center thickness of the removed valve lifters with a micrometer

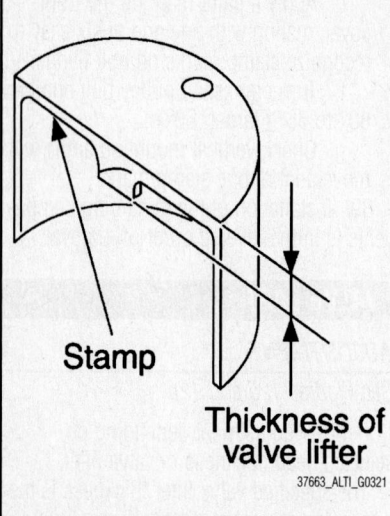

Fig. 126 Thickness of a new valve lifter can be identified by stamp marks on the reverse side (inside the cylinder)

➡**Available thickness of valve lifter: 26 sizes with a range of 0.3102 to 0.3299 inches (7.88 to 8.38 mm), in steps of 0.0008 inches (0.02 mm), when assembled at the factory.**

5. Install the selected valve lifter.
6. Install camshaft.
7. Manually turn crankshaft pulley a few turns.
8. Check that valve clearances for cold engine are within specifications, by referring to the specified values.
9. After completing the repair, check valve clearances again with the specifications for warmed engine. Use a feeler gauge to measure the clearance between the valve and camshaft. Make sure the values are within specifications.

INSPECTION

See Figures 127 through 129.

Perform this inspection as follows after removal, installation, or replacement of the camshaft or any valve related parts, or if there are any unusual engine conditions due to changes in valve clearance over time (starting, idling, and/or noise).

1. Warm up the engine, then stop it.
2. Remove the fender protector side cover RH.
3. Remove the rocker cover.
4. Turn crankshaft pulley in normal direction (clockwise when viewed from front) to align TDC identification mark (without paint mark) with timing indicator.
5. At this time, check that the both intake and exhaust cam lobes of No. 1 cylinder face outside.

➡**If they do not face outside, turn crankshaft pulley once more.**

6. Measure valve clearances with a feeler gauge at No. 1 INT and EXH, No. 2

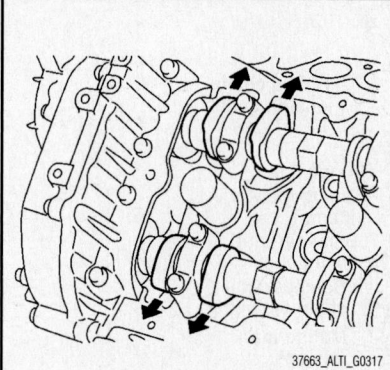

Fig. 127 Check that the both intake and exhaust cam lobes of No. 1 cylinder face outside

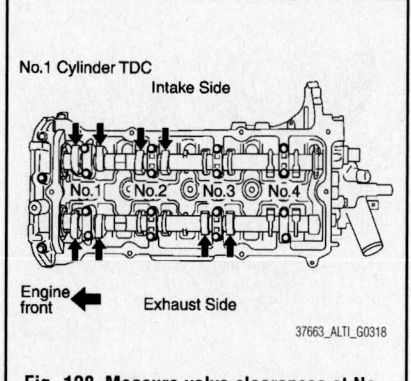

Fig. 128 Measure valve clearances at No. 1 INT and EXH, No. 2 INT, and No.3 EXH

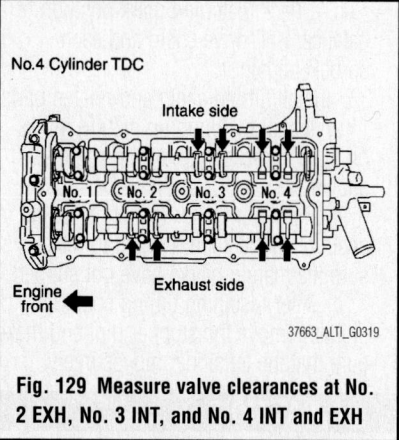

Fig. 129 Measure valve clearances at No. 2 EXH, No. 3 INT, and No. 4 INT and EXH

INT, and No.3 EXH with No.1 cylinder compression TDC.

   a. Use a feeler gauge to measure the clearance between valve and camshaft.

   b. Valve clearance standards:
- Cold: Intake: 0.009–0.013 inches (0.24–0.32 mm)
- Cold: Exhaust: 0.010–0.013 inches (0.26–0.34 mm)
- Hot: Intake: 0.012–0.016 inches (0.304–0.416 mm)
- Hot Exhaust: 0.012–0.017 inches (0.308–0.432 mm)

✳✳ CAUTION

If inspection was carried out with cold engine, check that values with fully warmed up engine are still within specifications.

7. Turn crankshaft one complete revolution (360°) and align mark on crankshaft pulley with pointer.
8. Measure valve clearances with a feeler gauge at No. 2 EXH, No. 3 INT, and No. 4 INT and EXH with No.4 cylinder compression TDC.
9. If out of specifications, make necessary adjustment.

ENGINE PERFORMANCE & EMISSION CONTROLS

CAMSHAFT POSITION (CMP) SENSOR

REMOVAL & INSTALLATION

1. Turn ignition switch OFF.
2. Loosen the fixing bolt of the sensor.
3. Disconnect camshaft position sensor (PHASE) harness connector.
4. Remove the sensor.
5. Installation is the reverse of removal.

CRANKSHAFT POSITION (CKP) SENSOR

REMOVAL & INSTALLATION

1. Turn ignition switch OFF.
2. Loosen the fixing bolt of the sensor.

3. Disconnect crankshaft position sensor (POS) harness connector.
4. Remove the sensor.
5. Installation is the reverse of removal.

ENGINE CONTROL MODULE (ECM)

LOCATION

The ECM is located in the engine compartment at the front driver's side.

REMOVAL & INSTALLATION

➡**The manufacturer does not provide a specific Removal and Installation pro-** cedure for this component. Refer to the graphic(s) when servicing this component.

ENGINE COOLANT TEMPERATURE (ECT) SENSOR

LOCATION

See Figure 130.

Refer to the accompanying illustration.

REMOVAL & INSTALLATION

1. Turn ignition switch OFF.
2. Disconnect engine coolant temperature sensor harness connector.

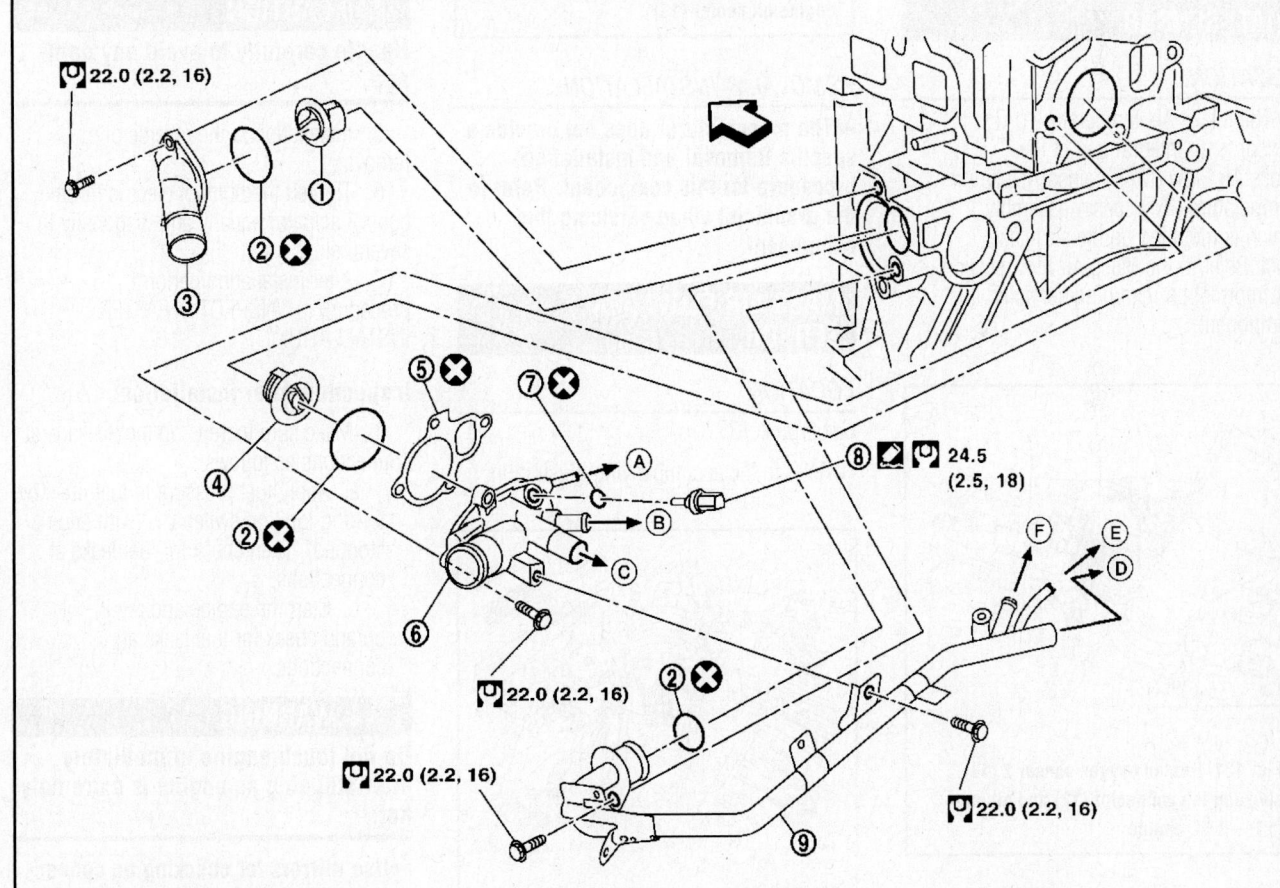

1. Thermostat
2. O-ring
3. Engine coolant inlet
4. Water control valve
5. Gasket
6. Engine coolant outlet
7. Copper washer
8. Engine coolant temperature sensor
9. Heater pipe
A. To electronic throttle control
B. To oil cooler
C. To heater
D. To heater
E. To electronic throttle control
F. To oil cooler

Fig. 130 Engine coolant temperature sensor (8)

37663_ALTH_G0020

3. Remove engine coolant temperature sensor.

4. Installation is the reverse of removal.

HEATED OXYGEN (HO2S) SENSOR

LOCATION

See Figure 131.

Refer to the accompanying illustration.

REMOVAL & INSTALLATION

➡**The manufacturer does not provide a specific Removal and Installation procedure for this component. Refer to the graphic(s) when servicing this component.**

INTAKE AIR TEMPERATURE (IAT) SENSOR

LOCATION

The Intake Air Temperature (IAT) sensor is an integral part of the Mass Air Flow (MAF) sensor/Intake Air Temperature (IAT) sensor assembly which is mounted on the air intake duct. Refer to the Mass Air Flow section for information regarding servicing this component.

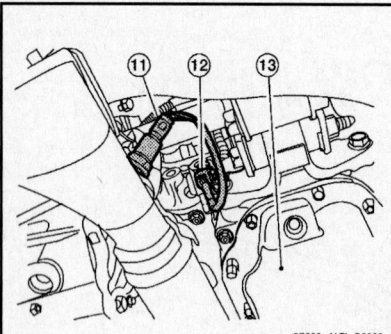

Fig. 131 Heated oxygen sensor 2 (11) showing the connector (12) and oil pan (13)—2.5L engine

REMOVAL & INSTALLATION

The Intake Air Temperature (IAT) sensor is an integral part of the Mass Air Flow (MAF) sensor/Intake Air Temperature (IAT) sensor assembly which is mounted on the air intake duct. Refer to the Mass Air Flow section for information regarding servicing this component.

KNOCK SENSOR (KS)

LOCATION

See Figure 132.

The knock sensor is attached to the cylinder block.

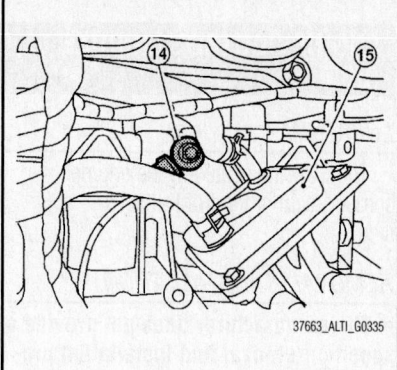

Fig. 132 Knock sensor (14) location near engine oil cooler (15)

REMOVAL & INSTALLATION

➡**The manufacturer does not provide a specific Removal and Installation procedure for this component. Refer to the graphic(s) when servicing this component.**

MASS AIR FLOW (MAF) SENSOR

LOCATION

See Figure 133.

Refer to the accompanying illustration.

Fig. 133 MAF sensor/IAT sensor location—2.5L engine

REMOVAL & INSTALLATION

➡**The manufacturer does not provide a specific Removal and Installation procedure for this component. Refer to the graphic(s) when servicing this component.**

THROTTLE CONTROL ACTUATOR (TAC)

LOCATION

See Figure 134.

The electrical throttle control actuator is located between the engine air intake duct and the intake manifold.

REMOVAL & INSTALLATION

1. Disconnect the engine air intake duct from the electric throttle control actuator.

2. Disconnect the electric throttle control actuator electrical connector.

3. Disconnect electric throttle control actuator coolant hoses.

4. Loosen bolts diagonally, and remove the electric throttle control actuator.

❄❄ CAUTION

Handle carefully to avoid any damage.

5. Installation is the reverse of removal.

6. Tighten the bolts of electric throttle control actuator equally and diagonally in several steps.

7. After installation perform procedure in "INSPECTION AFTER INSTALLATION".

Inspection After Installation

1. Make sure there is no fuel leakage at connections as follows:

   a. Apply fuel pressure to fuel lines by turning ignition switch ON (with engine stopped). Then check for fuel leaks at connections.

   b. Start the engine and rev it up and check for fuel leaks at connections.

❄❄ WARNING

Do not touch engine immediately after stopping as engine is extremely hot.

➡**Use mirrors for checking on connections out of the direct line of sight.**

THROTTLE POSITION SENSOR (TPS)

REMOVAL & INSTALLATION

Refer to Throttle Control Actuator section when servicing this component.

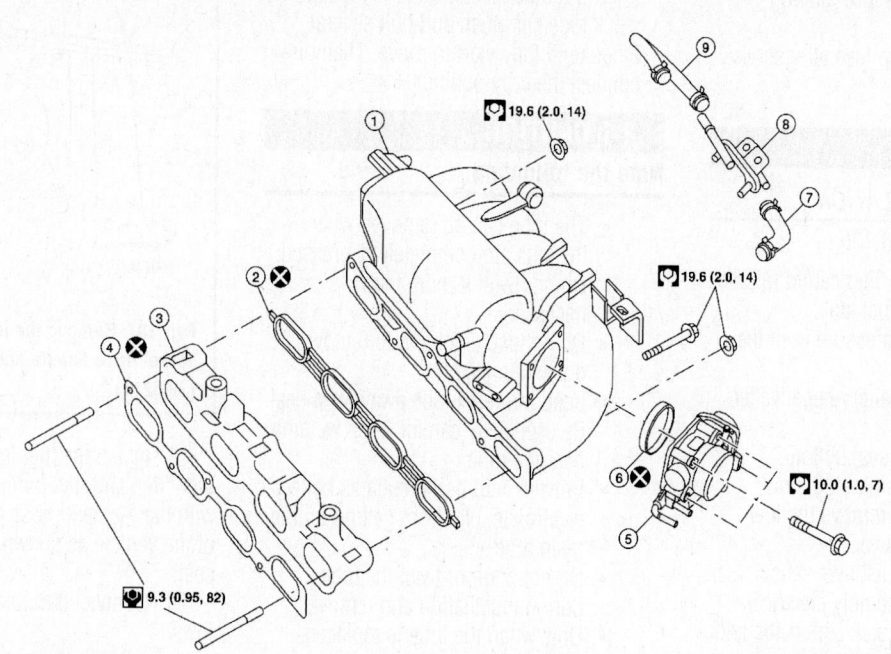

1. Intake manifold
4. Gasket
7. PCV hose

2. Rubber seal
5. Electric throttle control actuator
8. EVAP canister purge volume control solenoid valve

3. Intake manifold adapter
6. Rubber seal
9. EVAP hose

37663_ALTI_G0339

Fig. 134 Electric throttle control actuator (5) location—2.5L engine

---

FUEL   **GASOLINE FUEL INJECTION SYSTEM**

FUEL SYSTEM SERVICE PRECAUTIONS

Safety is the most important factor when performing not only fuel system maintenance, but any type of maintenance. Failure to conduct maintenance and repairs in a safe manner may result in serious personal injury or death. Work on a vehicle's fuel system components can be accomplished safely and effectively by adhering to the following rules and guidelines.

• To avoid the possibility of fire and personal injury, always disconnect the negative battery cable unless the repair or test procedure requires that battery voltage be applied.

• Always relieve the fuel system pressure prior to disconnecting any fuel system component (injector, fuel rail, pressure regulator, etc.) fitting or fuel line connection. Exercise extreme caution whenever relieving fuel system pressure to avoid exposing skin, face and eyes to fuel spray. Please be advised that fuel under pressure may penetrate the skin or any part of the body that it contacts.

• Always place a shop towel or cloth around the fitting or connection prior to loosening to absorb any excess fuel due to spillage. Ensure that all fuel spillage is quickly removed from engine surfaces. Ensure that all fuel-soaked cloths or towels are deposited into a flame-proof waste container with a lid.

• Always keep a dry chemical (Class B) fire extinguisher near the work area.

• Do not allow fuel spray or fuel vapors to come into contact with a spark or open flame.

• Always use a second wrench when loosening or tightening fuel line connection fittings. This will prevent unnecessary stress and torsion on fuel piping. Always follow the proper torque specifications.

• Always replace worn fuel fitting O-rings with new ones. Do not substitute fuel hose where rigid pipe is installed.

FUEL SYSTEM PRESSURE

RELIEVING

➡**If following procedure performed, a certain DTC may be detected.**

With CONSULT-III

1. Lift up the vehicle.
2. Turn ignition switch ON (READY).
3. Depress the accelerator pedal and keep it.
4. Shift the selector lever to N position with engine running.

❈❈ CAUTION

Never leave the selector lever in the N position for a long period of time. In the N position, the engine operates but electricity cannot be generated.

5. Perform "FUEL PRESSURE RELEASE" in "WORK SUPPORT" mode with CONSULT-III.
6. After engine stalls, turn ignition switch OFF.

Without CONSULT-III

1. Turn ignition switch OFF.
2. Remove fuel pump fuse located in IPDM E/R.
3. Turn ignition switch ON (READY).
4. Depress the accelerator pedal and keep it.

5. After engine stalls, turn ignition switch OFF.

6. Reinstall fuel pump fuse after servicing fuel system.

FUEL FILTER

REMOVAL & INSTALLATION

See Figures 135 through 138.

1. Unscrew the fuel filler cap to release the pressure inside the fuel tank.

2. Release the fuel pressure from the fuel lines.

3. Disconnect the battery negative terminal.

4. Remove the rear seat bottom.

5. Turn the four retainers 90° in a clockwise direction and remove the fuel pump inspection hole cover.

6. Disconnect the fuel level sensor, fuel filter, and fuel pump assembly electrical connector, EVAP hose quick connector, and the fuel feed hose quick connector from the fuel level sensor unit, fuel filter, and fuel pump assembly.

7. Remove the quick connector as follows:

   a. Hold the sides of the connector, push in tabs and pull out the tube.

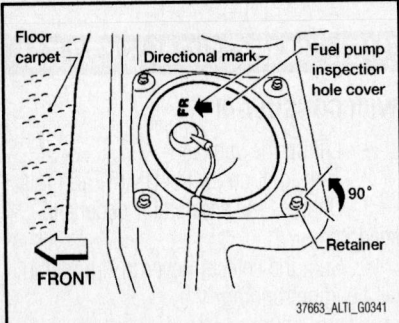

Fig. 135 Turn the four retainers 90° in a clockwise direction and remove the fuel pump inspection hole cover

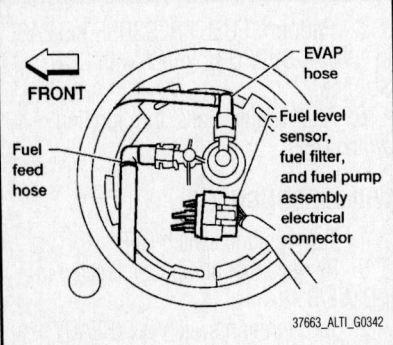

Fig. 136 Top view of fuel pump assembly

b. If the connector and the tube are stuck together, push and pull several times until they start to move. Then disconnect them by pulling.

✸✸ CAUTION

Note the following:

- The tube can be removed when the tabs are completely depressed. Do not twist it more than necessary.
- Do not use any tools to remove the quick connector.
- Keep the resin tube away from heat. Be especially careful when welding near the tube.
- Prevent acid liquid such as battery electrolyte, etc. from getting on the resin tube.
- Do not bend or twist the tube during installation and removal.
- Only when the tube is replaced, remove the remaining retainer on the tube or fuel level sensor, fuel filter, and fuel pump assembly.
- When the tube or fuel level sensor, fuel filter, and fuel pump assembly is replaced, also replace the retainer with a new one (green colored retainer).
- To keep the connecting portion clean and to avoid damage and foreign materials, cover them completely with plastic bags or something similar.

8. Remove the lock ring using a socket drive handle and Tool KV991J0090 (J-46214) as shown.

✸✸ CAUTION

Discard the lock ring, do not reuse the lock ring. Discard the ring seal, do not reuse the ring seal.

9. Remove the fuel level sensor, fuel filter, and fuel pump assembly.

✸✸ CAUTION

Do not bend the float arm during removal. Discard the ring seal, do not reuse the ring seal.

10. Inspect the fuel level sensor, fuel filter, and fuel pump for any defects and foreign materials. Replace as necessary.

To install:

11. Installation is in the reverse order of removal.

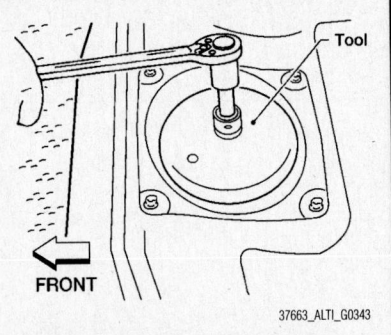

Fig. 137 Remove the lock ring using a socket drive handle and Tool KV991J0090 (J-46214)

12. Install the fuel level sensor, fuel filter, and fuel pump assembly with the fuel feed hose facing the front of the vehicle as shown. Use a new ring seal.

13. Connect the quick connector as follows:

   a. Check the connection for damage or any foreign materials.

   b. Align the connector with the tube, then insert the connector straight into the tube until a click is heard.

14. After the tube is connected, make sure the connection is secure by performing the following checks:

   a. Pull the tube and the connector to make sure they are securely connected.

   b. Visually confirm that the two retainer tabs are connected to the quick connector.

15. Turn the ignition switch to ON (without starting the engine) to apply fuel pressure to the fuel system, then check the connections for fuel leaks.

16. Start the engine and let it idle and check for fuel leaks at the fuel system connections.

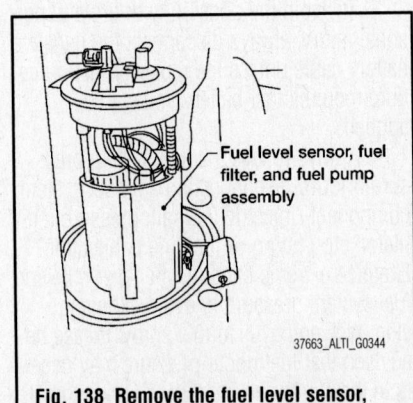

Fig. 138 Remove the fuel level sensor, fuel filter, and fuel pump assembly

FUEL LEVEL SENDING UNIT

LOCATION

The fuel level sensor unit is an integral part of the fuel pump assembly.

REMOVAL & INSTALLATION

See Figures 135 through 138.

1. Unscrew the fuel filler cap to release the pressure inside the fuel tank.
2. Release the fuel pressure from the fuel lines.
3. Disconnect the battery negative terminal.
4. Remove the rear seat bottom.
5. Turn the four retainers 90° in a clockwise direction and remove the fuel pump inspection hole cover.
6. Disconnect the fuel level sensor, fuel filter, and fuel pump assembly electrical connector, EVAP hose quick connector, and the fuel feed hose quick connector from the fuel level sensor unit, fuel filter, and fuel pump assembly.
7. Remove the quick connector as follows:
   a. Hold the sides of the connector, push in tabs and pull out the tube.
   b. If the connector and the tube are stuck together, push and pull several times until they start to move. Then disconnect them by pulling.

✵✵ CAUTION

Note the following:

- The tube can be removed when the tabs are completely depressed. Do not twist it more than necessary.
- Do not use any tools to remove the quick connector.
- Keep the resin tube away from heat. Be especially careful when welding near the tube.
- Prevent acid liquid such as battery electrolyte, etc. from getting on the resin tube.
- Do not bend or twist the tube during installation and removal.
- Only when the tube is replaced, remove the remaining retainer on the tube or fuel level sensor, fuel filter, and fuel pump assembly.
- When the tube or fuel level sensor, fuel filter, and fuel pump assembly is replaced, also replace the retainer with a new one (green colored retainer).
- To keep the connecting portion clean and to avoid damage and foreign materials, cover them completely with plastic bags or something similar.

8. Remove the lock ring using a socket drive handle and Tool KV991J0090 (J-46214) as shown.

✵✵ CAUTION

Discard the lock ring, do not reuse the lock ring. Discard the ring seal, do not reuse the ring seal.

9. Remove the fuel level sensor, fuel filter, and fuel pump assembly.

✵✵ CAUTION

Do not bend the float arm during removal. Discard the ring seal, do not reuse the ring seal.

10. Inspect the fuel level sensor, fuel filter, and fuel pump for any defects and foreign materials. Replace as necessary.

To install:

11. Installation is in the reverse order of removal.
12. Install the fuel level sensor, fuel filter, and fuel pump assembly with the fuel feed hose facing the front of the vehicle as shown. Use a new ring seal.
13. Connect the quick connector as follows:
    a. Check the connection for damage or any foreign materials.
    b. Align the connector with the tube, then insert the connector straight into the tube until a click is heard.
14. After the tube is connected, make sure the connection is secure by performing the following checks:
    a. Pull the tube and the connector to make sure they are securely connected.
    b. Visually confirm that the two retainer tabs are connected to the quick connector.
15. Turn the ignition switch to ON (without starting the engine) to apply fuel pressure to the fuel system, then check the connections for fuel leaks.
16. Start the engine and let it idle and check for fuel leaks at the fuel system connections.

FUEL PUMP ASSEMBLY

REMOVAL & INSTALLATION

See Figures 135 through 138.

1. Unscrew the fuel filler cap to release the pressure inside the fuel tank.

2. Release the fuel pressure from the fuel lines.
3. Disconnect the battery negative terminal.
4. Remove the rear seat bottom.
5. Turn the four retainers 90° in a clockwise direction and remove the fuel pump inspection hole cover.
6. Disconnect the fuel level sensor, fuel filter, and fuel pump assembly electrical connector, EVAP hose quick connector, and the fuel feed hose quick connector from the fuel level sensor unit, fuel filter, and fuel pump assembly.
7. Remove the quick connector as follows:
   a. Hold the sides of the connector, push in tabs and pull out the tube.
   b. If the connector and the tube are stuck together, push and pull several times until they start to move. Then disconnect them by pulling.

✵✵ CAUTION

Note the following:

- The tube can be removed when the tabs are completely depressed. Do not twist it more than necessary.
- Do not use any tools to remove the quick connector.
- Keep the resin tube away from heat. Be especially careful when welding near the tube.
- Prevent acid liquid such as battery electrolyte, etc. from getting on the resin tube.
- Do not bend or twist the tube during installation and removal.
- Only when the tube is replaced, remove the remaining retainer on the tube or fuel level sensor, fuel filter, and fuel pump assembly.
- When the tube or fuel level sensor, fuel filter, and fuel pump assembly is replaced, also replace the retainer with a new one (green colored retainer).
- To keep the connecting portion clean and to avoid damage and foreign materials, cover them completely with plastic bags or something similar.

8. Remove the lock ring using a socket drive handle and Tool KV991J0090 (J-46214) as shown.

✵✵ CAUTION

Discard the lock ring, do not reuse the lock ring. Discard the ring seal, do not reuse the ring seal.

9. Remove the fuel level sensor, fuel filter, and fuel pump assembly.

✳✳ CAUTION

Do not bend the float arm during removal. Discard the ring seal, do not reuse the ring seal.

10. Inspect the fuel level sensor, fuel filter, and fuel pump for any defects and foreign materials. Replace as necessary.

To install:

11. Installation is in the reverse order of removal.

12. Install the fuel level sensor, fuel filter, and fuel pump assembly with the fuel feed hose facing the front of the vehicle as shown. Use a new ring seal.

13. Connect the quick connector as follows:

   a. Check the connection for damage or any foreign materials.

   b. Align the connector with the tube, then insert the connector straight into the tube until a click is heard.

14. After the tube is connected, make sure the connection is secure by performing the following checks:

   a. Pull the tube and the connector to make sure they are securely connected.

   b. Visually confirm that the two retainer tabs are connected to the quick connector.

15. Turn the ignition switch to ON (without starting the engine) to apply fuel pressure to the fuel system, then check the connections for fuel leaks.

16. Start the engine and let it idle and check for fuel leaks at the fuel system connections.

FUEL RAIL AND INJECTOR

REMOVAL & INSTALLATION

See Figures 139 through 141.

1. Remove engine room cover.
2. Release the fuel pressure.
3. Remove the front air duct.
4. Disconnect the fuel hose quick connector at the fuel tube side.

✳✳ CAUTION

Prepare a container and cloth for catching any spilled fuel. This operation should be performed in a place that is free from any open flames. While hoses are disconnected seal their openings with vinyl bag or similar material to prevent foreign material from entering them.

5. Remove the intake manifold.
6. Disconnect sub-harness for injector at engine front side, and remove it from bracket.
7. Loosen the bolts in the reverse order shown, then remove fuel tube and fuel injectors as an assembly.
8. Remove the fuel injectors from the fuel tube.

   a. Release the clip and remove the fuel injector.

   b. Pull fuel injector straight out of the fuel tube.

✳✳ CAUTION

Be careful not to damage the nozzle. Avoid any impact, such as dropping the fuel injector. Do not disassemble or adjust the fuel injector.

To install:

9. Install new O-rings on the fuel injector, the fuel side black O-ring and the nozzle side green O-ring.

✳✳ CAUTION

Upper and lower O-rings are different. Be careful not to confuse them. Fuel tube side: black O-ring; Nozzle side: green O-ring.

✳✳ CAUTION

Note the following:

- Lubricate the O-rings lightly with new engine oil.
- Handle O-rings with bare hands only. Do not wear gloves.
- Do not clean O-rings with solvent.
- Make sure that O-ring and its mating part are free of foreign material.
- Be careful not to scratch O-rings during installation.
- Do not twist or stretch the O-ring. If the O-ring was stretched while it is attached, do not insert it into the fuel tube immediately.

10. Install the fuel injector into the fuel tube with the following procedure:

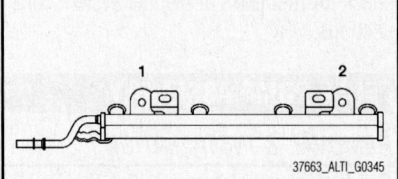

Fig. 139 Loosen the bolts in the reverse order shown

37663_ALTI_G0345

   a. Do not reuse the clip, replace it with a new one.

   b. Insert the new clip into the clip mounting groove on fuel injector.

   c. Insert the clip so that projection of fuel injector matches notch of the clip.

   d. Fuel tube side: black O-ring

   e. Nozzle side: green O-ring

11. Insert fuel injector into fuel tube with clip attached.

   a. Insert it while matching it to the axial center.

   b. Insert fuel injector so that projection of fuel tube matches notch of the clip.

   c. Make sure that fuel tube flange is securely fixed in flange fixing groove on the clip.

   d. Make sure that installation is complete by checking that fuel injector does not rotate or come off.

12. Install fuel tube assembly.

   a. Insert the tip of each fuel injector into intake manifold.

   b. Tighten the bolts in two steps in the numerical order as shown.

   c. Step 1: 7 ft. lbs. (10 Nm)

   d. Step 2: 16 ft. lbs. (22 Nm)

✳✳ CAUTION

After properly connecting fuel tube assembly to injector and fuel hose, check connection for fuel leakage.

13. Install the intake manifold.

14. Connect the fuel hose quick connector.

15. Installation of the remaining components is in the reverse order of removal.

16. Make sure there is no fuel leakage at connections as follows:

   a. Apply fuel pressure to fuel lines by turning ignition switch ON (with engine stopped). Then check for fuel leaks at connections.

   b. Start the engine and rev it up and check for fuel leaks at connections.

17. Perform procedures for "Throttle Valve Closed Position Learning" after finishing repairs.

18. If electric throttle control actuator is replaced, perform procedures for "Idle Air Volume Learning" after finishing repairs.

✳✳ WARNING

Do not touch engine immediately after stopping as engine is extremely hot.

➡Use mirrors for checking on connections out of the direct line of sight.

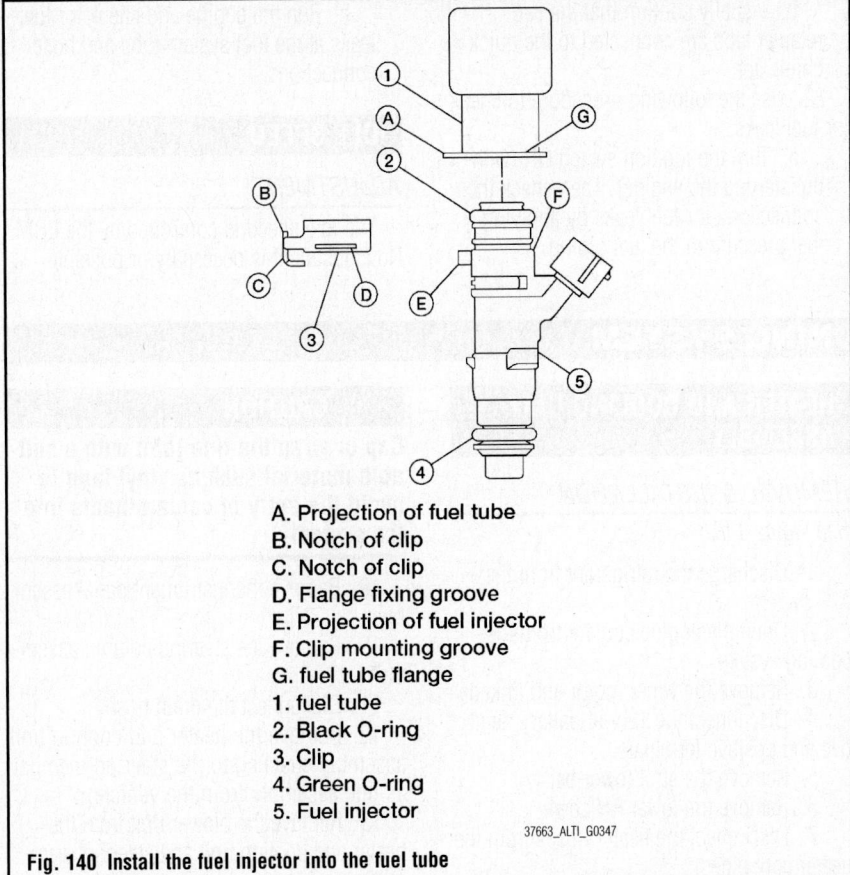

A. Projection of fuel tube
B. Notch of clip
C. Notch of clip
D. Flange fixing groove
E. Projection of fuel injector
F. Clip mounting groove
G. fuel tube flange
1. fuel tube
2. Black O-ring
3. Clip
4. Green O-ring
5. Fuel injector

37663_ALTI_G0347

Fig. 140 Install the fuel injector into the fuel tube

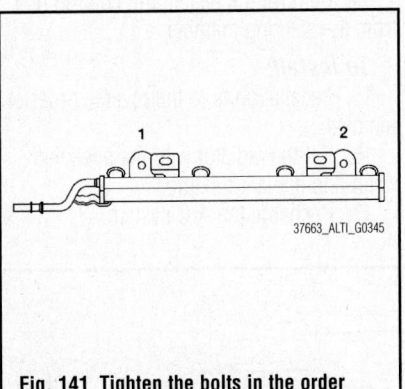

37663_ALTI_G0345

Fig. 141 Tighten the bolts in the order shown

FUEL TANK

REMOVAL & INSTALLATION

See Figures 135 and 136.

1. Disconnect the battery negative terminal.

2. Check the fuel level with the vehicle on a level surface. If the fuel gauge indicates more than the level (7/8 full), drain the fuel from the fuel tank until the fuel gauge indicates a level at or below (7/8 full).

3. In case the fuel pump does not operate, use the following procedure.

a. Insert fuel tubing of less than 0.98 inches (25 mm) diameter into the fuel filler tube through the fuel filler opening to drain fuel from the fuel filler tube.

b. Disconnect the fuel filler hose from the fuel filler tube.

c. Insert fuel tubing into the fuel tank through the fuel filler hose to drain fuel from the fuel tank.

d. As a guide, the fuel level reaches or is less than the level on the fuel gauge as shown, when approximately 2 ⅝ US gal. (10L) of fuel is drained from a full fuel tank.

4. Open the fuel filler cap to release the pressure inside the fuel tank.

5. Release fuel pressure from fuel line.

6. Remove rear seat bottom.

7. Turn the four retainers 90° in a clockwise direction and remove the fuel pump inspection hole cover.

8. Disconnect the fuel level sensor, fuel filter, and fuel pump assembly electrical connector, EVAP hose quick connector, and the fuel feed hose quick connector from the fuel level sensor unit, fuel filter, and fuel pump assembly.

9. Disconnect the quick connectors as follows:

a. Hold the sides of the connector, push in tabs and pull out the tube.

b. If the connector and the tube are stuck together, push and pull several times until they start to move. Then disconnect them by pulling.

☀☀ CAUTION

Note the following:

- The tube can be removed when the tabs are completely depressed. Do not twist it more than necessary.
- Do not use any tools to remove the quick connector.
- Keep the resin tube away from heat. Be especially careful when welding near the tube.
- Prevent acid liquid such as battery electrolyte, from getting on the resin tube.
- Do not bend or twist the tube during installation and removal.
- Only when the tube is replaced, remove the remaining retainer on the tube or fuel level sensor, fuel filter, and fuel pump assembly.
- When the tube or fuel level sensor, fuel filter, and fuel pump assembly is replaced, also replace the retainer with a new one (green colored retainer).
- To keep the connecting portion clean and to avoid damage and foreign materials, cover them completely with plastic bags or something similar.

10. Remove the center exhaust tube, with muffler(s).

11. Disconnect the fuel filler hose and the recirculation hose at the fuel tank side.

12. Disconnect the three parking brake cable mounting brackets on each cable and position the cables out of the way.

13. Remove the fuel tank protector.

14. Disconnect the fuel tank mounting straps while supporting the fuel tank.

15. Remove the fuel tank.

16. If replacing the fuel tank, remove the fuel level sensor, fuel filter and fuel pump assembly to transfer to the new fuel tank.

To install:

17. Install in the reverse order of removal paying attention to the following.

18. Before tightening the fuel tank mounting straps, temporarily install the filler hose and the recirculation hose.

19. Tighten all fuel tank mounting strap bolts to specification, then tighten the hose clamps.

20. Connect the quick connector as follows:

a. Check the connection for damage or any foreign materials.

b. Align the connector with the tube, then insert the connector straight into the tube until a click is heard.

21. After the tube is connected, make sure the connection is secure by performing the following checks:

a. Pull on the tube and the connector to make sure they are securely connected.

b. Visually confirm that the two retainer tabs are connected to the quick connector.

22. Use the following procedure to check for fuel leaks.

a. Turn the ignition switch ON (without starting the engine). Then check the connections for fuel leaks by applying fuel pressure to the fuel piping.

b. Run the engine and check for fuel leaks at the fuel system tube and hose connections.

IDLE SPEED

ADJUSTMENT

The idle speed is controlled by the ECM. No adjustment is necessary or possible.

HEATING & AIR CONDITIONING SYSTEM

BLOWER MOTOR

REMOVAL & INSTALLATION

See Figure 142.

1. Remove the glove box.
2. Disconnect the blower motor connector.
3. Remove the three blower motor screws and remove the blower motor from the blower unit.

To install:

4. Installation is in the reverse order of removal.

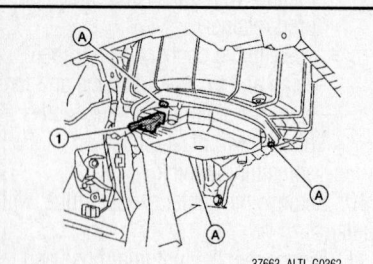

Fig. 142 Disconnect the blower motor connector (1); remove the three blower motor screws (A)

HEATER CORE

REMOVAL AND INSTALLATION

1. Remove the heater and cooling unit assembly.
2. Remove the heater grommet, heater pipe support and heater pipe cover.
3. Remove the heater and cooling unit foot duct LH.
4. Remove the heater core.

To install:

5. Installation is in the reverse order of removal.

➡**Make sure that the aspirator hose is securely attached to the aspirator on the heater and cooling unit foot duct LH.**

HEATER AND COOLING UNIT ASSEMBLY

REMOVAL & INSTALLATION

See Figure 143.

1. Discharge the refrigerant from the A/C system.
2. Drain the engine coolant from the cooling system.
3. Remove the wiper motor and linkage.
4. Disconnect the 12 volt battery negative and positive terminals.
5. Remove the strut tower bar.
6. Remove the lower RH cowl.
7. Disconnect the heater hoses from the heater core pipes.

❄ CAUTION

Cap or wrap the pipe joint with a suitable material such as vinyl tape to avoid the entry of contaminants into the system.

8. Disconnect the refrigerant lines from the expansion valve.

❄ CAUTION

Cap or wrap the line joint with a suitable material such as vinyl tape to avoid the entry of contaminants into the system.

9. Remove the instrument panel assembly.
10. Remove the steering column assembly.
11. Disconnect the drain hose.
12. Remove the heater and cooling unit assembly attached to the steering member as one assembly from the vehicle.
13. Remove the blower unit from the heater and cooling unit and steering member assembly.
14. Remove the heater and cooling unit from the steering member.

To install:

15. Installation is in the reverse order of removal.
16. Fill the radiator with the specified water and coolant mixture.
17. Recharge the A/C system.

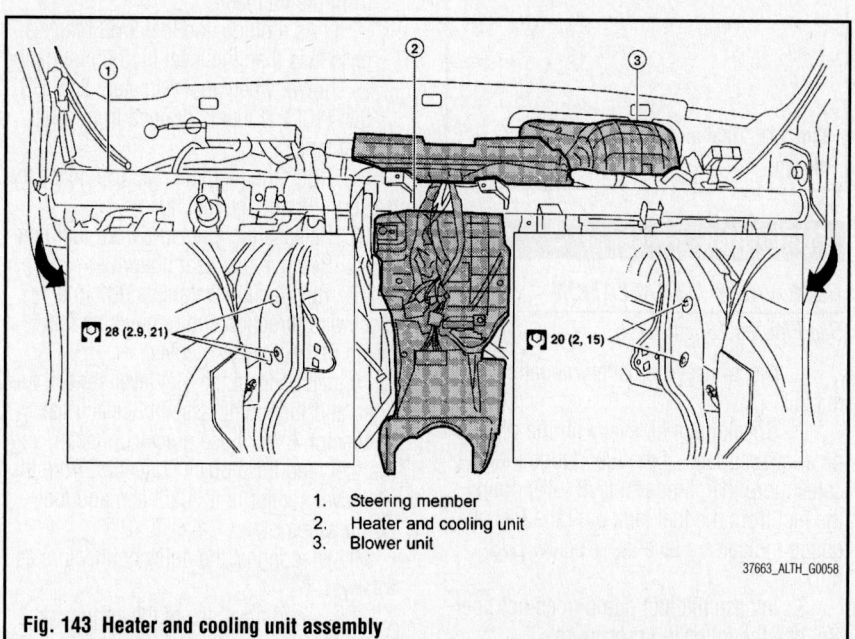

1. Steering member
2. Heater and cooling unit
3. Blower unit

Fig. 143 Heater and cooling unit assembly

STEERING

ELECTRIC POWER STEERING (EPS) CONTROL UNIT

REMOVAL & INSTALLATION

See Figures 144 and 145.

1. Remove the engine cover.
2. Remove the front wiper arm cover and wiper arm and blade assembly.
3. Remove the cowl top weatherstrip seal.
4. Remove the cowl top end caps.
5. Remove the cowl top finisher assembly.
6. Disconnect the washer hose.
7. Remove the strut brace.
8. Remove the wiper motor and connecting rod assembly.
9. Disconnect the 12-volt battery negative terminal.
10. Remove the left cowl extension.
11. Disconnect the MAF sensor connector.
12. Remove the air cleaner duct, blow-by hose and air cleaner duct hose.
13. Remove the fuse and fusible link box.
14. Disconnect the harness clips.
15. Disconnect the EPS DC/DC converter connector, EPS sensor connector, and EPS motor connector from the EPS ECU.

➡**For EPS DC/DC converter connector and EPS motor connector, perform the following:**

   a. Pull lock plate up until it stops.
   b. Turn the lock lever until it stops.
   c. Pull the connector to disconnect it.

16. Remove the EPS control unit nut and bolts and EPS control unit.

   To install:

17. Installation is in the reverse order of removal.

EPS DC/DC CONVERTER

REMOVAL & INSTALLATION

See Figure 146.

1. Pull the service plug to disconnect the high voltage battery.
2. Remove the rear seat.
3. Remove the EPS DC/DC converter cover nuts and remove the cover.

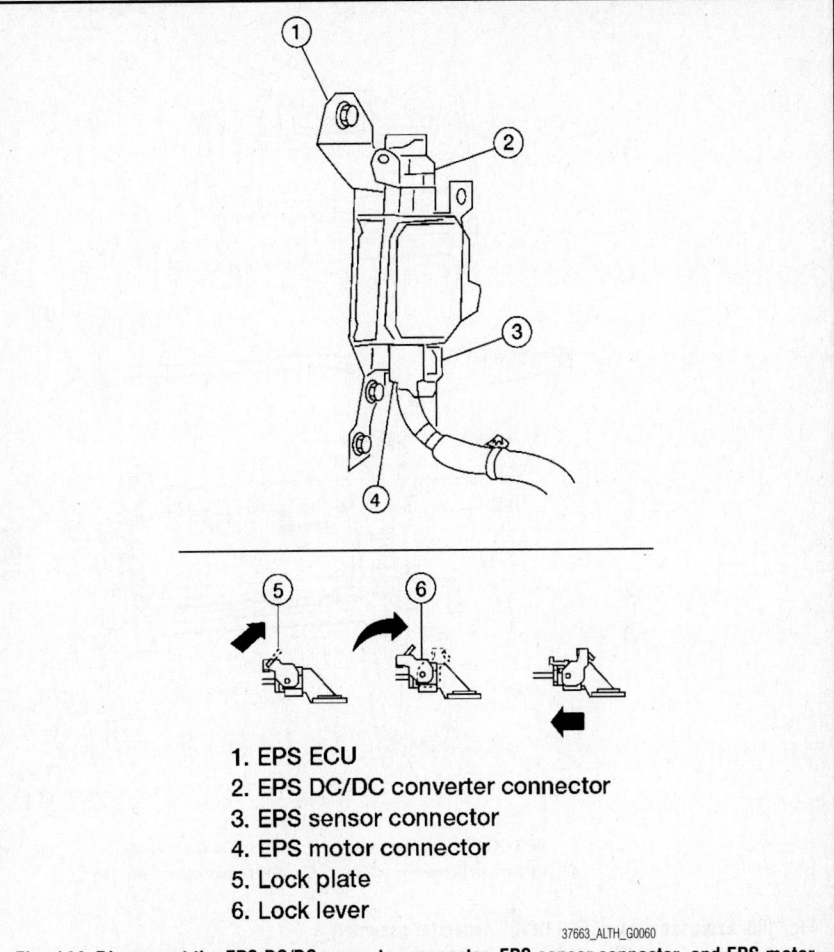

1. EPS ECU
2. EPS DC/DC converter connector
3. EPS sensor connector
4. EPS motor connector
5. Lock plate
6. Lock lever

37663_ALTH_G0060

Fig. 144 Disconnect the EPS DC/DC converter connector, EPS sensor connector, and EPS motor connector from the EPS ECU

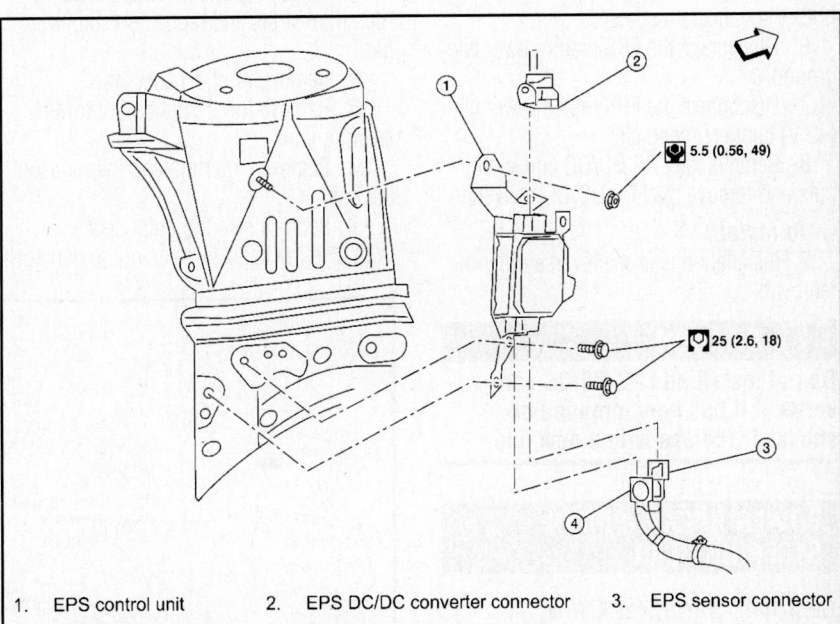

1. EPS control unit
2. EPS DC/DC converter connector
3. EPS sensor connector
4. EPS motor connector
⇐: Front

37663_ALTH_G0061

Fig. 145 EPS control unit location

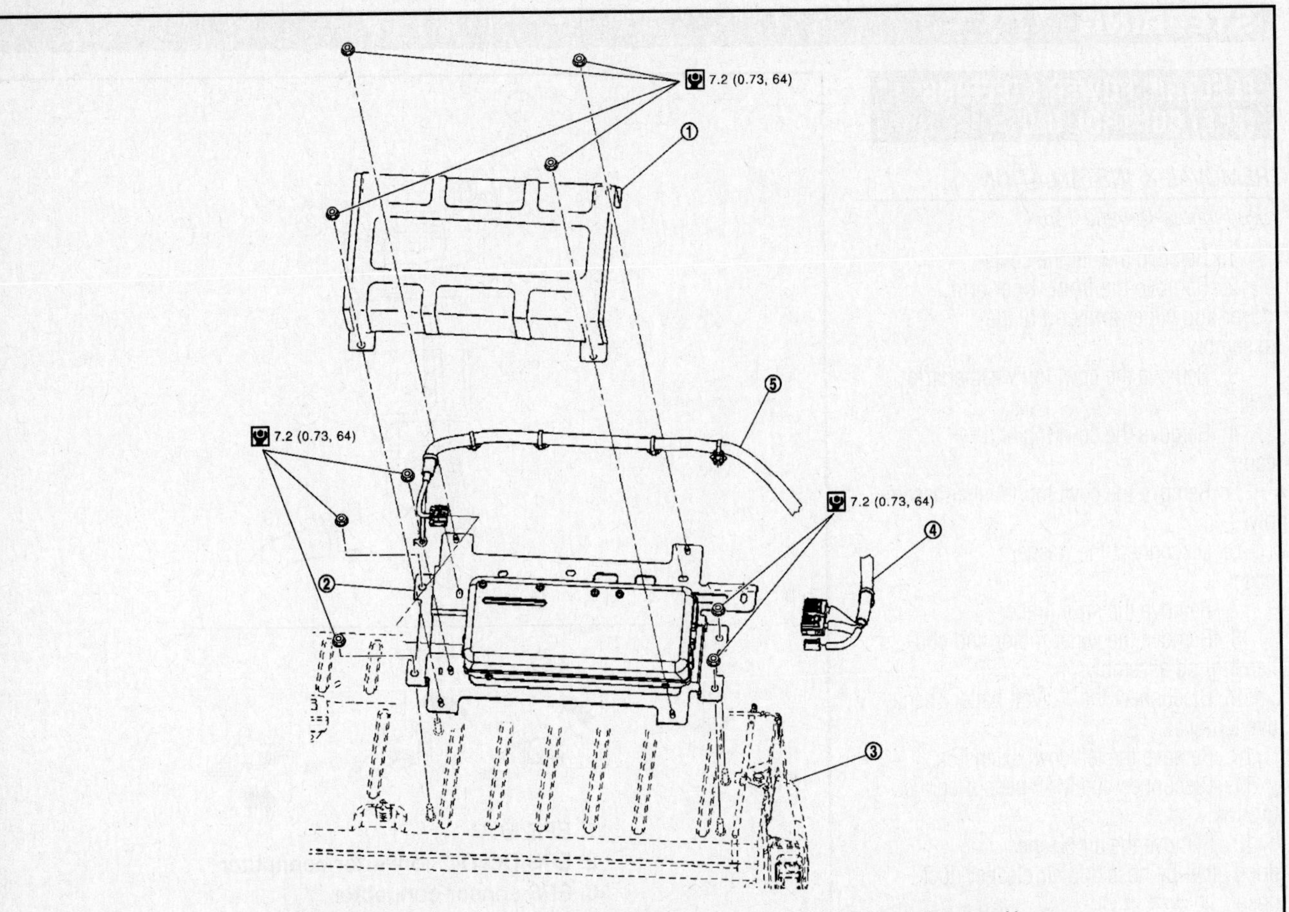

7.2 (0.73, 64)

① ⑤ ④ ③ ②

7.2 (0.73, 64)

7.2 (0.73, 64)

| 1. | EPS DC/DC converter cover | 2. | EPS DC/DC converter | 3. | High voltage battery assembly |
| 4. | EPS motor power line (42 V) | 5. | EPS motor power line (245 V) | | |

37663_ALTH_G0062

Fig. 146 Exploded view of EPS DC/DC converter assembly

4. Remove the shield earth nut.

5. Disconnect the EPS motor power line (245 V) connector and clip.

6. Disconnect the EPS motor power line ground.

7. Disconnect the EPS motor power line (42 V) connector and clip.

8. Remove the EPS DC/DC converter nuts and remove the EPS DC/DC converter.

To install:

9. Installation is in the reverse order of removal.

> ※※ **CAUTION**
>
> **Do not install an EPS DC/DC converter if it has been dropped or shocked, replace with a new one.**

POWER RACK & PINION STEERING GEAR

REMOVAL & INSTALLATION

See Figures 147 through 149.

1. Remove tires and wheels.

2. Remove engine undercover.

3. Remove stabilizer bar connecting rods from struts and reposition stabilizer bar.

4. Remove front exhaust tube.

5. Remove lower side bolt of lower steering joint.

6. Remove cotter pin and then loosen the nut.

7. Remove steering outer socket from steering knuckle so as not to damage

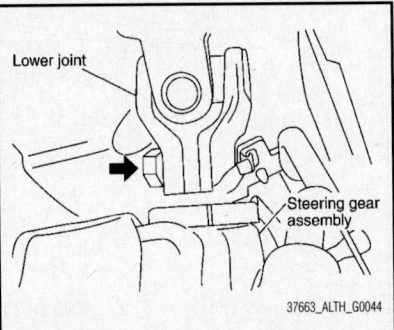

37663_ALTH_G0044

Fig. 147 Remove the lower side bolt of the lower steering joint

ball joint boot using Tool HT72520000 (J-25730-A) as shown.

> ※※ **CAUTION**
>
> **Temporarily tighten the nut to prevent damage to threads and to prevent the Tool from suddenly coming off.**

8. Disconnect EPS torque sensor harness connector, EPS motor angle sensor harness connector and EPS motor

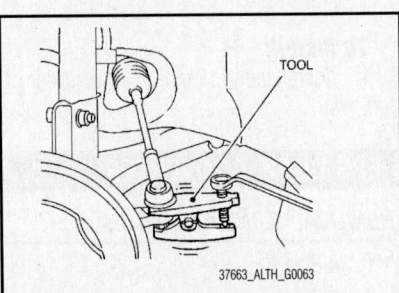

37663_ALTH_G0063

Fig. 148 Remove steering outer socket from steering knuckle so as not to damage ball joint boot using Tool HT72520000 (J-25730-A)

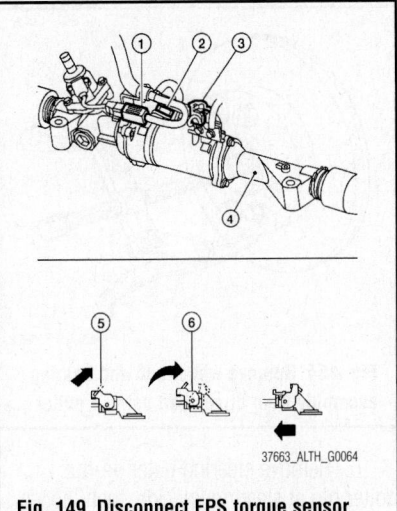

Fig. 149 Disconnect EPS torque sensor harness connector, EPS motor angle sensor harness connector and EPS motor power line connector from the steering gear

power line connector from the steering gear.

➡For EPS motor power line connector, perform the following;

  a. Pull lock plate up until it stops.
  b. Turn the lock lever until it stops.
  c. Pull EPS motor power line connector to disconnect it.

9. Remove ground bracket and harness bracket on steering gear.

10. Remove bolts and nuts of steering gear assembly, and then remove steering gear assembly from vehicle.

11. Check for damage to steering gear. If any exist, replace steering gear as an assembly.

To install:

12. Installation is the reverse order of removal.

✢✢ CAUTION

After connecting the EPS motor power line connector, make sure the connector is locked.

13. When installing lower joint to steering gear assembly, follow the procedure listed below.

  a. Set rack of steering gear in the neutral position.

➡To get the neutral position of rack, turn gear-sub assembly and measure the distance of inner socket, and then measure the intermediate position of the distance.

  b. Install slit part of lower shaft assembly joint aligning with the marking position of steering gear assembly input shaft. Make sure that the slit part of lower joint is aligned with the marking position of gear housing assembly input shaft.

14. Perform final tightening of nuts and bolts on each part under unladen conditions with tires on level ground when removing steering gear assembly. Check wheel alignment.

15. Make sure that steering wheel operates smoothly by turning several times from full left stop to full right stop.

SUSPENSION

LOWER CONTROL ARM (TRANSVERSE LINK)

REMOVAL AND & INSTALLATION

➡Nissan refers to the Lower Control Arm as Transverse Link. Refer to the exploded view of front suspension while servicing this component.

1. Remove wheel and tire.
2. Remove steering knuckle from transverse link.
3. Remove mounting nuts and washers on lower portion of stabilizer connecting rod.
4. Slightly loosen transverse link mounting bolts.
5. Remove transverse link bolts and nuts, and remove transverse link from suspension member.

To install:

6. Installation is in the reverse order of removal.
7. Refer to "Exploded View" for tightening torque.
8. Tighten transverse link bolts with vehicle unladen and all four tires on flat, level ground.
9. After installation, check wheel alignment.

STEERING KNUCKLE

REMOVAL & INSTALLATION

Refer to Wheel Hub, Steering Knuckle & Bearing (sealed unit) Section for Removal and Installation of the Steering Knuckle.

STRUT

REMOVAL & INSTALLATION

See Figure 150.

➡Refer to the exploded view of front suspension while servicing this component.

1. Remove wheel and tire.
2. Remove brake caliper and reposition aside using wire.

✢✢ CAUTION

Avoid depressing brake pedal with brake caliper removed.

3. Remove wheel sensor electrical harness from strut.
4. Remove brake hose lock plate.
5. Remove steering knuckle to strut bolts and nuts.
6. Remove bolt on strut tower bar then bolts on strut tower and remove strut from vehicle.

FRONT SUSPENSION

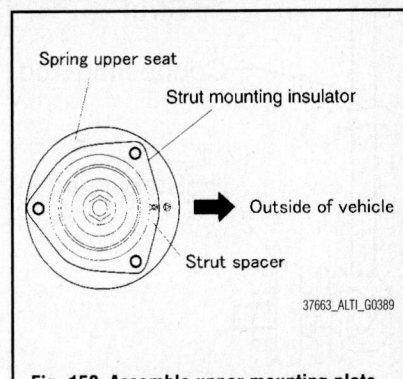

Fig. 150 Assemble upper mounting plate with its notch facing toward the outside

7. Check the strut for any oil leakage or other damage and replace as necessary.

To install:

8. Installation is in the reverse order of removal.
9. Refer to "Exploded View" for tightening torque.
10. Be sure arrows on strut mount insulator and spring upper seat are positioned as shown. Also be sure notch in strut spacer is positioned as shown. Then install strut.
11. Assemble upper mounting plate with its notch facing toward the outside.

STABILIZER BAR

REMOVAL & INSTALLATION

See Figures 151 through 153.

1. Remove steering gear.
2. Remove mounting nuts on upper portion of stabilizer connecting rod.
3. Remove stabilizer clamp bolts.
4. Remove stabilizer from the vehicle.
5. Check stabilizer, connecting rod, bushing and clamp for deformation, cracks and damage, and replace if necessary.

To install:

6. Installation is in the reverse order of removal.
7. Refer to "Exploded View" for tightening torque.
8. When installing stabilizer, make sure that the clamps are facing in the direction shown.

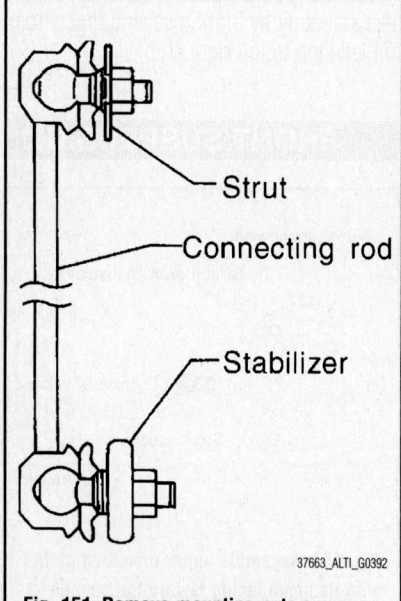

Fig. 151 Remove mounting nuts on upper portion of stabilizer connecting rod

Fig. 153 Position ball joint with case on pillow ball head parallel to stabilizer

9. Make sure the cut surface of the bushing faces the rear.
10. Stabilizer uses pillow ball type connecting rod. Position ball joint with case on pillow ball head parallel to stabilizer.

WHEEL HUB & BEARING

REMOVAL & INSTALLATION

See Figures 154 and 155.

1. Remove wheel and tire from vehicle.
2. Remove brake caliper, leaving brake caliper hydraulic lines connected. Reposition brake caliper aside with wire.

➡**Avoid depressing brake pedal while brake caliper is removed.**

3. Put alignment marks on disc rotor and wheel hub and bearing assembly, then remove disc rotor.
4. Remove wheel sensor from steering knuckle.

✲✲ CAUTION

Do not pull on wheel sensor harness.

5. Remove cotter pin, then remove lock nut from drive shaft.

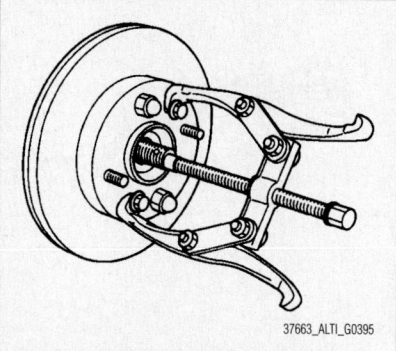

Fig. 154 Remove wheel hub and bearing assembly from drive shaft using a puller

6. Remove steering outer tie-rod cotter pin at steering knuckle, then loosen nut.
7. Disconnect the outer tie-rod end from steering knuckle using Tool HT2520000 (J-25730-A). Be careful not to damage ball joint boot.

✲✲ CAUTION

To prevent damage to threads and to prevent tool from coming off suddenly, temporarily tighten mounting nut.

8. Remove transverse link and steering knuckle pinch bolt and nut.
9. Remove wheel hub and bearing assembly from drive shaft using a puller or suitable tool.

✲✲ CAUTION

When removing wheel hub and bearing assembly, do not apply an excessive angle to drive shaft joint. Also be careful not to excessively extend slide joint. Support drive shaft when removing.

10. Remove wheel hub and bearing assembly bolts.

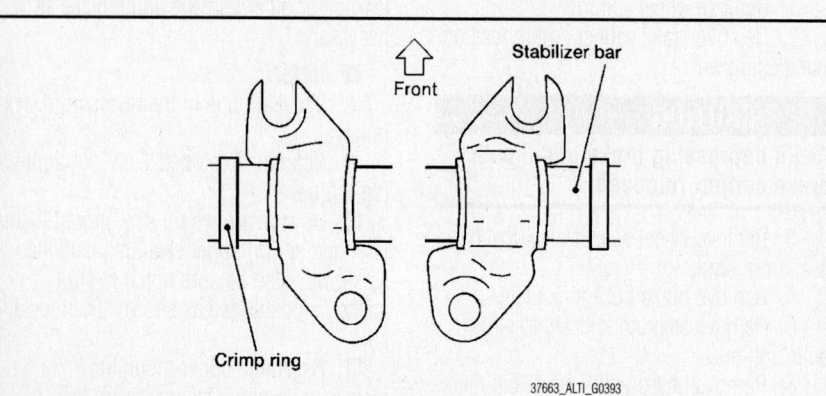

Fig. 152 When installing stabilizer, make sure that the clamps are facing in the direction shown

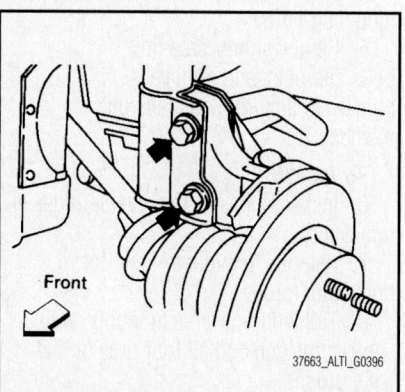

Fig. 155 Remove the lower strut bolts and nuts

11. Remove splash guard and wheel hub and bearing assembly from steering knuckle.

12. Remove the lower strut bolts and nuts.

13. Remove steering knuckle from vehicle.

14. Check for deformity, cracks and damage on each part, replace if necessary.

15. Check for boot breakage, axial looseness, and torque of transverse link ball joint and repair as necessary.

To install:

16. Installation is in the reverse order of removal.

❋❋ CAUTION

Do not reuse non-reusable parts.

17. When installing wheel hub and bearing assembly to steering knuckle, align cutout in toner ring cover with wheel sensor mounting hole in steering knuckle.

18. When installing disc rotor on wheel hub and bearing assembly, align the marks.

ADJUSTMENT

No adjustment is possible. If there is too much play, replace the bearing and/or hub.

SUSPENSION — REAR SUSPENSION

COIL SPRING

REMOVAL & INSTALLATION

See Figures 156 and 157.

1. Loosen the rear lower link bolt and nut from the suspension member side.

2. Support the rear lower link by placing a suitable jack under the knuckle.

3. Remove the rear lower link adjusting bolt and nut from the suspension member side.

➡**Do not reuse the adjusting nut, use a new adjusting nut for installation.**

4. Slowly lower the jack to lower the rear lower link and coil spring.

5. Remove the upper rubber seat, coil spring, and lower rubber seat from the rear lower link.

6. Remove rear lower link bolt and nut from the suspension member side.

7. Remove the rear lower link.

To install:

8. Installation is in the reverse order of removal.

➡**Do not reuse the adjusting nut, use a new adjusting nut for installation.**

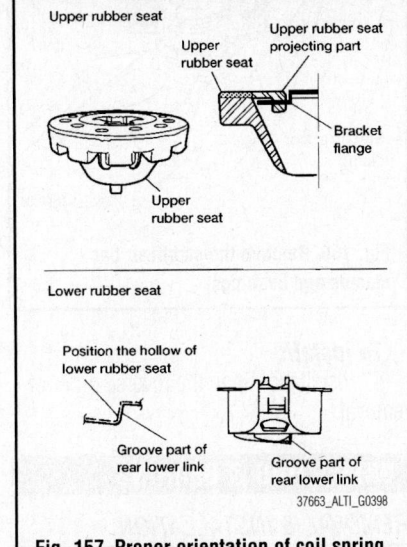

Fig. 157 Proper orientation of coil spring mounting components

9. Check that the projecting part inside the upper rubber seat and the bracket flange are attached as shown.

10. Check that the projection part outside the upper rubber seat is directed toward the front of the vehicle.

11. Position the hollow of the lower rubber seat with the groove part of the rear lower link.

12. Install the coil spring so that the side with the two paint markers is directed toward the lower side.

13. Check the rear wheel alignment and adjust if necessary.

CONTROL ARMS/LINKS

REMOVAL & INSTALLATION

Front Lower Link

See Figure 158.

1. Remove the front lower link nut and bolt from the knuckle side and the adjusting bolt and nut from the suspension member side.

➡**Do not reuse the adjusting nut, use a new adjusting nut for installation.**

2. Remove the front lower link.

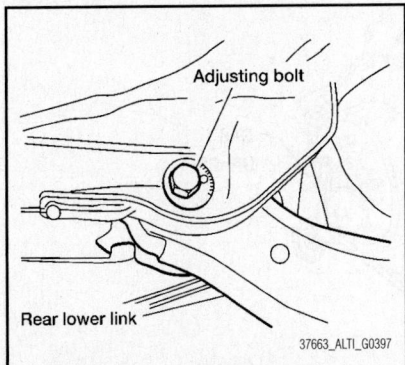

Fig. 156 Remove the rear lower link adjusting bolt and nut from the suspension member side

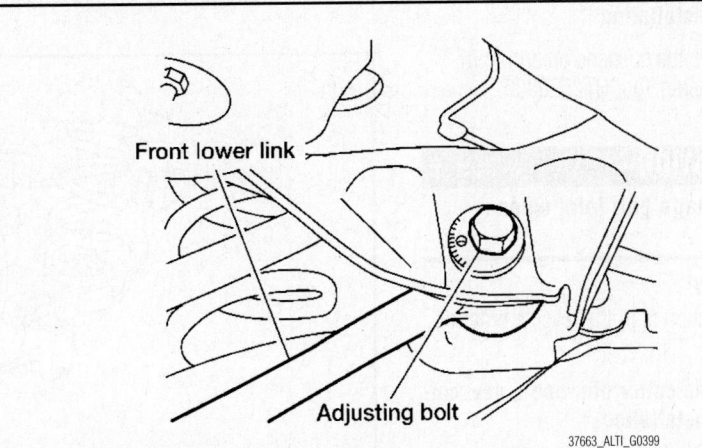

Fig. 158 Remove the front lower link nut and bolt from the knuckle side and the adjusting bolt and nut from the suspension member side

To install:

3. Installation is in the reverse order of removal.

➡ **Do not reuse the adjusting nut, use a new adjusting nut for installation.**

4. Check the rear wheel alignment and adjust if necessary.

Radius Rod

➡ **Refer to Exploded View of Rear Suspension while servicing this component.**

1. Remove the rear suspension assembly.

2. Remove the radius rod.

To install:

3. Installation is in the reverse order of removal.

4. Check the rear wheel alignment and adjust if necessary.

Rear Lower Link

Refer to Coil Spring Removal and Installation when servicing this component.

Suspension Arm

➡ **Refer to Exploded View of Rear Suspension while servicing this component.**

1. Remove the rear suspension assembly.

2. Remove the connecting rod bracket from the suspension arm.

3. Remove the two suspension arm bolts and nuts from the suspension member side of the suspension arm.

4. Remove the ball joint cotter pin and lock nut.

➡ **Discard the cotter pin, use a new cotter pin for installation.**

5. Remove the suspension arm from the knuckle using Tool HT2520000 (J-25730-A).

✷✷ CAUTION

Do not damage ball joint when removing.

To install:

6. Installation is in the reverse order of removal.

➡ **Discard the cotter pin, use a new cotter pin for installation.**

7. Check the rear wheel alignment and adjust if necessary.

STABILIZER BAR

REMOVAL & INSTALLATION

See Figure 159.

1. Disconnect the stabilizer bar from connecting rod.

2. Remove the stabilizer bar clamps and bushings.

3. Remove the stabilizer bar.

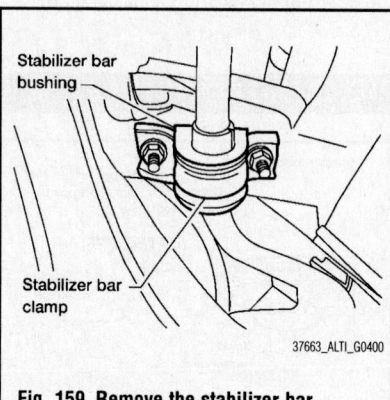

Fig. 159 Remove the stabilizer bar clamps and bushings

To install:

4. Installation is in the reverse order of removal.

WHEEL HUB & BEARING

REMOVAL & INSTALLATION

See Figure 160.

✷✷ CAUTION

Wheel hub assembly does not require maintenance. If any of the following symptoms are noted, replace the wheel hub assembly:

• A growling noise is emitted from the wheel hub assembly while driving.

• The wheel hub assembly drags or turns roughly.

1. Remove the rear wheel and tire.

2. Remove the brake caliper assembly.

a. The brake hose does not need to be disconnected from the brake caliper.

b. Suspend the brake caliper assembly using wire, do not stretch the brake hose.

c. Do not depress the brake pedal, or the caliper piston will pop out.

d. Do not twist the brake hose.

3. Remove the brake rotor.

4. Remove the rear ABS sensor, then move it away from the wheel hub assembly.

✷✷ CAUTION

Failure to remove the ABS sensor may result in damage to the sensor wires and the sensor being inoperative.

5. Remove the wheel hub assembly from knuckle.

6. Check for any deformity, cracks, or damage on the wheel hub assembly, replace if necessary.

To install:

7. Installation is in the reverse order of removal.

8. Check that the wheel bearings operate smoothly.

9. Check that the wheel hub bearing axial end play is within specification: 0.004 inches (0.1 mm).

ADJUSTMENT

No adjustment is possible. If there is too much play, replace the bearing and/or hub.

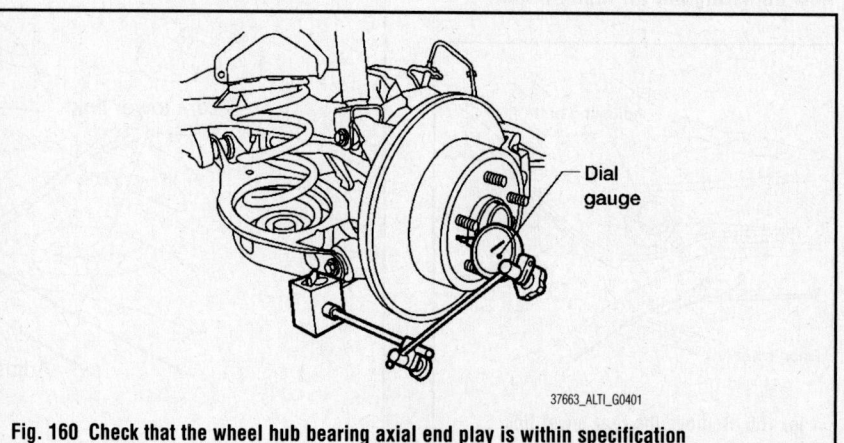

Fig. 160 Check that the wheel hub bearing axial end play is within specification

NISSAN

Armada

11

SPECIFICATIONS AND MAINTENANCE CHARTS

ENGINE AND VEHICLE IDENTIFICATION

Code ①	Engine Liters (cc)	Cu. In.	Cyl.	Fuel Sys.	Engine	Eng. Mfg.
VK56DE	5.6 (5552)	338.8	8	MFI	DOHC	Nissan

Model Year Code ②	Year
9	2009
A	2010

MFI: Multi-port Fuel Injection

DOHC: Double Overhead Camshafts

① Engine VIN: A. Engine VIN: B (FFV= flex fuel vehicle).

② 10th digit of the Vehicle Identification Number (VIN)

37663_ARMA_C0001

GENERAL ENGINE SPECIFICATIONS

Year	Model	Engine Displacement Liters	Engine ID	Net Horsepower @ rpm	Net Torque @ rpm (ft. lbs.)	Bore x Stroke (in.)	Com- pression Ratio	Oil Pressure @ rpm
2009	Armada	5.6	VK56DE	320@5200	393@3400	3.86X3.62	9.8:1	43@2000
2010	Armada	5.6	VK56DE	320@5200	393@3400	3.86X3.62	9.8:1	43@2000

37663_ARMA_C0002

ENGINE TUNE-UP SPECIFICATIONS

Year	Engine Displacement Liters	Engine ID	Spark Plug Gap (in.)	Ignition Timing	Fuel Pump (psi) ①	Idle Speed	Valve Clearance (in.) In.	Ex.
2009	5.6	VK56DE	0.043	②	51	600-700	0.010-0.013	0.011-0.015
2010	5.6	VK56DE	0.043	②	51	600-700	0.010-0.013	0.011-0.015

NOTE: The Vehicle Emission Control Information label often reflects specification changes made during production. The label figures must be used if they differ from those in this chart.

① At idle

② 15 degrees +/- 5 degrees BTDC

37663_ARMA_C0003

CAPACITIES

Year	Model	Engine Displacement Liters	Engine ID	Engine Oil with Filter (qts.)	Transmission (pts.)	Transfer Case (pts.)	Drive Axle Front (pts.)	Drive Axle Rear (pts.)	Fuel Tank (gal.)	Cooling System (qts.)
2009	Armada	5.6	VK56DE	6.50	22.50	6.36	3.380	3.75	28.0	15.25
2010	Armada	5.6	VK56DE	6.50	22.50	6.36	3.380	3.75	28.0	15.25

NOTE: All capacities are approximate. Add fluid gradually and check to be sure a proper fluid level is obtained.

37663_ARMA_C0004

FLUID SPECIFICATIONS

Year	Model	Engine Displ. Liters	Engine Oil	Auto. Trans.	Drive Axle Front	Drive Axle Rear	Transfer Case	Power Steering Fluid	Brake Master Cylinder	Cooling System
2009	Armada	5.6	SAE 5W-30	①	②	②	③	④	⑤	⑥
2010	Armada	5.6	SAE 5W-30	①	②	②	③	④	⑤	⑥

① Genuine Nissan Matic S ATF fluid. If not available genuine Nissan Matic J ATF may be used.

② Front: Nissan differential oil hypoid super GL-5 80W-90 or API GL-5 viscosity SAE 80W-90 gear oil.

Rear: Nissan differential oil API GL-5 synthetic 75W-90 gear oil.

③ Nissan Matic D ATF

④ Nissan Power Steering Fluid

⑤ Nissan Super Heavy Duty DOT 3

⑥ Nissan Long Life antifreeze or equivalent

37663_ARMA_C0014

VALVE SPECIFICATIONS

Year	Engine Displacement Liters	Engine ID	Seat Angle (deg.)	Face Angle (deg.)	Spring Test Pressure (lbs. @ in.)	Spring Installed Height (in.)	Stem-to-Guide Clearance (in.) Intake	Stem-to-Guide Clearance (in.) Exhaust	Stem Diameter (in.) Intake	Stem Diameter (in.) Exhaust
2009	5.6	VK56DE	45.15-45.45	45	37.0@1.457	1.9913	0.0008-0.0021	0.0012-0.0025	0.2348-0.2354	0.2344-0.2350
2010	5.6	VK56DE	45.15-45.45	45	37.0@1.457	1.9913	0.0008-0.0021	0.0012-0.0025	0.2348-0.2354	0.2344-0.2350

37663_ARMA_C0006

CAMSHAFT SPECIFICATIONS

All measurements are given in inches.

Year	Engine Displ. Liters	Engine ID/VIN	Journal Dia.	Brg. Oil Clearance	Shaft End-play	Runout	Journal Bore	Lobe Height Intake	Exhaust
2009	5.6	VK56DE	1.0217-1.0224	0.0012-0.0028	0.0045-0.0074	0.0008	1.0236-1.0244	1.7663-1.7738	1.7746-1.7821
2010	5.6	VK56DE	1.0278-1.0224	0.0012-0.0028	0.0045-0.0074	0.0008	1.0236-1.0244	1.7663-1.7738	1.7746-1.7821

37663_ARMA_C0007

CRANKSHAFT AND CONNECTING ROD SPECIFICATIONS

All measurements are given in inches.

Year	Engine Displ. Liters	Engine ID	Crankshaft Main Brg. Journal Dia.	Main Brg. Oil Clearance	Shaft End-play	Thrust on No.	Connecting Rod Journal Diameter	Oil Clearance	Side Clearance
2009	5.6	VK56DE	①	②	0.0039-0.0102	3	③	0.0008-0.0015	0.0079-0.0157
2010	5.6	VK56DE	①	②	0.0039-0.0102	3	③	0.0008-0.0015	0.0079-0.0157

① There are 24 different grades, ranging from 2.5182- 2.5174

② No. 1 and 5: 0.00004- 0.0004

　 No. 2, 3 and 4: 0.0003- 0.0007

③ There are 13 different grades, ranging from 2.2441- 2.2446. Specification is for rod bearing housing.

37663_ARMA_C0005

PISTON AND RING SPECIFICATIONS

All measurements are given in inches.

Year	Engine Displacement Liters	Engine ID	Piston Clearance	Ring Gap Top Comp.	Bottom Comp.	Oil Control	Ring Side Clearance Top Comp.	Bottom Comp.	Oil Control
2009	5.6	VK56DE	0.0004-0.0012	0.0091-0.0130	0.0098-0.0157	0.0079-0.0236	0.0014-0.0033	0.0012-0.0028	0.0006-0.0073
2010	5.6	VK56DE	0.0004-0.0012	0.0091-0.0130	0.0098-0.0157	0.0079-0.0236	0.0014-0.0033	0.0012-0.0028	0.0006-0.0073

37663_ARMA_C0008

TORQUE SPECIFICATIONS
All readings in ft. lbs.

Year	Engine Displacement Liters	Engine ID	Cylinder Head Bolts	Main Bearing Bolts	Rod Bearing Bolts	Crankshaft Damper Bolts	Flywheel Bolts	Manifold		Spark Plugs	Oil Pan Drain Plug
								Intake	Exhaust		
2009	5.6	VK56DE	①	②	③	④	65	NA	25	18	25
2010	5.6	VK56DE	①	②	③	④	65	NA	25	18	25

NA: not available

① Step 1: 33 ft. lbs

  Step 2: +70 degrees clockwise

  Step 3: loosen in reverse order of tightening sequence

  Step 4: 33 ft. lbs.

  Step 5: +60 degrees clockwise

  Step 6: +60 degrees clockwise

② Step 1: cap bolts in order 1-10: 29 ft. lbs.

  Step 2: cap sub bolts in order 11-20: 22 ft. lbs.

  Step 3: cap bolts in order 1-10: +40 degrees

  Step 4: cap sub bolts in order 11-20: +30 degrees

  Step 5: side bolts in order 21-30: 36 ft. lbs.

③ Step 1: 11 ft. lbs.

  Step 2: +90 degrees clockwise

④ Step 1: 69 ft. lbs.

  Step 2: +90 degrees

37663_ARMA_C0009

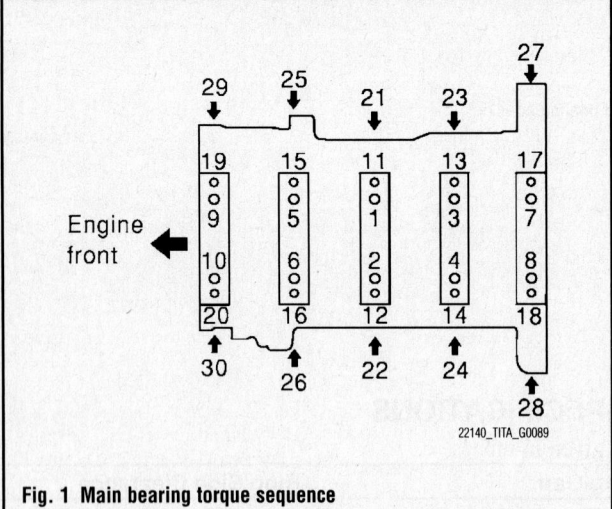

22140_TITA_G0089

Fig. 1 Main bearing torque sequence

WHEEL ALIGNMENT

Year	Model		Caster		Camber		Toe-in (in.)
			Range (+/-Deg.)	Preferred Setting (Deg.)	Range (+/-Deg.)	Preferred Setting (Deg.)	
2009	Armada	2WD	①	①	②	②	NA
		4WD	③	③	④	④	NA
2010	Armada	2WD	①	①	②	②	NA
		4WD	③	③	④	④	NA

NA - not available

① Minimum: 3 degrees 15' (3.25 degrees). Nominal: 4 degrees 0' (4.00 degrees). Maximum: 4 degrees 45' (4.75 degrees).

② Minimum: -0 degrees 51' (-0.85 degrees). Nominal: 0 degrees 6' (0.10 degrees). Maximum: 0 degree 39' (0.65 degree).

③ Minimum: 2 degrees 45' (2.75 degrees). Nominal: 3 degrees 30' (3.50 degrees). Maximum: 4 degrees 15' (4.25 degrees).

④ Minimum: -0 degrees 33' (-0.55 degrees). Nominal: 0 degrees 12' (0.20 degrees). Maximum: 0 degree 57' (0.95 degrees).

37663_ARMA_C0010

TIRE, WHEEL AND BALL JOINT SPECIFICATIONS

Year	Model	OEM Tires		Tire Pressures (psi)		Wheel Size	Ball Joint Inspection	Lug Nut Torque (ft. lbs.)
		Standard	Optional	Front	Rear			
2009	Armada	①	None	35	35	②	③	98
2010	Armada	①	None	35	35	②	③	98

OEM: Original Equipment Manufacturer

PSI: Pounds Per Square Inch

① SE: P265/70R18. LE: P275/60R20.

② SE: 18x8JJ. LE: 20x8JJ.

③ Axial play

Upper: 0

37663_ARMA_C0011

BRAKE SPECIFICATIONS

All measurements in inches unless noted

Year	Model		Brake Disc			Minimum Pad Thickness	Brake Caliper	
			Original Thickness	Minimum Thickness	Maximum Runout		Bracket Bolts (ft. lbs.)	Mounting Bolts (ft. lbs.)
2009	Armada	F	1.181	1.102	①	0.039	②	②
		R	0.551	0.472	①	0.039	③	③
2010	Armada	F	1.181	1.102	①	0.039	②	②
		R	0.551	0.472	①	0.039	③	③

① Maximum uneven wear measured at 8 positions: 0.0006. Runout limit, attached to vehicle: Front: 0.001, Rear: 0.002.

② Torque member mounting bolt: 110

Sliding pin bolt 53

③ Torque member mounting bolt: 76

37663_ARMA_C0012

SCHEDULED MAINTENANCE INTERVALS
NISSAN ARMADA

TO BE SERVICED	SERVICE	VEHICLE MILEAGE INTERVAL (x1000)												
		3.75	7.5	15	22.5	30	37.5	45	52.5	60	67.5	75	82.5	90
Engine oil & filter	R	✓	✓	✓	✓	✓	✓	✓	✓	✓	✓	✓	✓	✓
Brake lines & cables	I			✓		✓		✓		✓		✓		✓
Brake pads& rotors	L/I			✓		✓		✓		✓		✓		✓
Driveshaft boots & propeller shaft (4x4)	I					✓				✓				
Automatic transmission, final drive oil & transfer case	I				✓	✓		✓		✓				
LSD gear oil	I			✓		✓		✓		✓		✓		✓
Front wheel bearing grease (4x4)	R					✓				✓				
Air cleaner filter	R					✓				✓				✓
Engine coolant	R									✓				✓
Exhaust system	I						✓	✓	✓	✓				
Spark plugs	R	Replace every 105,000 miles												
Drive belt(s)	I			✓		✓		✓		✓		✓		
Cabin air filter	R							✓						✓
Exhaust system	I		✓			✓				✓				✓
Fuel lines	I		✓			✓				✓				
Steering gear (box) & linkage, axle & suspension parts	I					✓				✓				✓
Transfer case	I					✓				✓				✓
Tire rotation			✓	✓	✓	✓	✓	✓	✓	✓	✓	✓	✓	✓
Vapor lines	S/I					✓				✓				✓

R: Replace   S/I: Service or Inspect   L: Lubricate

FREQUENT OPERATION MAINTENANCE (SEVERE SERVICE)

If a vehicle is operated under any of the following conditions it is considered severe service:

- Extremely dusty areas.

- 50% or more of the vehicle operation is in 32°C (90°F) or higher temperatures, constant operation in temp. below 0°C (32°F).

- Prolonged idling (vehicle operation in stop and go traffic).

- Frequent short running periods (engine does not warm to normal operating temperatures).

- Police, taxi, delivery usage or trailer towing usage.

Oil & oil filter: replace every 3750 miles.

Brake pads, discs, drums & linings: service or inspect every 7500 miles.

Driveshaft boots & propeller shaft: service or inspect every 7500 miles.

Exhaust system: service or inspect every 7500 miles.

Final drive oil: Change every 30000 miles if towing a trailer.

Transfer case fluid: Change every 30000 miles if towing a trailer.

Steering gear (box) & linkage, (steering damper-4x4), axle & suspension parts: service or inspect every 7500 miles.

Steering linkage ball joints & front suspension ball joints: service or inspect every 7500 miles.

37663_ARMA_C0013

BRACES — INFORMATION AND PRECAUTIONS

ANTI-LOCK SYSTEMS

• Certain components within the ABS system are not intended to be serviced or repaired individually.

• Do not use rubber hoses or other parts not specifically specified for and ABS system. When using repair kits, replace all parts included in the kit. Partial or incorrect repair may lead to functional problems and require the replacement of components.

• Lubricate rubber parts with clean, fresh brake fluid to ease assembly. Do not use shop air to clean parts; damage to rubber components may result.

• Use only DOT 3 brake fluid from an unopened container.

• If any hydraulic component or line is removed or replaced, it may be necessary to bleed the entire system.

• A clean repair area is essential. Always clean the reservoir and cap thoroughly before removing the cap. The slightest amount of dirt in the fluid may plug an ori-fice and impair the system function. Perform repairs after components have been thoroughly cleaned; use only denatured alcohol to clean components. Do not allow ABS components to come into contact with any substance containing mineral oil; this includes used shop rags.

• The Anti-Lock control unit is a microprocessor similar to other computer units in the vehicle. Ensure that the ignition switch is **OFF** before removing or installing controller harnesses. Avoid static electricity discharge at or near the controller.

• If any arc welding is to be done on the vehicle, the control unit should be unplugged before welding operations begin.

DISC AND DRUM SYSTEMS

❋❋ CAUTION

Dust and dirt accumulating on brake parts during normal use may contain asbestos fibers from production or aftermarket brake linings. Breathing

excessive concentrations of asbestos fibers can cause serious bodily harm. Exercise care when servicing brake parts. Do not sand or grind brake lining unless equipment used is designed to contain the dust residue. Do not clean brake parts with compressed air or by dry brushing. Cleaning should be done by dampening the brake components with a fine mist of water, then wiping the brake components clean with a dampened cloth. Dispose of cloth and all residue containing asbestos fibers in an impermeable container with the appropriate label. Follow practices prescribed by the Occupational Safety and Health Administration (OSHA) and the Environmental Protection Agency (EPA) for the handling, processing, and disposing of dust or debris that may contain asbestos fibers.

BRAKES — BLEEDING THE BRAKE SYSTEM

BLEEDING PROCEDURE

BLEEDING PROCEDURE

➡**Be sure that the master cylinder is full of clean fresh brake fluid before starting the bleeding process. Use only the recommended brake fluid when bleeding the system. Do not allow brake fluid to spill on painted surfaces as damage will occur.**

1. Before servicing the vehicle, refer to the Precautions Section.

➡**If working near and/or around the SRS system and components, be sure to disable the SRS system. After disabling the system wait three minutes or more before servicing the vehicle.**

➡**Whenever the negative battery cable is disconnected the following components will require resetting. The Idle Air Volume Learning, Steering Angle Sensor Neutral Position, Sunroof Memory Reset/Initialization, Automatic Drive Positioner System, Audio presets and Navigation. Use the CONSULT-III diagnostic tool, or equivalent to perform the required resets.**

2. Disconnect the negative battery cable.

3. Turn the ignition switch OFF. Disconnect the ABS actuator and electric control unit connector.

4. Connect a vinyl tube to the rear right bleed valve. Be sure to have a catch pan handy to catch excess brake fluid.

5. Fully depress the brake pedal four or five times.

6. With the brake pedal depressed, loosen the bleed valve to let air out, then tighten it immediately.

7. Repeat the above steps until all air is removed from the system. Be sure to keep watch on the brake fluid level and replenish, as necessary.

8. Tighten the bleed valve.

9. Repeat the above steps at each wheel, with the master cylinder reservoir tank filled at least half way.

10. Bleed the remaining components in the following order: front left, rear left and front right.

11. Be sure to perform the reconnect/relearn procedures.

BRAKES

WHEEL SPEED SENSORS

REMOVAL & INSTALLATION

Front Sensor

See Figure 2.

1. Before servicing the vehicle, refer to the Precautions Section.

➡If working near and/or around the SRS system and components, be sure to disable the SRS system. After disabling the system wait three minutes or more before servicing the vehicle.

➡Whenever the negative battery cable is disconnected the following components will require resetting. The Idle Air Volume Learning, Steering Angle Sensor Neutral Position, Sunroof Memory Reset/Initialization, Automatic Drive Positioner System, Audio presets and Navigation. Use the CONSULT-III diagnostic tool, or equivalent to perform the required resets.

2. Disconnect the negative battery cable.
3. Raise and support the vehicle safely.
4. Remove the tire and wheel assembly.
5. Remove the wheel speed sensor mounting screw.

➡Remove the rotor to gain access to the wheel sensor mounting bolt.

6. Pull the sensor out, being careful to turn it as little as possible. Do not pull on the sensor harness.
7. Disconnect the wheel speed sensor electrical connector.
8. Remove the harness from it mount.

To install:

➡Be sure to use new fasteners, as required.

9. Inspect the sensor O-ring, replace as required.
10. Before installing the sensor be sure no foreign materials such as iron fragments are adhered to the pick-up part of the sensor or to the inside of the sensor mounting hole or on the rotor mounting surface.
11. Apply a thin coat of a suitable grease to the wheel sensor O-ring and mounting hole.
12. Tighten the sensor retaining bolt to 73 inch lbs.
13. Continue the installation in the reverse order of the removal procedure.
14. Be sure to perform the reconnect/relearn procedures.

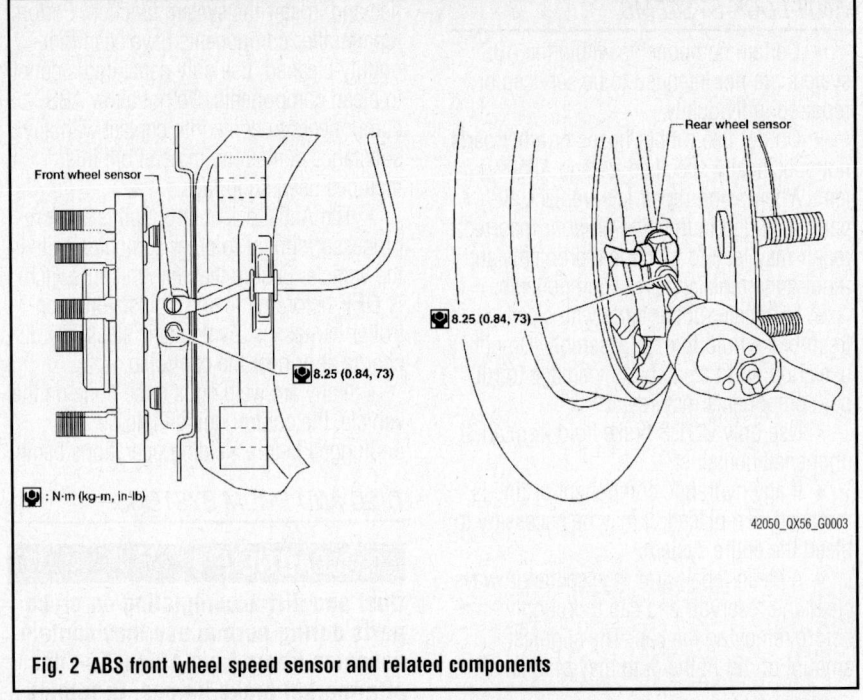

Fig. 2 ABS front wheel speed sensor and related components

Rear Sensor

See Figure 3.

1. Before servicing the vehicle, refer to the Precautions Section.

➡If working near and/or around the SRS system and components, be sure to disable the SRS system. After disabling the system wait three minutes or more before servicing the vehicle.

➡Whenever the negative battery cable is disconnected the following components will require resetting. The Idle Air Volume Learning, Steering Angle Sensor Neutral Position, Sunroof Memory Reset/Initialization, Automatic Drive Positioner System, Audio presets and Navigation. Use the CONSULT-III diagnostic tool, or equivalent to perform the required resets.

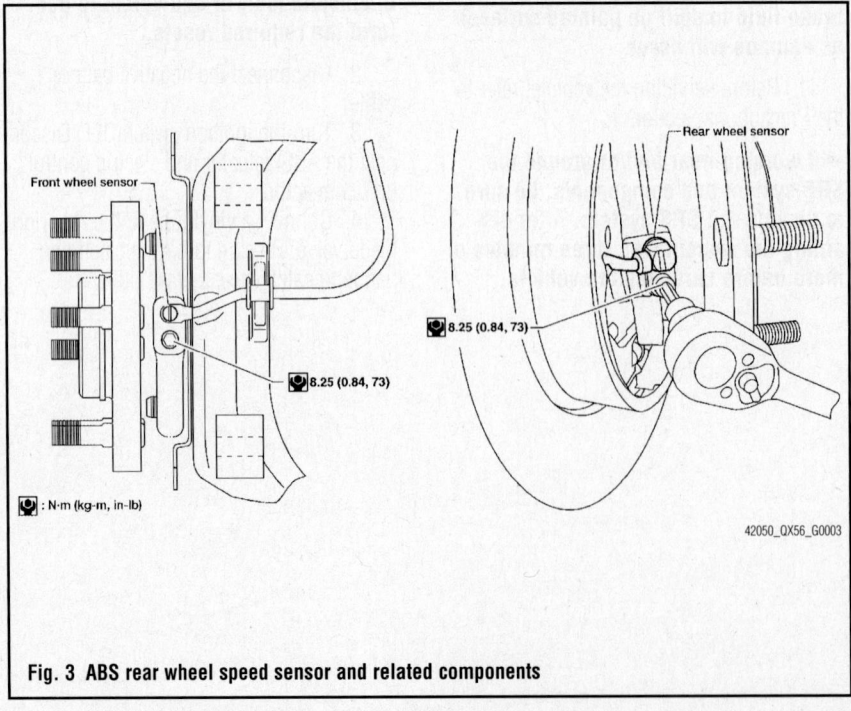

Fig. 3 ABS rear wheel speed sensor and related components

2. Disconnect the negative battery cable.

3. Raise and support the vehicle safely.

4. Remove the tire and wheel assembly.

5. Remove the wheel speed sensor mounting screw.

➡**Remove the rear hub and bearing assembly to gain access to the wheel sensor mounting bolt.**

6. Pull the sensor out, being careful to turn it as little as possible. Do not pull on the sensor harness.

7. Disconnect the wheel speed sensor electrical connector.

8. Remove the harness from it mount.

To install:

➡**Be sure to use new fasteners, as required.**

9. Inspect the sensor O-ring, replace as required.

10. Before installing the sensor be sure no foreign materials such as iron fragments are adhered to the pick-up part of the sensor or to the inside of the sensor mounting hole or on the rotor mounting surface.

11. Apply a thin coat of a suitable grease to the wheel sensor O-ring and mounting hole.

12. Tighten the sensor retaining bolt to 73 inch lbs.

13. Continue the installation in the reverse order of the removal procedure.

14. Be sure to perform the reconnect/relearn procedures.

BRAKES   **FRONT DISC BRAKES**

BRAKE CALIPER

REMOVAL & INSTALLATION

See Figure 4.

1. Before servicing the vehicle, refer to the Precautions Section.

➡**If working near and/or around the SRS system and components, be sure to disable the SRS system. After disabling the system wait three minutes or more before servicing the vehicle.**

➡**Whenever the negative battery cable is disconnected the following components will require resetting. The Idle Air Volume Learning, Steering Angle Sensor Neutral Position, Sunroof Memory Reset/Initialization, Automatic Drive Positioner System, Audio presets and Navigation. Use the CONSULT-III diagnostic tool, or equivalent to perform the required resets.**

2. Disconnect the negative battery cable.
3. Drain brake fluid as necessary.
4. Raise and safely support the vehicle.
5. Remove or disconnect the following:
   • Wheel and tire assembly
   • Union bolt
   • Disconnect and plug brake hose. Discard the washer.
   • Caliper-to-torque member slide pins, or remove the caliper and torque member as an assembly
   • Brake caliper

To install:

➡**Be sure to use new fasteners, as required.**

6. Install or connect the following:
   • Brake caliper, tighten torque member bolts to specification and the caliper slide pin to specification
   • Union bolt and tighten to 13 ft. lbs. (18 Nm)

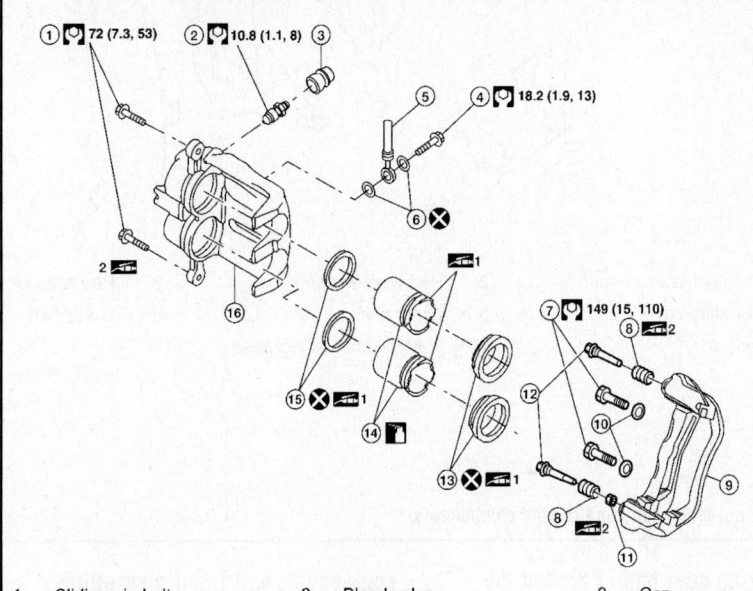

1.	Sliding pin bolt	2.	Bleed valve	3.	Cap
4.	Union bolt	5.	Brake hose	6.	Copper washer
7.	Torque member bolt	8.	Sliding pin boot	9.	Torque member
10.	Washers	11.	Bushing	12.	Sliding pin
13.	Piston boot	14.	Piston	15.	Piston seal
16.	Cylinder body		Brake fluid		1: Molykote M-77 grease

37663_QX56_G0018

Fig. 4 Front brake caliper and related components

7. Fill the master cylinder and bleed the brake system.

8. Install the wheel and tire assemblies.

9. Be sure to perform the reconnect/relearn procedures.

DISC BRAKE PADS

REMOVAL & INSTALLATION

See Figures 5 and 6.

1. Before servicing the vehicle, refer to the Precautions Section.

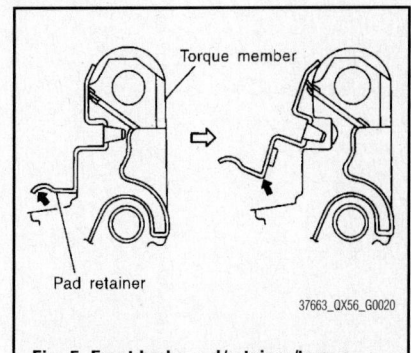

37663_QX56_G0020

Fig. 5 Front brake pad/retainer/torque member removal

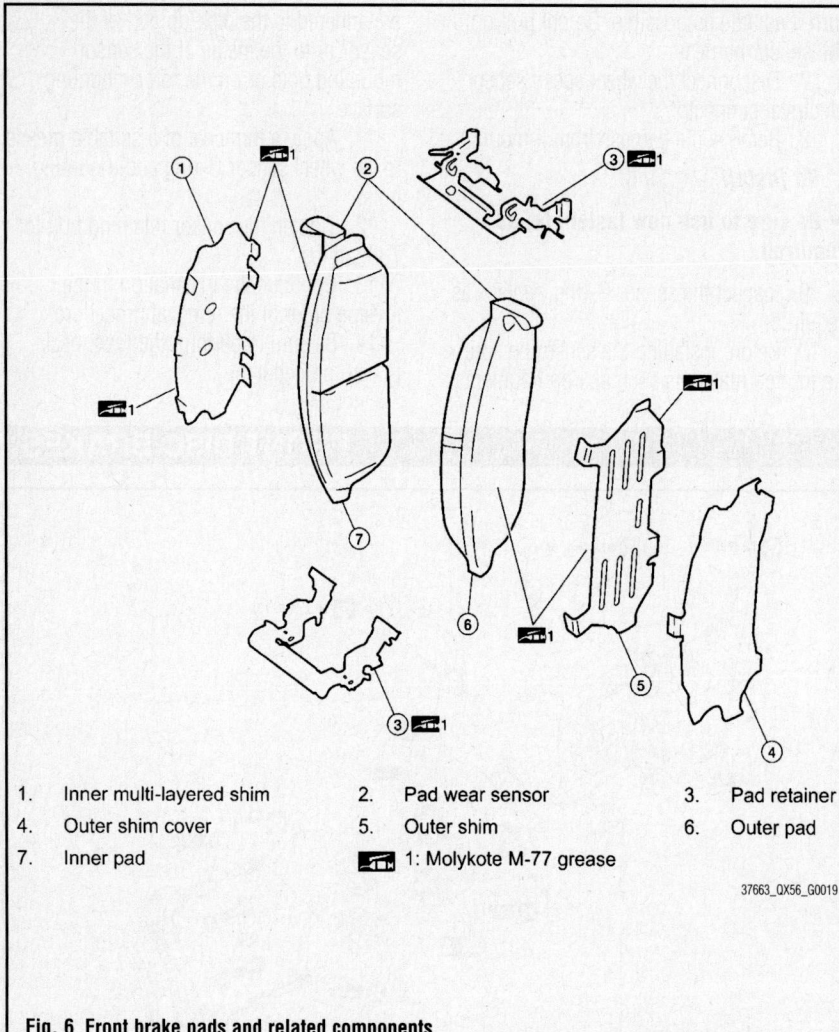

Fig. 6 Front brake pads and related components

1.	Inner multi-layered shim	2.	Pad wear sensor
4.	Outer shim cover	5.	Outer shim
7.	Inner pad		1: Molykote M-77 grease

3. Pad retainer

6. Outer pad

37663_QX56_G0019

➡ If working near and/or around the SRS system and components, be sure to disable the SRS system. After disabling the system wait three minutes or more before servicing the vehicle.

➡ Whenever the negative battery cable is disconnected the following components will require resetting. The Idle Air Volume Learning, Steering Angle Sensor Neutral Position, Sunroof Memory Reset/ Initialization, Automatic Drive Positioner System, Audio presets and Navigation. Use the CONSULT-III diagnostic tool, or equivalent to perform the required resets.

2. Disconnect the negative battery cable.

3. Drain brake fluid as necessary.

4. Raise and safely support the vehicle.

5. Remove the wheel and tire assembly.

6. Remove lower sliding pin bolt.

7. Suspend brake caliper with a remove and remove brake pad and shim from torque member.

➡ When removing the pad retainer from the torque member, lift it in the direction indicated by the arrow in the illustration.

To install:

➡ Be sure to use new fasteners, as required.

8. Push pistons in so that the pad is firmly installed, using a suitable tool.

9. Mount the brake caliper to torque member.

10. Attach pad retainer to torque member.

➡ Securely assemble the pad retainers so that they are not being lifted up from the torque member. Both inner and outer pads have a pad return system on the pad retainer. Install the pad return lever securely to the pad wear sensor.

11. Lubricate lower sliding pin bolt with a thin layer of silicone grease and install. Torque to specification.

12. Install the wheel and tire assembly.

13. Be sure to perform the reconnect/ relearn procedures.

BRAKES

REAR DISC BRAKES

BRAKE CALIPER

REMOVAL & INSTALLATION

See Figure 7.

1. Before servicing the vehicle, refer to the Precautions Section.

➡**If working near and/or around the SRS system and components, be sure to disable the SRS system. After disabling the system wait three minutes or more before servicing the vehicle.**

➡**Whenever the negative battery cable is disconnected the following components will require resetting. The Idle Air Volume Learning, Steering Angle Sensor Neutral Position, Sunroof Memory Reset/Initialization, Automatic Drive Positioner System, Audio presets and Navigation. Use the CONSULT-III diagnostic tool, or equivalent to perform the required resets.**

2. Disconnect the negative battery cable.
3. Drain brake fluid as necessary.
4. Raise and safely support the vehicle.
5. Drain brake fluid as necessary.
6. Remove or disconnect the following:
    - Wheel and tire assembly
    - Union bolt
    - Mounting bolts
    - Brake caliper assembly. Discard the copper washers.

To install:

➡**Be sure to use new fasteners, as required.**

7. Install or connect the following:
    - Brake caliper assembly and tighten mounting bolts to specification
    - Union bolt and tighten to 13 ft. lbs. (18 Nm)
8. Fill the master cylinder and bleed the brake system.
9. Install the wheel and tire assemblies.
10. Be sure to perform the reconnect/relearn procedures.

DISC BRAKE PADS

REMOVAL & INSTALLATION

See Figure 8.

1. Before servicing the vehicle, refer to the Precautions Section.

➡**If working near and/or around the SRS system and components, be sure to disable the SRS system. After disabling the system wait three minutes or more before servicing the vehicle.**

➡**Whenever the negative battery cable is disconnected the following components will require resetting. The Idle Air Volume Learning, Steering Angle Sensor Neutral Position, Sunroof Memory Reset/Initialization, Automatic Drive Positioner System, Audio presets and Navigation. Use the CONSULT-III diagnostic tool, or equivalent to perform the required resets.**

2. Disconnect the negative battery cable.
3. Drain brake fluid as necessary.
4. Raise and safely support the vehicle.
5. Remove the wheel and tire assembly.
6. Remove the sliding sleeves and pin bolts from the cylinder body
7. Support the cylinder body, with mechanics wire. Remove the pads, shims, cover and retainer.

To install:

8. Push pistons in so that the pad is firmly installed, using a suitable tool.
9. Install pads to the brake caliper.
10. Install top mounting bolt and tighten to specification.
11. Install the wheel and tire assembly.

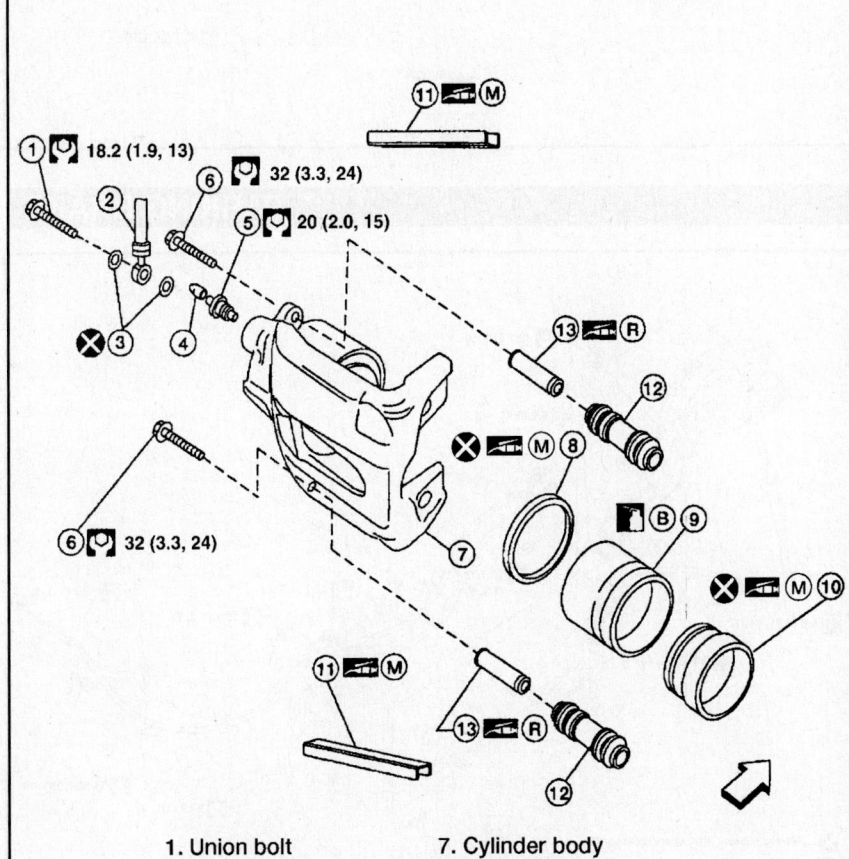

1. Union bolt
2. Brake hose
3. Washer
4. Cap
5. Bleed valve
6. Sliding pin bolt
7. Cylinder body
8. Piston seal
9. Piston
10. Piston boot
11. Knuckle side
12. Sliding sleeve bolt

37663_QX56_G0021

Fig. 7 Rear brake caliper and related components

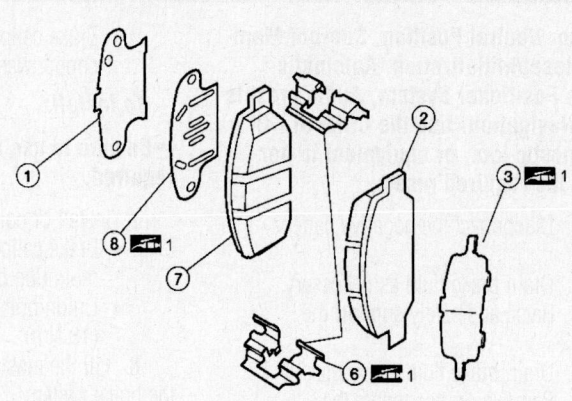
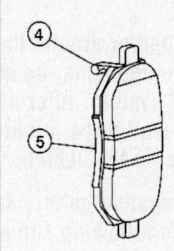

1.	Inner shim cover	2.	Outer pad	3.	Outer multi-layered shim
4.	Pad wear sensor	5.	Inner pad (RH)	6.	Pad retainer
7.	Inner pad	8.	Inner shim		

1: Molykote M-77 grease

37663_QX56_G0022

Fig. 8 Rear brake pads and related components

BRAKES   PARKING BRAKE

PARKING BRAKE CABLES

ADJUSTMENT

See Figure 9.

1. Before servicing the vehicle, refer to the Precautions Section.

➡ If working near and/or around the SRS system and components, be sure to disable the SRS system. After disabling the system wait three minutes or more before servicing the vehicle.

➡ Whenever the negative battery cable is disconnected the following components will require resetting. The Idle Air Volume Learning, Steering Angle Sensor Neutral Position, Sunroof Memory Reset/Initialization, Automatic Drive Positioner System, Audio presets and Navigation. Use the CONSULT-III diagnostic tool, or equivalent to perform the required resets.

2. Disconnect the negative battery cable.

3. Remove the lower instrument panel, driver's side.

4. Partially engage the parking brake pedal to access the adjusting nut.

5. Insert a deep socket wrench to rotate the adjusting nut and loosen the cable sufficiently.

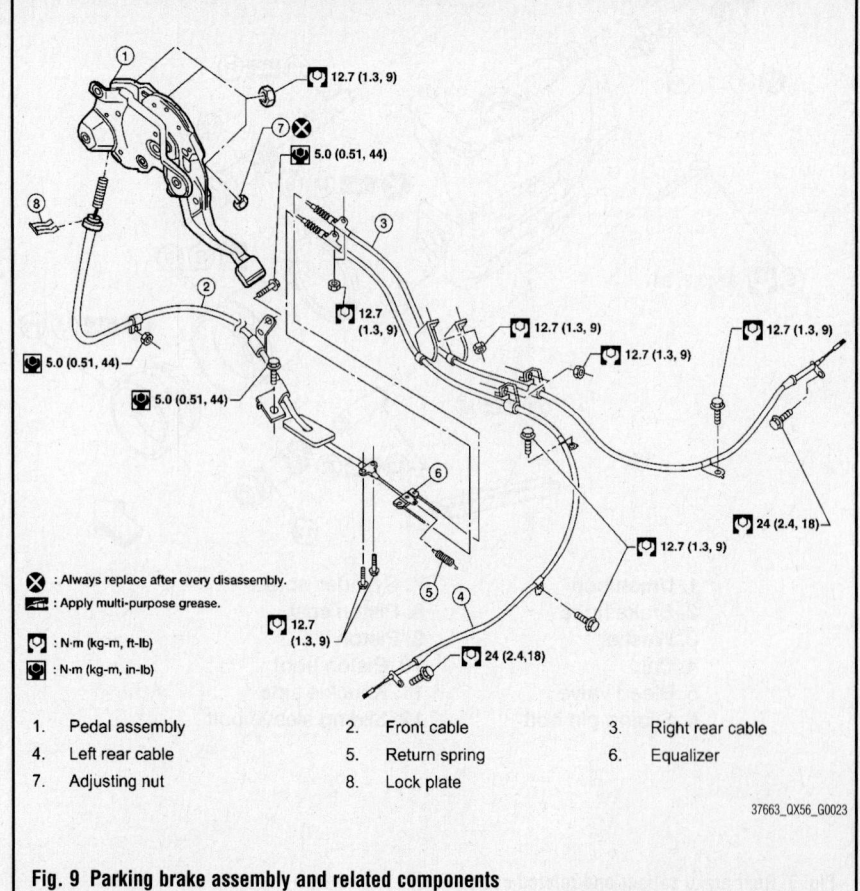

: Always replace after every disassembly.

: Apply multi-purpose grease.

: N·m (kg-m, ft-lb)

: N·m (kg-m, in-lb)

1.	Pedal assembly	2.	Front cable	3.	Right rear cable
4.	Left rear cable	5.	Return spring	6.	Equalizer
7.	Adjusting nut	8.	Lock plate		

37663_QX56_G0023

Fig. 9 Parking brake assembly and related components

6. Disengage the parking brake pedal.

7. Raise and support the vehicle safely.

8. Remove the tire and wheel assembly.

9. Remove the rotor. Measure the inner diameter at the widest point using tool J-21177A or equivalent.

10. Transfer the recorded measurement less 0.6 mm to the parking brake shoes and adjust accordingly.

11. Using wheel nuts, secure the rotor to the hub to prevent it from tilting.

12. Rotate the rotor to make sure that there is no drag.

13. To adjust the cable operate the pedal ten or more times with a force of 110 lbs.

14. Rotate the adjusting nut with a deep socket to adjust the pedal stroke to specification. Specification is 3–4 notches with a force of 44.1 lbs.

15. With the parking brake pedal completely disengaged, make sure there is no drag on the parking brake.

16. Reassemble and reinstall any removed components.

PARKING BRAKE SHOES

REMOVAL & INSTALLATION

See Figure 10.

1. Before servicing the vehicle, refer to the Precautions Section.

➡**If working near and/or around the SRS system and components, be sure to disable the SRS system. After disabling the system wait three minutes or more before servicing the vehicle.**

➡**Whenever the negative battery cable is disconnected the following components will require resetting. The Idle Air Volume Learning, Steering Angle Sensor Neutral Position, Sunroof Memory Reset/Initialization, Automatic Drive Positioner System, Audio presets and Navigation. Use the CONSULT-III diagnostic tool, or equivalent to perform the required resets.**

2. Disconnect the negative battery cable.

3. Remove the tire and wheel assembly.

4. Be sure that the parking brake lever is in the released position.

5. Remove the rear disc rotor.

6. Remove the return springs.

7. Remove the adjuster.

8. Disconnect the parking brake cable from the toggle lever.

9. Remove the retainers.

10. Remove the anti rattle pins and shoes.

To install:

➡**Be sure to use new fasteners, as required.**

11. Apply brake grease to the specified points during reassembly, see illustration for locating points.

12. Install the adjuster so that the threaded part expands when rotating it in the proper direction.

13. Continue the installation in the reverse order of the removal procedure.

14. Adjust the parking brake.

15. Perform the parking brake burnishing operation.

16. Be sure to perform the reconnect/relearn procedures.

ADJUSTMENT

See Figure 9.

Perform the parking brake burnishing operation by driving the vehicle forward under the following conditions: vehicle speed 25 mph forward direction, parking brake operating force 44.1 lbs set and apply time of 30 seconds. After parking brake burnishing operation, recheck parking brake adjustment, correct as required.

➡**To prevent the brake lining from getting to hot, allow a cool off period of five minutes between operations. Do not perform excessive break-in operations, because it may cause uneven or early wear of the lining.**

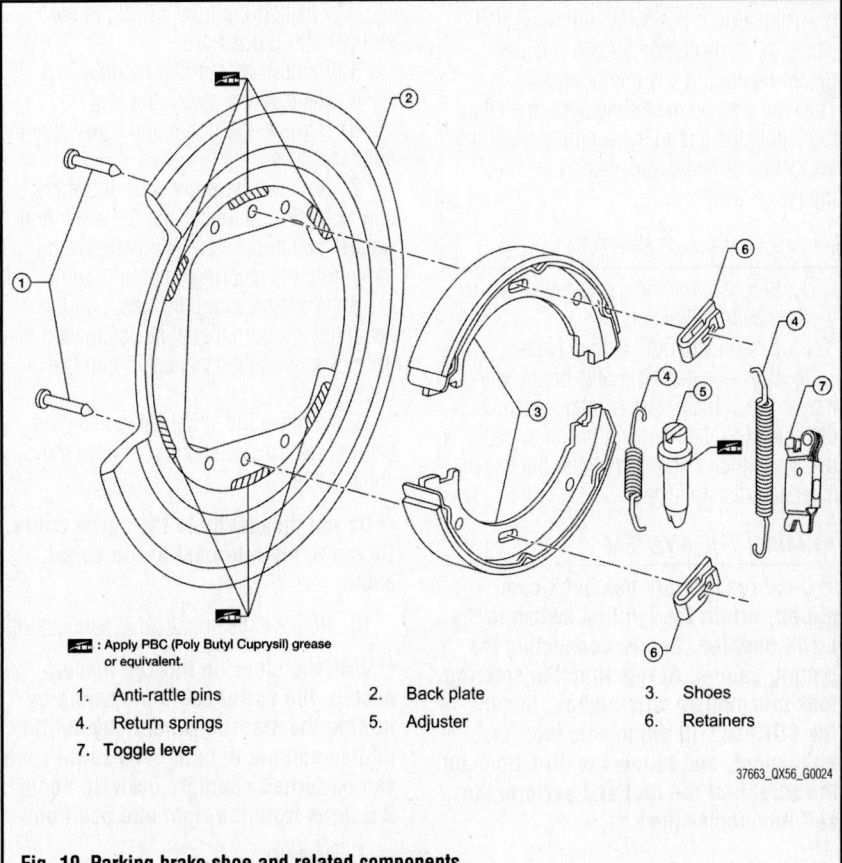

⬛ : Apply PBC (Poly Butyl Cuprysil) grease or equivalent.

1.	Anti-rattle pins	2.	Back plate	3.	Shoes
4.	Return springs	5.	Adjuster	6.	Retainers
7.	Toggle lever				

37663_QX56_G0024

Fig. 10 Parking brake shoe and related components

CHASSIS ELECTRICAL — AIR BAG (SUPPLEMENTAL RESTRAINT SYSTEM)

GENERAL INFORMATION

✳✳ CAUTION

These vehicles are equipped with an air bag system. The system must be disarmed before performing service on, or around, system components, the steering column, instrument panel components, wiring and sensors. Failure to follow the safety precautions and the disarming procedure could result in accidental air bag deployment, possible injury and unnecessary system repairs.

PRECAUTIONS

Disconnect and isolate the battery negative cable before beginning any airbag system component diagnosis, testing, removal, or installation procedures. Allow system capacitor to discharge for two minutes before beginning any component service. This will disable the airbag system. Failure to disable the airbag system may result in accidental airbag deployment, personal injury, or death.

DISARMING THE SYSTEM

1. Before servicing the vehicle, refer to the Precautions Section.
2. Disconnect both battery cables.
3. Wait at least 3 minutes before working on the vehicle. The air bag system is designed to retain enough power to deploy the air bag for a short time after the battery has been disconnected.

ARMING THE SYSTEM

➡Once repair work has been completed, return the ignition switch to the LOCK position, before connecting the battery cables. At this time the steering lock mechanism will engage. Install the CONSULT-III diagnostic tool, or equivalent, and follow the directions on the screen of the tool and perform the self diagnosis check.

CLOCKSPRING CENTERING

See Figure 11.

➡Before servicing, or working around, the SRS system, turn the ignition switch OFF, disconnect both battery cables and wait at least three minutes. When servicing, or working around, the SRS system do not work directly in front of the air bag module.

1. Before servicing the vehicle, refer to the Precautions Section.

➡If working near and/or around the SRS system and components, be sure to disable the SRS system. After disabling the system wait three minutes or more before servicing the vehicle.

➡Whenever the negative battery cable is disconnected the following components will require resetting. The Idle Air Volume Learning, Steering Angle Sensor Neutral Position, Sunroof Memory Reset/Initialization, Automatic Drive Positioner System, Audio presets and Navigation. Use the CONSULT-III diagnostic tool, or equivalent to perform the required resets.

2. Disconnect the negative battery cable. Disconnect the positive battery cable.
3. Position the front wheels in the straight ahead position.
4. Remove the air bag module.
5. Remove the steering wheel.
6. Remove the upper and lower steering column covers.
7. Remove the wiper washer switch connector. Pinch the tabs at the wiper and washer switch base and slide the switch away from the steering column to remove it.
8. While pressing the tabs, pull the headlight and turn signal switch toward the driver's door and disconnect it from the base.
9. Remove the spiral cable retaining screws, release the clip and remove the spiral cable.

➡Do not disassemble the spiral cable. Do not apply lubricant to the spiral cable.

10. Remove the spiral cable connectors.

➡With the steering linkage disconnected, the spiral cable may snap by turning the steering wheel beyond the limited number of turns. The spiral cable can be turned counterclockwise about 2.5 turns from the right end position.

To install:

➡Be sure to use new fasteners, as required.

11. Installation is the reverse of the removal procedure.
12. Be sure to align the spiral cable correctly when installing the steering wheel. Make sure that the spiral cable is in the neutral position.

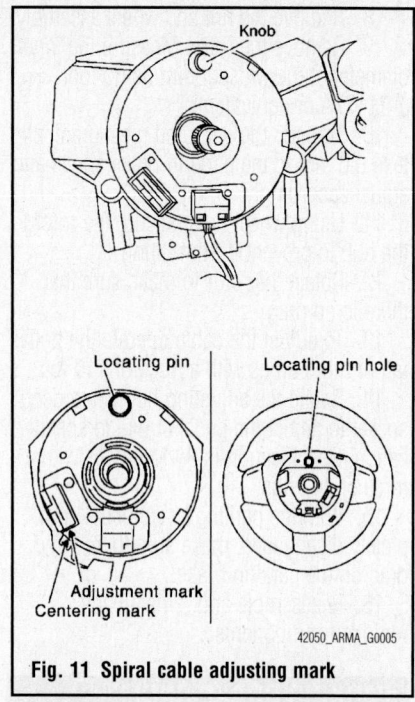

Fig. 11 Spiral cable adjusting mark

✳✳ WARNING

The neutral position is detected by turning to the left 2.5 revolutions from the right end position and ending with the knob at the top. The spiral cable may snap due to steering operation if the cable is installed incorrectly. Also, with the steering linkage disconnected the cable may snap by turning the steering wheel beyond the limited number of turns (2.5 from the neutral position to both the left and right).

13. If equipped with VDC adjust the steering angle sensor.
14. Use the CONSULT-III® tool and perform self diagnosis to ensure no malfunction is detected.

➡With the steering linkage disconnected, the spiral cable may snap by turning the steering wheel beyond the limited number of turns. The spiral cable can be turned counterclockwise about 2.5 turns from the right end position.

15. Tighten the steering wheel retaining nut to 25 ft. lbs. (33.9 Nm).
16. When reinstalling the air bag module, be sure to use new bolts. Tighten the bolts to 8 ft. lbs. (10.8 Nm).
17. Adjust the steering angle sensor neutral position, using the CONSULT-III diagnostic tool, or equivalent.
18. Be sure to perform the reconnect/relearn procedures.

DRIVE TRAIN

FRONT HALFSHAFT

REMOVAL & INSTALLATION

See Figure 12.

1. Before servicing the vehicle, refer to the Precautions Section.

➡️**If working near and/or around the SRS system and components, be sure to disable the SRS system. After disabling the system wait three minutes or more before servicing the vehicle.**

➡️**Whenever the negative battery cable is disconnected the following components will require resetting. The Idle Air Volume Learning, Steering Angle Sensor Neutral Position, Sunroof Memory Reset/Initialization, Automatic Drive Positioner System, Audio presets and Navigation. Use the CONSULT-III diagnostic tool, or equivalent to perform the required resets.**

2. Disconnect the negative battery cable.
3. Remove or disconnect the following:
   - Wheel and tire assembly
   - Engine splash guard
   - Wheel speed sensor harness from mount on knuckle, and reposition
   - Brake caliper, do not disconnect brake hose or allow caliper to hang by hose
   - Coil spring and shock absorber
   - Cotter pin and halfshaft nut
   - Halfshaft from front differential
   - Halfshaft from hub and bearing assembly

To install:

➡️**Be sure to use new fasteners, as required.**

4. Install or connect the following:
   - Halfshaft into hub
   - Halfshaft into front differential

- Halfshaft nut and tighten to 101 ft. lbs. (137 Nm) and replace cotter pin
- Wheel speed sensor harness
- Brake caliper
- Coil spring and shock absorber
- Engine splash guard
- Wheel and tire assembly

5. Be sure to perform the reconnect/relearn procedures.

FRONT PINION SEAL

REMOVAL & INSTALLATION

See Figures 13 through 16.

1. Before servicing the vehicle, refer to the Precautions Section.

➡️**If working near and/or around the SRS system and components, be sure to disable the SRS system. After disabling the system wait three minutes or more before servicing the vehicle.**

➡️**Whenever the negative battery cable is disconnected the following components will require resetting. The Idle Air Volume Learning, Steering Angle Sensor Neutral Position, Sunroof Memory Reset/Initialization, Automatic Drive Positioner System, Audio presets and Navigation. Use the CONSULT-III diagnostic tool, or equivalent to perform the required resets.**

2. Disconnect the negative battery cable.
3. Remove or disconnect the following:
   - Front driveshaft
   - Halfshafts

4. Measure and record the pinion bearing preload using special tool J-25765-A.

5. Loosen the pinion nut while holding the companion flange using special tool J-44195.

6. Remove the companion flange using a suitable tool.

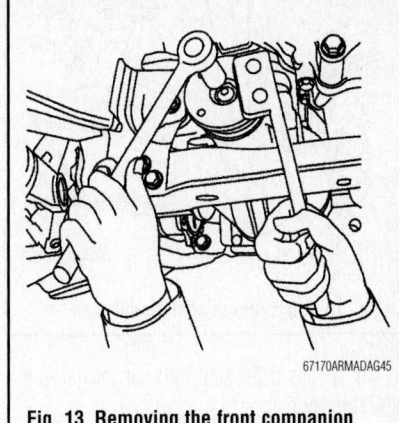

Fig. 13 Removing the front companion flange using Special Tool J-44195

7. Using a punch or drill, place a small hole in the case.

8. Remove the seal using special tool SP8P or equivalent.

To install:

➡️**Be sure to use new fasteners, as required.**

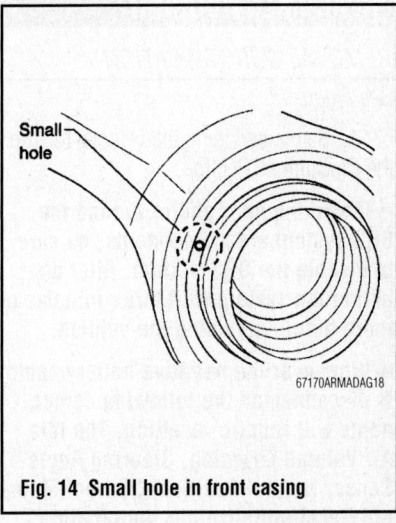

Fig. 14 Small hole in front casing

Small hole

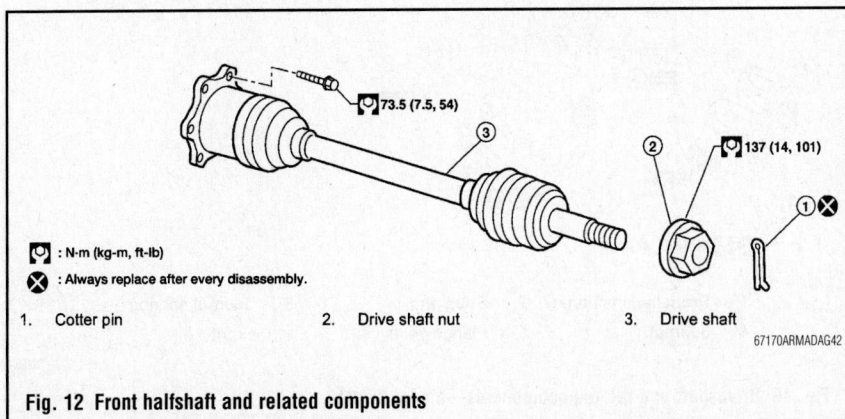

🔧 : N·m (kg·m, ft-lb)

❌ : Always replace after every disassembly.

1. Cotter pin
2. Drive shaft nut
3. Drive shaft

73.5 (7.5, 54)
137 (14, 101)

Fig. 12 Front halfshaft and related components

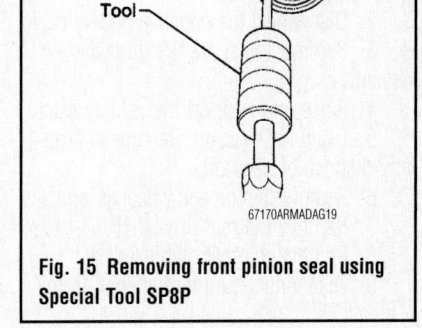

Tool

Fig. 15 Removing front pinion seal using Special Tool SP8P

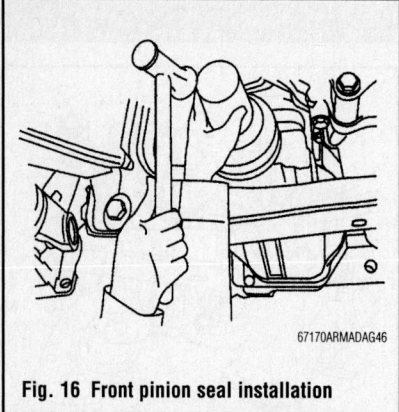

Fig. 16 Front pinion seal installation

9. Press front seal into carrier using a suitable tool.

10. Install companion flange and new pinion nut. Tighten pinion nut until there is no end play and until recorded pinion bearing preload is met plus an additional 5 inch lbs. (0.5 Nm).

11. Install or connect the following:
- Halfshafts
- Front driveshaft

12. Be sure to perform the reconnect/relearn procedures.

REAR AXLE HOUSING

REMOVAL & INSTALLATION

See Figure 17.

1. Before servicing the vehicle, refer to the Precautions Section.

➡**If working near and/or around the SRS system and components, be sure to disable the SRS system. After disabling the system wait three minutes or more before servicing the vehicle.**

➡**Whenever the negative battery cable is disconnected the following components will require resetting. The Idle Air Volume Learning, Steering Angle Sensor Neutral Position, Sunroof Memory Reset/Initialization, Automatic Drive Positioner System, Audio presets and Navigation. Use the CONSULT-III diagnostic tool, or equivalent to perform the required resets.**

2. Disconnect the negative battery cable.
3. Remove the spare tire and wheel assembly.
4. Raise and support the vehicle safely.
5. Drain the gear oil. Be sure to properly dispose of used oil.
6. Remove the tire and wheel assemblies.
7. Remove the rear driveshaft.
8. Remove the rear stabilizer bar.
9. Disconnect the rear halfshaft from

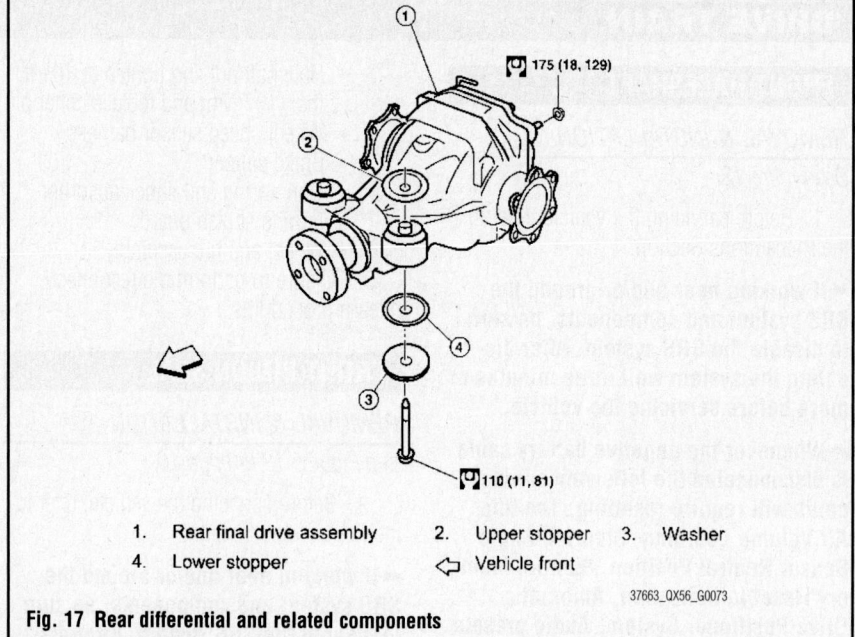

| 1. | Rear final drive assembly | 2. | Upper stopper | 3. | Washer |
| 4. | Lower stopper | | Vehicle front | | |

37663_QX56_G0073

Fig. 17 Rear differential and related components

the rear axle assembly. Position it aside using mechanics wire or equivalent.

10. Disconnect the breather hose from the axle cover.

11. Position a suitable jack under the assembly.

➡**Do not position the jack under the aluminum cover.**

12. Remove the rear axle assembly retaining bolts and nuts.

13. Carefully remove the assembly from the vehicle.

To install:

➡**Be sure to use new fasteners, as required.**

14. Installation is the reverse of the removal procedure.

15. Be sure to perform the reconnect/relearn procedures.

REAR DRIVESHAFT

REMOVAL & INSTALLATION

See Figures 18 through 20.

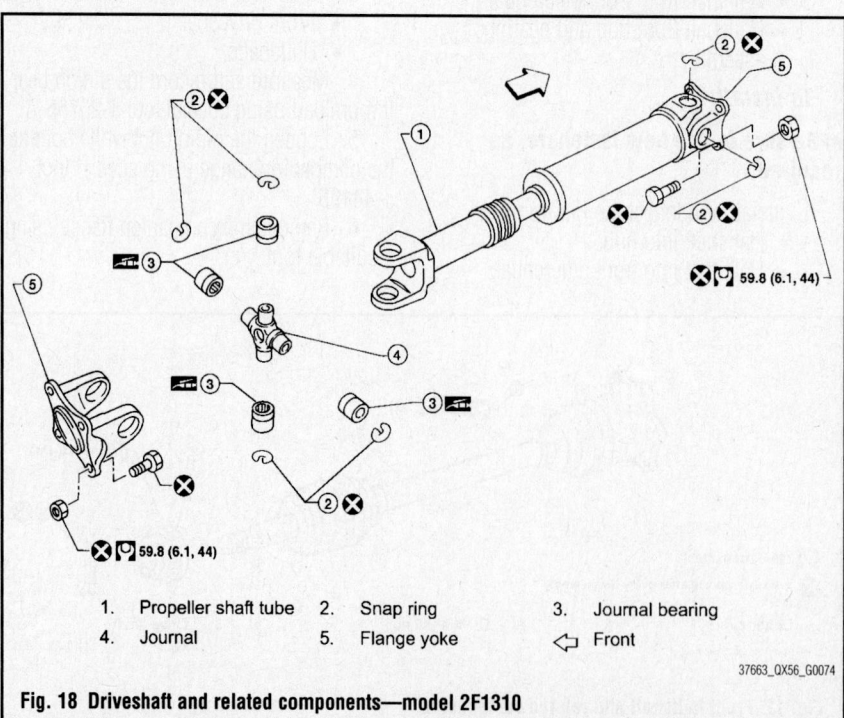

| 1. | Propeller shaft tube | 2. | Snap ring | 3. | Journal bearing |
| 4. | Journal | 5. | Flange yoke | | Front |

37663_QX56_G0074

Fig. 18 Driveshaft and related components—model 2F1310

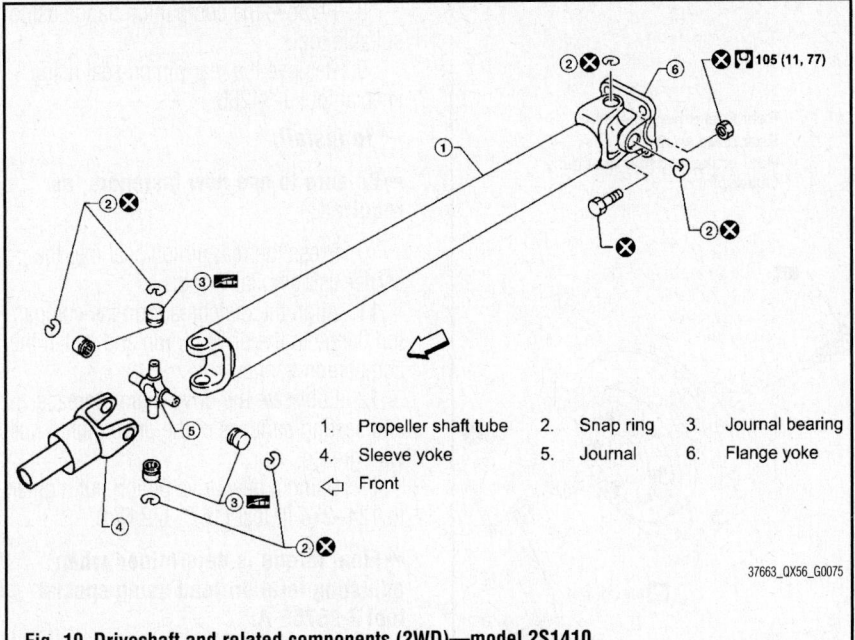

Fig. 19 Driveshaft and related components (2WD)—model 2S1410

1. Propeller shaft tube	2. Snap ring	3. Journal bearing
4. Sleeve yoke	5. Journal	6. Flange yoke
⇦ Front		

37663_QX56_G0075

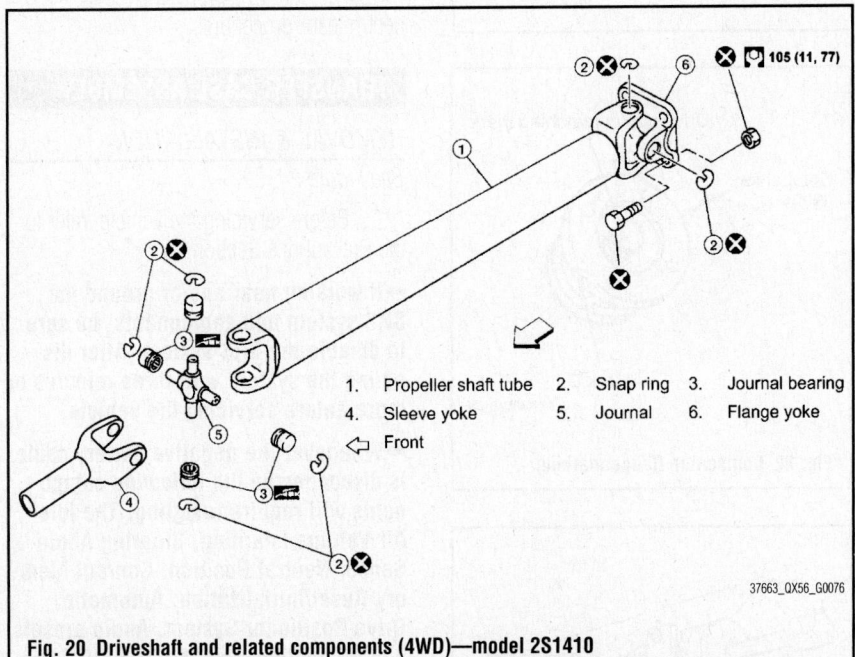

Fig. 20 Driveshaft and related components (4WD)—model 2S1410

1. Propeller shaft tube	2. Snap ring	3. Journal bearing
4. Sleeve yoke	5. Journal	6. Flange yoke
⇦ Front		

37663_QX56_G0076

1. Before servicing the vehicle, refer to the Precautions Section.

➡If working near and/or around the SRS system and components, be sure to disable the SRS system. After disabling the system wait three minutes or more before servicing the vehicle.

➡Whenever the negative battery cable is disconnected the following components will require resetting. The Idle Air Volume Learning, Steering Angle Sensor Neutral Position, Sunroof Memory Reset/Initialization, Automatic Drive Positioner System, Audio presets and Navigation. Use the CONSULT-III diagnostic tool, or equivalent to perform the required resets.

2. Disconnect the negative battery cable.
3. Position the selector lever in the N position.
4. Release the parking brake.
5. Raise and support the vehicle safely.
6. Matchmark the driveshaft and companion flange.
7. Remove the center support bearing bracket nuts, if equipped.

8. Remove the driveshaft. Discard the nuts.

To install:

➡Be sure to use new fasteners, as required.

9. Installation is the reverse of the removal procedure.
10. Check for vehicle vibration, correct as required.
11. Be sure to perform the reconnect/relearn procedures.

REAR HALFSHAFT

REMOVAL & INSTALLATION

See Figure 21.

1. Before servicing the vehicle, refer to the Precautions Section.

➡If working near and/or around the SRS system and components, be sure to disable the SRS system. After disabling the system wait three minutes or more before servicing the vehicle.

➡Whenever the negative battery cable is disconnected the following components will require resetting. The Idle Air Volume Learning, Steering Angle Sensor Neutral Position, Sunroof Memory Reset/Initialization, Automatic Drive Positioner System, Audio presets and Navigation. Use the CONSULT-III diagnostic tool, or equivalent to perform the required resets.

2. Disconnect the negative battery cable.
3. Remove or disconnect the following:
   - Wheel and tire assembly
   - Stabilizer bar clamp
   - Cotter pin and driveshaft nut
   - Bolts from the inside flange of the driveshaft

4. Separate the driveshaft from the wheel hub by lightly tapping the end with suitable hammer and wood block.
5. Remove the halfshaft.

✳✳ WARNING

Do not excessively extend the slide joint.

To install:

➡Be sure to use new fasteners, as required.

➡Do not reuse the halfshaft inside flange bolts and cotter pin.

6. Install or connect the following:

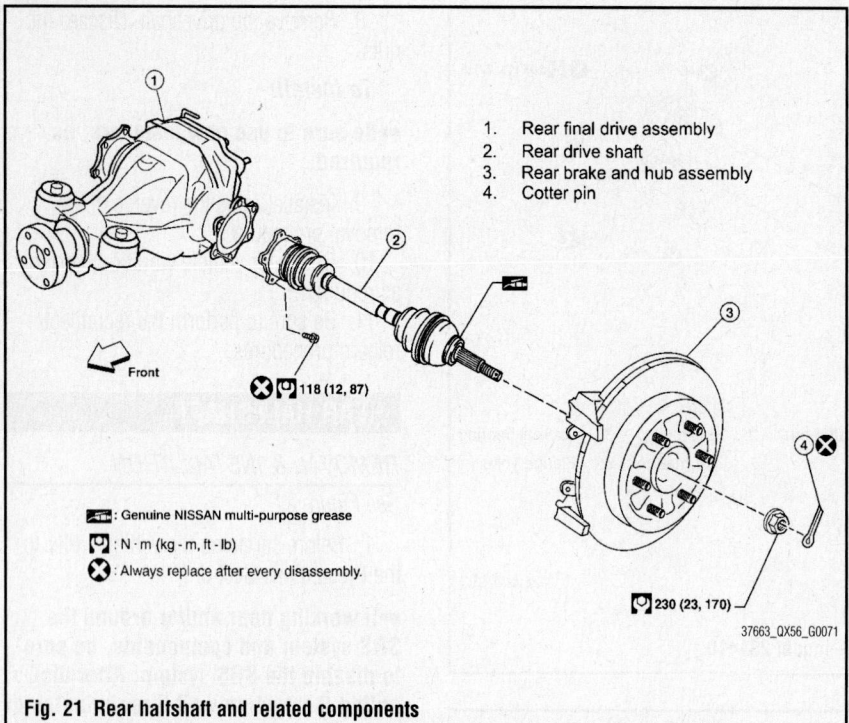

1. Rear final drive assembly
2. Rear drive shaft
3. Rear brake and hub assembly
4. Cotter pin

118 (12, 87)

230 (23, 170)

: Genuine NISSAN multi-purpose grease

N · m (kg - m, ft - lb)

: Always replace after every disassembly.

37663_QX56_G0071

Fig. 21 Rear halfshaft and related components

- Halfshaft
- Bolts for the inside flange and tighten to 87 ft. lbs. (118 Nm)
- Driveshaft nut and tighten nut to 101 ft. lbs. (137 Nm) and replace cotter pin
- Stabilizer bar clamp
- Wheel and tire assembly

7. Be sure to perform the reconnect/relearn procedures.

REAR PINION SEAL

REMOVAL & INSTALLATION

See Figures 22 and 23.

1. Before servicing the vehicle, refer to the Precautions Section.

➡ If working near and/or around the SRS system and components, be sure to disable the SRS system. After disabling the system wait three minutes or more before servicing the vehicle.

➡ Whenever the negative battery cable is disconnected the following components will require resetting. The Idle Air Volume Learning, Steering Angle Sensor Neutral Position, Sunroof Memory Reset/Initialization, Automatic Drive Positioner System, Audio presets and Navigation. Use the CONSULT-III diagnostic tool, or equivalent to perform the required resets.

2. Disconnect the negative battery cable.
3. Raise and safely support the vehicle.

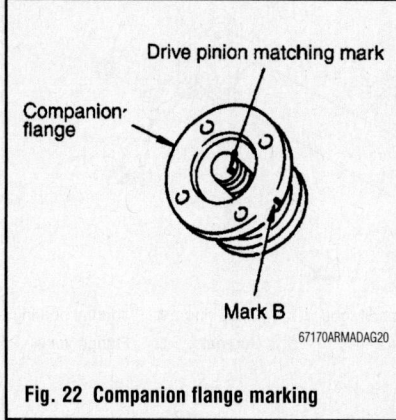

Drive pinion matching mark

Companion flange

Mark B

67170ARMADAG20

Fig. 22 Companion flange marking

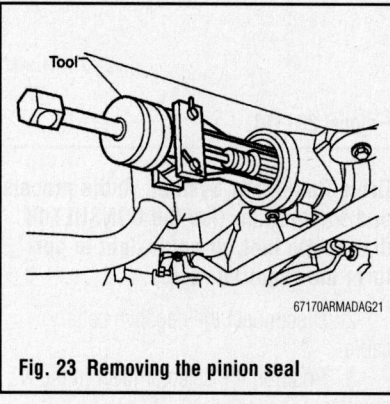

Tool

67170ARMADAG21

Fig. 23 Removing the pinion seal

4. Remove the rear driveshaft.
5. Measure and record the total preload.
6. Matchmark the drive pinion to position 'B' on the companion flange.
7. Remove the drive pinion nut using suitable tool.

8. Remove the companion flange using suitable tool.
9. Remove the rear pinion seal using special tool J-34286.

To install:

➡ Be sure to use new fasteners, as required.

10. Press the rear pinion seal into the carrier using suitable tool.
11. Align the matchmark on the companion flange to the drive pinion and install the companion flange.
12. Lubricate the drive pinion threads and seating surfaces of the drive pinion nut with grease.
13. Using a new drive pinion nut, tighten to 124–274 ft. lbs. (167–372 Nm).

➡ Final torque is determined when adjusting total preload using special tool J-25765-A.

14. Install rear driveshaft.
15. Be sure to perform the reconnect/relearn procedures.

TRANSFER CASE ASSEMBLY

REMOVAL & INSTALLATION

See Figure 24.

1. Before servicing the vehicle, refer to the Precautions Section.

➡ If working near and/or around the SRS system and components, be sure to disable the SRS system. After disabling the system wait three minutes or more before servicing the vehicle.

➡ Whenever the negative battery cable is disconnected the following components will require resetting. The Idle Air Volume Learning, Steering Angle Sensor Neutral Position, Sunroof Memory Reset/Initialization, Automatic Drive Positioner System, Audio presets and Navigation. Use the CONSULT-III diagnostic tool, or equivalent to perform the required resets.

2. Disconnect the negative battery cable.
3. Remove or disconnect the following:
   - Transmission splash guard
   - Center exhaust pipe and muffler
   - Front and rear driveshafts

➡ Plug rear oil seal after removing rear driveshaft.

   - Transmission assembly mounting bolts

4. Support the transmission assembly with a suitable jack and remove the cross-member.

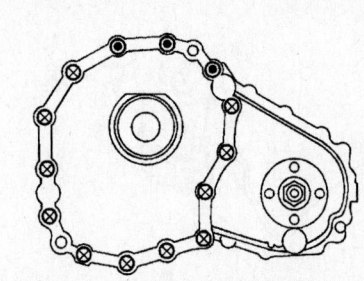

○ : Transfer → Automatic transmission
⊗ : Automatic transmission → Transfer

67170ARMADAG41

Fig. 24 Transfer case mounting bolt locations

5. Remove or disconnect the following:
- ATP switch, neutral 4LO switch, wait detection switch, transfer motor and transfer control device electrical connectors
- Breather hoses
- Shift actuator from the extension housing
- Transfer case to transmission assembly bolts
- Transfer case assembly

To install:

→ **Be sure to use new fasteners, as required.**

6. Install or connect the following:

- Transfer case to transmission assembly bolts tightening to 26 ft. lbs. (36 Nm)
- Shift actuator
- Breather hoses
- ATP switch, neutral 4LO switch, wait detection switch, transfer motor and transfer control device electrical connectors
- Support crossmember
- Transmission mounting bolts
- Driveshafts
- Muffler and center exhaust pipe
- Transmission splash guard

7. Be sure to perform the reconnect/relearn procedures.

ENGINE COOLING

ENGINE FAN

REMOVAL & INSTALLATION

Crankshaft Driven Type

See Figure 25.

→ **Never remove the radiator cap when the engine is hot. Serious burns could occur from high-pressure engine coolant escaping from the radiator.**

1. Before servicing the vehicle, refer to the Precautions Section.

→ **If working near and/or around the SRS system and components, be sure to disable the SRS system. After disabling the system wait three minutes or more before servicing the vehicle.**

→ **Whenever the negative battery cable is disconnected the following components will require resetting. The Idle Air Volume Learning, Steering Angle Sensor Neutral Position, Sunroof Memory Reset/Initialization, Automatic Drive Positioner System, Audio presets and Navigation. Use the CONSULT-III diagnostic tool, or equivalent to perform the required resets.**

2. Disconnect the negative battery cable.
3. Make sure the engine is cold.
4. Remove the air duct and resonator assembly.
5. Remove the engine front undercover.
6. Remove the lower radiator shroud.
7. Remove the drive belt.
8. Remove the cooling fan retaining bolts.
9. Remove the cooling fan from its mounting.

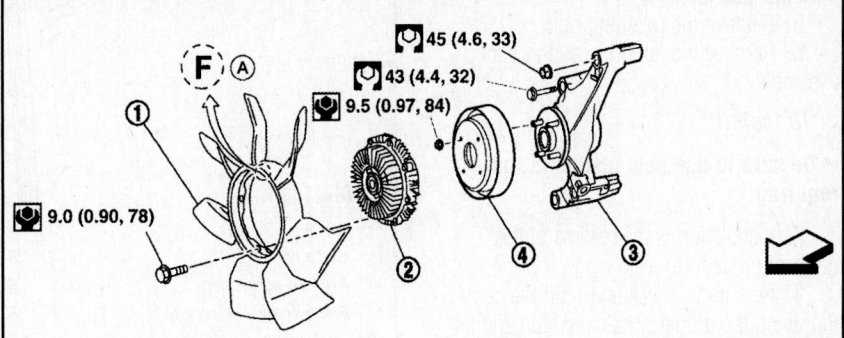

1. Cooling fan	2. Fan coupling	3. Fan bracket
4. Cooling fan pulley	A. Front mark	⇦ Engine front

37663_QX56_G0080

Fig. 25 Engine cooling fan and related components—crankshaft driven

To install:

→ **Be sure to use new fasteners, as required.**

10. Installation is the reverse of the removal procedure.
11. Be sure to install the fan with its front mark "F" facing the front of the engine.
12. Be sure to check and refill the cooling using the proper grade and type engine coolant, as required.
13. Start the engine and check for leaks.
14. Start the engine and allow it to reach operation temperature. Recheck the coolant level, fill as required.
15. Be sure to perform the reconnect/relearn procedures.

Motor Driven Type

See Figure 26.

→ **Never remove the radiator cap when the engine is hot. Serious burns could occur from high-pressure engine coolant escaping from the radiator.**

1. Before servicing the vehicle, refer to the Precautions Section.

→ **If working near and/or around the SRS system and components, be sure to disable the SRS system. After disabling the system wait three minutes or more before servicing the vehicle.**

→ **Whenever the negative battery cable is disconnected the following components will require resetting. The Idle Air Volume Learning, Steering Angle Sensor Neutral Position, Sunroof Memory Reset/Initialization, Automatic Drive Positioner System, Audio presets and Navigation. Use the CONSULT-III diagnostic tool, or equivalent to perform the required resets.**

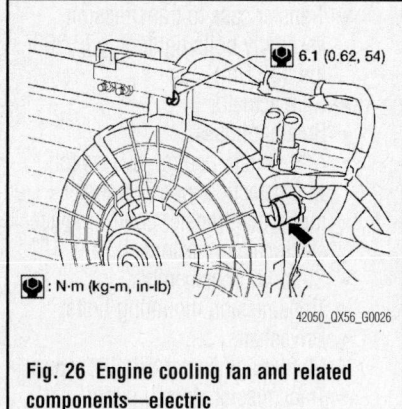

Fig. 26 Engine cooling fan and related components—electric

2. Disconnect the negative battery cable.

3. Remove the front bumper fascia.

4. Disconnect the harness connector from the fan motor.

5. Remove the retaining bolt.

6. Remove the fan grille and motor assembly.

To install:

➡**Be sure to use new fasteners, as required.**

7. Installation is the reverse of the removal procedure.

8. Be sure to check and refill the cooling using the proper grade and type engine coolant, as required.

9. Start the engine and check for leaks.

10. Start the engine and allow it to reach operation temperature. Recheck the coolant level, fill as required.

11. Be sure to perform the reconnect/relearn procedures.

RADIATOR

REMOVAL & INSTALLATION

See Figure 27.

✳✳ CAUTION

Never remove the radiator cap when the engine is hot. Serious burns could occur from high-pressure engine coolant escaping from the radiator.

1. Before servicing the vehicle, refer to the Precautions Section.

➡**If working near and/or around the SRS system and components, be sure to disable the SRS system. After disabling the system wait three minutes or more before servicing the vehicle.**

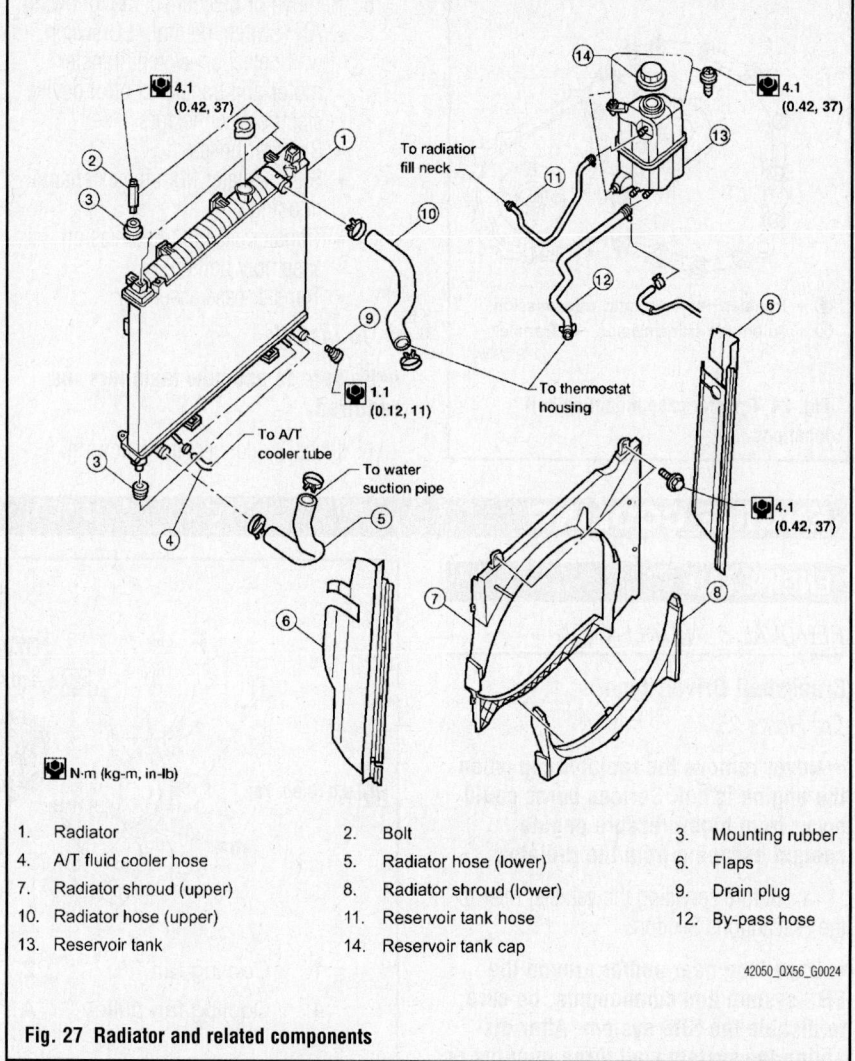

Fig. 27 Radiator and related components

1. Radiator	2. Bolt	3. Mounting rubber
4. A/T fluid cooler hose	5. Radiator hose (lower)	6. Flaps
7. Radiator shroud (upper)	8. Radiator shroud (lower)	9. Drain plug
10. Radiator hose (upper)	11. Reservoir tank hose	12. By-pass hose
13. Reservoir tank	14. Reservoir tank cap	

➡**Whenever the negative battery cable is disconnected the following components will require resetting. The Idle Air Volume Learning, Steering Angle Sensor Neutral Position, Sunroof Memory Reset/Initialization, Automatic Drive Positioner System, Audio presets and Navigation. Use the CONSULT-III diagnostic tool, or equivalent to perform the required resets.**

2. Disconnect the negative battery cable.

3. Make sure the engine is cold before removing the radiator.

4. Remove the engine room cover.

5. Remove the air cleaner and air duct assembly.

6. Drain the engine coolant. Be sure to properly dispose of the used coolant.

7. Disconnect and plug the automatic transmission fluid lines.

8. Disconnect the upper radiator hose. Do not allow coolant to contact the drive belts.

9. Disconnect the lower radiator hose. Do not allow coolant to contact the drive belts.

10. To remove the lower radiator shroud, release the tabs and pull the lower radiator shroud rearwards and down.

11. Remove the radiator shroud upper bolts and remove the upper radiator shroud.

12. Remove the air conditioning condenser bolts and brackets.

➡**Lift the condenser up and forward to remove it from the radiator.**

13. Remove the transmission fluid cooler bolts and the fluid cooler from the radiator. Position it to the side.

14. Lift up and remove the radiator. Be careful not to damage or scratch the air conditioning condenser and radiator core when removing the radiator.

To install:

➡**Be sure to use new fasteners, as required.**

15. Installation is the reverse of the removal procedure.

16. Be sure to refill the cooling using the proper grade and type engine coolant.

17. Start the engine and check for leaks.

18. Start the engine and allow it to reach operation temperature. Recheck the coolant level, fill as required.

19. Be sure to perform the reconnect/relearn procedures.

THERMOSTAT

REMOVAL & INSTALLATION

See Figures 28 and 29.

※ CAUTION

Never remove the radiator cap when the engine is hot. Serious burns could occur from high-pressure engine coolant escaping from the radiator.

1. Before servicing the vehicle, refer to the Precautions Section.

➡**If working near and/or around the SRS system and components, be sure to disable the SRS system. After disabling the system wait three minutes or more before servicing the vehicle.**

➡**Whenever the negative battery cable is disconnected the following components will require resetting. The Idle Air Volume Learning, Steering Angle Sensor Neutral Position, Sunroof Memory Reset/Initialization, Automatic Drive Positioner System, Audio presets and Navigation. Use the CONSULT-III diagnostic tool, or equivalent to perform the required resets.**

2. Disconnect the negative battery cable.

3. Make sure the engine is cold.

4. Remove the engine room cover.

5. Remove the air duct and resonator assembly.

6. Disconnect the water suction hose from the water inlet.

7. Remove the water inlet and thermostat.

➡**To remove the thermostat housing, water outlet and heater pipe, you will first have to remove the intake manifold.**

To install:

➡**Be sure to use new fasteners, as required.**

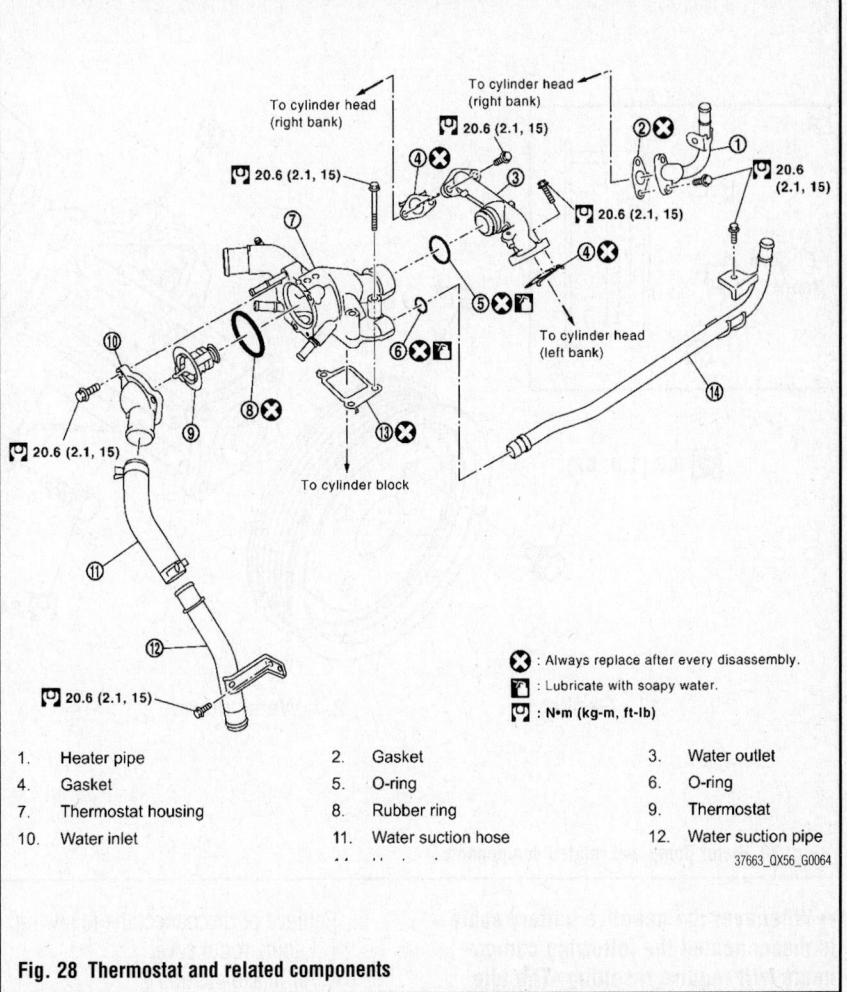

1.	Heater pipe	2.	Gasket	3.	Water outlet
4.	Gasket	5.	O-ring	6.	O-ring
7.	Thermostat housing	8.	Rubber ring	9.	Thermostat
10.	Water inlet	11.	Water suction hose	12.	Water suction pipe

❌ : Always replace after every disassembly.
🔲 : Lubricate with soapy water.
🔧 : N•m (kg-m, ft-lb)

37663_QX56_G0064

Fig. 28 Thermostat and related components

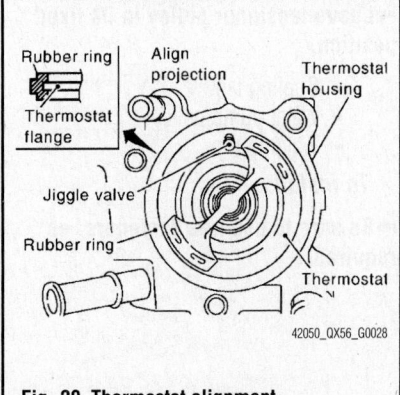

Fig. 29 Thermostat alignment

42050_QX56_G0028

8. Installation is the reverse of the removal procedure.

9. Be sure to use a new gasket.

10. Install the thermostat with the whole circumference of each flange part fitting securely inside the rubber ring, as shown in the illustration.

11. Be sure to perform the reconnect/relearn procedures.

12. Install the thermostat with the jiggle valve facing upward.

13. Be sure to refill the cooling using the proper grade and type engine coolant.

14. Start the engine and check for leaks.

15. Start the engine and allow it to reach operation temperature. Recheck the coolant level, fill as required.

WATER PUMP

REMOVAL & INSTALLATION

See Figure 30.

※ CAUTION

Never remove the radiator cap when the engine is hot. Serious burns could occur from high-pressure engine coolant escaping from the radiator.

1. Before servicing the vehicle, refer to the Precautions Section.

➡**If working near and/or around the SRS system and components, be sure to disable the SRS system. After disabling the system wait three minutes or more before servicing the vehicle.**

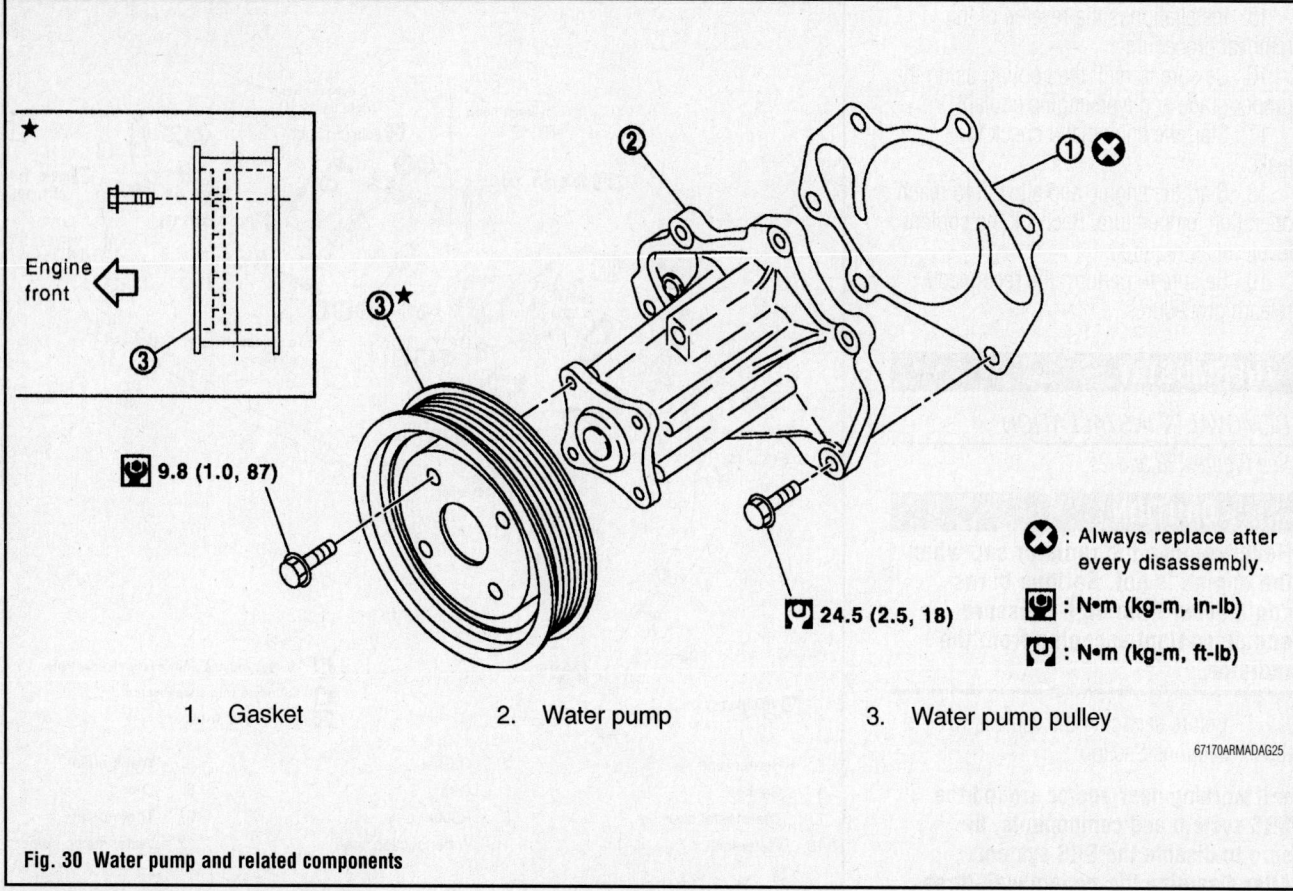

★

Engine front ←

9.8 (1.0, 87)

24.5 (2.5, 18)

⊗ : Always replace after every disassembly.

🔧 : N•m (kg-m, in-lb)

🔧 : N•m (kg-m, ft-lb)

1. Gasket   2. Water pump   3. Water pump pulley

67170ARMADAG25

Fig. 30 Water pump and related components

➡ **Whenever the negative battery cable is disconnected the following components will require resetting. The Idle Air Volume Learning, Steering Angle Sensor Neutral Position, Sunroof Memory Reset/Initialization, Automatic Drive Positioner System, Audio presets and Navigation. Use the CONSULT-III diagnostic tool, or equivalent to perform the required resets.**

2. Disconnect the negative battery cable.
3. Make sure the engine is cold.
4. Drain the cooling system.

5. Remove or disconnect the following:
    - Engine room cover
    - Air intake assembly
    - Accessory drive belt

➡ **Leave tensioner pulley in its fixed position.**

    - Cooling fan
    - Water pump pulley
    - Water pump

To install:

➡ **Be sure to use new fasteners, as required.**

6. Install or connect the following:
    - Water pump with a new gasket. Tighten bolts to 18 ft. lbs. (25 Nm)
    - Water pump pulley and tighten bolts to 87 inch lbs. (10 Nm)
    - Accessory drive belt
    - Air intake assembly
    - Engine splash guard
7. Refill the cooling system.
8. Start the engine and check for leaks.
9. Be sure to perform the reconnect/relearn procedures.

ENGINE ELECTRICAL   **CHARGING SYSTEM**

ALTERNATOR

REMOVAL & INSTALLATION

See Figure 31.

1. Before servicing the vehicle, refer to the Precautions Section.

➡ If working near and/or around the SRS system and components, be sure to disable the SRS system. After disabling the system wait three minutes or more before servicing the vehicle.

➡ Whenever the negative battery cable is disconnected the following components will require resetting. The Idle Air Volume Learning, Steering Angle Sensor Neutral Position, Sunroof Memory Reset/ Initialization, Automatic Drive Positioner System, Audio presets and

Navigation. Use the CONSULT-III diagnostic tool, or equivalent to perform the required resets.

2. Disconnect the negative battery cable.

3. Remove or disconnect the following:
   - Fan shroud
   - Drive belt
   - Lower alternator bracket
   - Alternator upper bolt
   - Alternator harness connectors
   - Alternator

To install:

➡ Be sure to use new fasteners, as required.

4. Install or connect the following:
   - Alternator
   - Alternator harness connectors
   - Upper bolt, tighten to 48 ft. lbs. (65 Nm)

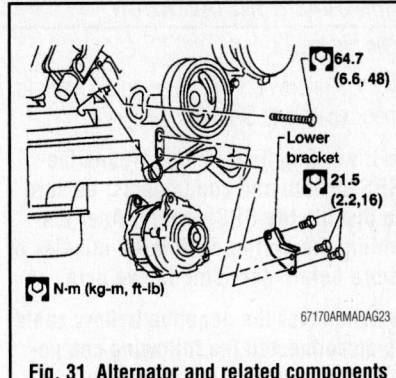

Fig. 31 Alternator and related components

   - Lower bracket, tighten to 16 ft. lbs (22 Nm)
   - Drive belt
   - Fan shroud
   - Negative battery cable

5. Be sure to perform the reconnect/ relearn procedures.

ENGINE ELECTRICAL   **IGNITION SYSTEM**

FIRING ORDER

1–8–7–3–6–5–4–2

IGNITION COIL

REMOVAL & INSTALLATION

See Figure 32.

1. Before servicing the vehicle, refer to the Precautions Section.

➡ If working near and/or around the SRS system and components, be sure to disable the SRS system. After disabling the system wait three minutes or more before servicing the vehicle.

➡ Whenever the negative battery cable is disconnected the following components will require resetting. The Idle Air Volume Learning, Steering Angle Sensor Neutral Position, Sunroof Memory Reset/Initialization, Automatic Drive Positioner System, Audio presets and Navigation. Use the CONSULT-III diagnostic tool, or equivalent to perform the required resets.

2. Disconnect the negative battery cable.

3. Remove the engine room cover. Remove the air cleaner assembly, as required.

4. Disconnect the harness connector from the ignition coil.

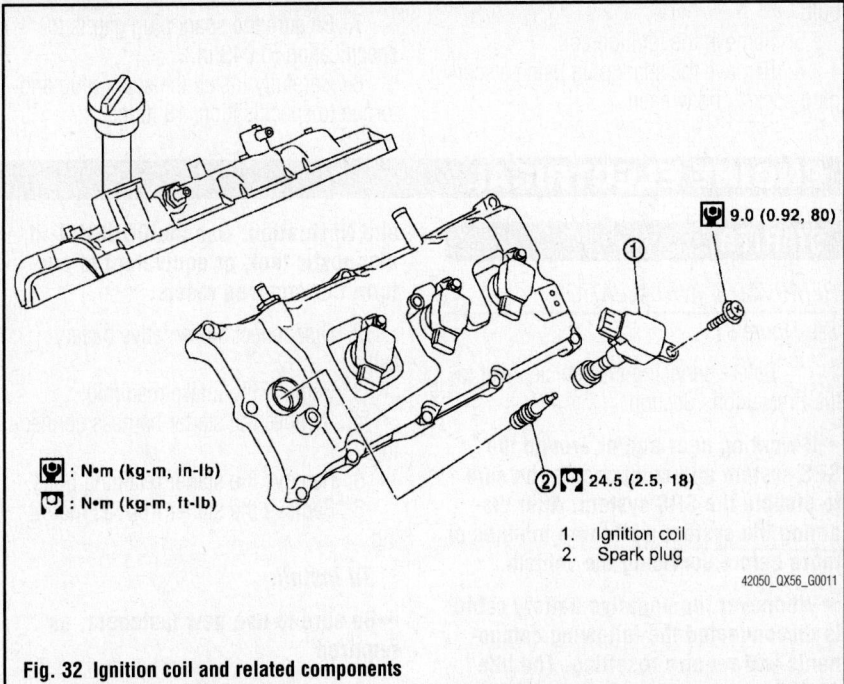

Fig. 32 Ignition coil and related components

5. Remove the ignition coil retaining bolt.

6. Remove the ignition coil.

To install:

➡ Be sure to use new fasteners, as required.

7. Install the ignition coil, torque the retaining bolt to 80 inch lbs. (9 Nm).

8. Connect the harness coil.

9. Connect the negative battery cable.

10. Be sure to perform the reconnect/ relearn procedures.

IGNITION TIMING

ADJUSTMENT

The ignition timing is controlled by the Powertrain Control Module (PCM). No adjustment is necessary or possible.

SPARK PLUGS

REMOVAL & INSTALLATION

See Figure 33.

1. Before servicing the vehicle, refer to the Precautions Section.

➡If working near and/or around the SRS system and components, be sure to disable the SRS system. After disabling the system wait three minutes or more before servicing the vehicle.

➡Whenever the negative battery cable is disconnected the following components will require resetting. The Idle Air Volume Learning, Steering Angle Sensor Neutral Position, Sunroof Memory Reset/Initialization, Automatic Drive Positioner System, Audio presets and Navigation. Use the CONSULT-III diagnostic tool, or equivalent to perform the required resets.

2. Disconnect the negative battery cable.
3. Disconnect the harness connector from the ignition coil.
4. Remove the ignition coil retaining bolt.
5. Remove the ignition coil.
6. Remove the spark plug using a spark plug socket and wrench.

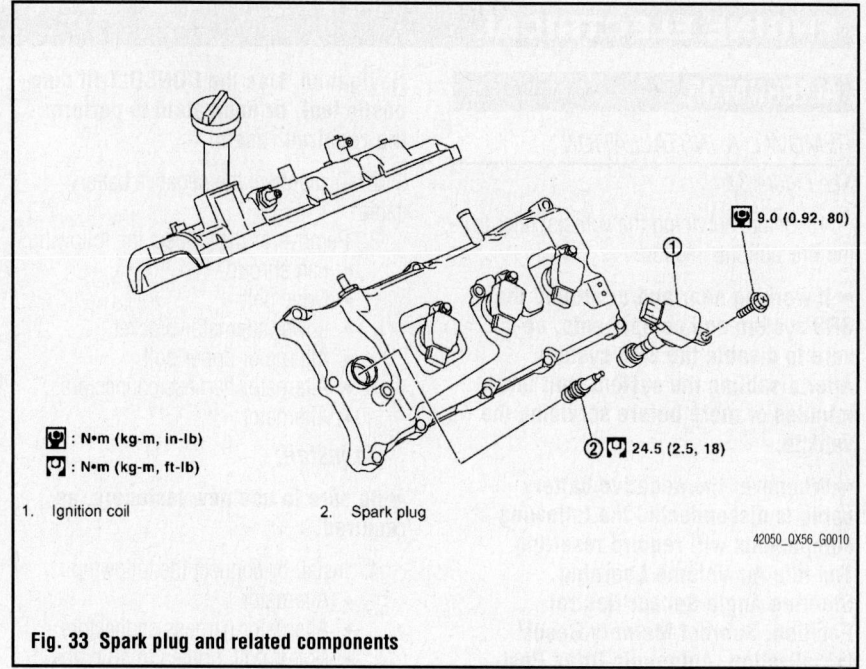

□ : N·m (kg-m, in-lb)
□ : N·m (kg-m, ft-lb)

1. Ignition coil
2. Spark plug

42050_QX56_G0010

Fig. 33 Spark plug and related components

To install:

➡Be sure to use new fasteners, as required.

7. Be sure the spark plug gap is to specification (0.043 in.).
8. Carefully install the spark plug and torque to specification, 18 ft. lbs.

9. Install the ignition coil, torque the retaining bolt to 80 inch lbs. (9 Nm).
10. Connect the harness coil.
11. Connect the negative battery cable.
12. Be sure to perform the reconnect/relearn procedures.

ENGINE ELECTRICAL   STARTING SYSTEM

STARTER

REMOVAL & INSTALLATION

See Figure 34.

1. Before servicing the vehicle, refer to the Precautions Section.

➡If working near and/or around the SRS system and components, be sure to disable the SRS system. After disabling the system wait three minutes or more before servicing the vehicle.

➡Whenever the negative battery cable is disconnected the following components will require resetting. The Idle Air Volume Learning, Steering Angle Sensor Neutral Position, Sunroof Memory Reset/Initialization, Automatic Drive Positioner System, Audio presets and Navigation. Use the CONSULT-III diagnostic tool, or equivalent to perform the required resets.

2. Disconnect the negative battery cable.
3. Remove the intake manifold.
4. Remove the starter harness connectors.
5. Remove the starter retaining bolts.
6. Remove the starter from its mounting.

To install:

➡Be sure to use new fasteners, as required.

7. Installation is the reverse of the removal procedure.
8. Tighten the retaining bolts to 34 ft. lbs. (46 Nm).

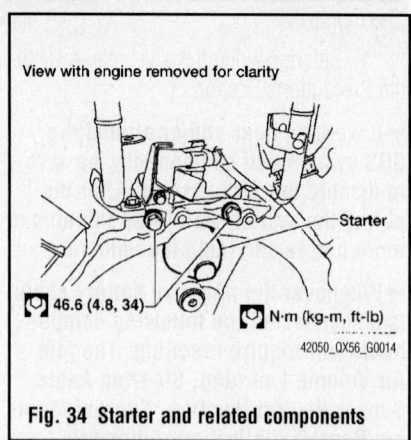

View with engine removed for clarity

Starter

□ 46.6 (4.8, 34)   □ N·m (kg-m, ft-lb)

42050_QX56_G0014

Fig. 34 Starter and related components

9. Tighten the terminal nut to 8 ft. lbs. (10.8 Nm).
10. Be sure to perform the reconnect/relearn procedures.

ENGINE MECHANICAL

ACCESSORY DRIVE BELTS

ACCESSORY BELT ROUTING

See Figure 35.

Refer to the accompanying illustration.

INSPECTION

Inspect the drive belt for signs of glazing or cracking. A glazed belt will be perfectly smooth from slippage, while a good belt will have a slight texture of fabric visible. Cracks will usually start at the inner edge of the belt and run outward. All worn or damaged drive belts should be replaced immediately.

ADJUSTMENT

Drive belt tension is not necessary, as it is automatically adjusted by the auto tensioner.

REMOVAL & INSTALLATION

See Figure 35.

1. Before servicing the vehicle, refer to the Precautions Section.

➡**If working near and/or around the SRS system and components, be sure to disable the SRS system. After disabling the system wait three minutes or more before servicing the vehicle.**

➡**Whenever the negative battery cable is disconnected the following components will require resetting. The Idle Air Volume Learning, Steering Angle Sensor Neutral Position, Sunroof Memory Reset/Initialization, Automatic Drive Positioner System, Audio presets and Navigation. Use the CONSULT-III diagnostic tool, or equivalent to perform the required resets.**

2. Disconnect the negative battery cable.
3. Remove the engine room cover.
4. Remove the air duct and resonator assembly.
5. Install special tool J-46535, or equivalent on the auto tensioner pulley bolt and move it upward.

➡**Avoid placing your hand in a location where pinching may occur if the holding tool accidentally comes off.**

6. Remove the drive belt from the vehicle.

To install:

➡**Be sure to use new fasteners, as required.**

7. Installation is the reverse of the removal procedure.
8. Be sure that the belt is securely installed around all pulleys.

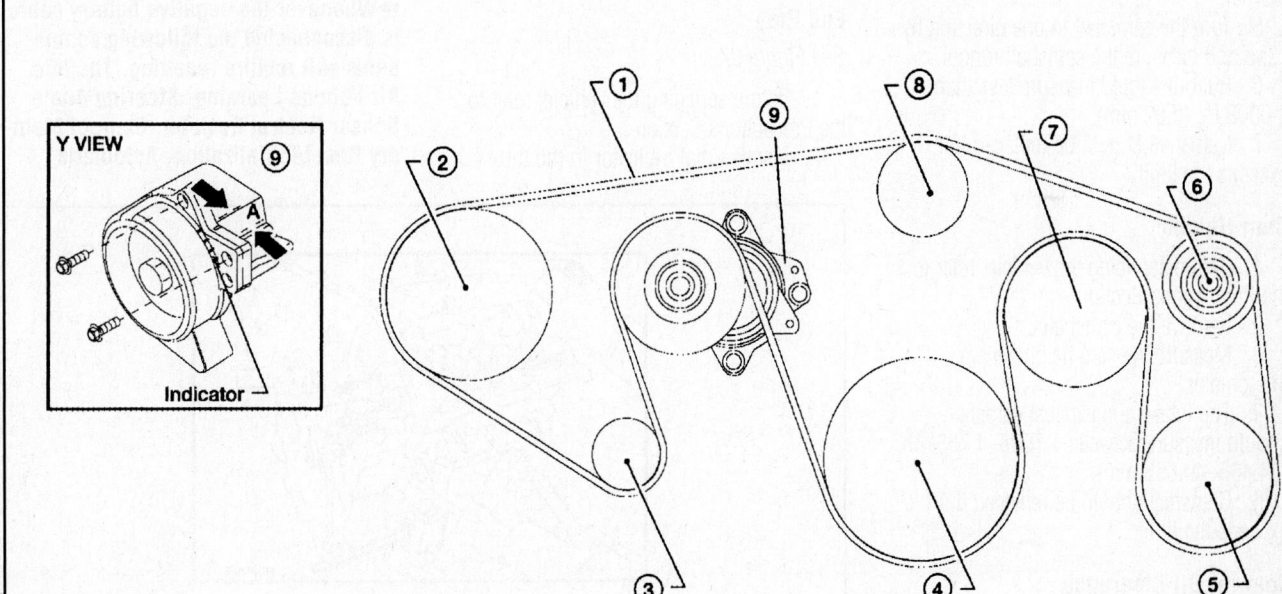

1.	Drive Belt	2.	Power Steering Pump Pulley	3.	Generator pulley
4.	Crankshaft Pulley	5.	A/C Compressor	6.	Idler Pulley
7.	Cooling Fan Pulley	8.	Water Pump Pulley	9.	Drive Belt Tensioner

67162-QX56-G47

Fig. 35 Accessory drive belt routing

9. Rotate the crankshaft several times clockwise to equalize belt tension between the pulleys.

10. Make sure that the belt tension is within the allowable working range, using the indicator notch on the auto tensioner.

11. Be sure to perform the reconnect/relearn procedures.

CAMSHAFT AND VALVE LIFTERS

INSPECTION

Runout

1. Before servicing the vehicle, refer to the Precautions Section.

2. Remove the camshafts.

3. Using a V-block on a precise flat table, support the No. 2 and 4 journals of the camshaft.

➡ Do not support journal No. 1 as it has a different diameter than the other locations.

4. Set the dial indicator to No. 3 journal.

5. Turn the camshaft to one direction by hand and measure the camshaft runout.

6. Runout should measure less than 0.0008 in. (0.02 mm).

7. Camshaft should be replaced if it exceeds the limit.

Cam Height

1. Before servicing the vehicle, refer to the Precautions Section.

2. Remove the camshafts.

3. Measure the cam height with a micrometer.

4. The intake and exhaust camshaft should measure between 1.7506–1.7581 in. (44.465–44.655 mm).

5. Camshaft should be replaced if it exceeds the limit.

Journal Oil Clearance

See Figure 36.

1. Before servicing the vehicle, refer to the Precautions Section.

2. Remove the camshafts.

3. Measure the outer diameter of camshaft journal with micrometer and record the result.

4. Reinstall the camshaft bearing caps in accordance to the installation procedure.

5. Measure the inner diameter of the camshaft bracket ("A") with a bore gauge and record the result.

6. Subtract the camshaft journal diameter from the camshaft bracket inner diame-

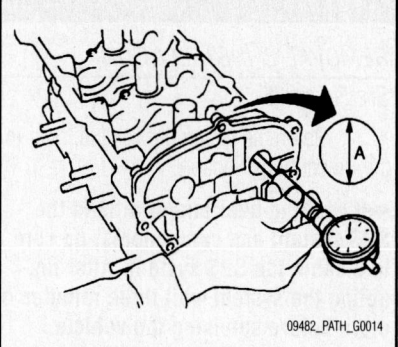

Fig. 36 Measuring the journal bore for oil clearance

ter. The difference should measure between 0.0012–0.0027 in. (0.030–0.068 mm).

7. The camshaft or camshaft bracket should be replaced if it exceeds the limit.

➡ The camshaft bracket cannot be replaced as an individual part, because it is machined together with the cylinder head. The entire cylinder head assembly must be replaced.

End Play

See Figure 37.

1. Before servicing the vehicle, refer to the Precautions Section.

2. Install a dial indicator in the thrust

direction on the front end of the camshaft. Measure the end play of the dial indicator when the camshaft is moved back and forth. The dial indicator should measure between 0.0045–0.0074 in. (0.115–0.188 mm).

3. Measure the No. 1 journal as shown. The distance ("A") should be 1.2008–1.2027 in. (30.500–30.548 mm).

4. Measure the No. 1 journal bearing as shown. The distance ("B") should be 1.1953–1.1963 in. (30.360–30.385 mm).

5. Replace either the camshaft or cylinder head assembly if the measurement is exceeded.

REMOVAL & INSTALLATION

See Figures 38 through 54.

1. Before servicing the vehicle, refer to the Precautions Section.

➡ If working near and/or around the SRS system and components, be sure to disable the SRS system. After disabling the system wait three minutes or more before servicing the vehicle.

➡ Whenever the negative battery cable is disconnected the following components will require resetting. The Idle Air Volume Learning, Steering Angle Sensor Neutral Position, Sunroof Memory Reset/Initialization, Automatic

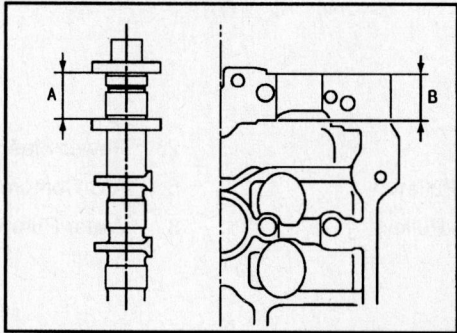

Fig. 37 Measuring for camshaft endplay

Drive Positioner System, Audio presets and Navigation. Use the CONSULT-III diagnostic tool, or equivalent to perform the required resets.

2. Disconnect the negative battery cable.
3. Remove the engine cover.
4. Remove the air cleaner assembly.
5. Remove the power steering reservoir tank bolts. Position the unit to the side.
6. Remove the valve covers.
7. Remove the spark plugs.
8. Remove the drive belt.
9. Be sure that the number one cylinder is at TDC on the compression stroke.

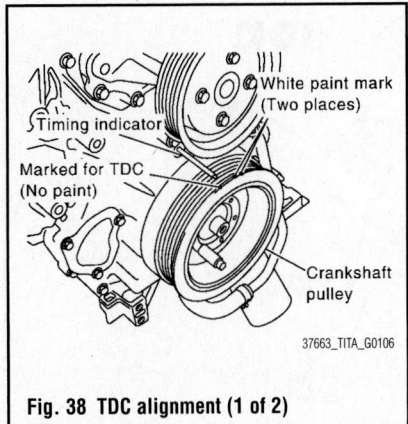

Fig. 38 TDC alignment (1 of 2)

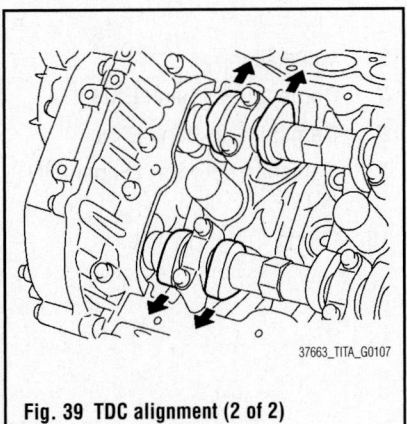

Fig. 39 TDC alignment (2 of 2)

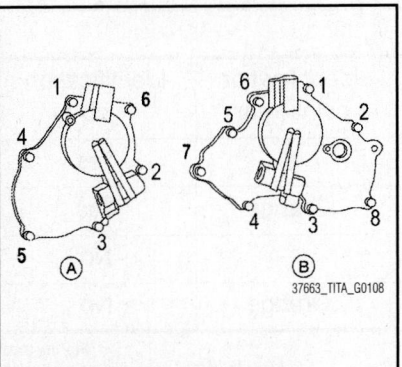

Fig. 40 Intake valve timing control solenoid cover tightening sequence

➡Turn the crankshaft pulley clockwise to align the TDC identification notch (without paint mark) with the timing indicator on the front cover. At this time make sure that the intake and exhaust cam lobes of the number one cylinder (top front on left bank) point outside. If not turn the crankshaft pulley once more. See illustration

10. Remove the CMP sensor.
11. Remove the intake valve timing control position sensors (right and left).
12. Remove the intake valve timing control solenoid valves (right and left).
13. Loosen and remove the intake valve timing control valve cover (right and left) bolts in the reverse order of the tightening sequence.
14. Paint alignment marks on the right bank (A) timing chain links (C) and left bank (B) timing chain links (D) and align with the camshaft sprocket alignment marks (E) and (F). See illustration.
15. To remove the left tensioner, squeeze the return proof clip ends using a suitable tool and push the plunger into the tensioner body. Secure the plunger using a stopper pin (hard wire 0.04 inch in diameter). Remove the bolts and the tensioner.

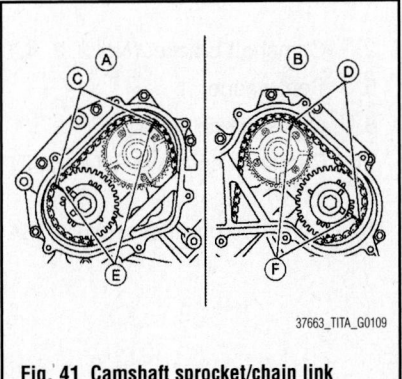

Fig. 41 Camshaft sprocket/chain link alignment

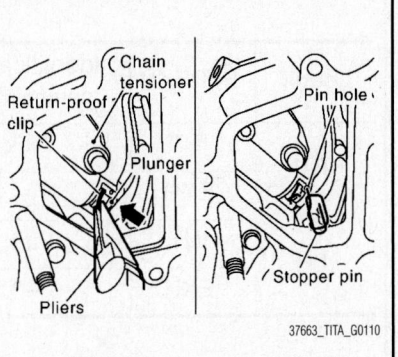

Fig. 42 Camshaft stopper pin installation

➡The plunger, spring and spring seat pop out when squeezing the return proof clip without holding the plunger head. It may cause serious injuries. Always hold the plunger head when removing.

➡Stop the plunger in the fully extended position using the return proof clip (1) if the stopper pin is removed. Push the plunger (2) into the chain tensioner body while squeezing the return proof clip (1). Secure it using a stopper pin (3). See illustration.

16. Remove the chain tensioner cover from the front cover, using tool KV10111100, or equivalent. Do not damage the mating surfaces.
17. To remove the left tensioner, squeeze the return proof clip ends using a suitable tool and push the plunger into the tensioner body. Secure the plunger using a stopper pin (hard wire 0.04 inch in diameter). Remove the bolts and the tensioner.

➡The plunger, spring and spring seat pop out when squeezing the return proof clip without holding the plunger head. It may cause serious injuries. Always hold the plunger head when removing.

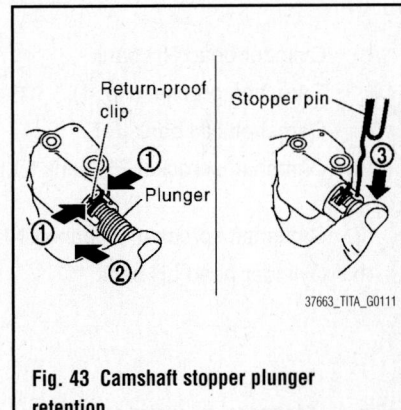

Fig. 43 Camshaft stopper plunger retention

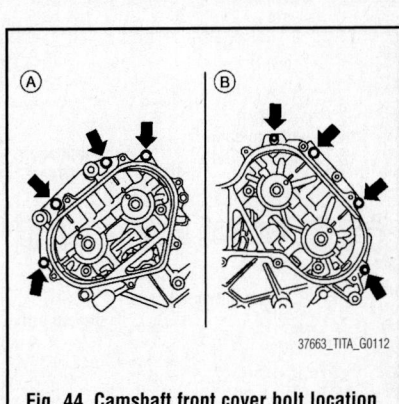

Fig. 44 Camshaft front cover bolt location (arrow)

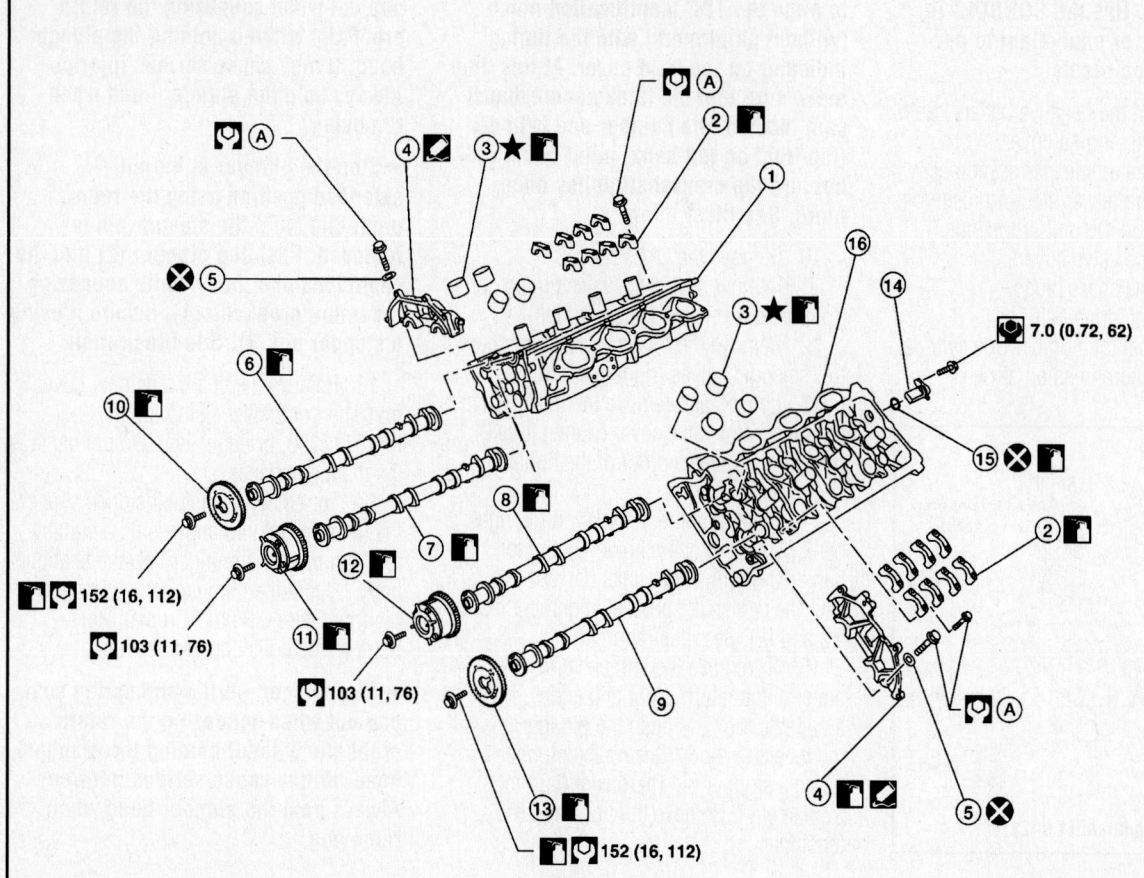

1. Cylinder head RH bank
2. Camshaft bracket (No. 2, 3, 4, 5)
3. Valve lifter
4. Camshaft bracket (No. 1)
5. Seal washer
6. Camshaft RH bank EXH
7. Camshaft RH bank INT
8. Camshaft LH bank INT
9. Camshaft LH bank EXH
10. Camshaft sprocket RH bank EXH
11. Camshaft sprocket RH bank INT (VTC)
12. Camshaft sprocket LH bank INT (VTC)
13. Camshaft sprocket LH bank EXH
14. Camshaft position sensor (PHASE)
15. O-ring
16. Cylinder head LH bank

37663_TITA_G0105

Fig. 45 Camshafts and related components

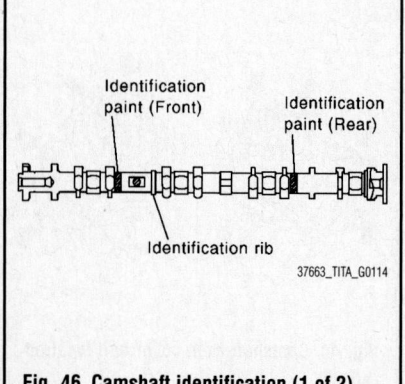

Fig. 46 Camshaft identification (1 of 2)

Bank	INT EXH	Identification paint (front)	Identification paint (rear)	Identification rib
RH	INT	Pink	—	Yes
	EXH	—	Orange	Yes
LH	INT	Pink	—	No
	EXH	—	Orange	No

37663_TITA_G0115

Fig. 47 Camshaft identification (2 of 2)

→If it is difficult to push the plunger on the tensioner, remove the plunger under the extended condition.

18. Loosen the camshaft sprocket bolts and remove the sprockets.

→To avoid interference between the valves and pistons, do not turn the crankshaft or camshaft with the timing chain disconnected.

19. Remove the front cover bolts. See illustration for location (arrow).

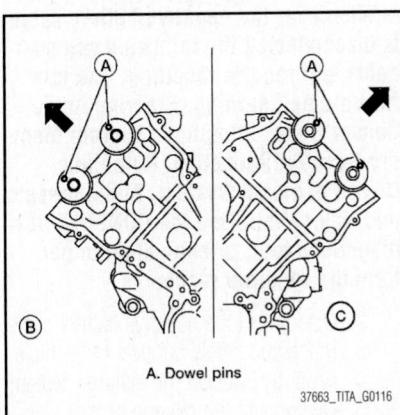

A. Dowel pins

37663_TITA_G0116

Fig. 48 Camshaft dowel pin installation

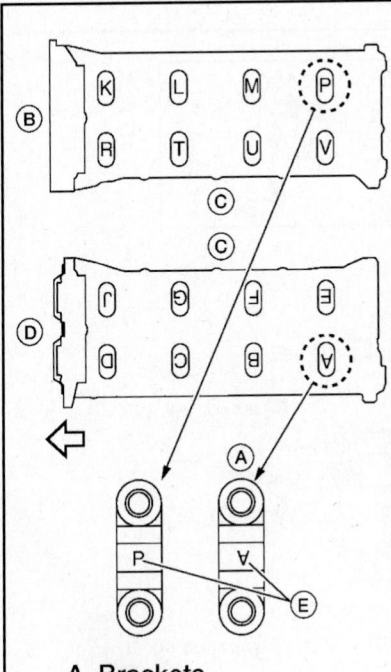

A. Brackets
B. Right
C. Intake manifold side
D. Left
E. Location mark

37663_TITA_G0117

Fig. 49 Camshaft bracket installation and identification

20. Remove the camshaft bracket bolts in the reverse order of the tightening sequence. Remove the number one camshaft bracket. The bottom of the front surface of the bracket will be stuck because of liquid gasket.

21. Remove the camshaft. Remove the lifters, as necessary.

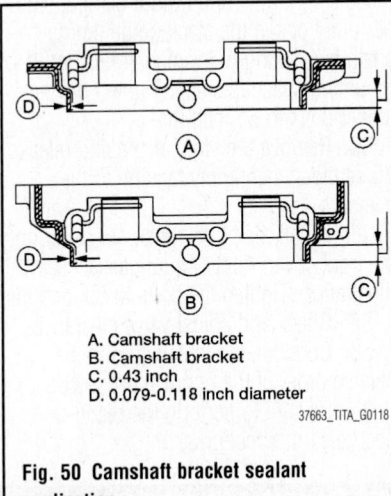

A. Camshaft bracket
B. Camshaft bracket
C. 0.43 inch
D. 0.079-0.118 inch diameter

37663_TITA_G0118

Fig. 50 Camshaft bracket sealant application

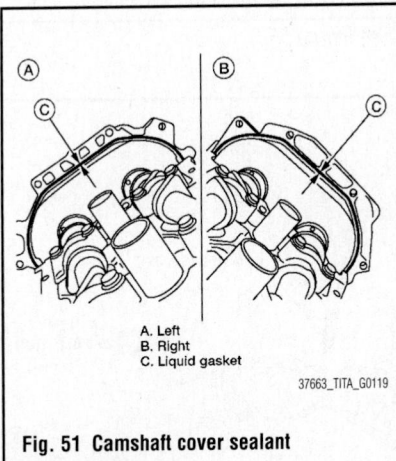

A. Left
B. Right
C. Liquid gasket

37663_TITA_G0119

Fig. 51 Camshaft cover sealant application (1 of 2)

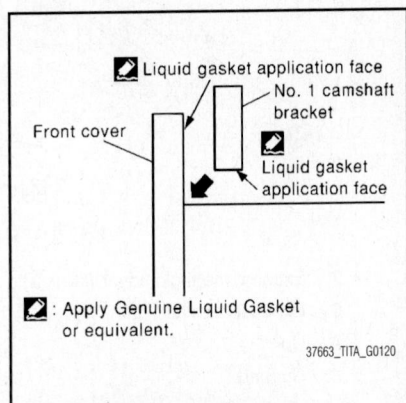

Liquid gasket application face
No. 1 camshaft bracket
Front cover
Liquid gasket application face

◢ : Apply Genuine Liquid Gasket or equivalent.

37663_TITA_G0120

Fig. 52 Camshaft cover sealant application (2 of 2)

To install:

→Be sure to use new fasteners, as required.

22. Install the camshafts. Be sure that the camshafts are properly identified. See illustrations.

23. Install the dowel pins at the front of the camshaft. See illustration for proper direction.

24. Install the camshaft brackets.

→Install by referring to the illustration location mark on the upper surface. Install so that the installation mark can be correctly read when viewed from the intake manifold side.

25. To install the number one camshaft bracket, apply liquid gasket as shown in the illustration. Be sure to wipe off any excessive gasket after installation.

26. Apply liquid gasket to the back side of the left front cover and the right front cover. Bead diameter should be 0.102–0.142 inch. Position the number one camshaft bracket close to the mounting position and then install it to prevent from touching gasket applied to each surface.

27. Temporarily tighten the right and left front cover bolts.

28. Tighten the camshaft bracket bolts to specification and in the proper sequence.

29. Tighten the right and left front cover bolts to 8 ft. lbs.

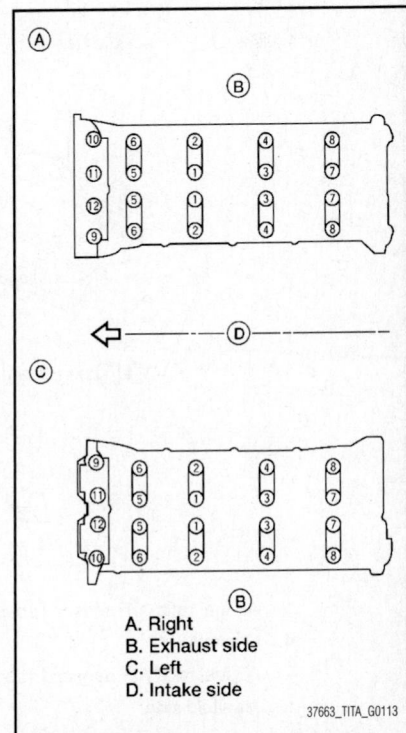

A. Right
B. Exhaust side
C. Left
D. Intake side

37663_TITA_G0113

Fig. 53 Camshaft bracket bolt tightening sequence

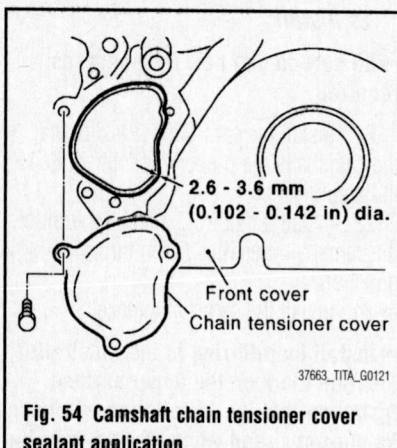

Fig. 54 Camshaft chain tensioner cover sealant application

2.6 - 3.6 mm (0.102 - 0.142 in) dia.

Front cover

Chain tensioner cover

37663_TITA_G0121

30. Install the camshaft sprockets aligning them with the matching marks painted on the timing chain and the camshaft sprockets, before removal. Align the sprocket key groove with the dowel pin on the camshaft front edge at the same time. Temporarily tighten the sprocket bolts.

31. Install the intake VTC and the exhaust side camshaft sprockets by selectively using the groove of the dowel pin according to the bank for the exhaust side camshaft sprockets, (common part used for both exhaust banks).

➡Use the groove marked "R" for right bank and "L" for left bank.

32. Lock the hex part of the camshaft in the same way as for removal. Tighten the sprocket bolts.

33. Check that the timing marks are properly aligned.

34. To install the chain tensioner, compress the plunger and hold it using a stopper pin. Loosen the slack guide timing chain by rotating the camshaft hex part if mounting space is small. Tighten the tensioner bolts to 61 inch lbs.

35. Remove the stopper pin and release the plunger, then apply tension to the chain.

36. Install the chain tensioner cover onto the front cover. Apply liquid gasket. See illustration. Tighten the bolts to 80 inch lbs.

37. Check and adjust valve clearances.

38. Continue the installation in the reverse order of the removal procedure.

39. Be sure to perform the reconnect/relearn procedures.

CATALYTIC CONVERTER

REMOVAL & INSTALLATION

See Figure 55.

At this time the manufacturer does not provide removal and installation procedures for this component. The following procedure is a guideline and may differ from the vehicle you are servicing.

1. Before servicing the vehicle, refer to the Precautions Section.

➡**If working near and/or around the SRS system and components, be sure to disable the SRS system. After disabling the system wait three minutes or more before servicing the vehicle.**

➡**Whenever the negative battery cable is disconnected the following components will require resetting. The Idle Air Volume Learning, Steering Angle Sensor Neutral Position, Sunroof Memory Reset/Initialization, Automatic Drive Positioner System, Audio presets and Navigation. Use the CONSULT-III diagnostic tool, or equivalent to perform the required resets.**

2. Disconnect the negative battery cable.

3. Raise and safely support the vehicle.

4. Properly support the exhaust system.

5. Disconnect the oxygen sensor electrical wires. As required, remove the sensor from its mounting.

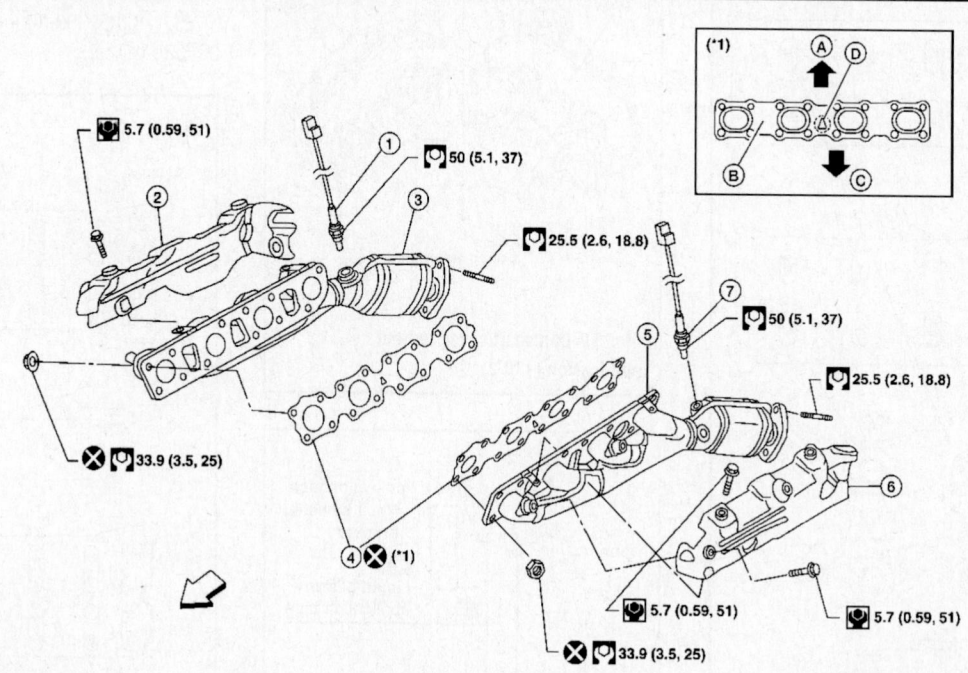

1. Air fuel ratio A/F sensor 1 (bank 2)
2. Exhaust manifold cover (bank 2)
3. Exhaust manifold (bank 2)
4. Gaskets
5. Exhaust manifold (bank 1)
6. Exhaust manifold cover (bank
7. Air fuel ratio A/F sensor 1 (bank 1)
A. Up
B. Coated face
C. Manifold side
D. Up mark
⇦ Front

37663_QX56_G0050

Fig. 55 Catalytic converters and related components

➡️**Be careful not to drop or damage the sensor. If a sensor has been dropped, it must be replaced.**

6. Remove the converter retaining bolts and nuts.

7. Remove the component from the vehicle.

8. Discard the gaskets.

To install:

➡️**Be sure to use new fasteners, as required.**

9. Installation is the reverse of the removal procedure.

10. Be sure to perform the reconnect/relearn procedures.

CRANKSHAFT DAMPER

REMOVAL & INSTALLATION

At this time the manufacturer does not provide removal and installation procedures for this component. The following procedure is a guideline and may differ from the vehicle you are servicing.

1. Before servicing the vehicle, refer to the Precautions Section.

➡️**If working near and/or around the SRS system and components, be sure to disable the SRS system. After disabling the system wait three minutes or more before servicing the vehicle.**

➡️**Whenever the negative battery cable is disconnected the following components will require resetting. The Idle Air Volume Learning, Steering Angle Sensor Neutral Position, Sunroof Memory Reset/Initialization, Automatic Drive Positioner System, Audio presets and Navigation. Use the CONSULT-III diagnostic tool, or equivalent to perform the required resets.**

2. Disconnect the negative battery cable.

3. Remove the engine cover, engine undercover and air cleaner assembly, as required for access.

4. Remove the drive belt.

5. Remove the necessary components to gain access to the crankshaft damper.

6. Remove the crankshaft pulley using suitable tool.

7. Set the bolts in the two bolt holes 0.04 inch (M6 x 1.0 mm) on the front surface.

8. Remove the crankshaft pulley from the crankshaft using tool.

To install:

9. Install the crankshaft damper pulley.

10. Tighten the crankshaft pulley bolt as follows:

- Step 1: 69 ft. lbs. (93 Nm)
- Step 2: Additional 90° (angle tightening)

11. Be sure to perform the reconnect/relearn procedures.

CRANKSHAFT FRONT SEAL

REMOVAL & INSTALLATION

See Figure 56.

1. Before servicing the vehicle, refer to the Precautions Section.

➡️**If working near and/or around the SRS system and components, be sure to disable the SRS system. After disabling the system wait three minutes or more before servicing the vehicle.**

➡️**Whenever the negative battery cable is disconnected the following components will require resetting. The Idle Air Volume Learning, Steering Angle Sensor Neutral Position, Sunroof Memory Reset/Initialization, Automatic Drive Positioner System, Audio presets and Navigation. Use the CONSULT-III diagnostic tool, or equivalent to perform the required resets.**

2. Disconnect the negative battery cable.

3. Remove the engine cover, engine undercover and air cleaner assembly, as required for access.

4. Remove the drive belt.

5. Remove the radiator.

6. Remove the necessary components to gain access to the crankshaft damper.

7. Remove the crankshaft pulley.

8. Remove the oil seal using a suitable tool.

To install:

9. Apply new engine oil to both the oil seal lip and dust seal lip of the new front oil seal.

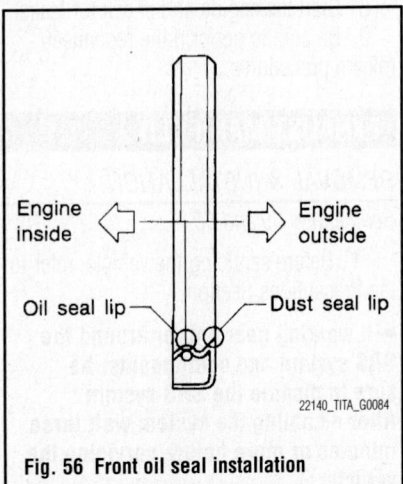

Fig. 56 Front oil seal installation

10. Install the front oil seal so that each seal lip is oriented as shown.

11. Install the crankshaft damper pulley.

12. Tighten the crankshaft pulley bolt as follows:

- Step 1: 69 ft. lbs. (93 Nm)
- Step 2: Additional 90° (angle tightening)

13. Be sure to fill the cooling system with the proper grade and type engine coolant.

14. Be sure to perform the reconnect/relearn procedures.

CYLINDER HEAD

REMOVAL & INSTALLATION

See Figures 57 and 58.

➡️**The engine must be removed from the vehicle to perform this procedure. Be sure that the engine is secured in a suitable holding fixture before performing this procedure.**

1. Before servicing the vehicle, refer to the Precautions Section.

➡️**If working near and/or around the SRS system and components, be sure to disable the SRS system. After disabling the system wait three minutes or more before servicing the vehicle.**

➡️**Whenever the negative battery cable is disconnected the following components will require resetting. The Idle Air Volume Learning, Steering Angle Sensor Neutral Position, Sunroof Memory Reset/Initialization, Automatic Drive Positioner System, Audio presets and Navigation. Use the CONSULT-III diagnostic tool, or equivalent to perform the required resets.**

2. Disconnect the negative battery cable.

3. Remove the engine room cover and air cleaner assembly.

4. Remove or disconnect the following:

- Engine assembly
- Belt tensioner
- Idler pulley
- Thermostat housing and hose
- Oil pan and strainer
- Fuel tube and injector assembly
- Intake manifold
- Ignition coil
- Rocker cover
- Crankshaft pulley
- Front engine cover
- Oil pump
- Timing chain
- Camshaft sprockets
- Camshafts

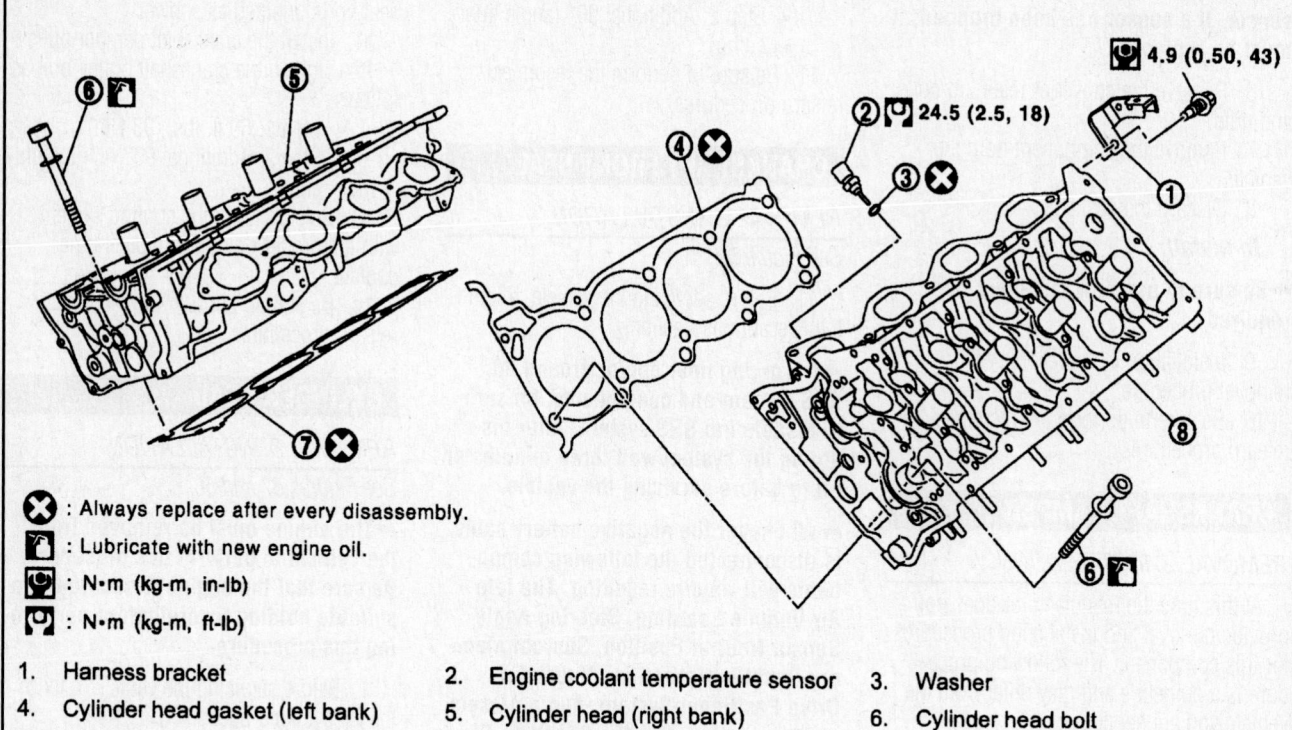

🅧 : Always replace after every disassembly.

🅰 : Lubricate with new engine oil.

🅞 : N·m (kg-m, in-lb)

🅟 : N·m (kg-m, ft-lb)

1. Harness bracket
4. Cylinder head gasket (left bank)
7. Cylinder head gasket (right bank)

2. Engine coolant temperature sensor
5. Cylinder head (right bank)
8. Cylinder head (left bank)

3. Washer
6. Cylinder head bolt

67170ARMADAG28

Fig. 57 Cylinder heads and related components

• Cylinder head, removing bolts in reverse order of installation sequence

To install:

➡ **Be sure to use new fasteners, as required.**

5. Install the cylinder head with a new gasket.

6. Tighten the bolts in sequence to specification.

7. Install or connect the following:
   • Camshaft
   • Camshaft sprockets

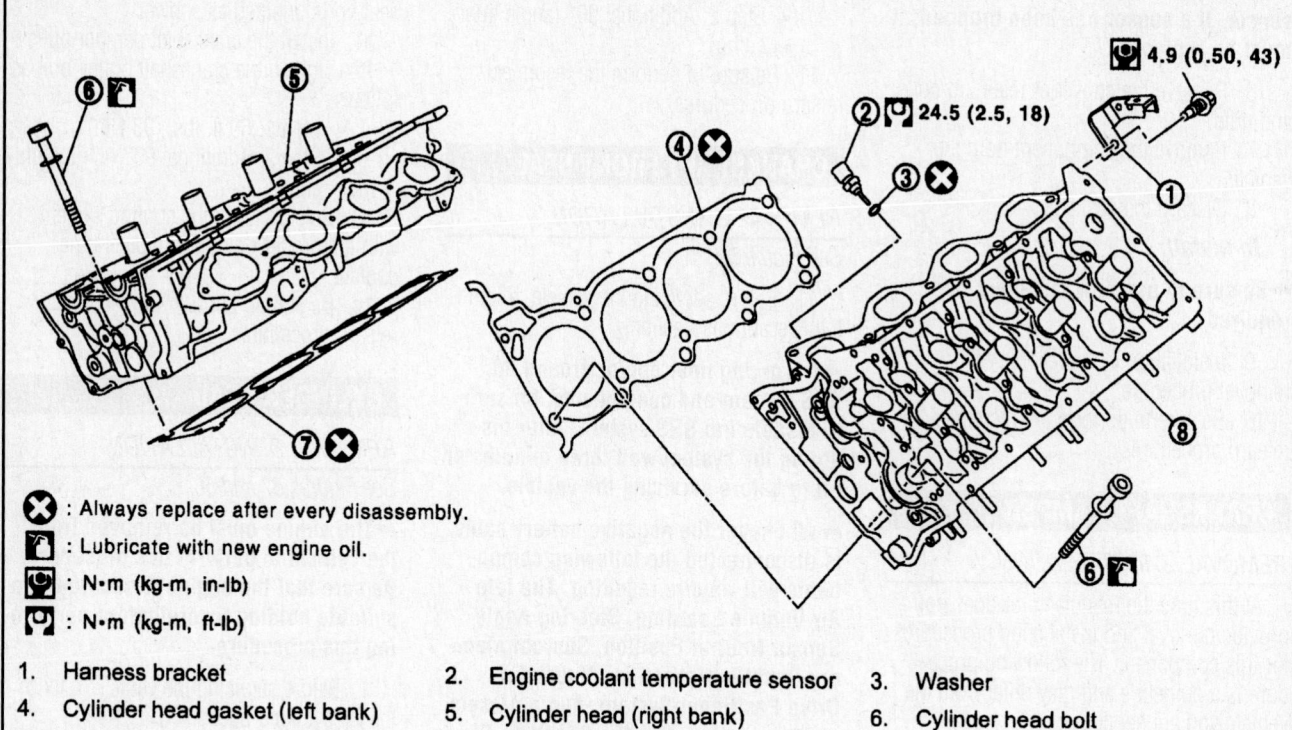

Right bank ⑦ ⑤ ② ④ ⑩
         ⑨ ③ ① ⑥ ⑧

◀ Engine front

Left bank ⑨ ③ ① ⑥ ⑧
         ⑦ ⑤ ② ④ ⑩

67170ARMADAG01

Fig. 58 Cylinder head bolt torque sequence

• Timing chain
• Oil pump
• Front engine cover
• Crankshaft pulley
• Rocker cover
• Ignition coil
• Intake manifold
• Fuel tube and injector assembly
• Oil pain and strainer
• Thermostat housing and hose
• Idler pulley
• Belt tensioner
• Engine assembly

8. Start the engine and check for leaks.

9. Be sure to perform the reconnect/relearn procedures.

EXHAUST MANIFOLD

REMOVAL & INSTALLATION

See Figures 59 and 60.

1. Before servicing the vehicle, refer to the Precautions Section.

➡ **If working near and/or around the SRS system and components, be sure to disable the SRS system. After disabling the system wait three minutes or more before servicing the vehicle.**

➡ **Whenever the negative battery cable is disconnected the following components will require resetting. The Idle Air Volume Learning, Steering Angle Sensor Neutral Position, Sunroof Memory Reset/Initialization, Automatic Drive Positioner System, Audio presets and Navigation. Use the CONSULT-III diagnostic tool, or equivalent to perform the required resets.**

2. Disconnect the negative battery cable.

3. Raise and support the vehicle safely.

4. Remove the engine under cover, if equipped.

5. Remove the front final drive assembly, if equipped.

6. Remove the main muffler and center exhaust tube.

7. Remove the front exhaust tubes.

8. Remove the tire and wheel assemblies.

9. Remove the fender protectors.

10. Remove the A/F sensors. Do not drop the sensors. If the sensor is dropped it must be replaced.

11. Properly support the engine.

12. Remove the engine mounting insulator.

13. Remove the exhaust manifold cover.

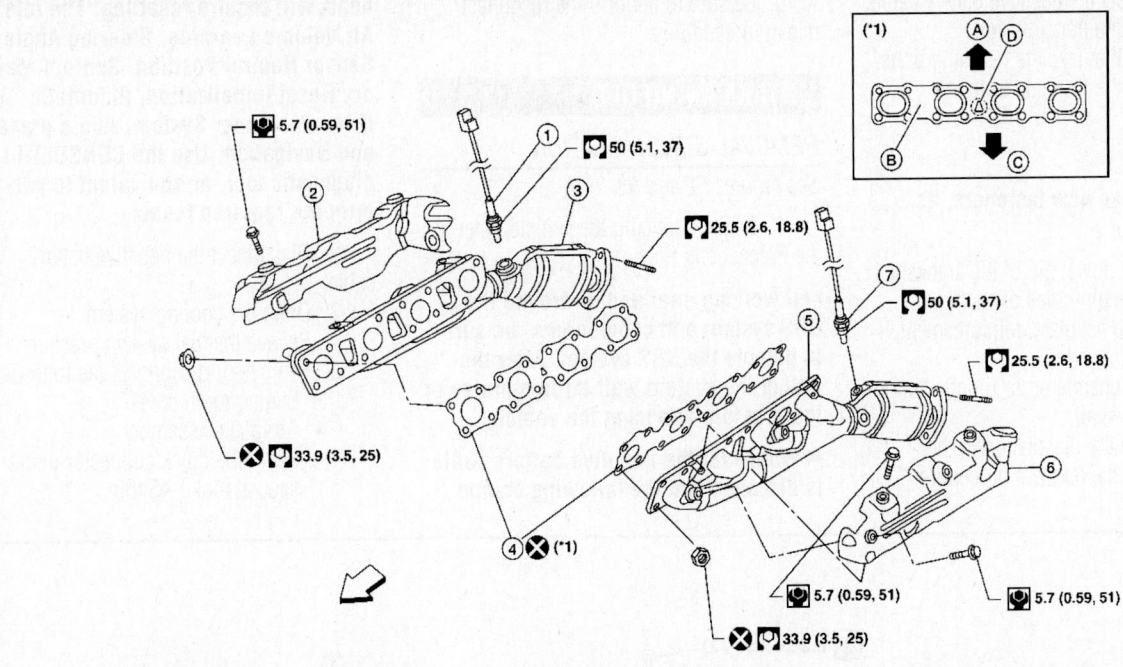

1. Air fuel ratio A/F sensor 1 (bank 2)
2. Exhaust manifold cover (bank 2)
3. Exhaust manifold (bank 2)
4. Gaskets
5. Exhaust manifold (bank 1)
6. Exhaust manifold cover (bank
7. Air fuel ratio A/F sensor 1 (bank 1)
A. Up
B. Coated face
C. Manifold side
D. Up mark
⟨▢ Front

37663_QX56_G0050

Fig. 59 Exhaust manifolds and related components

14. Remove the engine mounting bracket.

15. On right side remove the oil level dipstick.

16. Remove the left exhaust manifold nuts/bolts in the reverse order of the installation sequence.

17. Remove the exhaust manifold. Discard the gaskets.

To install:

➡**Be sure to use new fasteners, as required.**

18. Installation is the reverse of the removal procedure.

19. Install new gaskets with the top of the triangular UP mark on it facing up and its coated face (gray side) toward the exhaust manifold side.

20. Tighten the retaining nuts/bolts to specification and in the proper sequence.

21. Be sure to perform the reconnect/relearn procedures.

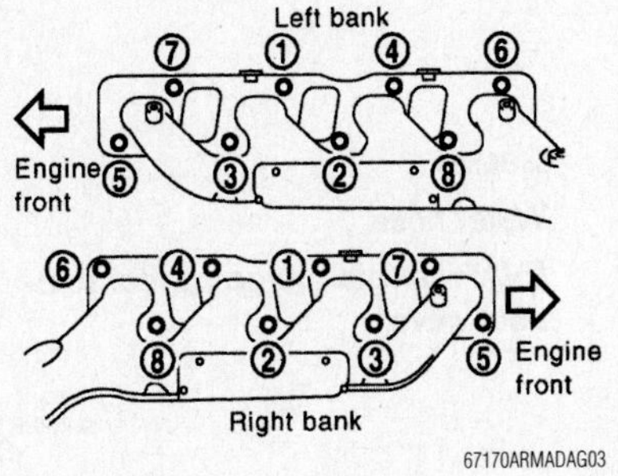

FLEXPLATE

REMOVAL & INSTALLATION

See Figure 61.

1. Before servicing the vehicle, refer to the Precautions Section.

➡**If working near and/or around the SRS system and components, be sure to disable the SRS system. After disabling the system wait three minutes or more before servicing the vehicle.**

➡**Whenever the negative battery cable is disconnected the following components will require resetting. The Idle Air Volume Learning, Steering Angle Sensor Neutral Position, Sunroof Memory Reset/Initialization, Automatic Drive Positioner System, Audio presets and Navigation. Use the CONSULT-III diagnostic tool, or equivalent to perform the required resets.**

67170ARMADAG03

Fig. 60 Exhaust manifold bolt torque sequence

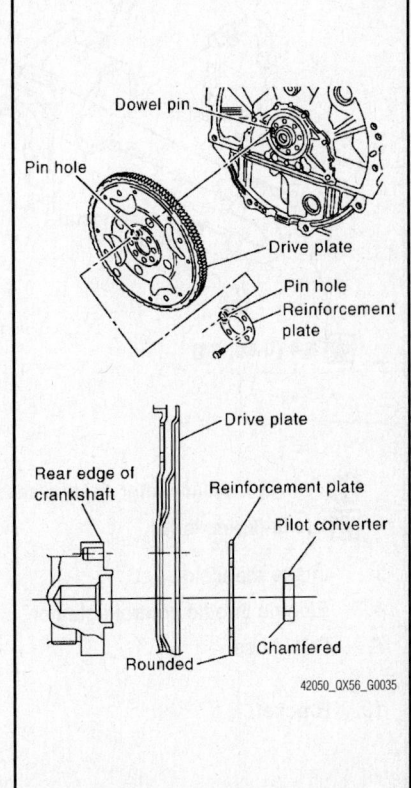

Fig. 61 Flexplate alignment and related components

2. Disconnect the negative battery cable.

3. Remove the transmission.

4. Remove the flexplate retaining bolts.

5. Remove the flexplate from the engine.

To install:

➡**Be sure to use new fasteners, as required.**

6. Align the dowel pin of the crankshaft rear end with the pin holes of each part.

7. Install the flexplate, reinforcement plate and pilot converter.

8. Face the chamfered or rounded edge side to the crankshaft.

9. Continue the installation in the reverse order of the removal procedure.

10. Be sure to perform the reconnect/relearn procedures.

INTAKE MANIFOLD

REMOVAL & INSTALLATION

See Figures 62 and 63.

1. Before servicing the vehicle, refer to the Precautions Section.

➡**If working near and/or around the SRS system and components, be sure to disable the SRS system. After disabling the system wait three minutes or more before servicing the vehicle.**

➡**Whenever the negative battery cable is disconnected the following compo-nents will require resetting. The Idle Air Volume Learning, Steering Angle Sensor Neutral Position, Sunroof Memory Reset/Initialization, Automatic Drive Positioner System, Audio presets and Navigation. Use the CONSULT-III diagnostic tool, or equivalent to perform the required resets.**

2. Disconnect the negative battery cable.

3. Drain the cooling system.

4. Relieve the fuel system pressure.

5. Remove or disconnect the following:
   - Engine room cover
   - Air intake assembly
   - Fuel tube quick connector using special tool J-45488

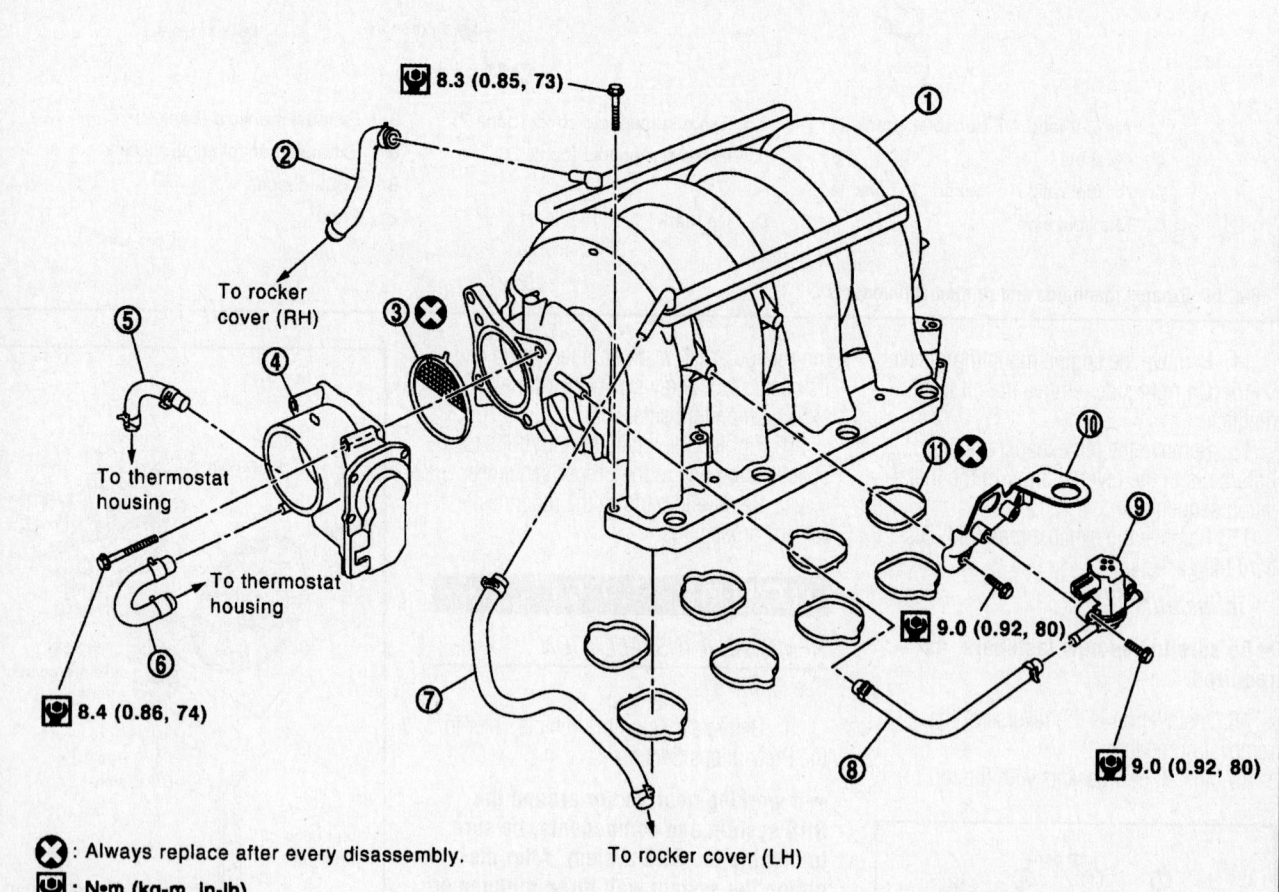

❌ : Always replace after every disassembly.

⚙ : N•m (kg-m, in-lb)

1. Intake manifold	2. PCV hose	3. Gasket
4. Electric throttle control actuator	5. Water hose	6. Water hose
7. PCV hose	8. EVAP hose	9. EVAP canister purge control solenoid valve
10. Bracket	11. Gasket	

67170ARMADAG29

Fig. 62 Intake manifold and related components

- Wiring harnesses and brackets from manifold
- Vacuum hoses
- PCV hose and tube
- Electric throttle control actuator, loosening bolts diagonally
- Fuel injectors
- Fuel tube assembly
- Intake manifold, removing bolts in reverse order of installation

To install:

→Be sure to use new fasteners, as required.

6. Install the intake manifold with new gaskets. Tighten the bolts in order as shown.

7. Install or connect the following:
- Fuel tube assembly
- Fuel injectors
- Electronic throttle control actuator, tightening the bolts in several steps
- PCV hose
- Vacuum hoses
- Wiring harnesses

8. Connect the fuel tube as follows:

a. Apply a thin layer of engine oil on the tube from tip end to spool end.

b. Insert tube into quick connector past the white identification mark.

c. Insert tube into quick connector until top spool is completely inside the connector and 2nd level spool is exposed right below the connector.

d. Pull slightly on the quick connector to ensure it is fully engaged.

e. Install quick connector cap on quick connector joint.

9. Install or connect the following:
- Air intake assembly
- Engine cover

10. Refill the cooling system.

11. Start engine and check for leaks.

12. Be sure to perform the reconnect/relearn procedures.

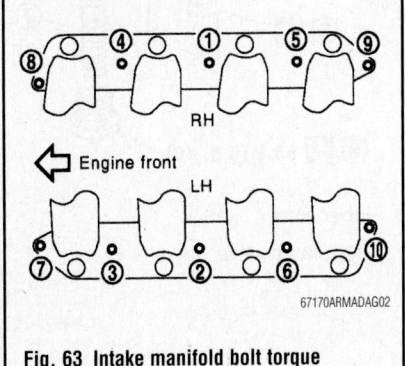

Fig. 63 Intake manifold bolt torque sequence

OIL PAN

REMOVAL & INSTALLATION

See Figures 64 through 68.

→The engine must be removed from the vehicle to perform this procedure. Be sure that the engine is secured in a suitable holding fixture before performing this procedure.

1. Before servicing the vehicle, refer to the Precautions Section.

→If working near and/or around the SRS system and components, be sure to disable the SRS system. After disabling the system wait three minutes or more before servicing the vehicle.

→Whenever the negative battery cable is disconnected the following components will require resetting. The Idle Air Volume Learning, Steering Angle Sensor Neutral Position, Sunroof Memory Reset/Initialization, Automatic Drive Positioner System, Audio presets and Navigation. Use the CONSULT-III diagnostic tool, or equivalent to perform the required resets.

2. Disconnect the negative battery cable.

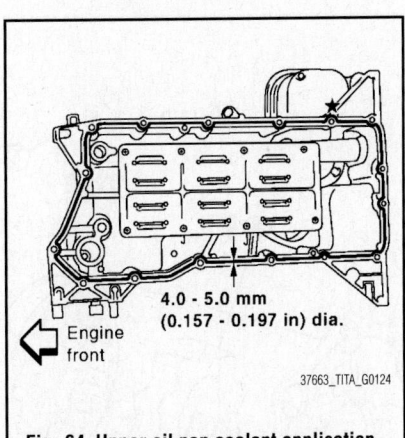

Fig. 64 Upper oil pan sealant application

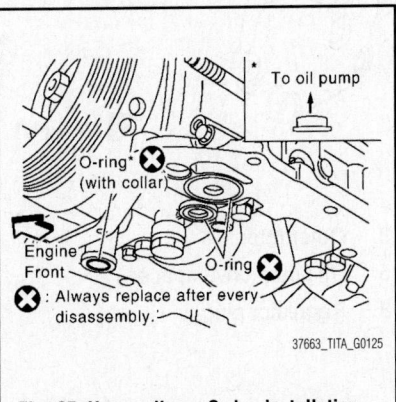

Fig. 65 Upper oil pan O-ring installation

3. Remove engine assembly and position it in a suitable holding fixture.

4. Remove lower oil pan, loosening bolts in reverse order of the installation sequence.

5. Remove oil strainer from upper oil pan.

6. Gently pry and remove upper oil pan from engine block.

To install:

→Be sure to use new fasteners, as required.

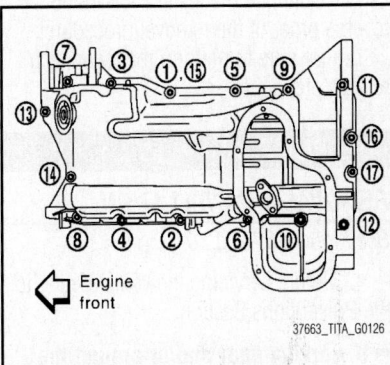

Fig. 66 Upper oil pan bolt tightening sequence

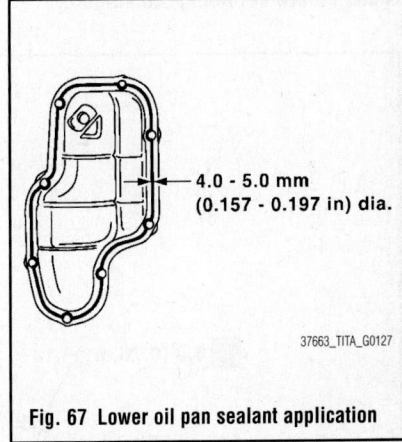

Fig. 67 Lower oil pan sealant application

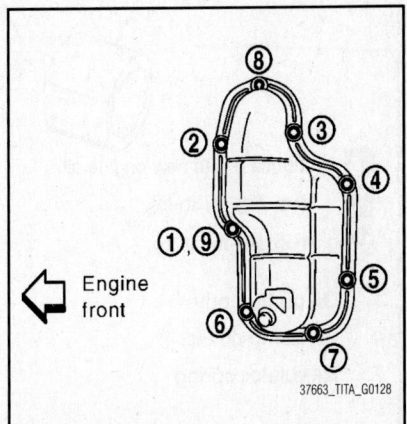

Fig. 68 Lower oil pan bolt tightening sequence

7. Apply liquid gasket to upper oil pan mating surfaces.

8. Install new O-rings to oil pump and front cover side.

9. Tighten upper oil pan bolts to specification and in the proper sequence.

10. Install or connect the following:
- Rear plate cover
- Oil strainer to upper oil pan
- Lower oil pan, tightening bolts in proper sequence and to specification

11. Continue the installation in the reverse order of the removal procedure.

12. Be sure to perform the reconnect/relearn procedures.

OIL PUMP

REMOVAL & INSTALLATION

See Figures 69 and 70.

1. Before servicing the vehicle, refer to the Precautions Section.

➡If working near and/or around the SRS system and components, be sure to disable the SRS system. After disabling the system wait three minutes or more before servicing the vehicle.

➡Whenever the negative battery cable is disconnected the following components will require resetting. The Idle Air Volume Learning, Steering Angle Sensor Neutral Position, Sunroof Memory Reset/Initialization, Automatic Drive Positioner System, Audio presets and Navigation. Use the CONSULT-III diagnostic tool, or equivalent to perform the required resets.

2. Disconnect the negative battery cable.

3. Remove or disconnect the following:
- Timing chain cover
- Oil pump drive spacer
- Oil pump

To install:

➡Be sure to use new fasteners, as required.

4. Install or connect the following:
- Oil pump
- Oil pump drive spacer
- Timing chain cover

➡When inserting the oil pump drive spacer, align the crankshaft key and the flat face of the inner rotor. If they are not aligned rotate the pump inner rotor by hand. Make sure that each part is

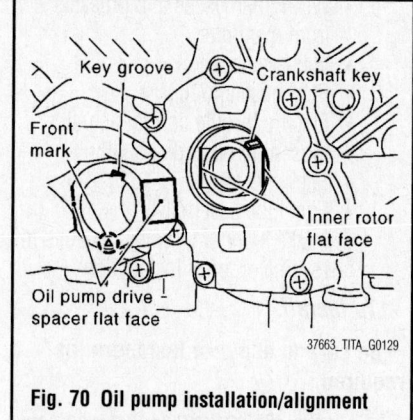

Fig. 70 Oil pump installation/alignment

aligned and tap lightly until it reaches the end.

5. Be sure to perform the reconnect/relearn procedures.

INSPECTION

See Figures 71 through 76.

1. Before servicing the vehicle, refer to the Precautions Section.

➡If working near and/or around the SRS system and components, be sure to disable the SRS system. After dis-

---

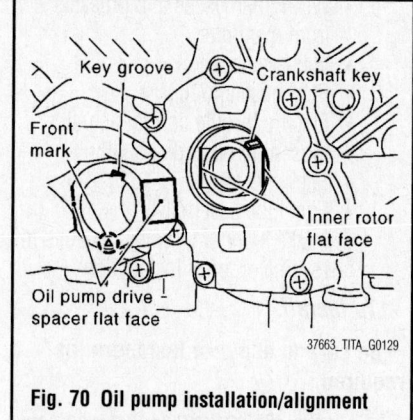

🔧 11.0 (1.1, 8)

🔧 6.9 (0.70, 61)

🔧 11.0 (1.1, 8)

⑧ 🔧 53.9 (5.5, 40)

🛢️ : Lubricate with new engine oil.

🔧 : N·m (kg-m, in-lb)

🔧 : N·m (kg-m, ft-lb)

1. Oil pump body
2. Outer rotor
3. Inner rotor
4. Oil pump cover
5. Oil pump drive spacer
6. Regulator valve
7. Regulator spring
8. Regulator plug

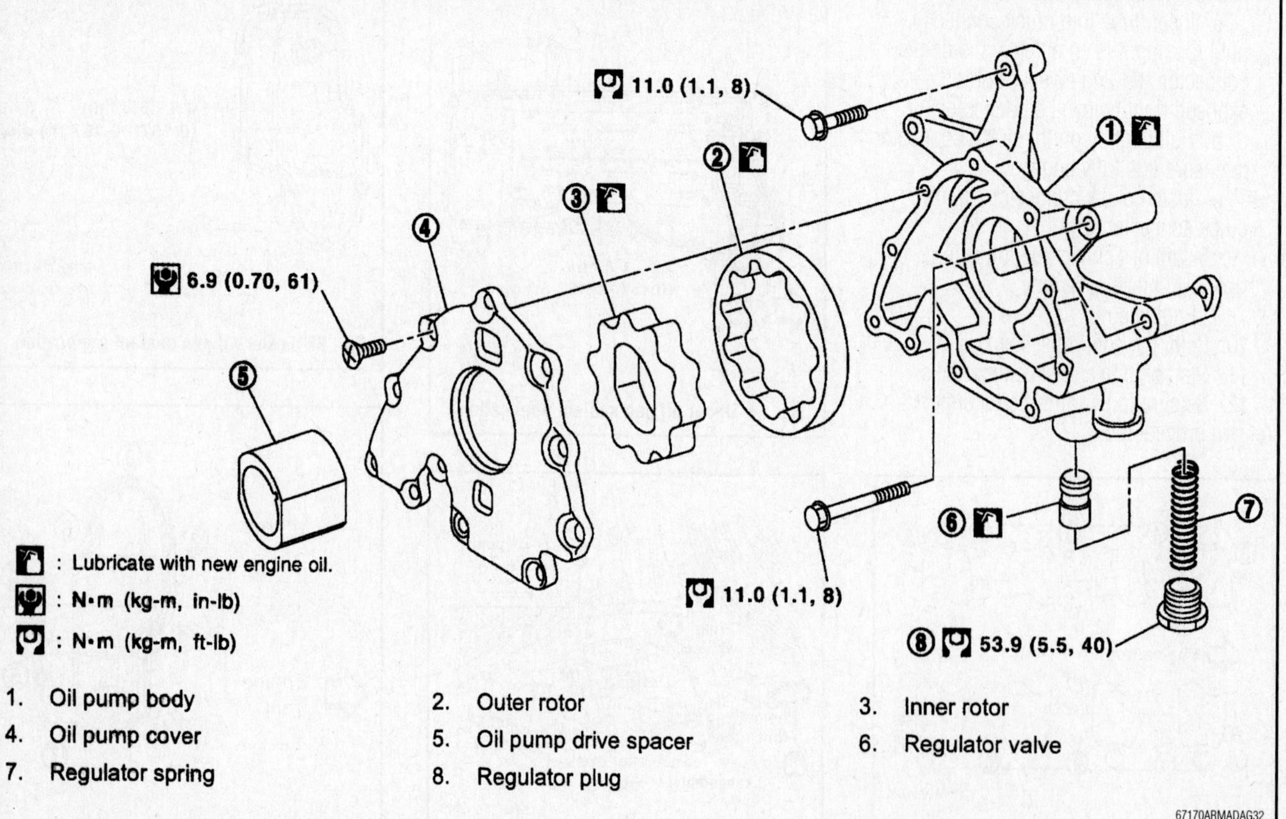

Fig. 69 Oil pump and related components

abling the system wait three minutes or more before servicing the vehicle.

➡ Whenever the negative battery cable is disconnected the following components will require resetting. The Idle Air Volume Learning, Steering Angle Sensor Neutral Position, Sunroof Memory Reset/Initialization, Automatic Drive Positioner System, Audio presets and Navigation. Use the CONSULT-III diagnostic tool, or equivalent to perform the required resets.

2. Disconnect the negative battery cable.

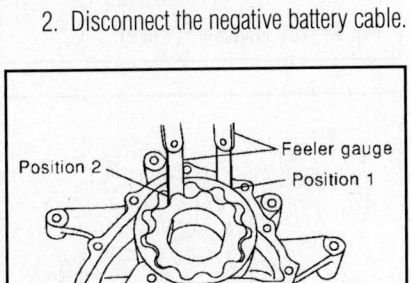

Fig. 71 Oil pump radial clearance measurement

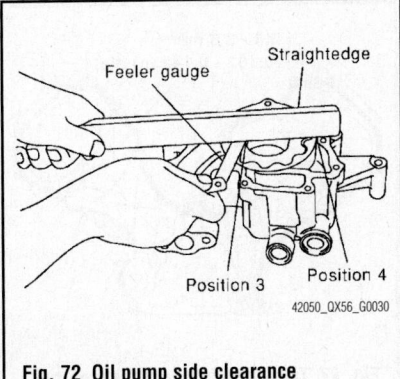

Fig. 72 Oil pump side clearance measurement

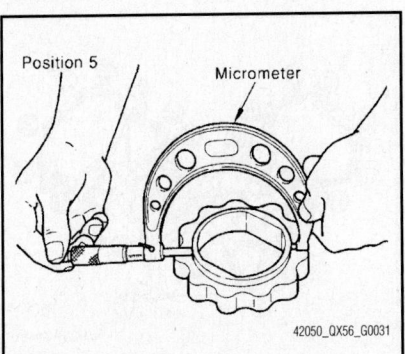

Fig. 73 Oil pump inner rotor and pump body clearance measurement

3. Remove the oil pump from the engine.
4. Remove the oil pump cover.
5. Remove the inner and outer rotors from the oil pump body.
6. Remove the regulator plug, regulator spring and regulator valve.
7. Measure the radial clearance using a feeler gauge. Body to outer rotor (position 1) should be 0.0045–0.0079 in. Inner rotor to outer tip (position 2) should be 0.0071 in.
8. Measure the side clearance using a feeler gauge. Body to inner rotor (position 3)

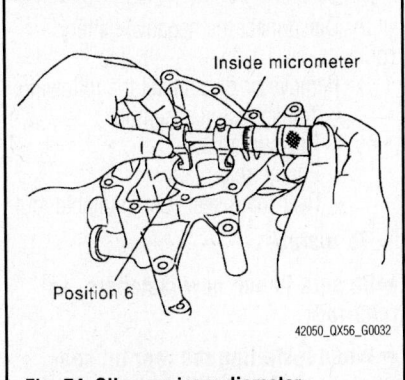

Fig. 74 Oil pump inner diameter clearance measurement

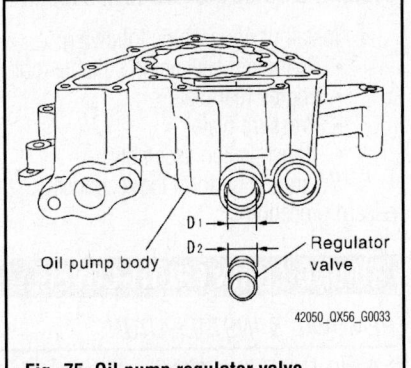

Fig. 75 Oil pump regulator valve clearance measurement

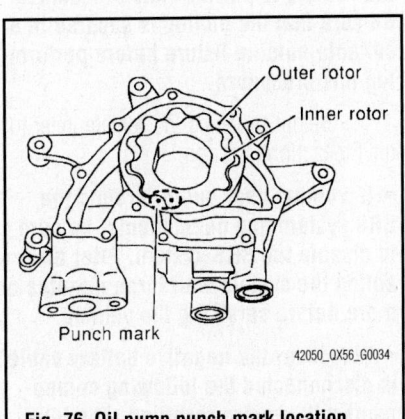

Fig. 76 Oil pump punch mark location

should be 0.0012–0.0028 in. Body to outer rotor (position 4) should be 0.0012–0.0035 in.

9. Calculate the clearance between the inner rotor and the oil pump body as follows:

a. Measure the outer diameter of the protruded portion of the inner rotor (position 5) using a feeler gauge. Measure the inner diameter of the oil pump body to the brazed portion (position 6) using a feeler gauge. Calculate the clearance using the following formula. Clearance=Inner diameter of oil pump body, minus outer diameter of inner rotor. Inner rotor to brazed portion of housing clearance specification is 0.0018–0.0036 in.

10. Check the regulator valve to oil pump cover clearance using the following formula: Clearance=D1 (valve hole diameter) minus D2 (outer diameter of valve). Regulator valve to oil pump specification should be 0.0016–0.0038 in.

➡ Coat the valve with clean engine oil. Check that it falls smoothly into the regulator valve hole by its own weight.

11. Assemble the oil pump in the reverse order.
12. Install the inner and outer rotor with the punched marks on the oil pump cover side.

To install:

➡ Be sure to use new fasteners, as required.

13. Installation is the reverse of the removal procedure.
14. Be sure to perform the reconnect/relearn procedures.

PISTON AND RING

POSITIONING

See Figures 77 and 78.

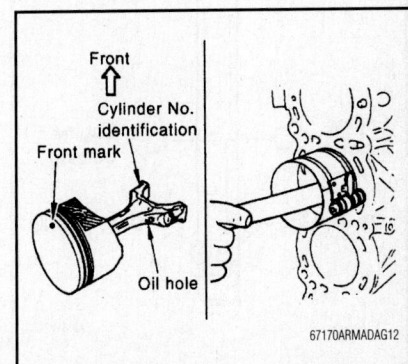

Fig. 77 Piston and rod positioning and identification

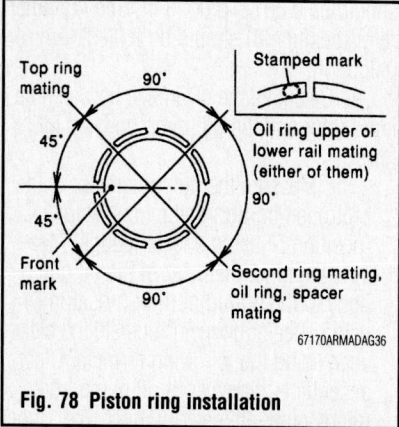

Fig. 78 Piston ring installation

REAR MAIN SEAL

REMOVAL & INSTALLATION

See Figures 79 and 80.

At this time the manufacturer does not provide removal and installation procedures for this component. The following procedure is a guideline and may differ from the vehicle you are servicing.

1. Before servicing the vehicle, refer to the Precautions Section.

➡If working near and/or around the SRS system and components, be sure to disable the SRS system. After disabling the system wait three minutes or more before servicing the vehicle.

➡Whenever the negative battery cable is disconnected the following components will require resetting. The Idle Air Volume Learning, Steering Angle Sensor Neutral Position, Sunroof Memory Reset/Initialization, Automatic Drive Positioner System, Audio presets and Navigation. Use the CONSULT-III diagnostic tool, or equivalent to perform the required resets.

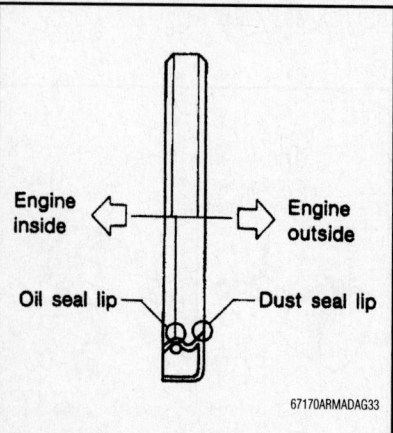

Fig. 79 Rear main seal installation positioning

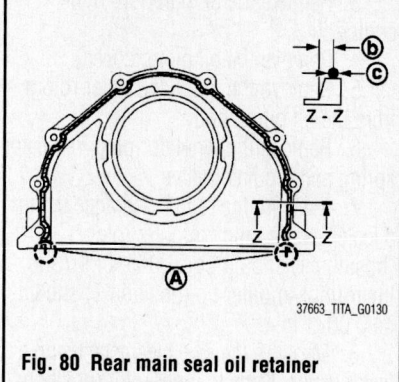

Fig. 80 Rear main seal oil retainer sealant application

2. Disconnect the negative battery cable.
3. Remove or disconnect the following:
   - Transmission assembly
   - Pressure plate
   - Engine rear plate
   - Rear main seal using suitable tool

To install:

➡Be sure to use new fasteners, as required.

➡When installing the rear oil seal retainer apply a continuous bead of sealant, as shown in the illustration. A= protrusion. B= 0.157–0.220 inch sealant. C= 0.134–0.173 inch sealant.

4. Install or connect the following:
   - Rear main seal using suitable tool
   - Engine rear plate
   - Pressure plate
   - Transmission assembly
5. Be sure to perform the reconnect/relearn procedures.

TIMING CHAIN FRONT COVER

REMOVAL & INSTALLATION

See Figures 81 through 86.

➡The engine must be removed from the vehicle to perform this procedure. Be sure that the engine is secured in a suitable holding fixture before performing this procedure.

1. Before servicing the vehicle, refer to the Precautions Section.

➡If working near and/or around the SRS system and components, be sure to disable the SRS system. After disabling the system wait three minutes or more before servicing the vehicle.

➡Whenever the negative battery cable is disconnected the following components will require resetting. The Idle

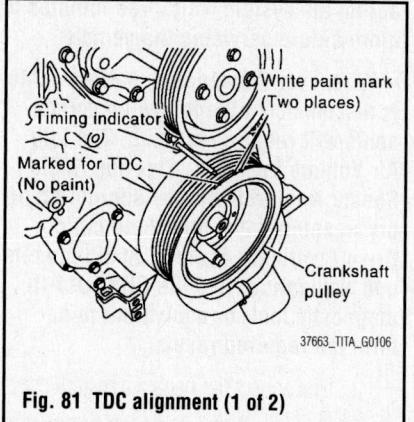

Fig. 81 TDC alignment (1 of 2)

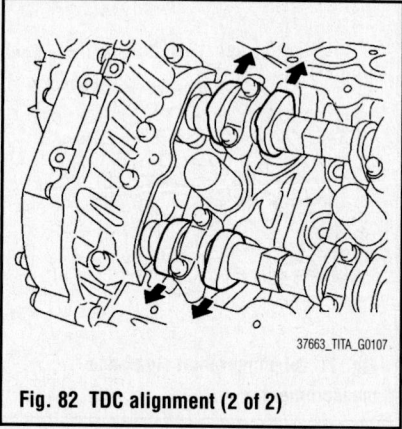

Fig. 82 TDC alignment (2 of 2)

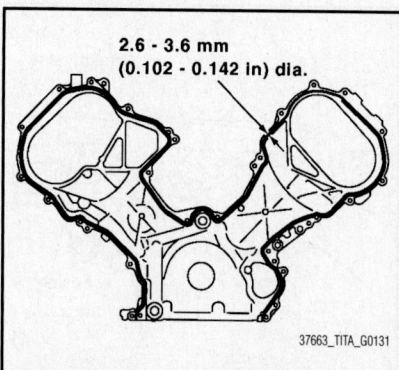

Fig. 83 Timing chain front cover sealant application

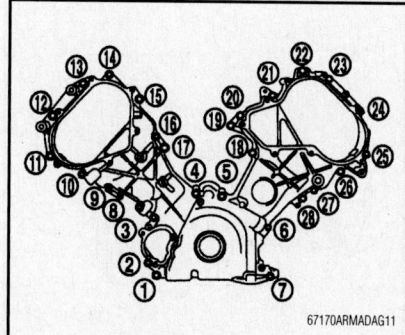

Fig. 84 Timing chain front cover bolt torque sequence

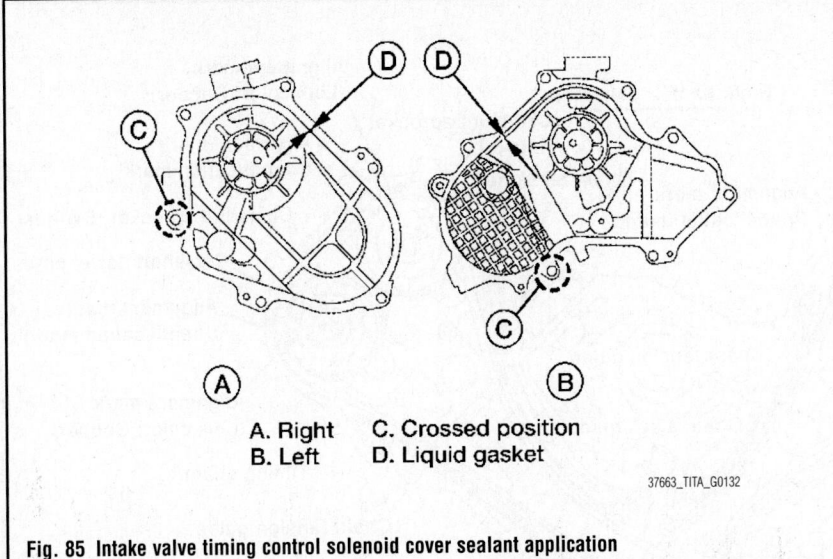

A. Right   C. Crossed position
B. Left   D. Liquid gasket

37663_TITA_G0132

Fig. 85 Intake valve timing control solenoid cover sealant application

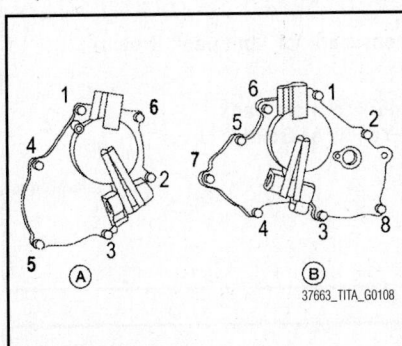

37663_TITA_G0108

**Fig. 86 Intake valve timing control
solenoid cover tightening sequence**

**Air Volume Learning, Steering Angle
Sensor Neutral Position, Sunroof Memory Reset/Initialization, Automatic
Drive Positioner System, Audio presets
and Navigation. Use the CONSULT-III
diagnostic tool, or equivalent to perform the required resets.**

2. Disconnect the negative battery
cable.

3. Remove or disconnect the following:
- Engine assembly
- Drive belt auto tensioner
- Idler pulley
- Thermostat housing and water hose
- Power steering pump bracket
- Oil pan (upper and lower)
- Oil strainer
- Alternator and bracket
- Rocker cover
- Water pump

4. Remove the CMP sensor.

5. Remove the intake valve timing control position sensors (right and left).

6. Remove the intake valve timing control solenoid valves (right and left).

7. Loosen and remove the intake valve
timing control valve cover (right and left)
bolts in the reverse order of the tightening
sequence.

8. Be sure that the number one cylinder
is at TDC on the compression stroke.

➡**Turn the crankshaft pulley clockwise
to align the TDC identification notch
(without paint mark) with the timing
indicator on the front cover. At this time
make sure that the intake and exhaust
cam lobes of the number one cylinder
(top front on left bank) point outside. If
not turn the crankshaft pulley once
more. See illustration**

9. Remove the crankshaft pulley.

10. Remove the front cover retaining
bolts. Remove the front cover. 11.
Discard the gasket.

To install:

➡**Be sure to use new fasteners, as
required.**

12. Installation is the reverse of the
removal procedure.

13. Apply sealant to the front cover. See
illustration.

14. Install retaining bolts and tighten to
specification and in the proper sequence.

15. Apply liquid gasket to the valve timing control solenoid covers.

➡**The start and end of the sealant
application should be crossed at position "C", see illustration, that cannot
be seen after attaching the intake valve
timing control solenoid valve cover.**

16. Install the intake valve timing control
solenoid covers. Tighten bolts in proper
sequence.

17. Continue the installation in the
reverse order of the removal procedure.

18. Be sure to perform the reconnect/
relearn procedures.

TIMING CHAIN & SPROCKETS

REMOVAL & INSTALLATION

See Figure 87.

➡**The engine must be removed from
the vehicle to perform this procedure.
Be sure that the engine is secured in a
suitable holding fixture before performing this procedure.**

1. Before servicing the vehicle, refer to
the Precautions Section.

➡**If working near and/or around the
SRS system and components, be sure
to disable the SRS system. After disabling the system wait three minutes or
more before servicing the vehicle.**

➡**Whenever the negative battery cable
is disconnected the following components will require resetting. The Idle
Air Volume Learning, Steering Angle
Sensor Neutral Position, Sunroof Memory Reset/Initialization, Automatic
Drive Positioner System, Audio presets
and Navigation. Use the CONSULT-III
diagnostic tool, or equivalent to perform the required resets.**

2. Disconnect the negative battery cable.

3. Remove the front cover.

4. Remove the oil pump drive spacer.

5. Remove the oil pump.
- Timing chain tensioner
- Chain tension guide and slack guide
- Timing chain
- Camshaft sprocket

To install:

➡**Be sure to use new fasteners, as
required.**

6. Ensure that the crankshaft key and
dowel pin of each camshaft are facing the
same direction.

7. Install or connect the following:
- Camshaft sprockets
- Timing chain
- Chain tension guide and slack guide
- Oil pump
- Oil pump drive spacer
- Front oil seal, using suitable tool

8. Continue the installation in the
reverse order of the removal procedure.

9. Be sure to perform the reconnect/
relearn procedures.

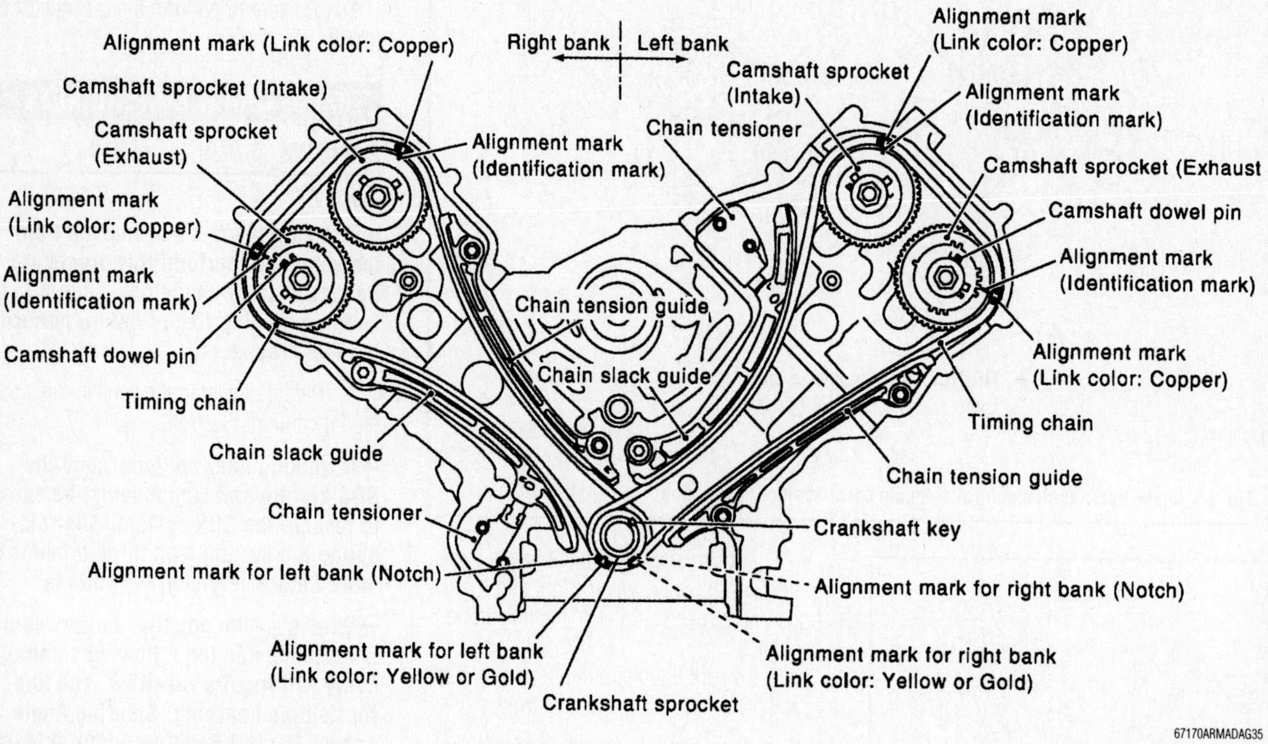

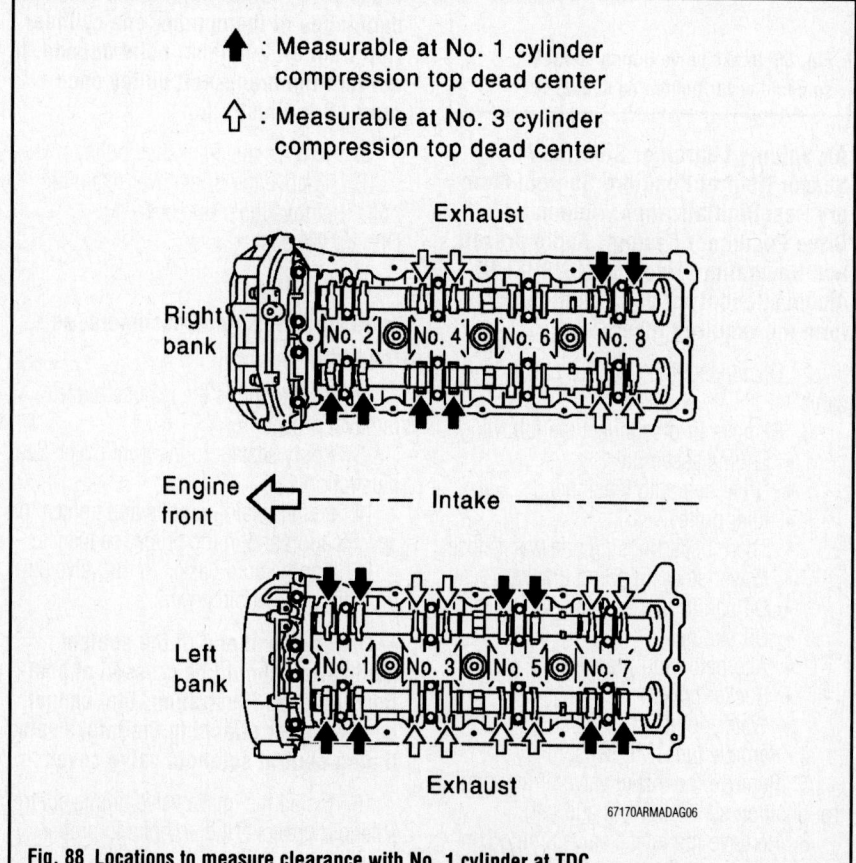

Fig. 87 Timing chains and related components

VALVE LASH

ADJUSTMENT

See Figures 88 and 89.

1. Before servicing the vehicle, refer to the Precautions Section.

➡ **Perform the following inspection after removal, installation or replacement of camshaft or valve-related parts, or if there are unusual engine conditions due to changes in valve clearance over time (starting, idling, and/or noise).**

2. Run engine to operating temperature.

3. Remove or disconnect the following:
   • Battery cover, if equipped
   • Engine room cover
   • Air intake assembly
   • Left and right rocker covers

4. Turn the crankshaft pulley clockwise to Top Dead Center (TDC) identification notch with timing indicator.

5. Ensure that both the intake and exhaust cam noses of the No. 1 cylinder face outside.

6. Measure the valve clearances at locations shown in figure.

Fig. 88 Locations to measure clearance with No. 1 cylinder at TDC

7. Turn the crankshaft pulley clockwise 270 degrees from the position of No. 1 cylinder compression to obtain No. 3 cylinder compression TDC.

8. Measure the valve clearances at locations shown in the figure.

9. Turn crankshaft pulley clockwise 90 degrees and measure the intake and exhaust valve clearance of No. 6 cylinder and exhaust valve clearance of No. 2 cylinder.

10. To adjust the valves, remove camshaft and valve lifter(s) out of specification.

11. Install replacement valve lifter(s).

12. Install the camshaft.

13. Manually turn the crankshaft pulley several turns.

14. Recheck valve clearances with engine at operating temperature.

INSPECTION

1. Before servicing the vehicle, refer to the Precautions Section.

2. Remove camshaft and valve lifter(s) out of specification.

3. Install replacement valve lifter(s).

4. Install the camshaft.

5. Manually turn the crankshaft pulley several turns.

6. Recheck valve clearances with engine at operating temperature.

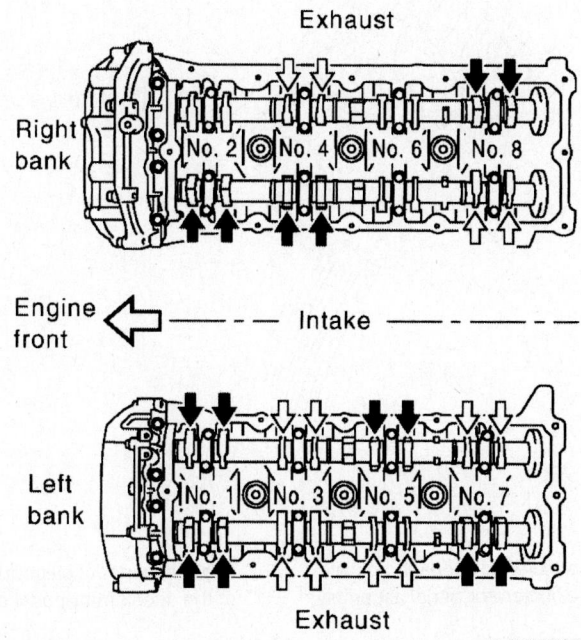

: Measurable at No. 1 cylinder compression top dead center

: Measurable at No. 3 cylinder compression top dead center

Fig. 89 Locations to measure clearance with No. 3 cylinder at TDC

ENGINE PERFORMANCE & EMISSION CONTROLS

ACCELERATOR PEDAL POSITION (APP) SENSOR

LOCATION

The Accelerator Pedal Position (APP) sensor is located on the upper end of the accelerator pedal assembly.

REMOVAL & INSTALLATION

See Figures 90 and 91.

1. Before servicing the vehicle, refer to the Precautions Section.

➡ If working near and/or around the SRS system and components, be sure to disable the SRS system. After disabling the system wait three minutes or more before servicing the vehicle.

➡ Whenever the negative battery cable is disconnected the following components will require resetting. The Idle Air Volume Learning, Steering Angle Sensor Neutral Position, Sunroof Memory Reset/Initialization, Automatic Drive Positioner System, Audio presets and Navigation. Use the CONSULT-III diagnostic tool, or equivalent to perform the required resets.

2. Disconnect the negative battery cable.

❉❉ CAUTION

Do not disassemble the accelerator pedal adjusting mechanism. Before removal and installation, the accelerator and brake pedals must be in the front most position. This is to align the base position of the accelerator and brake pedals. Do not disassemble the accelerator pedal assembly. Do not remove the Accelerator Pedal Position (APP) sensor from the accelerator pedal bracket. Avoid damage from dropping the accelerator pedal assembly during handling. Keep the accelerator pedal assembly away from water.

3. Move the accelerator and brake pedals to the front most position.

4. Turn the ignition switch **OFF** and disconnect the negative battery terminal.

5. Disconnect the adjustable brake pedal cable from the adjustable brake pedal. Unlock, then pull the adjustable brake pedal cable to disconnect it from the adjustable brake pedal.

6. Disconnect the adjustable pedal electric motor electrical connector.

7. Disconnect the adjustable pedal electric motor memory electrical connector, if equipped.

8. Disconnect APP sensor electrical connector, if equipped.

9. Disconnect the APP sensor electrical connector.

10. Remove the adjustable accelerator pedal assembly.

To install:

➡ Be sure to use new fasteners, as required.

11. Installation is the reverse of the removal procedure.

12. Be sure to perform the reconnect/relearn procedures.

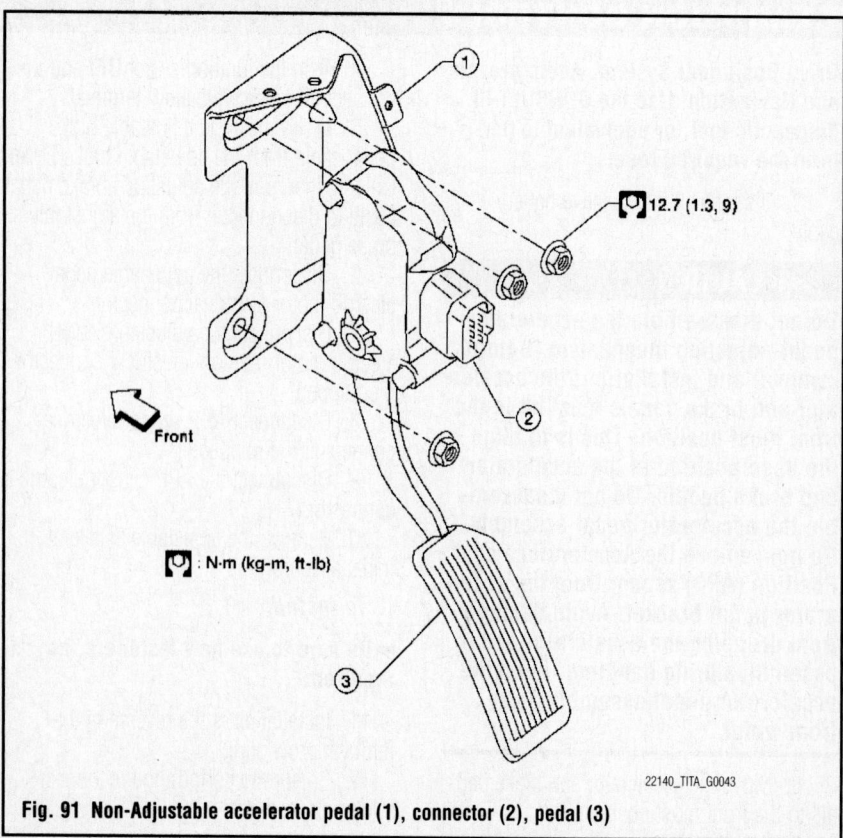

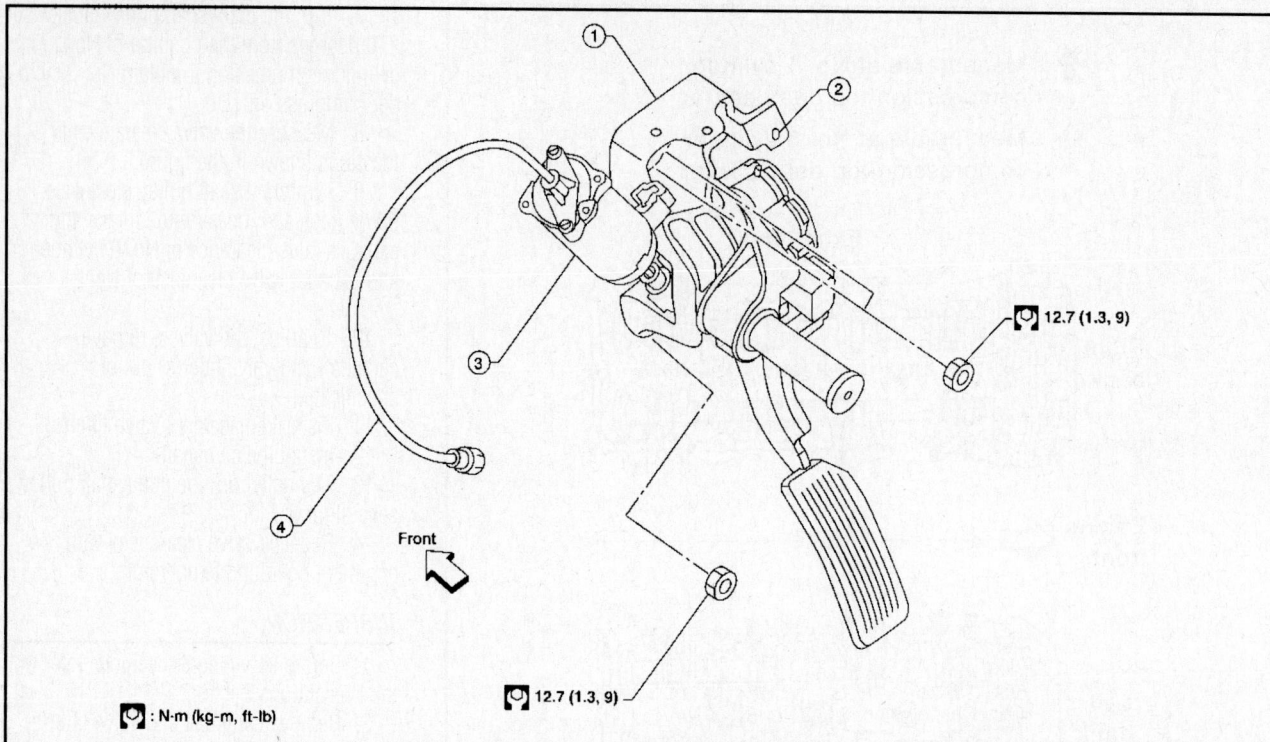

12.7 (1.3, 9)

12.7 (1.3, 9)

Front

: N·m (kg-m, ft-lb)

1. Adjustable accelerator pedal assembly
2. Adjustable accelerator pedal bracket (part of the accelerator pedal assembly)
3. Adjustable pedal electric motor (part of the accelerator pedal assembly)
4. Adjustable brake pedal cable (part of the accelerator pedal assembly)

22140_TITA_G0042

Fig. 90 Adjustable accelerator pedal assembly

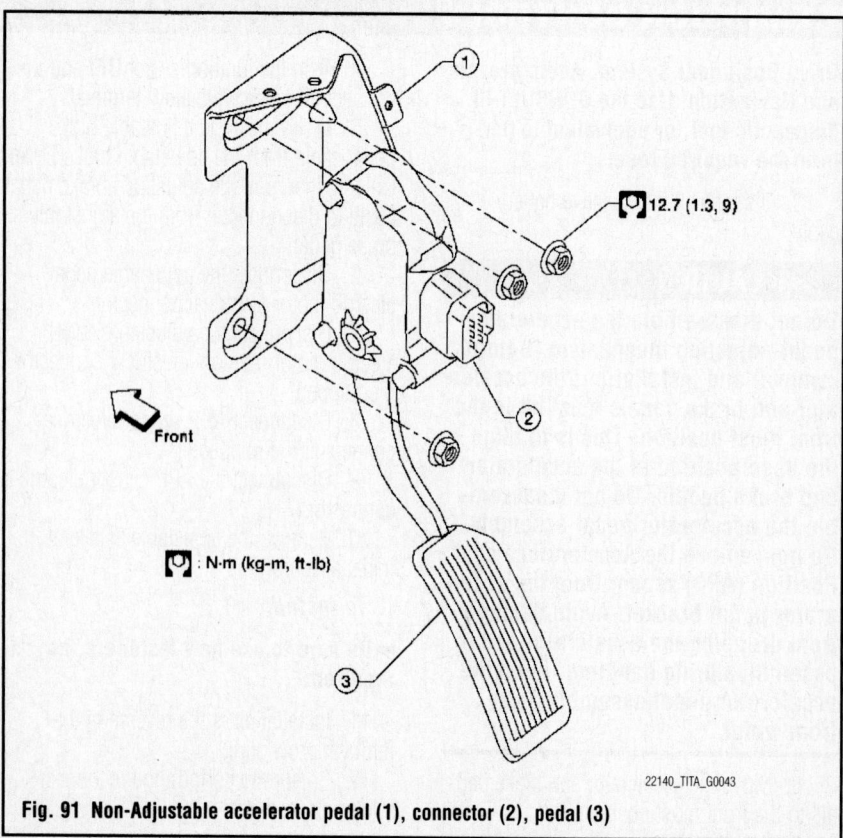

12.7 (1.3, 9)

Front

: N·m (kg-m, ft-lb)

22140_TITA_G0043

Fig. 91 Non-Adjustable accelerator pedal (1), connector (2), pedal (3)

CAMSHAFT POSITION (CMP) SENSOR

LOCATION

The Camshaft Position (CMP) sensor is located on the right front of the timing cover, facing the engine.

REMOVAL & INSTALLATION

1. Before servicing the vehicle, refer to the Precautions Section.

➡**If working near and/or around the SRS system and components, be sure to disable the SRS system. After disabling the system wait three minutes or more before servicing the vehicle.**

➡**Whenever the negative battery cable is disconnected the following components will require resetting. The Idle Air Volume Learning, Steering Angle Sensor Neutral Position, Sunroof Memory Reset/Initialization, Automatic Drive Positioner System, Audio presets and Navigation. Use the CONSULT-III diagnostic tool, or equivalent to perform the required resets.**

2. Disconnect the negative battery cable.

3. Remove the engine cover.

4. Remove air intake duct.

5. Disconnect the camshaft position sensor.

6. Remove the bolt and the Camshaft Position (CMP) sensor.

To install:

7. Install the CMP sensor and tighten the bolt.

8. Reconnect the camshaft electrical sensor.

9. Install the air intake duct.

10. Install the engine cover.

11. Be sure to perform the reconnect/relearn procedures.

CRANKSHAFT POSITION (CKP) SENSOR

LOCATION

The Crankshaft Position (CKP) sensor is located on the transmission assembly facing the gear teeth (cogs) of the signal plate.

REMOVAL & INSTALLATION

1. Before servicing the vehicle, refer to the Precautions Section.

➡If working near and/or around the SRS system and components, be sure to disable the SRS system. After disabling the system wait three minutes or more before servicing the vehicle.

➡Whenever the negative battery cable is disconnected the following components will require resetting. The Idle Air Volume Learning, Steering Angle Sensor Neutral Position, Sunroof Memory Reset/Initialization, Automatic Drive Positioner System, Audio presets and Navigation. Use the CONSULT-III diagnostic tool, or equivalent to perform the required resets.

2. Disconnect the negative battery cable.

3. Raise and support the vehicle safely.

4. Disconnect the Crankshaft Position (CKP) sensor connector.

5. Remove the mounting bolt and CKP sensor.

To install:

6. Install the CKP sensor and tighten the mounting bolt.

7. Reconnect the CKP sensor connector.

8. Lower the vehicle.

9. Be sure to perform the reconnect/relearn procedures.

ELECTRONIC CONTROL MODULE (ECM)

LOCATION

The Electronic Control Module (ECM) is located in the engine room passenger side behind battery.

REMOVAL & INSTALLATION

At this time the manufacturer does not provide removal and installation procedures for this component. The following procedure is a guideline and may differ from the vehicle you are servicing.

1. Before servicing the vehicle, refer to the Precautions Section.

➡If working near and/or around the SRS system and components, be sure to disable the SRS system. After disabling the system wait three minutes or more before servicing the vehicle.

➡Whenever the negative battery cable is disconnected the following components will require resetting. The Idle Air Volume Learning, Steering Angle Sensor Neutral Position, Sunroof Memory Reset/Initialization, Automatic Drive Positioner System, Audio presets and Navigation. Use the CONSULT-III diagnostic tool, or equivalent to perform the required resets.

2. Disconnect the negative battery cable.

3. Disconnect the positive battery cable and remove the battery, as required.

4. Carefully remove the Electronic Control Module (ECM) harness connectors.

5. Remove the ECM mounting bolts and the ECM.

To install:

6. Install the ECM and mounting bolts and tighten to 62 inch lbs. (7 Nm).

7. Carefully install the ECM harness connectors.

8. Install the battery.

9. Reconnect the battery cables.

10. Be sure to perform the reconnect/relearn procedures.

ENGINE COOLANT TEMPERATURE (ECT) SENSOR

LOCATION

The Engine Coolant Temperature (ECT) sensor is mounted in the front of the intake manifold. It is just to the right of the throttle body.

REMOVAL & INSTALLATION

1. Before servicing the vehicle, refer to the Precautions Section.

➡If working near and/or around the SRS system and components, be sure to disable the SRS system. After disabling the system wait three minutes or more before servicing the vehicle.

➡Whenever the negative battery cable is disconnected the following components will require resetting. The Idle Air Volume Learning, Steering Angle Sensor Neutral Position, Sunroof Memory Reset/Initialization, Automatic Drive Positioner System, Audio presets and Navigation. Use the CONSULT-III diagnostic tool, or equivalent to perform the required resets.

2. Disconnect the negative battery cable.

3. Remove the engine cover.

4. Remove the intake air duct.

5. Partially drain the cooling system.

6. Disconnect the harness connector.

7. Remove the Engine Coolant Temperature (ECT) sensor.

To install:

➡Be sure to use new fasteners, as required.

8. Install the ECT sensor and carefully tighten.

9. Reconnect the harness connector.

10. Install the intake air duct.

11. Install the engine cover.

12. Refill the engine coolant.

13. Be sure to perform the reconnect/relearn procedures.

EVAPORATIVE EMISSIONS (EVAP) CANISTER

LOCATION

See Figure 92.

This component is located under the vehicle near the fuel tank.

REMOVAL & INSTALLATION

See Figure 92.

At this time the manufacturer does not provide removal and installation procedures for this component. The following procedure is a guideline and may differ from the vehicle you are servicing.

1. Before servicing the vehicle, refer to the Precautions Section.

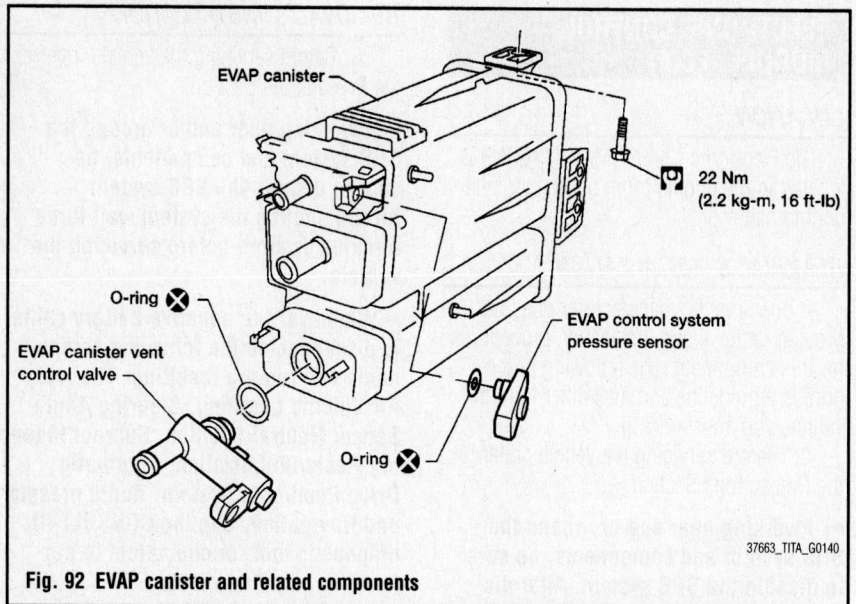

Fig. 92 EVAP canister and related components

→ If working near and/or around the SRS system and components, be sure to disable the SRS system. After disabling the system wait three minutes or more before servicing the vehicle.

→ Whenever the negative battery cable is disconnected the following components will require resetting. The Idle Air Volume Learning, Steering Angle Sensor Neutral Position, Sunroof Memory Reset/ Initialization, Automatic Drive Positioner System, Audio presets and Navigation. Use the CONSULT-III

diagnostic tool, or equivalent to perform the required resets.

2. Disconnect the negative battery cable.
3. Raise and support the vehicle safely.
4. Remove the fuel tank shield, as required.
5. Disconnect the component electrical connectors.
6. Disconnect the hoses.
7. Remove the mounting clips, nuts and or screws.
8. Remove the component from the vehicle.

To install:

→ Be sure to use new fasteners, as required.

9. Installation is the reverse of the removal procedure.
10. Be sure to perform the reconnect/ relearn procedures.

HEATED OXYGEN (HO2S) SENSOR

LOCATION

See Figure 93.

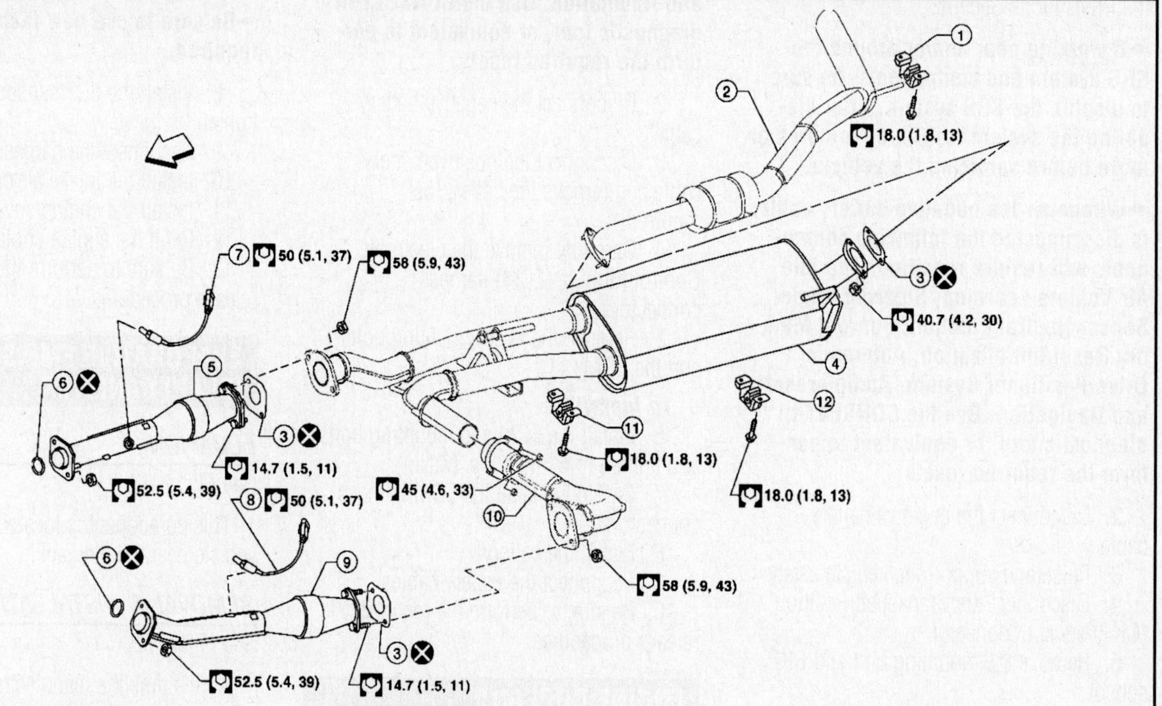

1. Tailpipe hanger bracket
4. Main muffler
7. Heated oxygen sensor 2 (bank 2)
10. Center exhaust tube
    Front

2. Tailpipe
5. Right front exhaust tube
8. Heated oxygen sensor 2 (bank 1)
11. Muffler hanger bracket front

3. Gasket
6. Ring gasket
9. Left front exhaust tube
12. Muffler hanger bracket rear

Fig. 93 Heated oxygen sensors and related components

The Heated Oxygen (HO2S) sensors are located after the exhaust manifold converter assembly, in the lower part of the exhaust system.

REMOVAL & INSTALLATION

1. Before servicing the vehicle, refer to the Precautions Section.

➡ **If working near and/or around the SRS system and components, be sure to disable the SRS system. After disabling the system wait three minutes or more before servicing the vehicle.**

➡ **Whenever the negative battery cable is disconnected the following components will require resetting. The Idle Air Volume Learning, Steering Angle Sensor Neutral Position, Sunroof Memory Reset/Initialization, Automatic Drive Positioner System, Audio presets and Navigation. Use the CONSULT-III diagnostic tool, or equivalent to perform the required resets.**

2. Disconnect the negative battery cable.
3. Raise and safely support the vehicle.
4. Remove the engine undercover, as needed.
5. Unplug the Heated Oxygen (HO2S) sensor harness.
6. Using an O2 wrench remove the HO2S sensor.

➡ **Lower the exhaust in needed.**

To install:

➡ **Be sure to use new fasteners, as required.**

7. Install the HO2S sensor and tighten to 37 ft. lbs. (50 Nm).
8. Install the harness connector.
9. Keep the harness connector and wiring away from exhaust system.
10. Be sure to perform the reconnect/relearn procedures.

INTAKE AIR TEMPERATURE (IAT) SENSOR

LOCATION

The Intake Air Temperature (IAT) sensor is integral to the mass air flow sensor, and is mounted on the air filter housing lid.

REMOVAL & INSTALLATION

See Figure 94.

1. Before servicing the vehicle, refer to the Precautions Section.

➡ **If working near and/or around the SRS system and components, be**

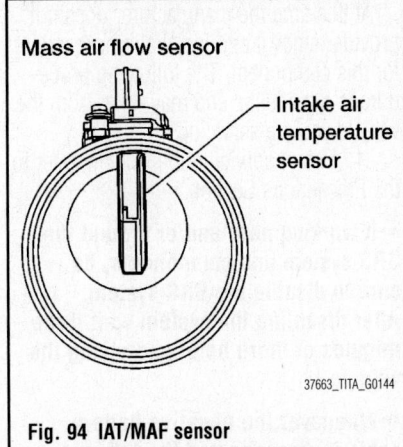

Mass air flow sensor

Intake air temperature sensor

37663_TITA_G0144

Fig. 94 IAT/MAF sensor

sure to disable the SRS system. After disabling the system wait three minutes or more before servicing the vehicle.

➡ **Whenever the negative battery cable is disconnected the following components will require resetting. The Idle Air Volume Learning, Steering Angle Sensor Neutral Position, Sunroof Memory Reset/Initialization, Automatic Drive Positioner System, Audio presets and Navigation. Use the CONSULT-III diagnostic tool, or equivalent to perform the required resets.**

2. Disconnect the negative battery cable.
3. Remove the engine room cover.
4. Remove the Intake Air Temperature (IAT/MAF) sensor harness.
5. Remove the mounting screws and the IAT/MAF sensor.

To install:

6. Install the IAT/MAF sensor.
7. Install the harness connector.
8. Install the engine room cover.
9. Be sure to perform the reconnect/relearn procedures.

KNOCK SENSOR (KS)

LOCATION

See Figure 95.

The Knock (KS) sensors are mounted under the intake manifold on the cylinder block.

REMOVAL & INSTALLATION

See Figure 95.

At this time the manufacturer does not provide removal and installation procedures for this component. The intake manifold will have to be removed to service this component.

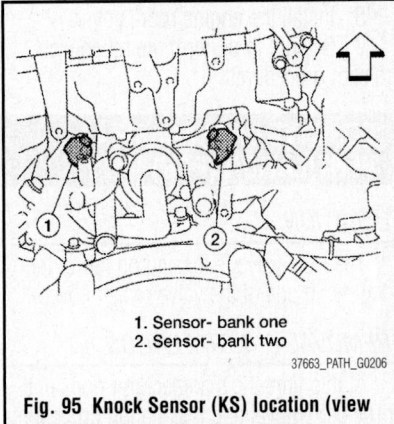

1. Sensor- bank one
2. Sensor- bank two

37663_PATH_G0206

Fig. 95 Knock Sensor (KS) location (view with engine removed from vehicle)

MALFUNCTION INDICATOR LIGHT (MIL)

RESET PROCEDURE

Clearing diagnostic trouble codes resets MIL.

MASS AIR FLOW (MAF) SENSOR

LOCATION

The Mass Air Flow (MAF) sensor is mounted on the air filter housing.

REMOVAL & INSTALLATION

See Figure 94.

1. Before servicing the vehicle, refer to the Precautions Section.

➡ **If working near and/or around the SRS system and components, be sure to disable the SRS system. After disabling the system wait three minutes or more before servicing the vehicle.**

➡ **Whenever the negative battery cable is disconnected the following components will require resetting. The Idle Air Volume Learning, Steering Angle Sensor Neutral Position, Sunroof Memory Reset/Initialization, Automatic Drive Positioner System, Audio presets and Navigation. Use the CONSULT-III diagnostic tool, or equivalent to perform the required resets.**

2. Disconnect the negative battery cable.
3. Remove the engine room cover.
4. Remove the Intake Air Temperature (IAT/MAF) sensor harness.
5. Remove the mounting screws and the IAT/MAF sensor.

To install:

6. Install the IAT/MAF sensor.
7. Install the harness connector.

8. Install the engine room cover.

9. Be sure to perform the reconnect/relearn procedures.

POSITIVE CRANKCASE VENTILATION (PCV) VALVE

LOCATION

The PCV valve is located on top of the engine, in one of the valve rocker covers.

REMOVAL & INSTALLATION

At this time the manufacturer does not provide removal and installation procedures for this component. The following procedure is a guideline and may differ from the vehicle you are servicing.

1. Before servicing the vehicle, refer to the Precautions Section.

➡If working near and/or around the SRS system and components, be sure to disable the SRS system. After disabling the system wait three minutes or more before servicing the vehicle.

➡Whenever the negative battery cable is disconnected the following components will require resetting. The Idle Air Volume Learning, Steering Angle Sensor Neutral Position, Sunroof Memory Reset/Initialization, Automatic Drive Positioner System, Audio presets and Navigation. Use the CONSULT-III diagnostic tool, or equivalent to perform the required resets.

2. Disconnect the negative battery cable.

3. Remove the necessary components in order to gain access to the component.

4. Disconnect the PCV hose.

5. Remove the valve from its mounting.

To install:

➡Be sure to use new fasteners, as required.

6. Installation is the reverse of the removal procedure.

7. Be sure to perform the reconnect/relearn procedures.

THROTTLE POSITION SENSOR (TPS)

LOCATION

The Throttle Position(TPS) sensor is integral to the electric Throttle Control actuator. The Throttle Control actuator is mounted at the front of the intake manifold.

REMOVAL & INSTALLATION

See Figures 96 and 97.

At this time the manufacturer does not provide removal and installation procedures for this component. The following procedure is a guideline and may differ from the vehicle you are servicing.

1. Before servicing the vehicle, refer to the Precautions Section.

➡If working near and/or around the SRS system and components, be sure to disable the SRS system. After disabling the system wait three minutes or more before servicing the vehicle.

➡Whenever the negative battery cable is disconnected the following components will require resetting. The Idle Air Volume Learning, Steering Angle Sensor Neutral Position, Sunroof Memory Reset/Initialization, Automatic Drive Positioner System, Audio presets and Navigation. Use the CONSULT-III diagnostic tool, or equivalent to perform the required resets.

2. Disconnect the negative battery cable.

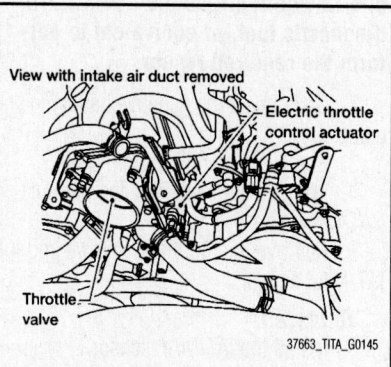

View with intake air duct removed

Electric throttle control actuator

Throttle valve

37663_TITA_G0145

Fig. 96 Throttle control actuator and related components

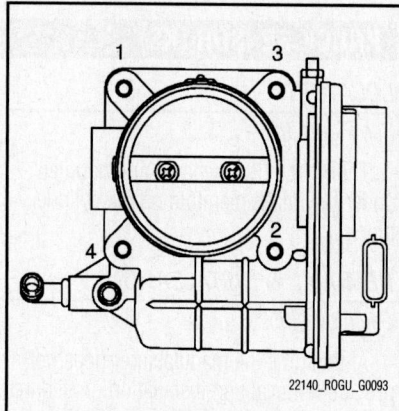

22140_ROGU_G0093

Fig. 97 Throttle body retaining bolt tightening sequence

3. Drain the cooling system, as required. Be sure to properly dispose of used engine coolant.

4. Remove the air intake duct.

5. Disconnect harness connector.

6. Disconnect water hoses.

7. Loosen the throttle body assembly mounting bolts in reverse order of the tightening sequence.

To install:

➡Be sure to use new fasteners, as required.

8. Install the throttle body assembly with a new gasket.

9. Tighten the mounting bolts in sequence to 74 inch lbs. (8.4 Nm).

10. Reconnect the water hose.

11. Reconnect the harness connector.

12. Reconnect the air intake duct.

13. Fill the cooling system with the proper grade and type engine coolant.

14. Be sure to perform the reconnect/relearn procedures.

VARIABLE CAMSHAFT TIMING OIL CONTROL SOLENOID

LOCATION

The Intake Valve Timing Control solenoid is located in the front timing chain cover.

REMOVAL & INSTALLATION

At this time the manufacturer does not provide removal and installation procedures for this component.

VEHICLE SPEED SENSOR (VSS)

LOCATION

The VSS sensor is located at the rear of the transmission case, under the tail shaft. On 4WD vehicles this component is located under the transfer case.

REMOVAL & INSTALLATION

1. Before servicing the vehicle, refer to the Precautions Section.

➡If working near and/or around the SRS system and components, be sure to disable the SRS system. After disabling the system wait three minutes or more before servicing the vehicle.

➡Whenever the negative battery cable is disconnected the following components will require resetting. The Idle Air Volume Learning, Steering Angle Sensor Neutral Position, Sunroof Memory Reset/Initialization, Automatic

Drive Positioner System, Audio presets and Navigation. Use the CONSULT-III diagnostic tool, or equivalent to perform the required resets.

2. Disconnect the negative battery cable.

3. Raise and safely support the vehicle.

4. Disconnect the sensor harness.

5. Remove the mounting bolt and the sensor.

To install:

➡**Be sure to use new fasteners, as required.**

6. Apply a small amount of transmission fluid to the sensor O-ring.

7. Install the Speed sensor and tighten the mounting bolt to 51 inch lbs. (5.8 Nm).

8. Be sure to perform the reconnect/relearn procedures.

FUEL   GASOLINE FUEL INJECTION SYSTEM

FUEL SYSTEM SERVICE PRECAUTIONS

Safety is the most important factor when performing not only fuel system maintenance, but any type of maintenance. Failure to conduct maintenance and repairs in a safe manner may result in serious personal injury or death. Work on a vehicle's fuel system components can be accomplished safely and effectively by adhering to the following rules and guidelines.

• To avoid the possibility of fire and personal injury, always disconnect the negative battery cable unless the repair or test procedure requires that battery voltage be applied.

• Always relieve the fuel system pressure prior to disconnecting any fuel system component (injector, fuel rail, pressure regulator, etc.) fitting or fuel line connection. Exercise extreme caution whenever relieving fuel system pressure to avoid exposing skin, face and eyes to fuel spray. Please be advised that fuel under pressure may penetrate the skin or any part of the body that it contacts.

• Always place a shop towel or cloth around the fitting or connection prior to loosening to absorb any excess fuel due to spillage. Ensure that all fuel spillage is quickly removed from engine surfaces. Ensure that all fuel-soaked cloths or towels are deposited into a flame-proof waste container with a lid.

• Always keep a dry chemical (Class B) fire extinguisher near the work area.

• Do not allow fuel spray or fuel vapors to come into contact with a spark or open flame.

• Always use a second wrench when loosening or tightening fuel line connection fittings. This will prevent unnecessary stress and torsion on fuel piping. Always follow the proper torque specifications.

• Always replace worn fuel fitting O-rings with new ones. Do not substitute fuel hose where rigid pipe is installed.

FUEL SYSTEM PRESSURE

RELIEVING

With CONSULT-III®

1. Turn ignition switch **ON**.

2. Perform "FUEL PRESSURE RELEASE" in "WORK SUPPORT" mode with CONSULT-III®.

3. Start engine.

4. After engine stalls, turn over the engine two or three times to release all fuel pressure.

5. Turn ignition switch **OFF**.

Without CONSULT-III®

See Figure 98.

1. Before servicing the vehicle, refer to the Precautions Section.

➡**If working near and/or around the SRS system and components, be sure to disable the SRS system. After disabling the system wait three minutes or more before servicing the vehicle.**

➡**Whenever the negative battery cable is disconnected the following components will require resetting. The Idle Air Volume Learning, Steering Angle Sensor Neutral Position, Sunroof Memory Reset/Initialization, Automatic Drive Positioner System, Audio presets and Navigation. Use the CONSULT-III diagnostic tool, or equivalent to perform the required resets.**

2. Remove fuel pump fuse located in IPDM E/R.

3. Start engine.

4. After engine stalls, turn over engine two or three times to release all fuel pressure.

5. Turn ignition switch **OFF**.

6. Disconnect the negative battery cable.

7. Reinstall fuel pump fuse after servicing fuel system.

8. Be sure to perform the reconnect/relearn procedures.

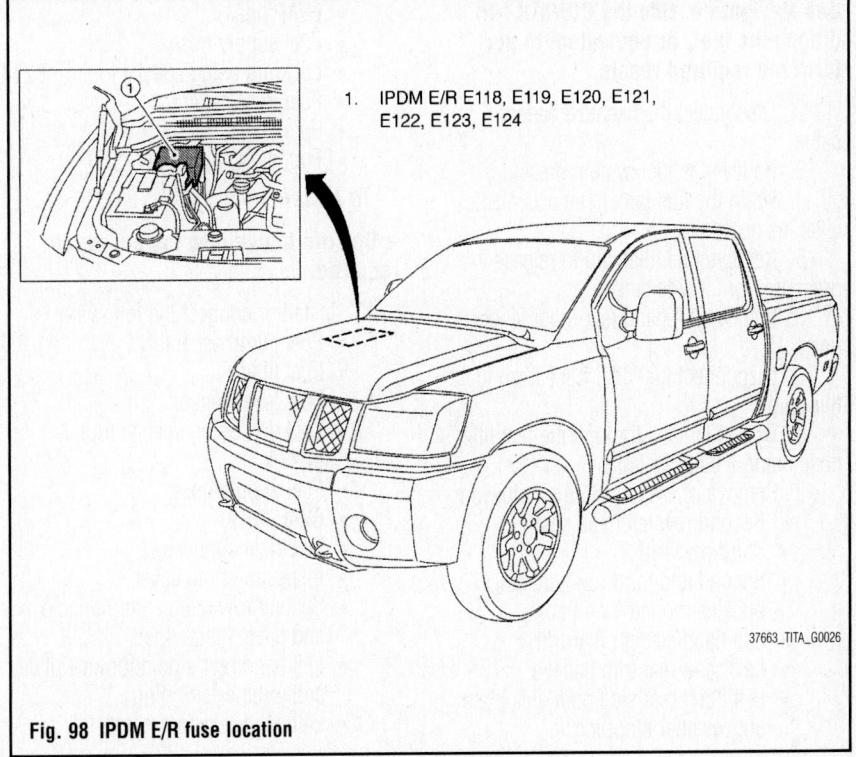

1. IPDM E/R E118, E119, E120, E121, E122, E123, E124

37663_TITA_G0026

Fig. 98 IPDM E/R fuse location

FUEL FILTER

REMOVAL & INSTALLATION

See Figure 99.

➡ **The fuel filter is part of the fuel pump assembly.**

FUEL LEVEL SENDING UNIT

REMOVAL & INSTALLATION

See Figure 99.

This component is located on the fuel pump module. This component is removed along with the fuel pump module.

FUEL PUMP MODULE

REMOVAL & INSTALLATION

See Figure 99.

1. Before servicing the vehicle, refer to the Precautions Section.

➡ **If working near and/or around the SRS system and components, be sure to disable the SRS system. After disabling the system wait three minutes or more before servicing the vehicle.**

➡ **Whenever the negative battery cable is disconnected the following components will require resetting. The Idle Air Volume Learning, Steering Angle Sensor Neutral Position, Sunroof Memory Reset/Initialization, Automatic Drive Positioner System, Audio presets and Navigation. Use the CONSULT-III diagnostic tool, or equivalent to perform the required resets.**

2. Disconnect the negative battery cable.

3. Relieve the fuel system pressure.

4. Drain the fuel tank to an acceptable level, as necessary.

5. Remove fuel filler cap to release pressure from inside tank.

6. Remove left hand rear inner fender liner.

7. Disconnect fuel filler hose from fuel filler pipe.

8. Drain fuel tank through the fuel filler hose using a suitable hose.

9. Remove or disconnect the following:
   - Second row left hand seat
   - Third row seat
   - Second and third row seat belt buckles mounted on floor
   - Left hand center pillar trim
   - Left hand rear trim panel
   - Left hand rear side door kick plate and weather stripping

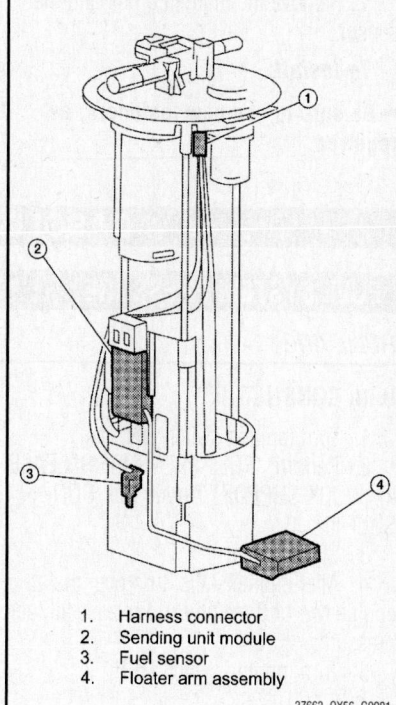

1. Harness connector
2. Sending unit module
3. Fuel sensor
4. Floater arm assembly

37663_QX56_G0081

Fig. 99 Fuel pump module and related components

   - Second row rear center console and base, if equipped
   - Inspection hole cover under carpet by turning retainers 90 degrees
   - Electrical connectors
   - EVAP hose
   - Fuel supply hose
   - Lockring using special tool J-46214
   - Fuel level sensor
   - Fuel filter
   - Fuel pump assembly

To install:

➡ **Be sure to use new fasteners, as required.**

10. Install or connect the following:
    - Fuel pump assembly
    - Fuel filter
    - Fuel level sensor
    - Lockring using special tool J-46214
    - Fuel supply hose
    - EVAP hose
    - Electrical connectors
    - Inspection hole cover
    - Second row rear center console and base, if equipped
    - Left hand rear side door kick plate and weather stripping
    - Left hand rear trim panel

    - Left hand center pillar trim
    - Second and third row seat belt buckles
    - Third row seat
    - Second row left hand seat
    - Fuel filler hose to fuel filler pipe
    - Left hand rear inner fender liner

11. Start the engine and check for leaks.

12. Be sure to perform the reconnect/relearn procedures.

FUEL RAIL AND INJECTOR

REMOVAL & INSTALLATION

See Figure 100.

1. Before servicing the vehicle, refer to the Precautions Section.

➡ **If working near and/or around the SRS system and components, be sure to disable the SRS system. After disabling the system wait three minutes or more before servicing the vehicle.**

➡ **Whenever the negative battery cable is disconnected the following components will require resetting. The Idle Air Volume Learning, Steering Angle Sensor Neutral Position, Sunroof Memory Reset/Initialization, Automatic Drive Positioner System, Audio presets and Navigation. Use the CONSULT-III diagnostic tool, or equivalent to perform the required resets.**

2. Disconnect the negative battery cable.

3. Remove engine cover. Remove the air cleaner assembly.

4. Relieve fuel system pressure.

5. Remove or disconnect the following:
   - Fuel injector harness connectors
   - Fuel hose assembly from right and left fuel rails
   - Fuel injectors with fuel rail as an assembly
   - Fuel injector from fuel rail

To install:

➡ **Be sure to use new fasteners, as required.**

6. Install or connect the following:

➡ **Always use a new O-ring when reinstalling the fuel injector to the fuel rail.**

   - New clip onto the fuel injector
   - Fuel injector to fuel rail
   - Fuel injectors and fuel rail as an assembly to the intake manifold. Tighten the bolts to 8 ft. lbs. (11 Nm).
   - Fuel hose assembly
   - Fuel injector harness connectors

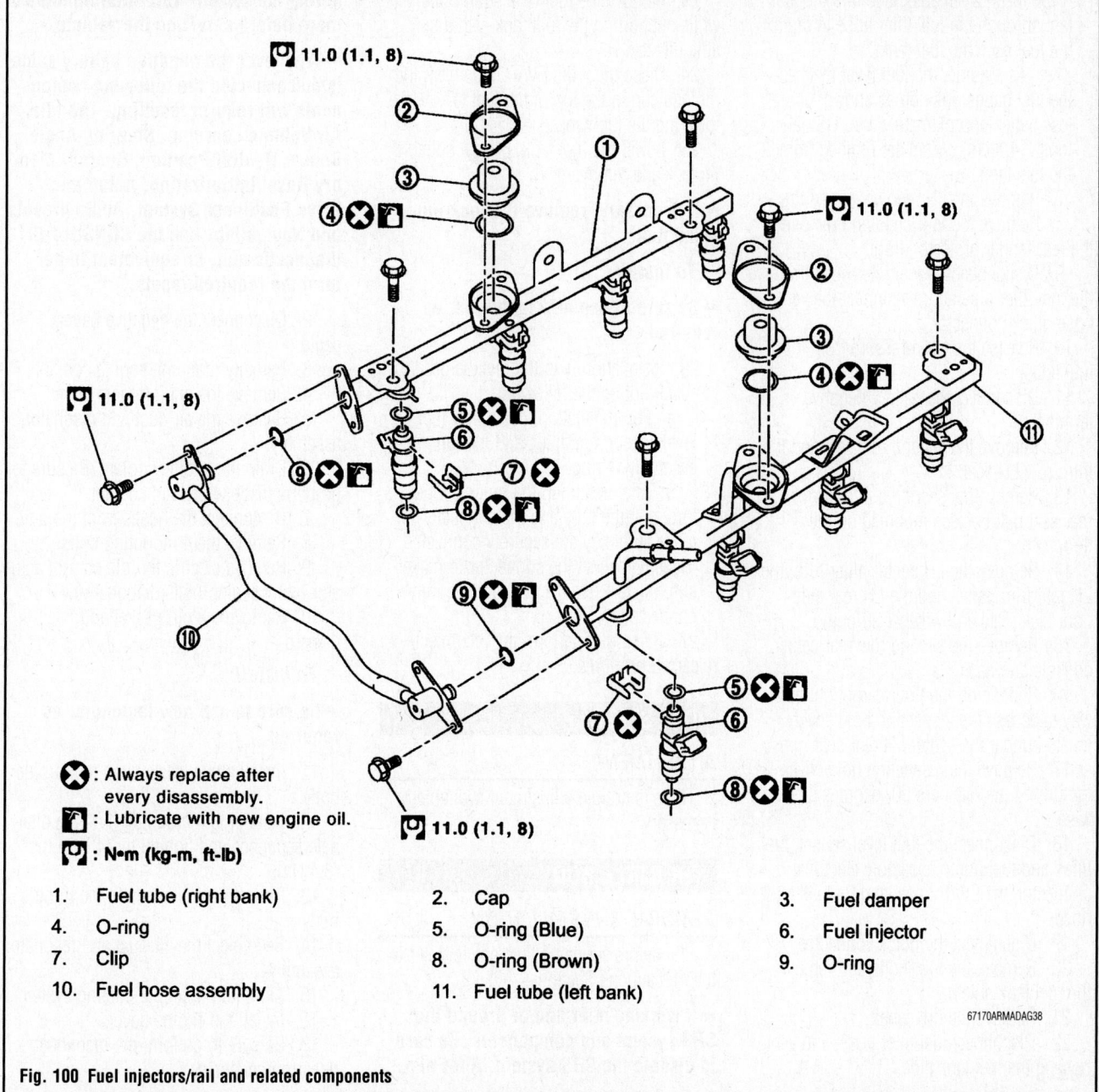

⊗ : Always replace after every disassembly.
▣ : Lubricate with new engine oil.
⬕ : N•m (kg-m, ft-lb)

1. Fuel tube (right bank)
2. Cap
3. Fuel damper
4. O-ring
5. O-ring (Blue)
6. Fuel injector
7. Clip
8. O-ring (Brown)
9. O-ring
10. Fuel hose assembly
11. Fuel tube (left bank)

67170ARMADAG38

Fig. 100 Fuel injectors/rail and related components

- Negative battery cable
- Engine cover

7. Start engine and check for leaks.

8. Be sure to perform the reconnect/relearn procedures.

FUEL TANK

REMOVAL & INSTALLATION

1. Before servicing the vehicle, refer to the Precautions Section.

➡**If working near and/or around the SRS system and components, be sure to disable the SRS system. After dis-abling the system wait three minutes or more before servicing the vehicle.**

➡**Whenever the negative battery cable is disconnected the following components will require resetting. The Idle Air Volume Learning, Steering Angle Sensor Neutral Position, Sunroof Memory Reset/Initialization, Automatic Drive Positioner System, Audio presets and Navigation. Use the CONSULT-III diagnostic tool, or equivalent to perform the required resets.**

2. Disconnect the negative battery cable.
3. Drain the fuel from the fuel tank, if necessary.

4. Remove the fuel filler cap to release the pressure from inside the fuel tank.

5. Check the fuel level on level gauge. If the fuel gauge indicates more than the level as shown (full or almost full), drain the fuel from the fuel tank until the fuel gauge indicates the level as shown, or less.

6. If the fuel pump does not operate, use the following procedure to drain the fuel to the specified level after disconnecting the fuel filler hose from the fuel filler pipe:

a. Insert a suitable hose of less than 15 mm (0.59 in.) diameter into the fuel filler pipe through the fuel filler opening to drain the fuel from fuel filler pipe.

b. Insert a suitable hose into the fuel tank through the fuel filler hose to drain the fuel from the fuel tank.

c. As a guide, the fuel level reaches the fuel gauge position as shown, or less, when approximately 3 ¾ US gallons (14 liters) of fuel are drained from the fuel tank.

7. Remove the LH rear wheel and tire.

8. Remove the four clips and remove the rear fender protector, front.

9. Disconnect the fuel filler hose from the fuel filler pipe and disconnect the vent hose quick connector.

10. Release the fuel pressure from the fuel lines.

11. Disconnect the battery negative terminal.

12. Remove the second row seat and the third row LH seat.

13. Remove the second and third row rear seat belt buckles mounted on the floor.

14. Remove the LH center pillar trim, the LH rear trim panel, and the LH rear side door kick plate and weather stripping.

15. Remove the second row rear center console and base.

16. Reposition the floor carpet out of the way to access the inspection hole cover, located under the center LH rear seat.

17. Remove the inspection hole cover by turning the retainers 90_degrees clockwise.

18. Disconnect the fuel level sensor, fuel filter, and fuel pump assembly electrical connector, the EVAP hose, and the fuel feed hose.

19. Disconnect the quick connector

20. Remove the four bolts and remove the fuel tank shield.

21. Remove the driveshaft.

22. Disconnect fuel filler hose, and vent hose at the fuel tank side.

23. Remove the fuel tank strap bolts while supporting the fuel tank with a suitable lift jack.

24. Disconnect the EVAP hose from the molded clip in the top of the fuel tank while lowering the fuel tank.

25. Lower the fuel tank using a suitable lift jack and remove it.

➡If necessary, remove the lockring using tool.

To install:

➡Be sure to use new fasteners, as required.

26. Installation is in the reverse order of removal, noting the following:

a. For installation, use a new fuel level sensor, fuel filter, and fuel pump assembly O-ring.

b. After installing the quick connectors, pull the tube and the connector to make sure they are securely connected. Visually inspect the connector to make sure the two retainer tabs are securely connected.

27. Be sure to perform the reconnect/relearn procedures.

IDLE SPEED

ADJUSTMENT

There is no idle adjustment available or necessary.

THROTTLE BODY

REMOVAL & INSTALLATION

1. Before servicing the vehicle, refer to the Precautions Section.

➡If working near and/or around the SRS system and components, be sure to disable the SRS system. After disabling the system wait three minutes or more before servicing the vehicle.

➡Whenever the negative battery cable is disconnected the following components will require resetting. The Idle Air Volume Learning, Steering Angle Sensor Neutral Position, Sunroof Memory Reset/Initialization, Automatic Drive Positioner System, Audio presets and Navigation. Use the CONSULT-III diagnostic tool, or equivalent to perform the required resets.

2. Disconnect the negative battery cable.

3. Partially drain the engine coolant.

4. Remove the engine room cover.

5. Remove the air duct and resonator assembly.

6. Drain the engine coolant. Be sure to properly dispose of used coolant.

7. Disconnect the hoses from the unit.

8. Remove the 4 mounting bolts.

9. Remove electric throttle control actuator by loosening bolts diagonally.

10. Remove the old gasket and discard it.

To install:

➡Be sure to use new fasteners, as required.

11. Install a new gasket and the throttle body.

12. Install the 4 mounting bolts in alternate sequence and tighten to 74 inch lbs. (8.4 Nm).

13. Reconnect the hoses to the throttle body.

14. Reconnect the air duct and resonator assembly.

15. As required, fill the cooling system.

16. Install the engine cover.

17. Be sure to perform the reconnect/relearn procedures.

HEATING & AIR CONDITIONING SYSTEM

BLOWER MOTOR

REMOVAL & INSTALLATION

See Figure 102.

1. Before servicing the vehicle, refer to the Precautions Section.

➡**If working near and/or around the SRS system and components, be sure to disable the SRS system. After disabling the system wait three minutes or more before servicing the vehicle.**

➡**Whenever the negative battery cable is disconnected the following components will require resetting. The Idle Air Volume Learning, Steering Angle Sensor Neutral Position, Sunroof Memory Reset/ Initialization, Automatic Drive Positioner System, Audio presets and Navigation. Use the CONSULT-III diagnostic tool, or equivalent to perform the required resets.**

2. Remove the glove box assembly.
3. Disconnect the front blower motor electrical connector.
4. Remove the blower retaining screws.
5. Remove the blower motor from its mounting.

To install:

➡**Be sure to use new fasteners, as required.**

6. Installation is the reverse of the removal procedure.
7. Be sure to perform the reconnect/relearn procedures.

HEATER/COOLING UNIT

REMOVAL & INSTALLATION

See Figures 103 through 106.

1. Before servicing the vehicle, refer to the Precautions Section.

➡**If working near and/or around the SRS system and components, be sure to disable the SRS system. After disabling the system wait three minutes or more before servicing the vehicle.**

➡**Whenever the negative battery cable is disconnected the following components will require resetting. The Idle Air Volume Learning, Steering Angle Sensor Neutral Position, Sunroof Memory Reset/ Initialization, Automatic Drive Positioner System, Audio presets and Navigation. Use the CONSULT-III diagnostic tool, or equivalent to perform the required resets.**

2. Position the front seats in the rearmost position.
3. Disconnect the negative battery cable.
4. Properly discharge the A/C system.
5. Drain the cooling system. Be sure to properly dispose of used coolant.
6. Disconnect the heater hoses at the heater core.
7. Disconnect and plug the refrigerant lines at the evaporator core.
8. Remove the instrument panel.
9. Remove the center console.
10. Disconnect the instrument panel wire harness at the right and left in-line connector brackets, and the fuse block (JB) electrical connectors.
11. Disconnect the steering member from each side of the vehicle body.
12. Remove the heater/cooling unit with it attached to the steering member from the vehicle.

➡**Use care not to damage the seats or interior trim panels.**

13. Remove the heater/cooling unit from the steering member.

To install:

➡**Be sure to use new fasteners, as required.**

14. Installation is the reverse of the removal procedure.
15. Be sure to use new O-rings coated with clean refrigerant oil, as required.
16. Fill the cooling system with the proper grade and the coolant.
17. Properly recharge the A/C system.
18. Start the engine and check for leaks, correct as required.
19. Be sure to perform the reconnect/relearn procedures.

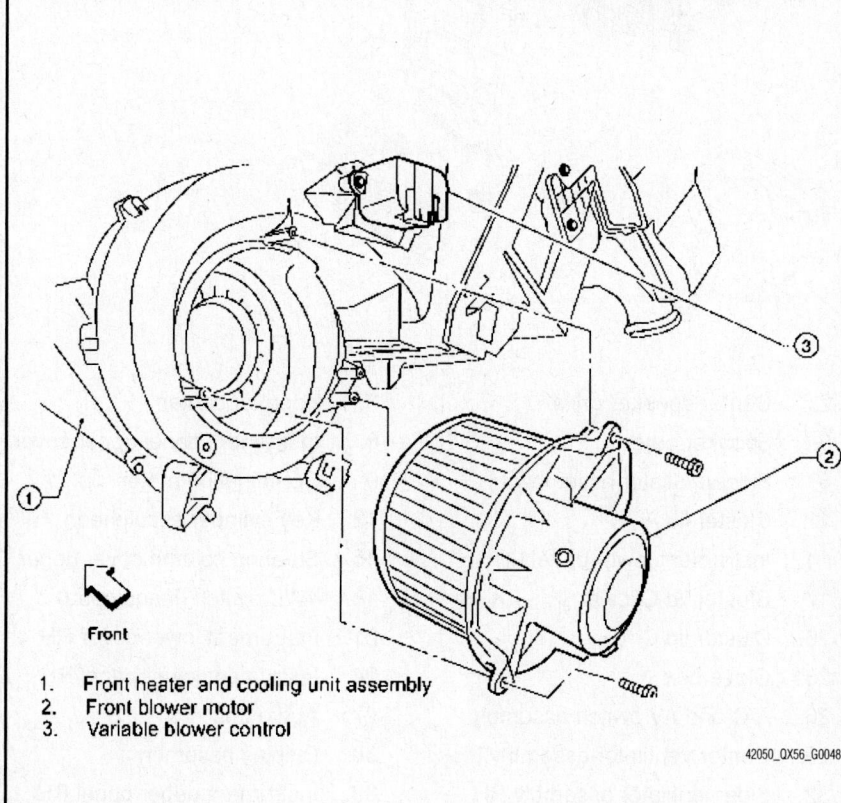

Front

1. Front heater and cooling unit assembly
2. Front blower motor
3. Variable blower control

42050_QX56_G0048

Fig. 102 Front blower motor and related components

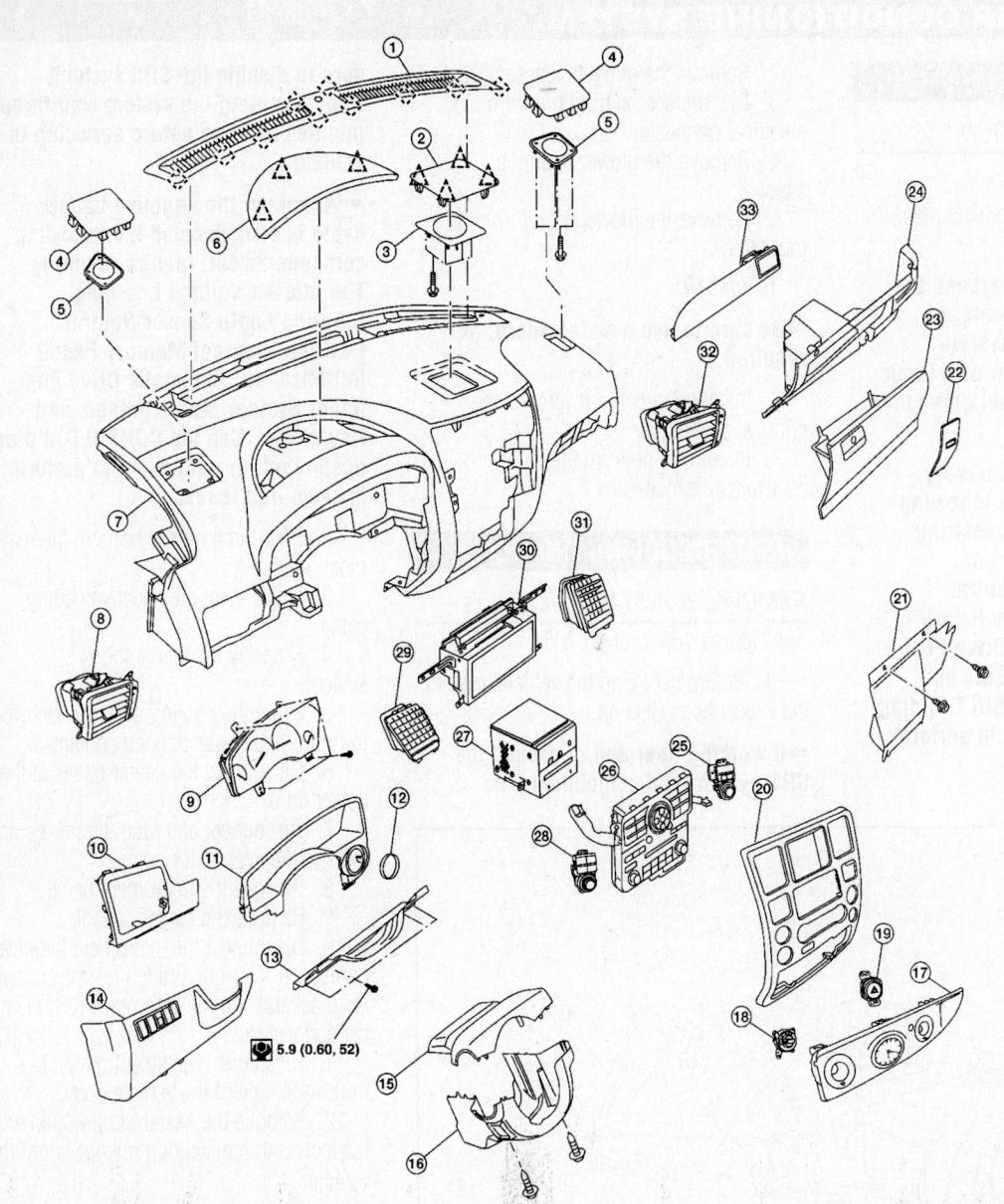

5.9 (0.60, 52)

1. Defroster grille	2. Center speaker grille	3. Speaker center
4. Speaker grille outer	5. Speaker outer	6. Instrument panel upper cover
7. Instrument panel and pad assembly	8. Side ventilator assembly LH	9. Combination meter
10. Instrument upper panel LH	11. Cluster lid A	12. Key cylinder escutcheon
13. Lower knee protector	14. Instrument lower panel LH	15. Steering column cover upper
16. Steering column cover lower	17. Cluster lid C lower	18. 4WD switch (if equipped)
19. Hazard switch	20. Cluster lid C	21. Instrument lower cover RH
22. Fuse access cover	23. Glove box	24. Instrument lower panel RH
25. Audio switch RH	26. A/C and AV switch assembly	27. Audio unit
28. Audio switch LH	29. Center ventilator assembly LH	30. Display assembly
31. Center ventilator assembly RH	32. Side ventilator assembly RH	33. Instrument upper panel RH
Metal clip	Clip	

37663_QX56_G0005

Fig. 103 Instrument panel and related components

HEATER CORE

REMOVAL & INSTALLATION

1. Before servicing the vehicle, refer to the Precautions Section.

➡ **If working near and/or around the SRS system and components, be sure to disable the SRS system. After disabling the system wait three minutes or more before servicing the vehicle.**

➡ **Whenever the negative battery cable is disconnected the following components will require resetting. The Idle Air Volume Learning, Steering Angle Sensor Neutral Position, Sunroof Memory Reset/Initialization, Automatic Drive Positioner System, Audio presets and Navigation. Use the CONSULT-III diagnostic tool, or equivalent to perform the required resets.**

2. Position the front seats in the rear-most position.
3. Disconnect the negative battery cable.
4. Properly discharge the A/C system.
5. Drain the cooling system. Be sure to properly dispose of used coolant.
6. Remove the heater/cooling unit.
7. Remove the heater core from the heater/cooling unit.

To install:

➡ **Be sure to use new fasteners, as required.**

8. Installation is the reverse of the removal procedure.
9. Be sure to use new O-rings coated with clean refrigerant oil, as required.
10. Fill the cooling system with the proper grade and the coolant.
11. Properly recharge the A/C system.
12. Start the engine and check for leaks, correct as required.
13. Be sure to perform the reconnect/relearn procedures.

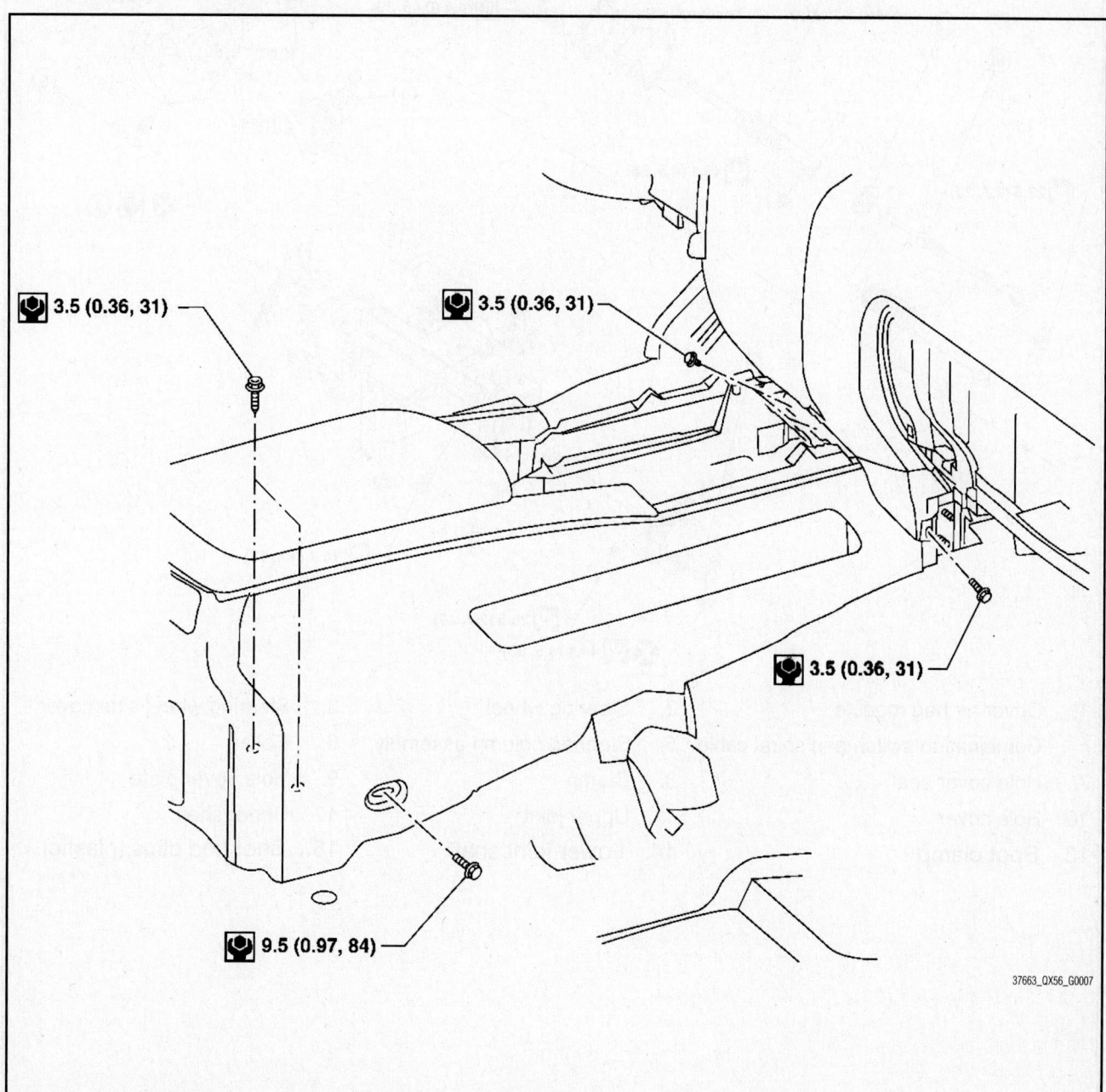

3.5 (0.36, 31)

3.5 (0.36, 31)

3.5 (0.36, 31)

9.5 (0.97, 84)

37663_QX56_G0007

Fig. 104 Front center console and related components

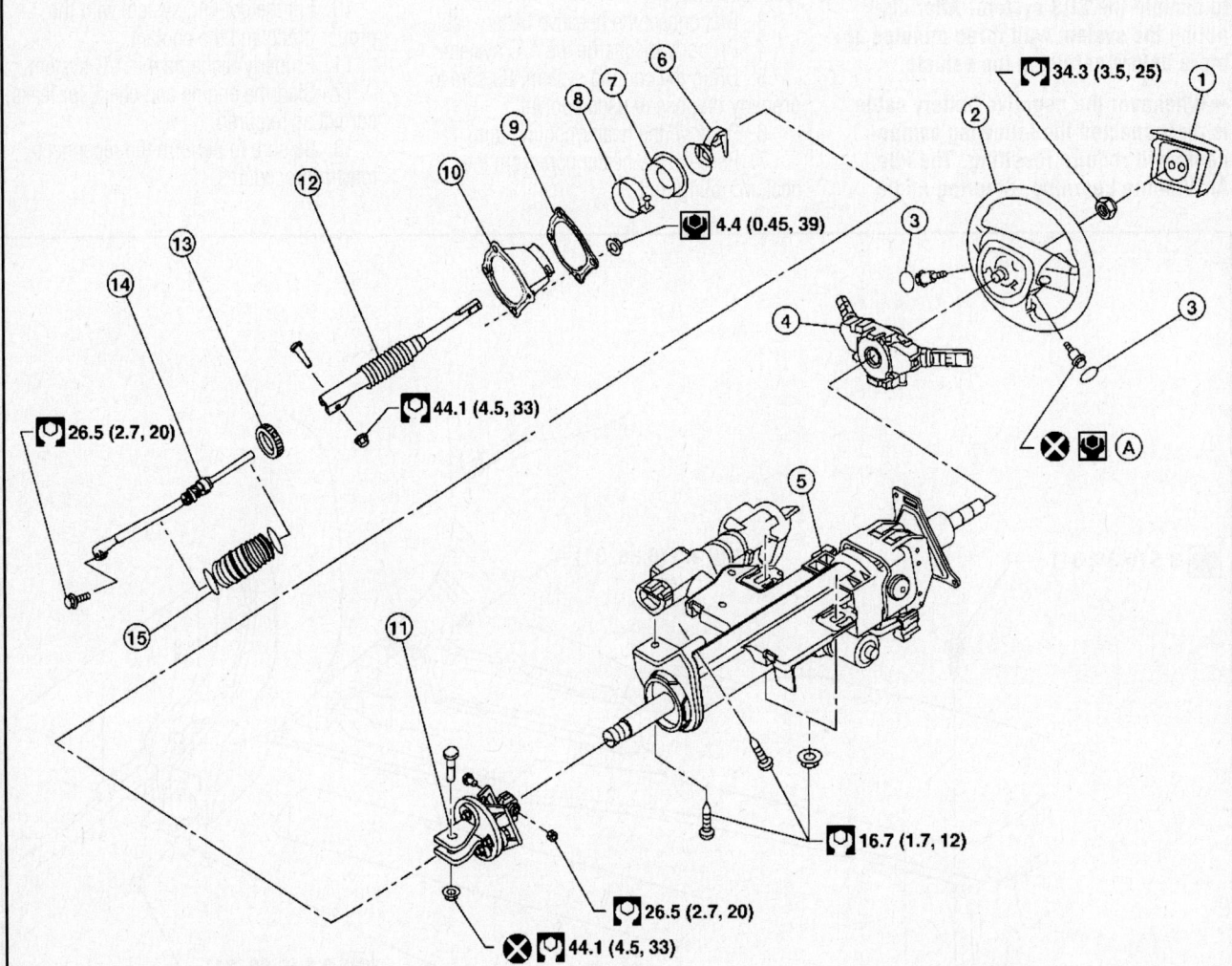

34.3 (3.5, 25)
4.4 (0.45, 39)
44.1 (4.5, 33)
26.5 (2.7, 20)
16.7 (1.7, 12)
26.5 (2.7, 20)
44.1 (4.5, 33)

1.	Driver air bag module	2.	Steering wheel	3.	Steering wheel side cover
4.	Combination switch and spiral cable	5.	Steering column assembly	6.	Collar
7.	Hole cover seal	8.	Clamp	9.	Hole cover plate
10.	Hole cover	11.	Upper joint	12.	Upper shaft
13.	**Boot clamp**	**14.**	**Lower joint shaft**	**15.**	**Boot and clips (plastic)**

37663_QX56_G0008

Fig. 105 Steering column and related components

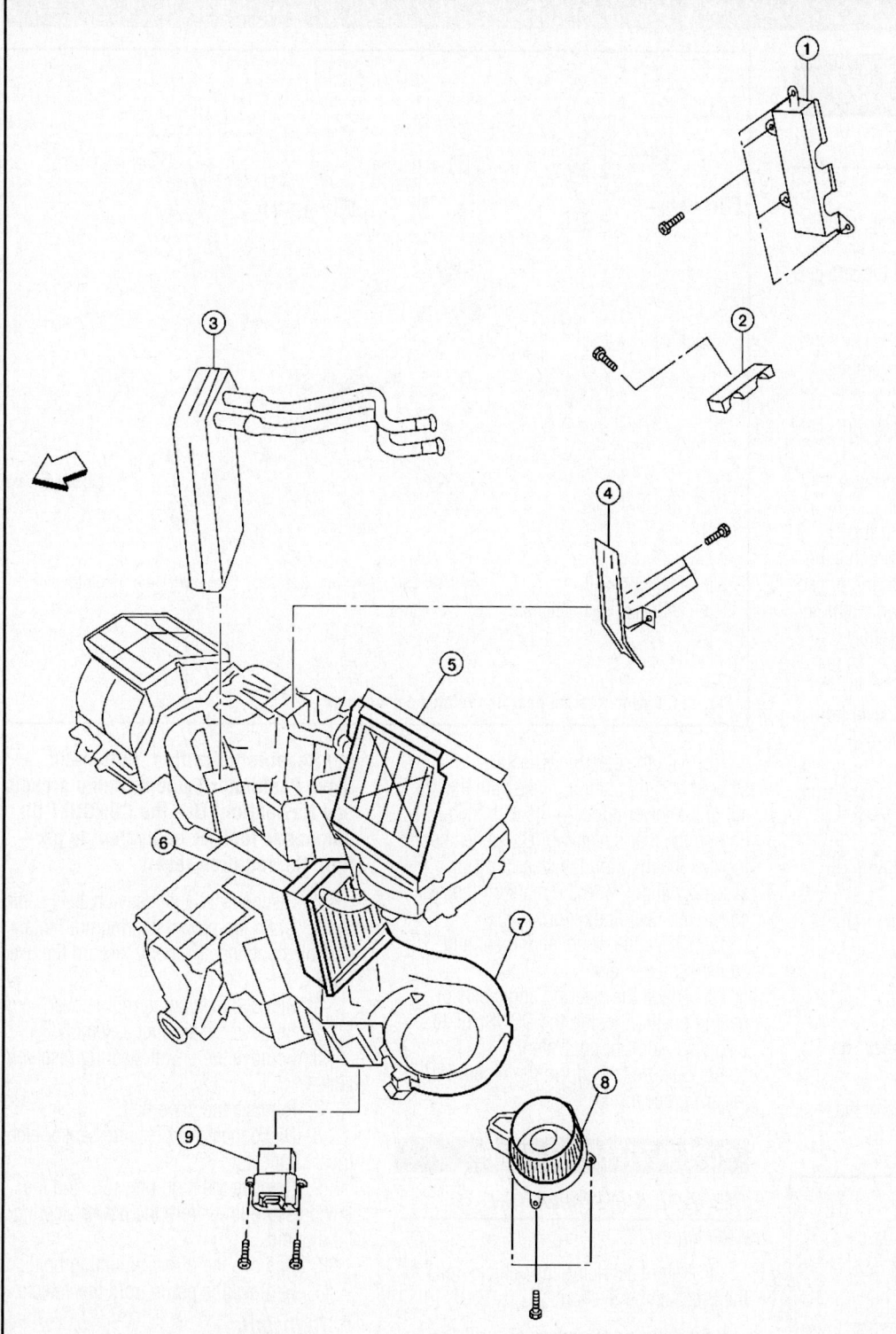

1. Heater core cover
2. Heater core pipe bracket
3. Heater core
4. Upper bracket
5. Upper heater and cooling unit case
6. A/C evaporator
7. Lower heater and cooling unit case
8. Blower motor
9. Variable blower control

⇦ Front

37663_QX56_G0089

Fig. 106 AC heater/cooling unit and related components

STEERING

POWER RACK & PINION STEERING GEAR

REMOVAL & INSTALLATION

See Figures 110 and 111.

1. Before servicing the vehicle, refer to the Precautions Section.
2. Ensure the wheels are in the straight-ahead position.
3. Remove or disconnect the following:
   - Wheels and tires
   - Engine splash guard
4. On 4WD, remove front final drive and support the halfshafts.
5. Remove cotter pin at steering outer socket and loosen mounting nut.
6. With the steering wheel in the straight ahead position, make sure that the slit of the lower joint (A) fits with the projection on the rear cover cap (B), while checking that the mark on the steering gear assembly aligns with the mark on the rear cover cap. See illustration.
7. Remove steering outer socket from steering knuckle using special tool J-25730-A.
8. Remove or disconnect the following:
   - Oil pipes from steering gear assembly
   - Lower joint mounting bolt from lower shaft
   - Mounting bolts and nuts from steering gear assembly
   - Steering gear assembly

To install:

➡**Be sure to use new fasteners, as required.**

9. Installation is the reverse of the removal procedure.

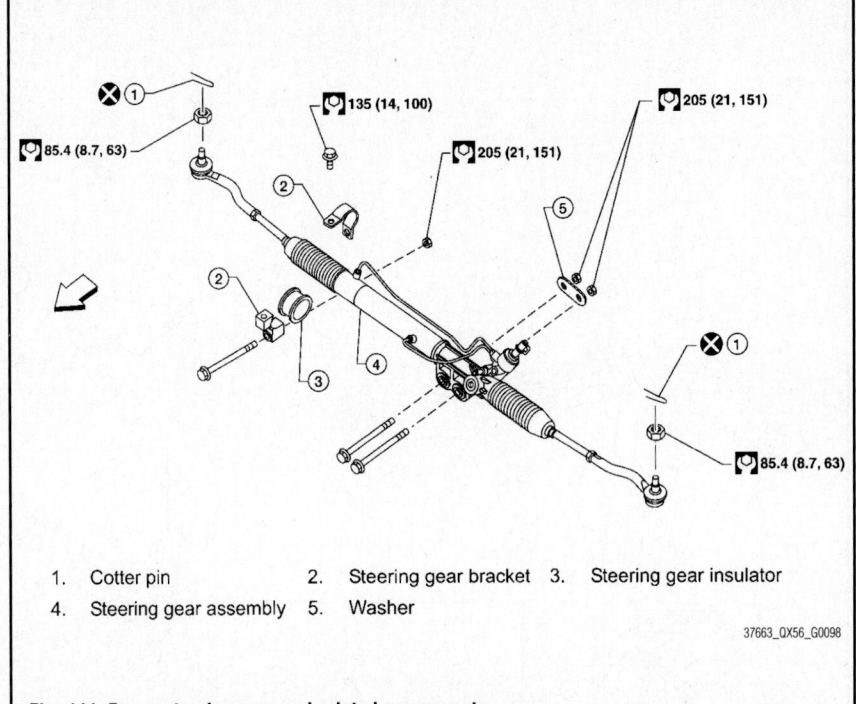

| 1. | Cotter pin | 2. | Steering gear bracket | 3. | Steering gear insulator |
| 4. | Steering gear assembly | 5. | Washer | | |

Fig. 111 Power steering gear and related components

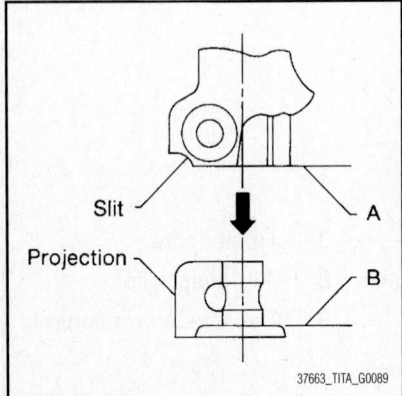

Fig. 110 Steering gear lower joint alignment

10. With the steering wheel in the straight ahead position, make sure that the slit of the lower joint (A) fits with the projection on the rear cover cap (B), while checking that the mark on the steering gear assembly aligns with the mark on the rear cover cap. See illustration.
11. Check the wheel alignment and adjust as necessary.
12. Adjust the steering angle sensor neutral position, using the CONSULT-III diagnostic tool, or equivalent.
13. Be sure to perform the reconnect/relearn procedures.

POWER STEERING PUMP

REMOVAL & INSTALLATION

See Figure 112.

1. Before servicing the vehicle, refer to the Precautions Section.

➡**If working near and/or around the SRS system and components, be sure to disable the SRS system. After disabling the system wait three minutes or more before servicing the vehicle.**

➡**Whenever the negative battery cable is disconnected the following components will require resetting. The Idle Air Volume Learning, Steering Angle Sensor Neutral Position, Sunroof Mem-ory Reset/Initialization, Automatic Drive Positioner System, Audio presets and Navigation. Use the CONSULT-III diagnostic tool, or equivalent to perform the required resets.**

2. Disconnect the negative battery cable.
3. Drain the power steering fluid into a suitable container. Properly discard the used fluid.
4. Remove the engine room cover.
5. Remove the air duct assembly.
6. Remove the power steering reservoir tank.
7. Remove the drive belt.
8. Disconnect the pressure sensor electrical connector.
9. Remove the high pressure and the low pressure lines from the power steering fluid pump.
10. Remove the pump mounting bolts.
11. Remove the pump from the vehicle.

To install:

➡**Be sure to use new fasteners, as required.**

12. Installation is the reverse of the removal procedure.
13. Bleed the power steering system.

➡**The drive belt tension is automatic and requires no adjustment.**

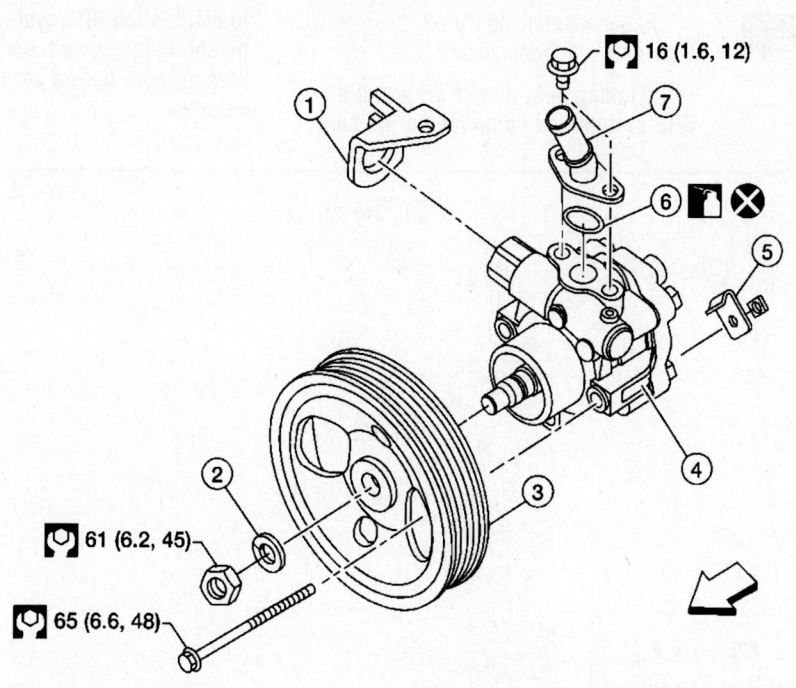

16 (1.6, 12)

61 (6.2, 45)

65 (6.6, 48)

1. Bracket	2. Spring washer	3. Pulley
4. Power steering pump	5. High pressure hose bracket	6. O-ring
7. Suction pipe	⟵ Front	

37663_QX56_G0100

Fig. 112 Power steering pump and related components

14. Be sure to perform the reconnect/relearn procedures.

BLEEDING

1. Before servicing the vehicle, refer to the Precautions Section.
2. Stop the engine.
3. Turn the steering wheel fully to the right and left several times.

➡**Do not allow the fluid level in the reservoir tank to go below the MIN level line. Check and add fluid as needed.**

4. Run the engine at idle speed. Turn the steering wheel fully to the right and then fully to the left. Hold for about three seconds. Check for fluid leakage.
5. Repeat the above step several times at three second intervals.

➡**Do not hold the steering wheel in the locked position for more than ten seconds.**

6. Check for air bubbles or cloudy fluid. If found, repeat the bleeding procedure.
7. Stop the engine and check the fluid level. Correct as required.

COIL SPRING

REMOVAL & INSTALLATION

See Figures 113 and 114.

1. Before servicing the vehicle, refer to the Precautions Section.

➥If working near and/or around the SRS system and components, be sure to disable the SRS system. After disabling the system wait three minutes or more before servicing the vehicle.

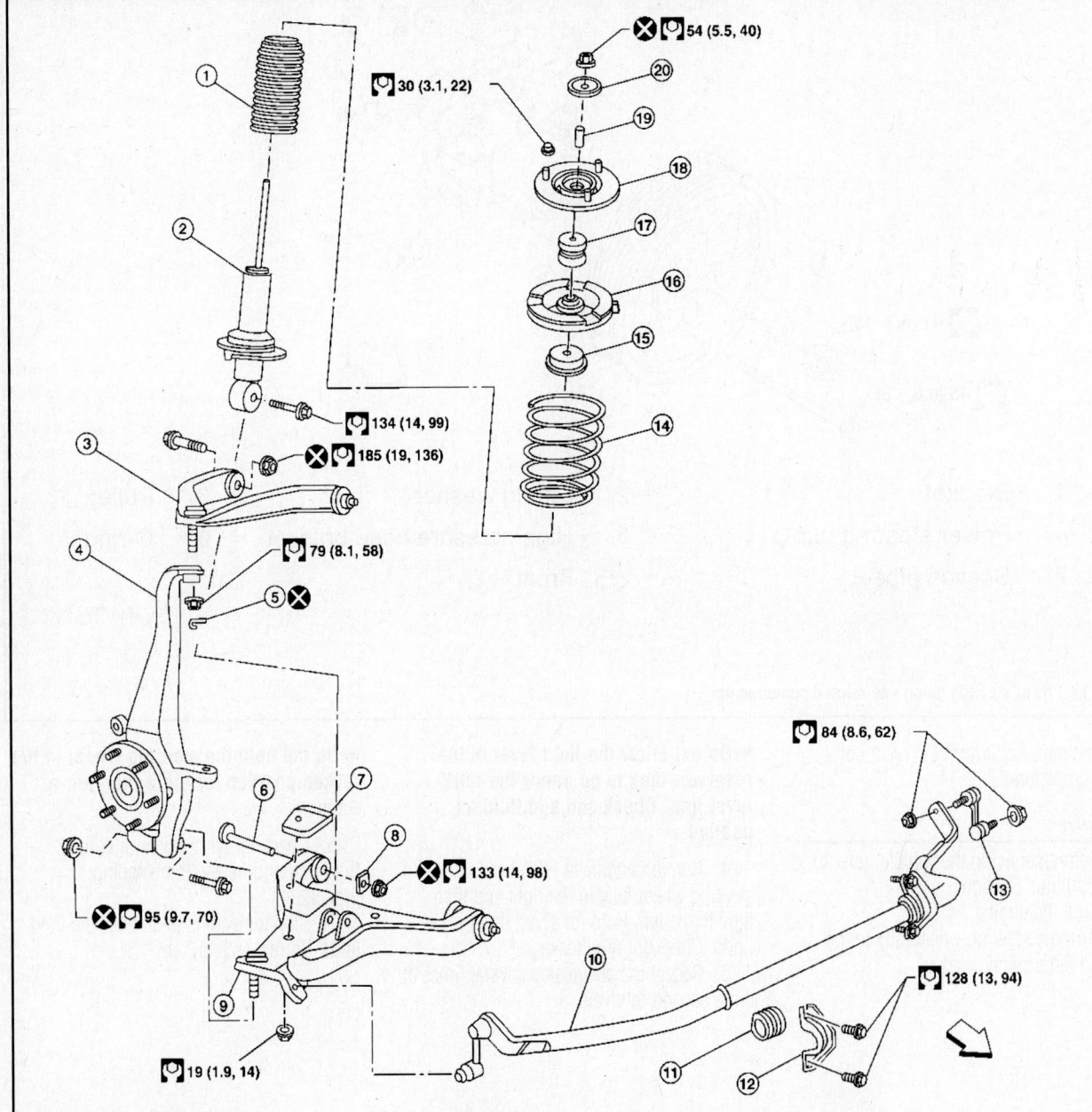

1.	Dust cover	
4.	Steering knuckle	
7.	Jounce bumper	
10.	Stabilizer bar	
13.	Connecting rod	
16.	Upper spring seat	
19.	Spacer	

2.	Shock absorber
5.	Cotter pin
8.	Washer
11.	Stabilizer bar bushing
14.	Coil spring
17.	Shock absorber bushing
20.	Washer

3.	Upper link
6.	Bolt
9.	Lower link
12.	Stabilizer bar mounting bracket
15.	Upper seat
18.	Shock absorber mounting insulator
⇦	Front

37663_QX56_G0104

Fig. 113 Front suspension and related components

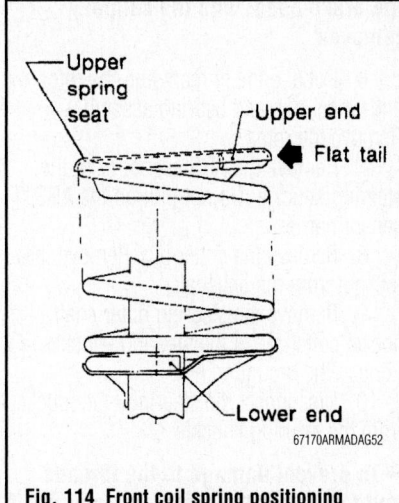

Fig. 114 Front coil spring positioning

➡**Whenever the negative battery cable is disconnected the following components will require resetting. The Idle Air Volume Learning, Steering Angle Sensor Neutral Position, Sunroof Memory Reset/Initialization, Automatic Drive Positioner System, Audio presets and Navigation. Use the CONSULT-III diagnostic tool, or equivalent to perform the required resets.**

2. Disconnect the negative battery cable.
3. Raise and safely support the vehicle.
4. Remove or disconnect the following:
   • Wheel and tire assembly
   • Lower shock absorber bolt
   • Upper shock absorber bolts
   • Coil spring and shock absorber assembly
5. Secure the shock absorber in a vice and loosen (without removing) the piston rod locknut.
6. Install a spring compressor and tighten until the shock absorber mounting insulator can be turned by hand.
7. Remove piston rod locknut and remove shock absorber from the coil spring.

To install:

➡**Be sure to use new fasteners, as required.**

8. Install upper mounting insulator in line with the lower shock absorber mount and step in shock absorber lower seat as shown in figure.
9. Tighten the new piston rod locknut to 40 ft. lbs. (54 Nm).
10. Install or connect the following:
    • Coil spring and shock absorber assembly
    • Upper shock absorber bolts and tighten to 22 ft. lbs (30 Nm)

• Lower shock absorber bolt and tighten to 99 ft. lbs. (134 Nm)
• Wheel and tire assembly
11. Check wheel alignment and adjust as necessary.
12. Be sure to perform the reconnect/relearn procedures.

LOWER BALL JOINT

REMOVAL & INSTALLATION

At this time the manufacturer does not provide removal and installation procedures for this component. The upper ball joint is part of the upper control arm assembly.

LOWER CONTROL ARM

REMOVAL & INSTALLATION
See Figure 115.

➡**Nissan/Infiniti refers to the lower control arm as a lower link.**

1. Before servicing the vehicle, refer to the Precautions Section.

➡**If working near and/or around the SRS system and components, be sure to disable the SRS system. After disabling the system wait three minutes or more before servicing the vehicle.**

➡**Whenever the negative battery cable is disconnected the following components will require resetting. The Idle Air Volume Learning, Steering Angle Sensor Neutral Position, Sunroof Memory Reset/Initialization, Automatic Drive Positioner System, Audio presets and Navigation. Use the CONSULT-III diagnostic tool, or equivalent to perform the required resets.**

2. Disconnect the negative battery cable.
3. Raise and safely support the vehicle.
4. Remove the tire and wheel assembly.
5. Remove the lower shock absorber retaining bolt.
6. Remove the stabilizer bar connecting rod lower nut.
7. Remove the pinch bolt from the steering knuckle, than separate the lower link ball joint from the steering knuckle.
8. Remove the lower link bolts and nuts.
9. Remove the lower link.

To install:

➡**Be sure to use new fasteners, as required.**

10. Installation is the reverse of the removal procedure.

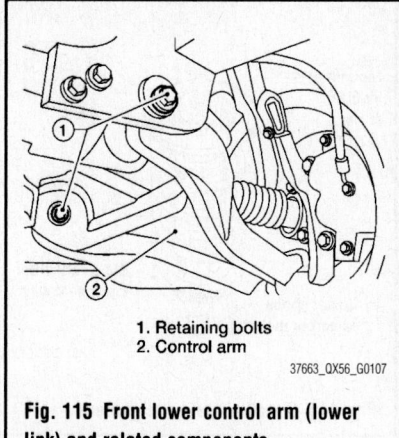

1. Retaining bolts
2. Control arm

37663_QX56_G0107

Fig. 115 Front lower control arm (lower link) and related components

11. Check and adjust alignment, as required.
12. Be sure to perform the reconnect/relearn procedures.

SHOCK ABSORBERS

REMOVAL & INSTALLATION
See Figures 113 and 116.

1. Before servicing the vehicle, refer to the Precautions Section.

➡**If working near and/or around the SRS system and components, be sure to disable the SRS system. After disabling the system wait three minutes or more before servicing the vehicle.**

➡**Whenever the negative battery cable is disconnected the following components will require resetting. The Idle Air Volume Learning, Steering Angle Sensor Neutral Position, Sunroof Memory Reset/Initialization, Automatic Drive Positioner System, Audio presets and Navigation. Use the CONSULT-III diagnostic tool, or equivalent to perform the required resets.**

2. Disconnect the negative battery cable.
3. Raise and safely support the vehicle.
4. Remove or disconnect the following:
   • Wheel and tire assembly
   • Lower shock absorber bolt
   • Upper shock absorber bolts
   • Coil spring and shock absorber assembly

To install:

➡**Be sure to use new fasteners, as required.**

5. Installation is the reverse of the removal procedure.
6. Install upper mounting insulator in line with the lower shock absorber mount

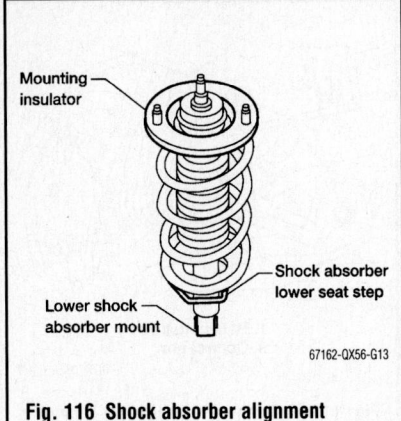

Fig. 116 Shock absorber alignment

and step in shock absorber lower seat as shown in figure.

7. Check wheel alignment and adjust as necessary.

8. Be sure to perform the reconnect/relearn procedures.

STEERING KNUCKLE

REMOVAL & INSTALLATION

See Figure 117.

1. Before servicing the vehicle, refer to the Precautions Section.

➡If working near and/or around the SRS system and components, be sure to disable the SRS system. After disabling the system wait three minutes or more before servicing the vehicle.

➡Whenever the negative battery cable is disconnected the following components will require resetting. The Idle Air Volume Learning, Steering Angle Sensor Neutral Position, Sunroof Memory Reset/Initialization, Automatic Drive Positioner System, Audio presets and Navigation. Use the CONSULT-III diagnostic tool, or equivalent to perform the required resets.

2. Disconnect the negative battery cable.

3. Raise and support the vehicle safely.

4. Remove the tire and wheel assembly.

5. Remove the brake caliper from its mounting and position it to the side.

➡Do not disconnect the hydraulic lines. It is not necessary to remove the bolts on the torque member and brake hose except for disassembly or replacement of the caliper. In this case hang the caliper to the side with mechanics wire so that the brake hose is not under tension. Avoid depressing

the brake pedal with the caliper removed.

6. Put alignment marks on the rotor and wheel hub and bearing assembly. Remove the rotor.

7. Remove the ABS sensor from the steering knuckle. Do not pull on the ABS sensor harness.

8. Remove the cotter pin. Remove the locknut from the halfshaft.

9. Remove the steering outer shaft socket cotter pin at the steering knuckle. Loosen the mounting nut.

10. Disconnect the steering outer socket from the steering knuckle.

➡To prevent damage to the threads and to prevent the tool from coming off suddenly, temporarily loosely install the mounting nut.

11. Remove the halfshaft.

12. Remove the wheel hub and bearing assembly bolts.

13. Remove the splash guard and wheel hub and bearing assembly from the steering knuckle.

14. Support the lower control arm assembly, using a suitable jack.

15. Remove the cotter pin and nut from the upper ball joint.

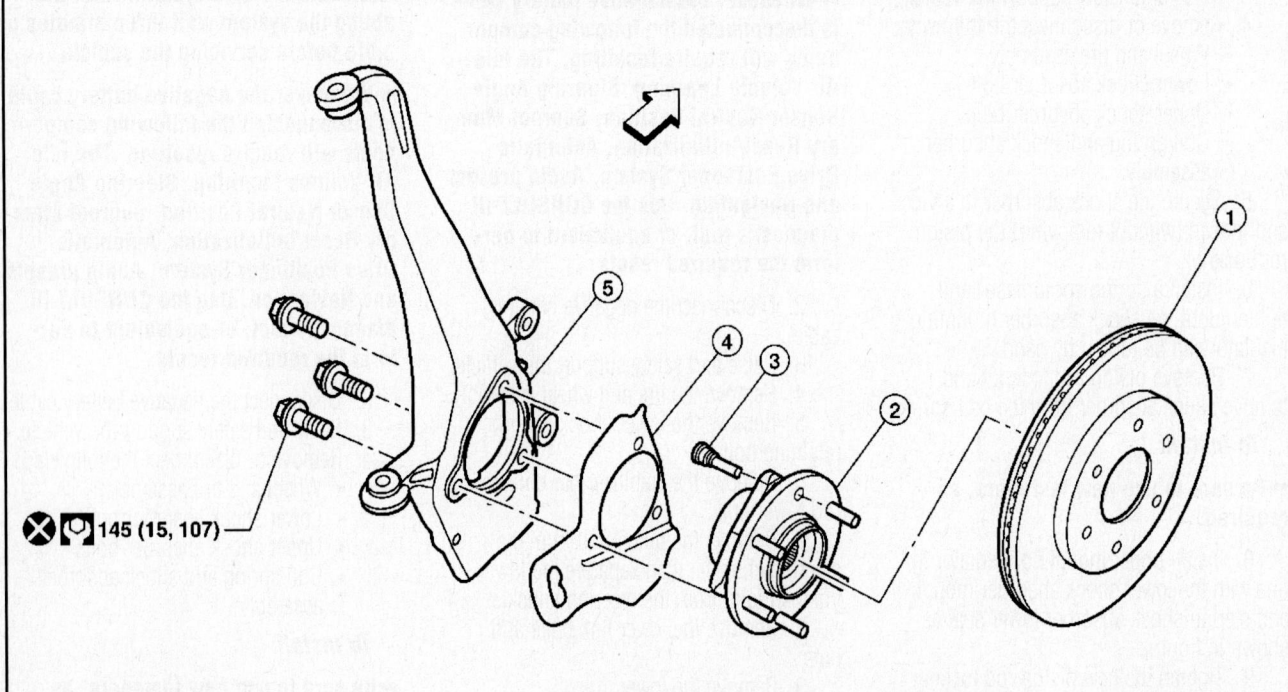

145 (15, 107)

| 1. Disc rotor | 2. Wheel hub and bearing assembly | 3. Wheel stud |
| 4. Splash guard | 5. Steering knuckle | 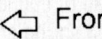 Front |

37663_QX56_G0108

Fig. 117 Front steering knuckle and related components

16. Separate the upper link ball joint from the steering knuckle using tool J-24319-01 or equivalent.

17. Remove the pinch bolt from the steering knuckle. Remove the steering knuckle from the lower control arm ball joint.

18. Remove the steering knuckle from the vehicle.

To install:

➡**Be sure to use new fasteners, as required.**

19. Installation is the reverse of the removal procedure.

20. Be sure to use the alignment marks made during the removal procedure when reinstalling removed components.

21. Check and adjust the front end alignment, as required.

22. Be sure to perform the reconnect/relearn procedures.

STABILIZER BAR

REMOVAL & INSTALLATION

See Figure 113.

1. Before servicing the vehicle, refer to the Precautions Section.

➡**If working near and/or around the SRS system and components, be sure to disable the SRS system. After disabling the system wait three minutes or more before servicing the vehicle.**

➡**Whenever the negative battery cable is disconnected the following components will require resetting. The Idle Air Volume Learning, Steering Angle Sensor Neutral Position, Sunroof Memory Reset/Initialization, Automatic Drive Positioner System, Audio presets and Navigation. Use the CONSULT-III diagnostic tool, or equivalent to perform the required resets.**

2. Disconnect the negative battery cable.

3. Raise and safely support the vehicle.

4. Remove the tire and wheel assembly.

5. Remove the engine under cover.

6. Remove the stabilizer bar mounting bracket retaining bolts and rubber bushings.

7. Remove the connecting rod nuts.

8. Remove the stabilizer bar from the vehicle.

To install:

➡**Be sure to use new fasteners, as required.**

9. Installation is the reverse of the removal procedure.

10. Be sure to perform the reconnect/relearn procedures.

UPPER BALL JOINT

REMOVAL & INSTALLATION

At this time the manufacturer does not provide removal and installation procedures for this component. The upper ball joint is part of the upper control arm assembly.

UPPER CONTROL ARM

REMOVAL & INSTALLATION

See Figures 118 and 119.

➡**Nissan/Infiniti refers to the upper control arm as a upper link.**

1. Before servicing the vehicle, refer to the Precautions Section.

➡**If working near and/or around the SRS system and components, be sure to disable the SRS system. After disabling the system wait three minutes or more before servicing the vehicle.**

➡**Whenever the negative battery cable is disconnected the following components will require resetting. The Idle Air Volume Learning, Steering Angle Sensor Neutral Position, Sunroof Memory Reset/Initialization, Automatic Drive Positioner System, Audio presets and Navigation. Use the CONSULT-III diagnostic tool, or equivalent to perform the required resets.**

2. Disconnect the negative battery cable.

3. Raise and safely support the vehicle.

4. Remove the tire and wheel assembly.

➡**Remove the fender protector to access the upper control arm.**

5. Remove or disconnect the following:
- Cotter pin and nut from upper ball joint

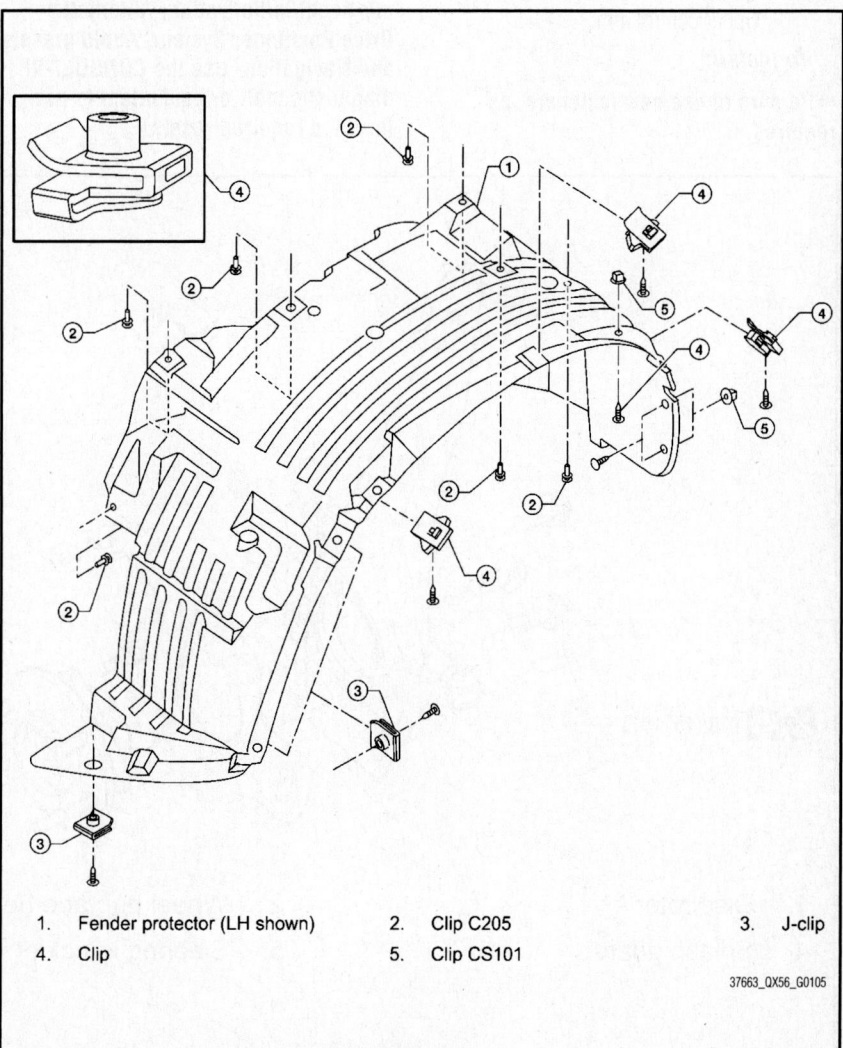

| 1. | Fender protector (LH shown) | 2. | Clip C205 | 3. | J-clip |
| 4. | Clip | 5. | Clip CS101 | | |

37663_QX56_G0105

Fig. 118 Front fender protector and related components

Fig. 119 Front upper control arm (upper link) and related components

6. Separate upper ball joint stud from steering knuckle using special tool J-24319-01.

7. Remove the following:
- Upper control arm mounting bolts. See illustration for bolt locations.
- Upper control arm

To install:

➡**Be sure to use new fasteners, as required.**

8. Installation is the reverse of the removal procedure.

9. Check and adjust alignment, as required.

10. Be sure to perform the reconnect/relearn procedures.

WHEEL HUB & BEARING

REMOVAL & INSTALLATION

See Figure 120.

1. Before servicing the vehicle, refer to the Precautions Section.

➡**If working near and/or around the SRS system and components, be sure to disable the SRS system. After disabling the system wait three minutes or more before servicing the vehicle.**

➡**Whenever the negative battery cable is disconnected the following components will require resetting. The Idle Air Volume Learning, Steering Angle Sensor Neutral Position, Sunroof Memory Reset/Initialization, Automatic Drive Positioner System, Audio presets and Navigation. Use the CONSULT-III diagnostic tool, or equivalent to perform the required resets.**

2. Disconnect the negative battery cable.

3. raise and safely support the vehicle.

4. Remove or disconnect the following:
- Wheel and tire assembly
- Engine splash guard
- Brake caliper without disconnecting the hydraulic lines, and reposition aside with wire

5. Matchmark the brake rotor to the wheel hub and remove the brake rotor.

6. Remove or disconnect the following:
- 4WD, cotter pin and locknut from halfshaft
- Halfshaft from wheel hub and bearing assembly
- ABS sensor
- Wheel hub and bearing assembly bolts
- Wheel hub and bearing assembly

To install:

➡**Be sure to use new fasteners, as required.**

7. Installation is the reverse of the removal procedure.

8. Check and adjust alignment, as required.

9. Be sure to perform the reconnect/relearn procedures.

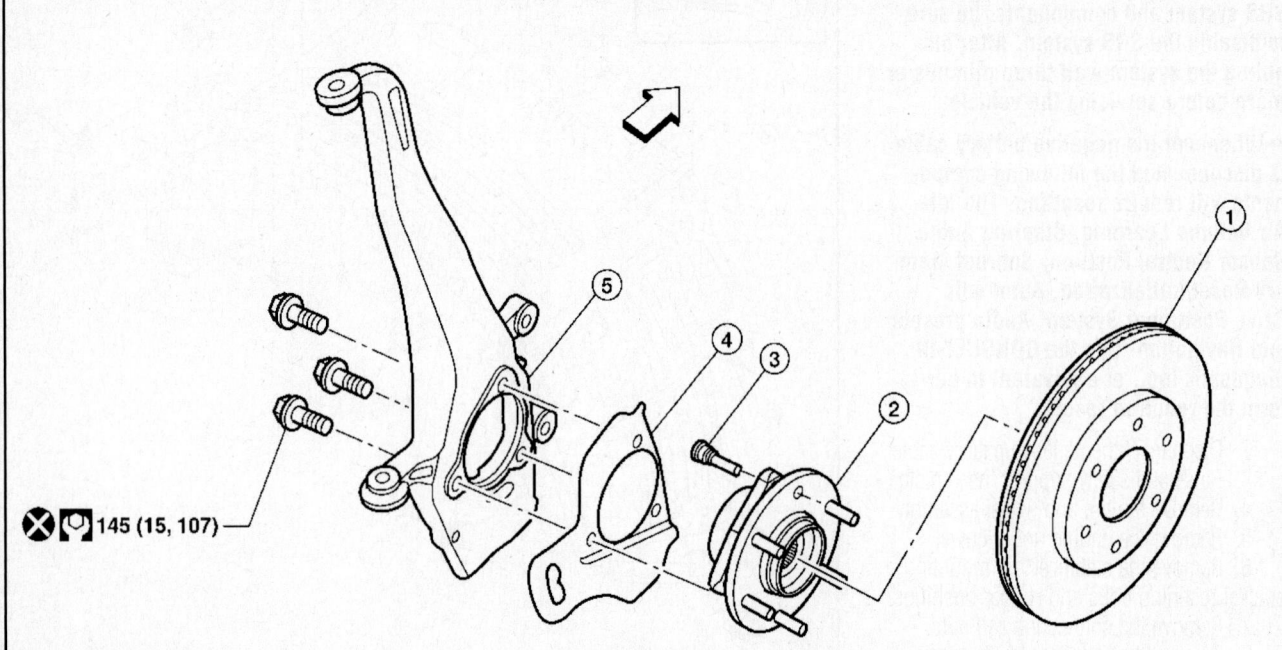

1. Disc rotor
2. Wheel hub and bearing assembly
3. Wheel stud
4. Splash guard
5. Steering knuckle
⇦ Front

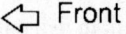

Fig. 120 Front hub/bearing assembly and related components

COIL SPRING

REMOVAL & INSTALLATION

See Figures 121 and 122.

1. Before servicing the vehicle, refer to the Precautions Section.

➡ If working near and/or around the SRS system and components, be sure to disable the SRS system. After disabling the system wait three minutes or more before servicing the vehicle.

➡ Whenever the negative battery cable is disconnected the following components will require resetting. The Idle Air Volume Learning, Steering Angle Sensor Neutral Position, Sunroof Memory Reset/Initialization, Automatic Drive Positioner System, Audio presets

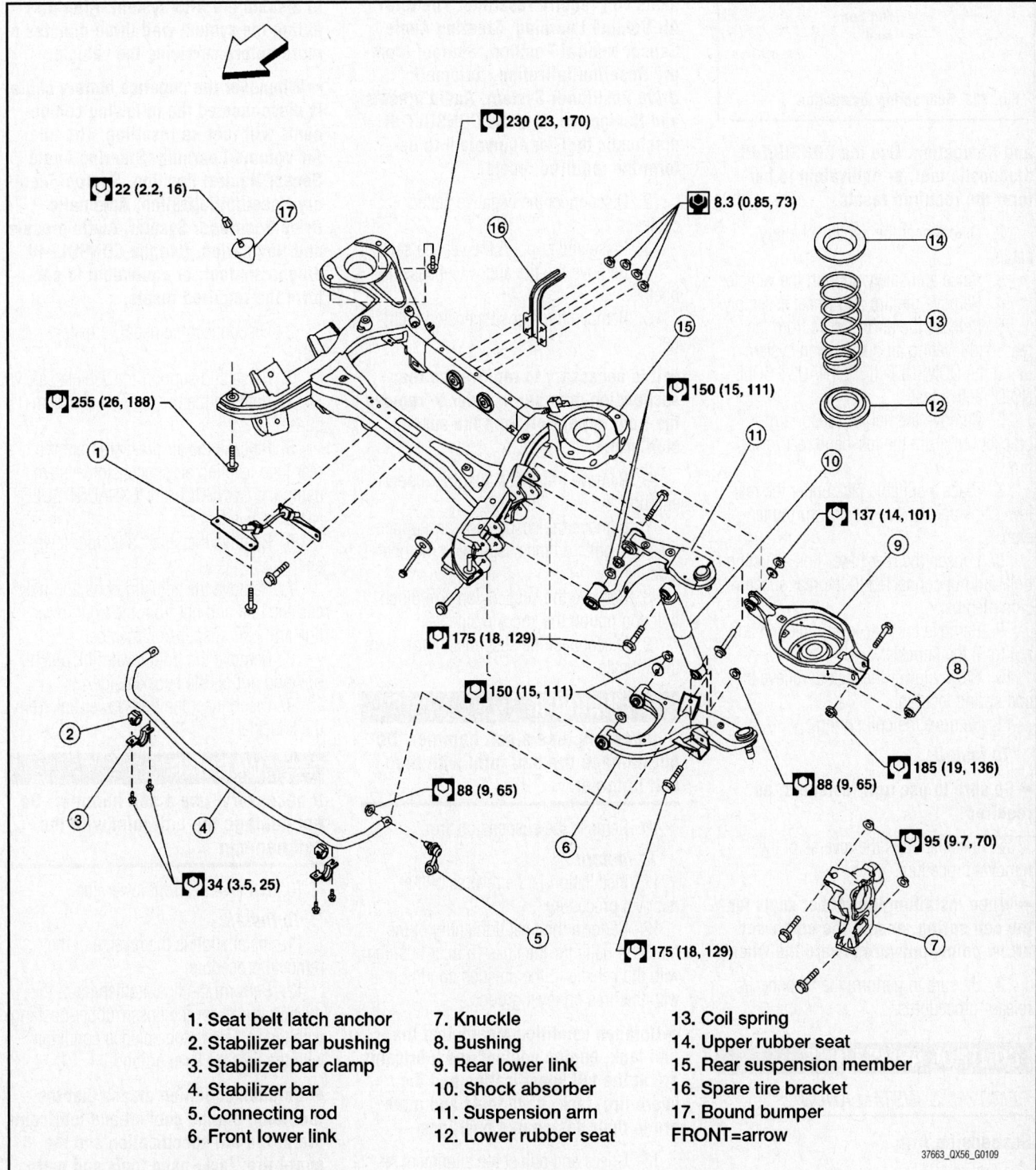

1. Seat belt latch anchor	7. Knuckle
2. Stabilizer bar bushing	8. Bushing
3. Stabilizer bar clamp	9. Rear lower link
4. Stabilizer bar	10. Shock absorber
5. Connecting rod	11. Suspension arm
6. Front lower link	12. Lower rubber seat

13. Coil spring	
14. Upper rubber seat	
15. Rear suspension member	
16. Spare tire bracket	
17. Bound bumper	
FRONT=arrow	

37663_QX56_G0109

Fig. 121 Rear suspension and related components

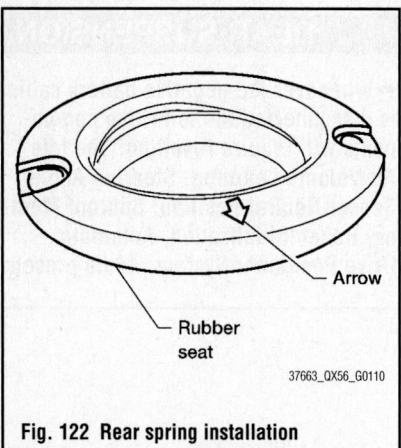

Arrow

Rubber
seat

37663_QX56_G0110

Fig. 122 Rear spring installation

and Navigation. Use the CONSULT-III diagnostic tool, or equivalent to perform the required resets.

2. Disconnect the negative battery cable.

3. Raise and safely support the vehicle.

4. Remove the tire and wheel assembly.

5. Release the air pressure from the rear load leveling air suspension system using the CONSULT-III® "EXHAUST SOLENOID" active test.

6. Remove the height sensor arm bracket bolt from the left-hand rear lower link.

7. Place a suitable jack under the rear lower link and relieve the coil spring tension.

8. Loosen the rear lower link adjusting bolt and nut connected to the rear suspension member.

9. Remove the rear lower link bolt and nut from the knuckle.

10. Slowly lower the jack to relieve the coil spring tension.

11. Remove the coil spring.

To install:

➡**Be sure to use new fasteners, as required.**

12. Installation is the reverse of the removal procedure.

➡**When installing the rubber seats for the coil spring, ensure the embossed arrow points outward toward the wheel.**

13. Be sure to perform the reconnect/relearn procedures.

CONTROL ARMS/LINKS

REMOVAL & INSTALLATION

Suspension Arm

See Figure 121.

1. Before servicing the vehicle, refer to the Precautions Section.

➡**If working near and/or around the SRS system and components, be sure to disable the SRS system. After disabling the system wait three minutes or more before servicing the vehicle.**

➡**Whenever the negative battery cable is disconnected the following components will require resetting. The Idle Air Volume Learning, Steering Angle Sensor Neutral Position, Sunroof Memory Reset/Initialization, Automatic Drive Positioner System, Audio presets and Navigation. Use the CONSULT-III diagnostic tool, or equivalent to perform the required resets.**

2. Disconnect the negative battery cable.

3. Raise and support the vehicle safely.

4. Remove the tire and wheel assemblies.

5. Remove the rear suspension member.

➡**It is necessary to remove the rear suspension member in order to remove the front upper bolt from the suspension arm.**

6. Remove the shock absorber upper end bolt.

7. Remove the suspension arm upper nuts and bolts on the suspension member side.

8. Remove the suspension arm pinch bolt and nut on the knuckle side.

9. Disconnect the suspension arm from the knuckle.

✳✳ WARNING

If necessary, use a soft hammer. Do not damage the ball joint with the soft hammer.

10. Remove the suspension arm.

To install:

11. Installation is the reverse of the removal procedure.

12. Perform the final tightening of the nuts and bolts for the links (rubber bushing) with the vehicle in the unladen condition with the tires on level ground.

➡**Unladen condition means that the fuel tank, engine coolant and lubricants are at the full specification and the spare tire, jack, hand tools and mats are in their designated positions.**

13. Check and adjust the alignment, as required.

14. Be sure to perform the reconnect/relearn procedures.

Front Lower Link

See Figure 121.

1. Before servicing the vehicle, refer to the Precautions Section.

➡**If working near and/or around the SRS system and components, be sure to disable the SRS system. After disabling the system wait three minutes or more before servicing the vehicle.**

➡**Whenever the negative battery cable is disconnected the following components will require resetting. The Idle Air Volume Learning, Steering Angle Sensor Neutral Position, Sunroof Memory Reset/Initialization, Automatic Drive Positioner System, Audio presets and Navigation. Use the CONSULT-III diagnostic tool, or equivalent to perform the required resets.**

2. Disconnect the negative battery cable.

3. Raise and support the vehicle safely.

4. Remove the tire and wheel assemblies.

5. Release the air pressure from the rear load leveling air suspension system using the CONSULT-III® "EXHAUST SOLENOID" active test.

6. Remove the shock absorber lower end bolt.

7. Remove the adjusting bolt and nut, and the bolt and nut from the front lower link and rear suspension member.

8. Remove the front lower link pinch bolt and nut on the knuckle side.

9. Disconnect the front lower link from the knuckle.

✳✳ WARNING

If necessary, use a soft hammer. Do not damage the ball joint with the soft hammer.

10. Remove the front lower link.

To install:

11. Installation is the reverse of the removal procedure.

12. Perform the final tightening of the nuts and bolts for the links (rubber bushing) with the vehicle in the unladen condition with the tires on level ground.

➡**Unladen condition means that the fuel tank, engine coolant and lubricants are at the full specification and the spare tire, jack, hand tools and mats are in their designated positions.**

13. Check and adjust the alignment, as required.

14. Be sure to perform the reconnect/relearn procedures.

Rear Lower Link and Spring

See Figure 121.

1. Before servicing the vehicle, refer to the Precautions Section.

➡If working near and/or around the SRS system and components, be sure to disable the SRS system. After disabling the system wait three minutes or more before servicing the vehicle.

➡Whenever the negative battery cable is disconnected the following components will require resetting. The Idle Air Volume Learning, Steering Angle Sensor Neutral Position, Sunroof Memory Reset/Initialization, Automatic Drive Positioner System, Audio presets and Navigation. Use the CONSULT-III diagnostic tool, or equivalent to perform the required resets.

2. Disconnect the negative battery cable.
3. Raise and support the vehicle safely.
4. Remove the tire and wheel assemblies.
5. Release the air pressure from the rear load leveling air suspension system using the CONSULT-III® "EXHAUST SOLENOID" active test.
6. Remove the height sensor arm bracket bolt from the left-hand rear lower link.
7. Place a suitable jack under the rear lower link and relieve the coil spring tension.
8. Loosen the rear lower link adjusting bolt and nut connected to the rear suspension member.
9. Remove the rear lower link bolt and nut from the knuckle.
10. Slowly lower the jack to relieve the coil spring tension.
11. Remove the coil spring.
12. Remove the upper rubber seat, coil spring and lower rubber seat from the rear lower link.
13. Remove the rear lower link adjusting bolt and nut from the rear suspension member.
14. Remove the rear lower link from its mounting.

To install:

15. Installation is the reverse of the removal procedure.
16. When installing the upper and lower rubber seats for the rear coil springs, the arrow embossed on the rubber seats must point out toward the wheel and tire assembly.

17. Tighten the rear lower link bolt to knuckle to 70 ft. lbs. (95 Nm).
18. Tighten the rear lower link adjusting bolt to rear suspension member to 101 ft. lbs. (137 Nm).
19. Tighten the height sensor arm bracket bolt to left-head rear lower link to 9 ft. lbs. (12 Nm).
20. Perform the final tightening of the nuts and bolts for the links (rubber bushing) with the vehicle in the unladen condition with the tires on level ground.

➡Unladen condition means that the fuel tank, engine coolant and lubricants are at the full specification and the spare tire, jack, hand tools and mats are in their designated positions.

21. Check and adjust the alignment, as required.
22. Be sure to perform the reconnect/relearn procedures.

SHOCK ABSORBER

REMOVAL & INSTALLATION

See Figure 123.

1. Before servicing the vehicle, refer to the Precautions Section.

➡If working near and/or around the SRS system and components, be sure to disable the SRS system. After disabling the system wait three minutes or more before servicing the vehicle.

➡Whenever the negative battery cable is disconnected the following components will require resetting. The Idle Air Volume Learning, Steering Angle Sensor Neutral Position, Sunroof Memory Reset/Initialization, Automatic Drive Positioner System, Audio presets and Navigation. Use the CONSULT-III diagnostic tool, or equivalent to perform the required resets.

2. Disconnect the negative battery cable.
3. Raise and safely support the vehicle.
4. Remove the tire and wheel assemblies.
5. Release the air pressure from the rear load leveling air suspension system using the CONSULT-III® "EXHAUST SOLENOID" active test.
6. Remove or disconnect the following:
   - Rear fender protector
   - Rear load leveling air suspension hose from the shock absorber
   - Shock absorber upper and lower end bolts
   - Shock absorber

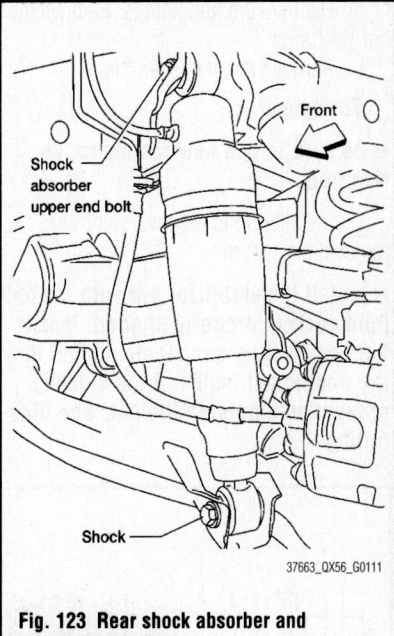

Fig. 123 Rear shock absorber and related components

Shock absorber upper end bolt

Front

Shock

37663_QX56_G0111

To install:

➡Be sure to use new fasteners, as required.

7. Installation is the reverse of the removal procedure.
8. Be sure to perform the reconnect/relearn procedures.

STABILIZER BAR

REMOVAL & INSTALLATION

See Figures 121 and 124.

1. Before servicing the vehicle, refer to the Precautions Section.

➡If working near and/or around the SRS system and components, be sure to disable the SRS system. After disabling the system wait three minutes or more before servicing the vehicle.

➡Whenever the negative battery cable is disconnected the following components will require resetting. The Idle Air Volume Learning, Steering Angle Sensor Neutral Position, Sunroof Memory Reset/Initialization, Automatic Drive Positioner System, Audio presets and Navigation. Use the CONSULT-III diagnostic tool, or equivalent to perform the required resets.

2. Disconnect the negative battery cable.
3. Raise and safely support the vehicle.
4. Disconnect the bar ends from the connecting rods.

5. Remove the bar clamps. Remove the bar bushings.

6. Remove the stabilizer bar.

To install:

➡ Be sure to use new fasteners, as required.

7. Installation is the reverse of the removal procedure.

➡ Install the stabilizer bar with the ball joint sockets properly aligned. Install the bar bushing and clamp so that they are positioned inside of the sideslip prevention clamp on the bar. See illustration

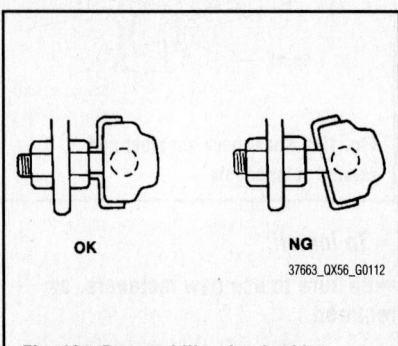

OK   NG

37663_QX56_G0112

Fig. 124 Rear stabilizer bar bushing positioning

8. Be sure to perform the reconnect/relearn procedures.

WHEEL HUB & BEARING (SEALED UNIT)

REMOVAL & INSTALLATION

See Figure 125.

1. Before servicing the vehicle, refer to the Precautions Section.

➡ If working near and/or around the SRS system and components, be sure to disable the SRS system. After disabling the system wait three minutes or more before servicing the vehicle.

➡ Whenever the negative battery cable is disconnected the following components will require resetting. The Idle Air Volume Learning, Steering Angle Sensor Neutral Position, Sunroof Memory Reset/Initialization, Automatic Drive Positioner System, Audio presets and Navigation. Use the CONSULT-III diagnostic tool, or equivalent to perform the required resets.

2. Disconnect the negative battery cable.
3. Raise and support the vehicle safely.
4. Remove or disconnect the following:

- Wheel and tire assembly
- Brake caliper without disconnecting the hydraulic lines, and reposition aside with wire
- Brake rotor
- Cotter pin and nut from driveshaft
- Driveshaft
- Wheel hub and bearing assembly bolts

5. Pulling out the wheel hub and bearing assembly slightly, remove the ABS sensor.

6. Remove the wheel hub and bearing assembly.

To install:

➡ Be sure to use new fasteners, as required.

7. Install or connect the following:
- ABS sensor
- Wheel hub and bearing assembly, using new bolts
- Driveshaft
- Lock nut and new cotter pin
- Brake rotor
- Brake caliper
- Wheel and tire assembly

8. Be sure to perform the reconnect/relearn procedures.

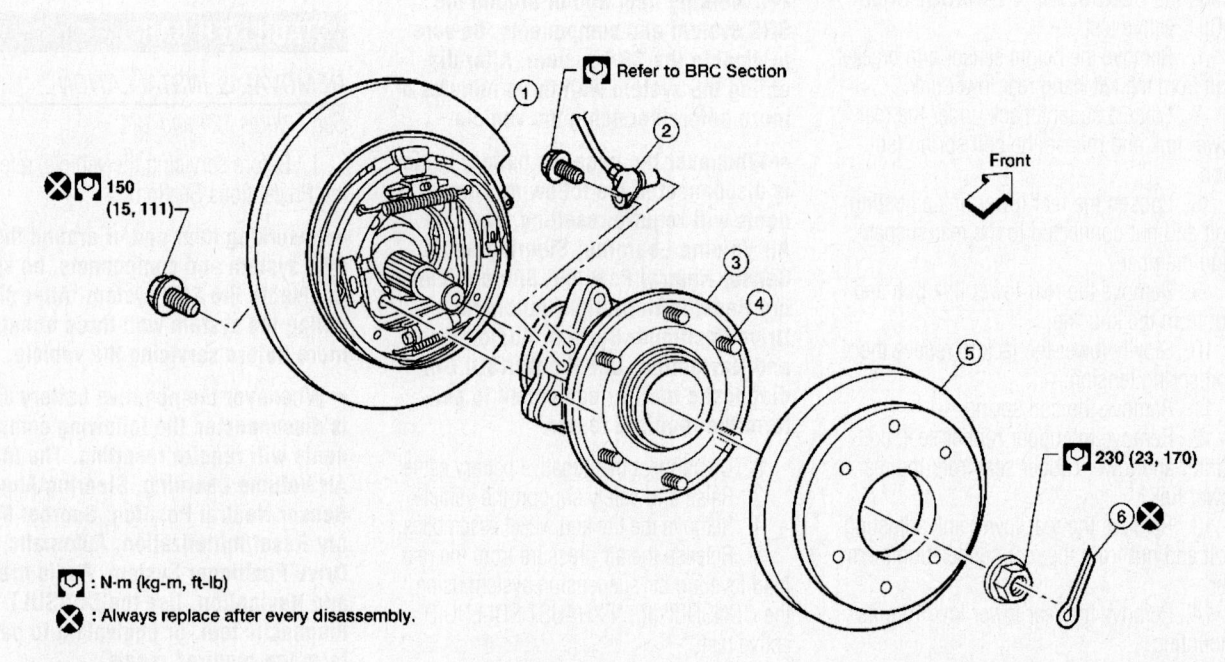

Refer to BRC Section

150 (15, 111)

Front

230 (23, 170)

🔧 : N·m (kg-m, ft-lb)

✖ : Always replace after every disassembly.

| 1. | Back plate | 2. | Rear ABS sensor | 3. | Wheel hub and bearing assembly |
| 4. | Wheel stud | 5. | Rear disc rotor | 6. | Cotter pin |

37663_QX56_G0113

Fig. 125 Rear hub/bearing assembly and related components

NISSAN

Cube

12

SPECIFICATIONS AND MAINTENANCE CHARTS

ENGINE AND VEHICLE IDENTIFICATION

		Engine						Model Year	
Code ①	Liters (cc)	Cu. In.	Cyl.	Fuel Sys.	Engine Type	Eng. Mfg.		Code ②	Year
MR18DE	1.8 (1797)	110	4	MFI	DOHC	Nissan		9	2009
								A	2010

MFI: Multi-port Fuel Injection

DOHC: Double Overhead Camshaft

① The Engine Code is stamped on the engine block near the starter.

② 10th position of the Vehicle Identification Number (VIN)

37663_CUBE_C0001

GENERAL ENGINE SPECIFICATIONS

Year	Model	Engine Displacement Liters	Engine Series (ID/VIN)	Net Horsepower @ rpm	Net Torque @ rpm (ft. lbs.)	Bore x Stroke (in.)	Compression Ratio	Oil Pressure @ rpm
2009	Cube	1.8	MR18DE	122@5200	127@4800	3.31X3.19	9.9:1	29@2000
2010	Cube	1.8	MR18DE	122@5200	127@4800	3.31X3.19	9.9:1	29@2000

37663_CUBE_C0002

ENGINE TUNE-UP SPECIFICATIONS

Year	Engine Displacement Liters	Engine ID/VIN	Spark Plug Gap (in.)	Ignition Timing (deg.) MT	Ignition Timing (deg.) AT	Fuel Pump (psi) ①	Idle Speed (rpm) MT	Idle Speed (rpm) AT ②	Valve Clearance Intake ③	Valve Clearance Exhaust ③
2009	1.8	MR18DE	0.043	13B	13B	51	650-750	650-750	0.010-0.013	0.011-0.015
2010	1.8	MR18DE	0.043	13B	13B	51	650-750	650-750	0.010-0.013	0.011-0.015

B: Before top dead center

① At idle

② Automatic transmission in neutral

③ Engine cold

37663_CUBE_C0003

CAPACITIES

Year	Model	Engine Displacement Liters	Engine ID/VIN	Engine Oil with Filter (qts.)	Transmission (pts.)		Fuel Tank (gal.)	Cooling System (qts.)
					5-Spd	Auto.		
2009	Cube	1.8	MR18DE	4.25	4.25	15.75	13.3	①
2010	Cube	1.8	MR18DE	4.25	4.25	15.75	13.3	①

① CVT models: 7.5 qts.
   M/T models: 7.25 qts.

37663_CUBE_C0004

FLUID SPECIFICATIONS

Year	Model	Engine Displ. Liters	Engine Oil	Man. Trans.	Auto. Trans.	Drive Axle		Transfer Case	Power Steering Fluid	Brake Master Cylinder	Cooling System
						Front	Rear				
2009	Cube	1.8	5W-30	75W-80	Nissan NS-2	NA	NA	NA	NA	DOT 3	N-LL
2010	Cube	1.8	5W-30	75W-80	Nissan NS-2	NA	NA	NA	NA	DOT 3	N-LL

NA: Not Applicable

N-LL: Nissan Long Life coolant

37663_CUBE_C0005

VALVE SPECIFICATIONS

Year	Engine ID/VIN	Engine Displacement Liters	Seat Angle (deg.)	Face Angle (deg.)	Spring Test Pressure (lbs. @ in.)	Spring Installed Height (in.)	Stem-to-Guide Clearance (in.)		Stem Diameter (in.)	
							Intake	Exhaust	Intake	Exhaust
2009	MR18DE	1.8	45.15-45.45	—	①	1.390	0.0008-0.0021	0.0012-0.0025	0.2152-0.2157	0.2148-0.2154
2010	MR18DE	1.8	45.15-45.45	—	①	1.390	0.0008-0.0021	0.0012-0.0025	0.2152-0.2157	0.2148-0.2154

① Intake: 34-39 lbs. @ 1.39 inches
   Exhaust: 31-36 lbs. @ 1.39 inches

37663_CUBE_C0006

CAMSHAFT SPECIFICATIONS

All measurements in inches unless noted

Year	Engine Displacement Liters	Engine Code/ID	Journal Dia.	Brg. Oil Clearance	Shaft End-play	Circle Runout	Lobe Height	
							Intake	Exhaust
2009	1.8	MR18DE	①	②	0.0030-0.0060	0.0008	1.7561-1.7636	1.6998-1.7073
2010	1.8	MR18DE	①	②	0.0030-0.0060	0.0008	1.7561-1.7636	1.6998-1.7073

① No. 1: 1.0998-1.1006
   All others: 0.98236-0.9831

② No. 1: 0.0018-0.0034 inches
   All others: 0.0012-0.0028 inches

37663_CUBE_C0007

CRANKSHAFT AND CONNECTING ROD SPECIFICATIONS

All measurements are given in inches.

Year	Engine Displacement Liters	Engine ID/VIN	Crankshaft				Connecting Rod		
			Main Brg. Journal Dia.	Main Brg. Oil Clearance	Shaft End-play	Thrust on No.	Journal Diameter	Oil Clearance	Side Clearance
2009	1.8	MR18DE	2.0457-2.0464	①	0.0039-0.0102	3	1.8504-1.8509	0.0002-0.0009	0.0079-0.0138
2010	1.8	MR18DE	2.0457-2.0464	①	0.0039-0.0102	3	1.8504-1.8509	0.0002-0.0009	0.0079-0.0138

① Nos. 1, 4, 5 : 0.0009-0.0013
   Nos. 2, 3 : 0.0005-0.0009

37663_CUBE_C0008

PISTON AND RING SPECIFICATIONS

All measurements are given in inches.

Year	Engine Displacement Liters	Engine ID/VIN	Piston Clearance	Ring Gap			Ring Side Clearance		
				Top Compression	Bottom Compression	Oil Control	Top Compression	Bottom Compression	Oil Control
2009	1.8	MR18DE	0.0008-0.0016	0.0079-0.0118	0.0197-0.0256	0.0059-0.0177	0.0016-0.0031	0.0012-0.0028	0.0060-0.0073
2010	1.8	MR18DE	0.0008-0.0016	0.0079-0.0118	0.0197-0.0256	0.0059-0.0177	0.0016-0.0031	0.0012-0.0028	0.0060-0.0073

37663_CUBE_C0009

TORQUE SPECIFICATIONS

All readings in ft. lbs.

Year	Engine Displacement Liters	Engine ID/VIN	Cylinder Head Bolts	Main Bearing Bolts	Rod Bearing Bolts	Crankshaft Damper Bolts	Flywheel Bolts	Manifold Intake	Manifold Exhaust	Spark Plugs	Oil Drain Plug
2009	1.8	MR18DE	①	②	③	④	80	20	25	14	25
2010	1.8	MR18DE	①	②	③	④	80	20	25	14	25

① Step 1: 30 ft. lbs.

   Step 2: Turn each bolt, in sequence, an additional 100 degrees

   Step3: Loosen completely, then retorque to 30 ft. lbs.

   Step 4: Turn each bolt, in sequence, an additional 100 degrees

   Step 5: Turn each bolt, in sequence, an additional 100 degrees

② Step 1: Tighten bolts 25 ft. lbs.

   Step 2: Tighten bolts an additional 60 degrees

③ Step 1: Tighten bolts to 20 ft. lbs.

   Step 2: Loosen all bolts

   Step 3: tighten bolts to 14 ft. lbs.

   Step 4: Tighten bolts an additional 60 degrees

④ Step 1: Tighten to 51 ft. lbs.

   Step 2: Loosen bolt completely

   Step 3: Tighten to 22 ft. lbs.

   Step 4: Tighten an additional 60 degrees

37663_CUBE_C0010

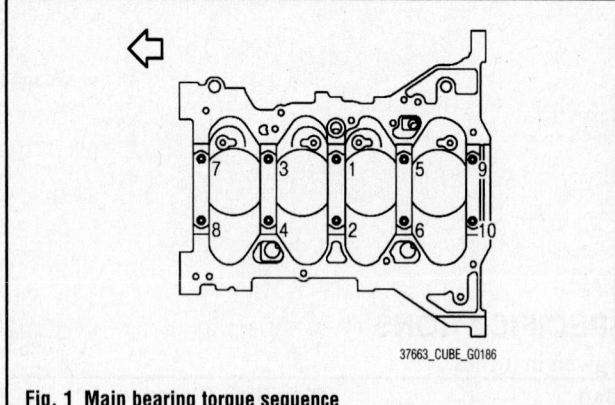

37663_CUBE_G0186

Fig. 1 Main bearing torque sequence

WHEEL ALIGNMENT

Year	Model		Caster		Camber		Toe-in (in.)
			Range (+/-Deg.)	Preferred Setting (Deg.)	Range (+/-Deg.)	Preferred Setting (Deg.)	
2009	Cube	F	0.75	4.67	0.75	-0.17	0.05 +0.02/- 0.05
		R	—	—	0.50	-1.51	0.118 +0.157/-0.079
2010	Cube	F	0.75	4.67	0.75	-0.17	0.05 +0.02/- 0.05
		R	—	—	0.50	-1.51	0.118 +0.157/-0.079

37663_CUBE_C0011

TIRE, WHEEL AND BALL JOINT SPECIFICATIONS

Year	Model	OEM Tires		Tire Pressures (psi)		Wheel Size	Lug Nut Torque (ft. lbs.)
		Standard	Optional	Front	Rear		
2009	Cube	P195/60R15	P195/55R16	33	33	6.5-JJ	80
2010	Cube	P195/60R15	P195/55R16	33	33	6.5-JJ	80

OEM: Original Equipment Manufacturer

PSI: Pounds Per Square Inch

37663_CUBE_C0012

BRAKE SPECIFICATIONS
All measurements in inches unless noted

Year	Model		Brake Disc			Brake Drum Diameter			Minimum Lining Thickness		Brake Caliper	
			Original Thickness	Minimum Thickness	Maximum Run-out	Original Inside Diameter	Max. Wear Limit	Maximum Machine Diameter	Front	Rear	Bracket Bolts (ft. lbs.)	Mounting Bolts (ft. lbs.)
2009	Cube	F	0.945	0.866	0.001	—	—	—	0.079	—	62	20
		R	0.354	0.315	0.002	9.00	9.06	—	—	0.079	—	—
2010	Cube	F	0.945	0.866	0.001	—	—	—	0.079	—	62	20
		R	0.354	0.315	0.002	9.00	9.06	—	—	0.079	—	—

37663_CUBE_C0013

SCHEDULED MAINTENANCE INTERVALS
Nissan—Cube

TO BE SERVICED	TYPE OF SERVICE	VEHICLE MILEAGE INTERVAL (x1000)												
		7.5	15	22.5	30	37.5	45	52.5	60	67.5	75	82.5	90	97.5
Engine oil & filter	R	✔	✔	✔	✔	✔	✔	✔	✔	✔	✔	✔	✔	✔
Brake lines & cables	S/I		✔		✔		✔		✔		✔		✔	
Brake pads, discs, drums & linings	S/I		✔		✔		✔		✔		✔		✔	
Driveshaft boots	S/I		✔		✔		✔		✔		✔		✔	
Exhaust system	S/I				✔				✔				✔	
Transaxle fluid	S/I		✔		✔		✔		✔		✔		✔	
Air cleaner filter	R				✔				✔				✔	
Spark plugs (except platinum)	R				✔				✔				✔	
Spark plugs (iridium and platinum)	R	Replace every 105,000 miles												
Steering gear & linkage, axle & suspension parts	S/I			✔					✔				✔	
Engine coolant	R	Replace every 60,000 miles, then every 30,000 miles												
Inverter coolant	R	Replace every 60,000 miles, then every 30,000 miles												
Drive belts	S/I								✔					
Fuel lines	S/I								✔					
Vapor lines	S/I								✔					
Cabin microfilter	R		✔		✔		✔		✔		✔		✔	
Valve adjustment	S/I	As needed												

R: Replace   S/I: Service or Inspect

FREQUENT OPERATION MAINTENANCE (SEVERE SERVICE)

If a vehicle is operated under any of the following conditions it is considered severe service:

- Extremely dusty areas.
- 50% or more of the vehicle operation is in 32°C (90°F) or higher temperatures, or constant operation in temperatures below 0°C (32°F).
- Prolonged idling (vehicle operation in stop and go traffic).
- Frequent short running periods (engine does not warm to normal operating temperatures).
- Police, taxi, delivery usage or trailer towing usage.

Oil & oil filter: change every 3750 miles.

Brake pads & discs: service or inspect every 7500 miles.

Driveshaft boots: service or inspect every 7500 miles.

Exhaust system: service or inspect every 7500 miles.

Steering gear & linkage, axle & suspension parts: service or inspect every 7500 miles.

Steering linkage ball joints & front suspension ball joints: service or inspect every 7500 miles.

Air cleaner filter: service or inspect every 15,000 miles.

37663_CUBE_C0014

BRAKES INFORMATION AND PRECAUTIONS

ANTI-LOCK SYSTEMS

• Certain components within the ABS system are not intended to be serviced or repaired individually.

• Do not use rubber hoses or other parts not specifically specified for and ABS system. When using repair kits, replace all parts included in the kit. Partial or incorrect repair may lead to functional problems and require the replacement of components.

• Lubricate rubber parts with clean, fresh brake fluid to ease assembly. Do not use shop air to clean parts; damage to rubber components may result.

• Use only DOT 3 brake fluid from an unopened container.

• If any hydraulic component or line is removed or replaced, it may be necessary to bleed the entire system.

• A clean repair area is essential. Always clean the reservoir and cap thoroughly before removing the cap. The slightest amount of dirt in the fluid may plug an orifice and impair the system function. Perform repairs after components have been thoroughly cleaned; use only denatured alcohol to clean components. Do not allow ABS components to come into contact with any substance containing mineral oil; this includes used shop rags.

• The Anti-Lock control unit is a microprocessor similar to other computer units in the vehicle. Ensure that the ignition switch is **OFF** before removing or installing controller harnesses. Avoid static electricity discharge at or near the controller.

• If any arc welding is to be done on the vehicle, the control unit should be unplugged before welding operations begin.

DISC AND DRUM SYSTEMS

✳✳ CAUTION

Dust and dirt accumulating on brake parts during normal use may contain asbestos fibers from production or aftermarket brake linings. Breathing excessive concentrations of asbestos fibers can cause serious bodily harm. Exercise care when servicing brake parts. Do not sand or grind brake lining unless equipment used is designed to contain the dust residue. Do not clean brake parts with compressed air or by dry brushing. Cleaning should be done by dampening the brake components with a fine mist of water, then wiping the brake components clean with a dampened cloth. Dispose of cloth and all residue containing asbestos fibers in an impermeable container with the appropriate label. Follow practices prescribed by the Occupational Safety and Health Administration (OSHA) and the Environmental Protection Agency (EPA) for the handling, processing, and disposing of dust or debris that may contain asbestos fibers.

BRAKES BLEEDING THE BRAKE SYSTEM

BLEEDING PROCEDURE

BLEEDING PROCEDURE

✳✳ CAUTION

Turn the ignition switch OFF and disconnect the ABS actuator and electric unit (control unit) connector or the battery negative terminal before performing the work. Monitor the fluid level in the reservoir tank while performing the air bleeding. Always use new brake fluid for refilling. Never reuse the drained brake fluid.

1. Connect a vinyl tube to the bleeder valve of the rear right brake.
2. Fully depress the brake pedal 4 to 5 times.
3. Loosen the bleeder valve and bleed air with the brake pedal depressed, and then quickly tighten the bleeder valve.

4. Repeat steps 2 and 3 until all of the air is out of the brake line.
5. Tighten the bleeder valve.
6. Perform steps 1 to 5 for the rear right brake, front left brake, rear left brake, and front right brake in that order.
7. Check that the fluid level in the reservoir tank is within the specified range after air bleeding.
8. Check each item of brake pedal. Adjust it if the measurement value is not the standard.

BRAKES ANTI-LOCK BRAKE SYSTEM (ABS)

WHEEL SPEED SENSORS

REMOVAL & INSTALLATION

Front

See Figure 2.

1. Remove the fender protector.
2. Remove the wheel sensor from steering knuckle.

✳✳ CAUTION

Never twist sensor harness as much as possible, when removing it. Pull wheel sensors out without pulling sensor harness.

3. Remove the wheel sensor harness from vehicle.

✳✳ CAUTION

Never twist sensor harness as much as possible, when removing it. Pull wheel sensors out without pulling sensor harness.

To install:

4. Note the following, and install in the reverse order of the removal.

a. Make sure there is no foreign material such as iron chips on and in the mounting hole of the wheel sensor.

b. Make sure no foreign material has been caught in the sensor rotor. Remove any foreign material and clean the mount. Replace the wheel sensor if necessary.

c. When installing wheel sensor, be sure to press rubber grommets in until they lock at locations shown above in the figure. When installed, harness must not be twisted.

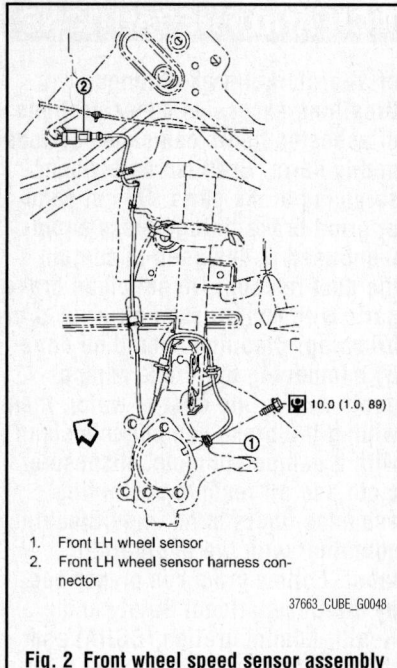

1. Front LH wheel sensor
2. Front LH wheel sensor harness connector

37663_CUBE_G0048

Fig. 2 Front wheel speed sensor assembly

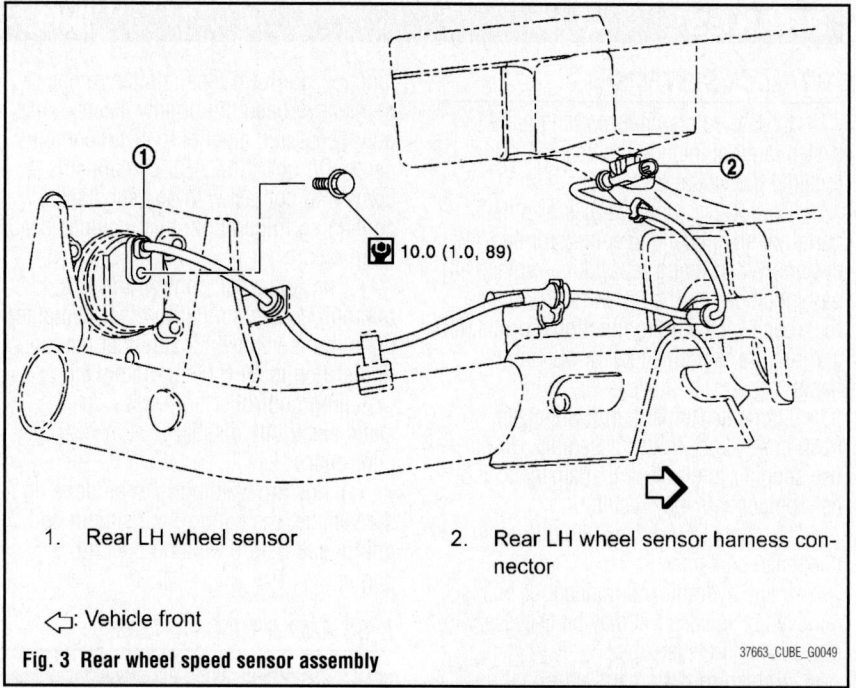

10.0 (1.0, 89)

| 1. Rear LH wheel sensor | 2. Rear LH wheel sensor harness con-
nector |

⇦: Vehicle front

Fig. 3 Rear wheel speed sensor assembly   37663_CUBE_G0049

Rear

See Figure 3.

1. Remove wheel sensor from wheel hub and bearing assembly.

❋❋ CAUTION

Never twist sensor harness as much as possible, when removing it. Pull wheel sensors out without pulling sensor harness.

2. Remove wheel sensor harness from vehicle.

❋❋ CAUTION

Never twist sensor harness as much as possible, when removing it. Pull wheel sensors out without pulling sensor harness.

To install:

3. Note the following, and install in the reverse order of the removal.

 a. Make sure there is no foreign material such as iron chips on and in the mounting hole of the wheel sensor.

 b. Make sure no foreign material has been caught in the sensor rotor. Remove any foreign material and clean the mount. Replace the wheel sensor if necessary.

 c. When installing wheel sensor, be sure to press rubber grommets in until they lock at locations shown above in the figure. When installed, harness must not be twisted.

WHEEL SPEED SENSOR RINGS (TOOTHED RINGS)

REMOVAL & INSTALLATION

❋❋ CAUTION

Sensor rotor cannot be disassembled. Remove the sensor rotor together with hub bearing assembly.

1. Remove the wheel hub and bearing assembly.

➡**Sensor rotor cannot be disassembled. Remove the sensor rotor together with hub bearing assembly.**

To install:

2. Install the wheel hub and bearing assembly.

BRAKES

BRAKE CALIPER

REMOVAL & INSTALLATION

See Figure 4.

❋❋ WARNING

Clean any dust from the brake caliper and brake pads with a vacuum dust collector. Never blow with compressed air.

❋❋ CAUTION

Never spill or splash brake fluid on painted surfaces. Brake fluid may seriously damage paint. Wipe it out immediately and wash with water if it gets on a protect surface.

1. Remove tires and wheels.
2. Fix the disc rotor using wheel nuts.
3. Drain brake fluid.

FRONT DISC BRAKES

❋❋ CAUTION

Never spill or splash brake fluid on the disc rotor.

4. Remove union bolt and copper washer, and disconnect brake hose from caliper assembly.

❋❋ CAUTION

Never depress the brake pedal. Brake fluid may splash while removing the brake hose.

5. Remove torque member mounting bolts, and remove brake caliper assembly.

✳ CAUTION

Never drop brake pad and caliper assembly.

6. Remove disc rotor.

✳ CAUTION

Put matching marks on the wheel hub and bearing assembly and the disc rotor before removing the disc rotor. Never drop disc rotor.

To install:

✳ WARNING

Clean any dust from the brake caliper and brake pads with a vacuum dust collector. Never blow with compressed air.

✳ CAUTION

Never depress the brake pedal. Brake fluid may splash while removing the brake hose. Never spill or splash brake fluid on painted surfaces. Brake fluid may seriously damage paint. Wipe it out immediately and wash with water if it gets on a protect surface.

7. Install disc rotor.

✳ CAUTION

Align the matching marks that have been made during removal when reusing the disc rotor.

8. Install the brake caliper assembly to the steering knuckle and tighten the torque member mounting bolts to the specified torque.

✳ CAUTION

Never spill or splash any grease and moisture on the brake caliper assembly mounting face, threads, mounting bolts and washers. Wipe out any grease and moisture.

9. Install brake hose and copper washers to brake caliper assembly, and tighten union bolts to the specified torque.

✳ CAUTION

Never reuse copper washer.

10. Refill with new brake fluid and perform the air bleeding.

✳ CAUTION

Never reuse drained brake fluid. Never spill or splash brake fluid on the disc rotor.

11. Check a drag of front disc brake. If any drag is found, inspect the installation.
12. Install tires.

DISC BRAKE PADS

REMOVAL & INSTALLATION
See Figure 5.

✳ WARNING

Clean any dust from the brake caliper and brake pads with a vacuum dust collector. Never blow with compressed air.

✳ CAUTION

Never depress the brake pedal while removing the brake pads because the piston may pop out. Never spill or splash brake fluid on the disc rotor.

1. Remove tires and wheels.
2. Remove lower sliding pin bolt.
3. Suspend the cylinder body with suitable wire so that the brake hose will not stretch. Then remove the brake pads, shims, shim covers and pad retainers from the torque member.

✳ CAUTION

Note the following:

- Never deform the pad retainer when removing the pad retainer from the torque member.
- Never damage the piston boot.
- Never drop the brake pads, shims, and the shim covers.
- Remember each position of the removed brake pads.

To install:

✳ WARNING

Clean any dust from the brake caliper and brake pads with a vacuum dust collector. Never blow with compressed air.

✳ CAUTION

Never depress the brake pedal while removing the brake pads or the cylinder body because the piston may pop out. Never spill or splash brake fluid on the disc rotor.

4. Install the pad retainers to the torque member if the pad retainers has been removed.

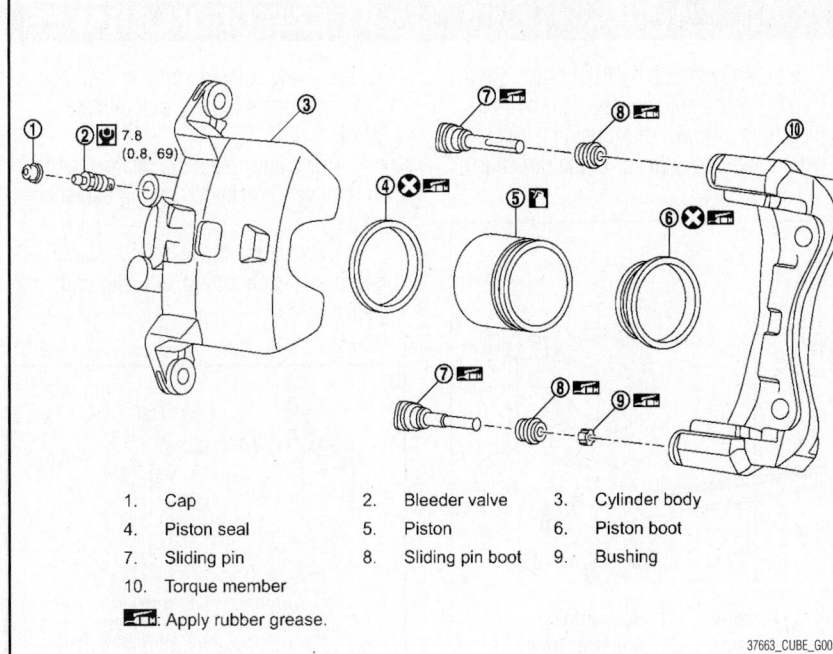

1. Cap
2. Bleeder valve
3. Cylinder body
4. Piston seal
5. Piston
6. Piston boot
7. Sliding pin
8. Sliding pin boot
9. Bushing
10. Torque member

▥: Apply rubber grease.

37663_CUBE_G0050

Fig. 4 Exploded view of front brake caliper

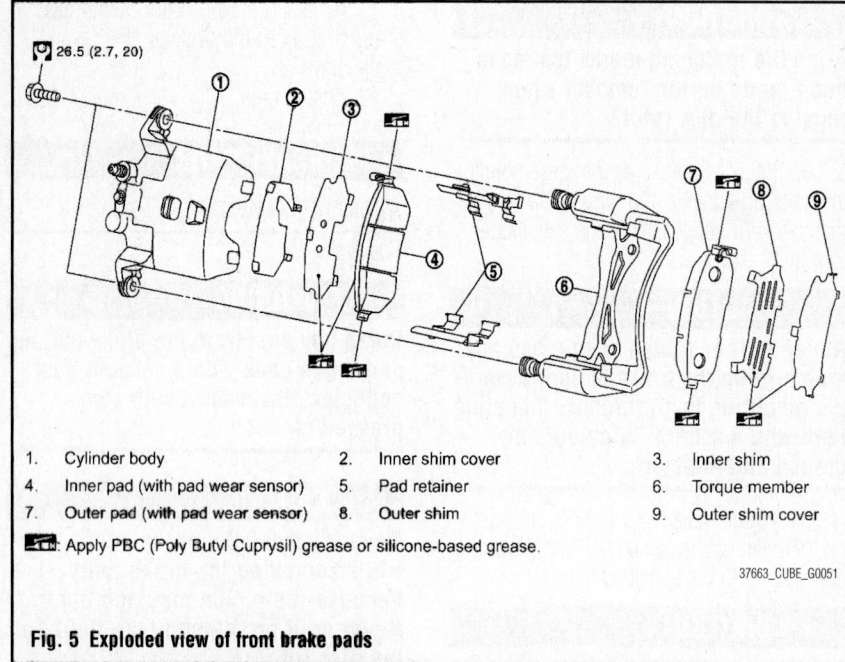

26.5 (2.7, 20)

1.	Cylinder body	2.	Inner shim cover	3.	Inner shim
4.	Inner pad (with pad wear sensor)	5.	Pad retainer	6.	Torque member
7.	Outer pad (with pad wear sensor)	8.	Outer shim	9.	Outer shim cover

▄▄: Apply PBC (Poly Butyl Cuprysil) grease or silicone-based grease.

37663_CUBE_G0051

Fig. 5 Exploded view of front brake pads

※※ CAUTION

Securely assemble the pad retainers so that it will not be lifted up from the torque member. Never deform the pad retainers.

5. Apply PBC (Poly Butyl Cuprysil) grease or silicone-based grease to the mating faces between the inner shim and the inner pad, and install them to the inner pad.

※※ CAUTION

Always replace the shim together with the shim cover when replacing the brake pad.

6. Apply PBC (Poly Butyl Cuprysil) grease or silicone-based grease to the mating faces between the outer shim and the outer shim cover, and install them to the outer pad.

※※ CAUTION

Always replace the shim together with the shim cover when replacing the brake pad.

7. Apply PBC (Poly Butyl Cuprysil) grease or silicone-based grease to the mating faces between the brake pads and the pad retainers, and Install the brake pads to the torque member.

8. Install cylinder body to torque member.

※※ CAUTION

Never damage the piston boot. When replacing brake pad with new one, check a brake fluid level in the reservoir tank because brake fluid returns to master cylinder reservoir tank when pressing piston in.

➡ **Use a disc brake piston tool to easily press piston.**

9. Install the lower sliding pin bolt and tighten it to the specified torque.

10. Depress the brake pedal several times to check that no drag feel is present for the front disc brake.

11. Install tires.

BRAKES

BRAKE DRUM

REMOVAL & INSTALLATION
See Figures 6 through 8.

※※ WARNING

Clean any dust from the brake caliper and brake pads with a vacuum dust collector. Never blow with compressed air.

※※ CAUTION

Never depress the brake pedal while removing the brake pads drum. Never drop the removed parts.

1. Remove tires and wheels.

2. Perform drain the brake fluid when remove or disassemble the wheel cylinder.

3. Remove the brake drum with the parking brake lever. If brake drum is difficult to the brake drum, remove according to the following procedure.

a. Remove the plug from brake drum.

b. Pull the adjuster lever from the plug hole of brake drum using a suitable wire, rotate the adjuster in the direction

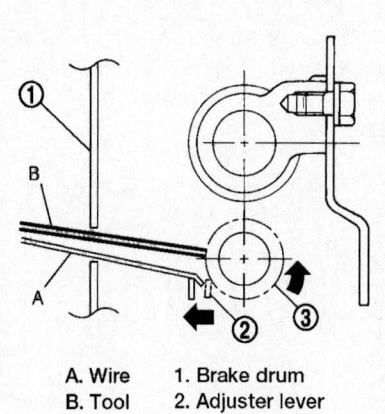

A. Wire	1. Brake drum
B. Tool	2. Adjuster lever

37663_CUBE_G0052

Fig. 6 Rotate the adjuster in the direction of the arrow using a suitable tool

REAR DRUM BRAKES

of the arrow using a suitable tool, and then compress the expanded brake shoe.

4. Press and rotate the retainer, and then remove the retainer, spring and shoe hold pin.

5. Remove the brake shoe assembly (brake shoe, each spring, adjuster and adjuster lever).

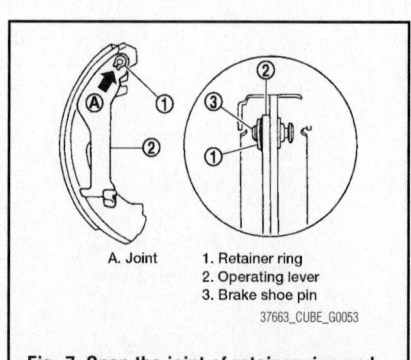

A. Joint	1. Retainer ring
	2. Operating lever
	3. Brake shoe pin

37663_CUBE_G0053

Fig. 7 Open the joint of retainer ring and remove the retainer ring to remove the operating lever from the brake shoe pin

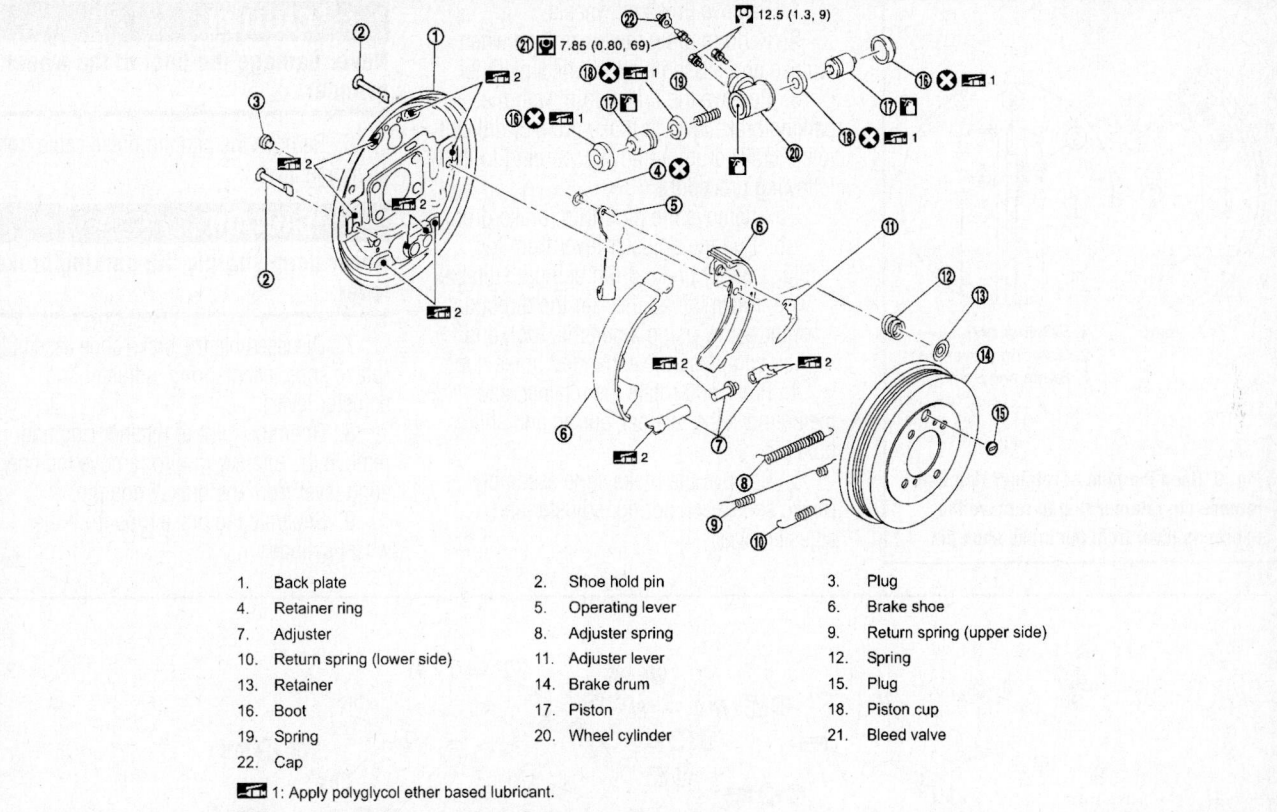

1. Back plate
2. Shoe hold pin
3. Plug
4. Retainer ring
5. Operating lever
6. Brake shoe
7. Adjuster
8. Adjuster spring
9. Return spring (upper side)
10. Return spring (lower side)
11. Adjuster lever
12. Spring
13. Retainer
14. Brake drum
15. Plug
16. Boot
17. Piston
18. Piston cup
19. Spring
20. Wheel cylinder
21. Bleed valve
22. Cap

1: Apply polyglycol ether based lubricant.

2: Apply PBC (Poly Butyl Cuprysil) grease or silicone-based grease.

: Apply brake fluid

Fig. 8 Exploded view of rear drum brake assembly

37663_CUBE_G0054

✳✳ CAUTION

Never damage the boot of the wheel cylinder.

6. Remove the parking brake cable from operating lever.

✳✳ CAUTION

Never bend sharply the parking brake lever.

7. Disassemble the brake shoe assembly (brake shoe, each spring, adjuster and adjuster lever).

8. Open the joint of retainer ring and remove the retainer ring to remove the operating lever from the brake shoe pin.

9. Separate the brake tube from the wheel cylinder.

10. Remove the wheel cylinder from back plate.

To install:

✳✳ WARNING

Clean any dust from the brake caliper and brake pads with a vacuum dust collector. Never blow with compressed air.

✳✳ CAUTION

Never depress the brake pedal while removing the brake drum. Never spill or splash brake fluid on the brake drum.

11. Note the following, and install in the reverse of removal.

a. After installing the retainer ring, close the joint of retainer ring until securely closed.

b. When disassembled adjuster, confirm the difference between left and right wheel for assemble.

c. Apply PBC (Poly Butyl Cuprysil) silicone-based grease to the adjuster screw.

d. Apply PBC (Poly Butyl Cuprysil) silicone-based grease to the mating faces between the adjuster and brake shoe.

e. Apply PBC (Poly Butyl Cuprysil) silicone-based grease to the mating faces between the back plate and brake shoe.

f. Shorten adjuster by rotating it.

g. Install the brake shoe assembly so that it does damage the wheel cylinder.

h. Check the component parts of

brake shoe assembly are installed properly.

i. Check the brake shoe sliding surface and brake drum inner surface for grease. Wipe it out any adheres to the surfaces.

j. Perform the air bleeding when removed or disassembled the wheel cylinder.

k. Adjust the brake shoe (parking brake lever stroke) after install and air bleeding.

BRAKE SHOES

REMOVAL & INSTALLATION
See Figures 6, 9 and 10.

✳✳ WARNING

Clean any dust from the brake caliper and brake pads with a vacuum dust collector. Never blow with compressed air.

✳✳ CAUTION

Never depress the brake pedal while removing the brake pads drum. Never drop the removed parts.

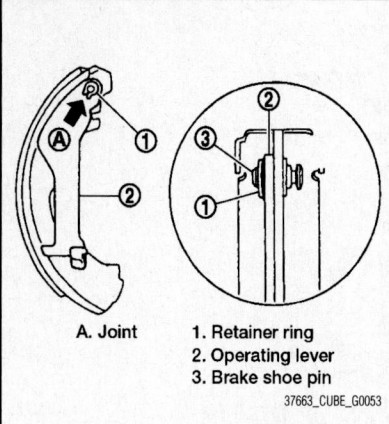

A. Joint

1. Retainer ring
2. Operating lever
3. Brake shoe pin

37663_CUBE_G0053

Fig. 9 Open the joint of retainer ring and remove the retainer ring to remove the operating lever from the brake shoe pin

1. Remove tires and wheels.
2. Perform drain the brake fluid when remove or disassemble the wheel cylinder.
3. Remove the brake drum with the parking brake lever. If brake drum is difficult to the brake drum, remove according to the following procedure.

   a. Remove the plug from brake drum.

   b. Pull the adjuster lever from the plug hole of brake drum using a suitable wire, rotate the adjuster in the direction of the arrow using a suitable tool, and then compress the expanded brake shoe.

4. Press and rotate the retainer, and then remove the retainer, spring and shoe hold pin.
5. Remove the brake shoe assembly (brake shoe, each spring, adjuster and adjuster lever).

❋❋ CAUTION

Never damage the boot of the wheel cylinder.

6. Remove the parking brake cable from operating lever.

❋❋ CAUTION

Never bend sharply the parking brake lever.

7. Disassemble the brake shoe assembly (brake shoe, each spring, adjuster and adjuster lever).
8. Open the joint of retainer ring and remove the retainer ring to remove the operating lever from the brake shoe pin.
9. Separate the brake tube from the wheel cylinder.

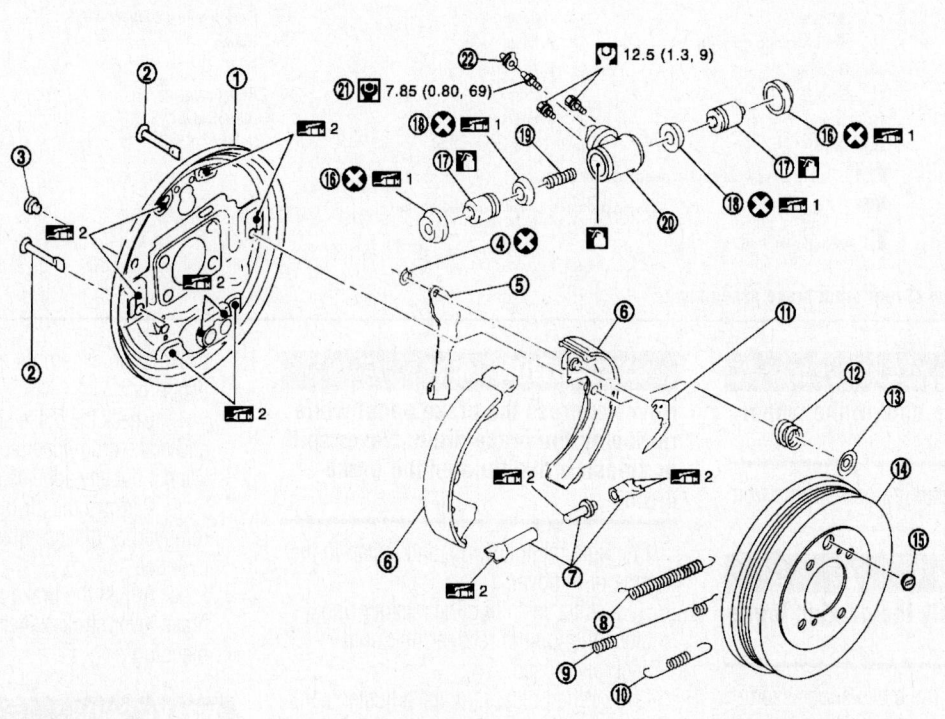

1. Back plate	2. Shoe hold pin	3. Plug
4. Retainer ring	5. Operating lever	6. Brake shoe
7. Adjuster	8. Adjuster spring	9. Return spring (upper side)
10. Return spring (lower side)	11. Adjuster lever	12. Spring
13. Retainer	14. Brake drum	15. Plug
16. Boot	17. Piston	18. Piston cup
19. Spring	20. Wheel cylinder	21. Bleed valve
22. Cap		

🛢️ 1: Apply polyglycol ether based lubricant.

🛢️ 2: Apply PBC (Poly Butyl Cuprysil) grease or silicone-based grease.

🛢️ : Apply brake fluid

37663_CUBE_G0054

Fig. 10 Exploded view of rear drum brake assembly

10. Remove the wheel cylinder from back plate.

To install:

11. Note the following, and install in the reverse of removal.

a. After installing the retainer ring, close the joint of retainer ring until securely closed.

b. When disassembled adjuster, confirm the difference between left and right wheel for assemble.

c. Apply PBC (Poly Butyl Cuprysil) silicone-based grease to the adjuster screw.

d. Apply PBC (Poly Butyl Cuprysil) silicone-based grease to the mating faces between the adjuster and brake shoe.

e. Apply PBC (Poly Butyl Cuprysil) silicone-based grease to the mating faces between the back plate and brake shoe.

f. Shorten adjuster by rotating it.

g. Install the brake shoe assembly so that it does damage the wheel cylinder.

h. Check the component parts of brake shoe assembly are installed properly.

i. Check the brake shoe sliding surface and brake drum inner surface for grease. Wipe it out any adheres to the surfaces.

j. Perform the air bleeding when removed or disassembled the wheel cylinder.

k. Adjust the brake shoe (parking brake lever stroke) after install and air bleeding.

BRAKES

PARKING BRAKE

PARKING BRAKE CABLES

ADJUSTMENT

Lever Stroke

1. Operate the parking brake lever with a force of 44 lbs. (196 N). Check that the lever stroke is within the specified number of notches. (Check it by listening to the clicks of the ratchet.)

2. When parking brake warning lamp turns ON, check that the lever stroke is within the specified number of notches. (Check it by listening to the clicks of the ratchet.)

Inspect Components

1. Check each component for installation condition such as looseness.

2. Check the parking brake lever assembly for bend, damage and cracks. Replace if necessary.

3. Check the cables and equalizer for wear, damage and cracks. Replace if necessary.

4. Check the parking brake switch, and replace it if necessary.

Adjustment

See Figure 11.

1. Remove the console mask.

2. Pull parking brake lever until a socket wrench can be inserted to adjusting nut.

3. Release the parking brake lever by turning the adjusting nut with a socket wrench and loosening the cable.

4. Depress the brake pedal with a force

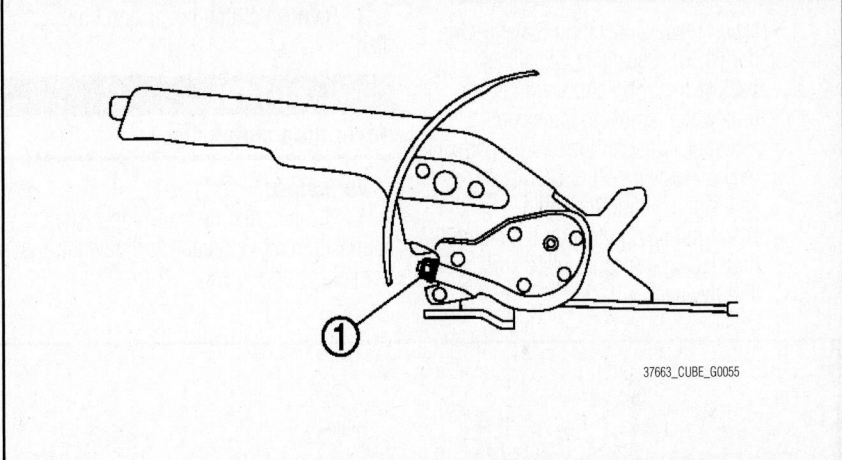

Fig. 11 Pull parking brake lever until a socket wrench can be inserted to adjusting nut (1)

of 44 lbs. (196 N) about 10 times and adjust the brake shoe clearance.

⁂ **CAUTION**

Make sure to securely depress the brake pedal.

5. Check a drag of rear drum brake.

6. Adjust the cable with the following procedure.

a. When replace parking brake cable, operate the parking brake lever with a force of 110 lbs. (490 N) for 10 strokes or more.

b. Adjust the parking brake lever stroke by turning the adjusting nut with a deep socket wrench.

⁂ **CAUTION**

Never reuse the adjusting nut if the nut is removed.

c. Operate the parking brake lever with a force of 44 lbs. (196 N). Check that the lever stroke is within the specified number of notches. (Check it by listening to the clicks of the ratchet.)

d. Rotate the brake drum with the parking brake lever released and check that there is no drag.

PARKING BRAKE SHOES

REMOVAL & INSTALLATION

See Rear Drum Brake.

CHASSIS ELECTRICAL   AIR BAG (SUPPLEMENTAL RESTRAINT SYSTEM)

GENERAL INFORMATION

DISARMING THE SYSTEM

Before servicing, turn ignition switch OFF, disconnect battery negative terminal and wait 3 minutes or more.

ARMING THE SYSTEM

Reconnect the battery cables. Wait several seconds to ensure that no system malfunction is detected by the air bag warning lamp.

DRIVE TRAIN

CLUTCH DRIVEN DISC & PRESSURE PLATE

REMOVAL & INSTALLATION
See Figures 12 and 13.

✳✳ CAUTION
Note the following:

- Never reuse Concentric Slave Cylinder (CSC). Because CSC slides back to the original position every time when removing transaxle assembly, dust on the sliding parts may damage a seal of CSC and may cause clutch fluid leakage.
- Never put any grease on the clutch disc facing, pressure plate surface and flywheel surface.

- When installing, be careful that grease applied to input shaft does not adhere to clutch disc.
- Never clean clutch disc using solvent.

1. Remove transaxle assembly from the engine.
2. Remove clutch cover mounting bolts.
3. Remove clutch cover and clutch disc.

✳✳ CAUTION
Never drop clutch disc.

To install:
4. Clean clutch disc and input shaft splines to remove grease and powder arisen from abrasion.

5. Apply recommended grease to clutch disc and input shaft splines.

✳✳ CAUTION
Be sure to apply grease to the points specified. Otherwise, noise, poor disengagement, or damage to the clutch may result. Excessive grease may cause slip or shudder. And if it adheres to seal of CSC, it cause clutch fluid leakage. wipe out excess grease. wipe out any grease oozing from the parts.

6. Install clutch disc, using a clutch aligner.
7. Install clutch cover and then temporarily tighten clutch cover mounting bolts.

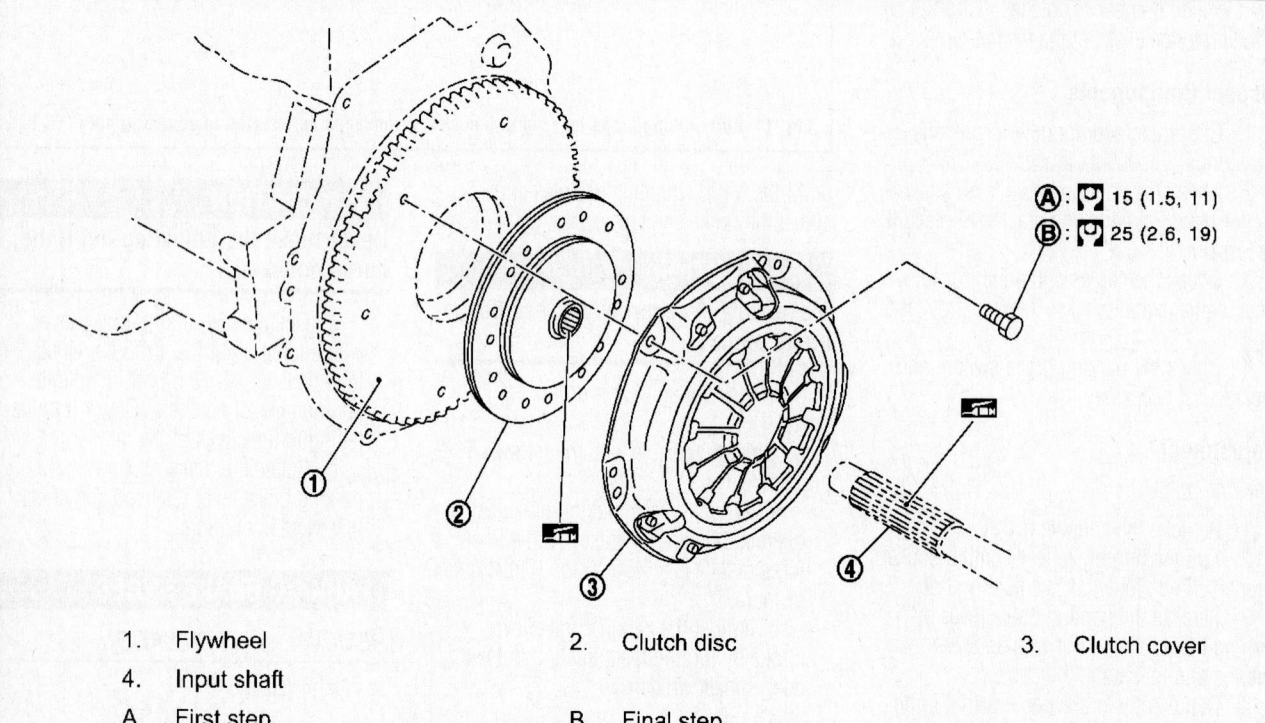

Ⓐ: 15 (1.5, 11)
Ⓑ: 25 (2.6, 19)

1. Flywheel
4. Input shaft
A. First step

2. Clutch disc
B. Final step

3. Clutch cover

⬛: Apply lithium-based grease including molybdenum disulphide.

37663_CUBE_G0103

Fig. 12 Exploded view of clutch assembly

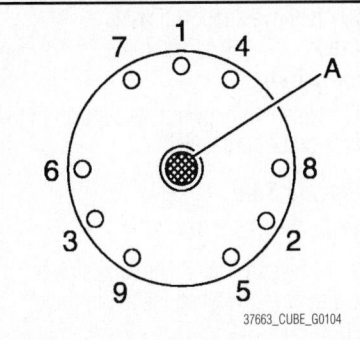

Fig. 13 Tighten clutch cover mounting bolts in two steps in the numerical order

8. Tighten clutch cover mounting bolts to the specified torque evenly in two steps in the numerical order as shown.

9. Install transaxle assembly to the engine.

HYDRAULIC SYSTEM BLEEDING

BLEEDING PROCEDURE

See Figures 14 and 15.

> ✳✳ **CAUTION**
>
> **Note the following:**

- Monitor clutch fluid level in reservoir tank so as not to empty the tank.
- Keep painted surface on the body or other parts free of clutch fluid. If it spills, wipe up immediately and wash the affected area with water.

➡**Do not use a vacuum assist or any other type of power bleeder on this system. Use of a vacuum assist or power bleeder will not purge all the air from the system.**

1. Fill reservoir tank with new clutch fluid.

> ✳✳ **CAUTION**
>
> **Never reuse drained clutch fluid.**

2. Connect a transparent vinyl hose to air bleeder of bleeding connector.

3. "Depress" and "release" the clutch pedal slowly and fully 15 times at an interval of 2 to 3 seconds and release the clutch pedal.

4. Press the lock pin into the bleeding connector and maintain the position.

> ✳✳ **CAUTION**
>
> **Since the inside of clutch tube is under hydraulic pressure, hold the tube to prevent it from getting disconnected.**

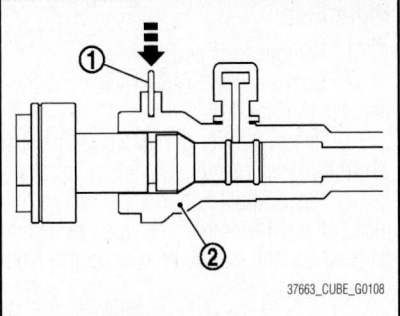

Fig. 14 Press the lock pin (1) into the bleeding connector (2), and maintain the position

tube to prevent it from getting disconnected.

5. Slide bleeding connector in the direction of the arrow as shown.

6. Depress the clutch pedal soon and hold it, and then bleed the air from the piping.

> ✳✳ **CAUTION**
>
> **Since the inside of clutch tube is under hydraulic pressure, hold the tube to prevent it from getting disconnected.**

7. Return clutch tube and lock pin in their original positions.

8. Release clutch pedal and wait for 5 seconds.

9. Repeat steps 3 to 8 until no bubbles are observed in the clutch fluid.

10. Check that the fluid level in the reservoir tank is within the specified range after air bleeding.

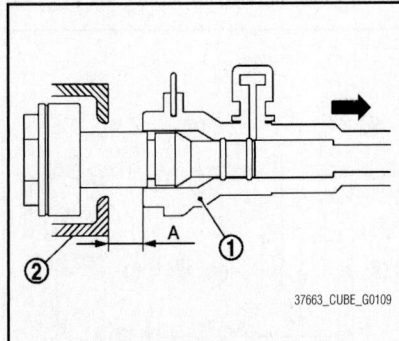

Fig. 15 Slide bleeding connector (1) in the direction of the arrow as shown

HALFSHAFTS

REMOVAL & INSTALLATION

Left Side

See Figure 16.

1. Remove tires and wheels.

2. Remove cotter pin, and then loosen wheel hub lock nut.

3. Patch wheel hub lock nut with a piece of wood. Hammer the wood to disengage wheel hub and bearing assembly from drive shaft.

> ✳✳ **CAUTION**
>
> **Never place drive shaft joint at an extreme angle. Also be careful not to overextend slide joint. Never allow drive shaft to hang down without support for joint sub-assembly, shaft and the other parts.**

➡**Use suitable puller, if wheel hub and bearing assembly and drive shaft cannot be separated even after performing the above procedure.**

4. Remove wheel hub lock nut.

5. Remove transverse link from steering knuckle.

6. Remove shaft assembly from wheel hub and bearing assembly.

> ✳✳ **CAUTION**
>
> **Be careful not to damage front wheel sensor and harness.**

7. Use the drive shaft attachment KV40107500 and a sliding hammer while inserting tip of the drive shaft attachment between shaft and transaxle assembly, and then remove drive shaft from transaxle assembly.

> ✳✳ **CAUTION**
>
> **Never place drive shaft joint at an extreme angle when removing drive shaft. Also be careful not to overextend slide joint. Confirm that the circular clip is attached to the drive shaft.**

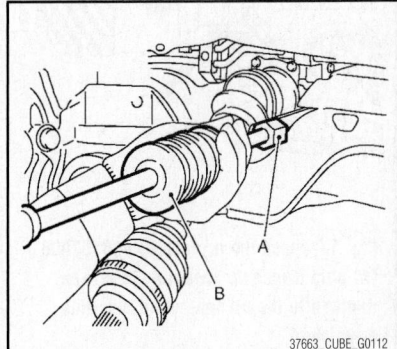

Fig. 16 Use the drive shaft attachment KV40107500 (A) and a sliding hammer (B), and then remove drive shaft from transaxle assembly

To install:

8. Note the following, and install in the reverse order of removal.

Transaxle Side

See Figure 17.

1. Always replace differential side oil seal with new one when installing drive shaft.

2. Place the protector KV38107900 onto transaxle assembly to prevent damage to the oil seal while inserting drive shaft. Slide drive shaft sliding joint and tap with a hammer to install securely.

✳✳ CAUTION

Check that circular clip is completely engaged.

3. Clean the matching surface of wheel hub lock nut and wheel hub and bearing assembly.

✳✳ CAUTION

Never apply lubricating oil to these matching surface.

4. Tighten the wheel hub lock nut to the specified torque.

✳✳ CAUTION

Never use a power tool to tighten the wheel hub lock nut.

5. Perform the final tightening of each of parts under unladen conditions, which were removed when removing wheel hub and bearing assembly and axle housing.

6. Never reuse cotter pin.

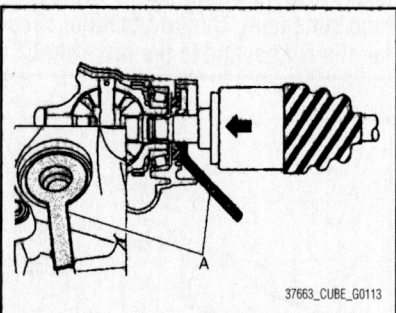

Fig. 17 Place the protector KV38107900 (A) onto transaxle assembly to prevent damage to the oil seal while inserting drive shaft

37663_CUBE_G0113

Right Side

1. Remove tires and wheels.
2. Remove wheel sensor and sensor harness if necessary.
3. Remove cotter pin, and then loosen wheel hub lock nut.
4. Patch wheel hub lock nut with a piece of wood. Hammer the wood to disengage wheel hub and bearing assembly from drive shaft.

✳✳ CAUTION

Never place drive shaft joint at an extreme angle. Also be careful not to overextend slide joint. Never allow drive shaft to hang down without support for joint sub-assembly, shaft and the other parts.

➡**Use suitable puller, if wheel hub and bearing assembly and drive shaft cannot be separated even after performing the above procedure.**

5. Remove wheel hub lock nut.
6. Remove transverse link from steering knuckle.
7. Remove drive shaft from wheel hub and bearing assembly.
8. Remove bearing housing plate bolts.
9. Remove drive shaft from transaxle assembly.

✳✳ CAUTION

Never place drive shaft joint at an extreme angle when removing drive shaft. Also be careful not to overextend slide joint.

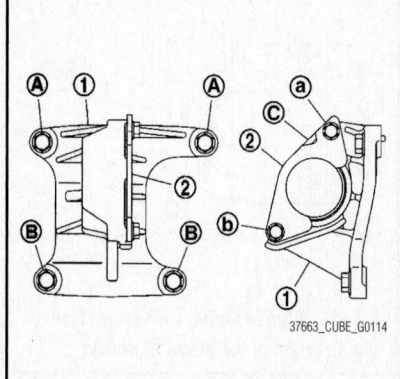

Fig. 18 Install support bearing bracket

37663_CUBE_G0114

10. Remove support bearing bracket.

To install:

11. Note the following, and install in the reverse order of removal.

Transaxle Side

See Figures 17 and 18.

1. Always replace differential side oil seal with new one when installing drive shaft.

2. Install support bearing bracket in following procedure:

   a. Temporarily tighten mounting bolts, then tighten them to specified torque.

   b. Set plate so that notch faces the upper side.

   c. Temporarily tighten mounting bolts, then tighten them.

✳✳ CAUTION

Never reuse plate.

3. Place the protector KV38107900 onto transaxle assembly to prevent damage to the oil seal while inserting drive shaft. Slide drive shaft sliding joint and tap with a hammer to install securely.

Wheel Side

1. Clean the matching surface of wheel hub lock nut and wheel hub and bearing assembly.

✳✳ CAUTION

Never apply lubricating oil to these matching surface.

2. Tighten the wheel hub lock nut to the specified torque.

✳✳ CAUTION

Never use a power tool to tighten the wheel hub lock nut.

3. Perform the final tightening of each of the parts, which were removed when removing wheel hub and bearing assembly and axle housing, under unladen conditions.

✳✳ CAUTION

Never reuse cotter pin.

ENGINE COOLING

ENGINE FAN

REMOVAL & INSTALLATION

See Figure 19.

1. Drain engine coolant from radiator.

> ☀ **CAUTION**
>
> **Perform this step engine is cold. Never spill engine coolant on drive belt.**

2. Remove air duct (inlet) and resonator assembly.
3. Remove reservoir tank as follows:
   a. Disconnect reservoir tank hose.
   b. Release the tab in the direction shown by the arrow.
   c. Lift up and remove the reservoir tank with the tab released.
4. Remove the upper radiator hose.
5. Disconnect harness connector from fan motor, and move harness to aside.
6. Remove cooling fan assembly.

> ☀ **CAUTION**
>
> **Be careful not to damage or scratch on radiator core when removing.**

> ***To install:***

7. Note the following, and install in the reverse order of removal.

> ☀ **CAUTION**
>
> **Only use genuine parts for fan shroud mounting bolt and observe the specified torque (to prevent radiator from being damaged).**

➟ Cooling fan is controlled by ECM.

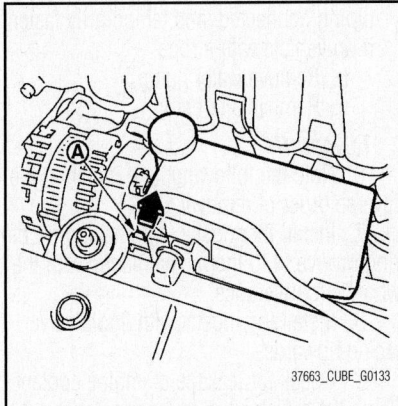

37663_CUBE_G0133

Fig. 19 Release the tab (A) in the direction shown by the arrow

RADIATOR

REMOVAL & INSTALLATION

See Figures 19 through 22.

> ☀ **WARNING**
>
> **Never remove radiator cap when engine is hot. Serious burns may occur from high-pressure engine coolant escaping from radiator. Wrap a thick cloth around the radiator cap. Slowly turn it a quarter of a turn to release built-up pressure. Then turn it all the way.**

1. Drain engine coolant from radiator.

> ☀ **CAUTION**
>
> **Perform this step when the engine is cold. Never spill engine coolant on drive belt.**

2. Remove air duct (inlet) and resonator assembly.
3. Remove reservoir tank as follows:
   a. Disconnect reservoir tank hose.
   b. Release the tab in the direction shown by the arrow.
   c. Lift up and remove the reservoir tank with tab released.
4. Remove radiator hose (upper and lower).
5. Disconnect harness connector from fan motor, and move harness aside.
6. Remove cooling fan assembly.

> ☀ **CAUTION**
>
> **Be careful not to damage or scratch the radiator core.**

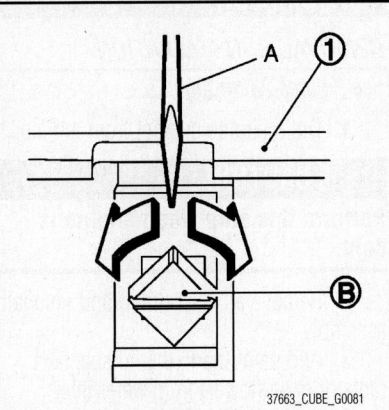

37663_CUBE_G0081

Fig. 20 Disengage front grille mounting clips (B) by rotating 45°using a flat-bladed screwdriver (A) through access hole of front grille (1) upper

7. Remove the front grille assembly.
   a. Remove front grille upper side fixing clips.
   b. Disengage front grille mounting clips by rotating 45°using a flat-bladed screwdriver through access hole of front grille upper while pulling front grill toward vehicle front.
8. Remove the front bumper fascia assembly.

> ☀ **CAUTION**
>
> **Bumper fascia is made of resin. Never apply strong force to it, and be careful to prevent contact with oil.**

   a. Fully open hood assembly.
   b. Remove bumper fascia upper fixing clips.
   c. Disengage front grille fixing pawls from back side of front grille while pull front grille horizontally to word vehicle front, and then remove front grille.
   d. Remove fender protector fixing clips and screws to access bumper fascia assembly fixing screw, and then remove bumper fascia assembly fixing screws (LH/RH).
   e. Remove bumper fascia assembly lower side fixing bolts and clips.
   f. Pull bumper fascia assembly side toward the vehicle side as shown by the arrows in the figure, and then disengage bumper fascia assembly from bumper side brackets (LH/RH).

> ☀ **CAUTION**
>
> **When removing bumper fascia, 2 workers are required so as to prevent it from dropping.**

   g. Disconnect front fog lamp harness connectors (LH/RH).

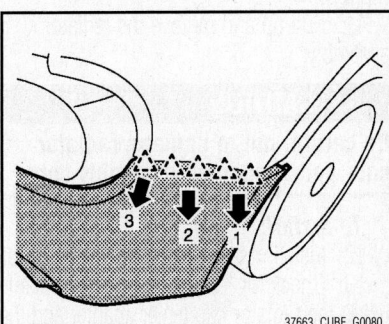

37663_CUBE_G0080

Fig. 21 Pull bumper fascia assembly side toward the vehicle side as shown by the arrows

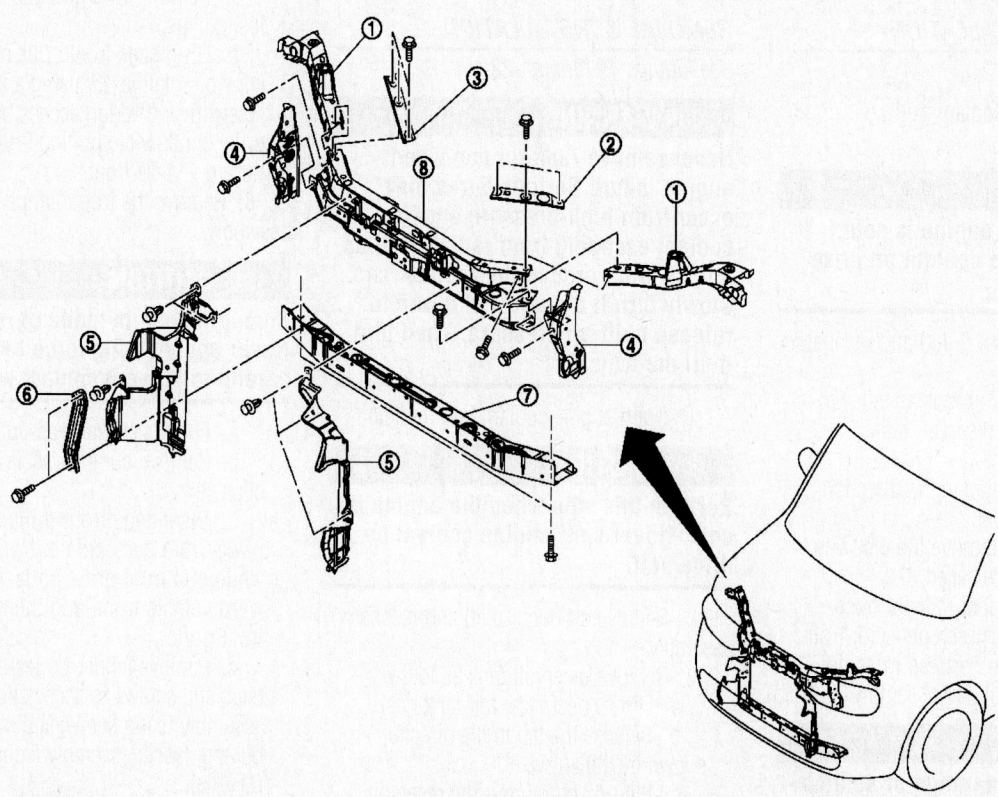

1.	Radiator core support side	2.	Radiator core support upper bracket (LH)	3.	Radiator core support upper bracket (RH)
4.	Radiator core reinforcement side	5.	Air guide	6.	Radiator core lower stay
7.	Radiator core support lower	8.	Radiator core support upper		

37663_CUBE_G0135

Fig. 22 Remove radiator core support (upper)

h. Remove bumper fascia assembly.

9. Remove the front combination lamp assembly (RH and LH).

10. Remove radiator core support (upper).

11. Pull up and remove the radiator assembly.

✳✳ CAUTION

Be careful not to damage radiator core and condenser assembly core.

To install:

12. Install in the reverse order of removal.

13. Check for leakage of engine coolant using the radiator cap tester adapter and the radiator cap tester.

14. Start and warm up the engine. Check visually that there is no leakage of engine coolant.

THERMOSTAT

REMOVAL & INSTALLATION

See Figures 23 through 25.

1. Drain engine coolant from radiator.

✳✳ CAUTION

Perform this step when engine is cold.

2. Remove air duct (inlet) and resonator assembly.

3. Add paint mark, then disconnect radiator hose (lower) from water inlet.

4. Remove water inlet and thermostat.

➡**Engine coolant leakage from cylinder block, so have a receptacle ready below.**

5. Remove thermostat housing with the following procedure:

a. Remove A/C compressor with A/C piping connected, and temporarily fasten it on vehicle with a rope.

b. Remove water pump.

c. Remove alternator.

To install:

6. Note the following, and install in the reverse order of removal.

7. Install thermostat with making rubber ring groove fit to thermostat flange with the whole circumference.

8. Install thermostat with jiggle valve facing upwards.

9. Check for leakage of engine coolant using the radiator cap tester adapter and the radiator cap tester.

10. Start and warm up the engine. Check

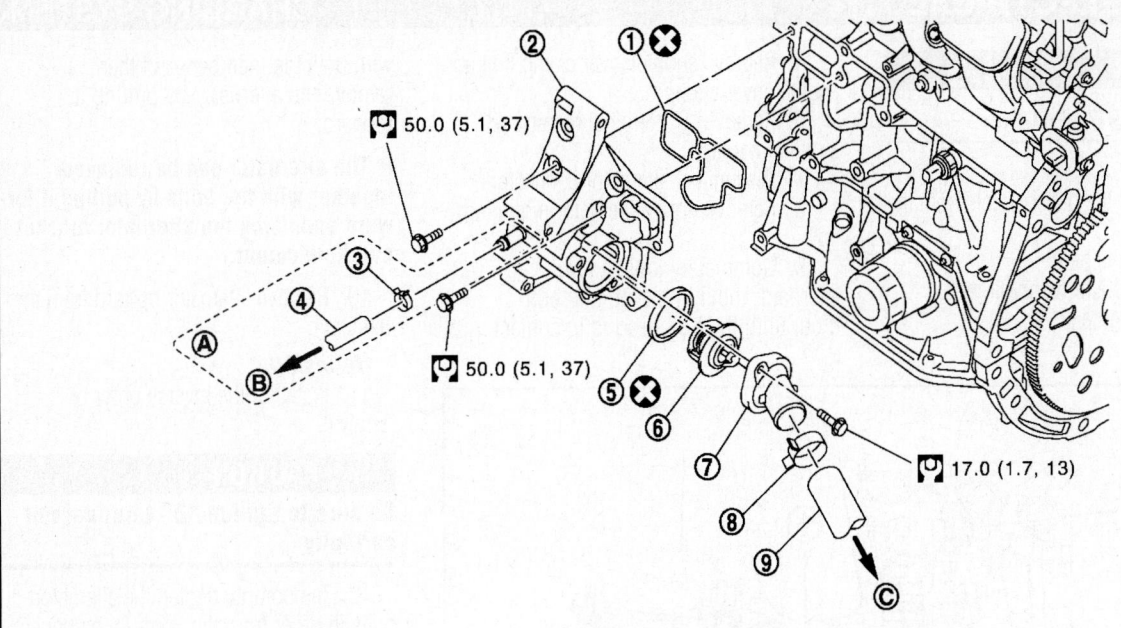

50.0 (5.1, 37)

50.0 (5.1, 37)

17.0 (1.7, 13)

1.	Gasket	2.	Thermostat housing	3.	Clamp
4.	Water hose	5.	Rubber ring	6.	Thermostat
7.	Water inlet	8.	Clamp	9.	Radiator hose (lower)
A.	CVT models	B.	To CVT fluid cooler	C.	To radiator

37663_CUBE_G0137

Fig. 23 Exploded view of thermostat housing assembly

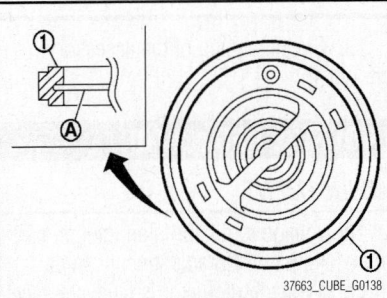

37663_CUBE_G0138

Fig. 24 Install thermostat with making rubber ring (1) groove fit to thermostat flange (A) with the whole circumference

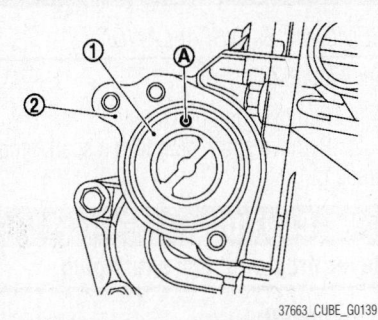

37663_CUBE_G0139

Fig. 25 Install thermostat (1) with jiggle valve (A) facing upwards; thermostat housing (2)

visually that there is no leakage of engine coolant.

WATER PUMP

REMOVAL & INSTALLATION
See Figure 26.

1. Drain engine coolant from radiator.

✳✳ CAUTION

Perform this step when the engine is cold. Never spill engine coolant on drive belt.

2. Remove front fender protector (RH).
3. Remove drive belt.
4. Remove water pump.

➡**Engine coolant leakage from cylinder block, so have a receptacle ready below.**

✳✳ CAUTION

Handle water pump vane so that it does not contact any other parts. Water pump cannot be disassembled and should be replaced as a unit.

5. Check visually that there is no significant dirt or rusting on water pump body and vane.

6. Check that there is no looseness in vane shaft, and that it turns smoothly when rotated by hand.

7. Replace water pump, if necessary.

To install:

8. Install in the reverse order of removal.

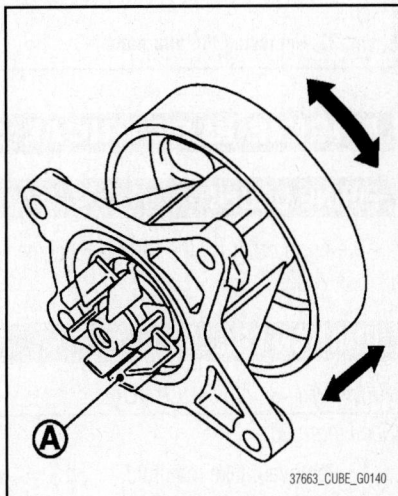

37663_CUBE_G0140

Fig. 26 Check visually that there is no significant dirt or rusting on water pump body and vane (A)

ENGINE ELECTRICAL
CHARGING SYSTEM

ALTERNATOR

REMOVAL & INSTALLATION

See Figure 27.

1. Disconnect the battery cable from the negative terminal.
2. Remove drive belt.
3. Remove radiator reservoir tank.
4. Remove engine cover.

5. Remove engine cover clamp bolt and engine cover clamp.
6. Remove "B" terminal nut and "B" terminal harness.
7. Disconnect alternator connector.
8. Remove upper alternator mounting bolt.
9. Completely loosen lower alternator mounting bolt, and pull it out until the bolt head is in contact

with the side member. And then, remove the alternator by pulling it forward.

➡The alternator can be removed together with the bolts by pulling it forward and using the alternator bracket bolt hole cutout.

10. Remove alternator upward from the vehicle.

To install:

11. Install in the reverse order of removal.

※※ **CAUTION**

Be sure to tighten "B" terminal nut carefully.

12. Temporarily tighten the alternator bolts in order from the lower to the upper, and then tighten them in order from the upper to the lower.

※※ **CAUTION**

For the alternator, the front side (pulley side) surface is the reference surface. Fit the reference surface to the alternator mounting part, and then tighten the bolts.

13. Check tension of the accessory drive belt.

VOLTAGE REGULATOR

ADJUSTMENT

The voltage regulator is an integral part of the alternator. No adjustment can be made. If voltage regulator is faulty, replace the component.

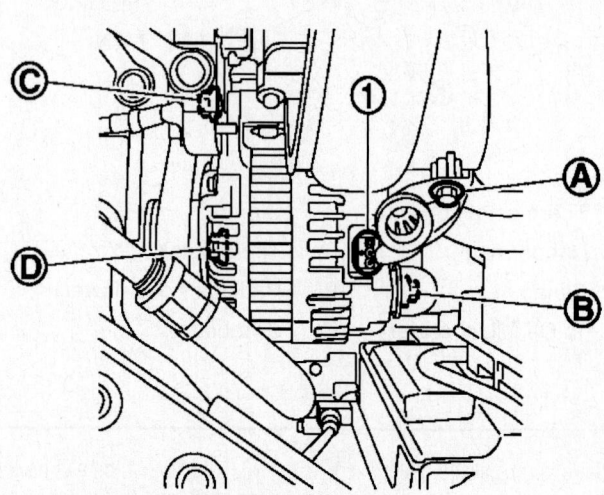

A. Engine cover clamp bolt
B. "B" terminal nut
C. Upper mounting bolt
D. Lower mounting bolt
1. Alternator connector

37663_CUBE_G0143

Fig. 27 Removing the alternator

ENGINE ELECTRICAL
IGNITION SYSTEM

FIRING ORDER

The firing order for the MR18DE engine is 1–3–4–2.

IGNITION COIL

REMOVAL & INSTALLATION

See Figure 28.

1. Remove intake manifold.
2. Remove ignition coil.

※※ **CAUTION**

**Never drop or shock ignition coil.
Never disassemble ignition coil.**

3. Installation is the reverse of removal.

IGNITION TIMING

INSPECTION

The ignition timing is 13° +/- 5° BTDC at idle speed.

ADJUSTMENT

The ignition timing is controlled by the ECM. No adjustment is necessary or possible.

SPARK PLUGS

REMOVAL & INSTALLATION

See Figure 28.

1. Remove ignition coil.
2. Remove spark plug with a spark plug wrench.

※※ **CAUTION**

Never drop or shock spark plug.

To install:

3. Install in the reverse order of removal.

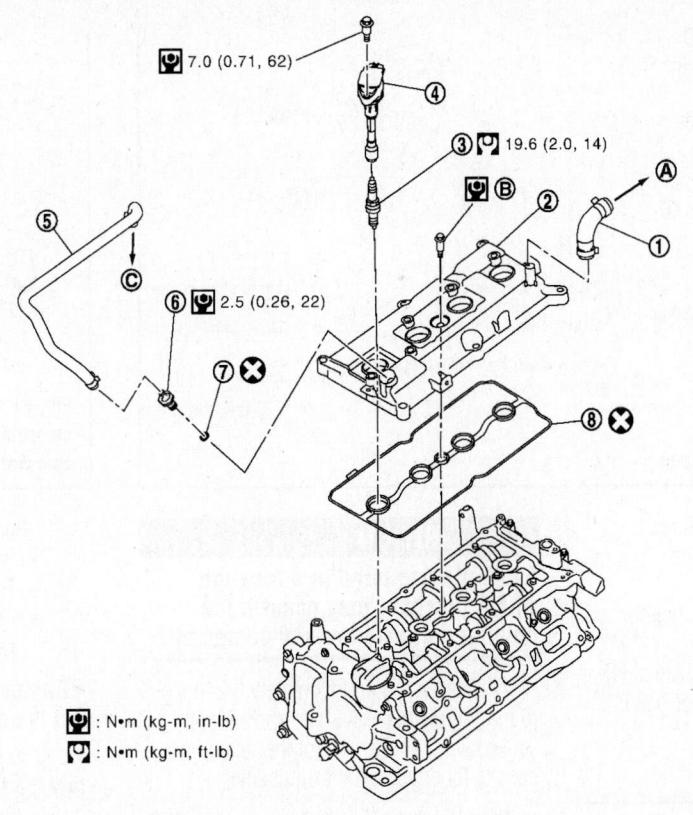

7.0 (0.71, 62)	④
	③ 19.6 (2.0, 14)
⑤	Ⓑ
	②
⑥ 2.5 (0.26, 22)	Ⓐ ①
⑦ ✕	⑧ ✕

🔧 : N•m (kg-m, in-lb)
🔧 : N•m (kg-m, ft-lb)

1. PCV hose	2. Rocker cover	3. Spark plug
4. Ignition coil	5. PCV hose	6. PCV valve
7. O-ring	8. Rocker cover gasket	
	Tightening must be done following	
A. To air duct assembly	B. the installation procedure. Refer to [ ].	C. To intake manifold

37663_CUBE_G0145

Fig. 28 Exploded view of ignition coil, spark plug, and rocker cover assembly

ENGINE ELECTRICAL

STARTER

REMOVAL & INSTALLATION

1. Disconnect the battery cable from the negative terminal.
2. Remove air duct (inlet).
3. Remove radiator reservoir tank.
4. Disconnect oil pressure switch connector.
5. Remove "B" terminal nut and "B" terminal harness.
6. Remove "S" terminal nut and "S" terminal harness.
7. Remove starter motor mounting bolts.
8. Remove starter motor upward from the vehicle.

STARTING SYSTEM

To install:
9. Install in the reverse order of removal.

✳✳ CAUTION

Be sure to tighten "B" terminal nut carefully.

ENGINE MECHANICAL

ACCESSORY DRIVE BELTS

ACCESSORY BELT ROUTING

See Figure 29.

Refer to the accompanying illustration.

INSPECTION

✳✳ WARNING

Perform this step when engine is stopped.

1. Check that the indicator (notch on fixed side) of drive belt auto-tensioner is within the possible use range.

➡**Note the following:**

- Check the drive belt auto-tensioner indication when the engine is cold.
- When new drive belt is installed,

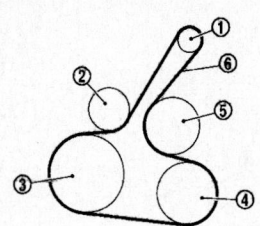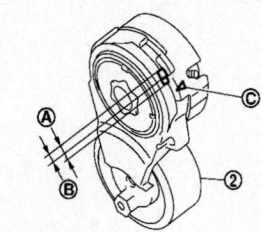

1. Alternator
2. Drive belt auto-tensioner
3. Crankshaft pulley
4. A/C compressor (models with A/C) Idler pulley (models without A/C)
5. Water pump
6. Drive belt
A. Possible use range
B. Range when new drive belt is installed
C. Indicator

37663_CUBE_G0148

Fig. 29 Accessory drive belt routing

the indicator (notch on fixed side) should be within the range.

2. Visually check entire drive belt for wear, damage or cracks.

3. If the indicator (notch on fixed side) is out of the possible use range or belt is damaged, replace drive belt.

ADJUSTMENT

Belt tension is not necessary, as it is automatically adjusted by drive belt auto-tensioner.

REMOVAL & INSTALLATION

See Figure 30.

1. Remove front wheel and tire (RH).
2. Remove front fender protector (RH).
3. Hold the hexagonal part of drive belt auto-tensioner with a wrench securely. Then move the wrench handle in the direction of arrow (loosening direction of tensioner).

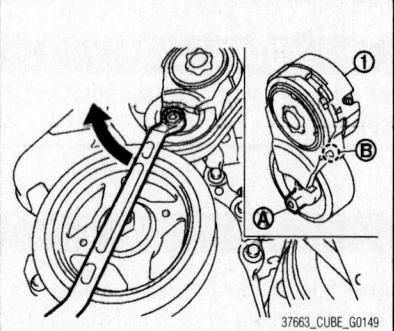

37663_CUBE_G0149

Fig. 30 Hold the hexagonal part (A) of drive belt auto-tensioner (1) with a wrench securely; then move the wrench handle in the direction of arrow; insert a rod into hole (B)

✳✳ CAUTION

Avoid placing hand in a location where pinching may occur if the holding tool accidentally comes off.

4. Insert a rod approximately 0.24 inches (6 mm) in diameter such as short-length screwdriver into the hole of the retaining boss to fix drive belt auto-tensioner.

➡**Keep drive belt auto-tensioner pulley arm locked after drive belt is removed.**

5. Remove drive belt.

To install:

6. Install drive belt.

✳✳ CAUTION

Confirm drive belt is completely set to pulleys. Check for engine oil, working fluid and engine coolant are not adhered to drive belt and each pulley groove.

7. Release drive belt auto-tensioner, and apply tension to drive belt.

8. Turn crankshaft pulley clockwise several times to equalize tension between each pulley.

9. Confirm tension of drive belt at indicator (notch on fixed side) is within the possible use range.

CAMSHAFT AND VALVE LIFTERS

REMOVAL & INSTALLATION

See Figures 31 through 39.

✳✳ CAUTION

The rotating direction in the text indicates all directions seen from the engine front.

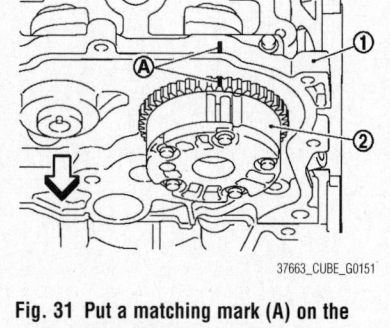

37663_CUBE_G0151

Fig. 31 Put a matching mark (A) on the camshaft sprocket (INT) (2) and the camshaft bracket (1)

1. Remove the following parts:
   - Intake manifold
   - Rocker cover
   - Front cover and timing chain related parts

➡**Removal of oil pump drive related part is not necessary.**

2. Remove camshaft position sensor (PHASE) from camshaft bracket.

✳✳ CAUTION

Note the following:

- Handle camshaft position sensor (PHASE) carefully and avoid impacts.
- Never disassemble camshaft position sensor (PHASE).
- Never place sensor where it is exposed to magnetism.

3. Put a matching mark on the camshaft sprocket (INT) and the camshaft bracket.

➡**It prevents the knock pin of the camshaft (INT) from engaging with the incorrect pin hole when installing the camshaft sprocket (INT).**

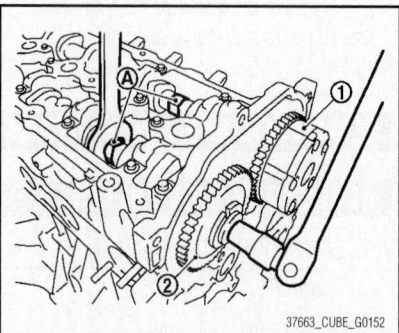

37663_CUBE_G0152

Fig. 32 Secure hexagonal part (A) of camshaft (INT) (1) (EXH) (2) with a wrench

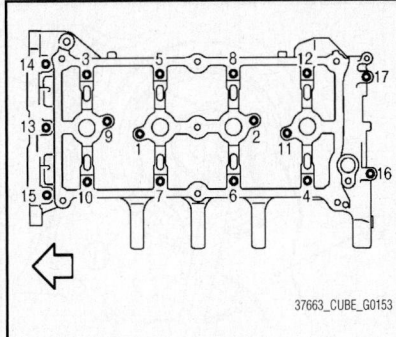

Fig. 33 Loosen mounting bolts in reverse order as shown

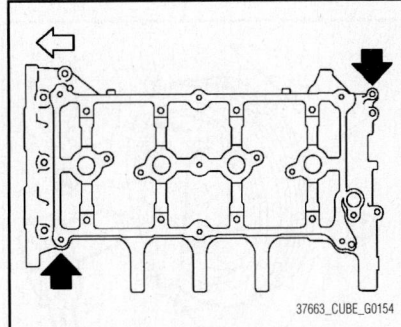

Fig. 34 Cut liquid gasket by prying at the positions indicated by the black arrows

4. Remove camshaft sprockets.
   a. Secure hexagonal part of camshaft with a wrench.
   b. Loosen camshaft sprocket mounting bolts and remove camshaft sprocket.

✲✲ CAUTION
Never rotate crankshaft or camshaft while timing chain is removed. It causes interference between valve and piston.

5. Remove camshaft bracket with the following procedure:
   a. Loosen mounting bolts in reverse order as shown.
   b. Cut liquid gasket by prying at the positions indicated by the black arrows, and then remove the camshaft bracket.
6. Remove camshafts.
7. Remove valve lifters.

✲✲ CAUTION
Identify installation positions, and store them in order.

8. Remove signal plate from camshaft (INT), if necessary.

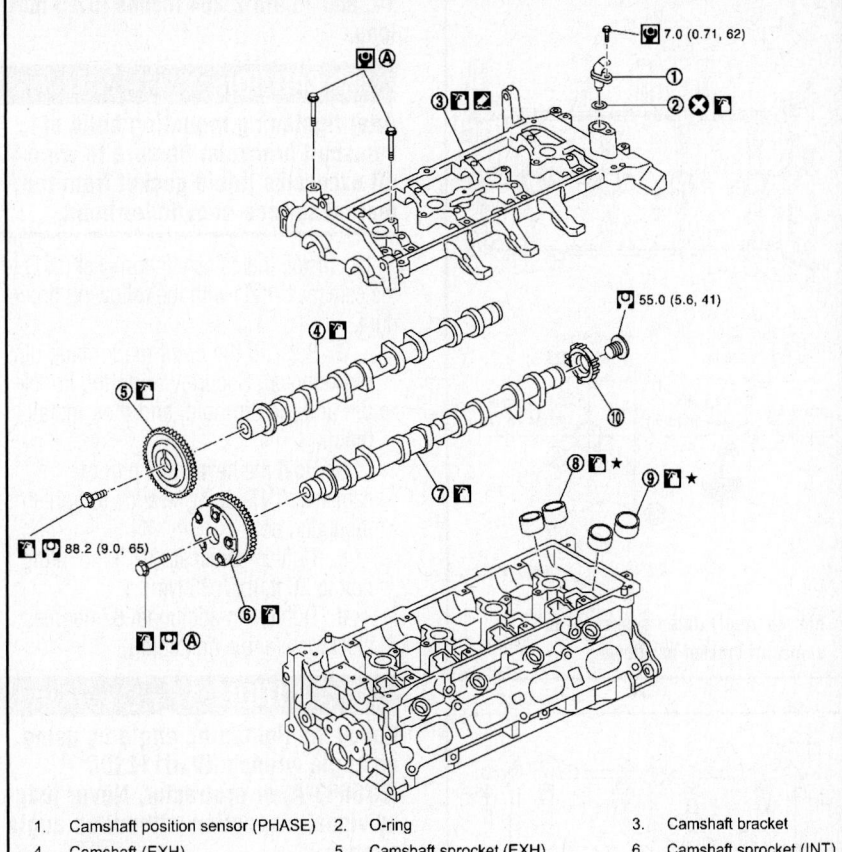

1. Camshaft position sensor (PHASE) 2. O-ring 3. Camshaft bracket
4. Camshaft (EXH) 5. Camshaft sprocket (EXH) 6. Camshaft sprocket (INT)
7. Camshaft (INT) 8. Valve lifter (EXH) 9. Valve lifter (INT)
10. Signal plate
A. Tightening must be done following the installation procedure.

Fig. 35 Exploded view of camshaft assembly

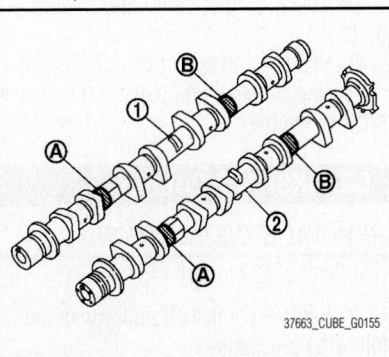

Fig. 36 Identifying intake and exhaust camshafts

To install:
9. Install valve lifters.

✲✲ CAUTION
Install them in their original positions.

10. Install camshafts.

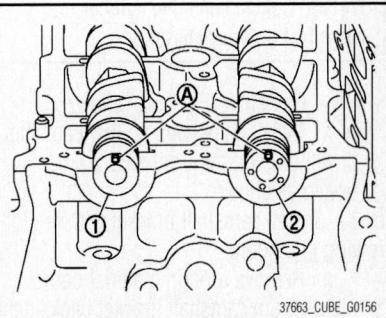

Fig. 37 Install camshafts so that camshaft dowel pins (A) on the front side are positioned as shown

a. Clean camshaft journal to remove any foreign material.
b. Distinguish between the intake and the exhaust by looking at the different shapes of the front and rear ends of the camshaft or using the identification colors.
• Camshaft (INT) A: Yellow

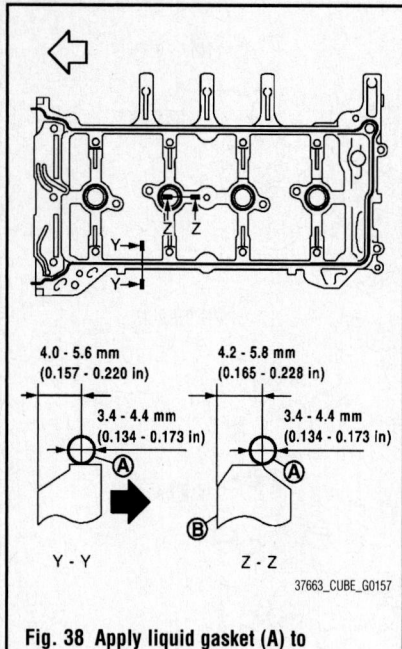

Fig. 38 Apply liquid gasket (A) to camshaft bracket as shown

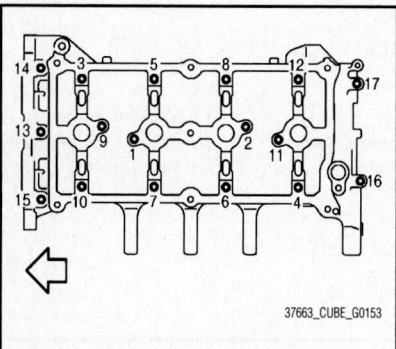

Fig. 39 Tighten mounting bolts in numerical order as shown

- Camshaft (EXH) B: Yellow

c. Install camshafts so that camshaft dowel pins on the front side are positioned as shown.

11: Install camshaft bracket with the following procedure:

 a. Remove foreign material completely from camshaft bracket backside and from cylinder head installation face.

 b. Apply liquid gasket to camshaft bracket as shown.

 c. Tighten mounting bolts of camshaft brackets in three steps, in numerical order as shown.
  - Step 1: 17 inch lbs. (2 Nm)
  - Step 2: 52 inch lbs. (6 Nm)
  - Step 3: 84 inch lbs. (9.5 Nm)

➡There are two types of mounting bolts. All bolt thread lengths, except numbers 13, 14, and 15, are 1.378 inches (35 mm) long. Bolt numbers 13,

14, and 15 are 2.264 inches (57.5 mm) long.

※※ CAUTION

After tightening mounting bolts of camshaft brackets, be sure to wipe off excessive liquid gasket from the mating surface of cylinder head.

12. Install the camshaft sprocket (INT) to the camshaft (INT) with the following procedure.

 a. Refer to the paint mark made during removal. Securely align the knock pin and the pin hole, and then install them.

 b. Hold the hexagonal part of camshaft (INT) using wrench to tighten mounting bolt.

 c. Tighten camshaft (INT) mounting bolt to 26 ft. lbs. (35 Nm).

 d. Tighten an additional 67 degrees clockwise (angle tightening).

※※ CAUTION

Check the tightening angle by using an angle wrench KV10112100 (BT8653-A) or protractor. Never judge by visual inspection without an angle wrench.

13. Install camshaft sprocket (EXH).

➡Secure the hexagonal part of camshaft (EXH) using wrench to tighten mounting bolt to 65 ft. lbs. (88 nm).

14. Install timing chain and related parts.

15. Inspect and adjust valve clearance.

16. Install remaining parts in the reverse order of removal.

CRANKSHAFT PULLEY

REMOVAL & INSTALLATION

See Figures 40 through 42.

1. Remove crankshaft pulley with the following procedure:

 a. Fix crankshaft pulley with a pulley holder, loosen crankshaft pulley bolt, and locate bolt seating surface at 0.39 inches (10 mm) from its original position.

※※ CAUTION

Never remove the crankshaft pulley bolt as they will be used as a supporting point for the pulley remover.

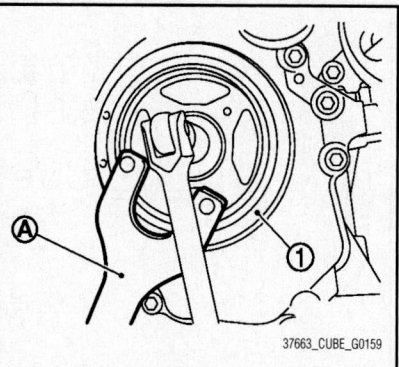

Fig. 40 Fix crankshaft pulley (1) with a pulley holder (A)

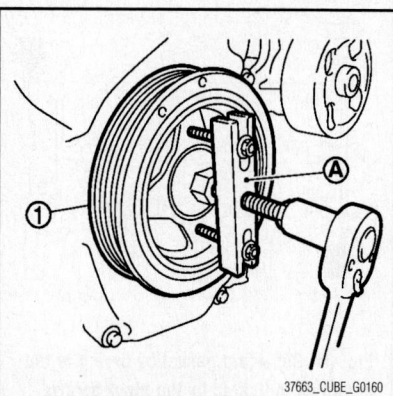

Fig. 41 Attach a pulley puller KV11103000 in the M6 thread hole on crankshaft pulley (1)

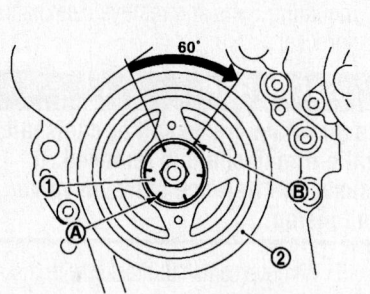

| A. Angle marks | 1. Crankshaft pulley bolt flange |
| B. Paint mark | 2. Crankshaft pulley |

Fig. 42 Put a paint mark on crankshaft pulley, matching with any one of six easy to recognize angle marks on crankshaft pulley bolt flange

 b. Attach a pulley puller KV11103000 in the M6 thread hole on crankshaft pulley and remove crankshaft pulley.

To install:

2. Install crankshaft pulley with the following procedure:

a. When inserting crankshaft pulley with a plastic hammer, tap on its center portion (not circumference).

✳✳ CAUTION

Never damage front oil seal lip section.

b. Secure crankshaft pulley with a pulley holder.

c. Apply new engine oil to thread and seat surfaces of crankshaft pulley bolt.

d. Tighten crankshaft pulley bolt to 51 ft. lbs. (69 Nm).

e. Completely loosen the bolt.

f. Tighten crankshaft pulley bolt 22 ft. lbs. (29 Nm).

g. Put a paint mark on crankshaft pulley, matching with any one of six easy to recognize angle marks on crankshaft pulley bolt flange.

h. Turn another 60 degrees clockwise (angle tightening). Check the tightening angle with movement of one angle mark.

i. Check that crankshaft rotates clockwise smoothly.

CRANKSHAFT FRONT SEAL

REMOVAL & INSTALLATION

See Figure 43.

1. Remove the following parts:
   - Front fender protector (RH)
   - Drive belt
   - Crankshaft pulley
2. Remove front oil seal with a suitable tool.

✳✳ CAUTION

Be careful not to damage front cover and crankshaft.

To install:

3. Apply new engine oil to new front oil seal joint surface and seal lip.

4. Install front oil seal so that each seal lip is oriented as shown.

5. Press-fit front oil seal using a suitable drift with outer diameter 2.24 inches (57 mm) and inner diameter 1.77 inches (45 mm).

✳✳ CAUTION

Press-fit oil seal straight to avoid causing burrs or tilting.

6. Install in the reverse order of removal, for the rest of parts.

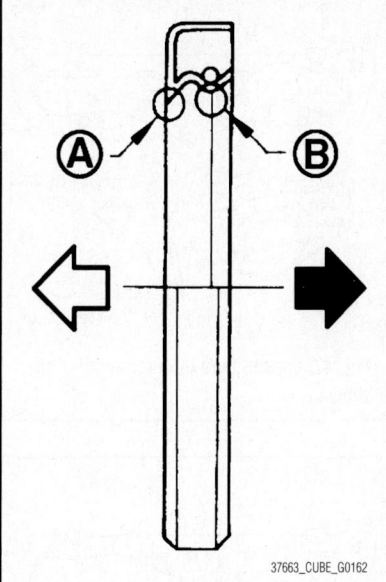

Fig. 43 Install front oil seal so that each seal lip is oriented as shown

CYLINDER HEAD

REMOVAL & INSTALLATION

See Figures 44 and 45.

1. Release fuel pressure.
2. Drain engine coolant and engine oil.
3. Remove the following components and related parts:
   - Exhaust manifold
   - Intake manifold
   - Fuel injector and fuel tube assembly
   - Water outlet
   - Rocker cover
   - Front cover, timing chain
   - Camshaft
4. Remove cylinder head.

a. Loosen cylinder head bolts in reverse order of the tightening sequence.

b. Using TORX socket, loosen cylinder head bolts.

5. Remove cylinder head gasket.

To install:

6. Install cylinder head gasket.

7. Install cylinder head, and tighten cylinder head bolts in numerical order as shown with the following procedure.

✳✳ CAUTION

If cylinder head bolts are reused, check their outer diameters before installation.

a. Apply new engine oil to threads and seating surface of mounting bolts.

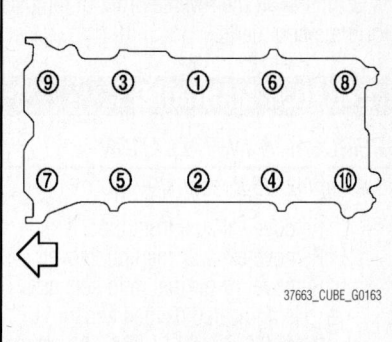

Fig. 44 Tighten cylinder head bolts in numerical order as shown

b. Tighten all cylinder head bolts to 30 ft. lbs. (40 Nm).

c. Turn all cylinder head bolts 100 degrees clockwise (angle tightening).

✳✳ CAUTION

Check and confirm the tightening angle by using an angle wrench KV10112100 (BT8653-A) or protractor. Never judge by visual inspection without the tool.

d. Completely loosen.

✳✳ CAUTION

In this step, loosen cylinder head bolts in reverse order of that indicated in the figure.

e. Tighten all cylinder head bolts to 30 ft. lbs. (40 Nm).

f. Turn all cylinder head bolts 100 degrees clockwise (angle tightening).

g. Turn all cylinder head bolts 100 degrees clockwise again (angle tightening).

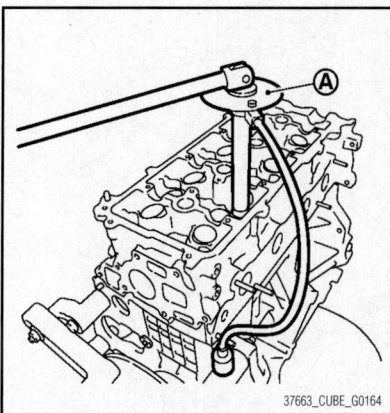

Fig. 45 Check and confirm the tightening angle by using an angle wrench (A) or protractor

8. Install in the reverse order of removal, for the rest of parts.

EXHAUST MANIFOLD

REMOVAL & INSTALLATION

See Figures 46 through 49.

1. Remove exhaust front tube.
2. Remove exhaust manifold cover.
3. Remove the air fuel ratio sensor 1.
   a. Using heated oxygen sensor wrench KV10117100 (J-3647-A), remove air fuel ratio sensor 1.

✳✳ CAUTION

Handle air fuel ratio sensor 1 carefully and avoid impacts.

➡The exhaust manifold can be removed and installed without removing the air fuel ratio sensor 1 (Disassembly of harness connector is necessary).

4. Remove exhaust manifold stay.
5. Remove exhaust manifold.
6. Loosen nuts in reverse order as shown.

➡Disregard Nos. 6 to 8 when loosening.

7. Remove gasket.

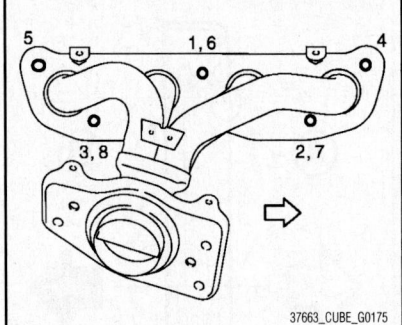

Fig. 47 Loosen nuts in reverse order as shown

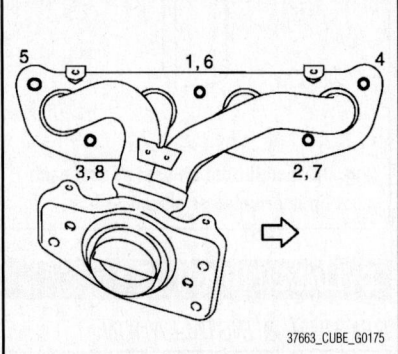

Fig. 48 Tighten nuts in numerical order as shown

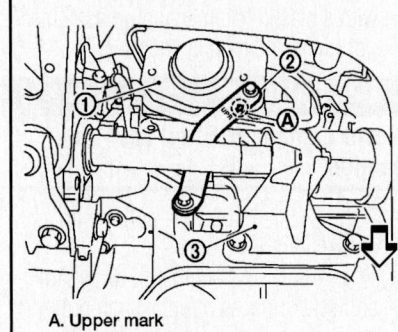

A. Upper mark
1. Exhaust manifold
2. Exhaust manifold stay
3. Drive shaft support bearing bracket

Fig. 49 Install exhaust manifold stay in the direction as shown

✳✳ CAUTION

Cover engine openings to avoid entry of foreign materials.

To install:

8. Install gasket to cylinder head.
9. Install exhaust manifold with the following procedure:
   a. Tighten nuts in numerical order as shown.

➡No. 6 to 8 mean double tightening of nuts No. 1 to 3.

   b. Install exhaust manifold stay in the direction as shown.
10. Install remaining parts in the reverse order of removal.

INTAKE MANIFOLD

REMOVAL & INSTALLATION

See Figures 50 through 52.

1. Remove engine cover.
2. Pull out oil level gauge.

✳✳ CAUTION

Cover the oil level gauge guide openings to avoid entry of foreign materials.

3. Disconnect PCV hose from intake manifold and rocker cover.
4. Remove air duct (inlet), resonator and air duct assembly.
5. Disconnect vacuum hose from intake manifold.
6. Disconnect water hoses from electric throttle control actuator.

➡Drain engine coolant from radiator or attach plug to prevent engine coolant

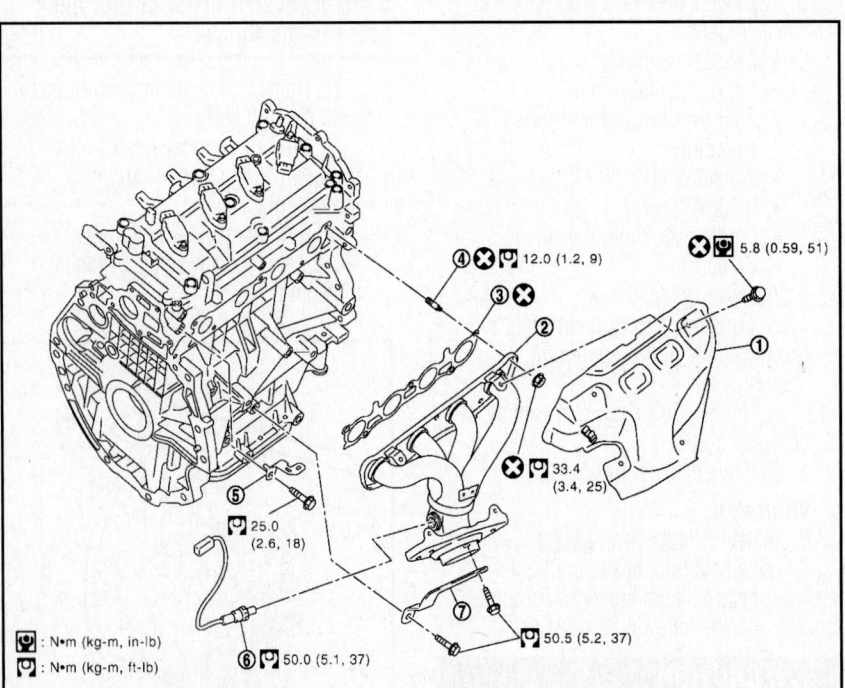

🔧 : N•m (kg-m, in-lb)
🔧 : N•m (kg-m, ft-lb)

1.	Exhaust manifold cover	2.	Exhaust manifold	3.	Gasket
4.	Stud bolt	5.	Harness bracket	6.	Air fuel ratio sensor 1
7.	Exhaust manifold stay				

Fig. 46 Exploded view of exhaust manifold assembly

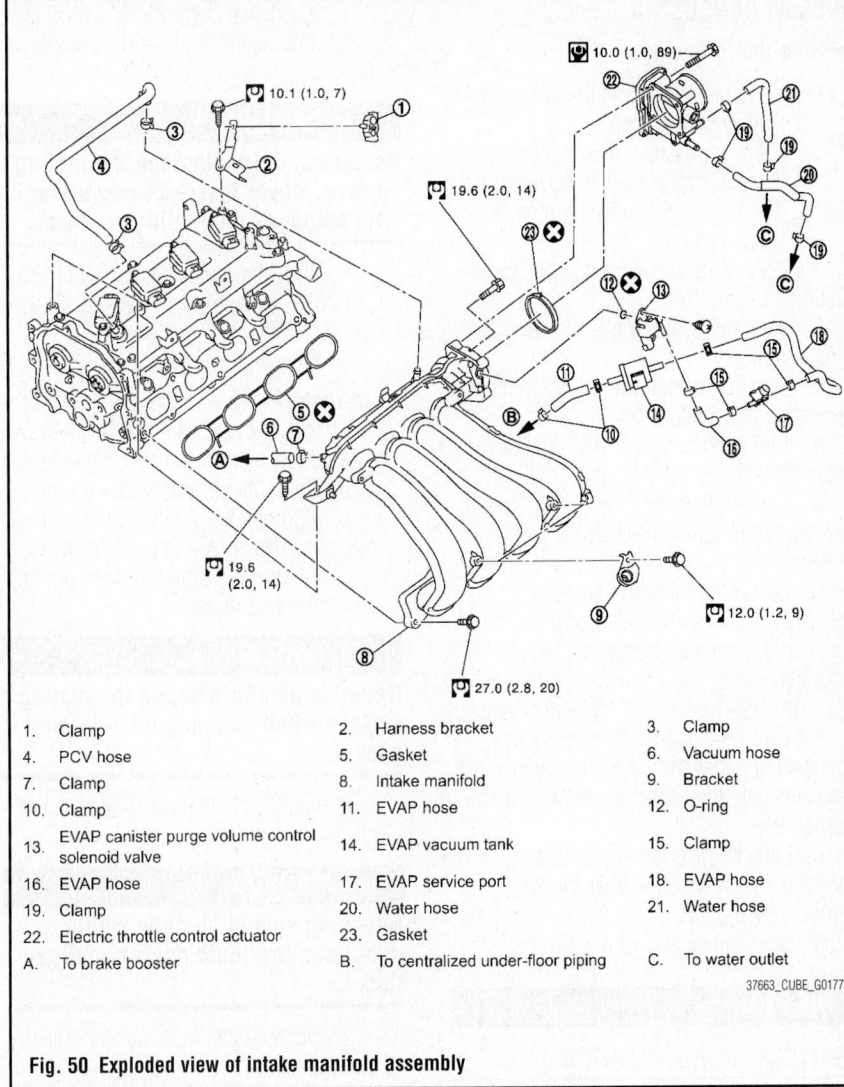

Fig. 50 Exploded view of intake manifold assembly

1.	Clamp	2.	Harness bracket	3.	Clamp
4.	PCV hose	5.	Gasket	6.	Vacuum hose
7.	Clamp	8.	Intake manifold	9.	Bracket
10.	Clamp	11.	EVAP hose	12.	O-ring
13.	EVAP canister purge volume control solenoid valve	14.	EVAP vacuum tank	15.	Clamp
16.	EVAP hose	17.	EVAP service port	18.	EVAP hose
19.	Clamp	20.	Water hose	21.	Water hose
22.	Electric throttle control actuator	23.	Gasket		
A.	To brake booster	B.	To centralized under-floor piping	C.	To water outlet

37663_CUBE_G0177

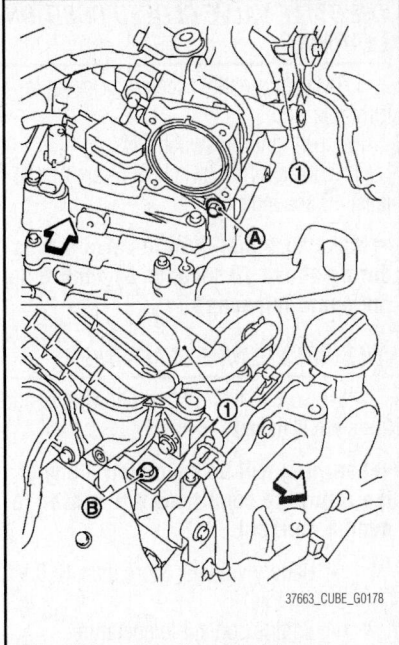

Fig. 51 Loosen and remove intake manifold (1) mounting bolts (A) and (B)

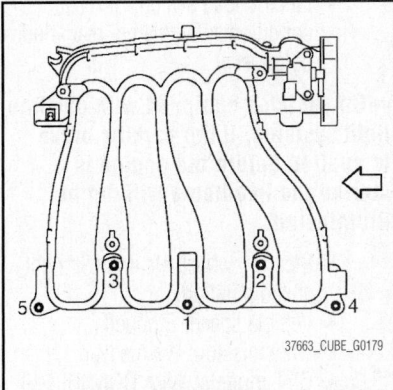

Fig. 52 Tighten intake manifold mounting bolts in numerical order

leakage when engine coolant is not drained.

✳✳ CAUTION

Perform this step when the engine is cold.

7. Remove electric throttle control actuator.

✳✳ CAUTION

Handle carefully to avoid any shock to electric throttle control actuator. Never disassemble electric throttle control actuator.

8. Remove intake manifold with the following procedure:
   a. Loosen and remove intake manifold mounting bolts.
   b. Remove harness clip from intake manifold side.
   c. Disconnect EVAP hose from intake manifold.

   d. Disconnect harness connector from EVAP canister purge volume control valve.
   e. Loosen mounting bolts in reverse order of the tightening sequence.

✳✳ CAUTION

Cover engine openings to avoid entry of foreign materials.

9. Remove brackets from intake manifold, if necessary.
10. Remove engine cover bracket, if necessary.
11. Remove EVAP canister purge volume control solenoid valve from intake manifold, if necessary.

To install:
12. Note the following, and install in the reverse order of removal.
13. Check if gasket is not dropped from the installation groove of intake manifold.

14. Install intake manifold with the following procedure:
   a. Tighten intake manifold mounting bolts in numerical order as shown.
   b. Tighten No. 1 bolt again.
   c. Tighten intake manifold mounting bolt A. Then tighten intake manifold mounting bolt B.
15. Tighten bolts of electric throttle control actuator equally and diagonally in several steps.
16. Perform "Throttle Valve Closed Position Learning" after repair when removing harness connector of the electric throttle control actuator.
17. Perform "Throttle Valve Closed Position Learning" and "Idle Air Volume Learning" after repair when replacing electric throttle control actuator.

THROTTLE VALVE CLOSED POSITION LEARNING

1. Make sure that accelerator pedal is fully released.
2. Turn ignition switch ON.
3. Turn ignition switch OFF and wait at least 10 seconds.

➡**Make sure that throttle valve moves during above 10 seconds by confirming the operating sound.**

IDLE AIR VOLUME LEARNING

Make sure that all of the following conditions are satisfied.

➡**Learning will be cancelled if any of the following conditions are missed for even a moment.**

- Battery voltage: More than 12.9 V (At idle)
- Engine coolant temperature: 158–212°F (70–100°C)
- Selector lever: P or N (CVT), Neutral (M/T)
- Electric load switch: OFF (Air conditioner, headlamp, rear window defogger)

➡**On vehicles equipped with daytime light systems, if the parking brake is applied before the engine is started the headlamp will not be illuminated.**

- Steering wheel: Neutral (Straight-ahead position)
- Vehicle speed: Stopped
- Transmission: Warmed-up
- CVT models: With CONSULT-III: Drive vehicle until "FLUID TEMP SE" in "DATA MONITOR" mode of "TRANSMISSION" system indicates less than 0.9 V.
- CVT models: Without CONSULT-III: Drive vehicle for 10 minutes.
- M/T models: Drive vehicle for 10 minutes.

With CONSULT-III

1. Perform Accelerator Pedal Released Position Learning.
2. Perform Throttle Valve Closed Position Learning.
3. Start engine and warm it up to normal operating temperature.
4. Select "IDLE AIR VOL LEARN" in "WORK SUPPORT" mode.
5. Touch "START" and wait 20 seconds.

Without CONSULT-III

➡**Note the following:**

- It is better to count the time accurately with a clock.
- It is impossible to switch the diagnostic mode when an accelerator pedal position sensor circuit has a malfunction.

1. Perform Accelerator Pedal Released Position Learning.
2. Perform Throttle Valve Closed Position Learning.
3. Start engine and warm it up to normal operating temperature.
4. Turn ignition switch OFF and wait at least 10 seconds.
5. Confirm that accelerator pedal is fully released, turn ignition switch ON and wait 3 seconds.
6. Repeat the following procedure quickly five times within 5 seconds.
   a. Fully depress the accelerator pedal.
   b. Fully release the accelerator pedal.
7. Wait 7 seconds, fully depress the accelerator pedal and keep it for approx. 20 seconds until the MIL stops blinking and turned ON.
8. Fully release the accelerator pedal within 3 seconds after the MIL turned ON.
9. Start engine and let it idle.

OIL PAN

REMOVAL & INSTALLATION

Lower Oil Pan

See Figures 53 and 54.

1. Drain engine oil.
2. Remove the lower oil pan with the following procedure:
   a. Loosen the lower oil pan mounting bolts in reverse order of the tightening sequence.

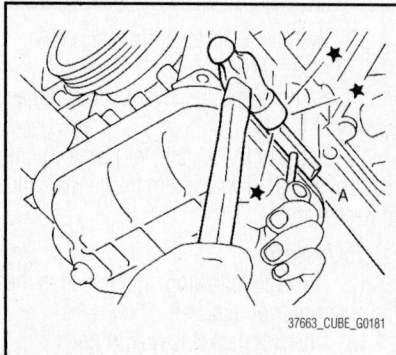

Fig. 53 Insert seal cutter (A) between the upper oil pan and the lower oil pan

b. Insert seal cutter KV10111100 (J-37228) between the upper oil pan and the lower oil pan.

❋❋ **CAUTION**

Be careful not to damage the mating surface. Never insert a screwdriver. This damages the mating surfaces.

c. Slide the seal cutter KV10111100 (J-37228) by tapping on the side of tool with a hammer.
d. Remove the lower oil pan.

To install:
3. Install the lower oil pan as follows:
   a. Use a scraper to remove old liquid gasket from mating surfaces.
   b. Also remove old liquid gasket from mating surface of the upper oil pan.
   c. Remove old liquid gasket from the bolt holes and threads.

❋❋ **CAUTION**

Never scratch or damage the mating surface when cleaning off old liquid gasket.

d. Apply a continuous bead of liquid gasket with a tube presser.

❋❋ **CAUTION**

Attaching should be done within 5 minutes after liquid gasket application.

e. Tighten bolts in numerical order as shown.
4. Install oil pan drain plug.
5. Install in the reverse order of removal after this step.
6. Clean oil strainer if any object attached.
7. Check the engine oil level and adjust engine oil.
8. Start engine, and check there is no leakage of engine oil.

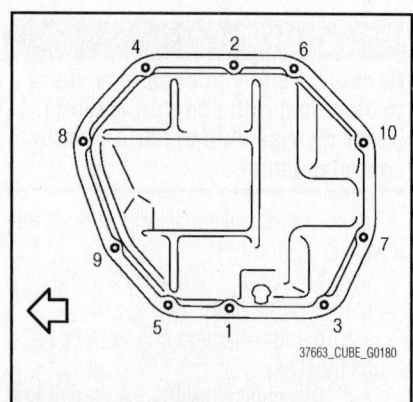

Fig. 54 Tighten bolts in numerical order

9. Stop engine and wait for 10 minutes.

10. Check the engine oil level again.

OIL PUMP

REMOVAL & INSTALLATION

See Figures 55 through 57.

1. Remove the lower oil pan.

2. Remove front cover, and other related parts.

3. Remove oil pump sprocket with the following procedure:

➡**Add matching mark if necessary for easier installation.**

a. Press stopper tab in the direction shown to push the oil pump chain tensioner slack guide toward oil pump chain tensioner.

➡**The oil pump chain tensioner slack guide is released by pressing the stopper tab. As the result, the oil pump chain tensioner slack guide can be moved.**

b. Insert a stopper pin into tensioner body hole to secure the oil pump chain tensioner slack guide.

➡**Use a hard metal pin with the diameter of approximately 0.47 inches (1.2 mm) as a stopper pin.**

c. Remove oil pump chain tensioner.

➡**When the holes on lever and tensioner body cannot be aligned, align these holes by slightly moving the oil pump chain tensioner slack guide.**

d. Hold the WAF part of oil pump shaft, and then loosen the oil pump sprocket bolt and remove it.

✳✳ CAUTION

Secure the oil pump shaft with the WAF part. Never loosen the oil pump sprocket bolt by tightening the oil pump drive chain.

e. Remove oil pump sprocket.

4. Remove oil pump. Loosen bolts in reverse order of the tightening sequence.

To install:

5. Note the following, and install in the reverse order of removal.

6. Tighten bolts in numerical order as shown.

7. Check the engine oil level.

8. Start the engine, and check that there is no leakage of engine oil.

9. Stop the engine and wait for 10 minutes.

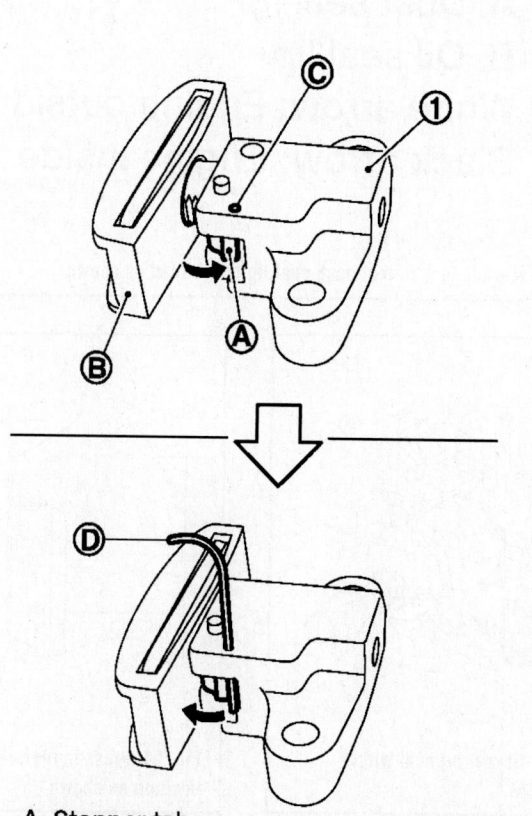

A. Stopper tab
B. Oil pump chain tensioner slack guide
C. Tensioner body hole
D. Stopper pin
1. Oil pump chain tensioner

37663_CUBE_G0183

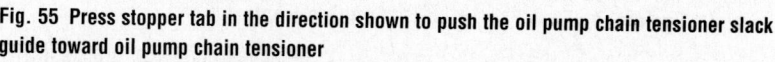

Fig. 55 Press stopper tab in the direction shown to push the oil pump chain tensioner slack guide toward oil pump chain tensioner

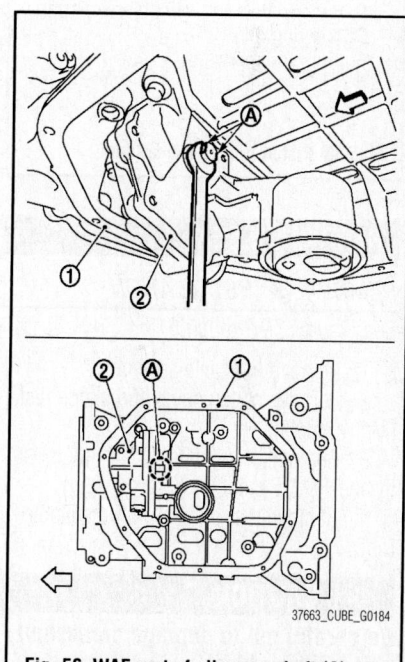

37663_CUBE_G0184

Fig. 56 WAF part of oil pump shaft (A), upper oil pan (1), and oil pump (2)

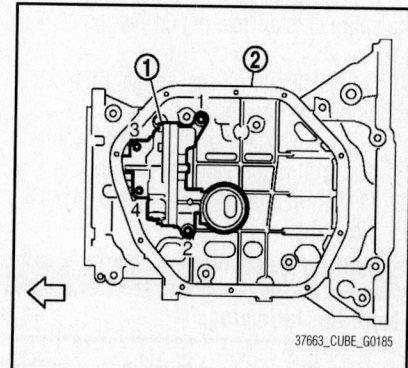

37663_CUBE_G0185

Fig. 57 Tighten bolts in numerical order

10. Check the engine oil level, and adjust the level.

PISTON AND RING

POSITIONING

See Figure 58.

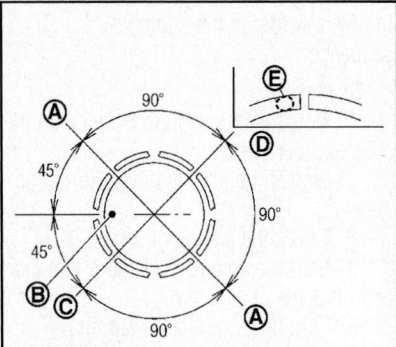

A. Oil ring upper or lower rail gap
B. Front mark
C. Second ring and oil ring spacer gap
D. Top ring gap
E. Stamped mark

37663_CUBE_G0187

Fig. 58 Piston ring gap locations

REAR MAIN SEAL

REMOVAL & INSTALLATION

See Figures 59 through 61.

1. Remove transaxle assembly.
2. Remove clutch cover and clutch disk (M/T models).
3. Remove drive plate (CVT models) or flywheel (M/T models).
4. Remove rear oil seal with a suitable tool.

❈❈ CAUTION

Be careful not to damage crankshaft and cylinder block.

To install:

5. Apply the liquid gasket lightly to entire outside area of new rear oil seal.
6. Install rear oil seal so that each seal lip is oriented as shown.

   a. Press-fit rear oil seal with a suitable drift outer diameter 4.53 inches (115 mm) and inner diameter 3.54 inches (90 mm).

❈❈ CAUTION

Note the following:

- Be careful not to damage crankshaft and cylinder block.

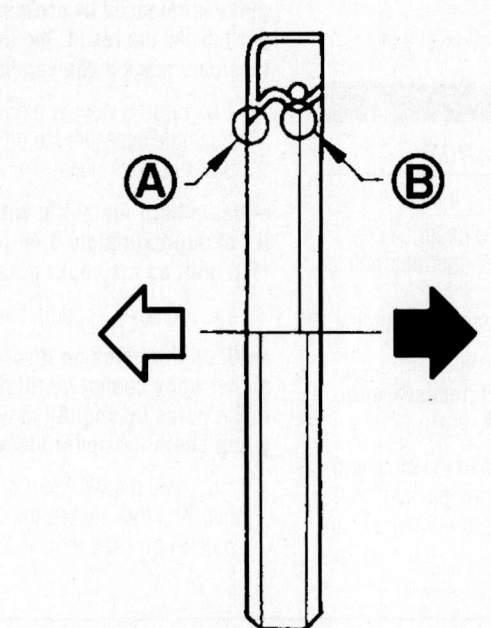

A. Dust seal lip
B. Oil seal lip
White arrow: Engine outside
Black arrow: Engine inside

37663_CUBE_G0188

Fig. 59 Install rear oil seal so that each seal lip is oriented as shown

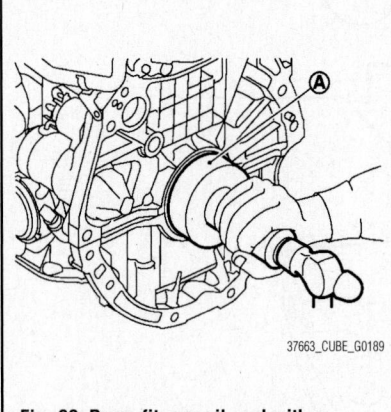

37663_CUBE_G0189

Fig. 60 Press-fit rear oil seal with a suitable drift (A)

- Press-fit oil seal straight to avoid causing burrs or tilting.
- Never touch grease applied onto oil seal lip.

   b. Press in rear oil seal to the position as shown.

7. Install in the reverse order of removal, for the rest of parts.

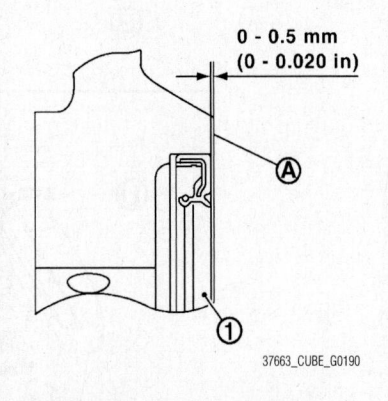

0 - 0.5 mm
(0 - 0.020 in)

37663_CUBE_G0190

Fig. 61 Press in rear oil seal (1) to the position as shown

TIMING CHAIN FRONT COVER

REMOVAL & INSTALLATION

See Figures 62 through 69.

➡ The rotating direction in the text indicates all directions viewed from the engine front.

1. Remove front fender protector (RH).

2. Drain engine oil.

✳✳ CAUTION

Perform this step when engine is cold.

3. Remove the following parts:
- Intake manifold
- Rocker cover
- Drive belt
- Ground cable (between front cover and radiator core support)

4. Set No. 1 cylinder at TDC on its compression stroke with the following procedure:

a. Rotate crankshaft pulley clockwise and align TDC mark (no paint) to timing indicator on front cover.

b. At the same time, check that the cam noses of the No. 1 cylinder are located as shown. If not, rotate crankshaft pulley one revolution (360 degrees) and align.

5. Remove crankshaft pulley with the following procedure:

a. Fix crankshaft pulley with a pulley holder, loosen crankshaft pulley bolt until the bolt seating surface is 0.39 inches (10 mm) from its original position.

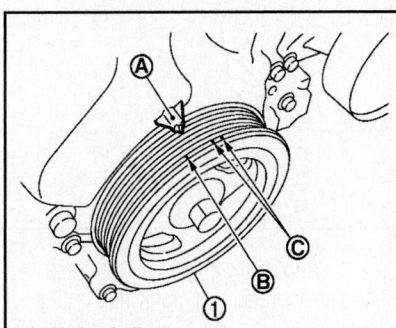

A. Timing indicator
B. TDC mark (no paint)
C. White paint mark (not used for service)
1. Crankshaft pulley

37663_CUBE_G0191

Fig. 62 Rotate crankshaft pulley clockwise and align TDC mark (no paint) to timing indicator on front cover

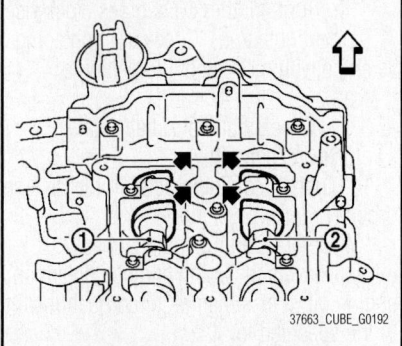

37663_CUBE_G0192

Fig. 63 Check that the cam noses of the No. 1 cylinder are located as shown

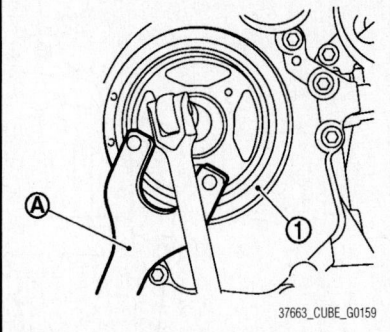

37663_CUBE_G0159

Fig. 64 Fix crankshaft pulley (1) with a pulley holder (A)

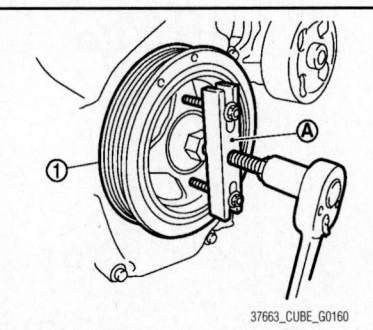

37663_CUBE_G0160

Fig. 65 Attach a pulley puller KV11103000 in the M6 thread hole on crankshaft pulley (1)

✳✳ CAUTION

Do not remove the crankshaft pulley bolt as it will be used as a supporting point for the pulley puller.

b. Attach a pulley puller KV11103000 in the M6 thread hole on crankshaft pulley, and remove crankshaft pulley.

6. Remove the lower oil pan.

➡ **If crankshaft sprocket and oil pump drive component are not removed, this step is unnecessary.**

7. Support the bottom surface of engine using a transmission jack, and then remove the engine mounting bracket (RH) and the engine mounting insulator (RH).

8. Remove intake valve timing control solenoid valve.

9. Remove drive belt auto-tensioner.

10. Remove front cover with the following procedure:

a. Remove the mounting bolts in reverse order as shown.

b. Cut liquid gasket by prying at the positions shown, and then remove the front cover.

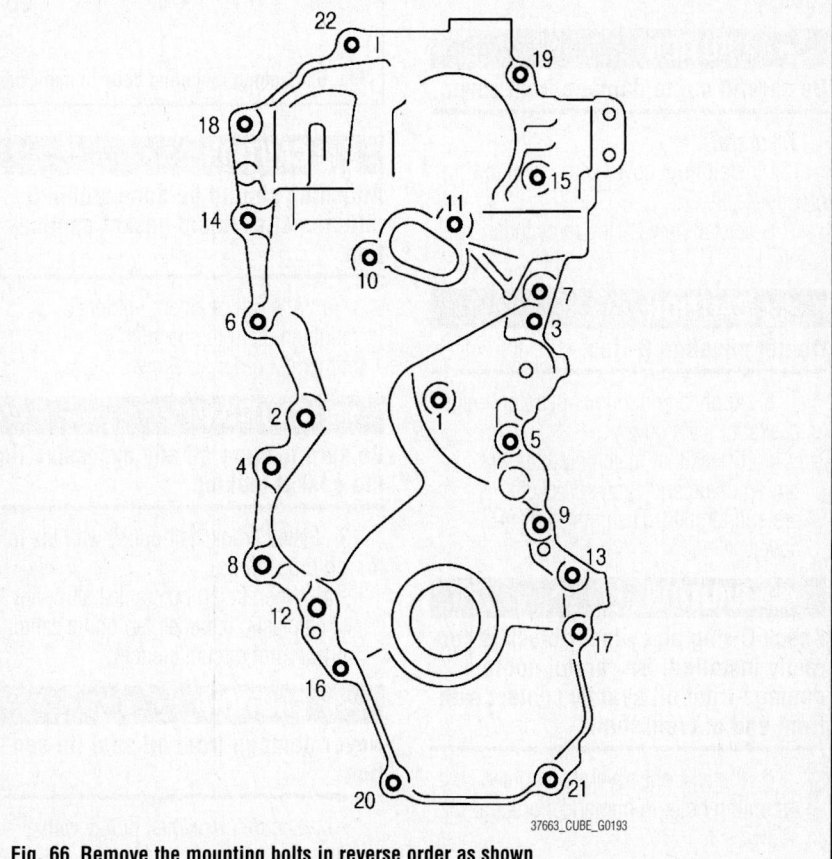

37663_CUBE_G0193

Fig. 66 Remove the mounting bolts in reverse order as shown

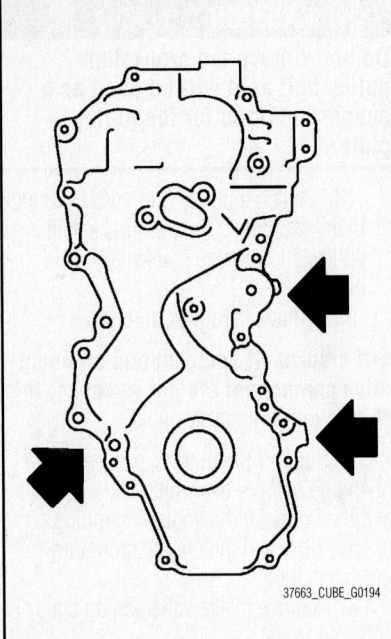

Fig. 67 Cut liquid gasket by prying at the positions shown

✳✳ CAUTION

Be careful not to damage the mating surface.

11. Remove front oil seal from front cover.

✳✳ CAUTION

Be careful not to damage front cover.

To install:

12. Install front cover with the following procedure:

   a. Install new O-ring to cylinder block.

✳✳ CAUTION

Do not misalign O-ring.

   b. Apply a continuous bead of liquid gasket to front cover.

   c. Check that matching marks of timing chain and each sprockets are still aligned. Then install front cover.

✳✳ CAUTION

Check O-ring on cylinder block is correctly installed. Be careful not to damage front oil seal by contact with front end of crankshaft.

   d. Install front cover, and tighten mounting bolts in numerical order as shown.

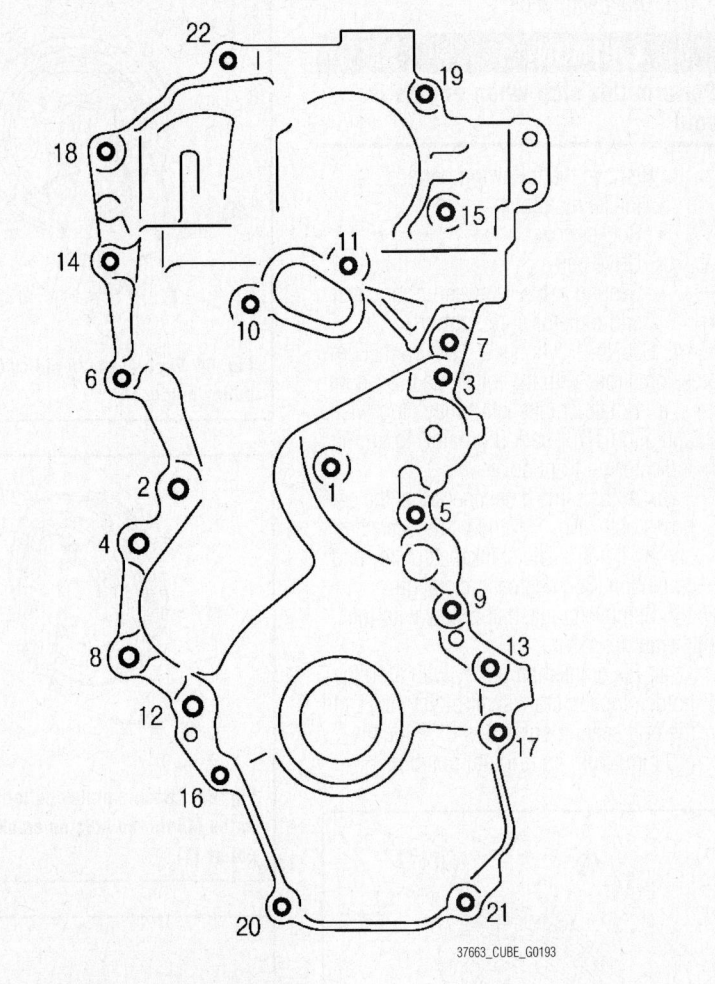

Fig. 68 Tighten mounting bolts in numerical order

✳✳ CAUTION

Attaching should be done within 5 minutes after liquid gasket application.

   e. After all bolts are tightened, retighten them to specified torque in numerical order as shown.

✳✳ CAUTION

Be sure to wipe off any excessive liquid gasket leaking.

13. Install crankshaft pulley with the following procedure:

   a. When inserting crankshaft pulley with a plastic hammer, tap on its center portion (not circumference).

✳✳ CAUTION

Never damage front oil seal lip section.

   b. Secure crankshaft pulley with a pulley holder.

   c. Apply new engine oil to thread and seat surfaces of crankshaft pulley bolt.

   d. Tighten crankshaft pulley bolt.

   e. Completely loosen.

   f. Tighten crankshaft pulley bolt.

   g. Put a paint mark on crankshaft pulley, matching with any one of six easy to recognize angle marks on crankshaft pulley bolt flange.

   h. Turn another 60 degrees clockwise (angle tightening). Check the tightening angle with movement of one angle mark.

   i. Check that crankshaft rotates clockwise smoothly.

14. Install remaining parts in the reverse order of removal.

15. Before starting engine, check oil/fluid levels including engine coolant and engine oil. If less than required quantity, fill to the specified level.

16. Use procedure below to check for fuel leakage.

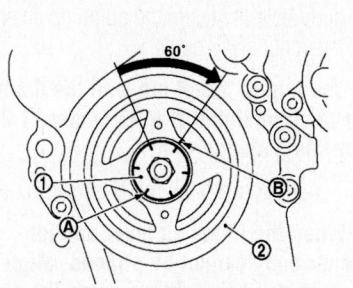

A. Angle marks
B. Paint mark
1. Crankshaft pulley bolt flange
2. Crankshaft pulley

37663_CUBE_G0161

Fig. 69 Put a paint mark on crankshaft pulley, matching with any one of six easy to recognize angle marks on crankshaft pulley bolt flange

a. Turn ignition switch "ON" (with engine stopped). With fuel pressure applied to fuel piping, check for fuel leakage at connection points.

b. Start engine. With engine speed increased, check again for fuel leakage at connection points.

17. Run engine to check for unusual noise and vibration.

➡If hydraulic pressure inside chain tensioner drops after removal/ installation, slack in guide may generate a pounding noise during and just after the engine start. However, this does not indicate an unusualness. Noise will stop after hydraulic pressure rises.

18. Warm up engine thoroughly to check there is no leakage of fuel, or any oil/fluids including engine oil and engine coolant.

19. Bleed air from lines and hoses of applicable lines, such as in cooling system.

20. After cooling down engine, again check oil/fluid levels including engine oil and engine coolant. Refill to the specified level, if necessary.

TIMING CHAIN & SPROCKETS

REMOVAL & INSTALLATION

See Figures 70 through 81.

➡The rotating direction in the text indicates all directions viewed from the engine front.

1. Remove front fender protector (RH).
2. Drain engine oil.

✳ CAUTION

Perform this step when engine is cold.

3. Remove the following parts:
- Intake manifold
- Rocker cover
- Drive belt
- Ground cable (between front cover and radiator core support)

4. Set No. 1 cylinder at TDC on its compression stroke with the following procedure:

a. Rotate crankshaft pulley clockwise and align TDC mark (no paint) to timing indicator on front cover.

b. At the same time, check that the cam noses of the No. 1 cylinder are located as shown. If not, rotate crankshaft pulley one revolution (360 degrees) and align.

5. Remove crankshaft pulley with the following procedure:

a. Fix crankshaft pulley with a pulley holder, loosen crankshaft pulley bolt until the bolt seating surface is 0.39 inches (10 mm) from its original position.

✳ CAUTION

Do not remove the crankshaft pulley bolt as it will be used as a supporting point for the pulley puller.

b. Attach a pulley puller KV11103000 in the M6 thread hole on crankshaft pulley, and remove crankshaft pulley.

6. Remove the lower oil pan.

➡If crankshaft sprocket and oil pump drive component are not removed, this step is unnecessary.

7. Support the bottom surface of engine using a transmission jack, and then remove the engine mounting bracket (RH) and the engine mounting insulator (RH).

8. Remove intake valve timing control solenoid valve.

9. Remove drive belt auto-tensioner.

10. Remove front cover with the following procedure:

a. Remove the mounting bolts in reverse order as shown.

b. Cut liquid gasket by prying at the positions shown, and then remove the front cover.

✳ CAUTION

Be careful not to damage the mating surface.

11. Remove front oil seal from front cover.

✳ CAUTION

Be careful not to damage front cover.

12. Remove timing chain tensioner with the following procedure:

a. Push in timing chain tensioner plunger.

b. Insert a stopper pin into the body hole, and then fix it with the plunger pushed in.

➡Use approximately 0.059 inches (1.5 mm) diameter. hard metal pin as a stopper pin.

c. Remove timing chain tensioner.

13. Remove slack guide, tension guide, and timing chain.

✳ CAUTION

Never rotate each crankshaft and camshaft individually while timing chain is removed. It causes interference between valve and piston.

➡If timing chain is difficult to remove, remove camshaft sprocket (EXH) first to remove timing chain.

14. Remove crankshaft sprocket and oil pump drive component with the following procedure:

a. Press stopper tab in the direction shown to push the oil pump chain tensioner slack guide toward oil pump chain tensioner.

➡The oil pump chain tensioner slack guide is released by pressing the stopper tab. As the result, the oil pump chain tensioner slack guide can be moved.

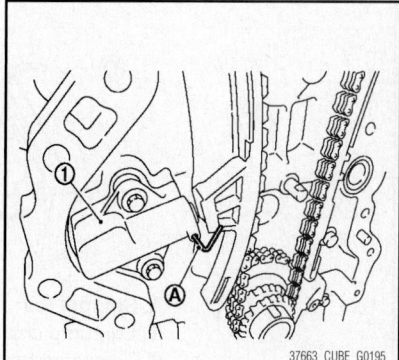

37663_CUBE_G0195

Fig. 70 Insert a stopper pin (A) into the body hole, and then fix it with the plunger pushed in; remove timing chain tensioner (1)

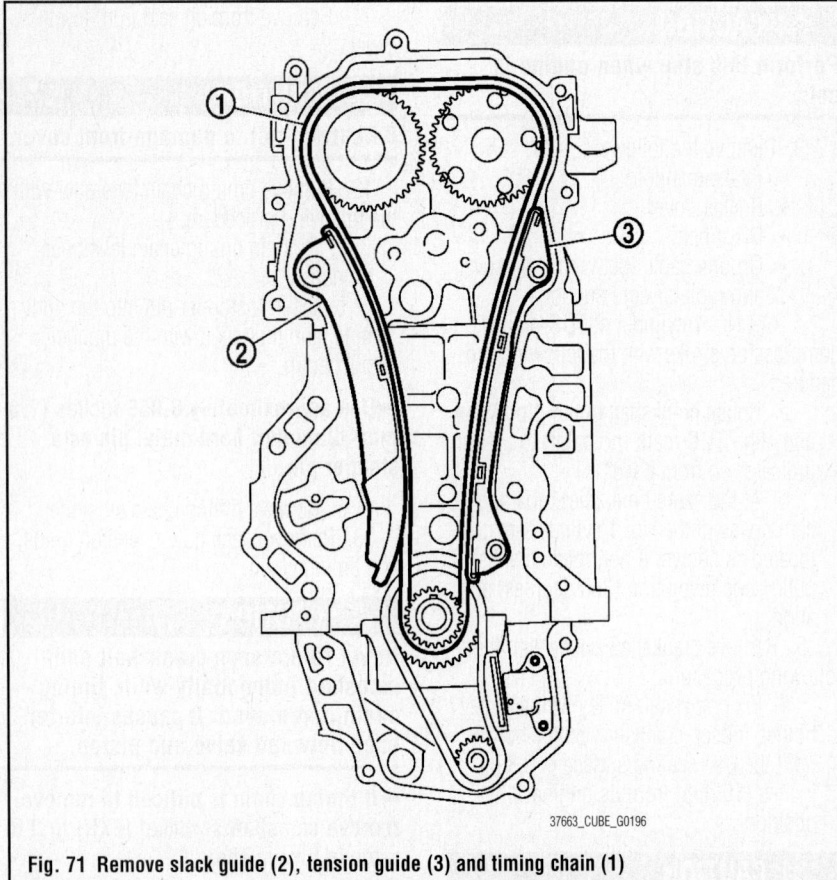

Fig. 71 Remove slack guide (2), tension guide (3) and timing chain (1)

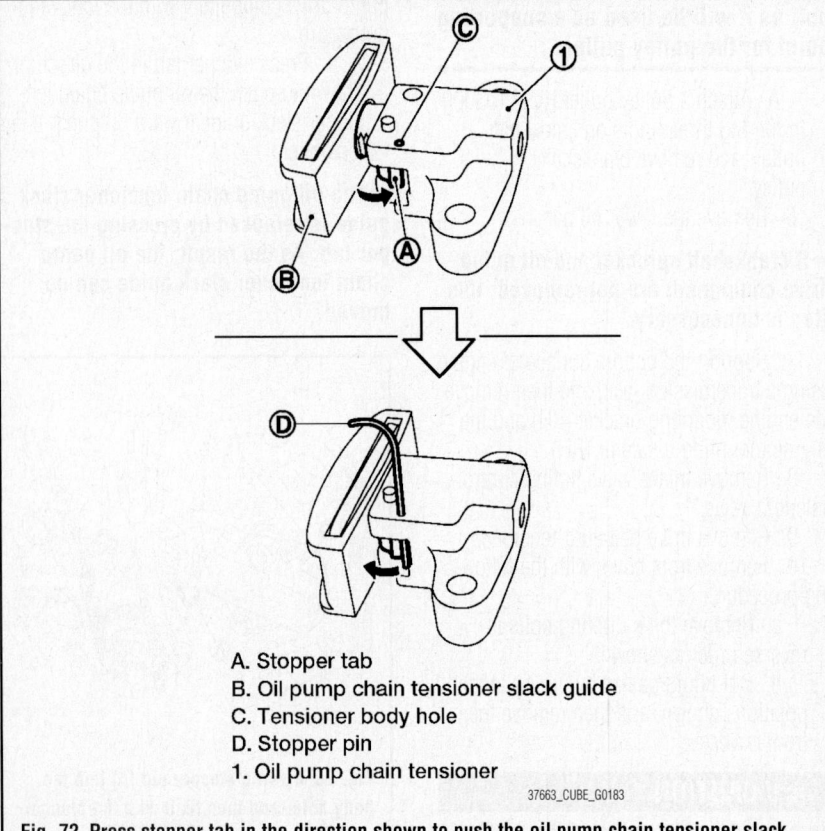

A. Stopper tab
B. Oil pump chain tensioner slack guide
C. Tensioner body hole
D. Stopper pin
1. Oil pump chain tensioner

Fig. 72 Press stopper tab in the direction shown to push the oil pump chain tensioner slack guide toward oil pump chain tensioner

b. Insert a stopper pin into tensioner body hole to secure the oil pump chain tensioner slack guide.

➡ Use a hard metal pin with the diameter of approximately 0.47 inches (1.2 mm) as a stopper pin.

c. Remove oil pump chain tensioner.

➡ When the holes on lever and tensioner body cannot be aligned, align these holes by slightly moving the oil pump chain tensioner slack guide.

d. Hold the WAF part of oil pump shaft, and then loosen the oil pump sprocket bolt and remove it.

✳✳ CAUTION

Secure the oil pump shaft with the WAF part. Never loosen the oil pump sprocket bolt by tightening the oil pump drive chain.

e. Remove oil pump sprocket.
15. Remove oil pump. Loosen bolts in reverse order as shown.

✳✳ CAUTION

Secure the oil pump shaft with the WAF part. Never loosen the oil pump sprocket bolt by tightening the oil pump drive chain.

a. Remove crankshaft sprocket, oil pump sprocket, and oil pump drive chain as a set.

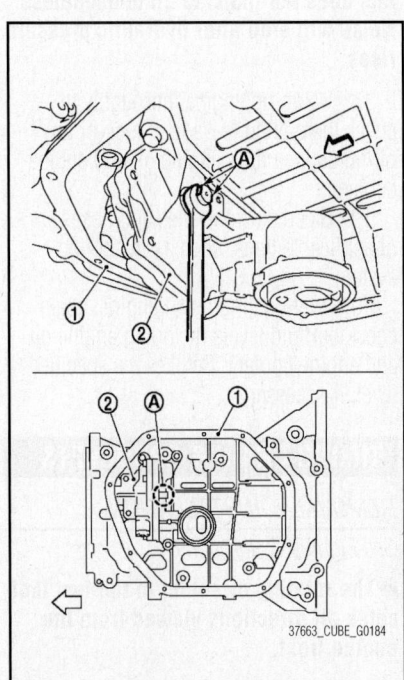

Fig. 73 WAF part of oil pump shaft (A), upper oil pan (1), and oil pump (2)

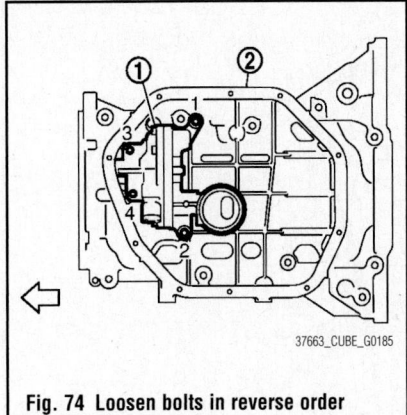

Fig. 74 Loosen bolts in reverse order

16. Remove tension guide (front cover side) from front cover, if necessary.

To install:

The figure shows the relationship between the matching mark on each timing chain and that on the corresponding sprocket, with the components installed.

17. Check that crankshaft key points straight up.

18. If the tension guide (front cover side) is removed, install it to the front cover.

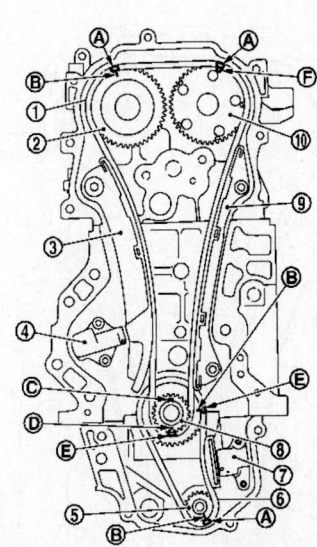

1. Timing chain
2. Camshaft sprocket (EXH)
3. Slack guide
4. Timing chain tensioner
5. Oil pump sprocket
6. Oil pump drive chain
7. Oil pump chain tensioner
8. Crankshaft sprocket
9. Tension guide
10. Camshaft sprocket (INT)

A. Matching mark (dark blue link)
B. Matching mark (stamping)
C. Crankshaft key position (straight up)
D. Matching mark (stamping)
E. Matching mark (orange link)
F. Matching mark (outer groove*)

*: There are two outer grooves in camshaft sprocket (INT).
  The wider one is a matching mark.

37663_CUBE_G0197

Fig. 75 Timing chain assembly

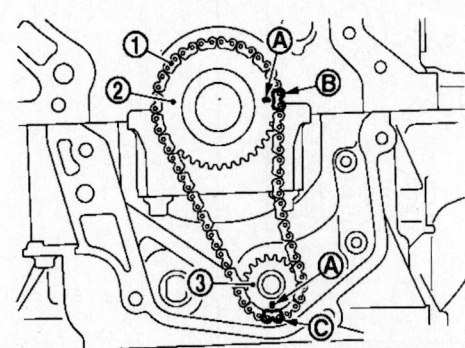

A. Matching mark (stamping)
B. Matching mark (orange link)
C. Matching mark (dark blue link)
1. Oil pump drive chain
2. Crankshaft sprocket
3. Oil pump sprocket

37663_CUBE_G0198

Fig. 76 Install crankshaft sprocket, oil pump sprocket, and oil pump drive chain

※※ **CAUTION**

Check the joint condition by sound or feeling.

19. Install crankshaft sprocket, oil pump sprocket, and oil pump drive chain.

a. Install it by aligning matching marks on each sprockets and oil pump drive chain.

b. If these matching marks are not aligned, rotate the oil pump shaft slightly to correct the position.

※※ **CAUTION**

Check matching mark position of each sprockets after installing the oil pump drive chain.

20. Hold the WAF part of oil pump shaft (A), and then tighten the oil pump shaft sprocket bolt.

※※ **CAUTION**

Secure the oil pump shaft with the WAF part. Never loosen the oil pump shaft sprocket bolt by tightening the oil pump drive chain.

21. Install oil pump chain tensioner.

a. Fix the plunger at the most compressed position using a stopper pin, and then install it.

b. Securely pull out the stopper pin after installing the oil pump chain tensioner.

c. Check matching mark position of oil pump drive chain and each sprockets again.

22. Align the matching marks of each sprocket with the matching marks of timing chain.

➡**If these matching marks are not aligned, rotate the camshaft slightly by holding the hexagonal portion to correct the position.**

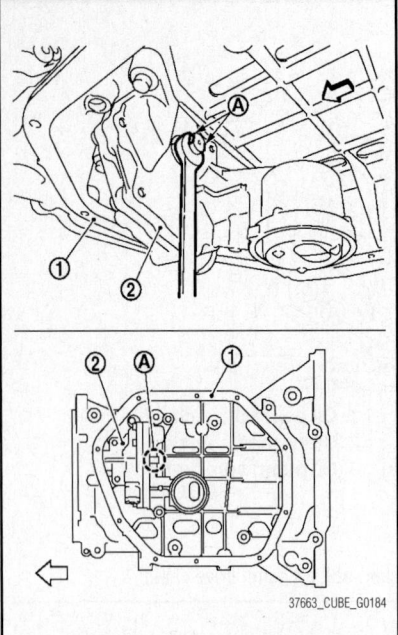

Fig. 77 WAF part of oil pump shaft (A), upper oil pan (1), and oil pump (2)

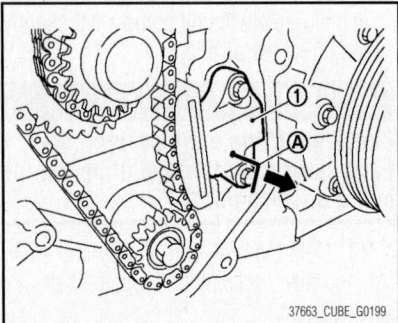

Fig. 78 Install oil pump chain tensioner (1); fix the plunger at the most compressed position using a stopper pin (A)

❊❊ CAUTION

Check matching mark position of each sprocket and timing chain again after installing the timing chain.

23. Install the slack guide and the tension guide.

24. Install timing chain tensioner.

    a. Fix the plunger at the most compressed position using a stopper pin, and then install it.

    b. Securely pull out the stopper pin after installing the timing chain tensioner.

25. Check matching mark position of timing chain and each sprockets again.

26. Install front oil seal.

27. Install front cover with the following procedure:

    a. Install new O-ring to cylinder block.

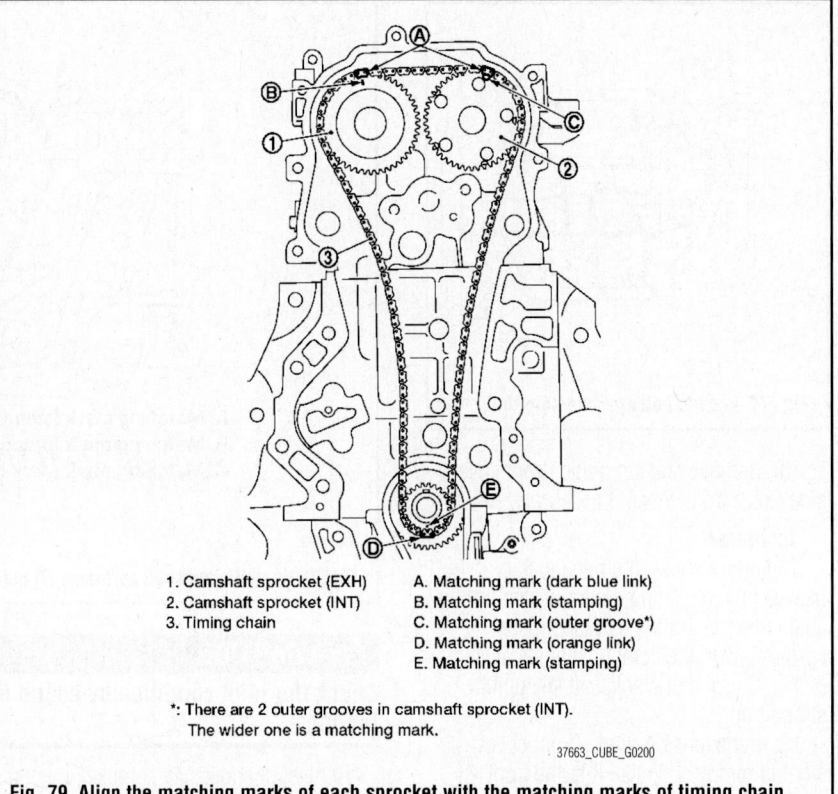

1. Camshaft sprocket (EXH)	A. Matching mark (dark blue link)
2. Camshaft sprocket (INT)	B. Matching mark (stamping)
3. Timing chain	C. Matching mark (outer groove*)
	D. Matching mark (orange link)
	E. Matching mark (stamping)

*: There are 2 outer grooves in camshaft sprocket (INT). The wider one is a matching mark.

Fig. 79 Align the matching marks of each sprocket with the matching marks of timing chain

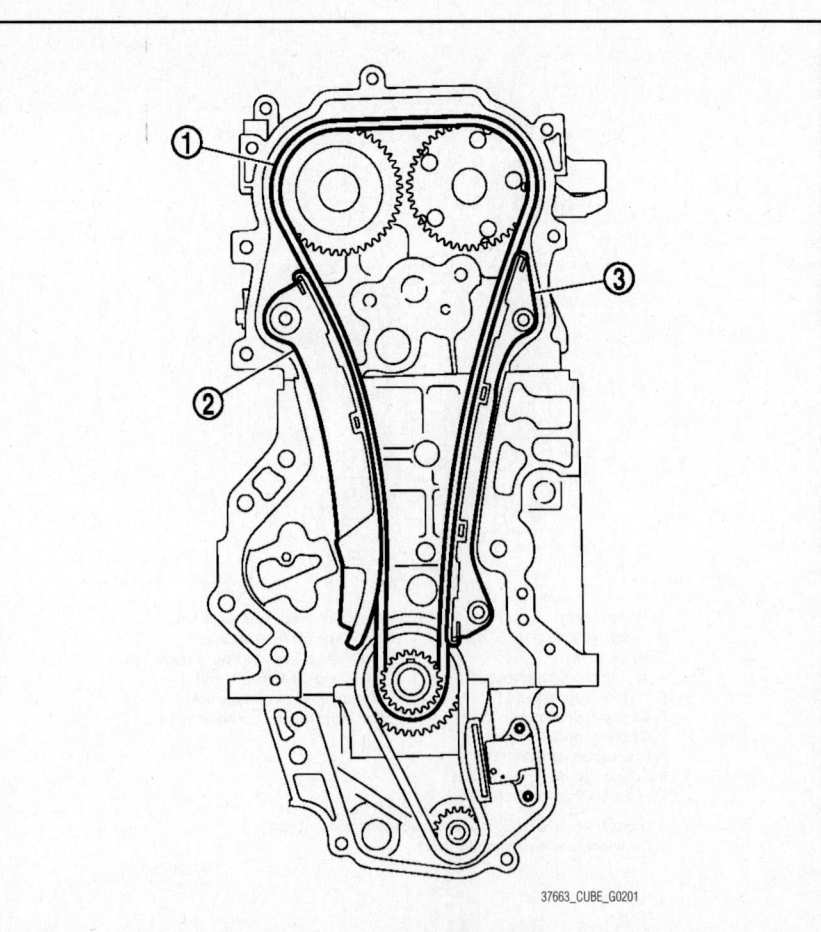

Fig. 80 Install the slack guide (2) and the tension guide (3)

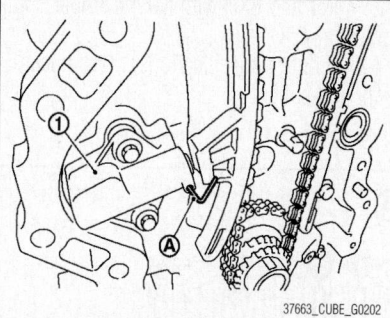

Fig. 81 Install timing chain tensioner (1); fix the plunger at the most compressed position using a stopper pin (A)

✳✳ CAUTION

Do not misalign O-ring.

b. Apply a continuous bead of liquid gasket to front cover.

c. Check that matching marks of timing chain and each sprockets are still aligned. Then install front cover.

✳✳ CAUTION

Check O-ring on cylinder block is correctly installed. Be careful not to damage front oil seal by contact with front end of crankshaft.

d. Install front cover, and tighten mounting bolts in numerical order as shown.

✳✳ CAUTION

Attaching should be done within 5 minutes after liquid gasket application.

e. After all bolts are tightened, retighten them to specified torque in numerical order as shown.

✳✳ CAUTION

Be sure to wipe off any excessive liquid gasket leaking.

28. Install crankshaft pulley with the following procedure:

a. When inserting crankshaft pulley with a plastic hammer, tap on its center portion (not circumference).

✳✳ CAUTION

Never damage front oil seal lip section.

b. Secure crankshaft pulley with a pulley holder.

c. Apply new engine oil to thread and seat surfaces of crankshaft pulley bolt.

d. Tighten crankshaft pulley bolt.

e. Completely loosen.

f. Tighten crankshaft pulley bolt.

g. Put a paint mark on crankshaft pulley, matching with any one of six easy to recognize angle marks on crankshaft pulley bolt flange.

h. Turn another 60 degrees clockwise (angle tightening). Check the tightening angle with movement of one angle mark.

i. Check that crankshaft rotates clockwise smoothly.

29. Install remaining parts in the reverse order of removal.

30. Before starting engine, check oil/fluid levels including engine coolant and engine oil. If less than required quantity, fill to the specified level.

31. Use procedure below to check for fuel leakage.

a. Turn ignition switch "ON" (with engine stopped). With fuel pressure applied to fuel piping, check for fuel leakage at connection points.

b. Start engine. With engine speed increased, check again for fuel leakage at connection points.

32. Run engine to check for unusual noise and vibration.

➡**If hydraulic pressure inside chain tensioner drops after removal/installation, slack in guide may generate a pounding noise during and just after the engine start. However, this does not indicate an unusualness. Noise will stop after hydraulic pressure rises.**

33. Warm up engine thoroughly to check there is no leakage of fuel, or any oil/fluids including engine oil and engine coolant.

34. Bleed air from lines and hoses of applicable lines, such as in cooling system.

35. After cooling down engine, again check oil/fluid levels including engine oil and engine coolant. Refill to the specified level, if necessary.

VALVE LASH

ADJUSTMENT

See Figures 82 and 83.

Perform adjustment depending on selected head thickness of valve lifter.

1. Remove camshaft.

2. Remove the valve lifters at the locations that are out of the standard.

3. Measure the center thickness of the removed valve lifters with a micrometer.

4. Use the equation below to calculate valve lifter thickness for replacement.

a. Valve lifter thickness calculation:

- $t = t1 + (C1 - C2)$
- t = Thickness of replacement valve lifter.
- $t1$ = Thickness of removed valve lifter.
- $C1$ = Measured valve clearance.
- $C2$ = Standard valve clearance.

b. Thickness of a new valve lifter can be identified by stamp marks on the reverse side (inside the cylinder). Stamp mark 302 indicates a thickness of 0.1189 inches (3.02 mm).

➡**Available thickness of valve lifter: 26 sizes with a range of 0.1181 to 0.1378 inches (3.00 to 3.50 mm), in steps of 0.0008 inches (0.02 mm), when assembled at the factory.**

5. Install the selected valve lifter.
6. Install camshaft.
7. Manually turn crankshaft pulley a few turns.

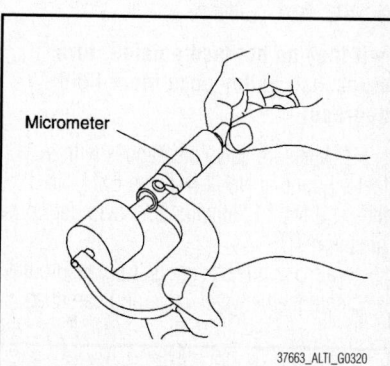

Fig. 82 Measure the center thickness of the removed valve lifters with a micrometer

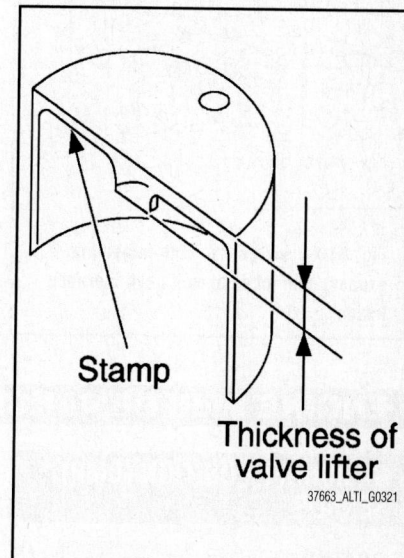

Fig. 83 Thickness of a new valve lifter can be identified by stamp marks on the reverse side (inside the cylinder)

8. Check that valve clearances for cold engine are within specifications, by referring to the specified values.

9. Install remaining parts in the reverse order of removal.

10. Warm up the engine, and check for unusual noise and vibration.

INSPECTION

See Figures 84 through 86.

Perform this inspection as follows after removal, installation, or replacement of the camshaft or any valve related parts, or if there are any unusual engine conditions due to changes in valve clearance.

1. Remove the rocker cover.

2. Turn crankshaft pulley in normal direction (clockwise when viewed from front) to align TDC identification mark (without paint mark) with timing indicator.

3. At this time, check that the both intake and exhaust cam lobes of No. 1 cylinder face inside.

➥ **If they do not face outside, turn crankshaft pulley once more (360 degrees).**

4. Measure valve clearances with a feeler gauge at No. 1 INT and EXH, No. 2 INT, and No.3 EXH with No.1 cylinder compression TDC.

   a. Use a feeler gauge to measure the clearance between valve and camshaft.

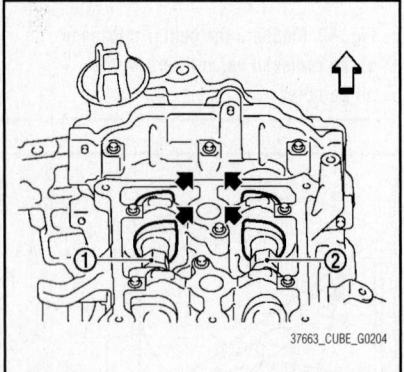

Fig. 84 Check that the both intake and exhaust cam lobes of No. 1 cylinder face inside

5. Turn crankshaft one complete revolution (360°) and align mark on crankshaft pulley with pointer.

6. Measure valve clearances with a feeler gauge at No. 2 EXH, No. 3 INT, and

No. 4 INT and EXH with No.4 cylinder compression TDC.

7. If out of specifications, make necessary adjustment.

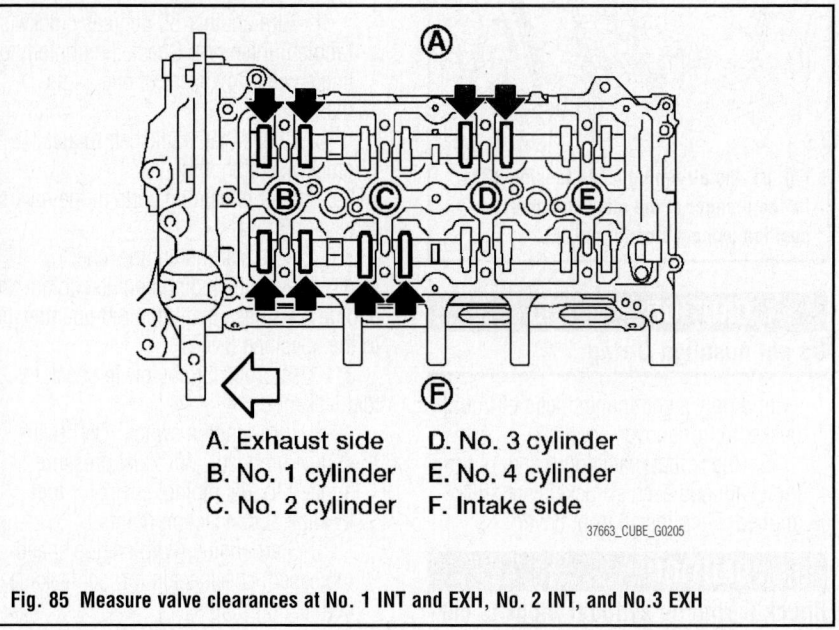

A. Exhaust side D. No. 3 cylinder
B. No. 1 cylinder E. No. 4 cylinder
C. No. 2 cylinder F. Intake side

37663_CUBE_G0205

Fig. 85 Measure valve clearances at No. 1 INT and EXH, No. 2 INT, and No.3 EXH

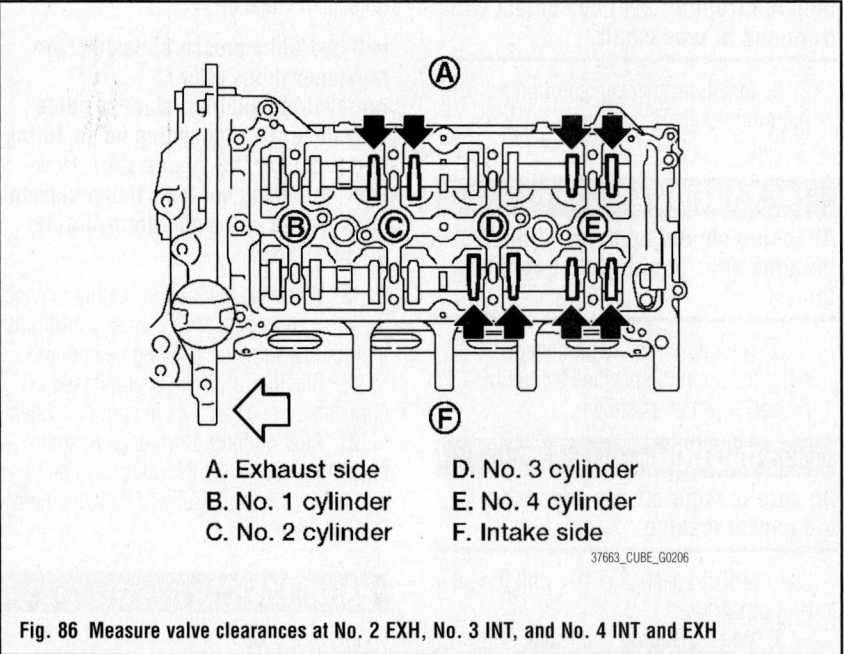

A. Exhaust side D. No. 3 cylinder
B. No. 1 cylinder E. No. 4 cylinder
C. No. 2 cylinder F. Intake side

37663_CUBE_G0206

Fig. 86 Measure valve clearances at No. 2 EXH, No. 3 INT, and No. 4 INT and EXH

ENGINE PERFORMANCE & EMISSION CONTROLS

CAMSHAFT POSITION (CMP) SENSOR

LOCATION

See Figure 87.

Refer to the accompanying illustration.

REMOVAL & INSTALLATION

See Figure 88.

1. Turn ignition switch OFF.

2. Loosen the fixing bolt of the sensor.

3. Disconnect camshaft position sensor (PHASE) harness connector.

4. Remove the sensor.

5. Installation is the reverse of removal.

7.0 (0.71, 62)
55.0 (5.6, 41)
88.2 (9.0, 65)
37663_CUBE_G0158

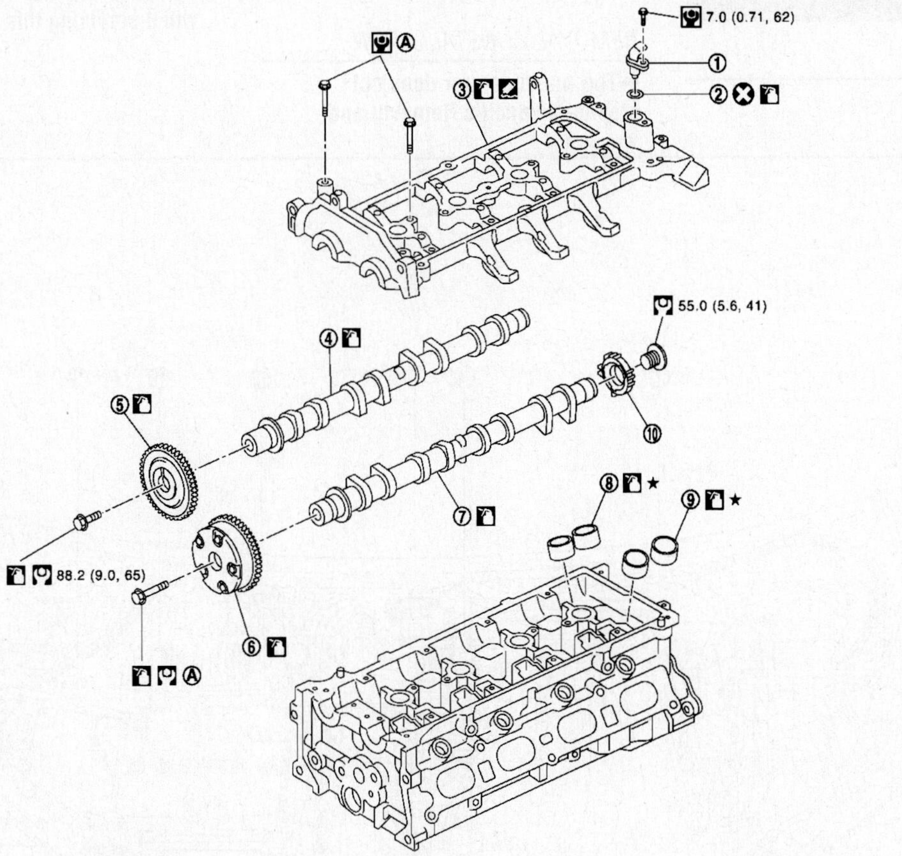

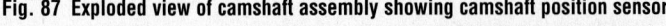

1.	Camshaft position sensor (PHASE)	2.	O-ring	3.	Camshaft bracket
4.	Camshaft (EXH)	5.	Camshaft sprocket (EXH)	6.	Camshaft sprocket (INT)
7.	Camshaft (INT)	8.	Valve lifter (EXH)	9.	Valve lifter (INT)
10.	Signal plate				
A.	Tightening must be done following the installation procedure.				

Fig. 87 Exploded view of camshaft assembly showing camshaft position sensor

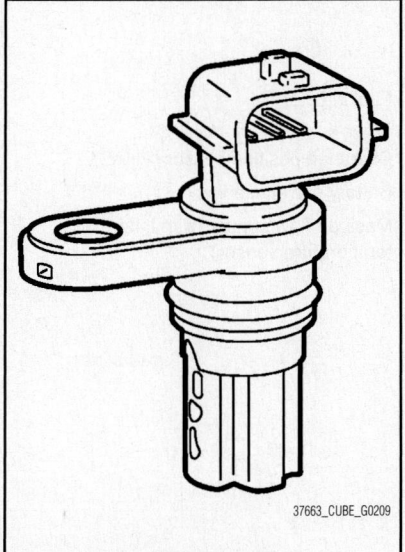

37663_CUBE_G0209

Fig. 88 Camshaft position sensor

CRANKSHAFT POSITION (CKP) SENSOR

LOCATION

The crankshaft position sensor (POS) is located on the oil pan facing the gear teeth (cogs) of the signal plate. It detects the fluctuation of the engine revolution.

REMOVAL & INSTALLATION

See Figure 89.

1. Turn ignition switch OFF.
2. Loosen the fixing bolt of the sensor.
3. Disconnect crankshaft position sensor (POS) harness connector.
4. Remove the sensor.
5. Installation is the reverse of removal.

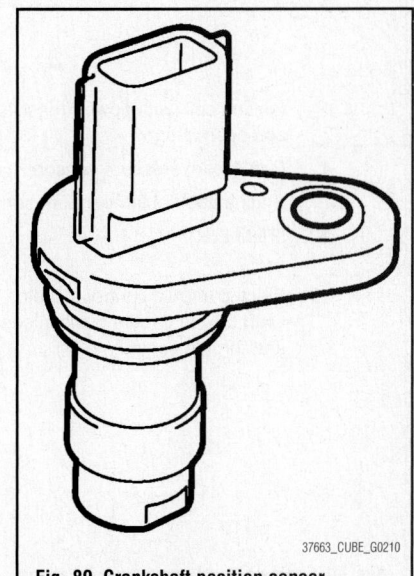

37663_CUBE_G0210

Fig. 89 Crankshaft position sensor

ENGINE CONTROL MODULE (ECM)

LOCATION

See Figure 90.

Refer to the accompanying illustration.

REMOVAL & INSTALLATION

➡**The manufacturer does not provide a specific Removal and**

Installation procedure for this component. Refer to the graphic(s) when servicing this component.

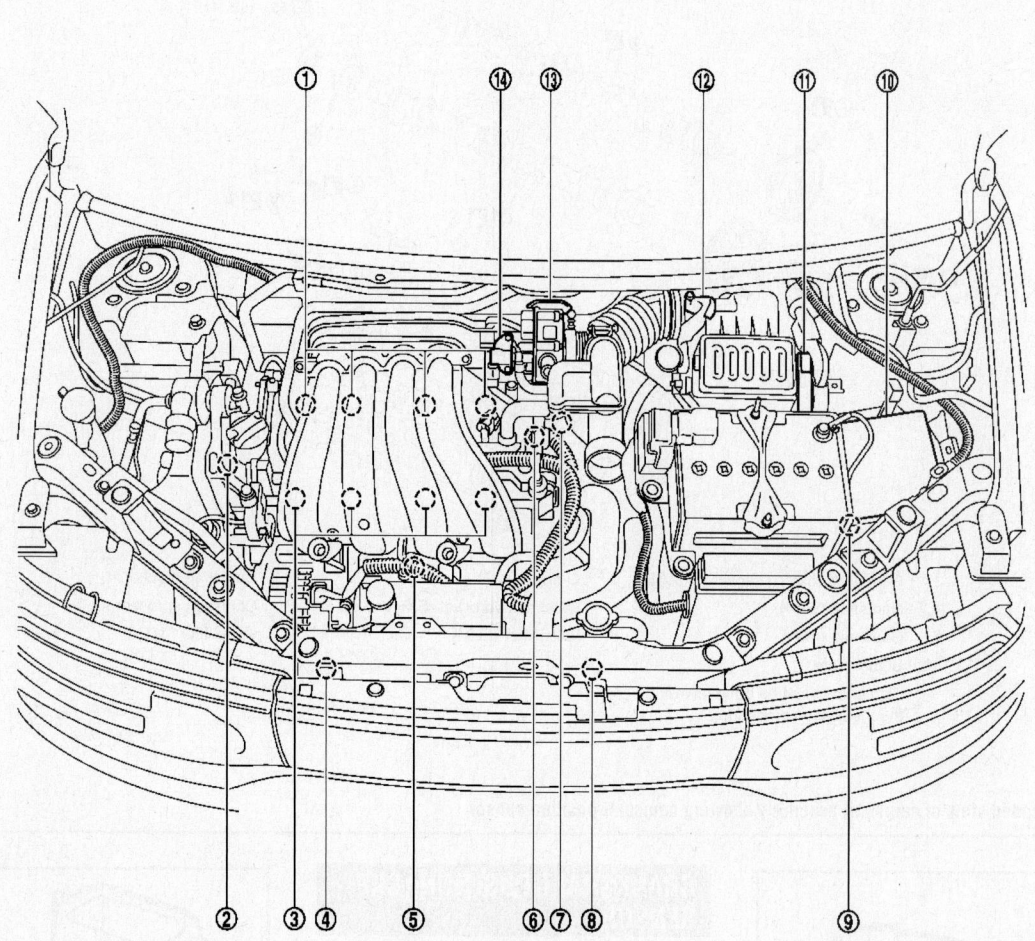

1. Ignition coil (with power transistor) and spark plug
2. Intake valve timing control solenoid valve
3. Fuel injector
4. Refrigerant pressure sensor
5. Knock sensor
6. Camshaft position sensor (PHASE)
7. Engine coolant temperature sensor
8. Cooling fan motor
9. Battery current sensor
10. IPDM E/R
11. ECM
12. Mass air flow sensor (with intake air temperature sensor)
13. Electric throttle control actuator (with built in throttle position sensor and throttle control motor)
14. EVAP canister purge volume control solenoid valve

37663_CUBE_G0207

Fig. 90 Engine control component locations

ENGINE COOLANT TEMPERATURE (ECT) SENSOR

LOCATION

See Figure 91.

Refer to the accompanying illustration.

REMOVAL & INSTALLATION

➡The manufacturer does not provide a specific Removal and Installation procedure for this component. Refer to the graphic(s) when servicing this component.

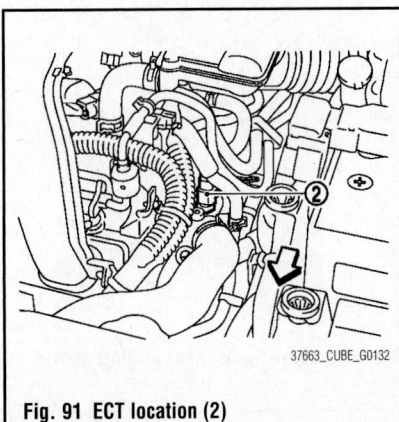

Fig. 91 ECT location (2)

HEATED OXYGEN (HO2S) SENSOR

LOCATION

See Figure 92.

Refer to the accompanying illustration.

REMOVAL & INSTALLATION

See Figure 93.

❋❋ WARNING

Allow the exhaust system to cool prior to removing the heated oxygen sensor 2 to avoid personal injury.

1. Remove heated oxygen sensor 2 with following procedure:
   a. Using heated oxygen sensor wrench KV10114400 (J-38365), remove the heated oxygen sensor 2.

❋❋ CAUTION

Be careful not to damage heated oxygen sensor 2.

To install:

2. Note the following, and install in the reverse order of removal.

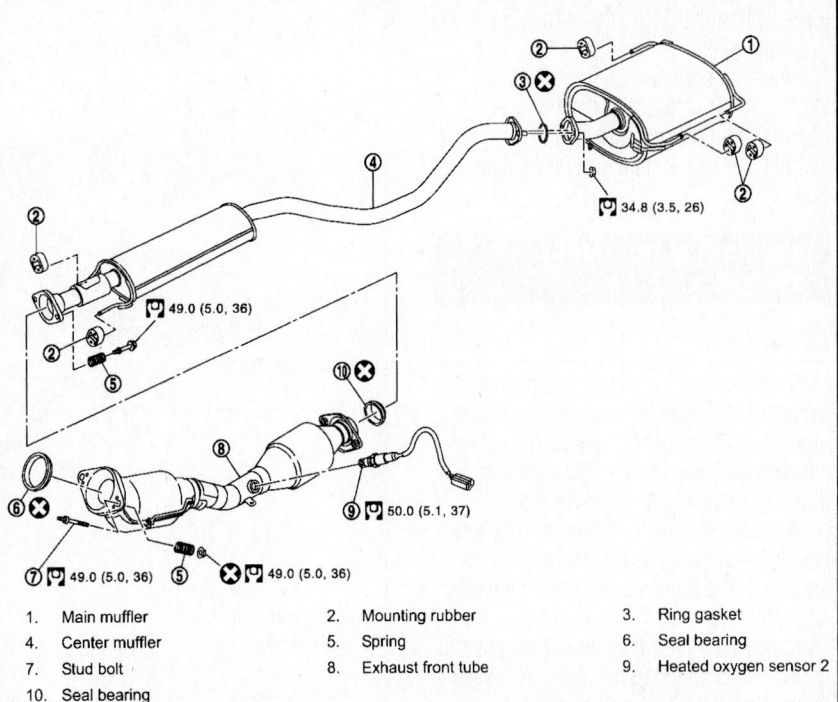

1. Main muffler
2. Mounting rubber
3. Ring gasket
4. Center muffler
5. Spring
6. Seal bearing
7. Stud bolt
8. Exhaust front tube
9. Heated oxygen sensor 2
10. Seal bearing

37663_CUBE_G0215

Fig. 92 Exploded view of exhaust system showing heated oxygen sensor 2 location

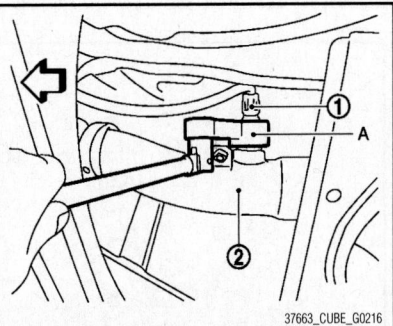

37663_CUBE_G0216

Fig. 93 Using heated oxygen sensor wrench (A), remove the heated oxygen sensor 2 (1)

❋❋ CAUTION

Note the following:

- Discard any heated oxygen sensor 2 which has been dropped onto a hard surface such as a concrete floor. Use a new one.
- Before installing a new heated oxygen sensor 2, clean exhaust system threads using the heated oxygen sensor thread cleaner (J-43897-18 or J-43897-12) and apply anti-seize lubricant.
- Never over torque heated oxygen sensor 2. Doing so may cause damage to the heated oxygen sen-

sor 2, resulting in the "MIL" coming on.

INTAKE AIR TEMPERATURE (IAT) SENSOR

LOCATION

The intake air temperature sensor is built-into mass air flow sensor (1). The sensor detects intake air temperature and transmits a signal to the ECM.

The temperature sensing unit uses a thermistor which is sensitive to the change in temperature. Electrical resistance of the thermistor decreases in response to the temperature rise.

REMOVAL & INSTALLATION

The Intake Air Temperature (IAT) sensor is an integral part of the Mass Air Flow (MAF) sensor/Intake Air Temperature (IAT) sensor assembly which is mounted on the air intake duct. Refer to the Mass Air Flow section for information regarding servicing this component.

MALFUNCTION INDICATOR LIGHT (MIL)

RESET PROCEDURE

The Malfunction Indicator Lamp (MIL) is located on the combination meter. The MIL will light up when the ignition switch is turned ON without the engine running. This is a bulb check.

When the engine is started, the MIL should go off. If the MIL remains on, the on board diagnostic system has detected an engine system malfunction.

MASS AIR FLOW (MAF) SENSOR

LOCATION

The mass air flow sensor is placed in the stream of intake air. It measures the intake flow rate by measuring a part of the entire intake flow. The mass air flow sensor controls the temperature of the hot wire to a certain amount. The heat generated by the hot wire is reduced as the intake air flows around it. The more air, the greater the heat loss.

Therefore, the electric current supplied to hot wire is changed to maintain the temperature of the hot wire as air flow increases. The ECM detects the air flow by means of this current change.

REMOVAL & INSTALLATION

See Figure 94.

➡ The manufacturer does not provide a specific Removal and Installation procedure for this component. Refer to the graphic(s) when servicing this component.

THROTTLE CONTROL ACTUATOR (TAC)

LOCATION

See Figure 95.

Refer to the accompanying illustration.

REMOVAL & INSTALLATION

➡ The manufacturer does not provide a specific Removal and Installation procedure for this component. Refer to the graphic(s) when servicing this component.

THROTTLE POSITION SENSOR (TPS)

LOCATION

The throttle position sensor is an integral part of the electric throttle control actuator.

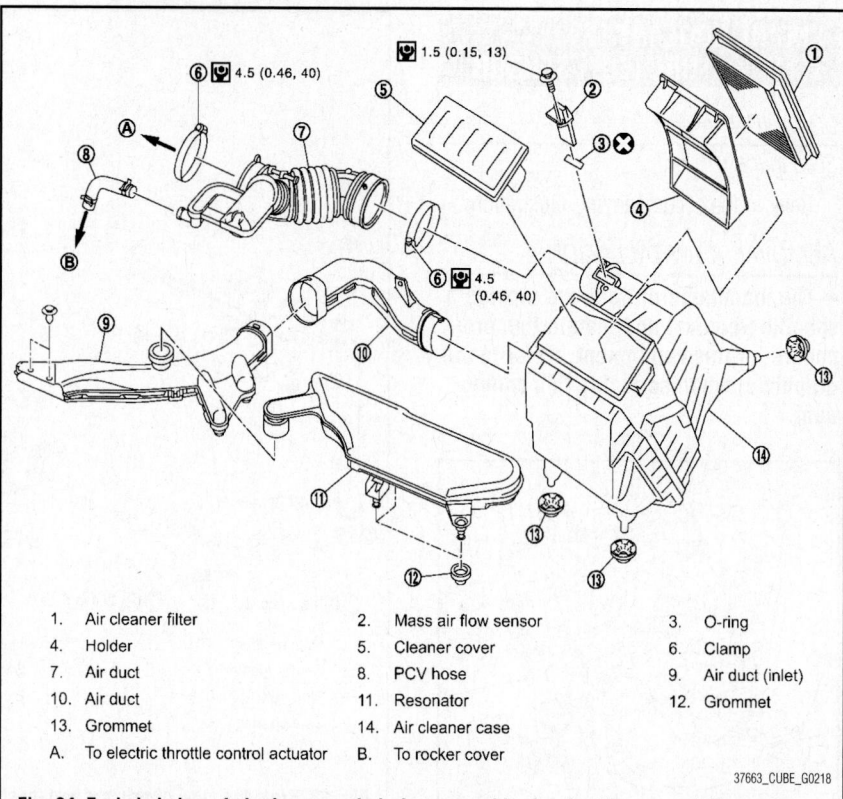

1.	Air cleaner filter	2.	Mass air flow sensor	3.	O-ring
4.	Holder	5.	Cleaner cover	6.	Clamp
7.	Air duct	8.	PCV hose	9.	Air duct (inlet)
10.	Air duct	11.	Resonator	12.	Grommet
13.	Grommet	14.	Air cleaner case		
A.	To electric throttle control actuator	B.	To rocker cover		

37663_CUBE_G0218

Fig. 94 Exploded view of air cleaner and air duct assembly showing the mass air flow sensor location

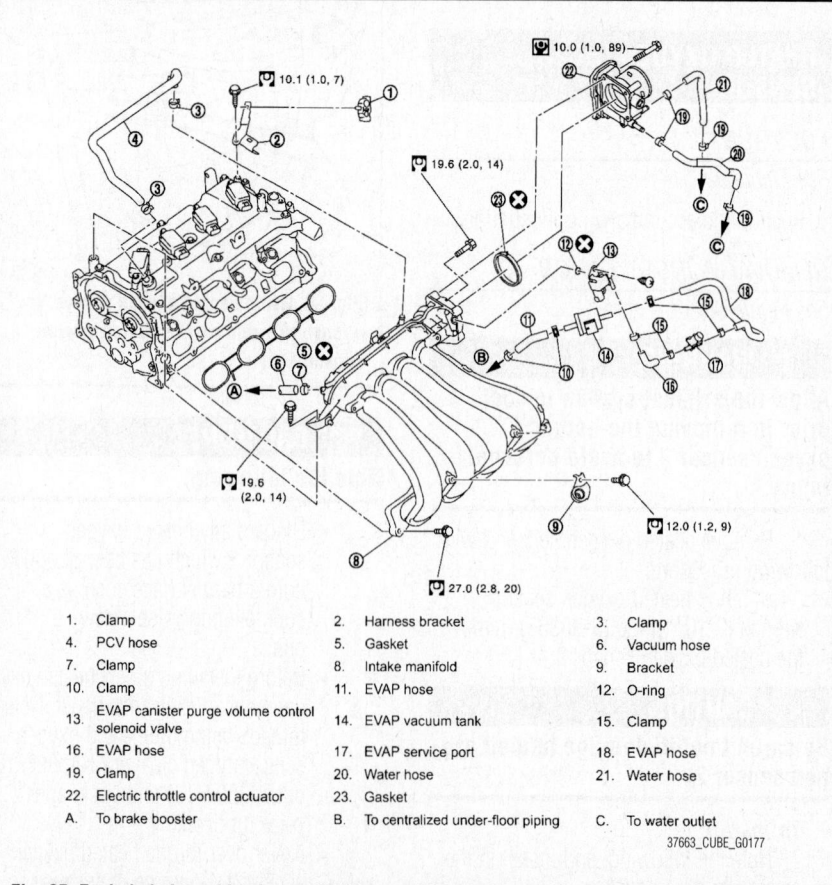

1.	Clamp	2.	Harness bracket	3.	Clamp
4.	PCV hose	5.	Gasket	6.	Vacuum hose
7.	Clamp	8.	Intake manifold	9.	Bracket
10.	Clamp	11.	EVAP hose	12.	O-ring
13.	EVAP canister purge volume control solenoid valve	14.	EVAP vacuum tank	15.	Clamp
16.	EVAP hose	17.	EVAP service port	18.	EVAP hose
19.	Clamp	20.	Water hose	21.	Water hose
22.	Electric throttle control actuator	23.	Gasket		
A.	To brake booster	B.	To centralized under-floor piping	C.	To water outlet

37663_CUBE_G0177

Fig. 95 Exploded view of intake manifold assembly showing electric throttle control actuator location

REMOVAL & INSTALLATION

Refer to the electric throttle control actuator section when servicing this component.

VEHICLE SPEED SENSOR (VSS)

LOCATION

See Figure 96.

Refer to the accompanying illustration.

REMOVAL & INSTALLATION

1. Remove battery.
2. Remove air duct (inlet), air duct and air cleaner case.
3. Remove battery bracket.
4. Remove control cable from manual lever.
5. Place manual lever to "L" position.
6. Disconnect primary speed sensor connector.
7. Remove primary speed sensor.
8. Remove O-ring from primary speed sensor.

To install:

9. Note the following, and install in the reverse order of removal.

❄❄ CAUTION

Never reuse O-ring. Apply CVT fluid to O-ring.

10. Check for CVT fluid leakage and check CVT fluid level.
11. Check the CVT position.

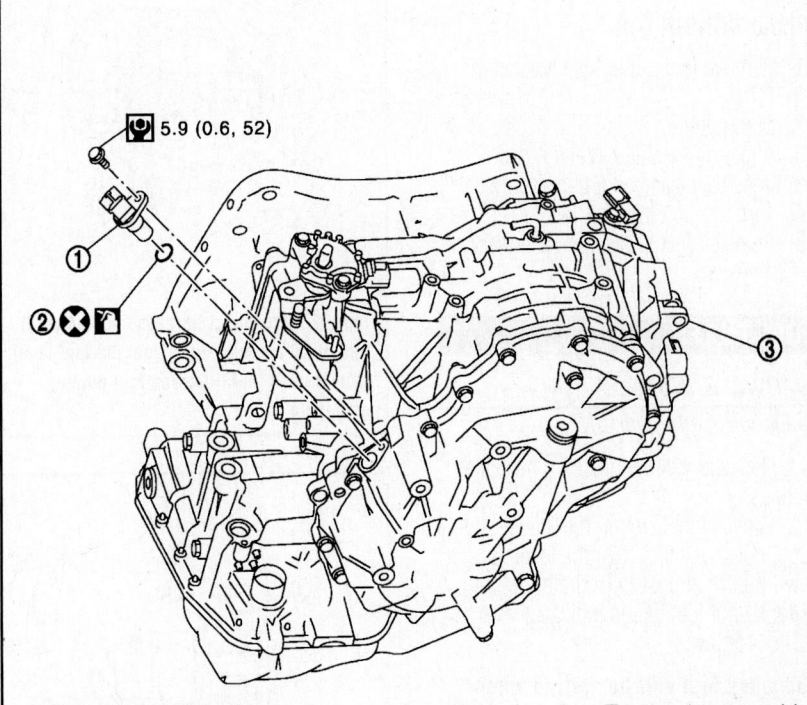

1. Primary speed sensor 2. O-ring 3. Transaxle assembly
: Apply CVT Fluid NS-2.

37663_CUBE_G0219

Fig. 96 Primary speed sensor location

FUEL   GASOLINE FUEL INJECTION SYSTEM

FUEL SYSTEM SERVICE PRECAUTIONS

Safety is the most important factor when performing not only fuel system maintenance, but any type of maintenance. Failure to conduct maintenance and repairs in a safe manner may result in serious personal injury or death. Work on a vehicle's fuel system components can be accomplished safely and effectively by adhering to the following rules and guidelines.

• To avoid the possibility of fire and personal injury, always disconnect the negative battery cable unless the repair or test procedure requires that battery voltage be applied.

• Always relieve the fuel system pressure prior to disconnecting any fuel system component (injector, fuel rail, pressure reg-

ulator, etc.) fitting or fuel line connection. Exercise extreme caution whenever relieving fuel system pressure to avoid exposing skin, face and eyes to fuel spray. Please be advised that fuel under pressure may penetrate the skin or any part of the body that it contacts.

• Always place a shop towel or cloth around the fitting or connection prior to loosening to absorb any excess fuel due to spillage. Ensure that all fuel spillage is quickly removed from engine surfaces. Ensure that all fuel-soaked cloths or towels are deposited into a flame-proof waste container with a lid.

• Always keep a dry chemical (Class B) fire extinguisher near the work area.

• Do not allow fuel spray or fuel vapors to come into contact with a spark or open flame.

• Always use a second wrench when loosening or tightening fuel line connection fittings. This will prevent unnecessary stress and torsion on fuel piping. Always follow the proper torque specifications.

• Always replace worn fuel fitting O-rings with new ones. Do not substitute fuel hose where rigid pipe is installed.

FUEL SYSTEM PRESSURE

RELIEVING

With CONSULT-III

1. Turn ignition switch ON.
2. Perform "FUEL PRESSURE RELEASE" in "WORK SUPPORT" mode with CONSULT-III.
3. Start engine.

4. After engine stalls, crank it two or three times to release all fuel pressure.

5. Turn ignition switch OFF.

Without CONSULT-III

1. Remove fuel pump fuse located in IPDM E/R.

2. Start engine.

3. After engine stalls, crank it two or three times to release all fuel pressure.

4. Turn ignition switch OFF.

5. Reinstall fuel pump fuse after servicing fuel system.

FUEL FILTER

REMOVAL & INSTALLATION

See Figures 97 through 101.

1. Release the fuel pressure from the fuel lines.

2. Check fuel level on fuel gauge. If fuel gauge indicates more than the level as shown (full or almost full), drain fuel from fuel tank until fuel gauge indicates level as shown or below.

→Because fuel will be spilled when removing fuel level sensor units for the top of the fuel is above the fuel level sensor units installation surface.

a. As a guide, drain approximately 2⅝ gal. (10L) of fuel.

b. In a case that fuel pump does not operate, perform the following procedure.

c. Insert hose of less than 1 inch (25 mm) in diameter into fuel filler tube through fuel filler opening to draw fuel from fuel filler tube.

d. Disconnect fuel filler hose from fuel filler tube.

e. Insert hose into fuel tank through fuel filler hose to draw fuel from fuel tank.

3. Open fuel filler lid.

4. Open filler cap and release the pressure inside fuel tank.

5. Remove rear seat.

6. Remove inspection hole cover. Using a screwdriver, remove it by turning clips clockwise by 90 degrees.

7. Disconnect harness connector and quick connector.

8. Remove quick connector in the following procedures.

a. Pinch quick connector square-part with your fingers, and pull out the quick connector by hand.

b. If quick connector and tube on sender unit are stuck, push and pull several times until they move, and pull out.

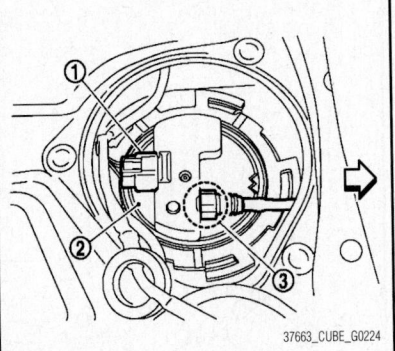

Fig. 97 Disconnect harness connector (1) and quick connector (3) from the fuel level sensor unit, fuel filter and fuel pump assembly (2)

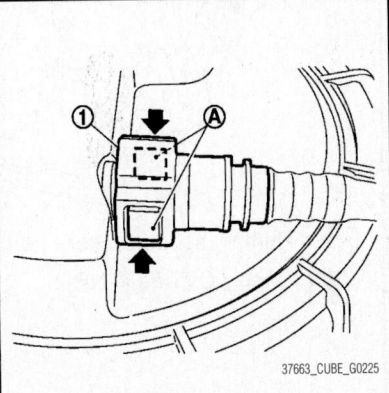

Fig. 98 Pinch quick connector square-part (A) with your fingers, and pull out the quick connector (1)

✳✳ CAUTION

Note the following:

- Quick connector can be removed when the tabs are completely depressed. Never twist it more than necessary.
- Never use any tools to disconnected quick connector.
- Keep resin tube away from heat. Be especially careful when welding near the resin tube.
- Prevent acid liquid such as battery electrolyte, etc. from getting on resin tube.
- Never bend or twist resin tube during installation and disconnection.
- To keep the connecting portion clean and to avoid damage and foreign materials, cover them completely with plastic bags or something similar.

- Never insert plug, preventing damage on O-ring in quick connector.

9. Remove lock ring for fuel level sensor unit, fuel filter and fuel pump assembly with fuel tank lock ring wrench KV991J0090 (J-46214) by turning counterclockwise.

10. Remove fuel level sensor unit, fuel filter and fuel pump assembly.

✳✳ CAUTION

Note the following:

- Never bend float arm during removal.
- Never pollute the inside by residue fuel. Draw out avoiding inclination by supporting with a cloth.
- Never cause impacts such by dropping when handling components.

To install:

11. Note to the following, and install in the reverse order of removal. Fuel Level Sensor Unit

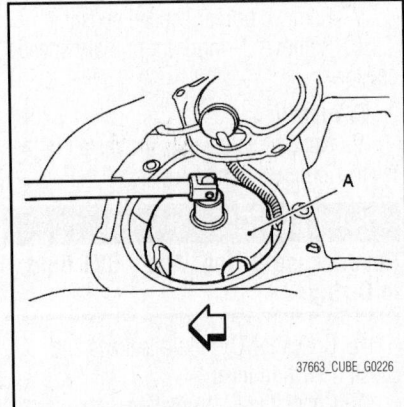

Fig. 99 Remove lock ring for fuel level sensor unit, fuel filter and fuel pump assembly with fuel tank lock ring wrench (A) by turning counterclockwise

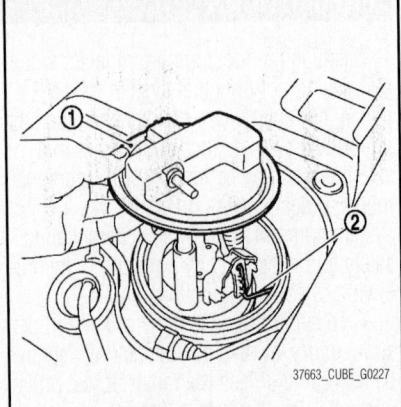

Fig. 100 Remove fuel level sensor unit, fuel filter and fuel pump assembly (1); float arm (2)

12. Install new O-ring to fuel tank without any twist.

13. Align A with B as shown. Install fuel level sensor unit, fuel filter and fuel pump assembly to fuel tank.

> ※※ **CAUTION**

Never allow O-ring to drop. Never bend float arm during installing.

14. Install lock ring for fuel level sensor unit, fuel filter and fuel pump assembly with lock ring wrench KV991J0090 (J-46214) by turning clockwise.

> ※※ **CAUTION**

Install lock ring horizontally.

15. Connect quick connector as follows:

a. Check the connection for damage or any foreign materials.

b. Align the connector with the tube, then insert the connector straight into the tube until a click sound is heard.

c. After connecting, check that the connection is secure by following method.

d. Visually confirm that the two tabs are connected to the connector.

e. Pull (A) the tube and the connector to check they are securely connected.

16. Turn ignition switch "ON" (with engine stopped), then check connections for leakage by applying fuel pressure to fuel piping.

17. Start engine and let it idle and check there are no fuel leakage at the fuel system connections.

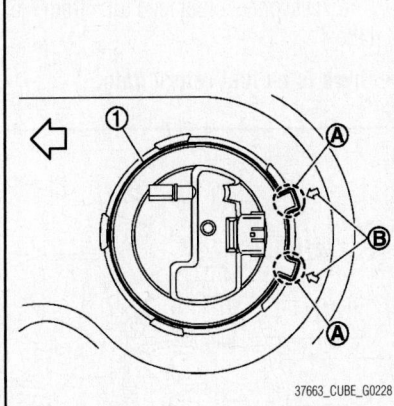

37663_CUBE_G0228

Fig. 101 Align (A) with (B) as shown; install fuel level sensor unit, fuel filter and fuel pump assembly (1)

FUEL LEVEL SENDING UNIT

REMOVAL & INSTALLATION

See Figures 97 through 101.

1. Release the fuel pressure from the fuel lines.

2. Check fuel level on fuel gauge. If fuel gauge indicates more than the level as shown (full or almost full), drain fuel from fuel tank until fuel gauge indicates level as shown or below.

➡**Because fuel will be spilled when removing fuel level sensor units for the top of the fuel is above the fuel level sensor units installation surface.**

a. As a guide, drain approximately 2⅝ gal. (10L) of fuel.

b. In a case that fuel pump does not operate, perform the following procedure.

c. Insert hose of less than 1 inch (25 mm) in diameter into fuel filler tube through fuel filler opening to draw fuel from fuel filler tube.

d. Disconnect fuel filler hose from fuel filler tube.

e. Insert hose into fuel tank through fuel filler hose to draw fuel from fuel tank.

3. Open fuel filler lid.

4. Open filler cap and release the pressure inside fuel tank.

5. Remove rear seat.

6. Remove inspection hole cover. Using a screwdriver, remove it by turning clips clockwise by 90 degrees.

7. Disconnect harness connector and quick connector.

8. Remove quick connector in the following procedures.

a. Pinch quick connector square-part with your fingers, and pull out the quick connector by hand.

b. If quick connector and tube on sender unit are stuck, push and pull several times until they move, and pull out.

> ※※ **CAUTION**

Note the following:

- Quick connector can be removed when the tabs are completely depressed. Never twist it more than necessary.
- Never use any tools to disconnected quick connector.
- Keep resin tube away from heat. Be especially careful when welding near the resin tube.
- Prevent acid liquid such as battery

electrolyte, etc. from getting on resin tube.

- Never bend or twist resin tube during installation and disconnection.
- To keep the connecting portion clean and to avoid damage and foreign materials, cover them completely with plastic bags or something similar.
- Never insert plug, preventing damage on O-ring in quick connector.

9. Remove lock ring for fuel level sensor unit, fuel filter and fuel pump assembly with fuel tank lock ring wrench KV991J0090 (J-46214) by turning counterclockwise.

10. Remove fuel level sensor unit, fuel filter and fuel pump assembly.

> ※※ **CAUTION**

Note the following:

- Never bend float arm during removal.
- Never pollute the inside by residue fuel. Draw out avoiding inclination by supporting with a cloth.
- Never cause impacts such by dropping when handling components.

To install:

11. Note to the following, and install in the reverse order of removal. Fuel Level Sensor Unit

12. Install new O-ring to fuel tank without any twist.

13. Align A with B as shown. Install fuel level sensor unit, fuel filter and fuel pump assembly to fuel tank.

> ※※ **CAUTION**

Never allow O-ring to drop. Never bend float arm during installing.

14. Install lock ring for fuel level sensor unit, fuel filter and fuel pump assembly with lock ring wrench KV991J0090 (J-46214) by turning clockwise.

> ※※ **CAUTION**

Install lock ring horizontally.

15. Connect quick connector as follows:

a. Check the connection for damage or any foreign materials.

b. Align the connector with the tube, then insert the connector straight into the tube until a click sound is heard.

c. After connecting, check that the connection is secure by following method.

d. Visually confirm that the two tabs are connected to the connector.

e. Pull (A) the tube and the connector to check they are securely connected.

16. Turn ignition switch "ON" (with engine stopped), then check connections for leakage by applying fuel pressure to fuel piping.

17. Start engine and let it idle and check there are no fuel leakage at the fuel system connections.

FUEL PUMP ASSEMBLY

REMOVAL & INSTALLATION

See Figures 97 through 101.

1. Release the fuel pressure from the fuel lines.

2. Check fuel level on fuel gauge. If fuel gauge indicates more than the level as shown (full or almost full), drain fuel from fuel tank until fuel gauge indicates level as shown or below.

➡**Because fuel will be spilled when removing fuel level sensor units for the top of the fuel is above the fuel level sensor units installation surface.**

a. As a guide, drain approximately 2⅝ gal. (10L) of fuel.

b. In a case that fuel pump does not operate, perform the following procedure.

c. Insert hose of less than 1 inch (25 mm) in diameter into fuel filler tube through fuel filler opening to draw fuel from fuel filler tube.

d. Disconnect fuel filler hose from fuel filler tube.

e. Insert hose into fuel tank through fuel filler hose to draw fuel from fuel tank.

3. Open fuel filler lid.

4. Open filler cap and release the pressure inside fuel tank.

5. Remove rear seat.

6. Remove inspection hole cover. Using a screwdriver, remove it by turning clips clockwise by 90 degrees.

7. Disconnect harness connector and quick connector.

8. Remove quick connector in the following procedures.

a. Pinch quick connector square-part with your fingers, and pull out the quick connector by hand.

b. If quick connector and tube on sender unit are stuck, push and pull several times until they move, and pull out.

⁕⁕ CAUTION
Note the following:

• Quick connector can be removed

when the tabs are completely depressed. Never twist it more than necessary.

• Never use any tools to disconnected quick connector.

• Keep resin tube away from heat. Be especially careful when welding near the resin tube.

• Prevent acid liquid such as battery electrolyte, etc. from getting on resin tube.

• Never bend or twist resin tube during installation and disconnection.

• To keep the connecting portion clean and to avoid damage and foreign materials, cover them completely with plastic bags or something similar.

• Never insert plug, preventing damage on O-ring in quick connector.

9. Remove lock ring for fuel level sensor unit, fuel filter and fuel pump assembly with fuel tank lock ring wrench KV991J0090 (J-46214) by turning counterclockwise.

10. Remove fuel level sensor unit, fuel filter and fuel pump assembly.

⁕⁕ CAUTION
Note the following:

• Never bend float arm during removal.

• Never pollute the inside by residue fuel. Draw out avoiding inclination by supporting with a cloth.

• Never cause impacts such by dropping when handling components.

To install:

11. Note to the following, and install in the reverse order of removal. Fuel Level Sensor Unit

12. Install new O-ring to fuel tank without any twist.

13. Align A with B as shown. Install fuel level sensor unit, fuel filter and fuel pump assembly to fuel tank.

⁕⁕ CAUTION
Never allow O-ring to drop. Never bend float arm during installing.

14. Install lock ring for fuel level sensor unit, fuel filter and fuel pump assembly with lock ring wrench KV991J0090 (J-46214) by turning clockwise.

⁕⁕ CAUTION
Install lock ring horizontally.

15. Connect quick connector as follows:
a. Check the connection for damage or any foreign materials.

b. Align the connector with the tube, then insert the connector straight into the tube until a click sound is heard.

c. After connecting, check that the connection is secure by following method.

d. Visually confirm that the two tabs are connected to the connector.

e. Pull (A) the tube and the connector to check they are securely connected.

16. Turn ignition switch "ON" (with engine stopped), then check connections for leakage by applying fuel pressure to fuel piping.

17. Start engine and let it idle and check there are no fuel leakage at the fuel system connections.

FUEL RAIL AND INJECTOR

REMOVAL & INSTALLATION

See Figures 102 through 107.

⁕⁕ CAUTION
Never remove or disassemble parts unless instructed.

⁕⁕ WARNING
Note the following:

• Put a "CAUTION: FLAMMABLE" sign in the workshop.

• Be sure to work in a well ventilated area and furnish workshop with a CO2 fire extinguisher.

• Never smoke while servicing fuel system. Keep open flames and sparks away from the work area.

1. Release the fuel pressure.

2. Remove intake manifold.

3. Disconnect quick connector with the following procedure.

a. Disconnect fuel feed tube from fuel tube.

➡**There is no fuel return path.**

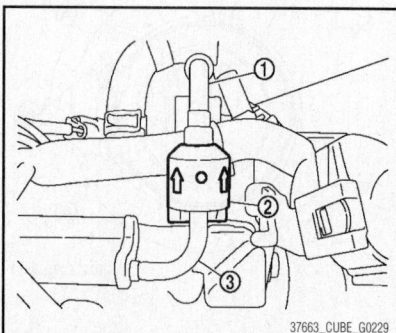

37663_CUBE_G0229

Fig. 102 Disconnect fuel feed tube (1) from fuel tube (3); remove quick connector cap (engine side) (2)

b. Remove quick connector cap (engine side) from quick connector connection.

c. With the sleeve side of quick connector release facing quick connector, install quick connector release onto fuel tube.

d. Insert quick connector release into quick connector until sleeve contacts and goes no further. Hold quick connector release on that position.

✳✳ CAUTION

Inserting quick connector release hard will not disconnect quick connector. Hold quick connector release where it contacts and goes no further.

e. Draw and pull out quick connector straight from fuel tube.

✳✳ CAUTION

Note the following:

- Pull quick connector at holding position.
- Never pull with lateral force applied. O-ring inside quick connector may be damaged.
- Prepare container and cloth beforehand as fuel will leakage out.
- Avoid fire and sparks.
- Keep parts away from heat source. Especially, be careful when welding is performed around them.

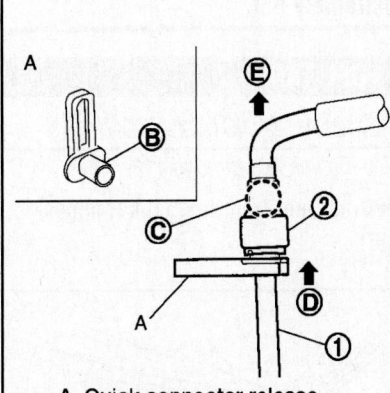

A. Quick connector release
B. Sleeve
C. Hold at this position
D. Insert and retain
E. Pull quick connector
1. Fuel tube
2. Quick connector

37663_CUBE_G0230

Fig. 103 Insert quick connector release into quick connector until sleeve contacts and goes no further

- Never expose parts to battery electrolyte or other acids.
- Never bend or twist connection between quick connector and fuel feed tube during installation/removal.
- To keep clean the connecting portion and to avoid damage and foreign materials, cover them completely with plastic bags, etc. or something similar.

4. Disconnect harness connector from fuel injector.

5. Remove fuel tube and fuel injector assembly.

6. Loosen mounting bolts in reverse order as shown.

✳✳ CAUTION

Note the following:

- When removing, be careful to avoid any interference with fuel injector.
- Use a shop cloth to absorb any fuel leakage from fuel tube.

7. Remove fuel injector from fuel tube with the following procedure:

a. Open and remove clip.

b. Remove fuel injector from fuel tube by pulling straight.

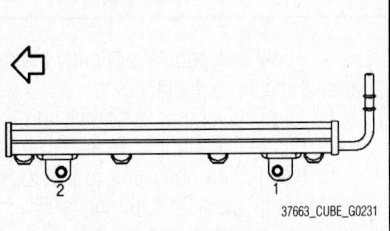

37663_CUBE_G0231

Fig. 104 Loosen mounting bolts in reverse order as shown

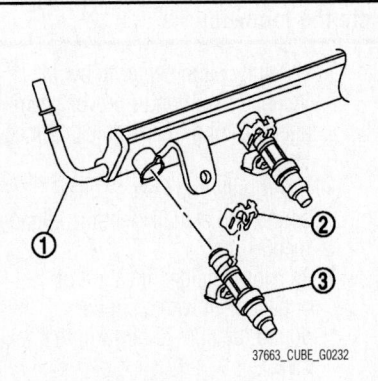

37663_CUBE_G0232

Fig. 105 Open and remove clip (2); remove fuel injector (3) from fuel tube (1) by pulling straight

✳✳ CAUTION

Note the following:

- Be careful with remaining fuel that may go out from fuel tube.
- Be careful not to damage fuel injector nozzle during removal.
- Never bump or drop fuel injector.
- Never disassemble fuel injector.

To install:

8. Note the following, and install O-rings to fuel injector.

✳✳ CAUTION

Note the following:

- Upper and lower O-rings are different. Be careful not to confuse them.
- Handle O-ring with bare hands. Never wear gloves.
- Lubricate O-ring with new engine oil.
- Never clean O-ring with solvent.
- Check that O-ring and its mating part are free of foreign material.
- When installing O-ring, be careful not to scratch it with tool or fingernails. Also be careful not to twist or stretch O-ring. If O-ring is stretched while installing, never insert it quickly into fuel tube.
- Insert O-ring straight into fuel tube. Never decenter or twist it.

9. Install fuel injector to fuel tube with the following procedure:

a. Insert clip into clip mounting groove on fuel injector.

b. Insert clip so that protrusion of fuel injector matches cutout of clip.

✳✳ CAUTION

Never reuse clip. Replace it with a new one. Be careful to keep clip from interfering with O-ring. If interference occurs, replace O-ring.

c. Insert fuel injector into fuel tube with clip attached.

d. Insert it while matching it to the axial center.

e. Insert fuel injector so that protrusion of fuel tube matches cut-out of clip.

f. Check that fuel tube flange is securely fixed in flange fixing groove on clip.

g. Check that installation is complete by checking that fuel injector does not rotate or come off.

10. Set fuel tube and fuel injector assembly at its position for installation on cylinder head.

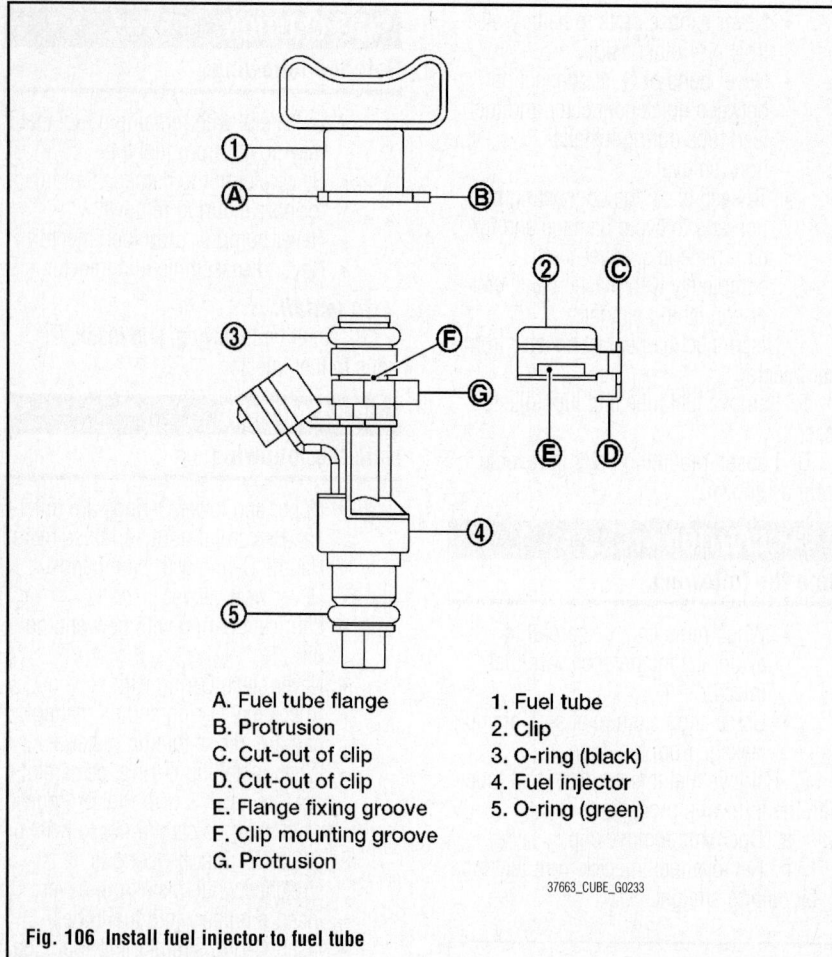

A. Fuel tube flange
B. Protrusion
C. Cut-out of clip
D. Cut-out of clip
E. Flange fixing groove
F. Clip mounting groove
G. Protrusion

1. Fuel tube
2. Clip
3. O-ring (black)
4. Fuel injector
5. O-ring (green)

37663_CUBE_G0233

Fig. 106 Install fuel injector to fuel tube

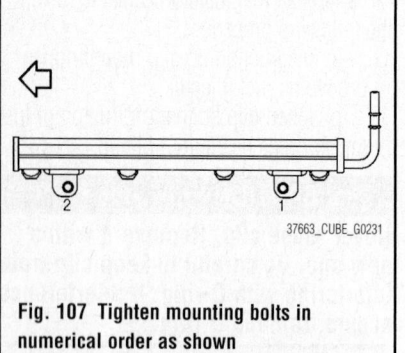

37663_CUBE_G0231

Fig. 107 Tighten mounting bolts in numerical order as shown

> ※※ **CAUTION**
>
> **For installation, be careful not to interfere with fuel injector nozzle.**

11. Install fuel tube and injector assembly onto cylinder.

12. Tighten mounting bolts in numerical order as shown.

13. Connect harness connector to fuel injector.

14. Connect fuel feed tube with the following procedure.

   a. Check for damage or foreign material on the fuel tube and quick connector.

   b. Apply new engine oil lightly to area around the top of fuel tube.

   c. Align center to insert quick connector straightly into fuel tube.

   d. Insert quick connector to fuel tube until the top spool on fuel tube is inserted completely and the 2nd level spool is positioned slightly below quick connector bottom end.

> ※※ **CAUTION**
>
> **Note the following:**
>
> - Carefully align center to avoid inclined insertion to prevent damage to O-ring inside quick connector.
> - Insert until you hear a "click" sound and actually feel the engagement.
> - To avoid misidentification of engagement with a similar sound, be sure to perform the next step.

   e. Pull quick connector hard by hand holding position. Check it is completely engaged (connected) so that it does not come out from fuel tube.

   f. Install quick connector cap (engine side) to quick connector connection.

   g. Install quick connector cap (engine side) with the side arrow facing quick connector side (fuel feed tube side).

> ※※ **CAUTION**
>
> **Note the following:**
>
> - Check that the quick connector and fuel tube are securely engaged with the quick connector cap (engine side) mounting groove.
> - Quick connector may not be connected correctly if quick connector cap (engine side) cannot be installed easily. Remove the quick connector cap (engine side), and then check the connection of quick connector again.

   h. Install fuel feed hose to hose clamp.

15. Install remaining parts in the reverse order of removal.

16. Turn ignition switch "ON" (with the engine stopped). With fuel pressure applied to fuel piping, check there are no fuel leakage at connection points.

➡ **Use mirrors for checking at points out of clear sight.**

17. Start the engine. With engine speed increased, check again that there are no fuel leakage at connection points.

> ※※ **CAUTION**
>
> **Never touch the engine immediately after stopped, as the engine becomes extremely hot.**

FUEL TANK

REMOVAL & INSTALLATION

See Figures 108 through 113.

➡ **Drain fuel from fuel tank if necessary.**

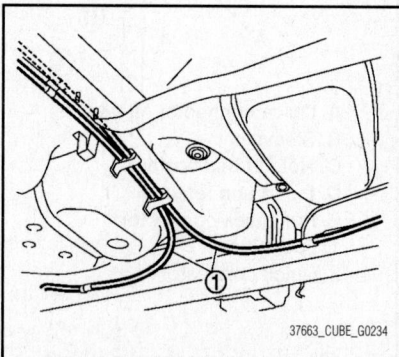

37663_CUBE_G0234

Fig. 108 Move parking brake cable (1) from the lower face of fuel tank

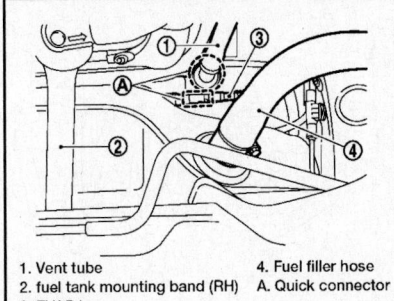

1. Vent tube
2. fuel tank mounting band (RH)
3. EVAP hose
4. Fuel filler hose
A. Quick connector

37663_CUBE_G0235

Fig. 109 Disconnect fuel filler hose at fuel tank side

※ **CAUTION**

Perform work on level place.

1. Perform steps 2 to 7 of "Removal" in "Fuel Level Sensor Unit, Fuel Filter And Fuel Pump Assembly" on fuel level sensor unit, fuel filter and fuel pump assembly.

2. Remove center muffler.

3. Remove insulator on vehicle side located above center and main mufflers.

4. Move parking brake cable from the lower face of fuel tank. Then remove clips for parking brake cable.

5. Remove brake tube protector.

6. Disconnect fuel filler hose at fuel tank side.

7. Disconnect EVAP hose and vent tube at the position shown.

➡ **Instruction for quick connector of EVAP hose and vent tube, refer to the following:**

※ **CAUTION**

Note the following:

- Quick connector can be disconnected when the tabs are depressed completely. Never twist it more than necessary.
- Never use any tools to disconnected quick connector.
- Keep resin tube away from heat. Be especially careful when welding near the resin tube.
- Prevent acid liquid such as battery electrolyte, etc. from getting on resin tube.
- Never bend or twist resin tube during installation and disconnection.
- Never remove the remaining retainer on hard tube (or the equivalent) except when resin tube or retainer is replaced.

- When resin tube or hard tube (or the equivalent) is replaced, also replace retainer with new green one.
- To keep the connecting portion clean and to avoid damage and foreign materials, cover them completely with plastic bags or something similar.
- Never insert plug, preventing damage on O-ring in quick connector.

8. Remove quick connector in the following procedures.

a. Pinch quick connector square-part with your fingers, and pull out the quick connector by hand.

b. If quick connector and tube on vehicle are stuck, push and pull several times until they move, and pull out.

※ **CAUTION**

Note the following:

- The tube can be removed when the tabs are completely depressed. Never twist it more than necessary.
- Never use any tools to disconnect quick connector.
- Keep the resin tube away from heat. Be especially careful when welding near the tube.
- Prevent acid liquid such as battery electrolyte, etc. from getting on the resin tube.
- Never bend or twist resin tube during installation and disconnection.

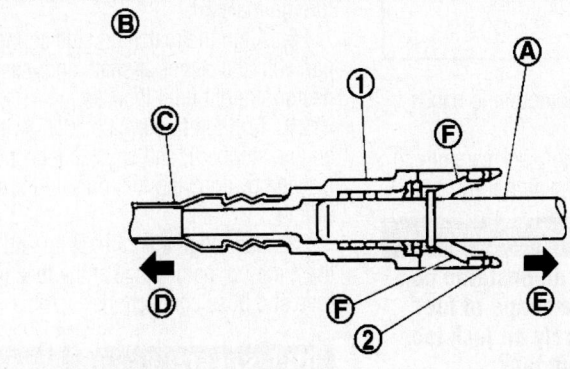

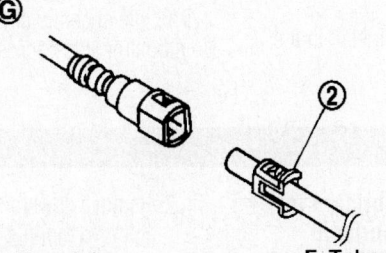

A. Hard tube
B. Connection (cross section)
C. Resin tube
D. To fuel tank
E. To fuel filler tube
F. Tabs
G. .Disconnection
1. Quick connector
2. Retainer

37663_CUBE_G0236

Fig. 110 Quick connector of EVAP hose and vent tube

37663_CUBE_G0237

Fig. 111 Pinch quick connector square-part (A) with your fingers, and pull out the quick connector (1) by hand

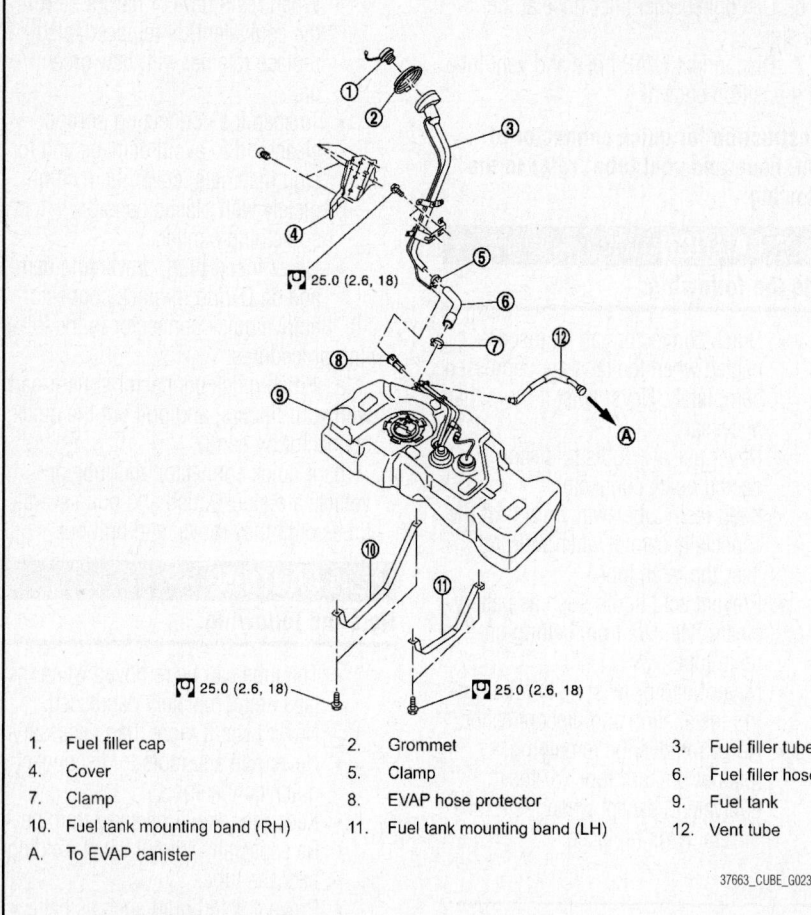

Fig. 112 Exploded view of the fuel tank assembly

1. Fuel filler cap
2. Grommet
3. Fuel filler tube
4. Cover
5. Clamp
6. Fuel filler hose
7. Clamp
8. EVAP hose protector
9. Fuel tank
10. Fuel tank mounting band (RH)
11. Fuel tank mounting band (LH)
12. Vent tube
A. To EVAP canister

🔧 25.0 (2.6, 18)

37663_CUBE_G0239

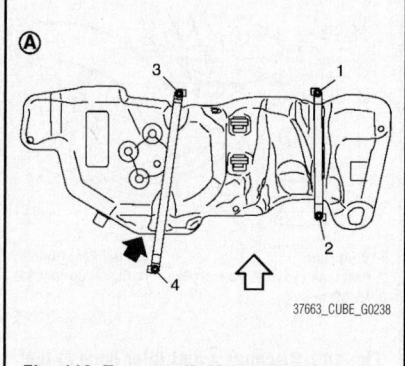

Fig. 113 Temporarily tighten bolts except #4 in numerical order as shown

37663_CUBE_G0238

• To keep the connecting portion clean and to avoid damage and foreign materials, cover them completely with plastic bags or something similar.
• Never insert plug, preventing damage on O-ring in quick connector.

9. Support the center part of fuel tank with transmission jack.

❋❋ CAUTION

Securely support the fuel tank with a piece of wood.

10. Remove EVAP canister filter.

11. Remove fuel tank mounting bands (RH and LH).

12. Lower transmission jack carefully to remove fuel tank while holding it by hand.

❋❋ CAUTION

Fuel tank may be in an unstable condition because of the shape of fuel tank bottom. Never rely on jack too much. Be sure to hold tank securely.

To install:

13. Note the following, and install in the reverse order of removal.

14. Temporarily tighten bolts except #4 in numerical order as shown.

15. Tighten bolt #4 to 18 ft. lbs. (25 Nm), pressing fuel tank in the direction shown by black arrow.

16. Tighten bolts 1, 2, and 3 to 18 ft. lbs. (25 Nm) in the reverse order as shown.

17. Insert the fuel filler tube into the fuel hose to the length of 1.38 inches (35 mm).

18. Securely clamp the hose and tube. Be sure hose clamp is not placed on swelled area of fuel filler tube.

19. Check connections for damage or foreign material.

20. Align the matching side connection part with the center of shaft, and insert connector straight until it clicks.

21. Turn ignition switch "ON" (with engine stopped), and check connections for leakage by applying fuel pressure to fuel piping.

22. Start engine and rev it up and check there are no fuel leakage at the fuel system tube and hose connections.

IDLE SPEED

ADJUSTMENT

The idle speed is controlled by the ECM. No adjustment is necessary or possible.

HEATING & AIR CONDITIONING SYSTEM

A/C UNIT

REMOVAL & INSTALLATION

See Figures 114 through 118.

❋❋ CAUTION

Perform lubricant return operation before each refrigeration system disassembly. However, if a large amount of refrigerant or lubricant is detected, never perform lubricant return operation.

1. Use a refrigerant collecting equipment (for HFC-134a) to discharge the refrigerant.

2. Drain engine coolant from cooling system.

3. Remove cowl top extension.

4. Remove mounting nut and lower dash insulator; position without the hindrance for work. (If equipped)

5. Remove mounting bolt, and then disconnect low-pressure flexible hose and high-pressure pipe from expansion valve. (If equipped)

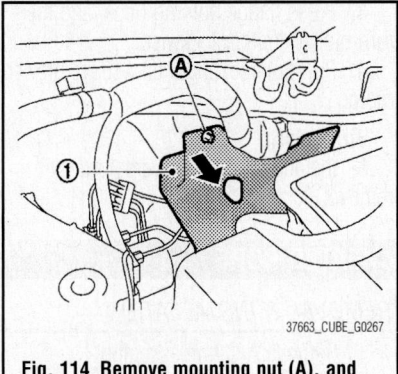

Fig. 114 Remove mounting nut (A), and lower dash insulator (1)

✳✳ CAUTION

Cap or wrap the joint of the A/C piping and expansion valve with suitable material such as vinyl tape to avoid the entry of air.

6. Remove clamps, and then disconnect heater hoses from A/C unit assembly.

✳✳ CAUTION

Some coolant may spill when heater hoses are disconnected. Close off the coolant inlet and outlet on the heater core (2 locations) with shop cloths.

7. Remove side ventilator duct.
8. Move steering column assembly to a position where it does not inhibit work.
9. Remove instrument stay.
10. Disconnect drain hose from A/C unit assembly.
11. Disconnect the harness connectors

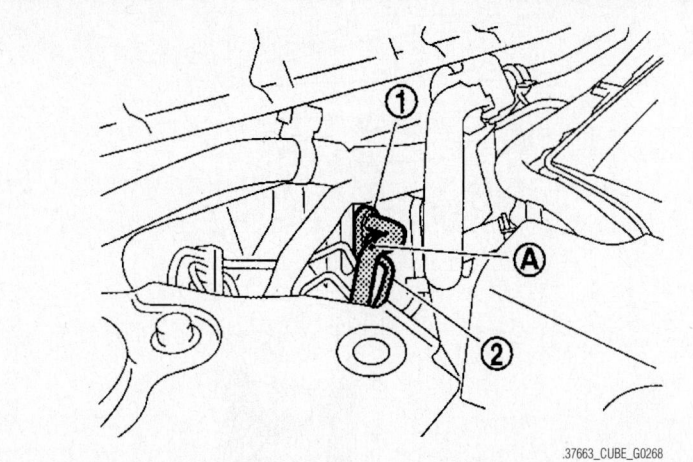

Fig. 115 Remove mounting bolt (A), and then disconnect low-pressure flexible hose (1) and high-pressure pipe (2) from expansion valve

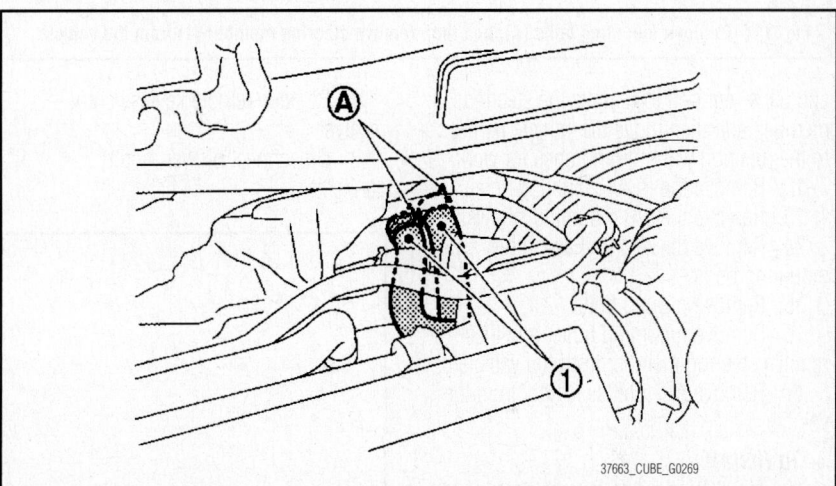

Fig. 116 Remove clamps (A), and then disconnect heater hoses (1) from A/C unit assembly

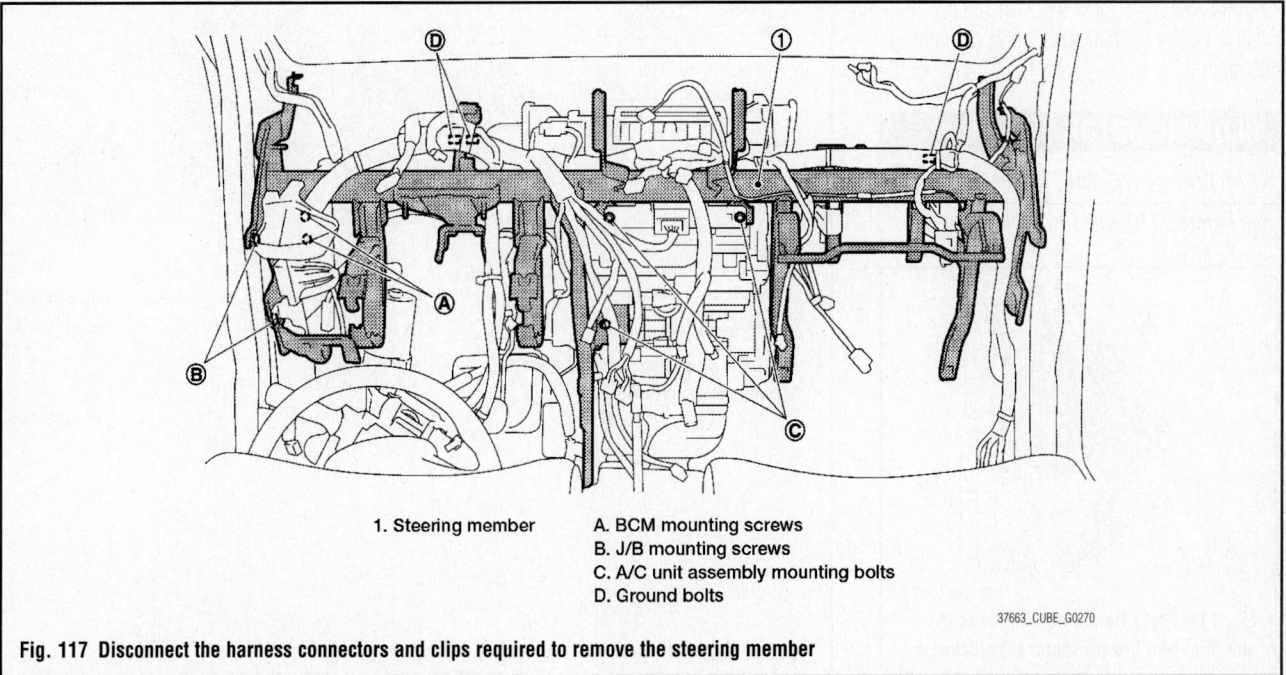

1. Steering member
A. BCM mounting screws
B. J/B mounting screws
C. A/C unit assembly mounting bolts
D. Ground bolts

Fig. 117 Disconnect the harness connectors and clips required to remove the steering member

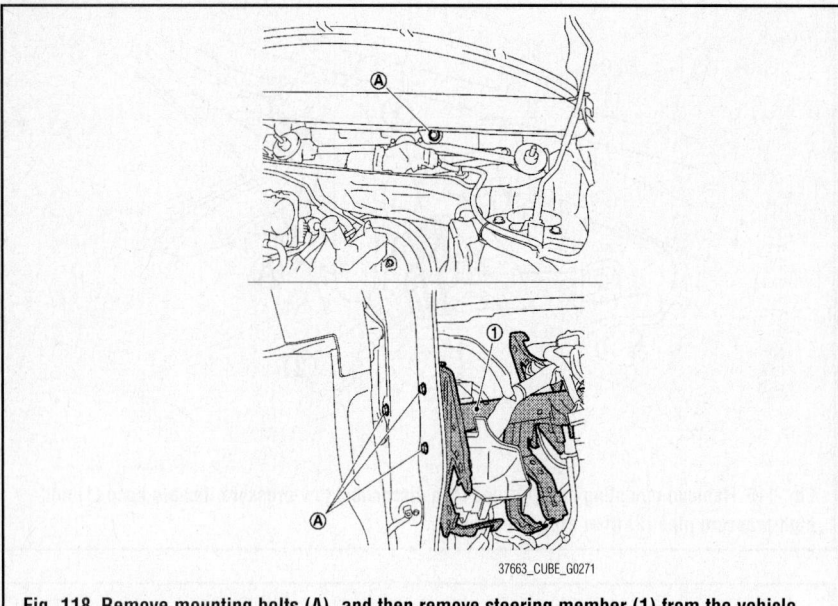

Fig. 118 Remove mounting bolts (A), and then remove steering member (1) from the vehicle

and clips required to remove the steering member, and then move the vehicle harness to the position without hindrance for work.

12. Remove the BCM mounting screws.

13. Remove the J/B mounting screws.

14. Remove the A/C unit assembly mounting bolts.

15. Remove ground bolts.

16. Remove mounting bolts, and then remove steering member from the vehicle.

17. Remove A/C unit assembly from the vehicle.

To install:

18. Installation is the reverse order of removal.

19. Replace O-rings with new ones. Apply compressor oil to them when installing.

20. Check for leakages when recharging refrigerant.

BLOWER MOTOR

REMOVAL & INSTALLATION

See Figures 119 and 120.

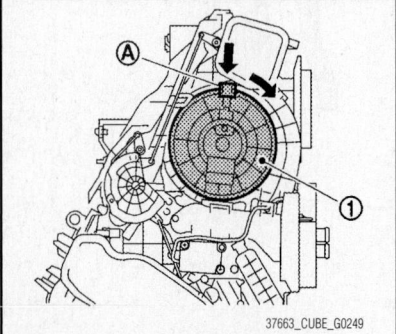

Fig. 119 Press flange holding hook (A), and then turn blower motor (1) clockwise

1. Remove remote keyless entry receiver.

2. Disconnect blower motor connector.

3. Press flange holding hook, and then turn blower motor clockwise.

4. Pull outside, and then remove blower motor.

To install:

5. Installation is the reverse order of removal.

HEATER CORE

REMOVAL & INSTALLATION

1. Remove A/C unit assembly.

2. Remove heater pipe grommet and heater pipe support from A/C unit assembly.

3. Remove foot duct LH.

4. Remove mounting screw, and then slide heater core to leftward.

5. Remove heater core from A/C unit assembly.

To install:

6. Installation is the reverse order of removal.

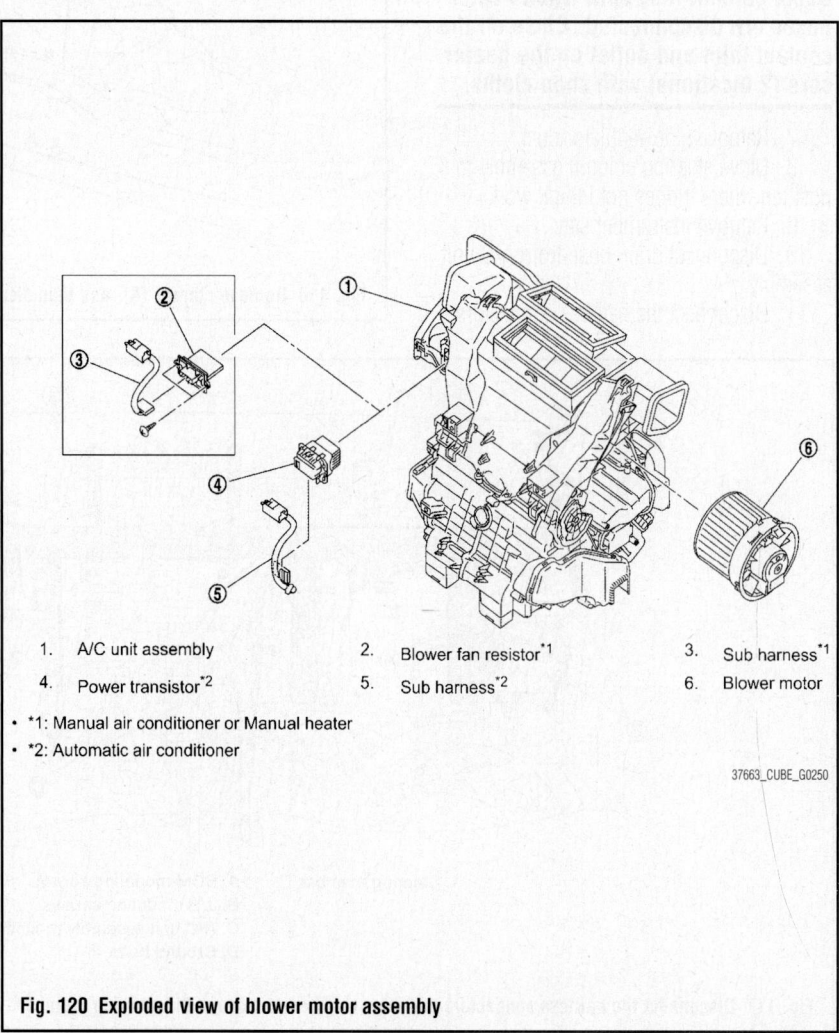

| 1. | A/C unit assembly | 2. | Blower fan resistor*1 | 3. | Sub harness*1 |
| 4. | Power transistor*2 | 5. | Sub harness*2 | 6. | Blower motor |

- *1: Manual air conditioner or Manual heater
- *2: Automatic air conditioner

Fig. 120 Exploded view of blower motor assembly

STEERING

EPS CONTROL UNIT

REMOVAL & INSTALLATION
See Figure 121.

✳✳ CAUTION
Note the following:

- Disconnect battery negative terminal before starting operations.
- Never shock EPS control unit, e.g. drop or hit.
- Never get EPS control unit wet with water or other liquid. Also, do not give EPS control unit a radical temperature change to avoid getting water drops.
- Never disassemble or remodel EPS control unit, EPS motor, torque sensor, harness and connectors.

1. Remove instrument lower panel LH.
2. Remove knee protector.
3. Disconnect EPS control unit connectors.

✳✳ CAUTION
Hold and pull the connector housing, not pulling harness, when disconnecting connectors. Also, do not grip, collapse or apply excessive force to the connector.

4. Remove EPS control unit from steering column assembly.

To install:
5. Note the following, and install in the reverse order of removal.
6. Check that harness is not damaged when installing EPS control unit. Also, check that EPS control unit is installed without trapping harness of foreign materials.
7. After installing steering column assembly, perform self-diagnosis with CONSULT-III to ensure correct operation.

POWER RACK & PINION STEERING GEAR

REMOVAL & INSTALLATION
See Figure 122.

1. Set vehicle to the straight-ahead position.
2. Remove bolt intermediate shaft (steering gear assembly side).

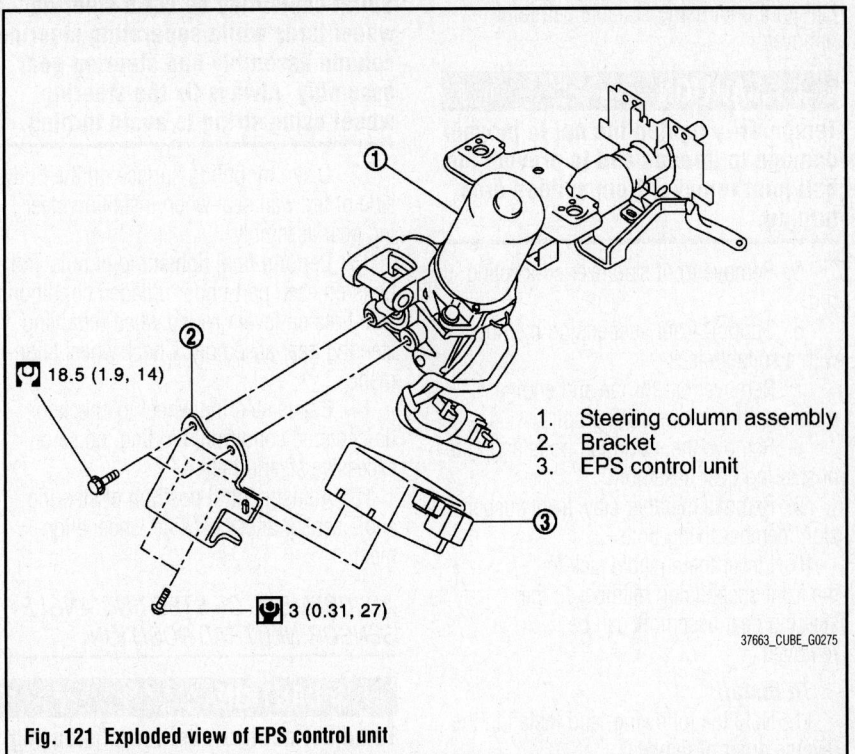

18.5 (1.9, 14)

3 (0.31, 27)

1. Steering column assembly
2. Bracket
3. EPS control unit

37663_CUBE_G0275

Fig. 121 Exploded view of EPS control unit

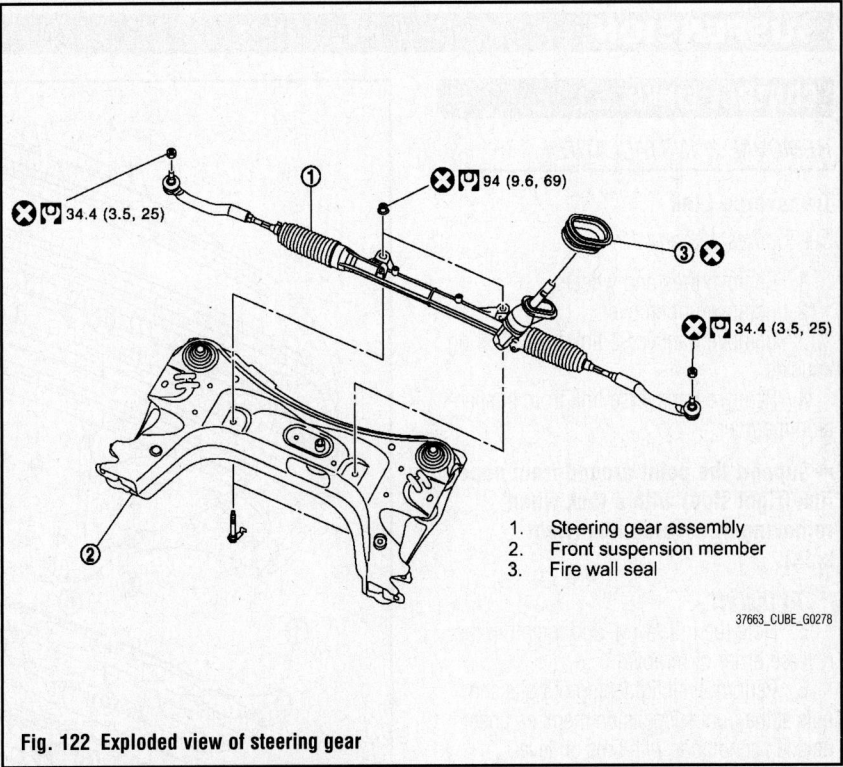

34.4 (3.5, 25)

94 (9.6, 69)

34.4 (3.5, 25)

1. Steering gear assembly
2. Front suspension member
3. Fire wall seal

37663_CUBE_G0278

Fig. 122 Exploded view of steering gear

✳✳ CAUTION
Spiral cable may be cut if steering wheel turns while separating steering column assembly and steering gear assembly. Always fix the steering wheel using string to avoid.

3. Remove tires and wheels.

4. Remove steering outer socket from steering knuckle so as not to damage ball joint boot using suitable ball joint remover.

✳✳ CAUTION

Temporarily tighten the nut to prevent damage to threads and to prevent the ball joint remover from sudden drop turning.

5. Remove front stabilizer connecting rod.

6. Support front suspension member with a suitable jack.

7. Remove rear torque and engine mounting bracket mounting bolts.

8. Remove the mounting bolts and nuts of steering gear assembly.

9. Remove member stay, front suspension member fixing bolts.

10. Lower the suitable jack for the front suspension member to the steering gear assembly can be removed.

To install:

11. Note the following, and install in the reverse order of removal.

✳✳ CAUTION

Spiral cable may be cut if steering wheel turns while separating steering column assembly and steering gear assembly. Always fix the steering wheel using string to avoid turning.

12. Clean mounting surface on the body side of fire wall seal when installing steering gear assembly.

13. Perform final tightening of nuts and bolts on each part under unladen conditions with tires on level ground when removing steering gear assembly. Check wheel alignment.

14. Rotate steering wheel to check for de-centered condition, binding, noise or excessive steering effort.

15. Adjust neutral position of steering angle sensor after checking wheel alignment.

ADJUSTMENT OF STEERING ANGLE SENSOR NEUTRAL POSITION

✳✳ CAUTION

To adjust neutral position of steering angle sensor, make sure to use CON-SULT-III. (Adjustment cannot be done without CONSULT-III.)

1. Stop vehicle with front wheels in straight-ahead position.

2. On the CONSULT-III screen, touch "WORK SUPPORT" and "ST ANGLE SENSOR ADJUSTMENT" in order.

3. Touch "START".

✳✳ CAUTION

Never touch steering wheel while adjusting steering angle sensor.

4. After approximately 10 seconds, touch "END".

➡**After approximately 60 seconds, it ends automatically.**

5. Turn the ignition switch OFF, then turn it ON again.

✳✳ CAUTION

Be sure to perform above operation.

SUSPENSION   FRONT SUSPENSION

CONTROL LINKS

REMOVAL & INSTALLATION

Transverse Link

See Figures 123 and 124.

1. Remove tires and wheels.
2. Remove undercover.
3. Remove transverse link from steering knuckle.
4. Remove transverse link from suspension member.

➡**Support the point around from upper link (right side) with a jack when removing transverse link (right side).**

To install:

5. Note the following, and install in the reverse order of removal.

6. Perform final tightening of bolts and nuts at the front suspension member, under unladen conditions with tires on level ground.

7. Check wheel sensor harness for proper connector.

8. Check wheel alignment.

9. Adjust neutral position of steering angle sensor.

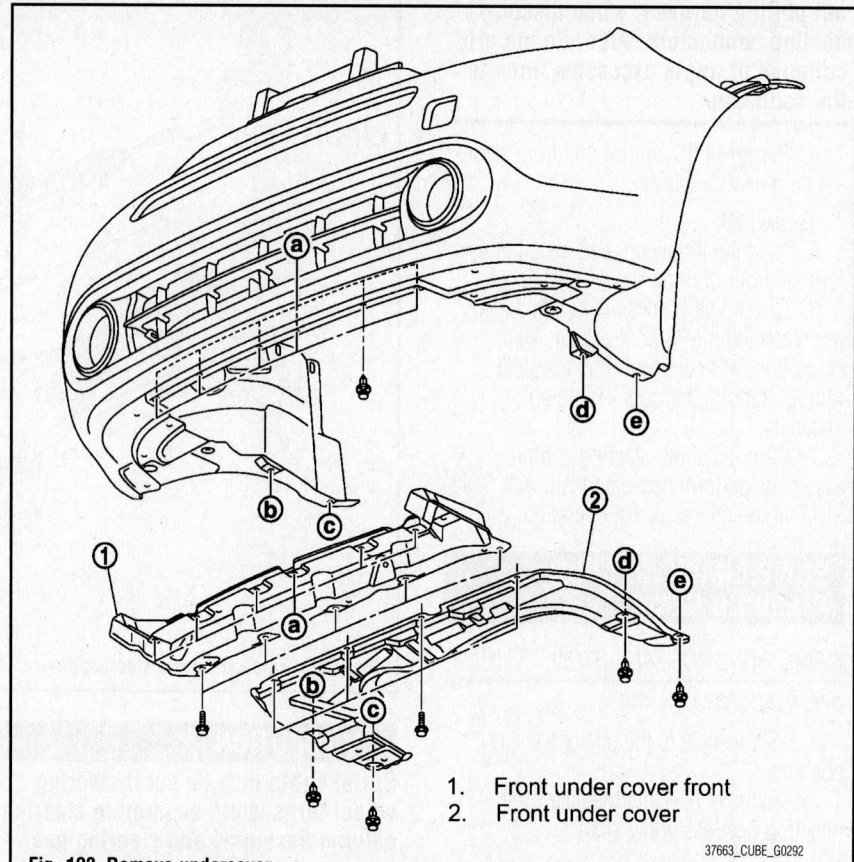

1. Front under cover front
2. Front under cover

Fig. 123 Remove undercover

37663_CUBE_G0292

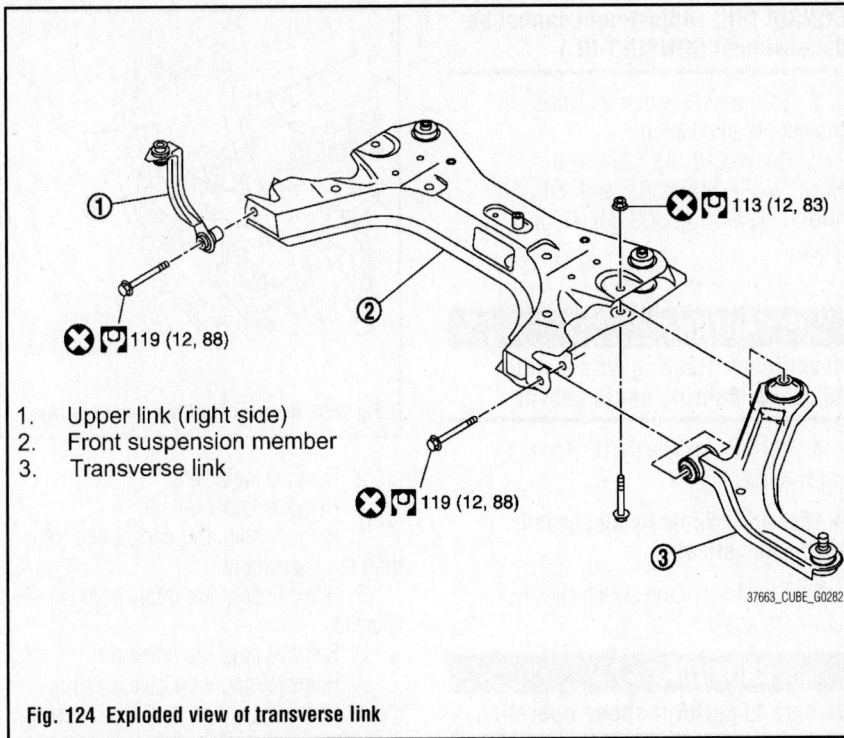

1. Upper link (right side)
2. Front suspension member
3. Transverse link

Fig. 124 Exploded view of transverse link

ADJUSTMENT OF STEERING ANGLE SENSOR NEUTRAL POSITION

✻✻ CAUTION

To adjust neutral position of steering angle sensor, make sure to use CONSULT-III. (Adjustment cannot be done without CONSULT-III.)

1. Stop vehicle with front wheels in straight-ahead position.
2. On the CONSULT-III screen, touch "WORK SUPPORT" and "ST ANGLE SENSOR ADJUSTMENT" in order.
3. Touch "START".

✻✻ CAUTION

Never touch steering wheel while adjusting steering angle sensor.

4. After approximately 10 seconds, touch "END".

➡After approximately 60 seconds, it ends automatically.

5. Turn the ignition switch OFF, then turn it ON again.

✻✻ CAUTION

Be sure to perform above operation.

STEERING KNUCKLE

REMOVAL & INSTALLATION

See Figure 125.

1. Remove tires and wheels.
2. Remove wheel sensor and sensor harness.
3. Remove lock plate from strut assembly.
4. Remove caliper assembly. Hang caliper assembly not to interfere with work.

✻✻ CAUTION

Never depress brake pedal while brake caliper is removed.

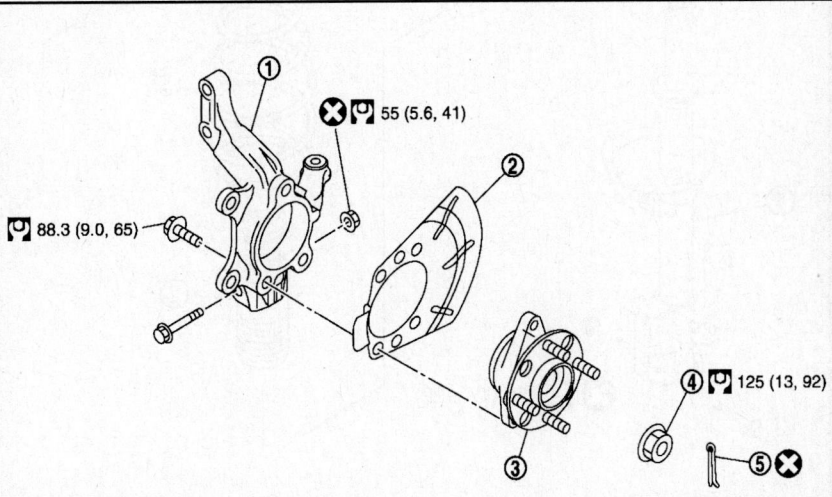

1. Steering knuckle 2. Splash guard 3. Wheel hub and bearing assembly
4. Wheel hub lock nut 5. Cotter pin

Fig. 125 Exploded view of steering knuckle assembly

5. Remove disc rotor.
6. Remove cotter pin, and then loosen wheel hub lock nut.
7. Patch wheel hub lock nut with a piece of wood. Hammer the wood to disengage wheel hub and bearing assembly from drive shaft.

✻✻ CAUTION

Note the following:

- Never place drive shaft joint at an extreme angle. Also be careful not to overextend slide joint.
- Never allow drive shaft to hang down without support for joint sub-assembly, shaft and the other parts.

➡Use suitable puller, if wheel hub and bearing assembly and drive shaft cannot be separated even after performing the above procedure.

8. Remove wheel hub lock nut.
9. Remove wheel hub and bearing assembly, and then remove splash guard.
10. Suspend the drive shaft with suitable wire.
11. Remove steering outer socket from steering knuckle.
12. Remove steering knuckle from transverse link.
13. Remove steering knuckle from strut assembly.

To install:

14. Note the following, and install in the reverse order of the removal.
   a. Clean the matching surface of

wheel hub lock nut and wheel hub and bearing assembly.

> ✴✴ **CAUTION**
>
> **Never apply lubricating oil to these matching surface.**

b. Tighten the wheel hub lock nut to 92 ft. lbs. (125 Nm).

> ✴✴ **CAUTION**
>
> **Never use a power tool to tighten the wheel hub lock nut.**

c. Perform the final tightening of each of parts under unladen conditions, which were removed when removing wheel hub and bearing assembly and axle housing.

d. Never reuse cotter pin.

15. Check wheel sensor harness for proper connection.

16. Check the wheel alignment.

17. Adjust neutral position of steering angle sensor.

ADJUSTMENT OF STEERING ANGLE SENSOR NEUTRAL POSITION

> ✴✴ **CAUTION**
>
> **To adjust neutral position of steering angle sensor, make sure to use**

CONSULT-III. (Adjustment cannot be done without CONSULT-III.)

1. Stop vehicle with front wheels in straight-ahead position.

2. On the CONSULT-III screen, touch "WORK SUPPORT" and "ST ANGLE SENSOR ADJUSTMENT" in order.

3. Touch "START".

> ✴✴ **CAUTION**
>
> **Never touch steering wheel while adjusting steering angle sensor.**

4. After approximately 10 seconds, touch "END".

➡ **After approximately 60 seconds, it ends automatically.**

5. Turn the ignition switch OFF, then turn it ON again.

> ✴✴ **CAUTION**
>
> **Be sure to perform above operation.**

STRUT

REMOVAL & INSTALLATION
See Figures 126 through 128.

1. Remove tires and wheels.

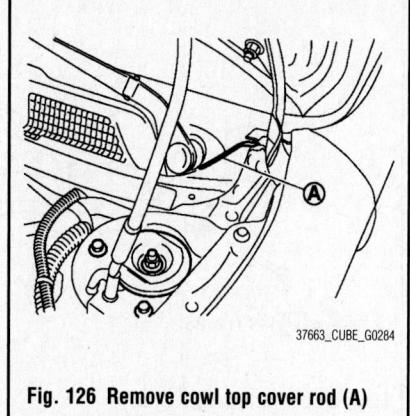

Fig. 126 Remove cowl top cover rod (A)

2. Remove lock plate.

3. Remove wheel sensor.

4. Remove stabilizer connecting rod from strut assembly.

5. Remove strut assembly from steering knuckle.

6. Remove cowl top cover rod.

7. Remove mounting bolts of strut mounting insulator, and then remove strut assembly.

To install:

8. Note the following, and install in the reverse order of removal.

a. Ensure the strut ID letters face the body front side.

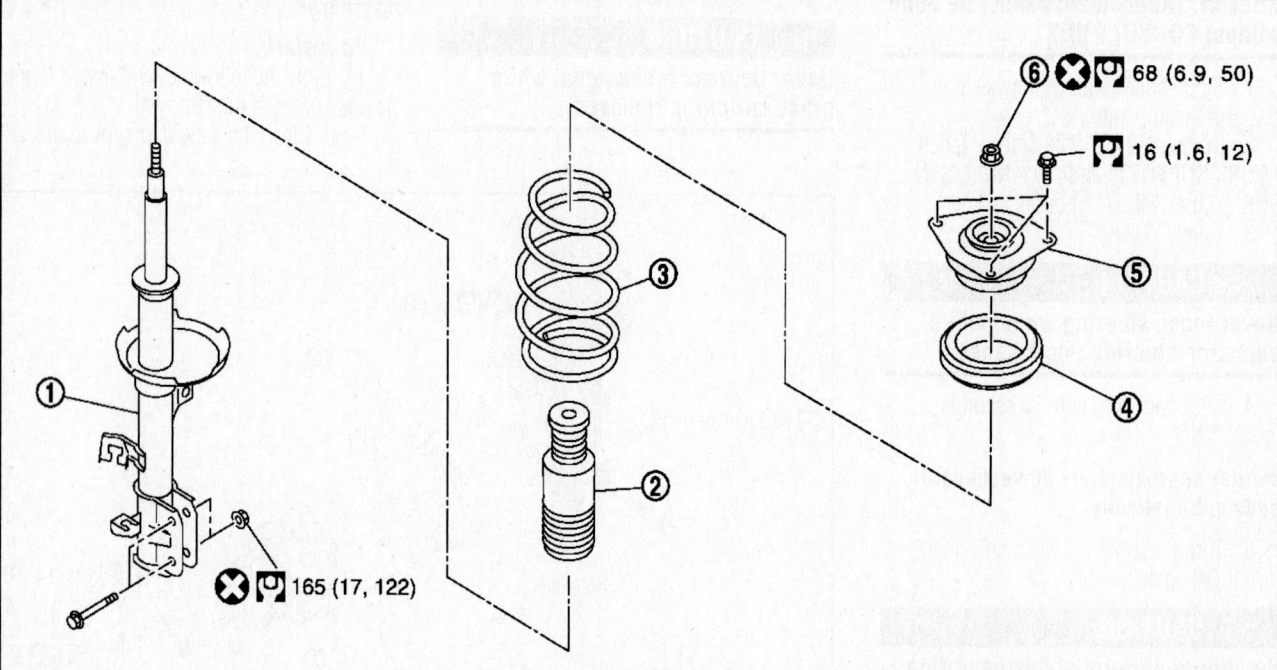

| 1. | Strut | 2. | Bound bumper | 3. | Coil spring |
| 4. | Strut mounting bearing | 5. | Strut mounting insulator | 6. | Piston rod lock nut |

Fig. 127 Exploded view of front coil spring and strut assembly

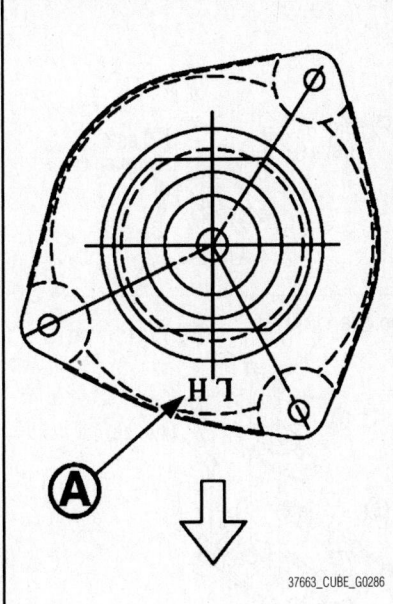

Fig. 128 Ensure the strut ID letters (A) face the body front side

b. Perform final tightening of bolts and nuts, under unladen conditions with tires on level ground.

9. Check wheel sensor harness for proper connection.

10. Check the wheel alignment.

11. Adjust neutral position of steering angle sensor.

ADJUSTMENT OF STEERING ANGLE SENSOR NEUTRAL POSITION

> ✳✳ **CAUTION**
>
> **To adjust neutral position of steering angle sensor, make sure to use CONSULT-III. (Adjustment cannot be done without CONSULT-III.)**

1. Stop vehicle with front wheels in straight-ahead position.

2. On the CONSULT-III screen, touch "WORK SUPPORT" and "ST ANGLE SENSOR ADJUSTMENT" in order.

3. Touch "START".

> ✳✳ **CAUTION**
>
> **Never touch steering wheel while adjusting steering angle sensor.**

4. After approximately 10 seconds, touch "END".

➡**After approximately 60 seconds, it ends automatically.**

5. Turn the ignition switch OFF, then turn it ON again.

> ✳✳ **CAUTION**
>
> **Be sure to perform above operation.**

STABILIZER BAR

REMOVAL & INSTALLATION

See Figures 129 and 130.

1. Remove tires and wheels.

2. Remove front suspension member.

3. Remove stabilizer connecting rod.

4. Remove mounting bolts from the stabilizer clamps, and then remove stabilizer clamps and stabilizer bushings from front suspension member.

5. Remove stabilizer bar.

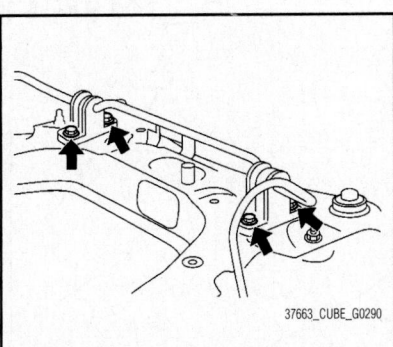

Fig. 129 Remove mounting bolts from the stabilizer clamps

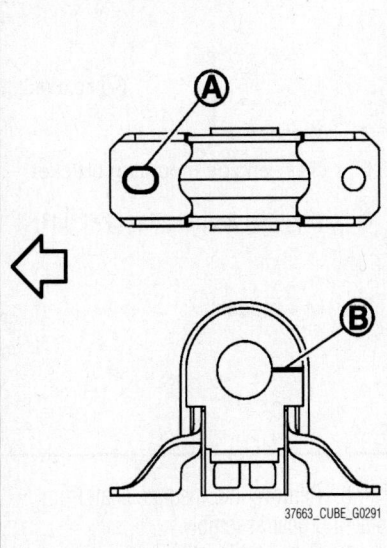

Fig. 130 Proper installation showing elongated hole (A) in clamp and slit (B) in bushing

To install:

6. Note the following, and install in the reverse order of removal.

a. Install stabilizer clamps so that the elongated hole faces toward the vehicle front.

b. Install stabilizer bushings so that the slit faces the vehicle rear side.

c. Tighten the mounting nut to the specified torque while holding a hexagonal part of stabilizer connecting rod side.

d. Perform final tightening of bolts and nuts at the vehicle installation position (rubber bushing), under unladen conditions with tires on level ground.

7. Check the wheel alignment.

8. Adjust neutral position of steering angle sensor.

ADJUSTMENT OF STEERING ANGLE SENSOR NEUTRAL POSITION

> ✳✳ **CAUTION**
>
> **To adjust neutral position of steering angle sensor, make sure to use CONSULT-III. (Adjustment cannot be done without CONSULT-III.)**

1. Stop vehicle with front wheels in straight-ahead position.

2. On the CONSULT-III screen, touch "WORK SUPPORT" and "ST ANGLE SENSOR ADJUSTMENT" in order.

3. Touch "START".

> ✳✳ **CAUTION**
>
> **Never touch steering wheel while adjusting steering angle sensor.**

4. After approximately 10 seconds, touch "END".

➡**After approximately 60 seconds, it ends automatically.**

5. Turn the ignition switch OFF, then turn it ON again.

> ✳✳ **CAUTION**
>
> **Be sure to perform above operation.**

SUSPENSION MEMBER

REMOVAL & INSTALLATION

See Figures 123, and 131 through 133.

1. Remove tires and wheels.

2. Remove undercover.

3. Remove wheel sensor.

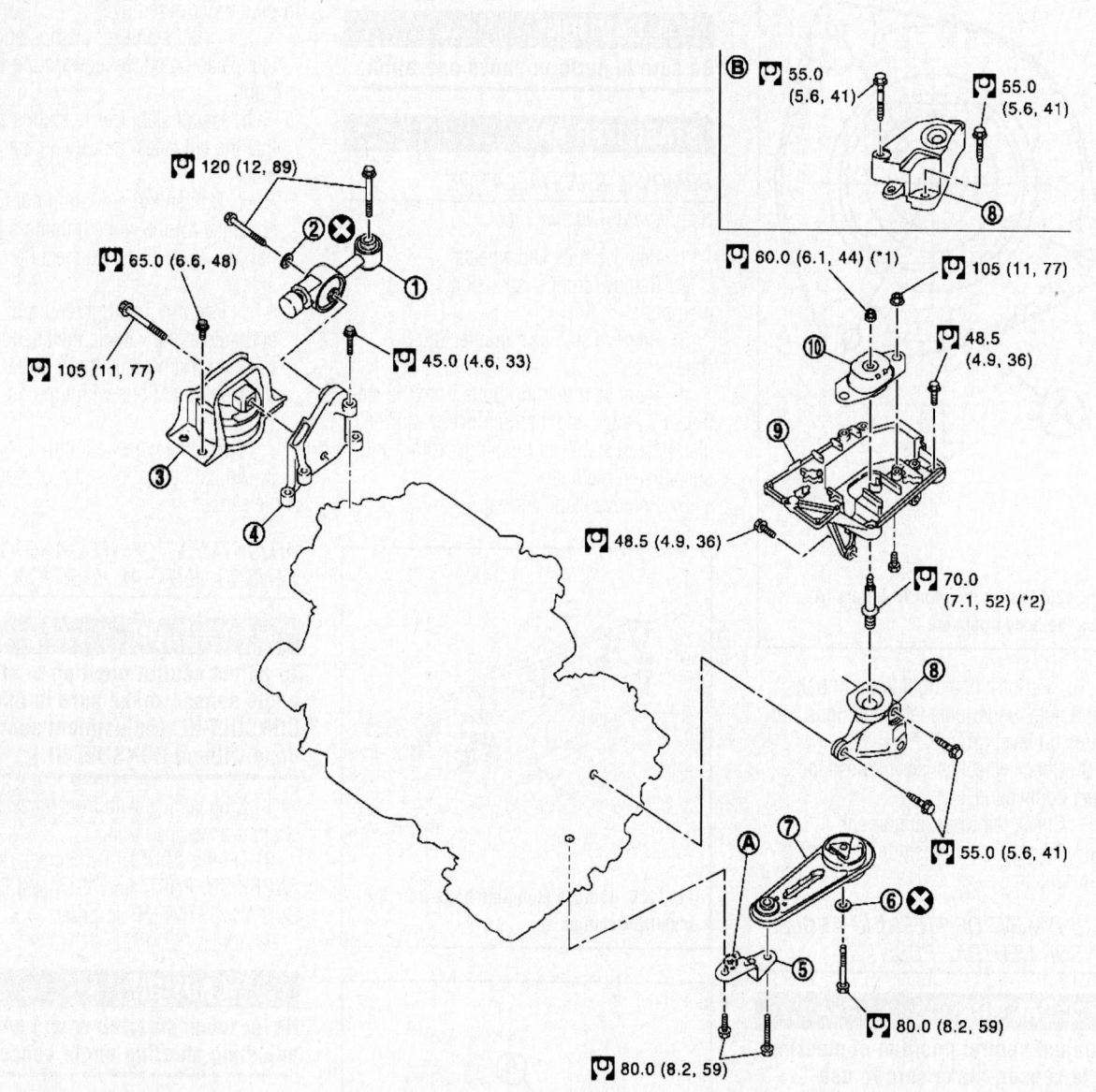

120 (12, 89)

65.0 (6.6, 48)

105 (11, 77)

45.0 (4.6, 33)

B 55.0 (5.6, 41)

55.0 (5.6, 41)

60.0 (6.1, 44) (*1)

105 (11, 77)

48.5 (4.9, 36)

48.5 (4.9, 36)

70.0 (7.1, 52) (*2)

55.0 (5.6, 41)

80.0 (8.2, 59)

80.0 (8.2, 59)

1. Upper torque rod (RH)
2. Washer
3. Engine mounting insulator (RH)
4. Engine mounting bracket (RH)
5. Rear engine mounting bracket
6. Washer
7. Rear torque rod
8. Engine mounting bracket (LH)
9. Engine mounting bracket support (LH)
10. Engine mounting insulator (LH)
A. Front mark
B. M/T models

37663_CUBE_G0165

Fig. 131 Exploded view of engine mount assemblies

4. Remove stabilizer connecting rod from strut assembly.

5. Remove rear torque rod.

6. Remove transverse link from steering knuckle.

7. Remove steering outer socket from steering knuckle.

8. Remove intermediate shaft from steering gear assembly.

9. Set suitable jack under front suspension member.

✳✳ CAUTION

Check the stable condition when using a jack.

10. Remove suspension member stay rear mounting bolts.

11. Remove upper link mounting bolts (right side: vehicle body side, left side: suspension member side), and suspension member mounting bolts.

12. Gradually lower the jack to remove front suspension member from vehicle body.

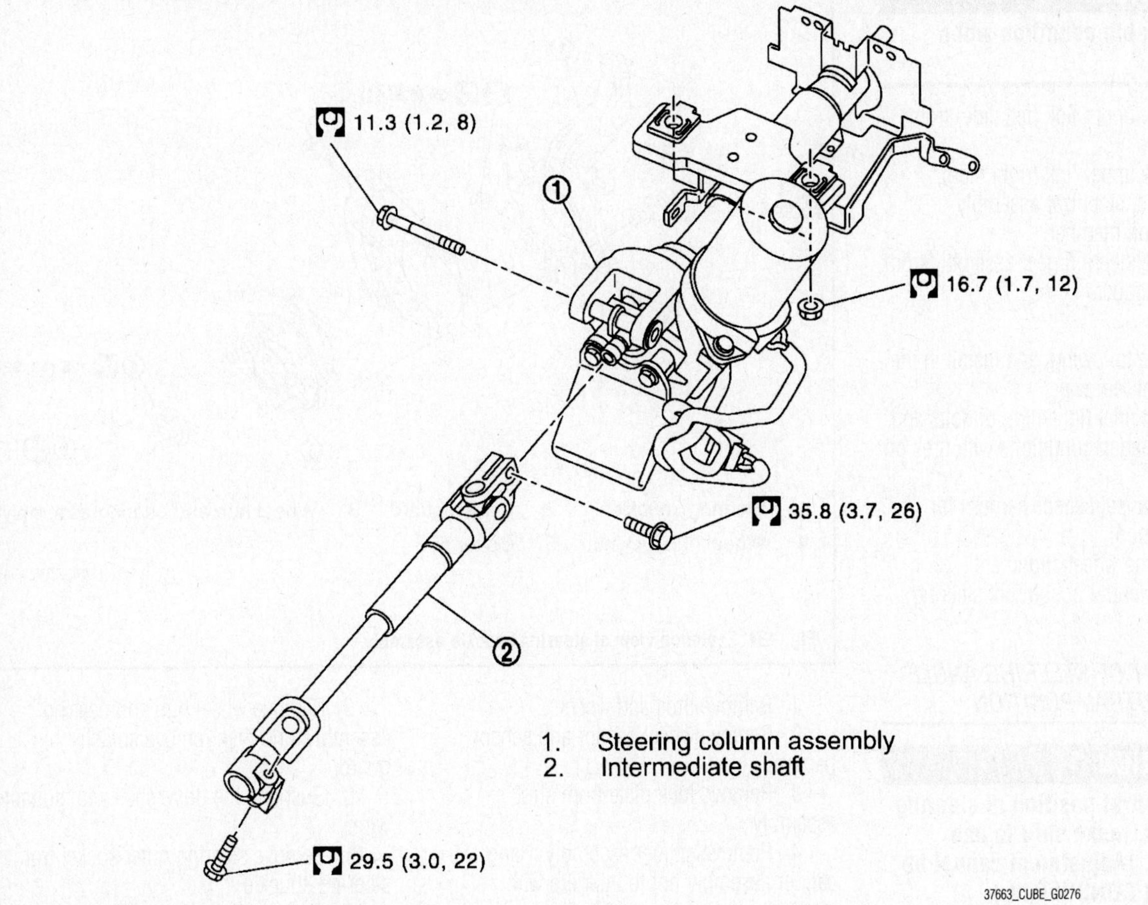

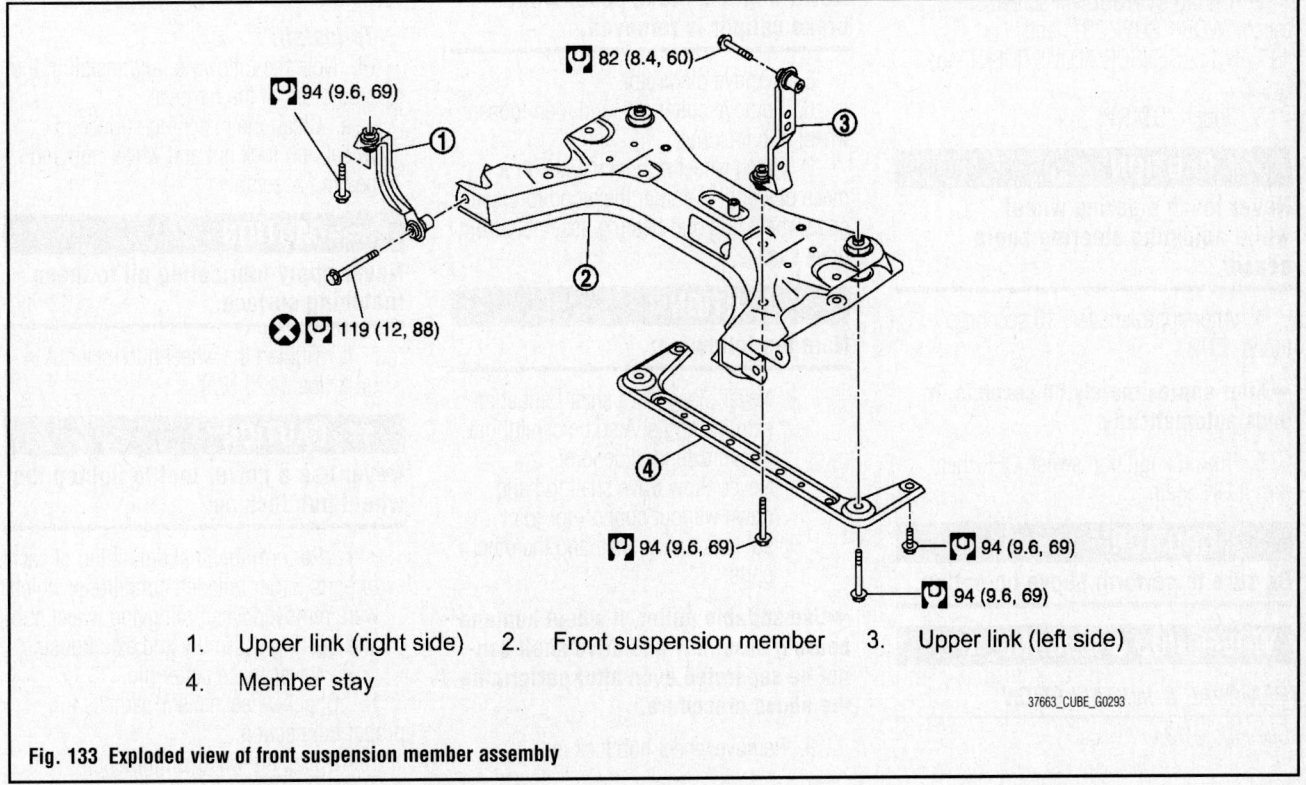

11.3 (1.2, 8)

16.7 (1.7, 12)

35.8 (3.7, 26)

29.5 (3.0, 22)

1. Steering column assembly
2. Intermediate shaft

37663_CUBE_G0276

Fig. 132 Exploded view of steering column

94 (9.6, 69)

82 (8.4, 60)

119 (12, 88)

94 (9.6, 69)

94 (9.6, 69)

94 (9.6, 69)

1. Upper link (right side)   2. Front suspension member   3. Upper link (left side)
4. Member stay

37663_CUBE_G0293

Fig. 133 Exploded view of front suspension member assembly

✳✳ CAUTION

Check the stable condition when using a jack.

13. Remove upper link (left side) from vehicle body.

14. Remove upper link (right side), transverse links, stabilizer assembly from suspension member.

15. Remove steering gear assembly from suspension member.

To install:

16. Note the following, and install in the reverse order of removal.

17. Perform final tightening of bolts and nuts, under unladen conditions with tires on level ground.

18. Check wheel sensor harness for proper connection.

19. Check the wheel alignment.

20. Adjust neutral position of steering angle sensor.

ADJUSTMENT OF STEERING ANGLE SENSOR NEUTRAL POSITION

✳✳ CAUTION

To adjust neutral position of steering angle sensor, make sure to use CONSULT-III. (Adjustment cannot be done without CONSULT-III.)

1. Stop vehicle with front wheels in straight-ahead position.

2. On the CONSULT-III screen, touch "WORK SUPPORT" and "ST ANGLE SENSOR ADJUSTMENT" in order.

3. Touch "START".

✳✳ CAUTION

Never touch steering wheel while adjusting steering angle sensor.

4. After approximately 10 seconds, touch "END".

➡**After approximately 60 seconds, it ends automatically.**

5. Turn the ignition switch OFF, then turn it ON again.

✳✳ CAUTION

Be sure to perform above operation.

WHEEL HUB & BEARING

REMOVAL & INSTALLATION

See Figure 134.

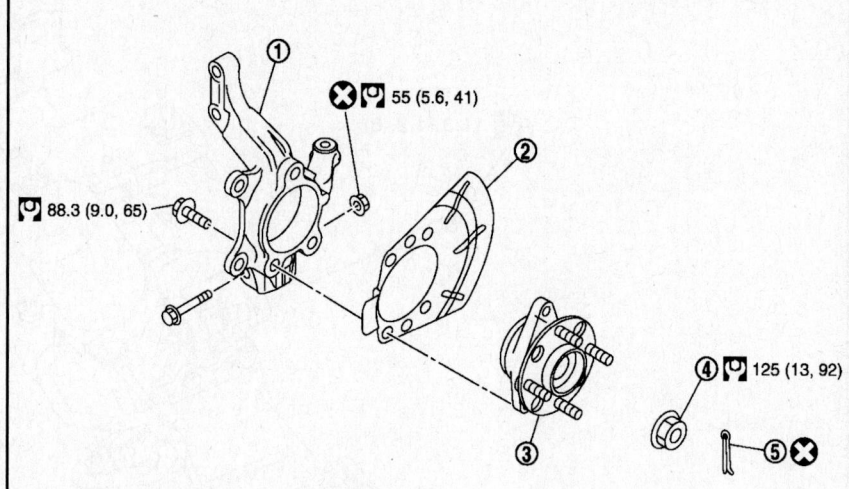

1. Steering knuckle
2. Splash guard
3. Wheel hub and bearing assembly
4. Wheel hub lock nut
5. Cotter pin

37663_CUBE_G0283

Fig. 134 Exploded view of steering knuckle assembly

1. Remove tires and wheels.

2. Remove wheel sensor and sensor harness.

3. Remove lock plate from strut assembly.

4. Remove caliper assembly. Hang caliper assembly not to interfere with work.

✳✳ CAUTION

Never depress brake pedal while brake caliper is removed.

5. Remove disc rotor.

6. Remove cotter pin, and then loosen wheel hub lock nut.

7. Patch wheel hub lock nut with a piece of wood. Hammer the wood to disengage wheel hub and bearing assembly from drive shaft.

✳✳ CAUTION

Note the following:

- Never place drive shaft joint at an extreme angle. Also be careful not to overextend slide joint.
- Never allow drive shaft to hang down without support for joint sub-assembly, shaft and the other parts.

➡**Use suitable puller, if wheel hub and bearing assembly and drive shaft cannot be separated even after performing the above procedure.**

8. Remove wheel hub lock nut.

9. Remove wheel hub and bearing assembly, and then remove splash guard.

10. Suspend the drive shaft with suitable wire.

11. Remove steering outer socket from steering knuckle.

12. Remove steering knuckle from transverse link.

13. Remove steering knuckle from strut assembly.

To install:

14. Note the following, and install in the reverse order of the removal.

a. Clean the matching surface of wheel hub lock nut and wheel hub and bearing assembly.

✳✳ CAUTION

Never apply lubricating oil to these matching surface.

b. Tighten the wheel hub lock nut to 92 ft. lbs. (125 Nm).

✳✳ CAUTION

Never use a power tool to tighten the wheel hub lock nut.

c. Perform the final tightening of each of parts under unladen conditions, which were removed when removing wheel hub and bearing assembly and axle housing.

d. Never reuse cotter pin.

15. Check wheel sensor harness for proper connection.

16. Check the wheel alignment.

17. Adjust neutral position of steering angle sensor.

ADJUSTMENT OF STEERING ANGLE SENSOR NEUTRAL POSITION

> ✳✳ **CAUTION**
>
> **To adjust neutral position of steering angle sensor, make sure to use CONSULT-III. (Adjustment cannot be done without CONSULT-III.)**

1. Stop vehicle with front wheels in straight-ahead position.

2. On the CONSULT-III screen, touch "WORK SUPPORT" and "ST ANGLE SENSOR ADJUSTMENT" in order.
3. Touch "START".

> ✳✳ **CAUTION**
>
> **Never touch steering wheel while adjusting steering angle sensor.**

4. After approximately 10 seconds, touch "END".

➡**After approximately 60 seconds, it ends automatically.**

5. Turn the ignition switch OFF, then turn it ON again.

> ✳✳ **CAUTION**
>
> **Be sure to perform above operation.**

ADJUSTMENT

No adjustments are possible. If the axle end play is greater than 0.002inches (0.05mm), replace the wheel bearing.

SUSPENSION

COIL SPRING

REMOVAL & INSTALLATION

See Figures 135 and 136.

1. Remove tires and wheels.
2. Set jack under rear suspension beam.

> ✳✳ **CAUTION**
>
> **Check the stable condition when using a jack.**

3. Remove right and left rear shock absorber mounting bolts (lower side).
4. Slowly lower jack, then remove upper rubber seat, coil spring and lower rubber seat from rear suspension beam.

> ✳✳ **CAUTION**
>
> **Check the stable condition when using a jack.**

5. Check lubber seat and coil spring for deformation, crack, and damage. Replace it if necessary.

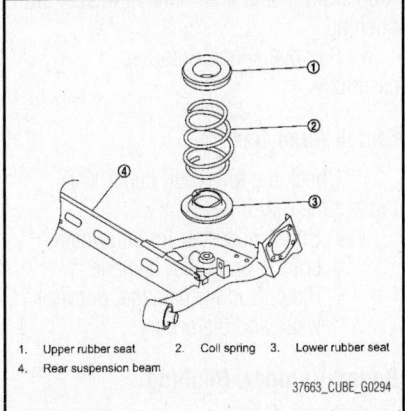

1. Upper rubber seat   2. Coil spring   3. Lower rubber seat
4. Rear suspension beam

37663_CUBE_G0294

Fig. 135 Exploded view of rear coil spring assembly

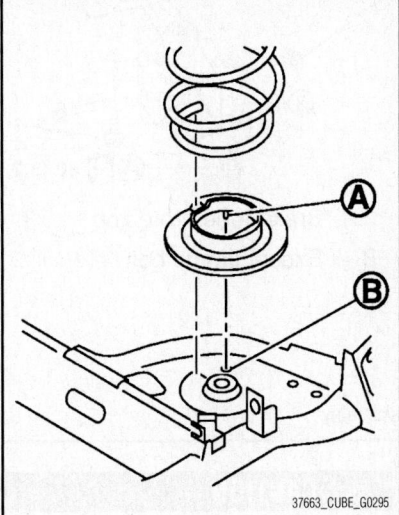

37663_CUBE_G0295

Fig. 136 Install the lower rubber seat with projection (A) as shown attached to rear suspension beam mounting hole (B)

To install:
6. Note the following, and install in the reverse order of removal.
    a. Install the lower rubber seat with projection as shown attached to rear suspension beam mounting hole.
    b. Match up lower rubber seat indentions and rear suspension beam grooves and attach.
7. Check wheel alignment.

REAR SUSPENSION BEAM

REMOVAL & INSTALLATION

See Figure 137.

1. Remove tires and wheels.
2. Drain brake fluid.
3. Remove parking brake cable and brake drum from rear suspension beam.

REAR SUSPENSION

4. Remove wheel sensor and sensor harness.
5. Set suitable jack under rear suspension beam.

> ✳✳ **CAUTION**
>
> **Check the stable condition when using a jack.**

6. Remove shock absorber mounting bolts (lower side).
7. Remove coil springs.
8. Separate brake hose and brake tube.
9. Remove suspension arm bracket mounting bolts.
10. Slowly lower jack, remove suspension arm bracket and rear suspension beam from vehicle body.

> ✳✳ **CAUTION**
>
> **Check the stable condition when using a jack.**

11. Remove wheel hub and bearing assembly.
12. Remove drum brake assembly.
13. Remove suspension arm bracket from rear suspension beam.
14. Remove brake tube protector from rear suspension beam.
15. Check rear suspension beam and rear suspension beam bracket for deformation, cracks or damage. Replace the part if necessary.

To install:
16. Note the following, and install in the reverse order of removal.
    a. Perform final tightening of rear suspension beam installation position (rubber bushing), under unladen conditions with tires on level ground.

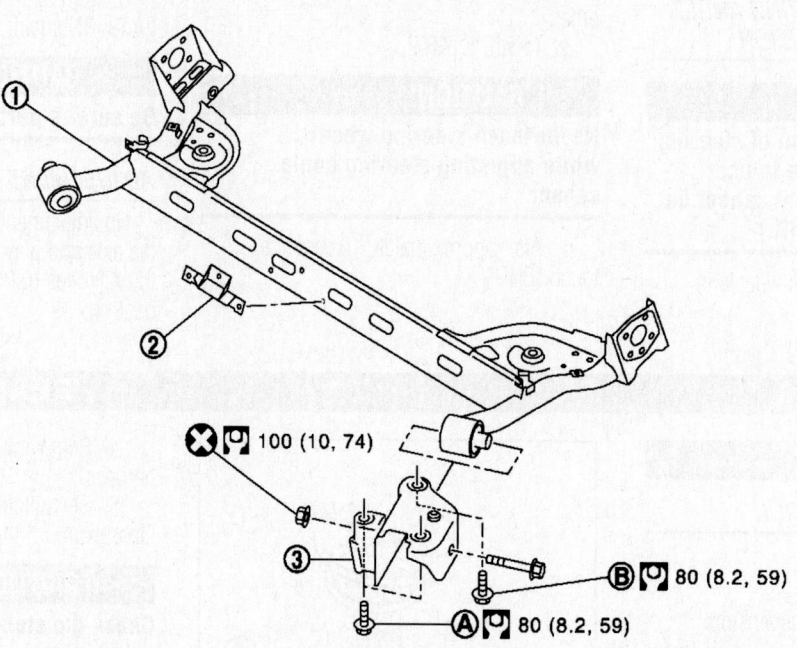

X [□] 100 (10, 74)

B [□] 80 (8.2, 59)

A [□] 80 (8.2, 59)

1. Rear suspension beam	2. Brake tube protector	3. Rear suspension arm bracket
A. Flange bolt	B. Except flange bolt	

37663_CUBE_G0296

Fig. 137 Exploded view of rear suspension beam assembly

b. Refill with new brake fluid and perform the air bleeding.

17. Check wheel sensor harness for proper connection.

18. Adjust parking brake.

19. Check the wheel alignment.

20. Adjust neutral position of steering angle sensor.

ADJUSTMENT OF STEERING ANGLE SENSOR NEUTRAL POSITION

✳✳ CAUTION

To adjust neutral position of steering angle sensor, make sure to use CONSULT-III. (Adjustment cannot be done without CONSULT-III.)

1. Stop vehicle with front wheels in straight-ahead position.

2. On the CONSULT-III screen, touch "WORK SUPPORT" and "ST ANGLE SENSOR ADJUSTMENT" in order.

3. Touch "START".

✳✳ CAUTION

Never touch steering wheel while adjusting steering angle sensor.

4. After approximately 10 seconds, touch "END".

➡**After approximately 60 seconds, it ends automatically.**

5. Turn the ignition switch OFF, then turn it ON again.

✳✳ CAUTION

Be sure to perform above operation.

SHOCK ABSORBER

REMOVAL & INSTALLATION

See Figures 138 through 140.

1. Remove tires and wheels.

2. Set suitable jack under rear suspension beam.

✳✳ CAUTION

Check the stable condition when using a jack.

3. Remove shock absorber mounting bolt (lower side).

4. Remove luggage lid from luggage side finisher.

5. Remove shock absorber mounting nut (upper side), and then remove washer and bushing.

6. Remove shock absorber assembly.

Shock Absorber

1. Check the following items, and replace the part if necessary.
- Shock absorber for deformation, cracks, and other damage.
- Piston rod for damage, uneven wear, and distortion.

Bound Bumper, Bushing

1. Check for cracks and damage. Replace it if necessary.

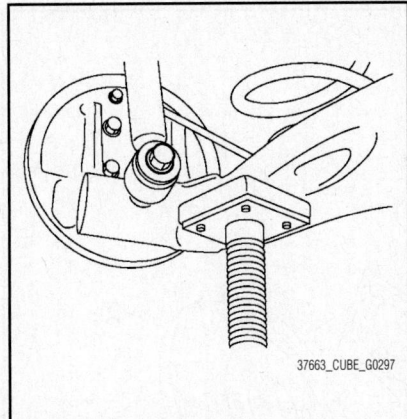

Fig. 138 Set suitable jack under rear suspension beam

Washer, Bound Bumper Cover, Distance Tube

1. Check for cracks and damage. Replace it if necessary.

To install:

2. Note the following, and install in the reverse order of removal.

  a. Perform final tightening of bolts and nuts at the shock absorber lower side (rubber bushing), under unladen conditions with tires on level ground.

  b. Install projection completely into the hole on the vehicle side as shown, when installing bushing.

  c. Hold the head of shock absorber

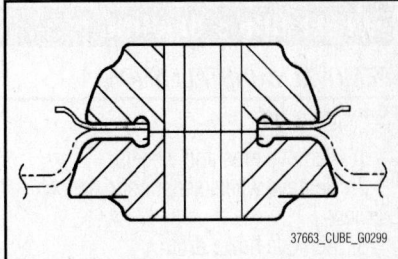

Fig. 140 Install projection completely into the hole on the vehicle side as shown

piston rod to prevent it from rotating, then tighten piston rod lock nut with a standard tightening torque value.

20 (2.0, 15)

124 (13, 91)

1.	Piston rod lock nut	2.	Washer	3.	Bushing
4.	Distance tube	5.	Bound bumper cover	6.	Bound bumper
7.	Shock absorber	8.	Rear suspension beam		

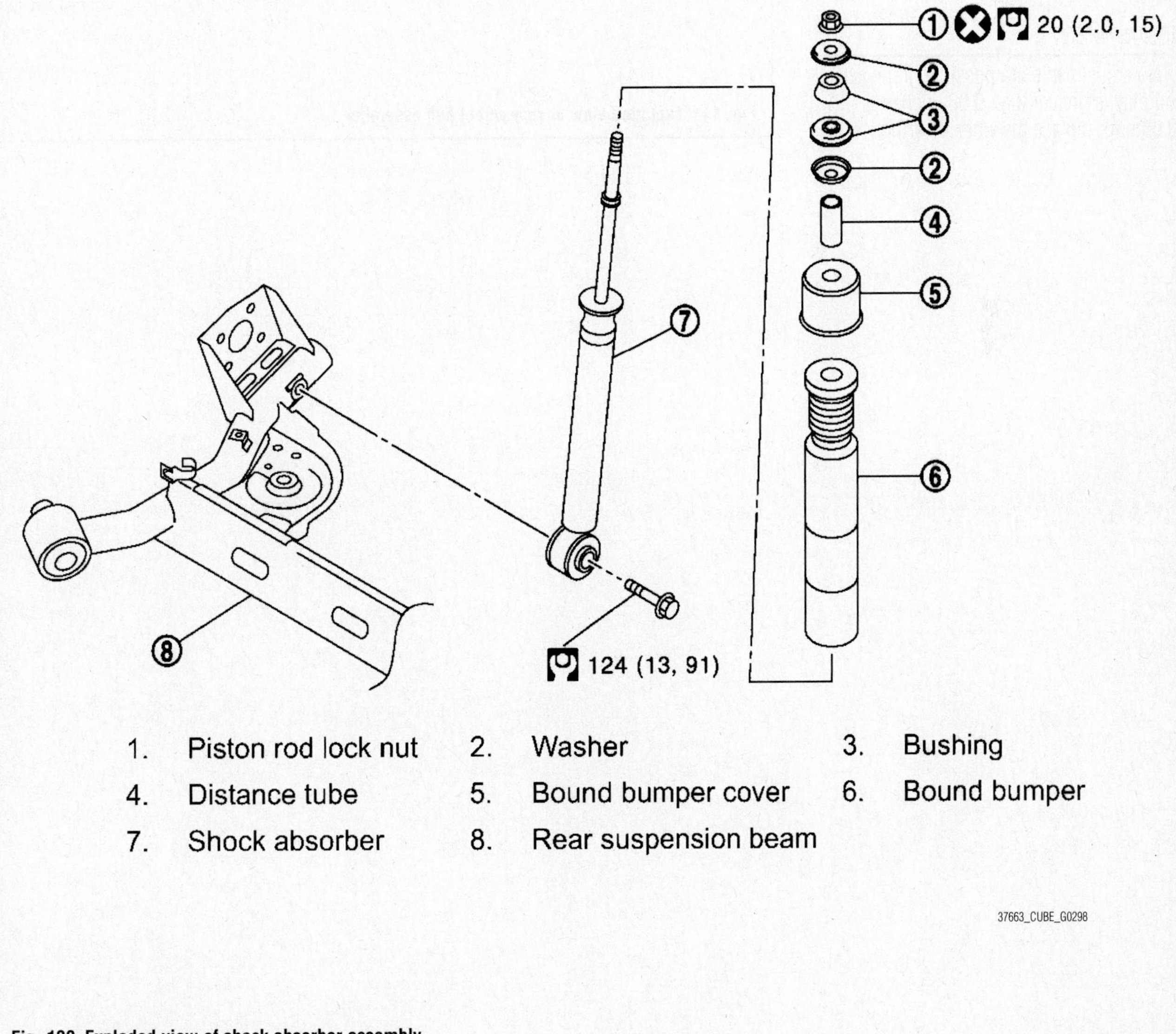

Fig. 139 Exploded view of shock absorber assembly

WHEEL HUB & BEARING

REMOVAL & INSTALLATION

See Figure 141.

1. Remove tires and wheels.
2. Remove wheel sensor and sensor harness.
3. Remove brake drum.

➡**Keep back plate and brake assembly not falling down without removing them.**

4. Remove wheel hub and bearing assembly.
5. Check the wheel hub and bearing assembly for wear, cracks, and damage. Replace if necessary.

To install:

6. Install in the reverse order of removal.
7. Check wheel sensor harness for proper connection.
8. Adjust parking brake operation (stroke).
9. Check wheel alignment.

ADJUSTMENT

No adjustments are possible. If the axle end play is greater than 0.002inches (0.05mm), replace the wheel bearing.

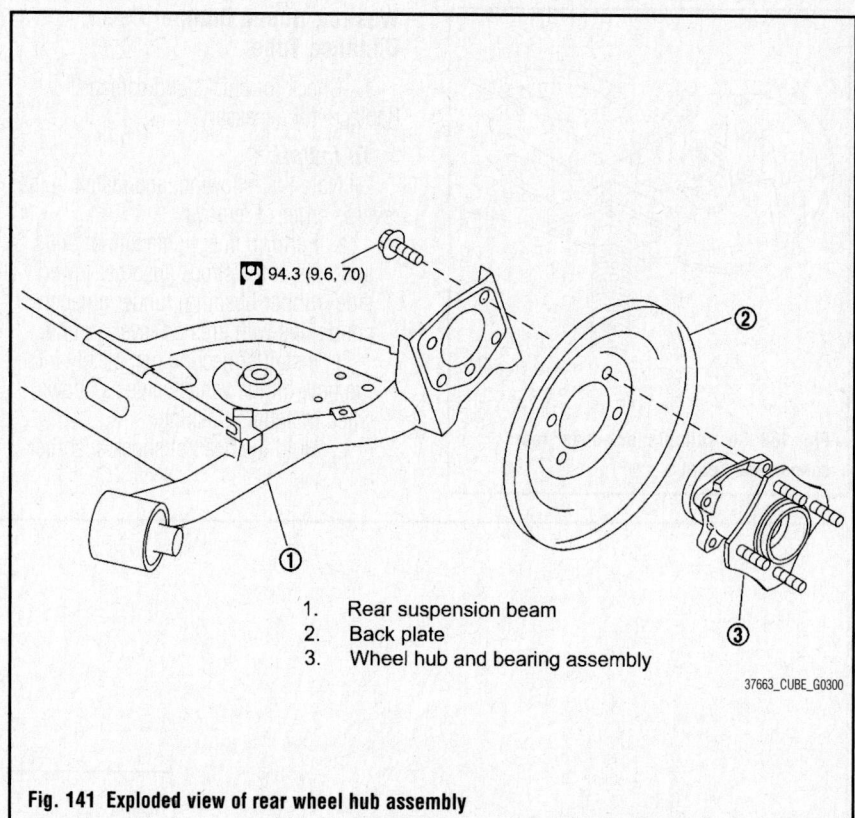

94.3 (9.6, 70)

1. Rear suspension beam
2. Back plate
3. Wheel hub and bearing assembly

37663_CUBE_G0300

Fig. 141 Exploded view of rear wheel hub assembly

SPECIFICATIONS AND MAINTENANCE CHARTS

ENGINE AND VEHICLE IDENTIFICATION

	Engine						Model Year	
ID/Code ①	Liters (cc)	Cu. In.	Cyl.	Fuel Sys.	Engine Type	Eng. Mfg.	Code ②	Year
QR25DE	2.5 (2488)	152	4	MFI	DOHC	Nissan	9	2009
VQ40DE	4.0 (3954)	241	6	MFI	DOHC	Nissan	A	2010

MFI: Multi-port Fuel Injection

SOHC: Single Overhead Camshaft

DOHC: Double Overhead Camshafts

① 4th digit of the Vehicle Identification Number (VIN)

② 10th digit of the Vehicle Identification Number (VIN)

37663_FRON_C0001

GENERAL ENGINE SPECIFICATIONS

Year	Model	Engine Displacement Liters	Engine ID	Net Horsepower @ rpm	Net Torque @ rpm (ft. lbs.)	Bore x Stroke (in.)	Compression Ratio	Oil Pressure @ rpm
2009	Frontier	2.5	QR25DE	154@5200	173@4400	3.50X3.94	9.5:1	43@2000
	Frontier	4.0	VQ40DE	265@5600	284@4000	3.76X3.62	9.7:1	43@2000
2010	Frontier	2.5	QR25DE	154@5200	173@4400	3.50X3.94	9.5:1	43@2000
	Frontier	4.0	VQ40DE	265@5600	284@4000	3.76X3.62	9.7:1	43@2000

37663_FRON_C0002

ENGINE TUNE-UP SPECIFICATIONS

Year	Engine Displ. Liters	Engine ID	Spark Plug Gap (in.)	Ignition Timing (deg.) MT	AT	Fuel Pump (psi)	Idle Speed (rpm) MT	AT ①	Valve Clearance (in.) In.	Ex.
2009	2.5	QR25DE	0.043	10-20B	10-20B	51 ②	575-675	650-750	③	④
	4.0	VQ40DE	0.043	10-20B	10-20B	51 ②	575-675	650-750	⑤	⑥
2010	2.5	QR25DE	0.043	10-20B	10-20B	51 ②	575-675	650-750	③	④
	4.0	VQ40DE	0.043	10-20B	10-20B	51 ②	575-675	650-750	⑤	⑥

NOTE: The Vehicle Emission Control Information label often reflects specification changes made during production. The label figures must be used
if they differ from those in this chart.

B: Before top dead center

HYD: Hydraulic

① Automatic transmission in Neutral

② At idle

③ 0.009-0.013 cold
   0.012-0.016 hot

④ 0.010-0.013 cold
   0.012-0.017 hot

⑤ 0.010-0.013 cold
   0.012-0.016 hot

⑥ 0.011-0.015 cold
   0.012-0.017 hot

37663_FRON_C0003

CAPACITIES

Year	Model	Engine Displacement Liters	Engine ID	Engine Oil with Filter (qts.)	Transmission (pts.) Manual	Transmission (pts.) Auto.	Transfer Case (pts.)	Drive Axle Front (pts.)	Drive Axle Rear (pts.)	Fuel Tank (gal.)	Cooling System (qts.)
2009	Frontier	2.5	QR25DE	5.15	6.15	21.50	—	—	3.3	21.2	10.0
	Frontier	4.0	VQ40DE	5.30	①	21.50	2.1	1.75	②	21.2	11.0
2010	Frontier	2.5	QR25DE	5.15	6.15	21.50	—	—	3.3	21.2	10.0
	Frontier	4.0	VQ40DE	5.15	①	21.50	2.1	1.75	②	21.2	11.0

NOTE: All capacities are approximate. Add fluid gradually and check to be sure a proper fluid level is obtained.

① 2WD: 8.3; 4WD: 8.9

② C200: 3.3; M226: 4.25

37663_FRON_C0004

VALVE SPECIFICATIONS

Year	Engine Displacement Liters	Engine ID	Seat Angle (deg.)	Face Angle (deg.)	Spring Test Pressure (lbs. @ in.)	Spring Installed Height (in.)	Stem-to-Guide Clearance (in.) Intake	Stem-to-Guide Clearance (in.) Exhaust	Stem Diameter (in.) Intake	Stem Diameter (in.) Exhaust
2009	2.5	QR25DE	①	②	③	1.390	0.0008-0.0021	0.0012-0.0025	0.2348-0.2354	0.2344-0.2350
	4.0	VQ40DE	①	②	④	1.456	0.0008-0.0021	0.0012-0.0025	0.2348-0.2354	0.2344-0.2350
2010	2.5	QR25DE	①	②	③	1.390	0.0008-0.0021	0.0012-0.0025	0.2348-0.2354	0.2344-0.2350
	4.0	VQ40DE	①	②	④	1.456	0.0008-0.0021	0.0012-0.0025	0.2348-0.2354	0.2344-0.2350

① 44 degrees 22 minutes to 45 degrees 8 minutes

② 45 degrees 15 minutes to 45 degrees 45 minutes

③ Installation: 37-42@1.457

  Valve open INT: 79-89@0.996

  Valve open EXH: 71-81@1.053

④ Installation: 37-42@1.457

  Valve open: 84-95@1.071

37663_FRON_C0005

CAMSHAFT SPECIFICATIONS CHART

All measurements are given in inches.

Year	Engine Displ. Liters	Engine VIN	Journal Dia.	Brg. Oil Clearance	Shaft End-play	Runout	Lobe Height	
							Intake	Exhaust
2009	2.5	QR25DE	①	0.0018-0.0034	0.0045-0.0074	0.0008	1.7722-1.7797	1.7313-1.7388
	4.0	VQ40DE	②	③	0.0045-0.0074	0.0010	1.7900-1.7921	1.7746-1.7821
2010	2.5	QR25DE	①	0.0018-0.0034	0.0045-0.0074	0.0008	1.7722-1.7797	1.7313-1.7388
	4.0	VQ40DE	②	③	0.0045-0.0074	0.0010	1.7900-1.7921	1.7746-1.7821

① No.1: 1.0998-1.1006. No's. 2, 3, 4, 5: 0.9226-0.9234

② No.1: 1.0211-1.0218. No's. 2, 3, 4: 0.9230-0.9238

③ No.1: 1.0018-0.0034. No's. 2, 3, 4: 0.0014-0.0030

37663_FRON_C0006

CRANKSHAFT AND CONNECTING ROD SPECIFICATIONS

All measurements are given in inches.

Year	Engine Displacement Liters	Engine ID	Crankshaft				Connecting Rod		
			Main Brg. Journal Dia.	Main Brg. Oil Clearance	Shaft End-play	Thrust on No.	Journal Diameter	Oil Clearance	Side Clearance
2009	2.5	QR25DE	①	②	0.0039-0.0102	NA	NA	0.0015-0.0022	0.0079-0.0138
	4.0	VQ40DE	③	0.0014-0.0018	0.0039-0.0098	NA	2.2441-2.2446	0.0013-0.0023	0.0079-0.0138
2010	2.5	QR25DE	①	②	0.0039-0.0102	NA	NA	0.0015-0.0022	0.0079-0.0138
	4.0	VQ40DE	③	0.0014-0.0018	0.0039-0.0098	NA	2.2441-2.2446	0.0013-0.0023	0.0079-0.0138

NA: Not Available

① There are 24 different grades, ranging from A (2.1645) to 7 (2.1636)

② No. 1, 3, 5: 0.0011-0.0017

No's. 2, 4: 0.0016-0.0022

③ There are 24 different grades, ranging from A (2.7549) to 7 (2.7540)

37663_FRON_C0007

PISTON AND RING SPECIFICATIONS

All measurements are given in inches.

Year	Engine Displacement Liters	Engine ID	Piston Clearance	Ring Gap			Ring Side Clearance		
				Top Comp.	Bottom Comp.	Oil Control	Top Comp.	Bottom Comp.	Oil Control
2009	2.5	QR25DE	0.0004-0.0012	0.0083-0.0122	0.0146-0.0205	0.0079-0.0236	0.0016-0.0031	0.0012-0.0028	0.0026-0.0053
	4.0	VQ40DE	0.0004-0.0012	0.0091-0.0130	0.0130-0.0189	0.0079-0.0197	0.0018-0.0031	0.0012-0.0028	0.0026-0.0053
2010	2.5	QR25DE	0.0004-0.0012	0.0083-0.0122	0.0146-0.0205	0.0079-0.0236	0.0016-0.0031	0.0012-0.0028	0.0026-0.0053
	4.0	VQ40DE	0.0004-0.0012	0.0091-0.0130	0.0130-0.0189	0.0079-0.0197	0.0018-0.0031	0.0012-0.0028	0.0026-0.0053

37663_FRON_C0008

TORQUE SPECIFICATIONS

All readings in ft. lbs.

Year	Engine Displacement Liters	Engine ID	Cylinder Head Bolts	Main Bearing Bolts	Rod Bearing Bolts	Crankshaft Damper Bolts	Flywheel Bolts	Manifold		Spark Plugs	Oil Pan Drain Plug
								Intake	Exhaust		
2009	2.5	QR25DE	①	②	③	④	80	⑤	⑥	14-22	25
	4.0	VQ40DE	⑦	⑧	14	⑨	65	⑩	⑪	14-22	25
2010	2.5	QR25DE	①	②	③	④	80	⑤	⑥	14-22	25
	4.0	VQ40DE	⑦	⑧	14	⑨	65	⑩	⑪	14-22	25

① Step 1: 37 ft. lbs.

Step 2: Plus 60 degrees clockwise

Step 3: loosen completely to 0 ft. lbs.

Step 4: 29 ft. lbs.

Step 5: Plus 75 degrees clockwise

Step 6: Plus 75 degrees clockwise

② Step 1: bolts 11-22 19 ft. lbs.

Step 2: bolts 1-10 29 ft. lbs.

Step 3: bolts 1-10 Plus 60-65 degrees

③ Step 1: 20 ft. lbs.

Step 2: loosen to 0 ft. lbs.

Step 3: 14 ft. lbs.

Step 4: Plus 85-95 degrees

④ Step 1: 31 ft. lbs.

Step 2: Plus 60 degrees

⑤ 14 ft. lbs.

⑥ Stud bolt: 11 ft. lbs.

Nuts: 22 ft. lbs.

⑦ Step 1: 72 ft. lbs.

Step 2: loosen completely to 0 ft. lbs.

Step 3: 29 ft. lbs.

Step 4: Plus 90 degrees clockwise

Step 5: Plus 90 degrees clockwise

⑧ Bolts: 17-24 (M8) 16 ft. lbs.

Install rear main seal

Bolts: 1-16 (M10) 26 ft. lbs.

Bolts: 1-16 (M10) Plus 90 degrees clockwise

⑨ Step 1: 33 ft. lbs.

Step 2: Plus 84-90 degrees clockwise

⑩ Intake manifold collector:

Bolts and nuts: 8 ft. lbs.

Stud bolts: 61 inch lbs.

⑪ Intake manifold:

Bolts and nuts: 5 ft. lbs. and

than to 21 ft. lbs.

Studs: 8 ft. lbs.

37663_FRON_C0009

WHEEL ALIGNMENT

Year	Model		Caster Range (+/-Deg.)	Caster Preferred Setting (Deg.)	Camber Range (+/-Deg.)	Camber Preferred Setting (Deg.)	Toe-in (in.)
2009	Frontier	2WD	①	①	②	②	③
		4WD	④	④	⑤	⑤	③
2010	Frontier	2WD	①	①	②	②	③
		4WD	④	④	⑤	⑤	③

NOTE: On 2009-2010 vehicles, fuel, coolant and engine oil must be full. Spare tire, jack, hand tools and mats must be in place.

Some 2009-2010 vehicles may be equipped with non adjustable lower link bolts and washers. In order to adjust caster and camber on these vehicles, first replace these bolts with adjustable cam bolts and washers.

① Minimum: 2 degrees 15' (2.25 degrees)
  Nominal: 3 degrees 0' (3.00 degrees)
  Maximum: 3 degrees 45' (3.75 degrees)
② Minimum: 2 degrees 0' (2.00 degrees)
  Nominal: 0 degrees 15' (0.25 degrees)
  Maximum: 1 degrees 0' (1.00 degrees)
③ Minimum: 0.08 inches
  Nominal 0.12 inches
  Maximum 0.16 inches

④ Minimum: 0 degrees 15' (-0.25 degrees)
  Nominal: 2 degrees 45' (2.75 degrees)
  Maximum: 3 degrees 30' (3.50 degrees)
⑤ Minimum: 0 degrees 15' (-0.25 degrees)
  Nominal: 0 degrees 30' (0.50 degrees)
  Maximum: 1 degrees 15' (1.25 degrees)
⑥ Minimum: 0.12 inches
  Nominal 0.16 inches
  Maximum 0.20 inches

37663_FRON_C0010

TIRE, WHEEL AND BALL JOINT SPECIFICATIONS

Year	Model	OEM Tires Standard	OEM Tires Optional	Tire Pressures (psi) Front	Tire Pressures (psi) Rear	Wheel Size	Ball Joint Inspection	Lug Nut Torque (ft. lbs.)
2009	Frontier XE	P235/75R15	None	①	①	7J	②	98
	Frontier SE	P265/70R15	None	①	①	7J	②	98
	Frontier NISMO off road	P265/75R16	None	①	①	7J	②	98
	Frontier LE	P265/65R17	None	①	①	7.5J	②	98
2010	Frontier XE	P235/75R15	None	①	①	7J	②	98
	Frontier SE (King Cab)	P265/70R16	None	①	①	7J	②	98
	Frontier SE (Club Cab)	P265/65R17	None	①	①	7.5J	②	98
	Frontier NISMO off road	P265/75R16	None	①	①	7J	②	98
	Frontier LE	P265/65R17	None	①	①	7.5J	②	98

OEM: Original Equipment Manufacturer

PSI: Pounds Per Square Inch

① See placard on vehicle

② Replace if any measurable movement is found.

37663_FRON_C0011

BRAKE SPECIFICATIONS

All measurements in inches unless noted

Year	Model	Brake Disc Original Thickness	Brake Disc Minimum Thickness	Brake Disc Maximum Runout	Brake Drum Diameter Original Inside Diameter	Brake Drum Diameter Max. Wear Limit	Brake Drum Diameter Maximum Machine Diameter	Minimum Lining Thickness Front	Minimum Lining Thickness Rear	Brake Caliper Bracket Bolts (ft. lbs.)	Brake Caliper Mounting Bolts (ft. lbs.)
2009	Frontier	①	②	0.002	③	④	—	0.079	0.079	⑤	⑥
2010	Frontier	①	②	0.002	③	④	—	0.079	0.079	⑤	⑥

① 4-cyl.: 0.710
6-cyl.: 1.100

② 4-cyl.: 0.630
6-cyl.: 1.024

③ Rear disc brakes: 0.710

④ Rear disc brakes: 0.630

⑤ Front: 136 ft. lbs.
Rear: 76 ft. lbs.

⑥ Front: 32 ft. lbs.
Rear: 19 ft. lbs.

37663_FRON_C0012

SCHEDULED MAINTENANCE INTERVALS
Nissan Frontier

TO BE SERVICED	TYPE OF SERVICE	VEHICLE MILEAGE INTERVAL (x1000)												
		7.5	15	22.5	30	37.5	45	52.5	60	67.5	75	82.5	90	97.5
Engine oil & filter	R	✓	✓	✓	✓	✓	✓	✓	✓	✓	✓	✓	✓	✓
Brake lines & cables	S/I		✓		✓		✓		✓		✓		✓	
Brake pads, discs, drums & linings	S/I		✓		✓		✓		✓		✓		✓	
Driveshaft boots & propeller shaft	S/I				✓				✓				✓	
Front wheel bearings (4X2)	S/I				✓				✓				✓	
Front wheel bearings (4X4)	S/I				✓				✓				✓	
Automatic & manual transmission, transfer & differential gear oil ①	S/I		✓		✓		✓		✓		✓		✓	
Air cleaner filter	R				✓				✓				✓	
Engine coolant	R								✓				✓	
Spark plugs (platinum)	R	replace every 105,000 miles												
Drive belt(s)	S/I				✓				✓				✓	
Exhaust system	S/I				✓				✓				✓	
Fuel lines	S/I				✓				✓				✓	
Steering gear (box) & linkage, axle & suspension parts	S/I				✓				✓				✓	
Vapor lines	S/I				✓				✓				✓	
Tires (rotate)	S/I	✓	✓	✓	✓	✓	✓	✓	✓	✓	✓	✓	✓	✓
Timing belt ②	R													

R: Replace   S/I: Service or Inspect

① Differential (w/limited-slip differential) oil: replace oil every 30,000 miles, 2007-2008 vehicles.

② Timing belt: replace at 105,000 miles.

FREQUENT OPERATION MAINTENANCE (SEVERE SERVICE)

If a vehicle is operated under any of the following conditions it is considered severe service:

- Extremely dusty areas.

- 50% or more of the vehicle operation is in 32°C (90°F) or higher temperatures, or constant operation in temperatures below 0°C (32°F).

- Prolonged idling (vehicle operation in stop and go traffic).

- Frequent short running periods (engine does not warm to normal operating temperatures).

- Police, taxi, delivery usage or trailer towing usage.

Oil & oil filter: replace every 3750 miles.

Brake pads, discs, drums & linings: service or inspect every 7500 miles.

Driveshaft boots & propeller shaft: service or inspect every 7500 miles.

Exhaust system: service or inspect every 7500 miles.

Steering gear (box) & linkage, (steering damper-4X4), axle & suspension parts: service or inspect every 7500 miles.

Steering linkage ball joints & front suspension ball joints: service or inspect every 7500 miles.

37663_FRON_C0013

BRAKES   INFORMATION AND PRECAUTIONS

ANTI-LOCK SYSTEMS

- Certain components within the ABS system are not intended to be serviced or repaired individually.
- Do not use rubber hoses or other parts not specifically specified for and ABS system. When using repair kits, replace all parts included in the kit. Partial or incorrect repair may lead to functional problems and require the replacement of components.
- Lubricate rubber parts with clean, fresh brake fluid to ease assembly. Do not use shop air to clean parts; damage to rubber components may result.
- Use only DOT 3 brake fluid from an unopened container.
- If any hydraulic component or line is removed or replaced, it may be necessary to bleed the entire system.
- A clean repair area is essential. Always clean the reservoir and cap thoroughly before removing the cap. The slightest amount of dirt in the fluid may

plug an orifice and impair the system function. Perform repairs after components have been thoroughly cleaned; use only denatured alcohol to clean components. Do not allow ABS components to come into contact with any substance containing mineral oil; this includes used shop rags.

- The Anti-Lock control unit is a microprocessor similar to other computer units in the vehicle. Ensure that the ignition switch is **OFF** before removing or installing controller harnesses. Avoid static electricity discharge at or near the controller.
- If any arc welding is to be done on the vehicle, the control unit should be unplugged before welding operations begin.

DISC AND DRUM SYSTEMS

❊❊ CAUTION

Dust and dirt accumulating on brake parts during normal use may contain

asbestos fibers from production or aftermarket brake linings. Breathing excessive concentrations of asbestos fibers can cause serious bodily harm. Exercise care when servicing brake parts. Do not sand or grind brake lining unless equipment used is designed to contain the dust residue. Do not clean brake parts with compressed air or by dry brushing. Cleaning should be done by dampening the brake components with a fine mist of water, then wiping the brake components clean with a dampened cloth. Dispose of cloth and all residue containing asbestos fibers in an impermeable container with the appropriate label. Follow practices prescribed by the Occupational Safety and Health Administration (OSHA) and the Environmental Protection Agency (EPA) for the handling, processing, and disposing of dust or debris that may contain asbestos fibers.

BRAKES   BLEEDING THE BRAKE SYSTEM

BLEEDING PROCEDURE

BLEEDING PROCEDURE

1. Before servicing the vehicle, refer to the Precautions Section.

❊❊ CAUTION

While bleeding the brake system, pay attention to the master cylinder fluid level.

2. Disconnect the negative battery cable.
3. Raise and safely support the vehicle.
4. Attach a vinyl tube to the right, rear bleeder valve.
5. Depress the brake pedal fully 4 or 5 times.
6. With the brake pedal depressed, loosen the bleeder valve to let the air out, then tighten it immediately.

7. Repeat steps 3 and 4 until no more air comes out.
8. Tighten the bleeder valve.
9. Fill the master cylinder reservoir.
10. Repeat the above steps for the left front, left rear, and the right front calipers, in that order.

BLEEDING THE ABS SYSTEM

See Bleeding Procedure.

BRAKES ANTI-LOCK BRAKE SYSTEM (ABS)

WHEEL SPEED SENSORS

REMOVAL & INSTALLATION

See Figure 1.

1. Before servicing the vehicle, refer to the Precautions Section.
2. Disconnect the negative battery cable.
3. Raise and support the vehicle safely.
4. Remove the tire and wheel assembly.
5. Remove the rotor.
6. Remove the wheel speed sensor mounting bolts and clip.
7. Pull the sensor out, being careful to turn it as little as possible. Do not pull on the sensor harness.

To install:

8. Before installing the sensor be sure no foreign materials such as iron fragments are adhered to the pick-up part of the sensor or to the inside of the sensor mounting hole or on the rotor mounting surface.
9. Be sure the harness is not twisted when installed.

10. Continue the installation in the reverse order of the removal procedure.

WHEEL SPEED SENSOR RINGS (TOOTHED RINGS)

REMOVAL & INSTALLATION

Front Wheel

The front sensor rotor cannot be disassembled. It is integrated with the hub and bearing assembly. To replace the sensor rotor, replace the front hub and knuckle assembly. The rear sensor rotor cannot be disassembled. It is integrated with the hub and bearing assembly. To replace the sensor rotor, replace the rear hub and bearing assembly.

Rear Wheel

C200 Axle

1. Disconnect the negative battery cable.
2. Raise and support the vehicle safely.
3. Remove the axle shaft.
4. Remove the sensor rotor using tool J25852-B or equivalent.

To install:

5. Installation is the reverse of the removal procedure.
6. Pay attention to the direction of the sensor rotor.

M226 Axle

1. Disconnect the negative battery cable.
2. Raise and support the vehicle safely.
3. Remove the axle shaft.

➡ **It is necessary to disassemble the rear axle to replace the sensor rotor.**

4. Pull the sensor off of the axle shaft using tool ST30031000, or equivalent, and a press.

To install:

5. Installation is the reverse of the removal procedure.
6. Make sure that the sensor rotor is fully seated.

➡ **Never reuse the old sensor rotor. Never reuse the old oil seal.**

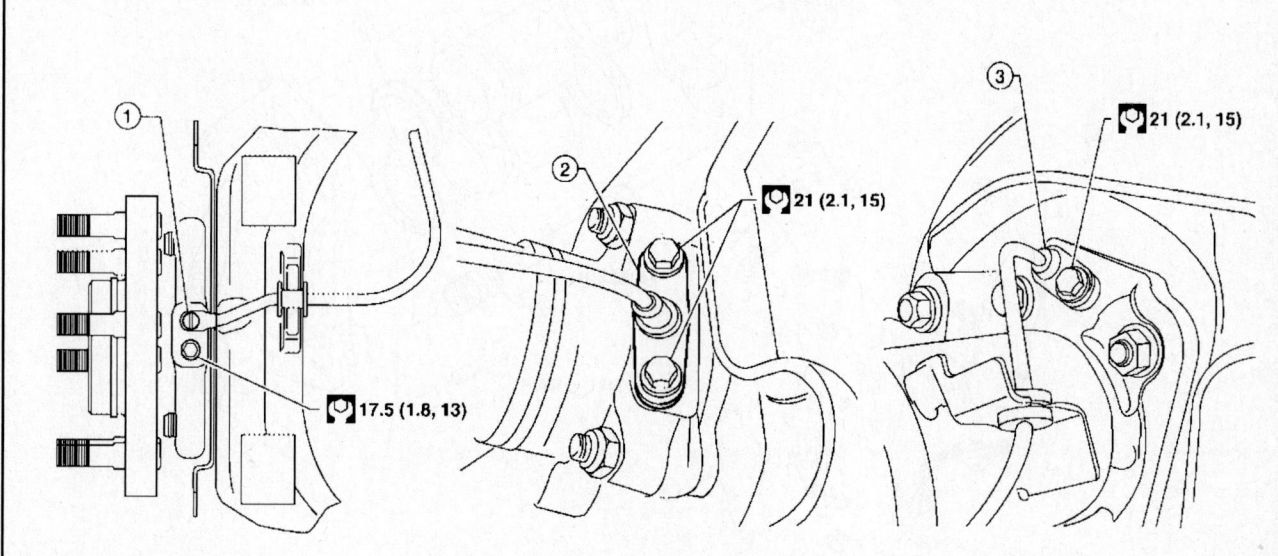

N·m (kg-m, ft-lb)

1. Front wheel sensor
2. Rear wheel sensor (C200)
3. Rear wheel sensor (M226)

42050_FRON_G0063

Fig. 1 Wheel speed sensor location

BRAKES   FRONT DISC BRAKES

BRAKE CALIPER

REMOVAL & INSTALLATION

See Figure 2.

1. Before servicing the vehicle, refer to the Precautions Section.
2. Drain the brake fluid, as necessary.
3. Raise the vehicle and support safely.
4. Remove the tire and wheel assembly.

5. Remove the bolt attaching the brake hose to the caliper. Plug the brake hose to prevent brake fluid loss.
6. Remove the caliper support mounting bolts and lift the caliper assembly from the knuckle.

To install:

7. Position the caliper assembly onto the knuckle and install the bolts. Make sure the rotor fits between the brake pads. Torque the bolts to specification.

8. Using new copper washers, connect the brake hose to the caliper. Torque the brake hose attaching bolt to specification.
9. Bleed the brake system.
10. Apply the brake pedal and inspect the system. Ensure proper operation and no leakage.
11. Install tire and wheel assembly. Lower the vehicle and road test.

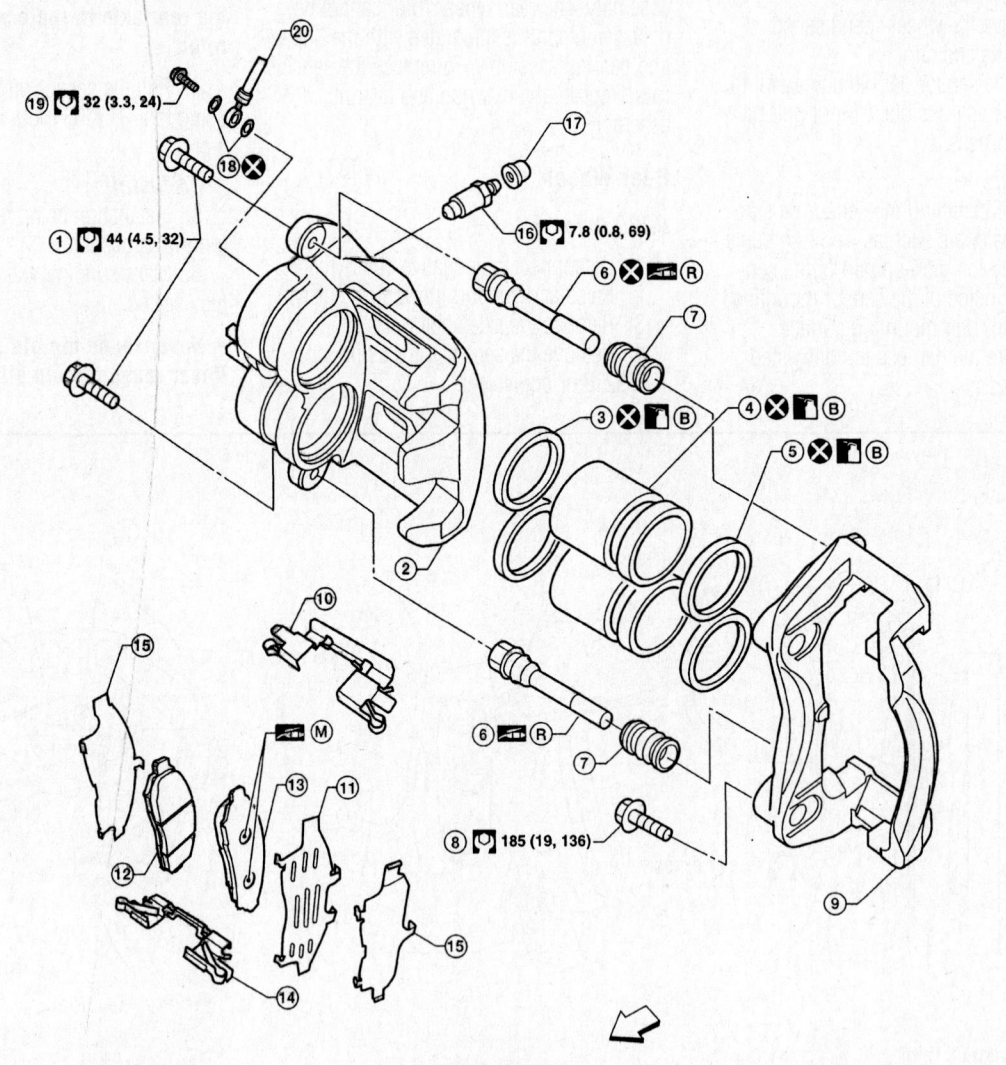

1.	Sliding pin bolt	2.	Cylinder body	3.	Piston seal
4.	Piston	5.	Piston boot	6.	Sliding pin
7.	Sliding pin boot	8.	Torque member bolt	9.	Torque member
10.	Pad retainer	11.	Inner shim	12.	Inner brake pad
13.	Outer brake pad	14.	Pad retainer	15.	Outer shim
16.	Bleed valve	17.	Cap	18.	Copper washers
19.	Union bolt	20.	Brake hose		

09482_FRON_G0121

Fig. 2 Front disc brake and related components

DISC BRAKE PADS

REMOVAL & INSTALLATION

1. Before servicing the vehicle, refer to the Precautions Section.
2. Drain the brake fluid, as necessary.

BRAKES

BRAKE CALIPER

REMOVAL & INSTALLATION

See Figure 3.

1. Before servicing the vehicle, refer to the Precautions Section.
2. Drain the brake fluid, as necessary.
3. Raise the vehicle and support safely.
4. Remove the tire and wheel assembly.
5. Remove the union bolt and brake hose. Remove the sliding pin bolts. Remove the caliper from the vehicle.

3. Raise the vehicle and support safely.
4. Remove the bottom pin from the caliper and swing the caliper cylinder body upward; support the caliper with a wire.
5. Remove the brake pad retainers, shims and the pads.

To install:
6. Installation is the reverse of the removal procedure.
7. Bleed the brake system.
8. Apply the brake pedal and inspect the system. Ensure proper operation and no leakage.
9. Install tire and wheel assembly. Lower the vehicle and road test.

DISC BRAKE PADS

REMOVAL & INSTALLATION

1. Before servicing the vehicle, refer to the Precautions Section.

To install:
6. Compress the piston of the disc brake caliper.
7. Install the brake pads and caliper assembly.

REAR DISC BRAKES

2. Drain the brake fluid, as necessary.
3. Raise the vehicle and support safely.
4. Remove the tire and wheel assembly.
5. Remove the top bolt from the caliper.
6. Swing the caliper open and remove the pads.

To install:
7. Compress the piston of the disc brake caliper.
8. Install the brake pads and caliper assembly. Tighten caliper bolts to 24 ft. lbs. (32 Nm).

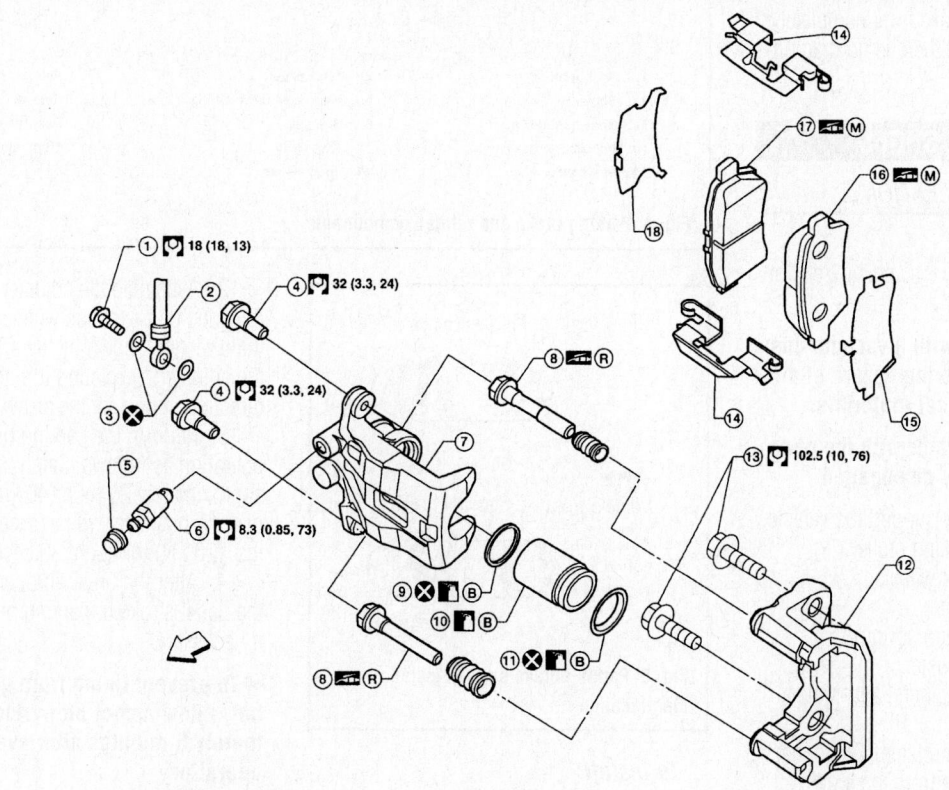

1.	Union bolt	2.	Brake hose	3.	Copper washers
4.	Sliding pin bolt	5.	Cap	6.	Bleed valve
7.	Cylinder body	8.	Sliding pin	9.	Piston seal
10.	Piston	11.	Piston boot	12.	Torque member
13.	Torque member bolt	14.	Pad retainer	15.	Outer shim
16.	Outer brake pad	17.	Inner brake pad	18.	Inner shim

09482_FRON_G0122

Fig. 3 Rear disc brake and related components

BRAKES | **PARKING BRAKE**

PARKING BRAKE CABLES

ADJUSTMENT

1. Remove rear half of the center console.

2. Rotate adjusting nut and loosen cable until tension is sufficiently released.

3. Remove the wheel and tire using power tool.

4. Remove the rotor and measure inner diameter at widest point.

5. Transfer measurement less 0.6 mm to the parking brake shoes and adjust accordingly.

6. Using wheel nuts, secure the disc to the hub to prevent it from tilting.

7. Rotate disc rotor to make sure there is no drag.

8. Operate parking brake lever 10 or more times with a force of 110 lbs. (490 N).

9. Rotate adjusting nut to adjust lever stroke to 6–8 notches with 44 lbs. (196 N) force.

10. With parking brake lever completely disengaged, make sure there is no drag on the parking brake.

PARKING BRAKE SHOES

REMOVAL & INSTALLATION

See Figures 4 and 5.

1. Before servicing the vehicle, refer to the Precautions Section.

➡ **Clean the brakes with a vacuum dust collector to minimize the hazard of airborne particles or other materials.**

➡ **Remove the disc rotor with the parking brake completely disengaged.**

2. Raise and safely support the vehicle.
3. Release the parking brake.
4. Remove the rear wheels.
5. Remove the rotor.
6. Remove the return springs.
7. Remove the adjuster.
8. Remove the, retainers, anti-rattle pins and shoes.
9. Remove the pin retainer. Disconnect the parking brake cable from the toggle lever.
10. Remove the back plate.

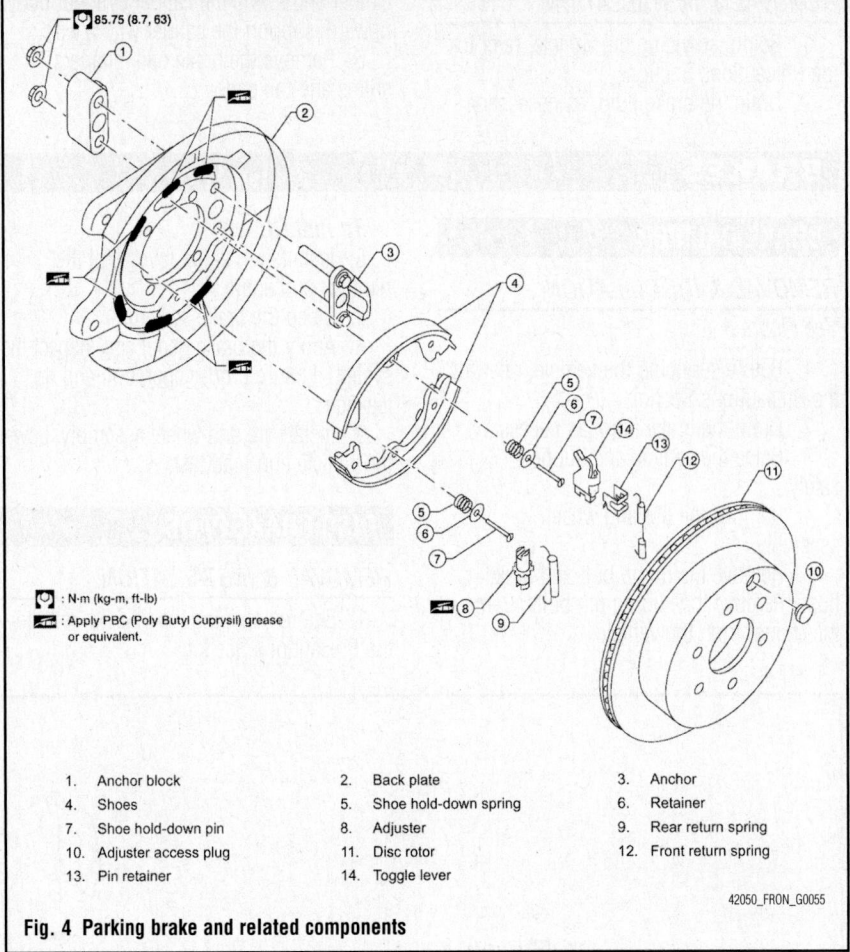

85.75 (8.7, 63)

: N·m (kg-m, ft-lb)

: Apply PBC (Poly Butyl Cuprysil) grease or equivalent.

1.	Anchor block	2.	Back plate	3.	Anchor
4.	Shoes	5.	Shoe hold-down spring	6.	Retainer
7.	Shoe hold-down pin	8.	Adjuster	9.	Rear return spring
10.	Adjuster access plug	11.	Disc rotor	12.	Front return spring
13.	Pin retainer	14.	Toggle lever		

42050_FRON_G0055

Fig. 4 Parking brake and related components

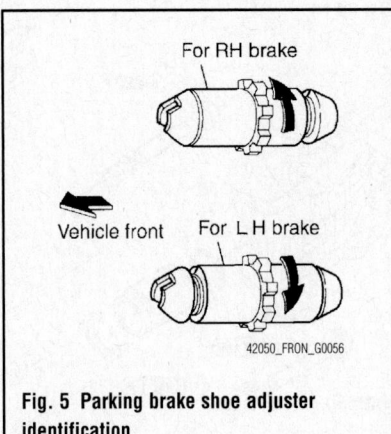

For RH brake

Vehicle front   For L H brake

42050_FRON_G0056

Fig. 5 Parking brake shoe adjuster identification

To install:

11. Installation is the reverse of the removal procedure.

12. Assemble the adjuster so that the threaded part expands when rotating it in the direction shown by the arrow. Shorten the adjuster by rotating it in the opposite direction shown by the arrow.

13. Perform the parking brake break-in operation as follows: Safely, drive forward at approximately 25 mph (40 km/h) with the parking brake set with a force of approx. 45 lbs. (200 N) for about 30 seconds.

14. After the break-in operation, check the pedal stroke of parking brake. Readjust if necessary.

➡ **To prevent lining from getting too hot, allow a cool off period of approximately 5 minutes after every break-in operation.**

15. Check and adjust the parking brake pedal stroke. Correct as required.

CHASSIS ELECTRICAL   AIR BAG (SUPPLEMENTAL RESTRAINT SYSTEM)

✳✳ CAUTION

These vehicles are equipped with an air bag system. The system must be disarmed before performing service on, or around, system components, the steering column, instrument panel components, wiring and sensors. Failure to follow the safety precautions and the disarming procedure could result in accidental air bag deployment, possible injury and unnecessary system repairs.

GENERAL INFORMATION

PRECAUTIONS

Disconnect and isolate the battery negative cable before beginning any airbag system component diagnosis, testing, removal, or installation procedures. Allow system capacitor to discharge for two minutes before beginning any component service. This will disable the airbag system. Failure to disable the airbag system may result in accidental airbag deployment, personal injury, or death.

DISARMING THE SYSTEM

To disarm the SRS system turn the ignition switch to the off position. Then, disconnect both battery cables starting with the negative cable first and wait at least 3 minutes after the cables are disconnected.

ARMING THE SYSTEM

To rearm the SRS system, turn the ignition switch to the off position. Connect both battery cables starting with the positive cable first.

CLOCKSPRING CENTERING

1. Be sure to align the spiral cable correctly when installing the steering wheel. Make sure that the spiral cable is in the neutral position.

➡The neutral position is detected by turning to the left 2.6 revolutions from the right end position and ending with the knob at the top. The spiral cable may snap due to steering operation if the cable is installed incorrectly. Also, with the steering linkage disconnected the cable may snap by turning the steering wheel beyond the limited number of turns (2.6 from the neutral position to both the left and right).

DRIVE TRAIN

CLUTCH DRIVEN DISC & PRESSURE PLATE

REMOVAL & INSTALLATION

See Figures 6 through 8.

1. Before servicing the vehicle, refer to the Precautions Section.
2. Remove or disconnect the following:
   - Negative battery cable
   - Transmission
   - Pressure plate. Loosen the bolts evenly in ½ turn steps.
   - Clutch disc

To install:
3. Install or connect the following:
   - Clutch disc and pressure plate.

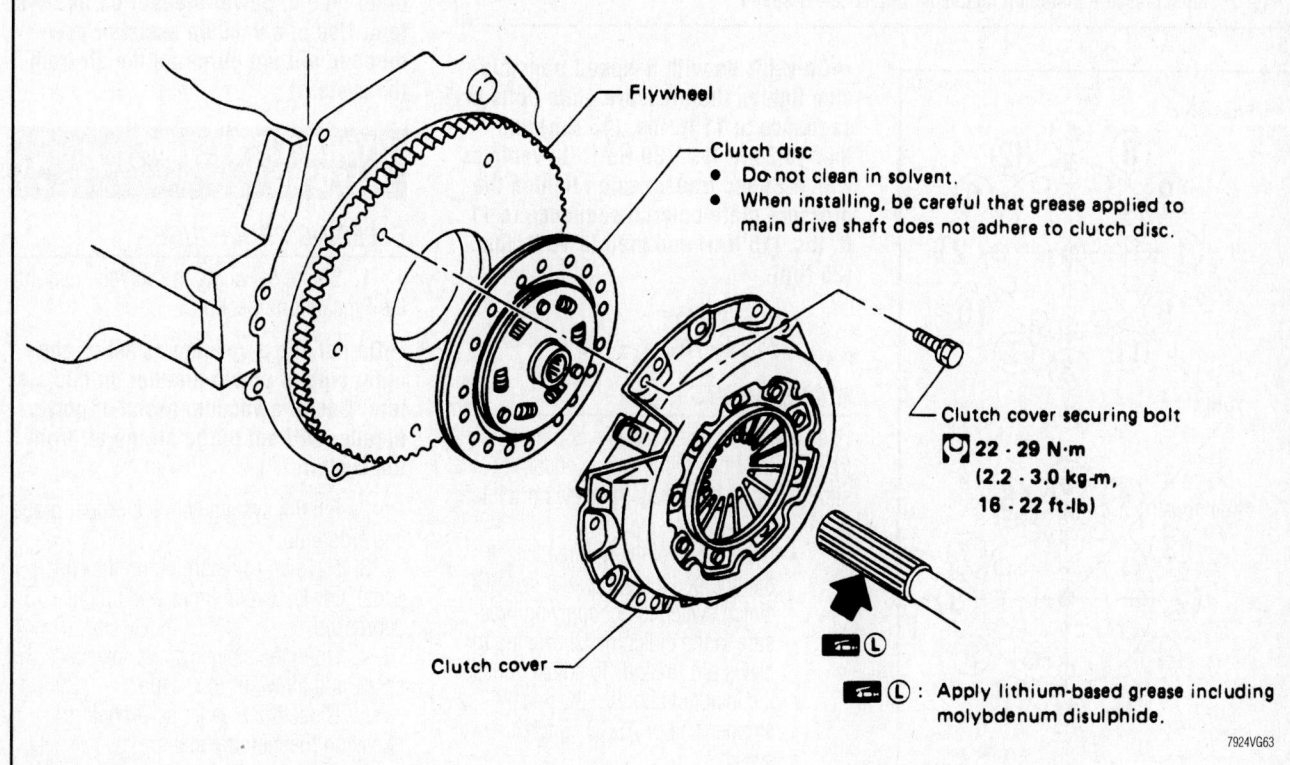

Fig. 6 Clutch and related components

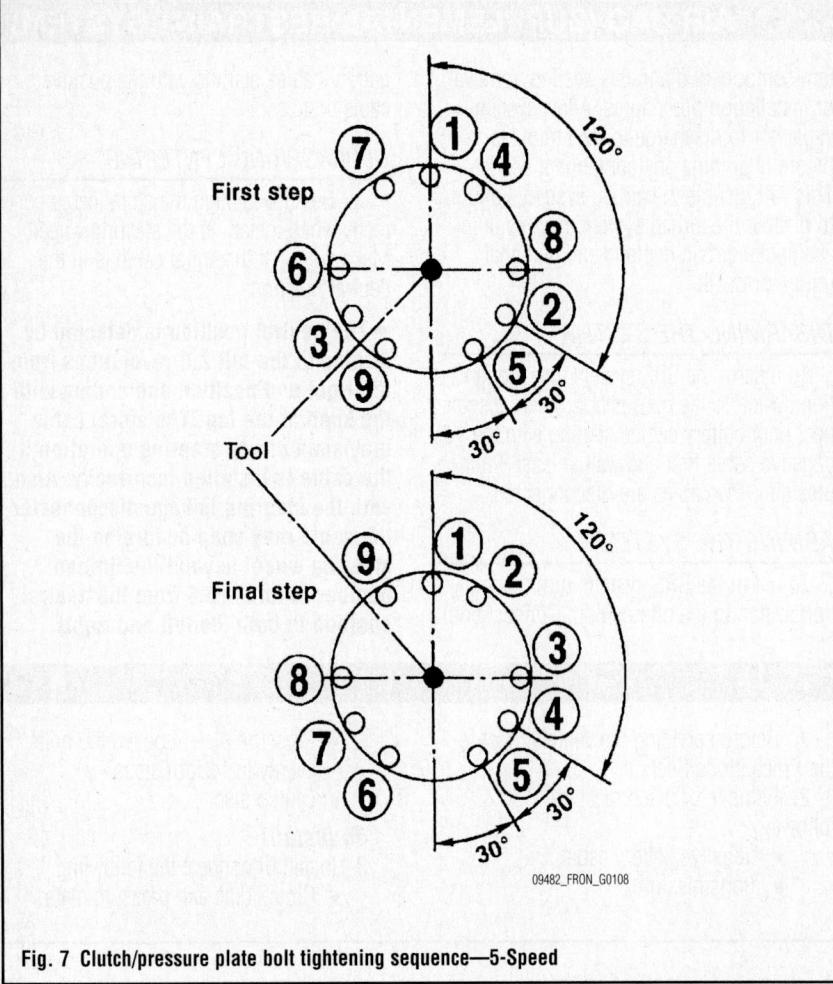

First step

Tool

Final step

09482_FRON_G0108

Fig. 7 Clutch/pressure plate bolt tightening sequence—5-Speed

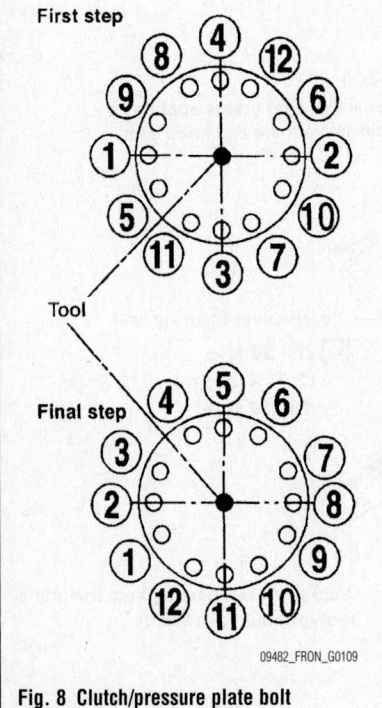

First step

Tool

Final step

09482_FRON_G0109

Fig. 8 Clutch/pressure plate bolt tightening sequence—6-Speed

➡On vehicles with 5–speed transmission tighten the pressure plate bolts in sequence to 11 ft. lbs. (15 Nm) and then to 29 ft. lbs. (39 Nm). On vehicles with 6-Speed transmission tighten the pressure plate bolts in sequence to 11 ft. lbs. (15 Nm) and then to 19 ft. lbs. (26 Nm).

- Transmission
- Negative battery cable

ADJUSTMENTS

1. Check to see if the clevis pin floats freely in the bore of the clutch pedal. It should not be bound by the clevis or clutch pedal.

- If the clevis pin is not free, check that the pedal stopper bolt or ASCD clutch switch is not applying pressure to the clutch pedal causing the clevis pin to bind. To adjust, loosen the lock nut and turn the pedal stopper bolt or ASCD clutch switch.
- Tighten the lock nut to 10 ft. lbs. (13 Nm)

- Verify that the clevis pin floats in the bore of the clutch pedal. It should not be bound by the clutch pedal.
- If the clevis pin is still not free, remove the clevis pin and check for deformation or damage. Replace clevis pin if necessary. Leave pin removed for step 2.

2. Check clutch pedal stroke for free range of movement.

- With the clevis pin removed, manually move the clutch pedal up and down to determine if it moves freely.
- If any sticking is noted, replace the assembly.

3. Adjust clearance "C" while depressing clutch pedal fully. (With clutch interlock switch).

4. Check clutch hydraulic and system components (clutch master cylinder, clutch operating cylinder, clutch withdrawal lever and clutch release bearing) for sticking or binding.

- If any sticking or binding noted, repair or replace related parts as necessary.
- If hydraulic system repair was necessary, bleed the clutch hydraulic system. Refer to bleeding.

➡Do not use a vacuum assist or any other type of power bleeder on this system. Use of a vacuum assist or power bleeder will not purge all the air from the system.

HYDRAULIC SYSTEM BLEEDING

BLEEDING PROCEDURE

1. Before servicing the vehicle, refer to the Precautions Section.

➡Do not use a vacuum assist or any other type of power bleeder on this system. Use of a vacuum assist or power bleeder will not purge all the air from the system.

2. Fill the system with the proper grade and type fluid.

3. Have an assistant pump the clutch pedal slowly several times and hold it depressed.

4. Open the slave cylinder bleeder screw and allow air to escape.

5. Close the bleeder screw before releasing the clutch pedal.

6. Repeat until all air is purged from the clutch hydraulic system.

7. Refill the reservoir to the full mark.

Do not spill brake fluid on painted surfaces. If it spills, wipe up immediately and wash the affected area with water.

➡**Do not use a vacuum assist or any other type of power bleeder on this system. Use of a vacuum assist or power bleeder will not purge all the air from the system. Monitor the fluid level in the reservoir tank to make sure it does not empty.**

8. Top off reservoir with new brake fluid.
9. Connect a transparent vinyl tube and container to the air bleeder valve on the clutch operating cylinder.
10. Fully depress the clutch pedal several times.
11. With the clutch pedal depressed, open the bleeder valve to release the air.
12. Close the bleeder valve.
13. Repeat steps 3 to 5 until clear brake fluid comes out of the air bleeder valve.
14. Tighten the air bleeder to 70 inch lbs. (7.9 Nm).

FRONT AXLE SHAFT, BEARING & SEAL

REMOVAL & INSTALLATION

1. Before servicing the vehicle, refer to the Precautions Section.
2. Raise and support the vehicle safely. Remove the tire and wheel assembly.
3. Remove the rear engine cover.
4. Remove the wheel sensor harness from the mount on the knuckle. Disconnect the harness connector.

➡**Do not pull on the wheel sensor harness.**

5. Remove the wheel hub and bearing assembly.

➡**It is not necessary to remove the wheel speed sensor from the wheel hub when the wheel hub is not being replaced. Carefully feed the sensor harness through the hole in the splash shield.**

6. Separate the upper link ball joint stud from the steering knuckle using tool ST29020001 (J-24319-01) or equivalent.
7. Remove the halfshaft assembly from the vehicle by prying the halfshaft from the front final drive using the proper tool.

To install:

8. Installation is the reverse of the removal procedure.

9. Be sure to use a new differential side oil seal.

REAR AXLE SHAFT, BEARING & SEAL

REMOVAL & INSTALLATION
See Figure 9.

1. Before servicing the vehicle, refer to the Precautions Section.
2. Remove or disconnect the following:
 • Rear wheel and tire assembly
 • Wheel speed sensor
 • Brake rotor
 • Brake caliper assembly
 • Parking brake cable
 • Brake fluid line
 • Bearing cage and backing plate bolts
 • Axle shaft assembly
 • Axle seal
 • Wheel speed sensor rotor
 • Snap ring and shim washer
 • Bearing ring retainer
 • back plate and torque member
 • Axle bearing studs
 • Wheel bearing
 • Grease catcher

To install:

➡**Use new seals, bearings, circlips and snap rings for assembly.**

3. Install grease catcher
4. Install the wheel studs through the grease catcher into the axle shaft using a suitable press.

➡**All six wheel studs must be pressed on at the same time and are flush with the grease catcher when installed.**

5. Position the axle bearing on the back plate and torque member.
6. Install the axle bearing studs using a suitable press to attach the axle bearing to the back plate and torque member.

➡**Always replace the axle bearing with a new one.**

7. Install the back plate and torque member, new axle bearing and new bearing ring retainer on the axle shaft using a suitable press. Do not exceed 11 tons force.
8. Press the new bearing ring retainer on the axle shaft with the taper side positioned toward press.

➡**Always replace the bearing ring retainer with a new one.**

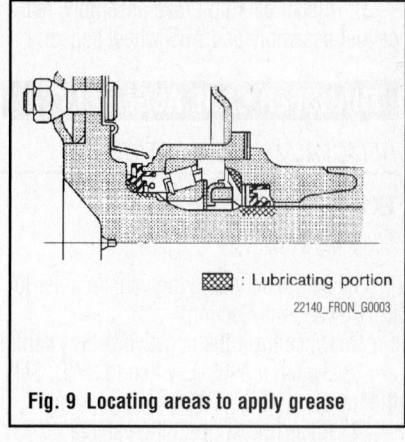

▨▨ : Lubricating portion

22140_FRON_G0003

Fig. 9 Locating areas to apply grease

9. Select the correct size shim washer. Select the size of shim washer so that the installed snap ring to shim washer clearance is 0.008 inch or less.
10. Install a new snap ring on the axle shaft.
11. Do not over spread the snap ring when installing, measure the outer diameter of the snap ring after installation and replace if the snap ring outer diameter exceeds 1.87 inch maximum.
12. Check the snap ring to shim washer clearance. Repeat previous steps as necessary.
13. Perform break-in rotation of the wheel bearing.
 • Rotate the wheel bearing in the forward direction for a minimum of 10 revolutions at 50–70 RPM.
 • Rotate the wheel bearing in the reverse direction for a minimum of 10 revolutions at 50–70 RPM.
14. Measure the rotational torque of the wheel bearing. Rotational torque should be 16 inch lbs. (1.8 Nm) at 8–12 RPM.
15. Inspect that the wheel bearing is free from axial play relative to the axle shaft.
16. Install a new ABS sensor rotor on the axle shaft with notch side away from press using a suitable press.

➡**Always replace the ABS sensor rotor with a new one.**

17. Install new axle seal in housing.
18. Apply multi-purpose grease to the recess of axle case end as shown in illustration.
19. Insert tool J-34296 into the new axle oil seal as a guide. Ensure tool ends do not overlap.
20. Insert the axle shaft assembly. Tighten the axle shaft nuts evenly in a criss-cross pattern to specification. Remove the tool when the axle shaft assembly is approximately 90 percent inserted to protect the new axle oil seal.

21. Install parking brake assembly, rear caliper assembly and ABS wheel sensor.

TRANSFER CASE ASSEMBLY

REMOVAL & INSTALLATION

TX15B

See Figure 10.

1. Before servicing the vehicle, refer to the Precautions Section.
2. Disconnect the negative battery cable.
3. Switch the 4WD switch to 2WD. Set the transfer case to 2WD.
4. Raise and support the vehicle safely.
5. Remove the undercovers. Drain the transfer case fluid.
6. Remove the center exhaust tube and main muffler.
7. Remove the front and rear driveshafts. Install plug in rear oil seal.
8. Remove the transmission-to-crossmember bolts. Properly support the transmission and transfer case assembly, using a suitable jack.
9. Remove the transmission crossmember.

➡**Support the transmission and transfer case using two suitable jacks while removing the transmission crossmember.**

10. Disconnect the ATP electrical connector, the 4LO switch connector, the wait detection switch and the transfer control device.
11. Disconnect each air breather hose from the transfer control device and the breather tube.
12. Remove the transfer case to transmission retaining bolts.

➡**Support the transmission and transfer case, using a suitable jack.**

13. Remove the transfer case from the vehicle.

To install:

14. Installation is the reverse of the removal procedure.

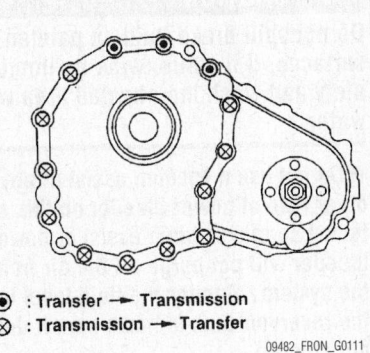

: Transfer → Transmission
: Transmission → Transfer

09482_FRON_G0111

Fig. 10 Transfer case bolt tightening sequence

15. Tighten the transfer case to transmission retaining bolts to specification and in the proper sequence. Specification is 27 ft. lbs. (36.6 Nm).
16. Start the engine and check for leaks, correct as required.

ENGINE COOLING

ENGINE FAN

REMOVAL & INSTALLATION

Electric

1. Remove air dam using power tool.
2. Remove engine front undercover using power tool.
3. Partially drain engine coolant from radiator.
4. Release the radiator shroud (lower) from the radiator shroud (upper) and position aside. Release the tabs, pull radiator shroud (lower) rearwards and down.
5. Remove air duct.
6. Remove reservoir tank hose from shroud.
7. Remove the radiator shroud (upper) bolts and remove the radiator shroud (upper).
8. Disconnect fan motor harness connector.
9. Remove bolt and fan grille and motor assembly.

To install:

10. Installation is the reverse of the removal procedure.
11. Start the engine and check for leaks, correct as required.

Mechanical (Belt Driven)

1. Remove air dam using power tool.

2. Remove engine front undercover using power tool.
3. Partially drain engine coolant from radiator.

✳✳ CAUTION

Perform this step when engine is cold. Do not spill engine coolant on drive belts.

4. Remove air duct.
5. Remove reservoir tank hose from shroud.
6. Removal radiator hose (upper) from radiator.

✳✳ CAUTION

Be careful not to allow engine coolant to contact drive belts.

7. Release the radiator shroud (lower) from the radiator shroud (upper) and position aside. Release the tabs, pull radiator shroud (lower) rearwards and down.
8. Remove the radiator shroud (upper) bolts and remove the radiator shroud (upper).
9. Remove the drive belt.
10. Remove the engine cooling fan.

To install:

11. Installation is the reverse of the removal procedure.

12. Start the engine and check for leaks, correct as required.

RADIATOR

REMOVAL & INSTALLATION

1. Remove air dam using power tool.
2. Remove engine undercover using power tool.
3. Drain engine coolant from radiator.

✳✳ CAUTION

Perform this step when engine is cold. Do not spill engine coolant on drive belts.

4. Remove air duct and air cleaner case assembly.
5. Remove reservoir tank hose.
6. Remove radiator hoses (upper and lower).

✳✳ CAUTION

Be careful not to allow engine coolant to contact drive belts.

7. Disconnect A/T fluid cooler hoses and plug.
8. Remove radiator shroud (lower).
9. Remove radiator shroud (upper).

10. Remove radiator cooling fan assembly.

11. Remove the radiator mounting bracket bolts.

12. Remove the two A/C condenser bolts.

13. Remove radiator as follows:

✳✳ CAUTION

Do not damage or scratch A/C condenser and radiator core when removing.

- With lifting and pulling radiator in a rear direction, disassemble mounting rubber (lower) from radiator core support center

- Because A/C condenser is attached to the front-lower portion of radiator, moving it in the rear direction should be at a minimum

- Lift A/C condenser up and remove radiator after disengaging the fitting as front-bottom surface

✳✳ CAUTION

Lifting A/C condenser should be minimum to prevent a load to A/C piping.

- After removing radiator, put A/C condenser on radiator core support center to prevent a load to A/C piping, and temporarily tie it in place

To install:

14. Installation is the reverse of the removal procedure.

15. Start the engine and check for coolant and transmission fluid leaks, correct as required.

THERMOSTAT

REMOVAL & INSTALLATION

2.5L Engine

See Figures 11 and 12.

➡**Never remove the radiator cap when the engine is hot. Serious burns could occur from high-pressure engine coolant escaping from the radiator.**

1. Be sure the engine is cold.
2. Disconnect the negative battery cable.
3. Drain the coolant. Properly disposed of used coolant.
4. Disconnect the lower radiator hose at the water inlet side (engine side).
5. Remove the water inlet retaining bolts.

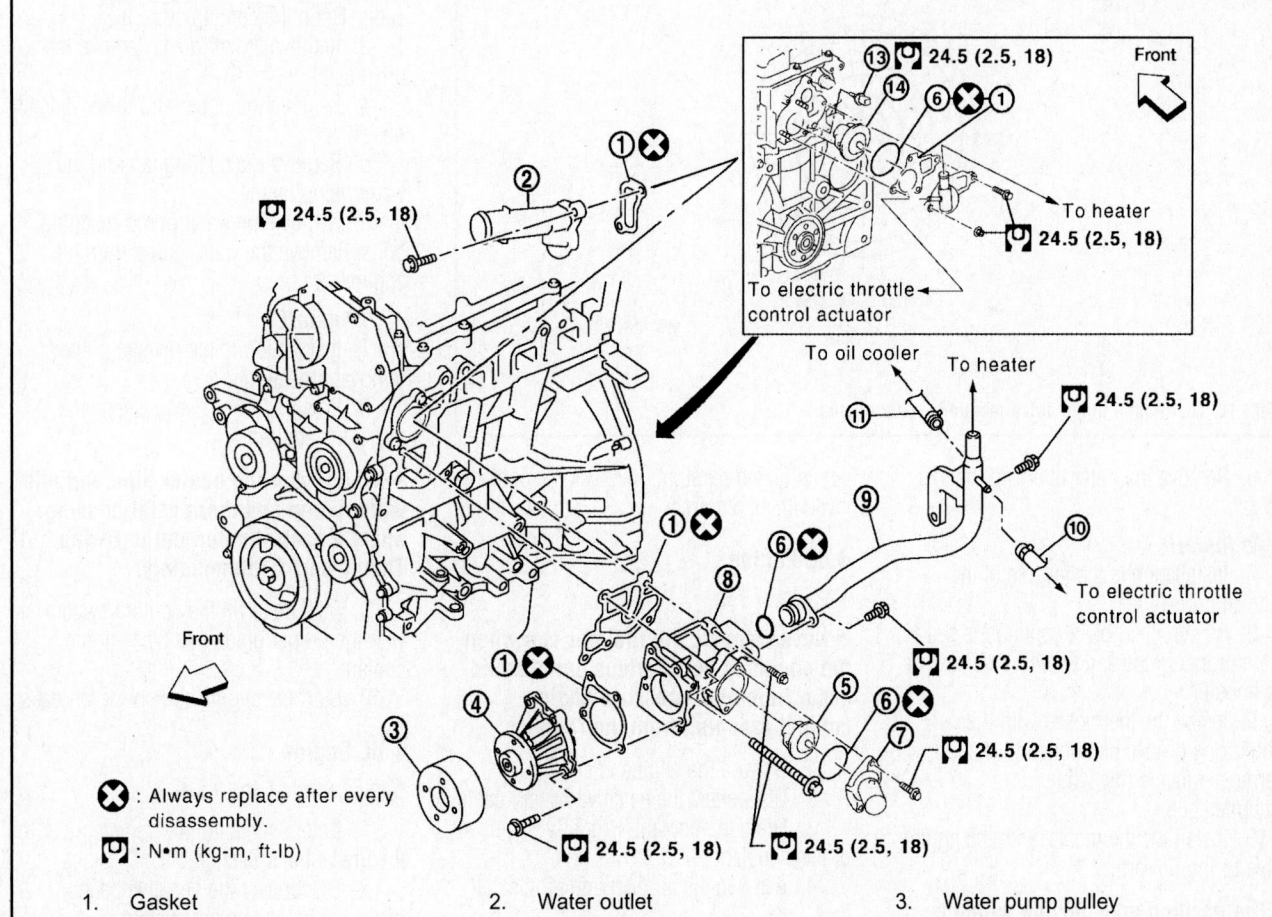

🔧 : Always replace after every disassembly.

🔩 : N•m (kg-m, ft-lb)

1. Gasket
2. Water outlet
3. Water pump pulley
4. Water pump
5. Thermostat
6. O-ring
7. Water inlet
8. Water pump and thermostat housing
9. Heater pipe
10. Water hose
11. Water hose
12. Heater outlet
13. Engine coolant temperature sensor
14. Water control valve

42050_FRON_G0030

Fig. 11 Thermostat assembly and related components—2.5L engine

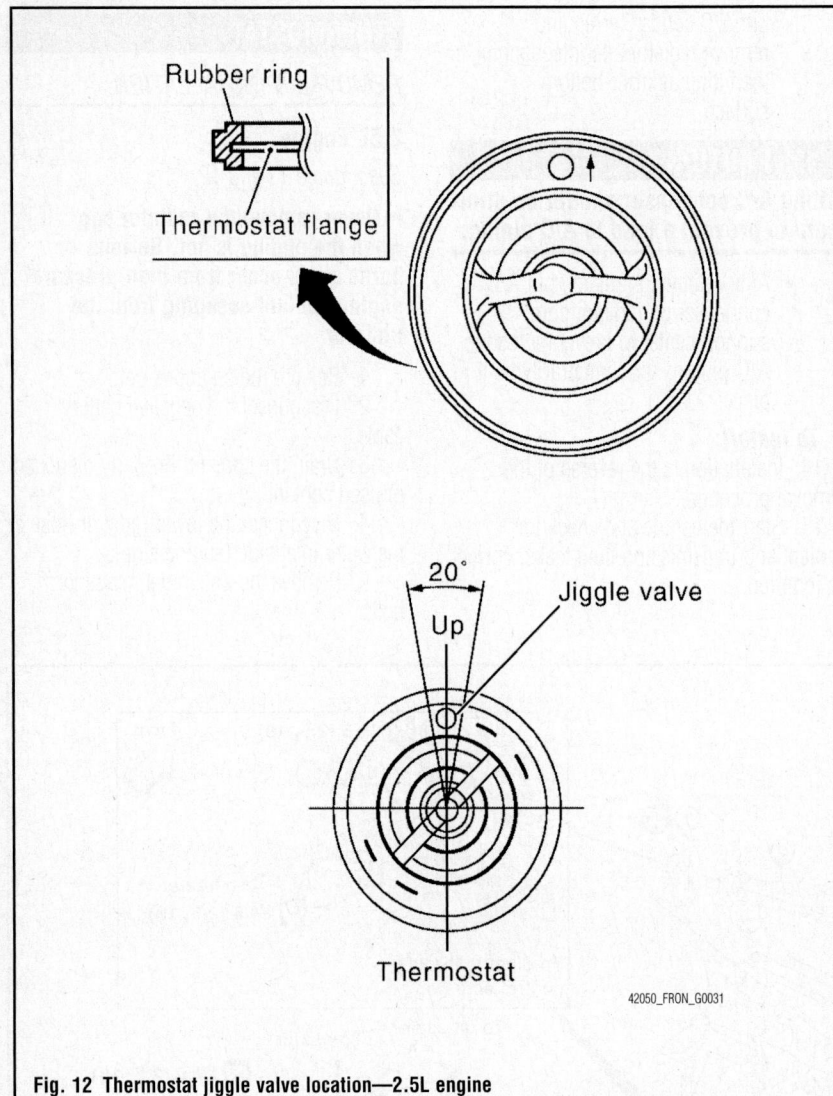

Fig. 12 Thermostat jiggle valve location—2.5L engine

6. Remove the water inlet and the thermostat.

To install:

7. Installation is the reverse of the removal procedure.

8. Be sure to apply a continuous bead of the proper grade and type RTV sealant to the housing.

9. Install the thermostat with the rubber ring groove positioned to fit the thermostat flange (the whole circumference).

10. Install the thermostat with the jiggle valve facing upward.

➡**The position may deviate within a range of 20 degrees.**

11. Be sure to refill the cooling using the proper grade and type engine coolant.

12. Start the engine and check for leaks.

13. Start the engine and allow it to reach operation temperature. Recheck the coolant level, fill as required.

4.0L Engine

See Figure 13.

➡**Never remove the radiator cap when the engine is hot. Serious burns could occur from high-pressure engine coolant escaping from the radiator.**

1. Be sure the engine is cold.
2. Disconnect the negative battery cable.
3. Drain the coolant. Properly disposed of used coolant.
4. Remove the air duct and air cleaner case.
5. Disconnect radiator hose (lower) and oil cooler hose from water inlet and thermostat assembly.
6. Remove the water inlet retaining bolts.
7. Remove the water inlet and the thermostat.

➡**Do not disassemble the water inlet and thermostat assembly. Replace them as a unit, if required.**

To install:

8. Installation is the reverse of the removal procedure.

9. Be sure to refill the cooling using the proper grade and type engine coolant.

10. Start the engine and check for leaks.

11. Start the engine and allow it to reach operation temperature. Recheck the coolant level, fill as required.

WATER PUMP

REMOVAL & INSTALLATION

2.5L Engine

See Figure 11.

1. Before servicing the vehicle, refer to the Precautions Section.
2. Disconnect the negative battery cable. Drain the cooling system.
3. Remove the air duct. Remove the drive belt.
4. Remove the upper and lower radiator hoses.
5. Remove the cooling fan and the water pump pulley.
6. Remove the water pump retaining bolts. Remove the water pump from the engine.

To install:

7. Installation is the reverse of the removal procedure.

8. Be sure to use new gaskets and O-rings, as required.

➡**When inserting heater pipe end into water pump and thermostat housing, apply a neutral detergent to O-ring. Then insert it immediately.**

9. Be sure to fill the cooling system with the proper grade and type engine coolant.

10. Start the engine and check for leaks.

4.0L Engine

See Figures 14 and 15.

1. Before servicing the vehicle, refer to the Precautions Section.
2. Disconnect the negative battery cable. Drain the cooling system.
3. Remove the air dam and undercover.
4. Remove air duct and resonator.
5. Remove the drive belts.
6. Remove the radiator upper hose. Remove the cooling fan.
7. Remove the chain tensioner cover and water pump cover from the front timing

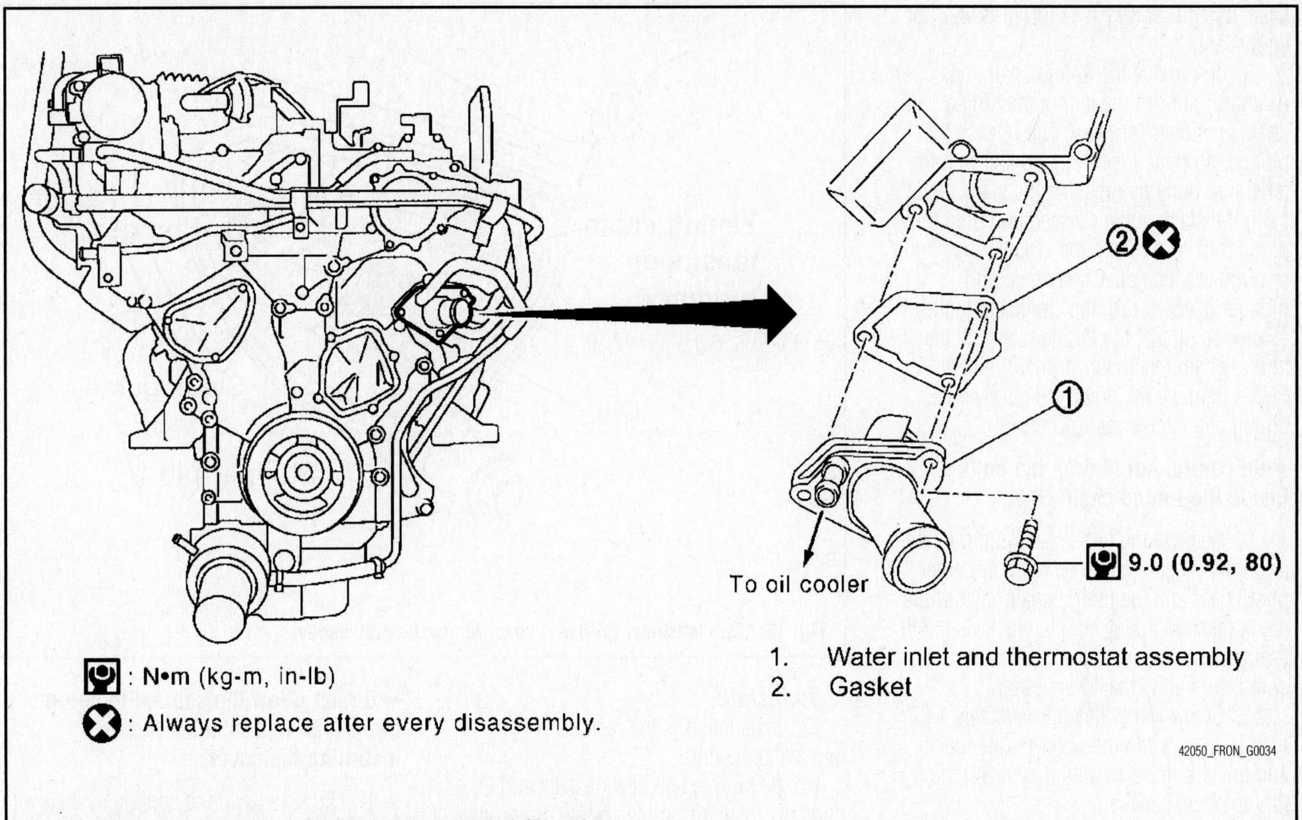

🔧 : N•m (kg-m, in-lb)

❌ : Always replace after every disassembly.

1. Water inlet and thermostat assembly
2. Gasket

To oil cooler

🔧 9.0 (0.92, 80)

42050_FRON_G0034

Fig. 13 Thermostat assembly and related components—4.0L engine

🔧 9.6 (0.98, 85)

🔧 8.1 (0.83, 72)

🛢 : Lubricate with new engine oil.

✏ : Apply Genuine RTV Silicone Sealant or equivalent. Refer to GI section.

❌ : Always replace after every disassembly.

🔧 : N•m (kg-m, in-lb)

🔧 : N•m (kg-m, ft-lb)

🔧 11.3 (1.2, 8)

④ ✏ 🔧 9.8 (1.0, 87)

⑤ ✏

🔧 11.3 (1.2, 8)

1. Water pump
2. Timing chain tensioner (primary)
3. Chain tensioner cover
4. Water drain plug (front)
5. Water pump cover
6. O-ring
7. O-ring

09482_FRON_G0016

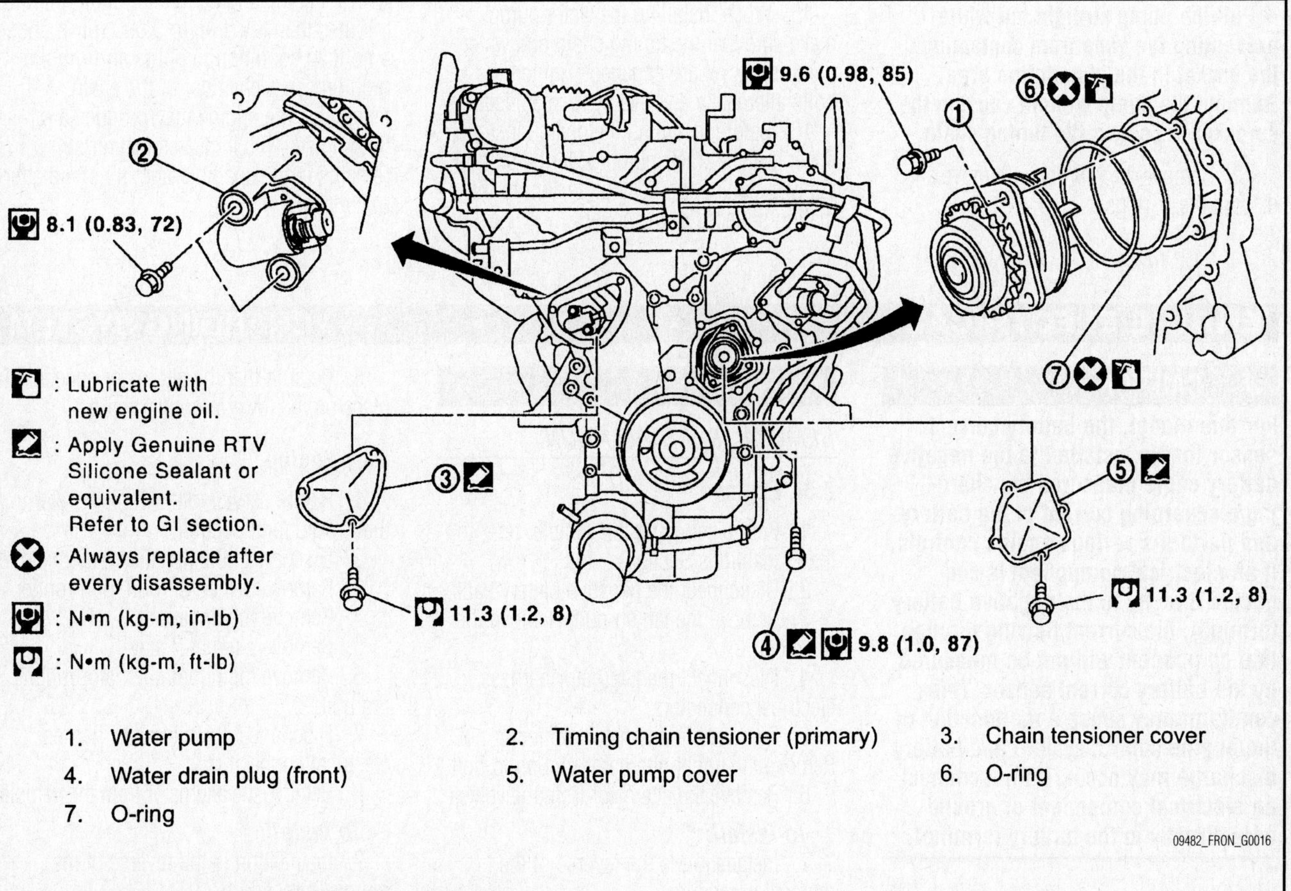

Fig. 14 Water pump and related components—4.0L engine

case, using tool KV10111100 (J-37228) or equivalent.

8. To remove the timing chain tensioner (primary), loosen the clip of the timing chain tensioner (primary) and release the plunger stopper. Insert the plunger into the tensioner body by pressing the slack guide. Keep the slack guide pressed and hold the plunger in by pushing the stopper pin through the tensioner body hole and plunger groove. Turn the crankshaft pulley clockwise so that the timing chain on the timing chain tensioner (primary) side is loose. Remove the bolts and remove the timing chain tensioner (primary).

➡**Be careful not to drop the bolts inside the timing chain case.**

9. Remove the three water pump retaining bolts. Secure a gap between the water pump gear and the timing chain, by turning the crankshaft pulley counterclockwise until timing chain looseness on the water pump sprocket is at its maximum point.

10. Screw M8 bolts approximately 1.97 inch in length into the water pumps upper and lower bolt holes until they reach the timing chain case.

11. Alternately tighten each bolt for a half turn and pull out the water pump.

➡**Pull the pump straight out while preventing the vane from contacting the socket in the installation area. Remove the pump without causing the sprocket to contact the timing chain.**

12. Remove the M8 bolts. Remove and discard the O-rings.

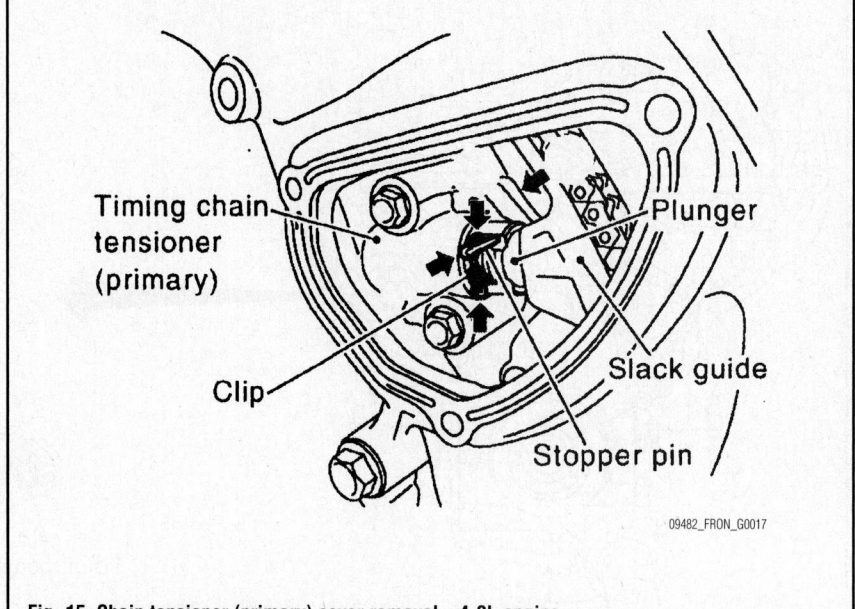

Fig. 15 Chain tensioner (primary) cover removal—4.0L engine

Labels: Timing chain tensioner (primary), Clip, Plunger, Slack guide, Stopper pin

09482_FRON_G0017

To install:

13. Installation is the reverse of the removal procedure.

14. Be sure to use new gaskets and O-rings, as required. Apply engine oil to new O-rings before installation. Locate O-ring with white paint mark in forward groove.

15. When installing the water pump make sure that the timing chain and water pump sprocket are engaged. Tighten the bolts alternately and evenly to specification.

16. Before installing the chain tensioner cover and the water pump cover be sure to apply a continuous bead of sealant to the mating surfaces of the covers.

➡**Do not allow the sealant to set for more than five minutes before installing the covers.**

17. Be sure to fill the cooling system with the proper grade and type engine coolant.

18. Start the engine and check for leaks.

19. Let the engine idle for about three minutes than rev it up to 3,000 rpm's under a no load condition to purge air from the high pressure chamber of the chain tensioner. The engine may produce a rattling noise. This indicates that air still remains in the chamber and is not a matter of concern.

ENGINE ELECTRICAL

✳✳ CAUTION

For this model, the battery current sensor that is installed to the negative battery cable measures the charging/discharging current of the battery and performs various engine controls. If an electrical component is connected directly to the negative battery terminal, the current flowing through that component will not be measured by the battery current sensor. This condition may cause a malfunction of the engine control system and battery discharge may occur. Do not connect an electrical component or ground wire directly to the battery terminal.

ALTERNATOR

REMOVAL & INSTALLATION

2.5L Engine

1. Before servicing the vehicle, refer to the Precautions Section.

2. Disconnect the negative battery cable.

3. Remove the fan shroud. Remove the drive belt.

4. Disconnect the alternator harness electrical connectors.

5. Remove the alternator mounting nut. Remove the upper alternator mounting bolt.

6. Remove the alternator from the vehicle.

To install:

7. Installation is the reverse of the removal procedure.

CHARGING SYSTEM

8. Be sure that the alternator spacer is in place on the lower mounting stud.

4.0L Engine

1. Before servicing the vehicle, refer to the Precautions Section.

2. Disconnect the negative battery cable.

3. Remove the fan shroud, on Frontier.

4. Remove the drive belt.

5. Remove alternator stay.

6. Remove the upper alternator mounting bolt.

7. Disconnect the alternator harness electrical connectors.

8. Remove the alternator from the vehicle.

To install:

9. Installation is the reverse of the removal procedure.

ENGINE ELECTRICAL   **IGNITION SYSTEM**

FIRING ORDER

See Figures 16 and 17.

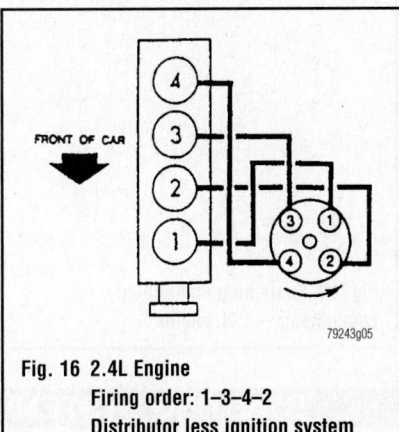

Fig. 16 2.4L Engine
Firing order: 1–3–4–2
Distributor less ignition system

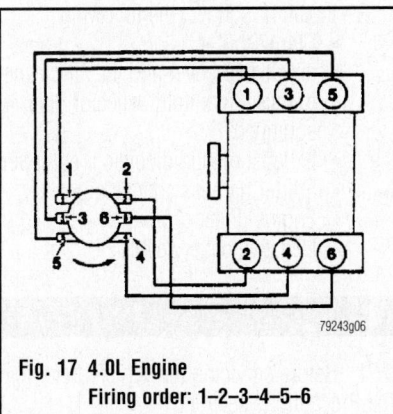

Fig. 17 4.0L Engine
Firing order: 1–2–3–4–5–6

IGNITION COIL

REMOVAL & INSTALLATION

2.5L Engine

1. Disconnect the negative battery cable.
2. Remove the intake manifold.

➡**If just removing the number one spark plug, it is not necessary to remove the intake manifold.**

3. Disconnect the harness connector from the ignition coil.
4. Remove the ignition coil.

To install:

5. Installation is the reverse of the removal procedure.

4.0L Engine

Left Bank

1. Disconnect the negative battery cable.
2. Remove the engine cover.
3. Remove the air cleaner case and air duct.
4. Move aside the harness, harness bracket and hoses which are located above the ignition coil.
5. Disconnect the harness connector from the ignition coil.
6. Remove the ignition coil.

To install:

7. Installation is the reverse of the removal procedure.

Right Bank

1. Disconnect the negative battery cable.
2. Remove the intake manifold collector.
3. Move aside the harness, harness bracket and hoses which are located above the ignition coil.
4. Disconnect the harness connector from the ignition coil.
5. Remove the ignition coil.

To install:

6. Installation is the reverse of the removal procedure.

IGNITION TIMING

INSPECTION

1. Before servicing the vehicle, refer to the Precautions Section.
2. Remove the number one ignition coil.
3. Connect the number one ignition coil and spark plug with a suitable high tension wire.
4. Attach the timing light clamp to the wire.
5. Check the ignition timing.

SPARK PLUGS

REMOVAL & INSTALLATION

2.5L Engine

See Figure 18.

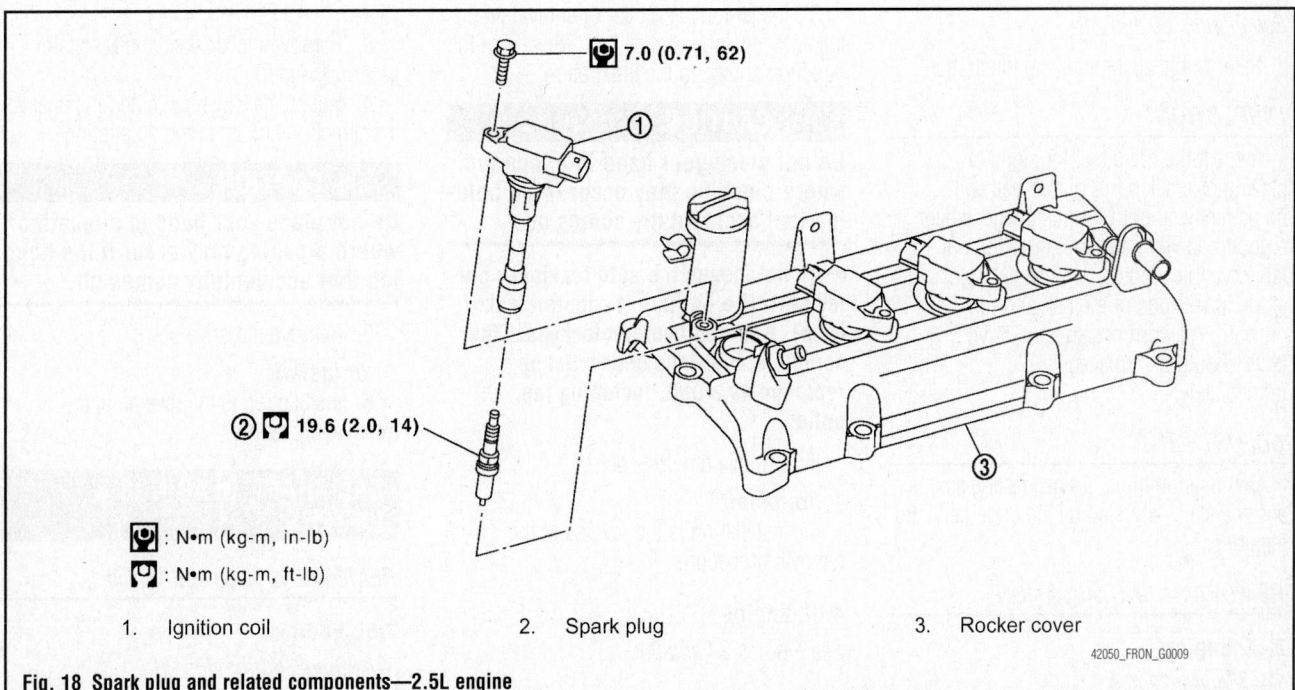

1. Ignition coil
2. Spark plug
3. Rocker cover

Fig. 18 Spark plug and related components—2.5L engine

1. Disconnect the negative battery cable.

2. Remove the intake manifold.

➡**If just removing the number one spark plug, it is not necessary to remove the intake manifold.**

3. Disconnect the harness connector from the ignition coil.

4. Remove the ignition coil.

5. Remove the spark plug using a spark plug socket and wrench.

To install:

6. Installation is the reverse of the removal procedure.

4.0L Engine

See Figure 19.

1. Disconnect the negative battery cable.

2. Remove the intake manifold collector.

3. Remove the ignition coil.

4. Remove the spark plug using a spark plug socket and wrench.

To install:

5. Installation is the reverse of the removal procedure.

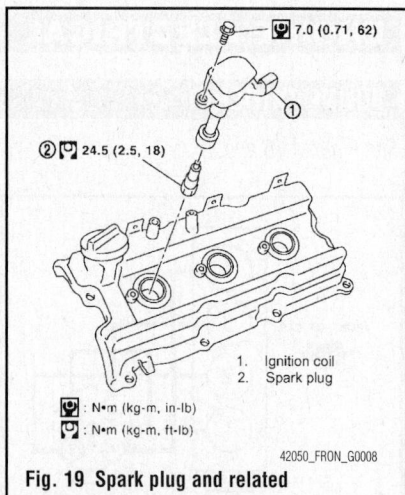

1. Ignition coil
2. Spark plug

⟨⟩ : N•m (kg-m, in-lb)
⟨⟩ : N•m (kg-m, ft-lb)

42050_FRON_G0008

Fig. 19 Spark plug and related components—4.0L engine

ENGINE ELECTRICAL

STARTER

REMOVAL & INSTALLATION

1. Before servicing the vehicle, refer to the Precautions Section.

2. Remove or disconnect the following:
- Negative battery cable
- Engine under cover
- On vehicles with 2.5L engine, remove the air cleaner cover and the air cleaner to intake manifold collector duct
- On vehicles with 4.0L engine, remove the exhaust manifold cover to gain access to the starter retaining bolts
- Starter harness connectors
- Starter bolts
- Starter motor

STARTING SYSTEM

To install:

3. Install or connect the following:
- Starter motor
- air cleaner cover and the air cleaner to intake manifold collector duct, if equipped
- Exhaust manifold cover, if equipped
- Starter harness connectors
- Engine under cover
- Negative battery cable

ENGINE MECHANICAL

ACCESSORY DRIVE BELTS

ACCESSORY BELT ROUTING

See Figures 20 and 21.

Refer to the accompanying illustration.

INSPECTION

Inspect the drive belt for signs of glazing or cracking. A glazed belt will be perfectly smooth from slippage, while a good belt will have a slight texture of fabric visible. Cracks will usually start at the inner edge of the belt and run outward. All worn or damaged drive belts should be replaced immediately.

ADJUSTMENT

Belt tensioning is not necessary, as it is automatically adjusted by the drive belt auto tensioner.

REMOVAL & INSTALLATION

2.5L Engine

See Figures 22 and 23.

1. Before servicing the vehicle, refer to the Precautions section.

2. Disconnect the negative battery cable.

3. Install tool J-46535, or equivalent on the auto tensioner pulley bolt. Move it in the direction shown in the illustration.

❊❊ **CAUTION**

Do not place your hand in a location where pinching may occur if the holding tool accidentally comes off.

➡**Do not loosen the auto tensioner pulley bolt. (Do not turn it counterclockwise). If turned counterclockwise, the complete auto tensioner must be replaced as a unit, including the pulley.**

4. Remove the drive belt.

To install:

5. Installation is the reverse of the removal procedure.

4.0L Engine

See Figures 24 and 25.

1. Before servicing the vehicle, refer to the Precautions section.

2. Disconnect the negative battery cable.

3. Remove the air duct and resonator assembly (inlet).

4. Rotate the drive belt auto tensioner in the direction shown in the illustration.

❊❊ **CAUTION**

Do not place your hand in a location where pinching may occur if the holding tool accidentally comes off.

5. Remove the drive belt.

To install:

6. Installation is the reverse of the removal procedure.

CAMSHAFT, BEARINGS & LIFTERS

REMOVAL & INSTALLATION

2.5L Engine

See Figures 26 through 36.

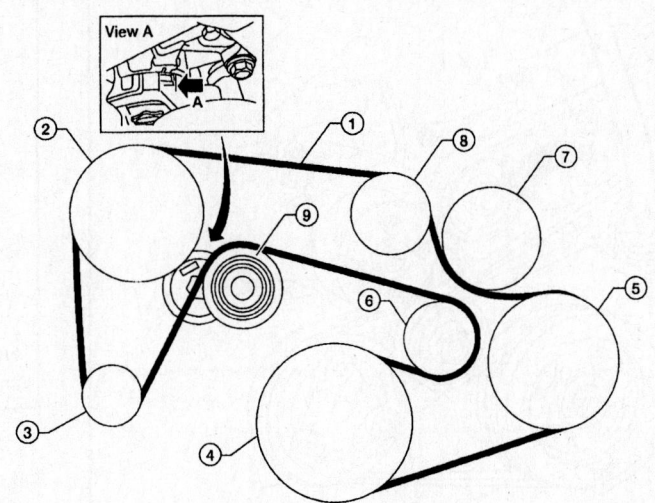

1. Drive belt
2. Power steering oil pump pulley
3. Generator pulley
4. Crankshaft pulley
5. A/C compressor (if equipped) or idler pulley
6. Idler pulley
7. Water pump
8. Idler pulley
9. Drive belt auto-tensioner

09482_FRON_G0005

Fig. 20 Accessory drive belt routing—2.5L engine

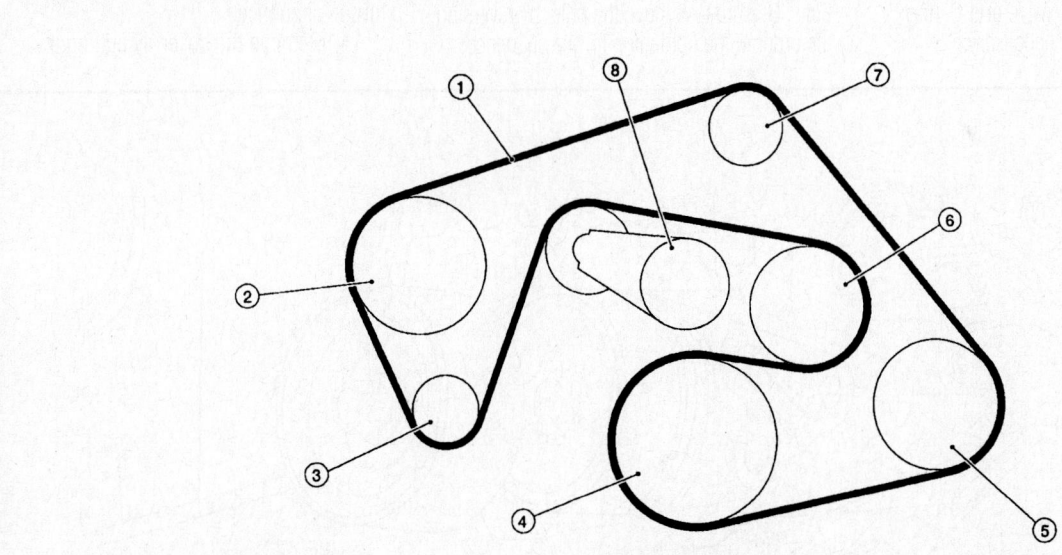

1. Drive belt
2. Power steering pump pulley
3. Generator pulley
4. Crankshaft pulley
5. A/C compressor
6. Cooling fan pulley
7. Idler pulley
8. Drive belt auto-tensioner

09482_FRON_G0006

Fig. 21 Accessory drive belt routing—4.0L engine

➡**The procedure below describes removal and installation of the camshaft without removing the front cover. If the front cover is removed** **refer to timing chain removal and installation.**

1. Before servicing the vehicle, refer to the Precautions Section.

2. Properly relieve the fuel system pressure.

3. Disconnect the negative battery cable. Drain the cooling system.

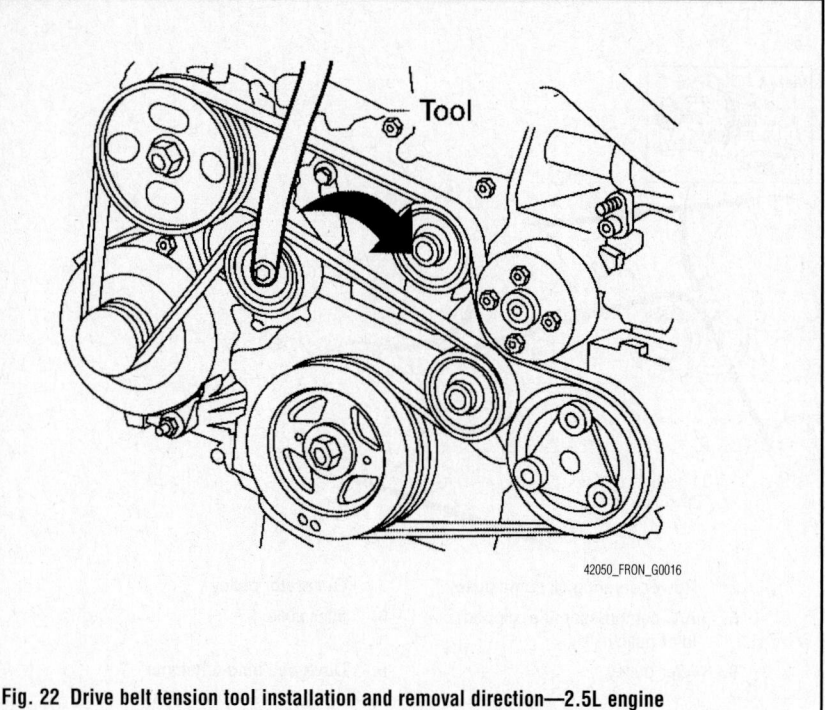

Fig. 22 Drive belt tension tool installation and removal direction—2.5L engine

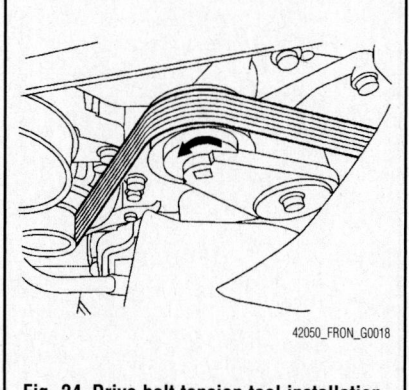

Fig. 24 Drive belt tension tool installation and removal direction—4.0L engine

4. Remove the intake manifold.

5. Disconnect the PCV hose from the rocker cover. Remove the ignition coil.

6. Remove the PCV valve and O-ring from the rocker cover, if necessary.

7. Remove the oil filler cap from the rocker cover, if necessary.

8. Remove the rocker cover retaining bolts. Be sure to remove the bolts by reversing the order of the tightening torque sequence.

9. Remove the rocker cover. Discard the gasket.

10. Remove the drive belt. Disconnect and remove the camshaft position sensor (PHASE).

11. Disconnect the IVT control solenoid electrical connector.

12. Disconnect the ground electrical connectors from the front cover.

13. Remove the IVT control solenoid retaining bolts. Be sure to remove the bolts by reversing the order of the tightening torque sequence.

14. Remove the cover by cutting the

21.6 (2.2, 16)

: N·m (kg-m, ft-lb)

Fig. 23 Drive belt auto tensioner and related components—2.5L engine

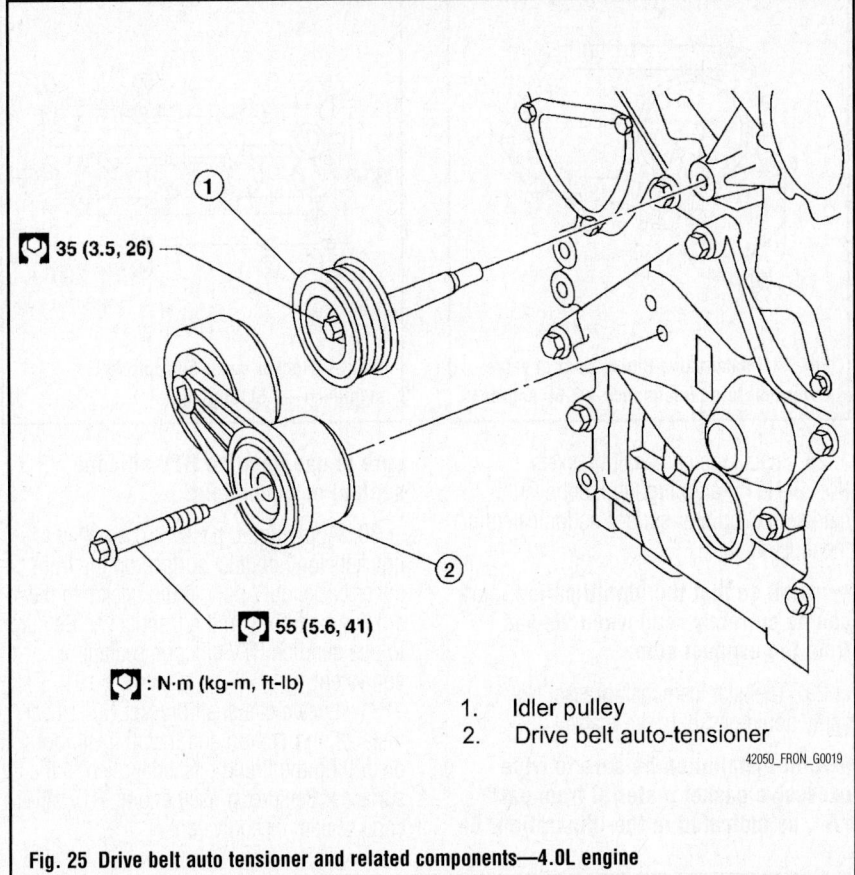

35 (3.5, 26)

55 (5.6, 41)

: N·m (kg-m, ft-lb)

1. Idler pulley
2. Drive belt auto-tensioner

42050_FRON_G0019

Fig. 25 Drive belt auto tensioner and related components—4.0L engine

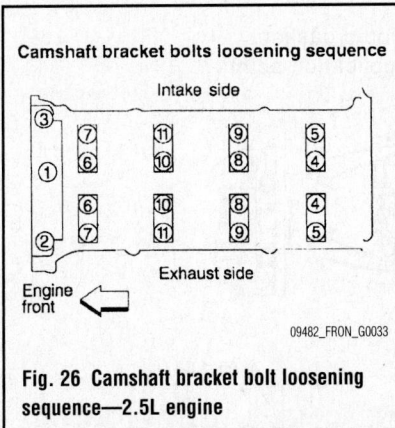

Camshaft bracket bolts loosening sequence

Intake side

Exhaust side

Engine
front

09482_FRON_G0033

Fig. 26 Camshaft bracket bolt loosening sequence—2.5L engine

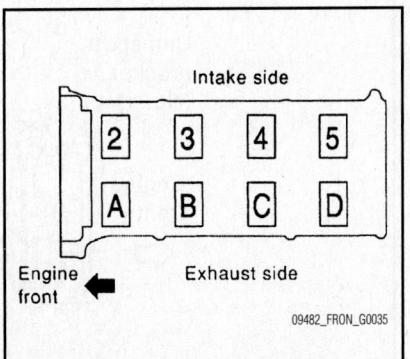

Intake side

2 3 4 5

A B C D

Engine
front

Exhaust side

09482_FRON_G0035

Fig. 28 Camshaft bracket identification—2.5L engine

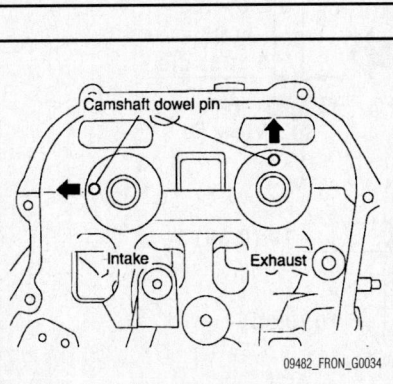

Camshaft dowel pin

Intake   Exhaust

09482_FRON_G0034

Fig. 27 Camshaft positioning—2.5L engine

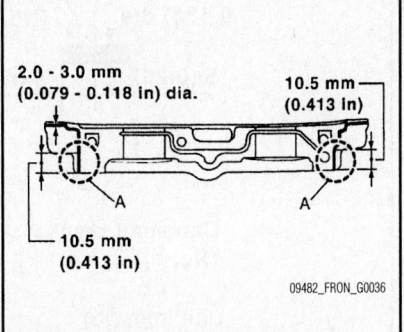

2.0 - 3.0 mm
(0.079 - 0.118 in) dia.

10.5 mm
(0.413 in)

A

A

10.5 mm
(0.413 in)

09482_FRON_G0036

Fig. 29 Camshaft bracket No. 1 sealant application point "A"—2.5L engine

sealant using tool KV10111100 (J-37228) or equivalent.

15. Position the number one cylinder on its compression stroke by rotating the crankshaft pulley clockwise. Align the mating marks for TDC with the timing indicator on the front cover.

16. At the same time make sure that the mating marks on the camshaft sprockets are lined up with the yellow links in the timing chain. If not rotate the crankshaft one more turn to line up the mating marks to the yellow links.

17. Pull the timing chain guide out between the camshaft sprockets through the front cover.

18. Line up the mating marks on the camshaft sprockets with the yellow links in the timing chain and paint an indelible mating mark on the sprocket and timing chain link plate.

➡ Do not rotate the crankshaft or the camshaft while the timing chain is removed.

➡ Chain tension holding work is not necessary. Crankshaft sprocket and timing chain do not disconnect structurally while the front cover is attached.

19. Push in the tensioner plunger and hold. Insert a stopper pin into the hole on the tensioner body to hold the chain tensioner. Remove the timing chain tensioner.

➡ Use a wire with 0.02 inch diameter for a stopper pin.

20. Secure the hexagonal part of the camshaft with a suitable tool. Loosen the camshaft sprocket bolts and remove the camshaft sprockets.

21. Loosen the camshaft bracket bolts. Be sure to remove the bolts by following the bolt removal sequence.

22. Remove the camshafts and brackets from the engine.

23. Remove the number one camshaft bracket by tapping lightly with a rubber mallet. Note the positions for installation.

24. As necessary, remove the valve lifters. Be sure to keep them in the proper order for installation.

To install:

25. Inspect the camshafts, replace as required.

26. Install the camshafts so that the camshaft dowel pins on the front side are positioned as indicated in the illustration.

27. Remove any foreign material from the camshaft bracket backside and from the cylinder head face.

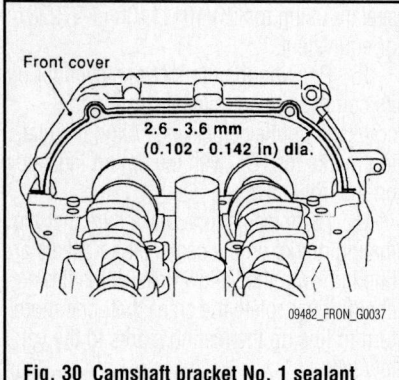

Fig. 30 Camshaft bracket No. 1 sealant application outside bolt hole—2.5L engine

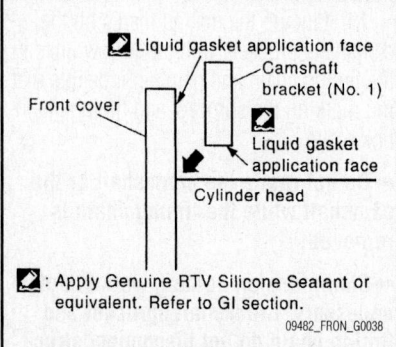

Fig. 31 Camshaft bracket No. 1 sealant application locating points—2.5L engine

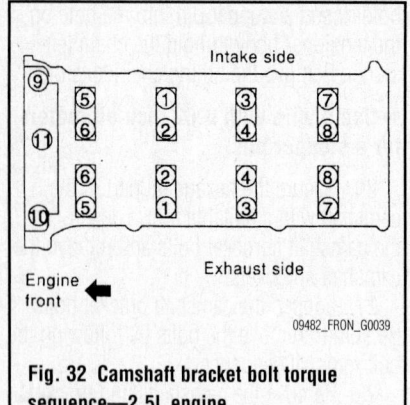

Fig. 32 Camshaft bracket bolt torque sequence—2.5L engine

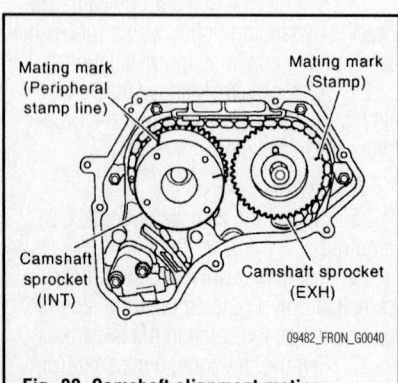

Fig. 33 Camshaft alignment mating marks—2.5L engine

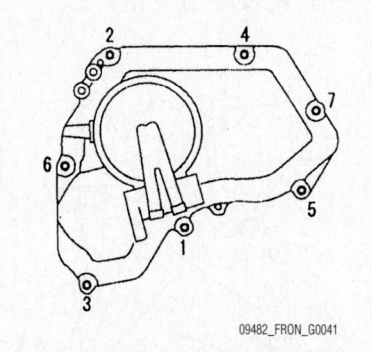

Fig. 34 Intake valve timing control valve cover bolt torque sequence—2.5L engine

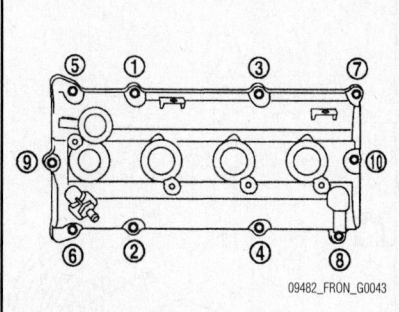

Fig. 36 Rocker cover bolt torque sequence—2.5L engine

28. Install the camshaft brackets (No. 2–No. 5) aligning the identification marks on the upper surface as indicated in the illustration.

➡**Install so that the identification mark can be correctly read when viewed from the exhaust side.**

29. To install camshaft bracket No. 1, apply liquid gasket to the bracket.

➡**After installation be sure to wipe excessive gasket material from part "A", as indicated in the illustration. Be** sure to use genuine RTV silicone sealant or equivalent.

30. Apply liquid gasket to camshaft bracket No. 1 contact surface on the front cover backside. Apply liquid gasket to the outside bolt hole on the front cover. Be sure to use genuine RTV silicone sealant or equivalent.

31. Locate camshaft bracket No. 1 near installation position and install it without disturbing the liquid gasket applied to the surfaces. Be sure to use genuine RTV silicone sealant or equivalent.

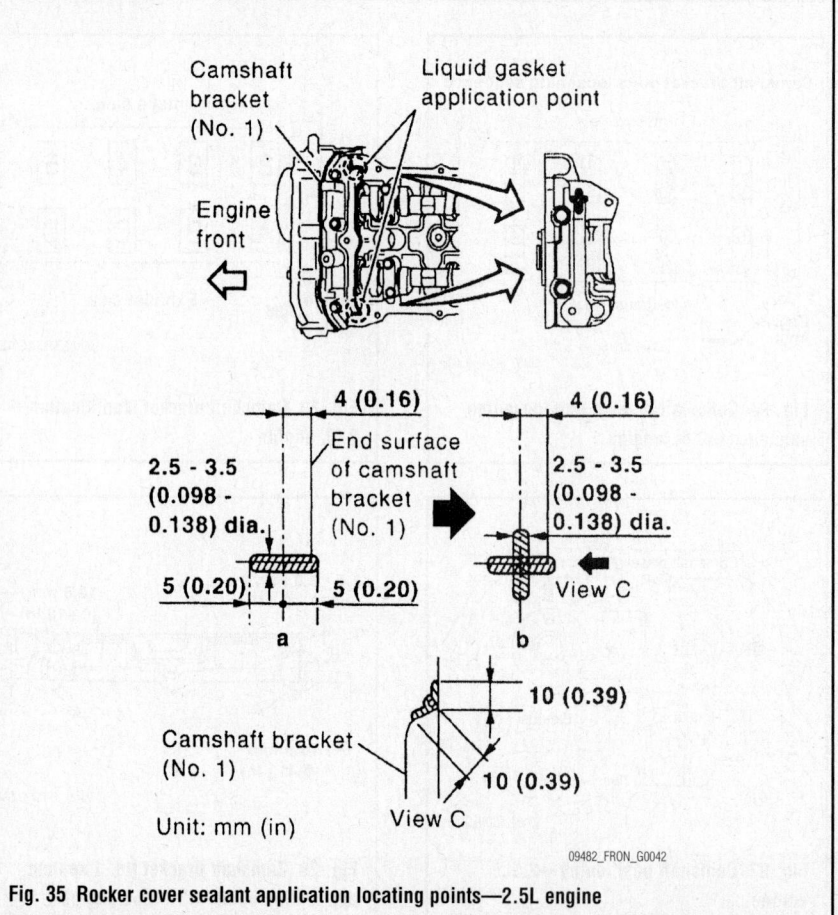

Fig. 35 Rocker cover sealant application locating points—2.5L engine

32. Tighten the camshaft bracket bolts in the proper sequence and to specification:

   a. Bolts 9–11 to 17 inch lbs. (1.24 to 1.92 Nm)

   b. Bolts 1–8 to 17 inch lbs. (0.90 to 1.92 Nm)

   c. Bolts 1–11 to 52 inch lbs. (1.24 to 5.88 Nm)

   d. Bolts 1–11 to 92 inch lbs. (1.24 to 10.39 Nm)

➡**After tightening the bolts be sure to wipe off any excessive liquid gasket. Be sure to use genuine RTV silicone sealant or equivalent.**

33. Install the camshaft position sensor. Install the camshaft sprockets.

➡**Install them by aligning the mating marks on each camshaft sprocket with the paint marks on the timing chain link plates, which were made during removal.**

✳✳ CAUTION

Aligned mating marks could slip. Therefore, after matching them, hold the timing chain in place by hand. Before and after installing the chain tensioner, make sure again that the mating marks have not slipped.

34. Install the chain tensioner. After installation, pull the stopper pin off completely, and make sure that the chain tensioner plunger is released.

➡**Before installation of the chain tensioner, it is possible to rematch the marks on the timing chain with new ones on each sprocket.**

35. Install the chain guide. Install oil rings to the camshaft sprocket (INT) insertion points on backside of intake valve timing control cover. Install the O-ring to the front cover.

36. Apply a 0.083–0.122 inch diameter bead of liquid gasket to the intake valve timing control cover. Be sure to use genuine RTV silicone sealant or equivalent.

37. Install the cover. Tighten the bolts in the proper sequence. Connect the ground cables and install the harness clip.

38. Check and adjust the valve clearance, as required.

➡**If hydraulic pressure inside the timing chain tensioner drops after removal/installation, slack in the guide may generate a pounding noise during and just after engine start. This is normal the noise will stop after hydraulic pressure rises.**

39. Continue the installation in the reverse order of the removal procedure.

40. Apply liquid gasket, be sure to use genuine RTV silicone sealant or equivalent, to the positions shown in the illustration. Refer to figure "a" to apply liquid gasket to joint part of camshaft bracket No. 1 and cylinder head. Refer to figure "b" to apply liquid gasket in 90 degrees to figure "b".

41. Install the rocker cover. Torque the retaining bolts to 18 inch lbs. (2.03 Nm) and then to 73 inch lbs. (8.25 Nm), in the proper sequence.

42. Inspect the camshaft sprocket (INT) oil groove.

➡**Perform this inspection only when DTC P0011 or DTC P0021 are detected in self diagnostic results of CONSULT-II.**

43. Be sure the engine is cold. Check and adjust oil level, as required.

44. Properly release the fuel system pressure. Disconnect the ignition coil and injector harness connectors.

➡**This is being done to prevent the engine from unintentionally being started while checking.**

45. Remove the intake valve timing control solenoid valve.

46. Crank the engine, and then make sure that engine oil comes out from camshaft bracket (No. 1) oil hole.

✳✳ WARNING

Be careful not to touch rotating parts, (drive belt, idler pulley, crankshaft pulley etc) as injury could result.

➡**Oil may squirt from the intake valve timing control solenoid valve installation hole during engine cranking. Use a shop towel to prevent oil from squirting on engine components.**

47. Clean the oil groove between the oil strainer and the intake timing control solenoid valve if engine oil does not come out from camshaft bracket (No. 1) oil hole.

48. Remove the components between the intake valve timing control solenoid valve and the camshaft sprocket (INT). Check each oil groove for clogging.

49. After inspection install any removed components.

4.0L Engine

See Figures 37 through 45.

1. Before servicing the vehicle, refer to the Precautions Section.

2. Properly relieve the fuel system pressure.

3. Disconnect the negative battery cable. Remove the engine cover.

Bank	INT/EXH	Dowel pin	Paint marks		Identification mark
			M1	M2	
RH	INT	No	Green	No	RE
	EXH	Yes	No	White	RE
LH	INT	No	Green	No	LH
	EXH	Yes	No	White	LH

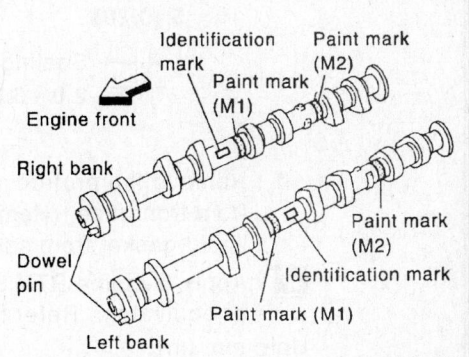

Fig. 37 Camshaft identification—4.0L engine

09482_FRON_G0044

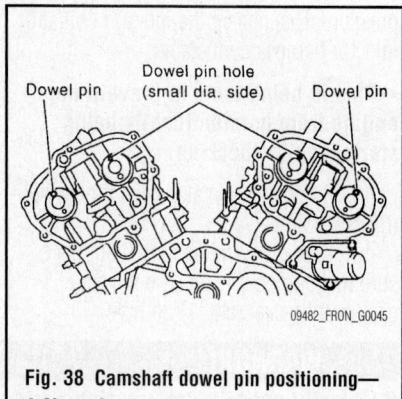

Fig. 38 Camshaft dowel pin positioning—4.0L engine

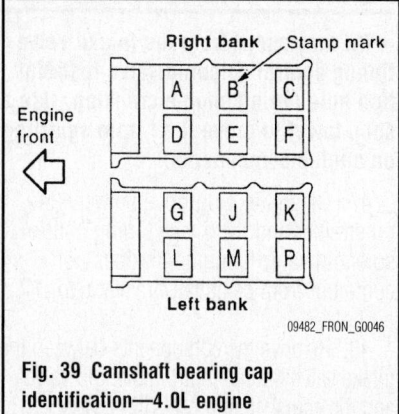

Fig. 39 Camshaft bearing cap identification—4.0L engine

4. Remove the front timing chain case, camshaft sprocket, timing chain and rear timing chain case.

5. Remove the camshaft position sensor (PHASE) from the cylinder head back side.

➡**Handle carefully to avoid dropping and shocks. Do not disassemble. Do not place in a location where the sensor can be exposed to magnetism.**

6. Remove the intake manifold collector.

7. Separate the engine harness and remove their brackets from the rocker covers. Remove the harness bracket from the cylinder head, if necessary.

8. Remove the ignition coil. Remove the PCV hoses. Remove the oil filler cap, if necessary.

9. Loosen the rocker cover retaining bolts, in the reverse order of the tightening sequence.

10. Remove the rocker covers from the engine.

11. Remove the intake valve timing control solenoid valves. Discard the gaskets.

12. Mark the camshaft brackets and bolts for reinstallation. Remove the camshaft bracket bolts. Be sure to remove

the bolts by reversing the order of the tightening torque sequence and in several steps.

13. Remove the camshafts.

14. If required, remove the valve lifters. Identify them for reinstallation in their original locations.

15. Remove the timing chain tensioner (secondary) from the cylinder head. Remove

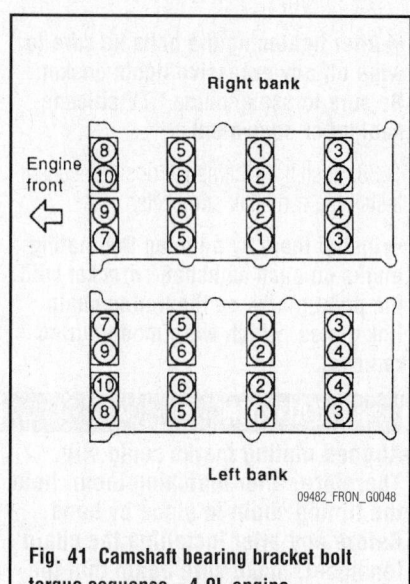

Fig. 41 Camshaft bearing bracket bolt torque sequence—4.0L engine

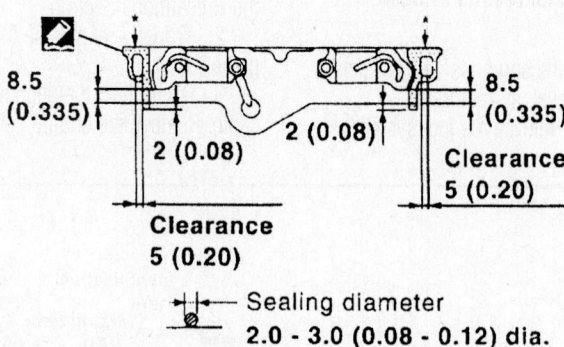

Camshaft bracket (No. 1)

8.5 (0.335)

8.5 (0.335)

2 (0.08)

2 (0.08)

Clearance 5 (0.20)

Clearance 5 (0.20)

Sealing diameter 2.0 - 3.0 (0.08 - 0.12) dia.

* : Remove the protruding liquid gasket from front face. (Remove the hardened liquid gasket from surface only.)

⬛ : Apply Genuine RTV Silicone Sealant or equivalent. Refer to GI section.

Unit: mm (in)

Fig. 40 Camshaft sealant application and location—4.0L engine

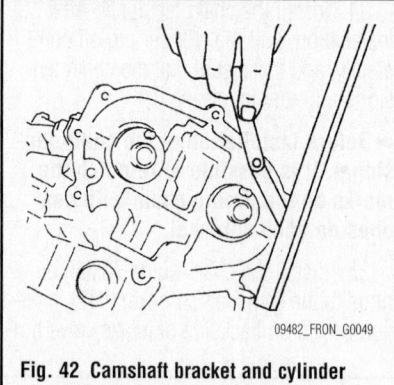

Fig. 42 Camshaft bracket and cylinder head measurement—4.0L engine

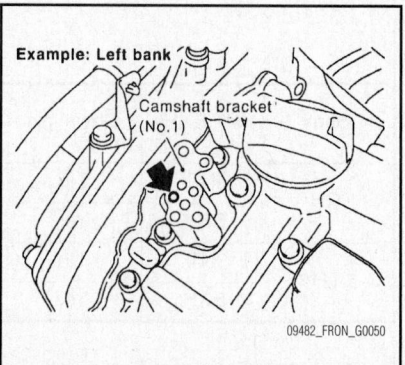

Fig. 43 Camshaft bracket (No. 1) oil hole location

the timing chain tensioner (secondary) with its stopper pin attached.

To install:

16. Inspect the camshafts, replace as required.

17. Install the timing chain tensioners (secondary) on both sides of the cylinder head. Be sure to use new O-rings.

➡ **Install the tensioner with its stopper pin attached. Install the tensioner with the sliding part facing downward on the right cylinder head and with the sliding part facing upward on the left cylinder head.**

18. Install the valve lifters, in their original bores.

19. Install the camshafts, with the dowel pin attached to its front end face on the exhaust side.

➡ **Follow the identification marks for proper placement and direction.**

20. Install the camshaft so that the dowel pin hole and dowel pin on the front end face are positioned as shown in the illustration (No. 1 piston at TDC on its compression stroke).

➡ **Large and small pin holes are located on the front end face of the camshaft (INT), at intervals of 180 degrees. Face small diameter side pin hole upward (in cylinder head upper face direction).**

➡ **Though the camshaft does not stop at the portion as shown, for placement of the cam nose, it is generally accepted that the camshaft is placed for the same direction as shown.**

21. Install the camshaft brackets in the same position that they were removed. Install brackets No. 2–No. 4 aligning the stamp marks as indicated in the illustration.

➡ **There are no identification marks indicating left or right for camshaft bracket No. 1.**

22. Apply liquid gasket to the mating surfaces of camshaft bracket No. 1 as shown in the illustration on both the left and right cylinder heads. Be sure to use genuine RTV sealant or equivalent.

23. Tighten the camshaft bracket bolts in the proper sequence and to specification:

    a. Bolts 7–10 to 17 inch lbs. (1.13 to 1.92 Nm)

    b. Bolts 1–6 to 17 inch lbs. (0.67 to 1.92 Nm)

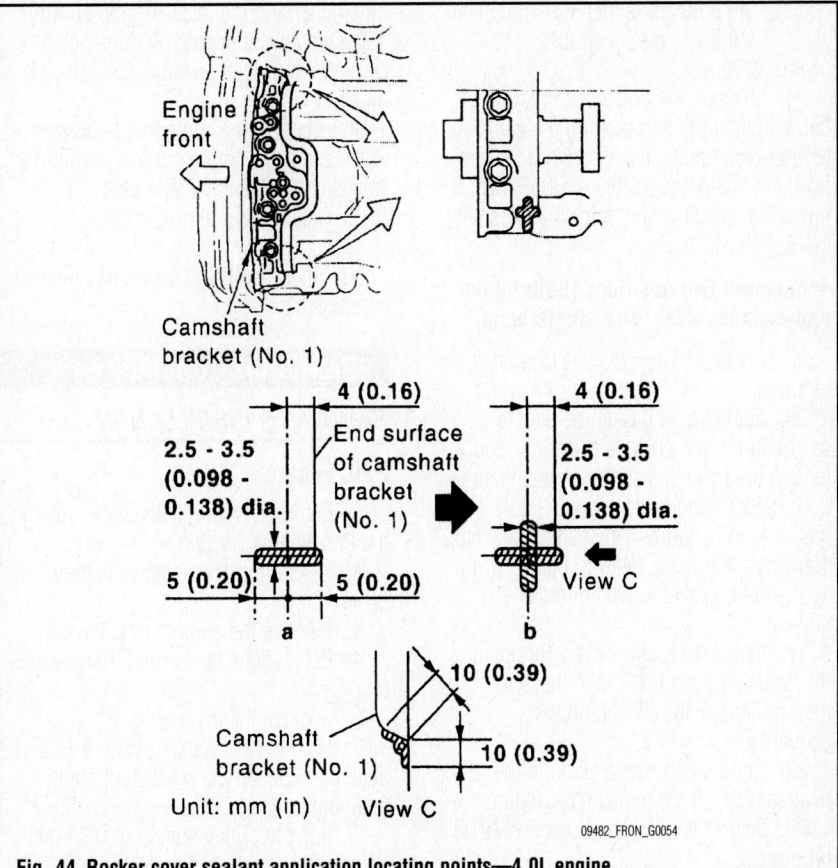

Fig. 44 Rocker cover sealant application locating points—4.0L engine

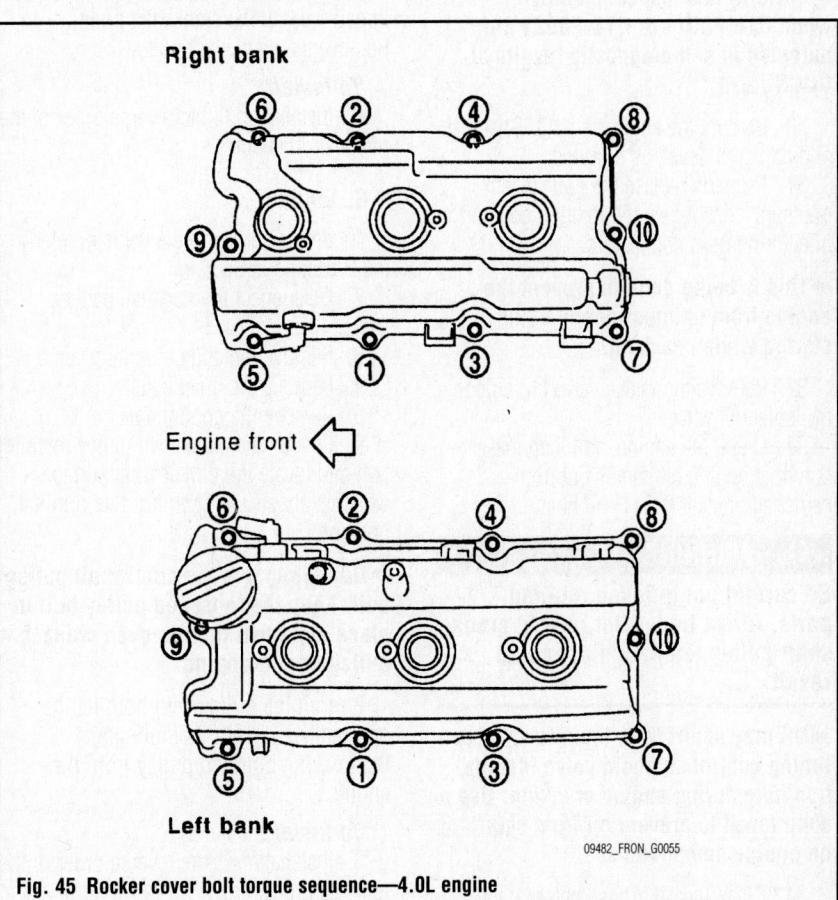

Fig. 45 Rocker cover bolt torque sequence—4.0L engine

c. All bolts to 52 inch lbs. (5.88 Nm)
d. All bolts to 92 inch lbs.
(10.39 Nm)

24. Measure the difference in levels between the front end faces of the camshaft bracket No. 1 and the cylinder head. Specification should be -0.0055–0.0055 inch. If not within specification, reinstall camshaft bracket No. 1.

➡**Measure two positions (both intake and exhaust side) for a single bank.**

25. Check and adjust valve clearance, as required.
26. Apply liquid gasket, be sure to use genuine RTV silicone sealant or equivalent, to the positions shown in the illustration. Refer to figure "a" to apply liquid gasket to joint part of camshaft bracket No. 1 and cylinder head. Refer to figure "b" to apply liquid gasket to the figure "a" squarely.
27. Install the rocker cover. Torque the retaining bolts to 17 inch lbs. and then to 74 inch lbs., in the proper sequence.
28. Continue the installation in the reverse order of the removal procedure.
29. Inspect the camshaft sprocket (INT) oil groove.

➡**Perform this inspection only when DTC P0011 or DTC P0021 are detected in self diagnostic results of CONSULT-II.**

30. Be sure the engine is cold. Check and adjust oil level, as required.
31. Properly release the fuel system pressure. Disconnect the ignition coil and injector harness connectors.

➡**This is being done to prevent the engine from unintentionally being started while checking.**

32. Remove the intake valve timing control solenoid valve.
33. Crank the engine, and then make sure that engine oil comes out from camshaft bracket (No. 1) oil hole.

❊❊ WARNING

Be careful not to touch rotating parts, (drive belt, idler pulley, crankshaft pulley, etc) as injury could result.

➡**Oil may squirt from the intake valve timing control solenoid valve installation hole during engine cranking. Use a shop towel to prevent oil from squirting on engine components.**

34. Clean the oil groove between the oil strainer and the intake timing control solenoid valve if engine oil does not come out from camshaft bracket (No. 1) oil hole.
35. Remove the components between the intake valve timing control solenoid valve and the camshaft sprocket (INT). Check each oil groove for clogging.
36. After inspection install any removed components.

CRANKSHAFT DAMPER

REMOVAL & INSTALLATION

2.5L Engine

1. Before servicing the vehicle, refer to the Precautions Section.
2. Disconnect the negative battery cable.
3. Remove the engine undercover.
4. Remove the fan shroud. Remove the cooling fan.
5. Remove the drive belts.
6. Hold the crankshaft pulley with a suitable tool. Loosen the crankshaft pulley retaining bolt.
7. Pull the pulley out about 0.39 inch. Remove the crankshaft pulley bolt.
8. Attach a pulley puller in the M6 thread hole on the crankshaft pulley. Remove the crankshaft pulley.

To install:
9. Installation is the reverse order of the removal procedure.

4.0L Engine

1. Before servicing the vehicle, refer to the Precautions Section.
2. Disconnect the negative battery cable.
3. Remove the engine undercover.
4. Remove the drive belts.
5. Remove the cooling fan.
6. Loosen the crankshaft pulley retaining bolt and locate the bolt seating surface, which is about 0.39 inch from its original position.

➡**Do not remove the crankshaft pulley bolt. Keep the loosened pulley bolt in place to protect the removed crankshaft pulley from dropping.**

7. Pull the pulley with both hands and remove it from its mounting. Remove the bolt and pulley from the engine.

To install:
8. Installation is the reverse order of the removal procedure.

CRANKSHAFT FRONT SEAL

REMOVAL & INSTALLATION

2.5L Engine

1. Before servicing the vehicle, refer to the Precautions Section.
2. Disconnect the negative battery cable.
3. Remove the engine undercover.
4. Remove the fan shroud. Remove the cooling fan.
5. Remove the drive belts.
6. Hold the crankshaft pulley with a suitable tool. Loosen the crankshaft pulley retaining bolt.
7. Pull the pulley out about 0.39 inch. Remove the crankshaft pulley bolt.
8. Attach a pulley puller in the M6 thread hole on the crankshaft pulley. Remove the crankshaft pulley.
9. Using a seal removal tool, remove the oil seal from its mounting.

➡**Be careful not to damage the front cover and/or the crankshaft.**

To install:
10. Installation is the reverse order of the removal procedure.
11. Press fit the seal until it is flush with the front end surface of the front cover, using the proper tools.

4.0L Engine

1. Before servicing the vehicle, refer to the Precautions Section.
2. Disconnect the negative battery cable.
3. Remove the engine undercover.
4. Remove the drive belts.
5. Remove the cooling fan.
6. Loosen the crankshaft pulley retaining bolt and locate the bolt seating surface, which is about 0.39 inch from its original position.

➡**Do not remove the crankshaft pulley bolt. Keep the loosened pulley bolt in place to protect the removed crankshaft pulley from dropping.**

7. Pull the pulley with both hands and remove it from its mounting. Remove the bolt and pulley from the engine.
8. Using a seal removal tool, remove the oil seal from its mounting.

➡**Be careful not to damage the front cover and/or the crankshaft.**

To install:
9. Installation is the reverse order of the removal procedure.
10. Press fit until the height of the front

oil seal is level with the mounting surface, using the proper tools.

CYLINDER HEAD

REMOVAL & INSTALLATION

2.5L Engine

See Figures 46 and 47.

1. Before servicing the vehicle, refer to the Precautions Section.
2. Properly relieve the fuel system pressure.
3. Disconnect the negative battery cable. Drain the cooling system. Drain the engine oil.
4. Remove the intake manifold and fuel tube assembly.
5. Remove the fuel injector and fuel tube assembly.
6. Remove the exhaust manifold and the three way catalyst.
7. Remove the water outlet. Remove the heater outlet.
8. Remove the front cover and the timing chain.

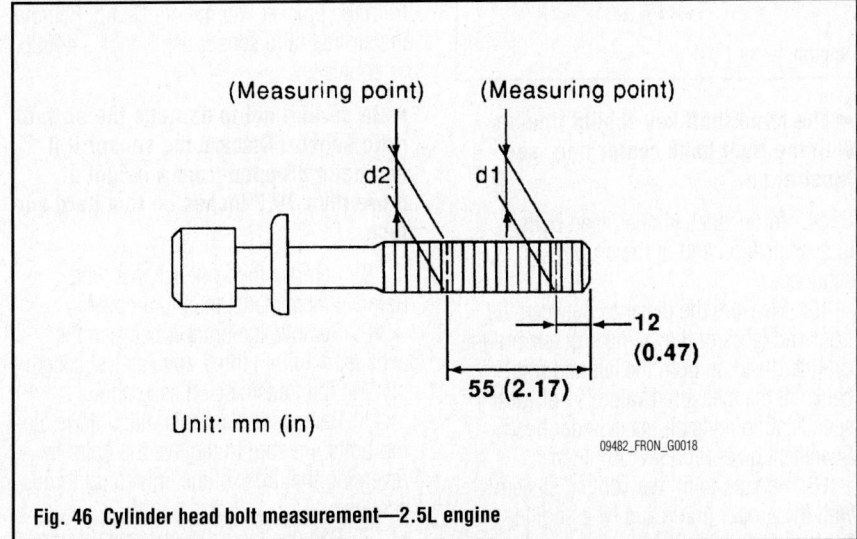

Fig. 46 Cylinder head bolt measurement—2.5L engine

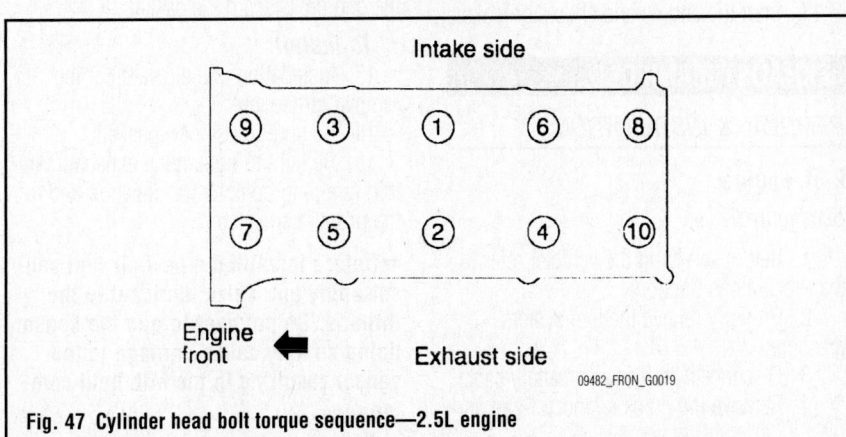

Fig. 47 Cylinder head bolt torque sequence—2.5L engine

9. Remove the camshafts.
10. Remove the cylinder head retaining bolts. Be sure to remove the bolts by reversing the order of the tightening torque sequence.
11. Remove the cylinder head from the engine. Discard the gasket.

To install:

12. Installation is the reverse of the removal procedure.
13. Be sure to inspect the cylinder head bolts. Replace as required.

➡**Head bolts are tightened by plastic zone tightening method. Whenever the size difference between "d1" and "d2" exceeds the limit, replace the bolt. "d1"-"d2" limit is 0.0091. If reduction of the outer diameter appears in a position other than "d2", use it the "d2" point.**

14. Install the new cylinder head gasket. Apply engine oil to cylinder head bolt threads and seating surfaces. Torque the cylinder head bolts to specification and in the proper sequence.

15. Be sure to fill the cooling system with the proper grade and type engine coolant.
16. Be sure to fill the engine with the proper grade and type motor oil.
17. Start the engine and check for leaks.

4.0L Engine

See Figures 48 through 51.

1. Before servicing the vehicle, refer to the Precautions Section.
2. Properly relieve the fuel system pressure.
3. Disconnect the negative battery cable. Drain the cooling system.
4. Remove the camshaft.
5. Remove the intake manifold.
6. Remove the exhaust manifold.
7. Remove the water inlet and thermostat assembly.
8. Remove the water outlet, water pipe and heater pipe.
9. Remove the cylinder head retaining bolts. Be sure to remove the bolts by reversing the order of the tightening torque sequence.
10. Remove the cylinder head from the engine. Discard the gasket.

To install:

11. Installation is the reverse of the removal procedure.
12. Be sure to inspect the cylinder head bolts. Replace as required.

➡**Head bolts are tightened by plastic zone tightening method. Whenever the size difference between "d1" and "d2" exceeds the limit, replace the bolt. "d1"-"d2" limit is 0.0043. If reduction of the outer diameter appears in a position other than "d2", use it the "d2" point.**

13. Install the new cylinder head gasket. Turn the crankshaft until the number one piston is at TDC.

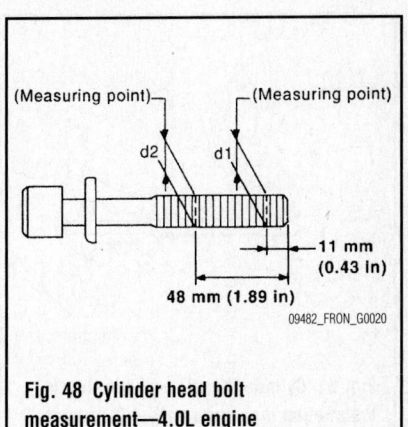

Fig. 48 Cylinder head bolt measurement—4.0L engine

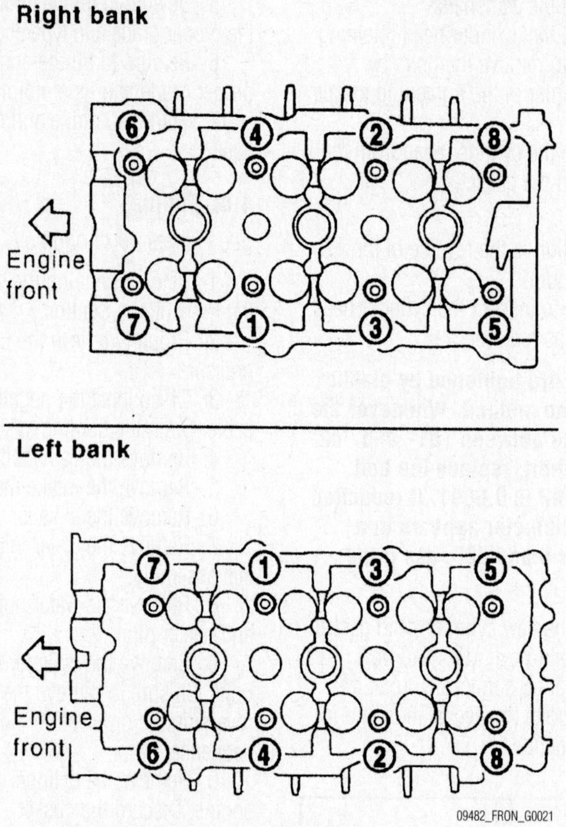

Right bank

Engine front →

Left bank

Engine front →

09482_FRON_G0021

Fig. 49 Cylinder head bolt torque sequence—4.0L engine

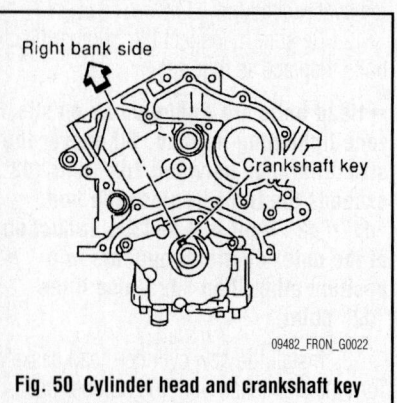

09482_FRON_G0022

Fig. 50 Cylinder head and crankshaft key alignment—4.0L engine

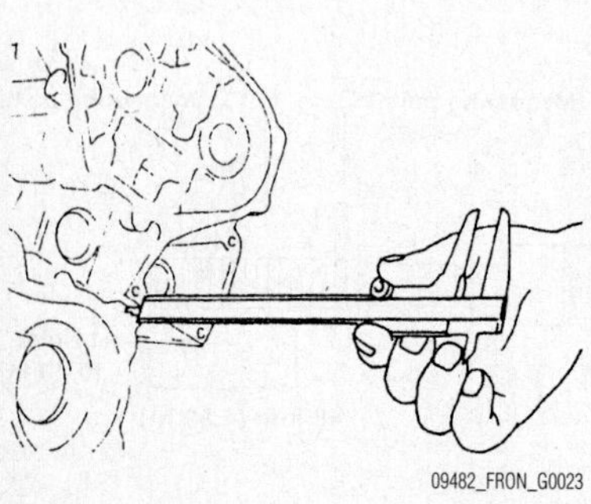

09482_FRON_G0023

Fig. 51 Cylinder head to cylinder block installation measurement—4.0L engine

➡ **The crankshaft key should line up with the right bank center line, see illustration.**

14. Torque the cylinder head bolts to specification and in the proper sequence.

15. Measure the distance between the front end faces of the cylinder block and the cylinder head on both the left and right banks. If the measured value is not within specification reinstall the cylinder head. Specification is 0.555–0.587 inch.

16. Be sure to fill the cooling system with the proper grade and type engine coolant.

17. Start the engine and check for leaks.

EXHAUST MANIFOLD

REMOVAL & INSTALLATION

2.5L Engine

See Figure 52.

1. Before servicing the vehicle, refer to the Precautions Section.

2. Properly relieve the fuel system pressure.

3. Disconnect the negative battery cable.

4. Remove the quick connector cap and

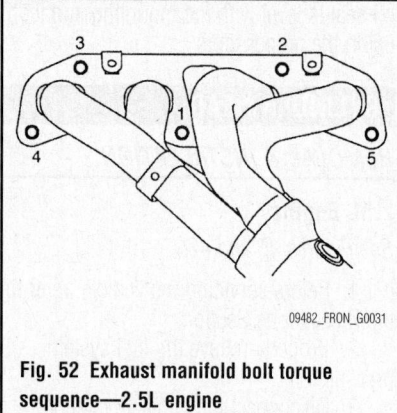

09482_FRON_G0031

Fig. 52 Exhaust manifold bolt torque sequence—2.5L engine

disconnect the quick connector at the engine side.

5. Remove the air duct and PCV hose.

6. Remove the electric throttle control actuator retaining bolts. Be sure to remove the bolts by reversing the order of the tightening torque sequence.

7. Remove the electric throttle control actuator and gasket.

8. Disconnect the harness connector of the air fuel ratio sensor and the harness from the bracket and middle clamp. Remove the air fuel ratio sensor using tool J-44626, or equivalent.

➡ **Be careful not to damage the air fuel ratio sensor. Discard the sensor if it has been dropped from a height of more than 19.7 inches on to a hard surface.**

9. Remove the front exhaust tube. Remove the exhaust manifold cover.

10. Remove the bracket between the exhaust manifold three way catalyst assembly and the transmission assembly.

11. Remove the exhaust manifold retaining bolts. Be sure to remove the bolts by reversing the order of the tightening torque sequence.

12. Remove the exhaust manifold from the engine. Discard the gasket.

To install:

13. Installation is the reverse of the removal procedure.

14. Be sure to use new gaskets.

15. Be sure to tighten the exhaust manifold retaining bolts to specification and in the proper sequence.

➡ **Before installing a new air fuel sensor apply anti seize lubricant to the threads. Do not over torque the sensor, doing so may cause damage to the sensor resulting in the MIL light coming on.**

➡ **See throttle valve closed position learning and idle air volume learning procedures, for relearning information.**

4.0L Engine

Left Side

See Figure 53.

1. Before servicing the vehicle, refer to the Precautions Section.
2. Disconnect the negative battery cable.
3. Remove air duct, PCV hose (between air duct and rocker cover) and electric throttle control actuator.
4. Disconnect the harness connector and remove the heated oxygen sensor.

➡ **Be careful not to damage the air fuel ratio sensor. Discard the sensor if it has been dropped from a height of more than 19.7 inches on to a hard surface.**

5. Remove the front exhaust tube.
6. Remove the exhaust manifold cover.
7. Remove bracket between exhaust manifold–three way catalyst assembly and transmission assembly.
8. Remove the exhaust manifold retaining bolts. Be sure to remove the bolts by reversing the order of the tightening torque sequence.
9. Remove the exhaust manifold from the engine. Discard the gasket.

To install:

10. Installation is the reverse of the removal procedure.
11. Be sure to use new gaskets.
12. Be sure to tighten the exhaust manifold retaining bolts to specification and in the proper sequence.

➡ **Before installing a new air fuel sensor and heated oxygen sensor, clean threads and apply anti seize lubricant to the threads. Do not over torque the sensor, doing so may cause damage to the sensor resulting in the MIL light coming on.**

Right Side

1. Before servicing the vehicle, refer to the Precautions Section.
2. Disconnect the negative battery cable.
3. Remove the engine from the vehicle. Position the assembly in a suitable holding fixture.
4. Remove the exhaust manifold retaining bolts. Be sure to remove the bolts by reversing the order of the tightening torque sequence.

➡ **Disregard the numerical order of No.7 and No.8 in the removal process.**

5. Discard the gaskets.

To install:

6. Installation is the reverse of the removal procedure.
7. Be sure to use new gaskets.
8. Be sure to tighten the exhaust manifold retaining bolts to specification and in the proper sequence.

➡ **Before installing a new air fuel sensor and heated oxygen sensor apply anti seize lubricant to the threads. Do not over torque the sensor, doing so may cause damage to the sensor resulting in the MIL light coming on.**

INTAKE MANIFOLD

REMOVAL & INSTALLATION

2.5L Engine

See Figures 54 and 55.

1. Before servicing the vehicle, refer to the Precautions Section.
2. Properly relieve the fuel system pressure.
3. Disconnect the negative battery cable. Drain the cooling system.
4. Remove the air cleaner case, air cleaner and air duct.
5. Disconnect and plug the water hoses from the electric throttle control actuator.
6. Remove the mass air flow sensor from the intake manifold. Remove the quick connector cap and disconnect the quick connector at the engine side.
7. Remove the electric throttle control actuator harness connector and retaining bolts. Be sure to remove the bolts by reversing the order of the tightening torque sequence.
8. Remove the electric throttle control actuator and gasket.
9. Disconnect the harness, vacuum hoses and PCV hoses from the intake manifold and position them to the side.
10. Remove intake manifold support.
11. Remove the intake manifold retaining bolts. Be sure to remove the bolts by reversing the order of the tightening torque sequence.
12. Remove the intake manifold, fuel tube protector and gasket from the engine.
13. As necessary, remove the EVAP canister purge volume solenoid valve and vacuum hose adapter from the intake manifold.

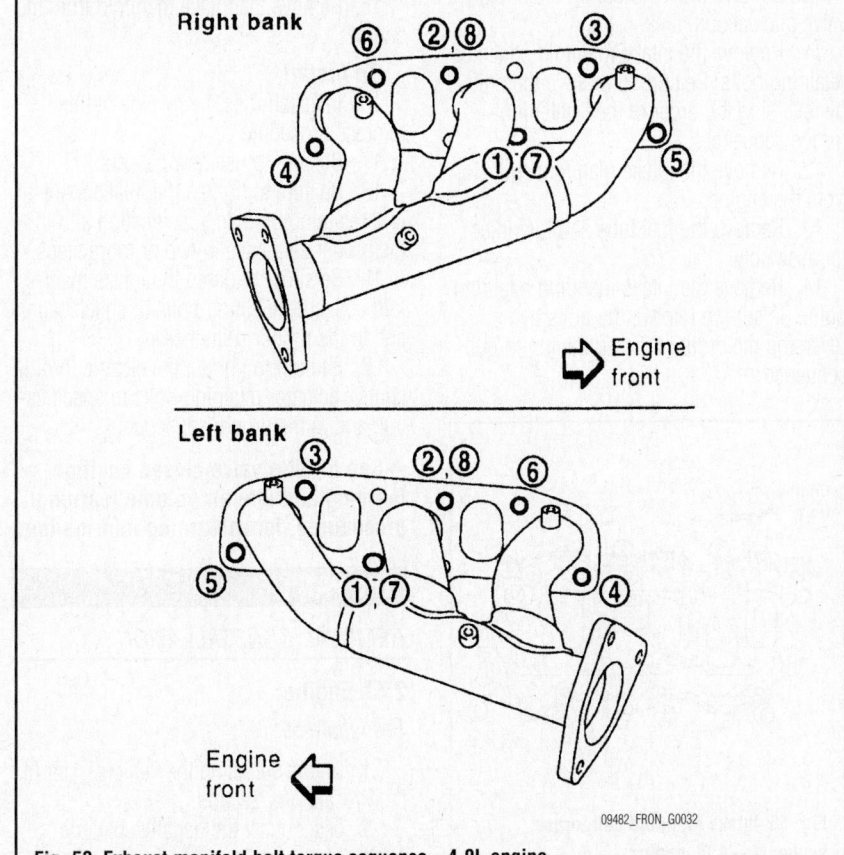

Right bank

Left bank

09482_FRON_G0032

Fig. 53 Exhaust manifold bolt torque sequence—4.0L engine

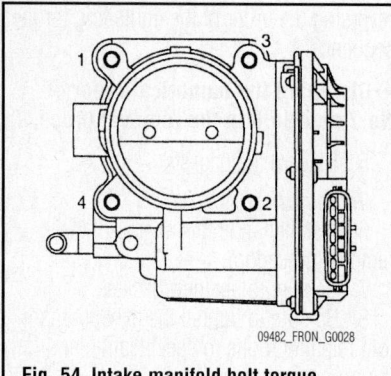

Fig. 54 Intake manifold bolt torque sequence—2.5L engine

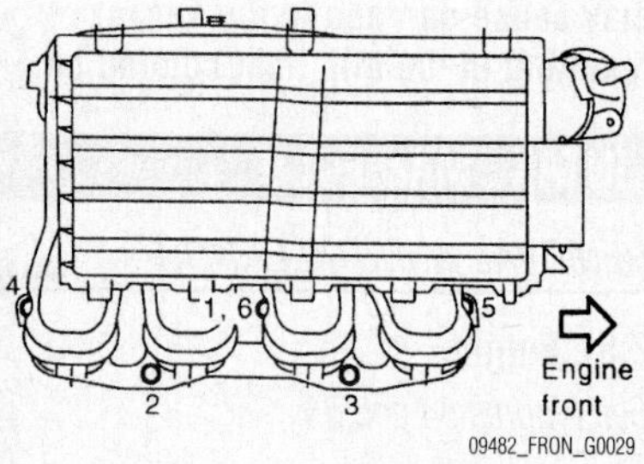

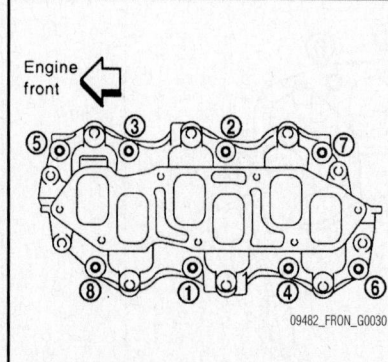

Fig. 55 Electric throttle control actuator bolt torque sequence—2.5L engine

14. Disconnect the sub frame harness from the fuel injectors. Remove the fuel tube and fuel injector assembly from the intake manifold, if required.

To install:

15. Installation is the reverse of the removal procedure.

16. Be sure to use new gaskets.

17. Be sure to tighten the intake manifold retaining bolts to specification and in the proper sequence.

➡**Refer to the torque sequence illustration, No. 6 means double tightening of bolt No. 1. M8xM38mm (1.50 inches) are green in color (No. 1, No.6). M8xM35 mm (1.38 inch) (No. 2, No. 3). Nut (No. 4, No. 5).**

18. Be sure to tighten the electric throttle control actuator retaining bolts to specification and in the proper sequence.

➡**See throttle valve closed position learning and idle air volume learning procedures, for relearning information.**

19. Be sure to fill the cooling system with the proper grade and type engine coolant.

20. Start the engine and check for leaks.

4.0L Engine

See Figures 56 through 58.

➡**Upper intake manifold is also referred to as intake manifold collector.**

1. Before servicing the vehicle, refer to the Precautions Section.

2. Properly relieve the fuel system pressure.

3. Disconnect the negative battery cable. Drain the cooling system.

4. Remove the engine cover. Remove the air cleaner case (upper) with the mass air flow sensor and air duct assembly.

5. Disconnect the water hoses from the electric throttle control actuator. Disconnect the harness connector.

6. Remove the electric throttle control actuator retaining bolts. Be sure to remove the bolts by reversing the order of the tightening torque sequence.

7. Remove the electric throttle control actuator.

8. Remove the brake booster vacuum hose and the PCV hose. Remove the intake manifold collector support.

9. Disconnect the EVAP hoses and harness connector from the EVAP canister purge volume control solenoid valve. Remove the EVAP canister purge volume control solenoid valve.

10. Remove the VIAS control solenoid valve and vacuum tank.

11. Remove the intake manifold collector retaining bolts. Be sure to remove the bolts by reversing the order of the tightening torque sequence.

12. Remove the intake manifold collector from the engine.

13. Remove the fuel tube and fuel injector assembly.

14. Remove the intake manifold retaining bolts. Be sure to remove the bolts by reversing the order of the tightening torque sequence.

Fig. 56 Intake manifold bolt torque sequence—4.0L engine

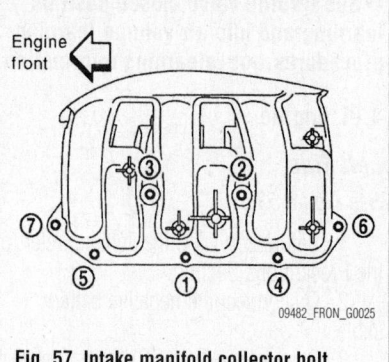

Fig. 57 Intake manifold collector bolt torque sequence—4.0L engine

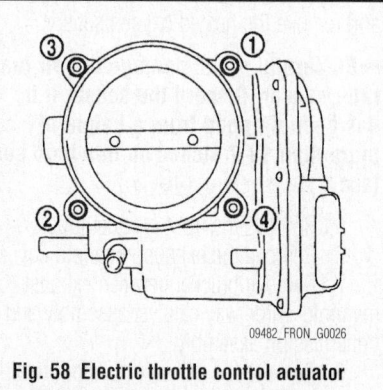

Fig. 58 Electric throttle control actuator bolt torque sequence—4.0L engine

15. Remove the intake manifold from the engine.

To install:

16. Installation is the reverse of the removal procedure.

17. Be sure to use new gaskets.

18. Be sure to tighten the intake manifold retaining bolts to specification and in the proper sequence in two or more steps.

19. Be sure to tighten the intake manifold collector retaining bolts to specification and in the proper sequence.

20. Be sure to tighten the electric throttle control actuator retaining bolts to specification and in the proper sequence.

➡**See throttle valve closed position learning and idle air volume learning procedures, for relearning information.**

OIL PAN

REMOVAL & INSTALLATION

2.5L Engine

See Figure 59.

1. Before servicing the vehicle, refer to the Precautions Section.

2. Disconnect the negative battery cable.

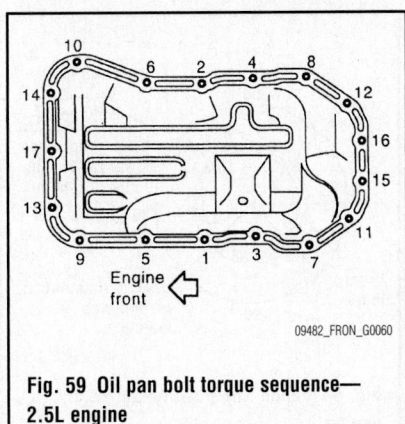

Fig. 59 Oil pan bolt torque sequence—2.5L engine

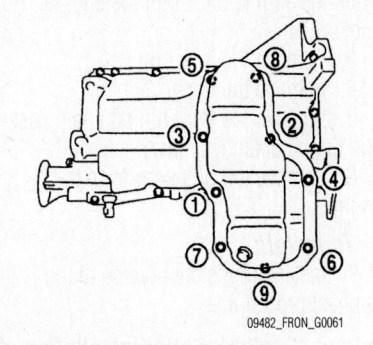

Fig. 60 Lower oil pan bolt torque sequence—4.0L engine

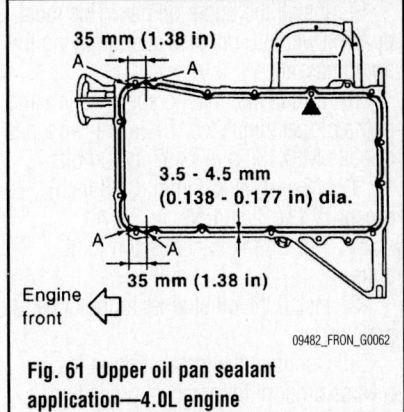

Fig. 61 Upper oil pan sealant application—4.0L engine

3. Remove the engine under cover. Drain the engine oil.

4. If equipped with automatic transmission, remove the fluid cooler tube.

5. Loosen the oil pan retaining bolts, in the reverse order of the installation sequence.

6. Insert a seal cutter tool between the oil pan and the cylinder block, and slide it by tapping on the side of the tool with a hammer.

7. Remove the oil pan from the engine.

To install:

8. Be sure to clean all the oil gasket material from both the oil pan and the cylinder block surfaces, using the proper tools.

9. Apply a continuous bead of sealant 0.138–0.177 inches (3.5–4.5 mm) to the oil pan mating surface.

10. Install the oil pan to the cylinder block. This must be done within 5 minutes after applying the liquid gasket.

11. Torque the bolts to specification and in the proper sequence.

12. Continue the installation in the reverse order of the removal procedure.

➡ **Wait 30 minutes after installation of the oil pan to allow the sealant to cure before adding oil.**

13. Fill the crankcase to the correct level.
14. Start the engine and check for leaks.

4.0L Engine

Lower

See Figure 60.

1. Before servicing the vehicle, refer to the Precautions Section.

2. Disconnect the negative battery cable.

3. Remove the engine under cover. Drain the engine oil.

4. Loosen the oil pan retaining bolts, in the reverse order of the installation sequence.

5. Insert a seal cutter tool between the oil pan and the cylinder block, and slide it by tapping on the side of the tool with a hammer.

6. Remove the oil pan from the engine.

To install:

7. Be sure to clean all the oil gasket material from both the oil pan and the cylinder block surfaces, using the proper tools.

8. Apply a continuous bead of sealant 0.138–0.177 inches (3.5–4.5 mm) to the oil pan mating surface. Be sure to use genuine RTV sealant or equivalent.

9. Install the oil pan to the cylinder block. This must be done within 5 minutes after applying the liquid gasket.

10. Torque the bolts to specification and in the proper sequence.

11. Continue the installation in the reverse order of the removal procedure.

➡ **Wait 30 minutes after installation of the oil pan to allow the sealant to cure before adding oil.**

12. Fill the crankcase to the correct level.
13. Start the engine and check for leaks.

Upper

See Figures 61 and 62.

1. Before servicing the vehicle, refer to the Precautions Section.

2. Disconnect the negative battery cable.

3. Remove the air duct. Remove the engine under cover.

4. Drain the engine oil. Drain the engine coolant.

5. Remove the final drive, if equipped with 4WD.

6. Disconnect the steering gear lower shaft joint bolt and steering gear nuts and bolts, position the assembly out of the way.

7. Remove the starter.

8. Disconnect the automatic transmission fluid cooler brackets, if equipped and position them out of the way.

9. Remove the oil filter, as necessary. Remove the oil cooler.

10. Remove the lower oil pan. Remove the oil strainer.

11. Remove the transmission joint bolts which pierce the oil pan.

12. Remove the rear cover plate.

13. Loosen the upper oil pan retaining bolts, in the reverse order of the installation sequence.

14. Insert a seal cutter tool between the oil pan and the cylinder block, and slide it by tapping on the side of the tool with a hammer.

15. Remove the oil pan from the engine. Remove the O-rings from the bottom lower cylinder block and oil pump.

To install:

16. Be sure to clean all the oil gasket material from both the oil pan and the cylinder block surfaces, using the proper tools.

17. Install new O-rings on the bottom lower cylinder block and oil pump.

18. Apply a continuous bead of sealant 0.138–0.177 inches (3.5–4.5 mm) to the lower cylinder block mating surfaces of the upper oil pan. Be sure to use genuine RTV sealant or equivalent.

➡ **For bolt holes marked with a solid black triangle, apply liquid gasket outside the hole. Apply a bead of sealant (0.177–0.217 inch diameter) to area "A".**

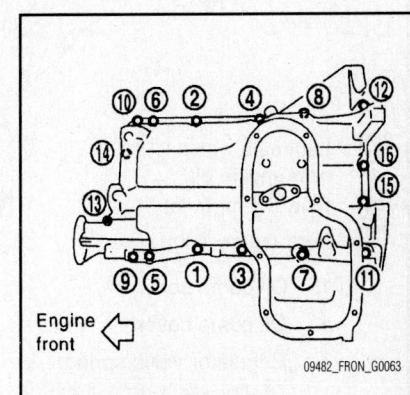

Fig. 62 Upper oil pan bolt torque sequence—4.0L engine

19. Install the upper oil pan. This must be done within 5 minutes after applying the liquid gasket.

20. Torque the bolts to specification and in the proper sequence. There are two types of bolts M8X100 mm (3.97 inch) bolts 7, 11, 12, 13 and M8X25 mm (0.98 inch) except 7, 11, 12 and 13.

21. Tighten the transmission joint bolts.

22. Install the oil strainer to the upper oil pan.

23. Continue the installation in the reverse order of the removal procedure.

➡**Wait 30 minutes after installation of the oil pan to allow the sealant to cure before adding oil.**

24. Fill the crankcase to the correct level.

25. Start the engine and check for leaks.

OIL PUMP

REMOVAL & INSTALLATION

4.0L Engine

See Figure 63.

1. Before servicing the vehicle, refer to the Precautions Section.

2. Disconnect the negative battery cable.

Drain the engine oil. Drain the engine coolant.

3. Remove the lower oil pan.

4. Remove the upper oil pan.

5. Remove the front timing chain case and timing chain (primary).

6. Remove the oil pump from the engine.

To install:

7. Installation is the reverse of the removal procedure.

➡**Wait 30 minutes after installation of the oil pan to allow the sealant to cure before adding oil.**

8. Fill the crankcase to the correct level.

9. Start the engine and check for leaks.

PISTON AND RING

POSITIONING

See Figures 64 and 65.

REAR MAIN SEAL

REMOVAL & INSTALLATION

1. Before servicing the vehicle, refer to the Precautions Section.

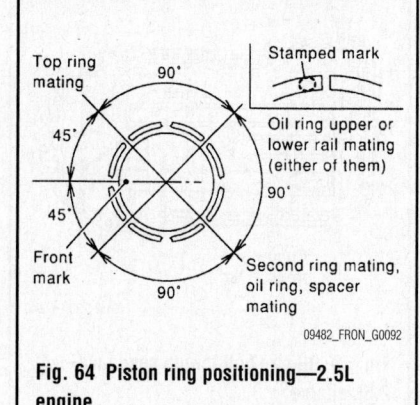

Fig. 64 Piston ring positioning—2.5L engine

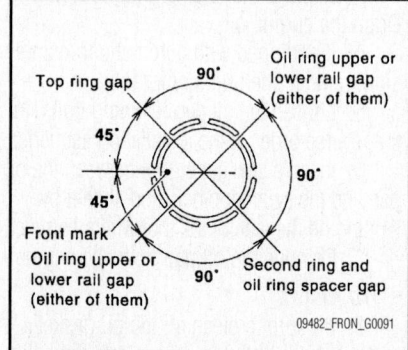

Fig. 65 Piston ring positioning—4.0L engine

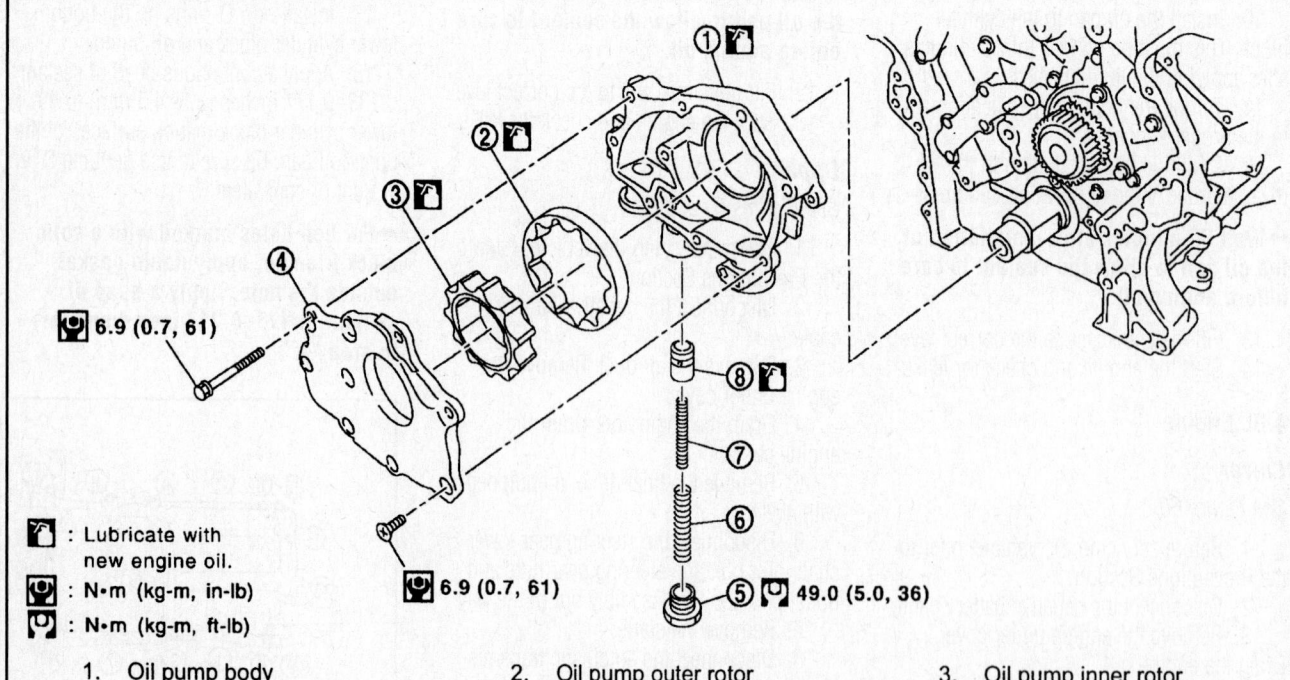

6.9 (0.7, 61)

6.9 (0.7, 61)

49.0 (5.0, 36)

: Lubricate with new engine oil.

: N•m (kg-m, in-lb)

: N•m (kg-m, ft-lb)

1. Oil pump body
2. Oil pump outer rotor
3. Oil pump inner rotor
4. Oil pump cover
5. Regulator valve plug
6. Regulator valve spring
7. Regulator valve spring
8. Regulator valve

Fig. 63 Oil pump and related components—4.0L engine

2. Remove or disconnect the following:
- Transmission
- Flywheel
- Clutch, if equipped
- Rear main seal

To install:

3. Install the seal so that it is flush with the retainer housing.

4. Install or connect the following:
- Flywheel.
- Transmission

TIMING CHAIN & SPROCKETS

REMOVAL & INSTALLATION

2.5L Engine

See Figures 66 through 73.

1. Before servicing the vehicle, refer to the Precautions Section.

2. Properly relieve the fuel system pressure.

3. Disconnect the negative battery cable.

4. Remove the air cleaner and the air duct assembly.

5. Remove the spark plugs.

6. Disconnect the PCV hose from the rocker cover. Remove the ignition coil.

7. Remove the PCV valve and O-ring from the rocker cover, if necessary.

8. Remove the oil filler cap from the rocker cover, if necessary.

9. Remove the rocker cover retaining bolts. Be sure to remove the bolts by reversing the order of the tightening torque sequence.

10. Remove the rocker cover. Discard the gasket.

11. Remove the coolant reservoir tank. Remove the auxiliary drive belt auto-tensioner.

12. Remove the alternator. Remove the strut tower brace.

13. Remove the air conditioning compressor and position it to the side. Do not disconnect the refrigerant lines.

14. Remove the power steering pump and reservoir tank; position the assembly to the side. Do not disconnect the fluid lines.

15. Remove the upper and lower oil pan. Remove the strainer.

16. Remove the IVT control cover bolts. Be sure to remove the bolts by reversing the order of the tightening torque sequence.

17. Remove the cover by cutting the sealant using tool KV10111100 (J-37228) or equivalent.

18. Position the number one cylinder on its compression stroke by rotating the crankshaft pulley clockwise. Align the mat- ing marks for TDC with the timing indicator on the front cover.

19. At the same time make sure that the mating marks on the camshaft sprockets are lined up as indicated in the illustration. If not rotate the crankshaft one more turn to line up the mating marks.

20. Hold the crankshaft pulley with a suitable tool. Loosen the crankshaft pulley retaining bolt.

21. Pull the pulley out about 0.39 inch. Remove the crankshaft pulley bolt.

22. Attach a pulley puller in the M6 thread hole on the crankshaft pulley. Remove the crankshaft pulley.

➡**Be careful not to damage the front cover and/or the crankshaft.**

23. Loosen the front cover retaining bolts, in the order indicated in the bolt loosening sequence illustration. Remove the front cover. Be careful not to damage the mating surfaces.

24. Using a seal removal tool, remove the oil seal, as required.

25. To remove the timing chain, push in on the chain tensioner plunger. Insert a stopper pin into the hole on the chain tensioner body to secure the chain tensioner plunger. Remove the chain tensioner. Remove the chain. Do not rotate the crankshaft with the chain removed.

➡**Use a 0.02 inch (approximate) metal pin as a stopper pin.**

26. Remove the camshaft sprockets.

27. Remove the timing chain slack guide, timing chain tensioner guide and spacer.

28. Remove the balancer unit timing chain tensioner by lifting the lever up and releasing the ratchet claw for return proof. Push the tensioner sleeve in and hold it. Matching the hole on the lever with the one on the body, insert a stopper pin to secure the tensioner sleeve. Remove the balancer unit timing chain tensioner.

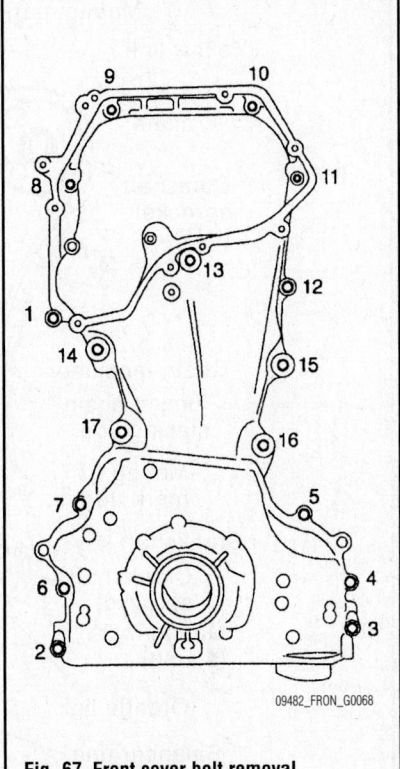

Fig. 67 Front cover bolt removal sequence—2.5L engine

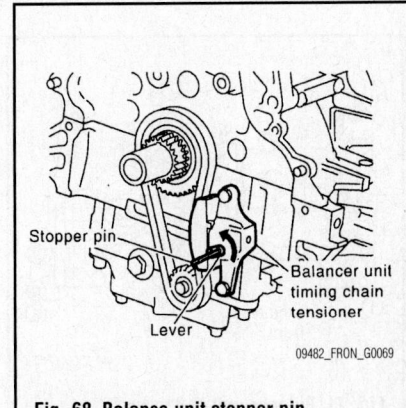

Fig. 68 Balance unit stopper pin installation—2.5L engine

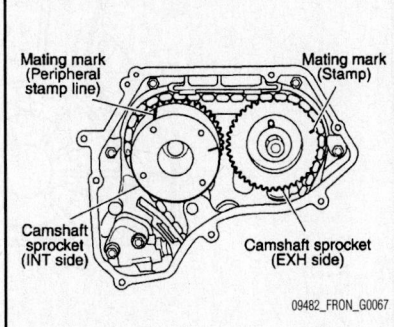

Fig. 66 Timing chain alignment marks— 2.5L engine

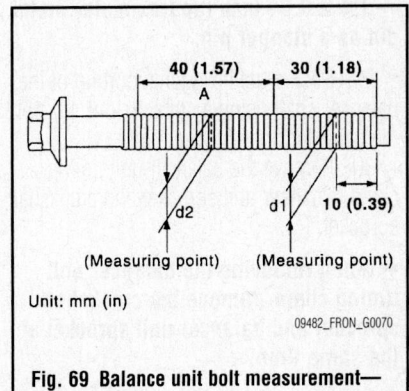

Fig. 69 Balance unit bolt measurement— 2.5L engine

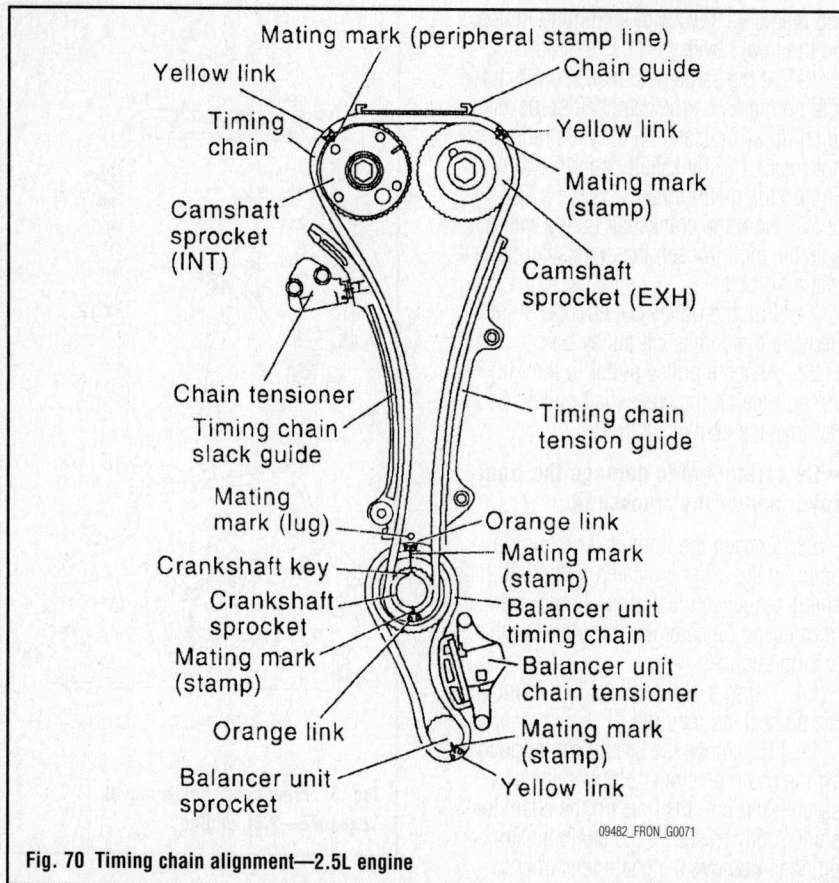

Fig. 70 Timing chain alignment—2.5L engine

bolts, in the order of the tightening sequence. Remove the balancer unit. Do not disassemble the balancer unit. Bolts one and four use a E14 Torx®head socket.

To install:

32. Check the chain for cracks and excessive wear, replace as required.

33. Measure the balancer unit bolt outer diameters ("d1" and "d2") at two positions, as shown in the illustration. If reduction appears in the "A" range, regard it as "d2". Specification is as follows: ("d1" - "d2"): 0.0059 inch. If it exceeds the specification (large difference in dimensions) replace the balancer unit bolt with a new one.

34. Measure the balancer bolt unit length. If it exceeds the specification replace the balancer unit bolt with a new one. Specification is 6.974 inch.

35. Make sure that the crankshaft key is pointing straight up. Install the O-ring to the balancer unit.

36. Install the balancer unit. Apply engine oil to the bolt threads and sealing surfaces. Torque the bolts to specification and in the proper sequence using tool KV10112100 (BT8653-A) or equivalent.
   a. Step 1: bolts 1–4 to 35 ft. lbs.
   b. Step 2: bolts 1–4 100 degrees clockwise

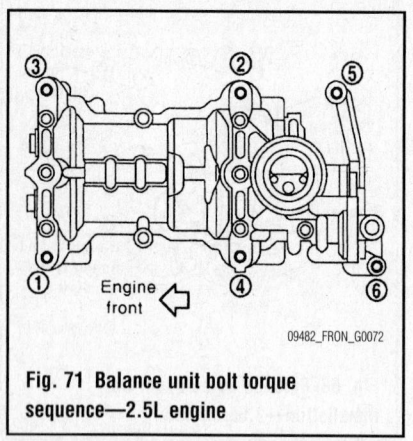

Fig. 71 Balance unit bolt torque sequence—2.5L engine

➡Use a 0.04 inch (approximate) metal pin as a stopper pin.

29. Secure the hexagonal portion of the balancer shaft using a suitable tool. Loosen the balancer unit sprocket bolt.

30. Remove the balancer unit timing chain, balancer unit sprocket and crankshaft sprocket.

➡When removing the balancer unit timing chain, remove the crankshaft sprocket and balancer unit sprocket at the same time.

31. Loosen the balancer unit mounting

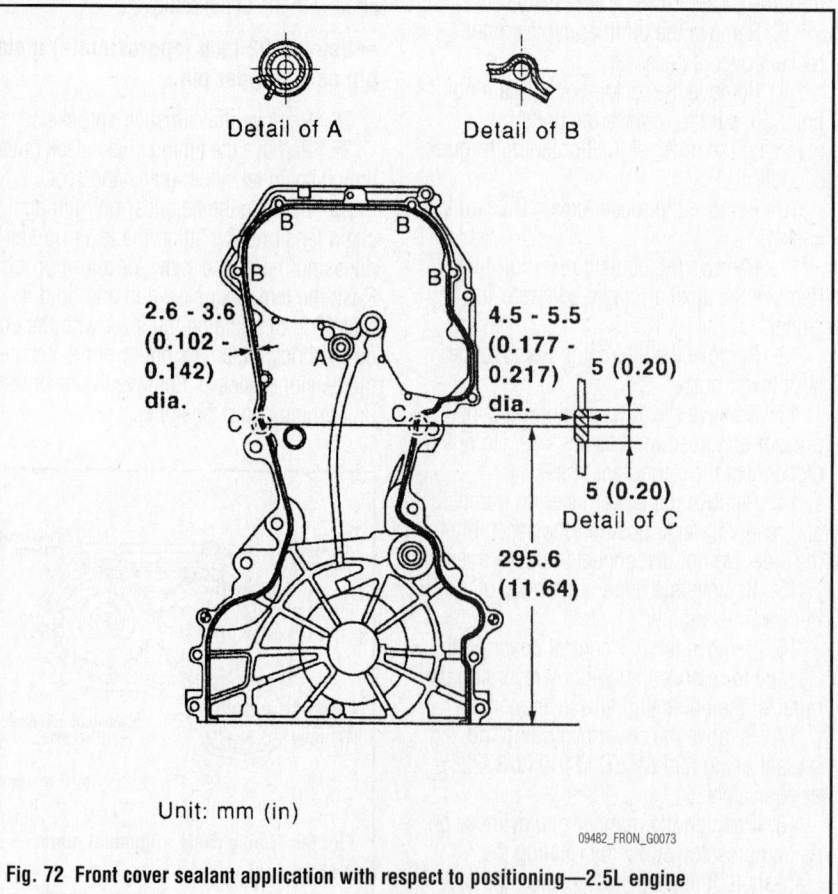

Unit: mm (in)

Fig. 72 Front cover sealant application with respect to positioning—2.5L engine

c. Step 3: bolts 1–4 loosen in the reverse order of the tightening sequence to zero

d. Step 4: bolts 1–4 to 35 ft. lbs.

e. Step 5: bolts 1–4 100 degrees clockwise

f. Step 6: bolts 5–6 to 22 ft. lbs.

➡Check the tightening angle using a tool or a protractor. Do not make a judgment by visual check alone.

37. Install the crankshaft sprocket, balancer unit sprocket and balancer timing chain.

38. Make sure that the crankshaft sprocket is positioned with the mating marks on the cylinder block and crankshaft sprocket meeting at the top.

39. Install it by aligning the mating marks on each sprocket and balancer unit timing chain.

40. Secure the hexagonal portion of the balancer shaft using a suitable tool. Tighten the balancer unit sprocket bolt to specification.

➡Install the crankshaft sprocket, balancer unit sprocket and balancer unit timing chain at the same time.

41. Install the balancer unit timing chain tensioner.

➡After installation, make sure that the mating marks have not slipped. Remove the stopper pin and release the tensioner sleeve.

42. Align the mating marks on each sprocket and timing chain. Install the timing chain and related parts.

43. Before and after installing the chain tensioner, check again to be sure that the mating marks have not slipped.

44. After installing the chain tensioner, remove the stopper pin. Make sure that the tensioner moves freely.

➡After the mating marks are aligned, keep them aligned by holding them with your hand. To avoid skipped teeth, do not rotate the crankshaft and camshaft until the cover is installed.

➡Before installing the chain tensioner, it is possible to change the position of the mating mark on the timing chain for that on each sprocket for alignment.

45. Install the front cover oil seal. Install O-rings to the cylinder head and the cylinder block.

46. Apply a continuous bead of liquid gasket to the front cover. Be sure to use genuine RTV sealant, or equivalent.

➡Sealant application instructions differ depending on position, refer to the illustration for positioning. Detail "A", cross over the start of the application and the end. Detail "B", apply liquid gasket outside of the bolt holes. For all bolt holes other than "B", apply to the inside. Detail "C", between here only, apply a bead of sealant 0.177–0.217 inch diameter.

47. Make sure that the mating marks of the chain and each sprocket are still aligned.

48. Install the front cover. Torque the retaining bolts to specification and in the proper sequence. Bolt position 5, 10, 14 and 17: 45 mm (1.77 inch). Except the above (except 1 to 4): 20mm (0.79 inch).

a. M6 bolts: 9 ft. lbs.

b. M10 bolts: 36 ft. lbs.

c. After all bolts are tightened, retighten them to specification and in the proper sequence.

➡Be sure to wipe off any excess liquid gasket leaking to the surface for installing the oil pan.

49. Install the chain guide between the camshaft sprockets.

50. Install the oil rings to the camshaft sprocket (INT) insertion points on backside of the intake valve timing control cover. Install the O-ring to the front cover.

51. Apply a continuous bead of liquid gasket, 0.122 inch in diameter, to the front cover. Be sure to use genuine RTV sealant or equivalent.

52. Install the cover. Tighten the bolts in the proper sequence to specification.

53. Install the intake valve timing control solenoid valve to the intake valve timing control cover, if removed.

54. Connect the ground cables, and install the harness clip.

55. Install the crankshaft pulley. Torque the retaining bolt to specification.

56. When installing the rocker cover, apply liquid gasket, be sure to use genuine RTV silicone sealant or equivalent, to the positions shown in the illustration. Refer to figure "a" to apply liquid gasket to joint part of camshaft bracket No. 1 and cylinder head. Refer to figure "b" to apply liquid gasket in 90 degrees to figure "b".

57. Install the rocker cover. Torque the retaining bolts to 18 inch lbs. and then to 73 inch lbs., in the proper sequence.

58. Continue the installation in the reverse order of the removal procedure.

➡If hydraulic pressure inside the timing chain tensioner drops after

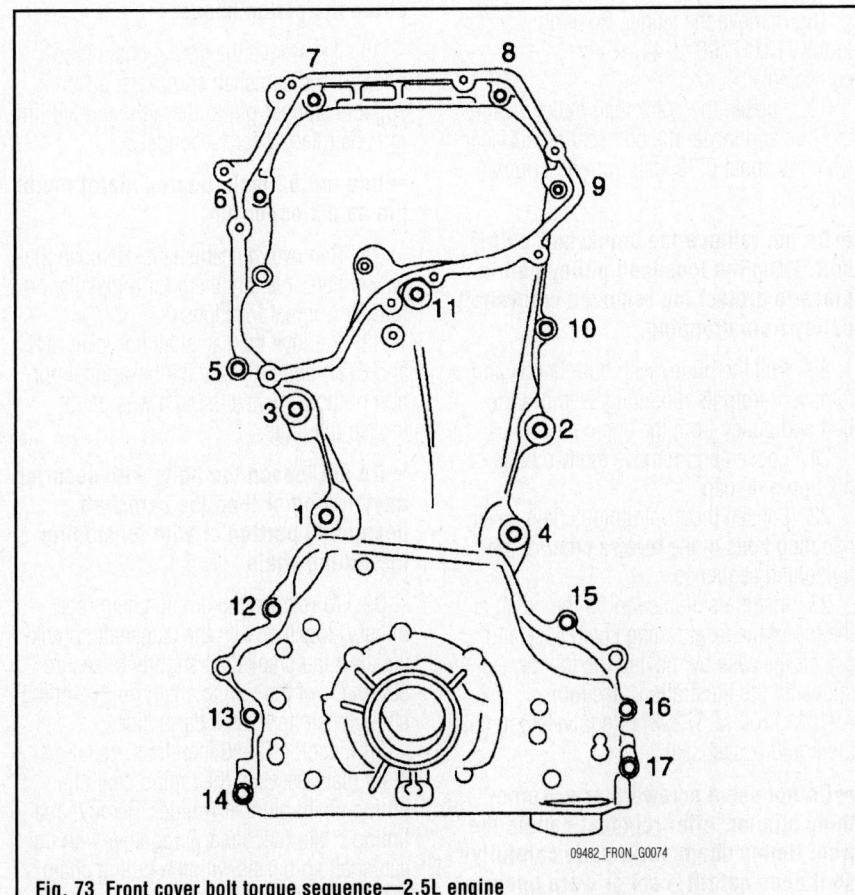

09482_FRON_G0074

Fig. 73 Front cover bolt torque sequence—2.5L engine

removal/installation, slack in the guide may generate a pounding noise during and just after engine start. This is normal the noise will stop after hydraulic pressure rises.

4.0L Engine

See Figures 73 through 88.

➡The procedure below describes the removal and installation of the front timing case and timing chain related parts and rear timing chain case, when the upper oil pan needs to be removed or installed. When only the timing chain (primary) is being removed it is not necessary to remove the rocker covers.

1. Before servicing the vehicle, refer to the Precautions Section.
2. Properly relieve the fuel system pressure.
3. Disconnect the negative battery cable.
4. Remove the engine cover. Drain the engine oil. Drain the engine coolant.
5. Remove the upper and lower oil pans.
6. Remove the radiator cooling fan assembly. Remove the drive belts.
7. Separate the engine wiring harnesses by removing their brackets from the front timing chain case.
8. Remove the power steering pump from the bracket with the fluid hoses attached. Position the assembly to the side. Do not disconnect the hoses. Remove the bracket.
9. Remove the alternator. Remove the water bypass hose, water hose clamp and idler pulley bracket from the front timing chain case.
10. Remove the left and right intake valve timing control covers. Loosen the bolts in the reverse order of the tightening sequence. Use tool KV10111100 (J-37228) or equivalent to cut the liquid gasket seal.

➡The shaft is internally jointed with the camshaft sprocket (INT) center hole. When removing, keep it horizontal until it is completely disconnected.

11. Remove the collared O-rings from the front timing chain case on both the left and right side.
12. Remove the intake manifold collector.
13. Separate the engine harness and remove their brackets from the rocker covers. Remove the harness bracket from the cylinder head, if necessary.

14. Remove the ignition coil. Remove the PCV hoses. Remove the oil filler cap, if necessary.
15. Loosen the rocker cover retaining bolts, in the reverse order of the tightening sequence.
16. Remove the rocker covers from the engine.

➡When only the timing chain (primary) is being removed it is not necessary to remove the rocker covers.

17. Set the No. 1 cylinder at TDC of its compression stroke by rotating the crankshaft pulley clockwise to align the timing mark (grooved line without color) with the timing indicator. Make sure that the intake and exhaust cam noses on No. 1 cylinder (engine front side on right bank) are in alignment as shown in the illustration. If not, rotate the crankshaft in the clockwise direction 360 degrees.

➡When only the timing chain (primary) is removed, the rocker cover does not need to be removed. To be sure that the No. 1 cylinder is set at TDC on the compression stroke, remove the front timing chain case cover first, then check the mating marks on the camshaft sprockets.

18. Remove the starter. Position tool KV10117700 (J-44716) or equivalent.
19. Loosen the crankshaft pulley retaining bolt and locate the bolt seating surface, which is about 0.39 inch from its original position.

➡Do not remove the crankshaft pulley bolt. Keep the loosened pulley bolt in place to protect the removed crankshaft pulley from dropping.

20. Pull the pulley with both hands and remove it from its mounting. Remove the bolt and pulley from the engine.
21. Loosen and remove the two bolts of the upper oil pan.
22. Loosen the front timing chain cover retaining bolts in the reverse order of the tightening sequence.
23. Insert a suitable tool in the notch at the top of the front timing chain case and pry off the case by moving the tool as shown in the illustration. Use tool KV10111100 (J-37228) or equivalent to cut the liquid gasket seal.

➡Do not use a screwdriver or something similar. After removal handle the front timing chain cover case carefully so it does not tilt, cant or warp under a load.

24. Remove the O-rings from the rear timing chain case.
25. Remove the water pump cover and chain tensioner cover from the front timing chain case cover, as required.
26. Remove the oil seal from the front timing chain case cover, as required.
27. Remove the timing chain tensioner (primary) by loosening the clip of the timing chain tensioner (primary) and release the plunger stopper. Insert the plunger into the tensioner body by pressing the slack guide. Keep the slack guide pressed and hold the plunger in by pushing the stopper pin through the tensioner body hole and the plunger groove. Remove the bolts and remove the timing chain tensioner (primary).
28. Remove the internal chain guide, tension guide and slack guide.

➡The tension guide can be removed after removing the timing chain (primary).

29. Remove the timing chain (primary) and the crankshaft sprocket.

➡After removing the timing chain (primary), do not turn the crankshaft and camshaft separately or the valves will strike the piston heads.

30. To remove the timing chain (secondary) and camshaft sprockets, attach a suitable stopper pin to the right and left timing chain tensioner (secondary).

➡Use a 0.02 inch (approximate) metal pin as a stopper pin.

31. Remove the camshafts. Remove the valve lifters. Identify them for reinstallation in their original locations.
32. Remove the camshaft sprocket (INT and EXH) bolts. Secure the hexagonal portion of the camshaft using a wrench to loosen the bolts.

➡Do not loosen the bolts with securing anything other than the camshaft hexagonal portion or with tensioning the timing chain.

33. To remove the timing chain (secondary) together with the camshaft sprockets, turn the crankshaft slightly to secure slackness of the timing chain on the timing chain tensioner (secondary) side.
34. Insert a 0.020 inch thick metal or resin plate between the timing chain and timing chain plunger (guide). Remove the timing chain (secondary) together with the camshaft sprockets with the timing chain loose from the guide groove.

✳✳ CAUTION

Be careful of the plunger coming off when removing the timing chain (secondary). This is because the plunger of the timing chain tensioner (secondary) moves during operation, leading to coming off its fixed stopper pin.

➡**The camshaft sprocket (INT) is a one piece integrated design sprocket for the timing chain (primary) and for the timing chain (secondary). When handling the sprocket avoid shock to the sprocket. Do not disassemble or loosen bolt "A", as shown in the illustration.**

35. Remove the water pump.
36. Remove the rear timing chain case cover bolts, in the reverse order of the tightening sequence. Using the proper tool, cut the liquid gasket sealant seal. Remove the cover.

➡**Do not remove the metal cover of the oil passage. After removal, handle the case carefully so it does not tilt, or warp under a load.**

37. Remove the O-rings from the cylinder head and No. 1 camshaft bracket. Remove the O-rings from the cylinder block.
38. If necessary, remove the timing chain tensioners (secondary) from the cylinder head by first removing the No. 1 camshaft bracket. Remove the timing chain tensioners (secondary) with the stopper pin attached.

To install:

39. Check the chain for cracks and excessive wear, replace as required.
40. Be sure to remove all old gasket material from bolts and bolt holes.
41. If removed install the timing chain tensioners (secondary) to the cylinder head.
42. Install camshaft brackets No. 1.
43. To install the rear timing chain case cover, first install new O-rings to the cylinder block, Install new O-rings to the cylinder head and camshaft bracket No. 1.
44. Apply liquid gasket sealant to the rear timing chain case back side, as shown in the illustration. Be sure to use genuine RTV sealant, or equivalent.

➡**For "A" in the figure, completely wipe out excessive liquid gasket extended on a portion touching at engine coolant. Apply liquid gasket on the installation position of the water pump and cylinder head very completely.**

45. Align the rear timing case with dowel pins (right and left) on the cylinder block. Install the rear timing chain case. Make sure that the O-rings stay in place during installation to the cylinder block, cylinder head and camshaft bracket No. 1.
46. Tighten the bolts to specification and in the proper sequence.
   a. Bolt length: 0.79 inch. Bolt position: 1,2,3,6,7,8,9,and 10.
   b. Bolt length: 0.63 inch. Bolt position: except 1,2,3,6,7,8,9,and 10.
   c. Torque bolts to 9 ft. lbs.
   d. After all bolts are tightened, retighten them to specification and in the proper sequence

➡**Be sure to wipe off any excess liquid gasket leaking to the surface for installing the oil pan.**

47. After installing the rear timing case, check the surface height deference between the rear timing chain case and the lower cylinder block. Specification should be -0.0094–0.0055 inch. If not within specification, repeat the installation procedure.
48. Install the water pump, using new O-rings.
49. Make sure that the dowel pin hole, dowel pin of camshaft and crankshaft key are located with number one piston at TDC on the compression stroke.

➡**Though the camshaft does not stop at the position, as shown in the illustration, for placement of the cam nose it is generally accepted that the camshaft is placed for the same direction as the illustration. Camshaft dowel pin hole (intake side): at the cylinder head upper face side in each bank. Camshaft dowel pin hole (exhaust side): at the cylinder head upper face side in each bank. Crankshaft key: at the cylinder head side of the right bank. Hole on the small diameter side must be used for the intake side dowel pin hole.**

50. To install the timing chains (secondary) and camshaft sprockets, push the plunger of the timing chain tensioner (secondary) and keep it pressed in with the stopper pin.

➡**Mating surfaces between the timing chain and sprockets slip easily. Confirm all mating mark positions repeatedly during the installation process.**

51. Install the timing chains (secondary) and camshaft sprockets (INT and EXH).
52. Align the mating marks on the timing chain (secondary) cooper color link, with the ones on the camshaft sprockets (INT and EXH) punched and install them.

➡**Mating marks for the camshaft sprocket (INT) are on the back side of the camshaft sprocket (secondary). There are two types of mating marks, circle and oval. They should be used for the right and the left banks, respectively. Right bank: circle type. Left bank: oval type.**

53. Align the dowel pin and pin hole on the camshafts with the groove and the dowel pin on the sprockets, and install them.
54. On the exhaust side, align the pin hole on the small diameter side of the camshaft front end with the dowel pin on the back side of the camshaft sprocket, and install them.
55. On the exhaust side, align the dowel pin on the camshaft front end with the pin groove on the camshaft sprocket, and install them.

➡**In case that the positions of each mating mark and each dowel pin will not fit on the mating marks, make a fine adjustment to the position holding the hexagonal portion on the camshaft with a wrench, or equivalent.**

➡**Bolts for the camshaft sprockets must be tightened. Tightening them by hand is enough to prevent the dislocation of the dowel pins. It may be difficult to visually check the dislocation of mating marks during and after installation. To make the matching easier, make a mating mark on the top of the sprocket teeth and its extended line in advance with paint.**

56. After confirming that the mating marks are aligned, tighten the camshaft sprocket bolts.
57. Pull the stopper pins out from the timing chain tensioners (secondary). Install the tension guide.
58. To install the timing chain (primary), install the crankshaft sprocket. Be sure that the mating marks on the crankshaft sprocket face the front of the engine.
59. Install the timing chain (primary).

➡**Install the timing chain (primary) so that the mating mark punched on the camshaft sprocket is aligned with the copper link on the timing chain, while the mating mark notched on the crankshaft sprocket is aligned with the yellow link on the timing chain, as shown in the illustration. If it is difficult to align mating marks with and with of the timing chain (primary) with each sprocket, gradually turn the camshaft using a wrench on the hexagonal portion to align it with the timing marks. During alignment be careful to prevent dislocation of the mating marks**

alignments of the timing chains (secondary). Note indicates the water pump.

60. Install the internal chain guide, slack guide and timing chain tensioner (primary).

➡ **Do not over tighten the slack guide bolts. It is normal for a gap to exist under the bolt seats when the bolts are tightened to specification.**

61. When installing the timing chain tensioner (primary), push in the plunger and keep it pressed in with the stopper pin. Remove any dirt on the surfaces. After

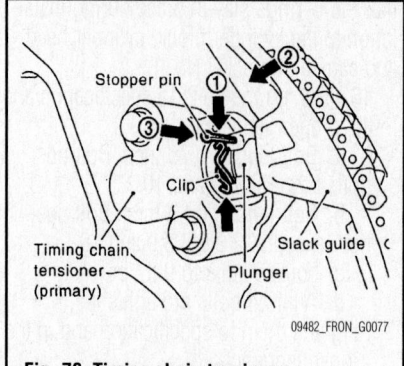

Fig. 76 Timing chain tensioner (primary)—4.0L engine

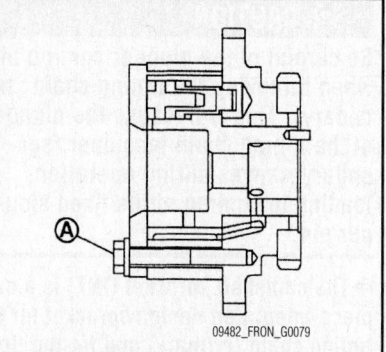

Fig. 78 Camshaft sprocket bolt location—4.0L engine

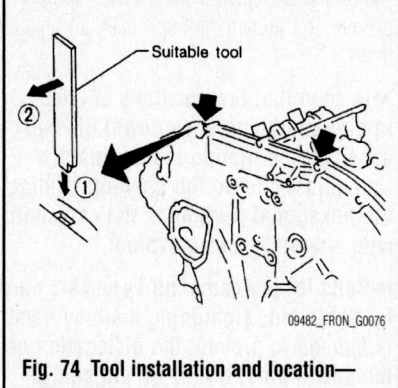

Fig. 74 Tool installation and location—4.0L engine

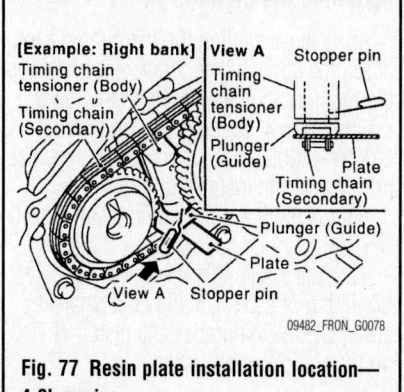

Fig. 77 Resin plate installation location—4.0L engine

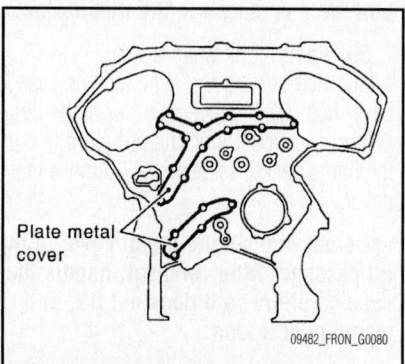

Fig. 79 Metal cover plate location on rear timing case cover—4.0L engine

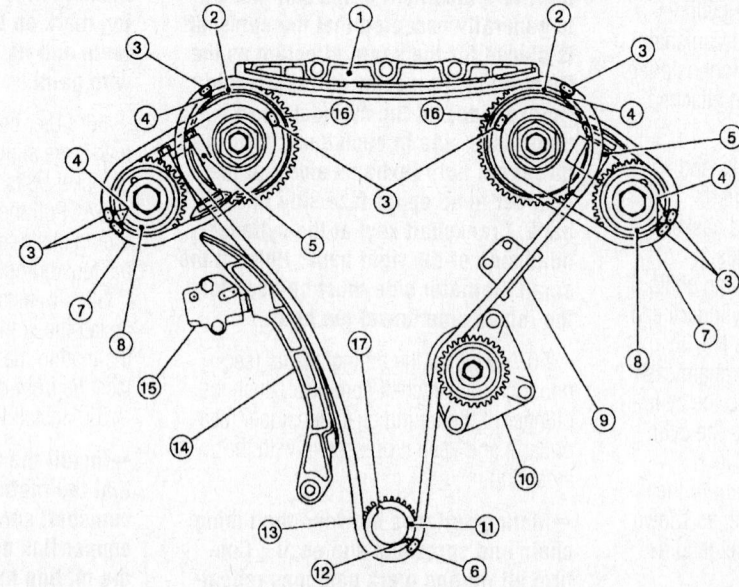

1. Internal chain guide	2. Camshaft sprocket (intake)	3. Mating mark (copper link)
4. Mating mark (punched)	5. Secondary timing chain tensioner	6. Mating mark (yellow link)
7. Secondary timing chain	8. Camshaft sprocket (exhaust)	9. Tensioner guide
10. Water pump	11. Crankshaft sprocket	12. Mating mark (notched)
13. Primary timing chain	14. Slack guide	15. Primary timing chain tensioner
16. Mating mark (back side)	17. Crankshaft key	

Fig. 75 Timing chain alignment—4.0L engine

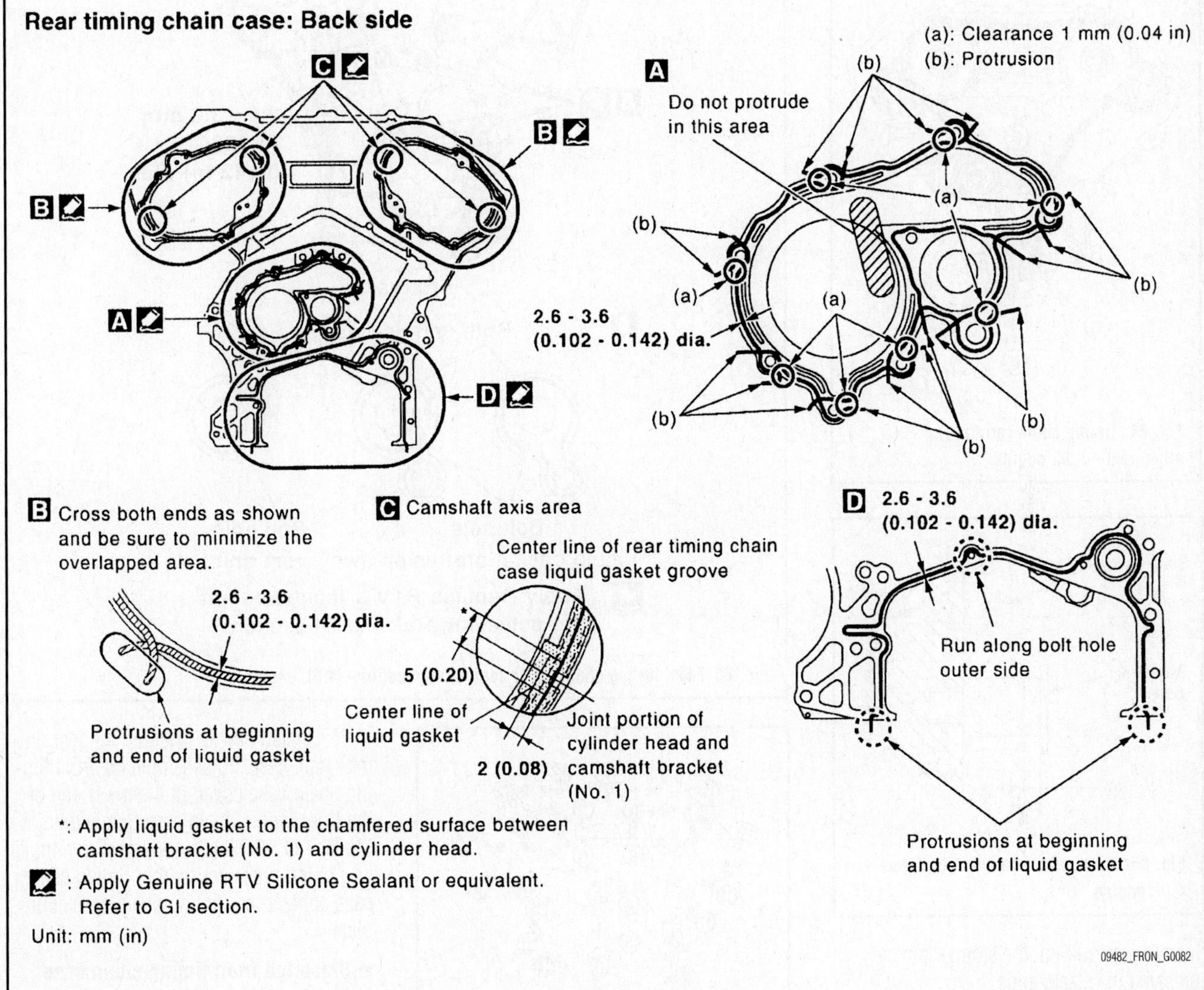

Rear timing chain case: Back side

C

B

B

A

D

(a): Clearance 1 mm (0.04 in)
(b): Protrusion

A

(b)

Do not protrude
in this area

(b)

(a)

(a)

2.6 - 3.6
(0.102 - 0.142) dia.

(a)

(b)

(b)

(b)

(b)

(b)

B Cross both ends as shown
and be sure to minimize the
overlapped area.

2.6 - 3.6
(0.102 - 0.142) dia.

Protrusions at beginning
and end of liquid gasket

C Camshaft axis area

Center line of rear timing chain
case liquid gasket groove

5 (0.20)

Center line of
liquid gasket

2 (0.08)

Joint portion of
cylinder head and
camshaft bracket
(No. 1)

*: Apply liquid gasket to the chamfered surface between
camshaft bracket (No. 1) and cylinder head.

: Apply Genuine RTV Silicone Sealant or equivalent.
Refer to GI section.

Unit: mm (in)

D 2.6 - 3.6
(0.102 - 0.142) dia.

Run along bolt hole
outer side

Protrusions at beginning
and end of liquid gasket

09482_FRON_G0082

Fig. 80 Rear timing chain cover sealant application—4.0L engine

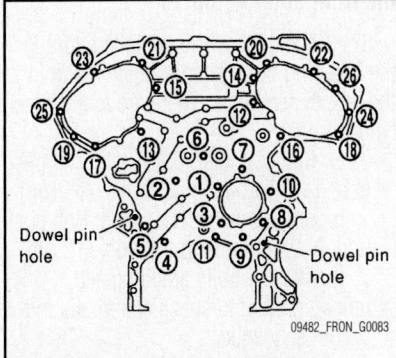

Dowel pin
hole

Dowel pin
hole

09482_FRON_G0083

**Fig. 81 Rear timing chain cover bolt
torque sequence—4.0L engine**

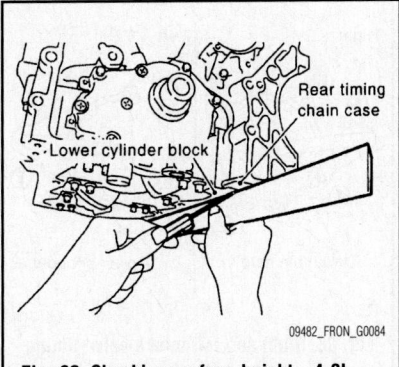

Rear timing
chain case

Lower cylinder block

09482_FRON_G0084

**Fig. 82 Checking surface height—4.0L
engine**

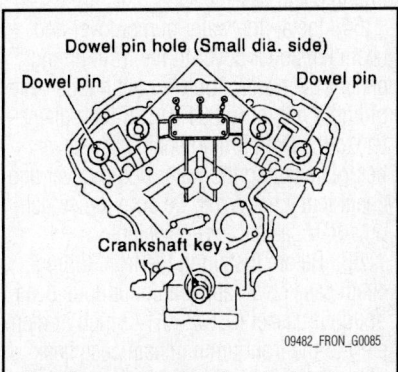

Dowel pin hole (Small dia. side)

Dowel pin

Dowel pin

Crankshaft key

09482_FRON_G0085

**Fig. 83 Dowel pin and crankshaft key
alignment—4.0L engine**

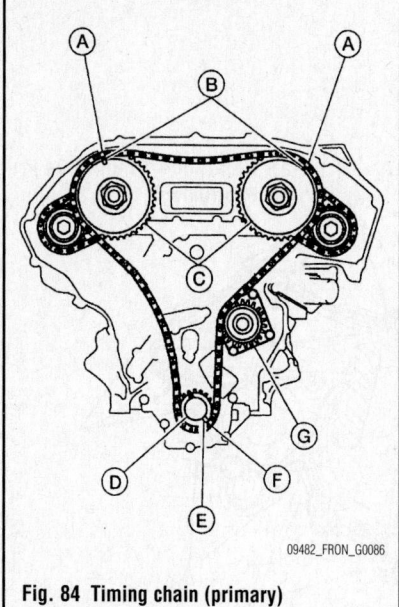

Fig. 84 Timing chain (primary) alignment—4.0L engine

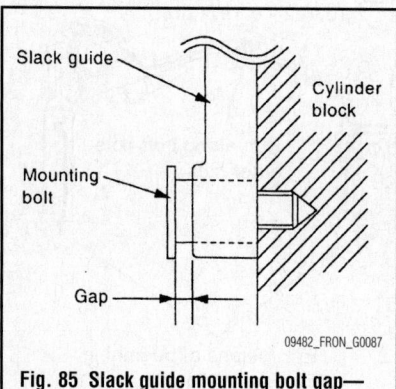

Fig. 85 Slack guide mounting bolt gap—4.0L engine

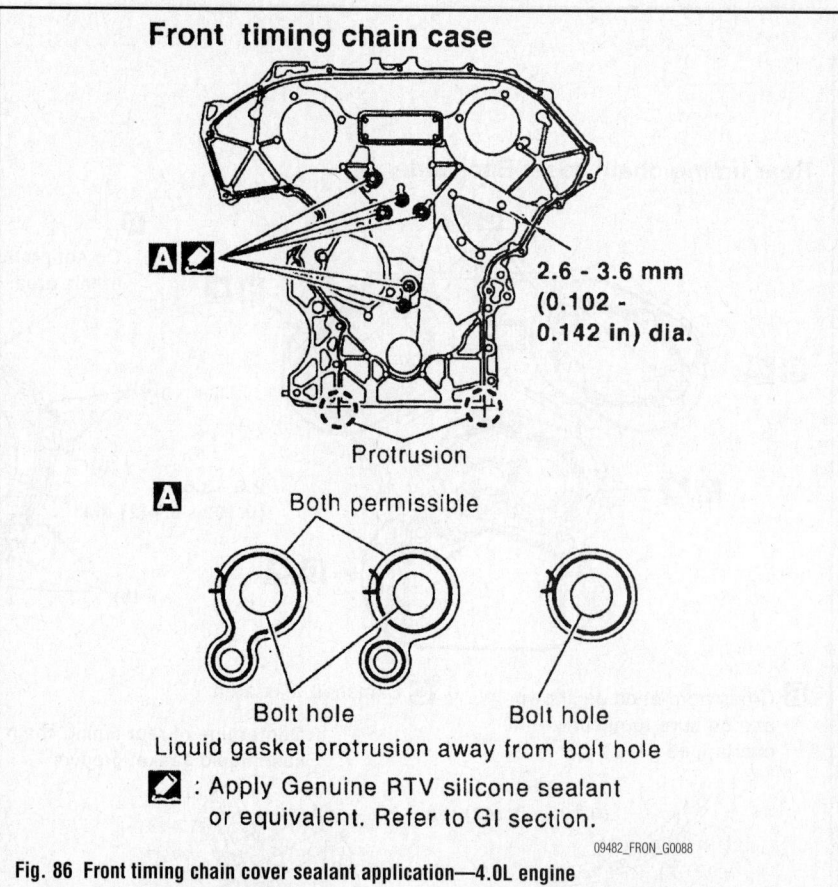

Front timing chain case

2.6 - 3.6 mm
(0.102 -
0.142 in) dia.

Protrusion

A

Both permissible

Bolt hole   Bolt hole

Liquid gasket protrusion away from bolt hole

: Apply Genuine RTV silicone sealant
or equivalent. Refer to GI section.

Fig. 86 Front timing chain cover sealant application—4.0L engine

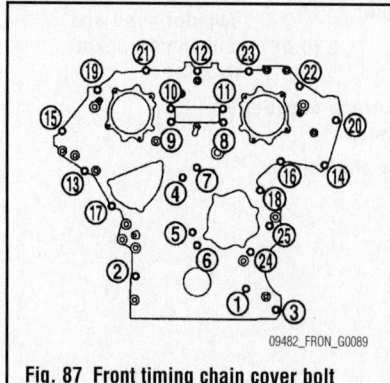

Fig. 87 Front timing chain cover bolt torque sequence—4.0L engine

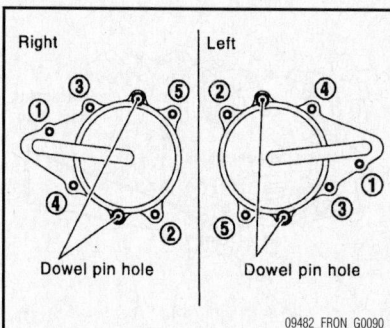

Right   Left

Dowel pin hole   Dowel pin hole

Fig. 88 Right and left intake valve timing control cover bolt torque sequence—4.0L engine

installation, pull out the stopper pin by pressing the slack guide.

62. Make sure, again, that the mating marks on the camshaft sprockets and timing chain have not slipped out of alignment. Install new O-rings on the rear timing chain case.

63. Install a new front seal in the front timing chain case cover.

64. Install the water pump cover and chain tensioner cover to the front timing chain case cover. Apply a continuous bead of liquid gasket (0.091–0.130 inch diameter) to the front timing chain case cover before installing the water pump cover and chain tensioner cover. Be sure to use genuine RTV sealant, or equivalent.

65. Before installing the front timing chain case cover apply a continuous bead of liquid gasket (0.102–0.142 inch in diameter) to the front timing chain case back side, as shown in the illustration. Be sure to use genuine RTV sealant or equivalent.

66. Install new O-rings on the rear timing chain case. To assemble the front timing chain case cover, fit the lower end of the front timing chain case tightly onto the top face of the oil pan (upper). From the fitting point, make entire front timing chain case contact rear timing chain case completely.

➡Since the front timing chain case cover is offset for difference of holt holes; tighten the bolts temporarily while holding the front timing chain case cover from the front and the top. Now insert a dowel pin while holding the front timing chain case cover from the front and the top.

67. Once the cover is installed, torque the retaining bolts to specification and in the proper sequence. There are four different types of bolts:

a. Bolt diameter: 0.39 inch. Bolt position: 1–5. Torque to 41 ft. lbs. (56 Nm).

b. Bolt diameter: 0.24 inch. Bolt position: 6–25. Torque to 9 ft. lbs. (12 Nm).

c. After all bolts are tightened, retighten them to specification and in the proper sequence.

68. Install the two bolts in the oil pan (upper). Torque to 16 ft. lbs.

69. Install new seal rings in the shaft grooves of the right and left intake valve timing control covers.

70. Apply a continuous bead of liquid gasket (0.083–0.122 inch in diameter) to the covers. Be sure to use genuine RTV sealant or equivalent.

71. Install new collared O-rings in the front timing chain case oil hole (left and right sides). Be careful not to move the seal ring from the installation groove, align the dowel pins on the front timing chain case with the holes to install the intake valve timing control covers.

72. Tighten the bolts in sequence and to specification.

73. Install the crankshaft pulley. Torque to specification.

74. Install the upper and lower oil pans.

75. Install the intake manifold collector.

76. Before installing the rocker cover, apply liquid gasket, be sure to use genuine RTV silicone sealant or equivalent, to the positions shown in the illustration. Refer to figure "a" to apply liquid gasket to joint part of camshaft bracket No. 1 and cylinder head. Refer to figure "b" to apply liquid gasket to the figure "a" squarely.

77. Install the rocker cover. Torque the retaining bolts to 17 inch lbs. and then to 74 inch lbs., in the proper sequence.

78. Continue the installation in the reverse order of the removal procedure.

➡ **If hydraulic pressure inside the timing chain tensioner drops after removal/installation, slack in the guide may generate a pounding noise during and just after engine start. This is normal the noise will stop after hydraulic pressure rises.**

VALVE LASH (CLEARANCE)

ADJUSTMENT

2.5L Engine

See Figures 89 through 91.

1. Before servicing the vehicle, refer to the Precautions Section.

2. Disconnect the negative battery cable. Drain the cooling system.

3. Remove the intake manifold.

4. Disconnect the PCV hose from the rocker cover. Remove the ignition coil.

5. Remove the PCV valve and O-ring from the rocker cover, if necessary.

6. Remove the oil filler cap from the rocker cover, if necessary.

7. Remove the rocker cover retaining bolts. Be sure to remove the bolts by reversing the order of the tightening torque sequence.

8. Remove the rocker cover. Discard the gasket.

9. Remove the undercover. Remove the lower radiator shroud.

10. Set the No. 1 cylinder at TDC of its

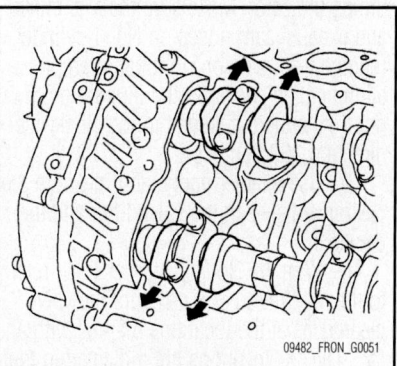

09482_FRON_G0051

Fig. 89 No. 1 cylinder at TDC (compression stroke)—2.5L engine

compression stroke by rotating the crankshaft pulley clockwise to align the TDC mark to the timing indicator on the front cover. At the same time make sure that both the intake and exhaust cam noses of the No. 1 cylinder face outward, as indicated by the arrows in the illustration. If not, rotate the crankshaft in the clockwise direction 360 degrees.

11. Use a feeler gauge and measure the clearance between the valve lifter and the camshaft.

12. With the No. 1 piston at TDC, refer to the illustration and measure the valve clearances at the locations marked with an "X". The "X" locations are indicated in the illustration with an arrow.

13. Rotate the crankshaft pulley clockwise 360 degrees and align the TDC mark to the timing indicator on the front cover.

14. With the No. 4 piston at TDC, refer to the illustration and measure the valve clearances at the locations marked with an "X". The "X" locations are indicated in the illustration with an arrow.

15. If measurements are not within specification, proceed to the next step.

16. Remove the camshaft. Remove the valve lifters that are not within specification.

17. Measure the center thickness of the removed lifters, using a micrometer.

18. Use the equation $(t=t1+(C1-C2)$ to calculate valve lifter thickness for replacement.

➡ **t= valve lifter thickness to be replaced. t1= removed valve lifter thickness. C1= measured valve clearance. C2= standard valve clearance.**

19. Thickness of the new valve lifter can be identified by the stamp mark on the reverse side (inside the cylinder). The stamp mark "696" indicates 6.96 mm (0.2740 inch) thickness.

Measuring position		No. 1 CYL.	No. 2 CYL.	No. 3 CYL.	No. 4 CYL.
No. 1 cylinder at compression TDC	INT	×	×		
	EXH	×		×	

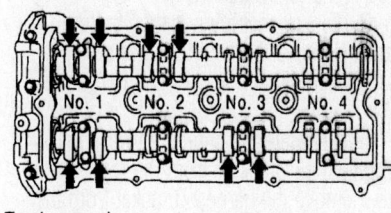

No. 1 cylinder compression TDC

Intake side

Engine front

Exhaust side

09482_FRON_G0052

Fig. 90 Valve adjustment measurement No. 1 cylinder at TDC (compression stroke)—2.5L engine

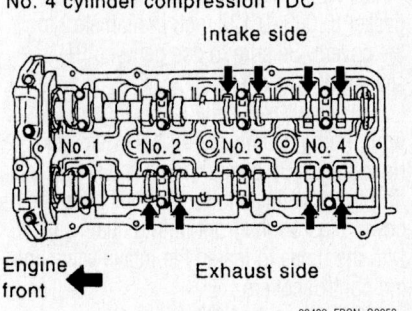

Measuring position		No. 1 CYL.	No. 2 CYL.	No. 3 CYL.	No. 4 CYL.
No. 4 cylinder at compression TDC	INT			×	×
	EXH		×		×

Fig. 91 Valve adjustment measurement No. 4 cylinder at TDC (compression stroke)—2.5L engine

➡**Available thickness of a valve lifter ranges from 6.96–7.46 mm (0.2740–0.2937 inch) in steps of 0.02 mm (0.0008 inch). There are 26 different sizes.**

20. Install the selected valve lifters.
21. Install the camshaft.
22. Manually rotate the crankshaft pulley in the clockwise direction a few rotations.
23. Check the valve clearance and be sure it is within specification.
24. When installing the rocker cover, apply liquid gasket, be sure to use genuine RTV silicone sealant or equivalent, to the positions shown in the illustration. Refer to figure "a" to apply liquid gasket to joint part of camshaft bracket No. 1 and cylinder head. Refer to figure "b" to apply liquid gasket in 90 degrees to figure "b".
25. Install the rocker cover. Torque the retaining bolts to 18 inch lbs. and then to 73 inch lbs., in the proper sequence.
26. Continue the installation in the reverse of the removal procedure.

4.0L Engine

See Figures 92 through 95.

1. Before servicing the vehicle, refer to the Precautions Section.
2. Disconnect the negative battery cable. Remove the engine under cover.
3. Remove the intake manifold collector.
4. Separate the engine harness and remove their brackets from the rocker covers. Remove the harness bracket from the cylinder head, if necessary.
5. Remove the ignition coil. Remove the PCV hoses. Remove the oil filler cap, if necessary.
6. Loosen the rocker cover retaining

bolts, in the reverse order of the tightening sequence.
7. Remove the rocker covers from the engine.
8. Set the No. 1 cylinder at TDC of its compression stroke by rotating the crankshaft pulley clockwise to align the timing mark (grooved line without color) with the timing indicator. Make sure that the intake and exhaust cam noses on No. 1 cylinder (engine front side on right bank) are in alignment as shown in the illustration. If not, rotate the crankshaft in the clockwise direction 360 degrees.
9. Use a feeler gauge and measure the clearance between the valve lifter and the camshaft.
10. With the No. 1 piston at TDC, refer to the illustration and measure the valve clearances at the locations marked with an "X". The "X" locations are indicated in the illustration with an arrow.
11. Rotate the crankshaft pulley clockwise 240 degrees (when viewed from the engine front) to align No. 3 cylinder at TDC on the compression stroke.

➡**The crankshaft pulley bolt flange has a stamped line every 60 degrees, which can be used as a guide to rotation angle.**

12. With the No. 3 piston at TDC, refer to the illustration and measure the valve clearances at the locations marked with an "X". The "X" locations are indicated in the illustration with an arrow.
13. Rotate the crankshaft pulley clockwise 240 degrees (when viewed from the engine front) to align No. 5 cylinder at TDC on the compression stroke.

➡**The crankshaft pulley bolt flange has a stamped line every 60 degrees,**

which can be used as a guide to rotation angle.

14. With the No. 5 piston at TDC, refer to the illustration and measure the valve clearances at the locations marked with an "X". The "X" locations are indicated in the illustration with an arrow.
15. If measurements are not within specification, proceed to the next step.
16. Remove the camshaft. Remove the valve lifters that are not within specification.
17. Measure the center thickness of the removed lifters, using a micrometer.
18. Use the equation (t=t1+(C1-C2) to calculate valve lifter thickness for replacement.

➡**t= valve lifter thickness to be replaced. t1= removed valve lifter thickness. C1= measured valve clearance. C2= standard valve clearance.**

19. Intake valve lifter thickness of the new valve lifter can be identified by the stamp mark on the reverse side (inside the cylinder). The stamp mark "788U" indicates 7.88 mm (0.3102 inch) thickness.

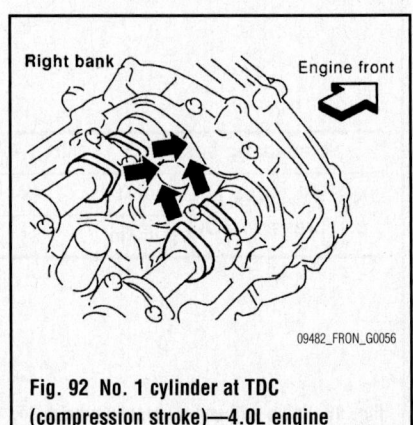

Fig. 92 No. 1 cylinder at TDC (compression stroke)—4.0L engine

Measuring position (right bank)		No. 1 CYL.	No. 3 CYL.	No. 5 CYL.
No. 1 cylinder at compression TDC	EXH		×	
	INT	×		
Measuring position (left bank)		No. 2 CYL.	No. 4 CYL.	No. 6 CYL.
No. 1 cylinder at compression TDC	INT			×
	EXH	×		

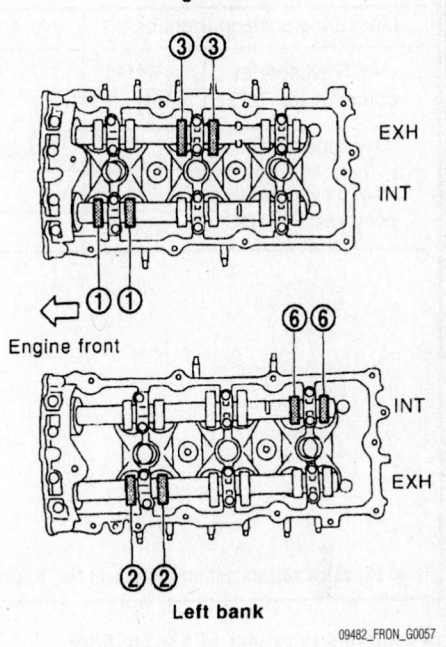

Fig. 93 Valve adjustment measurement No. 1 cylinder at TDC (compression stroke)—4.0L engine

Measuring position (right bank)		No. 1 CYL.	No. 3 CYL.	No. 5 CYL.
No. 3 cylinder at compression TDC	EXH			×
	INT		×	
Measuring position (left bank)		No. 2 CYL.	No. 4 CYL.	No. 6 CYL.
No. 3 cylinder at compression TDC	INT	×		
	EXH		×	

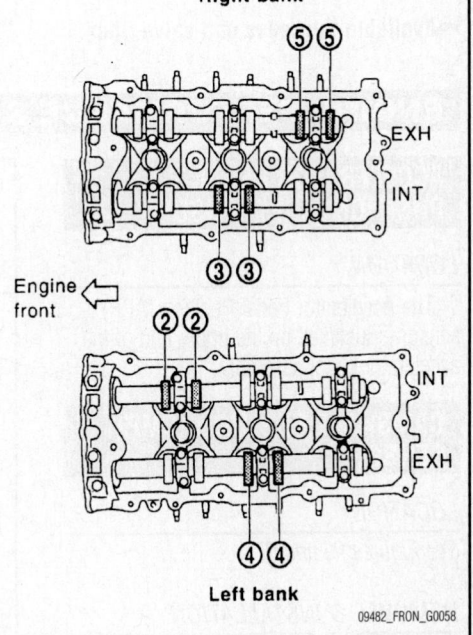

Fig. 94 Valve adjustment measurement No. 3 cylinder at TDC (compression stroke)—4.0L engine

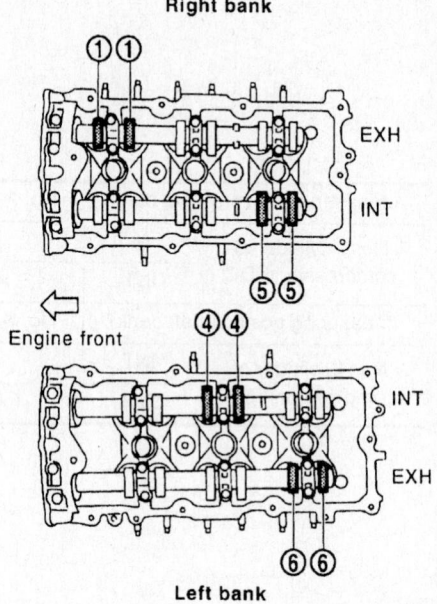

Measuring position (right bank)		No. 1 CYL.	No. 3 CYL.	No. 5 CYL.
No. 5 cylinder at compression TDC	EXH	×		
	INT			×
Measuring position (left bank)		No. 2 CYL.	No. 4 CYL.	No. 6 CYL.
No. 5 cylinder at compression TDC	INT		×	
	EXH			×

09482_FRON_G0059

Fig. 95 Valve adjustment measurement No. 5 cylinder at TDC (compression stroke)—4.0L engine

➡ **Available thickness of a valve lifter ranges from 7.88–8.40 mm (0.3102–0.3307 inch) in steps of 0.02 mm (0.0008 inch). There are 27 different sizes.**

20. Exhaust valve lifter thickness of the new valve lifter can be identified by the stamp mark on the reverse side (inside the cylinder). The stamp mark "N788" indicates 7.88 mm (0.3102 inch) thickness.

➡ **Available thickness of a valve lifter**

ranges from 7.88–8.36 mm (0.3102–0.3291 inch) in steps of 0.02 mm (0.0008 inch). There are 25 different sizes.

21. Install the selected valve lifters.
22. Install the camshaft.
23. Manually rotate the crankshaft pulley in the clockwise direction a few rotations.
24. Check the valve clearance and be sure it is within specification.
25. When installing the rocker cover,

apply liquid gasket, be sure to use genuine RTV silicone sealant or equivalent, to the positions shown in the illustration. Refer to figure "a" to apply liquid gasket to joint part of camshaft bracket No. 1 and cylinder head. Refer to figure "b" to apply liquid gasket to the figure "a" squarely.

26. Install the rocker cover. Torque the retaining bolts to 17 inch lbs. and then to 74 inch lbs., in the proper sequence.
27. Continue the installation in the reverse of the removal procedure.

ENGINE PERFORMANCE & EMISSION CONTROLS

ACCELERATOR PEDAL POSITION (APP) SENSOR

LOCATION

The Accelerator Pedal Position (APP) sensor is installed on the upper end of the accelerator pedal assembly.

CAMSHAFT POSITION (CMP) SENSOR

LOCATION

See Figures 96 and 97.

REMOVAL & INSTALLATION

1. Loosen the fixing bolt of the sensor.
2. Disconnect CMP sensor (PHASE) harness connector.
3. Remove the sensor.

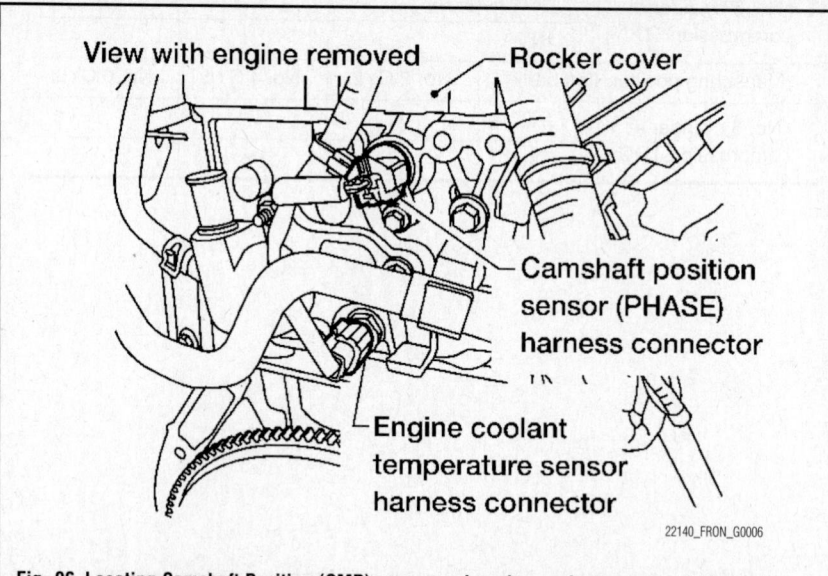

Fig. 96 Locating Camshaft Position (CMP) sensor and engine coolant sensor—2.5L engine

22140_FRON_G0006

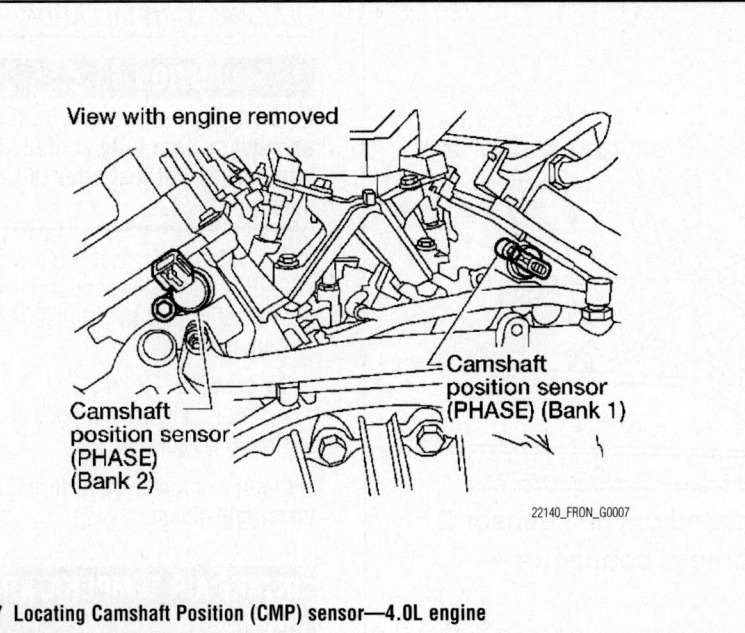

Fig. 97 Locating Camshaft Position (CMP) sensor—4.0L engine

To install:

4. Installation is the reverse of the removal procedure.

CRANKSHAFT POSITION (CKP) SENSOR

LOCATION

See Figures 98 and 99.

Refer to the accompanying illustrations.

REMOVAL & INSTALLATION

1. Loosen the fixing bolt of the sensor.
2. Disconnect CKP sensor (POS) harness connector.

3. Remove the sensor.

To install:

4. Installation is the reverse of the removal procedure.

ELECTRONIC CONTROL MODULE (ECM)

LOCATION

See Figure 100.

Refer to the accompanying illustration.

HEATED OXYGEN (HO2S) SENSOR

LOCATION

See Figures 101 and 102.

Refer to the accompanying illustrations.

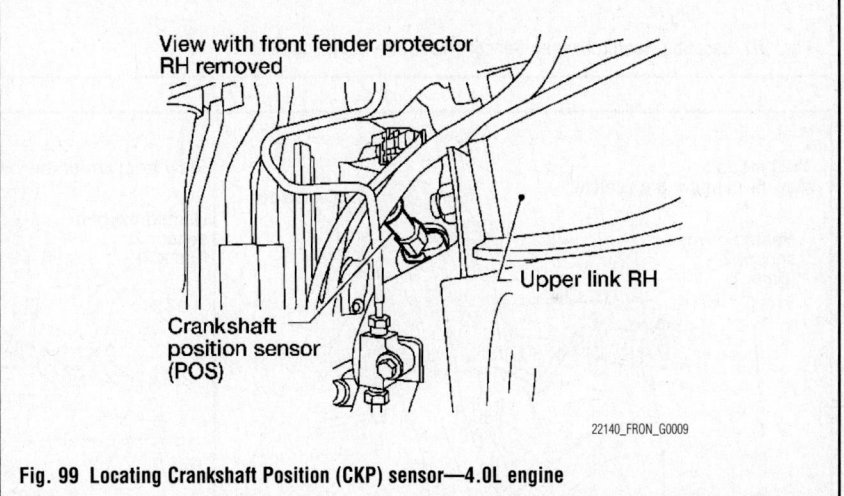

Fig. 99 Locating Crankshaft Position (CKP) sensor—4.0L engine

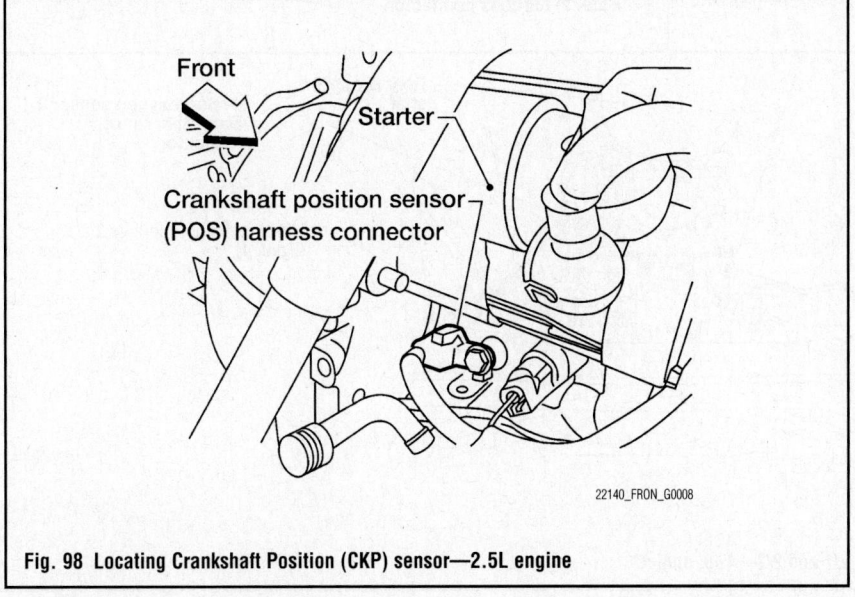

Fig. 98 Locating Crankshaft Position (CKP) sensor—2.5L engine

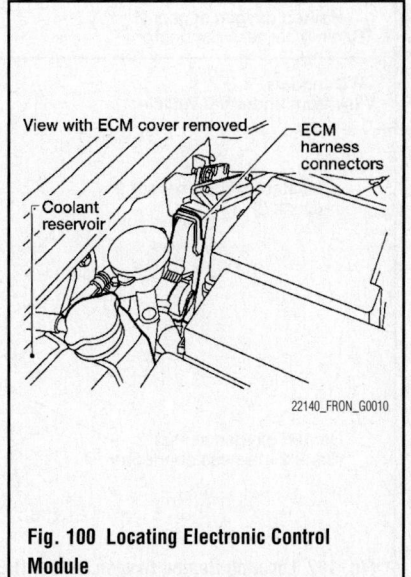

Fig. 100 Locating Electronic Control Module

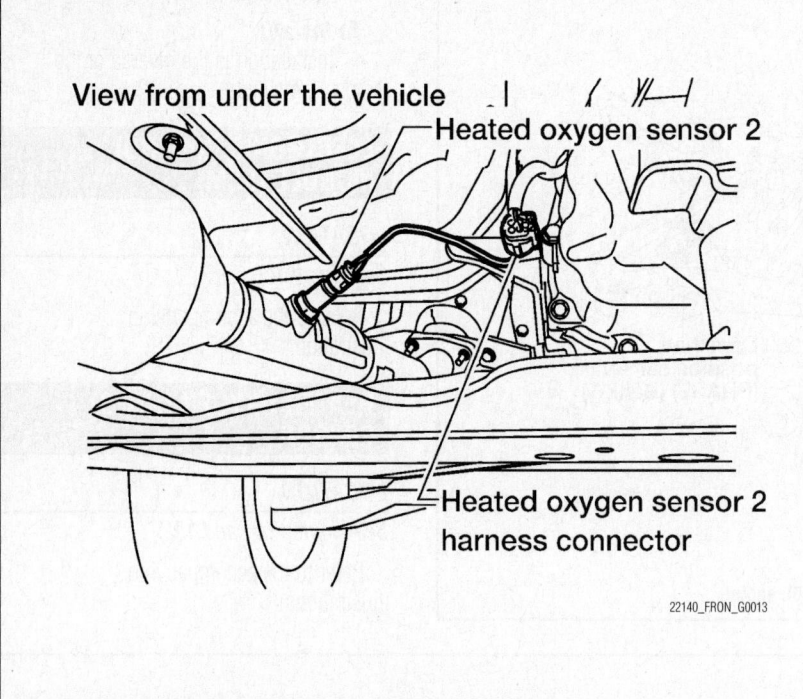

Fig. 101 Locating Heated Oxygen Sensor (HO2S) 2—2.5L engine

REMOVAL & INSTALLATION

✳✳ CAUTION

Perform the operation with the exhaust system fully cooled. The system will be hot just after the engine stops.

1. Disconnect sensor harness connector.
2. Remove the sensor using heated oxygen sensor wrench KV10114400 (J-38365) or equivalent.

To install:

3. Installation is the reverse of the removal procedure.

➡ Clean exhaust system threads before installing sensor.

INTAKE AIR TEMPERATURE (IAT) SENSOR

LOCATION

The Intake Air Temperature (IAT) sensor is built into Mass Air Flow (MAF) sensor.

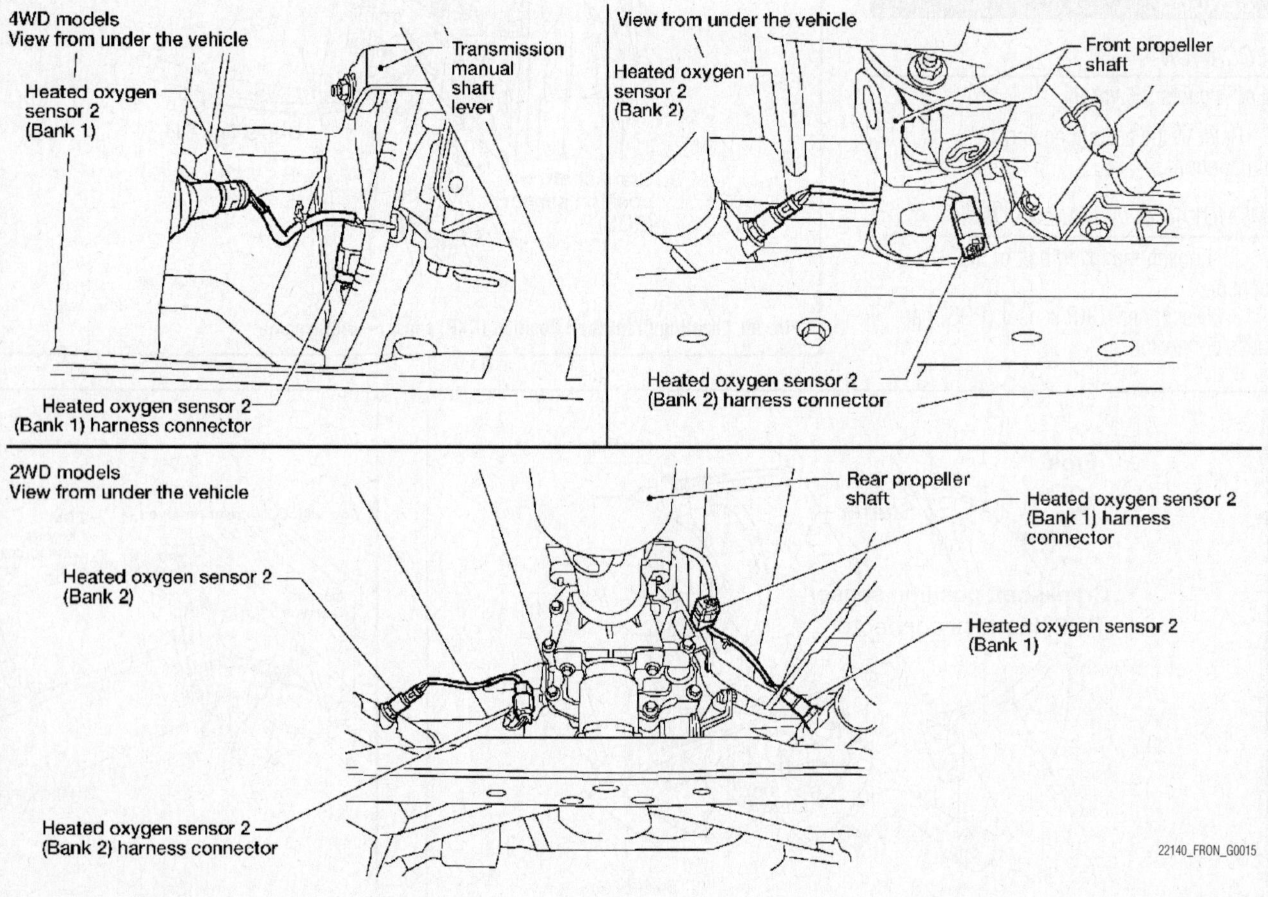

4WD models
View from under the vehicle

Heated oxygen sensor 2 (Bank 1)

Transmission manual shaft lever

Heated oxygen sensor 2 (Bank 1) harness connector

View from under the vehicle

Heated oxygen sensor 2 (Bank 2)

Front propeller shaft

Heated oxygen sensor 2 (Bank 2) harness connector

2WD models
View from under the vehicle

Heated oxygen sensor 2 (Bank 2)

Rear propeller shaft

Heated oxygen sensor 2 (Bank 1) harness connector

Heated oxygen sensor 2 (Bank 1)

Heated oxygen sensor 2 (Bank 2) harness connector

Fig. 102 Locating Heated Oxygen Sensor (HO2S) 2/1 and 2/2—4.0L engine

REMOVAL & INSTALLATION

See Mass Air Flow (MAF) sensor.

KNOCK SENSOR (KS)

LOCATION

See Figures 103 and 104.

Refer to the accompanying illustrations.

REMOVAL & INSTALLATION

1. On 4.0L engines, remove intake collector.

2. Remove sensor harness. Remove sensor.

To install:

3. Installation is the reverse of the removal procedure.

➡ **Use care when installing sensor. Do not use any knock sensors that have been dropped or physically damaged. Use only new ones. Torque sensor properly.**

MASS AIR FLOW (MAF) SENSOR

LOCATION

See Figures 105 and 106.

Refer to the accompanying illustrations.

REMOVAL & INSTALLATION

See Air Filter.

THROTTLE POSITION SENSOR (TPS)

REMOVAL & INSTALLATION

The Electric Throttle Control actuator must be replaced as a unit. See Intake Manifold.

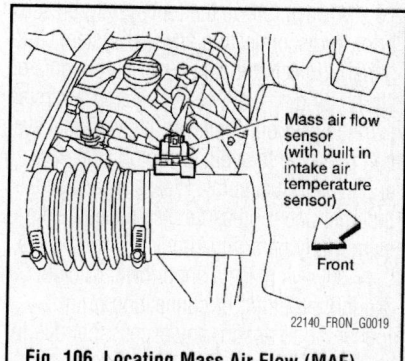

Fig. 106 Locating Mass Air Flow (MAF) sensor—4.0L engine

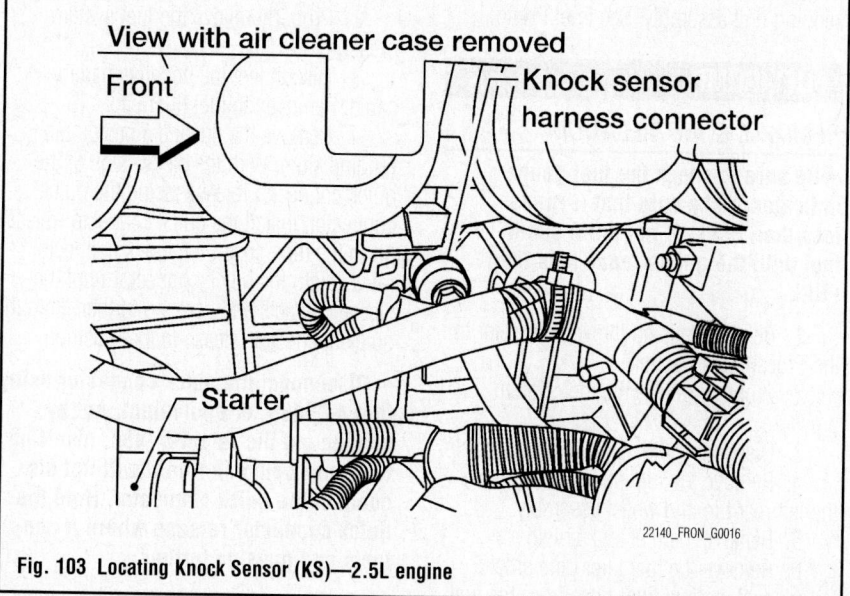

Fig. 103 Locating Knock Sensor (KS)—2.5L engine

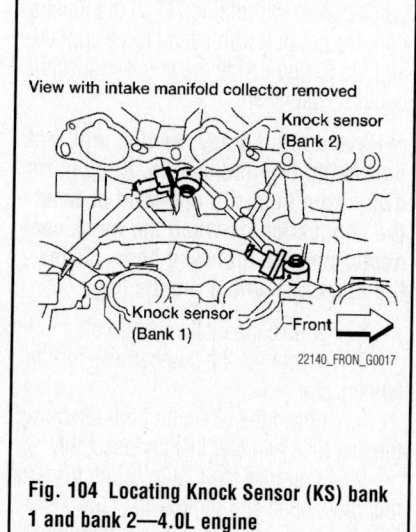

Fig. 104 Locating Knock Sensor (KS) bank 1 and bank 2—4.0L engine

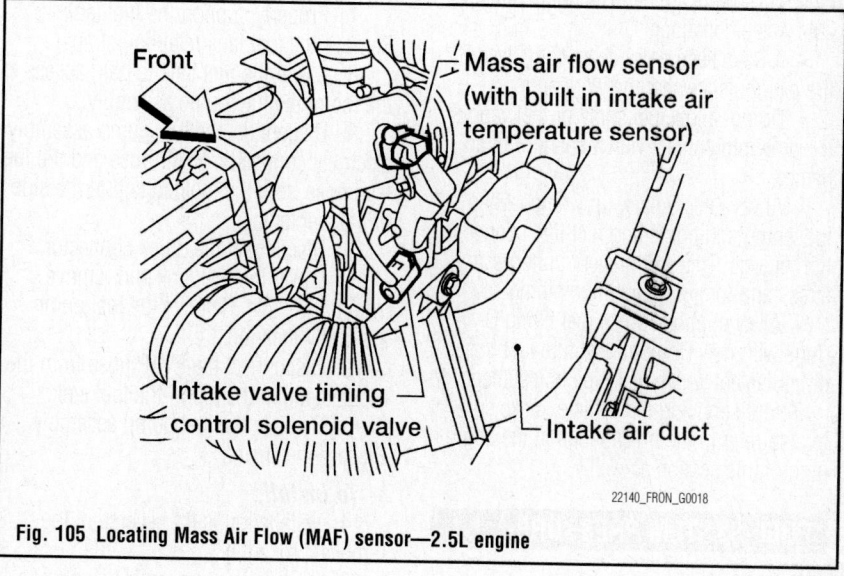

Fig. 105 Locating Mass Air Flow (MAF) sensor—2.5L engine

FUEL SYSTEM SERVICE PRECAUTIONS

Safety is the most important factor when performing not only fuel system maintenance but any type of maintenance. Failure to conduct maintenance and repairs in a safe manner may result in serious personal injury or death. Maintenance and testing of the vehicle's fuel system components can be accomplished safely and effectively by adhering to the following rules and guidelines.

• To avoid the possibility of fire and personal injury, always disconnect the negative battery cable unless the repair or test procedure requires that battery voltage be applied.

• Always relieve the fuel system pressure prior to disconnecting any fuel system component (injector, fuel rail, pressure regulator, etc.), fitting or fuel line connection. Exercise extreme caution whenever relieving fuel system pressure to avoid exposing skin, face and eyes to fuel spray. Please be advised that fuel under pressure may penetrate the skin or any part of the body that it contacts.

• Always place a shop towel or cloth around the fitting or connection prior to loosening to absorb any excess fuel due to spillage. Ensure that all fuel spillage (should it occur) is quickly removed from engine surfaces. Ensure that all fuel soaked cloths or towels are deposited into a suitable waste container.

• Always keep a dry chemical (Class B) fire extinguisher near the work area.

• Do not allow fuel spray or fuel vapors to come into contact with a spark or open flame.

• Always use a back-up wrench when loosening and tightening fuel line connection fittings. This will prevent unnecessary stress and torsion to fuel line piping.

• Always replace worn fuel fitting O-rings with new Do not substitute fuel hose or equivalent where fuel pipe is installed.

Before servicing the vehicle, make sure to also refer to the precautions in the beginning of this section as well.

FUEL SYSTEM PRESSURE

RELIEVING

1. Before servicing the vehicle, refer to the Precautions Section.
2. Remove the fuel pump fuse from the panel.

3. Start the engine and allow it to run until it stalls. Crank the engine for a few seconds to relieve additional fuel pressure.
4. Turn ignition switch off.
5. When repairs are complete, replace the fuel pump fuse and connect the negative battery cable.

FUEL FILTER

REMOVAL & INSTALLATION

Fuel Filter is serviced with fuel pump and sending unit assembly. See Fuel Pump.

FUEL PUMP

REMOVAL & INSTALLATION

➡**Be sure to check the fuel gauge indicator. Make sure that it reads less than FULL. If not drain some fuel until the gauge reads less than FULL.**

1. Before servicing the vehicle, refer to the Precautions Section.
2. Properly relieve the fuel system pressure.
3. Disconnect the negative battery cable.
4. Remove the fuel filler cap. Remove the left rear tire and wheel assembly.
5. Remove the fuel tank shield.
6. Remove the fuel filler pipe shield. Disconnect the fuel filler hose from the fuel filler pipe.
7. Properly support the fuel tank. Remove the fuel tank retaining straps.
8. Lower the fuel tank to gain access to the top of the fuel pump assembly.
9. Disconnect the fuel pump assembly electrical connector, EVAP hose and the fuel feed hose from the molded clip in the side of the fuel tank.
10. Disconnect the quick connector.
11. Lower the fuel tank and remove it from the vehicle. Remove the fuel pump assembly lockring.
12. Disconnect the EVAP hose from the molded clip in the top of the fuel tank.
13. Remove the fuel pump assembly. Discard the O-ring.

To install:

14. Installation is the reverse of the removal procedure.
15. Be sure to use a new O-ring upon installation.
16. Turn the ignition switch ON, but do not start the engine. Check the fuel lines and hose connections for leaks while applying fuel pressure to the system.

17. Start the engine and check for fuel leaks, correct as required.

FUEL RAIL (SUPPLY MANIFOLD) & INJECTOR

REMOVAL & INSTALLATION

2.5L Engine

See Figures 107 and 108.

1. Before servicing the vehicle, refer to the Precautions Section.
2. Properly relieve the fuel system pressure.
3. Disconnect the negative battery cable. Remove the fuel filler cap.
4. Remove the quick connector cap (engine side). With the sleeve side of the quick connector release facing the quick connector, install the quick connector release on to the tube. Insert the quick connector release into the quick connector until the sleeve contacts and goes no further. Hold the quick connector release in that position.

➡**Disconnect the quick connector using tool J-45488, or equivalent, not by picking out the retainer tabs. Inserting the quick connector hard will not disconnect the quick connector. Hold the quick connector release where it contacts and goes no further.**

5. Draw and pull out the quick connector straight from the fuel tube. Grasp the quick connector holding "A" in the illustration. Do not pull with lateral force applied and the O-ring inside the quick connector could be damaged.

➡**Have a cloth ready, as fuel will leak out. Avoid fire and sparks. Keep parts away from heat. Do not bend or twist the connection between the quick connector and the fuel feed hose. Cover the openings with a plastic bag.**

6. Remove the intake manifold.
7. Disconnect the sub harness for the fuel injector.
8. Loosen the retaining bolts. Remove the fuel tube and fuel injector assembly.
9. To remove the fuel injectors from the fuel tube, open and remove the clip. Remove the injector by pulling it straight out.

To install:

➡**Use new O-ring seals for assembly. Note that the upper and lower O-rings are different. Do not confuse them.**

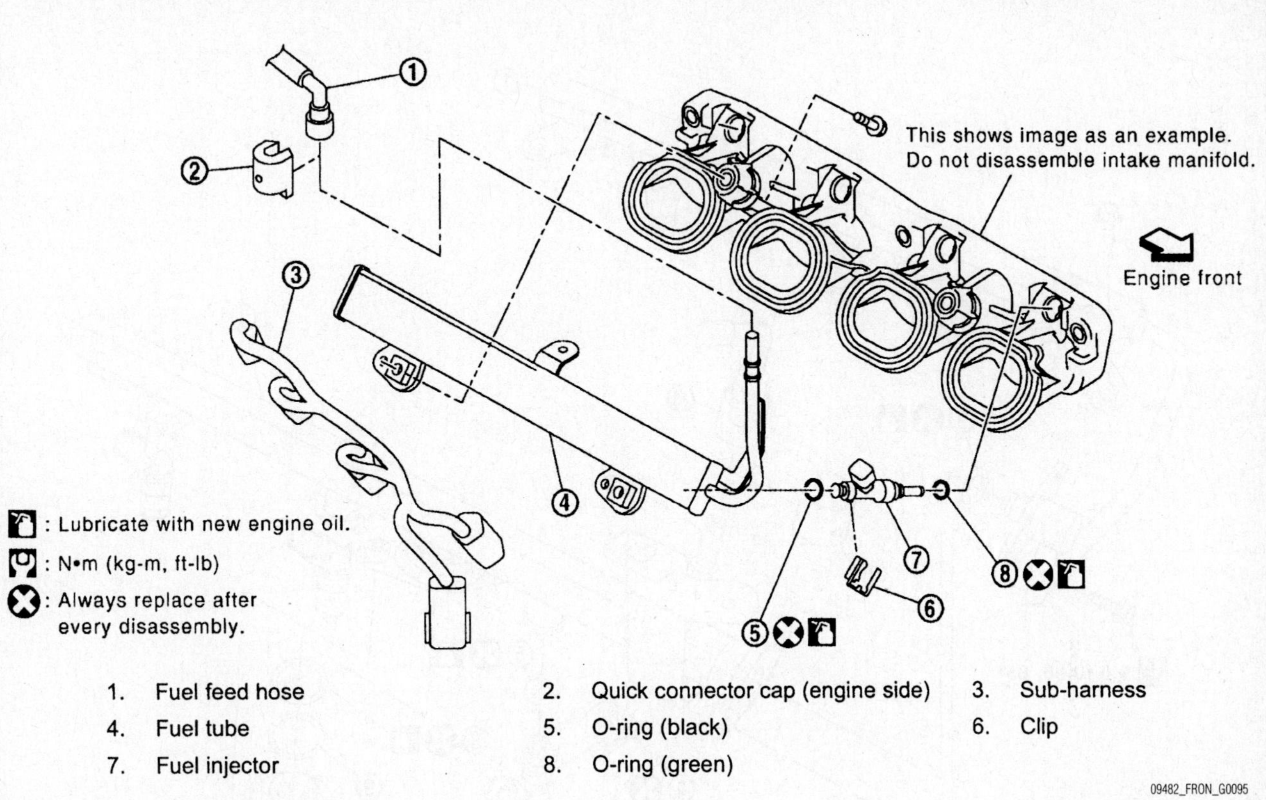

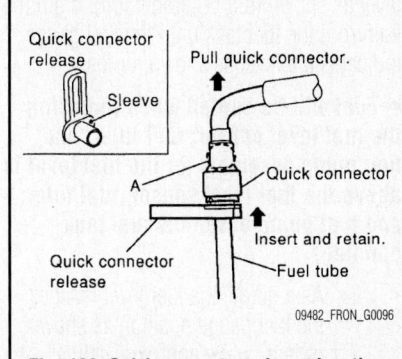

[oil symbol] : Lubricate with new engine oil.

[torque symbol] : N•m (kg-m, ft-lb)

[X symbol] : Always replace after every disassembly.

1.	Fuel feed hose	2.	Quick connector cap (engine side)	3.	Sub-harness
4.	Fuel tube	5.	O-ring (black)	6.	Clip
7.	Fuel injector	8.	O-ring (green)		

This shows image as an example.
Do not disassemble intake manifold.

Engine front

09482_FRON_G0095

Fig. 107 Fuel injector tube and related components—2.5L engine

Fuel tube side: Black. Nozzle side: Green.

10. Installation is the reverse of the removal procedure.

11. When installing the fuel feed tube be sure to torque the retaining bolts to 9 ft. lbs. (12 Nm) and then to 21 ft. lbs. (28 Nm) in an alternating order.

12. Turn the ignition switch ON, but do not start the engine. Check the fuel lines and hose connections for leaks while applying fuel pressure to the system.

13. Start the engine and check for fuel leaks, correct as required.

Quick connector release

Sleeve

Pull quick connector.

Quick connector

A

Quick connector release

Insert and retain.

Fuel tube

09482_FRON_G0096

Fig. 108 Quick connector release location "A"—2.5L engine

4.0L Engine

See Figure 109.

1. Before servicing the vehicle, refer to the Precautions Section.

2. Properly relieve the fuel system pressure.

3. Disconnect the negative battery cable. Remove the fuel filler cap.

4. Remove the intake manifold collector.

5. Remove the quick connector cap (engine side). With the sleeve side of the quick connector release facing the quick connector, install the quick connector release on to the tube. Insert the quick connector release into the quick connector until the sleeve contacts and goes no further. Hold the quick connector release in that position.

➡Disconnect the quick connector using tool J-45488, or equivalent, not by picking out the retainer tabs. Inserting the quick connector hard will not disconnect the quick connector. Hold the quick connector release where it contacts and goes no further.

6. Draw and pull out the quick connector straight from the fuel tube. Grasp the quick connector holding "A" in the illustration. Do not pull with lateral force applied and the O-ring inside the quick connector could be damaged.

➡Have a cloth ready, as fuel will leak out. Avoid fire and sparks. Keep parts away from heat. Do not bend or twist the connection between the quick connector and the fuel feed hose. Cover the openings with a plastic bag.

7. Remove the PCV hose between the rocker covers.

8. Disconnect the harness for the fuel injector.

9. Loosen the retaining bolts. Remove the fuel tube and fuel injector assembly. Remove the bolts which connect the left and right fuel tubes.

10. To remove the fuel injectors from the fuel tube, open and remove the clip. Remove the injector by pulling it straight out.

11. Disconnect the right fuel tube from the left fuel tube. Loosen the bolts, to remove the fuel damper cap and fuel damper, if necessary.

To install:

➡Use new O-ring seals for assembly. Note that the upper and lower O-rings are different. Do not confuse them.

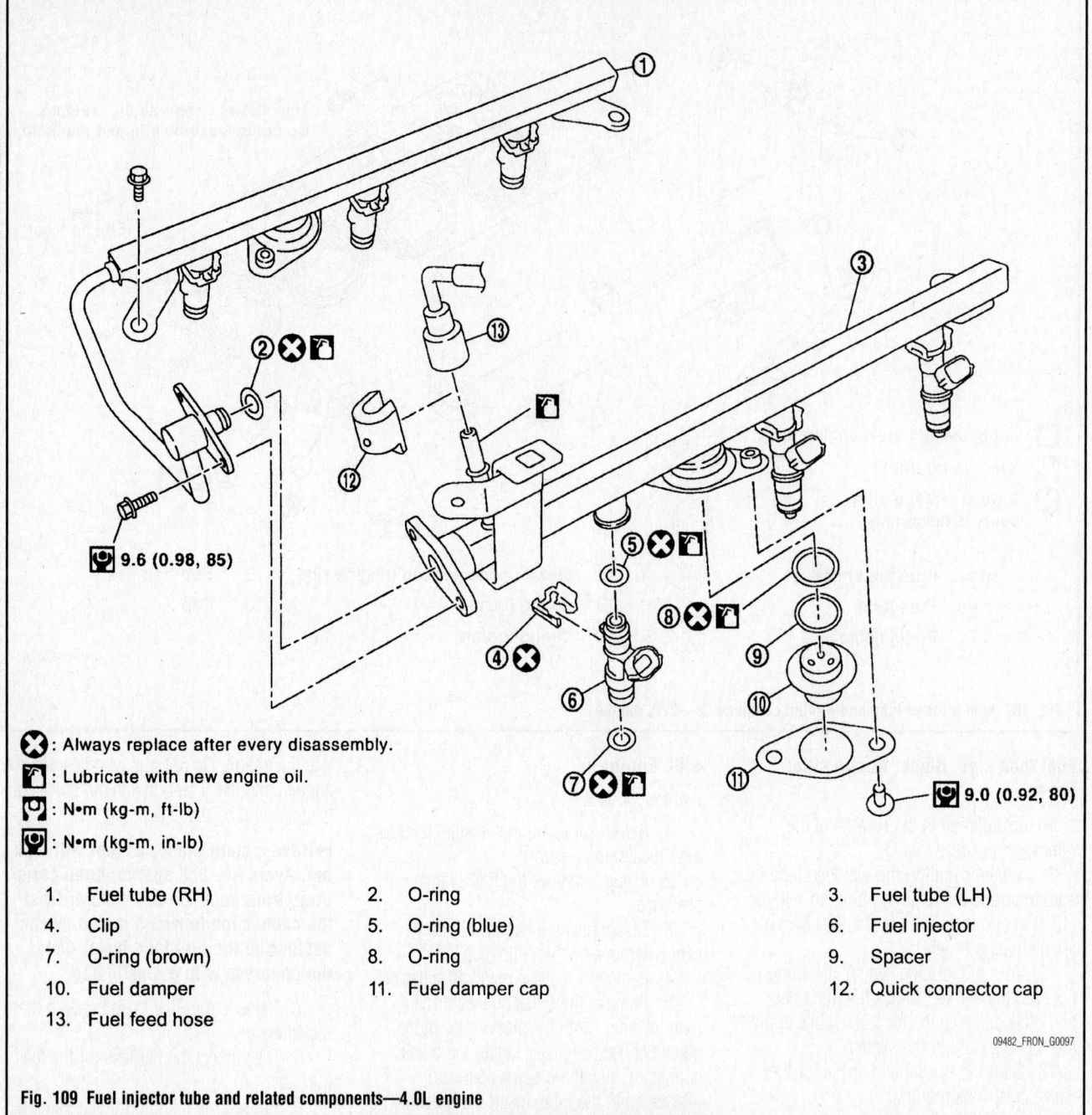

⊗ : Always replace after every disassembly.

🛢️ : Lubricate with new engine oil.

💪 : N•m (kg-m, ft-lb)

💪 : N•m (kg-m, in-lb)

9.6 (0.98, 85)

9.0 (0.92, 80)

1. Fuel tube (RH)	2. O-ring	3. Fuel tube (LH)
4. Clip	5. O-ring (blue)	6. Fuel injector
7. O-ring (brown)	8. O-ring	9. Spacer
10. Fuel damper	11. Fuel damper cap	12. Quick connector cap
13. Fuel feed hose		

09482_FRON_G0097

Fig. 109 Fuel injector tube and related components—4.0L engine

Fuel tube side: Blue. Nozzle side: Brown.

12. Installation is the reverse of the removal procedure.

13. When installing the fuel feed tube be sure to torque the retaining bolts to 7 ft. lbs and then to 16 ft. lbs. in an alternating order.

14. Turn the ignition switch ON, but do not start the engine. Check the fuel lines and hose connections for leaks while applying fuel pressure to the system.

15. Start the engine and check for fuel leaks, correct as required.

FUEL TANK

REMOVAL & INSTALLATION

❊❊ WARNING

Follow the "Fuel System Service Precautions" before working on the fuel system.

1. Remove the fuel filler cap to release the pressure from inside the fuel tank.
2. Remove the LH rear wheel and tire.
3. Check the fuel level on level gauge. If the fuel gauge indicates more than the level as shown (full or almost full), drain the fuel from the fuel tank until the fuel gauge indicates the level as shown, or less.

➡️**Fuel will be spilled when removing the fuel level sensor, fuel filter, and fuel pump assembly for the fuel level is above the fuel level sensor, fuel filter, and fuel pump assembly fuel tank opening.**

- As a guide, the fuel level reaches the fuel gauge position as shown, or less, when approximately 15 (4 US gal, 3 1/4 Imp gal) of fuel are drained from the fuel tank.

- If the fuel pump does not operate, use the following procedure to drain the fuel to the specified level.

  a. Insert a suitable hose of less than 15 mm (0.59 in) diameter into the fuel filler pipe through the fuel filler opening to drain the fuel from fuel filler pipe.

  b. Remove the fuel filler pipe shield.

  c. Disconnect the fuel filler hose from the fuel filler pipe.

  d. Insert a suitable hose into the fuel tank through the fuel filler hose to drain the fuel from the fuel tank.

4. Release the fuel pressure from the fuel lines.

5. Disconnect the battery negative terminal.

6. Remove the fuel tank shield.

7. Remove the fuel tank strap bolts while supporting the fuel tank with a suitable lift jack.

8. Lower the fuel tank using a suitable lift jack to access the top of the fuel level sensor, fuel filter, and fuel pump assembly.

9. Disconnect the fuel level sensor, fuel filter, and fuel pump assembly electrical connector, EVAP hose, and the fuel feed hose.

  a. Disconnect the fuel feed hose from the molded clip in the side of the fuel tank.

10. Disconnect the quick connector as follows:

  a. Hold the sides of the connector, push in tabs and pull out the tube.

  b. If the connector and the tube are stuck together, push and pull several times until they start to move. Then disconnect them by pulling.

❈❈ CAUTION
Note the following:

- The quick connector can be disconnected when the tabs are completely depressed. Do not twist the quick connector more than necessary.
- Do not use any tools to disconnect the quick connector.
- Keep the resin tube away from heat. Be especially careful when welding near the tube.
- Prevent any acid liquids such as battery electrolyte, from getting on the resin tube.
- Do not bend or twist the resin tube during connection.
- Do not remove the remaining retainer on the hard tube (or the equivalent) except when the resin tube or the retainer is replaced.
- When the resin tube or hard tube, or the equivalent, is replaced, also replace the retainer with a new one (semitransparent retainer).
- To keep the quick connector clean and to avoid damage and contamination from foreign materials, cover the quick connector with plastic bags or suitable material as shown.

11. Lower the fuel tank using a suitable lift jack and remove it from the vehicle.

12. Remove the lock ring using Tool as shown.

13. Disconnect the EVAP hose from the molded clip in the top of the fuel tank.

14. Remove the fuel level sensor, fuel fil-ter, and fuel pump assembly. Remove and discard the fuel level sensor, fuel filter, and fuel pump assembly O-ring.

❈❈ CAUTION
Note the following:

- Do not bend the float arm during removal.
- Avoid impacts such as dropping when handling the components.

To install:

15. Installation is the reverse of the removal procedure.

  a. For installation, use a new fuel level sensor, fuel filter, and fuel pump assembly O-ring.

  b. Connect the quick connector as follows:

- Check the connection for any damage or foreign materials.
- Align the connector with the pipe, then insert the connector straight into the pipe until a click is heard.
- After connecting the quick connector, make sure that the connection is secure by checking as follows:
- Pull the tube and the connector to make sure they are securely connected.
- Visually inspect the connector to make sure the two retainer tabs are securely connected.

IDLE SPEED

ADJUSTMENT

Idle speed is maintained by the Powertrain Control Module (PCM). No adjustment is necessary or possible.

HEATING & AIR CONDITIONING SYSTEM

BLOWER MOTOR

REMOVAL & INSTALLATION

See Figure 110.

➡**Before servicing, or working around, the SRS system, turn the ignition switch OFF, disconnect both battery cables and wait at least three minutes. When servicing, or working around, the SRS system do not work directly in front of the air bag module.**

1. Before servicing the vehicle, refer to the Precautions Section.

2. Disconnect the negative battery cable. Disconnect the positive battery cable.

3. Remove the lower glove box assembly.

4. Disconnect the blower motor electrical connector.

5. Remove the blower motor retaining screws.

6. Remove the blower motor from its mounting.

To install:

7. Installation is the reverse of the removal procedure.

HEATER CORE

REMOVAL & INSTALLATION

See Figures 111 and 112.

➡**Be sure to disarm the SRS system, prior to working on the vehicle. Turn the ignition switch OFF, disconnect both battery cables and wait at least three minutes before starting any work.**

1. Before servicing the vehicle, refer to the Precautions Section.

2. Position the front wheels in the straight ahead direction.

3. Disconnect the negative battery cable. Disconnect the positive battery cable.

4. Drain the cooling system.

5. Properly discharge the air conditioning system.

6. If equipped with the 4.0L engine, remove the right side heater core pipe nuts.

7. Disconnect the heater core hoses from the heater core.

8. Disconnect the air conditioning refrigerant lines from the expansion valve.

9. Position the front seats in the rearmost position on the seat tracks.

10. Remove the upper front pillar trim panel. Remove the steering lock

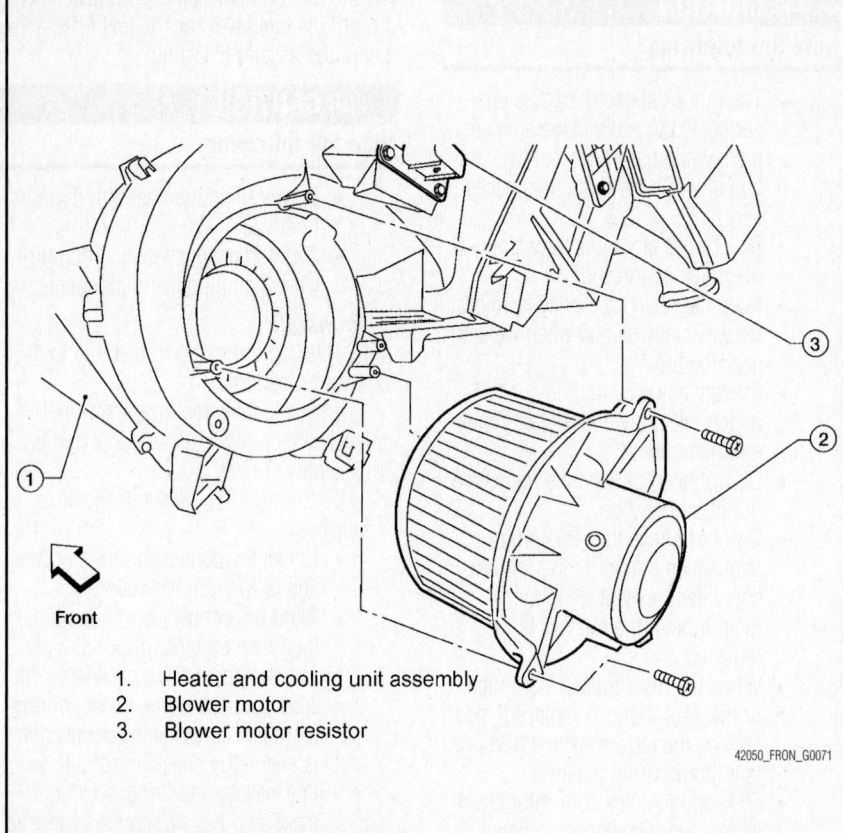

Front

1. Heater and cooling unit assembly
2. Blower motor
3. Blower motor resistor

42050_FRON_G0071

Fig. 110 Blower motor and related components

escutcheon. Remove the cluster lid "A". Remove the combination meter. Disconnect the electrical connections.

11. Remove the optical sensor. Remove the audio unit. Remove the cluster lid "D".

12. Remove the glove box. Remove the two bolts, through the glove box opening, retaining the front passenger's side air bag module to the steering member. Disconnect the air bag module connectors.

13. Remove the instrument stay right side and left side bolts. Remove the instrument panel.

14. Remove the two front floor ducts.

15. To remove the driver's side air bag module, locate the retaining clip access hole under the steering wheel. Insert a suitable blunt tool (4 mm–6 mm in size).

➡**Do not use sharp edged objects, such as a screwdriver, to release the driver's side airbag module from the steering wheel as SRS components may be unintentionally damaged.**

16. Press upward, toward the center of the steering wheel, on the retaining clip until the air bag module is released from the steering wheel.

17. Lift the air bag module from the steering wheel. Disconnect the electrical connectors. Remove the air bag module.

18. Disconnect the steering wheel switches. Remove the steering wheel center nut. Using a steering wheel removal tool, remove the steering wheel.

19. Remove the steering column upper and lower covers. Disconnect the wiper and washer switch connector. While pressing the tabs, pull the wiper and washer switch away from the spiral cable to remove it.

20. Disconnect the light and turn signal switch connector. While pressing the tabs, pull the light and turn signal switch toward the driver's door to remove it.

21. Remove the screws. While pressing the tab, pull the spiral cable away from the steering column assembly. Disconnect the electrical connectors.

➡**With the steering linkage disconnected, the spiral cable may snap by turning the steering wheel beyond the limited number of turns. The spiral cable can be turned counterclockwise about 2.5 turns from the neutral position.**

22. Remove the lower knee protector.

23. Remove the locknut and bolt from the upper joint and then separate the upper joint from the upper shaft.

24. Remove the three nuts and bolt from the steering column and then remove the steering column assembly from the steering member.

25. Remove the hole cover seal and clamp. Remove the hole cover nuts, remove the hole cover from the dash panel.

26. Remove the bolt from the lower joint of the lower joint shaft and remove the lower joint shaft from the vehicle.

27. Disconnect the instrument panel wire harness at the right and left in-line connector brackets, and the fuse block (SMJ) electrical connectors.

28. Remove the covers and then remove the three steering member bolts from each side to disconnect the steering member from the vehicle body.

29. Remove the heater/evaporator case assembly with it attached to the steering member from the vehicle.

30. Separate the steering member from the heater/evaporator unit.

31. Remove the heater cover retaining screws. Remove the cover.

32. Remove the heater core and the evaporator pipe bracket. Remove the heater core.

To install:

33. Installation is the reverse of the removal procedure.

➡**If the in-cabin microfilters are contaminated with coolant, replace them.**

34. Be sure to use new steering column retaining bolts and pinch bolt, as required.

➡**When installing the steering column, finger tighten all of the lower bracket and joint bolts and then tighten them to specification. Do not apply undue stress to the steering column.**

35. With the wheels in the straight ahead position align the slit of the lower joint with the projection on the dust cover. Insert the joint until surface "A" contacts surface "B".

36. Be sure to align the spiral cable correctly when installing the steering wheel. Make sure that the cable is in the neutral position. The neutral position is detected by turning left 2.6 revolutions from the right end position and ending with the locating pin at the top.

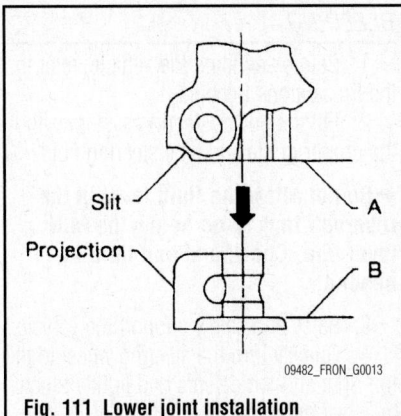

Fig. 111 Lower joint installation

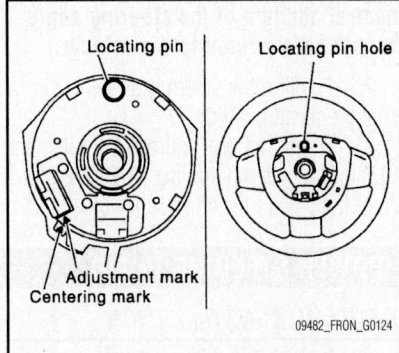

Fig. 112 Spiral cable installation and locating point

37. To adjust the steering angle sensor neutral position, position the steering wheel in the straight ahead position and rive the vehicle at 10 mph or more for ten minutes. When the procedure is complete, the SLP indicator lamp and the VDC OFF indicator lamp will turn off.

38. Be sure to fill the cooling system with the proper grade and type coolant.

39. Be sure to recharge the air conditioning system.

40. Check and adjust the front end alignment, as necessary.

STEERING

POWER RACK & PINION STEERING GEAR

REMOVAL & INSTALLATION
See Figures 113 and 114.

➡The spiral cable may snap due to steering operation if the steering column is separated from the steering gear assembly. Be sure to secure the steering wheel to avoid turning.

1. Before servicing the vehicle, refer to the Precautions Section.
2. Position the front wheels in the straight ahead position.

3. Disarm the SRS system.
4. Disconnect the negative battery cable.
5. Drain the power steering fluid.
6. Raise and support the vehicle safely. Remove the tire and wheel assemblies.
7. Remove the undercover.
8. If equipped with 4WD, remove the final drive, then support the halfshafts, using wire.
9. Remove the stabilizer bar brackets, and position the stabilizer bar aside.
10. Remove and discard the cotter pins at the steering outer sockets. Loosen the outer socket locknuts.

11. Remove the steering gear outer sockets from the steering knuckles, using tool HT72520000 (J-25730-A) or equivalent.
12. Disconnect and plug the power steering fluid lines at the steering gear.
13. Remove the bolt from the lower joint of the lower joint assembly. Separate the lower joint from the steering gear assembly. Be careful not to damage the lower joint.
14. Remove the steering gear retaining nuts and bolts. Remove the steering gear from the vehicle.

To install:
15. With the steering wheel in the straight ahead position, align the slit of the

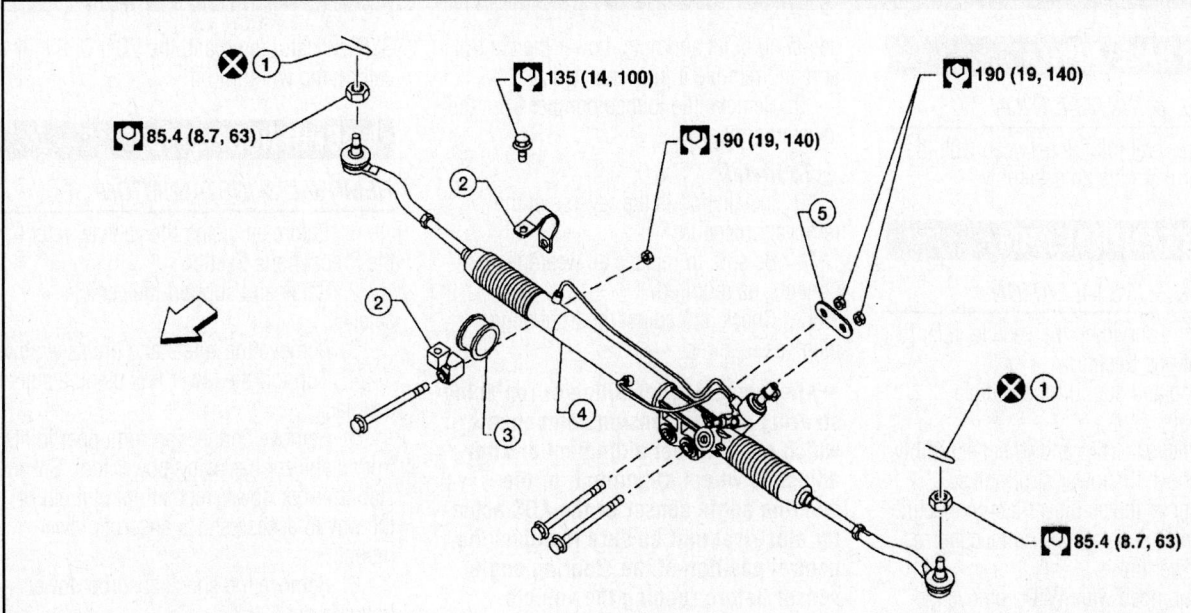

1. Cotter pin
4. Steering gear assembly
2. Mounting bracket
5. Washer
3. Mounting insulator
⇐ Front

Fig. 113 Power steering gear and related components

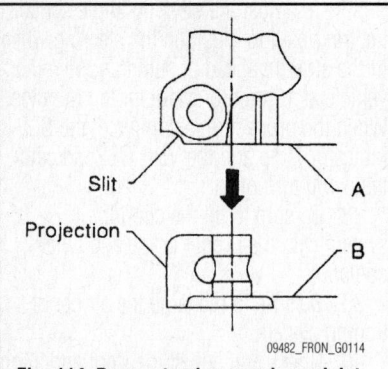

Slit — A

Projection — B

09482_FRON_G0114

Fig. 114 Power steering gear lower joint installation alignment

lower joint with the projection on the dust cover. Insert the joint until both surfaces contact each other.

16. Continue the installation in the reverse order of the removal procedure.

17. Check and adjust the front alignment, as required.

18. Bleed the power steering system.

19. Fill the power steering pump with the proper grade and type fluid.

➡ **After removing/installing or replacing steering and suspension components which effect wheel alignment or after adjusting wheel alignment, or the steering angle sensor or the ABS actuator electrical unit be sure to adjust the**

neutral position of the steering angle sensor before running the vehicle.

20. Position the steering wheel in the straight ahead position.

21. When this procedure is complete the SLP indicator lamp and the VDC OFF indicator lamp will turn off.

POWER STEERING PUMP

REMOVAL & INSTALLATION

1. Before servicing the vehicle, refer to the Precautions section.

2. Disconnect the negative battery cable.

3. Drain the power steering fluid from the reservoir tank. Properly dispose of used fluid.

4. On the 4.0L engine, remove the engine cover.

5. Remove the air duct assembly.

6. Remove the drive belt.

7. Disconnect the pressure sensor electrical connector.

8. Disconnect and plug the fluid lines.

9. Remove the pump retaining bolts.

10. Remove the pump from the vehicle.

To install:

11. Installation is the reverse of the removal procedure.

12. Bleed the power steering system.

BLEEDING

1. Before servicing the vehicle, refer to the Precautions Section.

2. Fill the power steering system with the proper grade and type steering fluid.

➡ **Do not allow the fluid level in the reservoir tank to go below the MIN level line. Check and add fluid as needed.**

3. Raise and safely support the vehicle.

4. Quickly turn the steering wheel to the full right and left detents and lightly touch the steering stoppers.

➡ **Do not hold the steering wheel in the locked position for more than ten seconds.**

5. Repeat this operation until the fluid level no longer decreases.

6. Start the engine.

7. Quickly turn the steering wheel to the full right and left detents and lightly touch the steering stoppers.

➡ **Do not hold the steering wheel in the locked position for more than ten seconds.**

8. Check for air bubbles or cloudy fluid. If found, repeat the bleeding procedure.

9. Stop the engine and check the fluid level. Correct as required.

SUSPENSION

LOWER BALL JOINT

REMOVAL & INSTALLATION

The lower ball joint is serviced with the lower control arm as an assembly.

LOWER CONTROL ARM

REMOVAL & INSTALLATION

1. Before servicing the vehicle, refer to the Precautions Section.

2. Raise and support the vehicle safely.

3. Remove the tire and wheel assembly.

4. Remove the lower strut bolt.

5. Remove the stabilizer bar connecting rod lower nut. Separate the connecting rod from the lower link.

6. If equipped with 4WD, remove the halfshaft.

7. Remove the pinch bolt from the steering knuckle. Separate the lower control arm ball joint stud from the steering knuckle, using the proper tool.

8. Remove the lower control arm

adjusting bolts and nuts. Lower the control arm and remove it from the vehicle.

9. Remove the jounce bumper from the lower control arm.

To install:

10. Installation is the reverse of the removal procedure.

11. Be sure to replace all wearable components, as required.

12. Check and adjust the front alignment, as required.

➡ **After removing/installing or replacing steering and suspension components which effect wheel alignment or after adjusting wheel alignment, or the steering angle sensor or the ABS actuator electrical unit be sure to adjust the neutral position of the steering angle sensor before running the vehicle.**

13. Position the steering wheel in the straight ahead position.

14. Drive the vehicle at 10 mph for more than 10 minutes.

15. When this procedure is complete the

FRONT SUSPENSION

SLP indicator lamp and the VDC OFF indicator lamp will turn off.

MACPHERSON STRUT

REMOVAL & INSTALLATION

1. Before servicing the vehicle, refer to the Precautions Section.

2. Raise and support the vehicle safely.

3. Remove the wheel and tire assembly.

4. Support the lower link using a suitable jack.

5. Remove connecting rod upper joints from stabilizer bar using power tool. Swing stabilizer bar down, repositioning it out of the way to access shock absorber lower mount.

6. Remove the shock absorber lower bolt and nut.

7. Remove the three shock absorber upper mounting nuts.

8. Remove the coil spring and shock absorber assembly. Turn steering knuckle out to gain enough clearance for removal.

To install:

9. Installation is the reverse of the removal procedure.

10. The step in the strut assembly lower seat faces outside of vehicle.

STEERING KNUCKLE

REMOVAL & INSTALLATION

1. Before servicing the vehicle, refer to the Precautions Section.

2. Raise and support the vehicle safely.

3. Remove the wheel and tire assembly.

4. Without disassembling the hydraulic lines, remove brake caliper. Reposition it aside with wire.

➡**Do not press the brake pedal while the brake caliper is removed.**

5. Put alignment marks on disc rotor and wheel hub and bearing assembly, then remove disc rotor.

6. Disconnect wheel sensor and remove bracket from steering knuckle.

7. On 4WD models, remove cotter pin, then remove lock nut from drive shaft.

8. Remove steering outer socket cotter pin at steering knuckle, then loosen mounting nut.

9. Remove the pinch bolt from the steering knuckle. Separate the lower control arm ball joint stud from the steering knuckle, using the proper tool.

10. Remove wheel hub and bearing assembly bolts.

11. Remove splash guard and wheel hub and bearing assembly from steering knuckle.

12. Remove cotter pin and nut from upper link ball joint.

13. Separate upper link ball joint from steering knuckle using Tool ST29020001 (J-24319-01).

14. Remove pinch bolt from steering knuckle, then separate lower link ball joint from steering knuckle.

15. Remove steering knuckle from vehicle.

To install:

16. Installation is the reverse of the removal procedure.

STABILIZER BAR & LINKS

REMOVAL & INSTALLATION

1. Before servicing the vehicle, refer to the Precautions Section.

2. Remove the front valance center.

3. Raise and support the vehicle safely.

4. Remove the engine undercover.

5. Remove the connecting rod nuts.

6. Loosen the top bolts for the stabilizer bar mounting brackets. Remove the lower bolts from the mounting brackets.

7. Remove the stabilizer bar from the vehicle.

8. Remove the bushings from the stabilizer bar.

To install:

9. Installation is the reverse of the removal procedure.

UPPER BALL JOINT

REMOVAL & INSTALLATION

The upper ball joint is serviced with the upper control arm as an assembly.

UPPER CONTROL ARM

REMOVAL & INSTALLATION

1. Before servicing the vehicle, refer to the Precautions Section.

2. Raise and support the vehicle safely.

3. Remove the tire and wheel assembly.

4. Using a suitable jack, support the lower control arm.

5. If working on the left side, remove the bolt from the lower joint of the lower joint shaft, then reposition the lower joint shaft out of the way. Do not damage the lower joint.

6. Remove the cotter pin and nut from the upper control arm ball joint.

7. Separate the upper control arm ball joint stud from the steering knuckle, using tool ST29020001 (J-24319-01) or equivalent.

8. Remove the upper control arm retaining bolts and nuts.

9. Remove the upper control arm from the vehicle.

To install:

10. Installation is the reverse of the removal procedure.

11. Be sure to replace all wearable components, as required.

12. Check and adjust the front alignment, as required.

➡After removing/installing or replacing steering and suspension components which effect wheel alignment or after adjusting wheel alignment, or the steering angle sensor or the ABS actuator electrical unit be sure to adjust the neutral position of the steering angle sensor before running the vehicle.

13. Position the steering wheel in the straight ahead position.

14. Drive the vehicle at 10 mph for more than 10 minutes.

15. When this procedure is complete the SLP indicator lamp and the VDC OFF indicator lamp will turn off.

WHEEL BEARINGS

REMOVAL & INSTALLATION
See Figure 115.

1. Before servicing the vehicle, refer to the Precautions Section.

2. Raise and support the vehicle safely.

3. Remove the tire and wheel assembly.

4. Remove the caliper and position it to the side with wire. Do not disconnect the brake fluid line.

➡**Do not press the brake pedal while the brake caliper is removed.**

5. Matchmark the brake rotor and the wheel hub. Remove the brake rotor.

6. Remove the cotter pin. Remove the lock nut.

7. Remove the halfshaft from the wheel hub and bearing assembly.

8. Remove the wheel sensor from the hub and bearing assembly. Do not pull on the wheel sensor harness.

9. Remove the wheel hub and bearing assembly bolts.

10. Remove the splash guard. Remove the wheel hub and bearing assembly from the steering knuckle.

➡**Carefully remove the wheel sensor and harness through the hole in the splash guard.**

To install:

11. Inspect the wheel sensor O-ring, replace the wheel speed sensor assembly, as required.

12. Installation is the reverse of the removal procedure.

13. Be sure to use new bolts when installing the wheel hub and bearing assembly.

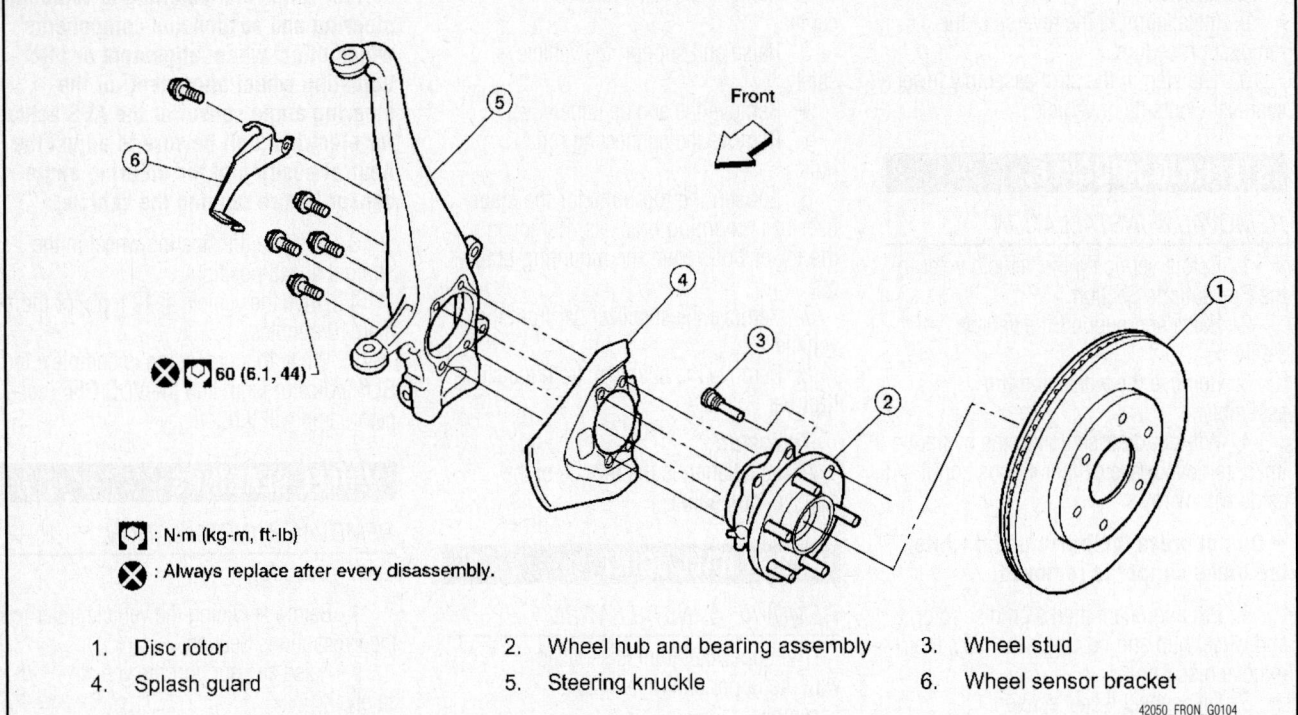

60 (6.1, 44)

⬡ : N·m (kg-m, ft-lb)

⊗ : Always replace after every disassembly.

1. Disc rotor
2. Wheel hub and bearing assembly
3. Wheel stud
4. Splash guard
5. Steering knuckle
6. Wheel sensor bracket

42050_FRON_G0104

Fig. 115 Hub and bearing assembly

SUSPENSION

LEAF SPRING

REMOVAL & INSTALLATION

1. Support rear axle assembly. Do not allow weight to be unsupported, but do not compress spring.
2. Remove lower end of shock absorber.
3. Remove u-bolt nuts and spring pad.
4. Remove rear shackle and bushings.
5. Remove front spring nut and bolt.
6. Remove spring.

To install:

7. Install front spring nut and bolt and rear shackle. Tighten finger tight.
8. Install U-bolts and bumper on axle. Install spring pad and nuts. Tighten nuts evenly to 53 ft. lbs. (72.5 Nm).
9. Install lower shock nut and bolt finger tight.
10. Lower vehicle and bounce to settle suspension. Tighten fasteners with vehicle on suspension.
11. Tighten lower shock absorber to 148 ft. lbs. (200 Nm).

REAR SUSPENSION

12. Tighten front spring nut and bolt to 165 ft. lbs. (190 Nm).
13. Tighten rear spring shackle nuts to 77 ft. lbs. (105 Nm)
14. Support rear axle assembly.
15. Remove fasteners and shock.

To install:

16. Install shock absorber.
17. Install upper nut and tighten to 33 ft. lbs. (45 Nm).
18. Install lower nut and bolt. Tighten to 148 ft. lbs. (200 Nm).
19. Remove support.

NISSAN

Maxima

14

SPECIFICATIONS AND MAINTENANCE CHARTS

ENGINE AND VEHICLE IDENTIFICATION

	Engine							Model Year	
Code ①	Liters (cc)	Cu. In.	Cyl.	Fuel Sys.	Engine Type	Eng. Mfg.		Code ②	Year
VQ35DE	3.5 (3498)	213	6	SFI	DOHC	Nissan		9	2009
								A	2010

SFI: Sequential Fuel Injection

DOHC: Double Overhead Camshaft

① The Engine Code is stamped on the engine block near the starter.

② 10th position of the Vehicle Identification Number (VIN)

37663_MAXI_C0001

GENERAL ENGINE SPECIFICATIONS

Year	Model	Engine Displacement Liters	Engine Series (ID/VIN)	Net Horsepower @ rpm	Net Torque @ rpm (ft. lbs.)	Bore x Stroke (in.)	Com- pression Ratio	Oil Pressure @ rpm
2009	Maxima	3.5	VQ35DE/A	290@6400	261@4400	3.76X3.20	10.6:1	43@2000
2010	Maxima	3.5	VQ35DE/A	290@6400	261@4400	3.76X3.20	10.6:1	43@2000

37663_MAXI_C0002

ENGINE TUNE-UP SPECIFICATIONS

Year	Model	Engine Displacement Liters	Engine ID/VIN	Spark Plug Gap (in.)	Ignition Timing (deg.) MT	AT	Fuel Pump (psi) ①	Idle Speed (rpm) MT	AT ②	Valve Clearance (in.) Intake ③	Exhaust ③
2009	Maxima	3.5	VQ35DE	0.043	NA	NA	51	—	600-700	0.012-0.016	0.012-0.017
2010	Maxima	3.5	VQ35DE	0.043	NA	NA	51	—	600-700	0.012-0.016	0.012-0.017

NOTE: The Vehicle Emission Control Information label often reflects specification changes made during production.

The label figures must be used if they differ from those in this chart.

NA: Not Available

B: Before top dead center

① System pressure at idle with vacuum hose connected; should increase to 43 psi when disconnected

② Automatic transmission in neutral

③ Engine cold

37663_MAXI_C0003

CAPACITIES

Year	Model	Engine ID/VIN	Engine Displacement Liters	Engine Oil with Filter (qts.)	Transmission (pts.) Man	Transmission (pts.) Auto.	Drive Axle Rear (pts.)	Fuel Tank (gal.)	Cooling System (qts.)
2009	Maxima	VQ35DE/A	3.5	4.6	—	21.2	—	20.0	8.6
2010	Maxima	VQ35DE/A	3.5	4.6	—	21.2	—	20.0	8.6

NOTE: All capacities are approximate. Add fluid gradually and check to be sure a proper fluid level is obtained.

37663_MAXI_C0004

FLUID SPECIFICATIONS

Year	Model	Engine Displ. Liters	Engine Oil	Man. Trans.	Auto. Trans.	Drive Axle Front	Drive Axle Rear	Transfer Case	Power Steering Fluid	Brake Master Cylinder	Cooling System
2009	Maxima	3.5	①	NA	②	NA	NA	NA	③	DOT 3	④
2010	Maxima	3.5	①	NA	②	NA	NA	NA	③	DOT 3	④

DOT: Department Of Transpotation

NA: Not Applicable

① Nissan Ester Engine Oil 5W-30

② Nissan CVT Fluid NS-2

③ Nissan PSF

④ Nissan Long Life Anti-Freeze Coolant

37663_MAXI_C0014

VALVE SPECIFICATIONS

Year	Engine ID/VIN	Engine Displacement Liters	Seat Angle (deg.)	Face Angle (deg.)	Spring Test Pressure (lbs. @ in.)	Spring Installed Height (in.)	Stem-to-Guide Clearance (in.) Intake	Stem-to-Guide Clearance (in.) Exhaust	Stem Diameter (in.) Intake	Stem Diameter (in.) Exhaust
2009	VQ35DE/A	3.5	45.15-45.45	45	37-42@ 1.4567	1.4567	0.0008-0.0021	0.0016-0.0029	0.2348-0.2354	0.2341-0.2346
2010	VQ35DE/A	3.5	45.15-45.45	45	37-42@ 1.4567	1.4567	0.0008-0.0021	0.0016-0.0029	0.2348-0.2354	0.2341-0.2346

37663_MAXI_C0005

CAMSHAFT AND BEARING SPECIFICATIONS

All measurements are given in inches.

Year	Engine Displacement Liters	Engine ID/VIN	Journal Diameter	Brg. Oil Clearance	Shaft End-play	Runout	Journal Bore	Lobe Lift Intake	Exhaust
2009	3.5	VQ35DE/A	①	②	0.0045-0.0074	0.0008	NA	NA	NA
2010	3.5	VQ35DE/A	①	②	0.0045-0.0074	0.0008	NA	NA	NA

NA: Not Available

① No 1: 1.0211 - 1.0218 inches

No 2, 3, 4: 0.9230 - 0.9238 inches

② No 1: 0.0018 - 0.0034 inches

No 2, 3, 4: 0.0014 - 0.0030 inches

37663_MAXI_C0013

CRANKSHAFT AND CONNECTING ROD SPECIFICATIONS

All measurements are given in inches.

Year	Engine Displacement Liters	Engine ID/VIN	Crankshaft Main Brg. Journal Dia.	Main Brg. Oil Clearance	Shaft End-play	Thrust on No.	Connecting Rod Journal Diameter	Oil Clearance	Side Clearance
2009	3.5	VQ35DE/A	2.3603-2.3612*	0.0014-0.0018	0.0039-0.0098	3	2.1654-2.16590	0.0008-0.0018	0.0079-0.0138
2010	3.5	VQ35DE/A	2.3603-2.3612*	0.0014-0.0018	0.0039-0.0098	3	2.1654-2.16590	0.0008-0.0018	0.0079-0.0138

* Based upon grade

37663_MAXI_C0008

PISTON AND RING SPECIFICATIONS

All measurements are given in inches.

Year	Engine Displ. Liters	Engine ID/VIN	Piston Clearance	Ring Gap Top Compression	Bottom Compression	Oil Control	Ring Side Clearance Top Compression	Bottom Compression	Oil Control
2009	3.5	VQ35DE/A	0.0004-0.0012	0.0091-0.0110	0.0130-0.0169	0.0079-0.0177	0.0018-0.0031	0.0012-0.0028	0.0018-0.0049
2010	3.5	VQ35DE/A	0.0004-0.0012	0.0091-0.0110	0.0130-0.0169	0.0079-0.0177	0.0018-0.0031	0.0012-0.0028	0.0018-0.0049

37663_MAXI_C0007

TORQUE SPECIFICATIONS

All readings in ft. lbs.

Year	Engine Displacement Liters	Engine ID/VIN	Cylinder Head Bolts	Main Bearing Bolts	Rod Bearing Bolts	Crankshaft Damper Bolts	Flywheel Bolts	Manifold Intake	Manifold Exhaust	Spark Plugs	Oil Drain Plug
2009	3.5	VQ35DE/A	①	②	③	④	65	⑤	23	15	25
2010	3.5	VQ35DE/A	①	②	③	④	65	⑤	23	15	25

① Step 1: 72 ft. lbs.

   Step 2: Loosen bolts completely

   Step 3: 29 ft. lbs.

   Step 4: Tighten an additional 103 degrees

   Step 5: Tighten an additional 103 degrees

② Step 1: 24 - 28 ft. lbs.

   Step 2: Tighten an additional 90 - 95 degrees

③ Step 1: 14 - 15 ft. lbs.

   Step 2: Tighten an additional 90 - 95 degrees

④ Step 1: 32 ft. lbs.

   Step 2: Tighten an additional 84 - 90 degrees

⑤ Step 1: 5 ft. lbs.

   Step 2: 20 ft. lbs.

37663_MAXI_C0006

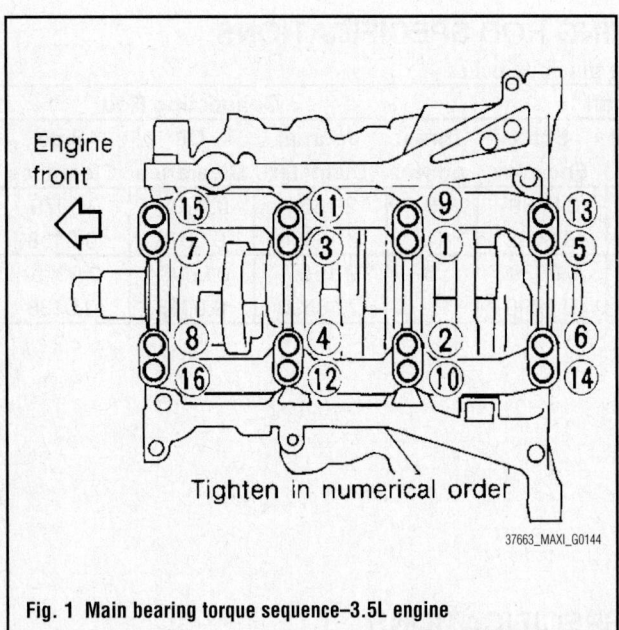

Fig. 1 Main bearing torque sequence–3.5L engine

WHEEL ALIGNMENT

Year	Model		Caster Range (+/-Deg.)	Caster Preferred Setting (Deg.)	Camber Range (+/-Deg.)	Camber Preferred Setting (Deg.)	Toe-in (in.)
2009	Maxima	F	0.55	+4.95	0.55	①	0.04 +/- 0.04
		R	—	—	NA	-0.42	0.0 +/- 0.08
2010	Maxima	F	0.55	+4.95	0.55	①	0.04 +/- 0.04
		R	—	—	NA	-0.42	0.0 +/- 0.08

NA: Not Available
① LH: -0.35
　 RH: -0.60

37663_MAXI_C0009

TIRE, WHEEL AND BALL JOINT SPECIFICATIONS

Year	Model	OEM Tires Standard	OEM Tires Optional	Tire Pressures (psi) Front	Tire Pressures (psi) Rear	Wheel Size	Lug Nut Torque (ft. lbs.)
2009	Maxima	P245/45R18	245/40R19	33/35	33/35	7-JJ/7.5-JJ	80
2010	Maxima	P245/45R18	245/40R19	33/35	33/35	7-JJ/7.5-JJ	80

OEM: Original Equipment Manufacturer
PSI: Pounds Per Square Inch

37663_MAXI_C0010

BRAKE SPECIFICATIONS
All measurements in inches unless noted

Year	Model		Brake Disc Original Thickness	Brake Disc Minimum Thickness	Brake Disc Maximum Run-out	Minimum Lining Thickness Front	Minimum Lining Thickness Rear	Brake Caliper Bracket Bolts (ft. lbs.)	Brake Caliper Mounting Bolts (ft. lbs.)
2009	Maxima	F	1.102	1.024	0.001	0.079	—	98	25
		R	0.650	0.551	0.002	—	0.039	62	32
2010	Maxima	F	1.102	1.024	0.001	0.079	—	98	25
		R	0.650	0.551	0.002	—	0.039	62	32

37663_MAXI_C0011

SCHEDULED MAINTENANCE INTERVALS
NISSAN—Maxima

TO BE SERVICED	TYPE OF SERVICE	VEHICLE MILEAGE INTERVAL (x1000)												
		7.5	15	22.5	30	37.5	45	52.5	60	67.5	75	82.5	90	97.5
Engine oil & filter	R	✓	✓	✓	✓	✓	✓	✓	✓	✓	✓	✓	✓	✓
Brake lines & cables	S/I		✓		✓		✓		✓		✓		✓	
Brake pads & discs	S/I		✓		✓		✓		✓		✓		✓	
Driveshaft boots	S/I		✓		✓		✓		✓		✓		✓	
Exhaust system	S/I				✓				✓				✓	
Transmission or transaxle fluid	S/I		✓		✓		✓		✓		✓		✓	
Air cleaner filter	R				✓				✓				✓	
Spark plugs (except platinum)	R				✓				✓				✓	
Spark plugs (platinum tip)	R								✓					
Steering gear & linkage, axle & suspension parts	S/I				✓				✓				✓	
Engine coolant	R								✓					
Drive belts	S/I								✓					
Fuel lines	S/I								✓					
Vapor lines	S/I								✓					

R: Replace   S/I: Service or Inspect

FREQUENT OPERATION MAINTENANCE (SEVERE SERVICE)

If a vehicle is operated under any of the following conditions it is considered severe service:

- Extremely dusty areas.

- 50% or more of the vehicle operation is in 32°C (90°F) or higher temperatures, or constant operation in temperatures below 0°C (32°F).

- Prolonged idling (vehicle operation in stop and go traffic).

- Frequent short running periods (engine does not warm to normal operating temperatures).

- Police, taxi, delivery usage or trailer towing usage.

Oil & oil filter: change every 3750 miles.

Brake pads & discs: service or inspect every 7500 miles.

Driveshaft boots: service or inspect every 7500 miles.

Exhaust system: service or inspect every 7500 miles.

Steering gear & linkage, axle & suspension parts: service or inspect every 7500 miles.

Steering linkage ball joints & front suspension ball joints: service or inspect every 7500 miles.

Air cleaner filter: service or inspect every 15,000 miles.

37663_MAXI_C0012

BRAKES INFORMATION AND PRECAUTIONS

ANTI-LOCK SYSTEMS

- Certain components within the ABS system are not intended to be serviced or repaired individually.
- Do not use rubber hoses or other parts not specifically specified for and ABS system. When using repair kits, replace all parts included in the kit. Partial or incorrect repair may lead to functional problems and require the replacement of components.
- Lubricate rubber parts with clean, fresh brake fluid to ease assembly. Do not use shop air to clean parts; damage to rubber components may result.
- Use only DOT 3 brake fluid from an unopened container.
- If any hydraulic component or line is removed or replaced, it may be necessary to bleed the entire system.
- A clean repair area is essential. Always clean the reservoir and cap thoroughly before removing the cap. The slightest amount of dirt in the fluid may plug an orifice and impair the system

function. Perform repairs after components have been thoroughly cleaned; use only denatured alcohol to clean components. Do not allow ABS components to come into contact with any substance containing mineral oil; this includes used shop rags.

- The Anti-Lock control unit is a microprocessor similar to other computer units in the vehicle. Ensure that the ignition switch is **OFF** before removing or installing controller harnesses. Avoid static electricity discharge at or near the controller.
- If any arc welding is to be done on the vehicle, the control unit should be unplugged before welding operations begin.

DISC AND DRUM SYSTEMS

✳ CAUTION

Dust and dirt accumulating on brake parts during normal use may contain asbestos fibers from production

or aftermarket brake linings. Breathing excessive concentrations of asbestos fibers can cause serious bodily harm. Exercise care when servicing brake parts. Do not sand or grind brake lining unless equipment used is designed to contain the dust residue. Do not clean brake parts with compressed air or by dry brushing. Cleaning should be done by dampening the brake components with a fine mist of water, then wiping the brake components clean with a dampened cloth. Dispose of cloth and all residue containing asbestos fibers in an impermeable container with the appropriate label. Follow practices prescribed by the Occupational Safety and Health Administration (OSHA) and the Environmental Protection Agency (EPA) for the handling, processing, and disposing of dust or debris that may contain asbestos fibers.**

BRAKES BLEEDING THE BRAKE SYSTEM

BLEEDING PROCEDURE

BLEEDING PROCEDURE

➡**Carefully monitor brake fluid level at master cylinder during bleeding operation. Fill reservoir with new brake fluid. Make sure it is full at all times while bleeding air out of system. Place a container under master cylinder to avoid spillage of brake fluid. Do not loosen the connecting portion of the actuator during air bleeding.**

1. Disconnect battery negative terminal.
2. Connect a transparent vinyl tube and container to air bleeder valve.
3. Fully depress brake pedal several times.
4. With brake pedal depressed, open air bleeder valve to release air.
5. Close air bleeder valve.
6. Release brake pedal slowly.
7. Tighten air bleeder valve to 69 inch lbs. (8 Nm).

8. Repeat steps 3 through 6 until no more air bubbles come out of air bleeder valve.
9. Bleed the brake hydraulic system air bleeder valves in the following order:
- Right rear brake
- Left front brake
- Left rear brake
- Right front brake

BLEEDING THE ABS SYSTEM

Refer to the Bleeding Procedure.

BRAKES

WHEEL SPEED SENSORS

REMOVAL & INSTALLATION

✳✳ CAUTION

Note the following:

- Be careful not to damage wheel sensor edge and sensor rotor teeth.
- When pulling out the wheel sensor, be careful to turn it as little as possible. Do not pull on the wheel sensor harness.
- Check if foreign objects such as iron fragments are adhered to the pick-up part of the sensor or to the inside of the hole for the wheel sensor, or if a foreign object is caught in the surface of the mating surface for the wheel sensor. Repair as necessary and then install the wheel sensor.

Front Wheel Sensor

See Figure 2.

1. Remove the front wheel and tire.
2. Partially remove front wheel fender protector and reposition out of the way.
3. Disconnect the wheel sensor harness connector.
4. Remove the wheel sensor harness from the brackets.
5. Remove the wheel sensor bolt and wheel sensor from the front hub assembly.

To install:

6. Installation is in the reverse order of removal.

Rear Wheel Sensor

1. Remove the rear wheel and tire.
2. Remove the stabilizer bar clamps and bushings using power tool, and reposition the stabilizer bar out of the way.
3. Disconnect the wheel sensor harness connector.
4. Remove the wheel sensor harness from the brackets.
5. Remove the wheel sensor bolt and wheel sensor from the rear hub assembly.

To install:

6. Installation is in the reverse order of removal.

WHEEL SPEED SENSOR ROTORS

REMOVAL & INSTALLATION

The front and rear wheel sensor rotors are an integral part of the wheel hub assemblies and cannot be disassembled. To replace the sensor rotor, replace the wheel hub assembly.

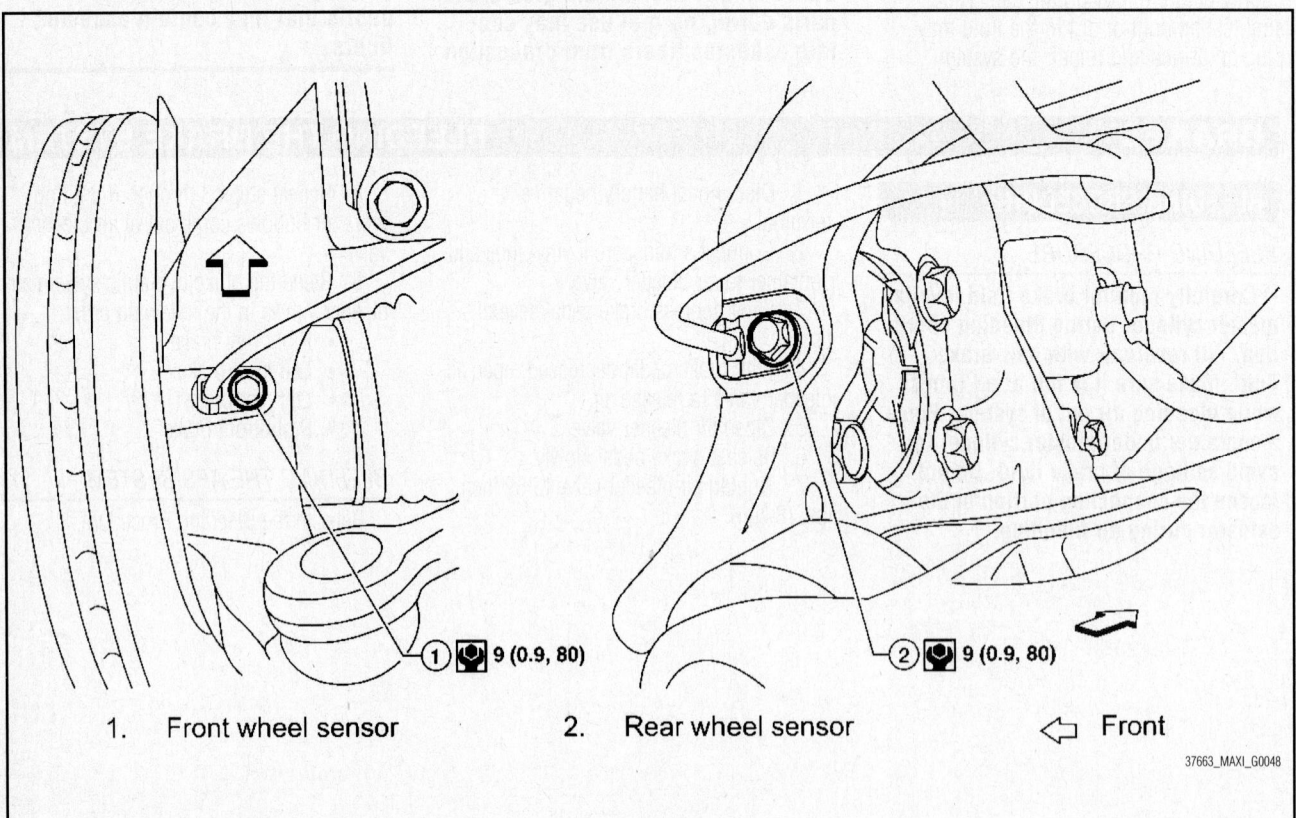

1. Front wheel sensor
2. Rear wheel sensor
⇐ Front

37663_MAXI_G0048

Fig. 2 Wheel sensor locations

BRAKES   **FRONT DISC BRAKES**

BRAKE CALIPER

REMOVAL & INSTALLATION

See Figures 3 through 6.

✳✳ WARNING

Clean dust on caliper and brake pad with a vacuum dust collector to minimize the hazard of air borne particles or other materials.

✳✳ CAUTION

When removing and installing the cylinder body, do not depress the brake pedal because the piston will pop out. Do not damage the piston boot. Keep the brake rotor free from grease and brake fluid. Refill the brake reservoir with new brake fluid only. Never reuse the drained brake fluid.

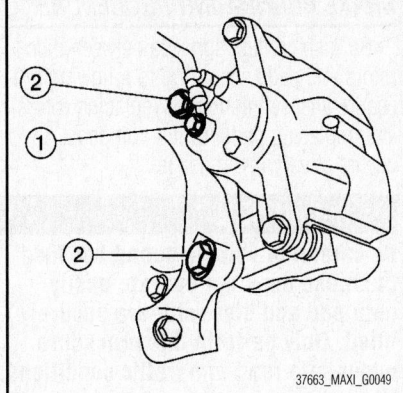

Fig. 3 Remove the union bolt (1) and then disconnect the brake hose from the caliper assembly; remove the torque member bolts (2)

1. Remove front wheel and tire.
2. Secure the disc rotor using a wheel nut.

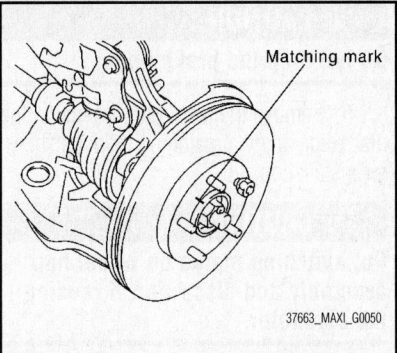

Fig. 4 If reusing the disc rotor, apply a matching mark as shown for installation

3. Drain the brake fluid.
4. Remove the union bolt and then disconnect the brake hose from the caliper assembly. Discard the copper washers.

✳✳ CAUTION

Do not reuse the copper washers.

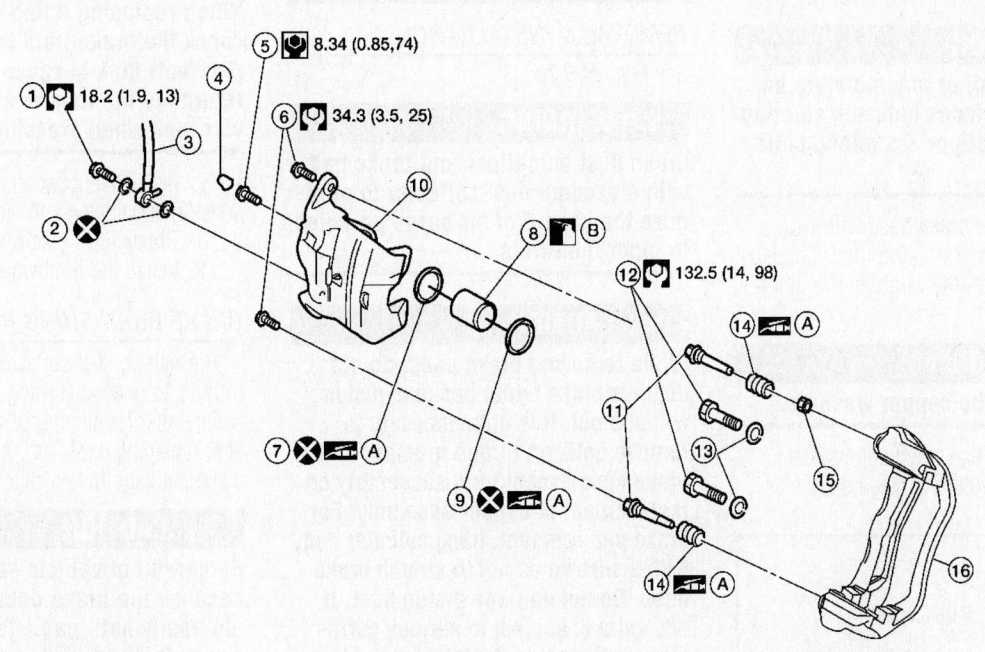

1.	Union bolt	2.	Copper washer	3.	Brake hose
4.	Cap	5.	Bleed valve	6.	Sliding pin bolt
7.	Piston seal	8.	Piston	9.	Piston boot
10.	Cylinder body	11.	Sliding pin	12.	Torque member bolt
13.	Washer	14.	Sliding pin boot	15.	Bushing
16.	Torque member	A.	Rubber grease	B.	Brake fluid

Fig. 5 Exploded view of front brake caliper assembly

5. Remove the torque member bolts and remove the brake caliper assembly.

> ✳✳ **CAUTION**
>
> **Do not drop the brake pads.**

6. Remove the disc rotor. If reusing the disc rotor, apply a matching mark as shown for installation.

> ✳✳ **CAUTION**
>
> **Put matching marks on wheel hub assembly and disc rotor, if reusing the disc rotor.**

To install:

7. Install the disc rotor. If reusing the disc rotor, align the matching mark on the disc rotor and wheel hub assembly for installation as shown.

> ✳✳ **CAUTION**
>
> **Align the matching marks on wheel hub assembly and disc rotor, if reusing the disc rotor.**

8. Install the brake caliper assembly, and tighten the torque member bolts to the specified torque.

> ✳✳ **CAUTION**
>
> **Do not allow oil or any moisture on all contact surfaces between steering knuckle and caliper assembly, bolts, and washer.**

9. Install the brake hose with two new copper washers, using the L-shaped pin for alignment, then tighten the union bolt.

> ✳✳ **CAUTION**
>
> **Do not reuse the copper washers.**

10. Refill the brake hydraulic system with new brake fluid and bleed out the air.

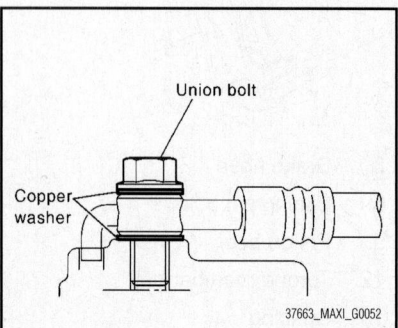

Fig. 6 Install the brake hose with two new copper washers, using the L-shaped pin for alignment

11. Check the front disc brakes for drag.
12. Install the front wheel and tire.

BRAKE BURNISHING PROCEDURE

Burnish contact surfaces between disc rotors and pads according to following procedure after refinishing or replacing rotors, after replacing pads, or if a soft pedal occurs at very low mileage.

> ✳✳ **CAUTION**
>
> **Be careful of vehicle speed because the brake does not operate easily until pad and disc rotor are securely fitted. Only perform this procedure under safe road and traffic conditions. Use extreme caution.**

1. Drive vehicle on straight, flatroad.
2. Depress brake pedal with the power to stop vehicle within 3 to 5 seconds until the vehicle stops.
3. Drive without depressing brake for a few minutes to cool the brake.
4. Repeat steps 1 to 3 until pad and disc rotor are securely fitted.

DISC BRAKE PADS

REMOVAL & INSTALLATION
See Figures 3 and 7.

> ✳✳ **WARNING**
>
> **Clean dust on caliper and brake pad with a vacuum dust collector to minimize the hazard of air borne particles or other materials.**

> ✳✳ **CAUTION**
>
> **While removing brake pads, do not depress brake pedal because piston will pop out. It is not necessary to remove bolts on torque member and brake hose except for disassembly or replacement of caliper assembly. For brake pad removal, hang cylinder body with a wire so as not to stretch brake hose. Do not damage piston boot. If any shim is subject to serious corrosion, replace it with a new one. Always replace shim and shim cover as a set when replacing brake pads. Keep rotor free from brake fluid. Burnish the brake pads and disc rotor contacting surfaces, after refinishing or replacing rotors, after replacing pads, or if a soft pedal occurs at very low mileage.**

1. Remove front wheel and tire.
2. Remove upper and lower sliding pin bolts without disconnecting the union bolt.

3. Hang cylinder body with a wire, and do not twist or stretch the brake hose.
4. Remove anti-rattle clips, pads, shims, and shim covers from torque member.

To install:

5. Apply Molykote M-77 grease or equivalent between the inner and outer shims and the back of the brake pads.
6. Install the inner and outer shims and shim covers to the inner pad and outer pad.

> ✳✳ **CAUTION**
>
> **Do not get grease on the inner and outer pad friction surfaces.**

7. Install the assembled inner and outer shims, shim covers, pads and anti-rattle clips to the torque member.

> ✳✳ **CAUTION**
>
> **Do not get grease on the inner and outer pad or rotor friction surfaces.**

8. Press piston into cylinder body using suitable tool, then install the cylinder body on the torque member.

> ✳✳ **CAUTION**
>
> **When replacing a pad with new one, check the brake fluid level in the reservoir tank because brake fluid returns to the master cylinder reservoir tank when pressing in the piston.**

9. Install the upper and lower sliding pin bolts (2) and tighten it to the specified torque.
10. Check front disc brakes for drag.
11. Install the front wheel and tire.

BRAKE BURNISHING PROCEDURE

Burnish contact surfaces between disc rotors and pads according to following procedure after refinishing or replacing rotors, after replacing pads, or if a soft pedal occurs at very low mileage.

> ✳✳ **CAUTION**
>
> **Be careful of vehicle speed because the brake does not operate easily until pad and disc rotor are securely fitted. Only perform this procedure under safe road and traffic conditions. Use extreme caution.**

1. Drive vehicle on straight, flat road.
2. Depress brake pedal with the power to stop vehicle within 3 to 5 seconds until the vehicle stops.
3. Drive without depressing brake for a few minutes to cool the brake.
4. Repeat steps 1 to 3 until pad and disc rotor are securely fitted.

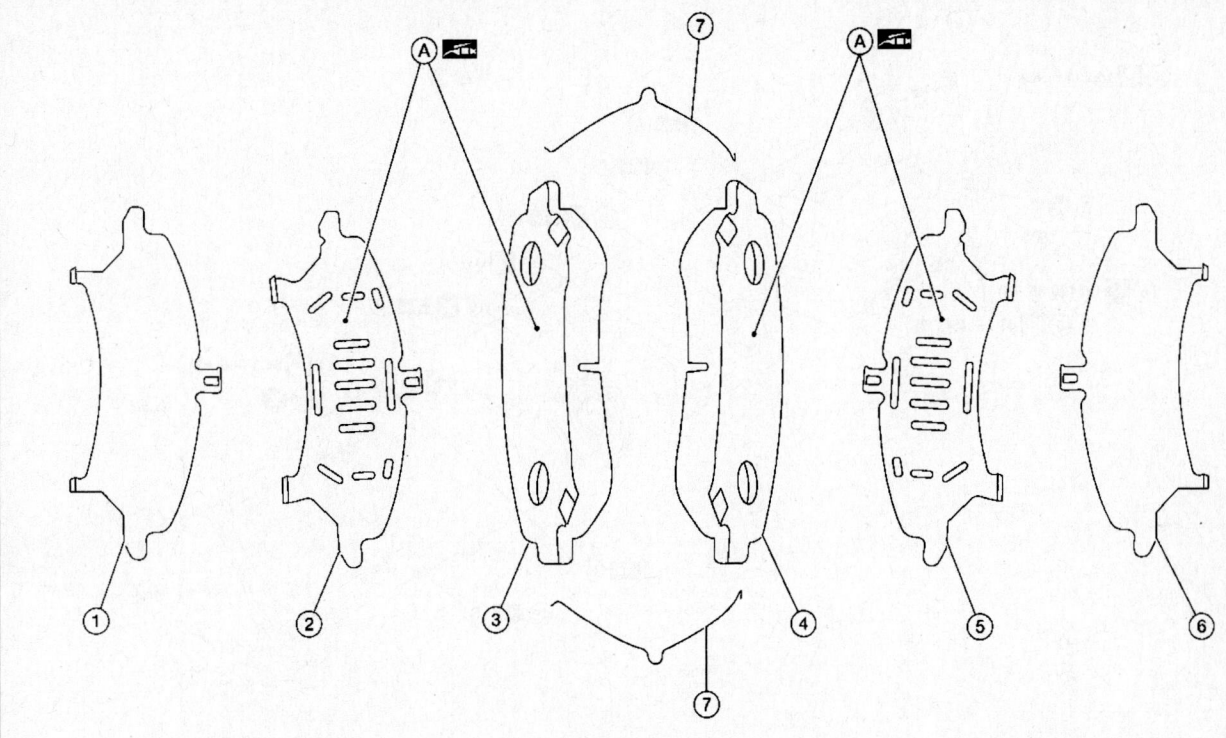

1. Inner shim cover	2. Inner shim	3. Inner pad
4. Outer pad	5. Outer shim	6. Outer shim cover
7. Anti-rattle clips	A. Molykote M-77 grease	

37663_MAXI_G0053

Fig. 7 Exploded view of brake pads

BRAKES

REAR DISC BRAKES

BRAKE CALIPER

REMOVAL & INSTALLATION

See Figures 8 through 10.

✳✳ WARNING

Clean dust on caliper and brake pad with a vacuum dust collector to minimize the hazard of air borne particles or other materials.

✳✳ CAUTION

While removing and installing the cylinder body, do not depress the brake pedal because the piston will pop out. Do not damage the piston boot. Keep rotor free from grease and brake fluid. Refill the brake reservoir with new brake fluid. Never reuse drained brake fluid.

1. Remove the rear wheel and tire.

2. Hold the disc rotor in place by installing a wheel nut.

3. Drain the brake fluid.

4. Remove the union bolt and copper washers, discard the copper washers. Then disconnect the brake hose from the cylinder body.

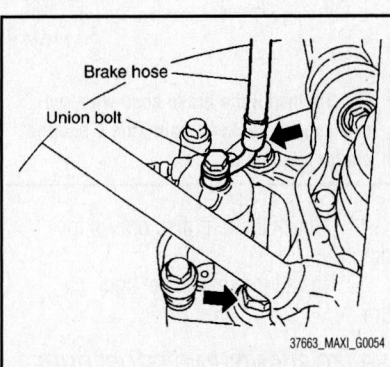

Brake hose

Union bolt

37663_MAXI_G0054

Fig. 8 Remove the union bolt and copper washers, discard the copper washers

✳✳ CAUTION

Do not reuse the copper washers.

5. Remove the torque member bolts, and remove the brake caliper assembly.

✳✳ CAUTION

Do not drop brake the pads.

6. Remove the disc rotor. If reusing the disc rotor, before removing the disc rotor apply matching mark.

✳✳ CAUTION

Put matching marks on wheel hub assembly and disc rotor, if reusing the disc rotor.

To install:

7. Install the disc rotor. If reusing the disc rotor, align the matching mark to position the disc rotor on the wheel hub assembly.

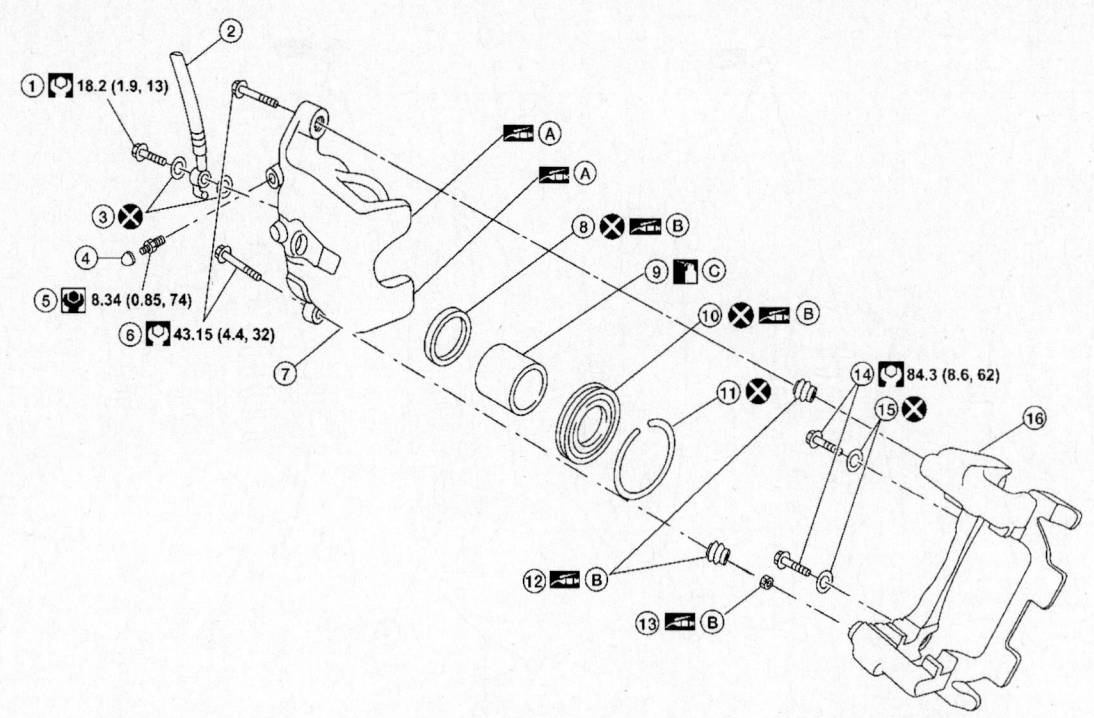

1. Union bolt	2. Brake hose	3. Copper washer
4. Cap	5. Bleed valve	6. Sliding pin bolt
7. Cylinder body	8. Piston seal	9. Piston
10. Piston boot	11. Retaining ring	12. Sliding pin boot
13. Bushing	14. Torque member bolt	15. Washer
16. Torque member	A. Molykote M-77 grease	B. Rubber grease
C. Brake fluid		

37663_MAXI_G0055

Fig. 9 Exploded view of rear brake caliper assembly

> **⁂ CAUTION**
>
> **Align the matching mark on wheel hub assembly and disc rotor, if reusing the disc rotor.**

8. Install the brake caliper assembly, and tighten the torque member bolts.

> **⁂ CAUTION**
>
> **Before installing caliper assembly, wipe off oil and moisture on all mounting surfaces of rear axle and caliper assembly and threads, bolts and washers.**

9. Install the brake hose with two new copper washers, using the L-shaped pin for alignment, then tighten the union bolt.

> **⁂ CAUTION**
>
> **Do not reuse the copper washers.**

10. Refill the brake hydraulic system with new brake fluid and bleed out the air.

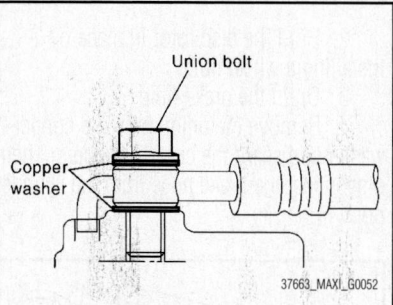

37663_MAXI_G0052

Fig. 10 Install the brake hose with two new copper washers, using the L-shaped pin for alignment

11. Check the rear disc brakes for drag.

12. Install the rear wheel and tire.

BRAKE BURNISHING PROCEDURE

Burnish contact surfaces between disc rotors and pads according to following procedure after refinishing or replacing rotors, after replacing pads, or if a soft pedal occurs at very low mileage.

> **⁂ CAUTION**
>
> **Be careful of vehicle speed because the brake does not operate easily until pad and disc rotor are securely fitted. Only perform this procedure under safe road and traffic conditions. Use extreme caution.**

1. Drive vehicle on straight, flat road.

2. Depress brake pedal with the power to stop vehicle within 3 to 5 seconds until the vehicle stops.

3. Drive without depressing brake for a few minutes to cool the brake.

4. Repeat steps 1 to 3 until pad and disc rotor are securely fitted.

DISC BRAKE PADS

REMOVAL & INSTALLATION

See Figures 11 and 12.

☀☀ WARNING

Clean dust on caliper and brake pad with a vacuum dust collector to minimize the hazard of air borne particles or other materials.

☀☀ CAUTION

While removing and installing cylinder body, do not depress brake pedal because piston will pop out. It is not necessary to remove bolts on torque member and brake hose except for disassembly or replacement of caliper assembly. For pad removal and installation, hang cylinder body with a wire so as not to stretch brake hose. Do not damage piston boot. If any shim is subject to serious corrosion, replace it with a new one. Always replace shim and shim covers as a set when replacing brake pads. Keep rotor free from brake fluid. Burnish the brake pads and disc rotor mutually contacting surfaces after refinishing or replacing rotors, after replacing pads, or if a soft pedal occurs at very low mileage.

1. Remove the rear wheel and tire.
2. Remove the upper sliding pin bolt and loosen the lower sliding pin bolt to swing the cylinder body down.
3. Remove the pads, pad retainers, shims, and shim covers from the torque member.

☀☀ CAUTION

Do not deform the pad retainers when removing them from the torque member.

To install:

4. Apply Molykote M-77 grease or equivalent to between shim covers and shims. Install inner shim, inner shim cover to inner pad. Install outer shim and outer shim cover to outer pad.
5. Apply Molykote M-77 grease or equivalent to between pad retainer and pad.

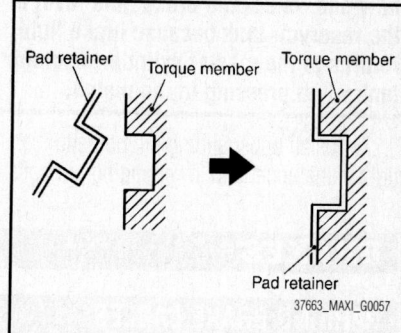

37663_MAXI_G0057

Fig. 12 Install pad retainers and pads to torque member

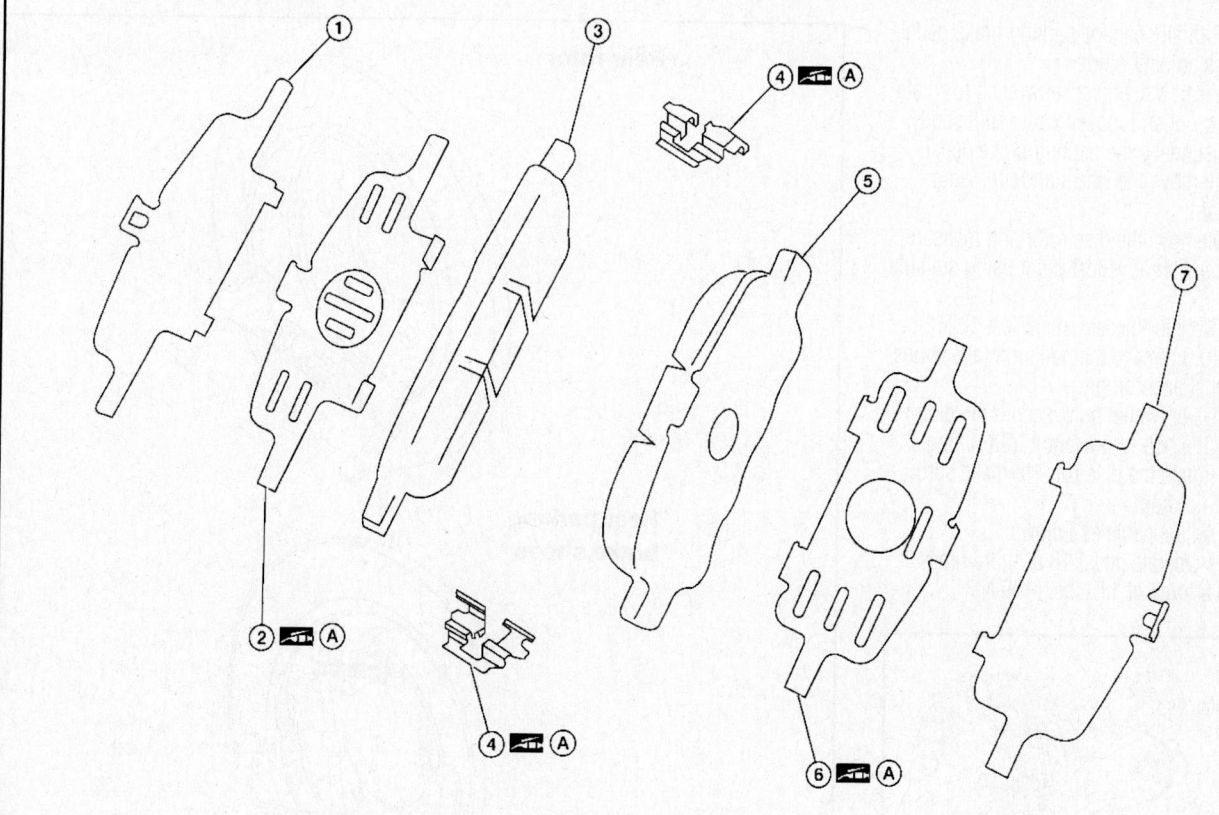

1.	Inner shim cover	2.	Inner shim	3.	Inner pad
4.	Pad retainer	5.	Outer pad	6.	Outer shim
7.	Outer shim cover	A.	Molykote M-77 grease		

37663_MAXI_G0056

Fig. 11 Exploded view of rear brake pads

Install pad retainers and pads to torque member.

6. Press in piston using suitable tool, until the pads can be installed, and then install the cylinder body in the torque member.

✳✳ CAUTION

In the case of replacing a pad with new one, check the brake fluid level in the reservoir tank because brake fluid returns to the master cylinder reservoir tank when pressing in the piston.

7. Install upper sliding pin bolt and tighten the upper and lower sliding pin bolts.

8. Check the rear disc brakes for drag.

9. Install the rear wheel and tire.

BRAKE BURNISHING PROCEDURE

Burnish contact surfaces between disc rotors and pads according to following procedure after refinishing or replacing rotors, after replacing pads, or if a soft pedal occurs at very low mileage.

✳✳ CAUTION

Be careful of vehicle speed because the brake does not oper- ate easily until pad and disc rotor are securely fitted. Only perform this procedure under safe road and traffic conditions. Use extreme caution.

1. Drive vehicle on straight, flat road.

2. Depress brake pedal with the power to stop vehicle within 3 to 5 seconds until the vehicle stops.

3. Drive without depressing brake for a few minutes to cool the brake.

4. Repeat steps 1 to 3 until pad and disc rotor are securely fitted.

BRAKES
PARKING BRAKE

PARKING BRAKE CABLES

ADJUSTMENT

See Figures 13 through 15.

1. Remove the lower instrument panel LH.

2. Partially engage parking brake pedal to access adjusting nut.

3. Insert a deep socket wrench to rotate adjusting nut and loosen cable sufficiently. Then, disengage the parking brake pedal.

4. Remove the wheel and tire using power tool.

5. Remove the disc rotor and measure inner diameter at widest point using suitable tool.

6. Transfer measurement less 0.24 inches (0.6 mm) to the parking brake shoes and adjust accordingly.

7. Using wheel nuts, secure the disc rotor to the hub to prevent it from tilting.

8. Rotate the disc rotor to make sure there is no drag.

9. Adjust cable as follows:

   a. Operate pedal 10 or more times with a force of 110 lbs. (490 N).

   b. Rotate adjusting nut with deep socket to adjust pedal stroke to specification of 4 to 5 clicks.

   c. With parking brake pedal completely disengaged, make sure there is no drag on the parking brake.

10. Install the disc rotor.

11. Install the wheel and tire using power tool.

12. Install the lower instrument panel LH.

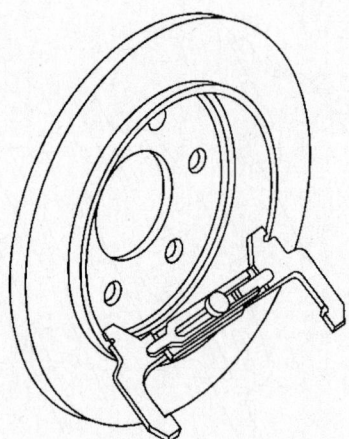

Rear rotor

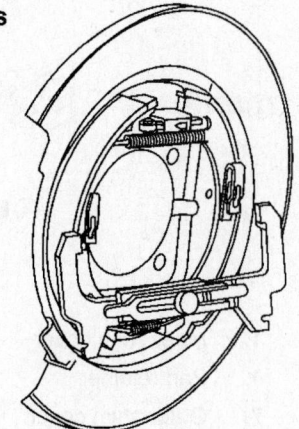

Rear parking brake shoes

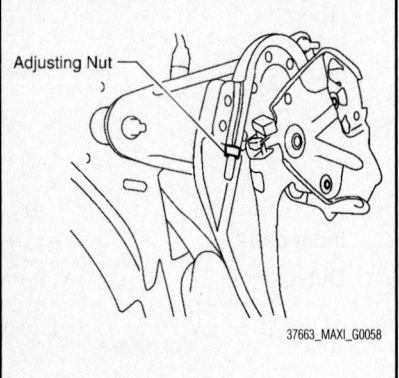

Fig. 13 Parking brake adjusting nut

Fig. 14 Measure inner diameter at widest point using suitable tool

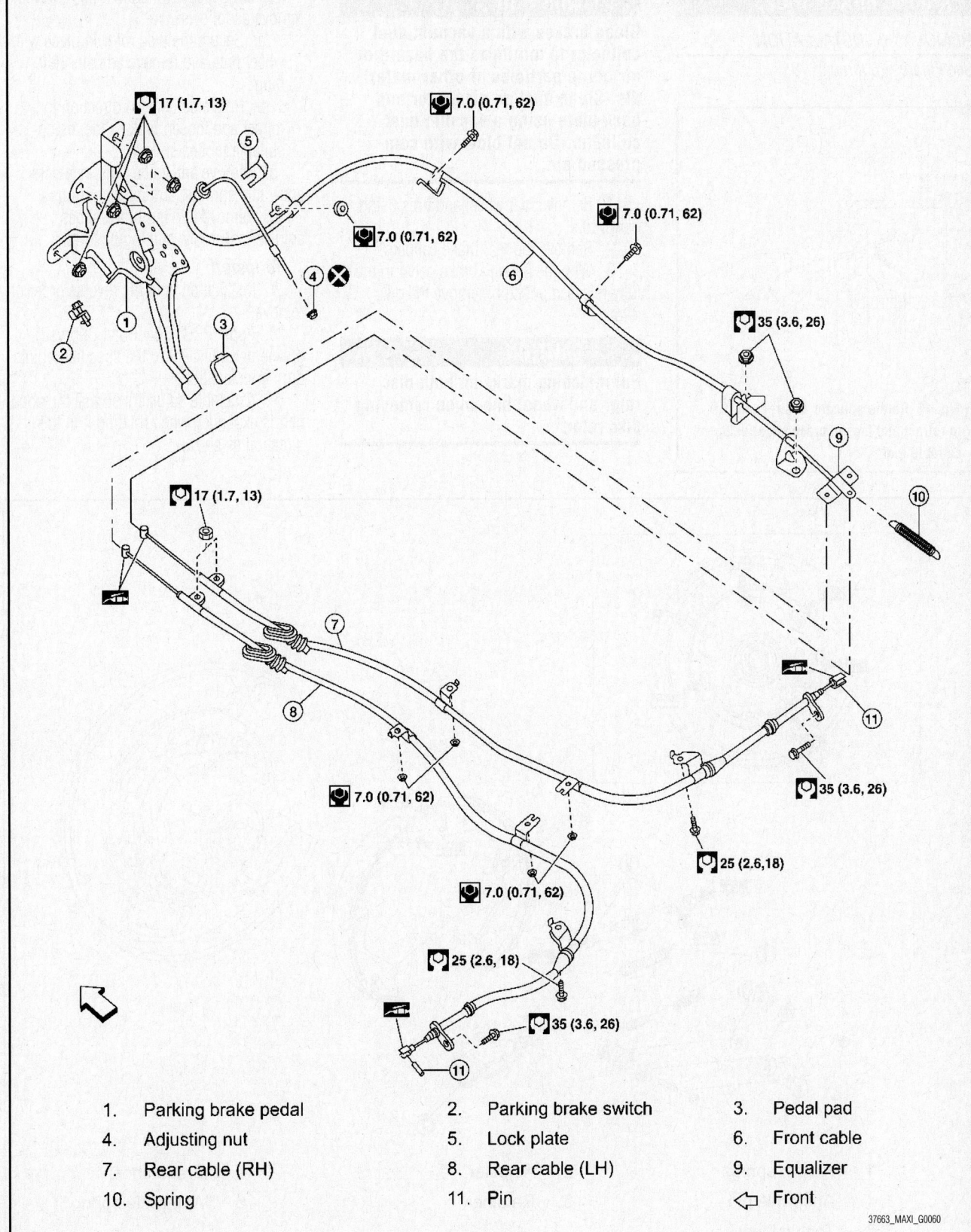

1. Parking brake pedal
2. Parking brake switch
3. Pedal pad
4. Adjusting nut
5. Lock plate
6. Front cable
7. Rear cable (RH)
8. Rear cable (LH)
9. Equalizer
10. Spring
11. Pin
⇦ Front

37663_MAXI_G0060

Fig. 15 Exploded view of parking brake control system

PARKING BRAKE SHOES

REMOVAL & INSTALLATION

See Figures 16 through 18.

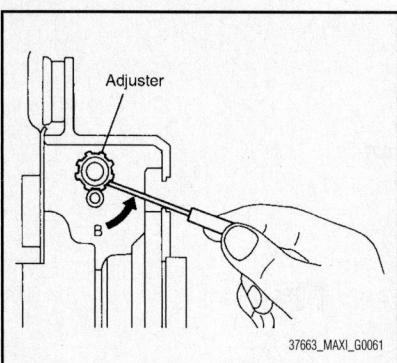

Fig. 16 Rotate adjuster in direction (B) to retract and loosen brake shoe, using suitable tool

❊❊ WARNING

Clean brakes with a vacuum dust collector to minimize the hazard of air borne particles or other materials. Clean dust on disc rotor and back plate using a vacuum dust collector. Do not blow with compressed air.

1. Remove rear wheel and tires using power tool.
2. Remove the rear brake calipers.
3. With the parking brake pedal in the fully released position, remove the disc rotor.

❊❊ CAUTION

Put matching marks on both disc rotor and wheel hub when removing disc rotor.

4. If the disc rotor cannot be removed, remove as follows:
   a. Secure the disc rotor in place with wheel nuts and remove adjuster hole plug.
   b. Rotate adjuster in direction to retract and loosen brake shoe, using suitable tool as shown.
5. Remove anti-rattle pins, retainers, anti-rattle springs, and return springs.
6. Remove parking brake shoes, adjuster assembly, and toggle lever.

To install:
7. Installation is in the reverse order of removal.
8. Apply PBC (Poly Butyl Cuprysil) grease or equivalent to the specified points during installation.
9. Assemble adjusters so that threaded part is expanded when rotating it in the direction as shown.

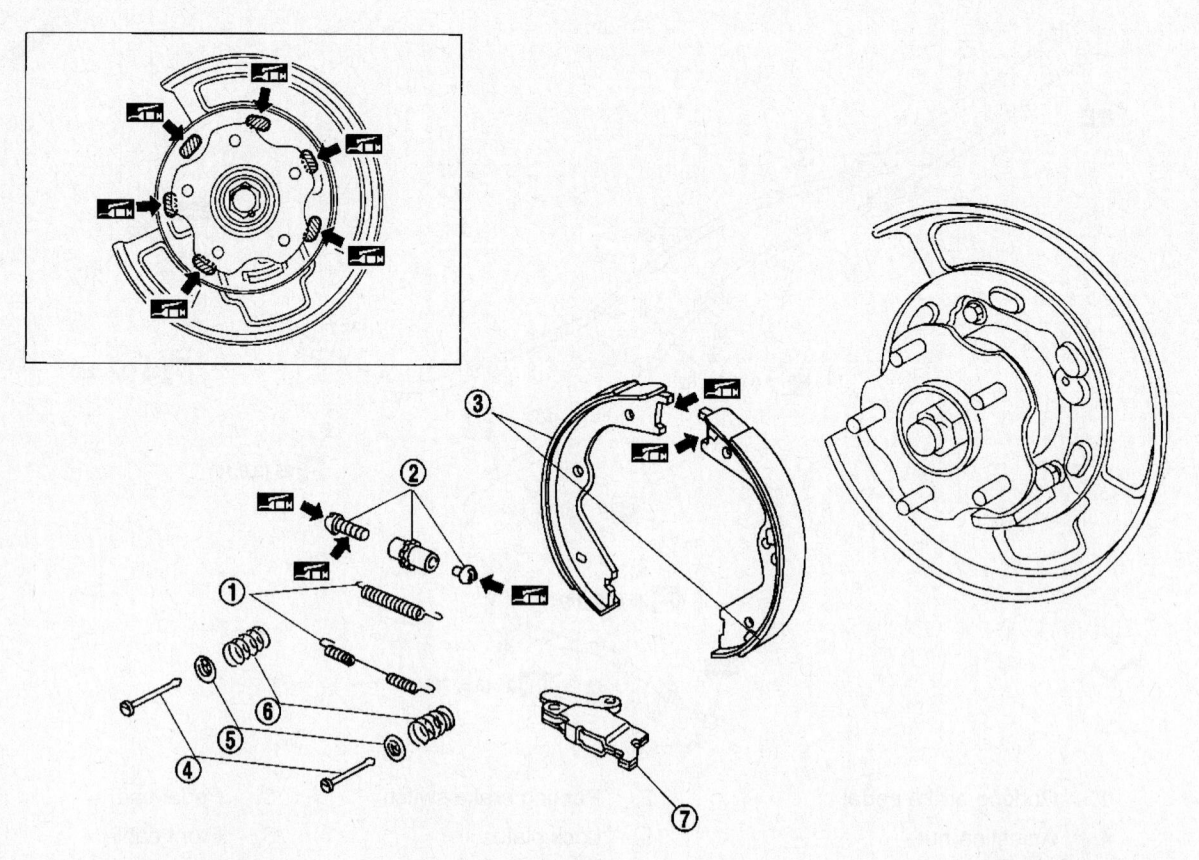

1. Return spring
2. Adjuster
3. Brake shoe
4. Anti-rattle pin
5. Retainer
6. Anti-rattle spring
7. Toggle lever
⬛ : PBC (Poly Butyl Cuprysil) grease or silicone-based grease

Fig. 17 Exploded view of parking assembly

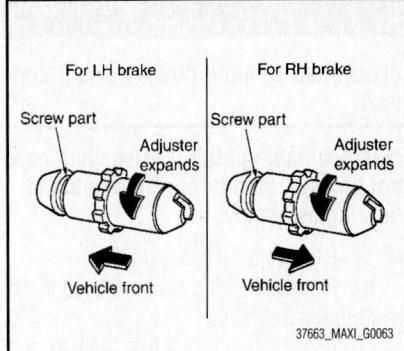

For LH brake

For RH brake

Screw part

Adjuster expands

Screw part

Adjuster expands

Vehicle front

Vehicle front

37663_MAXI_G0063

Fig. 18 Assemble adjusters so that threaded part is expanded when rotating it in the direction as shown

10. Shorten adjuster by rotating it the opposite as shown.

11. Check parking brake shoe sliding surface and drum inner surface for grease. Wipe off all grease adhering to the friction surfaces.

ADJUSTMENT

➡**After replacing the parking brake shoes or disc rotors, or if the parking brake does not function properly, perform the break-in operation as follows.**

1. Adjust parking brake pedal stroke to the specified amount of 4 to 5 notches.

2. Perform parking brake break-in (drag run) operation by driving and performing the following steps:

a. Drive forward at a constant speed of approximately 25 mph (40 km/h).

b. Apply the parking brake for approximately 10 seconds at an operating force at approximately 34–54 lbs. (150–199 N).

❋❋ **CAUTION**

To prevent lining from getting too hot, allow cool off period of approximately 5 minutes after every break-in operation. Do not perform excessive break-in operations, because it may cause uneven or early wear of lining.

3. After break-in operation, check that the parking brake pedal stroke is at specification of 4 to 5 clicks and adjust again as necessary.

CHASSIS ELECTRICAL

AIR BAG (SUPPLEMENTAL RESTRAINT SYSTEM)

GENERAL INFORMATION

❋❋ **CAUTION**

These vehicles are equipped with an air bag system. The system must be disarmed before performing service on, or around, system components, the steering column, instrument panel components, wiring and sensors. Failure to follow the safety precautions and the disarming procedure could result in accidental air bag deployment, possible injury and unnecessary system repairs.

SERVICE PRECAUTIONS

❋❋ **CAUTION**

Disconnect and isolate the battery negative cable before beginning any airbag system component diagnosis, testing, removal, or installation procedures. Wait at least 90 seconds after the ignition switch is turned off

and the negative (-) terminal cable is disconnected from the battery before starting the operation. The SRS is equipped with a backup power source, so if work is started within 90 seconds after disconnecting the negative (-) terminal cable from the battery, the SRS may be deployed. Failure to disable the airbag system may result in accidental airbag deployment, personal injury, or death.

DISARMING THE SYSTEM

All SRS electrical wiring harnesses and connectors are covered with YELLOW outer insulation. Do not use electrical test equipment on any circuit related to the SRS (air bag) sensors. When installing SRS components, always install with the arrow marks facing the front of the vehicle.

To disarm the SRS system turn the ignition switch to **OFF** position. Then, disconnect the both battery cables starting with the negative cable first and wait at least 10

minutes after the cables are disconnected. Be sure to insulate the battery terminal ends.

ARMING THE SYSTEM

To arm the SRS system turn the ignition switch to **OFF** position. Connect the both battery cables starting with the positive cable first.

The SRS or air bag system is equipped with a self-diagnostic operation. After turning the ignition key to the ON or START position, the AIR BAG warning lamp will illuminate for 7 seconds. After 7 seconds, the AIR BAG lamp will extinguish if no malfunction is detected. If the AIR BAG lamp does not extinguish after 7 seconds, check the SRS self-diagnostic system for a malfunction.

CLOCKSPRING CENTERING

Align spiral cable correctly when installing steering wheel. Make sure that the spiral cable is in the neutral position. The neutral position is detected by turning left 2.5 revolutions from the right end position and ending with the knob at the top.

DRIVE TRAIN

FRONT HALFSHAFTS

REMOVAL & INSTALLATION

Left Side

1. Remove wheel and tire from vehicle.
2. Remove wheel sensor from steering knuckle.

✳✳ CAUTION

Do not pull on wheel sensor harness.

3. Remove brake hose lock plate from strut assembly.
4. Remove brake caliper torque member bolts using power tool leaving brake hose attached, then remove disc rotor. Reposition caliper aside with wire.

➡️**Avoid depressing brake pedal while brake caliper is removed.**

5. Remove cotter pin, then loosen lock nut from drive shaft using power tool.
6. Remove lower strut bolts and nuts using power tool.
7. Using a piece of wood and a hammer, tap on lock nut to disengage drive shaft from wheel hub.

✳✳ CAUTION

Never place drive shaft joint at an extreme angle. Also be careful not to overextend slide joint. Never allow drive shaft to hang down without support.

➡️**Use suitable puller if drive shaft cannot be separated from wheel hub and bearing assembly.**

8. Remove drive shaft from transaxle assembly.
   a. Use Tool KV40107500 and a sliding hammer while inserting tip of tool between housing and transaxle assembly.

✳✳ CAUTION

Never place drive shaft joint at an extreme angle when removing drive shaft. Also be careful not to overextend slide joint.

To install:

9. Installation is in the reverse order of removal. Note the following:

✳✳ CAUTION

Do not reuse non-reusable parts.

- Install new circlip on drive shaft in the circular clip groove on transaxle side.

✳✳ CAUTION

Make sure the new circlip on the drive shaft is securely fastened.

- In order to prevent damage to differential side oil seal, place Tool KV38107900 onto oil seal before inserting drive shaft. Slide drive shaft into slide joint and tap with a hammer to install securely.

✳✳ CAUTION

Make sure that circlip is completely engaged.

Right Side

1. Remove wheel and tire from vehicle.
2. Remove wheel sensor from steering knuckle.

✳✳ CAUTION

Do not pull on wheel sensor harness.

3. Remove brake hose lock plate from strut assembly.
4. Remove brake caliper torque member bolts using power tool leaving brake hose attached, then remove disc rotor. Reposition caliper aside with wire.

➡️**Avoid depressing brake pedal while brake caliper is removed.**

5. Remove disc rotor.
6. Remove cotter pin, then loosen lock nut from drive shaft using power tool.
7. Remove lower strut bolts and nuts using power tool.
8. Using a piece of wood and a hammer, tap on lock nut to disengage drive shaft from wheel hub.

✳✳ CAUTION

Never place drive shaft joint at an extreme angle. Also be careful not to overextend slide joint. Never allow drive shaft to hang down without support.

➡️**Use suitable puller if drive shaft cannot be separated from wheel hub and bearing assembly.**

9. Remove front exhaust tube.
10. Remove bearing housing to support bearing bracket bolts.
11. Remove drive shaft from transaxle assembly.
    a. Use Tool KV40107500 and sliding hammer while inserting tip of tool between housing and transaxle assembly.

✳✳ CAUTION

Never place drive shaft joint at an extreme angle when removing drive shaft. Also be careful not to overextend slide joint.

12. If necessary, remove the support bearing bracket.

To install:

13. Installation is in the reverse order of removal. Note the following:

✳✳ CAUTION

Do not reuse non-reusable parts.

- Install new circlip on drive shaft in the circular clip groove on transaxle side.

✳✳ CAUTION

Make sure the new circlip on the drive shaft is securely fastened.

- In order to prevent damage to differential side oil seal, place Tool KV38107900 onto oil seal before inserting drive shaft as shown. Slide drive shaft into slide joint and tap with a hammer to install securely.

✳✳ CAUTION

Make sure that circlip is completely engaged.

- When installing support bearing bracket, temporarily tighten mounting bolts, then tighten to specified torque.

ENGINE COOLING

ENGINE FAN

REMOVAL & INSTALLATION

See Figure 19.

1. Drain engine coolant from radiator.

✳✳ CAUTION

Perform when engine is cold.

2. Remove ECM and transmission control module.
3. Remove battery tray.
4. Disconnect radiator upper hose.
5. Disconnect fan motor connectors.
6. Remove radiator cooling fan assembly.

To install:

7. Installation is in the reverse order of removal.

➡**Cooling fans are controlled by ECM.**

RADIATOR

REMOVAL & INSTALLATION

See Figure 20.

✳✳ WARNING

Never remove the radiator cap when the engine is hot. Serious burns could occur from high pressure coolant escaping from the radiator.

Wrap a thick cloth around the cap. Slowly turn it a quarter turn to allow built-up pressure to escape. Carefully remove the cap by turning it all the way.

1. Drain coolant.
2. Remove battery.
3. Remove Transmission Control Module (TCM).
4. Remove ECM and bracket.
5. Remove battery tray.
6. Remove front air duct.
7. Disconnect radiator upper hose and lower hose.
8. Remove front bumper fascia.
9. Remove A/C condenser.

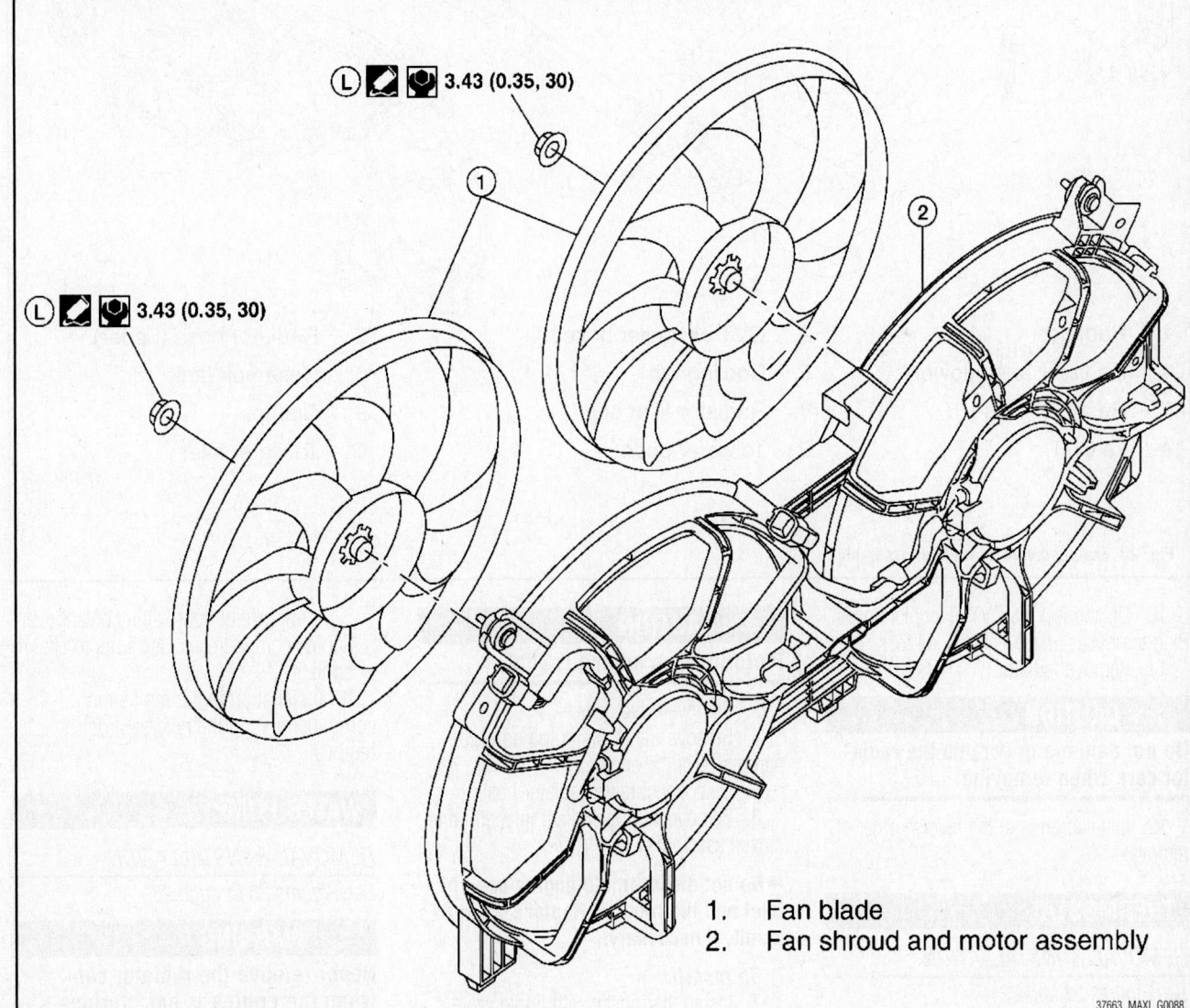

Ⓛ ✎ 🔧 3.43 (0.35, 30)

Ⓛ ✎ 🔧 3.43 (0.35, 30)

①
②

1. Fan blade
2. Fan shroud and motor assembly

37663_MAXI_G0088

Fig. 19 Exploded view of fan and shroud assembly

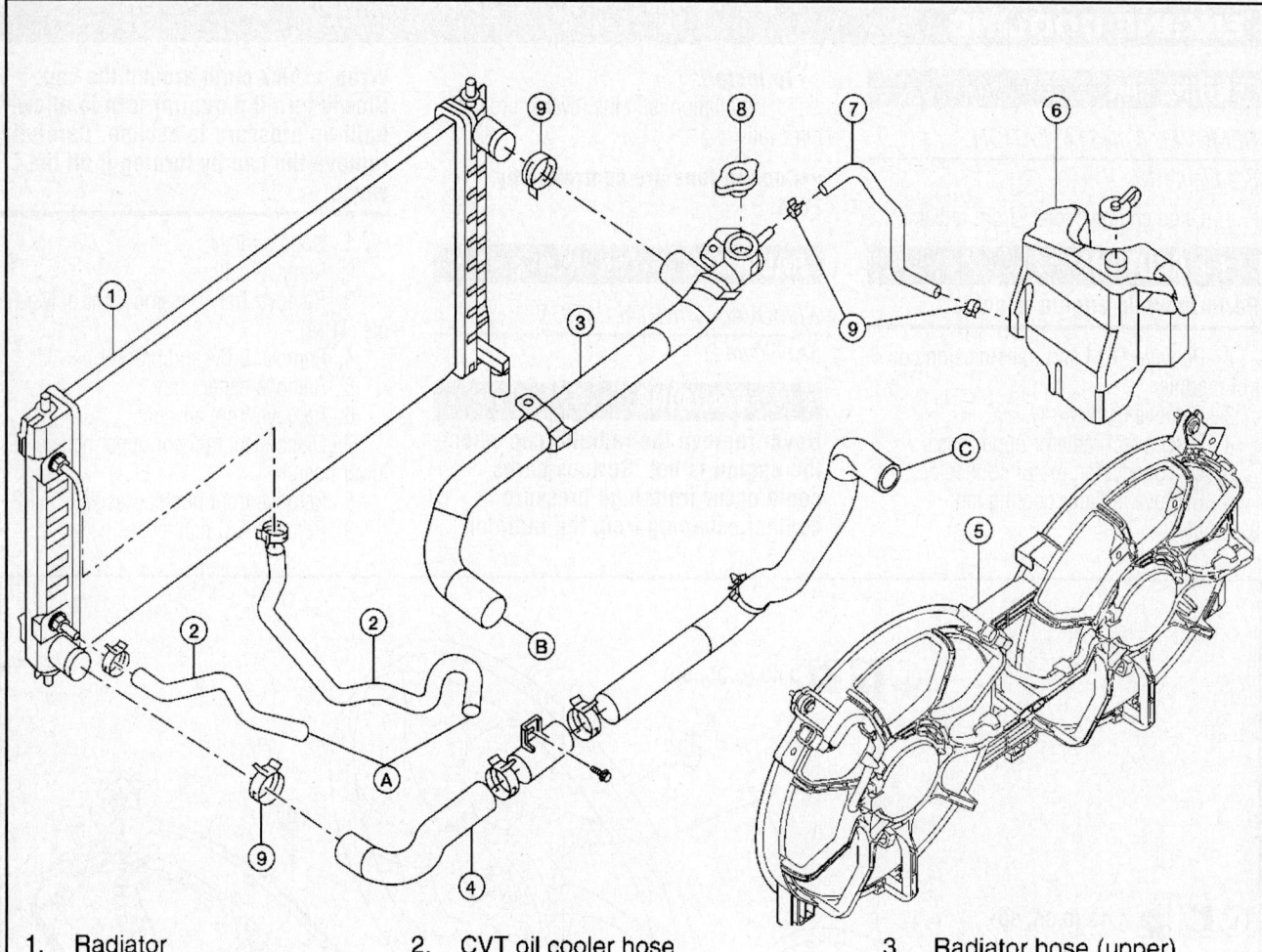

1. Radiator
2. CVT oil cooler hose
3. Radiator hose (upper)
4. Radiator hose (lower)
5. Cooling fan
6. Reservoir tank
7. Reservoir hose
8. Radiator filler cap
9. Clamps
A. To CVT
B. To water outlet
C. To water inlet

37663_MAXI_G0089

Fig. 20 Exploded view of radiator assembly

10. Disconnect the CVT oil cooler hoses. Plug the hoses to prevent CVT oil loss.
11. Remove radiator.

✼✼ CAUTION

Do not damage or scratch the radiator core when removing.

12. Installation is in the reverse order of removal.

THERMOSTAT

REMOVAL & INSTALLATION

See Figures 21 and 22.

1. Remove engine undercover using power tool.
2. Drain coolant from radiator.

✼✼ CAUTION

Perform when engine is cool.

3. Remove drive belts.
4. Remove water drain plug on water pump side of the engine.
5. Disconnect lower radiator hose.
6. Remove engine coolant inlet and thermostat assembly.

➡**Do not disassemble engine coolant inlet and thermostat. Replace them as a unit, if necessary.**

To install:

7. Install thermostat with jiggle valve facing upward.

   a. After installation, run engine for a few minutes, and check for leaks.

   b. Be careful not to spill coolant over engine compartment. Use a rag to absorb coolant.

8. Installation of the remaining components is in the reverse order of removal.

WATER PUMP

REMOVAL & INSTALLATION

See Figures 23 through 28.

✼✼ WARNING

Never remove the radiator cap when the engine is hot. Serious burns could occur from high pressure coolant escaping from the radiator.

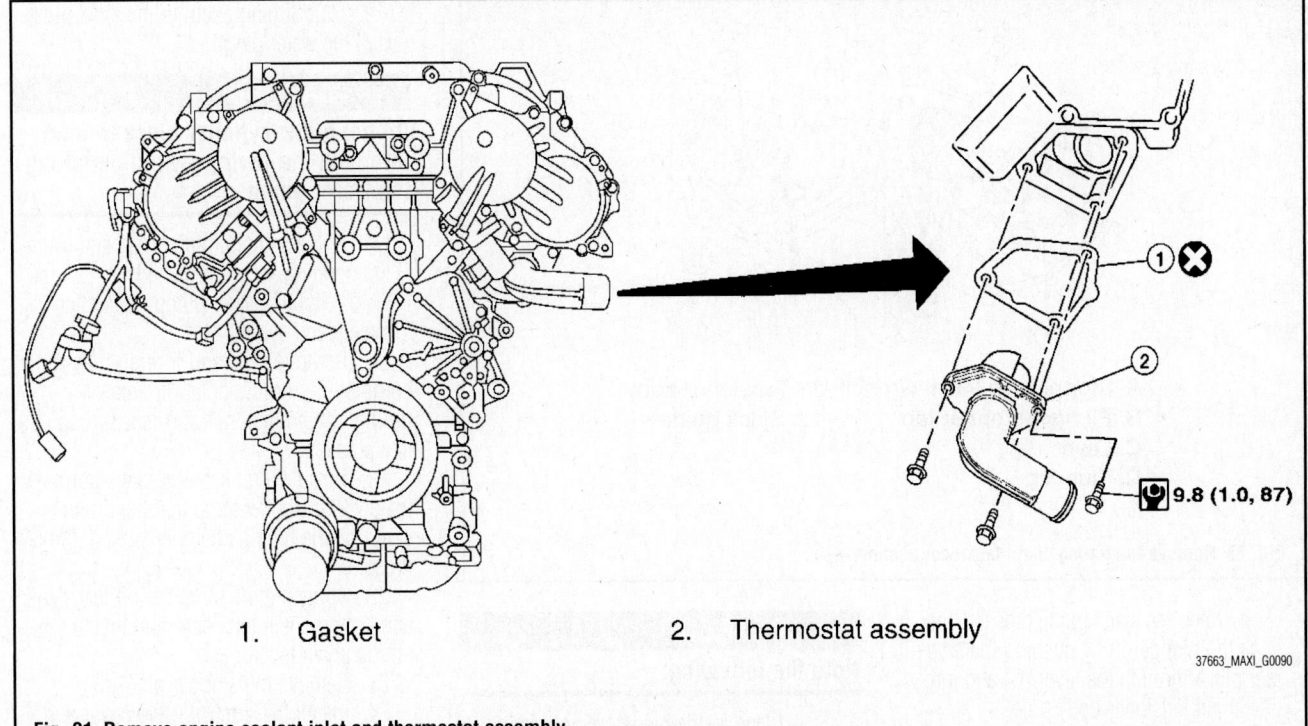

1. Gasket　　　　2. Thermostat assembly

37663_MAXI_G0090

Fig. 21 Remove engine coolant inlet and thermostat assembly

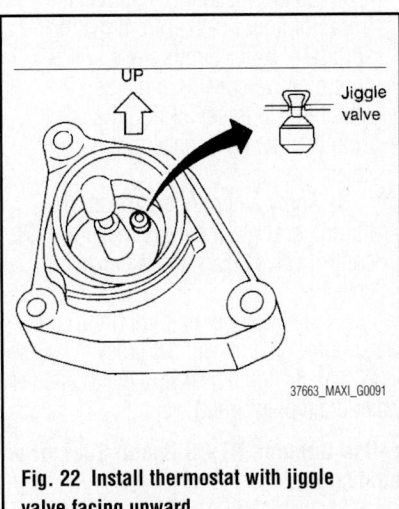

Fig. 22 Install thermostat with jiggle valve facing upward

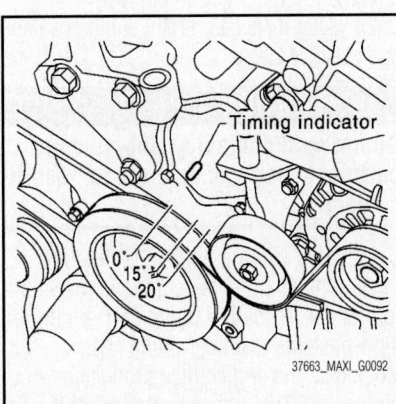

Fig. 23 Set No. 1 cylinder at TDC on its compression stroke

❊❊ CAUTION
Note the following:

- When removing water pump assembly, be careful not to get coolant on drive belt.
- Water pump cannot be disassembled and should be replaced as a unit.
- After installing water pump, connect hose and clamp securely, then check for leaks using radiator cap tester.

1. Drain engine coolant from the radiator.

❊❊ CAUTION
Perform when the engine is cold.

2. Remove RH wheel and tire.
3. Remove the fender protector side cover (RH).
4. Set No. 1 cylinder at TDC on its compression stroke.
5. Align pointer with TDC mark on crankshaft pulley.
6. Remove drive belt.
7. Remove the idler pulley and the A/C idler pulley.
8. Remove hood ledge cover (RH).
9. Remove water drain plug (front) on water pump side of cylinder block to drain engine coolant from engine.
10. Support engine and remove the front engine insulator and bracket.

11. Disconnect RH valve timing control connectors and remove RH IVT control valve cover.
12. Remove water pump cover.
13. Remove the timing chain tensioner assembly as follows:

　a. Pull the lever down to release the plunger stopper tab.

　b. Insert the stopper pin A into the tensioner body hole to hold the lever and keep the plunger stopper tab released.

➡**An Allen wrench {0.047 inches (1.2 mm)} is used for a stopper pin A as an example.**

　c. Compress the plunger into the tensioner body by pressing the slack guide.

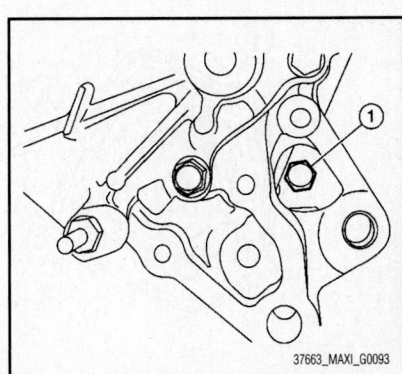

37663_MAXI_G0093

Fig. 24 Remove water drain plug (front) (1) on water pump side of cylinder block to drain engine coolant from engine

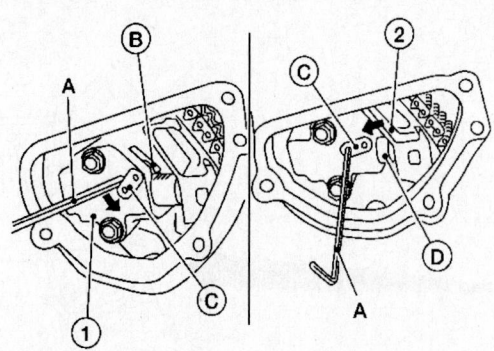

A. Stopper pin (Allen wrench)
B. Plunger stopper tab
C. Lever
D. Plunger

1. Tensioner body
2. Slack guide

37663_MAXI_G0094

Fig. 25 Remove the timing chain tensioner assembly

d. Keep the slack guide pressed and lock the plunger in by pushing the stopper pin A through the lever (C) and into the chain tensioner body hole.

e. Remove timing chain tensioner bolts and then remove the timing chain tensioner.

✳✳ CAUTION

Be careful not to drop timing chain tensioner bolts inside timing chain case.

14. Remove the three water pump bolts. Make a gap between water pump sprocket and timing chain, by carefully turning crankshaft pulley counterclockwise until timing chain loosens on water pump sprocket.

15. Screw M8 bolts into water pumps upper and lower bolt holes until they reach the timing chain case. Remove water pump.

✳✳ CAUTION

Note the following:

- Place a suitable shop cloth below the water pump housing to prevent any engine coolant from dripping into the timing chain case.
- Pull water pump straight out while preventing vane from contacting socket in installation area.
- Remove water pump without causing sprocket to contact timing chain.

16. Remove M8 bolts and O-rings from water pump.

To install:

17. Install new O-rings to water pump.

18. Apply engine oil and coolant to the O-rings.

➡**Locate the O-ring with white paint mark to engine front side.**

19. Hold timing chain to the side and install the water pump.

✳✳ CAUTION

Do not allow cylinder block to interfere with the O-rings when installing the water pump.

a. Check that timing chain and water pump sprocket are engaged.

b. Tighten water pump bolts alternately and evenly.

20. Remove dust and foreign material completely from installation area of timing chain tensioner and rear timing chain case.

21. Turn the crankshaft pulley approximately 20° clockwise so that the timing chain on the timing chain tensioner side is loose.

22. Apply engine oil to the oil feed hole and timing chain tensioner and install the timing chain tensioner.

23. Remove the stopper pin A.

24. Install IVT control valve cover and water pump cover.

a. Before installing, remove all traces of liquid gasket from mating surface of water pump cover and IVT control valve cover using a scraper. Also remove traces of liquid gasket from the mating surface of the front cover.

b. Apply a continuous bead of liquid gasket to mating surface of IVT control valve cover and water pump cover.

25. Install water drain plug (front) on water pump side of cylinder block.

26. Apply liquid gasket to the threads of water drain plug (front).

➡**Use Genuine RTV Silicone Sealant or equivalent.**

27. Installation of remaining components is in the reverse order of removal.

28. After installation refill engine coolant and check for leaks.

✳✳ CAUTION

Do not spill coolant in engine compartment. Use a shop cloth to absorb coolant.

29. After starting engine, let idle for three minutes, then rev engine up to 3,000 rpm under no load to purge air from the high-pressure chamber of the chain tensioner. The engine may produce a rattling noise. This indicates that air still remains in the chamber and is not a matter of concern.

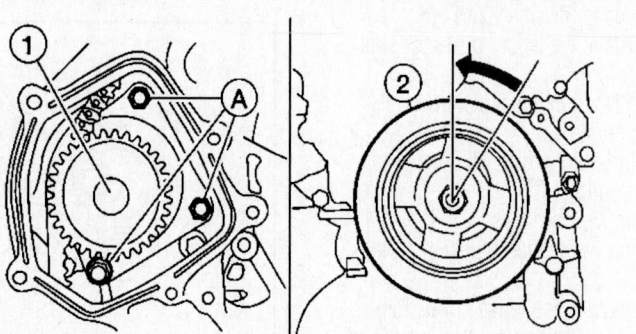

37663_MAXI_G0095

Fig. 26 Remove the three water pump bolts (A); make a gap between water pump sprocket (1) and timing chain, by carefully turning crankshaft pulley (2) counterclockwise

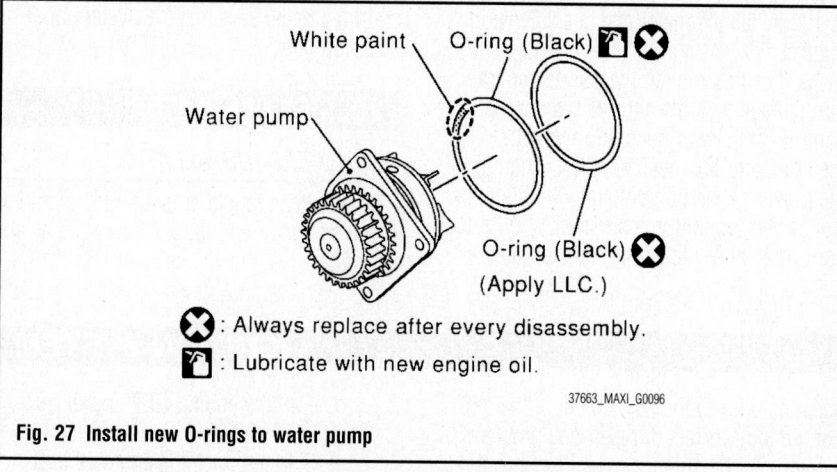

Fig. 27 Install new O-rings to water pump

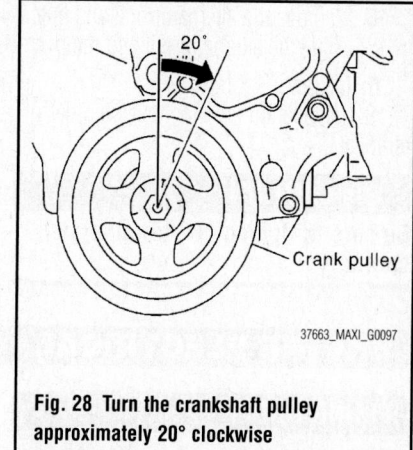

Fig. 28 Turn the crankshaft pulley approximately 20° clockwise

ENGINE ELECTRICAL

CHARGING SYSTEM

ALTERNATOR

REMOVAL & INSTALLATION

See Figure 29.

1. Disconnect the negative battery terminal.
2. Drain engine coolant.
3. Remove engine room cover.
4. Remove RH front wheel and tire assembly.
5. Remove front/right-side engine undercover.
6. Remove air cleaner and duct assembly.
7. Remove battery tray.
8. Remove cooling fan assembly.
9. Evacuate A/C system.
10. Remove the drive belt.
11. Remove the A/C compressor.
12. Remove idler pulley.
13. Remove A/C idler pulley.
14. Disconnect the oil pressure switch.
15. Disconnect the alternator harness connectors.

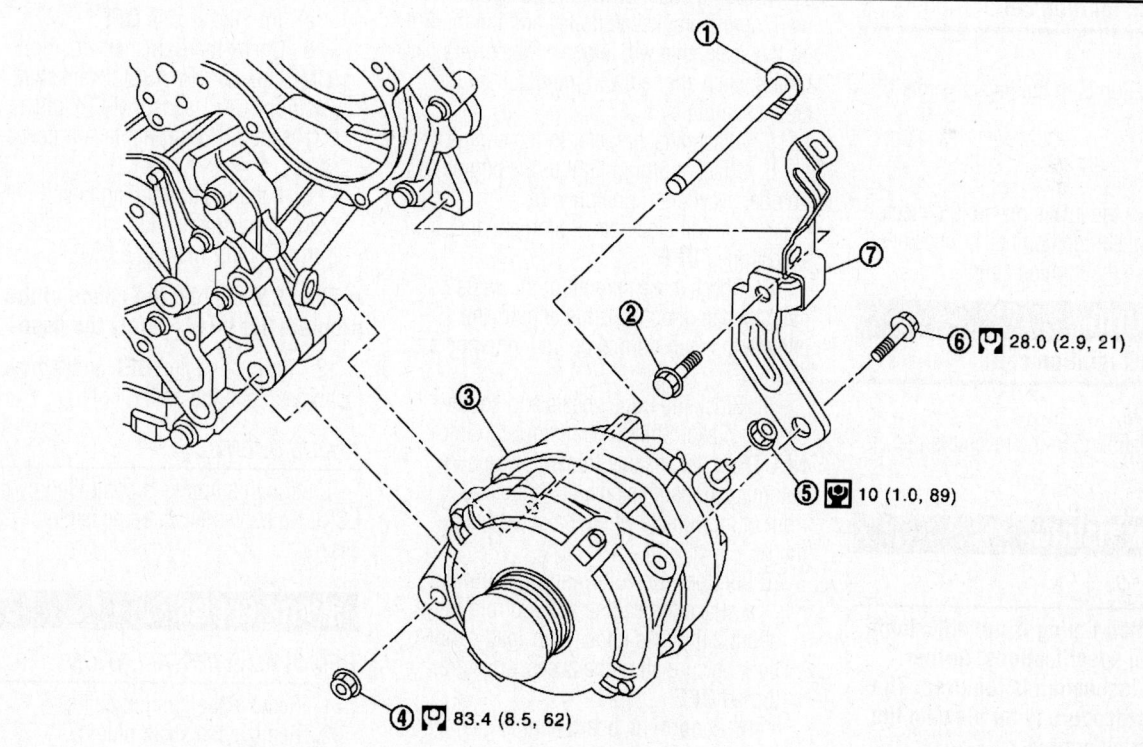

1.	Lower generator bolt	2.	Alternator stay bolt	3.	Alternator
4.	Lower generator nut	5.	"B" terminal nut	6.	Upper alternator bolt
7.	Alternator stay bracket				

Fig. 29 Exploded view of alternator assembly

16. Remove the alternator bolt and nuts.
17. Slide the alternator out and remove.

To install:

18. Installation is in the reverse order of removal.

✳✳ CAUTION

Be sure to tighten "B" terminal nut carefully.

19. Install alternator and check tension of belt.
20. For this model, the power generation voltage variable control system that controls the power generation voltage of the alternator has been adopted. Therefore, the power generation voltage variable control system operation inspection should be performed after replacing the alternator, and then make sure that the system operates normally.

VOLTAGE REGULATOR

REMOVAL & INSTALLATION

The voltage regulator is an integral part of the alternator and must be serviced as an assembly.

ENGINE ELECTRICAL

FIRING ORDER

Firing order for the engine is 1–2–3–4–5–6.

IGNITION COIL

REMOVAL & INSTALLATION

Left Side

1. Remove engine room cover.
2. Disconnect ignition coil connector.
3. Remove the ignition coil.

✳✳ CAUTION

Never shock ignition coil.

To install:

4. Installation is in the reverse order of removal.

Right Side

1. Remove the intake manifold collector.
2. Disconnect ignition coil connector.
3. Remove the ignition coil.

✳✳ CAUTION

Never shock ignition coil.

To install:

4. Installation is in the reverse order of removal.

IGNITION TIMING

INSPECTION

➡ **The ignition timing is not adjustable. If not within specifications, further diagnostic inspection is required. The following procedure is for viewing the ignition timing setting.**

Visually check the air cleaner, intake hoses, ducts, Exhaust Gas Recirculation (EGR) valve operation and electrical connections prior to the adjustment of the ignition timing. Correct or repair any problem as required. Be sure to inspect the throttle valve and Throttle Position (TP) sensor for proper operation.

1. Before servicing the vehicle, refer to the Precautions Section.
2. Locate the timing marks on the crankshaft pulley and the front of the engine.
3. Clean the timing marks.

➡ **The ignition timing specification is 15°, plus or minus 5° Before Top Dead Center (BTDC).**

4. Using chalk or white paint, color the mark on the crankshaft pulley and the mark on the scale, that will indicate the correct timing when aligned with the notch on the crankshaft pulley.
5. Attach a tachometer to the engine.
6. Attach a timing light to the engine to number 1 cylinder ignition wire.
7. Turn all electrical equipment and accessories **OFF**.
8. Check to be sure all of the wires clear the fan, then, start the engine and allow it to reach normal operating temperatures.
9. Block the front wheels and set the parking brake. Shift the transmission into **NEUTRAL** for manual transmission and automatic transmissions. Do not stand in front of the vehicle when making adjustments.
10. Perform the following procedures:
    a. Race the engine at 2000 rpm for about 2 minutes under a no-load condition; be sure all of the accessories are turned **OFF**.
    b. Perform on board engine diagnostics and repair any fault code.
    c. Race the engine at 2000 rpm for about 2 minutes under a no-load condition.

IGNITION SYSTEM

d. Turn the engine **OFF** and disconnect the TP sensor.
e. Start and race the engine 2 to 3 times under no-load, then run the engine at idle speed.

➡ **The ignition timing specification is 15°, plus or minus 5° BTDC.**

11. Aim the timing light at the timing marks. If the marks on the pulley and the engine are aligned when the light flashes, the timing is correct. Turn the engine **OFF** and remove the tachometer and the timing light. If the marks are not in alignment, proceed with the following steps:
    a. Turn the engine **OFF**.
    b. Check the Camshaft Position (CMP) sensor (PHASE), Crankshaft Position (CKP) sensor (REF) and CKP sensor (POS). Replace if necessary.
    c. If the ignition timing is still not correct, substitute a known good Electronic Control Module (ECM).

➡ **The ECM may be the cause of the problem but this is rarely the case.**

12. Turn the engine **OFF** and remove the tachometer and the timing light.

ADJUSTMENT

The ignition timing is controlled by the ECM. No adjustments are possible or necessary.

SPARK PLUGS

REMOVAL & INSTALLATION

1. Remove the ignition coil(s).
2. Remove the spark plug(s).
3. Install spark plug(s) and torque to 14–22 ft. lbs. (19–30 Nm).
4. Install the ignition coil(s).

ENGINE ELECTRICAL

STARTING SYSTEM

STARTER

REMOVAL & INSTALLATION

1. Disconnect the negative and positive battery terminals.
2. Remove the air cleaner assembly and air ducts.

3. Disconnect the following:
   - ECM
   - TCM
4. Remove the battery tray.
5. Disconnect the battery cable and starter harness connector.

6. Remove the starter bolts, then remove the starter.

To install:

7. Installation is in the reverse order of removal.

ENGINE MECHANICAL

ACCESSORY DRIVE BELTS

ACCESSORY BELT ROUTING

See Figure 30.

Refer to the accompanying illustration.

INSPECTION

✳✳ WARNING

Inspect and check the drive belts with the engine off.

1. Inspect belt for cracks, fraying, wear or oil adhesion. If necessary, replace with a new one.

2. Rotate the crankshaft pulley two times then check the belt tension using Tool BT-3373-F.

ADJUSTMENT

Belt tension is not manually adjustable, it is automatically adjusted by the drive belt auto-tensioner.

REMOVAL & INSTALLATION

See Figure 31.

1. Remove the front RH side cover.
2. While securely holding the hexagonal part in pulley center of drive belt auto-tensioner, move in the direction of arrow

(loosening direction of tensioner) using suitable tool.

✳✳ WARNING

Avoid placing hand in a location where pinching may occur if the holding tool accidentally comes off.

✳✳ CAUTION

Do not loosen the auto-tensioner pulley bolt. (Do not turn it counterclockwise. If turned counterclockwise, the complete auto-tensioner must be replaced as a unit, including pulley.)

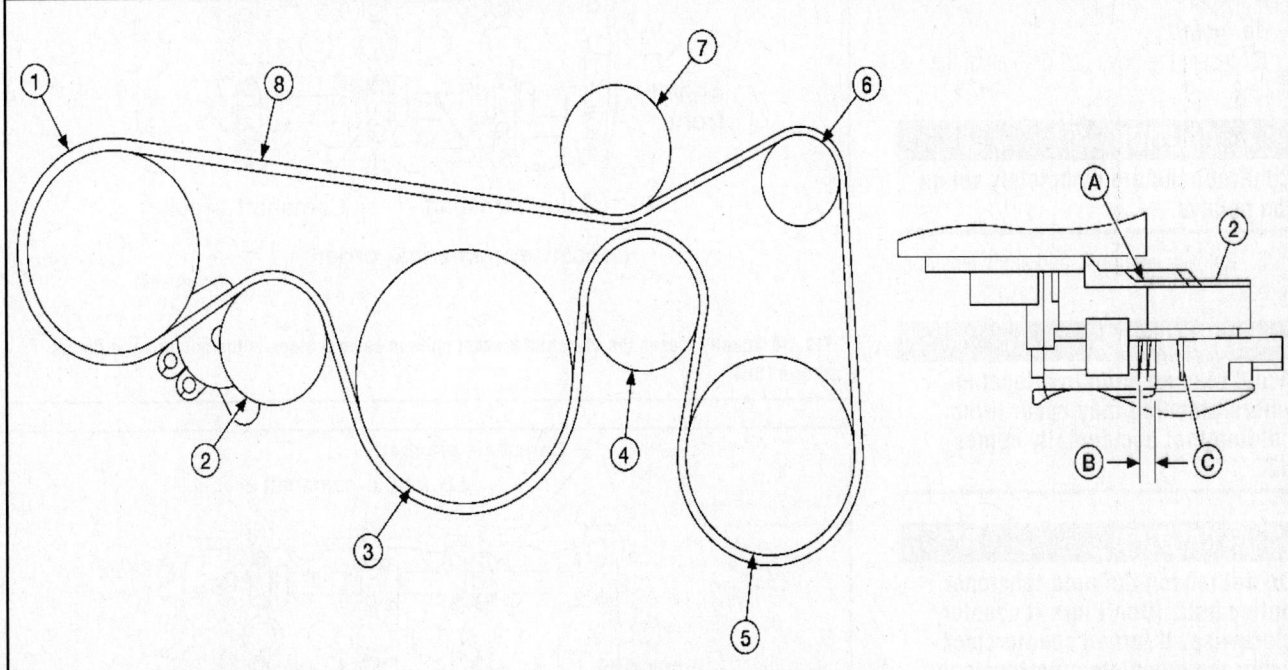

1.	Power steering pump	2.	Drive belt auto-tensioner	3.	Crankshaft
4.	Idler pulley	5.	A/C compressor pulley	6.	Generator pulley
7.	Idler pulley	8.	Drive belt	A.	Indicator
B.	Possible use range (for new belt)	C.	Belt replacement		

Fig. 30 Accessory drive belt routing

37663_MAXI_G0100

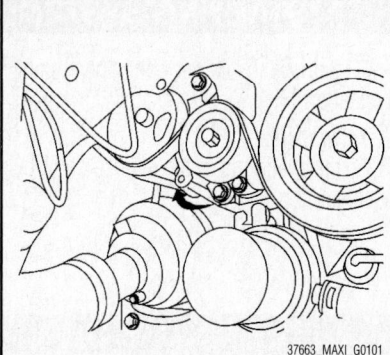

37663_MAXI_G0101

Fig. 31 While securely holding the hexagonal part in pulley center of drive belt auto-tensioner, move in the direction of arrow (loosening direction of tensioner) using suitable tool

3. Insert a rod approximately 0.24 inches (6 mm) in diameter through the rear of tensioner into retaining boss to lock tensioner pulley.

➡**Leave tensioner pulley arm locked until belt is installed again.**

4. Loosen auxiliary drive belt from water pump pulley and then remove it from the other pulleys.

To install:

5. Install the drive belt onto all of the pulleys.

✳✳ CAUTION

Confirm belts are completely set on the pulleys.

6. Release tensioner, and apply tensions to belt.

✳✳ WARNING

Avoid placing hand in a location where pinching may occur if the holding tool accidentally comes off.

✳✳ CAUTION

Do not loosen the auto-tensioner pulley bolt. (Don't turn it counterclockwise. If turned counterclockwise, the complete auto-tensioner must be replaced as a unit, including pulley.)

7. Turn crankshaft pulley clockwise several times to equalize tension between each pulley.

8. Confirm tensions of belt at indicator is within the allowable use range.

CAMSHAFT AND VALVE LIFTERS

REMOVAL & INSTALLATION

See Figures 32 through 43.

1. Remove the timing chains.

2. Remove camshaft position brackets (RH shown, LH similar).

3. Remove the intake and exhaust camshaft brackets and the camshafts.

a. Mark the camshafts, camshaft brackets, and bolts so they are placed in

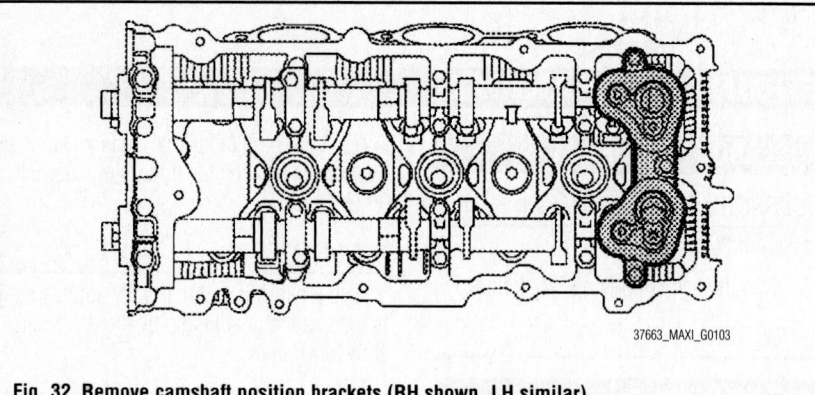

37663_MAXI_G0103

Fig. 32 Remove camshaft position brackets (RH shown, LH similar)

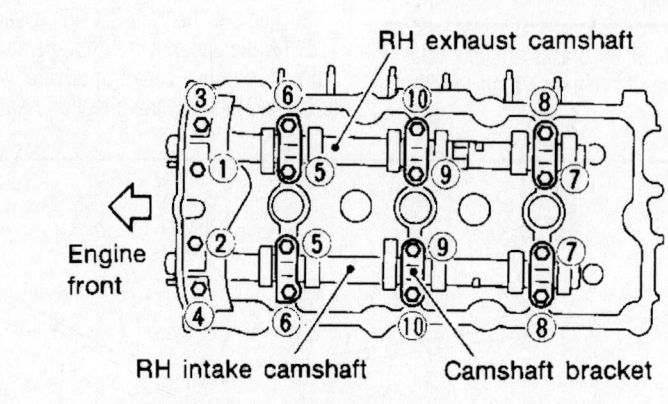

37663_MAXI_G0104

Fig. 33 Equally loosen the camshaft bracket bolts in several steps in the numerical order as shown (RH)

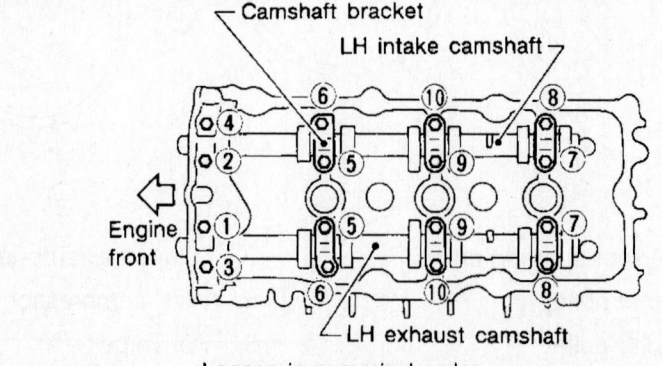

37663_MAXI_G0105

Fig. 34 Equally loosen the camshaft bracket bolts in several steps in the numerical order as shown (LH)

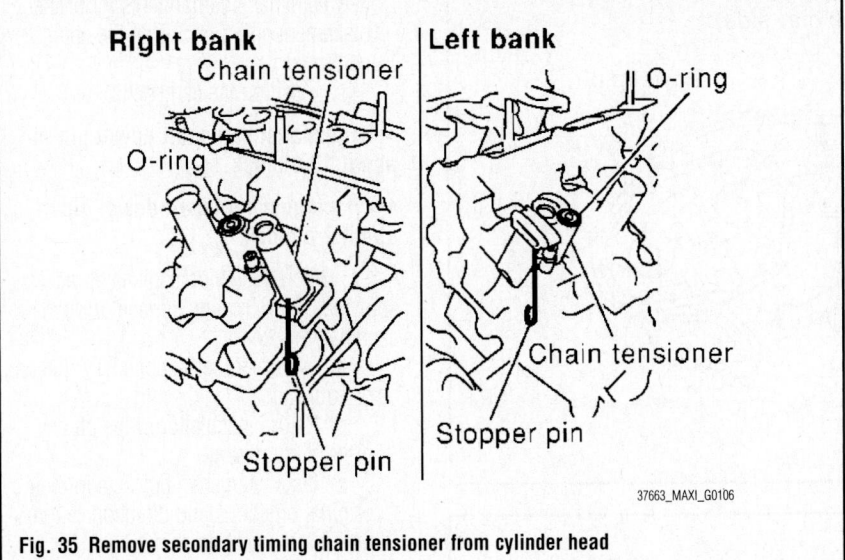

Fig. 35 Remove secondary timing chain tensioner from cylinder head

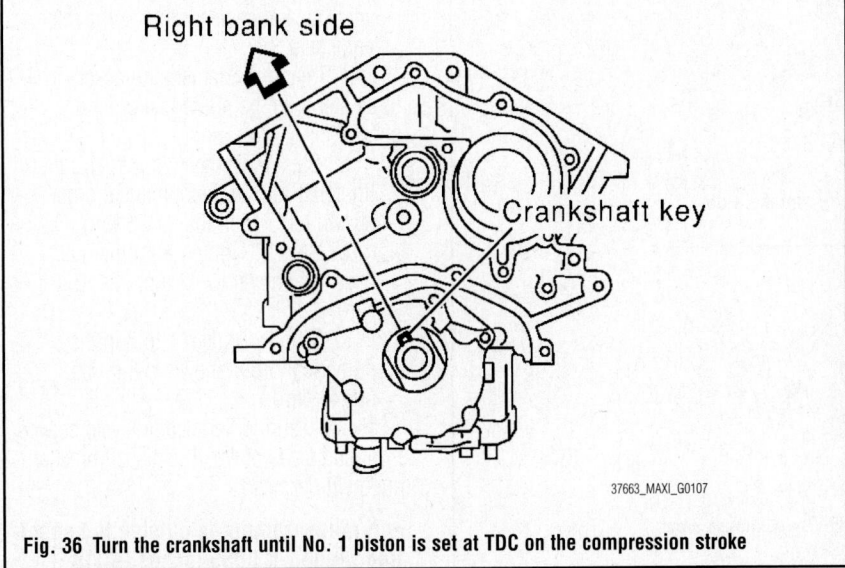

Fig. 36 Turn the crankshaft until No. 1 piston is set at TDC on the compression stroke

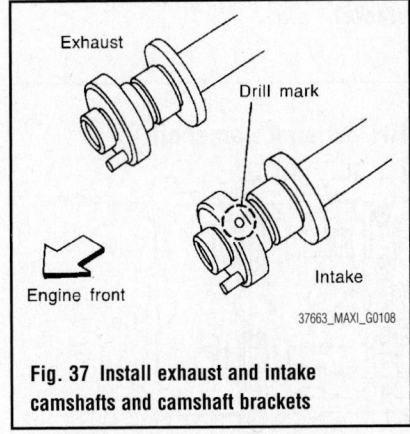

Fig. 37 Install exhaust and intake camshafts and camshaft brackets

4. Remove valve lifters, if necessary.

➡**Identify installation positions to ensure proper installation.**

5. Remove secondary timing chain tensioner from cylinder head.

6. Remove secondary tensioner with its stopper pin attached.

➡**Stopper pin was attached when secondary timing chain was removed.**

To install:

7. Before installation, remove any old Silicone RTV Sealant from component mating surfaces using a scraper.

➡**Remove the old Silicone RTV Sealant from the bolt holes and threads. Do not scratch or damage the mating surfaces.**

8. Before installing the front cam bracket, remove the old Silicone RTV Sealant from the mating surface using a scraper.

9. Turn the crankshaft until No. 1 piston is set at TDC on the compression stroke.

➡**The crankshaft key should line up with the right bank cylinder center line as shown.**

10. Install camshaft chain tensioners on both sides of cylinder head.

11. Install valve lifters, if removed.

➡**Install them in original positions.**

12. Install exhaust and intake camshafts and camshaft brackets.

   a. Intake camshaft has a drill mark on camshaft sprocket mounting flange.

   b. Follow your identification marks made during removal, or follow the identification marks that are present on the

the same position and direction for installation.

   b. Equally loosen the camshaft bracket bolts in several steps in the numerical order as shown.

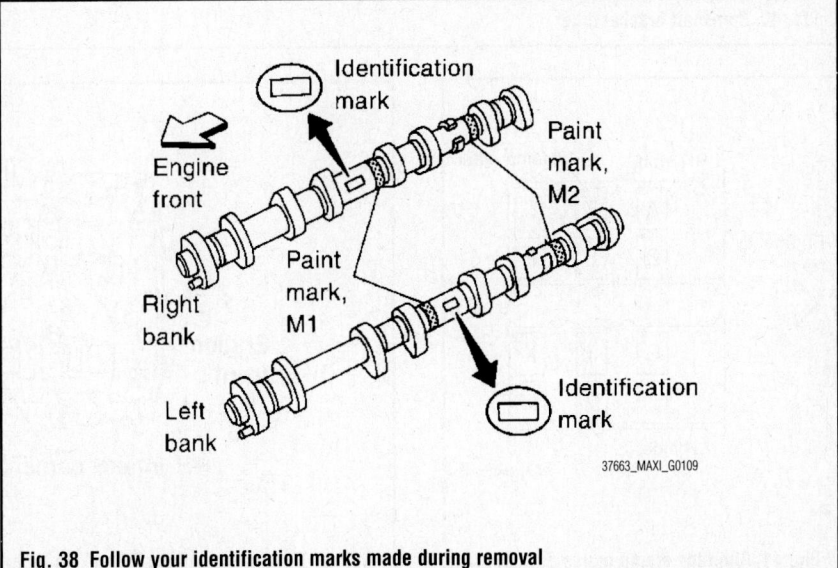

Fig. 38 Follow your identification marks made during removal

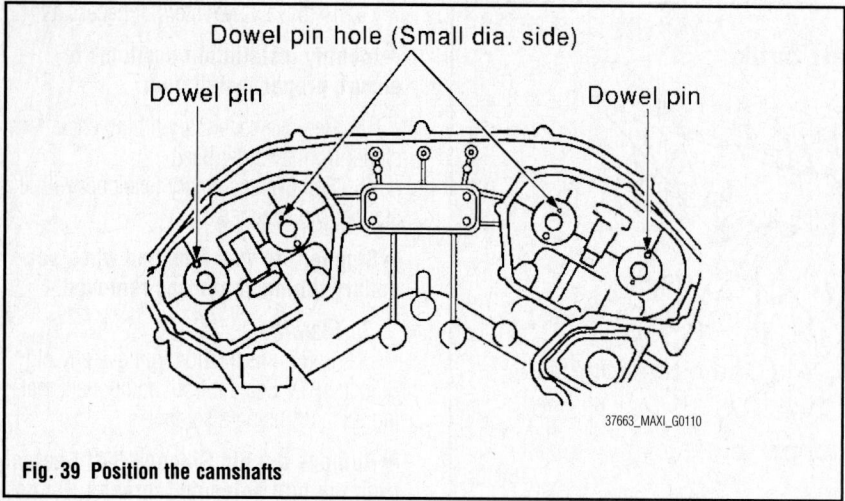

Fig. 39 Position the camshafts

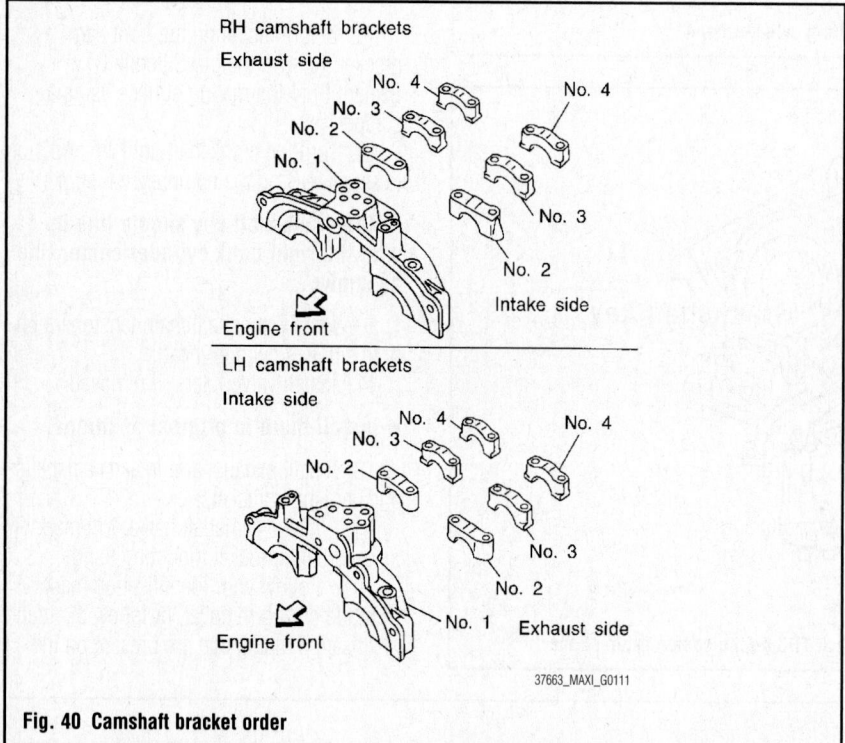

Fig. 40 Camshaft bracket order

new camshafts components for proper placement and direction of the components.

c. Position the camshafts.

➡**RH exhaust camshaft dowel pin at about 10 o'clock.**

➡**LH exhaust camshaft dowel pin at about 2 o'clock.**

13. Before installing camshaft brackets, apply sealant to mating surface of No. 1 camshaft bracket.

a. Use Genuine Silicone RTV Sealant, or equivalent.

b. Before installation, wipe off any protruding sealant.

c. Install camshaft brackets in their original positions and direction. Align the stamp marks as shown.

d. If checking and adjusting any part of valve assembly or camshaft, check valve clearance according to the reference data.

14. Tighten the camshaft brackets in three steps, in the numerical order as shown.

a. Step 1: Tighten No. 7 to 10, then tighten 1 to 6 in the numerical order shown to 17 inch lbs. (1.96 Nm).

b. Step 2: Tighten in numerical order as shown to 52 inch lbs. (5.88 Nm).

c. Step 3: Tighten 1 to 6 in the numerical order shown to 8 ft. lbs. (10.41 Nm).

15. Measure difference in levels between front end faces of No. 1 camshaft bracket and cylinder head.

➡**If measurement is outside the specified range {-0.0055 inches (-0.14 mm)}, re-install camshaft and camshaft bracket.**

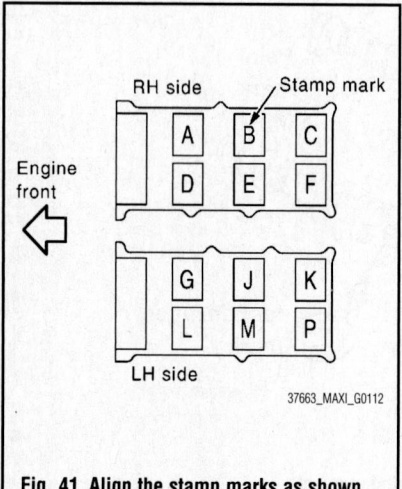

Fig. 41 Align the stamp marks as shown

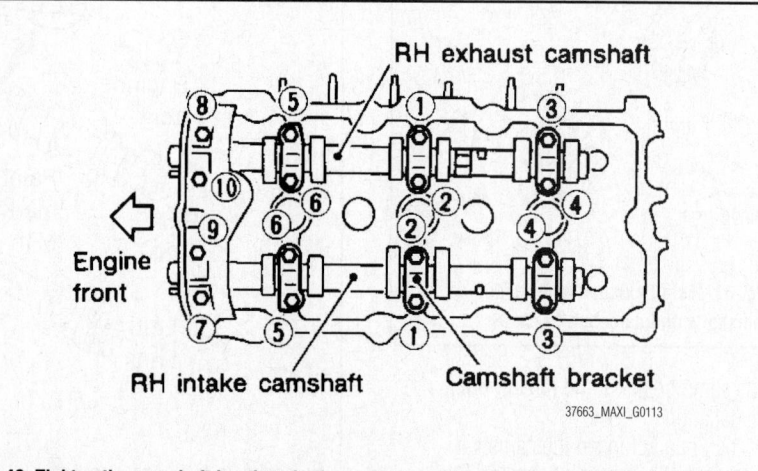

Fig. 42 Tighten the camshaft brackets in three steps, in the numerical order as shown (RH)

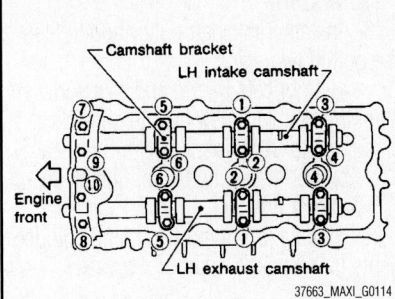

Fig. 43 Tighten the camshaft brackets in three steps, in the numerical order as shown (LH)

16. Install camshaft position sensors (PHASE) (RH and LH bank.)

17. Install the fuel rail and injectors.

18. Install the timing chains.

CATALYTIC CONVERTER

REMOVAL & INSTALLATION

See Figures 44 through 48.

❉❉ **WARNING**

Perform the work when the exhaust and cooling system have completely cooled down.

❉❉ **CAUTION**

When removing the front and rear engine mounting through bolts and nuts, lift the engine up slightly for safety.

1. Remove engine and CVT assembly.

2. Remove the RH and LH three way catalyst supports.

3. Remove heated oxygen sensor 2 (bank 1), heated oxygen sensor 2 (bank 2), Air Fuel Ratio (A/F) sensor 1 (bank 1) and air fuel ratio (A/F) sensor 1 (bank 2).

a. Remove harness connector of each sensor, and disconnect the harness from the bracket and middle clamp.

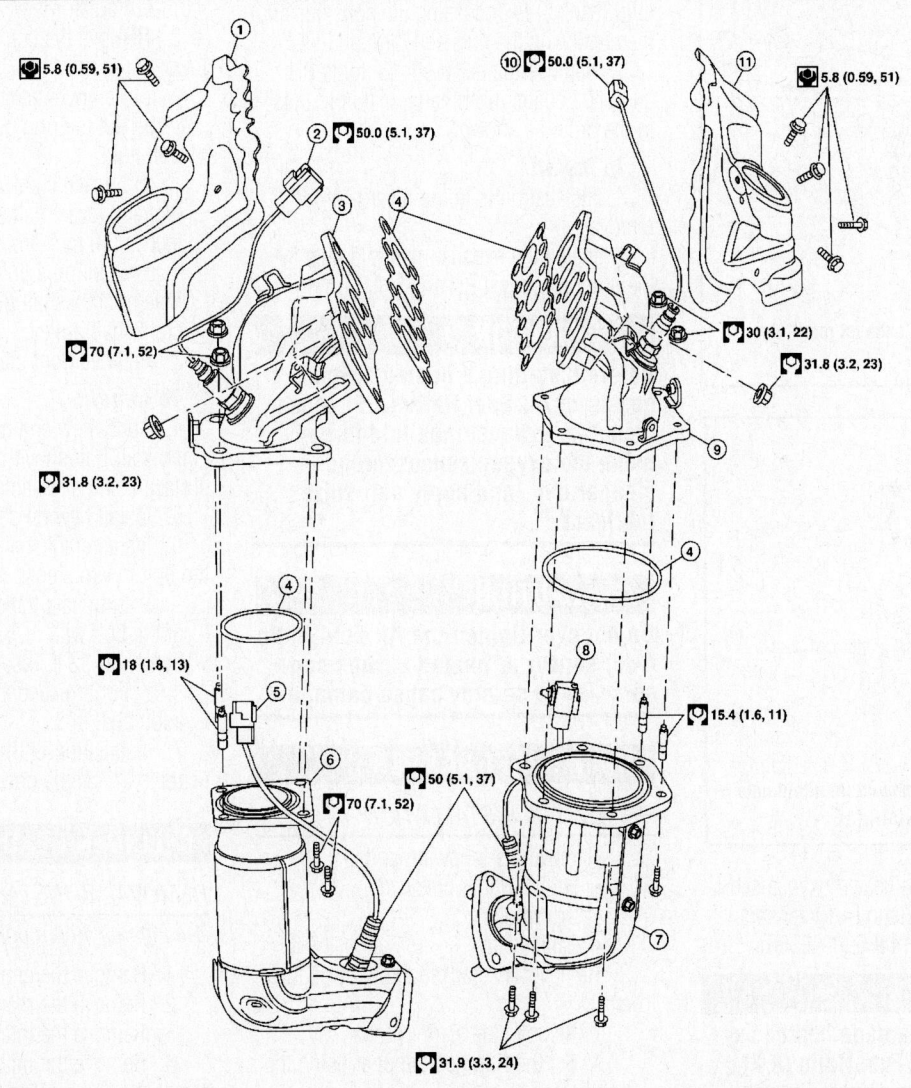

1. Exhaust manifold heat shield (RH)
2. Air fuel ratio (A/F) sensor 1 (bank 1)
3. Exhaust manifold (RH)
4. Gaskets
5. Heated oxygen sensor 2 (bank 1)
6. Three way catalyst (manifold) (bank 1)
7. Three way catalyst (manifold) (bank 2)
8. Heated oxygen sensor 2 (bank 2)
9. Exhaust manifold (LH)
10. Air fuel ratio (A/F) sensor 1 (bank 2)
11. Exhaust manifold heat shield (LH)

Fig. 44 Exploded view of exhaust manifold and three way catalyst

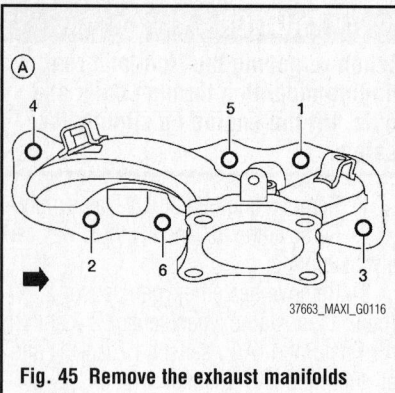

Fig. 45 Remove the exhaust manifolds RH (A)

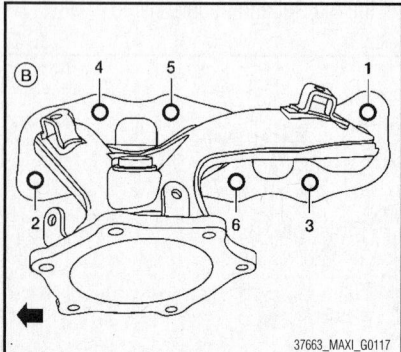

Fig. 46 Remove the exhaust manifolds LH (B)

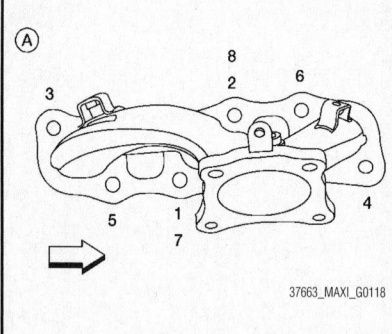

Fig. 47 Install the exhaust manifold nuts in the order as shown RH (A)

b. Remove both heated oxygen sensors and Air Fuel Ratio (A/F) sensors using Tool KV10114400 (J-38365).

✸✸ CAUTION

Be careful not to damage heated oxygen sensors or Air Fuel Ratio (A/F) sensors. Discard any heated oxygen sensor which has been dropped from a height of more than 19.7 inches (0.5 m) onto a hard surface such as a concrete floor; replace with a new sensor.

4. Remove exhaust manifold and three way catalyst heat shields.

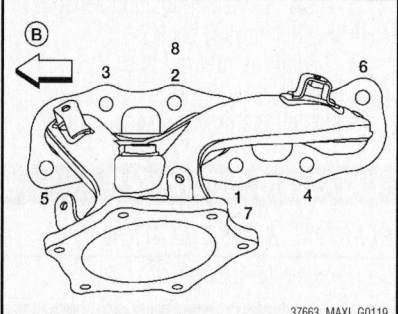

Fig. 48 Install the exhaust manifold nuts in the order as shown LH (B)

5. Remove the three way catalyst (manifold) (bank 1) and three way catalyst (manifold) (bank 2) by loosening the bolts first and then removing the nuts and through bolts.

6. Remove the exhaust manifolds RH and LH. Loosen the exhaust manifold nuts in the order as shown.

To install:

7. Installation is in the reverse order of removal.

8. Install the exhaust manifold nuts in the order as shown RH and LH.

✸✸ CAUTION

Before installing a heated oxygen sensor or Air Fuel Ratio (A/F) sensor, clean the exhaust manifold threads using the oxygen sensor thread cleaner tool, and apply anti-seize lubricant.

✸✸ CAUTION

Do not over-tighten the Air Fuel Ratio (A/F) sensor or heated oxygen sensors. Doing so may cause damage.

CRANKSHAFT DAMPER

REMOVAL & INSTALLATION

1. Remove the following parts:
   - Engine under cover
   - Drive belts
   - Radiator fan
2. Remove the crankshaft pulley as follows:
   a. Remove the starter motor.
   b. Set the ring gear stopper using the bolt hole.
   c. Loosen crankshaft pulley bolt using tool and locate bolt seating surface at 0.39 inches (10 mm) from its original position.
   d. Position a pulley puller at recess hole of crankshaft pulley to remove crankshaft pulley.

To install:

3. Install crankshaft pulley and tighten the bolt in two steps:
   a. Lubricate thread and seat surface of the bolt with new engine oil and tighten to 32 ft. lbs. (44 Nm).
   b. For the second step, tighten an additional 90°.
4. Installation of the remaining components is in reverse order of removal.

CRANKSHAFT FRONT SEAL

REMOVAL & INSTALLATION

1. Remove the following parts:
   - Engine under cover
   - Drive belts
   - Radiator fan
2. Remove the crankshaft pulley as follows:
   a. Remove the starter motor.
   b. Set the ring gear stopper using the bolt hole.
   c. Loosen crankshaft pulley bolt using tool and locate bolt seating surface at 0.39 inches (10 mm) from its original position.
   d. Position a pulley puller at recess hole of crankshaft pulley to remove crankshaft pulley.
3. Remove front oil seal from front cover.

To install:

4. Apply new engine oil to new oil seal and install it flush with front of mounting surface using a suitable tool.
5. Install new oil seal.
6. Install crankshaft pulley and tighten the bolt in two steps:
   a. Lubricate thread and seat surface of the bolt with new engine oil and tighten to 32 ft. lbs. (44 Nm).
   b. For the second step, tighten an additional 90°.
7. Installation of the remaining components is in reverse order of removal.

CYLINDER HEAD

REMOVAL & INSTALLATION

See Figures 49 through 54.

1. Remove the engine from the vehicle.
2. Remove the rear timing chain case.
3. Remove the intake manifold.
4. Remove the intake and exhaust camshafts.
5. Remove the coolant outlet housing.
6. Remove the RH and LH cylinder head bolts, with power tool.
   a. The bolts should be loosened gradually in three stages.
   b. Loosen the bolts in the numerical order as shown.

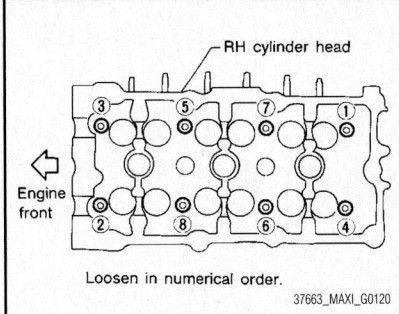

Fig. 49 Loosen the bolts in the numerical order as shown (RH)

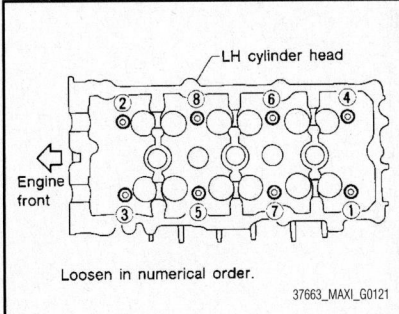

Fig. 50 Loosen the bolts in the numerical order as shown (LH)

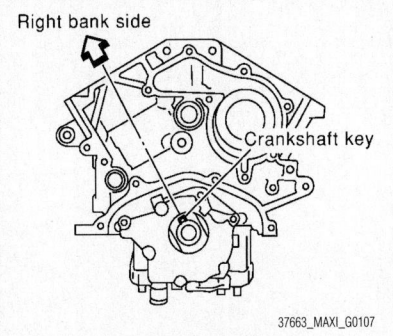

Fig. 51 Turn the crankshaft until No. 1 piston is set at TDC on the compression stroke

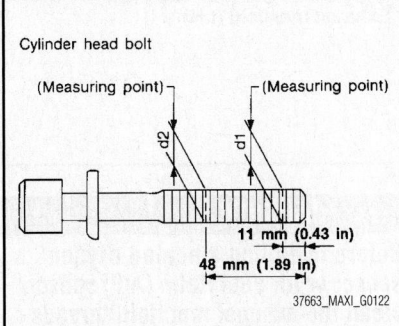

Fig. 52 Inspect the cylinder head bolts before installing the cylinder heads

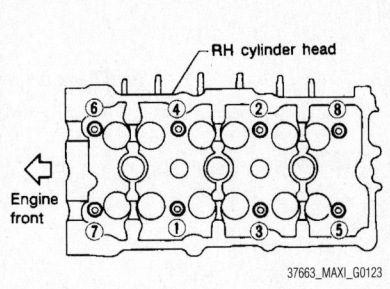

Fig. 53 RH cylinder head tightening sequence

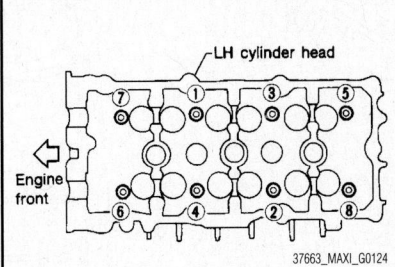

Fig. 54 LH cylinder head tightening sequence

   7. Remove cylinder heads and gaskets.
   8. Discard the cylinder head gaskets and use new gaskets for installation.

To install:
   9. Turn the crankshaft until No. 1 piston is set at TDC on the compression stroke.

➡**The crankshaft key should line up with the right bank cylinder center line as shown.**

   10. Install new gaskets on the cylinder heads.

✳✳ CAUTION

Do not rotate crankshaft and camshaft separately or valves will strike piston heads.

   11. Inspect the cylinder head bolts before installing the cylinder heads.

✳✳ CAUTION

Cylinder head bolts are tightened by degree rotation tightening method. Whenever the size difference between d1 and d2 exceeds the limit (0.0043 inches (0.11mm), replace the bolts with new ones.

➡**Lubricate threads and seat surfaces of the bolts with new engine oil.**

   12. Install the cylinder heads on the cylinder block. Tighten the cylinder head bolts in the five steps in the numerical order as shown using Tool KV10112100 (BT-8653-A).
     a. Step 1: 72 ft. lbs. (98 Nm)
     b. Step 2: Loosen in the reverse order of tightening.
     c. Step 3: 29 ft. lbs. (33 Nm)
     d. Step 4: 103° of rotation clockwise
     e. Step 5: 103° of rotation clockwise
   13. Installation of the remaining components is in the reverse order of removal.

DRIVE PLATE

REMOVAL & INSTALLATION

   1. Remove the transaxle.
   2. Remove drive plate.

To install:

   3. Installation is reverse of removal.
     a. Tighten bolts to 65 ft. lbs. (88 Nm).

EXHAUST MANIFOLD

REMOVAL & INSTALLATION

See Figures 55 through 59.

✳✳ WARNING

Perform the work when the exhaust and cooling system have completely cooled down.

✳✳ CAUTION

When removing the front and rear engine mounting through bolts and nuts, lift the engine up slightly for safety.

   1. Remove engine and CVT assembly.
   2. Remove the RH and LH three way catalyst supports.
   3. Remove heated oxygen sensor 2 (bank 1), heated oxygen sensor 2 (bank 2), Air Fuel Ratio (A/F) sensor 1 (bank 1) and air fuel ratio (A/F) sensor 1 (bank 2).
     a. Remove harness connector of each sensor, and disconnect the harness from the bracket and middle clamp.
     b. Remove both heated oxygen sensors and Air Fuel Ratio (A/F) sensors using Tool KV10114400 (J-38365).

✳✳ CAUTION

Be careful not to damage heated oxygen sensors or Air Fuel Ratio (A/F) sensors. Discard any heated oxygen sensor which has been dropped from a height of more than 19.7 inches (0.5 m) onto a hard surface such as a concrete floor; replace with a new sensor.

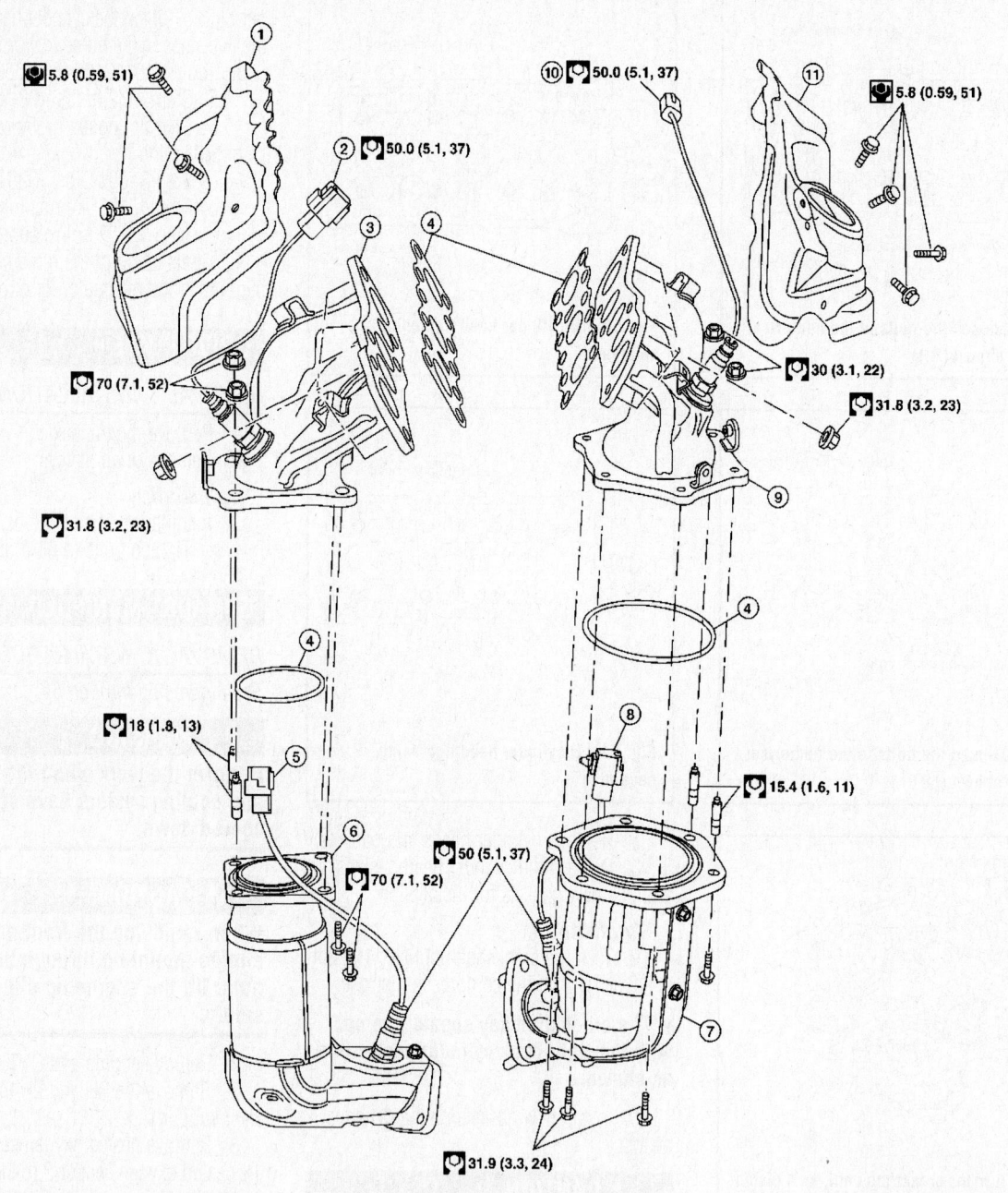

1. Exhaust manifold heat shield (RH)
4. Gaskets
7. Three way catalyst (manifold) (bank 2)
10. Air fuel ratio (A/F) sensor 1 (bank 2)

2. Air fuel ratio (A/F) sensor 1 (bank 1)
5. Heated oxygen sensor 2 (bank 1)
8. Heated oxygen sensor 2 (bank 2)
11. Exhaust manifold heat shield (LH)

3. Exhaust manifold (RH)
6. Three way catalyst (manifold) (bank 1)
9. Exhaust manifold (LH)

37663_MAXI_G0115

Fig. 55 Exploded view of exhaust manifold and three way catalyst

4. Remove exhaust manifold and three way catalyst heat shields.

5. Remove the three way catalyst (manifold) (bank 1) and three way catalyst (manifold) (bank 2) by loosening the bolts first and then removing the nuts and through bolts.

6. Remove the exhaust manifolds RH and LH. Loosen the exhaust manifold nuts in the order as shown.

To install:

7. Installation is in the reverse order of removal.

8. Install the exhaust manifold nuts in the order as shown RH and LH.

✳✳ CAUTION

Before installing a heated oxygen sensor or Air Fuel Ratio (A/F) sensor, clean the exhaust manifold threads using the oxygen sensor thread cleaner tool, and apply anti-seize lubricant.

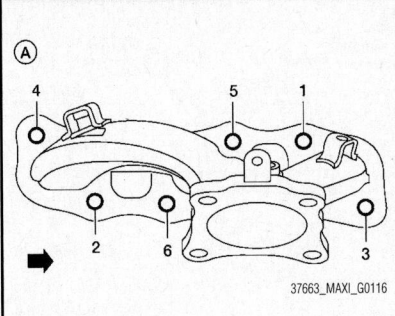

Fig. 56 Remove the exhaust manifolds RH (A)

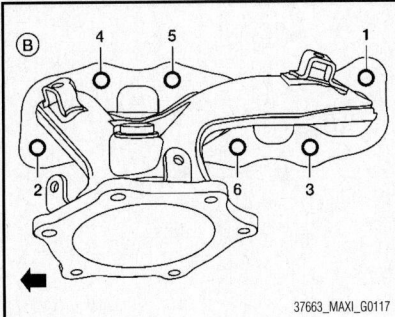

Fig. 57 Remove the exhaust manifolds LH (B)

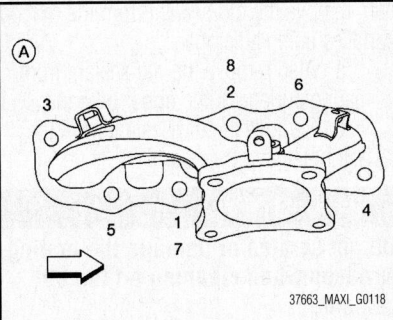

Fig. 58 Install the exhaust manifold nuts in the order as shown RH (A)

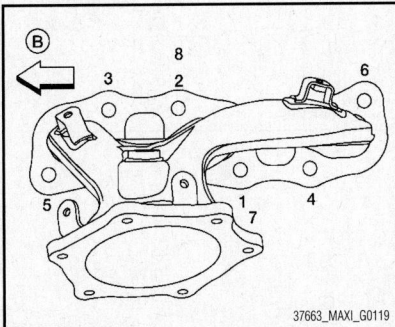

Fig. 59 Install the exhaust manifold nuts in the order as shown LH (B)

✳✳ CAUTION

Do not over-tighten the Air Fuel Ratio (A/F) sensor or heated oxygen sensors. Doing so may cause damage.

INTAKE MANIFOLD

REMOVAL & INSTALLATION

See Figures 60 through 63.

✳✳ WARNING

To avoid the danger of being scalded, never drain the coolant when the engine is hot.

1. Remove the engine cover.
2. Release the fuel pressure.
3. Remove intake manifold collector.
4. Disconnect fuel tube quick connector at vehicle piping side.
5. To remove the quick connector cap, hold the sides of the connector, push in the tabs and pull out the tube.

✳✳ CAUTION

Note the following:

- The tube can be removed when the tabs are completely depressed. Do not twist it more than necessary.
- Do not use any tools to remove the quick connector.
- Keep the resin tube away from heat.

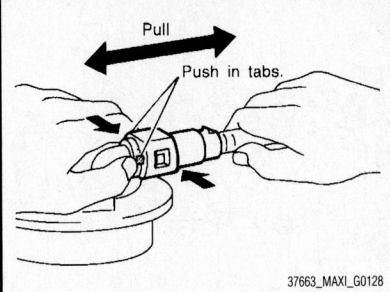

Fig. 61 To remove the quick connector cap, hold the sides of the connector, push in the tabs and pull out the tube

Be especially careful when welding near the tube.

- Prevent acid liquids such as battery electrolyte, etc. from getting on the resin tube.
- Do not bend or twist the tube during removal or installation.
- Do not remove the remaining retainer on the tube.
- When the tube is replaced, also replace the retainer with a new one.
- To keep the connecting portion clean and to avoid damage and foreign materials entering, cover the ends of the fuel tubes with plastic bags or something similar.

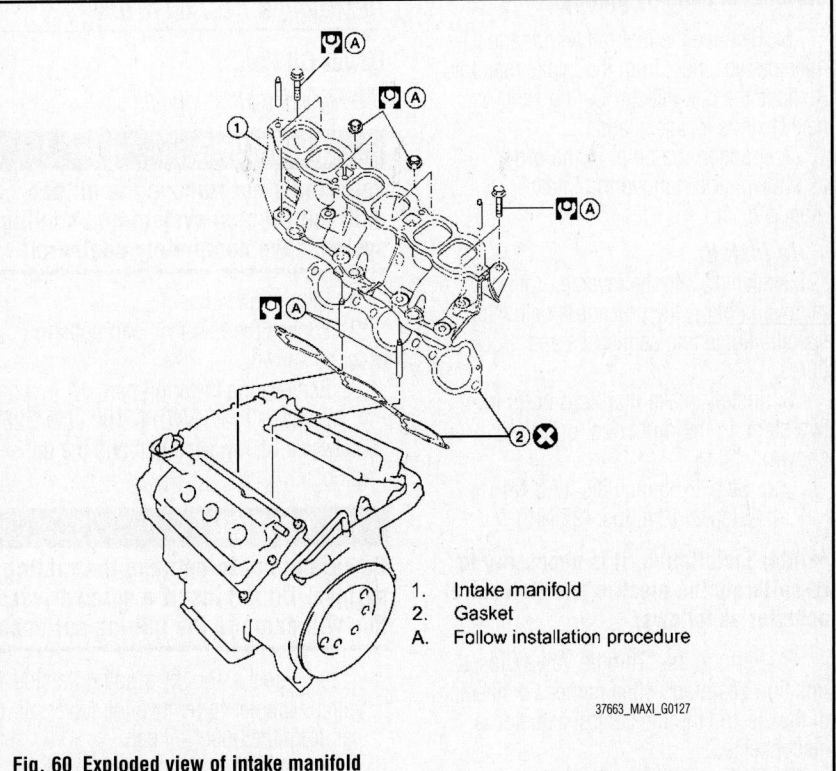

1. Intake manifold
2. Gasket
A. Follow installation procedure

Fig. 60 Exploded view of intake manifold

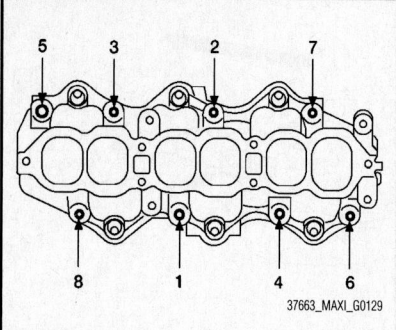

Fig. 62 Loosen the bolts in the order as shown

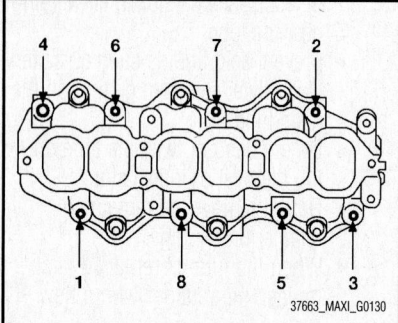

Fig. 63 Install intake manifold bolts in two steps in the numerical order

→**If the connector and the tube are stuck together, push and pull several times until they start to move. Then disconnect them by pulling.**

6. Remove the fuel rail with the fuel injectors attached, from the intake manifold. Remove the fuel injector O-rings and use new O-rings for installation.

7. Loosen the bolts in the order as shown, and remove the intake manifold.

To install:

Installation is in the reverse order of removal. Follow the procedure below for specific tightening sequences and procedures.

8. Install intake manifold bolts in two steps in the numerical order as shown.

    a. Step 1: 65 inch lbs. (7.3 Nm)
    b. Step 2: 19 ft. lbs. (25 Nm)

→**After installation, it is necessary to re-calibrate the electric throttle control actuator as follows:**

9. Perform the "Throttle Valve Closed Position Learning" when harness connector of the electric throttle control actuator is disconnected.

10. Perform the "Idle Air Volume Learn-

ing" when the electric throttle control actuator is replaced.

11. Install the quick connector as follows:

    a. Make sure no foreign substances are deposited in and around the fuel tube and quick connector and that there is no damage.

    b. Align the center to insert the quick connector straight onto the fuel tube.

    c. Insert the fuel tube until a click is heard.

    d. Install the quick connector cap on the quick connector joint. Align the arrow mark on the quick connector cap to the upper side.

    e. Install the fuel hose into the hose clamp.

12. Make sure there is no fuel leakage at connections as follows:

    a. Apply fuel pressure to fuel lines by turning ignition switch ON (with engine stopped). Then check for fuel leaks at connections.

    b. Start the engine and rev it up and check for fuel leaks at connections.

✳✳ CAUTION

Do not touch engine immediately after stopping as engine is extremely hot.

→**Use mirrors for checking on connections out of the direct line of sight.**

OIL PAN

REMOVAL & INSTALLATION

Lower Oil Pan

See Figures 64 through 68.

✳✳ WARNING

You should not remove the oil pan until the exhaust system and cooling system have completely cooled off.

1. Drain the engine oil.
2. Loosen the lower oil pan bolts in order as shown.
3. Remove the lower oil pan.
    a. Insert Tool KV10111100 (J-37228) between the lower oil pan and the upper oil pan.

✳✳ CAUTION

Be careful not to damage the mating surface. Do not insert a screwdriver, this will damage the mating surfaces.

    b. Slide the Tool by tapping its side with a hammer to remove the lower oil pan from the upper oil pan.
4. If re-installing the original lower oil

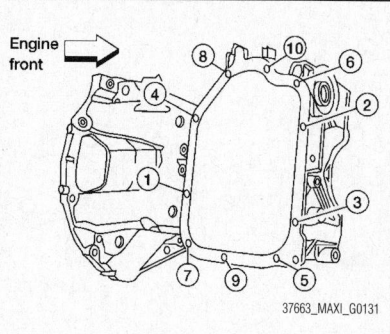

Fig. 64 Loosen the lower oil pan bolts in order

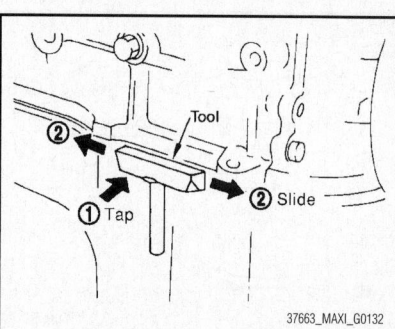

Fig. 65 Insert Tool KV10111100 (J-37228) between the lower oil pan and the upper oil pan

pan, remove the old sealant from the mating surfaces using a scraper.

    a. Also remove the old sealant from mating surface of the upper oil pan.

    b. Remove the old sealant from the bolt holes and threads.

✳✳ CAUTION

Do not scratch or damage the mating surfaces when cleaning off the old sealant.

5. Clean oil strainer if any object is attached.

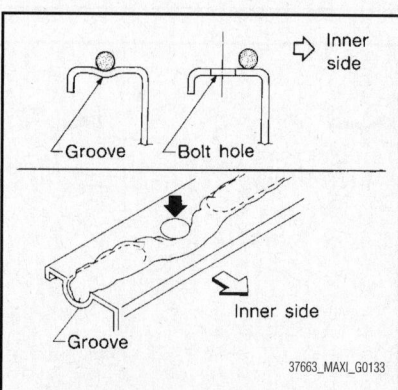

Fig. 66 Apply a continuous bead of sealant to the lower oil pan

To install:

6. Apply a continuous bead of sealant to the lower oil pan.
   a. Use Genuine Silicone RTV Sealant, or equivalent.
   b. Installation must be done within 5 minutes after applying sealant.
7. Install the lower oil pan. Tighten the lower oil pan bolts in order as shown.
   a. Wait at least 30 minutes before refilling the engine with oil.
   b. Start the engine and check for leaks.
   c. Inspect the engine oil level.

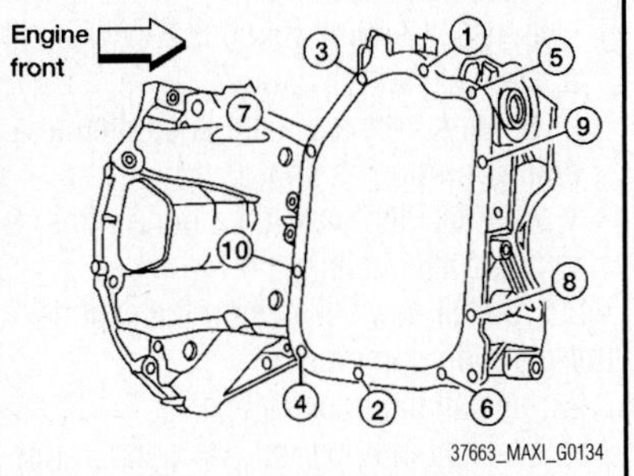

Fig. 67 Tighten the lower oil pan bolts in order

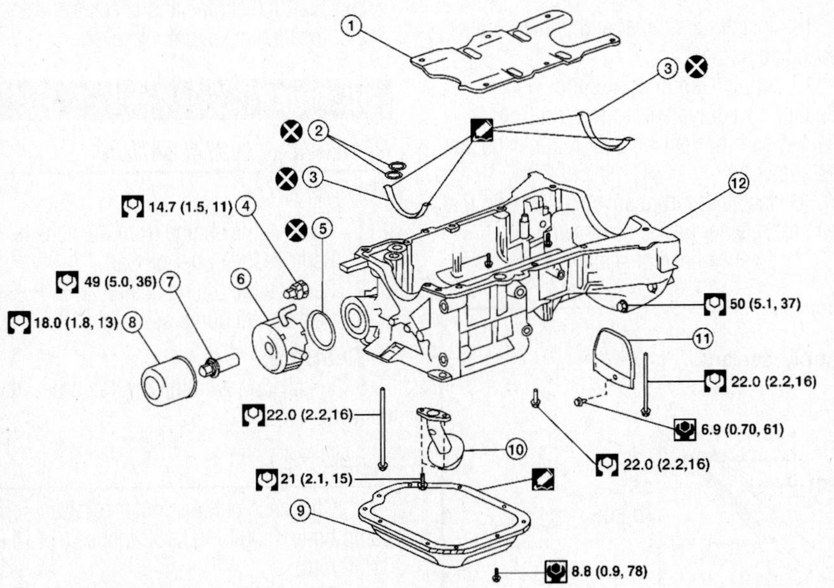

1.	Oil pan baffle	2.	O-ring	3.	Gasket
4.	Oil pressure switch	5.	Oil cooler gasket	6.	Oil cooler
7.	Oil cooler connection	8.	Oil filter	9.	Lower oil pan
10.	Oil strainer	11.	Rear plate cover	12.	Upper oil pan

37663_MAXI_G0135

Fig. 68 Exploded view of oil pan assembly

Upper Oil Pan

See Figures 69 through 76.

⁂ WARNING

You should not remove the oil pan until the exhaust system and cooling system have completely cooled off.

⁂ CAUTION

When removing the front and rear engine through bolts and nuts, lift the engine up slightly for safety.

⁂ CAUTION

When removing the upper oil pan from the engine, first remove the Crankshaft Position Sensor (CKP). Be careful not to damage sensor edges or signal plate teeth.

1. Remove the engine from the vehicle.
2. Drain the engine oil.
3. Remove the oil dipstick.
4. Remove the drive belt.
5. Disconnect the A/C compressor harness connector.
6. Remove the A/C compressor bolts and remove the A/C compressor.
7. Remove coolant pipe bolts.

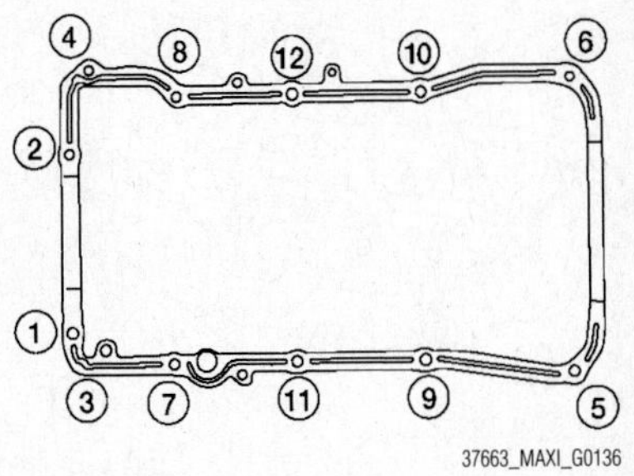

Fig. 69 Loosen the bolts in the order as shown

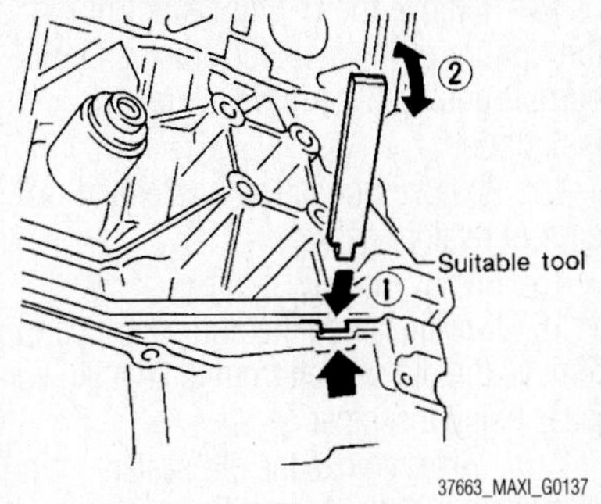

Fig. 70 Insert an appropriate size tool into the notch (1) of the upper oil pan

8. Disconnect the coolant lines from the engine oil cooler.
9. Remove the oil filter and engine oil cooler from the upper oil pan.
10. Remove the oil pressure switch, and the Crankshaft Position Sensor (CKP) from the upper oil pan.
11. Remove the lower oil pan.
12. Remove the upper oil pan.
    a. Loosen the bolts in the order as shown.
    b. Insert an appropriate size tool into the notch of the upper oil pan as shown.
    c. Pry off the upper oil pan by moving the tool up and down as shown.

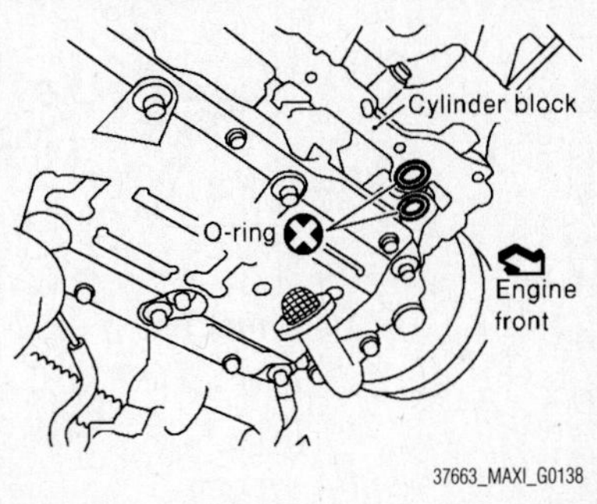

Fig. 71 Remove the O-ring seals from the bottom of the cylinder block

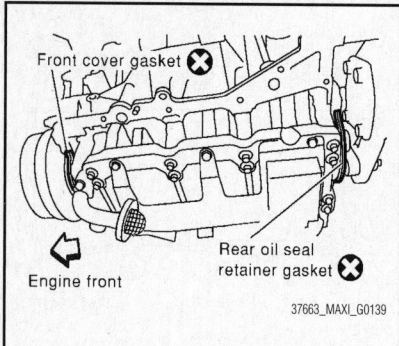

Fig. 72 Remove front cover gasket and rear oil seal retainer gasket

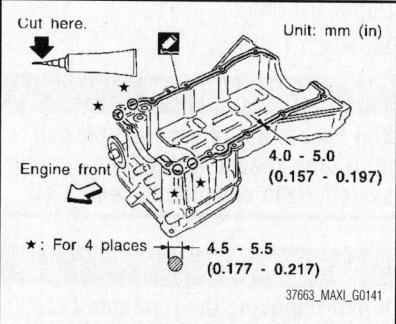

Fig. 74 Apply a bead of sealant to the cylinder block mating surface of the upper oil pan to a limited portion

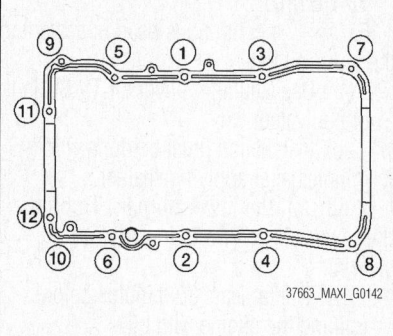

Fig. 76 Tighten upper oil pan bolts in the order as shown

13. Remove the O-ring seals from the bottom of the cylinder block and oil pump housing. Use new O-rings for installation.

14. Remove front cover gasket and rear oil seal retainer gasket.

15. Remove the oil strainer.

16. If re-installing the original oil pan, remove the old sealant from the mating surfaces using a scraper.

a. Also remove the old sealant from mating surface of the cylinder block.

b. Remove the old sealant from the bolt holes and threads.

✳✳ CAUTION

Do not scratch or damage the mating surfaces when cleaning off the old sealant.

17. Clean oil strainer if any object is attached.

To install:

✳✳ CAUTION

Wait at least 30 minutes before refilling the engine with oil.

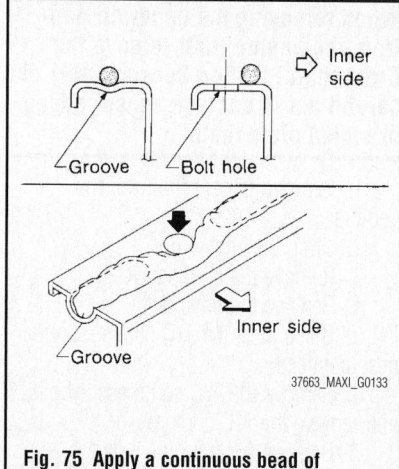

Fig. 75 Apply a continuous bead of sealant

18. Install oil strainer and tighten bolt to specified torque.

19. Apply Genuine Silicone RTV Sealant or equivalent, to the front cover gasket and the rear oil seal retainer gasket as shown.

20. Install the front cover gasket and rear oil seal retainer gasket as shown.

21. Apply a bead of sealant to the cylin-

der block mating surface of the upper oil pan to a limited portion as shown.

a. Use Genuine Silicone RTV Sealant, or equivalent.

b. Be sure the sealant is applied to a limited portion as shown.

c. Attaching should be done within 5 minutes after coating.

22. Install new O-rings on the cylinder block and oil pump body.

23. Install the upper oil pan.

a. Tighten upper oil pan bolts in the order as shown.

b. Wait at least 30 minutes before refilling the engine with oil.

24. Install the lower oil pan.

25. Installation of the remaining components is in the reverse order of removal.

26. Start the engine and check for leaks.

27. Inspect the engine oil level.

OIL PUMP

REMOVAL & INSTALLATION

See Figure 77.

1. Remove the engine from the vehicle.
2. Remove the upper oil pan.
3. Remove the timing chain.
4. Remove oil pump assembly.

To install:

5. Installation is in the reverse order of removal.

INSPECTION

1. Clearance between outer rotor and oil pump body: 0.0045–0.0102 inches (0.114–0.260 mm).

2. Tip clearance between inner rotor and outer rotor: Below 0.0071 inches (0.180 mm).

3. Side clearance with a straight-edge between inner rotor and oil pump body: 0.0012–0.0028 inches (0.030–0.070 mm).

4. Side clearance with a straight-edge between outer rotor and oil pump body: 0.0020–0.0043 inches (0.050–0.110 mm).

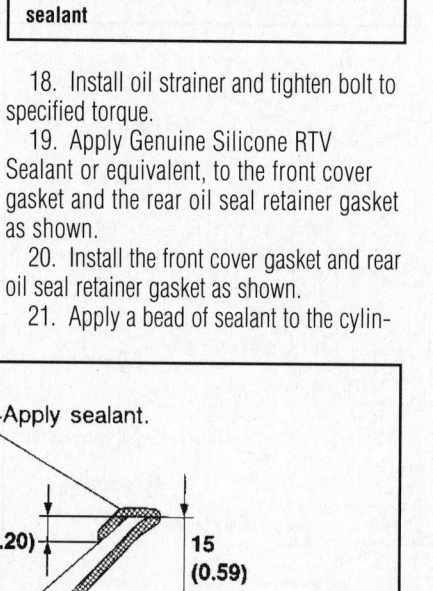

Fig. 73 Apply Genuine Silicone RTV Sealant or equivalent, to the front cover gasket and the rear oil seal retainer gasket

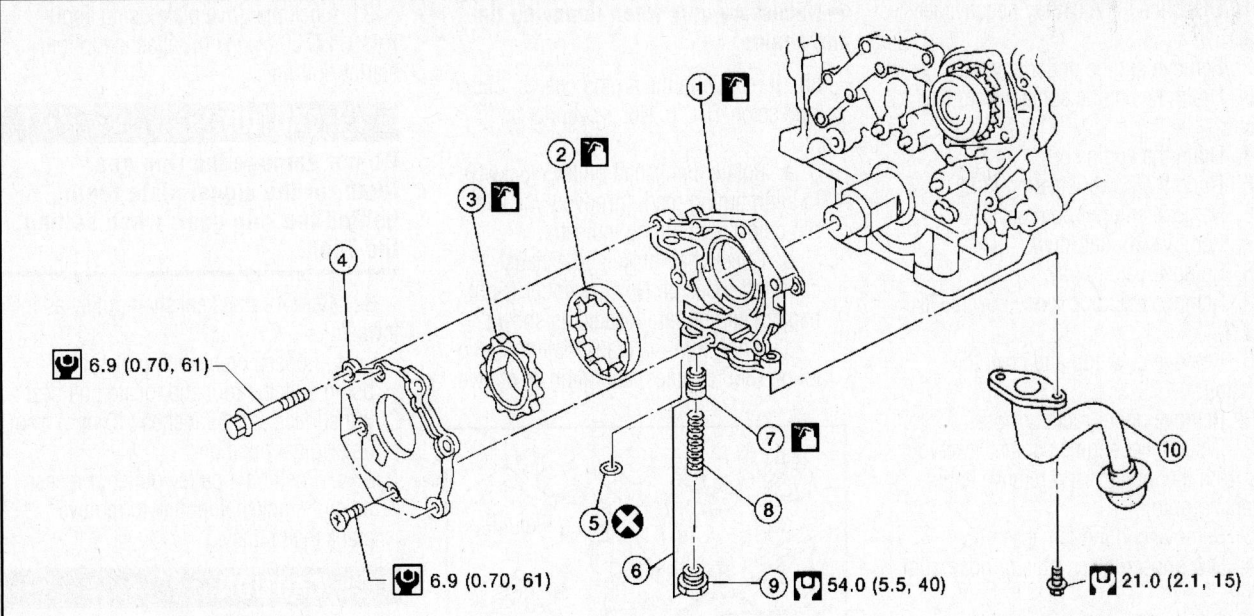

1. Oil pump housing
4. Oil pump cover
7. Regulator valve
10. Oil strainer

2. Outer rotor
5. O-ring
8. Spring

3. Inner rotor
6. Regulator valve set
9. Regulator plug

37663_MAXI_G0143

Fig. 77 Exploded view of oil pump assembly

PISTON AND RING

POSITIONING

See Figure 78.

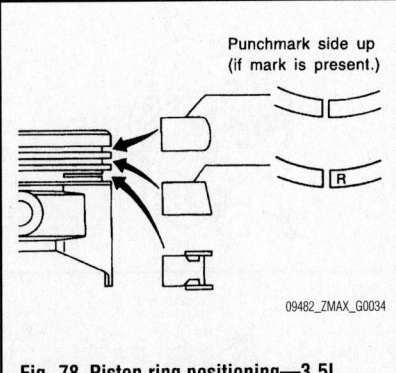

Punchmark side up
(if mark is present.)

R

09482_ZMAX_G0034

Fig. 78 Piston ring positioning—3.5L engine

REAR MAIN SEAL

REMOVAL & INSTALLATION

See Figure 79.

1. Remove engine and CVT assembly.
2. Separate the CVT from the engine assembly.
3. Remove drive plate.

4. Remove rear oil seal retainer using Tool KV10111100 (J-37228).

※※ CAUTION

Be careful not to damage mating surface. If rear oil retainer is removed, replace it with a new one.

To install:

5. Remove old liquid gasket material from mating surface of cylinder block and oil pan using a suitable scraper.
6. Apply liquid gasket to the new rear oil seal retainer using suitable tool.

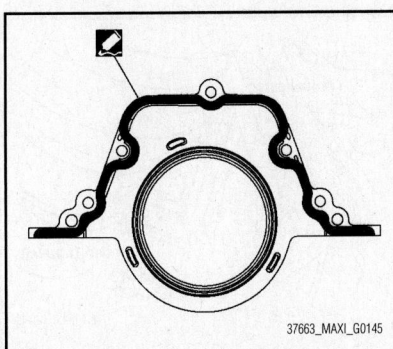

37663_MAXI_G0145

Fig. 79 Apply liquid gasket to the new rear oil seal retainer using suitable tool

a. Use Genuine Silicone RTV Sealant or equivalent.
b. Assembly should be completed within 5 minutes after coating.
7. Installation is in the reverse order of removal.

※※ CAUTION

When replacing an engine or transmission you must make sure the dowels are installed correctly during re-assembly. Improper alignment caused by missing dowels may cause vibration, oil leaks or breakage of drivetrain components.

TIMING CHAIN FRONT COVER

REMOVAL & INSTALLATION

See Figures 80 through 91.

➡This section describes the procedure for removal/installation of the front timing chain case without removing the oil pan (upper) from the vehicle. When rear timing chain case must be removed, remove the engine from the vehicle. Then remove front timing chain case, timing chain related parts, and rear timing chain case in this order, and install in reverse order of removal.

1. Disconnect the battery negative terminal.

2. Remove engine under cover.

3. Drain the engine coolant from the radiator.

4. Drain the engine oil.

5. Drain the power steering fluid.

6. Remove engine room cover.

7. Remove front air duct.

8. Remove battery tray.

9. Remove the hood ledge covers (RH and LH).

10. Remove cowl top and cowl top extension.

11. Remove upper radiator hose.

12. Disconnect engine coolant reservoir hose from the radiator and remove engine coolant reservoir.

13. Remove cooling fan assembly.

14. Disconnect lower radiator hose from engine.

15. Recover the A/C system R134a.

16. Remove the starter motor.

17. Disconnect the power steering fluid reservoir tank hose from the power steering pump and fluid cooler and remove the power steering fluid reservoir tank.

18. Remove the front RH wheel and tire.

19. Remove the engine side under cover.

20. Remove the drive belt.

21. Remove the power steering pump.

22. Remove the lower oil pan.

23. Remove upper oil pan bolts in reverse order as shown.

24. Remove the alternator.

25. Disconnect the A/C tubes from the A/C compressor and position aside.

26. Remove the A/C compressor bolts and remove the A/C compressor.

27. Remove the alternator bracket.

28. Support the engine with suitable jack and remove the RH engine insulator, mount and bracket.

29. Remove the rocker covers, if necessary.

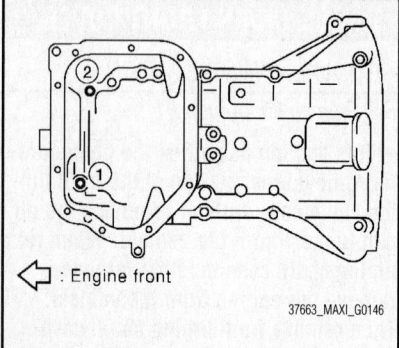

Fig. 80 Remove upper oil pan bolts in reverse order as shown

➡**Necessary only when removing timing chains.**

30. If removing the timing chains, obtain compression TDC of No. 1 cylinder as follows:

a. Rotate crankshaft pulley clockwise to align timing mark (grooved line without color) with timing indicator.

b. Check that intake and exhaust camshaft lobes on No. 1 cylinder (right bank of engine) are located as shown.

c. If not, turn the crankshaft one revolution (360 degrees) and align as shown.

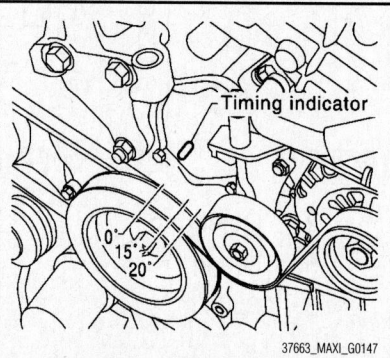

Fig. 81 Rotate crankshaft pulley clockwise to align timing mark (grooved line without color) with timing indicator

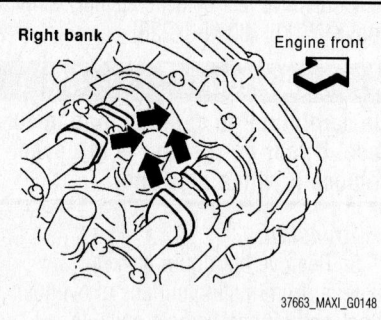

Fig. 82 Check that intake and exhaust camshaft lobes on No. 1 cylinder (right bank of engine) are located as shown

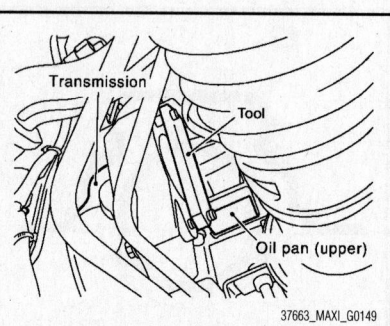

Fig. 83 Lock the drive plate using Tool KV10117700 (J-44716) attached to the starter bolt hole

31. Lock the drive plate using Tool KV10117700 (J-44716) attached to the starter bolt hole.

※※ CAUTION

Do not damage the ring gear teeth, or the signal plate teeth behind the ring gear, when setting the Tool.

32. Remove the crankshaft pulley as follows:

a. Loosen crankshaft pulley bolt using suitable tool and locate bolt seating surface at 0.39 inches (10 mm) from its original position.

b. Position a pulley puller at recess hole of crankshaft pulley to remove crankshaft pulley.

※※ CAUTION

Do not use a puller claw on crankshaft pulley periphery.

33. Remove engine oil cooler tube bolts and bracket.

34. Disconnect the oil pressure switch harness connector.

35. Disconnect valve timing control harness connector.

36. Remove the Bank 1 (RH) and Bank 2 (LH) IVT covers.

a. Loosen the IVT cover bolts in the reverse order as shown.

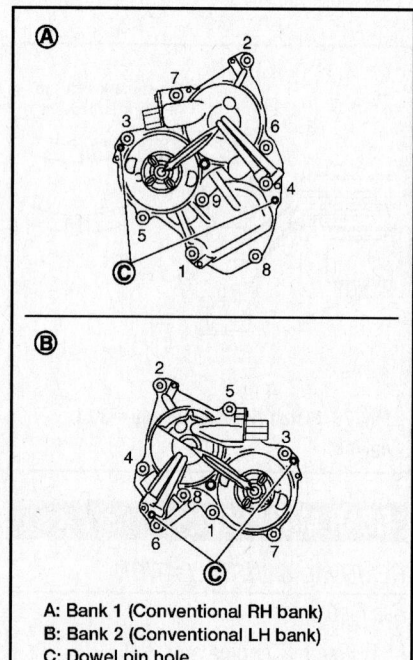

A: Bank 1 (Conventional RH bank)
B: Bank 2 (Conventional LH bank)
C: Dowel pin hole

Fig. 84 Loosen the IVT cover bolts in the reverse order as shown

❉❉ CAUTION

The shaft in the IVT cover is inserted into the center hole of the intake camshaft sprocket. Remove the IVT cover by pulling straight out until the IVT cover disengages from the camshaft sprocket.

b. Shaft is engaged with intake side camshaft sprocket center hole on inside. Pull straight out so as not to tilt until the joint is disengaged.

c. The mating surface of magnet retarder may be fitted with the exhaust side camshaft sprocket via engine oil. Open IVT cover carefully.

d. If the mating surface of the magnet retarder is fitted with the camshaft sprocket, open the IVT cover within the range that the load is not applied to the harness. Remove it so as to prevent magnet retarder from dropping.

❉❉ CAUTION

Be careful not to damage magnet retarder. When carrying IVT cover, face the magnet retarder side up to prevent the IVT cover from falling from magnet retarder. Never remove magnet retarder from IVT cover. (Disassembly prohibited)

37. Remove the A/C idler pulley and bracket and the drive belt auto-tensioner.

38. If necessary, remove the idler pulley and water pump cover.

39. Remove the front timing chain case.

a. Loosen the front timing chain case bolts in the order as shown.

b. Insert the appropriate size tool into the notch at the top of the front timing chain case.

c. Pry off the case by moving the suitable tool.

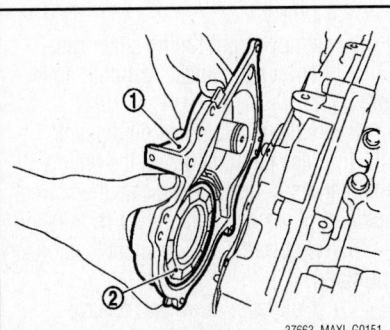

Fig. 85 The mating surface of magnet retarder (2) may be fitted with the exhaust side camshaft sprocket via engine oil; open IVT cover (1) carefully

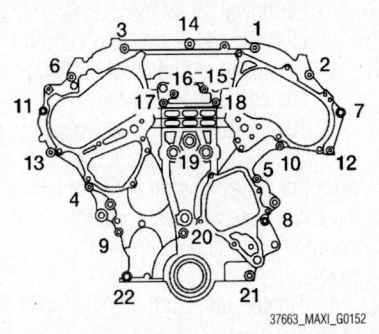

Fig. 86 Loosen the front timing chain case bolts in the order as shown

❉❉ CAUTION

Do not use a screwdriver or similar tool. After removal, handle carefully so it does not bend, or warp under a load.

40. Remove O-rings from rear timing chain case.

❉❉ CAUTION

Use new O-rings for installation.

41. Remove the front oil seal from the front timing chain case using a suitable tool.

❉❉ CAUTION

Do not damage the front cover.

42. Remove all old Silicone RTV Sealant from all the bolt holes and bolts.

❉❉ CAUTION

Do not damage the threads or mating surfaces.

43. Use a scraper to remove all of the old Silicone RTV Sealant from the front timing chain case and opposite mating surfaces.

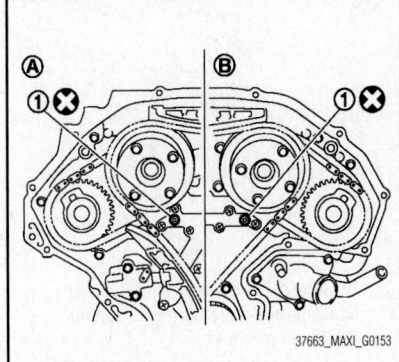

Fig. 87 Remove O-rings (1) from rear timing chain case

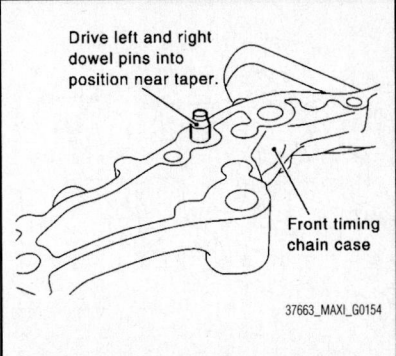

Fig. 88 Install dowel pins (right and left) into front timing chain case

❉❉ CAUTION

Do not damage the mating surfaces.

To install:

44. Install dowel pins (right and left) into front timing chain case up to a point close to taper in order to shorten protrusion length.

45. Install the new front oil seal on the front timing chain case. Apply new engine oil to the oil seal edges.

a. Install the new front oil seal so that it becomes flush with the face with front timing chain case using suitable drift.

❉❉ CAUTION

Press fit straight and avoid causing burrs or tilting the oil seal.

➡**Make sure the garter spring is in position and seal lip is not inverted.**

46. Install new O-rings on rear timing chain case.

❉❉ CAUTION

Use new O-rings for installation.

47. Apply Silicone RTV Sealant to front timing chain case.

a. Use Genuine Silicone RTV Sealant, or equivalent.

b. Before installation, wipe off the protruding sealant.

48. Install dowel pin on the rear timing chain case into dowel pin hole in front timing chain case.

49. Loosely install the front timing chain case bolts.

50. Tighten the front timing chain case bolts in the order as shown.

a. Retighten the front timing chain case bolts in the order as shown.

51. Install upper oil pan bolts.

52. Install lower oil pan.

53. Install IVT control valve covers.

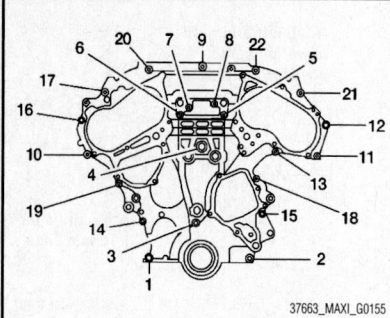

Fig. 89 Tighten the front timing chain case bolts in the order as shown

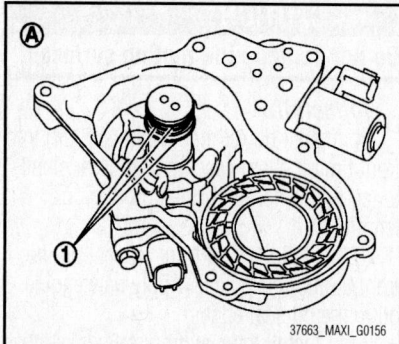

Fig. 90 Install new seal rings (1) in shaft grooves

a. Install new seal rings in shaft grooves.

> ※※ **CAUTION**
>
> **When replacing seal rings, replace all rings with new ones on both RH and LH IVT control valve covers.**

54. Install IVT covers with a new gasket to front timing chain case.

> ※※ **CAUTION**
>
> **Note the following:**

- Never face magnet retarder side down to prevent magnet retarder from dropping.
- Check the mating surface of the magnet retarder and the drum of the exhaust side camshaft sprocket for foreign materials.
- Align center of both shaft holes of the shaft and the intake side of the intake camshaft sprocket, and then insert them.
- Be careful not to drop the seal ring from the shaft groove.
- When setting the valve timing control cover in position by hand, if valve timing control cover is not centered with the valve timing case,

the dowel pin of the magnet retarder may not be aligned with the dowel pin holes of the cover. In this case, return to step "b".

a. Being careful not to move seal ring from the installation groove, align the dowel pins on the front timing chain case with the holes to install valve timing control covers.

b. Tighten bolts in the numerical order as shown.

55. Apply liquid gasket and install the water pump cover, if removed.

➡ **Use Genuine Silicone RTV Sealant or equivalent.**

56. Install crankshaft pulley and tighten the bolt in two steps.

a. Lubricate thread and seat surface of the bolt with new engine oil.

b. Apply a paint mark for the second step of angle tightening.

57. Rotate crankshaft pulley in normal direction (clockwise when viewed from front) to confirm it turns smoothly.

58. Installation of the remaining components is in reverse order of removal.

a. Before starting engine, check the levels of engine coolant, engine oil and working fluid. If less than

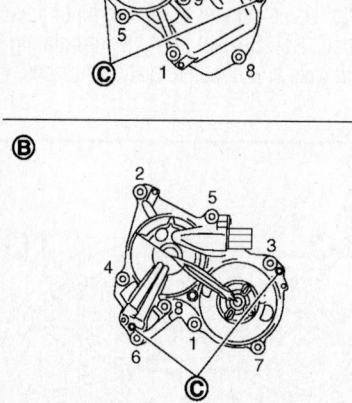

A: Bank 1 (Conventional RH bank)
B: Bank 2 (Conventional LH bank)
C: Dowel pin hole

Fig. 91 Tighten bolts in the numerical order as shown

required quantity, fill to the specified level.

b. Use procedure below to check for fuel leakage.

c. Turn ignition switch ON (with engine stopped). With fuel pressure applied to fuel piping, check for fuel leakage at connection points.

d. Start engine. With engine speed increased, check again for fuel leakage at connection points.

e. Run engine to check for unusual noise and vibration.

f. Warm up engine thoroughly to make sure there is no leakage of engine coolant, engine oil, working fluid, fuel and exhaust gas.

g. Bleed air from passages in pipes and tubes of applicable lines, such as in cooling system.

h. After cooling down engine, again check amounts of engine coolant, engine oil and working fluid. Refill to specified level, if necessary.

TIMING CHAIN & SPROCKETS

REMOVAL & INSTALLATION
See Figures 92 through 101.

> ※※ **CAUTION**
>
> **Note the following:**

- After removing timing chains, do not turn the crankshaft and camshaft separately, or the valves will strike the pistons.
- When installing camshafts, chain tensioners, oil seals, or other sliding parts, lubricate contacting surfaces with new engine oil.
- Apply new engine oil to bolt threads and seat surfaces when installing camshaft sprockets, camshaft brackets, and crankshaft pulley.

1. Remove front timing chain case.

2. Remove the intake manifold collector.

3. Remove the engine oil dipstick.

4. Place paint marks on the timing chain and sprockets to indicate the correct position of the components for installation.

5. Remove the timing chain tensioner (primary).

a. Pull lever down and release plunger stopper tab. Plunger stopper tab can be pushed up to release (coaxial structure with lever).

b. Insert stopper pin into timing chain tensioner (primary) body hole to

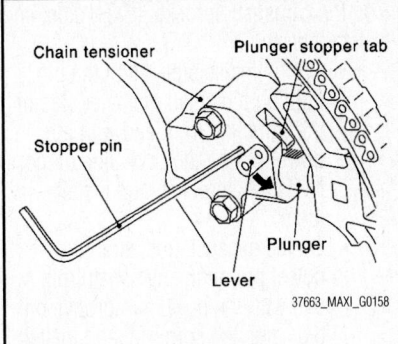

Fig. 92 Pull lever down and release plunger stopper tab

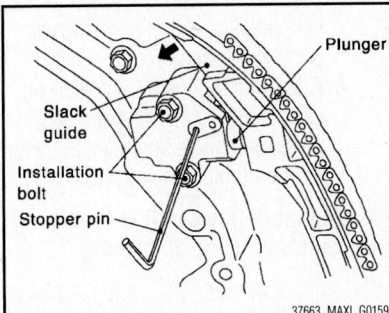

Fig. 93 Insert stopper pin into timing chain tensioner (primary) body hole to hold lever, and keep the tab released

hold lever, and keep the tab released. An Allen wrench [0.047 inches (1.2 mm)] is used for a stopper pin as an example.

c. Insert plunger into tensioner body by pressing the slack guide.

d. Keep the slack guide pressed and hold it by pushing the stopper pin through the lever hole and body hole.

e. Remove the bolts and remove the timing chain tensioner (primary).

6. Remove the internal chain guide, and slack guide.

7. Remove timing chain (primary) and crankshaft sprocket.

❊❊ CAUTION

After removing timing chains, do not turn the crankshaft and camshaft separately, or the valves will strike the pistons.

8. Attach a suitable stopper pin to the right and left timing chain tensioners (secondary).

9. Remove the timing chains (secondary) with camshaft sprockets (INT) and (EXH).

a. Insert metal or resin plate [0.020 inches (0.5 mm)] into guide between timing chain (secondary) and timing chain tensioner (secondary) plunger. Remove camshaft sprocket and timing

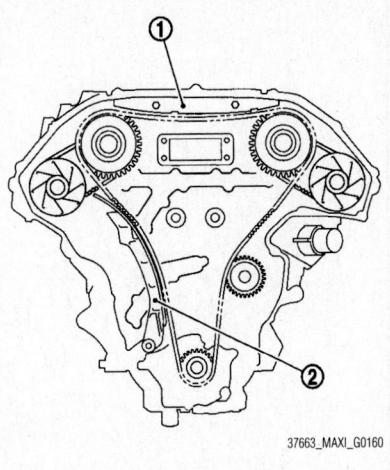

Fig. 94 Remove the internal chain guide (1), and slack guide (2)

chain (secondary) with timing chain removed from guide groove.

❊❊ CAUTION

Timing chain tensioner plunger can move while stopper pin is inserted in timing chain tensioner. Plunger can come out of tensioner when timing chain is removed. Use caution during removal.

➡ **Apply paint to the timing chain and camshaft sprockets for alignment during installation.**

b. Remove the camshaft sprocket (INT) and (EXH) bolts.

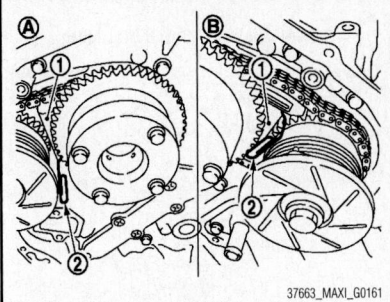

Fig. 95 Attach a suitable stopper pin (2) to the right and left timing chain tensioners (secondary) (1)

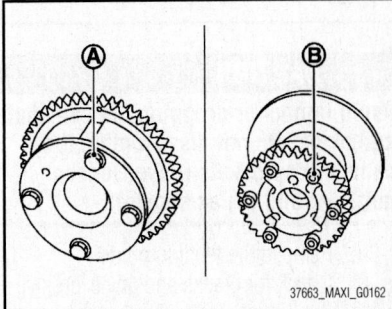

Fig. 96 Do not disassemble the camshaft sprockets (never loosen bolts (A) and (B) as shown)

c. Secure the hexagonal portion of the camshaft using a wrench to loosen the bolts.

d. Handle the sprockets as an assembly.

e. Remove timing chains (secondary).

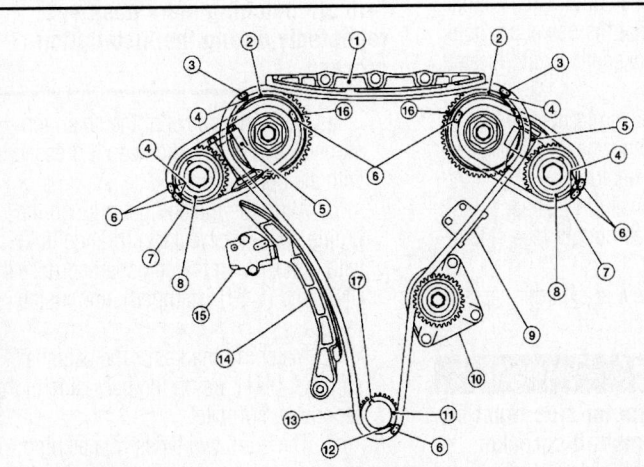

1.	Internal chain guide	2.	Camshaft sprocket (INT)	3.	Mating mark (pink link)
4.	Mating mark (punched)	5.	Timing chain tensioner (secondary)	6.	Mating mark (orange link)
7.	Timing chain (secondary)	8.	Camshaft sprocket (EXH)	9.	Tension guide
10.	Water pump	11.	Crankshaft sprocket	12.	Mating mark (notched)
13.	Timing chain (primary)	14.	Slack guide	15.	Timing chain tensioner (primary)
16.	Mating mark (back side)	17.	Crankshaft key		

Fig. 97 Timing chain assembly

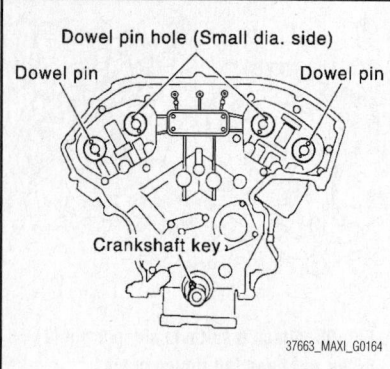

Fig. 98 Position the crankshaft so No. 1 piston is set at TDC on the compression stroke

✳✳ CAUTION

Avoid impact or dropping the camshaft sprockets. Do not disassemble the camshaft sprockets (never loosen bolts (A) and (B) as shown).

10. Remove the tension guide.
11. Check for cracks and any excessive wear of the timing chain. Replace the timing chain as necessary.

To install:

➡**The figure shows the relationship between the mating mark on each timing chain and that on the corresponding sprocket with the components installed.**

12. Install the tension guide.
13. Position the crankshaft so No. 1 piston is set at TDC on the compression stroke.
   a. Make sure that the dowel pin hole, dowel pin and crankshaft key are located as shown.
- Camshaft dowel pin hole (intake side): at cylinder head upper face side in each bank.
- Camshaft dowel pin (exhaust side): at cylinder head upper face side in each bank.
- Crankshaft key: at cylinder head side of RH bank.

✳✳ CAUTION

Hole on small diameter side must be used for intake camshaft sprocket dowel pin. Do not misidentify (ignore big diameter side).

14. Install the timing chains (secondary) and camshaft sprockets.

✳✳ CAUTION

Matching marks between the timing chain and sprockets slip easily. Con-

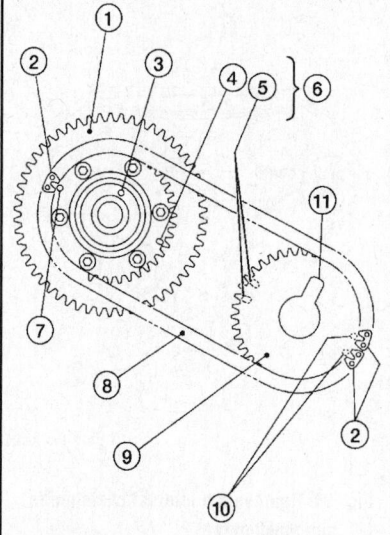

1. Camshaft sprocket (INT) side
2. Timing chain (orange link)
3. Dowel pin
4. Matching mark (oval type)
5. Matching mark (oval type)
6. LH bank
7. Matching mark (round type)
8. Timing chain (secondary)
9. Camshaft sprocket
10. Matching mark (round type)
11. Dowel pin groove

Fig. 99 Align the matching marks on the timing chain (secondary) (orange link) with the ones on the camshaft sprockets (INT) and (EXH) (stamped), and install them

firm all matching mark positions repeatedly during the installation process.

   a. Push the sleeve of the chain tensioner (secondary) and keep it pressed in with a stopper pin.
   b. Align the matching marks on the timing chain (secondary) (orange link) with the ones on the camshaft sprockets (INT) and (EXH) (stamped), and install them.
   c. Matching marks for the camshaft sprocket (INT) are on the back side of the secondary sprocket.
   d. There are two types of matching marks, round, oval types. They should be used for the RH and LH banks, respectively.
- RH bank: use round type.
- LH bank: use oval type.

   e. Align the dowel pin with and pin hole on the camshaft sprocket (INT) side, and dowel pin groove with the dowel pin

on the camshaft sprocket (EXH) side, and install them.
- On the intake side, align the pin hole on the small diameter side of the camshaft front end with the dowel pin on the back side of the camshaft sprocket, and install them.
- On the exhaust side, align the dowel pin on the camshaft front end with the dowel pin groove on the camshaft sprocket, and install them.
- Camshaft sprocket bolts must be tightened in the next step. Tightening them by hand is enough to prevent the dislocation of the dowel pins and dowel pin grooves.
- Check mating mark (punched) on each camshaft sprocket are positioned on the mating marks (orange link) on timing chain (secondary).

➡**Mating mark (punched) in the figure is for checking loose at this step.**

15. After confirming the mating marks are aligned, tighten the camshaft sprocket bolts.
   a. Secure the camshaft using a wrench at the hexagonal portion to tighten the bolts.
16. Pull the stopper pins out from the timing chain tensioners (secondary).
17. Install the crankshaft sprocket on the crankshaft.

➡**Make sure the mating marks on the crankshaft sprocket face the front of the engine.**

18. Install the timing chain (primary).
   a. Install timing chain (primary) so the mating mark (punched) on camshaft sprocket is aligned with the pink link on

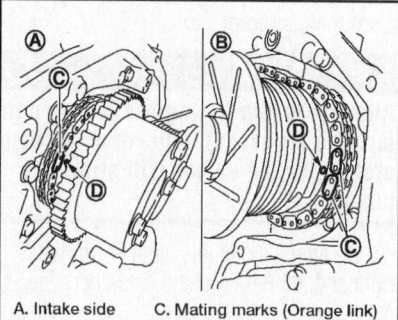

A. Intake side
B. Exhaust side
C. Mating marks (Orange link)
D. Mating mark (punched)

Fig. 100 Check mating mark (punched) on each camshaft sprocket are positioned on the mating marks (orange link) on timing chain (secondary)

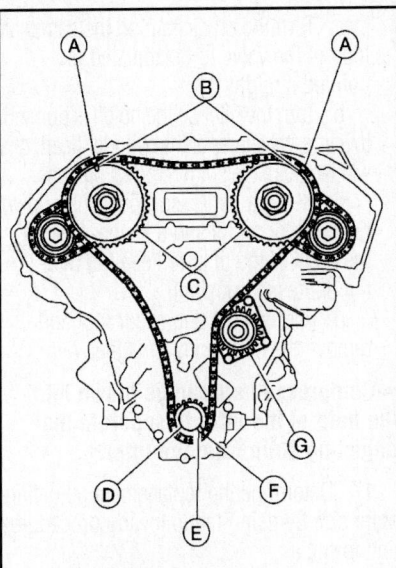

A. Pink link
B. Mating mark (punched)
C. Camshaft sprocket
D. Crankshaft sprocket
E. Mating mark (notched)
F. Orange link
G. Water pump

37663_MAXI_G0167

Fig. 101 Install the timing chain (primary)

the timing chain, while the mating mark (notched) on the crankshaft sprocket is aligned with the orange one on the timing chain, as shown.

b. When it is difficult to align mating marks of the timing chain (primary) with each sprocket, gradually turn the camshaft using a wrench on the hexagonal portion to align it with the mating marks.

c. During alignment, be careful to prevent dislocation of mating mark alignments of the secondary timing chains.

19. Install the internal chain guide (1), slack guide (2) and timing chain (primary).

✳✳ CAUTION

Do not overtighten the slack guide bolts. It is normal for a gap to exist under the bolt seats when the bolts are tightened to specification.

20. Install the timing chain tensioner (primary) for the slack guide.

a. When installing the timing chain tensioner (primary), push in the sleeve and keep it pressed in with the stopper pin.

b. Remove any dirt and foreign materials completely from the back and the mounting surfaces of the timing chain tensioner (primary).

c. After installation, pull out the stopper pin while pressing the slack guide.

21. Reconfirm that the matching marks on the sprockets and the timing chain have not slipped out of alignment.

22. Install the front timing chain case.

VALVE LASH

ADJUSTMENT

See Figures 102 through 104.

1. Remove the air duct with air cleaner case, collectors, hoses, wires, harnesses, and connectors.

2. Remove the intake manifold collectors.

3. Remove the ignition coils and spark plugs.

4. Remove the rocker covers.

5. Set the No. 1 cylinder at Top Dead Center (TDC) on its compression stroke. Align the pointer with the TDC mark on the crankshaft pulley. Check that the valve lifters on the No. 1 cylinder are loose and valve lifters on the No. 4 cylinder are tight. If not, turn the crankshaft 1 revolution (360 degrees) and align the pointer with the TDC mark on the crankshaft pulley.

6. Check the following valves:
- Both No. 1 intake valves
- Both No. 2 exhaust valves
- Both No. 3 exhaust valves
- Both No. 6 intake valves

7. Using a feeler gauge, measure the clearance between the valve lifter and the camshaft. Record any valve clearance measurements that are out of specification. Intake valve clearance (cold) is 0.010–0.013 inches (0.26–0.34 mm) and exhaust valve clearance (cold) is 0.011–0.015 inches (0.29–0.37 mm).

8. Turn the crankshaft 240 degrees and set the No. 3 cylinder to TDC of its compression stroke.

9. Check the following valves:
- Both No. 2 intake valves
- Both No. 3 intake valves
- Both No. 4 exhaust valves
- Both No. 5 exhaust valves

10. Using a feeler gauge, measure the clearance between the valve lifter and the camshaft. Record any valve clearance measurements that are out of specification.

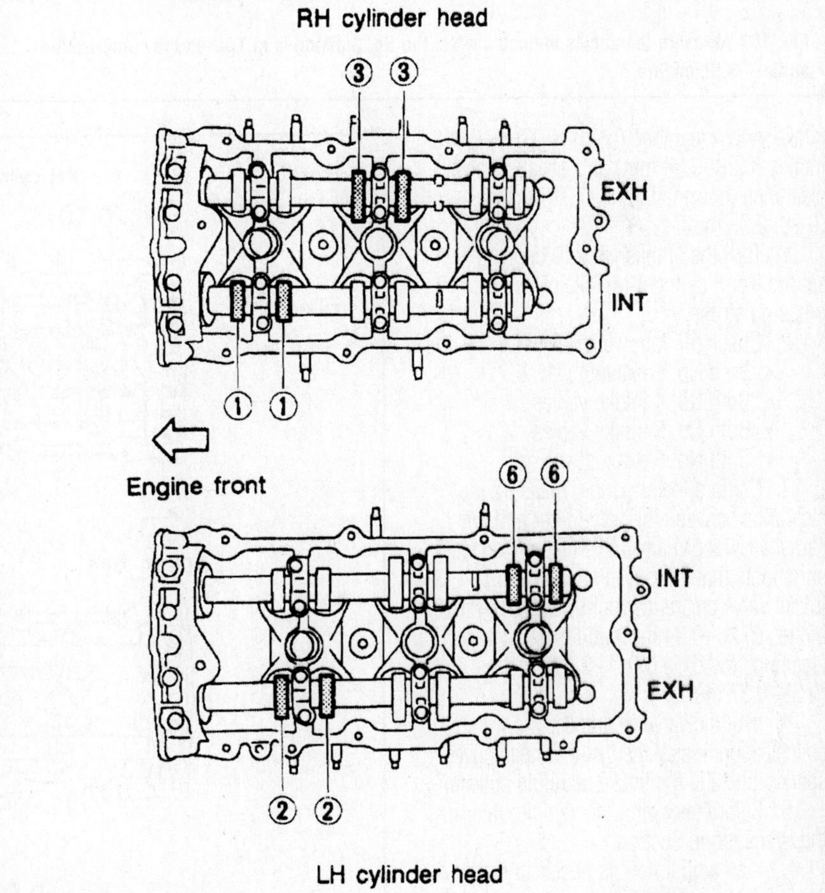

Fig. 102 Measure the valves indicated while the No. 1 piston is at TDC on the compression stroke—3.5L engine

09482_ZMAX_G0025

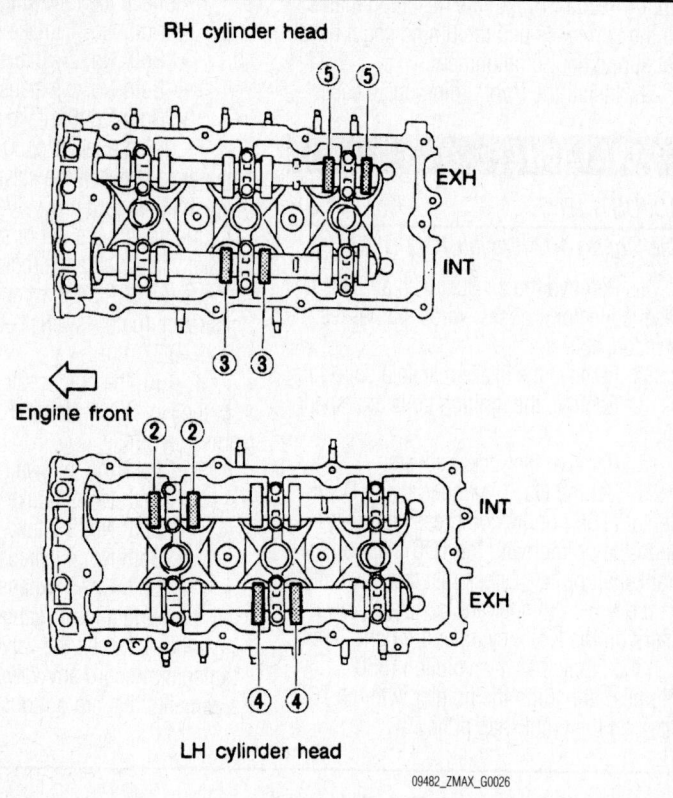

Fig. 103 Measure the valves indicated while the No. 3 piston is at TDC on the compression stroke—3.5L engine

a. Turn the crankshaft so the camshaft lobe of the valve to be adjusted is pointed straight up.

b. Turn the lifter so the notch is pointed towards the center of the cylinder head; this will facilitate the shim removal process.

c. Using a depressor tool, push down on the lifter and insert a keeper tool on the edge of the lifter to keep the lifter in the depressed position.

d. Remove the depressor tool and remove the shim with a magnet.

→**Compressed air can be blown into the hole of the lifter to separate the adjusting shim from the lifter.**

17. Determine the replacement adjusting shim size by using the following procedures and formula:

a. Using a micrometer determine thickness of the removed shim.

b. Calculate the thickness of a new adjusting shim so valve clearance is within the specified values.

c. R = thickness of the removed shim.

d. N = thickness of the new shim.

e. M = measured valve clearance.

• Intake shim determination formula: N = R + (M minus 0.0118 inches or 0.30 mm)

Intake valve clearance (cold) is 0.010–0.013 inches (0.26–0.34 mm) and exhaust valve clearance (cold) is 0.011–0.015 inches (0.29–0.37 mm).

11. Turn the crankshaft 240 degrees and set the No. 5 cylinder to TDC of its compression stroke.

12. Check the following valves:
• Both No. 1 exhaust valves
• Both No. 4 intake valves
• Both No. 5 intake valves
• Both No. 6 exhaust valves

13. Using a feeler gauge, measure the clearance between the valve lifter and the camshaft. Record any valve clearance measurements that are out of specification. Intake valve clearance (cold) is 0.010–0.013 inches (0.26–0.34 mm) and exhaust valve clearance (cold) is 0.011–0.015 inches (0.29–0.37 mm).

14. If all the valve clearances are within specification, install the cylinder head cover, spark plugs and the intake manifold collector.

15. Before servicing the vehicle, refer to the Precautions Section.

16. If an adjustment is necessary, adjust the valve clearance while engine is cold by removing the adjusting shim. The adjusting shim can be removed by using the following procedures:

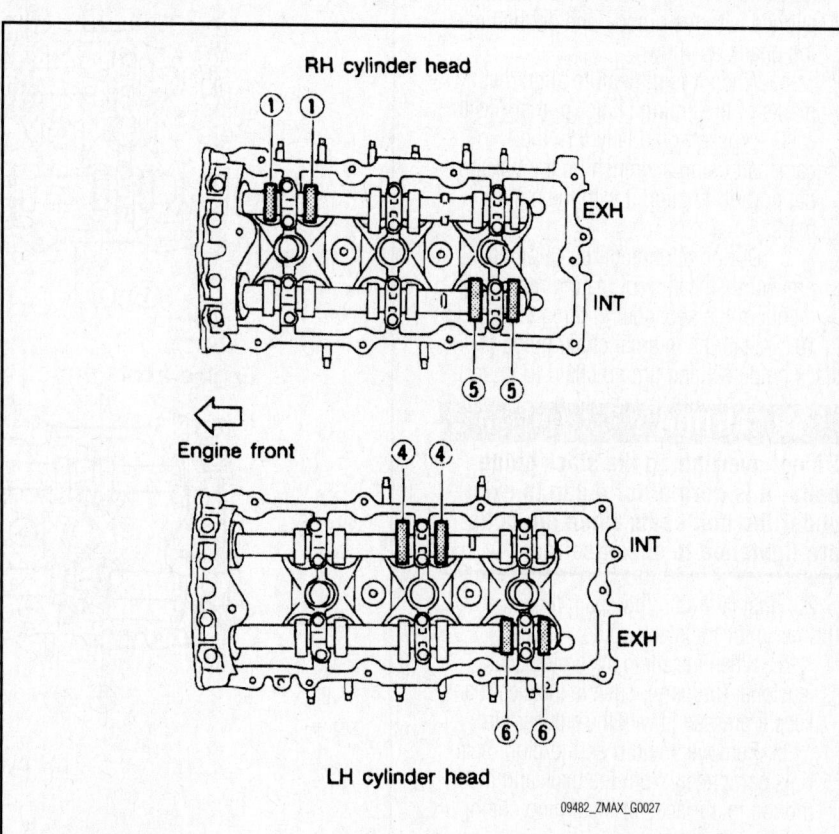

Fig. 104 Measure the valves indicated while the No. 5 piston is at TDC on compression—3.5L engine

- Exhaust shim determination formula: N = R + (M minus 0.0130 inches or 0.33 mm)

18. Shims are available in 64 sizes from 0.0913–0.1161 inches (2.32–2.95 mm) in steps of 0.004 inches (0.01 mm). The thickness is stamped on the shim; this side is always installed facing down. Select new shims with thickness as close as possible to calculated valve and install it in the lifter.

19. Install the new shim onto the lifter.

20. Depress the lifter and remove the keeper tool. Remove the depressor tool and recheck the valve clearance. Repeat this procedure for any other valves requiring adjustment.

21. When all valve adjustments are finished, install the cylinder head cover, spark plugs and the intake manifold collector.

➡️**Check and adjust the valve clearances while the engine is cold and not running.**

ENGINE PERFORMANCE & EMISSION CONTROLS

ACCELERATOR PEDAL POSITION (APP) SENSOR

LOCATION

See Figure 105.

The Accelerator Pedal Position (APP) sensor is an integral part of the accelerator pedal assembly.

REMOVAL & INSTALLATION

1. Disconnect the battery negative terminal.

2. Disconnect the accelerator position sensor electrical connector.

3. Remove the three accelerator pedal nuts.

4. Remove the accelerator pedal and accelerator position sensor assembly.

✳✳ CAUTION

Note the following:

- Do not disassemble the pedal assembly. Do not remove the accelerator pedal position sensor from the pedal assembly.
- Avoid impact from dropping during handling.
- Keep the pedal assembly away from water.

To install:

5. Installation is in the reverse order of removal.

6. Align and install accelerator pedal and accelerator position sensor assembly with locating pins in locating pin holes.

7. Check the accelerator pedal for smooth operation. There should be no binding or sticking when applying or releasing the accelerator pedal.

8. Check that the accelerator pedal moves through the full specified distance of pedal travel.

✳✳ CAUTION

When the harness connector of the accelerator pedal position sensor is disconnected, perform the "Accelerator pedal released position learning".

ACCELERATOR PEDAL RELEASED POSITION LEARNING

1. Check that accelerator pedal is fully released.

2. Turn ignition switch ON and wait at least 2 seconds.

3. Turn ignition switch OFF and wait at least 10 seconds.

4. Turn ignition switch ON and wait at least 2 seconds.

5. Turn ignition switch OFF and wait at least 10 seconds.

CAMSHAFT POSITION (CMP) SENSOR

LOCATION

The Camshaft Position (CMP) sensors are located at the rear of each cylinder head.

REMOVAL & INSTALLATION

1. Loosen the fixing bolt of the sensor.

2. Disconnect Camshaft Position (CMP) sensor harness connector.

3. Remove the sensor.

To install:

4. Installation is reverse of removal.

CRANKSHAFT POSITION (CKP) SENSOR

REMOVAL & INSTALLATION

1. Loosen the fixing bolt of the sensor.

2. Disconnect Crankshaft Position (CKP) sensor (POS) harness connector.

3. Remove the sensor.

To install:

4. Installation is reverse of removal.

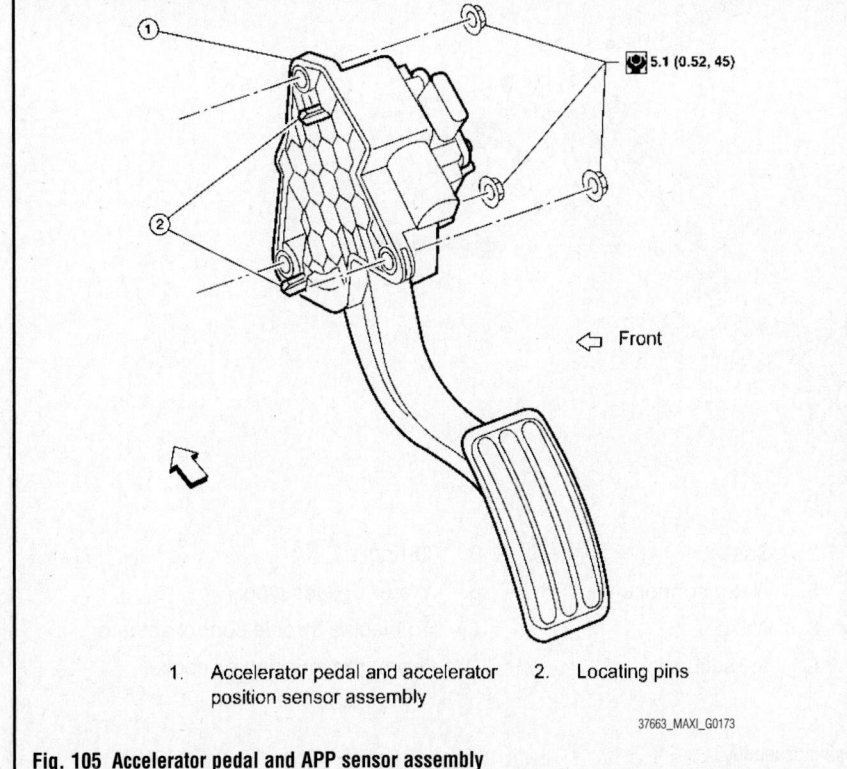

5.1 (0.52, 45)

⬅️ Front

1. Accelerator pedal and accelerator position sensor assembly
2. Locating pins

37663_MAXI_G0173

Fig. 105 Accelerator pedal and APP sensor assembly

ELECTRONIC CONTROL MODULE (ECM)

LOCATION

See Figure 106.

The ECM is located under the hood in the left front corner adjacent to the battery.

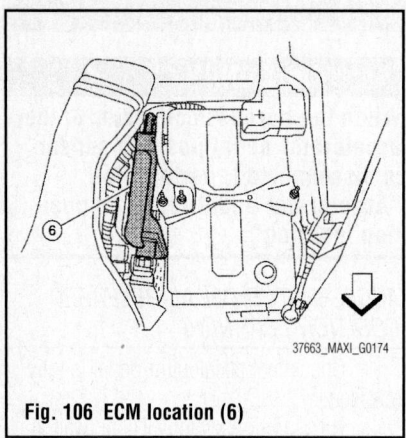

Fig. 106 ECM location (6)

REMOVAL & INSTALLATION

➡The manufacturer does not provide a specific Removal and Installation procedure for this component. Refer to the graphic(s) when servicing this component.

ENGINE COOLANT TEMPERATURE (ECT) SENSOR

LOCATION

See Figure 107.

The ECT sensor is located on the water outlet.

REMOVAL & INSTALLATION

➡The manufacturer does not provide a specific Removal and Installation procedure for this component. Refer to the graphic(s) when servicing this component.

EVAPORATIVE EMISSIONS (EVAP) CANISTER

REMOVAL & INSTALLATION

1. Lift up the vehicle.
2. Remove EVAP canister fixing bolt.
3. Remove EVAP canister.

➡The EVAP canister vent control valve and EVAP canister system pressure sensor can be removed without removing the EVAP canister.

To install:

4. Install in the reverse order of removal.

➡Tighten EVAP canister fixing bolt to the specified torque.

HEATED OXYGEN (HO2S) SENSOR

LOCATION

See Figure 108.

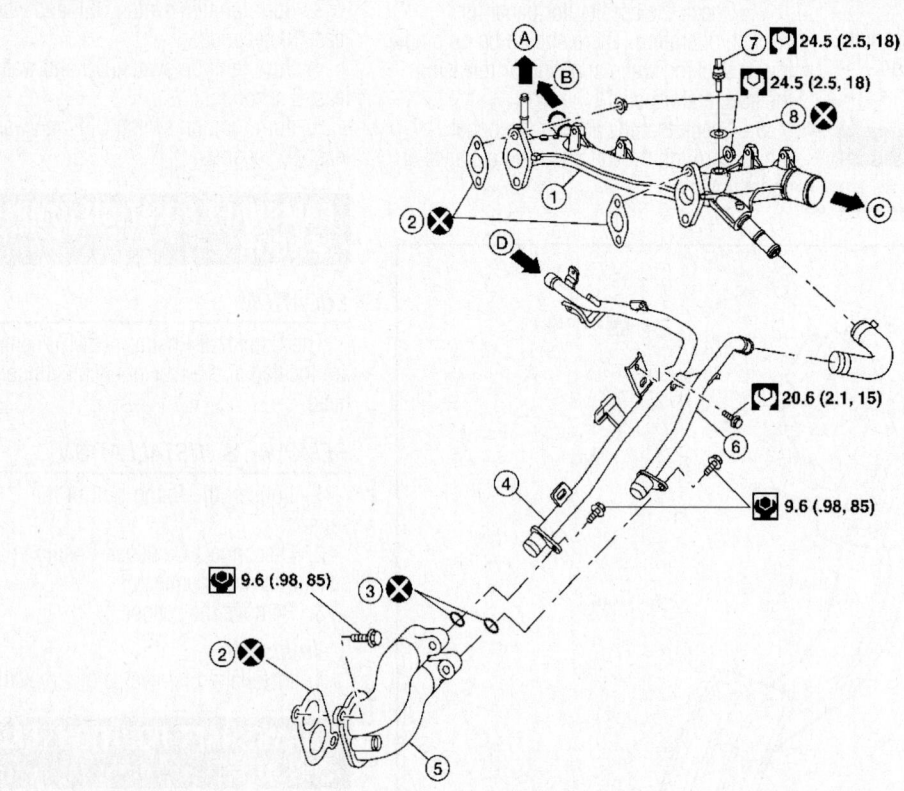

1.	Water outlet	2.	Gasket	3.	O-ring
4.	Heater pipe	5.	Water connector	6.	Water bypass pipe
7.	Engine coolant temperature sensor	8.	Washer	A.	To electric throttle control actuator
B.	To heater	C.	To radiator	D.	From transmission oil cooler

Fig. 107 Exploded view of water outlet and water piping assembly

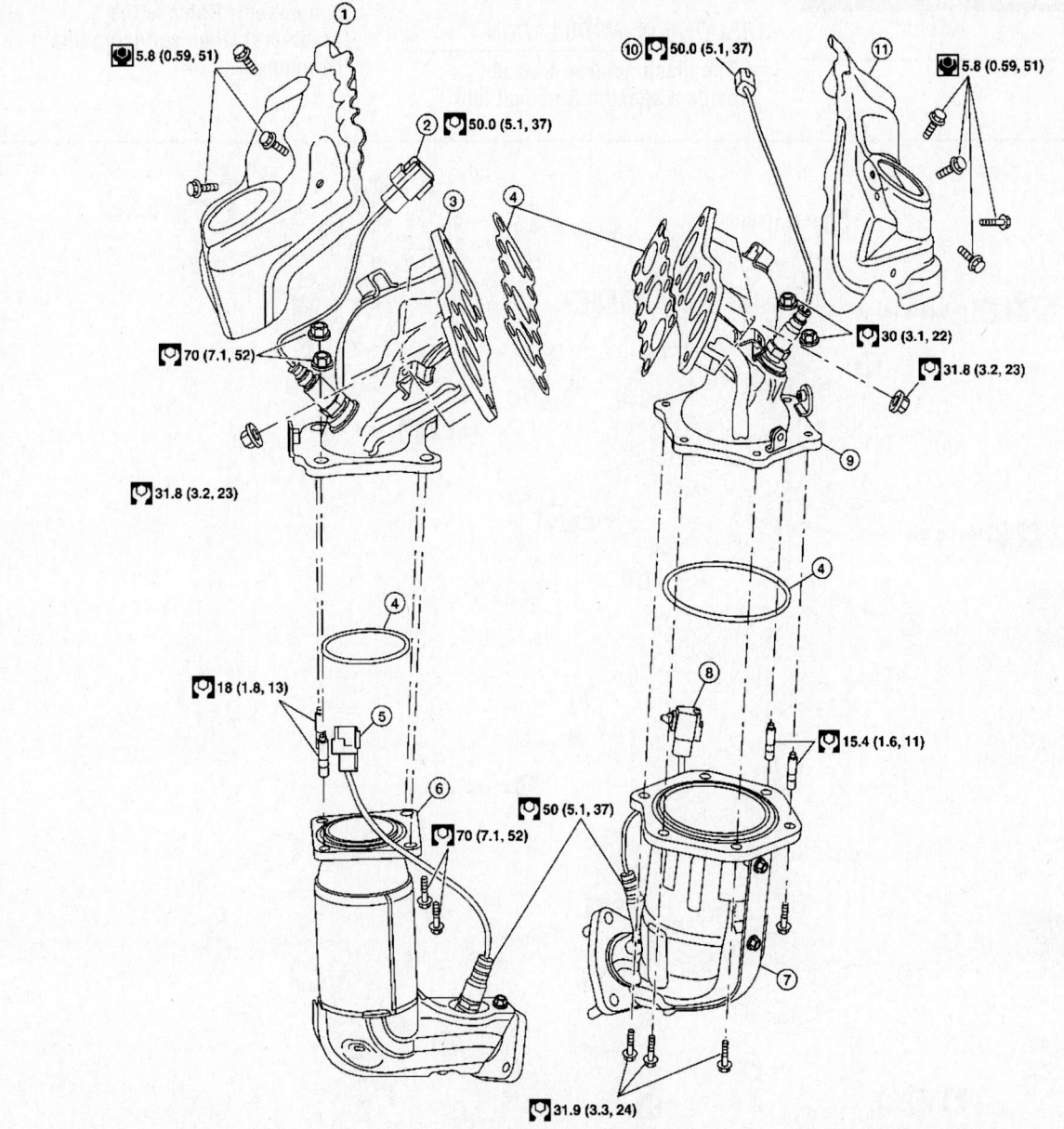

Fig. 108 Exploded view of exhaust manifold and three way catalyst assembly

1. Exhaust manifold heat shield (RH)
2. Air fuel ratio (A/F) sensor 1 (bank 1)
3. Exhaust manifold (RH)
4. Gaskets
5. Heated oxygen sensor 2 (bank 1)
6. Three way catalyst (manifold) (bank 1)
7. Three way catalyst (manifold) (bank 2)
8. Heated oxygen sensor 2 (bank 2)
9. Exhaust manifold (LH)
10. Air fuel ratio (A/F) sensor 1 (bank 2)
11. Exhaust manifold heat shield (LH)

The Heated Oxygen Sensors (HO2S) are located at the bottom end of the three way catalyst for each bank.

REMOVAL & INSTALLATION

1. Disconnect heated oxygen sensor harness connector.
2. Remove the sensor.

To install:

3. Installation is reverse of removal.

INTAKE AIR TEMPERATURE (IAT) SENSOR

LOCATION

The intake air temperature sensor is built-into the mass air flow sensor. The sensor detects intake air temperature and transmits a signal to the ECM.

The temperature sensing unit uses a thermistor which is sensitive to the change in temperature. Electrical resistance of the thermistor decreases in response to the rise in temperature.

REMOVAL & INSTALLATION

Refer to Mass Air Flow (MAF) sensor for Removal and Installation when servicing this component.

KNOCK SENSOR (KS)

LOCATION

See Figure 109.

Refer to the accompanying illustration.

REMOVAL & INSTALLATION

➡The manufacturer does not provide a specific Removal and

Installation procedure for this component. Refer to the graphic(s) when servicing this component.

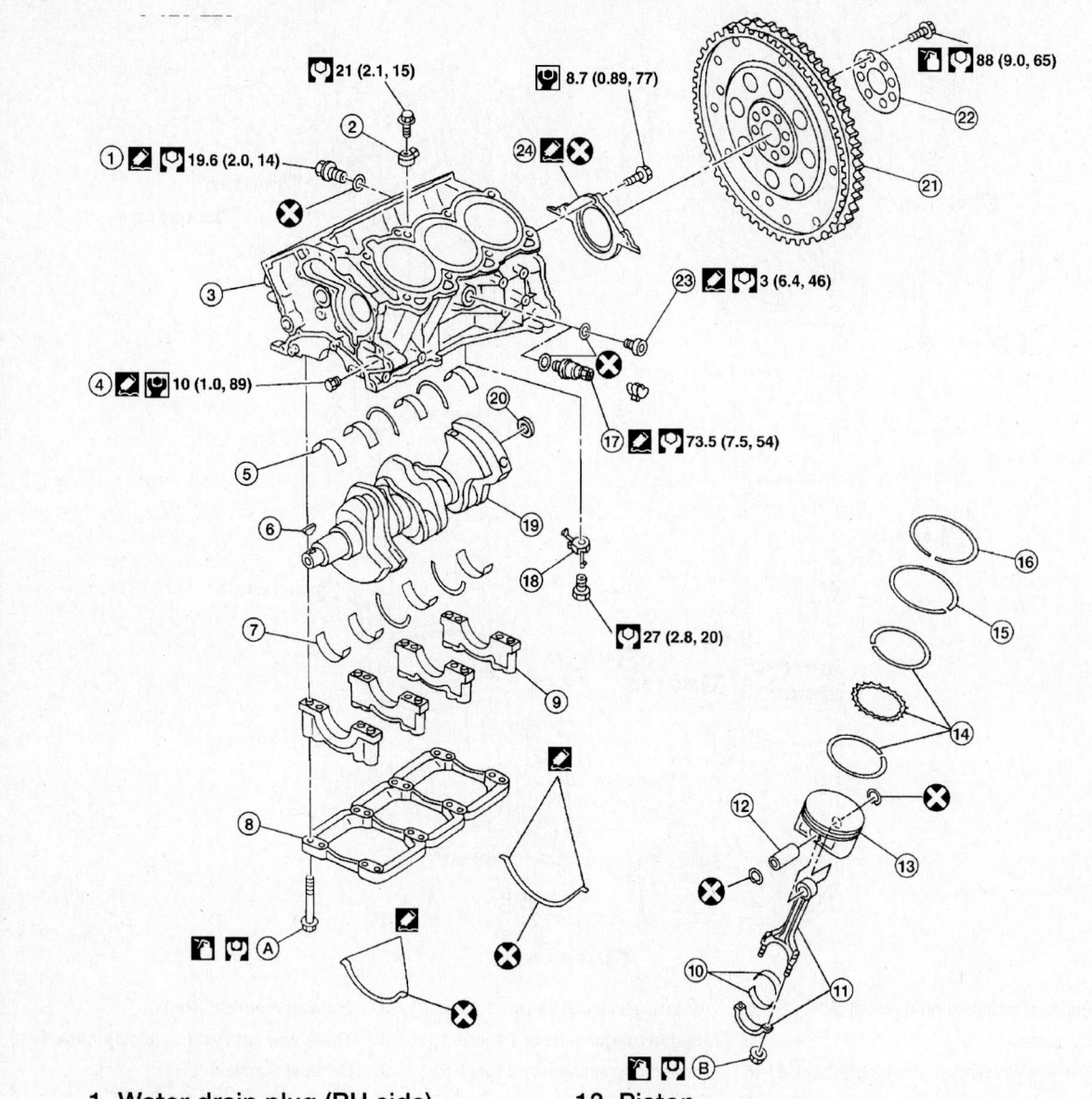

1. Water drain plug (RH side)
2. Knock sensor
3. Cylinder block
4. Water drain plug (water pump side)
5. Upper main bearing
6. Key
7. Lower main bearing
8. Main bearing beam
9. Main bearing cap
10. Connecting rod bearing
11. Connecting rod
12. Piston pin
13. Piston
14. Oil ring set
15. Second ring
16. Top ring
17. Cylinder block heater (Canada only)
18. Oil jet
19. Crankshaft
20. Pilot converter
21. Drive plate with signal plate
22. Drive plate reinforcement
23. Water drain plug (LH side)
24. Rear oil seal retainer

22140_350Z_G0006

Fig. 109 Knock sensor (KS) location (2)

MALFUNCTION INDICATOR LIGHT (MIL)

RESET PROCEDURE

Clearing diagnostic trouble codes resets the MIL.

MASS AIR FLOW (MAF) SENSOR

LOCATION

See Figure 110.

The Mass Air Flow (MAF) sensor is located in the intake air duct.

REMOVAL & INSTALLATION

➡The manufacturer does not provide a specific Removal and Installation procedure for this component. Refer to the graphic(s) when servicing this component. Also refer to air cleaner/air duct Removal and Installation.

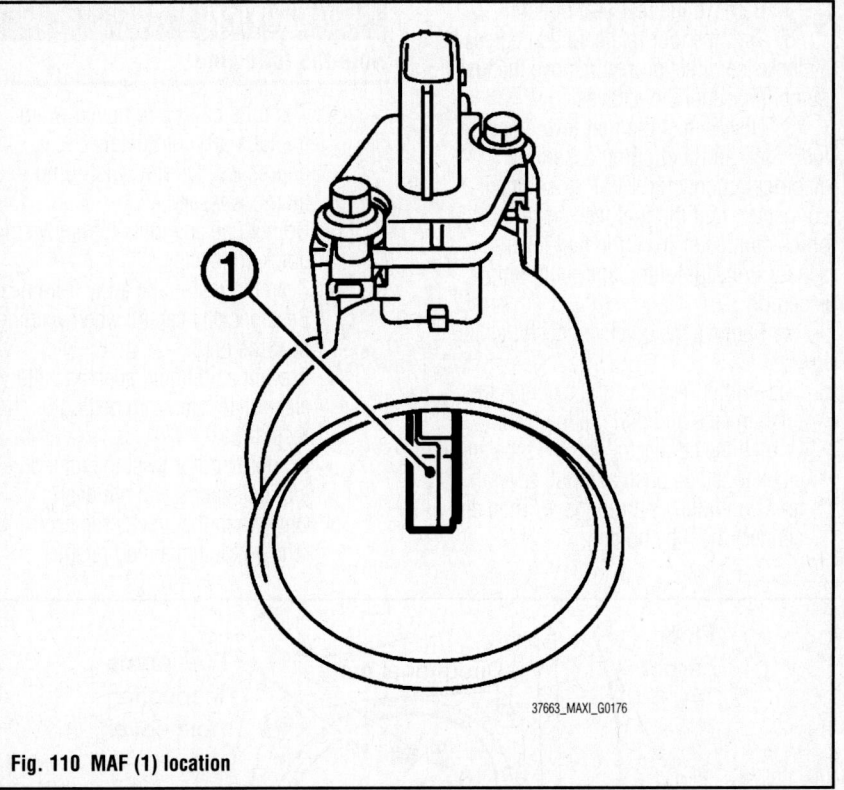

37663_MAXI_G0176

Fig. 110 MAF (1) location

FUEL   **GASOLINE FUEL INJECTION SYSTEM**

FUEL SYSTEM SERVICE PRECAUTIONS

Safety is the most important factor when performing not only fuel system maintenance, but any type of maintenance. Failure to conduct maintenance and repairs in a safe manner may result in serious personal injury or death. Work on a vehicle's fuel system components can be accomplished safely and effectively by adhering to the following rules and guidelines.

• To avoid the possibility of fire and personal injury, always disconnect the negative battery cable unless the repair or test procedure requires that battery voltage be applied.

• Always relieve the fuel system pressure prior to disconnecting any fuel system component (injector, fuel rail, pressure regulator, etc.) fitting or fuel line connection. Exercise extreme caution whenever relieving fuel system pressure to avoid exposing skin, face and eyes to fuel spray. Please be advised that fuel under pressure may penetrate the skin or any part of the body that it contacts.

• Always place a shop towel or cloth around the fitting or connection prior to loosening to absorb any excess fuel due to

spillage. Ensure that all fuel spillage is quickly removed from engine surfaces. Ensure that all fuel-soaked cloths or towels are deposited into a flame-proof waste container with a lid.

• Always keep a dry chemical (Class B) fire extinguisher near the work area.

• Do not allow fuel spray or fuel vapors to come into contact with a spark or open flame.

• Always use a second wrench when loosening or tightening fuel line connection fittings. This will prevent unnecessary stress and torsion on fuel piping. Always follow the proper torque specifications.

• Always replace worn fuel fitting O-rings with new ones. Do not substitute fuel hose where rigid pipe is installed.

FUEL SYSTEM PRESSURE

RELIEVING

The fuel pump fuse is located in the dash fuse box or in the engine compartment fuse box. Check the lid of the fuse box for exact location.

1. Remove the fuel pump fuse.
2. Start the engine.
3. Start the engine and run until the engine stalls.

4. After the engine stalls, try to restart the engine. If the engine will not start, the fuel pressure has been released.

5. Turn the ignition switch **OFF**. Reinstall the fuel pump fuse into the fuse block.

➡Do not crank the engine or turn the ignition switch ON after the fuel pump fuse has been reinstalled, or the fuel pressure will be re-established.

FUEL FILTER

REMOVAL & INSTALLATION

The fuel filter is an integral part of the fuel pump assembly. Refer to the Fuel Pump Assembly Removal and Installation when servicing this component.

FUEL PUMP ASSEMBLY

REMOVAL & INSTALLATION

See Figures 111 through 114.

1. Unscrew the fuel filler cap to release the pressure inside the fuel tank.
2. Release the fuel pressure from the fuel lines.
3. Disconnect the battery negative terminal.

4. Remove the rear seat bottom.

5. Turn the four retainers 90 degrees in a clockwise direction and remove the fuel pump inspection hole cover.

6. Disconnect the fuel level sensor, fuel filter, and fuel pump assembly electrical connector, EVAP hose quick connector, and the fuel feed hose quick connector from the fuel level sensor unit, fuel filter, and fuel pump assembly.

7. Remove the quick connector as follows:

   a. Hold the sides of the connector, push in tabs and pull out the tube.

   b. If the connector and the tube are stuck together, push and pull several times until they start to move. Then disconnect them by pulling.

✳✳ CAUTION

Note the following:

- The tube can be removed when the tabs are completely depressed. Do not twist it more than necessary.
- Do not use any tools to remove the quick connector.
- Keep the resin tube away from heat. Be especially careful when welding near the tube.
- Prevent acid liquid such as battery electrolyte, etc. from getting on the resin tube.
- Do not bend or twist the tube during installation and removal.
- Only when the tube is replaced, remove the remaining retainer on

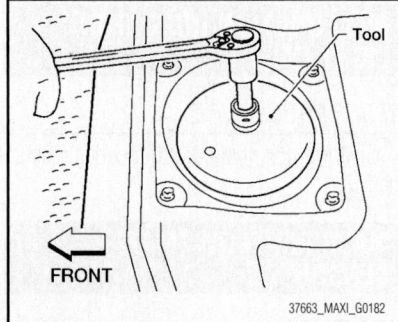

Fig. 113 Remove the lock ring using a socket drive handle and Tool KV991J0090 (J-46214) as shown

the tube or fuel level sensor, fuel filter, and fuel pump assembly.

- When the tube or fuel level sensor, fuel filter, and fuel pump assembly is replaced, replace the retainer with a new one (green colored retainer).
- To keep the connecting portion clean and to avoid damage and foreign materials, cover them completely with plastic bags or something similar.

8. Remove the lock ring using a socket drive handle and Tool KV991J0090 (J-46214) as shown.

✳✳ CAUTION

Discard the lock ring, do not reuse the lock ring. Discard the ring seal, do not reuse the ring seal.

9. Remove the fuel level sensor, fuel filter, and fuel pump assembly.

✳✳ CAUTION

Do not bend the float arm during removal. Discard the ring seal, do not reuse the ring seal.

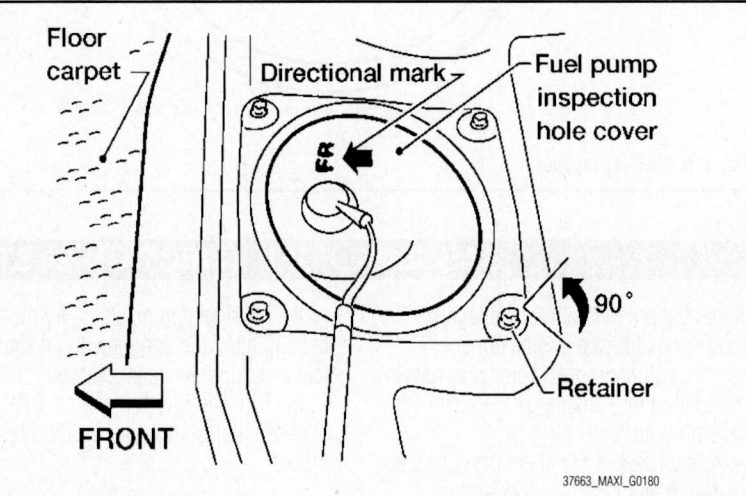

Fig. 111 Turn the four retainers 90 degrees in a clockwise direction and remove the fuel pump inspection hole cover

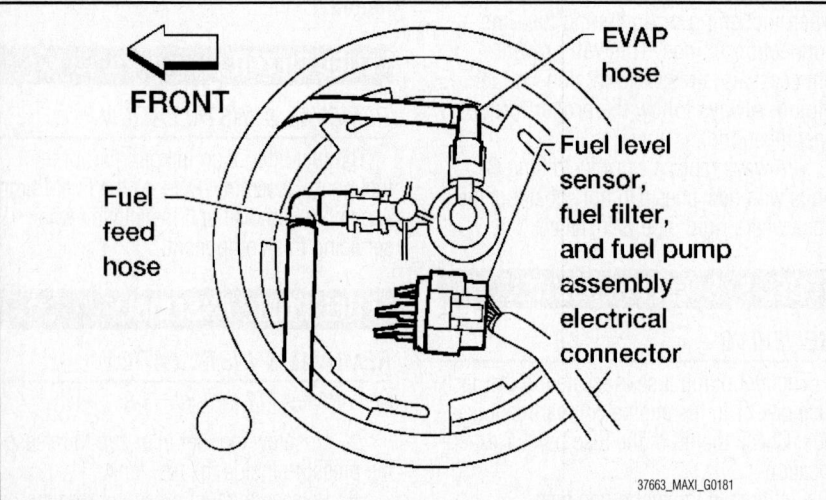

Fig. 112 Disconnect the fuel level sensor, fuel filter, and fuel pump assembly electrical connector, EVAP hose quick connector, and the fuel feed hose quick connector

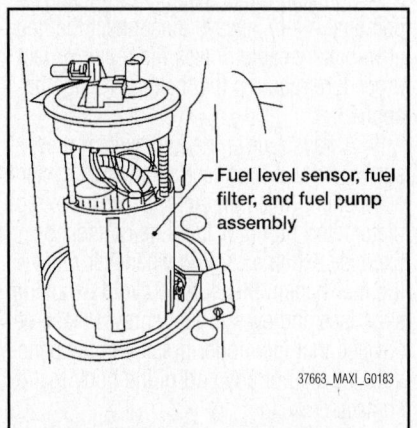

Fig. 114 Remove the fuel level sensor, fuel filter, and fuel pump assembly

10. Inspect the fuel level sensor, fuel filter, and fuel pump for any defects and foreign materials. Replace as necessary.

To install:

11. Installation is in the reverse order of removal.

12. Install the fuel level sensor, fuel filter, and fuel pump assembly with the fuel feed hose facing the front of the vehicle as shown. Use a new ring seal.

13. Connect the quick connector as follows:

a. Check the connection for damage or any foreign materials.

b. Align the connector with the tube, then insert the connector straight into the tube until a click is heard.

14. After the tube is connected, make sure the connection is secure by performing the following checks:

a. Pull the tube and the connector to make sure they are securely connected.

b. Visually confirm that the two retainer tabs are connected to the quick connector.

15. Use the following procedure to check for fuel leaks.

a. Turn the ignition switch to ON (without starting the engine) to apply fuel pressure to the fuel system, then check the connections for fuel leaks.

b. Start the engine and let it idle and check for fuel leaks at the fuel system connections.

FUEL RAIL AND INJECTOR

REMOVAL & INSTALLATION
See Figures 115 through 122.

✲✲ WARNING
Note the following:

- Be sure to work in a well ventilated area and furnish workshop with a CO2 fire extinguisher.
- Never smoke while servicing fuel system. Keep open flames and sparks away from the work area.
- To avoid the danger of being scalded, never drain engine coolant when engine is hot.

1. Remove engine cover.
2. Release the fuel pressure.
3. Remove front wiper arm and extension cowl top.
4. Remove the electric throttle control actuator bolts in the reverse order as shown and remove the electric throttle control actuator.

✲✲ CAUTION
Handle carefully to avoid any shock to the electric throttle control actuator. Do not disassemble.

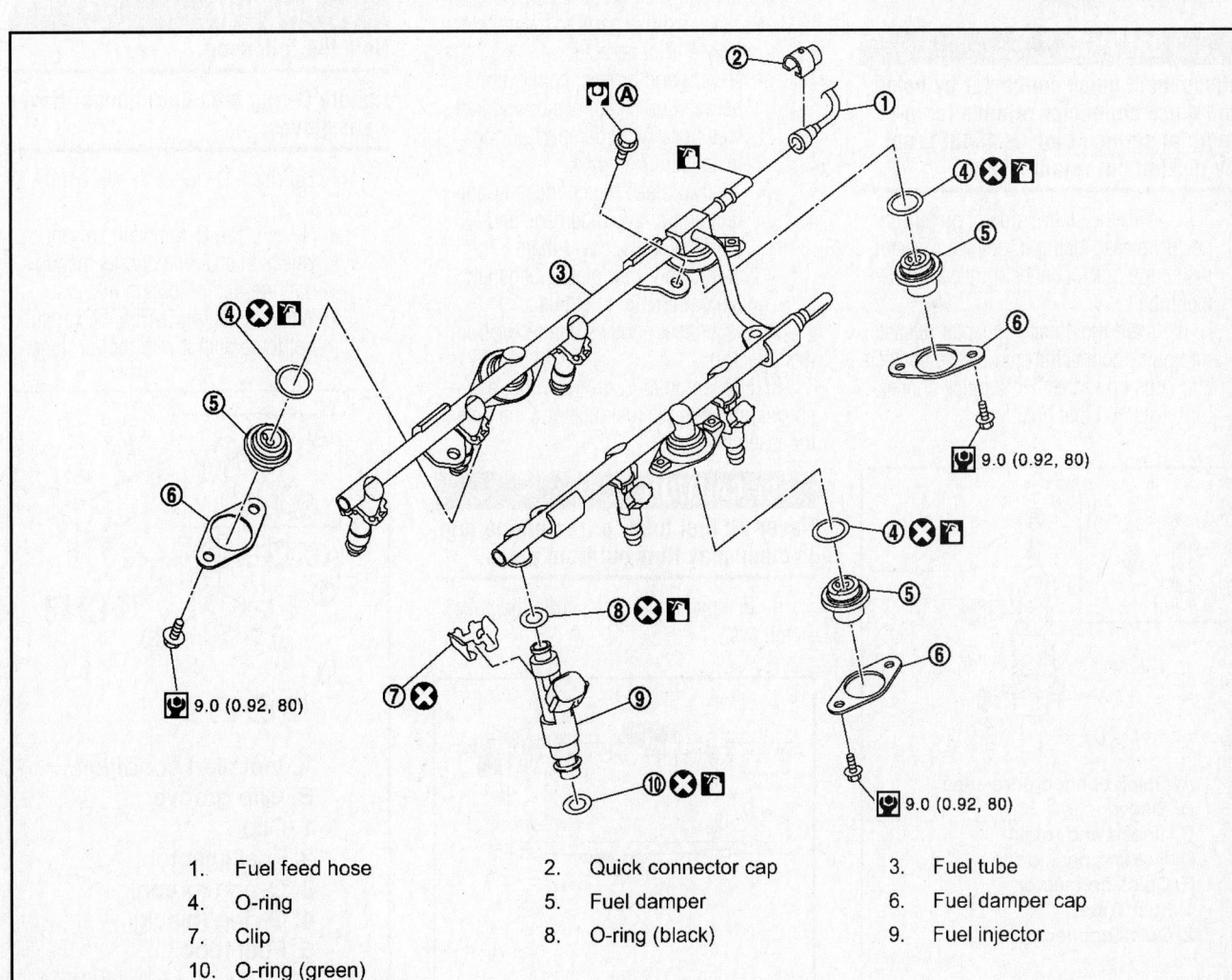

1. Fuel feed hose	2. Quick connector cap	3. Fuel tube
4. O-ring	5. Fuel damper	6. Fuel damper cap
7. Clip	8. O-ring (black)	9. Fuel injector
10. O-ring (green)		
A. Refer to installation		

37663_MAXI_G0186

Fig. 115 Exploded view of fuel injector and fuel tube assembly

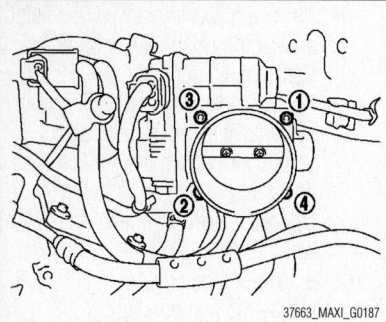

Fig. 116 Remove the electric throttle control actuator

5. Remove intake manifold collector.

6. When separating fuel feed hose and fuel tube connection, disconnect quick connector as follows:

   a. Remove quick connector cap from quick connector.

   b. Disconnect quick connector from fuel tube as follows:

✳ CAUTION

Disconnect quick connector by using the quick connector release (commercial service tool: J- 45488), not by picking out retainer tabs.

   c. With the sleeve side of quick connector release facing to quick connector, install the quick connector release onto fuel tube.

   d. Insert the quick connector release into quick connector until sleeve contacts and goes no further. Hold quick connector release on that position.

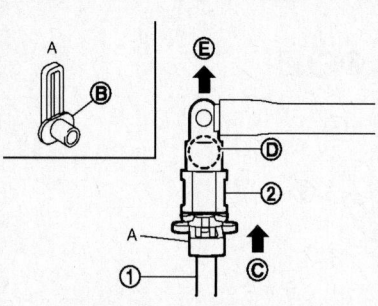

A. Quick connector release
B. Sleeve
C. "Insert and retain"
D. Holding position
E. Quick connector
1. Fuel Tube
2. Quick connector

Fig. 117 Insert the quick connector release into quick connector until sleeve contacts and goes no further

✳ CAUTION

Inserting quick connector release hard will not disconnect quick connector. Hold quick connector release where it contacts and goes no further.

   e. Draw and pull out quick connector straight from fuel tube.

✳ CAUTION

Note the following:

- Pull quick connector holding position as shown in the figure.
- Never pull with lateral force applied. O-ring inside quick connector may be damaged.
- Prepare container and cloth beforehand as fuel will leakage out.
- Avoid fire and sparks.
- Keep parts away from heat source. Especially, be careful when welding is performed around them.
- Never expose parts to battery electrolyte or other acids.
- Never bend or twist connection between quick connector and fuel feed hose (with damper) during installation/removal.
- To keep clean the connecting portion and to avoid damage and foreign materials, cover them completely with plastic bags, etc. (A) or something similar.

7. Disconnect harness connector from fuel injector.

8. Loosen bolts in reverse order as shown, and remove fuel tube and fuel injector assembly.

✳ CAUTION

Never tilt fuel tube, or remaining fuel in pipes may flow out from pipes.

9. Remove fuel injector from fuel tube as follows:

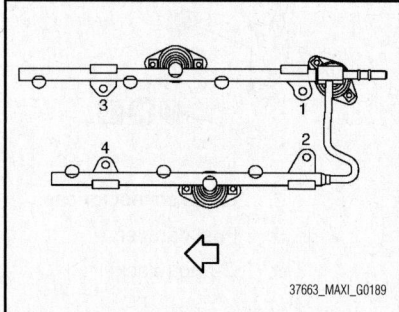

Fig. 118 Loosen bolts in reverse order as shown, and remove fuel tube and fuel injector assembly

   a. Open and remove clip.

   b. Remove fuel injector from fuel tube by pulling straight.

✳ CAUTION

Note the following:

- Be careful with remaining fuel that may go out from fuel tube.
- Be careful not to damage injector nozzle during removal.
- Never bump or drop fuel injector.
- Never disassemble fuel injector.

10. Remove fuel damper from fuel tube.

To install:

11. Install fuel damper as follows:

   a. Install new O-ring to fuel tube as shown. When handling new O-ring, be careful of the following caution:

✳ CAUTION

Note the following:

Handle O-ring with bare hands. Never wear gloves.

- Lubricate O-ring with new engine oil.
- Never clean O-ring with solvent.
- Check that O-ring and its mating part are free of foreign material.
- When installing O-ring, be careful not to scratch it with tool or finger-

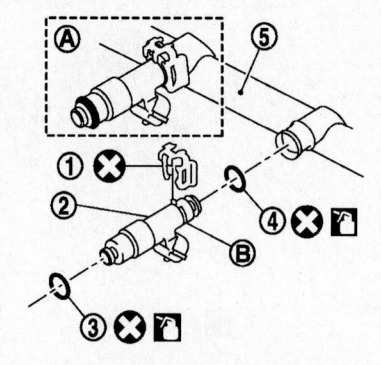

A. Installed condition
B. Clip groove
1. Clip
2. Fuel injector
3. O-ring (green)
4. O-ring (black)
5. Fuel tube

Fig. 119 Remove fuel injector from fuel tube

nails. Also be careful not to twist or stretch O-ring. If O-ring was stretched while it was being attached, never insert it quickly into fuel tube.

- Insert new O-ring straight into fuel tube. Never twist it.

b. Install spacer to fuel damper.

c. Insert fuel damper straight into fuel tube.

�֍✖ CAUTION

Note the following:

- Insert straight, checking that the axis is lined up.
- Never pressure-fit with excessive force.
- Insert fuel damper until is touching of fuel tube.

d. Tighten bolts evenly in turn.

e. After tightening bolts, check that there is no gap between fuel damper cap (5) and fuel tube.

12. Install new O-rings to fuel injector paying attention to the following.

✖✖ CAUTION

Note the following:

- Upper and lower O-ring are different. Be careful not to confuse them: Fuel tube side BLACK; Nozzle side GREEN.

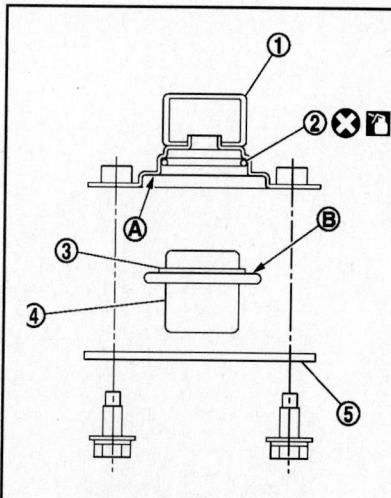

A. Touching fuel tube
B. Fuel damper
1. Fuel tube
2. O-ring
3. Spacer
4. Fuel damper
5. Fuel damper cap

37663_MAXI_G0191

Fig. 120 Install fuel damper

- Handle O-ring with bare hands. Never wear gloves.
- Lubricate O-ring with new engine oil.
- Never clean O-ring with solvent.
- Check that O-ring and its mating part are free of foreign material.
- When installing O-ring, be careful not to scratch it with tool or fingernails. Also be careful not to twist or stretch O-ring. If O-ring was stretched while it was being attached, never insert it quickly into fuel tube.
- Insert O-ring straight into fuel injector. Never decenter or twist it.

13. Install fuel injector to fuel tube as follows:

a. Insert clip into clip groove on fuel injector.

b. Insert clip so that protrusion of fuel injector matches cutout of clip.

✖✖ CAUTION

Never reuse clip. Replace it with new one. Be careful to keep clip from interfering with O-ring. If interference occurs, replace O-ring.

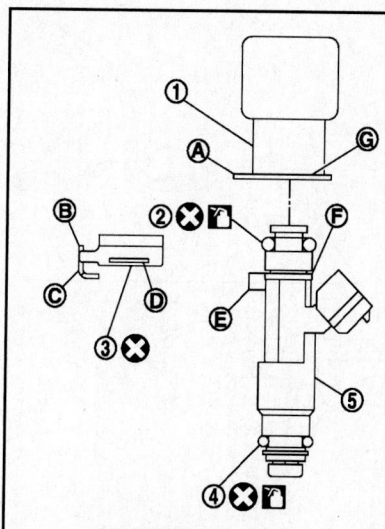

A. Protrusion
B. Cutout
C. Cutout
D. Flange fixing groove
E. Protrusion
F. Clip groove
G. Fuel tube flange
1. Fuel tube
2. O-ring (black)
3. Clip
4. O-ring (green)
5. Fuel injector

37663_MAXI_G0192

Fig. 121 Install fuel injector to fuel tube

c. Insert fuel injector into fuel tube with clip attached.

d. Insert it while matching it to the axial center.

e. Insert fuel injector so that protrusion of fuel tube matches cutout of clip.

f. Check that fuel tube flange is securely fixed in flange fixing groove on clip.

g. Check that installation is complete by checking that fuel injector does not rotate or come off.

h. Check that protrusions of fuel injectors and fuel tubes are aligned with cutouts of clips after installation.

14. Install fuel tube and fuel injector assembly to intake manifold.

✖✖ CAUTION

Be careful not to let tip of injector nozzle come in contact with other parts.

15. Tighten bolts in two steps in numerical order as shown in the figure.

a. Step 1: 7 ft. lbs. (10 Nm)
b. Step 2: 16 ft. lbs. (22 Nm)

16. Connect fuel injector harness.

17. Install intake manifold collector.

18. Connect quick connector between fuel feed hose and fuel tube connection with the following procedure:

a. Check no foreign substances are deposited in and around fuel tube and quick connector, and no damage on them.

b. Thinly apply new engine oil around fuel tube from tip end to spool end.

c. Align center to insert quick connector straightly into fuel tube.

d. Insert quick connector to fuel tube until top spool is completely inside quick connector, and 2nd level spool exposes right below quick connector.

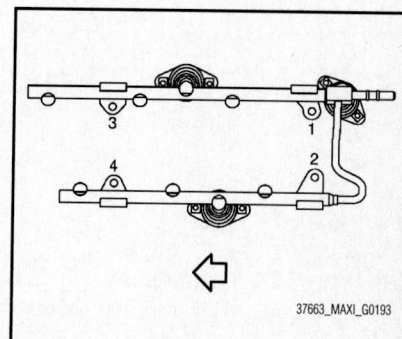

37663_MAXI_G0193

Fig. 122 Tighten bolts in two steps in numerical order

✳✳✳ CAUTION

Note the following:

- Hold position when inserting fuel tube into quick connector.
- Carefully align center to avoid inclined insertion to prevent damage to O-ring inside quick connector.
- Insert until you hear a "click" sound and actually feel the engagement.
- To avoid misidentification of engagement with a similar sound, be sure to perform the next step.

e. Pull quick connector by hand holding position. Check it is completely engaged (connected) so that it does not come out from fuel tube.

f. Install quick connector cap to quick connector.

g. Install quick connector cap with arrow on surface facing in direction of quick connector (fuel feed hose side).

✳✳✳ CAUTION

If quick connector cap cannot be installed smoothly, quick connector may have not been installed correctly. Check connection again.

h. Secure fuel feed hose to clamp of quick connector cap.

19. Installation is in the reverse order of removal.

20. Turn ignition switch "ON" (with the engine stopped). With fuel pressure applied to fuel piping, check there are no fuel leakage at connection points.

➡**Use mirrors for checking at points out of clear sight.**

21. Start the engine. With engine speed increased, check again that there are no fuel leakage at connection points.

✳✳✳ CAUTION

Never touch the engine immediately after stopped, as the engine becomes extremely hot.

FUEL TANK

DRAINING

1. Insert fuel tubing of less than 1 inch (25 mm) diameter into the fuel filler tube through the fuel filler opening to drain fuel from the fuel filler tube.

2. Disconnect the fuel filler hose from the fuel filler tube.

3. Insert fuel tubing into the fuel tank through the fuel filler hose to drain fuel from the fuel tank.

➡**As a guide, the fuel level reaches or is less than the level on the fuel gauge as shown, when approximately 2 ⅝ US gal (10 L) of fuel is drained from a full fuel tank.**

REMOVAL & INSTALLATION

See Figures 123 through 126.

1. Disconnect the battery negative terminal.

2. Open the fuel filler cap to release the pressure inside the fuel tank.

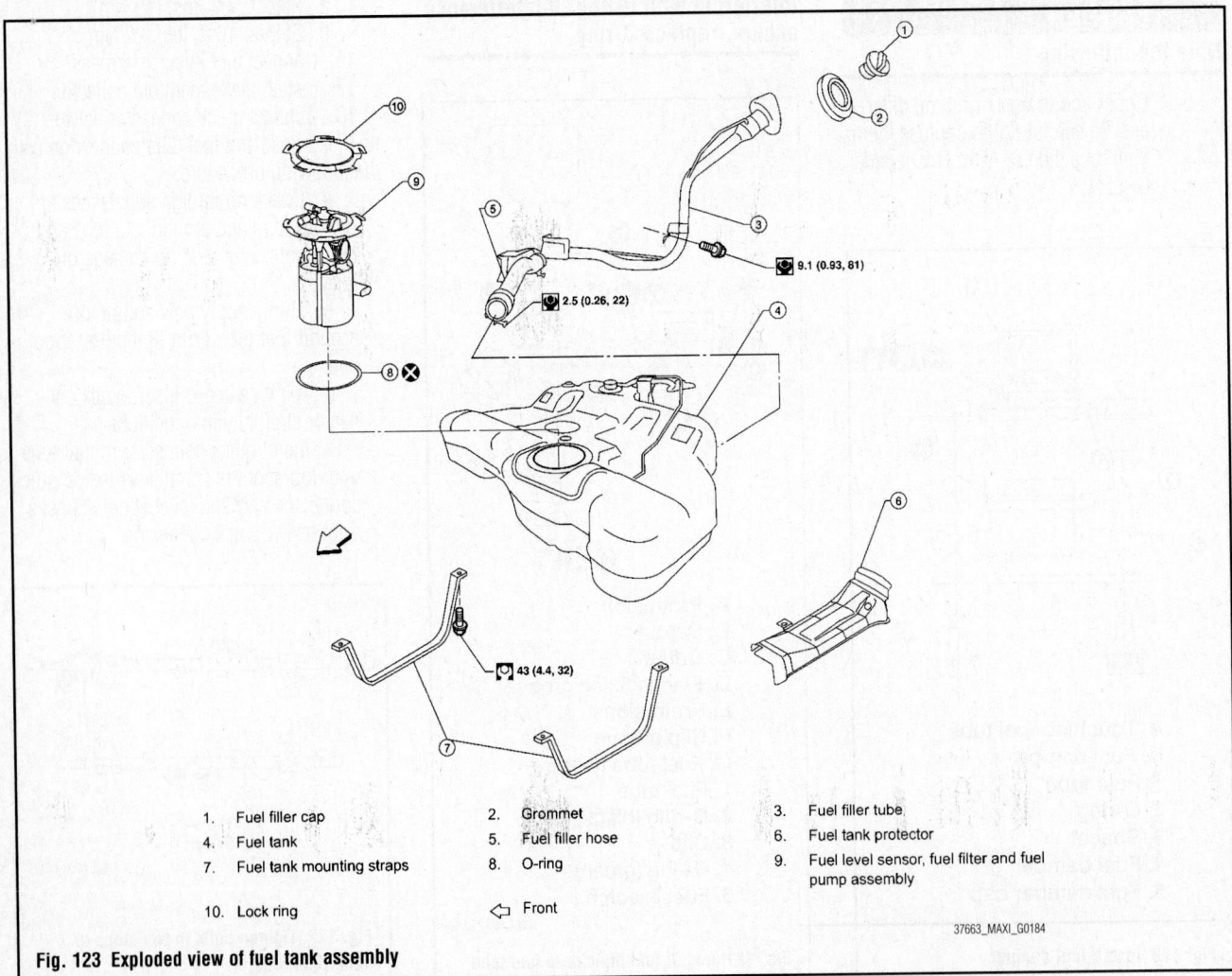

1. Fuel filler cap	2. Grommet	3. Fuel filler tube
4. Fuel tank	5. Fuel filler hose	6. Fuel tank protector
7. Fuel tank mounting straps	8. O-ring	9. Fuel level sensor, fuel filter and fuel pump assembly
10. Lock ring	⟨⟩ Front	

37663_MAXI_G0184

Fig. 123 Exploded view of fuel tank assembly

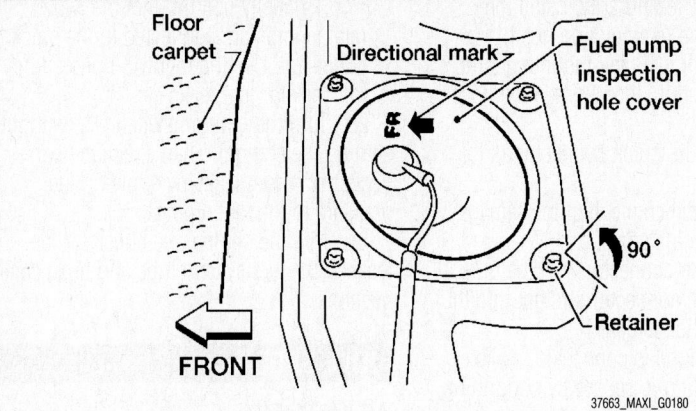

37663_MAXI_G0180

Fig. 124 Turn the four retainers 90 degrees in a clockwise direction and remove the fuel pump inspection hole cover

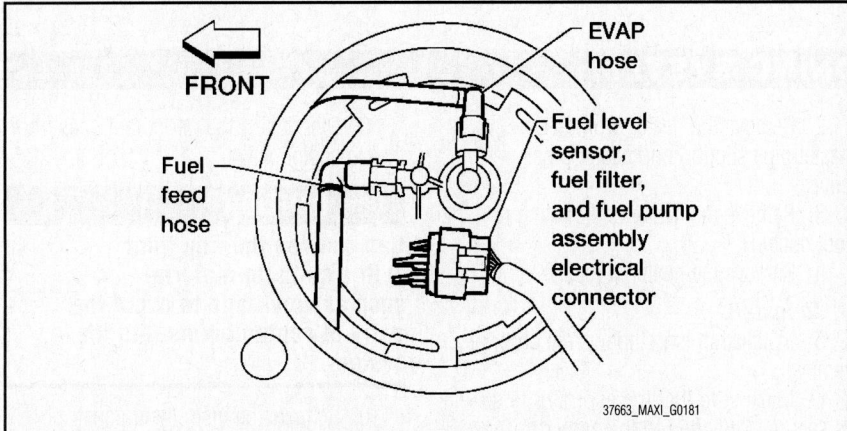

37663_MAXI_G0181

Fig. 125 Disconnect the fuel level sensor, fuel filter, and fuel pump assembly electrical connector, EVAP hose quick connector, and the fuel feed hose quick connector

3. Release fuel pressure from fuel line.

4. Check the fuel level with the vehicle on a level surface. If the fuel gauge indicates more than the level 7/8 full, drain the fuel from the fuel tank until the fuel gauge indicates a level at or below 7/8 full.

5. Remove rear seat bottom.

6. Turn the four retainers 90 degrees in a clockwise direction and remove the fuel pump inspection hole cover.

7. Disconnect the fuel level sensor, fuel filter, and fuel pump assembly electrical connector, EVAP hose quick connector, and fuel feed hose quick connector.

8. Disconnect the quick connectors as follows:

   a. Hold the sides of the connector, push in tabs and pull out the tube.

   b. If the connector and the tube are stuck together, push and pull several times until they start to move. Then disconnect them by pulling.

※※ **CAUTION**

Note the following:

- The tube can be removed when the tabs are completely depressed. Do not twist it more than necessary.
- Do not use any tools to remove the quick connector.
- Keep the resin tube away from heat. Be especially careful when welding near the tube.
- Prevent acid liquid such as battery electrolyte, from getting on the resin tube.
- Do not bend or twist the tube during installation and removal.
- Only when the tube is replaced, remove the remaining retainer on the tube or fuel level sensor, fuel filter, and fuel pump assembly.
- When the tube or fuel level sensor, fuel filter, and fuel pump assembly is replaced, also replace the retainer with a new one (green colored retainer).
- To keep the connecting portion clean and to avoid damage and foreign materials, cover them completely with plastic bags or something similar.

9. Remove the center exhaust tube, with muffler.

10. Disconnect the fuel filler hose and recirculation hose at the fuel tank side.

11. Disconnect the three parking brake cable mounting brackets on each cable and position the cables out of the way.

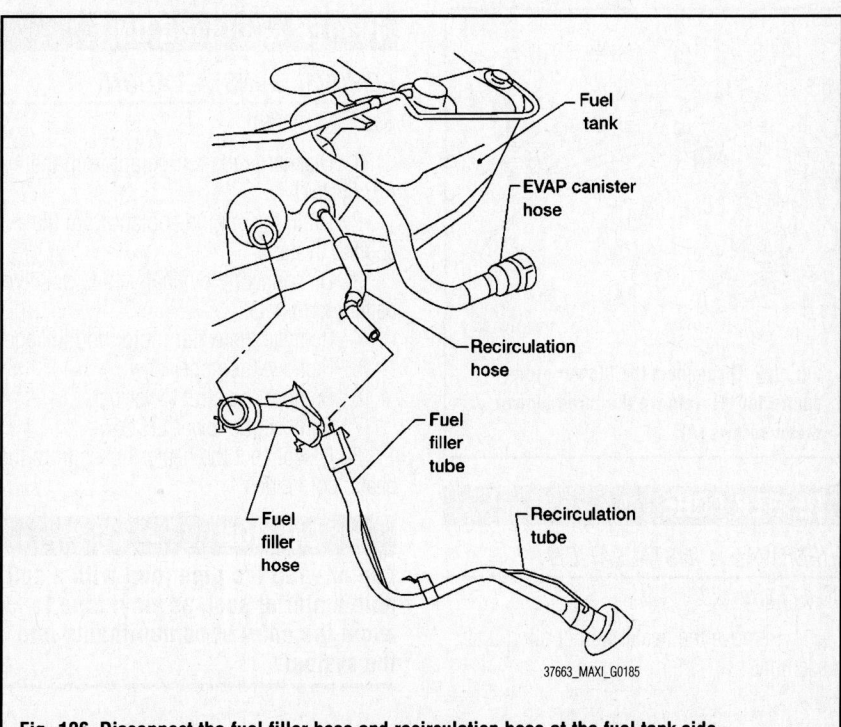

37663_MAXI_G0185

Fig. 126 Disconnect the fuel filler hose and recirculation hose at the fuel tank side

12. Remove rear stabilizer bar clamps, then allow stabilizer bar to hang.

13. Remove EVAP canister bolts. Then without disconnecting hoses, position EVAP canister aside.

14. Remove the fuel tank protector.

15. Remove the fuel tank mounting strap bolts and mounting straps while supporting the fuel tank with a suitable jack.

16. Remove the fuel tank.

17. If replacing the fuel tank, remove the fuel level sensor unit, fuel filter and fuel pump assembly to transfer to the new fuel tank.

To install:

18. Install in the reverse order of removal paying attention to the following.

19. Before tightening the fuel tank mounting straps, temporarily install the filler hose and the recirculation hose.

20. Tighten all fuel tank mounting strap bolts to specification, then tighten the hose clamps.

21. Connect the quick connector as follows:

a. Check the connection for damage or any foreign materials.

b. Align the connector with the tube, then insert the connector straight into the tube until a click is heard.

22. After the tube is connected, make sure the connection is secure by performing the following checks:

a. Pull on the tube and the connector to make sure they are securely connected.

b. Visually confirm that the two retainer tabs are connected to the quick connector. Use the following procedure to check for fuel leaks.

23. Turn the ignition switch ON (without starting the engine). Then check the connections for fuel leaks by applying fuel pressure to the fuel piping.

24. Run the engine and check for fuel leaks at the fuel system tube and hose connections.

IDLE SPEED

ADJUSTMENT

The idle speed is controlled by the ECM. No adjustment is necessary or possible.

HEATING & AIR CONDITIONING SYSTEM

BLOWER MOTOR

REMOVAL & INSTALLATION
See Figure 127.

1. Remove the glove box.
2. Disconnect the blower motor connector.
3. Remove the three blower motor screws and remove the blower motor from the blower unit.

To install:

4. Installation is in the reverse order of removal.

37663_MAXI_G0207

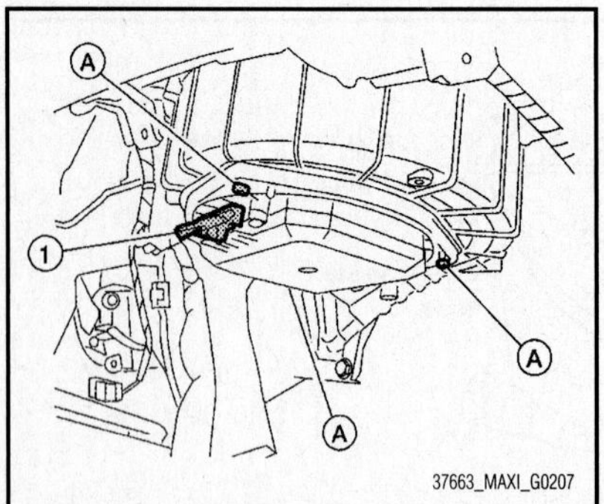

Fig. 127 Disconnect the blower motor connector (1); remove the three blower motor screws (A)

HEATER CORE

REMOVAL & INSTALLATION
See Figure 128.

1. Remove the heater and cooling unit assembly.

2. Remove the heater grommet, heater pipe support and heater pipe cover.

3. Remove the heater and cooling unit foot duct LH.

4. Remove the heater core.

To install:

5. Installation is in the reverse order of removal.

6. Make sure that the aspirator hose is securely attached to the aspirator on the heater and cooling unit foot duct LH.

HEATER &COOLING UNIT

REMOVAL & INSTALLATION
See Figure 129.

1. Discharge the refrigerant from the A/C system.

2. Drain the engine coolant from the cooling system.

3. Disconnect the negative and positive battery terminals.

4. Remove the wiper motor and linkage.

5. Remove the upper cowl.

6. Remove the strut tower bar.

7. Remove the lower LH cowl.

8. Disconnect the heater hoses from the heater core pipes.

✳✳ CAUTION

Cap or wrap the pipe joint with a suitable material such as vinyl tape to avoid the entry of contaminants into the system.

9. Disconnect the refrigerant lines from the expansion valve.

✳✳ CAUTION

Cap or wrap the line joint with a suitable material such as vinyl tape to avoid the entry of contaminants into the system.

10. Remove the instrument panel assembly.

11. Remove the steering column assembly.

12. Disconnect the drain hose.

13. Remove the LH, RH and center connector ducts.

14. Remove the steering member center stay.

15. Remove the heater and cooling unit assembly attached to the steering member as one assembly from the vehicle.

16. Remove the blower unit from the heater and cooling unit and steering member assembly.

17. Remove the heater and cooling unit from the steering member.

To install:

18. Installation is in the reverse order of removal.

19. Fill the radiator with the specified water and coolant mixture.

20. Recharge the A/C system.

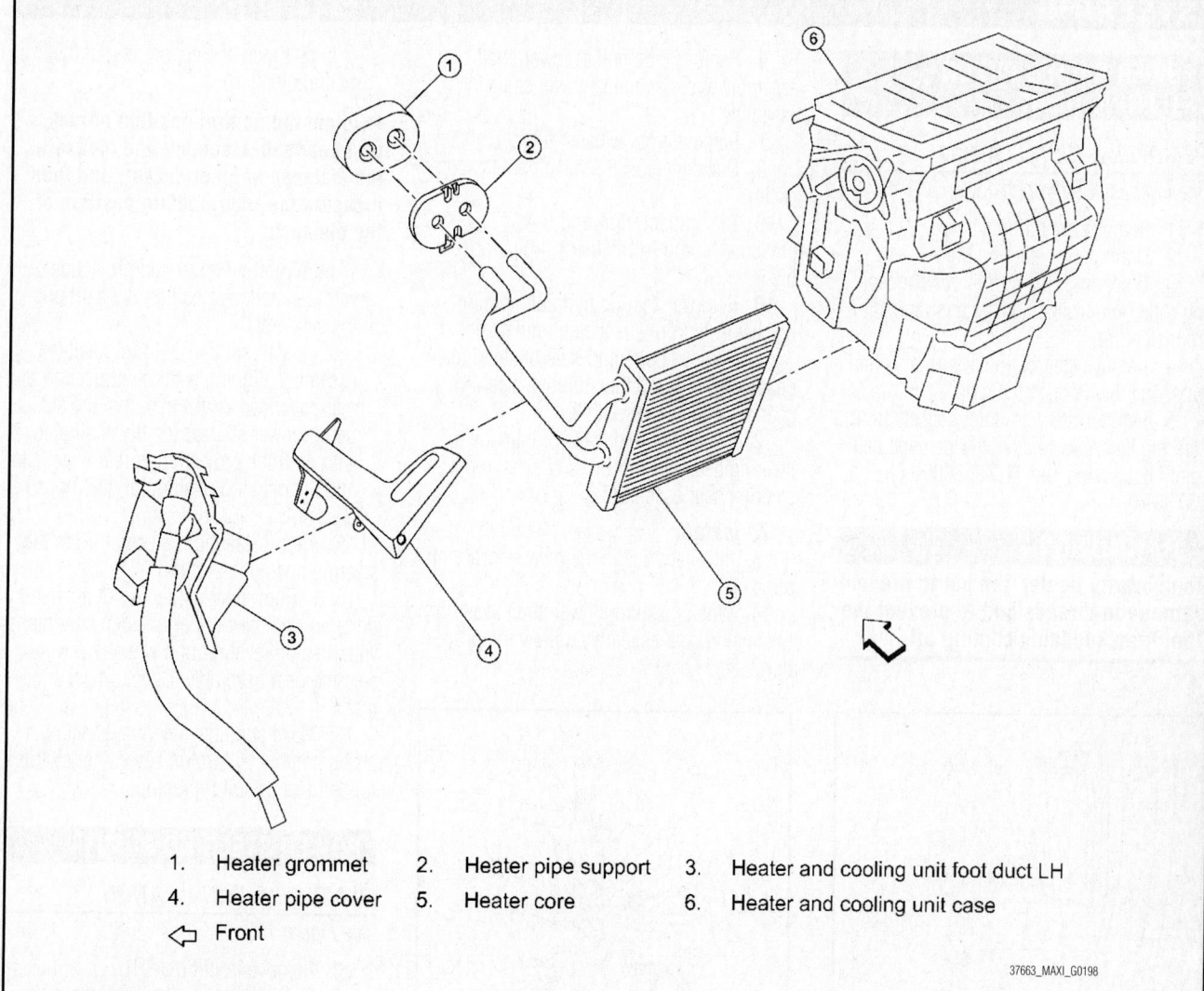

1. Heater grommet
2. Heater pipe support
3. Heater and cooling unit foot duct LH
4. Heater pipe cover
5. Heater core
6. Heater and cooling unit case
← Front

37663_MAXI_G0198

Fig. 128 Remove the heater core

1. Steering member
2. Heater and cooling unit
3. Blower unit

37663_MAXI_G0197

Fig. 129 Heater and cooling unit location

STEERING

POWER RACK & PINION STEERING GEAR

REMOVAL & INSTALLATION

See Figures 130 through 132.

1. Remove front tires.
2. Drain power steering fluid.
3. Disconnect front stabilizer connecting rods from front stabilizer and reposition front stabilizer.
4. Remove steering outer socket cotter pins, and then loosen the nuts.
5. Remove steering outer sockets from steering knuckles so as not to damage ball joint boots using Tool HT72520000 (J-25730-A).

✷✷ CAUTION

Temporarily tighten the nut to prevent damage to threads and to prevent the Tool from suddenly coming off.

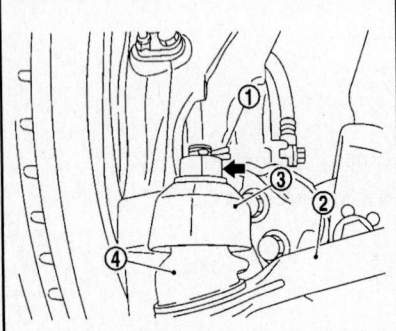

1. Steering outer socket cotter pins
2. Steering outer sockets
3. Steering knuckles
4. Ball joint boots

37663_MAXI_G0209

Fig. 130 Remove steering outer sockets from steering knuckles

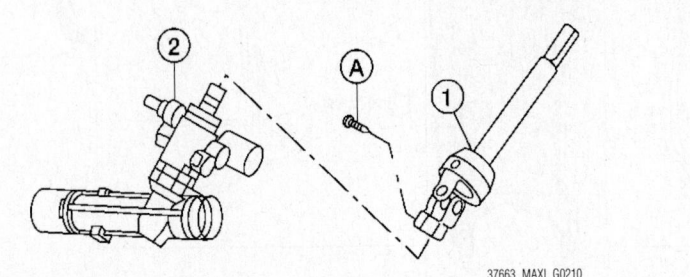

37663_MAXI_G0210

Fig. 131 Remove side bolt (A) of lower shaft assembly (1) and disconnect lower shaft assembly (2)

6. Remove side bolt of lower shaft assembly and disconnect lower shaft assembly.
7. Remove front exhaust tube.
8. Disconnect SSPS valve harness connector.
9. Disconnect high and low pressure piping from steering gear assembly.
10. Remove steering hydraulic piping bracket from front suspension member.
11. Remove bolts and nuts of steering gear assembly, and then remove steering gear assembly from vehicle.
12. Check for fluid leaks or damage to steering gear assembly. If any exist, replace steering gear assembly.

To install:

13. Installation is in the reverse order of removal.
14. When installing lower shaft assembly to steering gear assembly, follow the procedure listed below.

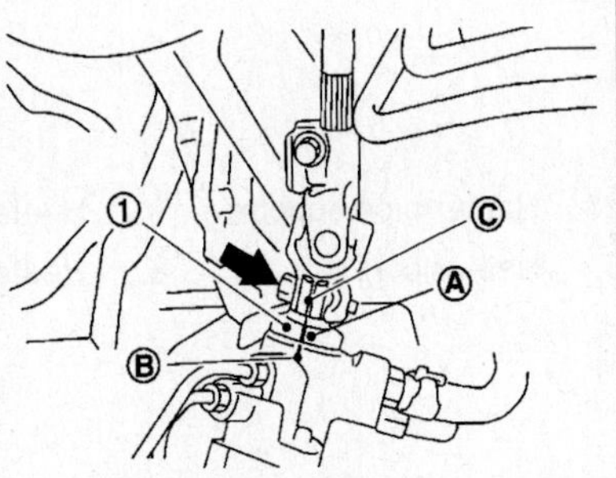

A. Rear cover cap projection
B. Marking position
C. Lower shaft assembly
1. Rear cover cap

37663_MAXI_G0211

Fig. 132 Align rear cover cap projection with the marking position of gear housing assembly

a. Set rack of steering gear in the neutral position.

➡**To get the neutral position of rack, turn gear sub-assembly and measure the distance of inner socket, and then measure the intermediate position of the distance.**

b. Align rear cover cap projection with the marking position of gear housing assembly.
c. Install slit part of lower shaft assembly aligning with the projection of rear cover cap. Make sure that the slit part of lower shaft assembly is aligned with both the projection of rear cover cap and the marking position of gear housing assembly.

15. After installation, bleed air from the steering hydraulic system.
16. Perform final tightening of nuts and bolts on each part under unladen conditions with tires on level ground when removing steering gear assembly. Check wheel alignment.
17. Make sure that steering wheel operates smoothly by turning several times from full left stop to full right stop.

POWER STEERING PUMP

REMOVAL & INSTALLATION

See Figure 133.

1. Remove front tire (RH).
2. Remove engine side undercover.
3. Remove hood ledge cover (RH).
4. Drain power steering fluid.
5. Disconnect high pressure piping and suction hose from power steering oil pump.
6. Loosen drive belt.
7. Remove drive belt from power steering oil pump pulley.
8. Remove power steering oil pump bolts, and then remove power steering oil pump.

To install:

9. Installation is in the reverse order of removal.
10. When installing power steering oil pump, install all bolts by hand initially, then tighten bolts to specification.
11. Perform the following procedures after installing.

a. Check belt tension.
b. Bleed air from power steering system.

AIR BLEEDING

➡**If air bleeding is not complete, the following symptoms can be observed:**

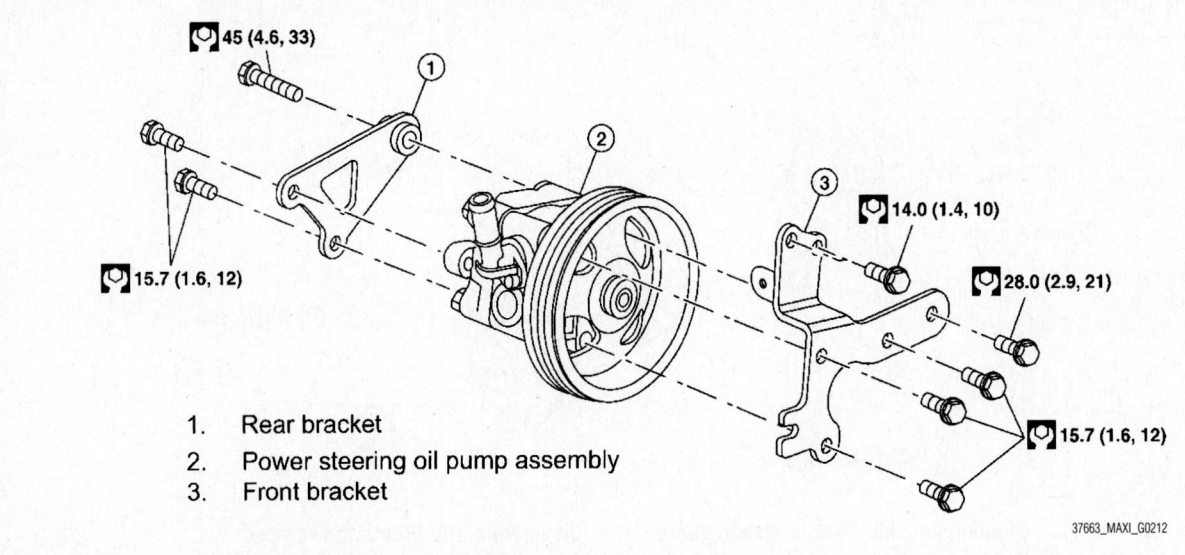

1. Rear bracket
2. Power steering oil pump assembly
3. Front bracket

37663_MAXI_G0212

Fig. 133 Exploded view of power steering oil pump assembly

- Bubbles are created in reservoir tank.
- Clicking noise can be heard from oil pump.
- Excessive buzzing in the oil pump.

➡Fluid noise may occur in the steering gear or oil pump. This does not affect performance or durability of the system.

1. Turn steering wheel several times from full left stop to full right stop with engine off.

⁂ CAUTION
Turn steering wheel while filling reservoir tank with fluid so as not to lower fluid level below the MIN line.

2. Start engine and hold steering wheel at each lock position for 3 seconds at idle to check for fluid leakage.
3. Repeat step 2 above several times at approximately 3 second intervals.

⁂ CAUTION
Do not hold the steering wheel in a locked position for more

than 10 seconds. (There is the possibility that oil pump may be damaged.)

4. Check fluid for bubbles and white contamination.
5. Stop engine if bubbles and white contamination do not drain out. Perform step 2 and 3 above after waiting until bubbles and white contamination drain out.
6. Stop the engine, and then check fluid level.

SUSPENSION   FRONT SUSPENSION

LOWER CONTROL ARM (TRANSVERSE LINK)

REMOVAL & INSTALLATION

1. Remove front tire and wheel.
2. Remove steering knuckle from transverse link using Tool HT7252000 (J-25730-B).
3. Remove mounting nuts and washers on lower portion of stabilizer connecting rod.
4. Slightly loosen transverse link mounting bolts.
5. Remove transverse link bolts and nuts, and remove transverse link from suspension member.

To install:
6. Installation is in the reverse order of removal.
7. Tighten transverse link bolts with vehicle unladen and all four tires on flat, level ground.

STEERING KNUCKLE

REMOVAL & INSTALLATION
See Figure 134.

1. Remove front wheel hub and bearing assembly.
2. Remove steering linkage from steering knuckle.
3. Remove steering knuckle lower pinch bolt.
4. Remove steering knuckle to strut bolts, then remove steering knuckle.

To install:
5. Installation is in the reverse order of removal.

⁂ CAUTION
Do not reuse non-reusable parts.

STRUT

REMOVAL & INSTALLATION
See Figure 135.

1. Remove tire using power tool.
2. Remove brake caliper and reposition aside using wire.

⁂ CAUTION
Avoid depressing brake pedal with brake caliper removed.

3. Remove wheel sensor electrical harness from strut.
4. Remove brake hose lock plate.
5. Remove steering knuckle to strut bolts and nuts.
6. Remove bolt on strut tower bar, then bolts on strut tower and remove strut from vehicle.

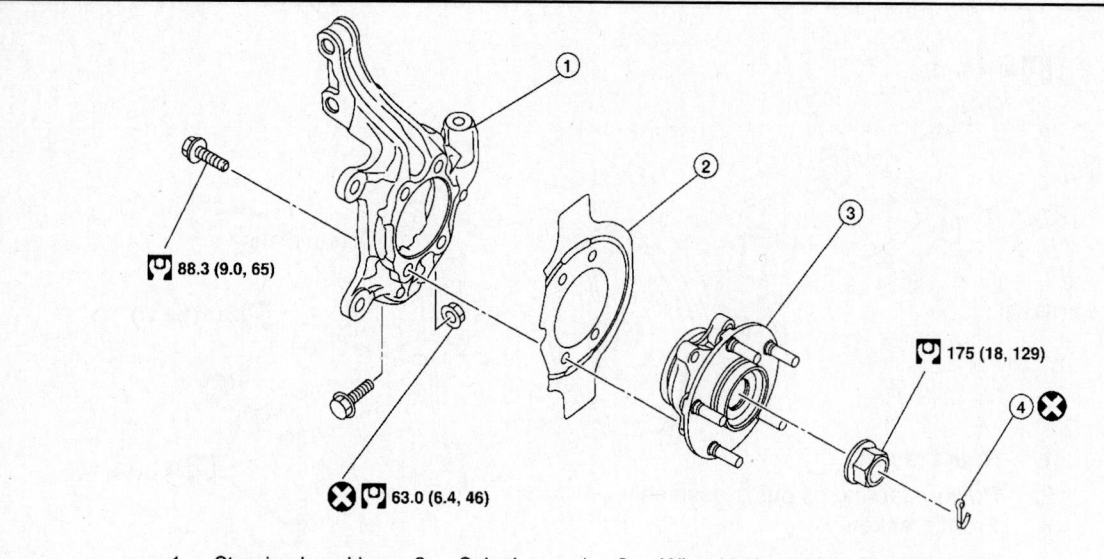

88.3 (9.0, 65)

63.0 (6.4, 46)

175 (18, 129)

1. Steering knuckle
2. Splash guard
3. Wheel hub and bearing assembly
4. Cotter pin

37663_MAXI_G0221

Fig. 134 Exploded view of the steering knuckle assembly

7. Check the strut for any oil leakage or other damage and replace as necessary.

To install:

8. Installation is in the reverse order of removal.

9. Be sure tab on strut mount insulator is positioned as shown.

OVERHAUL

Disassembly

See Figure 136.

1. Install Tool ST35652000 to strut and secure it in a vise.

✳ CAUTION

When installing Tool, wrap a shop cloth around strut to protect it from damage.

2. Slightly loosen piston rod lock nut.

✳ WARNING

Do not remove piston rod lock nut completely. If it is removed com-

pletely, the coil spring can jump out and may cause serious damage or injury.

3. Compress coil spring using a commercially available spring compressor.

✳ WARNING

Make sure that the pawls of the two spring compressors are firmly hooked on the spring. The spring compressors must be tightened alternately so as not to tilt the spring.

4. Making sure coil spring is free between upper and lower seats, then remove piston rod lock nut.

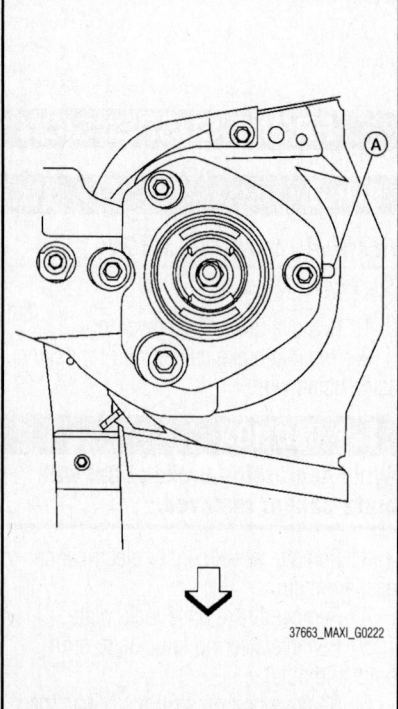

37663_MAXI_G0222

Fig. 135 Be sure tab (A) on strut mount insulator is positioned as shown

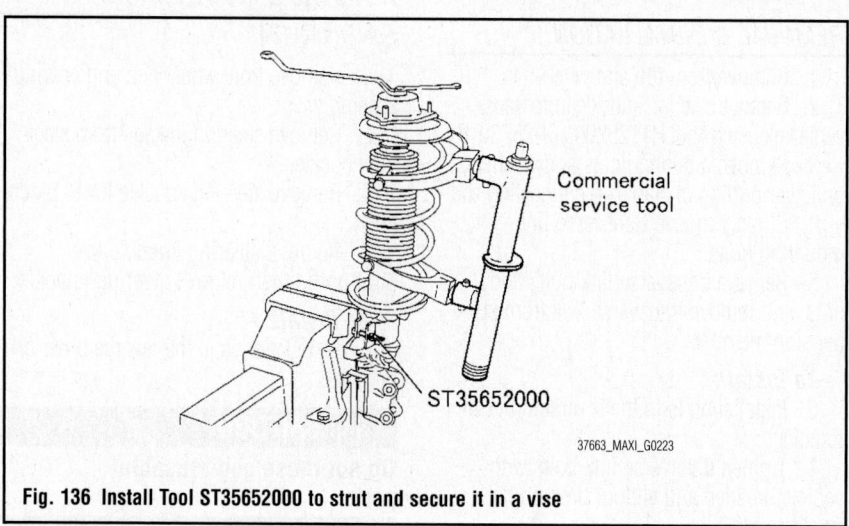

Commercial service tool

ST35652000

37663_MAXI_G0223

Fig. 136 Install Tool ST35652000 to strut and secure it in a vise

5. Remove small parts on strut; strut mount insulator, spring upper seat/strut bearing. Then remove coil spring.

6. Remove dust cover/jounce bumper from strut mount insulator.

7. Gradually release spring compressor (commercial service tool), and remove coil spring.

Assembly

See Figures 137 and 138.

1. Compress coil spring using a spring compressor (commercial service tool), and install it onto the strut.

> ❊❊ **CAUTION**
> Face tube side of coil spring downward. Align lower end to spring seat as shown.

> ❊❊ **WARNING**
> Be sure spring compressor is securely attached to coil spring. Compress coil spring.

2. Install dust cover/jounce bumper to strut mount insulator.

> ❊❊ **CAUTION**
> Be sure to install dust cover/jounce bumper to strut mount insulator securely. When installing dust cover/jounce bumper, use soapy water. Do not use machine oil or other lubricants.

3. Install small parts to the strut; spring upper seat/strut bearing and strut mount insulator. Temporarily install piston rod lock nut.

> ❊❊ **CAUTION**
> Do not reuse piston rod lock nut.

4. Be sure tab on strut mount insulator is positioned as shown.

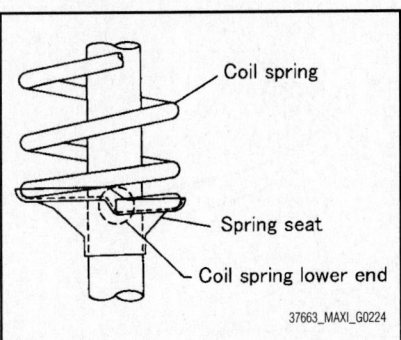

Fig. 137 Align lower end to spring seat as shown

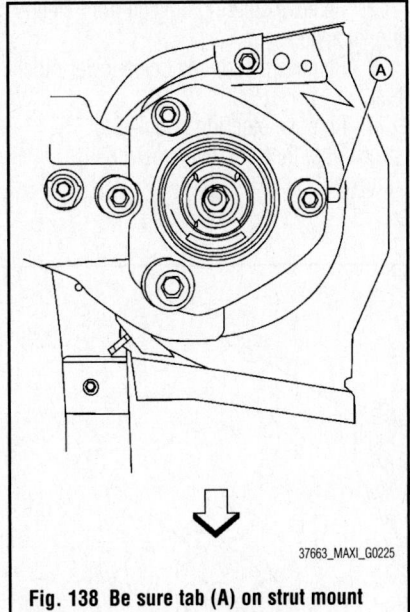

Fig. 138 Be sure tab (A) on strut mount insulator is positioned as shown

5. Be sure coil spring is properly set in spring rubber seat. Gradually release spring compressor.

> ❊❊ **CAUTION**
> Be sure upper rubber seat is properly aligned to spring upper seat and coil spring.

6. Tighten piston rod lock nut to the specified torque.

7. Remove Tool from strut.

Inspection After Disassembly

Strut

1. Check strut for deformation, cracks, and damage, and replace if necessary.

2. Check piston rod for damage, uneven wear, and distortion, and replace if necessary.

3. Check welded and sealed areas for oil leakage, and replace if necessary.

Insulator And Rubber Parts

1. Check strut mount insulator for cracks and rubber parts for wear. Replace them if necessary.

Coil Spring

1. Check for cracks, wear, and damage, and replace if necessary.

STABILIZER BAR

REMOVAL & INSTALLATION

See Figures 139 and 140.

1. Remove steering gear and linkage.

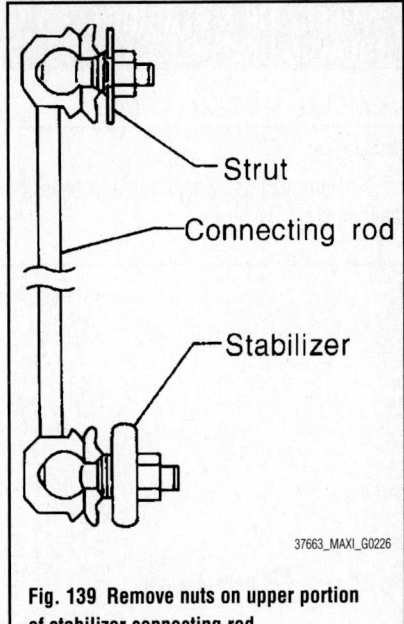

Fig. 139 Remove nuts on upper portion of stabilizer connecting rod

2. Remove nuts on upper portion of stabilizer connecting rod.

3. Remove stabilizer clamp bolts.

4. Remove stabilizer from the vehicle.

5. Check stabilizer, connecting rod, bushing and clamp for deformation, cracks and damage, and replace if necessary.

To install:

6. Installation is in the reverse order of removal.

7. When installing stabilizer, make sure that notch in stabilizer clips face front.

8. Make sure the slit in surface of stabilizer bushings face rear.

9. Stabilizer uses pillow ball type connecting rod. Position ball joint with case on pillow ball head parallel to stabilizer.

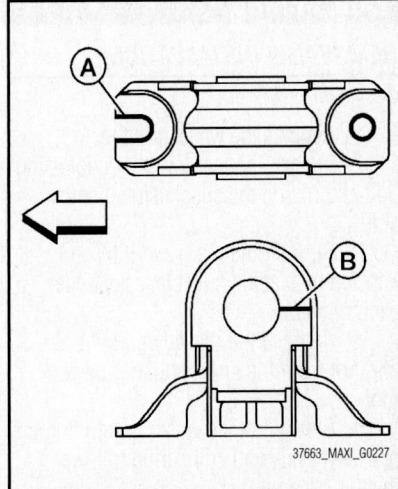

Fig. 140 Make sure that notch (A) in stabilizer clips face front and the slit (B) in surface of stabilizer bushings face rear

WHEEL HUB & BEARING (SEALED UNIT)

REMOVAL & INSTALLATION

See Figure 141.

1. Remove front wheel hub and bearing assembly.

2. Remove steering linkage from steering knuckle.

3. Remove steering knuckle lower pinch bolt.

4. Remove steering knuckle to strut bolts, then remove steering knuckle.

To install:

5. Installation is in the reverse order of removal.

✳✳ CAUTION

Do not reuse non-reusable parts.

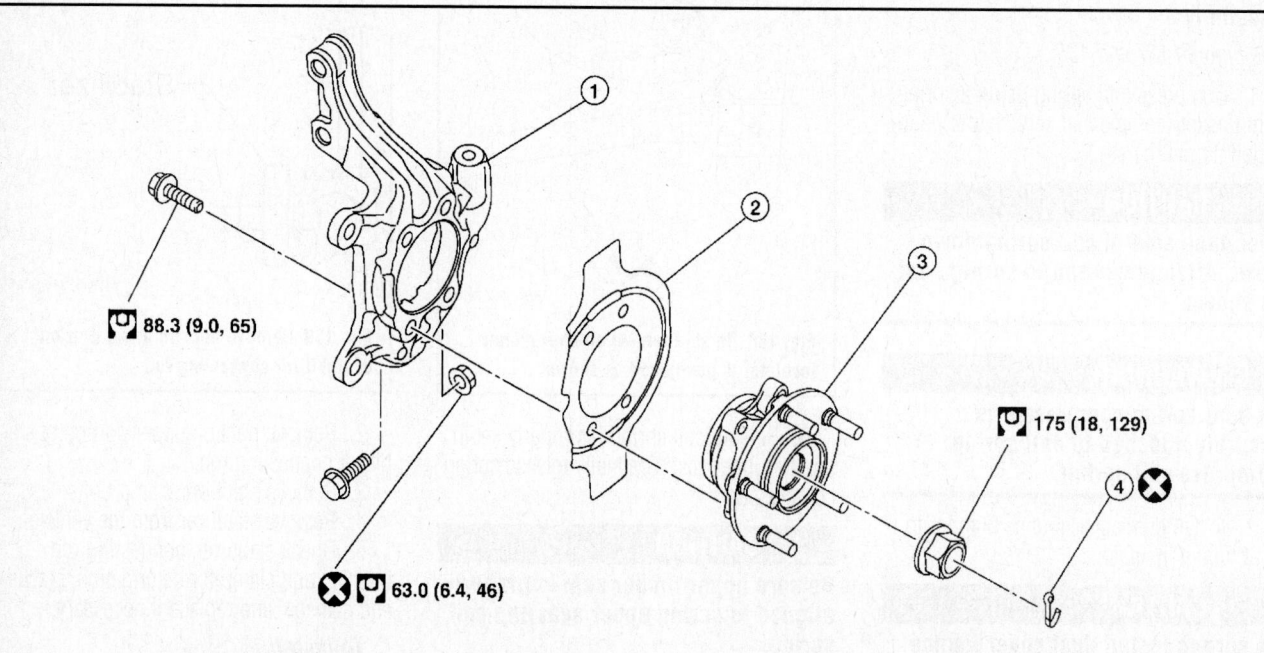

88.3 (9.0, 65)

175 (18, 129)

63.0 (6.4, 46)

1. Steering knuckle   2. Splash guard   3. Wheel hub and bearing assembly
4. Cotter pin

37663_MAXI_G0221

Fig. 141 Exploded view of the steering knuckle assembly

SUSPENSION

REAR SUSPENSION

COIL SPRING

REMOVAL & INSTALLATION

See Figures 142 and 143.

1. Remove the wheel and tire.

2. Loosen the rear lower link adjusting bolt and nut on the suspension member side.

3. Support the rear lower link and knuckle by placing suitable jacks under each of them.

4. Remove the rear lower link bolt and nut from the knuckle using power tool.

5. Slowly lower the jack supporting the rear lower link and coil spring to lower them.

6. Remove the upper rubber seat, coil spring, and lower rubber seat from the rear lower link.

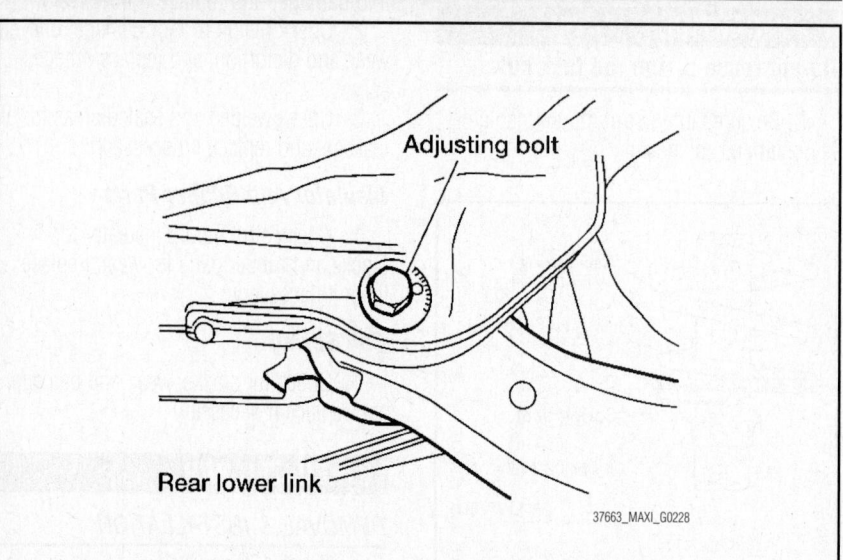

Adjusting bolt

Rear lower link

37663_MAXI_G0228

Fig. 142 Rear lower link adjusting bolt and nut on the suspension member side

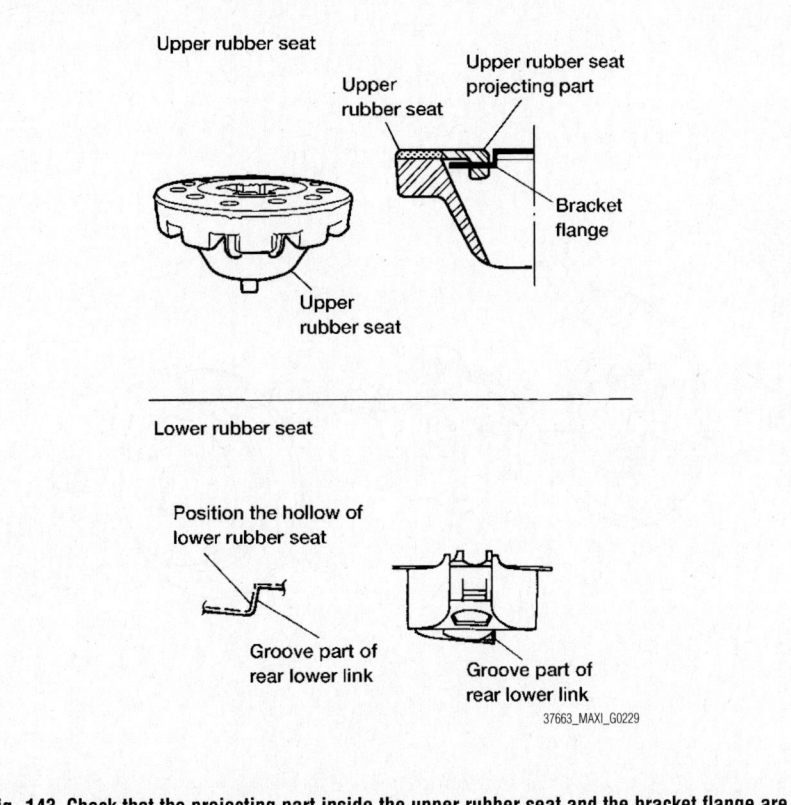

Fig. 143 Check that the projecting part inside the upper rubber seat and the bracket flange are attached as shown

7. Remove rear lower link adjusting bolt and nut from the suspension member side.

8. Remove the rear lower link.

To install:

9. Installation is in the reverse order of removal.

➡**Do not reuse the adjusting nut, use a new adjusting nut for installation.**

10. Check that the projecting part inside the upper rubber seat and the bracket flange are attached as shown.

11. Check that the projection part outside the upper rubber seat is directed toward the front of the vehicle.

12. Position the hollow of the lower rubber seat with the groove part of the rear lower link.

13. Install the coil spring so that the side with the two paint markers is directed toward the lower side.

14. Check the rear wheel alignment and adjust if necessary.

CONTROL ARMS/LINKS

REMOVAL & INSTALLATION

Front Lower Link

See Figure 144.

1. Remove the front lower link nut and bolt from the knuckle side and the adjusting bolt and nut from the suspension member side.

2. Remove the front lower link.

To install:

3. Installation is in the reverse order of removal.

➡**Do not reuse the adjusting nut, use a new adjusting nut for installation.**

4. Check the rear wheel alignment and adjust if necessary.

STABILIZER BAR

REMOVAL & INSTALLATION

1. Disconnect the stabilizer bar from connecting rod.

2. Remove the stabilizer bar clamps and bushings.

3. Remove the stabilizer bar.

To install:

4. Installation is in the reverse order of removal.

WHEEL HUB & BEARING (SEALED UNIT)

REMOVAL & INSTALLATION

See Figure 145.

✳✳ CAUTION

Wheel hub assembly does not require maintenance. If any of the following symptoms are noted, replace the wheel hub assembly.

- A growling noise is emitted from the wheel hub assembly while driving.
- The wheel hub assembly drags or turns roughly.

1. Remove the rear wheel and tire.

2. Remove the brake caliper assembly and brake rotor.

➡**The brake hose does not need to be disconnected from the brake caliper.**

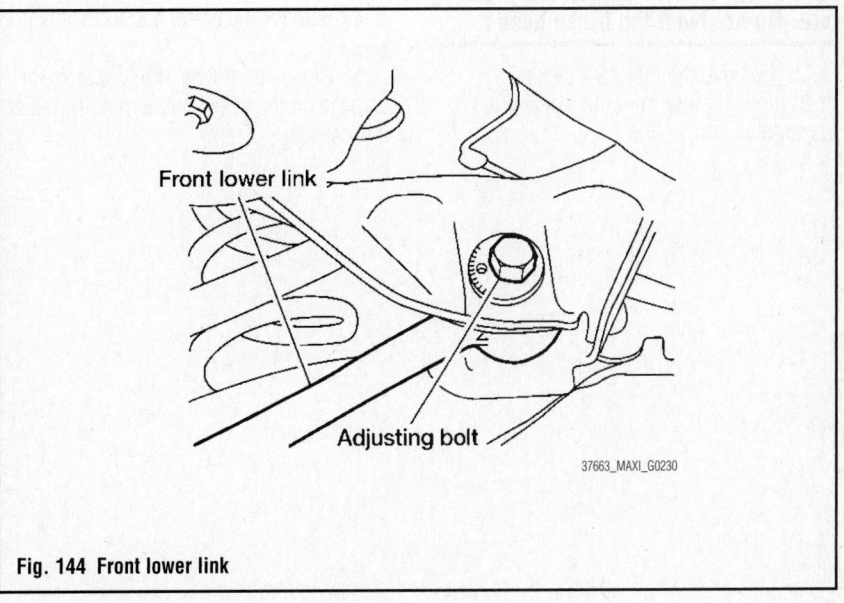

Fig. 144 Front lower link

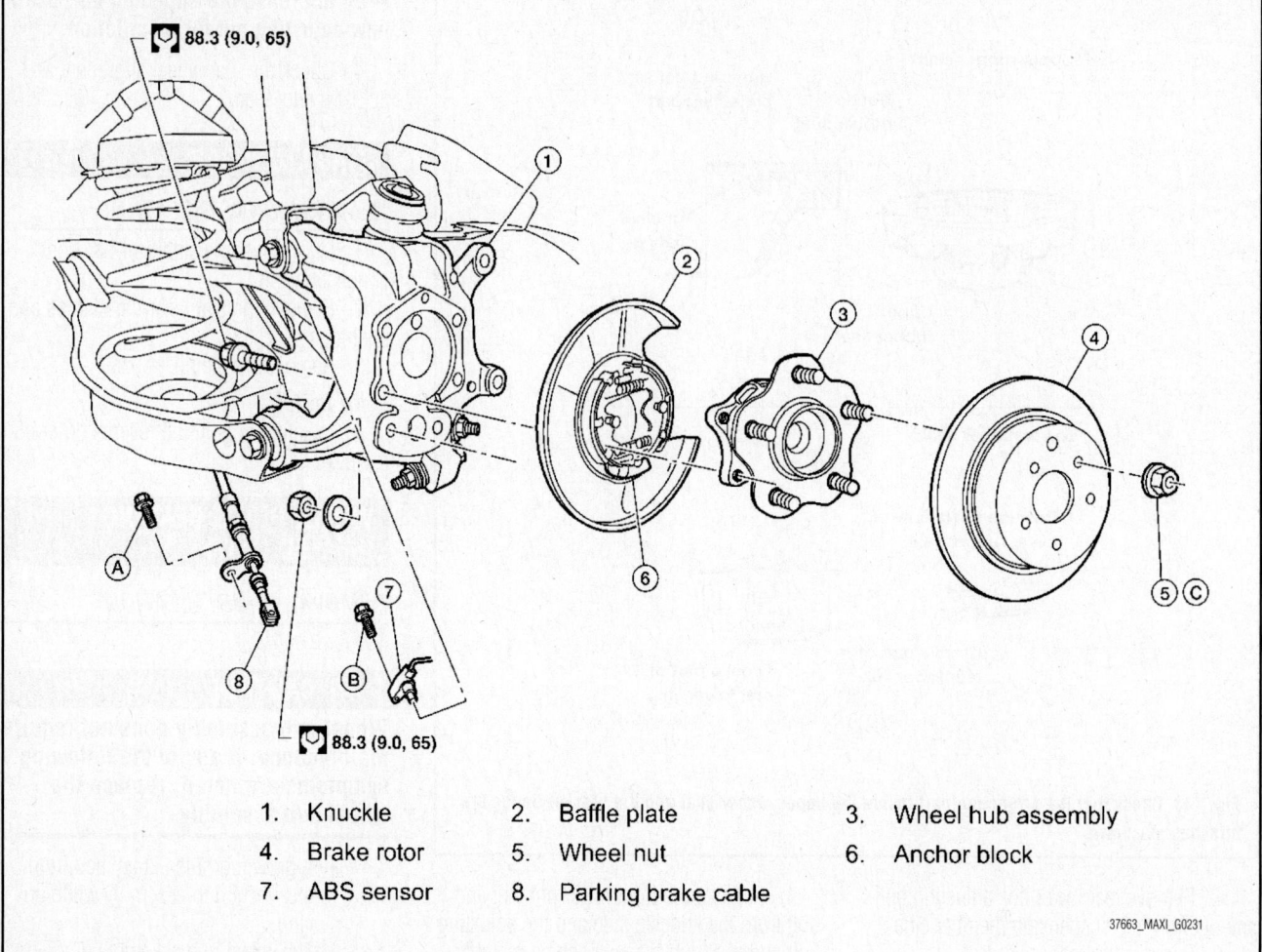

88.3 (9.0, 65)

88.3 (9.0, 65)

1. Knuckle
2. Baffle plate
3. Wheel hub assembly
4. Brake rotor
5. Wheel nut
6. Anchor block
7. ABS sensor
8. Parking brake cable

37663_MAXI_G0231

Fig. 145 Explode view of rear knuckle and hub assembly

❊❊ **CAUTION**

Suspend the brake caliper assembly using wire, do not stretch the brake hose. Do not depress the brake pedal, or the caliper piston will pop out. Do not twist the brake hose.

3. Remove the rear ABS sensor, then move it away from the wheel hub assembly.

❊❊ **CAUTION**

Failure to remove the ABS sensor may result in damage to the sensor wires and the sensor being inoperative.

4. Remove the wheel hub assembly from knuckle.

5. Check for any deformity, cracks, or damage on the wheel hub assembly, replace if necessary.

To install:

6. Installation is in the reverse order of removal.

7. Check that the wheel bearings operate smoothly.

8. Check that the wheel hub bearing axial end play is within specification: 0.004 inches (0.1 mm).

NISSAN

Murano

15

SPECIFICATIONS AND MAINTENANCE CHARTS

ENGINE AND VEHICLE IDENTIFICATION

	Engine						Model Year	
Code	Liters (cc)	Cu. In.	Cyl.	Fuel Sys.	Engine	Eng. Mfg.	Code	Year
VQ35DE	3.5 (3498)	213.45	6	MFI	DOHC	Nissan	9	2009
							A	2010

MFI: Multi-port Fuel Injection

DOHC: Double Overhead Camshaft

37663_MURA_C0001

GENERAL ENGINE SPECIFICATIONS

Year	Model	Engine Displacement Liters	Engine ID	Net Horsepower @ rpm	Net Torque @ rpm (ft. lbs.)	Bore x Stroke (in.)	Com- pression Ratio	Oil Pressure @ rpm
2009	Murano	3.5	VQ35DE	265@6000	248@4400	3.76X3.20	10.3:1	43@2000
2010	Murano	3.5	VQ35DE	265@6000	248@4400	3.76X3.20	10.3:1	43@2000

37663_MURA_C0002

ENGINE TUNE-UP SPECIFICATIONS

Year	Engine Displacement Liters	Engine ID	Spark Plug Gap (in.)	Ignition Timing (deg.)	Fuel Pump (psi)	Idle Speed RPM	Valve Clearance (in.) In.	Ex.
2009	3.5	VQ35DE	0.043	NA	51 ①	②	③	④
2010	3.5	VQ35DE	0.043	NA	51 ①	②	③	④

NA: Not Applicable

NOTE: The Vehicle Emission Control Information label often reflects specification changes made during production. The label figures must be used

① At idle

② Idle is computer controlled and is not adjustable.

③ 0.010-0.013 cold

0.012-0.016 hot

④ 0.011-0.015 cold

0.012-0.017 hot

37663_MURA_C0003

CAPACITIES

Year	Model	Engine Displacement Liters	Engine ID	Engine Oil with Filter (qts.)	Transmission (pts.)	Transfer Case (pts.)	Drive Axle Front (pts.)	Drive Axle Rear (pts.)	Fuel Tank (gal.)	Cooling System (qts.)
2009	Murano	3.5	VQ35DE	4.9	21.5	0.63	NA	1.1	21.7	9.9
2010	Murano	3.5	VQ35DE	4.9	21.5	0.63	NA	1.1	21.7	9.9

NA: Not Applicable

NOTE: All capacities are approximate. Add fluid gradually and check to be sure a proper fluid level is obtained.

37663_MURA_C0005

FLUID SPECIFICATIONS

Year	Model	Engine Displ. Liters (VIN)	Engine Oil	Man. Trans.	Auto. Trans.	Drive Axle Front	Drive Axle Rear	Trans. Case	Power Steering Fluid	Brake Master Cylinder	Cooling System
2009	Murano	VQ35DE	①	NA	②	③	③	③	NISSAN PSF	DOT 3	④
2010	Murano	VQ35DE	①	NA	②	③	③	③	NISSAN PSF	DOT 3	④

NA: Not Applicable

DOT: Department Of Transpotation

① API Service SM SAE 5W-30

② Nissan CVT Fluid NS-2 2

③ GL-5 80W90

④ Nissan Long Life Antifreeze/Coolant

37663_MURA_C0004

VALVE SPECIFICATIONS

Year	Engine Displacement Liters	Engine ID	Seat Angle (deg.)	Face Angle (deg.)	Spring Test Pressure (lbs. @ in.)	Spring Installed Height (in.)	Stem-to-Guide Clearance (in.) Intake	Stem-to-Guide Clearance (in.) Exhaust	Stem Diameter (in.) Intake	Stem Diameter (in.) Exhaust
2009	3.5	VQ35DE	45.15-45.45	45	42.3@1.467	1.457	0.0008-0.0021	0.0012-0.0022	0.2348-0.2354	0.2347-0.2350
2010	3.5	VQ35DE	45.15-45.45	45	45.4@1.457	1.457	0.0008-0.0021	0.0012-0.0022	0.2348-0.2354	0.2347-0.2350

37663_MURA_C0006

CAMSHAFT SPECIFICATIONS

All measurements are given in inches.

Year	Engine Displ. Liters	Engine VIN	Journal Dia.	Brg. Oil Clearance	Shaft End-play	Runout	Lobe Height Intake	Lobe Height Exhaust
2009	3.5	VQ35DE	①	②	0.0045-0.0074	③	1.7900-1.7974	1.7904-1.7978
2010	3.5	VQ35DE	①	②	0.0045-0.0074	③	1.7900-1.7974	1.7904-1.7978

① No.1: 1.0211-1.0218

No.2, No.3, No.4: 0.9230-0.9238

② No.1: 1.0018-1.0034

No.2, No.3, No.4: 0.0014-0.0030

③ Less then 0.001 (0.02 mm)

37663_MURA_C0007

CRANKSHAFT AND CONNECTING ROD SPECIFICATIONS

All measurements are given in inches.

Year	Engine Displacement Liters	Engine ID	Crankshaft Main Brg. Journal Dia.	Crankshaft Main Brg. Oil Clearance	Crankshaft Shaft End-play	Crankshaft Thrust on No.	Connecting Rod Journal Diameter	Connecting Rod Oil Clearance	Connecting Rod Side Clearance
2009	3.5	VQ35DE	①	0.0014-0.0018	0.0039-0.0098	4	②	0.0008-0.0018	0.0079-0.0138
2010	3.5	VQ35DE	①	0.0014-0.0018	0.0039-0.0098	4	②	0.0008-0.0018	0.0079-0.0138

① There are 24 different grades, ranging from A (2.3612) to 7 (2.3603)

② Grade 0: 2.0460-2.0462

Grade 1: 2.0457-2.0460

Grade 2: 2.0445-2.0457

37663_MURA_C0008

PISTON AND RING SPECIFICATIONS

All measurements are given in inches.

Year	Engine Displacement Liters	Engine ID	Piston Clearance	Ring Gap Top Comp.	Ring Gap Bottom Comp.	Ring Gap Oil Control	Ring Side Clearance Top Comp.	Ring Side Clearance Bottom Comp.	Ring Side Clearance Oil Control
2009	3.5	VQ35DE	0.0004-0.0012	0.0091-0.0130	0.0091-0.0130	0.0079-0.0177	0.0018-0.0031	0.0012-0.0028	0.0026-0.0049
2010	3.5	VQ35DE	0.0004-0.0012	0.0091-0.0130	0.0091-0.0130	0.0079-0.0177	0.0018-0.0031	0.0012-0.0028	0.0026-0.0049

37663_MURA_C0009

TORQUE SPECIFICATIONS
All readings in ft. lbs.

Year	Engine Displacement Liters	Engine ID	Cylinder Head Bolts	Main Bearing Bolts	Rod Bearing Bolts	Crankshaft Damper Bolts	Driveplate Bolts	Manifold Intake	Manifold Exhaust	Spark Plugs	Oil Pan Drain Plug
2009	3.5	VQ35DE	①	②	③	④	65	⑤	24	14-22	25
2010	3.5	VQ35DE	①	②	③	④	65	⑤	24	14-22	25

① Step 1: 72 ft. lbs.
  Step 2: Loosen all bolts completely
  Step 3: 29 ft. lbs.
  Step 4: +103 degrees
  Step 5: +103 degrees

② Step 1: 26 ft. lbs.
  Step 2: +90 degrees

③ Step 1: 14 ft. lbs.
  Step 2: +90 degrees

④ 33 ft. lbs. +90 degrees

⑤ Step 1: 5 ft. lbs
  Step 2: 19 ft. lbs.

37663_MURA_C0010

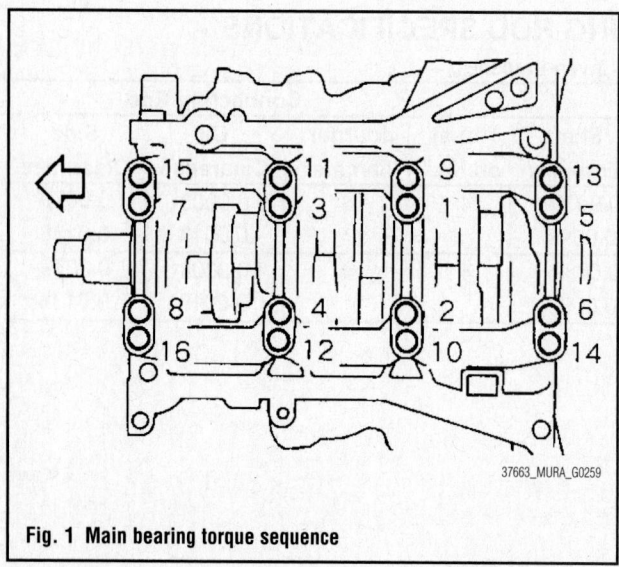

Fig. 1 Main bearing torque sequence

37663_MURA_G0259

WHEEL ALIGNMENT

Year	Model		Caster Range (+/-Deg.)	Caster Preferred Setting (Deg.)	Camber Range (+/-Deg.)	Camber Preferred Setting (Deg.)	Toe-in (in.)	Kingpin Inclination (Deg.)
2009	Murano	F	.75	+4.67	.75	-0.25	0.059+/-1.0	12.75+/-.75
		R	—	—	.50	-0.72	0.106+/-.71	—
2010	Murano	F	.75	+4.67	.75	-0.25	0.059+/-1.0	12.75+/-.75
		R	—	—	.50	-0.72	0.106+/-.71	—

Note: Specifications are taken with the following Fuel, radiator coolant and engine oil full. Spare tire, jack, hand tools and mats are in designated positions.

37663_MURA_C0011

TIRE, WHEEL AND BALL JOINT SPECIFICATIONS

| Year | Model | OEM Tires | | Tire Pressures (psi) | | Wheel Size | Ball Joint Inspection | Lug Nut Torque (ft. lbs.) |
		Standard	Optional	Front	Rear			
2009	Murano	P235/65R18	P235/55R20	33	33	7.5JJ	①	80
2010	Murano	P235/65R18	P235/55R20	33	33	7.5JJ	①	80

OEM: Original Equipment Manufacturer

PSI: Pounds Per Square Inch

① 0 (0mm) inches axial end play

37663_MURA_C0012

BRAKE SPECIFICATIONS

All measurements in inches unless noted

| Year | Model | | Brake Disc | | | Minimum Lining Thickness | Brake Caliper | |
			Original Thickness	Minimum Thickness	Maximum Runout		Bracket Bolts (ft. lbs.)	Mounting Bolts (ft. lbs.)
2009	Murano	F	1.102	1.024	0.0016	0.079	122	20
		R	0.630	0.551	0.0020	0.079	62	32
2010	Murano	F	1.102	1.024	0.0016	0.079	122	20
		R	0.630	0.551	0.0020	0.079	62	32

F: Front

R: Rear

37663_MURA_C0013

SCHEDULED MAINTENANCE INTERVALS (1)
Nissan—Murano

TO BE SERVICED	TYPE OF SERVICE	7.5	15	22.5	30	37.5	45	52.5	60
Engine oil & filter	R	every 3,750 miles							
Brake lines & cables	S/I		✓		✓		✓		✓
Brake pads, discs	I	✓	✓	✓	✓	✓	✓	✓	✓
Driveshaft boots & propeller shaft	L/I	✓	✓	✓	✓	✓	✓	✓	✓
CVT ①	I		✓		✓		✓		✓
Transfer case and differential fluid B	I		✓		✓		✓		✓
Air cleaner filter	R				✓				✓
Drive belt (s) ③	S/I								✓
Engine coolant ④	R								✓
Spark plugs	R	Platinum plugs, every 105,000 miles							
Cabin air filter	R		✓		✓		✓		✓
Exhaust system	I	✓	✓	✓	✓	✓	✓	✓	✓
Evap vapor lines	I				✓				✓
Fuel lines	S/I				✓				✓
Steering gear, linkage, axle & suspension parts	I		✓		✓		✓		✓
Tires (rotate)	S/I	every 5,000-6,000 miles							
Valve clearance ⑤	S/I				✓				✓

R: Replace   S/I: Service or Inspect   L: Lubricate

① If towing a trailer, using a camper or a car-top carrier, or driving on rough or muddy roads, change (not just inspect) oil at every 60,000 miles.

② If towing a trailer, using a camper or a car-top carrier, or driving on rough or muddy roads, change (not just inspect) oil at every 30,000 miles (48,000 km) or 24 months.

③ First at 60,000, then every 30,000 miles

④ After 60,000, replace every 30,000

⑤ Periodic maintenance not required, if valve noice increases, inspect valve clearance

Follow Periodic Maintenance Schedule 1 if the driving habits frequently include one or more of the following driving conditions:

  Repeated short trips of less than 5 miles (8 km).

  Repeated short trips of less than 10 miles (16 km) with outside temperatures remaining below freezing

  Operating in hot weather in stop-and-go "rush hour" traffic.

  Extensive idling and/or low speed driving for long distances, such as police, taxi or door-to-door delivery use

  Driving in dusty conditions.

  Driving on rough, muddy, or salt spread roads.

  Towing a trailer, using a camper or a car-top carrier.

Follow Periodic Maintenance Schedule 2 if none of driving conditions shown in Schedule 1 apply to the driving habits.

37663_MURA_C0015

SCHEDULED MAINTENANCE INTERVALS (2)
Nissan—Murano

TO BE SERVICED	SERVICE	7.5	15	22.5	30	37.5	45	52.5	60
Engine oil & filter	R	✓	✓	✓	✓	✓	✓	✓	✓
Brake lines & cables	S/I		✓		✓		✓		✓
Brake pads, discs	I	✓	✓	✓	✓	✓	✓	✓	✓
Driveshaft boots & propeller shaft	L/I		✓		✓		✓		✓
CVT ①	I		✓		✓		✓		✓
Transfer case and differential fluid ②	I		✓		✓		✓		✓
Air cleaner filter	R				✓				✓
Drive belt (s) ③	S/I								✓
Engine coolant ④	R								✓
Spark plugs	R	Platinum plugs, every 105,000 miles							
Cabin air filter	R		✓		✓		✓		✓
Exhaust system	I				✓				✓
Evap vapor lines	I				✓				✓
Fuel lines	S/I				✓				✓
Steering gear, linkage, axle & suspension parts	I				✓				✓
Tires (rotate)	S/I	every 5,000-6,000 miles							
Valve clearance ⑤	S/I				✓				✓

R: Replace   S/I: Service or Inspect   L: Lubricate

① If towing a trailer, using a camper or a car-top carrier, or driving on rough or muddy roads, change (not just inspect) oil at every 60,000 miles.

② If towing a trailer, using a camper or a car-top carrier, or driving on rough or muddy roads, change (not just inspect) oil at every 30,000 miles (48,000 km) or 24 months.

③ First at 60,000, then every 15,000 miles

④ After 60,000, replace every 30,000

⑤ Periodic maintenance not required, if valve noice increases, inspect valve clearance

Follow Periodic Maintenance Schedule 1 if the driving habits frequently include one or more of the following driving conditions:

   Repeated short trips of less than 5 miles (8 km).

   Repeated short trips of less than 10 miles (16 km) with outside temperatures remaining below freezing

   Operating in hot weather in stop-and-go "rush hour" traffic.

   Extensive idling and/or low speed driving for long distances, such as police, taxi or door-to-door delivery use

   Driving in dusty conditions.

   Driving on rough, muddy, or salt spread roads.

   Towing a trailer, using a camper or a car-top carrier.

Follow Periodic Maintenance Schedule 2 if none of driving conditions shown in Schedule 1 apply to the driving habits.

37663_MURA_C0014

BRAKES — INFORMATION AND PRECAUTIONS

ANTI-LOCK SYSTEMS

• Certain components within the ABS system are not intended to be serviced or repaired individually.

• Do not use rubber hoses or other parts not specifically specified for and ABS system. When using repair kits, replace all parts included in the kit. Partial or incorrect repair may lead to functional problems and require the replacement of components.

• Lubricate rubber parts with clean, fresh brake fluid to ease assembly. Do not use shop air to clean parts; damage to rubber components may result.

• Use only DOT 3 brake fluid from an unopened container.

• If any hydraulic component or line is removed or replaced, it may be necessary to bleed the entire system.

• A clean repair area is essential. Always clean the reservoir and cap thoroughly before removing the cap. The slightest amount of dirt in the fluid may plug an orifice and impair the system function. Perform repairs after components have been thoroughly cleaned; use only denatured alcohol to clean components. Do not allow ABS components to come into contact with any substance containing mineral oil; this includes used shop rags.

• The Anti-Lock control unit is a microprocessor similar to other computer units in the vehicle. Ensure that the ignition switch is **OFF** before removing or installing controller harnesses. Avoid static electricity discharge at or near the controller.

• If any arc welding is to be done on the vehicle, the control unit should be unplugged before welding operations begin.

DISC AND DRUM SYSTEMS

✳✳ CAUTION

Dust and dirt accumulating on brake parts during normal use may contain asbestos fibers from production or aftermarket brake linings. Breathing excessive concentrations of asbestos fibers can cause serious bodily harm. Exercise care when servicing brake parts. Do not sand or grind brake lining unless equipment used is designed to contain the dust residue. Do not clean brake parts with compressed air or by dry brushing. Cleaning should be done by dampening the brake components with a fine mist of water, then wiping the brake components clean with a dampened cloth. Dispose of cloth and all residue containing asbestos fibers in an impermeable container with the appropriate label. Follow practices prescribed by the Occupational Safety and Health Administration (OSHA) and the Environmental Protection Agency (EPA) for the handling, processing, and disposing of dust or debris that may contain asbestos fibers.

BRAKES — BLEEDING THE BRAKE SYSTEM

BLEEDING PROCEDURE

BLEEDING PROCEDURE

✳✳ WARNING

Monitor the fluid level in the sub tank during the air bleeding. Always use new brake fluid for refilling. Never reuse the drained brake fluid.

1. Turn the ignition switch OFF and disconnect the ABS actuator and electric unit (control unit) connector or the battery negative terminal before performing the work.

2. Connect a vinyl tube to the bleeder valve of the rear right brake.

3. Fully depress the brake pedal 4 to 5 times.

4. Loosen the bleeder valve and bleed air with the brake pedal depressed, and then quickly tighten the bleeder valve.

5. Repeat steps 3 and 4 until all of the air is out of the brake line.

6. Tighten the bleeder valve to the specified torque.

7. Perform steps 2 to 6 for the rear right brake, front left brake, rear left brake, and front right brake in order.

8. Check that the fluid level in the sub tank is within the specified range after air bleeding.

BLEEDING THE ABS SYSTEM

Refer to "Bleeding Procedure".

BRAKES — ANTI-LOCK BRAKE SYSTEM (ABS)

WHEEL SPEED SENSORS

REMOVAL & INSTALLATION

See Figures 2 and 3.

1. Before servicing the vehicle, refer to the Precautions Section.

2. Disconnect the negative battery cable.

3. Raise and support the vehicle safely.

4. Remove the tire and wheel assembly.

5. Remove the wheel speed sensor mounting bolts and clip.

6. Pull the sensor out, being careful to turn it as little as possible. Do not pull on the sensor harness.

To install:

7. Before installing the sensor be sure no foreign materials such as iron fragments are adhered to the pick-up part of the sensor or to the inside of the sensor mounting hole or on the rotor mounting surface.

8. Be sure the harness is not twisted when installed.

9. Continue the installation in the reverse order of the removal procedure.

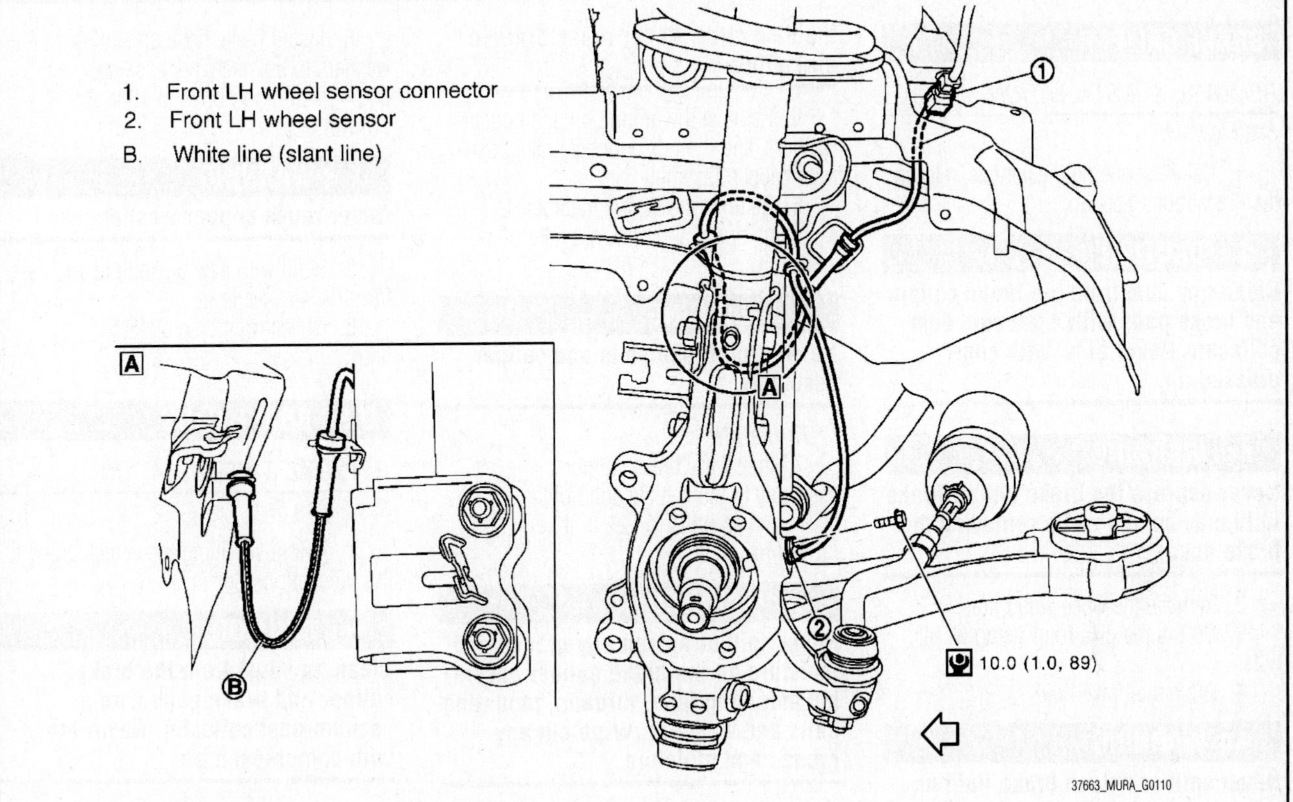

1. Front LH wheel sensor connector
2. Front LH wheel sensor
B. White line (slant line)

10.0 (1.0, 89)

37663_MURA_G0110

Fig. 2 Front wheel speed sensor

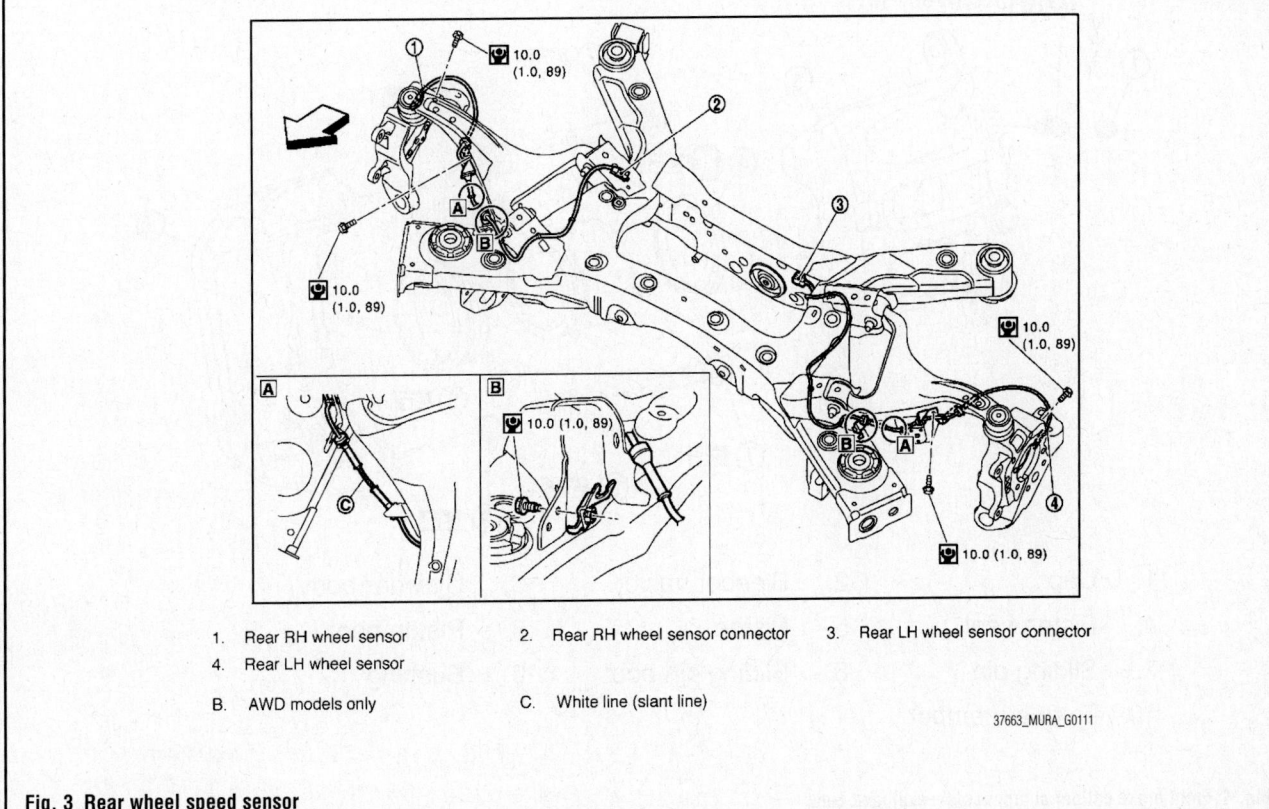

10.0
(1.0, 89)

10.0
(1.0, 89)

10.0
(1.0, 89)

10.0 (1.0, 89)

10.0 (1.0, 89)

1.	Rear RH wheel sensor	2.	Rear RH wheel sensor connector	3.	Rear LH wheel sensor connector
4.	Rear LH wheel sensor				
B.	AWD models only	C.	White line (slant line)		

37663_MURA_G0111

Fig. 3 Rear wheel speed sensor

BRAKES   FRONT DISC BRAKES

BRAKE CALIPER

REMOVAL & INSTALLATION
See Figure 4.

1. Before servicing the vehicle, refer to the Precautions Section.

> **⁂ CAUTION**
>
> **Clean any dust from the brake caliper and brake pads with a vacuum dust collector. Never blow with compressed air.**

> **⁂ WARNING**
>
> **Never depress the brake pedal. Brake fluid may splash while removing the brake hose.**

2. Remove the wheels and tires.
3. Secure the disc rotor using wheel nuts.
4. Drain the brake fluid.

> **⁂ WARNING**
>
> **Never spill or splash brake fluid on the disc rotor. Never reuse drained brake fluid.**

5. Remove the union bolt and copper washers, and then disconnect brake hose from caliper assembly.
6. Remove the torque member mounting bolts, and remove brake caliper assembly.

> **⁂ WARNING**
>
> **Never drop brake pads and caliper assembly.**

To install:

7. Install the brake caliper assembly to the vehicle and tighten the mounting bolts to 122 ft. lbs. (165 Nm).

> **⁂ WARNING**
>
> **Never spill or splash any grease and moisture on the brake caliper assembly mounting face, threads, mounting bolts and washers. Wipe out any grease and moisture.**

8. Install brake hose and copper washers to brake caliper assembly, and tighten union bolts to 13 ft. lbs. (18 Nm).

> **⁂ WARNING**
>
> **Never reuse copper washer.**

9. Refill with new brake fluid and perform the air bleeding.
10. Check front disc brake for drag.

DISC BRAKE PADS

REMOVAL & INSTALLATION
See Figure 5.

1. Before servicing the vehicle, refer to the Precautions Section.

> **⁂ CAUTION**
>
> **Clean any dust from the brake caliper and brake pads with a vacuum dust collector. Never blow with compressed air.**

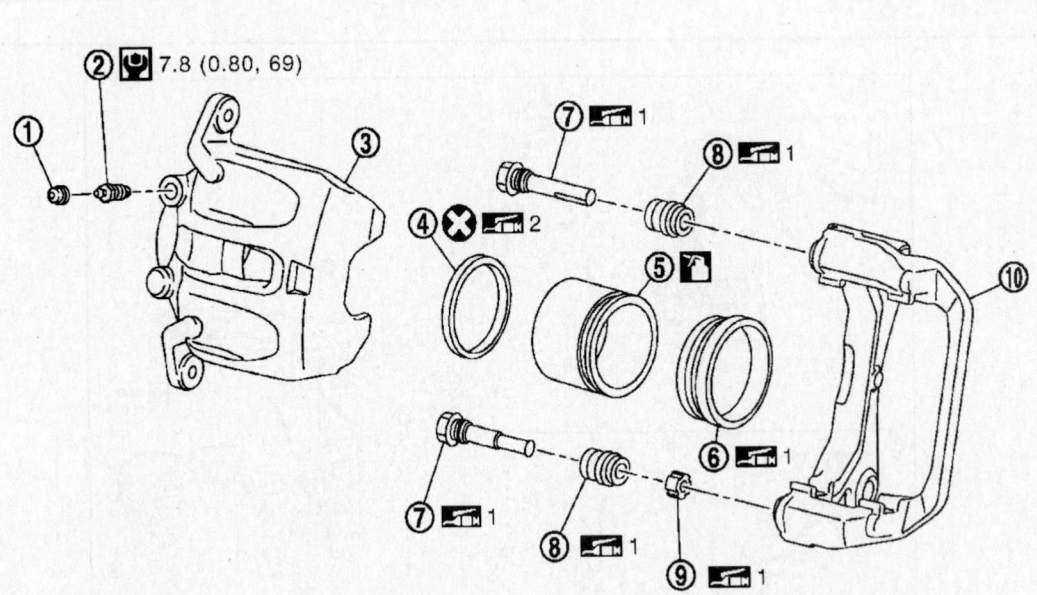

1. Cap	2. Bleeder valve	3. Cylinder body
4. Piston seal	5. Piston	6. Piston boot
7. Sliding pin	8. Sliding pin boot	9. Bushing
10. Torque member		

37663_MURA_G0099

Fig. 4 Front brake caliper components—exploded view

Never depress the brake pedal while removing the brake pads because the piston may pop out. Never spill or splash brake fluid on the disc rotor.

2. Remove the wheels and tires.
3. Remove the lower sliding pin bolt.
4. Suspend the cylinder body with suitable wire so that the brake hose will not stretch.
5. Remove the brake pads, shims and shim cover from the torque member.

Never reuse the pad retainers when removed the pad retainers from the torque member. Never damage the piston boot. Never drop the brake pads, shims, and the shim cover. Remember each position of the removed brake pads.

To install:

6. Install the pad retainer to the torque member if the pad retainer has been removed.

Never reuse the pad retainers. Securely assemble the pad retainers so that it will not be lifted up from the torque member. Never deform the pad retainers. Eliminate double-faced adhesive tape on torque member. Remove adhesive's protective liner on pad retainers.

7. Apply PBC (Poly Butyl Cuprysil) grease or silicone-based grease to the mating faces between the brake pads and pad retainers.
8. Apply copper based brake grease to the mating faces between the brake pads, shims and shim covers, and install them to the brake pad.

Always replace the shims together with the shim covers when replacing the brake pad.

9. Install the cylinder body and brake pads to the torque member.

Never damage the piston boot. When replacing a pad with new one, check a brake fluid level in the reservoir tank because brake fluid returns to master cylinder reservoir tank when pressing piston in.

➡**Use a disc brake piston tool to easily press the piston.**

10. Install the lower sliding pin bolt and tighten it to 20 ft. lbs. (27 Nm).
11. Depress the brake pedal several times to check that no drag feel is present for the front disc brake.

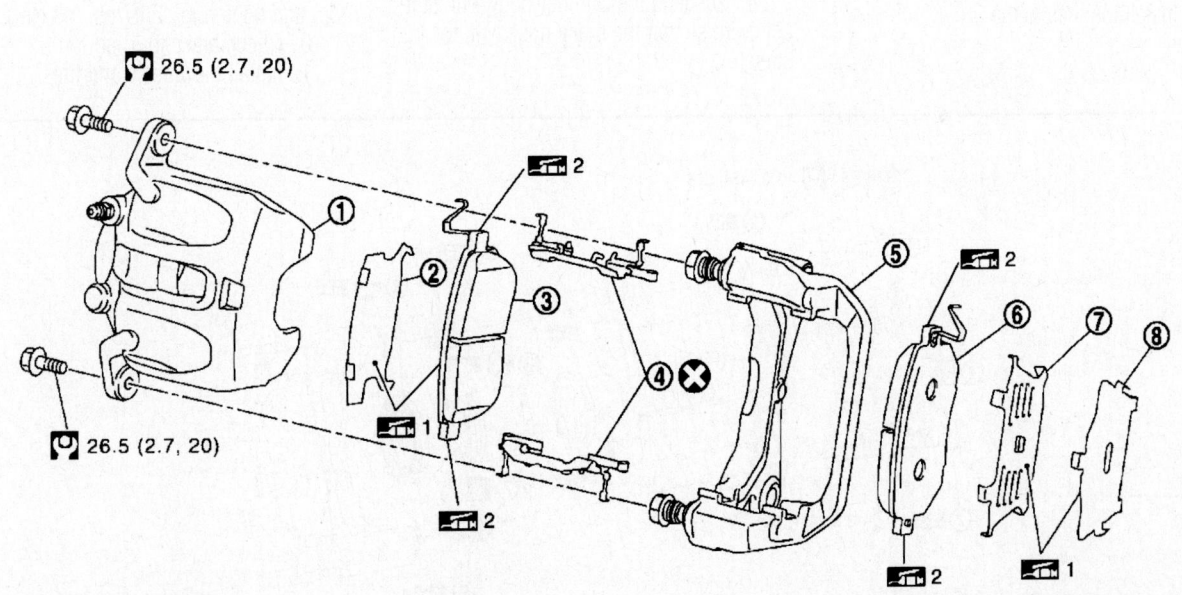

🔧 26.5 (2.7, 20)

🔧 26.5 (2.7, 20)

1. Cylinder body
2. Inner shim
3. Inner pad (with pad wear sensor)*
4. Pad retainer
5. Torque member
6. Outer pad
7. Outer shim
8. Outer shim cover

*: Some vehicles has pad wear sensor only for one side.

🔧1: Apply PBC (Poly Butyl Cuprysil) grease or silicone-based grease.

🔧2: Apply copper based brake grease.

37663_MURA_G0098

Fig. 5 Front brake pad—exploded view

BRAKES   **REAR DISC BRAKES**

BRAKE CALIPER

REMOVAL & INSTALLATION

See Figure 6.

1. Before servicing the vehicle, refer to the Precautions Section.
2. Raise and support the vehicle safely. Remove the tire and wheel assembly.
3. Drain brake fluid.
4. Remove union bolts and torque member bolts, and remove brake caliper assembly.

To install:

5. Install disc rotor.
6. Install caliper assembly to the vehicle, and tighten bolts to 62 ft. lbs. (84.3 Nm).
7. Install brake hose to caliper assembly and tighten union bolts to 13 ft. lbs. (18.2 Nm).

➡ **Do not reuse the copper washer for union bolts. Attach brake hose to the brake hose mounting boss.**

8. Refill new brake fluid.
9. Install the tires to the vehicle.

DISC BRAKE PADS

REMOVAL & INSTALLATION

See Figure 7.

1. Before servicing the vehicle, refer to the Precautions Section.

✳✳ CAUTION

Clean any dust from the brake caliper and brake pads with a vacuum dust collector. Never blow with compressed air.

✳✳ WARNING

Never depress the brake pedal while removing the brake pads because the piston may pop out. Never spill or splash brake fluid on the disc rotor.

2. Remove the wheels and tires.
3. Remove the upper sliding pin bolt.
4. Suspend the cylinder body with suitable wire so that the brake hose will not stretch.

5. Remove the brake pads, shims and shim cover from the torque member.

✳✳ WARNING

Never reuse the pad retainers when removed the pad retainers from the torque member. Never damage the piston boot. Never drop the brake pads, shims, and the shim cover. Remember each position of the removed brake pads.

To install:

6. Apply PBC (Poly Butyl Cuprysil) grease or silicon based grease to the rear of the pad and to both sides of the shim, and attach the inner shim and shim cover to the inner pad, and the outer shim and outer shim cover to the outer pad.
7. Attach the pad retainer and pad to the torque member.
8. Push the piston in so that the pad is attached and attach the cylinder body to the torque member.
9. Install the sliding pin bolt (one on top) and tighten to 32 ft. lbs. (43 Nm).
10. Check brake for drag.
11. Install the wheels and tires.

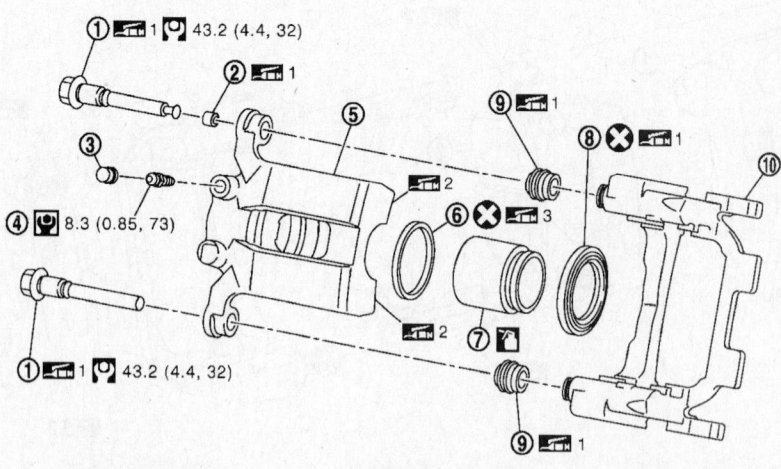

1. Sliding pin bolt
2. Bushing
3. Cap
4. Bleeder valve
5. Cylinder body
6. Piston seal
7. Piston
8. Piston boot
9. Sliding pin boot
10. Torque member

🔧1: Apply rubber grease.

🔧2: Apply PBC (Poly Butyl Cuprysil) grease or silicone-based grease.

🔧3: Apply polyglycol ether based lubricant.

🛢: Apply brake fluid.

37663_MURA_G0102

Fig. 6 Rear brake caliper components—exploded view

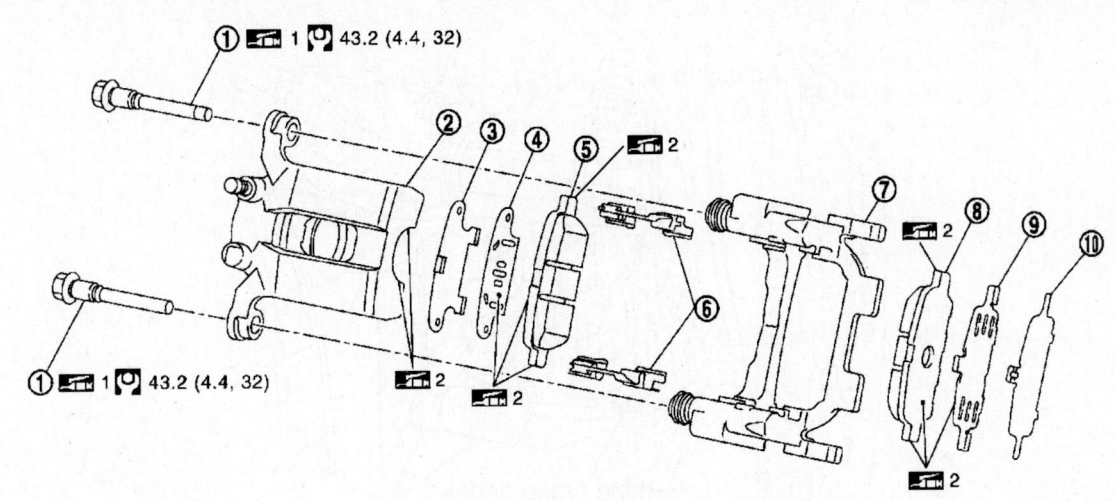

1. Sliding pin bolt
2. Cylinder body
3. Inner shim cover
4. Inner shim
5. Inner pad (with pad wear sensor)*
6. Pad retainer
7. Torque member
8. Outer pad
9. Outer shim
10. Outer shim cover

*: Some vehicles has pad wear sensor only for one side.

1: Apply rubber grease.

2: Apply PBC (Poly Butyl Cuprysil) grease or silicone-based grease.

37663_MURA_G0100

Fig. 7 Rear brake pad—exploded view

BRAKES   PARKING BRAKE

PARKING BRAKE CABLES

ADJUSTMENT

See Figure 8.

1. Before servicing the vehicle, refer to the Precautions Section.
2. Raise and safely support the vehicle.
3. Remove the rear wheels.
4. Insert a deep socket wrench to rotate the adjusting nut and loosen the cable. Then, return the pedal.
5. Using a couple of lug nuts, install them on the rotor to prevent it from tilting.
6. Remove the adjusting hole plug on the rotor. Using a flat bladed tool, turn the adjuster in direction "A" until the rotor is locked.
7. After locking the rotor turn the adjuster in the opposite direction (5–6 notches).
8. Rotate the rotor to make sure there is no drag. Install the adjusting plug cap.
9. Adjust the cable as follows. Operate the pedal 10 or more times with a force of 110 lbs.

10. Rotate the adjusting nut with a deep socket to adjust the pedal stroke.

➡**Do not reuse the adjusting nut after removing it.**

11. When the parking brake cable is operated with a force of 44 lbs, make sure the stroke is within the specified number of notches (5–6). Listen and count the ratcheting "click."
12. With the pedal completely returned, make sure there is no drag on the rear brake.
13. Correct as required.

PARKING BRAKE SHOES

REMOVAL & INSTALLATION

See Figures 9 and 10.

1. Before servicing the vehicle, refer to the Precautions Section.
2. Raise and safely support the vehicle.
3. Release the parking brake.
4. Remove the rear wheels.
5. Remove the rotor. If rotor cannot be removed:

a. Secure the rotor in place with lug nuts and remove adjuster hole plug.
b. Using a flat-bladed screwdriver, rotate the adjuster in direction "B" to retract and loosen the brake shoes.
6. Remove the anti-rattle pins, retainers, anti-rattle springs, and return springs.
7. Remove the parking brake shoes, adjuster assembly, adjuster spring and toggle lever.

To install:

8. Install in the reverse order of removal.
9. Apply PBC (Poly Butyl Cuprysil) grease or silicone-based grease to the back plate and brake shoe.

❊❊ WARNING

The parking brake shoes for the front side are made of different materials from those for the rear side. Never misidentify them when removing and replacing.

10. Assemble adjusters so that threaded part is expanded when rotating it in the direction shown by arrow.

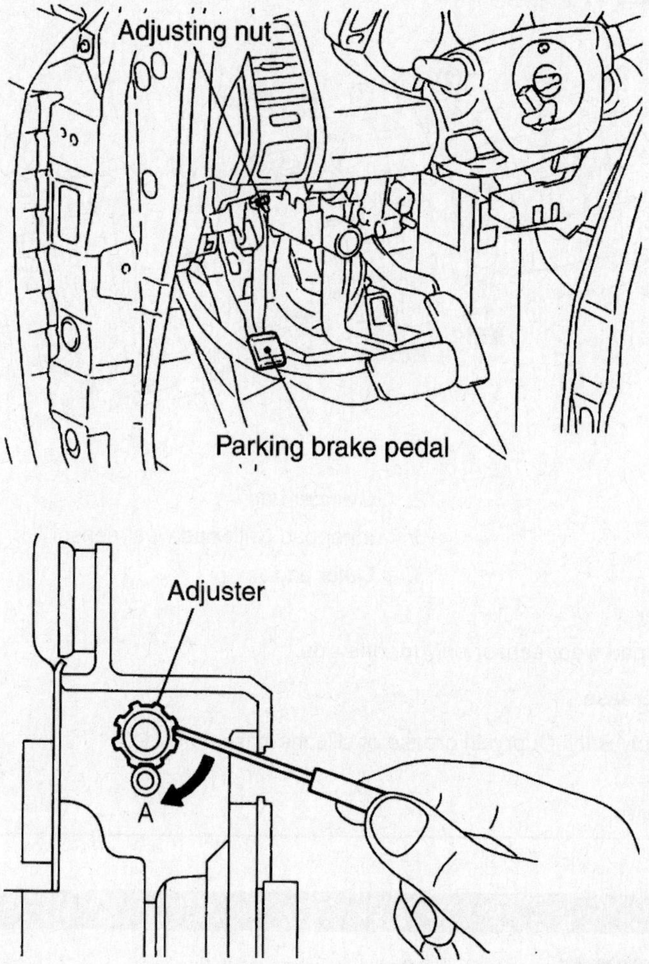

Adjusting nut

Parking brake pedal

Adjuster

A

42050_MURA_G0034

Fig. 8 Parking brake adjuster location

11. Shorten adjuster by rotating it.

12. When disassembling, apply PBC (Poly Butyl Cuprysil) grease or silicone-based grease to threads.

13. Check parking brake shoe

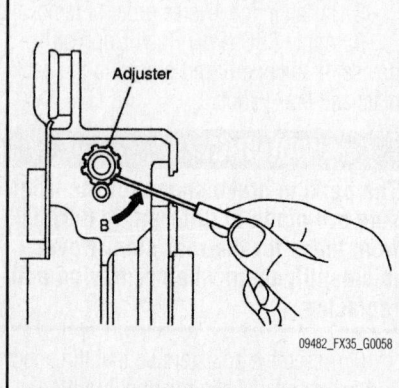

Adjuster

B

09482_FX35_G0058

Fig. 9 Backing off the brake shoe adjuster

sliding surface and drum inner surface for grease. Wipe it off if it adhere on the surfaces.

ADJUSTMENT

1. Adjust the parking brake pedal stroke to 4–5 clicks fully depressed. Insert a deep socket wrench to rotate the adjusting nut and loosen the cable sufficiently. Then, return pedal.

2. Remove the wheels.

3. Using a lug nut, secure the rotor to hub to prevent it from tilting.

4. Remove the adjusting hole plug. Using a flat-bladed screwdriver, turn the adjuster clockwise until the rotor is locked. After locking, turn the adjuster in the opposite direction by 5 or 6 notches.

5. Rotate the rotor to make sure that there is no drag. Then install the adjusting hole plug.

6. After adjusting the clearance of rear shoes, with no drag on rear brake, adjust the cable as follows:

7. Operate the pedal 10 or more times with a force of 490 N (110 lbs.).

8. Depress the pedal until a deep socket can be inserted. Insert the deep

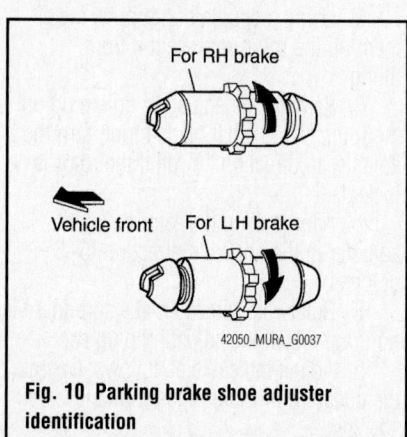

For RH brake

Vehicle front   For L H brake

42050_MURA_G0037

Fig. 10 Parking brake shoe adjuster identification

socket, and rotate the adjusting nut to adjust the pedal stroke.

➡**Do not reuse the adjusting nut.**

9. When the parking brake pedal is operated with a force of 196 N (44 lbs.), make sure the stroke is 5–6 notches. (Check it by listening and counting the ratchet clicks.)

10. With the parking brake pedal completely returned, make sure there is no drag on the rear brake.

11. Perform the parking brake break-in operation as follows: Safely, drive forward at approximately 40 km/h (25 MPH) with the parking brake set with a force of approx. 200 N (45 lbs.) for about 30 seconds.

12. After the break-in operation, check the pedal stroke of parking brake. Readjust if necessary.

➡**To prevent lining from getting too hot, allow a cool off period of approximately 5 minutes after every break-in operation.**

CHASSIS ELECTRICAL · AIR BAG (SUPPLEMENTAL RESTRAINT SYSTEM)

GENERAL INFORMATION

✻✻ CAUTION

These vehicles are equipped with an air bag system. The system must be disarmed before performing service on, or around, system components, the steering column, instrument panel components, wiring and sensors. Failure to follow the safety precautions and the disarming procedure could result in accidental air bag deployment, possible injury and unnecessary system repairs.

SERVICE PRECAUTIONS

✻✻ CAUTION

Disconnect and isolate the battery negative cable before beginning any airbag system component diagnosis, testing, removal, or installation procedures. Wait at least 90 seconds after the ignition switch is turned off and the negative (-) terminal cable is disconnected from the battery before starting the operation. The SRS is equipped with a backup power source, so if work is started within 90 seconds after disconnecting the negative (-) terminal cable from the battery, the SRS may be deployed. Failure to disable the airbag system may result in accidental airbag

deployment, personal injury, or death.

DISARMING THE SYSTEM

To disarm the **SRS** system turn the ignition switch to the **OFF** position. Then, disconnect both battery cables starting with the negative cable first and wait at least 3 minutes after the cables are disconnected.

ARMING THE SYSTEM

To rearm the **SRS** system, turn the ignition switch to the **OFF** position. Connect both battery cables starting with the positive cable first.

CLOCKSPRING CENTERING
See Figure 11.

✻✻ WARNING

The spiral cable may snap during steering operation if the cable is installed in an improper position.

1. Carefully turn the spiral cable clockwise to the end position.

2. Turn it counterclockwise (about 2 and a half turns) and stop turning at the mark "B" on which the stopper insertion holes are in the same position.

➡**The service part is installed in the neutral position by the stopper and can be set without adjusting after the stopper is removed.**

✻✻ WARNING

Never over turn the spiral cable or go beyond number of turns required. (This will cause the cable to snap.)

3. Adjust the spiral cable locating pin to the steering wheel locating pin hole.

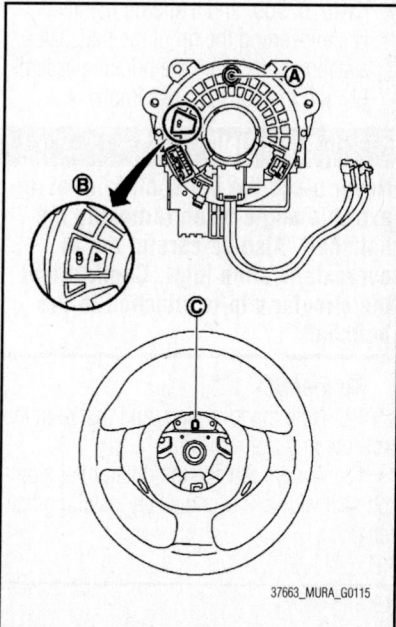

37663_MURA_G0115

Fig. 11 Adjust the spiral cable locating pin (A) to the steering wheel locating pin hole (C)

DRIVE TRAIN

FRONT HALFSHAFTS

REMOVAL & INSTALLATION

Left Side
See Figures 12 through 14.

1. Before servicing the vehicle, refer to the Precautions Section.

2. Remove the wheels and tires.

3. Remove the wheel sensor and sensor harness. Refer to Wheel Speed Sensor Removal & Installation in the ABS, Brake section.

4. Remove the lock plate from the strut assembly.

5. Remove the caliper assembly. Hang

caliper assembly aside. Refer to Front Disc Brakes, Caliper Removal & Installation in the Brake section.

✻✻ WARNING

Never depress brake pedal while brake caliper is removed.

6. Remove the disc rotor.

7. Remove the cotter pin, and then loosen the wheel hub lock nut.

8. Using suitable puller, separate the halfshaft from the wheel hub and bearing assembly.

❈❈ WARNING
Never place the halfshaft joint at an extreme angle. Also be careful not to overextend the slide joint. Never allow the halfshaft to hang down without support for joint sub-assembly, housing assembly and the other parts.

9. Remove the wheel hub lock nut.

10. Remove the strut assembly from steering knuckle. Refer to Strut Removal & Installation in the Front Suspension section.

11. Remove the halfshaft from the transaxle assembly.

   a. Use the halfshaft attachment (SST: KV40107500) and a sliding hammer while inserting the tip of the halfshaft attachment between the housing assembly and the transaxle assembly.

❈❈ WARNING
Never place the halfshaft joint at an extreme angle when removing the halfshaft. Also be careful not to overextend slide joint. Confirm that the circular clip is attached to the halfshaft.

To install:
12. Note the following, and install in the reverse order of removal.

13. Always replace the differential side oil seal with a new one when installing halfshaft.

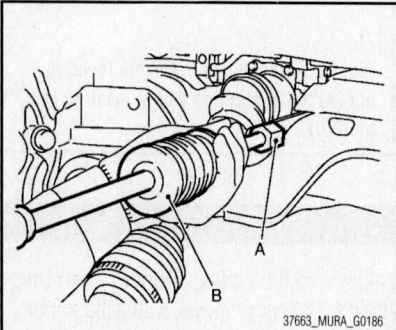

Fig. 12 Use the halfshaft attachment (A) (SST: KV40107500) and a sliding hammer (B) while inserting the tip of the halfshaft attachment between the housing assembly and the transaxle assembly

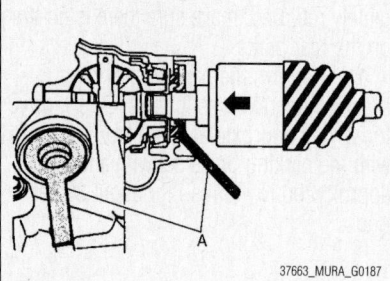

Fig. 13 Place the protector (A) (SST: KV38107900) onto the transaxle assembly

14. Place the protector (SST: KV38107900) onto the transaxle assembly to prevent damage to the oil seal while inserting the halfshaft. Slide the halfshaft sliding joint and tap with a hammer to install securely.

❈❈ WARNING
Make sure that the circular clip is completely engaged.

15. Clean the matching surfaces of the wheel hub lock nut and wheel hub and bearing assembly.

❈❈ WARNING
Never apply lubricating oil to the matching surfaces.

16. Clean the matching surfaces of the halfshaft and wheel hub and bearing assembly.

17. Apply paste (service part: 440037S000) to cover entire flat surface of the halfshaft joint sub assembly. Amount of paste: 0.008–0.035 oz. (0.2–1.0 g)

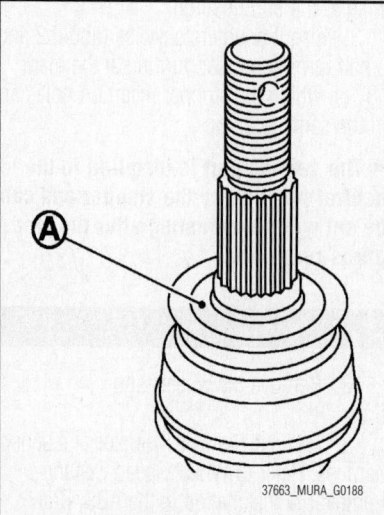

Fig. 14 Apply paste (service part: 440037S000) to the flat surface (A) of the halfshaft joint sub-assembly

❈❈ WARNING
Never use a power tool to tighten the wheel hub lock nut. Perform the final tightening of each of parts under unladen conditions. Never reuse the cotter pin.

Right Side

2WD Vehicles
See Figures 14 through 17.

1. Before servicing the vehicle, refer to the Precautions Section.

2. Remove the wheels and tires.

3. Remove the wheel sensor and sensor harness. Refer to Wheel Speed Sensor Removal & Installation in the ABS, Brake section.

4. Remove the lock plate from the strut assembly.

5. Remove the caliper assembly. Hang caliper assembly aside. Refer to Front Disc Brakes, Caliper Removal & Installation in the Brake section.

❈❈ WARNING
Never depress brake pedal while brake caliper is removed.

6. Remove the disc rotor.

7. Remove the cotter pin, and then loosen the wheel hub lock nut.

8. Using suitable puller, separate the halfshaft from the wheel hub and bearing assembly.

❈❈ WARNING
Never place the halfshaft joint at an extreme angle. Also be careful not to overextend the slide joint. Never allow the halfshaft to hang down without support for joint sub-assembly, housing assembly and the other parts.

9. Remove the wheel hub lock nut.

10. Remove the strut assembly from steering knuckle. Refer to Strut Removal & Installation in the Front Suspension section.

11. Remove the halfshaft from the wheel hub and bearing assembly.

12. Remove the bearing housing mounting bolts.

13. Remove the halfshaft from the transaxle assembly.

❈❈ WARNING
Never place the halfshaft joint at an extreme angle when removing the halfshaft. Also be careful not to overextend slide joint. Confirm that

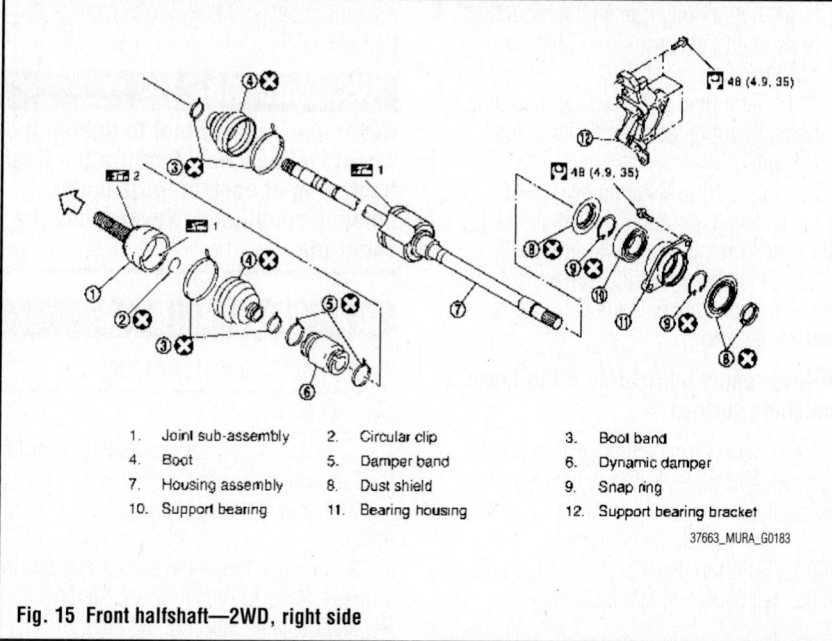

1. Joint sub-assembly
2. Circular clip
3. Boot band
4. Boot
5. Damper band
6. Dynamic damper
7. Housing assembly
8. Dust shield
9. Snap ring
10. Support bearing
11. Bearing housing
12. Support bearing bracket

37663_MURA_G0183

Fig. 15 Front halfshaft—2WD, right side

the circular clip is attached to the halfshaft.

14. Remove the support bearing bracket, follow the procedure described below.

    a. Remove the front exhaust tube.

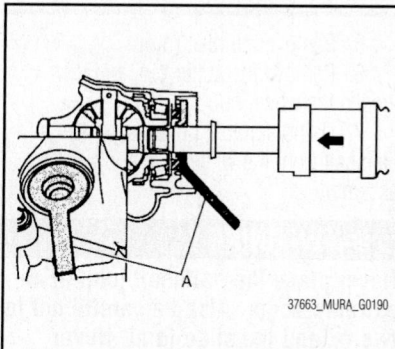

37663_MURA_G0190

Fig. 16 Place the protector (A) (SST: KV38107900) onto the transaxle assembly

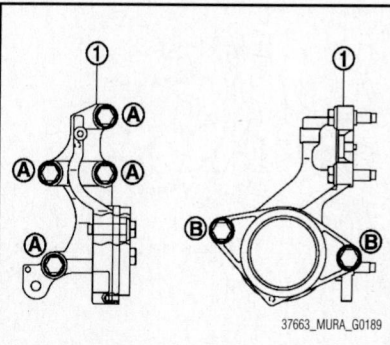

37663_MURA_G0189

Fig. 17 Install support bearing bracket (1) mounting bolts (A, B)

    b. Remove three way catalyst (Bank 1) and heated oxygen sensor harness bracket.

    c. Remove support bearing bracket.

To install:

15. Note the following, and install in the reverse order of removal.

16. Always replace the differential side oil seal with a new one when installing halfshaft.

17. Place the protector (SST: KV38107900) onto the transaxle assembly to prevent damage to the oil seal while inserting the halfshaft. Slide the halfshaft sliding joint and tap with a hammer to install securely.

18. Install support bearing bracket in following procedure,

    a. Temporarily tighten the mounting bolts, then tighten them to 35 ft. lbs. (48 Nm).

    b. Temporarily tighten the mounting bolts, then tighten them to 35 ft. lbs. (48 Nm).

19. Clean the matching surfaces of the halfshaft and wheel hub and bearing assembly.

20. Apply paste (service part: 440037S000) to cover entire flat surface of the halfshaft joint sub assembly. Amount of paste: 0.008–0.035 oz. (0.2–1.0 g)

✳✳ WARNING

Never use a power tool to tighten the wheel hub lock nut. Perform the final

tightening of each of parts under unladen conditions. Never reuse the cotter pin.

AWD Vehicles

See Figures 18 through 21.

1. Before servicing the vehicle, refer to the Precautions Section.

2. Remove the wheels and tires.

3. Remove the wheel sensor and sensor harness. Refer to Wheel Speed Sensor Removal & Installation in the ABS, Brake section.

4. Remove the lock plate from the strut assembly.

5. Remove the caliper assembly. Hang caliper assembly aside. Refer to Front Disc Brakes, Caliper Removal & Installation in the Brake section.

✳✳ WARNING

Never depress brake pedal while brake caliper is removed.

6. Remove the disc rotor.

7. Remove the cotter pin, and then loosen the wheel hub lock nut.

8. Using suitable puller, separate the halfshaft from the wheel hub and bearing assembly.

✳✳ WARNING

Never place the halfshaft joint at an extreme angle. Also be careful not to overextend the slide joint. Never allow the halfshaft to hang down without support for joint sub-assembly, housing assembly and the other parts.

9. Remove the wheel hub lock nut.

10. Remove the strut assembly from steering knuckle. Refer to Strut Removal & Installation in the Front Suspension section.

11. Remove the halfshaft from the wheel hub and bearing assembly.

12. Remove the halfshaft from link shaft.

    a. Use the halfshaft attachment (SST: KV40107500) and a sliding hammer while inserting the tip of the halfshaft attachment between the housing assembly and link shaft assembly.

✳✳ WARNING

Never place the halfshaft joint at an extreme angle when removing halfshaft. Also be careful not to overextend the slide joint.

13. Remove the bearing housing mounting bolts.

14. Remove the link shaft assembly from the support bearing bracket.

15. Remove the support bearing bracket, follow the procedure described below.

    a. Remove the front exhaust tube.

    b. Remove three way catalyst (Bank 1) and heated oxygen sensor harness bracket.

    c. Remove support bearing bracket.

To install:

16. Note the following, and install in the reverse order of removal.

17. Always replace the differential side oil seal with a new one when installing link shaft.

18. Place the protector (SST: KV38107900) onto the transaxle assembly to prevent damage to the oil seal while inserting the link shaft. Slide the link shaft sliding joint and tap with a hammer to install securely.

19. Install support bearing bracket in following procedure,

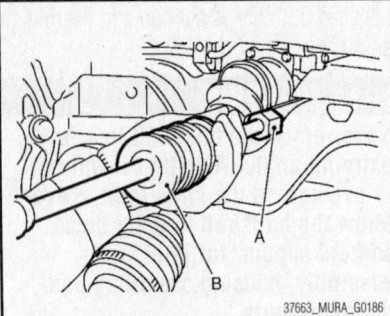

Fig. 18 Use the halfshaft attachment (A) (SST: KV40107500) and a sliding hammer (B) while inserting the tip of the halfshaft attachment between the housing assembly and the transaxle assembly

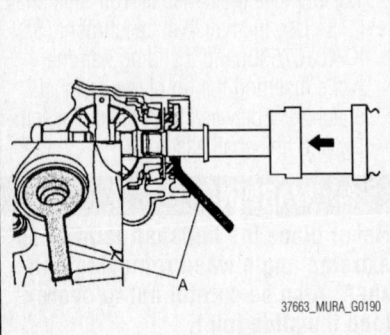

Fig. 19 Place the protector (A) (SST: KV38107900) onto the transaxle assembly

    a. Temporarily tighten the mounting bolts, then tighten them to 35 ft. lbs. (48 Nm).

    b. Temporarily tighten the mounting bolts, then tighten them to 35 ft. lbs. (48 Nm).

20. Apply NISSAN genuine grease to the halfshaft serration (link shaft side), and install the halfshaft onto link shaft.

21. Clean the matching surfaces of the wheel hub lock nut and wheel hub and bearing assembly.

➡**Never apply lubricating oil to these matching surface.**

22. Clean the matching surfaces of the halfshaft and wheel hub and bearing assembly.

23. Apply paste (service part: 440037S000) to cover entire flat surface of the halfshaft joint sub assembly.

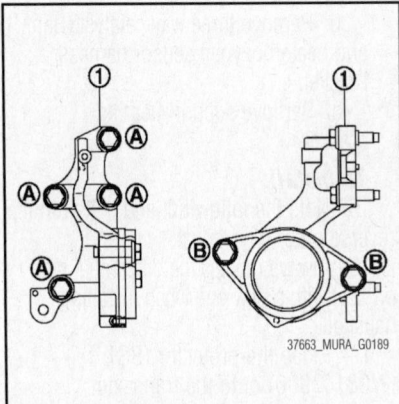

Fig. 20 Install support bearing bracket (1) mounting bolts (A, B)

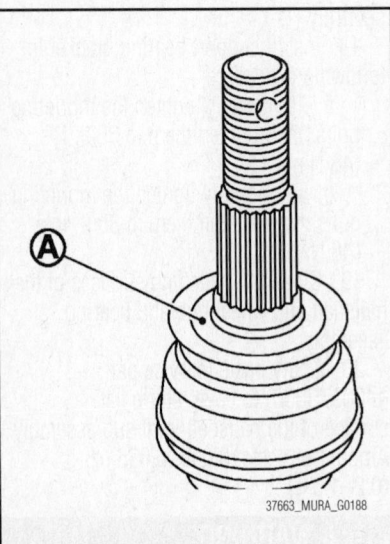

Fig. 21 Apply paste (service parts: 440037S000) to the flat surface (A) of the halfshaft joint sub-assembly

Amount of paste: 0.008–0.035 oz. (0.2–1.0 g)

> ✳✳ **WARNING**
>
> **Never use a power tool to tighten the wheel hub lock nut. Perform the final tightening of each of parts under unladen conditions. Never reuse the cotter pin.**

REAR HALFSHAFTS

REMOVAL & INSTALLATION

See Figures 22.

1. Before servicing the vehicle, refer to the Precautions Section.

2. Remove the wheels and tires.

3. Remove the wheel sensor and sensor harness. Refer to Wheel Speed Sensor Removal & Installation in the ABS, Brake section.

4. Remove the caliper assembly. Hang caliper assembly aside. Refer to Rear Disc Brakes, Caliper Removal & Installation in the Brake section.

> ✳✳ **WARNING**
>
> **Never depress brake pedal while brake caliper is removed.**

5. Remove the disc rotor.

6. Remove the cotter pin, and then loosen the wheel hub lock nut.

7. Using suitable puller, separate the halfshaft from the wheel hub and bearing assembly.

> ✳✳ **WARNING**
>
> **Never place the halfshaft joint at an extreme angle. Also be careful not to overextend the slide joint. Never**

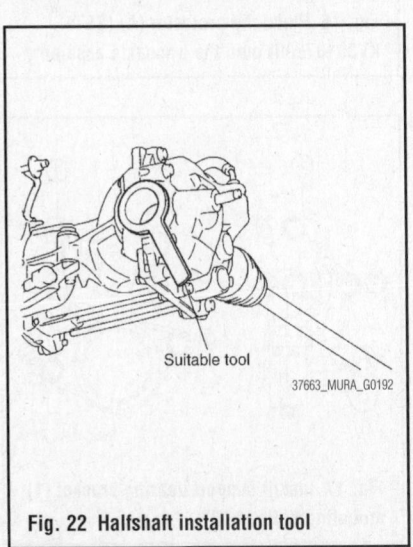

Fig. 22 Halfshaft installation tool

allow the halfshaft to hang down without support for joint sub-assembly, housing assembly and the other parts.

8. Remove the wheel hub lock nut.

9. Remove the wheel hub and bearing assembly.

10. Remove the hub cap.

11. Remove the halfshaft from the final drive assembly.

To install:

12. Note the following, and install in the reverse order of removal.

13. In order to prevent damage to the final drive of side oil seal, first fit to protector (suitable tool) onto side oil seal before inserting halfshaft. Slide halfshaft into slide joint tap with a hammer to install securely.

✳✳ WARNING

Never use a power tool to tighten the wheel hub lock nut. Perform the final tightening of each of parts under unladen conditions. Never reuse the cotter pin.

REAR PINION SEAL

REMOVAL & INSTALLATION

See Figures 23 through 26.

1. Before servicing the vehicle, refer to the Precautions Section.

2. Raise and support the vehicle safely.

3. Remove the propeller shaft. Refer to Propeller Shaft Removal & Installation.

4. Using paint, put a matchmark on the thread edge of the electric controlled coupling. The matchmark should be in line with the matching mark on the companion flange.

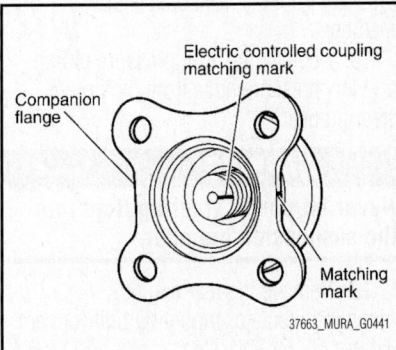

Fig. 23 Matchmark on the thread edge of the electric controlled coupling

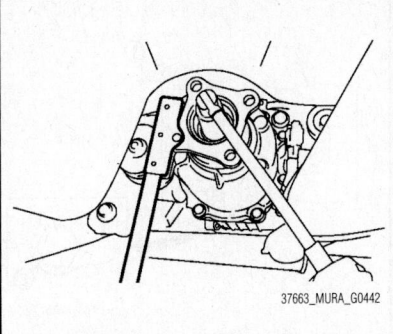

Fig. 24 Remove the companion flange lock nut and the companion flange

5. Using a flange wrench, remove the companion flange lock nut and remove the companion flange.

6. Using a suitable tool, remove the oil seal from the coupling cover. Be careful not to damage coupling cover.

To install:

7. Using the drifts install the oil seal until it becomes flush with the coupling cover end. Apply multi-purpose grease onto oil seal lips, and gear oil onto the circumference of oil seal.

✳✳ WARNING

Never reuse oil seal. When installing, never incline oil seal.

8. Align the matchmark of electric controlled coupling with the matchmark of companion flange, and install the companion flange.

9. Install the companion flange lock nut with a flange wrench, and tighten to 103 ft. lbs. (140 Nm). Never reuse the companion flange lock nut.

10. Install the propeller shaft.

11. Check the oil level.

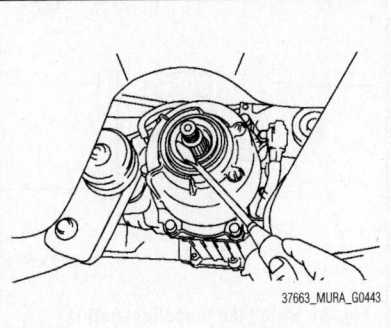

Fig. 25 Remove the oil seal from the coupling cover

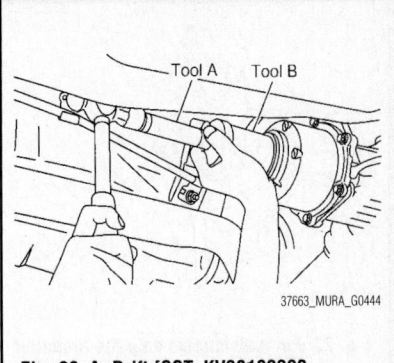

Fig. 26 A: Drift [SST: KV38100200 (J-26233)], B: Drift (SST: ST27861000)

REAR PROPELLER SHAFT

REMOVAL & INSTALLATION

See Figures 27 through 31.

1. Before servicing the vehicle, refer to the Precautions Section.

2. Disconnect the negative battery cable.

3. Shift the transaxle to the neutral position, and then release the parking brake.

4. Put matchmarks onto the propeller shaft flange yoke and the final drive and transfer companion flanges. Use paint for matchmarks.

5. Loosen the upper and lower center bearing mounting bracket mounting nuts. Tighten the mounting nuts temporarily.

6. Remove the propeller shaft assembly fixing bolts and nuts.

7. Remove the center bearing mounting bracket fixing nuts.

8. Remove the propeller shaft assembly.

✳✳ WARNING

If the constant velocity joint was bent during propeller shaft assembly removal, installation, or transportation, its boot may be damaged. Wrap boot interference area to metal part with shop cloth or rubber to protect boot from breakage.

9. Remove the clips and the center bearing mounting bracket (upper/lower).

To install:

10. Note the following, and install in the reverse order of removal.

11. Install the upper center bearing mounting bracket with its arrow mark facing forward.

12. Adjust the position of the center bearing mounting brackets, sliding back

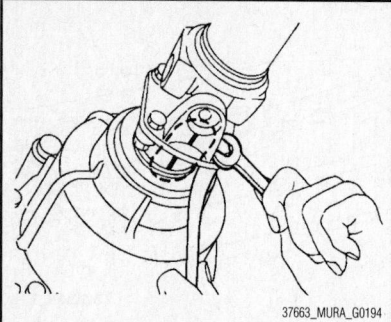

Fig. 27 Put matchmarks onto the propeller shaft flange yoke and the final drive and transfer companion flanges

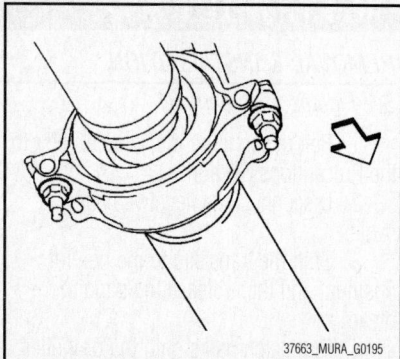

Fig. 28 Upper and lower center bearing mounting bracket mounting nuts

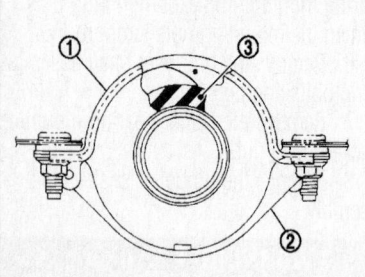

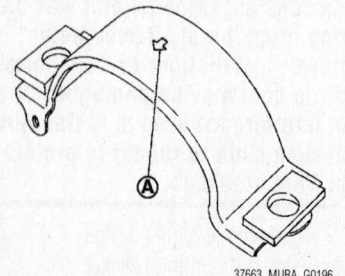

Fig. 29 Install the upper center bearing mounting bracket (1) with its arrow mark (A) facing forward and adjust the position of the upper center bearing mounting bracket and the lower center bearing mounting bracket (2), sliding back and forth to prevent play in thrust direction of center bearing insulator (3)

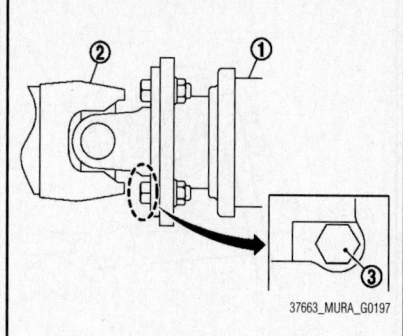

Fig. 30 Final drive assembly (1), propeller shaft assembly (2), and bolts (3)

and forth to prevent play in thrust direction of center bearing insulator. Install the upper and lower center bearing mounting brackets to the vehicle.

13. Align the matchmarks to install the propeller shaft assembly to the final drive and transfer companion flanges.

14. After assembly, perform a driving test to check propeller shaft vibration. If vibration occurs, separate the propeller shaft from final drive. Reinstall the companion flange after rotating it by 90, 180, 270 degrees. Perform the driving test and check propeller shaft vibration again at each point.

15. After tightening the bolts and nuts to 37 ft. lbs. (50 Nm), check that the bolts on the flange side are tightened as shown.

16. If the propeller shaft assembly or final drive assembly has been replaced, connect them as follows:

   a. Install the propeller shaft while aligning its matchmark with the matchmark of the final drive on the joint as close as possible.

   b. Tighten the mounting bolts and nuts of the propeller shaft and final drive to 37 ft. lbs. (50 Nm).

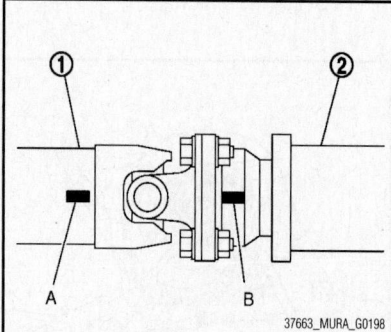

Fig. 31 Install the propeller shaft (1) while aligning its matchmark (A) with the matchmark (B) of the final drive (2) on the joint as close as possible

TRANSFER CASE ASSEMBLY

REMOVAL & INSTALLATION

See Figures 32 through 36.

1. Before servicing the vehicle, refer to the Precautions Section.
2. Disconnect the negative battery cable.
3. Remove the extension cowl top panel (lower and upper).
4. Remove the battery and battery tray.
5. Remove the air duct (inlet), air cleaner case (upper and lower) with mass air flow sensor and air duct assembly.
6. Remove the air fuel ratio sensor 1 (Bank 1).
7. Remove the front wheels and tires.
8. Remove the splash guards.
9. Remove the engine cover.
10. Remove the air guide.
11. Remove the right front halfshaft. Refer to Front Halfshaft Removal & Installation.

✸✸ WARNING

Be careful not to damage the gear ring oil seal inside of the transfer assembly.

12. Remove the exhaust front tube.
13. Separate the rear propeller shaft. Refer to Propeller Shaft Removal & Installation.
14. Disconnect the oxygen sensor 2 (Bank 1) harness connector.
15. Remove the heat insulator from the front suspension member.
16. Remove the three way catalyst (Bank 1).

✸✸ WARNING

Handle carefully to avoid any shock to the three way catalyst.

17. Remove the power steering tube brackets from the front suspension member.
18. Remove the high pressure piping and low pressure hose from the power steering gear.

✸✸ WARNING

Never let power steering fluid into the suspension member.

19. Remove the rear engine mounting insulator mounting bolt (upper side).
20. Support the transaxle assembly with a suitable jack.

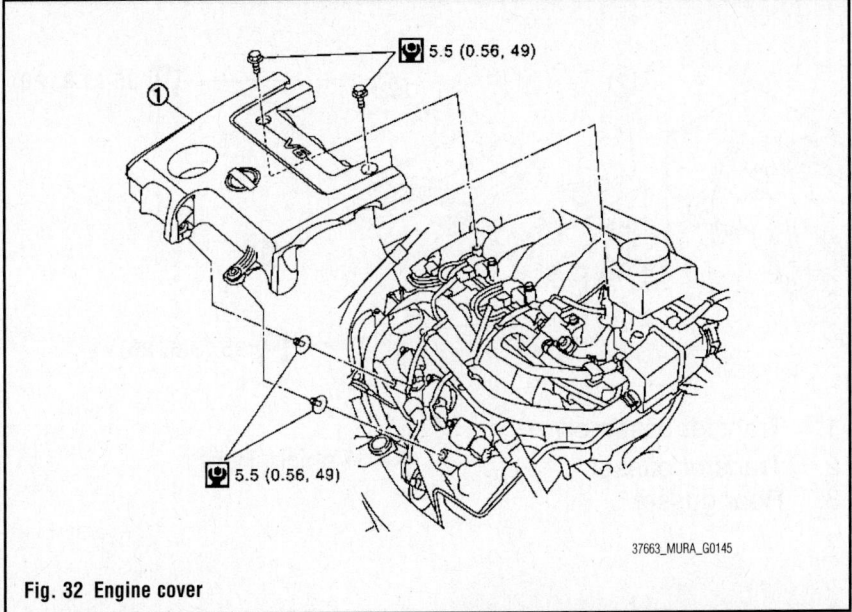

5.5 (0.56, 49)

5.5 (0.56, 49)

37663_MURA_G0145

Fig. 32 Engine cover

37663_MURA_G0146

Fig. 33 Remove the air guide (1)

✳✳ WARNING

When setting the transmission jack, be careful not to allow it to collide against the drain plug.

21. Support the front suspension member with a suitable jack.
22. Remove the rear gusset and the transfer gusset.
23. Remove the rear engine mounting insulator mounting bolt (lower side).
24. Disconnect the front and rear engine mounting insulator

harness connector and harness clip.

25. Move the rear engine mounting insulator to remove the rear engine mounting bracket.

✳✳ WARNING

Be careful not to damage the power steering gear boot.

26. Remove the rear engine mounting bracket.
27. Remove the left-hand and right-hand engine mounting brackets.
28. Remove the member stay, front suspension member mounting bolts and nuts.
29. Lower the jack for the front suspension member to the height where the transfer assembly can be removed.
30. Remove the transaxle assembly mounting bolts and the transfer assembly mounting bolts.
31. Remove the transfer assembly from the vehicle.

✳✳ WARNING

Be careful not to damage the air breather hose. After removing the transfer assembly from transaxle, be sure to replace differential side oil seal of the transaxle side with new one.

To install:

32. Note the following, and install in the reverse order of removal.

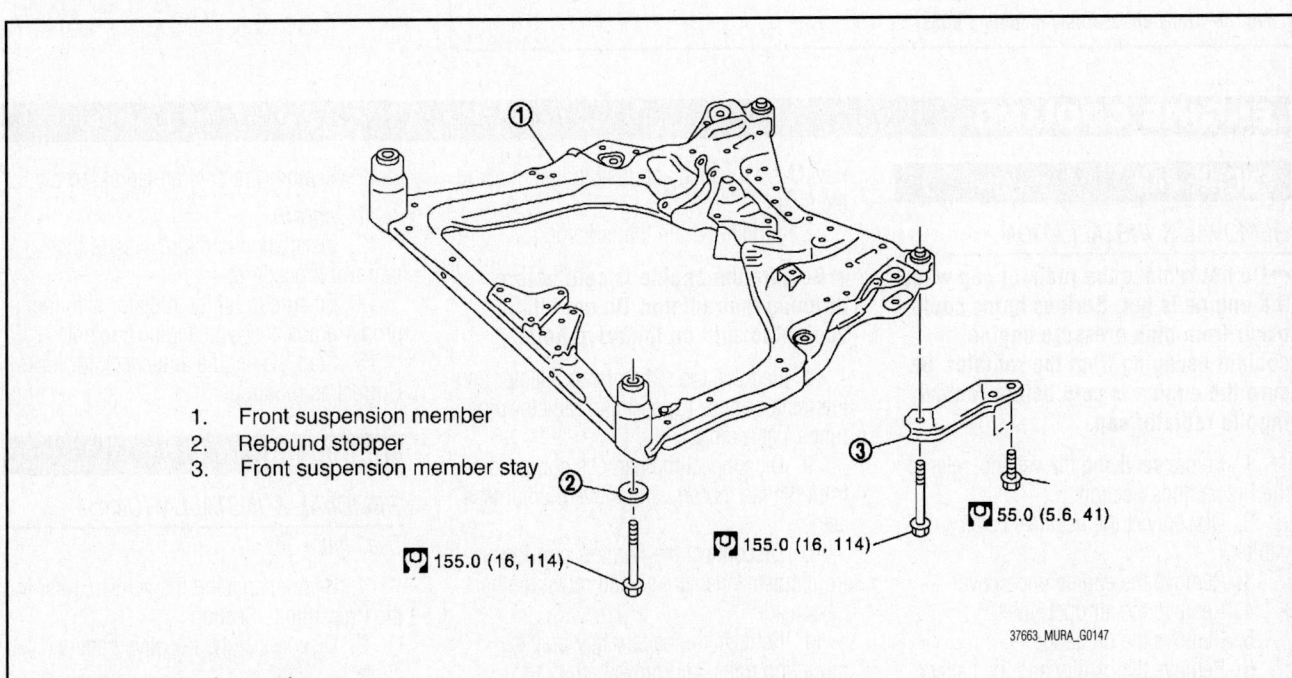

1. Front suspension member
2. Rebound stopper
3. Front suspension member stay

55.0 (5.6, 41)

155.0 (16, 114)

155.0 (16, 114)

37663_MURA_G0147

Fig. 34 Front suspension member

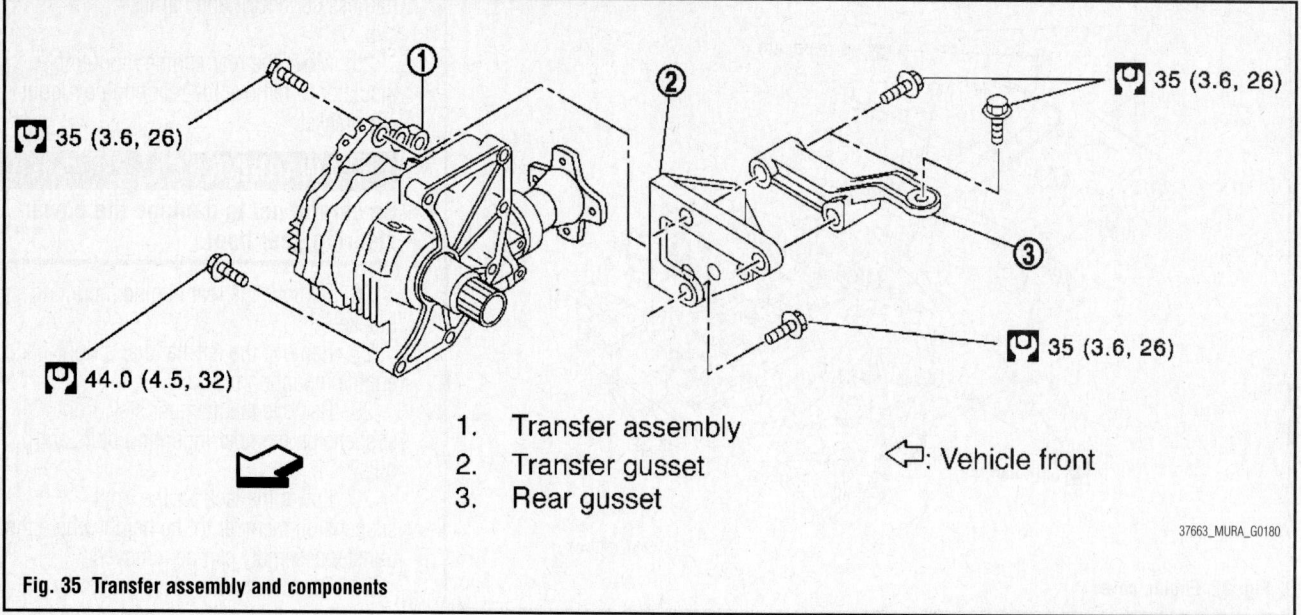

1. Transfer assembly
2. Transfer gusset
3. Rear gusset

⟵ : Vehicle front

37663_MURA_G0180

Fig. 35 Transfer assembly and components

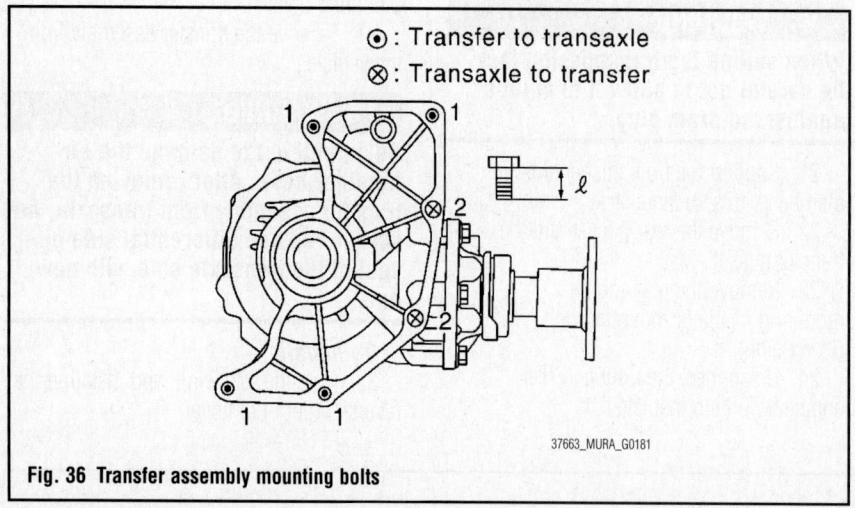

⊙ : Transfer to transaxle
⊗ : Transaxle to transfer

37663_MURA_G0181

Fig. 36 Transfer assembly mounting bolts

33. Install the mounting bolts:
- Bolt No. 1 (Qty: 4, length: 2.56 in. (65mm): Tighten to 32 ft. lbs. (44 Nm)
- Bolt No. 2 (Qty: 2, length: 1.57 in. (40mm): Tighten to 26 ft. lbs. (35 Nm)

✳✳ WARNING

Be careful not to damage the transaxle oil seal.

34. Check oil level and check for oil leakage after installation.

ENGINE COOLING

ENGINE FAN

REMOVAL & INSTALLATION

➡ **Do not remove the radiator cap when the engine is hot. Serious burns could occur from high pressure engine coolant escaping from the radiator. Be sure the engine is cold before removing the radiator cap.**

1. Before servicing the vehicle, refer to the Precautions Section.
2. Disconnect the negative battery cable.
3. Remove the engine undercover.
4. Remove the air duct (inlet).
5. Remove the oil gauge.
6. Remove the battery and the battery

tray. Move the fuse and fusible link block to the side.
7. Properly drain the radiator.

➡ **Be sure the engine is cold before draining the radiator. Do not allow coolant to spill on the drive belts.**

8. Remove the radiator cap adapter and the radiator hose (upper) and radiator pipe (upper) assembly.
9. Disconnect the harness connector from the fan motors, and move the harness aside.
10. Disconnect the harness connector from crash zone sensor, and move the harness aside.
11. Remove the battery tray bracket mounting bolts, and move battery tray bracket to aside.

12. Remove the cooling fan assembly.

To install:

13. Installation is the reverse of the removal procedure.
14. Be sure to fill the radiator with the proper grade and type engine coolant.
15. Start the engine and check for leaks. Correct as required.

RADIATOR

REMOVAL & INSTALLATION

See Figure 37.

1. Before servicing the vehicle, refer to the Precautions Section.
2. Disconnect the negative battery cable.

3. Remove the engine undercover.

4. Remove the right and left radiator core support covers.

5. Remove the air duct (inlet).

6. Remove the front grill.

7. Remove the horn.

8. Remove the hood lock.

9. Drain engine coolant from radiator.

❋❋ CAUTION

Perform this step when the engine is cold. Do not spill engine coolant on drive belt.

10. Disconnect the reservoir tank hose from the upper radiator pipe.

11. Disconnect the CVT fluid cooler hoses from the radiator. Install blind plug to avoid leakage of CVT fluid.

12. Remove the radiator cap adapter and the upper radiator hose and the upper radiator pipe assembly.

❋❋ WARNING

Be careful not to allow engine coolant to contact drive belt.

13. Disconnect the lower radiator hose from radiator.

14. Remove the condenser. Refer to Condenser Removal & Installation in the Heating & Air Conditioning section.

❋❋ WARNING

Be careful not to damage the condenser core.

15. Remove the radiator upper clips by pulling the tabs outside to release the lock and then remove the upper mounting rubbers.

❋❋ WARNING

Never pull the tabs outside excessively.

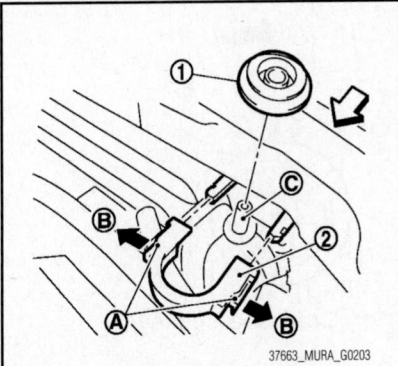

Fig. 37 Clip (2), tab (A), lock (B) upper mounting rubber (1), and mounting pin (C)

16. Lift up and remove the radiator from the front of radiator core support.

❋❋ WARNING

Be careful not to damage or scratch the radiator core.

To install:

17. Note the following, and install in the reverse order of removal.

18. Install the radiator upper clips on the radiator core connection as follows:

a. Install the upper mounting rubber on the radiator mounting pin.

b. Align the radiator upper clip with the radiator core connection, then insert each the radiator upper clip straight into the radiator core connections until a click is heard.

c. After connecting the radiator upper clips, visually confirm that the radiator upper clips are connected to the radiator core connections and move the radiator upper clips and the radiator forward and backward to check they are securely connected.

THERMOSTAT

REMOVAL & INSTALLATION

See Figures 38 and 39.

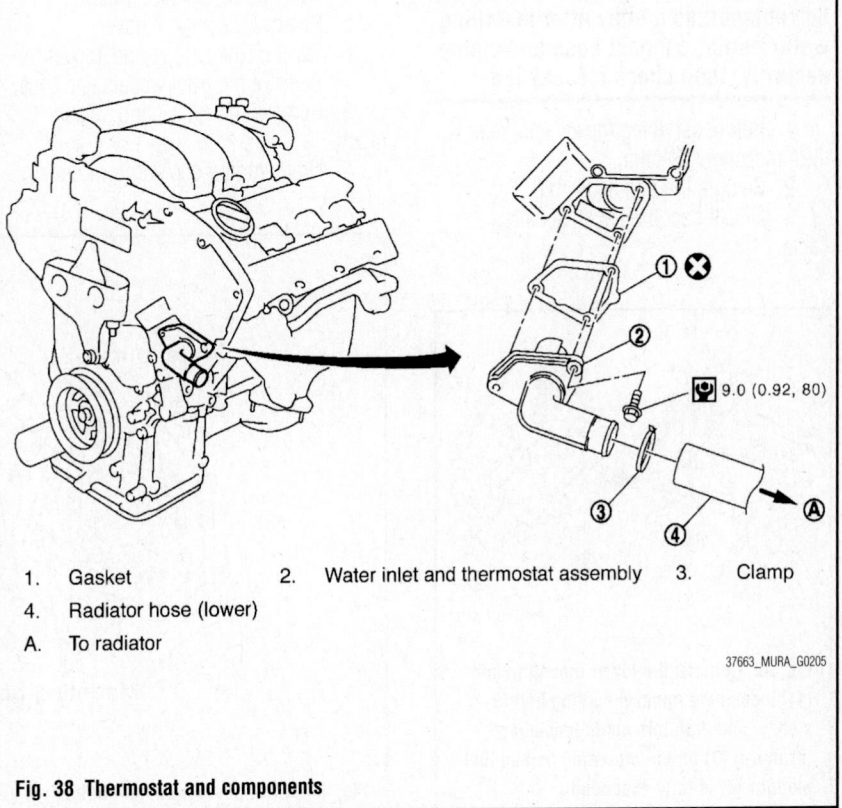

1. Gasket
2. Water inlet and thermostat assembly
3. Clamp
4. Radiator hose (lower)
A. To radiator

9.0 (0.92, 80)

37663_MURA_G0205

Fig. 38 Thermostat and components

➡**Never remove the radiator cap when the engine is hot. Serious burns could occur from high-pressure engine coolant escaping from the radiator.**

1. Before servicing the vehicle, refer to the Precautions Section.

2. Be sure the engine is cold.

3. Disconnect the negative battery cable.

4. Drain the engine coolant using the radiator drain plug and the water drain plug at the front of the cylinder block.

5. Remove the reservoir tank retaining bolts, and move it to the side.

6. Remove the intake valve timing control solenoid valve (Bank 2).

7. Disconnect the lower radiator hose from the water inlet and thermostat assembly.

8. Remove the water inlet and thermostat housing retaining bolts.

9. Remove the assembly from the engine.

➡**Do not disassemble the water inlet and thermostat assembly. Replace them as a unit, if required.**

To install:

10. Installation is the reverse of the removal procedure.

11. Be sure to refill the cooling

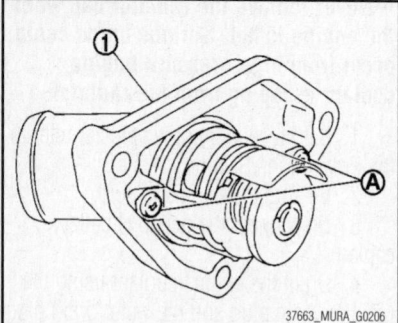

Fig. 39 Remove the water inlet and thermostat assembly (1) Never loosen these screws (A)

using the proper grade and type engine coolant.

12. Start the engine and check for leaks.

13. Start the engine and allow it to reach operation temperature. Recheck the coolant level, fill as required.

WATER PUMP

REMOVAL & INSTALLATION

See Figures 40 through 49.

✳✳ WARNING

When removing water pump assembly, be careful not to get engine coolant on drive belt. Water pump cannot be disassembled and should be replaced as a unit. After installing water pump, connect hose and clamp securely, then check for leakage.

1. Before servicing the vehicle, refer to the Precautions Section.

2. Be sure the engine is cold.

3. Disconnect the negative battery cable.

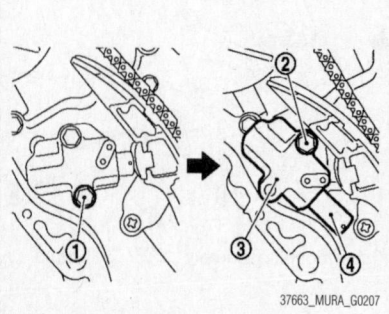

Fig. 40 Remove the lower mounting bolt (1), loosen the upper mounting bolt (2) slowly, and then turn chain tensioner (primary) (3) on the mounting bolt so that plunger (4) is fully expanded

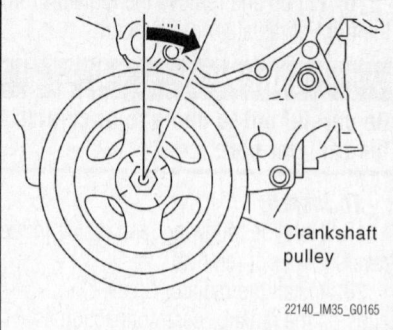

Fig. 41 Crankshaft pulleys clockwise rotation to loosen the timing chain (primary) side

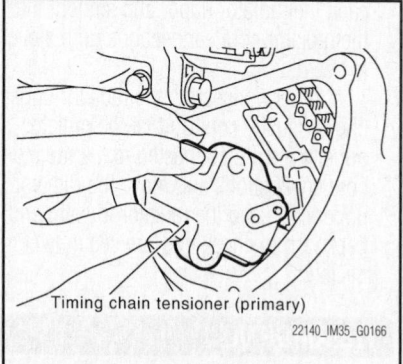

Fig. 42 Removal of the timing chain tensioner (primary)

4. Remove the air duct (inlet).

5. Remove the engine cover.

6. Remove the engine undercover.

7. Remove the front wheels and tires.

8. Remove the right-hand splash guard.

9. Drain engine coolant from radiator.

✳✳ CAUTION

Perform this step when the engine is cold. Do not spill engine coolant on drive belt.

10. Remove the drive belt. Refer to Accessory Drive Belt Removal & Installation in the Engine Mechanical section.

11. Remove the idler pulleys.

12. Remove the radiator reservoir tank.

13. Remove the reservoir tank of power steering oil pump with piping connected, and move it aside.

14. Support the oil pan (lower) bottom with transmission jack.

15. Remove the right-hand engine mounting insulator, right-hand engine mounting bracket and the upper torque rod.

16. Remove the water drain plug (front) on water pump side of cylinder block to drain engine coolant from the engine.

17. Remove the valve timing control cover (Bank 1) and the water pump cover from the front timing chain case. Cut the liquid gasket for removal.

18. Remove the timing chain tensioner (primary) as follows:

a. Remove the lower mounting bolt.

✳✳ WARNING

Be careful not to drop mounting bolt inside timing chain case.

b. Loosen the upper mounting bolt slowly, and then turn chain tensioner (primary) on the mounting bolt so that plunger is fully expanded.

➡**Even if plunger is fully expanded, it is not dropped from the body of timing chain tensioner (primary).**

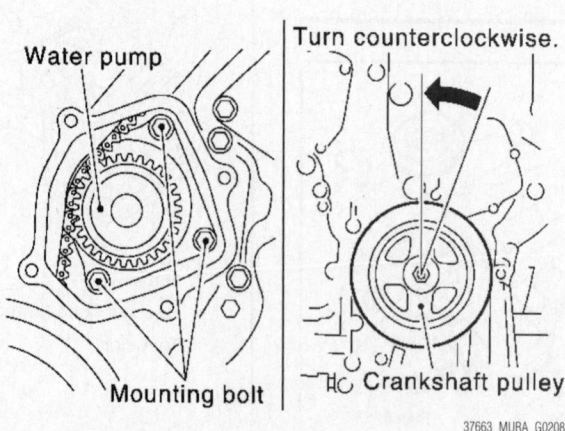

Fig. 43 Water pump mounting bolts

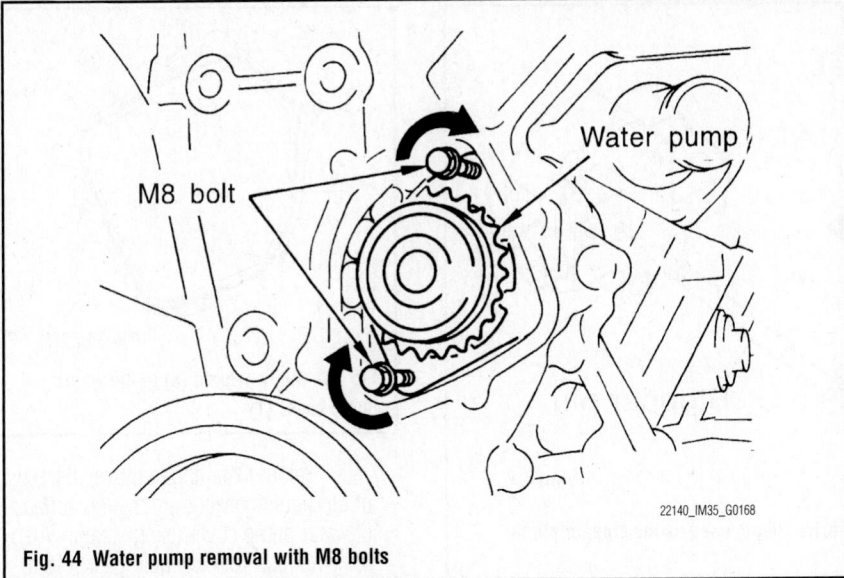

Water pump
M8 bolt
22140_IM35_G0168
Fig. 44 Water pump removal with M8 bolts

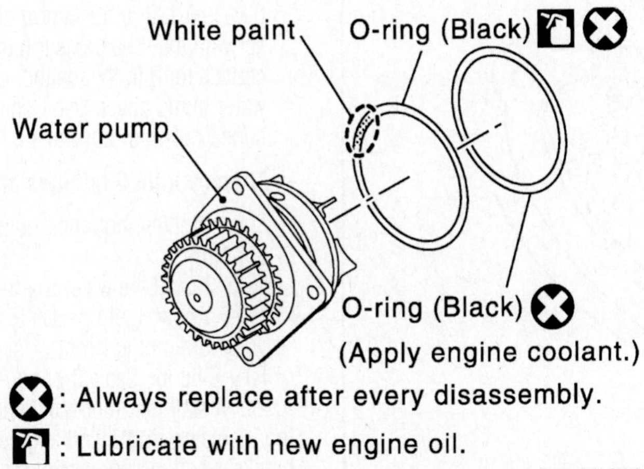

White paint
O-ring (Black)
Water pump
O-ring (Black)
(Apply engine coolant.)
: Always replace after every disassembly.
: Lubricate with new engine oil.
22140_IM35_G0169
Fig. 45 Water pump installation

c. Turn the crankshaft pulley clockwise so that the timing chain on the timing chain tensioner (primary) side is loose.

d. Remove upper mounting bolt, and then remove timing chain tensioner (primary).

19. Remove water pump as follows:

a. Remove the three water pump mounting bolts. Secure a gap between the water pump gear and timing chain, by turning the crankshaft pulley counterclockwise until timing chain water pump sprocket reaches maximum looseness.

b. Screw M8 bolts: pitch: 0.0492 in (1.25 mm) length: approx. 1.97 in (50 mm); into water pumps upper and lower mounting bolt holes until they reach timing chain case. Then, alternately tighten each bolt for a half turn, and pull out water pump.

c. Pull straight out while preventing vane from contacting socket in installation area. Remove water pump without causing sprocket to contact timing chain.

d. Remove M8 bolts and O-rings from water pump.

WARNING

Never disassemble water pump.

To install:

20. Install new O-rings to water pump.

a. Apply engine oil and engine coolant to O-rings as shown.

b. Locate O-ring with white paint mark to engine front side.

21. Install the water pump:

WARNING

Never allow cylinder block to nip O-rings when installing water pump.

a. Check that timing chain and water pump sprocket are engaged.

b. Install the water pump by tightening mounting bolts alternately and evenly to 85 inch lbs. (10 Nm).

22. Install timing chain tensioner (primary) as follows:

a. Turn crankshaft pulley clockwise so that timing chain on the timing chain tensioner (primary) side is loose.

b. Pull plunger stopper tab up (or turn lever downward) so as to remove plunger stopper tab from the ratchet of plunger.

22140_IM35_G0170
Fig. 46 Installing the timing chain to the water pump

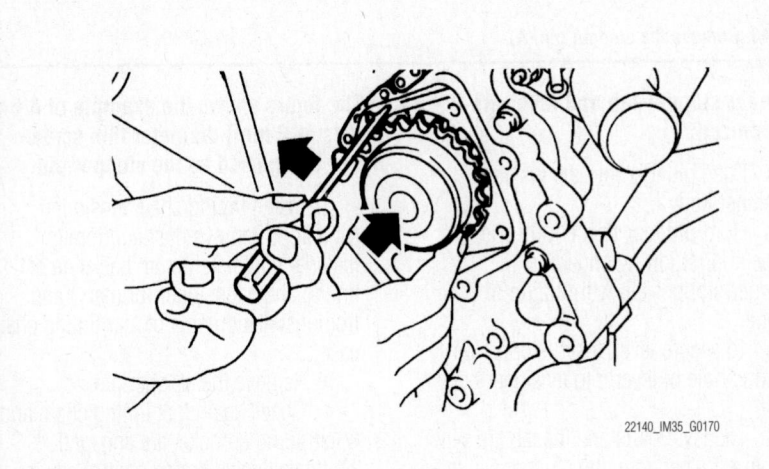

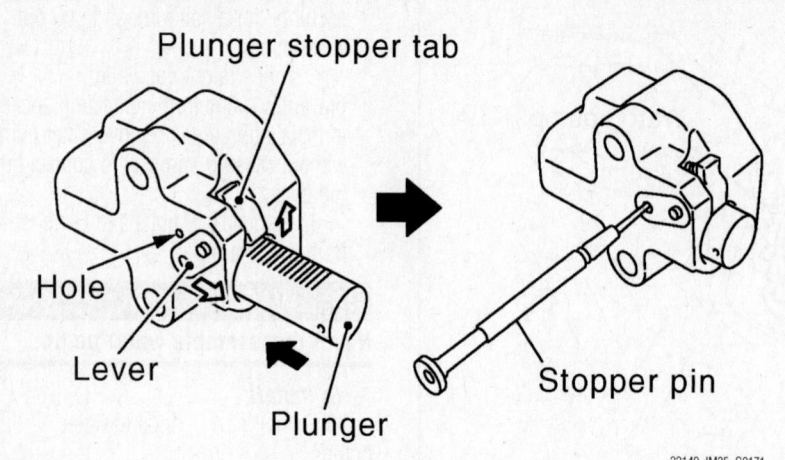

Fig. 47 0.047 inches (1.2 mm) diameter thin screwdriver being used as the stopper pin for the timing chain tensioner

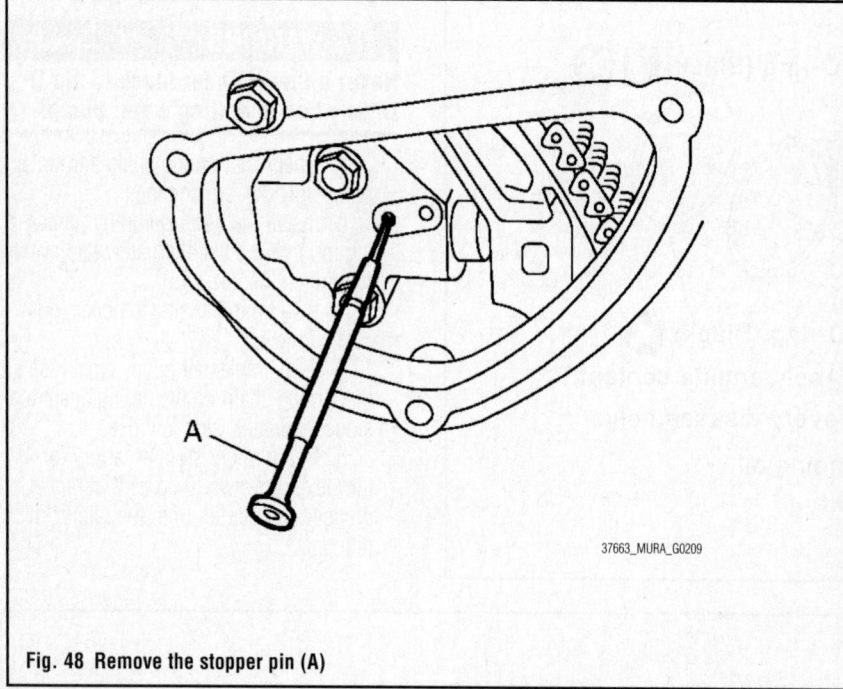

Fig. 48 Remove the stopper pin (A)

➡**Plunger stopper tab and lever are synchronized.**

c. Push plunger into the inside of tensioner body.

d. Hold plunger in the fully compressed position by engaging plunger stopper tab with the tip of ratchet.

e. To secure lever, insert stopper pin through hole of lever into tensioner body hole.

f. The lever parts and the tab are synchronized. Therefore, the plunger will be secured under this condition.

➡**The figure shows the example of 0.047 inches (1.2 mm) diameter thin screwdriver being used as the stopper pin.**

g. Install timing chain tensioner (primary). Remove dust and foreign material completely from backside of timing chain tensioner (primary) and from installation area of rear timing chain case.

h. Remove the stopper pin.

i. Check again that timing chain and water pump sprocket are engaged.

23. Install valve timing control cover (Bank 1) and water pump cover as follows:

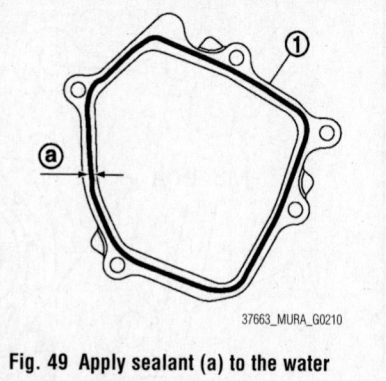

Fig. 49 Apply sealant (a) to the water pump cover (1)

a. Before installing, remove all traces of old liquid gasket from mating surface of water pump cover using scraper. Also remove traces of old liquid gasket from the mating surface of front timing chain case.

b. Apply a continuous bead 0.091–0.130 in. (2–3mm) of liquid gasket with the tube presser (commercial service tool) to the mating surface of water pump cover. Use Genuine RTV Silicone Sealant or equivalent.

➡**Attach within 5 minutes after coating.**

c. Tighten mounting bolts to 8 ft. lbs. (11 Nm).

d. Install the water drain plug.

e. Apply liquid gasket to the thread of water drain plug (front). Use Genuine RTV Silicone Sealant or equivalent.

24. Install the right-hand engine mounting insulator, right-hand engine mounting bracket and the upper torque rod.

25. Install the reservoir tank of power steering oil pump.

26. Install the radiator reservoir tank

27. Install the idler pulleys.

28. Install the drive belt.

29. Install the right-hand splash guard.

30. Install the front wheels and tires.

31. Install the engine undercover.

32. Install the engine cover.

33. Install the air duct (inlet).

34. Refill engine coolant.

35. Connect the negative battery cable.

36. After starting engine, let idle for three minutes, then rev engine up to 3,000 rpm under no load to purge air from the high-pressure chamber of chain tensioner. Engine may produce a rattling noise. This indicates that air still remains in the chamber and is not a matter of concern.

ENGINE ELECTRICAL

CHARGING SYSTEM

REMOVAL & INSTALLATION

See Figure 50.

1. Before servicing the vehicle, refer to the Precautions Section.
2. Disconnect the negative battery terminal.
3. Remove the engine cover.
4. Remove the front right-hand wheel and tire.
5. Remove the right-hand splash guard.
6. Remove the air cleaner and air duct assembly.
7. Remove the drive belt. Refer to Accessory Drive Belt Removal & Installation in the Engine Mechanical section.
8. Remove the A/C compressor. Refer to Compressor Removal & Installation in the Heating & Air Conditioning section.
9. Remove the idler pulley.
10. Disconnect the oil pressure switch.

11. Disconnect the alternator harness connectors.
12. Remove the alternator bolt and nuts, using power tools.
13. Slide the alternator out and remove.

To install:

14. Installation is in the reverse order of removal.
15. Temporarily tighten all of alternator bolt and nuts. Tighten them in numerical order shown.
    - Nut No. 1: 62 ft. lbs. (83 Nm)
    - Bolt No. 2, 3: 21 ft. lbs. (28 Nm)
16. Be sure to tighten the "B" terminal nut carefully.
17. Install the alternator and check tension of belt.
18. For this model, the power generation voltage variable control system that controls the power generation voltage of the alternator has been adopted. Therefore, the power

generation voltage variable control system operation inspection should be performed after replacing the alternator, and then make sure that the system operates normally.

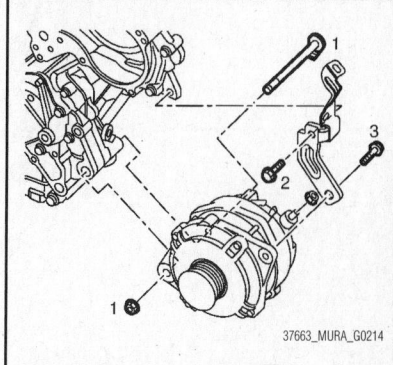

Fig. 50 Alternator bolt tightening sequence

ENGINE ELECTRICAL

IGNITION SYSTEM

FIRING ORDER

See Figure 51.

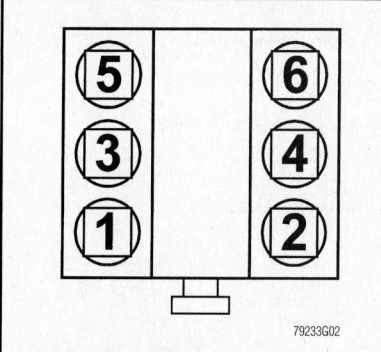

Fig. 51 3.5L Engines
Firing order: 1–2–3–4–5–6
Distributor less ignition system

IGNITION COIL

REMOVAL & INSTALLATION

See Figure 66.

1. Before servicing the vehicle, refer to the Precautions Section.
2. Disconnect the negative battery cable.
3. Remove the engine cover.
4. Remove the upper and lower air cleaner cases and the air duct assembly.

5. Remove the electric throttle control actuator.
6. Remove the intake manifold collector.
7. Move aside the wiring harness, wiring harness bracket, and hoses located above ignition coil.
8. Disconnect the wiring harness connector from the ignition coil.
9. Remove the ignition coil.

⁂ CAUTION

Do not subject the ignition coils to excessive shock or vibration.

To install:

10. Install the ignition coil on the engine.
11. Reconnect the wiring harness to the coil.
12. Reposition the wiring harness, bracket and hoses.
13. Install the air duct and the engine cover.

IGNITION TIMING

INSPECTION

1. Before servicing the vehicle, refer to the Precautions Section.
2. Remove the number one ignition coil.
3. Connect the number one ignition coil and spark plug with a suitable high tension wire.
4. Attach the timing light clamp to the wire.

5. Check the ignition timing.

ADJUSTMENT

The ignition timing is controlled by the ECM to maintain the best air-fuel ratio for every running condition of the engine. The ignition timing data is stored in the ECM.

SPARK PLUGS

REMOVAL & INSTALLATION

See Figure 52.

1. Before servicing the vehicle, refer to the Precautions Section.
2. Disconnect the negative battery cable.
3. Remove the engine cover.
4. Remove the upper and lower air cleaner cases and the air duct assembly.
5. Remove the electric throttle control actuator.
6. Remove the intake manifold collector.
7. Remove the ignition coil.
8. Remove the spark plug using a spark plug socket and wrench.

To install:

9. Install the ignition coil on the engine.
10. Reconnect the wiring harness to the coil.
11. Reposition the wiring harness, bracket and hoses.
12. Install the air duct and the engine cover.

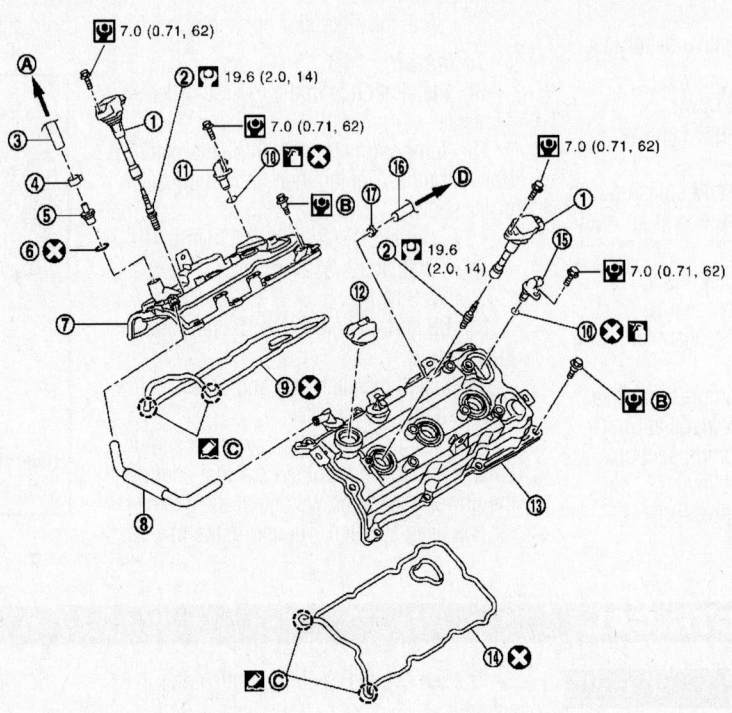

7.0 (0.71, 62)
19.6 (2.0, 14)
7.0 (0.71, 62)
7.0 (0.71, 62)
19.6 (2.0, 14)
7.0 (0.71, 62)

1.	Ignition coil	2.	Spark plug	3.	PCV hose
4.	Clamp	5.	PCV valve	6.	O-ring
7.	Rocker cover (bank 1)	8.	PCV hose	9.	Rocker cover gasket (bank 1)
10.	O-ring	11.	Camshaft position sensor (PHASE) (bank 1)	12.	Oil filler cap
13.	Rocker cover (bank 2)	14.	Rocker cover gasket (bank 2)	15.	Camshaft position sensor (PHASE) (bank 2)
16.	PCV hose	17.	Clamp		
A.	To intake manifold collector			C.	Camshaft bracket side
D.	To air duct assembly				

37663_MURA_G0218

Fig. 52 Ignition coil, spark plugs, and rocker cover components

ENGINE ELECTRICAL   STARTING SYSTEM

STARTER

REMOVAL & INSTALLATION

1. Before servicing the vehicle, refer to the Precautions Section.

2. Remove the battery.

3. Remove the air cleaner assembly and air ducts.

4. Disconnect the following unit connectors:
   - ECM
   - TCM
   - IPDM E/R

5. Remove the battery tray.

6. Disconnect the starter motor harness connectors.

7. Remove the starter motor mounting bolts, using power tools.

8. Remove the starter motor.

To install:

9. Installation is in the reverse order of removal. Tighten the mounting bolts to 41 ft. lbs. (55 Nm)

SOLENOID OR RELAY REPLACEMENT

The solenoid is not serviceable, the starter must be serviced as an assembly.

ENGINE MECHANICAL

ACCESSORY DRIVE BELTS

ACCESSORY BELT ROUTING

See Figure 53.

Refer to the accompanying illustration.

INSPECTION

See Figure 54.

➡**Check the drive belt auto-tensioner indication when the engine is cold.**

1. Check that the indicator of drive belt auto-tensioner is within the possible use range.

2. When new drive belt is installed, the indicator should be within the range in the figure.

3. Visually check entire drive belt for wear, damage or cracks.

4. If the indicator is out of the possible use range or belt is damaged, replace drive belt.

ADJUSTMENT

Belt tension is not necessary, as it is automatically adjusted by drive belt auto-tensioner.

REMOVAL & INSTALLATION

See Figure 55.

1. Before servicing the vehicle, refer to the Precautions Section.

2. Disconnect the negative battery cable.

3. Remove the right-hand front wheel and tire.

4. Remove the right-hand splash guard.

✴✴ CAUTION

Avoid placing hand in a location where pinching may occur if the holding tool accidentally comes off.

✴✴ WARNING

Never loosen the hexagonal part in center of drive belt auto-tensioner pulley (Never turn it counterclockwise). If turned counterclockwise, the complete drive belt auto-tensioner must be replaced as a unit, including the pulley.

5. Hold the hexagonal part in the center of the drive belt auto-tensioner pulley securely with a box wrench. Move the wrench handle in the direction of arrow (loosening direction of drive belt).

6. Insert a rod approximately 0.24 in. (6 mm) in diameter such as short-length screwdriver into the hole (A) of the retaining boss to fix the drive belt auto-tensioner pulley.

7. Loosen drive belt from water pump pulley in sequence, and remove it.

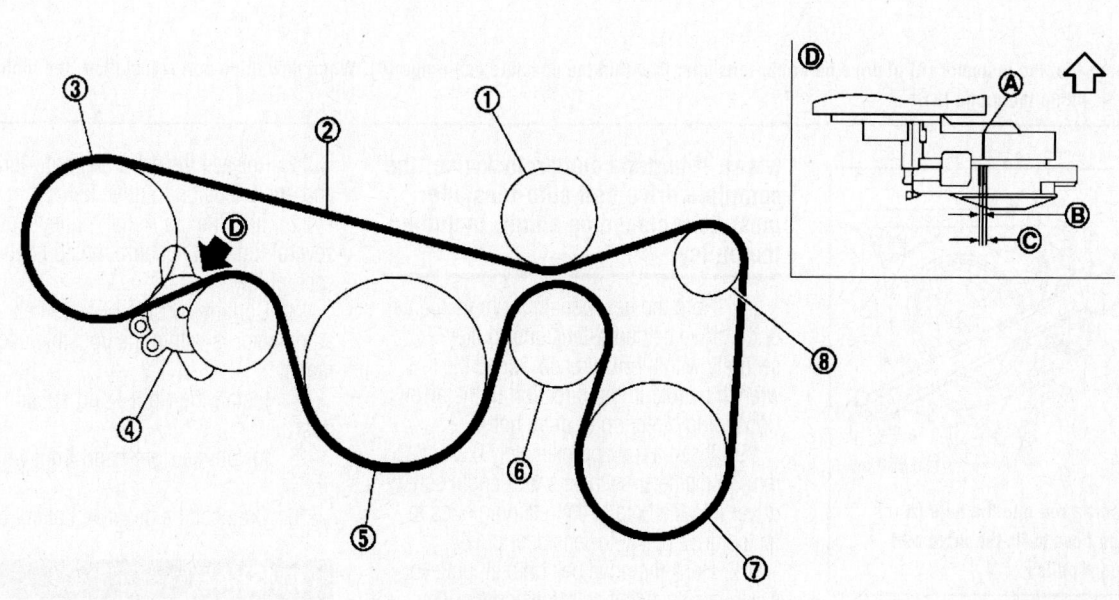

1.	Idler pulley	2.	Drive belt	3.	Power steering oil pump
4.	Drive belt auto-tensioner	5.	Crankshaft pulley	6.	Idler pulley
7.	A/C compressor	8.	Alternator	A.	Indicator	B.	Range when new drive belt is installed	C.	Possible use range
A.	Indicator	B.	Range when new drive belt is installed	C.	Possible use range
D.	View D				
⬅ : Engine front					

37663_MURA_G0220

Fig. 53 Accessory drive belt and components

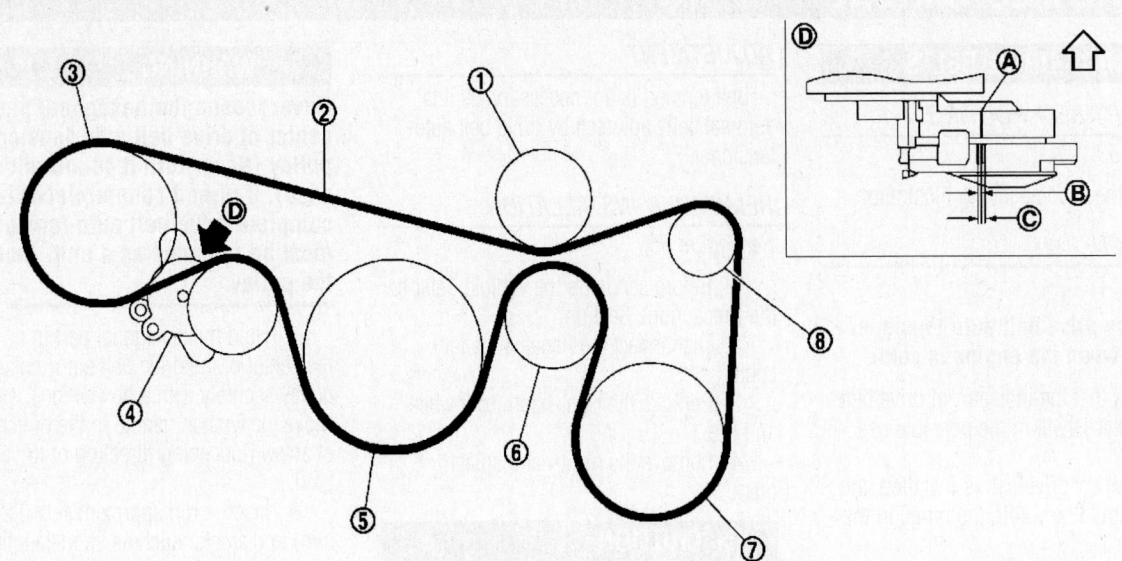

1. Idler pulley
4. Drive belt auto-tensioner
7. A/C compressor
A. Indicator
D. View D
⟵ : Engine front

2. Drive belt
5. Crankshaft pulley
8. Alternator
B. Range when new drive belt is installed

3. Power steering oil pump
6. Idler pulley

C. Possible use range

37663_MURA_G0220

Fig. 54 Check that the indicator (A) of drive belt auto-tensioner is within the possible use range (C). When new drive belt is installed, the indicator should be within the range (B).

37663_MURA_G0221

Fig. 55 Insert a rod into the hole (A) of the retaining boss to fix the drive belt auto-tensioner pulley

To install:

> ✳✳ **CAUTION**
>
> Avoid placing hand in a location where pinching may occur if the holding tool accidentally comes off.

> ✳✳ **WARNING**
>
> Never loosen the hexagonal part in center of drive belt auto-tensioner pulley (Never turn it counterclock-

wise). If turned counterclockwise, the complete drive belt auto-tensioner must be replaced as a unit, including the pulley.

8. Hold the hexagonal part in the center of the drive belt auto-tensioner pulley securely with a box wrench. Move the wrench handle in the direction of the arrow (loosening direction of drive belt).

9. Insert a rod approximately 0.24 in. (6 mm) in diameter such as short-length screwdriver into the hole of the retaining boss to fix the drive belt auto-tensioner pulley.

10. Hook the drive belt onto all pulleys except for drive belt auto-tensioner pulley, and onto drive belt auto-tensioner pulley last.

> ✳✳ **WARNING**
>
> Confirm drive belt is completely set to pulleys.

> ✳✳ **WARNING**
>
> Check for engine oil, working fluid and engine coolant are not adhered to drive belt and each pulley groove.

11. Release the drive belt auto-tensioner, and apply tension to drive belt.

12. Turn the crankshaft pulley clockwise several times to equalize tension between each pulley.

13. Confirm tension of drive belt at indicator is within the possible use range.

14. Install the right-hand splash guard.

15. Install the right-hand front wheel and tire.

16. Connect the negative battery cable.

CAMSHAFT

REMOVAL & INSTALLATION

See Figures 56 through 64.

1. Before servicing the vehicle, refer to the Precautions Section.

2. Disconnect the negative battery cable.

3. Drain the engine oil.

4. Drain the engine coolant from inside the engine.

5. Remove the front timing chain case, camshaft sprocket, timing chain and rear

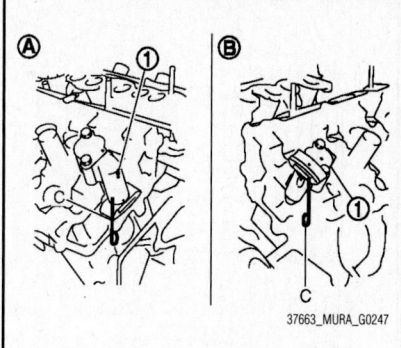

Fig. 56 Secondary timing chain tensioners (1), stopper pin (C), Bank 1 (A), Bank 2 (B)

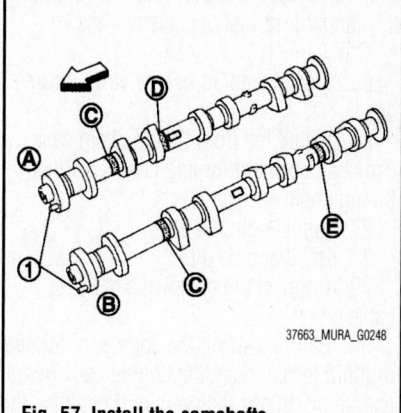

Fig. 57 Install the camshafts

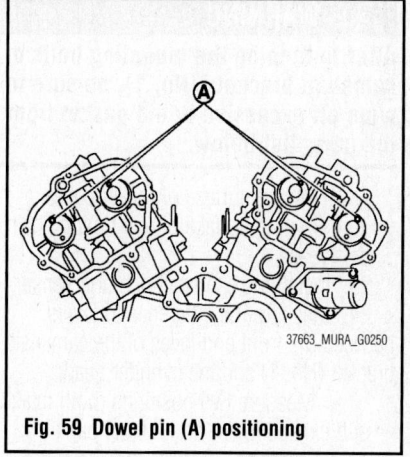

Fig. 59 Dowel pin (A) positioning

timing chain case. Refer to Timing Chain & Sprockets, Removal & Installation.

6. Remove the camshaft position sensor.

7. Remove the camshaft brackets.

a. Mark the camshafts, camshaft brackets and bolts so they are placed in the same position and direction for installation.

b. Equally loosen the camshaft bracket bolts in several steps in the reverse of the installation order shown below.

8. Remove the camshafts.

9. Remove the valve lifters.

a. Identify installation positions, and store them without mixing them up.

10. Remove the secondary timing chain tensioners from the cylinder head.

11. Remove the secondary timing chain tensioner with its stopper pin attached.

➡**Stopper pin should be attached when the secondary timing chain is removed.**

To install:

12. Install the secondary timing chain tensioners on both sides of the cylinder head.

a. Install the secondary timing chain tensioner with its stopper pin attached.

b. Install the secondary timing chain tensioner with sliding part facing downward on Bank 1 cylinder head, and with sliding part facing upward on Bank 2 cylinder head.

13. Install the valve lifters in the original position.

14. Install the camshafts.

15. Follow identification marks made during removal, or follow the identification marks that are present on new camshafts for proper placement and direction.

16. Install the camshaft so that the dowel pins on the front end face are positioned as shown. (No. 1 cylinder TDC on its compression stroke).

Bank	INT/EXH	Dowel pin (1)	Paint marks		
			M1 (D)	M2 (E)	M3 (C)
1	EXH (B)	Yes	No	Light blue	Light blue
	INT (A)	Yes	Pink	No	Light blue
2	INT (A)	Yes	Pink	No	Light blue
	EXH (B)	Yes	No	Light blue	Light blue

37663_MURA_G0249

Fig. 58 Camshaft identification marks

17. Though the camshaft does not stop at the portion as shown in the figure, for the placement of cam nose, it is generally accepted to position the camshaft in the same direction of the figure.

18. Install the camshaft brackets.

a. Remove the foreign material completely from the camshaft bracket backside and from the cylinder head installation face.

19. Install the camshaft bracket in the original position and direction as shown.

20. Install the camshaft brackets

No. 2 to 4, aligning the stamp marks as shown.

➡**There are no identification marks indicating left and right for the camshaft bracket No. 1.**

21. Apply liquid gasket to the mating surface of the No. 1 camshaft bracket as shown, on Bank 1 and Bank 2, as follows. Apply liquid gasket to rear timing chain case side. Use Genuine RTV Silicone Sealant or equivalent.

- 0.335 in. (8.5mm)
- 0.08 in. (2mm)
- Clearance 0.20 in. (5mm)
- 0.079–0.091 in. (2.0–2.3mm)

22. Tighten the camshaft bracket bolts in the following steps, in numerical order as shown.

a. Tighten No. 7 to 10 in numerical order as shown to 1 ft. lb. (1.96 Nm).

b. Tighten No. 1 to 6 in numerical order as shown to 1 ft. lb. (1.96 Nm).

c. Tighten No. 1 to 10 in numerical order as shown to 4 ft. lbs. (5.88 Nm).

d. Tighten No. 1 to 10 in numerical order as shown to 8 ft. lbs. (10.4 Nm).

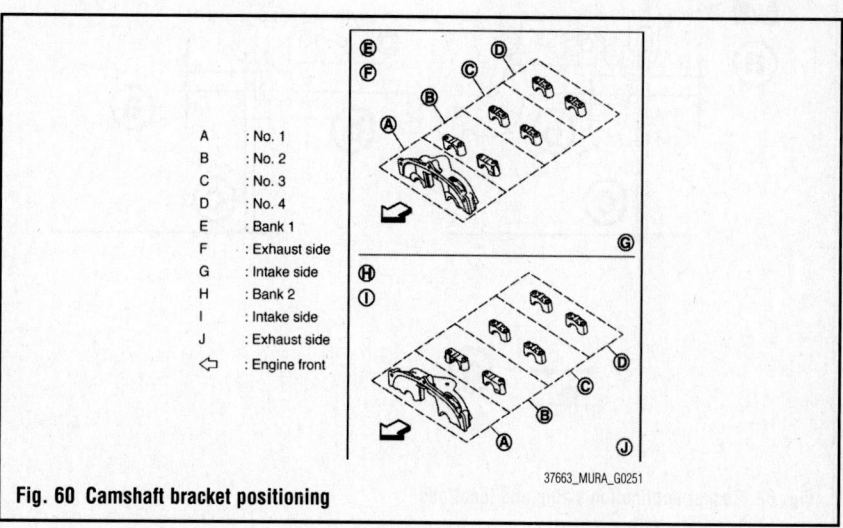

A : No. 1
B : No. 2
C : No. 3
D : No. 4
E : Bank 1
F : Exhaust side
G : Intake side
H : Bank 2
I : Intake side
J : Exhaust side
⟵ : Engine front

Fig. 60 Camshaft bracket positioning

After tightening the mounting bolts of camshaft brackets (No. 1), be sure to wipe off excessive liquid gasket from the parts list below.

- Mating surface of rocker cover
- Mating surface of rear timing chain case

23. Install the camshaft position sensor.

24. Measure the difference in levels between the front end faces of the camshaft bracket (No. 1) and the cylinder head.

   a. Measure two positions (both intake and exhaust side) for a single bank.
   - Standard: _0.0055 to 0.0055 in. (_0.14 to 0.14mm)

   b. If the measured value is out of the standard, re-install camshaft bracket (No. 1).

25. Inspect and adjust the valve clearance.

26. Install the front timing chain case, camshaft sprocket, timing chain and rear timing chain case.

27. Refill engine coolant.

28. Refill engine oil.

29. Connect the negative battery cable.

30. Before starting the engine, check the oil/fluid levels including engine coolant and engine oil. If less than required quantity, fill to the specified level.

31. Check for fluid leaks.

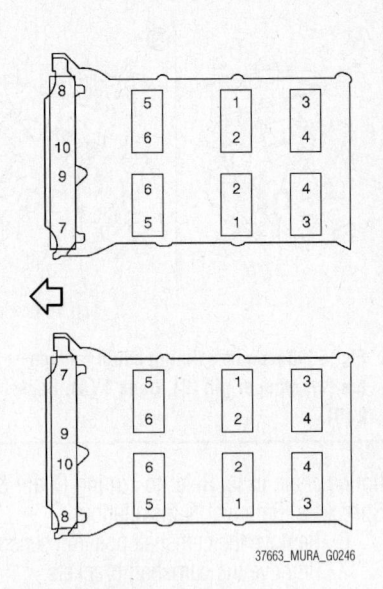

Fig. 63 Camshaft bracket bolts tightening sequence

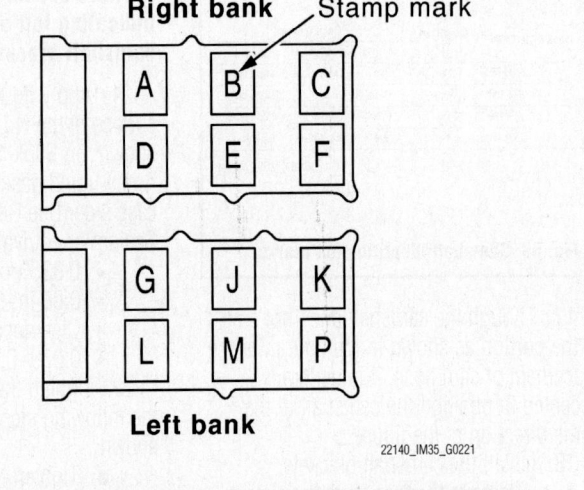

Fig. 61 Camshaft brackets No. 2 to 4 aligning sequence (Bank 1=Right Bank, Bank 2=Left Bank)

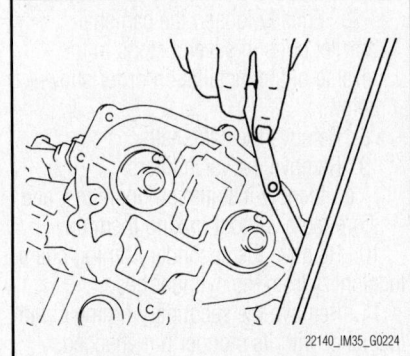

Fig. 64 Difference in levels between the front end faces of the camshaft bracket (No. 1) and the cylinder head

CATALYTIC CONVERTER

REMOVAL & INSTALLATION

See Combination Manifold.

COMBINATION MANIFOLD

REMOVAL & INSTALLATION

See Figures 65 through 67.

1. Before servicing the vehicle, refer to the Precautions Section.

2. Disconnect the negative battery cable.

3. Remove the radiator core support covers, air duct (inlet), air cleaner case (upper) with mass air flow sensor and air duct assembly.

4. Remove the engine cover.

5. Remove the front wiper arm.

Fig. 62 Gasket application sizing and locations

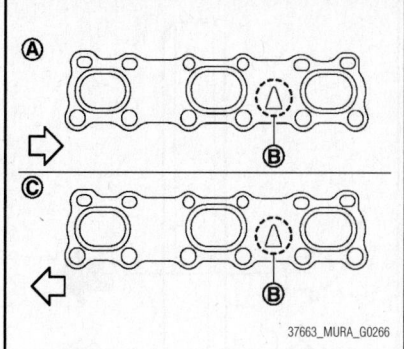

Fig. 65 Bank 1 (A), Bank 2 (C), Triangle press (B)

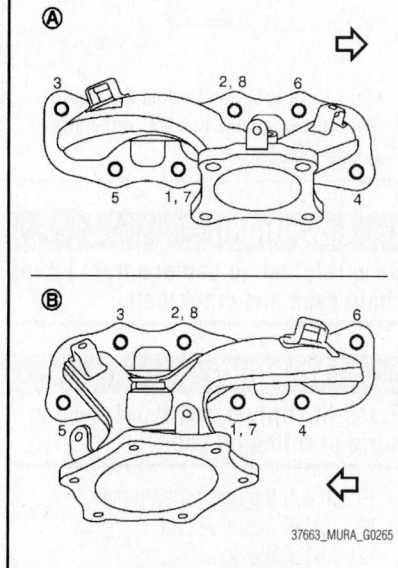

Fig. 66 Exhaust manifold bolt tightening sequence—reverse for bolt removal

6. Remove the extension cowl top.

7. Remove the front exhaust pipe.

8. Disconnect harness connector and remove air fuel ratio sensor 1 on both banks with the heated oxygen sensor wrench [SST: KV10117100 (J-3647-A)]. Put marks to identify installation positions of each air fuel ratio sensor 1. Be careful not to damage the sensor.

➡**Discard any sensor which has been dropped onto a hard surface, and replace with a new sensor.**

9. Disconnect harness connector and remove heated oxygen sensor 2 on both banks with the heated oxygen sensor wrench [SST: KV10114400 (J-38365)]. Put marks to identify installation positions of each heated oxygen sensor 2.

10. Remove exhaust manifold covers (Bank 1 and Bank 2).

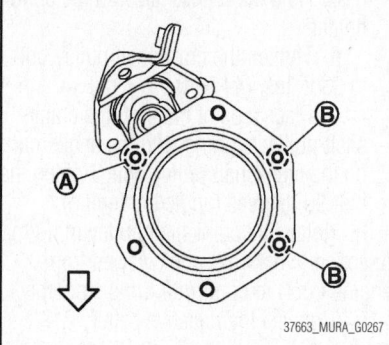

Fig. 67 Upper side bolts (B), lower side bolt (A)

11. Remove three way catalyst support mounting bolts (Bank 1 and Bank 2).

12. Remove three way catalysts by loosening bolts first and then removing nuts. (Stud bolt and flange bolt type) Handle carefully to avoid any shock to three way catalyst.

13. Loosen mounting nuts in the reverse of the installation order shown below.

14. Remove the exhaust manifolds (Bank 1 and Bank 2).

15. Remove the gaskets.

➡**Cover engine openings to avoid entry of foreign materials.**

To install:

16. Install the exhaust manifold gasket in the direction indicated in the figure.

17. If the stud bolts were removed, install them and tighten to 11 ft. lbs. (15 Nm).

18. Install the mounting nuts in numerical order as shown in the figure and tighten to 24 ft. lbs. (32 Nm). No. 7 and 8 mean double tightening of nuts No. 1 and 2.

19. Install the three way catalyst (Bank 2) flange type bolts:

    a. Temporarily tighten the upper side bolts and lower side bolt shown in the figure.

    b. Tighten the upper side bolts shown to 52 ft. lbs. (70 Nm).

    c. Tighten all the lower side bolts to 52 ft. lbs. (70 Nm).

20. Install the three way catalyst supports:

    a. Temporarily the tighten three way catalyst support mounting bolts.

    b. Tighten the three way catalyst support mounting bolts to oil pan (upper) to 16 ft. lbs. (22 Nm).

    c. Tighten the three way catalyst support mounting bolts to three way catalyst to 16 ft. lbs. (22 Nm).

➡**Before installing a new air fuel ratio sensor and a new heated oxygen sensor, clean exhaust system threads using oxygen sensor thread cleaner (commercial service tool: J-43897-18 or J-43897-12) and apply anti-seize lubricant.**

21. Install the air fuel ratio sensor 1 and heated oxygen sensor 2 in the original positions. If the installation positions cannot be identified, check the glass tube color (air fuel ratio sensor 1: black, heated oxygen sensor: white).

✳✳ WARNING

Never over-torque air fuel ratio sensor and heated oxygen sensor.

22. Install the exhaust manifold covers.

23. Install the front exhaust pipe.

24. Install the extension cowl top.

25. Install the front wiper arm.

26. Install the engine cover.

27. Install the radiator core support covers, air duct (inlet), air cleaner case (upper) with mass air flow sensor and air duct assembly.

28. Connect the negative battery cable.

CRANKSHAFT DAMPER

REMOVAL & INSTALLATION

See Figures 68 through 70.

1. Before servicing the vehicle, refer to the Precautions Section.

2. Disconnect the negative battery cable.

3. Remove the front right-hand wheel and tire.

4. Remove the front right-hand splash guard.

5. Remove the drive belt. Refer to Accessory Drive Belt Removal & Installation.

6. Remove the crankshaft pulley as follows:

    a. Hold the crankshaft with a pulley holder.

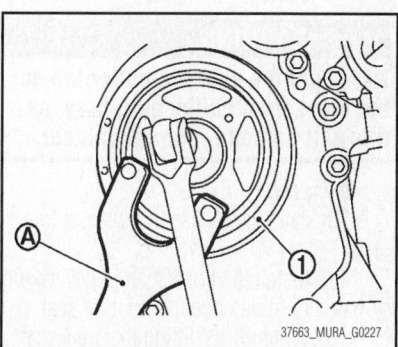

Fig. 68 Secure the crankshaft with a pulley holder (A)

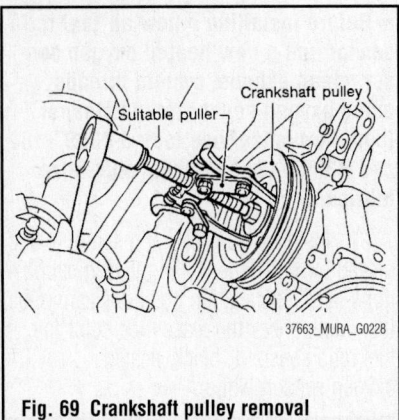

Fig. 69 Crankshaft pulley removal

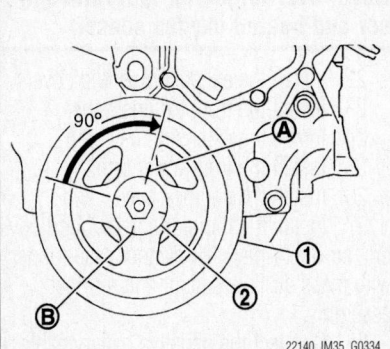

Fig. 70 Place a paint mark (A) on the crankshaft pulley (1) aligning with the angle mark (B) on the crankshaft pulley bolt (2)

b. Loosen the crankshaft pulley bolt and locate the bolt seating surface at 0.39 in. (10 mm) from its original position.

✳✳ WARNING

Never remove the crankshaft pulley bolt as it will be used as a supporting point for a suitable puller.

c. Place a suitable puller tab on the holes of crankshaft pulley, and pull the crankshaft pulley through.

✳✳ WARNING

Never put the suitable puller tab on the crankshaft pulley periphery, as this will damage internal damper.

To install:

7. Install the crankshaft pulley as follows:

a. Install the crankshaft pulley, taking care not to damage the front oil seal.

b. When press-fitting the crankshaft pulley with a plastic hammer, tap on its center portion (not circumference).

c. Hold the crankshaft with the pulley holder.

d. Tighten the crankshaft pulley bolt to 33 ft. lbs. (44 Nm).

e. Place a paint mark on the crankshaft pulley aligning with the angle mark on the crankshaft pulley bolt. Tighten the bolt 90 degrees (angle tightening).

8. Rotate the crankshaft pulley in normal direction (clockwise when viewed from engine front) to confirm it turns smoothly.

9. Connect the negative battery cable.

CRANKSHAFT FRONT SEAL

REMOVAL & INSTALLATION

See Figures 71 and 72.

1. Before servicing the vehicle, refer to the Precautions Section.

2. Disconnect the negative battery cable.

3. Remove the front right-hand wheel and tire.

4. Remove the front right-hand splash guard.

5. Remove the drive belt. Refer to Accessory Drive Belt Removal & Installation.

6. Remove the crankshaft pulley. Refer to Crankshaft Damper Removal & Installation.

7. Remove the front oil seal using a suitable tool. Be careful not to damage front timing chain case and crankshaft.

To install:

8. Apply new engine oil to both oil seal lip and dust seal lip of new front oil seal.

9. Install the front oil seal. Install so that each seal lip is oriented as shown.

a. Using a suitable drift, press-fit until the height of front oil seal is level with the mounting surface. Suitable drift:
- Outer diameter: 2.36 in. (60mm)
- Inner diameter: 1.97 in. (50mm)

b. Check the garter spring is in position and seal lips not inverted.

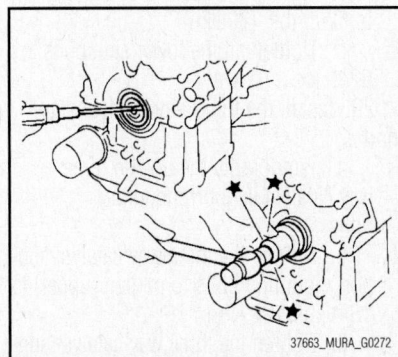

Fig. 71 Remove front oil seal

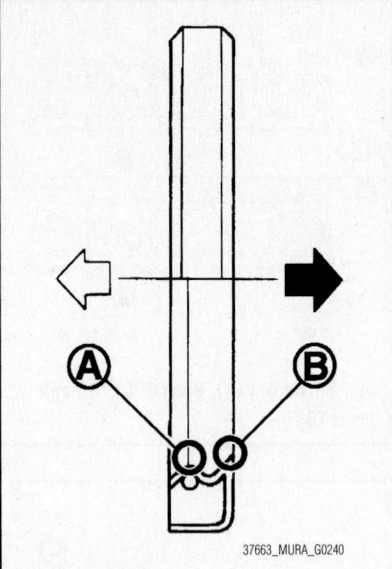

Fig. 72 Oil seal lip (A), dust seal lip (B), black arrow: engine outside, white arrow: engine inside

✳✳ WARNING

Be careful not to damage front timing chain case and crankshaft.

✳✳ WARNING

Press-fit straight and avoid causing burrs or tilting oil seal.

10. Install the crankshaft pulley.

11. Install the drive belt.

12. Install the splash guard.

13. Install the wheel and tire.

14. Refill fluids, as needed.

15. Connect the negative battery cable.

16. Check for fluid leaks.

CYLINDER HEAD

REMOVAL & INSTALLATION

See Figures 73 through 75.

1. Before servicing the vehicle, refer to the Precautions Section.

2. Disconnect the negative battery cable.

3. Remove the oil level gauge.

4. Remove the intake manifold collector.

5. Remove the rocker cover.

6. Remove the fuel tube and fuel injector assembly.

7. Remove the intake manifold. Refer to Intake manifold Removal & Installation.

8. Remove the exhaust manifold. Refer to Exhaust manifold Removal & Installation.

9. Remove the water inlet and thermo-

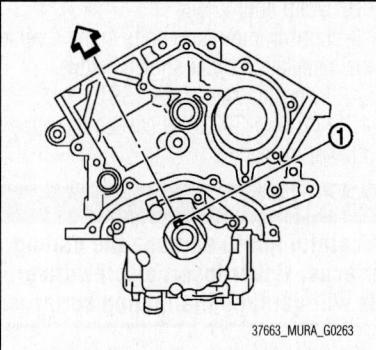

Fig. 73 The crankshaft key (1) should line up with the bank 1 cylinder center line

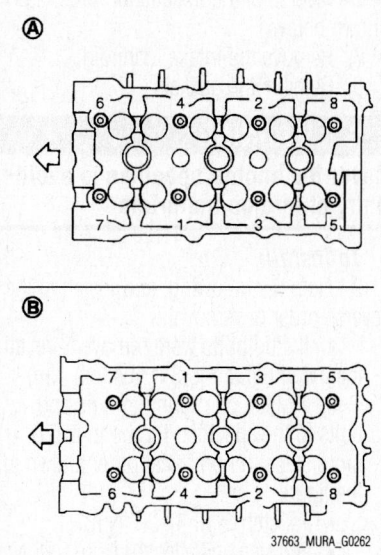

Fig. 74 Cylinder head bolt installation sequence, Bank 1 (A), Bank 2 (B)—reverse for bolt removal

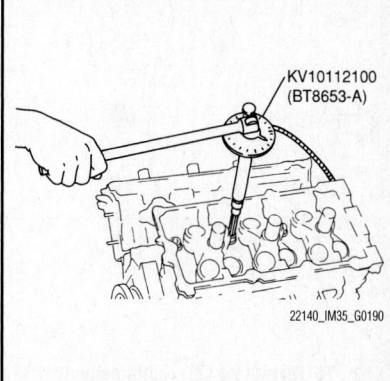

Fig. 75 Torque wrench (commercial service tool: J24239-01)

stat assembly. Refer to Thermostat Removal & Installation in the Engine Cooling section.

10. Remove the water outlet, water connector, water bypass pipe, and heater pipe.

11. Remove the timing chain and rear timing chain case. Refer to Timing Chain & Sprockets, Removal & Installation.

12. Remove the camshaft. Refer to Camshaft & Valve Lifters, Removal & Installation.

13. Remove the cylinder head bolts in the reverse of the installation order shown below.

14. Remove the cylinder heads (Bank 1 and Bank 2).

15. Remove the cylinder head gaskets.

To install:

16. Install new cylinder head gaskets.

17. Turn the crankshaft until No. 1 piston is set at TDC.

   a. The crankshaft key should line up with the bank 1 cylinder center line as shown.

18. Install the cylinder head: follow the steps below to tighten the cylinder head bolts in numerical order as shown in the figure, using a cylinder head bolts wrench (commercial service tool).

➡ **If cylinder head bolts are being reused, check their outer diameters before installation.**

➡ **Before installing the cylinder head, inspect cylinder head distortion.**

   a. Apply new engine oil to the threads and seat surfaces of the cylinder head bolts.

   b. Tighten all cylinder head bolts to 72 ft. lbs. (98 Nm).

   c. Completely loosen all cylinder head bolts, in the reverse order of that indicated in the figure.

   d. Tighten all cylinder head bolts to 29 ft. lbs. (39 Nm).

   e. Turn all cylinder head bolts 103 degrees clockwise (angle tightening). Check the tightening angle by using the angle wrench [SST: KV10112100 (BT8653-A)]. Check tightening angle indicated on the angle wrench indicator plate.

   f. Turn all cylinder head bolts 103 degrees clockwise again (angle tightening).

19. After installing the cylinder head, measure the distance between the front end faces of the cylinder block and the cylinder head (Bank 1 and Bank 2). If measured value is out of the standard, reinstall the cylinder head. Standard: 0.555–0.587 in. (14.1–14.9mm)

20. Install the camshaft.

21. Install the timing chain and rear timing chain case.

22. Install the water outlet, water connector, water bypass pipe, and heater pipe.

23. Install the water inlet and thermostat assembly.

24. Install the exhaust manifold.

25. Install the intake manifold.

26. Install the fuel tube and fuel injector assembly.

27. Install the rocker cover.

28. Install the intake manifold collector.

29. Install the oil level gauge.

30. Refill fluids, as needed.

31. Connect the negative battery cable.

32. Check for fluid leaks.

EXHAUST MANIFOLD

REMOVAL & INSTALLATION

See Combination Manifold.

FLEXPLATE/DRIVE PLATE

REMOVAL & INSTALLATION

See Figure 76.

1. Before servicing the vehicle, refer to the Precautions Section.

❊ WARNING

Never disassemble driveplate. Never place driveplate with signal plate facing down. When handling reinforcement plate, take care not to damage or scratch it. Handle reinforcement plate in a manner that prevents it from becoming magnetized.

2. Remove the transaxle.

3. Remove the driveplate retaining bolts.

4. Remove the driveplate from the engine.

To install:

5. When installing the driveplate be sure to correctly align the crankshaft side guide pin and drive side guide pin hole. If not correctly aligned the engine will run rough and turn on the MIL light.

6. Install the driveplate and reinforcement plate.

7. Hold the ring gear with the pulley stopper.

8. Tighten the mounting bolts to 35 ft. lbs. (88 Nm) and in a crosswise sequence in several passes.

➡ **Be sure that the dowel pin is installed at the rear end of the crankshaft.**

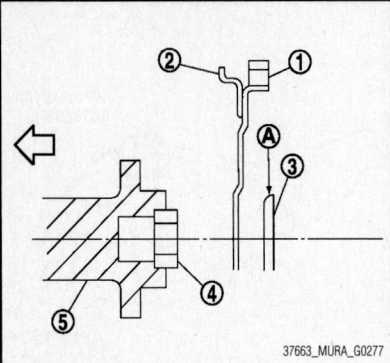

Fig. 76 Driveplate (2), reinforcement plate (3), ring gear (1), pilot converter (4), crankshaft (5) rounded (A)

INTAKE MANIFOLD

REMOVAL & INSTALLATION

See Figures 77 and 78.

1. Before servicing the vehicle, refer to the Precautions Section.
2. Disconnect the negative battery cable.
3. Release the fuel pressure.
4. Remove the intake manifold collector.
5. Remove the fuel tube and fuel injector assembly.

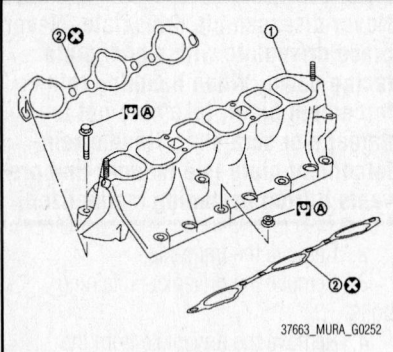

Fig. 77 Intake manifold (1) and gasket (2)

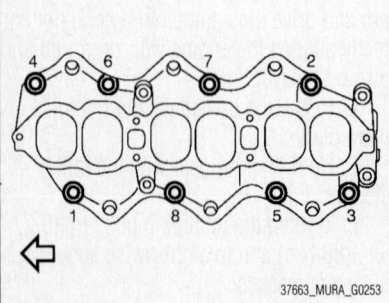

Fig. 78 Intake manifold bolt installation sequence

6. Loosen the mounting nuts and bolts in the reverse of the installation order as shown below.
7. Remove the intake manifold.
8. Remove the gaskets.

> ❋❋ **WARNING**
>
> **Cover the engine openings to avoid entry of foreign materials.**

To install:

9. Note the following, and install in the reverse order or removal.

  a. If stud bolts were removed, install them and tighten to 8 ft. lbs. (11 Nm).

  b. Tighten all mounting nuts and bolts to the specified torque in two or more steps in numerical order shown in the figure.
  - 1st step: 5 ft. lbs. (7 Nm)
  - 2nd step and after: 19 ft. lbs. (26 Nm)

OIL PAN

REMOVAL & INSTALLATION

Lower

See Figures 79 and 80.

1. Before servicing the vehicle, refer to the Precautions Section.

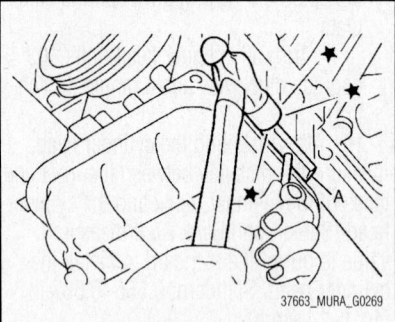

Fig. 79 Insert the seal cutter [SST: KV10111100 (J-37228)] (A) between the upper and lower oil pan

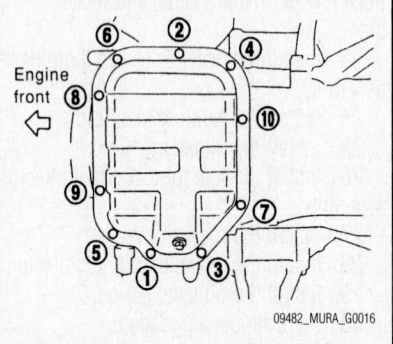

Fig. 80 Lower oil pan bolt torque sequence

2. Drain engine oil.
3. Loosen mounting bolts in the reverse of the installation order shown below.
4. Insert the seal cutter [SST: KV10111100 (J-37228)] between the upper and lower oil pan.

> ❋❋ **WARNING**
>
> **Be careful not to damage the mating surfaces. Never insert a screwdriver, this will damage the mating surfaces.**

5. Slide the seal cutter by tapping on the side of tool with a hammer.
6. Remove the lower oil pan.

To install:

7. Install the lower oil pan and tighten the mounting bolts in the order shown.
8. Use scraper to remove old liquid gasket from mating surfaces. Remove old liquid gasket from the bolt holes and thread.
9. Apply a continuous bead of liquid gasket to the lower oil pan. Use Genuine RTV Silicone Sealant or equivalent.

➡ **Attaching should be done within 5 minutes after coating.**

10. Install the lower oil pan. Install the mounting bolts in numerical order as shown in the figure and tighten to 78 inch lbs. (9 Nm).
11. Install oil pan drain plug.

➡ **Wait for at least 30 minutes after oil pan is installed to add engine oil.**

Upper

See Figures 81 through 84.

1. Before servicing the vehicle, refer to the Precautions Section.
2. Drain engine oil.
3. Drain engine coolant.
4. Remove the front wheels and tires.
5. Remove the front splash guards.

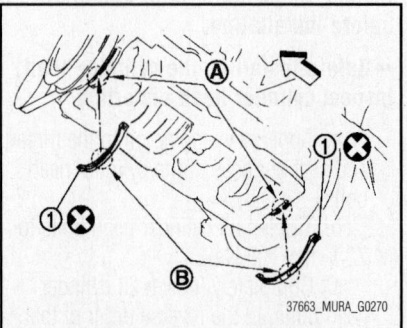

Fig. 81 Remove the O-rings (2) from bottom of cylinder block (1) and oil pump (3)

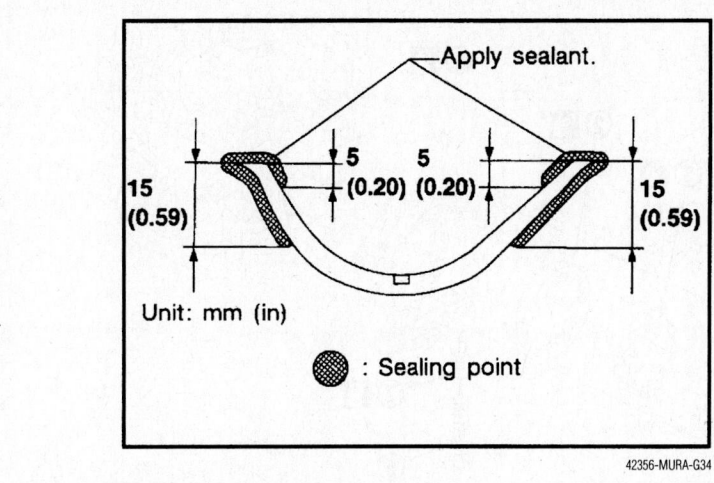

Fig. 82 RTV sealer application at the timing case

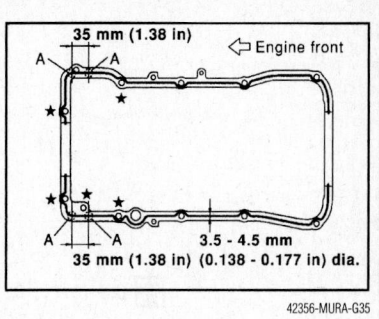

Fig. 84 RTV sealer application on the pan

6. Remove the drive belt. Refer to Accessory Drive Belt Removal & Installation.

7. Remove the oil level gauge.

8. Remove the right front halfshaft. Refer to Front Halfshaft Removal & Installation in the Drive Train section.

9. Remove the three way catalyst (Bank 1 and Bank 2) from exhaust manifolds (Bank 1 and Bank 2).

10. Remove the oil pressure switch.

11. Remove the oil filter.

12. Remove the oil cooler and water pipes.

13. Support transaxle assembly with a suitable jack. When setting the transmission jack, be careful not to allow it to collide against the drain plug.

14. Support front suspension member with a suitable jack.

15. Remove the rear engine mounting insulator.

16. Remove the left-hand engine mounting insulator mounting bolts from transaxle.

17. Remove the rear torque rod through bolts from rear torque rod bracket.

18. Remove the front suspension member stay and the mounting bolts and nuts.

19. Lower the jack for the front suspension member to the height.

20. Remove the transfer assembly. Refer to Transfer Case Removal & Installation in the Drive Train section.

21. Remove the lower oil pan.

22. Remove the oil strainer.

23. Remove the mounting bolts in reverse of the installation order shown below.

24. Insert the seal cutter [SST: KV10111100 (J-37228)] between the upper oil pan and cylinder block. Slide seal cutter

by tapping on the side of tool with a hammer.

✳✳ WARNING

Be careful not to damage the mating surfaces. Never insert a screwdriver, this will damage the mating surfaces.

25. Remove the upper oil pan.

26. Remove the O-rings from bottom of cylinder block and oil pump.

27. Remove the oil pan gaskets.

To install:

28. Use a scraper to remove old liquid gasket from mating surfaces. Remove old liquid gasket from mating surface of cylinder block. Remove old liquid gasket from the bolt holes and threads.

29. Install new oil pan gaskets.

30. Apply liquid gasket to new oil pan gaskets as shown in the figure. Use Genuine RTV Silicone Sealant or equivalent.

31. To install, align protrusion of oil pan gasket with notches of front timing chain case and rear oil seal retainer.

32. Install the oil pan gasket with smaller

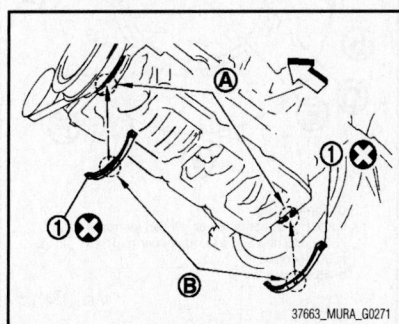

Fig. 83 Align protrusion (B) of oil pan gasket (1) with notches (A) of front timing chain case and rear oil seal retainer

arc to the front of the timing chain case side.

33. Install new O-rings on the bottom of cylinder block and oil pump.

34. Apply a continuous bead of sealant to the cylinder block mating surface of the upper oil pan to a limited portion shown. Use RTV silicone sealant or equivalent. For bolt holes with marks (5 locations), apply liquid gasket outside the holes. Apply a bead of 0.177–0.217 in. (4.5–5.5mm) diameter to area "A"

35. Install the upper oil pan. Avoid misalignment of both O-rings.

36. Tighten mounting bolts in numerical order as shown in the figure. There are three types of mounting bolts. Refer to the following for locating bolts:

- M8 _ 135 mm (5.31 in.): 11
- M8 _ 92 mm (3.62 in.): 5, 7, 8
- M8 _ 25 mm (0.98 in.): Except the above

37. Install the oil strainer to oil pump.

38. Install the lower oil pan.

39. Install the oil pan drain plug.

40. The remainder if installation is the reverse of removal.

➡**Wait for at least 30 minutes after oil pan is installed to add engine oil.**

OIL PUMP

REMOVAL & INSTALLATION

See Figure 85.

1. Before servicing the vehicle, refer to the Precautions Section.

2. Remove the upper and lower oil pans. Refer to Oil Pan Removal & Installation.

3. Remove the oil strainer. Refer to Oil Pan Removal & Installation.

4. Remove the front timing chain case and the primary timing chain. Refer to Timing Chain & Sprockets, Removal & Installation.

5. Remove oil pump assembly.

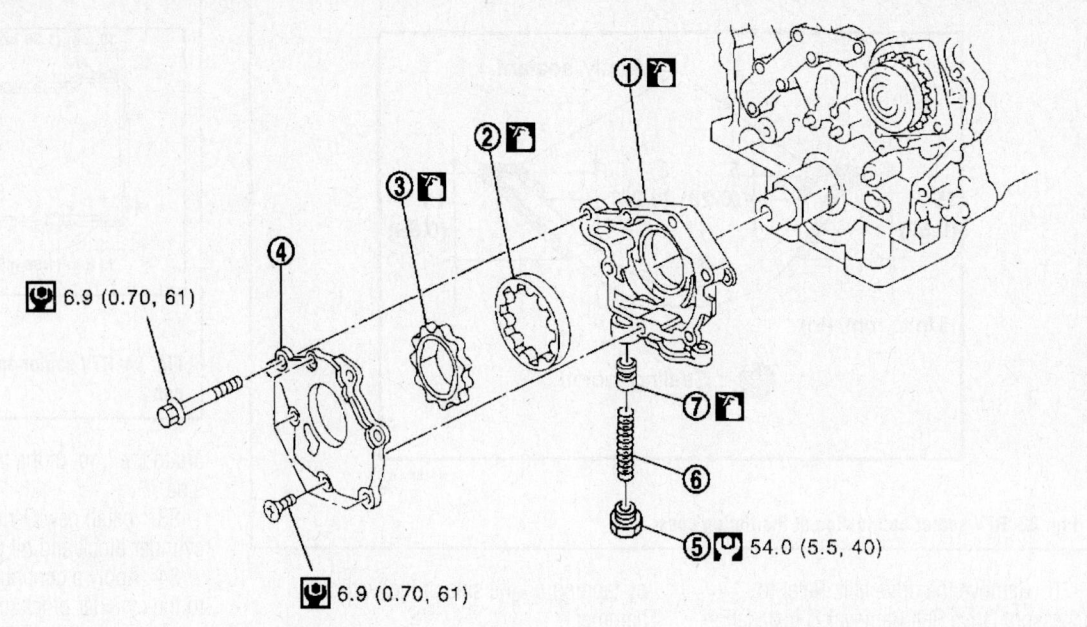

1. Oil pump body
2. Oil pump outer rotor
3. Oil pump inner rotor
4. Oil pump cover
5. Regulator valve plug
6. Regulator valve spring
7. Regulator valve

37663_MURA_G0254

Fig. 85 Oil pump

To install:

6. Installation is the reverse of the removal procedure.

7. Be sure to use new gaskets.

8. When installing, align crankshaft flat faces with inner rotor flat faces.

PISTON AND RING

POSITIONING

See Figures 86 and 87.

1. If there is stamped mark on ring, mount it with marked side up.

2. If there is no stamp on ring, no specific orientation is required for installation.

3. Stamped mark:
4. Top ring (A): —
5. Second ring (B): 2R
6. Position each ring with the gap as shown in the figure referring to the piston front mark.

REAR MAIN SEAL

REMOVAL & INSTALLATION

See Figure 88.

1. Remove the transaxle assembly.

2. Remove the drive plate. Refer to Flywheel/Drive Plate Removal & Installation.

3. Remove the oil pan (upper). Refer to Oil Pan Removal & Installation.

4. Using a special service tool, remove the liquid gasket and the rear oil seal retainer as an assembly. Be careful not to damage the mating surfaces.

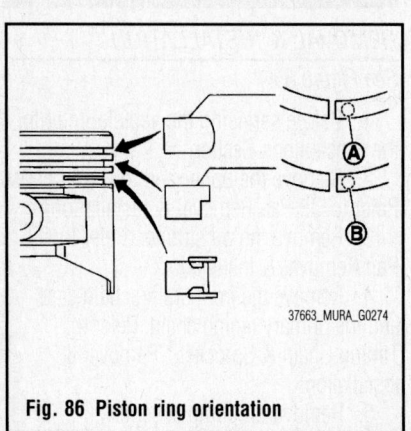

37663_MURA_G0274

Fig. 86 Piston ring orientation

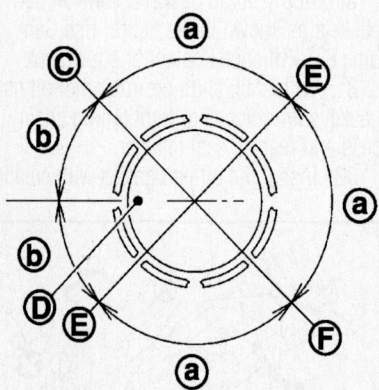

C: Top ring gap
E: Oil ring upper or lower rail gap (either of then)
F: Second ring and oil ring spacer gap
a: 90 degrees
b: 45 degrees

37663_MURA_G0275

Fig. 87 Position each ring with the gap as shown in the figure referring to the piston front mark (D)

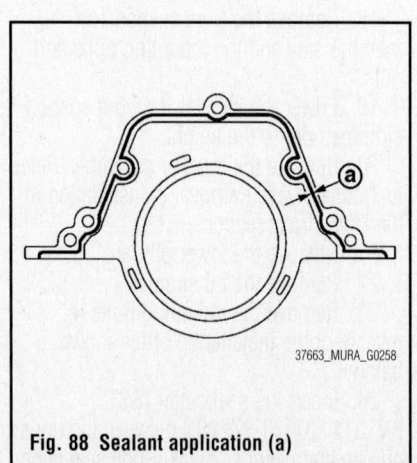

37663_MURA_G0258

Fig. 88 Sealant application (a)

To install:

5. Remove the old liquid gasket on the mating surfaces of the cylinder block and oil pan (upper).

6. Apply new engine oil to both the oil seal lip and the dust seal lip of the new oil seal retainer.

7. Apply a continuous bead of liquid gasket [0.08–0.12 in. (2.0–3.0mm)] to the rear oil seal retainer as shown. Use Genuine RTV Silicone Sealant or equivalent.

8. Install the rear oil seal retainer to the cylinder block. Ensure that the garter spring is in position and seal lips not inverted.

9. Install the oil pan (upper).

10. Install the drive plate.

11. Install the transaxle assembly.

TIMING CHAIN FRONT COVER

REMOVAL & INSTALLATION

See Timing Chain & Sprockets, Removal & Installation.

TIMING CHAIN & SPROCKETS

REMOVAL & INSTALLATION

See Figures 89 through 114.

1. Before servicing the vehicle, refer to the Precautions Section.

2. Disconnect the negative battery cable.

3. Drain the engine oil.

4. Drain the engine coolant from inside the engine.

5. Remove the intake manifold collector.

6. Remove the rocker covers (Bank 1 and Bank 2).

7. Remove the upper and lower oil pans and oil strainer.

8. Remove the drive belt, idler pulleys and bracket.

9. Remove the power steering oil pump with piping connected, and temporarily secure it aside.

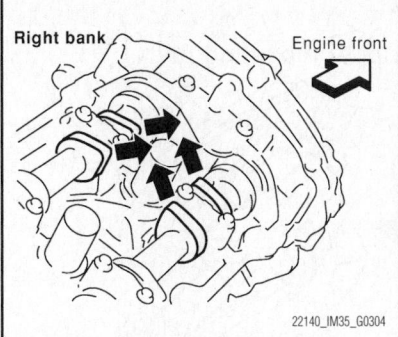

Fig. 90 Intake and exhaust cam noses on No. 1 cylinder (engine front side of Bank 1)

10. Separate the engine harness brackets from the front timing chain case.

11. Remove the valve timing control covers.

   a. Loosen the mounting bolts in the reverse of the installation order shown below.

> ✳✳ **WARNING**
>
> **Shaft is internally jointed with intake camshaft sprocket center hole. When removing, keep it horizontal until it is completely disconnected.**

12. Position the No. 1 cylinder at TDC of its compression stroke as follows:

   a. Rotate the crankshaft pulley clockwise to align the timing mark (grooved line without color) with the timing indicator.

13. Check that the intake and exhaust cam noses on the No. 1 cylinder (engine front side of Bank 1) are located as shown. If not, turn crankshaft one revolution (360 degrees) and align as shown.

14. Remove the crankshaft pulley as follows:

   a. Hold the crankshaft with a pulley holder.

   b. Loosen the crankshaft pulley bolt and locate the bolt seating surface at

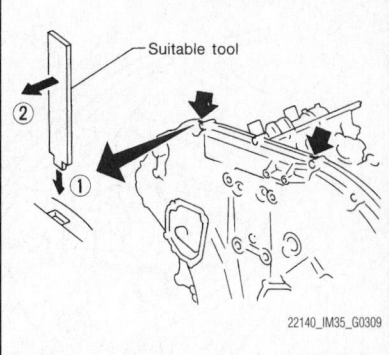

Fig. 92 Pry tool and location to pry front cover from motor

0.39 in. (10 mm) from its original position.

> ✳✳ **WARNING**
>
> **Never remove the crankshaft pulley bolt as it will be used as a supporting point for a suitable puller.**

   c. Place a suitable puller tab on the holes of crankshaft pulley, and pull the crankshaft pulley through.

> ✳✳ **WARNING**
>
> **Never put the suitable puller tab on the crankshaft pulley periphery, as this will damage internal damper.**

15. Remove the front timing chain case as follows:

   a. Loosen the mounting bolts in the reverse of the installation order shown below.

   b. Insert a suitable tool into the notch at the top of the front timing chain case as shown.

   c. Pry off the case by moving the tool as shown.

   d. Use the seal cutter [SST: KV10111100 (J-37228)] to cut the liquid gasket for removal.

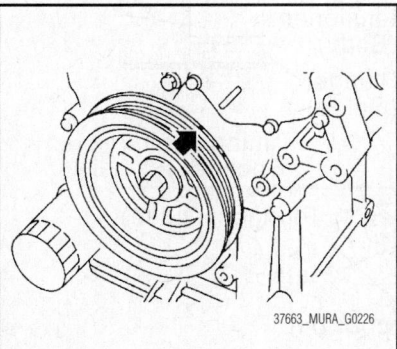

Fig. 89 Align the timing mark with timing indicator

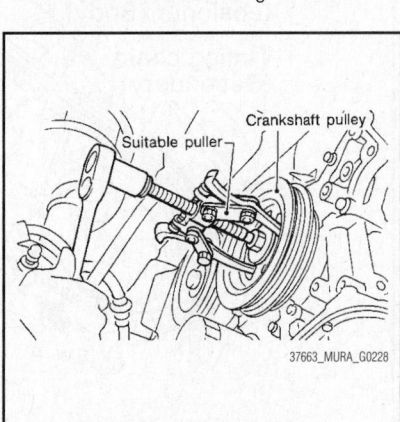

Fig. 91 Crankshaft pulley removal

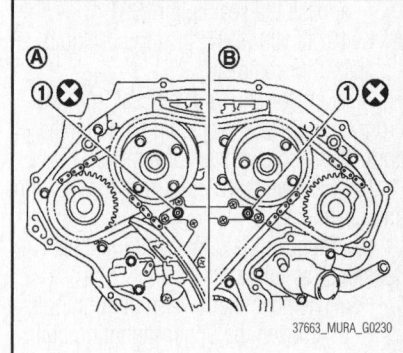

Fig. 93 O-ring (1), Bank 1 (A), Bank 2 (B)

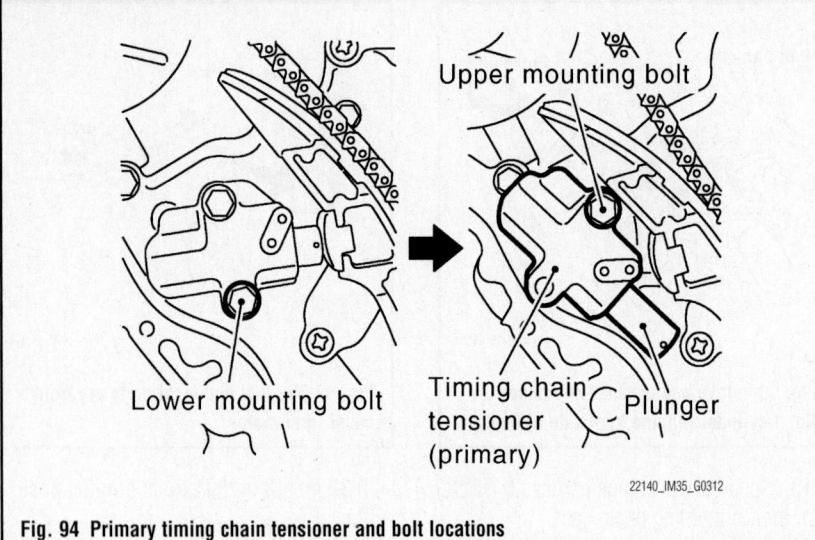

Fig. 94 Primary timing chain tensioner and bolt locations

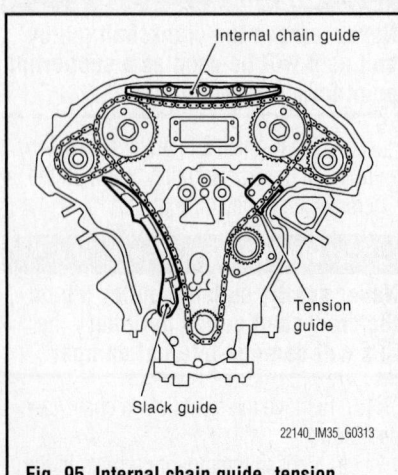

Fig. 95 Internal chain guide, tension guide and slack guide

※ WARNING

Never use screwdrivers or something similar. After removal, handle front timing chain case carefully so it does not tilt, cant, or warp under a load.

16. Remove the water pump cover from the front timing chain case.

a. Use the seal cutter [SST: KV10111100 (J-37228)] to cut liquid gasket for removal.

17. Remove the front oil seal from the front timing chain case using a suitable tool. Use a screwdriver for removal.

18. Remove the O-ring from the rear timing chain case.

19. Remove the primary timing chain tensioner as follows:

a. Remove the lower mounting bolt.

b. Loosen the upper mounting bolt slowly, and turn the primary timing chain tensioner on the mounting bolt so that the plunger is fully expanded.

➡ **Even if the plunger is fully expanded, it is not dropped from the body of the primary timing chain tensioner.**

c. Remove the upper mounting bolt, and remove the primary timing chain tensioner.

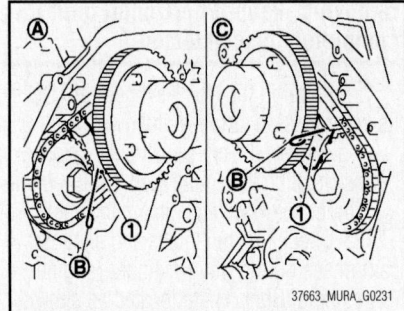

Fig. 96 Secondary timing chain tensioners (1), Bank 1 (A), Bank 2 (C), and stopper pin (B)

20. Remove the internal chain guide, tension guide and slack guide.

➡ **The tension guide can be removed after removing the primary timing chain.**

21. Remove the primary timing chain and the crankshaft sprocket.

※ WARNING

After removing the primary timing chain, never turn the crankshaft and camshaft separately, or valves will strike the piston heads.

22. Remove the secondary timing chain and camshaft sprockets as follows:

a. Attach a suitable stopper pin to the Bank 1 and Bank 2 the secondary timing chain tensioners.

➡ **Use an approximately 0.02 in. (0.5 mm) diameter hard metal pin as a stopper pin.**

b. For removal of the secondary timing chain tensioner, refer to Camshaft & Valve Lifters, Removal & Installation (removing the No. 1 camshaft bracket is required.)

c. Remove the camshaft sprocket mounting bolts.

d. Secure the hexagonal portion of the camshaft using a wrench to loosen the mounting bolts.

※ WARNING

Never loosen the mounting bolts with securing anything other than the camshaft hexagonal portion or with tensioning the timing chain.

e. Remove the secondary timing chain together with the camshaft sprockets.

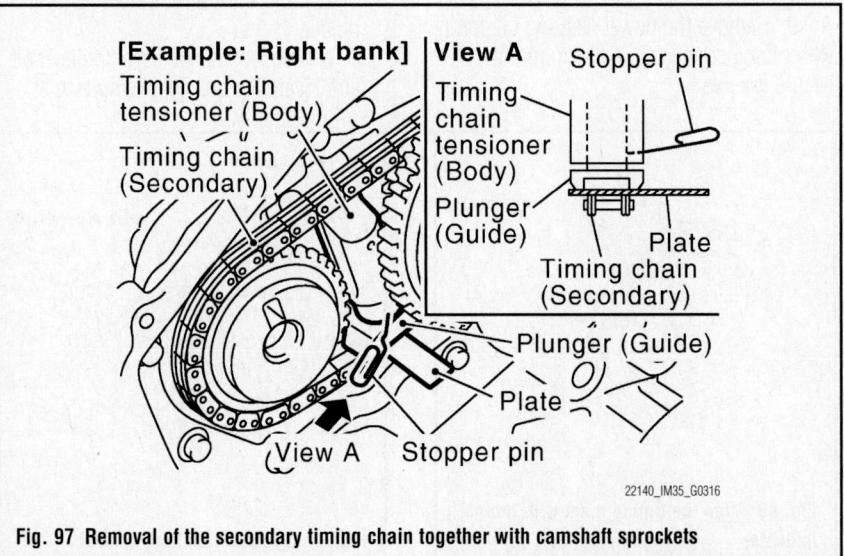

Fig. 97 Removal of the secondary timing chain together with camshaft sprockets

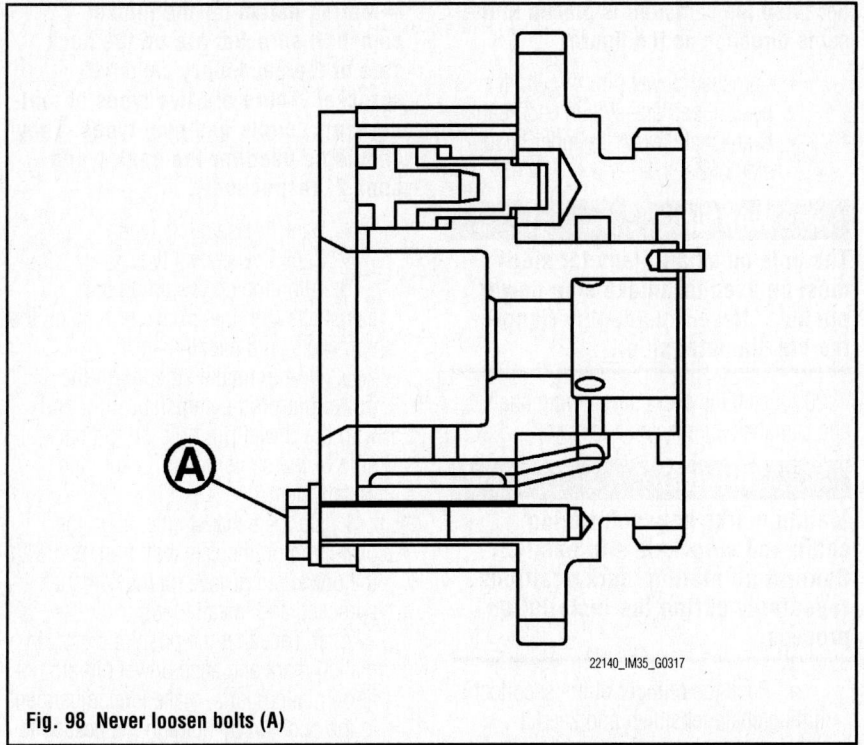

22140_IM35_G0317

Fig. 98 Never loosen bolts (A)

f. Turn the camshaft slightly to secure slackness of the timing chain on the secondary timing chain tensioner side.

g. Insert a 0.020 in. (0.5 mm) thick metal or resin plate between the timing chain and the timing chain tensioner plunger (guide). Remove the secondary timing chain together with the camshaft sprockets with the timing chain loose from the guide groove.

> ※※ **WARNING**
>
> **Be careful of the plunger coming off when removing the secondary timing chain.**

➡The intake camshaft sprocket is two-for-one structure of sprockets for the primary timing chain and for the secondary timing chain.

> ※※ **WARNING**
>
> **Handle the camshaft sprocket carefully to avoid any shock and never disassemble. Never loosen bolts as shown.**

23. Remove the secondary timing chain tensioners from the cylinder head as follows, if necessary.

a. Remove the secondary timing chain tensioners with a stopper pin attached.

24. Use a scraper to remove all traces of old liquid gasket from the front timing chain case, and opposite mating surfaces. Remove old liquid gasket from the bolt hole and thread.

25. Use a scraper to remove all traces of old liquid gasket from the water pump cover.

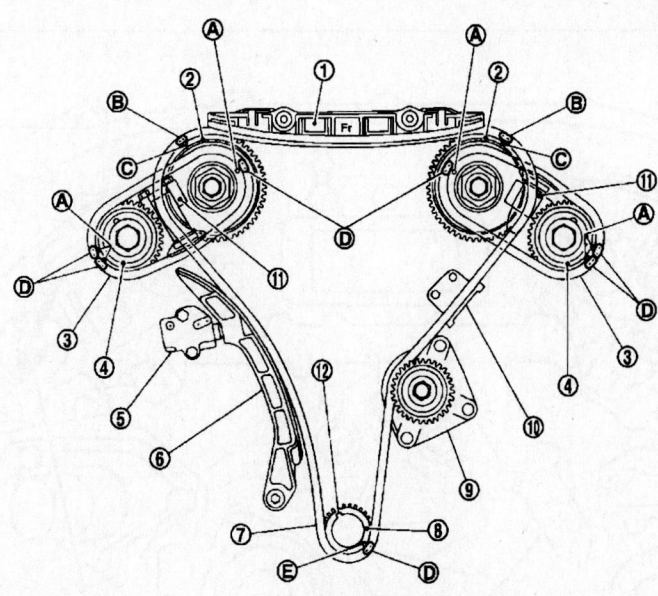

1. Internal chain guide	2. Camshaft sprocket (INT)	3. Timing chain (secondary)
4. Camshaft sprocket (EXH)	5. Timing chain tensioner (primary)	6. Slack guide
7. Timing chain (primary)	8. Crankshaft sprocket	9. Water pump
10. Tension guide	11. Timing chain tensioner (secondary)	12. Crankshaft key
A. Mating mark	B. Mating mark (pink link)	C. Mating mark (punched)
D. Mating mark (orange)	E. Mating mark (notched)	

37663_MURA_G0232

Fig. 99 Timing chain and sprockets mating marks

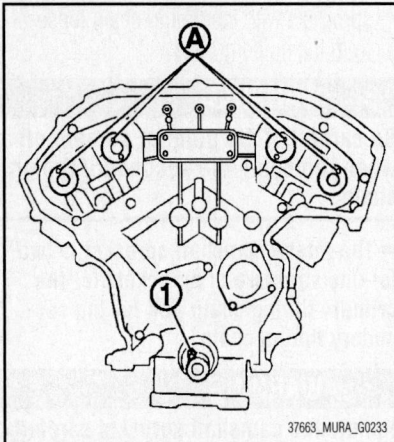

Fig. 100 Check that the dowel pin (A) and crankshaft key (1) are located as shown

37663_MURA_G0233

To install:

26. If removed, install the secondary timing chain tensioners to cylinder head as follows.

a. Install the secondary timing chain tensioners with a stopper pin attached and new O-ring.

27. Check that the dowel pin and crankshaft key are located as shown. (No. 1 cylinder at compression TDC)

➡Though camshaft does not stop at the position as shown in the figure, for the placement of cam nose, it is generally accepted the camshaft is placed in the same direction as the figure.

- Camshaft dowel pin: At cylinder head upper face side in each bank
- Crankshaft key: At cylinder head side of Bank 1

✳✳ WARNING

The hole on small diameter side must be used for intake side dowel pin hole. Never misidentify (ignore the big diameter side).

28. Install the secondary timing chain and camshaft sprockets as follows:

✳✳ WARNING

Mating marks between timing chain and sprockets slip easily. Confirm all mating mark positions repeatedly during the installation process.

a. Push the plunger of the secondary timing chain tensioner and keep it pressed in with stopper pin.

b. Install the secondary timing chain and camshaft sprockets.

c. Align the mating marks on the secondary timing chain (orange link) with the ones on the camshaft sprockets (punched), and install them.

➡Mating marks for the intake camshaft sprocket are on the back side of the secondary camshaft sprocket. There are two types of mating mark, circle and oval types. They should be used for the bank 1 and bank 2, respectively.

- Bank 1: Use circle type
- Bank 2: Use oval type

d. Align the dowel pin on the camshafts with the groove or hole on the sprockets, and install them.

e. On the intake side, align the dowel pin on the camshaft front end with the dowel pin hole on the back side of the camshaft sprocket, and install them.

f. On the exhaust side, align the dowel pin on the camshaft front end with the dowel pin groove on the camshaft sprocket, and install them.

g. In case that the positions of each mating mark and each dowel pin are not fit on mating parts, make fine adjustment to the position by holding the hexagonal portion on camshaft with wrench or equivalent.

h. Mounting bolts for camshaft sprockets must be tightened in the next step. Tightening them by hand is enough to prevent the dislocation of dowel pins.

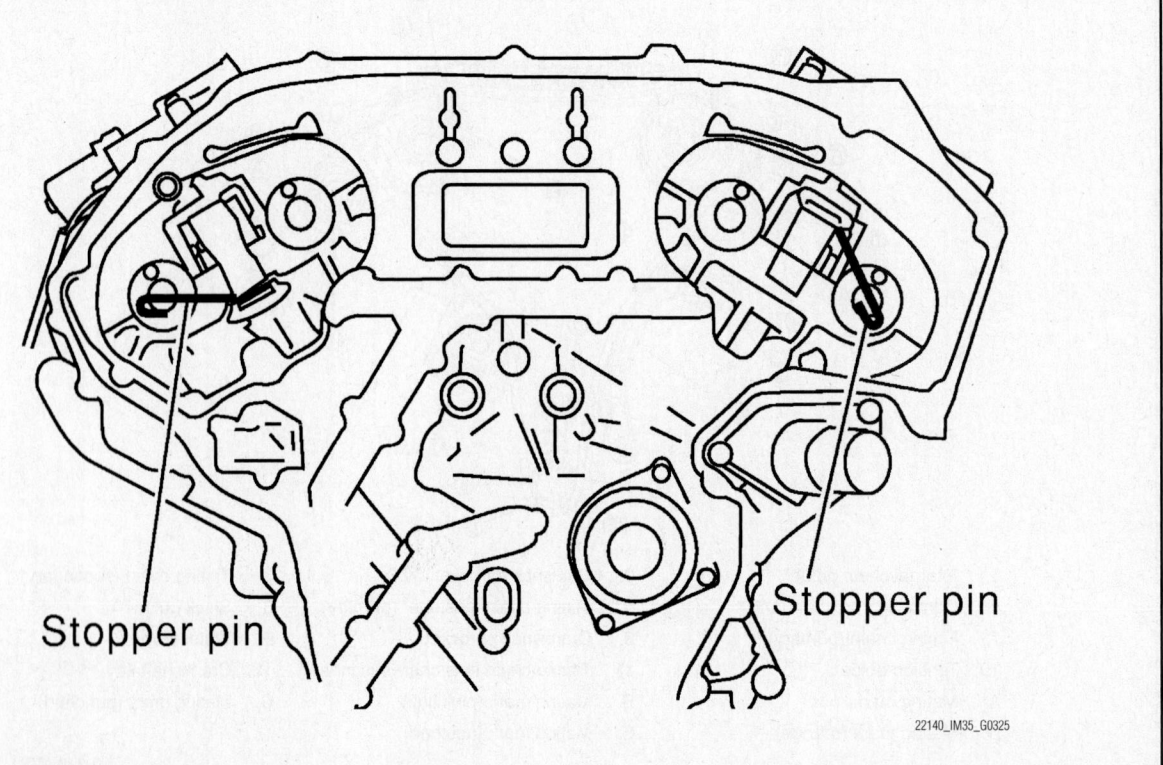

Stopper pin

Stopper pin

22140_IM35_G0325

Fig. 101 Locking the plunger of the secondary timing chain tensioner

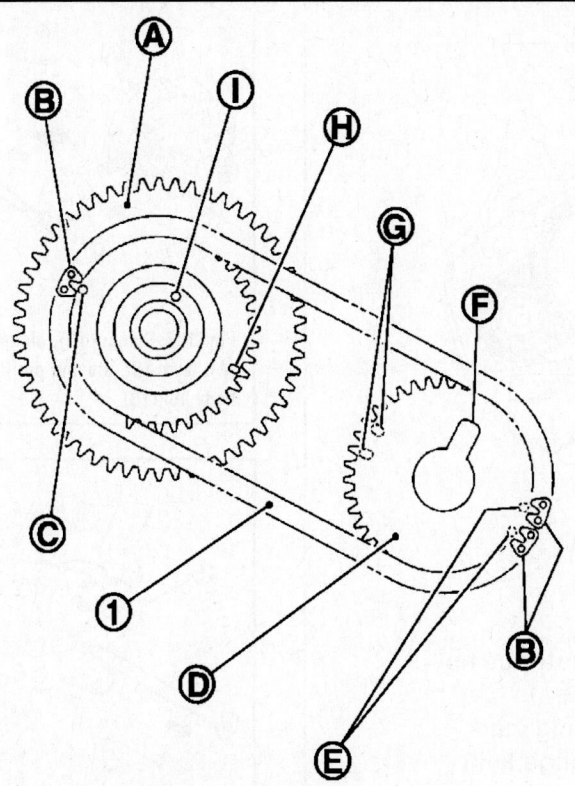

A. Intake camshaft sprocket back face
B. Orange link
C. Mating mark (Circle)
D. Exhaust camshaft sprocket back face
E. Mating mark (2 circle on front face)
F. Dowel pin groove
G. Mating mark (2 ovals on front face)
H. Mating mark (Oval)
I. Dowel pin hole

37663_MURA_G0234

Fig. 102 Install the secondary timing chain (1) and camshaft sprockets, figure shows Bank 1 (rear view)

i. It may be difficult to visually check the dislocation of mating marks during and after installation. To make the matching easier, make a mating mark on the top of sprocket teeth and its extended line in advance with paint.

j. After confirming the mating marks are aligned, tighten the camshaft

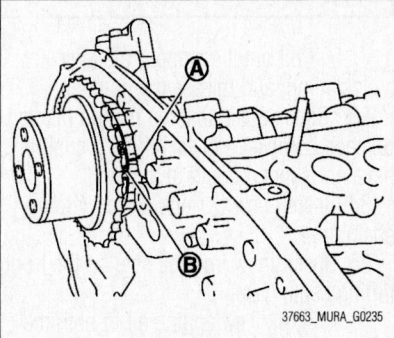

37663_MURA_G0235

Fig. 103 Mating mark (painted) (A), mating mark (orange link) (B)

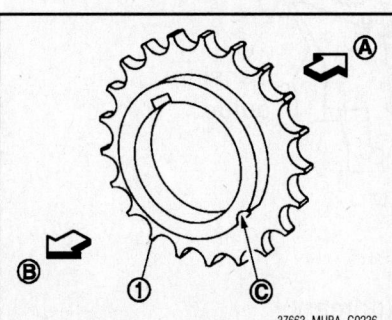

37663_MURA_G0236

Fig. 104 Install crankshaft sprocket (1): crankshaft side (A), engine front (B), mating mark (front side) (C)

sprocket mounting bolts. Secure the camshaft using a wrench at the hexagonal portion to tighten the mounting bolts.

k. Pull the stopper pins out from the secondary timing chain tensioners.

29. Install the tension guide.

30. Install the primary timing chain as follows:

a. Install crankshaft sprocket.

b. Check the mating marks on the crankshaft sprocket.

c. Install the primary timing chain.

d. Install the primary timing chain so the mating mark (punched) on the intake camshaft sprocket is aligned with the pink link on the timing chain, while the mating mark (notched) on the crankshaft sprocket is aligned with the orange link on the timing chain, as shown.

e. When it is difficult to align the mating marks of the primary timing chain with each sprocket, gradually turn the camshaft using a wrench on the hexagonal portion to align it with the mating marks.

f. During alignment, be careful to prevent dislocation of the mating mark alignments of the secondary timing chains.

31. Install the internal chain guide and slack guide.

✳✳ WARNING

Never over-tighten the slack guide mounting bolt. It is normal for a gap to exist under the bolt seat when the mounting bolt is tightened to specification.

32. Install the primary timing chain tensioner with the following procedure:

a. Pull the plunger stopper tab up (or turn lever downward) so as to remove the plunger stopper tab from the ratchet of plunger. Plunger stopper tab and lever are synchronized.

b. Push the plunger into the inside of the tensioner body.

c. Hold the plunger in the fully compressed position by engaging the plunger stopper tab with the tip of the ratchet.

d. To secure the lever, insert the stopper pin through the hole of the lever into the tensioner body hole. The lever parts and the tab are synchronized. Therefore, this will secure the plunger.

e. Install the primary timing chain tensioner. Remove any dirt and foreign materials completely from the back and the mounting surfaces of the primary timing chain tensioner.

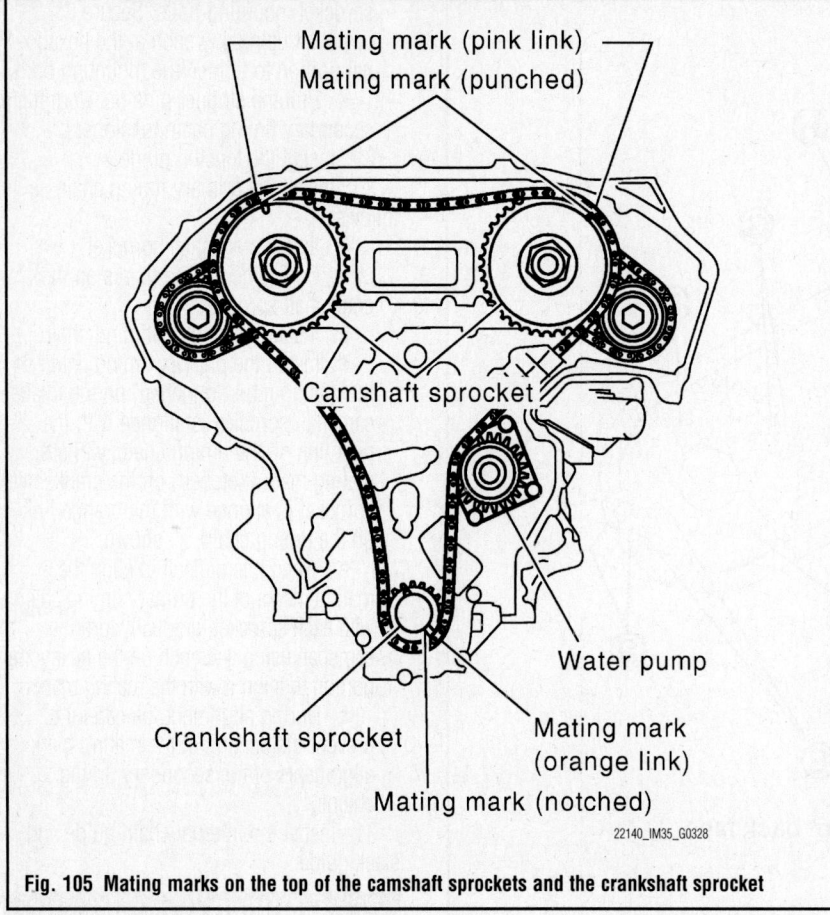

Fig. 105 Mating marks on the top of the camshaft sprockets and the crankshaft sprocket

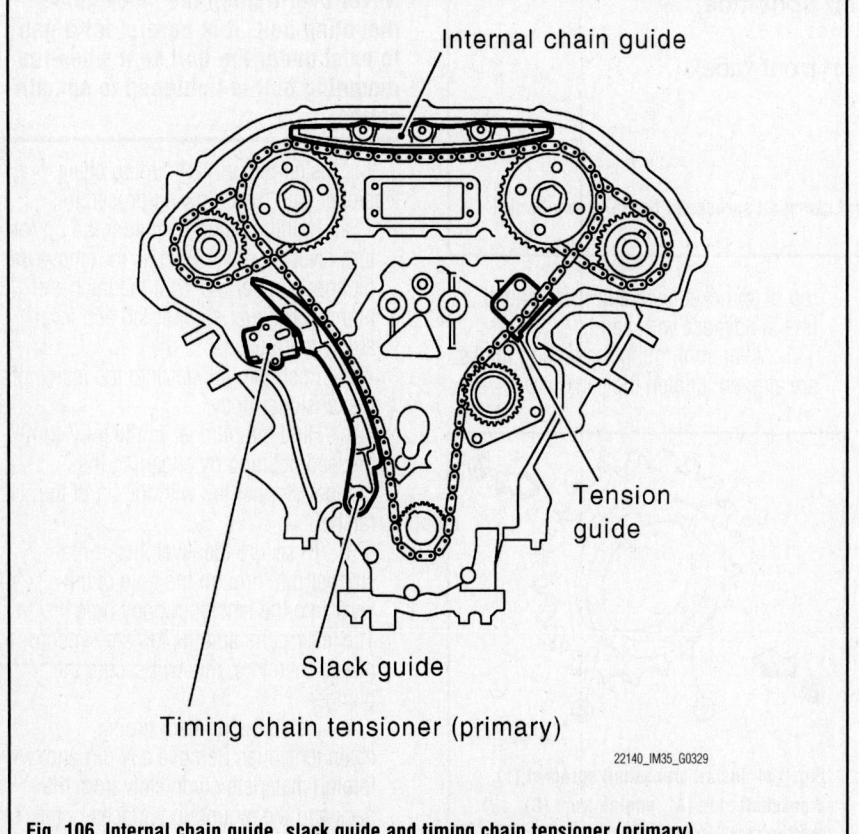

Fig. 106 Internal chain guide, slack guide and timing chain tensioner (primary)

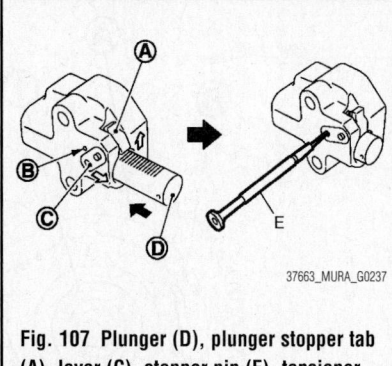

Fig. 107 Plunger (D), plunger stopper tab (A), lever (C), stopper pin (E), tensioner body hole (B)

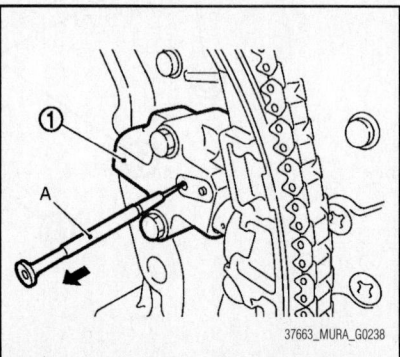

Fig. 108 Install the primary timing chain tensioner (1), pull out the stopper pin (A) after installing, and release the plunger

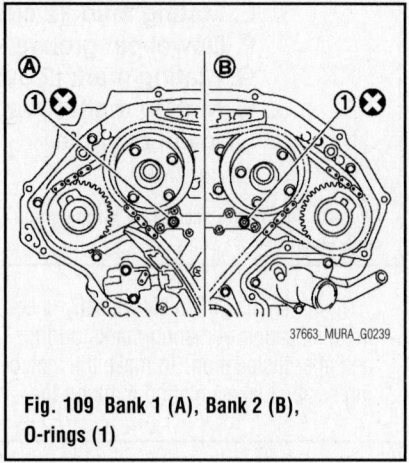

Fig. 109 Bank 1 (A), Bank 2 (B), O-rings (1)

f. Pull out the stopper pin after installing, and release the plunger.

33. Check again that the mating marks on each sprocket and each timing chain have not slipped out of alignment.

34. Install new O-rings on rear timing chain case.

35. Install new front oil seal on the front timing chain case.

a. Apply new engine oil to both the oil seal lip and the dust seal lip.

b. Install it so that each seal lip is oriented as shown.

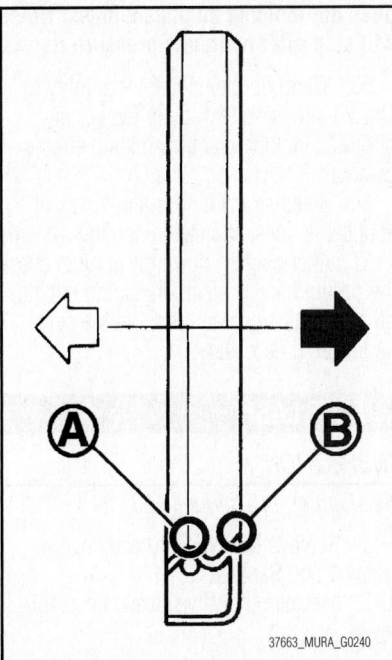

Fig. 110 Oil seal lip (A), dust seal lip (B), black arrow: engine outside, white arrow: engine inside

c. Using a suitable drift [outer diameter: 2.36 in. (60mm)], press-fit the oil seal until it becomes flush with the front timing chain case end face.

d. Check the garter spring is in position and the seal lip is not inverted.

36. Install the water pump cover to the front timing chain case.

a. Apply a continuous bead of liquid gasket to the water pump cover as shown. Use Genuine RTV Silicone Sealant or equivalent.

37. Install the front timing chain case as follows:

a. Apply a continuous bead of liquid gasket to the front timing chain case back side as shown. Use Genuine RTV Silicone Sealant or equivalent.

38. Install the front timing chain case as to fit its dowel pin holes together with the dowel pin on the rear timing chain case.

a. Tighten the mounting bolts to the specified torque in numerical order as shown. There are two types of mounting bolt. Refer to the following for locating the bolts:

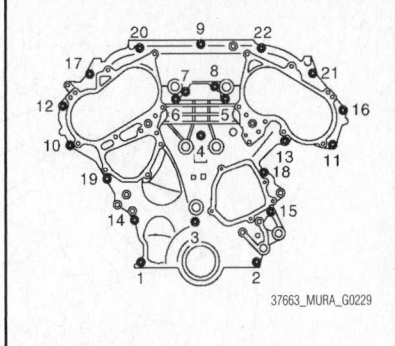

Fig. 112 Front timing chain case bolt tightening sequence

- M8 bolts (No. 1, 2): 21 ft. lbs. (28 Nm)
- M6 bolts (Except No. 1, 2): 9 ft. lbs. (13 Nm)

b. After all bolts are tightened, retighten them to the specified torque in numerical order as shown.

➡**Wipe off any excessive liquid gasket.**

c. After installing the front timing chain case, check the surface height difference between the following parts on the oil pan (upper) mounting surface: Front timing chain case to rear timing chain case: −0.006 to 0.006 in (−0.14 to 0.14 mm).

d. If not within the standard, repeat the installation procedure.

39. Install the intake valve timing control covers as follows:

a. Install new seal rings in shaft grooves.

b. Being careful not to move seal rings from the installation grooves, align the dowel pins on the front timing chain case with the holes to install the valve timing control covers.

c. Tighten the mounting bolts in numerical order as shown.

40. Install the crankshaft pulley as follows:

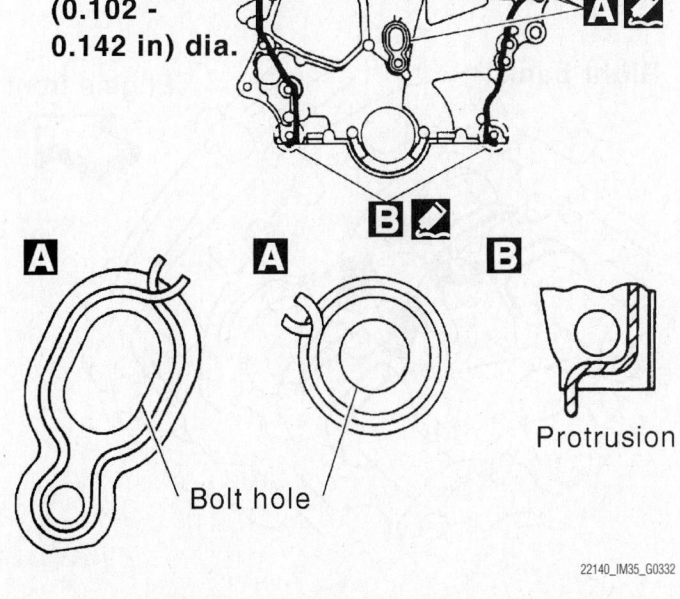

Front timing chain case

2.6 - 3.6 mm (0.102 - 0.142 in) dia.

Protrusion

Bolt hole

Fig. 111 Liquid gasket application chart for the front timing chain case back side

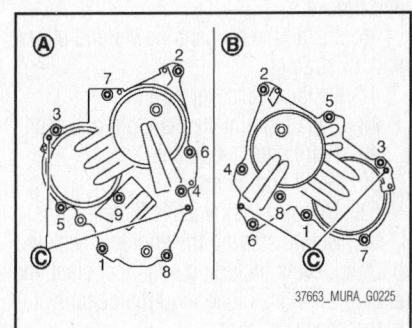

Fig. 113 Bank 1 (A), Bank 2 (B), dowel pin hole (C)

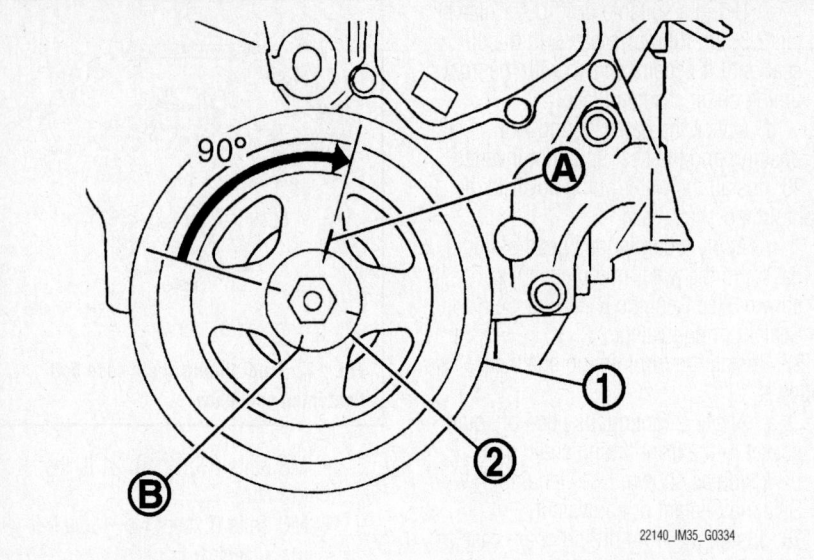

Fig. 114 Place a paint mark (A) on the crankshaft pulley (1) aligning with the angle mark (B) on the crankshaft pulley bolt (2)

a. Install the crankshaft pulley, taking care not to damage the front oil seal.

b. When press-fitting the crankshaft pulley with a plastic hammer, tap on its center portion (not circumference).

c. Hold the crankshaft with the pulley holder.

d. Tighten the crankshaft pulley bolt to 33 ft. lbs. (44 Nm).

e. Place a paint mark on the crankshaft pulley aligning with the angle mark on the crankshaft pulley bolt. Tighten the bolt 90 degrees (angle tightening).

41. Rotate the crankshaft pulley in normal direction (clockwise when viewed from engine front) to confirm it turns smoothly.

42. Install the valve timing control covers.

43. Install the engine harness brackets and connect the harnesses.

44. Install the power steering oil pump and piping.

45. Install the drive belt, idler pulleys and bracket.

46. Install the oil pans (lower and upper) and oil strainer.

47. Install the rocker covers.

48. Install the intake manifold collector.

49. Refill engine coolant.

50. Refill engine oil.

51. Connect the negative battery cable.

52. Before starting the engine, check the oil/fluid levels including engine coolant and engine oil. If less than required quantity, fill to the specified level.

53. Use the following procedure to check for fuel leakage:

a. Turn the ignition switch "ON" (with engine stopped). With fuel pressure applied to fuel piping, check for fuel leakage at connection points.

b. Start the engine. With engine speed increased, check again for fuel leakage at connection points.

54. Run the engine to check for unusual noise and vibration.

➡️If hydraulic pressure inside chain tensioner drops after removal/ installation, slack in guide may generate a pounding noise during and just after the engine start. However, this does not indicate an unusualness. Noise will stop after hydraulic pressure rises.

55. Warm up the engine thoroughly to check there is no leakage of fuel, or any oil/fluids including engine oil and engine coolant.

56. Bleed air from lines and hoses of applicable lines, such as in cooling system.

57. After cooling down the engine, check the oil/fluid levels again, including engine oil and engine coolant. Refill to the specified level, if necessary.

VALVE LASH

INSPECTION

See Figures 115 through 123.

1. Remove the cylinder head covers (Bank 1 and Bank 2).

2. Measure the valve clearance as follows:

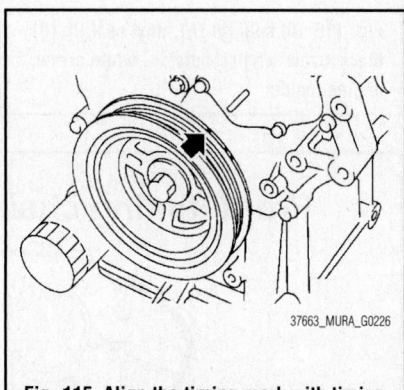

Fig. 115 Align the timing mark with timing indicator

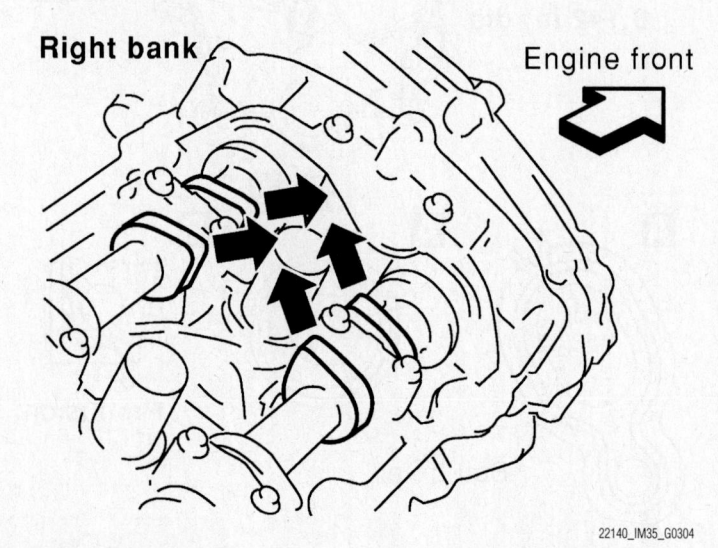

Fig. 116 Intake and exhaust cam noses on No. 1 cylinder (engine front side of Bank 1)

Measuring position [bank 1 (A)]		No. 1 CYL.	No. 3 CYL.	No. 5 CYL.
No. 1 cylinder at compression TDC	EXH (C)		× (B)	
	INT (D)	× (E)		
Measuring position [bank 2 (H)]		No. 2 CYL.	No. 4 CYL.	No. 6 CYL.
No. 1 cylinder at compression TDC	INT (D)			× (F)
	EXH (C)	× (G)		

37663_MURA_G0434

Fig. 117 Measure the valve clearances—No. 1 cylinder at compression TDC

a. Set the No. 1 cylinder at TDC of its compression stroke.

b. Rotate the crankshaft pulley clockwise to align the timing mark (grooved line without color) with the timing indicator.

3. Check that the intake and exhaust cam noses on the No. 1 cylinder (engine front side of Bank 1) are located as shown. If not, turn crankshaft one revolution (360 degrees) and align as shown.

a. Use a feeler gauge, measure the clearance between the valve lifter and the camshaft.

b. By referring to the figure, measure the valve clearances at locations marked

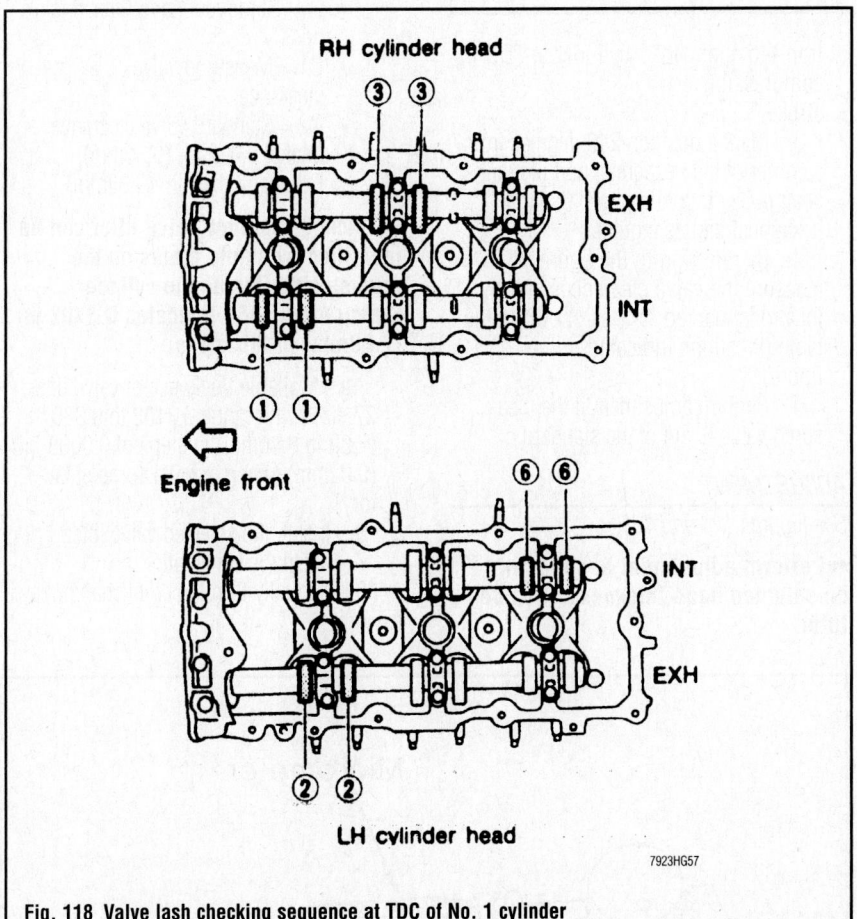

Fig. 118 Valve lash checking sequence at TDC of No. 1 cylinder

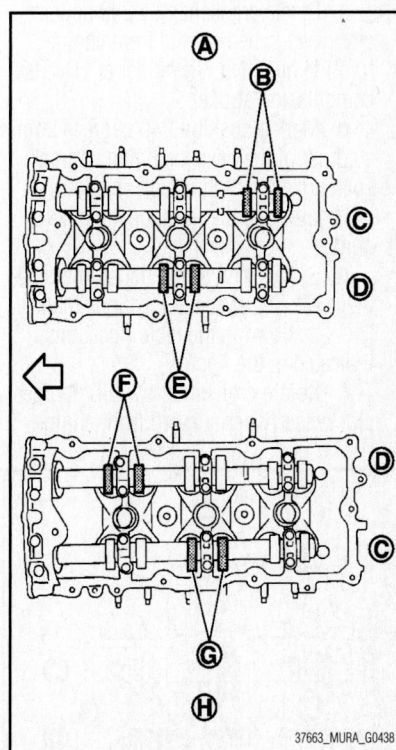

Fig. 121 Valve lash checking sequence at TDC of No. 3 cylinder

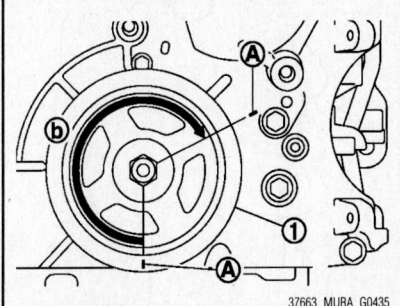

37663_MURA_G0435

Fig. 119 Mark (A) a position 240 degrees (b) from a corner of the hexagonal part of crankshaft pulley (1) mounting bolt

Measuring position [bank 1 (A)]		No. 1 CYL.	No. 3 CYL.	No. 5 CYL.
No. 3 cylinder at compression TDC	EXH (C)			× (B)
	INT (D)		× (E)	
Measuring position [bank 2 (H)]		No. 2 CYL.	No. 4 CYL.	No. 6 CYL.
No. 3 cylinder at compression TDC	INT (D)	× (F)		
	EXH (C)		× (G)	

37663_MURA_G0436

Fig. 120 Measure the valve clearances—No. 3 cylinder at compression TDC

Measuring position [bank 1 (A)]		No. 1 CYL.	No. 3 CYL.	No. 5 CYL.
No. 5 cylinder at compression TDC	EXH (C)	× (B)		
	INT (D)			× (E)
Measuring position [bank 2 (H)]		No. 2 CYL.	No. 4 CYL.	No. 6 CYL.
No. 5 cylinder at compression TDC	INT (D)		× (F)	
	EXH (C)			× (G)

37663_MURA_G0437

Fig. 122 Measure the valve clearances—No. 5 cylinder at compression TDC

"×" as shown in the table (locations indicated in the figure).

c. Rotate crankshaft by 240 degrees clockwise (when viewed from engine front) to align No. 3 cylinder at TDC its compression stroke.

d. Mark a position 240 degrees from a corner of the hexagonal part of crankshaft pulley mounting bolt as shown in the figure. Use the hexagonal part as a guide.

e. By referring to the figure, measure the valve clearances at locations marked "×" as shown in the table (locations indicated in the figure).

f. Rotate crankshaft by 240 degrees clockwise (when viewed from engine front) to align No. 5 cylinder at TDC its compression stroke.

g. Mark a position 240 degrees from a corner of the hexagonal part of crankshaft pulley mounting bolt. Use the hexagonal part as a guide.

h. By referring to the figure, measure the valve clearances at locations marked "×" as shown in the table (locations indicated in the figure).

i. Perform adjustment if the measured value is out of the standard.

ADJUSTMENT

See Figures 124 and 125.

➡**Perform adjustment depending on selected head thickness of valve lifter.**

1. Measure the valve clearance.
2. Remove the camshaft.
3. Remove the valve lifters at the locations that are out of the standard.
4. Measure the center thickness of the removed valve lifters with a micrometer.
5. Use the equation below to calculate valve lifter thickness for replacement.

- Valve lifter thickness calculation: $t = t1 + (C1 − C2)$
- t = Valve lifter thickness to be replaced
- $t1$ = Removed valve lifter thickness
- $C1$ = Measured valve clearance
- $C2$ = Standard valve clearance: Intake: 0.012 in. (0.30mm), Exhaust: 0.013 in. (0.33mm)

➡**Thickness of new valve lifter can be identified by stamp marks on the reverse side (inside the cylinder). Stamp mark 788P indicates 0.3102 in. (7.88mm) in thickness.**

a. Available thickness of valve lifter: 27 sizes with range 0.3102 to 0.3307 in. (7.88 to 8.40mm) in steps of 0.0008 in. (0.02mm) (when manufactured at factory).

6. Install the selected valve lifter.
7. Install the camshaft.
8. Manually turn the crankshaft pulley a few turns.

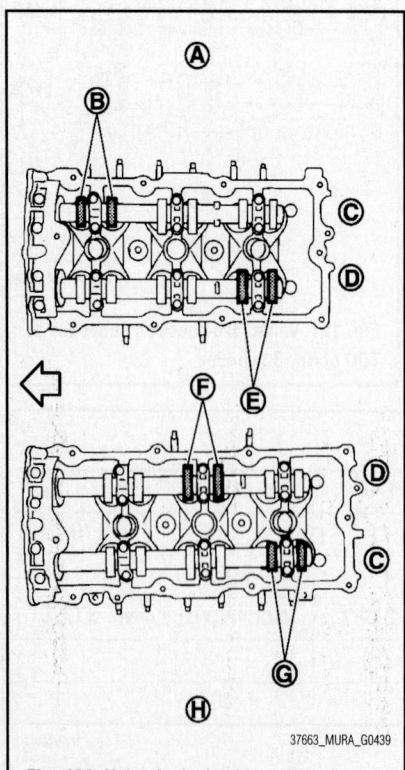

Fig. 123 Valve lash checking sequence at TDC of No. 5 cylinder

37663_MURA_G0439

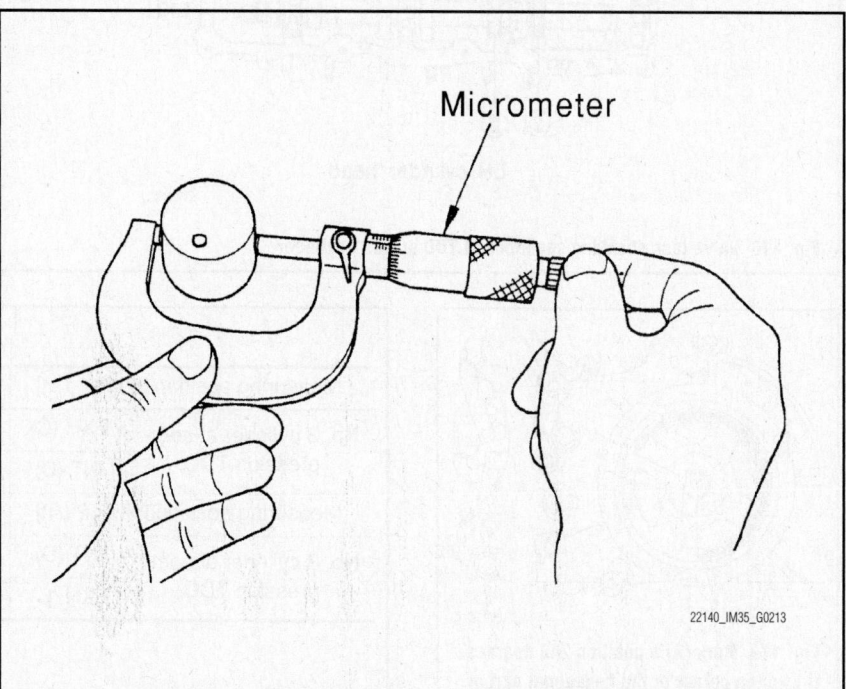

Micrometer

22140_IM35_G0213

Fig. 124 Valve lifter outer diameter inspection

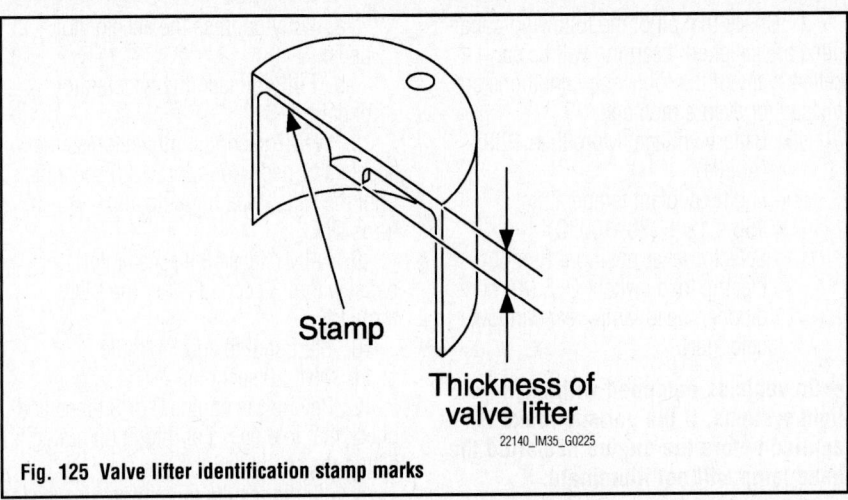

Stamp

Thickness of
valve lifter

22140_IM35_G0225

Fig. 125 Valve lifter identification stamp marks

9. Check that the valve clearances for cold engine are within the specifications by referring to the specified values.

a. Intake valve clearance (cold) is 0.010–0.013 in. (0.26–0.34mm) and exhaust valve clearance (cold) is 0.011–0.015 in. (0.29–0.37mm).

10. Install all removal parts in the reverse order of removal.

11. Warm up the engine, and check for unusual noise and vibration.

ENGINE PERFORMANCE & EMISSION CONTROLS

CAMSHAFT POSITION (CMP) SENSOR

LOCATION

See Figure 126.

Refer to the accompanying illustration.

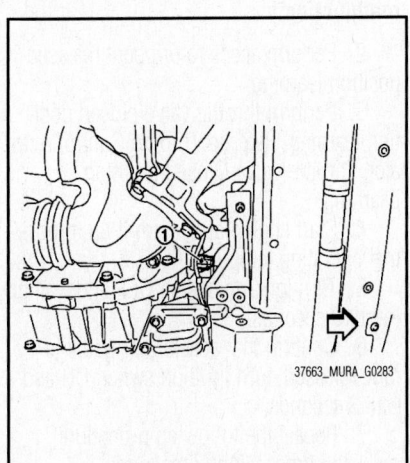

37663_MURA_G0282

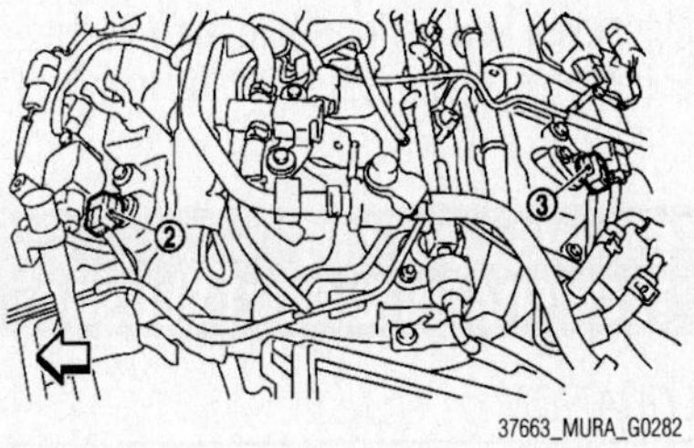

Fig. 126 Camshaft Position (CMP) sensor (PHASE) Bank 1 (2), Camshaft Position (CMP) sensor (PHASE) Bank 2 (3)

REMOVAL & INSTALLATION

1. Turn ignition switch OFF.
2. Loosen the sensor mounting bolt.
3. Disconnect the Camshaft Position (CMP) sensor (PHASE) harness connector.
4. Remove the sensor.

To install:

5. Install the CMP sensor.
6. Connect the CMP sensor connector.
7. Tighten the sensor mounting bolt.

CRANKSHAFT POSITION (CKP) SENSOR

LOCATION

See Figure 127.

Refer to the accompanying illustration.

37663_MURA_G0283

Fig. 127 Crankshaft Position (CKP) sensor

REMOVAL & INSTALLATION

1. Turn ignition switch OFF.
2. Loosen the sensor mounting bolt.
3. Disconnect the CKP sensor harness connector.
4. Remove the sensor.

To install:

5. Install the CKP sensor.
6. Connect the CKP sensor connector.
7. Tighten the sensor mounting bolt.

ELECTRONIC CONTROL MODULE (ECM)

LOCATION

See Figure 128.

Refer to the accompanying illustration.

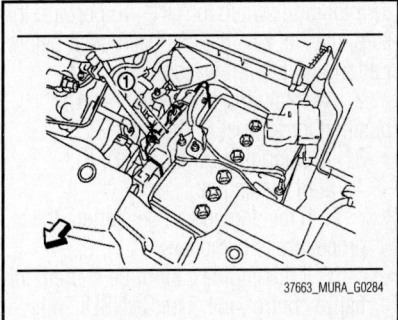

37663_MURA_G0284

Fig. 128 Electronic Control Module (ECM)

REMOVAL & INSTALLATION

See Figure 129.

1. Turn ignition switch OFF.
2. Loosen the sensor mounting bolt.
3. Disconnect the ECM harness connectors.
4. Remove the ECM.

To install:

5. Install the ECM.
6. Connect the ECM connectors.
7. Tighten the sensor mounting bolt.
8. When replacing the ECM:

a. Perform the initialization of NVIS (NATS) system and registration of all NVIS (NATS) ignition key IDs. See Reset, below.

b. Perform VIN registration. The CONSULT-III scan tool is required.

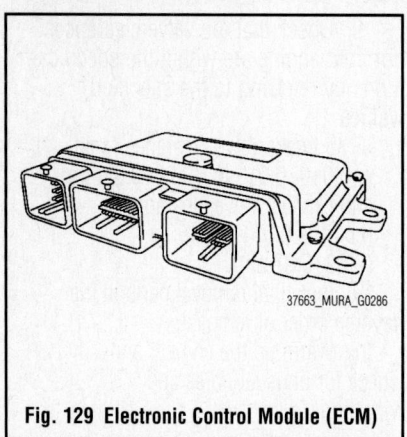

Fig. 129 Electronic Control Module (ECM)

c. Perform accelerator pedal released position learning.

d. Perform throttle valve closed position learning. Refer to Throttle Control Actuator, Throttle Valve Closed Position Learning.

e. Perform idle air volume learning. See Reset, below.

RESET

NVIS (NATS) System Initialization & Ignition Key ID Registration

1. Insert the registered Intelligent Key, turn ignition switch to "ON". To perform this step, use the key that has been used before performing ECM replacement.

2. Maintain ignition switch in "ON" position for at least 5 seconds.

3. Turn ignition switch to "OFF".

4. Start the engine.

a. If the engine can be started, the procedure is complete.

b. If the engine cannot be started, initialize control unit. The CONSULT-III scan tool is required.

Idle Air Volume Learning

See Figure 130.

1. Check that all of the following conditions are satisfied. Learning will be cancelled if any of the following conditions are missed for even a moment:

- Battery voltage: More than 12.9 V (At idle)
- Engine coolant temperature: 158–212°F (70–100°C)
- Selector lever position: P or N
- Electric load switch: OFF (Air conditioner, head lamp, rear window defogger)

➡ **On vehicles equipped with daytime light systems, if the parking brake is applied before the engine is started the head lamp will not illuminate.**

- Steering wheel: Neutral (Straight-ahead position)
- Vehicle speed: Stopped
- Transmission: Warmed-up
- Drive vehicle for 10 minutes

➡ **It is better to count the time accurately with a clock.**

➡ **It is impossible to switch the diagnostic mode when an accelerator pedal position sensor circuit has a malfunction.**

2. Perform accelerator pedal released position learning.

3. Perform throttle valve closed position learning. Refer to Throttle Control Actuator, Throttle Valve Closed Position Learning.

4. Start engine and warm it up to normal operating temperature.

5. Turn ignition switch OFF and wait at least 10 seconds.

6. Confirm that accelerator pedal is fully released, turn ignition switch ON and wait 3 seconds.

7. Repeat the following procedure quickly 5 times within 5 seconds.

a. Fully depress the accelerator pedal.

b. Fully release the accelerator pedal.

8. Wait 7 seconds, fully depress the accelerator pedal for approx. 20 seconds until the MIL stops blinking and turns ON.

9. Fully release the accelerator pedal within 3 seconds after the MIL turns ON.

10. Start engine and let it idle.

11. Wait 20 seconds.

12. Rev up the engine 2 or 3 times and check that idle speed and ignition timing are within the specifications.

a. If the results are normal, inspection in ended. If the results are not normal, check the following:

- Check that throttle valve is fully closed
- Check PCV valve operation
- Check that downstream of throttle valve is free from air leakage

b. If the results are not normal, repair or replace malfunctioning part. If the results are normal, engine component parts and their installation condition are questionable. Check and eliminate the cause of the incident. If any of the following conditions occur after the engine has started, eliminate the cause of the incident and perform Idle Air Volume Learning again: engine stalls, Incorrect idle.

ENGINE COOLANT TEMPERATURE (ECT) SENSOR

LOCATION

See Figure 131.

Refer to the accompanying illustration.

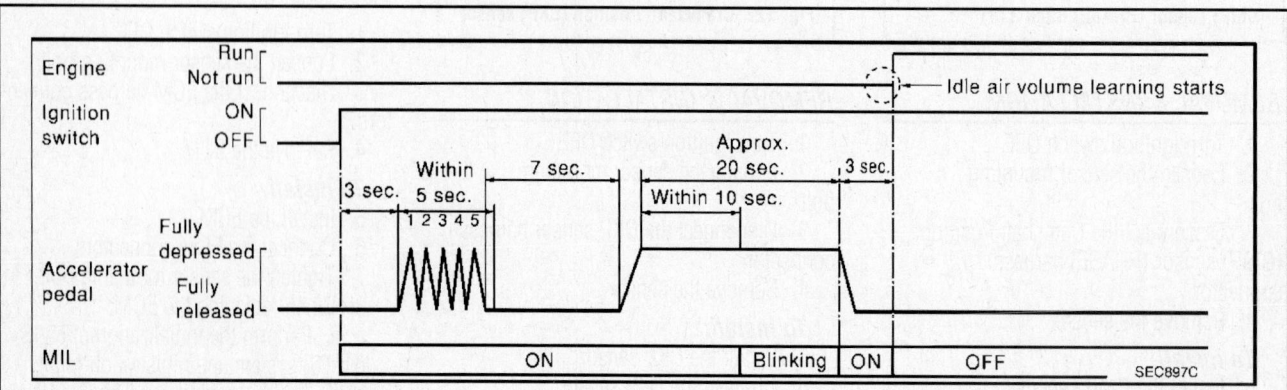

Fig. 130 Electronic Control Module (ECM)

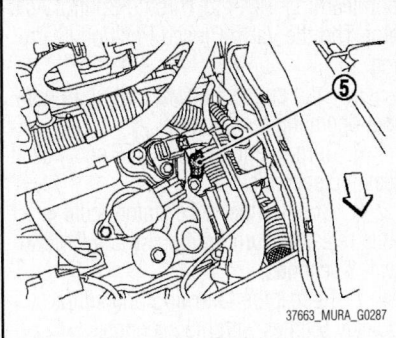

Fig. 131 Engine Coolant Temperature (ECT) Sensor

REMOVAL & INSTALLATION

1. Turn the ignition switch OFF.
2. Disconnect the engine coolant temperature sensor harness connector.
3. Remove the engine coolant temperature sensor.

To install:

4. Install the engine coolant temperature sensor.
5. Connect the sensor connector.

HEATED OXYGEN SENSOR (HO2S)

LOCATION

See Figure 132.

Refer to the accompanying illustration.

REMOVAL & INSTALLATION

1. Turn the ignition switch OFF.
2. Disconnect the heated oxygen sensor harness connector.

3. Remove the heated oxygen sensor.

To install:

4. Installation is the revere of removal, noting the following:

  a. Discard any heated oxygen sensor which has been dropped from a height of more than 0.5 m (19.7 in) onto a hard surface such as a concrete floor; use a new one.

  b. Before installing new oxygen sensor, clean exhaust system threads using Oxygen Sensor Thread Cleaner [commercial service tool (J-43897-18 or J-43897-12)] and approved anti-seize lubricant.

INTAKE AIR TEMPERATURE (IAT) SENSOR

LOCATION

See Figure 133.

Refer to the accompanying illustration.

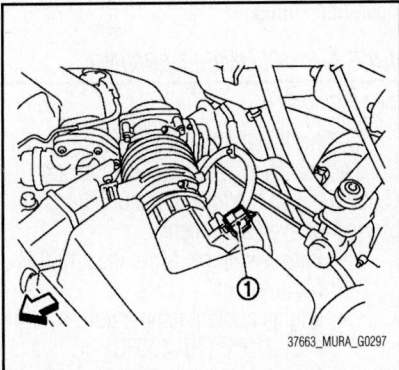

Fig. 133 MAF sensor and Intake Air Temperature (IAT) sensor

REMOVAL & INSTALLATION

1. Turn the ignition switch OFF.
2. Disconnect the IAT sensor harness connector.
3. Remove the IAT sensor.

To install:

4. Install in the reverse order of removal.

KNOCK SENSOR (KS)

LOCATION

See Figure 134.

Refer to the accompanying illustration.

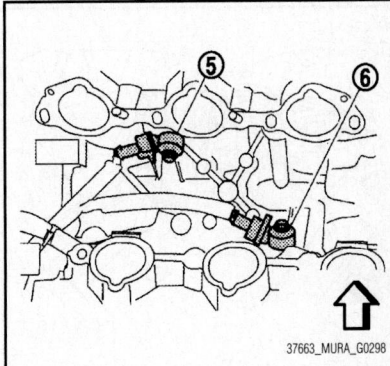

Fig. 134 Knock Sensor (KS), Bank 1 (6), Bank 2 (5)

REMOVAL & INSTALLATION

1. Turn the ignition switch OFF.
2. Disconnect the knock sensor harness connector.
3. Remove the knock sensor.

To install:

4. Install in the reverse order of removal.

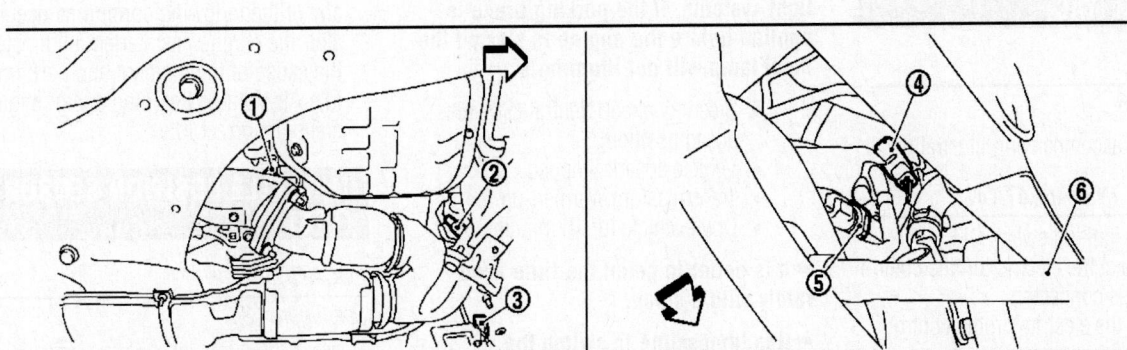

1. HO2S2 (bank 1)
2. HO2S2 (bank 2)
3. HO2S2 (bank 2) harness connector
4. HO2S2 (bank 1) harness connector
5. Power steering pressure sensor
6. Drive shaft (RH)

Fig. 132 Heated Oxygen Sensors (HO2S)

MALFUNCTION INDICATOR LIGHT (MIL)

RESET PROCEDURE

The MIL can be commanded off by the scan tool.

MASS AIR FLOW (MAF) SENSOR

LOCATION

See Figure 135.

Refer to the accompanying illustration.

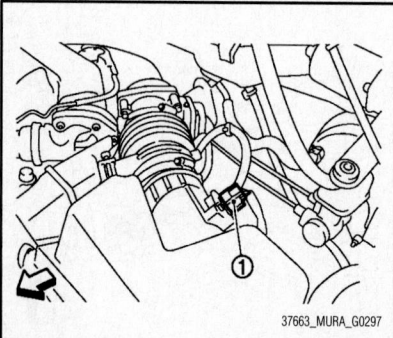

Fig. 135 Mass Air Flow (MAF) sensor and Intake Air Temperature (IAT) sensor

REMOVAL & INSTALLATION

1. Turn the ignition switch OFF.
2. Disconnect the MAF sensor harness connector.
3. Remove the MAF sensor.

To install:

4. Install in the reverse order of removal.

THROTTLE CONTROL ACTUATOR (TAC)

LOCATION

See Figure 136.

Refer to the accompanying illustration.

REMOVAL & INSTALLATION

1. Turn the ignition switch OFF.
2. Disconnect the electric throttle control actuator harness connector.
3. Remove the electric throttle control actuator.

To install:

4. Installation is the reverse of removal, noting the following:
   a. Perform throttle valve closed position learning.
   b. Perform idle air volume learning.

Fig. 136 Throttle Control Actuator (2)

THROTTLE VALVE CLOSED POSITION LEARNING

1. Check that accelerator pedal is fully released.
2. Turn ignition switch ON.
3. Turn ignition switch OFF and wait at least 10 seconds.
4. Check that throttle valve moves during the above 10 seconds by confirming the operating sound.

IDLE AIR VOLUME LEARNING

See Figure 170.

1. Check that all of the following conditions are satisfied. Learning will be cancelled if any of the following conditions are missed for even a moment:

- Battery voltage: More than 12.9 V (At idle)
- Engine coolant temperature: 158–212°F (70–100°C)
- Selector lever position: P or N
- Electric load switch: OFF (Air conditioner, head lamp, rear window defogger)

➡ **On vehicles equipped with daytime light systems, if the parking brake is applied before the engine is started the head lamp will not illuminate.**

- Steering wheel: Neutral (Straight-ahead position)
- Vehicle speed: Stopped
- Transmission: Warmed-up
- Drive vehicle for 10 minutes

➡ **It is better to count the time accurately with a clock.**

➡ **It is impossible to switch the diagnostic mode when an accelerator pedal position sensor circuit has a malfunction.**

2. Perform accelerator pedal released position learning.
3. Perform throttle valve closed posi-

tion learning. Refer to Throttle Control Actuator, Throttle Valve Closed Position Learning.

4. Start engine and warm it up to normal operating temperature.
5. Turn ignition switch OFF and wait at least 10 seconds.
6. Confirm that accelerator pedal is fully released, turn ignition switch ON and wait 3 seconds.
7. Repeat the following procedure quickly 5 times within 5 seconds.
   a. Fully depress the accelerator pedal.
   b. Fully release the accelerator pedal.
8. Wait 7 seconds, fully depress the accelerator pedal for approx. 20 seconds until the MIL stops blinking and turns ON.
9. Fully release the accelerator pedal within 3 seconds after the MIL turns ON.
10. Start engine and let it idle.
11. Wait 20 seconds.
12. Rev up the engine 2 or 3 times and check that idle speed and ignition timing are within the specifications.
    a. If the results are normal, inspection in ended. If the results are not normal, check the following:

    - Check that throttle valve is fully closed
    - Check PCV valve operation
    - Check that downstream of throttle valve is free from air leakage

    b. If the results are not normal, repair or replace malfunctioning part. If the results are normal, engine component parts and their installation condition are questionable. Check and eliminate the cause of the incident. If any of the following conditions occur after the engine has started, eliminate the cause of the incident and perform Idle Air Volume Learning again: engine stalls, incorrect idle.

THROTTLE POSITION SENSOR (TPS)

LOCATION

See Figure 136.

The Throttle Position Sensor (TPS) is part of the Throttle Control Actuator.

REMOVAL & INSTALLATION

See Throttle Control Actuator Removal & Installation.

VARIABLE CAMSHAFT TIMING OIL CONTROL SOLENOID

LOCATION

See Figure 137.

Refer to the accompanying illustration.

REMOVAL & INSTALLATION

1. Turn the ignition switch OFF.
2. Disconnect the intake valve timing control solenoid valve harness connector.
3. Remove the intake valve timing control solenoid valve.

To install:

4. Installation is the reverse of removal.

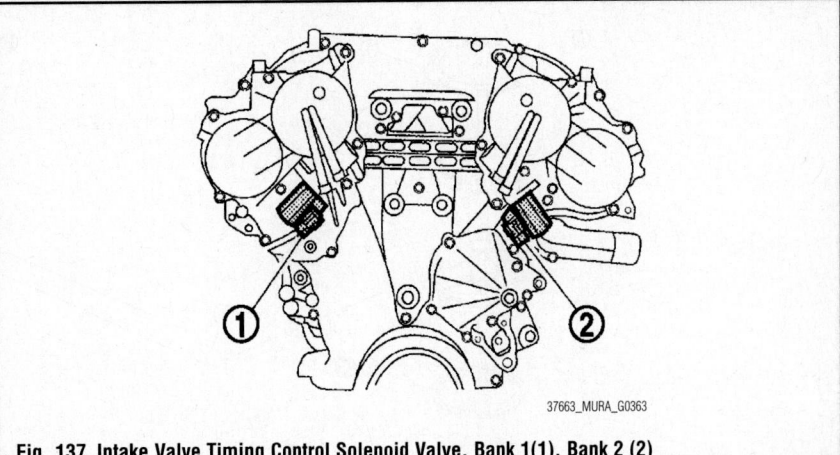

37663_MURA_G0363

Fig. 137 Intake Valve Timing Control Solenoid Valve, Bank 1(1), Bank 2 (2)

FUEL   GASOLINE FUEL INJECTION SYSTEM

FUEL SYSTEM SERVICE PRECAUTIONS

Safety is the most important factor when performing not only fuel system maintenance, but any type of maintenance. Failure to conduct maintenance and repairs in a safe manner may result in serious personal injury or death. Work on a vehicle's fuel system components can be accomplished safely and effectively by adhering to the following rules and guidelines.

• To avoid the possibility of fire and personal injury, always disconnect the negative battery cable unless the repair or test procedure requires that battery voltage be applied.

• Always relieve the fuel system pressure prior to disconnecting any fuel system component (injector, fuel rail, pressure regulator, etc.) fitting or fuel line connection. Exercise extreme caution whenever relieving fuel system pressure to avoid exposing skin, face and eyes to fuel spray. Please be advised that fuel under pressure may penetrate the skin or any part of the body that it contacts.

• Always place a shop towel or cloth around the fitting or connection prior to loosening to absorb any excess fuel due to spillage. Ensure that all fuel spillage is quickly removed from engine surfaces. Ensure that all fuel-soaked cloths or towels are deposited into a flame-proof waste container with a lid.

• Always keep a dry chemical (Class B) fire extinguisher near the work area.

• Do not allow fuel spray or fuel vapors to come into contact with a spark or open flame.

• Always use a second wrench when loosening or tightening fuel line connection fittings. This will prevent unnecessary stress and torsion on fuel piping. Always follow the proper torque specifications.

• Always replace worn fuel fitting O-rings with new ones. Do not substitute fuel hose where rigid pipe is installed.

FUEL SYSTEM PRESSURE

RELIEVING

See Figure 138.

1. Before servicing the vehicle, refer to the Precautions Section.
2. Remove the fuel pump fuse (2) located in IPDM E/R (1).
3. Start the engine.
4. After the engine stalls, crank it 2 or 3 times to release all fuel pressure.
5. Turn the ignition switch OFF.

6. Reinstall fuel pump fuse after servicing fuel system.

FUEL FILTER

REMOVAL & INSTALLATION

The filter is part of the pump and sender assembly. Service can only be done by removing the sender/pump assembly.

FUEL LEVEL SENDING UNIT

LOCATION

See Figures 139 and 140.

Refer to the accompanying illustrations.

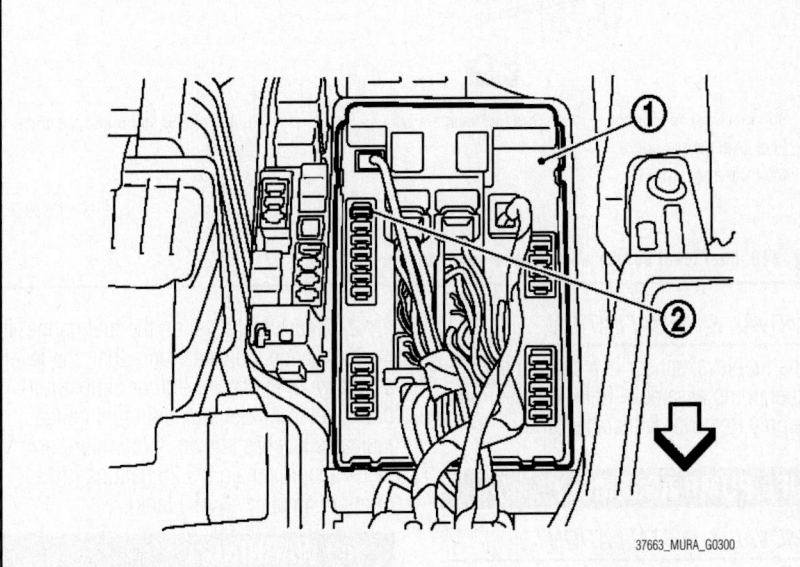

37663_MURA_G0300

Fig. 138 Remove the fuel pump fuse (2)

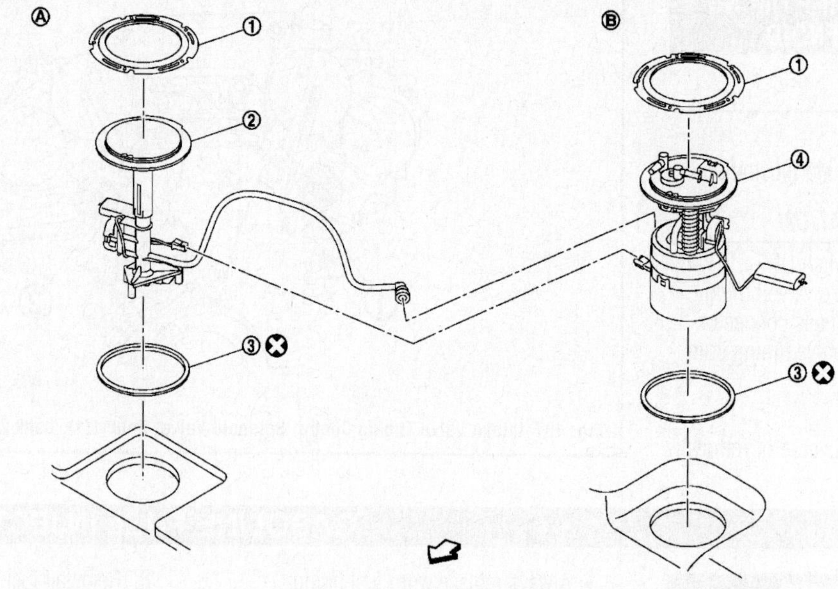

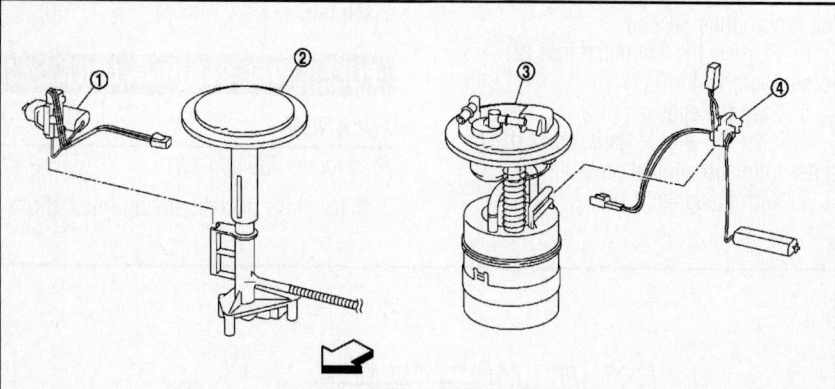

1. Retainer
2. Sub fuel level sensor unit
3. Seal packing
4. Main fuel level sensor unit, fuel filter and fuel pump assembly
A. Right side
B. Left side
⟨▭ : Vehicle front

37663_MURA_G0301

Fig. 139 Fuel pump assembly and related components

1. Sub fuel level sensor unit
2. Sub fuel level sensor unit base
3. Fuel filter and fuel pump assembly
4. Fuel level sensor unit
⟨▭ : Vehicle front

37663_MURA_G0302

Fig. 140 Fuel level sensor and sub fuel level sensor

REMOVAL & INSTALLATION

The fuel level sensor is a component of the fuel pump assembly. Refer to Fuel Pump Assembly Removal & Installation.

FUEL PUMP ASSEMBLY

REMOVAL & INSTALLATION

See Figures 141 through 147.

1. Before servicing the vehicle, refer to the Precautions Section.

2. Check fuel level on the fuel gauge. If the fuel gauge indicates more than the level as shown in the figure (full or almost full), drain fuel from fuel tank until fuel gauge indicates level as shown in the figure or below [approximately 5.25 gallons (20 L), or approximately ¾ of a tank].

❋❋ WARNING

Because fuel will be spilled when removing main and sub fuel level sensor units for the top of the fuel is above the main and sub fuel level sensor units installation surface. As a guide, fuel level becomes the position as shown in the figure or below when approximately 5.25 gallons (20 L) of fuel are drained from fuel tank.

3. Release the fuel pressure from the fuel lines.
4. Open the fuel filler lid.
5. Open the filler cap and release the pressure inside fuel tank.
6. Remove the rear seat cushion and reclining device assembly.

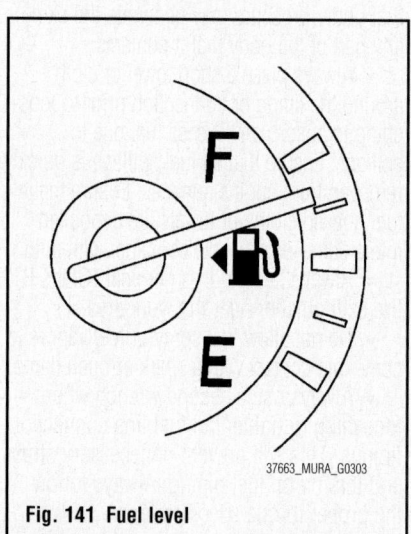

37663_MURA_G0303

Fig. 141 Fuel level

7. Peel off the floor carpet, then remove the inspection hole cover units by turning clips clockwise by 90 degrees.

8. Disconnect the harness connector, fuel feed tube and EVAP hose.

9. Disconnect the quick connector as follows:

a. Hold the sides of connector, push in the tabs and pull out the fuel feed tube.

10. Remove the retainer for the main fuel level sensor unit, fuel filter and fuel pump assembly and sub fuel level sensor unit with the fuel tank lock ring wrench (SST) by turning counterclockwise.

11. Remove the main fuel level sensor unit, fuel filter and fuel pump assembly, and sub fuel level sensor unit.

a. Raise the main fuel level sensor unit, fuel filter and fuel pump assembly, and disconnect the fuel hose connector (push in tabs and pull out) and sub fuel level sensor unit harness connector.

b. Raise and release the sub fuel level sensor unit to remove.

To install:

12. Note to the following, and install in the reverse order of removal.

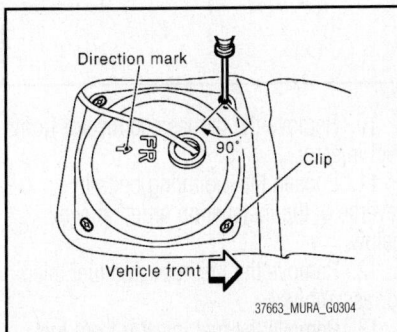

Fig. 142 Left side: Main fuel level sensor unit, fuel filter and fuel pump assembly, Right side: Sub fuel level sensor unit

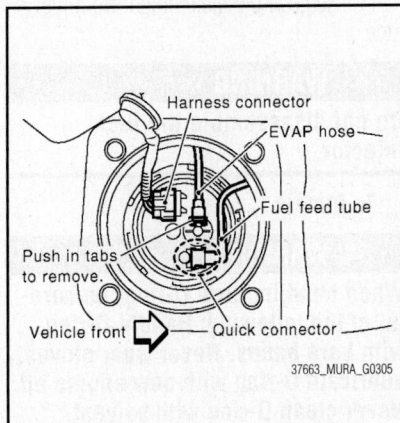

Fig. 143 Disconnect the harness connector, fuel feed tube and EVAP hose

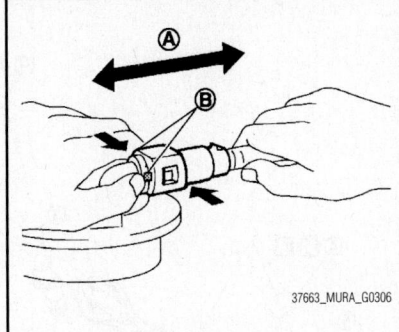

Fig. 144 Push in the tabs (B) and pull (A) out the fuel feed tube

13. Install the new seal packing to the fuel tank without any twist.

14. Connect the fuel hose connector (push in until it stops) and sub fuel level sensor unit harness connector. Insert the connector until you hear a click.

15. Align the direction mark on main fuel level sensor unit, fuel filter and fuel pump assembly and sub fuel level sensor unit with that on fuel tank as shown in the figure, and install them to fuel tank.

✷✷ WARNING

Never allow seal packing to drop. Never bend float arm during installing.

16. Install the retainer for the main fuel level sensor unit, fuel filter and fuel pump assembly and sub fuel level sensor unit with the fuel tank lock ring wrench (SST) by turning clockwise.

✷✷ WARNING

Install the retainer horizontally.

17. Connect the quick connector as follows:

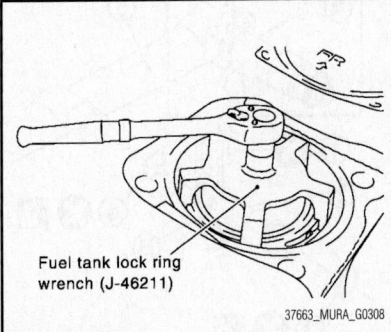

Fig. 145 Remove the retainer for the main fuel level sensor unit, fuel filter and fuel pump assembly and sub fuel level sensor unit

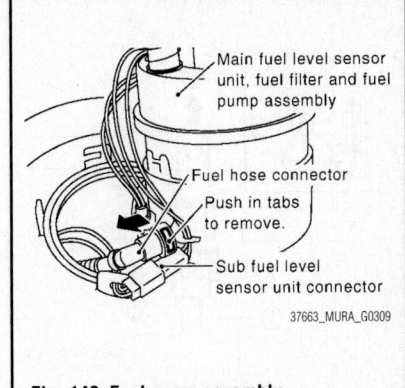

Fig. 146 Fuel pump assembly

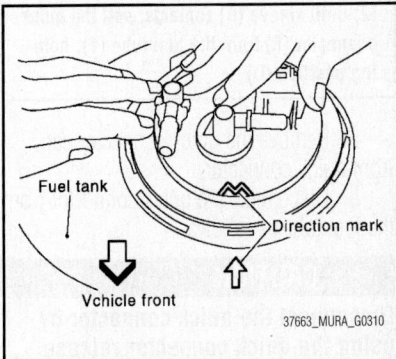

Fig. 147 Align the direction mark on main fuel level sensor unit, fuel filter and fuel pump assembly and sub fuel level sensor unit with that on fuel tank

a. Check the connection for damage or any foreign materials.

b. Align the connector with the tube, then insert the connector straight into the tube until a click sound is heard.

c. After connecting, check that the connection is secure by

FUEL RAIL AND INJECTOR

REMOVAL & INSTALLATION

See Figures 148 through 155.

1. Before servicing the vehicle, refer to the Precautions Section.

2. Remove the radiator core support covers, air duct (inlet), air cleaner cases (upper and lower) with mass air flow sensor and air duct assembly.

3. Remove the engine cover.

4. Properly release the fuel pressure.

5. Remove the front wiper arm and extension cowl top.

6. Drain the engine coolant, or when water hoses are disconnected, attach plug to prevent engine coolant leakage.

7. Remove the intake manifold collector.

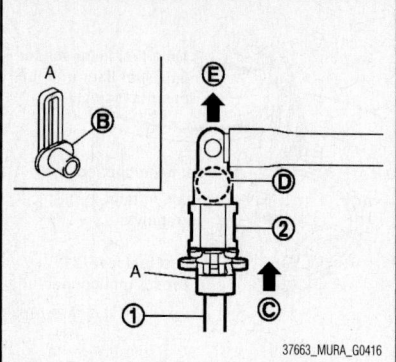

Fig. 148 Insert and retain (C) the quick connector release (A) into quick connector (2) until sleeve (B) contacts; pull the quick connector (E) from the fuel tube (1), holding position (D)

8. Remove the quick connector cap from quick connector.

9. Disconnect the quick connector from the fuel tube as follows:

※※ CAUTION

Disconnect the quick connector by using the quick connector release (commercial service tool: J-45488), not by picking out retainer tabs.

a. With the sleeve side of quick connector release facing toward the quick connector, install the quick connector release onto the fuel tube.

b. Insert the quick connector release into the quick connector until the sleeve contacts and goes no further. Hold the quick connector release in that position.

※※ WARNING

Inserting the quick connector release hard will not disconnect the quick connector. Hold the quick connector release where it contacts and goes no further.

c. Draw and pull out the quick connector straight from the fuel tube. Pull the quick connector holding position as shown.

※※ WARNING

Never pull with lateral force applied. Prepare container and cloth before as fuel will leak out. Never bend or twist connection between quick connector and fuel feed hose (with damper) during installation/removal. To keep the connecting portions clean and to avoid damage, cover them com-

pletely with plastic bags or something similar.

※※ CAUTION

Avoid fire and sparks. Keep parts away from heat source. Be careful if welding. Never expose parts to battery electrolyte or other acids.

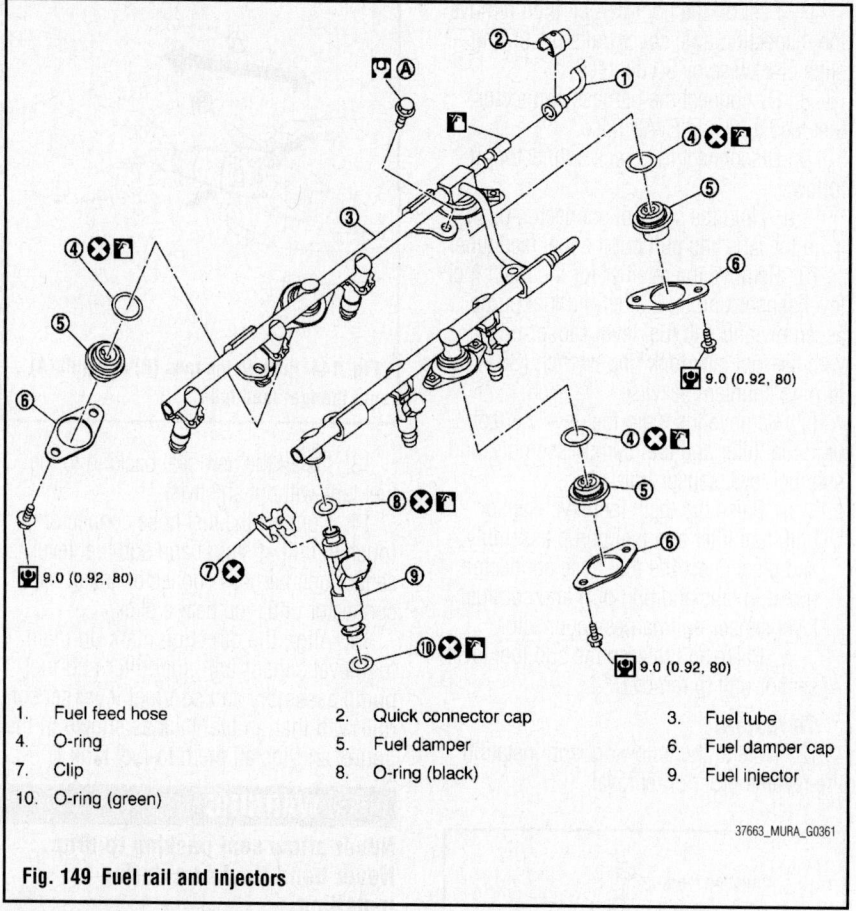

1. Fuel feed hose	2. Quick connector cap	3. Fuel tube
4. O-ring	5. Fuel damper	6. Fuel damper cap
7. Clip	8. O-ring (black)	9. Fuel injector
10. O-ring (green)		

37663_MURA_G0361

Fig. 149 Fuel rail and injectors

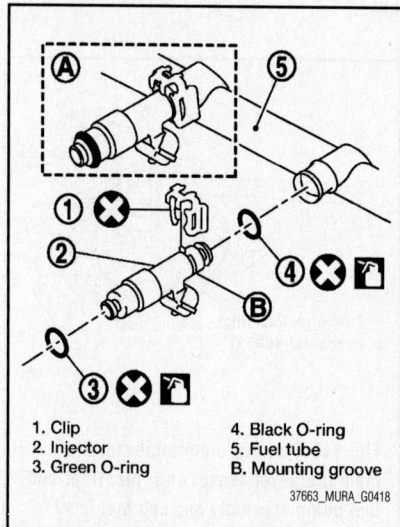

1. Clip	4. Black O-ring
2. Injector	5. Fuel tube
3. Green O-ring	B. Mounting groove

37663_MURA_G0418

Fig. 150 Installed position (A)

10. Remove the harness connector from fuel injector.

11. Loosen the mounting bolts in reverse of the installation order shown below.

12. Remove the fuel tube and fuel injector assembly.

13. Remove the fuel injector from fuel tube with following procedure.

a. Open and remove the clip.

b. Remove the fuel injector from the fuel tube by pulling straight.

14. Remove the fuel damper from fuel tube.

※※ WARNING

Do not disassemble the fuel injector.

To install:

※※ WARNING

When handling new O-ring, be careful of the following: Handle O-ring with bare hands. Never wear gloves. Lubricate O-ring with new engine oil. Never clean O-ring with solvent. Check that O-ring and its mating part are free of foreign material. When installing O-ring, be careful not to

scratch it with tool or fingernails. **Do not twist, de-center, or stretch O-ring. If O-ring was stretched while it was being attached, never insert it quickly into fuel tube. Insert new O-ring straight into fuel tube.**

15. Install the fuel damper as follows:
   a. Install a new O-ring to the fuel tube as shown.
   b. Install spacer to the fuel damper.
   c. Insert the fuel damper straight into the fuel tube. Insert straight, checking that the axis is lined up. Never pressure-fit with excessive force. Reference value: 29 lbs. (130 N)
   d. Tighten the bolts evenly in turn.
   e. After tightening the bolts, check that there is no gap between fuel damper cap and fuel tube.
16. Install new O-rings to the fuel injector. (Fuel tube side: black, nozzle side: green)
17. Install the fuel injector to the fuel tube as follows:
   a. Insert the clip into clip mounting groove on the fuel injector. Insert clip so that protrusion of fuel injector matches cutout of clip.

❋❋ **WARNING**

Never reuse the clip. Replace it with new one. Be careful to keep clip from interfering with O-ring. If interference occurs, replace O-ring.

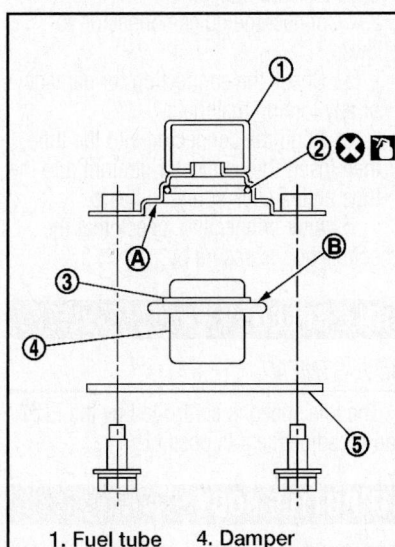

1. Fuel tube   4. Damper
2. O-ring   5. Fuel damper cap
3. Spacer

37663_MURA_G0419

Fig. 151 Insert fuel damper until (B) is touching (A) of fuel tube

   b. Insert the fuel injector into the fuel tube with clip attached. Insert it while matching it to the axial center. Insert the fuel injector so that protrusion of fuel tube matches cutout of clip.
   c. Check that fuel tube flange is securely fixed in flange fixing groove on clip.
   d. Check that installation is complete by checking that the fuel injector does not rotate or come off.
   e. Check that protrusions of the fuel

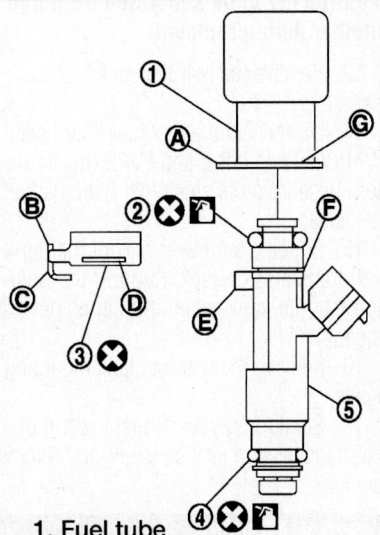

1. Fuel tube
2. Black O-ring
3. Clip
4. Green O-ring
5. Fuel injector
A. Fuel tube protrusion
B. Clip cutout
C. Clip cutout
D. Flange fixing groove
E. Fuel injector protrusion
F. Mounting groove
G. Fuel tube flange

37663_MURA_G0420

Fig. 152 Install fuel injector to fuel tube

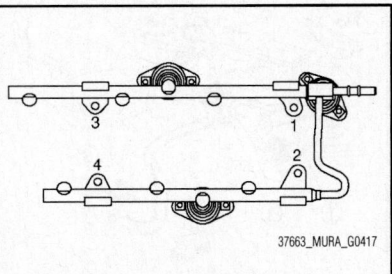

37663_MURA_G0417

Fig. 153 Fuel injector mounting bolt installation sequence—remove in reverse order

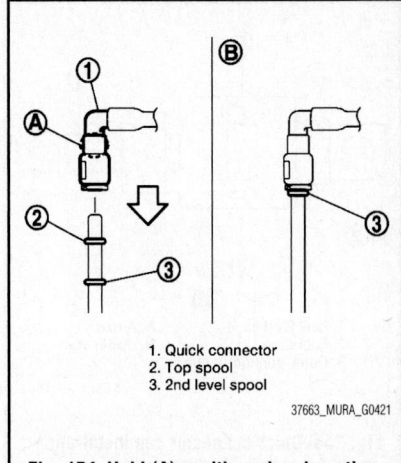

1. Quick connector
2. Top spool
3. 2nd level spool

37663_MURA_G0421

Fig. 154 Hold (A) position when inserting fuel tube; fitted (B) position

injectors and fuel tubes are aligned with cutouts of clips after installation.
18. Install the fuel tube and fuel injector assembly to the intake manifold. Tighten the mounting bolts in two steps in numerical order as shown.
   • Step 1: 7 ft. lbs. (10 Nm)
   • Step 2: 16 ft. lbs. (22 Nm)

❋❋ **WARNING**

Be careful not to let tip of injector nozzle come in contact with other parts.

19. Connect the fuel injector harness.
20. Install the intake manifold collector.
21. Connect the quick connector between fuel feed hose and fuel tube connection with the following procedure:
   a. Check no foreign substances are deposited in and around fuel tube and quick connector, and they are not damaged.
   b. Thinly apply new engine oil around the fuel tube from tip end to spool end.
   c. Align center to insert the quick connector straightly into fuel tube.
   d. Insert the quick connector to fuel tube until top spool is completely inside quick connector, and 2nd level spool exposes right below quick connector.
   e. Hold the position shown when inserting the fuel tube into quick connector.
   f. Carefully align center to avoid inclined insertion to prevent damage to O-ring inside quick connector.
   g. Insert until you hear a "click" sound and actually feel the engagement.
   h. To ensure engagement, pull the quick connector by hand, holding position. Ensure it is completely engaged

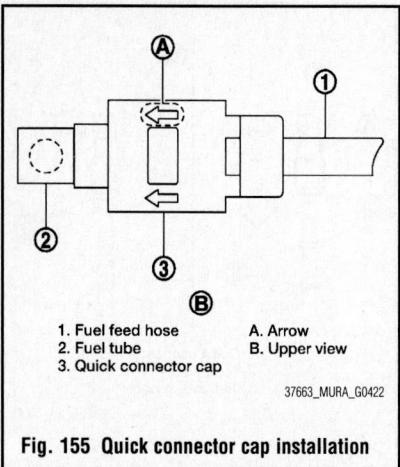

1. Fuel feed hose
2. Fuel tube
3. Quick connector cap
A. Arrow
B. Upper view

37663_MURA_G0422

Fig. 155 Quick connector cap installation

(connected) so that it does not come out from fuel tube.

i. Install the quick connector cap to quick connector.

j. Install the quick connector cap with the arrow the on surface facing in direction of quick connector (fuel feed hose side). If the quick connector cap cannot be installed smoothly, quick connector may have not been installed correctly. Check connection again.

k. Secure the fuel feed hose to the clamp of quick connector cap.

22. Install the front wiper arm and extension cowl top.

23. Install the engine cover.

24. Install the radiator core support covers, air duct (inlet), air cleaner cases (upper and lower) with mass air flow sensor and air duct assembly.

25. Add engine coolant, if necessary.

26. Turn the ignition switch ON (with engine stopped). With fuel pressure applied to fuel piping, check for fuel leakage at connection points.

27. Start the engine. With engine speed increased, check again for fuel leakage at connection points.

FUEL TANK

REMOVAL & INSTALLATION

See Figures 142 and 143, 156 through 158.

1. Before servicing the vehicle, refer to the Precautions Section.

2. Perform the work on level place.

3. Release the fuel pressure from the fuel lines.

4. Open the fuel filler lid.

5. Open the filler cap and release the pressure inside fuel tank.

6. Remove the rear seat cushion and reclining device assembly.

7. Peel off the floor carpet, then remove

the inspection hole cover units by turning clips clockwise by 90 degrees.

8. Disconnect the harness connector, fuel feed tube and EVAP hose.

9. Remove the center muffler.

10. Remove the propeller shaft. Refer to Propeller Shaft Removal & Installation in the Drive Train section.

11. Remove the rear parking brake cables.

12. Remove the rear suspension member assembly.

➡**For this service, halfshaft, final drive, and rear suspension member are required not to be separated from one another during removal.**

13. Remove the rear fuel tank protectors.

14. Disconnect the rear fuel filler hose, EVAP/Vent line hose, and EVAP (Recirculation) hose from the side other than the fuel tank side.

15. Support the lower part of fuel tank with transmission jack. Support the position that fuel tank mounting bands never engage.

16. Remove the rear fuel tank mounting bands.

17. Supporting with hands, lower the rear transmission jack carefully, and remove fuel tank.

❄ WARNING

Check that all connection points have been disconnected. Confirm there is no interference with vehicle.

18. Remove the rear fuel filler tube if necessary.

To install:

19. Note the following, and install in the reverse order of removal.

20. Securely clamp the fuel hoses and insert hose to the length below:

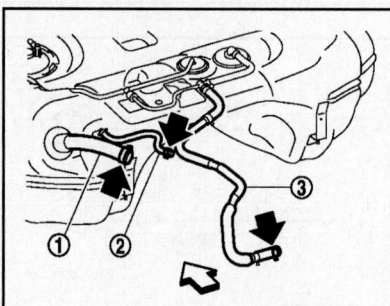

37663_MURA_G0312

Fig. 156 Disconnect fuel filler hose (1), EVAP/Vent line hose (3), and EVAP (Recirculation) hose (2)

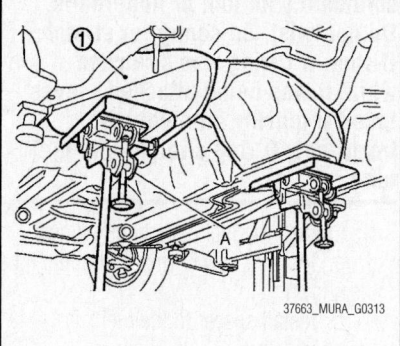

37663_MURA_G0313

Fig. 157 Support the lower part of fuel tank (1) with transmission jack (A)

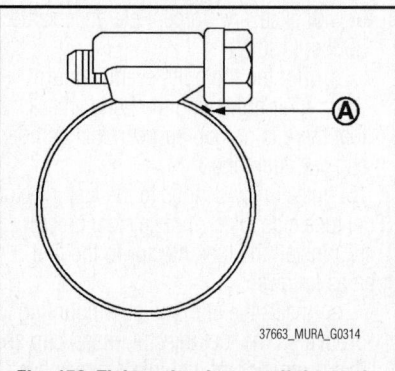

37663_MURA_G0314

Fig. 158 Tighten the clamp until the mark (A) is on the bolt head flange

- Fuel filler hose: 1.38 in. (35 mm)
- The other hoses: 0.98 in. (25 mm)

21. Be sure the rear hose clamp is not placed on swelled area of fuel tube.

22. Tighten the clamp until the mark is on the bolt head flange.

23. Connect the quick connector as follows:

a. Check the connection for damage or any foreign materials.

b. Align the connector with the tube, then insert the connector straight into the tube until a click sound is heard.

c. After connecting, check that the connection is secure by

IDLE SPEED

ADJUSTMENT

The Idle speed is controlled by the ECM and no adjustment is possible.

THROTTLE BODY

REMOVAL & INSTALLATION

See Figure 159.

1. Before servicing the vehicle, refer to the Precautions Section.

2. Remove the engine cover.

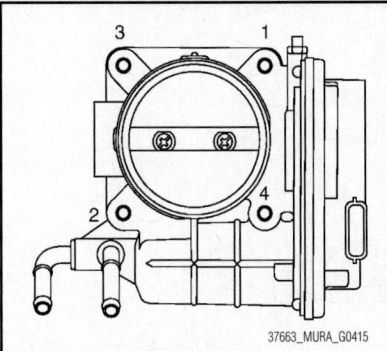

Fig. 159 Throttle body mounting bolt installation sequence—remove in reverse order

3. Remove the air cleaner cases (upper and lower) with mass air flow sensor and air duct assembly.

4. Drain the engine coolant, or when water hoses are disconnected, attach plug to prevent engine coolant leakage.

✳✳ CAUTION

Perform this step when the engine is cold.

5. Disconnect the water hoses the electric throttle control actuator. When engine coolant is not drained from the radiator, attach plug to water hoses to prevent engine coolant leakage.

6. Disconnect the electric throttle control actuator harness connector.

7. Loosen the electric throttle control actuator mounting bolts in reverse of the installation order shown.

➡**Never disassemble.**

To install:

8. Install the gasket with positioning no-protrusion surface upward or downward.

9. Tighten to 74 inch lbs. (8 Nm) in the order shown in above.

10. Perform the "Idle Air Volume Learn-ing" and "Throttle Valve Closed Position Learning".

THROTTLE VALVE CLOSED POSITION LEARNING

1. Check that accelerator pedal is fully released.

2. Turn ignition switch ON.

3. Turn ignition switch OFF and wait at least 10 seconds.

4. Check that throttle valve moves during the above 10 seconds by confirming the operating sound.

IDLE AIR VOLUME LEARNING

See Figure 130.

1. Check that all of the following conditions are satisfied. Learning will be cancelled if any of the following conditions are missed for even a moment:

- Battery voltage: More than 12.9 V (At idle)
- Engine coolant temperature: 158–212°F (70–100°C)
- Selector lever position: P or N
- Electric load switch: OFF (Air conditioner, head lamp, rear window defogger)

➡**On vehicles equipped with daytime light systems, if the parking brake is applied before the engine is started the head lamp will not illuminate.**

- Steering wheel: Neutral (Straight-ahead position)
- Vehicle speed: Stopped
- Transmission: Warmed-up
- Drive vehicle for 10 minutes

➡**It is better to count the time accurately with a clock.**

➡**It is impossible to switch the diagnostic mode when an accelerator pedal position sensor circuit has a malfunction.**

2. Perform accelerator pedal released position learning.

3. Perform throttle valve closed position learning. Refer to Throttle Control Actuator, Throttle Valve Closed Position Learning.

4. Start engine and warm it up to normal operating temperature.

5. Turn ignition switch OFF and wait at least 10 seconds.

6. Confirm that accelerator pedal is fully released, turn ignition switch ON and wait 3 seconds.

7. Repeat the following procedure quickly 5 times within 5 seconds.
   a. Fully depress the accelerator pedal.
   b. Fully release the accelerator pedal.

8. Wait 7 seconds, fully depress the accelerator pedal for approx. 20 seconds until the MIL stops blinking and turns ON.

9. Fully release the accelerator pedal within 3 seconds after the MIL turns ON.

10. Start engine and let it idle.

11. Wait 20 seconds.

12. Rev up the engine 2 or 3 times and check that idle speed and ignition timing are within the specifications.
   a. If the results are normal, inspection in ended. If the results are not normal, check the following:
   - Check that throttle valve is fully closed
   - Check PCV valve operation
   - Check that downstream of throttle valve is free from air leakage
   b. If the results are not normal, repair or replace malfunctioning part. If the results are normal, engine component parts and their installation condition are questionable. Check and eliminate the cause of the incident. If any of the following conditions occur after the engine has started, eliminate the cause of the incident and perform Idle Air Volume Learning again: engine stalls, Incorrect idle.

HEATING & AIR CONDITIONING SYSTEM

BLOWER MOTOR

REMOVAL & INSTALLATION

See Figure 160.

✳✳ CAUTION

Before servicing components near or affected by the SRS (air bag) system, read and observe all SRS Service Precautions. Refer to Supplemental Restraint System (SRS), in the Chassis Electrical section. Failure to observe all precautions may result in accidental airbag deployment, personal injury, or death.

1. Before servicing the vehicle, refer to the Precautions Section.

2. Disconnect the negative battery cable and wait at least 3 minutes for the SRS memory to drain.

3. Remove the lower right-hand instrument panel. Refer to Instrument Panel Removal & Installation in the Body Interior section.

4. Disconnect the blower motor electrical connector.

5. Remove the blower motor retaining screws.

6. Remove the blower motor.

To install:

7. Installation is the reverse of the removal procedure.

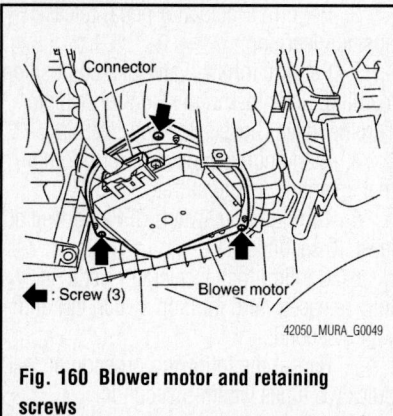

Fig. 160 Blower motor and retaining screws

BLOWER UNIT

REMOVAL & INSTALLATION

See Figures 161 through 163.

❋❋ CAUTION

Before servicing components near or affected by the SRS (air bag) system, read and observe all SRS Service Precautions. Refer to Supplemental Restraint System (SRS), in the Chassis Electrical section. Failure to observe all precautions may result in accidental airbag deployment, personal injury, or death.

1. Before servicing the vehicle, refer to the Precautions Section.
2. Disconnect the negative battery cable and wait at least 3 minutes for the SRS memory to drain.
3. Remove the instrument panel. Refer to Instrument Panel Removal & Installation in the Body Interior section.
4. Remove the mounting nuts and the instrument panel stay.
5. Disconnect the intake door motor and blower motor connectors.
6. Remove the heater and cooling unit assembly and blower unit mounting bolts.

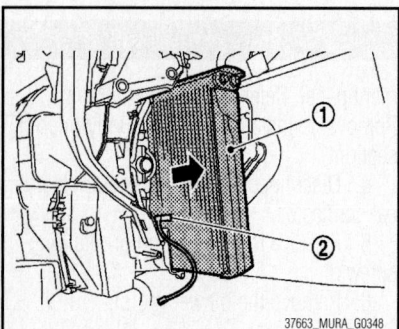

Fig. 161 Remove the mounting nuts (A) and the instrument panel stay (1)

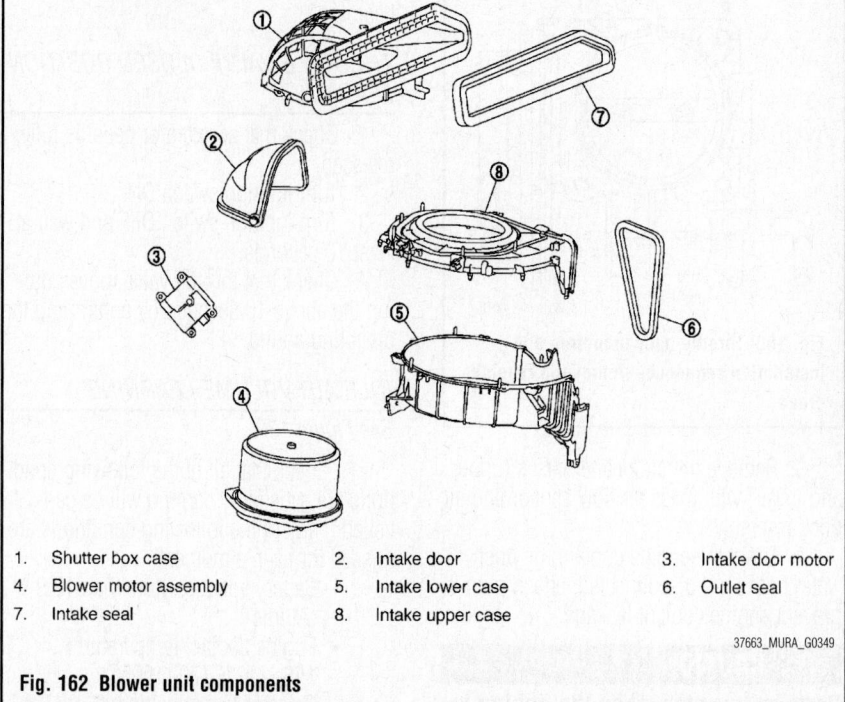

1.	Shutter box case	2.	Intake door	3.	Intake door motor
4.	Blower motor assembly	5.	Intake lower case	6.	Outlet seal
7.	Intake seal	8.	Intake upper case		

Fig. 162 Blower unit components

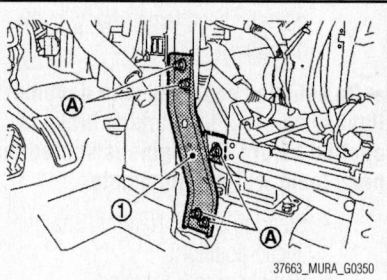

Fig. 163 Heater and cooling unit assembly and blower unit mounting bolts (A), steering member mounting bolts (B), blower unit (2) and steering member (1)

7. Remove the steering member mounting bolts (right).
8. Remove the blower unit while pulling the steering member to the front.

To install:

9. Installation is the reverse of removal procedure.

HEATER CORE

REMOVAL & INSTALLATION

See Figures 164 and 165.

❋❋ CAUTION

Before servicing components near or affected by the SRS (air bag) system, read and observe all SRS Service Precautions. Refer to Sup-

plemental Restraint System (SRS), in the Chassis Electrical section. Failure to observe all precautions may result in accidental airbag deployment, personal injury, or death.

1. Before servicing the vehicle, refer to the Precautions Section.
2. Disconnect the negative battery cable and wait at least 3 minutes for the SRS memory to drain.
3. Discharge the air conditioning system.
4. Remove the heater and cooling unit assembly. Refer to Heater & Cooling Unit Removal & Installation.
5. Remove the left foot duct.

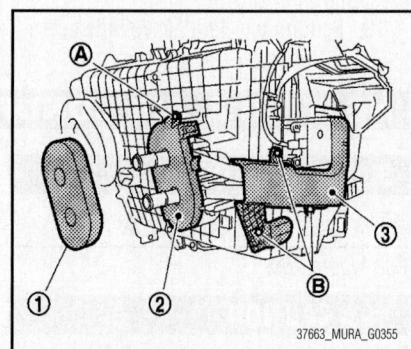

Fig. 164 Heater pipe grommet (1), heater pipe support (2) and mounting screw (A), heater pipe cover (3) and mounting screws (B)

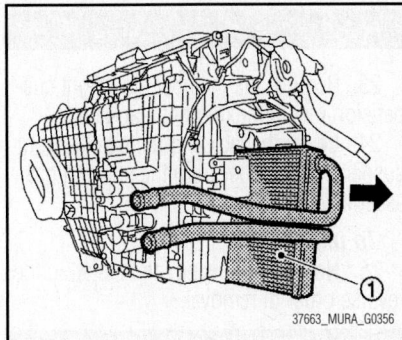

Fig. 165 Slide the heater core (1) in the direction shown by the arrow, and remove it

6. Remove the heater pipe grommet.

7. Remove the mounting screw, and remove the heater pipe support.

8. Remove the mounting screws, and remove the heater pipe cover.

9. Slide the heater core in the direction shown by the arrow, and remove it.

To install:

10. Installation is the reverse of the removal procedure.

   a. Replace O-rings with new ones. Apply compressor oil to the O-rings.

   b. Check for leakages when recharging refrigerant.

HEATER & COOLING UNIT

REMOVAL & INSTALLATION

See Figures 166 and 167.

❄❄ CAUTION

Before servicing components near or affected by the SRS (air bag) system, read and observe all SRS Service Precautions. Refer to Supplemental Restraint System (SRS), in the Chassis Electrical section. Failure to observe all precautions may result in accidental airbag deployment, personal injury, or death.

1. Before servicing the vehicle, refer to the Precautions Section.

2. Disconnect the negative battery cable and wait at least 3 minutes for the SRS memory to drain.

3. Discharge the air conditioning system.

4. Drain the engine coolant.

5. Remove the mounting bolt, and disconnect the low pressure pipe and high-pressure pipe from the expansion valve. Cap or wrap the joint of the A/C piping and expansion valve with suitable material such as vinyl tape to avoid the entry of air.

6. Remove the clamps and disconnect the heater hoses. Some coolant may spill. Close the coolant inlet/outlet on the heater core and heater hoses.

7. Remove the instrument panel assembly. Refer to Instrument Panel Removal & Installation in the Body Interior section.

8. Remove the mounting nuts and the instrument panel stay.

9. Remove the heater and cooling unit assembly and blower unit mounting bolts.

10. Remove the steering column mounting nuts.

11. Remove the ground bolts from the steering member.

12. Remove the harness clip from the steering member.

13. Disconnect the intake door motor and blower motor connectors.

14. Remove the steering member mounting bolts, and remove the steering member.

15. Remove the blower unit.

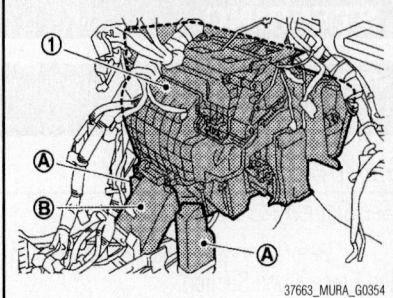

Fig. 167 Remove the rear foot duct 1 (left/right) (A) and rear ventilator duct 1 (B), and remove the heater and cooling unit assembly (1)

16. Disconnect the drain hose from the heater and cooling unit assembly.

17. Remove the rear foot duct 1 (left/right) and rear ventilator duct 1, and remove the heater and cooling unit assembly.

To install:

18. Installation is the reverse of the removal procedure.

   a. Replace O-rings with new ones. Apply compressor oil to the O-rings.

   b. Check for leakages when recharging refrigerant.

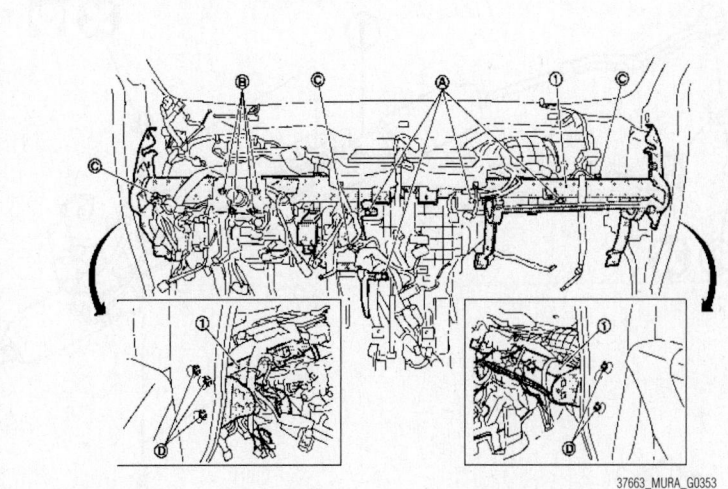

Fig. 166 Blower unit mounting bolts (A), steering column mounting nuts (B), ground bolts (C), steering member mounting bolts (D) and steering member (1)

STEERING

POWER RACK & PINION STEERING GEAR

REMOVAL & INSTALLATION

See Figures 168 through 172.

1. Before servicing the vehicle, refer to the Precautions Section.
2. Disarm the SRS system. Disconnect the negative battery cable.
3. Set the wheels in the straight ahead position.
4. Remove the front road wheel and tires.
5. Remove the splash guards.
6. Remove the engine under cover.
7. Remove the exhaust front pipe.
8. For AWD vehicles, separate the rear propeller shaft (front side).
9. Remove the heat insulator from the front floor.
10. Remove the cotter pin, and then loosen the nuts.
11. Using a ball joint remover, remove the steering outer socket from steering knuckle so as not to damage the ball joint boot.
12. Temporarily tighten the nut to prevent damage to the threads and to prevent the ball joint remover from suddenly coming off.

13. Remove the high pressure piping and low pressure hose of hydraulic piping, and then drain the power steering fluid.
14. Remove the steering hydraulic piping bracket from the front steering gear assembly.
15. Remove the power steering solenoid valve harness connector and harness clip.
16. Remove the lower joint fixing bolt (steering gear side).
17. Separate the lower shaft from the steering gear assembly by sliding the slide shaft.

> ⚠️ **WARNING**
>
> **Spiral cable may be cut if steering wheel turns while separating steering column assembly and steering gear assembly. Be sure to secure steering wheel using string to avoid turning.**

18. Remove the stabilizer assembly.
19. Support the front suspension member with a suitable jack.
20. Remove the engine mounting insulator (rear) mounting bolt (lower side).
21. Remove the left-hand engine mounting insulator.
22. Remove the steering gear assembly mounting bolts and nuts.

23. Remove the member stay, front suspension member fixing bolts and nuts.
24. Lower the jack and remove the front suspension member from the steering gear assembly.

To install:

25. Note the following, and install in the reverse order of removal.

> ⚠️ **WARNING**
>
> **Spiral cable may be cut if steering wheel turns while separating steering column assembly and steering gear assembly. Be sure to secure steering wheel using string to avoid turning.**

26. When installing low pressure hose, refer to the following:
    a. Never apply fluid to the hose and tube.
    b. Insert hose securely until it contacts spool of tube.
    c. Leave clearance when installing clamp. Standard: 0.12–0.31 in. (3–8mm)
27. When installing lower joint to steering gear assembly, follow the procedure listed below.
    a. Set rack of steering gear in the neutral position.

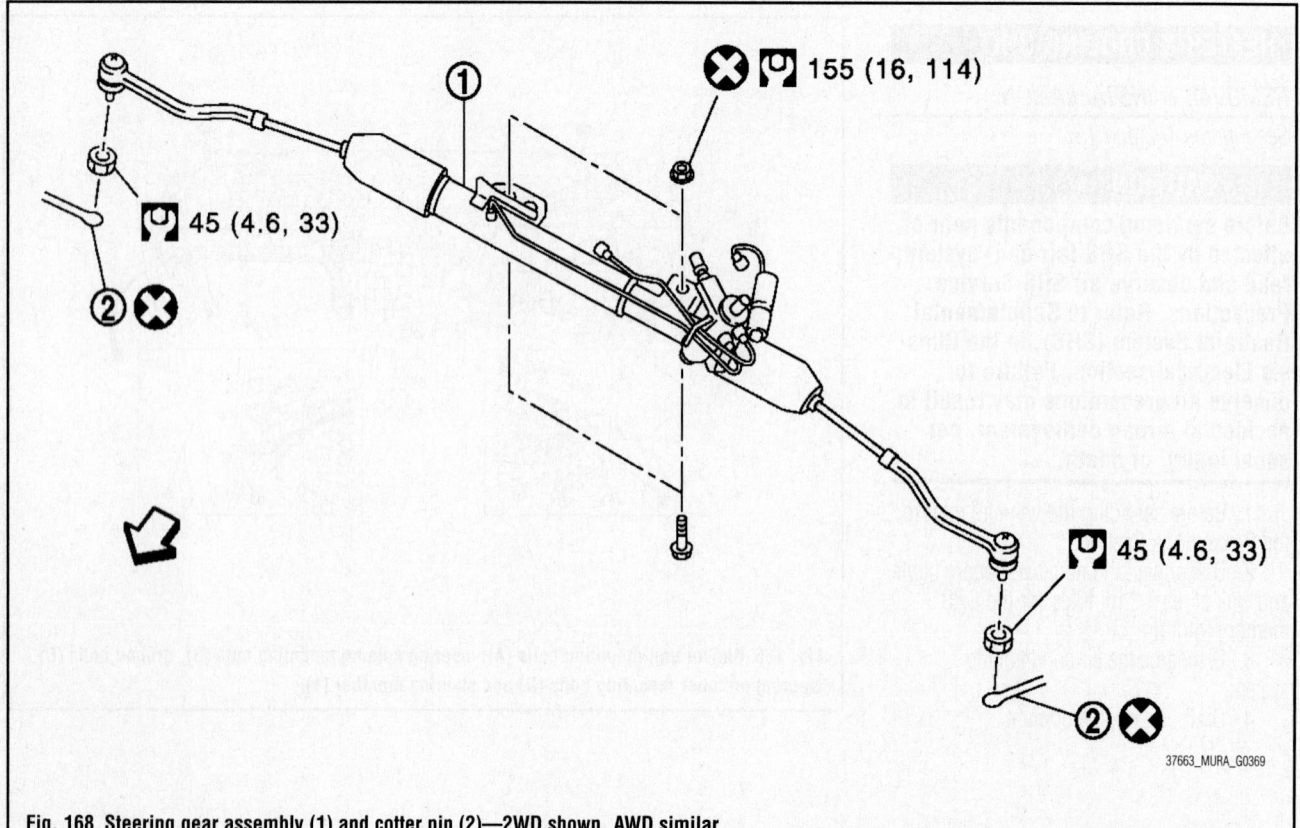

37663_MURA_G0369

Fig. 168 Steering gear assembly (1) and cotter pin (2)—2WD shown, AWD similar

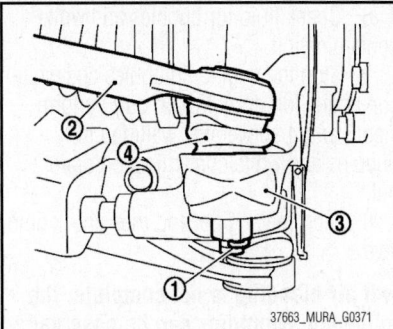

Fig. 169 Cotter pin (1), steering outer socket (2), steering knuckle (3), ball joint boot (4)

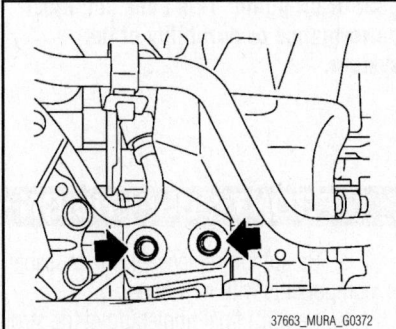

Fig. 170 Steering hydraulic piping bracket bolts

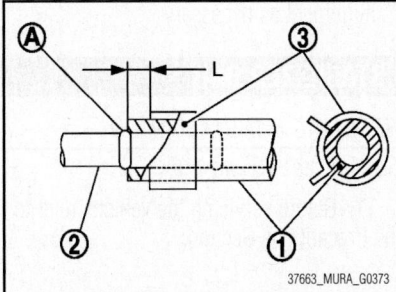

Fig. 171 Never apply fluid to the hose (1) and tube (2), insert hose until it contacts tube spool (A), and leave clearance (L) when installing clamp (3)

➡ **To get the neutral position of rack, turn gear-sub assembly and measure the distance of inner socket, and then measure the intermediate position of the distance.**

b. Align rear cover cap projection with the marking position of gear housing assembly.

c. Install slit part of lower joint aligning with the rear cover cap projection. Make sure that the slit part of lower joint is aligned with rear cover cap projection and the marking position of gear housing assembly.

28. After installation, bleed air from the steering hydraulic system.

29. Perform final tightening of nuts and bolts on each part under unladen conditions with tires on level ground when removing steering gear assembly. Check wheel alignment.

30. Adjust neutral position of steering angle sensor after checking wheel alignment.

POWER STEERING PUMP

REMOVAL & INSTALLATION

See Figures 171, 173 and 174.

1. Before servicing the vehicle, refer to the Precautions section.

2. Disconnect the negative battery cable.

3. Drain power steering fluid from the reservoir tank.

4. Remove the front wheels and tires.

5. Remove the splash guard.

6. Loosen the drive belt.

7. Remove the drive belt from the oil pump pulley.

8. Remove the copper washers and eye bolt (drain fluid).

9. Remove the suction hose (drain fluid).

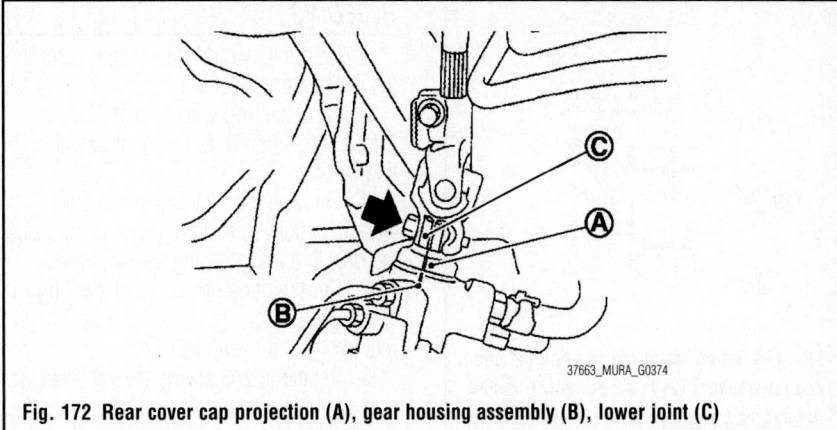

Fig. 172 Rear cover cap projection (A), gear housing assembly (B), lower joint (C)

10. Remove the oil pump mounting bolts, and then remove oil pump. Be careful not to damage halfshaft boot.

To install:

11. Note the following, and install in the reverse order of removal.

12. When installing suction hoses, refer to the following:

a. Never apply fluid to the hose and tube.

b. Insert hose securely until it contacts spool of tube.

c. Leave clearance when installing clamp. Standard: 0.12–0.31 in. (3–8mm)

13. When installing eye bolt and copper washers to oil pump, refer to the following:

a. Do not reuse copper washer.

b. Apply power steering fluid to around copper washer, and install eye bolt.

c. Install eye bolt with eye joint (assembled to high pressure hose) protrusion facing with pump side cutout, and then tighten it to the specified torque after tightening by hand.

d. Securely insert harness connector to pressure sensor.

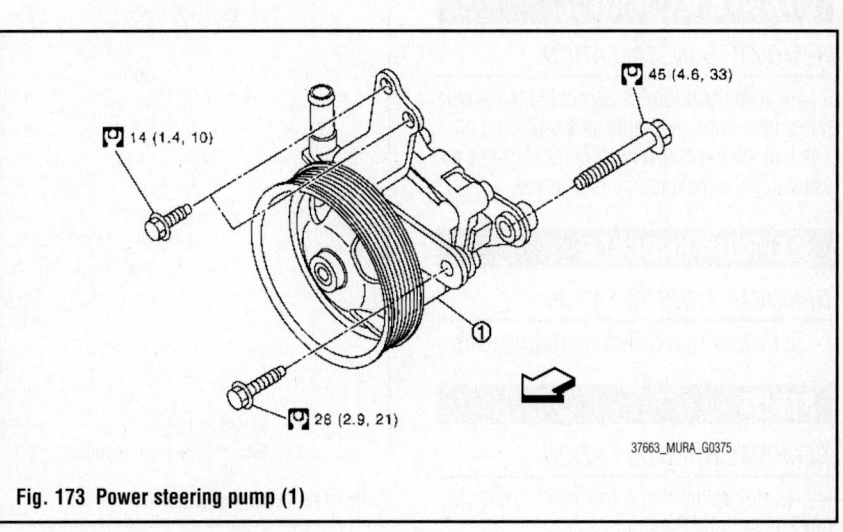

Fig. 173 Power steering pump (1)

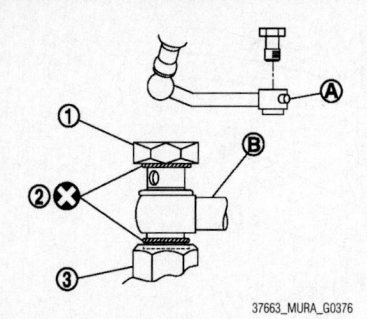

Fig. 174 When installing eye bolt (1) and copper washers (2) to oil pump (3), install the eye bolt with eye joint (B) protrusion (A) facing with pump side cutout

37663_MURA_G0376

14. Adjust belt tension.

15. Check fluid level, fluid leakage and air bleeding hydraulic system after the installation.

SUSPENSION

CONTROL LINKS

REMOVAL & INSTALLATION

1. Before servicing the vehicle, refer to the Precautions Section.

2. Raise and support the vehicle safely.

3. Remove the tire and wheel assembly.

4. Remove the power steering gear assembly.

5. Remove the stabilizer connecting rod lower nut.

6. Using the appropriate tools, separate the stabilizer bar and the stabilizer connecting rod.

To install:

7. Install in the reverse order of removal.

LOWER BALL JOINT

REMOVAL & INSTALLATION

The lower ball joint is part of the transverse link and is not serviceable as a separate unit. The ball joint and arm must be replaced as an assembly if a malfunction is detected.

STEERING KNUCKLE

REMOVAL & INSTALLATION

See Wheel Hub & Bearing (sealed unit).

STRUT

REMOVAL & INSTALLATION

1. Before servicing the vehicle, refer to the Precautions Section.

BLEEDING

1. Before servicing the vehicle, refer to the Precautions Section.

2. Turn steering wheel several times from full left stop to full right stop with engine off.

3. Fill reservoir tank with a sufficient amount of fluid so that fluid level is not below the MIN line while turning steering wheel.

4. Start the engine and hold steering wheel at each lock position for 3 second at idle to check for fluid leakage.

5. Repeat step 3 above several times at approximately 3 second intervals.

✳✳ WARNING

Never hold the steering wheel in a locked position for more than 10 seconds. (There is the possibility that oil pump may be damaged.)

2. Raise and support the vehicle safely.

3. Remove the wheels and tires.

4. Remove the lock plate.

5. Remove the wheel sensor.

6. Remove the stabilizer connecting rod from the strut assembly.

7. Remove the strut assembly from the steering knuckle.

8. Remove the cowl top cover.

9. Remove the strut mounting insulator mounting bolts, and remove the strut assembly.

To install:

10. Note the following, and install in the reverse order of the removal.

6. Check fluid for bubbles and white contamination.

7. Stop the engine if bubbles and white contamination do not drain out. Perform step 3 and 4 above after waiting until bubbles and white contamination drain out.

8. Stop the engine, and then check fluid level.

➡**If air bleeding is not complete, the following symptoms can be observed: bubbles in reservoir tank, clicking noise from oil pump, excessive buzzing in the oil pump.**

➡**Fluid noise may occur in the steering gear or oil pump. This does not affect performance or durability of the system.**

FRONT SUSPENSION

a. Be sure to replace all non reusable components with new ones.

b. Perform final tightening of the strut assembly lower side (rubber bushing) under unladen conditions with tires on level ground.

c. Check and adjust the front end alignment as necessary.

STABILIZER BAR

REMOVAL & INSTALLATION

See Figures 175 and 176.

1. Before servicing the vehicle, refer to the Precautions Section.

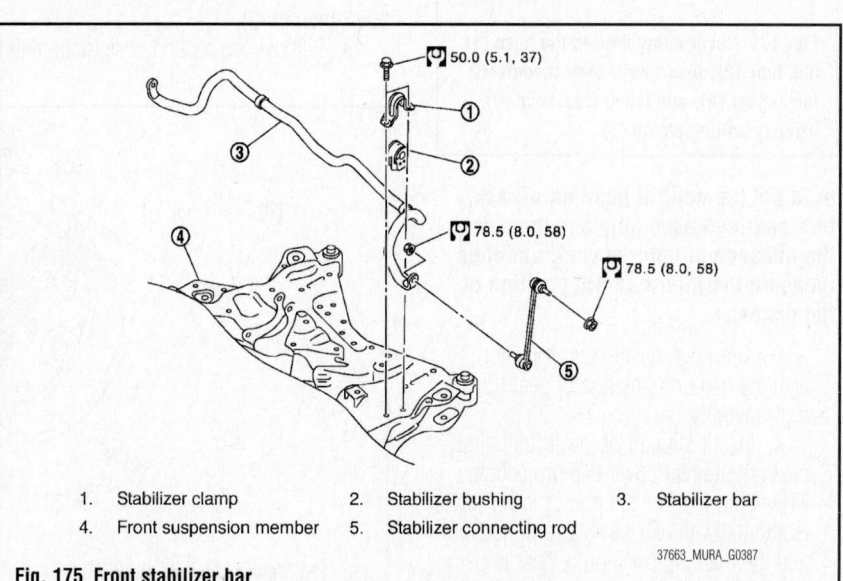

| 1. | Stabilizer clamp | 2. | Stabilizer bushing | 3. | Stabilizer bar |
| 4. | Front suspension member | 5. | Stabilizer connecting rod | | |

Fig. 175 Front stabilizer bar

37663_MURA_G0387

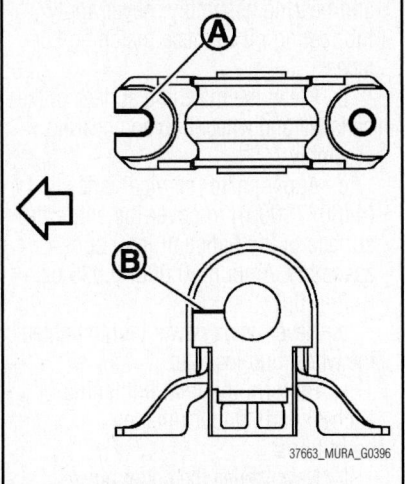

Fig. 176 Position the stabilizer clamp notch (A) and bushing slit (B)toward the front of the vehicle

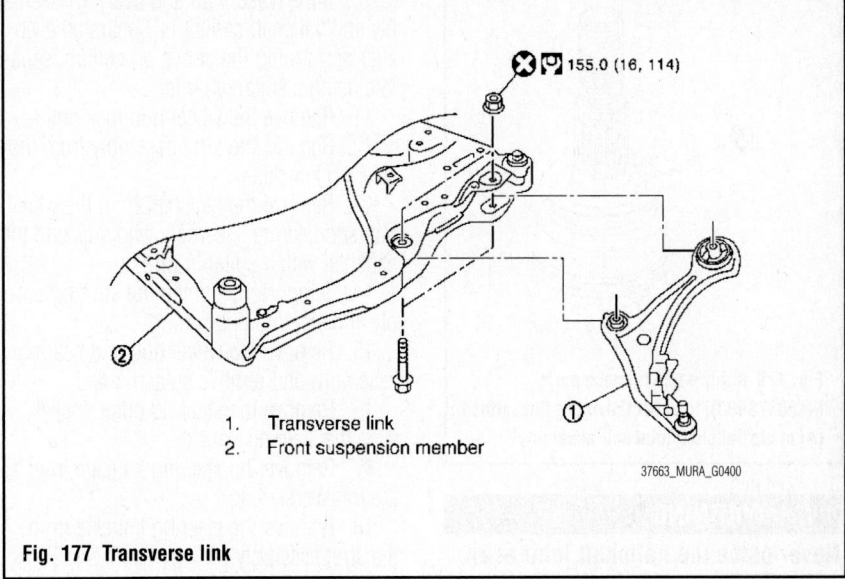

1. Transverse link
2. Front suspension member

Fig. 177 Transverse link

2. Raise and support the vehicle safely.

3. Remove the tire and wheel assembly.

4. Remove the front exhaust pipe.

5. Remove the rear propeller shaft from transfer, as applicable.

6. Remove the lock plate.

7. Remove the wheel sensor harness from the strut assembly.

8. Disconnect the power steering solenoid valve harness connector.

9. Remove the steering outer socket from the steering knuckle.

10. Remove the stabilizer connecting rod.

11. Remove the stabilizer clamp mounting bolts, and remove stabilizer clamp and stabilizer bushing from the front suspension.

12. Remove the stabilizer bar.

To install:

13. Install in the reverse order of removal, noting the following:

   a. Position the stabilizer clamp notch and bushing slit toward the front of the vehicle.

TRANSVERSE LINK

REMOVAL & INSTALLATION

See Figure 177.

1. Before servicing the vehicle, refer to the Precautions Section.

2. Raise and support the vehicle safely.

3. Remove the tire and wheel assembly.

4. Remove the halfshaft.

To install:

5. Note the following, and install in the reverse order of the removal.

a. Be sure to replace all non reusable components with new ones.

b. Perform final tightening of front suspension member installation position and strut assembly lower side (rubber bushing) under unladen conditions with tires on level ground.

c. Check the wheel alignment and adjust as necessary.

WHEEL HUB & BEARING

REMOVAL & INSTALLATION

See Figures 178 and 179.

1. Before servicing the vehicle, refer to the Precautions Section.

2. Raise and support the vehicle safely.

3. Remove the tire and wheel assembly.

4. Remove the wheel sensor and sensor harness.

5. Remove the lock plate from the strut assembly.

6. Remove the caliper assembly. Hang the caliper assembly aside.

➡ **Never depress the brake pedal while the brake caliper is removed.**

7. Remove the disc rotor.

8. Remove the cotter pin, and loosen the wheel hub lock nut.

9. Patch the wheel hub lock nut with a piece of wood. Hammer the wood to disengage wheel hub and bearing assembly from halfshaft.

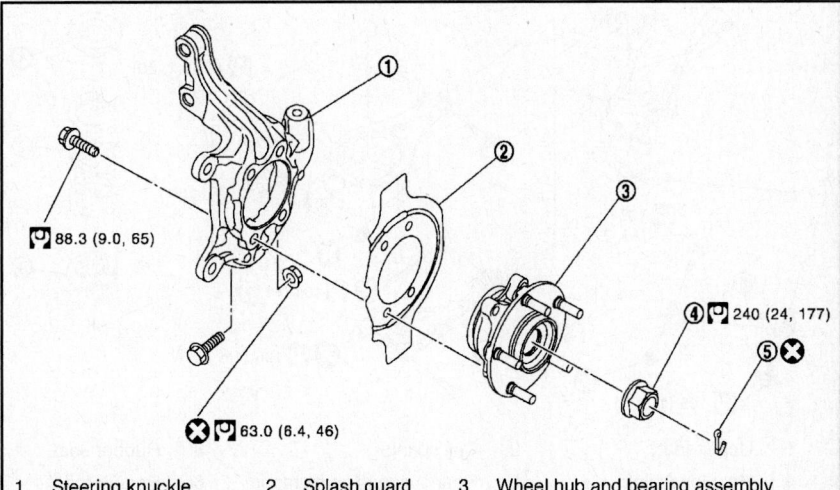

1. Steering knuckle
2. Splash guard
3. Wheel hub and bearing assembly
4. Wheel hub lock nut
5. Cotter pin

Fig. 178 Steering knuckle and bearing assembly

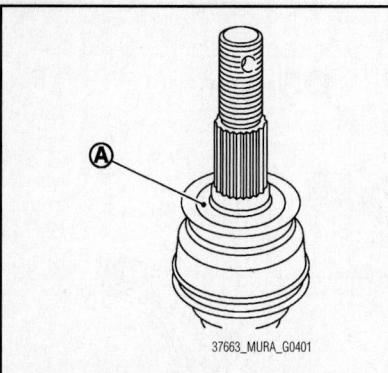

37663_MURA_G0401

Fig. 179 Apply paste [service parts (440037S000)] to cover the entire flat surface (A) of the halfshaft joint sub-assembly

✳✳ WARNING

Never place the halfshaft joint at an extreme angle. Also be careful not to overextend the slide joint. Never allow the halfshaft to hang down without support for the joint sub-assembly, shaft and the other parts.

10. If the wheel hub and bearing assembly and halfshaft cannot be separated even after performing the above procedure, separate using a suitable puller.

11. Remove the wheel hub lock nut.

12. Remove the strut assembly from the steering knuckle.

13. Remove the halfshaft from the wheel hub and bearing assembly, and suspend the halfshaft with a suitable wire.

14. Temporarily tighten the strut assembly and steering knuckle.

15. Remove the wheel hub and bearing assembly, and remove splash guard.

16. Remove the steering outer socket from the steering knuckle.

17. Remove the steering knuckle from the transverse link.

18. Remove the steering knuckle from the strut assembly.

To install:

19. Note the following, and install in the reverse order of the removal.

    a. Clean the matching surface of the wheel hub lock nut and wheel hub and bearing assembly; never apply lubricating oil to these matching surfaces.

    b. Clean the matching surface of the halfshaft and wheel hub and bearing assembly.

    c. Apply paste [service parts (440037S000)] to cover the entire flat surface of the halfshaft joint sub-assembly. Amount: 0.008–0.035 oz. (0.2–1.0g)

    d. Never use a power tool to tighten the wheel hub lock nut.

    e. Perform the final tightening of each of parts under unladen conditions.

    f. Never reuse the cotter pin.

ADJUSTMENT

The front wheel bearings are part of a unitized hub and are not adjustable. Move the wheel hub in the axial direction by hand. Make sure that there is no looseness of the wheel bearing. Axial end play is 0.002 inch or less.

SUSPENSION

REAR SUSPENSION

COIL SPRING

REMOVAL & INSTALLATION

See Figures 180 through 182.

1. Before servicing the vehicle, refer to the Precautions Section.

2. Raise and support the vehicle safely.

3. Remove the tire and wheel assembly.

4. Remove the stabilizer connecting rod (lower side).

5. Set a jack under the rear lower link.

6. Loosen the rear lower link mounting bolt and nut (rear suspension member

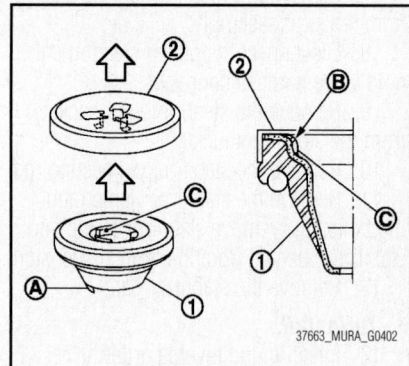

37663_MURA_G0402

Fig. 181 Upper seat (1), outside projection (A), inside projection (C), bracket (2), and tabs (B)

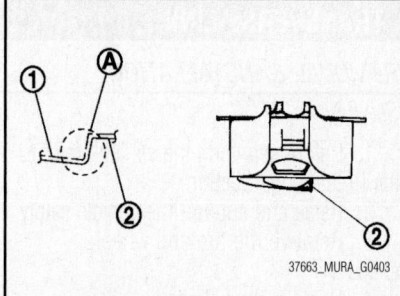

37663_MURA_G0403

Fig. 182 Align the lower end of the coil spring to the step (A) between the rubber seat (1) and the rear lower link (2)

✕ 🔧 105 (11, 77)

🔧 89 (9.1, 66)

🔧 105 (11, 77)

✕ 🔧 105 (11, 77)

✕ 🔧 105 (11, 77)

🔧 105 (11, 77)

① ② ③

1.	Upper seat	2.	Coil spring	3.	Rubber seat
4.	Rear lower link	5.	Rear suspension member	6.	Radius rod
7.	Front lower link				

37663_MURA_G0389

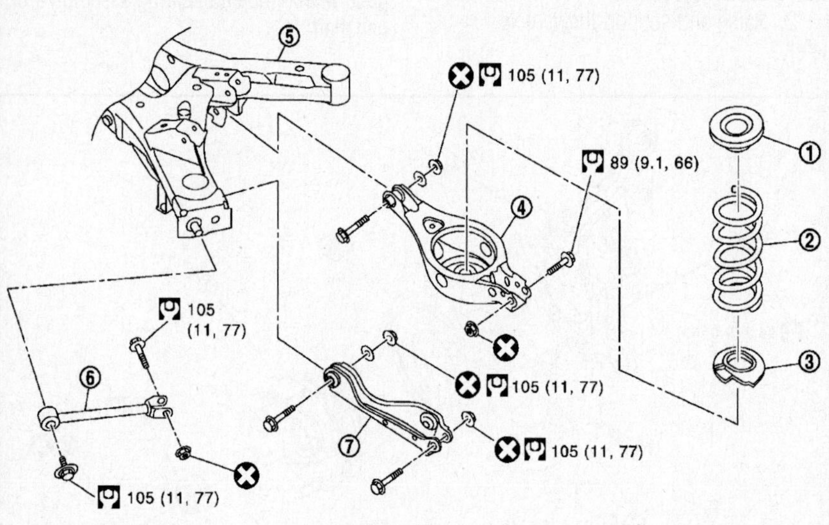

Fig. 180 Rear lower rear suspension assembly

side), and remove rear the lower link mounting bolt and nut (axle housing side).

7. Slowly lower the jack, and remove the upper seat, coil spring and rubber seat from the rear lower link.

8. Remove the rear lower link mounting bolt and nut (rear suspension member side), and remove the rear lower link.

To install:

9. Note the following, and install in the reverse order of removal.

a. Ensure that the upper seat is attached as removed.

b. Position the protrusion with the projection of the upper seat toward the outside of the vehicle (lateral direction).

c. Make sure that the projection on the inside of the upper seat is securely fitted on the bracket tabs.

d. Match up rubber seat indentions and rear lower link grooves and attach.

e. Install the coil spring by aligning the lower end of the coil spring to the step between the rubber seat and the rear lower link.

f. Set the coil spring so that its paint marks are aligned with the positions of 3.5 turns (2 places) and 4.5 turns (1 place) from the bottom end of the coil spring.

g. Perform the final tightening of rear suspension member and axle installation position (rubber bushing) under unladen condition with tires on level ground.

CONTROL ARMS/LINKS

REMOVAL & INSTALLATION

See Stabilizer Bar.

FRONT LOWER LINK

REMOVAL & INSTALLATION

See Figure 180.

1. Before servicing the vehicle, refer to the Precautions Section.

2. Raise and support the vehicle safely.

3. Remove the tire and wheel assembly.

4. Remove the wheel sensor and sensor harness.

5. Remove the rear lower link and coil spring. Refer to Coil Spring Removal & Installation.

6. Remove the mounting bolt in the lower side of the shock absorber.

7. Remove the stabilizer bushing and clamp.

8. Remove the front lower link mounting bolts and nuts (rear suspension member side).

9. Remove the front lower link mounting bolts and nuts (axle housing side).

10. Remove the front lower link from the vehicle.

To install:

11. Note the following, and install in the reverse order of the removal.

a. Be sure to replace all non reusable components with new ones.

b. Perform final tightening of the rear suspension member and axle installation position under unladen conditions with tires on level ground.

c. Check and adjust the rear alignment, as necessary.

RADIUS ARM

REMOVAL & INSTALLATION

See Figure 180.

1. Before servicing the vehicle, refer to the Precautions Section.

2. Raise and support the vehicle safely.

3. Remove the tire and wheel assembly.

4. Remove the wheel sensor and sensor harness.

5. Remove the rear lower link and coil spring. Refer to Coil Spring Removal & Installation.

6. Remove the mounting bolt in the lower side of the shock absorber.

7. Remove the front lower link mounting bolt and nut (axle housing side).

8. Loosen the front lower link mounting bolt and nut (suspension member side).

9. Remove the radius rod mounting bolts and nuts (axle housing side).

10. Remove the radius rod mounting bolt (rear suspension member side).

11. Remove the radius rod.

To install:

12. Note the following, and install in the reverse order of the removal.

a. Be sure to replace all non reusable components with new ones.

b. Perform final tightening of the rear suspension member and axle installation position (rubber bushing) under unladen conditions with tires on level ground.

c. Check and adjust the rear alignment, as necessary.

REAR SUSPENSION ARM

REMOVAL & INSTALLATION

See Figure 183.

1. Before servicing the vehicle, refer to the Precautions Section.

2. Raise and support the vehicle safely.

3. Remove the tire and wheel assembly.

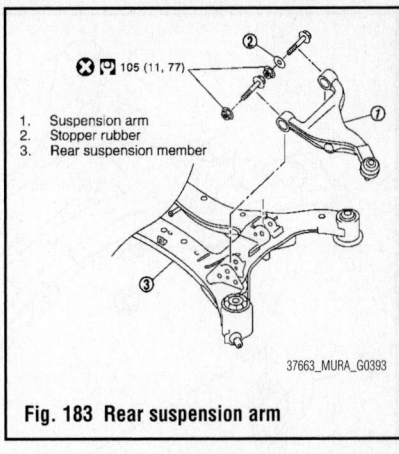

1. Suspension arm
2. Stopper rubber
3. Rear suspension member

37663_MURA_G0393

Fig. 183 Rear suspension arm

4. For AWD models, remove the suspension arm with the rear suspension member.

5. Remove the wheel sensor and sensor harness.

6. Remove the stabilizer connecting rod.

7. Remove the cotter pin of the suspension arm ball joint, and loosen the nut.

8. Use the ball joint remover to remove the suspension arm from the axle housing. Be careful not to damage the ball joint boot.

➡**Temporarily tighten the mounting nut to prevent damage to the threads and to prevent the ball joint remover from coming off.**

9. Remove the suspension arm.

To install:

10. Note the following, and install in the reverse order of removal.

a. Perform final tightening of rear suspension member installation position (rubber bushing), under unladen conditions with tires on level ground.

REAR SUSPENSION MEMBER

REMOVAL & INSTALLATION

See Figure 184.

1. Before servicing the vehicle, refer to the Precautions Section.

2. Raise and support the vehicle safely.

3. Remove the tire and wheel assembly.

4. Remove the caliper assembly. Hang the caliper assembly aside.

➡**Avoid depressing the brake pedal while the caliper assembly is removed.**

5. Remove the wheel sensor and sensor harness.

6. Remove the center muffler.

7. Remove the stabilizer bar. Refer to Stabilizer Bar Removal & Installation.

8. Remove the halfshaft (AWD models).

9. Remove the propeller shaft (AWD models).

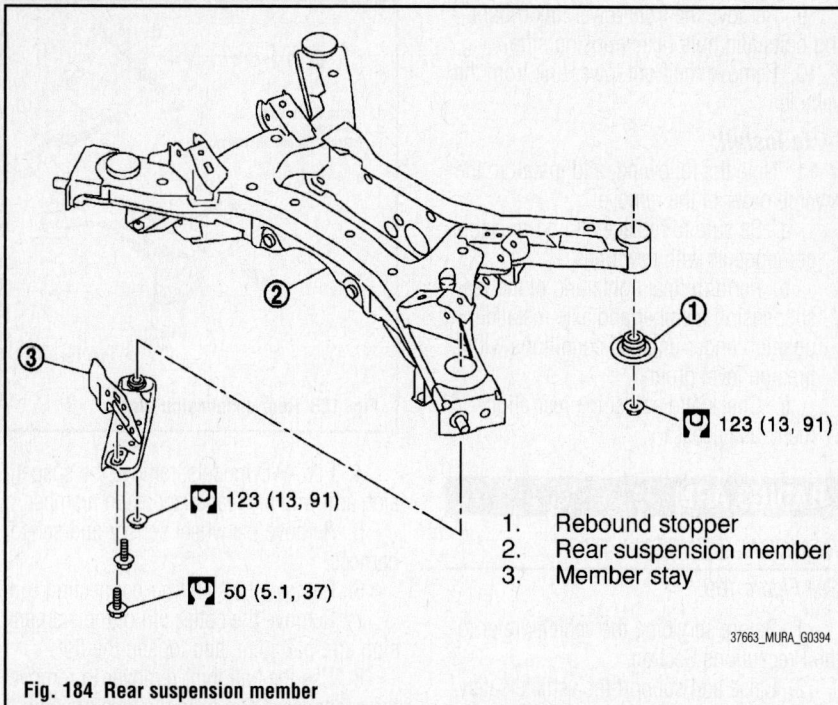

123 (13, 91)

123 (13, 91)

50 (5.1, 37)

1. Rebound stopper
2. Rear suspension member
3. Member stay

37663_MURA_G0394

Fig. 184 Rear suspension member

14. Remove the shock absorber (lower side).

15. Set a suitable jack under the rear suspension member.

16. Remove the member stay.

17. Remove the rear suspension member and rebound stopper.

18. Slowly lower the jack, and remove the rear suspension member, suspension arm, radius rod, front lower link and axle from the vehicle as a unit.

19. Remove the suspension arm.

20. Remove the radius rod.

21. Remove the front lower link.

To install:

22. Note the following, and install in the reverse order of the removal.

   a. Perform the final tightening of each part under unladen conditions.

   b. Check the wheel sensor harness for proper connection.

   c. Never reuse the cotter pin.

SHOCK ABSORBER

REMOVAL & INSTALLATION

See Figure 185.

1. Before servicing the vehicle, refer to the Precautions Section.

10. Remove the harness from the rear final drive and rear suspension member (AWD models).

11. Remove the rear final drive (AWD models).

12. Separate the attachment between the parking brake cable and the vehicle rear suspension member.

13. Remove the rear lower link and coil spring. Refer to Coil Spring Removal & Installation.

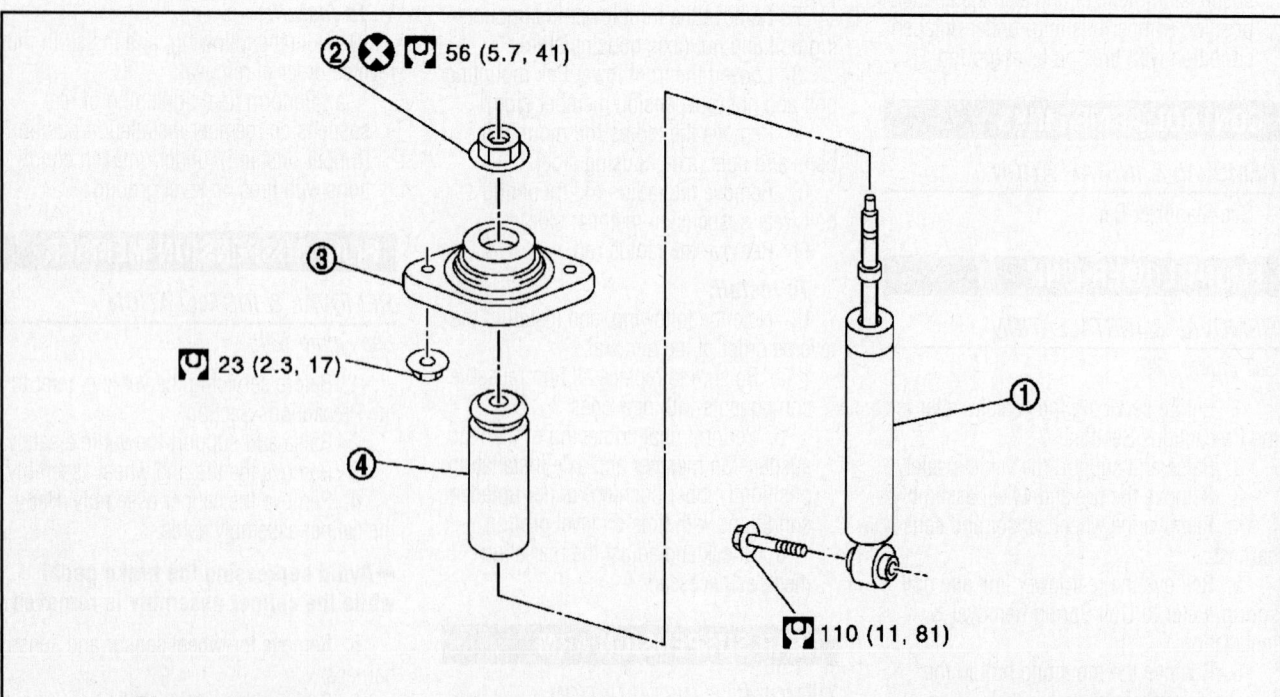

56 (5.7, 41)

23 (2.3, 17)

110 (11, 81)

1. Shock absorber
2. Piston rod lock nut
3. Shock absorber mounting bracket
4. Bound bumper

37663_MURA_G0390

Fig. 185 Shock absorber assembly

2. Raise and support the vehicle safely.

3. Remove the tire and wheel assembly.

4. Separate the stabilizer connecting rod (lower side).

5. Set a suitable jack under the axle housing to relieve the coil spring tension.

6. Remove the shock absorber mounting bolt (lower side).

7. Gradually lower the jack to remove it from front lower link.

8. Remove the shock absorber assembly mounting nuts (upper side) and remove the shock absorber assembly.

To install:

9. Note the following, and install in the reverse order of the removal.

   a. Be sure to replace all non reusable components with new ones.

   b. Perform final tightening of the shock absorber lower side (rubber bushing) under unladen conditions with tires on level ground.

TESTING

Check for oil leakage, damage and breakage of installation positions.

STABILIZER BAR

REMOVAL & INSTALLATION

See Figure 186.

1. Before servicing the vehicle, refer to the Precautions Section.

2. Raise and support the vehicle safely.

3. Remove the tire and wheel assembly.

4. Remove the stabilizer connecting rod mounting bracket.

5. Remove the stabilizer connecting rod. Apply matchmarks to identify installation position.

6. Remove the mounting bolts on the stabilizer clamp and remove the stabilizer bar.

To install:

7. Note the following, and install in the reverse order of removal.

   a. Check the matchmarks when installing.

   b. Tighten the mounting nut to 32 ft. lbs. (43 Nm) while holding a hexagonal part of the stabilizer connecting rod side.

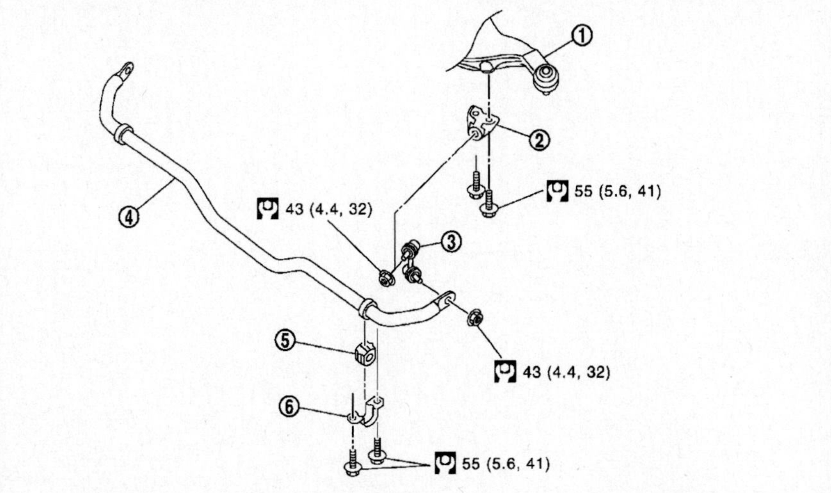

WHEEL HUB & BEARING (SEALED UNIT)

REMOVAL & INSTALLATION

See Figure 187.

1. Before servicing the vehicle, refer to the Precautions Section.

2. Raise and support the vehicle safely.

3. Remove the tire and wheel assembly.

4. Remove the wheel sensor and sensor harness.

5. Remove the caliper assembly. Hang caliper assembly aside.

✳✳ WARNING

Never depress the brake pedal while the caliper assembly is removed.

6. Remove the disc rotor.

7. Remove the cotter pin, and loosen the suspension arm mounting nut of the axle housing.

8. Remove the cotter pin, and loosen the wheel hub lock nut.

9. Patch the wheel hub lock nut with a piece of wood. Hammer the wood to disengage wheel hub and bearing assembly from halfshaft. Remove the wheel hub lock nut.

✳✳ WARNING

Never place halfshaft joint at an extreme angle. Also be careful not to overextend slide joint. Never allow the halfshaft to hang down without support for the housing (or joint sub-assembly), shaft and the other parts.

10. If the wheel hub and bearing assembly and halfshaft cannot be separated even after performing the above procedure, separate using suitable puller.

11. Remove the wheel hub and bearing assembly.

12. Remove the parking brake shoe and parking brake cable from the back plate.

13. Remove the anchor block mounting nuts, and remove the anchor block and back plate from the axle housing.

14. Remove the stabilizer connecting rod (upper side).

15. Remove the radius rod (axle housing side).

16. Remove the coil spring.

17. Set a suitable jack under axle housing.

18. Remove the front lower link (shock absorber side).

19. Remove the front lower link (axle housing side).

20. Using a using ball joint remover, separate the suspension arm from the axle housing so as not to damage the ball joint boot, and remove the axle housing from the vehicle.

21. Temporarily tighten the nuts to prevent damage to the threads and to prevent the ball joint remover from coming off.

✳✳ WARNING

Never place the halfshaft joint at an extreme angle. Also be careful not to overextend slide joint. Never allow halfshaft to hang down without support for counterpart such as joint sub-assembly, and other parts.

To install:

22. Note the following, and install in the reverse order of removal.

1.	Suspension arm	2.	Stabilizer connecting rod mount bracket	3.	Stabilizer connecting rod
4.	Stabilizer bar	5.	Stabilizer bushing	6.	Stabilizer clamp

37663_MURA_G0391

Fig. 186 Stabilizer bar assembly

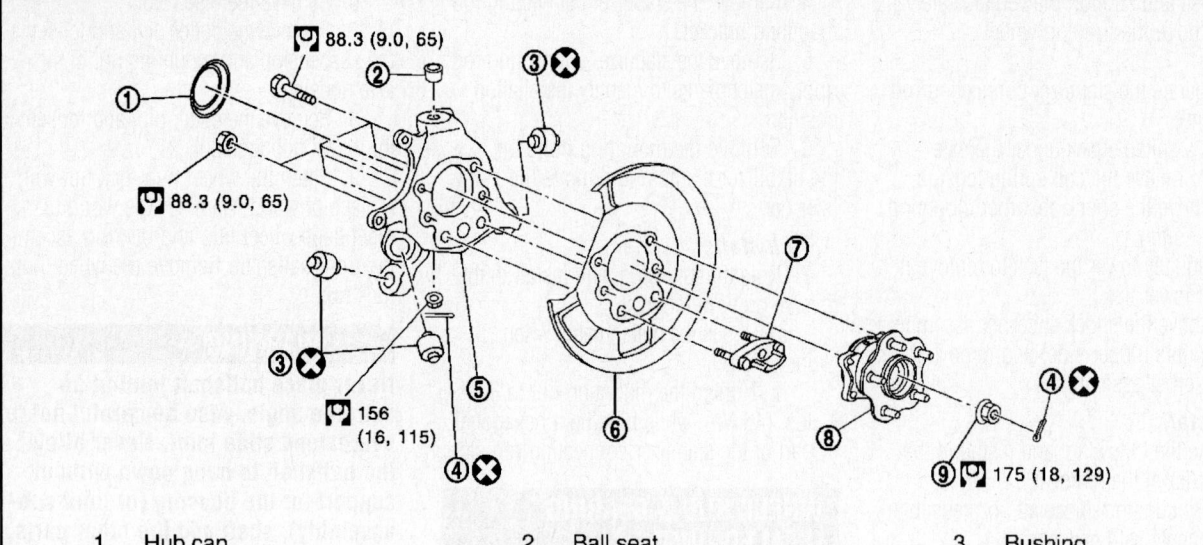

1. Hub cap
2. Ball seat
3. Bushing
4. Cotter pin
5. Axle housing
6. Back plate
7. Anchor block
8. Wheel hub and bearing assembly
9. Wheel hub lock nut

37663_MURA_G0392

Fig. 187 Rea wheel hub and bearing (sealed unit)

a. Never use a power tool to tighten the wheel hub lock nut.

b. Perform the final tightening of each of parts under un-laden conditions, which were removed when removing wheel hub and bearing assembly and axle housing.

c. Never reuse cotter pin.

ADJUSTMENT

The rear wheel bearings are part of a unitized hub and are not adjustable.

NISSAN

Pathfinder

16

SPECIFICATIONS AND MAINTENANCE CHARTS

ENGINE AND VEHICLE IDENTIFICATION

		Engine						Model Year	
Code ①	Liters (cc)	Cu. In.	Cyl.	Fuel Sys.	Engine	Eng. Mfg.		Code ②	Year
VQ40DE	4.0 (3954)	241	6	MFI	DOHC	Nissan		9	2009
VK56DE	5.6 (5552)	339	8	MFI	DOHC	Nissan		A	2010

MFI: Multi-port Fuel Injection

DOHC: Double Overhead Camshafts

① Fourth digit of the Vehicle Identification Number (VIN) is the engine ID, A: VQ40DE engine. B: VK56DE engine.

② Tenth digit of the Vehicle Identification Number (VIN)

37663_PATH_C0001

GENERAL ENGINE SPECIFICATIONS

Year	Model	Engine Displacement Liters (cc)	Engine ID/VIN	Fuel System Type	Net Horsepower @ rpm	Net Torque @ rpm (ft. lbs.)	Bore x Stroke (in.)	Com- pression Ratio	Oil Pressure @ rpm
2009	Pathfinder	4.0 (3954)	VQ40DE	MFI	266@5600	288@4000	3.76X3.62	9.7:1	43@2000
		5.6 (5552)	VQ56DE	MFI	310@5200	388@3400	3.86X3.62	9.8:1	43@2000
2010	Pathfinder	4.0 (3954)	VQ40DE	MFI	266@5600	288@4000	3.76X3.62	9.7:1	43@2000
		5.6 (5552)	VQ56DE	MFI	310@5200	388@3400	3.86X3.62	9.8:1	43@2000

MFI: Multi-port Fuel Injection

37663_PATH_C0002

ENGINE TUNE-UP SPECIFICATIONS

Year	Engine Displacement Liters (cc)	Engine ID/VIN	Spark Plug Gap (in.)	Ignition Timing (deg.) MT	Ignition Timing (deg.) AT	Fuel Pump (psi)	Idle Speed (rpm) MT	Idle Speed (rpm) AT	Valve Clearance (in.) In.	Valve Clearance (in.) Ex.
2009	4.0 (3954)	VQ40DE	0.043	NA	①	②	NA	575-675	HYD	HYD
	5.6 (5552)	VQ56DE	0.043	NA	①	②	NA	600-700	HYD	HYD
2010	4.0 (3954)	VQ40DE	0.043	NA	①	②	NA	575-675	HYD	HYD
	5.6 (5552)	VQ56DE	0.043	NA	①	②	NA	600-700	HYD	HYD

NOTE: The Vehicle Emission Control Information label often reflects specification changes made during production. The label figures must be used if they differ from those in this chart.

NA- Not Available

HYD: Hydraulic

① 15 degrees +/- 5 degrees

② 51 psi at idle

37663_PATH_C0003

CAPACITIES

Year	Model	Engine Displacement Liters	Engine ID/VIN	Engine Oil with Filter (qts.)	Transmission (pts.) 5-Spd	Transmission (pts.) Auto.	Transfer Case (pts.)	Drive Axle Front (pts.)	Drive Axle Rear (pts.)	Fuel Tank (gal.)	Cooling System (qts.)
2009	Pathfinder	4.0	VQ40DE	5.38	—	21.56	①	1.75	3.00	21.2	②
		5.6	VQ56DE	6.78	—	22.50	①	3.38	3.75	21.2	②
2010	Pathfinder	4.0	VQ40DE	5.38	—	21.56	①	1.75	3.00	21.2	②
		5.6	VQ56DE	6.78	—	22.50	①	3.38	3.75	21.2	②

NOTE: All capacities are approximate. Add fluid gradually and check to be sure a proper fluid level is obtained.

① Model ATX14B: 6.36 pts. Model TX15B: 4.36 pts.

② Without rear A/C: 10.8 qts. With rear A/C: 14.18 qts.

37663_PATH_C0004

FLUID SPECIFICATIONS

Year	Model	Engine Displ. Liters	Engine Oil	Auto. Trans.	Drive Axle Front	Drive Axle Rear	Transfer Case	Power Steering Fluid	Brake Master Cylinder	Cooling System
2009	Pathfinder	4.0	SAE 5W-30	①	②	②	③	④	⑤	⑥
		5.6	SAE 5W-30	①	②	②	③	④	⑤	⑥
2010	Pathfinder	4.0	SAE 5W-30	①	②	②	③	④	⑤	⑥
		5.6	SAE 5W-30	①	②	②	③	④	⑤	⑥

① Genuine Nissan Matic S ATF fluid. If not available genuine Nissan Matic J ATF may be used.

② Front: Nissan differential oil hypoid super GL-5 80W-90 or API GL-5 viscosity SAE 80W-90 gear oil.

Rear: Nissan differential oil synthetic 75W-90 or API GL-5 synthetic gear oil, viscosity SAE 75W-90.

③ Nissan Matic D ATF

④ Nissan Power Steering Fluid

⑤ Nissan Super Heavy Duty DOT 3

⑥ Nissan Long Life antifreeze or equivalent

37663_PATH_C0012

VALVE SPECIFICATIONS

Year	Engine Displacement Liters	Engine ID/VIN	Seat Angle (deg.)	Face Angle (deg.)	Spring Test Pressure (lbs. @ in.)	Spring Installed Height (in.)	Stem-to-Guide Clearance (in.) Intake	Stem-to-Guide Clearance (in.) Exhaust	Stem Diameter (in.) Intake	Stem Diameter (in.) Exhaust
2009	4.0	VQ40DE	45.15-45.45	45	37-42 @1.457	1.457	0.0008-0.0021	0.0012-0.0025	0.2348-0.2354	0.2344-0.2350
	5.6	VQ56DE	45.15-45.45	45	37-42 @1.457	1.457	0.0008-0.0021	0.0012-0.0025	0.2348-0.2354	0.2344-0.2350
2010	4.0	VQ40DE	45.15-45.45	45	37-42 @1.457	1.457	0.0008-0.0021	0.0012-0.0025	0.2348-0.2354	0.2344-0.2350
	5.6	VQ56DE	45.15-45.45	45	37-42 @1.457	1.457	0.0008-0.0021	0.0012-0.0025	0.2348-0.2354	0.2344-0.2350

37663_PATH_C0006

CAMSHAFT AND BEARING SPECIFICATIONS CHART

All measurements are given in inches.

Year	Engine Displacement Liters	Engine ID/VIN	Journal Diameter	Brg. Oil Clearance	Shaft End-play	Runout	Journal Bore	Lobe Height	
								Intake	Exhaust
2009	4.0	VQ40DE	①	②	0.0045-0.0074	0.0008	③	1.7900-1.7921	1.7746-1.7821
	5.6	VQ56DE	1.0217-1.0224	0.0012-0.0028	0.0045-0.0074	0.0008	1.0236-1.0244	1.7663-1.7738	1.7746-1.7821
2010	4.0	VQ40DE	①	②	0.0045-0.0074	0.001	③	1.7900-1.7921	1.7746-1.7821
	5.6	VQ56DE	1.0217-1.0224	0.0012-0.0028	0.0045-0.0074	0.0008	1.0236-1.0244	1.7663-1.7738	1.7746-1.7821

① No. 1: 1.0211-1.0218
   Nos. 2-4: 0.9230-0.9238
② No. 1: 0.0018-0.0034
   Nos. 2-4: 0.0014-0.0030
③ No. 1: 1.0236-1.0244
   Nos. 2-4: 0.9252-0.9260

37663_PATH_C0014

CRANKSHAFT AND CONNECTING ROD SPECIFICATIONS

All measurements are given in inches.

Year	Engine Displacement Liters	Engine ID/VIN	Crankshaft				Connecting Rod		
			Main Brg. Journal Dia.	Main Brg. Oil Clearance	Shaft End-play	Thrust on No.	Journal Diameter	Oil Clearance	Side Clearance
2009	4.0	VQ40DE	①	0.0014-0.0018	0.0039-0.0098	4	NA	0.0013-0.0023	0.0079-0.0138
	5.6	VQ56DE	②	③	0.0039-0.0102	4	NA	0.0008-0.0015	0.0079-0.0157
2010	4.0	VQ40DE	①	0.0014-0.0018	0.0039-0.0098	4	NA	0.0013-0.0023	0.0079-0.0138
	5.6	VQ56DE	②	③	0.0039-0.0102	4	NA	0.0008-0.0015	0.0079-0.0157

NA - Not Available
① There are 24 different grades, ranging from A (2.7549) to 7 (2.7540)
② There are 24 different grades, ranging from A (2.5182) to 2 (2.5174)
③ No. 1 and 5: 0.00004-0.0004
   No. 2, 3 and 4: 0.0003-0.0007

37663_PATH_C0005

PISTON AND RING SPECIFICATIONS
All measurements are given in inches.

Year	Engine Displacement Liters	Engine ID/VIN	Piston Clearance	Ring Gap			Ring Side Clearance		
				Top Comp.	Bottom Comp.	Oil Control	Top Comp.	Bottom Comp.	Oil Control
2009	4.0	VQ40DE	0.0004-0.0012	0.0091-0.0130	0.0130-0.0189	0.0079-0.0197	0.0018-0.0031	0.0012-0.0028	0.0026-0.0053
	5.6	VQ56DE	0.0004-0.0012	0.0091-0.0130	0.098-0.0157	0.0079-0.0236	0.0014-0.0033	0.0012-0.0028	0.0006-0.0020
2010	4.0	VQ40DE	0.0004-0.0012	0.0091-0.0130	0.0130-0.0189	0.0079-0.0197	0.0018-0.0031	0.0012-0.0028	0.0026-0.0053
	5.6	VQ56DE	0.0004-0.0012	0.0091-0.0130	0.098-0.0157	0.0079-0.0236	0.0014-0.0033	0.0012-0.0028	0.0006-0.0020

37663_PATH_C0007

TORQUE SPECIFICATIONS
All readings in ft. lbs.

Year	Engine Displacement Liters	Engine ID/VIN	Cylinder Head Bolts	Main Bearing Bolts	Rod Bearing Bolts	Crankshaft Damper Bolts	Flywheel Bolts	Manifold		Spark Plugs
								Intake	Exhaust	
2009	4.0	VQ40DE	①	②	③	④	65	⑤	22	18
	5.6	VQ56DE	⑥	⑦	⑧	⑨	65	NA	25	18
2010	4.0	VQ40DE	①	②	③	④	65	⑤	22	18
	5.6	VQ56DE	⑥	⑦	⑧	⑨	65	NA	25	18

NA - Not Available

① Step 1: 72 ft. lbs.

  Step 2: Loosen all bolts completely

  Step 3: 29 ft. lbs.

  Step 4: +90 degrees

  Step 5: +90 degrees

② Step 1: 26 ft. lbs.

  Step 2: +90 degrees

③ Step 1: 14 ft. lbs.

  Step 2: +90 degrees

④ 33 ft. lbs. +84 - 90 degrees

⑤ Step 1: 65 inch lbs.

  Step 2 and after: 21ft. lbs.

  Studs 8 ft. lbs.

⑥ Step 1: 72 ft. lbs.

  Step 2: Loosen all bolts completely

  Step 3: 33 ft. lbs.

  Step 4: +70 degrees

⑦ Step 1 Bolts 1-10 29 ft. lbs.

  Step 2: Bolts 11 - 20: 22 ft. lbs

  Step 3: Bolts 1 - 10: An additional 40 degrees

  Step 4: Bolts 11 - 20: an additional 30 degrees

  Side bolts in order of 21-30> 36 ft. lbs.

⑧ Step 1: 11 ft. lbs.

  Step 2: +90 degrees

⑨ 69 ft. lbs. +90 degrees

37663_PATH_C0008

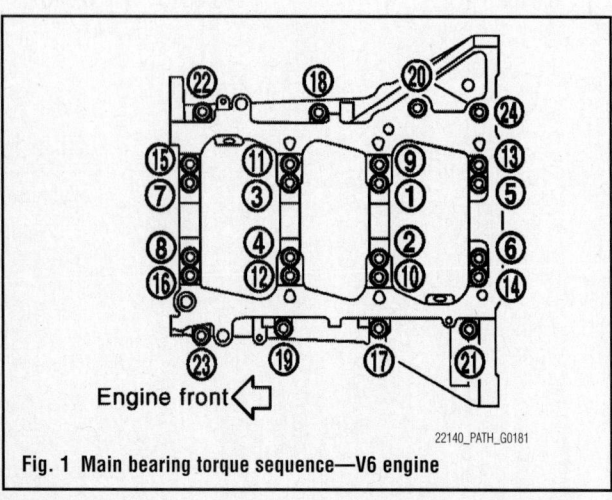

Fig. 1 Main bearing torque sequence—V6 engine

22140_PATH_G0181

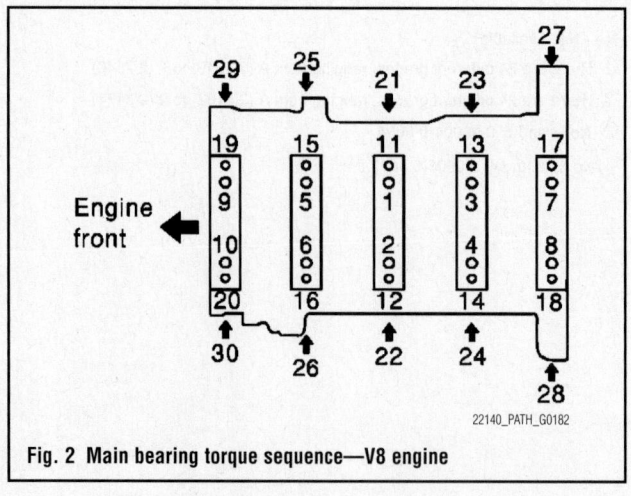

Fig. 2 Main bearing torque sequence—V8 engine

22140_PATH_G0182

WHEEL ALIGNMENT

Year	Model		Caster		Camber		Toe-in (in.)
			Range (+/-Deg.)	Preferred Setting (Deg.)	Range (+/-Deg.)	Preferred Setting (Deg.)	
2009	Pathfinder	2WD	①	①	②	②	NA
		4WD	③	③	④	②	NA
2010	Pathfinder	2WD	0.75	3.00	0.75	0.25	0.12+/-0.04
		4WD	0.75	2.75	0.75	0.50	0.12+/-0.04

NA - Not Available

① Minimum: 2 degrees 15' (2.25 degrees). Nominal: 3 degrees 0' (3.00 degrees). Maximum: 3 degrees 45' (3.75 degrees).

② Minimum: -0 degrees 30' (-0.50 degrees). Nominal: 0 degrees 15' (0.25 degrees). Maximum: 1 degree 00' (1.00 degree).

③ Minimum: 2 degrees 00' (2.00 degrees). Nominal: 2 degrees 45' (2.75 degrees). Maximum: 3 degrees 30' (3.50 degrees).

④ Minimum: -0 degrees 15' (-0.25 degrees). Nominal: 0 degrees 30' (0.50 degrees). Maximum: 1 degree 15' (1.25 degrees).

37663_PATH_C0009

TIRE, WHEEL AND BALL JOINT SPECIFICATIONS

Year	Model	OEM Tires		Tire Pressures (psi)		Wheel Size	Ball Joint Inspection	Lug Nuts (ft. lbs.)
		Standard	Optional	Front	Rear			
2009	Pathfinder	P245/75R16	none	35	35	16x7JJ	①	98
		P265/65R17	none	35	35	17x7.5JJ	①	98
		P265/60R18	none	35	35	18x8JJ	①	98
2010	Pathfinder	P245/75R16	none	35	35	16x7JJ	①	98
		P265/65R17	none	35	35	17x7.5JJ	①	98
		P265/60R18	none	35	35	18x8JJ	①	98

OEM: Original Equipment Manufacturer

PSI: Pounds Per Square Inch

① Axial play

Upper: 0

Lower: 0.008 in.

37663_PATH_C0010

BRAKE SPECIFICATIONS

All measurements in inches unless noted

Year	Model		Brake Disc			Minimum Lining Thickness		Brake Caliper	
			Original Thickness	Minimum Thickness	Maximum Runout	Front	Rear	Bracket Bolts (ft. lbs.)	Mounting Bolts (ft. lbs.)
2009	Pathfinder	F	1.102	1.024	0.0006	0.079	0.079	①	①
		R	0.079	0.630	0.0006	0.079	0.079	②	②
2010	Pathfinder	F	1.102	1.024	0.0006	0.079	0.079	①	①
		R	0.079	0.630	0.0006	0.079	0.079	②	②

NA: Not Available

① Torque member mounting bolt: 136 ft. lbs.

Sliding pin bolt 20 ft. lbs.

① Torque member mounting bolt: 76 ft. lbs.

Sliding pin bolt 20 ft. lbs.

37663_PATH_C0011

SCHEDULED MAINTENANCE INTERVALS
NISSAN PATHFINDER

TO BE SERVICED	TYPE OF SERVICE	VEHICLE MILEAGE INTERVAL (x1000)												
		3.75	7.5	15	22.5	30	37.5	45	52.5	60	67.5	75	82.5	90
Engine oil & filter	R	✓	✓	✓	✓	✓	✓	✓	✓	✓	✓	✓	✓	✓
Brake lines & cables	I			✓		✓		✓		✓		✓		✓
Brake pads& rotors	L/I			✓		✓		✓		✓		✓		✓
Driveshaft boots & propeller shaft (4x4)	I					✓				✓				
Automatic transmission, final drive oil & transfer case	I			✓		✓		✓		✓				
LSD gear oil	I			✓		✓		✓		✓		✓		✓
Front wheel bearing grease (4x4)	R					✓				✓				
Timing belt	R	Replace every 105,000 miles												
Air cleaner filter	R					✓				✓				✓
Engine coolant	R									✓				✓
Exhaust system	I						✓	✓	✓	✓				
Spark plugs	R	Replace every 105,000 miles												
Drive belt(s)	I			✓		✓		✓		✓		✓		✓
Cabin air filter	R							✓						✓
Exhaust system	I		✓			✓				✓				
Fuel lines	I		✓			✓				✓				
Steering gear (box) & linkage, axle & suspension parts	I					✓				✓				✓
Transfer case	I					✓				✓				✓
Tire rotation			✓	✓	✓	✓	✓	✓	✓	✓	✓	✓	✓	✓
Vapor lines	S/I					✓				✓				✓

R: Replace   S/I: Service or Inspect   L: Lubricate

FREQUENT OPERATION MAINTENANCE (SEVERE SERVICE)

If a vehicle is operated under any of the following conditions it is considered severe service:

- Extremely dusty areas.

- 50% or more of the vehicle operation is in 32°C (90°F) or higher temperatures, constant operation in temp. below 0°C (32°F).

- Prolonged idling (vehicle operation in stop and go traffic).

- Frequent short running periods (engine does not warm to normal operating temperatures).

- Police, taxi, delivery usage or trailer towing usage.

Oil & oil filter: replace every 3750 miles.

Brake pads, discs, drums & linings: service or inspect every 7500 miles.

Driveshaft boots & propeller shaft: service or inspect every 7500 miles.

Exhaust system: service or inspect every 7500 miles.

Final drive oil: Change every 30000 miles if towing a trailer.

Transfer case fluid: Change every 30000 miles if towing a trailer.

Steering gear (box) & linkage, (steering damper-4x4), axle & suspension parts: service or inspect every 7500 miles.

Steering linkage ball joints & front suspension ball joints: service or inspect every 7500 miles.

37663_PATH_C0013

BRAKES — INFORMATION AND PRECAUTIONS

ANTI-LOCK SYSTEMS

• Certain components within the ABS system are not intended to be serviced or repaired individually.

• Do not use rubber hoses or other parts not specifically specified for and ABS system. When using repair kits, replace all parts included in the kit. Partial or incorrect repair may lead to functional problems and require the replacement of components.

• Lubricate rubber parts with clean, fresh brake fluid to ease assembly. Do not use shop air to clean parts; damage to rubber components may result.

• Use only DOT 3 brake fluid from an unopened container.

• If any hydraulic component or line is removed or replaced, it may be necessary to bleed the entire system.

• A clean repair area is essential. Always clean the reservoir and cap thoroughly before removing the cap. The slightest amount of dirt in the fluid may plug an orifice and impair the system function. Perform repairs after components have been thoroughly cleaned; use only denatured alcohol to clean components. Do not allow ABS components to come into contact with any substance containing mineral oil; this includes used shop rags.

• The Anti-Lock control unit is a microprocessor similar to other computer units in the vehicle. Ensure that the ignition switch is **OFF** before removing or installing controller harnesses. Avoid static electricity discharge at or near the controller.

• If any arc welding is to be done on the vehicle, the control unit should be unplugged before welding operations begin.

DISC AND DRUM SYSTEMS

✳✳ CAUTION

Dust and dirt accumulating on brake parts during normal use may contain asbestos fibers from production or aftermarket brake linings. Breathing excessive concentrations of asbestos fibers can cause serious bodily harm. Exercise care when servicing brake parts. Do not sand or grind brake lining unless equipment used is designed to contain the dust residue. Do not clean brake parts with compressed air or by dry brushing. Cleaning should be done by dampening the brake components with a fine mist of water, then wiping the brake components clean with a dampened cloth. Dispose of cloth and all residue containing asbestos fibers in an impermeable container with the appropriate label. Follow practices prescribed by the Occupational Safety and Health Administration (OSHA) and the Environmental Protection Agency (EPA) for the handling, processing, and disposing of dust or debris that may contain asbestos fibers.

BRAKES — BLEEDING THE BRAKE SYSTEM

BLEEDING PROCEDURE

BLEEDING PROCEDURE

1. Before servicing the vehicle, refer to the Precautions Section.

➡**If working near and/or around the SRS system and components, be sure to disable the SRS system. After disabling the system wait three minutes or more before servicing the vehicle.**

➡**On this vehicle the battery current sensor that is installed to the negative battery cable measures the charging/discharging current of the battery and performs various engine controls. If an electrical component is connected directly to the negative battery cable, the current flowing through that component will not be measured by the battery current sensor. This condition may cause a malfunction of the engine control system and battery discharge may occur. Do not connect an electrical component or ground wire directly to the battery terminal.**

2. Disconnect the negative battery cable.

✳✳ CAUTION

While bleeding, monitor the master cylinder brake fluid level.

3. Turn ignition switch OFF.
4. Raise and support the vehicle safely, as required.
5. Connect a vinyl tube to the rear right bleed valve.
6. Fully depress brake pedal 4 to 5 times.
7. With brake pedal depressed, loosen bleed valve to let the air out, and then tighten it immediately.
8. Repeat steps 3 and 4 until no more air comes out.
9. Tighten bleed valve to the specified torque.
10. Perform steps 2 to 6 at each wheel, with master cylinder reservoir tank filled at least half way, bleed air from the front left, rear left, and front right bleed valve, in that order.
11. Be sure to perform the reconnect/relearn procedures.

BLEEDING THE ABS SYSTEM

Refer to "Bleeding the System".

BRAKES — ANTI-LOCK BRAKE SYSTEM (ABS)

WHEEL SPEED SENSORS

REMOVAL & INSTALLATION

See Figure 3.

1. Before servicing the vehicle, refer to the Precautions Section.

➡**If working near and/or around the SRS system and components, be sure to disable the SRS system. After disabling the system wait three minutes or more before servicing the vehicle.**

➡**On this vehicle the battery current sensor that is installed to the negative battery cable measures the charging/discharging current of the battery and performs various engine controls. If an electrical component is connected directly to the negative battery cable, the current flowing through that component will not be measured by the battery current sensor. This condition may cause a malfunction of the engine control system and battery discharge may** occur. Do not connect an electrical component or ground wire directly to the battery terminal.

2. Disconnect the negative battery cable.

3. Raise and support the vehicle safely.

4. Remove wheel and tire.

5. Remove wheel sensor mounting screw.

➡When removing front wheel sensor, first remove the disc rotor to gain access to the front wheel sensor mounting bolt. When removing rear wheel sensor, first remove spare tire.

6. Pull out the wheel sensor, being careful to turn it as little as possible.

✳✳ WARNING

Be careful not to damage sensor edge and sensor rotor teeth. Do not pull on the sensor harness.

7. Disconnect wheel sensor harness electrical connector, then remove harness from mounts.

To install:

➡Be sure to use new fasteners, as required.

8. Installation is the reverse of the removal procedure.

9. Be sure to perform the reconnect/relearn procedures.

✳✳ WARNING

Inspect wheel sensor O-ring, replace sensor assembly if damaged. Clean wheel sensor hole and mounting surface with brake cleaner and a lint-free shop rag. Be careful that dirt and debris do not enter the axle.

➡Apply a coat of suitable grease to the wheel sensor O-ring and mounting hole.

10. Tighten wheel sensor bolts to 73 inch lbs. (8.25 Nm).

WHEEL SPEED SENSOR RINGS (TOOTHED RINGS)

REMOVAL & INSTALLATION

Front Wheel

The wheel sensor rotors are built into the wheel hubs and are not removable. If damaged, replace wheel hub and bearing assembly.

Rear Wheel

1. Before servicing the vehicle, refer to the Precautions Section.

➡If working near and/or around the SRS system and components, be sure to disable the SRS system. After disabling the system wait three minutes or more before servicing the vehicle.

➡On this vehicle the battery current sensor that is installed to the negative battery cable measures the charging/discharging current of the battery and performs various engine controls. If an electrical component is connected directly to the negative battery cable, the current flowing through that component will not be measured by the battery current sensor. This condition may cause a malfunction of the engine control system and battery discharge may occur. Do not connect an electrical component or ground wire directly to the battery terminal.

2. Disconnect the negative battery cable.

3. Remove side flange from final drive assembly.

✳✳ CAUTION

Discard side oil seal.

4. Using tool and a suitable puller, remove sensor rotor from side flange.

To install:

➡Be sure to use new fasteners, as required.

5. Install new sensor rotor on side flange using a press.

6. Make sure sensor rotor is fully seated.

✳✳ CAUTION

Do not reuse the old sensor rotor.

7. Install side flange to final drive assembly.

✳✳ CAUTION

Do not reuse the side oil seal. The side oil seal must be replaced every time the side flange is removed from the final drive assembly.

8. Continue the installation in the reverse order of the removal procedure.

9. Be sure to perform the reconnect/relearn procedures.

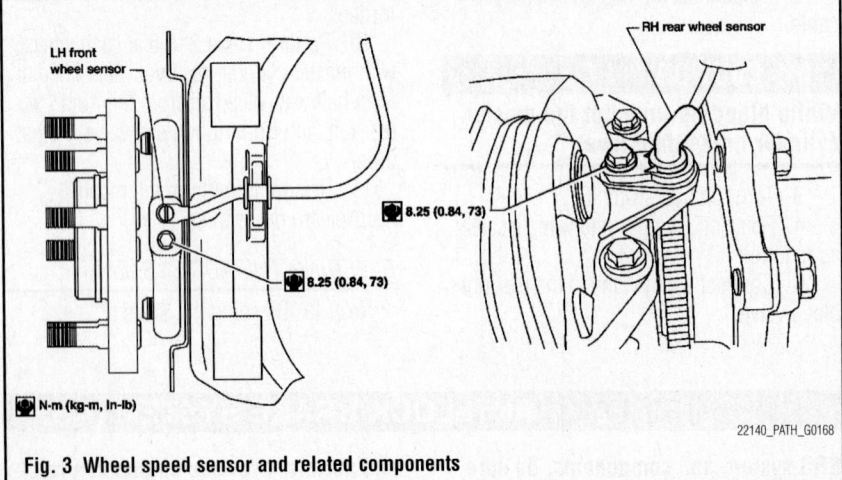

LH front wheel sensor

RH rear wheel sensor

8.25 (0.84, 73)

8.25 (0.84, 73)

N-m (kg-m, In-lb)

22140_PATH_G0168

Fig. 3 Wheel speed sensor and related components

BRAKES   **FRONT DISC BRAKES**

BRAKE CALIPER

REMOVAL & INSTALLATION

See Figure 4.

1. Before servicing the vehicle, refer to the Precautions Section.

➡**If working near and/or around the SRS system and components, be sure to disable the SRS system. After disabling the system wait three minutes or more before servicing the vehicle.**

➡**On this vehicle the battery current sensor that is installed to the negative**

battery cable measures the charging/discharging current of the battery and performs various engine controls. If an electrical component is connected directly to the negative battery cable, the current flowing through that component will not be measured by the battery current sensor. This condition may cause a malfunction of the engine control system and battery discharge may occur. Do not connect an electrical component or ground wire directly to the battery terminal.

2. Disconnect the negative battery cable.

3. Raise and support the vehicle safely. Remove wheel and tire.

4. Drain brake fluid as necessary.

➡**Do not remove union bolt unless removing cylinder body from vehicle.**

5. Remove union bolt as necessary and torque member bolts, then remove cylinder body from the vehicle.

6. Position cylinder body aside using suitable wire, as necessary.

7. When servicing brake caliper, remove sliding pin bolts and caliper from torque member.

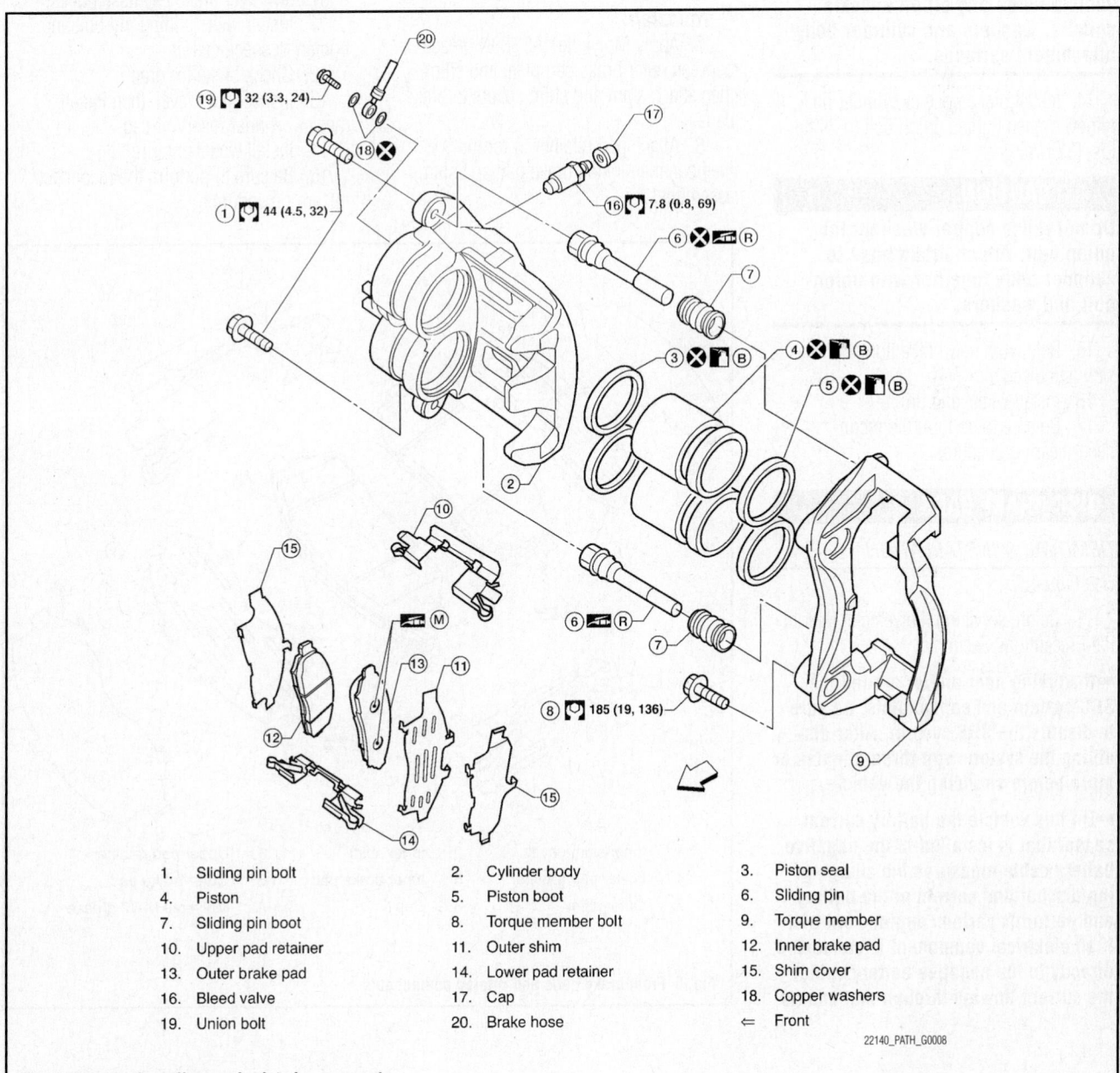

1. Sliding pin bolt	2. Cylinder body	3. Piston seal
4. Piston	5. Piston boot	6. Sliding pin
7. Sliding pin boot	8. Torque member bolt	9. Torque member
10. Upper pad retainer	11. Outer shim	12. Inner brake pad
13. Outer brake pad	14. Lower pad retainer	15. Shim cover
16. Bleed valve	17. Cap	18. Copper washers
19. Union bolt	20. Brake hose	⇐ Front

22140_PATH_G0008

Fig. 4 Front brake caliper and related components

8. Remove torque member.
9. Remove disc rotor.

To install:

※※ **CAUTION**

Refill with new brake fluid. Do not reuse drained brake fluid.

10. Install disc rotor.
11. Install torque member and tighten to specification.
12. Install sliding pin bolts, if removed.
13. Install cylinder body, then tighten sliding pin bolts to specification.

※※ **CAUTION**

When attaching cylinder body to the vehicle, wipe any oil off knuckle spindle, washers and cylinder body attachment surfaces.

14. Install brake hose to cylinder body, if removed, then tighten union bolt to 24 ft. lbs. (32 Nm).

※※ **CAUTION**

Do not reuse copper washers for union bolt. Attach brake hose to cylinder body together with union bolt and washers.

15. Refill with new brake fluid as necessary and bleed air.
16. Install wheel and tire.
17. Be sure to perform the reconnect/relearn procedures.

DISC BRAKE PADS

REMOVAL & INSTALLATION
See Figure 5.

1. Before servicing the vehicle, refer to the Precautions Section.

➡ **If working near and/or around the SRS system and components, be sure to disable the SRS system. After disabling the system wait three minutes or more before servicing the vehicle.**

➡ **On this vehicle the battery current sensor that is installed to the negative battery cable measures the charging/discharging current of the battery and performs various engine controls. If an electrical component is connected directly to the negative battery cable, the current flowing through that compo-** nent will not be measured by the battery current sensor. This condition may cause a malfunction of the engine control system and battery discharge may occur. Do not connect an electrical component or ground wire directly to the battery terminal.

2. Disconnect the negative battery cable.
3. Raise and support the vehicle safely. Remove wheel and tire.
4. Remove master cylinder reservoir cap.
5. Remove lower sliding pin bolt.
6. Suspend cylinder body with a wire and remove pads, shim, shim covers, and retainers from torque member.

To install:

7. Apply Molykote® AS880N grease between outer brake pad plate and shim, then attach shim and shim covers to brake pads.
8. Attach pad retainer to torque member, then install brake pad and shim assemblies.

※※ **CAUTION**

When attaching pad retainer, attach it firmly so that it is flush with torque member.

9. Push pistons into cylinder body.

➡ **Using a disc brake piston tool (commercial service tool), etc., makes it easier to push in piston.**

※※ **CAUTION**

By pushing in piston, brake fluid returns to master cylinder reservoir tank. Watch the level of the surface of reservoir tank.

10. Remove wire then swing cylinder body down over brake pad assemblies.
11. Install lower sliding pin bolt and tighten to specification.
12. Check brake for drag.
13. Inspect fluid level, then install master cylinder reservoir cap.
14. Install wheel and tire.
15. Be sure to perform the reconnect/relearn procedures.

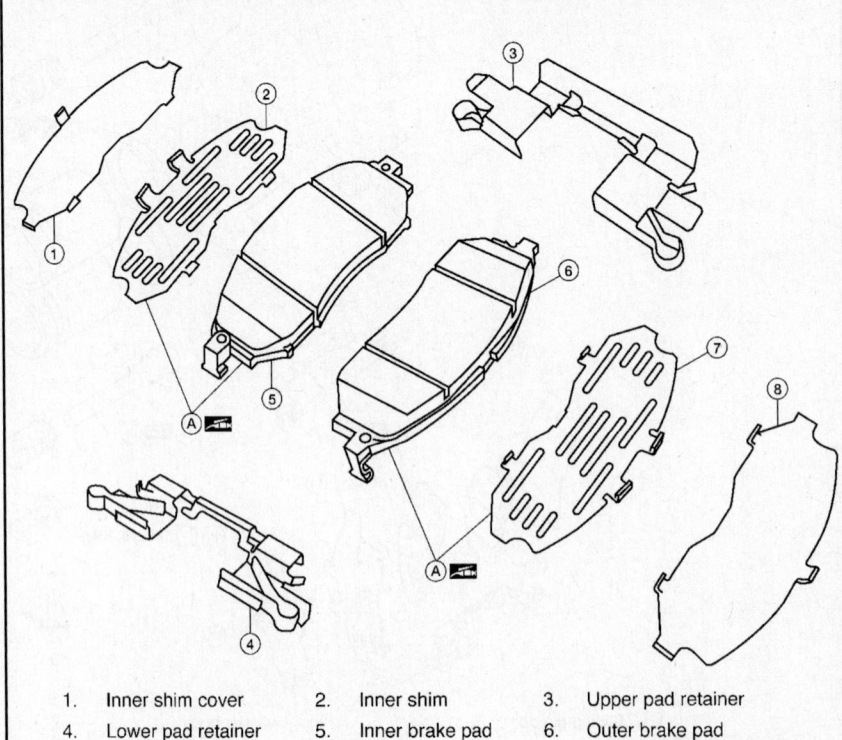

1.	Inner shim cover	2.	Inner shim
4.	Lower pad retainer	5.	Inner brake pad
7.	Outer shim	8.	Cap

3.	Upper pad retainer
6.	Outer brake pad
A.	Molykote M-77 grease

37663_PATH_G0025

Fig. 5 Front brake pads and related components

BRAKE CALIPER

REMOVAL & INSTALLATION

See Figure 6.

1. Before servicing the vehicle, refer to the Precautions Section.

➡If working near and/or around the SRS system and components, be sure to disable the SRS system. After disabling the system wait three minutes or more before servicing the vehicle.

➡On this vehicle the battery current sensor that is installed to the negative battery cable measures the charging/discharging current of the battery and performs various engine controls. If an electrical component is connected directly to the negative battery cable, the current flowing through that component will not be measured by the battery current sensor. This condition may cause a malfunction of the engine control system and battery discharge may occur. Do not connect an electrical component or ground wire directly to the battery terminal.

2. Disconnect the negative battery cable.
3. Raise and support the vehicle safely. Remove wheel and tire.
4. Drain brake fluid.
5. Remove union bolt and mounting bolts, and remove cylinder body.
6. Remove torque member.
7. Remove disc rotor.

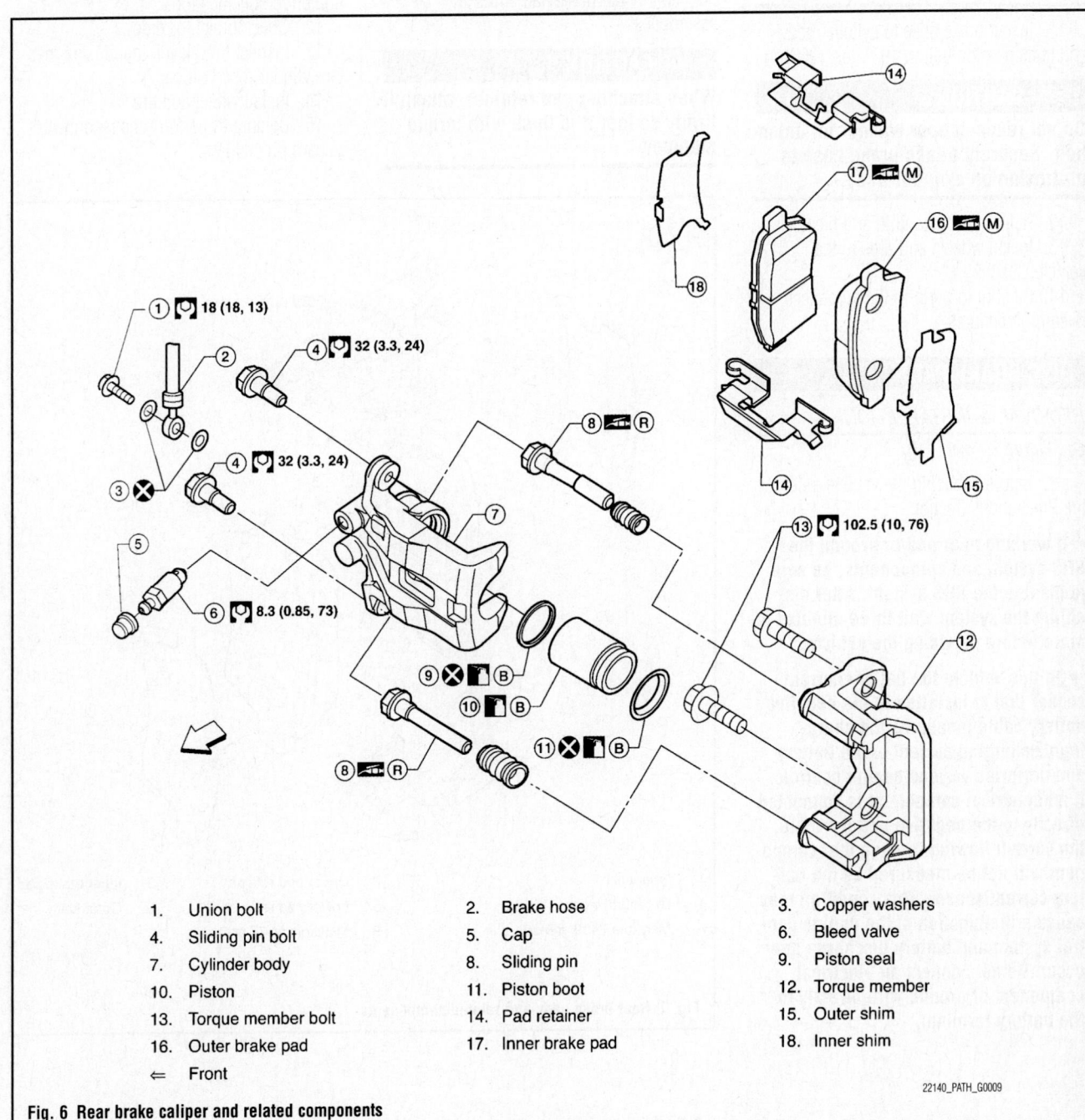

1. Union bolt	2. Brake hose	3. Copper washers
4. Sliding pin bolt	5. Cap	6. Bleed valve
7. Cylinder body	8. Sliding pin	9. Piston seal
10. Piston	11. Piston boot	12. Torque member
13. Torque member bolt	14. Pad retainer	15. Outer shim
16. Outer brake pad	17. Inner brake pad	18. Inner shim
⇐ Front		

Fig. 6 Rear brake caliper and related components

22140_PATH_G0009

To install:

> **✳✳ CAUTION**
>
> **Refill with new brake fluid. Do not reuse drained brake fluid.**

8. Install disc rotor.

9. Install torque member and tighten to specification.

10. Install cylinder body to the vehicle, and tighten mounting bolts to specification.

> **✳✳ CAUTION**
>
> **Before installing cylinder body to the vehicle, wipe off mounting surface of cylinder body.**

11. Install brake hose to cylinder body and tighten union bolt to 19 ft. lbs. (26 Nm).

> **✳✳ CAUTION**
>
> **Do not reuse copper washer for union bolt. Securely attach brake hose to protrusion on cylinder body.**

12. Refill new brake fluid and bleed air.

13. Install wheels and tires to the vehicle.

14. Be sure to perform the reconnect/relearn procedures.

DISC BRAKE PADS

REMOVAL & INSTALLATION

See Figure 7.

1. Before servicing the vehicle, refer to the Precautions Section.

➡**If working near and/or around the SRS system and components, be sure to disable the SRS system. After disabling the system wait three minutes or more before servicing the vehicle.**

➡**On this vehicle the battery current sensor that is installed to the negative battery cable measures the charging/discharging current of the battery and performs various engine controls. If an electrical component is connected directly to the negative battery cable, the current flowing through that component will not be measured by the battery current sensor. This condition may cause a malfunction of the engine control system and battery discharge may occur. Do not connect an electrical component or ground wire directly to the battery terminal.**

2. Disconnect the negative battery cable.

3. Raise and support the vehicle safely. Remove wheel and tire.

4. Remove master cylinder reservoir cap.

5. Remove lower sliding pin bolt.

6. Suspend cylinder body with a wire and remove pads, shim, shim covers, and retainers from torque member.

To install:

7. Apply Molykote® AS880N grease between outer brake pad plate and shim, then attach shim and shim covers to brake pads.

8. Attach pad retainer to torque member, then install brake pad and shim assemblies.

> **✳✳ CAUTION**
>
> **When attaching pad retainer, attach it firmly so that it is flush with torque member.**

9. Push pistons into cylinder body.

➡**Using a disc brake piston tool (commercial service tool), etc., makes it easier to push in piston.**

> **✳✳ CAUTION**
>
> **By pushing in piston, brake fluid returns to master cylinder reservoir tank. Watch the level of the surface of reservoir tank.**

10. Remove wire then swing cylinder body down over brake pad assemblies.

11. Install lower sliding pin bolt and tighten to specification.

12. Check brake for drag.

13. Inspect fluid level, then install master cylinder reservoir cap.

14. Install wheel and tire.

15. Be sure to perform the reconnect/relearn procedures.

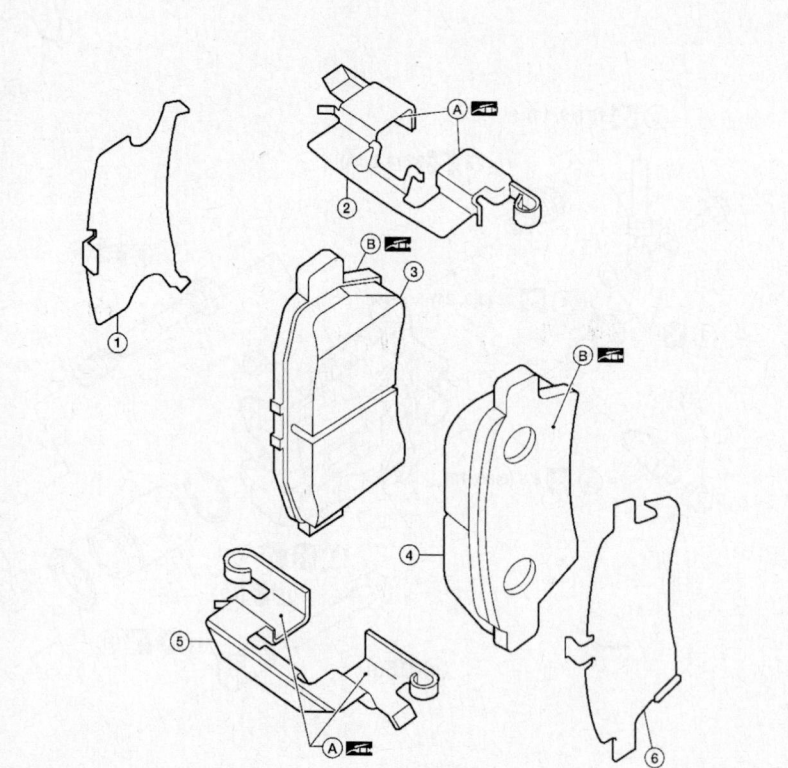

1.	Inner shim	2.	Upper pad retainer	3.	Inner brake pad
4.	Outer brake pad	5.	Lower pad retainer	6.	Outer shim
A.	Molykote 7439 grease	B.	Molykote M-77 grease		

37663_PATH_G0026

Fig. 7 Rear brake pads and related components

BRAKES

PARKING BRAKE

PARKING BRAKE CABLES

ADJUSTMENT

See Figures 8 and 9.

1. Before servicing the vehicle, refer to the Precautions Section.

➡If working near and/or around the SRS system and components, be sure to disable the SRS system. After disabling the system wait three minutes or more before servicing the vehicle.

➡On this vehicle the battery current sensor that is installed to the negative battery cable measures the charging/discharging current of the battery and performs various engine controls. If an electrical component is connected directly to the negative battery cable, the current flowing through that component will not be measured by the battery current sensor. This condition may cause a malfunction of the engine control system and battery discharge may occur. Do not connect an electrical component or ground wire directly to the battery terminal.

2. Disconnect the negative battery cable.
3. Remove front pillar lower finisher.
4. Remove lower instrument panel LH.
5. Pull rearward to release lower instrument panel LH.
6. Disconnect lower instrument panel LH harness connectors.
7. Partially engage parking brake pedal to access adjusting nut.
8. Insert a deep socket wrench to rotate adjusting nut and loosen cable until tension is sufficiently released. Then, disengage the parking brake pedal.
9. Remove the wheel and tire.

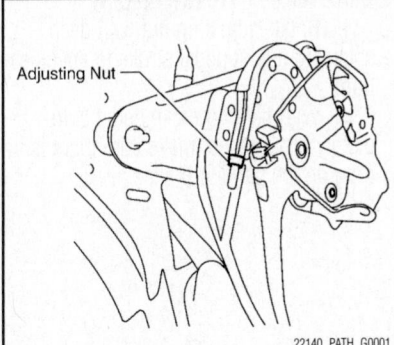

Fig. 8 Rotate adjusting nut and loosen cable until tension is sufficiently released

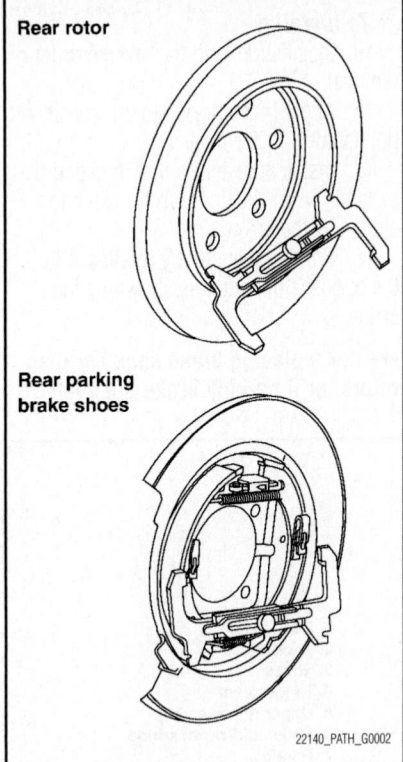

Rear rotor

Rear parking brake shoes

22140_PATH_G0002

Fig. 9 Measure inner diameter at widest point

10. Remove the rotor and measure inner diameter at widest point.
11. Transfer measurement less 0.6 mm to the parking brake shoes and adjust accordingly.
12. Using wheel nuts, secure the disc to the hub to prevent it from tilting.
13. Rotate disc rotor to make sure there is no drag.
14. Adjust cable as follows:
    a. Operate pedal 10 or more times with a force of 110 lbs. (490 N).
    b. Rotate adjusting nut with deep socket to adjust pedal stroke to 4 or 5 notches under a force of 44 lbs. (196 N).
    c. With parking brake pedal completely disengaged, make sure there is no drag on the parking brake.

PARKING BRAKE SHOES

REMOVAL & INSTALLATION

See Figure 10.

1. Before servicing the vehicle, refer to the Precautions Section.

➡If working near and/or around the SRS system and components, be sure to disable the SRS system. After dis-

abling the system wait three minutes or more before servicing the vehicle.

➡On this vehicle the battery current sensor that is installed to the negative battery cable measures the charging/discharging current of the battery and performs various engine controls. If an electrical component is connected directly to the negative battery cable, the current flowing through that component will not be measured by the battery current sensor. This condition may cause a malfunction of the engine control system and battery discharge may occur. Do not connect an electrical component or ground wire directly to the battery terminal.

2. Disconnect the negative battery cable.
3. Raise and support the vehicle safely. Remove wheel and tire.
4. Remove the rear brake caliper, without disconnecting the hydraulic hose.
5. Reposition the rear brake caliper aside using suitable wire.

➡Do not depress the brake pedal while the brake caliper is removed.

➡Remove the disc rotor only with the parking brake pedal completely disengaged.

6. Remove the rear disc rotor.
7. Remove the rear drive shaft.
    a. Move the A/T select lever to the N position and release the parking brake.
    b. Put matching marks on the rear propeller shaft flange yoke and the rear final drive companion flange as shown.

✳✳ WARNING

For matching marks, use paint. Never damage the rear propeller shaft flange yoke or the companion flange.

c. Remove the bolts, then remove the propeller shaft from the rear final drive and A/T or transfer.
8. Disconnect wheel sensor at harness connector. Then remove wheel sensor wire from grommet mounts.
9. Remove wheel hub and bearing assembly.
    a. Remove the cotter pin, then remove the rear drive shaft nut.
    b. Discard the cotter pin, use a new one for installation.

c. Remove the rear drive shaft.

d. Remove the four rear wheel hub and bearing assembly bolts.

e. Discard the four rear wheel hub and bearing assembly bolts, use new ones for installation.

f. Remove the rear wheel hub and bearing assembly.

➡**Withdraw wheel sensor harness through back plate when removing wheel hub and bearing assembly.**

10. Remove the return springs.

11. Remove the adjuster.

12. Remove the retainers, anti-rattle pins and shoes.

13. Disconnect the parking brake cable from the toggle lever.

14. Remove back plate.

To install:

15. Installation is in the reverse order of removal.

16. Apply brake grease to the specified points during assembly.

17. Install adjuster so that threaded part expands when rotating it in the direction shown by the arrow.

18. Shorten adjuster by rotating it in the opposite direction as shown by the arrow.

➡**After replacing brake shoes or disc rotors, or if parking brake does not function well, perform break-in operation as follows.**

19. Adjust parking brake pedal stroke.

20. Perform parking brake burnishing operation by driving the vehicle forward under the following conditions:

❊❊ CAUTION

To prevent lining from getting too hot, allow a cool off period of approximately 5 minutes after every break-in operation. Do not perform excessive break-in operations, because it may cause uneven or early wear of lining.

21. After burnishing operation, check parking brake pedal stroke. Readjust if it is now longer than the specified stroke.

22. Be sure to perform the reconnect/relearn procedures.

ADJUSTMENT

1. Before servicing the vehicle, refer to the Precautions Section.

2. Remove lower instrument panel LH.

3. Partially engage parking brake pedal to access adjusting nut.

4. Insert a deep socket wrench to rotate adjusting nut and loosen cable until tension is sufficiently released. Then, disengage the parking brake pedal.

5. Remove the wheel and tire.

6. Remove the rotor and measure inner diameter at widest point.

7. Transfer measurement less 0.6 mm to the parking brake shoes and adjust accordingly.

8. Using wheel nuts, secure the disc to the hub to prevent it from tilting.

9. Rotate disc rotor to make sure there is no drag.

10. Adjust cable as follows:

a. Operate pedal 10 or more times with a force of 110 lbs. (490 N).

b. Rotate adjusting nut with deep socket to adjust pedal stroke to specification.

c. With parking brake pedal completely disengaged, make sure there is no drag on the parking brake.

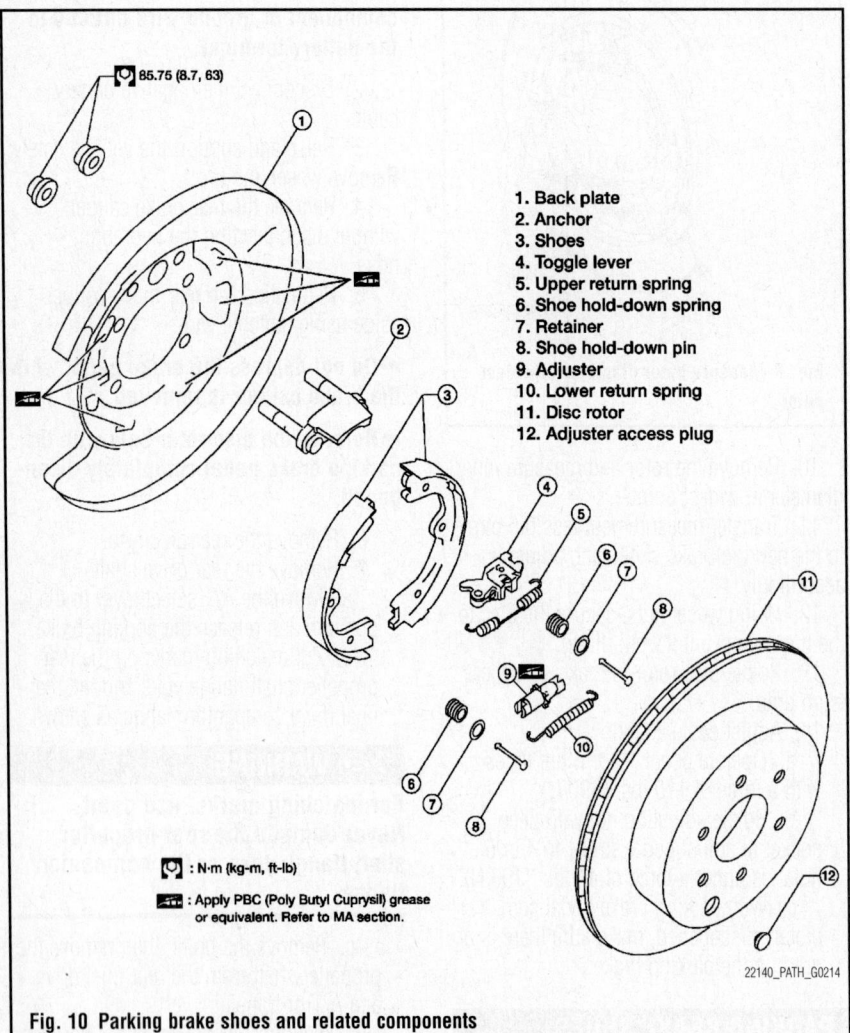

85.75 (8.7, 63)

1. Back plate
2. Anchor
3. Shoes
4. Toggle lever
5. Upper return spring
6. Shoe hold-down spring
7. Retainer
8. Shoe hold-down pin
9. Adjuster
10. Lower return spring
11. Disc rotor
12. Adjuster access plug

: N·m (kg-m, ft-lb)

: Apply PBC (Poly Butyl Cuprysil) grease or equivalent. Refer to MA section.

22140_PATH_G0214

Fig. 10 Parking brake shoes and related components

CHASSIS ELECTRICAL

AIR BAG (SUPPLEMENTAL RESTRAINT SYSTEM)

GENERAL INFORMATION

❋❋ CAUTION

These vehicles are equipped with an air bag system. The system must be disarmed before performing service on, or around, system components, the steering column, instrument panel components, wiring and sensors. Failure to follow the safety precautions and the disarming procedure could result in accidental air bag deployment, possible injury and unnecessary system repairs.

SERVICE PRECAUTIONS

❋❋ CAUTION

Disconnect and isolate the battery negative cable before beginning any airbag system component diagnosis, testing, removal, or installation procedures. Wait at least 90 seconds after the ignition switch is turned off and the negative (-) terminal cable is disconnected from the battery before starting the operation. The SRS is equipped with a backup power source, so if work is started within 90 seconds after disconnecting the negative (-) terminal cable from the battery, the SRS may be deployed. Failure to disable the airbag system may result in accidental airbag deployment, personal injury, or death.

DISARMING THE SYSTEM

❋❋ CAUTION

These vehicles are equipped with an air bag system. The system must be disarmed before performing service on, or around, system components, the steering column, instrument panel components, wiring and sensors. Failure to follow the safety precautions and the disarming procedure could result in accidental air bag deployment, possible injury and unnecessary system repairs.

❋❋ CAUTION

Before servicing the SRS system, turn ignition switch OFF, disconnect both battery cables and wait at least 3 minutes.

ARMING THE SYSTEM

➡Once repair work has been completed, return the ignition switch to the LOCK position, before connecting the battery cables. At this time the steering lock mechanism will engage. Install the CONSULT-III diagnostic tool, or equivalent, and follow the directions on the screen of the tool and perform the self diagnosis check.

DRIVE TRAIN

FRONT HALFSHAFTS

REMOVAL & INSTALLATION

See Figures 11 and 12.

1. Before servicing the vehicle, refer to the Precautions Section.

➡If working near and/or around the SRS system and components, be sure to disable the SRS system. After disabling the system wait three minutes or more before servicing the vehicle.

➡On this vehicle the battery current sensor that is installed to the negative battery cable measures the charging/discharging current of the battery and performs various engine controls. If an electrical component is connected directly to the negative battery cable, the current flowing through that component will not be measured by the battery current sensor. This condition may cause a malfunction of the engine control system and battery discharge may occur. Do not connect an electrical component or ground wire directly to the battery terminal.

2. Disconnect the negative battery cable.
3. Raise and support the vehicle safely.
4. Remove wheel and tire.

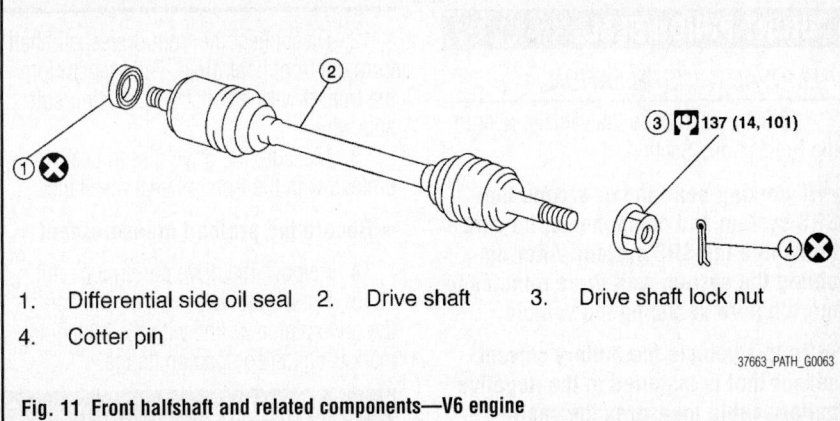

1. Differential side oil seal 2. Drive shaft 3. Drive shaft lock nut
4. Cotter pin

37663_PATH_G0063

Fig. 11 Front halfshaft and related components—V6 engine

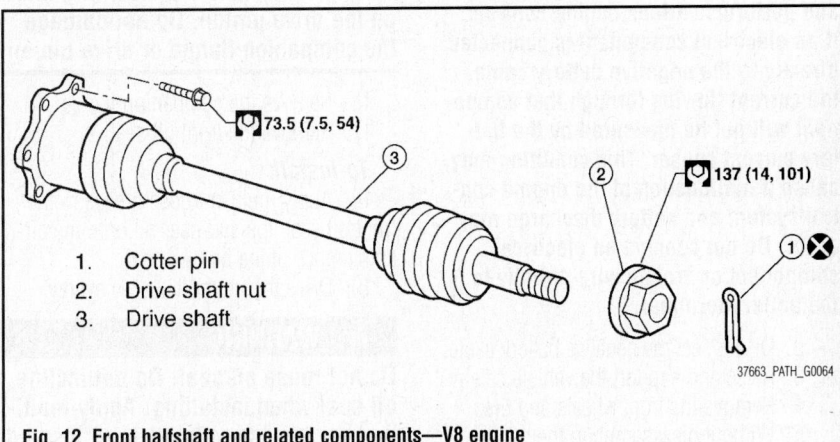

1. Cotter pin
2. Drive shaft nut
3. Drive shaft

37663_PATH_G0064

Fig. 12 Front halfshaft and related components—V8 engine

5. Remove rear engine under cover.
6. Remove wheel sensor harness from mount on knuckle, then disconnect wheel sensor harness connector.

✳✳ WARNING

Do not pull on wheel sensor harness.

7. Remove wheel hub and bearing assembly.

➡**It is not necessary to remove wheel sensor from wheel hub when wheel hub is not being replaced.**

8. Carefully feed wheel sensor harness through hole in splash shield.
9. Separate upper link ball joint stud from steering knuckle.
10. Support lower link with jack.
11. Remove halfshaft assembly.
   a. Pry halfshaft front final drive using suitable tool.

To install:

➡**Be sure to use new fasteners, as required.**

12. Installation is the reverse of the removal procedure.
13. Be sure to perform the reconnect/relearn procedures.

FRONT PINION SEAL

REMOVAL & INSTALLATION

1. Before servicing the vehicle, refer to the Precautions Section.

➡**If working near and/or around the SRS system and components, be sure to disable the SRS system. After disabling the system wait three minutes or more before servicing the vehicle.**

➡**On this vehicle the battery current sensor that is installed to the negative battery cable measures the charging/discharging current of the battery and performs various engine controls. If an electrical component is connected directly to the negative battery cable, the current flowing through that component will not be measured by the battery current sensor. This condition may cause a malfunction of the engine control system and battery discharge may occur. Do not connect an electrical component or ground wire directly to the battery terminal.**

2. Disconnect the negative battery cable.
3. Raise and support the vehicle safely.
4. Remove the front wheels and tires.
5. Without disassembling the hydraulic lines, remove the caliper torque member bolts.
6. Reposition the brake caliper aside using suitable wire.

✳✳ WARNING

Do not press the brake pedal while brake caliper is removed.

7. Remove the ABS sensor harness from the mount on the knuckle.

✳✳ CAUTION

Do not pull on the ABS sensor harness.

8. Support the lower link using a suitable jack.
9. Separate the upper link ball joint stud from the steering knuckle.

✳✳ WARNING

Support the lower link using a jack.

10. Remove the engine undercover.
11. Remove the RH and LH drive shafts from the front final drive.

✳✳ WARNING

Do not reuse the front final drive side oil seals.

12. Disconnect the front driveshaft shaft from the front final drive. Then reposition the front driveshaft shaft aside using suitable wire.
13. Measure the drive pinion bearing preload with the front oil seal resistance.

➡**Record the preload measurement.**

14. Remove the drive pinion lock nut.
15. Put a matching mark on the end of the drive pinion in line with the matching mark B on the companion flange.

✳✳ WARNING

Use paint to make the matching mark on the drive pinion. Do not damage the companion flange or drive pinion.

16. Remove the companion flange.
17. Remove the front oil seal.

To install:
18. Apply multi-purpose grease to the front oil seal lips and gear oil onto the circumference of the new oil seal.
19. Drive the front oil seal in evenly.

✳✳ WARNING

Do not reuse oil seal. Do not incline oil seal when installing. Apply multi-purpose grease onto oil seal lips and gear oil onto the circumference of oil seal.

20. Align the matching mark of the drive pinion with the matching mark B of the companion flange, then install the companion flange.
21. Apply gear oil on the threads of the drive pinion and the seating surface of the new drive pinion lock nut.
22. Install the new drive pinion lock nut. Tighten to the specified torque.

✳✳ WARNING

Do not reuse drive pinion lock nut.

23. Measure the drive pinion bearing preload with the front oil seal resistance.

➡**Drive pinion bearing preload should equal the measurement taken during removal plus an additional 5 inch lbs. (0.56 Nm). If the drive pinion bearing preload is low, tighten the drive pinion lock nut in 5ft. lbs (6.8 Nm) increments until the drive pinion preload is met.**

✳✳ WARNING

Never loosen the drive pinion nut to decrease drive pinion bearing preload. Do not exceed specified preload. If preload torque is exceeded a new collapsible spacer must be installed. If maximum torque is reached prior to reaching the required preload, the collapsible spacer may have been damaged. Replace the collapsible spacer.

24. Install new side oil seals into the front final drive assembly.
25. Install the RH and LH drive shafts to the front final drive.

✳✳ WARNING

When installing the drive shaft assembly into the front final drive assembly, do not damage the side oil seal.

26. Install the remaining components in the reverse order of removal.

✳✳ WARNING

Check the final drive gear oil level after installation.

27. Tighten the upper link ball joint stud nut to specifications.
28. Tighten the wheel nuts to specification.
29. Be sure to perform the reconnect/relearn procedures.

REAR HALFSHAFT

REMOVAL & INSTALLATION

See Figure 13.

1. Before servicing the vehicle, refer to the Precautions Section.

➡**If working near and/or around the SRS system and components, be sure to disable the SRS system. After disabling the system wait three minutes or more before servicing the vehicle.**

➡**On this vehicle the battery current sensor that is installed to the negative battery cable measures the charging/discharging current of the battery and performs various engine controls. If an electrical component is connected directly to the negative battery cable, the current flowing through that component will not be measured by the battery current sensor. This condition may cause a malfunction of the engine control system and battery discharge may occur. Do not connect an electrical component or ground wire directly to the battery terminal.**

2. Disconnect the negative battery cable.
3. Raise and support the vehicle safely. Remove the wheel and tire assembly.
4. Remove the cotter pin and discard, then remove the lock nut from the drive shaft.

✳✳ WARNING

Do not reuse the cotter pin, discard after removal and use a new cotter pin for installation.

5. Remove the six rear drive shaft bolts from the rear final drive assembly flange.

✳✳ WARNING

Do not reuse the rear drive shaft bolts, discard after removal and use new bolts for installation.

6. Separate the rear drive shaft from the rear wheel hub and bearing assembly by lightly tapping the end of the rear drive shaft with a suitable hammer and wood block. If it is difficult to separate, use a suitable puller.
7. Remove the rear drive shaft.

✳✳ WARNING

When removing the rear drive shaft, do not bend at an excessive angle to the rear drive shaft joint. Do not excessively extend the slide joint.

To install:

➡**Be sure to use new fasteners, as required.**

8. Installation is in the reverse order of removal.

✳✳ WARNING

Do not reuse the drive shaft inside flange bolts and washers, discard after removal and use new bolts and washers for installation.

✳✳ WARNING

Do not reuse the cotter pin, discard after removal and use a new cotter pin for installation.

REAR PINION SEAL

REMOVAL & INSTALLATION

1. Before servicing the vehicle, refer to the Precautions Section.

➡**If working near and/or around the SRS system and components, be sure to disable the SRS system. After disabling the system wait three minutes or more before servicing the vehicle.**

➡**On this vehicle the battery current sensor that is installed to the negative battery cable measures the charging/discharging current of the battery and performs various engine controls. If an electrical component is connected directly to the negative battery cable, the current flowing through that component will not be measured by the battery current sensor. This condition may cause a malfunction of the engine control system and battery discharge may occur. Do not connect an electrical component or ground wire directly to the battery terminal.**

2. Disconnect the negative battery cable.
3. Remove the rear driveshaft.
4. Put a matching mark on the end of the drive pinion in line with the matching mark B on the companion flange.

✳✳ WARNING

Use paint to make the matching mark on the drive pinion. Do not damage the companion flange or drive pinion.

➡**The matching mark B on the final drive companion flange indicates the maximum vertical run out position.**

5. Remove the drive pinion lock nut.
6. Remove the companion flange using suitable tool.
7. Remove the front oil seal.

To install:
8. Install the front oil seal.

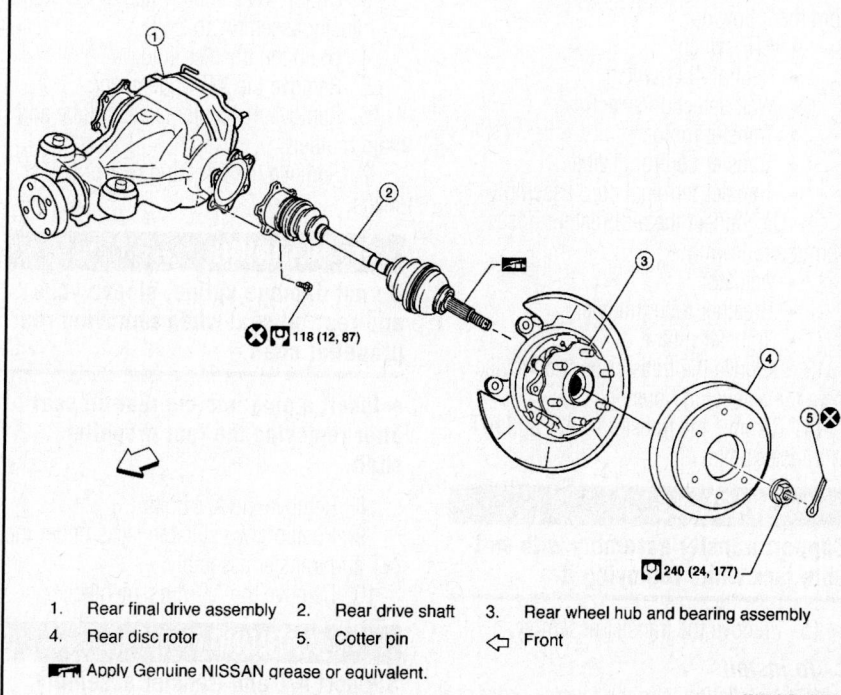

1. Rear final drive assembly   2. Rear drive shaft   3. Rear wheel hub and bearing assembly
4. Rear disc rotor   5. Cotter pin   ⟵ Front

▰ Apply Genuine NISSAN grease or equivalent.

37663_PATH_G0068

Fig. 13 Rear halfshaft and related components

WARNING

Do not reuse oil seal. Do not incline oil seal when installing. Apply multi-purpose grease onto oil seal lips, and gear oil onto the circumference of oil seal.

9. Align the matching mark of the drive pinion with the matching mark B of the companion flange, then install the companion flange.

10. Install the drive pinion lock nut.

WARNING

Do not reuse drive pinion lock nut.

11. Install the rear propeller shaft.

12. Be sure to perform the reconnect/relearn procedures.

TRANSFER CASE ASSEMBLY

REMOVAL & INSTALLATION

ATX14B

See Figure 14.

1. Before servicing the vehicle, refer to the Precautions Section.

➡️If working near and/or around the SRS system and components, be sure to disable the SRS system. After disabling the system wait three minutes or more before servicing the vehicle.

➡️On this vehicle the battery current sensor that is installed to the negative battery cable measures the charging/discharging current of the battery and performs various engine controls. If an electrical component is connected directly to the negative battery cable, the current flowing through that component will not be measured by the battery current sensor. This condition may cause a malfunction of the engine control system and battery discharge may occur. Do not connect an electrical component or ground wire directly to the battery terminal.

2. Disconnect the negative battery cable.

3. Set transfer state as 2WD when 4WD shift switch is at 2WD.

4. Remove the drain plug and gasket. Drain the fluid.

5. Remove the A/T undercover.

6. Remove the center exhaust tube and main muffler.

7. Remove the front and rear propeller shafts.

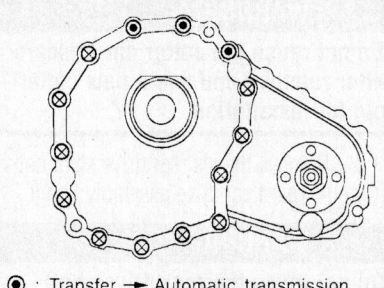

⊙ : Transfer → Automatic transmission
⊗ : Automatic transmission → Transfer

22140_PATH_G0011

Fig. 14 Transfer case bolt locations

WARNING

Do not damage spline, sleeve yoke and rear oil seal when removing rear propeller shaft.

➡️Insert a plug into the rear oil seal after removing the rear propeller shaft.

8. Remove the A/T nuts from the A/T crossmember.

9. Position two suitable jacks under the A/T and transfer assembly.

10. Remove the crossmember.

WARNING

Support A/T and transfer assembly using two suitable jacks while removing crossmember.

11. Disconnect the electrical connectors from the following:
- ATP switch
- Neutral 4LO switch
- Wait detection switch
- Transfer motor
- Transfer control device
- Transfer terminal cord assembly

12. Disconnect the air breather hoses from the following:
- Actuator
- Breather tube (transfer)
- Transfer motor

13. Remove the transfer control device from the extension housing.

14. Remove the transfer to A/T and A/T to transfer bolts.

WARNING

Support transfer assembly with suitable jack while removing it.

15. Remove the transfer assembly.

To install:

16. Installation is in the reverse order of removal.

17. Tighten the bolts to 27 ft. lbs. (36 Nm).

18. Fill the transfer with new fluid.

19. Check the transfer fluid.

20. Start the engine for one minute. Then stop the engine and recheck the transfer fluid.

21. Be sure to perform the reconnect/relearn procedures.

TX15B

See Figure 14.

1. Before servicing the vehicle, refer to the Precautions Section.

➡️If working near and/or around the SRS system and components, be sure to disable the SRS system. After disabling the system wait three minutes or more before servicing the vehicle.

➡️On this vehicle the battery current sensor that is installed to the negative battery cable measures the charging/discharging current of the battery and performs various engine controls. If an electrical component is connected directly to the negative battery cable, the current flowing through that component will not be measured by the battery current sensor. This condition may cause a malfunction of the engine control system and battery discharge may occur. Do not connect an electrical component or ground wire directly to the battery terminal.

2. Disconnect the negative battery cable.

3. Switch 4WD shift switch to 2WD and set transfer assembly to 2WD.

4. Drain the transfer fluid.

5. Remove the A/T undercover.

6. Remove the center exhaust tube and main muffler.

7. Remove the front and rear propeller shafts.

WARNING

Do not damage spline, sleeve yoke and rear oil seal when removing rear propeller shaft.

➡️Insert a plug into the rear oil seal after removing the rear propeller shaft.

8. Remove the A/T bolts.

9. Position two suitable jacks under the A/T and transfer assembly.

10. Remove the A/T crossmember.

WARNING

Support A/T and transfer assembly using two suitable jacks while removing A/T crossmember.

11. Disconnect the electrical connectors from the following:
- ATP switch
- 4LO switch
- Wait detection switch
- Transfer control device

12. Disconnect each air breather hose from the following:
- Transfer control device
- Breather tube (transfer)

13. Remove the transfer to A/T and A/T to transfer bolts.

> ⁂ **WARNING**
>
> **Support transfer assembly with suitable jack while removing it.**

14. Remove the transfer assembly.

> ⁂ **WARNING**
>
> **Do not damage rear oil seal (A/T).**

To install:

15. Installation is in the reverse order of removal.

16. Tighten the bolts to 27 ft. lbs. (36 Nm).

17. Fill the transfer with new fluid.

18. Check the transfer fluid.

19. Start the engine for one minute. Then stop the engine and recheck the transfer fluid.

20. After the installation, check the 4WD shift indicator pattern. If NG, adjust the position between the transfer assembly and transfer control unit.

21. Be sure to perform the reconnect/relearn procedures.

ENGINE COOLING

ENGINE FAN

REMOVAL & INSTALLATION

V6 Engine

Belt Driven

See Figure 15.

1. Before servicing the vehicle, refer to the Precautions Section.

➡ **If working near and/or around the SRS system and components, be sure to disable the SRS system. After disabling the system wait three minutes or more before servicing the vehicle.**

➡ **On this vehicle the battery current sensor that is installed to the negative battery cable measures the charging/discharging current of the battery and performs various engine controls. If an electrical component is connected directly to the negative battery cable, the current flowing through that component will not be measured by the battery current sensor. This condition may cause a malfunction of the engine control system and battery discharge may occur. Do not connect an electrical component or ground wire directly to the battery terminal.**

2. Disconnect the negative battery cable.

3. Remove the air dam.

4. Remove the engine under cover.

5. Drain the cooling system. Be sure to properly dispose of used coolant.

6. Remove air duct.
   a. Disconnect harness connector from Mass Air Flow (MAF) sensor.
   b. Disconnect PCV hose.
   c. Remove air cleaner case/mass air flow sensor assembly and air duct assembly disconnecting their joints.

7. Remove the reservoir tank hose from the radiator. Remove the upper radiator hose.

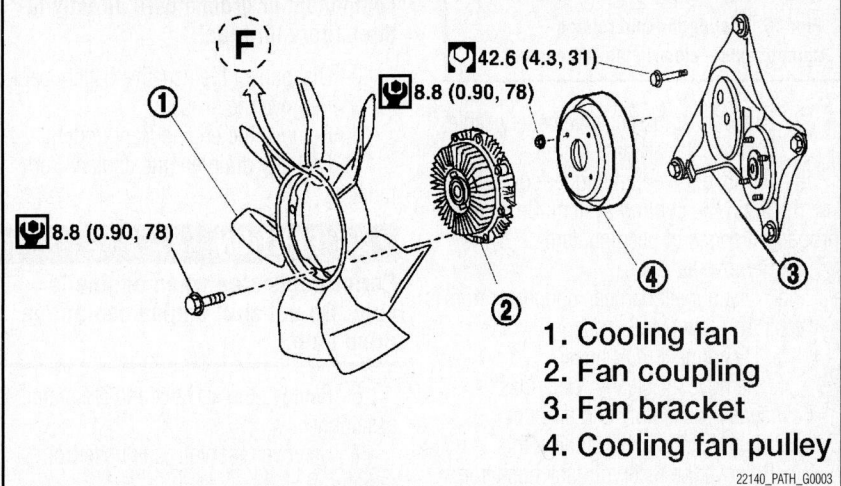

1. Cooling fan
2. Fan coupling
3. Fan bracket
4. Cooling fan pulley

22140_PATH_G0003

Fig. 15 Cooling fan and related components—belt driven fan

8. Remove the upper and lower radiator shrouds.

9. Remove drive belts.
   a. Rotate the drive belt auto tensioner in the direction of arrow (loosening direction of tensioner).

> ⁂ **CAUTION**
>
> **Avoid placing hand in a location where pinching may occur if the tool accidentally comes off.**

   b. Remove the drive belt.
10. Remove cooling fan.

To install:

➡ **Be sure to use new fasteners, as required.**

11. Installation is the reverse of the removal procedure.

12. Start and warm up the engine. Visually make sure that there are no leaks of the engine coolant.

13. Be sure to perform the reconnect/relearn procedures.

Electric

See Figure 16.

1. Before servicing the vehicle, refer to the Precautions Section.

➡ **If working near and/or around the SRS system and components, be sure to disable the SRS system. After disabling the system wait three minutes or more before servicing the vehicle.**

➡ **On this vehicle the battery current sensor that is installed to the negative battery cable measures the charging/discharging current of the battery and performs various engine controls. If an electrical component is connected directly to the negative battery cable, the current flowing through that component will not be measured by the battery current sensor. This condition may cause a malfunction of the engine control system and battery discharge may occur. Do not connect an electrical component or ground wire directly to the battery terminal.**

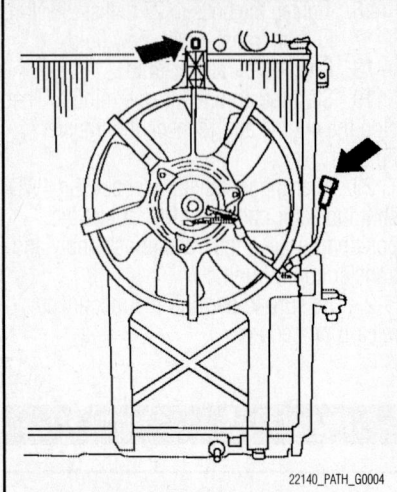

Fig. 16 Cooling fan and related components—electric fan (V6 engine)

2. Disconnect the negative battery cable.
3. Remove the air dam.
4. Remove the engine under cover.
5. Drain the cooling system. Be sure to properly dispose of used coolant.
6. Remove air duct.
   a. Disconnect harness connector from Mass Air Flow (MAF) sensor.
   b. Disconnect PCV hose.
   c. Remove air cleaner case/mass air flow sensor assembly and air duct assembly disconnecting their joints.
7. Remove the reservoir tank hose from the radiator. Remove the upper radiator hose.
8. Remove the upper and lower radiator shrouds.
9. Disconnect harness connector from fan motor.
10. Remove the bolt and remove the fan grille and motor assembly.

To install:

➡ **Be sure to use new fasteners, as required.**

11. Installation is the reverse of the removal procedure.
12. Start and warm up the engine. Visually make sure that there are no leaks of the engine coolant.
13. Be sure to perform the reconnect/relearn procedures.

V8 Engine

Belt Driven

See Figure 15.

1. Before servicing the vehicle, refer to the Precautions Section.

➡ **If working near and/or around the SRS system and components, be sure**
to disable the SRS system. After disabling the system wait three minutes or more before servicing the vehicle.

➡ On this vehicle the battery current sensor that is installed to the negative battery cable measures the charging/discharging current of the battery and performs various engine controls. If an electrical component is connected directly to the negative battery cable, the current flowing through that component will not be measured by the battery current sensor. This condition may cause a malfunction of the engine control system and battery discharge may occur. Do not connect an electrical component or ground wire directly to the battery terminal.

2. Disconnect the negative battery cable.
3. Remove the air dam.
4. Remove the engine front undercover.
5. Partially drain engine coolant from radiator.

❊❊ CAUTION

Perform this step when engine is cold. Do not spill engine coolant on drive belts.

6. Remove the air duct and resonator assembly.
7. Remove reservoir tank hose from radiator.
8. Remove reservoir tank hose from engine.
9. Removal radiator hose (upper) from radiator.

❊❊ WARNING

Do not spill engine coolant on drive belts.

10. Remove the radiator shroud (lower) and position aside.
    a. Release the tabs, pull radiator shroud (lower) rearwards and down to remove.
11. Remove the radiator shroud (upper) bolts and remove the radiator shroud (upper).
12. Remove the drive belt.
13. Remove the engine cooling fan.

To install:

14. Install cooling fan with its front mark "F" facing front of engine.
15. Check for leaks of the engine coolant.
16. Start and warm up the engine. Visually make sure that there are no leaks of the engine coolant.
17. Be sure to perform the reconnect/relearn procedures.

Electirc

See Figure 17.

1. Before servicing the vehicle, refer to the Precautions Section.

➡ **If working near and/or around the SRS system and components, be sure to disable the SRS system. After disabling the system wait three minutes or more before servicing the vehicle.**

➡ On this vehicle the battery current sensor that is installed to the negative battery cable measures the charging/discharging current of the battery and performs various engine controls. If an electrical component is connected directly to the negative battery cable, the current flowing through that component will not be measured by the battery current sensor. This condition may cause a malfunction of the engine control system and battery discharge may occur. Do not connect an electrical component or ground wire directly to the battery terminal.

2. Disconnect the negative battery cable.
3. Remove the air dam.
4. Remove the engine undercover.
5. Loosen the lower fan motor nuts.
6. Disconnect harness connector from fan motor.
7. Remove the upper fan motor bolts.
8. Remove the fan grille and motor assembly.

To install:

➡ **Be sure to use new fasteners, as required.**

9. Installation is the reverse of the removal procedure.
10. Start and warm up the engine. Visually make sure that there are no leaks of the engine coolant.

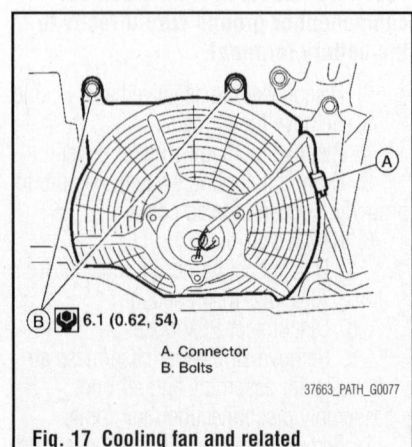

A. Connector
B. Bolts

Fig. 17 Cooling fan and related components—electric fan (V8 engine)

11. Be sure to perform the reconnect/relearn procedures.

RADIATOR

REMOVAL & INSTALLATION

1. Before servicing the vehicle, refer to the Precautions Section.

➡If working near and/or around the SRS system and components, be sure to disable the SRS system. After disabling the system wait three minutes or more before servicing the vehicle.

➡On this vehicle the battery current sensor that is installed to the negative battery cable measures the charging/discharging current of the battery and performs various engine controls. If an electrical component is connected directly to the negative battery cable, the current flowing through that component will not be measured by the battery current sensor. This condition may cause a malfunction of the engine control system and battery discharge may occur. Do not connect an electrical component or ground wire directly to the battery terminal.

2. Disconnect the negative battery cable.

✸✸ CAUTION

Do not remove radiator cap when engine is hot. Serious burns could occur from high-pressure engine coolant escaping from radiator. Wrap a thick cloth around the cap. Slowly turn it a quarter of a turn to release built-up pressure. Carefully remove radiator cap by turning it all the way.

3. Remove engine cover.
   a. Remove bolts.
   b. Lift up on engine cover firmly to dislodge snap fit mounts.
4. Drain engine coolant from radiator.

✸✸ CAUTION

Perform this step when engine is cold. Do not spill engine coolant on drive belts.

5. Remove air duct and air cleaner case assembly.
   a. Disconnect harness connector from mass air flow sensor.
   b. Disconnect PCV hose.
   c. Remove air cleaner case/mass

air flow sensor assembly and air duct assembly disconnecting their joints.
6. Remove reservoir tank hose.
7. Removal radiator hoses (upper and lower) and reservoir tank hose.

✸✸ WARNING

Be careful not to allow engine coolant to contact drive belts.

8. Remove radiator cooling fan assembly.
9. Disconnect A/T fluid cooler hoses.
   a. Install blind plug to avoid leakage of A/T fluid.
10. Remove the front grille.
11. Remove the upper mount bracket bolts.
12. Remove the two A/C condenser bolts.
13. Remove radiator as follows:

✸✸ WARNING

Do not damage or scratch A/C condenser and radiator core when removing.

   a. With lifting and pulling radiator in a rear direction, disassemble lower mount from radiator core support center.

✸✸ WARNING

Because A/C condenser is onto the front-lower portion of radiator, moving to rear direction should be at minimum.

   b. Lift A/C condenser up and remove radiator after disengaging the fitting as front-bottom surface.

✸✸ WARNING

Lifting A/C condenser should be minimum to prevent a load to A/C piping.

   c. After removing radiator, put A/C condenser on radiator core support center to prevent a load to A/C piping, and temporarily fix it with rope or similar means.

To install:

➡Be sure to use new fasteners, as required.

14. Installation is the reverse of the removal procedure.
15. Be sure to fill the cooling system with the proper grade and type engine coolant.
16. Be sure to perform the reconnect/relearn procedures.

THERMOSTAT

REMOVAL & INSTALLATION
See Figures 18 and 19.

1. Before servicing the vehicle, refer to the Precautions Section.

➡If working near and/or around the SRS system and components, be sure to disable the SRS system. After disabling the system wait three minutes or more before servicing the vehicle.

➡On this vehicle the battery current sensor that is installed to the negative battery cable measures the charging/discharging current of the battery and performs various engine controls. If an electrical component is connected directly to the negative battery cable, the current flowing through that component will not be measured by the battery current sensor. This condition may cause a malfunction of the engine control system and battery discharge may occur. Do not connect an electrical component or ground wire directly to the battery terminal.

2. Disconnect the negative battery cable.
3. Completely drain engine coolant.

✸✸ CAUTION

Perform this step when engine is cold. Do not spill engine coolant on drive belts.

4. Remove air duct and air cleaner case.
5. Disconnect radiator hose (lower) and oil cooler hose from water inlet and thermostat assembly.
6. Remove water inlet and thermostat assembly.

✸✸ WARNING

Do not disassemble water inlet and thermostat assembly. Replace them as a unit, if necessary.

To install:

➡Be sure to use new fasteners, as required.

7. Installation is the reverse of the removal procedure.

✸✸ WARNING

Be careful not to spill engine coolant over engine room. Use rag to absorb engine coolant.

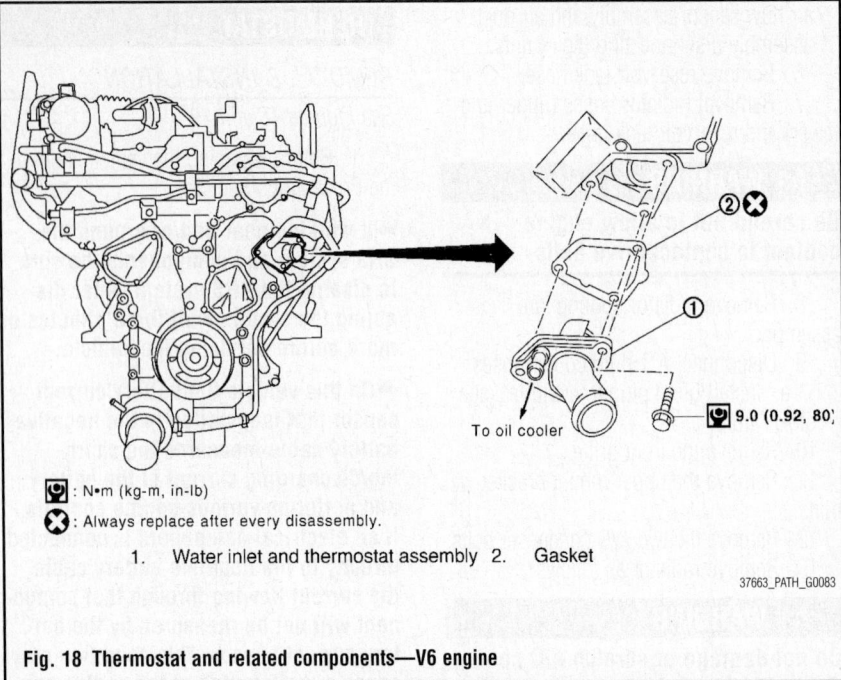

: N•m (kg-m, in-lb)

: Always replace after every disassembly.

1. Water inlet and thermostat assembly 2. Gasket

37663_PATH_G0083

Fig. 18 Thermostat and related components—V6 engine

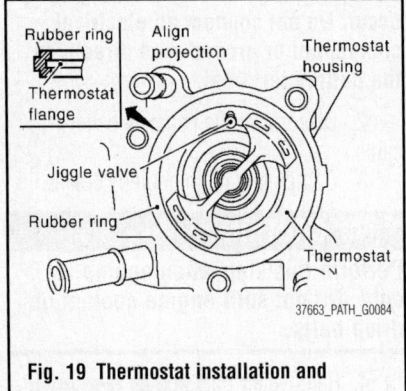

37663_PATH_G0084

Fig. 19 Thermostat installation and alignment—V8 engine

8. Start and warm up engine. Visually check there are no leaks of engine coolant.

9. Be sure to fill the cooling system with the proper grade and type engine coolant.

10. Be sure to perform the reconnect/relearn procedures.

WATER PUMP

REMOVAL & INSTALLATION

V6 Engine

See Figures 20 and 21.

✳✳ WARNING

When removing water pump assembly, be careful not to get engine coolant on timing chain and drive belts. Water pump cannot be disassembled and should be replaced as a unit. After installing water pump,

connect hose and clamp securely, then check for leaks.

1. Before servicing the vehicle, refer to the Precautions Section.

➡ If working near and/or around the SRS system and components, be sure to disable the SRS system. After disabling the system wait three minutes or more before servicing the vehicle.

➡ On this vehicle the battery current sensor that is installed to the negative battery cable measures the charging/discharging current of the battery and performs various engine controls. If an electrical component is connected directly to the negative battery cable, the current flowing through that component will not be measured by the battery current sensor. This condition may cause a malfunction of the engine control system and battery discharge may occur. Do not connect an electrical component or ground wire directly to the battery terminal.

2. Disconnect the negative battery cable.
3. Remove air dam.
4. Remove undercover.
5. Remove air duct and resonator.
6. Remove drive belts.
7. Drain engine coolant.

✳✳ CAUTION

Perform this step when engine is cold. Do not spill engine coolant on drive belts.

8. Remove radiator hoses (upper and lower) and cooling fan assembly.

9. Remove chain tensioner cover and water pump cover from front timing chain case.

10. Remove timing chain tensioner (primary) as follows:

   a. Loosen clip of timing chain tensioner (primary), and release plunger stopper (1).

   b. Insert plunger into tensioner body by pressing slack guide (2).

   c. Keep slack guide pressed and hold plunger in by pushing stopper pin through the tensioner body hole and plunger groove (3).

   d. Turn crankshaft pulley clockwise so that timing chain on the timing chain tensioner (primary) side is loose.

   e. Remove bolts and remove timing chain tensioner (primary).

✳✳ WARNING

Be careful not to drop bolts inside timing chain case.

11. Remove water pump as follows:

   a. Remove three water pump bolts. Secure a gap between water pump gear and timing chain, by turning crankshaft pulley counterclockwise until timing chain looseness on water pump sprocket becomes maximum.

   b. Screw M8 bolts [pitch: 1.25 mm (0.049 in.) length: approx. 50 mm (1.97 in.)] into water pumps upper and lower bolt holes until they reach timing chain case. Then, alternately tighten each bolt for a half turn, and pull out water pump.

✳✳ WARNING

Pull straight out while preventing vane from contacting socket in installation area. Remove water pump without causing sprocket to contact timing chain.

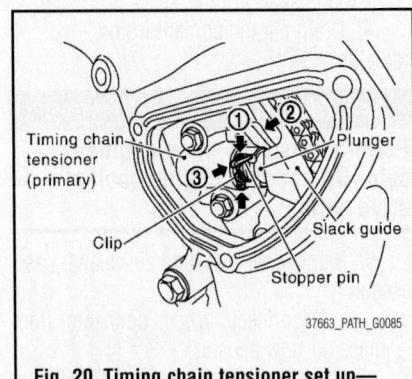

37663_PATH_G0085

Fig. 20 Timing chain tensioner set up—V6 engine

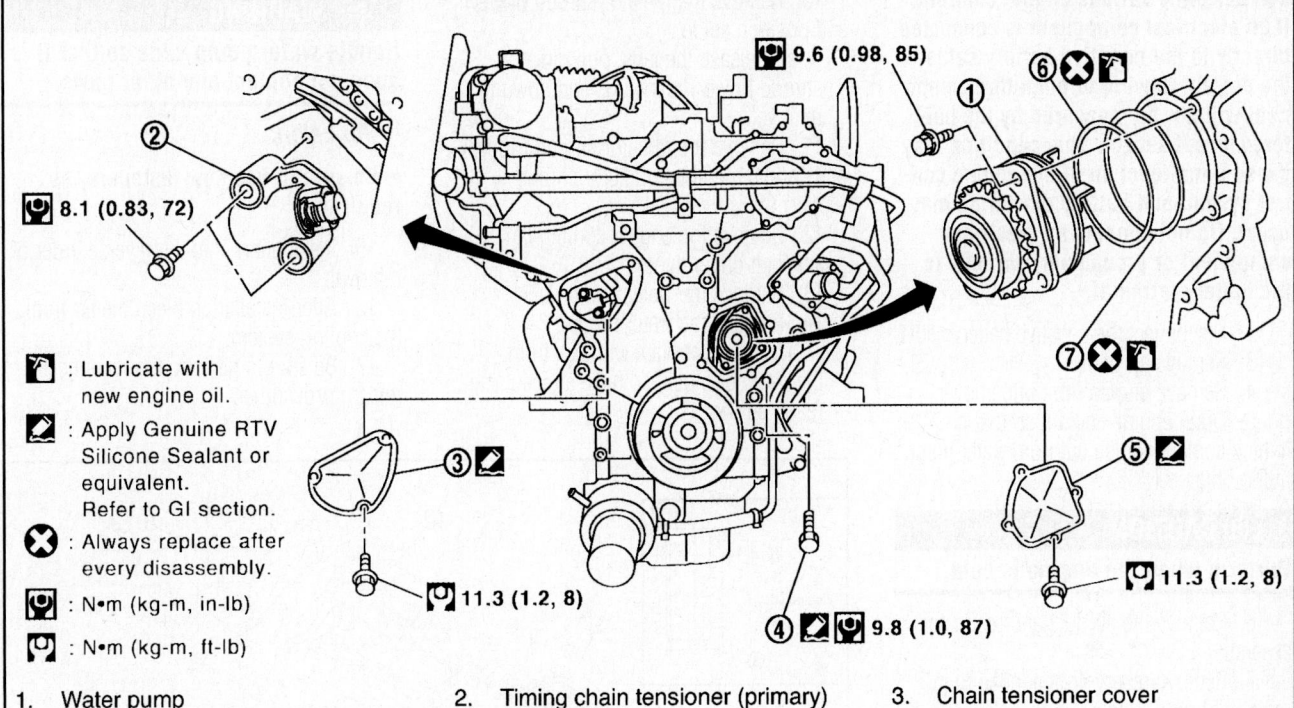

9.6 (0.98, 85)

2

8.1 (0.83, 72)

1

6

7

5

3

11.3 (1.2, 8)

4   9.8 (1.0, 87)

11.3 (1.2, 8)

: Lubricate with new engine oil.

: Apply Genuine RTV Silicone Sealant or equivalent. Refer to GI section.

: Always replace after every disassembly.

: N•m (kg-m, in-lb)

: N•m (kg-m, ft-lb)

1. Water pump
2. Timing chain tensioner (primary)
3. Chain tensioner cover
4. Water drain plug (front)
5. Water pump cover
6. O-ring
7. O-ring

22140_PATH_G0012

Fig. 21 Water pump and related components—V6 engine

c. Remove M8 bolts and O-rings from water pump.

❊❊ WARNING

Do not disassemble water pump.

➡**Do not reuse O-rings.**

To install:

➡**Be sure to use new fasteners, as required.**

12. Install new O-rings to water pump.

➡**Apply engine oil to O-rings. Locate O-ring with white paint mark to engine front side.**

13. Install water pump.

❊❊ WARNING

Do not allow timing chain case to pinch O-rings when installing water pump.

a. Make sure that timing chain and water pump sprocket are engaged.
b. Insert water pump by tightening bolts alternately and evenly.
14. Install timing chain tensioner (primary) as follows:
a. Remove dust and foreign material completely from backside of timing chain

tensioner (primary) and from installation area of rear timing chain case.
b. Turn crankshaft pulley clockwise so that timing chain on the timing chain tensioner (primary) side is loose.
c. Install timing chain tensioner (primary) with its stopper pin attached.

❊❊ WARNING

Be careful not to drop bolts inside timing chain case.

d. Remove stopper pin.
e. Make sure again that timing chain and water pump sprocket are engaged.
15. Install chain tensioner cover and water pump cover as follows:
a. Before installing, remove all traces of old liquid gasket from mating surface of water pump cover and chain tensioner cover using scraper. Also remove traces of old liquid gasket from the mating surface of front timing chain case.
b. Apply a continuous bead of liquid gasket, to mating surface of chain tensioner and water pump cover.

❊❊ WARNING

Attaching should be done within 5 minutes after coating.

c. Tighten bolts to specified torque.
16. Refill engine coolant system.
17. Installation of the remaining components is in the reverse order of removal after this step.
a. After starting engine, let idle for three minutes, then rev engine up to 3,000 rpm under no load to purge air from the high-pressure chamber of chain tensioner. Engine may produce a rattling noise. This indicates that air still remains in the chamber and is not a matter of concern.
18. Be sure to perform the reconnect/relearn procedures.

V8 Engine

See Figure 22.

1. Before servicing the vehicle, refer to the Precautions Section.

➡**If working near and/or around the SRS system and components, be sure to disable the SRS system. After disabling the system wait three minutes or more before servicing the vehicle.**

➡**On this vehicle the battery current sensor that is installed to the negative battery cable measures the charging/discharging current of the battery** and performs various engine controls. If an electrical component is connected directly to the negative battery cable, the current flowing through that component will not be measured by the battery current sensor. This condition may cause a malfunction of the engine control system and battery discharge may occur. Do not connect an electrical component or ground wire directly to the battery terminal.

2. Disconnect the negative battery cable.
3. Remove air dam.
4. Remove engine front undercover.
5. Drain engine coolant so that no engine coolant comes out from water pump fitting hole.

✳✳ CAUTION

Perform when the engine is cold.

6. Remove the air duct and resonator assembly.
7. Remove reservoir tank hose from radiator shroud (upper).
8. Remove reservoir tank hose from engine.
9. Remove radiator hose (upper) from radiator.

✳✳ WARNING

Be careful not to allow engine coolant to contact drive belts.

10. Remove the radiator shroud (lower) and position aside.
   a. Release the tabs, pull radiator shroud (lower) rearwards and down to remove.
11. Remove the radiator shroud (upper) bolts and remove the radiator shroud (upper) (A).
12. Remove the engine cooling fan (crankshaft driven type).
13. Remove the water pump pulley.
14. Remove the water pump.
   a. Engine coolant will leak from the cylinder block, so have a receptacle ready below.

✳✳ WARNING

Handle water pump vane so that it does not contact any other parts.

To install:

➡ **Be sure to use new fasteners, as required.**

15. Installation is in the reverse order of removal.
16. After installation bleed the air from the cooling system.
17. Be sure to perform the reconnect/relearn procedures.

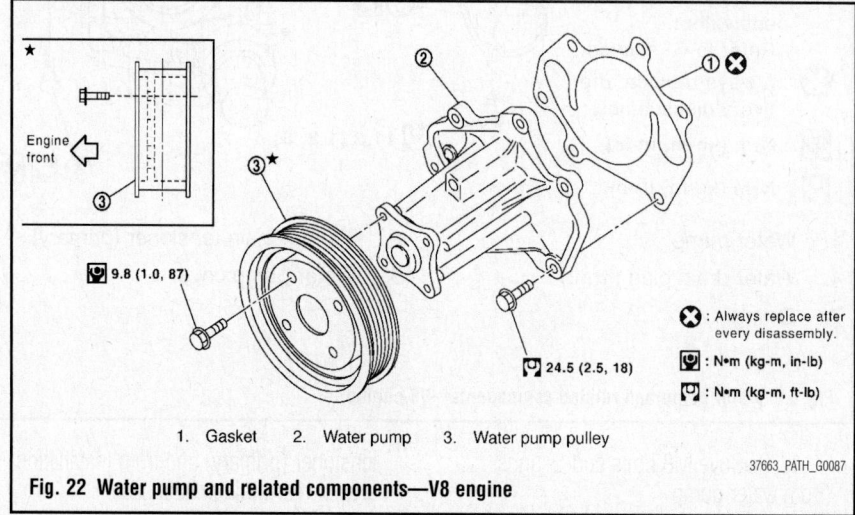

9.8 (1.0, 87)

24.5 (2.5, 18)

⊗ : Always replace after every disassembly.

[☉] : N•m (kg-m, in-lb)

[☉] : N•m (kg-m, ft-lb)

1. Gasket   2. Water pump   3. Water pump pulley

37663_PATH_G0087

Fig. 22 Water pump and related components—V8 engine

ENGINE ELECTRICAL   CHARGING SYSTEM

ALTERNATOR

REMOVAL & INSTALLATION
See Figure 23.

1. Before servicing the vehicle, refer to the Precautions Section.

➡ **If working near and/or around the SRS system and components, be sure to disable the SRS system. After disabling the system wait three minutes or more before servicing the vehicle.**

➡ **On this vehicle the battery current sensor that is installed to the negative battery cable measures the charging/discharging current of the battery and performs various engine controls. If an electrical component is connected directly to the negative battery cable, the current flowing through that component will not be measured by the battery current sensor. This condition may cause a malfunction of the engine con-**

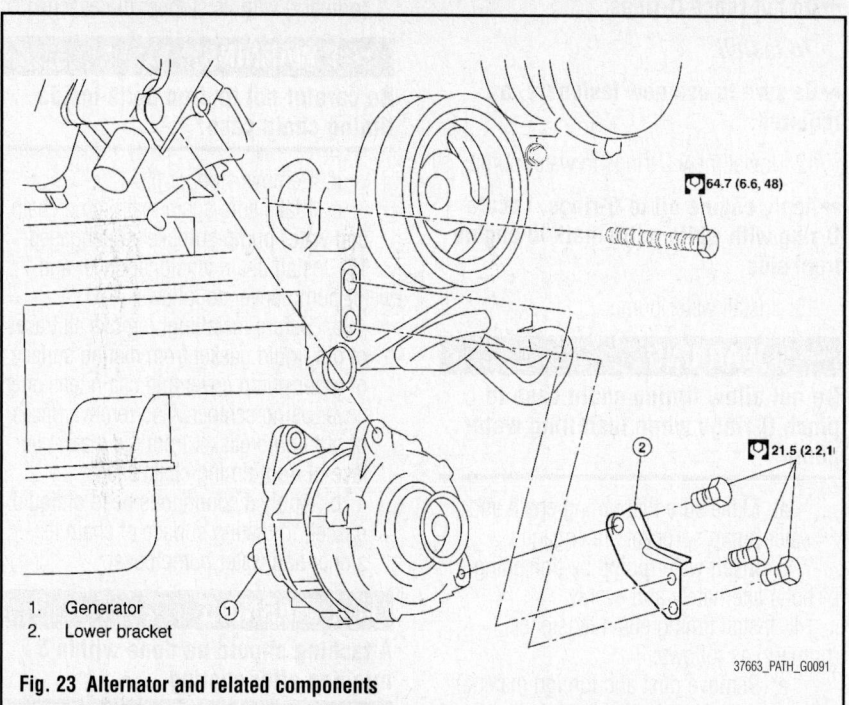

64.7 (6.6, 48)

21.5 (2.2, 1...

1. Generator
2. Lower bracket

37663_PATH_G0091

Fig. 23 Alternator and related components

trol system and battery discharge may occur. Do not connect an electrical component or ground wire directly to the battery terminal.

2. Disconnect the negative battery cable.

3. Disconnect the negative battery terminal.

4. Remove the fan shroud.

5. Remove the drive belt.

6. Remove alternator stay.

7. On V8 engine, remove the lower alternator bolt.

8. Remove the alternator upper bolt.

9. Disconnect the alternator harness connectors.

10. Remove the alternator.

To install:

11. Installation is in the reverse order of removal.

12. Install the alternator and check tension of drive belt.

➡The power generation variable voltage control system that controls the power generation voltage of the alternator has been adopted. Therefore, the power generation variable voltage control system inspection should be performed after replacing the alternator in order to ensure that the system operates normally.

13. Be sure to perform the reconnect/relearn procedures.

VOLTAGE REGULATOR

ADJUSTMENT

Adjustment of the regulator is not possible. If the regulator is defective, it must be replaced.

REMOVAL & INSTALLATION

The voltage regulator is an integral part of the alternator.

ENGINE ELECTRICAL

IGNITION COIL

REMOVAL & INSTALLATION

V6 Engine

Left Bank

See Figure 24.

1. Before servicing the vehicle, refer to the Precautions Section.

➡If working near and/or around the SRS system and components, be sure to disable the SRS system. After disabling the system wait three minutes or more before servicing the vehicle.

➡On this vehicle the battery current sensor that is installed to the negative battery cable measures the charging/discharging current of the battery and

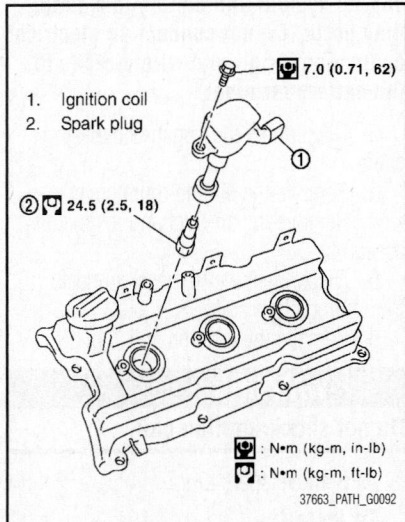

1. Ignition coil
2. Spark plug

⬚ 7.0 (0.71, 62)

② ⬚ 24.5 (2.5, 18)

⬚ : N•m (kg-m, in-lb)
⬚ : N•m (kg-m, ft-lb)

37663_PATH_G0092

Fig. 24 Ignition coil and related components—V6 engine

performs various engine controls. If an electrical component is connected directly to the negative battery cable, the current flowing through that component will not be measured by the battery current sensor. This condition may cause a malfunction of the engine control system and battery discharge may occur. Do not connect an electrical component or ground wire directly to the battery terminal.

2. Disconnect the negative battery cable.

3. Remove engine cover.

4. Remove air cleaner case and air duct. (At the left bank side, remove ignition coil).

5. Move aside harness, harness bracket, and hoses located above ignition coil.

6. Disconnect harness connector from ignition coil.

7. Remove ignition coil.

✳✳ WARNING

Do not shock it.

To install:

➡Be sure to use new fasteners, as required.

8. Installation is the reverse of the removal procedure.

9. Be sure to perform the reconnect/relearn procedures.

Right Bank

See Figure 24.

1. Before servicing the vehicle, refer to the Precautions Section.

➡If working near and/or around the SRS system and components, be sure to disable the SRS system. After dis-

IGNITION SYSTEM

abling the system wait three minutes or more before servicing the vehicle.

➡On this vehicle the battery current sensor that is installed to the negative battery cable measures the charging/discharging current of the battery and performs various engine controls. If an electrical component is connected directly to the negative battery cable, the current flowing through that component will not be measured by the battery current sensor. This condition may cause a malfunction of the engine control system and battery discharge may occur. Do not connect an electrical component or ground wire directly to the battery terminal.

2. Disconnect the negative battery cable.

3. Remove intake manifold collector.

4. Move aside harness, harness bracket, and hoses located above ignition coil.

5. Disconnect harness connector from ignition coil.

6. Remove ignition coil.

✳✳ WARNING

Do not shock it.

To install:

➡Be sure to use new fasteners, as required.

7. Installation is the reverse of the removal procedure.

8. Be sure to perform the reconnect/relearn procedures.

V8 Engine

See Figure 25.

1. Before servicing the vehicle, refer to the Precautions Section.

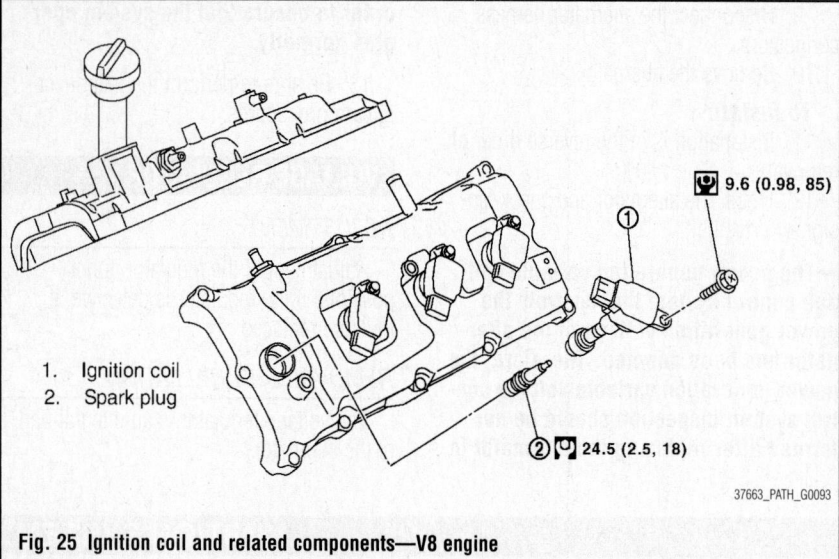

1. Ignition coil
2. Spark plug

9.6 (0.98, 85)

① ② 24.5 (2.5, 18)

37663_PATH_G0093

Fig. 25 Ignition coil and related components—V8 engine

➡If working near and/or around the SRS system and components, be sure to disable the SRS system. After disabling the system wait three minutes or more before servicing the vehicle.

➡On this vehicle the battery current sensor that is installed to the negative battery cable measures the charging/discharging current of the battery and performs various engine controls. If an electrical component is connected directly to the negative battery cable, the current flowing through that component will not be measured by the battery current sensor. This condition may cause a malfunction of the engine control system and battery discharge may occur. Do not connect an electrical component or ground wire directly to the battery terminal.

2. Disconnect the negative battery cable.
3. Remove the engine room cover.
4. Remove the air duct and resonator assembly.
5. Disconnect the harness connector from the ignition coil.
6. Remove the ignition coil.

✳✳ WARNING

Do not shock ignition coil.

To install:

➡Be sure to use new fasteners, as required.

7. Installation is the reverse of the removal procedure.

8. Be sure to perform the reconnect/relearn procedures.

IGNITION TIMING

ADJUSTMENT

Ignition timing is controlled by the ECM. No adjustment is necessary or possible.

SPARK PLUGS

REMOVAL & INSTALLATION

V6 Engine

See Figure 24.

1. Before servicing the vehicle, refer to the Precautions Section.

➡If working near and/or around the SRS system and components, be sure to disable the SRS system. After disabling the system wait three minutes or more before servicing the vehicle.

➡On this vehicle the battery current sensor that is installed to the negative battery cable measures the charging/discharging current of the battery and performs various engine controls. If an electrical component is connected directly to the negative battery cable, the current flowing through that component will not be measured by the battery current sensor. This condition may cause a malfunction of the engine control system and battery discharge may occur. Do not connect an electrical component or ground wire directly to the battery terminal.

2. Disconnect the negative battery cable.

3. Remove engine cover.
4. Remove air cleaner case and air duct.
5. Move aside harness, harness bracket, and hoses located above ignition coil.
6. Disconnect harness connector from ignition coil.
7. Remove ignition coil.
8. Remove the spark plug.

To install:

➡Be sure to use new fasteners, as required.

9. Installation is the reverse of the removal procedure.
10. Be sure to perform the reconnect/relearn procedures.

V8 Engine

See Figure 25.

1. Before servicing the vehicle, refer to the Precautions Section.

➡If working near and/or around the SRS system and components, be sure to disable the SRS system. After disabling the system wait three minutes or more before servicing the vehicle.

➡On this vehicle the battery current sensor that is installed to the negative battery cable measures the charging/discharging current of the battery and performs various engine controls. If an electrical component is connected directly to the negative battery cable, the current flowing through that component will not be measured by the battery current sensor. This condition may cause a malfunction of the engine control system and battery discharge may occur. Do not connect an electrical component or ground wire directly to the battery terminal.

2. Disconnect the negative battery cable.
3. Remove the engine room cover.
4. Remove the air duct and resonator assembly.
5. Disconnect the harness connector from the ignition coil.
6. Remove the ignition coil.

✳✳ WARNING

Do not shock ignition coil.

7. Remove spark plug.

To install:

➡Be sure to use new fasteners, as required.

8. Installation is the reverse of the removal procedure.

9. Be sure to perform the reconnect/relearn procedures.

ENGINE ELECTRICAL

REMOVAL & INSTALLATION

V6 Engine

1. Before servicing the vehicle, refer to the Precautions Section.

➡️**If working near and/or around the SRS system and components, be sure to disable the SRS system. After disabling the system wait three minutes or more before servicing the vehicle.**

➡️**On this vehicle the battery current sensor that is installed to the negative battery cable measures the charging/discharging current of the battery and performs various engine controls. If an electrical component is connected directly to the negative battery cable, the current flowing through that component will not be measured by the battery current sensor. This condition may cause a malfunction of the engine control system and battery discharge may occur. Do not connect an electrical component or ground wire directly to the battery terminal.**

2. Disconnect the negative battery cable.

3. Remove engine undercover.

4. Raise and safely support the vehicle, as necessary.

5. Remove exhaust manifold cover from exhaust manifold (right bank) to gain access to starter cover bolts.

6. Remove starter cover bolts and starter cover.

INSPECTION

Check spark plugs for the following:
1. Broken insulator
2. Worn electrode

7. Disconnect terminal "1" connector and terminal "2" nut.

8. Remove the two starter bolts.

9. Remove the starter.

To install:

➡️**Be sure to use new fasteners, as required.**

10. Installation is the reverse of the removal procedure.

11. Be sure to perform the reconnect/relearn procedures.

V8 Engine

See Figure 26.

1. Before servicing the vehicle, refer to the Precautions Section.

➡️**If working near and/or around the SRS system and components, be sure to disable the SRS system. After disabling the system wait three minutes or more before servicing the vehicle.**

➡️**On this vehicle the battery current sensor that is installed to the negative battery cable measures the charging/discharging current of the battery and performs various engine controls. If an electrical component is connected directly to the negative battery cable, the current flowing through that component will not be measured by the battery current sensor. This condition may cause a malfunction of the engine con-**

3. Carbon deposits
4. Damaged or broken gasket
5. Condition of the porcelain insulator at the tip of the spark plug.
6. Check the electrode gap.

STARTING SYSTEM

trol system and battery discharge may occur. Do not connect an electrical component or ground wire directly to the battery terminal.

2. Disconnect the negative battery cable.

3. Remove the intake manifold.

4. Disconnect terminal "1" connector and terminal "2" nut.

5. Remove the two starter bolts.

6. Remove the starter.

To install:

➡️**Be sure to use new fasteners, as required.**

7. Installation is the reverse of the removal procedure.

8. Be sure to perform the reconnect/relearn procedures.

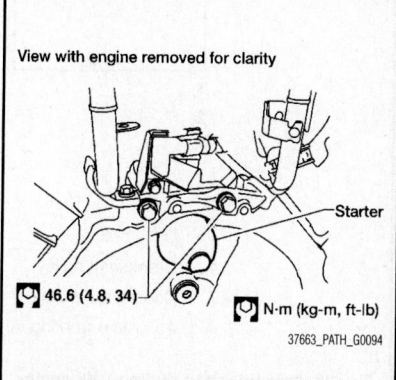

View with engine removed for clarity

46.6 (4.8, 34)   N·m (kg-m, ft-lb)

37663_PATH_G0094

Fig. 26 Starter retaining bolt locations—V8 engine

ENGINE MECHANICAL

ACCESSORY DRIVE BELTS

ACCESSORY BELT ROUTING
See Figures 27 and 28.

Refer to the accompanying illustration.

INSPECTION

Visually check entire belt for wear, damage or cracks.

ADJUSTMENT

Belt tensioning is not necessary, as it is

automatically adjusted by auto tensioner.

REMOVAL & INSTALLATION
See Figures 29 through 31.

1. Before servicing the vehicle, refer to the Precautions Section.

➡️**If working near and/or around the SRS system and components, be sure to disable the SRS system. After disabling the system wait three minutes or more before servicing the vehicle.**

➡️**On this vehicle the battery current sensor that is installed to the negative battery cable measures the charging/discharging current of the battery and performs various engine controls. If an electrical component is connected directly to the negative battery cable, the current flowing through that component will not be measured by the battery current sensor. This condition may cause a malfunction of the engine control system and battery discharge may**

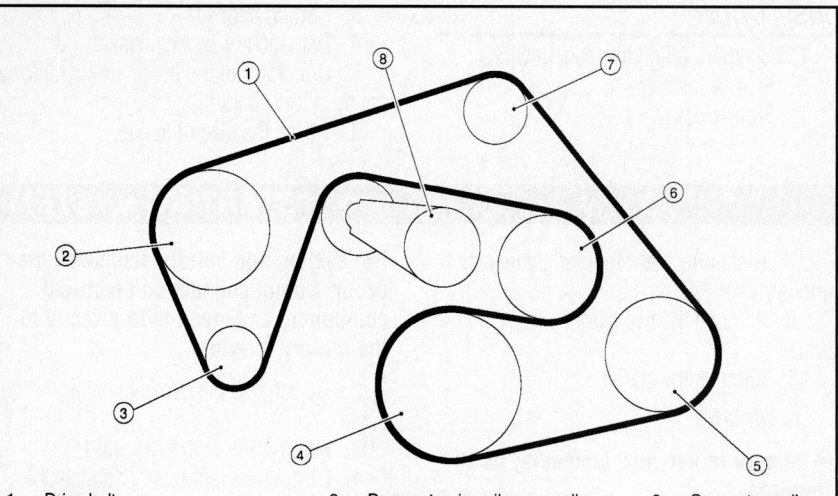

1. Drive belt
2. Power steering oil pump pulley
3. Generator pulley
4. Crankshaft pulley
5. A/C compressor
6. Cooling fan pulley
7. Idler pulley
8. Drive belt tensioner

22140_PATH_G0015

Fig. 27 Accessory belt routing—V6 engine

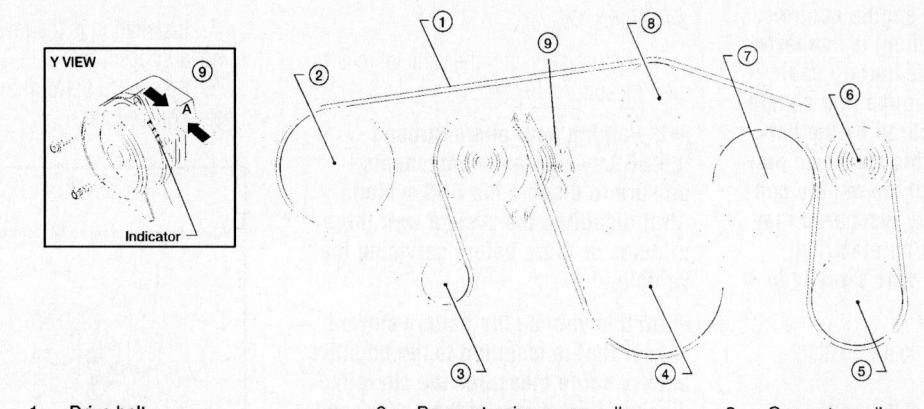

1. Drive belt
2. Power steering pump pulley
3. Generator pulley
4. Crankshaft pulley
5. A/C compressor
6. Idler pulley
7. Cooling fan pulley
8. Water pump pulley
9. Drive belt auto tensioner
A. Allowable working range

22140_PATH_G0082

Fig. 28 Accessory belt routing—V8 engine

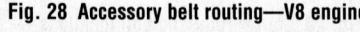

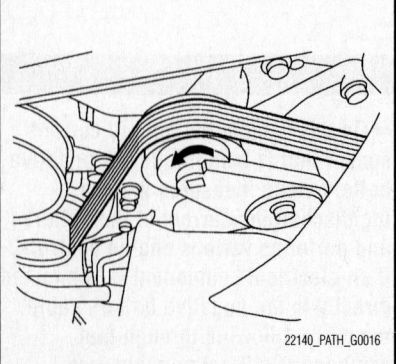

22140_PATH_G0016

Fig. 29 Rotate the drive belt auto tensioner in the direction of arrow

occur. Do not connect an electrical component or ground wire directly to the battery terminal.

2. Disconnect the negative battery cable.
3. Remove air duct and resonator assembly (inlet).
4. Rotate the drive belt auto tensioner in the direction of arrow (loosening direction of tensioner).

⁂ CAUTION

Avoid placing hand in a location where pinching may occur if the tool accidentally comes off.

5. Remove the drive belt.

To install:

➡Be sure to use new fasteners, as required.

6. Installation is the reverse of the removal procedure.

⁂ WARNING

Make sure belt is securely installed around all pulleys.

7. Be sure to perform the reconnect/relearn procedures.

CAMSHAFT AND VALVE LIFTERS

REMOVAL & INSTALLATION

V6 Engine

See Figures 32 through 35.

1. Before servicing the vehicle, refer to the Precautions Section.

➡If working near and/or around the SRS system and components, be sure to disable the SRS system. After disabling the system wait three minutes or more before servicing the vehicle.

➡On this vehicle the battery current sensor that is installed to the negative battery cable measures the charging/discharging current of the battery and performs various engine controls. If an electrical component is connected directly to the negative battery cable, the current flowing through that compo-

nent will not be measured by the battery current sensor. This condition may cause a malfunction of the engine control system and battery discharge may occur. Do not connect an electrical component or ground wire directly to the battery terminal.

2. Disconnect the negative battery cable.

3. Remove front timing chain case, camshaft sprocket, timing chain and rear timing chain case.

4. Remove Camshaft Position (CMP) sensor (PHASE) (right and left banks) from cylinder head back side.

✳✳ WARNING

Handle carefully to avoid dropping and shocks. Do not disassemble. Do not allow metal powder to adhere to magnetic part at sensor tip. Do not place sensors in a location where they are exposed to magnetism.

5. Remove intake valve timing control solenoid valves.

➡Discard intake valve timing control solenoid valve gaskets and use new gaskets for installation.

6. Remove camshaft brackets.
   a. Mark camshafts, camshaft brackets and bolts so they are placed in the same position and direction for installation.
   b. Equally loosen camshaft bracket bolts in several steps in reverse order of the tightening sequence.
7. Remove camshafts.
8. Remove valve lifters.

➡Identify installation positions, and store them without mixing them up.

9. Remove timing chain tensioner (secondary) from cylinder head with its stopper pin attached.

➡Stopper pin was attached when timing chain (secondary) was removed.

To install:

➡Be sure to use new fasteners, as required.

10. Install timing chain tensioners (secondary) on both sides of cylinder head.
    a. Install timing chain tensioner with its stopper pin attached.
    b. Install timing chain tensioner with sliding part facing downward on right-side cylinder head, and with sliding part facing upward on left-side cylinder head.
    c. Install new O-rings.
11. Install valve lifters.

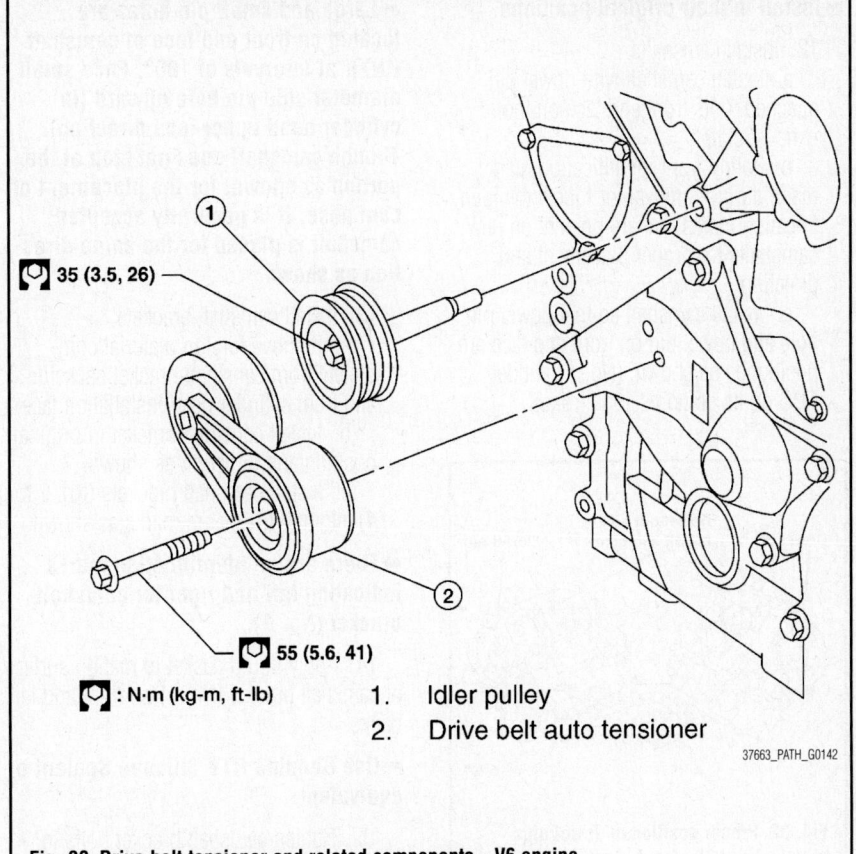

1. Idler pulley
2. Drive belt auto tensioner

Fig. 30 Drive belt tensioner and related components—V6 engine

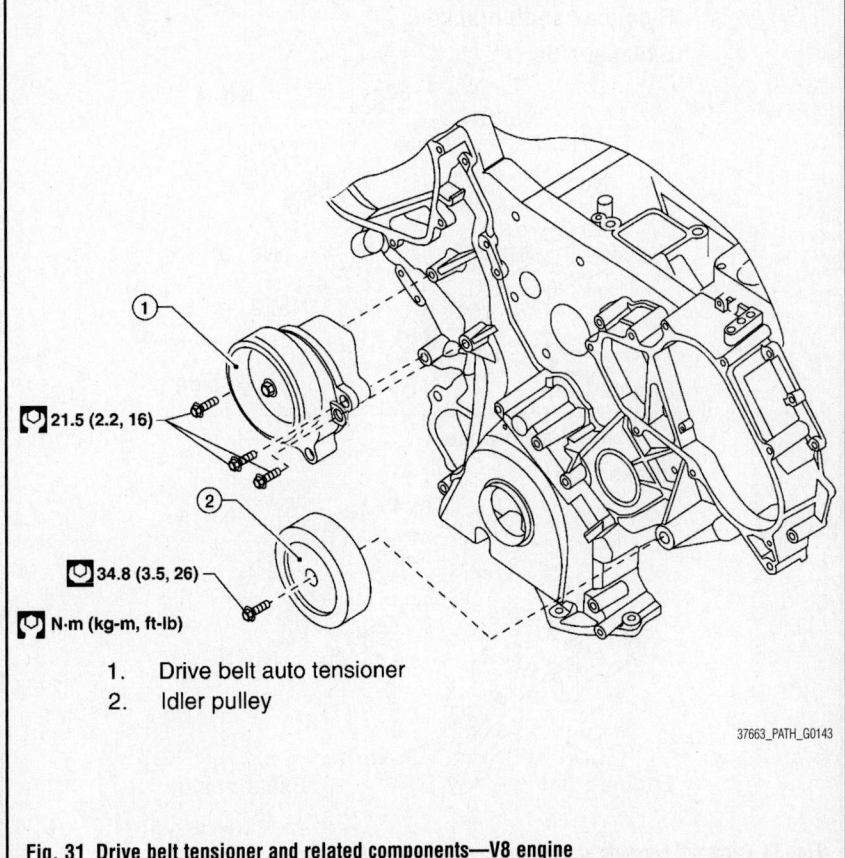

1. Drive belt auto tensioner
2. Idler pulley

Fig. 31 Drive belt tensioner and related components—V8 engine

➡️**Install in their original positions.**

12. Install camshafts.

a. Install camshaft with dowel pin attached to its front end face on the exhaust side.

b. Follow your identification marks made during removal, or follow the identification marks that are present on new camshafts for proper placement and direction.

c. Install camshaft so that dowel pin hole and dowel pin on front end face are positioned as shown. (No. 1 cylinder TDC on its compression stroke).

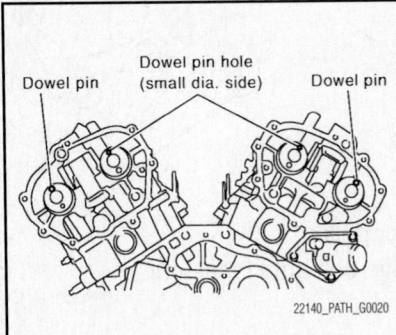

Fig. 32 Proper position of dowel pins during camshaft installation—V6 engine

➡️**Large and small pin holes are located on front end face of camshaft (INT), at intervals of 180°. Face small diameter side pin hole upward (in cylinder head upper face direction). Though camshaft does not stop at the portion as shown, for the placement of cam nose, it is generally accepted camshaft is placed for the same direction as shown.**

13. Install camshaft brackets.

a. Remove foreign material completely from camshaft bracket backside and from cylinder head installation face.

b. Install camshaft bracket in original position and direction as shown.

c. Install camshaft brackets (No. 2 to 4) aligning the stamp marks as shown.

➡️**There are no identification marks indicating left and right for camshaft bracket (No. 1).**

14. Apply liquid gasket to mating surface of camshaft bracket (No. 1) on right and left banks.

➡️**Use Genuine RTV Silicone Sealant or equivalent.**

15. Tighten camshaft bracket bolts in numerical order as shown.

16. Tighten camshaft bracket bolts in the following steps:

- Step 1: Bolts 7–10: 17 inch lbs. (1.96 Nm)
- Step 2: Bolts 1–6: 17 inch lbs. (1.96 Nm)
- Step 3: All bolts: 52 inch lbs. (5.88 Nm)
- Step 4: All bolts: 92 inch lbs. (10.4 Nm)

17. Measure the difference in levels between front end faces of camshaft bracket (No. 1) and cylinder head.

a. Standard: -0.0055 to 0.0055 in. (-0.14 to 0.14 mm)

b. Measure two positions (both intake and exhaust side) for a single bank.

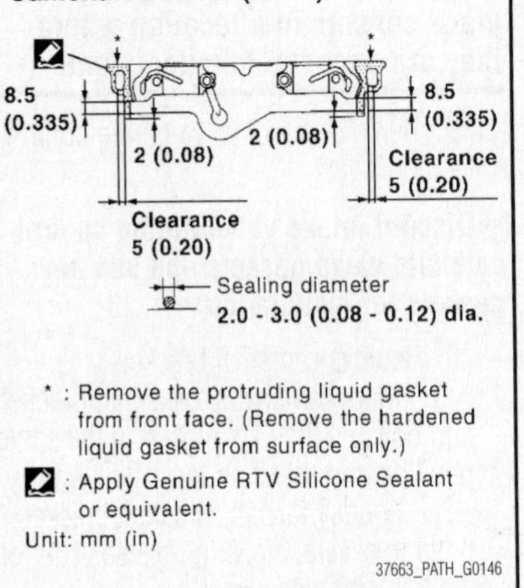

Camshaft bracket (No. 1)

8.5 (0.335)   8.5 (0.335)

2 (0.08)   2 (0.08)

Clearance 5 (0.20)   Clearance 5 (0.20)

Sealing diameter 2.0 - 3.0 (0.08 - 0.12) dia.

* : Remove the protruding liquid gasket from front face. (Remove the hardened liquid gasket from surface only.)

🔲 : Apply Genuine RTV Silicone Sealant or equivalent.

Unit: mm (in)

37663_PATH_G0146

Fig. 34 Camshaft bracket sealant application points—V6 engine

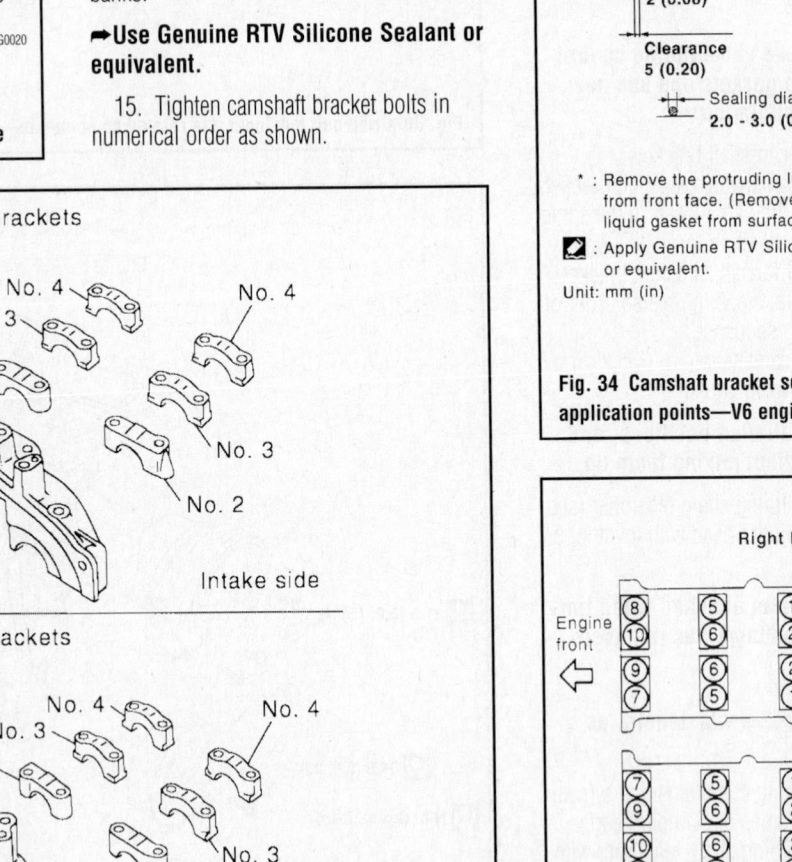

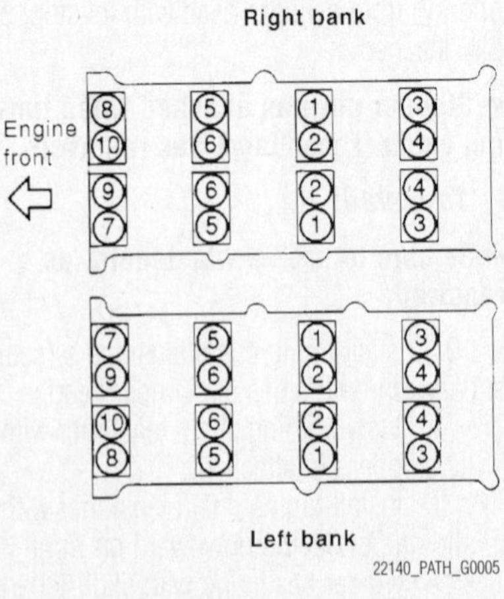

Right camshaft brackets
Exhaust side

No. 4
No. 4
No. 3
No. 2
No. 1
No. 3
No. 2

Engine front   Intake side

Left camshaft brackets
Intake side

No. 4
No. 4
No. 3
No. 2
No. 3
No. 2
No. 1

Engine front   Exhaust side

22140_PATH_G0021

Fig. 33 Camshaft brackets and related components—V6 engine

Right bank

Engine front

Left bank

22140_PATH_G0005

Fig. 35 Camshaft bracket bolt tightening sequence—V6 engine

c. If the measured value is out of the standard, re-install camshaft bracket (No. 1).

18. Check and adjust the valve clearance, as required.

19. Installation of the remaining components is in the reverse order of removal.

20. Be sure to perform the reconnect/relearn procedures.

V8 Engine

See Figures 36 through 51.

➡ **Do not remove the engine assembly to perform this procedure.**

1. Before servicing the vehicle, refer to the Precautions Section.

➡ **If working near and/or around the SRS system and components, be sure to disable the SRS system. After disabling the system wait three minutes or more before servicing the vehicle.**

➡ **On this vehicle the battery current sensor that is installed to the negative battery cable measures the charging/discharging current of the battery and performs various engine controls. If an electrical component is connected directly to the negative battery cable, the current flowing through that component will not be measured by the battery current sensor. This condition may cause a malfunction of the engine control system and battery discharge may occur. Do not connect an electrical component or ground wire directly to the battery terminal.**

2. Disconnect the negative battery cable.

3. Remove the RH bank and LH bank rocker covers.

4. Obtain compression TDC of No. 1 cylinder as follows:

a. Turn the crankshaft pulley clockwise to align the TDC identification notch (without paint mark) with the timing indicator on the front cover.

b. At this time, make sure both intake and exhaust cam lobes of No. 1 cylinder (top front on LH bank) point outside.

c. If they do not point outside, turn crankshaft pulley once more.

5. Remove the intake valve control solenoid cover RH bank (A) and intake valve control solenoid cover LH bank (B) as follows:

a. Loosen and remove the bolts, see illustration for proper removal sequence

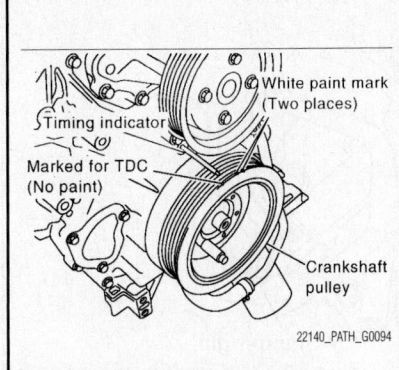

Fig. 36 Align the TDC identification notch—V8 engine

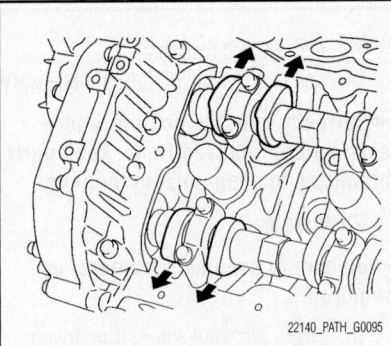

Fig. 37 Intake and exhaust cam lobes of No. 1 cylinder—V8 engine

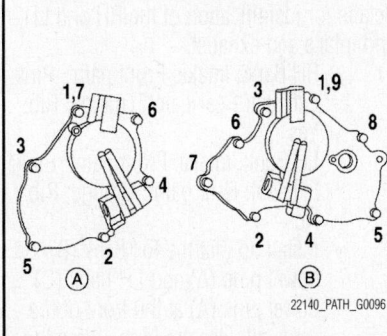

Fig. 38 Loosen and remove the bolts—V8 engine

b. Cut the liquid gasket and remove the covers.

✳✳ WARNING

Do not damage mating surfaces.

6. Paint alignment marks on the RH bank (A) timing chain links (C) and LH bank (B) timing chain links (D) and align with the camshaft sprocket alignment marks (E) and (F).

7. Remove the LH bank timing chain tensioner using the following steps.

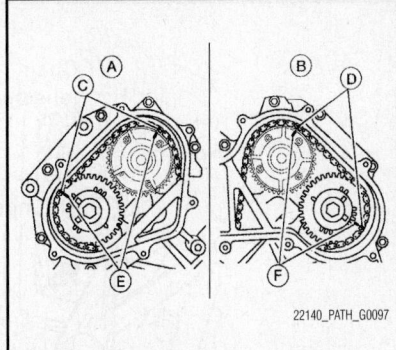

Fig. 39 Paint alignment marks on the RH bank (A) timing chain links (C) and LH bank (B) timing chain links—V8 engine

✳✳ CAUTION

Plunger, spring, and spring seat pop out when squeezing return-proof clip without holding plunger head. It may cause serious injuries. Always hold plunger head when removing.

a. Squeeze return-proof clip ends using suitable tool and push the plunger into the tensioner body.

b. Secure plunger using stopper pin.

➡ **Stopper pin is made from hard wire approximately 1 mm (0.04 in.) in diameter.**

c. Remove the bolts and the timing chain tensioner.

➡ **Stop plunger in the fully extended position using return-proof clip (1) if stopper pin is removed. Push the plunger (2) into the tensioner body while squeezing the return-proof clip (1). Secure it using stopper pin (3).**

8. Remove the RH bank timing chain tensioner cover from the front cover.

✳✳ WARNING

Do not damage mating surfaces.

9. Remove the RH bank timing chain tensioner using the following steps.

✳✳ CAUTION

Plunger, spring, and spring seat pop out when squeezing return-proof clip without holding plunger head. It may cause serious injuries. Always hold plunger head when removing.

a. Squeeze return-proof clip ends using suitable tool and push the plunger into the tensioner body.

b. Secure plunger using stopper pin.

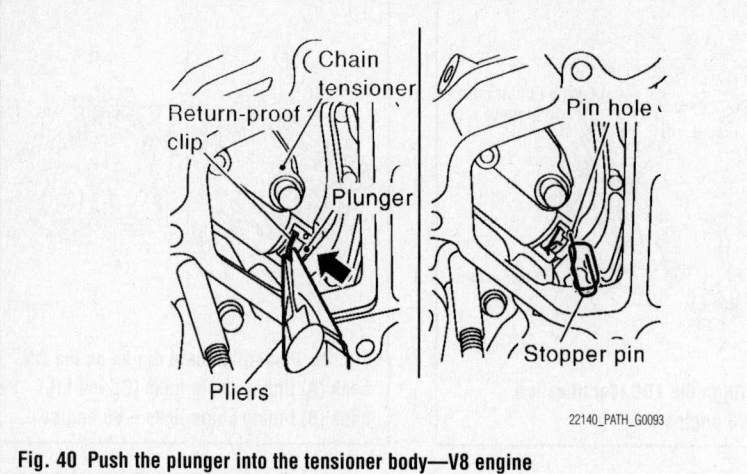

Fig. 40 Push the plunger into the tensioner body—V8 engine

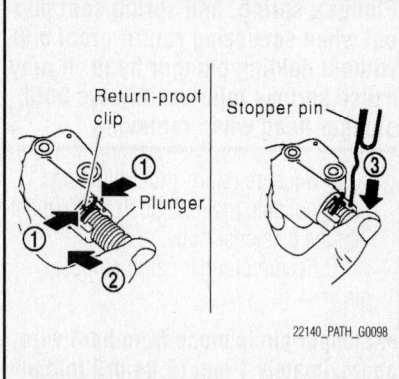

Fig. 41 Stop plunger in the fully extended position—V8 engine

c. Remove the bolts and the RH bank timing chain tensioner (A).

➡If it is difficult to push plunger on RH bank timing chain tensioner (A), remove the plunger under extended condition.

10. Loosen camshaft sprocket bolts as shown and remove camshaft sprockets.

⁂ WARNING

To avoid interference between valves and pistons, do not turn crankshaft or camshaft with timing chain disconnected.

11. Remove the RH (A) front cover bolts and LH (B) front cover bolts.
12. Remove RH (A) camshaft bracket bolts and LH (C) camshaft bracket bolts in the reverse order of the tightening sequence.
    a. Remove No. 1 camshaft bracket.

➡The bottom and front surface of bracket will be stuck because of liquid gasket.

13. Remove the camshaft.
14. Remove the valve lifters if necessary.

➡Correctly identify location where each part is removed from. Keep parts organized to avoid mixing them up.

To install:

➡Be sure to use new fasteners, as required.

15. Install the valve lifters if removed.

➡Install removed parts in their original locations.

16. Install the camshafts. Important details for identification of the RH and LH, and intake and exhaust.
   - RH Bank: Intake: Front paint: Pink; Exhaust: Rear paint: Orange; Rib: Yes
   - LH Bank: Intake: Front paint: Pink; Exhaust: Rear paint: Orange; Rib: No
   - Install so that the RH bank (B) dowel pins (A) and LH bank (C) dowel pins (A) at the front of the camshaft face are in the direction shown.

17. Install the RH bank (B) and LH bank (D) camshaft brackets (A).
    a. Install by referring to the installation location mark (E) on the upper surface.
    b. Install so that the installation location mark (E) can be correctly read when viewed from the intake manifold side (C).
    c. Install No. 1 camshaft bracket using the following procedure:
      - C: 0.43 in. (11 mm)
      - D: 0.079–0.118 in. (2.0–3.0 mm) diameter
      - Apply liquid gasket to No. 1 camshaft bracket (A) and (B) as shown.

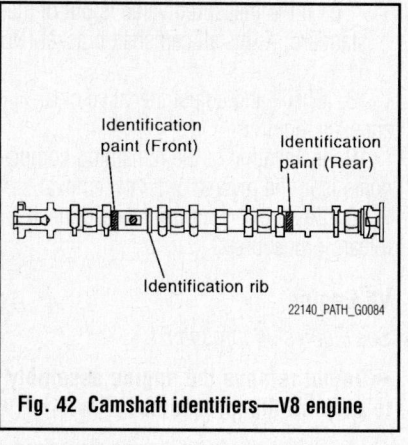

Fig. 42 Camshaft identifiers—V8 engine

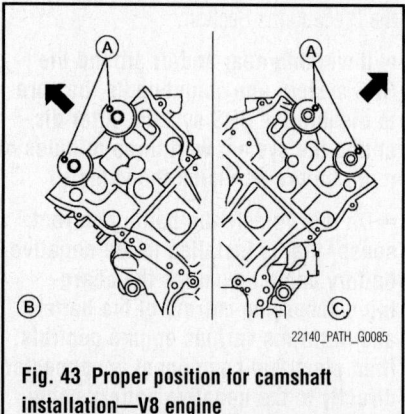

Fig. 43 Proper position for camshaft installation—V8 engine

⁂ WARNING

After installation, be sure to wipe off any excessive liquid gasket outside of application (C) and (D) both on RH and LH sides.

   - Remove completely any excess of liquid gasket inside bracket.
    d. Apply liquid gasket (C) to the back side of the LH (A) bank front cover and RH (B) bank front cover as shown.
    e. C: 0.102–0.142 in. (2.6–3.6 mm) diameter
    f. Position No. 1 camshaft bracket close to the mounting position, and then install it to prevent from touching liquid gasket applied to each surface.
    g. Temporarily tighten the RH (A) and LH (B) front cover bolts (4 for each bank) as shown.

18. Tighten the camshaft bracket bolts as follows:
    a. Step 1 (bolts 9–12): 17 inch lbs. (2.0 Nm)
    b. Step 2 (bolts 1–8): 17 inch lbs. (2.0 Nm)
    c. Step 3 (all bolts): 52 inch lbs. (5.9 Nm)
    d. Step 4 (all bolts): 92 inch lbs. (10.4 Nm)

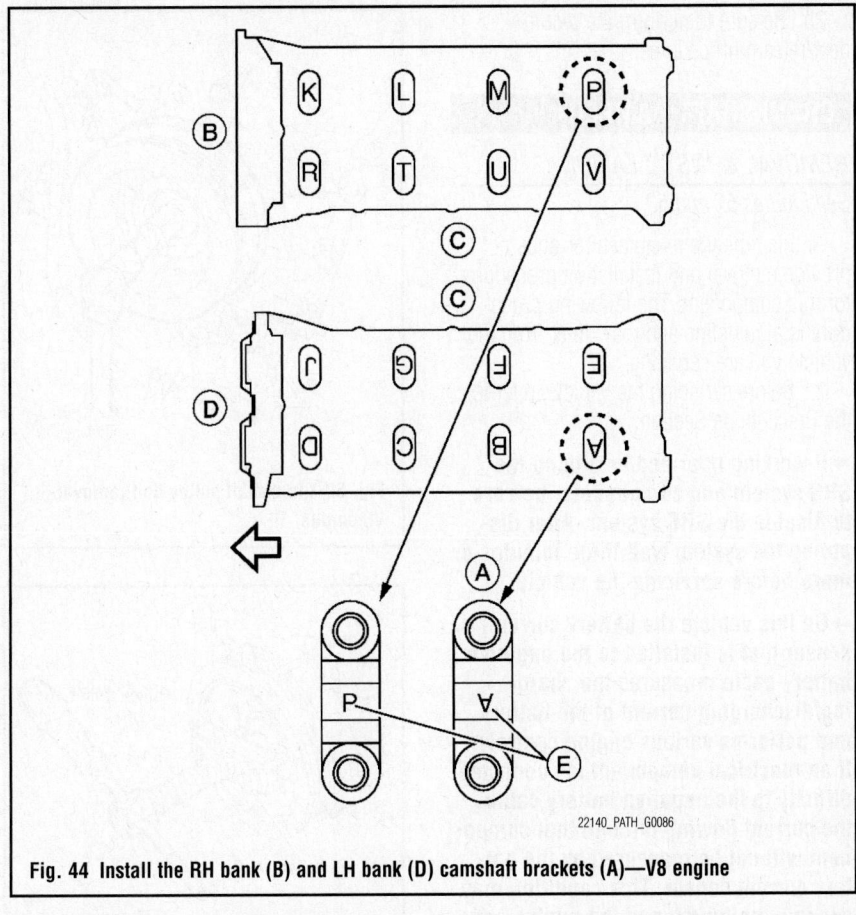

Fig. 44 Install the RH bank (B) and LH bank (D) camshaft brackets (A)—V8 engine

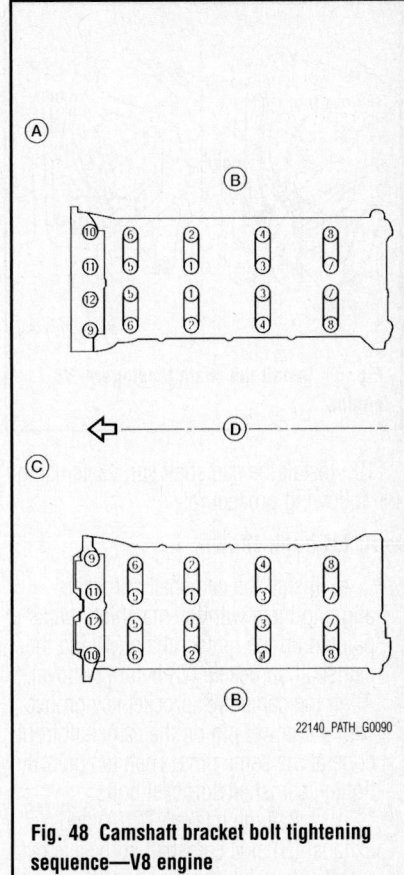

Fig. 48 Camshaft bracket bolt tightening sequence—V8 engine

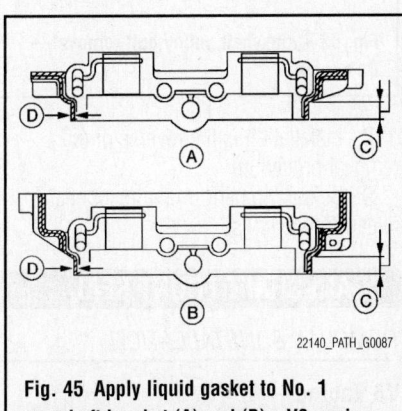

Fig. 45 Apply liquid gasket to No. 1 camshaft bracket (A) and (B)—V8 engine

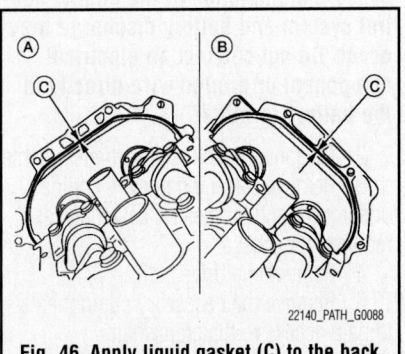

Fig. 46 Apply liquid gasket (C) to the back side of the LH (A) bank front cover and RH (B) bank front cover—V8 engine

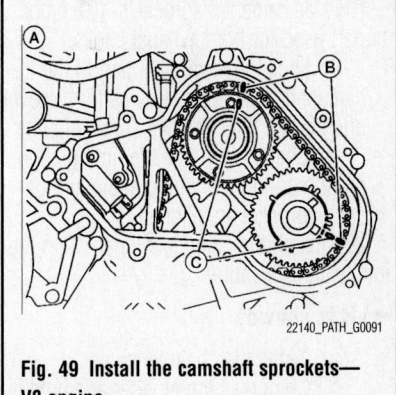

Fig. 49 Install the camshaft sprockets—V8 engine

✳✳ WARNING

After tightening the camshaft bracket bolts, be sure to wipe off excessive liquid gasket from the parts listed below.

- Mating surface of rocker cover
- Mating surface of front cover
- A: RH bank
- B: Exhaust side
- C: LH bank
- D: Intake side
- e. Tighten the RH (A) and LH (B) front cover bolts (4 for each bank) to 8 ft. lbs. (11.0 Nm)

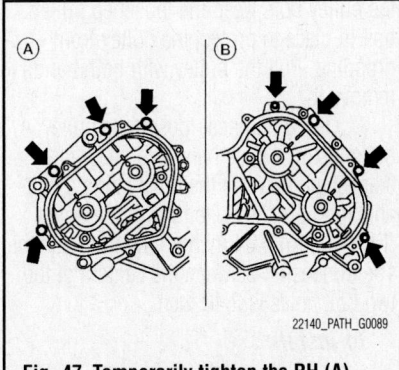

Fig. 47 Temporarily tighten the RH (A) and LH (B) front cover bolts—V8 engine

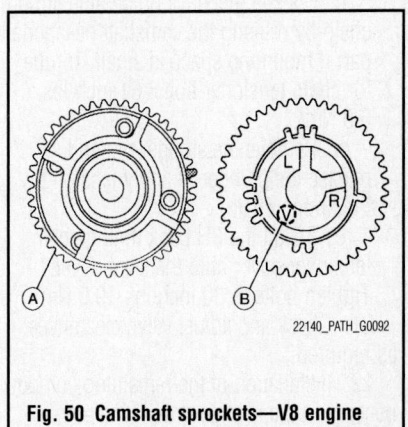

Fig. 50 Camshaft sprockets—V8 engine

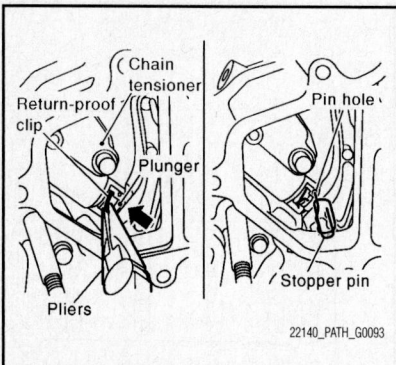

Fig. 51 Install the chain tensioner—V8 engine

19. Install the camshaft sprockets using the following procedure:

➡**A: LH bank shown.**

a. Install the camshaft sprockets aligning them with the matching marks painted on the timing chain (B) and the camshaft sprockets (C) before removal. Align the camshaft sprocket key groove with the dowel pin on the camshaft front edge at the same time. Then temporarily tighten camshaft sprocket bolts.

b. Install the intake VTC (A) and exhaust (B) side camshaft sprockets by selectively using the groove of the dowel pin according to the bank for the exhaust (B) side camshaft sprockets. (Common part used for both exhaust banks.)

c. Lock the hexagonal part of the camshaft in the same way as for removal, and tighten the camshaft sprocket bolts.

d. Check again that the timing alignment mark on the timing chain and on each sprocket are aligned.

20. Install the chain tensioner using the following procedure:

➡**LH is shown.**

a. Install the chain tensioner.

b. Compress the plunger and hold it using a stopper pin when installing.

c. Loosen the slack guide side timing chain by rotating the camshaft hexagonal part if mounting space is small. Torque for chain tensioner bolts: 61 inch lbs. (6.9 Nm).

d. Remove the stopper pin and release the plunger to apply tension to the timing chain.

e. Install the RH bank timing chain tensioner cover onto the front cover. Tighten bolts to 80 inch lbs. (9.0 Nm).

21. Check and adjust valve clearances, as required.

22. Installation of the remaining components is in the reverse order of removal.

23. Be sure to perform the reconnect/relearn procedures.

CRANKSHAFT DAMPER

REMOVAL & INSTALLATION

See Figures 52 and 53.

At this time the manufacturer does not provide removal and installation procedures for this component. The following procedure is a guideline and may differ from the vehicle you are servicing.

1. Before servicing the vehicle, refer to the Precautions Section.

➡**If working near and/or around the SRS system and components, be sure to disable the SRS system. After disabling the system wait three minutes or more before servicing the vehicle.**

➡**On this vehicle the battery current sensor that is installed to the negative battery cable measures the charging/discharging current of the battery and performs various engine controls. If an electrical component is connected directly to the negative battery cable, the current flowing through that component will not be measured by the battery current sensor. This condition may cause a malfunction of the engine control system and battery discharge may occur. Do not connect an electrical component or ground wire directly to the battery terminal.**

2. Disconnect the negative battery cable.

3. Remove the engine cover, engine undercover and air cleaner assembly, as required for access.

4. Remove the drive belt.

5. Remove the necessary components to gain access to the crankshaft damper.

6. On V6 engines, loosen the pulley bolt and locate the bolt seating surface 0.39 inch from its original position. Do not remove the pulley bolt. Keep the loosened pulley bolt in place to protect the pulley from dropping. Pull the pulley with both hands to remove it.

7. On V8 engines, loosen the pulley bolts, using a slide handle to secure the crankshaft. Remove the bolts using a suitable tool. Set the bolts in two bolt holes (M6x1.0mm) 0.04 inch on the front surface. The dimension between the centers of the two bolt holes is 2.40 inch.

To install:

➡**Be sure to use new fasteners, as required.**

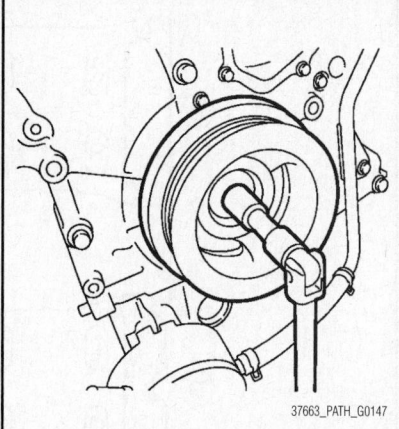

Fig. 52 Crankshaft pulley bolt removal—V6 engine

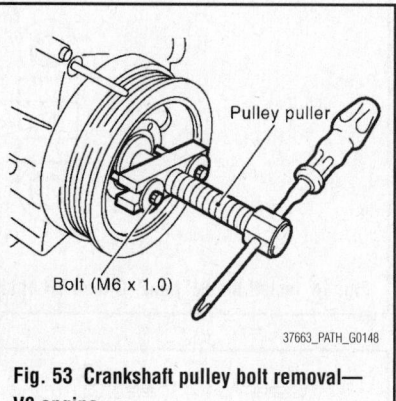

Fig. 53 Crankshaft pulley bolt removal—V8 engine

8. Installation is the reverse of the removal procedure.

9. Be sure to perform the reconnect/relearn procedures.

CRANKSHAFT FRONT SEAL

REMOVAL & INSTALLATION

V6 Engine

1. Before servicing the vehicle, refer to the Precautions Section.

➡**If working near and/or around the SRS system and components, be sure to disable the SRS system. After disabling the system wait three minutes or more before servicing the vehicle.**

➡**On this vehicle the battery current sensor that is installed to the negative battery cable measures the charging/discharging current of the battery and performs various engine controls. If an electrical component is connected directly to the negative battery cable, the current flowing through that compo-**

nent will not be measured by the battery current sensor. This condition may cause a malfunction of the engine control system and battery discharge may occur. Do not connect an electrical component or ground wire directly to the battery terminal.

2. Disconnect the negative battery cable.

3. Remove engine undercover.

4. Remove drive belts.

5. Remove engine cooling fan assembly.

6. Remove crankshaft pulley.

7. Remove front oil seal.

❄❄ WARNING
Be careful not to damage front timing chain case and crankshaft.

To install:

8. Apply new engine oil to both oil seal lip and dust seal lip of new front oil seal.

9. Install front oil seal.

a. Install front oil seal so that each seal lip is oriented as shown.

b. Press-fit until the height of front oil seal is level with the mounting surface using suitable tool.

c. Suitable drift: outer diameter 2.36 in. (60 mm), inner diameter 1.97 in. (50 mm).

❄❄ WARNING
Be careful not to damage front timing chain case and crankshaft. Press-fit straight and avoid causing burrs or tilting oil seal.

10. Installation is in the reverse order of removal after this step.

11. Be sure to perform the reconnect/relearn procedures.

V8 Engine
See Figure 54.

1. Before servicing the vehicle, refer to the Precautions Section.

➡If working near and/or around the SRS system and components, be sure to disable the SRS system. After disabling the system wait three minutes or more before servicing the vehicle.

➡On this vehicle the battery current sensor that is installed to the negative battery cable measures the charging/discharging current of the battery and performs various engine controls. If an electrical component is connected

---

directly to the negative battery cable, the current flowing through that component will not be measured by the battery current sensor. This condition may cause a malfunction of the engine control system and battery discharge may occur. Do not connect an electrical component or ground wire directly to the battery terminal.

2. Disconnect the negative battery cable.

3. Remove the air dam.

4. Remove the engine undercover.

5. Remove the air duct and resonator assembly and the air cleaner case (upper).

6. Remove the radiator assembly.

7. Remove the cooling fan (crankshaft driven type).

8. Remove the crankshaft pulley.

9. Remove the front oil seal using suitable tool.

❄❄ WARNING
Do not damage front cover and oil pump drive spacer.

To install:

10. Apply new engine oil to both the oil seal lip and dust seal lip of the new front oil seal.

11. Install the front oil seal.

a. Install the front oil seal so that each seal lip is oriented as shown.

b. Press-fit until the front oil seal is level with the front cover using suitable tool.

❄❄ WARNING
Do not damage front cover and crankshaft. Press-fit straight and avoid causing burrs or tilting oil seal.

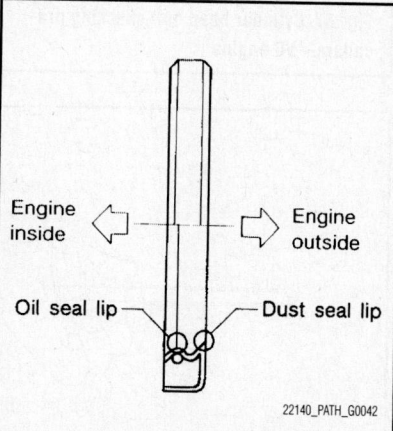

22140_PATH_G0042

Fig. 54 Install the front oil seal so that each seal lip is oriented as shown.

---

12. Installation of the remaining components is in the reverse order of removal.

13. Be sure to perform the reconnect/relearn procedures.

CYLINDER HEAD

REMOVAL & INSTALLATION

V6 Engine
See Figures 55 through 58.

1. Before servicing the vehicle, refer to the Precautions Section.

➡If working near and/or around the SRS system and components, be sure to disable the SRS system. After disabling the system wait three minutes or more before servicing the vehicle.

➡On this vehicle the battery current sensor that is installed to the negative battery cable measures the charging/discharging current of the battery and performs various engine controls. If an electrical component is connected directly to the negative battery cable, the current flowing through that component will not be measured by the battery current sensor. This condition may cause a malfunction of the engine control system and battery discharge may occur. Do not connect an electrical component or ground wire directly to the battery terminal.

2. Disconnect the negative battery cable.

3. Remove camshaft.

4. Remove intake manifold.

5. Remove exhaust manifold.

6. Remove water inlet and thermostat assembly.

7. Remove water outlet, water pipe and heater pipe.

8. Remove cylinder head bolts in reverse order of the tightening sequence.

9. Remove cylinder head gaskets. Discard the gaskets.

To install:

10. Install new cylinder head gasket.

11. Turn crankshaft until No. 1 piston is set at TDC.

➡Crankshaft key should line up with the right bank cylinder center line.

12. Install cylinder head and tighten cylinder head bolts in numerical order as shown.

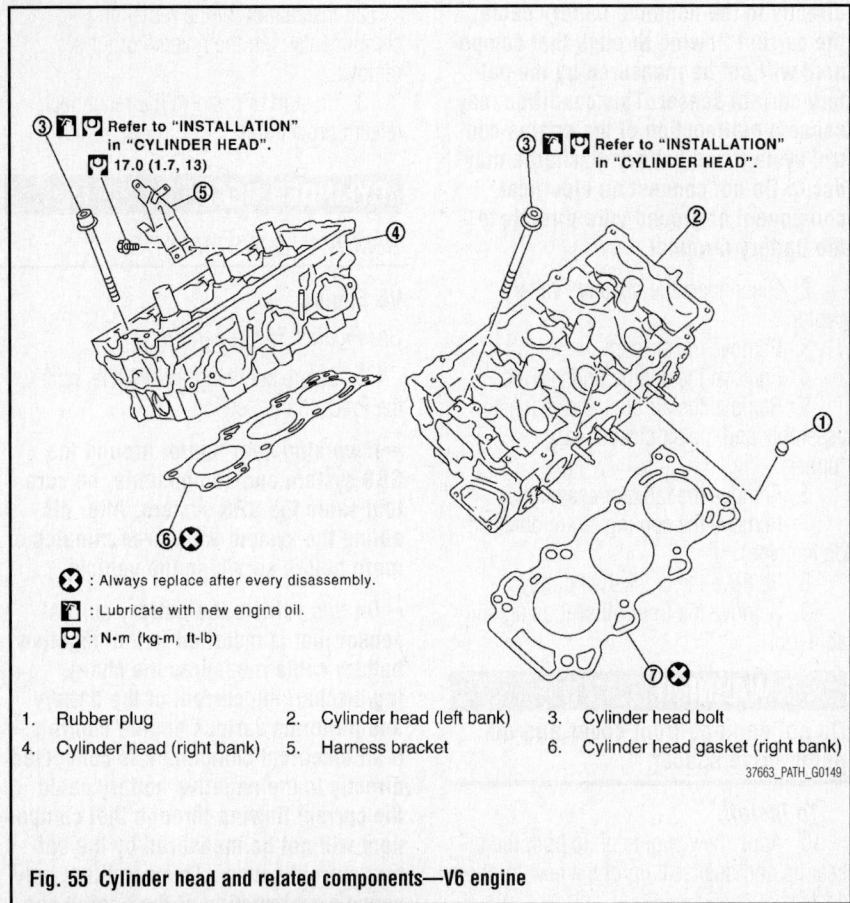

⊗ : Always replace after every disassembly.

⊡ : Lubricate with new engine oil.

⊙ : N•m (kg-m, ft-lb)

1. Rubber plug
2. Cylinder head (left bank)
3. Cylinder head bolt
4. Cylinder head (right bank)
5. Harness bracket
6. Cylinder head gasket (right bank)

37663_PATH_G0149

Fig. 55 Cylinder head and related components—V6 engine

❋❋ WARNING

If cylinder head bolts re-used, check their outer diameters before installation. These bolts are tightened using

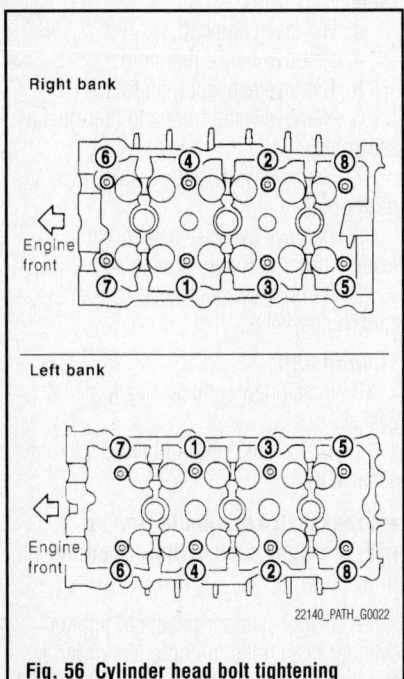

22140_PATH_G0022

Fig. 56 Cylinder head bolt tightening sequence—V6 engine

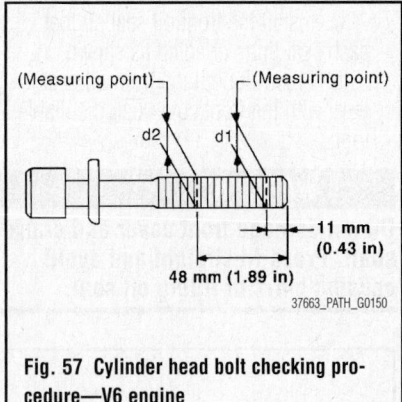

37663_PATH_G0150

Fig. 57 Cylinder head bolt checking procedure—V6 engine

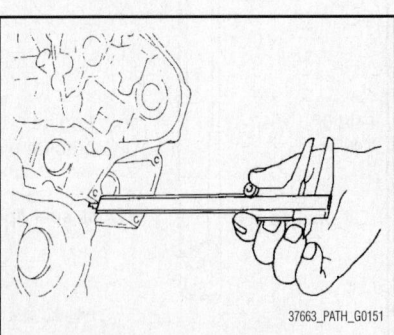

37663_PATH_G0151

Fig. 58 Cylinder head/cylinder block installation measurement point location—V6 engine

the plastic zone tightening method. Whenever the size difference between "d1" and "d2" exceeds the limit, replace the bolt. Specification: limit ("d1"-"d2") 0.0043 inch (0.11mm). If reduction of outer diameter appears in a position other than "d2", use it as "d2" point. See illustration.

13. Tighten cylinder head bolts using the following sequence:
- Step A: 72 ft. lbs. (98 Nm)
- Step B: Loosen bolts in the reverse order of tightening.
- Step C: 29 ft. lbs. (39.2 Nm)
- Step D: An additional 90° clockwise
- Step E: An additional 90° clockwise

14. After installing cylinder head, measure distance between front end faces of cylinder block and cylinder head (left and right banks). Specification should be 0.555–0.587 inch.

➡If the measured value is out of the standard, re-install cylinder head.

15. Installation of the remaining parts is in the reverse order of removal.

16. Be sure to perform the reconnect/relearn procedures.

V8 Engine

See Figures 59 through 60.

1. Before servicing the vehicle, refer to the Precautions Section.

➡If working near and/or around the SRS system and components, be sure to disable the SRS system. After disabling the system wait three minutes or more before servicing the vehicle.

➡On this vehicle the battery current sensor that is installed to the negative battery cable measures the charging/discharging current of the battery and performs various engine controls. If an electrical component is connected directly to the negative battery cable, the current flowing through that component will not be measured by the battery current sensor. This condition may cause a malfunction of the engine control system and battery discharge may occur. Do not connect an electrical component or ground wire directly to the battery terminal.

2. Disconnect the negative battery cable.

3. Remove the engine assembly from the vehicle. Position the assembly in a suitable holding fixture.

4. Remove the following components and related parts:

- Drive belt auto tensioner drive belts and idler pulley
- Thermostat housing and water piping
- Fuel tube and fuel injector assembly
- Starter
- Rocker cover
- Alternator and alternator bracket
- Oil pan and oil strainer

5. Remove the crankshaft pulley, front cover, oil pump, and timing chain.

6. Remove the camshaft sprockets and camshafts.

7. Remove the cylinder head bolts in the reverse order of the tightening sequence.

8. Remove the cylinder heads from the engine. Discard the gaskets.

To install:

9. Install a new cylinder head gasket.

10. Install the cylinder head. Follow the steps below to tighten the bolts in the numerical order shown.

※※ WARNING

If cylinder head bolts re-used, check their outer diameters before installation. These bolts are tightened using the plastic zone tightening method. Whenever the size difference between "d1" and "d2" exceeds the limit, replace the bolt. Specification: limit ("d1"-"d2") 0.0091 inch (0.23mm). If reduction of outer diameter appears in a position other than "d2", use it as "d2" point. See illustration.

a. Apply engine oil to threads and seating surface of the bolts.

b. Tighten cylinder head bolts as follows:

- Step A: 72 ft. lbs. (98.1 Nm)
- Step B: Loosen all bolts in the reverse order of tightening
- Step C: 33 ft. lbs. (44.1 Nm)

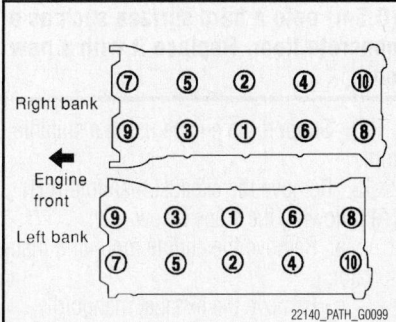

Fig. 59 Cylinder head bolt tightening sequence—V8 engine

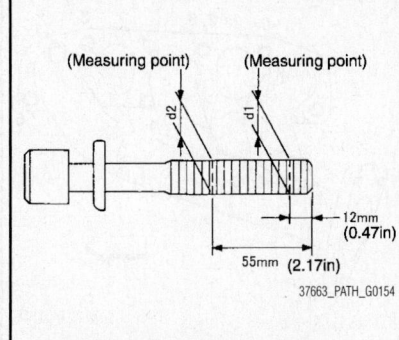

Fig. 60 Cylinder head bolt checking procedure—V8 engine

- Step D: 60° clockwise
- Step E: An additional 60° clockwise

c. Measure the tightening angle using Tool Number KV10112100 (BT8653A).

※※ WARNING

Measure the tightening angle using Tool. Do not measure visually.

11. Installation of the remaining components is in the reverse order of removal.

12. Be sure to perform the reconnect/relearn procedures.

EXHAUST MANIFOLD

REMOVAL & INSTALLATION

V6 Engine

Left Bank

See Figure 61.

1. Before servicing the vehicle, refer to the Precautions Section.

➡**If working near and/or around the SRS system and components, be sure to disable the SRS system. After disabling the system wait three minutes or more before servicing the vehicle.**

➡**On this vehicle the battery current sensor that is installed to the negative battery cable measures the charging/discharging current of the battery and performs various engine controls. If an electrical component is connected directly to the negative battery cable, the current flowing through that component will not be measured by the battery current sensor. This condition may cause a malfunction of the engine control system and battery discharge may occur. Do not connect an electrical component or ground wire directly to the battery terminal.**

2. Disconnect the negative battery cable.

3. Remove the engine room cover.

4. Remove the air cleaner assembly.

5. Remove the engine undercover.

6. Drain the engine coolant. Be sure to properly dispose of used coolant.

7. Raise and safely support the vehicle.

8. Remove the tire and wheel assembly.

9. Remove the exhaust manifold cover bolts.

10. Remove the center exhaust tube, main muffler and front exhaust tube.

11. Disconnect the oxygen sensor connector. Remove the sensor, as necessary.

12. Remove the three way catalyst nuts. Remove the catalyst.

13. Remove the exhaust manifold cover.

14. Remove the oil level gauge and gauge holder guide.

15. Disconnect the water hoses at the heater pipe.

16. Remove the heater pipe from the cylinder head.

17. Loosen the manifold retaining nuts in the reverse order of the tightening sequence.

➡**Discard the numeral order number 7 and 8 in the removal sequence.**

18. Remove the manifold retaining bolts and nuts.

19. Remove the component from the vehicle.

20. Discard the gasket. Discard the nuts.

To install:

➡**Be sure to use new fasteners, as required.**

21. Installation is the reverse of the removal procedure.

22. Tighten the retaining nuts to specification in two passes and in the proper sequence. Be sure to use new nuts.

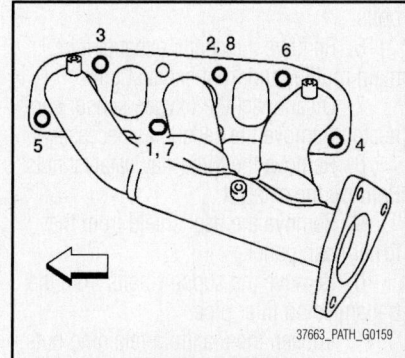

Fig. 61 Exhaust manifold tightening sequence (left side)—V6 engine

➡Tighten nuts number 1 and 2 in two steps. The numerical order of number 7 and 8 show second step.

✳✳ WARNING

Before installing a new air fuel ratio sensor 1 and heated oxygen sensor 2, clean exhaust system threads using oxygen sensor thread cleaner and apply anti-seize lubricant. _ Do not over torque air fuel ratio sensor 1 and heated oxygen sensor 2. Doing so may cause damage to air fuel ratio sensor 1 and heated oxygen sensor 2, resulting in the "MIL" coming on.

Right Bank
See Figure 62.

1. Before servicing the vehicle, refer to the Precautions Section.

➡If working near and/or around the SRS system and components, be sure to disable the SRS system. After disabling the system wait three minutes or more before servicing the vehicle.

➡On this vehicle the battery current sensor that is installed to the negative battery cable measures the charging/discharging current of the battery and performs various engine controls. If an electrical component is connected directly to the negative battery cable, the current flowing through that component will not be measured by the battery current sensor. This condition may cause a malfunction of the engine control system and battery discharge may occur. Do not connect an electrical component or ground wire directly to the battery terminal.

2. Disconnect the negative battery cable.
3. Raise and safely support the vehicle.
4. Remove the tire and wheel assembly.
5. Remove the exhaust manifold cover bolts.
6. Remove the center exhaust tube, main muffler and front exhaust tube.
7. Disconnect the oxygen sensor connector. Remove the sensor, as necessary.
8. Remove the three way catalyst nuts. Remove the catalyst.
9. Remove the heat shield from the lower dash panel.
10. Remove the support bolts from the transmission filler pipe.
11. Loosen the manifold retaining nuts in the reverse order of the tightening sequence.

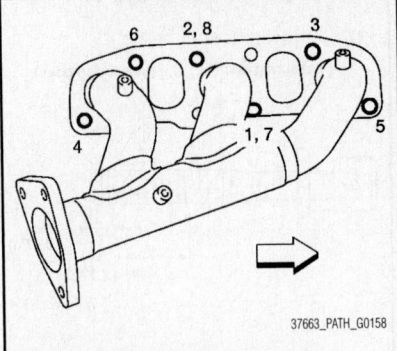

Fig. 62 Exhaust manifold tightening sequence (right side)—V6 engine

➡Discard the numeral order number 7 and 8 in the removal sequence.

12. Remove the manifold retaining bolts and nuts.
13. Remove the component from the vehicle.
14. Discard the gasket. Discard the nuts.

To install:

➡Be sure to use new fasteners, as required.

15. Installation is the reverse of the removal procedure.
16. Tighten the retaining nuts to specification in two passes and in the proper sequence. Be sure to use new nuts.

➡Tighten nuts number 1 and 2 in two steps. The numerical order of number 7 and 8 show second step.

✳✳ WARNING

Before installing a new air fuel ratio sensor 1 and heated oxygen sensor 2, clean exhaust system threads using oxygen sensor thread cleaner and apply anti-seize lubricant. Do not over torque air fuel ratio sensor 1 and heated oxygen sensor 2. Doing so may cause damage to air fuel ratio sensor 1 and heated oxygen sensor 2, resulting in the "MIL" coming on.

17. Be sure to perform the reconnect/relearn procedures.

V8 Engine
See Figure 63.

1. Before servicing the vehicle, refer to the Precautions Section.

➡If working near and/or around the SRS system and components, be sure to disable the SRS system. After dis-

abling the system wait three minutes or more before servicing the vehicle.

➡On this vehicle the battery current sensor that is installed to the negative battery cable measures the charging/discharging current of the battery and performs various engine controls. If an electrical component is connected directly to the negative battery cable, the current flowing through that component will not be measured by the battery current sensor. This condition may cause a malfunction of the engine control system and battery discharge may occur. Do not connect an electrical component or ground wire directly to the battery terminal.

2. Disconnect the negative battery cable.

✳✳ CAUTION

Perform the work when the exhaust and cooling system have cooled sufficiently.

3. Raise and safely support the vehicle.
4. Remove the air dam.
5. Remove the engine undercover.
6. Remove front final drive assembly (4WD).
7. Remove the main muffler assembly and center exhaust tube.
8. Remove the front exhaust tubes.
9. Remove front tires.
10. Remove fender protectors.
11. Remove the LH and RH air fuel ratio A/F sensors.
    a. Remove the harness connector of each air fuel ratio A/F sensor, and harness from bracket and middle clamp.

✳✳ WARNING

Do not damage the air fuel ratio A/F sensors. Discard any air fuel ratio A/F sensor which has been dropped from a height of more than 19.7 in. (0.5m) onto a hard surface such as a concrete floor. Replace it with a new one.

12. Support the engine using a suitable tool.
13. Remove the exhaust manifold (LH) (A) following the steps below.
    a. Remove the engine mounting insulator.
    b. Remove the exhaust manifold cover.
    c. Remove the engine mounting bracket.

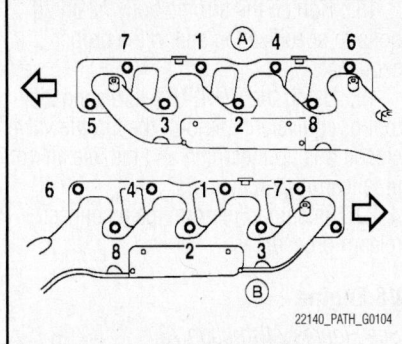

Fig. 63 Exhaust manifold retaining nut tightening sequence—V8 engine

d. Loosen the nuts LH side in reverse order of the tightening sequence.

e. Remove the exhaust manifold (LH).

14. Remove the exhaust manifold (RH) (B) following the steps below.

a. Remove the engine mounting insulator.

b. Remove the exhaust manifold cover.

c. Remove the engine mounting bracket.

d. Remove the oil level gauge guide.

e. Loosen the nuts RH side in reverse order of the tightening sequence.

f. Remove the exhaust manifold (RH).

15. Discard the gaskets. Discard the nuts.

To install:

➡**Be sure to use new fasteners, as required.**

16. Installation is in the reverse order of removal.

17. Install new exhaust manifold gasket with the top of the triangular up mark on it facing up and its coated face (gray side) toward the exhaust manifold side.

18. Tighten the retaining nuts to specification and in the proper sequence.

19. Do not over tighten the sensors. Doing so may interfere with sensor operation causing the MIL to come on.

20. Be sure to perform the reconnect/relearn procedures.

FLEXPLATE

REMOVAL & INSTALLATION

See Figures 64 through 67.

1. Before servicing the vehicle, refer to the Precautions Section.

➡**If working near and/or around the SRS system and components, be sure to disable the SRS system. After dis-**

abling the system wait three minutes or more before servicing the vehicle.

➡**On this vehicle the battery current sensor that is installed to the negative battery cable measures the charging/ discharging current of the battery and performs various engine controls. If an electrical component is connected directly to the negative battery cable, the current flowing through that component will not be measured by the battery current sensor. This condition may cause a malfunction of the engine control system and battery discharge may occur. Do not connect an electrical component or ground wire directly to the battery terminal.**

2. Disconnect the negative battery cable.

3. Raise and safely support the vehicle.

4. Remove the transmission assembly.

5. Remove the retaining bolts.

6. Remove the component from its mounting.

To install:

➡**Be sure to use new fasteners, as required.**

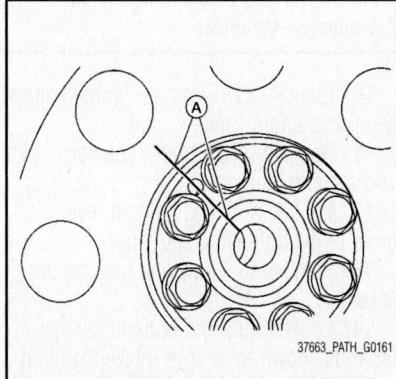

Fig. 64 Automatic transmission flexplate to transmission alignment—V6 engine

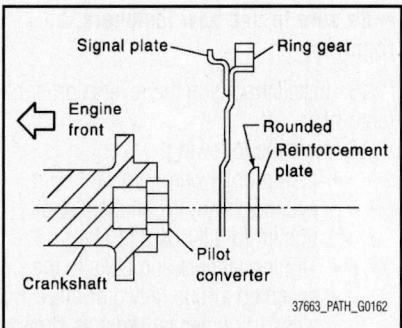

Fig. 65 Automatic transmission flexplate reinforcement plate installation direction—V6 engine

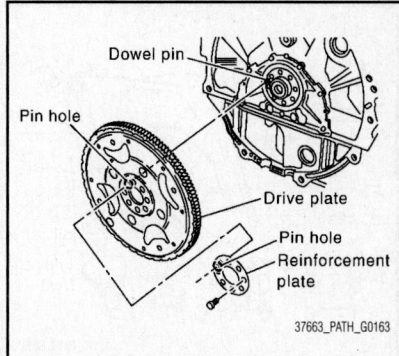

Fig. 66 Automatic transmission flexplate to transmission alignment—V8 engine

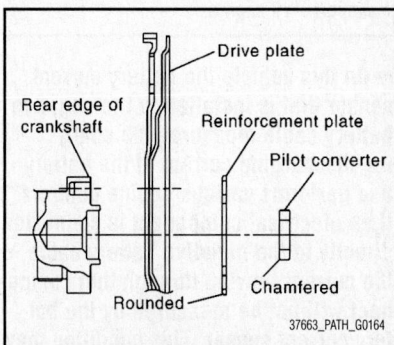

Fig. 67 Automatic transmission flexplate reinforcement plate installation direction—V8 engine

7. Be sure to correctly align the crankshaft side dowel pin and the flexplate side dowel pin hole (A).

➡**If not aligned correctly the MIL light will illuminate.**

8. Install the flexplate and reinforcement plate in the direction shown. See illustration.

9. Tighten the retaining bolts to specification in a crisscross pattern.

10. Continue the installation in the reverse order of the removal procedure.

11. Be sure to perform the reconnect/relearn procedures.

INTAKE MANIFOLD

REMOVAL & INSTALLATION

V6 Engine

See Figures 68 through 70.

1. Before servicing the vehicle, refer to the Precautions Section.

➡**If working near and/or around the SRS system and components, be sure to disable the SRS system. After disabling the system wait three minutes or more before servicing the vehicle.**

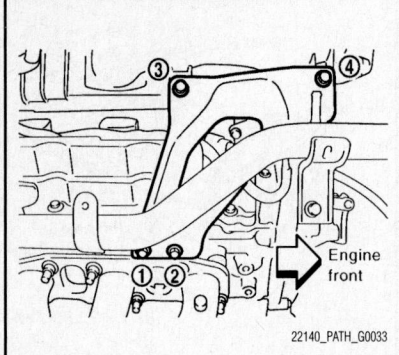

Fig. 68 Loosen bolts in reverse order as shown to remove intake manifold collector support—V6 engine

➡ On this vehicle the battery current sensor that is installed to the negative battery cable measures the charging/discharging current of the battery and performs various engine controls. If an electrical component is connected directly to the negative battery cable, the current flowing through that component will not be measured by the battery current sensor. This condition may cause a malfunction of the engine control system and battery discharge may occur. Do not connect an electrical component or ground wire directly to the battery terminal.

2. Disconnect the negative battery cable.
3. Properly relieve the fuel system pressure.
4. Remove engine cover.
5. Remove air cleaner case (upper) with mass air flow sensor and air duct assembly.
6. Remove electric throttle control actuator as follows:
    a. Drain engine coolant, or when water hoses are disconnected, attach plug to prevent engine coolant leakage.
    b. Disconnect water hoses from electric throttle control actuator.

➡ When engine coolant is not drained from radiator, attach plug to water hoses to prevent engine coolant leakage.

    c. Disconnect harness connector.
    d. Loosen bolts in reverse order of the tightening sequence.
7. Remove the following parts:
    • Vacuum hose (to brake booster)
    • PCV hose
8. Loosen bolts in reverse order as shown to remove intake manifold collector support.
9. Disconnect EVAP hoses and harness connector from EVAP canister purge volume control solenoid valve.

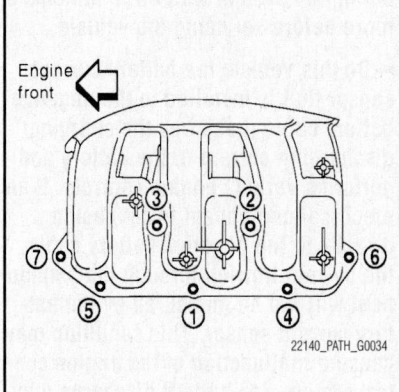

Fig. 69 Intake manifold bolt tightening sequence—V6 engine

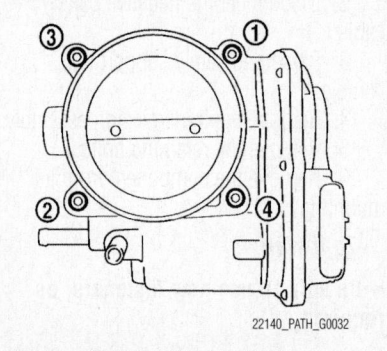

Fig. 70 Throttle body bolt tightening sequence—V6 engine

10. Remove EVAP canister purge volume control solenoid valve.
11. Remove VIAS control solenoid valve and vacuum tank.
12. Loosen nuts and bolts in reverse order of the tightening sequence.
13. Remove fuel tube and fuel injector assembly.
14. Loosen nuts and bolts in reverse order as shown to remove intake manifold.
15. Remove gaskets. Discard the gaskets.

To install:

➡ Be sure to use new fasteners, as required.

16. Installation is in the reverse order of removal.
17. Note the following:
    • If stud bolts were removed from cylinder head, install them and tighten to 8 ft. lbs. (11 Nm).
    • Tighten all nuts and bolts to the specified torque in two or more steps in numerical order as shown.
    • 1st step: 5 ft. lbs. (7.4 Nm)
    • 2nd step and after: 21 ft. lbs. (29.0 Nm)

18. Tighten the throttle body retaining bolts to specification and in the proper sequence.
19. Using the CONSULT-III diagnostic tool or equivalent, perform the throttle valve closed position learning and the idle air volume learning procedures.
20. Be sure to perform the reconnect/relearn procedures.

V8 Engine

See Figures 71 through 73.

1. Before servicing the vehicle, refer to the Precautions Section.

➡ If working near and/or around the SRS system and components, be sure to disable the SRS system. After disabling the system wait three minutes or more before servicing the vehicle.

➡ On this vehicle the battery current sensor that is installed to the negative battery cable measures the charging/discharging current of the battery and performs various engine controls. If an electrical component is connected directly to the negative battery cable, the current flowing through that component will not be measured by the battery current sensor. This condition may cause a malfunction of the engine control system and battery discharge may occur. Do not connect an electrical component or ground wire directly to the battery terminal.

2. Disconnect the negative battery cable.
3. Remove the air dam.
4. Remove the engine undercover.
5. Partially drain the engine coolant.

❄❄ CAUTION

To avoid the danger of being scalded, never drain the engine coolant when the engine is hot.

6. Remove the engine room cover.
7. Release the fuel pressure.
8. Remove the air duct and resonator assembly.
9. Disconnect the fuel tube quick connector on the engine side.
    a. Remove quick connector cap (engine side only).
    b. With the sleeve side of tool number 16641 6N210 (J45488) facing quick connector, install tool onto fuel tube.
    c. Insert Tool into quick connector until sleeve contacts and goes no further. Hold the Tool in that position.

d. Draw and pull out quick connector straight from fuel tube.

- Pull quick connector holding "A" position in illustration.
- Do not pull with lateral force applied. O-ring inside quick connector may be damaged.
- Prepare container and cloth beforehand as fuel will leak out.
- Avoid fire and sparks.
- Be sure to cover openings of disconnected pipes with plug or plastic bag to avoid fuel leakage and entry of foreign materials.

10. Remove or disconnect harnesses, brackets, vacuum hose, vacuum gallery and PCV hose and tube from intake manifold.

11. Remove electric throttle control actuator by loosening bolts diagonally.

12. Remove the fuel injectors and fuel tube assembly.

13. Loosen the bolts in reverse order of the tightening sequence.

14. Remove the intake manifold.

To install:

15. Installation is in the reverse order of removal.

16. Tighten the intake manifold bolts in numerical order as shown.

17. Install the EVAP canister purge control solenoid valve connector with it facing front of engine.

18. Tighten the electronic throttle control actuator bolts of the electric throttle control actuator equally and diagonally in several steps.

19. Install the water hose so that its overlap width for connection is between

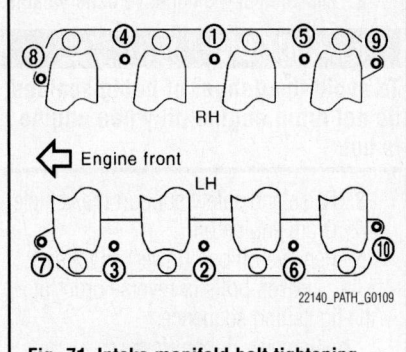

Fig. 71 Intake manifold bolt tightening sequence—V8 engine

1.06 in. (27 mm) and 1.26 in. (32 mm) (target: 1.06 in. or 27 mm).

20. Install quick connector as follows (the steps are the same for quick connectors on both engine side and vehicle side except for the quick connector cap).

21. Make sure no foreign substances are deposited in and around tube and quick connector, and they are not damaged.

22. Thinly apply new engine oil around the fuel tube from tip end to the spool end.

23. Align center to insert quick connector straight into fuel tube.

a. Insert until the paint mark for engagement identification (white) goes completely inside quick connector so that you cannot see it from the straight side of the connected part. Use a mirror to check this where it is not possible to view directly from the straight side, such as quick connector on vehicle side.

b. Insert fuel tube into quick connector until top spool is completely inside quick connector, and 2nd level spool exposes right below quick connector on engine side.

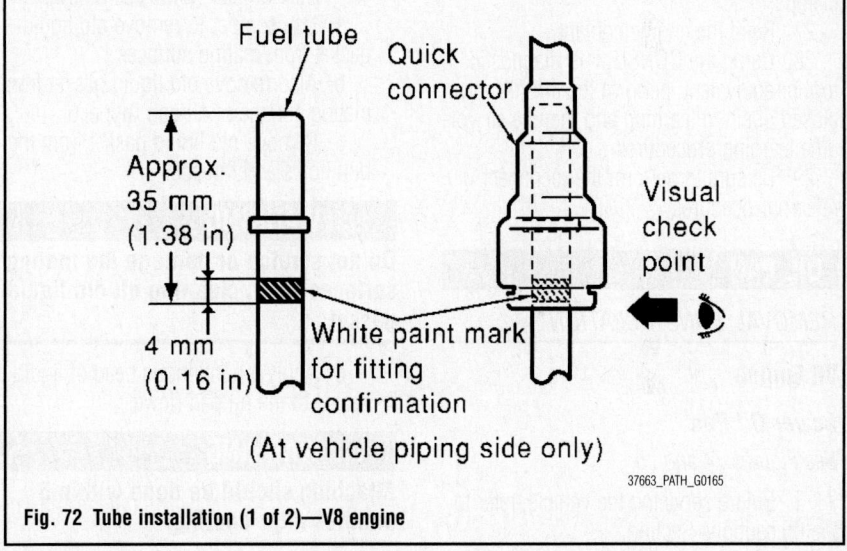

Fig. 72 Tube installation (1 of 2)—V8 engine

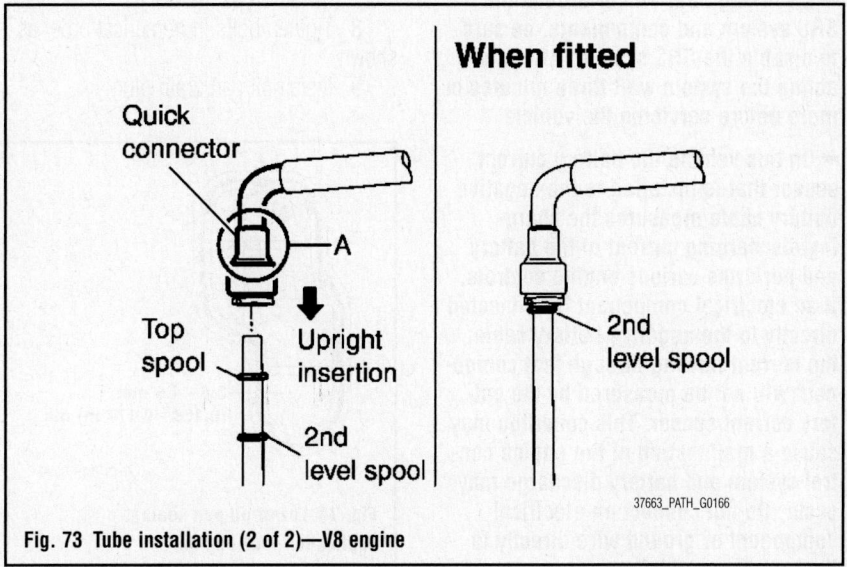

Fig. 73 Tube installation (2 of 2)—V8 engine

Hold "A" position in illustration when inserting fuel tube into quick connector. Carefully align center to avoid inclined insertion to prevent damage to O-ring inside quick connector. Insert until you hear a "click" sound and actually feel the engagement. To avoid misidentification of engagement with a similar sound, be sure to perform the next step.

24. Pull quick connector by hand holding "A" position. Make sure it is completely engaged (connected) so that it does not come out from fuel tube.

➡ Recommended pulling force is 50 N (5.1 kg, 11.2 lb).

25. Install the quick connector cap on the quick connector joint (on engine side only).
26. Install the fuel hose and tube to hose clamps.
27. Refill the engine coolant.
28. Using the CONSULT-III diagnostic tool or equivalent, perform the throttle valve closed position learning and the idle air volume learning procedures.
29. Be sure to perform the reconnect/relearn procedures.

OIL PAN

REMOVAL & INSTALLATION

V6 Engine

Lower Oil Pan

See Figures 74 and 75.

1. Before servicing the vehicle, refer to the Precautions Section.

➡ If working near and/or around the SRS system and components, be sure to disable the SRS system. After disabling the system wait three minutes or more before servicing the vehicle.

➡ On this vehicle the battery current sensor that is installed to the negative battery cable measures the charging/discharging current of the battery and performs various engine controls. If an electrical component is connected directly to the negative battery cable, the current flowing through that component will not be measured by the battery current sensor. This condition may cause a malfunction of the engine control system and battery discharge may occur. Do not connect an electrical component or ground wire directly to the battery terminal.

2. Disconnect the negative battery cable.

✳✳ CAUTION

To avoid the danger of being scalded, do not drain engine oil when engine is hot.

3. Raise and safely support the vehicle.
4. Drain engine oil.
5. Remove oil pan (lower) as follows:
   a. Loosen bolts in reverse order of the tightening sequence.
   b. Remove oil pan (lower).

✳✳ WARNING

Be careful not to damage the mating surfaces. Do not insert screwdriver, this will damage the mating surfaces.

➡ Slide seal cutter (1) by tapping on the side (2) of the tool with hammer.

To install:

6. Install oil pan (lower) as follows:
   a. Use scraper to remove old liquid gasket from mating surfaces.
   b. Also remove old liquid gasket from mating surface of oil pan (upper).
   c. Remove old liquid gasket from the bolt holes and threads.

✳✳ WARNING

Do not scratch or damage the mating surfaces when cleaning off old liquid gasket.

   d. Apply a continuous bead of liquid gasket to the oil pan (lower).

✳✳ WARNING

Attaching should be done within 5 minutes after coating.

7. Install oil pan (lower).
8. Tighten bolts in numerical order as shown.
9. Install oil pan drain plug.

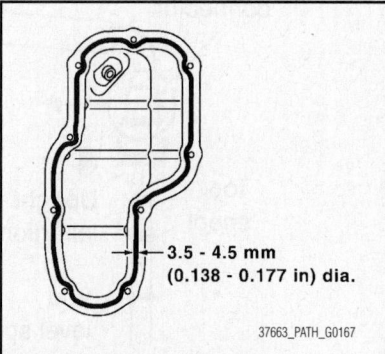

3.5 - 4.5 mm (0.138 - 0.177 in) dia.

37663_PATH_G0167

Fig. 74 Lower oil pan sealant application—V6 engine

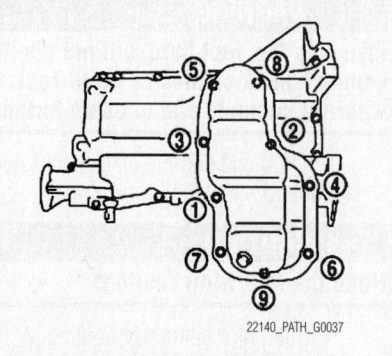

22140_PATH_G0037

Fig. 75 Lower oil pan bolt tightening sequence—V6 engine

➡ Wait at least 30 minutes after oil pan is installed before adding engine oil.

10. Check engine oil level and adjust engine oil.
11. Start engine, and check there is no leak of engine oil.
12. Stop engine and wait for 10 minutes.
13. Check engine oil level again.

Upper Oil Pan

See Figures 76 through 78.

1. Before servicing the vehicle, refer to the Precautions Section.

➡ If working near and/or around the SRS system and components, be sure to disable the SRS system. After disabling the system wait three minutes or more before servicing the vehicle.

➡ On this vehicle the battery current sensor that is installed to the negative battery cable measures the charging/discharging current of the battery and performs various engine controls. If an electrical component is connected directly to the negative battery cable, the current flowing through that component will not be measured by the battery current sensor. This condition may cause a malfunction of the engine control system and battery discharge may occur. Do not connect an electrical component or ground wire directly to the battery terminal.

2. Disconnect the negative battery cable.

✳✳ CAUTION

To avoid the danger of being scalded, do not drain engine oil when engine is hot.

3. Remove engine cover.
4. Remove air duct.
5. Raise and safely support the vehicle. Drain engine oil.

Perform this step when engine is cold. Do not spill engine oil on drive belts.

6. Drain engine coolant.

Perform this step when engine is cold. Do not spill engine coolant on drive belts.

7. Remove front final drive (4WD).

8. Disconnect steering gear lower joint shaft bolt and steering gear mounting nuts and bolts, position out of the way.

9. Remove the starter motor.

10. Disconnect A/T fluid cooler tube brackets and position out of the way.

11. Remove oil filter, as necessary.

12. Remove oil cooler.

13. Remove oil pan (lower).

14. Remove oil strainer.

15. Remove transmission joint bolts which pierce oil pan (upper).

16. Remove rear cover plate.

17. Loosen bolts in reverse order of the tightening sequence.

18. Remove O-rings from bottom of lower cylinder block and oil pump.

To install:

➥Be sure to use new fasteners, as required.

19. Install oil pan (upper) as follows:

a. Use scraper to remove old liquid gasket from mating surfaces.

b. Also remove the old liquid gasket from mating surface of lower cylinder block.

c. Remove old liquid gasket from the bolt holes and threads.

Do not scratch or damage the mating surfaces when cleaning off old liquid gasket.

d. Install new O-rings on the bottom of lower cylinder block and oil pump.

e. Apply a continuous bead of liquid gasket to the lower cylinder block mating surfaces of oil pan (upper).

f. For bolt holes with arrowhead mark, apply liquid gasket outside the hole.

g. Apply a bead of 4.5 to 5.5 mm (0.177 to 0.217 in) in diameter to area "A".

➥Attaching should be done within 5 minutes after coating.

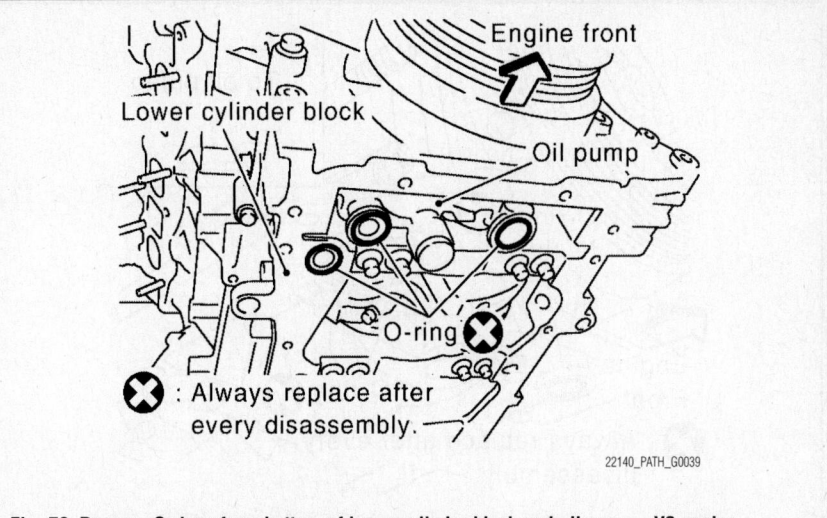

Fig. 76 Remove O-rings from bottom of lower cylinder block and oil pump—V6 engine

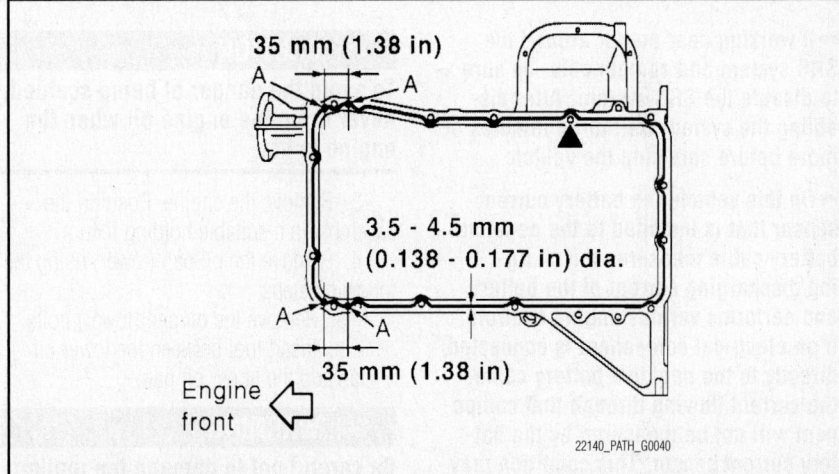

Fig. 77 Apply a continuous bead of liquid gasket to the lower cylinder block mating surfaces of oil pan (upper)—V6 engine

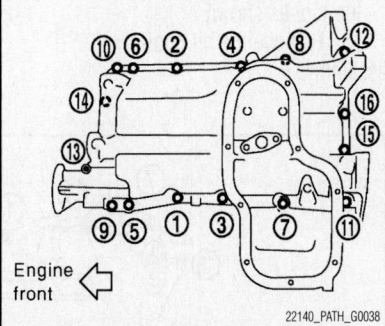

Fig. 78 Upper oil pan bolt tightening sequence—V6 engine

Install avoiding misalignment of both oil pan gaskets and O-rings.

20. Tighten bolts in numerical order as shown.

➥There are two types of bolts. Refer to the following for locating bolts.

• M8x4 in. (100 mm): holes 7, 11, 12, 13
• M8x1 in. (25 mm): All other holes

21. Tighten transmission joint bolts.

22. Install oil strainer to oil pan (upper).

23. Install oil pan (upper).

24. Check engine oil level and adjust engine oil.

25. Start engine, and check there is no leak of engine oil.

26. Stop engine and wait for 10 minutes.

27. Check engine oil level again.

28. Be sure to perform the reconnect/relearn procedures.

V8 Engine

See Figures 79 through 83.

1. Before servicing the vehicle, refer to the Precautions Section.

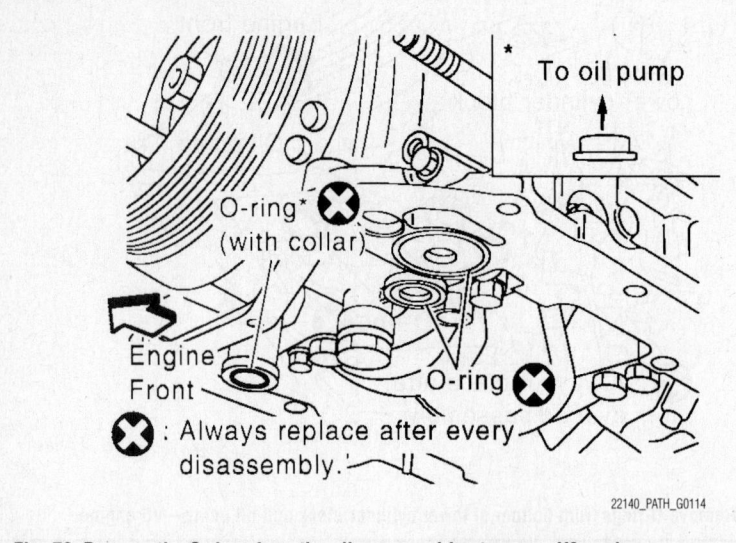

Fig. 79 Remove the O-rings from the oil pump and front cover—V8 engine

➡If working near and/or around the SRS system and components, be sure to disable the SRS system. After disabling the system wait three minutes or more before servicing the vehicle.

➡On this vehicle the battery current sensor that is installed to the negative battery cable measures the charging/discharging current of the battery and performs various engine controls. If an electrical component is connected directly to the negative battery cable, the current flowing through that component will not be measured by the battery current sensor. This condition may cause a malfunction of the engine control system and battery discharge may occur. Do not connect an electrical component or ground wire directly to the battery terminal.

2. Disconnect the negative battery cable.

✳✳ CAUTION

To avoid the danger of being scalded, never drain the engine oil when the engine is hot.

3. Remove the engine. Position the assembly in a suitable holding fixture.
4. Remove the oil pan (lower) using the following steps.
   a. Remove the oil pan (lower) bolts.
   b. Insert tool between the lower oil pan and the upper oil pan.

✳✳ WARNING

Be careful not to damage the mating surface.

   c. Tap seal cutter to insert it (1) and then slide it by tapping on the side (2) of the tool as shown.
5. Remove the oil cooler assembly.

6. Remove the oil strainer from the oil pan (upper).
7. Remove the oil pan (upper) using the following steps.
   a. Remove the oil pan (upper) bolts in the reverse order of the installation sequence.
   b. Remove the oil pan (upper) from the cylinder block by prying.

✳✳ WARNING

Do not damage mating surface.

8. Remove the O-rings from the oil pump and front cover.

➡Do not reuse O-rings.

To install:

➡Be sure to use new fasteners, as required.

9. Install the oil pan (upper) using the following steps.
   a. Apply liquid gasket thoroughly as shown.

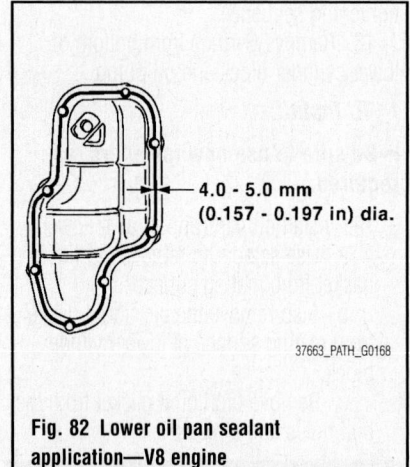

Fig. 82 Lower oil pan sealant application—V8 engine

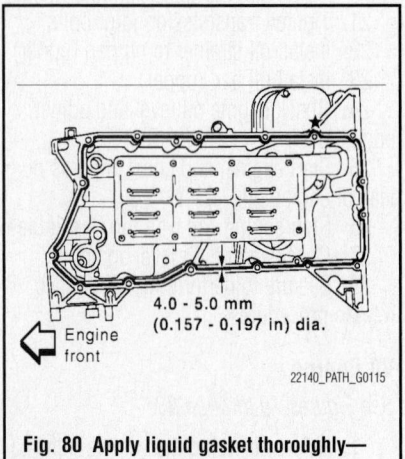

Fig. 80 Apply liquid gasket thoroughly—V8 engine

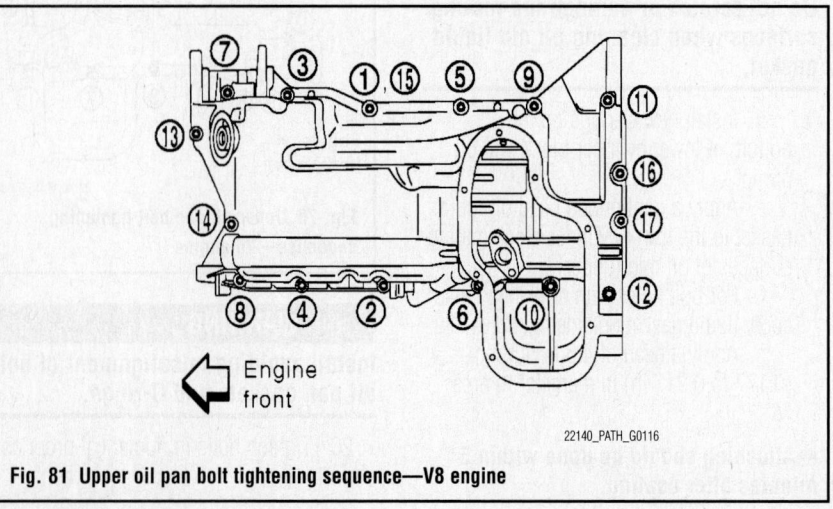

Fig. 81 Upper oil pan bolt tightening sequence—V8 engine

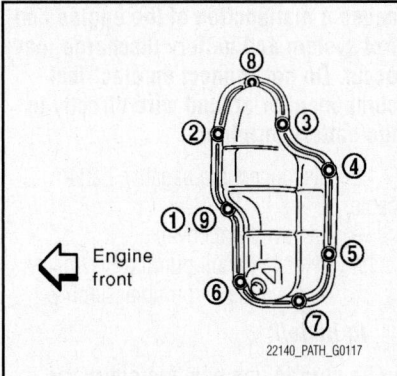

Fig. 83 Lower oil pan bolt tightening sequence—V8 engine

※※ WARNING

Apply liquid gasket to outside of bolt hole for the hole shown by star (*).

　b. Install new O-rings to the oil pump and front cover side.
　c. Tighten the bolts in numerical order as shown.

➡️**Bolt locations by size:**

- M6 _ 1.18 in. (30 mm): No. 15, 16
- M8 _ 0.98 in. (25 mm): No. 1, 3, 5, 7, 11, 13
- M8 _ 1.77 in. (45 mm): No. 2, 4, 6, 8, 10, 14
- M8 _ 4.84 in. (123 mm): No. 9, 12

　10. Install the oil strainer to the oil pan (upper).
　11. Install the oil pan (lower).
　　a. Apply liquid gasket thoroughly as shown.

※※ WARNING

Attaching should be done within 5 minutes after coating.

　b. Tighten the oil pan (lower) bolts in numerical order as shown.
　12. Install the oil pan drain plug.
　13. Install engine assembly.

➡️**Do not fill the engine with oil for at least 30 minutes after oil pan is installed.**

　14. Check engine oil level and add engine oil if necessary.

　15. Start the engine, and check for leaks of engine oil.
　16. Stop engine and wait for 10 minutes.
　17. Check engine oil level again.
　18. Be sure to perform the reconnect/relearn procedures.

OIL PUMP

REMOVAL & INSTALLATION

V6 Engine

See Figure 84.

　1. Before servicing the vehicle, refer to the Precautions Section.

➡️**If working near and/or around the SRS system and components, be sure to disable the SRS system. After disabling the system wait three minutes or more before servicing the vehicle.**

➡️**On this vehicle the battery current sensor that is installed to the negative battery cable measures the charging/discharging current of the battery and performs various engine controls.**

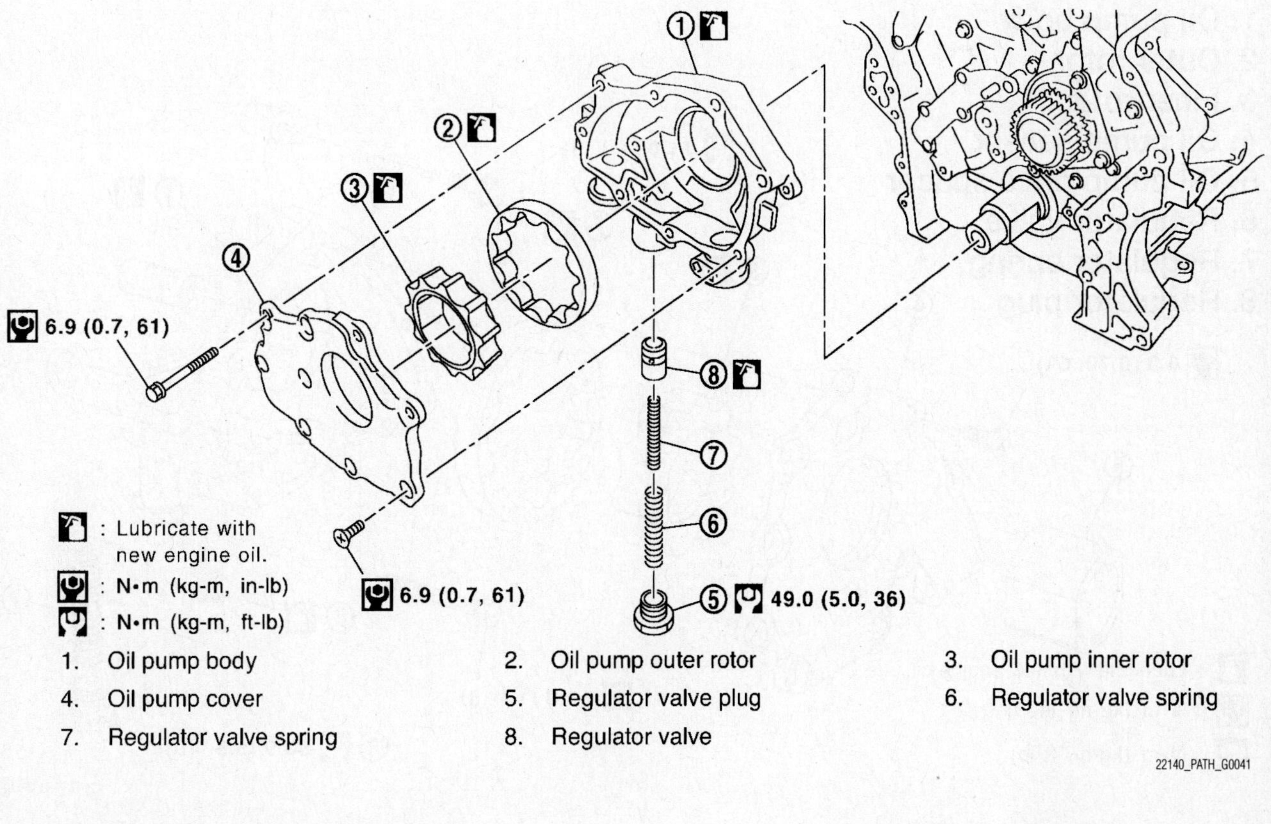

6.9 (0.7, 61)

🔧 : Lubricate with new engine oil.
⚙️ : N·m (kg-m, in-lb)
🔩 : N·m (kg-m, ft-lb)

6.9 (0.7, 61)

49.0 (5.0, 36)

1. Oil pump body
2. Oil pump outer rotor
3. Oil pump inner rotor
4. Oil pump cover
5. Regulator valve plug
6. Regulator valve spring
7. Regulator valve spring
8. Regulator valve

Fig. 84 Oil pump and related components—V6 engine

If an electrical component is connected directly to the negative battery cable, the current flowing through that component will not be measured by the battery current sensor. This condition may cause a malfunction of the engine control system and battery discharge may occur. Do not connect an electrical component or ground wire directly to the battery terminal.

2. Disconnect the negative battery cable.
3. Raise and safely support the vehicle. Drain the engine oil.
4. Remove oil pans (lower and upper).
5. Remove front timing chain case and timing chain (primary).
6. Remove oil pump assembly.

To install:

➡**Be sure to use new fasteners, as required.**

7. Installation is in the reverse order of removal, paying attention to the following:
8. When installing, align crankshaft flat faces with inner rotor flat faces.
9. Check the engine oil level.
10. Start engine, and check there are no leaks of engine oil.

11. Stop engine and wait for 10 minutes.
12. Check the engine oil level and add engine oil
13. Be sure to perform the reconnect/relearn procedures.

V8 Engine

See Figures 85 and 86.

1. Before servicing the vehicle, refer to the Precautions Section.

➡**If working near and/or around the SRS system and components, be sure to disable the SRS system. After disabling the system wait three minutes or more before servicing the vehicle.**

➡**On this vehicle the battery current sensor that is installed to the negative battery cable measures the charging/discharging current of the battery and performs various engine controls. If an electrical component is connected directly to the negative battery cable, the current flowing through that component will not be measured by the battery current sensor. This condition may** cause a malfunction of the engine control system and battery discharge may occur. Do not connect an electrical component or ground wire directly to the battery terminal.

2. Disconnect the negative battery cable.
3. Remove front cover.
4. Remove the oil pump drive spacer.
5. Remove the oil pump assembly.

To install:

➡**Be sure to use new fasteners, as required.**

6. Installation is in the reverse order of removal, paying attention of the following:
- When inserting the oil pump drive spacer, align the crankshaft key and the flat face of the inner rotor.
- If they are not aligned, rotate the oil pump inner rotor by hand.
- Make sure that each part is aligned and tap lightly until it reaches the end.
7. Check the engine oil level.
8. Start the engine and check for engine oil leaks.
9. Stop the engine and wait 10 minutes.

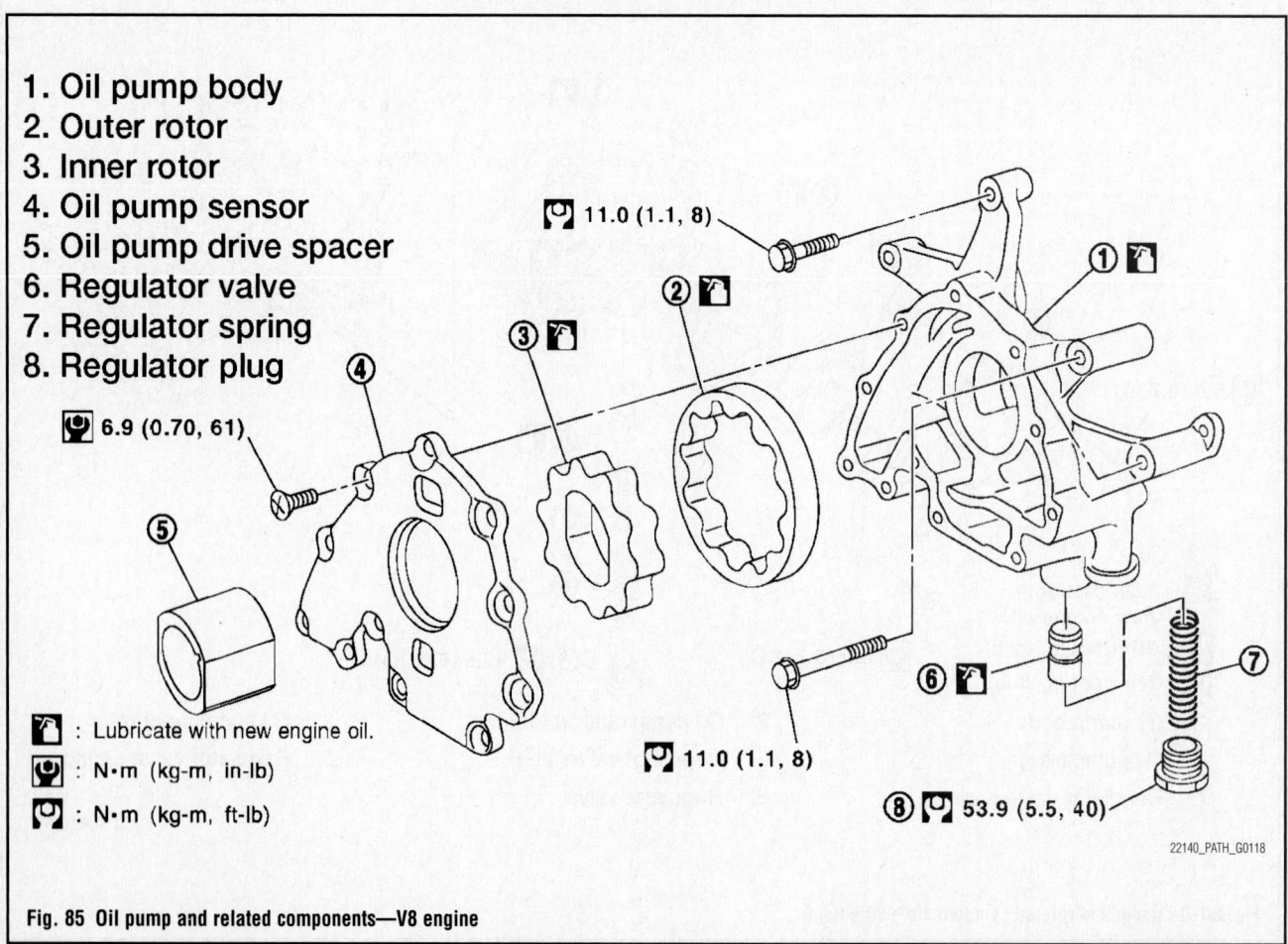

1. Oil pump body
2. Outer rotor
3. Inner rotor
4. Oil pump sensor
5. Oil pump drive spacer
6. Regulator valve
7. Regulator spring
8. Regulator plug

11.0 (1.1, 8)

6.9 (0.70, 61)

11.0 (1.1, 8)

8 53.9 (5.5, 40)

: Lubricate with new engine oil.

: N•m (kg-m, in-lb)

: N•m (kg-m, ft-lb)

22140_PATH_G0118

Fig. 85 Oil pump and related components—V8 engine

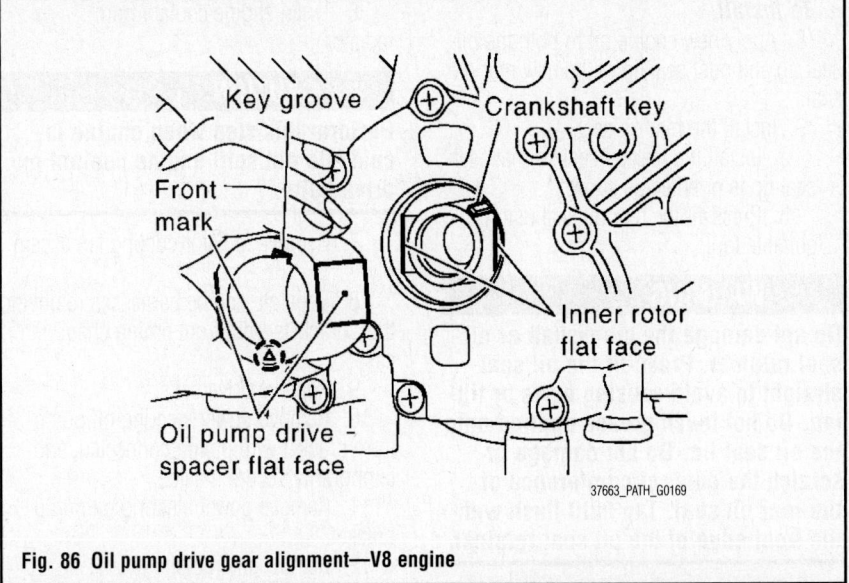

Fig. 86 Oil pump drive gear alignment—V8 engine

10. Check the engine oil level and adjust the engine oil level as required.

11. Be sure to perform the reconnect/relearn procedures.

PISTON AND RING

POSITIONING

See Figures 87 and 88.

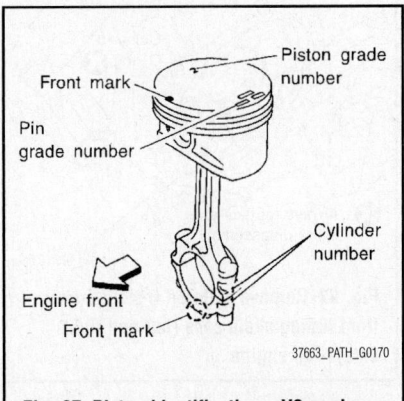

Fig. 87 Piston identification—V6 engine

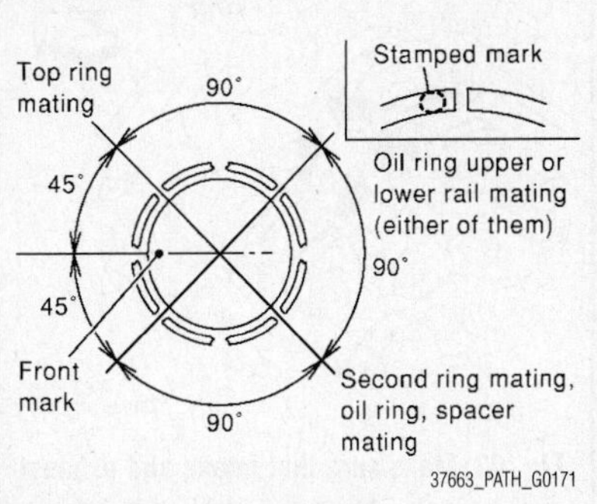

Fig. 88 Piston identification and alignment—V8 engine

REAR MAIN SEAL

REMOVAL & INSTALLATION

V6 Engine

See Figures 89 and 90.

1. Before servicing the vehicle, refer to the Precautions Section.

➡If working near and/or around the SRS system and components, be sure to disable the SRS system. After disabling the system wait three minutes or more before servicing the vehicle.

➡On this vehicle the battery current sensor that is installed to the negative battery cable measures the charging/discharging current of the battery and performs various engine controls. If an electrical component is connected directly to the negative battery cable, the current flowing through that component will not be measured by the battery current sensor. This condition may cause a malfunction of the engine control system and battery discharge may occur. Do not connect an electrical component or ground wire directly to the battery terminal.

2. Disconnect the negative battery cable.

3. Remove transmission assembly.

4. Matchmark and remove the flexplate.

5. Remove rear oil seal with a suitable tool.

✳✳ WARNING

Be careful not to damage crankshaft and cylinder block.

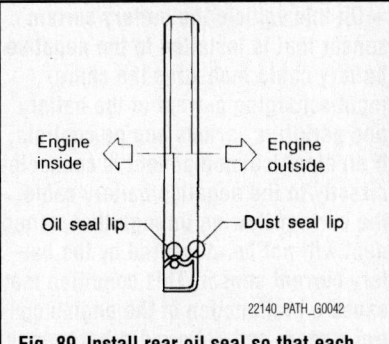

Fig. 89 Install rear oil seal so that each seal lip is oriented as shown—V6 & V8 engines

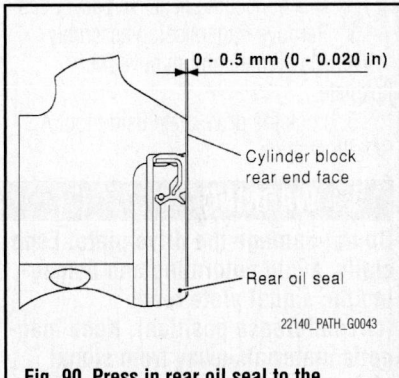

Fig. 90 Press in rear oil seal to the position as shown—V6 engine

To install:

6. Apply new engine oil to new rear oil seal joint surface and seal lip.

7. Install rear oil seal so that each seal lip is oriented as shown.

8. Press in rear oil seal to the position as shown.

✳✳ WARNING

Be careful not to damage crankshaft and cylinder block. Press-fit oil seal straight to avoid causing burrs or tilting. Do not touch grease applied onto oil seal lip.

9. Installation of the remaining components is in the reverse order of removal.

10. Be sure to perform the reconnect/relearn procedures.

V8 Engine

See Figures 54 and 91.

1. Before servicing the vehicle, refer to the Precautions Section.

➡If working near and/or around the SRS system and components, be sure to disable the SRS system. After disabling the system wait three minutes or more before servicing the vehicle.

→ On this vehicle the battery current sensor that is installed to the negative battery cable measures the charging/discharging current of the battery and performs various engine controls. If an electrical component is connected directly to the negative battery cable, the current flowing through that component will not be measured by the battery current sensor. This condition may cause a malfunction of the engine control system and battery discharge may occur. Do not connect an electrical component or ground wire directly to the battery terminal.

2. Disconnect the negative battery cable.
3. Remove transmission assembly.
4. Matchmark and remove the flexplate.
5. Lock the drive plate using tool A. See illustration.

✳✳ WARNING

Do not damage the drive plate. Especially, avoid deforming and damaging the signal plate teeth (circumference position). Keep magnetic materials away from signal plate.

6. Remove the drive plate.

✳✳ WARNING

Place the drive plate with the signal plate surface facing upward.

→ Remove the bolts diagonally.

7. Remove the rear oil seal using suitable tool.

✳✳ WARNING

Do not damage crankshaft or oil seal retainer surface.

Fig. 91 Lock the drive plate using tool A—V8 engine

To install:

8. Apply new engine oil to both the oil seal lip and dust seal lip of the new rear oil seal.
9. Install the rear oil seal.
  a. Install the rear oil seal so that each seal lip is oriented as shown.
  b. Press-fit the rear oil seal using suitable tool.

✳✳ WARNING

Do not damage the crankshaft or oil seal retainer. Press-fit the oil seal straight to avoid causing burrs or tilting. Do not touch grease applied onto the oil seal lip. Do not damage or scratch the outer circumference of the rear oil seal. Tap until flush with the front edge of the oil seal retainer.

10. Installation of the remaining components is in the reverse order of removal.

TIMING CHAIN FRONT COVER

REMOVAL & INSTALLATION

V6 Engine

See Figures 92 through 97.

1. Before servicing the vehicle, refer to the Precautions Section.

→ If working near and/or around the SRS system and components, be sure to disable the SRS system. After disabling the system wait three minutes or more before servicing the vehicle.

→ On this vehicle the battery current sensor that is installed to the negative battery cable measures the charging/discharging current of the battery and performs various engine controls. If an electrical component is connected directly to the negative battery cable, the current flowing through that component will not be measured by the battery current sensor. This condition may cause a malfunction of the engine control system and battery discharge may occur. Do not connect an electrical component or ground wire directly to the battery terminal.

2. Disconnect the negative battery cable.
3. Remove engine cover.
4. Release the fuel pressure.
5. Drain engine oil.

✳✳ CAUTION

Perform this step when engine is cold. Do not spill engine oil on drive belts.

6. Drain engine coolant from radiator.

✳✳ CAUTION

Perform this step when engine is cold. Do not spill engine coolant on drive belts.

7. Remove radiator cooling fan assembly.
8. Separate engine harnesses removing their brackets from front timing chain case.
9. Remove drive belts.
10. Remove power steering oil pump from bracket with piping connected, and temporarily secure it aside.
11. Remove power steering oil pump bracket.
12. Remove alternator.
13. Remove water bypass hose, water hose clamp and idler pulley bracket from front timing chain case.
14. Remove right and left intake valve timing control covers.
  a. Loosen bolts in reverse order of tightening sequence.
  b. Cut liquid gasket for removal.

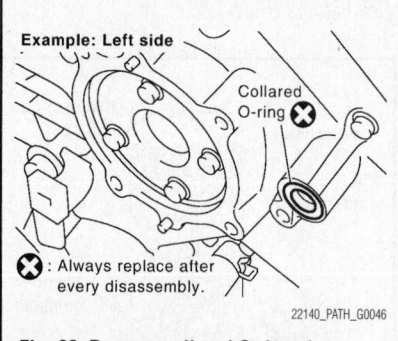

Fig. 92 Remove collared O-rings from front timing chain case (left and right side)—V6 engine

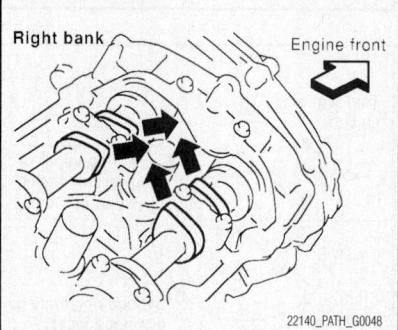

Fig. 93 Make sure that intake and exhaust cam noses on No. 1 cylinder (engine front side of right bank) are located as shown—V6 engine

⁂ WARNING

Shaft is internally jointed with camshaft sprocket (INT) center hole. When removing, keep it horizontal until it is completely disconnected.

15. Remove collared O-rings from front timing chain case (left and right side).

16. Remove rocker covers (right and left banks).

➡ **When only timing chain (primary) is removed, rocker cover does not need to be removed.**

17. Obtain No. 1 cylinder at TDC of its compression stroke as follows:

➡ **When timing chain is not removed/installed, this step is not required.**

a. Rotate crankshaft pulley clockwise to align timing mark (A) with timing indicator (B).

b. Make sure that intake and exhaust cam noses on No. 1 cylinder (engine front side of right bank) are located as shown.

c. If not, turn crankshaft one revolution (360°) and align as shown.

➡ **When only timing chain (primary) is removed, rocker cover does not need to be removed. To make sure that No. 1 cylinder is at its compression TDC, remove front timing chain case first. Then check mating marks on camshaft sprockets.**

18. Remove crankshaft pulley as follows:

a. Remove starter motor.

b. Loosen crankshaft pulley bolt and locate bolt seating surface as 0.39 in. (10 mm) from its original position.

⁂ WARNING

Do not remove crankshaft pulley bolt. Keep loosened crankshaft pulley bolt

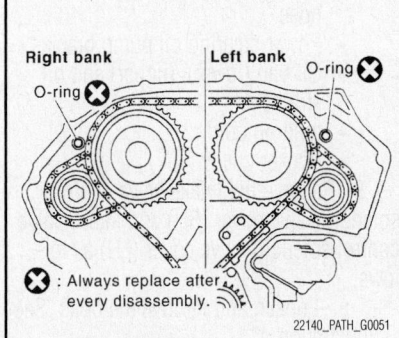

Right bank **Left bank**
O-ring ✖ O-ring ✖

✖ : Always replace after every disassembly.

22140_PATH_G0051

Fig. 94 Remove O-rings from rear timing chain case—V6 engine

in place protect removed crankshaft pulley from dropping.

c. Pull crankshaft pulley with both hands to remove it.

19. Loosen two bolts in front of oil pan (upper) in the reverse order of the tightening sequence.

20. Remove front timing chain case as follows:

a. Loosen bolts in reverse order of the tightening sequence.

b. Insert suitable tool into the notch at the top of the front timing chain case.

c. Pry off case by moving tool.

d. Cut liquid gasket for removal.

⁂ WARNING

Do not use screwdriver or something similar. After removal, handle front timing chain case carefully so it does not tilt, cant, or warp under a load.

21. Remove O-rings from rear timing chain case.

To install:

➡ **Be sure to use new fasteners, as required.**

22. Install front timing chain case as follows:

a. Apply a continuous bead of liquid gasket to front timing chain case back side as shown.

b. Install new O-rings on rear timing chain case.

c. Assemble front timing chain case as follows:

- Fit lower end of front timing chain case tightly onto top face of oil pan (upper). From the fitting point, make entire front timing chain case contact rear timing chain case completely.

- Since front timing chain case is off-set for difference of bolt holes, tighten bolts temporarily while holding front timing chain case from front and top.

- Same as the previous step, insert dowel pin while holding front timing chain case from front and top completely.

d. Tighten bolts to the specified torque in numerical order as shown.

➡ **There are two type of bolts. Refer to the following for locating bolts.**

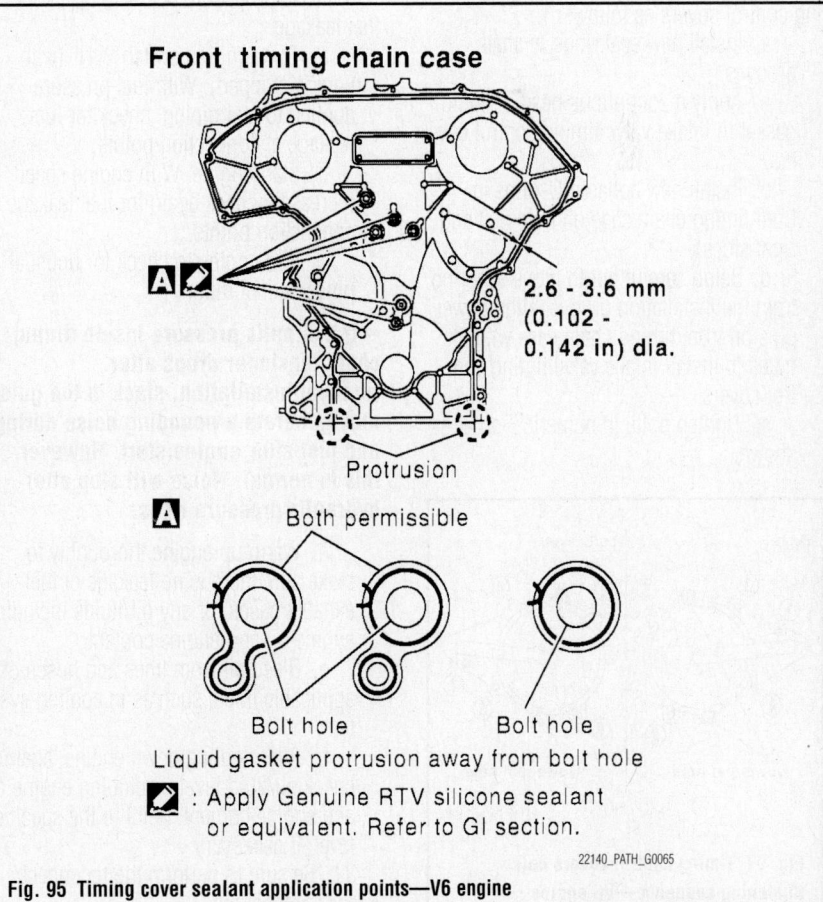

Front timing chain case

Ⓐ ✎

2.6 - 3.6 mm
(0.102 - 0.142 in) dia.

Protrusion

Ⓐ Both permissible

Bolt hole   Bolt hole

Liquid gasket protrusion away from bolt hole

✎ : Apply Genuine RTV silicone sealant or equivalent. Refer to GI section.

22140_PATH_G0065

Fig. 95 Timing cover sealant application points—V6 engine

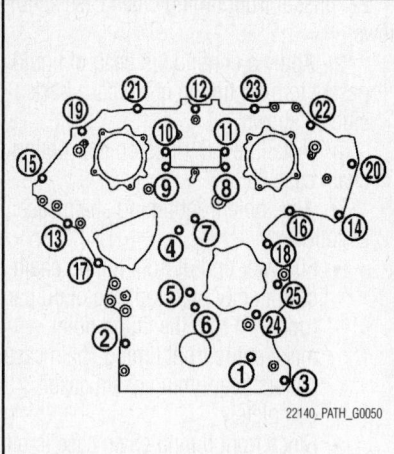

Fig. 96 Timing cover bolt tightening sequence—V6 engine

- 1–5: 0.39 in. (10 mm)
- 6–25: 0.24 in. (6 mm)
- Bolt position torque specification: 1–5: 41 ft. lbs. (55.0 Nm); 6–25: 9 ft. lbs. (12.7 Nm)

e. After all bolts tightened, retighten them to the specified torque in numerical order as shown.

23. Install two bolts in front of oil pan (upper).

24. Install right and left intake valve timing control covers as follows:

a. Install new seal rings in shaft grooves.

b. Apply a continuous bead of liquid gasket to intake valve timing control covers.

c. Install new collared O-rings in front timing chain case oil hole (left and right sides).

d. Being careful not to move seal ring from the installation groove, align dowel pins on front timing chain case with the holes to install intake valve timing control covers.

e. Tighten bolts in numerical order as shown.

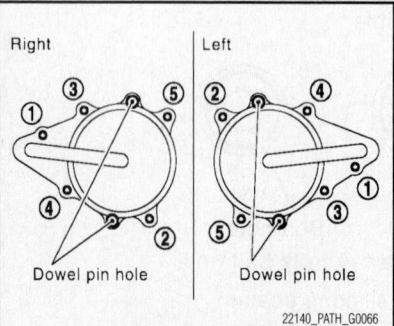

Fig. 97 Timing control covers bolt tightening sequence—V6 engine

25. Install crankshaft pulley as follows:

a. Install crankshaft pulley, taking care not to damage front oil seal.

➡ **When press-fitting crankshaft pulley with plastic hammer, tap on its center portion (not circumference).**

b. Tighten crankshaft pulley bolt. Crankshaft bolt torque: 33ft. lbs. (44.1 Nm)

c. Put a paint mark on crankshaft pulley aligning with angle mark on crankshaft pulley bolt. Then, further retighten bolt by 60°(equivalent to one graduation).

26. Rotate crankshaft pulley in normal direction (clockwise when viewed from front) to confirm it turns smoothly.

27. Install oil pans (upper and lower).

28. Install rocker covers (right and left banks).

29. Installation of the remaining components is in the reverse order of removal after this step.

30. The following are procedures for checking fluid leaks, lubricant leaks and exhaust gases leaks.

a. Before starting engine, check oil/fluid levels including engine coolant and engine oil. If less than required quantity, fill to the specified level.

31. Use procedure below to check for fuel leakage.

a. Turn ignition switch "ON" (with engine stopped). With fuel pressure applied to fuel piping, check for fuel leakage at connection points.

b. Start engine. With engine speed increased, check again for fuel leakage at connection points.

c. Run engine to check for unusual noise and vibration.

➡ **If hydraulic pressure inside timing chain tensioner drops after removal/installation, slack in the guide may generate a pounding noise during and just after engine start. However, this is normal. Noise will stop after hydraulic pressure rises.**

d. Warm up engine thoroughly to make sure there is no leakage of fuel, exhaust gases, or any oil/fluids including engine oil and engine coolant.

e. Bleed air from lines and hoses of applicable lines, such as in cooling system.

f. After cooling down engine, again check oil/fluid levels including engine oil and engine coolant. Refill to the specified level, if necessary.

32. Be sure to perform the reconnect/relearn procedures.

V8 Engine

See Figures 98 through 101.

1. Before servicing the vehicle, refer to the Precautions Section.

➡ **If working near and/or around the SRS system and components, be sure to disable the SRS system. After disabling the system wait three minutes or more before servicing the vehicle.**

➡ **On this vehicle the battery current sensor that is installed to the negative battery cable measures the charging/discharging current of the battery and performs various engine controls. If an electrical component is connected directly to the negative battery cable, the current flowing through that component will not be measured by the battery current sensor. This condition may cause a malfunction of the engine control system and battery discharge may occur. Do not connect an electrical component or ground wire directly to the battery terminal.**

2. Disconnect the negative battery cable.

➡ **To remove timing chain and associated parts, start with those on the LH bank. The procedure for removing parts on the RH bank is omitted because it is the same as that for removal on the LH bank.**

To install timing chain and associated parts, start with those on the RH bank. The procedure for installing parts on the LH bank is omitted because it is the same as that for installation on the RH bank.

3. Remove the engine from the vehicle. Position the assembly in a suitable holding fixture.

4. Remove the following components and related parts:

- Drive belt auto tensioner and idler pulley.
- Thermostat housing and water hose.
- Power steering oil pump bracket.
- Oil pan (lower), (upper) and oil strainer.
- Ignition coil.
- Rocker cover.

5. Remove the Intake valve control solenoid valve cover (RH) and Intake valve control solenoid valve cover (LH) as follows:

a. Loosen and remove the bolts. See illustration.

b. Cut the liquid gasket and remove the covers.

※※ **WARNING**

Do not damage mating surfaces.

6. Obtain compression TDC of No. 1 cylinder as follows:

a. Turn the crankshaft pulley clockwise to align the TDC identification notch (without paint mark) with the timing indicator on the front cover.

b. At this time, make sure both intake and exhaust cam lobes of No. 1 cylinder (top front on LH bank) point outside. If they do not point outside, turn crankshaft pulley once more.

7. Remove the crankshaft pulley.

a. Loosen the crankshaft pulley bolts using a hammer handle to secure the crankshaft.

b. Remove the crankshaft pulley from the crankshaft.

c. Remove the crankshaft pulley.

➡**The dimension between the centers of the two bolt holes is 61 mm (2.40 in.).**

8. Remove the front cover.

a. Loosen and remove the bolts in the reverse order of the tightening sequence.

b. Cut the liquid gasket and remove the covers.

※※ **WARNING**

Do not damage mating surfaces.

9. Remove the front oil seal using suitable tool.

※※ **WARNING**

Do not damage front cover.

To install:

➡**Be sure to use new fasteners, as required.**

10. Install the front oil seal using suitable tool.

※※ **WARNING**

Do not scratch or make burrs on the circumference of the oil seal.

11. Install the chain tensioner cover.

12. Install the front cover as follows:

a. Install a new O-ring on the cylinder block.

b. Apply liquid gasket as shown.

c. Check again that the timing alignment marks on the timing chain and on each sprocket are aligned. Then install the front cover.

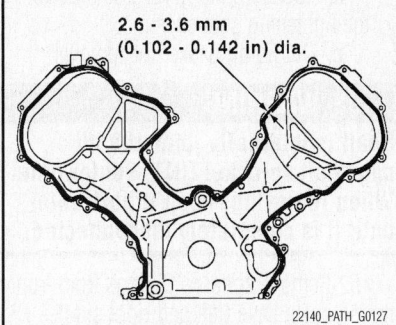

Fig. 98 Timing cover sealant application—V8 engine

d. Install the bolts in the numerical order shown.

e. After tightening, re-tighten to the specified torque.

※※ **WARNING**

Be sure to wipe off any excessive liquid gasket leaking onto surface mating with oil pan.

13. Install the Intake valve control solenoid valve cover (RH) (A) and Intake valve control solenoid valve cover (LH) (B) as follows:

a. Cross mark (C) that cannot be seen after assembly.

b. Apply liquid gasket (D) as shown.

※※ **WARNING**

The start and end of the application of the liquid gasket should be crossed at a position that cannot be seen after attaching the Intake valve control solenoid valve cover.

c. Install the bolts in the numerical order shown.

14. Install the crankshaft pulley.

a. Install the key of the crankshaft.

b. Insert the pulley by lightly tapping it.

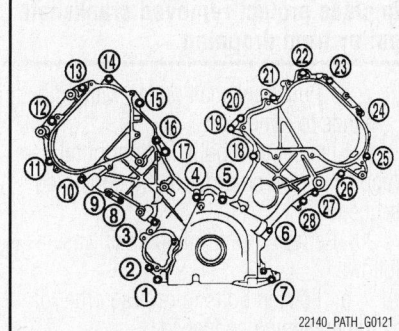

Fig. 99 Timing cover bolt tightening sequence—V8 engine

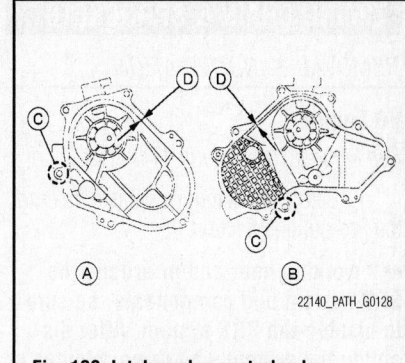

Fig. 100 Intake valve control valve sealant application (D)—V8 engine

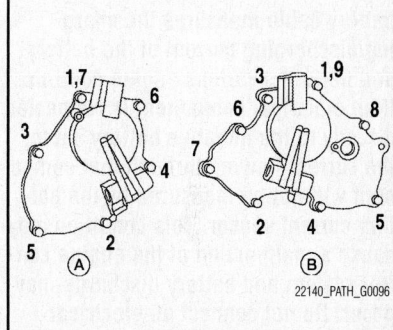

Fig. 101 Intake valve control solenoid bolt tightening sequence—V8 engine

※※ **WARNING**

Do not tap pulley on the side surface where the belt is installed (outer circumference).

15. Tighten the crankshaft pulley bolt.

a. Lock the crankshaft using suitable tool, then tighten the bolt.

b. Perform the following steps for angular tightening:

c. Apply engine oil onto the threaded parts of the bolt and seating area.

d. Select the one most visible notch of the four on the bolt flange. Corresponding to the selected notch, put a alignment mark (such as paint) on the crankshaft pulley.

16. Rotate the crankshaft pulley in normal direction (clockwise when viewed from engine front) to check for parts interference.

17. Installation of the remaining components is in the reverse of order of removal.

18. Be sure to perform the reconnect/relearn procedures.

TIMING CHAIN & SPROCKETS

REMOVAL & INSTALLATION

V6 Engine

See Figures 102 through 114.

1. Before servicing the vehicle, refer to the Precautions Section.

➡If working near and/or around the SRS system and components, be sure to disable the SRS system. After disabling the system wait three minutes or more before servicing the vehicle.

➡On this vehicle the battery current sensor that is installed to the negative battery cable measures the charging/discharging current of the battery and performs various engine controls. If an electrical component is connected directly to the negative battery cable, the current flowing through that component will not be measured by the battery current sensor. This condition may cause a malfunction of the engine control system and battery discharge may occur. Do not connect an electrical component or ground wire directly to the battery terminal.

2. Disconnect the negative battery cable.
3. Remove engine cover.
4. Release the fuel pressure.
5. Drain engine oil.

✳✳ CAUTION

Perform this step when engine is cold. Do not spill engine oil on drive belts.

6. Drain engine coolant from radiator.

✳✳ CAUTION

Perform this step when engine is cold. Do not spill engine coolant on drive belts.

7. Remove radiator cooling fan assembly.
8. Separate engine harnesses removing their brackets from front timing chain case.
9. Remove drive belts.
10. Remove power steering oil pump from bracket with piping connected, and temporarily secure it aside.
11. Remove power steering oil pump bracket.
12. Remove alternator.
13. Remove water bypass hose, water hose clamp and idler pulley bracket from front timing chain case.
14. Remove right and left intake valve timing control covers.

a. Loosen bolts in reverse order of the tightening sequence.
b. Cut liquid gasket for removal.

✳✳ WARNING

Shaft is internally jointed with camshaft sprocket (INT) center hole. When removing, keep it horizontal until it is completely disconnected.

15. Remove collared O-rings from front timing chain case (left and right side).
16. Remove rocker covers (right and left banks).

➡When only timing chain (primary) is removed, rocker cover does not need to be removed.

17. Obtain No. 1 cylinder at TDC of its compression stroke as follows:

➡When timing chain is not removed/installed, this step is not required.

a. Rotate crankshaft pulley clockwise to align timing mark (grooved line without color) with timing indicator.
b. Make sure that intake and exhaust cam noses on No. 1 cylinder (engine front side of right bank) are located as shown.
c. If not, turn crankshaft one revolution (360°) and align as shown.

➡When only timing chain (primary) is removed, rocker cover does not need to be removed. To make sure that No. 1 cylinder is at its compression TDC, remove front timing chain case first. Then check mating marks on camshaft sprockets.

18. Remove crankshaft pulley as follows:
a. Remove starter motor.
b. Loosen crankshaft pulley bolt and locate bolt seating surface as 0.39 in. (10 mm) from its original position.

✳✳ WARNING

Do not remove crankshaft pulley bolt. Keep loosened crankshaft pulley bolt in place protect removed crankshaft pulley from dropping.

c. Pull crankshaft pulley with both hands to remove it.
19. Loosen two bolts in front of oil pan (upper) in reverse order of the tightening sequence.
20. Remove front timing chain case as follows:
a. Loosen bolts in reverse order of the tightening sequence.
b. Insert suitable tool into the notch at the top of the front timing chain case.

c. Pry off case by moving tool.
d. Cut liquid gasket for removal.

✳✳ WARNING

Do not use screwdriver or something similar. After removal, handle front timing chain case carefully so it does not tilt, cant, or warp under a load.

21. Remove O-rings from rear timing chain case.
22. Remove water pump cover and chain tensioner cover from front timing chain case, if necessary.
23. Remove front oil seal from front timing chain case using suitable tool.

✳✳ WARNING

Be careful not to damage front timing chain case.

24. Use a scraper to remove all traces of old liquid gasket from front and rear timing chain cases and oil pan (upper), and liquid gasket mating surfaces.

✳✳ WARNING

Be careful not to allow gasket fragments to enter oil pan.

25. Remove old liquid gasket from bolt holes and threads.
26. Use a scraper to remove all traces of old liquid gasket from water pump cover, chain tensioner cover and intake valve timing control covers.
27. Remove timing chain tensioner (primary) as follows:
a. Loosen clip of timing chain tensioner (primary), and release plunger stopper (1).
b. Insert plunger into tensioner body by pressing slack guide (2).
c. Keep slack guide pressed and hold plunger in by pushing stopper pin

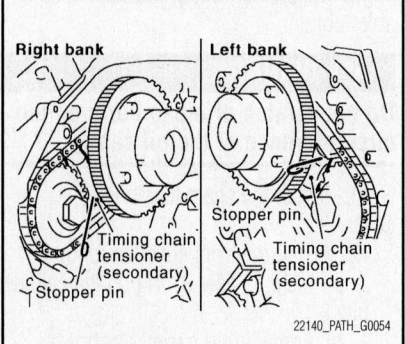

Fig. 102 Attach suitable stopper pin to the right and left timing chain tensioners (secondary)—V6 engine

through the tensioner body hole and plunger groove (3).

d. Remove bolts and remove timing chain tensioner (primary).

28. Remove internal chain guide, tension guide and slack guide.

➡**Tension guide can be removed after removing timing chain (primary).**

29. Remove timing chain (primary) and crankshaft sprocket.

❈❈ WARNING

After removing timing chain (primary), do not turn crankshaft and camshaft separately, or valves will strike the piston heads.

30. Remove timing chain (secondary) and camshaft sprockets as follows:

a. Attach suitable stopper pin to the right and left timing chain tensioners (secondary).

➡**Use approximately 0.02 in. (0.5 mm) diameter hard metal pin as a stopper pin.**

b. Remove camshaft sprocket (INT and EXH) bolts.

➡**Secure the hexagonal portion of camshaft using wrench to loosen bolts.**

❈❈ WARNING

Do not loosen bolts with securing anything other than the camshaft hexagonal portion or with tensioning the timing chain.

c. Remove timing chain (secondary) together with camshaft sprockets.

• Turn camshaft slightly to secure slackness of timing chain on timing chain tensioner (secondary) side.

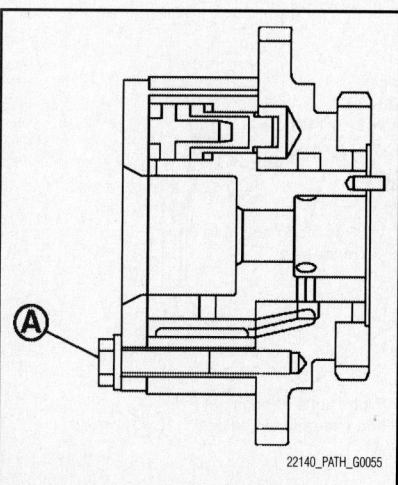

22140_PATH_G0055

Fig. 103 Do not loosen bolts "A" as shown—V6 engine

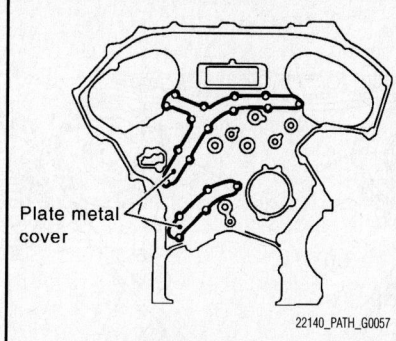

22140_PATH_G0057

Fig. 104 Do not remove plate metal cover of oil passage—V6 engine

• Insert 0.5 mm (0.020 in) thick metal or resin plate between timing chain and timing chain tensioner plunger (guide). Remove timing chain (secondary) together with camshaft sprockets with timing chain loose from guide groove.

❈❈ WARNING

Be careful of plunger coming off when removing timing chain (secondary). This is because plunger of timing chain tensioner (secondary) moves during operation, leading to coming off of fixed stopper pin.

➡**Camshaft sprocket (INT) is a one piece integrated design sprockets for timing chain (primary) and for timing chain (secondary).**

❈❈ WARNING

When handling camshaft sprocket (INT), be careful of the following: Handle carefully to avoid any shock to camshaft sprocket. Do not disassemble. (Do not loosen bolts "A" as shown).

31. Remove water pump.

32. Remove rear timing chain case as follows:

a. Loosen and remove bolts in reverse order of the tightening sequence.

b. Cut liquid gasket and remove rear timing chain case.

❈❈ WARNING

Do not remove plate metal cover of oil passage.

❈❈ WARNING

After removal, handle rear timing chain case carefully so it does not tilt, cant, or warp under a load.

33. Remove O-rings from cylinder head and camshaft bracket (No. 1).

34. Remove O-rings from cylinder block.

35. Remove timing chain tensioners (secondary) from cylinder head if necessary.

a. Remove camshaft brackets (No. 1).

b. Remove timing chain tensioners (secondary) with stopper pin attached.

36. Use scraper to remove all traces of old liquid gasket from front and rear timing chain cases, and opposite mating surfaces. Remove old liquid gasket from bolt hole and thread.

37. Use scraper to remove all traces of liquid gasket from water pump cover, chain tensioner cover and intake valve timing control covers.

38. Check for cracks and any excessive wear at link plates and roller links of timing chain.

39. Replace timing chain as necessary.

To install:

➡**Be sure to use new fasteners, as required.**

➡**The figure below shows the relationship between the mating mark on each timing chain and that on the corresponding sprocket, with the components installed.**

40. Install timing chain tensioners (secondary) to cylinder head if removed.

a. Install timing chain tensioners (secondary) with stopper pin attached and new O-ring.

b. Install camshaft brackets (No. 1).

41. Install rear timing chain case as follows:

a. Install new O-rings onto cylinder block.

b. Install new O-rings to cylinder head and camshaft bracket (No. 1).

c. Apply liquid gasket to rear timing chain case back side as shown.

❈❈ WARNING

For "A" in the figure, completely wipe out liquid gasket extended on a portion touching at engine coolant. Apply liquid gasket on installation position of water pump and cylinder head very completely

d. Align rear timing chain case with dowel pins (right and left) on cylinder block and install rear timing chain case.

➡**Make sure O-rings stay in place during installation to cylinder block, cylinder head and camshaft bracket (No. 1).**

e. Tighten bolts in numerical order as shown.

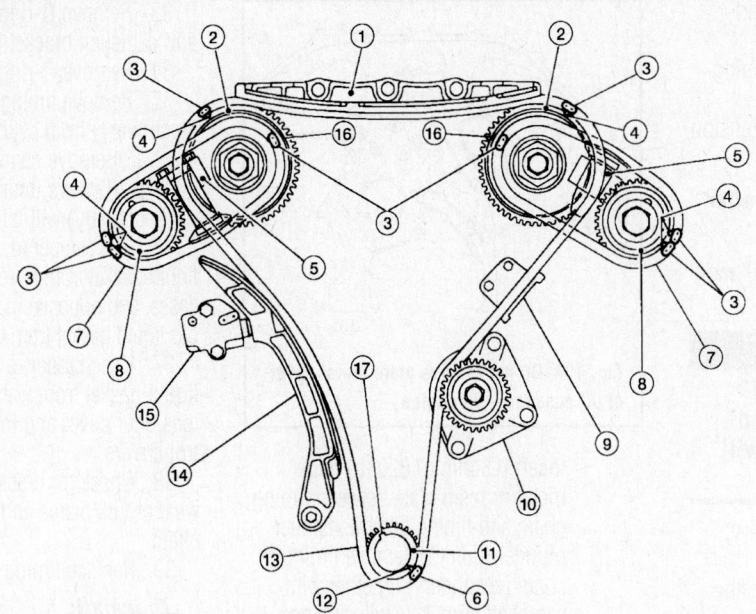

1. Internal chain guide
2. Camshaft sprocket (intake)
3. Mating mark (copper link)
4. Mating mark (punched)
5. Secondary timing chain tensioner
6. Mating mark (yellow link)
7. Secondary timing chain
8. Camshaft sprocket (exhaust)
9. Tensioner guide
10. Water pump
11. Crankshaft sprocket
12. Mating mark (notched)
13. Primary timing chain
14. Slack guide
15. Primary timing chain tensioner
16. Mating mark (back side)
17. Crankshaft key

22140_PATH_G0068

Fig. 105 Timing chain assembly with mating marks—V6 engine

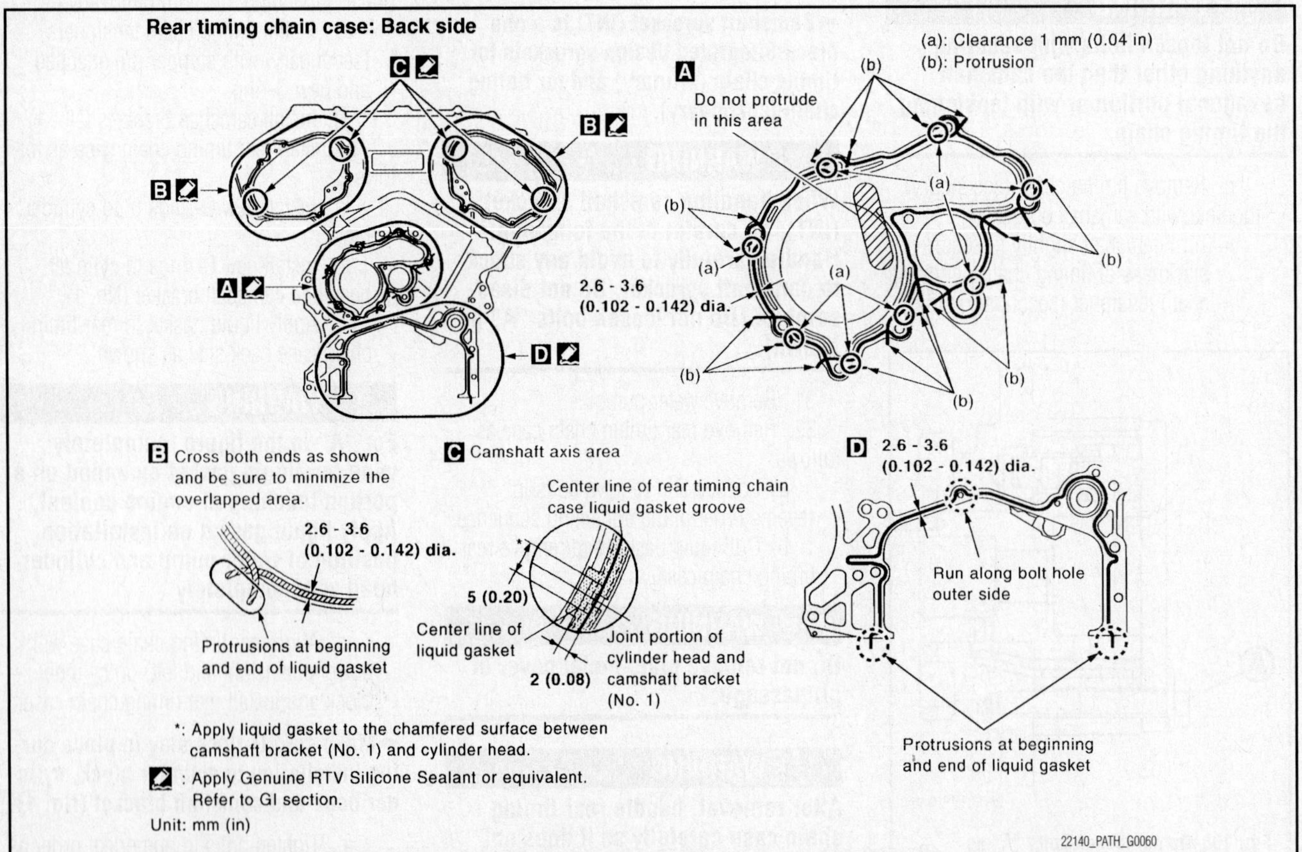

Rear timing chain case: Back side

(a): Clearance 1 mm (0.04 in)
(b): Protrusion

Do not protrude in this area

2.6 - 3.6 (0.102 - 0.142) dia.

B Cross both ends as shown and be sure to minimize the overlapped area.

2.6 - 3.6 (0.102 - 0.142) dia.

Protrusions at beginning and end of liquid gasket

C Camshaft axis area

Center line of rear timing chain case liquid gasket groove

5 (0.20)

Center line of liquid gasket

2 (0.08)

Joint portion of cylinder head and camshaft bracket (No. 1)

D 2.6 - 3.6 (0.102 - 0.142) dia.

Run along bolt hole outer side

Protrusions at beginning and end of liquid gasket

*: Apply liquid gasket to the chamfered surface between camshaft bracket (No. 1) and cylinder head.

: Apply Genuine RTV Silicone Sealant or equivalent. Refer to GI section.

Unit: mm (in)

22140_PATH_G0060

Fig. 106 Apply liquid gasket to rear timing chain case back side as shown—V6 engine

➥**There are two type of bolts. Refer to the following for locating bolts.**

- 0.79 in. (20 mm): 1, 2, 3, 6, 7, 8, 9, 10
- 0.63 in. (16 mm): Except the above
- Rear timing case bolt torque: 9 ft lbs. (12.7 Nm)

f. After all bolts are tightened, retighten them to the specified torque in numerical order as shown.

g. If liquid gasket protrudes, wipe it off immediately.

h. After installing rear timing chain case, check the surface height difference between following parts on oil pan (upper) mounting surface. If not within the standard, repeat the installation procedure.

i. Standard: Rear timing chain case to lower cylinder block: -0.0094–0.0055 in. (-0.24–0.14 mm)

42. Install water pump with new O-rings.

43. Make sure that dowel pin hole, dowel pin of camshaft and crankshaft key are located as shown. (No. 1 cylinder at compression TDC).

➥**Though camshaft does not stop at the position as shown, for the placement of cam nose, it is generally accepted camshaft is placed for the same direction of the figure.**

- Camshaft dowel pin hole (intake side): At cylinder head upper face side in each bank.
- Camshaft dowel pin (exhaust side): At cylinder head upper face side in each bank.
- Crankshaft key: At cylinder head side of right bank.

※※ WARNING

Hole on small diameter side must be used for intake side dowel pin hole.

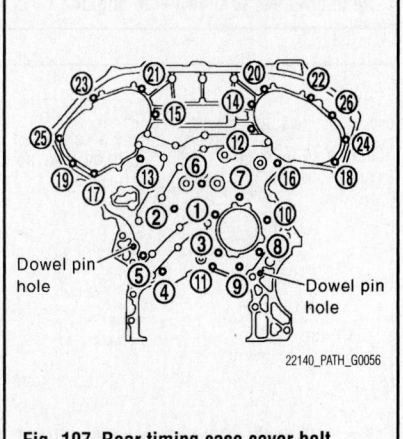

Fig. 107 Rear timing case cover bolt tightening sequence—V6 engine

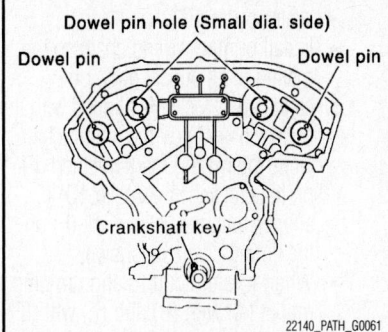

Fig. 108 Make sure that dowel pin hole, dowel pin of camshaft and crankshaft key are located as shown. (No. 1 cylinder at compression TDC)—V6 engine

Do not misidentify (ignore big diameter side).

44. Install timing chains (secondary) and camshaft sprockets as follows:

※※ WARNING

Mating marks between timing chain and sprockets slip easily. Confirm all mating mark positions repeatedly during the installation process.

a. Push plunger of timing chain tensioner (secondary) and keep it pressed in with stopper pin.

b. Install timing chains (secondary) and camshaft sprockets (INT and EXH).

c. Align the mating marks on timing chain (secondary) (copper color link) with the ones on camshaft sprockets (INT and EXH) (punched), and install them.

➥**Mating marks for camshaft sprocket (INT) are on the back side of camshaft sprocket (secondary).**

➥**There are two types of mating marks, circle and oval types.**

- Right bank: Use circle type.
- Left bank: Use oval type.
- They should be used for the right and left banks, respectively.

d. Align dowel pin and pin hole on camshafts with the groove and dowel pin on sprockets, and install them.

e. On the intake side, align pin hole on the small diameter side of the camshaft front end with dowel pin on the back side of camshaft sprocket, and install them.

f. On the exhaust side, align dowel pin on camshaft front end with pin groove on camshaft sprocket, and install them.

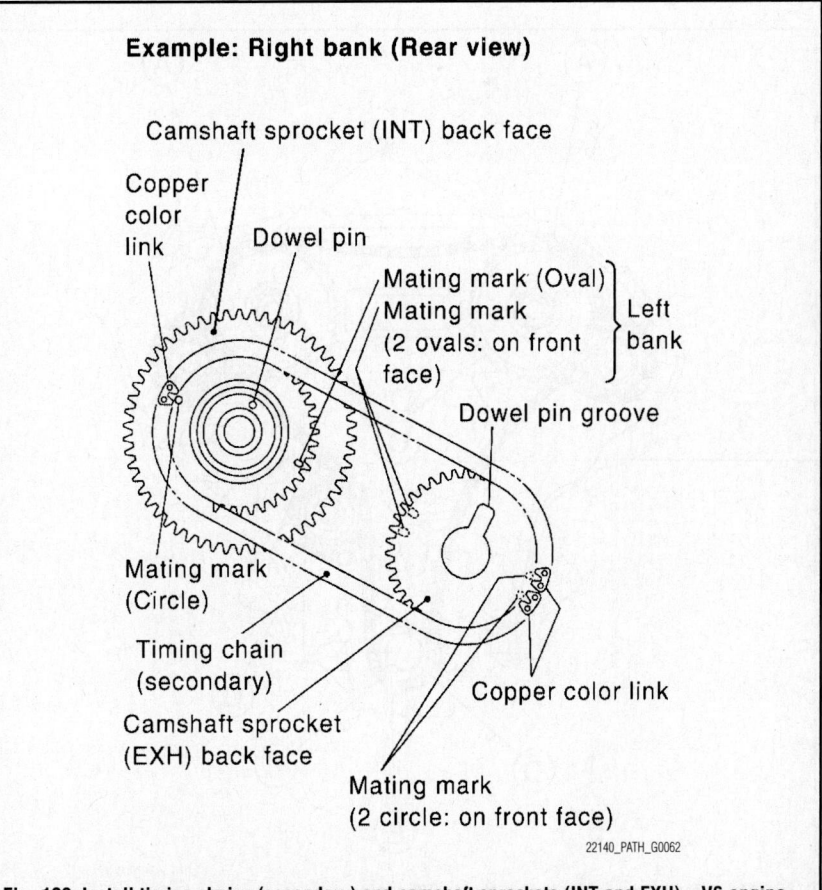

Fig. 109 Install timing chains (secondary) and camshaft sprockets (INT and EXH)—V6 engine

g. In case that positions of each mating mark and each dowel pin are not fit on mating parts, make fine adjustment to the position holding the hexagonal portion on camshaft with wrench or equivalent.

h. Bolts for camshaft sprockets must be tightened in the next step. Tightening them by hand is enough to prevent the dislocation of dowel pins.

i. It may be difficult to visually check the dislocation of mating marks during and after installation. To make the matching easier, make a mating mark on the top of sprocket teeth and its extended line in advance with paint.

j. After confirming the mating marks are aligned, tighten camshaft sprocket bolts.

➡ **Secure camshaft using wrench at the hexagonal portion to tighten bolts.**

k. Pull stopper pins out from timing chain tensioners (secondary).

45. Install tension guide.

46. Install timing chain (primary) as follows:

a. Install crankshaft sprocket.

➡ **Make sure the mating marks on crankshaft sprocket face the front of engine.**

b. Install the primary timing chain.
• Water pump (G).
• Install primary timing chain so the mating mark punched (B) on camshaft sprocket is aligned with the copper link (A) on the timing chain, while the mating mark notched (E) on the crankshaft sprocket (D) is aligned with the yellow link (F) on the timing chain, as shown.
• When it is difficult to align mating marks (A) with (B) and (E) with (F) of the primary timing chain with each sprocket, gradually turn the camshaft using a wrench on the hexagonal portion to align it with the mating marks.
• During alignment, be careful to prevent dislocation of mating mark alignments of the secondary timing chains.

47. Install internal chain guide, slack guide and timing chain tensioner (primary).

☀☀ **WARNING**

Do not overtighten slack guide bolts. It is normal for a gap to exist under the bolt seats when bolts are tightened to specification.

• When installing timing chain tensioner (primary), push in plunger and keep it pressed in with stopper pin.
• Remove any dirt and foreign materials completely from the back and the mounting surfaces of timing chain tensioner (primary).
• After installation, pull out stopper pin by pressing slack guide.

48. Make sure again that the mating marks on camshaft sprockets and timing chain have not slipped out of alignment.

49. Install new O-rings on rear timing chain case.

50. Install new front oil seal on front timing chain case.

a. Apply new engine oil to both oil seal lip and dust seal lip.

b. Install it so that each seal lip is oriented as shown.

c. Press-fit oil seal until it becomes flush with front timing chain case end face using suitable drift with outer diameter: 2.36 in. (60 mm).

d. Make sure the garter spring is in position and seal lip is not inverted.

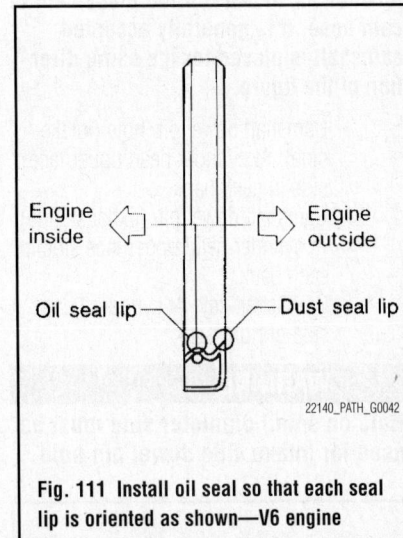

Fig. 111 Install oil seal so that each seal lip is oriented as shown—V6 engine

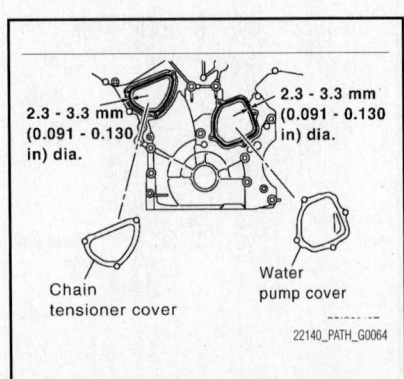

Fig. 112 Apply a continuous bead of liquid gasket to front timing chain—V6 engine

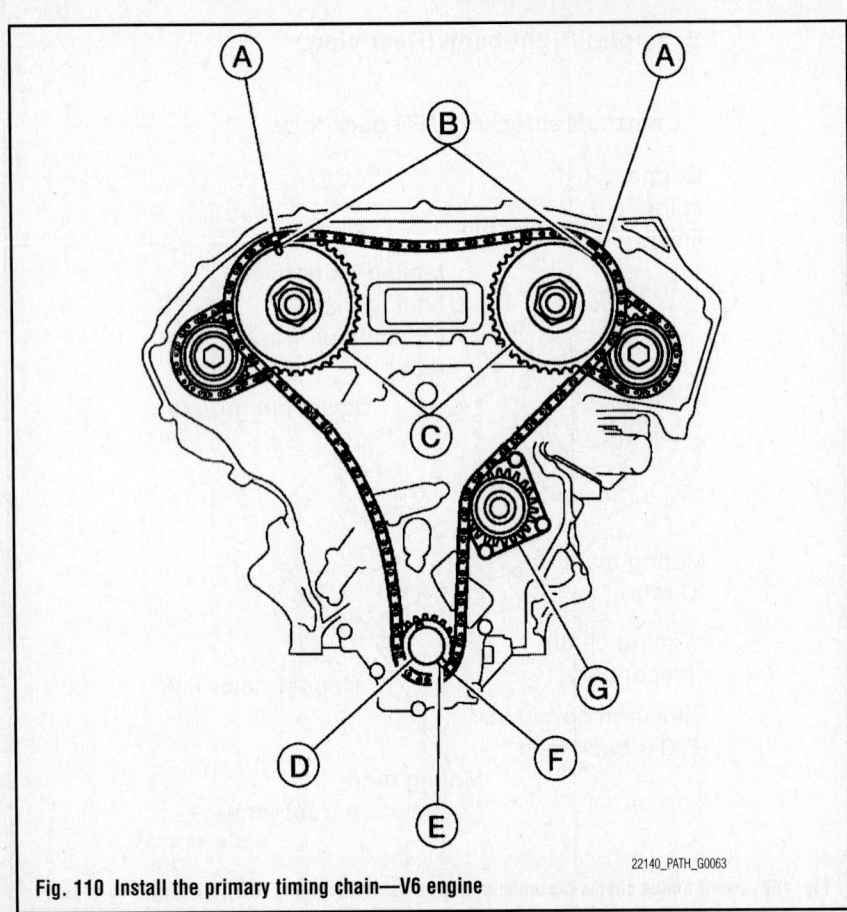

22140_PATH_G0063

Fig. 110 Install the primary timing chain—V6 engine

51. Install water pump cover and chain tensioner cover to front timing chain case.

a. Apply a continuous bead of liquid gasket to front timing chain case as shown.

52. Install front timing chain case as follows:

a. Apply a continuous bead of liquid gasket to front timing chain case back side as shown.

b. Install new O-rings on rear timing chain case.

c. Assemble front timing chain case as follows:

- Fit lower end of front timing chain case tightly onto top face of oil pan (upper). From the fitting point, make entire front timing chain case contact rear timing chain case completely.
- Since front timing chain case is offset for difference of bolt holes, tighten bolts temporarily while holding front timing chain case from front and top.
- Same as the previous step, insert dowel pin while holding front timing chain case from front and top completely.

d. Tighten bolts to the specified torque in numerical order as shown.

➡ **There are two type of bolts. Refer to the following for locating bolts.**

- 1–5: 0.39 in. (10 mm)
- 6–25: 0.24 in. (6 mm)
- Bolt position torque specification: 1–5: 41 ft. lbs. (55.0 Nm); 6–25: 9 ft. lbs. (12.7 Nm)

e. After all bolts tightened, retighten them to the specified torque in numerical order as shown.

53. Install two bolts in front of oil pan (upper).

54. Install right and left intake valve timing control covers as follows:

a. Install new seal rings in shaft grooves.

b. Apply a continuous bead of liquid gasket to intake valve timing control covers.

c. Install new collared O-rings in front timing chain case oil hole (left and right sides).

d. Being careful not to move seal ring from the installation groove, align dowel pins on front timing chain case with the holes to install intake valve timing control covers.

e. Tighten bolts in numerical order as shown.

55. Install crankshaft pulley as follows:

a. Install crankshaft pulley, taking care not to damage front oil seal.

➡ **When press-fitting crankshaft pulley with plastic hammer, tap on its center portion (not circumference).**

b. Tighten crankshaft pulley bolt. Crankshaft bolt torque: 33 ft. lbs. (44.1 Nm)

c. Put a paint mark on crankshaft pulley aligning with angle mark on crankshaft pulley bolt. Then, further retighten bolt by 60° (equivalent to one graduation).

56. Rotate crankshaft pulley in normal direction (clockwise when viewed from front) to confirm it turns smoothly.

57. Install oil pans (upper and lower).

58. Install rocker covers (right and left banks).

59. Installation of the remaining components is in the reverse order of removal after this step.

60. The following are procedures for checking fluid leaks, lubricant leaks and exhaust gases leaks.

a. Before starting engine, check oil/fluid levels including engine coolant and engine oil. If less than required quantity, fill to the specified level.

61. Use procedure below to check for fuel leakage.

a. Turn ignition switch "ON" (with engine stopped). With fuel pressure applied to fuel piping, check for fuel leakage at connection points.

b. Start engine. With engine speed increased, check again for fuel leakage at connection points.

c. Run engine to check for unusual noise and vibration.

➡ **If hydraulic pressure inside timing chain tensioner drops after removal/installation, slack in the guide may generate a pounding noise during and just after engine start. However, this is**

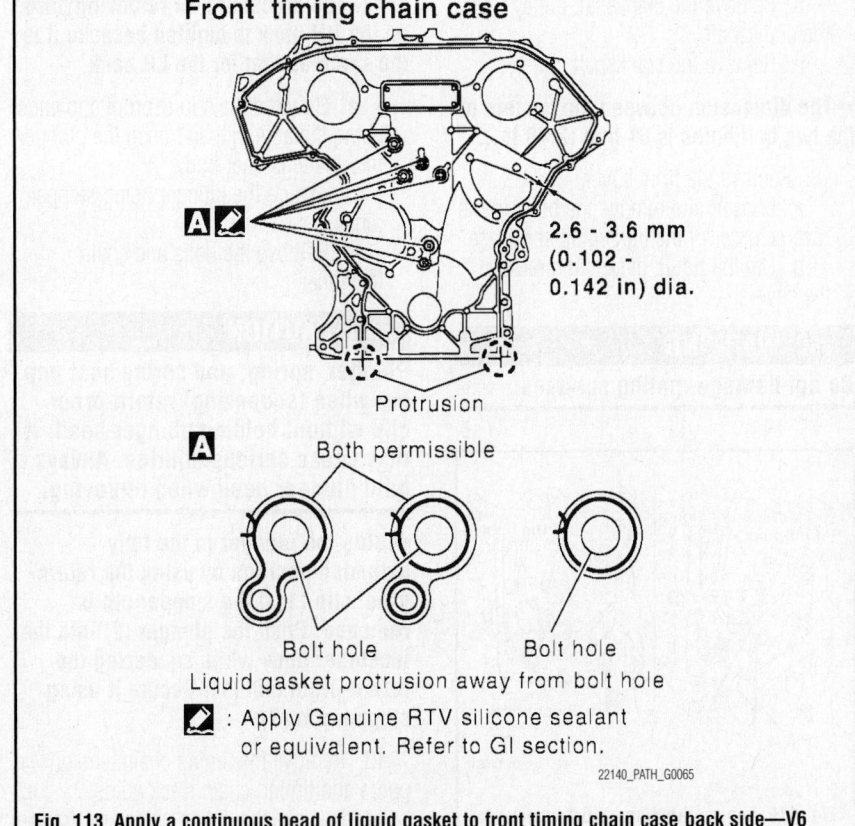

Front timing chain case

2.6 - 3.6 mm
(0.102 - 0.142 in.) dia.

Protrusion

Ⓐ Both permissible

Bolt hole   Bolt hole

Liquid gasket protrusion away from bolt hole

✎ : Apply Genuine RTV silicone sealant or equivalent. Refer to GI section.

22140_PATH_G0065

Fig. 113 Apply a continuous bead of liquid gasket to front timing chain case back side—V6 engine

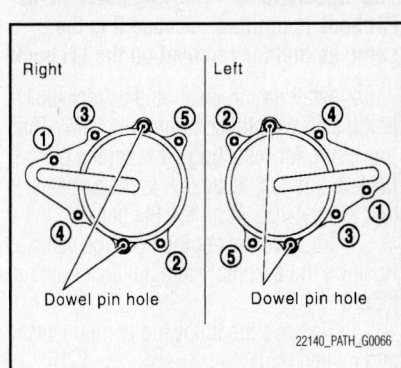

Right   Left

Dowel pin hole   Dowel pin hole

22140_PATH_G0066

Fig. 114 Valve timing control bolt tightening sequence—V6 engine

normal. **Noise will stop after hydraulic pressure rises.**

d. Warm up engine thoroughly to make sure there is no leakage of fuel, exhaust gases, or any oil/fluids including engine oil and engine coolant.

e. Bleed air from lines and hoses of applicable lines, such as in cooling system.

f. After cooling down engine, again check oil/fluid levels including engine oil and engine coolant. Refill to the specified level, if necessary.

62. Be sure to perform the reconnect/relearn procedures.

V8 Engine

See Figures 115 through 120.

1. Before servicing the vehicle, refer to the Precautions Section.

➡ **If working near and/or around the SRS system and components, be sure to disable the SRS system. After disabling the system wait three minutes or more before servicing the vehicle.**

➡ **On this vehicle the battery current sensor that is installed to the negative battery cable measures the charging/discharging current of the battery and performs various engine controls. If an electrical component is connected directly to the negative battery cable, the current flowing through that component will not be measured by the battery current sensor. This condition may cause a malfunction of the engine control system and battery discharge may occur. Do not connect an electrical component or ground wire directly to the battery terminal.**

2. Disconnect the negative battery cable.

➡ **To remove timing chain and associated parts, start with those on the LH bank. The procedure for removing parts on the RH bank is omitted because it is the same as that for removal on the LH bank.**

To install timing chain and associated parts, start with those on the RH bank. The procedure for installing parts on the LH bank is omitted because it is the same as that for installation on the RH bank.

3. Remove the engine from the vehicle. Position the assembly in a suitable holding fixture.

4. Remove the following components and related parts:

- Drive belt auto tensioner and idler pulley

- Thermostat housing and water hose
- Power steering oil pump bracket
- Oil pan (lower), (upper) and oil strainer
- Ignition coil
- Rocker cover

5. Remove the Intake valve control solenoid valve cover (RH) (A) and Intake valve control solenoid valve cover (LH) (B) as follows:

a. Loosen and remove the bolts in the order of the tightening sequence.

b. Cut the liquid gasket and remove the covers.

❋❋ WARNING

Do not damage mating surfaces.

6. Obtain compression TDC of No. 1 cylinder as follows:

a. Turn the crankshaft pulley clockwise to align the TDC identification notch (without paint mark) with the timing indicator on the front cover.

b. At this time, make sure both intake and exhaust cam lobes of No. 1 cylinder (top front on LH bank) point outside. If they do not point outside, turn crankshaft pulley once more.

7. Remove the crankshaft pulley.

a. Loosen the crankshaft pulley bolts using a hammer handle to secure the crankshaft.

b. Remove the crankshaft pulley from the crankshaft.

c. Remove the crankshaft pulley.

➡ **The dimension between the centers of the two bolt holes is 61 mm (2.40 in.).**

8. Remove the front cover.

a. Loosen and remove the bolts in the reverse order of the tightening sequence.

b. Cut the liquid gasket and remove the covers.

❋❋ WARNING

Do not damage mating surfaces.

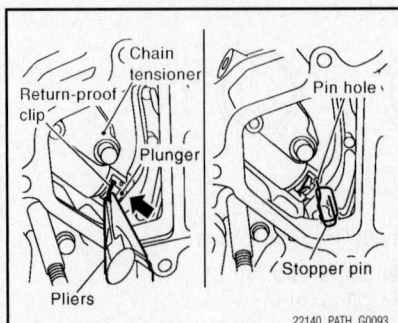

Fig. 115 Secure the plunger using stopper pin—V8 engine

22140_PATH_G0093

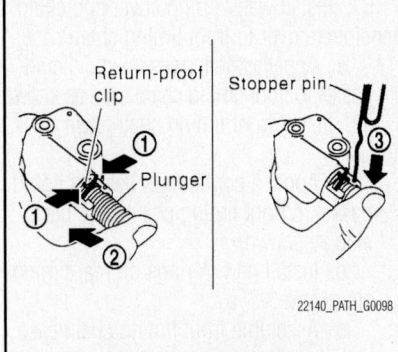

22140_PATH_G0098

Fig. 116 Stop the plunger in the fully extended position—V8 engine

9. Remove the front oil seal using suitable tool.

❋❋ WARNING

Do not damage front cover.

10. Remove the oil pump drive spacer.

➡ **Hold and remove the flat space of the oil pump drive spacer by pulling it forward.**

11. Remove the oil pump.

12. Remove the chain tensioner on the LH bank using the following steps.

➡ **To remove the timing chain and associated parts, start with those on the LH bank. The procedure for removing parts on the RH bank is omitted because it is the same as that for the LH bank.**

a. Squeeze the return-proof clip ends using suitable tool and push the plunger into the tensioner body.

b. Secure the plunger using stopper pin.

c. Remove the bolts and chain tensioner.

❋❋ CAUTION

Plunger, spring, and spring seat pop out when (squeezing) return-proof clip without holding plunger head. It may cause serious injuries. Always hold plunger head when removing.

➡ **Stop the plunger in the fully extended position by using the return-proof clip (1) if the stopper pin is removed. Push the plunger (2) into the tensioner body while squeezing the return-proof clip (1). Secure it using stopper pin (3).**

13. Remove the timing chain tension guide and timing chain slack guide.

14. Remove the timing chain and crankshaft sprocket.

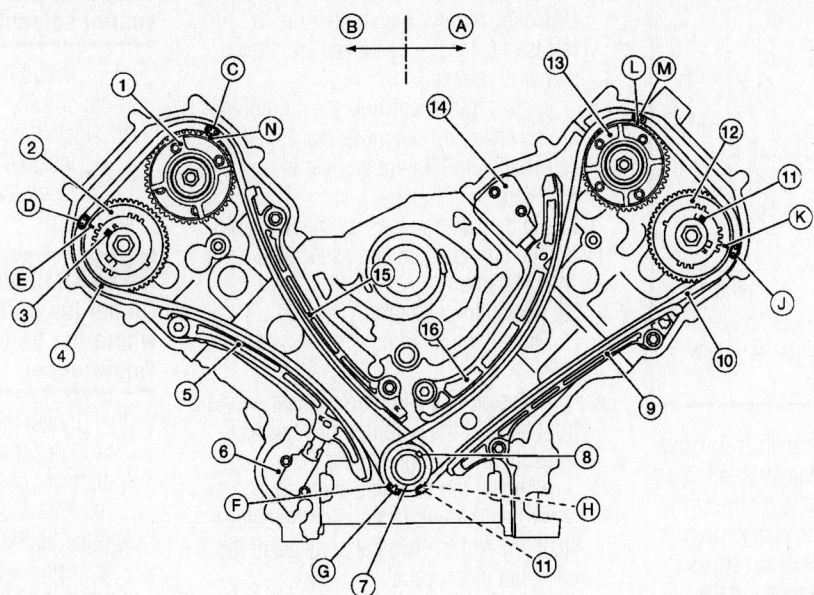

1. RH bank Camshaft sprocket (INT) (VTC)
2. RH bank Camshaft sprocket (EXH)
3. RH bank camshaft dowel pin
4. Timing chain
5. RH bank Timing chain slack guide
6. Primary timing chain tensioner
7. Crankshaft sprocket
8. Crankshaft key
9. LH Timing chain tension guide
10. Timing chain
11. LH Camshaft dowel pin
12. LH bank Camshaft sprocket (EXH)
13. LH bank Camshaft sprocket (INT) (VTC)
14. Secondary timing chain tensioner
15. RH bank timing chain tension guide
16. LH timing chain slack guide
A. LH bank
B. RH bank
C. Alignment mark (Link color: copper)
D. Alignment mark (Link color: copper)
E. Alignment mark (Identification mark)
F. Alignment mark for LH bank (Notch)
G. Alignment mark for LH bank (Link color: Yellow)
H. Alignment mark for RH bank (Link color: Yellow)
J. Alignment mark (Link color: copper)
K. Alignment mark (Identification mark)
L. Alignment mark (Identification mark)
M Alignment mark (Link color: copper)
N. Alignment mark (Identification mark)

22140_PATH_G0123

Fig. 117 Timing chains and related components—V8 engine

15. Loosen the camshaft sprocket bolts as shown and remove the camshaft sprocket.

✳✳ WARNING

To avoid interference between valves and pistons, do not turn crankshaft or camshaft when timing chain is disconnected.

16. Repeat the same procedure to remove the RH timing chain and associated parts.

To install:

➡Be sure to use new fasteners, as required.

➡The above figure shows the relationship between the mating mark on each timing chain and that of the corresponding sprocket, with the components installed.

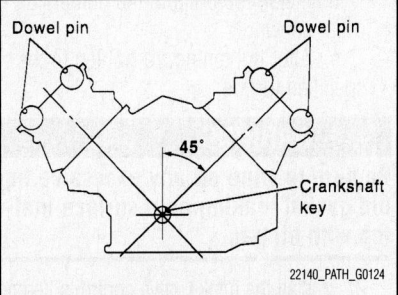

22140_PATH_G0124

Fig. 118 Make sure the crankshaft key and RH bank camshaft dowel pin and LH bank camshaft dowel pin are facing in the direction—V8 engine

To install the timing chain and associated parts, start with those on the RH bank. The procedure for installing parts on the LH bank is omitted because it is the same as that for installation on the RH bank.

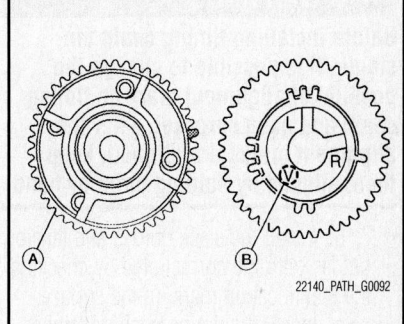

22140_PATH_G0092

Fig. 119 Install the camshaft sprockets—V8 engine

17. Make sure the crankshaft key and RH bank camshaft dowel pin and LH bank camshaft dowel pin are facing in the direction shown.

18. Install the camshaft sprockets.

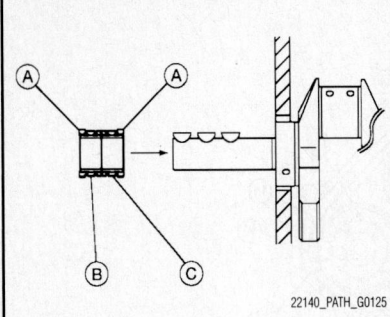

Fig. 120 Install the crankshaft sprockets for both banks—V8 engine

a. Install the intake camshaft sprocket (VTC) (A) and exhaust camshaft sprockets (B) by selectively using the groove of the dowel pin according to the bank. (Common part used for both exhaust banks.)

b. Lock the hexagonal part of the camshaft in the same way as for removal, and tighten the bolts.

• A: Intake
• B = V: Exhaust

19. Install the crankshaft sprockets for both banks.

a. Install LH bank crankshaft sprocket (B) and RH bank crankshaft sprocket (C) so that their flange side (A) (the larger diameter side without teeth) faces in the direction shown.

➡The same parts are used but facing directions are different.

20. Install the timing chains and associated parts.

a. Align the alignment mark on each sprocket and the timing chain for installation.

❈❈ WARNING

Before installing timing chain tensioner, it is possible to change the position of alignment mark on timing chain and each sprocket. After the alignment marks are aligned, keep them aligned by holding them by hand.

b. Install the slack guides and tension guides onto the correct side by checking the identification mark on the surface.

c. Install the timing chain tensioner with the plunger locked in with the stopper pin.

❈❈ WARNING

Before and after the installation of the timing chain tensioner, make sure that the alignment mark on the timing chain is not out of alignment.

d. After installing the timing chain tensioner, remove the stopper pin to release the tensioner. Make sure the tensioner is released.

e. To avoid chain-link skipping of the timing chain, do not move crankshaft or camshafts until the front cover is installed.

21. In the same way as for the RH bank, install the timing chain and associated parts on the LH bank.

22. Install the oil pump.

23. Install the oil pump drive spacer as follows:

a. Install so that the front mark on the front edge of the oil pump drive spacer faces the front of the engine.

b. Insert the oil pump drive spacer according to the directions of the crankshaft key and the two flat surfaces of the oil pump inner rotor.

c. If the positional relationship does not allow the insertion, rotate the oil pump inner rotor to allow the oil pump drive spacer to be inserted.

24. Install the front oil seal using suitable tool.

❈❈ WARNING

Do not scratch or make burrs on the circumference of the oil seal.

25. Install the chain tensioner cover.

26. Install the front cover as follows:

a. Install a new O-ring on the cylinder block.

b. Apply liquid gasket as shown.

c. Check again that the timing alignment marks on the timing chain and on each sprocket are aligned. Then install the front cover.

d. Install the bolts in the numerical order shown.

e. After tightening, re-tighten to the specified torque.

❈❈ WARNING

Be sure to wipe off any excessive liquid gasket leaking onto surface mating with oil pan.

27. Install the Intake valve control solenoid valve cover (RH) (A) and Intake valve control solenoid valve cover (LH) (B) as follows:

a. Cross mark (C) that cannot be seen after assembly.

b. Apply liquid gasket (D) as shown.

❈❈ WARNING

The start and end of the application of the liquid gasket should be crossed at a position that cannot be

seen after attaching the Intake valve control solenoid valve cover.

c. Install the bolts in the numerical order shown.

28. Install the crankshaft pulley.

a. Install the key of the crankshaft.

b. Insert the pulley by lightly tapping it.

❈❈ WARNING

Do not tap pulley on the side surface where the belt is installed (outer circumference).

29. Tighten the crankshaft pulley bolt.

a. Lock the crankshaft using suitable tool, then tighten the bolt.

b. Perform the following steps for angular tightening:

c. Apply engine oil onto the threaded parts of the bolt and seating area.

d. Select the one most visible notch of the four on the bolt flange. Corresponding to the selected notch, put a alignment mark (such as paint) on the crankshaft pulley.

30. Rotate the crankshaft pulley in normal direction (clockwise when viewed from engine front) to check for parts interference.

31. Installation of the remaining components is in the reverse of order of removal.

32. Be sure to perform the reconnect/relearn procedures.

VALVE LASH

ADJUSTMENT

V6 Engine

See Figures 121 through 129.

1. Before servicing the vehicle, refer to the Precautions Section.

➡If working near and/or around the SRS system and components, be sure to disable the SRS system. After disabling the system wait three minutes or more before servicing the vehicle.

➡On this vehicle the battery current sensor that is installed to the negative battery cable measures the charging/discharging current of the battery and performs various engine controls. If an electrical component is connected directly to the negative battery cable, the current flowing through that component will not be measured by the battery current sensor. This condition may cause a malfunction of the engine control system and battery discharge may occur. Do not connect an electrical

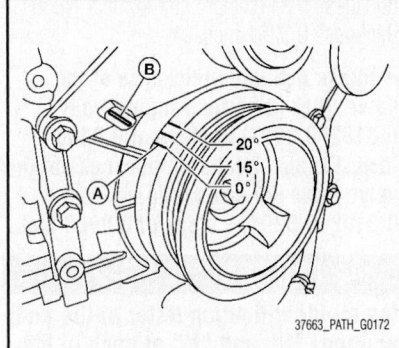

Fig. 121 Rotate crankshaft pulley clockwise to align timing mark—V6 engine

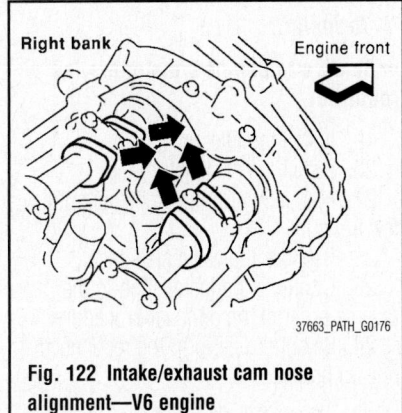

Fig. 122 Intake/exhaust cam nose alignment—V6 engine

component or ground wire directly to the battery terminal.

2. Disconnect the negative battery cable.
3. Remove the engine room cover.
4. Remove the air cleaner assembly.

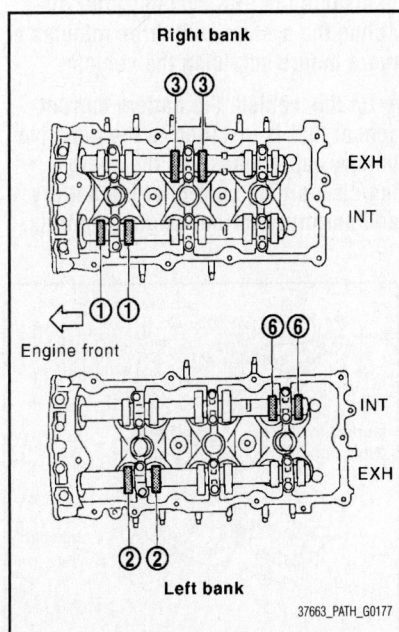

Fig. 123 Clearance measurement number 1 cylinder at TDC (1 of 2)—V6 engine

Measuring position (RH bank)		No. 1 CYL.	No. 3 CYL.	No. 5 CYL.
No. 1 cylinder at compression TDC	EXH		×	
	INT	×		
Measuring position (LH bank)		No. 2 CYL.	No. 4 CYL.	No. 6 CYL.
No. 1 cylinder at compression TDC	INT			×
	EXH	×		

Fig. 124 Clearance measurement number 1 cylinder at TDC (2 of 2)—V6 engine

5. Remove the valve covers.
6. Position the number 1 cylinder at TDC on the compression stroke.

➡**To accomplish this, rotate the crankshaft pulley clockwise to align the timing mark (A) with the timing indicator**

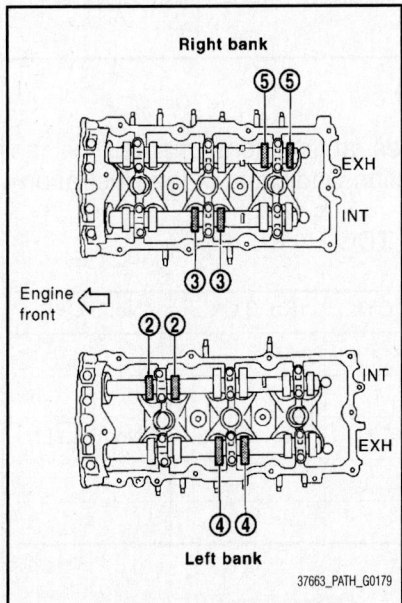

Fig. 125 Clearance measurement number 3 cylinder at TDC (1 of 2)—V6 engine

(B). Make sure that the intake and exhaust cam noses of the number 1 cylinder (right bank) are located as shown in the illustration. If not rotate the crankshaft 360 degrees.

7. Use a feeler gauge and check the clearance between the valve lifter and camshaft.
8. Specification should be Cold: Intake 0.010–0.013, Exhaust 0.011–0.015. Hot: Intake 0.012–0.016, Exhaust 0.012–0.017.
9. Measure the valve clearance at locations marked "X" as shown in the illustrations using a feeler gauge, for number 1 cylinder at TDC.
10. Rotate the crankshaft 240 degrees clockwise (viewed from engine front) to align number 3 cylinder at TDC on the compression stroke.

➡**Crankshaft pulley bolt flange has a stamped line every 60 degrees. Markings can be used as a guide.**

11. Measure the valve clearance at locations marked "X" as shown in the illustrations using a feeler gauge, for number 3 cylinder at TDC.
12. Rotate the crankshaft 240 degrees clockwise (viewed from engine front) to align number 5 cylinder at TDC on the compression stroke.

Measuring position (RH bank)		No. 1 CYL.	No. 3 CYL.	No. 5 CYL.
No. 3 cylinder at compression TDC	EXH			×
	INT		×	
Measuring position (LH bank)		No. 2 CYL.	No. 4 CYL.	No. 6 CYL.
No. 3 cylinder at compression TDC	INT	×		
	EXH		×	

Fig. 126 Clearance measurement number 3 cylinder at TDC (2 of 2)—V6 engine

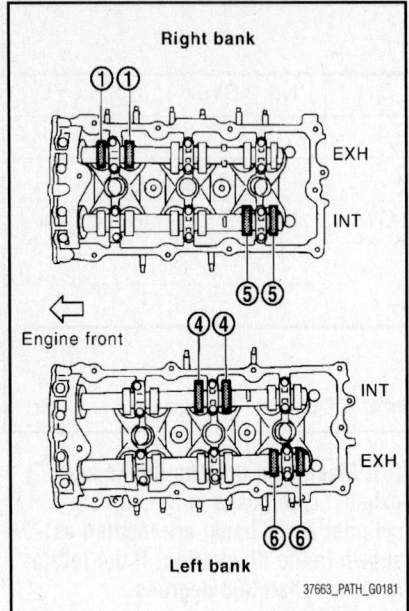

Fig. 127 Clearance measurement number 5 cylinder at TDC (1 of 2)—V6 engine

➡Crankshaft pulley bolt flange has a stamped line every 60 degrees. Markings can be used as a guide.

13. Measure the valve clearance at locations marked "X" as shown in the illustrations using a feeler gauge, for number 5 cylinder at TDC.

14. If not within specification, remove the camshaft.

15. Remove valve lifters at locations that are out of standard.

16. Measure center thickness with a micrometer. Replace lifter as required after performing replacement calculation. See illustration.

17. Thickness of the new lifter can be identified by stamp marks on the reverse side (inside the cylinder).

18. Intake: stamp mark 788U thickness 0.3102 inch, stamp mark 790U thickness 0.3110 inch, stamp mark 840U thickness 0.3307 inch. Exhaust: stamp mark N788 thickness 0.3102 inch, stamp mark N790 thickness 0.3110 inch, stamp mark N836 thickness 0.3291 inch.

➡Intake available thickness of the valve lifter (27 sizes ranging from 0.3102–0.3307) in steps of 0.0008 inch. Exhaust available thickness of the valve lifter (25 sizes ranging from 0.3102–0.3291 in steps of 0.0008 inch.

✱✱ CAUTION

Install identification letter at the end and top, "U" and "N" at each of the proper positions. Be careful of incorrect installation between intake and exhaust.

To install:

➡Be sure to use new fasteners, as required.

19. Install the lifter.
20. Install the camshaft.
21. Manually turn the crankshaft pulley a few turns.
22. Recheck the valve clearance.
23. Continue the installation in the reverse order of the removal procedure.
24. Be sure to perform the reconnect/relearn procedures.

V8 Engine

See Figures 130 through 137.

1. Before servicing the vehicle, refer to the Precautions Section.

➡If working near and/or around the SRS system and components, be sure to disable the SRS system. After disabling the system wait three minutes or more before servicing the vehicle.

➡On this vehicle the battery current sensor that is installed to the negative battery cable measures the charging/discharging current of the battery and performs various engine controls.

• Measure the valve clearances at locations marked "×" as shown in the table below (locations indicated in the illustration) with feeler gauge.
• No. 5 cylinder at compression TDC

Measuring position (RH bank)		No. 1 CYL.	No. 3 CYL.	No. 5 CYL.
No. 5 cylinder at compression TDC	EXH	×		
	INT			×
Measuring position (LH bank)		No. 2 CYL.	No. 4 CYL.	No. 6 CYL.
No. 5 cylinder at compression TDC	INT		×	
	EXH			×

37663_PATH_G0182

Fig. 128 Clearance measurement number 5 cylinder at TDC (2 of 2)—V6 engine

---

Valve lifter thickness calculation: $t = t_1 + (C_1 - C_2)$

t = Valve lifter thickness to be replaced

t_1 = Removed valve lifter thickness

C_1 = Measured valve clearance

C_2 = Standard valve clearance:

Intake : 0.26 - 0.34 mm (0.010 - 0.013 in)*

Exhaust : 0.29 - 0.37 mm (0.011 - 0.015 in)*

*: Approximately 20°C (68°F)

37663_PATH_G0183

Fig. 129 Valve lifter replacement thickness measurement

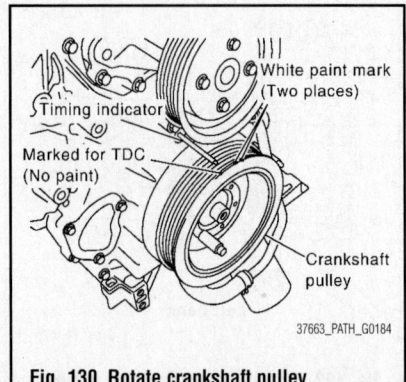

Fig. 130 Rotate crankshaft pulley clockwise to align timing mark—V8 engine

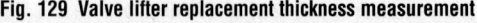

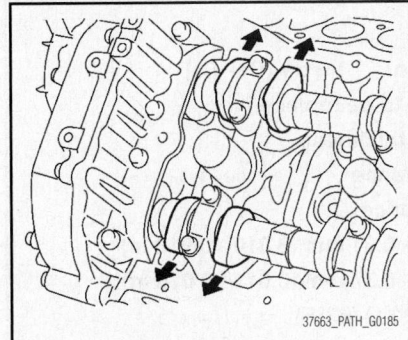

37663_PATH_G0185

Fig. 131 Intake/exhaust cam nose alignment—V8 engine

If an electrical component is connected directly to the negative battery cable, the current flowing through that component will not be measured by the battery current sensor. This condition may cause a malfunction of the engine control system and battery discharge may occur. Do not connect an electrical component or ground wire directly to the battery terminal.

2. Disconnect the negative battery cable.
3. Remove the engine room cover.
4. Remove the air cleaner assembly.
5. Remove the valve covers.
6. Position the number 1 cylinder at TDC on the compression stroke.

➡To accomplish this, rotate the crankshaft pulley (clockwise when viewed from engine front) to align the TDC identification notch (without paint mark) with the timing indicator.). Make sure that the intake and exhaust cam noses of the number 1 cylinder (top front on left bank) face outside as

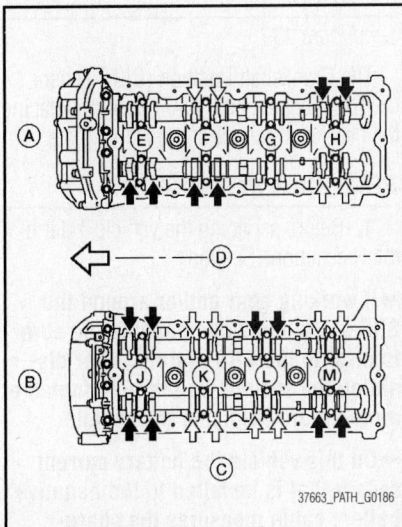

37663_PATH_G0186

Fig. 132 Clearance measurement (1 of 5)—V8 engine

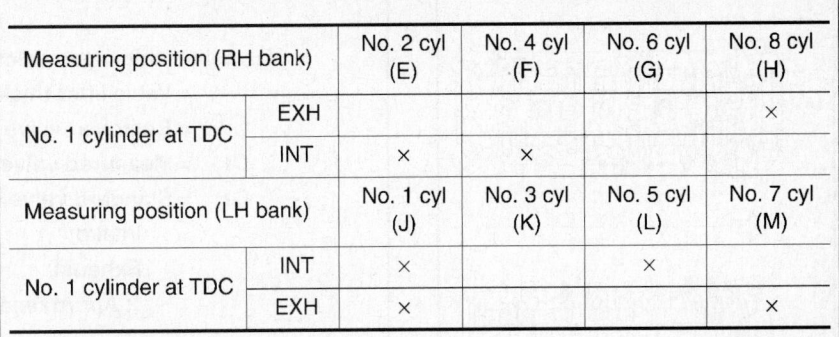

Measuring position (RH bank)		No. 2 cyl (E)	No. 4 cyl (F)	No. 6 cyl (G)	No. 8 cyl (H)
No. 1 cylinder at TDC	EXH				×
	INT	×	×		
Measuring position (LH bank)		No. 1 cyl (J)	No. 3 cyl (K)	No. 5 cyl (L)	No. 7 cyl (M)
No. 1 cylinder at TDC	INT	×		×	
	EXH	×			×

37663_PATH_G0187

Fig. 133 Clearance measurement (2 of 5)—V8 engine

shown in the illustration. If not rotate the crankshaft 360 degrees.

7. Measure the valve clearance at locations marked "X" as shown in the illustrations (locations indicated with black arrow), using a feeler gauge.

➡White arrow: engine front. Black arrow: measurable at number 1 cylin-

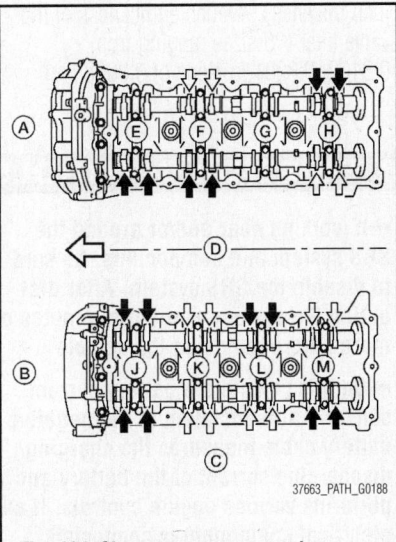

37663_PATH_G0188

Fig. 134 Clearance measurement (3 of 5)—V8 engine

der at TDC compression stroke. White arrow: measurable at number 3 cylinder at TDC compression stroke. A: right side. B: left side. C: exhaust. D: intake.

8. Rotate the crankshaft 270 degrees clockwise to align number 3 cylinder at TDC on the compression stroke.

9. Measure the valve clearance at locations marked "X" as shown in the illustrations (locations indicated with black arrow), using a feeler gauge.

➡White arrow: engine front. Black arrow: measurable at number 1 cylinder at TDC compression stroke. White arrow: measurable at number 3 cylinder at TDC compression stroke. A: right side. B: left side. C: exhaust. D: intake.

10. Rotate the crankshaft pulley 90 degrees from the position of number 3 cylinder at TDC on the compression stroke (clockwise by 360 degrees from the position of number 1 cylinder at TDC on the compression stroke) to measure the intake and exhaust valve clearances of the number 6 cylinder and the exhaust valve clearance of the number 2 cylinder.

➡White arrow: engine front. A: right side. B: left side. C: exhaust. D: intake.

Measuring position (RH bank)		No. 2 cyl (E)	No. 4 cyl (F)	No. 6 cyl (G)	No. 8 cyl (H)
No. 3 cylinder at TDC	EXH		×		
	INT				×
Measuring position (LH bank)		No. 1 cyl (J)	No. 3 cyl (K)	No. 5 cyl (L)	No. 7 cyl (M)
No. 3 cylinder at TDC	INT		×		×
	EXH		×	×	

37663_PATH_G0189

Fig. 135 Clearance measurement (4 of 5)—V8 engine

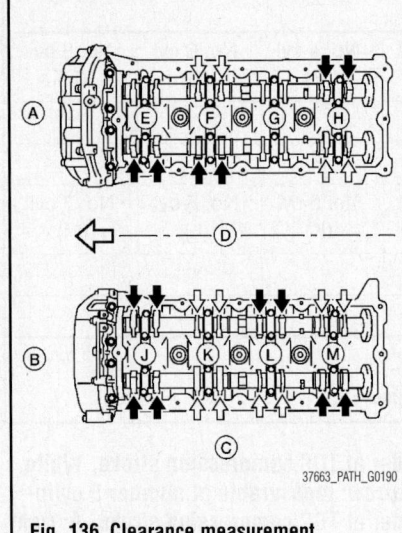

Fig. 136 Clearance measurement (5 of 5)—V8 engine

E: number 2 cylinder. F: number 4 cylinder. G: number 6 cylinder. H: number 8 cylinder. J: number 1 cylinder. K: number 3 cylinder. L: number 5 cylinder. M: number 7 cylinder.

11. If not within specification, remove the camshaft.

Valve lifter thickness calculation: $t = t_1 + (C_1 - C_2)$

t = Valve lifter thickness to be replaced

t_1 = Removed valve lifter thickness

C_1 = Measured valve clearance

C_2 = Standard valve clearance:

    Intake   : 0.26 - 0.34 mm (0.010 - 0.013 in)*

    Exhaust : 0.29 - 0.37 mm (0.011 - 0.015 in)*

    *: Approximately 20°C (68°F)

37663_PATH_G0183

Fig. 137 Valve lifter replacement thickness measurement

---

12. Remove valve lifters at locations that are out of standard.

13. Measure center thickness with a micrometer. Replace lifter as required after performing replacement calculation. See illustration.

14. Thickness of the new lifter can be identified by stamp marks on the reverse side (inside the cylinder).

15. Stamp mark N788 indicates 0.3102 inch thickness. Available thickness of the valve lifter (25 sizes ranging from 0.3102–0.3291 in steps of 0.0008 inch.

To install:

➡Be sure to use new fasteners, as required.

16. Install the lifter.

17. Install the camshaft.

18. Manually turn the crankshaft pulley a few turns.

19. Recheck the valve clearance.

20. Continue the installation in the reverse order of the removal procedure.

21. Be sure to perform the reconnect/relearn procedures.

ENGINE PERFORMANCE & EMISSION CONTROLS

CAMSHAFT POSITION (CMP) SENSOR

LOCATION

See Figure 138.

The Camshaft Position (CMP) sensors are located on the right and left bank cylinder heads at the back side.

REMOVAL & INSTALLATION

1. Before servicing the vehicle, refer to the Precautions Section.

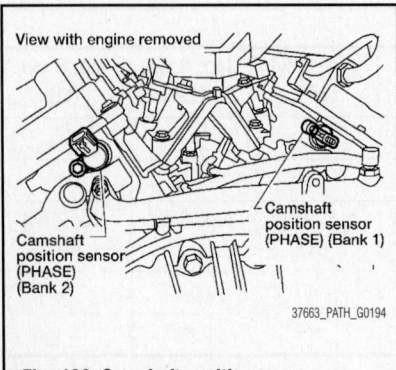

Fig. 138 Camshaft position sensor location—V6 engine

➡If working near and/or around the SRS system and components, be sure to disable the SRS system. After disabling the system wait three minutes or more before servicing the vehicle.

➡On this vehicle the battery current sensor that is installed to the negative battery cable measures the charging/discharging current of the battery and performs various engine controls. If an electrical component is connected directly to the negative battery cable, the current flowing through that component will not be measured by the battery current sensor. This condition may cause a malfunction of the engine control system and battery discharge may occur. Do not connect an electrical component or ground wire directly to the battery terminal.

2. Disconnect the negative battery cable.

3. Loosen the fixing bolt of the sensor.

4. Disconnect the electrical connector.

5. Remove the bolt securing the sensor.

To install:

➡Be sure to use new fasteners, as required.

---

6. Installation is the reverse of the removal procedure.

7. Be sure to perform the reconnect/relearn procedures.

CRANKSHAFT POSITION (CKP) SENSOR

LOCATION

See Figure 139.

The Crankshaft Position (CKP) sensor (POS) is located on the A/T assembly facing the gear teeth (cogs) of the signal plate.

REMOVAL & INSTALLATION

1. Before servicing the vehicle, refer to the Precautions Section.

➡If working near and/or around the SRS system and components, be sure to disable the SRS system. After disabling the system wait three minutes or more before servicing the vehicle.

➡On this vehicle the battery current sensor that is installed to the negative battery cable measures the charging/discharging current of the battery and performs various engine controls.

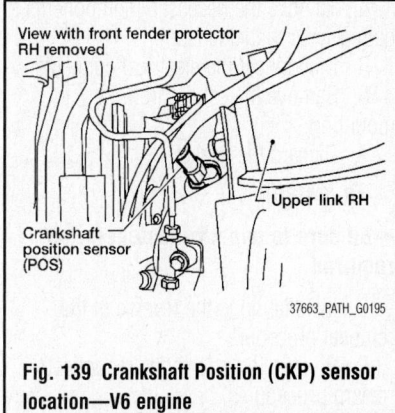

Fig. 139 Crankshaft Position (CKP) sensor location—V6 engine

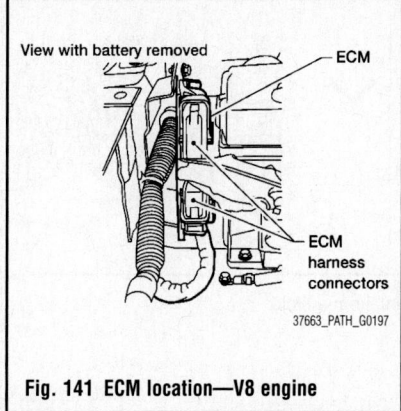

Fig. 141 ECM location—V8 engine

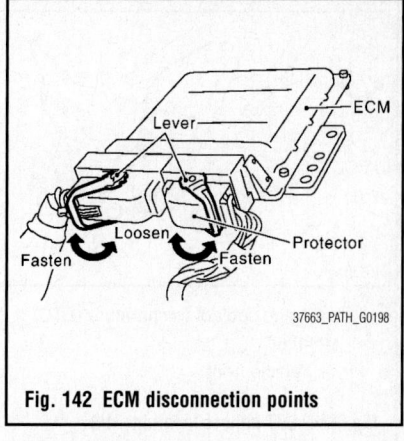

Fig. 142 ECM disconnection points

If an electrical component is connected directly to the negative battery cable, the current flowing through that component will not be measured by the battery current sensor. This condition may cause a malfunction of the engine control system and battery discharge may occur. Do not connect an electrical component or ground wire directly to the battery terminal.

2. Disconnect the negative battery cable.
3. Loosen the fixing bolt of the sensor.
4. Disconnect Crankshaft Position (CKP) sensor (POS) harness connector.
5. Remove the sensor.
6. Visually check the sensor for chipping.

To install:

➡Be sure to use new fasteners, as required.

7. Installation is the reverse of the removal procedure.
8. Be sure to perform the reconnect/relearn procedures.

ELECTRONIC CONTROL MODULE (ECM)

LOCATION

See Figures 140 and 141.

On the V6 engine the Electronic Control Module (ECM) is located in the engine

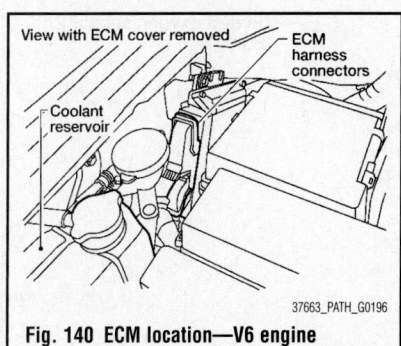

Fig. 140 ECM location—V6 engine

room passenger side behind reservoir tank . On the V8 engine the Electronic Control Module (ECM) is located in the engine room passenger side behind the battery .

REMOVAL & INSTALLATION

See Figures 140 through 142.

At this time the manufacturer does not provide removal and installation procedures for this component. The following procedure is a guideline and may differ from the vehicle you are servicing.

1. Before servicing the vehicle, refer to the Precautions Section.

➡If working near and/or around the SRS system and components, be sure to disable the SRS system. After disabling the system wait three minutes or more before servicing the vehicle.

➡On this vehicle the battery current sensor that is installed to the negative battery cable measures the charging/discharging current of the battery and performs various engine controls. If an electrical component is connected directly to the negative battery cable, the current flowing through that component will not be measured by the battery current sensor. This condition may cause a malfunction of the engine control system and battery discharge may occur. Do not connect an electrical component or ground wire directly to the battery terminal.

2. Disconnect the negative battery cable.
3. Remove the engine cover, if required.
4. Remove the air cleaner assembly, if required.
5. Remove the necessary components to gain access to the ECM.
6. Disconnect the electrical connectors.
7. Remove the ECM retaining screws/bolts/clips etc.
8. Remove the component from its mounting.

To install:

➡Be sure to use new fasteners, as required.

9. Installation is the reverse of the removal procedure.
10. Be sure to perform the reconnect/relearn procedures.

ENGINE COOLANT TEMPERATURE (ECT) SENSOR

LOCATION

See Figures 143 and 144.

Refer to the accompanying illustrations.

REMOVAL & INSTALLATION

At this time the manufacturer does not provide removal and installation procedures for this component. The following procedure is a guideline and may differ from the vehicle you are servicing.

1. Before servicing the vehicle, refer to the Precautions Section.

➡If working near and/or around the SRS system and components, be sure to disable the SRS system. After disabling the system wait three minutes or more before servicing the vehicle.

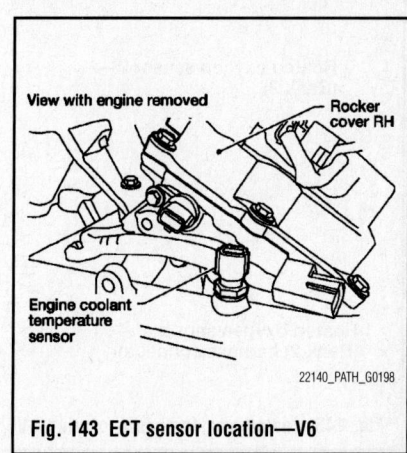

Fig. 143 ECT sensor location—V6

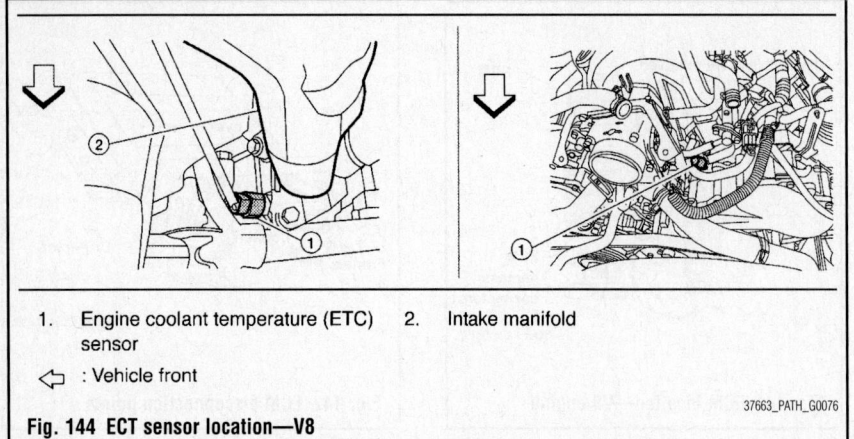

1. Engine coolant temperature (ETC) sensor
2. Intake manifold

⟵ : Vehicle front

37663_PATH_G0076

Fig. 144 ECT sensor location—V8

➡On this vehicle the battery current sensor that is installed to the negative battery cable measures the charging/discharging current of the battery and performs various engine controls. If an electrical component is connected directly to the negative battery cable, the current flowing through that component will not be measured by the battery current sensor. This condition may cause a malfunction of the engine control system and battery discharge may occur. Do not connect an electrical component or ground wire directly to the battery terminal.

2. Disconnect the negative battery cable.
3. Drain the coolant to an acceptable level, below the sensor. Be sure to properly dispose of used coolant.

4. Remove the necessary components to gain access to the sensor.
5. Disconnect the electrical connector.
6. Remove the sensor from its mounting.
7. Discard the gasket.

To install:

➡**Be sure to use new fasteners, as required.**

8. Installation is the reverse of the removal procedure.
9. Be sure to perform the reconnect/relearn procedures.

HEATED OXYGEN SENSOR (HO2S)

LOCATION

See Figures 145 and 146.

Refer to the accompanying illustrations.

REMOVAL & INSTALLATION

At this time the manufacturer does not provide removal and installation procedures

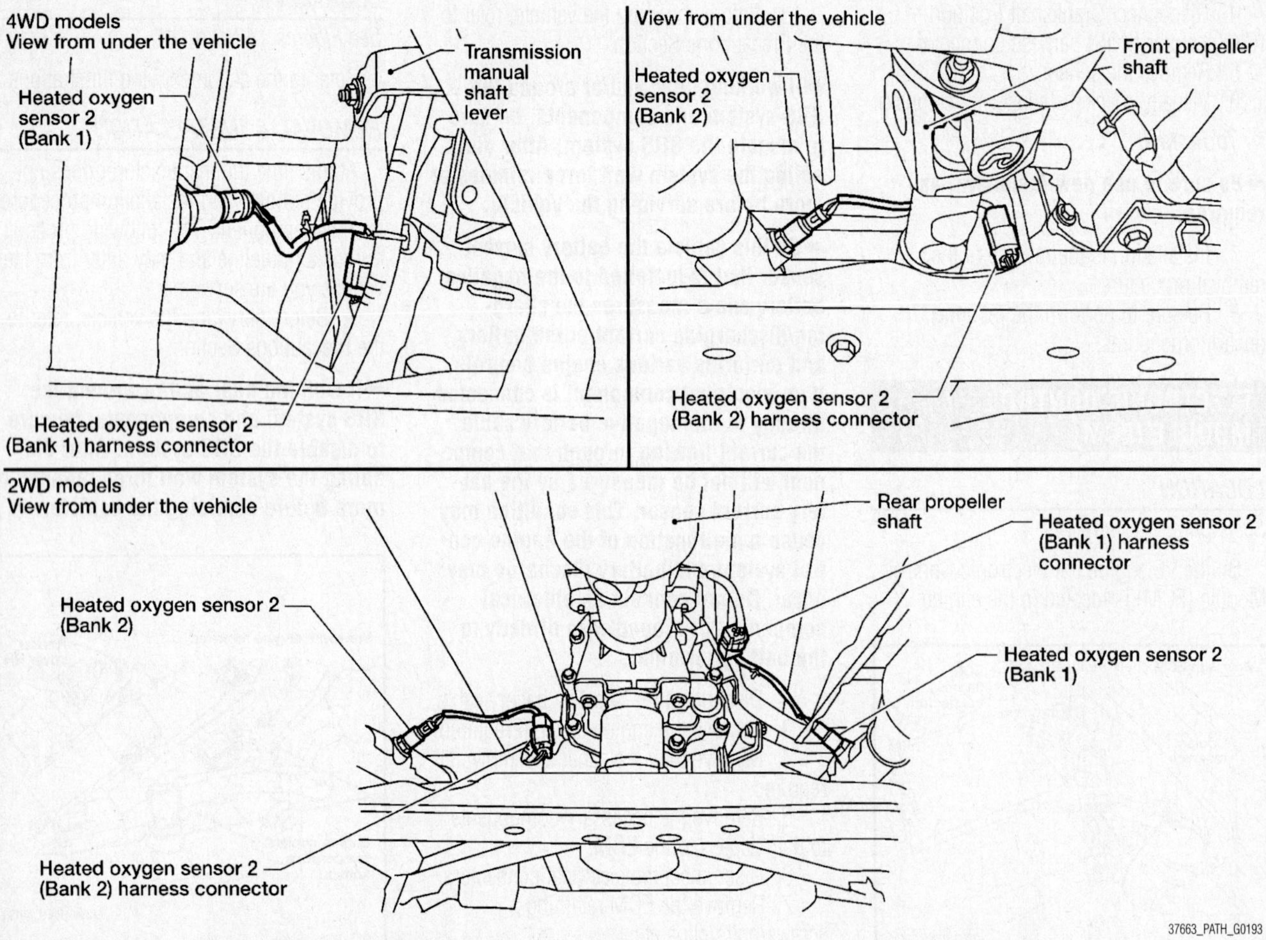

4WD models
View from under the vehicle

Heated oxygen sensor 2 (Bank 1)

Transmission manual shaft lever

Heated oxygen sensor 2 (Bank 1) harness connector

View from under the vehicle

Heated oxygen sensor 2 (Bank 2)

Front propeller shaft

Heated oxygen sensor 2 (Bank 2) harness connector

2WD models
View from under the vehicle

Heated oxygen sensor 2 (Bank 2)

Rear propeller shaft

Heated oxygen sensor 2 (Bank 1) harness connector

Heated oxygen sensor 2 (Bank 1)

Heated oxygen sensor 2 (Bank 2) harness connector

37663_PATH_G0193

Fig. 145 Heated oxygen sensor location—V6 engine

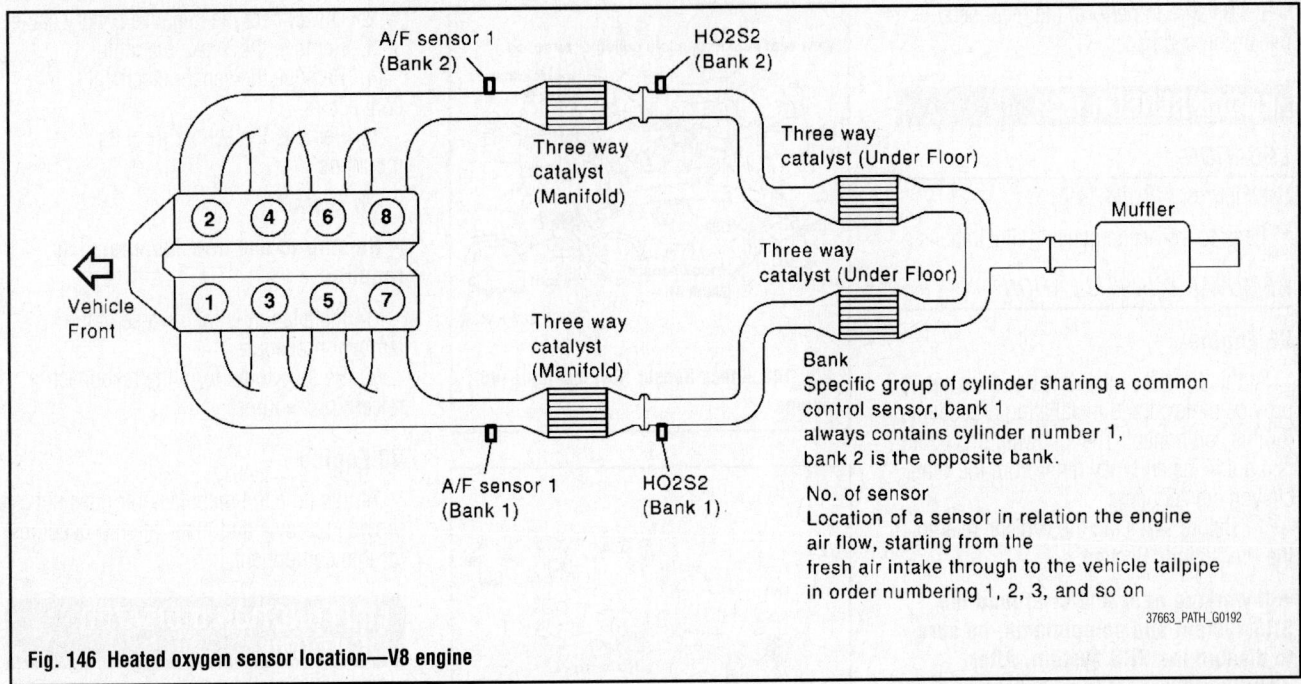

Fig. 146 Heated oxygen sensor location—V8 engine

for this component. The following procedure is a guideline and may differ from the vehicle you are servicing.

1. Before servicing the vehicle, refer to the Precautions Section.

➡️If working near and/or around the SRS system and components, be sure to disable the SRS system. After disabling the system wait three minutes or more before servicing the vehicle.

➡️On this vehicle the battery current sensor that is installed to the negative battery cable measures the charging/discharging current of the battery and performs various engine controls. If an electrical component is connected directly to the negative battery cable, the current flowing through that component will not be measured by the battery current sensor. This condition may cause a malfunction of the engine control system and battery discharge may occur. Do not connect an electrical component or ground wire directly to the battery terminal.

2. Disconnect the negative battery cable.
3. Remove air cleaner case and air duct.
4. Remove engine undercover.
5. Raise and support the vehicle, as required.
6. Disconnect harness connector.
7. Remove the sensor from its mounting.

To install:

➡️Be sure to use new fasteners, as required.

8. Installation is the reverse of the removal procedure.

9. Be sure to perform the reconnect/relearn procedures.

INTAKE AIR TEMPERATURE (IAT) SENSOR

LOCATION

See Figure 147.

The Intake Air Temperature (IAT) sensor is built-into Mass Air Flow (MAF) sensor, which is positioned in the air cleaner assembly.

REMOVAL & INSTALLATION

1. Before servicing the vehicle, refer to the Precautions Section.

➡️If working near and/or around the SRS system and components, be sure to disable the SRS system. After dis-

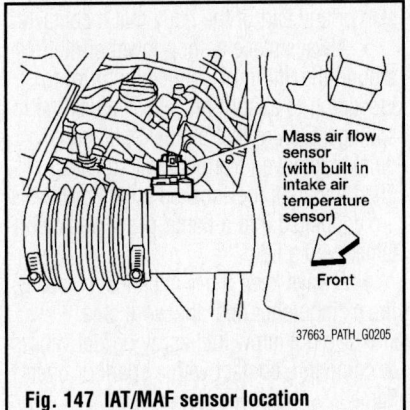

Fig. 147 IAT/MAF sensor location

abling the system wait three minutes or more before servicing the vehicle.

➡️On this vehicle the battery current sensor that is installed to the negative battery cable measures the charging/discharging current of the battery and performs various engine controls. If an electrical component is connected directly to the negative battery cable, the current flowing through that component will not be measured by the battery current sensor. This condition may cause a malfunction of the engine control system and battery discharge may occur. Do not connect an electrical component or ground wire directly to the battery terminal.

2. Disconnect the negative battery cable.
3. Disconnect harness connector from mass air flow sensor.
4. Disconnect PCV hose.
5. Remove air cleaner case/Mass Air Flow (MAF) sensor assembly and air duct assembly disconnecting their joints. Add marks as necessary for easier installation.

✴️ WARNING

Handle MAF sensor with care. Do not shock it. Do not disassemble it. Do not touch its sensor.

To install:

➡️Be sure to use new fasteners, as required.

6. Installation is the reverse of the removal procedure.

7. Be sure to perform the reconnect/relearn procedures.

KNOCK SENSOR (KS)

LOCATION

See Figures 148 and 149.

Refer to the accompanying illustrations.

REMOVAL & INSTALLATION

V6 Engine

At this time the manufacturer does not provide removal and installation procedures for this component. The following procedure is a guideline and may differ from the vehicle you are servicing.

1. Before servicing the vehicle, refer to the Precautions Section.

➡ **If working near and/or around the SRS system and components, be sure to disable the SRS system. After disabling the system wait three minutes or more before servicing the vehicle.**

➡ **On this vehicle the battery current sensor that is installed to the negative battery cable measures the charging/discharging current of the battery and performs various engine controls. If an electrical component is connected directly to the negative battery cable, the current flowing through that component will not be measured by the battery current sensor. This condition may cause a malfunction of the engine control system and battery discharge may**

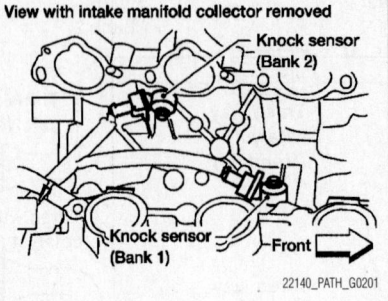

View with intake manifold collector removed

Knock sensor (Bank 2)

Knock sensor (Bank 1)   Front

22140_PATH_G0201

Fig. 148 Knock Sensor (KS) location—V6 engine

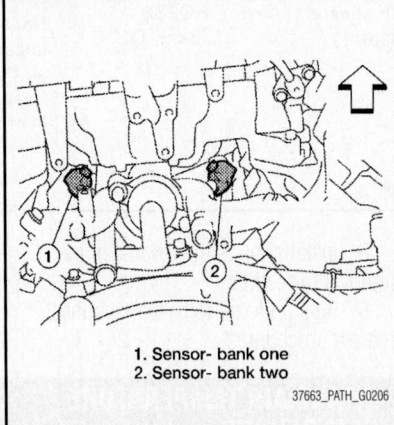

1. Sensor- bank one
2. Sensor- bank two

37663_PATH_G0206

Fig. 149 Knock Sensor (KS) location (view with engine removed from vehicle)—V8 engine

occur. Do not connect an electrical component or ground wire directly to the battery terminal.

2. Disconnect the negative battery cable.
3. Remove the intake manifold.
4. Remove the sensor electrical connectors.
5. Remove the sensor from its mounting.

To install:

➡ **Be sure to use new fasteners, as required.**

6. Installation is the reverse of the removal procedure.
7. Be sure to perform the reconnect/relearn procedures.

V8 Engine

At this time the manufacturer does not provide removal and installation procedures for this component.

MALFUNCTION INDICATOR LIGHT (MIL)

RESET PROCEDURE

Clearing diagnostic trouble codes resets the MIL.

MASS AIR FLOW (MAF) SENSOR

LOCATION

Refer to Intake Air Temperature (IAT) Sensor.

REMOVAL & INSTALLATION

Refer to Intake Air Temperature (IAT) Sensor.

FUEL

GASOLINE FUEL INJECTION SYSTEM

FUEL SYSTEM SERVICE PRECAUTIONS

Safety is the most important factor when performing not only fuel system maintenance, but any type of maintenance. Failure to conduct maintenance and repairs in a safe manner may result in serious personal injury or death. Work on a vehicle's fuel system components can be accomplished safely and effectively by adhering to the following rules and guidelines.

• To avoid the possibility of fire and personal injury, always disconnect the negative battery cable unless the repair or test procedure requires that battery voltage be applied.
• Always relieve the fuel system pressure prior to disconnecting any fuel system component (injector, fuel rail, pressure regulator,

etc.) fitting or fuel line connection. Exercise extreme caution whenever relieving fuel system pressure to avoid exposing skin, face and eyes to fuel spray. Please be advised that fuel under pressure may penetrate the skin or any part of the body that it contacts.

• Always place a shop towel or cloth around the fitting or connection prior to loosening to absorb any excess fuel due to spillage. Ensure that all fuel spillage is quickly removed from engine surfaces. Ensure that all fuel-soaked cloths or towels are deposited into a flame-proof waste container with a lid.
• Always keep a dry chemical (Class B) fire extinguisher near the work area.
• Do not allow fuel spray or fuel vapors to come into contact with a spark or open flame.

• Always use a second wrench when loosening or tightening fuel line connection fittings. This will prevent unnecessary stress and torsion on fuel piping. Always follow the proper torque specifications.
• Always replace worn fuel fitting O-rings with new ones. Do not substitute fuel hose where rigid pipe is installed.

FUEL SYSTEM PRESSURE

RELIEVING

See Figures 150 and 151.

1. Before servicing the vehicle, refer to the Precautions Section.

➡ **If working near and/or around the SRS system and components, be sure to disable the SRS system. After dis-**

abling the system wait three minutes or more before servicing the vehicle.

2. Remove fuel pump fuse located in IPDM E/R.

3. Start engine.

4. After engine stalls, crank it two or three times to release all fuel pressure.

5. Turn ignition switch OFF.

6. Reinstall fuel pump fuse after servicing fuel system.

➡**On this vehicle the battery current sensor that is installed to the negative battery cable measures the charging/discharging current of the battery and performs various engine controls. If an electrical component is connected directly to the negative battery cable, the current flowing through that component will not be measured by the battery current sensor. This condition may cause a malfunction of the engine control system and battery discharge may occur. Do not connect an electrical component or ground wire directly to the battery terminal.**

7. Disconnect the negative battery cable.

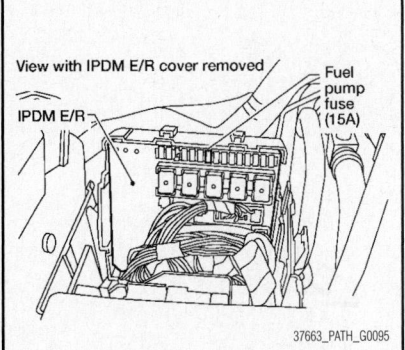

Fig. 150 Fuel system fuse location—V6 engine

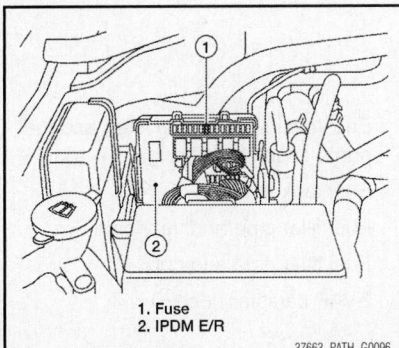

1. Fuse
2. IPDM E/R

Fig. 151 Fuel system fuse location—V8 engine

FUEL FILTER

REMOVAL & INSTALLATION

The fuel filter is part of the fuel pump module.

FUEL LEVEL SENDING UNIT

REMOVAL & INSTALLATION

The fuel level sending unit is part of the fuel pump module.

FUEL PUMP MODULE

REMOVAL & INSTALLATION

See Figures 152 through 154.

1. Before servicing the vehicle, refer to the Precautions Section.

➡**If working near and/or around the SRS system and components, be sure to disable the SRS system. After disabling the system wait three minutes or more before servicing the vehicle.**

➡**On this vehicle the battery current sensor that is installed to the negative battery cable measures the charging/discharging current of the battery and performs various engine controls. If an electrical component is connected directly to the negative battery cable, the current flowing through that component will not be measured by the battery current sensor. This condition may cause a malfunction of the engine control system and battery discharge may occur. Do not connect an electrical component or ground wire directly to the battery terminal.**

2. Disconnect the negative battery cable.

3. Remove the fuel filler cap to release the pressure from inside the fuel tank.

4. Remove the LH rear wheel and tire.

5. Check the fuel level on level gauge. If the fuel gauge indicates more than the level as shown (full or almost full), drain the fuel from the fuel tank until the fuel gauge indicates the level as shown, or less.

➡**Fuel will be spilled when removing the fuel level sensor, fuel filter, and fuel pump assembly for the fuel level is above the fuel level sensor, fuel filter, and fuel pump assembly fuel tank opening.**

• As a guide, the fuel level reaches the fuel gauge position as shown,

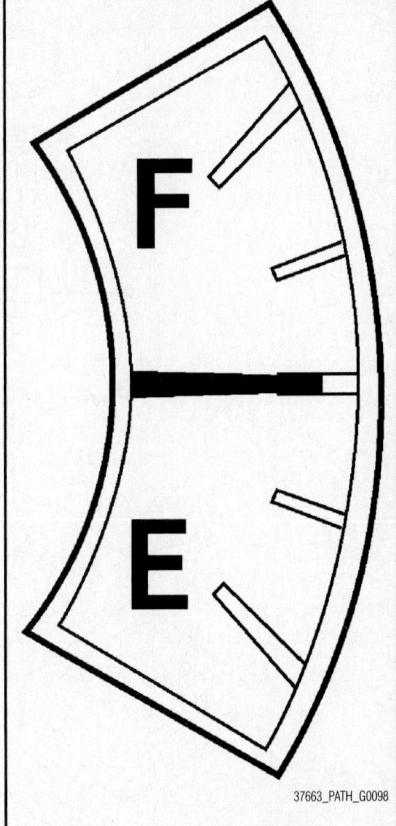

Fig. 152 Fuel gauge level draining position

or less, when approximately 4 US gal (15L) of fuel are drained from the fuel tank.

• If the fuel pump does not operate, use the following procedure to drain the fuel to the specified level.

a. Insert a suitable hose of less than 15 mm (0.59 in.) diameter into the fuel filler pipe through the fuel filler opening to drain the fuel from fuel filler pipe.

b. Remove the fuel filler pipe shield.

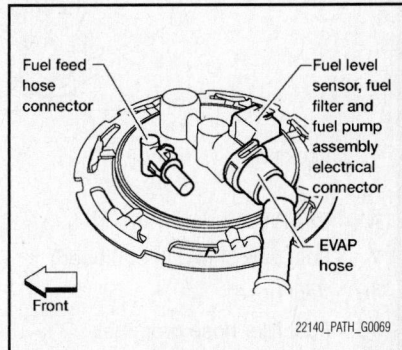

Fig. 153 Disconnect the fuel level sensor, fuel filter, and fuel pump assembly electrical connector, and the fuel feed hose

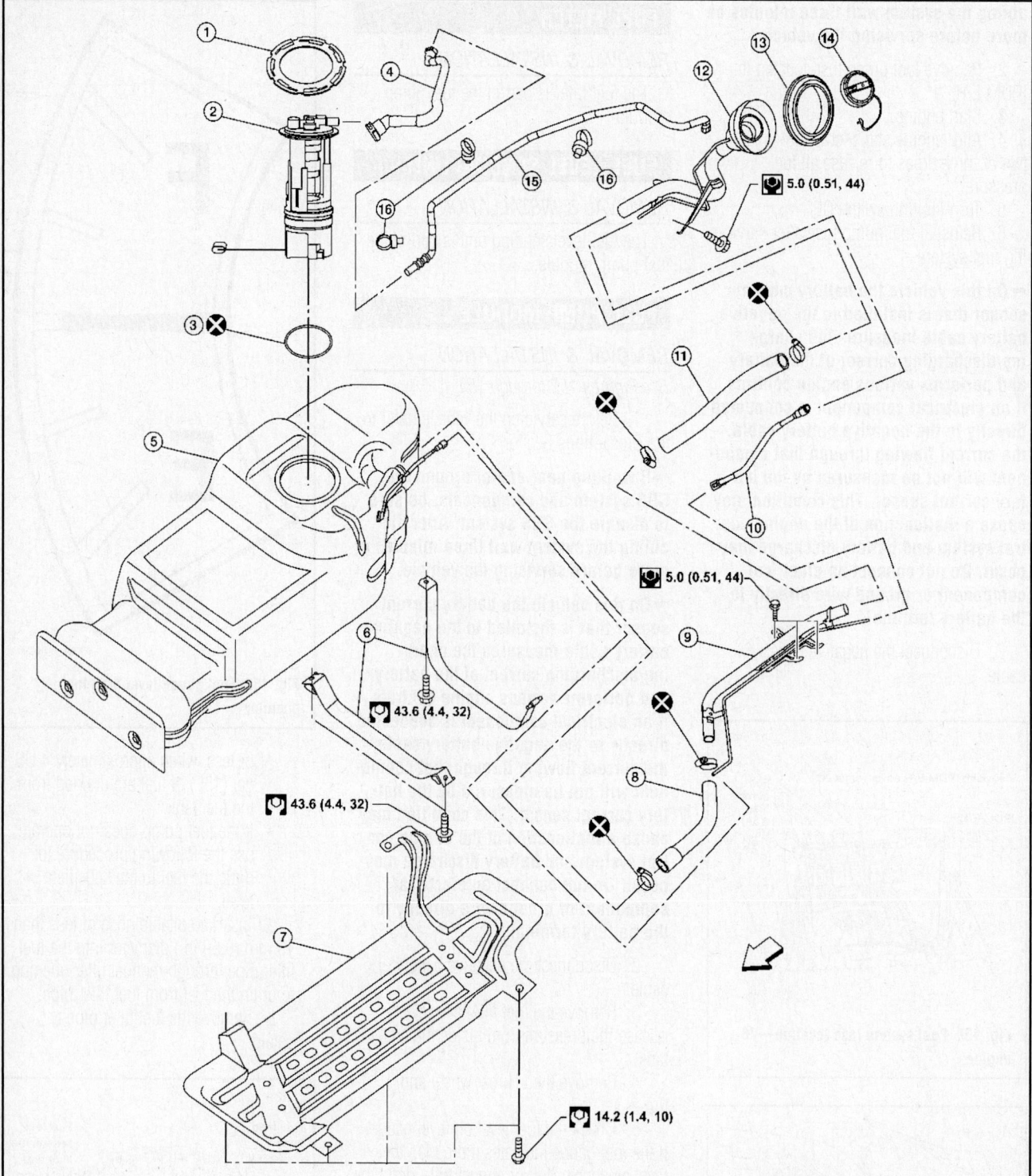

5.0 (0.51, 44)

5.0 (0.51, 44)

43.6 (4.4, 32)

43.6 (4.4, 32)

14.2 (1.4, 10)

1. Lock ring	2. Fuel level sensor, fuel filter, and fuel pump assembly	3. Fuel level sensor, fuel filter, and fuel pump assembly O-ring
4. EVAP hose	5. Fuel tank	6. Fuel tank straps
7. Fuel tank shield (if equipped)	8. Lower fuel filler hose	9. Fuel filler pipe and vent pipe
10. Vent hose	11. Upper fuel filler hose	12. Fuel filler pipe and cup
13. Fuel filler hose grommet	14. Fuel filler cap	15. EVAP canister hose
16. clamp	⇐ Front	

37663_PATH_G0097

Fig. 154 Fuel pump module and related components

c. Disconnect the fuel filler hose from the fuel filler pipe.

d. Insert a suitable hose into the fuel tank through the fuel filler hose to drain the fuel from the fuel tank.

6. Release the fuel pressure from the fuel lines.

7. Disconnect the lower fuel filler hose from the fuel tank, the EVAP hose, and the vent pipe quick connector.

a. Disconnect the fuel feed hose from the molded clip in the side of the fuel tank.

- Disconnect the quick connector as follows:
- Hold the sides of the connector, push in the tabs and pull out the tube.
- If the connector and the tube are stuck together, push and pull several times until they start to move. Then disconnect them by pulling.

❈❈ WARNING

The quick connector can be disconnected when the tabs are completely depressed. Do not twist the quick connector more than necessary. Do not use any tools to disconnect the quick connector. Keep the resin tube away from heat. Be especially careful when welding near the tube. Do not bend or twist the resin tube during connection.

8. Remove the four bolts and remove the fuel tank shield.

9. Remove the driveshaft shaft.

10. Support the fuel tank using a suitable lift jack.

11. Remove the three fuel tank strap bolts while supporting the fuel tank with a suitable lift jack.

12. Remove the fuel tank straps and slowly lower the fuel tank to access the top of the fuel level sensor, fuel filter and fuel pump assembly.

❈❈ CAUTION

Do not lower the fuel tank too far to prevent damage to the fuel feed hose and the fuel level sensor, fuel filter and fuel pump assembly connector.

13. Disconnect the fuel level sensor, fuel filter, and fuel pump assembly electrical connector, and the fuel feed hose.

a. Disconnect the quick connector as follows:

- Hold the sides of the connector, push in the tabs and pull out the tube.

- If the connector and the tube are stuck together, push and pull several times until they start to move. Then disconnect them by pulling.

❈❈ WARNING

The quick connector can be disconnected when the tabs are completely depressed. Do not twist the quick connector more than necessary. Do not use any tools to disconnect the quick connector. Do not bend or twist the resin tube during connection.

14. Lower the fuel tank using a suitable lift jack and remove the fuel tank.

15. Disconnect the EVAP hose from the fuel pump and remove the EVAP hose from the molded clip in the top of the fuel tank.

16. Remove the lock ring.

17. Remove the fuel level sensor, fuel filter, and fuel pump assembly.

❈❈ WARNING

Do not bend the float arm during removal. Avoid impacts such as dropping when handling the components.

To install:

18. Installation is in the reverse order of removal.

19. Connect the quick connector as follows:

- Check the connection for any damage or foreign materials.
- Align the connector with the pipe, then insert the connector straight into the pipe until a click is heard.
- Pull the tube and the connector to make sure they are securely connected.
- Visually inspect the connector to make sure the two retainer tabs are securely connected.

❈❈ WARNING

Do not bend the float arm during installation. Avoid impacts such as dropping when handling the components.

20. Turn the ignition switch ON but do not start engine, then check the fuel pipe and hose connections for leaks while applying fuel pressure.

21. Start the engine and rev it above idle, then check that there are no fuel leaks at any of the fuel pipe and hose connections.

22. Be sure to perform the reconnect/relearn procedures.

FUEL RAIL AND INJECTOR

REMOVAL & INSTALLATION

V6 Engine

See Figures 155 through 157.

1. Before servicing the vehicle, refer to the Precautions Section.

➡ If working near and/or around the SRS system and components, be sure to disable the SRS system. After disabling the system wait three minutes or more before servicing the vehicle.

➡ On this vehicle the battery current sensor that is installed to the negative battery cable measures the charging/discharging current of the battery and performs various engine controls. If an electrical component is connected directly to the negative battery cable, the current flowing through that component will not be measured by the battery current sensor. This condition may cause a malfunction of the engine control system and battery discharge may occur. Do not connect an electrical component or ground wire directly to the battery terminal.

2. Disconnect the negative battery cable.

3. Remove intake manifold collector.

❈❈ CAUTION

Perform this step when engine is cold.

4. Disconnect the fuel quick connector on the engine side.

a. Using Tool No. 45488, perform the following steps to disconnect the quick connector.

b. Remove quick connector cap.

c. With the sleeve side of tool facing quick connector, install tool onto fuel tube.

d. Insert tool into quick connector until sleeve contacts and goes no further. Hold the tool on that position.

❈❈ WARNING

Inserting the tool hard will not disconnect quick connector. Hold tool where it contacts and goes no further.

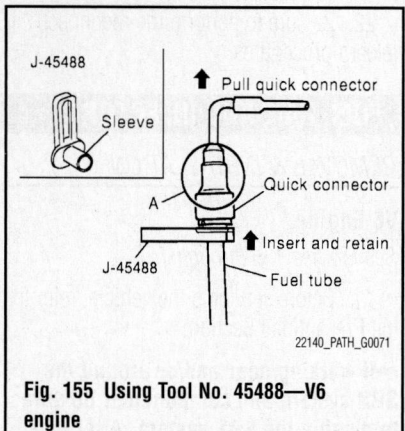

Fig. 155 Using Tool No. 45488—V6 engine

e. Pull the quick connector straight out from the fuel tube.

> ※ **CAUTION**
>
> **Pull quick connector holding it at the A position. Do not pull with lateral force applied. O-ring inside quick connector may be damaged. Prepare container and cloth beforehand as fuel will leak out. Avoid fire and**

sparks. Be sure to cover openings of disconnected pipes with plug or plastic bag to avoid fuel leakage and entry of foreign materials.

5. Remove PCV hose between rocker covers (right and left banks).
6. Disconnect harness connector from fuel injector.
7. Loosen bolts in reverse order as shown, and remove fuel tube and fuel injector assembly.

> ※ **CAUTION**
>
> **Do not tilt it, or remaining fuel in pipes may flow out from pipes.**

8. Remove bolts which connects fuel tube (RH) and fuel tube (LH).
9. Remove fuel injector from fuel tube as follows:
    a. Carefully open and remove clip.
    b. Remove fuel injector from fuel tube by pulling straight.

> ※ **CAUTION**
>
> **Be careful with remaining fuel that may go out from fuel tube. Be careful not to damage injector nozzles during removal. Do not bump or drop fuel injector. Do not disassemble fuel injector.**

10. Disconnect fuel tube (RH) from fuel tube (LH).
11. Loosen bolts, to remove fuel damper cap and fuel damper, if necessary.

To install:
12. Install fuel damper as follows:
    a. Install new O-ring to fuel tube.

➡When handling new O-rings, be careful of the following caution:

> ※ **WARNING**
>
> **Handle O-ring with bare hands. Do not wear gloves. Lubricate O-ring**

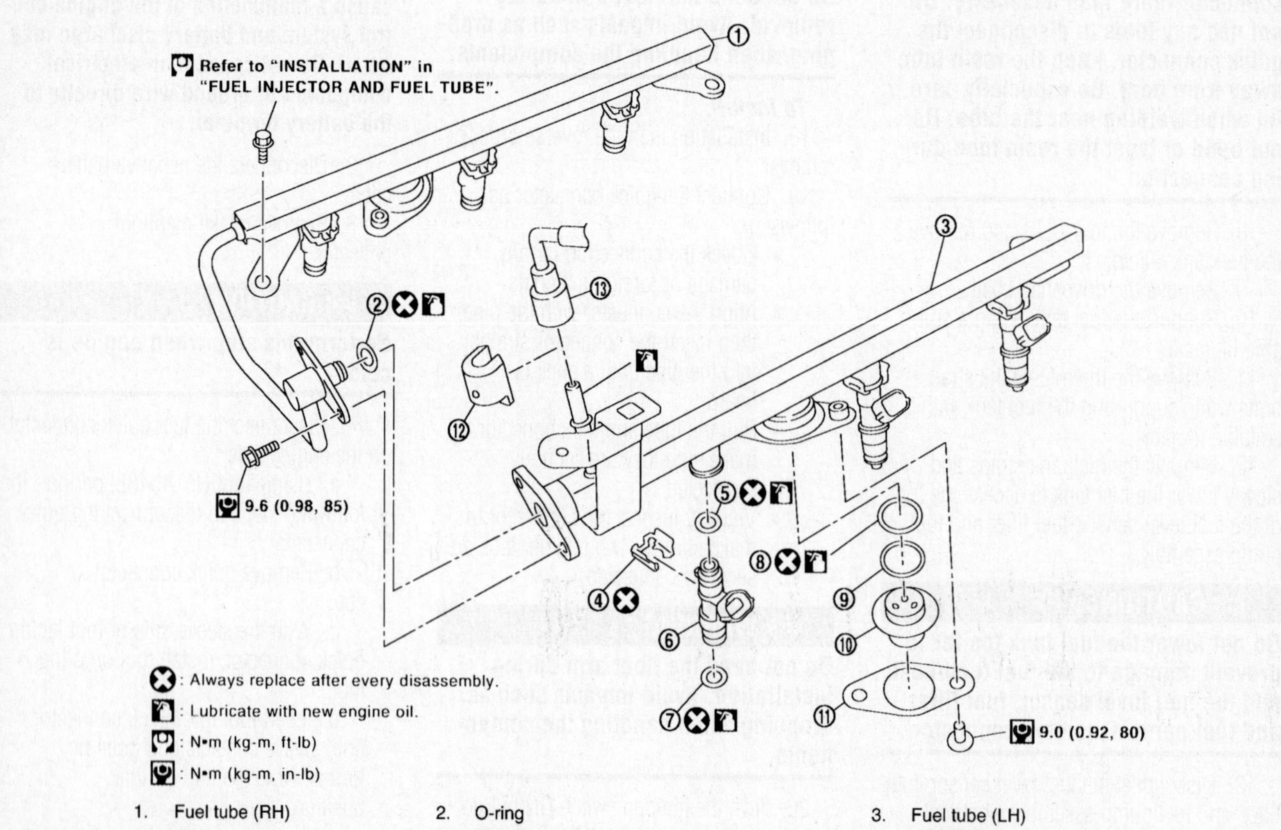

Refer to "INSTALLATION" in "FUEL INJECTOR AND FUEL TUBE".

9.6 (0.98, 85)

9.0 (0.92, 80)

❌ : Always replace after every disassembly.

🛢 : Lubricate with new engine oil.

🔧 : N•m (kg-m, ft-lb)

🔧 : N•m (kg-m, in-lb)

1. Fuel tube (RH)	2. O-ring	3. Fuel tube (LH)
4. Clip	5. O-ring (blue)	6. Fuel injector
7. O-ring (brown)	8. O-ring	9. Spacer
10. Fuel damper	11. Fuel damper cap	12. Quick connector cap
13. Fuel feed hose		

Fig. 156 Fuel injector rail and related components—V6 Engine

with new engine oil. Do not clean O-ring with solvent. Make sure that O-ring and its mating part are free of foreign material. When installing O-ring, be careful not to scratch it with tool or fingernails. Also be careful not to twist or stretch O-ring. If O-ring was stretched while it was being attached, do not insert it quickly into fuel tube. Insert new O-ring straight into fuel tube. Do not twist it.

   b. Install spacer to fuel damper.
   c. Insert fuel damper straight into fuel tube.

✱✱ WARNING

Insert straight, making sure that the axis is lined up. Do not pressure-fit with excessive force.

   d. Tighten bolts evenly in turn.

➡ **After tightening bolts, make sure that there is no gap between fuel damper cap and fuel tube.**

   13. Install new O-rings to fuel injector, paying attention to the following.

✱✱ WARNING

Upper and lower O-ring are different. Be careful not to confuse them.

- Fuel tube side: Blue
- Nozzle side: Brown

✱✱ WARNING

Handle O-ring with bare hands. Do not wear gloves. Lubricate O-ring with new engine oil. Do not clean O-ring with solvent. Make sure that O-ring and its mating part are free of foreign material. When installing O-ring, be careful not to scratch it with tool or fingernails. Also be careful not to twist or stretch O-ring. If O-ring was stretched while it was being attached, do not insert it quickly into fuel tube. Insert O-ring straight into fuel injector. Do not twist it.

   14. Install fuel injector to fuel tube as follows:
   a. Insert clip into clip mounting groove on fuel injector.

✱✱ WARNING

Do not reuse clip. Replace it with a new one. Be careful to keep clip from interfering with O-ring. If interference occurs, replace O-ring.

   b. Insert fuel injector into fuel tube with clip attached.
   c. Make sure that installation is complete by checking that fuel injector does not rotate or come off.
   d. Make sure that protrusions of fuel injectors are aligned with cutouts of clips after installation.
   15. Connect fuel tube (RH) to fuel tube (LH), and tighten bolts temporarily.
   a. Tighten bolts with the specified torque after installing fuel tube and fuel injector assembly.

✱✱ WARNING

Handle O-ring with bare hands. Do not wear gloves. Lubricate O-ring with new engine oil. Do not clean O-ring with solvent. Make sure that O-ring and its mating part are free of foreign material. When installing O-ring, be careful not to scratch it with tool or fingernails. Also be careful not to twist or stretch O-ring. If O-ring was stretched while it was being attached, do not insert it quickly into fuel tube. Insert new O-ring straight into fuel tube. Do not twist it.

   16. Install fuel tube and fuel injector assembly to intake manifold.

✱✱ WARNING

Be careful not to let tip of injector nozzle come in contact with other parts.

   17. Tighten bolts in two steps in numerical order as shown.
   a. Fuel injector tube assembly bolts:
- 1st step: 7 ft. lbs. (10.1 Nm)
- 2nd step: 16 ft. lbs. (22.0 Nm)

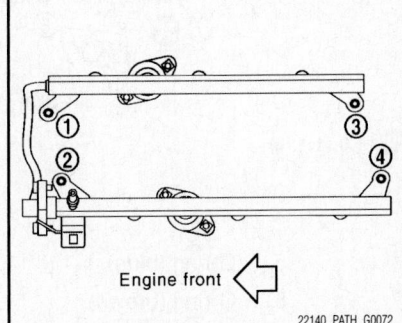

Engine front

22140_PATH_G0072

Fig. 157 Tighten bolts in two steps in numerical order as shown—V6 engine

   18. Tighten bolts which connects fuel tube (RH) and fuel tube (LH) with the specified torque.
   19. Connect fuel injector harness connector.
   20. Install intake manifold collector.
   21. Installation of the remaining components is in the reverse order of removal.
   22. Turn ignition switch "ON" (with engine stopped). With fuel pressure applied to fuel piping, check for fuel leakage at connection points.

➡ **Use mirrors for checking at points out of clear sight.**

   23. Start engine. With engine speed increased, check again for fuel leakage at connection points.
   24. Be sure to perform the reconnect/relearn procedures.

V8 Engine

See Figures 158 through 162.

   1. Before servicing the vehicle, refer to the Precautions Section.

➡ **If working near and/or around the SRS system and components, be sure to disable the SRS system. After disabling the system wait three minutes or more before servicing the vehicle.**

➡ **On this vehicle the battery current sensor that is installed to the negative battery cable measures the charging/discharging current of the battery and performs various engine controls. If an electrical component is connected directly to the negative battery cable, the current flowing through that component will not be measured by the battery current sensor. This condition may cause a malfunction of the engine control system and battery discharge may occur. Do not connect an electrical component or ground wire directly to the battery terminal.**

   2. Disconnect the negative battery cable.
   3. Remove the engine room cover.
   4. Release the fuel pressure.
   5. Remove the air duct and resonator assembly.
   6. Disconnect the fuel injector harness connectors.
   7. Disconnect the fuel hose assembly from the fuel tubes (RH and LH).

✱✱ CAUTION

While hoses are disconnected, plug them to prevent fuel from draining. Do not separate the fuel connector and fuel hose.

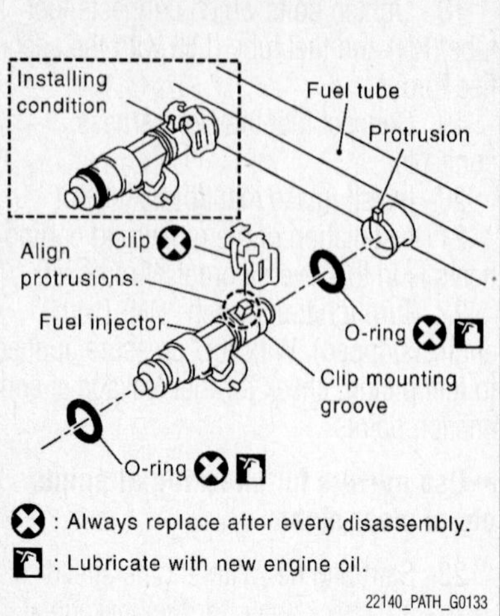

Installing condition
Fuel tube
Protrusion
Clip
Align protrusions.
Fuel injector
O-ring
Clip mounting groove
O-ring
: Always replace after every disassembly.
: Lubricate with new engine oil.
22140_PATH_G0133

Fig. 158 Remove the fuel injector from the fuel tube—V8 engine

8. Remove the fuel injectors with the fuel tube assembly.
9. Remove the fuel injector from the fuel tube using the following steps:
   a. Spread open and remove the clip.
   b. Remove the fuel injector from the fuel tube by pulling straight out.

CAUTION

Be careful with remaining fuel that may leak out from fuel tube. Do not damage injector nozzles during removal. Do not bump or drop fuel injectors. Do not disassemble fuel injectors.

10. Remove the fuel damper from each fuel tube.

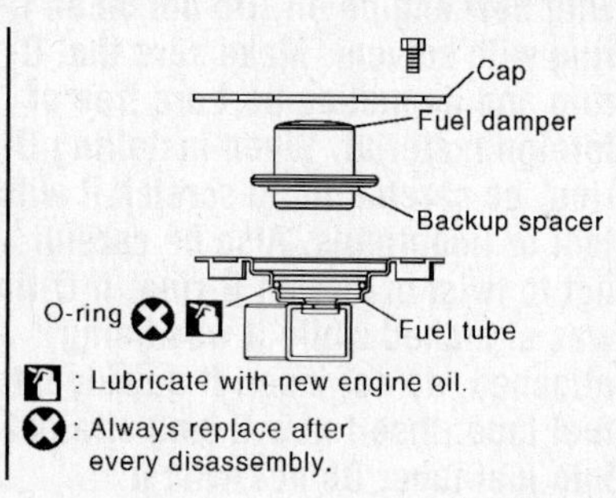

Cap
Fuel damper
Backup spacer
O-ring
Fuel tube
: Lubricate with new engine oil.
: Always replace after every disassembly.
22140_PATH_G0134

Fig. 160 Install the fuel damper to each fuel tube—V8 engine

To install:

11. Install the fuel damper to each fuel tube using the following steps:
   a. Apply engine oil to the new O-ring and set it into the cup of the fuel tube.

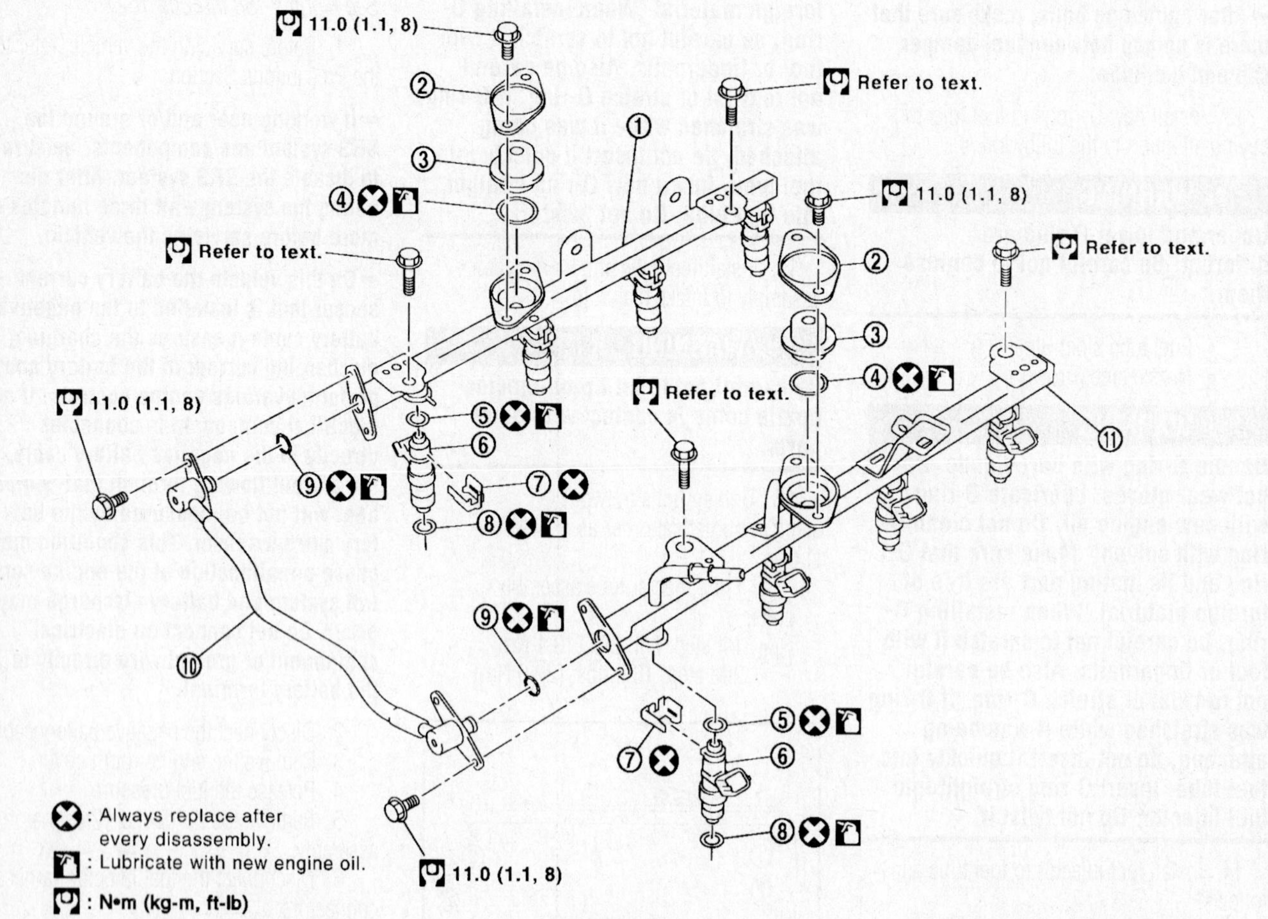

11.0 (1.1, 8)
Refer to text.
11.0 (1.1, 8)
Refer to text.
Refer to text.
11.0 (1.1, 8)
Refer to text.
11.0 (1.1, 8)
: Always replace after every disassembly.
: Lubricate with new engine oil.
: N•m (kg-m, ft-lb)

1. Fuel tube (RH)
2. Cap
3. Fuel damper
4. O-ring
5. O-ring (blue)
6. Fuel injector
7. Clip
8. O-ring (brown)
9. O-ring
10. Fuel hose assembly
11. Fuel tube (LH)

22140_PATH_G0132

Fig. 159 Fuel injector rail and related components—V8 engine

Handle O-ring with bare hands. Never wear gloves. Lubricate new O-ring with new engine oil. Do not clean O-ring with solvent. Make sure that O-ring and its mating part are free of foreign material. When installing O-ring, do not scratch it with tool or fingernails. Do not twist or stretch the O-ring.

b. Make sure that the backup spacer is in the O-ring connecting surface of the fuel damper.

➡ The backup spacer is part of the fuel damper assembly.

c. Insert the fuel damper until it seats on the fuel tube.

Insert straight, making sure that the axis is lined up. Do not pressure-fit with excessive force. Install the cap, and then tighten the bolts evenly. After tightening the bolts, make sure that there is no gap between the cap and fuel tube.

12. Install new O-rings to the fuel injector paying attention to the items below.

Upper and lower O-rings are different colors. Handle O-ring with bare hands. Never wear gloves. Lubricate new O-ring with new engine oil. Do not clean O-ring with solvent. Make sure that O-ring and its mating part are free of foreign material. When installing O-ring, be careful not to scratch it with tool or fingernails. Also be careful not to twist or stretch O-ring. If O-ring was stretched while it was being attached, do not insert it quickly into fuel tube. Insert O-ring straight into fuel tube. Do not angle or twist it.

13. Install the fuel injector to the fuel tube using the following steps.

a. Insert new clip into clip mounting groove on the fuel injector.
• Insert clip so that lug A of fuel injector matches notch A of the clip.

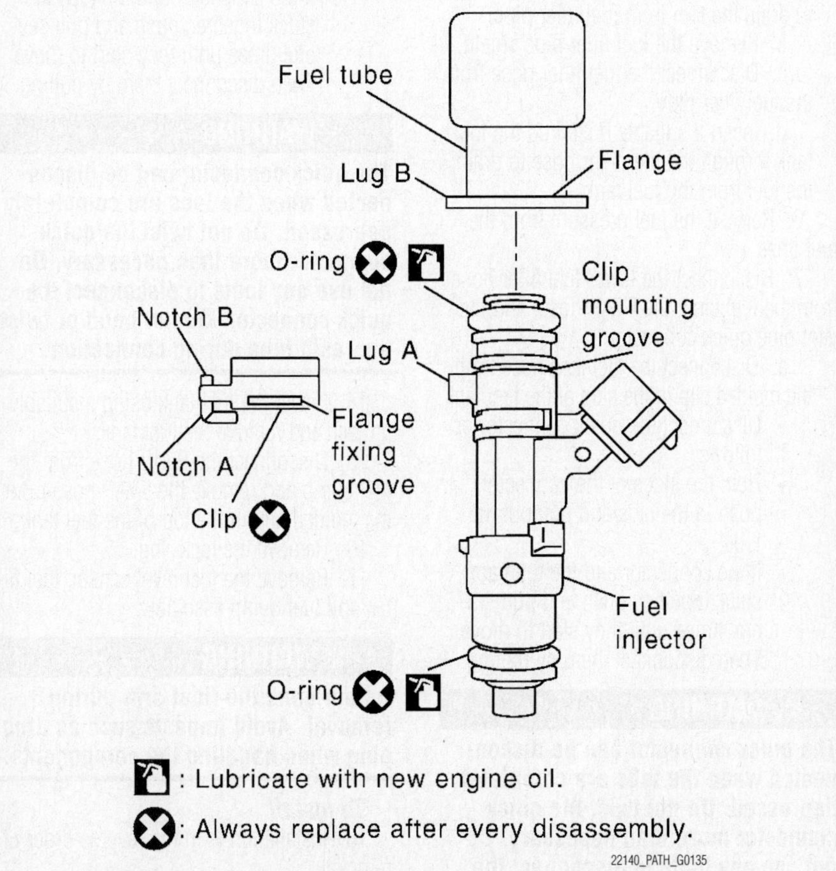

Do not reuse clip. Replace it with a new one. Do not allow the clip to interfere with the O-ring. If interference occurs, replace the O-ring.

b. Insert the fuel injector into the fuel tube with the clip attached.
• Insert it while matching it to the axial center.
• Insert fuel injector so that lug B of fuel tube matches notch B of the clip.
• Make sure that the fuel tube flange is securely seated in the flange fixing groove on the clip.

c. Make sure that installation is complete by checking that the fuel injector does not rotate or come off.
• Make sure that the protrusions of the fuel injectors are aligned with the cutouts of the clips after installation.

14. Install the fuel tube and fuel injector assembly to the intake manifold.

Do not let the tip of the injector nozzle come in contact with other parts.

a. Tighten fuel tube assembly bolts A to B in two steps:
• Step 1: 9 ft. lbs. (12.8 Nm)
• Step 2: 18 ft. lbs. (24.5 Nm)

15. Install the fuel hose assembly.
a. Insert connector's straight, making sure that the axis is lined up with fuel tube side to prevent O-ring from being damaged.
b. Tighten bolts evenly in several steps.
c. Make sure that there is no gap between the flange and fuel tube after tightening the bolts.

Fuel tube

Lug B — Flange

O-ring ❌ 🛢

Notch B

Lug A — Clip mounting groove

Flange fixing groove

Notch A

Clip ❌

Fuel injector

O-ring ❌ 🛢

🛢 : Lubricate with new engine oil.

❌ : Always replace after every disassembly.

22140_PATH_G0135

Fig. 161 Install the fuel injector to the fuel tube—V8 engine

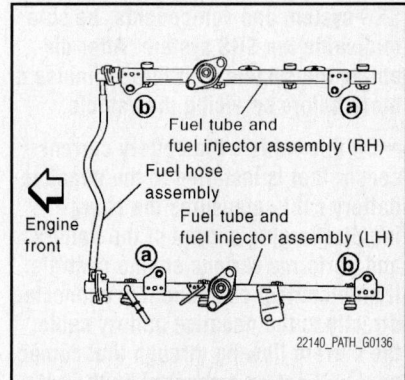

Engine front

Fuel tube and fuel injector assembly (RH)
Fuel hose assembly
Fuel tube and fuel injector assembly (LH)

22140_PATH_G0136

Fig. 162 Injector rail retaining bolt tightening sequence—V8 engine

Handle O-ring with bare hands. Do not wear gloves. Lubricate O-ring with new engine oil. Do not clean O-ring with solvent. Make sure that O-ring and its mating part are free of foreign material. When installing O-ring, be careful not to scratch it with tool or fingernails. Also be careful not to twist or stretch O-ring. If O-ring was stretched while it was being attached, do not insert it quickly into fuel tube. Insert new O-ring straight into fuel tube. Do not twist it.

16. Installation of the remaining components is in the reverse order of removal.

17. After installing the fuel tubes, make sure there are no fuel leaks at the connections using the following steps.

18. Apply fuel pressure to the fuel lines by turning ignition switch ON (with engine stopped). Then check for fuel leaks at the connections.

➡ **Use mirrors for checking on hidden points.**

19. Start the engine and rev it up and check for fuel leaks at the connections.

Do not touch the engine immediately after stopping, as engine becomes extremely hot.

20. Be sure to perform the reconnect/relearn procedures.

FUEL TANK

REMOVAL & INSTALLATION
See Figure 154.

1. Before servicing the vehicle, refer to the Precautions Section.

➡ **If working near and/or around the SRS system and components, be sure to disable the SRS system. After disabling the system wait three minutes or more before servicing the vehicle.**

➡ **On this vehicle the battery current sensor that is installed to the negative battery cable measures the charging/discharging current of the battery and performs various engine controls. If an electrical component is connected directly to the negative battery cable, the current flowing through that component will not be measured by the battery current sensor. This condition may cause a malfunction of the engine con-**

trol system and battery discharge may occur. Do not connect an electrical component or ground wire directly to the battery terminal.

2. Disconnect the negative battery cable.

3. Remove the fuel filler cap to release the pressure from inside the fuel tank.

4. Remove the LH rear wheel and tire.

5. Check the fuel level on level gauge. If the fuel gauge indicates more than the level as shown (full or almost full), drain the fuel from the fuel tank until the fuel gauge indicates the level as shown, or less.

➡ **Fuel will be spilled when removing the fuel level sensor, fuel filter, and fuel pump assembly for the fuel level is above the fuel level sensor, fuel filter, and fuel pump assembly fuel tank opening.**

- As a guide, the fuel level reaches the fuel gauge position as shown, or less, when approximately 4 US gal (15L) of fuel are drained from the fuel tank.
- If the fuel pump does not operate, use the following procedure to drain the fuel to the specified level.

a. Insert a suitable hose of less than 15 mm (0.59 in.) diameter into the fuel filler pipe through the fuel filler opening to drain the fuel from fuel filler pipe.

b. Remove the fuel filler pipe shield.

c. Disconnect the fuel filler hose from the fuel filler pipe.

d. Insert a suitable hose into the fuel tank through the fuel filler hose to drain the fuel from the fuel tank.

6. Release the fuel pressure from the fuel lines.

7. Disconnect the lower fuel filler hose from the fuel tank, the EVAP hose, and the vent pipe quick connector.

a. Disconnect the fuel feed hose from the molded clip in the side of the fuel tank.

- Disconnect the quick connector as follows:
- Hold the sides of the connector, push in the tabs and pull out the tube.
- If the connector and the tube are stuck together, push and pull several times until they start to move. Then disconnect them by pulling.

The quick connector can be disconnected when the tabs are completely depressed. Do not twist the quick connector more than necessary. Do not use any tools to disconnect the quick connector. Keep the resin tube away from heat. Be especially care-

ful when welding near the tube. Do not bend or twist the resin tube during connection.

8. Remove the four bolts and remove the fuel tank shield.

9. Remove the driveshaft shaft.

10. Support the fuel tank using a suitable lift jack.

11. Remove the three fuel tank strap bolts while supporting the fuel tank with a suitable lift jack.

12. Remove the fuel tank straps and slowly lower the fuel tank to access the top of the fuel level sensor, fuel filter and fuel pump assembly.

Do not lower the fuel tank too far to prevent damage to the fuel feed hose and the fuel level sensor, fuel filter and fuel pump assembly connector.

13. Disconnect the fuel level sensor, fuel filter, and fuel pump assembly electrical connector, and the fuel feed hose.

a. Disconnect the quick connector as follows:

- Hold the sides of the connector, push in the tabs and pull out the tube.
- If the connector and the tube are stuck together, push and pull several times until they start to move. Then disconnect them by pulling.

The quick connector can be disconnected when the tabs are completely depressed. Do not twist the quick connector more than necessary. Do not use any tools to disconnect the quick connector. Do not bend or twist the resin tube during connection.

14. Lower the fuel tank using a suitable lift jack and remove the fuel tank.

15. Disconnect the EVAP hose from the fuel pump and remove the EVAP hose from the molded clip in the top of the fuel tank.

16. Remove the lock ring.

17. Remove the fuel level sensor, fuel filter, and fuel pump assembly.

Do not bend the float arm during removal. Avoid impacts such as dropping when handling the components.

To install:

18. Installation is in the reverse order of removal.

19. Connect the quick connector as follows:

- Check the connection for any damage or foreign materials.
- Align the connector with the pipe, then insert the connector straight into the pipe until a click is heard.
- Pull the tube and the connector to make sure they are securely connected.
- Visually inspect the connector to make sure the two retainer tabs are securely connected.

❊❊ WARNING

Do not bend the float arm during installation. Avoid impacts such as dropping when handling the components.

20. Turn the ignition switch ON but do not start engine, then check the fuel pipe and hose connections for leaks while applying fuel pressure.

21. Start the engine and rev it above idle, then check that there are no fuel leaks at any of the fuel pipe and hose connections.

22. Be sure to perform the reconnect/relearn procedures.

IDLE SPEED

ADJUSTMENT

Idle speed is controlled by the ECM. No Adjustment is necessary or possible.

THROTTLE BODY

REMOVAL & INSTALLATION

V6 Engine

See Figures 163 and 164.

1. Before servicing the vehicle, refer to the Precautions Section.

➡ **If working near and/or around the SRS system and components, be sure to disable the SRS system. After disabling the system wait three minutes or more before servicing the vehicle.**

➡ **On this vehicle the battery current sensor that is installed to the negative battery cable measures the charging/discharging current of the battery and performs various engine controls. If an electrical component is connected** directly to the negative battery cable, the current flowing through that component will not be measured by the battery current sensor. This condition may cause a malfunction of the engine control system and battery discharge may occur. Do not connect an electrical component or ground wire directly to the battery terminal.

2. Disconnect the negative battery cable.
3. Remove the air dam.

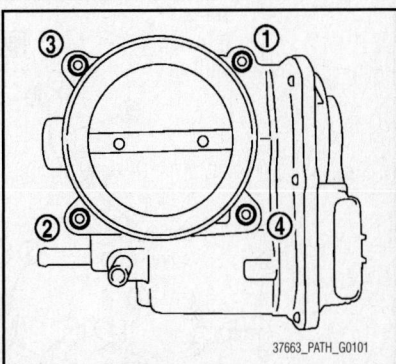

37663_PATH_G0101

Fig. 164 Throttle body retaining bolt tightening sequence—V6 engine

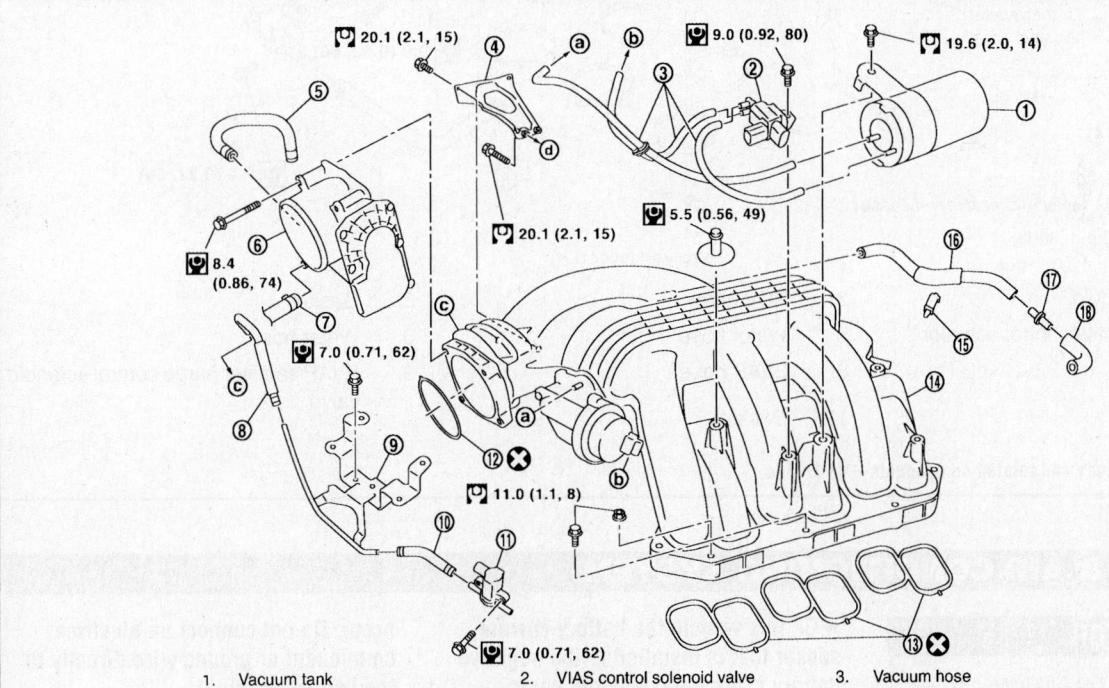

1. Vacuum tank	2. VIAS control solenoid valve	3. Vacuum hose
4. Intake manifold collector support	5. Water hose	6. Electric throttle control actuator
7. Water hose	8. EVAP hose	9. Bracket
10. EVAP hose	11. EVAP canister purge volume control solenoid valve	12. Gasket
13. Gasket	14. Intake manifold collector	15. Clip
16. PCV hose	17. Connector	18. PCV hose
a. To intake manifol collector	b. To power valve	c. To throttle body
d. To cylinder head (RH bank)		

22140_PATH_G0031

Fig. 163 Throttle body and related components—V6 Engine

4. Remove the engine undercover.

5. Remove the engine room cover,

6. Remove the air cleaner assembly.

7. Drain the engine coolant. Be sure to properly dispose of used coolant.

8. Disconnect the hoses from the unit.

9. Disconnect the electrical connectors.

10. Remove the retaining bolts in the reverse order of the tightening sequence.

11. Remove the component from its mounting. Discard the gasket.

To install:

➡ **Be sure to use new fasteners, as required.**

12. Installation is the reverse of the removal procedure.

13. Tighten the retaining bolts to specification and in the proper sequence.

14. Using the CONSULT-III diagnostic tool or equivalent, perform the throttle valve closed position learning and the idle air volume learning procedures.

15. Be sure to perform the reconnect/relearn procedures.

V8 Engine

See Figure 165.

At this time the manufacturer does not provide removal and installation procedures for this component.

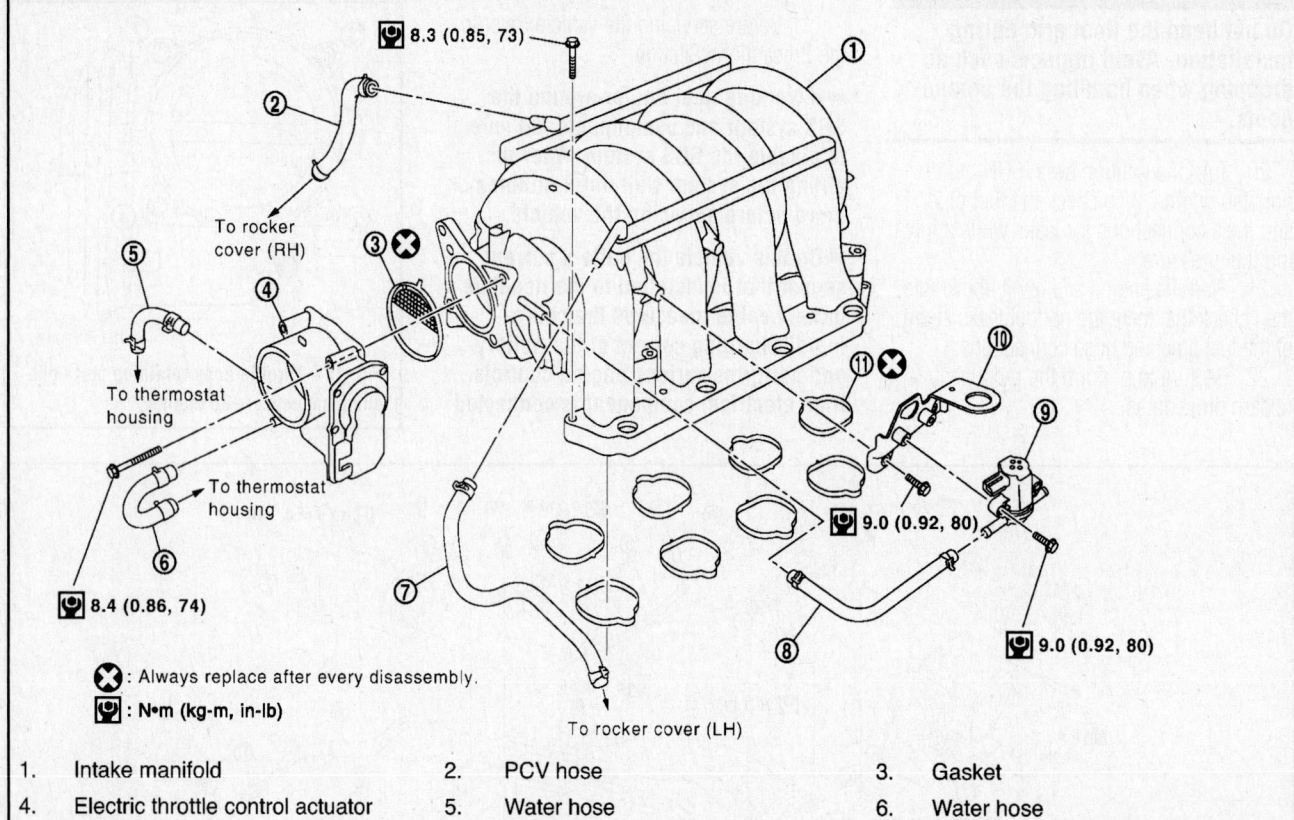

⊗ : Always replace after every disassembly.

🔧 : N•m (kg-m, in-lb)

1.	Intake manifold	2.	PCV hose	3.	Gasket
4.	Electric throttle control actuator	5.	Water hose	6.	Water hose
7.	PCV hose	8.	EVAP hose	9.	EVAP canister purge control solenoid valve
10.	Bracket	11.	Gasket		

Fig. 165 Throttle body and related components—V8 Engine

22140_PATH_G0105

HEATING & AIR CONDITIONING SYSTEM

BLOWER MOTOR

REMOVAL & INSTALLATION

See Figures 166 and 167.

1. Before servicing the vehicle, refer to the Precautions Section.

➡ **If working near and/or around the SRS system and components, be sure to disable the SRS system. After disabling the system wait three minutes or more before servicing the vehicle.**

➡ **On this vehicle the battery current sensor that is installed to the negative battery cable measures the charging/discharging current of the battery and performs various engine controls. If an electrical component is connected directly to the negative battery cable, the current flowing through that component will not be measured by the battery current sensor. This condition may cause a malfunction of the engine control system and battery discharge may occur. Do not connect an electrical component or ground wire directly to the battery terminal.**

2. Disconnect the negative battery cable.

3. Remove the lower glove box assembly.

4. Disconnect the front blower motor electrical connector.

5. Remove the three screws and remove the front blower motor.

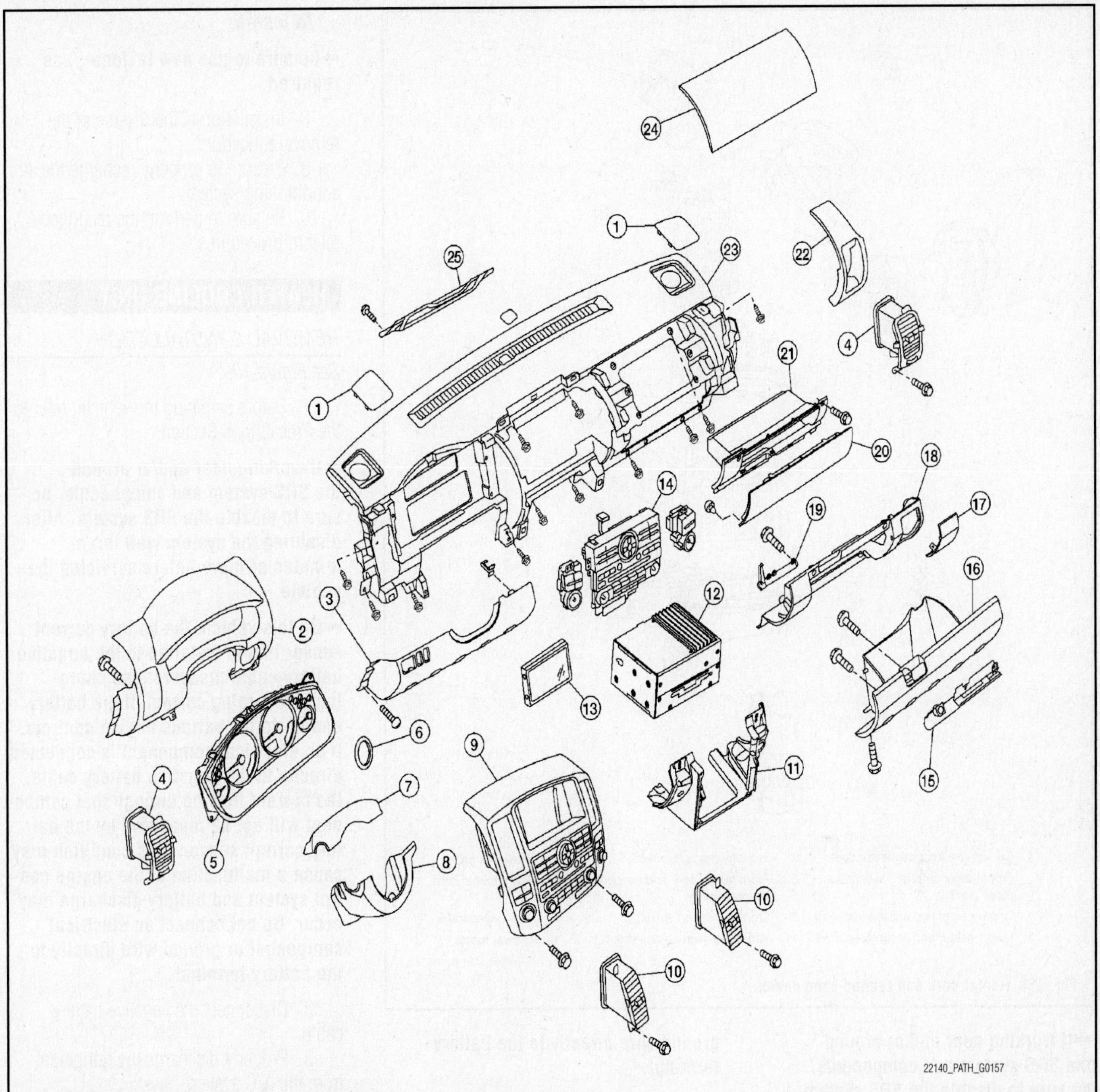

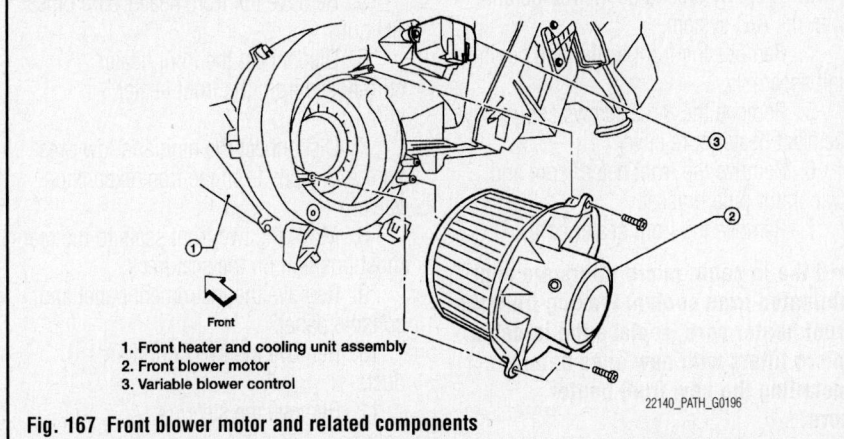

Fig. 166 Instrument panel and related components

22140_PATH_G0157

Fig. 167 Front blower motor and related components

Front

1. Front heater and cooling unit assembly
2. Front blower motor
3. Variable blower control

22140_PATH_G0196

To install:

➡**Be sure to use new fasteners, as required.**

6. Installation is the reverse of the removal procedure.

7. Be sure to perform the reconnect/relearn procedures.

HEATER CORE

REMOVAL & INSTALLATION

See Figure 168.

1. Before servicing the vehicle, refer to the Precautions Section.

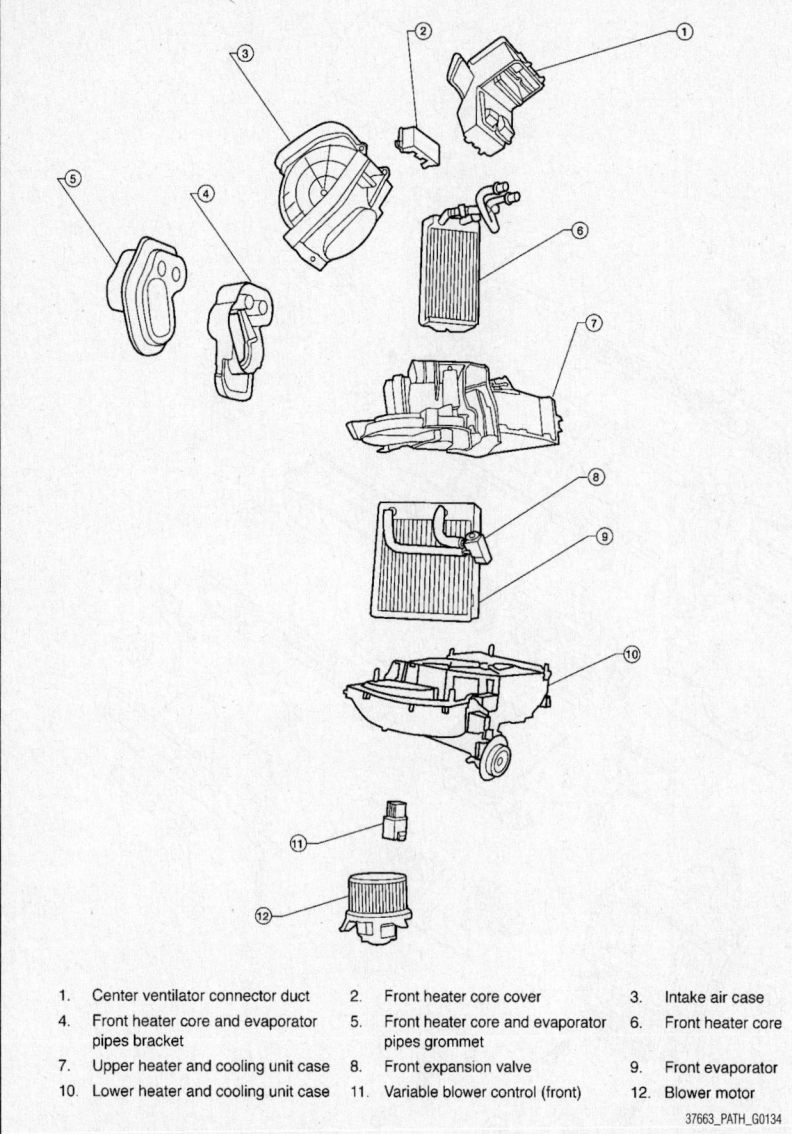

1.	Center ventilator connector duct	2.	Front heater core cover	3.	Intake air case
4.	Front heater core and evaporator pipes bracket	5.	Front heater core and evaporator pipes grommet	6.	Front heater core
7.	Upper heater and cooling unit case	8.	Front expansion valve	9.	Front evaporator
10.	Lower heater and cooling unit case	11.	Variable blower control (front)	12.	Blower motor

37663_PATH_G0134

Fig. 168 Heater core and related components

➡️ **If working near and/or around the SRS system and components, be sure to disable the SRS system. After disabling the system wait three minutes or more before servicing the vehicle.**

➡️ **On this vehicle the battery current sensor that is installed to the negative battery cable measures the charging/discharging current of the battery and performs various engine controls. If an electrical component is connected directly to the negative battery cable, the current flowing through that component will not be measured by the battery current sensor. This condition may cause a malfunction of the engine control system and battery discharge may occur. Do not connect an electrical component or ground wire directly to the battery terminal.**

2. Disconnect the negative battery cable.

3. Properly discharge the refrigerant from the A/C system.

4. Remove the front heater and cooling unit assembly.

5. Remove the three screws and remove the front heater core cover.

6. Remove the front heater core and evaporator pipe bracket.

7. Remove the front heater core.

➡️ **If the in-cabin micro filters are contaminated from coolant leaking from the front heater core, replace the in-cabin micro filters with new ones before installing the new front heater core.**

To install:

➡️ **Be sure to use new fasteners, as required.**

8. Installation is the reverse of the removal procedure.

9. Be sure to properly recharge the air conditioning system.

10. Be sure to perform the reconnect/relearn procedures.

HEATER/COOLING UNIT

REMOVAL & INSTALLATION

See Figure 169.

1. Before servicing the vehicle, refer to the Precautions Section.

➡️ **If working near and/or around the SRS system and components, be sure to disable the SRS system. After disabling the system wait three minutes or more before servicing the vehicle.**

➡️ **On this vehicle the battery current sensor that is installed to the negative battery cable measures the charging/discharging current of the battery and performs various engine controls. If an electrical component is connected directly to the negative battery cable, the current flowing through that component will not be measured by the battery current sensor. This condition may cause a malfunction of the engine control system and battery discharge may occur. Do not connect an electrical component or ground wire directly to the battery terminal.**

2. Disconnect the negative battery cable.

3. Properly discharge the refrigerant from the A/C system.

4. Drain the coolant from the engine cooling system.

5. Remove the front heater core pipes RH nut.

6. Disconnect the front heater core hoses from the front heater core.

7. Disconnect the high and low pressure A/C pipes from the front expansion valve.

8. Move the two front seats to the rearmost position on the seat track.

9. Remove the instrument panel and console panel.

10. Remove the two front floor ducts.

11. Remove the steering column.

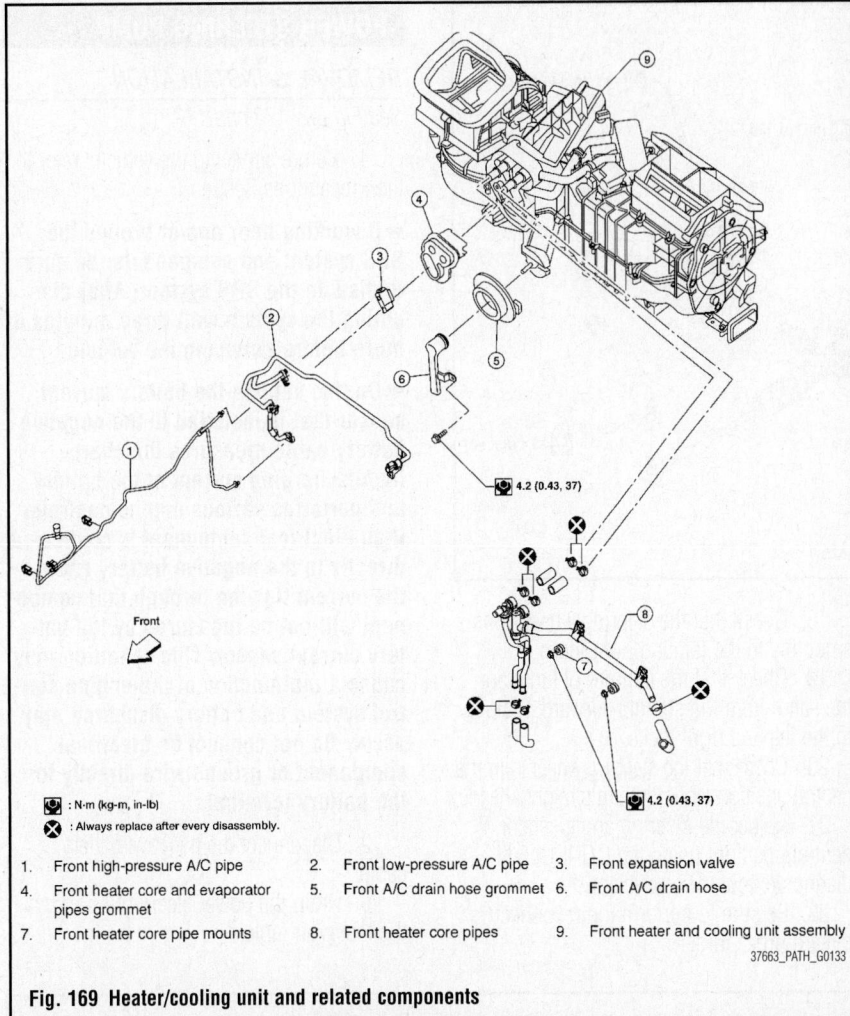

Key:
⬛ : N·m (kg-m, in-lb)
⊗ : Always replace after every disassembly.

1. Front high-pressure A/C pipe
2. Front low-pressure A/C pipe
3. Front expansion valve
4. Front heater core and evaporator pipes grommet
5. Front A/C drain hose grommet
6. Front A/C drain hose
7. Front heater core pipe mounts
8. Front heater core pipes
9. Front heater and cooling unit assembly

37663_PATH_G0133

Fig. 169 Heater/cooling unit and related components

12. Disconnect the instrument panel wire harness at the RH and LH in-line connector brackets, and the fuse block (SMJ) electrical connectors.

13. Remove the covers then remove the three steering member bolts from each side to disconnect the steering member from the vehicle body.

14. Remove the front heater and cooling unit assembly with it attached to the steering member, from the vehicle.

❊❊ CAUTION

Use care not to damage the seats and interior trim panels when removing the front heater and cooling unit assembly with it attached to the steering member. Use suitable plugs on the heater core pipes to prevent coolant leakage.

15. Remove the front heater and cooling unit assembly from the steering member.

To install:

➡**Be sure to use new fasteners, as required.**

16. Installation is the reverse of the removal procedure.

17. Be sure to properly recharge the air conditioning system.

18. Be sure to perform the reconnect/relearn procedures.

STEERING

POWER RACK & PINION STEERING GEAR

REMOVAL & INSTALLATION

See Figures 170 and 171.

❊❊ CAUTION

Spiral cable may snap due to steering operation if the steering column is separated from the steering gear assembly. Therefore secure the steering wheel to avoid turning.

1. Before servicing the vehicle, refer to the Precautions Section.

➡**If working near and/or around the SRS system and components, be sure to disable the SRS system. After disabling the system wait three minutes or more before servicing the vehicle.**

➡**On this vehicle the battery current sensor that is installed to the negative**

battery cable measures the charging/discharging current of the battery and performs various engine controls. If an electrical component is connected directly to the negative battery cable, the current flowing through that component will not be measured by the battery current sensor. This condition may cause a malfunction of the engine control system and battery discharge may occur. Do not connect an electrical component or ground wire directly to the battery terminal.

2. Disconnect the negative battery cable.

3. Set front wheels in the straight ahead position.

4. Remove the front tires from the vehicle.

5. Remove the undercover.

6. On 4WD models, remove the front final drive, then support the drive shafts, using suitable wire.

7. Remove the stabilizer bar brackets and reposition the stabilizer bar.

8. Remove the cotter pins at the steering outer sockets.

❊❊ WARNING

Do not reuse the cotter pins.

9. Loosen the outer socket nuts.

10. Remove the steering outer sockets from the steering knuckles, then remove the nuts.

❊❊ WARNING

Do not damage the outer socket boots. Do not damage the outer socket threads. Thread the ball joint nut onto the end of the outer socket during removal.

11. Remove the high pressure and low pressure piping from the steering gear assembly, then drain the fluid from the piping.

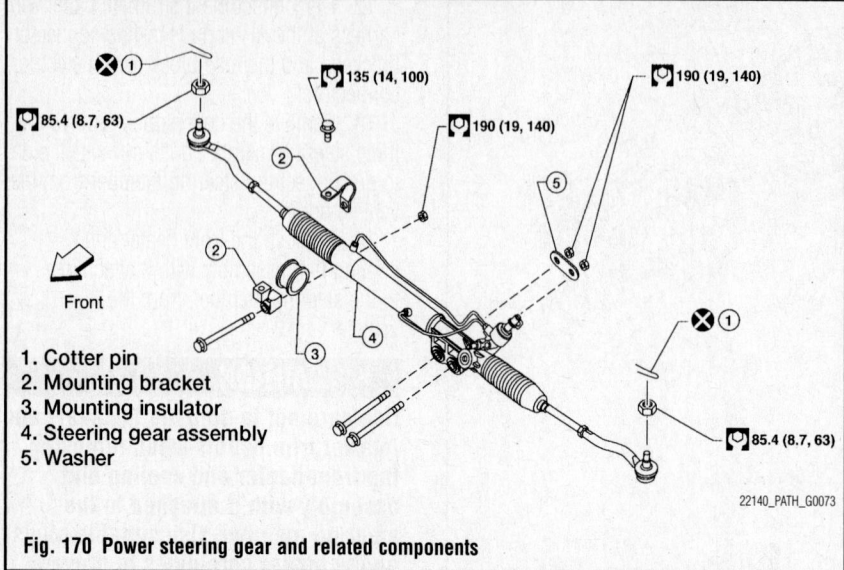

Fig. 170 Power steering gear and related components

1. Cotter pin
2. Mounting bracket
3. Mounting insulator
4. Steering gear assembly
5. Washer

Front

135 (14, 100)
85.4 (8.7, 63)
190 (19, 140)
190 (19, 140)
85.4 (8.7, 63)

22140_PATH_G0073

12. Remove the bolt from the lower joint of the lower joint shaft, then separate the lower joint from the steering gear assembly.

✳✳ WARNING

Do not damage the lower joint.

13. Remove the nuts and bolts of the steering gear assembly, then remove the steering gear assembly from the vehicle.

To install:

14. Installation is in the reverse order of removal.

15. With the steering wheel in the straight ahead position, align the slit of the lower joint with the projection on the dust cover. Insert the joint until surface "A" contacts surface "B".

16. After removing/installing or replacing steering components, check wheel alignment.

17. Bleed the air from the steering hydraulic system.

18. Check that the steering wheel turns smoothly to the left and right locks.

19. Check that the number of turns are the same from the straight-forward position to the left and right locks.

20. Check that the steering wheel is in the neutral position when driving straight ahead.

21. Adjust the steering angle sensor neutral position, using the CONSULT-III diagnostic tool, or equivalent.

22. Be sure to perform the reconnect/relearn procedures.

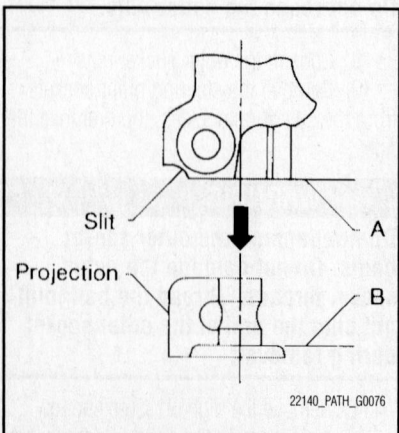

Slit
Projection
A
B

22140_PATH_G0076

Fig. 171 Insert the joint until surface "A" contacts surface "B"

POWER STEERING PUMP

REMOVAL & INSTALLATION

See Figures 172 and 173.

1. Before servicing the vehicle, refer to the Precautions Section.

➡**If working near and/or around the SRS system and components, be sure to disable the SRS system. After disabling the system wait three minutes or more before servicing the vehicle.**

➡**On this vehicle the battery current sensor that is installed to the negative battery cable measures the charging/discharging current of the battery and performs various engine controls. If an electrical component is connected directly to the negative battery cable, the current flowing through that component will not be measured by the battery current sensor. This condition may cause a malfunction of the engine control system and battery discharge may occur. Do not connect an electrical component or ground wire directly to the battery terminal.**

2. Disconnect the negative battery cable.

3. Drain the power steering fluid from the reservoir tank.

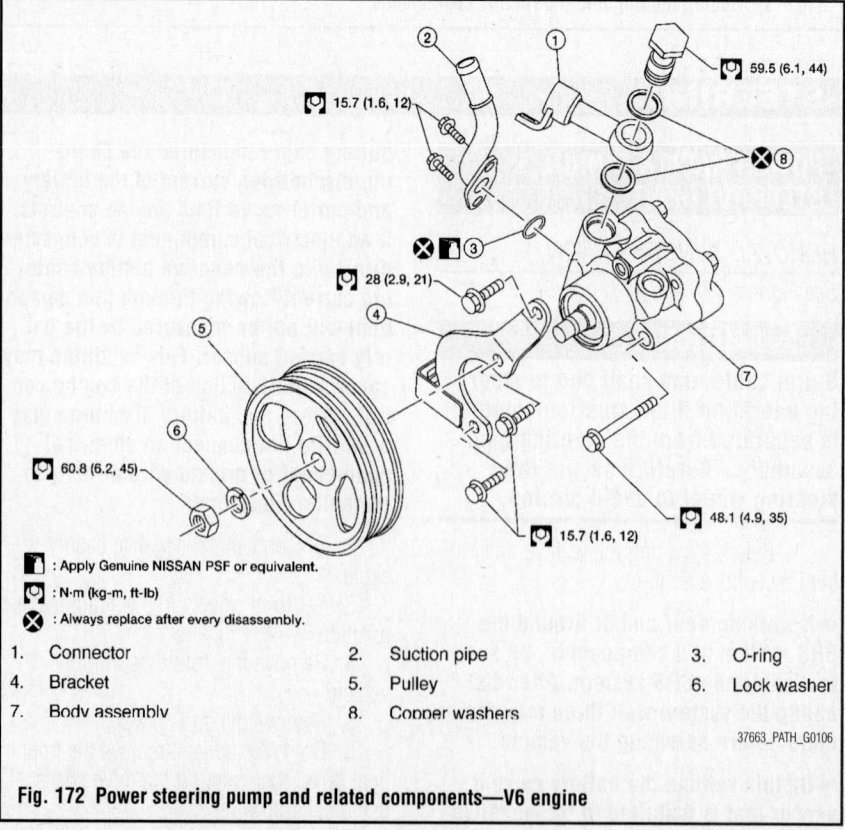

15.7 (1.6, 12)
59.5 (6.1, 44)
28 (2.9, 21)
60.8 (6.2, 45)
48.1 (4.9, 35)
15.7 (1.6, 12)

🔧 : Apply Genuine NISSAN PSF or equivalent.
🔧 : N·m (kg-m, ft-lb)
✖ : Always replace after every disassembly.

1. Connector
2. Suction pipe
3. O-ring
4. Bracket
5. Pulley
6. Lock washer
7. Body assembly
8. Copper washers

37663_PATH_G0106

Fig. 172 Power steering pump and related components—V6 engine

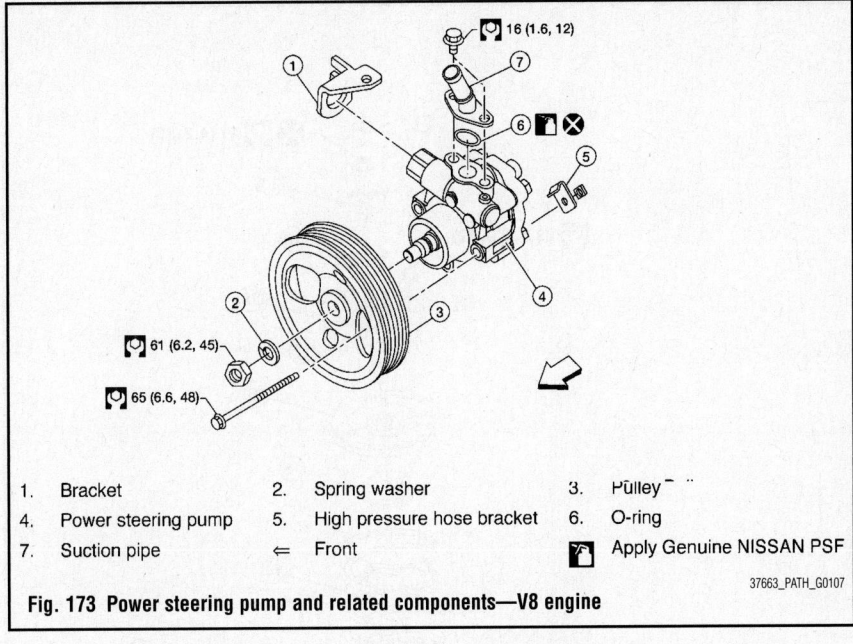

16 (1.6, 12)

61 (6.2, 45)

65 (6.6, 48)

1.	Bracket	2.	Spring washer	3. Pulley
4.	Power steering pump	5.	High pressure hose bracket	6. O-ring
7.	Suction pipe	⇐	Front	Apply Genuine NISSAN PSF

37663_PATH_G0107

Fig. 173 Power steering pump and related components—V8 engine

4. Remove the engine room cover.

5. Remove the air duct assembly.

6. Remove the serpentine drive belt from the auto tensioner and power steering oil pump.

7. Disconnect the pressure sensor electrical connector.

8. Remove the high pressure and low pressure piping from the power steering oil pump.

9. Remove the power steering oil pump bolts, then remove the power steering pump.

To install:

➡Be sure to use new fasteners, as required.

10. Installation is the reverse of the removal procedure.

11. After installation, bleed the air from the hydraulic circuit thoroughly.

12. Be sure to perform the reconnect/relearn procedures.

BLEEDING

1. Before servicing the vehicle, refer to the Precautions Section.

➡When the vehicle is stationary or while the steering wheel is being turned slowly, some noise may be heard from the oil pump or gear. This noise is normal and does not affect any system.

2. Check for fluid leakage.

3. Start the engine and turn the steering wheel fully to the right and left several times.

✳✳ CAUTION

Do not allow steering fluid reservoir tank to go below the MIN level line. Check tank frequently and add fluid as needed.

4. Run the engine at idle speed. Hold the steering wheel at each "locked" position for three seconds.

✳✳ WARNING

Do not hold steering wheel in the locked position for more than 10 seconds. (There is the possibility that oil pump may be damaged.)

5. Repeat step 3 several times at about three second intervals.

6. Check for air bubbles, cloudy fluid and fluid leakage.

7. If air bubbles or cloudiness exists, perform steps 3 and 4 again until air bubbles and cloudiness do not exist.

8. Stop the engine and check fluid level.

SUSPENSION

FRONT SUSPENSION

CONTROL LINKS

REMOVAL & INSTALLATION

Lower Link

See Figure 174.

1. Before servicing the vehicle, refer to the Precautions Section.

➡If working near and/or around the SRS system and components, be sure to disable the SRS system. After disabling the system wait three minutes or more before servicing the vehicle.

➡On this vehicle the battery current sensor that is installed to the negative battery cable measures the charging/discharging current of the battery and performs various engine controls. If an electrical component is connected directly to the negative battery cable, the current flowing through that component will not be measured by the battery current sensor. This condition may cause a malfunction of the engine control system and battery discharge may occur. Do not connect an electrical component or ground wire directly to the battery terminal.

2. Disconnect the negative battery cable.

3. Raise and support the vehicle safely.

4. Remove the tire and wheel assembly.

5. Remove the lower shock absorber bolt and nut.

6. Remove the stabilizer bar connecting rod lower nut. Separate the connecting rod from the lower link.

7. On 4WD vehicles, remove the driveshaft.

8. Remove the pinch bolt from the steering knuckle. Separate the lower link ball joint from the steering knuckle.

9. Remove the lower link adjusting bolts and nuts. Remove the lower link.

➡Some vehicles may be equipped with straight (non adjustable) lower link bolts and washers. In order to adjust camber and caster on these vehicles first replace the lower link bolts and washers with adjustable (cam) bolts and washers.

To install:

➡Be sure to use new fasteners, as required.

10. Installation is the reverse of the removal procedure.

11. Check and adjust alignment.

12. Be sure to perform the reconnect/relearn procedures.

Upper Link

See Figure 174.

1. Before servicing the vehicle, refer to the Precautions Section.

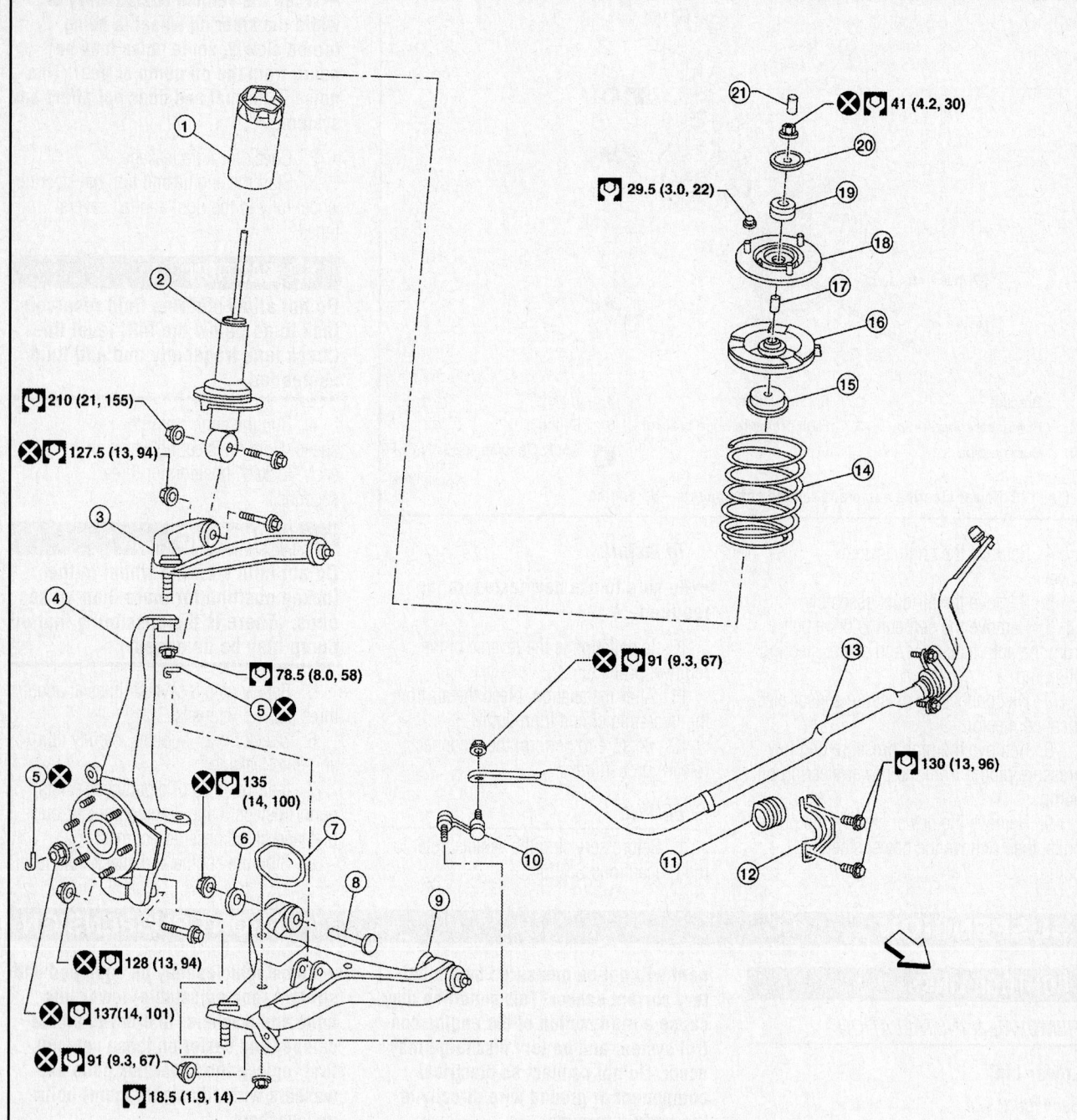

1. Dust cover
2. Shock absorber
3. Upper link
4. Steering knuckle
5. Cotter pin
6. Washer
7. Jounce bumper
8. Bolt
9. Lower link
10. Connecting rod
11. Stabilizer bar bushing
12. Stabilizer bar bracket
13. Stabilizer bar
14. Coil spring
15. Dust cover cap
16. Upper spring seat
17. Spacer
18. Shock absorber mounting insulator
19. Spacer
20. Washer
21. Cap
⇐ Front

37663_PATH_G0110

Fig. 174 Front suspension and related components

➡️If working near and/or around the SRS system and components, be sure to disable the SRS system. After disabling the system wait three minutes or more before servicing the vehicle.

➡️On this vehicle the battery current sensor that is installed to the negative battery cable measures the charging/discharging current of the battery and performs various engine controls. If an electrical component is connected directly to the negative battery cable, the current flowing through that component will not be measured by the battery current sensor. This condition may cause a malfunction of the engine control system and battery discharge may occur. Do not connect an electrical component or ground wire directly to the battery terminal.

2. Disconnect the negative battery cable.
3. Raise and support the vehicle safely.
4. Remove the tire and wheel assembly.
5. Support the lower link using the proper jack.
6. For left side, remove the bolt from the lower joint of the lower joint shaft. Reposition the lower joint out of the way. Do not damage the lower joint.
7. Remove the cotter pin and nut from the upper link ball joint. Discard the pin.
8. Separate the upper link ball joint stud from the knuckle using tool ST2902001, or equivalent.
9. Remove the upper link bolts and nuts.
10. Remove the component from its mounting.

To install:

➡️Be sure to use new fasteners, as required.

11. Installation is the reverse of the removal procedure.
12. Check and adjust alignment.
13. Be sure to perform the reconnect/relearn procedures.

LOWER BALL JOINT

REMOVAL & INSTALLATION

➡️The lower ball joint is part of the lower link (control arm).

LOWER CONTROL ARM

REMOVAL & INSTALLATION

See Figure 174.

1. Before servicing the vehicle, refer to the Precautions Section.

➡️If working near and/or around the SRS system and components, be sure to disable the SRS system. After disabling the system wait three minutes or more before servicing the vehicle.

➡️On this vehicle the battery current sensor that is installed to the negative battery cable measures the charging/discharging current of the battery and performs various engine controls. If an electrical component is connected directly to the negative battery cable, the current flowing through that component will not be measured by the battery current sensor. This condition may cause a malfunction of the engine control system and battery discharge may occur. Do not connect an electrical component or ground wire directly to the battery terminal.

2. Disconnect the negative battery cable.
3. Raise and support the vehicle safely.
4. Remove the tire and wheel assembly.
5. Remove the lower shock absorber bolt and nut.
6. Remove the stabilizer bar connecting rod lower nut. Separate the connecting rod from the lower link.
7. On 4WD vehicles, remove the driveshaft.
8. Remove the pinch bolt from the steering knuckle. Separate the lower link ball joint from the steering knuckle.
9. Remove the lower link adjusting bolts and nuts. Remove the lower link.

➡️Some vehicles may be equipped with straight (non adjustable) lower link bolts and washers. In order to adjust camber and caster on these vehicles first replace the lower link bolts and washers with adjustable (cam) bolts and washers.

To install:

➡️Be sure to use new fasteners, as required.

10. Installation is the reverse of the removal procedure.
11. Check and adjust alignment.
12. Be sure to perform the reconnect/relearn procedures.

MACPHERSON STRUT

REMOVAL & INSTALLATION

See Figure 174.

1. Before servicing the vehicle, refer to the Precautions Section.

➡️If working near and/or around the SRS system and components, be sure to disable the SRS system. After disabling the system wait three minutes or more before servicing the vehicle.

➡️On this vehicle the battery current sensor that is installed to the negative battery cable measures the charging/discharging current of the battery and performs various engine controls. If an electrical component is connected directly to the negative battery cable, the current flowing through that component will not be measured by the battery current sensor. This condition may cause a malfunction of the engine control system and battery discharge may occur. Do not connect an electrical component or ground wire directly to the battery terminal.

2. Disconnect the negative battery cable.
3. Raise and support the vehicle safely. Remove the wheel and tire.
4. Support the lower link using a suitable jack.
5. Remove connecting rod upper joints from stabilizer bar.
6. Swing stabilizer bar down, repositioning it out of the way to access shock absorber lower mount.
7. Remove the shock absorber lower bolt and nut.
8. Remove the three shock absorber upper mounting nuts.
9. Remove the coil spring and shock absorber assembly.
10. Turn steering knuckle out to gain enough clearance for removal.

To install:

➡️Be sure to use new fasteners, as required.

11. Installation is the reverse of the removal procedure.
12. Be sure to perform the reconnect/relearn procedures.

STEERING KNUCKLE

REMOVAL & INSTALLATION

See Figure 175.

1. Before servicing the vehicle, refer to the Precautions Section.

➡️If working near and/or around the SRS system and components, be sure to disable the SRS system. After disabling the system wait three minutes or more before servicing the vehicle.

➡ On this vehicle the battery current sensor that is installed to the negative battery cable measures the charging/discharging current of the battery and performs various engine controls. If an electrical component is connected directly to the negative battery cable, the current flowing through that component will not be measured by the battery current sensor. This condition may cause a malfunction of the engine control system and battery discharge may occur. Do not connect an electrical component or ground wire directly to the battery terminal.

2. Disconnect the negative battery cable.
3. Raise and safely support the vehicle.
4. Remove wheel and tire from vehicle.
5. Without disassembling the hydraulic lines, remove brake caliper. Reposition it aside with wire.

➡ Avoid depressing brake pedal while brake caliper is removed.

6. Put alignment marks on disc rotor and wheel hub and bearing assembly, then remove disc rotor.
7. Disconnect wheel sensor and remove bracket from steering knuckle.

✳✳ WARNING
Do not pull on wheel sensor harness.

8. On 4WD models, remove cotter pin, then remove lock nut from drive shaft.
9. Remove steering outer socket cotter pin at steering knuckle, then loosen nut.
10. Disconnect steering outer socket from steering knuckle. Be careful not to damage outer socket boot.

✳✳ WARNING
To prevent damage to threads and to prevent Tool from coming off suddenly, temporarily tighten nut.

11. Remove wheel hub and bearing assembly bolts.
12. Remove splash guard and wheel hub and bearing assembly from steering knuckle.

✳✳ WARNING
Do not pull on wheel sensor harness.

13. Remove cotter pin and nut from upper link ball joint.
14. Separate upper link ball joint from steering knuckle.
15. Remove pinch bolt from steering knuckle.
16. Separate lower link ball joint from steering knuckle.
17. Remove steering knuckle from vehicle.

To install:

➡ Be sure to use new fasteners, as required.

18. Installation is the reverse of the removal procedure.

✳✳ WARNING
Always replace drive shaft lock nut and cotter pin.

➡ When installing disc rotor on wheel hub and bearing assembly, align the marks.

19. Perform wheel alignment.
20. Be sure to perform the reconnect/relearn procedures.

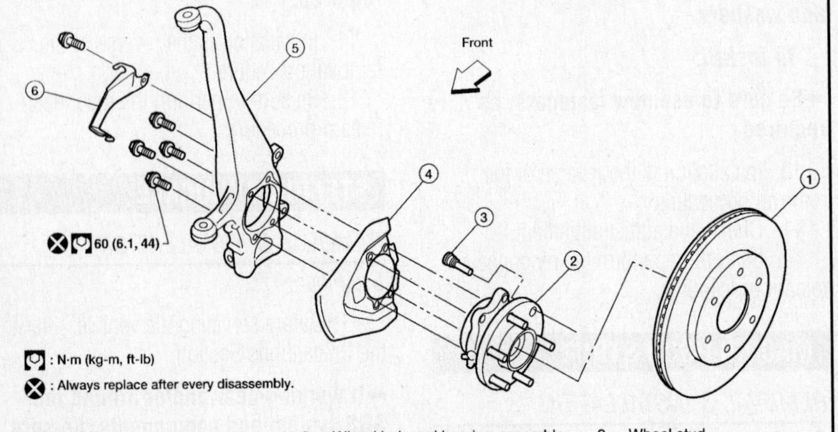

STABILIZER BAR

REMOVAL & INSTALLATION
See Figures 174 and 176.

1. Before servicing the vehicle, refer to the Precautions Section.

➡ If working near and/or around the SRS system and components, be sure to disable the SRS system. After disabling the system wait three minutes or more before servicing the vehicle.

➡ On this vehicle the battery current sensor that is installed to the negative battery cable measures the charging/discharging current of the battery and performs various engine controls. If an electrical component is connected directly to the negative battery cable, the current flowing through that component will not be measured by the battery current sensor. This condition may cause a malfunction of the engine control system and battery discharge may occur. Do not connect an electrical component or ground wire directly to the battery terminal.

2. Disconnect the negative battery cable.
3. Raise and safely support the vehicle.
4. Remove the front valance center.
5. Remove engine undercover.
6. Remove connecting rod nuts.
7. Loosen top bolts for stabilizer bar brackets, then remove lower bolts from brackets and remove stabilizer bar.
8. Remove bushings from stabilizer bar.

To install:

➡ Be sure to use new fasteners, as required.

9. Installation is the reverse of the removal procedure.

✪ 🔧 60 (6.1, 44)

🔧 : N·m (kg-m, ft-lb)
✪ : Always replace after every disassembly.

| 1. Disc rotor | 2. Wheel hub and bearing assembly | 3. Wheel stud |
| 4. Splash guard | 5. Steering knuckle | 6. Wheel sensor bracket |

22140_PATH_G0079

Fig. 175 Steering knuckle and related components

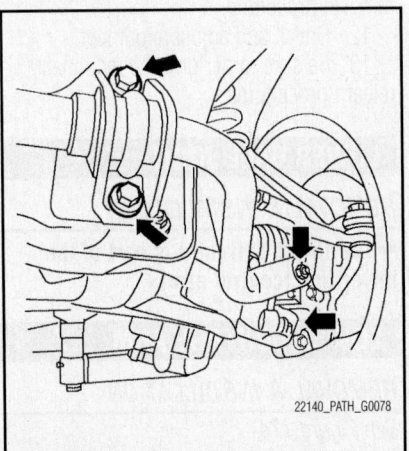

22140_PATH_G0078

Fig. 176 Loosen top bolts for stabilizer bar brackets

10. Tighten all nuts and bolts to specification.

11. Be sure to perform the reconnect/relearn procedures.

UPPER BALL JOINT

REMOVAL & INSTALLATION

➡ **The upper ball joint is part of the upper link (control arm).**

UPPER CONTROL ARM

REMOVAL & INSTALLATION

See Figure 174.

1. Before servicing the vehicle, refer to the Precautions Section.

➡ **If working near and/or around the SRS system and components, be sure to disable the SRS system. After disabling the system wait three minutes or more before servicing the vehicle.**

➡ **On this vehicle the battery current sensor that is installed to the negative battery cable measures the charging/discharging current of the battery and performs various engine controls. If an electrical component is connected directly to the negative battery cable, the current flowing through that component will not be measured by the battery current sensor. This condition may cause a malfunction of the engine control system and battery discharge may occur. Do not connect an electrical component or ground wire directly to the battery terminal.**

2. Disconnect the negative battery cable.
3. Raise and support the vehicle safely.
4. Remove the tire and wheel assembly.
5. Support the lower link using the proper jack.
6. For left side, remove the bolt from the lower joint of the lower joint shaft. Reposition the lower joint out of the way. Do not damage the lower joint.
7. Remove the cotter pin and nut from the upper link ball joint. Discard the pin.
8. Separate the upper link ball joint

stud from the knuckle using tool ST2902001, or equivalent.

9. Remove the upper link bolts and nuts.
10. Remove the component from its mounting.

To install:

➡ **Be sure to use new fasteners, as required.**

11. Installation is the reverse of the removal procedure.
12. Check and adjust alignment.
13. Be sure to perform the reconnect/relearn procedures.

WHEEL HUB & BEARING

REMOVAL & INSTALLATION

See Figure 175.

1. Before servicing the vehicle, refer to the Precautions Section.

➡ **If working near and/or around the SRS system and components, be sure to disable the SRS system. After disabling the system wait three minutes or more before servicing the vehicle.**

➡ **On this vehicle the battery current sensor that is installed to the negative battery cable measures the charging/discharging current of the battery and performs various engine controls. If an electrical component is connected directly to the negative battery cable, the current flowing through that component will not be measured by the battery current sensor. This condition may cause a malfunction of the engine control system and battery discharge may occur. Do not connect an electrical component or ground wire directly to the battery terminal.**

2. Disconnect the negative battery cable.
3. Raise and safely support the vehicle.
4. Remove wheel and tire.
5. Without disassembling the hydraulic lines, remove caliper torque member bolts.
6. Reposition brake caliper aside with wire.

✳✳ WARNING

Do not press brake pedal while brake caliper is removed.

7. Put alignment mark on disc rotor and wheel hub and bearing assembly, then remove disc rotor.
8. Remove cotter pin, then remove lock nut from drive shaft.
9. Remove driveshaft from wheel hub and bearing assembly.
10. Remove wheel sensor from wheel hub and bearing assembly.
    a. Inspect the wheel sensor O-ring, replace the wheel sensor assembly if damaged.
    b. Clean the wheel sensor hole and mounting surface with a suitable brake cleaner and clean lint-free shop rag. Be careful that dirt and debris do not enter the axle bearing area.
    c. Apply a coat of suitable grease to the wheel sensor O-ring and mounting hole.

✳✳ WARNING

Do not pull on the wheel sensor harness.

11. Remove wheel hub and bearing assembly bolts.
12. Remove splash guard and wheel hub and bearing assembly from steering knuckle.
13. Carefully remove wheel sensor and harness through hole in splash guard.

To install:

➡ **Be sure to use new fasteners, as required.**

14. Installation is the reverse of the removal procedure.
15. Use new bolts when installing the wheel hub and bearing assembly.
16. When installing disc rotor on wheel hub and bearing assembly, position the disc rotor according to alignment mark.
17. Check and adjust alignment.
18. Be sure to perform the reconnect/relearn procedures.

SUSPENSION

REAR SUSPENSION

COIL SPRING

REMOVAL & INSTALLATION

See Figure 177.

1. Before servicing the vehicle, refer to the Precautions Section.

➡ **If working near and/or around the SRS system and components, be sure to disable the SRS system. After disabling the system wait three minutes or more before servicing the vehicle.**

➡ **On this vehicle the battery current sensor that is installed to the negative**

battery cable measures the charging/discharging current of the battery and performs various engine controls. If an electrical component is connected directly to the negative battery cable, the current flowing through that component will not be measured by the battery

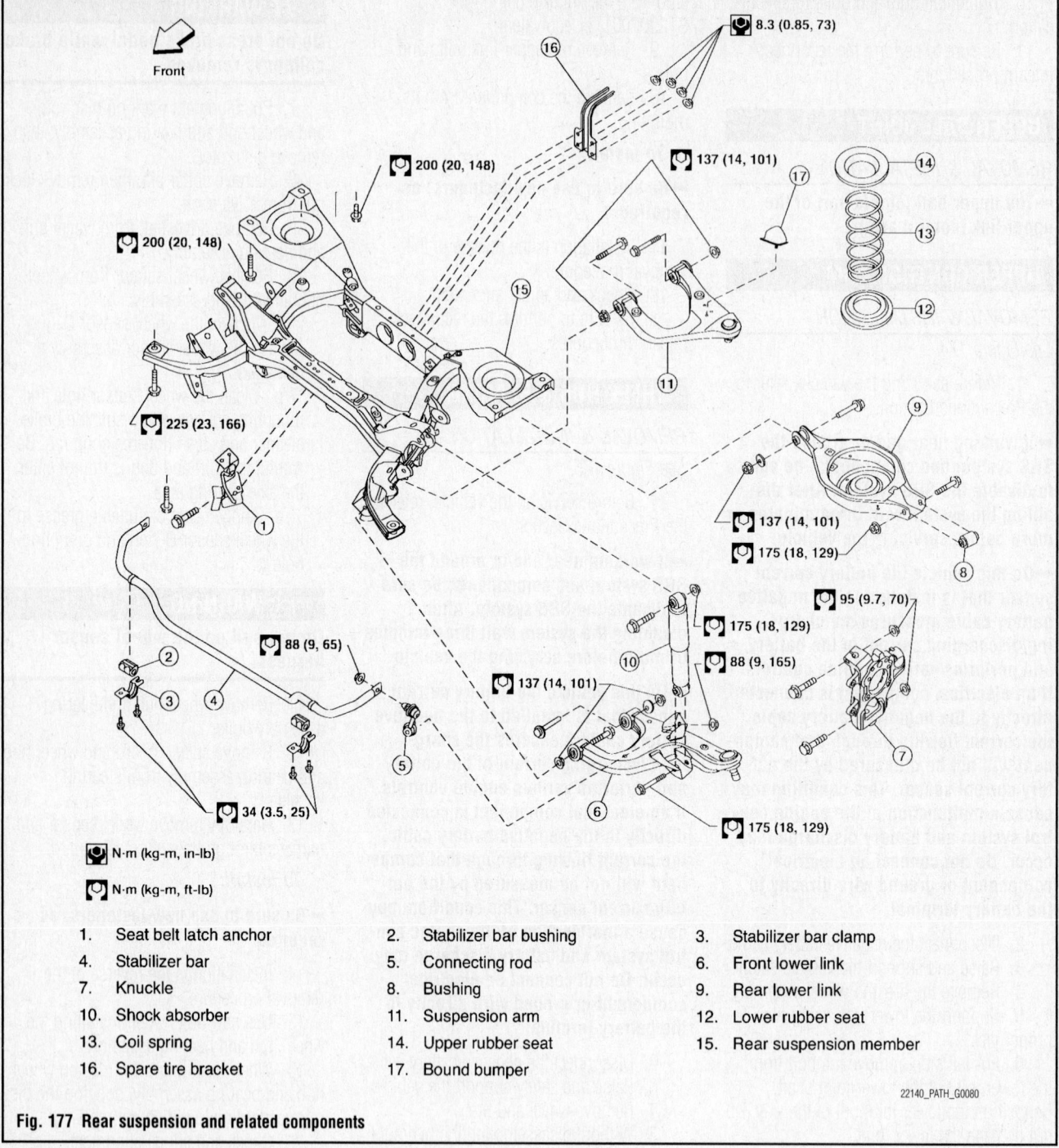

Front

8.3 (0.85, 73)

200 (20, 148)

137 (14, 101)

200 (20, 148)

225 (23, 166)

137 (14, 101)

175 (18, 129)

95 (9.7, 70)

175 (18, 129)

88 (9, 65)

137 (14, 101)

88 (9, 165)

34 (3.5, 25)

175 (18, 129)

N·m (kg-m, in-lb)

N·m (kg-m, ft-lb)

1.	Seat belt latch anchor	2.	Stabilizer bar bushing	3.	Stabilizer bar clamp
4.	Stabilizer bar	5.	Connecting rod	6.	Front lower link
7.	Knuckle	8.	Bushing	9.	Rear lower link
10.	Shock absorber	11.	Suspension arm	12.	Lower rubber seat
13.	Coil spring	14.	Upper rubber seat	15.	Rear suspension member
16.	Spare tire bracket	17.	Bound bumper		

22140_PATH_G0080

Fig. 177 Rear suspension and related components

current sensor. This condition may cause a malfunction of the engine control system and battery discharge may occur. Do not connect an electrical component or ground wire directly to the battery terminal.

2. Disconnect the negative battery cable.
3. Raise and support the vehicle safely.
4. Remove the wheel and tire assembly.
5. If removing the LH rear lower link and coil spring, remove the spare wheel and tire assembly.

6. Set a suitable jack to relieve the coil spring tension and support the rear lower link.
7. Loosen the rear lower link adjusting bolt and nut connected to the rear suspension member without removing the adjusting bolt and nut.
8. Remove the rear lower link pinch bolt and nut from the knuckle.
9. Slowly lower the rear lower link using the suitable jack to release the coil spring tension.
10. Remove the upper rubber seat, coil

spring and lower rubber seat from the rear lower link.

To install:

➡**Be sure to use new fasteners, as required.**

11. Installation is the reverse of the removal procedure.
    a. When installing the upper and lower rubber seats for the rear coil springs, the arrow embossed on the rubber seats must point out toward the wheel and tire assembly.

b. Perform the final tightening of the rear lower link nuts and bolts (with rubber bushings) under no-load conditions with tires on level ground.

c. Tighten the nuts and bolts to specification.

d. Check the wheel alignment.

12. Be sure to perform the reconnect/relearn procedures.

CONTROL ARMS/LINKS

REMOVAL & INSTALLATION

Front Lower Link

See Figure 177.

1. Before servicing the vehicle, refer to the Precautions Section.

➡**If working near and/or around the SRS system and components, be sure to disable the SRS system. After disabling the system wait three minutes or more before servicing the vehicle.**

➡**On this vehicle the battery current sensor that is installed to the negative battery cable measures the charging/discharging current of the battery and performs various engine controls. If an electrical component is connected directly to the negative battery cable, the current flowing through that component will not be measured by the battery current sensor. This condition may cause a malfunction of the engine control system and battery discharge may occur. Do not connect an electrical component or ground wire directly to the battery terminal.**

2. Disconnect the negative battery cable.

3. Raise and safely support the vehicle.

4. Remove the wheel and tire assembly.

5. Remove the stabilizer bar.

6. Set a suitable jack under the rear lower link to relieve the coil spring tension.

7. Remove the shock absorber lower end bolt.

8. Remove the adjusting bolt and nut, and the bolt and nut, from the front lower link and rear suspension member.

9. Remove the front lower link pinch bolt and nut on the knuckle side.

10. Disconnect the front lower link from the knuckle using a soft hammer.

✷✷ WARNING

Do not damage the ball joint with the soft hammer.

11. Remove the front lower link.

To install:

➡**Be sure to use new fasteners, as required.**

12. Installation is the reverse of the removal procedure.

a. Tighten the nuts and bolts to specification.

b. Perform the final tightening of the front lower link nuts and bolts (with rubber bushings) under no-load conditions with tires on level ground.

c. Check the wheel alignment.

13. Be sure to perform the reconnect/relearn procedures.

Rear Lower Control Link

See Figure 177.

1. Before servicing the vehicle, refer to the Precautions Section.

➡**If working near and/or around the SRS system and components, be sure to disable the SRS system. After disabling the system wait three minutes or more before servicing the vehicle.**

➡**On this vehicle the battery current sensor that is installed to the negative battery cable measures the charging/discharging current of the battery and performs various engine controls. If an electrical component is connected directly to the negative battery cable, the current flowing through that component will not be measured by the battery current sensor. This condition may cause a malfunction of the engine control system and battery discharge may occur. Do not connect an electrical component or ground wire directly to the battery terminal.**

2. Disconnect the negative battery cable.

3. Raise and safely support the vehicle.

4. Remove the wheel and tire assembly.

5. If removing the LH rear lower link and coil spring, remove the spare wheel and tire assembly.

6. Set a suitable jack to relieve the coil spring tension and support the rear lower link.

7. Loosen the rear lower link adjusting bolt and nut connected to the rear suspension member without removing the adjusting bolt and nut.

8. Remove the rear lower link pinch bolt and nut from the knuckle.

9. Slowly lower the rear lower link using the suitable jack to release the coil spring tension.

10. Remove the upper rubber seat, coil spring and lower rubber seat from the rear lower link.

11. Remove the rear lower link adjusting bolt and nut from the rear suspension member.

12. Remove the rear lower link.

To install:

➡**Be sure to use new fasteners, as required.**

13. Installation is the reverse of the removal procedure.

a. When installing the upper and lower rubber seats for the rear coil springs, the arrow embossed on the rubber seats must point out toward the wheel and tire assembly.

b. Perform the final tightening of the rear lower link nuts and bolts (with rubber bushings) under no-load conditions with tires on level ground.

c. Tighten the nuts and bolts to specification.

d. Check the wheel alignment.

14. Be sure to perform the reconnect/relearn procedures.

SHOCK ABSORBER

REMOVAL & INSTALLATION

See Figure 177.

1. Before servicing the vehicle, refer to the Precautions Section.

➡**If working near and/or around the SRS system and components, be sure to disable the SRS system. After disabling the system wait three minutes or more before servicing the vehicle.**

➡**On this vehicle the battery current sensor that is installed to the negative battery cable measures the charging/discharging current of the battery and performs various engine controls. If an electrical component is connected directly to the negative battery cable, the current flowing through that component will not be measured by the battery current sensor. This condition may cause a malfunction of the engine control system and battery discharge may occur. Do not connect an electrical component or ground wire directly to the battery terminal.**

2. Disconnect the negative battery cable.

3. Raise and safely support the vehicle.

4. Remove the wheel and tire assembly.

5. Position a suitable jack under the front lower link to support the shock absorber.

6. Remove the shock absorber upper and lower end bolts.

7. Remove the shock absorber.

To install:

➡ **Be sure to use new fasteners, as required.**

8. Installation is the reverse of the removal procedure.

   a. Tighten the nuts and bolts to specification.

   b. Check the wheel alignment.

9. Be sure to perform the reconnect/relearn procedures.

STABILIZER BAR

REMOVAL & INSTALLATION

See Figures 177 and 178.

1. Before servicing the vehicle, refer to the Precautions Section.

➡ **If working near and/or around the SRS system and components, be sure to disable the SRS system. After disabling the system wait three minutes or more before servicing the vehicle.**

➡ **On this vehicle the battery current sensor that is installed to the negative battery cable measures the charging/discharging current of the battery and performs various engine controls. If an electrical component is connected directly to the negative battery cable, the current flowing through that component will not be measured by the battery current sensor. This condition may cause a malfunction of the engine control system and battery discharge may occur. Do not connect an electrical component or ground wire directly to the battery terminal.**

2. Disconnect the negative battery cable.

3. Raise and support the vehicle safely. Remove the tire and wheel assemblies.

4. Disconnect the stabilizer bar ends from the connecting rods.

5. Remove the stabilizer bar clamps, and remove the stabilizer bar bushings.

6. Remove the stabilizer bar.

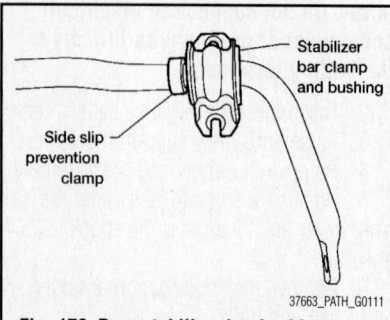

Fig. 178 Rear stabilizer bar bushing positioning

To install:

➡ **Be sure to use new fasteners, as required.**

7. Installation is the reverse of the removal procedure.

8. Install the stabilizer bar bushings and clamps so they are positioned outside of the sideslip prevention clamp on the stabilizer bar.

9. Be sure to perform the reconnect/relearn procedures.

WHEEL HUB & BEARING (SEALED UNIT)

REMOVAL & INSTALLATION

See Figures 177 and 179.

1. Before servicing the vehicle, refer to the Precautions Section.

➡ **If working near and/or around the SRS system and components, be sure to disable the SRS system. After disabling the system wait three minutes or more before servicing the vehicle.**

➡ **On this vehicle the battery current sensor that is installed to the negative battery cable measures the charging/discharging current of the battery and performs various engine controls. If an electrical component is connected directly to the negative battery cable, the current flowing through that component will not be measured by the battery current sensor. This condition may cause a malfunction of the engine control system and battery discharge may occur. Do not connect an electrical component or ground wire directly to the battery terminal.**

2. Disconnect the negative battery cable.

3. Raise and support the vehicle safely.

4. Remove the wheel and tire assembly.

5. Remove the rear brake caliper, without disconnecting the hydraulic hose.

6. Reposition the rear brake caliper aside using suitable wire.

✳✳ WARNING

Do not depress the brake pedal while the brake caliper is removed.

7. Remove the rear disc rotor.

8. Remove the cotter pin, then remove the rear drive shaft nut.

➡ **Discard the cotter pin, use a new one for installation.**

9. Remove the rear drive shaft.

10. Remove the four rear wheel hub and bearing assembly bolts.

➡ **Discard the four rear wheel hub and bearing assembly bolts, use new ones for installation.**

11. Remove the rear wheel hub and bearing assembly.

To install:

➡ **Be sure to use new fasteners, as required.**

12. Installation is the reverse of the removal procedure.

   a. Use a new cotter pin for installation.

   b. Use new rear wheel hub and bearing assembly bolts for installation.

13. Be sure to perform the reconnect/relearn procedures.

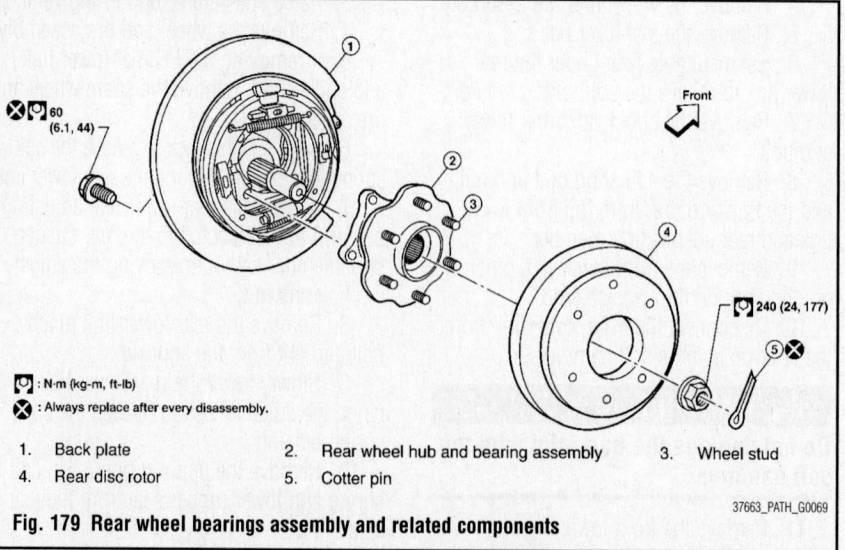

Fig. 179 Rear wheel bearings assembly and related components

SPECIFICATIONS AND MAINTENANCE CHARTS

VEHICLE AND ENGINE IDENTIFICATION CHART

Engine Code							Model Year	
Code	Liters (cc)	Cu. In.	Cyl.	Fuel Sys.	Engine Type	Eng. Mfg.	Code ①	Year
VQ35DE	3.5 (3498)	213	6	MFI	DOHC	Nissan	7	2007
							8	2008
MFI: Multi-port Fuel Injection							9	2009

37663_QUES_C0001

GENERAL ENGINE SPECIFICATIONS

Year	Engine Displacement Liters (VIN)	Net Horsepower @ rpm	Net Torque @ rpm (ft. lbs.)	Bore x Stroke (in.)	Compression Ratio	Oil Pressure @ rpm
2007	3.5 (VQ35DE)	235@5800	240@4400	3.76X3.20	10.0:1	43@2000
2008	3.5 (VQ35DE)	235@5800	240@4400	3.76X3.20	10.0:1	43@2000
2009	3.5 (VQ35DE)	235@5800	240@4400	3.76X3.20	10.0:1	43@2000

37663_QUES_C0002

ENGINE TUNE-UP SPECIFICATIONS

Year	Engine Displacement Liters (VIN)	Spark Plug Gap (in.)	Ignition Timing (deg.) MT	Ignition Timing (deg.) AT	Fuel Pump (psi)	Idle Speed (rpm) MT	Idle Speed (rpm) AT	Valve Clearance In.	Valve Clearance Ex.
2007	3.5 (VQ35DE)	0.043	NA	10-20 ①	51 ②	NA	625-725	HYD	HYD
2008	3.5 (VQ35DE)	0.043	NA	10-20 ①	51 ②	NA	625-725	HYD	HYD
2009	3.5 (VQ35DE)	0.043	NA	10-20 ①	51 ②	NA	625-725	HYD	HYD

NA: Not Applicable

NOTE: The Vehicle Emission Control Information label must be used if they differ from those in this chart.

HYD: Hydraulic

① Before top dead center

② System pressure at idle

37663_QUES_C0003

CAPACITIES

Year	Model	Engine Displacement Liters (VIN)	Engine Oil with Filter (qts.)	Transmission (pts.)			Drive Axle		Fuel Tank (gal.)	Cooling System (qts.) ①
				4-Spd	5-Spd	Auto.	Front (pts.)	Rear (pts.)		
2007	Quest	3.5 (VQ35DE)	3.75	NA	NA	16.5	NA	NA	20	11.2
2008	Quest	3.5 (VQ35DE)	3.75	NA	NA	16.5	NA	NA	20	11.2
2009	Quest	3.5 (VQ35DE)	3.75	NA	NA	16.5	NA	NA	20	11.2

NA: Not Applicable

NOTE: All capacities are approximate. Add fluid gradually and check to be sure a proper fluid level is obtained.

① Includes reservoir tank.

37663_QUES_C0005

FLUID SPECIFICATIONS

Year	Model	Engine Displ. Liters (VIN)	Engine Oil	Man. Trans.	Auto. Trans.	Drive Axle		Trans. Case	Power Steering Fluid	Brake Master Cylinder	Cooling System
						Front	Rear				
2007	Quest	VQ35DE	①	NA	②	NA	NA	NA	Dextrol VI	DOT 3	③
2008	Quest	VQ35DE	①	NA	②	NA	NA	NA	Dextrol VI	DOT 3	③
2009	Quest	VQ35DE	①	NA	②	NA	NA	NA	Dextrol VI	DOT 3	③

NA: Not Applicable

DOT: Department Of Transpotation

① API Service SM SAE 5W-30

② NISSAN Matic K

③ Nissan Long Life Antifreeze/Coolant

37663_QUES_C0004

VALVE SPECIFICATIONS

Year	Engine Displacement Liters (VIN)	Seat Angle (deg.)	Face Angle (deg.)	Spring Test Pressure (lbs. @ in.)	Spring Installed Height (in.)	Stem-to-Guide Clearance (in.)		Stem Diameter (in.)	
						Intake	Exhaust	Intake	Exhaust
2007	3.5 (VQ35DE)	45.15-45.45	45	37.3-42.3@ 1.457	1.457	0.0008-0.0021	0.0012-0.0025	0.2348-0.2354	0.2341-0.2346
2008	3.5 (VQ35DE)	45.15-45.45	45	37.3-42.3@ 1.457	1.457	0.0008-0.0021	0.0012-0.0025	0.2348-0.2354	0.2341-0.2346
2009	3.5 (VQ35DE)	45.15-45.45	45	37.3-42.3@ 1.457	1.457	0.0008-0.0021	0.0012-0.0025	0.2348-0.2354	0.2341-0.2346

37663_QUES_C0006

CAMSHAFT SPECIFICATIONS

All measurements are given in inches.

Year	Engine Displacement Liters	Engine VIN	Journal Dia.	Brg. Oil Clearance	Shaft End-play	Runout	Lobe Height Intake	Lobe Height Exhaust
2007	3.5	VQ35DE	①	②	0.0045-0.0074	③	1.7663-1.7738	1.7663-1.7738
2008	3.5	VQ35DE	①	②	0.0045-0.0074	③	1.7663-1.7738	1.7663-1.7738
2009	3.5	VQ35DE	①	②	0.0045-0.0074	③	1.7663-1.7738	1.7663-1.7738

① No.1: 1.0211-1.0218

　No.2, No.3, No.4: 0.9230-0.9238

② No.1: 1.0018-1.0034

　No.2, No.3, No.4: 0.0014-0.0030

③ Less then 0.0008 (0.02 mm)

37663_QUES_C0007

CRANKSHAFT AND CONNECTING ROD SPECIFICATIONS

All measurements are given in inches.

Year	Engine Displacement Liters (VIN)	Crankshaft Main Brg. Journal Dia.	Crankshaft Main Brg. Oil Clearance	Crankshaft Shaft End-play	Crankshaft Thrust on No.	Connecting Rod Journal Diameter	Connecting Rod Oil Clearance	Connecting Rod Side Clearance
2007	3.5 (VQ35DE)	①	0.0014-0.0018	0.0118	4	②	0.0013-0.0023	0.0079-0.0138
2008	3.5 (VQ35DE)	①	0.0014-0.0018	0.0118	4	②	0.0013-0.0023	0.0079-0.0138
2009	3.5 (VQ35DE)	①	0.0014-0.0018	0.0118	4	②	0.0013-0.0023	0.0079-0.0138

① There are 24 different grades, ranging from A (2.3612) to 7 (2.3603)

② Grade 0: 2.0460-2.0462

　Grade 1: 2.0457-2.0460

　Grade 2: 2.0445-2.0457

37663_QUES_C0008

PISTON AND RING SPECIFICATIONS

All measurements are given in inches.

Year	Engine Displacement Liters (VIN)	Piston Clearance	Ring Gap Top Compression	Ring Gap Bottom Compression	Ring Gap Oil Control	Ring Side Clearance Top Compression	Ring Side Clearance Bottom Compression	Ring Side Clearance Oil Control
2007	3.5 (VQ35DE)	0.0004-0.0012	0.0091-0.0130	0.0130-0.0189	0.0079-0.0197	0.0018-0.0031	0.0012-0.0028	0.0026-0.0053
2008	3.5 (VQ35DE)	0.0004-0.0012	0.0091-0.0130	0.0130-0.0189	0.0079-0.0197	0.0018-0.0031	0.0012-0.0028	0.0026-0.0053
2009	3.5 (VQ35DE)	0.0004-0.0012	0.0091-0.0130	0.0130-0.0189	0.0079-0.0197	0.0018-0.0031	0.0012-0.0028	0.0026-0.0053

37663_QUES_C0009

TORQUE SPECIFICATIONS

All readings in ft. lbs.

Year	Engine Displacement Liters (VIN)	Cylinder Head Bolts	Main Bearing Bolts	Rod Bearing Bolts	Crankshaft Damper Bolts	Flywheel Bolts	Manifold		Spark Plugs	Oil Pan Drain Plug
							Intake	Exhaust		
2007	3.5 (VQ35DE)	①	②	③	④	61-69	14	23	18	25
2008	3.5 (VQ35DE)	①	②	③	④	61-69	14	23	18	25
2009	3.5 (VQ35DE)	①	②	③	④	61-69	14	23	18	25

① Step 1: 72 ft. lbs.

   Step 2: Loosen all bolts completely

   Step 3: 29 ft. lbs.

   Step 4: +90 degrees

   Step 5: +90 degrees

② Step 1: 26 ft. lbs.

   Step 2: +90-95 degrees

③ Step 1: 15 ft. lbs.

   Step 2: +90-95 degrees

④ 32 ft. lbs. +60-65 degrees

37663_QUES_C0010

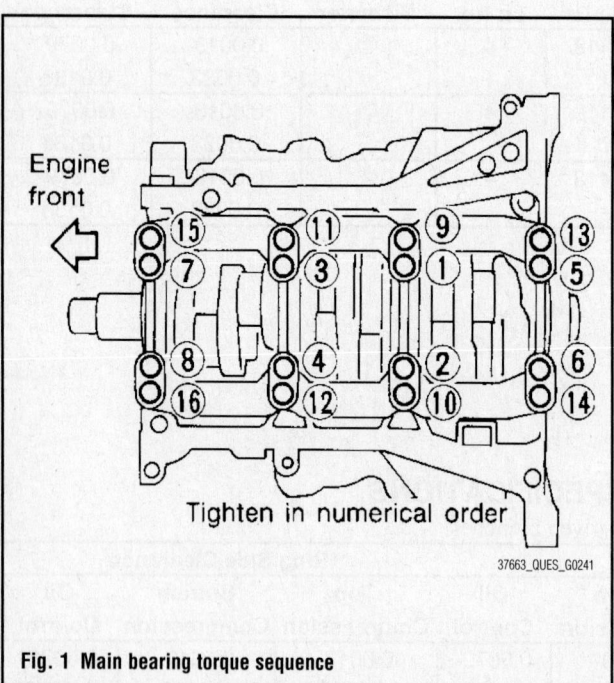

Fig. 1 Main bearing torque sequence

37663_QUES_G0241

WHEEL ALIGNMENT

Year	Model		Caster Range (Deg.)	Caster Preferred Setting (Deg.)	Camber Range (Deg.)	Camber Preferred Setting (Deg.)	Toe-in (in.)
2007	Quest ① ②	F	③	③	④	④	0.01 +/- 0.04
		R	NA	NA	⑤	⑤	0.13 +/- 0.07
2008	Quest ① ②	F	③	③	④	④	0.01 +/- 0.04
		R	NA	NA	⑤	⑤	0.13 +/- 0.07
2009	Quest ① ②	F	③	③	④	④	0.01 +/- 0.04
		R	NA	NA	⑤	⑤	0.13 +/- 0.07

NA: Not Applicable

① Vehicle unladen

② Specifications are decimal degrees

③ Minimum: 1.95

  Nominal: 2.70

  Maximum: 3.45

  Left and right difference 0.75 or less

④ Minimum: - 1.25

  Nominal: - 0.50

  Maximum: 0.25

  Left and right difference 0.75 or less

⑤ Minimum: - 1.05

  Nominal: - 0.55

  Maximum: - 0.05

37663_QUES_C0012

TIRE, WHEEL AND BALL JOINT SPECIFICATIONS

Year	Model	OEM Tires Standard	OEM Tires Optional	Tire Pressures (psi) Front	Tire Pressures (psi) Rear	Wheel Size	Ball Joint Inspection	Lug Nut (ft. lbs.)
2007	Quest	P225/65R16	P225/60R17	35	35	NA	①	83
2008	Quest	P225/65R16	P225/60R17	35	35	NA	①	83
2009	Quest	P225/65R16	P225/60R17	35	35	NA	①	83

NA: Not Available

OEM: Original Equipment Manufacturer

PSI: Pounds Per Square Inch

① Replace if any measurable movement is found.

37663_QUES_C0013

BRAKE SPECIFICATIONS
All measurements in inches unless noted

Year	Model		Brake Disc Original Thickness	Brake Disc Minimum Thickness	Brake Disc Maximum Runout	Minimum Lining Thickness	Brake Caliper Bracket-to-Hub Bolt (ft. lbs.)	Brake Caliper Mounting Pin or Bolt (ft. lbs.)
2007	Quest	F	1.100	1.02	0.0016	0.079	112	55
		R	0.630	0.551	0.0020	0.079	62	32
2008	Quest	F	1.100	1.02	0.0016	0.079	112	55
		R	0.630	0.551	0.0020	0.079	62	32
2009	Quest	F	1.100	1.02	0.0016	0.079	112	55
		R	0.630	0.551	0.0020	0.079	62	32

NOTE: Due to changes made during production, refer to the manufacturer's specifications if they differ from those in this chart

F: Front

R: Rear

SCHEDULED MAINTENANCE INTERVALS (1)
Nissan—Quest

TO BE SERVICED	TYPE OF SERVICE	7.5	15	22.5	30	37.5	45	52.5	60
Engine oil & filter	R	every 3,750 miles							
Brake lines & cables	S/I		✓		✓		✓		✓
Brake pads, discs	I	✓	✓	✓	✓	✓	✓	✓	✓
Driveshaft boots	I	✓	✓	✓	✓	✓	✓	✓	✓
Automatic transmission ①	I		✓		✓		✓		✓
Air cleaner filter	R				✓				✓
Drive belt (s) ②	S/I								✓
Engine coolant ③	R								✓
Spark plugs	R	Platinum plugs, every 105,000 miles							
Cabin air filter	R		✓		✓		✓		✓
Exhaust system	I	✓	✓	✓	✓	✓	✓	✓	✓
Evap vapor lines	I				✓				✓
Fuel lines	S/I				✓				✓
Steering gear, linkage, axle & suspension parts	I	✓	✓	✓	✓	✓	✓	✓	✓
Tires (rotate)	S/I	✓	✓	✓	✓	✓	✓	✓	✓
Valve clearance ④	S/I				✓				✓

R: Replace   S/I: Service or Inspect   L: Lubricate   I: Inspect

① If towing a trailer, using a camper or a car-top carrier, or driving on rough or muddy roads, change (not just inspect) oil at every 30,000 miles.

② First at 60,000, inspect every 15,000 miles and replace as necessary

③ After 60,000, replace every 30,000

④ Periodic maintenance not required, if valve noise increases, inspect valve clearance

Follow Periodic Maintenance Schedule 1 if the driving habits frequently include one or more of the following driving conditions:

Repeated short trips of less than 5 miles (8 km).

Repeated short trips of less than 10 miles (16 km) with outside temperatures remaining below freezing

Operating in hot weather in stop-and-go "rush hour" traffic.

Extensive idling and/or low speed driving for long distances, such as police, taxi or door-to-door delivery use

Driving in dusty conditions.

Driving on rough, muddy, or salt spread roads.

Towing a trailer, using a camper or a car-top carrier.

Follow Periodic Maintenance Schedule 2 if none of driving conditions shown in Schedule 1 apply to the driving habits.

SCHEDULED MAINTENANCE INTERVALS (2)
Nissan—Quest

TO BE SERVICED	TYPE OF	7.5	15	22.5	30	37.5	45	52.5	60
Engine oil & filter	R	✓	✓	✓	✓	✓	✓	✓	✓
Brake lines & cables	S/I		✓		✓		✓		✓
Brake pads, discs	I		✓		✓		✓		✓
Driveshaft boots	I		✓		✓		✓		✓
Automatic transmission	I		✓		✓		✓		✓
Air cleaner filter	R				✓				✓
Drive belt (s) ①	S/I								✓
Engine coolant ②	R								✓
Spark plugs	R	Platinum plugs, every 105,000 miles							
Cabin air filter	R		✓		✓		✓		✓
Exhaust system	I				✓				✓
Evap vapor lines	I				✓				✓
Fuel lines	S/I				✓				✓
Steering gear, linkage, axle & suspension parts	I				✓				✓
Tires (rotate)	S/I	✓	✓	✓	✓	✓	✓	✓	✓
Valve clearance ③	S/I				✓				✓

R: Replace   S/I: Service or Inspect   L: Lubricate   I: Inspect

① First at 60,000, then every 15,000 miles

② After 60,000, replace every 30,000

③ Periodic maintenance not required, if valve noice increases, inspect valve clearance

Follow Periodic Maintenance Schedule 1 if the driving habits frequently include one or more of the following driving conditions:

Repeated short trips of less than 5 miles (8 km).

Repeated short trips of less than 10 miles (16 km) with outside temperatures remaining below freezing

Operating in hot weather in stop-and-go "rush hour" traffic.

Extensive idling and/or low speed driving for long distances, such as police, taxi or door-to-door delivery use

Driving in dusty conditions.

Driving on rough, muddy, or salt spread roads.

Towing a trailer, using a camper or a car-top carrier.

Follow Periodic Maintenance Schedule 2 if none of driving conditions shown in Schedule 1 apply to the driving habits.

37663_QUES_C0015

BRAKES — INFORMATION AND PRECAUTIONS

ANTI-LOCK SYSTEMS

• Certain components within the ABS system are not intended to be serviced or repaired individually.

• Do not use rubber hoses or other parts not specifically specified for and ABS system. When using repair kits, replace all parts included in the kit. Partial or incorrect repair may lead to functional problems and require the replacement of components.

• Lubricate rubber parts with clean, fresh brake fluid to ease assembly. Do not use shop air to clean parts; damage to rubber components may result.

• Use only DOT 3 brake fluid from an unopened container.

• If any hydraulic component or line is removed or replaced, it may be necessary to bleed the entire system.

• A clean repair area is essential. Always clean the reservoir and cap thoroughly before removing the cap. The slightest amount of dirt in the fluid may plug an ori-fice and impair the system function. Perform repairs after components have been thoroughly cleaned; use only denatured alcohol to clean components. Do not allow ABS components to come into contact with any substance containing mineral oil; this includes used shop rags.

• The Anti-Lock control unit is a microprocessor similar to other computer units in the vehicle. Ensure that the ignition switch is **OFF** before removing or installing controller harnesses. Avoid static electricity discharge at or near the controller.

• If any arc welding is to be done on the vehicle, the control unit should be unplugged before welding operations begin.

DISC AND DRUM SYSTEMS

✷✷ CAUTION
Dust and dirt accumulating on brake parts during normal use may contain asbestos fibers from production or aftermarket brake linings. Breathing excessive concentrations of asbestos fibers can cause serious bodily harm. Exercise care when servicing brake parts. Do not sand or grind brake lining unless equipment used is designed to contain the dust residue. Do not clean brake parts with compressed air or by dry brushing. Cleaning should be done by dampening the brake components with a fine mist of water, then wiping the brake components clean with a dampened cloth. Dispose of cloth and all residue containing asbestos fibers in an impermeable container with the appropriate label. Follow practices prescribed by the Occupational Safety and Health Administration (OSHA) and the Environmental Protection Agency (EPA) for the handling, processing, and disposing of dust or debris that may contain asbestos fibers.

BRAKES — BLEEDING THE BRAKE SYSTEM

BLEEDING PROCEDURE

BLEEDING PROCEDURE

✷✷ WARNING
Carefully monitor brake fluid level at the sub tank during bleeding operation.

✷✷ WARNING
Fill the sub tank with new brake fluid. Make sure it is full at all times while bleeding the air out of system.

➡Place a container under the sub tank to avoid spilling brake fluid.

➡Do not loosen the line fittings at the ABS actuator during air bleeding.

1. Turn ignition switch OFF and disconnect ABS actuator and control unit connector or negative battery terminal.
2. Connect a transparent vinyl tube and container to the air bleeder valve.
3. Fully depress the brake pedal several times.
4. With the brake pedal depressed, open the air bleeder valve to release air.
5. Close the air bleeder valve.

6. Release the brake pedal slowly.
7. Tighten the air bleeder valve to 71 inch lbs. (8 Nm).
8. Repeat steps 2 through 7 until no more air bubbles come out of the air bleeder valve.
9. Bleed the brake hydraulic system air bleeder valves in the following order:
• Right rear brake
• Left front brake
• Left rear brake
• Right front brake

BLEEDING THE ABS SYSTEM
Refer to the Bleeding Procedure.

BRAKES — ANTI-LOCK BRAKE SYSTEM (ABS)

WHEEL SPEED SENSORS

REMOVAL & INSTALLATION

Front
See Figure 2.

✷✷ WARNING
Be careful not to damage the sensor edge and sensor rotor teeth. When removing the front or rear wheel hub assembly, first remove the wheel sensor from the assembly. Failure to do so may result in damage to the sensor wires, making the sensor inoperative. Pull out the sensor, be careful to turn it as little as possible. Do not pull on the sensor harness.

1. Remove the wheel and tire.
2. Partially remove the front wheel fender protector.
3. Remove the wheel sensor bolt and wheel sensor.
4. Remove the harness wire from mounts and disconnect the wheel sensor harness connector.

To install:
5. Installation is the reverse of removal, noting the following:

✷✷ WARNING
Before installing the wheel sensor, make sure there are no foreign materials (such as iron fragments) adhered to the pick-up part of the wheel sensor, to the inside of the wheel sensor mounting hole or on the rotor mounting surface.

Rear

See Figure 2.

> ❋❋ **WARNING**
>
> Be careful not to damage the sensor edge and sensor rotor teeth. When removing the front or rear wheel hub assembly, first remove the wheel sensor from the assembly. Failure to do so may result in damage to the sensor wires, making the sensor inoperative. Pull out the sensor, be careful to turn it as little as possible. Do not pull on the sensor harness.

1. Remove the wheel and tire.
2. Remove the wheel sensor bolt and wheel sensor from the rear wheel hub and bearing assemblies.
3. Remove the harness wire from mounts and harness wire clips from the rear suspension member.
4. Disconnect the wheel sensor harness connector.

To install:

5. Installation is the reverse of removal, noting the following:

> ❋❋ **WARNING**
>
> Before installing the wheel sensor, make sure there are no foreign materials (such as iron fragments) adhered to the pick-up part of the wheel sensor, to the inside of the wheel sensor mounting hole oron the rotor mounting surface.

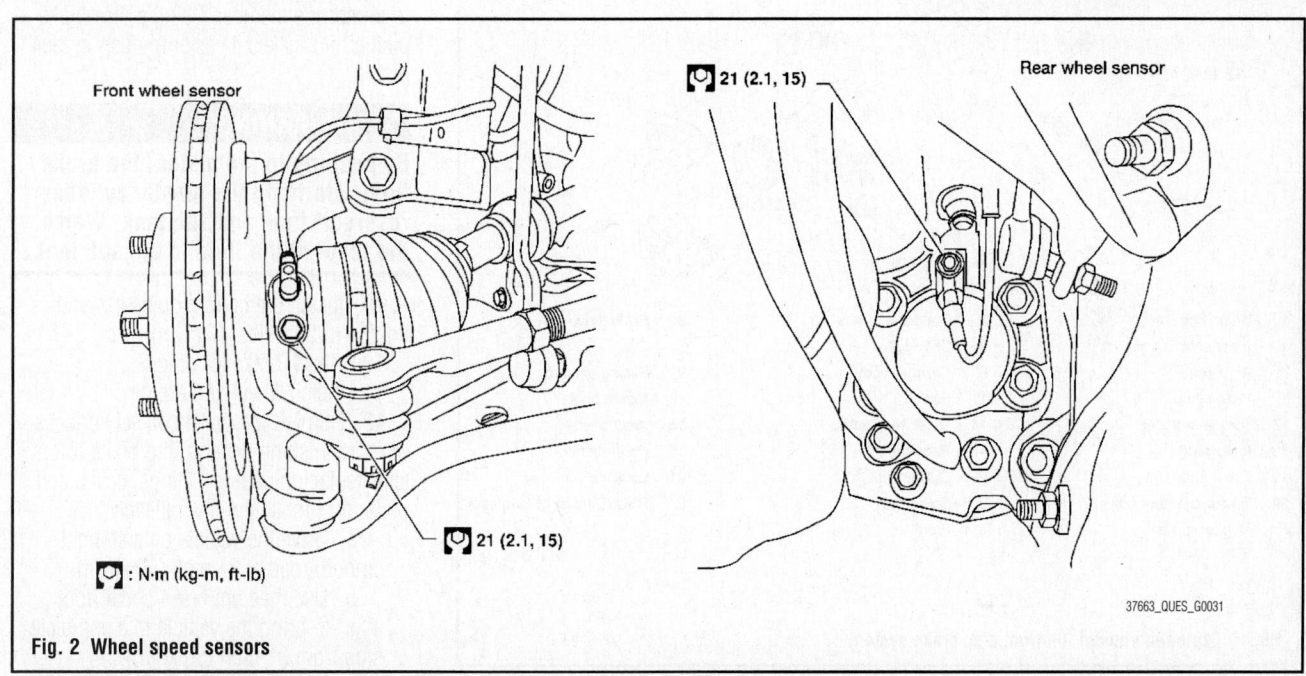

Front wheel sensor

21 (2.1, 15)

: N·m (kg-m, ft-lb)

21 (2.1, 15)

Rear wheel sensor

37663_QUES_G0031

Fig. 2 Wheel speed sensors

BRAKES

FRONT DISC BRAKES

BRAKE CALIPER

REMOVAL & INSTALLATION

See Figures 3 and 4.

1. Before servicing the vehicle, refer to the Precautions section.
2. Remove the wheel and tire.
3. Drain the brake fluid.
4. Remove the caliper bracket mounting bolts and the caliper hose bolt and remove the caliper assembly.

To install:

5. Install caliper assembly, and tighten caliper bracket mounting bolts to 112 ft. lbs. (152 Nm).
6. Install brake hose to the caliper assembly using a new copper washer, and tighten the bolt to 13 ft. lbs. (18 Nm).
7. Refill the system with brakes.
8. Install the wheel and tire.

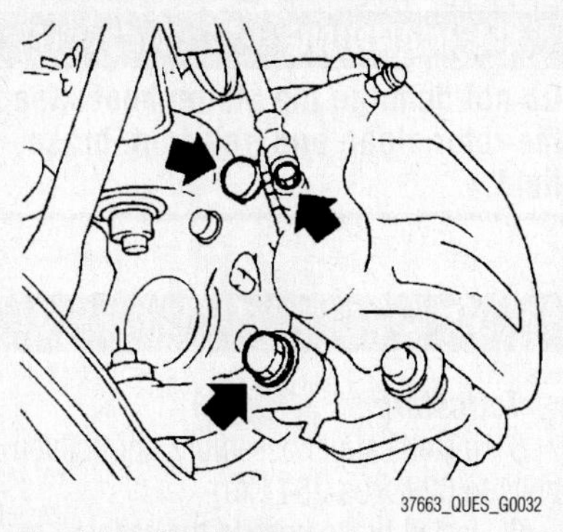

Fig. 3 Front caliper

37663_QUES_G0032

DISC BRAKE PADS

REMOVAL & INSTALLATION

See Figure 4.

> ❋❋ **CAUTION**
>
> Clean brake pads with a vacuum dust collector to minimize a hazard of airborne particles or other materials.

> ❋❋ **WARNING**
>
> While removing pad assemblies, do not depress brake pedal because piston will pop out. Be careful not to damage piston boot or get oil on disc rotor. Always replace shims when replacing pads. If shims are rusted or show peeling of rubber coat, replace them with new shims. Carefully monitor brake fluid level because brake fluid will return to reservoir when pushing back piston. Burnish the brake pads (or linings) and disc rotor mutually contacting surfaces after refinishing or replacing drums or rotors, after replacing pads or linings, or if a soft pedal occurs at a very low mileage.

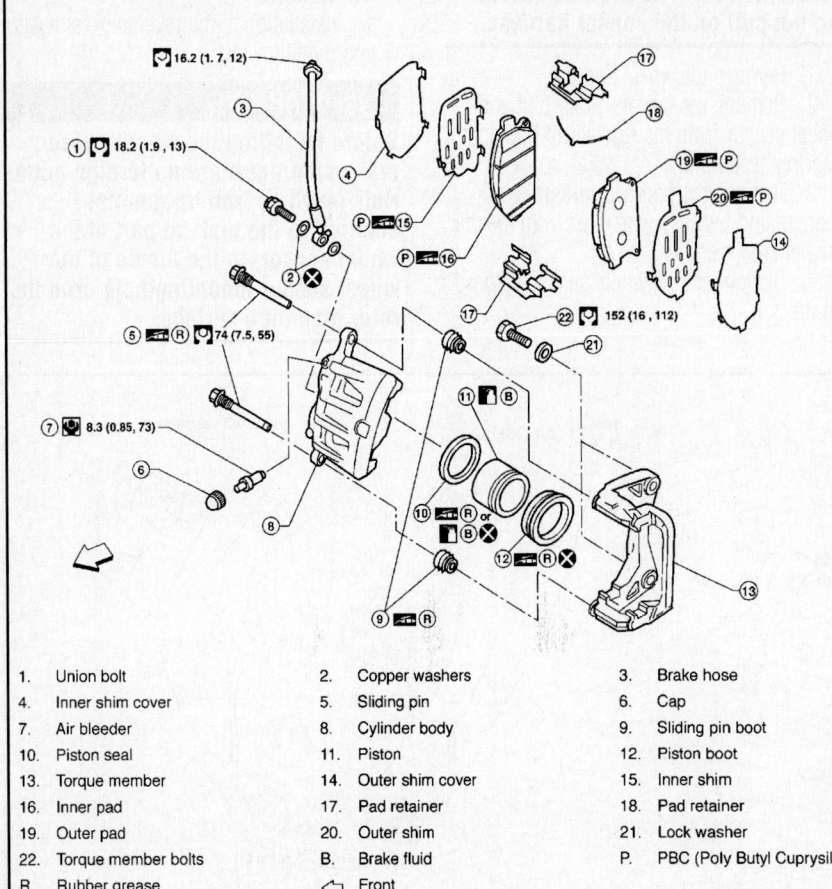

1. Union bolt
2. Copper washers
3. Brake hose
4. Inner shim cover
5. Sliding pin
6. Cap
7. Air bleeder
8. Cylinder body
9. Sliding pin boot
10. Piston seal
11. Piston
12. Piston boot
13. Torque member
14. Outer shim cover
15. Inner shim
16. Inner pad
17. Pad retainer
18. Pad retainer
19. Outer pad
20. Outer shim
21. Lock washer
22. Torque member bolts
B. Brake fluid
P. PBC (Poly Butyl Cuprysil)
R. Rubber grease
⇦ Front

37663_QUES_G0033

Fig. 4 Exploded view of the front disc brake system

1. Before servicing the vehicle, refer to the Precautions section.
2. Remove the wheel and tire.
3. Remove the bottom guide pin from the caliper and swing the caliper cylinder body upward; support the caliper with a wire.
4. Remove the brake pad retainers and the pads. When removing the pad retainer from torque member, carefully lift the pad retainer upward.

✳✳ WARNING

Do not damage the piston boot. Keep the rotor clean and free from brake fluid.

To install:

5. Apply PBC (Poly Butyl Cuprysil) or equivalent, to the pad back plate and inner shim sides.
6. Install inner shims and outer shim covers to inner and outer pads.
7. Install pad retainers and pad assemblies to the torque member. When attaching the pad retainer, attach it firmly so that it does not move higher than the torque member.
8. Push the piston in so that the pad is firmly attached, and install the cylinder body to the torque member. Using a disc brake piston tool makes it easier to push in the piston.

✳✳ WARNING

By pushing in the piston, the brake fluid returns to the master cylinder reservoir tank and sub tank. Watch the level of the fluid in the sub tank.

9. Install the bottom guide pin and tighten to 55 ft. lbs. (74 Nm).
10. Check brake for drag.
11. Install wheel and tire.
12. Burnish the brake contact surfaces when refinishing or replacing brake rotors, after replacing pads or linings, or if a soft pedal occurs at very low mileage:
   a. Drive the vehicle on a straight smooth road at 31 mph (50 km/h).
   b. Use medium brake pedal /foot effort to bring the vehicle to a complete stop from 31 mph (50 km/h).
   c. Adjust brake pedal /foot pressure such that vehicle stopping time equals 3 to 5 seconds.
   d. To cool the brake system, drive the vehicle at 31 mph (50 km/h) for 1 minute without stopping.
   e. Repeat the first 3 steps, 10 times or more to complete the burnishing procedure.

BRAKES

BRAKE CALIPER

REMOVAL & INSTALLATION

See Figure 5.

1. Before servicing the vehicle, refer to the precautions section.
2. Remove the wheel and tire.
3. Drain the brake fluid.
4. Remove the caliper bracket mounting bolts and the caliper hose bolt and remove the caliper assembly.

To install:

5. Install caliper assembly, and tighten bolts to 62 ft. lbs. (84 Nm).
6. Install brake hose to the caliper assembly using a new copper washer, and tighten the bolt to 13 ft. lbs. (18 Nm).
7. Refill the system with new brake fluid and bleed the brakes.
8. Adjust the parking brake.

REAR DISC BRAKES

9. Install the wheel and tire.

DISC BRAKE PADS

REMOVAL & INSTALLATION

See Figure 5.

✳✳ CAUTION

Clean brake pads with a vacuum dust collector to minimize a hazard of airborne particles or other materials.

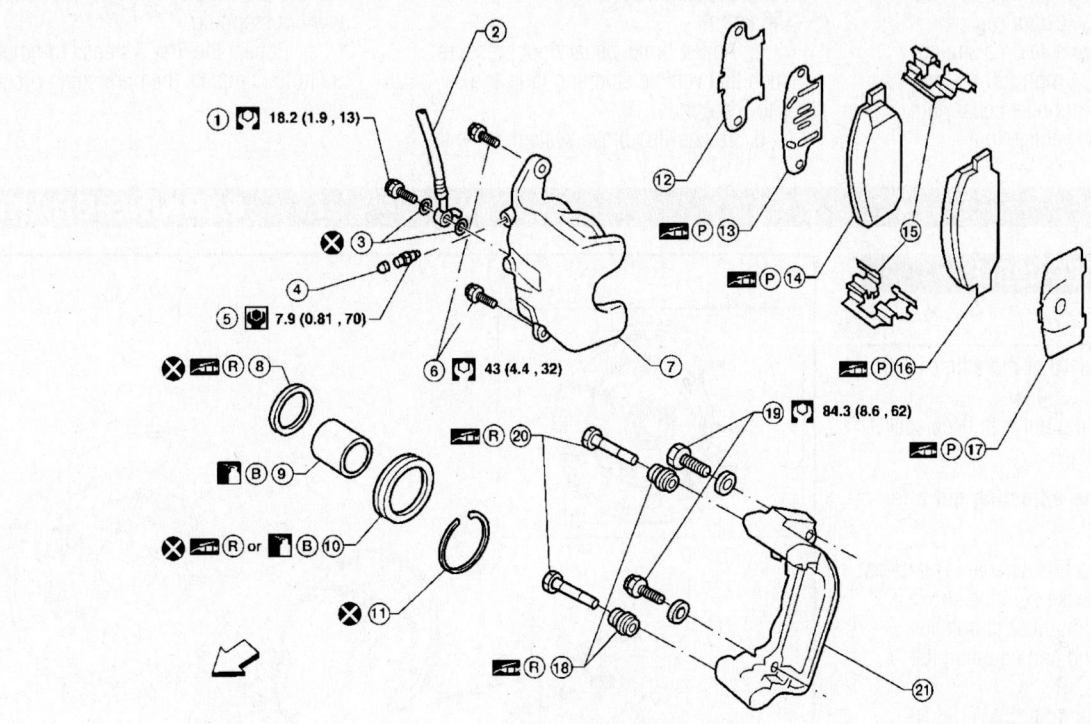

1.	Union bolt	2.	Brake hose	3.	Copper washer
4.	Cap	5.	Air bleeder	6.	Sliding pin bolt
7.	Cylinder body	8.	Piston seal	9.	Piston
10.	Piston boot	11.	Retaining ring	12.	Inner shim cover
13.	Inner shim	14.	Inner pad	15.	Pad retainer
16.	Outer pad	17.	Outer shim	18.	Sliding pin boot
19.	Torque member bolts	20.	Sliding pin	21.	Torque member
B.	Brake fluid	P.	PBC (Poly Butyl Cuprysil)	R.	Rubber grease
⇦	Front				

37663_QUES_G0035

Fig. 5 Exploded view of the rear disc brake system

> **※※ WARNING**
>
> While removing pad assemblies, do not depress brake pedal because piston will pop out. Be careful not to damage piston boot or get oil on disc rotor. Always replace shims when replacing pads. If shims are rusted or show peeling of rubber coat, replace them with new shims. Carefully monitor brake fluid level because brake fluid will return to reservoir when pushing back piston. Burnish the brake pads (or linings) and disc rotor mutually contacting surfaces after refinishing or replacing drums or rotors, after replacing pads or linings, or if a soft pedal occurs at a very low mileage.

1. Before servicing the vehicle, refer to the Precautions section.
2. Remove the wheel and tire.

3. Remove the bottom guide pin from the caliper and swing the caliper cylinder body upward; support the caliper with a wire.
4. Remove the brake pad retainers and the pads and shims. When removing the pad retainer from torque member, carefully lift the pad retainer upward.

> **※※ WARNING**
>
> Do not damage the piston boot. Keep the rotor clean and free from brake fluid.

To install:

5. Apply PBC (Poly Butyl Cuprysil) or equivalent, to the pad back plate and inner shim sides.
6. Install inner shims and outer shim covers to inner and outer pads.
7. Install pad retainers and pad assemblies to the torque member. When attaching the pad retainer, attach it firmly so that it does not move higher than the torque member.
8. Push the piston in so that the pad is firmly attached, and install the cylinder body to the torque member. Using a disc brake piston tool makes it easier to push in the piston.

> **※※ WARNING**
>
> By pushing in the piston, the brake fluid returns to the master cylinder reservoir tank and sub tank. Watch the level of the fluid in the sub tank.

9. Install the bottom guide pin and tighten to 32 ft. lbs. (43 Nm).
10. Check brake for drag.
11. Install wheel and tire.
12. Burnish the brake contact surfaces when refinishing or replacing brake rotors,

after replacing pads or linings, or if a soft pedal occurs at very low mileage:

a. Drive the vehicle on a straight smooth road at 31 mph (50 km/h).

b. Use medium brake pedal /foot effort to bring the vehicle to a

complete stop from 31 mph (50 km/h).

c. Adjust brake pedal /foot pressure such that vehicle stopping time equals 3 to 5 seconds.

d. To cool the brake system, drive the

vehicle at 31 mph (50 km/h) for 1 minute without stopping.

e. Repeat the first 3 steps, 10 times or more to complete the burnishing procedure.

BRAKES   PARKING BRAKE

PARKING BRAKE CABLES

ADJUSTMENT

1. Operate pedal 10 or more times with a force of 66 ft. lbs. (294 N).

2. Rotate adjusting nut with deep socket to adjust pedal stroke.

➡**Do not reuse the adjusting nut after removing it.**

3. When parking brake pedal is operated at specified force, make sure the stroke is within the specified number of notches. (Check it by listening and counting the ratchet clicks.)

• Pedal force: 44 ft. lbs. (196 N)
• Pedal stroke: 5 to 6 notches

4. With the pedal completely returned, make sure there is no drag on the rear brake.

PARKING BRAKE SHOES

REMOVAL & INSTALLATION

See Figures 6 through 9.

1. Before servicing the vehicle, refer to the precautions section.

2. Remove wheel and tire.

3. Remove the disc rotor with the parking brake pedal completely in the released position.

4. If disc rotor cannot be removed, remove as follows:

a. Secure the disc rotor in place with wheel nuts and remove disc rotor plug.

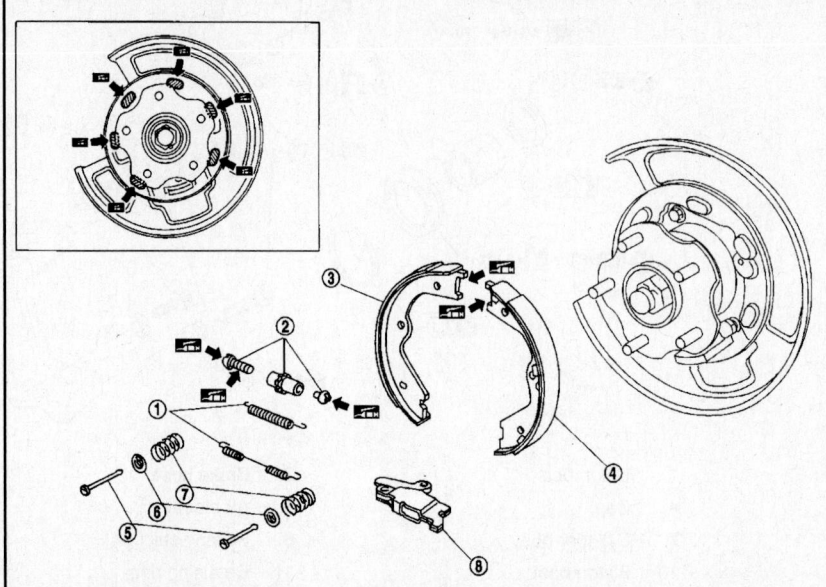

: PBC (Poly Butyl Cuprysil) grease or silicone-based grease point

1.	Return spring	2.	Adjuster assembly	3.	Shoe (secondary)
4.	Shoe (primary)	5.	Anti-rattle pins	6.	Retainers
7.	Anti-rattle springs	8.	Toggle lever		

37663_QUES_G0039

Fig. 7 Parking brake shoes

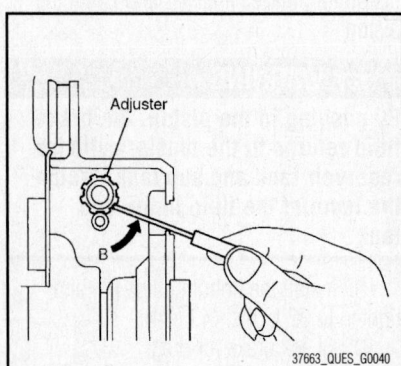

Adjuster

37663_QUES_G0040

Fig. 6 Rotate the adjuster in the direction shown (B)

b. Using a suitable tool, rotate the adjuster in the direction shown to retract and loosen brake shoes.

5. Remove the anti-rattle pins, retainers, anti-rattle springs, and return springs.

6. Remove the parking brake shoes, adjuster assembly, and toggle lever.

To install:

7. To install, reverse removal procedure, noting the following:

a. Apply brake grease to the specified points during assembly.

b. There is a difference between the adjusters orientation from left and right. Assemble the adjuster so the threaded part expands when rotating it in the direction shown.

c. Shorten the adjuster by rotating it.

d. When disassembling the adjuster, apply PBC (Poly Butyl Cuprysil)

grease or silicone based grease to the threads.

8. After replacing brake shoes or disc rotors, or if brakes do not function well, adjust the parking brake pedal stroke to the specified stroke, (Pedal force: 44 ft. lbs. (196 N), Pedal stroke: 5 to 6 notches) and perform break-in operation as follows:

a. Drive the vehicle forward.

b. Maintain vehicle speed at approximately 25 mph (40 km/h), keeping it constant in forward direction.

c. Apply the parking brake at an operating force of approximately 33 ft. lbs. (147 N) with constantforce.

d. Release the parking brake after approximately 10 seconds.

9. After break-in operation, check the lever stroke of the parking brake. Readjust if it is no longer at the specified stroke.

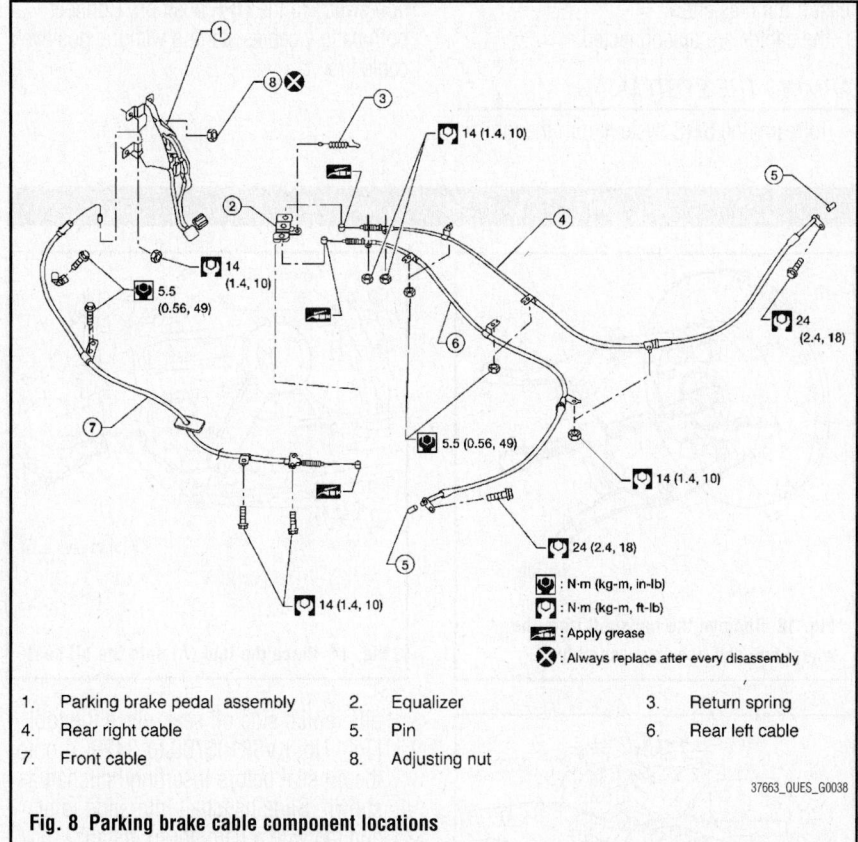

1. Parking brake pedal assembly
2. Equalizer
3. Return spring
4. Rear right cable
5. Pin
6. Rear left cable
7. Front cable
8. Adjusting nut

37663_QUES_G0038

Fig. 8 Parking brake cable component locations

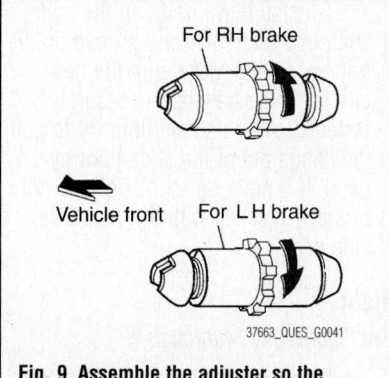

37663_QUES_G0041

Fig. 9 Assemble the adjuster so the threaded part expands when rotating it in the direction shown

a. To prevent the lining from getting too hot, allow a cool off period of approximately 5 minutes after every break-in operation.

b. Do not perform excessive break-in operations, because it may cause uneven or early wear of the lining.

ADJUSTMENT

See Figure 10.

1. Make sure the components are attached properly (check for looseness, backlash, etc.). Check parking brake pedal assembly for bend, damage and cracks, and replace

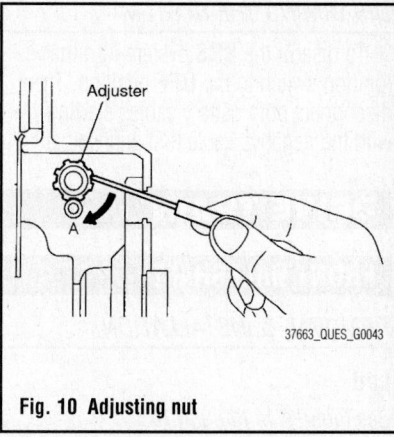

37663_QUES_G0043

Fig. 10 Adjusting nut

if necessary. Check cable for wear and damage, and replace if necessary. Check parking brake warning lamp switch for malfunction, and replace if necessary.

2. Remove lower left-hand instrument panel. Refer to Instrument Panel Removal & Installation in the Body Exterior section.

3. Insert a deep socket wrench to rotate adjusting nut and loosen the cable sufficiently. Then, return the pedal to the free height.

4. Remove wheel and tire using power tool.

5. Using wheel nuts, secure the disc to the hub and prevent it from tilting.

6. Remove adjusting hole plug installed on the disc rotor. Rotate the adjuster in the direction shown, until the disc rotor is locked, using a suitable tool. After locking, turn the adjuster in the opposite direction by 5 or 6 notches.

7. Rotate the disc rotor to make sure there is no drag. Install the adjusting hole plug.

8. Adjust the cable as necessary.

CHASSIS ELECTRICAL   AIR BAG (SUPPLEMENTAL RESTRAINT SYSTEM)

GENERAL INFORMATION

✻✻ CAUTION

These vehicles are equipped with an air bag system. The system must be disarmed before performing service on, or around, system components, the steering column, instrument panel components, wiring and sensors. Failure to follow the safety precautions and the disarming

procedure could result in accidental air bag deployment, possible injury and unnecessary system repairs.

SERVICE PRECAUTIONS

✻✻ CAUTION

Disconnect and isolate the battery negative cable before beginning any airbag system component diagnosis, testing, removal, or installation procedures. Wait at least 90 seconds after

the ignition switch is turned off and the negative (-) terminal cable is disconnected from the battery before starting the operation. The SRS is equipped with a backup power source, so if work is started within 90 seconds after disconnecting the negative (-) terminal cable from the battery, the SRS may be deployed. Failure to disable the airbag system may result in accidental airbag deployment, personal injury, or death.

DISARMING THE SYSTEM

To disarm the **SRS** system turn the ignition switch to the **OFF** position. Then, disconnect both battery cables starting with the negative cable first and wait at least 3 minutes after the cables are disconnected.

ARMING THE SYSTEM

To rearm the **SRS** system, turn the ignition switch to the **OFF** position. Connect both battery cables starting with the positive cable first.

DRIVE TRAIN

FRONT HALFSHAFT

REMOVAL & INSTALLATION

Left

See Figures 11 through 14.

1. Before servicing the vehicle, refer to the Precautions Section.
2. Remove the wheel and tire.
3. Remove the wheel sensor from the steering knuckle.
4. Remove the cotter pin, and remove the lock nut from the halfshaft.
5. Remove the brake hose lock plate, and remove the brake hose from the strut.
6. Remove the lower ball joint pinch bolt, and separate the lower ball joint from the steering knuckle.
7. Using a puller or suitable tool, remove the halfshaft from the wheel hub and bearing assembly.

✳✳ WARNING

When removing the halfshaft, do not apply an excessive angle to the halfshaft joint. Also be careful not to excessively extend the slide joint.

8. Remove the halfshaft from the transaxle using a suitable tool and the halfshaft puller (Tool No. KV40107500).

   a. Set the appropriate tool (Tool No. KV40107500) and a halfshaft puller or suitable tool between the halfshaft (slide joint side) and transaxle as shown and remove the halfshaft.

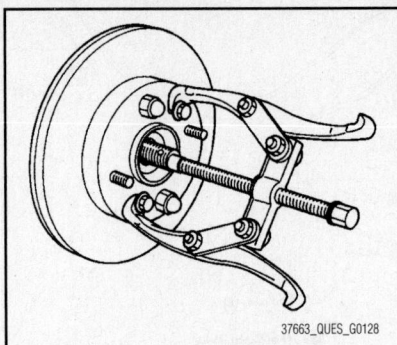

Fig. 12 Remove the halfshaft from the wheel hub and bearing assembly

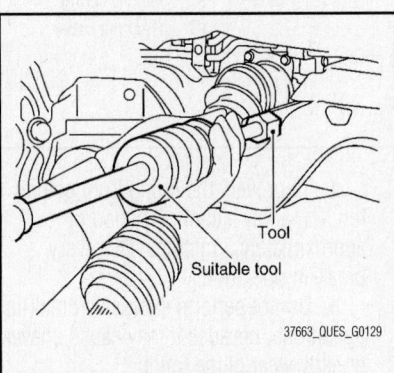

Fig. 13 Remove the halfshaft from the transaxle

To install:

9. Installation is in the reverse order of removal, noting the following:

   a. Tighten the axle nut to 101 ft. lbs. (137 Nm).

   b. In order to prevent damage to the

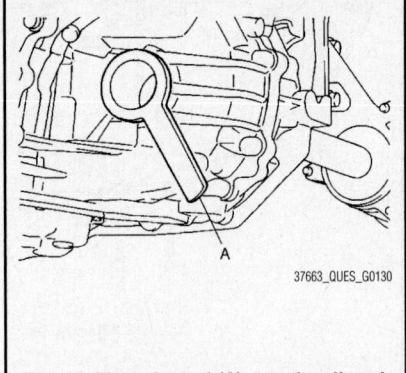

Fig. 14 Place the tool (A) onto the oil seal

differential side oil seal, place the tool [Tool No. KV38106700 (J-34296)] onto the oil seal before inserting halfshaft as shown. Slide halfshaft into slide joint and tap with a hammer to install securely.

   c. Install new circlip on the halfshaft in the circular clip groove on the transaxle side. Make sure the new circlip on the halfshaft is securely fastened. After its insertion, try to pull the flange out of the slide joint by hand. If it pulls out, the circlip is not properly meshed with the transaxle side gear.

Right

See Figures 12, 14 through 16.

1. Before servicing the vehicle, refer to the Precautions Section.
2. Remove the wheel and tire.
3. Remove the wheel sensor from the steering knuckle.
4. Remove the cotter pin, and remove the lock nut from the halfshaft.
5. Remove the brake hose lock plate, and remove the brake hose from the strut.
6. Remove the lower ball joint pinch bolt, and separate the lower ball joint from the steering knuckle.
7. Using a puller or suitable tool, remove the halfshaft from the wheel hub and bearing assembly.

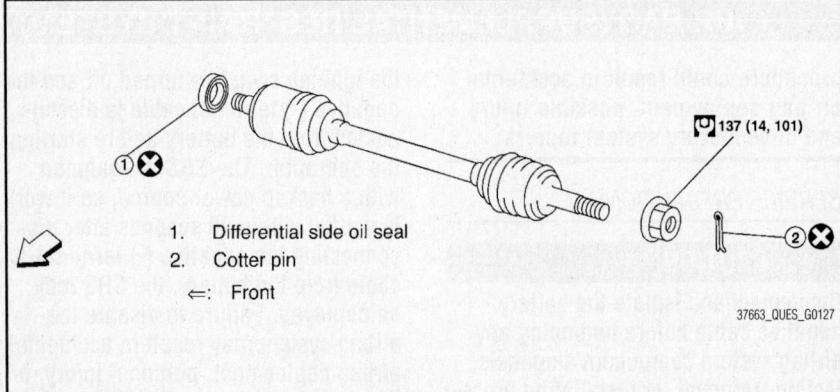

1. Differential side oil seal
2. Cotter pin
⇐: Front

Fig. 11 Exploded view of the left front halfshaft

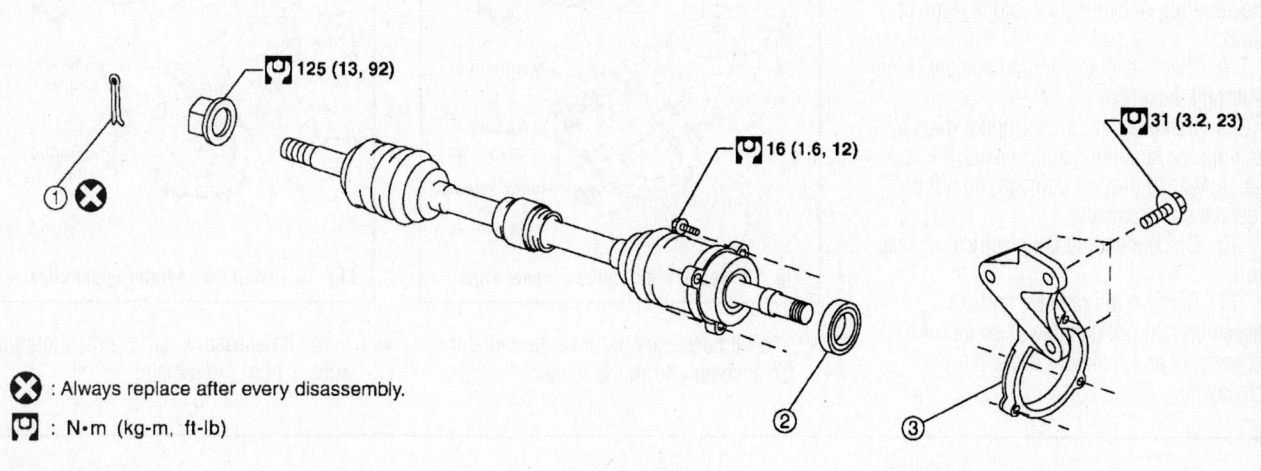

: Always replace after every disassembly.

: N•m (kg-m, ft-lb)

1. Cotter pin
2. Differential side oil seal
3. Support bearing bracket

67170-NIQU-G82

Fig. 15 Exploded view of the right front halfshaft—2004–06 models

---

❋❋ WARNING

When removing the halfshaft, do not apply an excessive angle to the halfshaft joint. Also be careful not to excessively extend the slide joint.

8. Remove the support bearing bolts.
9. Remove the halfshaft from the transaxle using a suitable pry tool.

To install:

10. Installation is in the reverse order of removal, noting the following:

    a. Tighten the axle nut to 92 ft. lbs. (125 Nm) , and the support bearing bolts to 23 ft. lbs. (31 Nm).

    b. In order to prevent damage to the differential side oil seal, place the tool [Tool No. KV38106800 (J-34297)] onto the oil seal before inserting halfshaft as shown. Slide halfshaft into slide joint and tap with a hammer to install securely.

    c. Install new circlip on the halfshaft in the circular clip groove on the transaxle side. Make sure the new circlip on the halfshaft is securely fastened. After its insertion, try to pull the flange out of the slide joint by hand. If it pulls out, the circlip is not properly meshed with the transaxle side gear.

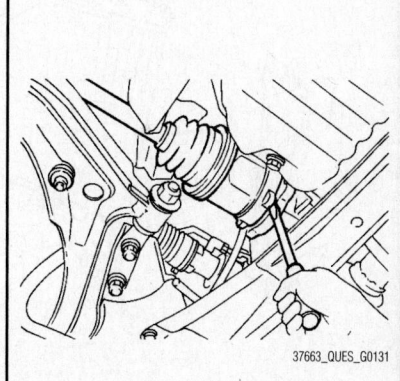

37663_QUES_G0131

Fig. 16 Remove the halfshaft from the transaxle using a suitable pry tool

---

ENGINE COOLING

ENGINE FAN

REMOVAL & INSTALLATION

See Figure 17.

1. Before servicing the vehicle, refer to the Precautions Section.
2. Disconnect the negative battery cable.
3. Remove the radiator. Refer to Radiator Removal & Installation.
4. Remove radiator cooling fan assembly from radiator.

To install:

5. Installation is the reverse of removal. Tighten the mounting bolts to 37 inch lbs. (4 Nm).

RADIATOR

REMOVAL & INSTALLATION

See Figures 17 through 19.

❋❋ CAUTION

Never remove the radiator cap when the engine is hot. Serious burns could occur from high pressure-coolant escaping from the radiator.

1. Before servicing the vehicle, refer to the Precautions Section.
2. Disconnect the negative battery cable.
3. Drain the engine coolant.
4. Partially drain A/T fluid.
5. Remove the fresh air duct. Refer to Air Cleaner Removal & Installation in the Engine Mechanical section.

6. Disconnect the upper and lower radiator hoses.

7. Disconnect the A/T fluid cooler hoses. Plug A/T lines to avoid leakage of fluid.

8. Disconnect the engine coolant reservoir tank hose.

9. Remove the radiator upper clips by pulling the tabs outside to release the lock, as shown. To prevent damage, do not pull lock tabs excessively.

10. Disconnect the fan electrical connectors.

11. Remove the radiator and fan assembly. Do not damage or scratch A/C condenser and radiator core when removing.

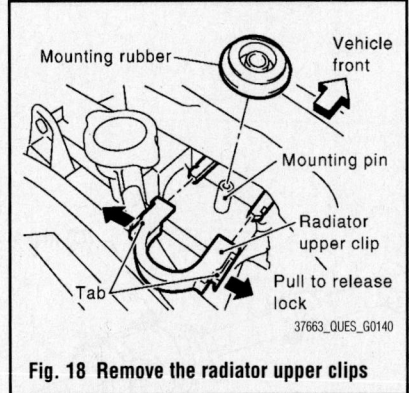

Fig. 18 Remove the radiator upper clips

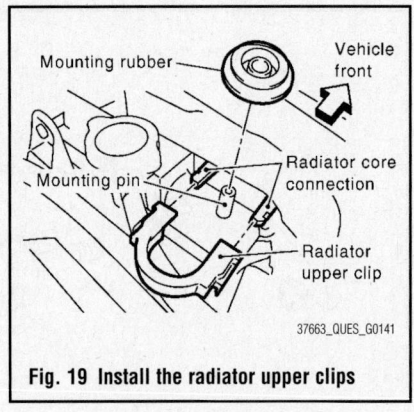

Fig. 19 Install the radiator upper clips

a. If necessary, remove the radiator fan assembly from the radiator.

b. If necessary, remove the mounting rubber from the radiator.

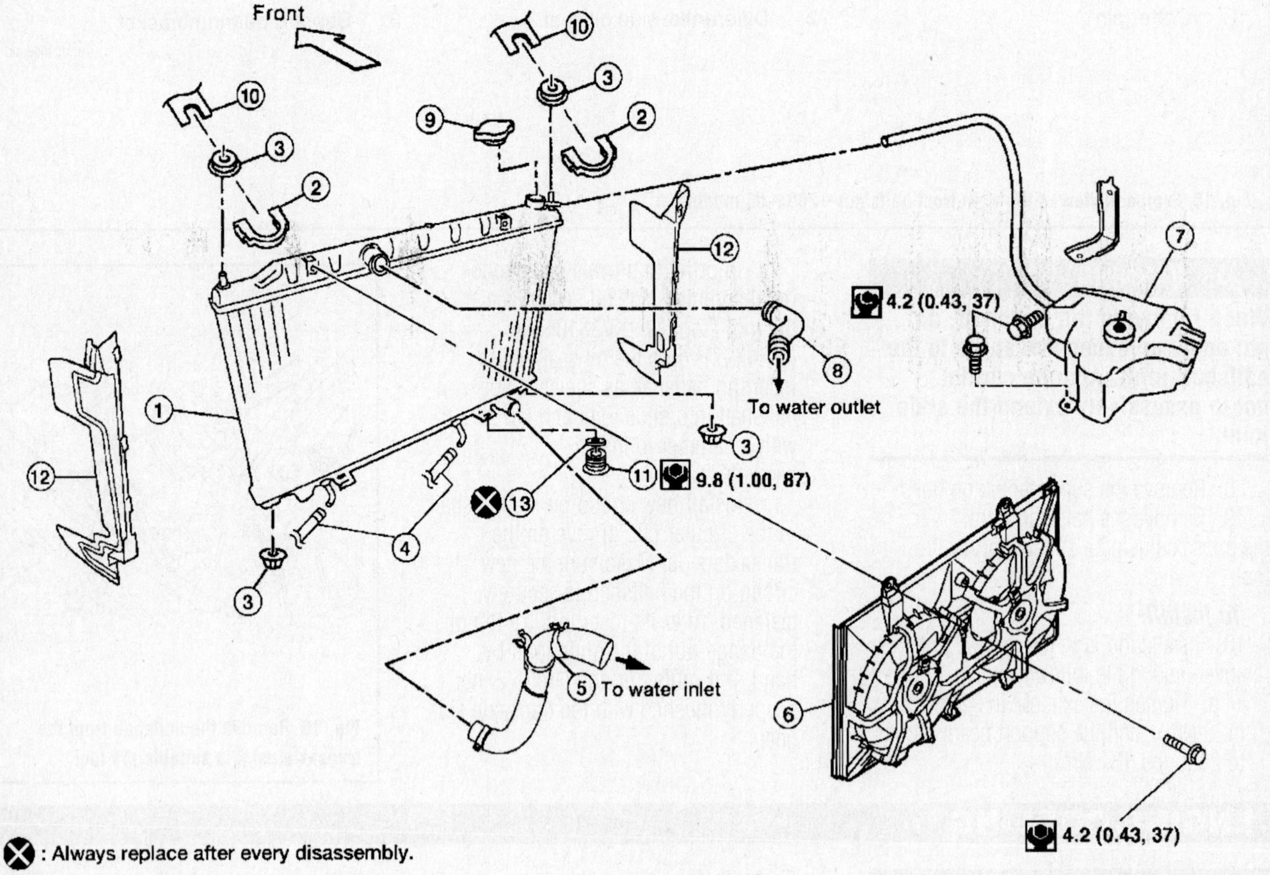

✖ : Always replace after every disassembly.

⚙ : N·m (kg-m, in-lb)

1. Radiator	2. Radiator upper clip	3. Mounting rubber
4. A/T fluid cooler hose (if equipped)	5. Radiator hose (lower)	6. Radiator fan assembly
7. Engine coolant reservoir tank	8. Radiator hose (upper)	9. Radiator filler cap
10. Radiator core connection	11. Radiator drain plug	12. Air guide
13. O-ring		

Fig. 17 Exploded view of radiator and components

To install:

12. To install, reverse removal procedure, noting the following:

13. Fill the radiator with coolant.

14. Install radiator upper clip:

a. Install the rubber on the mounting pin of the radiator core.

b. Align the radiator upper clip with the radiator core connector, and insert the radiator upper clip straight into the radiator core connections until a click is heard.

c. Visually confirm that the two radiator upper clips are connected to the radiator core connections.

d. Move the radiator upper clip and the radiator core forward and backward to make sure they are securely connected.

15. Check for engine coolant leaks.

16. Start and warm up the engine. Visually make sure that there are no engine coolant leaks or A/T fluid leaks.

THERMOSTAT

REMOVAL & INSTALLATION

See Figure 20.

1. Before servicing the vehicle, refer to the Precautions Section.

2. Disconnect the negative battery cable.

3. Drain the engine coolant.

4. Remove the IPDM E/R bolts and position aside.

5. Disconnect the lower radiator hose.

6. Disconnect the oil cooler line.

7. Remove the engine coolant inlet and thermostat assembly.

➡**Do not disassemble engine coolant inlet and thermostat assembly. Replace them as a unit, if necessary.**

To install:

8. Install the engine coolant inlet and thermostat assembly.

9. Connect the oil cooler line.

10. Connect the lower radiator hose.

11. Install the IPDM E/R bolts and position aside.

12. Refill engine coolant. Do not spill coolant in engine compartment. Use a shop cloth to absorb coolant.

13. Connect the negative battery cable.

14. Check for engine coolant leaks.

15. Start and warm up the engine. Visually make sure that there are no engine coolant leaks or A/T fluid leaks.

WATER PUMP

REMOVAL & INSTALLATION

See Figures 21 through 24.

1. Before servicing the vehicle, refer to the Precautions Section.

2. Disconnect the negative battery cable.

3. Drain the engine coolant from the radiator.

4. Remove the engine coolant reservoir tank.

5. Remove the IPDM E/R bolts and position aside.

6. Remove the right-hand wheel and tire assembly.

7. Remove the right-hand fender protector.

8. Remove the drive belts. Refer to Accessory Drive Belt Removal & Installation in the Engine Mechanical section.

9. Remove the idler pulley and idler pulley bracket.

10. Support the engine and remove the front engine insulator and bracket.

11. Remove the water drain plug on the water pump side of the cylinder block.

12. Remove the chain tensioner cover and water pump cover.

13. Remove the timing chain tensioner assembly:

a. Pull the lever down and release the plunger stopper tab.

b. Insert the stopper pin into the

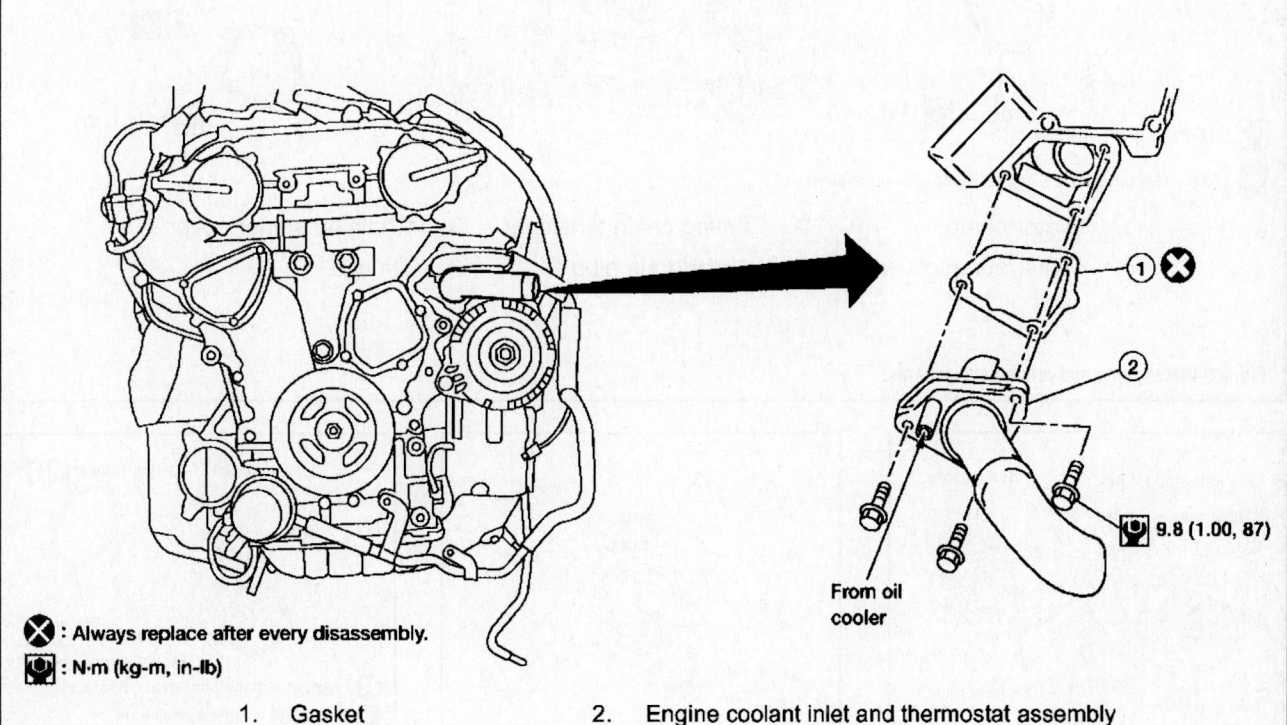

⊗ : Always replace after every disassembly.

🔧 : N·m (kg-m, in-lb)

1. Gasket

2. Engine coolant inlet and thermostat assembly

🔧 9.8 (1.00, 87)

From oil cooler

42050_QUES_G0012

Fig. 20 Removing thermostat from thermostat housing

tensioner body hole to hold the lever and keep the stopper tab released.

c. Insert the plunger into the tensioner body by pressing the timing chain slack guide.

d. Keep the slack guide pressed and hold the plunger stopper tab in by pushing the stopper pin deeper through the lever and into the timing chain tensioner body hole.

e. Make a gap between water pump gear and timing chain by turning the crankshaft pulley approximately 20 degrees clockwise.

f. Remove the chain tensioner.

14. Remove the 3 water pump bolts. Make a maximum gap between the water pump gear and the timing chain by turning the crankshaft pulley counterclockwise until the timing chain loosens on the water pump sprocket.

15. Screw M8 bolts (pitch: 0.049 in. (1.25mm), length: Approx. 1.97 in. (50mm) into the water pump's upper and lower mounting bolt holes until they reach the timing chain case. Alternately tighten each bolt for a half turn, and pull out water pump.

16. Pull straight out while preventing the vane from contacting socket in the installation area. Remove the water pump without the sprocket contacting the timing chain.

17. Remove the M8 bolts and O-rings from water pump.

✻✻ WARNING

When removing water pump assembly, be careful not to get coolant on drive belt.

✻✻ WARNING

Water pump cannot be disassembled and should be replaced as a unit.

To install:
18. Install new O-rings to water pump.

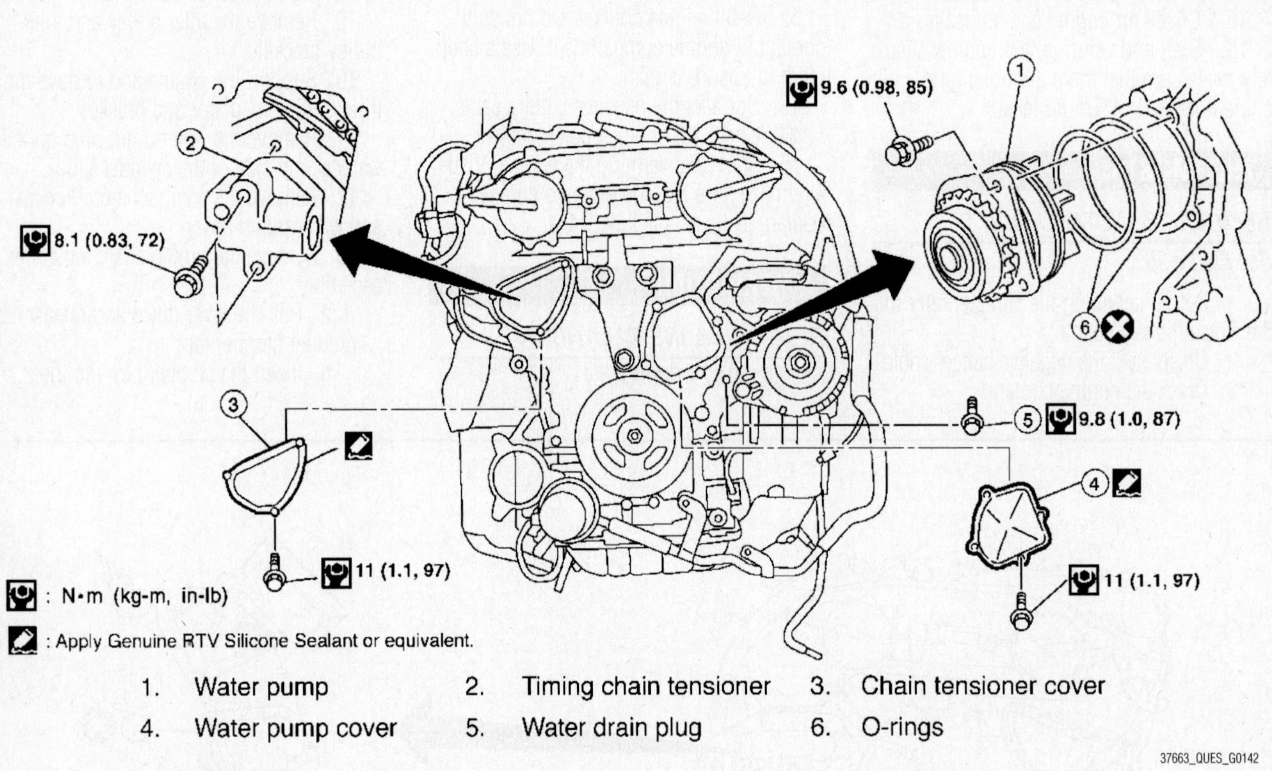

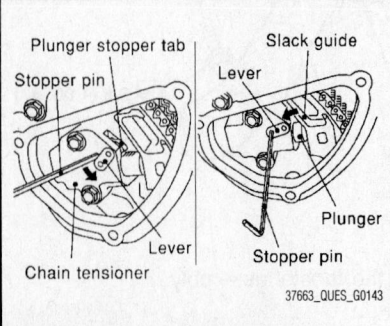

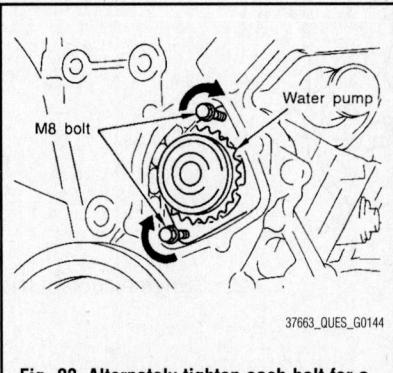

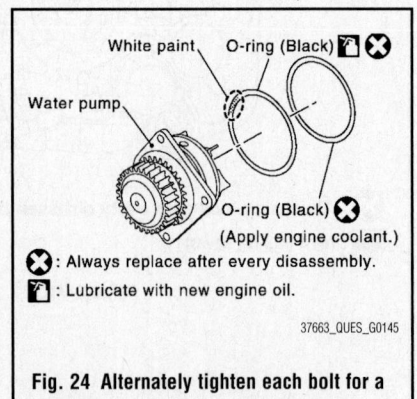

: N·m (kg-m, in-lb)

: Apply Genuine RTV Silicone Sealant or equivalent.

| 1. Water pump | 2. Timing chain tensioner | 3. Chain tensioner cover |
| 4. Water pump cover | 5. Water drain plug | 6. O-rings |

37663_QUES_G0142

Fig. 21 Water pump and related components

37663_QUES_G0143

Fig. 22 Insert the stopper pin into the tensioner body hole to hold the lever and keep the stopper tab released

37663_QUES_G0144

Fig. 23 Alternately tighten each bolt for a half turn, and pull out water pump

37663_QUES_G0145

Fig. 24 Alternately tighten each bolt for a half turn, and pull out water pump

19. Apply engine oil and engine coolant to the O-rings as shown. Locate the O-ring with white paint mark to engine front side.

20. Install the water pump.

➡ **Do not allow cylinder block to interfere with the O-rings when installing the water pump.**

21. Check that timing chain and water pump sprocket are engaged.

22. Insert the water pump by tightening the mounting bolts alternately and evenly.

23. Remove the dust and foreign material completely from the back side of the chain tensioner and from the installation area of the rear timing chain case.

24. Turn the crankshaft pulley clockwise so that the timing chain on the timing chain tensioner side is loose (approximately 20 degrees).

➡ **When installing the timing chain tensioner, engine oil should be applied to the oil hole and tensioner.**

25. Install the timing chain tensioner.

26. Remove the stopper pin.

27. Install chain tensioner cover and water pump cover:

a. Before installing, remove all traces of sealant from mating surface of water pump cover and chain tensioner cover. Also remove traces of sealant from the mating surface of the front cover.

b. Apply a continuous bead of RTV Silicone Sealant, or equivalent, to the mating surface of the chain tensioner cover and the water pump cover.

28. Install the water drain plug on the water pump side of the cylinder block, and tighten to 76 inch lbs. (10 Nm).

29. Install the idler pulley and idler pulley bracket. Tighten idler pulley bolts to 26 ft. lbs. (35 Nm).

30. Install the front engine insulator and bracket.

31. Install the drive belts.

32. Install the right-hand fender protector.

33. Install the right-hand wheel and tire assembly.

34. Install the IPDM E/R bolts and position aside.

35. Install the engine coolant reservoir tank.

36. Connect the negative battery cable.

37. Refill engine coolant.

38. After starting the engine, let it idle for three minutes, then rev the engine up to 3,000 rpm under no load to purge the air from the high-pressure chamber of the chain tensioner. The engine may produce a rattling noise. This indicates that air still remains in the chamber and is not a matter of concern.

ENGINE ELECTRICAL — CHARGING SYSTEM

ALTERNATOR

REMOVAL & INSTALLATION

See Figure 25.

1. Before servicing the vehicle, refer to the precautions section.

2. Disconnect the negative battery terminal.

3. Remove radiator. Refer to Radiator Removal & Installation in the Engine Cooling section.

4. Remove the right-hand wheel and tire assembly.

5. Remove the right-hand fender protector.

6. Remove the drive belt. Refer to Accessory Drive Belt Removal & Installation in the Engine Mechanical section.

7. Remove idler pulley.

8. Remove the alternator adjustable top mount.

9. Remove the alternator lower bolt and nut.

10. Disconnect the alternator harness connectors.

11. Remove the alternator upper bolt.

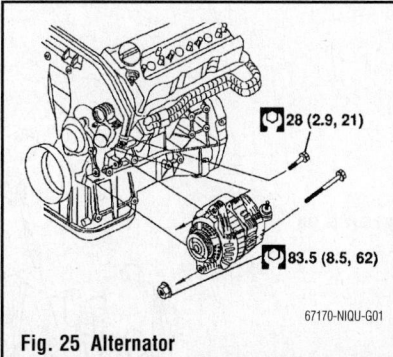

Fig. 25 Alternator

67170-NIQU-G01

12. Remove the alternator.

To install:

13. Install the alternator. Tighten the upper bolt to 21 ft. lbs. (28 Nm).

14. Connect the alternator harness connectors.`

15. Tighten the alternator lower bolt and nut to 62 ft. lbs. (84 Nm).

16. Install the alternator adjustable top mount.

17. Install idler pulley.

18. Install the drive belt. Check the drive belt tension.

19. Install the right-hand fender protector.

20. Install the right-hand wheel and tire assembly.

21. Install radiator.

22. Connect the negative battery terminal.

23. Be sure to tighten B terminal nut carefully to 8 ft. lbs. (11 Nm).

ENGINE ELECTRICAL — IGNITION SYSTEM

FIRING ORDER

See Figure 26.

IGNITION COIL

REMOVAL & INSTALLATION

See Figure 28.

1. Before servicing the vehicle, refer to the Precautions Section.

2. Disconnect the negative battery cable.

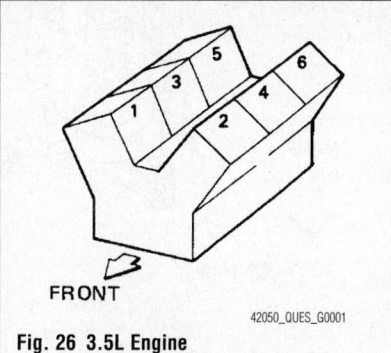

FRONT

42050_QUES_G0001

Fig. 26 3.5L Engine
Firing Order: 1–2–3–4–5–6

3. Remove the engine cover.

4. Remove the intake manifold collector.

5. Disconnect ignition coil connector.

6. Remove the ignition coil. Do not shock ignition coil.

To install:

7. Installation is in the reverse order of removal.

IGNITION TIMING

INSPECTION

See Figure 27.

A
0°
15°
20°
37663_QUES_G0156

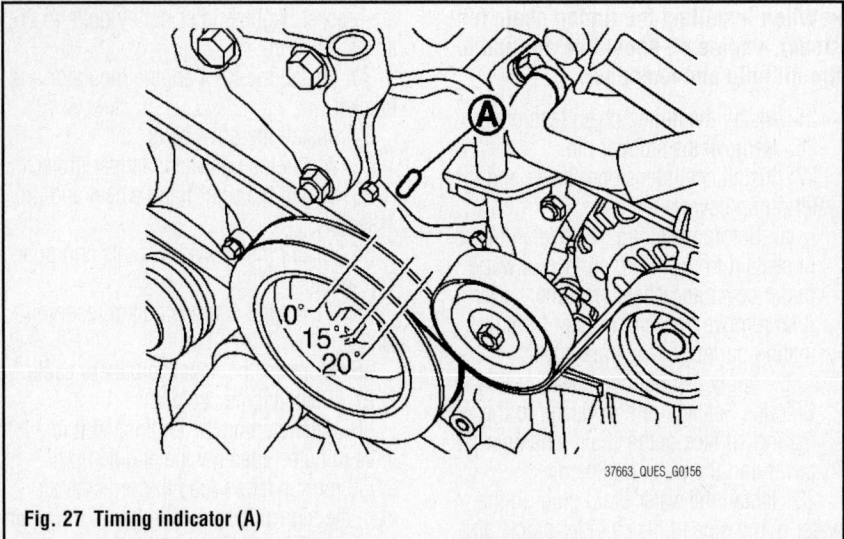

Fig. 27 Timing indicator (A)

1. Before servicing the vehicle, refer to the Precautions Section.
2. Remove the number one ignition coil.
3. Connect the number one ignition coil and spark plug with a suitable high tension wire.
4. Attach the timing light clamp to the wire.
5. Check the ignition timing.

ADJUSTMENT

The ignition timing is controlled by the Powertrain Control Module (PCM). No adjustment is necessary or possible.

SPARK PLUGS

REMOVAL & INSTALLATION

See Figure 28.

1. Before servicing the vehicle, refer to the Precautions Section.
2. Disconnect the negative battery cable.
3. Remove the engine cover.
4. Remove the ignition coils. Refer to Ignition Coil Removal & Installation.
5. Remove the spark plug using a suitable tool.

To install:

6. Install the spark plug using a suitable tool. Use standard type spark plug for normal driving conditions.
   a. The hot type spark plug is suitable when fouling occurs with the standard type spark plug under conditions such as:
      - Frequent engine starts
      - Low ambient temperatures
   b. The cold type spark plug is suitable when engine spark knock occurs with the standard type spark plug under conditions such as:
      - Extended highway driving
      - Frequent high engine revolution
7. Install the ignition coils.
8. Install the engine cover.
9. Connect the negative battery cable.

7 (0.7, 62)
1
2
24.5 (2.5, 18)
1
2
24.5 (2.5, 18)
3
4
: N·m (kg-m, in-lb)
: N·m (kg-m, ft-lb)
Engine front
RH
LH
1. Ignition coil
2. Spark plug
3. Rocker cover (RH)
4. Rocker cover (LH)
42050_QUES_G0002

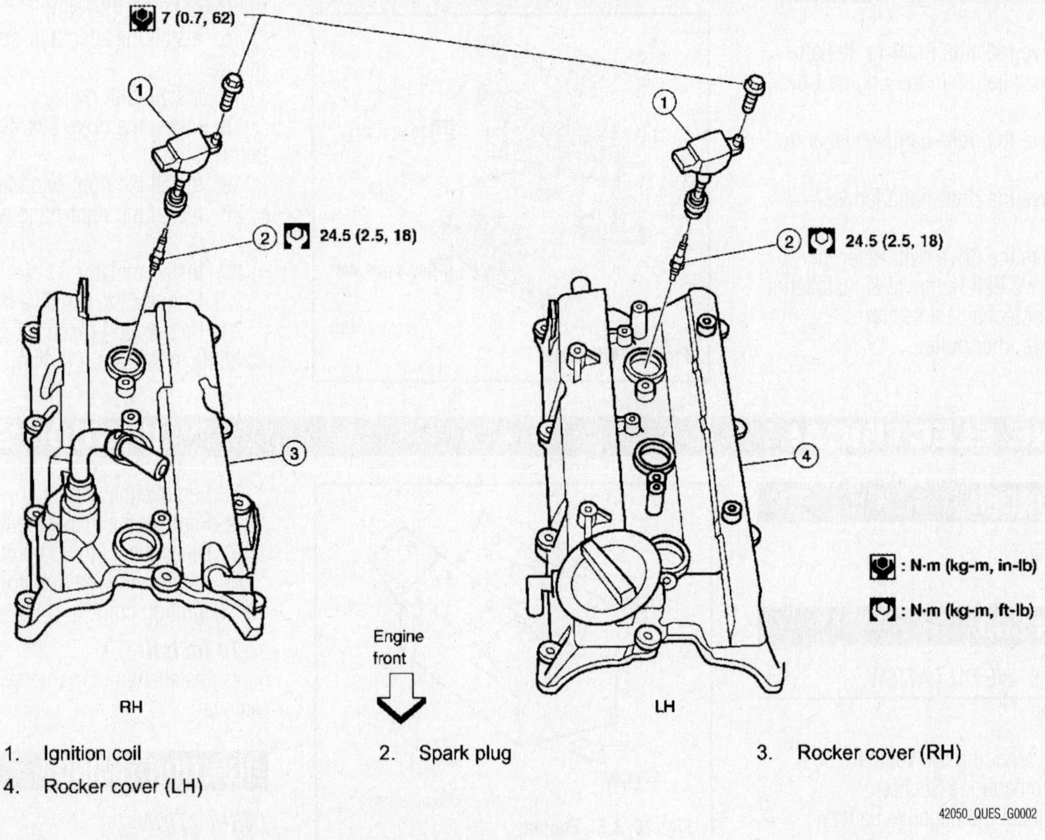

Fig. 28 Ignition coil and spark plugs

ENGINE ELECTRICAL

STARTER

REMOVAL & INSTALLATION

See Figures 29 and 30.

1. Before servicing the vehicle, refer to the Precautions Section.
2. Disconnect the negative battery terminal.
3. Remove the starter insulator.

4. Remove the harness bracket and harness protector from the starter engine room harness.
5. Disconnect the starter harness connectors.
6. Remove the two starter mounting bolts.
7. Remove the starter.

To install:

8. Installation is the reverse of removal.

a. Tighten the starter mounting bolts to 33 ft. lbs. (45 Nm).
b. Tighten the starter mounting bolts to 80 inch lbs. (95 Nm).

SOLENOID OR RELAY REPLACEMENT

The solenoid is not serviceable, the starter must be serviced as an assembly.

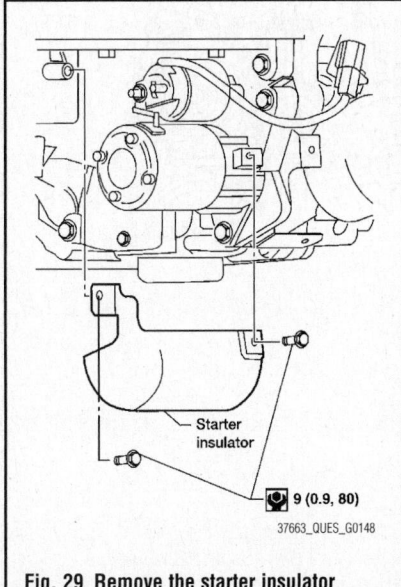

Fig. 29 Remove the starter insulator

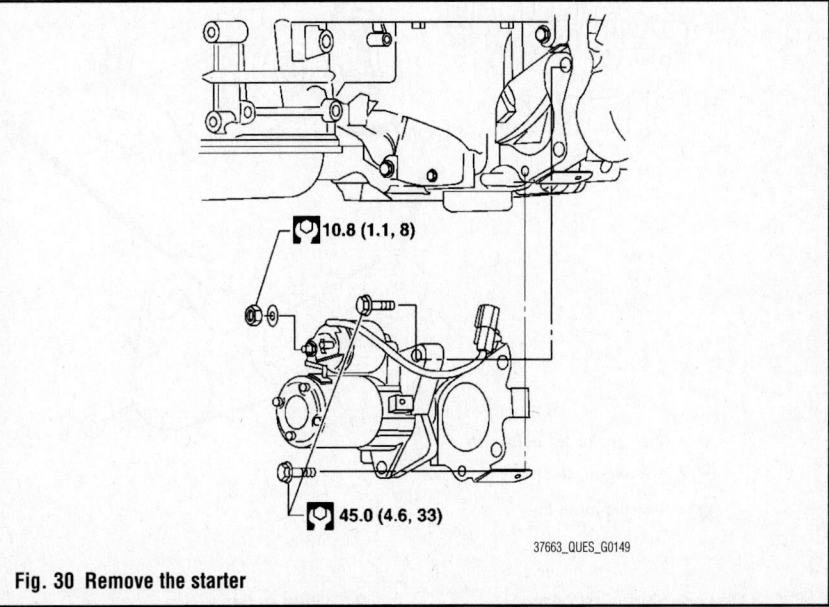

Fig. 30 Remove the starter

ENGINE MECHANICAL

ACCESSORY DRIVE BELTS

ACCESSORY BELT ROUTING

See Figure 31.

Refer to the accompanying illustrations.

INSPECTION

See Figures 31 and 32.

❋❋ CAUTION

Be sure to perform inspection when the engine is not running.

1. Before servicing the vehicle, refer to the Precautions Section.
2. Inspect belt for cracks, fraying, wear or oil adhesion. If necessary, replace with a new one.
3. Inspect drive belt deflections by applying 22 ft. lbs. (98 N) pressure on the belt midway between pulleys at the points shown.
4. Measure the belt tension using the appropriate tool (Tool No. BT-3373-F) at the locations shown. Adjust if belt deflections exceed the limit or if belt tension is not within specifications.
   a. Tighten idler pulley lock nut by hand and measure deflection or tension without looseness.

➡Inspect drive belt deflection or tension when engine is cold.

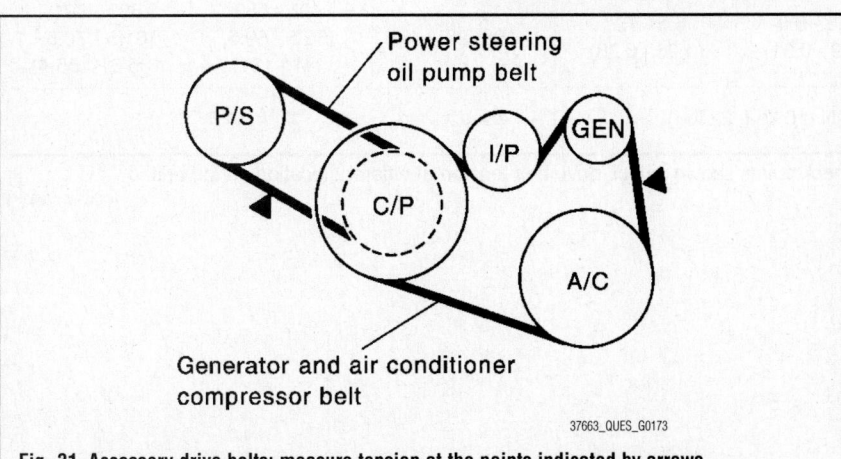

Fig. 31 Accessory drive belts; measure tension at the points indicated by arrows

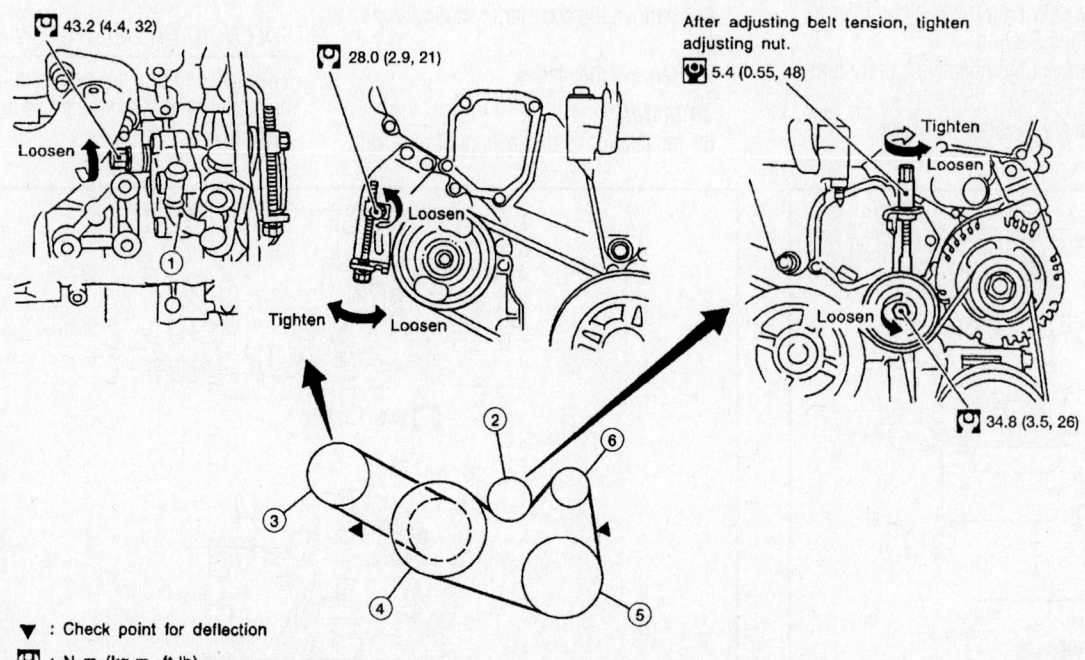

Check point for deflection symbol: ▼ : Check point for deflection

N·m (kg-m, ft-lb) symbol

N·m (kg-m, in-lb) symbol

1. Power steering oil pump
2. Idler pulley
3. Power steering oil pump
4. Crankshaft pulley
5. Air conditioner compressor
6. Generator

	Deflection adjustment		Unit: mm (in)	Tension adjustment*		Unit: N (kg-f, lb-f)
	Used belt		New belt	Used belt		New belt
	Limit	After adjustment		Limit	After adjustment	
Generator and air conditioner compressor	7 (0.28)	4.2 - 4.6 (0.17 - 0.18)	3.7 - 4.1 (0.15 - 0.16)	294 (30, 66)	730 - 818 (74.5 - 83.5, 164 - 184)	838 - 926 (85.5 - 94.5, 188 - 208)
Power steering oil pump	11 (0.43)	7.3 - 8.0 (0.29 - 0.31)	6.5 - 7.2 (0.26 - 0.28)	196 (20, 44)	495 - 583 (50.5 - 59.5, 111 - 131)	603 - 691 (61.5 - 70.5, 135.6 - 155.4)
Applied pushing force	98 N (10 kg-f, 22 lb-f)			—		

*: If belt tension gauge cannot be installed at check points shown, check drive belt tension at different location on the belt.

37663_QUES_G0174

Fig. 32 Drive belt service information

※※ WARNING

When checking belt deflection or tension immediately after installation, first adjust it to the specification value. Then, after turning the crankshaft two turns or more, readjust to the specified value to avoid variation in deflection between pulleys.

ADJUSTMENT

Generator & Air Conditioner Compressor Belt

See Figures 32 and 33.

1. Before servicing the vehicle, refer to the Precautions Section.

➡When belt is replaced with a new one, adjust it to value for _New belt_ to accommodate for insufficient adaptability with pulley grooves.

➡When deflection or tension of belt being used exceeds _Used belt limit_ adjust it to value for _After

adjustment_ of _Used belt_. When checking belt deflection or tension immediately after installation, first adjust it to the specification value. Then, after turning the crankshaft two turns or more, readjust to the specified value to avoid variation in deflection between pulleys.

➡When installing belt, make sure that it is correctly engaged with pulley groove. Keep oil and water away from belt. Do not twist or bend belt excessively.

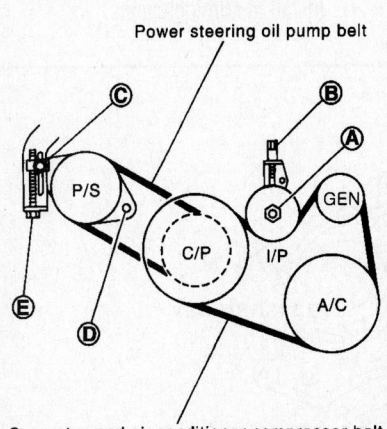

Power steering oil pump belt

Generator and air conditioner compressor belt

37663_QUES_G0175

Fig. 33 Idler pulley lock nut (A) and adjusting bolt (B); power steering adjusting bolts (C, E) and power steering oil pump bolt (D)

2. Remove engine undercover.
3. Loosen idler pulley lock nut and adjust the belt by turning adjusting bolt.
4. Tighten lock nut to 26 ft. lbs. (35 Nm).
5. Tighten adjusting bolt to 48 inch lbs. (6 Nm).

Power Steering Oil Pump Belt

See Figures 32 and 33.

1. Before servicing the vehicle, refer to the Precautions Section.

➡When belt is replaced with a new one, adjust it to value for _New belt_ to accommodate for insufficient adaptability with pulley grooves.

➡When deflection or tension of belt being used exceeds _Used belt limit_ adjust it to value for _After

adjustment_ of _Used belt_. When checking belt deflection or tension immediately after installation, first adjust it to the specification value. Then, after turning the crankshaft two turns or more, readjust to the specified value to avoid variation in deflection between pulleys.

➡When installing belt, make sure that it is correctly engaged with pulley groove. Keep oil and water away from belt. Do not twist or bend belt excessively.

2. Remove engine undercover.
3. Loosen adjusting bolt.
4. Loosen power steering oil pump bolt (bolt head is at the engine rear side).
5. Adjust the belt by turning the adjusting bolt (bolt is loosened with counterclockwise rotation).
6. Tighten adjusting bolt.
7. Tighten power steering oil pump bolt.

REMOVAL & INSTALLATION

1. Before servicing the vehicle, refer to the Precautions Section.
2. Disconnect the negative battery terminal.
3. Remove engine undercover.
4. Fully loosen each belt. Remove generator and air conditioner compressor belt and then power steering oil pump belt.

※※ WARNING

Grease is applied to idler pulley adjusting bolt. Be careful to keep grease away from the belts.

To install:

5. Installation is in the reverse order of removal, noting the following:
   a. Make sure belts are correctly engaged with the pulley groove.
   b. Clean off any oil and coolant on belts and each pulley groove.
   c. Adjust belt tension.

CAMSHAFT AND VALVE LIFTERS

REMOVAL & INSTALLATION

See Figures 34 through 43.

1. Before servicing the vehicle, refer to the Precautions Section.
2. Disconnect the negative battery terminal.
3. Drain the cooling system.
4. Remove the front timing chain case, camshaft sprocket, timing chain and the rear timing chain case.
5. If necessary, remove both the Camshaft Position (CMP) sensors from cylinder head back side.

➡Handle carefully to avoid dropping and shocks. Do not disassemble. Do not allow metal powder to adhere to magnetic part at sensor tip. Do not place sensors in a location where they are exposed to magnetism.

6. Remove the Intake Valve Timing (IVT) control solenoid valves. Discard the IVT control solenoid valve gaskets and use new gaskets for installation.
7. Remove the intake and exhaust camshaft brackets. Mark the camshafts, camshaft brackets, and bolts so they are placed in the same position and direction for installation.
8. Equally loosen the camshaft bracket bolts in several steps in the numerical order shown.
9. Remove the camshaft.
10. Remove the valve lifters. Identify installation positions, and store them without mixing them up.

To install:

11. Using a suitable tool, remove any old Silicone RTV Sealant from component mating surfaces, bolt holes and threads, and the No. 1 camshaft bracket mating surface. Do not scratch or damage the surfaces.

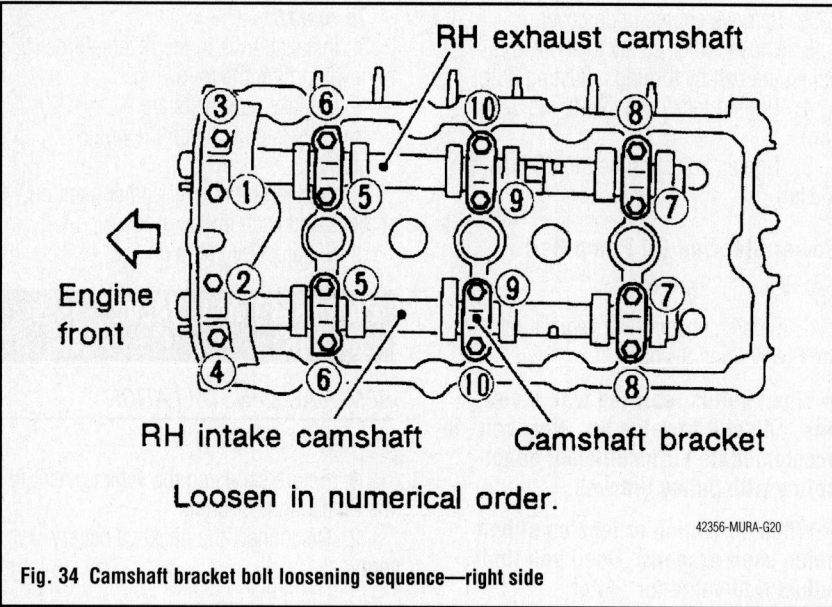

Fig. 34 Camshaft bracket bolt loosening sequence—right side

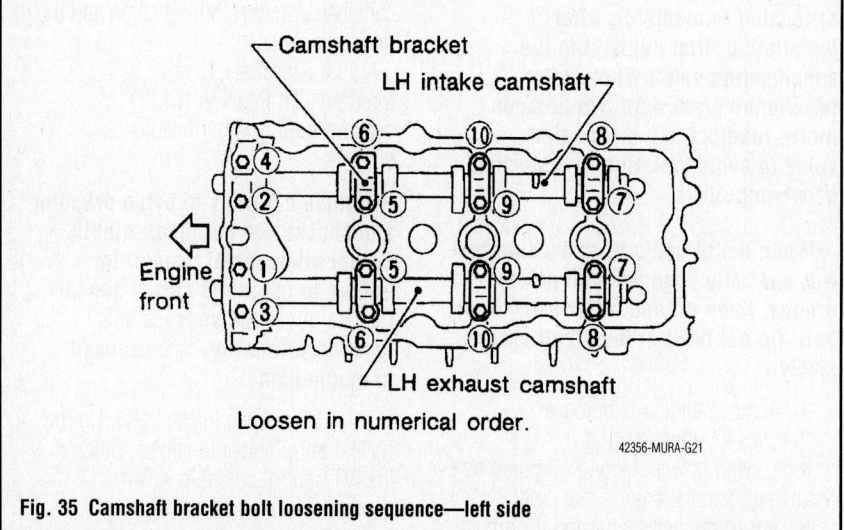

Fig. 35 Camshaft bracket bolt loosening sequence—left side

12. Turn the crankshaft until No. 1 piston is set at TDC on the compression stroke. The crankshaft key should line up with the right bank cylinder center line as shown.

13. Install the valve lifters in original positions.

14. Install the exhaust and intake camshafts and camshaft brackets.

   a. The intake camshaft has a drill mark on camshaft sprocket mounting flange.

   b. Follow identification marks made during removal, or follow the identification marks that are present on the new camshafts components for proper placement and direction of the components.

   c. Position the camshafts:

- Right-hand exhaust camshaft dowel pin at about 10 o'clock.
- Left-hand exhaust camshaft dowel pin at about 2 o'clock.

15. Before installing No. 1 camshaft brackets, apply sealant to mating surface of No. 1 camshaft bracket. Use Genuine Silicone RTV Sealant, or equivalent. Before installation, wipe off any protruding sealant.

   a. Install remaining camshaft brackets in their original positions and direction. Align the stamp marks as shown.

   b. If checking and adjusting any part of valve assembly or camshaft, check valve clearance according to the reference data.

16. Tighten the camshaft brackets in the four steps, as shown. Camshaft bracket bolts:

- Step 1 (bolts 7–10): 17 inch lbs. (2 Nm)
- Step 2 (bolts 1–6): 17 inch lbs. (2 Nm)
- Step 3: 52 inch lbs. (6 Nm)
- Step 4: 92 inch lbs. (10 Nm)

17. Measure the difference in levels between the front end faces of the No.1 camshaft bracket and the cylinder head. Standard: - 0.14 (- 0.0055 in.)

   a. If measurement is outside the specified range, re-install camshaft and camshaft bracket.

18. Install the IVT control solenoid valves with new gaskets.

19. If applicable, install the camshaft position sensors.

20. Install the timing chains.

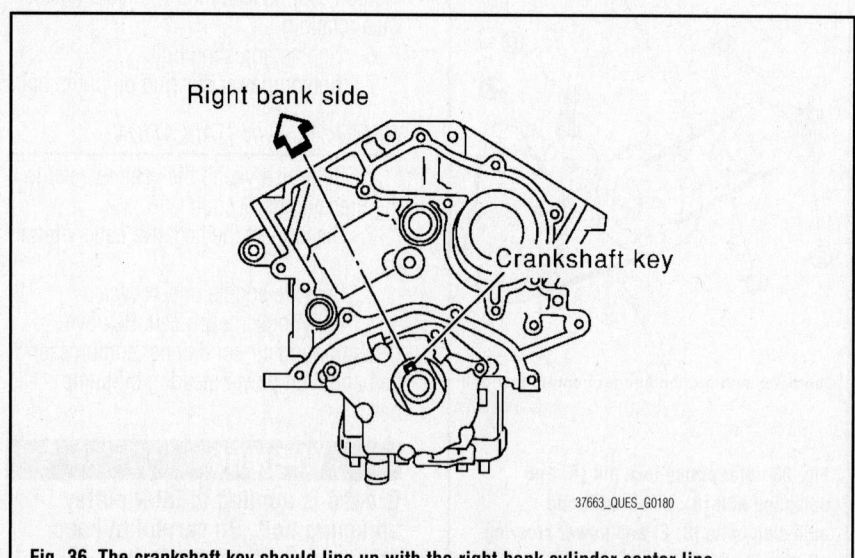

Fig. 36 The crankshaft key should line up with the right bank cylinder center line

Bank	INT/EXH	ID mark	Drill mark	Paint marks	
				M1	M2
RH	INT	RE	Yes	Yes	No
	EXH	RE	No	No	Yes
LH	INT	LH	Yes	Yes	No
	EXH	LH	No	No	Yes

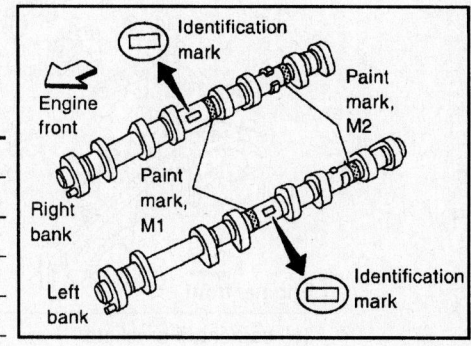

67170-NIQU-G69

Fig. 37 Camshaft identification

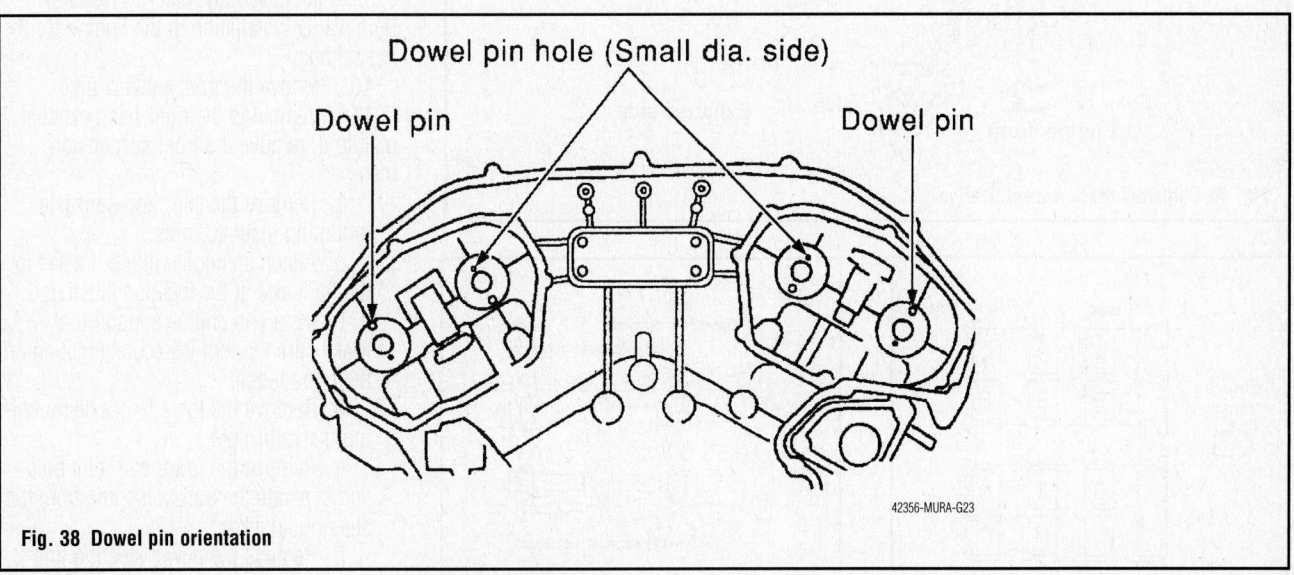

42356-MURA-G23

Fig. 38 Dowel pin orientation

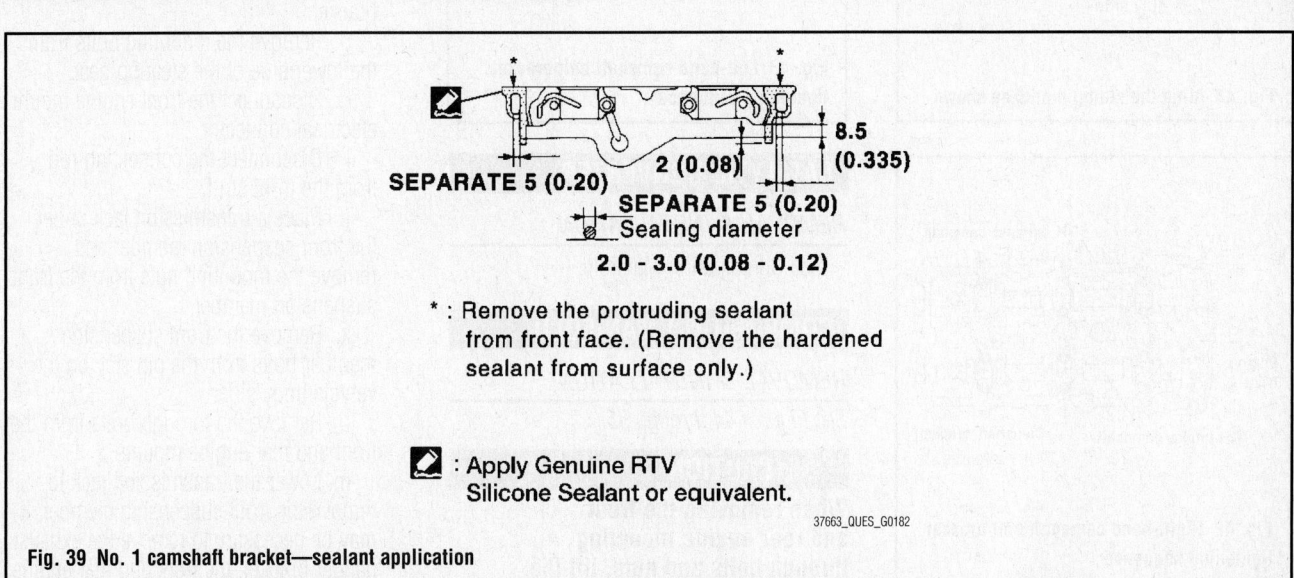

SEPARATE 5 (0.20)

2 (0.08)

SEPARATE 5 (0.20)

8.5 (0.335)

Sealing diameter
2.0 - 3.0 (0.08 - 0.12)

* : Remove the protruding sealant
from front face. (Remove the hardened
sealant from surface only.)

: Apply Genuine RTV
Silicone Sealant or equivalent.

37663_QUES_G0182

Fig. 39 No. 1 camshaft bracket—sealant application

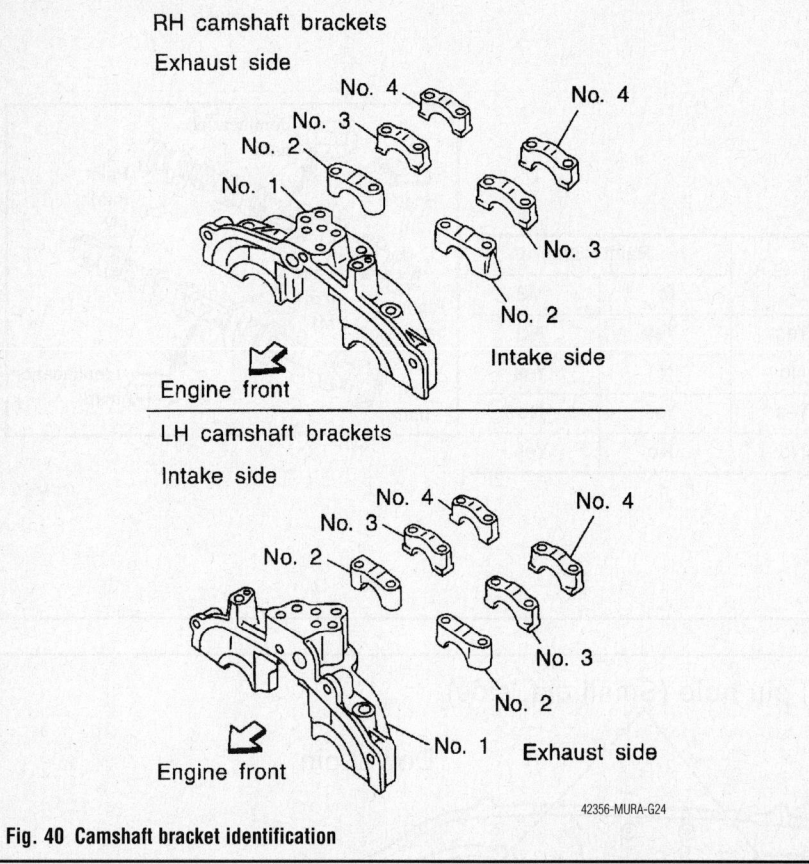

RH camshaft brackets

Exhaust side

No. 4
No. 3
No. 2
No. 1

No. 4

No. 3

No. 2

Intake side

Engine front

LH camshaft brackets

Intake side

No. 4
No. 3
No. 2

No. 4

No. 1

No. 3

No. 2

Engine front

Exhaust side

42356-MURA-G24

Fig. 40 Camshaft bracket identification

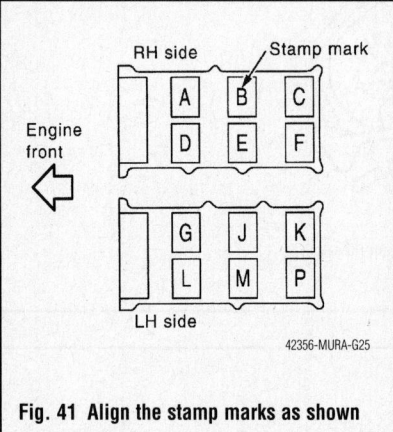

RH side   Stamp mark

| A | B | C |
| D | E | F |

Engine front

| G | J | K |
| L | M | P |

LH side

42356-MURA-G25

Fig. 41 Align the stamp marks as shown

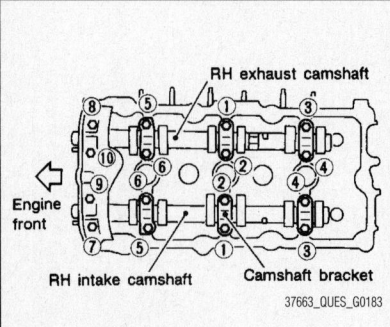

RH exhaust camshaft

Engine front

RH intake camshaft   Camshaft bracket

37663_QUES_G0183

Fig. 42 Right-hand camshaft bolt bracket tightening sequence

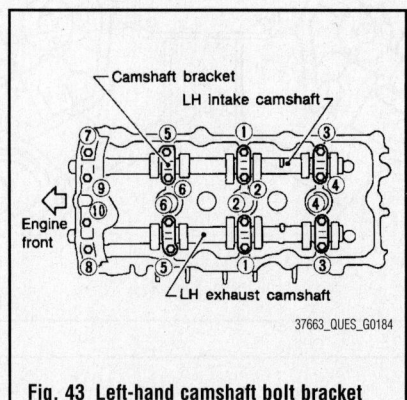

Camshaft bracket
LH intake camshaft

Engine front

LH exhaust camshaft

37663_QUES_G0184

Fig. 43 Left-hand camshaft bolt bracket tightening sequence

CATALYTIC CONVERTER

REMOVAL & INSTALLATION

See Combination Manifold.

COMBINATION MANIFOLD

REMOVAL & INSTALLATION

See Figures 44 through 55.

✳✳ CAUTION

When removing the front and rear engine mounting through bolts and nuts, lift the

engine up slightly for safety.

1. Before servicing the vehicle, refer to the Precautions Section.
2. Disconnect the negative battery terminal.
3. Remove cowl top. Refer to Cowl Top Removal & Installation in the Body Exterior section.
4. Disconnect air fuel ratio sensor 1 (bank 2) connector.
5. Remove the front wheel and tires.
6. Remove the engine undercover.
7. Remove the inner wheel well splash shields.
8. If removing only the air fuel ratio sensor 1 (bank 2) do so at this time through the wheel well opening,
   using Tool No. KV991J0050 (J-44626).
9. If removing the left-hand bank exhaust manifold, remove the radiator and cooling fan assembly. Refer to Radiator Removal & Installation in the Engine Cooling section.
10. Remove the front exhaust pipe.
11. If removing the right-hand exhaust manifold, remove the front suspension member.

   a. Remove the left-hand transaxle mounting insulator nuts.
   b. Attach an engine lifting bracket to the transaxle at the location illustrated.
   c. Install an engine support tool. Make sure the tool is securely resting on the hood ledge.
   d. Remove the three transaxle mounting insulator nuts.
   e. Remove the lower ball joint bolt and separate the transverse link from the steering knuckle.
   f. Remove the power steering line bracket.
   g. Remove the mounting bolts from the lower side of the steering gear.
   h. Disconnect the front engine mount electrical connector.
   i. Disconnect the connecting rod from the front strut.
   j. Place a transmission jack under the front suspension member and remove the mounting nuts from the front suspension member.
   k. Remove the front suspension member bolts from the pin stay on the vehicle body side.
   l. Remove the through bolts from the front and rear engine mounts.
   m. Lower the transmission jack to remove the front suspension member. It may be necessary to remove the exhaust hanger bracket, the front and rear engine

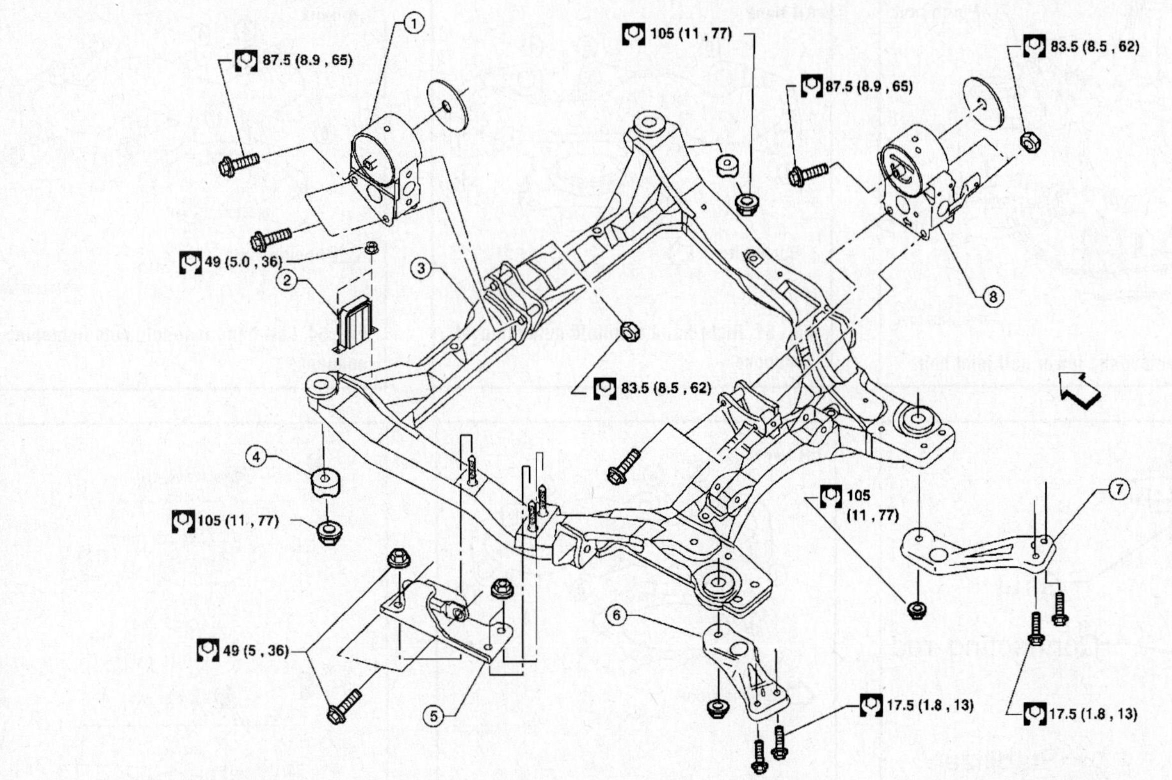

87.5 (8.9 , 65)
1
105 (11 , 77)
83.5 (8.5 , 62)
87.5 (8.9 , 65)
49 (5.0 , 36)
2
3
8
83.5 (8.5 , 62)
4
105 (11 , 77)
105
(11 , 77)
7
6
49 (5 , 36)
5
17.5 (1.8 , 13)
17.5 (1.8 , 13)

1. Front engine mount
2. Dynamic damper
3. Front suspension member
4. Cup
5. LH transaxle mounting insulator
6. Member pin stay, LH
7. Member pin stay, RH
8. Rear engine mount

⇦ Front

37663_QUES_G0187

Fig. 44 Front suspension member

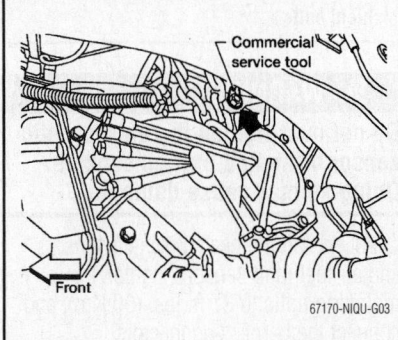

Commercial
service tool
Front
67170-NIQU-G03

Fig. 45 Attach a engine lifting bracket to the transaxle

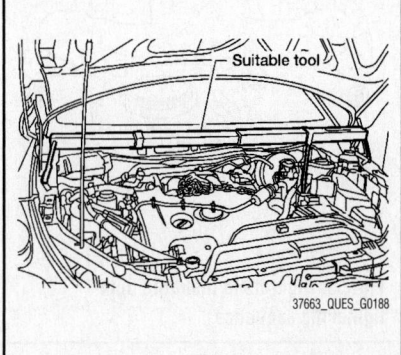

Suitable tool
37663_QUES_G0188

Fig. 46 Install an engine support tool

LH transaxle
mounting insulator
37663_QUES_G0189

Fig. 47 Install an engine support tool

mounts, the transverse link, and the dynamic damper to enable removal.

12. Remove the bank 1 and bank 2 three way catalyst manifolds support bolts in the order as shown.

13. Remove heated oxygen sensor 2 (bank 1), heated oxygen sensor 2 (bank 2), air fuel ratio sensor 1 (bank 1) and air fuel ratio (A/F) sensor 1 (bank 2).

a. Remove harness connector of each sensor, and disconnect the harness from the bracket and middle clamp.

b. Remove both heated oxygen sensors and air fuel ratio (A/F) sensors using Tool No. KV10114400 (J-38365).

✱✱ WARNING

Discard and replace any heated oxygen sensor which has been dropped from a height of more than 19.7 in (0.5 m) onto a hard surface.

14. Remove exhaust manifold and three way catalyst manifold heat shields.

15. Remove the three way catalyst (manifold) (bank 1) and three way catalyst (manifold) (bank 2) by loosening the bolts first and then removing the nuts and through bolts.

16. Remove the right-hand and left-hand bank exhaust manifold nuts in the order shown.

17. Remove the right-hand and left-hand bank exhaust manifold.

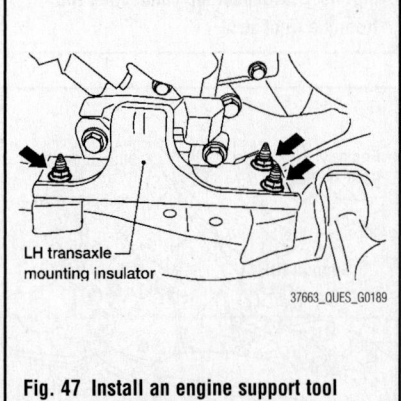

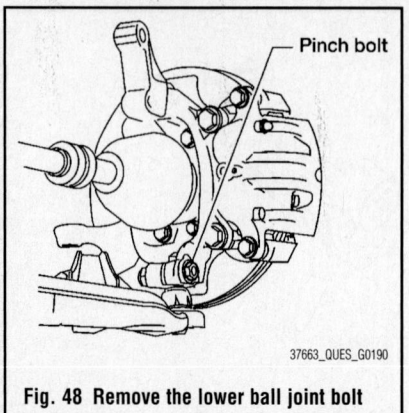

Fig. 48 Remove the lower ball joint bolt

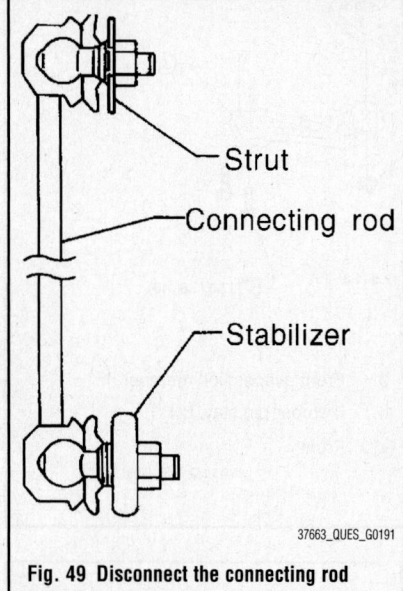

Fig. 49 Disconnect the connecting rod from the front strut

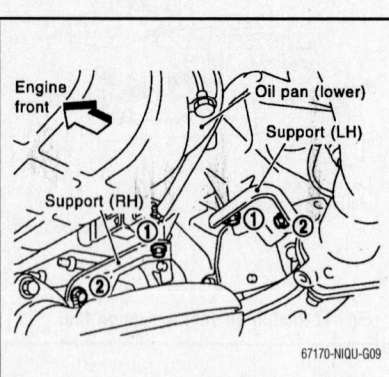

Fig. 50 Three way catalyst support bolts removal and tightening sequence

To install:

18. Install the exhaust manifolds and tighten the right-hand and left-hand bank exhaust manifold nuts in the order shown to 23 ft. lbs. (31 Nm).

19. Install the right-hand and left-hand bank three way catalyst support bolts in the

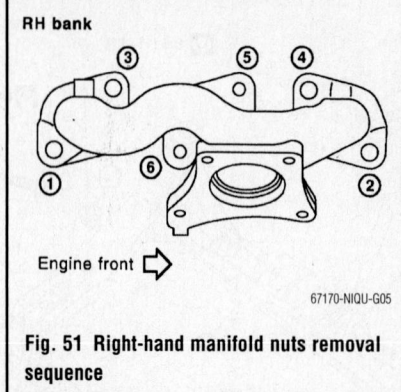

Fig. 51 Right-hand manifold nuts removal sequence

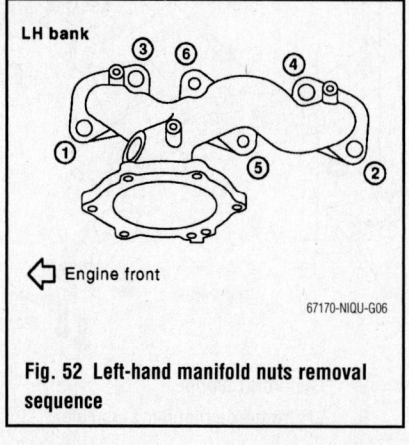

Fig. 52 Left-hand manifold nuts removal sequence

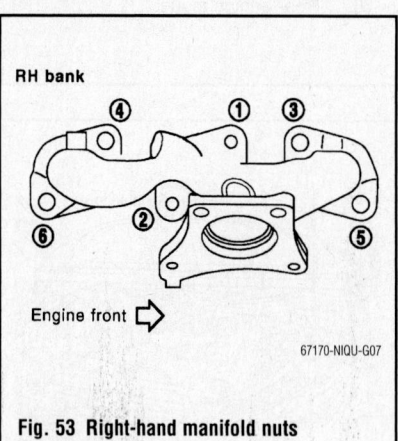

Fig. 53 Right-hand manifold nuts tightening sequence

order shown and tighten to 16 ft. lbs. (22 Nm).

20. Install the exhaust manifold heat shield bolts and tighten to 51 inch lbs. (6 Nm).

21. Install the three way catalyst manifolds heat shield bolts and tighten to 73 inch lbs. (8 Nm).

22. Before installing a heated oxygen sensor or air fuel ratio sensor, clean the exhaust manifold threads using the appropriate tool (Tool No. J-43897-18, J-43897-12), and apply anti-seize lubricant.

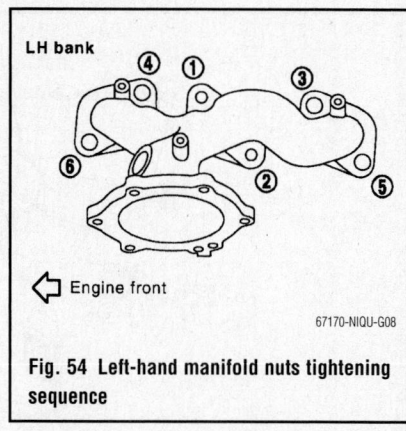

Fig. 54 Left-hand manifold nuts tightening sequence

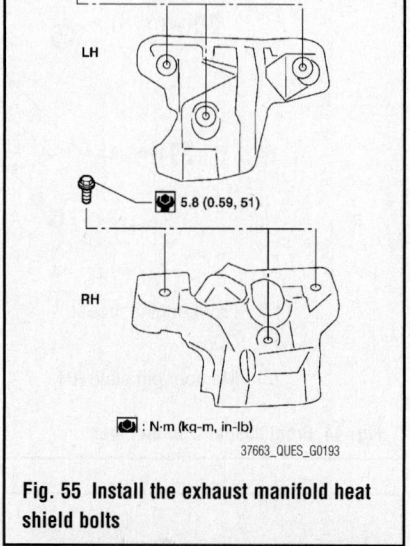

Fig. 55 Install the exhaust manifold heat shield bolts

✳✳ WARNING

Do not over-tighten the air fuel ratio sensor or heated oxygen sensors. Doing so may cause damage.

23. Install the heated oxygen sensors and air fuel ratio sensors, tighten the mounting bolts to 37 ft. lbs. (50 Nm), and connect the harness connectors.

24. Install the front support member in the reverse order of removal and tighten the retainers to the specifications shown in the front suspension member mounting exploded view illustration.

25. Install the front exhaust pipe.

26. Install the radiator and cooling fan assembly.

27. Install the air fuel ratio sensor 1 (bank 2), as necessary.

28. Install the inner wheel well splash shields.

29. Install the engine undercover.

30. Install the front wheel and tires.

31. Install the cowl top.

32. Connect the negative battery cable.

CRANKSHAFT DAMPER

REMOVAL & INSTALLATION

See Figures 56 and 57.

1. Before servicing the vehicle, refer to the Precautions Section.

2. Disconnect the negative battery terminal.

3. Remove the engine undercover.

4. Remove the drive belts. Refer to Accessory Drive Belt Removal & Installation.

5. Remove the radiator fan. Refer to Engine Cooling Fan Removal & Installation in the Engine Cooling section.

6. Remove the starter motor. Refer to Starter Removal & Installation in the Engine Electrical section.

7. Lock the ring gear using the appropriate tool [Tool No. KV10117700 (J-44716)] attached to the starter bolt hole.

✳✳ WARNING

Do not damage the ring gear teeth, or the signal plate teeth behind the ring gear when setting the tool.

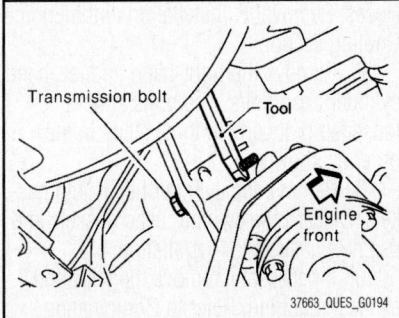

Fig. 56 Lock the ring gear using the appropriate tool [Tool No. KV10117700 (J-44716)] attached to the starter bolt hole

Fig. 57 Position a pulley puller at recess hole of crankshaft pulley

8. Loosen crankshaft pulley bolt and locate bolt seating surface at 0.39 in. (10mm) from its original position.

9. Position a pulley puller at recess hole of crankshaft pulley and remove the crankshaft pulley.

✳✳ WARNING

Do not use a puller claw on crankshaft pulley periphery.

To install:

10. Install the crankshaft pulley and tighten the bolt in two steps.

   a. Lubricate thread and seat surface of the bolt with new engine oil.

   b. For the second step of angle tightening use the appropriate tool [Tool No. KV10112100 (BT-8653-A)].

   • Step 1: 32 ft. lbs. (44 Nm)

   • Step 2: 84–90 degrees clockwise

11. Remove the tool attached to the starter bolt hole.

12. Rotate crankshaft pulley in normal direction (clockwise when viewed from front) to confirm it turns smoothly.

13. Install the radiator fan.

14. Install the drive belts.

15. Install the engine undercover.

16. Connect the negative battery cable.

CRANKSHAFT FRONT SEAL

REMOVAL & INSTALLATION

See Figures 58 through 60.

1. Before servicing the vehicle, refer to the Precautions Section.

2. Disconnect the negative battery terminal.

3. Remove the engine undercover.

4. Remove the drive belts. Refer to Accessory Drive Belt Removal & Installation.

5. Remove the radiator fan. Refer to Engine Cooling Fan Removal & Installation in the Engine Cooling section.

6. Remove the crankshaft pulley. Refer to Crankshaft Damper Removal & Installation.

7. Remove the front oil seal from the front timing chain case.

To install:

8. Install the front oil seal on the front timing chain case using a suitable tool. Apply clean engine oil to the oil seal edges.

   a. Install it so that each seal lip is oriented as shown. Suitable drift:

   • Outer diameter: 2.32 in. (59mm)

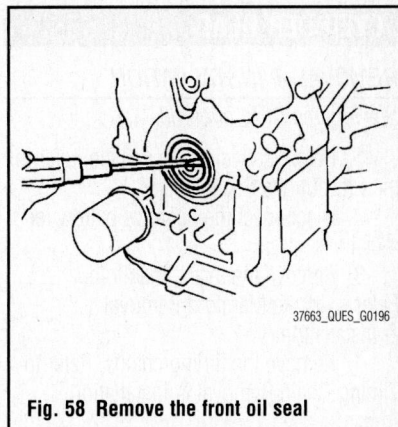

Fig. 58 Remove the front oil seal

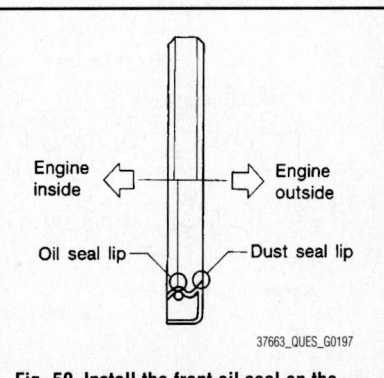

Fig. 59 Install the front oil seal on the front timing chain case

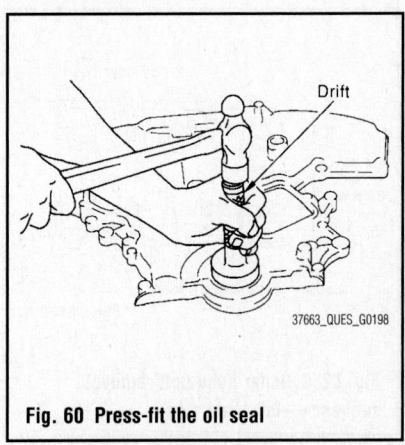

Fig. 60 Press-fit the oil seal

   • Inner diameter: 1.93 in. (49mm)

   b. Press-fit the oil seal until it becomes flush with timing chain case end face, using a suitable drift. Press fit straight and avoid causing burrs or tilting the oil seal.

   c. Make sure the garter spring in the oil seal is in position and seal lip is not inverted.

9. Install the crankshaft pulley.

10. Install the radiator fan.

11. Install the drive belts.

12. Install the engine undercover.

13. Connect the negative battery cable.

CYLINDER HEAD

REMOVAL & INSTALLATION

See Figures 61 through 68.

1. Before servicing the vehicle, refer to the Precautions Section.

2. Disconnect the negative battery terminal.

3. Remove the intake manifold. Refer to Intake Manifold Removal & Installation.

4. Remove the timing chains. Refer to Timing Chain Removal & Installation.

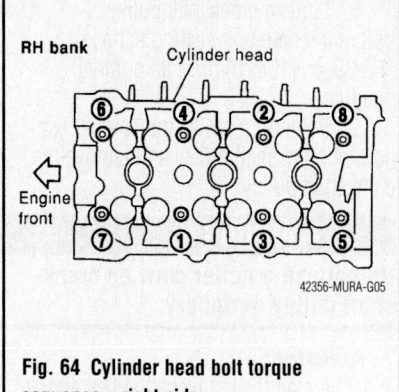

Fig. 64 Cylinder head bolt torque sequence—right side

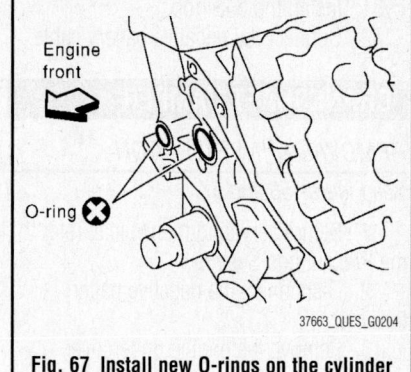

Fig. 67 Install new O-rings on the cylinder block

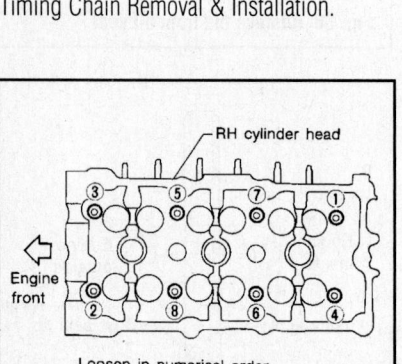

Fig. 61 Cylinder head bolt removal sequence—right-hand side

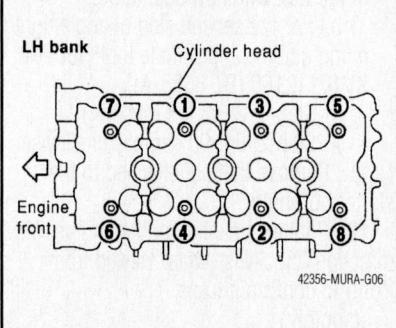

Fig. 65 Cylinder head bolt torque sequence—left side

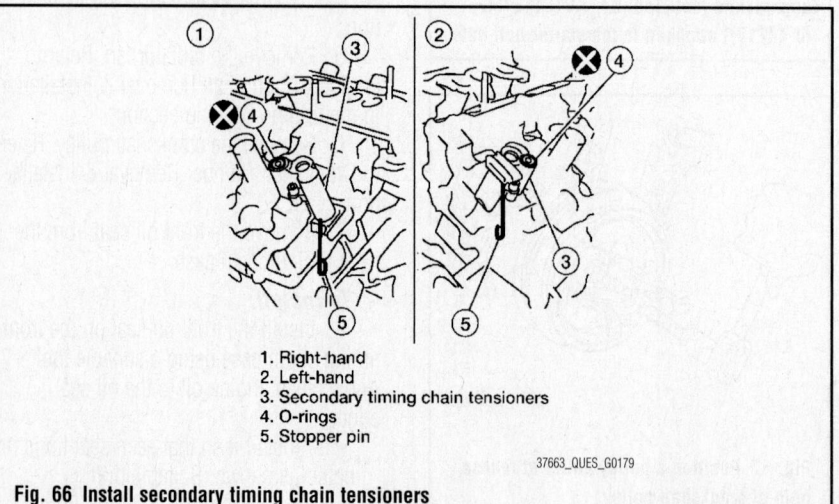

Fig. 68 Install (1) right-hand and left-hand (2) new O-rings (3) on the cylinder heads

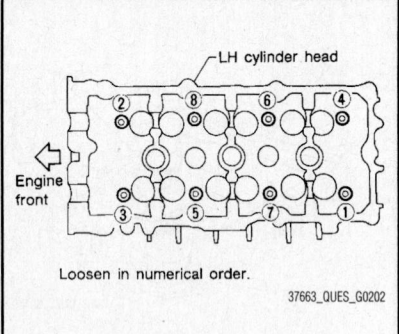

Fig. 62 Cylinder head bolt removal sequence—left-hand side

5. Remove the three way catalysts (manifolds), exhaust manifold heat shields and right-hand exhaust manifold. Refer to Combination Manifold Removal & Installation.

6. Remove the intake and exhaust camshafts and the camshaft brackets. Mark the camshaft brackets so they are placed in the same position and direction for installation. Refer to Camshaft & Valve Lifters, Removal & Installation.

7. Remove the water outlet. Refer to

Hoses, Removal & Installation in the Engine Cooling section.

8. Remove the right-hand and left-hand cylinder head bolts. The bolts should be loosened gradually in three steps in the order as shown.

9. Remove the cylinder heads and gaskets. Discard the cylinder head gaskets and use new gaskets for installation.

10. If necessary, remove the left-hand exhaust manifold. Refer to Combination Manifold Removal & Installation.

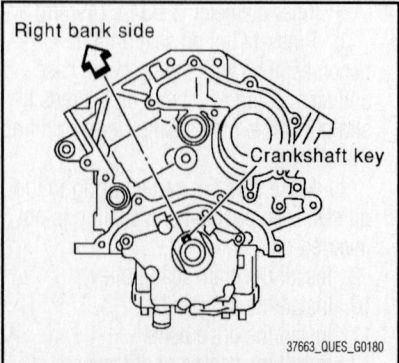

Fig. 63 The crankshaft key should line up with the right bank cylinder center line

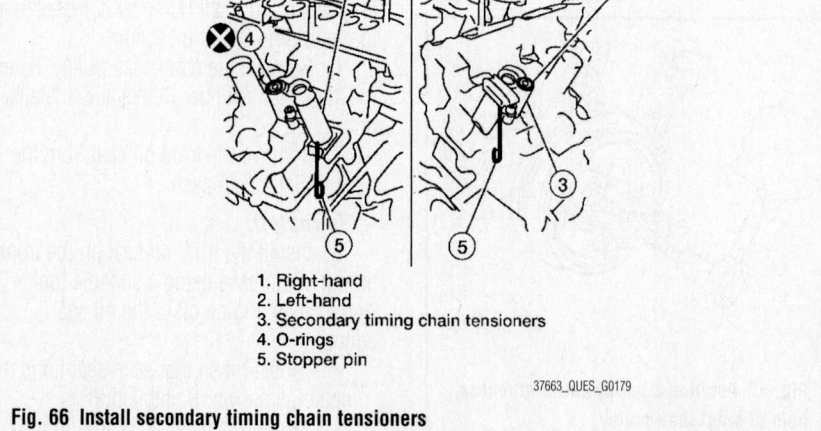

1. Right-hand
2. Left-hand
3. Secondary timing chain tensioners
4. O-rings
5. Stopper pin

Fig. 66 Install secondary timing chain tensioners

To install:

11. Turn the crankshaft until No. 1 piston is set at TDC on the compression stroke. The crankshaft key should line up with the right bank cylinder center line as shown.

12. Install new gaskets on the cylinder heads.

✳✳ CAUTION

Do not rotate crankshaft and camshaft separately or valves will strike piston heads.

13. If applicable, install the left-hand exhaust manifold.

14. Install the cylinder heads on the cylinder block. Tighten the cylinder head bolts in five steps in the order as shown using the appropriate tool [Tool No. KV10112100 (BT-8653-A)].
- Step a: 72 ft. lbs. (98 Nm)
- Step b: Loosen in the reverse order of tightening.
- Step c: 29 ft. lbs. (39.2 Nm)
Step d: 90° clockwise
Step e: 90° clockwise

15. Install the water outlet.

16. Install secondary timing chain tensioners on the right-hand and left-hand cylinder heads.

17. Install the exhaust and intake camshafts and camshaft brackets.

18. Install the IVT control solenoid valves with new gaskets.

19. Install new O-rings on the cylinder block.

20. Install right-hand and left-hand new O-rings on the cylinder heads.

21. Install the right-hand exhaust manifold, three way catalysts (manifolds) and exhaust manifold heat shields.

22. Install the timing chain.

23. Install the intake manifold.

24. Connect the negative battery cable.

EXHAUST MANIFOLD

REMOVAL & INSTALLATION

See Combination Manifold.

FLYWHEEL/DRIVE PLATE

REMOVAL & INSTALLATION

See Figures 69 and 70.

1. Before servicing the vehicle, refer to the precautions section.

2. Disconnect the negative and positive battery cables.

3. Remove the engine assembly and separate the transaxle.

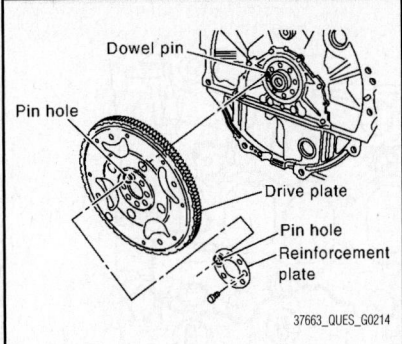

Fig. 69 Drive plate and reinforcement plate

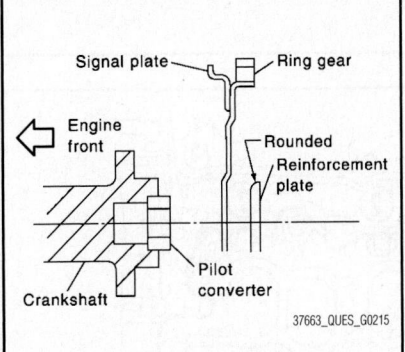

Fig. 70 Install the drive plate and reinforcement plate in the direction shown

4. Matchmark the drive plate position to the crankshaft dowel pin prior to removal to assist in installation.

5. Remove the drive plate. Secure the drive plate using the appropriate tool [Tool No. KV10117700 (J-44716)], and remove drive plate bolts. Loosen the bolts in diagonal order.

✳✳ WARNING

Do not disassemble the drive plate. Never place the drive plate with the signal plate facing down. When handling the signal plate, take care not to damage or scratch it. Handle the signal plate in a manner that prevents it from becoming magnetized.

To install:

6. Install the drive plate and reinforcement plate, aligning the matchmark and tighten the bolts to 65 ft. lbs. (88 Nm).
a. Ensure the dowel pin is installed in the crankshaft.
b. When installing the drive plate to crankshaft, be sure to correctly align crankshaft side dowel pin and drive plate side dowel pin hole.
c. Install the drive plate and reinforcement plate in the direction shown.

7. Secure the drive plate using the appropriate tool [Tool No. KV10117700 (J-44716)]. Tighten the drive plate bolts crosswise several times using a suitable tool.

8. Install the engine assembly.

9. Refill fluids and check for leaks.

INTAKE MANIFOLD

REMOVAL & INSTALLATION

See Figures 71 through 73.

1. Before servicing the vehicle, refer to the precautions section.

2. Relieve the fuel system pressure.

3. Remove the intake manifold upper and lower collector.

4. Remove the fuel rail with the fuel injectors.

5. Loosen the intake manifold nuts and bolts in the sequence illustrated.

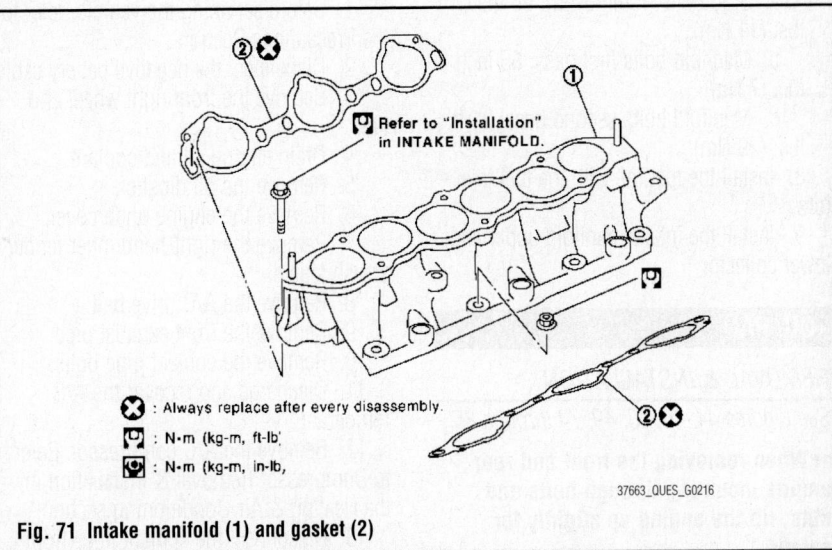

Fig. 71 Intake manifold (1) and gasket (2)

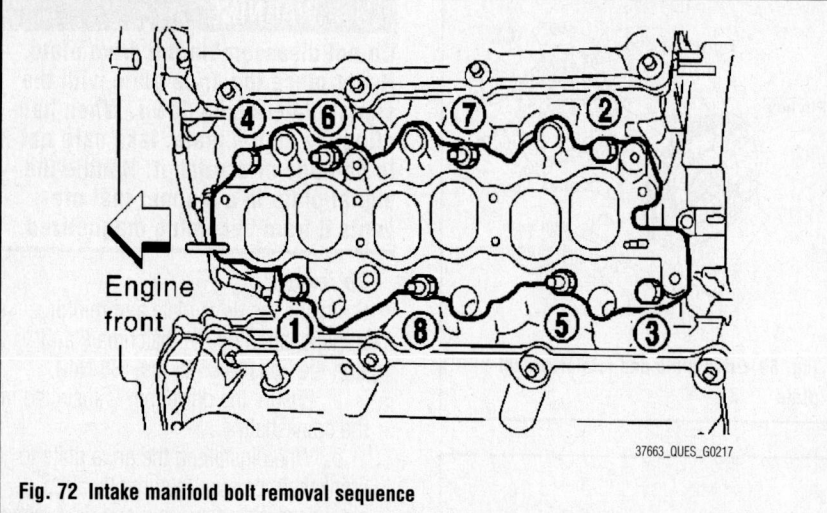

Fig. 72 Intake manifold bolt removal sequence

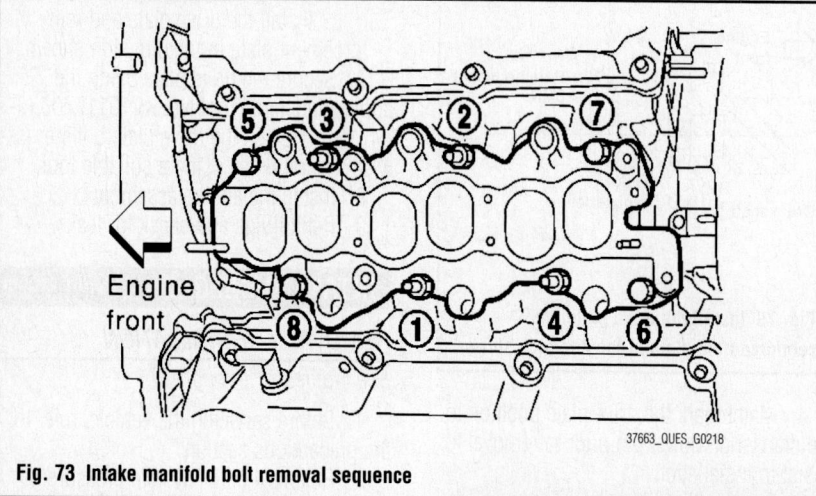

Fig. 73 Intake manifold bolt removal sequence

6. Remove the intake manifold.

To install:

7. Install the intake manifold. Tighten the intake manifold nuts and bolts in the sequence illustrated in the following steps:_

    a. Stud bolts (if removed): 96 inch lbs. (11 Nm)

    b. Manifold bolts first pass: 65 inch lbs. (7 Nm)

    c. Manifold bolts second pass: 21 ft. lbs. (29 Nm)

8. Install the fuel rail with the fuel injectors.

9. Install the intake manifold upper and lower collector.

OIL PAN

REMOVAL & INSTALLATION

See Figures 44, 46, 48, 49, 74 through 85.

➡**When removing the front and rear engine mounting through bolts and nuts, lift the engine up slightly for safety.**

✳✳ CAUTION

When removing the upper oil pan from the engine, first remove the Crankshaft Position (CKP) sensor.

1. Before servicing the vehicle, refer to the Precautions Section.

2. Disconnect the negative battery cable.

3. Remove the front right wheel and tire.

4. Drain engine oil and coolant.

5. Remove the oil dipstick.

6. Remove the engine undercover.

7. Remove the right-hand inner fender splash shield.

8. Remove the A/C drive belt.

9. Remove the front exhaust pipe.

10. Remove the coolant pipe bolts.

11. Discharge and recover the A/C refrigerant.

12. Remove the A/C compressor. Refer to Compressor Removal & Installation in the Heating & Air Conditioning section.

13. Disconnect the coolant lines from the engine oil cooler and plug the lines them to prevent coolant loss.

14. Remove the oil filter and the engine oil cooler from the upper oil pan.

15. Remove the oil pressure switch and the CKP sensor from the upper oil pan.

16. Remove the front halfshafts. Refer to Halfshaft Removal & Installation in the Drive Train section.

17. Remove the front suspension member as follows:

    a. Remove the left-hand transaxle mounting insulator nuts.

    b. Attach an engine lifting bracket to the transaxle at the location illustrated.

    c. Install an engine support tool. Make sure the tool is securely resting on the hood ledge.

    d. Remove the three transaxle mounting insulator nuts.

    e. Remove the lower ball joint bolt and separate the transverse link from the steering knuckle.

    f. Remove the power steering line bracket.

    g. Remove the mounting bolts from the lower side of the steering gear.

    h. Disconnect the front engine mount electrical connector.

    i. Disconnect the connecting rod from the front strut.

    j. Place a transmission jack under the front suspension member and remove the mounting nuts from the front suspension member.

    k. Remove the front suspension member bolts from the pin stay on the vehicle body side.

    l. Remove the through bolts from the front and rear engine mounts.

    m. Lower the transmission jack to remove the front suspension member. It may be necessary to remove the exhaust hanger bracket, the front and rear engine mounts, the transverse link, and the dynamic damper to enable removal.

18. Disconnect the Heated Oxygen Sensors (HO2S) and Air Flow Ratio (AFR) sensors and remove the two three way catalysts from the exhaust manifolds. Refer to Combination Manifold Removal & Installation.

19. Remove the rear cover plate from the upper oil pan.

20. Loosen the lower oil pan bolts in the order shown.

21. Remove the lower oil pan:

    a. Insert a removal tool [Tool No. KV10111100 (J-37228)] between the lower oil pan and the upper oil pan.

    b. After removing the bolts and nuts, separate the mating surface using the removal tool and remove the sealant.

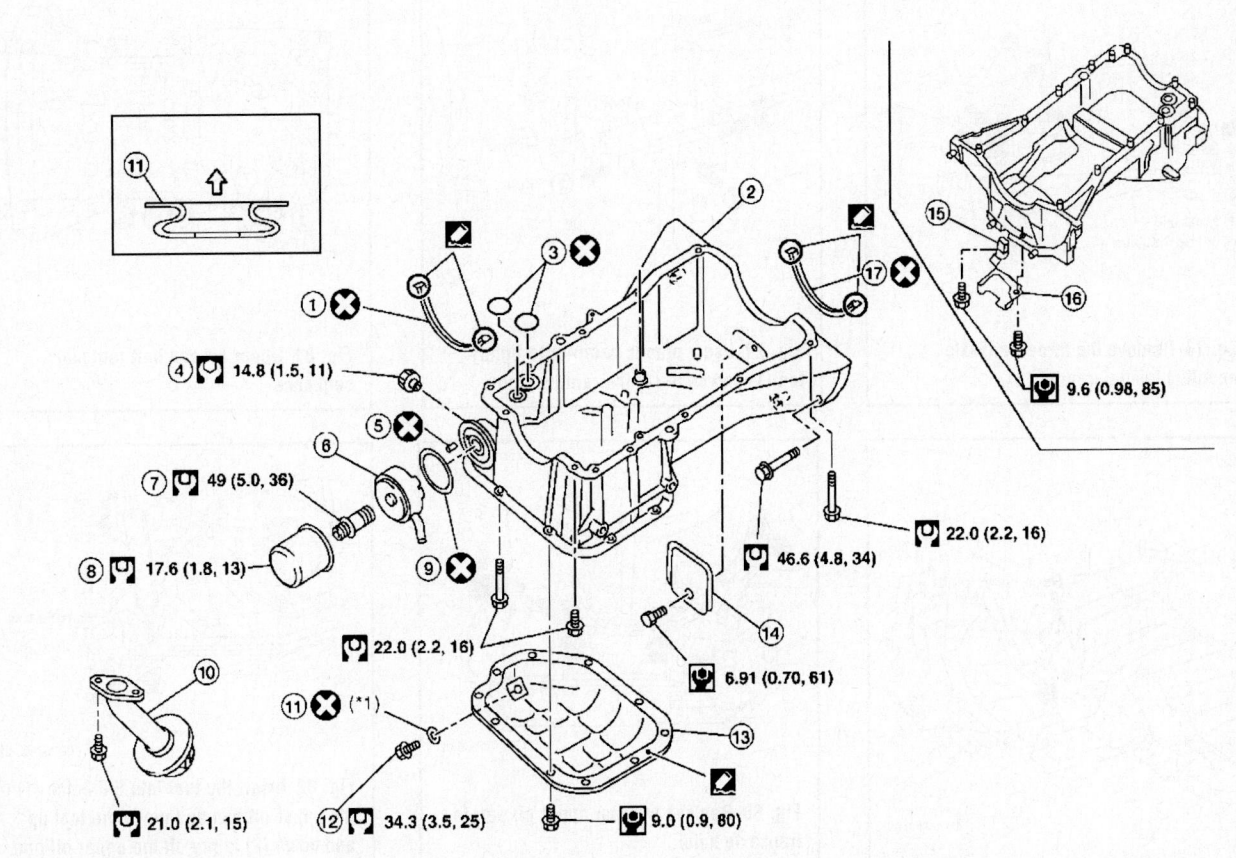

1. Front cover gasket
4. Oil pressure switch
7. Oil cooler connection
10. Oil strainer (connects to oil pump body)
13. Lower oil pan
16. Crankshaft position sensor (POS) shield

2. Upper oil pan
5. Relief valve
8. Oil filter
11. Gasket (⇐ oil pan side)
14. Rear cover plate
17. Rear oil seal retainer gasket

3. O-ring
6. Oil cooler
9. Gasket
12. Drain plug
15. Crankshaft position sensor (POS)
⇐ Oil pan side

37663_QUES_G0219

Fig. 74 Oil pan and components

c. In areas where the removal tool is difficult to use, use a plastic hammer to lightly tap the removal tool where the Silicone RTV Sealant is applied. Use a plastic hammer to slide the removal tool by tapping on the side.

※※ WARNING

If for some unavoidable reason a tool such as a flat blade screwdriver is used, be careful not to damage the mating surfaces.

22. Remove the four upper oil pan-to-transaxle bolts.
23. Remove the upper oil pan:
   a. Loosen the bolts in the order shown.
   b. Insert an appropriate size tool into the notch of the upper oil pan as shown.
   c. Pry off the upper oil pan by moving the tool up and down as illustrated.
24. If re-installing the original oil pan, remove the old sealant from the oil pan, cylinder block, and bolt holes and threads.

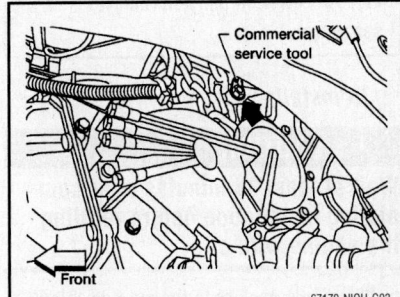

67170-NIQU-G03

Fig. 75 Attach a engine lifting bracket to the transaxle

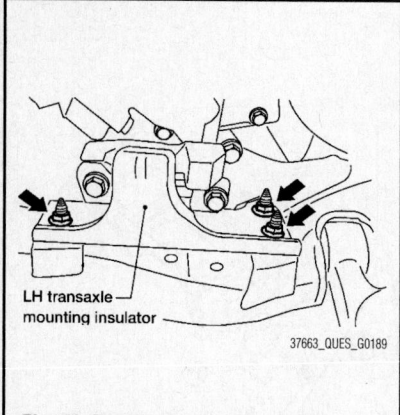

Fig. 76 Remove the three transaxle mounting insulator nuts

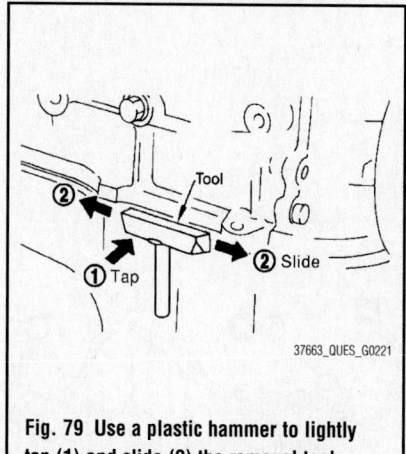

Fig. 79 Use a plastic hammer to lightly tap (1) and slide (2) the removal tool

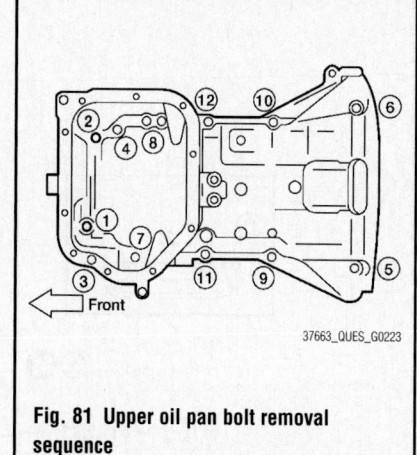

Fig. 81 Upper oil pan bolt removal sequence

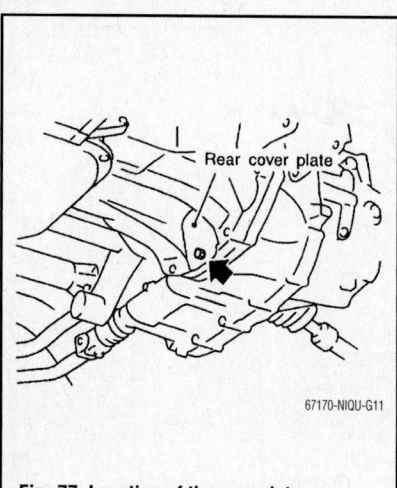

Fig. 77 Location of the rear plate

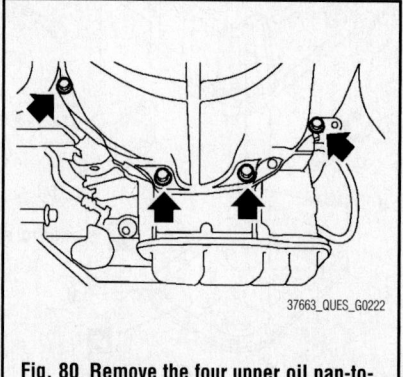

Fig. 80 Remove the four upper oil pan-to-transaxle bolts

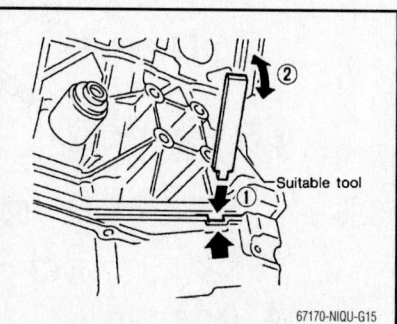

Fig. 82 Insert the tool into the notch (1) of the upper oil pan and move the tool up and down (2) to pry off the upper oil pan

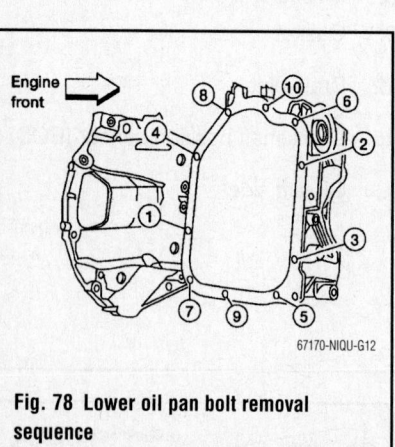
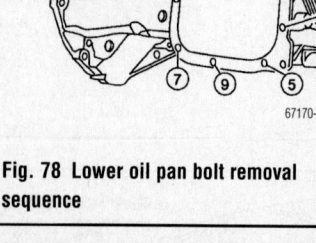

Fig. 78 Lower oil pan bolt removal sequence

To install:

❈❈ CAUTION

Wait at least 30 minutes after completing installation before refilling the engine with oil.

25. Apply sealant to the cylinder block mating surface of the upper oil pan as illustrated. Attach the pan within 5 minutes of applying the sealant.

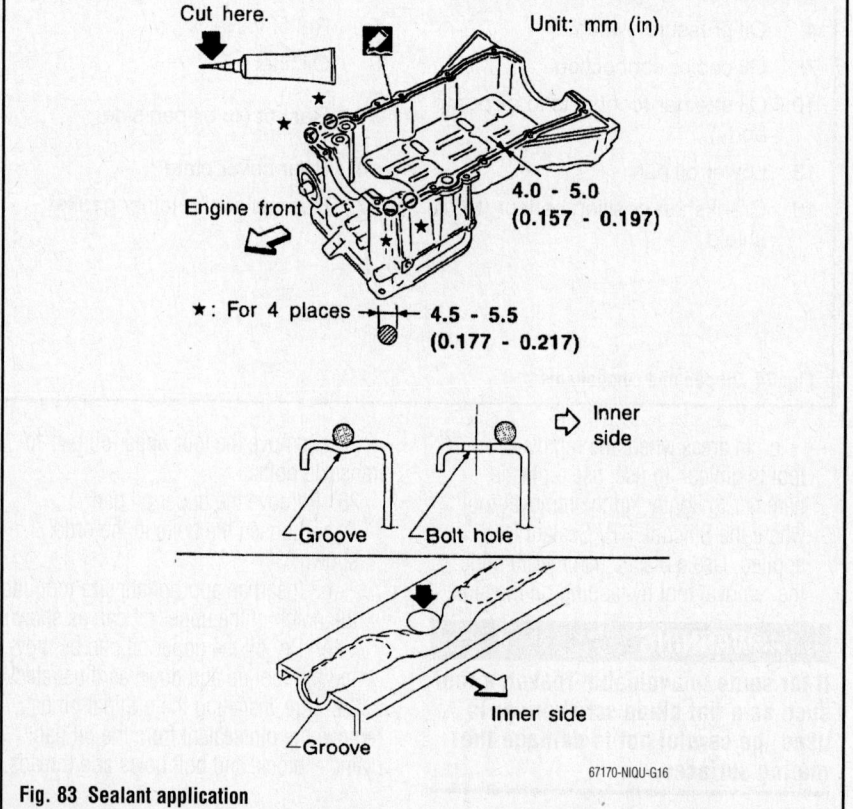

Fig. 83 Sealant application

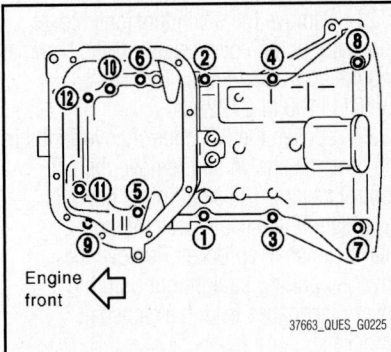

Fig. 84 Upper oil pan bolt tightening sequence

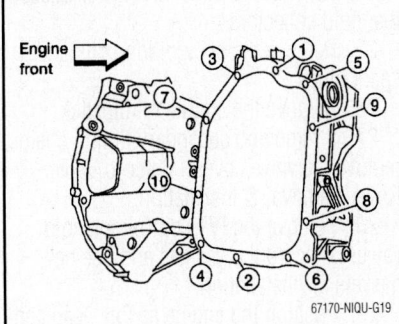

Fig. 85 Lower oil pan bolt tightening sequence

26. Install new O-rings on the cylinder block and oil pump body.

27. Install the upper oil pan. Tighten upper oil pan bolts in the order shown.

28. Install the four upper oil pan to transaxle bolts.

29. Apply sealant to the lower oil pan. Attach the pan within 5 minutes of applying the sealant.

30. Install the lower oil pan. Tighten the lower oil pan bolts in the sequence shown.

31. Install the rear plate cover.

32. Connect the heated oxygen sensors and air flow ratio (A/F) sensors and install the two three way catalysts manifold from the exhaust manifolds.

33. Install the front suspension member.

34. Install the front halfshafts.

35. Install the oil pressure switch and the CKP sensor.

36. Install the oil filter and the engine oil cooler and connect the coolant lines.

37. Install the A/C compressor.

38. Install the coolant pipe bolts.

39. Install the front exhaust pipe.

40. Install the A/C drive belt.

41. Install the right-hand inner fender splash shield.

42. Install the engine undercover.

43. Install the oil dipstick.

44. Install the wheel and tire.

❊❊ WARNING

Wait at least 30 minutes before refilling the engine with oil.

45. Refill engine oil and coolant and recharge the A/C refrigerant.

46. Connect the negative battery cable.

47. Start the engine and check for leaks.

48. Inspect the engine oil level.

OIL PUMP

REMOVAL & INSTALLATION

See Figure 86.

1. Before servicing the vehicle, refer to the Precautions Section.

2. Remove the engine.

3. Remove the timing chain. Refer to Timing Chain & Sprockets, Removal & Installation.

4. Remove the oil strainer. Refer to Oil Pan Removal & Installation.

5. Remove oil pump assembly.

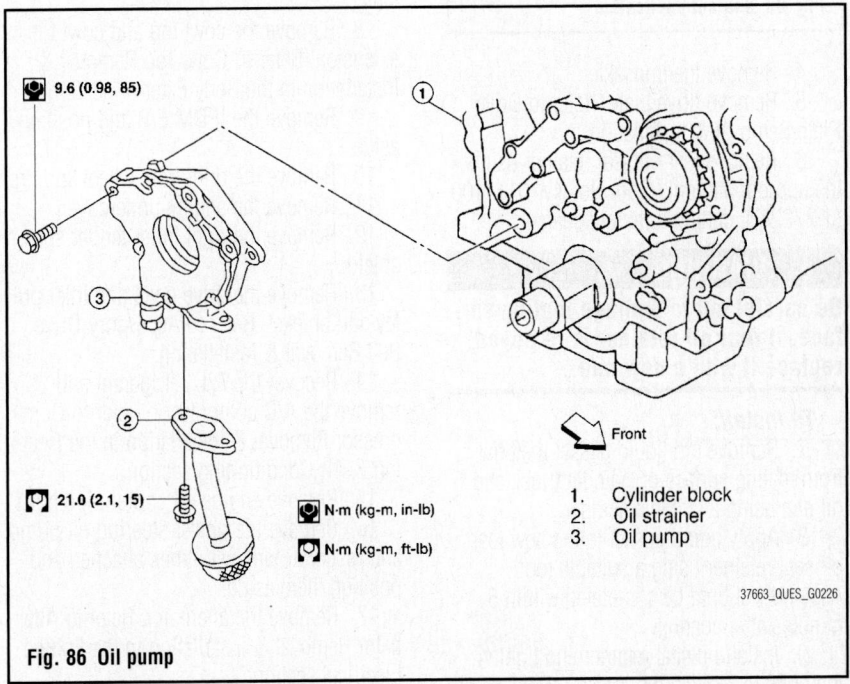

Fig. 86 Oil pump

To install:

6. Installation is the reverse of the removal procedure.

7. Be sure to use new gaskets.

8. When installing, align crankshaft flat faces with inner rotor flat faces.

PISTON AND RING

POSITIONING

See Figure 87.

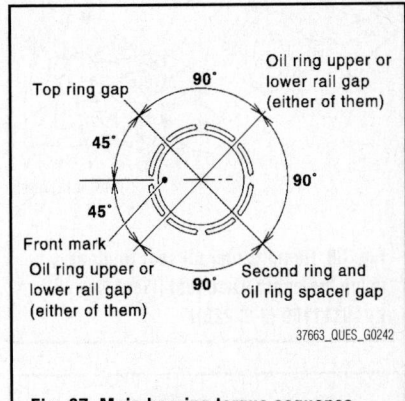

Fig. 87 Main bearing torque sequence

REAR MAIN SEAL

REMOVAL & INSTALLATION

See Figures 88 and 89.

➡**Rear oil seal is replaced with the rear oil seal retainer and must be replaced as an assembly.**

1. Before servicing the vehicle, refer to the Precautions Section.

2. Disconnect the negative battery terminal.

3. Remove upper oil pan. Refer to Oil Pan Removal & Installation.

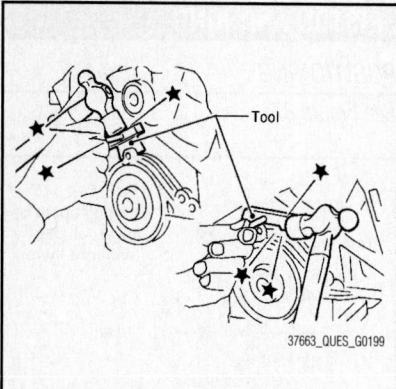

Fig. 88 Remove rear oil seal retainer using the appropriate tool [(Tool No. KV10111100 (J-37228)]

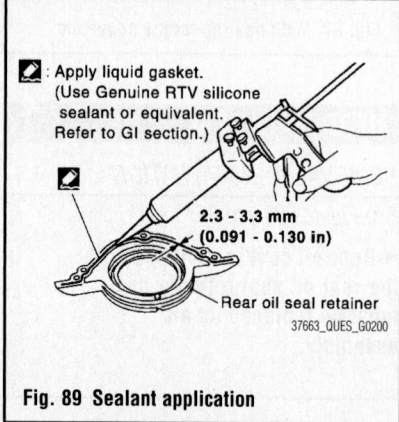

Fig. 89 Sealant application

4. Remove the transaxle.

5. Remove drive plate. Refer to Drive Plate Removal & Installation.

6. Remove rear oil seal retainer using the appropriate tool [(Tool No. KV10111100 (J-37228)], cutting away old sealant.

✳✳ WARNING

Be careful not to damage mating surface. If rear oil retainer is removed, replace it with a new one.

To install:

7. Remove old liquid gasket material from mating surface of cylinder block and oil pan using a suitable scraper.

8. Apply liquid gasket to the new rear oil seal retainer using a suitable tool. Assembly should be completed within 5 minutes after coating.

9. Install oil seal retainer and tighten the bolts to 78 inch lbs. (9 Nm). Replace oil seal and retainer as an assembly.

10. Install the drive plate.

11. Install the transaxle.

12. Install the upper oil pan.

13. Connect the negative battery cable.

TIMING CHAIN FRONT COVER

REMOVAL & INSTALLATION

See Figures 90 through 108.

This section describes procedures for removal/installation procedure of the front timing chain case and timing chain related parts without removing the upper oil pan from the vehicle.

When upper oil pan needs to be removed or installed, or when rear timing chain case is removed or
installed, remove upper and lower oil pans first. Then remove front timing chain case, timing chain related
parts, and rear timing chain case in this order, and install in reverse order of removal.

1. Before servicing the vehicle, refer to the Precautions Section.

2. Disconnect the negative battery cable.

3. Drain the cooling system.

4. Drain the engine oil.

5. Remove the engine cover.

6. Remove the upper air cleaner case, mass air flow sensor and air cleaner to electric throttle control actuator tube. Refer to Air Cleaner Removal & Installation.

7. Remove the engine coolant reservoir tank.

8. Remove the cowl top and cowl top extension. Refer to Cowl Top Removal & Installation in the Body Exterior section.

9. Remove the IPDM E/R and position aside.

10. Remove the right front wheel and tire.

11. Remove the engine undercover.

12. Remove the right inner fender splash shield.

13. Remove the drive belts and idler pulley and bracket. Refer to Accessory Drive Belt Removal & Installation.

14. Recover the A/C refrigerant and remove the A/C compressor. Refer to Compressor Removal & Installation in the Heating & Air Conditioning section.

15. Remove engine oil cooler pipe bolts.

16. Remove the power steering oil pump and reservoir tank with lines attached and position them aside.

17. Remove the alternator. Refer to Alternator Removal & Installation in the Engine Electrical section.

18. Disconnect the engine harness and position aside.

19. Remove the A/C low-pressure flexible hose.

20. Support the engine and remove the right-hand engine mounting insulator, mount and bracket.

21. Remove the chain tensioner cover and water pump cover using plastic hammer and the appropriate tool (Tool No. KV10111100 (J-37228).

22. Loosen the IVT control cover bolts in the order as shown and remove the IVT control covers. The shaft in the cover is inserted into the center hole of the intake camshaft sprocket. Remove the cover by pulling straight out until the cover disengages from the camshaft sprocket.

23. Remove the starter motor. Refer to Starter Removal & Installation in the Engine Electrical section.

24. Remove the upper and lower intake manifold collectors.

25. Remove the six ignition coils and spark plugs.

26. Remove the engine oil dipstick.

27. If removing secondary timing chains, remove the valve covers. Refer to Valve Cover Removal & Installation.

28. Remove the IVT control solenoid valves. Discard the gaskets and use new gaskets for installation.

29. Position the engine at Top Dead center (TDC) of No. 1 cylinder as follows:

a. Rotate crankshaft pulley clockwise to align the timing mark (grooved line

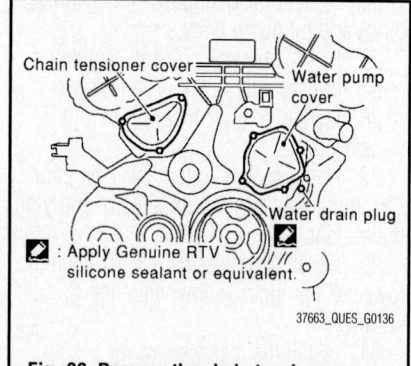

Fig. 90 Remove the chain tensioner cover and water pump cover

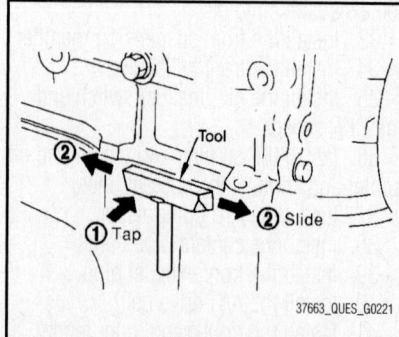

Fig. 91 Use a plastic hammer to lightly tap (1) and slide (2) the removal tool

without color) with the timing indicator.

b. Check that intake and exhaust camshaft lobes on No. 1 cylinder (right bank of engine) are located as shown. If not, turn the crankshaft one revolution (360 degrees) and align as illustrated.

30. Lock the ring gear using the appropriate tool [Tool No. KV10117700 (J-44716)] attached to the starter bolt hole.

✳✳ WARNING

Do not damage the ring gear teeth, or the signal plate teeth behind the ring gear when setting the tool.

31. Remove the crankshaft pulley. Refer to Crankshaft Damper Removal & Installation.

32. Remove the lower oil pan. Refer to Oil Pan Removal & Installation.

33. Loosen the upper oil pan front bolts in the order as shown.

34. Temporarily install lower oil pan.

35. Support the front of engine under oil pan using a jack.

36. Remove the front timing chain case:

a. Loosen the front timing chain case bolts in the order as shown.

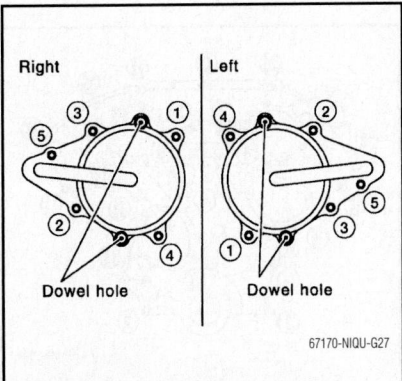

Fig. 92 IVT control cover bolt removal sequence

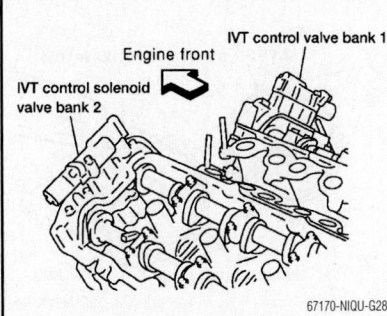

Fig. 93 Remove the IVT control solenoid valves

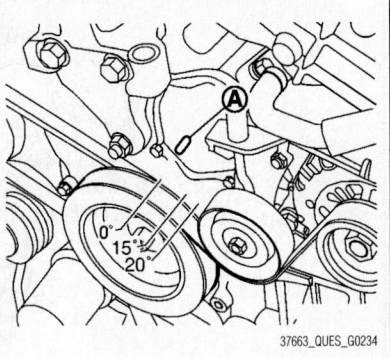

Fig. 94 Rotate crankshaft pulley clockwise to align the timing mark (grooved line without color) with the timing indicator (A)

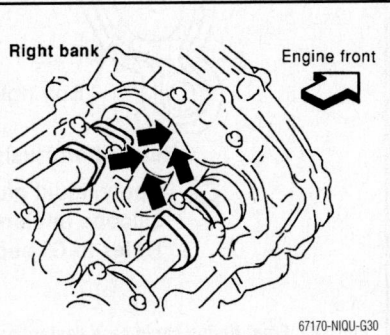

Fig. 95 Intake and exhaust camshaft lobes positioning

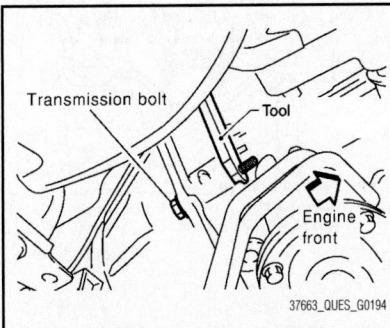

Fig. 96 Lock the ring gear using the appropriate tool [Tool No. KV10117700 (J-44716)] attached to the starter bolt hole

b. Insert an appropriate tool into the notch at the top of the front timing chain case to pry it loose, as shown.

✳✳ WARNING

Do not use a screwdriver or similar tool. After removal, handle carefully so it does not bend or warp under a load.

37. Remove the water pump cover and chain tensioner cover from the front timing

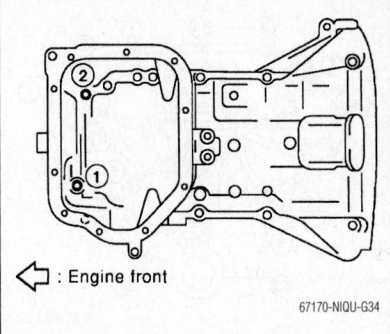

Fig. 97 Upper oil pan bolt removal sequence

chain case using the appropriate tool [Tool No. KV10111100 (J-37228)]. Do not insert a screwdriver, this will damage the mating surfaces.

38. Remove the front oil seal from the front timing chain case using a suitable tool being careful not to damage the front cover.

39. If necessary, remove timing chain and related parts.

40. Remove O-rings and seal rings from the front and rear timing chain case.

41. Use a scraper to carefully remove all of the old sealant from the front timing chain mating surfaces.

To install:

42. Install O-rings and seals to the front and rear timing chain case.

43. Install the timing chain and related parts, if removed.

44. Install the left and right dowel pins into the front timing chain case up to a point close to taper in order to shorten protrusion length.

45. Install the front oil seal on the front timing chain case. Refer to Crankshaft Front Seal Removal & Installation

46. Apply sealant to front timing chain case as shown. Make sure to wipe off the protruding sealant prior to installation.

47. Install the dowel pin on the rear timing chain case into the dowel pin hole in the front timing chain case.

48. Apply sealant to the top surface of the upper oil pan as illustrated.

49. Install the front timing chain case:

a. Install the lower end of the front timing chain case tightly onto the top surface of the upper oil pan. Make sure that the oil pan gasket is in place.

b. While pressing the front timing chain case from its front and top as shown, hammer the dowel pin until the outer end becomes flush with the surface. Install dowel pin on the rear timing chain case into dowel pin hole in front timing chain case.

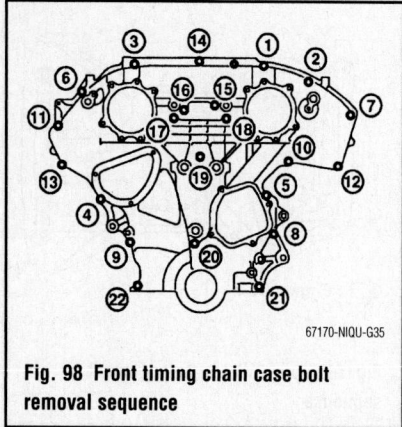

Fig. 98 Front timing chain case bolt removal sequence

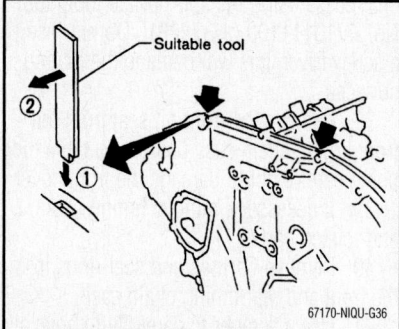

Fig. 99 Insert an appropriate tool into the notch (1) at the top of the front timing chain case to pry (2) it loose

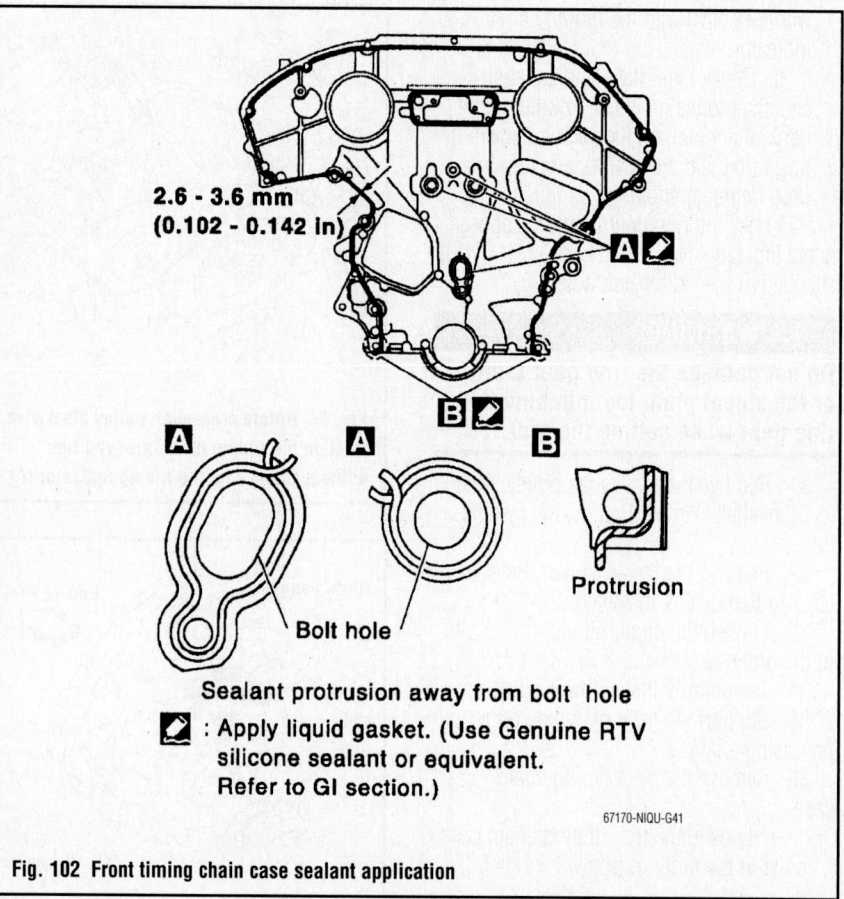

2.6 - 3.6 mm (0.102 - 0.142 in)

Bolt hole

Sealant protrusion away from bolt hole

Protrusion

: Apply liquid gasket. (Use Genuine RTV silicone sealant or equivalent. Refer to GI section.)

Fig. 102 Front timing chain case sealant application

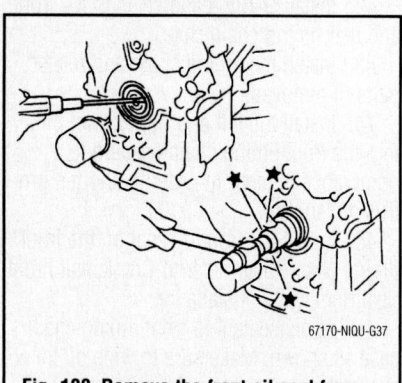

Fig. 100 Remove the front oil seal from the front timing chain case

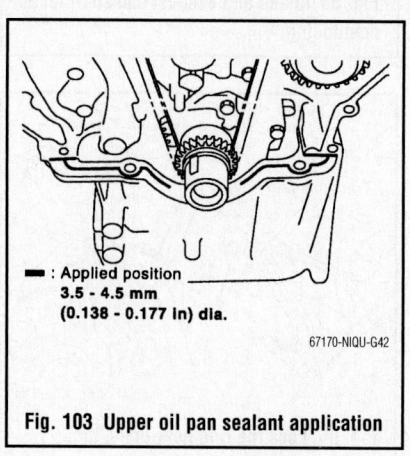

: Applied position
3.5 - 4.5 mm
(0.138 - 0.177 in) dia.

Fig. 103 Upper oil pan sealant application

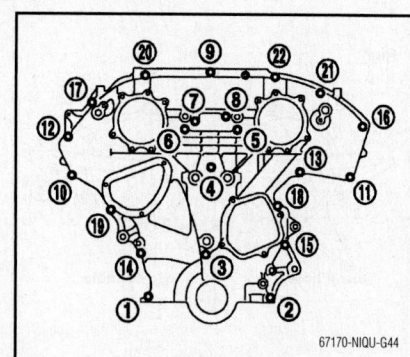

Fig. 105 Front timing chain case bolt tightening sequence

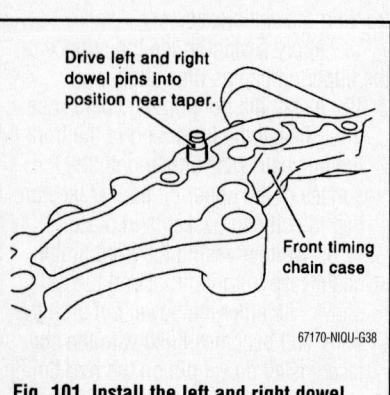

Drive left and right dowel pins into position near taper.

Front timing chain case

Fig. 101 Install the left and right dowel pins into the front timing chain case

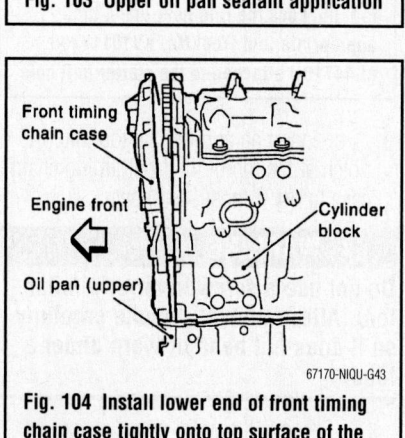

Front timing chain case

Engine front

Oil pan (upper)

Cylinder block

Fig. 104 Install lower end of front timing chain case tightly onto top surface of the upper oil pan

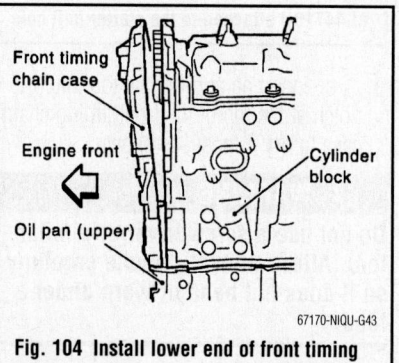

2.1 - 3.1 mm (0.083 - 0.122 in) dia.

Seal ring

Identification code

Seal ring

Fig. 106 IVT control cover sealant application

50. Loosely install the front timing chain case bolts.

51. Tighten the front timing chain case bolts in the order shown.
- Bolt 1–2, diameter 0.31 in. (8mm): 21 ft. lbs. (29 Nm)
- Bolt 3–22, diameter 0.24 in. (6mm): 9 ft. lbs. (13 Nm)

52. Install the lower oil pan.

53. Install the upper oil pan front bolts and tighten to 13 ft. lbs. (17 Nm).

54. Install the IVT control solenoid valves and new gaskets.

55. Install the IVT control valve covers as follows:

a. Install new collared O-rings in the front cover oil hole on both sides.

b. Install new seal rings on the IVT control covers.

c. Apply sealant to the IVT control covers as shown.

d. Being careful not to move the seal ring from the installation groove, align the dowel pins on the chain case with the holes to install the IVT control covers.

e. Tighten the intake valve timing control cover bolts in the order illustrated to 100 inch lbs. (11 Nm).

56. Apply sealant to the water pump cover and the chain tensioner cover and

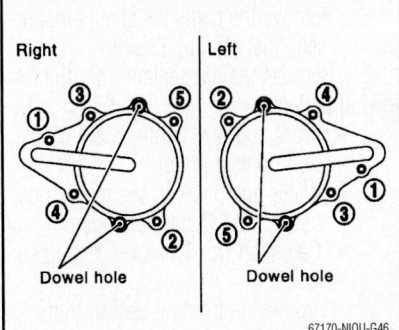

Fig. 107 Intake valve timing control cover bolt tightening sequence

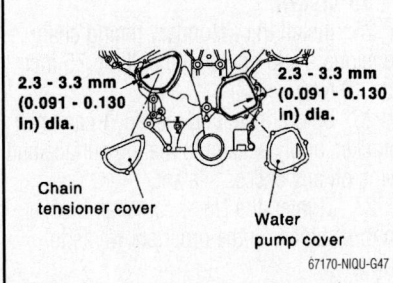

Fig. 108 Water pump cover and the chain tensioner cover sealant application

install the covers. Tighten the bolts on the pump and cover to 97 inch lbs. (11 Nm).

57. Install the crankshaft pulley.

58. Remove the tool that was used to lock the ring gear.

59. Install the valve covers, if removed.

60. Install the engine dipstick.

61. Install the ignition coils and spark plugs.

62. Install the upper and lower intake manifold collectors.

63. Install the starter.

64. Install the right-hand engine mounting insulator, mount and bracket.

65. Install the A/C low-pressure flexible hose.

66. Connect the engine harness.

67. Install the alternator.

68. Install the power steering oil pump and reservoir tank

69. Install the engine oil cooler pipe bolts.

70. Install the A/C compressor.

71. Install the drive belts and idler pulley.

72. Install the right inner fender splash shield.

73. Install the engine undercover.

74. Install the right front wheel and tire.

75. Install the IPDM E/R.

76. Install the cowl top and cowl top extension.

77. Install the engine coolant reservoir tank.

78. Install the upper air cleaner case, mass air flow sensor and air cleaner to electric throttle control actuator tube.

79. Install the engine cover.

80. Refill engine coolant and engine oil.

81. Recharge A/C refrigerant.

82. Connect the negative battery cable.

83. Start the engine and check for leaks.

84. Inspect the engine oil level.

TIMING CHAIN & SPROCKETS

REMOVAL & INSTALLATION

See Figures 109 through 129.

1. Before servicing the vehicle, refer to the Precautions Section.

2. Disconnect the negative battery cable.

3. Before removing the upper oil pan, remove the crankshaft position sensor.

❋❋ WARNING

Be careful not to damage sensor edges. Do not spill engine oil or coolant on drive belts.

4. Remove the front timing chain case. Refer to Front Timing Chain Cover Removal & Installation.

5. If removing rear timing chain case, remove the engine.

6. If removing rear timing chain case, remove the upper and lower oil pans. Refer to Oil Pan Removal & Installation.

7. If removing timing chain, remove the lower oil pan and loosen the front 2 bolts from the upper oil pan. Refer to Oil Pan Removal & Installation.

8. Remove the Valve Covers. Refer to Valve Cover Removal & Installation.

9. Remove the internal chain guide.

10. Remove the timing chain tensioner and slack guide. Place matchmarks on the timing chain and sprockets to indicate the correct position of the components for installation.

a. Pull lever down and release plunger stopper tab. Plunger stopper tab can be pushed up to release (coaxial structure with lever).

b. Insert stopper pin into tensioner body hole to hold lever, and keep the tab released.

c. Insert plunger into tensioner body hole by pressing the slack guide.

d. Keep the slack guide pressed and hold it by pushing the stopper pin through the lever hole and body hole as shown.

e. Remove the timing chain tensioner installation bolts and remove the timing chain tensioner.

f. Remove slack guide installation bolt and the slack guide.

11. Remove primary timing chain and crankshaft sprocket.

❋❋ WARNING

After removing timing chain, do not turn the crankshaft and camshaft separately, or the valves will strike the pistons.

12. Attach a suitable stopper pin to the right and left secondary timing chain tensioners.

13. Remove the intake and exhaust camshaft sprocket bolts.

a. Matchmark the timing chain and camshaft sprockets for alignment during installation.

14. Remove the secondary timing chains with camshaft sprockets.

a. Rotate camshaft slightly, and loosen secondary timing chain on secondary timing chain tensioner side.

b. Insert metal or resin plate [0.020 in. (0.5mm)] into plunger between

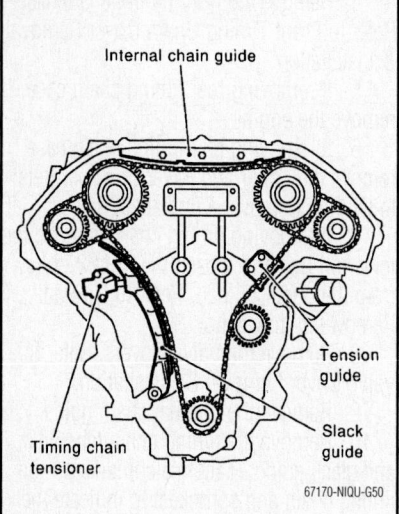

Fig. 109 Location of the internal chain guide

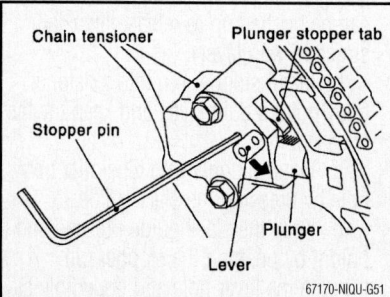

Fig. 110 Insert a stopper pin into the tensioner body hole to hold the chain tensioner lever—Allen wrench used as stopper pin for example

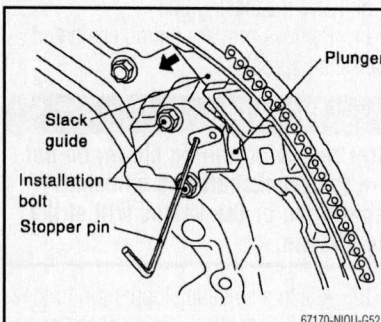

Fig. 111 Keep the slack side chain guide pressed and hold it by pushing a stopper pin

secondary timing chain and secondary timing chain tensioner plunger. Remove camshaft sprocket and secondary timing chain with secondary timing chain removed from plunger groove. Secondary timing chain tensioner plunger can move while stopper pin is inserted in timing chain tensioner. Plunger can come out of tensioner when

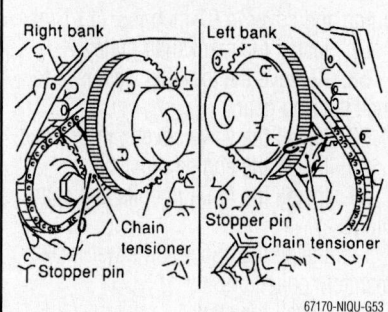

Fig. 112 Attach a suitable stopper pin to the right and left camshaft chain tensioners

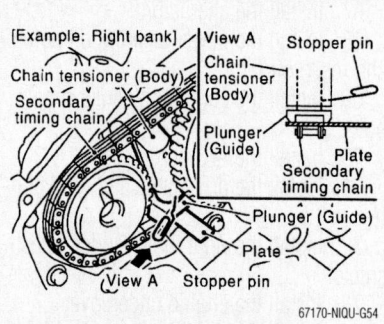

Fig. 113 Insert a 0.020 in. (0.5 mm) metal or resin plate into the guide between timing chain and chain tensioner plunger

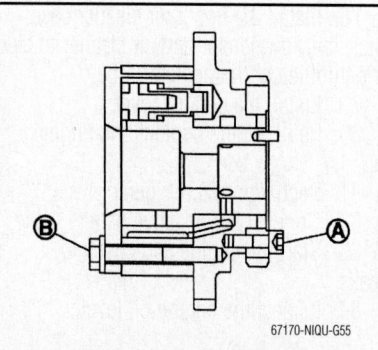

Fig. 114 Do not disassemble the intake sprockets (never loosen bolts A and B)

timing chain is removed. Use caution during removal.

➡ Intake camshaft sprocket is two-for-one structure of primary and secondary sprockets.

➡ Handle the intake sprockets as an assembly.

✳✳ WARNING

Avoid impact or dropping the intake sprockets.

✳✳ WARNING

Do not disassemble the intake sprockets (never loosen bolts A and B as shown).

15. Loosen the No. 1 camshaft bracket bolts in several steps in the order as shown and remove the No. 1 camshaft brackets.

16. Remove the timing chain tension guide.

17. Remove the rear timing chain case, if necessary. Do not remove the plate metal covers for the oil passage.

✳✳ WARNING

After removing the chain case, do not apply any load to the case that might bend it.

a. Loosen and remove the rear timing chain case bolts in the order shown.

b. Carefully remove the sealant using the appropriate tool [Tool No. KV10111100 (J-37228)], and remove the rear timing chain case.

18. Remove the engine coolant inlet and thermostat assembly.

19. Remove O-rings on the right-hand and left-hand No. 1 camshaft bracket, cylinder head and cylinder block.

20. If necessary, remove the water pump.

21. Remove the camshaft chain tensioners (for secondary timing chains).

22. Remove the old sealant from the following surfaces:
- Front and rear timing chain case
- Bolt holes and bolts
- Water pump cover, chain tensioner cover and IVT control covers
- Camshaft No. 1 bracket mating surface

23. Remove the front oil seal from the front timing chain case using an appropriate tool.

24. Check for cracks and any excessive wear at the roller links of the timing chain. Replace the timing chain as necessary.

To install:

25. Install the secondary timing chain tensioners and tighten the bolts to 75 inch lbs. (9 Nm).

26. Before installing the No. 1 camshaft bracket, apply sealant to mating surface and wipe off any excess sealant.

27. Tighten the No. 1 camshaft bracket in three steps, in the order shown as follows:
- Step 1: 17 inch lbs. (2 Nm).
- Step 2: 52 inch lbs. (6 Nm).
- Step 3: 92 inch lbs. (10 Nm).

28. Install the thermostat, gasket and

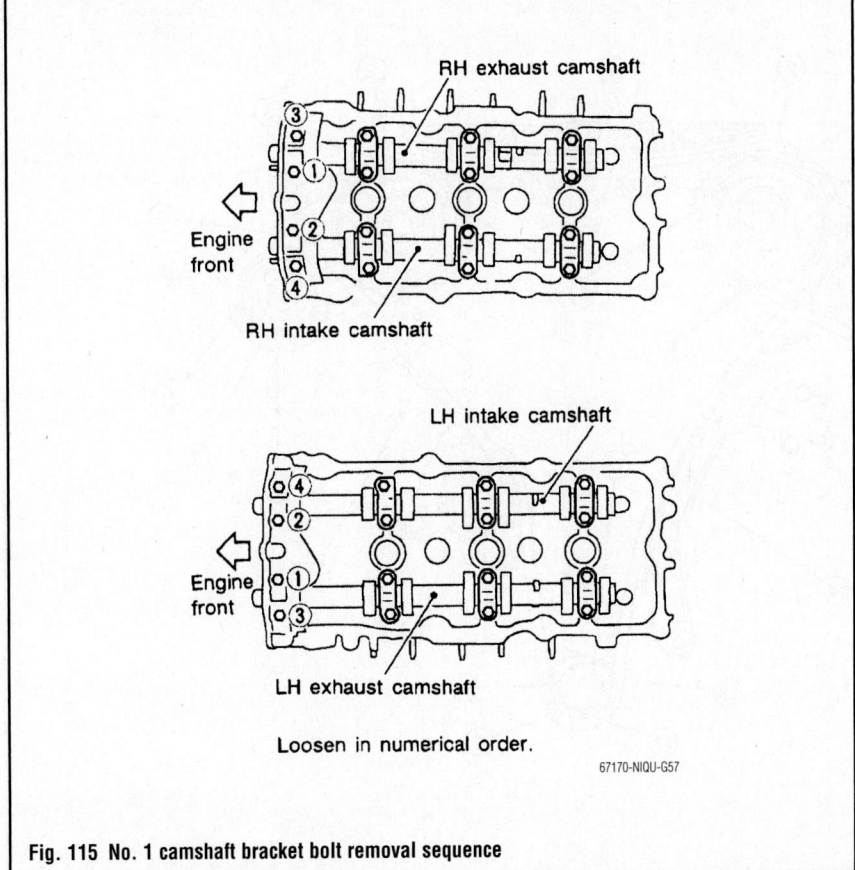

RH exhaust camshaft

Engine front

RH intake camshaft

LH intake camshaft

Engine front

LH exhaust camshaft

Loosen in numerical order.

67170-NIQU-G57

Fig. 115 No. 1 camshaft bracket bolt removal sequence

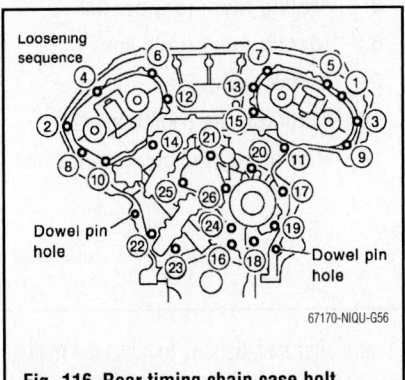

Loosening sequence

Dowel pin hole

Dowel pin hole

67170-NIQU-G56

Fig. 116 Rear timing chain case bolt removal sequence

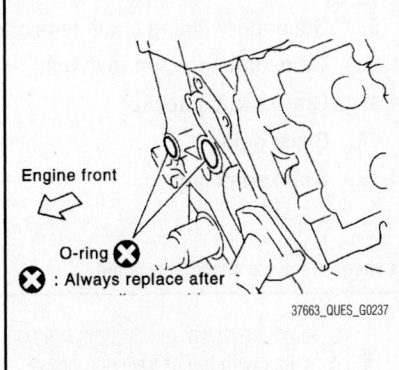

Engine front

O-ring

⊗ : Always replace after

37663_QUES_G0237

Fig. 118 Remove O-rings

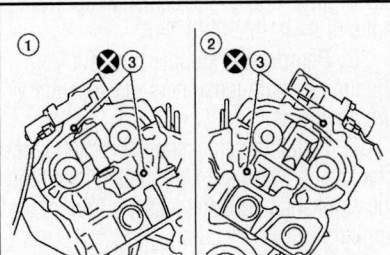

37663_QUES_G0236

Fig. 117 Remove O-rings (3) on the right-hand (1) and left-hand (2) No. 1 camshaft bracket, cylinder head and cylinder block

e. Tighten the rear timing chain case bolts in the order shown. After all the bolts are initially tightened, retighten them to 9 ft. lbs. (13 Nm). Bolt lengths:
- Bolts 1, 2, 3, 6, 7, 8, 9, 10: 0.79 inch (20 mm)
- Bolts 4, 5, 11–26: 0.63 inch (16 mm)

f. After installing the rear timing chain case, check the surface height difference between the rear timing chain case to cylinder block. The measurement should be 0.0094–0.0055 in. (0.24–0.14 mm). If not within specification, repeat the cover installation procedure.

30. Install the timing chain tension guide. Tighten the bolts to 16 ft. lbs. (22 Nm).

31. Position the crankshaft so the No. 1 piston is set at TDC on the compression stroke.

32. Make sure that the dowel pin hole, dowel pin and crankshaft key are located as shown.
- The camshaft dowel pin hole (intake side): at cylinder head upper face side in each bank
- The camshaft dowel pin (exhaust side): at cylinder head upper face side in each bank
- The crankshaft key: at cylinder head side of right-hand bank

✸✸ WARNING

The hole on small diameter side must be used for the intake camshaft sprocket dowel pin. Do not misidentify (ignore the big diameter side).

33. Install the secondary timing chains and camshaft sprockets as follows:

✸✸ WARNING

Matchmarks between the timing chain and sprockets can slip easily. Check all matchmark positions repeatedly during the installation process.

a. Push the sleeve of the secondary chain tensioner and keep it pressed in with a stopper pin.

b. Align the matchmarks on the secondary timing chain (gold link) with the ones on the intake and exhaust sprockets (stamped), and install them.

➡**Matchmarks for the intake sprocket are on the back side of the secondary sprocket. There are two types of matchmarks, round and oval types. The right hand banks use round type and the left hand bank use oval type.**

coolant inlet housing. Tighten the bolts to 87 inch lbs. (10 Nm).

29. Install rear timing chain case as follows:

a. Install new O-rings on the cylinder block and cylinder head.

b. Install the water pump, if removed.

c. Apply silicone sealant to the rear timing chain case as shown, cleaning off the protruding sealant prior to installation.

d. Align the rear timing chain case with the dowel pins on the cylinder block and install the case. Make sure the O-rings stay in place during installation.

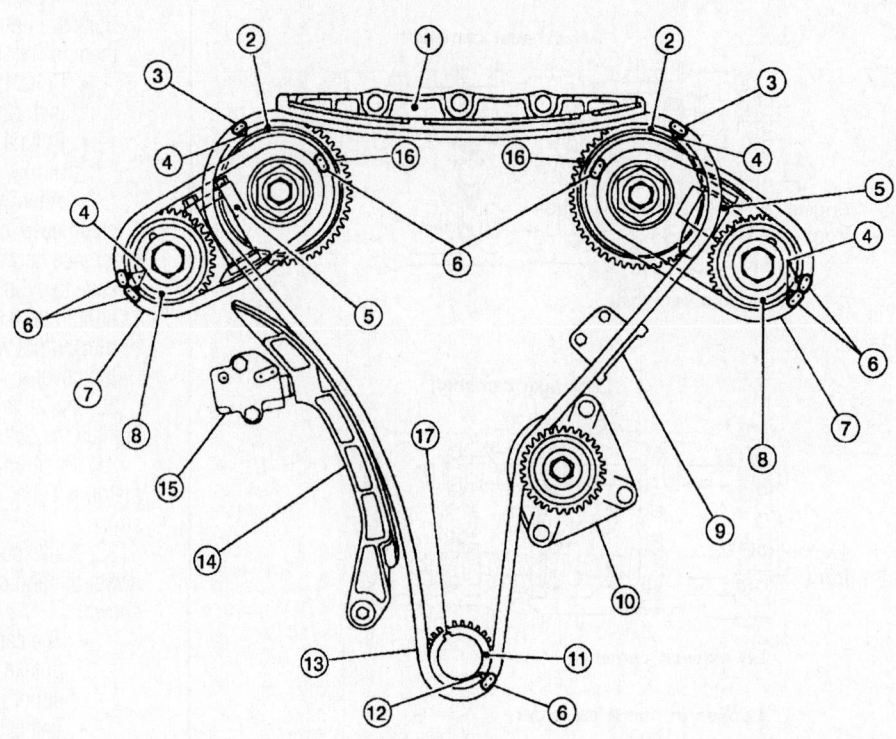

1. Internal chain guide	2. Camshaft sprocket (intake)	3. Mating mark (copper link)
4. Mating mark (punched)	5. Secondary timing chain tensioner	6. Mating mark (gold link)
7. Secondary timing chain	8. Camshaft sprocket (exhaust)	9. Tensioner guide
10. Water pump	11. Crankshaft sprocket	12. Mating mark (notched)
13. Primary timing chain	14. Slack guide	15. Primary timing chain tensioner
16. Mating mark (back side)	17. Crankshaft key	

67170-NIQU-G58

Fig. 119 Make sure all timing marks are aligned as shown when the chain is installed

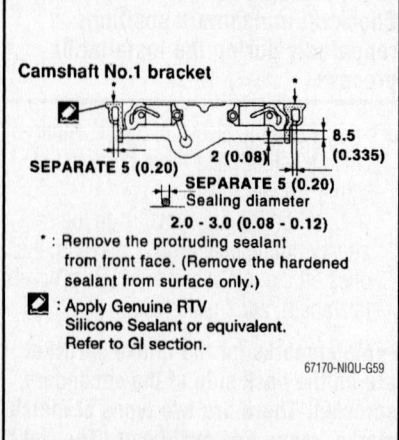

Camshaft No.1 bracket

SEPARATE 5 (0.20)
2 (0.08)
8.5 (0.335)
SEPARATE 5 (0.20)
Sealing diameter
2.0 - 3.0 (0.08 - 0.12)

* : Remove the protruding sealant from front face. (Remove the hardened sealant from surface only.)

: Apply Genuine RTV Silicone Sealant or equivalent. Refer to GI section.

67170-NIQU-G59

Fig. 120 Apply sealant to the No. 1 camshaft bracket

c. Align the dowel pin and pin hole on the camshaft with the groove and dowel pin on the sprocket, and install them.

d. On the intake side, align the pin hole on the small diameter side of the camshaft front end with the dowel pin on the back side of the camshaft sprocket, and install them.

e. On the exhaust side, align the dowel pin on the camshaft front end with the pin groove on the camshaft sprocket, and install them.

f. Tighten the camshaft sprocket bolts by hand to prevent the dislocation of dowel pins.

34. Tighten the timing chain tension guide bolts to 16 ft. lbs. (22 Nm).

a. It may be difficult to visually check the dislocation of mating marks during

and after installation. To make the matching easier, make a matchmark on the sprocket teeth in advance with paint.

35. After confirming the mating marks are aligned, tighten the camshaft sprocket bolts to 91 ft. lbs. (123 Nm).

36. Remove the stopper pins out from the timing chain tensioners, on secondary timing chains.

37. Install the crankshaft sprocket on the crankshaft. Make sure the mating marks on the crankshaft sprocket face the front of the engine.

38. Install the primary timing chain as follows:

a. Install the primary timing chain so the mating mark (punched) on the camshaft sprocket is aligned with the copper link on the timing chain, while the

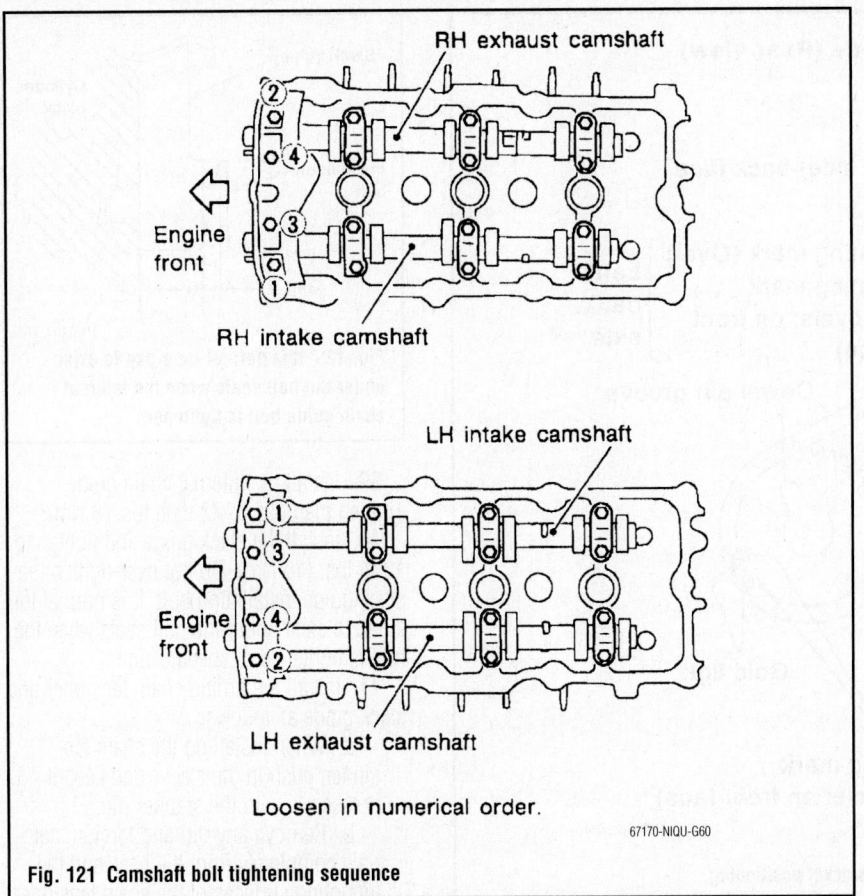

Fig. 121 Camshaft bolt tightening sequence

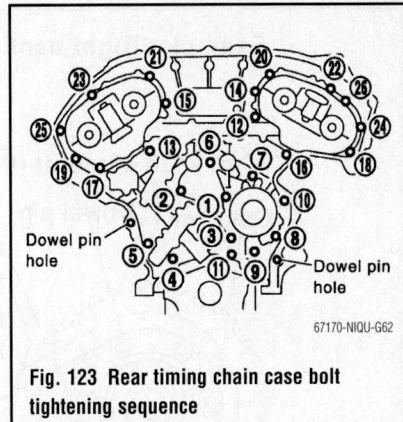

Fig. 123 Rear timing chain case bolt tightening sequence

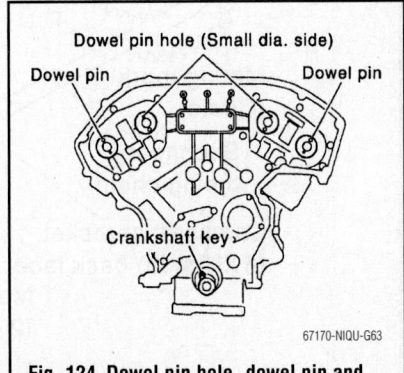

Fig. 124 Dowel pin hole, dowel pin and crankshaft key positioning

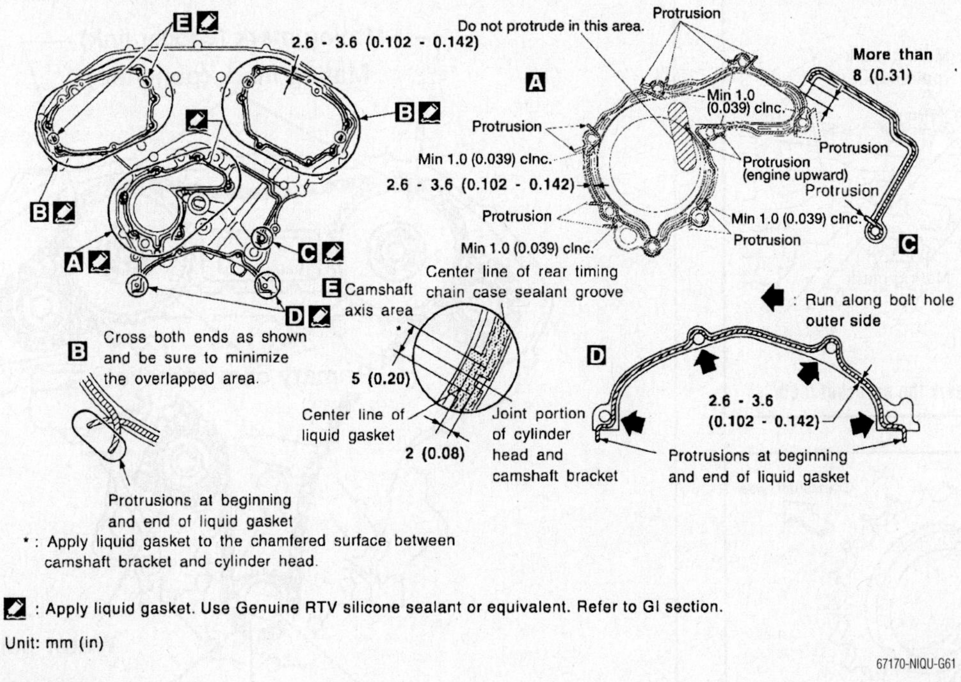

Fig. 122 Sealant application

mating mark (notched) on the crankshaft sprocket is aligned with the gold link on the timing chain, as shown.

b. When it is difficult to align mating marks of the primary timing chain with each sprocket, gradually turn the camshaft using a wrench on the hexagonal portion to align it with the mating marks.

c. During alignment, be careful to prevent dislocation of mating mark alignments of the secondary timing chains.

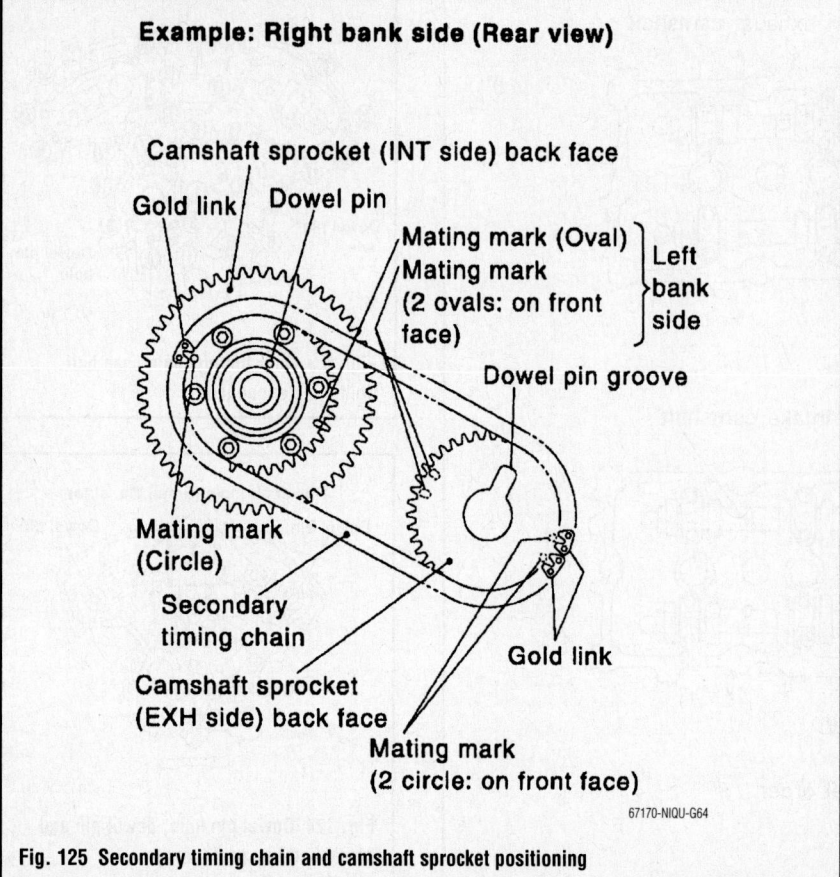

Example: Right bank side (Rear view)

Camshaft sprocket (INT side) back face
Gold link / Dowel pin
Mating mark (Oval)
Mating mark (2 ovals: on front face) } Left bank side
Dowel pin groove
Mating mark (Circle)
Secondary timing chain
Gold link
Camshaft sprocket (EXH side) back face
Mating mark (2 circle: on front face)

67170-NIQU-G64

Fig. 125 Secondary timing chain and camshaft sprocket positioning

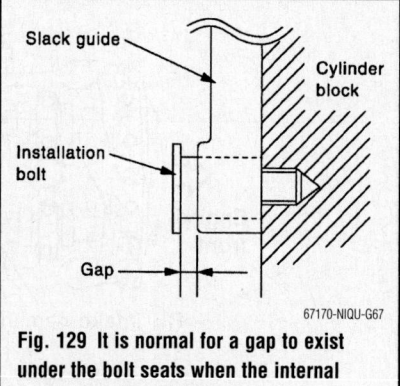

Slack guide
Cylinder block
Installation bolt
Gap

67170-NIQU-G67

Fig. 129 It is normal for a gap to exist under the bolt seats when the internal chain guide bolt is tightened

39. Install the internal chain guide. Tighten the bolts to 72 inch lbs. (8 Nm).

40. Install the slack guide and tighten to 12 ft. lbs. (16 Nm). Do not over-tighten the slack guide installation bolt. It is normal for a gap to exist under the bolt seats when the bolt is tightened to specification.

41. Install the timing chain tensioner and slack guide as follows:

a. When installing the chain tensioner, push in the sleeve and keep it pressed in with the stopper pin.

b. Remove any dirt and foreign materials completely from the back and the mounting surfaces of the chain tensioner.

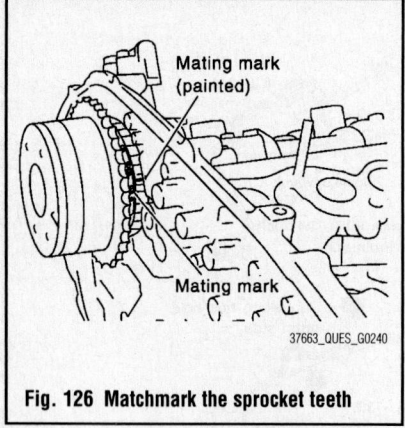

Mating mark (painted)
Mating mark

37663_QUES_G0240

Fig. 126 Matchmark the sprocket teeth

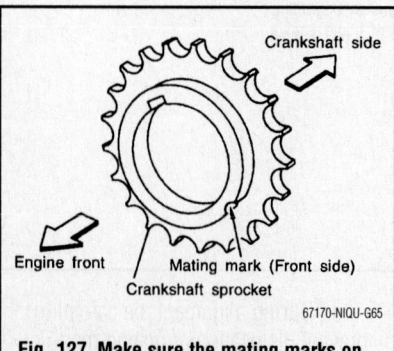

Crankshaft side
Engine front
Mating mark (Front side)
Crankshaft sprocket

67170-NIQU-G65

Fig. 127 Make sure the mating marks on the crankshaft sprocket face the front of the engine

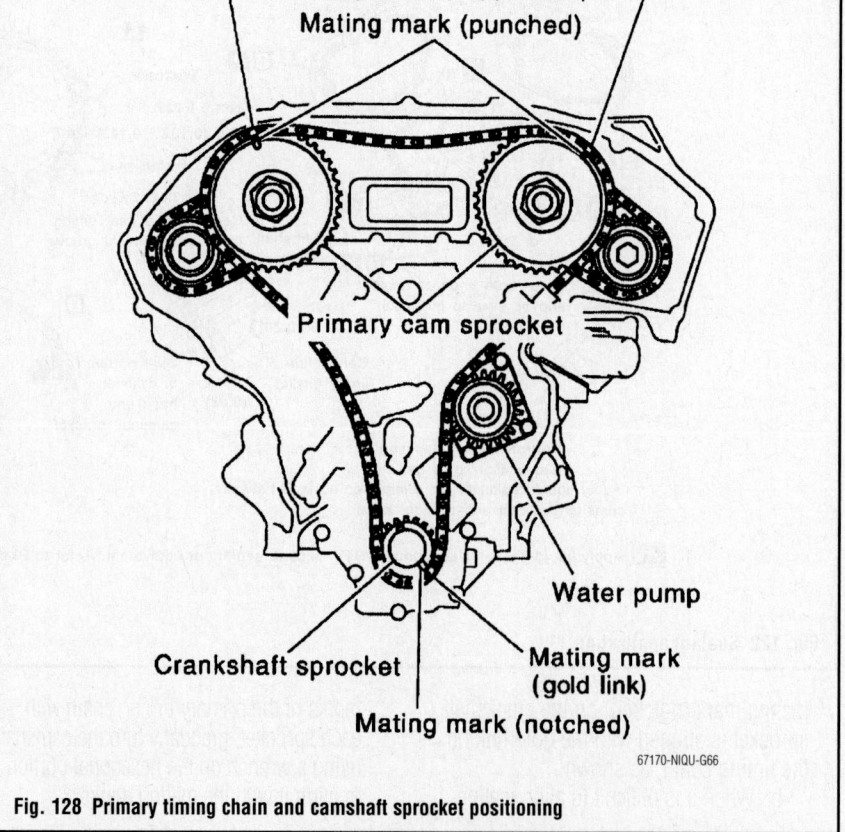

Mating mark (copper link)
Mating mark (punched)
Primary cam sprocket
Water pump
Crankshaft sprocket
Mating mark (gold link)
Mating mark (notched)

67170-NIQU-G66

Fig. 128 Primary timing chain and camshaft sprocket positioning

c. Install the timing chain tensioner and install the timing chain tensioner installation bolts.

d. Keep the slack guide pressed and hold it by pushing the stopper pin through the lever hole and body hole as shown.

e. Insert plunger into tensioner body hole by pressing the slack guide.

f. Insert stopper pin into tensioner body hole to hold lever, and keep the tab released.

g. Pull lever down and release plunger stopper tab. Plunger stopper tab can be pushed up to release (coaxial structure with lever).

h. After installation, pull out the stopper pin by pressing the slack guide.

42. Confirm that the match marks on the sprockets and the timing chain have not slipped out of alignment.

43. Install new O-rings on the rear timing chain case.

44. Install the front timing cover case and related components. Follow the Timing Chain Front Cover installation instructions to install all the remaining components.

45. Install the valve covers and oil pans, as necessary.

46. Install the engine, if removed.

47. Refill the engine with oil and coolant. Wait at least 30 minutes for the Silicone RTV Sealant to set before filling the engine with fluids to avoid leaks.

48. Activate the fuel system. Check for any leaks when the system is repressurized and correct as necessary.

49. Start the engine and check all systems for leaks or improper operation. Correct as necessary.

50. After starting engine, keep idling for three minutes. Then rev engine up to 3,000 rpm under no load to purge air from the high-pressure oil chamber of the chain tensioners. The engine may produce a rattling noise. This indicates that air still remains in the chamber and is not a matter of concern.

51. Warm up engine thoroughly to make sure there is no leakage of fuel, exhaust gases, or any oil/fluids including engine oil and engine coolant.

52. Bleed air from lines and hoses of applicable lines, such as in cooling system.

53. After cooling down engine, again check oil/fluid levels including engine oil and engine coolant. Refill to the specified level, if necessary.

VALVE LASH

INSPECTION

See Figures 130 through 139.

Perform inspection as follows after removal, installation or replacement of camshaft or valve related parts, or if there is unusual engine conditions regarding valve clearance. Check valve clearance while engine is cold and not running.

1. Set the No. 1 cylinder at TDC on its compression stroke.

a. Align pointer with TDC mark on crankshaft pulley.

b. Check that the valve lifters on No. 1 cylinder are loose and valve lifters on No. 4 are tight. If not, turn the crankshaft one full revolution (360°) and align as shown.

2. Check only the valves as shown.

a. Using a feeler gauge, measure the clearance between the valve lifter and camshaft.

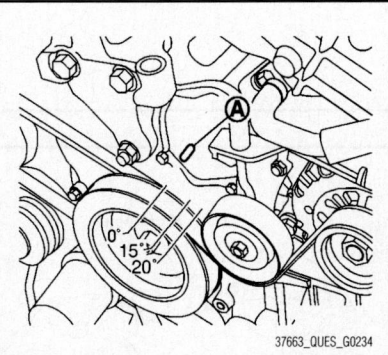

Fig. 130 Align pointer (A) with TDC mark on crankshaft pulley

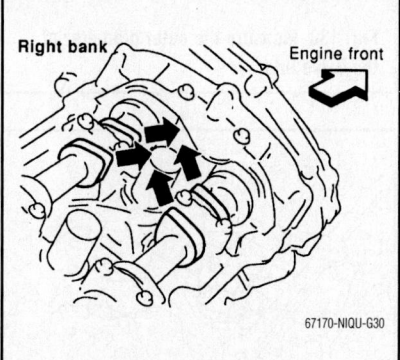

Fig. 131 Valve lifter alignment

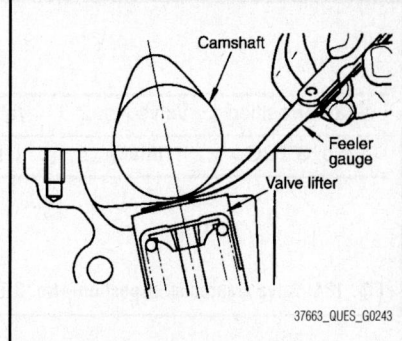

Fig. 133 Measure the clearance between the valve lifter and camshaft

b. Record any valve clearance measurements which are out of specification. They will be used later to determine the required replacement lifter size. Valve Clearance for Checking (cold):

- Intake: 0.010–0.013 in. (0.26–0.34 mm)
- Exhaust: 0.011– 0.015 in. (0.29–0.37 mm)

3. Turn the crankshaft 240°.

4. Set No. 3 cylinder at TDC on its compression stroke.

5. Turn the crankshaft 240° and align as above.

6. Set No. 5 cylinder at TDC on its compression stroke.

7. If the valve clearances are out of specification, adjust the valve clearances.

Adjust valve clearance while engine is cold. Perform adjustment by selecting the correct head thickness of the valve lifter (adjusting shims are not used). The specified valve lifter thickness is the dimension at normal temperatures. Ignore dimensional differences caused by temperature. Use specifications for hot engine condition to confirm valve clearances.

8. Remove the camshaft.

9. Remove the valve lifter that was measured as being outside the standard specifications.

10. Measure the center thickness of the removed lifter with a micrometer.

11. Use the equation below to calculate

Crank Position	Valve No. 1	Valve No. 2	Valve No. 3	Valve No. 6
No. 1 TDC	Intake	Exhaust	Exhaust	Intake

67170-NIQU-G70

Fig. 132 Valve clearance inspection—No. 1 TDC

Crank Position	Valve No. 2	Valve No. 3	Valve No. 4	Valve No. 5
No. 3 TDC	Intake	Intake	Exhaust	Exhaust

67170-NIQU-G71

Fig. 134 Valve clearance inspection—No. 3 TDC

Crank Position	Valve No. 1	Valve No. 4	Valve No. 5	Valve No. 6
No. 5 TDC	Exhaust	Intake	Intake	Exhaust

67170-NIQU-G72

Fig. 135 Valve clearance inspection—No. 5 TDC

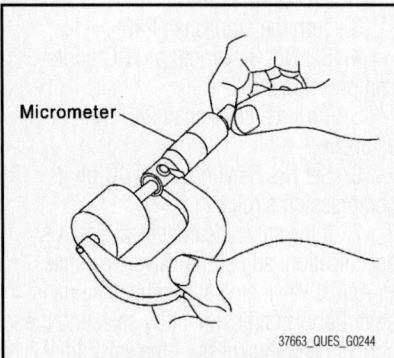

37663_QUES_G0244

Fig. 136 Measure the center thickness of the removed lifter with a micrometer

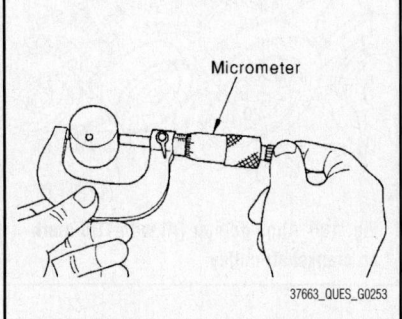

37663_QUES_G0253

Fig. 138 Measure the outer diameter of the valve lifter

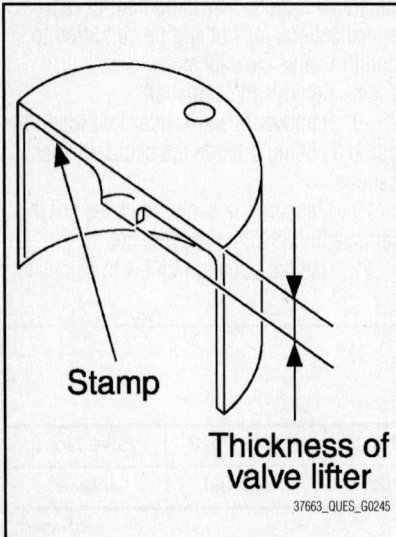

37663_QUES_G0245

Fig. 137 New valve lifter thickness

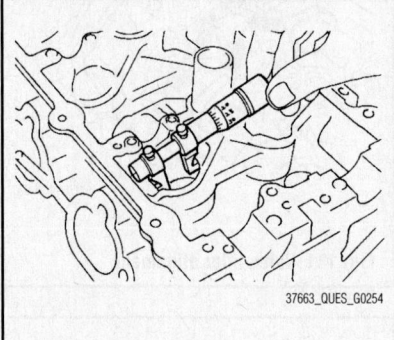

37663_QUES_G0254

Fig. 139 Measure the diameter of the valve lifter bore

valve lifter thickness for replacement. Valve lifter thickness calculation:

- Thickness of replacement valve lifter = t1 + (C1 - C2)

- t1 = Thickness of removed valve lifter
- C1 = Measured valve clearance
- C2=Standard valve clearance:

Thickness of a new valve lifter can be identified by stamp marks on the reverse side (inside the cylinder). Stamp mark 788U indicates 0.3102 in. (7.88 mm) in thickness. Stamp mark 840U indicates 0.3307 in. (8.40 mm) in thickness. Install identification letter at the end and top, (U) and (N), at each of proper positions. (Be careful of mis-installation between intake and exhaust.)

Available thickness of valve lifter: 27 sizes with range 0.3102–0.3307 in. (7.88–8.40 mm) in steps of 0.0008 in. (0.02 mm) (when manufactured at factory).

12. Install the selected replacement valve lifter.

13. Install the camshaft.

14. Manually rotate the crankshaft a few turns.

15. Confirm that the valve clearances are within specification.

16. After the engine has been run to full operating temperature, confirm that the valve clearances are within specification.

  a. Cold valve clearance:
- Intake: 0.010–0.013 in. (0.26–0.34mm)
- Exhaust: 0.011–0.015 in. (0.29–0.37mm)

  b. Hot valve clearance approximately 176°F (80°C)]:
- Intake: 0.012–0.016 in. (0.304–0.416 mm)
- Exhaust: 0.012–0.017 in. (0.308–0.432 mm)

17. Inspect the Valve Lifter Clearance:

  a. Measure the outer diameter of the valve lifter. If out of the specified range, replace the valve lifter:
- 1.3377–1.3381 in. (33.977–33.987 mm)

18. Inspect the Valve Lifter Bore Diameter:

  a. Using inside micrometer, measure diameter of the valve lifter bore of cylinder head. If out of the specified range, replace the cylinder head assembly:
- 1.3386–1.3392 in. (34.000–34.016 mm)

19. Calculate the Valve Lifter Clearance:

  a. (Valve lifter clearance) = (hole diameter for valve lifter) minus (outer diameter of valve lifter). If out of specified range, replace either or both valve lifter and cylinder head assembly:
- 0.0005–0.0015 in. (0.013–0.039 mm)

ENGINE PERFORMANCE & EMISSION CONTROLS

CAMSHAFT POSITION (CMP) SENSOR

Refer to the accompanying illustration.

LOCATION

See Figure 140.

CRANKSHAFT POSITION (CKP) SENSOR

LOCATION

See Figure 141.

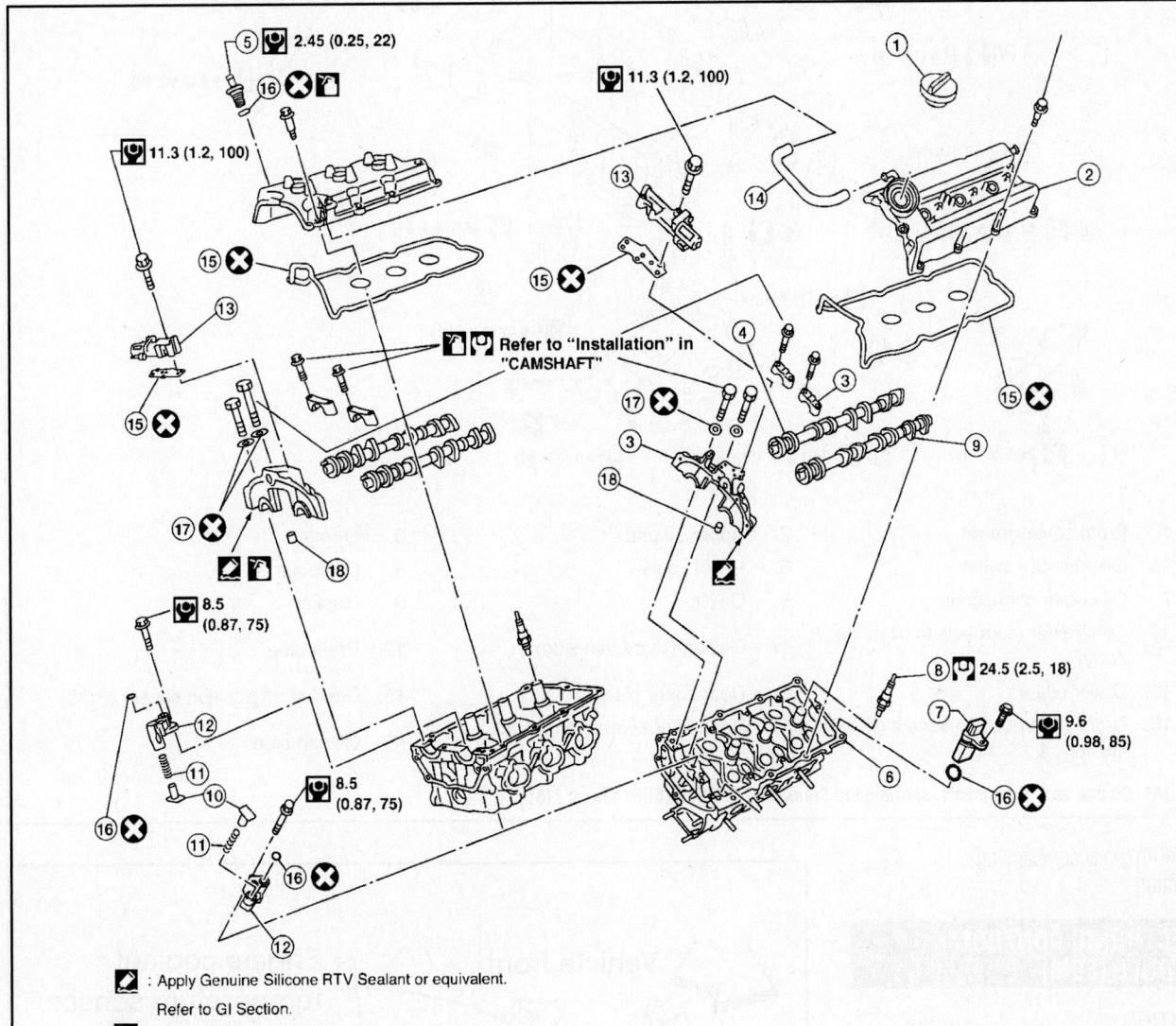

5 : 2.45 (0.25, 22)

11.3 (1.2, 100)

11.3 (1.2, 100)

8.5 (0.87, 75)

8.5 (0.87, 75)

24.5 (2.5, 18)

9.6 (0.98, 85)

: Apply Genuine Silicone RTV Sealant or equivalent.
 Refer to GI Section.

: Lubricate with engine oil

: N·m (kg-m, ft-lb)

: N·m (kg-m, in-lb)

: Always replace after every disassembly.

1. Oil filler cap	2. Rocker cover (LH)	3. Camshaft bracket (LH)
4. Camshaft (INT)	5. PCV valve	6. Cylinder head (LH)
7. Camshaft position sensor (PHASE)	8. Spark plug	9. Camshaft (EXH)
10. Tensioner sleeve	11. Tensioner spring	12. Secondary camshaft chain tensioner
13. IVT control solenoid valve	14. PCV hose	15. Gasket
16. O-ring	17. Seal washer	18. Dowel pin

37663_QUES_G0178

Fig. 140 Exploded view of camshaft components, showing the Camshaft Position (CMP) sensor (7)

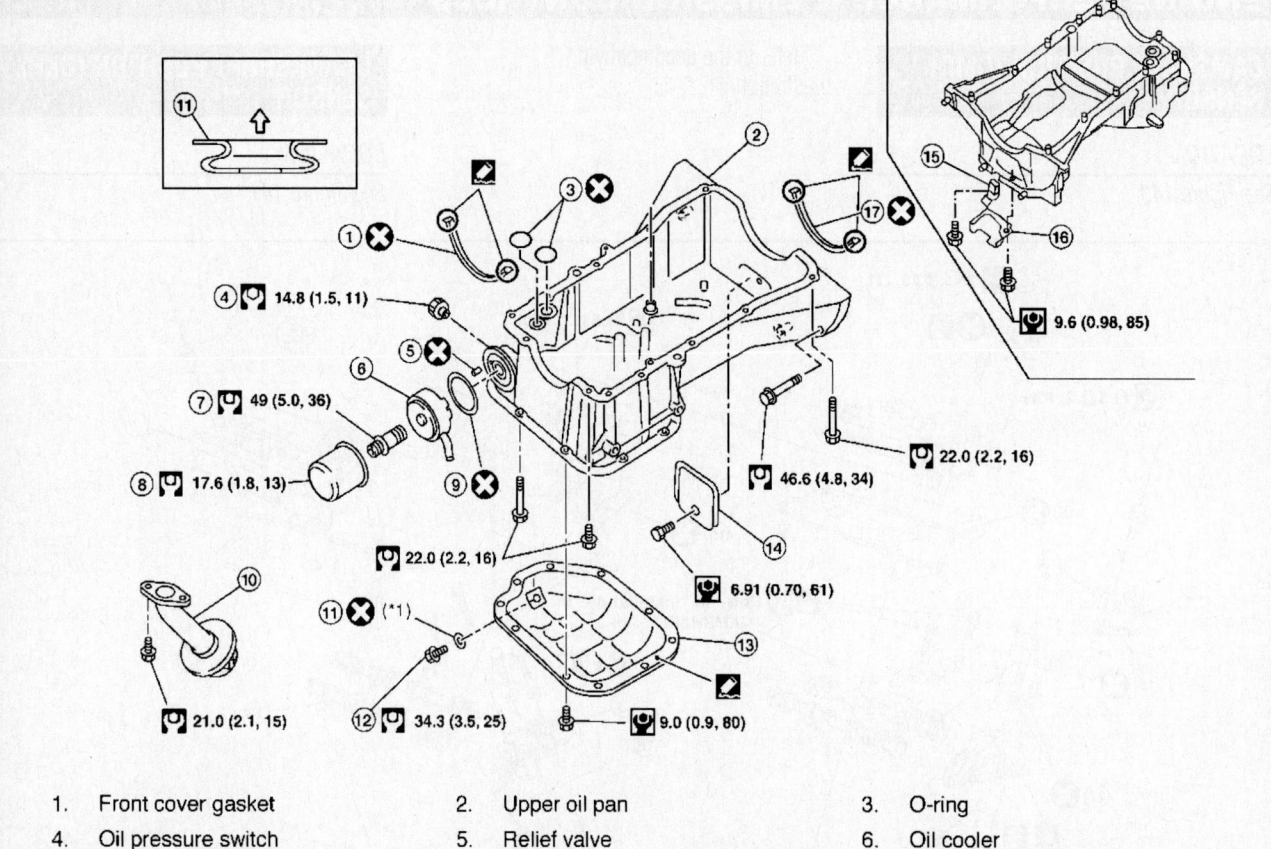

1. Front cover gasket
2. Upper oil pan
3. O-ring
4. Oil pressure switch
5. Relief valve
6. Oil cooler
7. Oil cooler connection
8. Oil filter
9. Gasket
10. Oil strainer (connects to oil pump body)
11. Gasket (⇐ oil pan side)
12. Drain plug
13. Lower oil pan
14. Rear cover plate
15. Crankshaft position sensor (POS)
16. Crankshaft position sensor (POS) shield
17. Rear oil seal retainer gasket
⇐ Oil pan side

37663_QUES_G0219

Fig. 141 Oil pan and components, showing the Crankshaft Position (CKP) sensor (15)

Refer to the accompanying illustration.

ELECTRONIC CONTROL MODULE (ECM)

LOCATION

The ECM is located in the right side of the cowl top (behind the strut tower).

ENGINE COOLANT TEMPERATURE (ECT) SENSOR

LOCATION

See Figure 142.

Refer to the accompanying illustration.

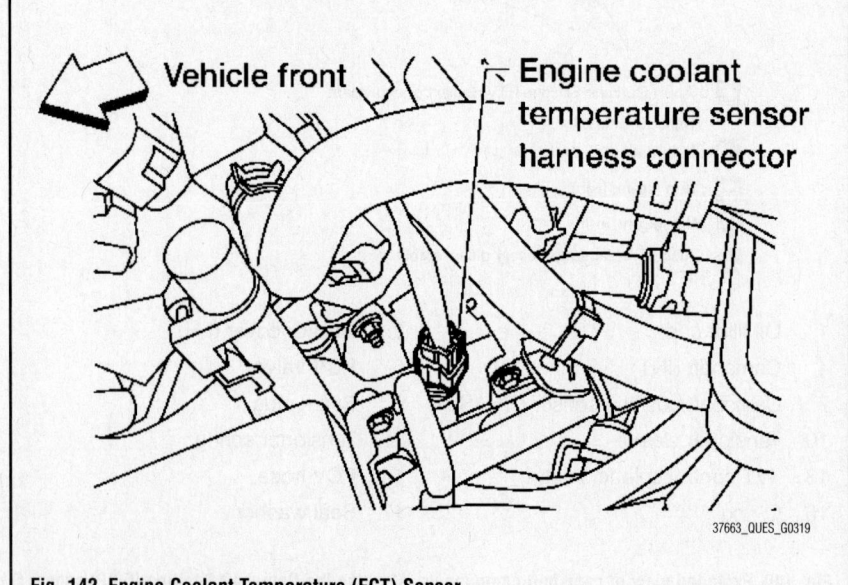

Vehicle front

Engine coolant temperature sensor harness connector

37663_QUES_G0319

Fig. 142 Engine Coolant Temperature (ECT) Sensor

HEATED OXYGEN (HO2S) SENSOR

LOCATION

See Figure 143.

Refer to the accompanying illustration.

INTAKE AIR TEMPERATURE (IAT) SENSOR

LOCATION

See Figure 144.

Refer to the accompanying illustration.

KNOCK SENSOR (KS)

LOCATION

See Figure 145.

Refer to the accompanying illustration.

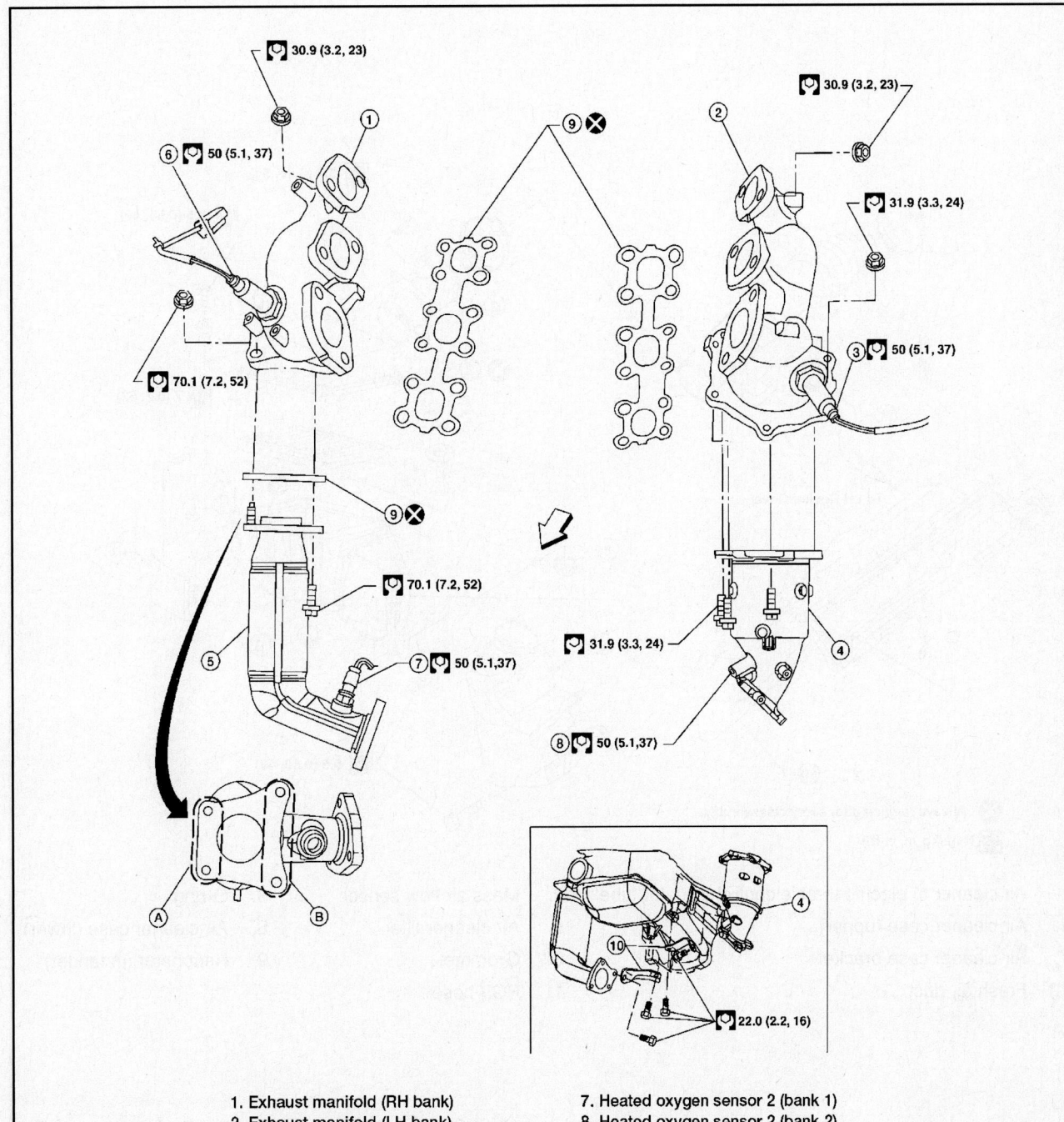

1. Exhaust manifold (RH bank)
2. Exhaust manifold (LH bank)
3. Air fuel ratio (A/F) sensor 1 (bank 2)
4. Three way catalyst (manifold) (bank 2)
5. Three way catalyst (manifold) (bank 1)
6. Air fuel ratio (A/F) sensor 1 (bank 1)
7. Heated oxygen sensor 2 (bank 1)
8. Heated oxygen sensor 2 (bank 2)
9. Gasket
10. Three way catalyst supports
A. Stud
B. Bolt

37663_QUES_G0192

Fig. 143 Combination manifold system, showing Heated Oxygen Sensor (HO2S) 2, Bank 1 (7), and Bank 2 (8)

MASS AIR FLOW (MAF) SENSOR

Refer to Intake Air Temperature (IAT) Sensor.

LOCATION

See Figure 144.

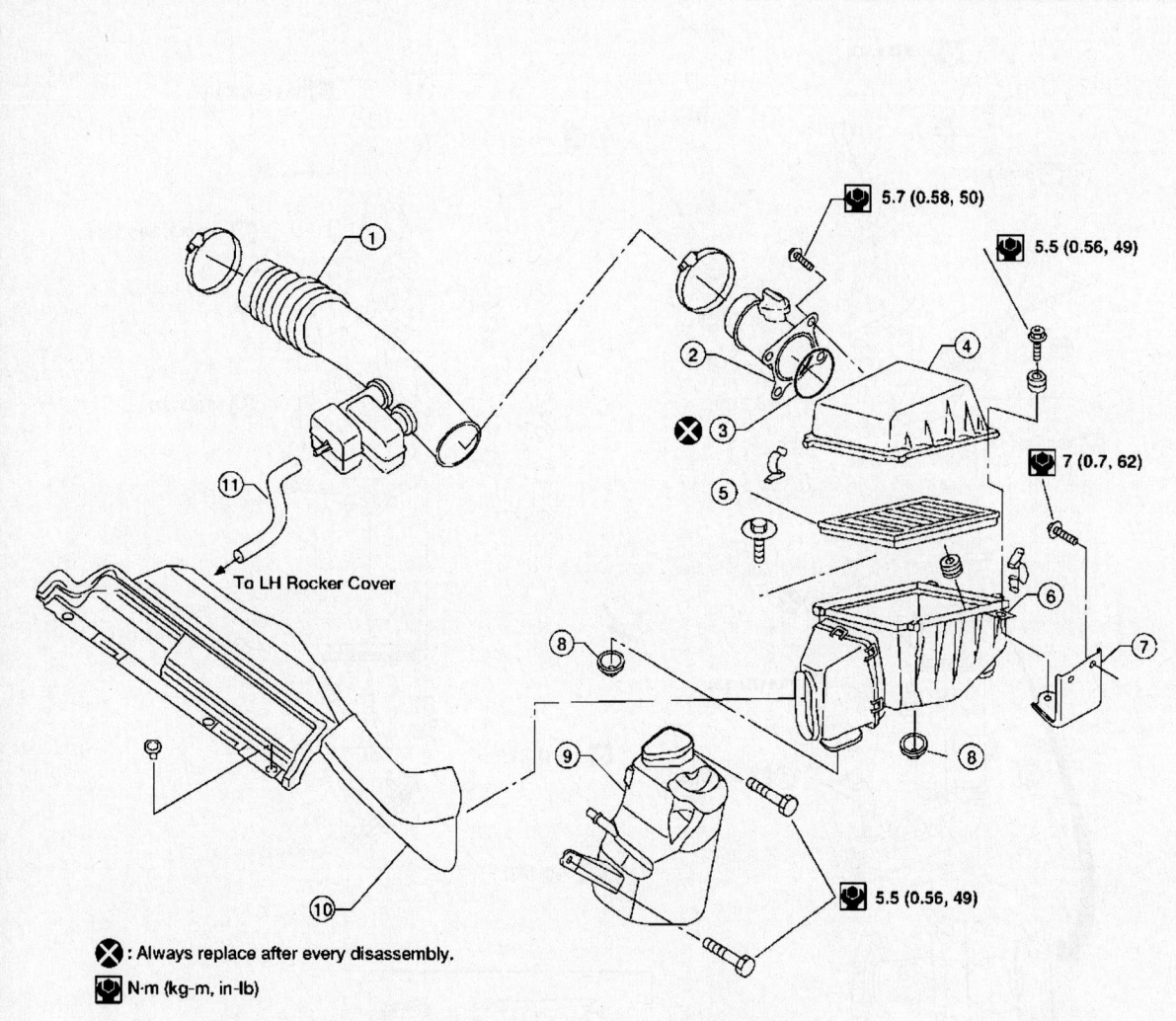

5.7 (0.58, 50)

5.5 (0.56, 49)

7 (0.7, 62)

To LH Rocker Cover

5.5 (0.56, 49)

⊗ : Always replace after every disassembly.

N·m (kg-m, in-lb)

1. Air cleaner to electric throttle control actuator tube	2. Mass air flow sensor	3. O-ring
4. Air cleaner case (upper)	5. Air cleaner filter	6. Air cleaner case (lower)
7. Air cleaner case bracket	8. Grommet	9. Resonator (in fender)
10. Fresh air duct	11. PCV hose	

37663_QUES_G0176

Fig. 144 Air cleaner and air duct components, showing Mass Air Flow (MAF)/Intake Air Temperature (IAT) sensors (2)

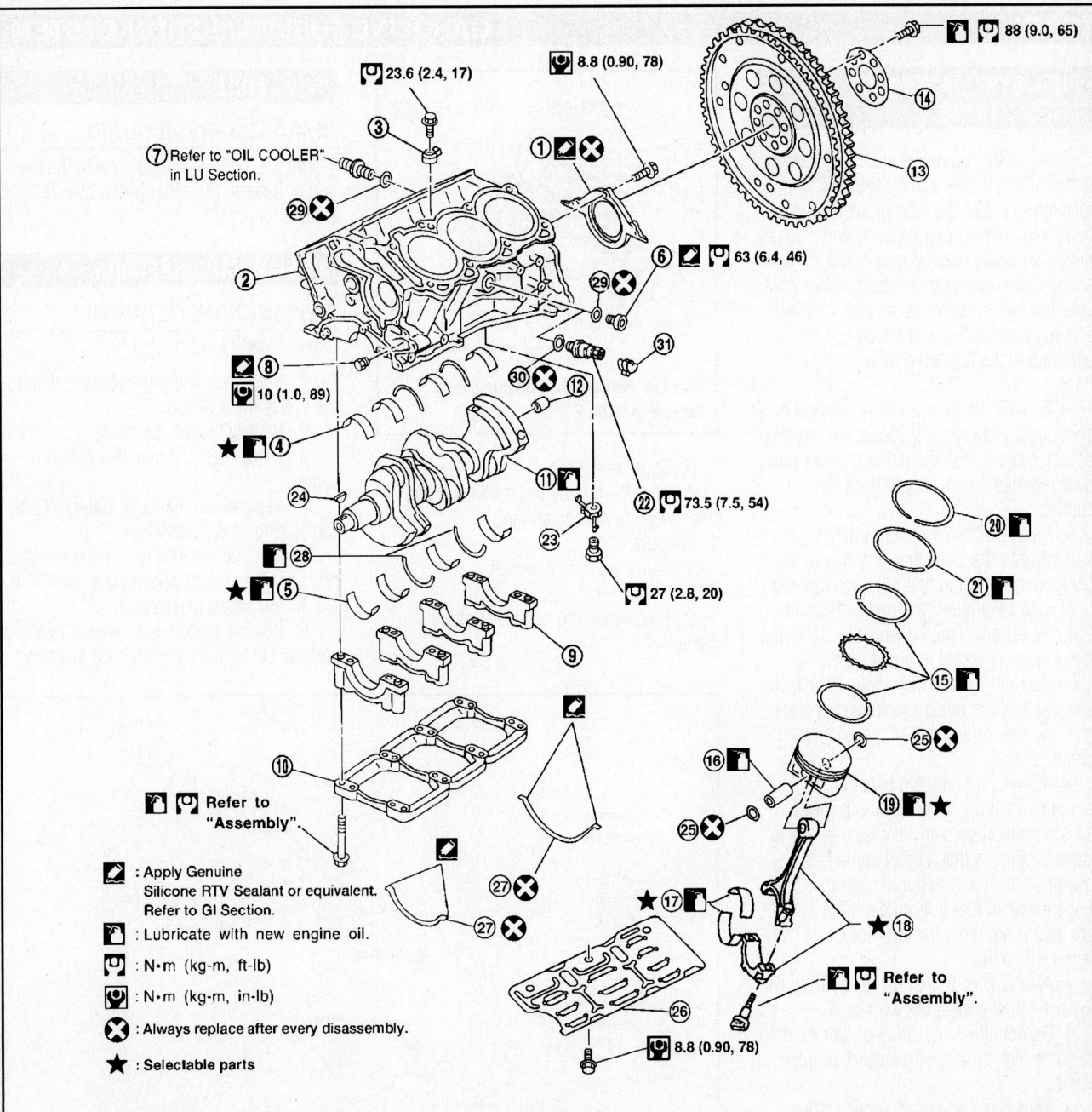

Fig. 145 Exploded view of cylinder block, showing Knock Sensor (3)

1. Rear oil seal retainer	2. Cylinder block	3. Knock sensor
4. Upper main bearing	5. Lower main bearing	6. Water drain plug (LH side)
7. Water connector (RH side)	8. Water drain plug (water pump side)	9. Main bearing cap
10. Main bearing beam	11. Crankshaft	12. Pilot converter
13. Drive plate with signal plate	14. Drive plate reinforcement	15. Oil ring set
16. Piston pin	17. Connecting rod bearing	18. Connecting rod
19. Piston	20. Top ring	21. Second ring
22. Cylinder block heater (Canada only)	23. Oil jet	24. Key
25. Snap ring	26. Baffle plate	27. Gasket
28. Thrust bearing	29. Gasket	30. O-ring
31. Connector protector cap		

37663_QUES_G0213

FUEL SYSTEM SERVICE PRECAUTIONS

Safety is the most important factor when performing not only fuel system maintenance, but any type of maintenance. Failure to conduct maintenance and repairs in a safe manner may result in serious personal injury or death. Work on a vehicle's fuel system components can be accomplished safely and effectively by adhering to the following rules and guide-lines.

• To avoid the possibility of fire and per-sonal injury, always disconnect the negative battery cable unless the repair or test proce-dure requires that battery voltage be applied.

• Always relieve the fuel system pres-sure prior to disconnecting any fuel system component (injector, fuel rail, pressure reg-ulator, etc.) fitting or fuel line connection. Exercise extreme caution whenever relieving fuel system pressure to avoid exposing skin, face and eyes to fuel spray. Please be advised that fuel under pressure may pene-trate the skin or any part of the body that it contacts.

• Always place a shop towel or cloth around the fitting or connection prior to loosening to absorb any excess fuel due to spillage. Ensure that all fuel spillage is quickly removed from engine surfaces. Ensure that all fuel-soaked cloths or towels are deposited into a flame-proof waste con-tainer with a lid.

• Always keep a dry chemical (Class B) fire extinguisher near the work area.

• Do not allow fuel spray or fuel vapors to come into contact with a spark or open flame.

• Always use a second wrench when loosening or tightening fuel line connection fittings. This will prevent unnecessary stress and torsion on fuel piping. Always follow the proper torque specifications.

• Always replace worn fuel fitting O-rings with new ones. Do not substitute fuel hose where rigid pipe is installed.

FUEL SYSTEM PRESSURE

RELIEVING

See Figure 146.

1. Before servicing the vehicle, refer to the Precautions Section.
2. Remove the fuel pump fuse located in IPDM E/R.

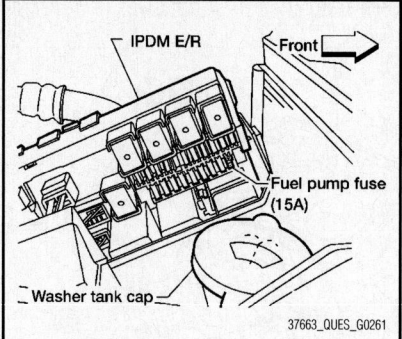

Fig. 146 Remove the fuel pump fuse located in IPDM E/R

3. Start the engine.
4. After the engine stalls, crank it two or three times to release all fuel pressure.
5. Turn the ignition switch OFF.
6. Disconnect the negative battery cable.

FUEL FILTER

REMOVAL & INSTALLATION

The fuel filter is integrated with the fuel pump. Refer to Fuel Pump Removal & Installation.

FUEL PUMP ASSEMBLY

REMOVAL & INSTALLATION

See Figures 147 through 151.

1. Before servicing the vehicle, refer to the Precautions Section.
2. Relieve the fuel pressure.
3. Disconnect the negative battery cable.
4. Remove the fuel tank. Refer to Fuel Tank Removal & Installation.
5. Remove the lock ring using a socket drive handle and the appropriate tool [Tool No. KV991J0090 (J-46214)].
6. Remove the fuel level sensor, fuel filter, and fuel pump assembly from the fuel tank.

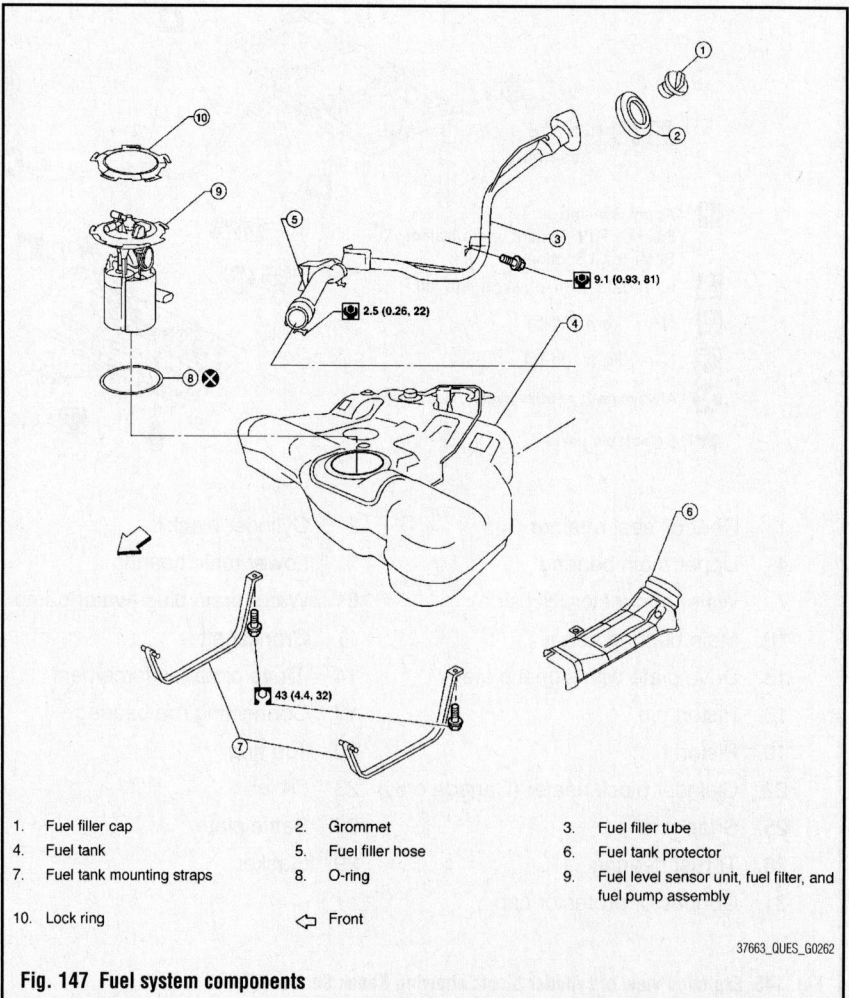

1.	Fuel filler cap	2.	Grommet	3.	Fuel filler tube
4.	Fuel tank	5.	Fuel filler hose	6.	Fuel tank protector
7.	Fuel tank mounting straps	8.	O-ring	9.	Fuel level sensor unit, fuel filter, and fuel pump assembly
10.	Lock ring	⇦	Front		

Fig. 147 Fuel system components

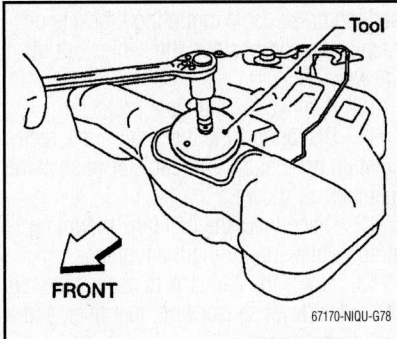

Fig. 148 Remove the lock ring using a socket drive handle and tool J-16214

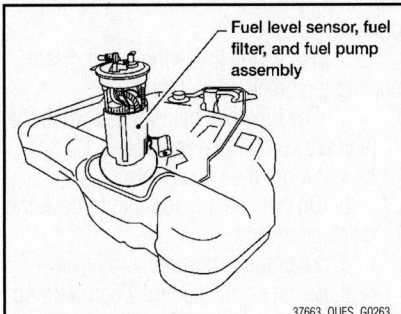

Fig. 149 Remove the fuel level sensor, fuel filter, and fuel pump assembly from the fuel tank

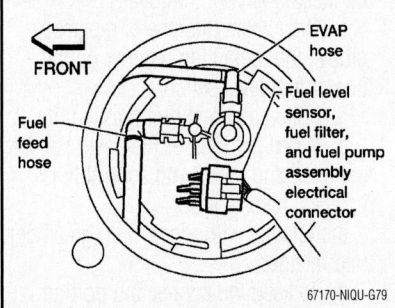

Fig. 150 Install the fuel level sensor, fuel filter, and fuel pump assembly with the fuel feed hose facing the front of the vehicle

✳✳ CAUTION

Do not bend the float arm during removal.

To install:

7. Make sure the fuel level sensor, fuel filter, and fuel pump is free from defects and foreign materials.

8. Install the fuel level sensor, fuel filter, and fuel pump assembly with the fuel feed hose facing the front of the vehicle.

9. Turn the lock ring until the lock ring

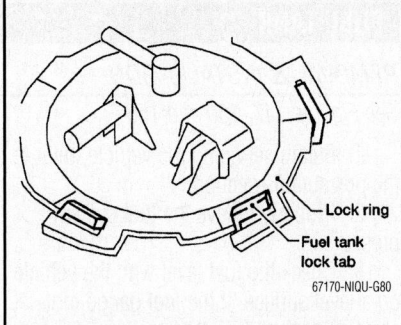

Fig. 151 Turn the lock ring until the lock ring is fully rotated into the fuel tank lock tabs

is fully rotated into the fuel tank lock tabs as shown.

10. Install the fuel tank.

11. Turn the ignition switch to ON with the engine OFF to check the connections for fuel leaks with the electric fuel pump applying fuel pressure to the fuel lines.

12. Start the engine and let it idle to check that there are no fuel leaks at the fuel system tube and hose connections.

FUEL RAIL AND INJECTOR

REMOVAL & INSTALLATION

See Figures 152 through 155.

1. Before servicing the vehicle, refer to the Precautions Section.

2. Relieve the fuel pressure.

3. Remove the upper and lower intake manifold collector.

4. Disconnect the fuel quick connector on the engine side, using the appropriate tool No. J-45488, as follows:

   a. Remove the quick connector cap.

   b. With the sleeve side of tool facing quick connector, install the tool on to fuel tube.

   c. Insert the tool into quick connector until sleeve contacts and goes no further.

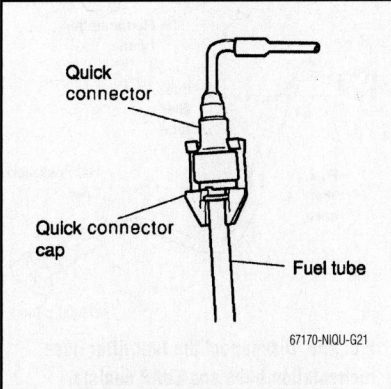

Fig. 152 Quick connector components

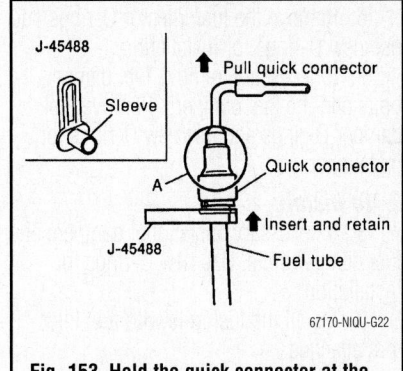

Fig. 153 Hold the quick connector at the "A" position

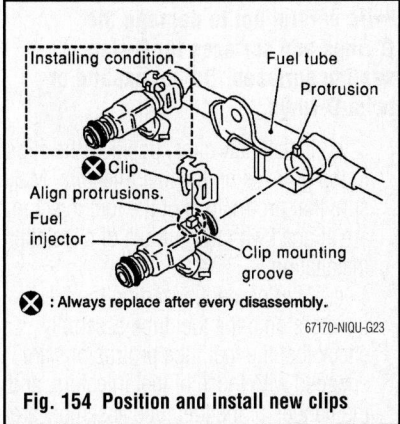

Fig. 154 Position and install new clips

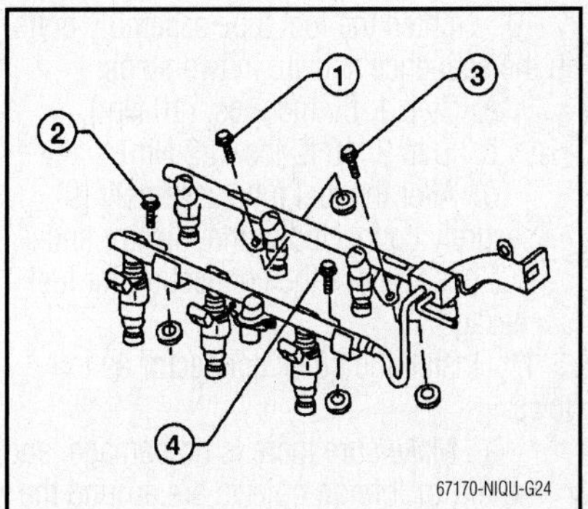

Fig. 155 Fuel tube assembly bolts tightening sequence

Hold the tool in that position. Inserting the tool hard will not disconnect quick connector. Hold tool where it contacts and will go no further.

   d. Pull the quick connector straight out from the fuel tube. _Pull the quick connector holding it at the position shown.

➡**Do not pull with lateral force applied as the O-ring inside the quick connector may be damaged.**

5. Remove the fuel rail, with the fuel injectors attached, from the intake manifold.

6. Remove the fuel injector O-rings and use new O-rings for installation.

7. If necessary, remove fuel damper bolts and the fuel dampers. Remove fuel damper O-rings and use new O-rings for installation.

To install:

8. If necessary, install fuel dampers and fuel damper bolts. Use new O-rings for installation.

9. Install the fuel rails with fuel injectors attached.

   a. Install new O-rings. Lubricate the O-rings by lightly coating with new engine oil.

➡**Be careful not to damage the O-rings and surfaces for O-ring sealing surfaces. Do not expand or twist O-rings.**

   b. Install new clips, position the clips in the grooves on the fuel injectors. Make sure that protrusions of the fuel injectors are aligned with the cutouts of clips after installation.

   c. After properly inserting the fuel injectors onto the fuel tube assembly, check that the fuel tube protrusions are engaged with those of fuel injectors, and the flanges of the fuel tube assembly are fully engaged with the clips.

10. Tighten the fuel tube assembly bolts in the sequence shown, in two steps:

   a. Step 1: 89 inch lbs. (10 Nm).

   b. Step 2: 16 ft. lbs. (22 Nm).

   c. After the fuel tube assembly is securely connected to the injector and fuel hose, check the connection for fuel leakage.

11. Install the quick connector as follows:

   a. Make sure there is no damage, and no dirt or foreign objects are around the tube and quick connector.

   b. Align the center to insert the quick connector straight onto the fuel tube.

   c. Insert the fuel tube until a click is heard.

   d. Install the quick connector cap.

12. Install the upper and lower intake manifold collector.

13. Make sure there is no fuel leakage at connections as follows:

   a. Apply fuel pressure to fuel lines by turning ignition switch **ON** with the engine **OFF** and check for fuel leaks at connections.

   b. Start the engine and rev it up and check for fuel leaks at connections. Use mirrors for checking on connections out of the direct line of sight.

FUEL TANK

REMOVAL & INSTALLATION

See Figures 147, 150 and 156.

1. Before servicing the vehicle, refer to the precautions section.

2. Properly relieve the fuel system pressure.

3. Check the fuel level with the vehicle on a level surface. If the fuel gauge indicates more than ⅞ full, drain the fuel from the fuel tank until the fuel gauge indicates a level at

or below ⅞ full.

4. In case the fuel pump does not operate, use the following procedure:

   a. Insert fuel tubing of less than 0.98 in. (25 mm) diameter into the fuel filler tube through the fuel filler opening to drain fuel from the fuel filler tube.

   b. Disconnect the fuel filler hose from the fuel filler tube.

   c. Insert fuel tubing into the fuel tank through the fuel filler hose to drain fuel from the fuel tank.

5. Open the fuel filler lid and unscrew the fuel filler cap to release the pressure inside the fuel tank.

6. Release the fuel pressure from the fuel lines.

7. Disconnect the negative battery cable.

8. Remove the center exhaust tube, with mufflers.

9. Disconnect the parking brake cables from the equalizer, then disconnect the three

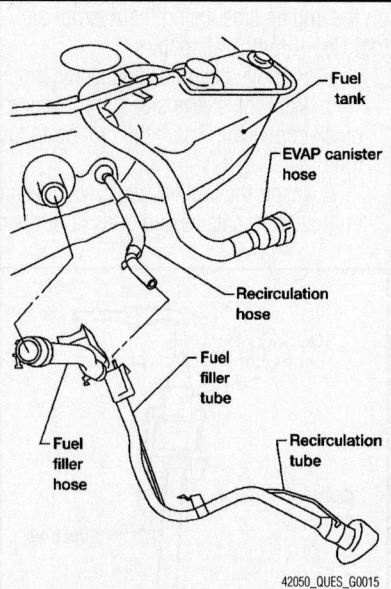

Fig. 156 Disconnect the fuel filler hose, recirculation hose and EVAP canister hose at the fuel tank

42050_QUES_G0015

parking brake cable mounting brackets on each cable and position the cables out of the way.

10. Remove the fuel tank protector.

11. Disconnect the fuel filler hose, recirculation hose and EVAP canister hose at the fuel tank as shown.

12. Disconnect the fuel tank mounting straps while supporting the fuel tank.

13. Lower the fuel tank to access the top of the fuel level sensor unit, fuel filter, and fuel pump assembly.

14. Disconnect the fuel level sensor unit, fuel filter, and fuel pump assembly electrical connector, and the fuel feed hose from the fuel level sensor unit, fuel filter, and fuel pump assembly.

15. Observe the following when disconnecting the quick-connectors:

   a. The tube can be removed when the tabs are completely depressed. Do not twist it more than necessary.

   b. Do not use any tools to remove the quick connector.

   c. Keep the resin tube away from heat. Be especially careful when welding near the tube.

   d. Prevent liquid acids, such as battery electrolyte, from getting on the resin tube.

   e. Do not bend or twist the tube during installation and removal.

   f. Only when the tube is replaced, remove the remaining retainer on the tube or fuel level sensor,

fuel filter, and fuel pump assembly.

   g. When the tube or fuel level sensor, fuel filter, and fuel pump assembly is replaced, also replace

the retainer with a new one (green colored retainer).

   h. To keep the connecting portion clean and to avoid damage and foreign materials, cover them completely with plastic bags or something similar.

16. Disconnect the quick-connectors by performing the following:

   a. Hold the sides of the connector, push in tabs and pull out the tube.

   b. If the connector and the tube are stuck together, push and pull several times until they start to move. Then disconnect them by pulling.

17. Remove the fuel tank.

18. If replacing the fuel tank, disconnect the EVAP hose and remove the fuel level sensor unit, fuel filter, and fuel pump assembly to transfer to the new fuel tank.

To install:

19. To install, reverse removal procedure.

20. Before tightening the fuel tank mounting straps, temporarily install the fuel filler hose, recirculation hose and EVAP canister hose. Tighten the straps to 31 ft. lbs. (41 Nm).

21. Connect the quick-connectors by performing the following:

　a. Check the connection for damage or any foreign materials

　b. Align the connector with the tube, then insert the connector straight into the tube until a click is heard

　c. After the tube is connected, make sure the connection is secure by pulling on the tube and the connector to make sure they are securely connected

22. Visually confirm that the two retainer tabs are connected to the quick connector.

23. Check the parking brake for proper operation.

24. Turn the ignition switch to ON with the engine OFF to check the connections for fuel leaks with the

electric fuel pump applying fuel pressure to the fuel lines.

25. Start the engine and let it idle to check that there are no fuel leaks at the fuel system tube and hose connections.

IDLE SPEED

ADJUSTMENT

Idle speed is maintained by the Powertrain Control Module (PCM). No adjustment is necessary or possible.

HEATING & AIR CONDITIONING SYSTEM

BLOWER MOTOR

REMOVAL & INSTALLATION

See Figure 157.

❄❄ CAUTION

Before servicing components near or affected by the SRS (air bag) system, read and observe all SRS Service Precautions. Refer to Supplemental Restraint System (SRS), in the Chassis Electrical section. Failure to observe all precautions may result in accidental airbag deployment, personal injury, or death.

1. Before servicing the vehicle, refer to the Precautions Section.

2. Disconnect the negative battery cable and wait at least 3 minutes for the SRS memory to drain.

3. Remove the instrument panel.

4. Remove the center ventilator duct.

5. Remove the front blower motor.

To install:

6. To install, reverse removal procedure.

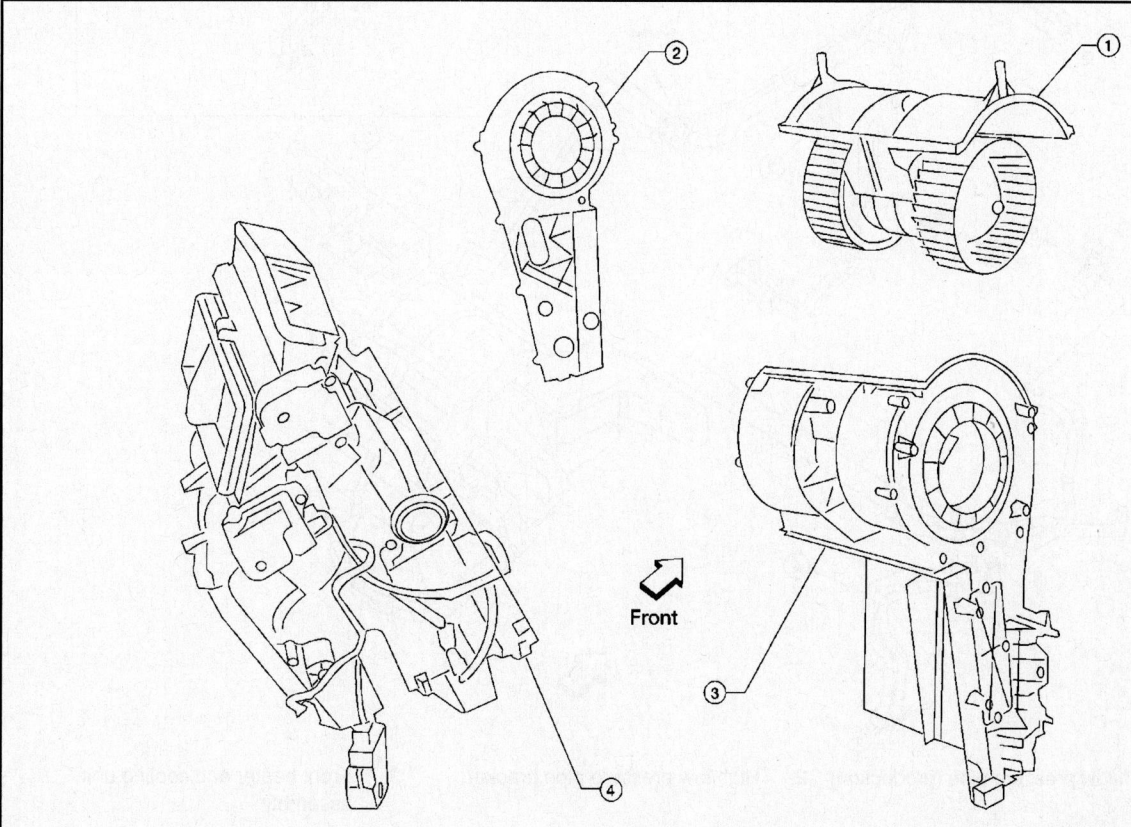

1. Front blower motor
2. Blower motor side cover
3. Blower motor case
4. Heater core and evaporator case

Front

37663_QUES_G0267

Fig. 157 Front blower motor

HEATER CORE

REMOVAL & INSTALLATION

See Figure 157.

✳✳ CAUTION

Before servicing components near or affected by the SRS (air bag) system, read and observe all SRS Service Precautions. Refer to Supplemental Restraint System (SRS), in the Chassis Electrical section. Failure to observe all precautions may result in accidental airbag deployment, personal injury, or death.

1. Before servicing the vehicle, refer to the Precautions Section.
2. Disconnect the negative battery cable and wait at least 3 minutes for the SRS memory to drain.
3. Discharge the refrigerant from the A/C system.
4. Drain the engine cooling system.
5. Remove the front heater and cooling unit assembly.

To install:

6. Installation is the reverse of removal.

HEATER & A/C UNIT

REMOVAL & INSTALLATION

See Figure 158.

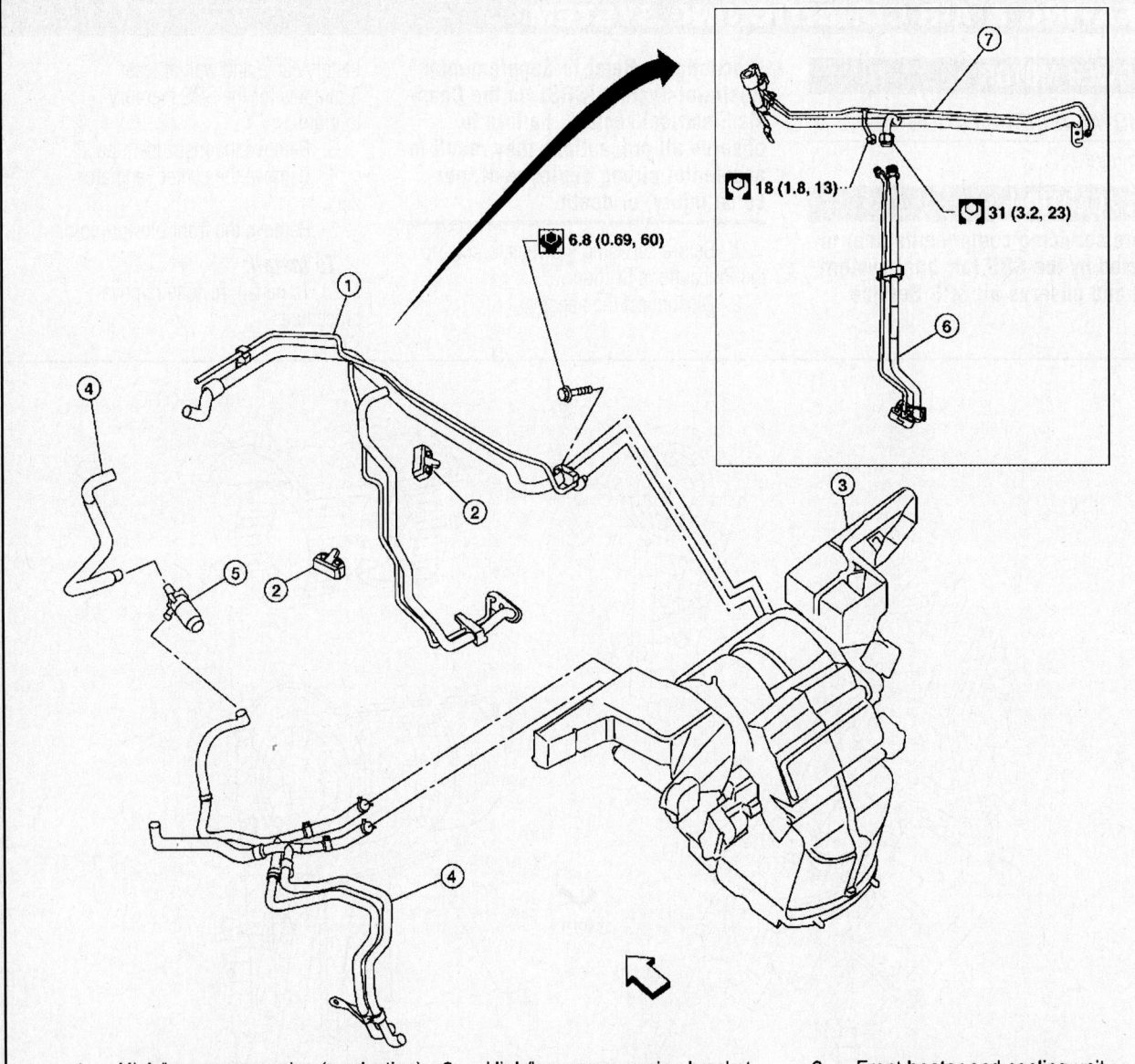

1. High/low pressure pipe (production)
2. High/low pressure pipe bracket
3. Front heater and cooling unit assembly
4. Front heater core pipe and hose assembly
5. Heater pump
6. High/low pressure pipe - lower (service)
7. High/low pressure pipe - upper (service)
⇐ Front

42050_QUES_G0044

Fig. 158 Heater and cooling unit

1. Before servicing the vehicle, refer to the Precautions Section.
2. Disconnect the negative battery cable and wait at least 3 minutes for the SRS memory to drain.
3. Discharge the refrigerant from the A/C system.

4. Drain the engine cooling system.
5. Remove the cowl top extension. Refer to Cowl Top Removal & Installation in the Body Exterior section.
6. Disconnect the front heater hoses from the front heater core.
7. Disconnect the high/low pressure pipe from the front expansion valve.
8. Move the two front seats to the rearmost position on the seat track.
9. Remove the instrument panel. Refer to Instrument Panel Removal & Installation in the Body Interior section.
10. Disconnect the instrument panel wire harness at the right-hand and left-hand in-line connector brackets, and the fuse block electrical connectors.

11. Disconnect the steering member from each side of the vehicle body.
12. Carefully remove the front heater and cooling unit assembly with it attached to the steering member.
13. Remove the front heater and cooling unit assembly from the steering member.

To install:
14. Installation is the reverse of removal, noting the following:
15. Replace the O-ring of the low-pressure flexible hose and high-pressure flexible hose with a new

one, and apply compressor oil to it when installing it.
16. Fill the engine cooling system.
17. Recharge the A/C system and check for leaks.

STEERING

POWER RACK & PINION STEERING GEAR

REMOVAL & INSTALLATION
See Figures 159 through 163.

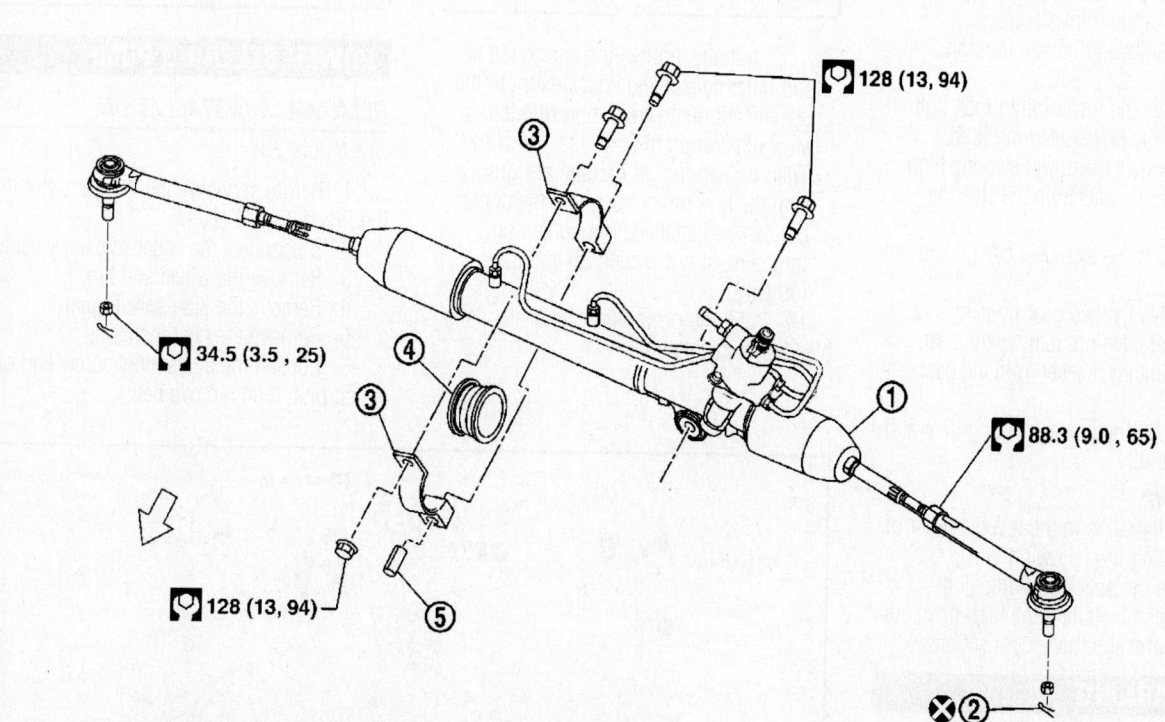

128 (13, 94)
34.5 (3.5 , 25)
128 (13, 94)
88.3 (9.0 , 65)

1. Steering gear assembly
2. Cotter pin
3. Gear housing mounting bracket
4. Gear housing insulator
5. Sleeve
⇐ Front

37663_QUES_G0292

Fig. 159 Power steering gear

1. Before servicing the vehicle, refer to the Precautions Section.

2. Disconnect the negative battery cable and wait at least 3 minutes for the SRS memory to drain.

※※ WARNING

The rotation of the driver air bag spiral cable is limited. If the steering gear must be removed, set the front wheels in the straight-ahead direction. Do not rotate the steering column while the steering gear is removed. The steering wheel and spiral cable must be removed before removing the steering lower joint to avoid damaging the supplemental restraint system spiral cable.

3. Remove the steering wheel and spiral cable. Refer to Steering Wheel Removal & Installation.

4. Remove the two front wheels.

5. Remove the cotter pins and nuts, and disconnect the outer tie-rod ends using an appropriate tool [Tool No. HT72520000 (J-25730-A)].

6. Disconnect the outer stabilizer bar ends from the connecting rods.

7. Remove the front stabilizer bar bracket rear bolts and loosen the front bolts.

8. Remove the lower joint pinch bolt.

9. Drain the power steering fluid.

10. Disconnect the power steering high and low pressure lines from the steering gear.

11. Position the stabilizer bar up and out of the way.

12. Remove the two gear housing mounting bolts. Do not remove the gear housing mounting bracket from the gear housing.

13. Remove the power steering gear and linkage assembly.

To install:

14. Installation is in the reverse order of removal, noting the following:

a. Use the specified tightening torque when installing the high pressure and low-pressure hose connections.

※※ WARNING

Excessive tightening will damage threads of connection or O-ring.

b. The O-ring in low-pressure hose connector is larger than that in high-pressure connector. Take care to install the proper O-ring.

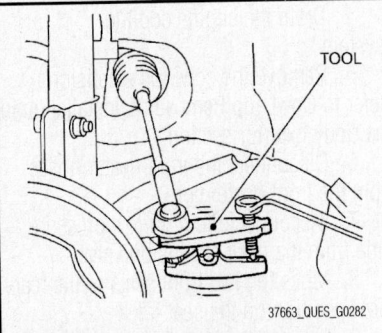

Fig. 160 Disconnect the outer tie-rod end from the steering knuckle using the appropriate tool

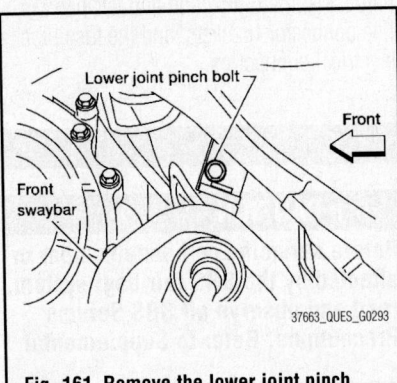

Fig. 161 Remove the lower joint pinch bolt

c. Initially, tighten the nut on the tie-rod outer socket and knuckle arm to the specification shown in the exploded view illustration of 25 ft. lbs. (34 Nm). Then tighten further to align nut groove with the first pin hole so that the cotter pin can be installed. The tightening torque must not exceed 36 ft. lbs. (49 Nm).

15. Refill the power steering system and bleed after installation.

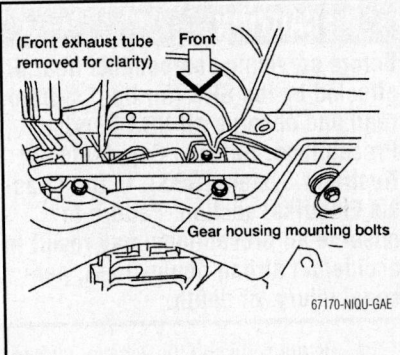

Fig. 162 Steering gear housing mounting bolts

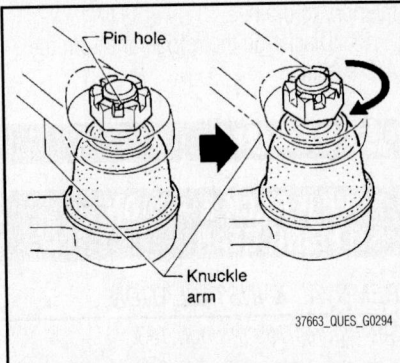

Fig. 163 Align nut groove with the first pin hole

POWER STEERING PUMP

REMOVAL & INSTALLATION

See Figure 164.

1. Before servicing the vehicle, refer to the Precautions Section.

2. Disconnect the negative battery cable.

3. Remove the wheel and tire.

4. Remove the side splash guard.

5. Remove the heat insulator.

6. Loosen the adjustment screw and oil pump bolt, then remove belt.

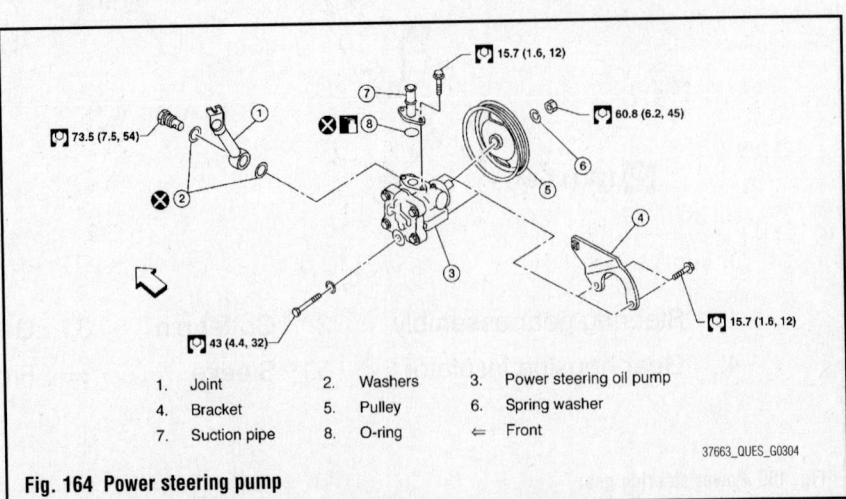

1. Joint	2. Washers	3. Power steering oil pump
4. Bracket	5. Pulley	6. Spring washer
7. Suction pipe	8. O-ring	⇐ Front

Fig. 164 Power steering pump

7. Drain the power steering fluid.

8. Remove the oil pump union bolts and hose.

9. Remove the oil pump bracket bolts.

10. Remove the power steering pump.

To install:

11. Installation is in the reverse order of removal.

12. Adjust belt tension.

13. Refill power steering system and bleed air after installation.

BLEEDING

1. Safely raise the front end of vehicle until the wheels are clear of the ground.

2. Add Genuine NISSAN PSF or equivalent into the steering fluid reservoir tank to the specified level. Then quickly turn the steering wheel fully to right and left and lightly touch steering

stoppers. Repeat steering wheel operation until the fluid level no longer decreases.

3. Start the engine then repeat step 2 above. Incomplete air bleeding will cause the following symptoms:

- Air bubbles in reservoir tank
- Clicking noise in oil pump
- Excessive buzzing in oil pump

4. If this happens, bleed out the air repeating step 2 above. Fluid noise may occur in the valve or oil pump. This is common when the vehicle is stationary or while turning the steering wheel slowly. This does not affect the performance or durability of the system.

SUSPENSION   FRONT SUSPENSION

CONTROL LINKS

REMOVAL & INSTALLATION

See Stabilizer Bar Removal & Installation.

STEERING KNUCKLE

REMOVAL & INSTALLATION

See Wheel Hub & Bearing (sealed unit).

STRUT

REMOVAL & INSTALLATION

See Figures 165 through 167.

1. Before servicing the vehicle, refer to the Precautions Section.

2. Remove the wheel and tire.

3. Remove the cowl top and cowl top extension. Refer to Cowl Top Removal & Installation in the Body Exterior section.

4. Disconnect the wheel sensor wire and front brake hose from the brackets on the front shock absorber (strut).

5. Disconnect the connecting rod upper link.

6. Support the wheel hub and steering knuckle assembly with a suitable wire.

7. Remove the lower shock absorber (strut) bolts and nuts.

8. Remove the three upper strut mounting nuts.

❊❊ WARNING

Do not remove the piston rod lock nut on vehicle.

9. Remove the coil spring and shock absorber (strut) assembly.

To install:

10. Installation of the assembly is in the reverse order of removal, noting the following:

a. Make sure the front strut spacer is positioned as illustrated.

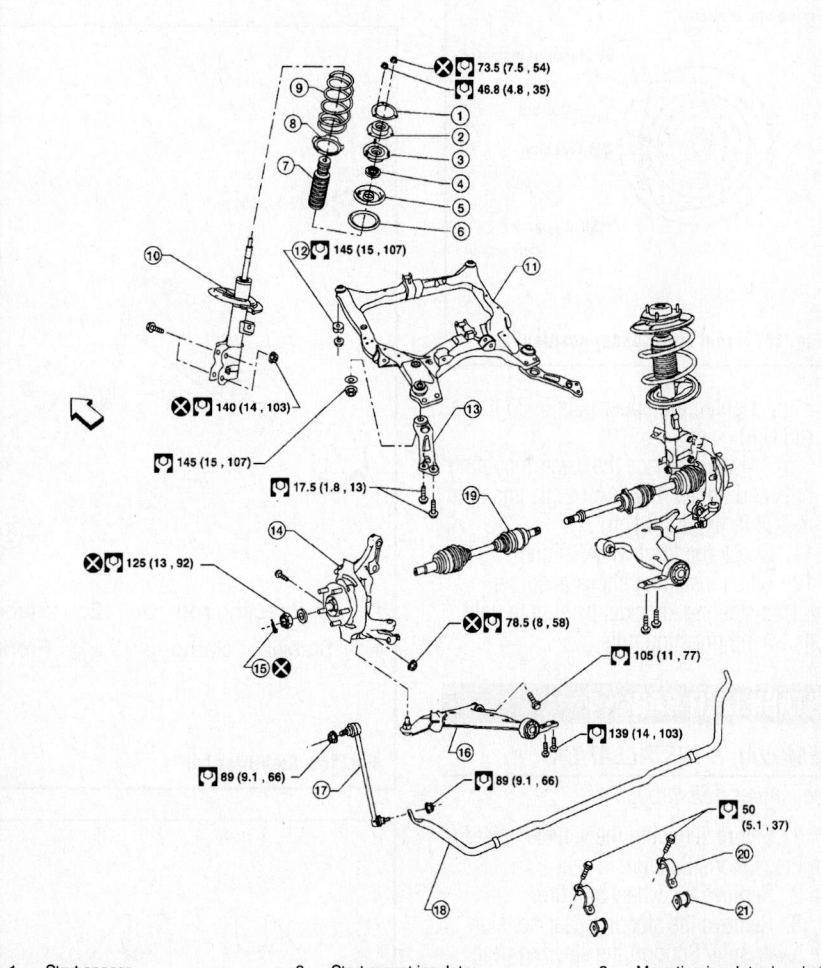

1.	Strut spacer	2.	Strut mount insulator	3.	Mounting insulator bracket
4.	Thrust bearing	5.	Upper spring seat	6.	Upper rubber seat
7.	Dust cover	8.	Lower rubber seat	9.	Coil spring
10.	Shock absorber (Strut)	11.	Front suspension member	12.	Cup
13.	Member pin stay	14.	Wheel hub and steering knuckle assembly	15.	Cotter pin
16.	Transverse link	17.	Connecting rod	18.	Stabilizer bar
19.	Drive shaft	20.	Stabilizer clamp	21.	Stabilizer bushing

37663_QUES_G0274

Fig. 165 Exploded view of the front suspension assembly

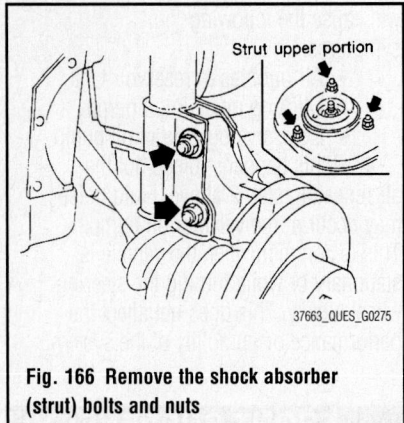

Fig. 166 Remove the shock absorber (strut) bolts and nuts

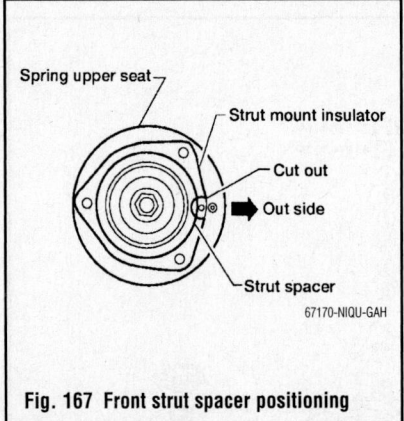

Fig. 167 Front strut spacer positioning

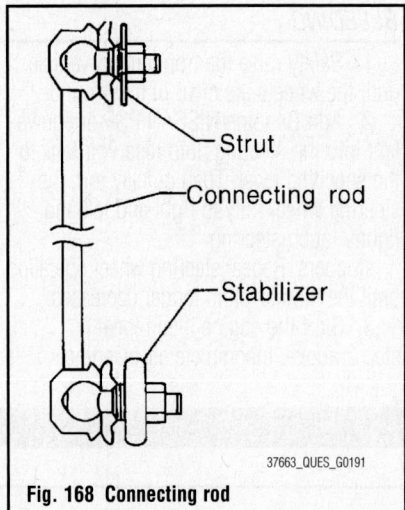

Fig. 168 Connecting rod

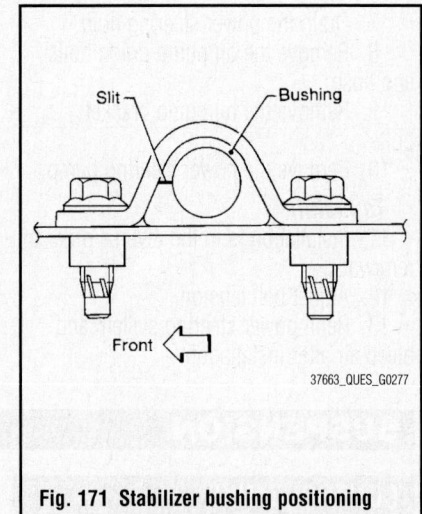

Fig. 171 Stabilizer bushing positioning

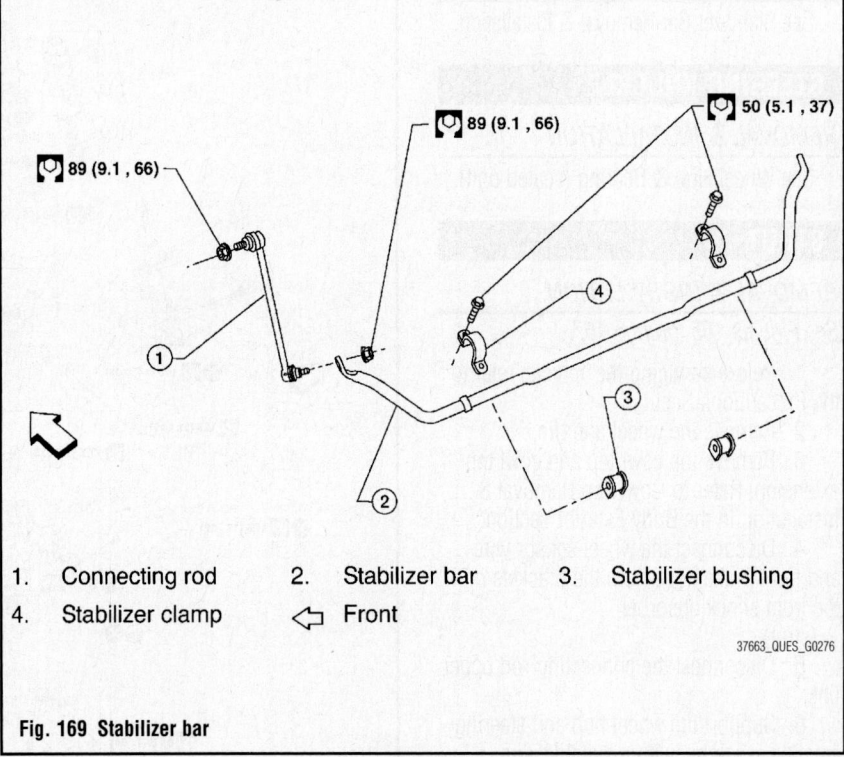

1. Connecting rod   2. Stabilizer bar   3. Stabilizer bushing
4. Stabilizer clamp   ⇦ Front

Fig. 169 Stabilizer bar

b. Tighten the upper nuts to 35 ft. lbs. (46 Nm)

c. Always replace the lower mounting nuts and tighten the lower bolts and nuts to 103 ft. lbs. (140 Nm).

11. Check the front wheel alignment.

12. When installing shock absorber (strut) to steering knuckle, be sure to hold bolts when tightening nuts.

STABILIZER BAR

REMOVAL & INSTALLATION

See Figures 168 through 172.

1. Before servicing the vehicle, refer to the Precautions Section.

2. Remove the wheel and tire.

3. Remove the steering gear bolts on the lower side. Support the steering gear.

4. Disconnect the connecting rod end at the stabilizer bar. Prevent the connecting rod from turning by inserting a hex wrench into the end of the ball stud, then remove nut.

5. Remove the two stabilizer bar clamps from the front suspension member.

6. Remove the stabilizer bar by withdrawing from the side.

7. Remove the two stabilizer bushings as necessary.

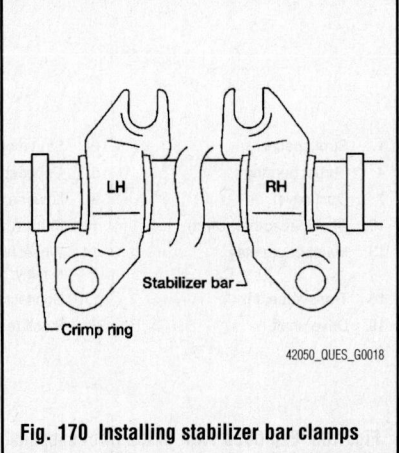

Fig. 170 Installing stabilizer bar clamps

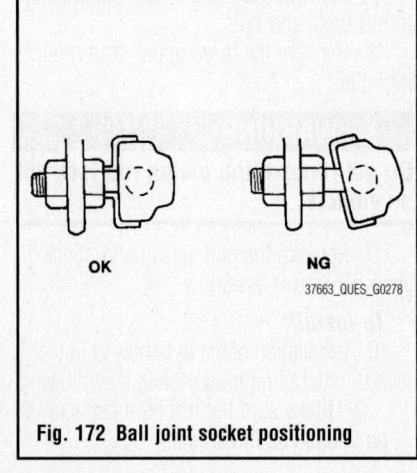

Fig. 172 Ball joint socket positioning

To install:

8. To install, reverse removal procedure.

9. When installing stabilizer bar, make sure that the stabilizer clamps are facing in the correct direction shown.

10. Make sure that the slit in the stabilizer bushing is in the position as shown.

11. Lubricate the inner and outer surfaces of the stabilizer bushing using a silicone lubricant.

12. Install stabilizer bar with ball joint socket properly placed as shown.

TRANSVERSE LINK

REMOVAL & INSTALLATION

See Figures 165 and 173.

1. Before servicing the vehicle, refer to the Precautions Section.

2. Remove the wheel and tire.

3. Remove the lower ball joint pinch bolt and nut.

4. Separate the transverse link from the wheel hub and steering knuckle assembly.

5. Remove the two transverse link pivot bolts and transverse link bolt using power tool.

6. Remove the transverse link.

To install:

7. To install, reverse removal procedure.

➡**During installation, final tightening must be carried out at curb weight with wheels on the ground.**

8. After installation, check wheel alignment.

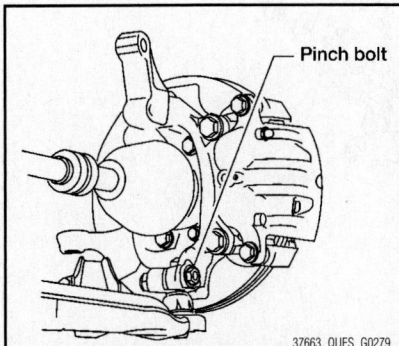

Fig. 173 Remove the lower ball joint pinch bolt and nut

WHEEL HUB & BEARING

REMOVAL & INSTALLATION

See Figures 174 through 176.

1. Before servicing the vehicle, refer to the Precautions Section.

2. Remove the wheel and tire.

3. Remove the brake caliper, leaving the line attached and suspend the caliper using a piece of wire.

4. Place alignment marks on the brake rotor and the wheel hub and bearing assembly, and remove the rotor.

5. Remove the wheel sensor from the steering knuckle.

6. Remove the cotter pin and lock nut from the drive shaft.

7. Remove the steering outer tie-rod cotter pin at the steering knuckle, and loosen the mounting nut.

8. Disconnect the outer tie-rod end from the steering knuckle using the appropriate tool [Tool No. HT72520000 (J-25730-A)]. Be careful not to damage ball joint boot.

9. To prevent damage to the threads and to prevent the tool from coming off suddenly, temporarily tighten mounting nut.

10. Remove the transverse link and steering knuckle pinch bolt and nut.

11. Remove the wheel hub and bearing assembly from the drive shaft using a suitable puller.

➡**When removing the wheel hub and bearing assembly, do not apply an excessive angle to drive shaft joint. Be careful not to excessively extend the slide joint. Support the drive shaft when removing.**

12. Remove the wheel hub and bearing assembly bolts.

13. Remove the splash guard and the wheel hub and bearing assembly from the steering knuckle.

14. Remove the lower strut bolts and nuts using power tool.

15. Remove steering knuckle from vehicle.

To install:

16. Installation is in the reverse order of removal, noting the following:

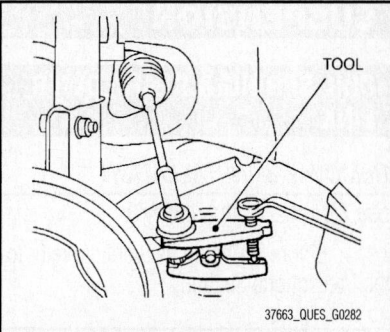

Fig. 175 Disconnect the outer tie-rod end from the steering knuckle using the appropriate tool

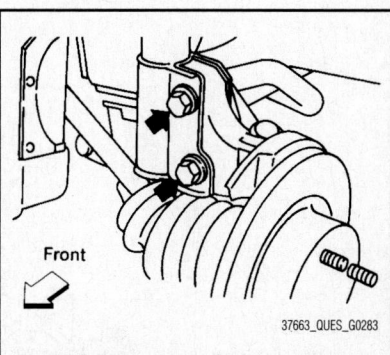

Fig. 176 Remove the wheel hub and bearing assembly bolts

a. When installing the wheel hub and bearing assembly to steering knuckle, align cutout in toner ring cover with wheel sensor mounting hole in steering knuckle.

b. When installing the rotor on the wheel hub and bearing assembly, align the marks made prior to removal.

c. A front end alignment is recommended.

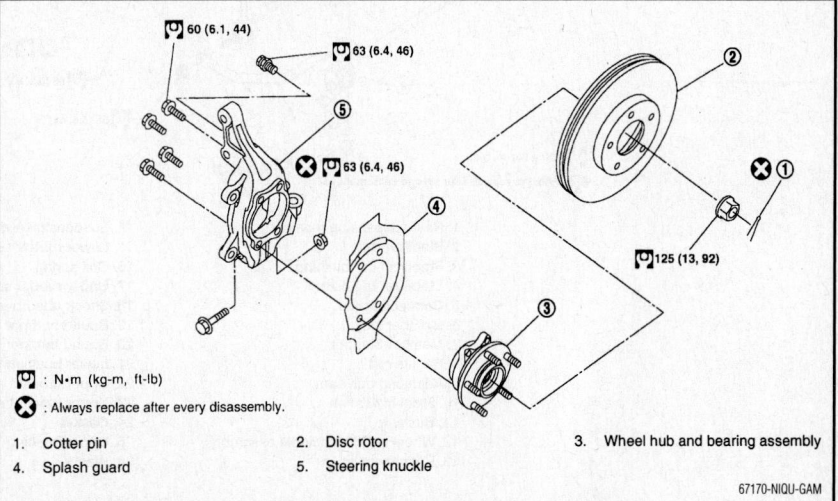

Fig. 174 Exploded view of the front wheel hub and bearing components

1. Cotter pin
2. Disc rotor
3. Wheel hub and bearing assembly
4. Splash guard
5. Steering knuckle

COIL SPRING & REAR LOWER LINK

REMOVAL & INSTALLATION

See Figures 177 through 179.

1. Before servicing the vehicle, refer to the Precautions Section.

2. Remove the wheel and tire assembly.

3. Position a transmission jack or suitable tool to relieve the coil spring tension and support the rear lower link.

4. Loosen the rear lower link adjusting bolt and nut connected to the rear suspension member.

5. Remove the rear lower link bolt and nut from the wheel hub and spindle assembly.

6. Slowly lower the transmission jack to release the coil spring tension. Remove upper rubber seat, coil spring, and lower rubber seat from the rear lower link.

7. Remove the rear lower link adjusting

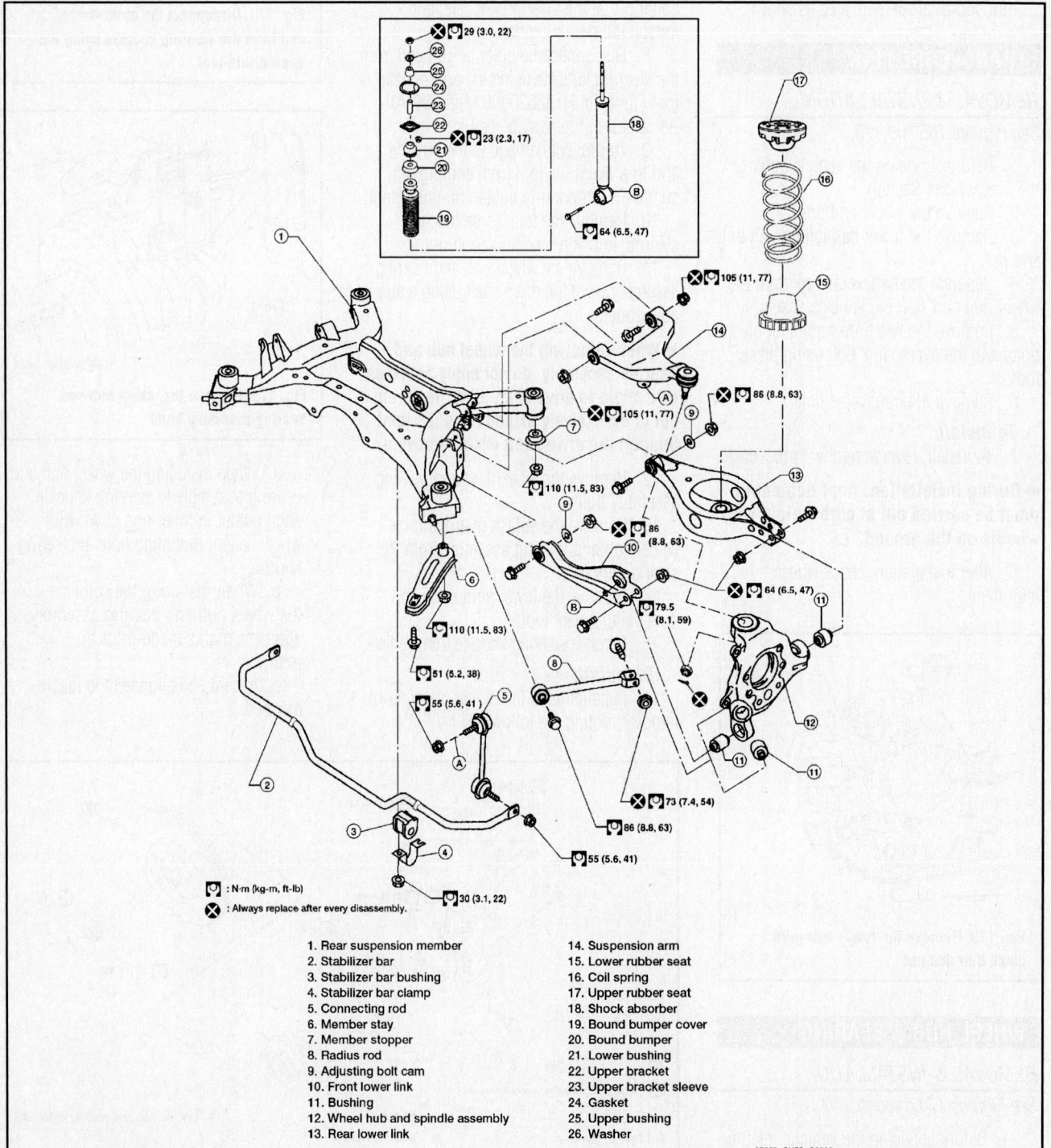

☐ : N·m (kg-m, ft-lb)

✕ : Always replace after every disassembly.

1. Rear suspension member
2. Stabilizer bar
3. Stabilizer bar bushing
4. Stabilizer bar clamp
5. Connecting rod
6. Member stay
7. Member stopper
8. Radius rod
9. Adjusting bolt cam
10. Front lower link
11. Bushing
12. Wheel hub and spindle assembly
13. Rear lower link
14. Suspension arm
15. Lower rubber seat
16. Coil spring
17. Upper rubber seat
18. Shock absorber
19. Bound bumper cover
20. Bound bumper
21. Lower bushing
22. Upper bracket
23. Upper bracket sleeve
24. Gasket
25. Upper bushing
26. Washer

37663_QUES_G0280

Fig. 177 Rear suspension assembly

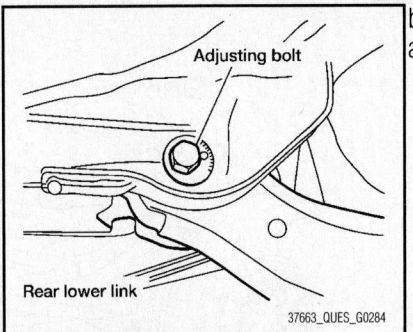

Fig. 178 Loosen the rear lower link adjusting bolt and nut connected to the rear suspension member

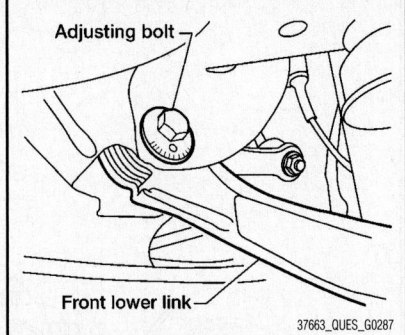

Fig. 180 Remove the front lower link adjusting bolt

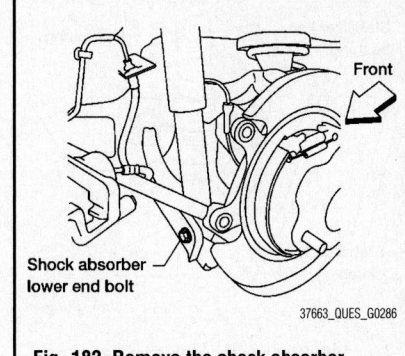

Fig. 182 Remove the shock absorber lower end bolt

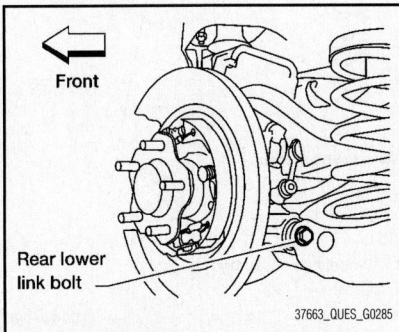

Fig. 179 Remove the rear lower link bolt and nut from the wheel hub and spindle assembly

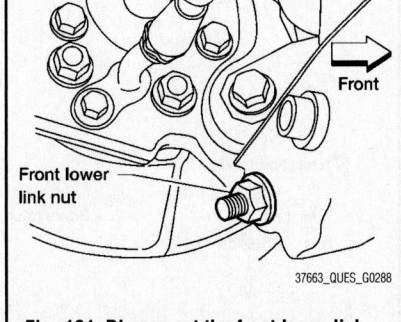

Fig. 181 Disconnect the front lower link nut and bolt

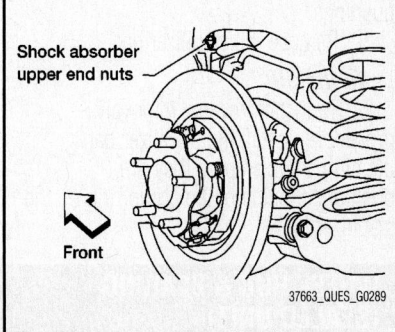

Fig. 183 Remove the shock absorber upper end nuts

bolt and nut from the rear suspension member using power tool.

8. Remove the rear lower link.

To install:

9. Installation is the reverse of removal.
10. Check the wheel alignment and adjust if necessary.

FRONT LOWER LINK

REMOVAL & INSTALLATION

See Figures 177, 180 and 181.

1. Before servicing the vehicle, refer to the Precautions Section.
2. Raise and safely support the vehicle.
3. Remove the wheel and tire assembly.
4. Position a transmission jack or suitable tool to relieve the coil spring tension and support the front lower link.
5. Remove the shock absorber lower end bolt.
6. Remove the front lower link adjusting bolt.
7. Disconnect the front lower link nut and bolt from the wheel hub and spindle assembly.
8. Remove the front lower link.

To install:

9. Installation is the reverse of removal.

10. Check the wheel alignment and adjust if necessary.

SHOCK ABSORBER

REMOVAL & INSTALLATION

See Figures 177, 182 and 183.

1. Before servicing the vehicle, refer to the Precautions Section.
2. Raise and safely support the vehicle.
3. Remove the wheel and tire assembly.
4. Position a transmission jack or suitable tool under the rear lower link to relieve the coil spring tension.
5. Remove the shock absorber lower end bolt.
6. Remove the transmission jack supporting the rear lower link.
7. Remove the shock absorber upper end nuts.
8. Remove the shock absorber from the vehicle.

To install:

9. Installation is the reverse of removal.

INSPECTION

1. Check for smooth operation through a full stroke, both compression and extension.

2. Check for oil leakage.
3. Check piston rod for cracks, deformation or other damage and replace if necessary.

STABILIZER BAR

REMOVAL & INSTALLATION

See Figures 177, 184 and 185.

1. Before servicing the vehicle, refer to the precautions section.
2. Disconnect the stabilizer bar ends from the connecting rods.
3. Remove the stabilizer bar clamps.

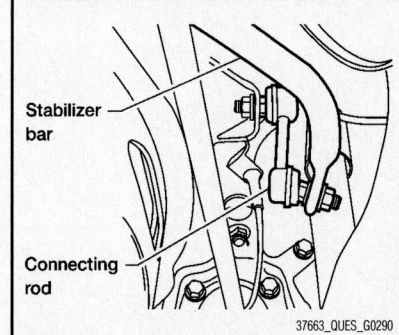

Fig. 184 Disconnect the stabilizer bar ends from the connecting rods

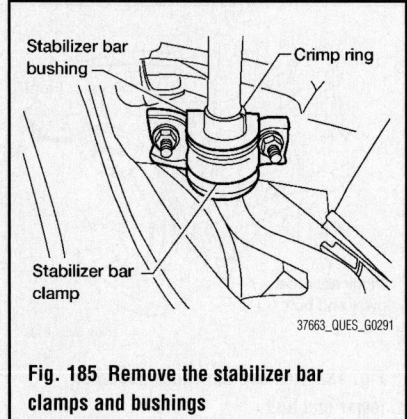

Fig. 185 Remove the stabilizer bar clamps and bushings

4. Remove the stabilizer bar bushings.

5. Remove the stabilizer bar.

To install:

6. To install, reverse removal procedure. Install the stabilizer bar bushing and clamp so they are positioned inside of the crimp ring on the stabilizer bar.

WHEEL HUB & BEARING (SEALED UNIT)

REMOVAL & INSTALLATION

See Figure 186.

1. Before servicing the vehicle, refer to the precautions section.

2. Remove the rear wheel.

3. Remove the brake caliper assembly without disconnecting the hydraulic line and suspend the caliper aside using wire.

4. Release the parking brake and remove the rotor.

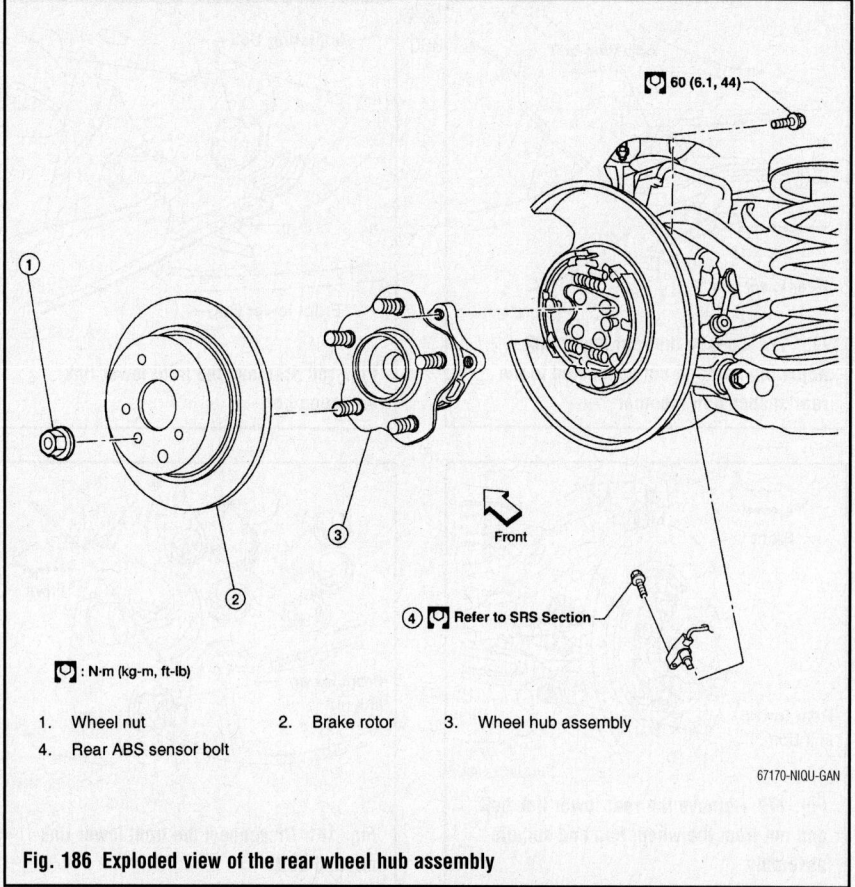

| : N·m (kg-m, ft-lb)

1. Wheel nut   2. Brake rotor   3. Wheel hub assembly
4. Rear ABS sensor bolt

Fig. 186 Exploded view of the rear wheel hub assembly

5. Remove the rear ABS sensor, and position it aside using wire.

6. Remove the wheel hub assembly from the knuckle.

7. Check for cracks, and damage on the wheel hub assembly and replace as necessary.

To install:

8. Installation is in the reverse order of removal. Tighten the wheel hub bolts to 44 ft. lbs. (60 Nm).

9. Check that the wheel bearing operates smoothly.

10. Check that the wheel hub bearing axial end play is within 0.002 in. (0.05mm) or less using a dial indicator.

NISSAN

Rogue

18

SPECIFICATIONS AND MAINTENANCE CHARTS

ENGINE AND VEHICLE IDENTIFICATION

Engine							Model Year	
Code ①	Liters (cc)	Cu. In.	Cyl.	Fuel Sys.	Engine Type	Eng. Mfg.	Code ②	Year
QR25DE	2.5 (2488)	151.82	4	MPI	DOHC	Nissan	9	2009
							10	2010

SFI: Sequential Fuel Injection

DOHC: Double Overhead Camshaft

① Stamped on the left side of the engine block

② 10th digit of the Vehicle Identification Number (VIN)

37663_ROGU_C0001

GENERAL ENGINE SPECIFICATIONS

Year	Model	Engine Displacement Liters	Engine Series ID	Net Horsepower @ rpm	Net Torque @ rpm (ft. lbs.)	Bore x Stroke (in.)	Com-pression Ratio	Oil Pressure @ rpm
2009	Rogue	2.5	QR25DE	170@6000	175@4400	3.50x3.94	9.6:1	43@2000
2010	Rogue	2.5	QR25DE	170@6000	175@4400	3.50x3.94	9.6:1	43@2000

37663_ROGU_C0002

ENGINE TUNE-UP SPECIFICATIONS

Year	Engine Disp. Liters	Engine ID	Spark Plug Gap (in.)	Ignition Timing (deg.)	Fuel Pump (psi)	Idle Speed (rpm)	Valve Clearance Cold Intake	Valve Clearance Cold Exhaust
2009	2.5	QR25DE	0.043	10 BTDC ①	51	700+/-50 ①	0.009-0.013	0.010-0.013
2010	2.5	QR25DE	0.043	10 BTDC ①	51	700+/-50 ①	0.009-0.013	0.010-0.013

NOTE: The Vehicle Emission Control Information label often reflects specification changes made during production.

The label figures must be used if they differ from those in this chart.

① Component is computer controlled and is not adjustible.

37663_ROGU_C0003

CAPACITIES

Year	Model	Engine Displacement Liters	Engine ID	Engine Oil with Filter (qts.)	Auto Trans. (qts.)	Transfer Case (pts)	Rear Drive Axle (pts.) ①	Fuel Tank (gal.)	Cooling System (qts.)
2009	Rogue	2.5	QR25DE	4.9	②	1.2	0.6	15.8	7.9 ③
2010	Rogue	2.5	QR25DE	4.9	②	1.2	0.6	15.8	7.9 ③

① Synthetic GL-5 (75W-90) or equivalent

② 2WD (8qts) and AWD (9 qts)

③ The use of genuine Nissan engine coolant is recommended or similar ethylene glycol based non-silicate, non-amine, non- nitrite, and non- borat coolant

37663_ROGU_C0005

FLUID SPECIFICATIONS

Year	Model	Engine Displacement Liters	Engine ID/VIN	Engine Oil	Auto. Trans. ①	Drive Axle ②	Power Steering Fluid	Brake Master Cylinder	Engine Coolant ③
2009	Rogue	2.5	QR25DE	5W-30	Nissan	80W-90	NA	DOT 3	Nissan
2010	Rogue	2.5	QR25DE	5W-30	Nissan	80W-90	NA	DOT 3	Nissan

NA: Not Available

DOT: Department Of Transpotation

① Using trasmission fluid other than genuine Nissan CVT fluid NS-2 will damage the CVT

② GL-5 (80W-90) or equivalent

③ The use of genuine Nissan engine coolant is recommended

37663_ROGU_C0004

VALVE SPECIFICATIONS

Year	Engine Displacement Liters	Engine ID	Seat Angle (deg.)	Face Angle (deg.)	Inner Spring free length (in.)	Spring Installed Height (in.)	Stem-to-Guide Clearance (in.) Intake	Stem-to-Guide Clearance (in.) Exhaust	Stem Diameter (in.) Intake	Stem Diameter (in.) Exhaust
2009	2.5	QR25DE	45	NS	①	1.390	0.0008-0.0021	0.0012-0.0025	0.234-0.235	0.234-0.235
2010	2.5	QR25DE	45	NS	①	1.390	0.0008-0.0021	0.0012-0.0025	0.234-0.235	0.234-0.235

NS: Not Supplied

① Intake (1.7213-1.7291) and Exhuast (1.7831-1.7909)

37663_ROGU_C0006

CAMSHAFT AND BEARING SPECIFICATIONS CHART

All measurements are given in inches.

Year	Engine Displ. Liters	Engine ID/VIN	Journal Dia.	Brg. Oil Clearance	Shaft End-play	Runout	Journal Bore	Lobe Height	
								Intake	Exhaust
2009	2.5	QR25DE	①	0.0018-0.0034	0.0045-0.0074	0.0008	NA	②	1.7313-1.7388
2010	2.5	QR25DE	①	0.0018-0.0034	0.0045-0.0074	0.0008	NA	②	1.7313-1.7388

NA: Not Available

① Journal 1: 1.0998-1.1006
   All Others: 0.9226-0.9234

② Non-California: 1.7644-1.7718
   California: 1.7722-1.7797

37663_ROGU_C0008

CRANKSHAFT AND CONNECTING ROD SPECIFICATIONS

All measurements are given in inches.

Year	Engine Displacement Liters	Engine ID	Crankshaft				Connecting Rod		
			Main Brg. Journal Dia.	Main Brg. Oil Clearance	Shaft End-play	Thrust on No.	Journal Diameter	Oil Clearance	Side Clearance
2009	2.5	QR25DE	2.1636-2.1645	①	0.0039-0.0102	3	1.7699-1.7706	0.0014-0.0018	0.0079-0.0138
2010	2.5	QR25DE	2.1636-2.1645	①	0.0039-0.0102	3	1.7699-1.7706	0.0014-0.0018	0.0079-0.0138

① Nos. 1, 3 and 5: 0.0005-0.0009 in.
   Nos. 2 and 4: 0.0007-0.0011

37663_ROGU_C0007

PISTON AND RING SPECIFICATIONS

All measurements are given in inches.

Year	Engine Displ. Liters	Engine ID	Piston Clearance	Ring Gap			Ring Side Clearance		
				Top Comp.	Bottom Comp.	Oil Control	Top Comp.	Bottom Comp.	Oil Control
2009	2.5	QR25DE	0.0004-0.0012	0.0091-0.0130	0.0130-0.0189	0.0079-0.0177	0.0016-0.0031	0.0012-0.0028	0.0018-0.0049
2010	2.5	QR25DE	0.0004-0.0012	0.0091-0.0130	0.0130-0.0189	0.0079-0.0177	0.0016-0.0031	0.0012-0.0028	0.0018-0.0049

37663_ROGU_C0009

TORQUE SPECIFICATIONS
All readings in ft. lbs.

Year	Engine Displacement Liters	Engine ID	Cylinder Head Bolts	Main Bearing Bolts	Rod Bearing Bolts	Crankshaft Damper Bolts	Flywheel Bolts	Manifold Intake	Manifold Exhaust	Spark Plugs	Oil Pan Drain Plug
2009	2.5	QR25DE	①	②	③	④	76-83	13-15	29-32	14	25
2010	2.5	QR25DE	①	②	③	④	76-83	13-15	29-32	14	25

① Step 1: 37 ft. lbs.

   Step 2: 60-66 degrees

   Step 3: Loosen bolts completely, then retorque to 26-32 ft. lbs.

   Step 4: Turn each bolt, in sequence, an additional 75-80 degrees

   Step 5: Repeat, turn each bolt, in sequence, an additional 75-80 degrees

② Bolt Nos. 1-10:

   Step 1: 27-31 ft. lbs.

   Step 2: Torque an additional 60-65 degrees

   Bolt Nos. 11-14: Torque last, to 17-20 ft. lbs.

③ Step 1: 20 ft. lbs.

   Step 2: Loosen bolts completely

   Step 3: 14 ft. lbs.

   Step 4: 85-95 degrees

④ Step 1: 29-36 ft. lbs.

   Step 2: 60-66 degrees

37663_ROGU_C0010

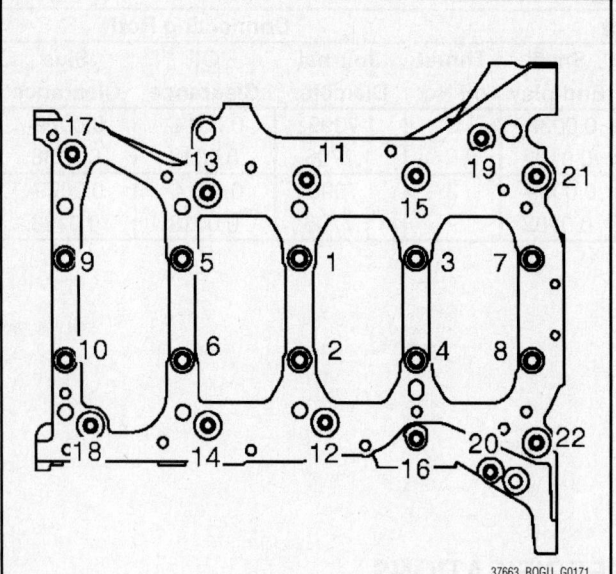

Fig. 1 Main Bearing Torque Sequence

37663_ROGU_G0171

WHEEL ALIGNMENT

Year	Model		Caster Range (+/-Deg.)	Caster Preferred Setting (Deg.)	Camber Range (+/-Deg.)	Camber Preferred Setting (Deg.)	Toe-in (in.)
2009	Rogue	F	0.75	+4.75	0.75	①	0.08+/-0.04
		R	NA	NA	0.50	-0.92	0.08+/-0.08
2010	Rogue	F	0.75	+4.75	0.75	①	0.08+/-0.04
		R	NA	NA	0.50	-0.92	0.08+/-0.08

NA: Not Applicable

F: Front

R: Rear

① Left side front camber -0.25

   Right side front front camber -0.50

37663_ROGU_C0011

TIRE, WHEEL AND BALL JOINT SPECIFICATIONS

Year	Model	OEM Tires Standard	OEM Tires Optional	Tire Pressures (psi) Front	Tire Pressures (psi) Rear	Wheel Size	Ball Joint Inspection	Lug Nut Torque (ft. lbs.)
2009	Rogue	P215/70R16	225/60R17	33	33	6.5-JJ and 7-JJ	①	80
2010	Rogue	P215/70R16	225/60R17	33	33	6.5-JJ and 7-JJ	①	80

OEM: Original Equipment Manufacturer

PSI: Pounds Per Square Inch

STD: Standard

OPT: Optional

① Replace if any measurable movement is found.

37663_ROGU_C0012

BRAKE SPECIFICATIONS

All measurements in inches unless noted

Year	Model		Brake Disc Original Thickness	Brake Disc Minimum Thickness	Brake Disc Maximum Runout	Minimum Lining Thickness	Brake Caliper Bracket Bolts (ft. lbs.)	Brake Caliper Mounting Bolts (ft. lbs.)
2009	Rogue	F	1.024	0.945	0.0014	0.079	122	25
		R	0.630	0.551	0.0028	0.059	62	32
2010	Rogue	F	1.024	0.945	0.0014	0.079	122	25
		R	0.630	0.551	0.0028	0.059	62	32

F: Front

R: Rear

37663_ROGU_C0013

SCHEDULED MAINTENANCE INTERVALS (2)
Nissan—Rogue

TO BE SERVICED	SERVICE	7.5	15	22.5	30	37.5	45	52.5	60
Engine oil & filter	R	✓	✓	✓	✓	✓	✓	✓	✓
Brake lines & cables	S/I		✓		✓		✓		✓
Brake pads, discs	I		✓		✓		✓		✓
Driveshaft boots & propeller shaft	L/I		✓		✓		✓		✓
CVT ①	I		✓		✓		✓		✓
Transfer case and differential fluid ②	I		✓		✓		✓		✓
Air cleaner filter	R				✓				✓
Drive belt (s) ③	S/I								✓
Engine coolant ④	R								✓
Spark plugs	R	Platinum plugs, every 105,000 miles							
Cabin air filter	R		✓		✓		✓		✓
Exhaust system	I				✓				✓
Evap vapor lines	I				✓				✓
Fuel lines	S/I				✓				✓
Steering gear, linkage, axle & suspension parts	I	✓	✓	✓	✓	✓	✓	✓	✓
Tires (rotate)	S/I	every 5,000-6,000 miles							
Valve clearance ⑤	S/I								

R: Replace   S/I: Service or Inspect   L: Lubricate

① If towing a trailer, using a camper or a car-top carrier, or driving on rough or muddy roads, change (not just inspect) oil at every 60,000 miles.

② If towing a trailer, using a camper or a car-top carrier, or driving on rough or muddy roads, change (not just inspect) oil at every 30,000 miles (48,000 km) or 24 months.

③ First at 60,000, then every 15,000 miles

④ After 60,000, replace every 30,000

⑤ Periodic maintenance not required, if valve noice increases, inspect valve clearance

Follow Periodic Maintenance Schedule 1 if the driving habits frequently include one or more of the following driving conditions:

Repeated short trips of less than 5 miles (8 km).

Repeated short trips of less than 10 miles (16 km) with outside temperatures remaining below freezing

Operating in hot weather in stop-and-go "rush hour" traffic.

Extensive idling and/or low speed driving for long distances, such as police, taxi or door-to-door delivery use

Driving in dusty conditions.

Driving on rough, muddy, or salt spread roads.

Towing a trailer, using a camper or a car-top carrier.

Follow Periodic Maintenance Schedule 2 if none of driving conditions shown in Schedule 1 apply to the driving habits.

37663_ROGU_C0014

SCHEDULED MAINTENANCE INTERVALS (1)
Nissan—Rogue

TO BE SERVICED	TYPE OF SERVICE	7.5	15	22.5	30	37.5	45	52.5	60
Engine oil & filter	R	every 3,750 miles							
Brake lines & cables	S/I		✓		✓		✓		✓
Brake pads, discs	I	✓	✓	✓	✓	✓	✓	✓	✓
Driveshaft boots & propeller shaft	L/I	✓	✓	✓	✓	✓	✓	✓	✓
CVT ①	I		✓		✓		✓		✓
Transfer case and differential fluid ②	I		✓		✓		✓		✓
Air cleaner filter	R					✓			✓
Drive belt (s) ③	S/I								✓
Engine coolant ④	R								✓
Spark plugs	R	Platinum plugs, every 105,000 miles							
Cabin air filter	R		✓		✓		✓		✓
Exhaust system	I	✓	✓	✓	✓	✓	✓	✓	✓
Evap vapor lines	I				✓				✓
Fuel lines	S/I				✓				✓
Steering gear, linkage, axle & suspension parts	I	✓	✓	✓	✓	✓	✓	✓	✓
Tires (rotate)	S/I	every 5,000-6,000 miles							
Valve clearance ⑤	S/I								

R: Replace   S/I: Service or Inspect   L: Lubricate

① If towing a trailer, using a camper or a car-top carrier, or driving on rough or muddy roads, change (not just inspect) oil at every 60,000 miles.

② If towing a trailer, using a camper or a car-top carrier, or driving on rough or muddy roads, change (not just inspect) oil at every 30,000 miles (48,000 km) or 24 months.

③ First at 60,000, then every 30,000 miles

④ After 60,000, replace every 30,000

⑤ Periodic maintenance not required, if valve noice increases, inspect valve clearance

Follow Periodic Maintenance Schedule 1 if the driving habits frequently include one or more of the following driving conditions:

Repeated short trips of less than 5 miles (8 km).

Repeated short trips of less than 10 miles (16 km) with outside temperatures remaining below freezing

Operating in hot weather in stop-and-go "rush hour" traffic.

Extensive idling and/or low speed driving for long distances, such as police, taxi or door-to-door delivery use

Driving in dusty conditions.

Driving on rough, muddy, or salt spread roads.

Towing a trailer, using a camper or a car-top carrier.

Follow Periodic Maintenance Schedule 2 if none of driving conditions shown in Schedule 1 apply to the driving habits.

37663_ROGU_C0015

BRAKES · INFORMATION AND PRECAUTIONS

ANTI-LOCK SYSTEMS

• Certain components within the ABS system are not intended to be serviced or repaired individually.

• Do not use rubber hoses or other parts not specifically specified for and ABS system. When using repair kits, replace all parts included in the kit. Partial or incorrect repair may lead to functional problems and require the replacement of components.

• Lubricate rubber parts with clean, fresh brake fluid to ease assembly. Do not use shop air to clean parts; damage to rubber components may result.

• Use only DOT 3 brake fluid from an unopened container.

• If any hydraulic component or line is removed or replaced, it may be necessary to bleed the entire system.

• A clean repair area is essential. Always clean the reservoir and cap thoroughly before removing the cap. The slightest amount of dirt in the fluid may plug an orifice and impair the system function. Perform repairs after components have been thoroughly cleaned; use only denatured alcohol to clean components. Do not allow ABS components to come into contact with any substance containing mineral oil; this includes used shop rags.

• The Anti-Lock control unit is a microprocessor similar to other computer units in the vehicle. Ensure that the ignition switch is **OFF** before removing or installing controller harnesses. Avoid static electricity discharge at or near the controller.

• If any arc welding is to be done on the vehicle, the control unit should be unplugged before welding operations begin.

DISC AND DRUM SYSTEMS

✴✴ CAUTION

Dust and dirt accumulating on brake parts during normal use may contain asbestos fibers from production or aftermarket brake linings. Breathing excessive concentrations of asbestos fibers can cause serious bodily harm. Exercise care when servicing brake parts. Do not sand or grind brake lining unless equipment used is designed to contain the dust residue. Do not clean brake parts with compressed air or by dry brushing. Cleaning should be done by dampening the brake components with a fine mist of water, then wiping the brake components clean with a dampened cloth. Dispose of cloth and all residue containing asbestos fibers in an impermeable container with the appropriate label. Follow practices prescribed by the Occupational Safety and Health Administration (OSHA) and the Environmental Protection Agency (EPA) for the handling, processing, and disposing of dust or debris that may contain asbestos fibers.

BRAKES · BLEEDING THE BRAKE SYSTEM

BLEEDING PROCEDURE

BLEEDING PROCEDURE

✴✴ WARNING

Carefully monitor brake fluid level in the reservoir tank during bleeding operation.

✴✴ WARNING

Fill the sub tank with new brake fluid. Make sure it is full at all times while bleeding the air out of system.

1. Turn ignition switch OFF and disconnect ABS actuator and control unit connector or negative battery terminal.

2. Connect a vinyl tube to the bleeder valve.

3. Fully depress the brake pedal 4 to 5 times.

4. Loosen the bleeder valve and bleed air with the brake pedal depressed, and then quickly tighten the bleeder valve.

5. Repeat the above steps until all of the air is out of the brake line.

6. Tighten the bleeder valve to 73 inch lbs. (8 Nm).

7. Bleed the brake hydraulic system air bleeder valves in the following order:
• Right rear brake
• Left front brake
• Left rear brake
• Right front brake

BLEEDING THE ABS SYSTEM

Refer to the Bleeding Procedure.

BRAKES · ANTI-LOCK BRAKE SYSTEM (ABS)

WHEEL SPEED SENSORS

REMOVAL & INSTALLATION

Front Sensor

1. Before servicing the vehicle, refer to the Precautions Section.

2. Disconnect the negative battery cable.

✴✴ WARNING

Avoiding twisting the sensor harness as much as possible when removing it. Pull sensors out without pulling front or rear wheel hub. This is to avoid damage to sensor wiring and loss of sensor function. When you see the harness of the wheel sensor from the front side of the vehicle ensure that the yellow lines are not twisted. Avoiding twisting the sensor harness as much as possible when removing it. Pull sensors out without pulling on the sensor harness.

3. Disconnect the wheel speed sensor harness.

4. Remove the mounting bolt.

5. Remove the front speed sensor.

To install:

6. Install the front speed sensor.

7. Install the mounting bolt and tighten to 7 ft. lbs. (10 Nm).

8. Reconnect the wheel speed sensor harness.

9. Connect the negative battery cable.

Rear Sensor

1. Before servicing the vehicle, refer to the Precautions Section.

2. Disconnect the negative battery cable.

✳✳ WARNING

Avoiding twisting the sensor harness as much as possible when removing it. Pull sensors out without pulling front or rear wheel hub. This is to

avoid damage to sensor wiring and loss of sensor function. When you see the harness of the wheel sensor from the front side of the vehicle ensure that the yellow lines are not twisted. Avoiding twisting the sensor harness as much as possible, when removing it. Pull sensors out without pulling on the sensor harness.

3. Disconnect the wheel speed sensor harness.

4. Remove the mounting bolt.

5. Remove the front speed sensor.

To install:

6. Install the front speed sensor.

7. Install the mounting bolt and tighten to 7 ft. lbs. (10 Nm).

8. Reconnect the wheel speed sensor harness.

9. Connect the negative battery cable.

BRAKES   FRONT DISC BRAKES

BRAKE CALIPER

REMOVAL & INSTALLATION

See Figures 2 and 3.

1. Before servicing the vehicle, refer to the Precautions Section.

2. Remove the front wheels and tires.

3. Secure the disc rotor using wheel nuts.

4. Drain the brake fluid.

5. Remove the union bolt and

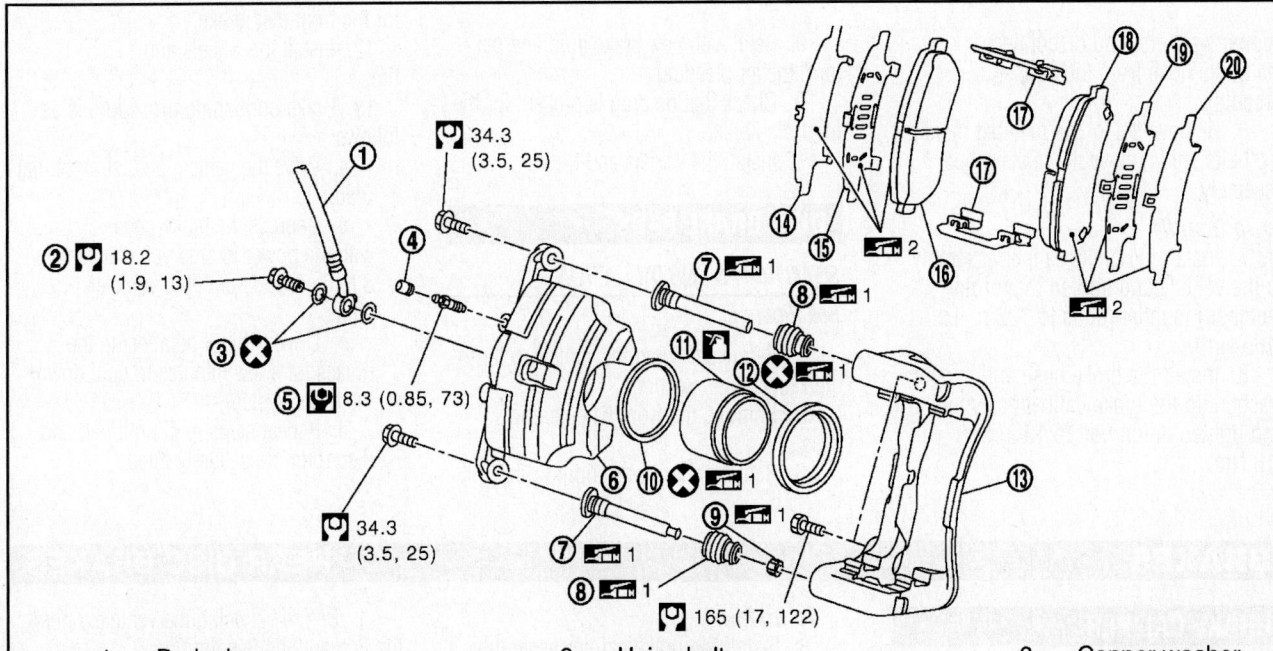

1.	Brake hose	2.	Union bolt	3.	Copper washer
4.	Cap	5.	Bleeder valve	6.	Cylinder body
7.	Sliding pin	8.	Sliding pin boot	9.	Bushing
10.	Piston seal	11.	Piston	12.	Piston boot
13.	Torque member	14.	Inner shim cover	15.	Inner shim
16.	Inner pad (only RH side with pad wear sensor)	17.	Pad retainer	18.	Outer pad
19.	Outer shim	20.	Outer shim cover		

⬛1: Apply rubber grease.

⬛2: Apply copper based brake grease.

⬛: Apply brake fluid.

22140_ROGU_G0020

Fig. 2 Front disc brake components

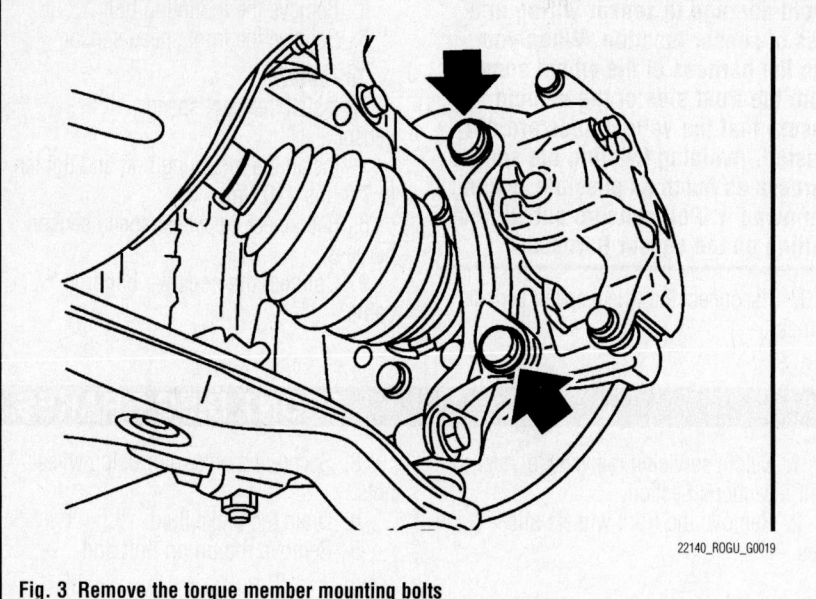

22140_ROGU_G0019

Fig. 3 Remove the torque member mounting bolts

copper washers, and disconnect the brake hose from the caliper assembly.

6. Remove the torque member mounting bolts, and remove the brake caliper assembly.

To install:

7. Install the brake caliper assembly to the vehicle and tighten the torque member mounting bolts to 122 ft. lbs. (165 Nm).

8. Install the brake hose and copper washers to the brake caliper assembly, and tighten union bolt to 13 ft. lbs. (18 Nm).

9. Refill with new brake fluid and perform the air bleeding.

10. Check that no drag is present for the front disc brake.

11. Install the wheels and tires.

DISC BRAKE PADS

REMOVAL & INSTALLATION

See Figure 2.

1. Before servicing the vehicle, refer to the Precautions Section.

2. Remove the front wheels and tires.

3. Remove the lower sliding pin bolt.

4. Suspend the cylinder body with suitable wire so that the brake hose will not stretch.

5. Remove the brake pad from the torque member.

To install:

6. Install the pad retainer to the torque member if the pad retainers have been removed.

7. Apply copper based brake grease and install them to the brake pad.

8. Install the cylinder body and brake pads to the torque member.

9. Use a disc brake piston tool to easily press the piston back into the caliper.

10. Install the lower sliding pin bolt and tighten.

11. Depress the brake pedal several times to check that no drag feel is present for the front disc brake.

12. Install the wheels and tires.

13. Brake burnishing procedure is as follows:

   a. Drive the vehicle on a straight, flat road.

   b. Depress the brake pedal with the power to stop vehicle within 3 to 5 seconds until the vehicle stops.

   c. Drive without depressing the brakes for a few minutes to cool down the brake system.

   d. Repeat steps A–C until pad and disc rotor are securely fitted.

BRAKES

BRAKE CALIPER

REMOVAL & INSTALLATION

See Figures 4 and 5.

1. Before servicing the vehicle, refer to the Precautions Section.

2. Clean any dust from the brake caliper and brake pads with a vacuum dust collector. Never blow with compressed air.

3. Remove the rear wheels and tires.

4. Secure the disc rotor using wheel nuts.

5. Drain brake fluid.

6. Remove the union bolt and copper washers and disconnect the brake hose from the caliper assembly.

7. Remove the torque member mounting bolts, and remove the brake caliper assembly.

To install:

8. Install the brake caliper assembly to the vehicle and tighten the torque member mounting bolts to 62 ft. lbs. (84 Nm).

9. Install brake hose and copper washers to brake caliper assembly, and tighten union bolt to 13 ft. lbs. (18 Nm).

10. Refill with new brake fluid and perform the air bleeding.

11. Check that no drag is present for the rear disc brake.

12. Install the wheels and tires.

DISC BRAKE PADS

REMOVAL & INSTALLATION

See Figure 4.

REAR DISC BRAKES

1. Before servicing the vehicle, refer to the Precautions Section.

2. Remove the rear wheels and tires.

3. Remove lower sliding pin bolt.

4. Suspend the cylinder body with suitable wire so that the brake hose will not stretch.

5. Remove the brake pad from the torque member.

To install:

6. Install the pad retainer to the torque member if the pad retainers have been removed.

7. Apply copper based brake grease and install them to the brake pad.

8. Install the cylinder body and brake pads to the torque member.

9. Use a disc brake piston tool to easily press the piston back into the caliper.

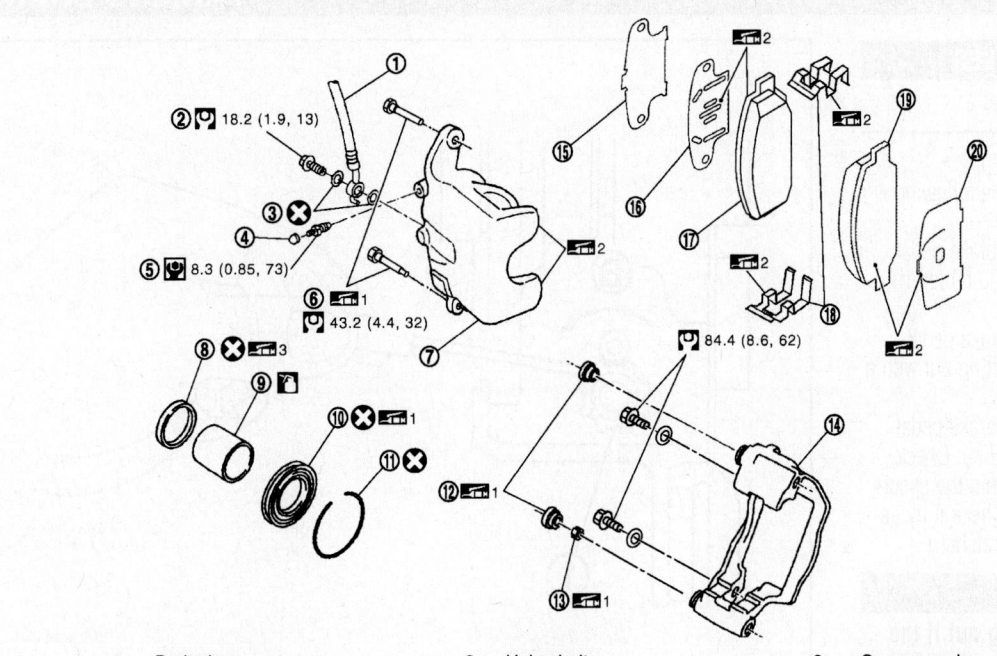

1. Brake hose
2. Union bolt
3. Copper washer
4. Cap
5. Bleeder valve
6. Sliding pin bolt
7. Cylinder body
8. Piston seal
9. Piston
10. Piston boot
11. Retaining ring
12. Sliding pin boot
13. Bushing
14. Torque member
15. Inner shim cover
16. Inner shim
17. Inner pad
18. Pad wear sensor (RH inner pad only)
19. Pad retainer
20. Outer pad
21. Outer shim

1: Apply rubber grease.

2: Apply copper based brake grease.

3: Apply polyglycol ether based lubricant.

: Apply brake fluid.

22140_ROGU_G0022

Fig. 4 Rear disc brake components

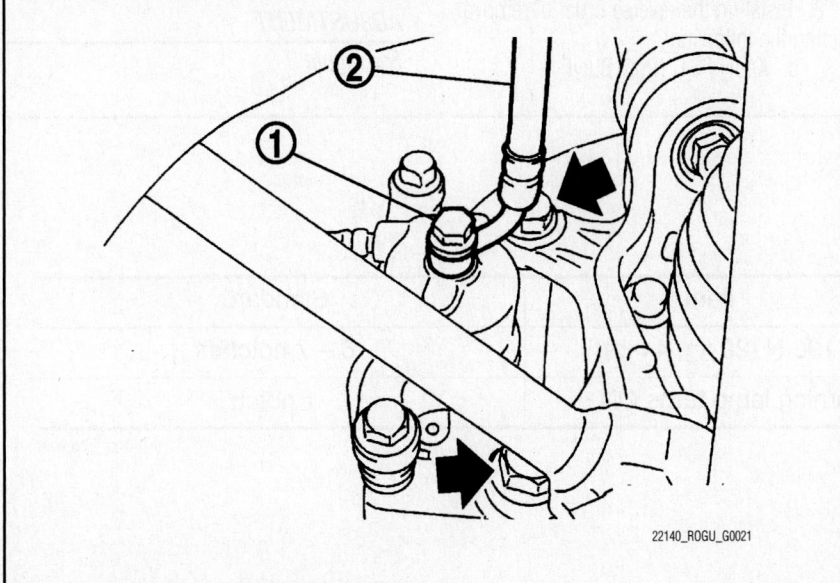

22140_ROGU_G0021

Fig. 5 Union bolt and mounting bolts shown

10. Install the lower sliding pin bolt and tighten it to 32 ft. lbs. (43 Nm).

11. Depress the brake pedal several times to check that no drag feel is present for the front disc brake.

12. Install the wheels and tires.

13. Brake burnishing procedure is as follows:

a. Drive the vehicle on a straight, flat road.

b. Depress the brake pedal with the power to stop vehicle within 3 to 5 seconds until the vehicle stops.

c. Drive without depressing the brakes for a few minutes to cool down the brake system.

d. Repeat steps A—C until pad and disc rotor are securely fitted.

BRAKES   **PARKING BRAKE**

PARKING BRAKE CABLES

ADJUSTMENT

See Figure 6.

1. Adjust the cable with the following procedure:

   a. Operate the parking brake pedal with a force of 110 lbs. (490 N) for 10 strokes or more.

   b. Adjust the parking brake pedal stroke by turning the adjusting nut with a deep socket wrench.

   c. Operate the parking brake pedal with a force of 44 lbs. (196 N). Check that the pedal stroke is within the specified number of notches. (Check it by listening to the clicks of the ratchet.)

✳✳ WARNING

Never reuse the adjusting nut if the nut is removed.

2. Install the rear tires with power tool.

PARKING BRAKE SHOES

REMOVAL & INSTALLATION

See Figures 7 and 8.

1. Before servicing the vehicle, refer to the Precautions Section.

2. Remove the rear wheels and tires.

3. Remove the disc rotor with parking brake completely in the released position.

4. Put matchmarks on the disc rotor and the wheel hub and bearing assembly when reusing the disc rotor.

5. If disc rotor cannot be removed, remove as follows:

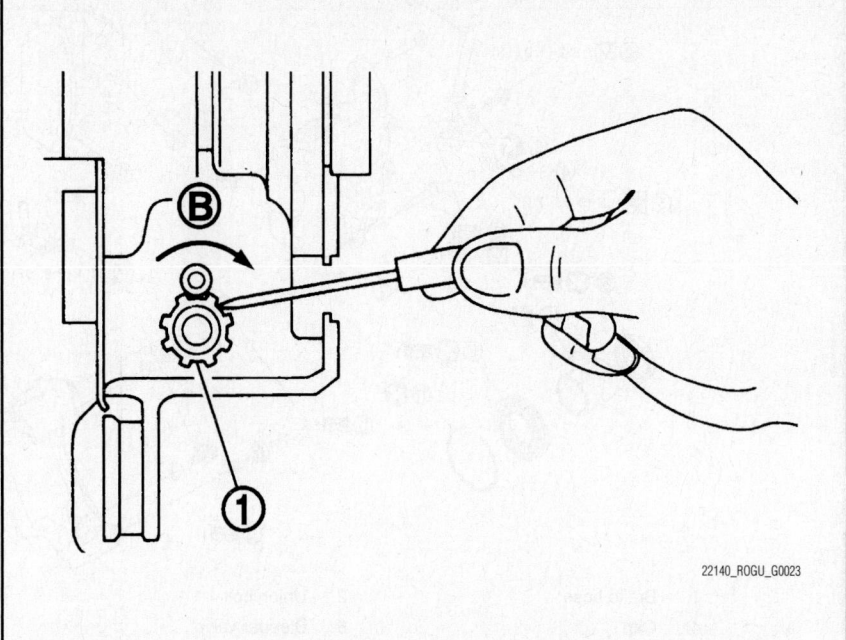

22140_ROGU_G0023

Fig. 7 Using suitable tool, rotate adjuster (1) in direction (B) to retract and loosen parking brake shoe

   a. Secure the disc rotor with wheel nuts and remove the adjusting hole plug.

   b. Using suitable tool, rotate the adjuster in the direction shown to retract and loosen parking brake shoe.

6. Remove anti-rattle pins, springs, and return springs.

7. Remove brake strut, adjuster, parking brake shoes and toggle lever.

To install:

8. Install in the reverse order of removal, noting the following:

   a. Apply PBC (Poly Butyl Cuprysil) grease or silicone-based grease to the back plate and brake shoe.

   b. Assemble adjusters so that threaded part is expanded when rotated.

   c. Shorten the adjuster by rotating it.

   d. When disassembling, apply PBC (Poly Butyl Cuprysil) grease or silicone-based grease to threads.

ADJUSTMENT

See Figure 9.

Item	Standard
Number of notches [under force of 196 N (20 kg, 44 lb)]	6 – 7 notches
Number of notches when brake warning lamp turns ON	1 notch

37663_ROGU_G0049

Fig. 6 Parking brake adjustment

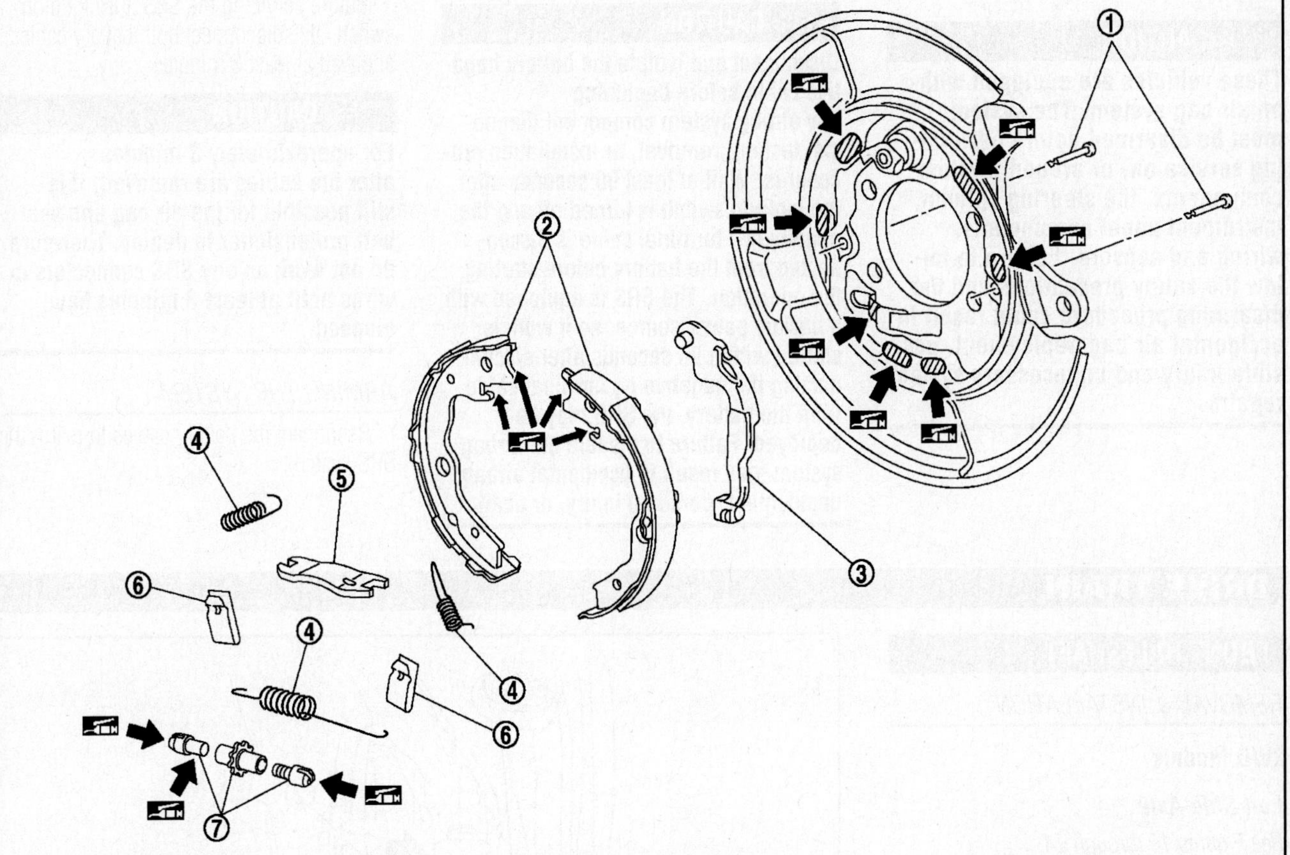

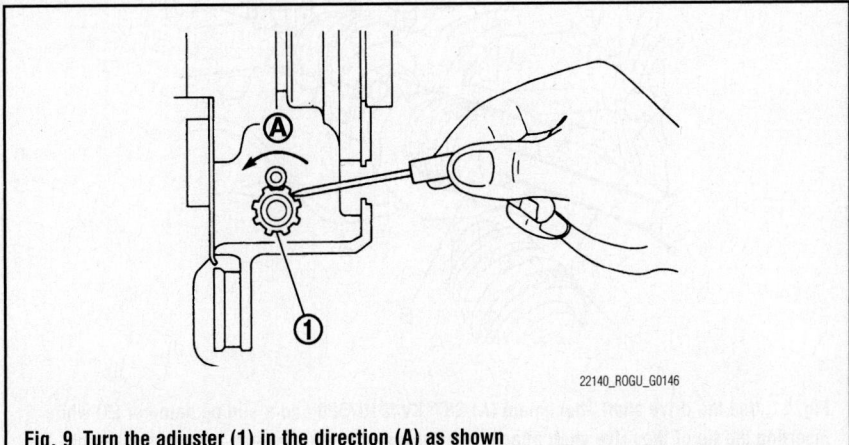

1. Anti-rattle pin
4. Return spring
7. Adjuster

2. Parking brake shoe
5. Brake strut

3. Toggle lever
6. Spring

▄▟ᵀᴴ: Apply PBC (Poly Butyl Cuprysil) grease or silicone-based grease.

22140_ROGU_G0024

Fig. 8 Exploded view of the parking brake shoes

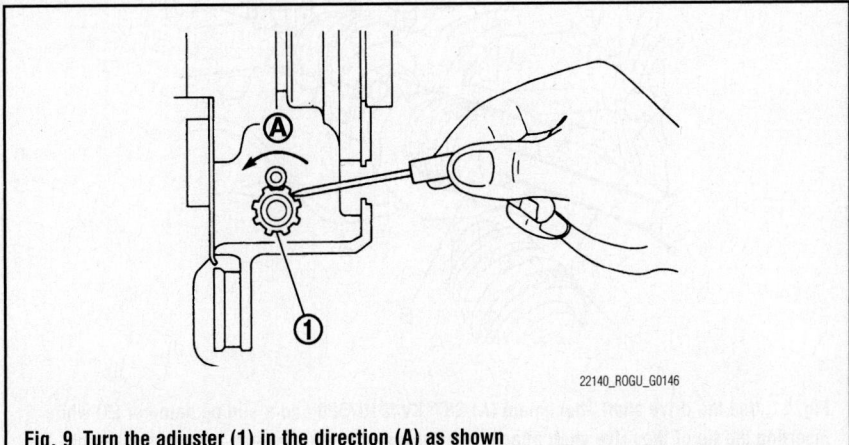

22140_ROGU_G0146

Fig. 9 Turn the adjuster (1) in the direction (A) as shown

1. Before servicing the vehicle, refer to the Precautions Section.

2. Remove the rear wheels and tires.

3. Remove disc rotor with parking brake completely in the released position.

4. Remove the adjusting hole plug from the disc rotor. Turn the adjuster in the direction as shown in the figure using a suitable tool until the disc rotor is locked.

5. Turn back the adjuster 7 notches from the locked position.

6. Rotate the disc rotor to check that there is no drag. Install the adjusting hole plug.

CHASSIS ELECTRICAL

AIR BAG (SUPPLEMENTAL RESTRAINT SYSTEM)

GENERAL INFORMATION

> ❋❋ **CAUTION**
>
> These vehicles are equipped with an air bag system. The system must be disarmed before performing service on, or around, system components, the steering column, instrument panel components, wiring and sensors. Failure to follow the safety precautions and the disarming procedure could result in accidental air bag deployment, possible injury and unnecessary system repairs.

SERVICE PRECAUTIONS

> ❋❋ **CAUTION**
>
> Disconnect and isolate the battery negative cable before beginning any airbag system component diagnosis, testing, removal, or installation procedures. Wait at least 90 seconds after the ignition switch is turned off and the negative (-) terminal cable is disconnected from the battery before starting the operation. The SRS is equipped with a backup power source, so if work is started within 90 seconds after disconnecting the negative (-) terminal cable from the battery, the SRS may be deployed. Failure to disable the airbag system may result in accidental airbag deployment, personal injury, or death.

DISARMING THE SYSTEM

Before servicing the SRS, turn ignition switch OFF, disconnect both battery cables, and wait at least 3 minutes.

> ❋❋ **CAUTION**
>
> For approximately 3 minutes after the cables are removed, it is still possible for the air bag and seat belt pretensioner to deploy. Therefore, do not work on any SRS connectors or wires until at least 3 minutes have elapsed.

ARMING THE SYSTEM

Reconnect the battery cables to rearm the SRS system.

DRIVE TRAIN

FRONT HALFSHAFT

REMOVAL & INSTALLATION

2WD Models

Left Side Axle

See Figures 10 through 14.

1. Before servicing the vehicle, refer to the Precautions Section.
2. Remove the wheels and tires.
3. Remove the wheel sensor from the steering knuckle.

> ❋❋ **WARNING**
>
> Never pull on the wheel sensor harness.

4. Remove the lock plate from the strut assembly.
5. Remove the torque member mounting bolts. Hang torque member aside.
6. Remove the brake disc rotor. Put matchmarks on the wheel hub and bearing assembly and the disc rotor before removing the disc rotor. Never drop the brake disc rotor.
7. Remove the cotter pin and loosen the hub lock nut.
8. Patch the hub lock nut with a piece of wood. Hammer the wood to disengage the wheel hub and bearing assembly from the drive shaft.

> ❋❋ **WARNING**
>
> Never place drive shaft joint at an extreme angle. Also, be careful

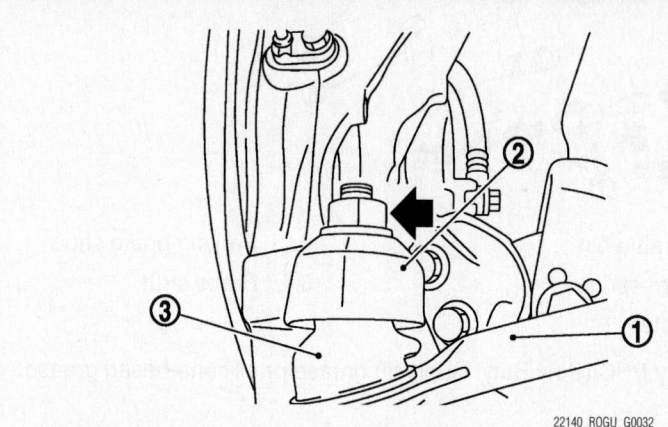

Fig. 10 Remove steering outer socket (1) from steering knuckle (2) using the ball joint remover so as not to damage ball joint boot (3)

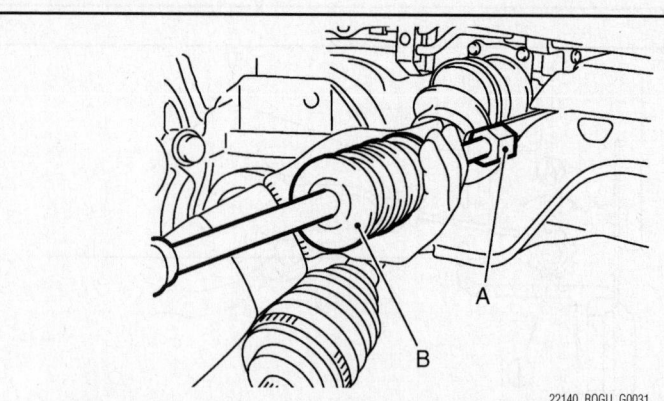

Fig. 11 Use the drive shaft attachment (A) SST: KV40107500 and a sliding hammer (B) while inserting the tip of the drive shaft attachment between the housing and transaxle assembly

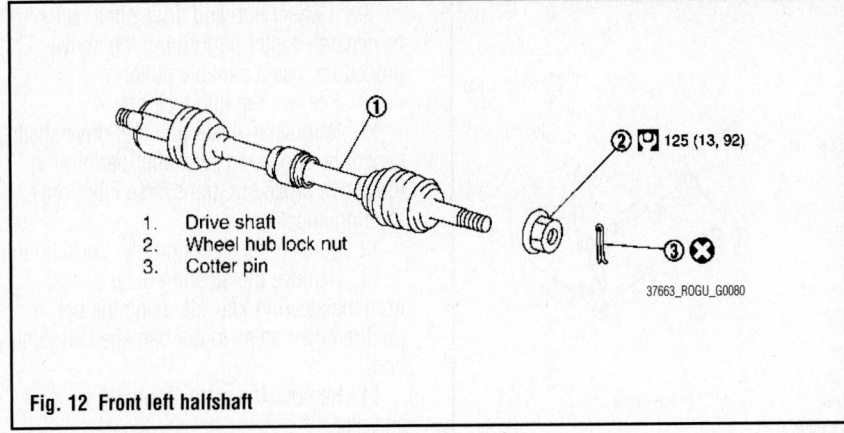

1. Drive shaft
2. Wheel hub lock nut
3. Cotter pin

37663_ROGU_G0080

Fig. 12 Front left halfshaft

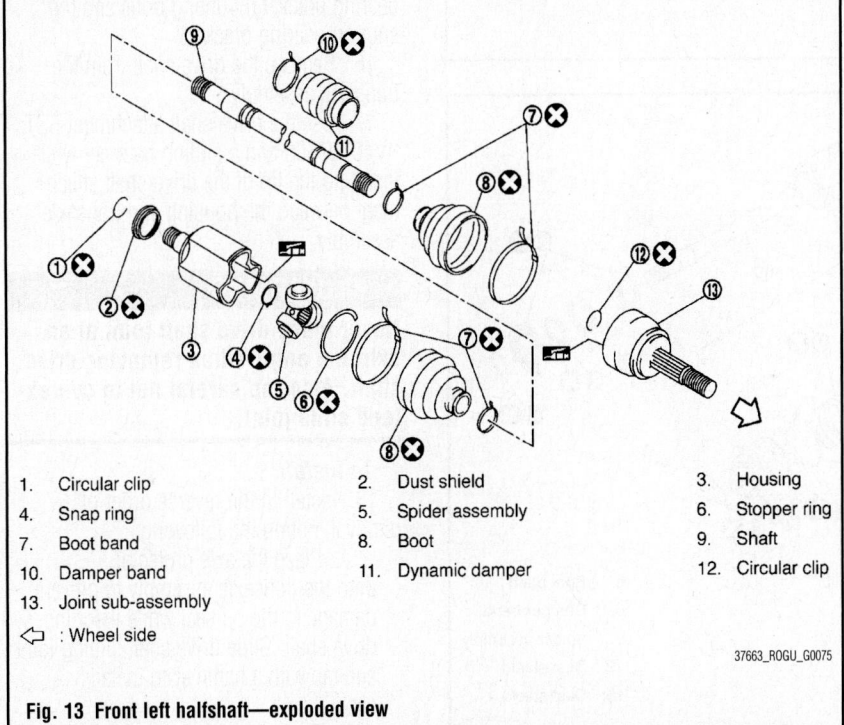

1.	Circular clip	2.	Dust shield	3.	Housing
4.	Snap ring	5.	Spider assembly	6.	Stopper ring
7.	Boot band	8.	Boot	9.	Shaft
10.	Damper band	11.	Dynamic damper	12.	Circular clip
13.	Joint sub-assembly				

⇦ : Wheel side

37663_ROGU_G0075

Fig. 13 Front left halfshaft—exploded view

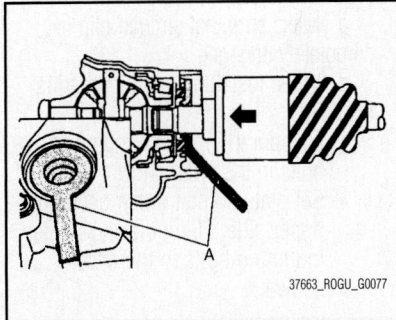

37663_ROGU_G0077

Fig. 14 Place the axle protector (A) SST: KV38107900 onto transaxle assembly

not to overextend slide joint. Never allow drive shaft to hang down without support for housing (or joint

sub-assembly), shaft and the other parts.

9. If wheel hub and drive shaft cannot be separated after performing the above procedure, use a suitable puller.

10. Remove the hub lock nut.

11. Remove if wheel hub and drive shaft cannot be separated even after performing the above procedure transverse link from steering knuckle.

12. Loosen the steering outer socket nut.

13. Remove the steering outer socket from the steering knuckle using the ball joint remover so as to not damage ball joint boot.

14. Remove drive shaft from transaxle assembly.

15. Use the drive shaft attachment SST: KV40107500 and a sliding hammer while inserting the tip of the drive shaft attachment between the housing and transaxle assembly.

➡**Never place drive shaft joint at an extreme angle when removing drive shaft. Also, be careful not to overextend slide joint. Confirm that the circular clip is attached to the drive shaft.**

To install:

16. Install in the reverse order of removal, noting the following:

a. Place the axle protector SST: KV38107900 onto transaxle assembly to prevent damage to the oil seal while inserting drive shaft. Slide drive shaft sliding joint and tap with a hammer to install securely.

❋❋ WARNING

Make sure that circular clip is completely engaged.

b. Tighten the axle nut to 92 ft. lbs. (125 Nm). Install the cotter pin.

Rigth Side Axle

See Figures 10, 15 through 18.

1. Before servicing the vehicle, refer to the Precautions Section.

2. Remove the wheels and tires.

3. Remove the wheel sensor from the steering knuckle.

❋❋ WARNING

Never pull on the wheel sensor harness.

4. Remove the lock plate from the strut assembly.

5. Remove the torque member mounting bolts. Hang torque member aside.

6. Remove the brake disc rotor. Put matchmarks on the wheel hub and bearing assembly and the disc rotor before removing the disc rotor. Never drop the brake disc rotor.

7. Remove the cotter pin and loosen the hub lock nut.

8. Patch the hub lock nut with a piece of wood. Hammer the wood to disengage the wheel hub and bearing assembly from the drive shaft.

❋❋ WARNING

Never place drive shaft joint at an extreme angle. Also, be careful not to overextend slide joint. Never allow drive shaft to hang down without support for housing (or joint sub-assembly), shaft and the other parts.

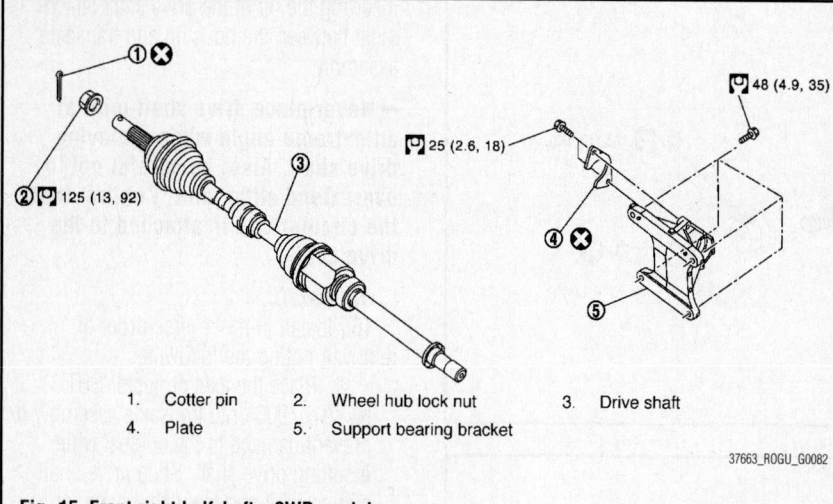

1.	Cotter pin	2.	Wheel hub lock nut	3.	Drive shaft
4.	Plate	5.	Support bearing bracket		

37663_ROGU_G0082

Fig. 15 Front right halfshaft—2WD models

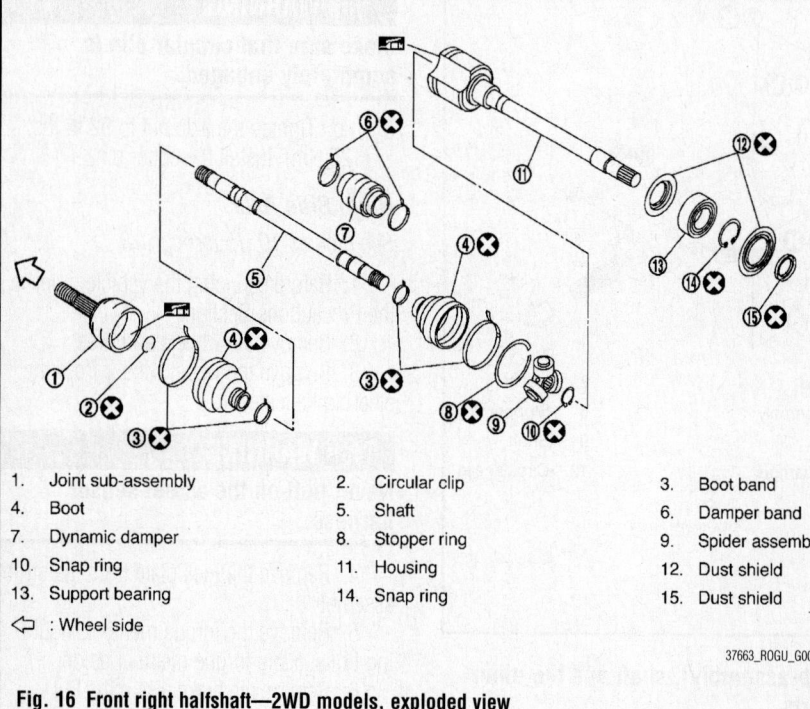

1.	Joint sub-assembly	2.	Circular clip	3.	Boot band
4.	Boot	5.	Shaft	6.	Damper band
7.	Dynamic damper	8.	Stopper ring	9.	Spider assembly
10.	Snap ring	11.	Housing	12.	Dust shield
13.	Support bearing	14.	Snap ring	15.	Dust shield
⇦ : Wheel side					

37663_ROGU_G0076

Fig. 16 Front right halfshaft—2WD models, exploded view

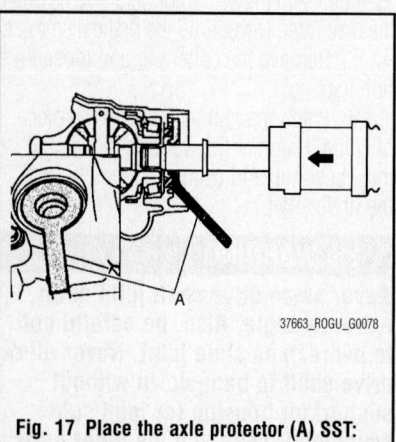

37663_ROGU_G0078

Fig. 17 Place the axle protector (A) SST: KV38107900 onto transaxle assembly

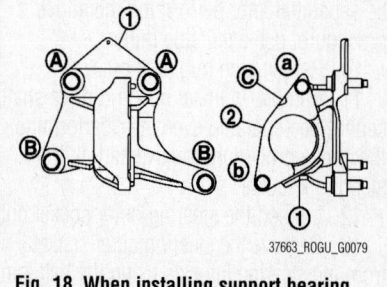

37663_ROGU_G0079

Fig. 18 When installing support bearing bracket (1), temporarily tighten mounting bolts in the order of (A), (B). Set plate (2) so that notch (C) becomes upper side. Temporarily tighten mounting bolts in the order of (a), (b)

9. If wheel hub and drive shaft cannot be separated after performing the above procedure, use a suitable puller.

10. Remove the hub lock nut.

11. Remove if wheel hub and drive shaft cannot be separated even after performing the above procedure transverse link from steering knuckle.

12. Loosen the steering outer socket nut.

13. Remove the steering outer socket from the steering knuckle using the ball joint remover so as to not damage ball joint boot.

14. Remove the plate mounting bolts and plate.

15. If necessary, remove the support bearing bracket mounting bolts and the support bearing bracket.

16. Remove the drive shaft from the transaxle assembly.

17. Use the drive shaft attachment SST: KV40107500 and a sliding hammer while inserting the tip of the drive shaft attachment between the housing and transaxle assembly.

❋❋ WARNING

Never place drive shaft joint at an extreme angle when removing drive shaft. Also, be careful not to overextend slide joint.

To install:

18. Install in the reverse order of removal, noting the following:

   a. Place the axle protector onto the transaxle assembly to prevent damage to the oil seal while inserting drive shaft. Slide drive shaft sliding joint and tap with a hammer to install securely.

   b. Tighten the axle nut to 92 ft. lbs. (125 Nm). Install the cotter pin.

   c. Make sure that circular clip is completely engaged.

   d. When installing support bearing bracket:

   • (Temporarily) tighten mounting bolts in the order shown.
   • Set plate so that notch becomes upper side. (Temporarily) tighten mounting bolts in the order shown.

❋❋ WARNING

Never reuse plate.

AWD Models

Left Side Axle

See Figures 10 through 13 and 19.

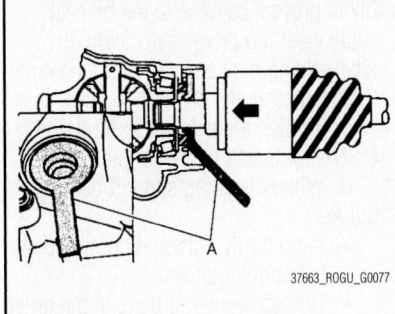

Fig. 19 Place the axle protector (A) onto transaxle assembly

1. Before servicing the vehicle, refer to the Precautions Section.
2. Remove the wheel and tires.
3. Remove the wheel sensor from the steering knuckle.

> ❋❋ **WARNING**
>
> **Never pull on the wheel speed sensor harness.**

4. Remove the lock plate from the strut assembly.
5. Remove the torque member mounting bolts. Hang the torque member aside.
6. Remove the brake disc rotor. Put matchmarks on the wheel hub and bearing assembly and the disc rotor before removing the disc rotor. Never drop brake disc rotor.
7. Remove cotter pin, and loosen hub lock nut.
8. Patch hub lock nut with a piece of wood. Hammer the wood to disengage wheel hub and bearing assembly from drive shaft.

> ❋❋ **WARNING**
>
> **Never place drive shaft joint at an extreme angle. Also, be careful not to overextend slide joint. Never allow drive shaft to hang down without support for housing (or joint sub-assembly), shaft and the other parts.**

9. If the wheel hub and drive shaft cannot be separated after performing the above procedure, use a suitable puller.
10. Remove the hub lock nut.
11. Remove the transverse link from the steering knuckle.
12. Loosen the steering outer socket steering outer socket nut.
13. Remove the steering outer socket from the steering knuckle using the ball joint remover so as not to damage ball joint boot.
14. Temporarily tighten the nut to prevent

damage to threads and to prevent the ball joint remover from suddenly coming off.
15. Remove the drive shaft from the transaxle assembly.
16. Use the drive shaft attachment (A) SST: KV40107500 and a sliding hammer (B) while inserting tip of the drive shaft attachment between housing and transaxle assembly.

> ❋❋ **WARNING**
>
> **Never place drive shaft joint at an extreme angle when removing drive shaft. Also be careful not to overextend slide joint. Confirm that the circular clip is attached to the drive shaft.**

To install:

17. Install in the reverse order of removal, noting the following:
   a. Always replace the differential side oil seal with new one when installing drive shaft.
   b. Place the protector (SST: KV38107900) onto transaxle assembly to prevent damage to the oil seal while inserting drive shaft. Slide drive shaft sliding joint and tap with a hammer to install securely.

> ❋❋ **WARNING**
>
> **Make sure that circular clip is completely engaged.**

   c. Tighten the axle nut to 92 ft. lbs. (125 Nm). Install the cotter pin.

Right Side Axle

See Figures 10, 20 through 23.

1. Before servicing the vehicle, refer to the Precautions Section.
2. Remove the wheel and tires.
3. Remove the wheel sensor from the steering knuckle.

> ❋❋ **WARNING**
>
> **Never pull on the wheel speed sensor harness.**

4. Remove the lock plate from the strut assembly.
5. Remove the torque member mounting bolts. Hang the torque member aside.
6. Remove the brake disc rotor. Put matchmarks on the wheel hub and bearing assembly and the disc rotor before removing the disc rotor. Never drop brake disc rotor.
7. Remove cotter pin, and loosen hub lock nut.
8. Patch hub lock nut with a piece of wood. Hammer the wood to disengage wheel hub and bearing assembly from drive shaft.

> ❋❋ **WARNING**
>
> **Never place drive shaft joint at an extreme angle. Also, be careful not to overextend slide joint. Never allow drive shaft to hang down without support for housing (or joint sub-assembly), shaft and the other parts.**

9. If the wheel hub and drive shaft cannot be separated after performing the above procedure, use a suitable puller.
10. Remove the hub lock nut.
11. Remove the transverse link from the steering knuckle.

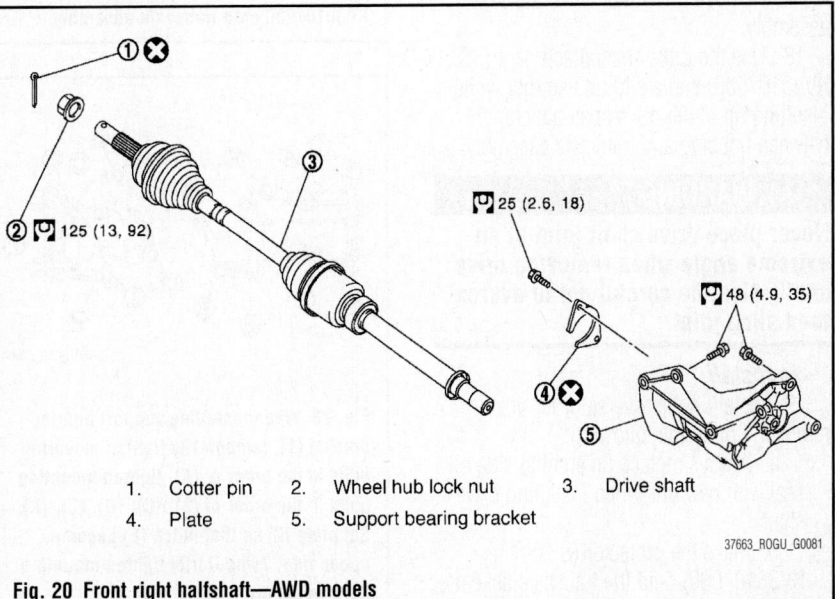

1. Cotter pin
2. Wheel hub lock nut
3. Drive shaft
4. Plate
5. Support bearing bracket

Fig. 20 Front right halfshaft—AWD models

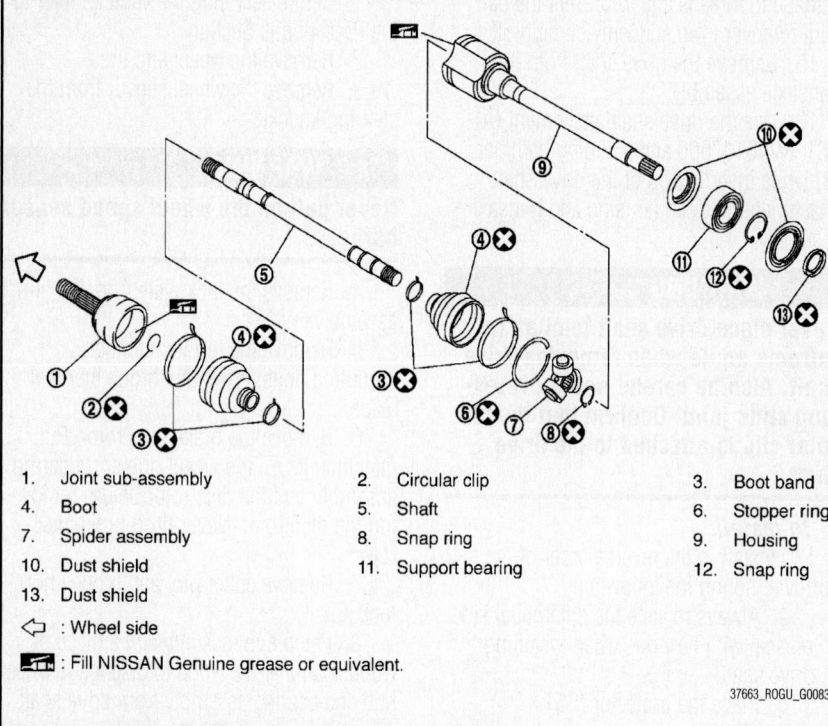

1. Joint sub-assembly
2. Circular clip
3. Boot band
4. Boot
5. Shaft
6. Stopper ring
7. Spider assembly
8. Snap ring
9. Housing
10. Dust shield
11. Support bearing
12. Snap ring
13. Dust shield

◁ : Wheel side

⬛ : Fill NISSAN Genuine grease or equivalent.

37663_ROGU_G0083

Fig. 21 Front right halfshaft—AWD models, exploded view

12. Loosen the steering outer socket steering outer socket nut.

13. Remove the steering outer socket from the steering knuckle using the ball joint remover so as not to damage ball joint boot.

14. Temporarily tighten the nut to prevent damage to threads and to prevent the ball joint remover from suddenly coming off.

15. Remove plate mounting bolts and plate.

16. If necessary, remove the support bearing bracket mounting bolts and the support bearing bracket.

17. Remove drive shaft from transaxle assembly.

18. Use the drive shaft attachment (SST: KV40107500) and a sliding hammer while inserting tip of the drive shaft attachment between housing and transaxle assembly.

⁕⁕ WARNING

Never place drive shaft joint at an extreme angle when removing drive shaft. Also, be careful not to overextend slide joint.

To install:

19. Install in the reverse order of removal, noting the following:

a. Always replace differential side oil seal with new one when installing drive shaft.

b. Place the protector (SST: KV38107900) onto the transaxle assem-

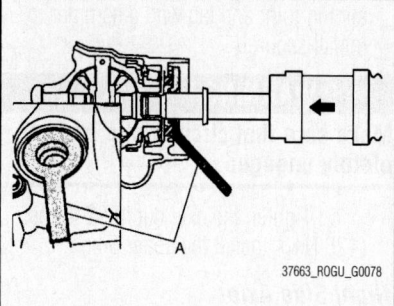

37663_ROGU_G0078

Fig. 22 Place the protector (A) SST: KV38107900 onto transaxle assembly

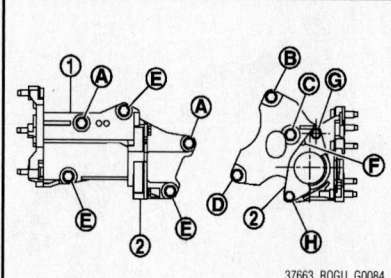

37663_ROGU_G0084

Fig. 23 When installing support bearing bracket (1), temporarily tighten mounting bolts in the order of (A), tighten mounting bolts in the order of (B), (C), (D), (E), (A). Set plate (2) so that notch (F) becomes upper side. Temporarily tighten mounting bolts in the order of (G), (H)

bly to prevent damage to the oil seal while inserting drive shaft. Slide drive shaft sliding joint and tap with a hammer to install securely.

c. Tighten the axle nut to 92 ft. lbs. (125 Nm). Install the cotter pin.

d. When installing support bearing bracket.

• Temporarily tighten mounting bolts in the order shown.

• Tighten mounting bolts in the order shown.

• Set plate so that notch becomes upper side. (Temporarily) tighten mounting bolts in the order shown.

⁕⁕ WARNING

Never reuse plate.

REAR HALFSHAFT

REMOVAL & INSTALLATION

See Figures 24 and 25.

1. Before servicing the vehicle, refer to the Precautions Section.

2. Remove the wheels and tires.

3. Remove the torque member mounting bolts. Hang torque member aside.

4. Remove the disc rotor.

5. Remove the cotter pin and loosen the hub lock nut.

6. Patch the hub lock nut with a piece of wood. Hammer the wood to disengage the wheel hub and bearing assembly from the drive shaft.

⁕⁕ WARNING

Never place drive shaft joint at an extreme angle. Also be careful not to overextend slide joint. Never allow drive shaft to hang down without support for housing (or joint sub-assembly), shaft and the other parts.

7. If the wheel hub and bearing assembly and drive shaft cannot be separated after performing the above procedure, use a suitable puller.

8. Remove the hub lock nut.

9. Remove the wheel sensor from the axle housing.

10. Remove the stabilizer link.

11. Set a suitable jack under the suspension arm.

12. Remove the shock absorber from the suspension arm.

13. Remove the upper link from the suspension arm.

14. Remove the lower link from suspension arm.

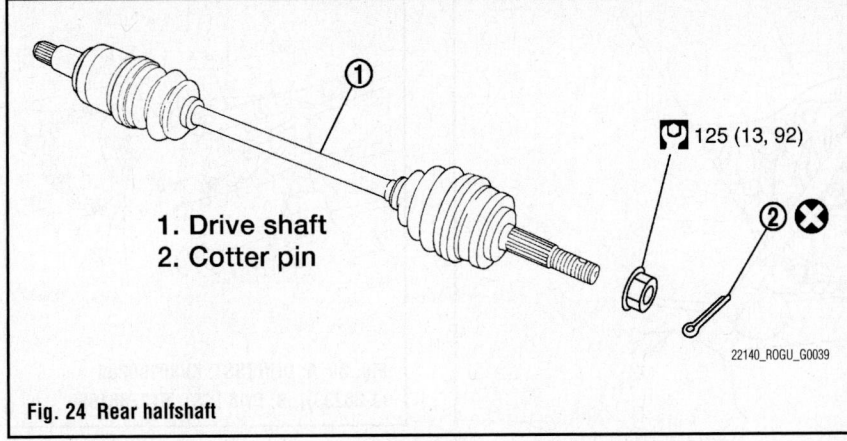

1. Drive shaft
2. Cotter pin

125 (13, 92)

22140_ROGU_G0039

Fig. 24 Rear halfshaft

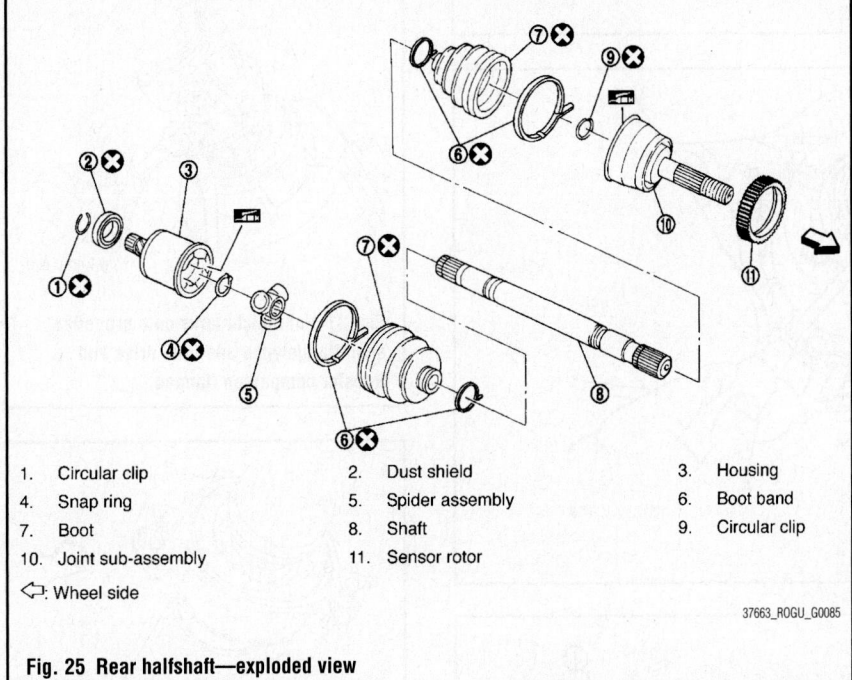

1.	Circular clip	2.	Dust shield	3.	Housing
4.	Snap ring	5.	Spider assembly	6.	Boot band
7.	Boot	8.	Shaft	9.	Circular clip
10.	Joint sub-assembly	11.	Sensor rotor		

⇦: Wheel side

37663_ROGU_G0085

Fig. 25 Rear halfshaft—exploded view

15. Remove the drive shaft from the final drive assembly.

To install:

16. Install in the reverse order of removal, noting the following:

a. Align the matchmarks made during removal, if reusing the disc rotor.

b. Tighten the axle nut to 92 ft. lbs. (125 Nm).

c. Perform final tightening of bolts and nuts at suspension arm (rubber bushing), under relaxed conditions with tires on level ground.

REAR PINION SEAL

REMOVAL & INSTALLATION

See Figures 26 through 30.

1. Before servicing the vehicle, refer to the Precautions Section.

matching mark should be in line with the matching mark on companion flange.

4. Remove the companion flange lock nut, using a flange wrench (commercial service tool).

5. Remove the companion flange.

6. Remove the front oil seal from coupling cover, using a flat-bladed screwdriver.

To install:

7. Apply multi-purpose grease onto oil seal lips, and gear oil onto the circumference of oil seal.

8. Install front oil seal until it becomes flush with the coupling cover. Never reuse the oil seal. Never incline the oil the seal.

9. Align the matchmark of the electric controlled coupling with the matchmark of companion flange, and install the companion flange.

10. Install the companion flange lock nut with a flange wrench (commercial service tool) and tighten to 103 ft. lbs. (140 Nm).

✳✳ WARNING

Never reuse companion flange lock nut.

11. Install the rear propeller shaft.

12. If oil leaks while removing, check oil level after the installation.

PROPELLER SHAFT

REMOVAL & INSTALLATION

See Figures 31 through 35.

1. Shift the transaxle to the neutral position, and then release the parking brake.

2. Remove the following:
   • Muffler assembly
   • Exhaust center pipe

2. Remove rear propeller shaft.

3. Put matching mark on the thread edge of electric controlled coupling. The

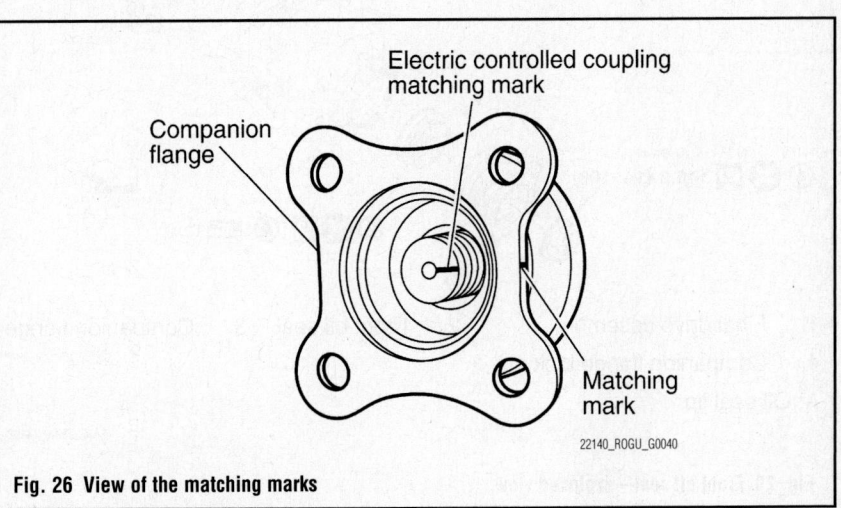

Electric controlled coupling matching mark

Companion flange

Matching mark

22140_ROGU_G0040

Fig. 26 View of the matching marks

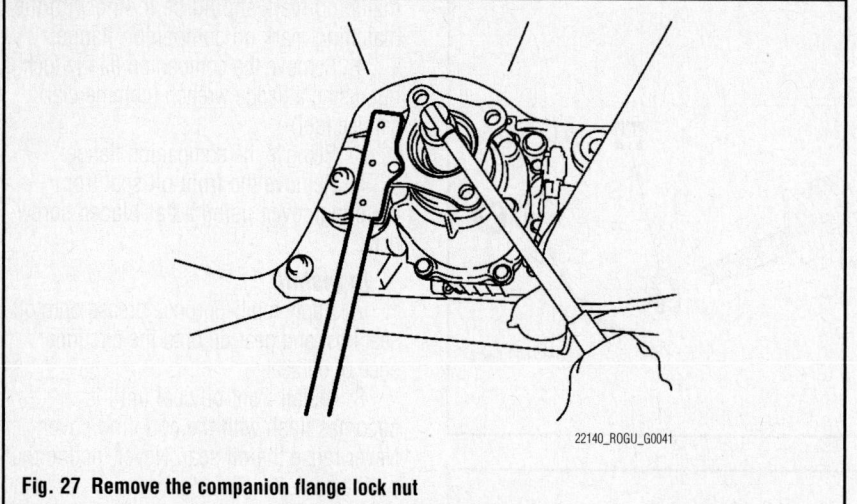

Fig. 27 Remove the companion flange lock nut

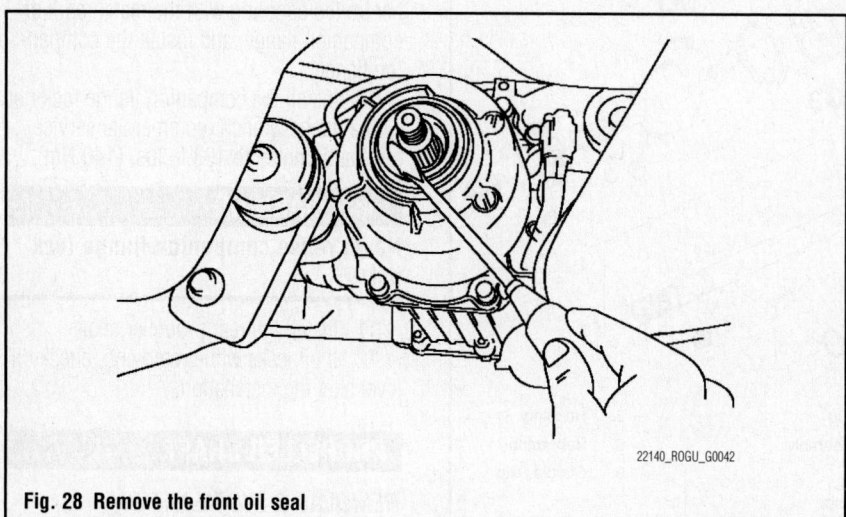

Fig. 28 Remove the front oil seal

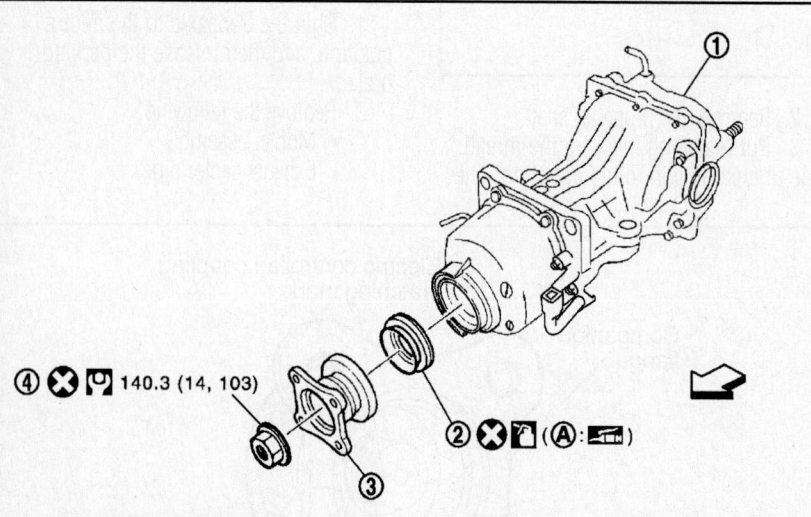

④ ❌ 🔧 140.3 (14, 103)

② ❌ 🔧 (A: ▱)

1. Final drive assembly
2. Front oil seal
3. Companion flange
4. Companion flange lock nut

A: Oil seal lip

Fig. 29 Front oil seal—exploded view

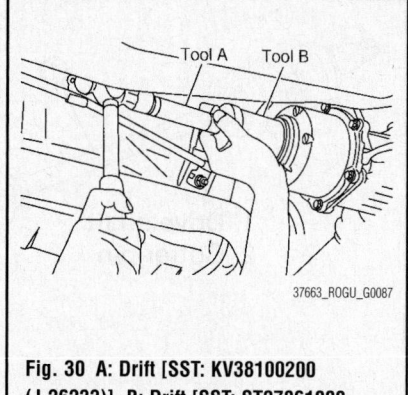

Fig. 30 A: Drift [SST: KV38100200 (J-26233)], B: Drift [SST: ST27861000

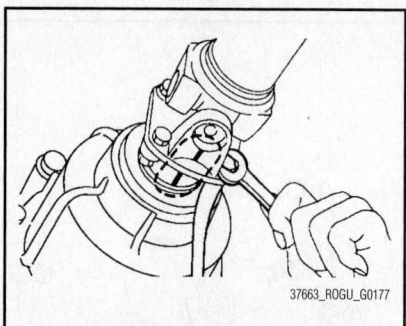

Fig. 31 Put matchmarks onto propeller shaft flange yoke and final drive and transfer companion flanges

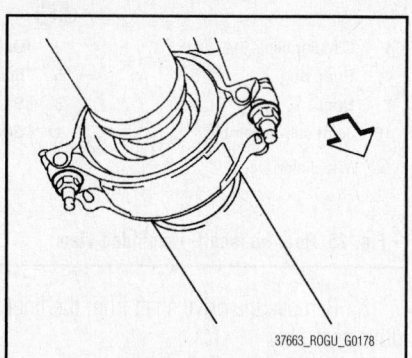

Fig. 32 Loosen mounting nuts of center bearing mounting brackets (upper/lower)

3. Put matchmarks onto propeller shaft flange yoke and final drive and transfer companion flanges. For matching marks, use paint. Never damage propeller shaft flange yoke and transfer companion flange.

4. Loosen mounting nuts of center bearing mounting brackets (upper/lower). Tighten mounting nuts temporarily.

5. Remove propeller shaft assembly fixing bolts and nuts.

6. Remove center bearing mounting bracket fixing nuts.

7. Remove propeller shaft assembly.

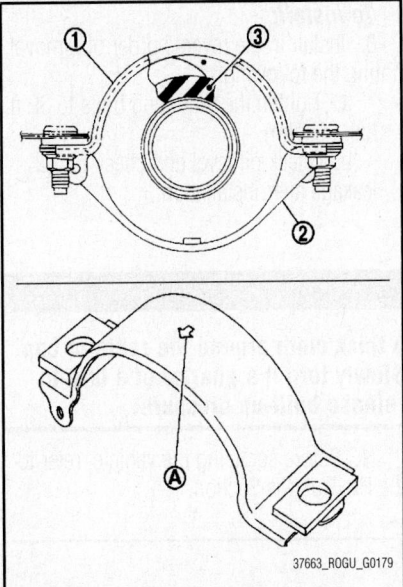

Fig. 33 Install center bearing mounting bracket (upper) (1) with its arrow mark (A) facing forward, and adjust position of center bearing mounting bracket (upper), center bearing mounting bracket (lower) (2) sliding back and forth to prevent play in thrust direction of center bearing insulator (3)

☀☀ WARNING

If constant velocity joint was bent during propeller shaft assembly removal, installation, or transportation, its boot may be damaged. Wrap boot interference area to metal part with shop cloth or rubber to protect boot from breakage.

8. Remove clips and center bearing mounting bracket (upper/lower).

To install:

9. Note the following, and install in the reverse order of removal.

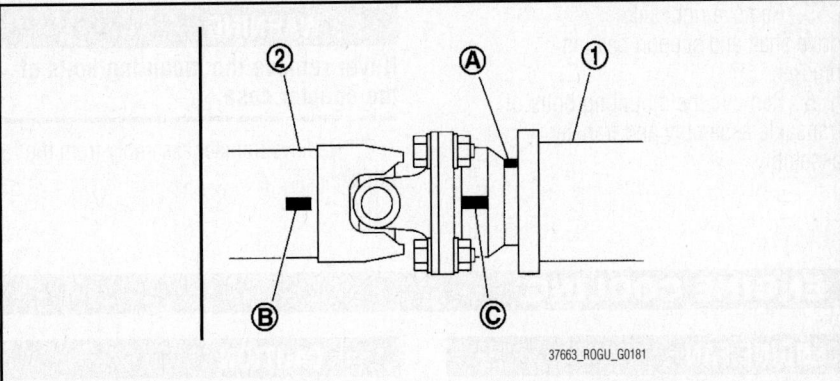

Fig. 35 With the mark (A) faced upward (1), couple the propeller shaft and the final drive so that the matching mark (B) of propeller shaft (2) can be positioned as closest as possible with the matching mark (C)

a. Install center bearing mounting bracket (upper) with its arrow mark facing forward.

b. Adjust position of center bearing mounting bracket (upper), center bearing mounting bracket (lower) sliding back and forth to prevent play in thrust direction of center bearing insulator. Install center bearing mounting bracket (upper/lower) to vehicle.

c. Align matching marks to install propeller shaft assembly to final drive and transfer companion flanges.

d. After assembly, perform a driving test to check propeller shaft vibration. If vibration occurred, separate propeller shaft from final drive. Reinstall companion flange after rotating it by 90, 180, 270 degrees. Then perform driving test and check propeller shaft vibration again at each point.

e. After tightening the bolts and nuts to the specified torque, make sure that the bolts on the flange side is tightened as shown.

f. If propeller shaft assembly or final

drive assembly has been replaced, connect them as follows:

- Face the companion flange mark of the final drive upward. With the mark faced upward, couple the propeller shaft and the final drive so that the matching mark of propeller shaft can be positioned as closest as possible with the matching mark of the final drive companion flange.
- Tighten mounting bolts and nuts of propeller shaft and final drive to the specified torque.

TRANSFER CASE ASSEMBLY

REMOVAL & INSTALLATION

See Figure 36.

1. Before servicing the vehicle, refer to the Precautions Section.
2. Remove the exhaust front tube.
3. Remove the exhaust center tube.
4. Separate the rear propeller shaft.

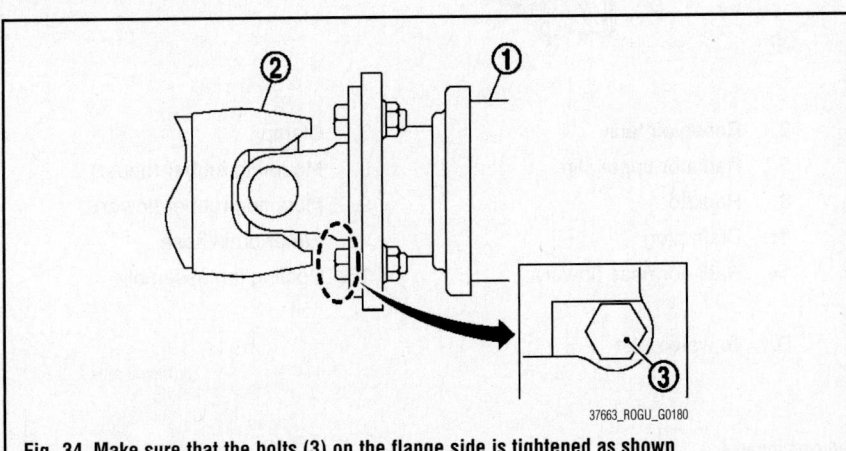

Fig. 34 Make sure that the bolts (3) on the flange side is tightened as shown

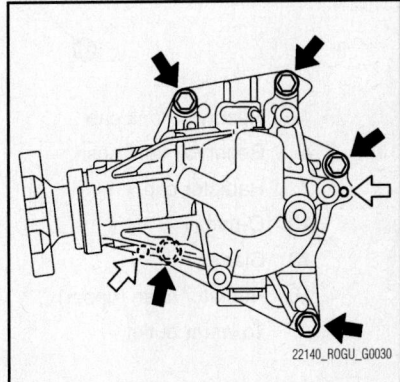

Fig. 36 Transfer case assembly mounting bolts

5. Remove right side drive shaft and support bearing bracket.

6. Remove the mounting bolts of transaxle assembly and transfer assembly.

✳✳ WARNING

Never remove the mounting bolts of the adapter case.

7. Remove transfer assembly from the vehicle.

To install:

8. Install in the reverse order of removal, noting the following:

a. Tighten the mounting bolts to 32 ft. lbs. (44 Nm).

b. Check oil level and check for oil leakage after installation.

ENGINE COOLING

ENGINE FAN

REMOVAL & INSTALLATION

See Figure 37.

✳✳ CAUTION

Never remove radiator cap when engine is hot. Serious burns may occur from high-pressure engine coolant escaping from radiator. Wrap a thick cloth around the radiator cap. Slowly turn it a quarter of a turn to release built-up pressure.

1. Before servicing the vehicle, refer to the Precautions Section.

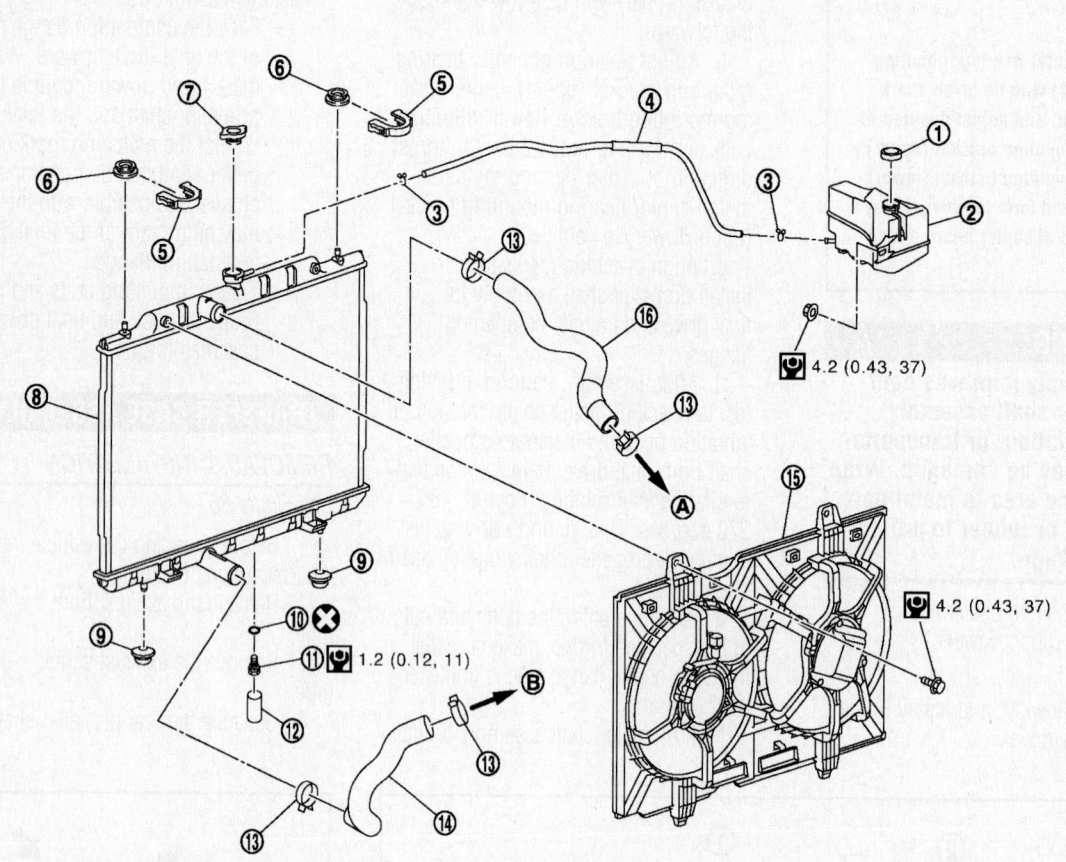

1. Reservoir tank cap	2. Reservoir tank	3. Clamp
4. Reservoir tank hose	5. Radiator upper clip	6. Mounting rubber (upper)
7. Radiator cap	8. Radiator	9. Mounting rubber (lower)
10. O-ring	11. Drain plug	12. Water drain hose
13. Clamp	14. Radiator hose (lower)	15. Cooling fan assembly
16. Radiator hose (upper)		
A. To water outlet	B. To water inlet	

22140_ROGU_G0159

Fig. 37 Exploded view of the radiator fan and related components

2. Remove engine under cover.

3. Drain engine coolant from radiator.

4. Remove air duct (inlet).

5. Remove radiator hose (upper) and reservoir tank hose.

6. Disconnect harness connector from fan motor, and move harness to aside.

7. Remove cooling fan assembly.

❋❋ WARNING

Be careful not to damage or scratch on radiator core when removing.

8. Remove cooling fan mounting nuts, and then remove the left-hand and right-hand cooling fans.

9. Remove fan motor cover and left-hand and right-hand fan motors.

To install:

10. Install in the reverse order of removal and note the following:

a. Add engine coolant and bleed the cooling system.

b. Pressure test engine and check for leaks.

c. Check the fan operation.

RADIATOR

REMOVAL & INSTALLATION

See Figure 37.

❋❋ CAUTION

Never remove radiator cap when engine is hot. Serious burns may occur from high-pressure engine coolant escaping from radiator. Wrap a thick cloth around the radiator cap. Slowly turn it a quarter of a turn to release built-up pressure.

1. Before servicing the vehicle, refer to the Precautions Section.

2. Remove engine under cover.

3. Drain engine coolant from radiator.

4. Remove air duct (inlet).

5. Remove radiator hose (upper) and reservoir tank hose.

6. Disconnect harness connector from fan motor, and move harness to aside.

7. Remove cooling fan assembly.

8. Remove the lower radiator hose.

9. Remove radiator upper clips by pulling the tabs outside to release the lock.

10. Remove the radiator.

❋❋ WARNING

Be careful not to damage or scratch on radiator core when removing.

To install:

11. Install in the reverse order of removal and note the following:

a. Add engine coolant and bleed the cooling system.

b. Pressure test engine and check for leaks.

THERMOSTAT

REMOVAL & INSTALLATION

See Figures 38 and 39.

❋❋ CAUTION

Never remove radiator cap when engine is hot. Serious burns may occur from high-pressure engine coolant escaping from radiator. Wrap a thick cloth around the radiator cap. Slowly turn it a quarter of a turn to release built-up pressure.

1. Before servicing the vehicle, refer to the Precautions Section.

2. Remove the battery.

3. Disconnect engine room harness connectors at unit sides TCM and ECM, and then move it aside.

4. Remove the battery tray.

5. Remove the air duct and air cleaner case assembly.

6. Drain the engine coolant.

7. Disconnect the lower radiator hose at water inlet side.

8. Disconnect water hose at water inlet side (Type 1).

9. Remove water inlet and thermostat.

10. Remove the water control valve with the following procedure:

a. Disconnect radiator hose (upper) at water control valve housing (water outlet) side.

b. Disconnect harness connector from engine coolant temperature sensor.

c. Remove the CVT fluid level gauge and CVT fluid charging pipe.

d. Disconnect water hoses.

e. Disconnect the air fuel ratio sensor 1 and heated oxygen sensor 2 harness connectors, and remove harness clips from heater pipe.

f. Remove the heater pipe and heater hose.

g. After removing the water control valve housing (water outlet), remove water control valve.

To install:

11. Install in the reverse order of removal, noting the following:

a. Refer to the exploded view for tightening specifications.

b. Install thermostat and water control valve with making rubber ring groove fit to thermostat flange and water control valve flange.

➡**The same procedure is applied for installation of thermostat.**

c. Install the thermostat with jiggle valve facing upwards. (The position deviation may be within the range of 20 degrees as shown in the figure.)

d. Install water control valve with the arrow facing up and the frame center part facing upwards. (The position deviation may be within the range of 20 degrees as shown in the figure.)

WATER PUMP

REMOVAL & INSTALLATION

See Figure 40.

❋❋ CAUTION

Never remove radiator cap when engine is hot. Serious burns may occur from high-pressure engine coolant escaping from radiator. Wrap a thick cloth around the radiator cap. Slowly turn it a quarter of a turn to release built-up pressure.

1. Before servicing the vehicle, refer to the Precautions Section.

2. Drain the engine coolant.

3. Remove the following parts:

- Drive belt
- Drive belt auto-tensioner
- Alternator
- Water pump

4. Engine coolant will leak from the cylinder block, so have a receptacle ready below.

5. Remove water pump housing with the following procedure:

a. Remove exhaust manifold cover.

b. Remove oil level gauge and oil level gauge guide.

c. Remove the mounting bolts for water pipe.

d. Remove water pump housing.

e. Remove exhaust manifold and three way catalyst assembly.

f. Remove the water pipe.

To install:

6. Install in the reverse order of removal, noting the following:

a. When inserting water pipe end into cylinder block, apply a neutral detergent to O-ring, and then insert it immediately.

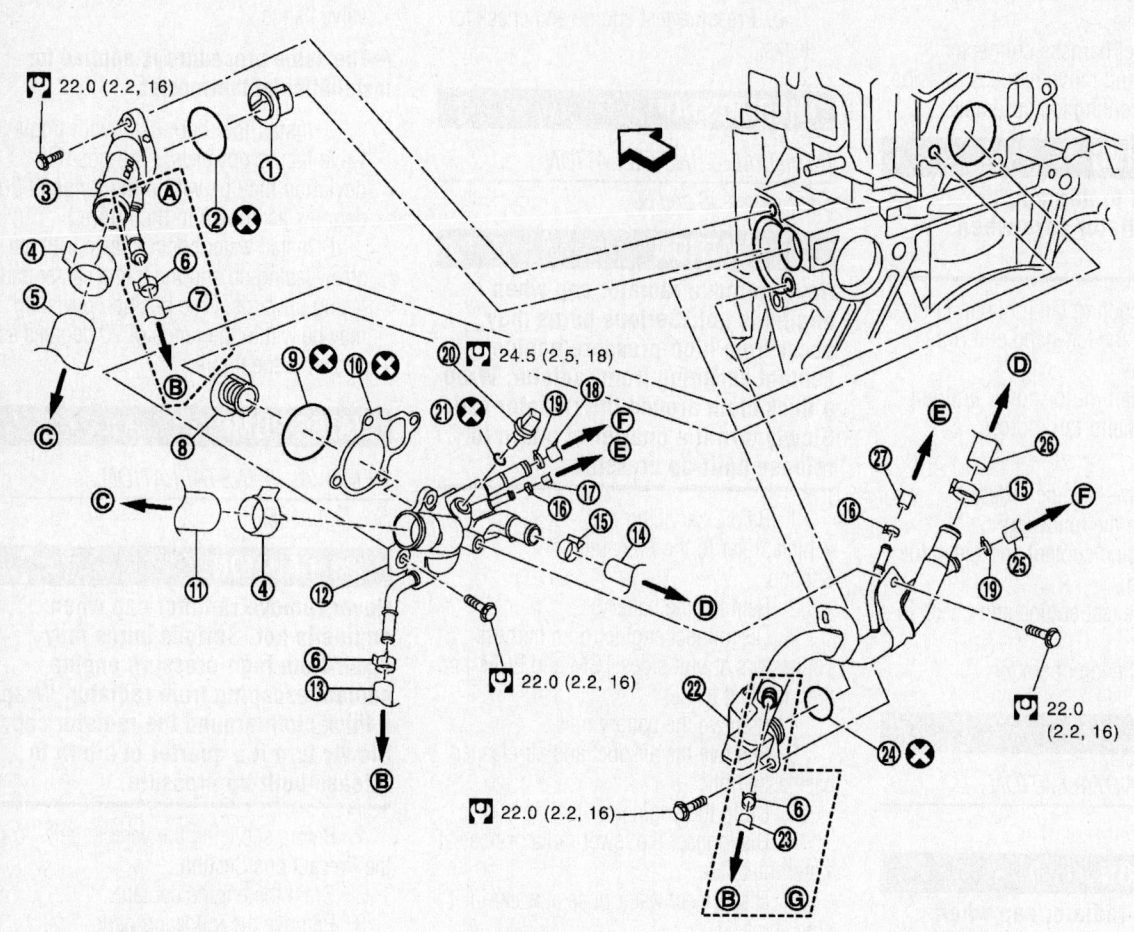

1. Thermostat
2. O-ring
3. Water inlet
4. Clamp
5. Radiator hose (lower)
6. Clamp
7. Water hose
8. Water control valve
9. O-ring
10. Gasket
11. Radiator hose (upper)
12. Water control valve housing (water outlet)
13. Water hose
14. Heater hose
15. Clamp
16. Clamp
17. Water hose
18. Water hose
19. Clamp
20. Engine coolant temperature sensor
21. Washer
22. Heater pipe
23. Water hose
24. O-ring
25. Water hose
26. Heater hose
27. Water hose
A. Type 1
B. To CVT fluid cooler
C. To radiator
D. To heater
E. To electric throttle control actuator
F. To oil cooler
G. Type 2
◁ : Engine front

22140_ROGU_G0043

Fig. 38 Exploded view of the thermostat, water control valve and related components

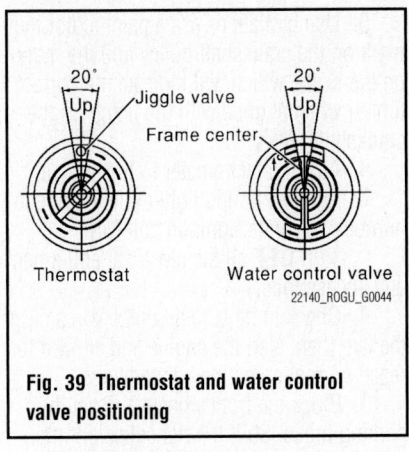

Fig. 39 Thermostat and water control valve positioning

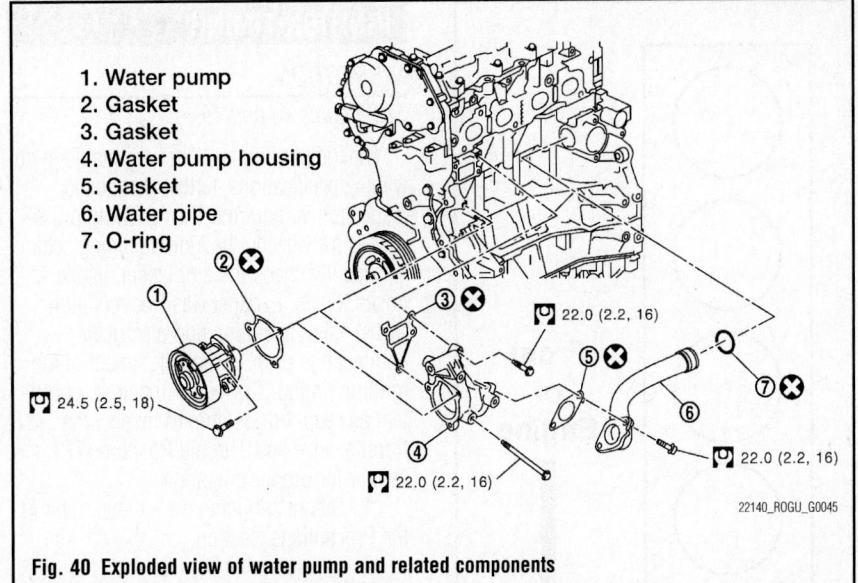

1. Water pump
2. Gasket
3. Gasket
4. Water pump housing
5. Gasket
6. Water pipe
7. O-ring

24.5 (2.5, 18)

22.0 (2.2, 16)

22.0 (2.2, 16)

22.0 (2.2, 16)

22.0 (2.2, 16)

22140_ROGU_G0045

Fig. 40 Exploded view of water pump and related components

   b. Tighten water pump mounting bolts to 18 ft. lbs. (25 Nm).

   c. Refer to the exploded water pump view for additional tightening specification.

ENGINE ELECTRICAL

CHARGING SYSTEM

ALTERNATOR

REMOVAL & INSTALLATION

See Figure 41.

   1. Before servicing the vehicle, refer to the Precautions Section.

   2. Disconnect the battery cable from the negative terminal.

   3. Remove drive belt.

   4. Disconnect the alternator connector.

   5. Remove the "B" terminal nut and "B" terminal harness.

   6. Remove harness bracket.

   7. Remove the upper alternator mounting bolt.

   8. Remove lower alternator mounting bolt.

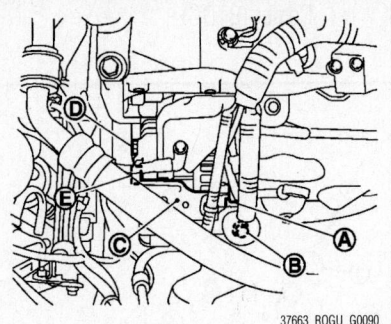

37663_ROGU_G0090

Fig. 41 Disconnect alternator connector (A), remove "B" terminal nut (B), "B" terminal harness, harness bracket (C), upper alternator mounting bolt (D), lower alternator mounting bolt (E)

   9. Remove the alternator upward from the vehicle.

To install:

   10. Install in the reverse order of removal, noting the following:

   a. Install the alternator and tighten the mounting bolts to 48 ft. lbs. (65 Nm).

   b. Install the terminal nut "B" and tighten to 7 ft. lbs. (10 Nm).

   c. Connect the negative battery cable.

   d. Check tension of the drive belt.

VOLTAGE REGULATOR

REMOVAL & INSTALLATION

   The voltage regulator is integral to the alternator.

ENGINE ELECTRICAL

IGNITION SYSTEM

FIRING ORDER

See Figure 42.

IGNITION COIL

REMOVAL & INSTALLATION

See Figure 43.

   1. Before servicing the vehicle, refer to the Precautions Section.

   2. Remove air duct and resonator assembly.

   3. Remove the electric throttle control actuator without disconnecting water hose.

   4. Loosen the intake manifold mounting bolts and nuts.

   5. Remove the intake manifold.

   6. Disconnect harness connector from ignition coil.

   7. Support the bottom surface of engine using a transmission jack.

   8. Remove ground cable and harness from the right-hand engine mounting bracket.

   9. Remove the ignition coil.

✳✳ WARNING

Never drop or shock the ignition coil. Never disassemble the ignition coil.

   10. Disconnect PCV hose from rocker cover.

   11. Remove the right-hand engine mounting insulator.

   12. Remove the right-hand engine mounting bracket.

   13. Remove PCV valve and O-ring from rocker cover, if necessary.

   14. Remove oil filler cap from rocker cover if needed.

To install:

   15. Install in the reverse order of removal, noting the following:

   a. Tighten the ignition coil mounting bolt to 62 inch lbs. (7 Nm).

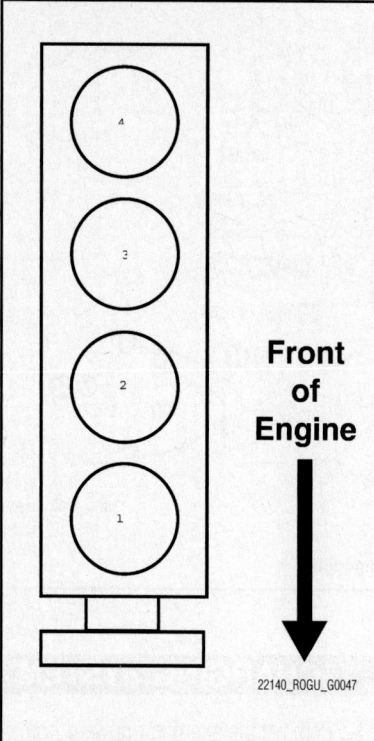

Front of Engine

22140_ROGU_G0047

Fig. 42 Engine Firing order: 1—3—4—2

IGNITION TIMING

INSPECTION

See Figures 44 and 45.

The ignition timing is not adjustable. If not within specifications, further diagnostic inspection is required. The following procedure is for viewing the ignition timing setting.

Visually check the air cleaner, intake hoses, ducts, Exhaust Gas Recirculation (EGR) valve operation and electrical connections prior to the adjustment of the ignition timing. Correct or repair any problem as required. Be sure to inspect the throttle valve and Throttle Position (TP) sensor for proper operation.

1. Before servicing the vehicle, refer to the Precautions Section.

2. Locate the timing marks on the crankshaft pulley and the front of the engine.

3. Clean the timing marks.

4. The ignition timing specifications are as follows:

- 10–20 degrees Before Top Dead Center (BTDC)

5. Using chalk or white paint, color the mark on the crankshaft pulley and the mark on the scale, which will indicate the correct timing when aligned with the notch on the crankshaft pulley.

6. Attach a tachometer to the engine.

7. Attach a timing light to the engine to number 1 cylinder ignition coil wire.

8. Turn **OFF** all the electrical equipment and accessories.

9. Check to be sure all of the wires clear the fan, then, start the engine and allow it to reach normal operating temperatures.

10. Block the front wheels and set the parking brake. Shift the transmission into **NEUTRAL**.

11. Perform the following procedures:

a. Race the engine at 2000 rpm for about 2 minutes under a no-load condition; be sure all of the accessories are turned **OFF**.

b. Perform on board engine diagnostics and repair any fault code.

c. Race the engine at 2000 rpm for about 2 minutes under a no-load condition.

d. Turn the engine **OFF** and disconnect the TP sensor.

e. Start and race the engine 2–3 times under no-load, then run the engine at idle speed.

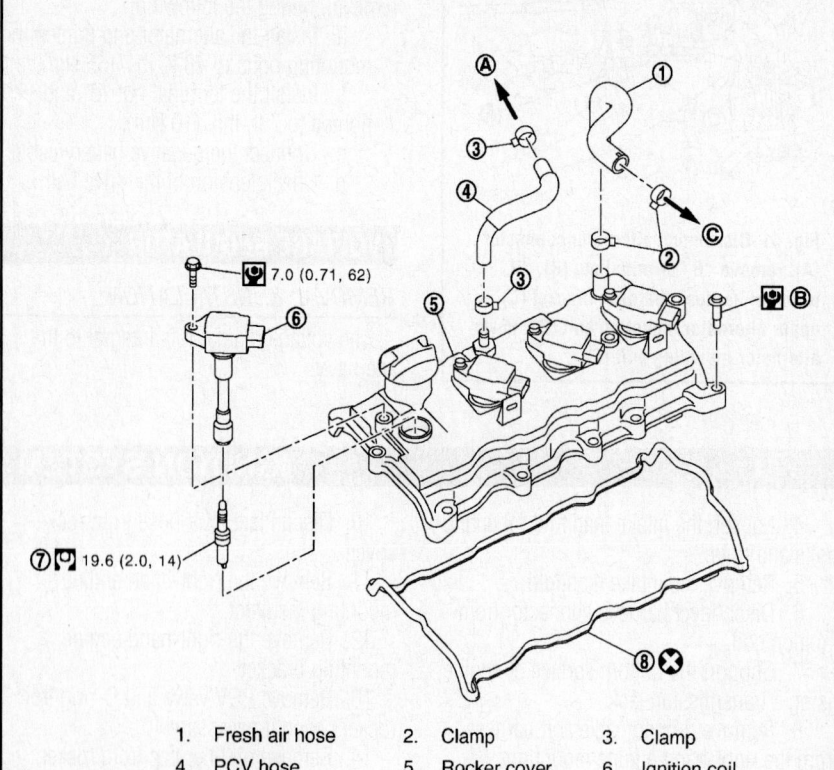

1. Fresh air hose
2. Clamp
3. Clamp
4. PCV hose
5. Rocker cover
6. Ignition coil
7. Spark plug
A. To intake manifold
C. To air duct

22140_ROGU_G0048

Fig. 43 Ignition coil and related components

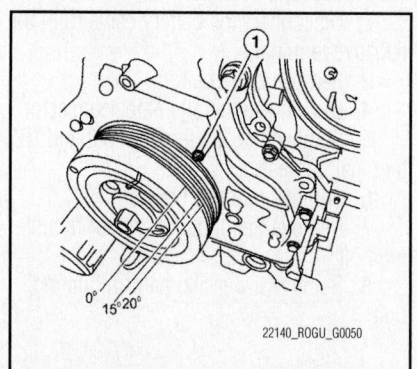

22140_ROGU_G0050

Fig. 44 Locate the timing marks on the crankshaft pulley

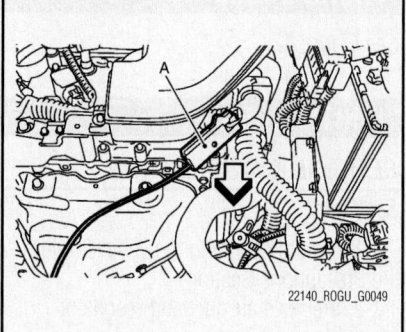

22140_ROGU_G0049

Fig. 45 Attach a timing light to the engine to number 1 cylinder ignition coil wire

12. Aim the timing light at the timing marks. If the marks on the pulley and the engine are aligned when the light flashes, the timing is correct. Turn the engine **OFF** and remove the tachometer and the timing light. If the marks are not in alignment, proceed with the following steps:

a. Turn the engine **OFF**.

b. Check the Camshaft Position (CMP) sensor (PHASE), Crankshaft Position (CKP) sensor (REF) and CKP sensor (POS). Replace if necessary.

c. Check that all the timing chain and gears are correctly aligned.

d. If the ignition timing is still not correct, substitute a known good Electronic Control Module (ECM).

➡**The ECM may be the cause of the problem, but this is rarely the case.**

e. Turn the engine **OFF** and remove the tachometer and the timing light.

ADJUSTMENT

The ignition timing is not adjustable.

SPARK PLUGS

REMOVAL & INSTALLATION

1. Before servicing the vehicle, refer to the Precautions Section.

2. Remove air duct and resonator assembly.

3. Remove the electric throttle control actuator without disconnecting water hose.

4. Loosen the intake manifold mounting bolts and nuts.

5. Remove the intake manifold.

6. Disconnect harness connector from ignition coil.

7. Support the bottom surface of engine using a transmission jack.

8. Remove ground cable and harness from the right-hand engine mounting bracket.

9. Remove the ignition coils.

10. Remove the spark plugs.

To install:

11. Install in the reverse order of removal, noting the following:

a. Tighten the spark plugs to 14 ft. lbs. (20 Nm).

b. Tighten the ignition coil mounting bolts to 62 inch lbs. (7 Nm).

ENGINE ELECTRICAL

STARTING SYSTEM

STARTER

REMOVAL & INSTALLATION

2WD Models

See Figures 46 and 47.

1. Before servicing the vehicle, refer to the Precautions Section.

2. Disconnect the battery cable from the negative terminal.

3. Remove the terminal nut and terminal harness.

4. Disconnect "S" connector.

5. Remove the starter motor mounting bolts.

6. Remove the starter motor downward from the vehicle.

To install:

7. Install in the reverse order of removal, noting the following:

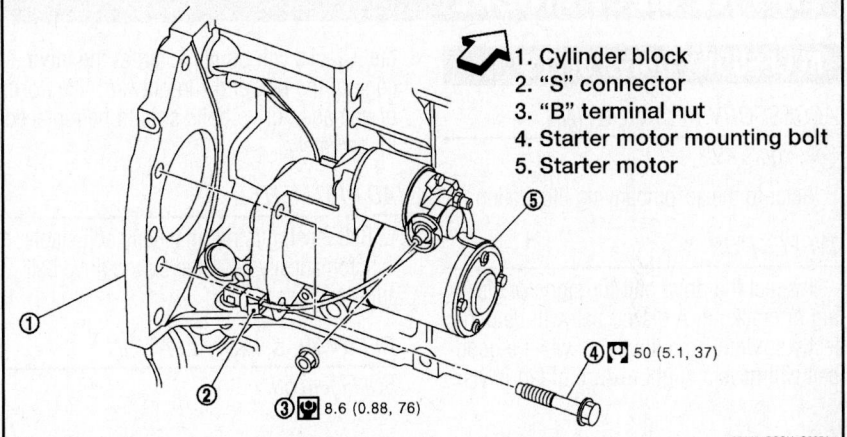

1. Cylinder block
2. "S" connector
3. "B" terminal nut
4. Starter motor mounting bolt
5. Starter motor

③ 8.6 (0.88, 76)
④ 50 (5.1, 37)

22140_ROGU_G0051

Fig. 47 View of starter motor mounting

a. Tighten the starter mounting bolts to 37 ft. lbs. (50 Nm).

b. Tighten terminal nut "B" carefully to 76 inch lbs. (8.6 Nm).

8. Connect the negative battery cable.

AWD Models

See Figures 47 and 48.

1. Before servicing the vehicle, refer to the Precautions Section.

2. Disconnect the battery cable from the negative terminal.

3. Remove the front wheel and tire.

4. Remove the "B" terminal nut and "B" terminal harness.

5. Disconnect "S" connector.

6. Remove the starter motor mounting bolts.

7. Slide the alternator out and remove.

To install:

8. Install in the reverse order of removal, noting the following:

a. Tighten the starter mounting bolts to 37 ft. lbs. (50 Nm).

b. Tighten the terminal nut "B" carefully to 76 inch lbs. (9 Nm).

c. Connect the negative battery cable.

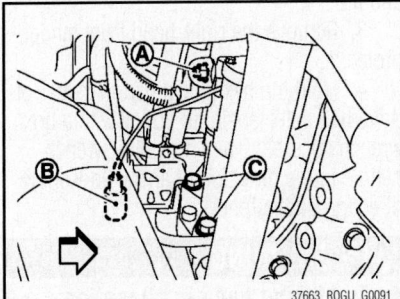

37663_ROGU_G0091

Fig. 46 Remove "B" terminal nut (A) and "B" terminal harness, disconnect "S" connector (B), and remove starter motor mounting bolts (C)

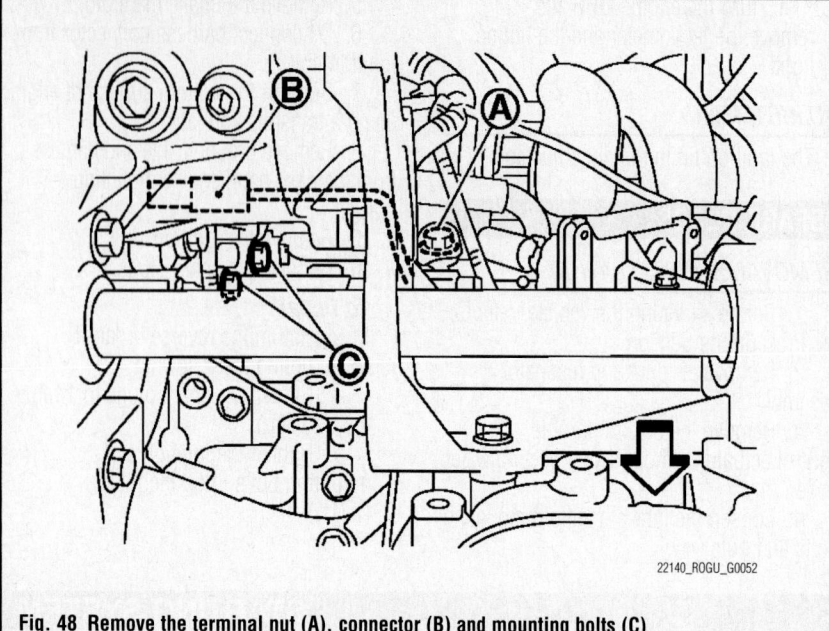

Fig. 48 Remove the terminal nut (A), connector (B) and mounting bolts (C)

SOLENOID OR RELAY REPLACEMENT

1. Before servicing the vehicle, refer to the Precautions Section.
2. Disconnect the negative battery cable.
3. Remove the starter motor.
4. Remove the screws from the motor gear case.
5. Keep the adjustment shims for installation.
6. Remove the solenoid.

To install:

7. Install the solenoid with adjustment shims.
8. Tighten the mounting screws to 52 inch lbs. (6 Nm).
9. Test starter drive for proper operation.
10. Install the starter motor.
11. Connect the negative battery cable.

ENGINE MECHANICAL

ACCESSORY DRIVE BELTS

ACCESSORY BELT ROUTING

See Figure 49.

Refer to the accompanying illustration.

INSPECTION

Inspect the drive belt for signs of glazing or cracking. A glazed belt will be perfectly smooth from slippage, while a good belt will have a slight texture of fabric visible. Cracks will usually start at the inner edge of the belt and run outward. All worn or damaged drive belts should be replaced immediately.

ADJUSTMENT

Belt tension is not manually adjustable, it is automatically adjusted by the drive belt auto-tensioner.

REMOVAL & INSTALLATION

See Figure 50.

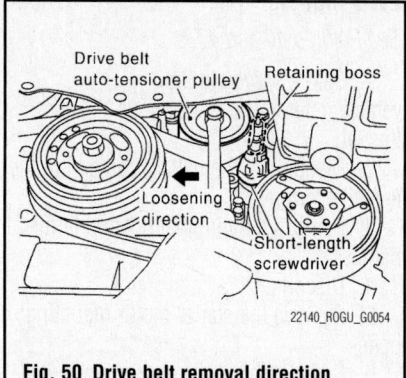

Fig. 50 Drive belt removal direction shown

1. Before servicing the vehicle, refer to the Precautions Section.
2. Remove the right-hand front wheel and tire.
3. Remove the right-hand front fender protector.
4. Hold the hexagonal part in center of drive belt auto-tensioner pulley with a box wrench securely. Then move the wrench handle in the direction of arrow (loosening direction of tensioner).

✳✳ WARNING

Avoid placing hand in a location where pinching may occur if the holding tool accidentally comes off. Never loosen the hexagonal part in center of drive belt auto-tensioner pulley (Never turn it counterclock-

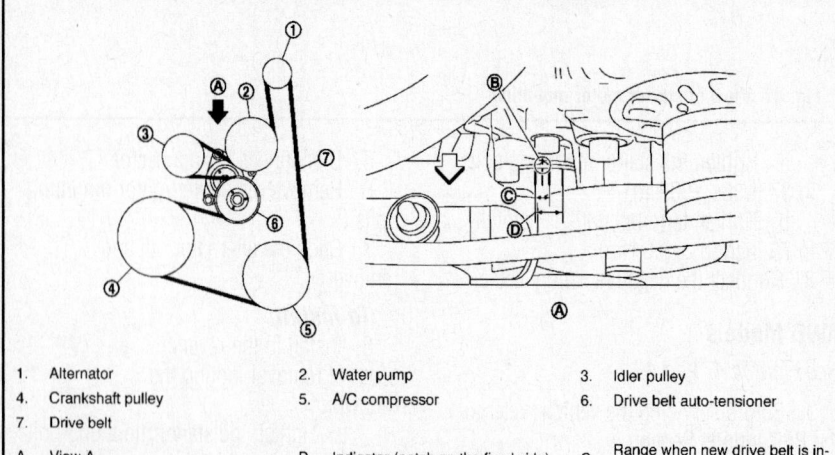

1.	Alternator	2.	Water pump	3.	Idler pulley
4.	Crankshaft pulley	5.	A/C compressor	6.	Drive belt auto-tensioner
7.	Drive belt				
A.	View A	B.	Indicator (notch on the fixed side)	C.	Range when new drive belt is installed
D.	Possible use range				
⟨⟩	: Engine front				

Fig. 49 Accessory drive belt

wise). If turned counterclockwise, the complete drive belt auto-tensioner must be replaced as a unit, including the pulley.

5. Insert a rod approximately 0.24 in. (6 mm) in diameter such as short-length screwdriver into the hole of the retaining boss to fix drive belt auto-tensioner pulley.

6. Loosen drive belt from water pump pulley in sequence, and remove it.

To install:

7. Hold the hexagonal part in center of drive belt auto-tensioner pulley with a box wrench securely. Then move the wrench handle in the direction of arrow (loosening direction of tensioner).

8. Insert a rod approximately 0.24 in. (6 mm) in diameter such as short-length screwdriver into the hole of retaining boss to fix drive belt auto-tensioner pulley.

9. Hook drive belt onto all pulleys except for water pump, and then onto water pump pulley finally.

10. Confirm drive belt is completely set to pulleys.

11. Release the drive belt auto-tensioner, and apply tension to drive belt.

12. Turn the crankshaft pulley clockwise several times to equalize tension between each pulley.

13. Confirm tension of drive belt at indicator (notch on fixed side) is within the possible use range.

CAMSHAFT AND VALVE LIFTERS

REMOVAL & INSTALLATION

See Figures 51 through 57.

1. Before servicing the vehicle, refer to the Precautions Section.

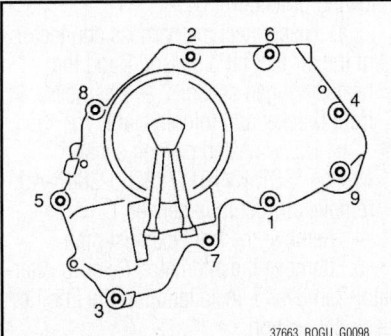

Fig. 51 Intake valve timing control cover bolt tightening sequence—reverse the sequence for removal

2. Disconnect the negative battery cable.

3. Relieve the fuel system pressure.

4. Drain the coolant from the engine and radiator.

5. Remove the intake manifold. Refer to Intake Manifold Removal & Installation.

6. Remove the valve cover.

7. Remove the camshaft position sensor.

8. Remove the camshaft position sensor bracket.

9. Remove the intake valve timing control cover, as follows:

a. Disconnect the intake valve timing control solenoid valve harness connector.

b. Remove the intake valve timing control solenoid valve, if necessary.

c. Loosen the bolts in the reverse order as shown in the figure.

d. Use a seal cutter [SST: KV10111100 (J-37228)] or equivalent tool to cut liquid gasket for removal.

10. Pull the chain guide between camshaft sprockets out through front cover.

11. Set the No. 1 cylinder at TDC on its compression stroke with the following procedure:

a. Open splash guard on the right-hand undercover.

b. Rotate the crankshaft pulley clockwise and align TDC mark to the timing indicator on the front cover.

c. At the same time, check that the mating marks on camshaft sprockets are located as shown. If not, rotate crankshaft pulley one more turn to align mating marks to the positions in the figure.

12. Remove the camshaft sprockets with the following procedure:

a. Line up the mating marks on camshaft sprockets, and paint indelible mating marks on timing chain link plate.

b. Push in the chain tensioner plunger. Insert a stopper pin into the hole on the chain tensioner body to secure the chain tensioner plunger and remove chain tensioner.

➡**Use approximately 0.020 in. (0.5 mm) diameter hard metal pin as a stopper pin.**

c. Secure the hexagonal part of the camshaft with a wrench. Loosen the camshaft sprocket mounting bolts and remove the camshaft sprockets.

❈❈ WARNING

Never rotate the crankshaft or camshaft while timing chain is removed. It causes interference between valve and piston.

➡**Chain tension holding work is not necessary. Crankshaft sprocket and timing chain do not disconnect structurally while front cover is attached.**

13. Loosen the mounting bolts in the reverse order as shown in the figure, and remove camshaft brackets and camshafts.

a. Remove camshaft bracket (No. 1) by slightly tapping it with a plastic hammer.

14. Remove the valve lifters.

a. Identify installation positions, and store them without mixing them up.

To install:

15. Install the valve lifters in the original positions.

16. Install the camshafts.

a. Distinction between intake and exhaust camshafts is determined by the different shapes of the rear end.

17. Install camshafts so that camshaft dowel pins on the front side are positioned as shown.

18. Install camshaft brackets with the following procedure:

a. Remove foreign material completely from camshaft bracket backside and from cylinder head installation face.

b. Install camshaft brackets (No. 2 to 5) aligning the identification marks on upper surface as shown in the figure. Install so

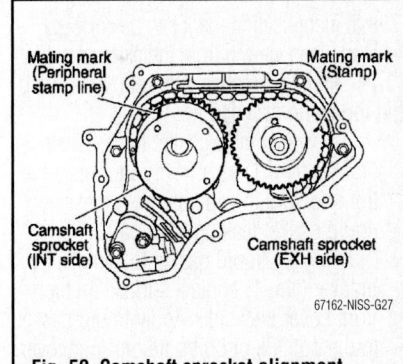

Fig. 52 Camshaft sprocket alignment marks

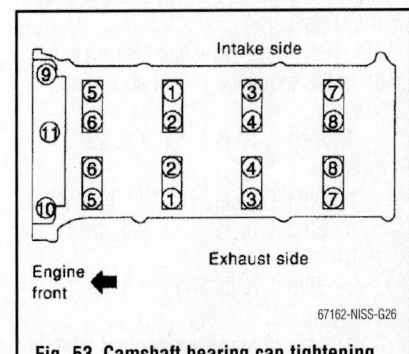

Fig. 53 Camshaft bearing cap tightening sequence—reverse the sequence for removal

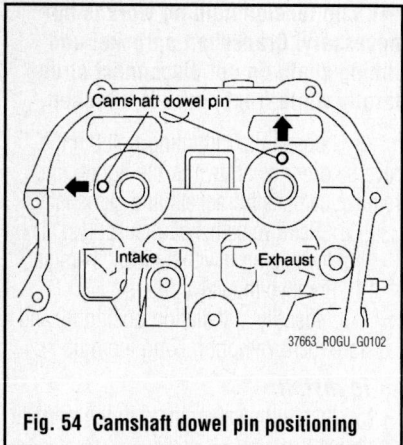

Fig. 54 Camshaft dowel pin positioning

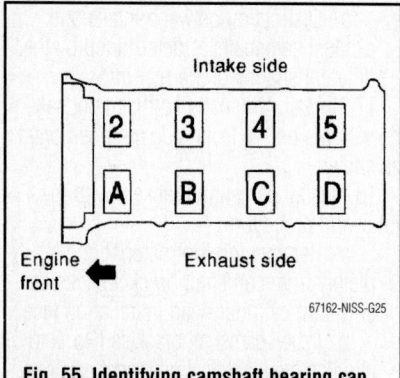

Fig. 55 Identifying camshaft bearing cap installation marks

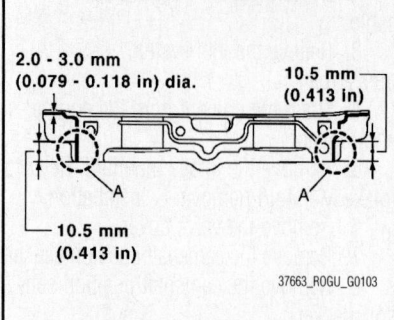

Fig. 56 Apply liquid gasket to camshaft bracket (No. 1) and be sure to wipe off any excessive liquid gasket leaking from part "A"

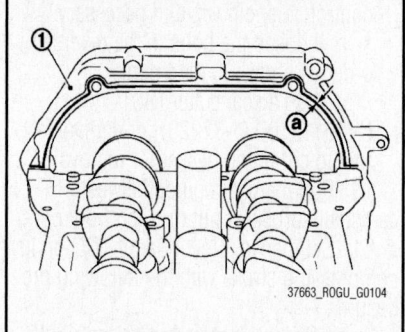

Fig. 57 1: Front Cover, a: 0.134–0.173 in. (3.4–4.4mm)

that identification mark can be correctly read when viewed from the exhaust side.

19. Install camshaft bracket (No. 1) with the following procedure:

a. Apply liquid gasket to camshaft bracket (No. 1) as shown. After installation, be sure to wipe off any excessive liquid gasket leaking from part "A".

b. Apply liquid gasket to camshaft bracket (No. 1) contact surface on the front cover backside. Apply liquid gasket to the outside of bolt hole on front cover.

c. For camshaft bracket (No. 1) near installation position, and install it without disturbing the liquid gasket applied to the surfaces.

20. Tighten the camshaft bearing caps bolts in the sequence shown above, and as follows:

a. Step 1, bolts 9–11: 1 ft. lbs (2 Nm).

b. Step 2, bolts 1–8: 1 ft. lbs (2 Nm).

c. Step 3, bolts 1–11: 4 ft. lbs. (6 Nm).

d. Step 4, bolts 1–11: 8 ft. lbs. (10 Nm).

e. After tightening mounting bolts of camshaft brackets, be sure to wipe off excessive liquid gasket from the following:

• Mating surface of rocker cover
• Mating surface of front cover (When installed without front cover)

21. Install the camshaft sprockets.

a. Align the mating marks on each camshaft sprocket with the ones painted on timing chain link plate during removal.

b. Aligned mating marks could slip. Therefore, after matching them, hold the timing chain in place by hand. Before and after installing chain tensioner, check again that mating marks have not slipped. Before installation of chain tensioner, it is possible to re-match the marks on timing chain with the ones on each sprocket.

22. Install the chain tensioner.

a. After installation, pull the stopper pin off completely, and check that chain tensioner plunger is released.

23. Install the chain guide.

24. Install the intake valve timing control cover with the following procedure:

a. Install the intake valve timing control solenoid valve to intake valve timing control cover if removed.

b. Install the oil rings to the camshaft sprocket (INT) insertion points on backside of intake valve timing control cover.

c. Install new O-ring to front cover.

d. Apply liquid gasket with a tube presser to intake valve timing control cover.

➡**Attaching should be done within 5 minutes after liquid gasket application.**

e. Tighten the mounting bolts in numerical order as shown above.

25. Install the camshaft position sensor bracket.

a. Apply liquid gasket with a tube presser the camshaft position sensor bracket. After installation, be sure to wipe off any excessive liquid gasket leaking from part "b".

➡**Attaching should be done within 5 minutes after liquid gasket application.**

b. Tighten mounting bolts in numerical order as shown above.

26. Install camshaft position sensor.

27. Inspect and adjust valve clearance.

28. Install the valve cover.

29. Install the intake manifold.

30. Connect the negative battery cable.

31. Fill the cooling system.

32. Start the vehicle, check for leaks and repair if necessary.

CATALYTIC CONVERTER

REMOVAL & INSTALLATION

See Combination Manifold.

COMBINATION MANIFOLD

REMOVAL & INSTALLATION

See Figures 58 and 59.

1. Before servicing the vehicle, refer to the Precautions Section.

2. Disconnect the negative battery cable.

3. Remove the air fuel ratio sensor 1 and the heated oxygen sensor 2 with the following procedure:

a. Disconnect the harness connector of the air fuel ratio sensor 1 and the heated oxygen sensor 2 and harness from bracket and middle clamp.

b. Using heated oxygen sensor wrench [SST: KV10117100 (J-3647-A)], remove air fuel ratio sensor 1.

4. Remove the front exhaust pipe.

5. Remove the alternator. Refer to Alternator Removal & Installation in the Engine Electrical section.

6. Remove the exhaust manifold cover (upper).

7. Loosen the nuts in reverse order as shown in the figure to remove exhaust man-

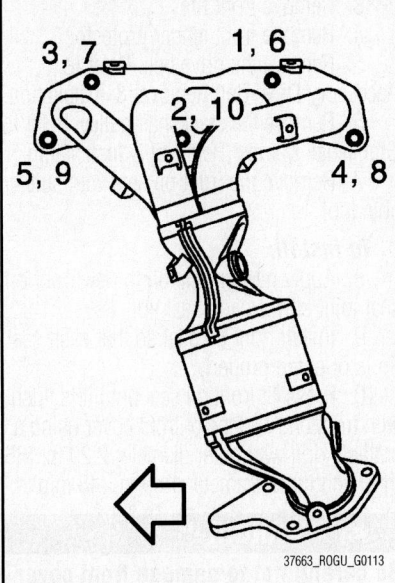

Fig. 58 Exhaust manifold and three way catalyst assembly tightening sequence—reverse the sequence for removal

ifold and three way catalyst assembly. Disregard No. 6 to 10 when loosening.

8. Remove the gasket. Cover engine openings to avoid entry of foreign materials.

9. Remove the exhaust manifold cover (lower) and three way catalyst cover from exhaust manifold and three way catalyst assembly.

To install:

10. Install the lower cover and three way catalyst cover.

11. Clean all gasket mounting surfaces and install new gaskets.

12. Install the stud bolts, if removed, and tighten to 11 ft. lbs. (15 Nm).

13. Tighten nuts in numerical order as shown in the figure, above. No. 6 to 10 refer to double tightening of bolts No. 1 and 5.

14. Install the upper cover.

15. Install the alternator.

16. Install the front exhaust pipe.

17. Before installing a new air fuel ratio sensor 1 and heated oxygen sensor 2, clean the exhaust system threads using heated oxygen sensor thread cleaner and apply anti-seize lubricant (commercial service tool: J-43897-18 or J-43897-12).

18. Install the air fuel ratio sensor 1 and the heated oxygen sensor 2

19. Never over torque the air fuel ratio sensor 1 and heated oxygen sensor 2. Doing so may cause damage to the air fuel ratio sensor 1 and heated

oxygen sensor 2, resulting in the "MIL" coming on.

20. Connect the negative battery cable.

CRANKSHAFT DAMPER

REMOVAL & INSTALLATION

See Figures 60 through 62.

1. Before servicing the vehicle, refer to the Precautions Section.

2. Remove engine under cover.

3. Remove front tire.

4. Remove front fender protector.

5. Fix crankshaft pulley with a pulley holder (commercial service tool), loosen crankshaft pulley bolt, and locate bolt seating surface at 0.39 inch (10 mm) from its original position.

6. Attach a pulley puller (commercial service tool) in the M 6 thread hole on crankshaft pulley, and then remove crankshaft pulley.

To install:

7. Secure crankshaft pulley with a pulley holder (commercial service tool), and tighten crankshaft pulley.

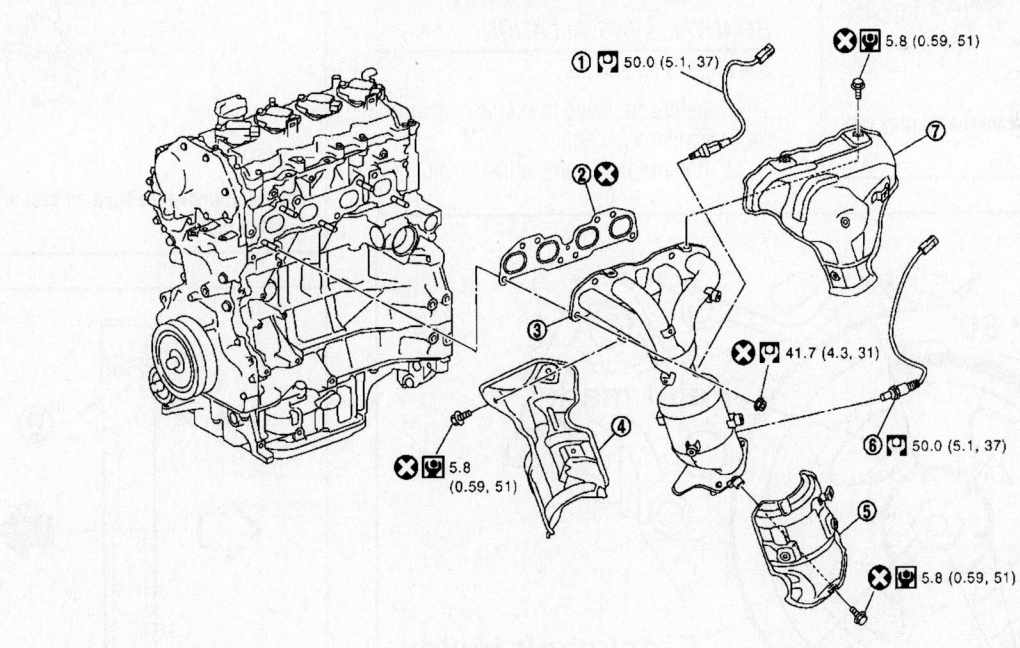

1. Air fuel ratio sensor 1	2. Gasket	3. Exhaust manifold and three way catalyst assembly
4. Three way catalyst cover	5. Exhaust manifold cover (lower)	6. Heated oxygen sensor 2
7. Exhaust manifold cover (upper)		

Fig. 59 Exhaust manifold and three way catalyst

8. Install front oil seal so that each seal lip is oriented properly.

9. Press-fit front oil seal until it is flush with front end surface of front cover using a suitable drift with outer diameter 2.20 inches (56 mm) and inner diameter 1.89 inches (48 mm).

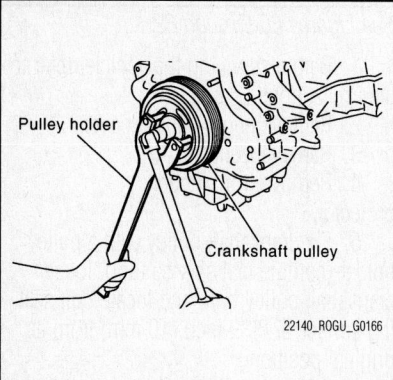

Fig. 60 Loosen the crankshaft pulley bolt

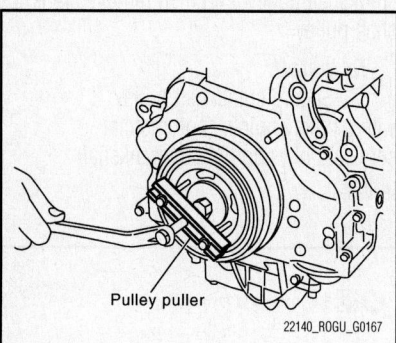

Fig. 61 Remove the crankshaft pulley with a puller

❊❊ WARNING

Be careful not to damage front cover and crankshaft. Press-fit oil seal straight to avoid causing burrs or tilting.

10. Insert crankshaft pulley by aligning with crankshaft key.

11. When inserting crankshaft pulley with a plastic hammer, tap on its center portion (not circumference).

12. Secure crankshaft pulley with a pulley holder (commercial service tool).

13. Perform angle tightening with the following procedure:

 a. Apply new engine oil to thread and seat surfaces of crankshaft pulley bolt.

 b. Tighten crankshaft pulley bolt to 31 ft. lbs. (42 Nm).

 c. Put a paint mark on crankshaft pulley, mating with any one of six easy to recognize angle marks on bolt flange.

 d. Turn another 60 degrees clockwise (angle tightening).

14. Check the tightening angle with movement of one angle mark.

15. Install all removed parts in the reverse order of removal.

CRANKSHAFT FRONT SEAL

REMOVAL & INSTALLATION

See Figures 63 and 64.

1. Before servicing the vehicle, refer to the Precautions Section.

2. Remove the engine under cover.

3. Remove front tire.

4. Remove front fender protector.

5. Remove the drive belt. Refer to Accessory Drive Belt Removal & Installation.

6. Remove the crankshaft pulley. Refer to Crankshaft Damper Removal & Installation.

7. Remove the front oil seal with suitable tool.

To install:

8. Apply new engine oil to new front oil seal joint surface and seal lip.

9. Install front oil seal so that each seal lip is oriented properly.

10. Press-fit front oil seal until it is flush with front end surface of front cover using a suitable drift with outer diameter 2.20 in. (56 mm) and inner diameter 1.89 in. (48 mm).

❊❊ WARNING

Be careful not to damage front cover and crankshaft. Press-fit oil seal straight to avoid causing burrs or tilting.

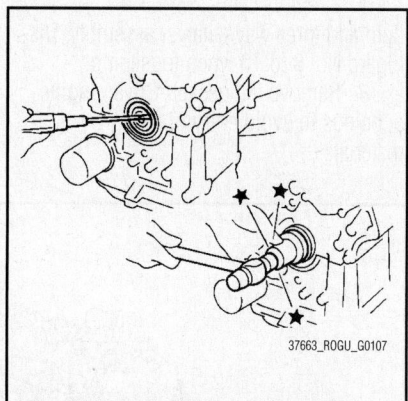

Fig. 63 Remove the front oil seal with suitable tool

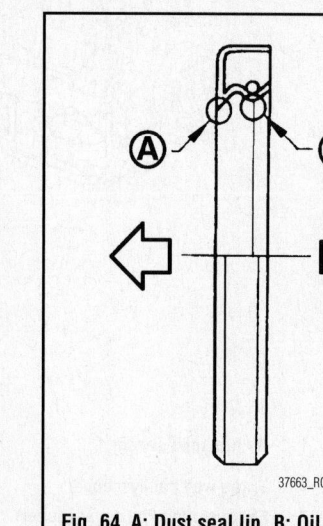

Fig. 64 A: Dust seal lip, B: Oil seal lip, White arrow: Engine outside, Black arrow: Engine inside.

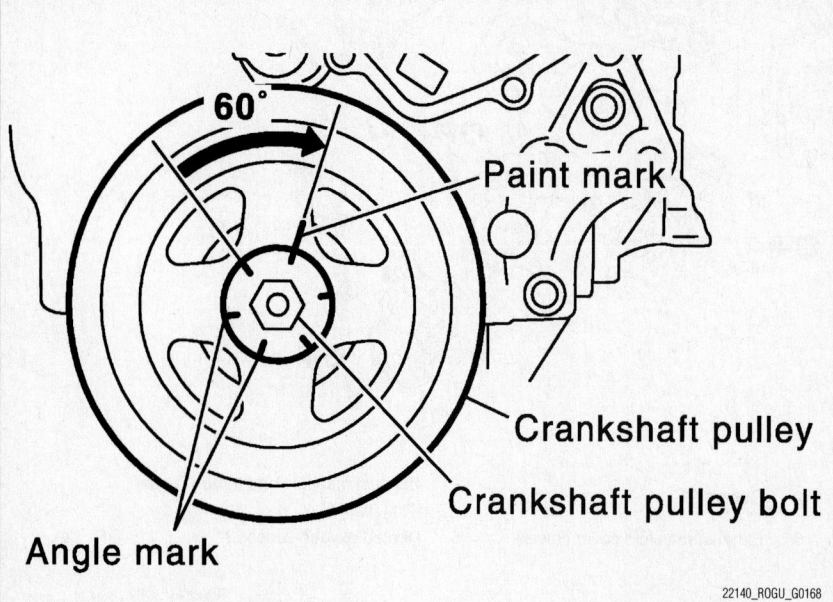

Fig. 62 Turn another 60 degrees clockwise

11. Install all removed parts in the reverse order of removal.

CYLINDER HEAD

REMOVAL & INSTALLATION

See Figures 65 through 68.

1. Before servicing the vehicle, refer to the Precautions Section.
2. Release fuel pressure.
3. Drain the engine coolant and engine oil.
4. Remove the following components and related parts:
   - Exhaust manifold and three way catalyst assembly
   - Intake manifold and fuel tube assembly
   - Water control valve and water control valve housing (water outlet)
5. Remove the front cover and timing chain.
6. Remove the camshafts.
7. Securely support bottom of cylinder block with a jack or equivalent tool, and release the hoist that was supporting it.
8. Remove the cylinder head, loosening the bolts in the reverse order of the tightening sequence.
9. Using TORX® socket (size E20), loosen cylinder head bolts.
10. Remove the cylinder head gasket.

To install:

11. Carefully clean the engine block and cylinder head, check to see if cylinder head surface is warped.
12. At each of several locations on bottom surface of cylinder head, measure the distortion in six directions.
13. Cylinder head bolts are tightened by plastic zone tightening method. Whenever the size difference between (A) and (B)

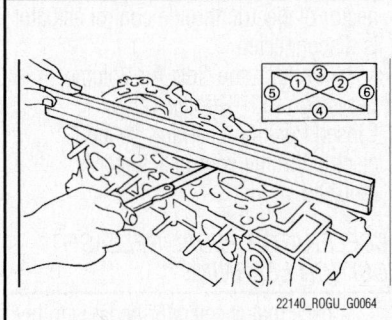

Fig. 65 Cylinder head distortion measurement locations

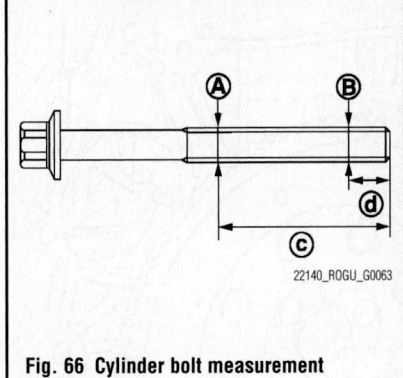

Fig. 66 Cylinder bolt measurement location view

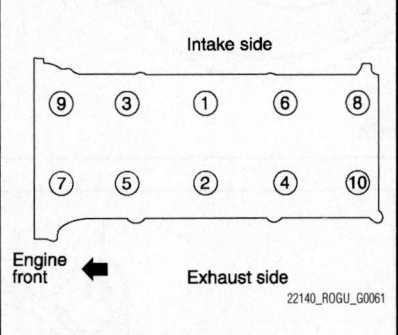

Fig. 67 Tighten cylinder head bolts in numerical order as shown

exceeds the limit, replace them with new one.
14. Limits are as follows:
   - Limit ("B" minus "A") : 0.0091 in. (0.23 mm)
   - c: 2.165 in. (55 mm)
   - d: 0.472 in. (12 mm)
15. Install the cylinder head gasket.
16. Tighten cylinder head bolts in numerical order as shown in figure with the following procedure, and install cylinder head.

❋❋ WARNING

If cylinder head bolts are reused, check their outer diameters before installation.

17. Apply new engine oil to threads and seating surface of mounting bolts.
18. Tighten all bolts to 37 ft. lbs. (50 Nm).
19. Turn all bolts 60 degrees clockwise (angle tightening).
20. Completely loosen.
21. In this step, loosen bolts in reverse order of that indicated in the figure.

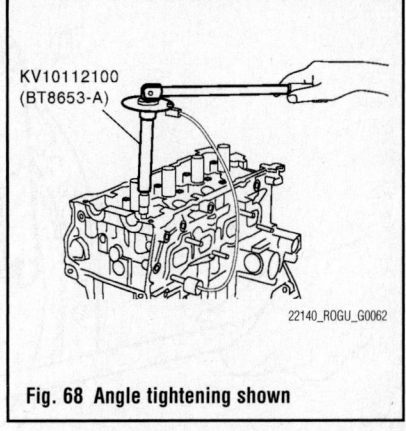

Fig. 68 Angle tightening shown

22. Tighten all bolts as follows to 29 ft. lbs. (39 Nm):
   a. Turn all bolts 75 degrees clockwise (angle tightening).
   b. Turn all bolts 75 degrees clockwise again (angle tightening).

❋❋ WARNING

Check and confirm the tightening angle by using an angle wrench (SST) or protractor. Avoid judgment by visual inspection without the tool.

23. Installation is in the reverse order of removal procedure after this step.

EXHAUST MANIFOLD

REMOVAL & INSTALLATION

See Combination Manifold.

FLYWHEEL

REMOVAL & INSTALLATION

See Figure 69.

1. Before servicing the vehicle, refer to the Precautions Section.
2. Disconnect the negative battery cable.
3. Remove the transaxle assembly. Refer to Transaxle Removal & Installation in the Drive Train section.
4. Secure the flywheel and remove the mounting bolts.
5. Remove the flywheel.

To install:

6. Install the flywheel.
7. Secure the flywheel and tighten in sequence the mounting bolts to 80 ft. lbs. (108 Nm).
8. Install the transaxle assembly.
9. Connect the negative battery cable.

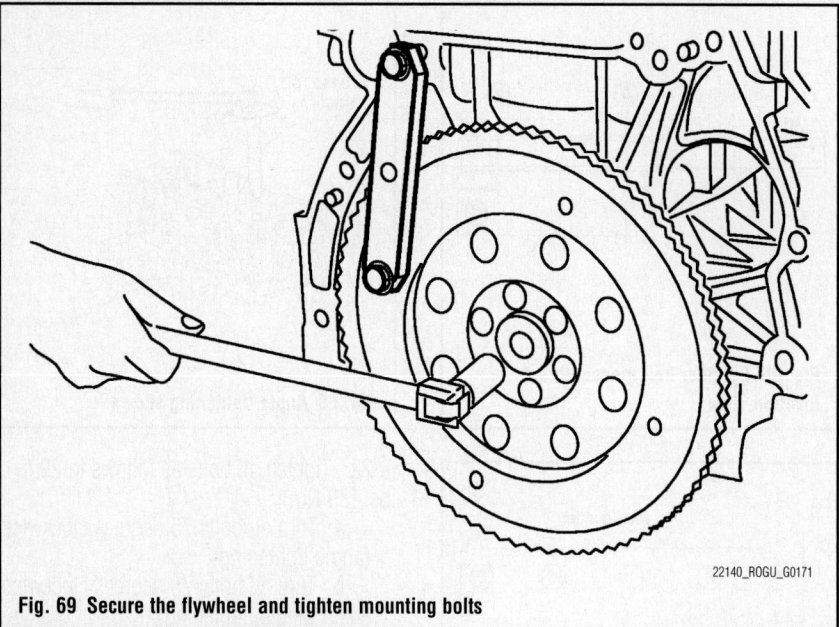

Fig. 69 Secure the flywheel and tighten mounting bolts

INTAKE MANIFOLD

REMOVAL & INSTALLATION

See Figures 70 through 71.

1. Before servicing the vehicle, refer to the Precautions Section.

2. Release the fuel pressure.

3. Remove the cowl top cover.

4. Remove the air cleaner case and mass air flow sensor assembly and air duct and resonator assembly.

5. Remove the electric throttle control actuator with the following procedure:

    a. Disconnect the harness connector.

    b. Loosen the mounting bolts in reverse order as shown in the figure, and remove electric throttle control actuator and gasket.

❈❈ WARNING

Handle carefully to avoid any shock to electric throttle control actuator. Never disassemble.

➡**When removing only the intake manifold, move electric throttle control actuator without disconnecting the water hose.**

6. Disconnect harness, vacuum hose and PCV hose from intake manifold, and move them aside.

7. Remove the intake manifold support.

8. Disconnect harness connector from tumble control valve motor (For California).

9. Loosen mounting bolts and nuts in reverse order as shown in the figure, and remove intake manifold and gasket. Disregard No. 6 when loosening.

10. Disconnect the sub-harness from fuel injector.

11. Remove the fuel tube and fuel injector assembly from intake manifold adaptor.

12. Remove the EVAP canister purge volume control solenoid valve from intake manifold, if necessary.

To install:

13. Install in the reverse order of removal, noting the following:

    a. If the stud bolts were removed, install them and tighten to 83 inch lbs. (9 Nm).

    b. No. 6 refers to double tightening of bolt No. 1.

    c. Tighten the mounting bolts equally to 14 ft. lbs. (20 Nm), and diagonally in several steps and in numerical order in the reverse order of removal.

    d. Tighten the electric throttle body to 74 inch lbs. (8 Nm).

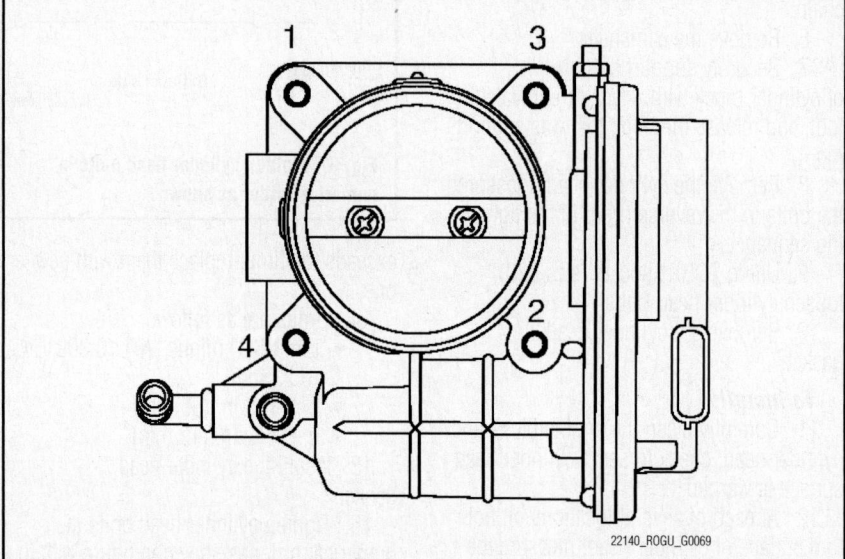

Fig. 70 Loosen mounting bolts in reverse order as shown in the figure, and remove electric throttle control actuator and gasket

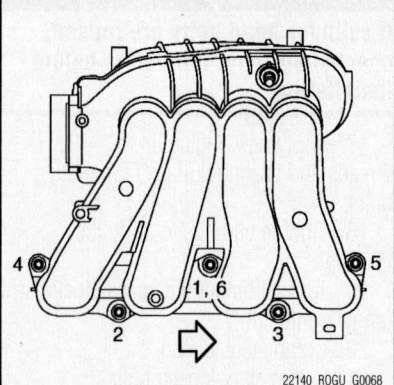

Fig. 71 Loosen mounting bolts and nuts in reverse order as shown in the figure

    e. Perform the "Throttle Valve Closed Position Learning" when harness connector of electric throttle control actuator is disconnected.

    f. Perform the "Idle Air Volume Learning" and "Throttle Valve Closed Position Learning" when electric throttle control actuator is replaced.

ACCELERATOR PEDAL RELEASED POSITION LEARNING

1. Check that accelerator pedal is fully released.

2. Turn ignition switch ON and wait at least 2 seconds.

3. Turn ignition switch OFF and wait at least 10 seconds.

4. Turn ignition switch ON and wait at least 2 seconds.

5. Turn ignition switch OFF and wait at least 10 seconds.

IDLE AIR VOLUME LEARNING

See Figures 72 and 73.

The Idle Air Volume Learning is a function of ECM to learn the idle air volume that keeps each engine idle speed within the specific range. It must be performed under the following conditions:

• Each time the electric throttle control actuator or ECM is replaced

• Idle speed or ignition timing is out of the specification

1. Check that all of the following conditions are satisfied. Learning will be cancelled if any of the following conditions are missed for even a moment:

• Battery voltage: More than 12.9 V (At idle)

• Engine coolant temperature: 158–212°F (70–100°C)
• Selector lever: P or N
• Electric load switch: OFF (Air conditioner, headlamp, rear window defogger)
• On vehicles equipped with daytime light systems, if the parking brake is applied before the engine is started, the headlamp will not illuminate.
• Steering wheel: Neutral (Straight-ahead position)

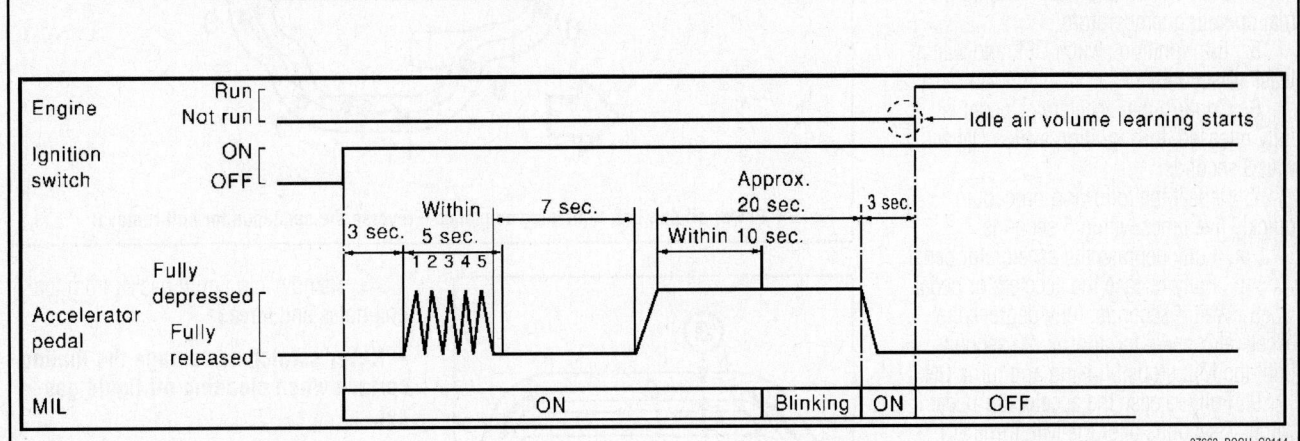

Fig. 72 Idle Air Volume Learning

Idle Speed

Transmission	Condition	Specification
CVT	No load* (in P or N position)	700 ± 50 rpm

*: Under the following conditions
• A/C switch: OFF
• Electric load: OFF (Lights, heater fan & rear window defogger)
• Steering wheel: Kept in straight-ahead position

Ignition Timing

Transmission	Condition	Specification
CVT	No load* (in P or N position)	10 ± 5° BTDC

*: Under the following conditions
• A/C switch: OFF
• Electric load: OFF (Lights, heater fan & rear window defogger)
• Steering wheel: Kept in straight-ahead position

37663_ROGU_G0115

Fig. 73 Idle Speed and Ignition Timing

- Vehicle speed: Stopped
- Transmission: Warmed-up
- With CONSULT-III: Drive vehicle until "ATF TEMP SEN" in "DATA MONITOR" mode of "CVT" system indicates less than 0.9V.
- Without CONSULT-III: Drive vehicle for 10 minutes.

2. If a CONSULT-III scan tool is not available, perform Accelerator Pedal Released Position Learning.

3. Perform Throttle Valve Closed Position Learning.

4. Start engine and warm it up to normal operating temperature.

5. Turn ignition switch OFF and wait at least 10 seconds.

6. Confirm that accelerator pedal is fully released, turn ignition switch ON and wait 3 seconds.

7. Repeat the following procedure quickly five times within 5 seconds.
   a. Fully depress the accelerator pedal.
   b. Fully release the accelerator pedal.

8. Wait 7 seconds, fully depress the accelerator pedal for approx. 20 seconds until the MIL stops blinking and turns ON.

9. Fully release the accelerator pedal within 3 seconds after the MIL turns ON.

10. Start engine and let it idle.

11. Wait 20 seconds.

12. Rev up the engine two or three times and check that idle speed and ignition timing are within the specifications. If the inspection result is not normal, find the malfunctioning part.

THROTTLE VALVE CLOSED POSITION LEARNING

The Throttle Valve Closed Position Learning is a function of ECM to learn the fully closed position of the throttle valve by monitoring the throttle position sensor output signal. It must be performed each time harness connector of electric throttle control actuator or ECM is disconnected.

1. Check that accelerator pedal is fully released.

2. Turn ignition switch ON.

3. Turn ignition switch OFF and wait at least 10 seconds.

4. Check that throttle valve moves during the above 10 seconds by confirming the operating sound.

OIL PAN

REMOVAL & INSTALLATION

Lower Oil Pan

See Figures 74 through 78.

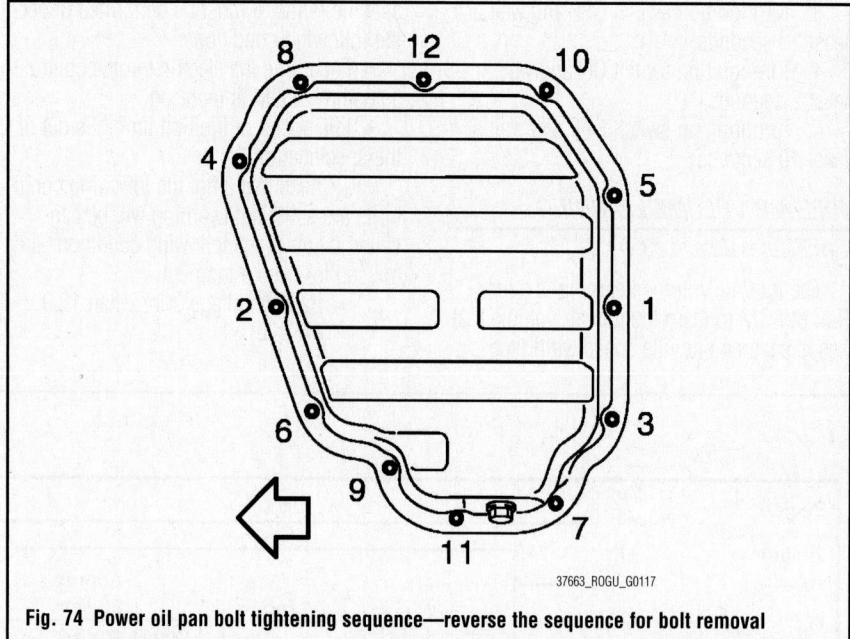

Fig. 74 Power oil pan bolt tightening sequence—reverse the sequence for bolt removal

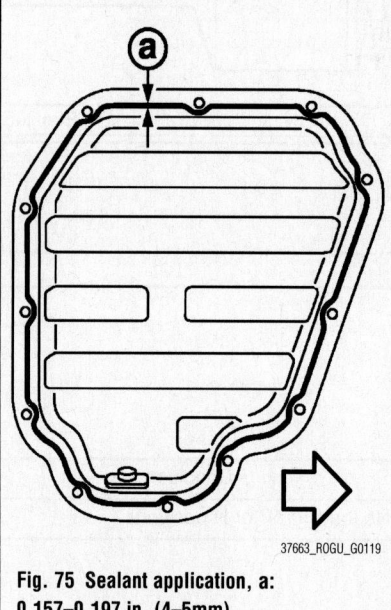

Fig. 75 Sealant application, a: 0.157–0.197 in. (4–5mm)

1. Before servicing the vehicle, refer to the Precautions Section.

2. Drain the engine oil.

3. Loosen the bolts in reverse of tightening sequence, shown.

4. Insert seal cutter (SST) between oil pan (upper) and oil pan (lower). Be careful not to damage the mating surface.

5. Remove the lower oil pan.

To install:

6. Use a scraper to remove old liquid gasket from mating surfaces.

7. Remove old liquid gasket from mating surface of oil pan (upper).

8. Remove old liquid gasket from the bolt holes and threads.

➡**Never scratch or damage the mating surface when cleaning off liquid gasket.**

9. Apply a continuous bead of liquid gasket with a tube presser (commercial service tool).

10. Attaching should be done within 5 minutes after liquid gasket application.

11. Tighten bolts in numerical order as shown above.

12. Install oil pan drain plug.

13. Pour engine oil at least 30 minutes after oil pan is installed.

14. Check engine oil level and adjust engine oil.

15. Start engine, and check there is no leaks of engine oil.

16. Stop engine and wait for 10 minutes.

17. Check engine oil level again.

Upper Oil Pan

See Figures 76 through 78.

1. Before servicing the vehicle, refer to the Precautions Section.

2. Remove the undercover.

3. Drain the engine oil.

4. Remove the oil pan (lower).

5. Remove the oil level gauge and guide.

6. Disconnect the steering lower joint at steering gear assembly side, and release the steering lower shaft.

7. Disconnect the steering outer sockets from steering knuckle.

12
11
6
4
2
10
13
16
17
9
14
1
15
8
7
5
3
22140_ROGU_G0072

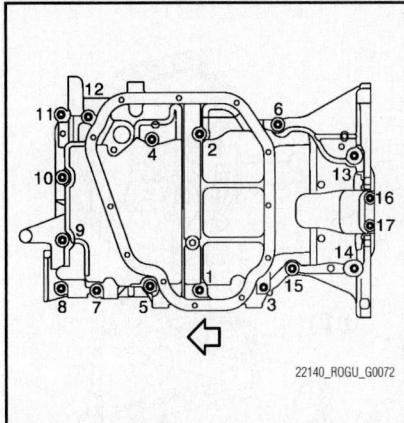

Fig. 76 Bolt tightening sequence, loosen bolts in reverse order

a
37663_ROGU_G0116

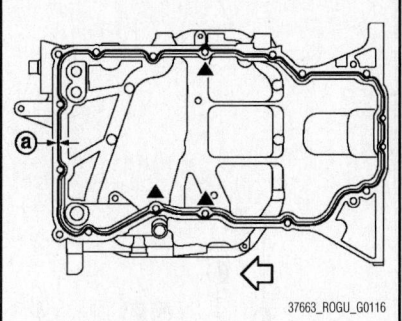

Fig. 77 Sealant application: for bolt holes with triangle marks (3 locations), apply sealant outside the holes, a: 0.157–0.197 in. (4–5mm)

8. Remove the rear torque rod.
9. Remove the stabilizer connecting rod.
10. Remove the front suspension member.
11. Remove the A/C compressor without disconnecting A/C piping, and temporarily fasten it on vehicle with a rope.
12. Remove the oil strainer.
13. Loosen the bolts in reverse of tightening sequence, shown.
14. Insert the seal cutter (SST) between oil pan (upper) and lower cylinder block, and slide it by tapping on the side of the tool with a hammer.

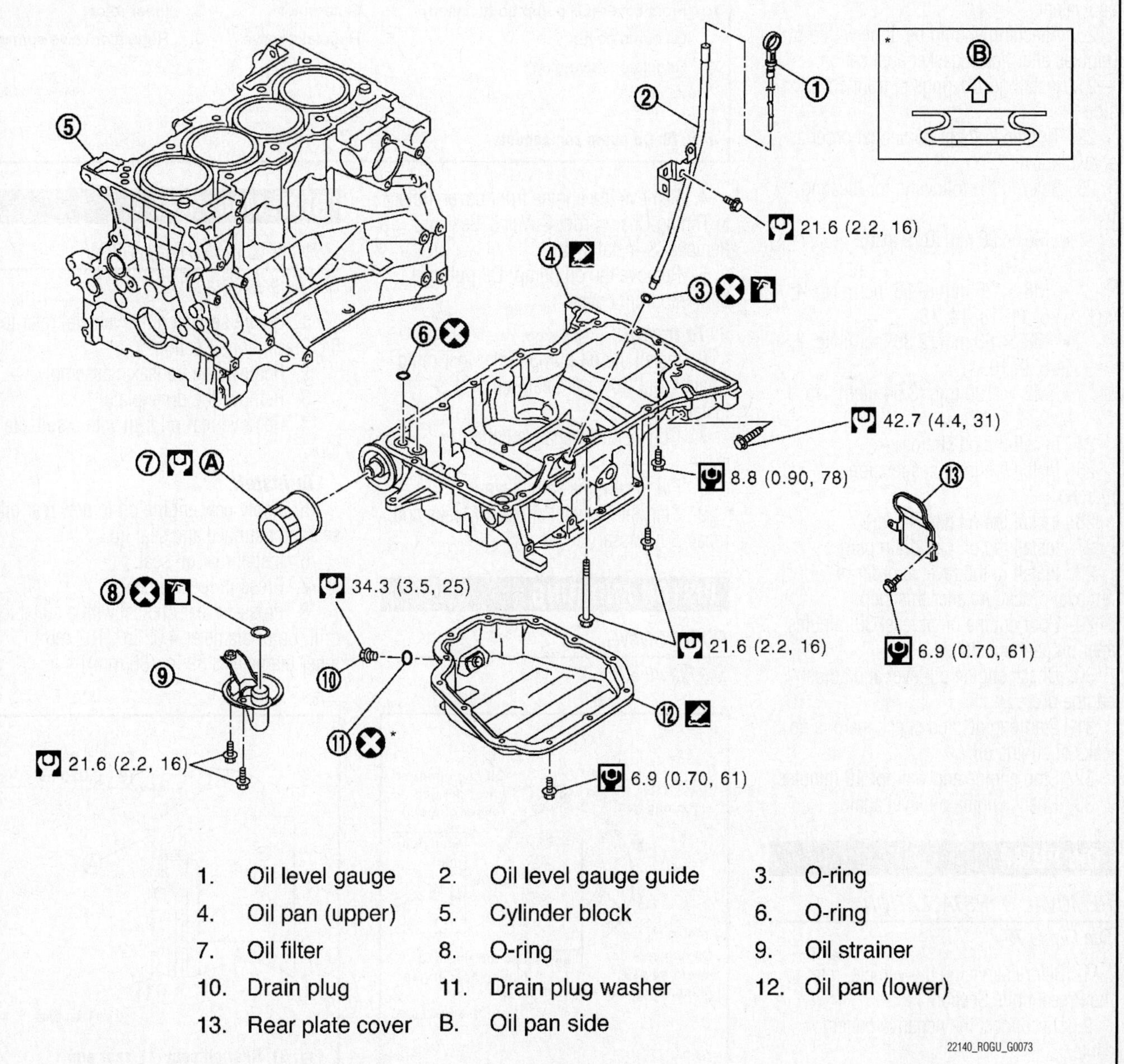

*
B
21.6 (2.2, 16)
42.7 (4.4, 31)
8.8 (0.90, 78)
34.3 (3.5, 25)
21.6 (2.2, 16)
6.9 (0.70, 61)
21.6 (2.2, 16)
6.9 (0.70, 61)
A
1. Oil level gauge
2. Oil level gauge guide
3. O-ring
4. Oil pan (upper)
5. Cylinder block
6. O-ring
7. Oil filter
8. O-ring
9. Oil strainer
10. Drain plug
11. Drain plug washer
12. Oil pan (lower)
13. Rear plate cover
B. Oil pan side
22140_ROGU_G0073

Fig. 78 Exploded view of upper and lower oil pan and mounting bolt tightening specifications

➡**Be careful not to damage the mating surface.**

15. Remove the O-rings at front cover side.

To install:

16. Use a scraper to remove old liquid gasket from mating surfaces.

17. Remove the old liquid gasket from mating surface of cylinder block.

18. Remove old liquid gasket from the bolt holes and threads.

➡**Never scratch or damage the mating surfaces when cleaning off old liquid gasket.**

19. Apply a continuous bead of liquid gasket with a tube presser outside the holes. Use Genuine RTV Silicone Sealant or equivalent.

20. Attaching should be done within 5 minutes after liquid gasket application.

21. Install new O-rings at front cover side.

22. Tighten bolts in numerical order as shown above.

23. Refer to the following for locating bolts:

- M6 × 20 mm (0.79 inch): No. 16, 17
- M8 × 25 mm (0.98 inch): No. 4, 6, 11, 13, 14, 15
- M8 × 60 mm (2.36 inch): No. 7, 8, 9, 10
- M8 × 100 mm (3.94 inch): No. 1, 2, 3, 5, 12

24. Install the oil strainer.

25. Install the front suspension member.

26. Install the oil pan (lower).

27. Install the oil pan drain plug.

28. Install in the reverse order of removal procedure after this step.

29. Pour engine oil at least 30 minutes after oil pan is installed.

30. Check engine oil level and adjust engine oil.

31. Start engine, and check there is no leaks of engine oil.

32. Stop engine and wait for 10 minutes.

33. Check engine oil level again.

OIL PUMP

REMOVAL & INSTALLATION

See Figure 79.

1. Before servicing the vehicle, refer to the Precautions Section.

2. Disconnect the negative battery cable.

3. Drain the engine oil.

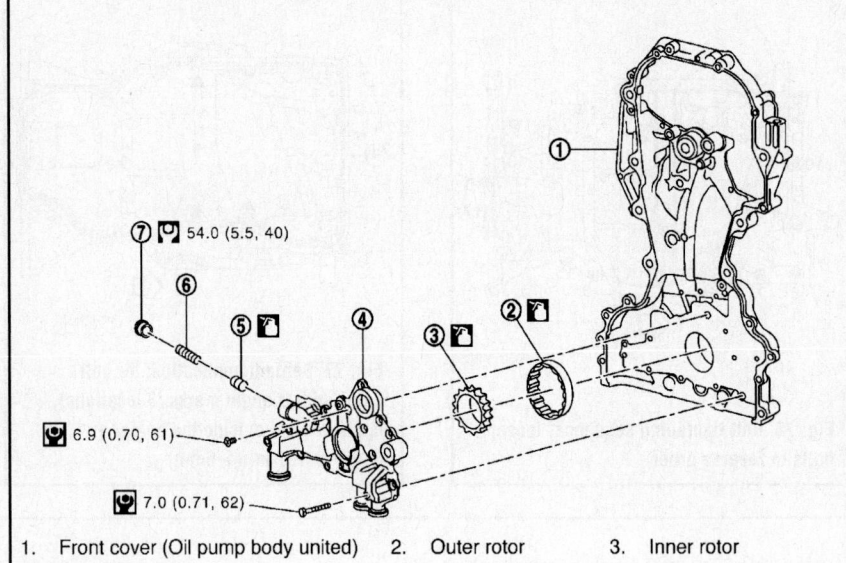

1. Front cover (Oil pump body united)
2. Outer rotor
3. Inner rotor
4. Oil pump cover
5. Regulator valve
6. Regulator valve spring
7. Regulator valve plug

37663_ROGU_G0120

Fig. 79 Oil pump components

4. Remove the engine front cover. Refer to Timing Chain Front Cover & Seal Removal & Installation.

5. Remove the oil pump. Oil pump is built into front cover.

To install:

6. Install the oil pump. Refer to Timing Chain Front Cover & Seal Removal & Installation.

7. Connect the negative battery cable.

8. Fill the engine with clean oil.

9. Start the vehicle; check for leaks and repair if necessary.

PISTON AND RING

POSITIONING

See Figure 80.

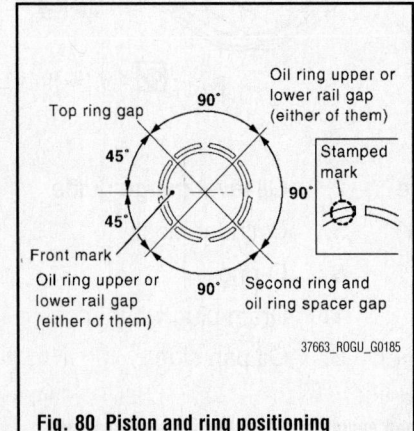

37663_ROGU_G0185

Fig. 80 Piston and ring positioning

REAR MAIN SEAL

REMOVAL & INSTALLATION

See Figures 81 and 82.

1. Before servicing the vehicle, refer to the Precautions Section.

2. Remove the transaxle assembly.

3. Remove the drive plate.

4. Remove rear oil seal with a suitable tool.

To install:

5. Apply new engine oil to new rear oil seal joint surface and seal lip.

6. Install rear oil seal.

7. Press in rear oil seal.

8. Press-fit rear oil seal with a suitable drift [outer diameter 4.02 in. (102 mm), inner diameter 3.39 in. (86 mm)].

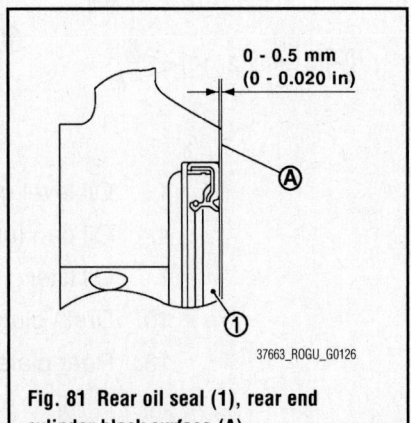

37663_ROGU_G0126

Fig. 81 Rear oil seal (1), rear end cylinder block surface (A)

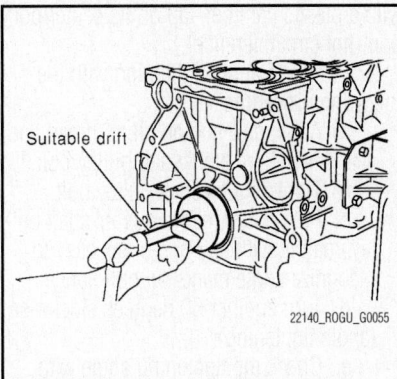

Fig. 82 Rear main seal installation shown

⁂ **WARNING**

Be careful not to damage crankshaft and cylinder block. Press-fit oil seal straight to avoid causing burrs or tilting. Never touch grease applied onto oil seal lip.

9. Install the drive plate.
10. Install the transaxle assembly.

TIMING CHAIN FRONT COVER

REMOVAL & INSTALLATION

See Figures 83 through 89.

1. Before servicing the vehicle, refer to the Precautions Section.
2. Disconnect the negative battery cable.
3. Remove PCV hose.
4. Remove intake manifold. Refer to Intake Manifold Removal & Installation.
5. Remove ignition coil.
6. Remove drive belt. Refer to Accessory Drive Belt Removal & Installation.
7. Remove drive belt auto-tensioner.
8. Remove engine mounting bracket.

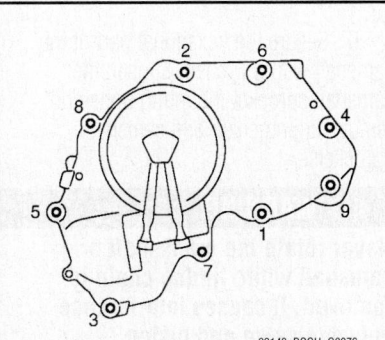

Fig. 83 Intake valve timing control cover bolt tightening sequence—reverse the sequence for removal

9. Remove rocker cover. Refer to Valve Cover Removal & Installation.
10. Remove oil pan (lower). Refer to Oil Pan Removal & Installation.
11. Remove the oil pan (upper), and oil strainer. Refer to Oil Pan Removal & Installation.
12. Remove intake valve timing control cover.
13. Loosen bolts in reverse order as shown in the figure.
14. Use a seal cutter tool to cut liquid gasket for removal.

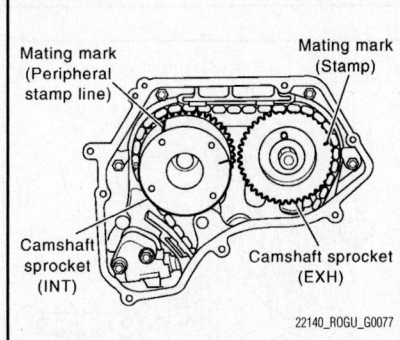

Fig. 84 Check that the mating marks on camshaft sprockets are located as shown

⁂ **WARNING**

Be careful not to damage mounting surface.

15. Pull chain guide between camshaft sprockets out through front cover.
16. Set the No. 1 cylinder at TDC on its compression stroke with the following procedure:
    a. Rotate the crankshaft pulley clockwise and align TDC mark to timing indicator on front cover.
    b. At the same time, check that the mating marks on camshaft sprockets are located as shown.
    c. If not, rotate crankshaft pulley one more turn to align mating marks to the positions shown.
17. Remove the crankshaft pulley. Refer to Crankshaft Damper Removal & Installation.
18. Remove the front cover with the following procedure:
    a. Remove the mounting bolts in reverse order as shown.
    b. Use a seal cutter to cut liquid gasket for removal.
19. If front oil seal needs to be replaced, lift it with a suitable tool, and remove it.

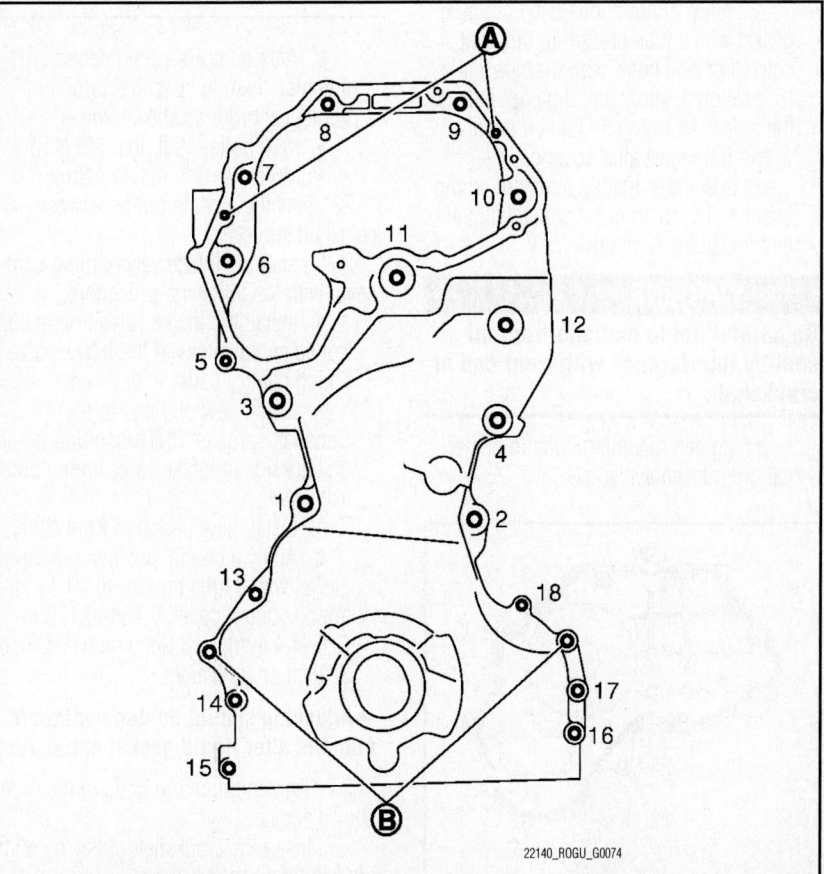

Fig. 85 Front cover tightening sequence with dowel alignment shown (A) and (B)—reverse the sequence for removal

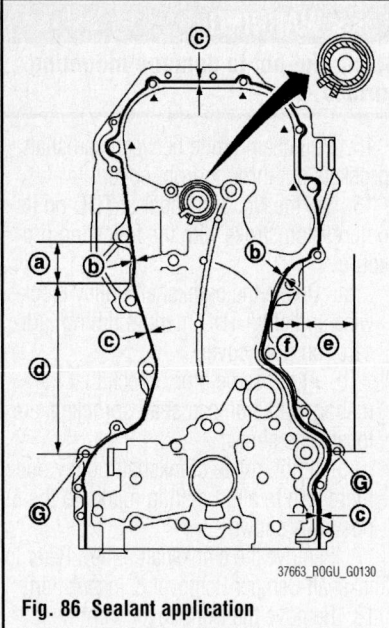

Fig. 86 Sealant application

To install:

20. Install the front oil seal to front cover.

21. Install the front cover with the following procedure:

a. Install O-rings to cylinder head and cylinder block.

b. Apply a continuous bead of liquid gasket with a tube presser to the front cover. For bolt holes with triangle marks (5 locations), apply liquid gasket outside the holes. Attaching should be done within 5 minutes after coating.

c. Check that mating marks of timing chain and each sprocket are still aligned, and install the front cover.

> **❊❊ WARNING**
>
> **Be careful not to damage front oil seal by interference with front end of crankshaft.**

d. Tighten mounting bolts in numerical order as shown above.

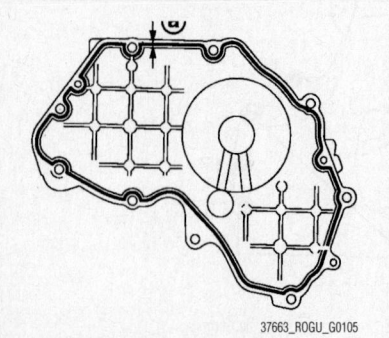

Fig. 87 Intake valve timing control cover sealant application

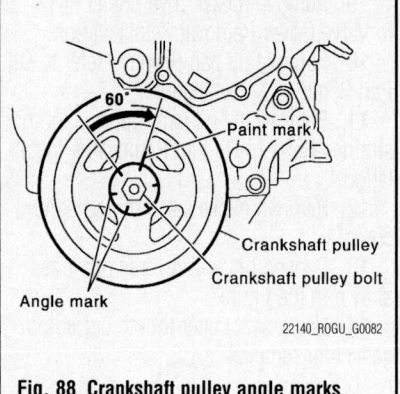

Fig. 88 Crankshaft pulley angle marks

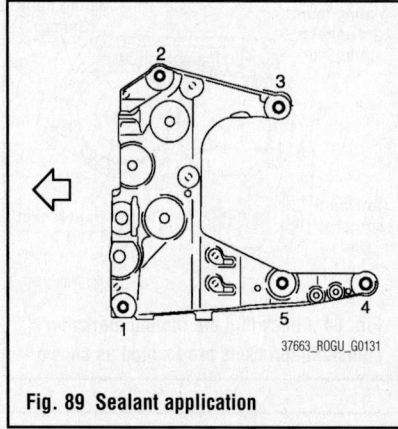

Fig. 89 Sealant application

e. After all bolts are tightened, retighten them to specified torque in numerical order as shown above.
- M10 bolts: 36 ft. lbs. (49 Nm)
- M6 bolts: 9 ft. lbs. (13 Nm)

22. Install the chain guide between camshaft sprockets.

23. Install the intake valve timing control cover with the following procedure:

a. Install the intake valve timing control solenoid valves to the intake valve timing control cover if removed.

b. Install new oil rings to the camshaft sprocket (INT) insertion points on backside of intake valve timing control cover.

c. Install new O-ring to front cover.

d. Apply a continuous bead of liquid gasket with a tube presser to intake valve timing control cover, 0.134–0.173 in. (3.4–4.4 mm). Use Genuine RTV Silicone Sealant or equivalent.

➡ **Attaching should be done within 5 minutes after liquid gasket application.**

24. Tighten mounting bolts in the order shown above.

25. Insert the crankshaft pulley by aligning with crankshaft key.

26. When inserting crankshaft pulley

with a plastic hammer, tap on its center portion (not circumference).

27. Perform angle tightening with the following procedure:

a. Apply new engine oil to thread and seat surfaces of crankshaft pulley bolt.

b. Tighten crankshaft pulley bolt.

c. Put a paint mark on crankshaft pulley, mating with any one of six easy to recognize angle marks on bolt flange.

d. Turn another 60 degrees clockwise (angle tightening).

e. Check the tightening angle with movement of one angle mark.

28. Install the oil pans and oil strainer.

29. Install the valve covers.

30. Install the engine mounting bracket.

a. Tighten the bolts No. 3, 5 as shown in the figure. (temporarily)

b. Tighten the bolts in numerical order as shown in the figure.

31. Connect the negative battery cable.

32. Install all removed parts in the reverse order of removal.

TIMING CHAIN & SPROCKETS

REMOVAL & INSTALLATION

See Figures 90 through 94.

1. Before servicing the vehicle, refer to the Precautions Section.

2. Disconnect the negative battery cable.

3. Remove the timing chain front cover. Refer to Timing Chain Front Cover Removal & Installation.

4. Remove timing chain and camshaft sprockets with the following procedure:

5. Push in the chain tensioner plunger. Insert a stopper pin into hole on chain tensioner body to secure chain tensioner plunger and remove chain tensioner. Use approximately 0.02 in. (0.5 mm) diameter hard metal pin as a stopper pin.

6. Secure the hexagonal part of the camshaft with a wrench. Loosen the camshaft sprocket mounting bolts and remove timing chain and camshaft sprockets.

> **❊❊ WARNING**
>
> **Never rotate the crankshaft or camshaft while timing chain is removed. It causes interference between valve and piston.**

7. Remove timing chain slack guide, timing chain tension guide and oil pump drive spacer.

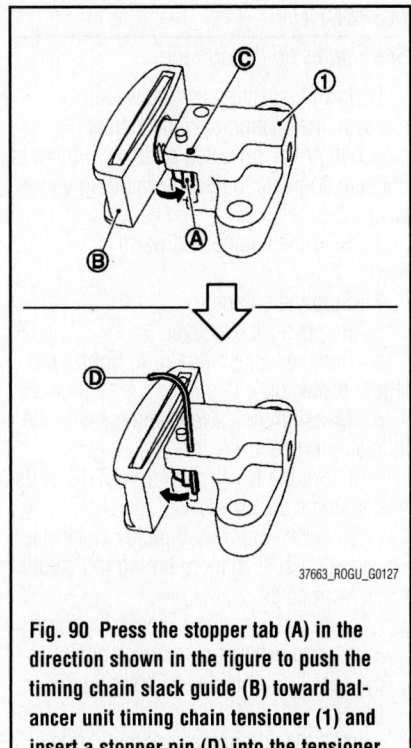

Fig. 90 Press the stopper tab (A) in the direction shown in the figure to push the timing chain slack guide (B) toward balancer unit timing chain tensioner (1) and insert a stopper pin (D) into the tensioner body hole (C)

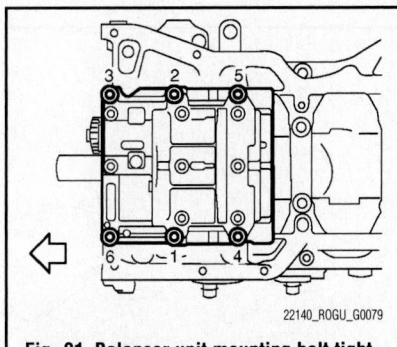

Fig. 91 Balancer unit mounting bolt tightening sequence—reverse the sequence for removal

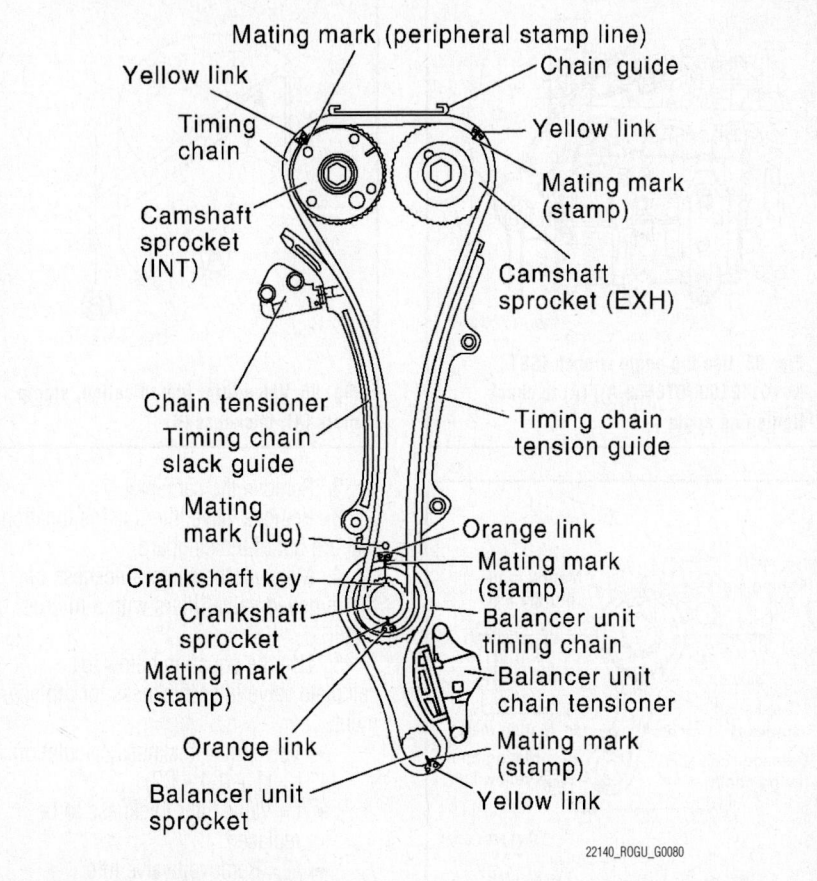

Fig. 92 The figure shows the relationship between the mating mark on each timing chain and that on the corresponding sprocket

8. Remove the balancer unit timing chain tensioner with the following procedure:

a. Press the stopper tab in the direction shown in the figure to push the timing chain slack guide toward balancer unit timing chain tensioner. The slack guide is released by pressing the stopper tab. As the result, the slack guide can be moved.

b. Insert a stopper pin into tensioner body hole to secure the timing chain slack guide. Use a hard metal pin with the diameter of approximately 0.047 in. (1.2 mm) as a stopper pin.

c. Remove the balancer unit timing chain tensioner.

d. When the holes on the lever and tensioner body cannot be aligned, align these holes by slightly moving the slack guide.

9. Remove the balancer unit timing chain and crankshaft sprocket.

10. Remove the mounting bolts in reverse order as shown in the figure, and remove balancer unit.

⁂ WARNING

Never disassemble balancer unit.

11. The figure shows the relationship between the mating mark on each timing chain and that on the corresponding sprocket, with the components installed.

To install:

12. Check that crankshaft key points straight up.

13. Tighten mounting bolts in numerical order as shown above, and install the balancer unit:

a. Apply new engine oil to threads and seat surfaces of mounting bolts.

b. Tighten No. 1–5 bolts to 31 ft. lbs. (42 Nm).

c. Tighten the No. 6 bolt to 27 ft. lbs. (36 Nm).

d. Turn No. 1–5 bolts 120 degrees clockwise (angle tightening).

➡**Use the angle wrench [SST: KV10112100 (BT8653-A)] to check tightening angle. Never make judgment by visual inspection.**

e. Turn No. 6 bolt 90 degrees clockwise (angle tightening).

f. Completely loosen all bolts.

g. In this step, loosen bolts in reverse order.

h. Repeat the steps b to e.

14. Install the crankshaft sprocket and balancer unit timing chain.

a. Check that crankshaft sprocket is positioned with mating marks on cylinder block and crankshaft sprocket meeting at the top.

b. Install it by aligning mating marks on each sprocket and balancer unit timing chain.

15. Install the balancer unit timing chain tensioner. Be careful not to let mating marks of each sprocket and timing chain slip.

16. After installation, check the mating marks have not slipped, then remove

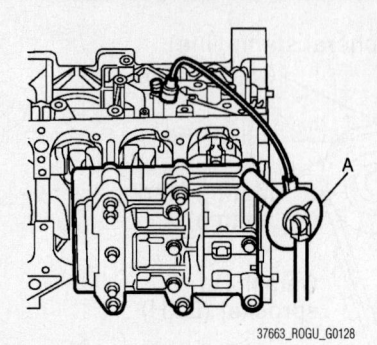

Fig. 93 Use the angle wrench [SST: KV10112100 (BT8653-A)] (A) to check tightening angle

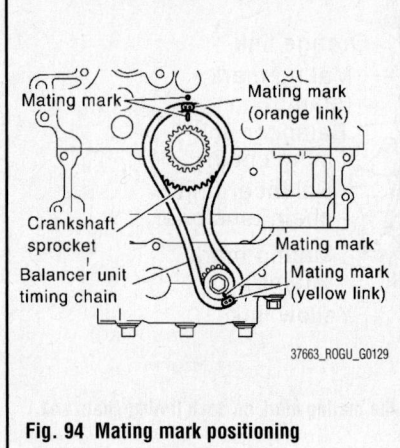

Fig. 94 Mating mark positioning

stopper pin and release tensioner sleeve.

17. Install the timing chain and related parts.

　a. Install by aligning mating marks on each sprocket and timing chain as shown above.

　b. Before and after installing chain tensioner, check again to ensure that mating marks have not slipped.

18. After installing chain tensioner, remove stopper pin, and check that tensioner moves freely.

➤**After the mating marks are aligned, keep them aligned by holding them by hand. To avoid skipped teeth, never rotate crankshaft and camshaft until front cover is installed.**

19. Install the timing chain front cover.

VALVE LASH

ADJUSTMENT

See Figure 95.

1. Perform adjustments depending on selected head thickness of valve lifter.

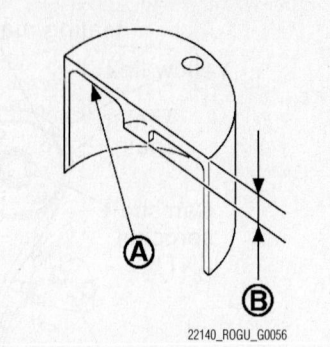

Fig. 95 Valve lifter identification, stamp mark (A), thickness (B)

2. Remove the camshaft.

3. Remove valve lifters at the locations that are out of the standard.

4. Measure the center thickness of the removed valve lifters with a micrometer.

5. Use the equation below to calculate valve lifter thickness for replacement:

- Valve lifter thickness calculation: $t = t1 + (C1 - C2)$
- t = Valve lifter thickness to be replaced
- $t1$ = Removed valve lifter thickness
- $C1$ = Measured valve clearance
- $C2$ = Standard valve clearance: Intake: 0.011 in. (0.28 mm), Exhaust: 0.012 in. (0.30 mm)

6. Thickness of new valve lifter can be identified by stamp mark on the reverse side (inside the cylinder).

7. Stamp mark "788" indicates 0.3102 inch (7.88 mm) in thickness.

➤**Available thickness of valve lifter: 26 sizes range 0.3102 to 0.3299 inch (7.88 to 8.38 mm) in steps of 0.0008 inch (0.02 mm) (when manufactured at factory).**

8. Install the selected valve lifter.

9. Install the camshaft.

10. Manually rotate the crankshaft pulley a few rotations.

11. Check that the valve clearances for cold engine are within specifications by referring to the specified values.

12. Install all removed parts in the reverse order of removal.

13. Warm up the engine, and check for unusual noise and vibration.

INSPECTION

See Figures 96 through 98.

Perform inspection as follows after removal, installation or replacement of camshaft or valve-related parts, or if there is unusual engine conditions regarding valve clearance.

1. Start the engine and warm it up.

2. Stop the engine.

3. Remove rocker cover.

4. Remove splash guard on right-hand fender protector.

5. Measure the valve clearance with the following procedure:

　a. Set the No. 1 cylinder at TDC of its compression stroke.

　b. Rotate crankshaft pulley clockwise and align TDC mark to timing indicator on front cover.

　c. At the same time, check that both intake and exhaust cam noses of No. 1 cylinder face outside as shown in the figure.

　d. If they do not face outside, rotate crankshaft pulley once more (360 degrees).

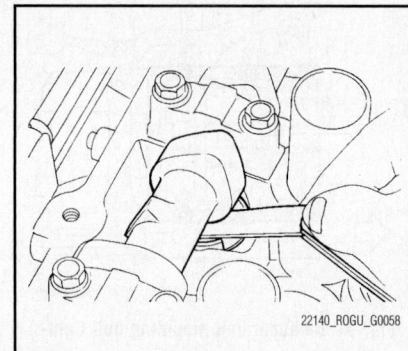

Fig. 96 Using a feeler gauge, measure the clearance between valve lifter and camshaft

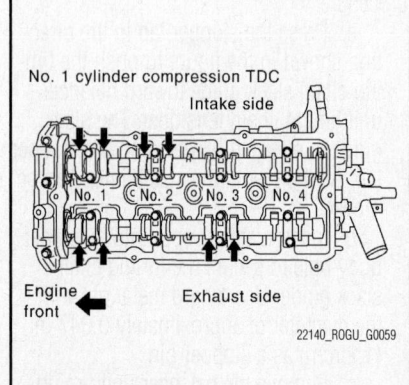

Fig. 97 Measure the valve clearances at locations marked (X)

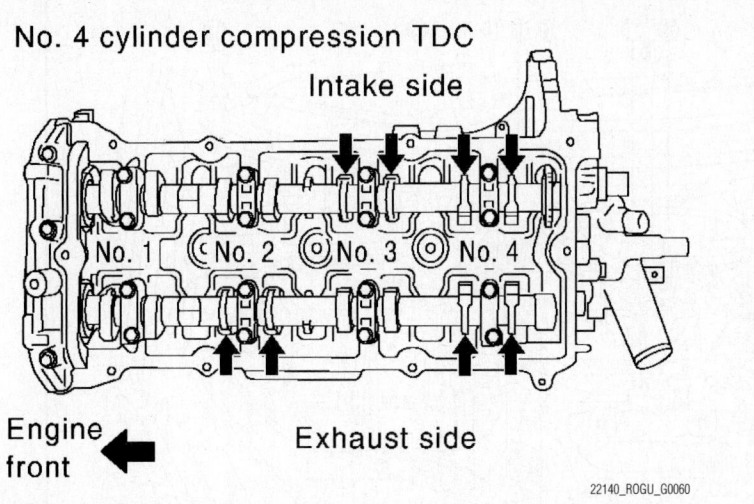

No. 4 cylinder compression TDC

Intake side

No. 1   No. 2   No. 3   No. 4

Engine front

Exhaust side

22140_ROGU_G0060

Fig. 98 Measure the valve clearance at locations marked (X) as shown in the table below (locations indicated with black arrow in the figure) with a feeler gauge.

e. Using a feeler gauge, measure the clearance between valve lifter and camshaft.

f. By referring to the figure, measure the valve clearances at locations marked (X) as shown in the table below (locations indicated with black arrow in the figure) with a feeler gauge.

g. With No. 1 cylinder compression TDC, rotate crankshaft pulley one revolution (360 degrees) and align TDC mark to timing indicator on front cover (No. 4 cylinder compression TDC).

h. By referring to the figure, measure the valve clearance at locations marked (X) as shown in the table below (locations indicated with black arrow in the figure) with a feeler gauge.

6. If out of standard, perform adjustment.

ENGINE PERFORMANCE & EMISSION CONTROLS

CAMSHAFT POSITION (CMP) SENSOR

LOCATION

See Figure 99.

Refer to the accompanying illustration.

REMOVAL & INSTALLATION

1. Before servicing the vehicle, refer to the Precautions Section.
2. Disconnect the negative battery cable.
3. Remove the air cleaner duct and resonator.
4. Remove the Camshaft Position (CMP) sensor mounting bolt.
5. Remove the CMP sensor harness connector.
6. Remove the CMP sensor.

To install:

7. Installation is the reverse of removal.

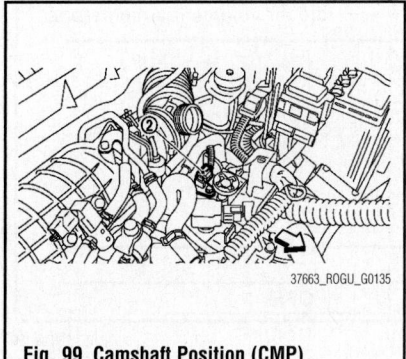

37663_ROGU_G0135

Fig. 99 Camshaft Position (CMP) Sensor (2)

CRANKSHAFT POSITION (CKP) SENSOR

LOCATION

See Figure 100.

Refer to the accompanying illustration.

REMOVAL & INSTALLATION

1. Before servicing the vehicle, refer to the Precautions Section.
2. Disconnect the negative battery cable.
3. Disconnect the Crankshaft Position (CKP) sensor harness connector.
4. Remove the CKP sensor mounting bolt.
5. Remove the CKP sensor.

To install:

6. Installation is the reverse of removal.

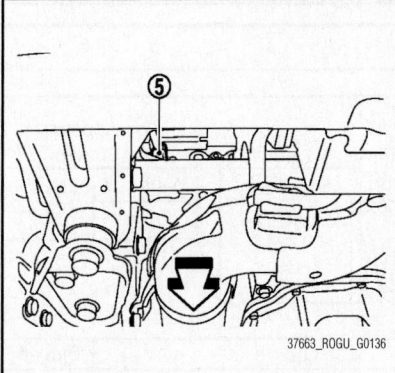

37663_ROGU_G0136

Fig. 100 Crankshaft Position (CMP) Sensor (5)

ELECTRONIC CONTROL MODULE (ECM)

LOCATION

See Figure 101.

Refer to the accompanying illustration.

RESET

The CONSULT-III scan tool is required.

ACCELERATOR PEDAL RELEASED POSITION LEARNING

1. Check that accelerator pedal is fully released.
2. Turn ignition switch ON and wait at least 2 seconds.
3. Turn ignition switch OFF and wait at least 10 seconds.
4. Turn ignition switch ON and wait at least 2 seconds.
5. Turn ignition switch OFF and wait at least 10 seconds.

IDLE AIR VOLUME LEARNING

See Figures 102 and 103.

The Idle Air Volume Learning is a function of ECM to learn the idle air volume that keeps each engine idle speed within the specific range. It must be performed under the following conditions:

• Each time the electric throttle control actuator or ECM is replaced

• Idle speed or ignition timing is out of the specification

1. Check that all of the following

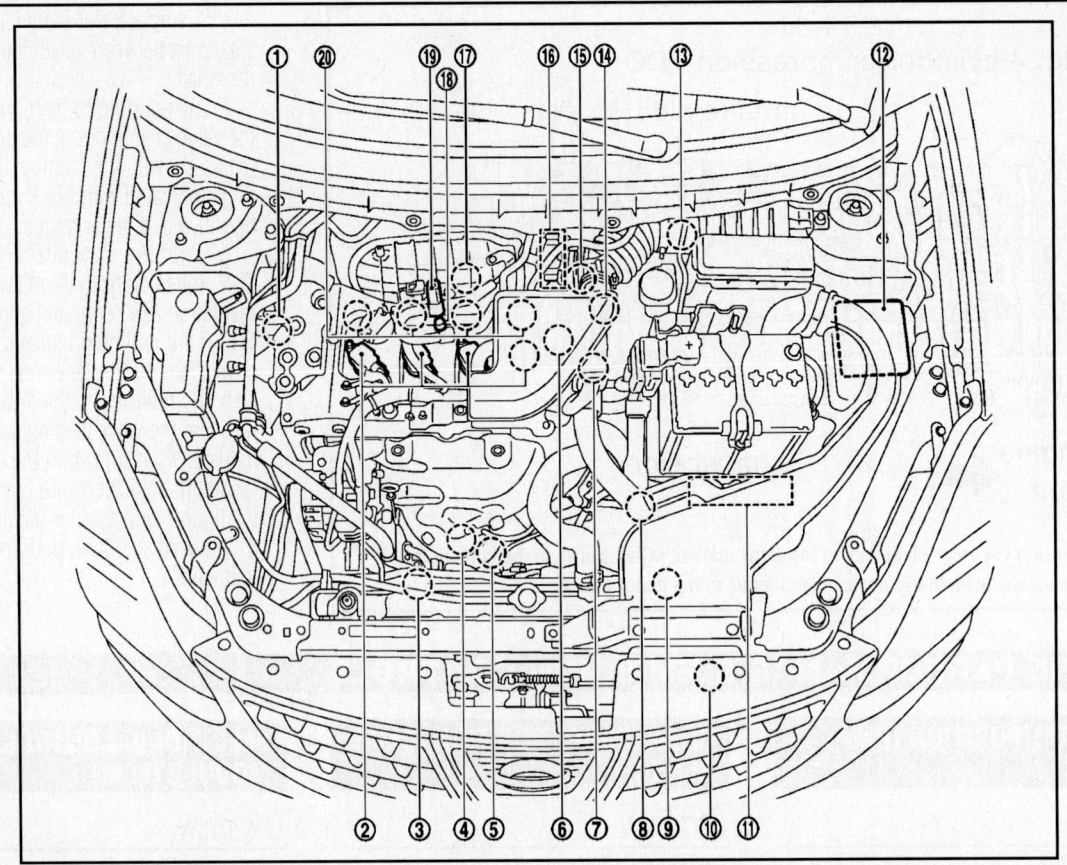

1. Intake valve timing control solenoid valve
2. Ignition coil (with power transistor) and spark plug
3. Cooling fan motor-2
4. Air fuel ratio (A/F) sensor 1
5. Heated oxygen sensor 2
6. Camshaft position sensor (PHASE)
7. Engine coolant temperature sensor
8. Transmission range switch
9. Cooling fan motor-1
10. Refrigerant pressure sensor
11. ECM
12. IPDM E/R
13. Mass air flow sensor (with intake air temperature sensor)
14. Tumble control valve actuator
15. Crankshaft position sensor (POS)
16. Electric throttle control actuator (with built in throttle position sensor and throttle control motor)
17. Knock sensor
18. EVAP service port
19. EVAP canister purge volume control solenoid valve
20. Fuel injector

37663_ROGU_G0132

Fig. 101 View of the emission control component locations

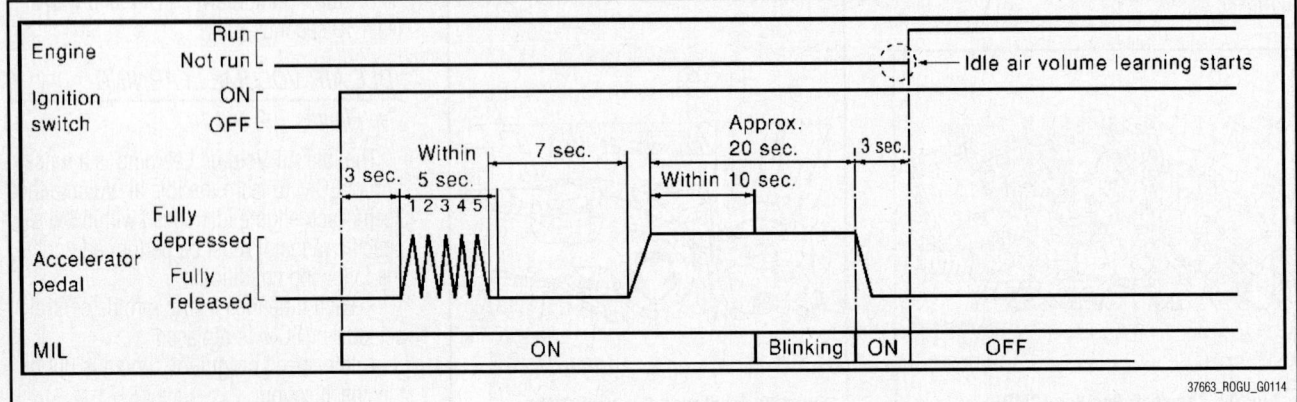

37663_ROGU_G0114

Fig. 102 Idle Air Volume Learning

Idle Speed

Transmission	Condition	Specification
CVT	No load* (in P or N position)	700 ± 50 rpm

*: Under the following conditions
- A/C switch: OFF
- Electric load: OFF (Lights, heater fan & rear window defogger)
- Steering wheel: Kept in straight-ahead position

Ignition Timing

Transmission	Condition	Specification
CVT	No load* (in P or N position)	10 ± 5° BTDC

*: Under the following conditions
- A/C switch: OFF
- Electric load: OFF (Lights, heater fan & rear window defogger)
- Steering wheel: Kept in straight-ahead position

37663_ROGU_G0115

Fig. 103 Idle Speed and Ignition Timing

conditions are satisfied. Learning will be cancelled if any of the following conditions are missed for even a moment:
- Battery voltage: More than 12.9 V (At idle)
- Engine coolant temperature: 158–212°F (70–100°C)
- Selector lever: P or N
- Electric load switch: OFF (Air conditioner, headlamp, rear window defogger)
- On vehicles equipped with daytime light systems, if the parking brake is applied before the engine is
started, the headlamp will not illuminate.
- Steering wheel: Neutral (Straight-ahead position)
- Vehicle speed: Stopped
- Transmission: Warmed-up
- With CONSULT-III: Drive vehicle until "ATF TENP SEN" in "DATA MONITOR" mode of "CVT" system indicates less than 0.9V.
- Without CONSULT-III: Drive vehicle for 10 minutes.

2. If a CONSULT-III scan tool is not available, perform Accelerator Pedal Released Position Learning.

3. Perform Throttle Valve Closed Position Learning.

4. Start engine and warm it up to normal operating temperature.

5. Turn ignition switch OFF and wait at least 10 seconds.

6. Confirm that accelerator pedal is fully released, turn ignition switch ON and wait 3 seconds.

7. Repeat the following procedure quickly five times within 5 seconds.
   a. Fully depress the accelerator pedal.
   b. Fully release the accelerator pedal.

8. Wait 7 seconds, fully depress the accelerator pedal for approx. 20 seconds until the MIL stops blinking and turns ON.

9. Fully release the accelerator pedal within 3 seconds after the MIL turns ON.

10. Start engine and let it idle.

11. Wait 20 seconds.

12. Rev up the engine two or three times and check that idle speed and ignition timing are within the specifications. If the inspection result is not normal, find the malfunctioning part.

ENGINE COOLANT TEMPERATURE (ECT) SENSOR

LOCATION
See Figure 104.

Refer to the accompanying illustration.

REMOVAL & INSTALLATION

1. Before servicing the vehicle, refer to the Precautions Section.

2. Disconnect the negative battery cable.

3. Disconnect the Engine Coolant Temperature (ECT) sensor harness connector.

4. Remove the ECT sensor.

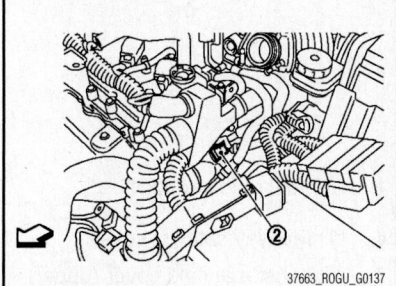

37663_ROGU_G0137

Fig. 104 Engine Coolant Temperature (ECT) Sensor (2)

To install:

5. Installation is the reverse of removal.

HEATED OXYGEN (HO2S) SENSOR

LOCATION
See Figure 105.

Refer to the accompanying illustration.

REMOVAL & INSTALLATION

1. Raise the vehicle.

2. Remove the front engine under cover.

3. Remove the Heated Oxygen (HO2S) sensor harness connector.

4. Remove the HO2S sensor.

To install:

5. Installation is the reverse of removal.

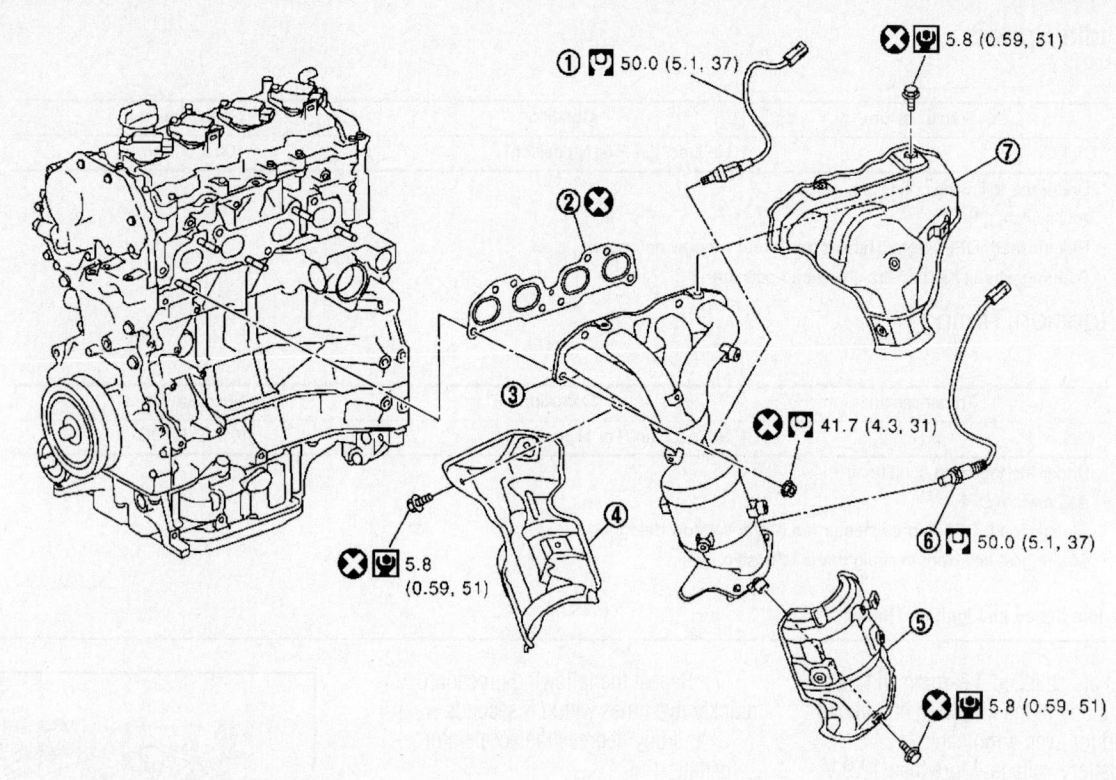

1. Air fuel ratio sensor 1
2. Gasket
3. Exhaust manifold and three way catalyst assembly
4. Three way catalyst cover
5. Exhaust manifold cover (lower)
6. Heated oxygen sensor 2
7. Exhaust manifold cover (upper)

37663_ROGU_G0112

Fig. 105 Exhaust manifold and three way catalyst

INTAKE AIR TEMPERATURE (IAT) SENSOR

LOCATION

See Figures 106 and 107.

Refer to the accompanying illustrations.

REMOVAL & INSTALLATION

1. Remove the Mass Air Flow (MAF) sensor (with Intake Air Temperature sensor) harness connector.

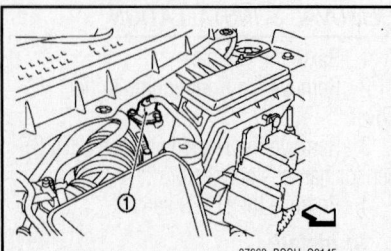

37663_ROGU_G0145

Fig. 106 Mass Air Flow (MAF) sensor and Intake Air Temperature (IAT) sensor (1)

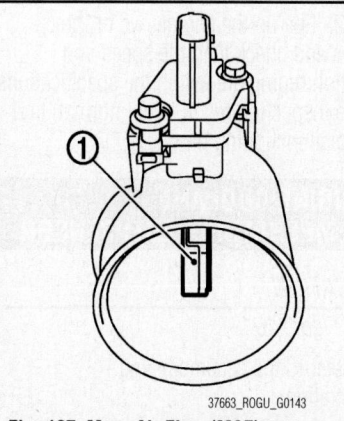

37663_ROGU_G0143

Fig. 107 Mass Air Flow (MAF) sensor and Intake Air Temperature (IAT) sensor (1)

2. Remove the retaining screws and remove the sensor.

To install:

3. Install the sensor and tighten the retaining screws.

4. Install the MAF sensor (with IAT sensor) harness connector.

KNOCK SENSOR (KS)

LOCATION

See Figure 101.

Refer to the accompanying illustration.

MALFUNCTION INDICATOR LIGHT (MIL)

RESET PROCEDURE

See Figure 108.

Install the CONSULT-III® or Commercial Scanner to the diagnostic connector, and reset the DTC code.

MASS AIR FLOW (MAF) SENSOR

LOCATION

See Figures 106 and 107.

Refer to the Intake Air Temperature (IAT) Sensor.

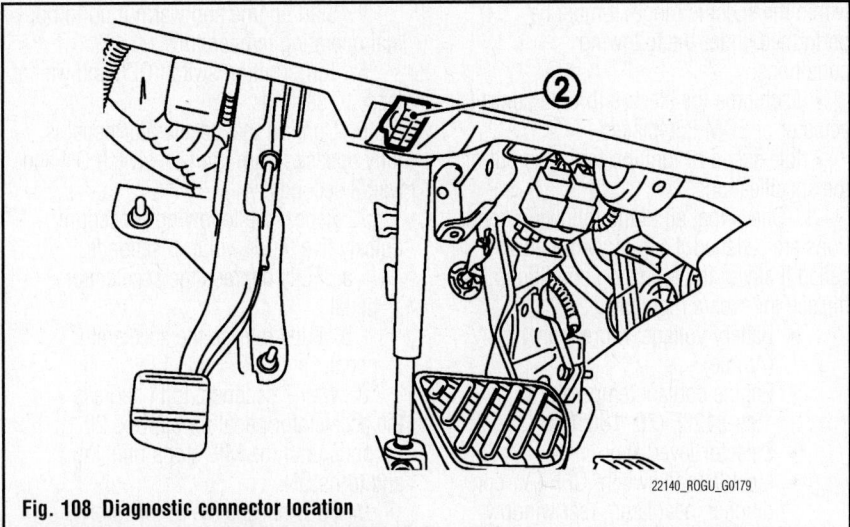

Fig. 108 Diagnostic connector location

22140_ROGU_G0179

POSITIVE CRANKCASE VENTILATION (PCV) VALVE

LOCATION

See Figure 109.

Refer to the accompanying illustration.

REMOVAL & INSTALLATION

1. Remove PCV hose.
2. Remove PCV valve from valve cover.

To install:

3. Installation is reverse of removal.

THROTTLE CONTROL ACTUATOR (TAC)

LOCATION

See Figure 110.

Refer to the accompanying illustration.

37663_ROGU_G0144

Fig. 110 Throttle Control Actuator (TAC) (1)

REMOVAL & INSTALLATION

See Figure 111.

1. Before servicing the vehicle, refer to the Precautions Section.
2. Release the fuel pressure.
3. Remove the cowl top cover.
4. Remove the air cleaner case and mass air flow sensor assembly and air duct and resonator assembly.
5. Remove the electric throttle control actuator with the following procedure:
    a. Disconnect the harness connector.
    b. Loosen the mounting bolts in reverse order as shown in the figure, and remove electric throttle control actuator and gasket.

❋❋ WARNING

Handle carefully to avoid any shock to electric throttle control actuator. Never disassemble.

To install:

6. Install in the reverse order of removal, noting the following:
    a. Tighten the mounting bolts equally to 14 ft. lbs. (20 Nm), and diagonally in several steps and in numerical order in the reverse order of removal.

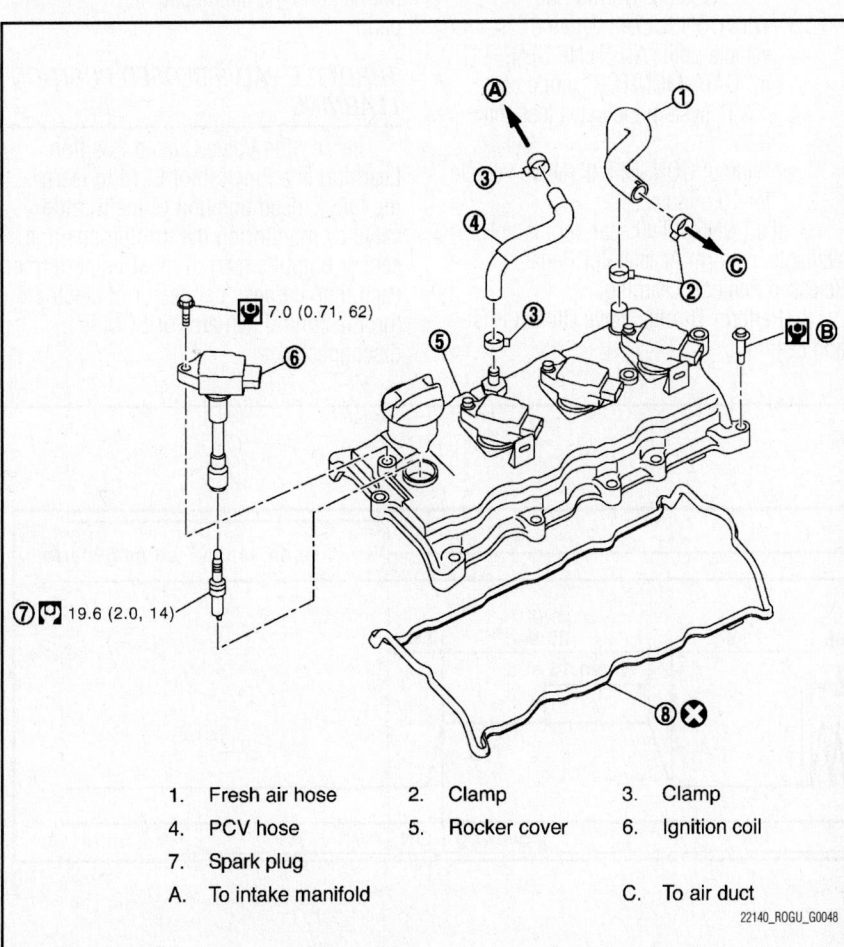

7.0 (0.71, 62)

19.6 (2.0, 14)

1.	Fresh air hose	2.	Clamp	3.	Clamp
4.	PCV hose	5.	Rocker cover	6.	Ignition coil
7.	Spark plug				
A.	To intake manifold			C.	To air duct

22140_ROGU_G0048

Fig. 109 Valve covers and related components

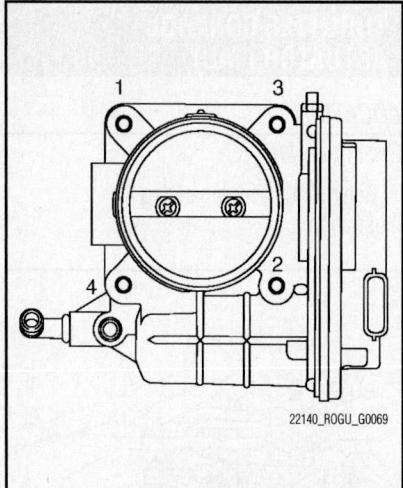

22140_ROGU_G0069

Fig. 111 Loosen mounting bolts in reverse order as shown in the figure, and remove electric throttle control actuator and gasket

b. Tighten the electric throttle body to 74 inch lbs. (8 Nm).

c. Perform the "Throttle Valve Closed Position Learning" when harness connector of electric throttle control actuator is disconnected.

d. Perform the "Idle Air Volume Learning" and "Throttle Valve Closed Position Learning" when electric throttle control actuator is replaced.

IDLE AIR VOLUME LEARNING

See Figures 112 and 113.

The Idle Air Volume Learning is a function of ECM to learn the idle air volume that keeps each engine idle speed within the specific range. It must be performed under the following conditions:

- Each time the electric throttle control actuator or ECM is replaced
- Idle speed or ignition timing is out of the specification

1. Check that all of the following conditions are satisfied. Learning will be cancelled if any of the following conditions are missed for even a moment:

- Battery voltage: More than 12.9 V (At idle)
- Engine coolant temperature: 158–212°F (70–100°C)
- Selector lever: P or N
- Electric load switch: OFF (Air conditioner, headlamp, rear window defogger)
- On vehicles equipped with daytime light systems, if the parking brake is applied before the engine is started, the headlamp will not illuminate.
- Steering wheel: Neutral (Straight-ahead position)
- Vehicle speed: Stopped
- Transmission: Warmed-up
- With CONSULT-III: Drive vehicle until "ATF TENP SEN" in "DATA MONITOR" mode of "CVT" system indicates less than 0.9V.
- Without CONSULT-III: Drive vehicle for 10 minutes.

2. If a CONSULT-III scan tool is not available, perform Accelerator Pedal Released Position Learning.

3. Perform Throttle Valve Closed Position Learning.

4. Start engine and warm it up to normal operating temperature.

5. Turn ignition switch OFF and wait at least 10 seconds.

6. Confirm that accelerator pedal is fully released, turn ignition switch ON and wait 3 seconds.

7. Repeat the following procedure quickly five times within 5 seconds.

a. Fully depress the accelerator pedal.

b. Fully release the accelerator pedal.

8. Wait 7 seconds, fully depress the accelerator pedal for approx. 20 seconds until the MIL stops blinking and turns ON.

9. Fully release the accelerator pedal within 3 seconds after the MIL turns ON.

10. Start engine and let it idle.

11. Wait 20 seconds.

12. Rev up the engine two or three times and check that idle speed and ignition timing are within the specifications. If the inspection result is not normal, find the malfunctioning part.

THROTTLE VALVE CLOSED POSITION LEARNING

The Throttle Valve Closed Position Learning is a function of ECM to learn the fully closed position of the throttle valve by monitoring the throttle position sensor output signal. It must be performed each time harness connector of electric throttle control actuator or ECM is disconnected.

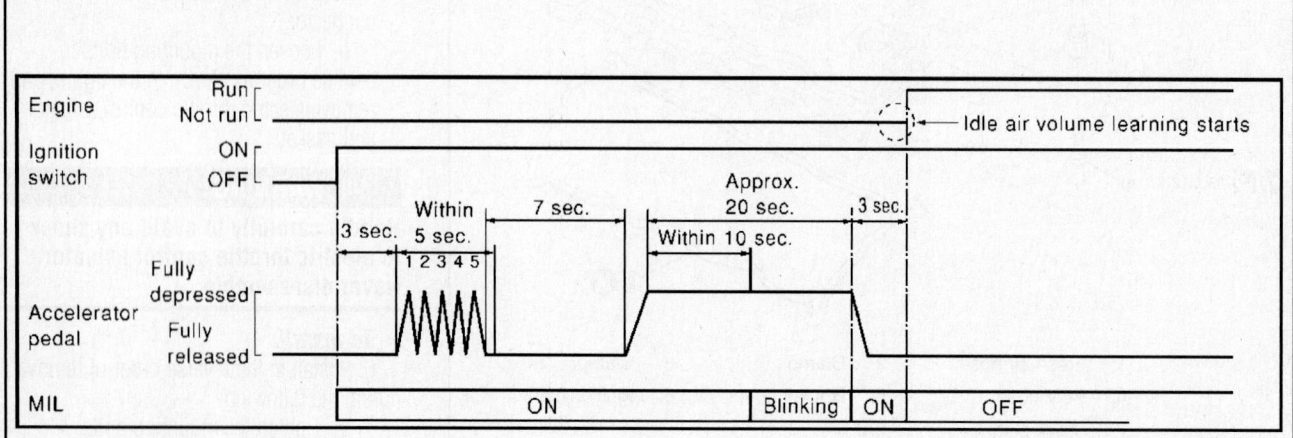

Fig. 112 Idle Air Volume Learning

37663_ROGU_G0114

Idle Speed

Transmission	Condition	Specification
CVT	No load* (in P or N position)	700 ± 50 rpm

*: Under the following conditions
- A/C switch: OFF
- Electric load: OFF (Lights, heater fan & rear window defogger)
- Steering wheel: Kept in straight-ahead position

Ignition Timing

Transmission	Condition	Specification
CVT	No load* (in P or N position)	10 ± 5° BTDC

*: Under the following conditions
- A/C switch: OFF
- Electric load: OFF (Lights, heater fan & rear window defogger)
- Steering wheel: Kept in straight-ahead position

37663_ROGU_G0115

Fig. 113 Idle Speed and Ignition Timing

1. Check that accelerator pedal is fully released.
2. Turn ignition switch ON.
3. Turn ignition switch OFF and wait at least 10 seconds.
4. Check that throttle valve moves during the above 10 seconds by confirming the operating sound.

THROTTLE POSITION SENSOR (TPS)

LOCATION

See Throttle Control Actuator (TAC).

REMOVAL & INSTALLATION

See Throttle Control Actuator (TAC).

VARIABLE CAMSHAFT TIMING OIL CONTROL SOLENOID

LOCATION

See Figures 114 and 115.

Refer to the accompanying illustrations.

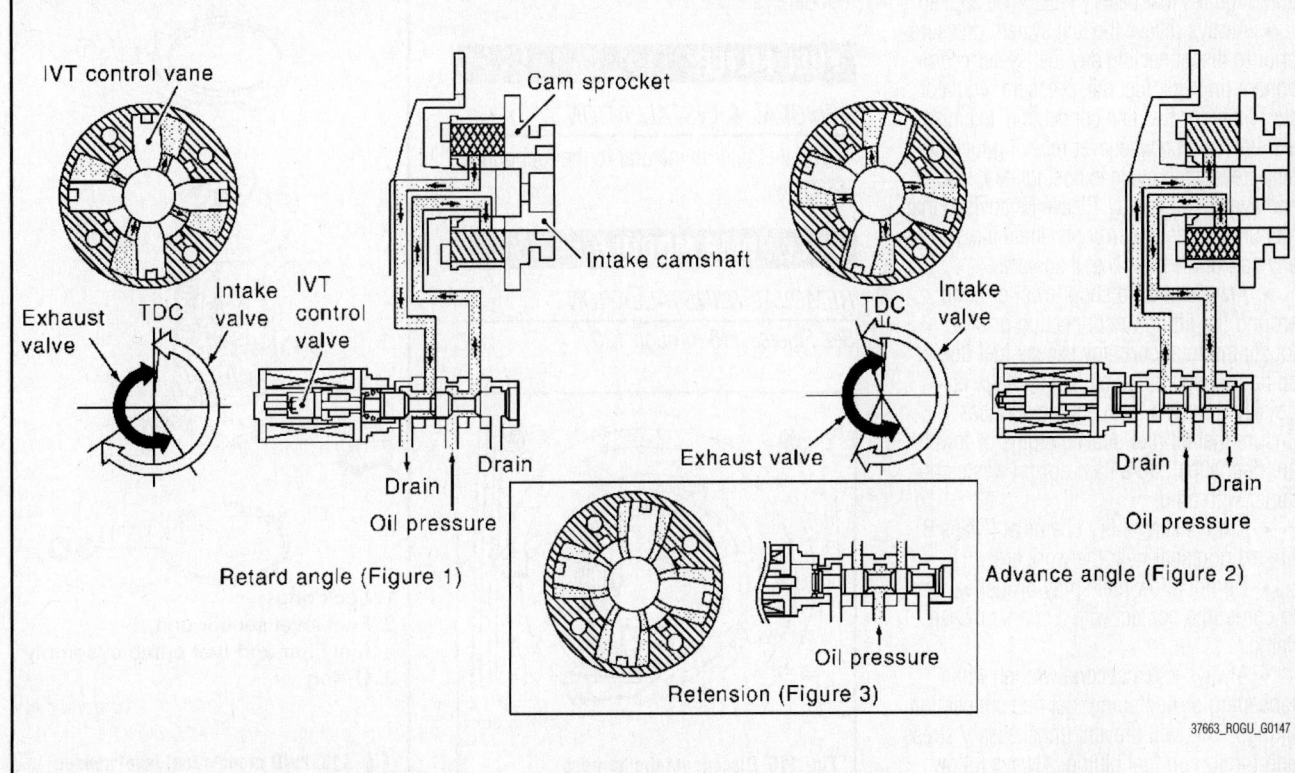

37663_ROGU_G0147

Fig. 114 Intake Valve Timing Control System Diagram

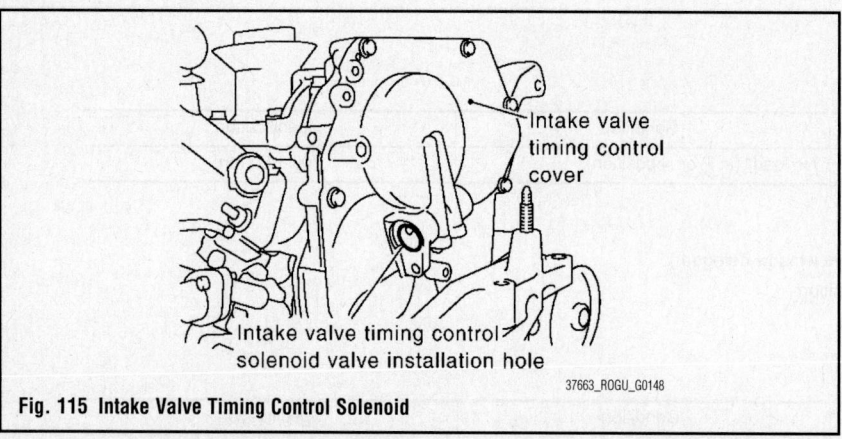

Fig. 115 Intake Valve Timing Control Solenoid

FUEL SYSTEM SERVICE PRECAUTIONS

Safety is the most important factor when performing not only fuel system maintenance, but any type of maintenance. Failure to conduct maintenance and repairs in a safe manner may result in serious personal injury or death. Work on a vehicle's fuel system components can be accomplished safely and effectively by adhering to the following rules and guidelines.

• To avoid the possibility of fire and personal injury, always disconnect the negative battery cable unless the repair or test procedure requires that battery voltage be applied.

• Always relieve the fuel system pressure prior to disconnecting any fuel system component (injector, fuel rail, pressure regulator, etc.) fitting or fuel line connection. Exercise extreme caution whenever relieving fuel system pressure to avoid exposing skin, face and eyes to fuel spray. Please be advised that fuel under pressure may penetrate the skin or any part of the body that it contacts.

• Always place a shop towel or cloth around the fitting or connection prior to loosening to absorb any excess fuel due to spillage. Ensure that all fuel spillage is quickly removed from engine surfaces. Ensure that all fuel-soaked cloths or towels are deposited into a flame-proof waste container with a lid.

• Always keep a dry chemical (Class B) fire extinguisher near the work area.

• Do not allow fuel spray or fuel vapors to come into contact with a spark or open flame.

• Always use a second wrench when loosening or tightening fuel line connection fittings. This will prevent unnecessary stress and torsion on fuel piping. Always follow the proper torque specifications.

• Always replace worn fuel fitting O-rings with new ones. Do not substitute fuel hose where rigid pipe is installed.

FUEL SYSTEM PRESSURE

RELIEVING

1. Remove fuel pump fuse located in IPDM E/R. or unplug the fuel pump connector.
2. Start the engine.
3. After engine stalls, crank it two or three times to release all fuel pressure.
4. Turn ignition switch OFF.
5. Reinstall fuel pump fuse after servicing fuel system.

FUEL FILTER

REMOVAL & INSTALLATION

The fuel filter is integral to the fuel pump assembly.

FUEL PUMP MODULE

REMOVAL & INSTALLATION

See Figures 116 through 120.

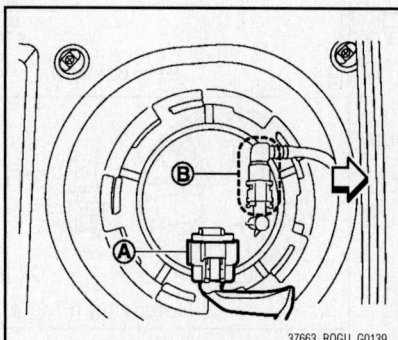

Fig. 116 Disconnect the harness connector (A) and the quick connector (B)

REMOVAL & INSTALLATION

1. Before servicing the vehicle, refer to the Precautions Section.
2. Disconnect the negative battery cable.
3. Disconnect the connector.
4. Remove the retaining bolts.
5. Remove the Intake Valve Timing Control solenoid valve.

To install:

6. Installation is the reverse of removal.

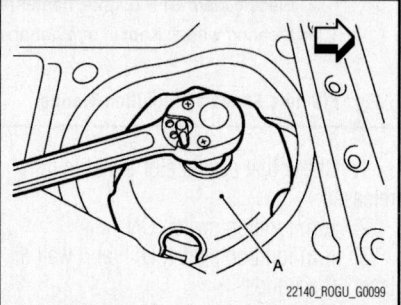

Fig. 117 Remove the lock ring with a lock ring wrench (A)

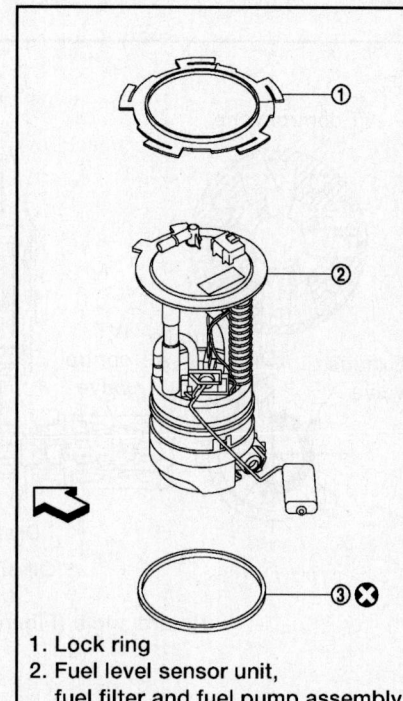

1. Lock ring
2. Fuel level sensor unit, fuel filter and fuel pump assembly
3. O-ring

Fig. 118 2WD models fuel level sensor assembly

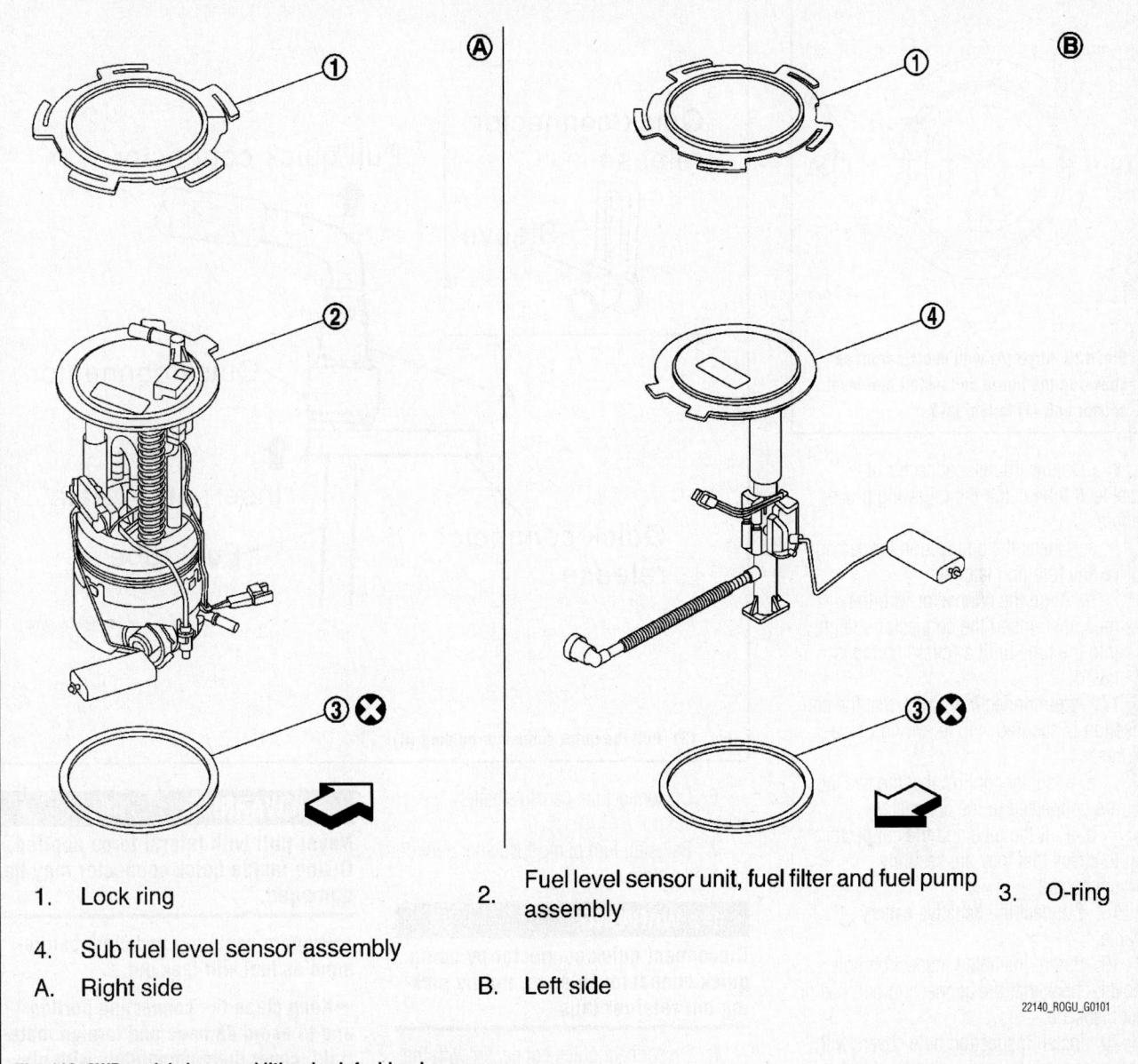

1. Lock ring
2. Fuel level sensor unit, fuel filter and fuel pump assembly
3. O-ring
4. Sub fuel level sensor assembly
A. Right side
B. Left side

22140_ROGU_G0101

Fig. 119 AWD models has an additional sub-fuel level sensor

1. Before servicing the vehicle, refer to the Precautions Section.

2. Disconnect the negative battery cable.

3. Check fuel level on fuel gauge. If fuel gauge indicates more than ¾ of a tank, drain fuel tank to around a half a tank.

4. In case fuel pump does not operate, perform the following procedure:

   a. Insert hose of less than 0.79 in. (20 mm) diameter into fuel filler tube through fuel filler opening to draw fuel from fuel filler tube.

   b. Disconnect fuel filler hose from fuel filler tube.

   c. Insert hose into fuel tank through fuel filler hose to draw fuel from fuel tank.

5. Release the fuel pressure from the fuel lines.

6. Open the fuel filler lid.

7. Open fuel filler cap and release the pressure inside the fuel tank.

8. Remove rear seat cushion.

9. Remove inspection hole cover.

10. Using a screwdriver, remove it by turning clips clockwise by 90 degrees.

11. Disconnect the harness connector and fuel line quick connector.

12. Using lock ring wrench [SST: KV991J0090 (J-46214)], remove lock ring.

➡**For reference when installing, put a matching mark on lock ring, fuel pump assembly and fuel tank.**

13. Raise fuel level sensor unit, fuel filter and fuel pump assembly, and disconnect fuel tube and harness connector.

❋❋ WARNING

Never bend float arm during removal. Never pollute the inside by residue fuel. Draw out avoiding inclination by supporting with a cloth. Never cause impact such as dropping when handling components.

To install:

14. Install O-ring to fuel tank without any twist.

15. Align with vehicle front as shown in the figure. Install fuel level sensor unit (1) to fuel tank.

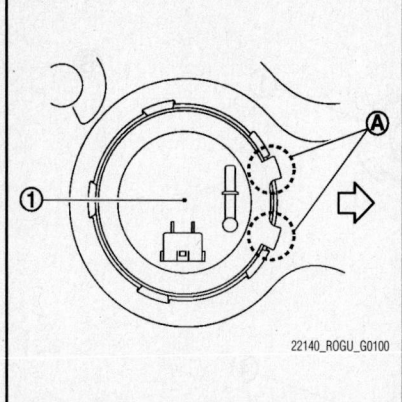

Fig. 120 Align (A) with vehicle front as shown in the figure and install fuel level sensor unit (1) to fuel tank

16. Connect quick connector of fuel feed tube using the following procedures:

a. Check the connection for damage or any foreign materials.

b. Align the connector with the tube, then insert the connector straight into the tube until a "click" sound is heard.

17. After connecting, check that the connection is secured with following procedures:

a. Visually confirm that the two tabs are connected to the connector.

b. Pull the tube and the connector to check that they are securely connected.

18. Connect the negative battery cable.

19. Before installing inspection hole cover, check that the connecting part has no fuel leakage.

20. Install inspection hole covers with the front mark (arrow) facing front of vehicle.

21. Lock clips by turning counterclockwise.

22. Install the rear seat cushion.

23. Install the fuel cap.

FUEL RAIL AND INJECTOR

REMOVAL & INSTALLATION

See Figures 121 through 123.

1. Before servicing the vehicle, refer to the Precautions Section.

2. Remove fuel pump fuse located in IPDM E/R.

3. Start the engine.

4. After engine stalls, crank it two or three times to release all fuel pressure.

5. Turn ignition switch OFF.

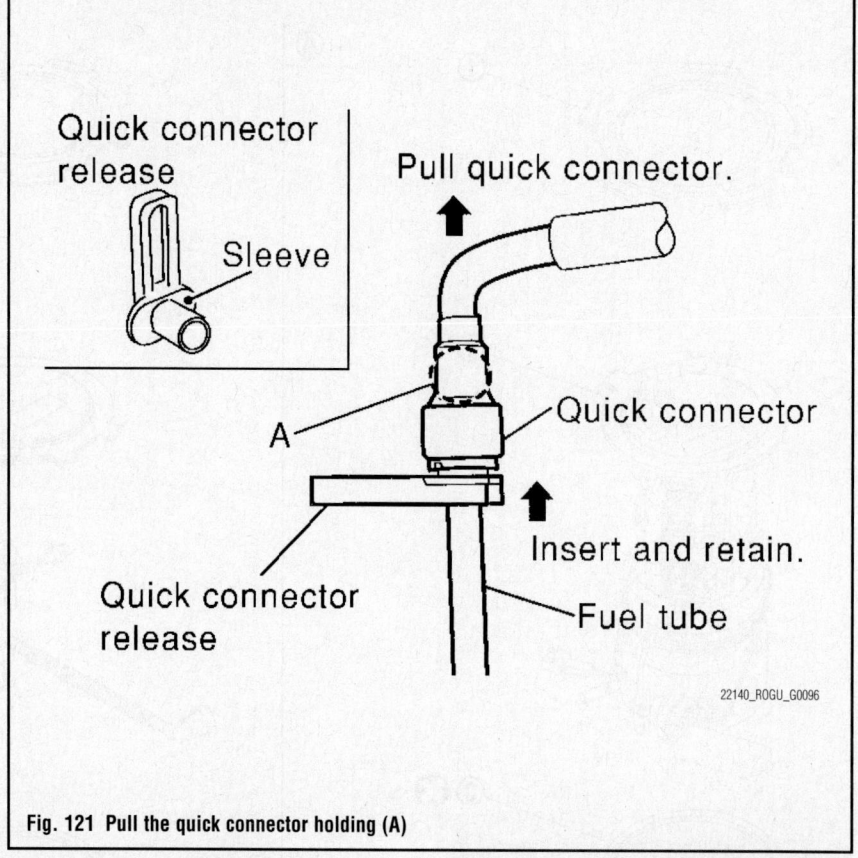

Fig. 121 Pull the quick connector holding (A)

6. Disconnect the negative battery cable.

7. Reinstall fuel pump fuse after servicing fuel system.

❋❋ WARNING

Disconnect quick connector by using quick connector release, not by picking out retainer tabs.

8. Disconnect quick connector with the following procedure:

a. Remove the quick connector cap.

b. With the sleeve side of quick connector release facing quick connector, install quick connector release onto fuel tube.

c. Insert quick connector release into quick connector until sleeve contacts and goes no further. Hold quick connector release on that position.

➡**Inserting quick connector release hard will not disconnect quick connector. Hold quick connector release where it contacts and goes no further.**

d. Draw and pull out quick connector straight from fuel tube.

e. Pull the quick connector holding position in the figure.

❋❋ WARNING

Never pull with lateral force applied. O-ring inside quick connector may be damaged.

➡**Prepare container and cloth beforehand as fuel will leak out.**

➡**Keep clean the connecting portion and to avoid damage and foreign materials, cover them completely with plastic bags or something similar.**

9. Remove the intake manifold.

10. Disconnect sub-harness for fuel injector.

11. Remove the fuel tube and fuel injector assembly.

12. Loosen the mounting bolts and in reverse order as shown.

13. Remove fuel injector from fuel tube with the following procedure:

a. Open and remove clip.

b. Remove fuel injector from fuel tube by pulling straight.

To install:

14. Note the following, and install O-rings to fuel injector as follows:

➡**Upper and lower O-rings are different. Be careful not to confuse them.**

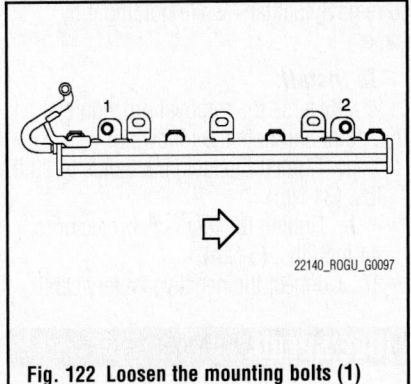

Fig. 122 Loosen the mounting bolts (1) and (2) in reverse order

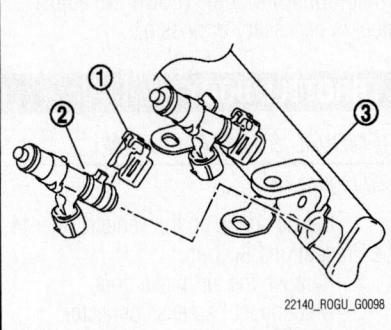

Fig. 123 Open and remove clip (1) remove fuel injector (2) from fuel tube (3)

- Except for California: Fuel tube side—Blue, Nozzle side—Brown
- For California: Fuel tube side—Black, Nozzle side—Green
- Handle O-ring with bare hands. Never wear gloves.
- Lubricate O-ring with new engine oil.
- Never clean O-ring with solvent.
- Check that O-ring and its mating part are free of foreign material.
- When installing O-ring, be careful not to scratch it with tool or fingernails. Also be careful not to twist or stretch O-ring.
- If O-ring was stretched while it was being attached, never insert it quickly into fuel tube.
- Insert O-ring straight into fuel tube. Never decenter or twist it.

15. Install fuel injector to fuel tube with the following procedure:

   a. Insert clip into clip mounting groove on fuel injector.

   b. Insert clip so that protrusion of fuel injector matches cutout of clip.

✷✷ WARNING

Never reuse clip. Replace it with a new one. Be careful to keep clip from

interfering with O-ring. If interference occurs, replace O-ring.

   c. Insert fuel injector into fuel tube with clip attached.

   d. Insert it while matching it to the axial center.

   e. Insert fuel injector so that protrusion of fuel tube matches cutout of clip.

   f. Check that fuel tube flange is securely fixed in flange fixing groove on clip.

   g. Check that installation is complete by making sure that fuel injector does not rotate or come off.

16. Install fuel tube and fuel injector assembly with the following procedure:

   a. Insert the tip of each fuel injector into intake manifold adapter.

   b. Tighten mounting bolts in numerical order

17. Connect sub-harness for fuel injector.

18. Install the intake manifold.

19. Note the following, and connect quick connector to install fuel feed hose:

   a. Check the connection for foreign material and damage.

   b. Align center to insert quick connector straight into fuel tube. Insert fuel tube into quick connector until the top spool on fuel tube is inserted completely and the second level spool is positioned slightly below quick connector bottom end.

   c. Hold position when inserting fuel tube into quick connector.

   d. Carefully align center to avoid inclined insertion to prevent damage to O-ring inside quick connector.

   e. Insert until you hear a (click sound) and actually feel the engagement.

20. To avoid misidentification of engagement with a similar sound, be sure to perform the next step:

   a. Before clamping fuel feed hose with hose clamps, pull quick connector hard by hand holding position.

   b. Check it is completely engaged (connected) so that it does not come out from fuel feed tube.

21. Install the quick connector cap to quick connector connection.

22. Install so that the arrow mark on the side faces up.

23. Install fuel feed hose to hose clamp.

24. Install in the reverse order of removal procedure after this step.

FUEL TANK

REMOVAL & INSTALLATION

2WD Models

See Figure 124.

1. Before servicing the vehicle, refer to the Precautions Section.

2. Disconnect the negative battery cable.

3. Check the fuel level on the fuel gauge. If fuel gauge indicates more than ¾ of a tank, drain fuel tank to around a half a tank.

4. In case fuel pump does not operate, perform the following procedure:

   a. Insert hose of less than 0.79 inch (20 mm) diameter into fuel filler tube through fuel filler opening to draw the fuel from fuel filler tube.

   b. Disconnect fuel filler hose from fuel filler tube.

   c. Insert hose into fuel tank through fuel filler hose to draw fuel from fuel tank.

5. Release the fuel pressure from the fuel lines.

6. Open the fuel filler lid.

7. Open fuel filler cap and release the pressure inside the fuel tank.

8. Remove rear seat cushion.

9. Remove inspection hole cover.

10. Disconnect fuel lines and harness connectors.

11. Remove the muffler assembly.

12. Remove the protector from the fuel tank.

13. Remove the vent hose at rear side of fuel tank.

14. Disconnect the EVAP tube at rear side of fuel tank.

15. Remove fuel filler hose at fuel filler tube side.

16. Remove the suspension bar.

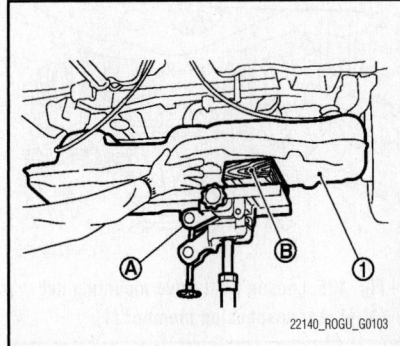

Fig. 124 Fuel tank (1) supported with transmission jack (A) and block of wood (B)

17. Remove parking brake cable mounting bolts and separate the parking brake cable from suspension arm.

18. Support center of fuel tank with transmission jack.

19. Securely support the fuel tank with a piece of wood.

20. Remove fuel tank band (RH and LH).

21. Lower the transmission jack carefully to remove fuel tank while holding it by hand.

To install:

22. Reverse the removal at this point and note the following:

   a. Tighten the fuel tank bands to 23 ft. lbs. (31 Nm).

   b. Tighten the fuel tank protector to 44 inch lbs. (5 Nm).

23. Connect the negative battery cable.

AWD Models

See Figures 125 and 126.

1. Before servicing the vehicle, refer to the Precautions Section.

2. Disconnect the negative battery cable.

3. Check the fuel level on the fuel gauge. If fuel gauge indicates more than ¾ of a tank, drain fuel tank to around a half a tank.

4. In case fuel pump does not operate, perform the following procedure:

   a. Insert hose of less than 0.79 in. (20 mm) diameter into fuel filler tube through fuel filler opening to draw the fuel from fuel filler tube.

   b. Disconnect fuel filler hose from fuel filler tube.

   c. Insert hose into fuel tank through fuel filler hose to draw fuel from fuel tank.

5. Release the fuel pressure from the fuel lines.

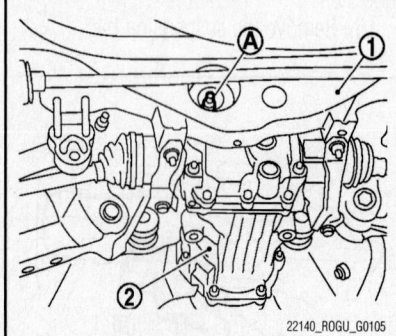

Fig. 125 Loosen final drive mounting nut (A) at rear suspension member (1)

6. Open the fuel filler lid.

7. Open fuel filler cap and release the pressure inside the fuel tank.

8. Remove rear seat cushion.

9. Remove inspection hole cover.

10. Disconnect fuel lines and harness connectors.

11. Remove the muffler assembly.

12. Remove the propeller shaft.

13. Remove the protector from the fuel tank.

14. Remove vent hose at rear side of fuel tank.

15. Disconnect the EVAP tube at rear side of fuel tank.

16. Remove fuel filler hose at fuel filler tube side.

17. Remove the suspension bar.

18. Remove parking brake cable mounting bolts and separate parking brake cable from suspension arm.

19. Disconnect the AWD solenoid harness connector and harness clip.

20. Loosen final drive mounting nut at rear suspension member.

➡**Never remove final drive mounting nut.**

21. Remove final drive mounting bolts from final drive mounting bracket (3) to tilt final drive assembly.

➡**Final drive assembly does not have to be removed from the vehicle.**

22. Support center of fuel tank with transmission jack.

23. Securely support the fuel tank with a piece of wood.

24. Remove fuel tank band (RH and LH).

25. Lower the transmission jack carefully

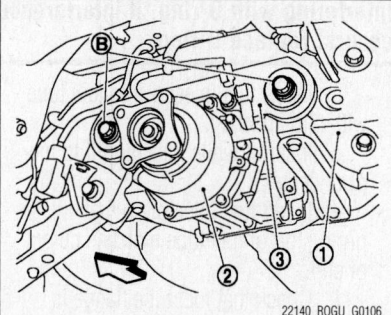

Fig. 126 Remove final drive mounting bolts (B) from final drive mounting bracket (3) to tilt final drive assembly (2)

to remove fuel tank while holding it by hand.

To install:

26. Reverse the removal procedure at this point and note the following:

   a. Tighten the fuel tank bands to 23 ft. lbs. (31 Nm).

   b. Tighten the fuel tank protector to 44 inch lbs. (5 Nm).

27. Connect the negative battery cable.

IDLE SPEED

ADJUSTMENT

Idle speed is maintained by the Electronic Control Module (ECM). No adjustment is necessary or possible.

THROTTLE BODY

REMOVAL & INSTALLATION

See Figure 127.

1. Before servicing the vehicle, refer to the Precautions Section.

2. Remove the air intake duct.

3. Disconnect harness connector.

4. Disconnect water hose.

5. Loosen the throttle body assembly mounting bolts in reverse order as shown in the figure.

To install:

6. Install the throttle body assembly with a new gasket.

7. Tighten the mounting bolts in sequence to 74 inch lbs. (8.4 Nm)

8. Reconnect the water hose.

9. Reconnect the harness connector.

10. Reconnect the air intake duct.

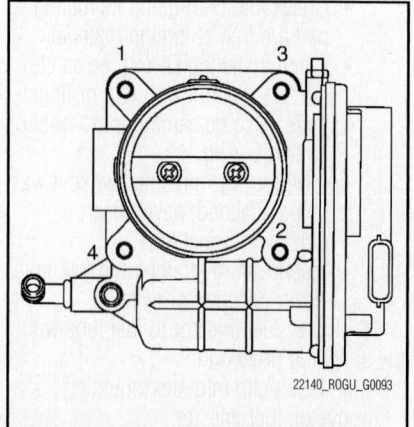

Fig. 127 Loosen mounting bolts in reverse order as shown

HEATING & AIR CONDITIONING SYSTEM

BLOWER MOTOR

REMOVAL & INSTALLATION

See Figures 128 and 129.

1. Before servicing the vehicle, refer to the Precautions Section.
2. Remove instrument driver lower cover. Refer to Instrument Panel Removal & Installation in the Body Interior section.
3. Remove combination meter.
4. Remove knee protector.
5. Remove foot duct LH.
6. Remove mode door motor.
7. Remove brake pedal assembly and accelerator pedal.
8. Disconnect blower motor connector.
9. Press flange holding hook. Turn blower motor counterclockwise.
10. Pull outside and remove blower motor.

❊❊ WARNING

The balance is adjusted when blower fan and blower motor are assembled, so do not replace the individual parts.

To install:
11. Installation is the reverse of removal.

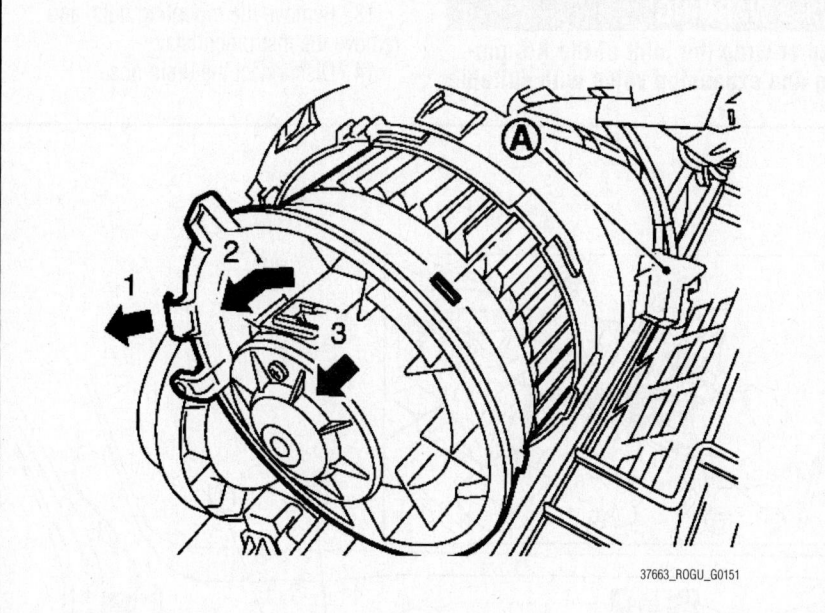

Fig. 128 Disconnect blower motor connector (A), press flange holding hook (1), turn blower motor counterclockwise (2), and pull outside (3) and remove blower motor

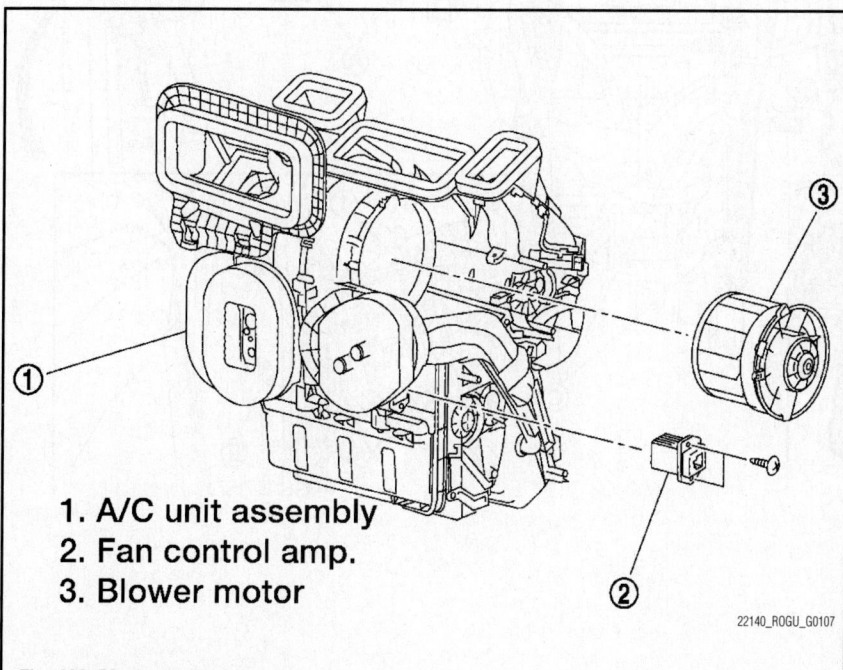

1. A/C unit assembly
2. Fan control amp.
3. Blower motor

Fig. 129 Blower motor removal

HEATER CORE

REMOVAL & INSTALLATION

See Figures 130 and 133.

1. Before servicing the vehicle, refer to the Precautions Section.
2. Remove the A/C unit assembly.
3. Remove heater packing.
4. Remove heater pipe flange.
5. Remove mounting screw, and then remove heater pipe clamp.
6. Slide heater core to leftward (as shown in the figure).

To install:
7. Installation is the reverse order of removal.

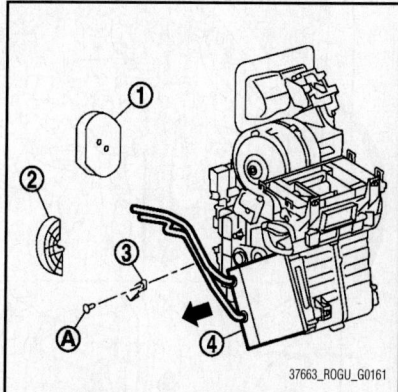

Fig. 130 Remove heater packing (1), heater pipe flange (2), mounting screw (A), heater pipe clamp (3), and heater core (4)

HEATER UNIT

REMOVAL & INSTALLATION

See Figures 131 through 133.

1. Before servicing the vehicle, refer to the Precautions Section.
2. Disconnect the negative battery cable.

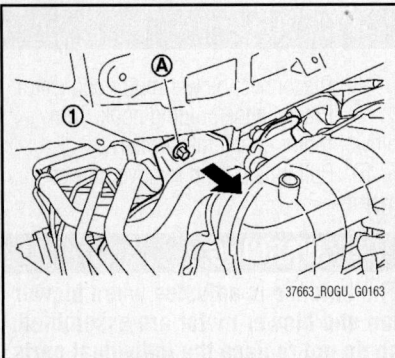

Fig. 131 Remove mounting nut (A), and move lower dash insulator (1) aside

37663_ROGU_G0163

3. Use a refrigerant collecting equipment (for HFC-134a) to discharge the refrigerant.

4. Drain engine coolant from the cooling system.

5. Remove the cowl top cover.

6. Remove the mounting nut, and position the lower dash insulator aside.

7. Remove the mounting bolt, and disconnect the low-pressure flexible hose and high-pressure pipe from the expansion valve.

❋❋ WARNING

Cap or wrap the joint of the A/C piping and expansion valve with suitable material such as vinyl tape to avoid the entry of air.

8. Remove the clamps and disconnect the heater hoses from the heater core.

9. Remove the instrument panel. Refer to Instrument Panel Removal & Installation in the Body Interior section.

10. Remove the center ventilator duct.

11. Remove the rear foot duct 1.

12. Remove the rear foot duct 2.

13. Remove the mounting nuts, and remove the instrument stay.

14. Disconnect the drain hose.

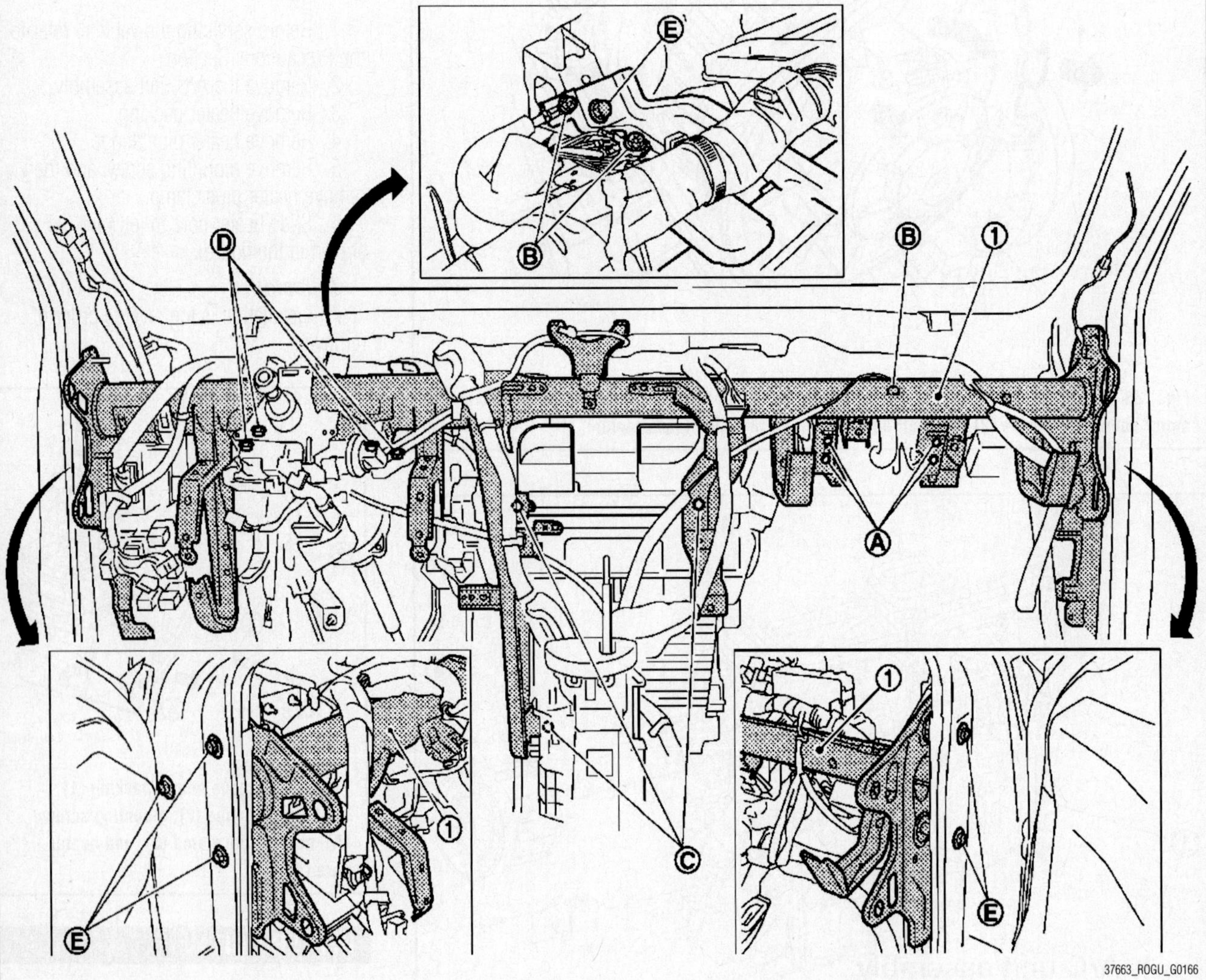

Fig. 132 Bolts, screws and nut locations

37663_ROGU_G0166

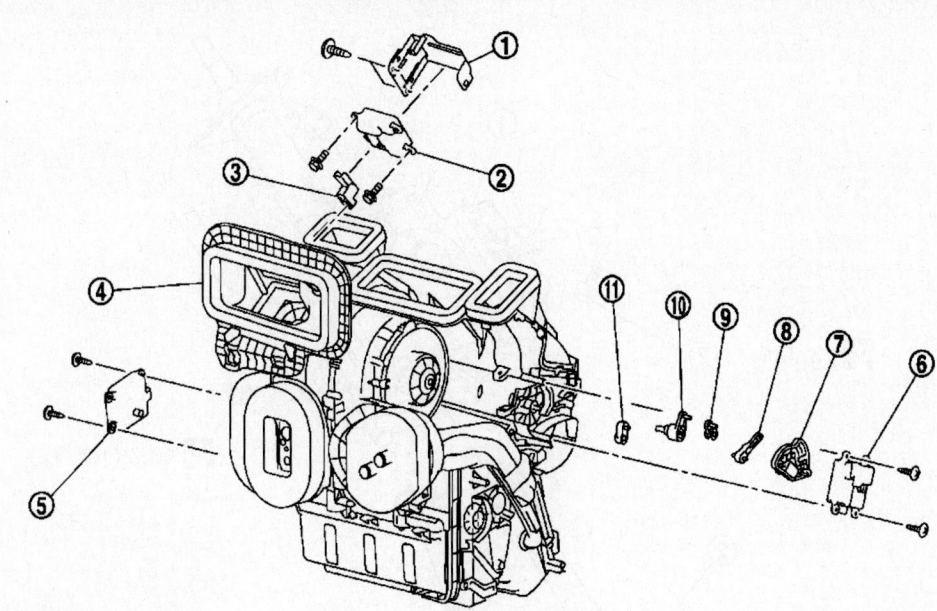

1.	Intake door motor bracket	2.	Intake door motor	3.	Intake door lever
4.	A/C unit assembly	5.	Air mix door motor	6.	Mode door motor
7.	Main link	8.	Foot door link	9.	Foot door lever
10.	Ventilator door lever	11.	Defroster door lever		

37663_ROGU_G0152

Fig. 133 A/C unit assembly (4)

15. Remove the mounting screws, and remove the BCM with bracket attached.

16. Remove the ground bolts from the steering member.

17. Remove the mounting screws from the A/C unit assembly.

18. Remove the steering column mounting nuts.

19. Remove the steering member mounting bolts, and remove the steering member.

20. Remove the A/C unit assembly.

To install:

21. Installation is the reverse order of removal.

a. Replace O-rings with new ones. Then apply compressor oil to the O-rings.

b. Check for leakages when recharging refrigerant.

STEERING

POWER RACK & PINION STEERING GEAR

REMOVAL & INSTALLATION

See Figures 134 and 135.

1. Before servicing the vehicle, refer to the Precautions Section.

2. Set vehicle to the straight-ahead position.

3. Disconnect the negative battery cable.

4. Remove the upper cover.

5. Remove dash seal.

6. Remove hole cover.

7. Remove the bolt of intermediate shaft (lower side), and then remove intermediate shaft from steering gear pinion shaft.

8. Remove tires with a power tool.

9. Remove steering outer socket from steering knuckle so as not to damage ball joint boot using suitable ball joint remover.

10. Temporarily tighten the nut to prevent damage to threads and to prevent the ball joint remover from suddenly coming off.

11. Remove front suspension member.

12. Remove the steering gear assembly.

To install:

13. Note the following, and install in the reverse order of removal:

a. Tighten the steering gear mounting nuts to 109 ft. lbs. (147.5 Nm).

b. Tighten the tie rod end mounting nuts to 25 ft. lbs. (34.4 Nm).

14. Spiral cable may be cut if steering wheel turns while separating steering column assembly and steering gear assembly. Be sure to secure steering wheel using string to avoid turning.

15. Check each part of dash seal for damage or other malfunctions. Replace if necessary.

16. Perform final tightening of nuts and bolts on each part under relaxed conditions with tires on level ground when removing steering gear assembly.

17. Check wheel alignment.

18. Adjust the neutral position of the steering angle sensor as follows:

a. Stop the vehicle with front wheels in straight-ahead position.

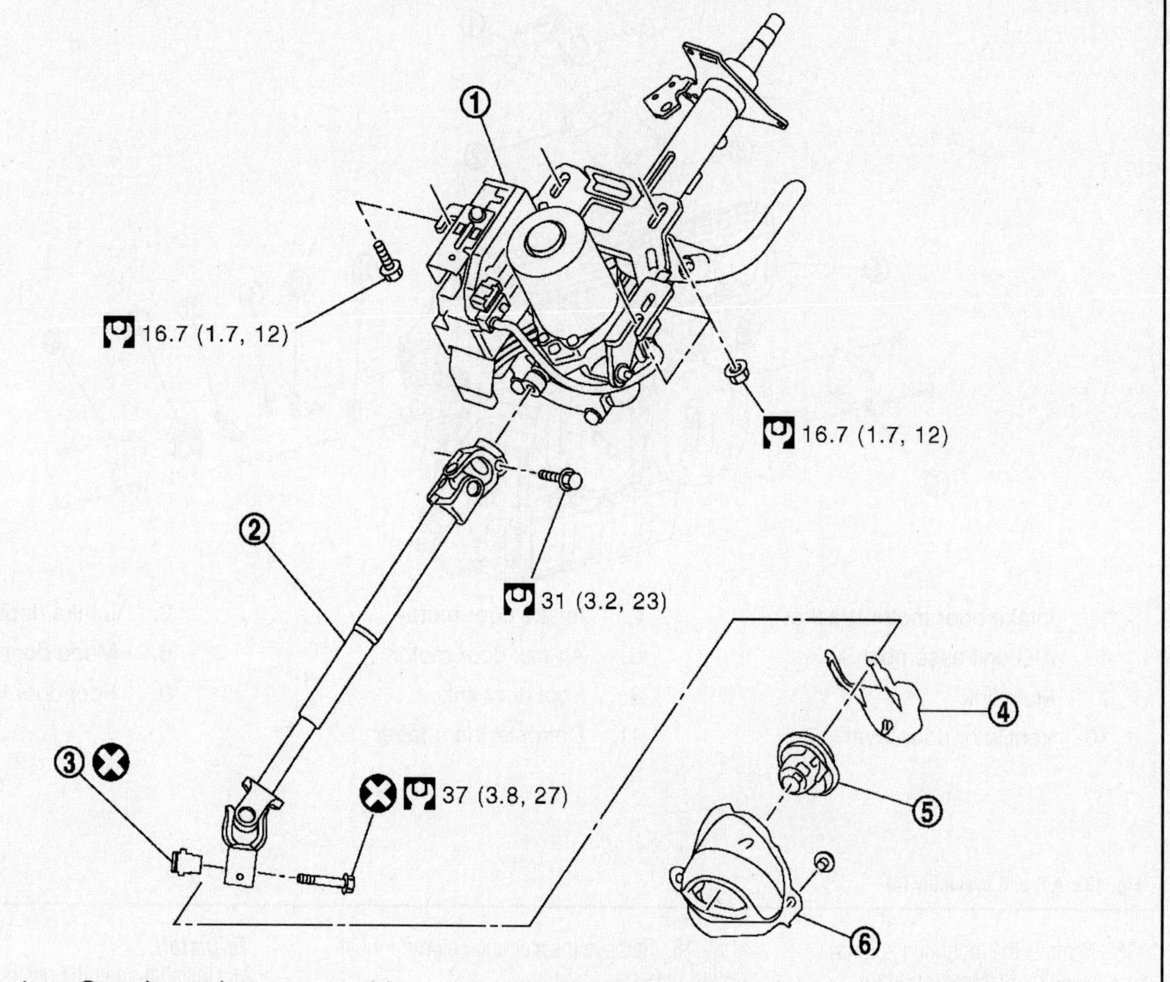

[🔧] 16.7 (1.7, 12)

[🔧] 16.7 (1.7, 12)

[🔧] 31 (3.2, 23)

③ ⊗

⊗ [🔧] 37 (3.8, 27)

1. Steering column assembly   2. Intermediate shaft   3. Cam nut
4. Upper cover   5. Dash seal   6. Hole cover

22140_ROGU_G0111

Fig. 134 Remove the bolt of intermediate shaft (lower side)

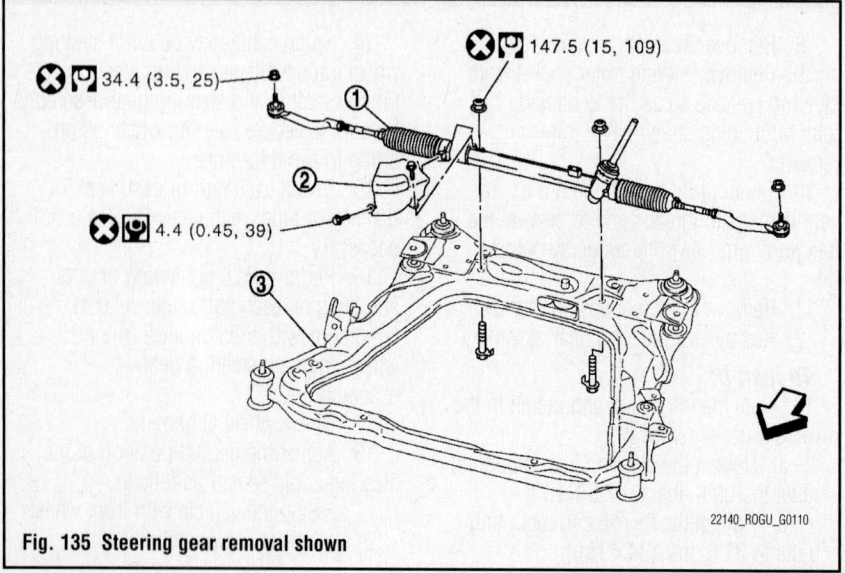

⊗ [🔧] 34.4 (3.5, 25)

⊗ [🔧] 147.5 (15, 109)

⊗ [🔧] 4.4 (0.45, 39)

22140_ROGU_G0110

Fig. 135 Steering gear removal shown

b. On the CONSULT-III® screen, touch "WORK SUPPORT" and "ST ANG SEN ADJUSTMENT" in order.
  c. Touch "START".

➡**Do not touch steering wheel while adjusting steering angle sensor.**

  d. After approximately 10 seconds, touch "END".
  e. After approximately 60 seconds, it ends automatically.
  f. Turn ignition switch OFF, then turn it ON again.
  g. Run vehicle with front wheels in straight-ahead position, then stop.
  h. Select "DATA MONITOR". Then make sure "STR ANGLE SIG" is within 0 plus or minus 2.5°.

SUSPENSION   **FRONT SUSPENSION**

CONTROL LINKS

REMOVAL & INSTALLATION

1. Remove tires with power tool.
2. Remove under cover from vehicle.
3. Remove upper and lower retaining nuts from stabilizer connecting rod.
4. Remove stabilizer connecting rod.

To install:

5. Installation is the reverse order of removal.

   a. Tighten retaining nut to 62 ft. lbs. (84 Nm)

LOWER BALL JOINT

REMOVAL & INSTALLATION

The lower ball joint is integral to the control arm. The ball is serviced with the control arm assembly.

LOWER CONTROL ARM

REMOVAL AND & INSTALLATION

See Figure 136.

1. Before servicing the vehicle, refer to the Precautions Section.
2. Raise the vehicle.
3. Remove the tire.
4. Remove the lower control arm from the steering knuckle.
5. Remove the lower control arm from the suspension member.

➡**The transverse link cannot be pulled out because the mounting bolt of transverse link at the rear of the mounting area located on the front side of vehicle hits against the stabilizer bar. Therefore, get stabilizer bar out of the way to remove the transverse link.**

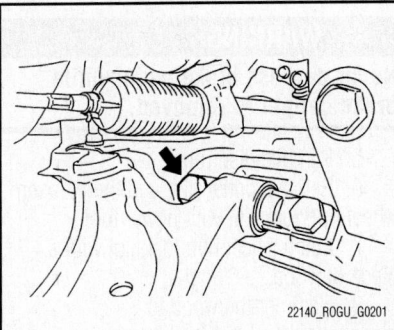

22140_ROGU_G0201

Fig. 136 Lower control arm to suspension member mounting bolts

To install:

6. Note the following, and install in the reverse order of removal:

   a. Perform final tightening of bolts and nuts at the front suspension member, under relaxed conditions with tires on level ground.

   b. Tighten the front control arm mounting bolts to 126 ft. lbs. (171 Nm).

   c. Tighten the rear control arm bolt and nut to 104 ft. lbs. (142 Nm).

STEERING KNUCKLE

REMOVAL & INSTALLATION

See Figure 137.

1. Before servicing the vehicle, refer to the Precautions Section.
2. Remove tires with power tool.
3. Remove wheel sensor from steering knuckle.

✳✳ WARNING

Never pull on the wheel sensor harness.

4. Remove lock plate from strut assembly.
5. Remove torque member mounting bolts with power tool. Hang torque member so as not to interfere with work.
6. Remove disc rotor. Put matching marks on the wheel hub and bearing assembly and the disc rotor before removing the disc rotor.
7. Remove cotter pin, and then loosen hub lock nut with power tool and disengage wheel hub and bearing assembly from drive shaft.
8. Remove the hub lock nut.

➡**Use suitable puller, if wheel hub and bearing assembly and driveshaft can-**

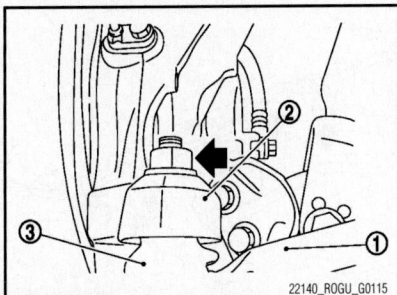

22140_ROGU_G0115

Fig. 137 Remove steering outer socket (1) from steering knuckle (2) using the ball joint remover so as not to damage ball joint boot (3)

not be separated even after performing the above procedure.

9. Remove wheel hub and bearing assembly, and then remove splash guard.
10. Remove transverse link from steering knuckle.
11. Remove the steering knuckle from strut assembly.
12. Loosen the nut of steering outer socket.
13. Remove steering outer socket from steering knuckle using the ball joint remover so as not to damage ball joint boot.
14. Temporarily tighten the nut to prevent damage to threads and to prevent the ball joint remover from suddenly coming off.
15. Remove the steering knuckle from vehicle.

To install:

16. Note the following, and install in the reverse order of the removal:

   a. Align the matching marks made during removal when reusing the disc rotor.

   b. Install removed wheel hub and bearing assembly and steering knuckle and perform the final tightening of each part under relaxed conditions on the level surface.

   c. Tighten the stabilizer link mounting nuts to 62 ft. lbs. (84 Nm).

   d. Tighten the lower strut knuckle bolt to 104 ft. lbs. (141 Nm).

   e. Tighten the tie rod end mounting nuts to 25 ft. lbs. (34 Nm).

   f. Tighten the axle nut to 92 ft. lbs. (125 Nm).

   g. Install the front wheels, lower the vehicle and perform a front end alignment.

STRUT

REMOVAL & INSTALLATION

See Figure 138.

1. Before servicing the vehicle, refer to the Precautions Section.
2. Raise the vehicle.
3. Remove the tire.
4. Remove the brake hose lock plate clip.
5. Remove cap and mounting nut on the upper side of stabilizer connecting rod, and then remove stabilizer connecting rod from strut assembly with power tool.
6. Separate steering knuckle from strut assembly.
7. Remove mounting bolts of strut

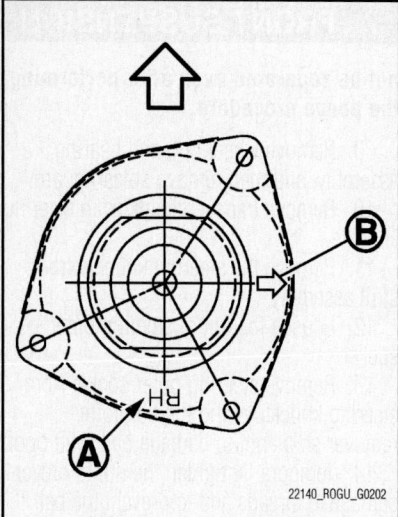

Fig. 138 Identification mark (A), Arrow mark (B) and Large arrow front of the vehicle

mounting insulator, and then remove strut assembly with power tool.

To install:

8. Note the following, and install in the reverse order of removal:

  a. Perform final tightening of bolts and nuts, under relaxed conditions with tires on level ground.

  b. Tighten the strut plate nuts to 14 ft. lbs. (19 Nm).

  c. Tighten the strut to steering knuckle mounting nuts to 104 ft. lbs. (142 Nm).

  d. Tighten the stabilizer link nut to 62 ft. lbs. (84 Nm).

OVERHAUL

1. Before servicing the vehicle, refer to the Precautions Section.
2. Raise the vehicle.
3. Remove the tire.
4. Remove the strut assembly.
5. Install the strut assembly into a spring compressor service tool.
6. Compress the coil spring. Make sure coil spring with a spring compressor between strut mounting bearing and lower rubber seat (strut assembly) is free.
7. Remove piston rod lock nut while securing the piston rod tip so that piston rod does not turn.
8. Remove strut mounting insulator and strut mounting bearing, and bound bumper from strut.
9. After remove coil spring with a spring compressor, and then gradually release a spring compressor.
10. Remove the strut.

To install:

11. Install the strut.
12. Install lower rubber seat.
13. Install bound bumper onto strut mounting insulator.
14. Compress coil spring using a spring compressor (commercial service tool), and install it onto strut assembly.
15. Install strut mounting bearing and strut mounting insulator with bound bumper to strut.
16. Secure piston rod tip so that piston rod does not turn, then tighten to 58 ft. lbs. (79 Nm).
17. Gradually remove the spring pressure.
18. Remove the strut assembly from service tool.
19. Install the strut assembly to the vehicle.
20. Install the tire and lower the vehicle.

STABILIZER BAR

REMOVAL & INSTALLATION

See Figures 139 and 140.

1. Remove tires power tool.
2. Remove under cover from vehicle.
3. Remove steering outer socket from steering knuckle.
4. Remove stabilizer connecting rod.
5. Remove rear torque rod.
6. Separate intermediate shaft from steering gear.
7. Set suitable jack under front suspension member.
8. Remove front suspension member stay from vehicle.
9. Gradually lower jack front suspension member in order to remove stabilizer mounting bolts.
10. Remove mounting bolts of stabilizer clamp, and then remove stabilizer clamp

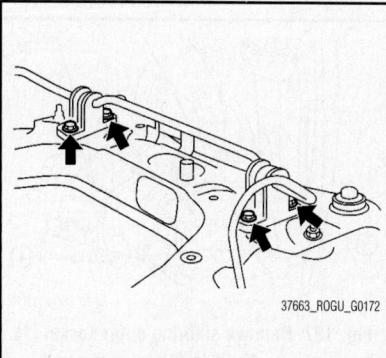

Fig. 139 Remove the stabilizer clamp mounting

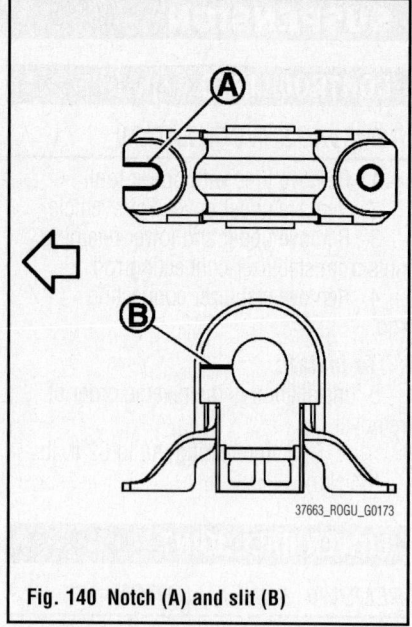

Fig. 140 Notch (A) and slit (B)

and stabilizer bushing from front suspension member.

11. Remove stabilizer bar.

To install:

12. Note the following, and install in the reverse order of removal.

  a. Install stabilizer clamp that notch is toward vehicle front side.

  b. Install stabilizer bushing that slit is toward vehicle front side.

WHEEL HUB & BEARING

REMOVAL & INSTALLATION

See Figure 141.

1. Remove tires with power tool.
2. Remove wheel sensor from steering knuckle.
3. Remove lock plate from strut assembly.
4. Remove torque member mounting bolts with power tool. Hang torque member not to interfere with work.

❄❄ WARNING

Never depress brake pedal while brake caliper is removed.

5. Remove disc rotor.
6. Remove cotter pin, and then loosen wheel hub lock nut with power tool.
7. Patch wheel hub lock nut with a piece of wood.
8. Hammer the wood to disengage wheel hub and bearing assembly from drive shaft. Remove the wheel hub lock nut.

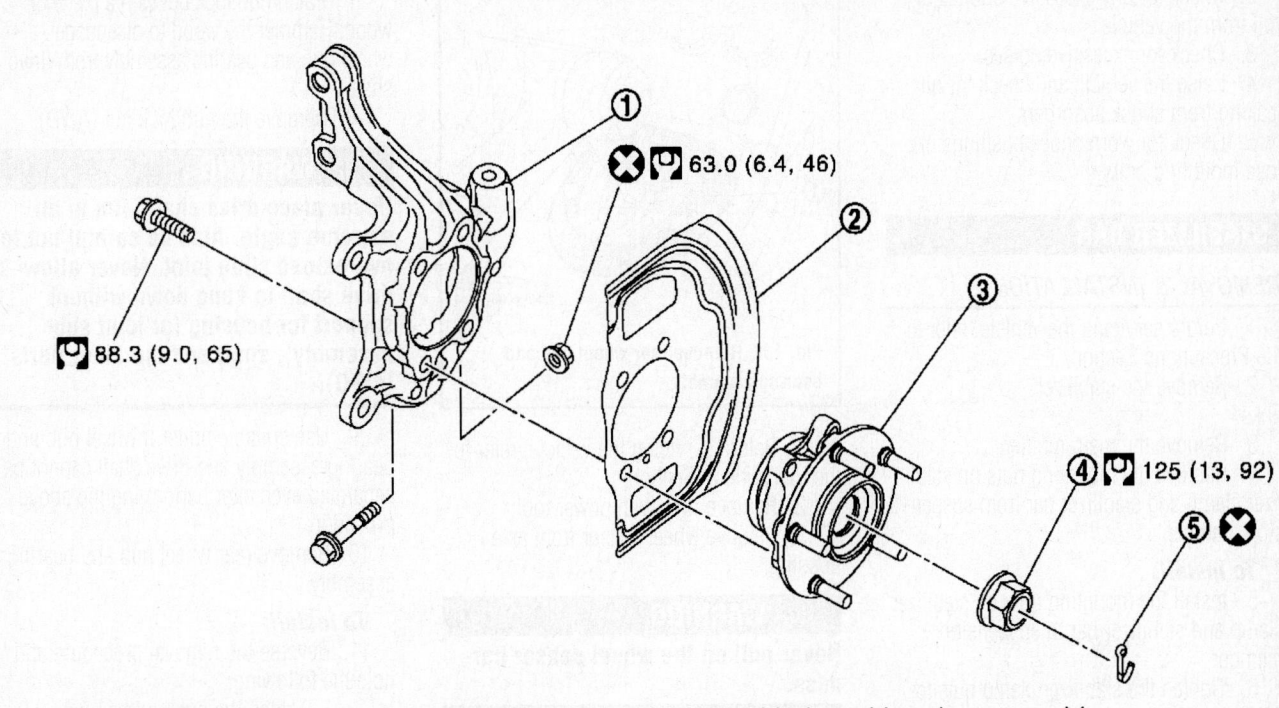

1. Steering knuckle   2. Splash guard   3. Wheel hub and bearing assembly
4. Wheel hub lock nut   5. Cotter pin

37663_ROGU_G0174

Fig. 141 Wheel hub and bearing assembly

✳✳ WARNING

Never place drive shaft joint at an extreme angle. Also be careful not to overextend slide joint. Never allow drive shaft to hang down without support for

housing (or joint sub-assembly), shaft and the other parts.

➡ Use suitable puller, if wheel hub and bearing assembly and drive shaft cannot be separated even after performing the above procedure.

9. Remove wheel hub and bearing assembly.

To install:
10. Installation is the reverse of removal.

ADJUSTMENT

No adjustment is necessary or possible.

SUSPENSION

COIL SPRING

REMOVAL & INSTALLATION

Refer to Control Arms/Links for Removal & Installation.

CONTROL ARMS/LINKS

REMOVAL & INSTALLATION

1. Before servicing the vehicle, refer to the Precautions Section.
2. Remove tires with power tool.
3. Set suitable jack under suspension arm to relieve the coil spring tension.
4. Remove the shock absorber.
5. Remove the coil spring.

To install:

➡**Perform final tightening of the bolts**

and nuts at the shock absorber lower side (rubber bushing), under relaxed conditions with tires on level ground.

6. Install the coil spring.
7. Raise the suspension arm to line up shock mounting holes.
8. Install the shock absorber.
9. Tighten the top mounting nut and bolt to 89 ft. lbs. (120 Nm).
10. Tighten the bottom mounting nut and bolt to 89 ft. lbs. (120 Nm).

SHOCK ABSORBER

REMOVAL & INSTALLATION

1. Before servicing the vehicle, refer to the Precautions Section.
2. Remove tires with power tool.

REAR SUSPENSION

3. Set suitable jack under suspension arm to relieve the coil spring tension.
4. Remove the shock absorber.

To install:

➡**Perform final tightening of the bolts and nuts at the shock absorber lower side (rubber bushing), under relaxed conditions with tires on level ground.**

5. Raise the suspension arm to line up shock mounting holes.
6. Install the shock absorber.
7. Tighten the top mounting nut and bolt to 89 ft. lbs. (120 Nm).
8. Tighten the bottom mounting nut and bolt to 89 ft. lbs. (120 Nm).

TESTING

1. Road test the vehicle.

2. Check for any excessive bounce or roll from the vehicle.

3. Check for excessive noises.

4. Raise the vehicle and check for oil leaking from shock absorbers.

5. Check for worn shock bushings or lose mounting bolts.

STABILIZER BAR

REMOVAL & INSTALLATION

1. Before servicing the vehicle, refer to the Precautions Section.

2. Remove the stabilizer links.

3. Remove the main muffler.

4. Remove the mounting nuts on stabilizer clamp and stabilizer bar from suspension member.

To install:

5. Install the mounting nuts on stabilizer clamp and stabilizer bar to suspension member.

6. Tighten the stabilizer clamp nuts to 26 ft. lbs. (36 Nm).

7. Install the main muffler.

8. Tighten the stabilizer link mounting bolts to 81 ft. lbs. (110 Nm).

WHEEL HUB & BEARING

REMOVAL & INSTALLATION

See Figure 142.

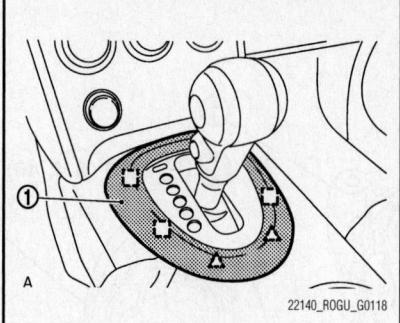

Fig. 142 Remove rear wheel hub and bearing assembly

1. Before servicing the vehicle, refer to the Precautions Section.

2. Remove tires with power tool.

3. Remove wheel sensor from axle housing.

✷✷ WARNING

Never pull on the wheel sensor harness.

4. Remove torque member mounting bolts. Hang torque member not to interfere with work.

5. Remove disc rotor. Put matching marks on the wheel hub and bearing assembly and the disc rotor before removing the disc rotor.

6. Remove cotter pin, and then loosen hub lock nut with power tool (AWD).

7. Patch hub lock nut with a piece of wood. Hammer the wood to disengage wheel hub and bearing assembly from drive shaft.

8. Remove the hub lock nut (AWD).

✷✷ WARNING

Never place drive shaft joint at an extreme angle. Also be careful not to overextend slide joint. Never allow drive shaft to hang down without support for housing (or joint sub-assembly), shaft and the other parts (AWD).

9. Use suitable puller if wheel hub and bearing assembly and drive shaft cannot be separated even after performing the above procedure.

10. Remove rear wheel hub and bearing assembly.

To install:

11. Reverse the removal procedure and note the following:

a. Tighten the hub and bearing assembly mounting bolts to 65 ft. lbs. (88.3 Nm).

b. Tighten the axle locking nut to 92 ft. lbs. (125 Nm) (AWD).

c. Install new cotter pin (AWD).

ADJUSTMENT

No adjustment is necessary or possible.

SPECIFICATIONS AND MAINTENANCE CHARTS

ENGINE AND VEHICLE IDENTIFICATION

	Engine							Model Year	
Code ①	Liters (cc)	Cu. In.	Cyl.	Fuel Sys.	Engine Type	Eng. Mfg.		Code ②	Year
MR20DE	2.0 (1997)	122	4	MFI	DOHC	Nissan		9	2009
QR25DE	2.5 (2488)	152	4	MFI	DOHC	Nissan		A	2010

MFI: Multi-port Fuel Injection

DOHC: Double Overhead Camshaft

① The Engine Code is stamped on the engine block near the starter.

② 10th position of the Vehicle Identification Number (VIN)

37663_SENT_C0001

GENERAL ENGINE SPECIFICATIONS

Year	Model	Engine Displacement Liters	Engine Series (ID/VIN)	Net Horsepower @ rpm	Net Torque @ rpm (ft. lbs.)	Bore x Stroke (in.)	Com-pression Ratio	Oil Pressure @ rpm
2009	Sentra	2.0	MR20DE	140@5100	147@4800	3.31X3.55	10.2:1	29@2000
	Sentra SE-R	2.5	QR25DE	150@5600	154@5600	3.50x3.94	9.5:1	60@3000
2010	Sentra	2.0	MR20DE	140@5100	147@4800	3.31X3.55	10.2:1	29@2000
	Sentra SE-R	2.5	QR25DE	150@5600	154@5600	3.50x3.94	9.5:1	60@3000

37663_SENT_C0002

ENGINE TUNE-UP SPECIFICATIONS

Year	Engine Displacement Liters	Engine ID/VIN	Spark Plug Gap (in.)	Ignition Timing (deg.) MT	AT	Fuel Pump (psi) ①	Idle Speed (rpm) MT	AT ②	Valve Clearance Intake ③	Exhaust ③
2009	2.0	MR20DE	0.043	6B	6B	51	625-725	650-750	0.010-0.013	0.011-0.015
	2.5	QR25DE	0.043	15B	15B	51	650-750	650-750	0.009-0.013	0.010-0.013
2010	2.0	MR20DE	0.043	6B	6B	51	625-725	650-750	0.010-0.013	0.011-0.015
	2.5	QR25DE	0.043	15B	15B	51	650-750	650-750	0.009-0.013	0.010-0.013

B: Before top dead center

① At idle

② Automatic transmission in neutral

③ Engine cold

37663_SENT_C0003

CAPACITIES

Year	Model	Engine ID/VIN	Engine Displacement Liters	Engine Oil with Filter (qts.)	Transmission (pts.) 5-Spd	Auto.	Fuel Tank (gal.)	Cooling System (qts.)
2009	Sentra	MR20DE	2.0	4.0	4.25	17.5	14.5	8.0
	Sentra SE-R	QR25DE	2.5	4.2	4.90	18.0	13.2	6.4
2010	Sentra	MR20DE	2.0	4.0	4.25	17.5	14.5	8.0
	Sentra SE-R	QR25DE	2.5	4.2	4.90	18.0	13.2	6.4

① Engine coolant, 8 qts.: Inverter coolant, 3 qts.

37663_SENT_C0004

FLUID SPECIFICATIONS

Year	Model	Engine Displacement Liters	Engine Oil	Man. Trans.	Auto. Trans.	Brake Master Cylinder	Cooling System
2009	Sentra	2.0	5W-30	75W-80	Nissan NS-2	DOT 3	N-LL
	Sentra SE-R	2.5	5W-30	75W-80	Nissan NS-2	DOT 3	N-LL
2010	Sentra	2.0	5W-30	75W-80	Nissan NS-2	DOT 3	N-LL
	Sentra SE-R	2.5	5W-30	75W-80	Nissan NS-2	DOT 3	N-LL

N-LL: Nissan Long Life coolant

37663_SENT_C0005

VALVE SPECIFICATIONS

Year	Engine ID/VIN	Engine Displacement Liters	Seat Angle (deg.)	Face Angle (deg.)	Spring Test Pressure (lbs. @ in.)	Spring Installed Height (in.)	Stem-to-Guide Clearance (in.) Intake	Exhaust	Stem Diameter (in.) Intake	Exhaust
2009	MR20DE	2.0	45.15-45.45	—	34-39@ 1.39	1.390	0.0008-0.0021	0.0012-0.0025	0.2152 0.2157	0.2148 0.2154
	QR25DE	2.5	45.15-45.45	—	34-39@ 1.39	1.390	0.0008-0.0021	0.0012-0.0025	0.2348-0.2354	0.2344-0.2350
2010	MR20DE	2.0	45.15-45.45	—	34-39@ 1.39	1.390	0.0008-0.0021	0.0012-0.0025	0.2152 0.2157	0.2148 0.2154
	QR25DE	2.5	45.15-45.45	—	34-39@ 1.39	1.390	0.0008-0.0021	0.0012-0.0025	0.2348-0.2354	0.2344-0.2350

37663_SENT_C0006

CAMSHAFT SPECIFICATIONS

All measurements in inches unless noted

Year	Engine Displacement Liters	Engine Code/ID	Journal Dia.	Brg. Oil Clearance	Shaft End-play	Circle Runout	Lobe Height	
							Intake	Exhaust
2009	2.0	MR20DE	①	② 0.0018	0.0030-0.0060	0.0008	1.7560-1.7635	1.6997-1.7072
	2.5	QR25DE	③	⑤ 0.0034	0.0045-0.0074	0.0016	1.7644-1.7718	1.7313-1.7388
2010	2.0	MR20DE	①	② 0.0018	0.0030-0.0060	0.0008	1.7560-1.7635	1.6997-1.7072
	2.5	QR25DE	③	⑤ 0.0034	0.0045-0.0074	0.0016	1.7644-1.7718	1.7313-1.7388

NA: Not Available

① No. 1: 1.0098-1.1006
  All others: 0.9823-0.9831
② No. 1: 0.0018-0.0034
  All others: 0.0012-0.0028
③ No. 1: 1.0998-1.1006
  All others: 0.9926-0.9234

④ No. 1: 1.0211-1.0218
  All others: 0.9230-0.9238
⑤ No. 1: 0.0018-0.0034
  All others: 0.0014-0.0030

37663_SENT_C0007

CRANKSHAFT AND CONNECTING ROD SPECIFICATIONS

All measurements are given in inches.

Year	Engine Displacement Liters	Engine ID/VIN	Crankshaft				Connecting Rod		
			Main Brg. Journal Dia.	Main Brg. Oil Clearance	Shaft End-play	Thrust on No.	Journal Diameter	Oil Clearance	Side Clearance
2009	2.0	MR20DE	2.0457-2.0464	①	0.0039-0.1020	3	1.7305-1.7311	0.0015-0.0019	0.0079-0.0138
	2.5	QR25DE	2.1636-2.1645	②	0.0039-0.0102	3	1.7699-1.7706	0.0014-0.0018	0.0039-0.0102
2010	2.0	MR20DE	2.0457-2.0464	①	0.0039-0.1020	3	1.7305-1.7311	0.0015-0.0019	0.0079-0.0138
	2.5	QR25DE	2.1636-2.1645	②	0.0039-0.0102	3	1.7699-1.7706	0.0014-0.0018	0.0039-0.0102

① Nos. 1, 4, 5 : 0.0009-0.0013
  Nos. 2, 3 : 0.0005-0.0009
② Nos. 1, 3, 5 : 0.0005-0.0009
  Nos. 2, 4 : 0.0007-0.0011

37663_SENT_C0008

PISTON AND RING SPECIFICATIONS

All measurements are given in inches.

Year	Engine Displacement Liters	Engine ID/VIN	Piston Clearance	Ring Gap			Ring Side Clearance		
				Top Compression	Bottom Compression	Oil Control	Top Compression	Bottom Compression	Oil Control
2009	2.0	MR20DE	0.0004-0.0012	0.0080-0.0120	0.0200-0.0260	0.0060-0.0180	0.0020-0.0030	0.0010-0.0030	0.0010-0.0070
	2.5	QR25DE	0.0004-0.0012	0.0083-0.0122	0.0146-0.0205	0.0079-0.0177	0.0016-0.0031	0.0012-0.0028	0.0018-0.0049
2010	2.0	MR20DE	0.0004-0.0012	0.0080-0.0120	0.0200-0.0260	0.0060-0.0180	0.0020-0.0030	0.0010-0.0030	0.0010-0.007
	2.5	QR25DE	0.0004-0.0012	0.0083-0.0122	0.0146-0.0205	0.0079-0.0177	0.0016-0.0031	0.0012-0.0028	0.0018-0.0049

37663_SENT_C0009

TORQUE SPECIFICATIONS

All readings in ft. lbs.

Year	Engine Displacement Liters	Engine ID/VIN	Cylinder Head Bolts	Main Bearing Bolts	Rod Bearing Bolts	Crankshaft Damper Bolts	Flywheel Bolts	Manifold		Spark Plugs	Oil Drain Plug
								Intake	Exhaust		
2009	2.0	MR20DE	①	②	③	④	80	20	25	14	25
	2.5	QR25DE	⑤	⑥	⑦	④	76-83	13-15	29-32	18	25
2010	2.0	MR20DE	①	②	③	④	80	20	25	14	25
	2.5	QR25DE	⑤	⑥	⑦	④	76-83	13-15	29-32	18	25

① Step 1: Tighten all in sequence to 30 ft. lbs.

   Step 2: Tighten an additional 100 degrees

   Step 3: Loosen all bolts in reverse order

   Step 4: Repeat steps 1 and 2

   Step 5: Tighten an additional 100 degrees

② Step 1: 25 ft. lbs.

   Step 2: Tighten an additional 60 degrees

③ Step 1: 20 ft. lbs.

   Step 2: loosen bolts

   Step 3: 14 ft. lbs.

④ Step 1: 29-36 ft. lbs.

   Step 2: 60-66 degrees

⑤ Step 1: 72 ft. lbs.

   Step 2: Loosen completely, then retorque to 26-32 ft. lbs.

   Step 3: Turn each bolt, in sequence, an additional 75-80 degrees

   Step 4: Turn each bolt, in sequence, an additional 75-80 degrees

⑥ Tighten bolts 11-22 to 19 ft. lbs.

   Step 2: Tighten bolts 1-10 to 29 ft. lbs.

   Step 3: Tighten bolts 1-10 an additional 60 degrees

⑦ Step 1: Tighten bolts to 20 ft. lbs.

   Step 2: Loosen all bolts

   Step 3: tighten bolts to 14 ft. lbs.

   Step 4: Tighten bolts an additional 90 degrees

37663_SENT_C0010

WHEEL ALIGNMENT

Year	Model		Caster		Camber		Toe-in (in.)
			Range (+/-Deg.)	Preferred Setting (Deg.)	Range (+/-Deg.)	Preferred Setting (Deg.)	
2009	Sentra	F	0.75	+4.92	0.75	0.17	0.00 +/- 0.04
		R	—	—	0.50	-1.50	0.0197 +/- 0.56
2010	Sentra	F	0.75	+4.92	0.75	0.17	0.00 +/- 0.04
		R	—	—	0.50	-1.50	0.0197 +/- 0.56

37663_SENT_C0011

TIRE, WHEEL AND BALL JOINT SPECIFICATIONS

Year	Model	OEM Tires		Tire Pressures (psi)		Wheel Size	Lug Nut Torque (ft. lbs.)
		Standard	Optional	Front	Rear		
2009	Sentra	P205/60HR15	P205/55HR16	33	30	6.5-JJ	80
	Sentra SE-R	P215/45ZR17	None	33	33	7-JJ	80
2010	Sentra	P205/60HR15	P205/55HR16	33	30	6.5-JJ	80
	Sentra SE-R	P215/45ZR17	None	33	33	7-JJ	80

OEM: Original Equipment Manufacturer

PSI: Pounds Per Square Inch

37663_SENT_C0012

BRAKE SPECIFICATIONS

All measurements in inches unless noted

Year	Model		Brake Disc			Brake Drum Diameter		Minimum Lining Thickness		Brake Caliper	
			Original Thickness	Minimum Thickness	Maximum Run-out	Original Inside Diameter	Max. Wear Limit	Front	Rear	Bracket Bolts (ft. lbs.)	Mounting Bolts (ft. lbs.)
2009	Sentra	F	0.945	0.866	0.001		—	0.079	—	122	20
		R	—	—	—	9.000	9.079	—	0.059	—	—
	Sentra SE-R	F	1.181	1.118	0.002	—	—	0.079	—	112	—
		R	0.350	0.315	0.003	—	—	—	0.059	28-38	16-23
2010	Sentra	F	0.945	0.866	0.001		—	0.079	—	122	20
		R	—	—	—	9.000	9.079	—	0.059	—	—
	Sentra SE-R	F	1.181	1.118	0.002	—	—	0.079	—	112	—
		R	0.350	0.315	0.003	—	—	—	0.059	28-38	16-23

37663_SENT_C0013

SCHEDULED MAINTENANCE INTERVALS
Nissan—Sentra

TO BE SERVICED	TYPE OF SERVICE	VEHICLE MILEAGE INTERVAL (x1000)												
		7.5	15	22.5	30	37.5	45	52.5	60	67.5	75	82.5	90	97.5
Engine oil & filter	R	✓	✓	✓	✓	✓	✓	✓	✓	✓	✓	✓	✓	✓
Brake lines & cables	S/I		✓		✓		✓		✓		✓		✓	
Brake pads, discs, drums & linings	S/I		✓		✓		✓		✓		✓		✓	
Driveshaft boots	S/I		✓		✓		✓		✓		✓		✓	
Exhaust system	S/I				✓				✓				✓	
Transaxle fluid	S/I		✓		✓		✓		✓		✓		✓	
Air cleaner filter	R				✓				✓				✓	
Spark plugs (except platinum)	R				✓				✓				✓	
Spark plugs (iridium and platinum)	R	Replace every 105,000 miles												
Steering gear & linkage, axle & suspension parts	S/I				✓				✓				✓	
Engine coolant	R	Replace every 60,000 miles, then every 30,000 miles												
Inverter coolant	R	Replace every 60,000 miles, then every 30,000 miles												
Drive belts	S/I								✓					
Fuel lines	S/I								✓					
Vapor lines	S/I								✓					
Cabin microfilter	R		✓		✓		✓		✓		✓		✓	
Valve adjustment	S/I	As needed												

R: Replace   S/I: Service or Inspect

FREQUENT OPERATION MAINTENANCE (SEVERE SERVICE)

If a vehicle is operated under any of the following conditions it is considered severe service:

- Extremely dusty areas.

- 50% or more of the vehicle operation is in 32°C (90°F) or higher temperatures, or constant operation in temperatures below 0°C (32°F).

- Prolonged idling (vehicle operation in stop and go traffic).

- Frequent short running periods (engine does not warm to normal operating temperatures).

- Police, taxi, delivery usage or trailer towing usage.

Oil & oil filter: change every 3750 miles.

Brake pads & discs: service or inspect every 7500 miles.

Driveshaft boots: service or inspect every 7500 miles.

Exhaust system: service or inspect every 7500 miles.

Steering gear & linkage, axle & suspension parts: service or inspect every 7500 miles.

Steering linkage ball joints & front suspension ball joints: service or inspect every 7500 miles.

Air cleaner filter: service or inspect every 15,000 miles.

37663_SENT_C0014

BRAKES — INFORMATION AND PRECAUTIONS

ANTI-LOCK SYSTEMS

- Certain components within the ABS system are not intended to be serviced or repaired individually.
- Do not use rubber hoses or other parts not specifically specified for and ABS system. When using repair kits, replace all parts included in the kit. Partial or incorrect repair may lead to functional problems and require the replacement of components.
- Lubricate rubber parts with clean, fresh brake fluid to ease assembly. Do not use shop air to clean parts; damage to rubber components may result.
- Use only DOT 3 brake fluid from an unopened container.
- If any hydraulic component or line is removed or replaced, it may be necessary to bleed the entire system.
- A clean repair area is essential. Always clean the reservoir and cap thoroughly before removing the cap. The slightest amount of dirt in the fluid may plug an orifice and impair the system function. Perform repairs after components have been thoroughly cleaned; use only denatured alcohol to clean components. Do not allow ABS components to come into contact with any substance containing mineral oil; this includes used shop rags.
- The Anti-Lock control unit is a microprocessor similar to other computer units in the vehicle. Ensure that the ignition switch is **OFF** before removing or installing controller harnesses. Avoid static electricity discharge at or near the controller.
- If any arc welding is to be done on the vehicle, the control unit should be unplugged before welding operations begin.

DISC AND DRUM SYSTEMS

✳✳ CAUTION

Dust and dirt accumulating on brake parts during normal use may contain asbestos fibers from production or aftermarket brake linings. Breathing excessive concentrations of asbestos fibers can cause serious bodily harm. Exercise care when servicing brake parts. Do not sand or grind brake lining unless equipment used is designed to contain the dust residue. Do not clean brake parts with compressed air or by dry brushing. Cleaning should be done by dampening the brake components with a fine mist of water, then wiping the brake components clean with a dampened cloth. Dispose of cloth and all residue containing asbestos fibers in an impermeable container with the appropriate label. Follow practices prescribed by the Occupational Safety and Health Administration (OSHA) and the Environmental Protection Agency (EPA) for the handling, processing, and disposing of dust or debris that may contain asbestos fibers.

BRAKES — BLEEDING THE BRAKE SYSTEM

BLEEDING PROCEDURE

✳✳ CAUTION

Be careful not to splash brake fluid on painted areas; it may cause paint damage. If brake fluid is splashed on painted areas, wash it away with water immediately. All hoses must be free from excessive bending, twisting and pulling.

1. While bleeding, pay attention to master cylinder fluid level.
2. Disconnect ABS actuator and the hydraulic electric unit or disconnect the negative (-) battery cable.
3. Connect a vinyl tube to the rear right bleed valve.
4. Fully depress brake pedal 4 to 5 times.
5. With brake pedal depressed, loosen bleed valve to let the air out, and then tighten it immediately.
6. Repeat until no more air comes out.
7. Tighten bleed to 61–78 inch lbs. (7–9 Nm).
8. Following these steps, with master cylinder reservoir filled at least half way, bleed the remaining brake cylinders in order: front left, rear left, and front right.

BRAKES   ANTI-LOCK BRAKE SYSTEM (ABS)

WHEEL SPEED SENSORS

REMOVAL & INSTALLATION

✳✳ WARNING

As much as possible, avoid rotating wheel sensor and avoid pulling on sensor harness. Take care to avoid damaging wheel sensor edges or rotor teeth. Remove wheel sensor before removing wheel hub.

1. Raise and safely support the vehicle and remove the wheel.

2. Disconnect wiring and remove bolt.

3. Pull sensor straight out.

To install:

4. Make sure there is no rust or foreign material such on or in the mounting hole of the wheel sensor. Make sure no foreign material has been caught in the sensor rotor. Remove any foreign material and clean the mount.

5. Press rubber grommets of strut bracket and body all the way in until they lock. Harness should not be twisted after

installation. Install with harness paint mark on body side grommet facing front of vehicle, and the strut side grommet facing outside of vehicle.

6. When installing rear wheel sensor, press rubber grommets of suspension arm bracket and harness of side member all the way in until they lock Harness should not be twisted after installation. (Aim the paint mark upward towards vehicle.)

7. Torque front sensor bolts to 80 inch lbs. (9 Nm).

8. Torque rear sensor bolts to 8 ft. lbs. (11 Nm).

BRAKES   FRONT DISC BRAKES

BRAKE CALIPER

REMOVAL & INSTALLATION

See Figure 1.

1. Raise and safely support vehicle and remove tires.

2. Secure disc rotor using wheel nuts.

✳✳ WARNING

Put matching marks on wheel hub assembly and disc rotor if it is necessary to remove rotor.

3. Attach a tube to the brake bleeder and drain brake fluid from caliper.

4. Remove union bolt, and then remove brake hose from caliper assembly.

5. Remove mounting bolts from torque member and remove caliper assembly from vehicle.

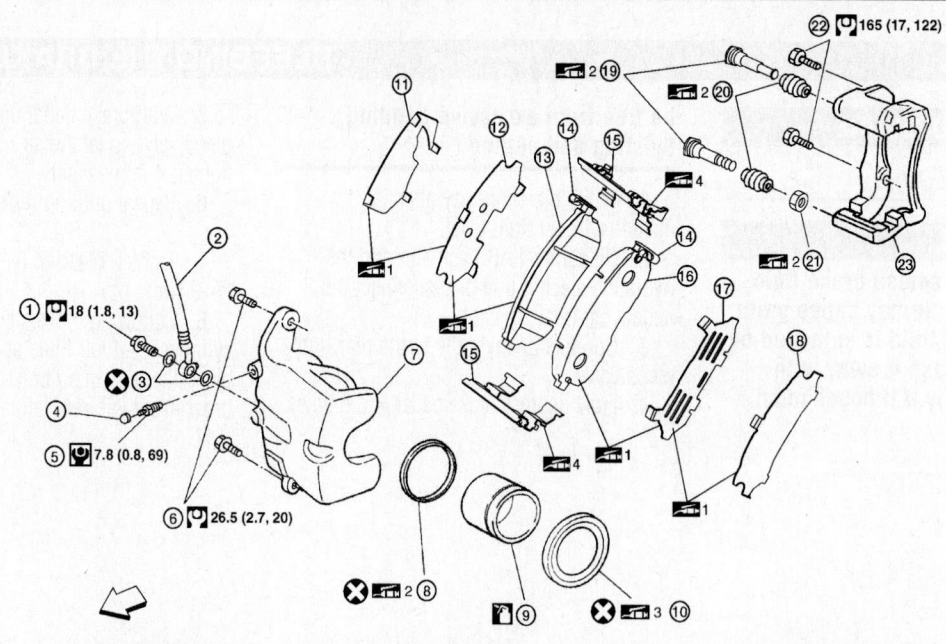

1.	Union bolt	2.	Brake hose	3.	Copper washer
4.	Cap	5.	Bleed valve	6.	Sliding pin bolt
7.	Cylinder body	8.	Piston seal	9.	Piston
10.	Piston boot	11.	Inner shim cover	12.	Inner shim
13.	Inner pad	14.	Pad wear sensor	15.	Pad retainer
16.	Outer pad	17.	Outer shim	18.	Outer shim cover
19.	Sliding pin	20.	Sliding pin boot	21.	Bushing
22.	Torque member mounting bolt	23.	Torque member	⇐	: Front
🅱	: Brake fluid	⎰1: M-77 grease		⎰2: Rubber grease

⎰3: Polyglycol ether based lubricant   ⎰4: M-7439 grease

Fig. 1 Front brake caliper assembly

22140_ALTI_G0001

To install:

> ※※ **WARNING**
>
> **Before installing torque member, wipe oil and grease from mounting surface of steering knuckle and torque member.**

6. Install torque member. On Sentra, tighten torque member mounting bolts to 122 ft. lbs. (165 Nm).
7. Install brake hose to caliper assembly.
8. Refill with new brake fluid and bleed air.
9. Check front disc brake for drag.
10. Install tires to the vehicle.

DISC BRAKE PADS

REMOVAL & INSTALLATION

1. Raise and safely support vehicle and remove front wheels.

2. Remove lower sliding pin bolt.
3. Remove caliper and hang from body with a wire. Do not let the cylinder hang by the hose.
4. Remove pads, shims and pad retainers from torque member.

> ※※ **WARNING**
>
> **When removing pad retainer from torque member, lift pad retainer in the direction shown by arrow, so as not to deform it.**

To install:

5. Apply Molykote M-77 grease or equivalent to the shims. Install shims to pads.

> ※※ **WARNING**
>
> **Securely install shims according to mounting direction of pads.**

6. Apply grease to pad contact surface on pad retainers. Install pad retainers and pads to the torque member.

> ※※ **WARNING**
>
> **When installing pad retainer, attach it firmly so that it is not lifted up from torque member.**

7. Check the brake fluid level in the reservoir because fluid level will rise when pressing piston in.
8. Use a disc brake piston tool (commercial service tool) to easily press to piston in.
9. Install cylinder body to torque member.
10. Install lower sliding pin bolt (lower side), and tighten it to 20 ft. lbs. (26 Nm).
11. Pump the brake pedal and check the master cylinder reservoir. Add fluid if necessary.

BRAKES   REAR DRUM BRAKES

BRAKE DRUM

REMOVAL & INSTALLATION

1. Raise and safely support vehicle and remove rear wheel.
2. Slide the drum off the studs. If the drum is difficult to remove:
   - Press up on adjuster lever with a wire or thin rod through adjuster plug hole on the back plate.
   - Turn adjuster with a flat bladed screwdriver to retract the brake shoes.
3. After installing brake drum, adjust brake shoes.

BRAKE SHOES

REMOVAL & INSTALLATION

See Figure 2.

1. Raise and safely support vehicle and remove rear wheels.
2. With the parking brake lever released, remove the brake drum.
3. While pushing and rotating the retainer, pull out shoe hold pin and remove shoe assembly.

> ※※ **WARNING**
>
> **Do not damage the wheel cylinder boot.**

4. Remove the parking brake cable from the operating lever. Do not bend the parking brake cable.
5. Disassemble the shoe assembly (shoe, springs, adjuster, adjuster lever).

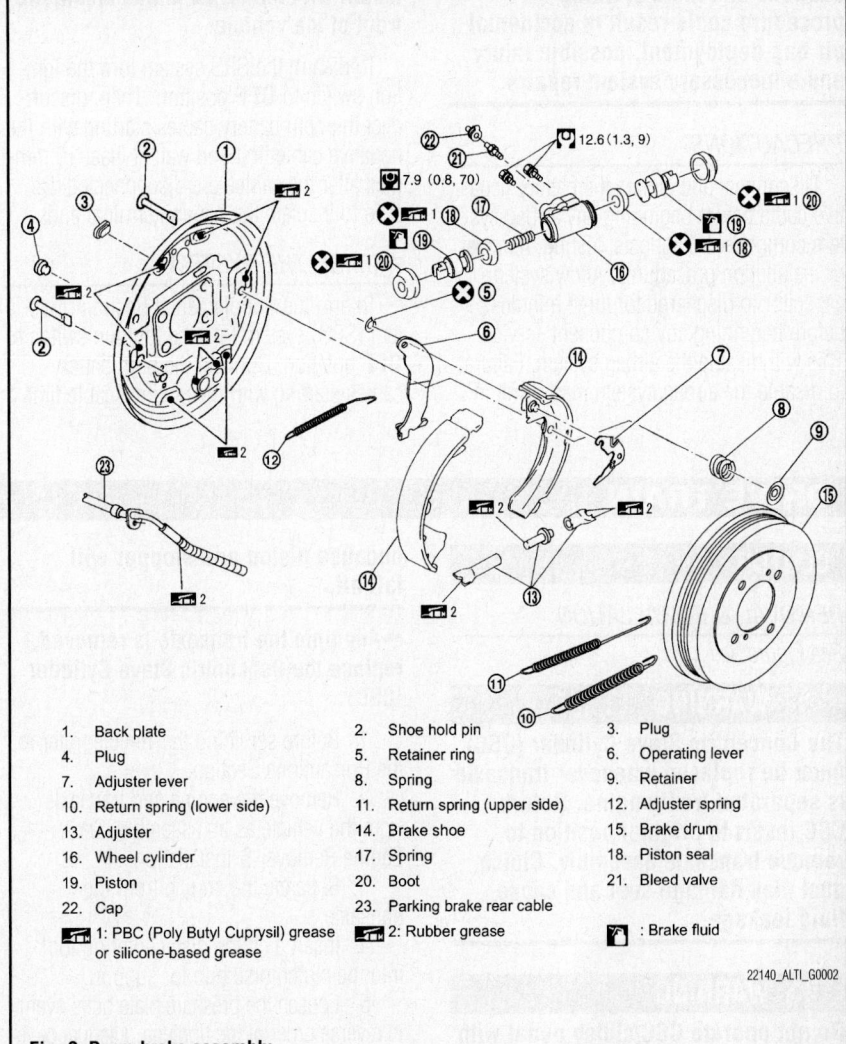

1. Back plate	2. Shoe hold pin	3. Plug
4. Plug	5. Retainer ring	6. Operating lever
7. Adjuster lever	8. Spring	9. Retainer
10. Return spring (lower side)	11. Return spring (upper side)	12. Adjuster spring
13. Adjuster	14. Brake shoe	15. Brake drum
16. Wheel cylinder	17. Spring	18. Piston seal
19. Piston	20. Boot	21. Bleed valve
22. Cap	23. Parking brake rear cable	

🔳1: PBC (Poly Butyl Cuprysil) grease or silicone-based grease   🔳2: Rubber grease   🔳 : Brake fluid

22140_ALTI_G0002

Fig. 2 Drum brake assembly

6. Remove retainer ring to separate operating lever from brake shoe.

To install:

7. If parking brake operating lever is removed, install operating lever to brake shoe. Install and crimp retainer ring.

8. Apply small amount of brake grease to brake shoe sliding surfaces (the worn areas) on the back plate.

9. Apply brake grease to adjuster screw and assemble the screw. (Screws are different for right and left wheels.)

10. Assemble the shoe, adjuster, adjuster lever and springs to the shoe assembly.

11. Connect the parking brake rear cable to the operating lever.

12. Install the shoe assembly. After assembly, be sure that each part is installed properly.

13. Install the brake drum.

14. Depress brake pedal several times.

15. Adjust clearance of brake shoe. See "Parking Brake Adjustment."

16. Install wheels.

CHASSIS ELECTRICAL

AIR BAG (SUPPLEMENTAL RESTRAINT SYSTEM)

GENERAL INFORMATION

✳ CAUTION

These vehicles are equipped with an air bag system. The system must be disarmed before performing service on, or around, system components, the steering column, instrument panel components, wiring and sensors. Failure to follow the safety precautions and the disarming procedure could result in accidental air bag deployment, possible injury and unnecessary system repairs.

PRECAUTIONS

Disconnect and isolate the battery negative cable before beginning any airbag system component diagnosis, testing, removal, or installation procedures. Allow system capacitor to discharge for three minutes before beginning any component service. This will disable the airbag system. Failure to disable the airbag system may result in

accidental airbag deployment, personal injury, or death.

DISARMING THE SYSTEM

➡All Supplemental Restraint System (SRS) electrical wiring harnesses and connectors are covered with YELLOW outer insulation. Do not use electrical test equipment on any circuit related to the SRS (air bag) sensors. When installing SRS components, always install with the arrow marks facing the front of the vehicle.

To disarm the SRS system turn the ignition switch to **OFF** position. Then, disconnect the both battery cables starting with the negative cable first and wait at least 10 minutes after the cables are disconnected. Be sure to insulate the battery terminal ends.

ARMING THE SYSTEM

To arm the Supplemental Restraint System (SRS) system turn the ignition switch to **OFF** position. Connect the both battery cables starting with the positive cable first.

➡The SRS or air bag system is equipped with a self-diagnostic operation. After turning the ignition key to the ON or START position, the AIR BAG warning lamp will illuminate for 7 seconds. After 7 seconds, the AIR BAG lamp will extinguish if no malfunction is detected. If the AIR BAG lamp does not extinguish after 7 seconds, check the SRS self-diagnostic system for a malfunction.

CLOCKSPRING CENTERING

1. Slowly turn the clockspring (spiral spring) clockwise till it stops.

2. Turn it counterclockwise about 2.0 turns, then stop turning at the point where the alignment arrows are directly across from each other.

3. Rotate the clockspring slightly as needed so the locating pin is positioned at the top.

➡A new clockspring comes in the neutral position with a stopper clip in place.

DRIVE TRAIN

CLUTCH

REMOVAL & INSTALLATION

See Figure 3.

✳ WARNING

The Concentric Slave Cylinder (CSC) must be replaced whenever transaxle is separated from engine. Return CSC insert to original position to remove transaxle assembly. Clutch dust may damage seal and cause fluid leakage.

✳ WARNING

Do not operate CSC/clutch pedal with transaxle and engine separated

because piston and stopper will fall off.

➡Any time the transaxle is removed, replace the Concentric Slave Cylinder (CSC).

1. Before servicing the vehicle, refer to the Precautions Section.

2. Remove the engine and transaxle from the vehicle as an assembly. See Engine Removal & Installation.

3. Separate the engine from the transaxle.

4. Insert a clutch disc centering tool into the clutch disc hub for support.

5. Loosen the pressure plate bolts evenly in reverse order of the tightening sequence, a little at a time to prevent distortion

6. Remove clutch assembly.

7. Remove CSC from clutch housing.

To install:

8. Install a new CSC and torque the bolts to 15 ft. lbs. (20 Nm).

9. Apply a light coating of chassis lube to the splines on the transaxle input shaft.

✳ WARNING

Keep grease off CSC to prevent fluid leakage.

10. Fit the clutch disc in place using the centering tool.

11. Install the pressure plate and all bolts finger tight.

12. Torque the pressure plate bolts

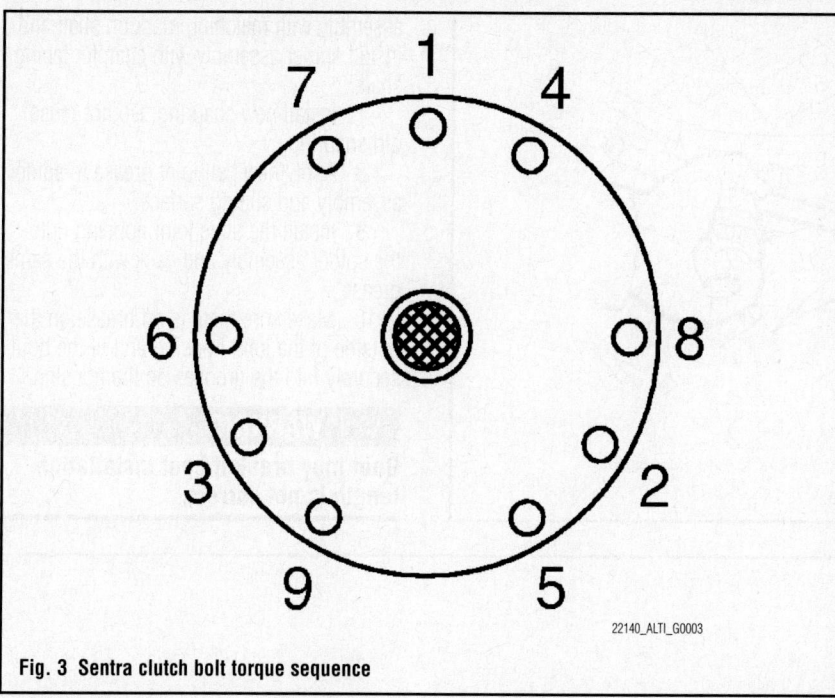

Fig. 3 Sentra clutch bolt torque sequence

in a crisscross pattern, first to 11 ft. lbs. (15 Nm), then to 19 ft. lbs. (25 Nm).
- Remove the clutch disc centering tool.
- Attach transaxle to engine. Carefully fit input shaft through clutch disc to avoid distorting/damaging clutch cover springs. Torque bolts to 46 ft. lbs. (62 Nm).

➡**DO NOT draw the transaxle to the engine with the bolts. This may damage the clutch and/or transaxle.**

13. Install or connect the following:
14. Install Engine/Transaxle assembly into the vehicle.
15. Bleed the clutch cylinders.

BLEEDING

See Figure 4.

❈❈ **WARNING**

Two people are required to bleed the clutch cylinder. Do not use vacuum assist or any type of power bleeder. It will not purge all the air from this system.

1. Fill master cylinder reservoir with new brake fluid.
2. Connect a clear tube to the bleeding connector on the Concentric Slave Cylinder (CSC).
3. Push and release the clutch pedal slowly and fully 15 times, waiting 3 seconds between each cycle.
4. Push in the lock pin on the bleeding connector and hold it in.

❈❈ **WARNING**

Hold the lock pin in to prevent the bleeding connector from separating when fluid pressure is applied.

5. Slide the bleeding connector out ³⁄₁₆ (5mm), then press the clutch pedal and hold it down.
6. Push the bleeding connector back in and release the clutch pedal.
7. Repeat until no bubbles are observed in the fluid flow.

FRONT HALFSHAFT

REMOVAL & INSTALLATION

See Figures 5 and 6.

1. Raise and safely support the vehicle and remove the wheel.

2. Remove wheel speed sensor from steering knuckle.

❈❈ **WARNING**

Do not pull on sensor wiring harness.

3. Remove cotter pin, then loosen hub lock nut. Temporarily leave the nut installed to prevent damage to threads.
4. Remove nuts and bolts securing steering knuckle to strut assembly. Alternately, remove the bolt to separate the lower ball joint from the steering knuckle.
5. Separate the halfshaft from the wheel hub and bearing assembly by lightly tapping the end of the shaft using a hammer and brass drift or wood block, then remove hub lock nut.

➡**Use a suitable puller if hub and shaft cannot be separated using the above procedure.**

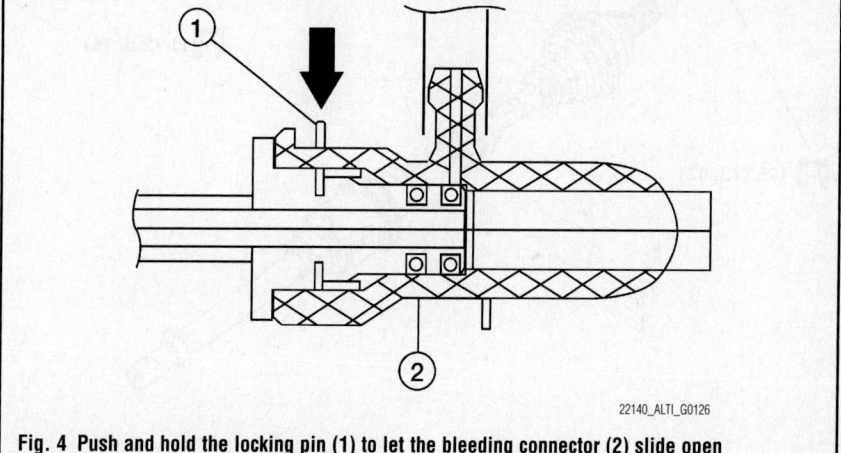

Fig. 4 Push and hold the locking pin (1) to let the bleeding connector (2) slide open

6. Remove the halfshaft from the wheel hub and bearing assembly.

❈❈ **WARNING**

Do not apply an excessive angle to halfshaft joint when removing from the wheel hub and bearing assembly. Do not excessively extend inner joint or allow the shaft to hang by the joint. Support the entire halfshaft.

7. To remove the right side halfshaft, remove the bolts and plate from the support bearing bracket.
8. Carefully pry off halfshaft from transaxle assembly.
9. Installation is the reverse of removal. Torque the castle nut to 92 ft. lbs. (125 Nm) and install a new cotter pin. Torque the

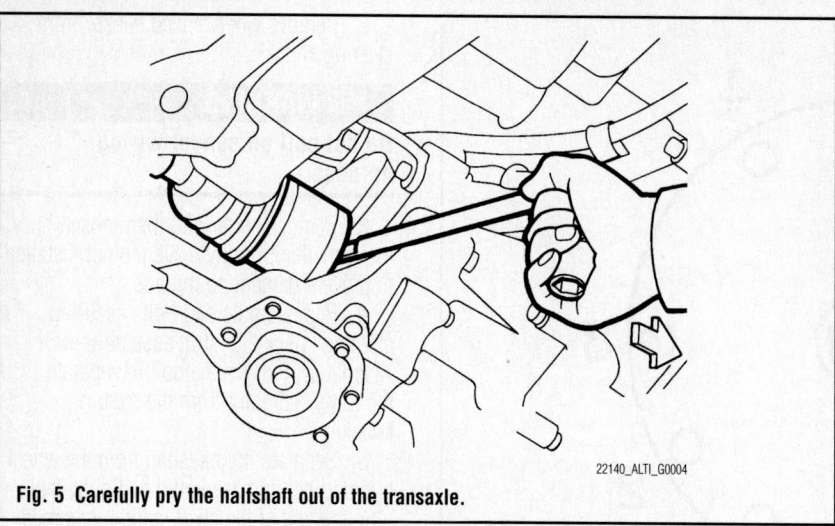

Fig. 5 Carefully pry the halfshaft out of the transaxle.

assembly with matching mark on shaft and install spider assembly with chamfer facing shaft.

7. Install new snapring. Do not reuse old snapring.

8. Apply new ball joint grease to spider assembly and sliding surface.

9. Install the slide joint housing onto the spider assembly and pack with the same grease.

10. Make sure there is no grease on the outside of the joint housing and fit the boot securely into the grooves on the housing.

✴✴ WARNING

Boot may break if boot installation length is not correct.

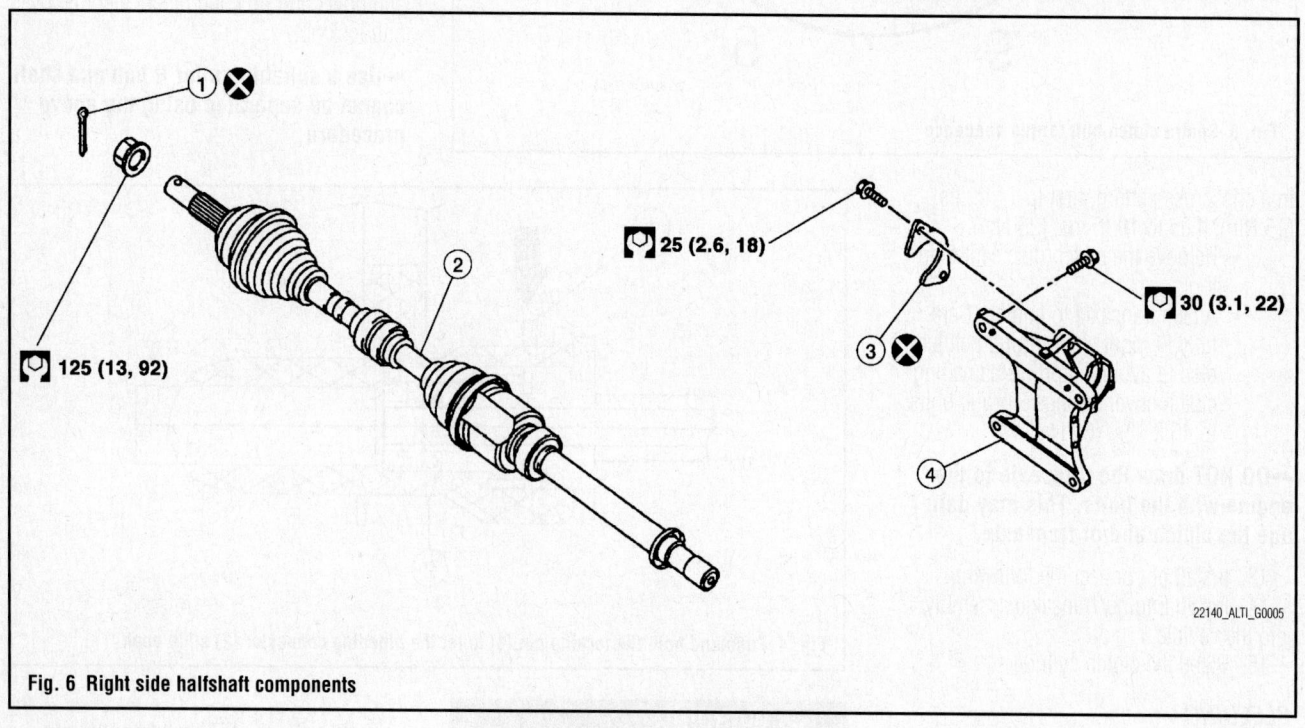

Fig. 6 Right side halfshaft components

steering knuckle bolts to 110 ft. lbs. (150 Nm). Torque the ball joint pinch bolt to 41 ft. lbs. (55 Nm).

CV-JOINT OVERHAUL

Inner Joint

See Figures 7 and 8.

1. Mount halfshaft in a vise.

✴✴ WARNING

When mounting shaft in a vise, always use copper or aluminum plates between vise and shaft.

2. To remove inner joint:
  • Remove boot bands and slide the boot back.

  • Put matching marks on slide joint housing and shaft, then pull out shaft from slide joint housing.
  • Put matching marks on spider assembly and shaft.
  • Remove snap ring, then remove spider assembly from shaft.
  • Remove boot from shaft.

To install:

3. Cover halfshaft spline with tape to prevent damage to boot during installation.

4. Install new boot and new small boot band on shaft. Do not reuse boot or boot band.

5. Remove protective tape from halfshaft spline.

6. Align matching mark on spider

11. Make sure boot installation length "L" is correct.
  • With CVT: L = 7 in. (178 mm)
  • With manual transmission: L = 6.6 in. (168 mm)

12. Insert a flat-bladed screwdriver into the large end of boot to bleed air from boot to prevent boot deformation.

13. Secure large end of boot with new boot band. Do not reuse old boot band.

Outer Joint

1. Remove boot bands and slide the boot back.

2. Screw a slide hammer 1.2 in (30 mm) or more into threaded part of joint. Pull joint straight off of shaft.

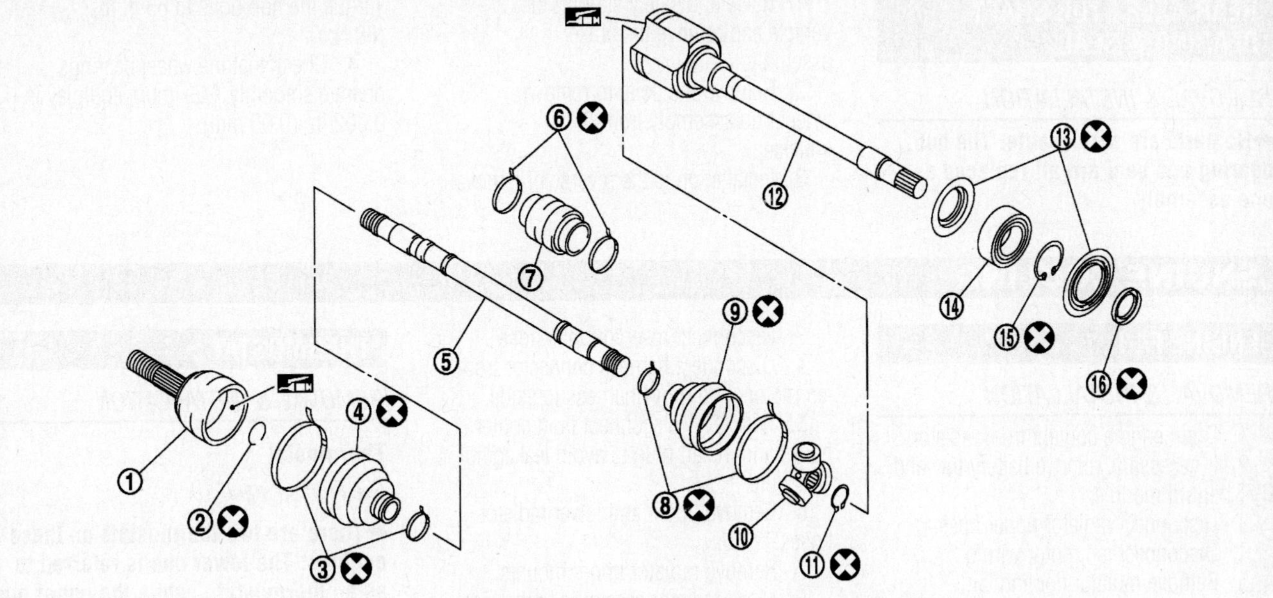

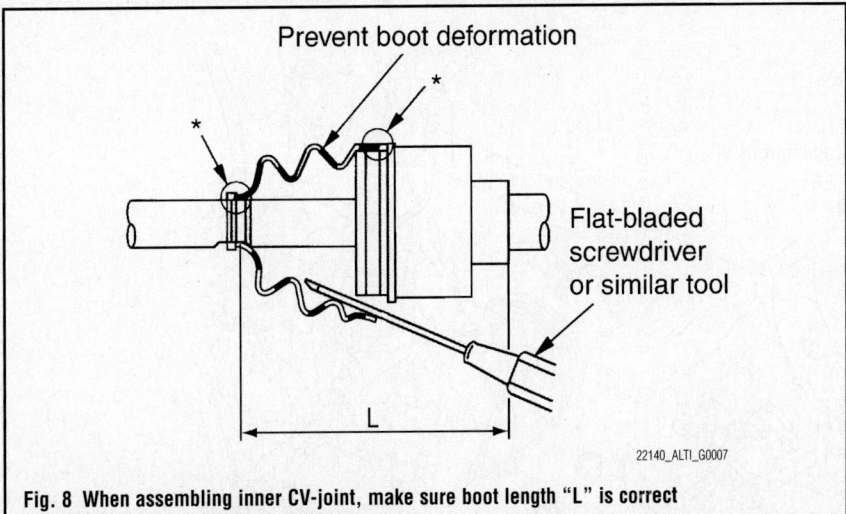

1. Joint sub-assembly
2. Circlip
3. Boot bands
4. Boot
5. Shaft
6. Damper bands
7. Damper
8. Boot band
9. Boot
10. Spider assembly
11. Snap ring
12. Slide joint housing
13. Dust shield
14. Bearing
15. Snap ring
16. Dust shield

22140_ALTI_G0006

Fig. 7 Halfshaft components: inner joint can be rebuilt. Right side halfshaft shown.

Prevent boot deformation

Flat-bladed
screwdriver
or similar tool

L

22140_ALTI_G0007

Fig. 8 When assembling inner CV-joint, make sure boot length "L" is correct

❋❋ WARNING

If joint assembly cannot be removed after five or more unsuccessful attempts, replace the entire halfshaft assembly.

3. Remove circlip from shaft.
4. Remove boot from shaft.

To install:

5. Cover halfshaft spline with tape to prevent damage to the boot.

6. Install a new small boot band on shaft, then the new boot. Do not reuse old boot or boot band.

7. Remove protective tape.

8. Install a new circlip to shaft, making sure it fits securely into the groove. Do not reuse old circlip.

9. Thread castle nut onto joint and fit joint onto shaft.

10. Drive joint home over the circlip with a mallet.

11. Pack the joint with new CV-joint grease. Pack the remainder of the grease into boot.

12. Make sure there is no grease on the outside of the joint housing and fit the boot securely into the grooves on the housing.

❋❋ WARNING

Boot may break if boot installation length is not correct.

13. Make sure boot installation length from groove on the joint to the groove on the shaft is 5.6 in. (142 mm). Insert a flat-bladed screwdriver into the large end of boot to bleed air from boot to prevent boot deformation.

14. Fit new boot bands and tighten securely using the proper tool.

Halfshaft Support Bearing

1. Pry outer dust shield from slide joint assembly using a suitable tool.

2. Remove snap ring.

3. Press support bearing assembly off shaft.

4. Remove dust shield.

5. Installation is the reverse of removal. Use new dust shields and snap ring.

REAR AXLE STUB SHAFT BEARING & SEAL

REMOVAL & INSTALLATION

➥**No parts are serviceable. The hub, bearing and seal are all replaced as one assembly.**

ENGINE COOLING

ENGINE FAN

REMOVAL & INSTALLATION

1. Drain engine coolant from radiator.
2. If necessary, remove battery tray and CVT control module.
3. Disconnect radiator upper hose.
4. Disconnect fan motor wiring.
5. Remove radiator cooling fan assembly.
6. Installation is the reverse of removal.

RADIATOR

REMOVAL & INSTALLATION

1. Drain engine coolant from radiator.
2. Remove air intake duct.

1. Raise and safely support the vehicle and remove the brake assembly.
2. Remove the bolts to remove wheel hub assembly from knuckle.
3. Installation is the reverse of removal.

3. Disconnect reservoir tank hose.
4. Disconnect harness connector from fan motors and move harness to aside.
5. With CVT, disconnect fluid cooler hoses and install plug to avoid leakage of CVT fluid.
6. Remove upper and lower radiator hoses.
7. Remove radiator upper mounts.
8. Move radiator assembly to the rear-ward direction of vehicle, and then lift it upward to remove.

✳✳ WARNING

Do not damage or scratch A/C condenser and radiator core when removing.

9. Installation is the reverse of removal.

Torque the hub bolts to 65 ft. lbs. (88 Nm).
4. Check that the wheel bearings operate smoothly. Maximum end play is 0.002 in. (0.05 mm).

THERMOSTAT

REMOVAL & INSTALLATION

Thermostat

See Figures 9 and 10.

➥**There are two thermostats on these engines. The lower one is referred to as a "thermostat," while the upper one is referred to as a "water control valve."**

1. Before servicing the vehicle, refer to the Precautions Section.
2. Remove front air duct and engine undercover as necessary.
3. Drain the cooling system.

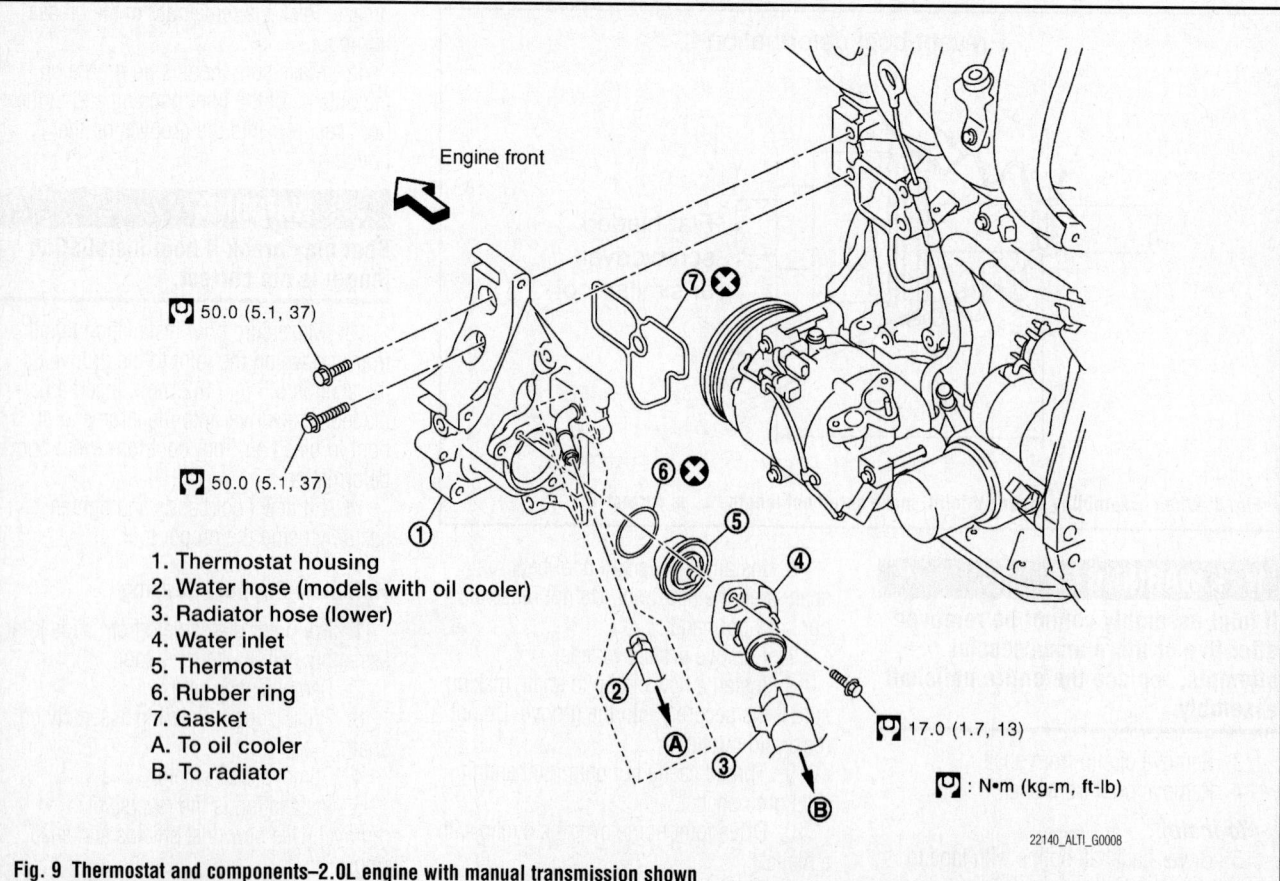

Engine front

50.0 (5.1, 37)

50.0 (5.1, 37)

17.0 (1.7, 13)

1. Thermostat housing
2. Water hose (models with oil cooler)
3. Radiator hose (lower)
4. Water inlet
5. Thermostat
6. Rubber ring
7. Gasket
A. To oil cooler
B. To radiator

: N•m (kg-m, ft-lb)

22140_ALTI_G0008

Fig. 9 Thermostat and components–2.0L engine with manual transmission shown

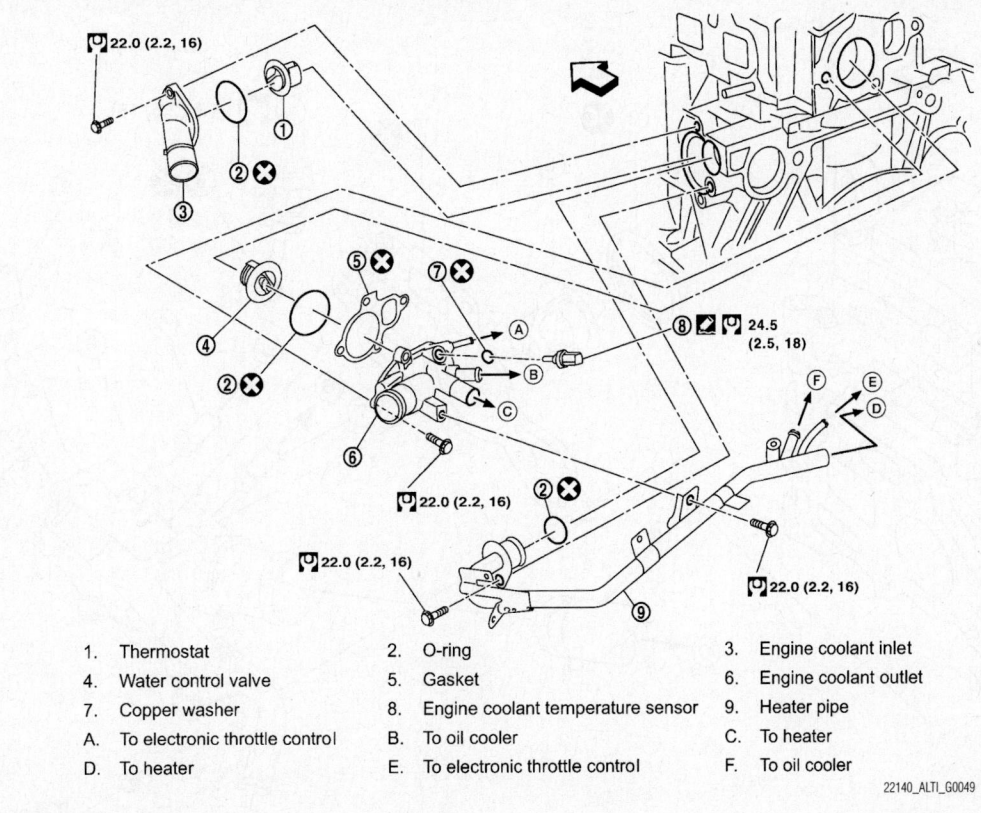

1.	Thermostat	2.	O-ring	3.	Engine coolant inlet
4.	Water control valve	5.	Gasket	6.	Engine coolant outlet
7.	Copper washer	8.	Engine coolant temperature sensor	9.	Heater pipe
A.	To electronic throttle control	B.	To oil cooler	C.	To heater
D.	To heater	E.	To electronic throttle control	F.	To oil cooler

22140_ALTI_G0049

Fig. 10 Thermostat (1) and water valve (4) housings–2.5L engine

4. Remove radiator lower hose.

5. Remove engine coolant inlet and thermostat.

To install:

6. Fit a new rubber ring on the thermostat, making sure the flange seats properly inside the ring.

7. Install the thermostat with the jiggle valve facing upwards. The position deviation may be within the range of +/- 10°.

8. To complete installation, reverse remaining removal procedure

9. Fill the cooling system.

10. After installation, run engine for a few minutes and check for leaks.

Water control Valve

See Figure 11.

❄❄ **WARNING**

Never remove the radiator cap when the engine is hot. Serious burns could occur from high pressure coolant escaping from the radiator.

1. Before servicing the vehicle, refer to the Precautions Section.

2. Drain the cooling system.

3. Remove the air duct and air cleaner.

4. Remove the upper radiator hose, heater hoses and throttle body hoses.

5. Remove the water outlet and remove the water control valve from the cylinder head.

To install:

6. To install, fit a new rubber ring on the water control valve.

7. Fit the valve into the cylinder head with arrow on outer face pointing up. The valve frame should be vertical.

8. Fit a new gasket and install the outlet onto the cylinder head.

9. Reconnect the hoses using new O-rings, gaskets and copper washers as needed.

10. Fill the cooling system.

11. After installation, run engine for a few minutes and check for leaks.

WATER PUMP

REMOVAL & INSTALLATION

2.0L Engine

1. Before servicing the vehicle, refer to the precautions in the beginning of this section.

2. Drain the cooling system.

3. To remove the drive belt tensioner:
- Working underneath the vehicle, place a wrench on the tensioner idler pulley nut.
- Push the wrench clockwise and insert a short screwdriver in the hole that appears to the right of the tensioner to hold the tensioner in place.
- Remove the drive belt

4. Loosen the water pump bolts and remove the pump.

5. Remove all traces of gasket material from sealing surfaces.

To install:

6. Install a new gasket:

7. Install the water pump and torque the bolts to 18 ft. lbs. (25 Nm).

8. Install the drive belt and hold the tensioner with a wrench to remove the screwdriver.

9. Fill the cooling system.

10. Start the vehicle, check for leaks and repair if necessary.

2.5L Engine

1. Before servicing the vehicle, refer to the precautions in the beginning of this section.

2. Drain the cooling system.

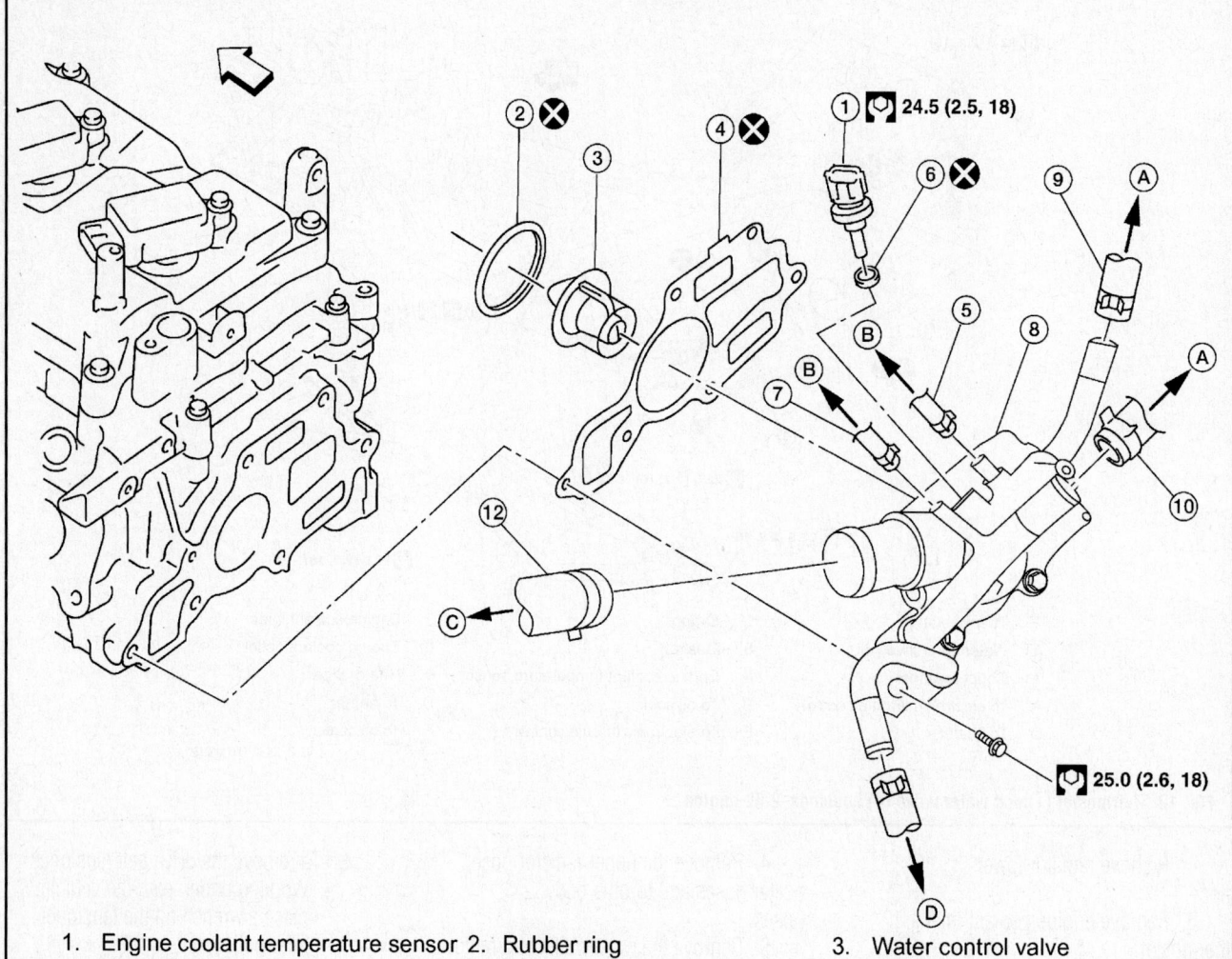

1. Engine coolant temperature sensor
2. Rubber ring
3. Water control valve
4. Gasket
5. Water hose
6. Gasket
7. Water hose
8. Water outlet
9. Heater hose
10. Heater hose
11. Water hose (CVT fluid cooler)
12. Radiator hose (upper)
⟸ Front
A. To heater
B. To electric throttle control actuator
C. To radiator
D. To CVT fluid cooler

22140_ALTI_G0009

Fig. 11 Exploded view of water valve–2.0L engine with CVT shown

3. Remove the engine undercover.
4. A special tool is needed to remove the drive belt tensioner. To remove the tensioner:
   - Working underneath the vehicle, place tool J-46535 on the tensioner idler pulley and push clockwise.

⁂⁂ WARNING

Do not loosen the belt tensioner pulley bolt or turn the tensioner counterclockwise. If the bolt is loosened, the tensioner assembly must be replaced.

   - Insert a short screwdriver into the tensioner retaining boss to lock the tensioner in place
   - Remove the drive belt
5. Remove the coolant reservoir.
6. Remove the Intelligent Power Distribution Module (IPDM) by unlocking the pawls and unplugging the connector.
7. Raise and safely support the vehicle and remove the right front wheel.
8. Remove the inner fender.
9. Remove the ground strap.

➡The alternator and exhaust system may interfere with removal of the water pipe.

10. Loosen the water pump bolts and remove the pump.
11. Remove all traces of gasket material from sealing surfaces.

To install:
12. Install a new gasket:
13. Install the water pump and torque the bolts to 20 ft. lbs. (28 Nm).
14. Install the drive belt and hold the tensioner with the special tool to remove the screwdriver.
15. Reinstall the remaining components.
16. Fill the cooling system.
17. Start the engine, check for leaks and repair if necessary.

ENGINE ELECTRICAL | CHARGING SYSTEM

ALTERNATOR

REMOVAL & INSTALLATION

2.0L Engine

1. Before servicing the vehicle, refer to the precautions in the beginning of this section.
2. Disconnect the negative battery cable
3. Working underneath the vehicle, place a wrench on the tensioner idler pulley nut.
4. Push the wrench clockwise and insert a short screwdriver in the hole that appears to the right of the tensioner to hold the tensioner in place.
5. Remove the drive belt
6. Unplug the alternator connector and disconnect the large B+ cable
7. Remove the bolts to remove the alternator

To install:

8. Install the alternator and torque the bolts to 18 ft. lbs. (25 Nm).

9. Connect the wiring. Tighten the B+ cable carefully.
10. Install the drive belt and hold the tensioner with the wrench to remove the screwdriver.
11. Connect the battery cable.

2.5L Engine

1. Before servicing the vehicle, refer to the Precautions Section.
2. Disconnect negative battery cable
3. Remove engine undercover
4. A special tool is needed to remove the drive belt tensioner. To remove the tensioner:

   - Working underneath the vehicle, place tool J-46535 on the tensioner idler pulley and push clockwise.

> #### ☀ WARNING
>
> **Do not loosen the belt tensioner pulley bolt or turn the tensioner**

counterclockwise. **If the bolt is loosened, the tensioner assembly must be replaced.**

   - Insert a short screwdriver into the tensioner retaining boss to lock the tensioner in place
   - Remove the drive belt

5. Unplug the connector from the alternator and disconnect the B+ cable.
6. Remove the alternator mounting bolts and remove the alternator.

To install:

7. Install the alternator and torque the bolts to 18 ft. lbs. (25 Nm)
8. Connect the wiring. Tighten the B+ cable carefully.
9. Install the drive belt and hold the tensioner with the special tool to remove the screwdriver.
10. Install the engine undercover
11. Connect the negative battery cable

ENGINE ELECTRICAL | IGNITION SYSTEM

FIRING ORDER

2.0L and 2.5L Engines; Firing order: 1–3–4–2
Distributorless ignition system (one coil on each cylinder)

IGNITION COIL

REMOVAL & INSTALLATION

2.0L and 3.5L Engines

1. Remove the intake manifold.
2. Disconnect coil wiring.
3. Remove bolt to remove ignition coil.
4. Installation is reverse of removal.

2.5L Engine

1. Disconnect the ignition coil wiring as required.
2. Remove ignition coil.
3. Installation is the reverse of removal. Make sure coil seals the opening in the rocker cover.

IGNITION TIMING

ADJUSTMENT

2.0L Engine

See Figures 12 and 13.

1. There are two different ways to connect a standard timing light. If equipped with a loop wire as shown, simply attach the pick-up clamp and check ignition timing at the crankshaft pulley.
2. If no loop wire is installed:

   - Remove No. 4 ignition coil
   - Make up a high-tension extension wire using a suitable sparkplug wire.
   - Connect No. 4 ignition coil to No. 4 spark plug with

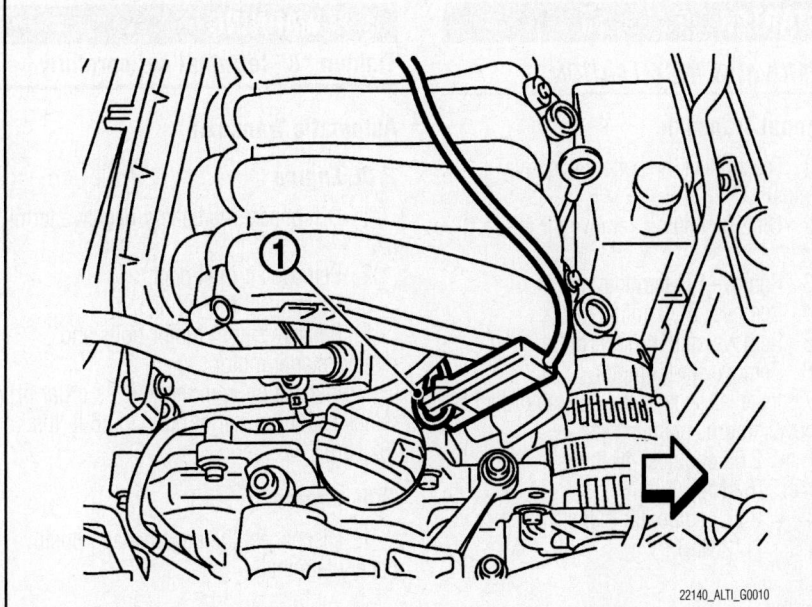

22140_ALTI_G0010

Fig. 12 Some vehicles have a timing loop for connecting a standard timing light

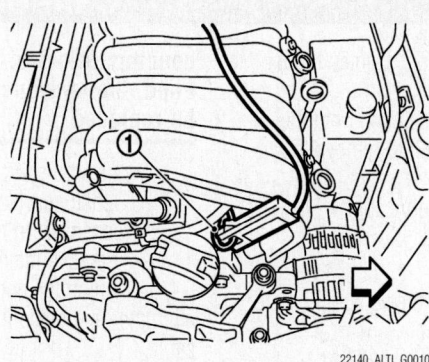

Fig. 13 Add an extension wire between the No. 4 ignition coil and spark plug to connect a timing light

high-tension extension wire and attach timing light clamp to this wire.

3. Check ignition timing at the crankshaft pulley. Timing should be $6° ± 5°$ BTDC (in Neutral), but there is no adjustment.

2.5L Engine

See Figure 14.

1. Slide the wiring harness protector of ignition coil No.1 back (2) to reach the wires.

2. Attach timing light to the ignition coil No.1 wires as shown in the figure.

3. On 3.5L engine, ignition timing should be $18° ± 5°$ BTDC at idle. There is no adjustment.

4. On 2.5L engine, ignition timing should be $15° ± 5°$ BTDC at idle. There is no adjustment.

SPARK PLUGS

REMOVAL & INSTALLATION

1. On 2.0L engines, remove intake manifold

2. Remove ignition coils

3. Remove spark plugs.

4. Installation is the reverse of removal.

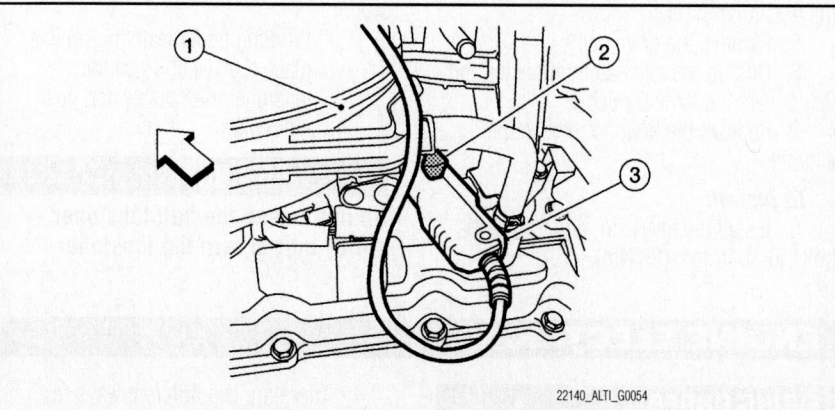

Fig. 14 On 3.5L engine, a timing light pickup can be connected to the No.1 ignition coil on the rear cylinder bank below the intake manifold—3.5L engine

ENGINE ELECTRICAL

STARTING SYSTEM

STARTER

REMOVAL & INSTALLATION

Manual Transaxle

1. Disconnect the battery negative terminal.

2. On 2.0L engine, remove air inlet duct.

3. Remove "S" terminal nut.

4. Remove "B" terminal nut.

5. Remove starter motor bolts.

6. Remove starter motor.

7. Installation is in the reverse order of removal. Torque starter bolts:
- 2.0L engines: 46 ft. lbs. (62 Nm).
- 2.5L engine, 83 ft. lbs. (112 Nm).

✳✳ WARNING

Tighten "B" terminal nut carefully.

Automatic Transaxle

2.0L Engine

1. Disconnect the battery negative terminal.

2. Remove air inlet duct.

3. Disconnect the wiring.

4. Remove starter motor bolts and remove starter motor.

5. Installation is in the reverse order of removal. Torque starter bolts to 46 ft. lbs. (62 Nm).

2.5L Engine

1. Disconnect the negative and positive battery terminal.

2. Remove the air cleaner assembly and air ducts.

3. On 3.5L engine, remove the following:
- ECM
- CVT control unit
- IPDM/ER (fuel/relay box)

4. Remove the battery tray.

5. Disconnect the starter wiring.

6. Remove the two starter bolts and remove the starter.

To install:

7. Fit starter into place and install bolts. Torque to 45 ft. lbs. (62 Nm).

8. Connect the wiring, torque the large cable nut to 86 inch lbs. (10 Nm).

9. Install remaining components and connect the battery.

ENGINE MECHANICAL

ACCESSORY DRIVE BELTS

ACCESSORY BELT ROUTING

See Figures 15 and 16.

Refer to the accompanying illustrations.

INSPECTION

1. Check the drive belt auto-tensioner indicator when the engine is cold.

2. Make sure that the indicator (notch on fixed side) of drive belt auto-tensioner points to range A.

3. Visually check entire drive belt for wear, damage or cracks. If the indicator points outside of range A or the belt is cracked or damaged, replace drive belt.

4. When a new drive belt is installed, the indicator should point to range B.

ADJUSTMENT

The spring-loaded belt tensioner adjusts automatically as required.

REMOVAL & INSTALLATION

1. Working underneath the vehicle, place a wrench on the tensioner idler pulley nut.

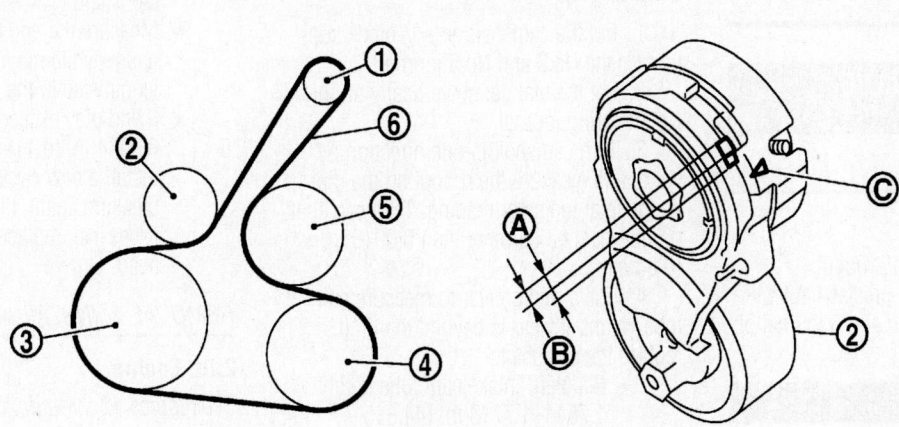

1. Alternator
2. Drive belt auto–tensioner
3. Crankshaft pulley
4. A/C compressor (models with A/C)
   Idler pulley (models without A/C)
5. Water pump
6. Drive belt
A. Possible use range
B. Range when new drive belt is installed
C. Indicator

22140_ALTI_G0012

Fig. 15 Accessory drive belt and automatic tensioner–2.0L engine

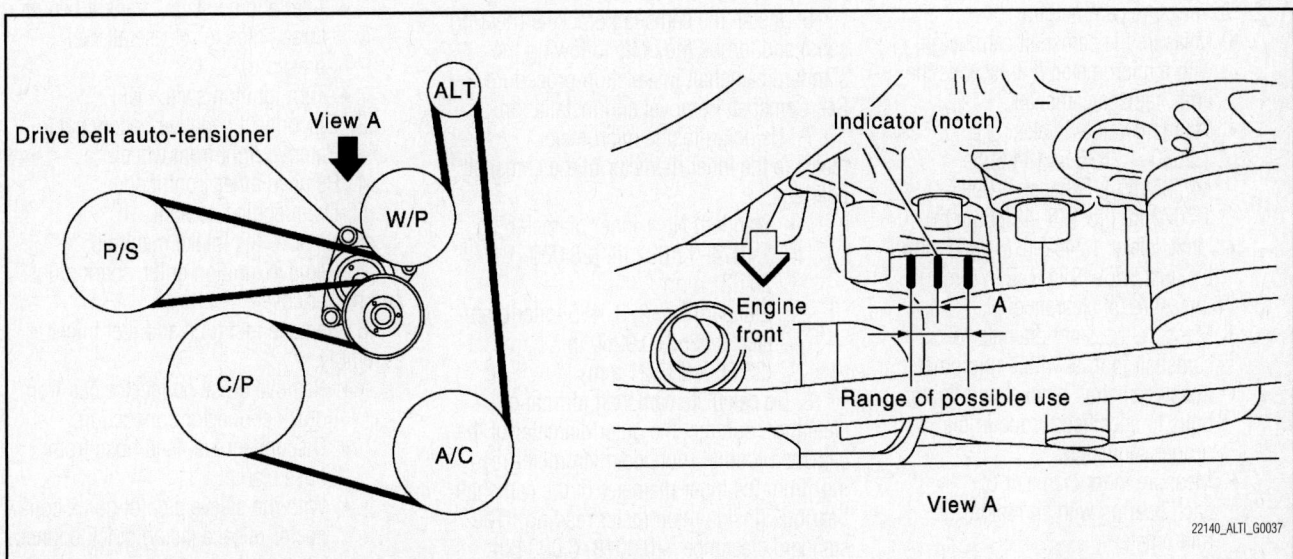

22140_ALTI_G0037

Fig. 16 Accessory drive belt and automatic tensioner–2.5L engine

2. Push the wrench clockwise and insert a short screwdriver in the hole that appears in the tensioner to hold it in place.

3. Remove the drive belt

4. Install the new drive belt and hold the tensioner with the wrench to remove the screwdriver.

✳✳ WARNING

Do not loosen the auto-tensioner pulley bolt. (Do not turn it counterclockwise.) If turned counterclockwise, the complete auto-tensioner must be replaced as a unit, including pulley.

CAMSHAFT AND VALVE LIFTERS

INSPECTION

2.0L Engine

1. To measure camshaft runout:
   • Put V-block on a precise flat table, and support No. 2 and 5 journal of camshaft.

✳✳ WARNING

Never support No. 1 journal (on the side of camshaft sprocket) because it has a different diameter from the other four locations.

   • Set dial indicator (A) vertically to No. 3 journal.
   • Turn camshaft to one direction with hands, and measure the camshaft runout on dial indicator.
   • If runout exceeds 0.0020 in. (0.05mm), replace camshaft.
2. To measure cam height:
   • Measure the camshaft cam height with a micrometer. If it exceeds the limit, replace camshaft.
   • Standard value: Intake, 1.7560–1.7635 in. (44.605–44.795 mm): Exhaust, 1.6997–1.7072 in. (43.175–43.365 mm)
   • Limit: Intake, 1.7482 in. (44.405 mm): Exhaust, 1.6919 in. (42.975 mm)
3. To measure oil clearance:
   • Measure the outer diameter of camshaft journal with a micrometer.
   • Install camshaft bracket and tighten bolts to specified torque in the proper sequence.
   • Measure inner diameter of each bearing with an inside micrometer.
   • Standard value: Bearing No. 1, 0.0018–0.0034 in. (0.045–0.086

mm): Remaining bearings, 0.0012–0.0028 in. (0.030–0.071 mm)
   • Limit: 0.0059 in (0.15 mm)

➡**If oil clearance is not correct, the cylinder head must be replaced.**

4. To measure endplay
   • Camshaft endplay can be measured with the valve tappets removed or installed. Standard value, 0.0030–0.0060 in. (0.075–0.153 mm): Limit, 0.0094 in. (0.24 mm)

2.5L Engine

1. Put the camshaft on a V-block supporting the No.2 and No.5 journals.
2. Set the dial gauge vertically on the No.3 bearing journal.
3. Turn camshaft in one direction by hand, and measure the runout on the dial gauge total indicator reading. Total camshaft runout must be no more than 0.0016 in. (0.04 mm).
4. Use a micrometer to measure cam lobe height. If wear is beyond the limit, replace the camshaft.
   • Standard intake cam lobe height: 1.7644–1.7718 in. (44.815–45.005 mm)
   • Standard exhaust cam lobe height: 1.7313–1.7388 in. (43.975–44.165 mm)
5. Use a micrometer to measure the camshaft bearing journal.
   • Standard No.1 bearing journal diameter: 1.0998–1.1006 in. (27.935–27.955 mm)
   • Standard No. 2, 3, 4, 5 bearing journal diameter: 0.9226–0.9234 in. (23.435–23.455 mm)
6. Install the camshaft brackets (bearing caps) and torque the bolts following the standard camshaft installation procedure. See Camshaft Removal and Installation.
7. Using an inside micrometer, measure the inner diameter of the camshaft bearings.
   • Standard No.1 inner diameter: 1.1024–1.1032 in. (28.000–28.021 mm)
   • Standard No. 2, 3, 4, 5 inner diameter: 0.9252–0.9260 in. (23.500–23.521 mm)
8. To calculate camshaft journal oil clearance, subtract the outer diameter of the camshaft journal (outside micrometer reading) from the inner diameter of the camshaft bearings (inside micrometer reading). The standard clearance is 0.0018–0.0034 in. (0.045–0.086 mm)
9. If clearance is out of the specified

range, replace either or both the camshaft and the cylinder head assembly.

➡**Inner diameter of the camshaft bearings is manufactured together with the cylinder head. If the camshaft oil clearance is out of specification, replace the whole cylinder head assembly.**

10. To measure camshaft end play:
    • Install the camshafts with the valve lifters removed and torque the bearing cap bolts as specified.
    • Set a dial gauge in the thrust direction on the front end of the camshaft.
    • Measure the end play with the dial gauge while moving the camshaft lengthwise in the bearings.
    • If end play exceeds 0.0045–0.0074 in. (0.115–0.188 mm), install a new camshaft and measure again. If end play is still excessive, replace the cylinder head.

REMOVAL & INSTALLATION

2.0L Engine

See Figures 17 through 20.

✳✳ CAUTION

Be sure to work in a well ventilated area and keep a CO2 fire extinguisher handy. Do not smoke while servicing fuel system. Keep open flames and sparks away from the work area.

1. Release the fuel pressure.
   • Remove fuel pump fuse.
   • Start engine.
   • After engine stalls, crank it two or three times to release all fuel pressure.
   • Turn ignition switch OFF.
2. Disconnect negative battery cable.
3. Remove right front wheel.
4. Remove inner front fender.
5. Drain engine coolant.
6. Remove the intake manifold:
7. Remove ignition coils, spark plugs and rocker cover.
8. Remove fuel tube and fuel injector assembly
   • Remove quick connector cap from quick connector connection.
   • Disconnect fuel feed hose from hose clamp.
   • With the sleeve side of quick connector release facing quick connector, install quick connector release tool onto fuel tube.

- Insert quick connector release into quick connector until sleeve contacts and goes no further. Hold quick connector release on that position. Inserting quick connector release hard will not disconnect quick connector. Hold quick connector release where it contacts and goes no further.
- Draw and pull out quick connector straight from fuel tube.

➡**Valves are adjusted by replacing the lifters. Now is the time to measure and record valve clearances.**

9. Remove the rocker cover.
10. Remove the timing cover, timing chain and related parts.
- Remove Camshaft Position (CMP) sensor from camshaft bracket.

※※ **WARNING**

Handle camshaft sensor carefully to avoid dropping and shocks. Never

disassemble. Never allow metal powder to adhere to magnetic part at sensor tip. Never place sensor in a location where it is exposed to magnetism.

11. Align the match marks (A) on the intake camshaft sprocket (2) and the camshaft bracket (1) as shown. This prevents the knock pin of the camshaft from engaging with the incorrect pin hole when installing the camshaft sprocket.
12. Hold the camshaft with a wrench and loosen camshaft sprocket bolts to remove camshaft sprocket.

※※ **WARNING**

Never rotate crankshaft or camshaft while timing chain is removed. It causes interference between valve and piston. Never loosen the sprocket bolts without holding the camshaft securely with a wrench.

13. Loosen bolts in reverse of tightening order.
14. Cut liquid gasket by prying at the right front and left rear corners of the camshaft bracket, then remove the camshaft bracket.

※※ **WARNING**

Be careful not to damage the mating surface. A more adhesive liquid gasket is applied compared to previous types when shipped, so it should not be forced off the position not specified.

15. Carefully lift out the camshafts.

To install:

16. Install valve lifters in their original positions.
17. Install camshafts. Position the dowel pins at the sprocket end at the 12 o'clock position.

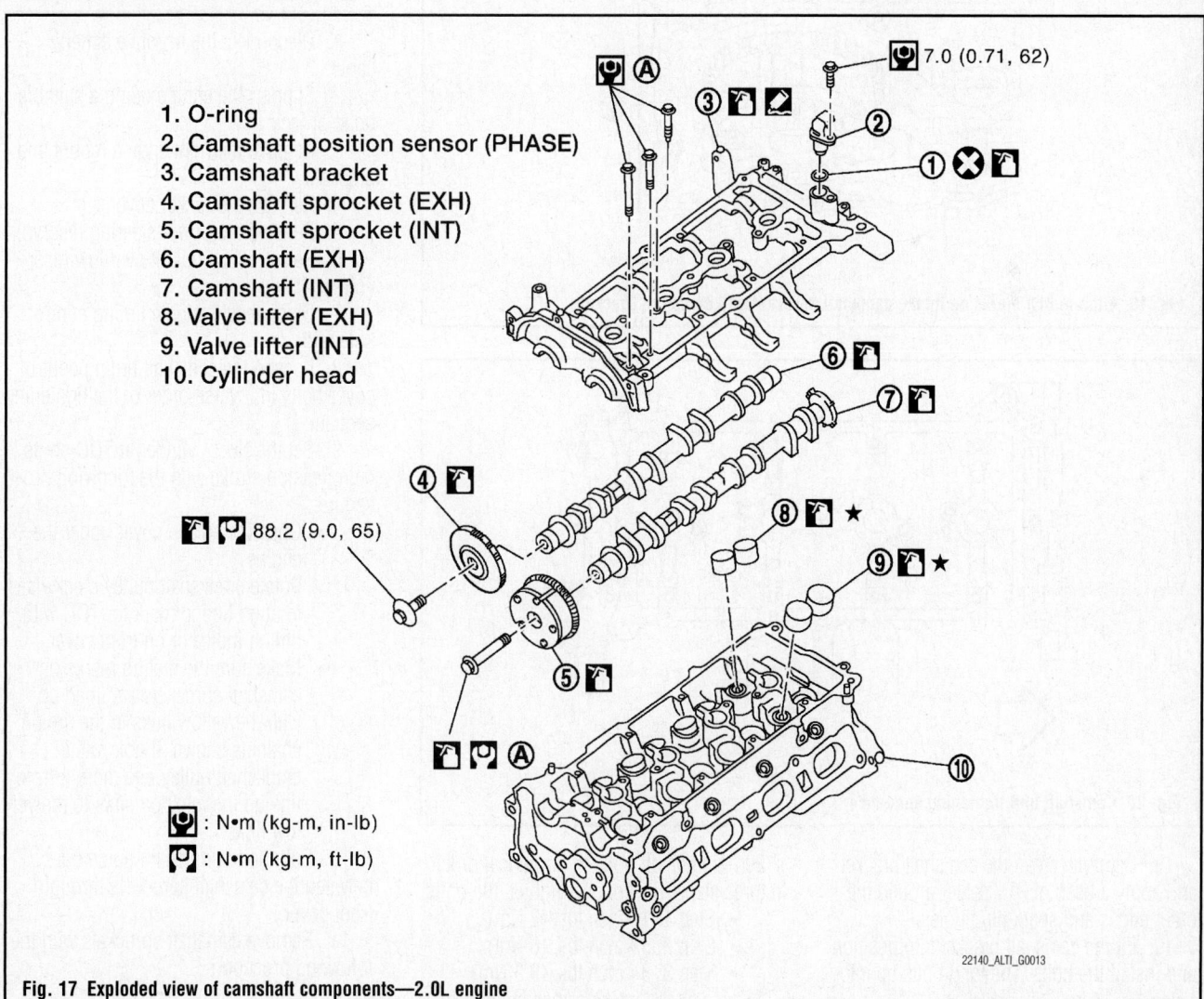

1. O-ring
2. Camshaft position sensor (PHASE)
3. Camshaft bracket
4. Camshaft sprocket (EXH)
5. Camshaft sprocket (INT)
6. Camshaft (EXH)
7. Camshaft (INT)
8. Valve lifter (EXH)
9. Valve lifter (INT)
10. Cylinder head

7.0 (0.71, 62)

88.2 (9.0, 65)

: N•m (kg-m, in-lb)

: N•m (kg-m, ft-lb)

22140_ALTI_G0013

Fig. 17 Exploded view of camshaft components—2.0L engine

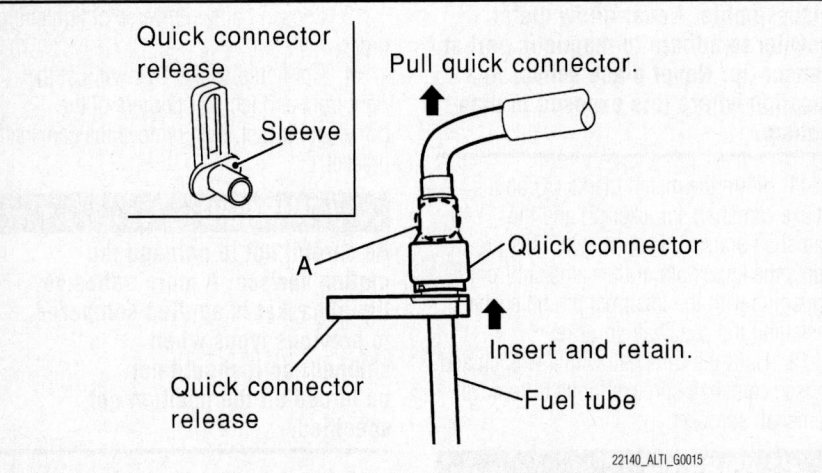

Fig. 18 Hold the quick connector release tool up while pulling fuel line straight off. Don't twist the fuel line.

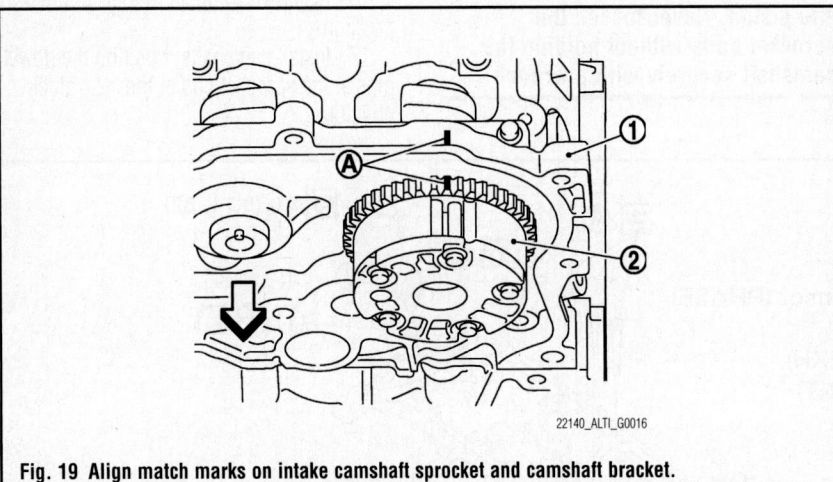

Fig. 19 Align match marks on intake camshaft sprocket and camshaft bracket.

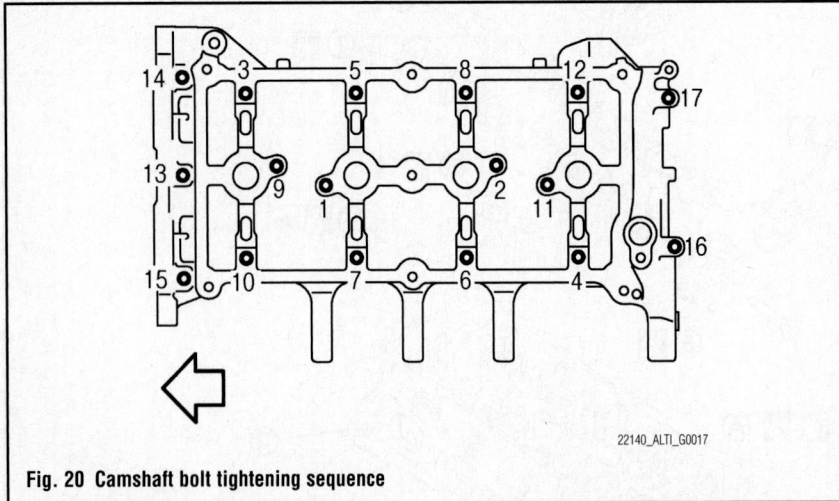

Fig. 20 Camshaft bolt tightening sequence

18. Carefully clean the camshaft bracket and apply a bead of RTV sealer around the outer edges and spark plug holes.

19. Fit the camshaft bracket into position and install the bolts. The long bolts go into holes 13, 14 and 15.

20. Tighten all bolts in numerical order in three steps. Do not over tighten the bolts.
- Step 1: 17 inch lbs. (2 Nm)
- Step 2: 52 inch lbs. (6 Nm)
- Step 3: 84 inch lbs. (9.5 Nm)

21. Install the intake camshaft sprocket, making sure to align the match marks. Hold the camshaft with a wrench and tighten the bolt to 26 ft. lbs. (35 Nm) plus 67°.

✳✳ WARNING

Use an angle-measuring wrench.

22. Install the exhaust camshaft sprocket, hold the camshaft with a wrench and torque the bolt to 65 ft. lbs. (88 Nm).

23. Install timing chain and related parts.

24. Inspect and adjust valve clearance.

25. Installation of the remaining components is in the reverse order of removal.

2.5L Engine

See Figures 21 through 26.

✳✳ CAUTION

Be sure to work in a well ventilated area and keep a CO2 fire extinguisher handy. Do not smoke while servicing fuel system. Keep open flames and sparks away from the work area.

1. Disconnect the negative battery cable.

2. Support the engine using a suitable hoist or jack.

3. Remove the right engine mount and brackets.

4. Remove the rocker cover.

5. Remove the power steering reservoir.

6. Remove the coolant overflow reservoir.

7. Disconnect variable timing control solenoid.

8. Loosen the camshaft timing control cover bolts in reverse order of the tightening sequence.

9. Set the No.1 cylinder at TDC on its compression stroke with the following procedure:
- Open the splash cover under the engine.
- Rotate crankshaft pulley clockwise to align timing mark for TDC with timing indicator on front cover.
- Make sure the mating marks on camshaft sprockets are lined up with the yellow links in the timing chain as shown. If not, rotate crankshaft pulley one more turn to line up the mating marks to the yellow links.

10. Pull the timing chain guide out between the camshaft sprockets through front cover.

11. Remove camshaft sprockets with the following procedure.

※※ **WARNING**

Do not rotate the crankshaft or camshaft while the timing chain is removed. This is an interference engine.

➡**Chain tension holding work is not necessary. Crankshaft sprocket and timing chain do not disconnect while front cover is attached.**

- Make sure marks on camshaft sprockets are aligned

with the yellow links in the timing chain
- Paint an indelible mating mark on the sprocket and timing chain link plate.
- Push in the chain tensioner plunger and insert a 0.020 in. (0.5 mm) pin into the hole on tensioner body to hold the chain tensioner.
- Remove the timing chain tensioner.

12. Hold the hexagonal part of camshaft with a wrench and loosen the camshaft sprocket mounting bolts to remove the camshaft sprockets.

13. Loosen the camshaft bearing cap bolts in reverse of the order and remove the caps. A rubber mallet may be needed.

14. Remove the camshafts.

15. When removing the valve lifters, keep them in order so they can be installed in the same locations.

To install:

16. Install the valve lifters.

17. Lubricate and carefully fit the camshafts into place. The back of the intake camshaft has the signal plate for the camshaft position sensor. Make sure the

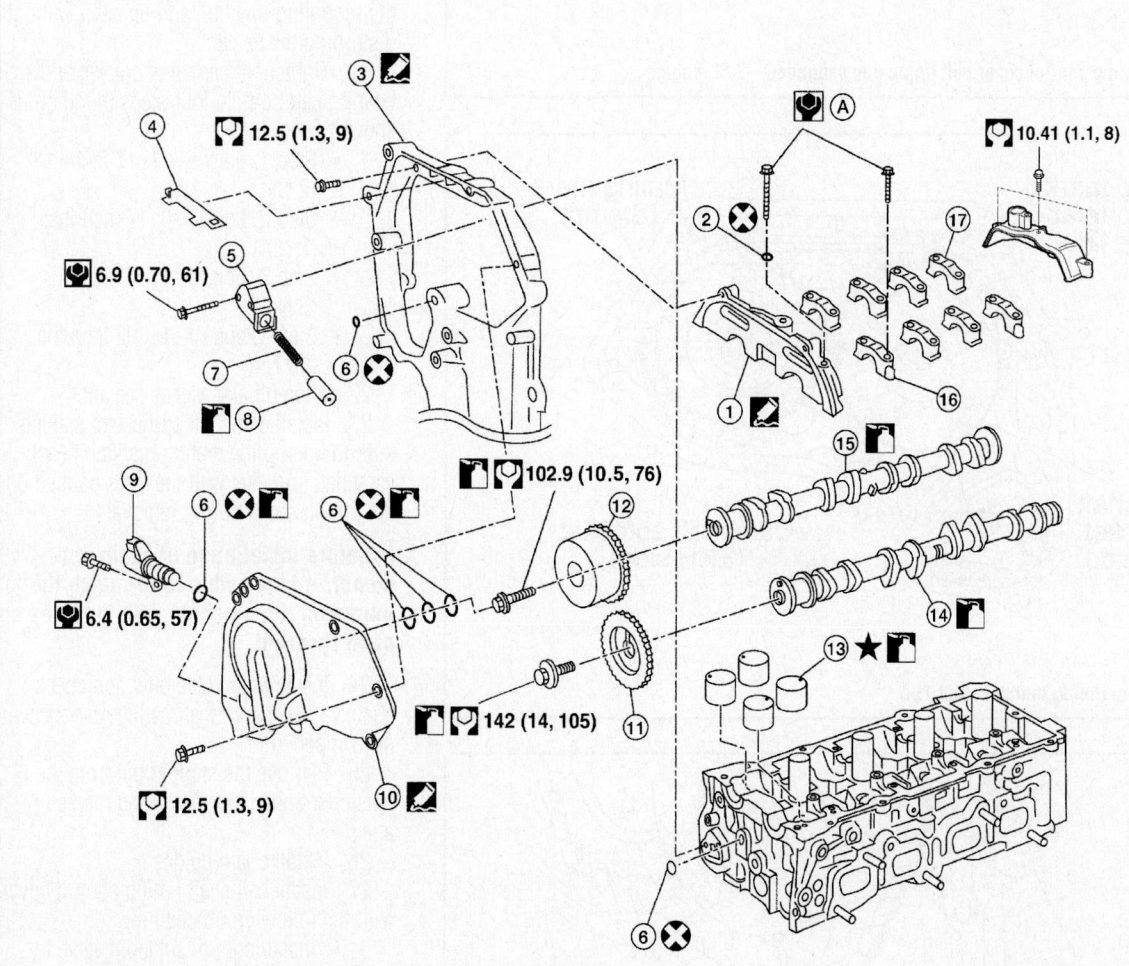

1.	Camshaft bracket (No.1)	2.	Washer	3.	Front cover (partial view)
4.	Chain guide	5.	Chain tensioner	6.	O-ring(s)
7.	Chain tensioner spring	8.	Chain tensioner plunger	9.	IVT control solenoid valve
10.	IVT control cover	11.	Camshaft sprocket (EXH)	12.	Camshaft sprocket (INT)
13.	Valve lifter	14.	Camshaft (EXH)	15.	Camshaft (INT)
16.	Camshaft bracket (No. 2)	17.	Camshaft bracket (No. 5)	A.	Follow installation procedure

22140_ALTI_G0056

Fig. 21 Exploded view of camshaft components—2.5L engine

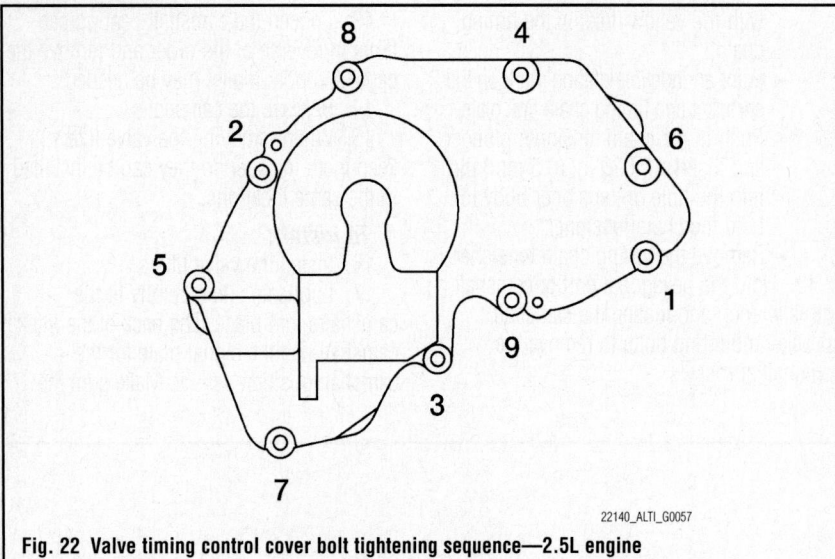

Fig. 22 Valve timing control cover bolt tightening sequence—2.5L engine

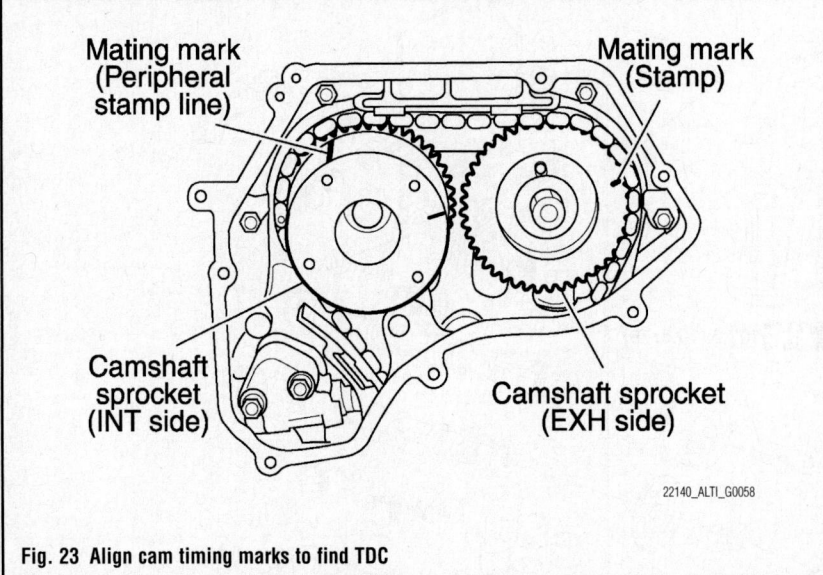

Fig. 23 Align cam timing marks to find TDC

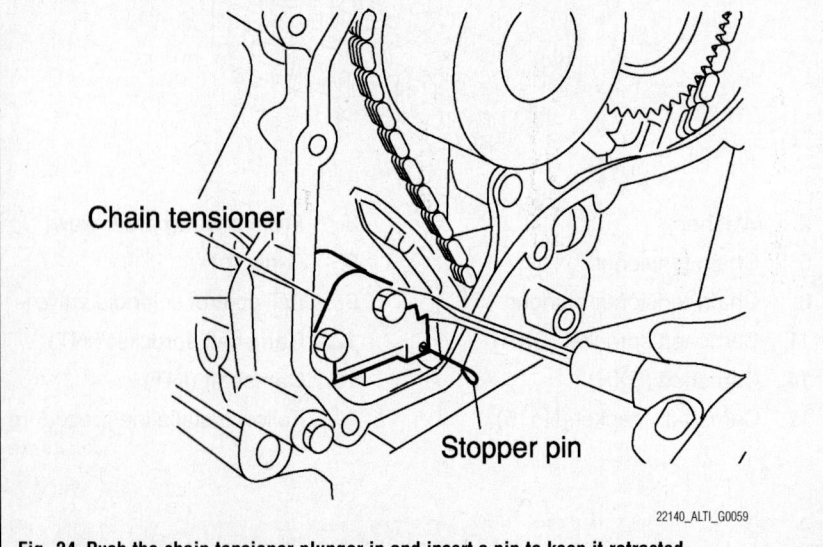

Fig. 24 Push the chain tensioner plunger in and insert a pin to keep it retracted

dowel pins on the front side are positioned as shown.

18. Install the camshaft bearing caps so the numbers or letters can be read from the exhaust side of the cylinder head. Caps 2, 3, 4 and 5 go on the intake camshaft, front to rear of engine. Caps A, B, C and D are front-to-rear exhaust camshaft caps. Make the bolts only finger tight at this time.

19. Make sure all sealing surfaces are clean and apply RTV sealant to the bracket where it contacts the cylinder head and the front cover. The sealant bead should be outside the bolt holes.

20. Position the camshaft bracket near the mounting position and install it without disturbing the sealant.

21. Tighten all camshaft bracket and bearing cap bolts in four steps in the order shown.

- Step 1, bolts 9–11: 17 inch lbs. (2 Nm)
- Step 2, bolts 1–8: 17 inch lbs. (2 Nm)
- Step 3, bolts 1–11 52 inch lbs. (6 Nm)
- Step 4, bolts 1–11: 92 inch lbs. (10 Nm)

22. Wipe off any excess sealant.

23. Install camshaft sprockets, making sure to line up the mating marks on each camshaft sprocket with the ones painted on the timing chain during removal.

➡**Before installation of chain tensioner, it is possible to re-match the marks on timing chain with the ones on each sprocket.**

24. Install chain tensioner and check again to make sure that mating marks have not slipped.

25. Remove the stopper pin from the tensioner and check the timing marks again.

26. Install chain guide.

27. Install camshaft timing control cover with the following procedure.

- Install control solenoid valve to intake valve timing control cover.
- Install new O-ring to front cover side.

28. Apply RTV sealant to the cover. The bead should be inside the bolt holes.

- Install the control cover and tighten the bolts in numerical order to 9 ft. lbs. (12 Nm).

29. Check and adjust valve clearances.

30. Installation of the remaining components is in the reverse order.

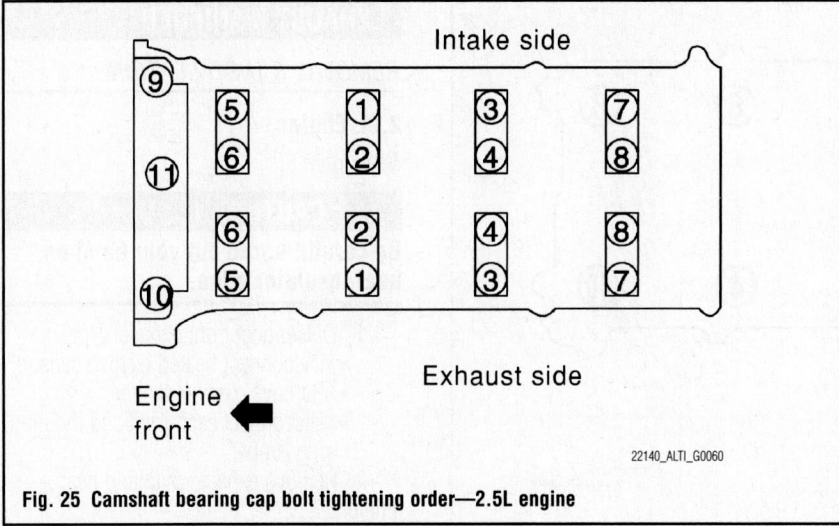

Fig. 25 Camshaft bearing cap bolt tightening order—2.5L engine

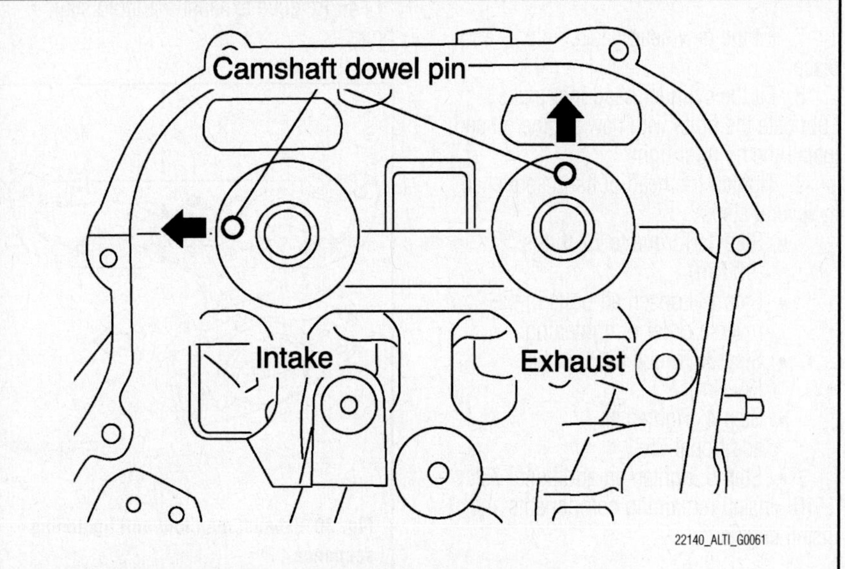

22140_ALTI_G0060

Camshaft dowel pin

Intake   Exhaust

22140_ALTI_G0061

Fig. 26 Make sure camshaft dowel pins are positioned as shown.

CRANKSHAFT FRONT SEAL

REMOVAL & INSTALLATION

See Figure 27.

2.0L Engine

1. Remove front timing cover.
2. Carefully pry out old seal.
3. Using a suitable tool, press-fit the new seal until it is flush with the front of the cover. Make sure the seal is straight and not curled.

2.5L Engine

1. Raise and safely support the vehicle and remove the right front wheel.
2. Remove the engine under cover.
3. Remove the drive belts.
4. Remove the crankshaft pulley.
5. Carefully pry the front oil seal from

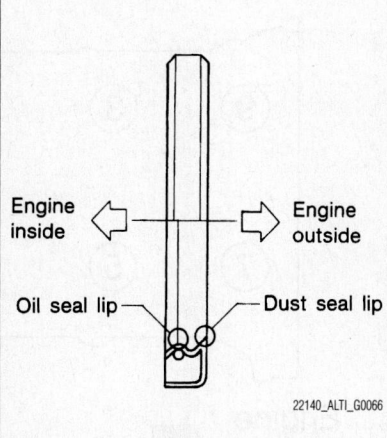

Engine inside   Engine outside

Oil seal lip   Dust seal lip

22140_ALTI_G0066

Fig. 27 Front and rear main seals are installed with the longer lip towards the outside of the engine—all engines

the front cover. Be careful not to scratch front cover or crankshaft.

To install:

6. Apply new engine oil to new oil seal and fit it into place with the longer lip towards the outside of the engine.
7. Press the seal straight in using an appropriately sized drift. Make sure the garter spring in the oil seal is in position and seal lip is not inverted.
8. Install crankshaft pulley and toque the bolts to 31 ft. lbs. plus an additional 60°.
9. Installation of the remaining components is in reverse order of removal.

CYLINDER HEAD

REMOVAL & INSTALLATION

2.0 Engine

See Figure 28.

> ✳✳ **CAUTION**
>
> **Be sure to work in a well ventilated area and keep a CO2 fire extinguisher handy. Do not smoke while servicing fuel system. Keep open flames and sparks away from the work area.**

1. Release the fuel pressure.
   - Remove fuel pump fuse.
   - Start engine.
   - After engine stalls, crank it two or three times to release all fuel pressure.
   - Turn ignition switch OFF.
2. Disconnect negative battery cable.
3. Drain engine coolant and engine oil.
4. Remove right front wheel and inner front fender.
5. Remove drive belt.
6. Remove intake and exhaust manifolds.
7. Remove fuel tube and fuel injector assembly
8. Remove water outlet assembly.
9. Remove ignition coils, spark plugs and rocker cover.
10. Remove front cover and timing chain.
11. Remove camshafts.
12. Use a TORX socket size E18 and carefully loosen cylinder head bolts in reverse of installation order.
13. Remove bolts and remove cylinder head.
14. Remove cylinder head gasket.

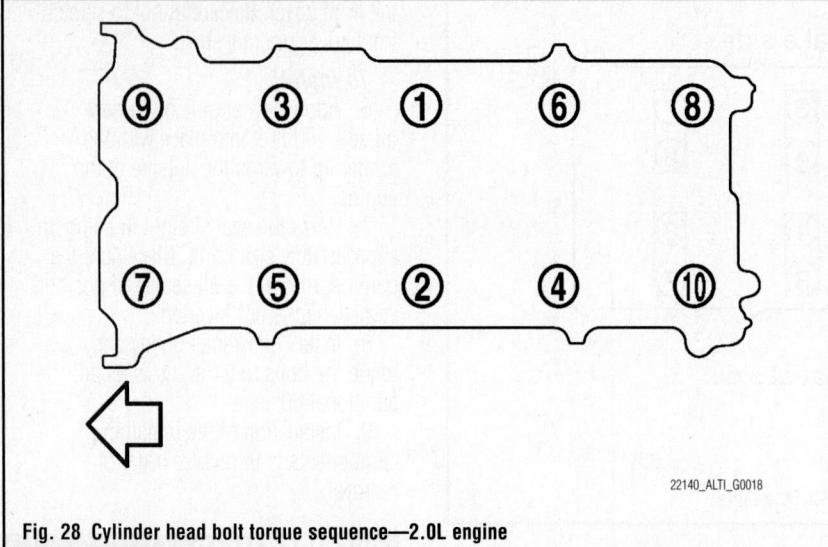

Fig. 28 Cylinder head bolt torque sequence—2.0L engine

To install:

➡**Use new cylinder head bolts.**

15. Clean the bolt hole threads and all sealing surfaces.

16. Install a new cylinder head gasket.

17. Apply new engine oil to threads and seating surface of bolts.

18. Install cylinder head and tighten bolts in numerical order in the following sequence.

- Step a: 30 ft. lbs. (40 Nm)
- Step b: 100° clockwise
- Step c: Loosen in the reverse order of tightening sequence.
- Step d: 30 ft. lbs. (40 Nm)
- Step e: 100° clockwise
- Step f: 100° clockwise

✳✳ WARNING

Check and confirm the tightening angle by using an angle torque wrench or protractor. Never judge by visual inspection without the tool.

19. Install the remaining components in the reverse order of removal.

2.5L Engine

See Figure 29.

1. Remove the engine and transaxle assembly.

2. Remove the timing chain.

3. Remove the camshafts.

4. Remove the spark plugs.

5. Loosen the cylinder head bolts in reverse order of tightening sequence and remove the cylinder head.

To install:

6. Clean all gasket surfaces. Use new cylinder head bolts.

7. Fit the new head gasket into place.

8. Fit the cylinder head into place. Lubricate the bolts with new engine oil and install them finger tight.

9. Tighten the head bolts in sequence in several steps:

- Step 1: Torque to 72 ft. lbs. (98 Nm)
- Step 2: Loosen all bolts in the reverse order of tightening.
- Step 3: Torque to 29 ft. lbs. (39 Nm)
- Step 4: Tighten an additional 75°
- Step 5: tighten an additional 75°

10. Install remaining components and install engine.

EXHAUST MANIFOLD

REMOVAL & INSTALLATION

2.0L Engine

See Figure 30.

✳✳ CAUTION

Be careful not to cut your hand on heat insulator edge.

1. Disconnect front exhaust pipe.
- Disconnect heated oxygen sensor.
- Remove oxygen sensor.
- Disconnect each joint and mounting rubber.

2. Remove exhaust manifold heat shield.

3. Remove the A/F sensor.

4. Remove exhaust manifold stay bolt.

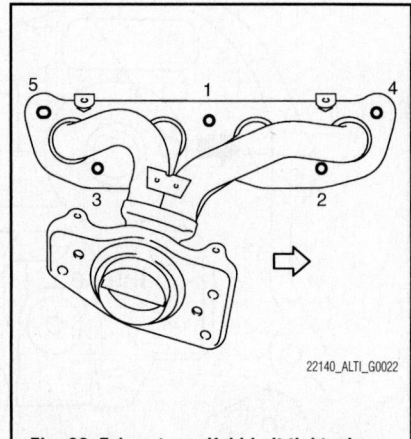

Fig. 30 Exhaust manifold bolt tightening sequence

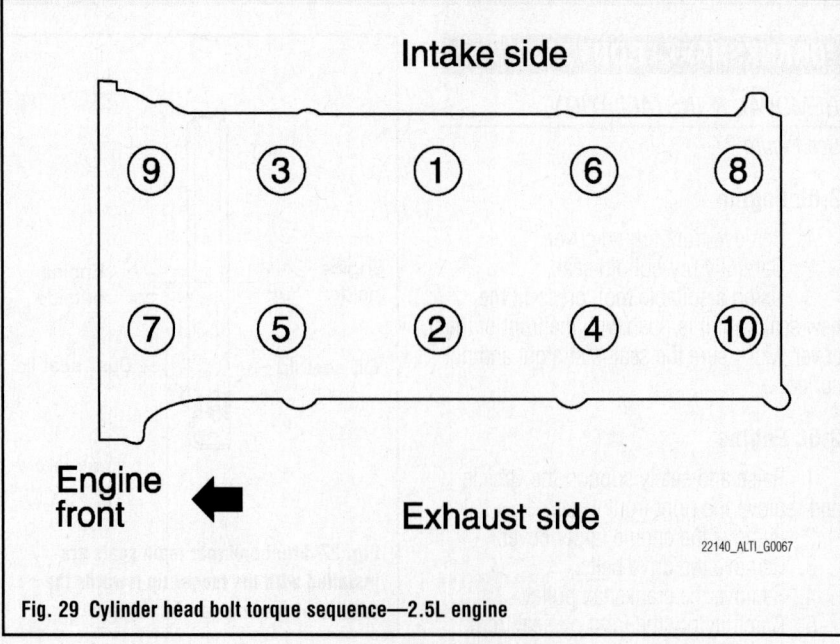

Fig. 29 Cylinder head bolt torque sequence—2.5L engine

5. Loosen nuts in reverse order and remove exhaust manifold.

To install:

6. Install a new gasket.

7. Fit manifold into place and loosely install all nuts. The upper end of the stay is marked "up."

8. Torque nuts in sequence in two steps to 25 ft. lbs. (33 Nm).

9. To connect exhaust pipe:

- Securely insert new seal bearing into exhaust manifold. Be careful not to damage seal bearing surface when installing.
- Install spring and tighten nut. Be careful that the stud bolt nut does not interfere with the flanged area.
- Make sure the spring sits properly on the flange surface by aligning it to the locator dimples.

10. Install the A/F and oxygen sensors. Do not over tighten or the MIL may turn on.

2.5L Engine

See Figure 31.

1. Remove the engine undercover.

2. Disconnect the air fuel ratio (A/F) sensor, unhook the harness from the bracket and remove the sensor.

3. Remove the lower exhaust manifold covers.

4. Remove the front exhaust pipe.

5. Remove the upper exhaust manifold cover.

6. Loosen the manifold nuts and remove the exhaust manifold and three way catalyst. Discard the gasket.

To install:

7. Install a new gasket and fit the manifold/catalyst into place.

8. Install the nuts and torque in sequence to 11 ft. lbs. (15 Nm).

9. Install the exhaust pipe and torque the bolts to 11 ft. lbs. (15 Nm).

10. Install the A/F sensor and reconnect the wiring.

INTAKE MANIFOLD

REMOVAL & INSTALLATION

2.0L Engine

See Figure 32.

1. Drain engine coolant or disconnect water hoses from electronic throttle control actuator and plug the hoses.

2. Disconnect wiring from electronic throttle control actuator.

✳✳ WARNING

Handle carefully to avoid any shock to electric throttle control actuator. Never disassemble.

3. Remove dipstick and cover the opening to avoid entry of foreign materials.

4. Loosen and install intake manifold bolts in reverse order.

5. Remove manifold and cover intake ports.

6. Make sure all sealing surfaces are clean and fit a new manifold gasket into place.

7. Install manifold and loosely install all nuts and bolts.

8. Torque the manifold-to-cylinder head bolts in sequence to 20 ft. lbs. (27 Nm).

9. Connect all hoses and wiring.

10. Check/refill coolant level.

11. Run the engine to check for leaks.

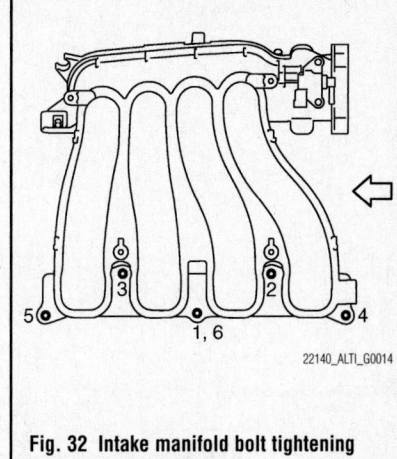

Fig. 32 Intake manifold bolt tightening sequence: 2.0L engine

2.5L Engine

1. Release the fuel pressure.

2. Drain coolant.

3. Disconnect the Mass Air Flow (MAF) sensor.

4. Remove air cleaner and air duct assembly.

5. Remove windshield wipers and cowl.

6. Disconnect the following components from the intake manifold:

- PCV hose
- EVAP hose and EVAP canister purge volume control solenoid
- Electric throttle control actuator
- Brake booster vacuum hose

7. Disconnect the fuel line quick connector on the engine side.

8. Disconnect throttle control actuator coolant hoses.

9. Remove the bolts and nuts and remove the intake manifold assembly. Cover engine openings to avoid entry of foreign materials.

To install:

10. Install a new manifold gasket and fit the manifold into place. Working from the center towards the ends, torque the nuts and bolts to 14 ft. lbs. (20 Nm), then repeat the torque sequence.

11. Reconnect all wiring, hoses and vacuum lines.

12. Refill the cooling system and run the engine to check for leaks.

13. Reinstall the cowl and wiper arms.

OIL PAN

REMOVAL & INSTALLATION

2.0L Engine

See Figures 33 through 35.

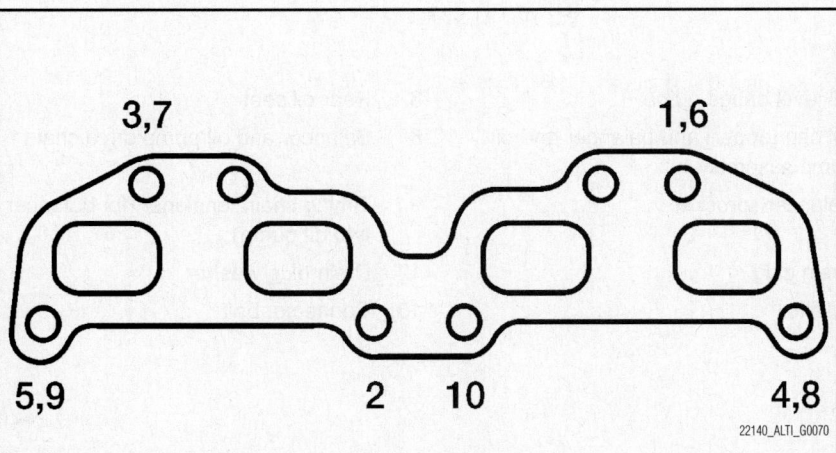

Fig. 31 Exhaust manifold bolt torque sequence—2.5L engine

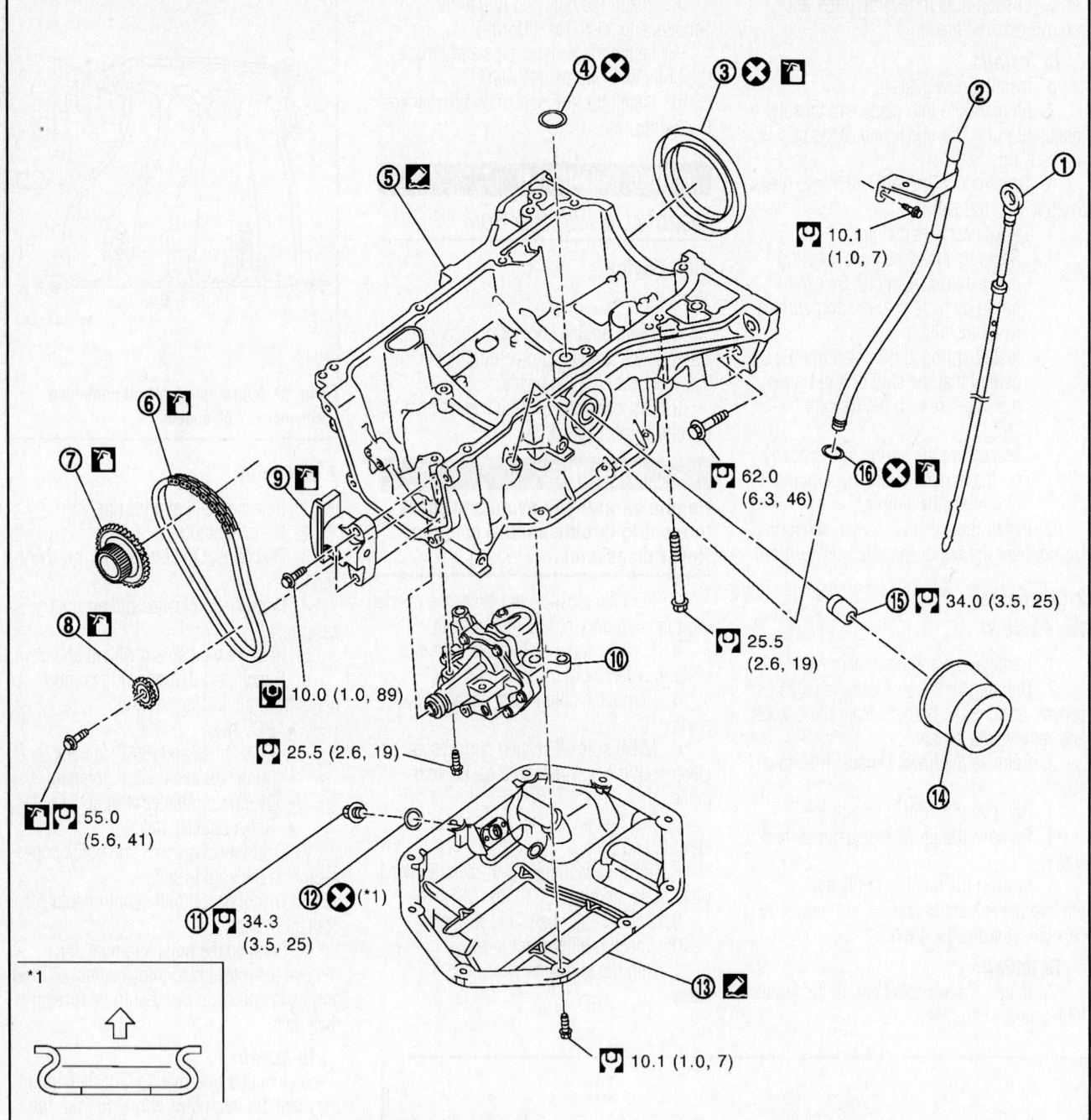

1. Oil level gauge
2. Oil level gauge guide
3. Rear oil seal
4. O-ring
5. Oil pan (upper) and balancer and oil pump assembly
6. Balancer and oil pump drive chain
7. Crankshaft sprocket
8. Balancer sprocket
9. Timing chain tensioner (for balancer and oil pump)
10. Balancer and oil pump
11. Drain plug
12. Drain plug washer
13. Oil pan (lower)
14. Oil filter
15. Connector bolt
16. O-ring
⇦ : Oil pan side

22140_ALTI_G0023

Fig. 33 Oil pan assembly—2.0L engine

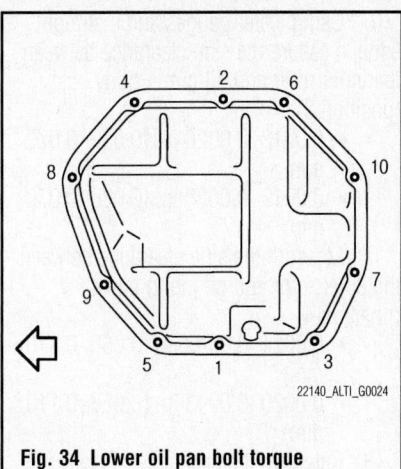

Fig. 34 Lower oil pan bolt torque sequence

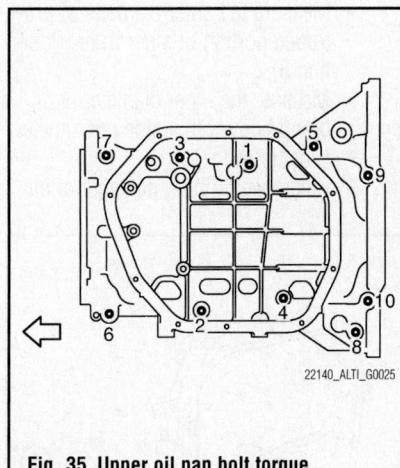

Fig. 35 Upper oil pan bolt torque sequence

1. Drain engine oil.
2. Remove engine and transaxle assembly.
3. Remove oil filter.
4. Remove lower oil pan bolts in reverse order.
5. After removing the pan, clean off the sealant being careful not to damage the mating surfaces.
6. Remove the flywheel.
7. Remove the front cover, timing chain, and the oil pump drive chain
8. Remove oil pan (lower) bolts in reverse order as shown.

To install:

✳✳ WARNING

The rear oil seal should be installed within 5 minutes after installing upper oil pan. Always replace rear oil seal with new one. Never touch oil seal lip.

9. Carefully scrape old liquid gasket from mating surfaces on oil pan and cylinder block.

10. Remove old liquid gasket from the bolt holes and threads.
11. Apply RTV liquid gasket to the oil pan. The bead should be outside the four center bolt holes and inside the four corner bolt holes.
12. Install new oil filter passage O-ring on the cylinder block. Make sure it's properly aligned.
13. Fit the pan into place on the block and tighten bolts in numerical order to 19 ft. lbs. (25 Nm).
14. Install rear oil seal with the following procedure.

- Wipe off liquid gasket protruding from the block/pan assembly.
- Apply engine oil to entire outside area of rear oil seal.
- Press-fit the rear oil seal using a press tool with outer diameter 4.53 in. (115mm) and inner diameter 3.54 in. (90 mm) Press-fit straight until seal is flush with engine block, making sure that rear oil seal does not curl or tilt.

15. Install oil pump sprocket, oil pump drive chain and other related parts if removed.
16. Use a scraper (A) to remove old liquid gasket from mating surfaces on lower oil pan.
17. Remove old liquid gasket from the bolt holes and threads.
18. Apply a bead of RTV sealant, staying inside all the bolt holes.
19. Tighten bolts in numerical order to 7 ft. lbs. (10 Nm).
20. Install oil filter, hand tight only.
21. Installation of the remaining components is in the reverse order of removal.

2.5L Engine

See Figure 36.

1. Raise and safely support the vehicle and drain the oil.
2. Remove the front exhaust pipe.
3. Remove power steering cooler hose bracket from suspension member.
4. Remove the front subframe for clearance to remove the oil pan.

- Remove nut on lower portion of stabilizer bar connecting rod from lower suspension arm.
- Remove suspension arm from subframe and swing it outward.
- Remove front exhaust pipe.
- Support engine or transmission with a jack.
- Remove steering gear bolts. Remove steering gear and power steering tube bracket from suspension member. Hang steering gear.
- Set a jack under subframe and remove the nuts.

5. Remove the lower oil pan bolts and remove the pan.
6. Remove the oil strainer.
7. Remove rear plate cover and four engine-to transaxle bolts.
8. Loosen the upper oil pan bolts and remove the upper oil pan. Note the different bolt lengths.

To install:

9. Carefully clean all sealing surfaces and apply a bead of RTV sealant to the upper oil pan. Install new O-rings and fit the pan into place. Start the bolts finger tight.
10. Torque the oil pan bolts in sequence to 16 ft. lbs. (22 Nm).

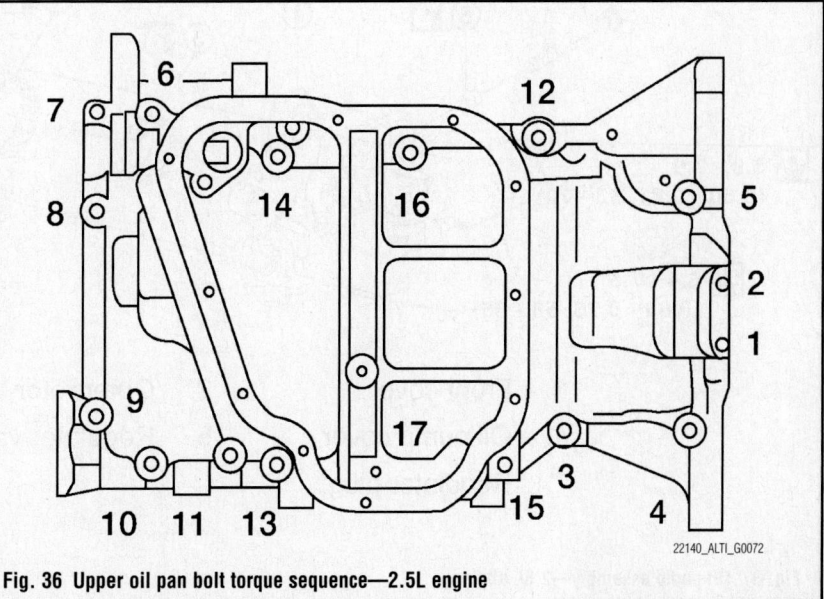

Fig. 36 Upper oil pan bolt torque sequence—2.5L engine

11. Torque the pan-to-transaxle bolts to 31 ft. lbs. (46 Nm).

12. Install the strainer and torque the bolts

13. Apply RTV to the lower oil pan to 16 ft. lbs. (22 Nm).

14. Install the subframe:
   • Raise the subframe into place and start all fasteners.
   • Torque the large subframe bolts to 107 ft. lbs. (145 Nm).
   • Torque the suspension arm bolts to 114 ft. lbs. (155 Nm).
   • Reconnect steering gear and stabilizer bar.

15. Install the remaining components and refill the engine with oil.

16. Run the engine to check for leaks.

OIL PUMP

REMOVAL & INSTALLATION

See Oil Pan Removal and Installation

INSPECTION

2.0L Engine

The oil pump and balance shafts are all one assembly and cannot be disassembled or repaired.

2.5L and 3.5L Engines

See Figures 37 through 42.

1. Remove the oil pump cover.

2. Remove inner rotor and outer rotor from front cover.

3. After removing regulator plug, remove regulator spring and regulator valve.

4. Using a feeler gauge, measure the clearance between outer rotor and oil pump body (position 1).
   • 0.0045–0.0070 in. (0.114–0.179 mm)

5. Measure the clearance between the inner rotor and outer rotor (position 2).
   • 0.0067–0.0087 in. (0.170–0.220 mm)

6. Using feeler gauges and a straight edge, measure the side clearance between the inner rotor and oil pump body (position 3)
   • 0.0012–0.0028 in. (0.030–0.070 mm.
   • 0.0012–0.0028 in. (0.030–0.070 mm.

7. Measure the side clearance between the outer rotor and oil pump body (position 4)
   • 0.0024–0.0043 in. (0.060–0.110 mm)
   • 0.0020–0.0043 in. (0.050–0.110 mm)

8. Calculate the clearance between inner rotor and oil pump body as follows:
   • Measure the outer diameter of protruded portion of inner rotor (Position 5).
   • Measure the inner diameter of oil pump body with inside micrometer (Position 6).
   • Clearance = (Inner diameter of oil

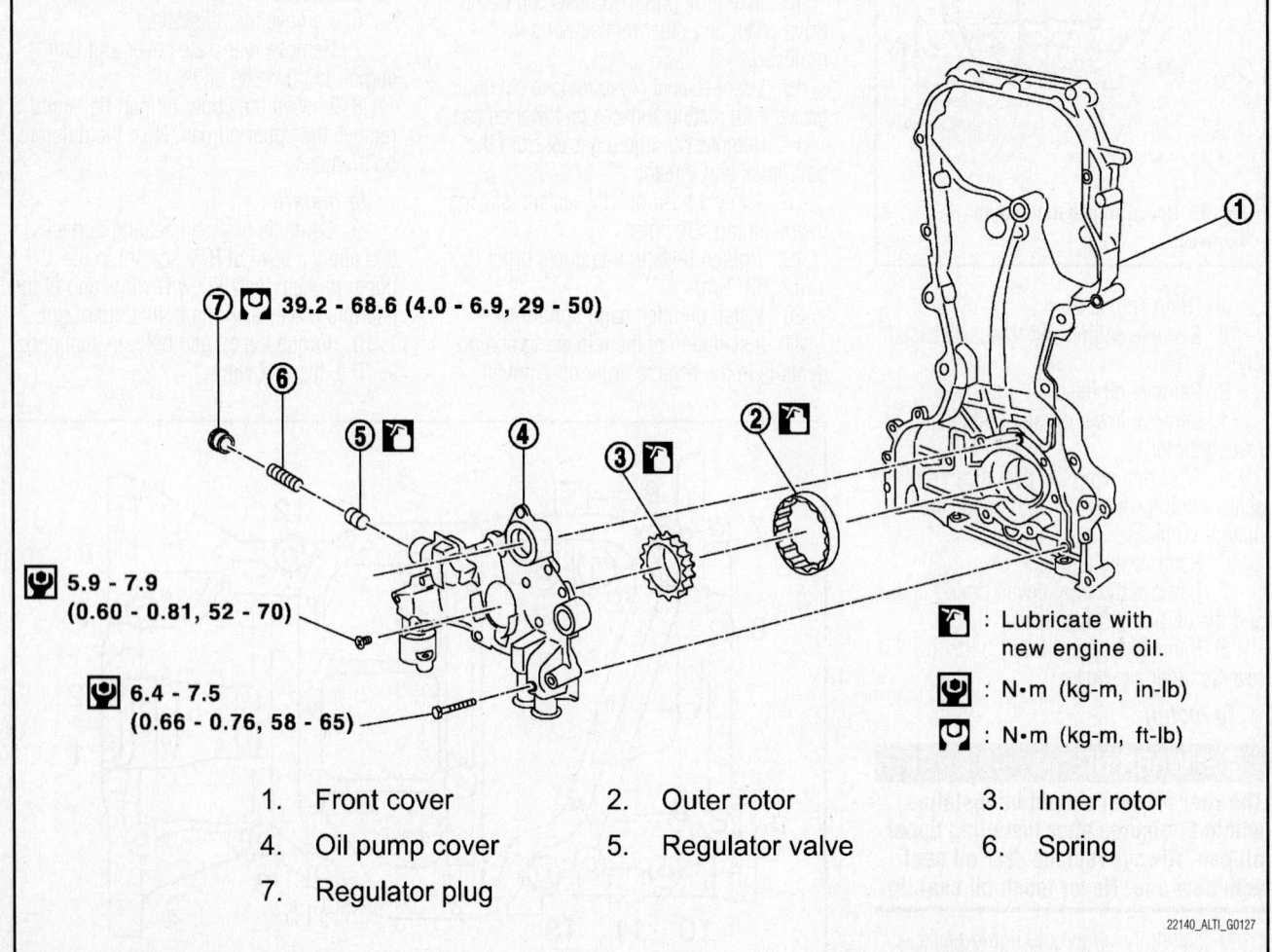

7 [torque] 39.2 - 68.6 (4.0 - 6.9, 29 - 50)

5.9 - 7.9 (0.60 - 0.81, 52 - 70)

6.4 - 7.5 (0.66 - 0.76, 58 - 65)

: Lubricate with new engine oil.

[torque symbol] : N•m (kg-m, in-lb)

[torque symbol] : N•m (kg-m, ft-lb)

1.	Front cover	2.	Outer rotor	3.	Inner rotor
4.	Oil pump cover	5.	Regulator valve	6.	Spring
7.	Regulator plug				

22140_ALTI_G0127

Fig. 37 Oil pump assembly—2.5L engine

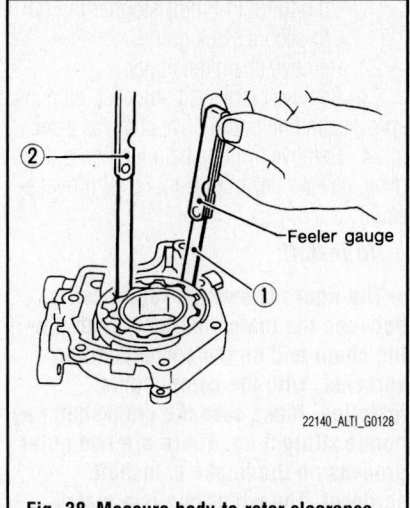

Fig. 38 Measure body-to-rotor clearance (1) and inner/outer rotor clearances (2)

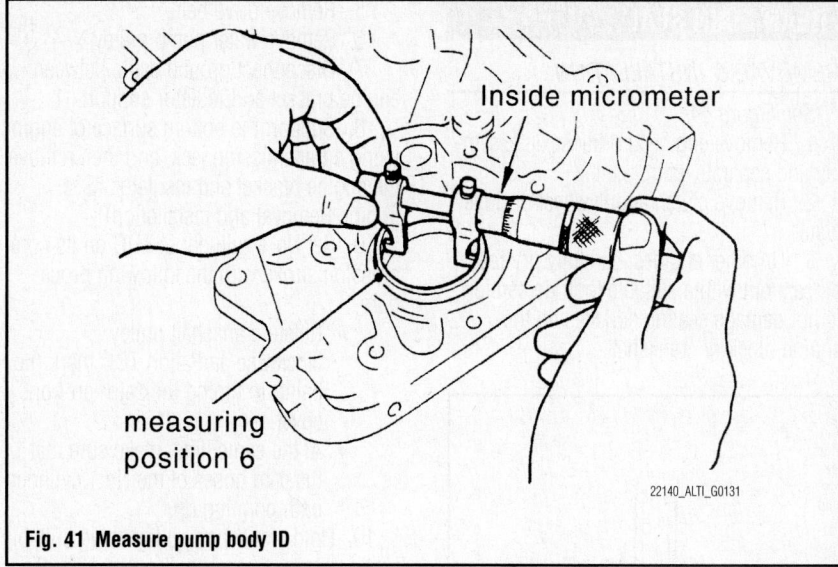

Fig. 41 Measure pump body ID

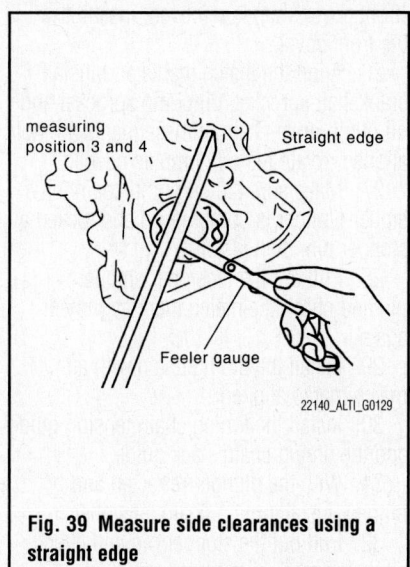

Fig. 39 Measure side clearances using a straight edge

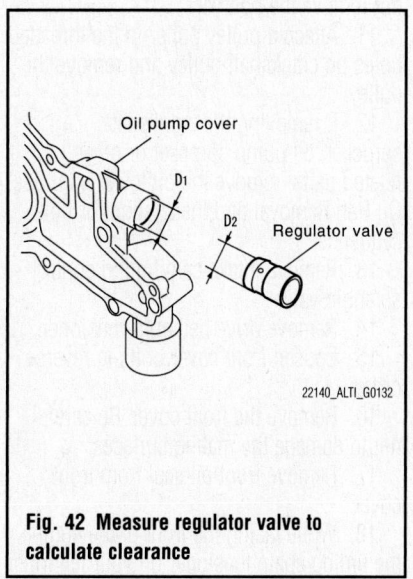

Fig. 42 Measure regulator valve to calculate clearance

pump body) – (Outer diameter of inner rotor).
  - 0.0016–0.0038 in. (0.040–0.097 mm)
9. Calculate regulator valve clearance using the same technique.
  - 0.0014–0.0028 in. (0.035–0.070 mm).

❊❊ WARNING

Coat regulator valve with engine oil. Check that it falls smoothly into the valve hole by its own weight.

PISTON AND RING

POSITIONING

See Figure 43.

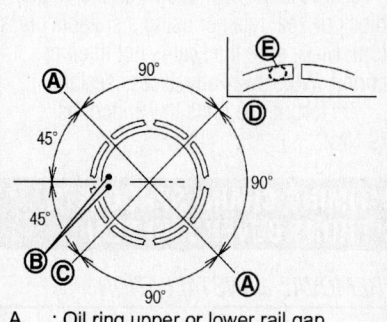

A : Oil ring upper or lower rail gap
B : Front mark
C : Second ring and oil ring spacer gap
D : Top ring gap
E : Stamped mark

Fig. 43 Piston ring positioning—all engines

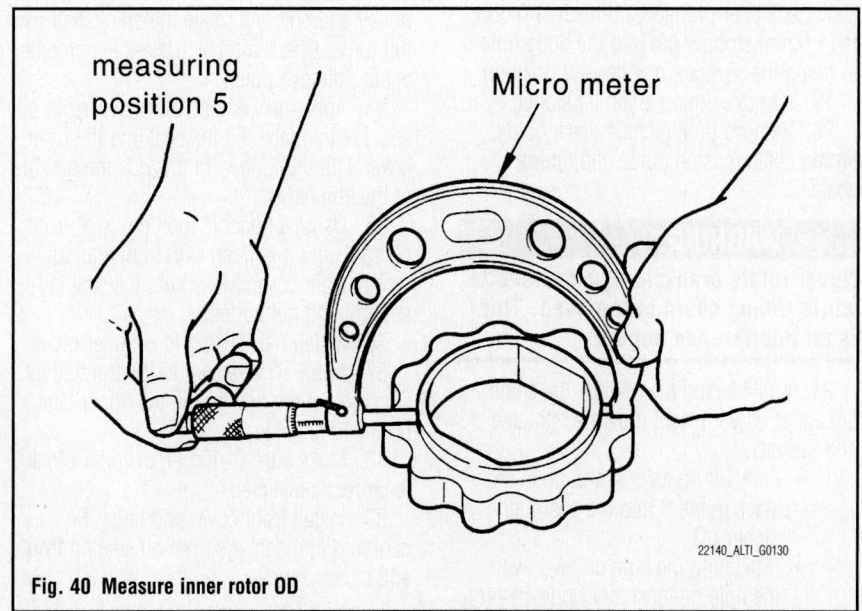

Fig. 40 Measure inner rotor OD

REAR MAIN SEAL

REMOVAL & INSTALLATION

See Figure 44.

1. Remove engine and transaxle assembly and separate them.

2. Remove clutch and flywheel or drive plate.

3. On other engines, carefully pry rear oil seal out with a suitable tool. Be careful to not damage sealing surfaces on the engine block or crankshaft.

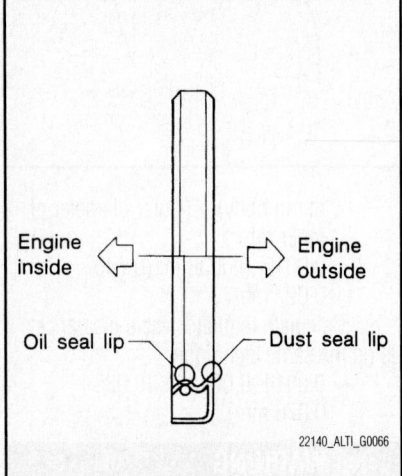

Fig. 44 Front and rear main seals are installed with the longer lip towards the outside of the engine—all engines

To install:

4. Apply RTV sealant to entire outside area of new rear oil seal.

5. Fit the seal into place with the longer lip towards the outside of the engine.

6. Press the seal flush with the engine block or seal retainer using a suitable press tool. Make sure the seal is not tilted or curled. Wipe away any excess sealant.

7. Torque the bolts to 78 inch lbs. (9 Nm).

TIMING CHAIN, SPROCKETS, FRONT COVER AND SEAL

REMOVAL & INSTALLATION

2.0L Engine

See Figures 45 through 49.

1. Remove right front wheel and inner fender

2. Drain engine oil.

3. Remove the intake manifold:

4. Remove ignition coils, spark plugs and rocker cover.

5. Remove drive belt.

6. Remove water pump pulley.

7. Disconnect ground cable between engine bracket and radiator support.

8. Support the bottom surface of engine using a transmission jack, and then remove the engine bracket and insulator. (See Engine Removal and Installation)

9. Set No.1 cylinder at TDC on its compression stroke with the following procedure:

- Rotate crankshaft pulley clockwise and align TDC mark (no paint) to timing indicator on front cover.
- At the same time, make sure that the cam noses of the No.1 cylinder both pointing up.

10. Hold crankshaft pulley using suitable tool and loosen crankshaft pulley bolt. Do not remove the bolt yet.

11. Attach a pulley puller in the threaded holes on crankshaft pulley and remove the pulley.

12. If removing the crankshaft sprocket, oil pump sprocket or other related parts, remove the oil lower pan. See Oil Pan Removal and Installation for bolt sequence.

13. Remove intake cam timing control solenoid valve.

14. Remove drive belt auto-tensioner.

15. Loosen front cover bolts in reverse order.

16. Remove the front cover. Be careful not to damage the mating surfaces.

17. Remove front oil seal from front cover.

18. While facing the front of the engine, the timing chain tensioner on your left must be retracted and locked. Push in timing chain tensioner plunger and insert a 0.060 in (1.5mm) stopper pin into the body hole to retain the plunger in collapsed position.

19. Remove timing chain tensioner.

20. Remove timing chain slack guide, timing chain tension guide and timing chain.

❊❊ WARNING

Never rotate crankshaft or camshafts while timing chain is removed. This is an interference engine.

21. While facing the engine, the chain tensioner on your right must be retracted and locked.

- Fully lift up lever A and push the slack guide B into the chain tensioner (1).
- Matching the hole on lever with the hole on tensioner body, insert a

0.040 in. (1.0mm) stopper pin (C) to secure slack guide.

22. Remove chain tensioner.

23. Remove crankshaft sprocket, oil pump sprocket and oil pump drive chain as a set.

24. Remove timing chain tension guide (front cover side) from front cover if necessary.

To install:

➡ **The figure shows the relationship between the match mark on each timing chain and on the corresponding sprocket, with the components installed. Make sure the crankshaft key points straight up. There are two outer grooves on the intake camshaft sprocket. The wider one is a match mark.**

25. If the timing chain tension guide (front cover side) is removed, install it to the front cover.

26. Align the match marks and install crankshaft sprocket, oil pump sprocket and oil pump drive chain. If these marks are not aligned, rotate the oil pump as needed.

27. Make sure the oil pump chain tensioner plunger is compressed and locked a stopper pin, then install it.

28. Pull out the tensioner stopper pin and check the match mark alignment again.

29. Install the timing chain with all match marks aligned.

30. Install the timing chain tension guide and the timing chain slack guide.

31. With the plunger retracted and pinned, install timing chain tensioner.

32. Pull out the stopper pin and check all match mark alignments again.

33. Temporarily install the crankshaft pulley and bolt and rotate the crankshaft two full turns. Check all match mark alignments again. Remove pulley.

34. Apply new engine oil to new front oil seal joint surface. Fit the seal into the front cover with the longer lip towards the outside of the engine.

35. Using a suitable tool, press-fit front oil seal until it is flush with front end surface of front cover. Make sure the seal is straight and not curled.

36. Install new O-ring to cylinder block.

37. Apply RTV sealant to the front cover. Don't forget the opening in the upper center (bolts 10 and 11).

38. Make sure O-ring on cylinder block is correctly installed.

39. Install front cover and bolts. Be careful not to damage front oil seal on front end of crankshaft.

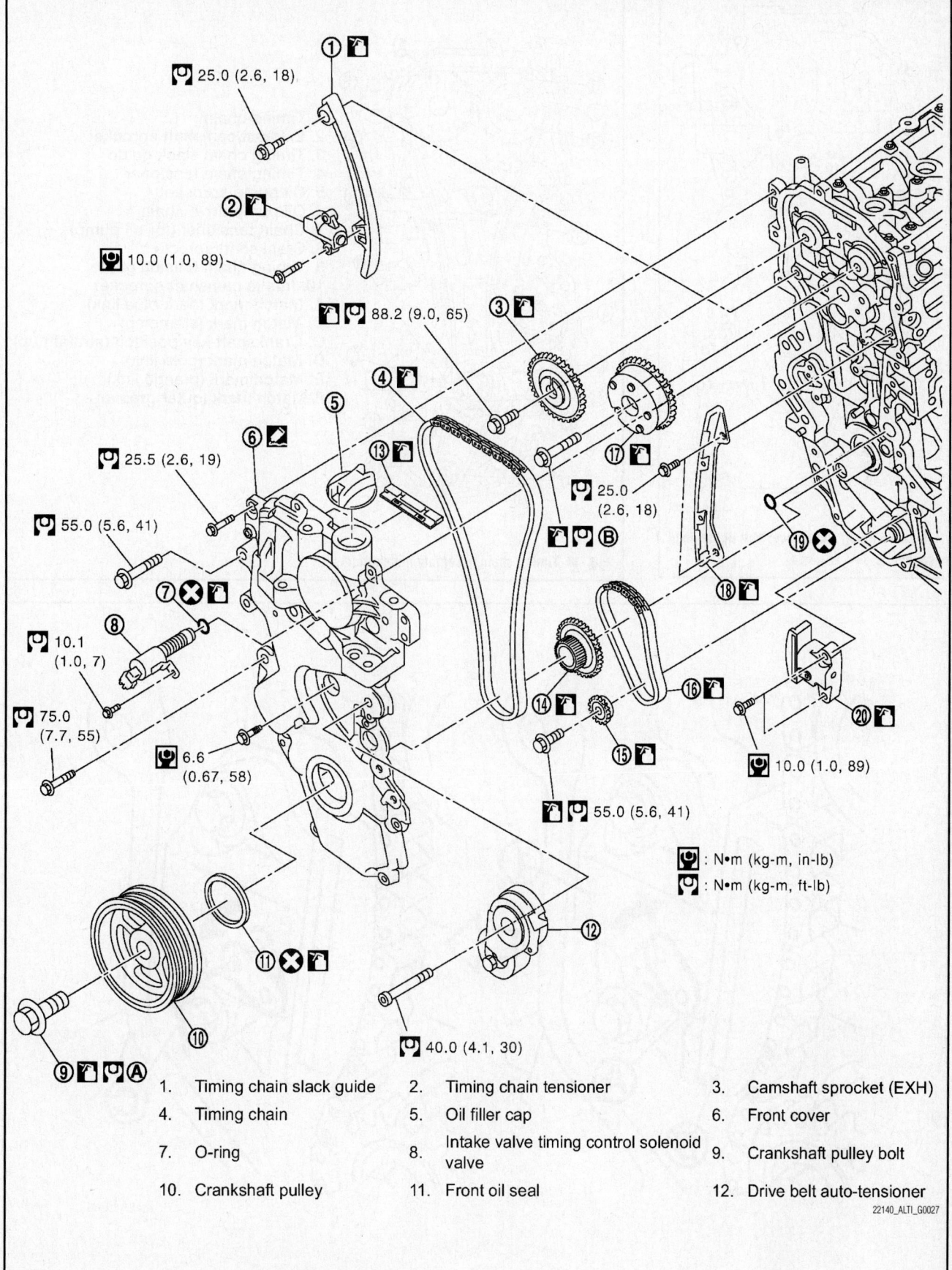

1. Timing chain slack guide
2. Timing chain tensioner
3. Camshaft sprocket (EXH)
4. Timing chain
5. Oil filler cap
6. Front cover
7. O-ring
8. Intake valve timing control solenoid valve
9. Crankshaft pulley bolt
10. Crankshaft pulley
11. Front oil seal
12. Drive belt auto-tensioner

22140_ALTI_G0027

Fig. 45 Exploded view of timing cover and timing chain

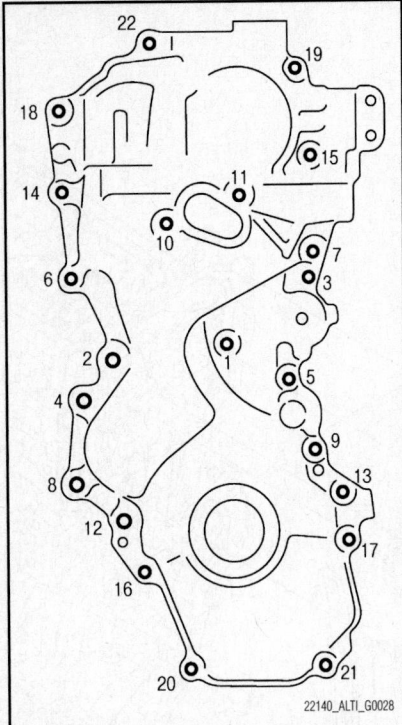

22140_ALTI_G0028

Fig. 46 Timing chain cover bolt tightening sequence

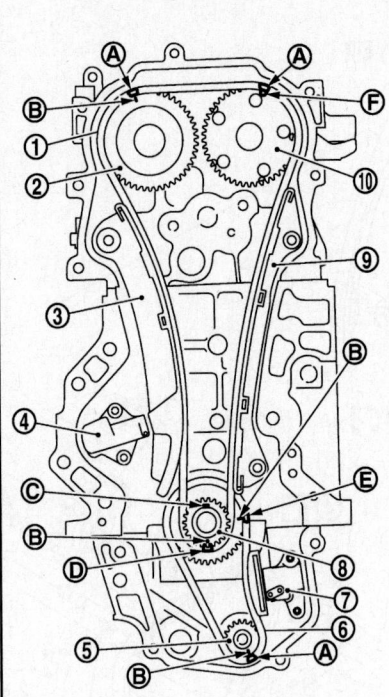

1. Timing chain
2. Exhaust camshaft sprocket
3. Timing chain slack guide
4. Timing chain tensioner
5. Oil pump sprocket
6. Oil pump drive chain
7. Chain tensioner (for oil pump)
8. Crankshaft sprocket
9. Timing chain tension guide
10. Intake camshaft sprocket
A. Match mark (dark blue link)
B. Match mark (stamping)
C. Crankshaft key position (straight up)
D. Match mark (gold link)
E. Match mark (orange link)
F. Match mark (outer groove)

22140_ALTI_G0030

Fig. 48 Timing chain assembly match marks.

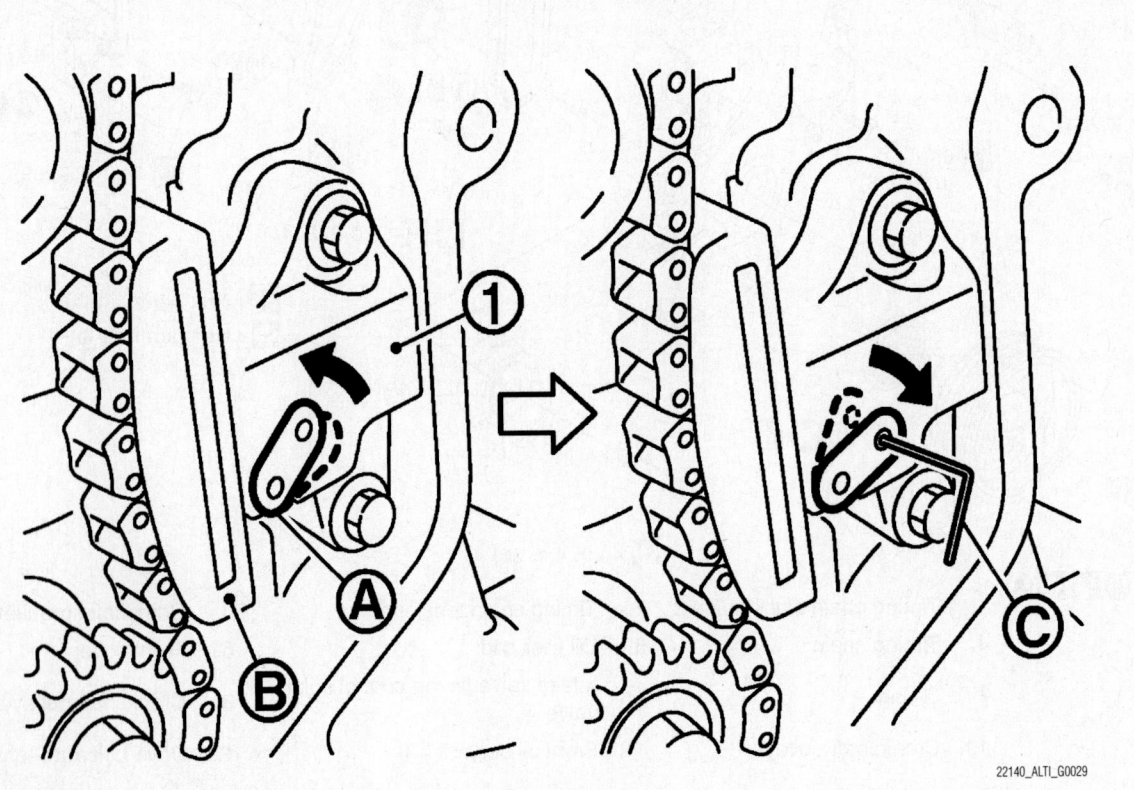

22140_ALTI_G0029

Fig. 47 Timing chain tensioner removal

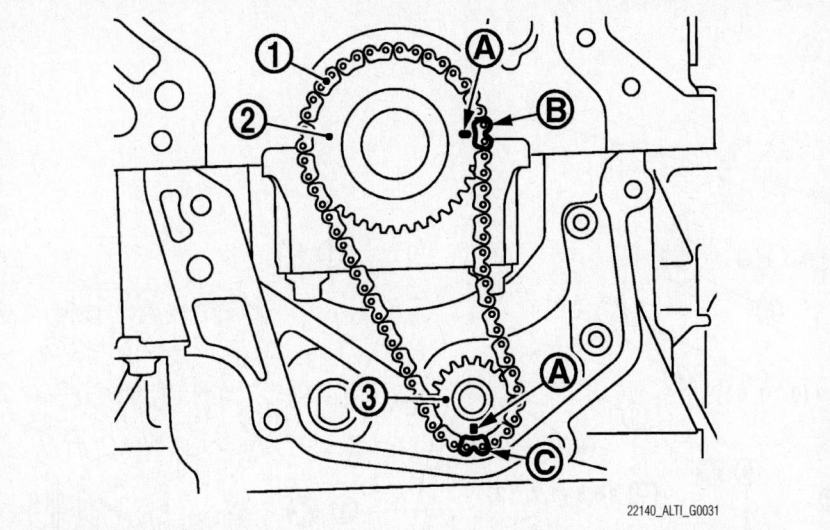

Fig. 49 The oil pump must be indexed to the crankshaft by aligning the marks on the chain and sprockets.

40. Tighten bolts in numerical order in two steps to specified torque.

41. Wipe off any excess liquid gasket.

42. Install crankshaft pulley and apply new oil to the pulley bolt.

43. Hold crankshaft pulley using the proper tool.

44. Tighten crankshaft pulley bolt in four steps.

- 51 ft. lbs. (68.6 Nm)
- 0 ft. lbs. (0 Nm)
- 22 ft. lbs. (29.4 Nm)
- 60°

45. Make sure crankshaft rotates clockwise smoothly.

46. Installation of the remaining components is in the reverse order of removal.

2.5L Engine

See Figures 50 through 54.

1. Raise and safely support the vehicle.

2. Remove the oil pan.

3. Remove the alternator.

4. Disconnect variable timing control solenoid harness connector.

5. Remove the coolant overflow reservoir.

6. Position the engine compartment fuse and relay box aside.

7. Remove the right engine mount and bracket.

8. Remove the camshaft timing control (IVT) cover using.

9. Pull the chain guide (between camshaft sprockets) out through front cover.

10. Set the No.1 cylinder at TDC on the compression stroke. Make sure the timing marks on the camshaft sprockets are aligned. See Camshaft Removal and Installation.

11. Hold the crankshaft pulley and loosen but do not remove the crankshaft pulley bolt.

12. Remove the crankshaft pulley with a bolt-on puller. Do not use a claw type puller.

13. Remove the timing cover bolts in reverse of the tightening sequence.

14. Push in the chain tensioner plunger and secure it with a stopper pin in the hole on the tensioner body. Remove chain tensioner.

15. Remove the timing chain.

16. If necessary, hold the hexagonal part of the camshaft with a wrench and loosen the camshaft sprocket bolt and remove the camshaft sprocket.

✷✷ WARNING

Do not rotate the crankshaft or camshafts while the timing chain is removed. This is an interference engine.

17. To remove the balance shaft chain, compress the chain guide tensioner and remove the tensioner, guide, timing chain, and oil pump drive spacer.

➡**The balance shaft bolts must be replaced once removed.**

18. Secure the left balancer shaft with a wrench and loosen the sprocket bolt.

19. Remove balancer unit timing chain, balancer unit sprocket and crankshaft sprocket.

To install:

➡**There may be two color variations of the link marks (link colors) on the timing chain. There are 26 links between the gold/yellow marks on the timing chain; and 64 links between the camshaft sprocket gold/yellow link and the crankshaft sprocket orange/blue link on the timing chain side without the tensioner.**

20. Make sure the crankshaft key points straight up.

21. Use new bolts for the balance shaft assembly and oil the threads before installation. Install the balancer unit and tighten the bolts in numerical in the following steps:

- Step 1: Bolts 1-5, 31 ft. lbs. (42 Nm): Bolt 6, 27 ft. lbs. (36 Nm)
- Step 2: Bolts 1-5, 120°: Bolt 6, 90°
- Step 3: Loosen in reverse order or tightening sequence
- Step 4: Bolts 1-5, 31 ft. lbs. (42 Nm): Bolt 6, 27 ft. lbs. (36 Nm)
- Step 5: Bolts 1-5, 120°: Bolt 6, 90°

22. Install the crankshaft sprocket and timing chain for the balancer unit. Make sure the crankshaft sprocket mark is at the top to align with the mark on the block. The orange or blue link on the chain aligns with this same mark, while the gold or yellow link aligns with the mark on the balance shaft sprocket.

23. Install balancer timing chain tensioner and remove the stopper pin. Make sure the timing marks are still aligned.

24. Install the camshaft timing chain and related parts.

25. After installing timing chain tensioner, remove the stopper pin and make sure that the tensioner moves freely.

26. Rotate the crankshaft two full turns and make sure all timing marks still line up.

27. Install a new front oil seal

28. Carefully clean all sealing surfaces and apply RTV sealant to the timing cover. Install the cover and torque bolts A and D to 36 ft. lbs. (49 Nm) and bolts B and C to 9 ft. lbs. (13 Nm).

29. Carefully clean all sealing surfaces and apply RTV sealant to the rocker cover. Install the cover

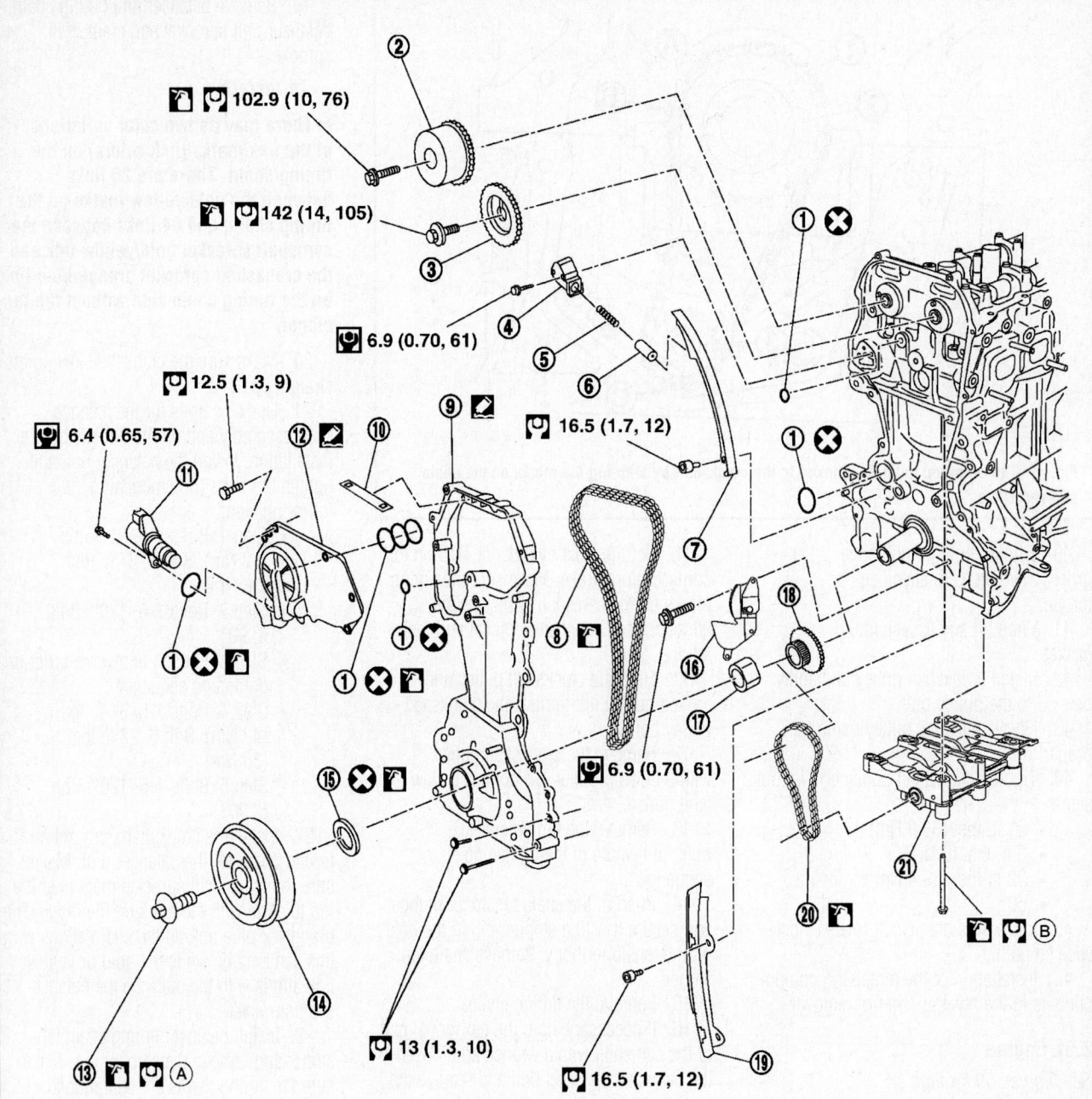

102.9 (10, 76)

142 (14, 105)

6.9 (0.70, 61)

12.5 (1.3, 9)

6.4 (0.65, 57)

16.5 (1.7, 12)

6.9 (0.70, 61)

13 (1.3, 10)

16.5 (1.7, 12)

1. O-rings	2. Camshaft sprocket (INT)	3. Camshaft sprocket (EXH)
4. Chain tensioner	5. Spring	6. Chain tensioner plunger
7. Timing chain slack guide	8. Timing chain	9. Front cover
10. Chain guide	11. IVT solenoid valve	12. IVT cover
13. Crankshaft pulley bolt	14. Crankshaft pulley	15. Front oil seal
16. Balancer unit timing chain tensioner	17. Oil pump drive spacer	18. Crankshaft sprocket
19. Timing chain tension guide	20. Balancer unit timing chain	21. Balancer unit
A. Follow installation procedure	B. Follow installation procedure	

22140_ALTI_G0083

Fig. 50 Timing chain and balance shaft chain components—2.5L engine

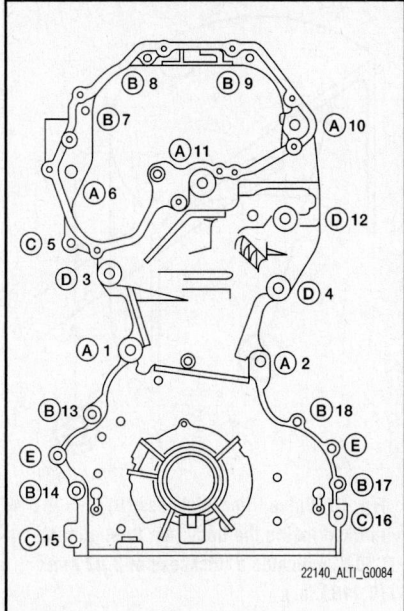

Fig. 51 Timing chain cover bolt torque sequence—2.5L engine

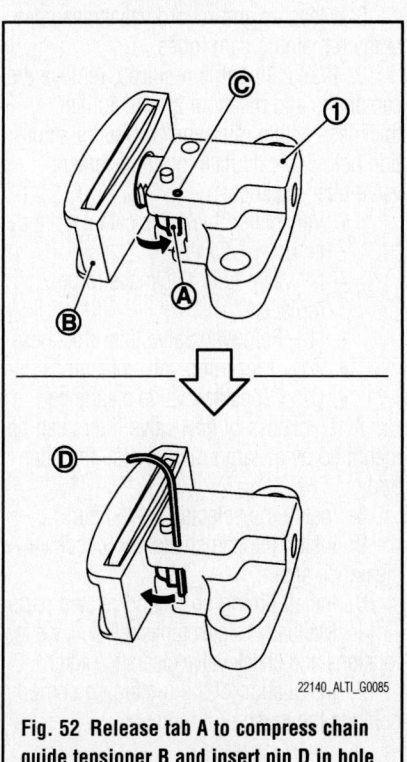

Fig. 52 Release tab A to compress chain guide tensioner B and insert pin D in hole C to lock it in place—2.5L engine

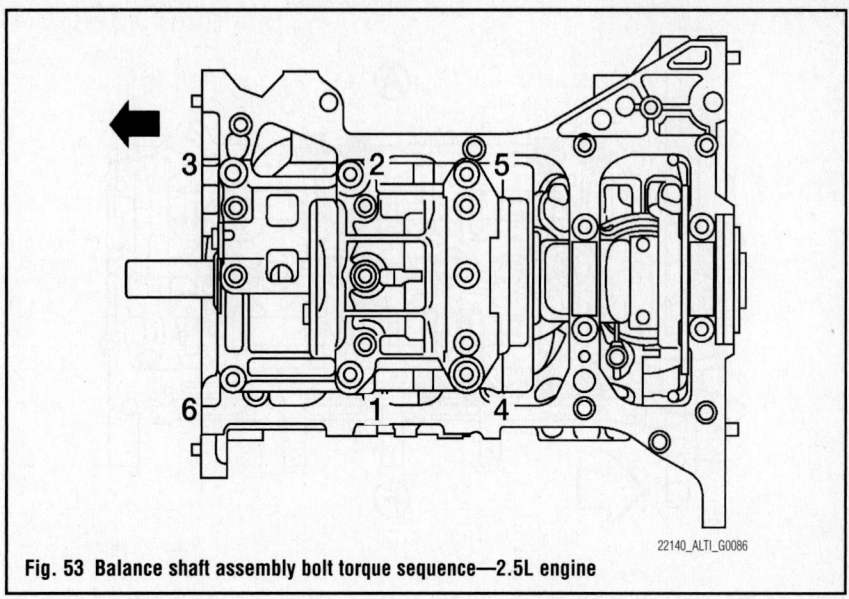

Fig. 53 Balance shaft assembly bolt torque sequence—2.5L engine

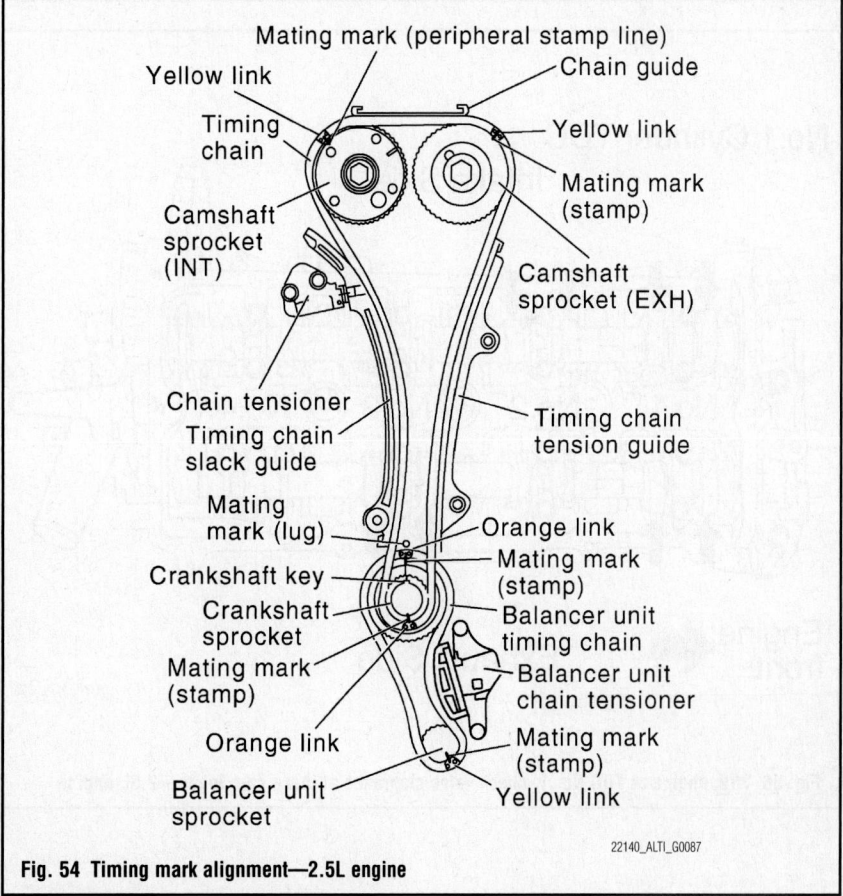

Fig. 54 Timing mark alignment—2.5L engine

30. Install IVT solenoid valve to IVT cover with a new O-ring.

31. Apply RTV sealant to the IVT cover and install it. Torque the bolts in order to 9 ft. lbs. (13 Nm).

32. Install crankshaft pulley and torque bolt to 31 ft. lbs. (42 Nm) plus 60°.

33. Install remaining components.

VALVE LASH

ADJUSTMENT

See Figures 55 through 57.

➡**Valves are adjusted by changing lifters. There are 26 different thicknesses of Nissan valve lifter available in sizes ranging from 3.00 to 3.50 mm**

(0.1181 to 0.1378 in) in steps of 0.02 mm (0.0008 in.).

1. Remove the rocker cover.

2. Rotate crankshaft to TDC on No.1 cylinder. Align TDC mark (no paint) to timing indicator on front cover.

3. Measure and record valve clearance at the cam lobes indicated.

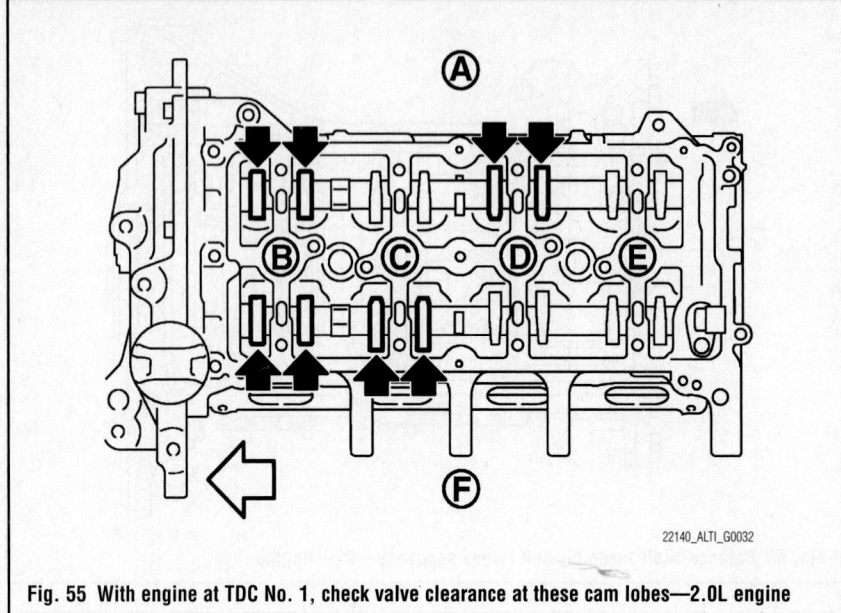

Fig. 55 With engine at TDC No. 1, check valve clearance at these cam lobes—2.0L engine

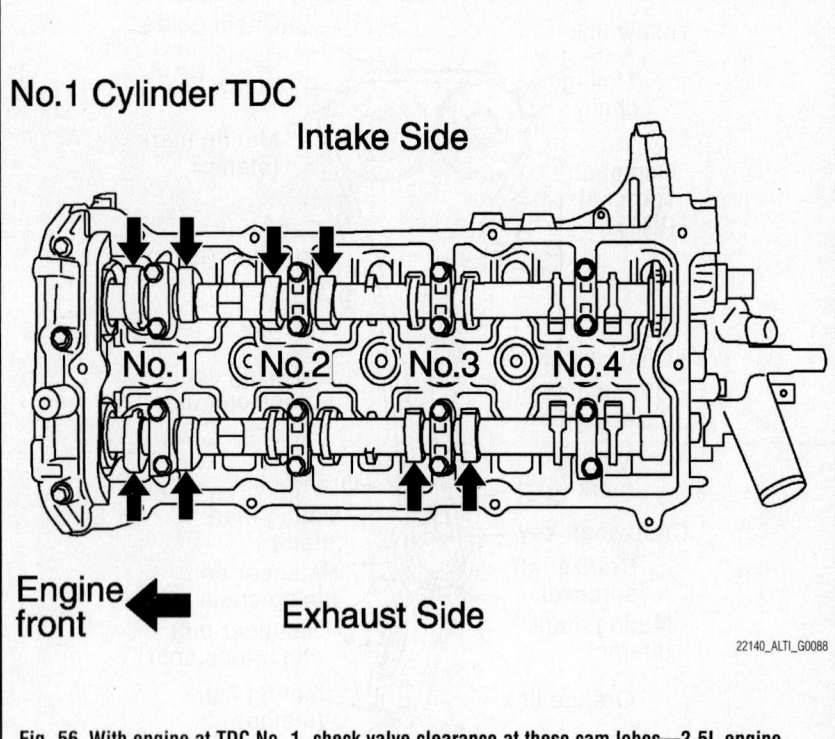

No.1 Cylinder TDC
Intake Side

No.1 No.2 No.3 No.4

Engine front

Exhaust Side

22140_ALTI_G0088

Fig. 56 With engine at TDC No. 1, check valve clearance at these cam lobes—2.5L engine

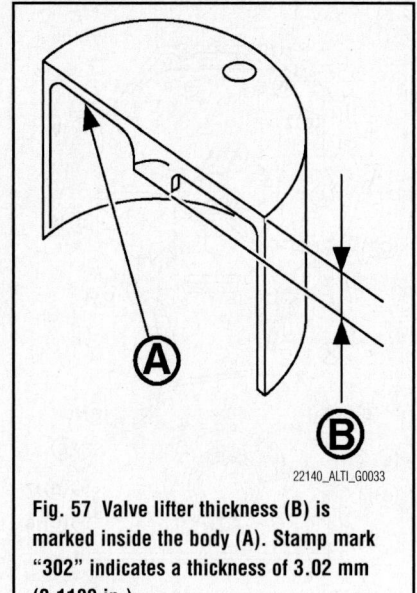

22140_ALTI_G0033

Fig. 57 Valve lifter thickness (B) is marked inside the body (A). Stamp mark "302" indicates a thickness of 3.02 mm (0.1189 in.).

4. Rotate the crankshaft one full turn to TDC No. 4.

5. Measure and record valve clearance at the remaining cam lobes.

6. If adjustment is required, remove the camshafts and measure the valve lifter thickness with a micrometer. Use the equation below to calculate the replacement valve lifter thickness.

- Valve lifter thickness calculation: $t = t_1 + (C_1 - C_2)$
- t = Valve lifter thickness to be replaced
- t_1 = Removed valve lifter thickness
- C_1 = Measured valve clearance
- C_2 = Specified valve clearance

7. Thickness of new valve lifters can be identified by a stamp mark inside the lifter body.

8. Install the selected valve lifters.

9. Install the camshafts and check valve clearance again.

10. Install timing chain and related parts.

11. Manually rotate crankshaft pulley a few rotations and check valve clearance again.

12. Installation of the remaining components is the reverse of removal.

ENGINE PERFORMANCE & EMISSION CONTROLS

CAMSHAFT POSITION (CMP) SENSOR

LOCATION

See Figures 58 and 59.

Refer to the accompanying illustrations.

REMOVAL & INSTALLATION

1. Unplug the connector.
2. Remove the bolt to remove the sensor.
3. During installation, use a new O-ring and make sure the sensor is clean and free of metal filings on the magnet end.

CRANKSHAFT POSITION (CKP) SENSOR

LOCATION

See Figures 60 and 61.

Refer to the accompanying illustrations.

REMOVAL & INSTALLATION

1. Unplug the connector.
2. Remove the bolt to remove the sensor.

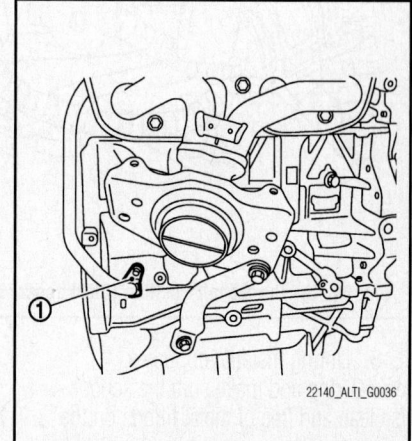

Fig. 60 Crankshaft Position (CKP) sensor is on the rear of the cylinder block—2.0L engine

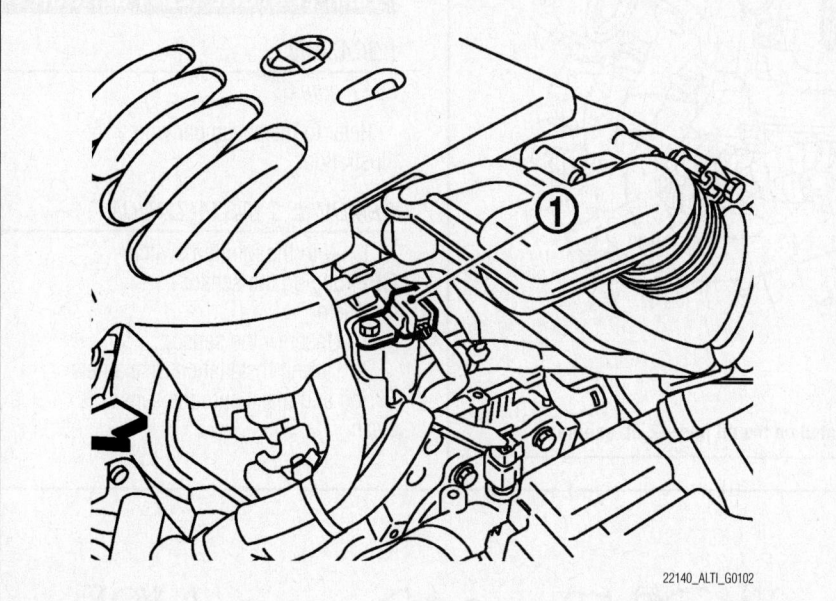

Fig. 58 The Camshaft Position (CMP) sensor is on top of the engine near the ignition coil.—2.0L engine

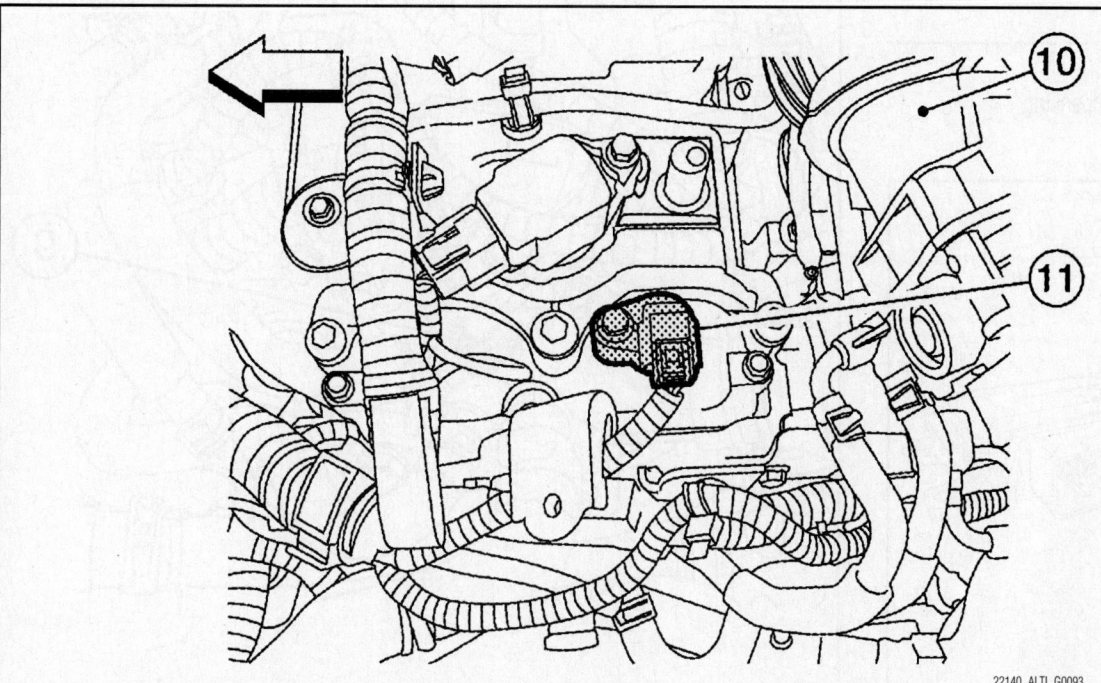

Fig. 59 The Camshaft Position (CMP) sensor is located on top of the engine at the flywheel end.—2.5L engine

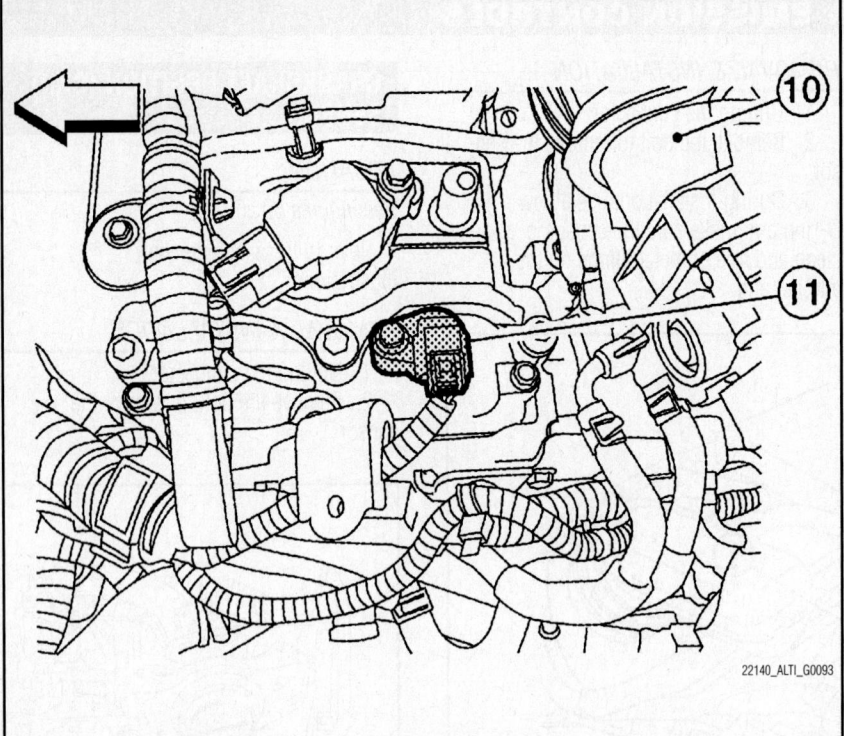

Fig. 61 The Crankshaft Position (CKP) sensor is located on the oil pan—2.5L engine

3. During installation, use a new O-ring and make sure the sensor is clean and free of metal filings on the magnet end.

ELECTRONIC CONTROL MODULE (ECM)

LOCATION

See Figure 62.

Refer to the accompanying illustration.

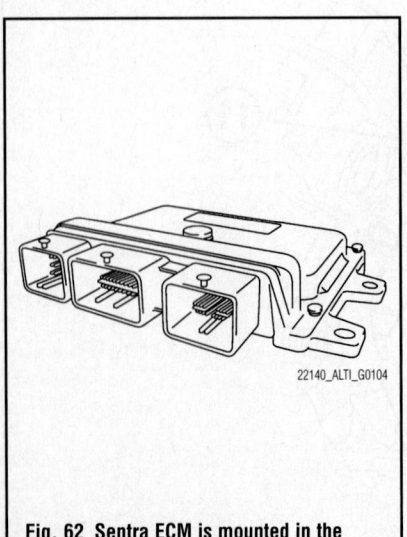

Fig. 62 Sentra ECM is mounted in the engine compartment.

REMOVAL & INSTALLATION

1. Disconnect both battery cables.
2. Remove the connector cover and remove the ECM.
3. Unplug the connector.
4. When installing, take care not to bend the pins in the connector.

ENGINE COOLANT TEMPERATURE (ECT) SENSOR

LOCATION

See Figure 63.

Refer to the accompanying illustration.

REMOVAL & INSTALLATION

1. With the ignition switch **OFF**, unplug the sensor connector.
2. Unscrew the sensor.
3. During installation, use a new O-ring and make sure the sensor is clean.

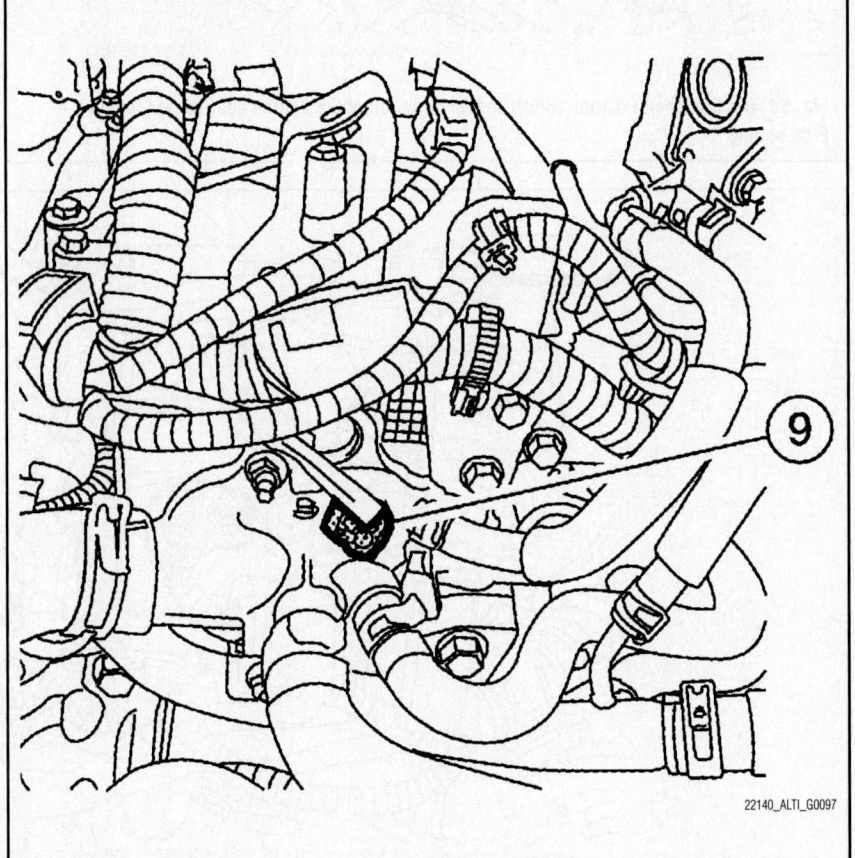

Fig. 63 The ECT is in the water outlet assembly near the upper radiator hose connection.

HEATED OXYGEN (HO2S) SENSOR

LOCATION

See Figure 64.

Refer to the accompanying illustration.

REMOVAL & INSTALLATION

1. Make sure the exhaust manifold is cool.

2. Unplug the connector and remove the sensor with an oxygen sensor wrench or socket.

3. When installing, lightly coat the threads with anti-seize and torque to 23 ft. lbs. (31 Nm). Do not over tighten or the Malfunction Indicator Light (MIL) may turn on.

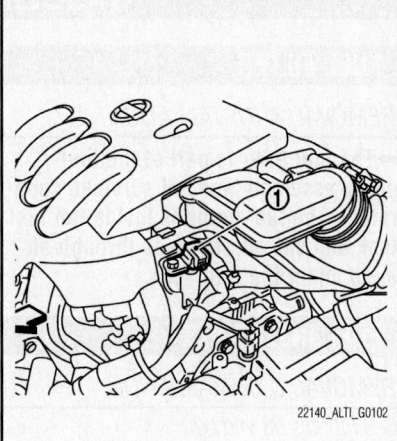

22140_ALTI_G0102

Fig. 65 The Knock sensor is mounted on the engine block below the intake manifold—2.0L engine

22140_ALTI_G0101

Fig. 67 The Mass Air Flow (MAF) sensor is in the air filter housing—2.5L engine shown, 2.0L engine similar

2. During installation, do not over tighten the bolt or the sensor will not function properly.

MASS AIR FLOW (MAF) SENSOR

LOCATION

See Figure 67.

Refer to the accompanying illustration.

REMOVAL & INSTALLATION

The Mass Air Flow (MAF) sensor has a thin wire extended into the air stream flowing into the engine. The sensor uses electric current to heat the wire to a specific temperature. As air

flowing past the wire cools it, more current is required to maintain the predetermined temperature. The ECM detects the air flow by measuring this current change.

VEHICLE SPEED SENSOR (VSS)

LOCATION

On vehicles with ABS, vehicle speed is calculated by the ABS control unit using inputs from all four wheel speed sensors. There is no dedicated vehicle speed sensor.

On vehicles with a manual transaxle and without an Anti-lock Brake System (ABS), a speedometer drive gear is located on the transaxle housing.

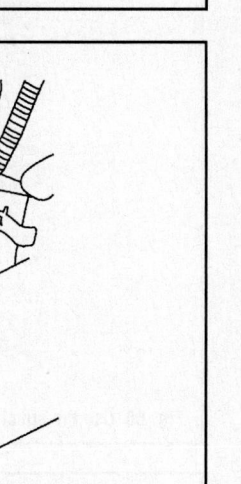

22140_ALTI_G0100

Fig. 64 The primary oxygen sensor (air/fuel ratio sensor) is in the intake manifold below the heat shield—2.5L engine shown.

INTAKE AIR TEMPERATURE (IAT) SENSOR

LOCATION

The intake air temperature sensor is part of the Mass Air Flow (MAF) sensor.

KNOCK SENSOR (KS)

LOCATION

See Figures 65 and 66.

Refer to the accompanying illustrations.

REMOVAL & INSTALLATION

1. Remove the bolt to remove the sensor from the engine block.

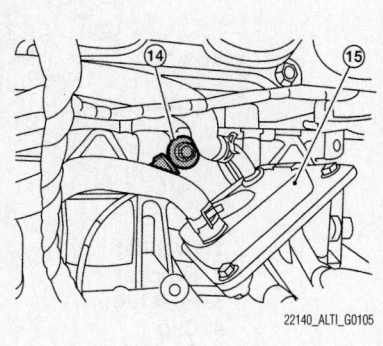

22140_ALTI_G0105

Fig. 66 The knock sensor is on the front of the engine block near the oil cooler—2.5L engine

FUEL SYSTEM SERVICE PRECAUTIONS

Safety is the most important factor when performing not only fuel system maintenance, but any type of maintenance. Failure to conduct maintenance and repairs in a safe manner may result in serious personal injury or death. Work on a vehicle's fuel system components can be accomplished safely and effectively by adhering to the following rules and guidelines.

• To avoid the possibility of fire and personal injury, always disconnect the negative battery cable unless the repair or test procedure requires that battery voltage be applied.

• Always relieve the fuel system pressure prior to disconnecting any fuel system component (injector, fuel rail, pressure regulator, etc.) fitting or fuel line connection. Exercise extreme caution whenever relieving fuel system pressure to avoid exposing skin, face and eyes to fuel spray. Please be advised that fuel under pressure may penetrate the skin or any part of the body that it contacts.

• Always place a shop towel or cloth around the fitting or connection prior to loosening to absorb any excess fuel due to spillage. Ensure that all fuel spillage is quickly removed from engine surfaces. Ensure that all fuel-soaked cloths or towels are deposited into a flame-proof waste container with a lid.

• Always keep a dry chemical (Class B) fire extinguisher near the work area.

• Do not allow fuel spray or fuel vapors to come into contact with a spark or open flame.

• Always use a second wrench when loosening or tightening fuel line connection fittings. This will prevent unnecessary stress and torsion on fuel piping. Always follow the proper torque specifications.

• Always replace worn fuel fitting O-rings with new ones. Do not substitute fuel hose where rigid pipe is installed.

FUEL SYSTEM PRESSURE

RELIEVING

1. Remove the fuel pump fuse.
2. Start the engine.
3. After the engine stalls, crank it three times to release all remaining fuel pressure.
4. Turn ignition switch OFF.

FUEL FILTER

REMOVAL & INSTALLATION

➡**The fuel filter is part of the fuel pump assembly and not serviced separately. The fuel pump is inside the fuel tank and can be accessed through an open under the rear seat.**

FUEL INJECTORS

REMOVAL & INSTALLATION

See Figures 68 and 69.

1. Release the fuel pressure.
2. Remove quick connector cap from quick connector connection.

3. Disconnect fuel feed hose from hose clamp.

• With the sleeve side of quick connector release facing quick connector, install quick connector release onto fuel tube.

• Insert quick connector release into quick connector until sleeve contacts and goes no further. Hold quick connector release on that position.

✳✳ WARNING

Inserting quick connector release hard will not disconnect quick connector. Hold quick connector release where it contacts and goes no further.

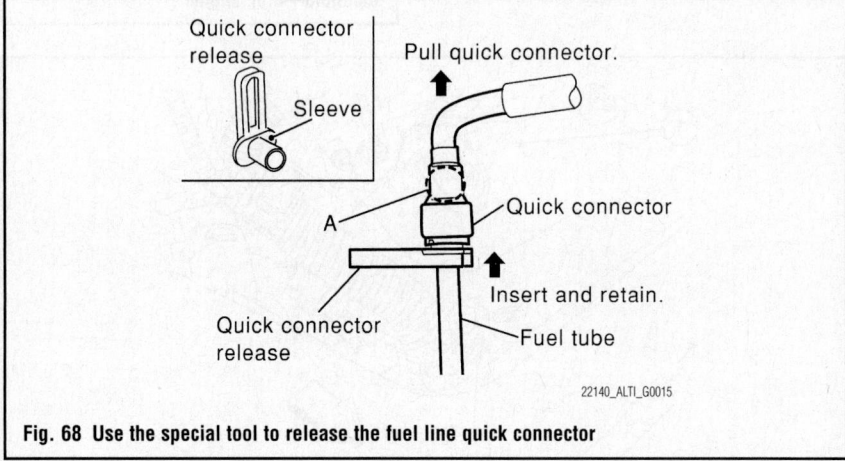

Fig. 68 Use the special tool to release the fuel line quick connector

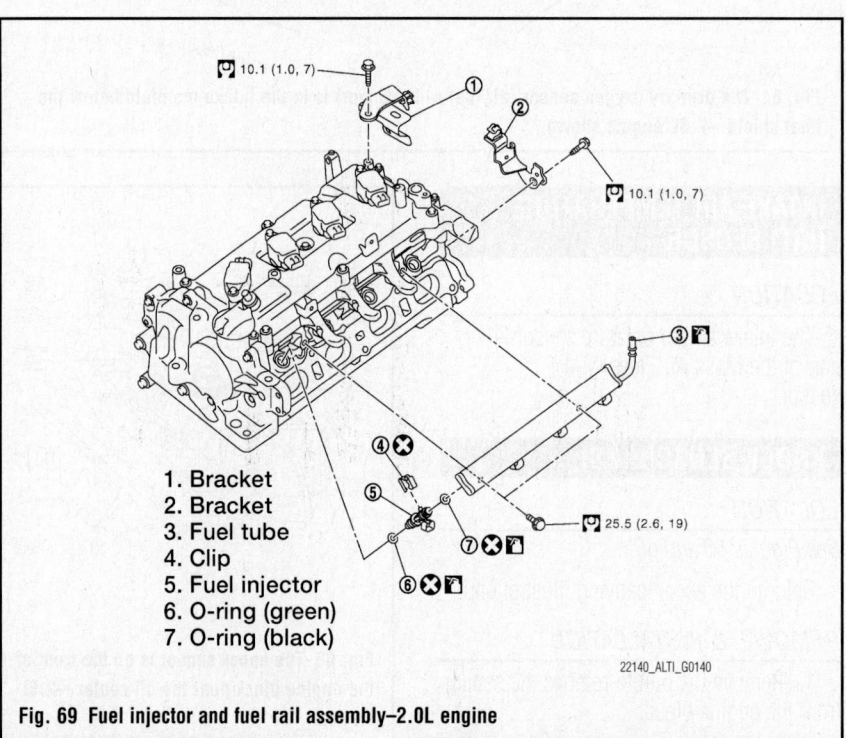

1. Bracket
2. Bracket
3. Fuel tube
4. Clip
5. Fuel injector
6. O-ring (green)
7. O-ring (black)

Fig. 69 Fuel injector and fuel rail assembly–2.0L engine

- Draw and pull out quick connector straight from fuel tube.

❄❄ WARNING

Pull quick connector holding "A" position. Do not pull with lateral force applied. O-ring inside quick connector may be damaged. Do not bend or twist connection between quick connector and fuel feed hose.

- To keep clean the connecting portion and to avoid damage and foreign materials, cover them completely with plastic bags or something similar.
4. Remove intake manifold.
5. Remove fuel tube bolts.
6. Remove the fuel tube and fuel injector assembly.

❄❄ WARNING

When removing, be careful to avoid any interference with fuel injector.

7. Remove fuel injector from fuel tube with the following procedure:
   - Open and remove clip.
   - Remove fuel injector from fuel tube by pulling straight. Be careful not to damage fuel injector nozzle during removal.

To install:

➡ **Upper and lower injector O-rings are different. Be careful not to confuse them. Handle O-ring with bare hands. Never wear gloves.**

8. Lubricate O-ring with new engine oil and fit them onto the injector. The engine-side O-ring is green, the fuel rail O-ring is black. Be careful not to twist or stretch O-rings.
9. Slowly insert injector straight into fuel tube. Never twist it.
10. Fit fuel injectors into the cylinder head and torque the bolts to 19 ft. lbs. (25 Nm).
11. Connect the fuel hose quick connector and pressurize the fuel system to check for leaks. The pump will run for two seconds each time the ignition switch is turned **ON**.
12. Install the intake manifold.

FUEL PUMP

REMOVAL & INSTALLATION

See Figure 70.

➡ **Fuel will be spilled when removing fuel pump assembly if the tank is full.**

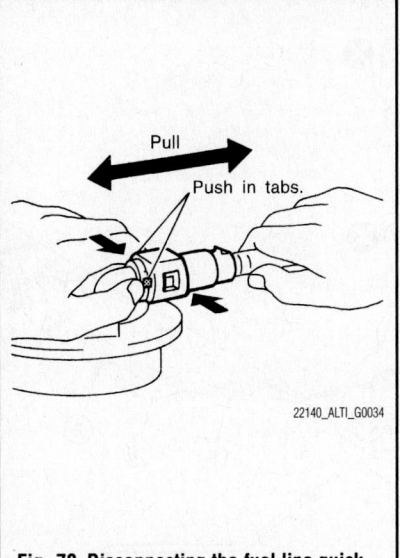

Fig. 70 Disconnecting the fuel line quick connector

If the fuel gauge indicates more than (⅞ full), drain at least 3 ⅛ gallons (12L) from the fuel tank.

1. Open fuel door and unscrew the fuel filler cap to release the pressure inside the fuel tank.
2. Release fuel system pressure.
3. Remove rear seat bottom.
4. Turn the three cover retainers 90° in counterclockwise and remove the fuel pump inspection hole cover.
5. Disconnect the wiring and fuel hose quick connectors. To remove the quick connector, hold the sides of the connector, push in the tabs and pull the tube straight out. The tube can be removed only when the tabs are completely depressed. Do not twist or use any tools.
6. To keep the connectors clean and to avoid damage, cover them completely with plastic bags or something similar. Do not insert plugs to prevent damage to O-ring.
7. Remove the locking ring using the correct tool and carefully lift the pump/filter/sending unit out of the tank. Take care not to bend the float arm.

To install:

8. Carefully fit the fuel pump assembly into the tank.
9. Reconnect the hoses and wiring.
10. Connect the battery and turn the ignition switch ON three or four times to run the pump, pressurize the system and check for leaks.

FUEL TANK

REMOVAL & INSTALLATION

See Figure 71.

➡ **Fuel will be spilled when removing fuel pump assembly if the tank is full. If the fuel gauge indicates more than (⅞ full), drain at least 3 ⅛ gallons (12L) from the fuel tank.**

1. Check fuel level with the vehicle on a level surface. If the fuel gauge indicates more than (⅞ full), drain at least 3 ⅛ gallons (12L) from the fuel tank.
2. Siphon fuel from fuel tank if necessary.
3. Open fuel door and unscrew the fuel filler cap to release the pressure inside the fuel tank.
4. Release the fuel pressure.
5. Remove rear seat bottom.
6. Turn the three retainers 90° in a counterclockwise direction and remove the fuel pump inspection hole cover.
7. Disconnect wiring and fuel feed hoses. To keep the connectors clean and to avoid damage, cover them completely with plastic bags or something similar. Do not insert plugs to prevent damage to O-ring.
8. Remove center exhaust pipe.
9. Remove fuel tank protector.
10. Loosen fuel filler hose clamp and remove fuel filler hose.

❄❄ WARNING

Do not remove fuel filler hose from fuel filler tube. Mark components for alignment.

11. Remove vent hose and EVAP hose at rear of fuel tank.
12. Support center of fuel tank with transmission jack.
13. Remove fuel tank bands. If they are not marked "R" and "L," mark them now.
14. Lower transmission jack carefully to remove fuel tank while supporting it by hand.

❄❄ WARNING

Fuel tank may be in an unstable position because of the shape of fuel tank bottom. Be sure to support tank securely.

To install:

15. Secure tank on a transmission jack and carefully raise it into position
16. Fit the fuel tank bands (marked "R" and "L") and tighten the bolts to 23 ft. lbs. (31 Nm).
17. Connect the filler and vent hoses and tighten the clamps.

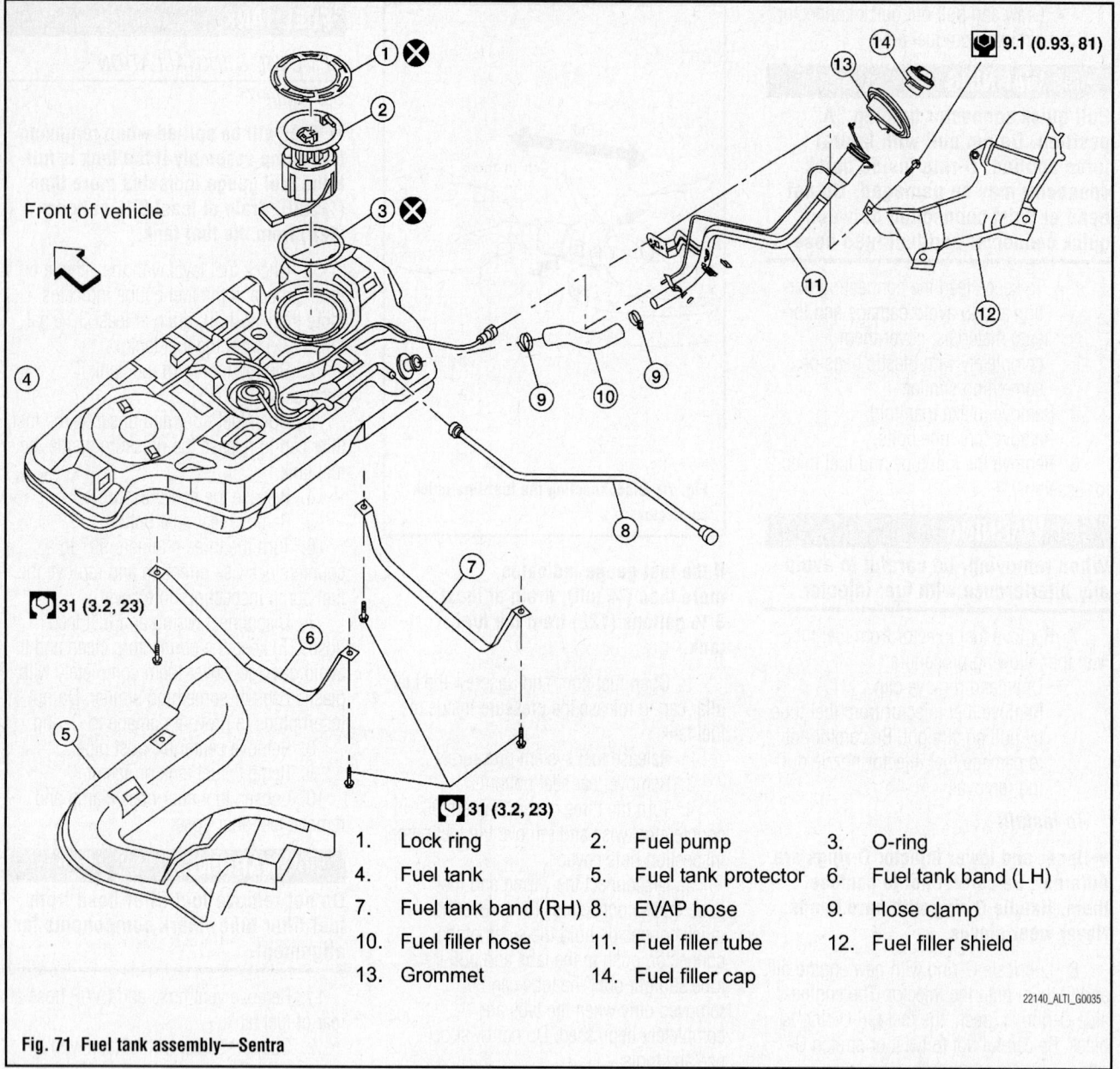

Front of vehicle

31 (3.2, 23)

31 (3.2, 23)

9.1 (0.93, 81)

1.	Lock ring	2.	Fuel pump	3.	O-ring
4.	Fuel tank	5.	Fuel tank protector	6.	Fuel tank band (LH)
7.	Fuel tank band (RH)	8.	EVAP hose	9.	Hose clamp
10.	Fuel filler hose	11.	Fuel filler tube	12.	Fuel filler shield
13.	Grommet	14.	Fuel filler cap		

22140_ALTI_G0035

Fig. 71 Fuel tank assembly—Sentra

18. Connect remaining tubes, wiring and fuel lines.

19. With some fuel in the tank, turn the ignition switch ON to pressurize the system and check for leaks.

20. Install remaining components.

THROTTLE BODY

REMOVAL & INSTALLATION

1. Remove engine cover.

2. Remove air intake ducts and air filter housing as required.

3. Disconnect throttle wiring.

4. Disconnect coolant hoses and plug them immediately to prevent coolant leaks.

5. Remove the bolts to remove throttle assembly.

6. Installation is the reverse of removal. Torque the bolts to 7 ft. lbs. (10 Nm). Do not over tighten these bolts or the throttle may not work properly.

7. A throttle relearn procedure is required.

- Make sure that accelerator pedal is fully released.
- Turn ignition switch ON.
- Turn ignition switch OFF. During the next 10 seconds, you should hear the throttle moving as the ECM learns the "closed throttle" position.

HEATING & AIR CONDITIONING SYSTEM

BLOWER MOTOR

REMOVAL & INSTALLATION

1. Remove the instrument panel. See Heater Core Removal & Installation

2. Disconnect the wiring and remove the blower motor.

3. Installation is the reverse of removal.

HEATER CORE

REMOVAL & INSTALLATION

See Figure 72.

➡ **Heater core removal requires removing the dashboard.**

1. Before servicing the vehicle, refer to the Precautions Section.

2. Position the steering wheel in the straight-ahead position.

3. Turn the ignition switch OFF.

4. Disconnect the negative (() battery cable; then, the positive (+) battery cable.

➡ **Wait at least 10 minutes after disconnecting the battery cables for the charge in the air bag circuit to dissipate before working on the air bag module(s).**

5. Properly discharge and recover the refrigerant from the A/C system.

6. Drain the cooling system.

7. Install the steering column by removing the following:

- Lower dash cover
- Airbag module: store it face up out of the way
- Steering wheel
- Steering column covers
- Combination switch and spiral cable
- Disconnect all wiring
- Install the bolt to disconnect the upper end of the intermediate shaft from the power steering motor.
- Install the nuts to lower the power steering assembly from the vehicle.

8. Install the center console. There are two screws under the front panel (below the brake handle) and two more at the back (open the lid).

9. Install the lower instrument panel covers and disengage the diagnostic connector and hood release handle.

10. Install the uppermost part of the center cluster trim, then install four screws to

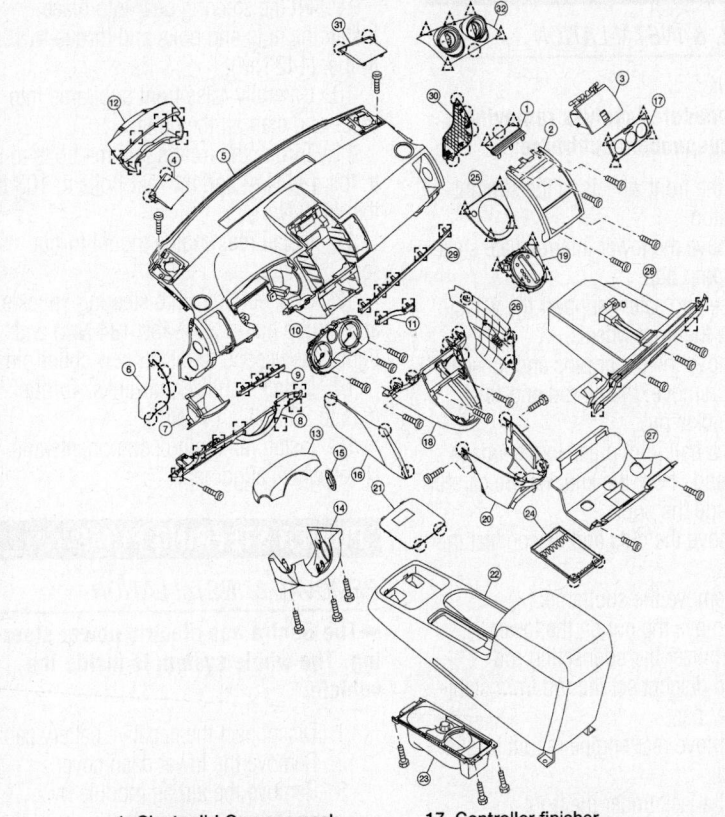

1. Cluster lid C upper mask
2. Cluster lid C
3. Cluster lid C storage bin
4. Speaker grille (LH)
5. Instrument panel
6. Instrument side mask (LH)
7. Fuse block lid storage bin
8. Instrument lower finisher
9. Instrument panel trim (LH)
10. Combination meter
11. Instrument panel trim center
12. Cluster lid A
13. Steering column cover upper
14. Steering column cover lower
15. Steering lock escutcheon
16. Instrument lower cover (LH)
17. Controller finisher
18. Instrument upper cover (center)
19. CVT finisher
20. Instrument lower cover (center)
21. Center console mat
22. Center console
23. Center console cup holder
24. Center console tray
25. MT finisher
26. Instrument lower cover (RH)
27. Glove box lower finisher
28. Glove box assembly
29. Instrument panel trim (RH)
30. Instrument side mask (RH)
31. Speaker grille (RH)

22140_ALTI_G0038

Fig. 72 Exploded view of Sentra dashboard

install the whole center cluster. Unplug wiring connectors as needed.

11. Install the glove box. There are two screws below the door and four inside at the top.

12. Disengage three clips to install the instrument cluster cover, then install three screws to install the instrument cluster. Unplug the connectors as necessary.

13. Install the passenger side airbag module from the steering member. Store it face up out of the way.

14. Install the speakers at each front corner of the dashboard and install the bolts securing the dashboard.

15. Install the screws at each upper and lower corner of the dashboard and in the instrument cluster opening.

16. Disconnect the antenna and install the dashboard.

17. Disconnect the refrigerant lines from the evaporator.

18. Disconnect the hoses from the heater core.

STEERING

POWER STEERING GEAR

REMOVAL & INSTALLATION

See Figure 73.

➡**This procedure involves removing the front suspension subframe.**

1. Set the front wheels to the straight-ahead position.

2. Remove the lower intermediate steering shaft clamp bolt.

3. Raise and safely support the vehicle and remove the front wheels.

4. Remove the cotter pins and loosen (but do not remove) the tie rod end-to-steering knuckle nuts.

5. Use a ball joint press to disengage the tie rod ends from the knuckle. Be careful not to damage the boot.

6. Remove the nuts and disconnect the tie rod ends.

7. To remove the subframe:

- Remove the nut on the lower stabilizer bar connecting rod and disconnect the rod from stabilizer bar.
- Remove rear engine mount torque rod.
- Set a jack under the front subframe, then remove the subframe bolts.
- Gradually lower the subframe far enough to remove the steering gear.

8. Remove the bolts and nuts to remove the steering gear assembly.

To install:

9. Fit the steering gear into place, install the nuts and bolts and torque to 105 ft. lbs. (142 Nm).

10. Carefully raise front subframe into place and start all the bolts.

11. Torque the front subframe bolts to 69 ft. lbs. (94 Nm) and the rear bolts to 103 ft. lbs. (140 Nm).

12. Install rear engine mount torque rod.

13. Connect tie rod to steering knuckle and torque nut to 25 ft. lbs. (34 Nm) and tighten as needed to install new cotter pins.

14. Install stabilizer bar links. Torque nuts to 30 ft. lbs. (41 Nm).

15. Install remaining components and check wheel alignment.

POWER STEERING PUMP

REMOVAL & INSTALLATION

➡**The Sentra has electric power steering. The whole system is inside the vehicle.**

1. Disconnect the negative battery cable.

2. Remove the lower dash cover.

3. Remove the airbag module by removing the two tamper-proof bolts at the back of the steering wheel. Store the airbag module face up out of the way.

4. Remove the steering wheel using a puller.

5. Remove the steering column covers.

6. Remove the turn signal and wiper switches by pressing the tabs and pulling straight out.

7. Remove the spiral cable and carefully disconnect the wiring.

8. Remove the bolt to disconnect the upper end of the intermediate shaft from the power steering motor.

9. Remove the nuts to lower the power steering assembly from the vehicle.

To install:

10. Fit the power steering assembly into place, install the nuts and connect the wiring.

11. Connect the intermediate shaft and install the bolt.

12. Install the switches and connect the wiring.

13. When installing the spiral cable, make sure the centering marks (at the 7 o'clock position) are aligned with the cable in the neutral position. To find neutral, turn the locating pin (that engages the steering wheel) all the way to the right, then back 2.6 turns.

14. Install the steering wheel and torque the nut to 25 ft. lbs. (31 Nm).

15. Install the airbag module and torque the screws to 8 ft. lbs. (10 Nm). Do not over tighten these screws or the airbag may not deploy properly.

16. Make sure no one is inside the vehicle when connecting the battery.

BLEEDING

1. Run engine until the fluid temperature reaches 50–80° C (122–176° F) in reservoir tank, and keep engine speed idle.

2. Turn steering wheel several times from full left stop to full right stop.

3. Hold steering wheel at each lock position for five seconds and carefully, check for fluid leakage.

❋❋ WARNING

Do not hold the steering wheel in a locked position for more than 10 seconds. There is the possibility that oil pump may be damaged.

4. If fluid leakage at connections is noticed, then loosen flare nut and then retighten. Do not overtighten connector as this can damage O-ring, washer and connector.

5. Check steering gear boots for accumulation of fluid indicating leaks in the steering gear.

6. Check steering fluid level.

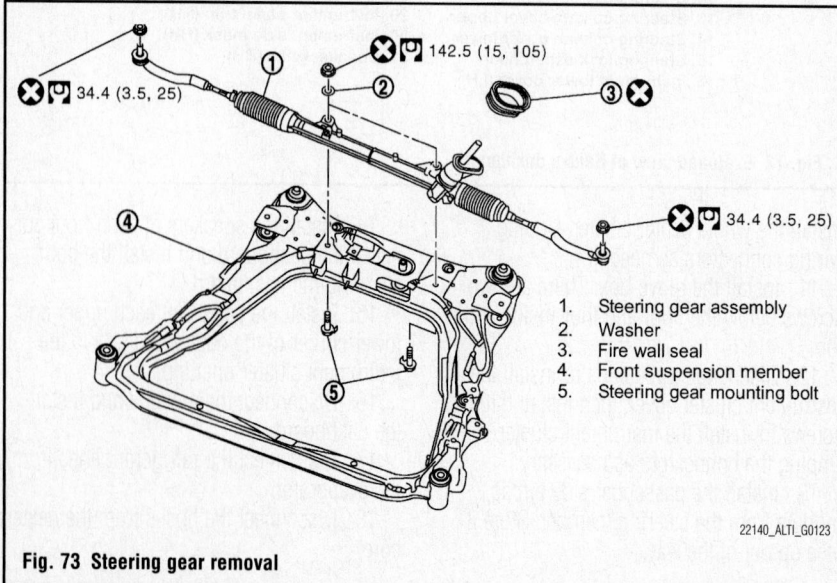

34.4 (3.5, 25)
142.5 (15, 105)
34.4 (3.5, 25)

1. Steering gear assembly
2. Washer
3. Fire wall seal
4. Front suspension member
5. Steering gear mounting bolt

22140_ALTI_G0123

Fig. 73 Steering gear removal

LOWER BALL JOINT

REMOVAL & INSTALLATION

The lower ball joint is part of the lower control arm and cannot be replaced separately.

LOWER CONTROL ARM

REMOVAL & INSTALLATION

See Figure 74.

1. Raise and safely support vehicle and remove front wheels.

2. Disconnect the stabilizer bar.

3. Remove ball joint pinch bolt and disengage ball joint from steering knuckle with a ball joint separator tool. Be careful not to damage the boot.

4. Remove the bolts and remove the control arm.

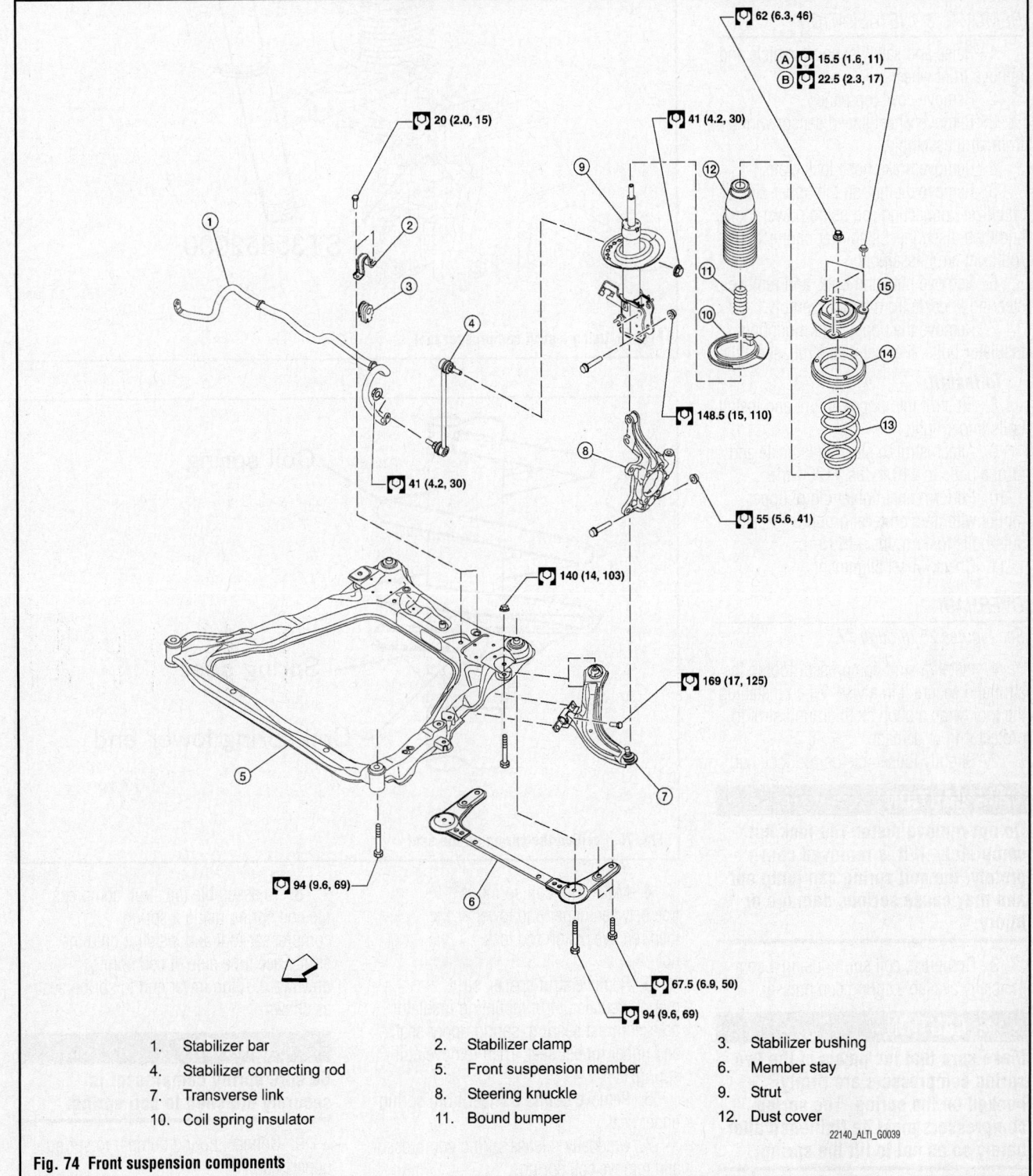

1.	Stabilizer bar	2.	Stabilizer clamp	3.	Stabilizer bushing
4.	Stabilizer connecting rod	5.	Front suspension member	6.	Member stay
7.	Transverse link	8.	Steering knuckle	9.	Strut
10.	Coil spring insulator	11.	Bound bumper	12.	Dust cover

22140_ALTI_G0039

Fig. 74 Front suspension components

To install:

5. Installation is the reverse of removal, noting the following:
- Torque front control arm bolts to 125 ft. lbs. (169 Nm) and rear bolt to 103 ft. lbs. (140 Nm). Torque ball joint pinch bolt to 41 ft. lbs. (55 Nm).

MACPHERSON STRUT

REMOVAL & INSTALLATION

1. Raise and safely support vehicle and remove front wheels.
2. Remove cowl top panel.
3. Remove wheel speed sensor wiring from strut assembly.
4. Remove brake hose lock plate.
5. Remove the nut on the upper side of stabilizer connecting rod using power tool, and then disconnect stabilizer connecting rod from strut assembly.
6. Remove nuts and bolts and remove steering knuckle from strut assembly.
7. Remove the upper strut mounting insulator bolts, then remove strut assembly.

To install:

8. Fit strut into upper mount and install bolts finger tight.
9. Attach strut to steering knuckle and torque bolts to 110 ft. lbs. (120 Nm).
10. Perform final tightening of upper mount with tires on level ground. Torque nuts/bolts to 11 ft. lbs. (15 Nm).
11. Check wheel alignment.

OVERHAUL

See Figures 75 through 77.

1. Install a strut compressor tool to the strut and secure it in a vise. When installing the tool, wrap a shop cloth around strut to protect it from damage.
2. Slightly loosen piston rod lock nut.

✳✳ CAUTION

Do not remove piston rod lock nut completely. If it is removed completely, the coil spring can jump out and may cause serious damage or injury.

3. Compress coil spring using a commercially available spring compressor.

✳✳ CAUTION

Make sure that the pawls of the two spring compressors are firmly hooked on the spring. The spring compressors must be tightened alternately so as not to tilt the spring.

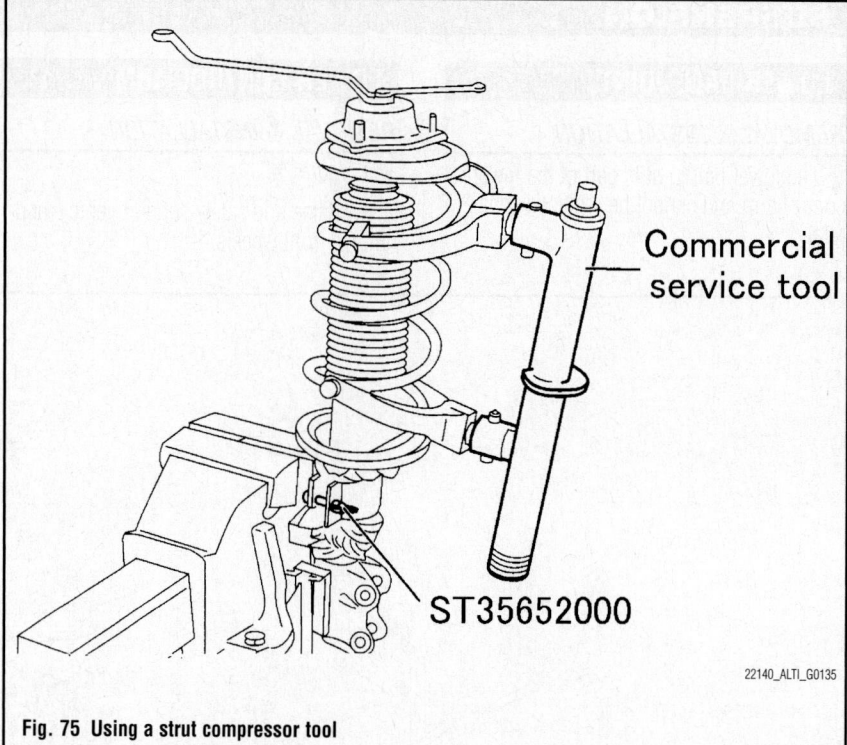

Fig. 75 Using a strut compressor tool

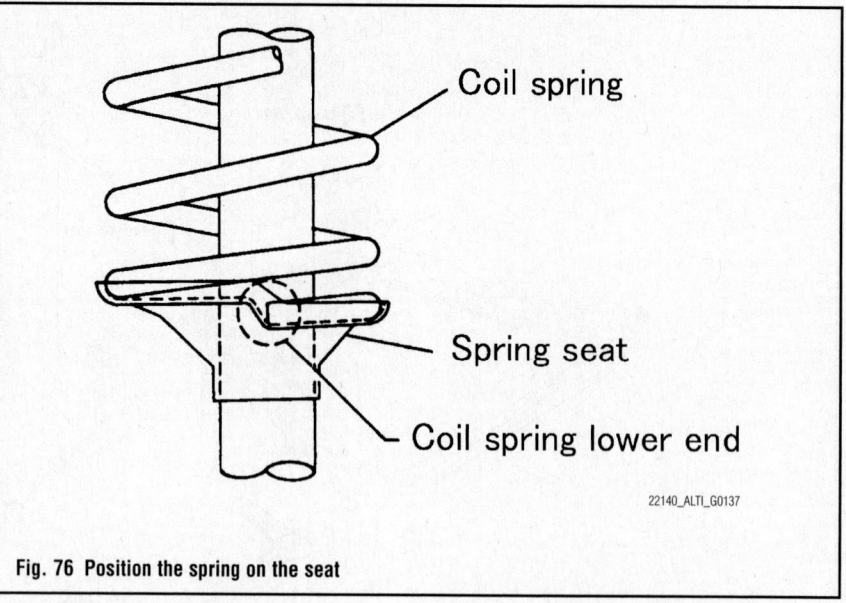

Fig. 76 Position the spring on the seat

4. Make sure coil spring is free between upper and lower seats, then remove piston rod lock nut.
5. Remove strut spacer, strut mount insulator, strut mounting insulator bracket thrust bearing, spring upper seat, and upper rubber seat. Then remove coil spring.
6. Remove bound bumper from spring upper seat.
7. Gradually release spring compressor and remove coil spring.

8. To assemble the strut, compress the coil spring using a spring compressor tool and install it onto the strut. Face tube side of coil spring downward. Align lower end to spring seat as shown.

✳✳ CAUTION

Be sure spring compressor is securely attached to coil spring.

9. Connect bound bumper to spring upper seat.

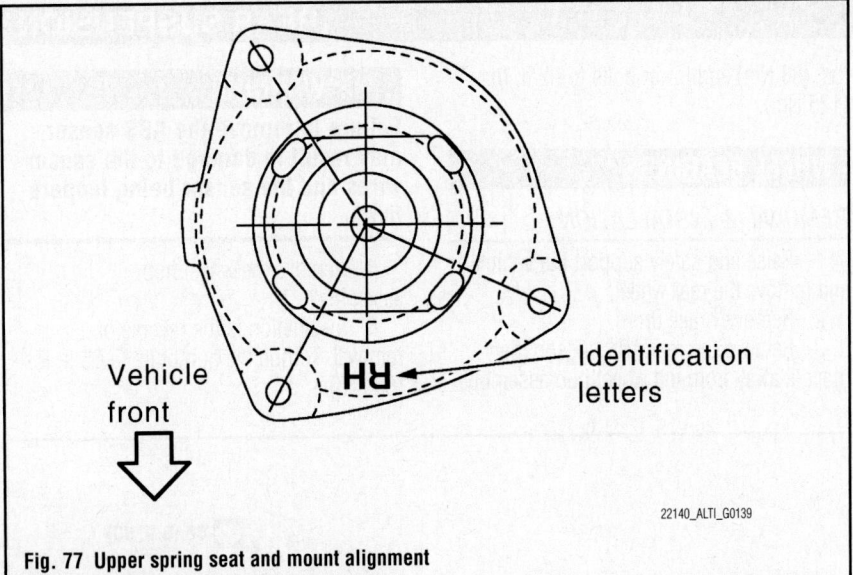

Vehicle front

Identification letters

22140_ALTI_G0139

Fig. 77 Upper spring seat and mount alignment

※※ **WARNING**

Be sure to install bound bumper to spring upper seat securely. When installing bound bumper, use soapy water. Do not use machine oil or other lubricants.

10. Install the upper rubber seat, spring upper seats, thrust bearing, strut mount insulator, and strut spacer. Install a new lock nut but don't tighten it yet.

11. Be sure the strut mount insulator and spring seat are positioned properly.

12. Torque the new piston rod lock nut to 46 ft. lbs. (62 Nm).

13. Carefully remove the spring compressor.

STEERING KNUCKLE

REMOVAL & INSTALLATION

1. Raise and safely support vehicle and remove front wheels.

2. Remove brake rotor and brake caliper without disconnecting hydraulic line. Hang caliper from body with wire.

3. Remove cotter pin, then loosen half-shaft lock nut. Temporarily leave the nut installed to prevent damage to threads.

4. Separate the halfshaft from the wheel hub and bearing assembly by tapping the end of the shaft using a hammer and brass drift or wood block, then remove hub lock nut.

➡Use a suitable puller if hub and shaft cannot be separated using the above procedure.

5. Loosen tie rod nut and disengage tie rod from steering knuckle with a ball joint separator. Be careful not to damage the boot.

6. Remove the nut and separate tie rod from steering knuckle.

7. Remove lower ball joint pinch bolt from steering knuckle.

8. Remove nuts and bolts and remove steering knuckle from strut assembly.

To install:

9. Fit steering knuckle to strut and install bolts finger tight.

10. Fit halfshaft into hub and install nut finger tight.

11. Fit ball joint into steering knuckle:
   - Torque bolt to 41 ft. lbs. (55 Nm).
   - Torque knuckle/strut bolts to 110 ft. lbs. (120 Nm).

12. Fit tie rod end into steering knuckle. Torque nut to 25 ft. lbs. (35 Nm) and tighten as needed to install a new cotter pin.

13. Torque halfshaft nut to 92 ft. lbs. (125 Nm) and install a new cotter pin.

14. Install remaining components and check wheel alignment.

STABILIZER BAR

REMOVAL & INSTALLATION

➡This procedure involves lowering the whole front suspension subframe.

1. Separate intermediate shaft from steering gear pinion shaft.

2. Raise and safely support the vehicle and remove the front wheels.

3. Remove the nut on the lower side of stabilizer bar connecting rod and disconnect the rod from stabilizer bar. If necessary remove stabilizer bar connecting rod upper nut.

4. Loosen tie rod joint nut and disengage tie rod from steering knuckle using a ball joint separator. Be careful not to damage the boot. Remove the nut and disconnect the tie rod joint from the steering knuckle.

5. Remove rear engine mount torque rod.

6. Set a jack under the front subframe, then remove the subframe bolts.

7. Gradually lower subframe in order to remove stabilizer bar bolts.

8. Remove the stabilizer clamp bolts, then remove stabilizer clamps and stabilizer bushing.

9. Remove stabilizer bar.

To install:

10. Fit stabilizer bar into place and install the clamps. Torque bolts to 15 ft. lbs. (20 Nm).

11. Carefully raise front subframe into place and start all the bolts.

12. Torque the rear subframe bolts to 103 ft. lbs. (140 Nm) and front bolts to 69 ft. lbs. (94 Nm).

13. Install rear engine mount torque rod.

14. Connect tie rod to steering knuckle and torque nut to 25 ft. lbs. (34 Nm) and tighten as needed to install new cotter pins.

15. Install stabilizer bar links. Torque nuts to 30 ft. lbs. (41 Nm).

16. Install remaining components and check wheel alignment.

WHEEL BEARINGS

REMOVAL & INSTALLATION

1. Remove the steering knuckle.

2. Remove the bolts to remove the hub and bearing assembly from the steering knuckle.

3. Installation is the reverse of removal.
   - Torque the hub assembly bolts to 65 ft. lbs. (88 Nm).

COIL SPRING

REMOVAL & INSTALLATION

See Figure 78.

1. Set jack under rear suspension beam.
2. Remove both lower shock absorber bolts.
3. Carefully lower the suspension and remove the spring.
4. Installation is the reverse of removal. Torque upper shock absorber bolts to 50 ft.

lbs. (88 Nm) and lower bolts to 92 ft. lbs. (125 Nm).

WHEEL BEARINGS

REMOVAL & INSTALLATION

1. Raise and safely support the vehicle and remove the rear wheel.
2. Remove brake drum.
3. Remove the rear ABS sensor, then move it away from the wheel hub assembly.

✳✳ WARNING

Failure to remove the ABS sensor may result in damage to the sensor wires and the sensor being inoperative.

4. Remove the wheel hub assembly.
5. Installation is the reverse of removal. Torque the hub bolts to 65 ft. lbs. (88 Nm).

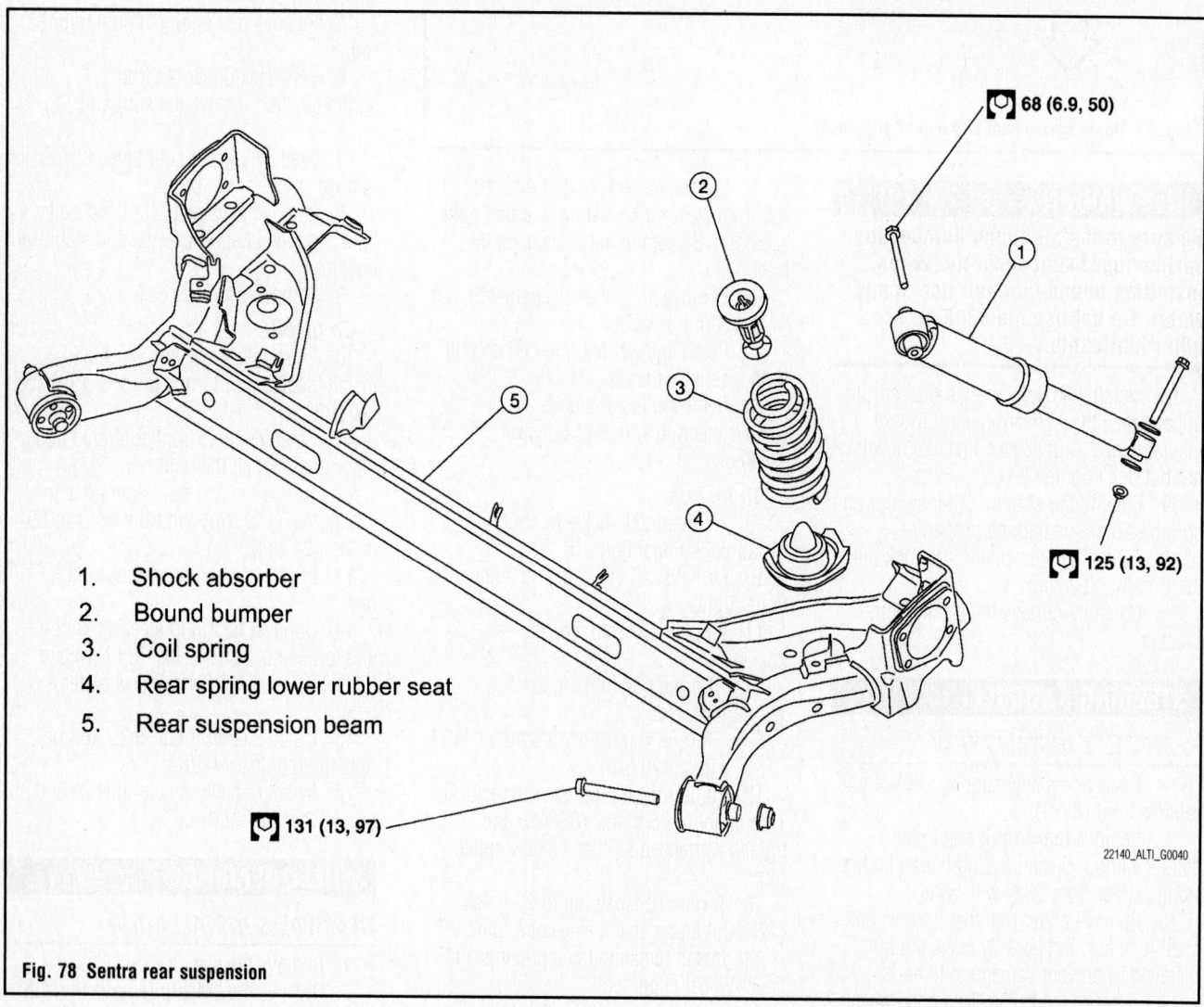

1. Shock absorber
2. Bound bumper
3. Coil spring
4. Rear spring lower rubber seat
5. Rear suspension beam

68 (6.9, 50)

125 (13, 92)

131 (13, 97)

22140_ALTI_G0040

Fig. 78 Sentra rear suspension

NISSAN

Titan

20

SPECIFICATIONS AND MAINTENANCE CHARTS

ENGINE AND VEHICLE IDENTIFICATION

Engine							Model Year	
Code ①	Liters (cc)	Cu. In.	Cyl.	Fuel Sys.	Engine	Eng. Mfg.	Code ②	Year
VK56DE	5.6 (5552)	338.8	8	MFI	DOHC	Nissan	9	2009
							A	2010

MFI: Multi-port Fuel Injection

DOHC: Double Overhead Camshafts

① 4th digit is engine type: A= VK56DE. B= VK56DE (flex fuel).

② 10th digit of the Vehicle Identification Number (VIN)

37663_TITA_C0001

GENERAL ENGINE SPECIFICATIONS

Year	Model	Engine Displacement Liters	Engine ID	Net Horsepower @ rpm	Net Torque @ rpm (ft. lbs.)	Bore x Stroke (in.)	Com- pression Ratio	Oil Pressure @ rpm
2009	Titan	5.6	VK56DE	317@5200	385@3400	3.86X3.62	9.8:1	43@2000
2010	Titan	5.6	VK56DE	317@5200	385@3400	3.86X3.62	9.8:1	43@2000

37663_TITA_C0002

ENGINE TUNE-UP SPECIFICATIONS

Year	Engine Displacement Liters	Engine ID	Spark Plug Gap (in.)	Ignition Timing	Fuel Pump (psi) ①	Idle Speed ②	Valve Clearance (in.) In.	Valve Clearance (in.) Ex.
2009	5.6	VK56DE	0.043	③	51	④	0.010-0.013	0.011-0.015
1010	5.6	VK56DE	0.043	③	51	④	0.010-0.013	0.011-0.015

NOTE: The Vehicle Emission Control Information label often reflects specification changes made during production. The label figures must be used if they differ from those in this chart.

B: Before top dead center

① At idle

② Automatic transmission in Neutral

③ 10-20 degrees BTDC

④ 600-700

37663_TITA_C0003

CAPACITIES

Year	Model	Engine Displacement Liters	Engine ID	Engine Oil with Filter (qts.)	Transmission (pts.)	Transfer Case (pts.)	Drive Axle Front (pts.)	Drive Axle Rear (pts.)	Fuel Tank (gal.) ①	Cooling System (qts.)
2009	Titan	5.6	VK56DE	6.78	22.5	4.36	3.38	4.25	28.0	12.18
2010	Titan	5.6	VK56DE	6.78	22.5	4.36	3.38	4.25	28.0	12.18

NOTE: All capacities are approximate. Add fluid gradually and check to be sure a proper fluid level is obtained.

① Long wheel base (37)

37663_TITA_C0005

FLUID SPECIFICATIONS

Year	Model	Engine Displ. Liters	Engine ID	Engine Oil	Auto Trans.	Transfer Case	Drive Axle Front	Drive Axle Rear	Power Steering Fluid	Engine Coolant	Brake Fluid
2009	Titan	5.6	VK56DE	5W30	①	②	③	④	⑤	⑥	DOT 3
2010	Titan	5.6	VK56DE	5W30	①	②	③	④	⑤	⑥	DOT 3

DOT: Department Of Transpotation

① Nissan Matic S ATF. If not available Nissan Matic J ATF may be used.

② Nissan Matic D ATF.

③ Nissan differential oil hypoid super GL-5 80W-90 or API GL-5 viscosity 80W-90. For hot climates, viscosity SAE 90 is suitable above ambient temperature 32 degrees F.

④ Nissan differential oil synthe 75W-140 or API GL-5 synthetic gear oil viscosity SAE 75W-140.

⑤ Nissan PSF. Dexron VI type ATF may also be used.

⑥ Nissan long life antifreeze

37663_TITA_C0004

VALVE SPECIFICATIONS

Year	Engine Displacement Liters	Engine ID	Seat Angle (deg.)	Face Angle (deg.)	Spring Test Pressure (lbs. @ in.)	Spring Installed Height (in.)	Stem-to-Guide Clearance (in.) Intake	Stem-to-Guide Clearance (in.) Exhaust	Stem Diameter (in.) Intake	Stem Diameter (in.) Exhaust
2009	5.6	VK56DE	45.15-45.45	45	37.0@1.457	1.9913	0.0008-0.0021	0.0012-0.0025	0.2348-0.2354	0.2344-0.2350
2010	5.6	VK56DE	45.15-45.45	45	37.0@1.457	1.9913	0.0008-0.0021	0.0012-0.0025	0.2348-0.2354	0.2344-0.2350

37663_TITA_C0006

CAMSHAFT AND BEARING SPECIFICATIONS CHART

All measurements are given in inches.

Year	Engine Displ. Liters	Engine ID/VIN	Journal Dia.	Brg. Oil Clearance	Shaft End-play	Runout	Journal Bore	Lobe Height	
								Intake	Exhaust
2009	5.6	VK56DE	1.0217-1.0224	0.0012-0.0028	0.0045-0.0074	0.0008	1.0236-1.0244	1.7663-1.7738	1.7746-1.7821
2010	5.6	VK56DE	1.0217-1.0224	0.0012-0.0028	0.0045-0.0074	0.0008	1.0236-1.0244	1.7663-1.7738	1.7746-1.7821

37663_TITA_C0007

CRANKSHAFT AND CONNECTING ROD SPECIFICATIONS

All measurements are given in inches.

Year	Engine Displ. Liters	Engine ID	Crankshaft				Connecting Rod		
			Main Brg. Journal Dia.	Main Brg. Oil Clearance	Shaft End-play	Thrust on No.	Journal Diameter	Oil Clearance	Side Clearance
2009	5.6	VK56DE	①	②	0.0118	3	③	0.0008-0.0015	0.0079-0.0157
2010	5.6	VK56DE	①	②	0.0118	3	③	0.0008-0.0015	0.0079-0.0157

① There are 24 different grades, ranging from G (2.5183) to 9 (2.5173)

② No. 1 and 5: 0.00004-0.0004

   No. 2, 3 and 4: 0.0003-0.0007

③ Grade 0: 2.2441-2.2441

   Grade 1: 2.2441-2.2442

   Grade 2: 2.2442-2.2442

   Grade 3: 2.2442-2.2443

   Grade 4: 2.2443-2.2443

   Grade 5: 2.2443-2.2443

   Grade 6: 2.2443-2.2444

   Grade 7: 2.2444-2.2444

   Grade 8: 2.2444-2.2444

   Grade 9: 2.2444-2.2445

   Grade A: 2.2445-2.2445

   Grade B: 2.2445-2.22446

   Grade C: 2.2446-2.2446

37663_TITA_C0008

PISTON AND RING SPECIFICATIONS

All measurements are given in inches.

Year	Engine Displacement Liters	Engine ID	Piston Clearance	Ring Gap			Ring Side Clearance		
				Top Comp.	Bottom Comp.	Oil Control	Top Comp.	Bottom Comp.	Oil Control
2009	5.6	VK56DE	0.0004-0.0012	0.0091-0.0130	0.0098-0.0157	0.0079-0.0236	0.0014-0.0033	0.0012-0.0028	0.0006-0.0073
2010	5.6	VK56DE	0.0004-0.0012	0.0091-0.0130	0.0098-0.0157	0.0079-0.0236	0.0014-0.0033	0.0012-0.0028	0.0006-0.0073

37663_TITA_C0009

TORQUE SPECIFICATIONS
All readings in ft. lbs.

Year	Engine Displacement Liters	Engine ID	Cylinder Head Bolts	Main Bearing Bolts	Rod Bearing Bolts	Crankshaft Damper Bolts	Flywheel Bolts	Manifold		Spark Plugs	Oil Pan Drain Plug
								Intake	Exhaust		
2009	5.6	VK56DE	①	②	③	④	65	73	25	18	25
2010	5.6	VK56DE	①	②	③	④	65	73	25	18	25

① Step 1: 72 ft. lbs.

    Step 2: Loosen all bolts completely

    Step 3: 33 ft. lbs.

    Step 4: +70 degrees

② Step 1 Bolts 1-10 29 ft. lbs.

    Step 2: Bolts 11 - 20: 22 ft. lbs

    Step 3: Bolts 1 - 10: An additional 40 degrees

    Step 4: Bolts 11 - 20: an additional 30 degrees

    Side bolts in order of 21-30> 36 ft. lbs.

③ Step 1: 11 ft. lbs.

    Step 2: +90 degrees

④ Step 1: 69 ft. lbs.

    Step 2: +90 degrees

37663_TITA_C0010

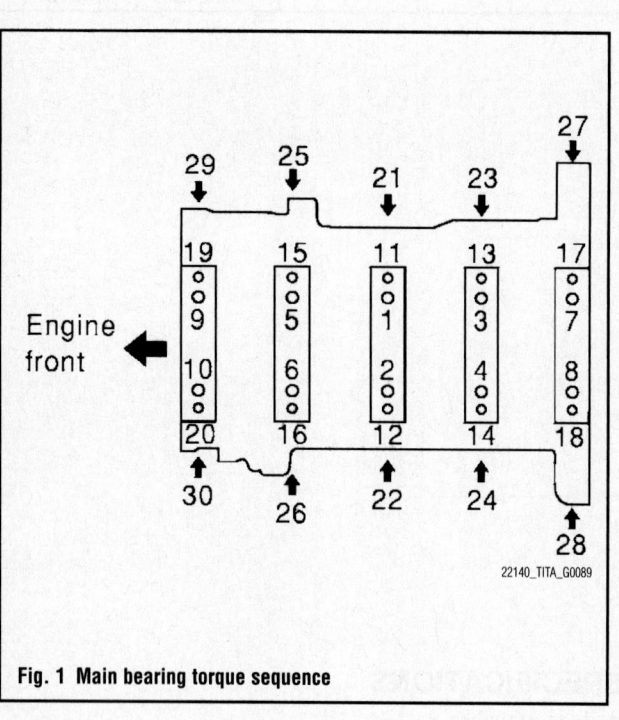

Fig. 1 Main bearing torque sequence

22140_TITA_G0089

WHEEL ALIGNMENT

Year	Model		Caster Range (+/-Deg.)	Caster Preferred Setting (Deg.)	Camber Range (+/-Deg.)	Camber Preferred Setting (Deg.)	Toe-in (in.)
2009	Titan	2WD	①	①	②	②	NA
		4WD	③	③	④	②	NA
2010	Titan	2WD	①	①	②	②	NA
		4WD	③	③	④	②	NA

NA - Not Available

① Minimum: 2 degrees 15' (2.25 degrees). Nominal: 3 degrees 0' (3.00 degrees). Maximum: 3 degrees 45' (3.75 degrees).

② Minimum: -0 degrees 57' (-0.95 degrees). Nominal: 0 degrees 12' (0.20 degrees). Maximum: 0 degree 33' (0.55 degree).

③ Minimum: 1 degrees 27' (1.45 degrees). Nominal: 2 degrees 12' (2.20 degrees). Maximum: 2 degrees 57' (2.95 degrees).

④ Minimum: -0 degrees 27' (-0.45 degrees). Nominal: 0 degrees 18' (0.30 degrees). Maximum: 1 degree 03' (1.05 degrees).

37663_TITA_C0011

TIRE, WHEEL AND BALL JOINT SPECIFICATIONS

Year	Model	OEM Tires Standard	OEM Tires Optional	Tire Pressures (psi) Front	Tire Pressures (psi) Rear	Wheel Size	Ball Joint Inspection	Lug Nut Torque (ft. lbs.)
2009	Titan	①	①	35	35	②	③	98
2010	Titan	①	①	35	35	②	③	98

OEM: Original Equipment Manufacturer

PSI: Pounds Per Square Inch

① P275/70R18

   P275/60R20

② 18 and 20

③ Axial play

   Upper: 0

37663_TITA_C0012

BRAKE SPECIFICATIONS

All measurements in inches unless noted

Year	Model		Brake Disc Original Thickness	Brake Disc Minimum Thickness	Brake Disc Maximum Runout	Minimum Lining Thickness	Brake Caliper Bracket Bolts (ft. lbs.)	Brake Caliper Mounting Bolts (ft. lbs.)
2009	Titan	F	1.181	1.102	0.001	0.039	①	①
		R	0.551	0.492	0.002	0.039	NA	②
2010	Titan	F	1.181	1.102	0.001	0.039	①	①
		R	0.551	0.492	0.002	0.039	NA	②

NA: Not applicable

① Sliding pin bolt: 53

   Torque member bolt: 110

② Sliding pin bolt: 24

37663_TITA_C0013

SCHEDULED MAINTENANCE INTERVALS
Nissan Titan

TO BE SERVICED	TYPE OF SERVICE	7.5	15	22.5	30	37.5	45	52.5	60
Engine oil & filter	R	✓	✓	✓	✓	✓	✓	✓	✓
Brake lines & cables	S/I		✓		✓		✓		✓
Brake pads and rotors	I	✓	✓	✓	✓	✓	✓	✓	✓
Driveshaft boots & propeller shaft (4x4)	L/I		✓		✓		✓		✓
Transmission, transfer & differential gear oil	I		✓		✓		✓		✓
Air cleaner filter	R								✓
Engine coolant	R	Replace after 60,000miles. Inspect every 15,000 miles							
Spark plugs (Platinum)	R	Replace every 105,000 miles							
Drive belt(s)	S/I	Replace after 60,000miles. Inspect every 15,000 miles							
Cabin air filter	R		✓		✓		✓		✓
Exhaust system	I	✓	✓	✓	✓	✓	✓	✓	✓
Fuel lines	S/I				✓				✓
Fuel filter		Maintenance free item							
Steering gear (box) & linkage, axle & suspension parts	I				✓				✓
Tire Rotation	S	✓	✓	✓	✓	✓	✓	✓	✓
Vapor lines	S/I				✓				✓

R: Replace   S/I: Service or Inspect   L: Lubricate

FREQUENT OPERATION MAINTENANCE (SEVERE SERVICE)

If a vehicle is operated under any of the following conditions it is considered severe service:

- **Extremely dusty areas.**
- **Rough, muddy, or salt spread roads.**
- **50% or more of the vehicle constant operation is in 32°C (90°F) or higher temperatures, or temperatures below 0°C (32°F).**
- **Prolonged idling (vehicle operation in stop and go traffic).**
- **Frequent short running periods (engine does not warm to normal operating temperatures).**
- **Police, taxi, delivery usage or trailer towing usage.**

Oil & oil filter: replace every 3750 miles.

Brake pads, discs, drums & linings: service or inspect every 7500 miles.

Driveshaft boots & propeller shaft: service or inspect every 7500 miles.

Exhaust system: service or inspect every 7500 miles.

Steering gear (box) & linkage, (steering damper-4x4), axle & suspension parts: service or inspect every 7500 miles.

Steering linkage ball joints & front suspension ball joints: service or inspect every 7500 miles.

Transfer case fluid, transmission fluid and differential gear oil: Change every 30000 miles if towing a trailer.

37663_TITA_C0014

BRAKES — INFORMATION AND PRECAUTIONS

ANTI-LOCK SYSTEMS

- Certain components within the ABS system are not intended to be serviced or repaired individually.
- Do not use rubber hoses or other parts not specifically specified for and ABS system. When using repair kits, replace all parts included in the kit. Partial or incorrect repair may lead to functional problems and require the replacement of components.
- Lubricate rubber parts with clean, fresh brake fluid to ease assembly. Do not use shop air to clean parts; damage to rubber components may result.
- Use only DOT 3 brake fluid from an unopened container.
- If any hydraulic component or line is removed or replaced, it may be necessary to bleed the entire system.
- A clean repair area is essential. Always clean the reservoir and cap thoroughly before removing the cap. The slightest amount of dirt in the fluid may plug an orifice and impair the system function. Perform repairs after components have been thoroughly cleaned; use only denatured alcohol to clean components. Do not allow ABS components to come into contact with any substance containing mineral oil; this includes used shop rags.

- The Anti-Lock control unit is a microprocessor similar to other computer units in the vehicle. Ensure that the ignition switch is **OFF** before removing or installing controller harnesses. Avoid static electricity discharge at or near the controller.
- If any arc welding is to be done on the vehicle, the control unit should be unplugged before welding operations begin.

DISC AND DRUM SYSTEMS

✳✳ CAUTION

Dust and dirt accumulating on brake parts during normal use may contain asbestos fibers from production or aftermarket brake linings. Breathing excessive concentrations of asbestos fibers can cause serious bodily harm. Exercise care when servicing brake parts. Do not sand or grind brake lining unless equipment used is designed to contain the dust residue. Do not clean brake parts with compressed air or by dry brushing. Cleaning should be done by dampening the brake components with a fine mist of water, then wiping the brake components clean with a dampened cloth. Dispose of cloth and all residue containing asbestos fibers in an impermeable container with the appropriate label. Follow practices prescribed by the Occupational Safety and Health Administration (OSHA) and the Environmental Protection Agency (EPA) for the handling, processing, and disposing of dust or debris that may contain asbestos fibers.

BRAKES — BLEEDING THE BRAKE SYSTEM

BLEEDING PROCEDURE

✳✳ CAUTION

Carefully monitor brake fluid level at the sub tank during bleeding operation.

✳✳ CAUTION

Fill the sub tank with new brake fluid. Make sure it is full at all times while bleeding the air out of system.

➡Place a container under the sub tank to avoid spilling brake fluid.

➡Do not loosen the line fittings at the ABS actuator during air bleeding.

1. Before servicing the vehicle, refer to the Precautions Section.
2. Turn ignition switch OFF and disconnect ABS actuator and control unit connector or negative battery terminal.
3. Connect a transparent vinyl tube and container to air bleeder valve.
4. Fully depress brake pedal several times.
5. With brake pedal depressed, open air bleeder valve to release air.
6. Close air bleeder valve.
7. Release brake pedal slowly.
8. Tighten air bleeder valve to 71 inch lbs. (8 Nm).
9. Repeat steps 2 through 7 until no more air bubbles come out of air bleeder valve.
10. Bleed the brake hydraulic system air bleeder valves in the following order:
- Right rear brake
- Left front brake
- Left rear brake
- Right front brake

BLEEDING THE ABS SYSTEM

✳✳ CAUTION

Carefully monitor brake fluid level at the sub tank during bleeding operation.

✳✳ CAUTION

Fill the sub tank with new brake fluid. Make sure it is full at all times while bleeding the air out of system.

➡Place a container under the sub tank to avoid spilling brake fluid.

➡Do not loosen the line fittings at the ABS actuator during air bleeding.

1. Before servicing the vehicle, refer to the Precautions Section.
2. Turn ignition switch OFF and disconnect ABS actuator and control unit connector or negative battery terminal.
3. Connect a transparent vinyl tube and container to air bleeder valve.
4. Fully depress brake pedal several times.
5. With brake pedal depressed, open air bleeder valve to release air.
6. Close air bleeder valve.
7. Release brake pedal slowly.
8. Tighten air bleeder valve to 71 inch lbs. (8 Nm).
9. Repeat steps 2 through 7 until no more air bubbles come out of air bleeder valve.
10. Bleed the brake hydraulic system air bleeder valves in the following order:
- Right rear brake
- Left front brake
- Left rear brake
- Right front brake

WHEEL SPEED SENSORS

REMOVAL & INSTALLATION
See Figure 2.

✳✳ WARNING

Be careful not to damage sensor edge and sensor rotor teeth. Do not pull on the sensor harness.

➡ **When removing the front wheel sensor, first remove the disc rotor to gain access to the front wheel sensor bolt.**

1. Before servicing the vehicle, refer to the Precautions Section.

➡ **If working near and/or around the SRS system and components, be sure to disable the SRS system. After dis-** abling the system wait three minutes or more before servicing the vehicle.

➡ **Whenever the negative battery cable is disconnected the following components will require resetting. The Idle Air Volume Learning, Steering Angle Sensor Neutral Position, Sunroof Memory Reset/Initialization, Automatic Drive Positioner System, Audio presets and Navigation. Use the CONSULT-III diagnostic tool, or equivalent to perform the required resets.**

2. Disconnect the negative battery cable.

3. Remove wheel sensor bolt.

4. Pull out the sensor, being careful to turn it as little as possible.

5. Disconnect wheel sensor harness electrical connector, then remove harness from mounts.

6. Inspect wheel sensor O-ring, replace sensor assembly if damaged.

7. Clean wheel sensor hole and mounting surface with brake cleaner and a lint-free shop rag. Be careful that dirt and debris do not enter the axle.

8. Apply a coat of suitable grease to the wheel sensor O-ring and mounting hole.

To install:

➡ **Be sure to use new fasteners, as required.**

9. Installation is the reverse of the removal procedure.

10. Be sure to perform the reconnect/relearn procedures.

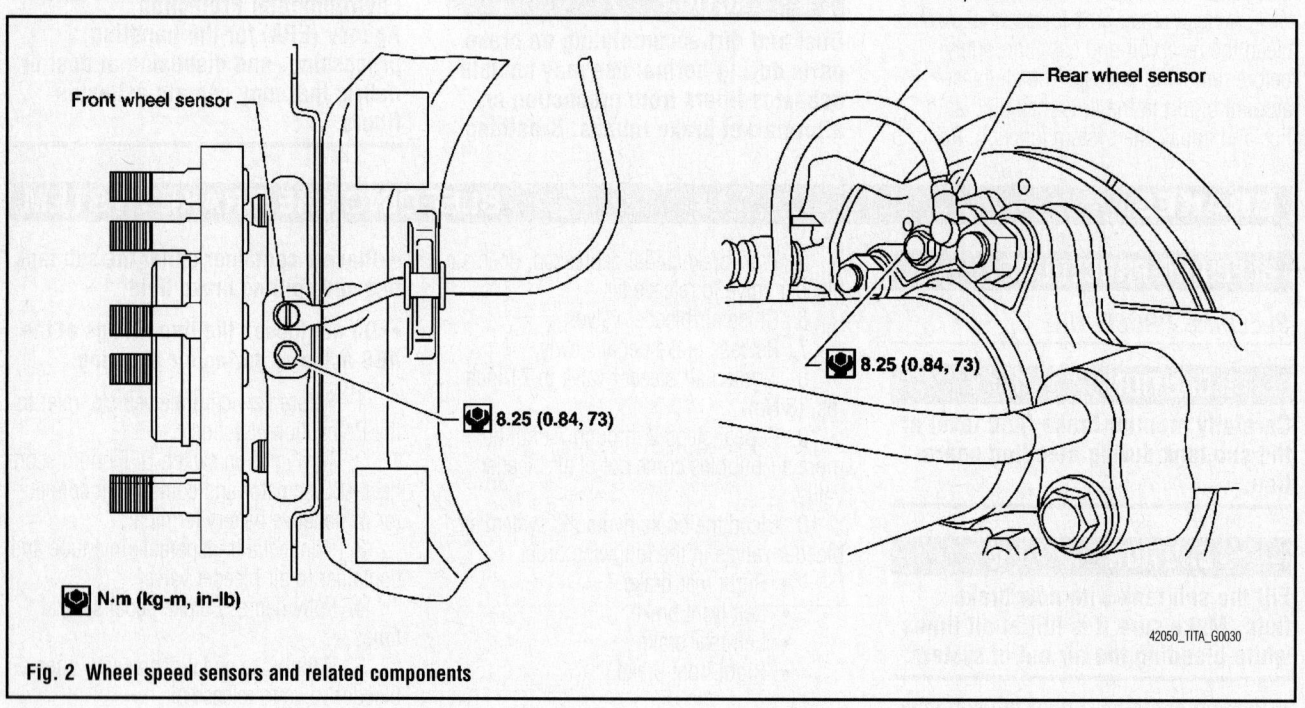

Fig. 2 Wheel speed sensors and related components

42050_TITA_G0030

BRAKES   **FRONT DISC BRAKES**

BRAKE CALIPER

REMOVAL & INSTALLATION

See Figure 3.

1. Before servicing the vehicle, refer to the Precautions Section.

➡ **If working near and/or around the SRS system and components, be sure to disable the SRS system. After disabling the system wait three minutes or more before servicing the vehicle.**

➡ **Whenever the negative battery cable is disconnected the following components will require resetting. The Idle Air Volume Learning, Steering Angle Sensor Neutral Position,** Sunroof Memory Reset/Initialization, Automatic Drive Positioner System, Audio presets and Navigation. **Use the CONSULT-III diagnostic tool, or equivalent to perform the required resets.**

2. Disconnect the negative battery cable.
3. Drain brake fluid as necessary.
4. Raise and safely support the vehicle.
5. Remove or disconnect the following:
   - Wheel and tire assembly
   - Union bolt, discard copper washers
   - Caliper-to-torque member slide pins, or remove the caliper and torque member as an assembly
   - Brake caliper

To install:

➡ **Be sure to use new fasteners, as required.**

6. Install the brake caliper, tighten torque member bolts to specification, and caliper slide pins to specification.
7. Install the union bolt and tighten to 13 ft. lbs. (18 Nm).
8. Fill the master cylinder and bleed the brake system.
9. Install the wheels.
10. Be sure to perform the reconnect/relearn procedures.

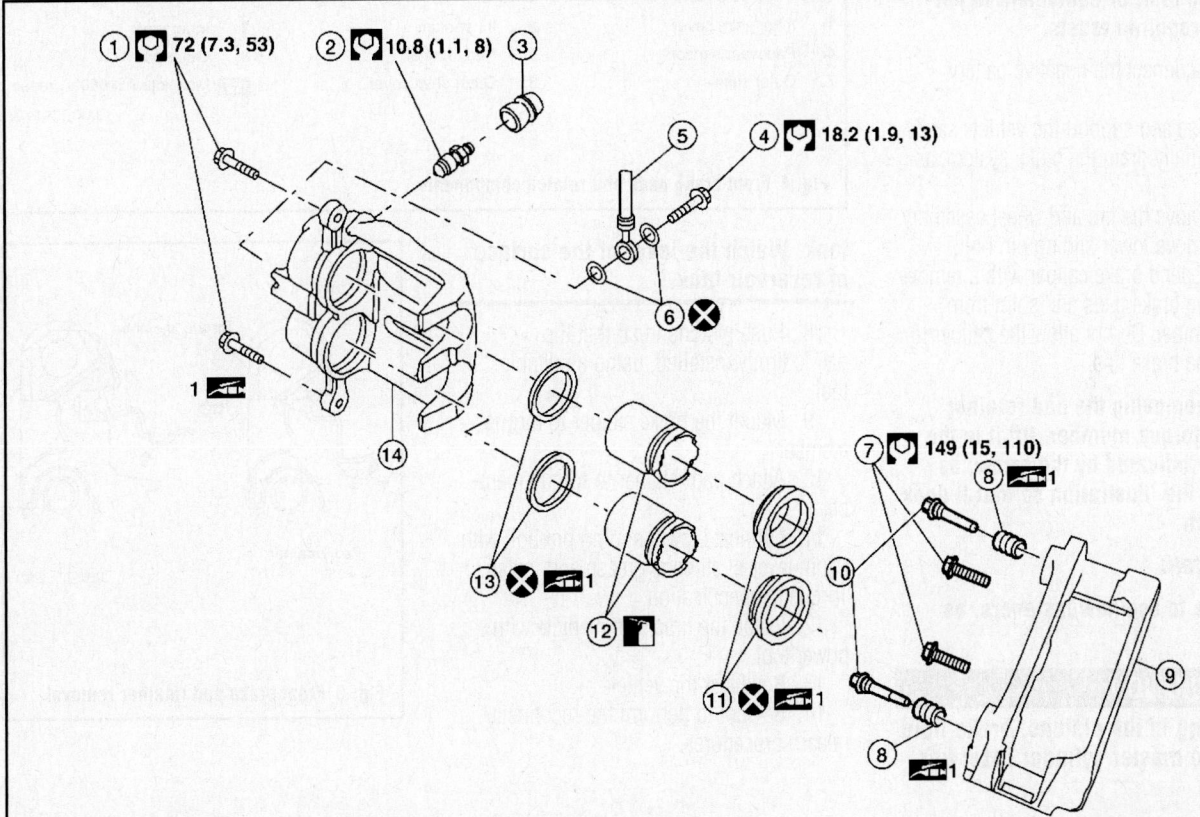

1. Sliding pin bolt	2. Bleed valve
4. Union bolt	5. Brake hose
7. Torque member bolt	8. Sliding pin boot
10. Sliding pin	11. Piston boot
13. Piston seal	14. Cylinder body

3. Cap
6. Copper washer
9. Torque member
12. Piston
 Brake fluid

37663_TITA_G0019

Fig. 3 Front caliper and related components

DISC BRAKE PADS

REMOVAL & INSTALLATION

See Figures 4 and 5.

1. Before servicing the vehicle, refer to the Precautions Section.

➡ **If working near and/or around the SRS system and components, be sure to disable the SRS system. After disabling the system wait three minutes or more before servicing the vehicle.**

➡ **Whenever the negative battery cable is disconnected the following components will require resetting. The Idle Air Volume Learning, Steering Angle Sensor Neutral Position, Sunroof Memory Reset/Initialization, Automatic Drive Positioner System, Audio presets and Navigation. Use the CONSULT-III diagnostic tool, or equivalent to perform the required resets.**

2. Disconnect the negative battery cable.

3. Raise and support the vehicle safely.

4. Partially drain the brake system, as necessary.

5. Remove the tire and wheel assembly.

6. Remove lower sliding pin bolt.

7. Suspend brake caliper with a remove and remove brake pads and shim from torque member. Do not allow the caliper to hang by the brake line.

➡ **When removing the pad retainer from the torque member, lift it in the direction indicated by the arrow, as shown in the illustration so that it does not deform.**

To install:

➡ **Be sure to use new fasteners, as required.**

✳✳ CAUTION

By pushing in the pistons, brake fluid returns to master cylinder reservoir

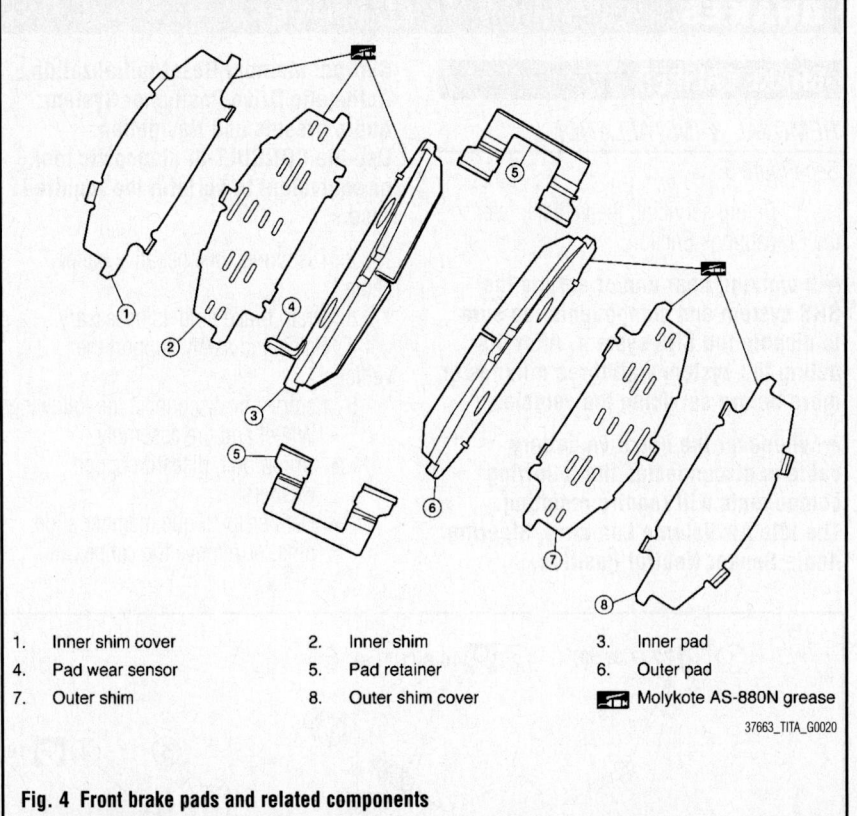

1.	Inner shim cover	2.	Inner shim	3.	Inner pad
4.	Pad wear sensor	5.	Pad retainer	6.	Outer pad
7.	Outer shim	8.	Outer shim cover		Molykote AS-880N grease

37663_TITA_G0020

Fig. 4 Front brake pads and related components

tank. Watch the level of the surface of reservoir tank.

8. Push pistons in so that the pad is firmly installed, using a suitable tool.

9. Mount the brake caliper to torque member.

10. Attach pad retainer to torque member.

11. Lubricate lower sliding pin bolt with a thin layer of silicone grease and install. Torque to specification

12. Install the tires from vehicle with power tool.

13. Road test the vehicle.

14. Be sure to perform the reconnect/relearn procedures.

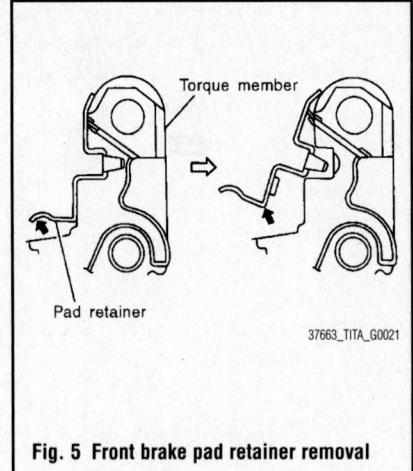

37663_TITA_G0021

Fig. 5 Front brake pad retainer removal

BRAKES

BRAKE CALIPER

REMOVAL & INSTALLATION
See Figure 6.

1. Before servicing the vehicle, refer to the Precautions Section.

➡**If working near and/or around the SRS system and components, be sure to disable the SRS system. After disabling the system wait three minutes or more before servicing the vehicle.**

➡**Whenever the negative battery cable is disconnected the following components will require resetting. The Idle Air Volume Learning, Steering Angle Sensor Neutral Position, Sunroof Memory Reset/Initialization, Automatic Drive Positioner System, Audio presets and Navigation. Use the CONSULT-III diagnostic tool, or equivalent to perform the required resets.**

2. Disconnect the negative battery cable.
3. Raise and safely support the vehicle.
4. Remove the tire and wheel assembly.
5. Drain the brake fluid, as necessary.
6. Remove the brake hose mounting bolt and brake hose.
7. Remove the brake caliper assembly. Discard the copper washer. Do not reuse.

To install:

➡**Be sure to use new fasteners, as required.**

8. Install the brake caliper assembly and tighten mounting bolts to specification.
9. Install brake hose and tighten to 13 ft. lbs. (18 Nm).
10. Bleed the air from the brake caliper.
11. Install the tires from vehicle with power tool.
12. Road test the vehicle.
13. Be sure to perform the reconnect/relearn procedures.

DISC BRAKE PADS

REMOVAL & INSTALLATION
See Figure 7.

1. Before servicing the vehicle, refer to the Precautions Section.

➡**If working near and/or around the SRS system and components, be sure to disable the SRS system. After disabling the system wait three minutes or more before servicing the vehicle.**

➡**Whenever the negative battery cable is disconnected the following compo-**

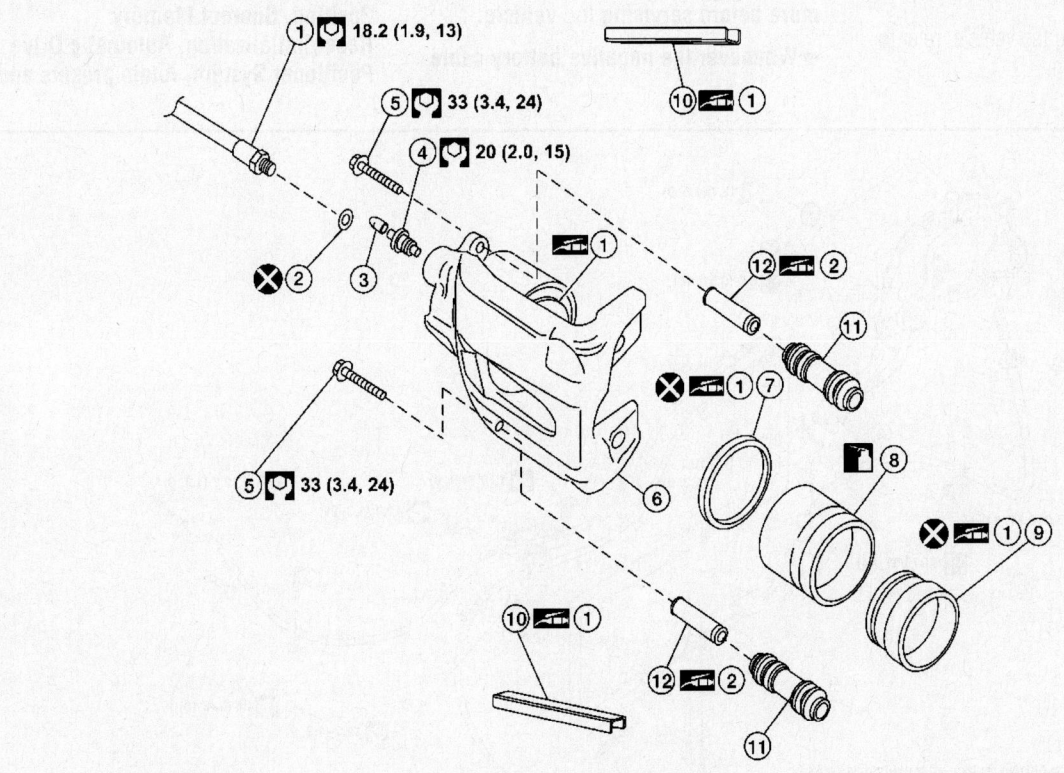

1.	Brake hose	2.	Copper washer	3.	Cap
4.	Bleed valve	5.	Sliding pin bolt	6.	Cylinder body
7.	Piston seal	8.	Piston	9.	Piston boot
10.	Slipper	11.	Sliding sleeve boot	12.	Sliding sleeve
	Brake fluid		1: Molykote M-77 grease		2: Rubber grease

37663_TITA_G0022

Fig. 6 Rear caliper and related components

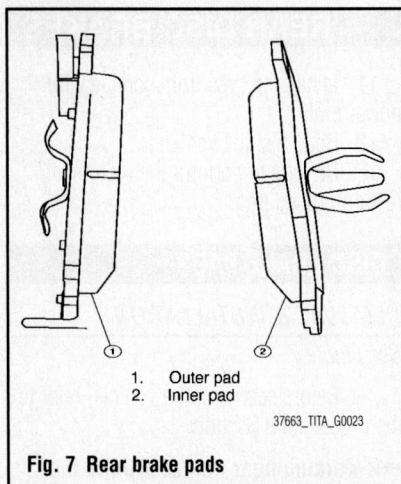

1. Outer pad
2. Inner pad

37663_TITA_G0023

Fig. 7 Rear brake pads

nents will require resetting. The Idle Air Volume Learning, Steering Angle Sensor Neutral Position, Sunroof Memory Reset/Initialization, Automatic Drive Positioner System, Audio presets and Navigation. Use the CONSULT-III diagnostic tool, or equivalent to perform the required resets.

2. Disconnect the negative battery cable.

3. Raise and safely support the vehicle.

4. Remove the tire and wheel assembly.

5. Drain the brake fluid, as necessary.

6. Remove the sliding pin.

7. Remove the cylinder body and remove the pads.

To install:

➡ Be sure to use new fasteners, as required.

8. Apply Molykote® (M-77) grease to the knuckle slide where the brake pads contact.

9. Install pads to cylinder body.

10. Install top mounting bolt and tighten to specification.

11. Check brake for drag.

12. Install tires to the vehicle.

13. Be sure to perform the reconnect/relearn procedures.

BRAKES   PARKING BRAKE

PARKING BRAKE CABLES

ADJUSTMENT

See Figure 8.

1. Before servicing the vehicle, refer to the Precautions Section.

➡ If working near and/or around the SRS system and components, be sure to disable the SRS system. After disabling the system wait three minutes or more before servicing the vehicle.

➡ Whenever the negative battery cable

is disconnected the following components will require resetting. The Idle Air Volume Learning, Steering Angle Sensor Neutral Position, Sunroof Memory Reset/Initialization, Automatic Drive Positioner System, Audio presets and

---

12.7 (1.3, 9)

5.0 (0.51, 44)

5.0 (0.51, 44)

5.0 (0.51, 44)

12.7 (1.3, 9)

12.7 (1.3, 9)

12.7 (1.3, 9)

12.7 (1.3, 9)

12.7 (1.3, 9)

12.7 (1.3, 9)

24 (2.4, 18)

12.7 (1.3, 9)

24 (2.4, 18)

✖ : Always replace after every disassembly.

▭ : Apply multi-purpose grease.

▣ : N·m (kg-m, ft-lb)

▣ : N·m (kg-m, in-lb)

1. Pedal assembly	2. Front cable	3. Right rear cable
4. Left rear cable	5. Return spring	6. Equalizer
7. Adjusting nut	8. Lock plate	

37663_TITA_G0024

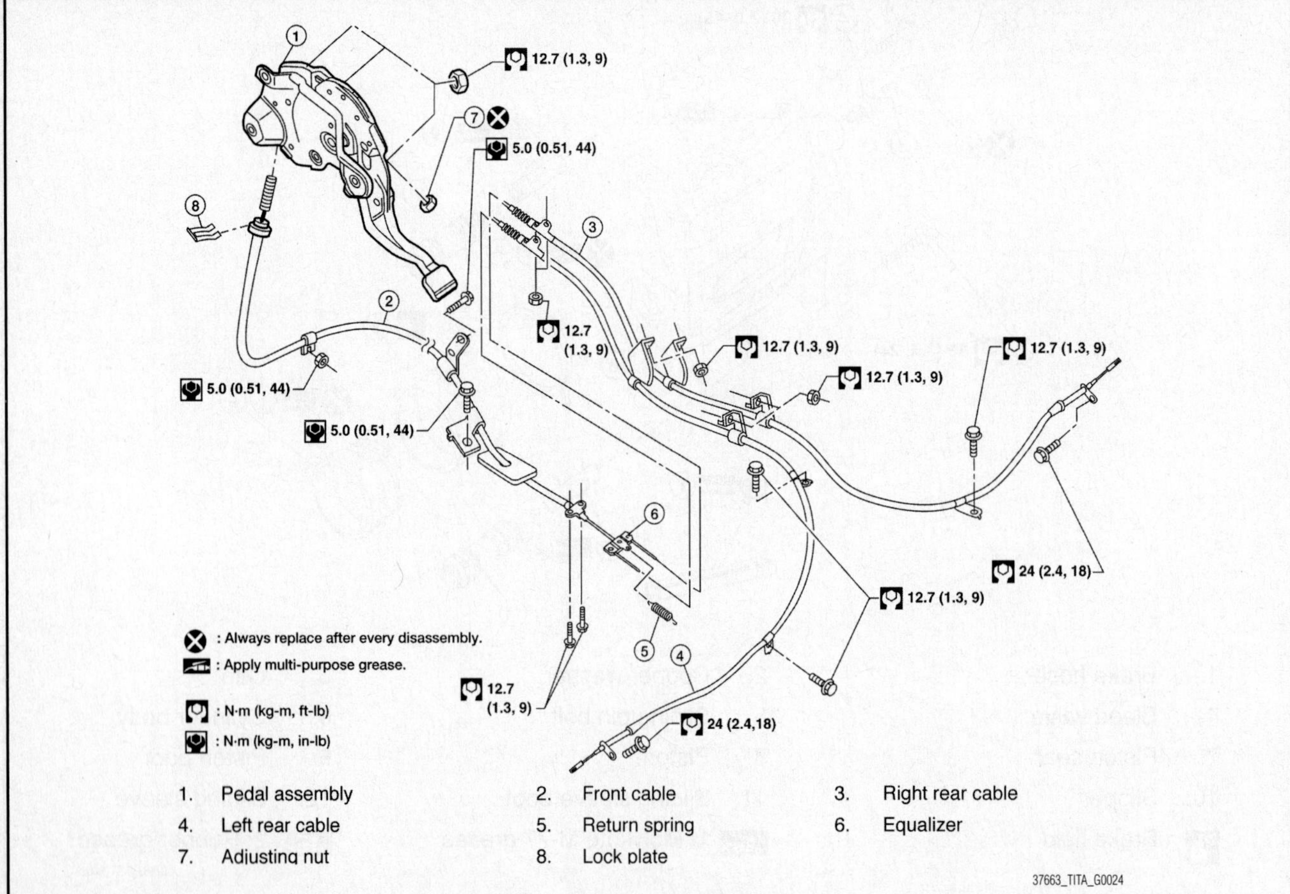

Fig. 8 Parking brake system and related components

Navigation. Use the CONSULT-III diagnostic tool, or equivalent to perform the required resets.

2. Disconnect the negative battery cable.
3. Remove the left-hand lower instrument panel.
4. Partially engage parking brake pedal to access adjusting nut.
5. Insert a deep socket wrench to rotate adjusting nut and loosen cable sufficiently. Then, disengage the parking brake pedal.
6. Remove the wheel and tire.
7. Remove the disc rotor and measure inner diameter at widest point using tool J-21177-A.
8. Transfer measurement less 0.6 mm to the parking brake shoes and adjust accordingly.
9. Using wheel nuts, secure the disc rotor to the hub to prevent it from tilting.
10. Rotate the disc rotor to make sure there is no drag.
11. Adjust cable as follows:
    a. Operate pedal 10 or more times with a force of 110 ft. lbs. (490 Nm).
    b. Rotate adjusting nut with deep socket to adjust pedal stroke to specification: 3 to 4 notches; under force of 44.1 lbs. (196 N).
    c. With parking brake pedal completely disengaged, make sure there is no drag on the parking brake.
12. Be sure to perform the reconnect/relearn procedures.

PARKING BRAKE SHOES

REMOVAL & INSTALLATION

See Figure 9.

CAUTION

Clean the brakes with a vacuum dust collector to minimize the hazard of airborne particles or other materials.

➡Remove the disc rotor only with the parking brake pedal completely in the released position.

1. Before servicing the vehicle, refer to the Precautions Section.

➡If working near and/or around the SRS system and components, be sure to disable the SRS system. After disabling the system wait three minutes or more before servicing the vehicle.

➡Whenever the negative battery cable is disconnected the following components will require resetting. The Idle Air Volume Learning, Steering Angle Sensor Neutral Position, Sunroof Memory Reset/Initialization, Automatic Drive Positioner System, Audio presets and Navigation. Use the CONSULT-III diagnostic tool, or equivalent to perform the required resets.

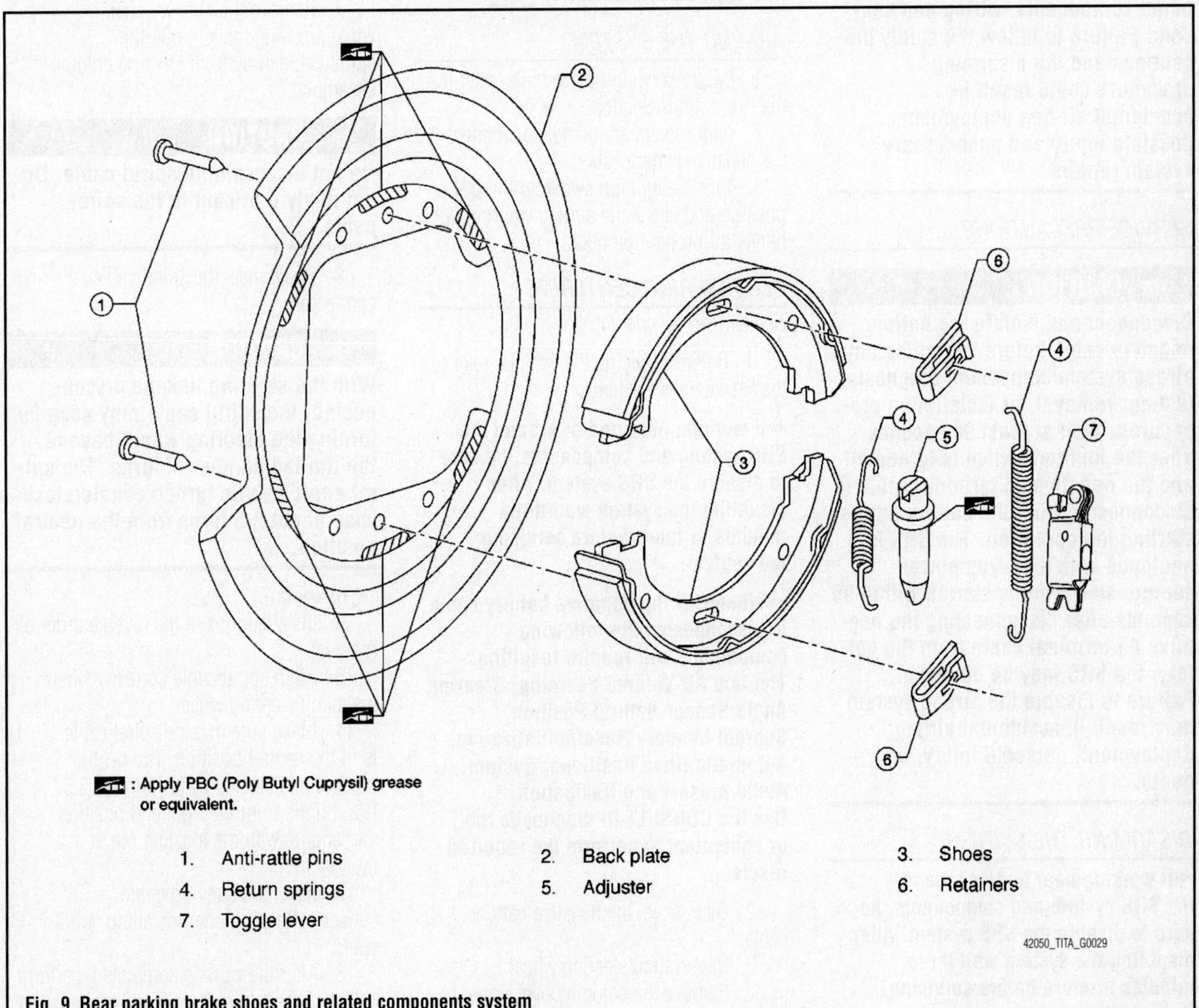
1
2
3
4
5
6
7
: Apply PBC (Poly Butyl Cuprysil) grease or equivalent.
1. Anti-rattle pins
2. Back plate
3. Shoes
4. Return springs
5. Adjuster
6. Retainers
7. Toggle lever
42050_TITA_G0029

Fig. 9 Rear parking brake shoes and related components system

2. Disconnect the negative battery cable.
3. Remove or disconnect the following:
- Rear disc rotor
- Return springs
- Adjuster
- Rear cable from the toggle lever, if necessary
- Retainers, anti-rattle pins and shoes
4. Check shoe sliding surface on back plate for excessive wear and damage.
5. Check anti-rattle pins for excessive wear and corrosion.
6. Check the return springs for sagging.
7. Check the adjuster for rough operation.

To install:

➡ Be sure to use new fasteners, as required.

8. To install, reverse removal procedure.

➡ **Apply brake grease to the specified points during assembly.**

➡ **There is a difference between the adjuster's orientation from left and right. Assemble the adjuster so the threaded part expands when rotating it.**

9. Be sure to perform the reconnect/relearn procedures.

CHASSIS ELECTRICAL

AIR BAG (SUPPLEMENTAL RESTRAINT SYSTEM)

GENERAL INFORMATION

※※ CAUTION

These vehicles are equipped with an air bag system. The system must be disarmed before performing service on, or around, system components, the steering column, instrument panel components, wiring and sensors. Failure to follow the safety precautions and the disarming procedure could result in accidental air bag deployment, possible injury and unnecessary system repairs.

SERVICE PRECAUTIONS

※※ CAUTION

Disconnect and isolate the battery negative cable before beginning any airbag system component diagnosis, testing, removal, or installation procedures. Wait at least 90 seconds after the ignition switch is turned off and the negative (-) terminal cable is disconnected from the battery before starting the operation. The SRS is equipped with a backup power source, so if work is started within 90 seconds after disconnecting the negative (-) terminal cable from the battery, the SRS may be deployed. Failure to disable the airbag system may result in accidental airbag deployment, personal injury, or death.

DISARMING THE SYSTEM

➡ If working near and/or around the SRS system and components, be sure to disable the SRS system. After disabling the system wait three minutes or more before servicing the vehicle.

1. Before servicing the vehicle, refer to the Precautions Section.
2. Disconnect both battery cables.
3. Wait at least 3 minutes before working on the vehicle.

➡ **The air bag system is designed to retain enough power to deploy the air bag for a short time after the battery has been disconnected.**

ARMING THE SYSTEM

1. Before servicing the vehicle, refer to the Precautions Section.
2. After repairs are complete, connect the negative battery cable.
3. Turn the ignition switch to the **ON** position and check the air bag warning light blinks for proper operation.

CLOCKSPRING CENTERING

See Figures 10 and 11.

1. Before servicing the vehicle, refer to the Precautions Section.

➡ If working near and/or around the SRS system and components, be sure to disable the SRS system. After disabling the system wait three minutes or more before servicing the vehicle.

➡ Whenever the negative battery cable is disconnected the following components will require resetting. The Idle Air Volume Learning, Steering Angle Sensor Neutral Position, Sunroof Memory Reset/Initialization, Automatic Drive Positioner System, Audio presets and Navigation. Use the CONSULT-III diagnostic tool, or equivalent to perform the required resets.

2. Disconnect the negative battery cable.
3. Remove the steering wheel.
4. Remove the column cover upper and lower.

5. Disconnect wiper and washer switch connector. Then while pressing tabs, pull wiper and washer switch away from spiral cable to remove.
6. Disconnect lighting and turn signal switch connector. Then while pressing tabs, pull lighting and turn signal switch toward driver door to remove.
7. Remove the screws. Then while pressing the tab, pull the spiral cable away from steering column assembly.

※※ CAUTION

Do not disassemble spiral cable. Do not apply lubricant to the spiral cable.

8. Disconnect the spiral cable connectors.

※※ CAUTION

With the steering linkage disconnected, the spiral cable may snap by turning the steering wheel beyond the limited number of turns. The spiral cable can be turned counterclockwise about 2.5 turns from the neutral position.

To install:

9. Installation is in the reverse order of removal.
10. Align spiral cable correctly when installing steering wheel.
11. Make sure that the spiral cable is in the neutral position. The neutral position is detected by turning left 2.5 revolutions from the right end position and ending with the locating pin at the top.
12. Reset the steering angle sensor calibration after installing spiral cable.
13. After the work is completed, perform self-diagnosis to make sure no malfunction is detected.

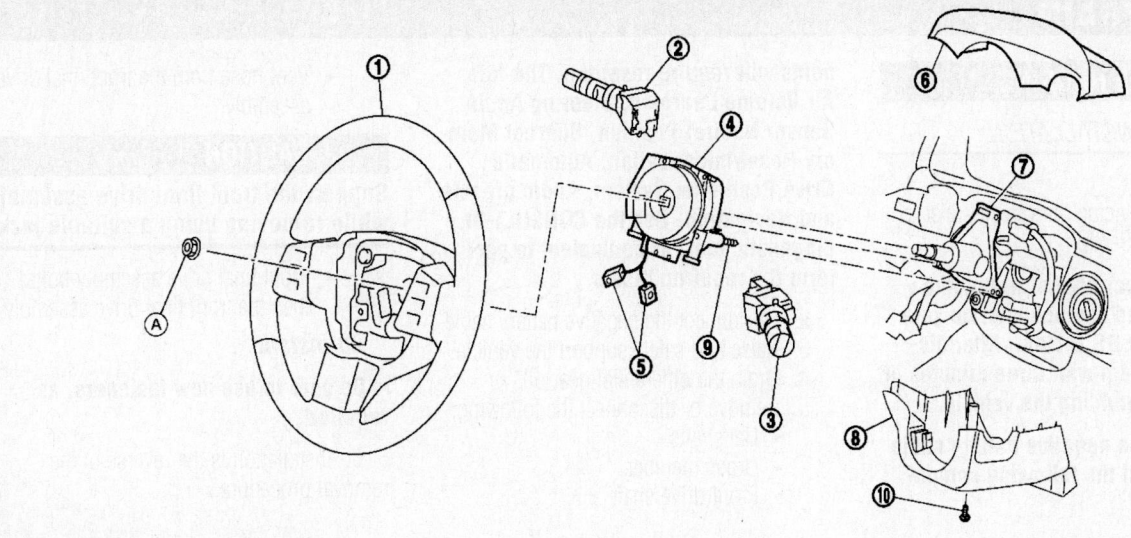

1. Steering wheel
2. Lighting and turn signal switch
3. Wiper and washer switch
4. Spiral cable
5. Driver air bag module connector
6. Column cover upper
7. Column assembly
8. Column cover lower
9. Screw (Do not remove)
10. Screw

37663_TITA_G0032

Fig. 10 Spiral cable and related components

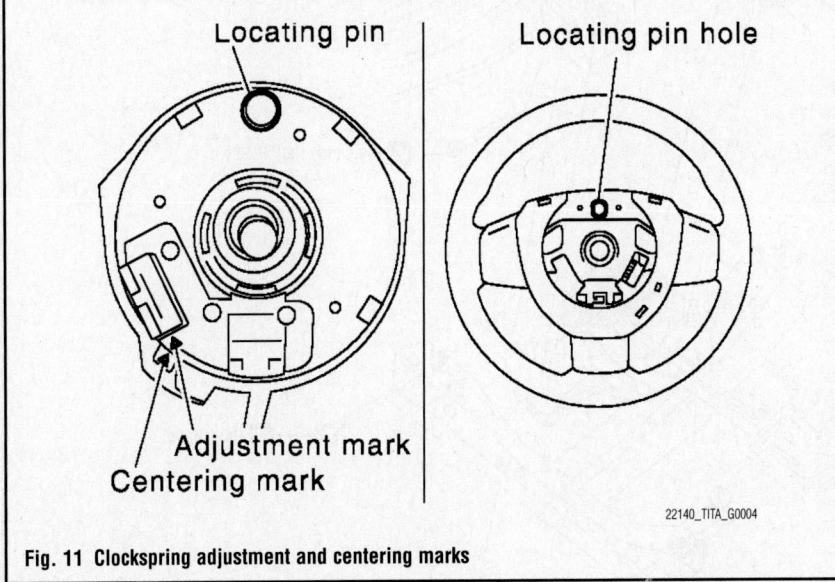

Locating pin

Locating pin hole

Adjustment mark
Centering mark

22140_TITA_G0004

Fig. 11 Clockspring adjustment and centering marks

✲✲ CAUTION

The spiral cable may snap due to steering operation if the cable is not installed in the correct position. With the steering linkage disconnected, the cable may snap by turning the steering wheel beyond the limited number of turns. The spiral cable can be turned counterclockwise about 2.5 turns from the neutral position.

14. Adjust the steering angle sensor neutral position, using the CONSULT-III diagnostic tool, or equivalent.

15. Be sure to perform the reconnect/relearn procedures.

DRIVE TRAIN

FRONT AXLE HOUSING

REMOVAL & INSTALLATION

See Figure 12.

1. Before servicing the vehicle, refer to the Precautions Section.

→If working near and/or around the SRS system and components, be sure to disable the SRS system. After disabling the system wait three minutes or more before servicing the vehicle.

→Whenever the negative battery cable is disconnected the following components will require resetting. The Idle Air Volume Learning, Steering Angle Sensor Neutral Position, Sunroof Memory Reset/Initialization, Automatic Drive Positioner System, Audio presets and Navigation. Use the CONSULT-III diagnostic tool, or equivalent to perform the required resets.

2. Disconnect the negative battery cable.
3. Raise and safely support the vehicle.
4. Drain the differential gear oil.
5. Remove or disconnect the following:
   - Halfshafts
   - Cross member
   - Front drive shaft
   - Vent hose from the front final drive assembly

※ CAUTION

Support the front final drive assembly while removing using a suitable jack.

   - Front final drive assembly bolts, then the front final drive assembly

To install:

→Be sure to use new fasteners, as required.

6. Installation is the reverse of the removal procedure.

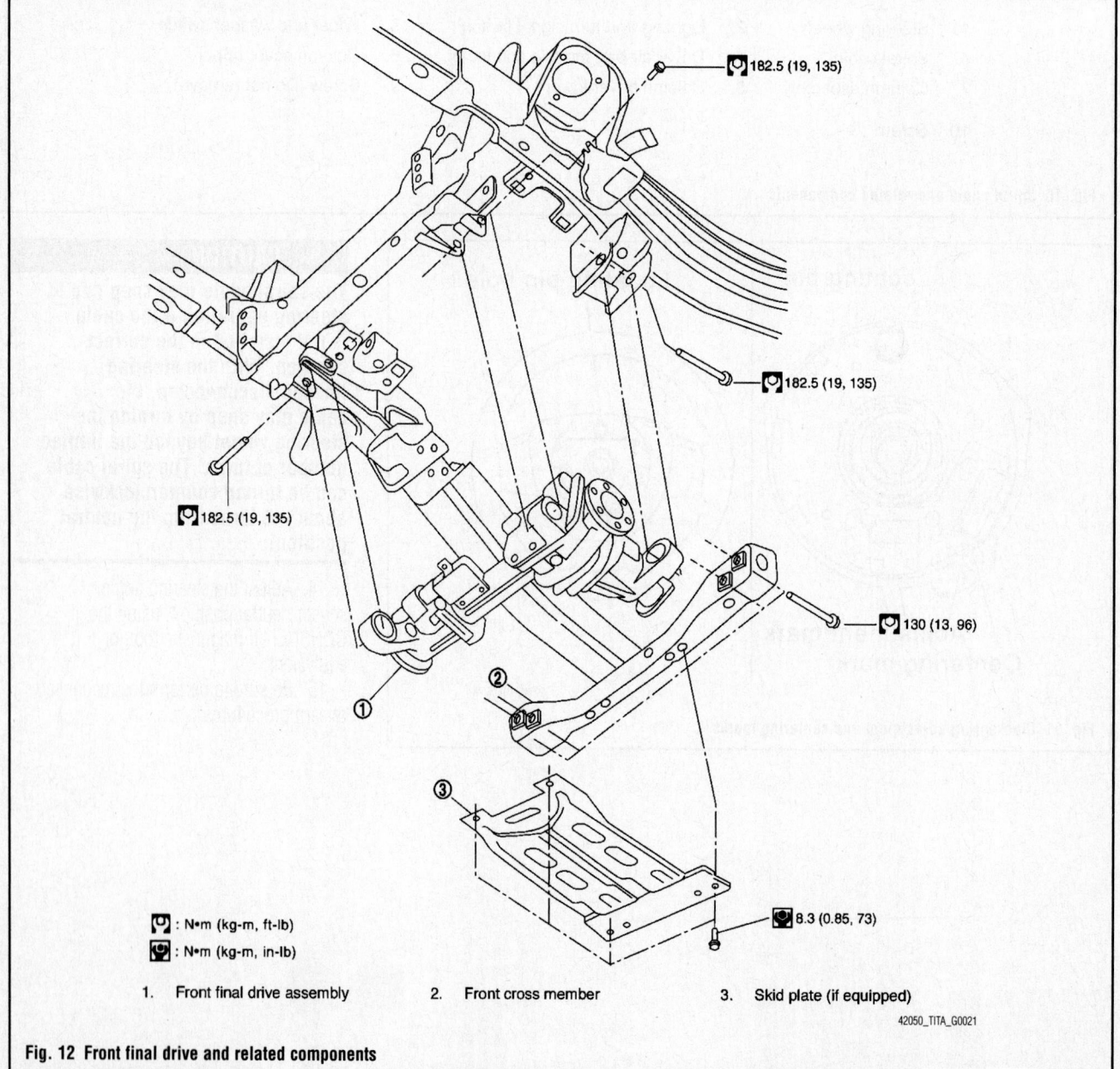

182.5 (19, 135)

182.5 (19, 135)

182.5 (19, 135)

130 (13, 96)

8.3 (0.85, 73)

🔧 : N•m (kg-m, ft-lb)
🔧 : N•m (kg-m, in-lb)

| 1. Front final drive assembly | 2. Front cross member | 3. Skid plate (if equipped) |

42050_TITA_G0021

Fig. 12 Front final drive and related components

➡Make sure there are no pinched or restricted areas on the breather hose caused by folding or bending when installing it.

7. Fill the front final drive assembly with differential gear oil after installation.

8. Be sure to perform the reconnect/relearn procedures.

FRONT AXLE SEAL

REMOVAL & INSTALLATION

See Figure 12.

1. Before servicing the vehicle, refer to the Precautions Section.

➡If working near and/or around the SRS system and components, be sure to disable the SRS system. After disabling the system wait three minutes or more before servicing the vehicle.

➡Whenever the negative battery cable is disconnected the following components will require resetting. The Idle Air Volume Learning, Steering Angle Sensor Neutral Position, Sunroof Memory Reset/Initialization, Automatic Drive Positioner System, Audio presets and Navigation. Use the CONSULT-III diagnostic tool, or equivalent to perform the required resets.

2. Disconnect the negative battery cable.

3. Raise and safely support the vehicle.

4. Remove front final drive assembly.

5. Remove differential side shaft and differential side flange using suitable tool.

6. Using a punch or drill, place a small hole in the case.

7. Remove the seal using special tool SP8P.

To install:

➡Be sure to use new fasteners, as required.

❄❄ CAUTION

Do not re-use seal. Do not incline the new side oil seal when installing. Apply multi-purpose grease to the lips of the new side oil seal.

8. Apply multi-purpose grease to the lips of the new side oil seal. Then drive the new side oil seal in evenly to the gear carrier using suitable tool.

9. To complete installation, reverse remaining removal procedure.

10. Check the differential gear oil level after installation.

11. Be sure to perform the reconnect/relearn procedures.

HALFSHAFTS

REMOVAL & INSTALLATION

See Figure 13.

1. Before servicing the vehicle, refer to the Precautions Section.

➡If working near and/or around the SRS system and components, be sure to disable the SRS system. After disabling the system wait three minutes or more before servicing the vehicle.

➡Whenever the negative battery cable is disconnected the following components will require resetting. The Idle Air Volume Learning, Steering Angle

Sensor Neutral Position, Sunroof Memory Reset/Initialization, Automatic Drive Positioner System, Audio presets and Navigation. Use the CONSULT-III diagnostic tool, or equivalent to perform the required resets.

2. Disconnect the negative battery cable.

3. Raise and support the vehicle safely.

4. Remove or disconnect the following:
   • Wheel and tire assembly
   • Engine splash guard
   • ABS sensor harness on knuckle
   • Brake caliper and suspend it aside
   • Coil spring and shock absorber assembly

5. Separate upper ball joint stud from steering knuckle using special tool J-24319-01.

6. Remove or disconnect the following:
   • Cotter pin and half shaft nut
   • Halfshaft from front differential
   • Hallfshaft from hub and bearing assembly

To install:

➡Be sure to use new fasteners, as required.

7. Install or connect the following:
   • Half shaft into hub
   • Halfshaft into front differential
   • Halfshaft nut and tighten to 101 ft. lbs. and replace cotter pin
   • Upper ball joint to steering knuckle
   • Coil spring and shock absorber assembly
   • Brake caliper
   • ABS sensor
   • Engine splash guard
   • Wheel

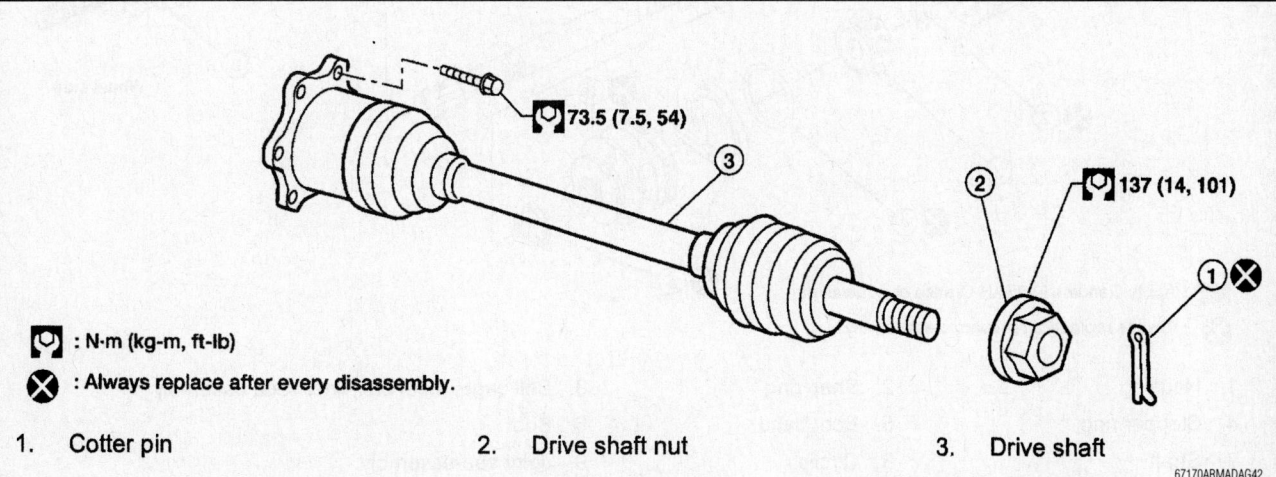

🔧 : N·m (kg-m, ft-lb)

✖ : Always replace after every disassembly.

1. Cotter pin

2. Drive shaft nut

3. Drive shaft

67170ARMADAG42

Fig. 13 Front halfshaft and related components

8. Be sure to perform the reconnect/relearn procedures.

CV-BOOTS OVERHAUL

Inner CV-Joint

See Figure 14.

1. Before servicing the vehicle, refer to the Precautions Section.
2. Remove the halfshaft from the vehicle.
3. Mount the halfshaft in a vise.
4. Remove the dust boot bands.
5. Remove the stopper ring with a flat-bladed tool.
6. Remove the snapring.
7. Disassemble the cage, ball and inner race assembly and dust boot for cleaning and inspection.

To install:

➡ **Discard old dust boot, dust boot bands and snapring and use new ones for assembly.**

8. Wrap the serrated part of the halfshaft with tape.
9. Install new dust boot and band onto halfshaft.
10. Remove tape from serrated part of halfshaft.

11. Install the cage, ball and inner race assembly.
12. Install the new snapring.
13. Insert 4.50–5.3 oz of genuine NISSAN grease or equivalent onto the housing and also onto halfshaft.
14. Install the stopper ring onto the housing.
15. Install the dust boot into the grooves on joint sub-assembly.
16. Secure the big and small ends of the dust boot using new boot bands.

Outer CV-Joint

See Figure 15.

1. Before servicing the vehicle, refer to the Precautions Section.
2. Remove the halfshaft from the vehicle.
3. Mount halfshaft in a vise.
4. Remove the dust boot bands and dust boot from joint sub-assembly.
5. Insert a suitable puller into the threaded part of the halfshaft. Pull the joint sub-assembly off of the halfshaft as shown in figure.
6. Remove dust boot and circlip from halfshaft for cleaning and inspection.

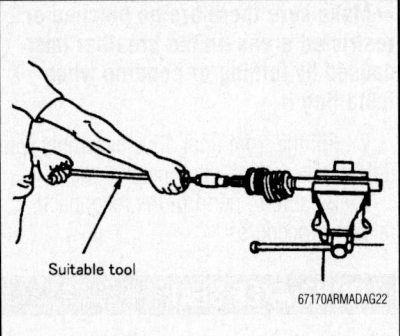

Suitable tool

67170ARMADAG22

Fig. 15 Using a suitable puller to remove joint sub-assembly.

To install:

➡ **Discard old dust boot, dust boot bands and circlip and use new ones for assembly.**

7. Insert genuine NISSAN grease or equivalent into the joint sub-assembly until grease oozes from the ball groove and serration hole.
8. Wrap the serrated part of the halfshaft with tape.
9. Install new dust boot and band onto halfshaft.
10. Remove tape from serrated part of the halfshaft.

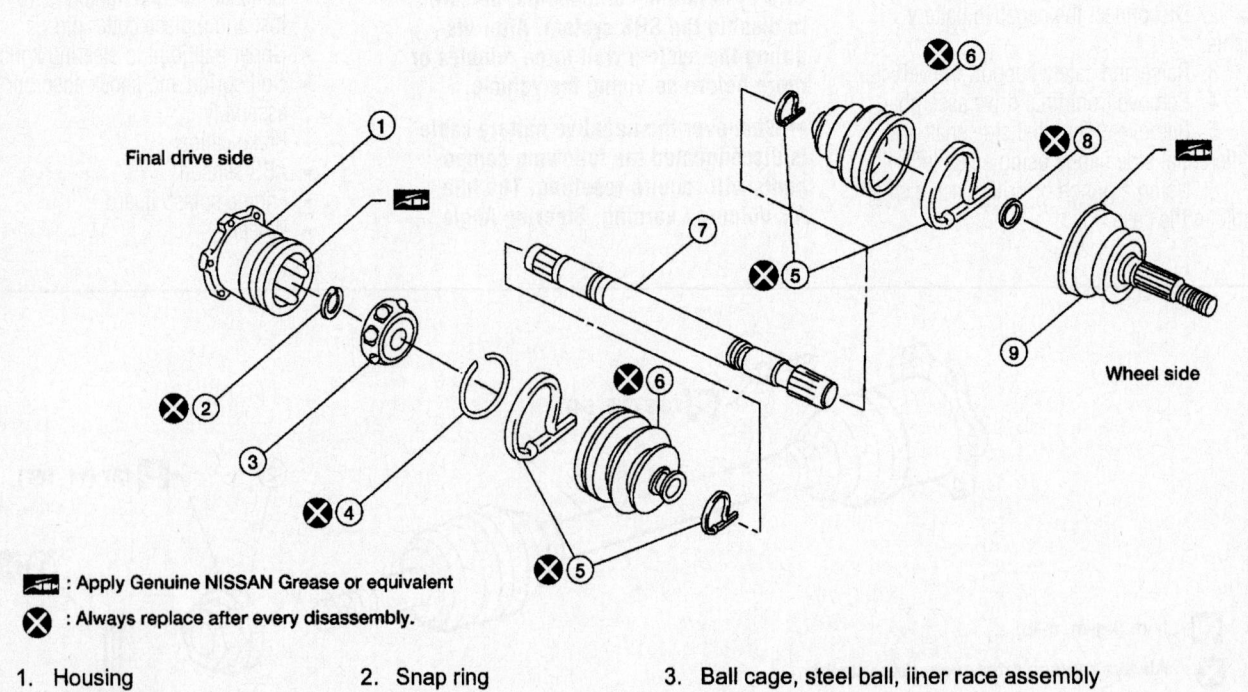

Final drive side

Wheel side

⊗ : Apply Genuine NISSAN Grease or equivalent

⊗ : Always replace after every disassembly.

1. Housing
2. Snap ring
3. Ball cage, steel ball, iiner race assembly
4. Stopper ring
5. Boot band
6. Boot
7. Shaft
8. Circlip
9. Joint sub-assembly

67170ARMADAG43

Fig. 14 Front halfshaft—exploded view

11. Press-fit the new circlip to the half-shaft.

12. Insert 5.1–5.8 oz of genuine NISSAN grease or equivalent into the joint sub-assembly and large end of boot.

13. Install the dust boot into the grooves on the joint sub-assembly.

14. Secure the big and small ends of the dust boot using new boot bands.

FRONT PINION SEAL

REMOVAL & INSTALLATION

See Figures 16 through 19.

1. Before servicing the vehicle, refer to the Precautions Section.

➡If working near and/or around the SRS system and components, be sure to disable the SRS system. After disabling the system wait three minutes or more before servicing the vehicle.

➡Whenever the negative battery cable is disconnected the following components will require resetting. The Idle Air Volume Learning, Steering Angle Sensor Neutral Position, Sunroof Memory Reset/Initialization, Automatic Drive Positioner System, Audio presets and Navigation. Use the CONSULT-III diagnostic tool, or equivalent to perform the required resets.

2. Disconnect the negative battery cable.

3. Raise and safely support the vehicle.

4. Remove front driveshaft.

5. Remove halfshafts.

6. Measure and record the pinion bearing preload using special tool J-25765-A.

7. Loosen the pinion nut while holding the companion flange using special tool J-44195.

8. Remove the companion flange using a suitable tool.

9. Using a punch or drill, place a small hole in the case.

10. Remove the seal using special tool SP8P.

To install:

➡Be sure to use new fasteners, as required.

11. Press front seal into carrier using a suitable tool.

12. Install companion flange and new pinion nut. Tighten pinion nut until there is no endplay and until recorded pinion bearing preload is met plus an additional 5 inch lbs. (0.5 Nm).

13. Install halfshafts.

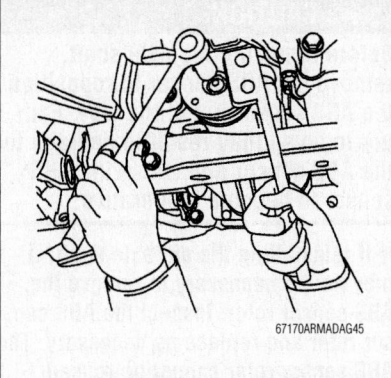
Fig. 16 Removing the companion flange

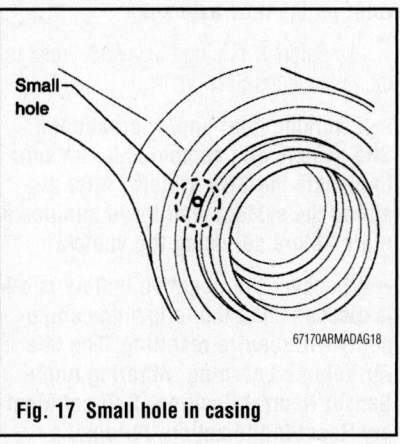
Fig. 17 Small hole in casing

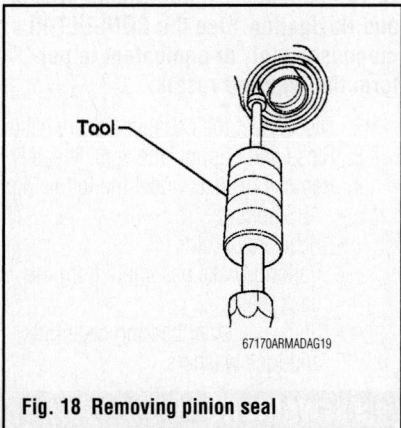
Fig. 18 Removing pinion seal

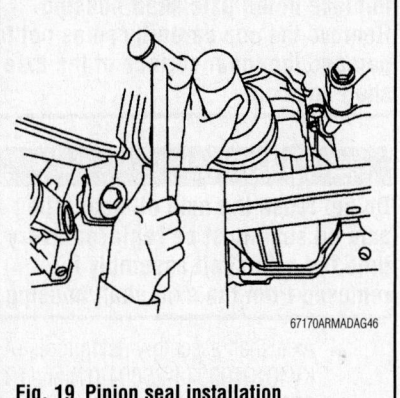
Fig. 19 Pinion seal installation

14. Install front driveshaft.

15. Be sure to perform the reconnect/relearn procedures.

REAR AXLE HOUSING

REMOVAL & INSTALLATION

See Figures 20 and 21.

✲✲ CAUTION

Do not damage spline, companion flange and front oil seal when removing propeller shaft. Before removing final drive assembly or rear axle assembly, disconnect ABS sensor harness connector from the assembly and move it away from final drive/rear axle assembly area. Failure to do so may result in sensor wires being damaged and sensor becoming inoperative.

1. Before servicing the vehicle, refer to the Precautions Section.

➡If working near and/or around the SRS system and components, be sure to disable the SRS system. After disabling the system wait three minutes or more before servicing the vehicle.

➡Whenever the negative battery cable is disconnected the following components will require resetting. The Idle Air Volume Learning, Steering Angle Sensor Neutral Position, Sunroof Memory Reset/Initialization, Automatic Drive Positioner System, Audio presets and Navigation. Use the CONSULT-III diagnostic tool, or equivalent to perform the required resets.

2. Disconnect the negative battery cable.

3. Raise and support the vehicle safely.

4. Drain the differential gear oil.

5. Disconnect or disconnect the following:

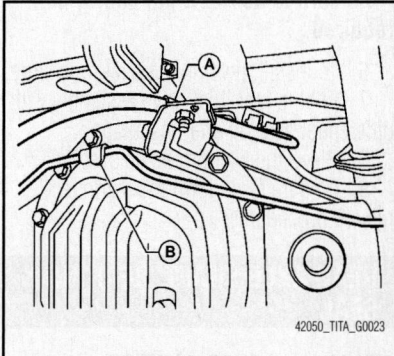
Fig. 20 Disconnecting parking brake cable (A) and brake tube (B)

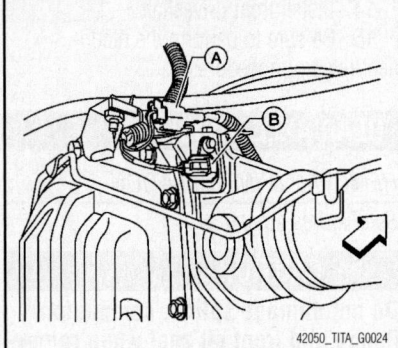

Fig. 21 Disconnecting differential lock position switch harness connector (A) and differential lock solenoid harness connector (B)

- Rear driveshaft
- Axle shaft
- Brake tube block connectors
- ABS sensor wire harness
- Parking brake cable (A)
- Brake tube (B)
- If equipped with electronic locking differential, differential lock position switch harness connector (A)
- If equipped with electronic locking differential, differential lock solenoid harness connector (B)
- Brake hose from brake tube at the mounting clip on top of rear final drive assembly. Then remove the metal clip to disconnect brake line from the mounting clip on top of the rear final drive assembly.

➡Support rear final drive assembly using a suitable jack. Secure rear final drive assembly to the jack while removing it.

- Rear shock absorber lower bolts
- Leaf spring U-bolt nuts
- Rear final drive assembly

To install:

➡Be sure to use new fasteners, as required.

6. To install, reverse removal procedure.
7. Fill the rear final drive assembly with differential gear oil after installation.
8. Bleed the air from brake system.
9. Be sure to perform the reconnect/relearn procedures.

REAR AXLE SHAFT, BEARING & SEAL

REMOVAL & INSTALLATION
See Figure 22.

✳✳ CAUTION

Before removing the axle shaft, remove the ABS sensor to reposition the ABS sensor out of the way. Failure to do so may result in damage to the ABS sensor and cause the ABS sensor to become inoperative.

➡If reinstalling the old axle shaft, it may not be necessary to remove the ABS sensor rotor. Inspect the ABS sensor rotor and replace as necessary. The ABS sensor rotor cannot be reused after it is removed. If replacing the axle shaft, install a new ABS sensor rotor on the new axle shaft.

1. Before servicing the vehicle, refer to the Precautions Section.

➡If working near and/or around the SRS system and components, be sure to disable the SRS system. After disabling the system wait three minutes or more before servicing the vehicle.

➡Whenever the negative battery cable is disconnected the following components will require resetting. The Idle Air Volume Learning, Steering Angle Sensor Neutral Position, Sunroof Memory Reset/Initialization, Automatic Drive Positioner System, Audio presets and Navigation. Use the CONSULT-III diagnostic tool, or equivalent to perform the required resets.

2. Disconnect the negative battery cable.
3. Raise and support the vehicle safely.
4. Remove or disconnect the following:
   - ABS sensor
   - Rear brake rotor
   - Parking brake assembly from the back plate
   - Four axle shaft bearing cage nuts and lock washers

✳✳ WARNING

The axle shaft bearing cup may stay in place in the axle shaft housing. Remove the cup carefully so as not to damage the inner surface of the axle shaft housing.

✳✳ CAUTION

Do not reuse the axle oil seal. The axle oil seal must be replaced every time the axle shaft assembly is removed from the axle shaft housing.

- Axle shaft assembly using tools (A) KV40101000 (J-25604-01) and (B) ST36230000 (J-25840-A)

- Snap ring from the axle shaft using suitable snapring pliers

✳✳ WARNING

Mount the axle shaft using a soft jaw vise to avoid damaging the axle shaft. Do not drill all the way through the bearing ring retainer, the drill may damage the axle shaft surface.

5. Strike the bearing ring retainer using a suitable chisel and hammer, with the chisel positioned across the drilled hole. Break the bearing ring retainer to remove it.

✳✳ WARNING

Do not heat or cut the axle shaft bearing or bearing ring retainer with a torch during removal, doing so will damage the axle shaft.

6. Remove or disconnect the following:
   - Axle shaft bearing cage studs using a suitable hammer or press

✳✳ WARNING

Do not tighten the tool against the axle shaft. Do not heat or cut the axle shaft bearing or bearing ring retainer with a torch during removal, doing so will damage the axle shaft.

- Axle shaft bearing off of the axle shaft using tool ST30031000 and a suitable press

✳✳ CAUTION

Do not reuse the axle oil seal. The axle oil seal must be replaced every time the axle shaft assembly is removed from the axle shaft housing.

- Axle oil seal and discard
- Wheel bearing cage

7. Clean and remove all nicks and burrs.
8. Check for straightness and distortion. Replace if necessary.
9. Inspect machined surfaces for evidence of overheating, damage and wear. Replace if necessary.
10. Measure the bearing ring retainer axle journal diameter: 1.5640 inches (39.726 mm). Replace if necessary.
11. Check that the axle shaft bearing and cup roll freely and are free from noise, cracks, pitting and wear. Replace if necessary.
12. Check the axle shaft bearing cage for deformation and cracks. Replace if necessary.

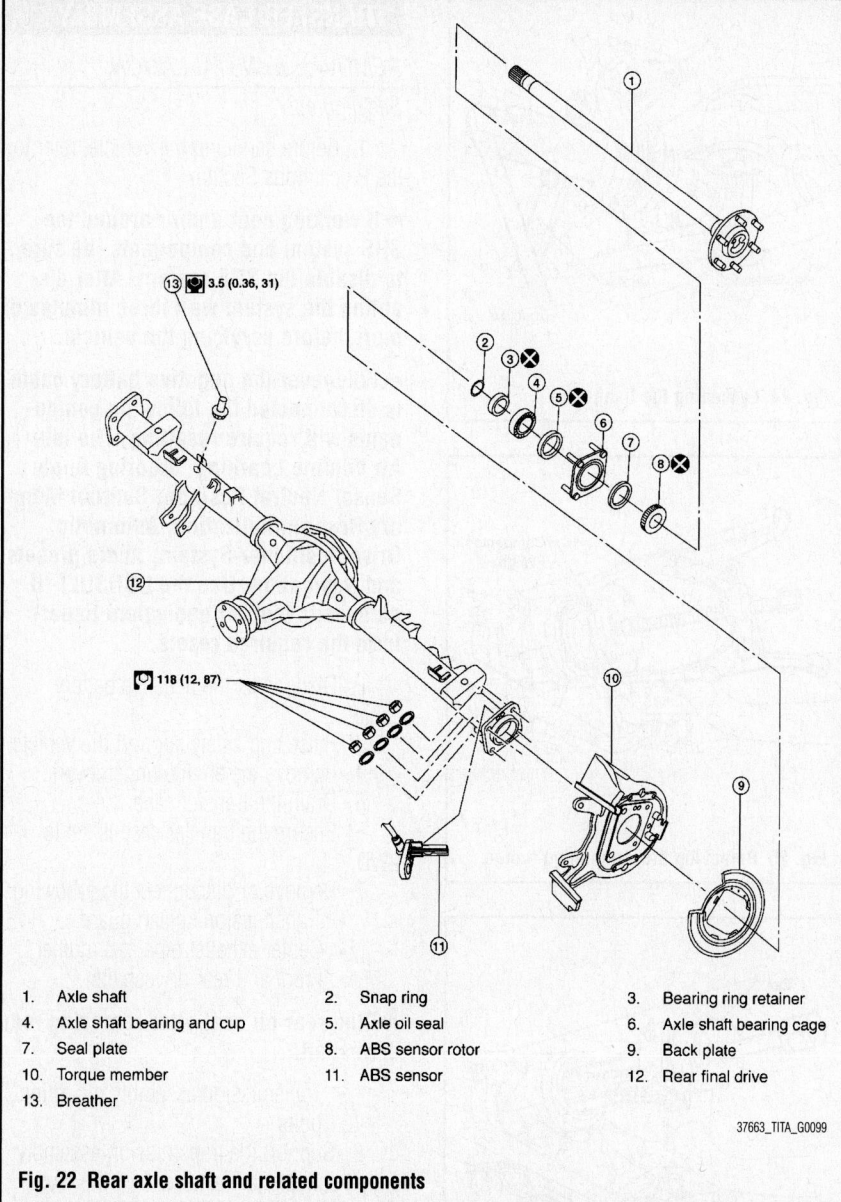

13 ⊡ 3.5 (0.36, 31)

118 (12, 87)

1.	Axle shaft	2.	Snap ring	3.	Bearing ring retainer
4.	Axle shaft bearing and cup	5.	Axle oil seal	6.	Axle shaft bearing cage
7.	Seal plate	8.	ABS sensor rotor	9.	Back plate
10.	Torque member	11.	ABS sensor	12.	Rear final drive
13.	Breather				

37663_TITA_G0099

Fig. 22 Rear axle shaft and related components

13. Check axle shaft housing exterior and inner machined surfaces for deformation and cracks. Replace if necessary.

To install:

➡**Be sure to use new fasteners, as required.**

➡**Do not reuse the old ABS sensor rotor.**

14. If installing a new axle shaft, install a new ABS sensor rotor onto the new axle shaft.

15. Install the axle shaft bearing cage.

☀ CAUTION

Do not reuse the axle oil seal. The axle oil seal must be replaced every time the axle shaft assembly is removed from the axle shaft housing.

16. Install the axle shaft bearing and cup on the axle shaft by performing the following:

a. Prepare an installer tool from a steel tube measuring 30 inches (762 mm) long with an outside diameter of 2.125 inches (53.98 mm) and an inside diameter of 1.625 inches (41.28 mm).

b. Press the axle shaft bearing and cup onto the axle shaft using a suitable press and the installer tool, until a .0015 inch (0.038 mm) feeler gauge does not fit in between the axle shaft bearing cup and seat.

☀ CAUTION

Make sure the axle shaft bearing and cup, axle oil seal, and axle shaft bearing cage are installed facing in the correct direction.

17. Install the bearing ring retainer onto the axle shaft by pressing the bearing ring retainer onto the axle shaft with a minimum force of 6992 lbs. (3172 kg, 31,100 N) until a .0015 inch (0.038 mm) feeler gauge does not fit between the bearing inner race and the bearing ring retainer in at least one point.

18. To complete installation, reverse remaining removal procedure.

19. Be sure to perform the reconnect/relearn procedures.

DRIVESHAFT

REMOVAL & INSTALLATION

1. Before servicing the vehicle, refer to the Precautions Section.

➡**If working near and/or around the SRS system and components, be sure to disable the SRS system. After disabling the system wait three minutes or more before servicing the vehicle.**

➡**Whenever the negative battery cable is disconnected the following components will require resetting. The Idle Air Volume Learning, Steering Angle Sensor Neutral Position, Sunroof Memory Reset/Initialization, Automatic Drive Positioner System, Audio presets and Navigation. Use the CONSULT-III diagnostic tool, or equivalent to perform the required resets.**

2. Disconnect the negative battery cable.

3. Position the selector lever in the N position.

4. Release the parking brake.

5. Raise and support the vehicle safely.

6. Matchmark the driveshaft and companion flange.

7. Remove the center support bearing bracket nuts, if equipped.

8. Remove the driveshaft. Discard the nuts.

To install:

➡**Be sure to use new fasteners, as required.**

9. Installation is the reverse of the removal procedure.

10. Check for vehicle vibration, correct as required.

11. Be sure to perform the reconnect/relearn procedures.

REAR PINION SEAL

REMOVAL & INSTALLATION

See Figures 23 through 26.

1. Before servicing the vehicle, refer to the Precautions Section.

➡ If working near and/or around the SRS system and components, be sure to disable the SRS system. After disabling the system wait three minutes or more before servicing the vehicle.

➡ Whenever the negative battery cable is disconnected the following components will require resetting. The Idle Air Volume Learning, Steering Angle Sensor Neutral Position, Sunroof Memory Reset/Initialization, Automatic Drive Positioner System, Audio presets and Navigation. Use the CONSULT-III diagnostic tool, or equivalent to perform the required resets.

2. Disconnect the negative battery cable.
3. Raise and safely support the vehicle.
4. Remove the rear driveshaft.
5. Measure and record the total preload.
6. Matchmark the drive pinion to position 'B' on the companion flange.
7. Remove the drive pinion nut using suitable tool.
8. Remove the companion flange using suitable tool.
9. Remove the rear pinion seal using special tool J-34286.

To install:

➡ Be sure to use new fasteners, as required.

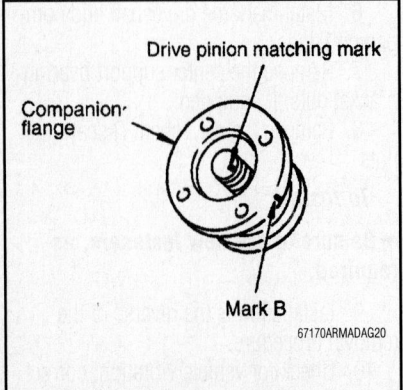

Fig. 23 Companion flange marking

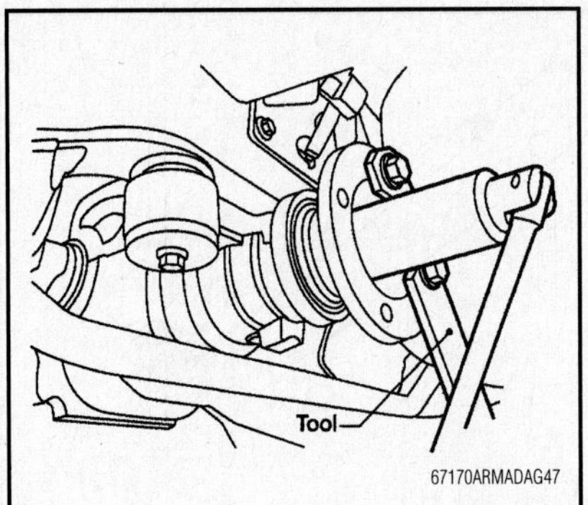

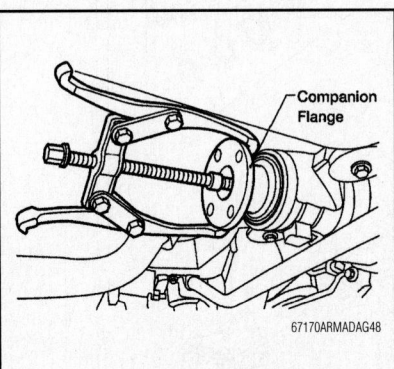

Fig. 24 Loosening the flange nut

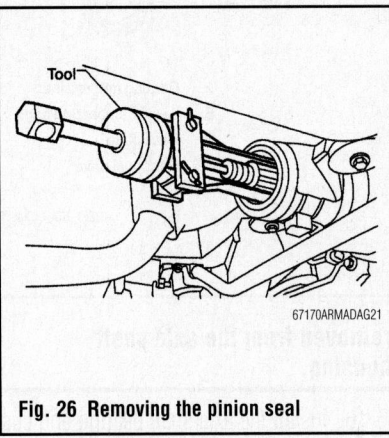

Fig. 25 Removing the companion flange

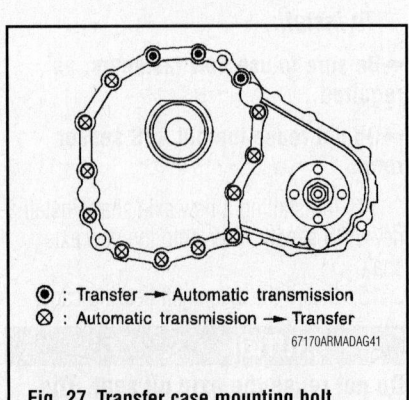

Fig. 26 Removing the pinion seal

10. Press the rear pinion seal into the carrier using suitable tool.
11. Align the matchmark on the companion flange to the drive pinion and install the companion flange.
12. Lubricate the drive pinion threads and seating surfaces of the drive pinion nut with grease.
13. Using a new drive pinion nut, tighten to 124–274 ft. lbs. (167–372 Nm).
14. Install rear driveshaft.
15. Be sure to perform the reconnect/relearn procedures.

TRANSFER CASE ASSEMBLY

REMOVAL & INSTALLATION

See Figure 27.

1. Before servicing the vehicle, refer to the Precautions Section.

➡ If working near and/or around the SRS system and components, be sure to disable the SRS system. After disabling the system wait three minutes or more before servicing the vehicle.

➡ Whenever the negative battery cable is disconnected the following components will require resetting. The Idle Air Volume Learning, Steering Angle Sensor Neutral Position, Sunroof Memory Reset/Initialization, Automatic Drive Positioner System, Audio presets and Navigation. Use the CONSULT-III diagnostic tool, or equivalent to perform the required resets.

2. Disconnect the negative battery cable.
3. Raise and safely support the vehicle.
4. Remove the engine undercovers.
5. Drain the transfer case.
6. Ensure the transfer case is set to 2WD.
7. Remove or disconnect the following:
   • Transmission splash guard
   • Center exhaust pipe and muffler
   • Front and rear driveshafts

➡ Plug rear oil seal after removing rear driveshaft.

   • Transmission assembly mounting bolts

8. Support the transmission assembly with a suitable jack and remove the crossmember.
9. Remove or disconnect the following:
   • ATP switch, neutral 4LO switch, wait detection switch, transfer

Fig. 27 Transfer case mounting bolt locations

motor and transfer control device electrical connectors
- Breather hoses
- Shift actuator from the extension housing
- Transfer case to transmission assembly bolts
- Transfer case assembly

To install:
10. Install or connect the following:
- Transfer case to transmission assembly bolts, tightening to 27 ft. lbs. (36 Nm)
- Shift actuator
- Breather hoses
- ATP switch, neutral 4LO switch, wait detection switch, transfer

motor and transfer control device electrical connectors
- Support crossmember
- Transmission mounting bolts
- Drive shafts
- Muffler and center exhaust pipe
- Transmission splash guard
11. Be sure to perform the reconnect/relearn procedures.

ENGINE COOLING

ENGINE FAN

REMOVAL & INSTALLATION

See Figure 28.

1. Before servicing the vehicle, refer to the Precautions Section.

➡️**If working near and/or around the SRS system and components, be sure to disable the SRS system. After disabling the system wait three minutes or more before servicing the vehicle.**

➡️**Whenever the negative battery cable is disconnected the following components will require resetting. The Idle Air Volume Learning, Steering Angle Sensor Neutral Position, Sunroof Memory Reset/Initialization, Automatic Drive Positioner System, Audio presets and Navigation. Use the CONSULT-III diagnostic tool, or equivalent to perform the required resets.**

2. Disconnect the negative battery cable.
3. Remove or disconnect the following:
- Air duct and resonator assembly

- Engine cover
- Lower radiator shroud
- Drive belt
- Cooling fan

To install:

➡️**Be sure to use new fasteners, as required.**

4. To install, reverse removal procedure.
5. Install cooling fan with its front mark "F" facing front of engine.
6. Be sure to perform the reconnect/relearn procedures.

RADIATOR

REMOVAL & INSTALLATION

See Figure 29.

✳✳ CAUTION

Never remove the radiator cap when the engine is hot. Serious burns could occur from high-pressure engine coolant escaping from the radiator. Perform this procedure when engine is cold.

1. Before servicing the vehicle, refer to the Precautions Section.

➡️**If working near and/or around the SRS system and components, be sure to disable the SRS system. After disabling the system wait three minutes or more before servicing the vehicle.**

➡️**Whenever the negative battery cable is disconnected the following components will require resetting. The Idle Air Volume Learning, Steering Angle Sensor Neutral Position, Sunroof Memory Reset/Initialization, Automatic Drive Positioner System, Audio presets and Navigation. Use the CONSULT-III diagnostic tool, or equivalent to perform the required resets.**

2. Disconnect the negative battery cable.
3. Drain the cooling system.
4. Remove or disconnect the following:
- Engine splash guard
- Air intake assembly
- A/T fluid cooler hoses, install blind plug to avoid leakage of A/T fluid

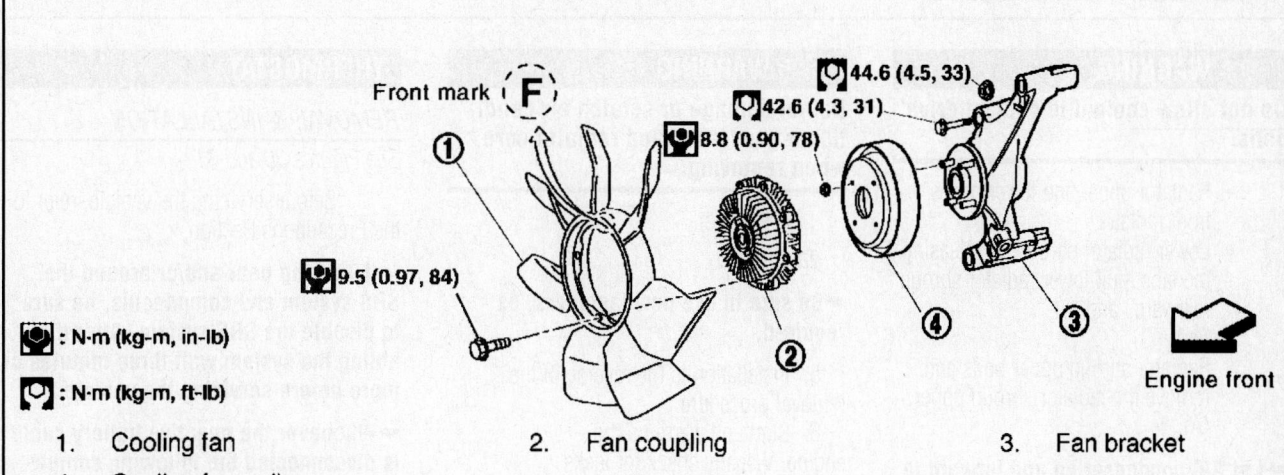

Front mark (F)

44.6 (4.5, 33)
42.6 (4.3, 31)
8.8 (0.90, 78)

9.5 (0.97, 84)

① ② ③ ④

Engine front

⚙ : N·m (kg-m, in-lb)

🔧 : N·m (kg-m, ft-lb)

1. Cooling fan
4. Cooling fan pulley
2. Fan coupling
3. Fan bracket

42050_TITA_G0015

Fig. 28 Cooling fan and related components

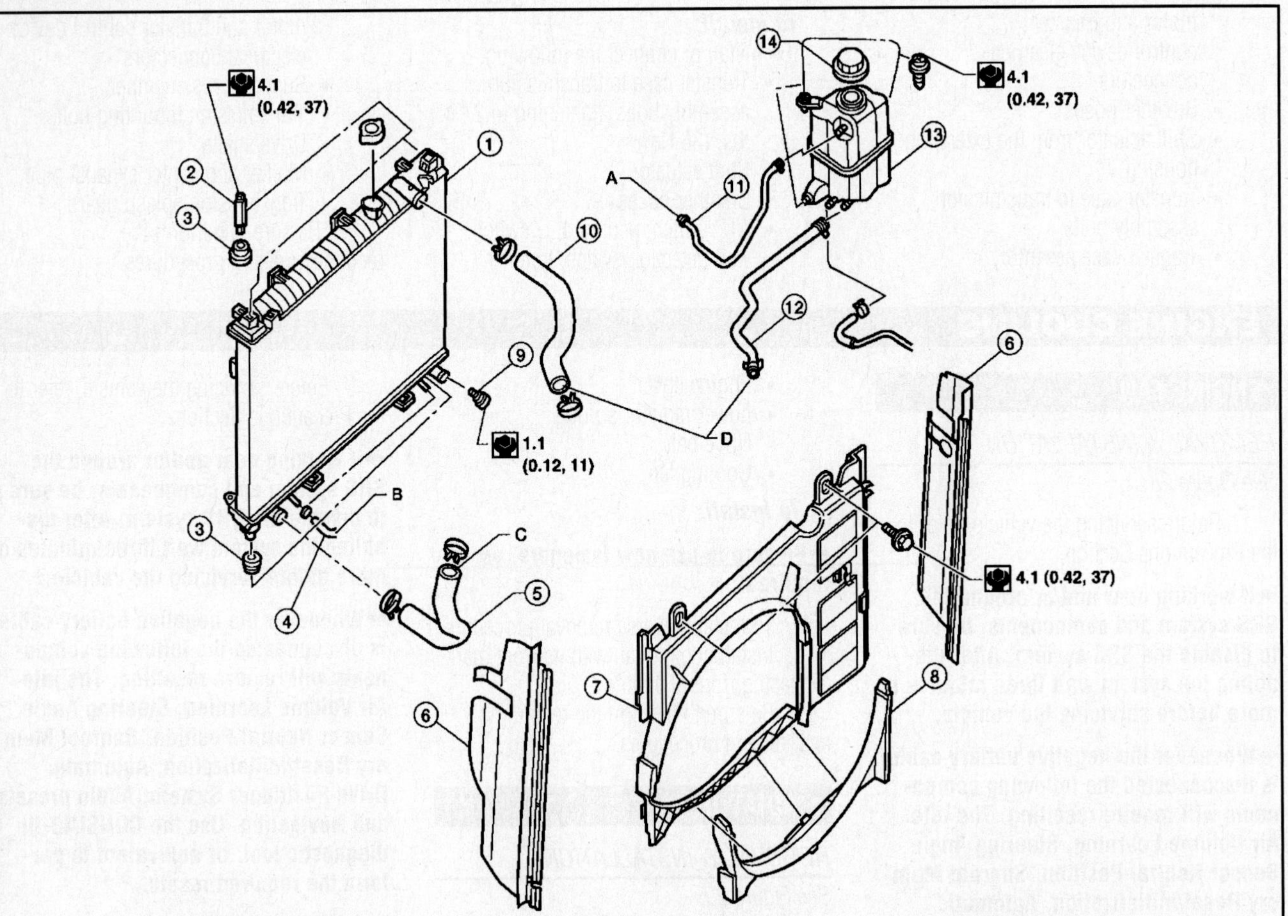

1. Radiator	2. Bolt	3. Mounting rubber
4. A/T fluid cooler hose	5. Radiator hose (lower)	6. Flaps
7. Radiator shroud (upper)	8. Radiator shroud (lower)	9. Drain plug
10. Radiator hose (upper)	11. Reservoir tank hose	12. By-pass hose
13. Reservoir tank	14. Reservoir tank cap	A. To radiator fill neck
B. To A/T cooler tube	C. To water suction pipe	D. To thermostat housing

42050_TITA_G0010

Fig. 29 Radiator and related components

⁂ WARNING

Do not allow coolant to contact drive belts.

- Radiator upper and lower hoses from radiator
- Lower radiator shroud by releasing the tabs, pull lower radiator shroud rearwards and

down

- Radiator shroud upper bolts and remove the radiator shroud upper (A)

➡Lift A/C condenser up and forward to remove from radiator.

- A/C condenser bolts and brackets
- A/T oil cooler bolts and oil cooler from radiator and position aside

⁂ WARNING

Do not damage or scratch air conditioner condenser and radiator core when removing.

- Radiator

To install:

➡Be sure to use new fasteners, as required.

5. Installation is the reverse of the removal procedure.
6. Start and warm up the engine. Visually check for leaks of the engine coolant and A/T fluid.
7. Be sure to perform the reconnect/relearn procedures.

THERMOSTAT

REMOVAL & INSTALLATION

See Figures 30 and 31.

1. Before servicing the vehicle, refer to the Precautions Section.

➡If working near and/or around the SRS system and components, be sure to disable the SRS system. After disabling the system wait three minutes or more before servicing the vehicle.

➡Whenever the negative battery cable is disconnected the following components will require resetting. The Idle Air Volume Learning, Steering Angle Sensor Neutral Position, Sunroof Memory Reset/Initialization, Automatic

Drive Positioner System, Audio presets and Navigation. Use the CONSULT-III diagnostic tool, or equivalent to perform the required resets.

2. Disconnect the negative battery cable.

3. Drain the cooling system.

4. Remove or disconnect the following:
- Air duct and resonator assembly
- Engine cover
- Water suction hose from the water inlet
- Water inlet and thermostat

To install:

➡**Be sure to use new fasteners, as required.**

5. To install, reverse removal procedure.

6. Install the thermostat with the whole circumference of each flange part fit securely inside the rubber ring as shown.

7. Install the thermostat with the jiggle valve facing upwards.

8. Tighten the thermostat mounting bolts to 15 ft. lbs. (20.6 Nm).

9. Install engine coolant and bleed the system.

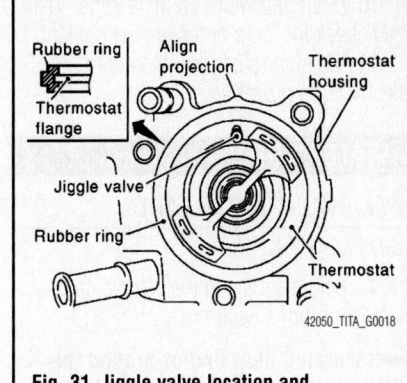

Fig. 31 Jiggle valve location and installation

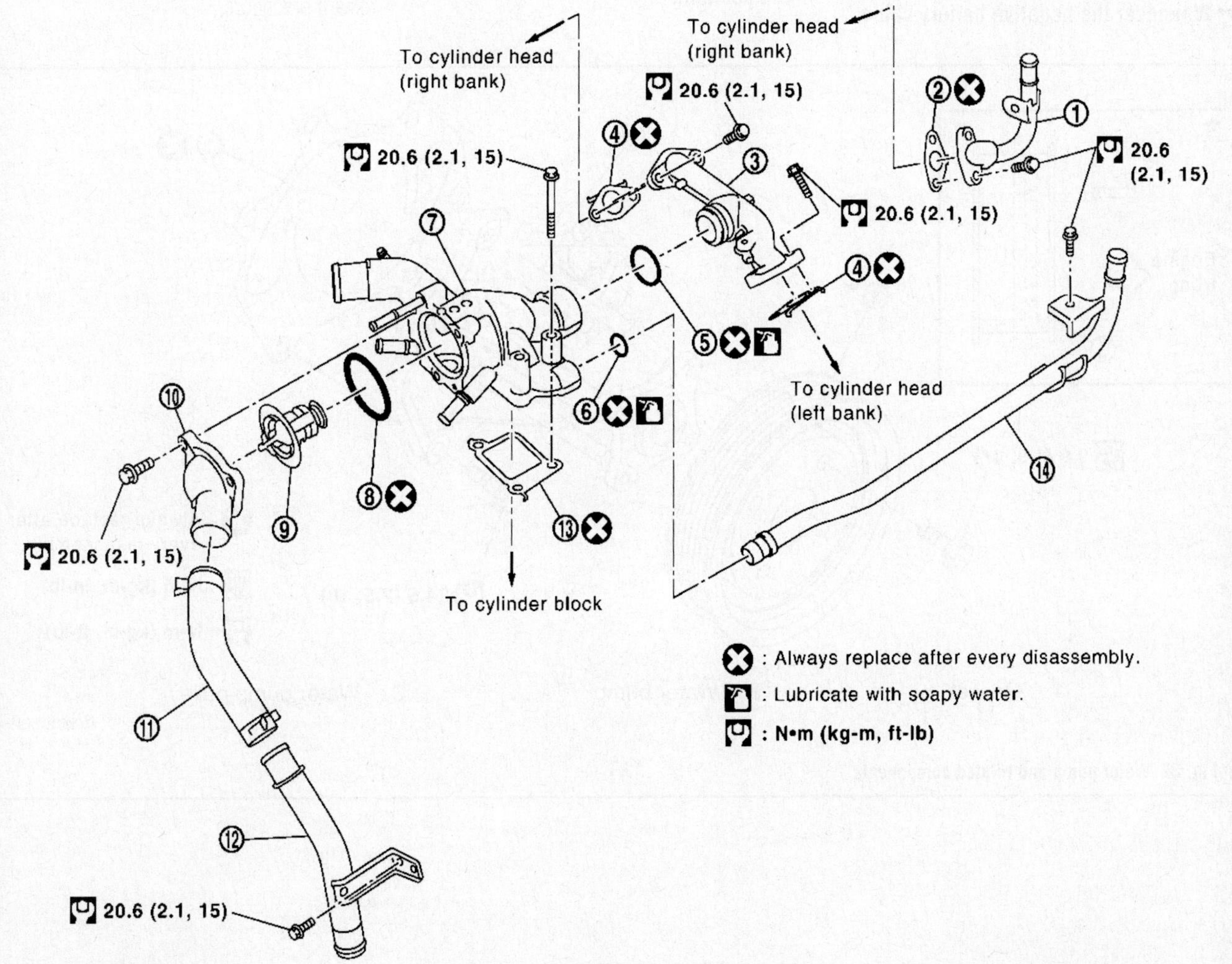

To cylinder head (right bank)

To cylinder head (right bank)

To cylinder head (left bank)

To cylinder block

⊗ : Always replace after every disassembly.

🔧 : Lubricate with soapy water.

🔩 : N•m (kg-m, ft-lb)

1.	Heater pipe	2.	Gasket	3.	Water outlet
4.	Gasket	5.	O-ring	6.	O-ring
7.	Thermostat housing	8.	Rubber ring	9.	Thermostat
10.	Water inlet	11.	Water suction hose	12.	Water suction pipe
13.	Gasket	14.	Heater pipe		

Fig. 30 Thermostat and related components

10. Start and warm up the engine. Visually check for leaks of the engine coolant.

11. Be sure to perform the reconnect/relearn procedures.

WATER PUMP

REMOVAL & INSTALLATION

See Figure 32.

1. Before servicing the vehicle, refer to the Precautions Section.

➡**If working near and/or around the SRS system and components, be sure to disable the SRS system. After disabling the system wait three minutes or more before servicing the vehicle.**

➡**Whenever the negative battery cable** is disconnected the following components will require resetting. The Idle Air Volume Learning, Steering Angle Sensor Neutral Position, Sunroof Memory Reset/Initialization, Automatic Drive Positioner System, Audio presets and Navigation. Use the **CONSULT-III** diagnostic tool, or equivalent to perform the required resets.

2. Disconnect the negative battery cable.
3. Drain the cooling system.
4. Remove or disconnect the following:
   - Engine splash guard
   - Air intake assembly
   - Accessory drive belt

➡**Leave the tensioner pulley in its fixed position.**

- Water pump pulley
- Water pump

To install:

➡**Be sure to use new fasteners, as required.**

5. Install or connect the following:
   - Water pump with a new gasket, tighten bolts to 18 ft. lbs. (25 Nm)
   - Water pump pulley, tighten bolts to 87 inch lbs. (10 Nm)
   - Accessory drive belt
   - Air intake assembly
   - Engine splash guard
6. Refill the cooling system.
7. Start the engine and check for leaks.
8. Be sure to perform the reconnect/relearn procedures.

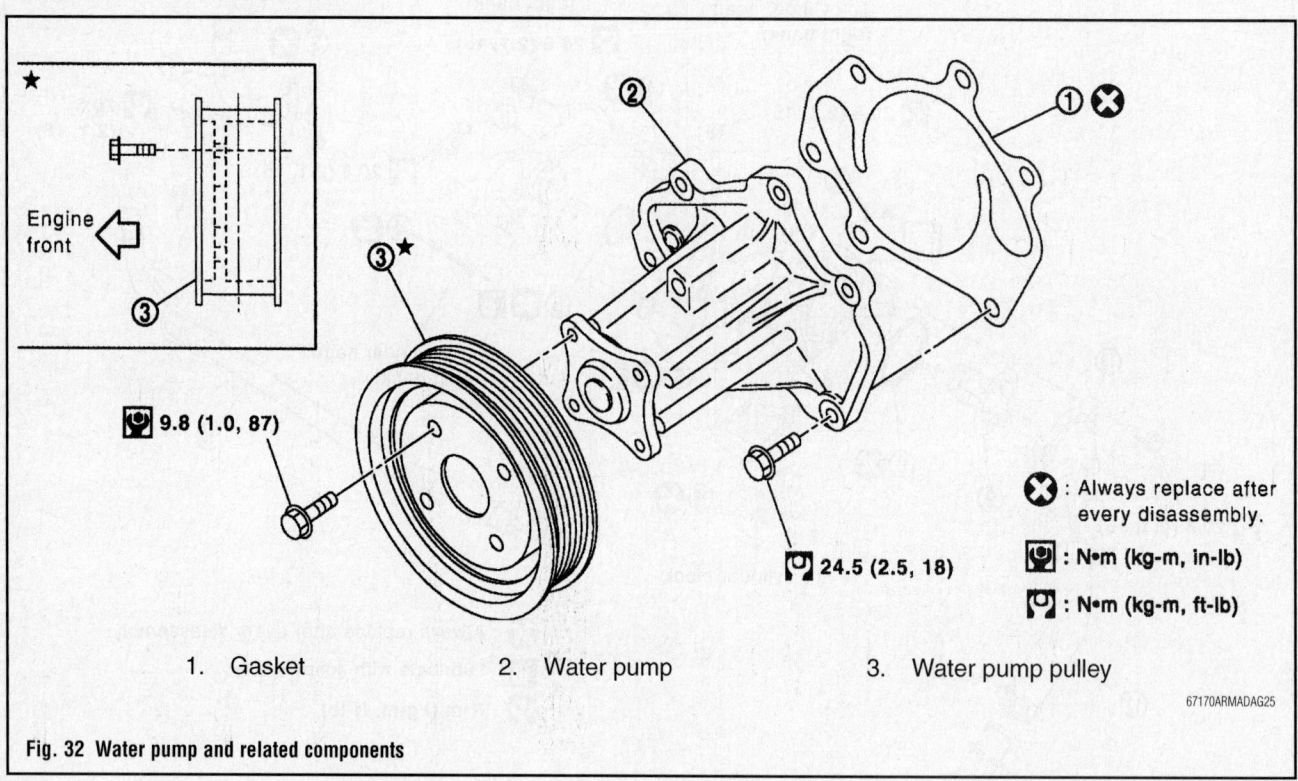

9.8 (1.0, 87)

24.5 (2.5, 18)

⊗ : Always replace after every disassembly.

: N•m (kg-m, in-lb)

: N•m (kg-m, ft-lb)

1. Gasket   2. Water pump   3. Water pump pulley

Engine front

67170ARMADAG25

Fig. 32 Water pump and related components

ENGINE ELECTRICAL

CHARGING SYSTEM

ALTERNATOR

REMOVAL & INSTALLATION

See Figure 33.

1. Before servicing the vehicle, refer to the Precautions Section.

➡ If working near and/or around the SRS system and components, be sure to disable the SRS system. After disabling the system wait three minutes or more before servicing the vehicle.

➡ Whenever the negative battery cable is disconnected the following components will require resetting. The Idle Air Volume Learning, Steering Angle Sensor Neutral Position, Sunroof Memory Reset/Initialization, Automatic Drive Positioner System, Audio presets and Navigation. Use the CONSULT-III diagnostic tool, or equivalent to perform the required resets.

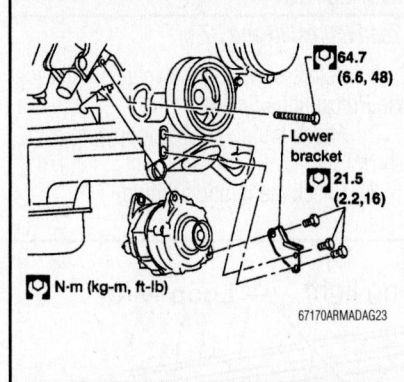

Fig. 33 Alternator and related components

2. Disconnect the negative battery cable.
3. Remove or disconnect the following:
   - Fan shroud
   - Drive belt
   - Lower alternator bracket
   - Alternator upper bolt
   - Alternator harness connectors
   - Alternator

To install:

➡ Be sure to use new fasteners, as required.

4. Install or connect the following:
   - Alternator
   - Alternator harness connectors
   - Upper bolt, tighten to 48 ft. lbs. (65 Nm)
   - Lower bracket, tighten to 16 ft. lbs (22 Nm)
   - Drive belt
   - Fan shroud
   - Negative battery cable
5. Be sure to perform the reconnect/relearn procedures.

VOLTAGE REGULATOR

REMOVAL & INSTALLATION

The voltage regulator is integral to the alternator assembly.

ENGINE ELECTRICAL

IGNITION SYSTEM

FIRING ORDER

See Figure 34.

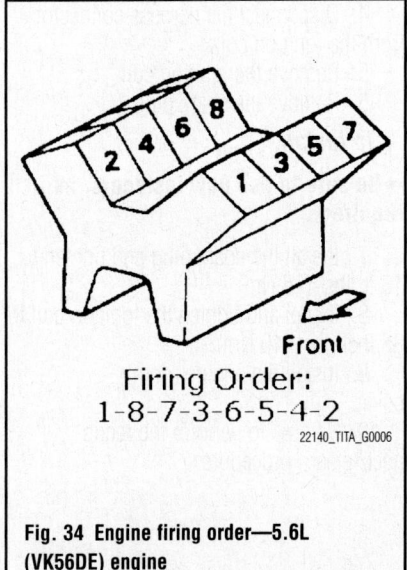

Firing Order:
1-8-7-3-6-5-4-2

22140_TITA_G0006

Fig. 34 Engine firing order—5.6L (VK56DE) engine

IGNITION COIL

REMOVAL & INSTALLATION

See Figure 35.

1. Before servicing the vehicle, refer to the Precautions Section.

➡ If working near and/or around the SRS system and components, be sure to disable the SRS system. After disabling the system wait three minutes or more before servicing the vehicle.

➡ Whenever the negative battery cable is disconnected the following components will require resetting. The Idle Air Volume Learning, Steering Angle Sensor Neutral Position, Sunroof Memory Reset/Initialization, Automatic

Drive Positioner System, Audio presets and Navigation. Use the CONSULT-III diagnostic tool, or equivalent to perform the required resets.

2. Disconnect the negative battery cable.
3. Remove the engine room cover using power tool.
4. Disconnect the harness connector from the ignition coil.
5. Remove the ignition coil.

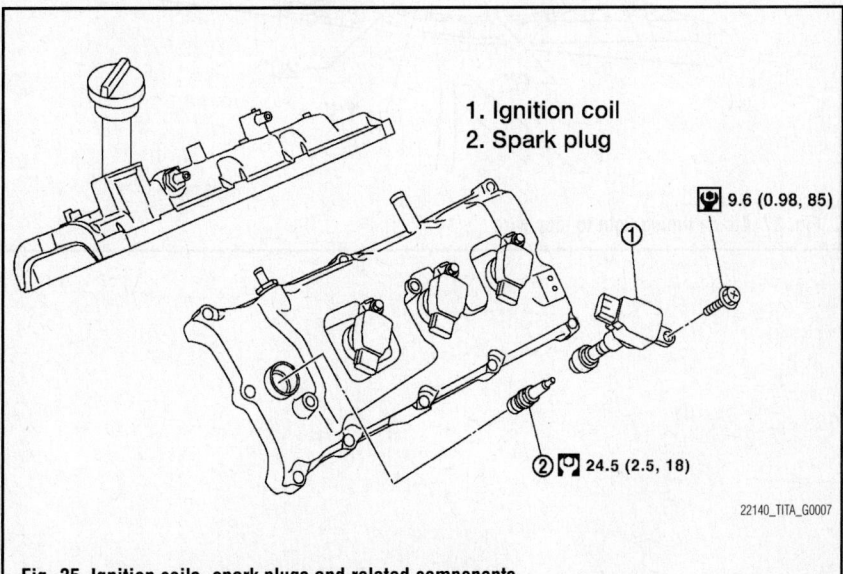

1. Ignition coil
2. Spark plug

Fig. 35 Ignition coils, spark plugs and related components

To install:

➡ **Be sure to use new fasteners, as required.**

6. Install and tighten the ignition coil to 85 inch lbs. (10 Nm).

7. Install the engine room cover.

8. Be sure to perform the reconnect/relearn procedures.

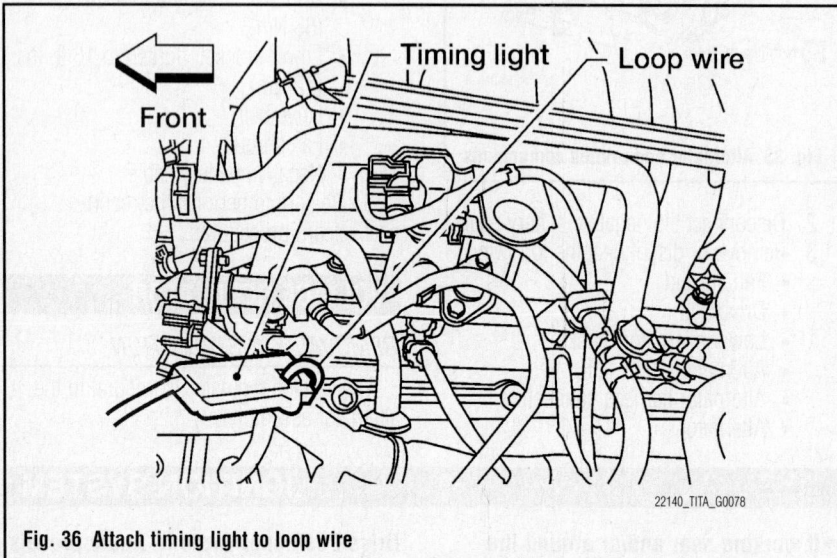

Fig. 36 Attach timing light to loop wire

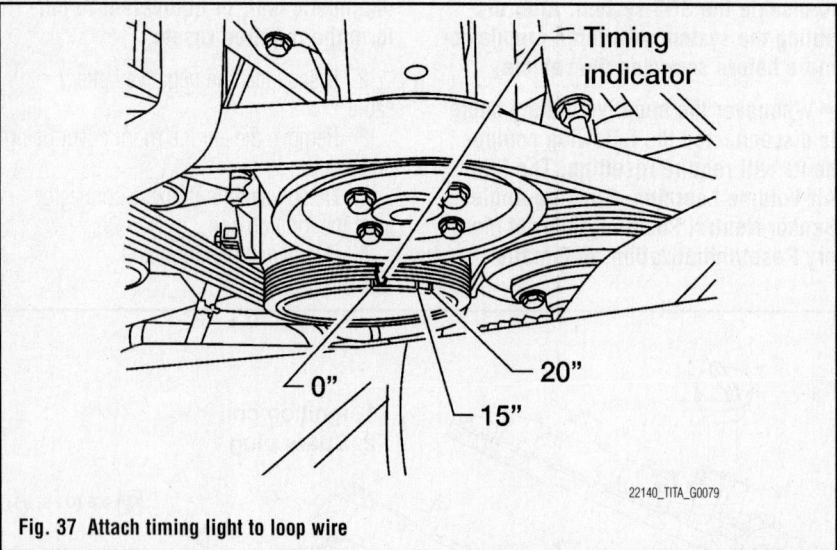

Fig. 37 Attach timing light to loop wire

IGNITION TIMING

INSPECTION

See Figures 36 and 37.

1. Before servicing the vehicle, refer to the Precautions Section.

2. Attach timing light to loop wire as shown.

3. Check the ignition timing.

ADJUSTMENT

The ignition timing is controlled by the Engine Control Module (ECM). No adjustment is necessary or possible.

SPARK PLUGS

REMOVAL & INSTALLATION

See Figure 35.

1. Before servicing the vehicle, refer to the Precautions Section.

➡ **If working near and/or around the SRS system and components, be sure to disable the SRS system. After disabling the system wait three minutes or more before servicing the vehicle.**

➡ **Whenever the negative battery cable is disconnected the following components will require resetting. The Idle Air Volume Learning, Steering Angle Sensor Neutral Position, Sunroof Memory Reset/Initialization, Automatic Drive Positioner System, Audio presets and Navigation. Use the CONSULT-III diagnostic tool, or equivalent to perform the required resets.**

2. Disconnect the negative battery cable.

3. Remove the engine room cover using power tool.

4. Disconnect the harness connector from the ignition coil.

5. Remove the ignition coil.

6. Remove the spark plug.

To install:

➡ **Be sure to use new fasteners, as required.**

7. Install the spark plug and tighten to 18 ft. lbs. (25 Nm).

8. Install and tighten the ignition coil to 85 inch lbs. (10 Nm).

9. Install the engine room cover.

10. Be sure to perform the reconnect/relearn procedures.

ENGINE ELECTRICAL   **STARTING SYSTEM**

STARTER

REMOVAL & INSTALLATION

See Figure 38.

1. Before servicing the vehicle, refer to the Precautions Section.

➡**If working near and/or around the SRS system and components, be sure to disable the SRS system. After disabling the system wait three minutes or more before servicing the vehicle.**

➡**Whenever the negative battery cable is disconnected the following components will require resetting. The Idle Air Volume Learning, Steering Angle Sensor Neutral Position, Sunroof Memory Reset/Initialization, Automatic Drive Positioner System, Audio presets and Navigation. Use the CONSULT-III diagnostic tool, or equivalent to perform the required resets.**

2. Disconnect the negative battery cable.
3. Remove or disconnect the following:
   • Intake manifold

   • Starter harness connectors
   • Two starter bolts using power tools
   • Starter

To install:

➡**Be sure to use new fasteners, as required.**

4. To install, reverse removal procedure and note the following:
   a. Tighten the starter mounting bolts to 34 ft. lbs. (47 Nm).
   b. Connect the negative battery cable.
5. Be sure to perform the reconnect/relearn procedures.

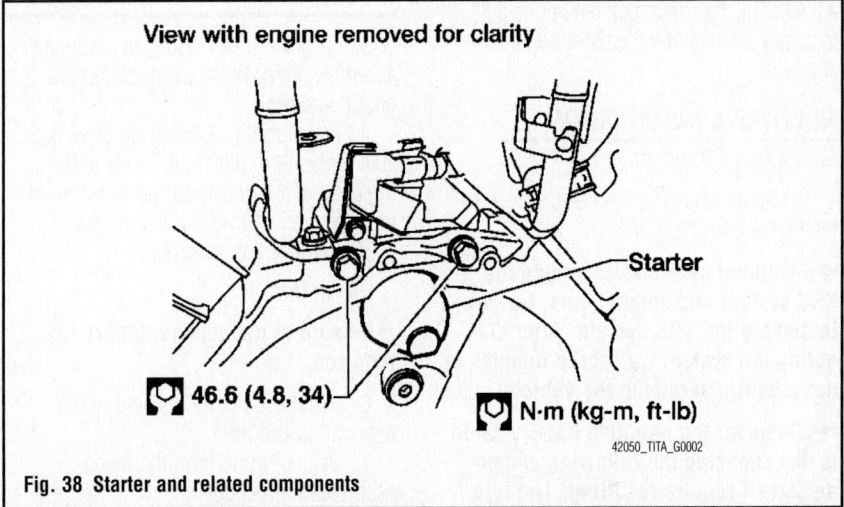

View with engine removed for clarity

Starter

46.6 (4.8, 34)   N·m (kg-m, ft-lb)

42050_TITA_G0002

Fig. 38 Starter and related components

ENGINE MECHANICAL

ACCESSORY DRIVE BELTS

ACCESSORY BELT ROUTING

See Figure 39.

Refer to the accompanying illustration.

INSPECTION

See Figure 39.

Remove air duct and resonator assembly when inspecting drive belt. Make sure that indicator (single line notch) of each auto

tensioner is within the allowable working range "A" (between three line notches). The indicator notch is located on the moving side of the drive belt auto tensioner. Inspect the drive belt for signs of glazing or cracking. A glazed belt will be perfectly smooth

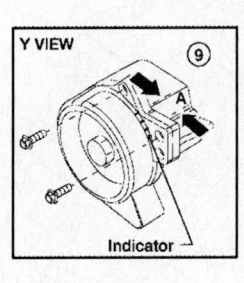

Y VIEW

Indicator

1. Drive belt	2. Power steering pump pulley	3. Generator pulley
4. Crankshaft pulley	5. A/C compressor	6. Idler pulley
7. Cooling fan pulley	8. Water pump pulley	9. Drive belt auto tensioner

42050_TITA_G0004

Fig. 39 Drive belt routing & working range "A"

from slippage, while a good belt will have a slight texture of fabric visible. Cracks will usually start at the inner edge of the belt and run outward. All worn or damaged drive belts should be replaced immediately. If the indicator is out of allowable working range or belt is damaged, replace the belt

ADJUSTMENT

There is no manual drive belt tension adjustment. The drive belt tension is automatically adjusted by the drive belt auto tensioner.

REMOVAL & INSTALLATION

See Figures 39 and 40.

1. Before servicing the vehicle, refer to the Precautions Section.

➡ If working near and/or around the SRS system and components, be sure to disable the SRS system. After disabling the system wait three minutes or more before servicing the vehicle.

➡ Whenever the negative battery cable is disconnected the following components will require resetting. The Idle Air Volume Learning, Steering Angle Sensor Neutral Position, Sunroof Memory Reset/Initialization, Automatic

Drive Positioner System, Audio presets and Navigation. Use the CONSULT-III diagnostic tool, or equivalent to perform the required resets.

2. Disconnect the negative battery cable.

✷✷ CAUTION

Avoid placing hand in a location where pinching may occur if the holding tool accidentally comes off.

3. Remove the air duct and resonator assembly. Remove the air duct and resonator assembly.
4. Install tool (J-46535) on drive belt auto tensioner pulley bolt, move in the direction of arrow (loosening direction of tensioner) as shown in illustration.
5. Remove the drive belt.

To install:

➡ Be sure to use new fasteners, as required.

6. Installation is the reverse of the removal procedure.
7. Be sure to perform the reconnect/relearn procedures.

➡ Make sure belt is securely installed around all pulleys.

8. Rotate the crankshaft pulley several turns clockwise to equalize belt tension between pulleys.
9. Make sure belt tension is within the allowable working range, using the indicator notch on the drive belt auto tensioner.

CAMSHAFT AND VALVE LIFTERS

INSPECTION

Runout

1. Before servicing the vehicle, refer to the Precautions Section.
2. Remove the camshafts.
3. Using a V-block on a precise flat table, support the No. 2 and No. 4 journals of the camshaft.

➡ Do not support journal No. 1 as it has a different diameter than the other locations.

4. Set the dial indicator to No. 3 journal.
5. Turn the camshaft to one direction by hand and measure the camshaft runout.
6. Runout should measure less than 0.0008 inches (0.02 mm).
7. Camshaft should be replaced if it exceeds the limit.

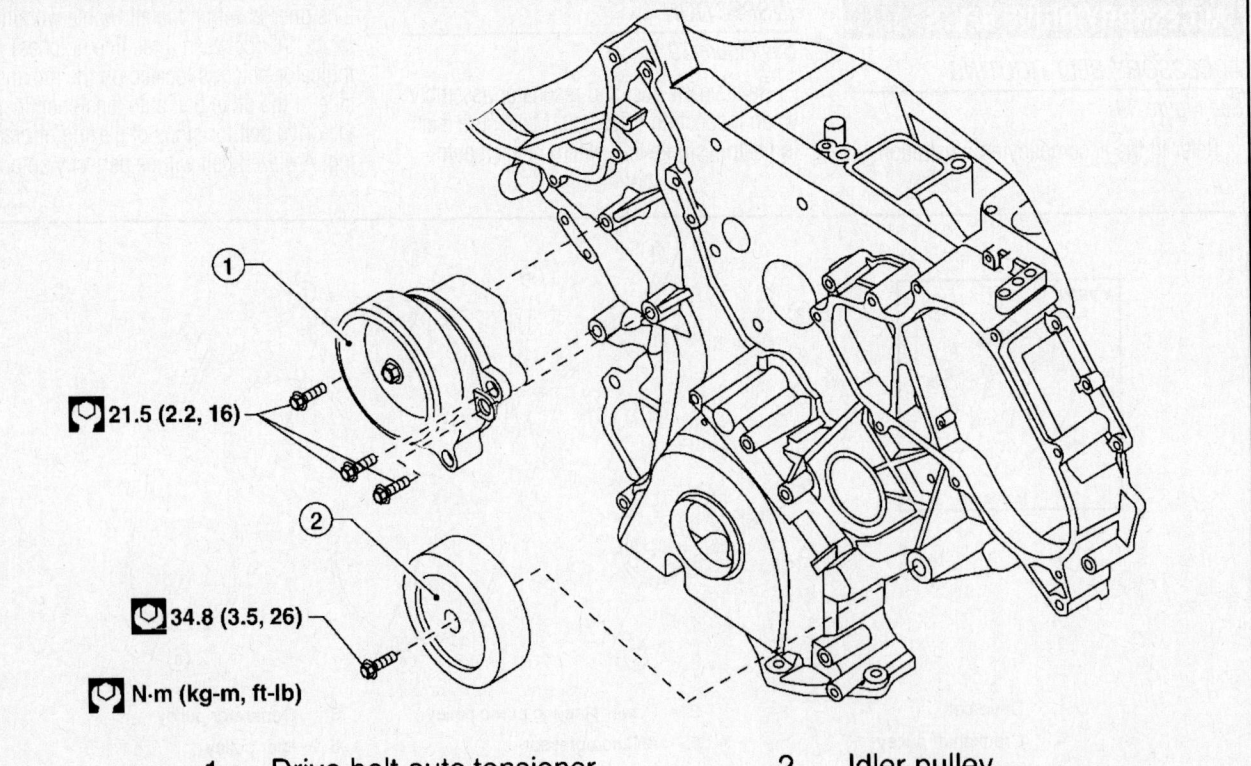

21.5 (2.2, 16)

34.8 (3.5, 26)

N·m (kg-m, ft-lb)

1. Drive belt auto tensioner   2. Idler pulley

37663_TITA_G0102

Fig. 40 Drive belt auto tensioner and idler pulley

Cam Height

1. Before servicing the vehicle, refer to the Precautions Section.
2. Remove the camshafts.
3. Measure the cam height with a micrometer.
4. The intake and exhaust camshaft should measure between 1.7506–1.7581 inches (44.465–44.655 mm).
5. Camshaft should be replaced if it exceeds the limit.

Journal Oil Clearance

1. Before servicing the vehicle, refer to the Precautions Section.
2. Remove the camshafts.
3. Measure the outer diameter of camshaft journal with micrometer and record the result.
4. Reinstall the camshaft bearing caps in accordance to the installation procedure.
5. Measure the inner diameter of the camshaft bracket ("A") with a bore gauge and record the result.
6. Subtract the camshaft journal diameter from the camshaft bracket inner diameter. The difference should measure between 0.0012–0.0027 inches (0.030–0.068 mm).
7. The camshaft or camshaft bracket should be replaced if it exceeds the limit.

➡**The camshaft bracket cannot be replaced as an individual part, because it is machined together with the cylinder head. The entire cylinder head assembly must be replaced.**

End Play

See Figure 41.

1. Before servicing the vehicle, refer to the Precautions Section.
2. Install a dial indicator in the thrust direction on the front end of the camshaft. Measure the end play of the dial indicator when the camshaft is moved back and forth. The dial indicator should measure between 0.0045–0.0074 inches (0.115–0.188 mm).
3. Measure the No. 1 journal as shown. The distance ("A") should be 1.2008–1.2027 inches (30.500–30.548 mm).
4. Measure the No. 1 journal bearing as shown. The distance ("B") should be 1.1953–1.1963 inches (30.360–30.385 mm).
5. Replace either the camshaft or cylinder head assembly if the measurement is exceeded.

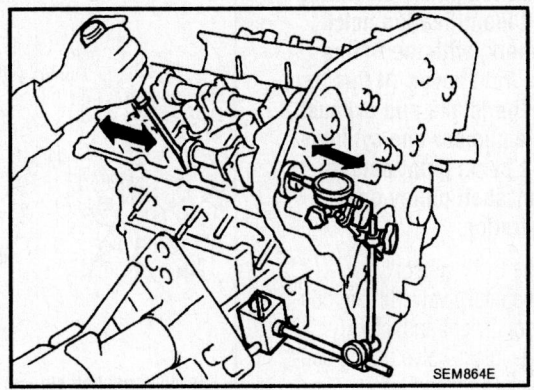

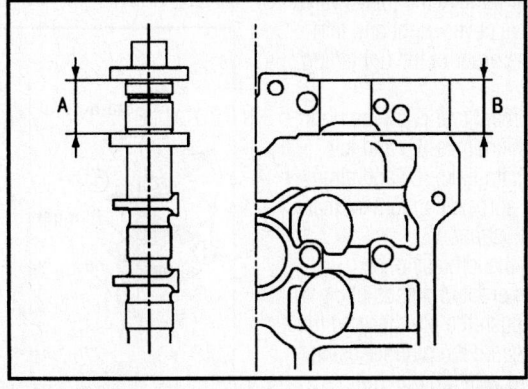

Fig. 41 Measuring for camshaft endplay

REMOVAL & INSTALLATION

See Figures 42 through 58.

1. Before servicing the vehicle, refer to the Precautions Section.

➡**If working near and/or around the SRS system and components, be sure to disable the SRS system. After disabling the system wait three minutes or more before servicing the vehicle.**

➡**Whenever the negative battery cable is disconnected the following components will require resetting. The Idle Air Volume Learning, Steering Angle Sensor Neutral Position, Sunroof Memory Reset/Initialization, Automatic Drive Positioner System, Audio presets and Navigation. Use the CONSULT-III diagnostic tool, or equivalent to perform the required resets.**

2. Disconnect the negative battery cable.
3. Remove the power steering reservoir tank bolts. Position the unit to the side.
4. Remove the valve covers.
5. Remove the spark plugs.
6. Remove the drive belt.
7. Be sure that the number one cylinder is at TDC on the compression stroke.

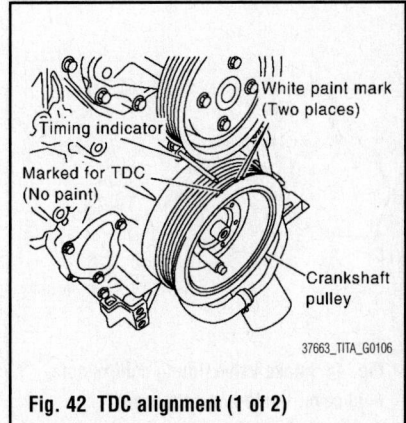

Fig. 42 TDC alignment (1 of 2)

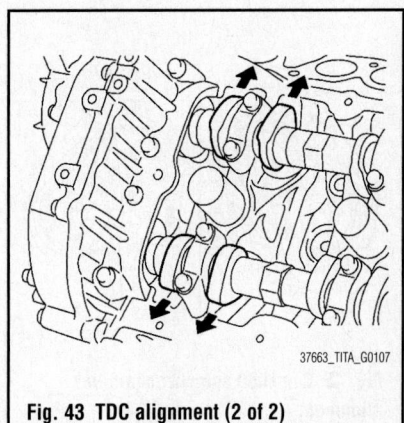

Fig. 43 TDC alignment (2 of 2)

➡Turn the crankshaft pulley clockwise to align the TDC identification notch (without paint mark) with the timing indicator on the front cover. At this time make sure that the intake and exhaust cam lobes of the number one cylinder (top front on left bank) point outside. If not turn the crankshaft pulley once more. See illustration

8. Remove the CMP sensor.

9. Remove the intake valve timing control position sensors (right and left).

10. Remove the intake valve timing control solenoid valves (right and left).

11. Loosen and remove the intake valve timing control valve cover (right and left) bolts in the reverse order of the tightening sequence.

12. Paint alignment marks on the right bank (A) timing chain links (C) and left bank (B) timing chain links (D) and align with the camshaft sprocket alignment marks (E) and (F). See illustration.

13. To remove the left tensioner, squeeze the return proof clip ends using a suitable tool and push the plunger into the tensioner body. Secure the plunger using a stopper pin (hard wire 0.04 inch in diameter). Remove the bolts and the tensioner.

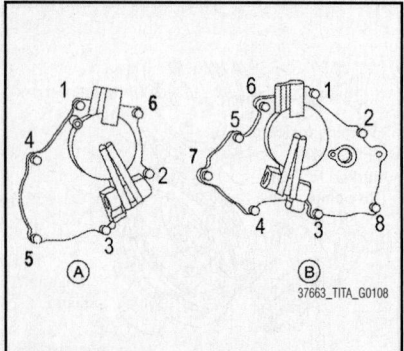

Fig. 44 Intake valve timing control solenoid cover tightening sequence

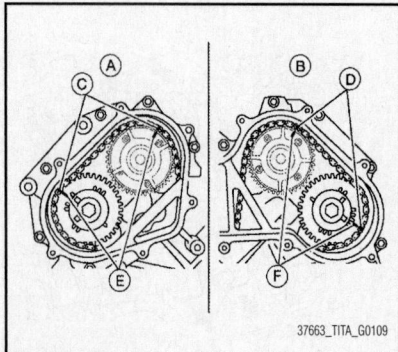

Fig. 45 Camshaft sprocket/chain link alignment

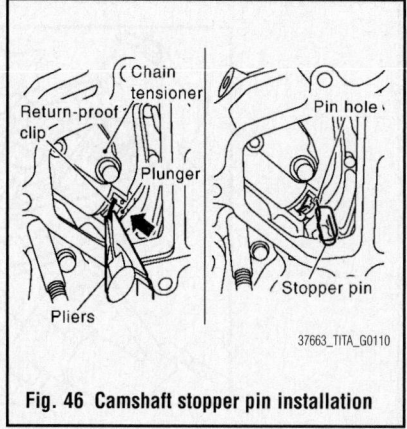

Fig. 46 Camshaft stopper pin installation

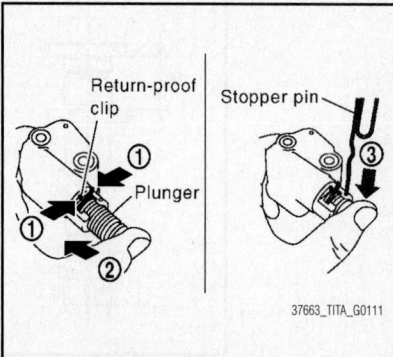

Fig. 47 Camshaft stopper plunger retention

➡The plunger, spring and spring seat pop out when squeezing the return proof clip without holding the plunger head. It may cause serious injuries. Always hold the plunger head when removing.

➡Stop the plunger in the fully extended position using the return proof clip (1) if the stopper pin is removed. Push the plunger (2) into the chain tensioner body while squeezing the return proof clip (1). Secure it using a stopper pin (3). See illustration.

14. Remove the chain tensioner cover from the front cover, using tool KV10111100, or equivalent. Do not damage the mating surfaces.

15. To remove the left tensioner, squeeze the return proof clip ends using a suitable tool and push the plunger into the tensioner body. Secure the plunger using a stopper pin (hard wire 0.04 inch in diameter). Remove the bolts and the tensioner.

➡The plunger, spring and spring seat pop out when squeezing the return proof clip without holding the plunger head. It may cause serious injuries. Always hold the plunger head when removing.

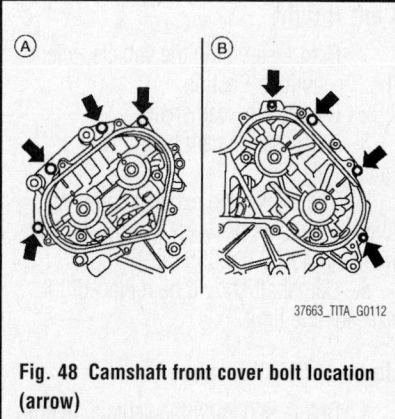

Fig. 48 Camshaft front cover bolt location (arrow)

➡If it is difficult to push the plunger on the tresioner, remove the plunger under the extended condition.

16. Loosen the camshaft sprocket bolts and remove the sprockets.

➡To avoid interference between the valves and pistons, do not turn the crankshaft or camshaft with the timing chain disconnected.

17. Remove the front cover bolts. See illustration for location (arrow).

18. Remove the camshaft bracket bolts in the reverse order of the tightening sequence. Remove the number one camshaft bracket. The bottom of the front surface of the bracket will be stuck because of liquid gasket.

19. Remove the camshaft. Remove the lifters, as necessary.

To install:

➡Be sure to use new fasteners, as required.

20. Install the camshafts. Be sure that the camshafts are properly identified. See illustrations.

21. Install the dowel pins at the front of the camshaft. See illustration for proper direction.

22. Install the camshaft brackets.

➡Install by referring to the illustration location mark on the upper surface. Install so that the installation mark can be correctly read when viewed from the intake manifold side.

23. To install the number one camshaft bracket, apply liquid gasket as shown in the illustration. Be sure to wipe off any excessive gasket after installation.

24. Apply liquid gasket to the back side of the left front cover and the right front cover. Bead diameter should be

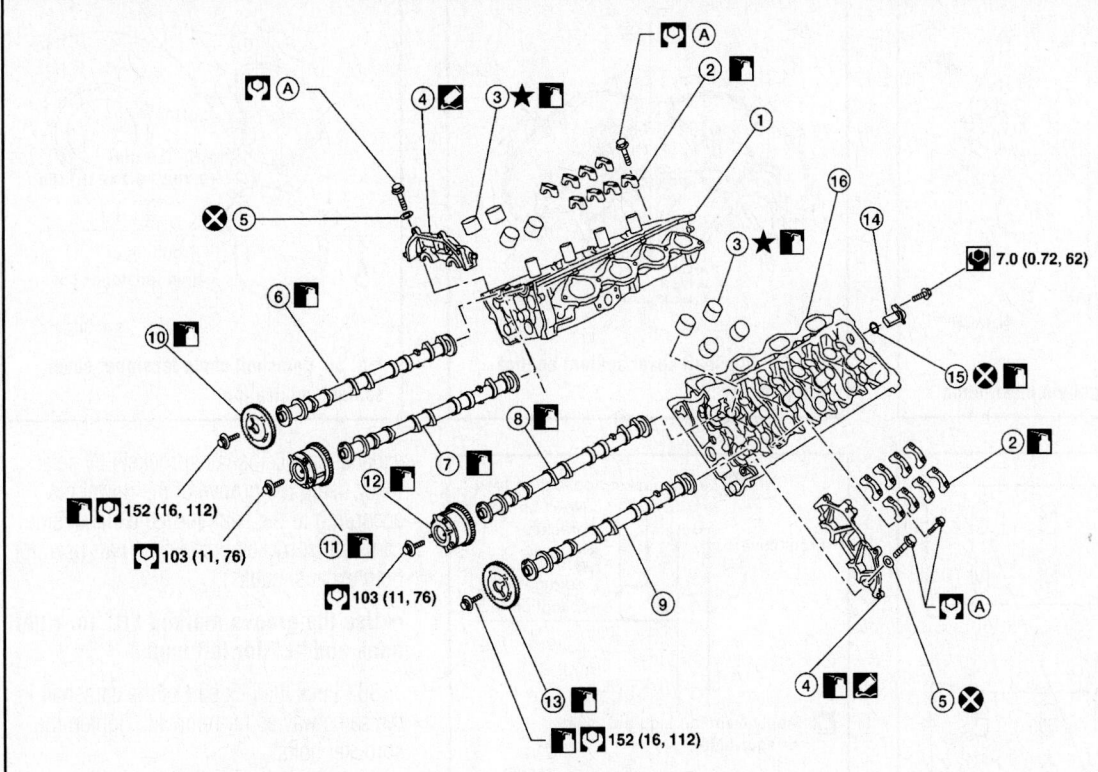

1. Cylinder head RH bank
4. Camshaft bracket (No. 1)
7. Camshaft RH bank INT
10. Camshaft sprocket RH bank EXH
13. Camshaft sprocket LH bank EXH
16. Cylinder head LH bank

2. Camshaft bracket (No. 2, 3, 4, 5)
5. Seal washer
8. Camshaft LH bank INT
11. Camshaft sprocket RH bank INT (VTC)
14. Camshaft position sensor (PHASE)

3. Valve lifter
6. Camshaft RH bank EXH
9. Camshaft LH bank EXH
12. Camshaft sprocket LH bank INT (VTC)
15. O-ring

37663_TITA_G0105

Fig. 49 Camshafts and related components

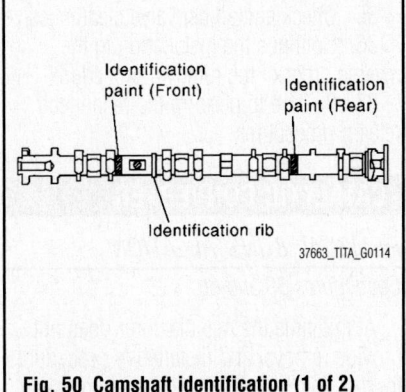

37663_TITA_G0114

Fig. 50 Camshaft identification (1 of 2)

Bank	INT EXH	Identification paint (front)	Identification paint (rear)	Identification rib
RH	INT	Pink	—	Yes
	EXH	—	Orange	Yes
LH	INT	Pink	—	No
	EXH	—	Orange	No

37663_TITA_G0115

Fig. 51 Camshaft identification (2 of 2)

0.102–0.142 inch. Position the number one camshaft bracket close to the mounting position and then install it to prevent from touching gasket applied to each surface.

25. Temporarily tighten the right and left front cover bolts.

26. Tighten the camshaft bracket bolts to specification and in the proper sequence.

27. Tighten the right and left front cover bolts to 8 ft. lbs.

28. Install the camshaft sprockets aligning them with the matching marks painted on the timing chain and the camshaft sprockets, before removal. Align the sprocket key groove with the dowel pin on the camshaft front edge at the same time. Temporarily tighten the sprocket bolts.

29. Install the intake VTC and the

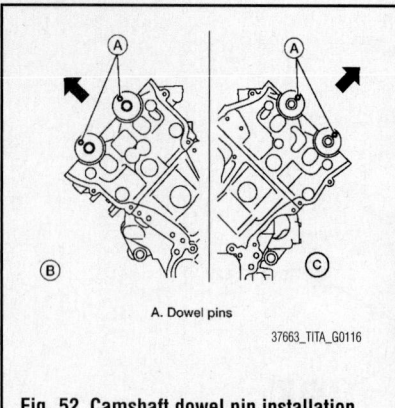

A. Dowel pins

37663_TITA_G0116

Fig. 52 Camshaft dowel pin installation

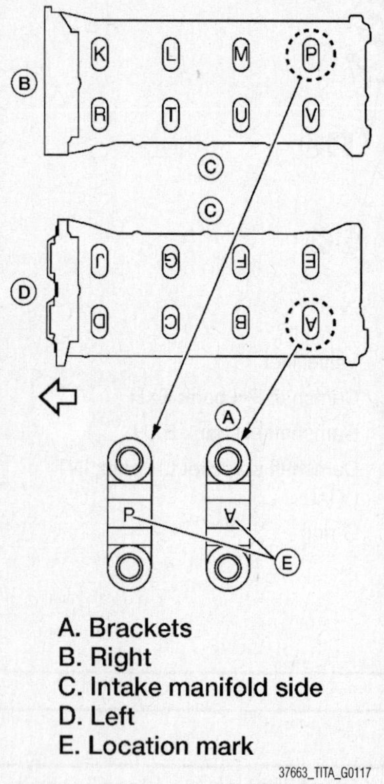

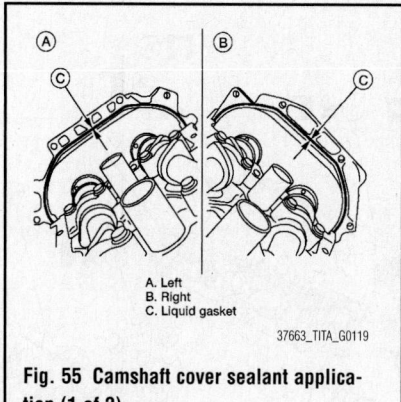

A. Left
B. Right
C. Liquid gasket

37663_TITA_G0119

Fig. 55 Camshaft cover sealant application (1 of 2)

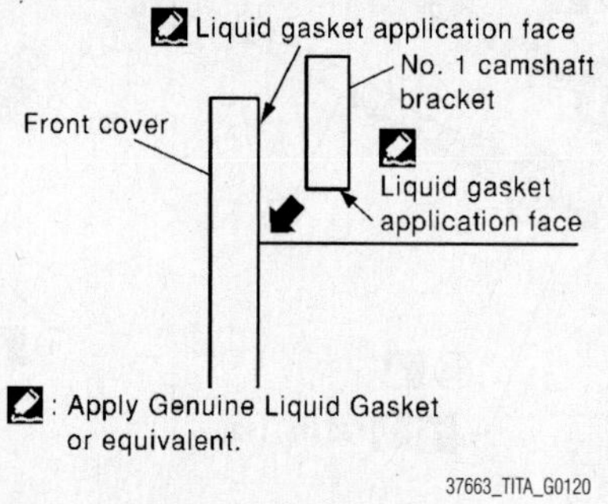

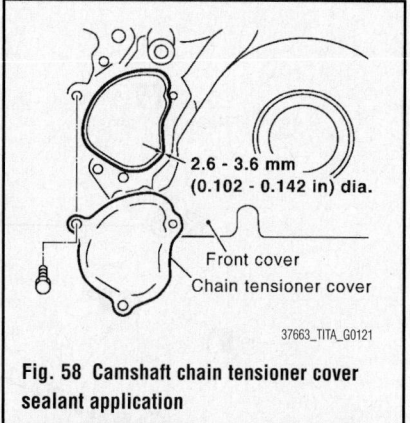

2.6 - 3.6 mm
(0.102 - 0.142 in) dia.

Front cover
Chain tensioner cover

37663_TITA_G0121

Fig. 58 Camshaft chain tensioner cover sealant application

A. Brackets
B. Right
C. Intake manifold side
D. Left
E. Location mark

37663_TITA_G0117

Fig. 53 Camshaft bracket installation and identification

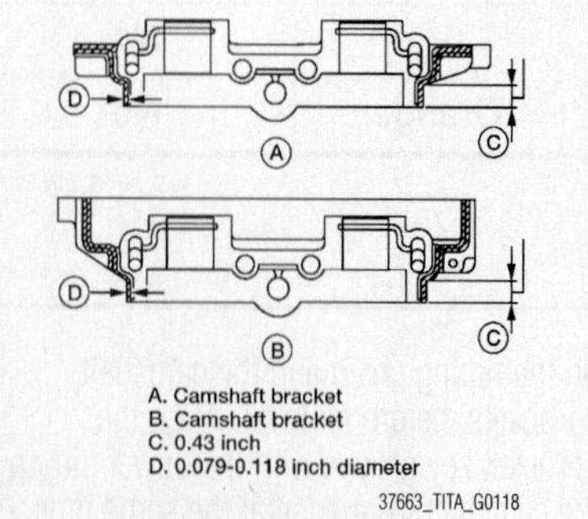

Liquid gasket application face
No. 1 camshaft bracket
Front cover
Liquid gasket application face

⬛ : Apply Genuine Liquid Gasket or equivalent.

37663_TITA_G0120

Fig. 56 Camshaft cover sealant application (2 of 2)

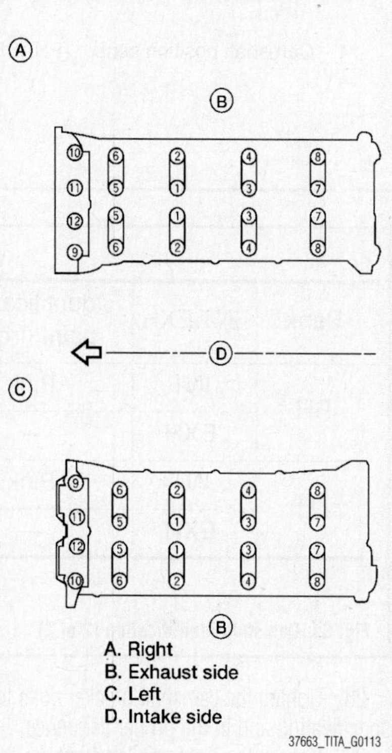

A. Camshaft bracket
B. Camshaft bracket
C. 0.43 inch
D. 0.079-0.118 inch diameter

37663_TITA_G0118

Fig. 54 Camshaft bracket sealant application

A. Right
B. Exhaust side
C. Left
D. Intake side

37663_TITA_G0113

Fig. 57 Camshaft bracket bolt tightening sequence

exhaust side camshaft sprockets by selectively using the groove of the dowel pin according to the bank for the exhaust side camshaft sprockets, (common part used for both exhaust banks).

➡**Use the groove marked "R" for right bank and "L" for left bank.**

30. Lock the hex part of the camshaft in the same way as for removal. Tighten the sprocket bolts.

31. Check that the timing marks are properly aligned.

32. To install the chain tensioner, compress the plunger and hold it using a stopper pin. Loosen the slack guide timing chain by rotating the camshaft hex part if mounting space is small. Tighten the tensioner bolts to 61 inch lbs.

33. Remove the stopper pin and release the plunger, then apply tension to the chain.

34. Install the chain tensioner cover onto the front cover. Apply liquid gasket. See illustration. Tighten the bolts to 80 inch lbs.

35. Check and adjust valve clearances.

36. Continue the installation in the reverse order of the removal procedure.

37. Be sure to perform the reconnect/relearn procedures.

CATALYTIC CONVERTER

REMOVAL & INSTALLATION

See Figures 59 and 60.

At this time the manufacturer does not provide removal and installation procedures for this component. The following procedure is a guideline and may differ from the vehicle you are servicing.

1. Before servicing the vehicle, refer to the Precautions Section.

➡**If working near and/or around the SRS system and components, be sure to disable the SRS system. After dis-**

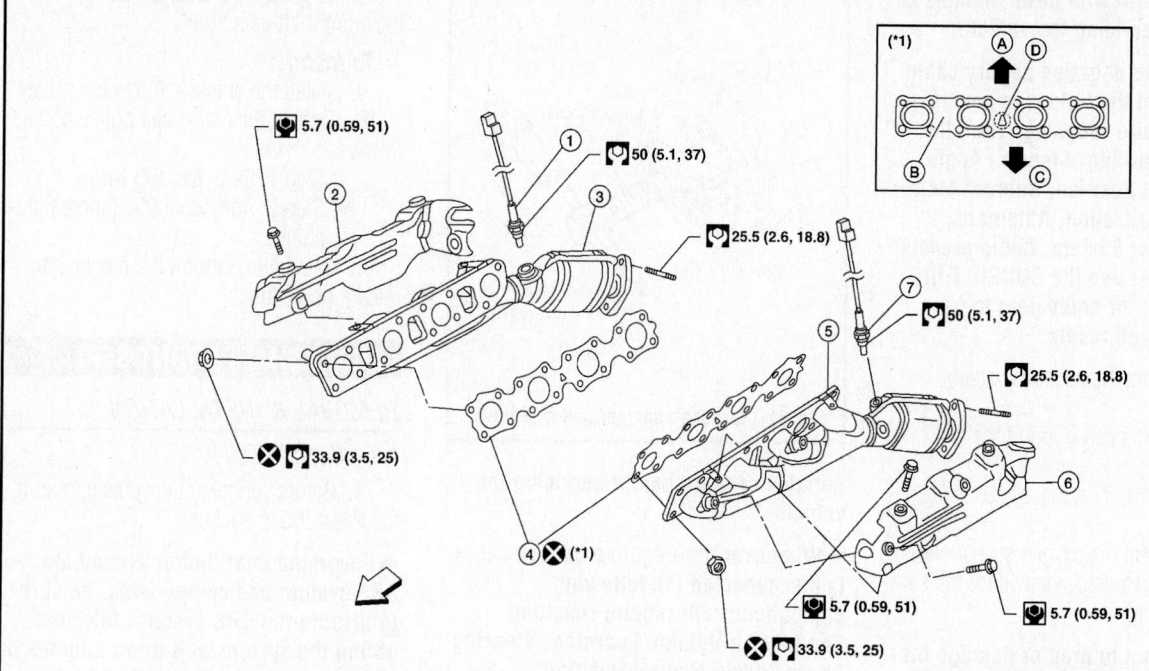

(*1)
A
B
C
D
1
2
3
4
5
6
7
5.7 (0.59, 51)
50 (5.1, 37)
25.5 (2.6, 18.8)
50 (5.1, 37)
25.5 (2.6, 18.8)
33.9 (3.5, 25)
(*1)
5.7 (0.59, 51)
5.7 (0.59, 51)
33.9 (3.5, 25)

1. Air fuel ratio A/F sensor 1 (bank 2)
2. Exhaust manifold cover (bank 2)
3. Exhaust manifold (bank 2)
4. Gaskets
5. Exhaust manifold (bank 1)
6. Exhaust manifold cover (bank 1)
7. Air fuel ratio A/F sensor 1 (bank 1)

A. Up
B. Coated face
C. Manifold side
D. Up mark

⇦ Front

37663_TITA_G0122

Fig. 59 Catalytic converters and related components (1 of 2)

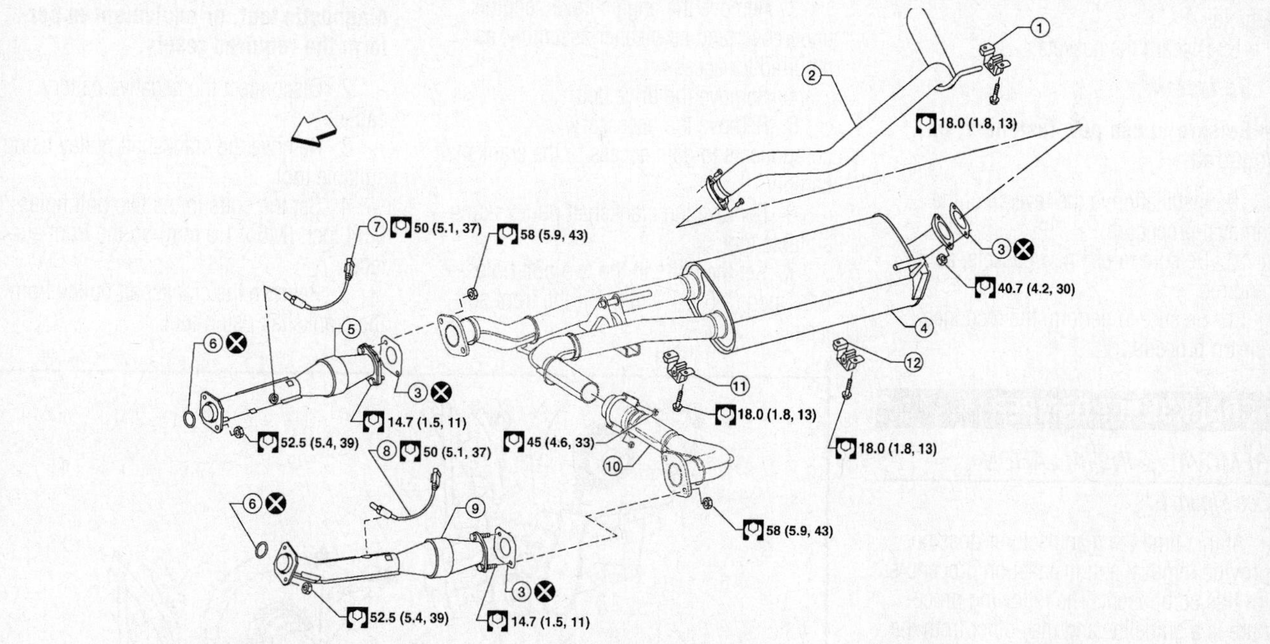

1
2
18.0 (1.8, 13)
3
40.7 (4.2, 30)
4
7
50 (5.1, 37)
58 (5.9, 43)
5
6
3
14.7 (1.5, 11)
52.5 (5.4, 39)
8
50 (5.1, 37)
45 (4.6, 33)
11
18.0 (1.8, 13)
12
18.0 (1.8, 13)
10
6
9
58 (5.9, 43)
3
52.5 (5.4, 39)
14.7 (1.5, 11)

1. Tailpipe hanger bracket
2. Tailpipe
3. Gasket
4. Main muffler
5. Right front exhaust tube
6. Ring gasket
7. Heated oxygen sensor 2 (bank 2)
8. Heated oxygen sensor 2 (bank 1)
9. Left front exhaust tube
10. Center exhaust tube
11. Muffler hanger bracket front
12. Muffler hanger bracket rear

⇦ Front

37663_TITA_G0123

Fig. 60 Catalytic converters and related components (2 of 2)

abling the system wait three minutes or more before servicing the vehicle.

➡ Whenever the negative battery cable is disconnected the following components will require resetting. The Idle Air Volume Learning, Steering Angle Sensor Neutral Position, Sunroof Memory Reset/Initialization, Automatic Drive Positioner System, Audio presets and Navigation. Use the CONSULT-III diagnostic tool, or equivalent to perform the required resets.

2. Disconnect the negative battery cable.

3. Raise and safely support the vehicle.

4. Properly support the exhaust system.

5. Disconnect the oxygen sensor electrical wires. As required, remove the sensor from its mounting.

➡ Be careful not to drop or damage the sensor. If a sensor has been dropped, it must be replaced.

6. Remove the converter retaining bolts and nuts.

➡ Depending upon the vehicle configuration the exhaust manifold may have to be removed along with the converter.

7. Remove the component from the vehicle.

8. Discard the gaskets.

To install:

➡ Be sure to use new fasteners, as required.

9. Installation is the reverse of the removal procedure.

10. Be sure to use new gaskets, as required.

11. Be sure to perform the reconnect/relearn procedures.

CRANKSHAFT DAMPER

REMOVAL & INSTALLATION

See Figure 61.

At this time the manufacturer does not provide removal and installation procedures for this component. The following procedure is a guideline and may differ from the vehicle you are servicing.

1. Before servicing the vehicle, refer to the Precautions Section.

➡ If working near and/or around the SRS system and components, be sure to disable the SRS system. After disabling the system wait three

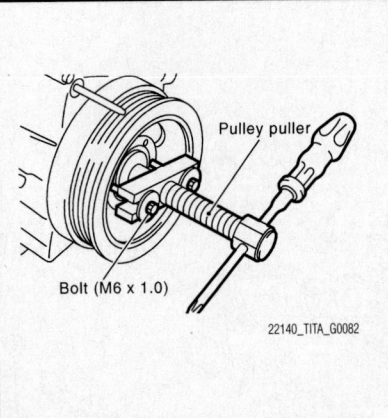

Fig. 61 Crankshaft damper pulley removal

minutes or more before servicing the vehicle.

➡ Whenever the negative battery cable is disconnected the following components will require resetting. The Idle Air Volume Learning, Steering Angle Sensor Neutral Position, Sunroof Memory Reset/Initialization, Automatic Drive Positioner System, Audio presets and Navigation. Use the CONSULT-III diagnostic tool, or equivalent to perform the required resets.

2. Disconnect the negative battery cable.

3. Remove the engine cover, engine undercover and air cleaner assembly, as required for access.

4. Remove the drive belt.

5. Remove the necessary components to gain access to the crankshaft damper.

6. Remove the crankshaft pulley using suitable tool.

7. Set the bolts in the two bolt holes 0.04 inch (M6 x 1.0 mm) on the front surface.

8. Remove the crankshaft pulley from the crankshaft using tool.

To install:

9. Install the crankshaft damper pulley.

10. Tighten the crankshaft pulley bolt as follows:
   - Step 1: 69 ft. lbs. (93 Nm)
   - Step 2: Additional 90° (angle tightening)

11. Be sure to perform the reconnect/relearn procedures.

CRANKSHAFT FRONT SEAL

REMOVAL & INSTALLATION

See Figures 62 and 63.

1. Before servicing the vehicle, refer to the Precautions Section.

➡ If working near and/or around the SRS system and components, be sure to disable the SRS system. After disabling the system wait three minutes or more before servicing the vehicle.

➡ Whenever the negative battery cable is disconnected the following components will require resetting. The Idle Air Volume Learning, Steering Angle Sensor Neutral Position, Sunroof Memory Reset/Initialization, Automatic Drive Positioner System, Audio presets and Navigation. Use the CONSULT-III diagnostic tool, or equivalent to perform the required resets.

2. Disconnect the negative battery cable.

3. Remove the crankshaft pulley using suitable tool.

4. Set the bolts in the two bolt holes 0.04 inch (M6 x 1.0 mm) on the front surface.

5. Remove the crankshaft pulley from the crankshaft using tool.

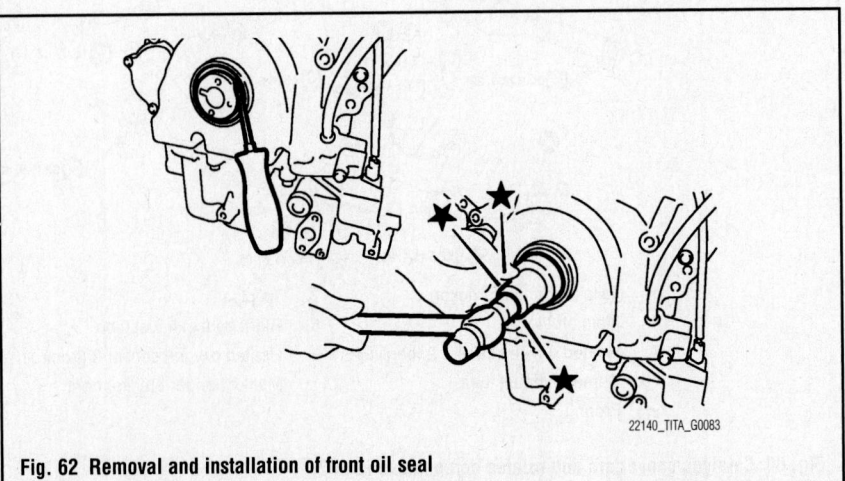

Fig. 62 Removal and installation of front oil seal

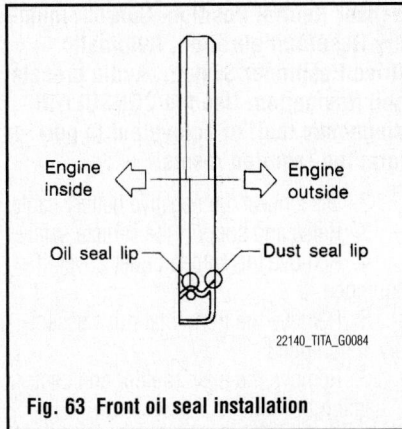

Fig. 63 Front oil seal installation

6. Remove the oil seal using a suitable tool.

To install:

7. Apply new engine oil to both the oil seal lip and dust seal lip of the new front oil seal.

8. Install the front oil seal so that each seal lip is oriented as shown.

9. Install the crankshaft damper pulley.

10. Tighten the crankshaft pulley bolt as follows:
- Step 1: 69 ft. lbs. (93 Nm)
- Step 2: Additional 90° (angle tightening)

11. Be sure to perform the reconnect/relearn procedures.

CYLINDER HEAD

REMOVAL & INSTALLATION

See Figures 64 and 65.

➡ **The engine must be removed from the vehicle to perform this procedure. Be sure that the engine is secured in a suitable holding fixture before performing this procedure.**

1. Before servicing the vehicle, refer to the Precautions Section.

➡ **If working near and/or around the SRS system and components, be sure to disable the SRS system. After disabling the system wait three minutes or more before servicing the vehicle.**

➡ **Whenever the negative battery cable is disconnected the following components will require resetting. The Idle Air Volume Learning, Steering Angle Sensor Neutral Position, Sunroof Memory Reset/Initialization, Automatic Drive Positioner System, Audio presets and Navigation. Use the CONSULT-III**

diagnostic tool, or equivalent to perform the required resets.

2. Disconnect the negative battery cable.

3. Remove or disconnect the following:
- Engine assembly
- Belt tensioner
- Idler pulley
- Thermostat housing and hose
- Oil pan and strainer
- Fuel tube and injector assembly
- Intake manifold
- Ignition coil
- Rocker cover
- Crankshaft pulley
- Front engine cover
- Oil pump
- Timing chain
- Camshaft sprockets
- Camshafts
- Cylinder head, removing bolts in reverse order of installation sequence

To install:

➡ **Be sure to use new fasteners, as required.**

4. Install the cylinder head with a new gasket.

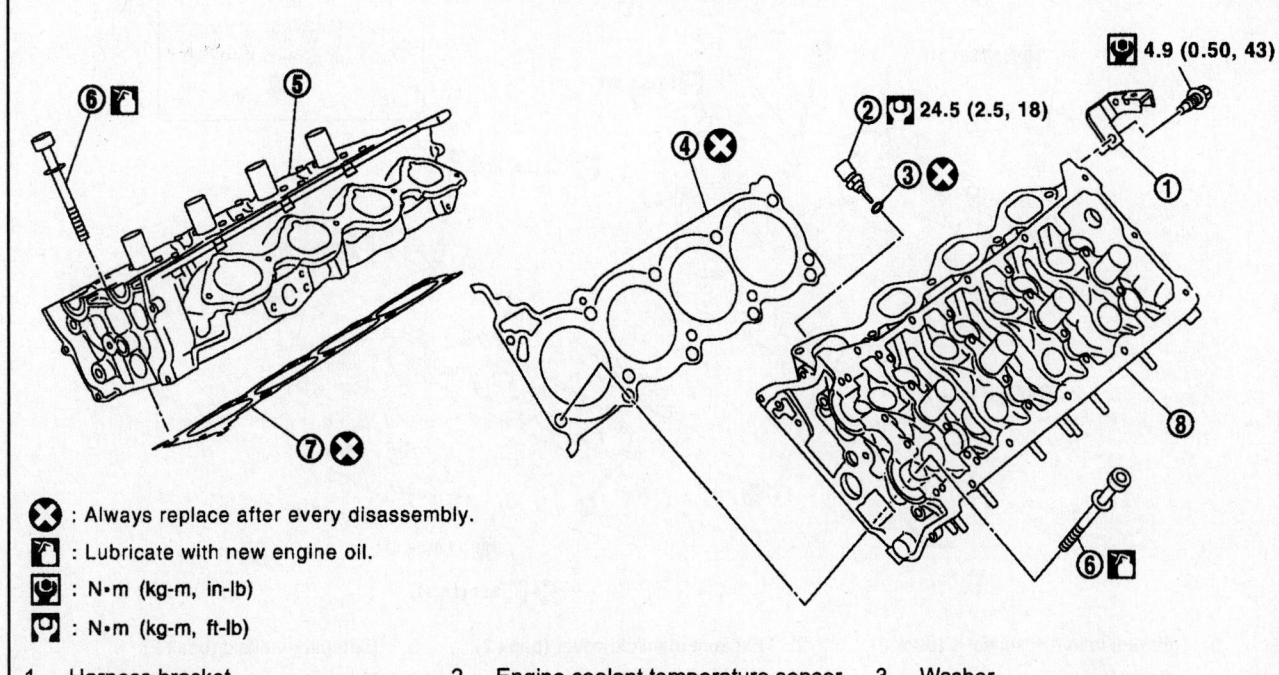

⊗ : Always replace after every disassembly.

🛢 : Lubricate with new engine oil.

⊙ : N•m (kg-m, in-lb)

⊡ : N•m (kg-m, ft-lb)

1. Harness bracket	2. Engine coolant temperature sensor	3. Washer
4. Cylinder head gasket (left bank)	5. Cylinder head (right bank)	6. Cylinder head bolt
7. Cylinder head gasket (right bank)	8. Cylinder head (left bank)	

Fig. 64 Cylinder heads and related components

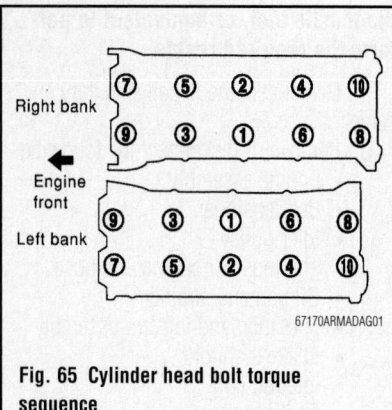

Fig. 65 Cylinder head bolt torque sequence

5. Tighten the bolts in sequence to specification.

6. Install or connect the following:
- Camshaft
- Camshaft sprockets
- Timing chain
- Oil pump
- Front engine cover
- Crankshaft pulley
- Rocker cover
- Ignition coil
- Intake manifold

- Fuel tube and injector assembly
- Oil pain and strainer
- Thermostat housing and hose
- Idler pulley
- Belt tensioner
- Engine assembly

7. Start the engine and check for leaks.

8. Be sure to perform the reconnect/relearn procedures.

EXHAUST MANIFOLD

REMOVAL & INSTALLATION

See Figures 66 and 67.

1. Before servicing the vehicle, refer to the Precautions Section.

➡**If working near and/or around the SRS system and components, be sure to disable the SRS system. After disabling the system wait three minutes or more before servicing the vehicle.**

➡**Whenever the negative battery cable is disconnected the following components will require resetting. The Idle Air Volume Learning, Steering Angle Sensor Neutral Position, Sunroof Memory Reset/Initialization, Automatic Drive Positioner System, Audio presets and Navigation. Use the CONSULT-III diagnostic tool, or equivalent to perform the required resets.**

2. Disconnect the negative battery cable.

3. Raise and support the vehicle safely.

4. Remove the engine under cover, if equipped.

5. Remove the front final drive assembly, if equipped.

6. Remove the main muffler and center exhaust tube.

7. Remove the front exhaust tubes.

8. Remove the tire and wheel assemblies.

9. Remove the fender protectors.

10. Remove the A/F sensors. Do not drop the sensors. If the sensor is dropped it must be replaced.

11. Properly support the engine.

12. Remove the engine mounting insulator.

13. Remove the exhaust manifold cover.

14. Remove the engine mounting bracket.

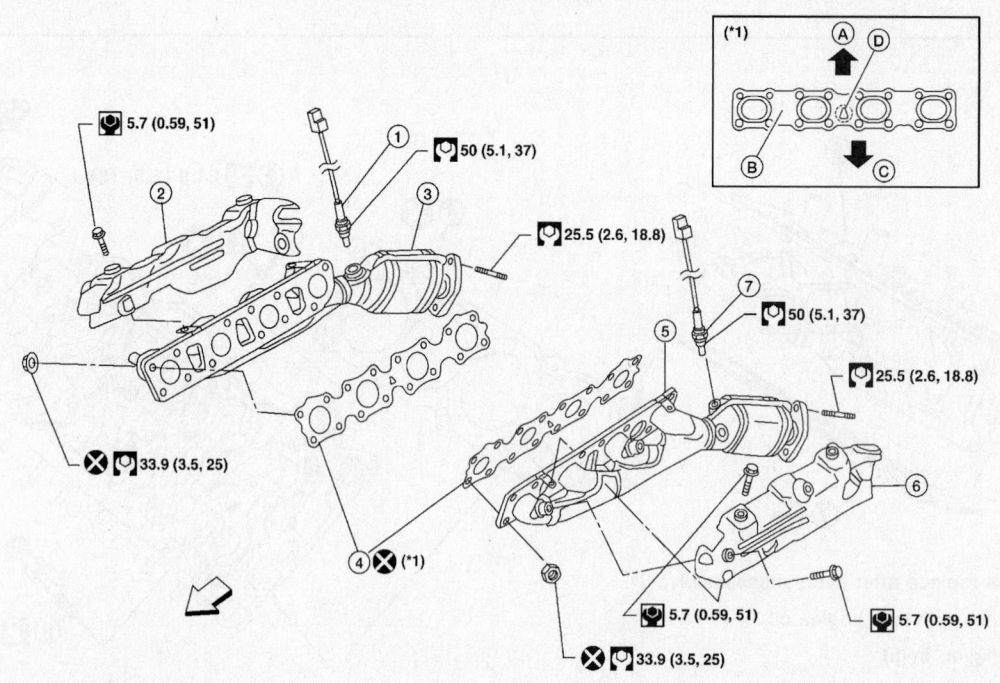

1. Air fuel ratio A/F sensor 1 (bank 2)	2. Exhaust manifold cover (bank 2)	3. Exhaust manifold (bank 2)
4. Gaskets	5. Exhaust manifold (bank 1)	6. Exhaust manifold cover (bank 1)
7. Air fuel ratio A/F sensor 1 (bank 1)	A. Up	B. Coated face
C. Manifold side	D. Up mark	⇦ Front

Fig. 66 Exhaust manifolds and related components

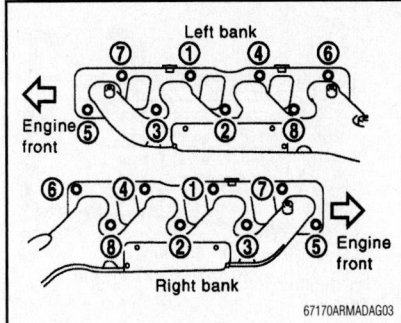

Fig. 67 Exhaust manifold bolt torque sequence

15. On right side remove the oil level dipstick.
16. Remove the left exhaust manifold nuts/bolts in the reverse order of the installation sequence.
17. Remove the exhaust manifold. Discard the gaskets.

To install:

➡Be sure to use new fasteners, as required.

18. Installation is the reverse of the removal procedure.
19. Install new gaskets with the top of the triangular UP mark on it facing up and its coated face (gray side) toward the exhaust manifold side.
20. Tighten the retaining nuts/bolts to specification and in the proper sequence.
21. Be sure to perform the reconnect/relearn procedures.

FLEXPLATE

REMOVAL & INSTALLATION

1. Before servicing the vehicle, refer to the Precautions Section.

➡If working near and/or around the SRS system and components, be sure to disable the SRS system. After disabling the system wait three minutes or more before servicing the vehicle.

➡Whenever the negative battery cable is disconnected the following components will require resetting. The Idle Air Volume Learning, Steering Angle Sensor Neutral Position, Sunroof Memory Reset/Initialization, Automatic Drive Positioner System, Audio presets and Navigation. Use the CONSULT-III diagnostic tool, or equivalent to perform the required resets.

2. Disconnect the negative battery cable.
3. Remove the transmission assembly.
4. Lock the flexplate.
5. Remove the flexplate diagonally.
6. Remove the flexplate.

To install:

➡Be sure to use new fasteners, as required.

7. Install the flexplate.
8. Install the flexplate bolts diagonally.
9. Lock the flexplate and tighten the bolts to specification
10. Install the transmission assembly.
11. Be sure to perform the reconnect/relearn procedures.

INTAKE MANIFOLD

REMOVAL & INSTALLATION

See Figures 68 and 69.

1. Before servicing the vehicle, refer to the Precautions Section.

➡If working near and/or around the SRS system and components, be sure to disable the SRS system. After disabling the system wait three minutes or more before servicing the vehicle.

Ⓧ: Always replace after every disassembly.
: N•m (kg-m, in-lb)

1.	Intake manifold	2.	PCV hose	3.	Gasket
4.	Electric throttle control actuator	5.	Water hose	6.	Water hose
7.	PCV hose	8.	EVAP hose	9.	EVAP canister purge control solenoid valve
10.	Bracket	11.	Gasket		

67170ARMADAG29

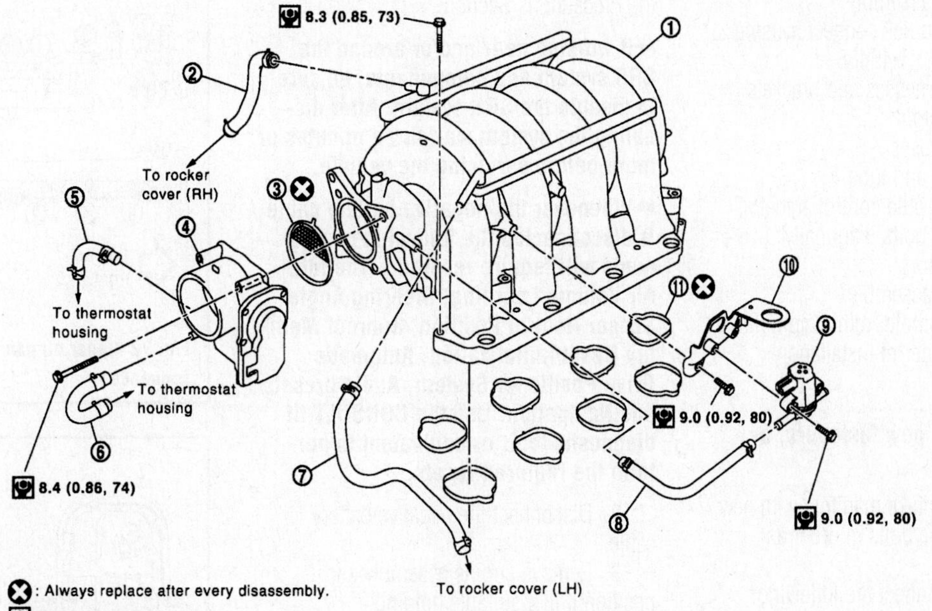

Fig. 68 Intake manifold and related components

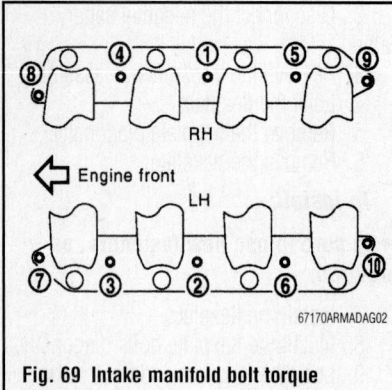

Fig. 69 Intake manifold bolt torque sequence

➡Whenever the negative battery cable is disconnected the following components will require resetting. The Idle Air Volume Learning, Steering Angle Sensor Neutral Position, Sunroof Memory Reset/Initialization, Automatic Drive Positioner System, Audio presets and Navigation. Use the CONSULT-III diagnostic tool, or equivalent to perform the required resets.

2. Disconnect the negative battery cable.
3. Drain the cooling system.
4. Relieve the fuel system pressure.
5. Remove or disconnect the following:
- Engine cover
- Air intake assembly
- Fuel tube quick connector using special tool J-45488
- Wiring harnesses and brackets from manifold
- Vacuum hoses
- PCV hose and tube
- Electric throttle control actuator, loosening bolts diagonally
- Fuel injectors
- Fuel tube assembly
- Intake manifold, removing bolts in reverse order of installation

To install:

➡Be sure to use new fasteners, as required.

6. Install the intake manifold with new gaskets. Tighten the bolts in order as shown.
7. Install or connect the following:
- Fuel tube assembly
- Fuel injectors
- Electronic throttle control actuator, tightening the bolts in several steps
- PCV hose
- Vacuum hoses
- Wiring harnesses
8. Connect the fuel tube as follows:

a. Apply a thin layer of engine oil on the tube from tip end to spool end.
b. Insert tube into quick connector past the white identification mark.
c. Insert tube into quick connector until top spool is completely inside the connector and 2nd level spool is exposed right below the connector.
d. Pull slightly on the quick connector to ensure it is fully engaged.
e. Install quick connector cap on quick connector joint.
9. Install or connect the following:
- Air intake assembly
- Engine cover
10. Refill the cooling system.
11. Start engine and check for leaks.
12. Be sure to perform the reconnect/relearn procedures.

OIL PAN

REMOVAL & INSTALLATION

See Figures 70 through 74.

➡The engine must be removed from the vehicle to perform this procedure. Be sure that the engine is secured in a suitable holding fixture before performing this procedure.

1. Before servicing the vehicle, refer to the Precautions Section.

➡If working near and/or around the SRS system and components, be sure to disable the SRS system. After disabling the system wait three minutes or more before servicing the vehicle.

➡Whenever the negative battery cable is disconnected the following components will require resetting. The Idle Air Volume Learning, Steering Angle Sensor Neutral Position, Sunroof Memory Reset/Initialization, Automatic Drive Positioner System, Audio presets and Navigation. Use the CONSULT-III diagnostic tool, or equivalent to perform the required resets.

2. Disconnect the negative battery cable.
3. Remove engine assembly and position it in a suitable holding fixture.
4. Remove lower oil pan, loosening bolts in reverse order of the installation sequence.
5. Remove oil strainer from upper oil pan.
6. Gently pry and remove upper oil pan from engine block.

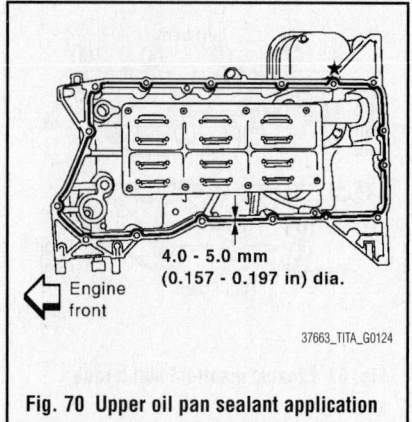

4.0 - 5.0 mm
(0.157 - 0.197 in) dia.

Fig. 70 Upper oil pan sealant application

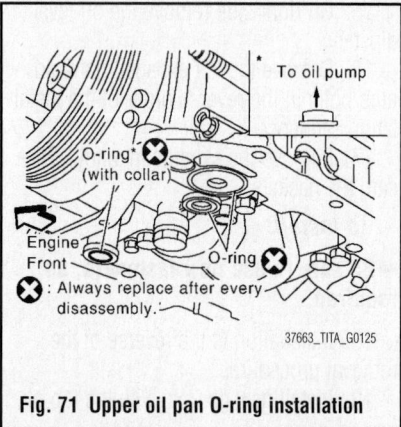

Fig. 71 Upper oil pan O-ring installation

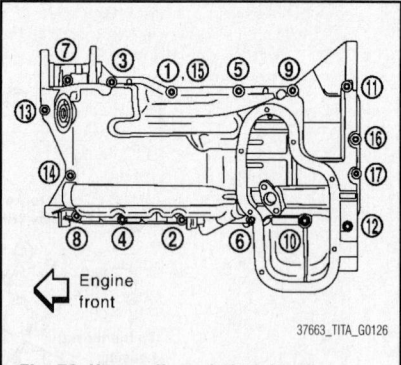

Fig. 72 Upper oil pan bolt tightening sequence

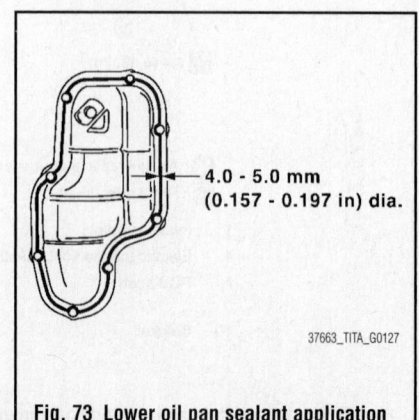

4.0 - 5.0 mm
(0.157 - 0.197 in) dia.

Fig. 73 Lower oil pan sealant application

Engine front
37663_TITA_G0128

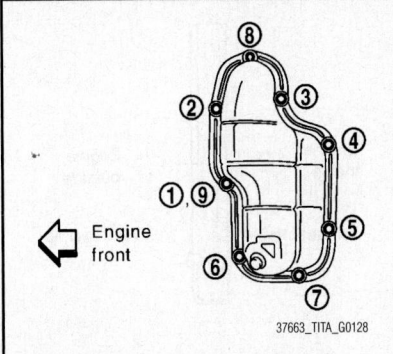

Fig. 74 Lower oil pan bolt tightening sequence

To install:

➡**Be sure to use new fasteners, as required.**

7. Apply liquid gasket to upper oil pan mating surfaces.
8. Install new O-rings to oil pump and front cover side.
9. Tighten upper oil pan bolts to specification and in the proper sequence.
10. Install or connect the following:
    - Rear plate cover
    - Oil strainer to upper oil pan
    - Lower oil pan, tightening bolts in proper sequence and to specification
11. Be sure to perform the reconnect/relearn procedures.

OIL PUMP

REMOVAL & INSTALLATION

See Figures 75 and 76.

1. Before servicing the vehicle, refer to the Precautions Section.

➡**If working near and/or around the SRS system and components, be sure to disable the SRS system. After disabling the system wait three minutes or more before servicing the vehicle.**

➡**Whenever the negative battery cable is disconnected the following components will require resetting. The Idle Air Volume Learning, Steering Angle Sensor Neutral Position, Sunroof Memory Reset/Initialization, Automatic Drive Positioner System, Audio presets and Navigation. Use the CONSULT-III diagnostic tool, or equivalent to perform the required resets.**

2. Disconnect the negative battery cable.
3. Remove or disconnect the following:
    - Timing chain cover
    - Oil pump drive spacer
    - Oil pump

To install:

➡**Be sure to use new fasteners, as required.**

4. Install or connect the following:
    - Oil pump
    - Oil pump drive spacer
    - Timing chain cover

➡**When inserting the oil pump drive spacer, align the crankshaft key and the flat face of the inner rotor.**

Key groove
Crankshaft key
Front mark
Inner rotor flat face
Oil pump drive spacer flat face
37663_TITA_G0129

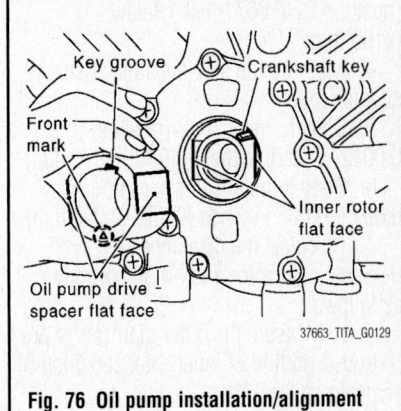

Fig. 76 Oil pump installation/alignment

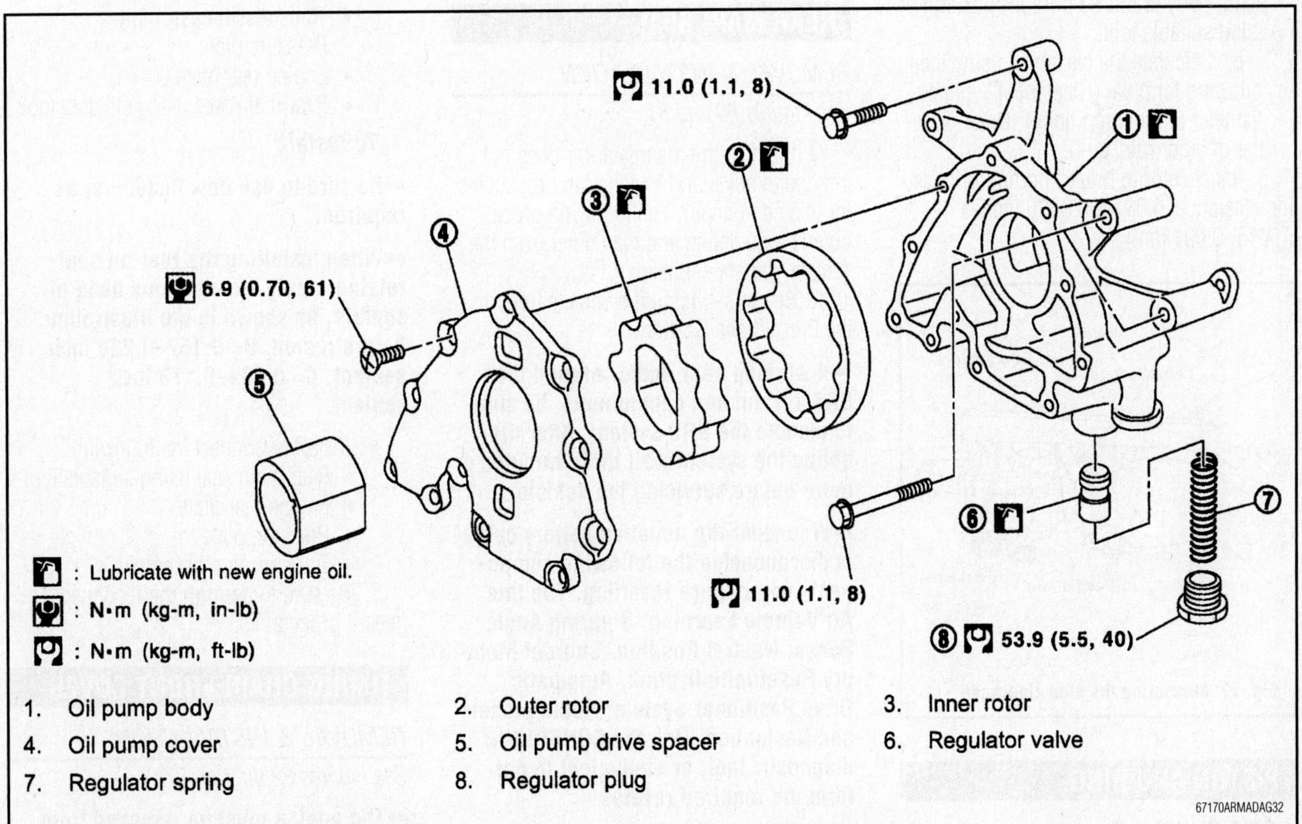

11.0 (1.1, 8)
6.9 (0.70, 61)
11.0 (1.1, 8)
53.9 (5.5, 40)
: Lubricate with new engine oil.
: N•m (kg-m, in-lb)
: N•m (kg-m, ft-lb)
1. Oil pump body
2. Outer rotor
3. Inner rotor
4. Oil pump cover
5. Oil pump drive spacer
6. Regulator valve
7. Regulator spring
8. Regulator plug
67170ARMADAG32

Fig. 75 Oil pump and related components

If they are not aligned rotate the pump inner rotor by hand. Make sure that each part is aligned and tap lightly until it reaches the end.

5. Be sure to perform the reconnect/relearn procedures.

INSPECTION

See Figures 77.

1. Measure the radial clearance using a suitable tool.
2. Body to outer rotor (position 1): 0.0045–0.0079 inches (0.114–0.200 mm).
3. Inner rotor to outer rotor tip (position 2): 0.0071 inch (Below 0.180 mm).
4. Measure the side clearance using a suitable tool.
5. Body to inner rotor (position 3): 0.0012–0.0028 inches (0.030–0.070 mm)
6. Body to outer rotor (position 4): 0.0012–0.0035 inches (0.030–0.090 mm)
7. Calculate the clearance between inner rotor and oil pump body as follows:
   a. Measure the outer diameter of protruded portion of inner rotor (position 5) using suitable tool.
   b. Measure the inner diameter of oil pump body to brazed portion (position 6) using suitable tool.
   c. Calculate the clearance using the following formula: (Clearance) = (Inner diameter of oil pump body) (Outer diameter of inner rotor).
8. Inner rotor to brazed portion of housing clearance: 0.0018–0.0036 inches (0.045–0.091 mm).

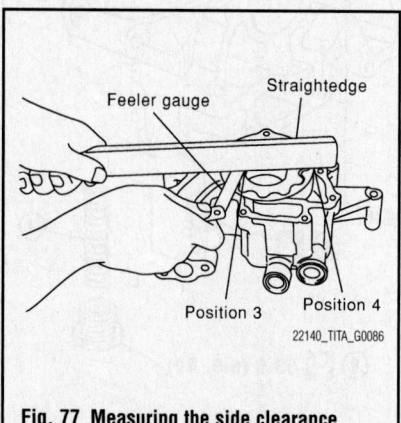

Fig. 77 Measuring the side clearance

PISTON AND RING

POSITIONING

See Figures 78 and 79.

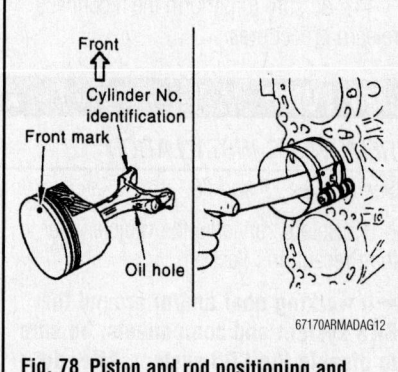

Fig. 78 Piston and rod positioning and identification

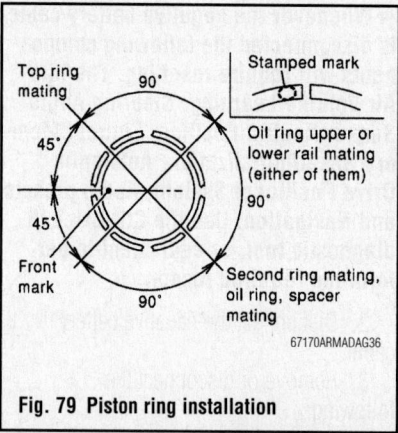

Fig. 79 Piston ring installation

REAR MAIN SEAL

REMOVAL & INSTALLATION

See Figures 80 and 81.

At this time the manufacturer does not provide removal and installation procedures for this component. The following procedure is a guideline and may differ from the vehicle you are servicing.

1. Before servicing the vehicle, refer to the Precautions Section.

➡ If working near and/or around the SRS system and components, be sure to disable the SRS system. After disabling the system wait three minutes or more before servicing the vehicle.

➡ Whenever the negative battery cable is disconnected the following components will require resetting. The Idle Air Volume Learning, Steering Angle Sensor Neutral Position, Sunroof Memory Reset/Initialization, Automatic Drive Positioner System, Audio presets and Navigation. Use the CONSULT-III diagnostic tool, or equivalent to perform the required resets.

2. Disconnect the negative battery cable.
3. Remove or disconnect the following:

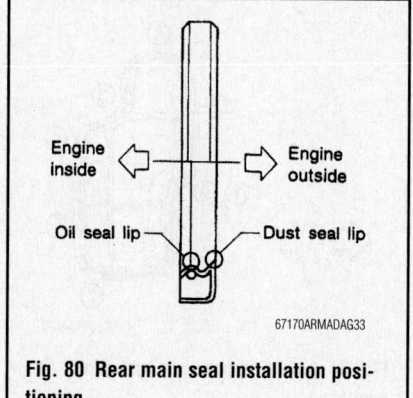

Fig. 80 Rear main seal installation positioning

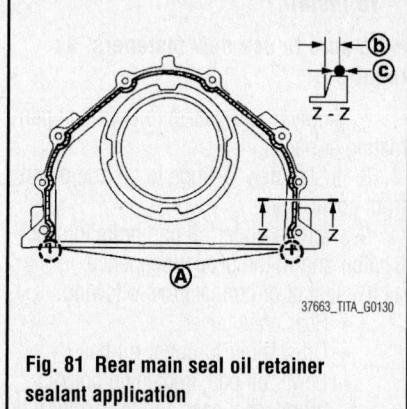

Fig. 81 Rear main seal oil retainer sealant application

- Transmission assembly
- Pressure plate
- Engine rear plate
- Rear main seal using suitable tool

To install:

➡ Be sure to use new fasteners, as required.

➡ When installing the rear oil seal retainer apply a continuous bead of sealant, as shown in the illustration. A= protrusion. B= 0.157–0.220 inch sealant. C= 0.134–0.173 inch sealant.

4. Install or connect the following:
   - Rear main seal using suitable tool
   - Engine rear plate
   - Pressure plate
   - Transmission assembly
5. Be sure to perform the reconnect/relearn procedures.

TIMING CHAIN FRONT COVER

REMOVAL & INSTALLATION

See Figures 82 through 87.

➡ The engine must be removed from the vehicle to perform this procedure. Be sure that the engine is secured in a suitable holding fixture before performing this procedure.

1. Before servicing the vehicle, refer to the Precautions Section.

➡ If working near and/or around the SRS system and components, be sure to disable the SRS system. After disabling the system wait three minutes or more before servicing the vehicle.

➡ Whenever the negative battery cable is disconnected the following components will require resetting. The Idle Air Volume Learning, Steering Angle Sensor Neutral Position, Sunroof Memory Reset/Initialization, Automatic Drive Positioner System, Audio presets and Navigation. Use the CONSULT-III diagnostic tool, or equivalent to perform the required resets.

2. Disconnect the negative battery cable.
3. Remove or disconnect the following:
   • Engine assembly
   • Drive belt auto tensioner
   • Idler pulley
   • Thermostat housing and water hose
   • Power steering pump bracket
   • Oil pan (upper and lower)
   • Oil strainer

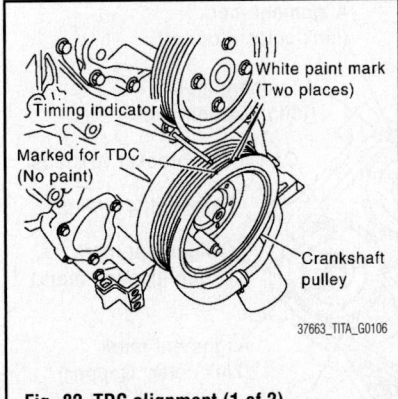

Fig. 82 TDC alignment (1 of 2)

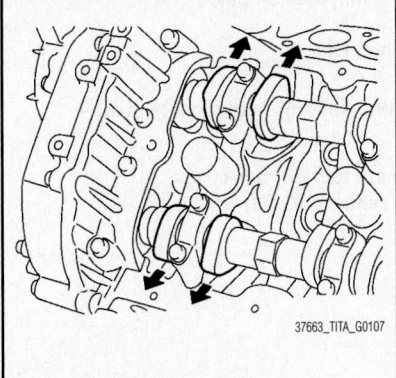

Fig. 83 TDC alignment (2 of 2)

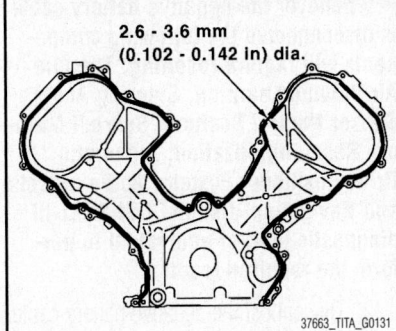

Fig. 84 Timing chain front cover sealant application

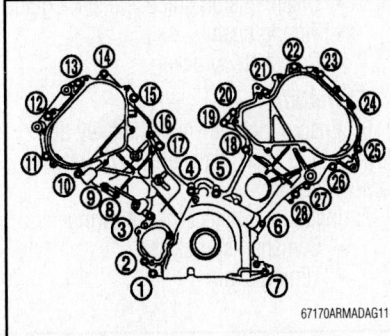

Fig. 85 Timing chain front cover bolt torque sequence

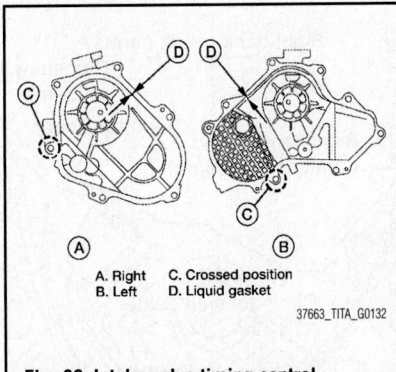

Fig. 86 Intake valve timing control solenoid cover sealant application

• Alternator and bracket
• Rocker cover
• Water pump

4. Remove the CMP sensor.
5. Remove the intake valve timing control position sensors (right and left).
6. Remove the intake valve timing control solenoid valves (right and left).
7. Loosen and remove the intake valve timing control valve cover (right and left) bolts in the reverse order of the tightening sequence.
8. Be sure that the number one cylinder is at TDC on the compression stroke.

➡ Turn the crankshaft pulley clockwise to align the TDC identification notch (without paint mark) with the timing indicator on the front cover. At this time make sure that the intake and exhaust cam lobes of the number one cylinder (top front on left bank) point outside. If not turn the crankshaft pulley once more. See illustration

9. Remove the crankshaft pulley.
10. Remove the front cover retaining bolts. Remove the front cover.
11. Discard the gasket.

To install:

➡ Be sure to use new fasteners, as required.

12. Installation is the reverse of the removal procedure.
13. Apply sealant to the front cover. See illustration.
14. Install retaining bolts and tighten to specification and in the proper sequence.
15. Apply liquid gasket to the valve timing control solenoid covers.

➡ The start and end of the sealant application should be crossed at position "C", see illustration, that cannot be seen after attaching the intake valve timing control solenoid valve cover.

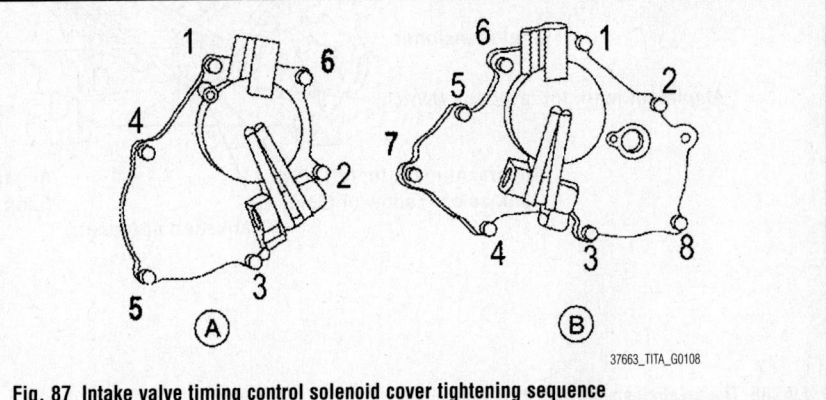

Fig. 87 Intake valve timing control solenoid cover tightening sequence

16. Install the intake valve timing control solenoid covers. Tighten bolts in proper sequence.

17. Continue the installation in the reverse order of the removal procedure.

18. Be sure to perform the reconnect/relearn procedures.

TIMING CHAIN & SPROCKETS

REMOVAL & INSTALLATION

See Figure 88.

➡The engine must be removed from the vehicle to perform this procedure. Be sure that the engine is secured in a suitable holding fixture before performing this procedure.

1. Before servicing the vehicle, refer to the Precautions Section.

➡If working near and/or around the SRS system and components, be sure to disable the SRS system. After disabling the system wait three minutes or more before servicing the vehicle.

➡Whenever the negative battery cable is disconnected the following components will require resetting. The Idle Air Volume Learning, Steering Angle Sensor Neutral Position, Sunroof Memory Reset/Initialization, Automatic Drive Positioner System, Audio presets and Navigation. Use the CONSULT-III diagnostic tool, or equivalent to perform the required resets.

2. Disconnect the negative battery cable.
3. Remove the front cover.
4. Remove the oil pump drive spacer.
5. Remove the oil pump.
   • Timing chain tensioner
   • Chain tension guide and slack guide
   • Timing chain
   • Camshaft sprocket

To install:

6. Ensure that the crankshaft key and dowel pin of each camshaft are facing the same direction.
7. Install or connect the following:
   • Camshaft sprockets
   • Timing chain

   • Chain tension guide and slack guide
   • Oil pump
   • Oil pump drive spacer
   • Front oil seal, using suitable tool

8. Continue the installation in the reverse order of the removal procedure.

9. Be sure to perform the reconnect/relearn procedures.

VALVE LASH

ADJUSTMENT

1. Before servicing the vehicle, refer to the Precautions Section.
2. Remove camshaft and valve lifter(s) out of specification.
3. Install replacement valve lifter(s).
4. Install the camshaft.
5. Manually turn the crankshaft pulley several turns.
6. Recheck valve clearances with engine at operating temperature.

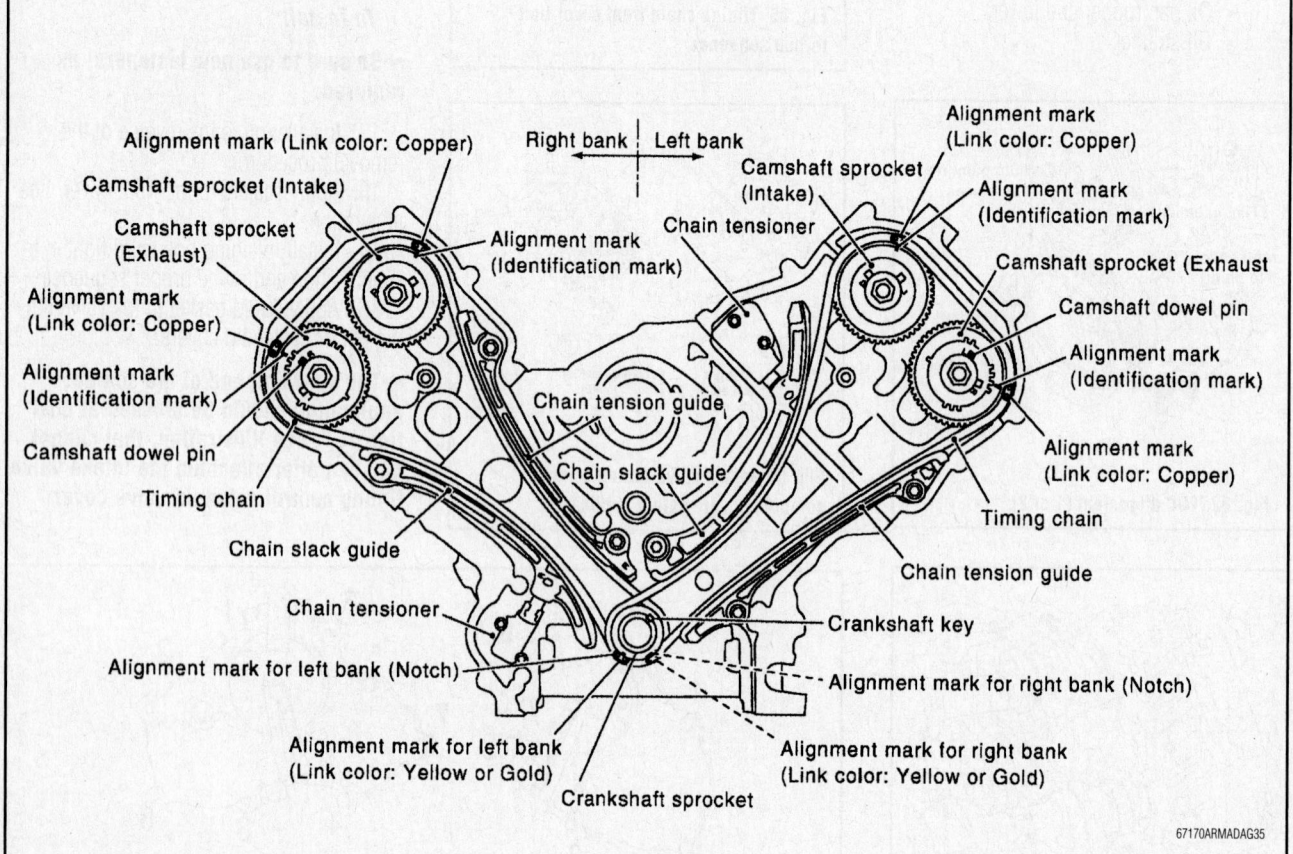

Fig. 88 Timing chains and related components

INSPECTION

See Figures 89 and 90.

1. Before servicing the vehicle, refer to the Precautions Section.

2. Run engine to operating temperature.

3. Remove or disconnect the following:
   - Engine cover
   - Battery cover
   - Air intake assembly
   - Left and right rocker covers

4. Turn the crankshaft pulley clockwise to Top Dead Center (TDC) identification notch with timing indicator.

5. Ensure that both the intake and exhaust cam noses of the No. 1 cylinder face outside.

6. Measure the valve clearances at locations marked 'x' shown in figure.

7. Turn the crankshaft pulley clockwise 270 degrees from the position of No. 1 cylinder compression to obtain No. 3 cylinder compression TDC.

8. Measure the valve clearances at locations marked 'x' shown in next figure.

9. Turn crankshaft pulley clockwise 90 degrees and measure the intake and exhaust valve clearance of No. 6 cylinder and exhaust valve clearance of No. 2 cylinder.

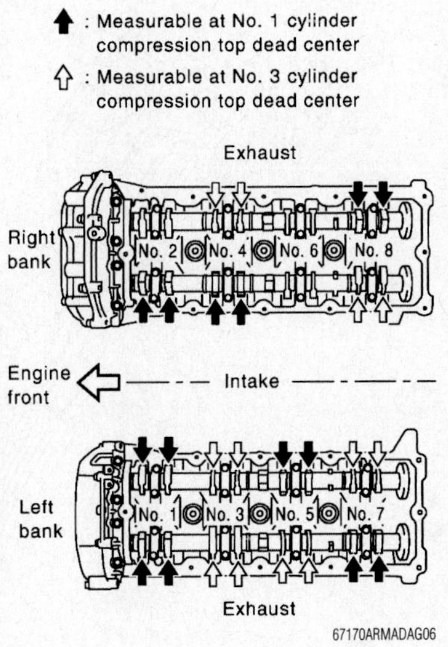

Fig. 89 Locations to measure clearance with No. 1 cylinder at TDC

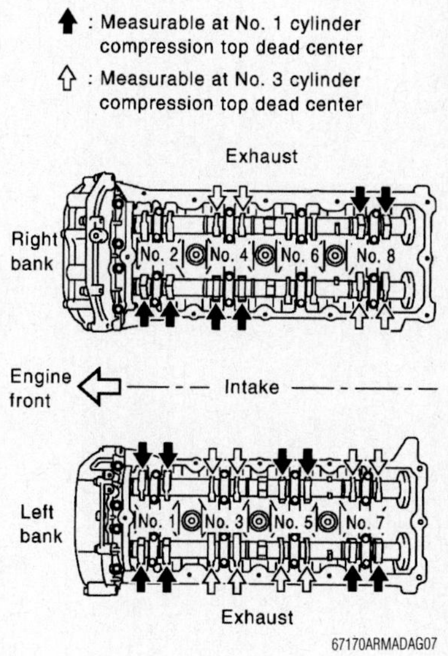

Fig. 90 Locations to measure clearance with No. 3 cylinder at TDC

ENGINE PERFORMANCE & EMISSION CONTROLS

ACCELERATOR PEDAL POSITION (APP) SENSOR

LOCATION

The Accelerator Pedal Position (APP) sensor is located on the upper end of the accelerator pedal assembly.

REMOVAL & INSTALLATION

See Figures 91 and 92.

1. Before servicing the vehicle, refer to the Precautions Section.

➡**If working near and/or around the SRS system and components, be sure to disable the SRS system. After disabling the system wait three minutes or more before servicing the vehicle.**

➡**Whenever the negative battery cable is disconnected the following components will require resetting. The Idle Air Volume Learning, Steering Angle Sensor Neutral Position, Sunroof Memory Reset/Initialization, Automatic Drive Positioner System, Audio presets and Navigation. Use the CONSULT-III diagnostic tool, or equivalent to perform the required resets.**

2. Disconnect the negative battery cable.

✳✳ CAUTION

Do not disassemble the accelerator pedal adjusting mechanism. Before removal and installation, the accelerator and brake pedals must be in the front most position. This is to align the base position of the accelerator and brake pedals. Do not disassemble the accelerator pedal assembly. Do not remove the Accelerator Pedal Position (APP) sensor from the accelerator pedal bracket. Avoid damage from dropping the accelerator pedal assembly during handling. Keep the accelerator pedal assembly away from water.

3. Move the accelerator and brake pedals to the front most position.

4. Turn the ignition switch **OFF** and disconnect the negative battery terminal.

5. Disconnect the adjustable brake pedal cable from the adjustable brake pedal. Unlock, then pull the adjustable brake pedal cable to disconnect it from the adjustable brake pedal.

6. Disconnect the adjustable pedal electric motor electrical connector.

7. Disconnect the adjustable pedal electric motor memory electrical connector, if equipped.

8. Disconnect APP sensor electrical connector, if equipped.

9. Disconnect the APP sensor electrical connector.

10. Remove the adjustable accelerator pedal assembly.

To install:

➡**Be sure to use new fasteners, as required.**

11. Installation is the reverse of the removal procedure.

12. Be sure to perform the reconnect/relearn procedures.

CAMSHAFT POSITION (CMP) SENSOR

LOCATION

The Camshaft Position (CMP) sensor is located on the right front of the timing cover, facing the engine.

REMOVAL & INSTALLATION

1. Before servicing the vehicle, refer to the Precautions Section.

➡**If working near and/or around the SRS system and components, be sure to disable the SRS system. After**

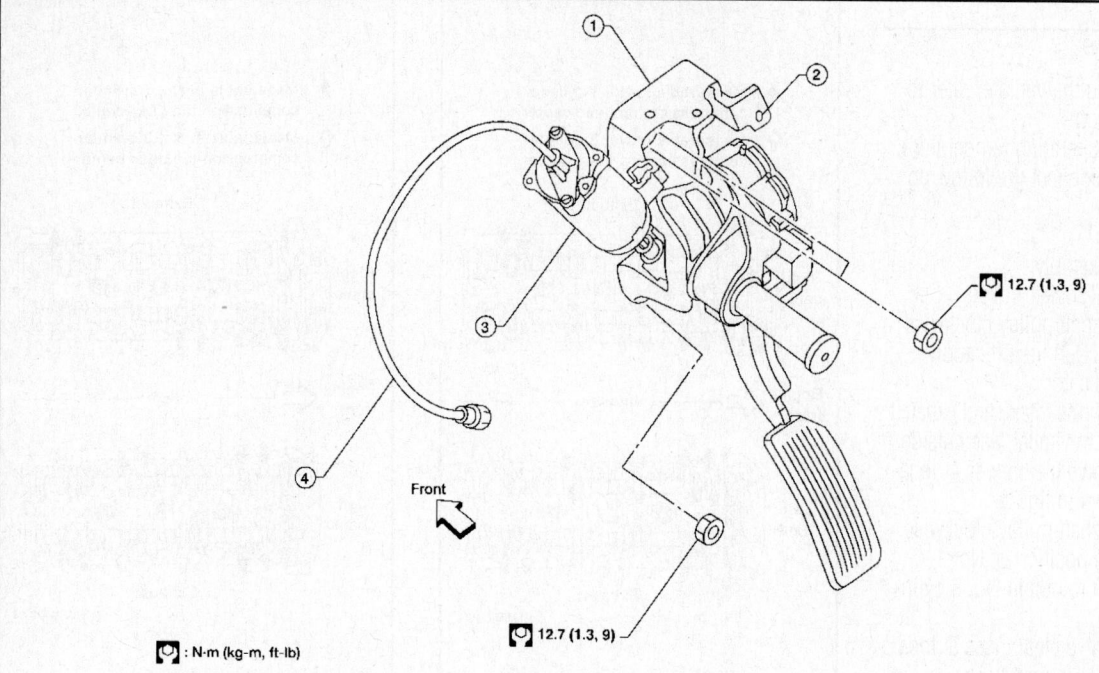

1. Adjustable accelerator pedal assembly
2. Adjustable accelerator pedal bracket (part of the accelerator pedal assembly)
3. Adjustable pedal electric motor (part of the accelerator pedal assembly)
4. Adjustable brake pedal cable (part of the accelerator pedal assembly)

: N·m (kg-m, ft-lb)

22140_TITA_G0042

Fig. 91 Adjustable accelerator pedal assembly

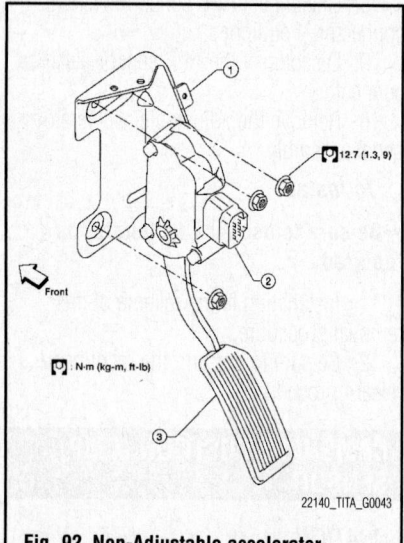

22140_TITA_G0043

Fig. 92 Non-Adjustable accelerator pedal (1), connector (2), pedal (3)

disabling the system wait three minutes or more before servicing the vehicle.

➡Whenever the negative battery cable is disconnected the following components will require resetting. The Idle Air Volume Learning, Steering Angle Sensor Neutral Position, Sunroof Memory Reset/Initialization, Automatic Drive Positioner System, Audio presets and Navigation. Use the CONSULT-III diagnostic tool, or equivalent to perform the required resets.

2. Disconnect the negative battery cable.
3. Remove the engine cover.
4. Remove air intake duct.
5. Disconnect the camshaft position sensor.
6. Remove the bolt and the Camshaft Position (CMP) sensor.

To install:

7. Install the CMP sensor and tighten the bolt.
8. Reconnect the camshaft electrical sensor.
9. Install the air intake duct.
10. Install the engine cover.
11. Be sure to perform the reconnect/relearn procedures.

CRANKSHAFT POSITION (CKP) SENSOR

LOCATION

The Crankshaft Position (CKP) sensor is located on the transmission assembly facing the gear teeth (cogs) of the signal plate.

REMOVAL & INSTALLATION

1. Before servicing the vehicle, refer to the Precautions Section.

➡If working near and/or around the SRS system and components, be sure to disable the SRS system. After disabling the system wait three minutes or more before servicing the vehicle.

➡Whenever the negative battery cable is disconnected the following components will require resetting. The Idle Air Volume Learning, Steering Angle Sensor Neutral Position, Sunroof Memory Reset/Initialization, Automatic Drive Positioner System, Audio presets and Navigation. Use the CONSULT-III diagnostic tool, or equivalent to perform the required resets.

2. Disconnect the negative battery cable.
3. Raise and support the vehicle safely.
4. Disconnect the Crankshaft Position (CKP) sensor connector.
5. Remove the mounting bolt and CKP sensor.

To install:

6. Install the CKP sensor and tighten the mounting bolt.
7. Reconnect the CKP sensor connector.

8. Lower the vehicle.

9. Be sure to perform the reconnect/relearn procedures.

ELECTRONIC CONTROL MODULE (ECM)

LOCATION

The Electronic Control Module (ECM) is located in the engine room passenger side behind battery.

REMOVAL & INSTALLATION

At this time the manufacturer does not provide removal and installation procedures for this component. The following procedure is a guideline and may differ from the vehicle you are servicing.

1. Before servicing the vehicle, refer to the Precautions Section.

➡**If working near and/or around the SRS system and components, be sure to disable the SRS system. After disabling the system wait three minutes or more before servicing the vehicle.**

➡**Whenever the negative battery cable is disconnected the following components will require resetting. The Idle Air Volume Learning, Steering Angle Sensor Neutral Position, Sunroof Memory Reset/Initialization, Automatic Drive Positioner System, Audio presets and Navigation. Use the CONSULT-III diagnostic tool, or equivalent to perform the required resets.**

2. Disconnect the negative battery cable.

3. Disconnect the positive battery cable and remove the battery, as required..

4. Carefully remove the Electronic Control Module (ECM) harness connectors.

5. Remove the ECM mounting bolts and the ECM.

To install:

6. Install the ECM and mounting bolts and tighten to 62 inch lbs. (7 Nm).

7. Carefully install the ECM harness connectors.

8. Install the battery.

9. Reconnect the battery cables.

10. Be sure to perform the reconnect/relearn procedures.

ENGINE COOLANT TEMPERATURE (ECT) SENSOR

LOCATION

The Engine Coolant Temperature (ECT) sensor is mounted in the front of the intake manifold. It is just to the right of the throttle body.

REMOVAL & INSTALLATION

1. Before servicing the vehicle, refer to the Precautions Section.

➡**If working near and/or around the SRS system and components, be sure to disable the SRS system. After disabling the system wait three minutes or more before servicing the vehicle.**

➡**Whenever the negative battery cable is disconnected the following components will require resetting. The Idle Air Volume Learning, Steering Angle Sensor Neutral Position, Sunroof Memory Reset/Initialization, Automatic Drive Positioner System, Audio presets and Navigation. Use the CONSULT-III diagnostic tool, or equivalent to perform the required resets.**

2. Disconnect the negative battery cable.

3. Remove the engine cover.

4. Remove the intake air duct.

5. Partially drain the cooling system.

6. Disconnect the harness connector.

7. Remove the Engine Coolant Temperature (ECT) sensor.

To install:

➡**Be sure to use new fasteners, as required.**

8. Install the ECT sensor and carefully tighten.

9. Reconnect the harness connector.

10. Install the intake air duct.

11. Install the engine cover.

12. Refill the engine coolant.

13. Be sure to perform the reconnect/relearn procedures.

EVAPORATIVE EMISSIONS (EVAP) CANISTER

LOCATION

This component is located under the vehicle near the fuel tank.

REMOVAL & INSTALLATION

See Figure 93.

At this time the manufacturer does not provide removal and installation procedures for this component. The following procedure is a guideline and may differ from the vehicle you are servicing.

1. Before servicing the vehicle, refer to the Precautions Section.

➡**If working near and/or around the SRS system and components, be sure to disable the SRS system. After disabling the system wait three minutes or more before servicing the vehicle.**

➡**Whenever the negative battery cable is disconnected the following components will require resetting. The Idle Air Volume Learning, Steering Angle Sensor Neutral Position, Sunroof Memory Reset/Initialization, Automatic Drive Positioner System, Audio presets and Navigation. Use the CONSULT-III diagnostic tool, or equivalent to perform the required resets.**

2. Disconnect the negative battery cable.

3. Raise and support the vehicle safely.

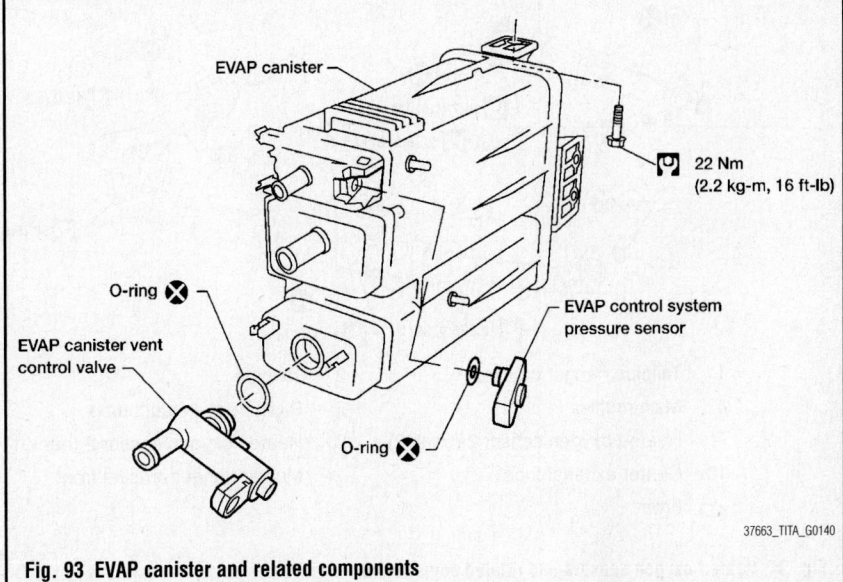

Fig. 93 EVAP canister and related components

37663_TITA_G0140

4. Remove the fuel tank shield, as required.

5. Disconnect the component electrical connectors.

6. Disconnect the hoses.

7. Remove the mounting clips, nuts and or screws.

8. Remove the component from the vehicle.

To install:

➡**Be sure to use new fasteners, as required.**

9. Installation is the reverse of the removal procedure.

10. Be sure to perform the reconnect/relearn procedures.

HEATED OXYGEN (HO2S) SENSOR

LOCATION

The Heated Oxygen (HO2S) sensors are located after the exhaust manifold converter assembly, in the lower part of the exhaust system.

REMOVAL & INSTALLATION

See Figure 94.

1. Before servicing the vehicle, refer to the Precautions Section.

➡**If working near and/or around the SRS system and components, be sure to disable the SRS system. After disabling the system wait three minutes or more before servicing the vehicle.**

➡**Whenever the negative battery cable is disconnected the following components will require resetting. The Idle Air Volume Learning, Steering Angle Sensor Neutral Position, Sunroof Memory Reset/Initialization, Automatic Drive Positioner System, Audio presets and Navigation. Use the CONSULT-III diagnostic tool, or equivalent to perform the required resets.**

2. Disconnect the negative battery cable.

3. Raise and safely support the vehicle.

4. Remove the engine undercover, as needed.

5. Unplug the Heated Oxygen (HO2S) sensor harness.

6. Using an O2 wrench remove the HO2S sensor.

➡**Lower the exhaust in needed.**

To install:

➡**Be sure to use new fasteners, as required.**

7. Install the HO2S sensor and tighten to 37 ft. lbs. (50 Nm).

8. Install the harness connector.

9. Keep the harness connector and wiring away from exhaust system.

10. Be sure to perform the reconnect/relearn procedures.

INTAKE AIR TEMPERATURE (IAT) SENSOR

LOCATION

The Intake Air Temperature (IAT) sensor is integral to the Mass Air Flow (MAF) sensor, and is mounted on the air filter housing lid.

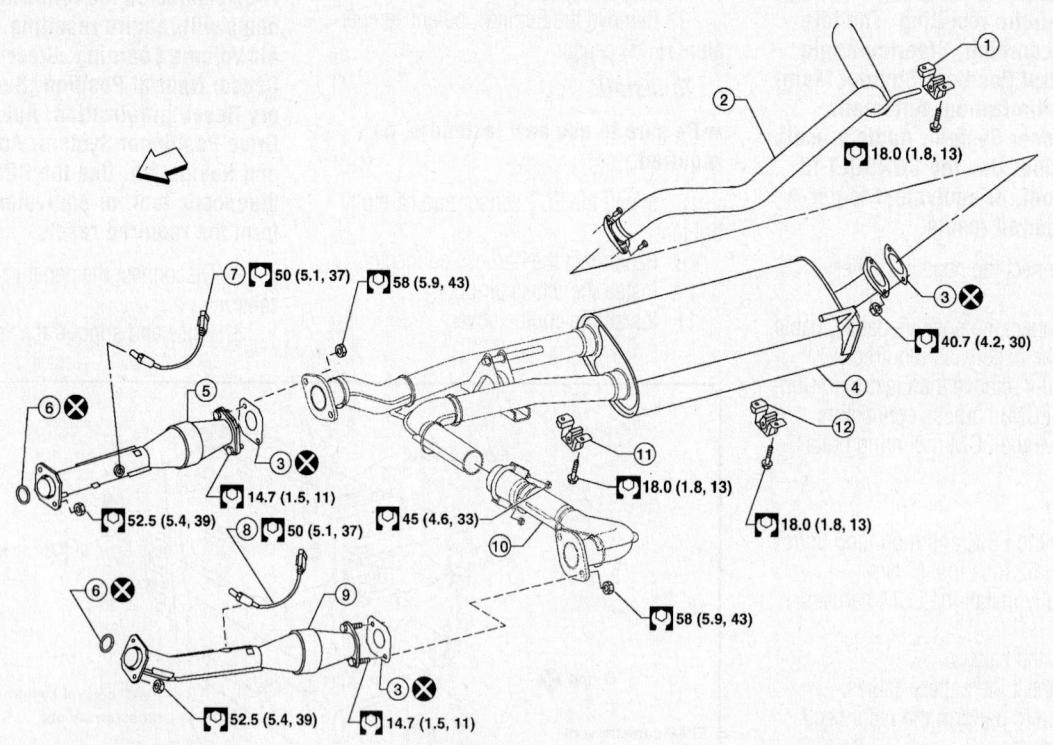

1. Tailpipe hanger bracket
2. Tailpipe
3. Gasket
4. Main muffler
5. Right front exhaust tube
6. Ring gasket
7. Heated oxygen sensor 2 (bank 2)
8. Heated oxygen sensor 2 (bank 1)
9. Left front exhaust tube
10. Center exhaust tube
11. Muffler hanger bracket front
12. Muffler hanger bracket rear
⇦ Front

37663_TITA_G0123

Fig. 94 Heated oxygen sensors and related components

REMOVAL & INSTALLATION

See Figure 95.

1. Before servicing the vehicle, refer to the Precautions Section.

➡ If working near and/or around the SRS system and components, be sure to disable the SRS system. After disabling the system wait three minutes or more before servicing the vehicle.

➡ Whenever the negative battery cable is disconnected the following components will require resetting. The Idle Air Volume Learning, Steering Angle Sensor Neutral Position, Sunroof Memory Reset/Initialization, Automatic Drive Positioner System, Audio presets and Navigation. Use the CONSULT-III diagnostic tool, or equivalent to perform the required resets.

2. Disconnect the negative battery cable.
3. Remove the engine room cover.
4. Remove the Intake Air Temperature (IAT/MAF) sensor harness.
5. Remove the mounting screws and the IAT/MAF sensor.

To install:

6. Install the IAT/MAF sensor.
7. Install the harness connector.
8. Install the engine room cover.
9. Be sure to perform the reconnect/relearn procedures.

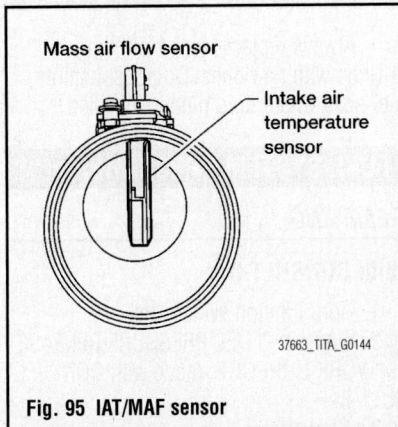

Mass air flow sensor

Intake air temperature sensor

37663_TITA_G0144

Fig. 95 IAT/MAF sensor

KNOCK SENSOR (KS)

LOCATION

See Figure 96.

The Knock (KS) sensors are mounted under the intake manifold on the cylinder block.

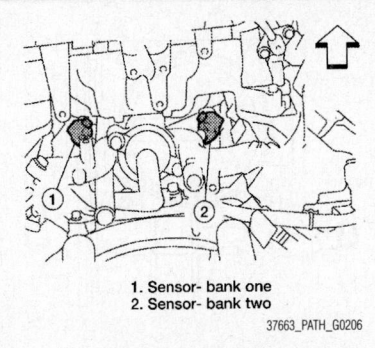

1. Sensor- bank one
2. Sensor- bank two

37663_PATH_G0206

Fig. 96 Knock Sensor (KS) location (view with engine removed from vehicle)

REMOVAL & INSTALLATION

At this time the manufacturer does not provide removal and installation procedures for this component. The intake manifold will have to be removed to service this component.

MALFUNCTION INDICATOR LIGHT (MIL)

RESET PROCEDURE

Clearing diagnostic trouble codes resets the MIL.

MASS AIR FLOW (MAF) SENSOR

LOCATION

Refer to the Intake Air Temperature (IAT) sensor.

REMOVAL & INSTALLATION

Refer to the Intake Air Temperature (IAT) sensor.

POSITIVE CRANKCASE VENTILATION (PCV) VALVE

LOCATION

The PCV valve is located on top of the engine, in one of the valve rocker covers.

REMOVAL & INSTALLATION

At this time the manufacturer does not provide removal and installation procedures for this component. The following procedure is a guideline and may differ from the vehicle you are servicing.

1. Before servicing the vehicle, refer to the Precautions Section.

➡ If working near and/or around the SRS system and components, be sure to disable the SRS system. After disabling the system wait three minutes or more before servicing the vehicle.

➡ Whenever the negative battery cable is disconnected the following components will require resetting. The Idle Air Volume Learning, Steering Angle Sensor Neutral Position, Sunroof Memory Reset/Initialization, Automatic Drive Positioner System, Audio presets and Navigation. Use the CONSULT-III diagnostic tool, or equivalent to perform the required resets.

2. Disconnect the negative battery cable.
3. Remove the necessary components in order to gain access to the component.
4. Disconnect the PCV hose.
5. Remove the valve from its mounting.

To install:

➡ Be sure to use new fasteners, as required.

6. Installation is the reverse of the removal procedure.
7. Be sure to perform the reconnect/relearn procedures.

THROTTLE POSITION SENSOR (TPS)

LOCATION

The Throttle Position(TPS) sensor is integral to the electric Throttle Control actuator. The Throttle Control actuator is mounted at the front of the intake manifold.

REMOVAL & INSTALLATION

See Figures 97 and 98.

At this time the manufacturer does not provide removal and installation procedures for this component. The following procedure is a guideline and may differ from the vehicle you are servicing.

1. Before servicing the vehicle, refer to the Precautions Section.

➡ If working near and/or around the SRS system and components, be sure to disable the SRS system. After disabling the system wait three minutes or more before servicing the vehicle.

➡ Whenever the negative battery cable is disconnected the following components will require resetting. The Idle Air Volume Learning, Steering Angle Sensor Neutral Position, Sunroof Memory Reset/Initialization, Automatic Drive Positioner System, Audio presets and Navigation. Use the CONSULT-III diagnostic tool, or equivalent to perform the required resets.

2. Disconnect the negative battery cable.

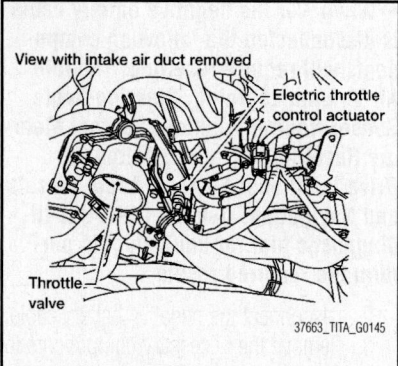

Fig. 97 Throttle control actuator and related components

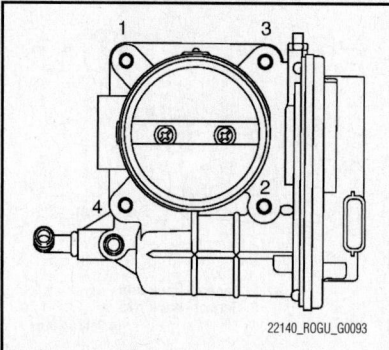

Fig. 98 Throttle body retaining bolt tightening sequence

3. Drain the cooling system, as required. Be sure to properly dispose of used engine coolant.
4. Remove the air intake duct.
5. Disconnect harness connector.
6. Disconnect water hoses.
7. Loosen the throttle body assembly mounting bolts in reverse order of the tightening sequence.

To install:

➡ Be sure to use new fasteners, as required.

8. Install the throttle body assembly with a new gasket.
9. Tighten the mounting bolts in sequence to 74 inch lbs. (8.4 Nm).

10. Reconnect the water hose.
11. Reconnect the harness connector.
12. Reconnect the air intake duct.
13. Fill the cooling system with the proper grade and type engine coolant.
14. Be sure to perform the reconnect/relearn procedures.

VEHICLE SPEED SENSOR (VSS)

LOCATION

The VSS sensor is located at the rear of the transmission case, under the tail shaft. On 4WD vehicles this component is located under the transfer case.

REMOVAL & INSTALLATION

1. Before servicing the vehicle, refer to the Precautions Section.

➡ If working near and/or around the SRS system and components, be sure to disable the SRS system. After disabling the system wait three minutes or more before servicing the vehicle.

➡ Whenever the negative battery cable is disconnected the following components will require resetting. The Idle Air Volume Learning, Steering Angle Sensor Neutral Position, Sunroof Memory Reset/Initialization, Automatic Drive Positioner System, Audio presets and Navigation. Use the CONSULT-III diagnostic tool, or equivalent to perform the required resets.

2. Disconnect the negative battery cable.
3. Raise and safely support the vehicle.
4. Disconnect the sensor harness.
5. Remove the mounting bolt and the sensor.

To install:

➡ Be sure to use new fasteners, as required.

6. Apply a small amount of transmission fluid to the sensor O-ring.
7. Install the Speed sensor and tighten the mounting bolt to 51 inch lbs. (5.8 Nm).
8. Be sure to perform the reconnect/relearn procedures.

FUEL

GASOLINE FUEL INJECTION SYSTEM

FUEL SYSTEM SERVICE PRECAUTIONS

Safety is the most important factor when performing not only fuel system maintenance, but any type of maintenance. Failure to conduct maintenance and repairs in a safe manner may result in serious personal injury or death. Work on a vehicle's fuel system components can be accomplished safely and effectively by adhering to the following rules and guidelines.

• To avoid the possibility of fire and personal injury, always disconnect the negative battery cable unless the repair or test procedure requires that battery voltage be applied.

• Always relieve the fuel system pressure prior to disconnecting any fuel system component (injector, fuel rail, pressure regulator, etc.) fitting or fuel line connection. Exercise extreme caution whenever relieving fuel system pressure to avoid exposing skin, face and eyes to fuel spray. Please be advised that fuel under pressure may penetrate the skin or any part of the body that it contacts.

• Always place a shop towel or cloth around the fitting or connection prior to loosening to absorb any excess fuel due to spillage. Ensure that all fuel spillage is quickly removed from engine surfaces. Ensure that all fuel-soaked cloths or towels are deposited into a flame-proof waste container with a lid.

• Always keep a dry chemical (Class B) fire extinguisher near the work area.

• Do not allow fuel spray or fuel vapors to come into contact with a spark or open flame.

• Always use a second wrench when loosening or tightening fuel line connection fittings. This will prevent unnecessary stress and torsion on fuel piping. Always follow the proper torque specifications.

• Always replace worn fuel fitting O-rings with new ones. Do not substitute fuel hose where rigid pipe is installed.

FUEL SYSTEM PRESSURE

RELIEVING

With CONSULT-II®

1. Turn ignition switch **ON**.
2. Perform "FUEL PRESSURE RELEASE" in "WORK SUPPORT" mode with CONSULT-II®.
3. Start engine.
4. After engine stalls, turn over the engine two or three times to release all fuel pressure.
5. Turn ignition switch **OFF**.

Without CONSULT-II®

See Figure 99.

1. Before servicing the vehicle, refer to the Precautions Section.

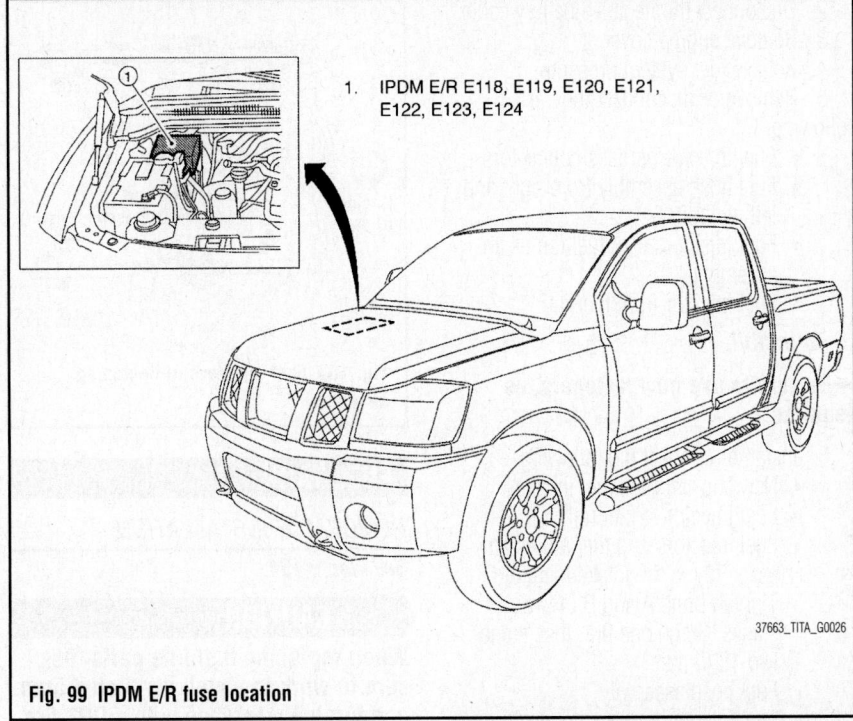

1. IPDM E/R E118, E119, E120, E121, E122, E123, E124

Fig. 99 IPDM E/R fuse location

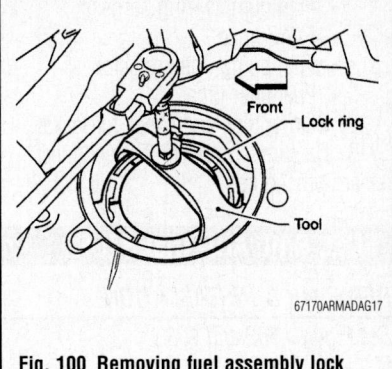

Fig. 100 Removing fuel assembly lock ring

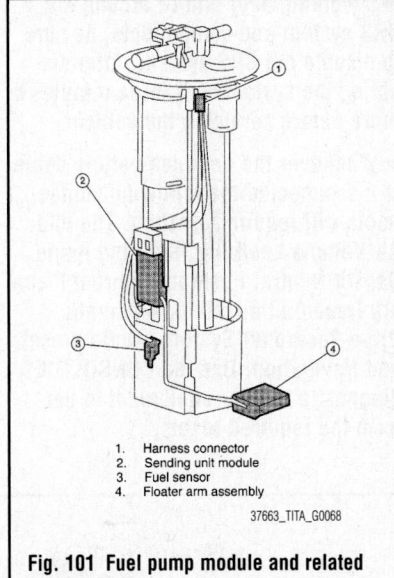

1. Harness connector
2. Sending unit module
3. Fuel sensor
4. Floater arm assembly

Fig. 101 Fuel pump module and related components

➡ If working near and/or around the SRS system and components, be sure to disable the SRS system. After disabling the system wait three minutes or more before servicing the vehicle.

➡ Whenever the negative battery cable is disconnected the following components will require resetting. The Idle Air Volume Learning, Steering Angle Sensor Neutral Position, Sunroof Memory Reset/Initialization, Automatic Drive Positioner System, Audio presets and Navigation. Use the CONSULT-III diagnostic tool, or equivalent to perform the required resets.

2. Remove fuel pump fuse located in IPDM E/R.
3. Start engine.
4. After engine stalls, turn over engine two or three times to release all fuel pressure.
5. Turn ignition switch **OFF**.
6. Disconnect the negative battery cable.
7. Reinstall fuel pump fuse after servicing fuel system.
8. Be sure to perform the reconnect/relearn procedures.

FUEL FILTER

REMOVAL & INSTALLATION

The fuel filter is part of the fuel pump module.

FUEL PUMP MODULE

REMOVAL & INSTALLATION
See Figures 100 and 101.

1. Before servicing the vehicle, refer to the Precautions Section.

➡ If working near and/or around the SRS system and components, be sure to disable the SRS system. After disabling the system wait three minutes or more before servicing the vehicle.

➡ Whenever the negative battery cable is disconnected the following components will require resetting. The Idle Air Volume Learning, Steering Angle Sensor Neutral Position, Sunroof Memory Reset/Initialization, Automatic Drive Positioner System, Audio presets and Navigation. Use the CONSULT-III diagnostic tool, or equivalent to perform the required resets.

2. Disconnect the negative battery cable.
3. Relieve the fuel system pressure.
4. Remove fuel filler cap to release pressure from inside tank.
5. Disconnect fuel filler hose from fuel filler pipe.
6. Drain fuel tank through the fuel filler opening using a suitable hose.
7. Disconnect the following:
   • Fuel pump line protector
   • EVAP hose
   • Fuel level sensor

• Fuel filter
• Fuel pump wiring harness
• Fuel supply hose
8. Using a suitable jack to support the fuel tank, remove the strap bolts and remove the fuel tank from the vehicle.
9. Remove the lock ring using special tool J-46536, or equivalent.
10. Remove the following:
   • Fuel level sensor
   • Fuel filter
   • Fuel pump assembly

To install:

➡ Be sure to use new fasteners, as required.

11. Install or connect the following:
   • Fuel pump assembly, using new O-ring
   • Fuel filter, using new filter
   • Fuel level sensor, using new sensor
   • Fuel pump assembly lock ring
   • Fuel tank
   • Fuel supply hose

- Fuel pump wiring harness
- EVAP hose
- Fuel pump line protector
- Fuel filler pipe

12. Start engine and check for leaks.
13. Be sure to perform the reconnect/relearn procedures.

FUEL RAIL AND INJECTOR

REMOVAL & INSTALLATION

See Figures 102 and 103.

1. Before servicing the vehicle, refer to the Precautions Section.

➡**If working near and/or around the SRS system and components, be sure to disable the SRS system. After disabling the system wait three minutes or more before servicing the vehicle.**

➡**Whenever the negative battery cable is disconnected the following components will require resetting. The Idle Air Volume Learning, Steering Angle Sensor Neutral Position, Sunroof Memory Reset/Initialization, Automatic Drive Positioner System, Audio presets and Navigation. Use the CONSULT-III diagnostic tool, or equivalent to perform the required resets.**

2. Disconnect the negative battery cable.
3. Remove engine cover.
4. Relieve fuel system pressure.
5. Remove or disconnect the following:
   - Fuel injector harness connectors
   - Fuel hose assembly from right and left fuel rails
   - Fuel injectors with fuel rail as an assembly
   - Fuel injector from fuel rail

To install:

➡**Be sure to use new fasteners, as required.**

6. Install or connect the following:
   - New clip onto the fuel injector
   - Fuel injector to fuel rail
   - Fuel injectors and fuel rail as an assembly to the intake manifold. Tighten bolts A and B in two stages. Stage one 9 ft. lbs., stage two 18 ft. lbs.
   - Fuel hose assembly
   - Fuel injector harness connectors
   - Negative battery cable
   - Engine cover
7. Start engine and check for leaks.
8. Be sure to perform the reconnect/relearn procedures.

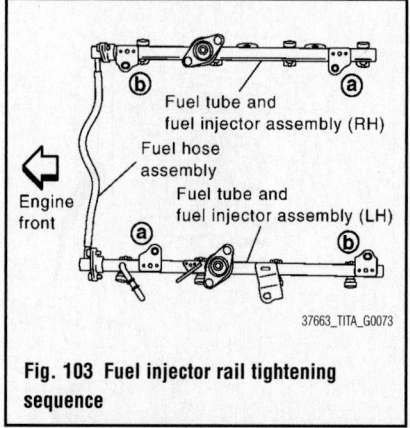

37663_TITA_G0073

Fig. 103 Fuel injector rail tightening sequence

FUEL TANK

REMOVAL & INSTALLATION

See Figure 104.

✳✳ CAUTION

When replacing fuel line parts, be sure to work in a well ventilated area and furnish workshop with a CO_2 fire extinguisher. Do not smoke while servicing fuel system. Keep open flames and sparks away from the work area.

✳✳ CAUTION

Always replace O-rings and clamps with new ones. Do not kink or twist hoses when they are being installed. Do not tighten hose clamps excessively to avoid damaging hoses. Tighten high-pressure rubber hose clamp so that clamp end is 0.12 inches (3 mm) from hose end. Tightening torque specifications are the same for all rubber hose clamps. Ensure that screw does not contact adjacent parts.

1. Before servicing the vehicle, refer to the Precautions Section.

➡**If working near and/or around the SRS system and components, be sure to disable the SRS system. After disabling the system wait three minutes or more before servicing the vehicle.**

➡**Whenever the negative battery cable is disconnected the following components will require resetting. The Idle Air Volume Learning, Steering Angle Sensor Neutral Position, Sunroof Memory Reset/Initialization, Automatic Drive Positioner System, Audio presets and Navigation. Use the CONSULT-III diagnostic tool, or equivalent to perform the required resets.**

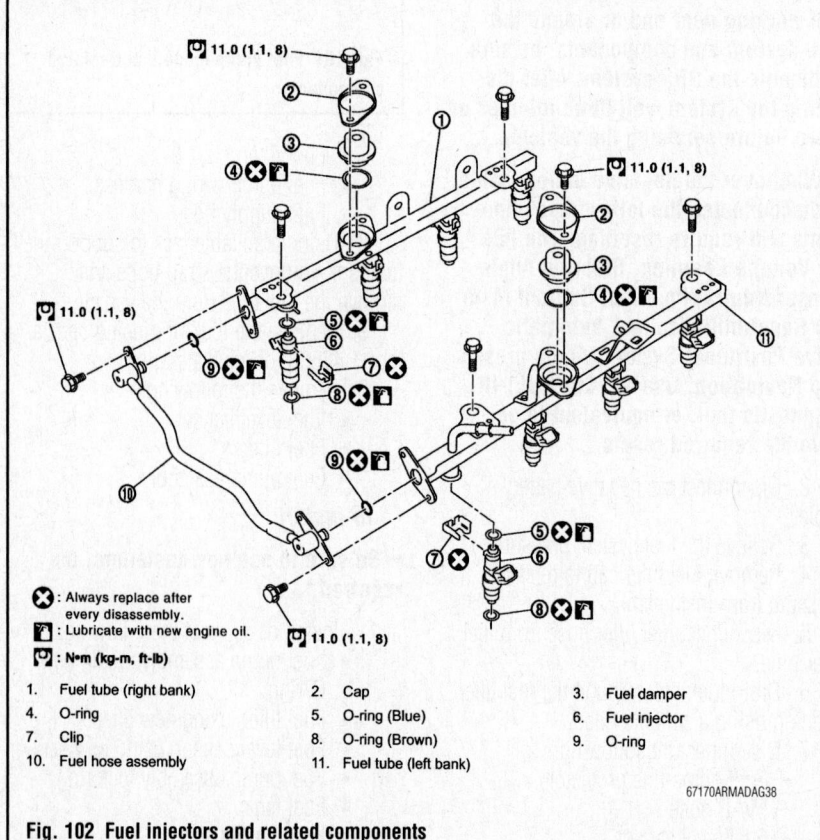

✗ : Always replace after every disassembly.
▯ : Lubricate with new engine oil.
▭ : N•m (kg-m, ft-lb)

1. Fuel tube (right bank)	2. Cap	3. Fuel damper
4. O-ring	5. O-ring (Blue)	6. Fuel injector
7. Clip	8. O-ring (Brown)	9. O-ring
10. Fuel hose assembly	11. Fuel tube (left bank)	

67170ARMADAG38

Fig. 102 Fuel injectors and related components

2. Remove the fuel filler cap to release the pressure from inside the fuel tank.

3. Check the fuel level on level gauge. If the fuel gauge indicates more than the level as full or almost full, drain the fuel from the fuel tank until the fuel gauge indicates the level as less.

4. If the fuel pump does not operate, use the following procedure to drain the fuel:

a. Insert a suitable hose of less than 0.59 inches (15 mm) diameter into the fuel filler pipe through the fuel filler opening to drain the fuel from fuel filler pipe.

b. Remove the left-hand rear wheel and tire.

c. Remove the fuel filler pipe shield.

d. Disconnect the fuel filler hose from the fuel filler pipe and disconnect the vent hose quick connector.

e. Insert a suitable hose into the fuel tank through the fuel filler hose to drain the fuel from the fuel tank.

5. Release the fuel pressure from the fuel lines.

6. Disconnect the negative battery cable.

7. Remove the three nuts and remove fuel line pump protector.

8. Disconnect the EVAP hose at the EVAP canister.

9. Disconnect the fuel level sensor, fuel filter, and fuel pump assembly electrical connector, and the fuel feed hose.

✳✳ CAUTION

Observe the following when disconnecting the quick-connectors:

- The tube can be removed when the tabs are completely depressed. Do not twist it more than necessary.
- Do not use any tools to remove the quick connector.
- Keep the resin tube away from heat. Be especially careful when welding near the tube.
- Prevent liquid acids, such as battery electrolyte, from getting on the resin tube.
- Do not bend or twist the tube during installation and removal.
- Only when the tube is replaced, remove the remaining retainer on the tube or fuel level sensor, fuel filter, and fuel pump assembly.
- When the tube or fuel level sensor, fuel filter, and fuel pump assembly is replaced, also replace

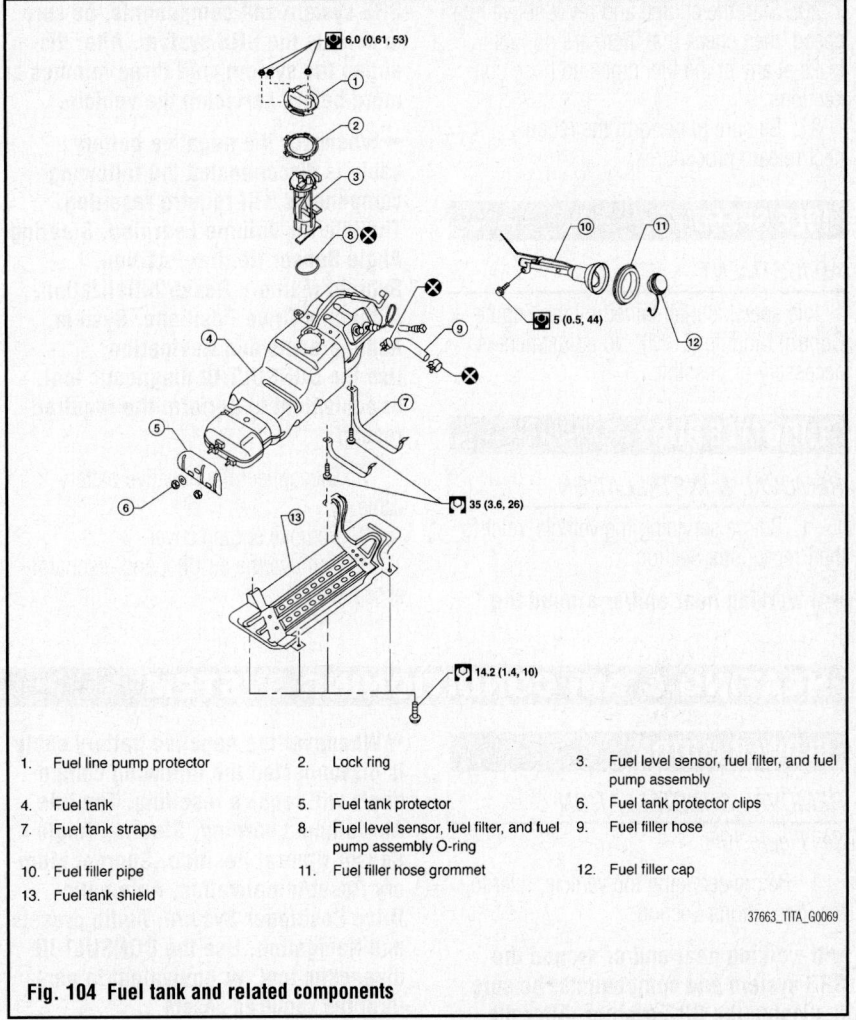

1.	Fuel line pump protector	2.	Lock ring	3.	Fuel level sensor, fuel filter, and fuel pump assembly
4.	Fuel tank	5.	Fuel tank protector	6.	Fuel tank protector clips
7.	Fuel tank straps	8.	Fuel level sensor, fuel filter, and fuel pump assembly O-ring	9.	Fuel filler hose
10.	Fuel filler pipe	11.	Fuel filler hose grommet	12.	Fuel filler cap
13.	Fuel tank shield				

37663_TITA_G0069

Fig. 104 Fuel tank and related components

the retainer with a new one (green colored retainer).
- To keep the connecting portion clean and to avoid damage and foreign materials, cover them completely with plastic bags or something similar.

10. Disconnect the quick-connectors by performing the following:

a. Hold the sides of the connector, push in tabs and pull out the tube.

b. If the connector and the tube are stuck together, push and pull several times until they start to move. Then disconnect them by pulling.

11. Remove the four bolts and remove the fuel tank shield using power tool.

12. Disconnect fuel filler hose at the fuel tank side.

13. Remove the fuel tank strap bolts while supporting the fuel tank with a suitable lift jack.

14. Lower the fuel tank using a suitable lift jack and remove it from the vehicle.

15. If necessary, remove the lock ring using tool No. J-46536.

16. If necessary, remove the fuel level sensor, fuel filter, and fuel pump assembly. Discard the fuel level sensor, fuel filter, and fuel pump assembly O-ring.

To install:

➡**Be sure to use new fasteners, as required.**

17. To install, reverse removal procedure.

18. Connect the quick-connectors by performing the following:

a. Check the connection for damage or any foreign materials.

b. Align the connector with the tube, then insert the connector straight into the tube until a click is heard.

c. After the tube is connected, make sure the connection is secure by pulling on the tube and the connector to make sure they are securely connected.

19. Turn the ignition switch ON but do not start engine, then check the fuel pipe and hose connections for leaks while applying fuel pressure to the system.

20. Start the engine and rev it above idle speed, then check that there are no fuel leaks at any of the fuel pipe and hose connections.

21. Be sure to perform the reconnect/relearn procedures.

IDLE SPEED

ADJUSTMENT

Idle speed is maintained by the Engine Control Module (ECM). No adjustment is necessary or possible.

THROTTLE BODY

REMOVAL & INSTALLATION

1. Before servicing the vehicle, refer to the Precautions Section.

➡️If working near and/or around the

SRS system and components, be sure to disable the SRS system. After disabling the system wait three minutes or more before servicing the vehicle.

➡️Whenever the negative battery cable is disconnected the following components will require resetting. The Idle Air Volume Learning, Steering Angle Sensor Neutral Position, Sunroof Memory Reset/Initialization, Automatic Drive Positioner System, Audio presets and Navigation. Use the CONSULT-III diagnostic tool, or equivalent to perform the required resets.

2. Disconnect the negative battery cable.

3. Remove engine cover.

4. Remove the air duct and resonator assembly.

5. Drain the engine coolant. Be sure to properly dispose of used coolant.

6. Disconnect the hoses from the unit.

7. Remove the 4 mounting bolts.

8. Remove the old gasket and discard it

To install:

➡️Be sure to use new fasteners, as required.

9. Install a new gasket and the throttle body.

10. Install the 4 mounting bolts in alternate sequence and tighten to 74 inch lbs. (8.4 Nm).

11. Reconnect the hoses to the throttle body.

12. Reconnect the air duct and resonator assembly.

13. Install the engine cover.

14. Be sure to perform the reconnect/relearn procedures.

HEATING & AIR CONDITIONING SYSTEM

BLOWER MOTOR

REMOVAL & INSTALLATION

See Figure 105.

1. Before servicing the vehicle, refer to the Precautions Section.

➡️If working near and/or around the SRS system and components, be sure to disable the SRS system. After disabling the system wait three minutes or more before servicing the vehicle.

➡️Whenever the negative battery cable is disconnected the following components will require resetting. The Idle Air Volume Learning, Steering Angle Sensor Neutral Position, Sunroof Memory Reset/Initialization, Automatic Drive Positioner System, Audio presets and Navigation. Use the CONSULT-III diagnostic tool, or equivalent to perform the required resets.

2. Disconnect the negative battery cable.

3. Disconnect or remove the following:
   • Glove box assembly

• Blower motor electrical connector
• Three screws and blower motor

To install:

➡️Be sure to use new fasteners, as required.

4. Installation is the reverse of the removal procedure.

5. Be sure to perform the reconnect/relearn procedures.

HEATER/COOLING UNIT

REMOVAL & INSTALLATION

See Figures 106 through 110.

1. Before servicing the vehicle, refer to the Precautions Section.

➡️If working near and/or around the SRS system and components, be sure to disable the SRS system. After disabling the system wait three minutes or more before servicing the vehicle.

➡️Whenever the negative battery cable is disconnected the following components will require resetting. The Idle Air Volume Learning, Steering Angle Sensor Neutral Position, Sunroof Memory Reset/Initialization, Automatic Drive Positioner System, Audio presets and Navigation. Use the CONSULT-III diagnostic tool, or equivalent to perform the required resets.

2. Position the front seats in the rearmost position.

3. Disconnect the negative battery cable.

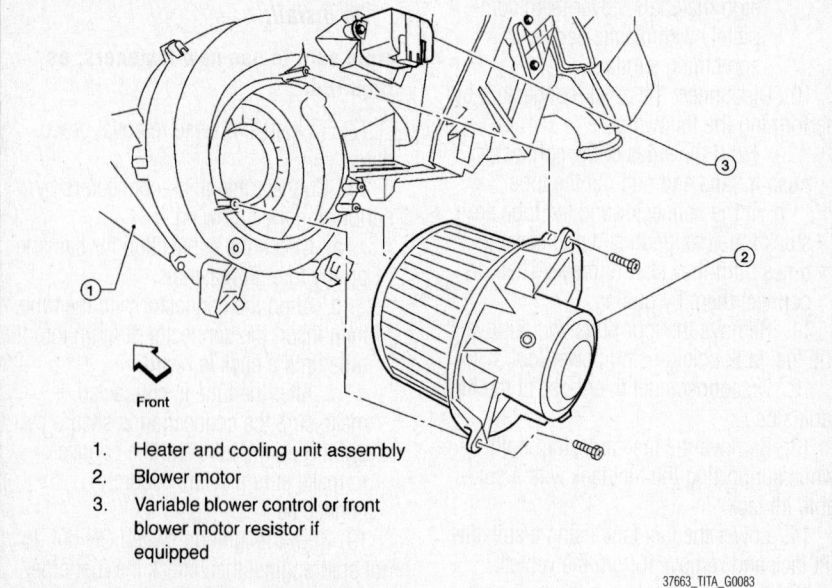

Front

1. Heater and cooling unit assembly
2. Blower motor
3. Variable blower control or front blower motor resistor if equipped

37663_TITA_G0083

Fig. 105 Blower motor and related components

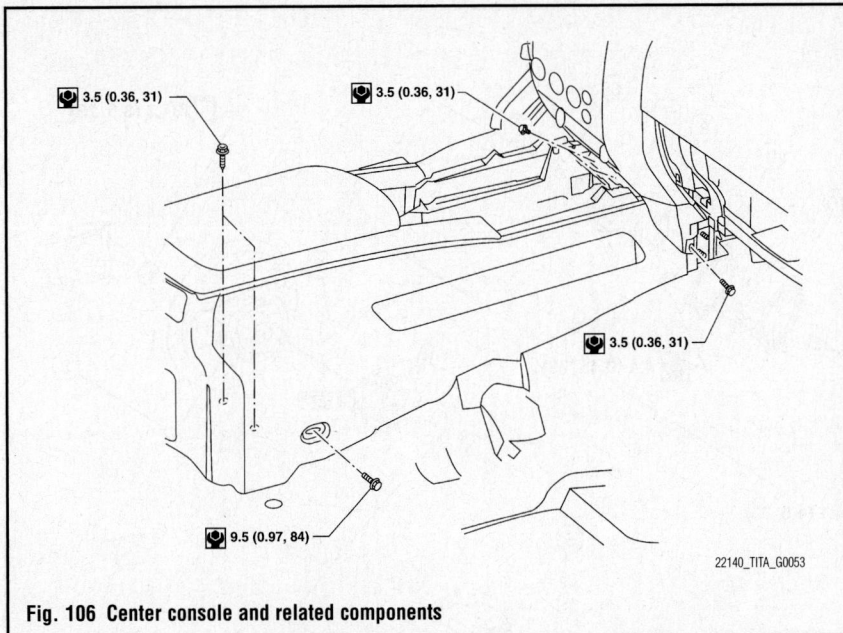

3.5 (0.36, 31)

3.5 (0.36, 31)

3.5 (0.36, 31)

9.5 (0.97, 84)

22140_TITA_G0053

Fig. 106 Center console and related components

4. Properly discharge the A/C system.

5. Drain the cooling system. Be sure to properly dispose of used coolant.

6. Disconnect the heater hoses at the heater core.

7. Disconnect and plug the refrigerant lines at the evaporator core.

8. Remove the instrument panel lower cover.

9. Remove the center console.

10. Remove the steering column.

11. Remove the combination meter.

12. Remove the audio unit.

13. Remove the display unit, if equipped.

14. Remove the lower knee protector.

15. Remove the defroster grille. Disconnect the optical sensor electrical connector.

16. Remove the left and right side ventilator assembly.

17. Remove the left and right side assist grip and windshield garnish.

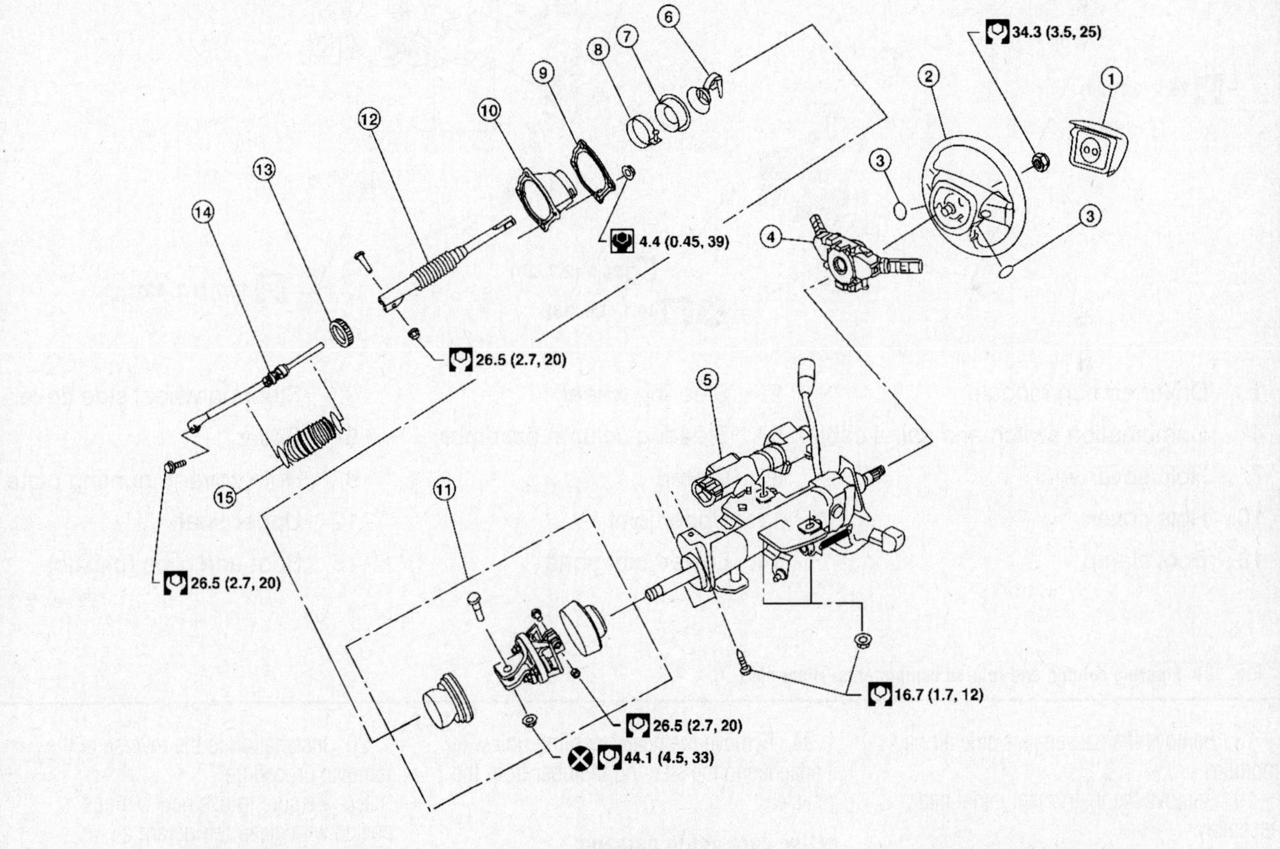

34.3 (3.5, 25)

4.4 (0.45, 39)

26.5 (2.7, 20)

26.5 (2.7, 20)

16.7 (1.7, 12)

26.5 (2.7, 20)

44.1 (4.5, 33)

1.	Driver air bag module	2.	Steering wheel	3.	Steering wheel side cover
4.	Combination switch and spiral cable	5.	Steering column assembly	6.	Collar
7.	Hole cover seal	8.	Clamp	9.	Hole cover mounting plate
10.	Hole cover	11.	Upper joint	12.	Upper shaft
13.	Boot clamp	14.	Lower joint shaft	15.	Boot and clips (plastic)

37663_TITA_G0007

Fig. 107 Steering column and related components—column shift

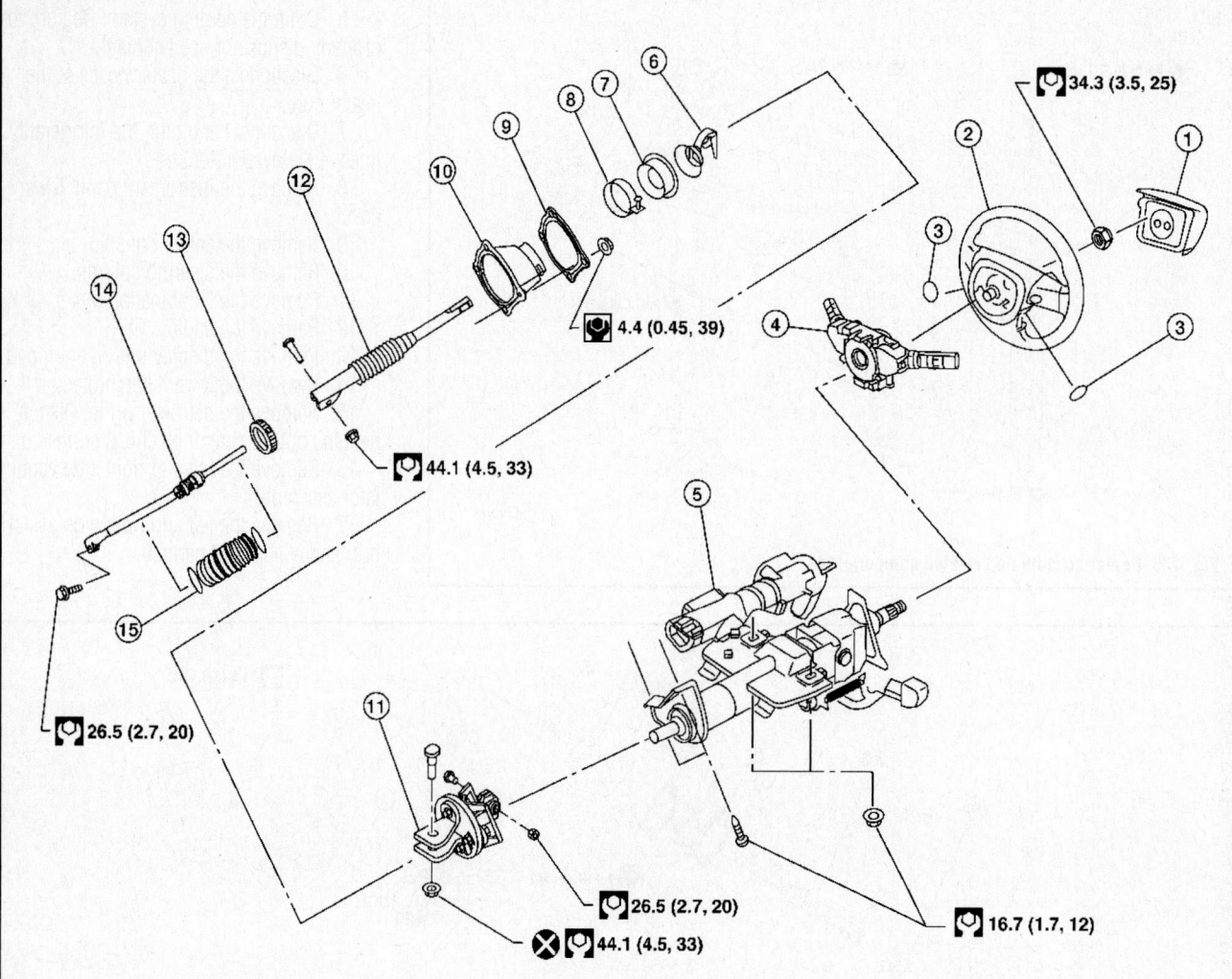

1. Driver air bag module
2. Steering wheel
3. Steering wheel side cover
4. Combination switch and spiral cable
5. Steering column assembly
6. Collar
7. Hole cover seal
8. Clamp
9. Hole cover mounting plate
10. Hole cover
11. Upper joint
12. Upper shaft
13. Boot clamp
14. Lower joint shaft
15. Boot and clips (plastic)

37663_TITA_G0008

Fig. 108 Steering column and related components—floor shift

18. Remove the passenger's side air bag module.

19. Remove the instrument panel pad assembly.

20. Disconnect the remaining electrical connectors. Remove the retaining bolts.

21. Carefully remove the instrument panel.

22. Disconnect the instrument panel wire harness at the right and left in-line connector brackets, and the fuse block (JB) electrical connectors.

23. Disconnect the steering member from each side of the vehicle body.

24. Remove the heater/cooling unit with it attached to the steering member from the vehicle.

➡**Use care not to damage the seats or interior trim panels.**

25. Remove the heater/cooling unit from the steering member.

To install:

➡**Be sure to use new fasteners, as required.**

26. Installation is the reverse of the removal procedure.

27. Be sure to use new O-rings coated with clean refrigerant oil, as required.

28. Fill the cooling system with the proper grade and the coolant.

29. Properly recharge the A/C system.

30. Start the engine and check for leaks, correct as required.

31. Be sure to perform the reconnect/relearn procedures.

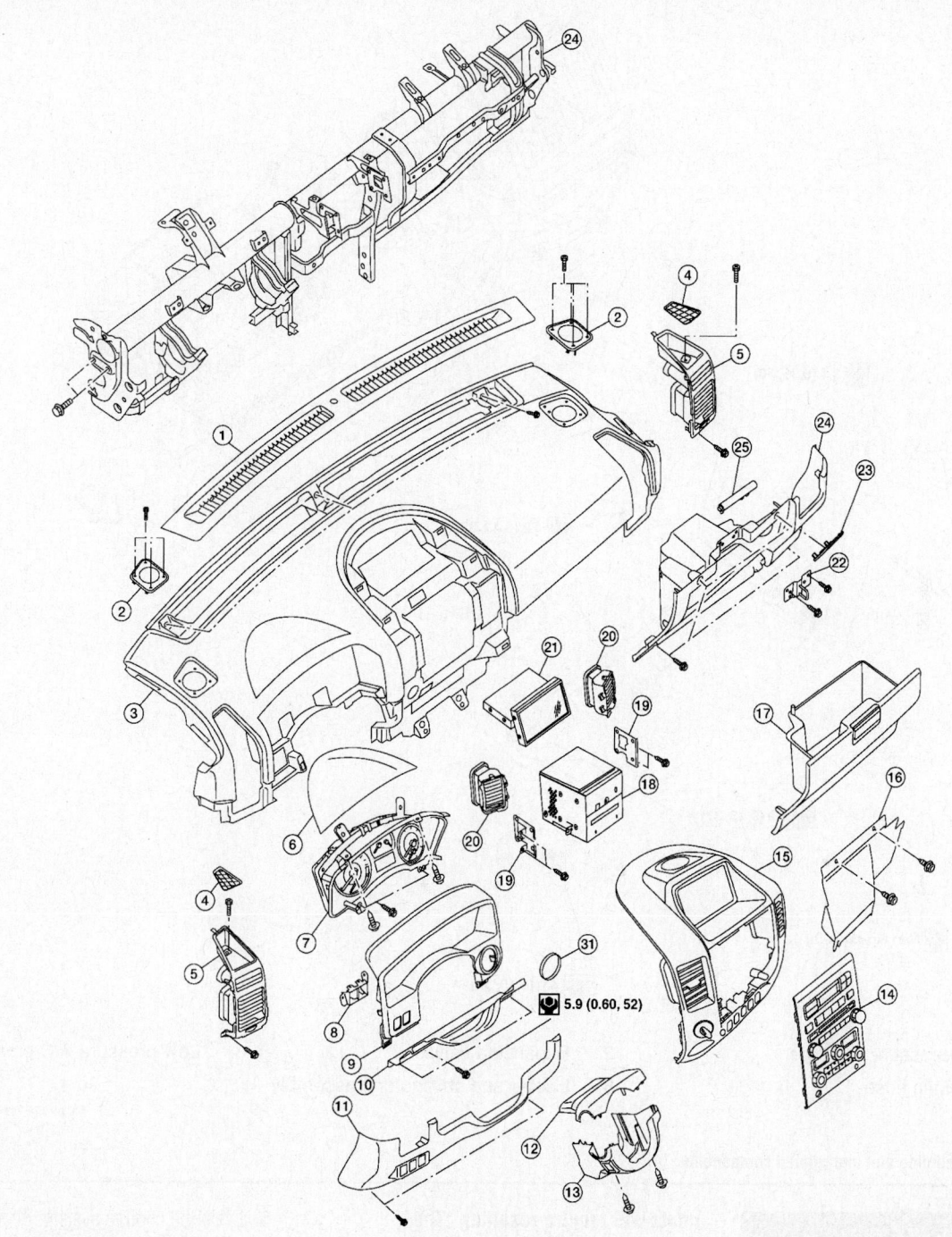

1.	Defroster grille	2.	Speaker grille RH/LH	3.	Instrument panel and pad assembly
4.	Deck pocket mat RH/LH	5.	Side ventilator assembly RH/LH	6.	Combination meter cover
7.	Combination meter	8.	Switch assembly	9.	Cluster lid A
10.	Lower knee protector	11.	Lower instrument panel LH	12.	Steering column cover upper
13.	Steering column cover lower	14.	Clluster lid D	15.	Cluster lid C
16.	Instrument lower cover RH	17.	Glove box	18.	Audio unit
19.	Audio bracket RH/LH	20.	Center ventilator assembly RH/LH	21.	Display assembly (if equipped)
22.	Glove box lid striker	23.	Fuse block cover	24.	Lower instrument panel RH
25.	Glove box damper (if equipped)	26.	Steering member		

Fig. 109 Instrument panel and related components (1 of 2)

37663_TITA_G0005

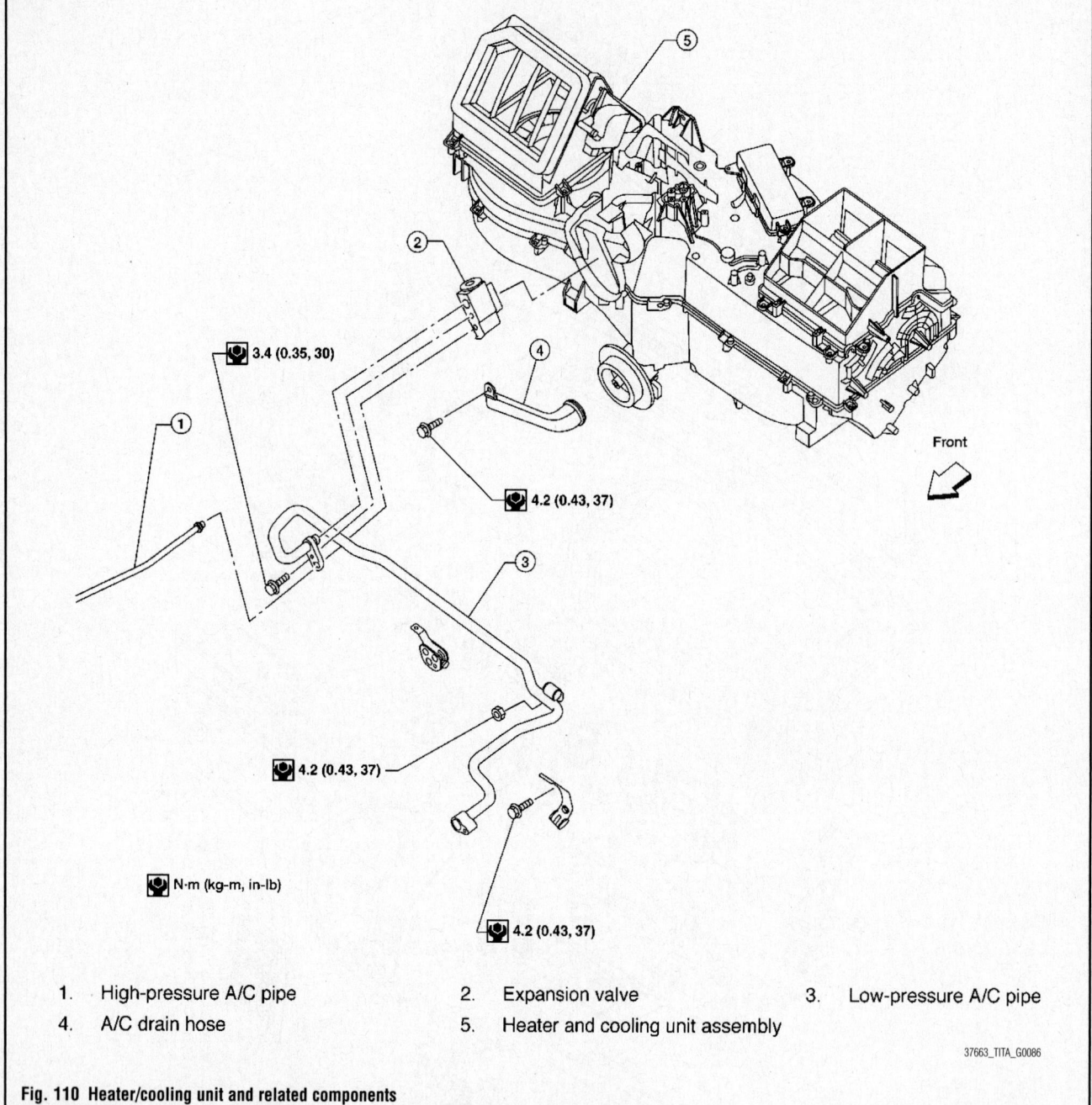

3.4 (0.35, 30)

4.2 (0.43, 37)

4.2 (0.43, 37)

N·m (kg-m, in-lb)

4.2 (0.43, 37)

Front

1.	High-pressure A/C pipe	2.	Expansion valve	3.	Low-pressure A/C pipe
4.	A/C drain hose	5.	Heater and cooling unit assembly		

37663_TITA_G0086

Fig. 110 Heater/cooling unit and related components

HEATER CORE

REMOVAL & INSTALLATION

See Figures 109 through 111.

1. Before servicing the vehicle, refer to the Precautions Section.

➡**If working near and/or around the SRS system and components, be sure to disable the SRS system. After disabling the system wait three minutes or more before servicing the vehicle.**

➡**Whenever the negative battery cable is disconnected the following compo-** nents will require resetting. The **Idle Air Volume Learning, Steering Angle Sensor Neutral Position, Sunroof Memory Reset/Initialization, Automatic Drive Positioner System, Audio presets and Navigation. Use the CONSULT-III diagnostic tool, or equivalent to perform the required resets.**

2. Position the front seats in the rear-most position.

3. Disconnect the negative battery cable.

4. Properly discharge the A/C system.

5. Drain the cooling system. Be sure to properly dispose of used coolant.

6. Disconnect the heater hoses at the heater core.

7. Disconnect and plug the refrigerant lines at the evaporator core.

8. Remove the instrument panel.

9. Remove the heater/cooling unit.

10. Remove the four screws and remove the upper bracket.

11. Remove the four screws and remove the heater core cover.

12. Remove the core pipe bracket.

13. Remove the heater core from its mounting.

➡**Be sure to replace the in cabin micro-filter.**

To install:

➡**Be sure to use new fasteners, as required.**

14. Installation is the reverse of the removal procedure.

15. Be sure to use new O-rings coated with clean refrigerant oil, as required.

16. Fill the cooling system with the proper grade and the coolant.

17. Properly recharge the A/C system.

18. Start the engine and check for leaks, correct as required.

19. Be sure to perform the reconnect/relearn procedures.

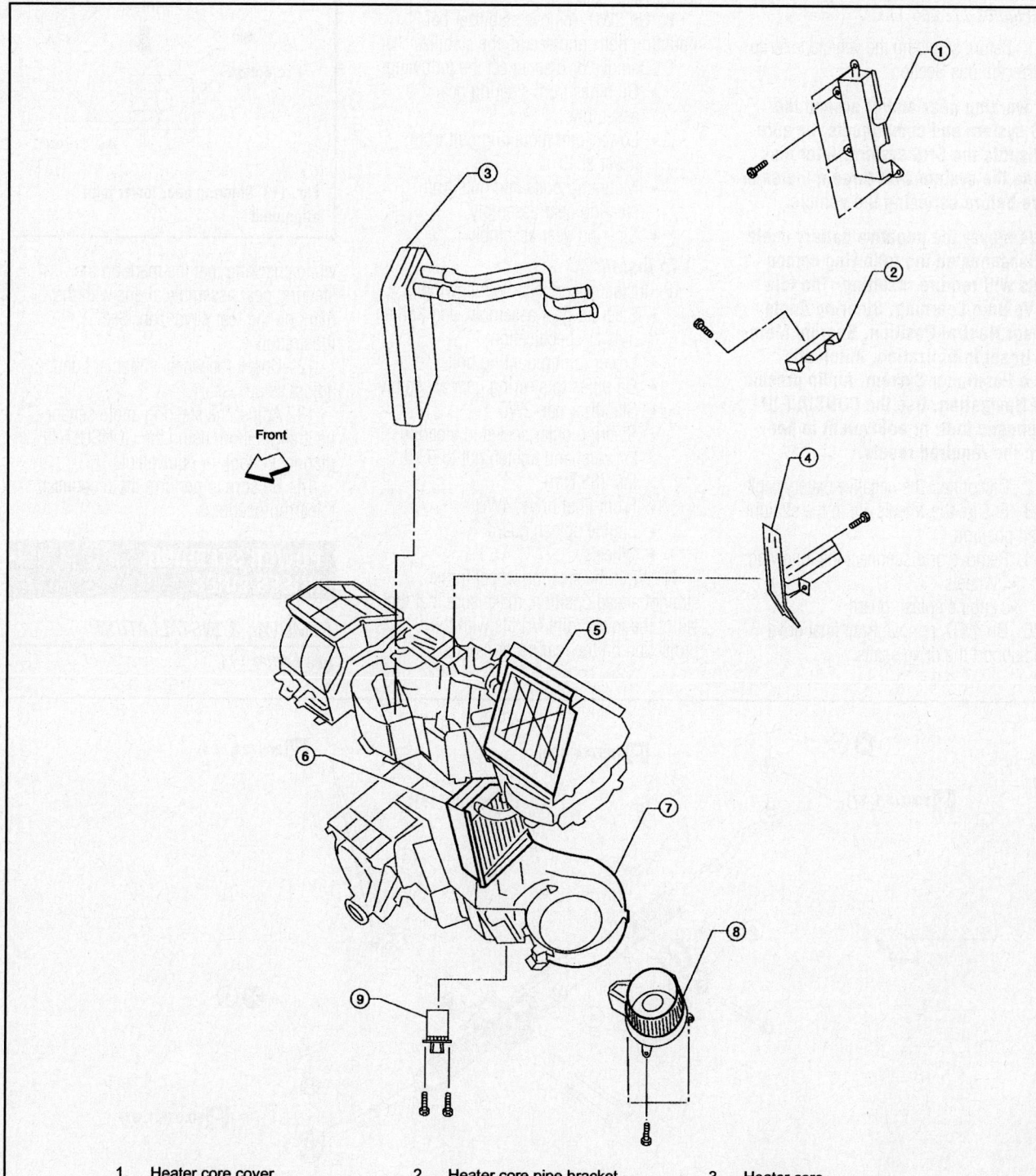

Front

1. Heater core cover	2. Heater core pipe bracket	3. Heater core
4. Upper bracket	5. Upper heater and cooling unit case	6. A/C evaporator
7. Lower heater and cooling unit case	8. Blower motor	9. Variable blower control

67170ARMADAG26

Fig. 111 Heater core and related components

STEERING

POWER RACK & PINION STEERING GEAR

REMOVAL & INSTALLATION

See Figures 112 and 113.

1. Before servicing the vehicle, refer to the Precautions Section.

➡ **If working near and/or around the SRS system and components, be sure to disable the SRS system. After disabling the system wait three minutes or more before servicing the vehicle.**

➡ **Whenever the negative battery cable is disconnected the following components will require resetting. The Idle Air Volume Learning, Steering Angle Sensor Neutral Position, Sunroof Memory Reset/Initialization, Automatic Drive Positioner System, Audio presets and Navigation. Use the CONSULT-III diagnostic tool, or equivalent to perform the required resets.**

2. Disconnect the negative battery cable.
3. Ensure the wheels are in the straight-ahead position.
4. Remove or disconnect the following:
   • Wheels
   • Engine splash guard
5. On 4WD, remove front final drive and support the drive shafts.

6. Remove cotter pin at steering outer socket and loosen mounting nut.
7. Remove steering outer socket from steering knuckle using special tool J-25730-A.
8. On 2WD, remove stabilizer bar mounting bolts and secure the stabilizer bar.
9. Remove or disconnect the following:
   • Oil pipes from steering gear assembly
   • Lower joint mounting bolt from lower shaft
   • Mounting bolts and nuts from steering gear assembly
   • Steering gear assembly

To install:

10. Install or connect the following:
   • Steering gear assembly, and tighten nuts to specification
   • Lower joint mounting bolt
   • Oil pipes to steering gear assembly
   • Stabilizer bar, 2WD
   • Steering outer socket to steering knuckle, and tighten nut to 63 ft. lbs. (86 Nm)
   • Front final drive, 4WD
   • Engine splash guard
   • Wheels

11. With the steering wheel in the straight ahead position, make sure that the slit of the lower joint (A) fits with the projection on the rear cover cap (B),

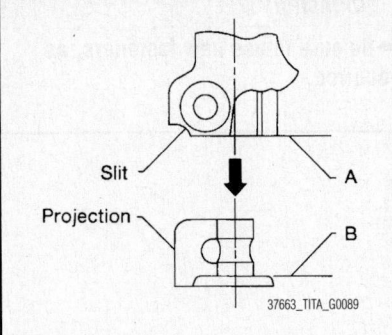

Fig. 113 Steering gear lower joint alignment

while checking that the mark on the steering gear assembly aligns with the mark on the rear cover cap. See illustration.

12. Check the wheel alignment and adjust as necessary.
13. Adjust the steering angle sensor neutral position, using the CONSULT-III diagnostic tool, or equivalent.
14. Be sure to perform the reconnect/relearn procedures.

POWER STEERING HOSES/LINES

REMOVAL & INSTALLATION

See Figure 114.

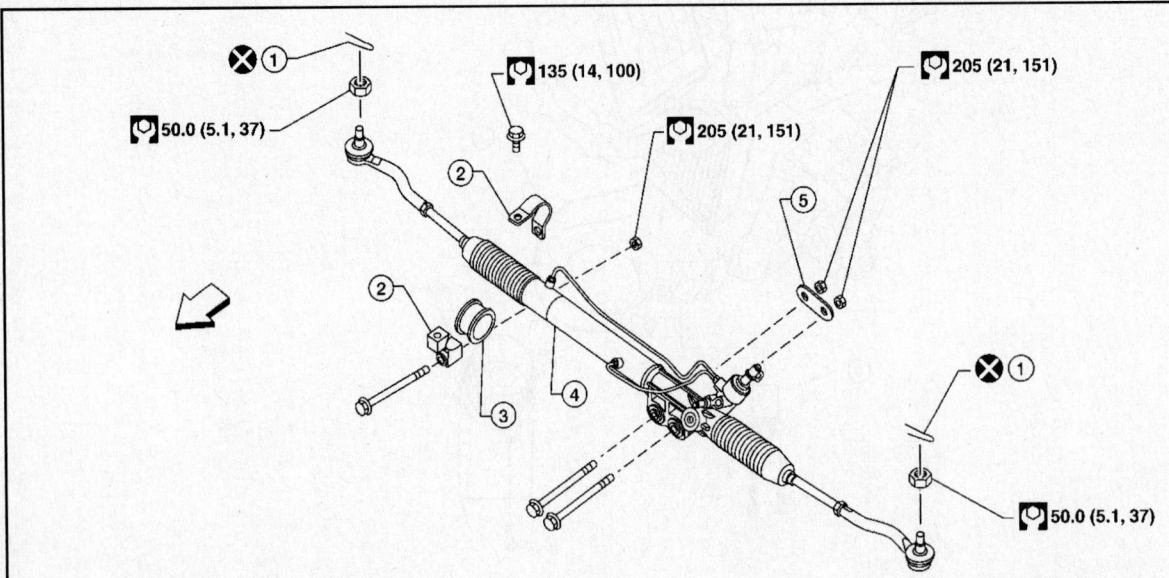

| 1. | Cotter pin | 2. | Steering gear bracket | 3. | Steering gear insulator |
| 4. | Steering gear assembly | 5. | Washer | ⬅ | Front |

Fig. 112 Steering gear and related components

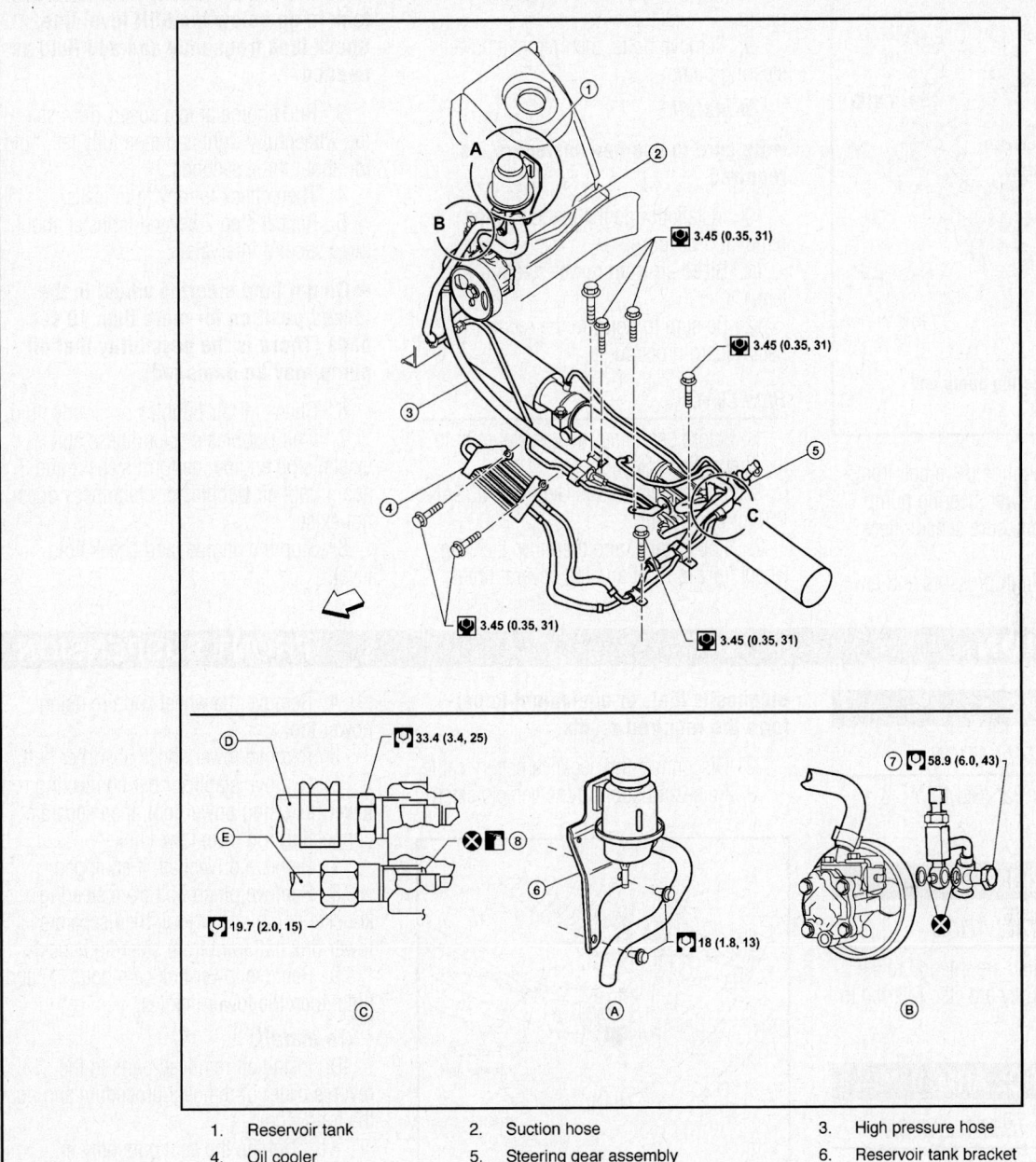

Fig. 114 Power steering hoses and related components

1.	Reservoir tank	2.	Suction hose	3.	High pressure hose
4.	Oil cooler	5.	Steering gear assembly	6.	Reservoir tank bracket
7.	Eye bolt	8.	O-rings	⇐	Front
D.	Low pressure piping	E.	High pressure piping		

37663_TITA_G0090

At this time the manufacturer does not provide removal and installation procedures for this component, refer to the illustration as required.

POWER STEERING PUMP

REMOVAL & INSTALLATION

See Figure 115.

1. Before servicing the vehicle, refer to the Precautions Section.

➡ **If working near and/or around the SRS system and components, be sure to disable the SRS system. After disabling the system wait three minutes or more before servicing the vehicle.**

➡ **Whenever the negative battery cable is disconnected the following components will require resetting. The Idle Air Volume Learning, Steering Angle Sensor Neutral Position, Sunroof Memory Reset/Initialization, Automatic**

Drive Positioner System, Audio presets and Navigation. Use the CONSULT-III diagnostic tool, or equivalent to perform the required resets.

2. Disconnect the negative battery cable.

3. Drain power steering fluid from reservoir tank.

4. Remove air duct assembly.

5. Remove power steering reservoir tank.

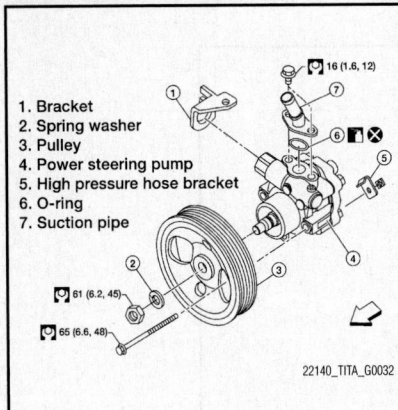

1. Bracket
2. Spring washer
3. Pulley
4. Power steering pump
5. High pressure hose bracket
6. O-ring
7. Suction pipe

16 (1.6, 12)

61 (6.2, 45)

65 (6.6, 48)

22140_TITA_G0032

Fig. 115 Power steering pump and related components

6. Remove serpentine drive belt from auto tensioner and power steering pump.

7. Disconnect pressure sensor electrical connector.

8. Remove the high pressure and low pressure piping from power steering oil pump.

9. Remove bolts, then remove power steering pump.

To install:

➡**Be sure to use new fasteners, as required.**

10. Installation is the reverse of the removal procedure.

11. Bleed air from power steering system.

12. Be sure to perform the reconnect/relearn procedures.

BLEEDING

1. Before servicing the vehicle, refer to the Precautions Section.
   Recommended fluid is Genuine NISSAN PSF or equivalent.

2. Stop engine, and then turn steering wheel fully to right and left several times.

➡**Do not allow steering fluid reservoir tank to go below the MIN level line. Check tank frequently and add fluid as needed.**

3. Run engine at idle speed. Turn steering wheel fully right and then fully left, hold for about three seconds.

4. Then check for any fluid leaks.

5. Repeat step 2 several times at about three second intervals.

➡**Do not hold steering wheel in the locked position for more than 10 seconds (There is the possibility that oil pump may be damaged).**

6. Check for air bubbles or cloudy fluid.

7. If air bubbles or cloudiness still exists, stop engine, perform steps 2 and 3 again until air bubbles or cloudiness does not exist.

8. Stop the engine, and check fluid level.

SUSPENSION

CONTROL LINKS

REMOVAL & INSTALLATION

Refer to Stabilizer Bar (REMOVAL & INSTALLATION).

LOWER BALL JOINT

REMOVAL & INSTALLATION

The lower ball joints are integral to the upper control arm. They are also referred to as an upper link.

LOWER CONTROL ARM

REMOVAL AND & INSTALLATION

See Figures 116 through 118.

1. Before servicing the vehicle, refer to the Precautions Section.

➡**If working near and/or around the SRS system and components, be sure to disable the SRS system. After disabling the system wait three minutes or more before servicing the vehicle.**

➡**Whenever the negative battery cable is disconnected the following components will require resetting. The Idle Air Volume Learning, Steering Angle Sensor Neutral Position, Sunroof Memory Reset/Initialization, Automatic Drive Positioner System, Audio presets and Navigation. Use the CONSULT-III**

diagnostic tool, or equivalent to perform the required resets.

2. Disconnect the negative battery cable.

3. Raise and support the vehicle safely.

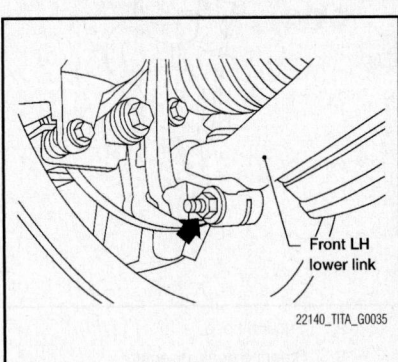

Front LH lower link

22140_TITA_G0035

Fig. 116 Remove pinch bolt from steering knuckle

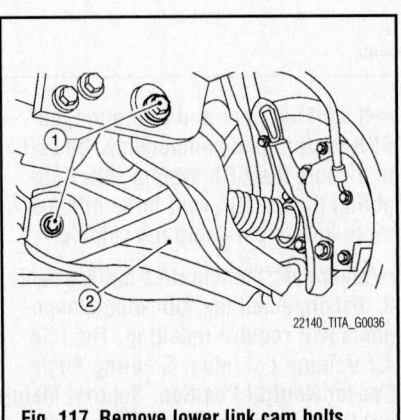

22140_TITA_G0036

Fig. 117 Remove lower link cam bolts

FRONT SUSPENSION

4. Remove the wheel and tire using power tool.

5. Remove lower shock absorber bolt.

6. Remove stabilizer bar connecting rod lower nut using power tool, then separate connecting rod from lower link.

7. Remove driveshaft, if equipped.

8. Remove pinch bolt from steering knuckle using power tool, then separate lower link ball joint from steering knuckle.

9. Remove lower link cam bolts (1) and nuts, then the lower link (2).

To install:

10. Install all removed parts in the reverse order of removal procedure and note the following:

   a. Tighten the cam nuts only in a relaxed position to 98 ft. lbs. (133 Nm). Cam bolts are for adjustment.

   b. Tighten the steering knuckle pinch bolt to 70 ft. lbs. (95 (Nm).

   c. After installation, check that the front wheel alignment is within specification.

➡**Some vehicles may be equipped with straight (non adjustable) lower link bolts and washers. In order to adjust camber and caster on these vehicles, first replace the lower link bolts and washers with adjustable ones.**

11. Be sure to perform the reconnect/relearn procedures.

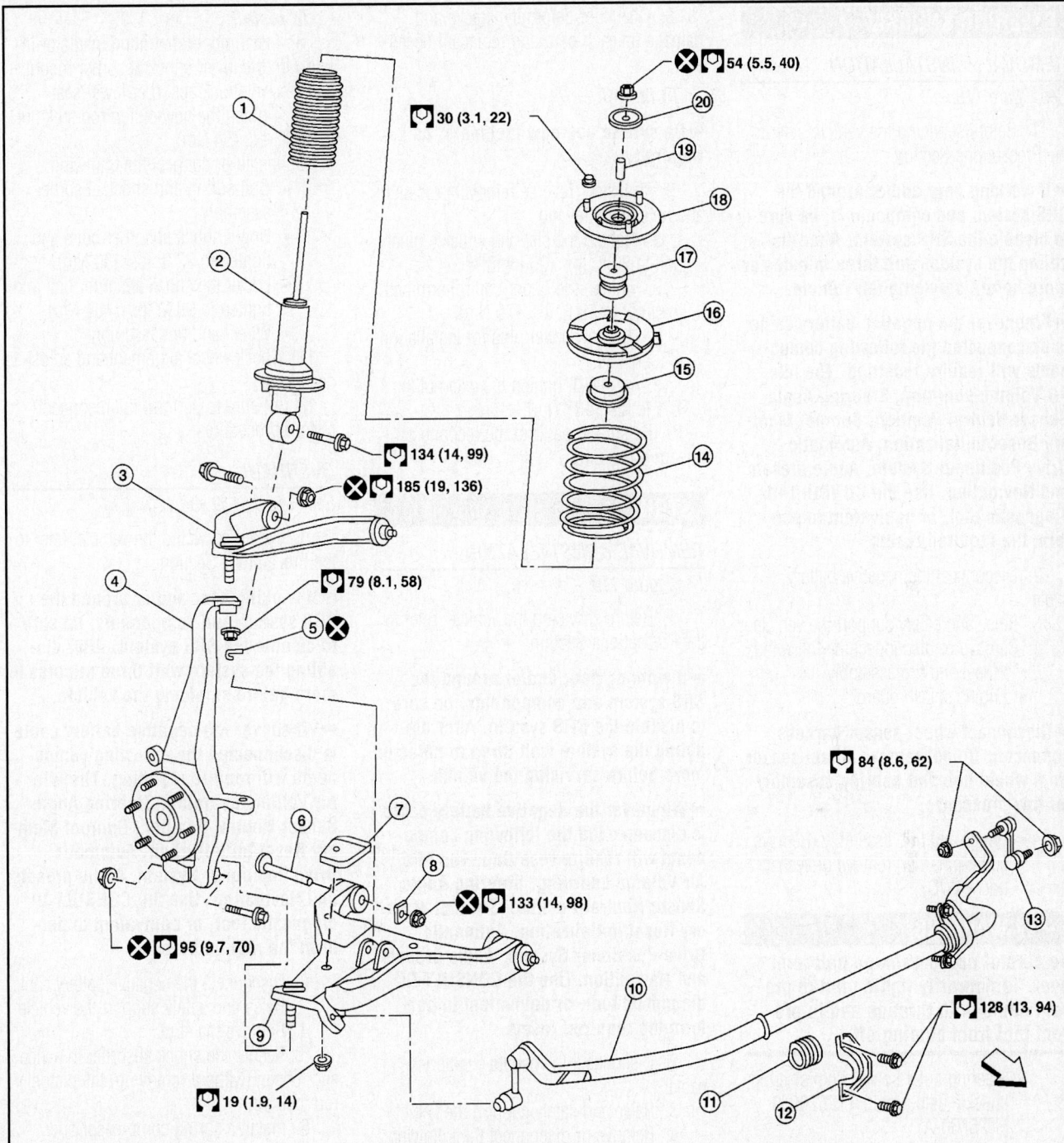

1.	Dust cover	2.	Shock absorber	3. Upper link
4.	Steering knuckle	5.	Cotter pin	6. Bolt
7.	Jounce bumper	8.	Washer	9. Lower link
10.	Stabilizer bar	11.	Stabilizer bar bushing	12. Stabilizer bar mounting bracket
13.	Connecting rod	14.	Coil spring	15. Upper seat
16.	Upper spring seat	17.	Shock absorber bushing	18. Shock absorber mounting insulator
19.	Spacer	20.	Washer	⇦ Front

37663_TITA_G0094

Fig. 118 Front suspension components

STEERING KNUCKLE

REMOVAL & INSTALLATION

See Figure 118.

1. Before servicing the vehicle, refer to the Precautions Section.

➡**If working near and/or around the SRS system and components, be sure to disable the SRS system. After disabling the system wait three minutes or more before servicing the vehicle.**

➡**Whenever the negative battery cable is disconnected the following components will require resetting. The Idle Air Volume Learning, Steering Angle Sensor Neutral Position, Sunroof Memory Reset/Initialization, Automatic Drive Positioner System, Audio presets and Navigation. Use the CONSULT-III diagnostic tool, or equivalent to perform the required resets.**

2. Disconnect the negative battery cable.

3. Raise and safely support the vehicle.

4. Remove or disconnect the following:
   - Wheel and tire assembly
   - Engine splash guard

➡**Disconnect wheel sensor harness connector. Do not remove wheel sensor from wheel hub and bearing assembly for this procedure.**

   - Wheel and hub assembly (Remove cotter pin, then remove driveshaft nut—4WD)

✳✳ WARNING

Be careful not to damage ball joint boot. Temporarily tighten nut to prevent damage to threads and to prevent tool from coming off.

   - Steering outer socket from steering knuckle using tool HT72520000 (J-25730-A)
   - Coil spring and shock absorber assembly

➡**Support the lower link using a suitable jack.**

   - Cotter pin and nut from upper link ball joint and discard the cotter pin

5. Separate the upper link ball joint from steering knuckle using tool ST29020001 (J-24319-01).

6. Remove pinch bolt from steering knuckle using power tool, then separate lower link ball joint from steering knuckle.

7. Remove the steering knuckle from vehicle.

8. Check for deformity, cracks and damage on each part, and replace if necessary.

To install:

➡**Be sure to use new fasteners, as required.**

9. To install, reverse removal procedure and note the following:

   a. Tighten the steering knuckle pinch bolt to 70 ft. lbs. (95 (Nm).

   b. Tighten the upper control arm ball joint nut to 58 ft. lbs. (79 Nm).

   c. Use new cotter pins for installation of lock nuts.

   d. For 4WD, tighten the axle nut to 101 ft. lbs. (137 Nm).

10. Be sure to perform the reconnect/relearn procedures.

STRUT

REMOVAL & INSTALLATION

See Figure 118.

1. Before servicing the vehicle, refer to the Precautions Section.

➡**If working near and/or around the SRS system and components, be sure to disable the SRS system. After disabling the system wait three minutes or more before servicing the vehicle.**

➡**Whenever the negative battery cable is disconnected the following components will require resetting. The Idle Air Volume Learning, Steering Angle Sensor Neutral Position, Sunroof Memory Reset/Initialization, Automatic Drive Positioner System, Audio presets and Navigation. Use the CONSULT-III diagnostic tool, or equivalent to perform the required resets.**

2. Disconnect the negative battery cable.

3. Raise and safely support the vehicle.

4. Remove or disconnect the following:
   - Wheel and tire assembly
   - Lower shock absorber bolt
   - Upper shock absorber bolts
   - Coil spring and shock absorber assembly

5. Secure the shock absorber in a vice and loosen (without removing) the piston rod lock nut.

6. Install a spring compressor and tighten until the shock absorber mounting insulator can be turned by hand.

7. Remove piston rod lock nut and remove shock absorber from the coil spring.

To install:

8. Install upper mounting insulator in line with the lower shock absorber mount and step in shock absorber lower seat.

9. Tighten the new piston rod lock nut to 40 ft. lbs. (54 Nm).

10. Install or connect the following:
    - Coil spring and shock absorber assembly
    - Upper shock absorber bolts and tighten to 22 ft. lbs. (30 Nm)
    - Lower the shock absorber bolt and tighten to 99 ft. lbs. (134 Nm)
    - Wheel and tire assembly

11. Check wheel alignment and adjust as necessary.

12. Be sure to perform the reconnect/relearn procedures.

OVERHAUL

See Figures 119 and 120.

1. Before servicing the vehicle, refer to the Precautions Section.

➡**If working near and/or around the SRS system and components, be sure to disable the SRS system. After disabling the system wait three minutes or more before servicing the vehicle.**

➡**Whenever the negative battery cable is disconnected the following components will require resetting. The Idle Air Volume Learning, Steering Angle Sensor Neutral Position, Sunroof Memory Reset/Initialization, Automatic Drive Positioner System, Audio presets and Navigation. Use the CONSULT-III diagnostic tool, or equivalent to perform the required resets.**

2. Disconnect the negative battery cable.

3. Raise and safely support the vehicle.

4. Remove the strut.

5. Secure the shock absorber in a vice and loosen (without removing) the piston rod lock nut.

6. Install a spring compressor and tighten until the shock absorber mounting insulator can be turned by hand.

7. Remove piston rod lock nut and remove shock absorber from the coil spring.

To install:

8. Install upper mounting insulator in line with the lower shock absorber mount and step in shock absorber lower seat as shown in figure.

9. Tighten the new piston rod lock nut to 40 ft. lbs. (54 Nm).

10. Install or connect the following:
    - Coil spring and shock absorber assembly

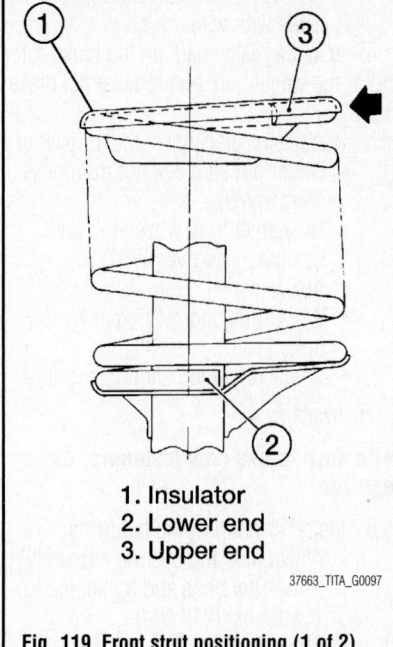

1. Insulator
2. Lower end
3. Upper end

37663_TITA_G0097

Fig. 119 Front strut positioning (1 of 2)

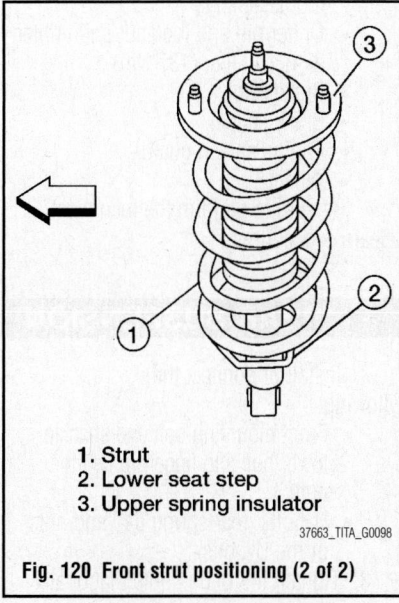

1. Strut
2. Lower seat step
3. Upper spring insulator

37663_TITA_G0098

Fig. 120 Front strut positioning (2 of 2)

- Upper shock absorber bolts and tighten to 22 ft. lbs (30 Nm)
- Lower the shock absorber bolt and tighten to 99 ft. lbs. (134 Nm)
- Wheel and tire assembly
11. Check wheel alignment and adjust as necessary.

STABILIZER BAR

REMOVAL & INSTALLATION
See Figures 118 and 121.

1. Before servicing the vehicle, refer to the Precautions Section.

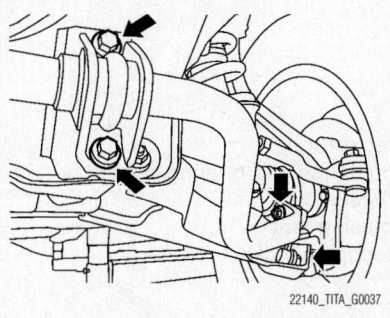

22140_TITA_G0037

Fig. 121 Stabilizer bar mounting bracket bolts and connecting rod nuts

➡If working near and/or around the SRS system and components, be sure to disable the SRS system. After disabling the system wait three minutes or more before servicing the vehicle.

➡Whenever the negative battery cable is disconnected the following components will require resetting. The Idle Air Volume Learning, Steering Angle Sensor Neutral Position, Sunroof Memory Reset/Initialization, Automatic Drive Positioner System, Audio presets and Navigation. Use the CONSULT-III diagnostic tool, or equivalent to perform the required resets.

2. Disconnect the negative battery cable.
3. Raise and safely support the vehicle.
4. Remove or disconnect the following:
   - Engine undercover
   - Stabilizer bar mounting bracket bolts and connecting rod nuts using power tool
   - Bushings from stabilizer bar
5. Check stabilizer bar for twist and deformation. Replace if necessary. Check rubber bushing for cracks, wear and deterioration. Replace if necessary.

To install:
6. To install, reverse removal procedure and note the following:
   a. Tighten stabilizer bar bracket mounting bolts to 94 ft. lbs. (128 Nm).
   b. Tighten the connecting rod nuts to 62 ft. lbs. (84 Nm).
7. Be sure to perform the reconnect/relearn procedures.

UPPER BALL JOINT

REMOVAL & INSTALLATION
See Figure 118.

The upper ball joints are integral to the upper control arm. They are also referred to as an upper link.

UPPER CONTROL ARM

REMOVAL & INSTALLATION
See Figure 118.

1. Before servicing the vehicle, refer to the Precautions Section.

➡If working near and/or around the SRS system and components, be sure to disable the SRS system. After disabling the system wait three minutes or more before servicing the vehicle.

➡Whenever the negative battery cable is disconnected the following components will require resetting. The Idle Air Volume Learning, Steering Angle Sensor Neutral Position, Sunroof Memory Reset/Initialization, Automatic Drive Positioner System, Audio presets and Navigation. Use the CONSULT-III diagnostic tool, or equivalent to perform the required resets.

2. Disconnect the negative battery cable.
3. Remove or disconnect the following:
   - Wheel and tire assembly
   - Fender protector to access the upper link
   - Cotter pin and nut from upper ball joint
4. Separate upper ball joint stud from steering knuckle using special tool J-24319-01.
5. Remove the upper control arm mounting bolts.
6. Remove the upper arm.

To install:
7. Install or connect the following:
   - Upper control arm and tighten bolts to 107 ft. lbs. (145 Nm)
   - Upper ball joint with new cotter pin and tighten nut to 58 ft. lbs. (79 Nm)
   - Fender protector
   - Wheel and tire assembly
8. Check front end alignment, adjust as required.
9. Be sure to perform the reconnect/relearn procedures.

WHEEL HUB & BEARING

REMOVAL & INSTALLATION
See Figure 122.

1. Before servicing the vehicle, refer to the Precautions Section.

➡If working near and/or around the SRS system and components, be sure to disable the SRS system. After dis-

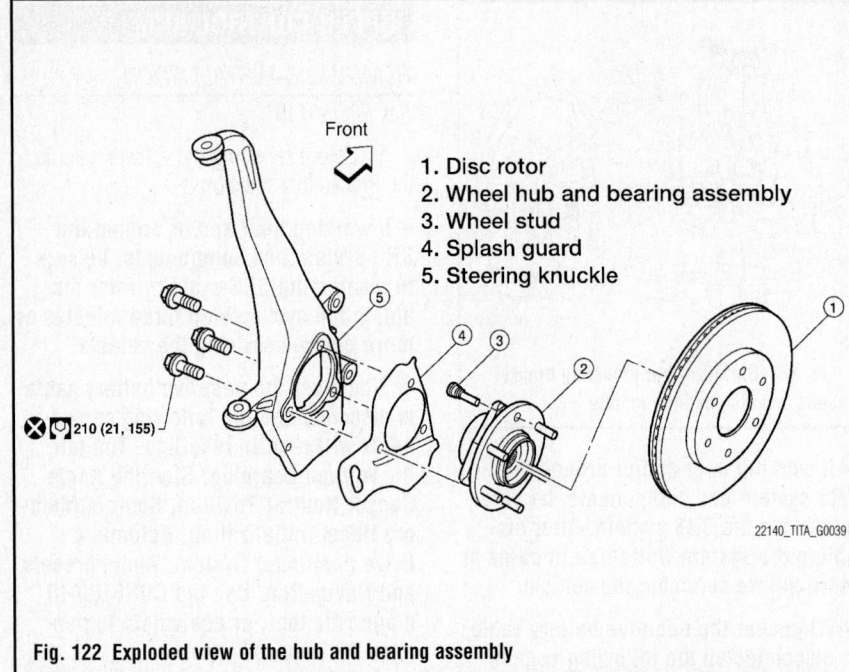

1. Disc rotor
2. Wheel hub and bearing assembly
3. Wheel stud
4. Splash guard
5. Steering knuckle

210 (21, 155)

22140_TITA_G0039

Fig. 122 Exploded view of the hub and bearing assembly

abling the system wait three minutes or more before servicing the vehicle.

➡Whenever the negative battery cable is disconnected the following components will require resetting. The Idle Air Volume Learning, Steering Angle Sensor Neutral Position, Sunroof Memory Reset/Initialization, Automatic Drive Positioner System, Audio presets and Navigation. Use the CONSULT-III diagnostic tool, or equivalent to perform the required resets.

2. Disconnect the negative battery cable.
3. Remove or disconnect the following:
   • Wheel and tire assembly
   • Engine undercover
   • Brake caliper without disconnecting

the hydraulic lines, and reposition aside with wire

4. Install a matchmark on the brake rotor and to the wheel hub, and remove the brake rotor.
5. Remove or disconnect the following:
   • Cotter pin and lock nut from drive shaft (4WD)
   • Driveshaft from wheel hub and bearing assembly (4WD)
   • ABS sensor
   • Wheel hub and bearing assembly bolts
   • Wheel hub and bearing assembly

To install:

➡Be sure to use new fasteners, as required.

6. Install or connect the following:
   • Wheel hub and bearing assembly, using new bolts and tightening to 155 ft. lbs. (210 Nm)
   • ABS sensor
   • Driveshaft to wheel hub and bearing assembly
   • Cotter pin and lock nut and tighten to 101 ft. lbs. (137 Nm)
   • Brake rotor
   • Brake caliper
   • Engine splash guard
   • Wheel
7. Be sure to perform the reconnect/relearn procedures.

SUSPENSION

REAR SUSPENSION

LEAF SPRING

REMOVAL & INSTALLATION

See Figure 123.

1. Before servicing the vehicle, refer to the Precautions Section.

➡If working near and/or around the SRS system and components, be sure to disable the SRS system. After disabling the system wait three minutes or more before servicing the vehicle.

➡Whenever the negative battery cable is disconnected the following components will require resetting. The Idle Air Volume Learning, Steering Angle Sensor Neutral Position, Sunroof Memory Reset/Initialization, Automatic Drive Positioner System, Audio presets and Navigation. Use the CONSULT-III diagnostic tool, or equivalent to perform the required resets.

2. Disconnect the negative battery cable.
3. Raise and support the vehicle safely.
4. Remove the tire and wheel assembly.
5. Support the rear differential with a suitable jack to relieve the tension from the leaf spring.
6. Remove or disconnect the following:
   • Shock absorber lower mounting bolt
   • Spring clip U-bolt nuts
   • Spring pad
   • Storage box, if equipped
   • Rear shackle lower bolt
   • Leaf spring front mounting bolt
   • Leaf spring

To install:

➡Be sure to use new fasteners, as required.

7. Install or connect the following:
   • Front mounting bolt and shackle lower bolt and finger tighten the nuts
   • U-bolts, rear spring pad and nuts or the U-bolts
8. Tighten the U-bolt nuts diagonally and evenly to specification.
9. Install the shock absorber and finger tighten the nuts.
10. Remove the jack supporting the rear differential and bounce the rear of the vehicle to stabilize the suspension.
11. Tighten the front mount bolt to specification.
12. Tighten the rear shackle lower bolt to specification.
13. Tighten the shock absorber lower mounting bolt to specification.
14. Be sure to perform the reconnect/relearn procedures.

SHOCK ABSORBER

REMOVAL & INSTALLATION

See Figures 123 through 124.

1. Before servicing the vehicle, refer to the Precautions Section.

➡If working near and/or around the SRS system and components, be sure to disable the SRS system. After disabling the system wait three minutes or more before servicing the vehicle.

➡Whenever the negative battery cable is disconnected the following components will require resetting. The Idle Air Volume Learning, Steering Angle Sensor Neutral Position, Sunroof Memory Reset/Initialization, Automatic Drive Positioner System,

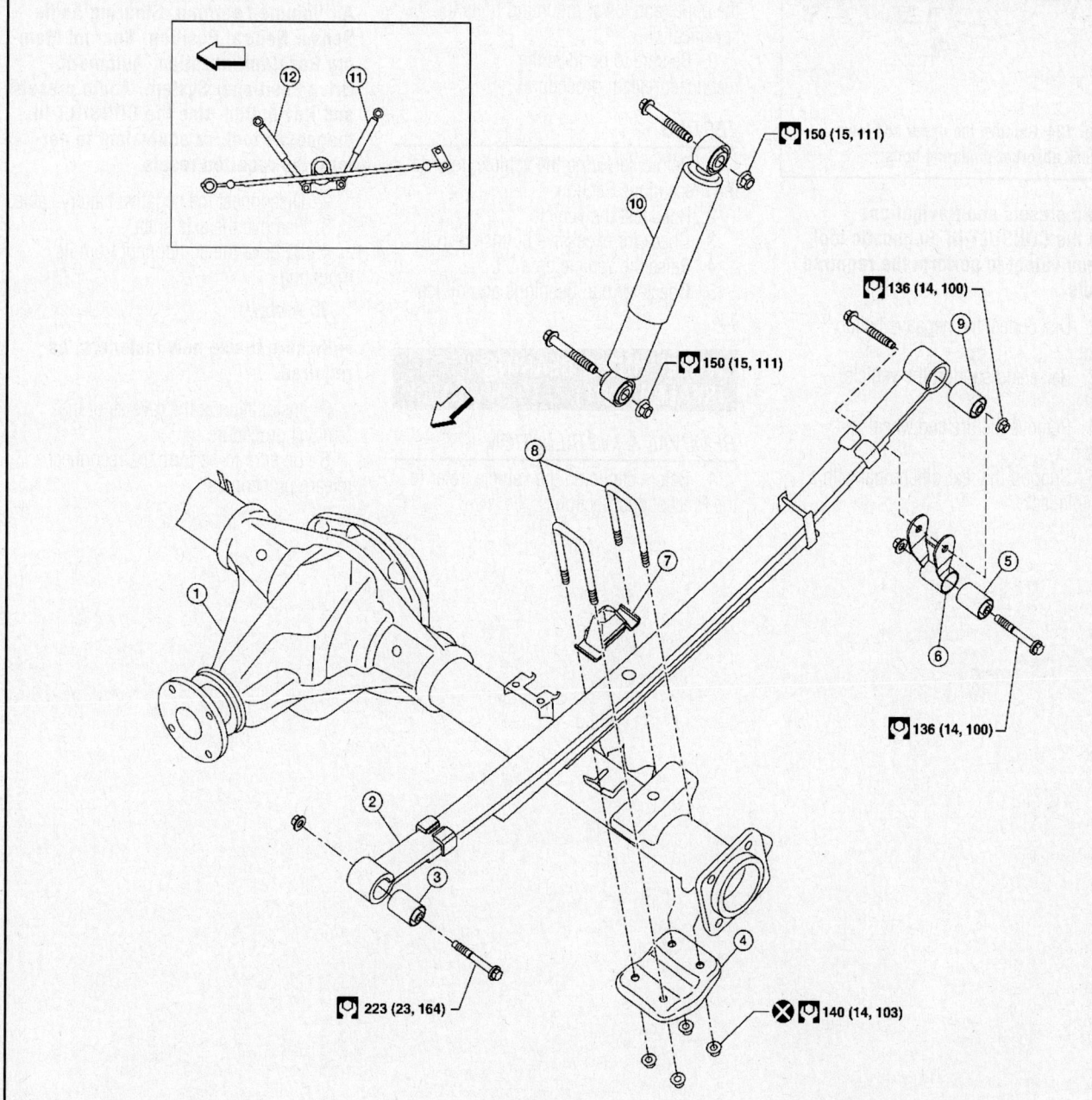

1.	Rear final drive	2.	Rear leaf spring	3.	Rear spring bushing (front)
4.	Rear spring pad	5.	Rear spring shackle bushing	6.	Rear spring shackle
7.	Bumper	8.	Rear spring clip U-bolts	9.	Rear spring bushing (rear)
10.	Shock absorber	11.	Shock absorber (left side)	12.	Shock absorber (right side)
⬅	Front				

37663_TITA_G0095

Fig. 123 Rear suspension components

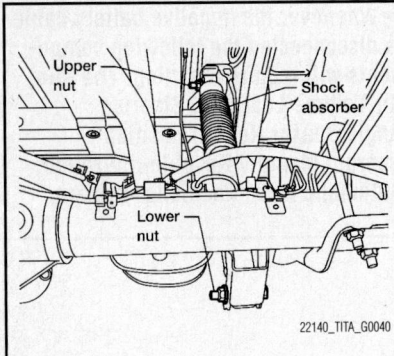

Fig. 124 Remove the upper and lower shock absorber mounting bolts

Audio presets and Navigation. Use the CONSULT-III diagnostic tool, or equivalent to perform the required resets.

2. Disconnect the negative battery cable.

3. Raise and support the vehicle safely.

4. Remove the tire and wheel assembly.

5. Support the rear differential with a suitable jack.

6. Remove the upper and lower shock absorber mounting bolts.

7. Remove the shock absorber.

To install:

➡ Be sure to use new fasteners, as required.

8. Install the shock absorber and tighten the upper and lower mounting bolts to specification.

9. Be sure to perform the reconnect/relearn procedures.

TESTING

1. Before servicing the vehicle, refer to the Precautions Section.

2. Road test the vehicle.

3. Check for excessive bounce or roll.

4. Raise the vehicle on a lift.

5. Check for bad bushings and oil leakage.

WHEEL HUB & BEARING (SEALED UNIT)

REMOVAL & INSTALLATION

1. Before servicing the vehicle, refer to the Precautions Section.

➡ If working near and/or around the SRS system and components, be sure to disable the SRS system. After disabling the system wait three minutes or more before servicing the vehicle.

➡ Whenever the negative battery cable is disconnected the following components will require resetting. The Idle Air Volume Learning, Steering Angle Sensor Neutral Position, Sunroof Memory Reset/Initialization, Automatic Drive Positioner System, Audio presets and Navigation. Use the CONSULT-III diagnostic tool, or equivalent to perform the required resets.

2. Disconnect the negative battery cable.

3. Remove the axle shaft.

4. Remove the component from its mounting.

To install:

➡ Be sure to use new fasteners, as required.

5. Installation is the reverse of the removal procedure.

6. Be sure to perform the reconnect/relearn procedures.

SPECIFICATIONS AND MAINTENANCE CHARTS

ENGINE AND VEHICLE IDENTIFICATION

			Engine				Model Year	
Code ①	Liters (cc)	Cu. In.	Cyl.	Fuel Sys.	Engine Type	Eng. Mfg.	Code ②	Year
HR16DE	1.6 (1598)	97.51	4	MPFI	DOHC	Nissan	9	2009
MR18DE	1.8 (1797)	109.65	4	SFI D4S	DOHC	Nissan	10	2010

SFI: Sequential Fuel Injection

DOHC: Double Overhead Camshaft

NA: Information not available

① Stamped on the left side of the engine block

② 10th digit of the Vehicle Identification Number (VIN)

37663_VERS_C0001

GENERAL ENGINE SPECIFICATIONS

Year	Model	Engine Displacement Liters	Engine Series ID	Net Horsepower @ rpm	Net Torque @ rpm (ft. lbs.)	Bore x Stroke (in.)	Com- pression Ratio	Oil Pressure @ rpm
2009	Versa	1.6	HR16DE	110@6000	111@4600	3.07x3.29	10.7:1	39 plus@2000
	Versa	1.8	MR18DE	122@5200	127@4800	3.31x3.19	9.9:1	29 plus@2000
2010	Versa	1.6	HR16DE	110@6000	111@4600	3.07x3.29	10.7:1	39 plus@2000
	Versa	1.8	MR18DE	122@5200	127@4800	3.31x3.19	9.9:1	29 plus@2000

37663_VERS_C0002

ENGINE TUNE-UP SPECIFICATIONS

Year	Engine Displacement Liters	Engine ID	Spark Plug Gap (in.)	Ignition Timing (deg.)*	Fuel Pump (psi)	Idle Speed (rpm)	Valve Clearance (Cold) Intake	Valve Clearance (Cold) Exhaust
2009	1.6	HR16DE	0.043	1-11B	N/A	600-800	0.010-0.013	0.011-0.015
	1.8	MR18DE	0.039-0.043	8-18B	N/A	650-750	0.010-0.013	0.011-0.015
2010	1.6	HR16DE	0.043	1-11B	N/A	600-800	0.010-0.013	0.011-0.015
	1.8	MR18DE	0.039-0.043	8-18B	N/A	650-750	0.010-0.013	0.011-0.015

NOTE: The Vehicle Emission Control Information label often reflects specification changes made during production.

The label figures must be used if they differ from those in this chart.

37663_VERS_C0003

CAPACITIES

Year	Model	Engine Displacement Liters	Engine ID	Engine Oil with Filter (qts.)	Transmission (pts.)			Drive Axle		Fuel Tank (gal.)	Cooling System (qts.)
					Manual	Auto.	CVT	Front (pts.)	Rear (pts.)		
2009	Versa	1.6	HR16DE	3.2	①	②	17.5	NA	NA	13.7	6.6
	Versa	1.8	MR18DE	4.1	4.25	16.75	17.5	NA	NA	13.7	7.25
2010	Versa	1.6	HR16DE	3.2	①	②	17.5	NA	NA	13.7	6.6
	Versa	1.8	MR18DE	4.1	4.25	16.75	17.5	NA	NA	13.7	7.25

NA: Not Applicable

① 5 speed: 5.5

  6 Speed: 4.25

② 1.6L: 16.25

  1.8L: 16.75

FLUID SPECIFICATIONS

Year	Model	Engine Displ. Liters	Engine Oil	5 spd Trans.	6 spd Trans.	Auto. Trans.	CVT Trans.	Trans. Case	Power Steering Fluid	Brake Master Cylinder	Cooling System
2009	Versa	1.6L/1.8L	5W-30	①	②	③	④	NA	Dextrol VI	DOT 3	⑤
2010	Versa	1.6L/1.8L	5W-30	①	②	③	④	NA	Dextrol VI	DOT 3	⑤

NA: Not Applicable

DOT: Department Of Transpotation

① GL-4 75W-85

② XT 4447M+ 75W-80

③ Nissan Matic D ATF

④ Nissan CVT NS-2

⑤ Nissan Long Life Antifreeze

VALVE SPECIFICATIONS

Year	Engine Displacement Liters	Engine ID	Seat Angle (deg.)	Face Angle (deg.)	Spring Test Pressure (lbs. @ in.)	Spring Free-Length (in.)	Stem-to-Guide Clearance (in.)		Stem Diameter (in.)	
							Intake	Exhaust	Intake	Exhaust
2009	1.6	HR16DE	45	45	59-67@ 0.9433	1.6638	0.0008- 0.0021	0.0012- 0.0025	0.1955 0.1961	0.1951 0.1957
	1.8	MR18DE	45	45	①	②	0.0008- 0.0021	0.0012- 0.0025	0.2152 0.2157	0.2148 0.2154
2010	1.6	HR16DE	45	45	59-67@ 0.9433	1.6638	0.0008- 0.0021	0.0012- 0.0025	0.1955 0.1961	0.1951 0.1957
	1.8	MR18DE	45	45	①	②	0.0008- 0.0021	0.0012- 0.0025	0.2152 0.2157	0.2148 0.2154

① Intake: 75-85@1.0377

  Exhaust: 60-67@1.0944

② Intake: 1.7677-1.7755

  Exhaust: 1.8007-1.8086

CAMSHAFT SPECIFICATIONS

All measurements are given in inches.

Year	Engine Displ. Liters	Engine VIN	Journal Dia.	Brg. Oil Clearance	Shaft End-play	Runout	Lobe Height Intake	Lobe Height Exhaust
2009	1.6	HR16DE	①	②	0.0030-0.0060	③	1.6419-1.6494	1.5817-1.5892
	1.8	MR18DE	①	②	0.0030-0.0060	③	1.7560-1.7635	1.6997-1.7072
2010	1.6	HR16DE	①	②	0.0030-0.0060	③	1.6419-1.6494	1.5817-1.5892
	1.8	MR18DE	①	②	0.0030-0.0060	③	1.7560-1.7635	1.6997-1.7072

① No.1: 1.0998-1.1006

　No.2, No.3, No.4, No.5: 0.9823-0.9831

② No.1: 0.0018-0.0034

　No.2, No.3, No.4, No.5: 0.0012-0.0028

③ Less then 0.0008 (0.02 mm)

37663_VERS_C0007

CRANKSHAFT AND CONNECTING ROD SPECIFICATIONS

All measurements are given in inches.

Year	Engine Displacement Liters	Engine ID	Crankshaft Main Brg. Journal Dia.	Crankshaft Main Brg. Oil Clearance	Crankshaft Shaft End-play	Crankshaft Thrust on No.	Connecting Rod Journal Diameter	Connecting Rod Oil Clearance	Connecting Rod Side Clearance
2009	1.6	HR16DE	1.8881-1.8889*	0.0009-0.0013	0.0039-0.0102	3	1.5729-1.5736*	0.0011-0.0015	0.0079-0.0139
	1.8	MR18DE	2.0456-2.0464*	①	0.0039-0.0102	3	1.7304-1.7311*	0.0015-0.0019	0.0079-0.0138
2010	1.6	HR16DE	1.8881-1.8889*	0.0009-0.0013	0.0039-0.0102	3	1.5729-1.5736*	0.0011-0.0015	0.0079-0.0139
	1.8	MR18DE	2.0456-2.0464*	①	0.0039-0.0102	3	1.7304-1.7311*	0.0015-0.0019	0.0079-0.0138

* Based upon grade

① Journal No. 1, 4 and 5: 0.0009 - 0.0013 inch

　Remaining journals: 0.0005 - 0.0009 inch

37663_VERS_C0008

PISTON AND RING SPECIFICATIONS
All measurements are given in inches.

Year	Engine Displ. Liters	Engine ID	Piston Clearance	Ring Gap			Ring Side Clearance		
				Top Comp.	Bottom Comp.	Oil Control	Top Comp.	Bottom Comp.	Oil Control
2009	1.6	HR16DE	0.0008-0.0020	0.0079-0.0118	0.0138-0.0197	0.0079-0.0236	0.0016-0.0031	0.0012-0.0028	0.0018-0.0049
	1.8	MR18DE	0.0008-0.0016	0.008-0.012	0.020-0.026	0.006-0.018	0.002-0.003	0.001-0.003	0.001-0.007
2010	1.6	HR16DE	0.0008-0.0020	0.0079-0.0118	0.0138-0.0197	0.0079-0.0236	0.0016-0.0031	0.0012-0.0028	0.0018-0.0049
	1.8	MR18DE	0.0008-0.0016	0.008-0.012	0.020-0.026	0.006-0.018	0.002-0.003	0.001-0.003	0.001-0.007

37663_VERS_C0009

TORQUE SPECIFICATIONS
All readings in ft. lbs.

Year	Engine Displacement Liters	Engine ID	Cylinder Head Bolts	Main Bearing Bolts	Rod Bearing Bolts	Crankshaft Damper Bolts	Flywheel Bolts	Manifold		Spark Plugs	Oil Pan Drain Plug
								Intake	Exh		
2009	1.6	HR16DE	①	②	③	④	80	20	25	13	25
	1.8	MR18DE	①	⑤	③	⑥	80	20	25	13	25
2010	1.6	HR16DE	①	②	③	④	80	20	25	13	25
	1.8	MR18DE	①	⑤	③	⑥	80	20	25	13	25

① Step 1: 30 ft. lbs.
Step 2: Tighten an additional 100 degrees
Step 3: Loosen to 0 ft. lbs.
Step 4: 30 ft. lbs.
Step 5: Tighten an additional 100 degrees
Step 6: Tighten an additional 100 degrees

② Step 1: 24 ft. lbs.
Step 2: Plus 60 degrees

③ Step 1: 14 ft. lbs.
Step 2: Tighten an additional 60 degrees

④ Step 1: 26 ft. lbs.
Step 2: Plus 60 degrees

⑤ Step 1: 25 ft. lbs.
Step 2: Plus 60 degrees

⑥ Step 1: 22 ft. lbs.
Step 2: Plus 60 degrees

37663_VERS_C0010

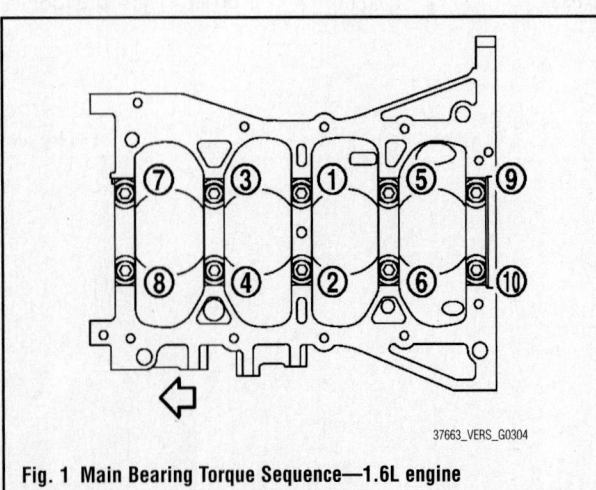

37663_VERS_G0304

Fig. 1 Main Bearing Torque Sequence—1.6L engine

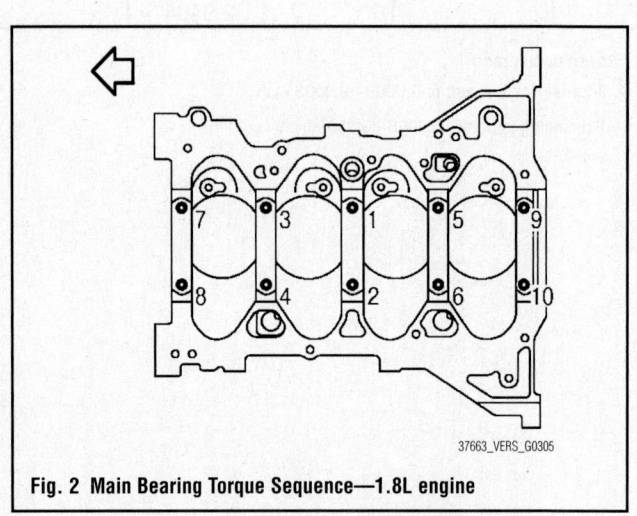

37663_VERS_G0305

Fig. 2 Main Bearing Torque Sequence—1.8L engine

WHEEL ALIGNMENT

Year	Model		Caster Range (+/-Deg.)	Caster Preferred Setting (Deg.)	Camber Range (+/-Deg.)	Camber Preferred Setting (Deg.)	Toe-in (in.)	Steering Axis Inclination (Deg.)
2009	Versa	RF	0.75	+4.83	0.75	-0.17	0.04+/-0.04	NA
		LF	0.75	+4.67	0.75	-0.17		
		Rear	NA	NA	0.50	-1.52	0.08+/-0.18	
2010	Versa	RF	0.75	+4.83	0.75	-0.17	0.04+/-0.04	NA
		LF	0.75	+4.67	0.75	-0.17		
		Rear	NA	NA	0.50	-1.52	0.08+/-0.18	

NA: Not Applicable

37663_VERS_C0011

TIRE, WHEEL AND BALL JOINT SPECIFICATIONS

Year	Model	OEM Tires Standard	OEM Tires Optional	Tire Pressures (psi) Front	Tire Pressures (psi) Rear	Wheel Size	Ball Joint Inspection	Lug Nut Torque (ft. lbs.)
2009	Versa	P185/65R14	P185/65R15	33	33	5.5	①	83
	Versa	P195/55R16	N/A	33	33	5.5	①	83
2010	Versa	P185/65R14	P185/65R15	33	33	5.5	①	83
	Versa	P195/55R16	N/A	33	33	5.5	①	83

OEM: Original Equipment Manufacturer

PSI: Pounds Per Square Inch

① Replace if any measurable movement is found.

37663_VERS_C0012

BRAKE SPECIFICATIONS

All measurements in inches unless noted

Year	Model		Brake Disc Original Thickness	Brake Disc Minimum Thickness	Brake Disc Maximum Runout	Minimum Pad Thickness	Drum Brake Inner Diameter	Drum Brake Lining Thickness	Brake Caliper Bracket Bolts (ft. lbs.)	Brake Caliper Mounting Bolts (ft. lbs.)
2009	Versa	F	0.374	0.079	0.0016	0.866	NA	NA	62	20
		R	NA	NA	NA	NA	①	0.059	NA	NA
2010	Versa	F	0.374	0.079	0.0016	0.866	NA	NA	62	20
		R	NA	NA	NA	NA	①	0.059	NA	NA

NA: Not Applicable

F: Front

R: Rear

① LT20 System: 8.051 in.

　 LT23 System: 9.055 in.

37663_VERS_C0013

SCHEDULED MAINTENANCE INTERVALS (1)
Nissan—Versa

TO BE SERVICED	TYPE OF SERVICE	7.5	15	22.5	30	37.5	45	52.5	60
Engine oil & filter	R	Every 3,750 miles							
Brake lines & cables	S/I		✓		✓		✓		✓
Brake pads, drums, discs	I	✓	✓	✓	✓	✓	✓	✓	✓
Driveshaft boots	I	✓	✓	✓	✓	✓	✓	✓	✓
Automatic & CVT transmission * ①	I		✓		✓		✓		✓
Manual transmission	I		✓		✓		✓		✓
Air cleaner filter	R				✓				✓
Drive belt (s) ②	S/I								✓
Engine coolant ③	R								✓
Spark plugs	R	Platinum plugs, every 105,000 miles							
Cabin air filter	R		✓		✓		✓		✓
Exhaust system	I	✓	✓	✓	✓	✓	✓	✓	✓
Evap vapor lines	I				✓				✓
Fuel lines	S/I				✓				✓
Fuel Filter	R	Maintenance free item							
Steering gear, linkage, axle & suspension parts	I	✓	✓	✓	✓	✓	✓	✓	✓
Tires (rotate)	S/I	✓	✓	✓	✓	✓	✓	✓	✓
Valve clearance ④	S/I				✓				✓

R: Replace   S/I: Service or Inspect   L: Lubricate   I: Inspect

* Using transmission fluid other than Genuine NISSAN CVT Fluid NS-2 will damage the CVT

① If towing a trailer, using a camper or a car-top carrier, or driving on rough or muddy roads, change (not just inspect) oil at every 30,000 miles.

② First at 60,000, inspect every 15,000 miles and replace as necessary

③ After 60,000, replace every 30,000

④ Periodic maintenance not required, if valve noise increases, inspect valve clearance

Follow Periodic Maintenance Schedule 1 if the driving habits frequently include one or more of the following driving conditions:

Repeated short trips of less than 5 miles (8 km).

Repeated short trips of less than 10 miles (16 km) with outside temperatures remaining below freezing

Operating in hot weather in stop-and-go "rush hour" traffic.

Extensive idling and/or low speed driving for long distances, such as police, taxi or door-to-door delivery use

Driving in dusty conditions.

Driving on rough, muddy, or salt spread roads.

Towing a trailer, using a camper or a car-top carrier.

Follow Periodic Maintenance Schedule 2 if none of driving conditions shown in Schedule 1 apply to the driving habits.

37663_VERS_C0014

SCHEDULED MAINTENANCE INTERVALS (2)
Nissan—Versa

TO BE SERVICED	SERVICE	7.5	15	22.5	30	37.5	45	52.5	60
Engine oil & filter	R	✓	✓	✓	✓	✓	✓	✓	✓
Brake lines & cables	S/I		✓		✓		✓		✓
Brake pads, drums, discs	I		✓		✓		✓		✓
Driveshaft boots	I		✓		✓		✓		✓
Automatic & CVT transmission *	I		✓		✓		✓		✓
Manual transmission	I		✓		✓		✓		✓
Air cleaner filter	R				✓				✓
Drive belt (s) ①	S/I								✓
Engine coolant ②	R								✓
Spark plugs	R	Platinum plugs, every 105,000 miles							
Cabin air filter	R		✓		✓		✓		✓
Exhaust system	I				✓				✓
Evap vapor lines	I				✓				✓
Fuel lines	S/I				✓				✓
Fuel Filter	R	Maintenance free item							
Steering gear, linkage, axle & suspension parts	I				✓				✓
Tires (rotate)	S/I	✓	✓	✓	✓	✓	✓	✓	✓
Valve clearance ③	S/I				✓				✓

R: Replace   S/I: Service or Inspect   L: Lubricate   I: Inspect

* Using transmission fluid other than Genuine NISSAN CVT Fluid NS-2 will damage the CVT

① First at 60,000, then every 15,000 miles

② After 60,000, replace every 30,000

③ Periodic maintenance not required, if valve noice increases, inspect valve clearance

Follow Periodic Maintenance Schedule 1 if the driving habits frequently include one or more of the following driving conditions:

Repeated short trips of less than 5 miles (8 km).

Repeated short trips of less than 10 miles (16 km) with outside temperatures remaining below freezing

Operating in hot weather in stop-and-go "rush hour" traffic.

Extensive idling and/or low speed driving for long distances, such as police, taxi or door-to-door delivery use

Driving in dusty conditions.

Driving on rough, muddy, or salt spread roads.

Towing a trailer, using a camper or a car-top carrier.

Follow Periodic Maintenance Schedule 2 if none of driving conditions shown above apply to the driving habits.

37663_VERS_C0015

BRAKES   INFORMATION AND PRECAUTIONS

ANTI-LOCK SYSTEMS

- Certain components within the ABS system are not intended to be serviced or repaired individually.

- Do not use rubber hoses or other parts not specifically specified for and ABS system. When using repair kits, replace all parts included in the kit. Partial or incorrect repair may lead to functional problems and require the replacement of components.

- Lubricate rubber parts with clean, fresh brake fluid to ease assembly. Do not use shop air to clean parts; damage to rubber components may result.

- Use only DOT 3 brake fluid from an unopened container.

- If any hydraulic component or line is removed or replaced, it may be necessary to bleed the entire system.

- A clean repair area is essential. Always clean the reservoir and cap thoroughly before removing the cap. The slightest amount of dirt in the fluid may plug an orifice and impair the system function. Perform repairs after components have been thoroughly cleaned; use only denatured alcohol to clean components. Do not allow ABS components to come into contact with any substance containing mineral oil; this includes used shop rags.

- The Anti-Lock control unit is a microprocessor similar to other computer units in the vehicle. Ensure that the ignition switch is **OFF** before removing or installing controller harnesses. Avoid static electricity discharge at or near the controller.

- If any arc welding is to be done on the vehicle, the control unit should be unplugged before welding operations begin.

DISC AND DRUM SYSTEMS

✳✳ CAUTION

Dust and dirt accumulating on brake parts during normal use may contain asbestos fibers from production or aftermarket brake linings. Breathing excessive concentrations of asbestos fibers can cause serious bodily harm. Exercise care when servicing brake parts. Do not sand or grind brake lining unless equipment used is designed to contain the dust residue. Do not clean brake parts with compressed air or by dry brushing. Cleaning should be done by dampening the brake components with a fine mist of water, then wiping the brake components clean with a dampened cloth. Dispose of cloth and all residue containing asbestos fibers in an impermeable container with the appropriate label. Follow practices prescribed by the Occupational Safety and Health Administration (OSHA) and the Environmental Protection Agency (EPA) for the handling, processing, and disposing of dust or debris that may contain asbestos fibers.

BRAKES   BLEEDING THE BRAKE SYSTEM

BLEEDING PROCEDURE

BLEEDING PROCEDURE

1. Before servicing the vehicle, refer to the Precautions Section.

➡**Before working, disconnect connectors of ABS actuator and electric unit (control unit) or the battery cable from the negative terminal.**

➡**While bleeding, pay attention to master cylinder fluid level.**

2. Connect a vinyl tube to the rear right bleed valve.
3. Fully depress brake pedal 4 to 5 times.
4. With brake pedal depressed, loosen bleed valve to let the air out, and then tighten it immediately.
5. Repeat the above 2 steps until no more air comes out.
6. Tighten bleed valve to 74 inch lbs. (8 Nm) for front disc brakes and 70 inch lbs. (8 Nm) for rear drum brakes.
7. Following the above 5 steps, with master cylinder reservoir tank filled at least half way, bleed air from the rear right, front left, rear left, and front right brake, in that order.

BLEEDING THE ABS SYSTEM

See Brake Bleeding procedure.

BRAKES   ANTI-LOCK BRAKE SYSTEM (ABS)

WHEEL SPEED SENSORS

REMOVAL & INSTALLATION

1. Before servicing the vehicle, refer to the Precautions Section.

✳✳ WARNING

As much as possible, avoid rotating wheel sensor when removing it. Pull wheel sensors out without pulling on sensor harness.

✳✳ WARNING

Take care to avoid damaging wheel sensor edges or rotor teeth.

Remove wheel sensor first before removing front or rear wheel hub. This is to avoid damage to wheel sensor wiring and loss of sensor function.

2. Remove the wheel and tire.
3. For front wheel speed sensors, remove the front fender protector.
4. Disconnect wheel sensor connector.
5. Remove wheel sensor.

   To install:
6. Installation is the reverse of removal, noting the following:

✳✳ WARNING

When installing, make sure there is no foreign material such as iron chips on and in the mounting hole of the wheel sensor. Make sure no foreign material has been caught in the sensor rotor. Remove any foreign material and clean the mount.

✳✳ WARNING

When installing front wheel sensor, press rubber grommets of strut bracket and body all the way in until they get locked, and be careful not to apply a twist to harness. Harness should not be twisted after installation. (Install it with harness paint mark on body side grommet facing front of vehicle, and the strut side grommet facing outside of vehicle.)

✳✳ **WARNING**

When installing rear wheel sensor, press rubber grommets of suspension arm bracket and harness of side member all the way in until they get locked, and be careful not to apply a twist to harness. Harness should not be twisted after installation. (Aim the paint mark upward of vehicle.)

BRAKES
FRONT DISC BRAKES

BRAKE CALIPER

REMOVAL & INSTALLATION

See Figures 3 and 4.

1. Before servicing the vehicle, refer to the Precautions Section.
2. Remove wheel and tire.
3. Secure disc rotor using wheel nuts. Put matching marks on wheel hub assembly and disc rotor, if it is necessary to remove disc rotor.
4. Drain brake fluid.
5. Remove the union bolt (A) and discard the copper washers.
6. Remove the brake hose.

➡**Do not reuse the copper washers.**

7. Remove torque member mounting bolts from torque member, and remove caliper assembly from vehicle.

To install:

➡**Before installing torque member to vehicle, wipe oil and grease on mounting surface of steering knuckle and torque member.**

8. Install caliper assembly to vehicle, and tighten mounting bolts to 62 ft. lbs (84 Nm).
9. Install brake hose to caliper assembly.
10. Refill with new brake fluid and bleed air.
11. Check front disc brake for drag.
12. Install tires to the vehicle.

DISC BRAKE PADS

REMOVAL & INSTALLATION

See Figures 4 and 5.

1. Before servicing the vehicle, refer to the Precautions Section.
2. Partially drain brake fluid reservoir.
3. Remove wheel and tire.
4. Remove sliding pin bolt (lower side).
5. Hang cylinder body with a wire, and remove pad return spring, pads, shims and pad retainers from torque member.

➡**When removing pad retainer from torque member, lift pad retainer, so as not to deform it.**

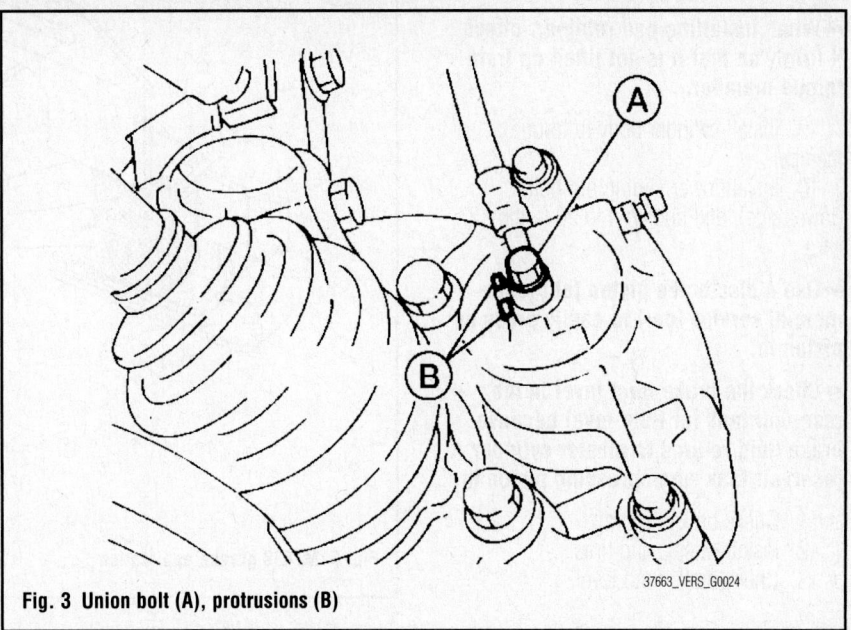

Fig. 3 Union bolt (A), protrusions (B)

37663_VERS_G0024

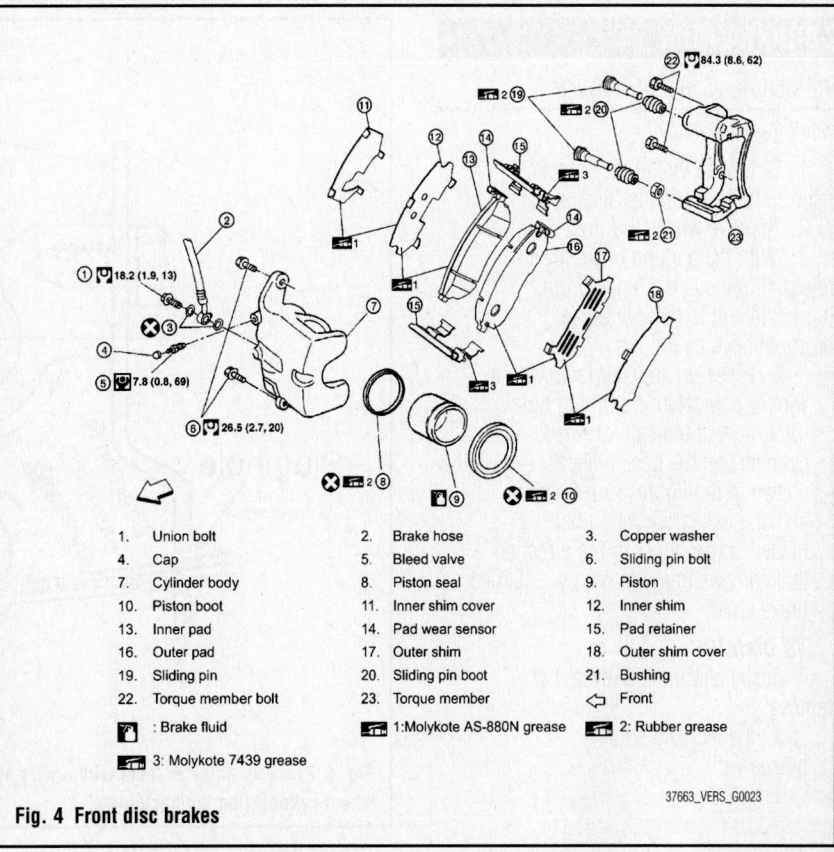

1.	Union bolt	2.	Brake hose	3.	Copper washer
4.	Cap	5.	Bleed valve	6.	Sliding pin bolt
7.	Cylinder body	8.	Piston seal	9.	Piston
10.	Piston boot	11.	Inner shim cover	12.	Inner shim
13.	Inner pad	14.	Pad wear sensor	15.	Pad retainer
16.	Outer pad	17.	Outer shim	18.	Outer shim cover
19.	Sliding pin	20.	Sliding pin boot	21.	Bushing
22.	Torque member bolt	23.	Torque member	⇦	Front

🛢 : Brake fluid ⊞ 1:Molykote AS-880N grease ⊞ 2: Rubber grease

⊞ 3: Molykote 7439 grease

Fig. 4 Front disc brakes

37663_VERS_G0023

To install:

6. Apply AS-880N grease or equivalent to the shims. Install shims to pads. Securely install shims according to mounting direction of pads.

7. Apply M7439 grease or equivalent to pad contact surface on pad retainers.

8. Install pad retainers, pad return spring, shims and pads to the torque member.

➡**When installing pad retainer, attach it firmly so that it is not lifted up from torque member.**

9. Install cylinder body to torque member.

10. Install lower sliding pin bolt (lower side), and tighten it to 20 ft. lbs. (27 Nm).

➡**Use a disc brake piston tool (commercial service tool) to easily press to piston in.**

➡**Check the brake fluid level in the reservoir tank for fluid level because brake fluid returns to master cylinder reservoir tank when pressing piston in.**

11. Check brake for drag.
12. Install wheels and tires.
13. Check brake fluid level.

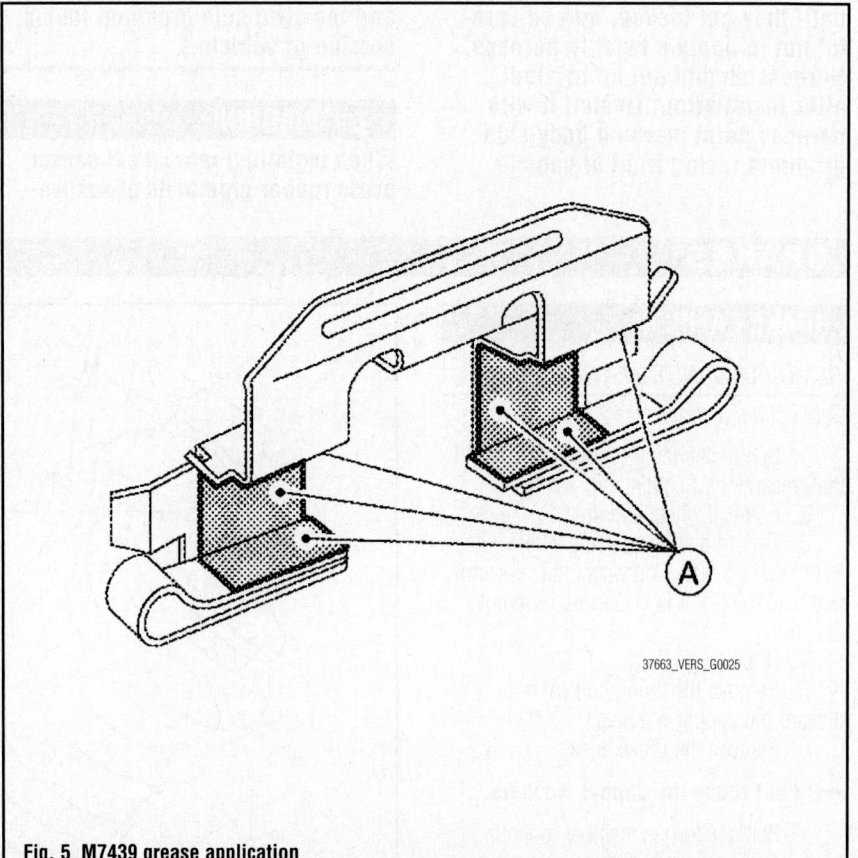

37663_VERS_G0025

Fig. 5 M7439 grease application

BRAKES

REAR DRUM BRAKES

BRAKE DRUM

REMOVAL & INSTALLATION

See Figures 6 and 7.

1. Before servicing the vehicle, refer to the Precautions Section.

2. Remove wheel and tire.

3. With the parking brake lever released, remove the brake drum. If it is difficult to remove brake drum, remove as follows:

a. Press up adjuster lever with a wire or equivalent from plug hole (plug hole at the side of wheel cylinder) on the back plate as shown in the figure. Turn frame of adjuster assembly with a flat bladed screw driver in the direction that narrows frame to narrow enlarged brake shoe.

To install:

4. Installation is the reverse of removal.

a. Inspect and adjust brakes.

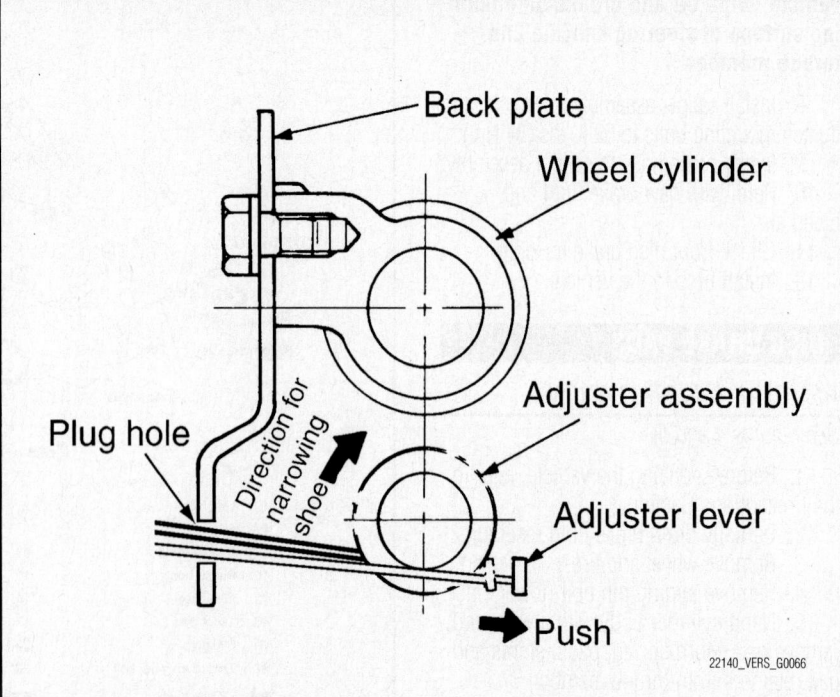

Back plate
Wheel cylinder
Plug hole
Direction for narrowing shoe
Adjuster assembly
Adjuster lever
➡ Push

22140_VERS_G0066

Fig. 6 Press up adjuster lever with a wire or equivalent from plug hole (plug hole at the side of wheel cylinder) on the back plate

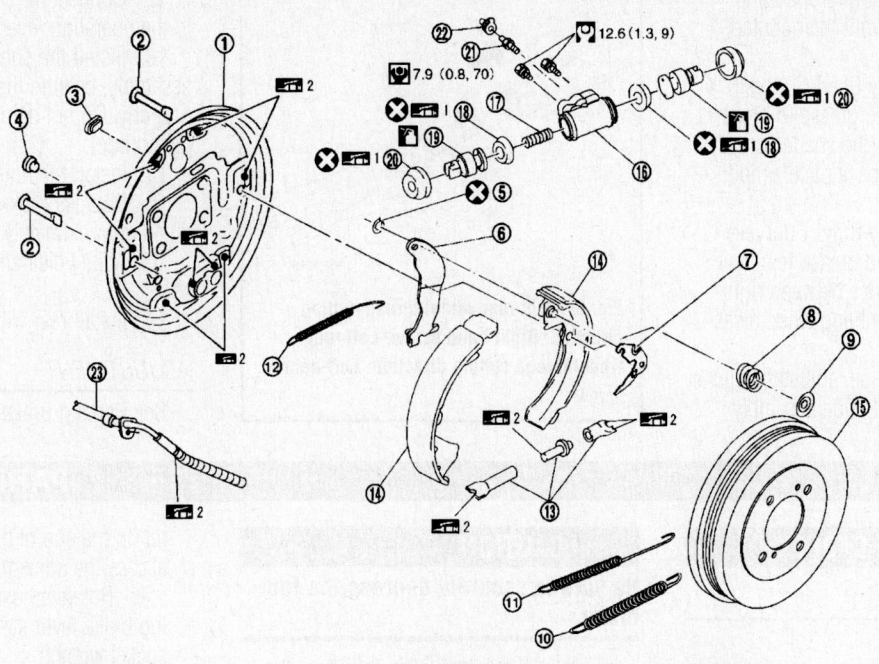

1. Back plate
2. Shoe hold pin
3. Plug
4. Plug
5. Retainer ring
6. Operating lever
7. Adjuster lever
8. Spring
9. Retainer
10. Return spring (lower side)
11. Return spring (upper side)
12. Adjuster spring
13. Adjuster
14. Brake shoe
15. Brake drum
16. Wheel cylinder
17. Spring
18. Piston seal
19. Piston
20. Boot
21. Bleed valve
22. Cap
23. Parking brake rear cable

1: Rubber grease
2: PBC (Poly Butyl Cuprysil) grease or silicone-based grease
: Brake fluid

37663_VERS_G0026

Fig. 7 Rear brake assembly

BRAKE SHOES

REMOVAL & INSTALLATION

See Figures 7 through 9.

1. Before servicing the vehicle, refer to the Precautions Section.

2. Remove brake drum. Refer to Brake Drum Removal & Installation.

3. While pushing and rotating the retainer, pull out shoe hold pin, and then remove shoe assembly. Do not damage the wheel cylinder boot.

4. Remove the parking brake rear cable from the operating lever. Do not bend the parking brake cable.

5. Disassemble the shoe assembly (shoe, springs, adjuster, adjuster lever).

6. Remove retainer ring with a tool to separate operating lever from brake shoe.

To install:

7. Install operating lever to brake shoe, if necessary.

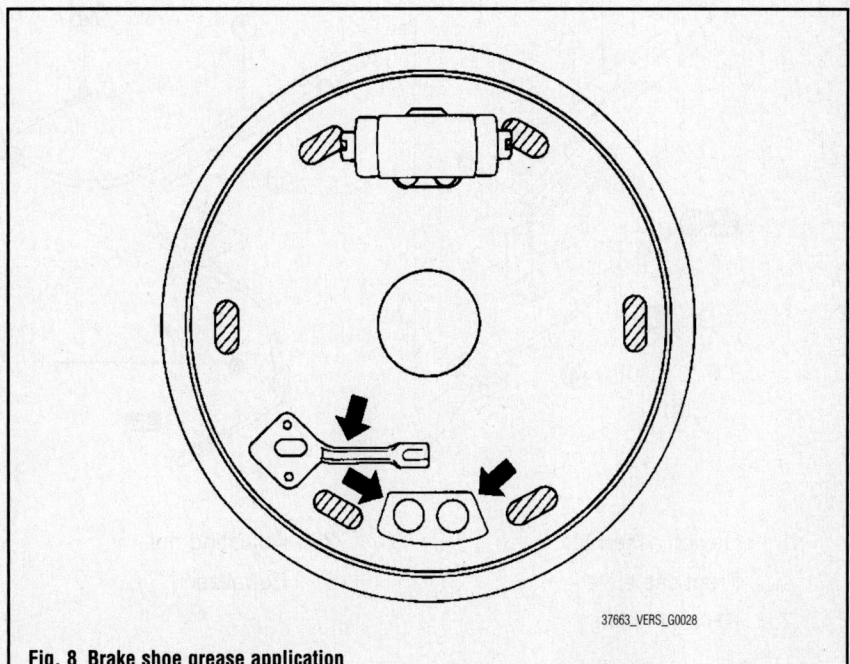

37663_VERS_G0028

Fig. 8 Brake shoe grease application

8. Install retainer ring to operating lever, and crimp them until their contact points are met.

9. Apply PBC (Poly Butyl Cuprysil) grease or silicone based grease to brake shoes sliding surfaces (the shaded areas) and other parts on the back plate as indicated by arrows.

10. Apply PBC (Poly Butyl Cuprysil) grease or silicone based grease to screw and confirm the difference between right and left wheel for assembling when disassembled.

11. Assemble the shoe, adjuster, adjuster lever and springs to the shoe assembly.

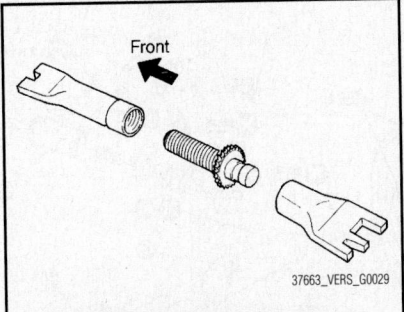

Front

37663_VERS_G0029

Fig. 9 Right rear wheel thread cutting direction: Right-hand screw, Left rear wheel thread cutting direction: Left-hand screw

12. Connect the parking brake rear cable to the operating lever.

13. Install the shoe assembly. After assembly, be sure that each part is installed properly. Do not damage the wheel cylinder piston boot.

14. Install the brake drum.

15. Depress brake pedal for several times (approximately 2 or 3 times).

16. Adjust clearance of brake shoe.

17. Install rear wheel and tire.

ADJUSTMENT

See Parking Brake Cable Adjustment.

BRAKES

PARKING BRAKE

PARKING BRAKE CABLES

ADJUSTMENT

See Figure 10.

1. Remove console mask cover.

2. Engage parking brake lever, then lift up the end of the trim on the lever to access the adjusting nut.

3. Insert a deep socket wrench onto adjusting nut. Rotate adjusting nut to fully loosen cable, and then release parking brake lever.

4. Depress the foot brake about 10 times and adjust the rear shoe clearance.

☀️ CAUTION

Be sure to securely depress the foot brake.

5. Rotate brake drum to make sure that there is no drag.

6. Adjust parking brake cable with the following procedure:

a. When replace parking brake cable, operate parking brake lever with a force of 110 ft. lbs. (490 Nm) about 10 times.

b. Engage parking brake lever, then

lift up the end of the trim on the lever to access the adjusting nut.

c. Rotate adjusting nut to adjust parking brake lever stroke using a deep socket wrench.

d. Operate parking brake lever with a force of 44 ft. lbs. (196 Nm), make sure the parking brake lever stroke is within 8 to 9 notches. (Check it by listening and counting ratchet clicks.)

e. Make sure that there is no drag on rear brake with parking brake lever completely released.

7. Install console mask.

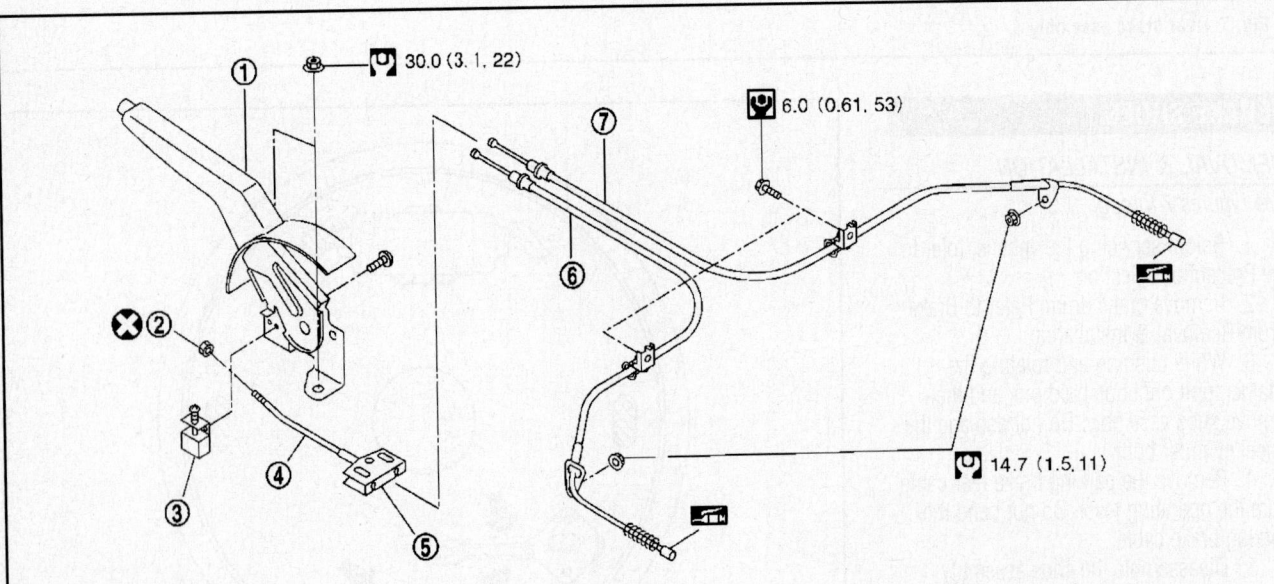

30.0 (3.1, 22)

6.0 (0.61, 53)

14.7 (1.5, 11)

1. Device assembly	2. Adjusting nut	3. Parking brake switch
4. Front cable	5. Equalizer	6. LH rear cable
7. RH rear cable		

37663_VERS_G0030

Fig. 10 Parking brake components

GENERAL INFORMATION

✳✳ CAUTION

These vehicles are equipped with an air bag system. The system must be disarmed before performing service on, or around, system components, the steering column, instrument panel components, wiring and sensors. Failure to follow the safety precautions and the disarming procedure could result in accidental air bag deployment, possible injury and unnecessary system repairs.

SERVICE PRECAUTIONS

✳✳ CAUTION

Disconnect and isolate the battery negative cable before beginning any airbag system component diagnosis, testing, removal, or installation procedures. Wait at least 90 seconds after the ignition switch is turned off and the negative (-) terminal cable is disconnected from the battery before starting the operation. The SRS is equipped with a backup power source, so if work is started within 90 seconds after disconnecting the negative (-) terminal cable from the battery, the SRS may be deployed. Failure to disable the airbag system may result in accidental airbag deployment, personal injury, or death.

DISARMING THE SYSTEM

1. Before servicing the vehicle, refer to the Precautions Section.
2. Disconnect the negative and positive battery cables, then wait at least 3 minutes. For approximately 3 minutes after the cables are removed, it is still possible for the air bag and seat belt pre-tensioner to deploy.

ARMING THE SYSTEM

1. Before servicing the vehicle, refer to the Precautions Section.
2. Connect the negative and positive battery cables.

CLOCKSPRING CENTERING

See Figure 11.

✳✳ CAUTION

When servicing the SRS, do not work from directly in front of air bag module.

✳✳ CAUTION

Before servicing the SRS, turn ignition switch OFF, disconnect both battery cables and wait at least 3 minutes.

✳✳ WARNING

Do not tap or bump the steering wheel.

✳✳ WARNING

Do not disassemble the spiral cable.

✳✳ WARNING

Do not allow oil, grease or water to come in contact with the spiral cable.

✳✳ WARNING

Replace the spiral cable if it has been dropped or sustained an impact.

1. Before servicing the vehicle, refer to the Precautions Section.
2. Make sure spiral cable alignment and centering marks are matched and in the neutral position. The neutral position is detected by turning left 2.5 revolutions from the right end position, ending with the locating pin at the top. Place steering wheel in straight ahead position, then install it with the locating pin hole directly over spiral cable locating pin.

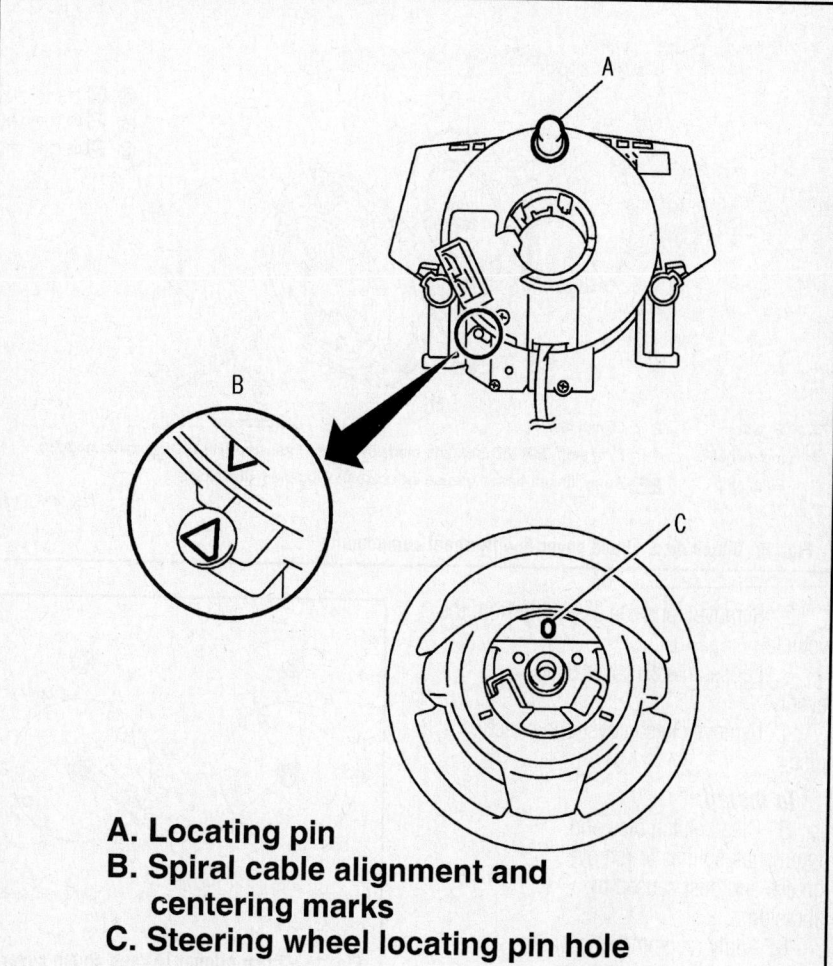

A. Locating pin
B. Spiral cable alignment and centering marks
C. Steering wheel locating pin hole

22140_VERS_G0060

Fig. 11 Spiral cable alignment

DRIVE TRAIN

CLUTCH DRIVEN DISC & PRESSURE PLATE

REMOVAL & INSTALLATION

See Figures 12 and 13.

1. Before servicing the vehicle, refer to the Precautions Section.

✳✳ WARNING

Return CSC insert to original position to remove transaxle assembly. Dust on clutch disc sliding parts may damage CSC seal and may cause clutch fluid leakage. DO NOT apply any grease to the clutch disc facing, pressure plate surface and flywheel surface. When installing, be careful that grease applied to input shaft does not adhere to clutch disc. Never clean clutch disc using solvent.

✳✳ WARNING

Be sure to apply grease to the points specified. Otherwise, noise, poor disengagement, or damage to the clutch may result. Excessive grease may cause slip or shudder. If it adheres to CSC seal, it will cause clutch fluid leakage. Wipe off excess grease.

7. Install clutch disc using suitable clutch aligner.
8. Install clutch cover. Pre-tighten clutch cover bolts.
9. Tighten clutch cover bolts evenly in two steps:
   - HR16DE engine models first step: 11 ft. lbs. (15 Nm)
   - MR18DE engine models first step: 14 ft. lbs. (20 Nm)
   - Final step: 19 ft. lbs. (25.5 Nm)
10. Install transaxle assembly.

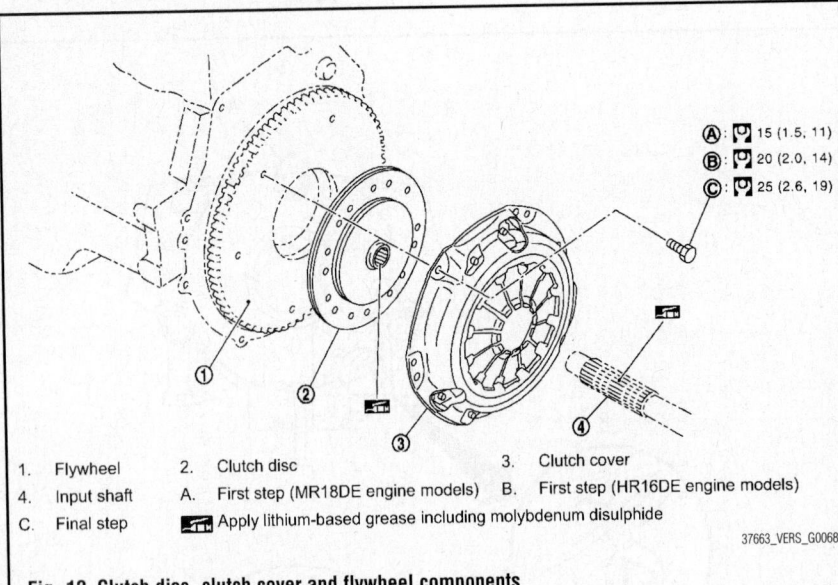

A: ⬚ 15 (1.5, 11)
B: ⬚ 20 (2.0, 14)
C: ⬚ 25 (2.6, 19)

1. Flywheel	2. Clutch disc	3. Clutch cover
4. Input shaft	A. First step (MR18DE engine models)	B. First step (HR16DE engine models)
C. Final step	Apply lithium-based grease including molybdenum disulphide	

37663_VERS_G0068

Fig. 12 Clutch disc, clutch cover and flywheel components

2. Remove transaxle assembly from the vehicle.
3. Loosen clutch cover bolts evenly.
4. Remove clutch cover and clutch disc.

To install:

5. Clean clutch disc and input shaft splines to remove grease and dust caused by abrasion.
6. Apply recommended grease to clutch disc and input shaft splines.

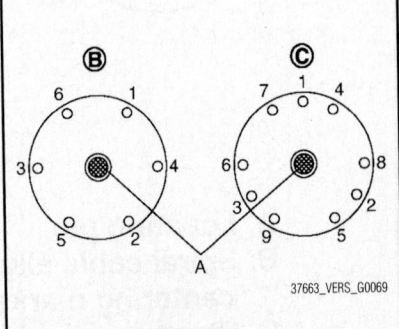

37663_VERS_G0069

Fig. 13 Clutch aligner (A) and clutch cover tightening sequence: 1.6L engines (B), 1.8L engines (C)

HYDRAULIC SYSTEM BLEEDING

BLEEDING PROCEDURE

See Figures 14 through 18.

1. Before servicing the vehicle, refer to the Precautions Section.

✳✳ WARNING

Do not use a vacuum assist or any other type of power bleeder on this system. Use of a vacuum assist or power bleeder will not purge all the air from the system. Carefully monitor fluid level in reservoir tank during bleeding operation. Keep painted surface of body and other parts free of clutch fluid. If it spills, wipe up immediately and wash the affected area with water.

2. Fill master cylinder reservoir tank with new clutch fluid.
3. Connect a transparent vinyl tube and container to the bleeding connector on the slave cylinder (CSC).
4. Depress and release the clutch pedal slowly and fully 15 times at an interval of 2 to 3 seconds and release the clutch pedal.
5. Push in the lock pin of the bleeding connector and maintain the position. Hold it to prevent the bleeding connector from separating when fluid pressure is applied.
6. For 5 speed transaxles, slide clutch tube in the direction of the arrow as shown to the dimension A [0.20 inch (5 mm)].
7. For 6 speed transaxles, slide bleeding connector in the direction of the arrow as shown to the dimension A [0.39 inch (10 mm)].
8. Depress the clutch pedal soon and

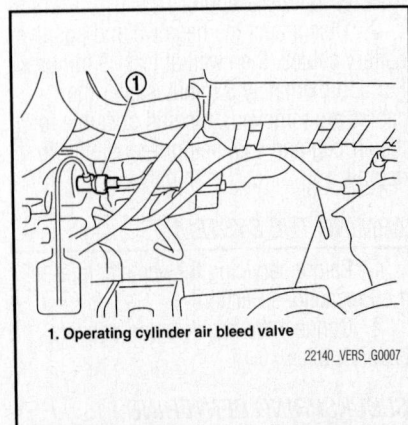

1. Operating cylinder air bleed valve

22140_VERS_G0007

Fig. 14 Slave cylinder bleeding connector

hold it, and then bleed air from the piping. wait for 5 seconds. Hold the clutch pedal down to prevent air from getting back into the clutch system.

9. Return clutch tube and lock pin to their original positions.

10. Release clutch pedal and wait for 5 seconds.

11. Repeat steps 4 to 10 until no bubbles are observed in the clutch fluid.

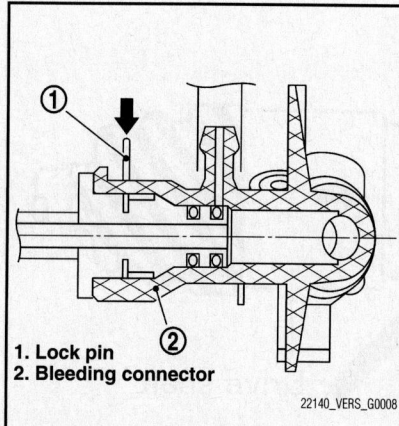

1. Lock pin
2. Bleeding connector

22140_VERS_G0008

Fig. 15 Bleeding connector lock pin installation—5 speed transaxles

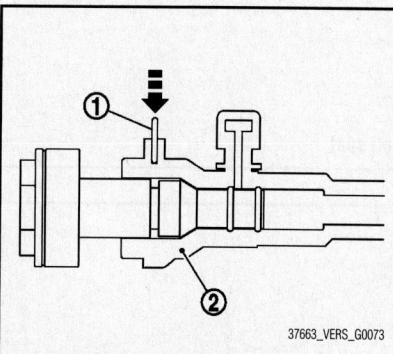

37663_VERS_G0073

Fig. 16 Bleeding connector lock pin installation—6 speed transaxles

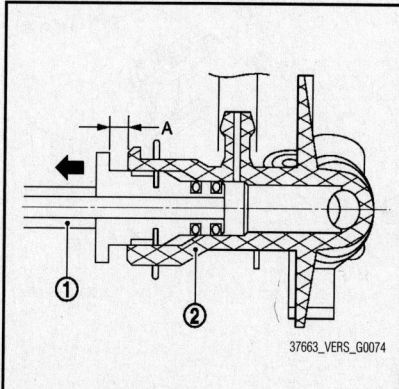

37663_VERS_G0074

Fig. 17 Clutch tube (1) and bleeding connector (2)

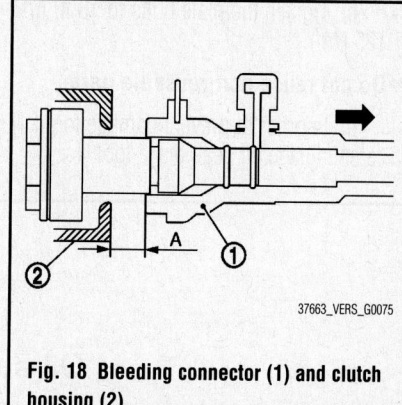

37663_VERS_G0075

Fig. 18 Bleeding connector (1) and clutch housing (2)

FRONT HALFSHAFT

REMOVAL & INSTALLATION

Left

See Figures 19 through 21.

1. Before servicing the vehicle, refer to the Precautions Section.

2. Remove wheel and tire.

3. Remove wheel sensor from steering knuckle. Do not pull on wheel sensor harness.

4. Remove transverse link ball joint nut and bolt. Then, remove transverse link from steering knuckle.

5. Remove cotter pin, and loosen hub lock nut.

6. Temporarily leave the hub lock nut installed to prevent damage to threads.

7. Separate the drive shaft from the wheel hub and bearing assembly by lightly tapping the end of the drive shaft using a

hammer or suitable tool and wood block, and then remove hub lock nut.

8. If wheel hub and bearing assembly and drive shaft cannot be separated after performing the above procedure, use a suitable puller.

9. Remove the drive shaft from the wheel hub and bearing assembly.

➡**Do not apply an excessive angle to drive shaft joint when removing from the wheel hub and bearing assembly. Do not excessively extend slide joint. Do not allow drive shaft to hang down. Support the entire drive shaft.**

10. Pry off drive shaft from transaxle assembly side as shown. Make sure that circlip is attached on the edge.

To install:

11. Installation is in the reverse order of removal, noting the following:

a. Tighten the drive shaft nut to 83 ft. lbs. (113 Nm).

➡**Do not reuse non-reusable parts.**

b. In order to prevent damage to differential side oil seal, place tool No. KV38105500 (J-33904) onto oil seal before inserting drive shaft as shown. Slide drive shaft into slide joint and tap with a hammer to install securely.

c. Install new circlip on drive shaft in the circlip groove on transaxle side. Make sure the new circlip on the drive shaft is securely fastened.

d. After its insertion, try to pull the flange out of the slide joint by hand. If it

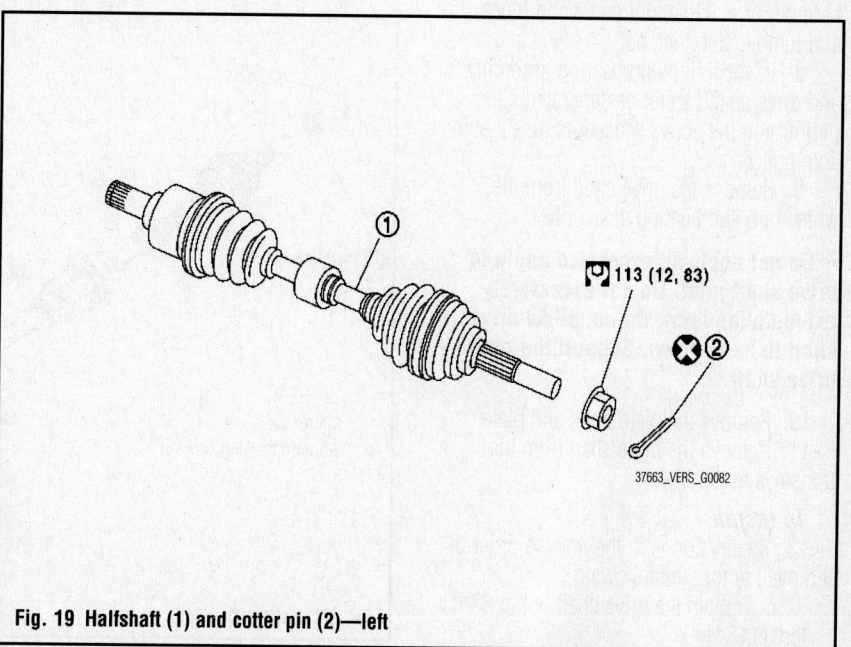

113 (12, 83)

37663_VERS_G0082

Fig. 19 Halfshaft (1) and cotter pin (2)—left

Fig. 20 Drive shaft removal

pulls out, the circlip is not properly meshed with the transaxle side gear.

e. Check transaxle fluid level.

Right

See Figures 21 and 22.

1. Before servicing the vehicle, refer to the Precautions Section.

2. Remove wheel and tire.

3. Remove wheel sensor from steering knuckle. Do not pull on wheel sensor harness.

4. Remove transverse link ball joint nut and bolt. Remove transverse link from steering knuckle.

5. Remove cotter pin and loosen hub lock nut.

6. Temporarily leave the hub lock nut installed to prevent damage to threads.

7. Separate the drive shaft from the wheel hub and bearing assembly by lightly tapping the end of the drive shaft using a hammer or suitable tool and wood block, and remove hub lock nut.

8. If wheel hub and bearing assembly and drive shaft cannot be separated after performing the above procedure, use a suitable puller.

9. Remove the drive shaft from the wheel hub and bearing assembly.

➡**Do not apply an excessive angle to drive shaft joint. Do not excessively extend slide joint. Do not allow drive shaft to hang down. Support the entire drive shaft.**

10. Remove the plate bolts and plate.

11. Remove the drive shaft from the transaxle assembly.

To install:

12. Installation is in the reverse order of removal, noting the following:

a. Tighten the drive shaft nut to 83 ft. lbs. (113 Nm).

b. Tighten the plate bolts to 18 ft. lbs. (25 Nm).

➡**Do not reuse non-reusable parts.**

c. In order to prevent damage to differential side oil seal, place tool No. KV38106700 (J-34296) onto oil seal before inserting drive shaft as shown. Slide drive shaft into slide joint and tap with a hammer to install securely.

d. Check transaxle fluid level.

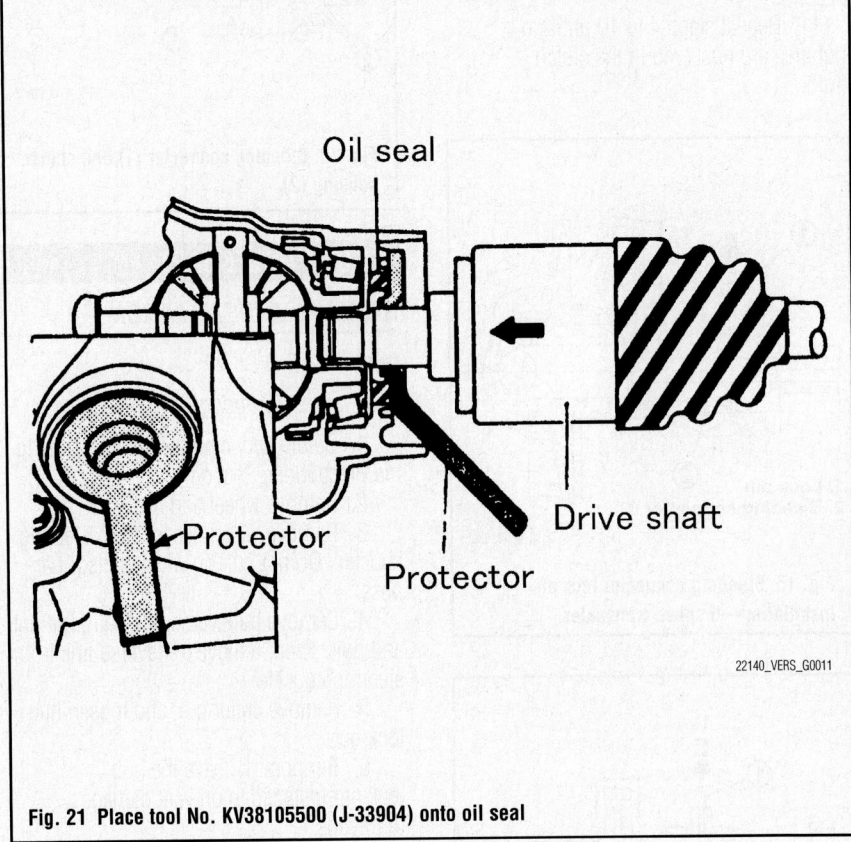

Fig. 21 Place tool No. KV38105500 (J-33904) onto oil seal

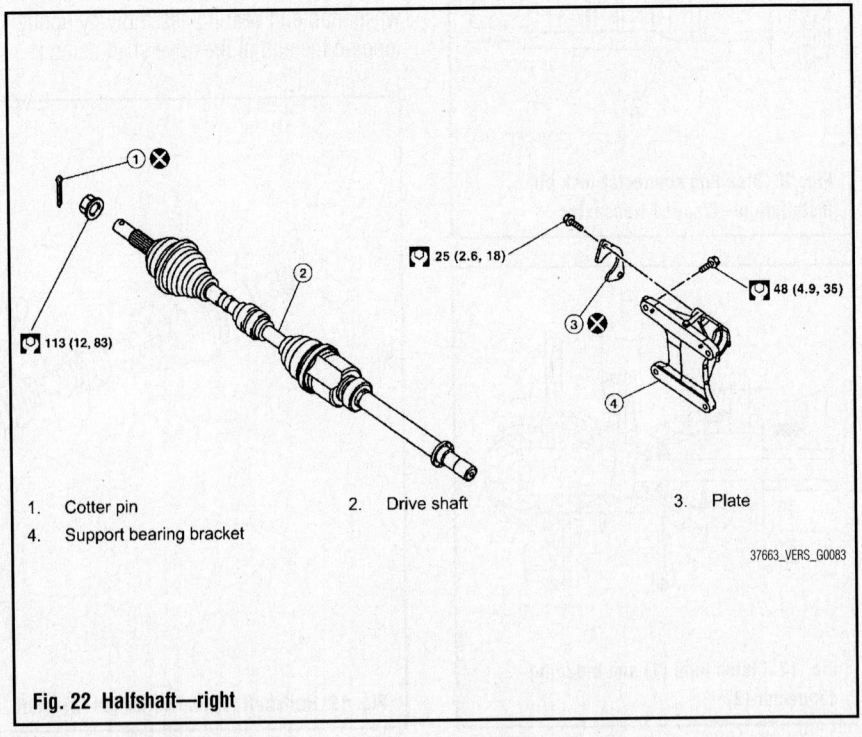

1. Cotter pin
2. Drive shaft
3. Plate
4. Support bearing bracket

Fig. 22 Halfshaft—right

CV-JOINT OVERHAUL

Inner—Transaxle Side

See Figures 23 through 30.

1. Before servicing the vehicle, refer to the precautions section.

2. Remove the front halfshaft from the vehicle.

3. Mount front drive shaft in a vise. When mounting shaft in a vise, always use copper or aluminum plates between vise and shaft.

4. Remove boot bands and slide the boot back.

5. Remove circlip and dust shield from slide joint housing.

6. Put matchmarks on slide joint housing and shaft and pull out shaft from slide joint housing.

7. Put matchmarks on spider assembly and shaft.

8. Remove snap ring using a suitable tool, and remove spider assembly from shaft.

9. Remove boot from shaft.

10. Clean the old grease off of the slide joint assembly.

To install:

11. Install new boot and new small boot

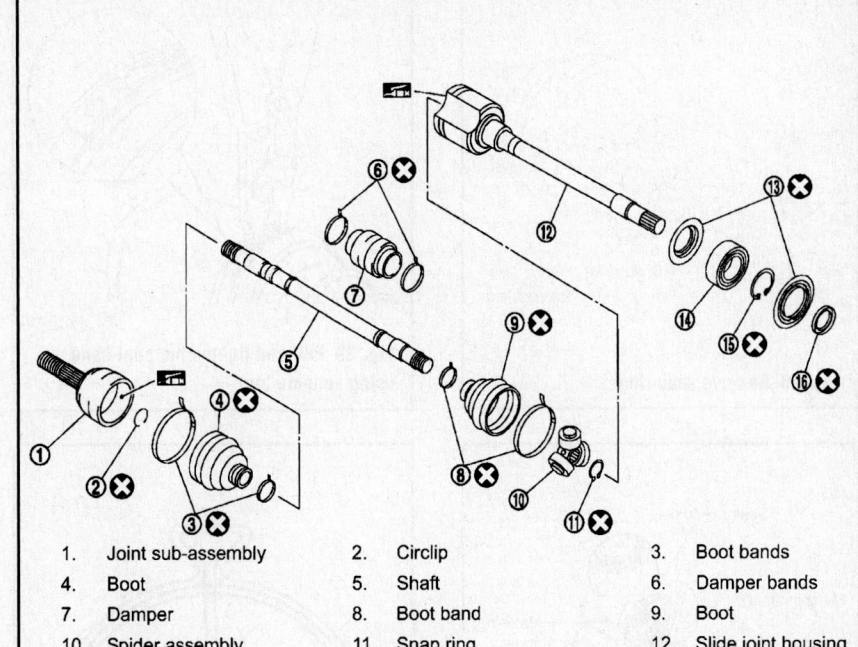

1.	Joint sub-assembly	2.	Circlip	3.	Boot bands
4.	Boot	5.	Shaft	6.	Damper bands
7.	Damper	8.	Boot band	9.	Boot
10.	Spider assembly	11.	Snap ring	12.	Slide joint housing
13.	Dust shield	14.	Bearing	15.	Snap ring
16.	Dust shield				

37663_VERS_G0085

Fig. 24 Right front halfshaft—exploded view

band on shaft. Cover drive shaft serration with tape to prevent damage to boot during installation. Discard old boot and boot band; replace with new ones.

12. Remove protective tape wound around serrated part of shaft.

13. Align matching mark on spider assembly with matching mark on shaft and install spider assembly with chamfer facing shaft.

14. Install new snap ring using a suitable tool. Do not reuse snap ring.

15. Apply recommended grease (Genuine NISSAN Grease or equivalent) to spider assembly and sliding surface.

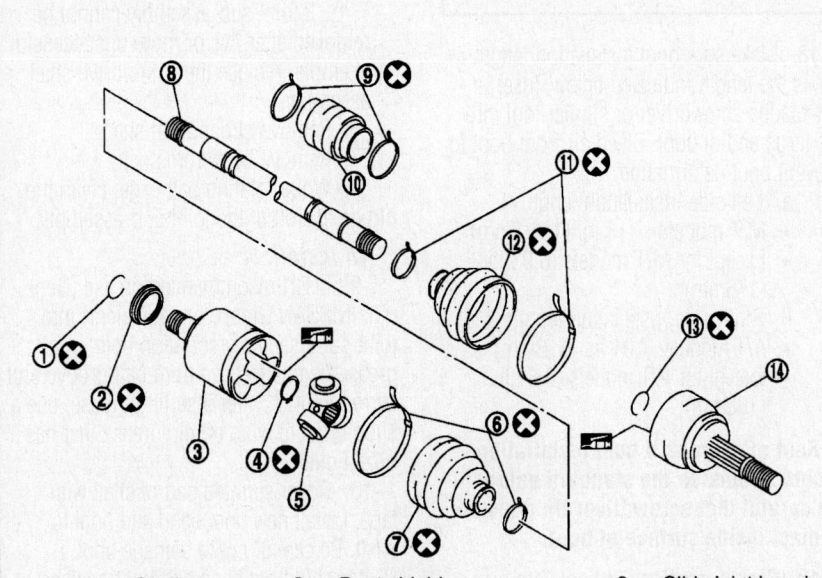

1.	Circlip	2.	Dust shield	3.	Slide joint housing
4.	Snap ring	5.	Spider assembly	6.	Boot band
7.	Boot	8.	Shaft	9.	Damper band
10.	Damper	11.	Boot band	12.	Boot
13.	Circlip	14.	Joint sub-assembly			

37663_VERS_G0084

Fig. 23 Left front halfshaft—exploded view

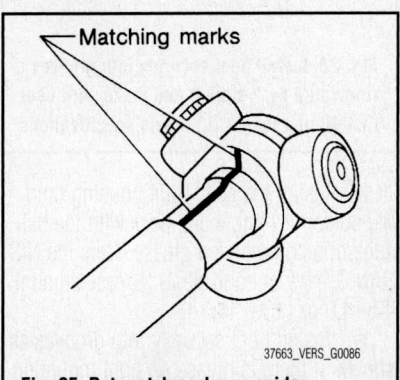

Matching marks

37663_VERS_G0086

Fig. 25 Put matchmarks on spider assembly and shaft

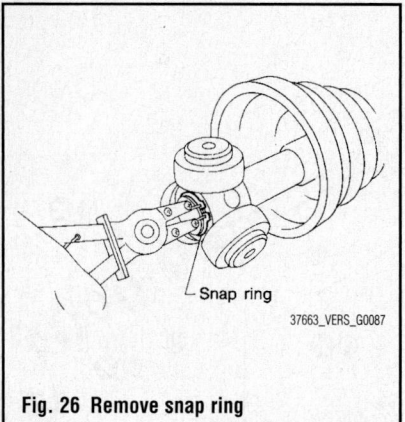

Fig. 26 Remove snap ring

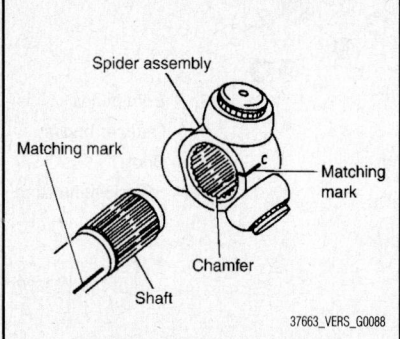

Fig. 27 Align matching mark on spider assembly with matching mark on shaft and install spider assembly with chamfer facing shaft

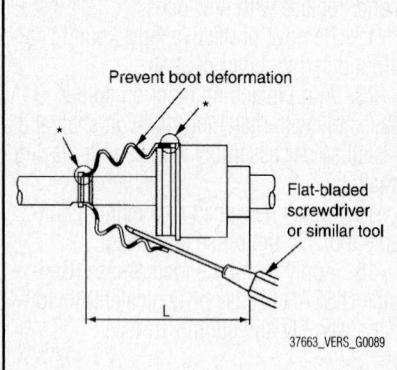

Fig. 28 Install boot securely into grooves (indicated by * marks) and make sure boot installation length (L) meets specifications

16. Install the slide joint housing onto the spider assembly and pack with the balance of recommended grease (Genuine NISSAN Grease or equivalent). Grease amount: 4.6–4.9 oz (130–140 g)

17. Install boot securely into grooves as shown. If there is grease on boot mounting surfaces of shaft and housing, boot may come off. Clean all grease from surfaces.

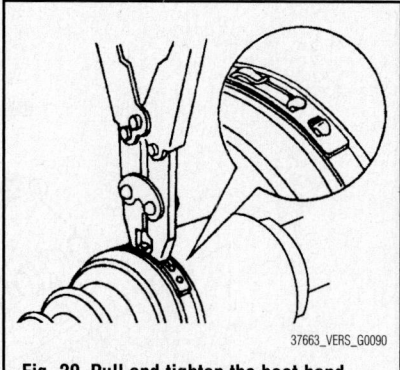

Fig. 29 Pull and tighten the boot band using suitable tool

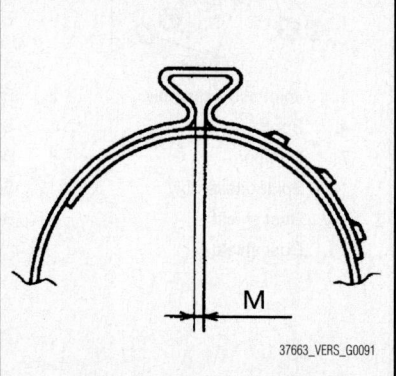

Fig. 30 Install new small boot band and secure so that dimension "M" meets specification

18. Make sure boot installation length "L" is the length indicated below. Insert a flat-bladed screwdriver or similar tool into the large end of boot. Bleed air from boot to prevent boot deformation.

   a. Left side installation length "L":
- M/T models: 7.13 in. (181.2mm)
- Except for M/T models: 6.93 in. (176mm)

   b. Right side installation length "L":
- A/T models: 7.01 in. (178mm)
- Except for A/T models: 6.61 in. (168mm)

➡**Boot may break if boot installation length is outside the standard value. Be careful that screwdriver tip does not contact inside surface of boot.**

19. Secure large end of boot with new boot band:

   a. Set boot band onto boot and insert first pawl on boot band into the first groove on opposite end of boot band.

   b. Pull and tighten the boot band using suitable tool until both pawls of boot band are secured in the boot band grooves. Do not reuse boot band.

20. Install new small boot band securely using tool No. KV40107300. Do not reuse boot band. Secure boot band so that dimension "M" meets specification [0.039–0.157 in. (1.0–4.0mm)].

21. Rotate the slide joint and confirm that the boot position is correct. If boot position is not correct, remove the boot bands, reposition the boot and install new boot bands.

22. Install new dust shield to slide joint housing.

Outer—Wheel Side

See Figures 23, 24 and 29 through 31.

1. Before servicing the vehicle, refer to the precautions section.

2. Remove the front halfshaft from the vehicle.

3. Mount front drive shaft in a vise. When mounting shaft in a vise, always use copper or aluminum plates between vise and shaft.

4. Remove boot bands and slide the boot back.

5. Screw a sliding hammer or suitable tool 1.18 in. (30mm) or more into threaded part of joint sub-assembly. Pull joint sub-assembly out of shaft.

   a. Align sliding hammer or suitable tool and drive shaft then remove joint sub-assembly by pulling directly.

   b. If joint sub-assembly cannot be removed after five or more unsuccessful attempts, replace the entire drive shaft assembly.

6. Remove circlip from shaft.

7. Remove boot from shaft.

8. While rotating ball cage, clean the old grease off of the joint sub-assembly.

To install:

9. Insert recommended grease (Genuine NISSAN Grease or equivalent) into joint sub-assembly serration hole until grease begins to ooze from ball groove and serration hole. After inserting grease, use a shop cloth to wipe off old grease that has oozed out.

10. Cover serrated part of shaft with tape. Install new boot band and boot to shaft. Be careful not to damage boot. Discard old boot band and boot; replace with new one.

11. Remove protective tape wound around serrated part of shaft.

12. Attach new circlip to shaft. Thread nut onto end of joint subassembly and press fit using suitable tool. The circlip must fit securely into shaft groove. Discard old circlip and replace with new one.

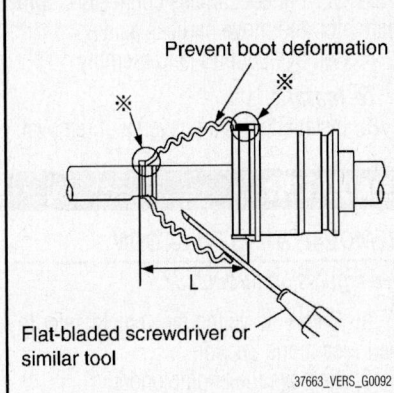

Fig. 31 Install boot securely into grooves (indicated by * marks) and make sure boot installation length (L) meets specifications

13. Insert the balance of new grease (2.5–3.2 oz (70–90 g) into housing from large end of boot.

14. Install boot securely into grooves as shown. If there is grease on boot mounting surfaces of shaft and housing, boot may come off. Remove all grease from surfaces.

15. Make sure boot installation length "L" is the specified length indicated below. Insert a flat-bladed screwdriver or similar tool into the large end of boot. Bleed air from boot to prevent boot deformation.

- Boot installation length "L": 5.59 in. (142mm)

➡**Boot may break if boot installation length is outside the standard value. Be careful that screwdriver tip does not contact inside surface of boot.**

16. Install new large and small boot bands securely using the appropriate tool (Tool No. KV40107300). Do not reuse boot bands.

a. Secure boot band so that dimension "M" meets specification (Dimension "M": 0.039–0.157 in. (1.0–4.0 mm).

17. After installing housing and shaft, rotate boot to check whether or not the actual position is correct. If boot position is not correct, remove old boot bands then reposition the boot and secure with new boot bands.

Damper

See Figure 32.

1. Remove the damper bands, and remove damper from the shaft.

To install:

2. Secure with new damper bands.
- CVT models "A": 8.86–9.09 in. (225–231mm)
- Except for CVT models "A": 9.06–9.29 in. (230–236mm)
- "B": 2.76 in. (70mm)

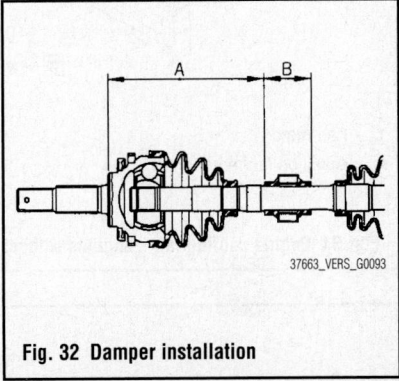

Fig. 32 Damper installation

ENGINE COOLING

ENGINE FAN

REMOVAL & INSTALLATION

See Figures 33 and 34.

1. Before servicing the vehicle, refer to the Precautions Section.

2. Partially drain engine coolant from radiator.

3. Remove air duct (inlet).
4. Remove reservoir tank.
5. Disconnect radiator hose (upper) at radiator side.

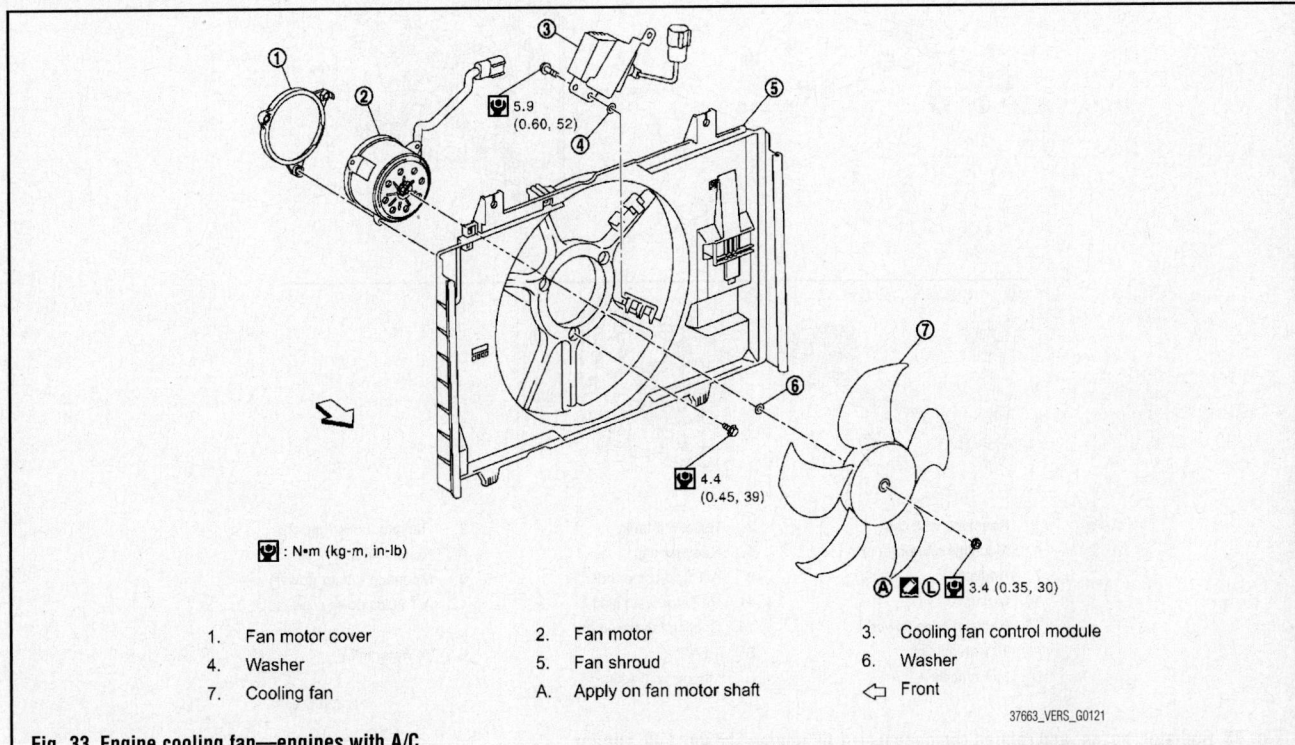

1.	Fan motor cover	2.	Fan motor	3.	Cooling fan control module
4.	Washer	5.	Fan shroud	6.	Washer
7.	Cooling fan	A.	Apply on fan motor shaft	⬅	Front

Fig. 33 Engine cooling fan—engines with A/C

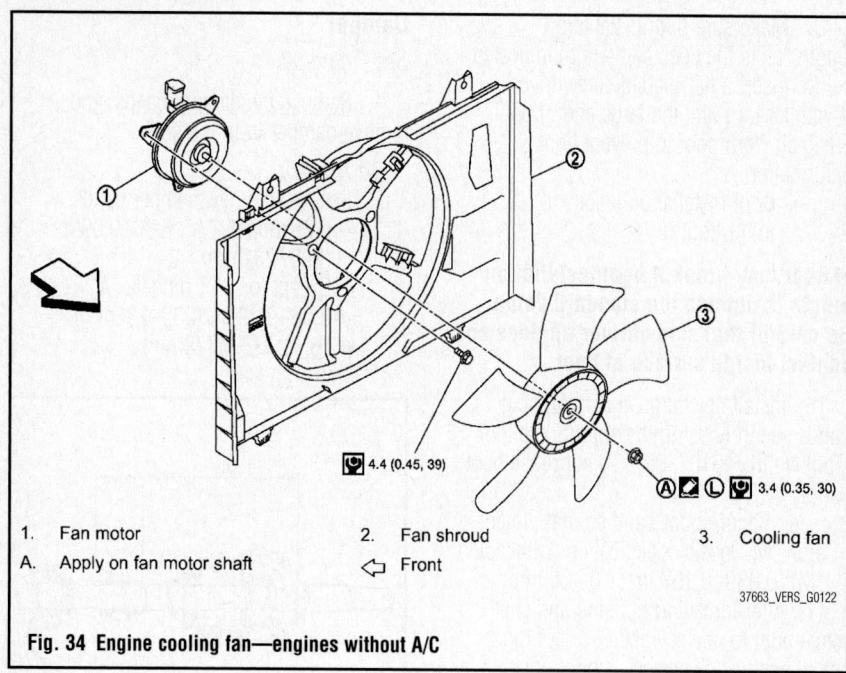

1. Fan motor
A. Apply on fan motor shaft
2. Fan shroud
◁ Front
3. Cooling fan

4.4 (0.45, 39)

A L 3.4 (0.35, 30)

37663_VERS_G0122

Fig. 34 Engine cooling fan—engines without A/C

6. Disconnect harness connectors from fan motor and move harness aside.
7. Remove cooling fan assembly.

To install:
8. Installation is the reverse of removal.

RADIATOR

REMOVAL & INSTALLATION

See Figures 35 through 37.

1. Before servicing the vehicle, refer to the Precautions Section.
2. Remove the engine under cover.
3. Drain engine coolant from radiator.
4. Remove air duct (inlet).
5. Remove reservoir tank as follows:
a. Disconnect reservoir tank hose.
b. Release the tab in the direction shown by the arrow.

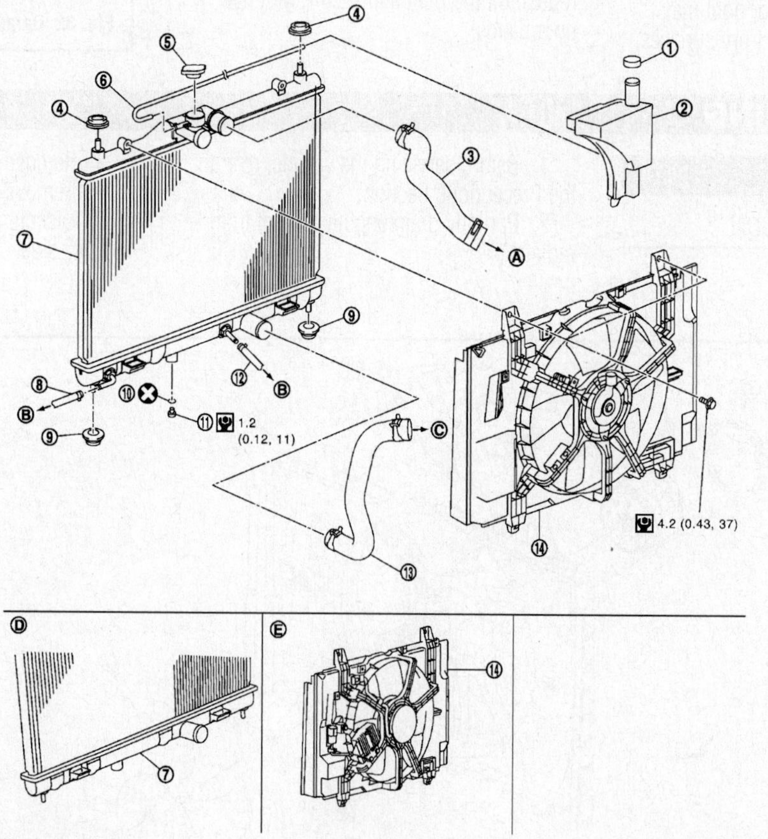

1.2 (0.12, 11)

4.2 (0.43, 37)

1. Reservoir tank cap	2. Reservoir tank	3. Radiator hose (upper)
4. Mounting rubber (upper)	5. Radiator cap	6. Reservoir tank hose
7. Radiator	8. A/T fluid cooler hose	9. Mounting rubber (lower)
10. O-ring	11. Radiator drain plug	12. A/T fluid cooler hose
13. Radiator hose (lower)	14. Cooling fan assembly	
A. To water outlet	B. To A/T	C. To water inlet
D. M/T models	E. Models with A/C	

37663_VERS_G0109

Fig. 35 Radiator, hoses, and related components—1.6L engine shown, 1.8L similar

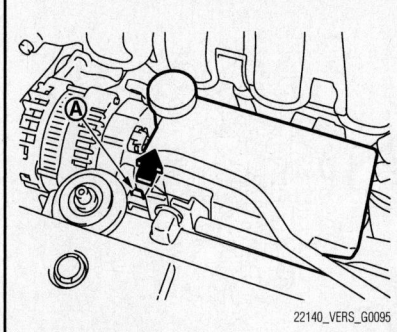

Fig. 36 Release the tab in the direction shown

c. Lift up while removing the reservoir tank hose, and remove it.

6. Disconnect the harness connector from the fan motor and move the harness aside.

7. Disconnect CVT or A/T fluid cooler hoses, if equipped. Install plug to avoid leakage of CVT or A/T fluid, if necessary.

8. Remove radiator hoses (upper and lower).

9. Remove radiator core support cover.

10. Remove the cooling fan assembly.

11. Remove the radiator core support (upper) bolts, the bolts of the stationary part on the radiator core support side and clip. Lift the radiator from the radiator (upper) mount part of the radiator core support (upper).

12. Move radiator assembly to the rearward direction of vehicle, and then lift it upward to remove.

To install:

13. Installation is the reverse of removal.

THERMOSTAT

REMOVAL & INSTALLATION

1.6L Engine

See Figures 38 through 41.

1. Before servicing the vehicle, refer to the Precautions Section.

2. Drain engine coolant from radiator.

3. Remove air duct (inlet).

4. Remove the reservoir tank.

5. Add paint mark, then disconnect radiator hose (lower) from water inlet.

6. Remove water inlet and thermostat. Engine coolant will leak from cylinder block, so have a receptacle ready below.

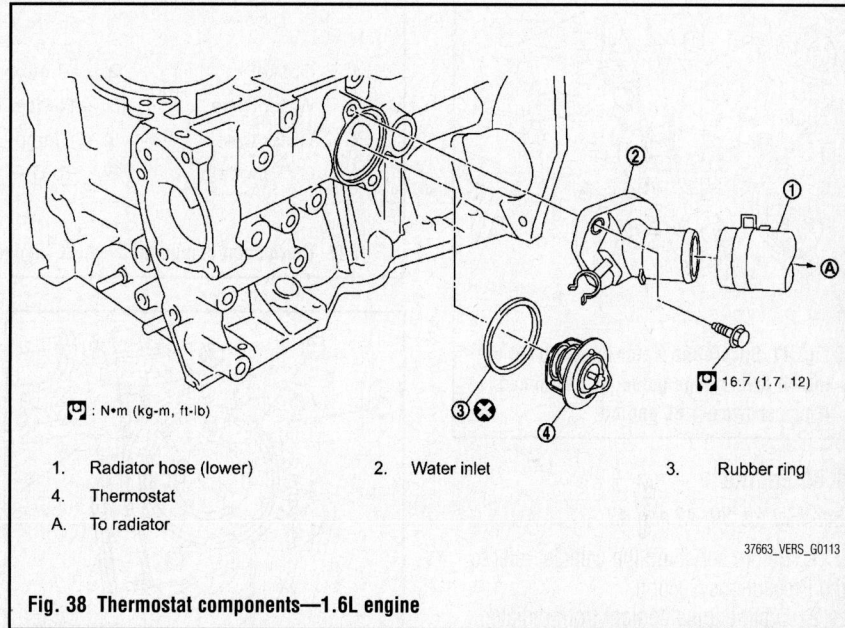

: N•m (kg-m, ft-lb)

16.7 (1.7, 12)

1.	Radiator hose (lower)	2.	Water inlet	3.	Rubber ring
4.	Thermostat				
A.	To radiator				

Fig. 38 Thermostat components—1.6L engine

To install:

7. Installation is in the reverse order of removal, noting the following:

a. Replace the rubber ring with a new one.

b. Install thermostat while making the rubber ring groove fit securely to the thermostat flange.

c. Install the thermostat to the cylinder block with jiggle valve facing upwards.

d. After installation, secure the water inlet clip on the oil level gauge guide positioned as shown.

e. Check that the reservoir tank cap is tightened. Check for leaks of engine coolant.

f. Start and warm up the engine. Visually make sure that there is no leaks of engine coolant.

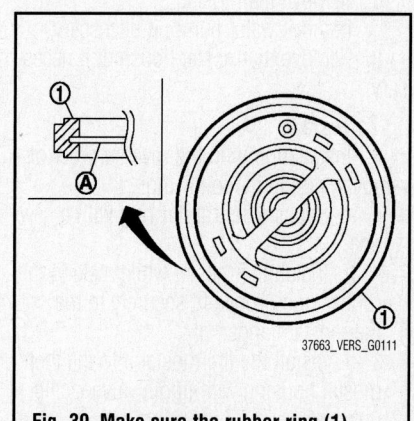

Fig. 39 Make sure the rubber ring (1) groove fits securely to the thermostat flange (A)—1.8L engine shown, 1.6L engine similar

45 N•m
(4.6 kg-m, 33 ft-lb)

1.	Radiator core support upper	2.	Radiator core support lower	3.	Radiator core support lower stay
4.	Radiator core support side stay	5.	Air guide			

Fig. 37 Radiator core support components

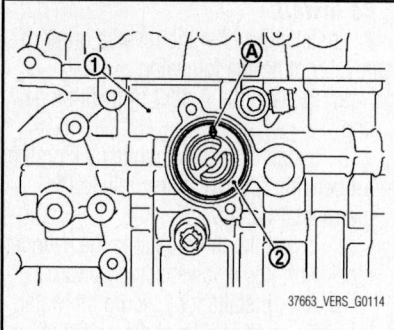

Fig. 40 Install the thermostat (2) to the cylinder block (1) with jiggle valve (A) facing upwards—1.6L engine

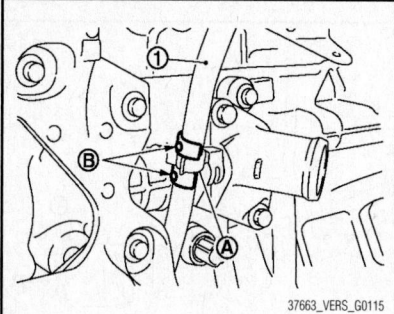

Fig. 41 Secure the water inlet clip (A) on the oil level gauge guide (1) positioned (B) as shown—1.6L engine

1.8L Engine

See Figures 39, 42 and 43.

1. Before servicing the vehicle, refer to the Precautions Section.
2. Drain engine coolant from radiator.
3. Remove air duct (inlet).
4. Remove the radiator hose (lower) from the engine.
5. Remove water inlet.
6. Remove thermostat.
7. Remove water pump, if necessary.
8. Remove thermostat housing, if necessary.

To install:

9. Installation is in the reverse order of removal, noting the following:

   a. Replace the rubber ring with a new one.

   b. Install thermostat while making the rubber ring groove fit securely to the thermostat flange.

   c. Install the thermostat into the thermostat housing with jiggle valve facing upwards.

   d. If installing the thermostat housing, securely insert the rubber ring into the mating groove of thermostat housing and

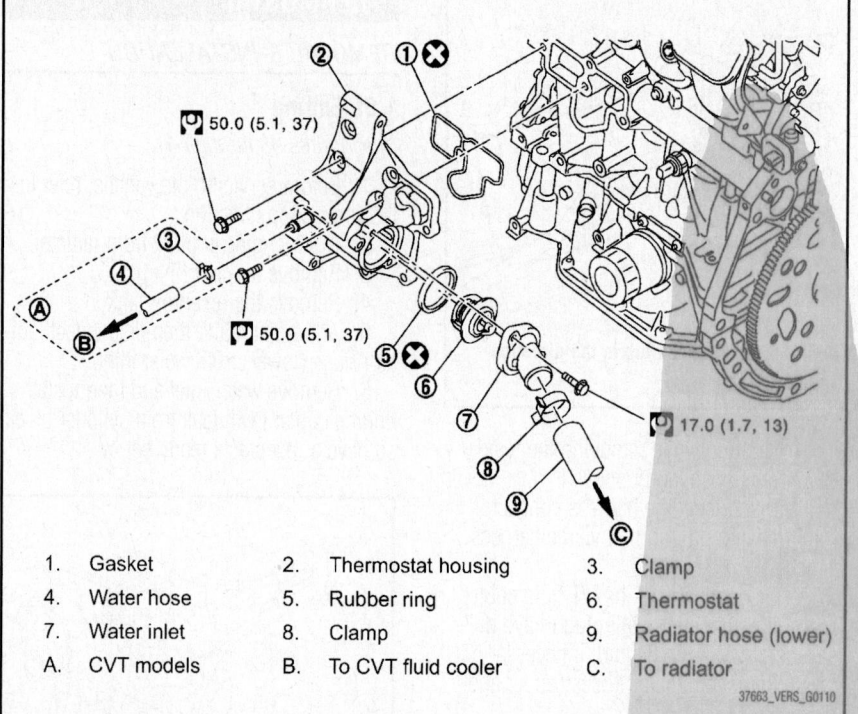

1.	Gasket	2.	Thermostat housing	3.	Clamp
4.	Water hose	5.	Rubber ring	6.	Thermostat
7.	Water inlet	8.	Clamp	9.	Radiator hose (lower)
A.	CVT models	B.	To CVT fluid cooler	C.	To radiator

Fig. 42 Thermostat components—1.8L engine

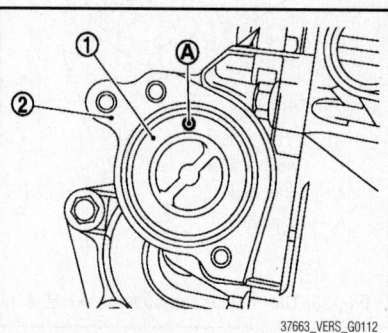

Fig. 43 Install the thermostat (1) into the thermostat housing (2) with jiggle valve (A) facing upwards—1.8L engine

install it. Replace the rubber ring with a new one. Install the thermostat housing to the cylinder block without displacing the gasket from the gasket position.

   e. Check that the reservoir tank cap is tightened. Check for leaks of engine coolant.

   f. Start and warm up the engine. Visually make sure that there is no leaks of engine coolant.

WATER PUMP

REMOVAL & INSTALLATION

1.6L Engine

See Figures 44 and 45.

1. Before servicing the vehicle, refer to the Precautions Section.
2. Disconnect the negative battery cable.
3. Drain engine coolant from radiator.
4. Partially remove the right-hand front fender protector.
5. Loosen mounting bolts of water pump pulley before loosening belt tension of drive belt.
6. Remove drive belt. Refer to Accessory Drive Belt Removal & Installation in the Engine Mechanical section.
7. Remove water pump pulley.
8. Loosen mounting bolts in reverse of the order shown, and remove the water pump.

➡**Handle water pump vane so that it does not contact any other parts. Water pump cannot be disassembled and should be replaced as a unit.**

To install:

9. Installation is the reverse of removal, noting the following:

   a. Tighten the water pump bolts as shown above.

   b. When installing the water pump pulley, never install mounting bolts to oblong holes.

   c. Check that the reservoir tank cap is tightened. Check for leaks of engine coolant.

   d. Start and warm up the engine. Visually make sure that there is no leaks of engine coolant.

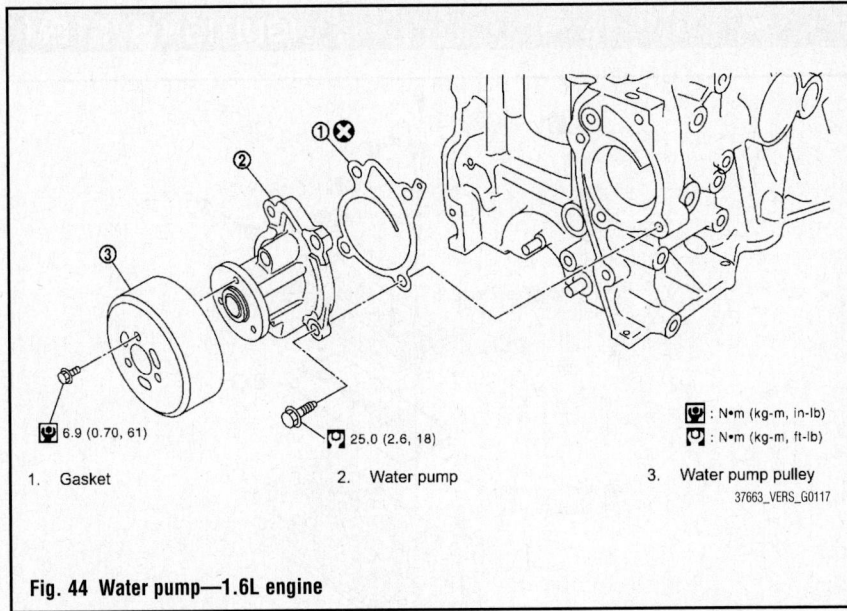

Fig. 44 Water pump—1.6L engine

1. Gasket
2. Water pump
3. Water pump pulley

6.9 (0.70, 61)
25.0 (2.6, 18)

: N•m (kg-m, in-lb)
: N•m (kg-m, ft-lb)

37663_VERS_G0117

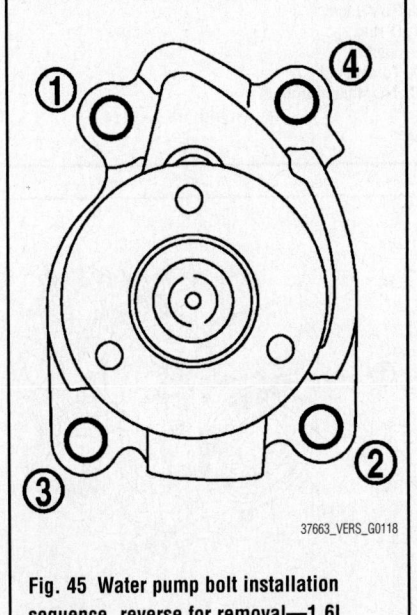

Fig. 45 Water pump bolt installation sequence, reverse for removal—1.6L engine

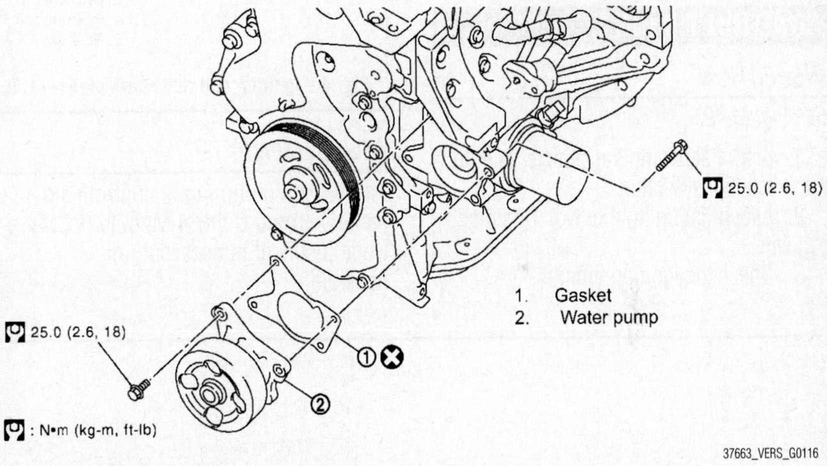

37663_VERS_G0118

1.8L Engine

See Figure 46.

Fig. 46 Water pump—1.8L engine

25.0 (2.6, 18)
25.0 (2.6, 18)

: N•m (kg-m, ft-lb)

1. Gasket
2. Water pump

37663_VERS_G0116

5. Drain engine coolant from radiator.

6. Remove the right-hand front fender protector.

7. Remove drive belt. Refer to Accessory Drive Belt Removal & Installation in the Engine Mechanical section.

8. Remove alternator. Refer to Alternator Removal & Installation in the Engine Electrical section.

9. Remove radiator hose (lower).

10. Remove water pump.

➡ **Handle water pump vane so that it does not contact any other parts. Water pump cannot be disassembled and should be replaced as a unit.**

To install:

11. Installation is the reverse of removal, noting the following:

a. Check that the reservoir tank cap is

1. Before servicing the vehicle, refer to the Precautions Section.

2. Disconnect the negative battery cable.

3. Remove the radiator.

4. Remove reservoir tank.

tightened. Check for leaks of engine coolant.

b. Start and warm up the engine. Visually make sure that there is no leaks of engine coolant.

ENGINE ELECTRICAL

ALTERNATOR

REMOVAL & INSTALLATION

1. Before servicing the vehicle, refer to the Precautions Section.

2. Disconnect the battery cable from the negative terminal.

3. Remove drive belt. Refer

to Accessory Drive Belt Removal & Installation in the Engine Mechanical section.

4. Remove radiator reservoir tank.

5. Disconnect alternator connector.

6. Remove "B" terminal nut.

7. Remove alternator bolts.

CHARGING SYSTEM

8. Remove alternator assembly from the vehicle.

To install:

9. Installation is in the reverse order of removal.

➡ **Be sure to tighten "B" terminal nut carefully.**

FIRING ORDER

Firing Order: 1–3–4–2

IGNITION COIL

REMOVAL & INSTALLATION

See Figures 47 and 48.

1. Before servicing the vehicle, refer to the Precautions Section.
2. Disconnect the negative battery cable.
3. Remove intake manifold. Refer to Intake Manifold Removal & Installation in the Engine Mechanical section.
4. Remove ignition coil.

To install:
5. Install ignition coil.
6. Install intake manifold.
7. Connect the negative battery cable.

IGNITION TIMING

INSPECTION

See Figure 49.

1. Before servicing the vehicle, refer to the Precautions Section.
2. Attach timing light to No. 1 ignition coil wire.
3. Check the ignition timing.

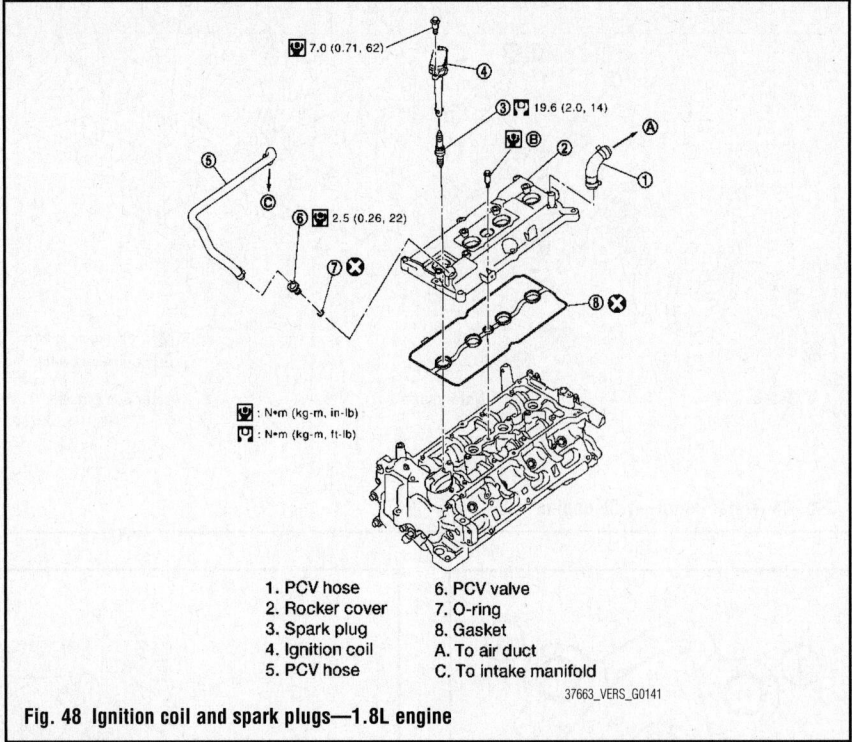

1. PCV hose
2. Rocker cover
3. Spark plug
4. Ignition coil
5. PCV hose
6. PCV valve
7. O-ring
8. Gasket
A. To air duct
C. To intake manifold

☒ : N•m (kg-m, in-lb)
☒ : N•m (kg-m, ft-lb)

37663_VERS_G0141

Fig. 48 Ignition coil and spark plugs—1.8L engine

ADJUSTMENT

The ignition timing is controlled by the Powertrain Control Module (PCM). No adjustment is necessary or possible.

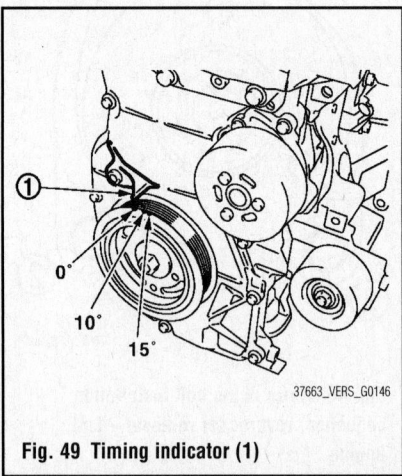

37663_VERS_G0146

Fig. 49 Timing indicator (1)

SPARK PLUGS

REMOVAL & INSTALLATION

See Figures 47 and 48.

1. Before servicing the vehicle, refer to the Precautions Section.
2. Disconnect the negative battery cable.
3. Remove ignition coil.
4. Remove spark plug using suitable tool.

To install:
5. Installation is the reverse of removal.

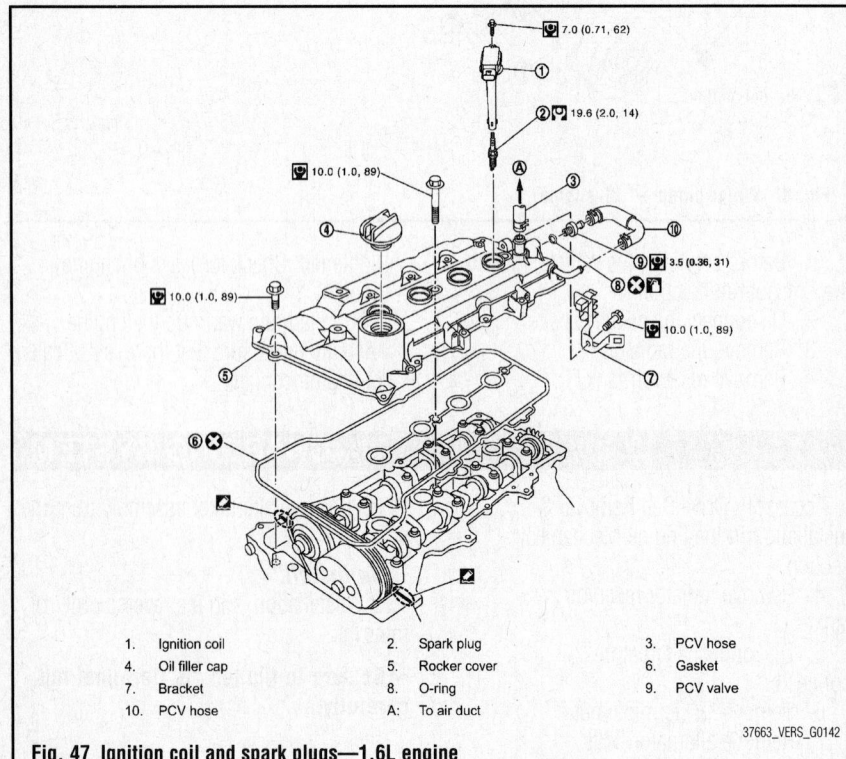

1. Ignition coil
2. Spark plug
3. PCV hose
4. Oil filler cap
5. Rocker cover
6. Gasket
7. Bracket
8. O-ring
9. PCV valve
10. PCV hose
A: To air duct

37663_VERS_G0142

Fig. 47 Ignition coil and spark plugs—1.6L engine

ENGINE ELECTRICAL

STARTER

REMOVAL & INSTALLATION

1. Before servicing the vehicle, refer to the Precautions Section.

2. Disconnect the battery negative terminal.
3. Remove air duct (inlet).
4. Remove reservoir tank.
5. Remove "S" terminal nut.
6. Remove "B" terminal nut.
7. Remove starter motor bolts.

STARTING SYSTEM

8. Remove starter motor.

To install:

9. Installation is in the reverse order of removal.

➡**Be sure to tighten terminal nuts carefully.**

ENGINE MECHANICAL

ACCESSORY DRIVE BELTS

ACCESSORY BELT ROUTING

See Figures 50 and 51.

Refer to the accompanying illustrations.

INSPECTION

1.6L Engine

1. Before servicing the vehicle, refer to the Precautions Section.

✳✳ **CAUTION**

Be sure to perform this step when the engine is stopped.

2. Visually check belts for wear, damage, and cracks on inside and edges.
3. Turn crankshaft pulley two times clockwise, and make sure tension on all pulleys is equal before doing the test.

1.8L Engine

1. Make sure that the indicator (notch on fixed side) of drive belt auto-tensioner is within the possible use range.

➡**Check the drive belt auto-tensioner indication when the engine is cold.**

➡**When new drive belt is installed, the indicator (notch on fixed side) should be within the range.**

2. Visually check entire drive belt for wear, damage or cracks.
3. If the indicator (notch on fixed side) is out of the possible use range or belt is damaged, replace drive belt.

ADJUSTMENT

1.6L Engine

See Figure 52.

➡**When belt is replaced with new one, adjust belt tension to the value for "New belt", because new belt will not fully seat in the pulley groove. When tension of the belt being used exceeds "Limit", adjust it to the value for "After adjusted". When installing a belt, make sure it is correctly engaged with the pulley groove.**

✳✳ **WARNING**

Never allow oil or engine coolant to get on the belt. Never twist or bend the belt strongly.

1. Loosen the idler pulley lock nut from the tightening position with the specified torque by 45 degrees.

   a. When the lock nut is loosened excessively, the idler pulley tilts and the correct tension adjustment cannot be performed. Never loosen it excessively (more than 45 degrees).

   b. Put a matching mark on the lock nut, and check turning angle with a protractor. Never visually check the tightening angle.

2. Adjust the belt tension by turning the adjusting bolt.

   a. When checking immediately after installation, first adjust it to the specified value. Then, after turning crankshaft two turns or more, re-adjust to the specified value to avoid variation in deflection between pulleys.

   b. When the tension adjustment is performed, the lock nut should be in the condition at step"2". If the tension adjustment is performed when the lock nut is loosened more than the standard, the idler pulley tilts and the correct tension adjustment cannot be performed.

3. Tighten the idler pulley lock nut to 26 ft. lbs. (35 Nm).

1.8L Engine

Belt tension adjustment is not necessary, as it is automatically adjusted by drive belt auto-tensioner.

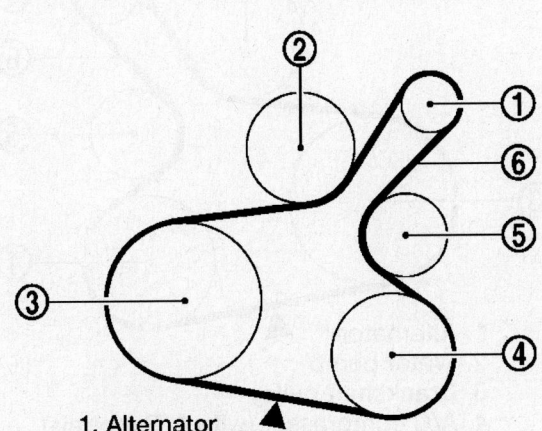

1. Alternator
2. Water pump
3. Crankshaft pulley
4. A/C compressor (with A/C models) or Idler pulley (without A/C models)
5. Idler pulley
6. Drive belt
A. Idler pulley lock nut
B. Adjusting bolt

37663_VERS_G0147

Fig. 50 Accessory belt routing—1.6L engine

1. Alternator
2. Drive belt auto-tensioner
3. Crankshaft pulley
4. A/C compressor (models with A/C)
   /Idler pulley (models without A/C)
5. Water pump
6. Drive belt

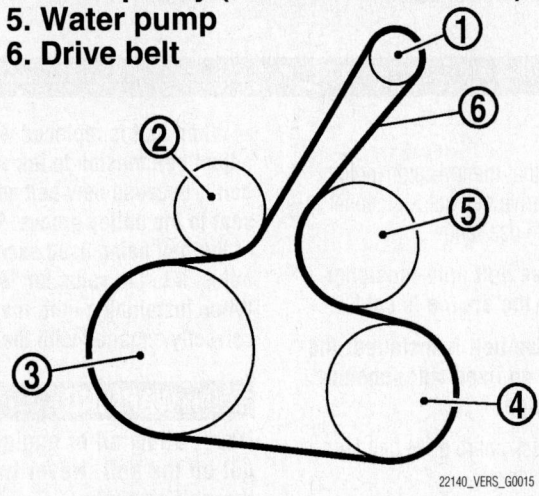

22140_VERS_G0015

Fig. 51 Accessory belt routing—1.8L engine

REMOVAL & INSTALLATION

1.6L Engine

See Figure 52.

1. Before servicing the vehicle, refer to the Precautions Section.

2. Loosen the idler pulley lock nut, and then adjust the belt tension by turning the adjusting bolt.

3. Remove drive belt.

To install:

4. Pull the idler pulley in the loosening direction, and then temporarily tighten the idler pulley lock nut to 39 inch lbs. (4 Nm).

5. Do not move the lock nut from the tightened position. Proceed to the next step.

6. Install the drive belt to each pulley. Make sure that there is no oil, grease, or coolant, etc. in pulley grooves. Make sure that the belt is securely inside the groove on each pulley.

7. Adjust drive belt tension by turning the adjusting bolt.

   a. Perform the belt tension adjustment with the lock nut temporarily tightened at the step "1" so as not to tilt the idler pulley.

   b. When checking immediately after installation, first adjust it to the specified value. Then, after turning crankshaft two

turns or more, re-adjust to the specified value to avoid variation in deflection between pulleys.

8. Tighten the idler pulley lock nut to 26 ft. lbs. (35 Nm).

9. Make sure that belt tension of each belt is within the standard.

1.8L Engine

See Figure 53.

1. Before servicing the vehicle, refer to the Precautions Section.

2. Remove the right-hand fender protector.

3. Hold the hexagonal part of drive belt auto-tensioner with a wrench securely. Then move the wrench handle in the direction of arrow (loosening direction of tensioner).

> ❊❊ **CAUTION**
>
> **Never place hand in a location where pinching may occur if the holding tool accidentally comes off.**

> ❊❊ **WARNING**
>
> **Do not loosen the auto-tensioner pulley bolt. (Do not turn it counterclockwise.) If turned counterclockwise, the**

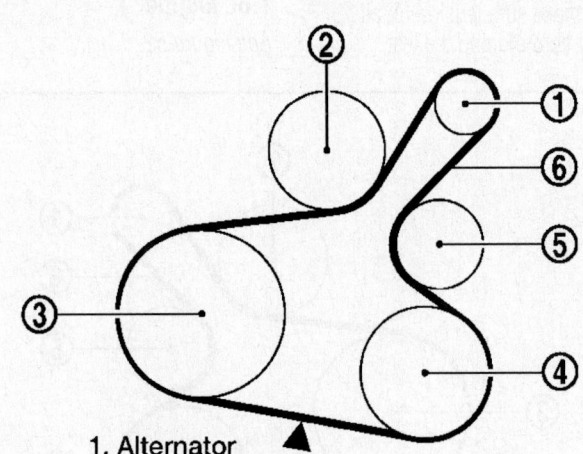

1. Alternator
2. Water pump
3. Crankshaft pulley
4. A/C compressor (with A/C models)
   or Idler pulley (without A/C models)
5. Idler pulley
6. Drive belt
A. Idler pulley lock nut
B. Adjusting bolt

37663_VERS_G0147

Fig. 52 Accessory drive belt routing—1.6L engine

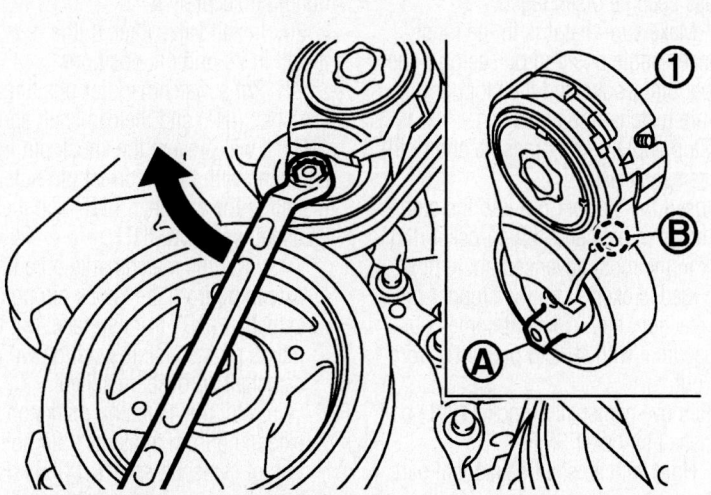

A. Hexagonal part of drive belt auto-tensioner
1. Drive belt auto-tensioner
B. Retaining rod insertion hole

22140_VERS_G0014

Fig. 53 Hold the hexagonal part (A) of drive belt auto-tensioner (1) with a wrench, loosen, and insert a rod into the hole (B) of the retaining boss

complete auto-tensioner must be replaced as a unit, including pulley.

4. Insert a rod such as short-length screwdriver approximately 0.24 in. (6 mm) in diameter into the hole of the retaining boss to hold the drive belt auto-tensioner.

5. Remove drive belt.

To install:

6. Install drive belt. Confirm drive belt is completely set to pulleys.

7. Release drive belt auto-tensioner, and apply tension to drive belt.

8. Check for engine oil, working fluid and engine coolant are not adhered to drive belt and each pulley groove.

9. Turn crankshaft pulley clockwise several times to equalize tension between each pulley.

10. Confirm tension of drive belt at indicator (notch on fixed side) is within the possible use range.

11. Install the right-hand fender protector.

CAMSHAFT AND VALVE LIFTERS

REMOVAL & INSTALLATION

1.6L Engine
See Figures 54 through 65.

37663_VERS_G0242

Fig. 54 Camshaft sprocket (EXH) (1), hexagonal part (A)

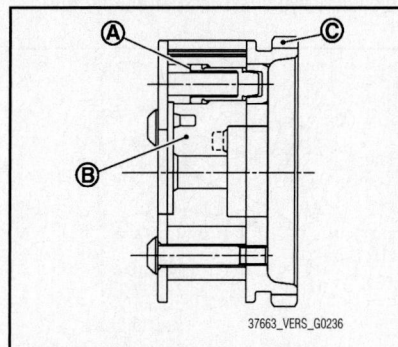

37663_VERS_G0236

Fig. 55 Sprocket (C), vane (camshaft coupling) (B), lock pin (A)

1. Remove timing chain. Refer to Timing Chain and Sprockets Removal and Installation.

2. Remove camshaft position sensor (PHASE) from rear end of cylinder head. Handle it carefully and avoid impacts.

3. Remove the exhaust camshaft sprocket (EXH) bolt and camshaft sprocket (EXH).

a. Hold the camshaft hexagonal part, and secure the camshaft. Never rotate crankshaft and camshaft separately, so as not to contact valve with piston in the following steps.

4. Turn the camshaft sprocket (INT) to the most advanced position. Installation and removal of the camshaft sprocket (INT) must be done in the most advanced position for the following reasons:

a. The sprocket and vane (camshaft coupling) are designed to spin and move within the range of a certain angle.

b. With the engine stopped and the vane in the most retarded angle, it will not spin because it is locked to the sprocket side by the internal lock pin.

c. If the camshaft sprocket bolts are turned in the situation described above, the lock pin will become damaged and cause malfunctions because of the increased horizontal load (cutting force) on the lock pin.

5. Remove camshaft bracket (No. 1) by loosening the bolts in several steps, and removing them.

6. Apply air pressure (44 psi or more) to the No. 1 journal oil hole of camshaft (INT) using an air gun.

The air pressure is used to move the lock pin into the disengage position.

a. Apply and maintain the air pressure into the oil hole on the second groove from the front of camshaft thrust.

b. Attach the rubber nozzle narrowed to the top of the air gun to prevent air leakage from the oil hole. Securely apply the air pressure to the oil hole.

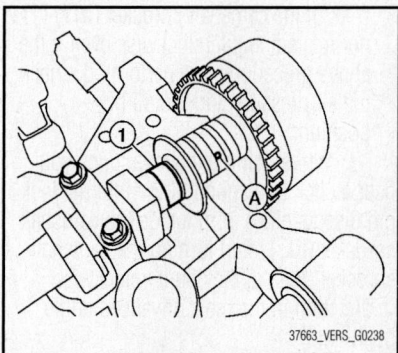

37663_VERS_G0238

Fig. 56 Apply air pressure to the No. 1 journal oil hole of camshaft (INT)

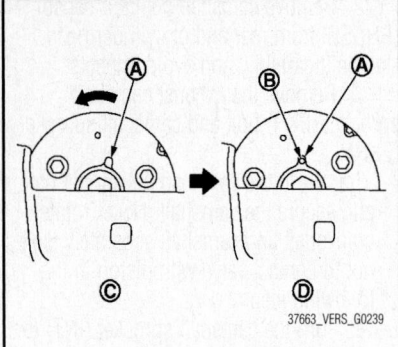

Fig. 57 Most retarded angle (lock pin engaged) (C), most advanced angle (D), stopper pin groove (A) and stopper pin hole (B)

✳✳ WARNING

There are other oil holes in the side grooves. Never use the incorrect oil holes. Be sure not to damage the oil path with the tip of the air gun. Wipe all the oil off the air gun to prevent oil from being blown all over along with the air, and the area around the air gun should be wiped with a rag when applying air pressure.

✳✳ CAUTION

Eye protection should be worn as needed.

7. Hold the camshaft sprocket (INT) with hands, and then apply the power counterclockwise/clockwise alternatively.

   a. Finally rotate the sprocket of the camshaft sprocket (INT) counterclockwise.

   b. Perform the work while applying the air pressure to the oil hole.

   c. If the lock pin is not released by hand, tap the camshaft sprocket (INT) (1) lightly with a plastic hammer.

   d. If the camshaft sprocket (INT) (1) is not rotated counterclockwise even if the above procedures are performed, check the air pressure and the oil hole position.

8. While doing the above, once you hear a click (the sound of the internal lock pin disengaging) from inside the camshaft sprocket (INT), start turning the camshaft sprocket (INT) in the counterclockwise direction in the most advanced angle position.

   a. Keep the air pressure on.

   b. If there is no click, as soon as the vane-side (camshaft side) starts moving

independently of the sprocket, the lock pin has become disengaged.

   c. Make sure that it is in the most advanced angle position by seeing if the stopper pin groove and the stopper pin hole are matched up as shown.

9. Stop applying air pressure and holding the camshaft (INT).

10. Insert the stopper pin into the stopper pin holes in the camshaft sprocket (INT) and lock in the most advanced angle position. No load is exerted on the stopper pin (spring reaction, etc.). Since it comes out easily, secure it with tape to prevent it from coming out.

11. Remove camshaft sprocket (INT) bolt and camshaft sprocket (INT).

   a. Hold the camshaft hexagonal part, and then secure the camshaft. Never rotate crankshaft and camshaft separately, so as not to contact valve with piston in the following steps.

12. Remove camshaft brackets (No. 2 to 5). Loosen bolts in several steps in the reverse of the order shown. The camshaft bracket (No. 1) has been already removed.

13. Remove camshaft (EXH).

14. Remove camshaft (INT).

15. Remove valve lifter. Identify installation positions, and store them without mixing them up.

16. Remove intake valve timing control solenoid valve.

17. Remove the alternator and bracket, if necessary, then remove the plug, washer and oil filter.
Discard the washer, do not reuse.

To install:

18. Install the oil filter and new washer.

   a. Attach the oil filter to the plug, and install it to the cylinder head.

19. Install intake valve timing control solenoid valve. Insert it straight into the

cylinder head, and tighten bolts after positioning it securely.

20. Install valve lifter. If it is reused, install in its original positions.

21. Put a matchmark for positioning the camshaft (INT) and the camshaft sprocket (INT). It will prevent the knock pin from engaging with the incorrect pin hole after installing the camshaft (INT) and the camshaft sprocket (INT).

   a. Put the matchmarks on a line extending from the knock pin position of camshaft (INT) front surface. Put the marks on the visible position with the camshaft sprocket installed.

   b. Put the matching marks on a line extending from the knock pin hole position of camshaft sprocket (INT). Put the marks on the visible position with it installed to the camshaft.

22. Install camshafts.

   a. Distinction between camshaft (INT and EXH) is performed by identifying the different shapes of the rear end.

   b. Install camshafts to the cylinder head so that knock pins on front end are positioned as shown.

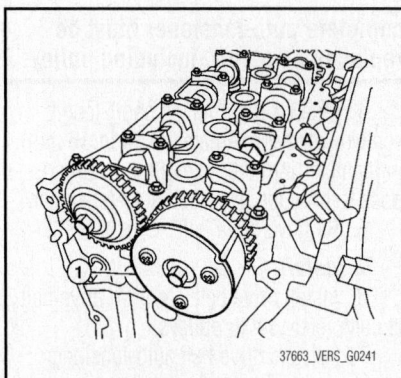

Fig. 59 Camshaft sprocket (INT) (1), hexagonal part (A)

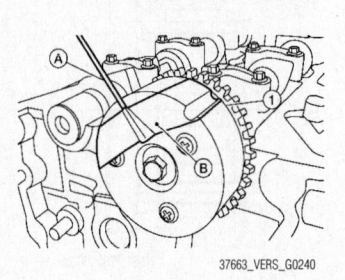

Fig. 58 Insert the stopper pin (A) into the stopper pin holes in the camshaft sprocket (INT) (1) and lock in the most advanced angle position and secure stopper pin with tape (B)

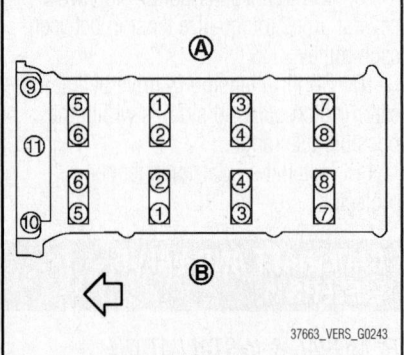

Fig. 60 Camshaft bracket bolt installation sequence, EXH side (A), INT side (B)— reverse for removal

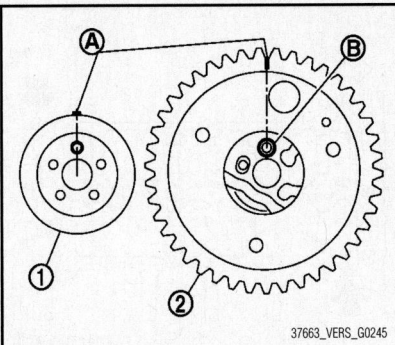

Fig. 61 Matchmarks (A, B), knock pin position (1) and knock pin hole (2)

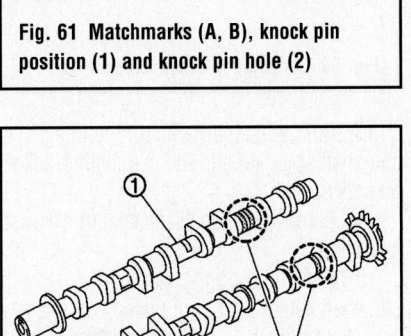

Fig. 62 Camshaft (EXH) (1), camshaft (INT) (2), identification mark (A)

➡**Though camshaft does not stop at the position shown, for the placement of cam nose, it is generally accepted camshaft is placed in the same direction.**

23. Install camshaft brackets (No. 2 to 5), aligning the identification marks on upper surface as shown. Install so that identification mark can be correctly read when viewed from the INT side.

24. Tighten camshaft bracket bolts in the following steps, in numerical order as shown above.

a. Tighten No. 9 to 11 in numerical order to 17 inch lbs. (2 Nm).

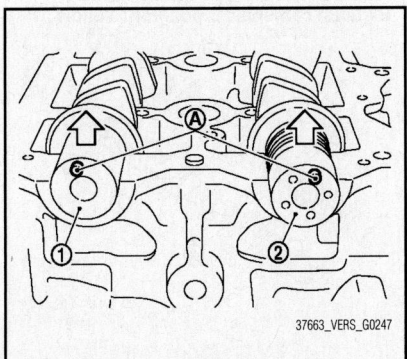

Fig. 63 Knock pin (A) positioning

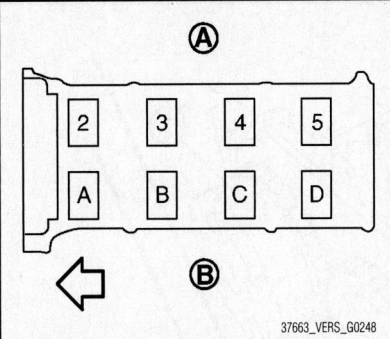

Fig. 64 Identification mark alignment, EXH side (A),

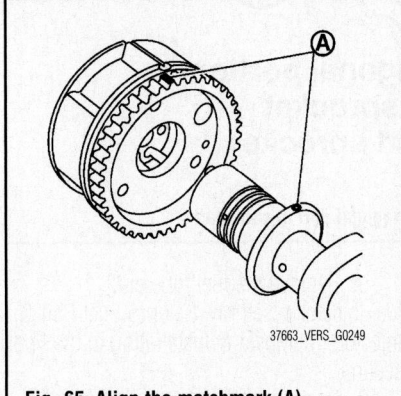

Fig. 65 Align the matchmark (A)

b. Tighten No. 1 to 8 in numerical order to 17 inch lbs. (2 Nm).

c. Tighten all bolts in numerical order to 52 inch lbs. (6 Nm).

d. Tighten all bolts in numerical order to 8 ft. lbs. (10 Nm).

25. Install the camshaft sprocket (INT) to the camshaft (INT):

a. Align the matchmark. Securely align the knock pin and the pin hole, and then install them.

b. Temporarily tighten the camshaft sprocket (INT) bolt on the front side of camshaft sprocket (INT).

c. Hold the camshaft hexagonal part to secure the camshaft and tighten the bolt.

26. Install the camshaft sprocket (EXH) to the camshaft (EXH) while aligning the matching mark and the matchmark of camshaft sprocket (EXH).

a. If the positions of knock pin and pin groove are not aligned, move the camshaft (EXH) slightly to correct these positions.

b. Hold the camshaft hexagonal part to secure the camshaft and tighten the bolt.

c. Make sure that the matchmark and each camshaft sprocket matchmark are in the correct location.

27. Install the timing chain.

28. Install the camshaft position sensor (PHASE) to the rear end of cylinder head. Tighten bolts with it seated completely.

29. Check and adjust valve clearance.

30. Installation of the remaining components is in the reverse order of removal.

1.8L Engine

See Figures 66 through 73.

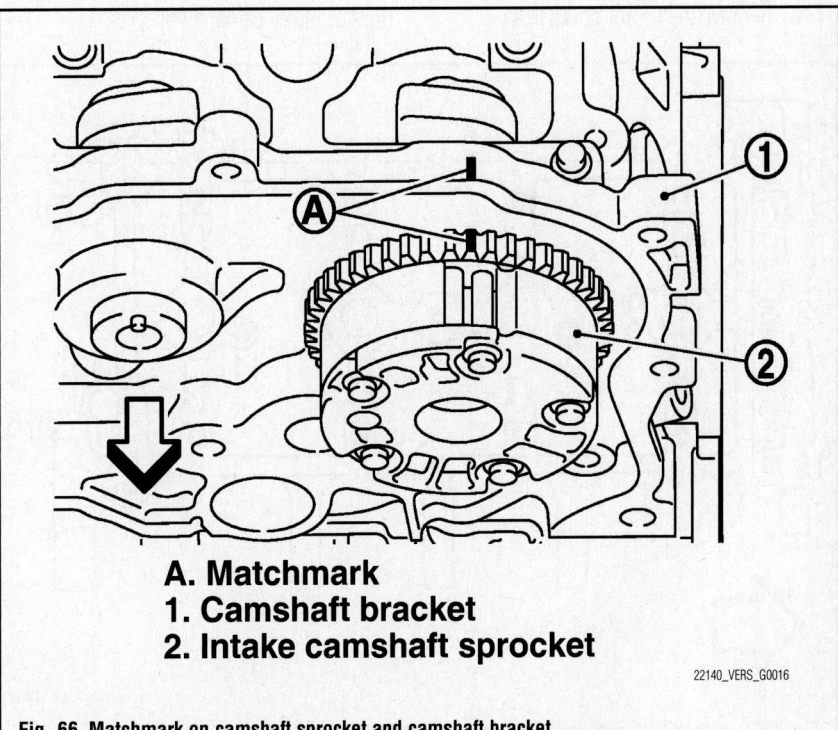

A. Matchmark
1. Camshaft bracket
2. Intake camshaft sprocket

Fig. 66 Matchmark on camshaft sprocket and camshaft bracket

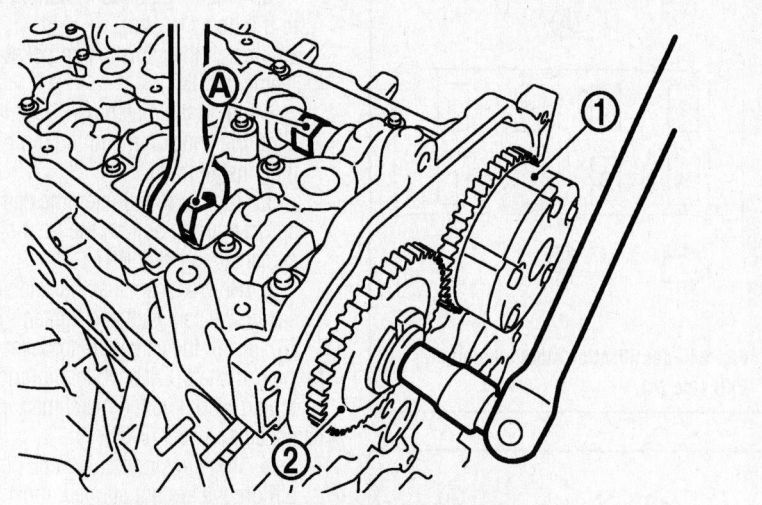

A. Camshaft - hexagonal portion
1. Intake camshaft sprocket
2. Exhaust camshaft sprocket

22140_VERS_G0017

Fig. 67 Remove camshaft sprockets and secure hexagonal part of camshaft

1. Before servicing the vehicle, refer to the Precautions Section.
2. Release the fuel pressure.
3. Disconnect negative battery cable.
4. Remove the right front wheel.
5. Remove the right front fender protector.
6. Drain engine coolant.
7. Remove the intake manifold. Refer to Intake Manifold Removal & Installation.
8. Remove the rocker cover.

9. Remove the fuel tube and fuel injector assembly. Refer to Fuel Rail & Injectors, Removal & Installation in the Fuel section.
10. Remove the front cover, timing chain and related parts. Refer to Timing Chain Removal & Installation.
11. Remove the alternator. Refer to Alternator Removal & Installation in the Engine Electrical section.
12. Remove camshaft position sensor (PHASE) from camshaft bracket.

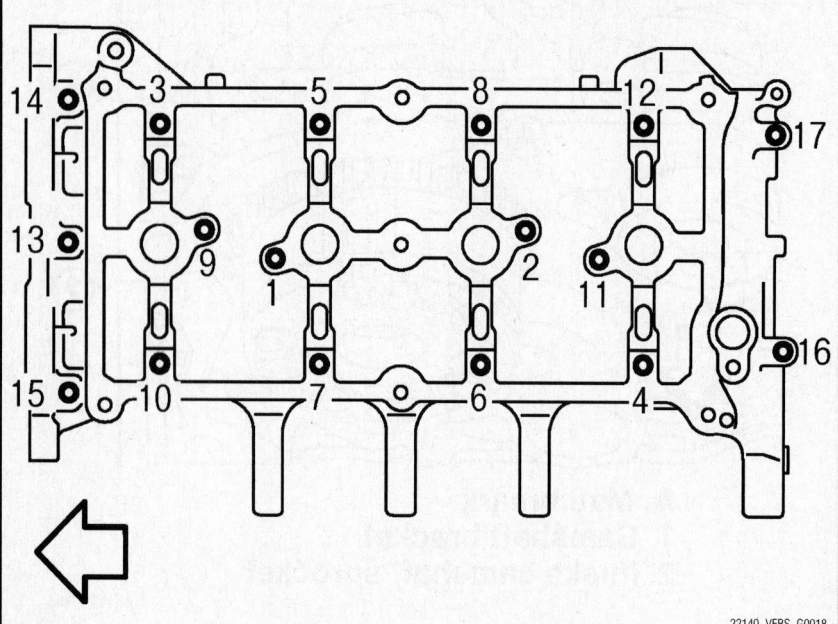

22140_VERS_G0018

Fig. 68 Camshaft bolt installation sequence—reverse for removal

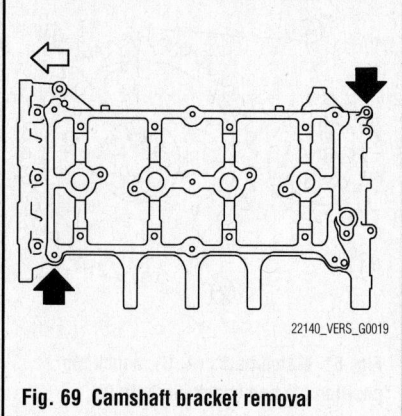

22140_VERS_G0019

Fig. 69 Camshaft bracket removal

13. Put the matchmarks on the intake camshaft sprocket and the camshaft bracket as shown.
14. Remove camshaft intake and exhaust sprockets.
15. Secure hexagonal part of camshaft with a wrench. Loosen camshaft sprocket bolts and remove the camshaft sprocket.

➡**Never rotate crankshaft or camshaft while timing chain is removed. It causes interference between valve and piston.**

➡**Never loosen the bolts with securing anything other than the camshaft hexagonal part or with tensioning the timing chain.**

16. Loosen bolts in reverse order as shown.
17. Cut liquid gasket by prying the position shown, and then remove the camshaft bracket.

➡**Be careful not to damage the mating surface. A more adhesive liquid gasket is applied compared to**

1. Exhaust camshaft
2. Intake camshaft
Intake camshaft A position: Yellow
Exhaust camshaft B position: Yellow

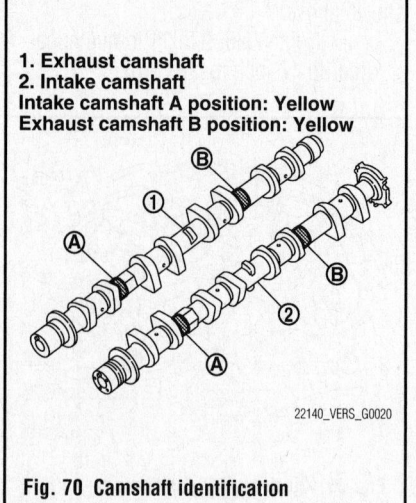

22140_VERS_G0020

Fig. 70 Camshaft identification

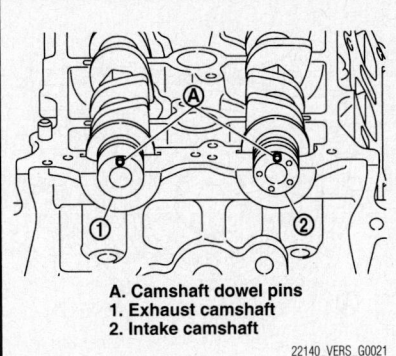

Fig. 71 Camshaft dowel pin installation positioning

A. Camshaft dowel pins
1. Exhaust camshaft
2. Intake camshaft

22140_VERS_G0021

previous types when shipped, so it should not be forced off the position not specified.

18. Remove the camshafts.
19. Remove the valve lifters.

➡️Identify installed positions, and store them without mixing them up.

To install:
20. Install valve lifters in the original positions.
21. Install camshafts.
   a. Clean camshaft journal to remove any foreign material.

➡️Distinguish between the intake and the exhaust by looking at the different shapes of the front and rear ends of the camshaft or using the identification colors.

22. Install camshafts so that camshaft dowel pins on the front side are positioned as shown.

➡️Though camshaft does not stop at the positions as shown, for the placement of cam nose, it is generally accepted camshaft is placed for the same direction as shown.

23. Remove foreign material completely from camshaft bracket backside and from cylinder head installation face.
24. Apply liquid gasket to camshaft bracket as shown.

➡️Use Genuine Silicone RTV Sealant (Tool No. WS39930000) or equivalent.

25. Install camshaft bracket bolts numerically as shown above, noting the following:
   a. Note the 2 types of M6 bolts: bolts No. 13, No. 14, and No. 15 in the figure have a thread length of 57.5 mm (2.264 inch), and the remaining bolts have a thread length of 35 mm (1.378 inch).
   b. Tighten all the bolts in 3 steps:

- Step 1: 17 inch lbs. (1.96 Nm)
- Step 2: 52 inch lbs. (5.88 Nm)
- Step 3: 84 inch lbs. (9.5 Nm)

26. Install the intake camshaft sprocket to the intake camshaft.

➡️When the installing the intake camshaft sprocket, refer to the match-mark. Securely align the knock pin and the pin hole, and then install.

27. Tighten the intake camshaft sprocket bolt to 26 ft. lbs. (35 Nm). Secure the hexagonal part of the intake camshaft using a wrench to tighten the bolt.
28. Turn 67 degrees clockwise (angle tightening) using Tool No. KV10112100 (BT-8653-A).

❄❄ CAUTION

Never judge by visual inspection without an angle wrench.

29. Install the exhaust camshaft sprocket and tighten the bolt to 65 ft. lbs. (88.2 Nm). Secure the hexagonal part of the camshaft using a wrench to tighten the bolt.
30. Install timing chain and related parts.
31. Inspect and adjust valve clearance.
32. Installation of the remaining components is in the reverse order of removal.

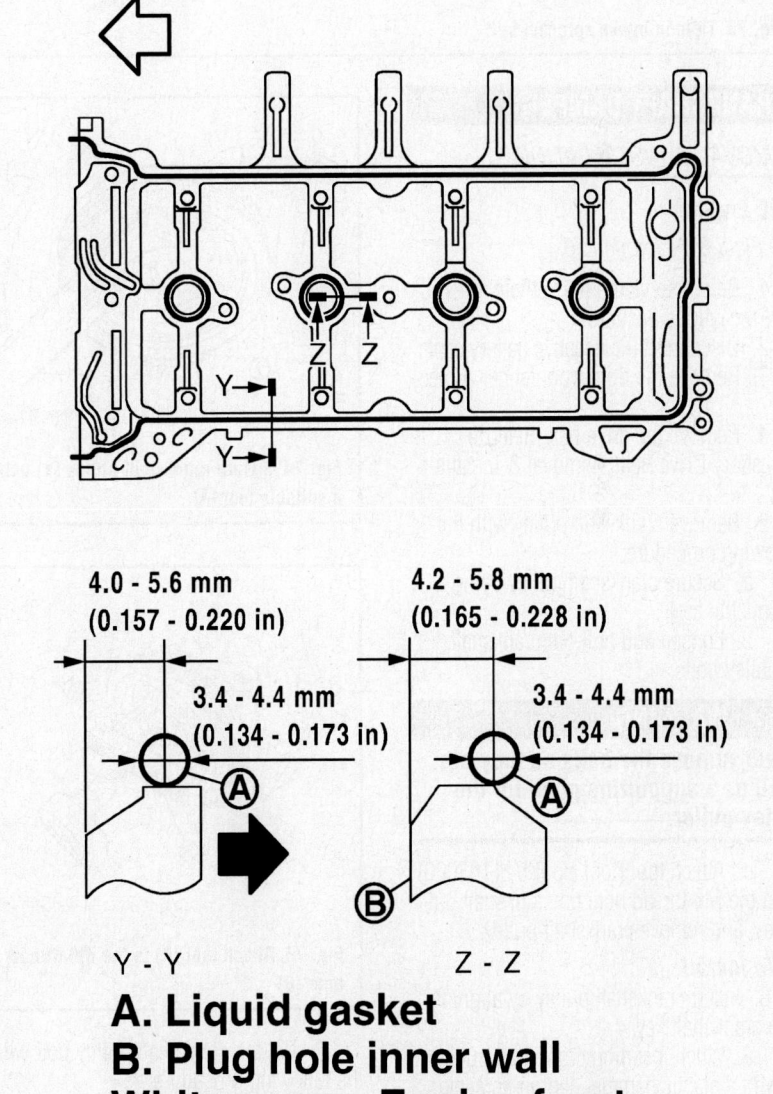

4.0 - 5.6 mm
(0.157 - 0.220 in)

3.4 - 4.4 mm
(0.134 - 0.173 in)

4.2 - 5.8 mm
(0.165 - 0.228 in)

3.4 - 4.4 mm
(0.134 - 0.173 in)

Y - Y

Z - Z

A. Liquid gasket
B. Plug hole inner wall
White arrow: Engine front
Black arrow: Engine outside

22140_VERS_G0022

Fig. 72 Sealant application

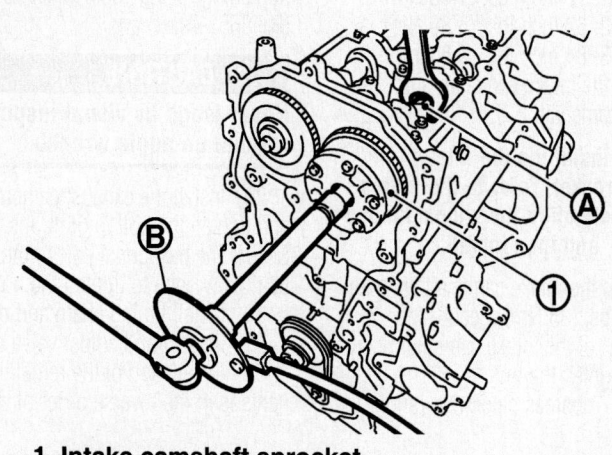

1. Intake camshaft sprocket
A. Intake camshaft - hexagonal portion

22140_VERS_G0024

Fig. 73 Tighten intake sprocket bolt

CRANKSHAFT DAMPER

REMOVAL & INSTALLATION

1.6L Engine

See Figures 74 through 76.

1. Before servicing the vehicle, refer to the Precautions Section.
2. Disconnect the negative battery cable.
3. Remove the right front fender protector.
4. Remove the drive belt. Refer to Accessory Drive Belts Removal & Installation.
5. Remove crankshaft pulley with the following procedure:
   a. Secure crankshaft pulley using a suitable tool.
   b. Loosen and pull out crankshaft pulley bolts.

> ※※ **WARNING**
>
> **Never remove the bolts as they are used as a supporting point for the pulley puller.**

   c. Attach tool (tool No. KV11103000) in the M6 thread hole on crankshaft pulley, and remove crankshaft pulley.

To install:

6. Insert crankshaft pulley by aligning with crankshaft key.
   a. When inserting crankshaft pulley with a plastic hammer, tap on its center portion (not circumference).

> ※※ **WARNING**
>
> **Never damage front oil seal lip section.**

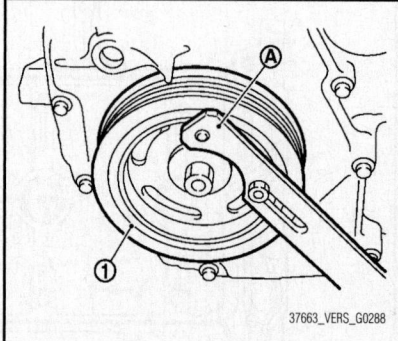

37663_VERS_G0288

Fig. 74 Secure crankshaft pulley (1) using a suitable tool (A)

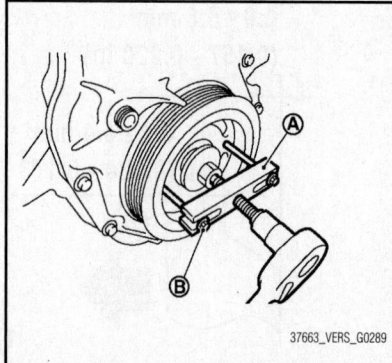

37663_VERS_G0289

Fig. 75 Attach tool (A) in the M6 thread hole (B)

7. Tighten crankshaft pulley bolt with the following procedure:
   a. Secure crankshaft pulley with a suitable tool, and tighten crankshaft pulley bolt.
   b. Apply new engine oil to thread and seat surfaces of crankshaft pulley bolt.
   c. Tighten crankshaft pulley bolt to 26 ft. lbs. (35 Nm).

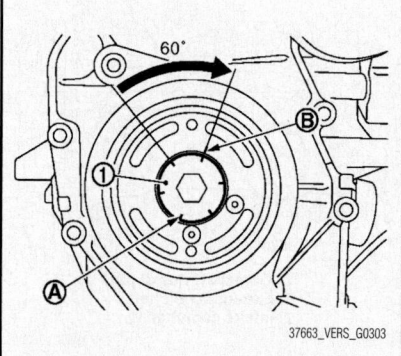

37663_VERS_G0303

Fig. 76 Put a paint mark (B) on crankshaft pulley, mating with any one of six easy to recognize angle marks (A) on crankshaft bolt flange (1)

   d. Put a paint mark on crankshaft pulley, mating with any one of six easy to recognize angle marks on crankshaft bolt flange.
   e. Turn another 60 degrees clockwise (angle tightening). Check the tightening angle with movement of one angle mark.
8. Make sure that crankshaft turns smoothly by rotating by hand clockwise.
9. Install the drive belt.
10. Install the front fender protector (RH).
11. Connect the negative battery cable.

1.8L Engine

See Figures 77 through 79.

1. Before servicing the vehicle, refer to the Precautions Section.
2. Disconnect the negative battery cable.
3. Remove the right front fender protector.
4. Remove the drive belt. Refer to Accessory Drive Belts Removal & Installation.
5. Hold crankshaft pulley using suitable

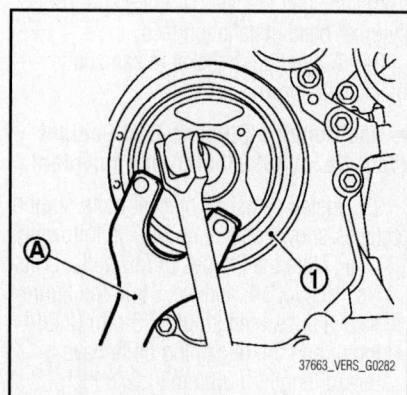

37663_VERS_G0282

Fig. 77 Crankshaft pulley angle tightening

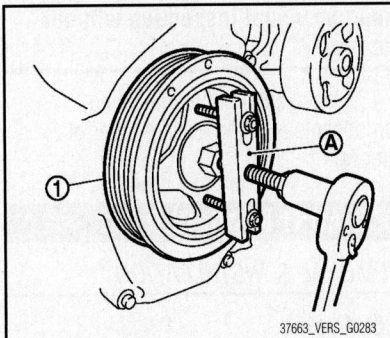

Fig. 78 Crankshaft pulley (1) and tool No. KV11103000 (A)

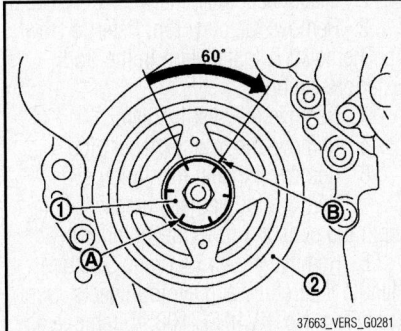

Fig. 79 Crankshaft pulley angle tightening

tool (tool No. KV11103000), and loosen crankshaft pulley bolt. Locate bolt seating surface at 0.39 in. (10 mm) from its original position. Never remove the crankshaft pulley bolt as it will be used as a supporting point for the pulley puller.

6. Attach a pulley puller in the M6 thread hole on crankshaft pulley, and remove crankshaft pulley.

To install:

7. Install crankshaft pulley as follows:

✴✴ WARNING

Never damage front oil seal lip section. If needed use a plastic hammer, tap on its center portion (not circumference) to seat crankshaft pulley.

a. Apply new engine oil to thread and seat surfaces of crankshaft pulley bolt.

b. Secure crankshaft pulley using tool No. KV10109300.

c. Tighten crankshaft pulley bolt in two steps:
- Step 1: 22 ft. lbs. (29 Nm)
- Step 2: 60° clockwise

d. For angle tightening, put a paint mark on crankshaft pulley matching with any one of six easy to recognize angle marks on crankshaft pulley bolt flange.

e. Turn 60 degrees clockwise (angle tightening).

f. Check the tightening angle with movement of one angle mark.

g. Make sure that crankshaft rotates clockwise smoothly.

8. Installation of the remaining components is in the reverse order of removal.

CRANKSHAFT FRONT SEAL

REMOVAL & INSTALLATION

See Figures 80 and 81.

1. Before servicing the vehicle, refer to the Precautions Section.

2. Remove the right front fender protector.

3. Remove the drive belt. Refer to Accessory Drive Belt Removal & Installation.

4. Remove the crankshaft pulley. Refer to Crankshaft Damper Removal & Installation.

5. Remove front oil seal using a suitable tool.

➡**Be careful not to damage front cover and crankshaft.**

To install:

6. Apply new engine oil to new front oil seal joint surface and seal lip.

7. Install front oil seal so that each seal lip is oriented as shown in the figure.

8. Using a suitable drift, press-fit until the height of front oil seal is level with the mounting surface:
- 1.8L engine outer diameter: 57 mm (2.24 inch)
- 1.8L engine inner diameter: 45 mm (1.77 inch)

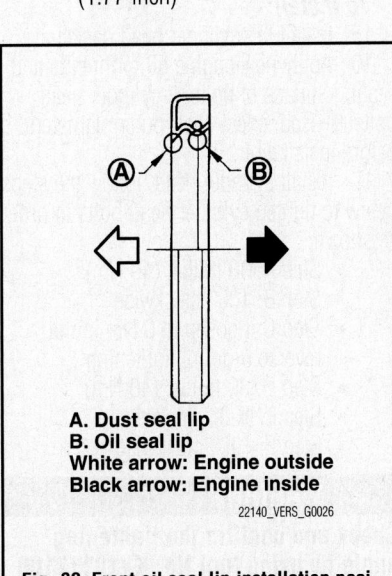

A. Dust seal lip
B. Oil seal lip
White arrow: Engine outside
Black arrow: Engine inside

22140_VERS_G0026

Fig. 80 Front oil seal lip installation positioning

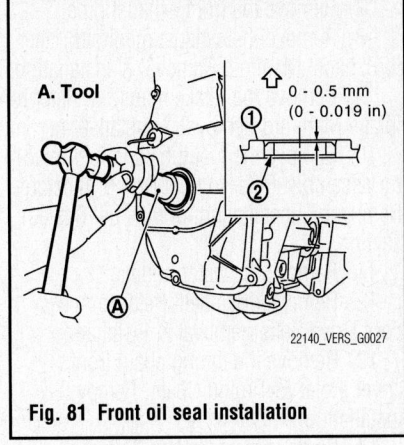

A. Tool

0 - 0.5 mm (0 - 0.019 in)

22140_VERS_G0027

Fig. 81 Front oil seal installation

- 1.6L engine outer diameter: 50 mm (1.97 inch)
- 1.6L engine inner diameter: 44 mm (1.73 inch)

➡**Press-fit oil seal straight to avoid causing burrs or tilting.**

➡**Do not touch grease applied on oil seal lip.**

9. Installation of the remaining components is in the reverse order of removal.

CYLINDER HEAD

REMOVAL & INSTALLATION

1.6L Engine

See Figures 82.

1. Before servicing the vehicle, refer to the Precautions Section.

2. Release the fuel pressure.

3. Disconnect the negative battery cable.

4. Drain engine coolant and engine oil.

5. Remove the right front fender protector.

6. Remove alternator. Refer to Alternator Removal & Installation in the Engine Electrical section.

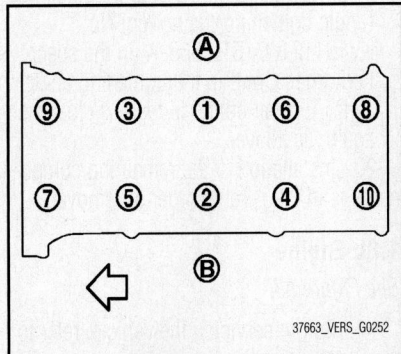

37663_VERS_G0252

Fig. 82 Cylinder head bolt installation sequence—reverse for removal

7. Remove the front exhaust pipe.

8. Remove the exhaust manifold. Refer to Exhaust Manifold Removal & Installation.

9. Remove the intake manifold. Refer to Intake Manifold Removal & Installation.

10. Remove the fuel tube and fuel injector assembly. Refer to Fuel Rail & Injectors, Removal & Installation, in the Fuel System section.

11. Remove the water outlet.

12. Remove drive belt. Refer to Accessory Drive Belts Removal & Installation.

13. Remove the timing chain front cover. Refer to Timing Chain, Removal & Installation.

14. Remove the camshaft. Refer to Camshaft & Valve Lifters, Removal & Installation.

15. Remove the air cleaner. Refer to Air Cleaner Removal & Installation.

16. Remove cylinder head bolts in reverse order as shown.

17. Remove the cylinder head gasket.

To install:

18. Install the cylinder head gasket.

19. Tighten cylinder head bolts in numerical order shown above, with the following procedure. to install cylinder head.

a. Apply new engine oil to threads and seating surface of bolts. If cylinder head bolts re-used, check their outer diameters before installation.

b. Tighten all bolts in the specified order to 30 ft. lbs. (40 Nm).

c. Turn all bolts 60 degrees clockwise (angle tightening) using tool No. KV10112100 (BT-8653-A) in the specified order.

d. Check and confirm the tightening angle by using tightening tool or protractor. Avoid judgment by visual inspection without the tool.

e. Completely loosen all bolts in reverse order.

f. Retighten all bolts in the specified order to 30 ft. lbs. (40 Nm).

g. Turn all bolts 75 degrees clockwise (angle tightening) using tool No. KV10112100 (BT-8653-A) in the specified order. Confirm the tightening angle.

h. Turn all bolts 75 degrees clockwise again, as above.

20. Installation of the remaining components is in the reverse order of removal.

1.8L Engine

See Figure 83.

1. Before servicing the vehicle, refer to the Precautions Section.

2. Release the fuel pressure.

3. Drain engine coolant and engine oil.

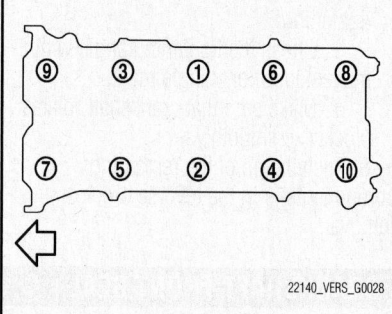

Fig. 83 Cylinder head bolt installation sequence—reverse for removal

4. Remove the right front fender protector.

5. Remove drive belt. Refer to Accessory Drive Belts Removal & Installation.

6. Remove the exhaust manifold. Refer to Exhaust Manifold Removal & Installation.

7. Remove the intake manifold. Refer to Intake Manifold Removal & Installation.

8. Remove the water outlet.

9. Remove the fuel tube and fuel injector assembly. Refer to Fuel Rail & Injectors, Removal & Installation, in the Fuel System section.

10. Remove the rocker cover.

11. Remove the timing chain front cover. Refer to Timing Chain, Removal & Installation.

12. Remove the camshaft. Refer to Camshaft & Valve Lifters, Removal & Installation.

13. Using a TORX® socket (size E18), remove cylinder head, loosening the bolts in reverse order as shown.

14. Remove the cylinder head gasket.

To install:

15. Install the cylinder head gasket.

16. Apply new engine oil to threads and seating surface of bolts. If cylinder head bolts re-used, check their outer diameters before installation.

17. Install cylinder head, follow the steps below to tighten cylinder head bolts in order as shown:

- Step A: 30 ft. lbs. (40 Nm)
- Step B: 100° clockwise
- Step C: Loosen to 0 Nm in the reverse order of tightening
- Step D: 30 ft. lbs. (40 Nm)
- Step E: 100° clockwise
- Step F: 100° clockwise

✳✳ WARNING

Check and confirm the tightening angle by using tool No. KV10112100 (BT-8653-A) or protractor. Never judge by visual inspection without the tool.

18. Installation of the remaining components is in the reverse order of removal.

EXHAUST MANIFOLD

REMOVAL & INSTALLATION

1.6L Engine

See Figure 84.

1. Before servicing the vehicle, refer to the Precautions Section.

2. Disconnect the negative battery cable.

3. Remove the cowl top. Refer to Cowl Top Removal & Installation in the Body Exterior section.

4. Remove the heat insulator.

5. Remove the front exhaust pipe.

6. Remove exhaust manifold cover.

7. Remove the harness bracket of air fuel ratio sensor 1 from the cylinder head.

8. Remove the air fuel ratio sensor 1. Handle it carefully and avoid impacts. Use Tool (Tool No. KV10117100) to remove air fuel ratio sensor 1.

➡ **The exhaust manifold can be removed and installed without removing the air fuel ratio sensor 1 (Disassembly of harness connector is necessary).**

9. Remove exhaust manifold side bolt of exhaust manifold stay.

10. Remove exhaust manifold. Loosen nuts in the reverse of the order shown.

11. Remove gasket.

12. Cover engine openings to avoid entry of foreign materials.

13. Remove exhaust manifold cover from back of exhaust manifold.

To install:

14. Installation is in the reverse order of removal, noting the following:

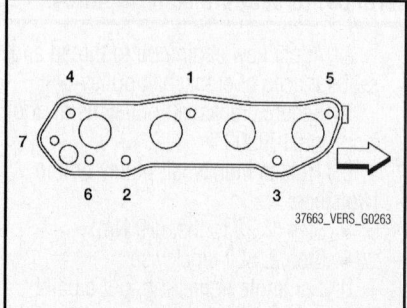

Fig. 84 Exhaust manifold bolt installation sequence—reverse for removal

15. Tighten exhaust manifold nuts to specification in two stages in the numerical order shown above.

16. Before installing a new air fuel ratio sensor 1, clean the exhaust tube threads using suitable tool and approved anti-seize lubricant. Use Tool (Tool No. KV10117100) to install air fuel ratio sensor 1. Do not over-tighten.

1.8L Engine

See Figures 85 and 86.

1. Before servicing the vehicle, refer to the Precautions Section.

2. Disconnect the negative battery cable.

3. Remove the cowl top. Refer to Cowl Top Removal & Installation in the Body Exterior section.

4. Remove the front exhaust pipe.

5. Remove exhaust manifold cover.

6. Remove the A/F sensor 1, using tool No. KV991J0050 (J-44626). Handle it carefully and avoid impacts.

7. Remove exhaust manifold side bolt of exhaust manifold stay.

8. Loosen nuts in reverse order as shown and remove exhaust manifold.

9. Remove gasket.

10. Cover engine openings to avoid entry of foreign materials.

To install:

11. Install exhaust manifold gasket.

12. Tighten exhaust manifold nuts to specification in two stages in the numerical order as shown above.

13. Install exhaust manifold stay.

14. Before installing a new A/F ratio sensor, clean the exhaust tube threads using suitable tool and approved anti-seize lubricant.

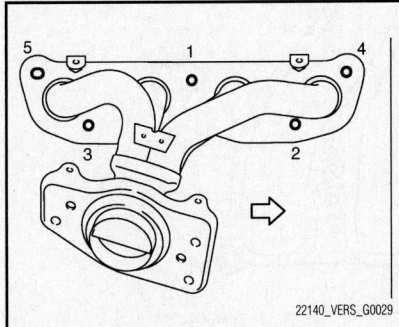

Fig. 85 Exhaust manifold bolt installation sequence—reverse for removal

Fig. 86 Install exhaust manifold stay (2) in the direction shown

15. Install the A/F ratio sensor 1, using tool No. KV991J0050 (J-44626). Do not over-tighten.

16. Installation of the remaining parts is in the reverse order of removal.

FLYWHEEL/DRIVE PLATE

REMOVAL & INSTALLATION

See Figures 87 and 88.

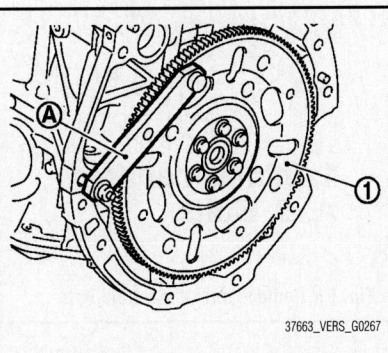

Fig. 87 Drive plate (1) and tool (A)—A/T or CVT models

1. Before servicing the vehicle, refer to the Precautions Section.

2. Remove transaxle assembly.

3. Remove clutch cover and clutch disk (M/T models).

4. Remove drive plate (A/T or CVT models) or flywheel (M/T models).

   a. Secure flywheel (M/T models) or drive plate (A/T or CVT models) using Tool No. KV 11105210 (J-44716), and remove bolts. Be careful not to damage or scratch drive plate or contact surface of flywheel clutch disc.

To install:

5. Installation is the reverse of removal.

INTAKE MANIFOLD

REMOVAL & INSTALLATION

1.6L Engine

See Figures 89 and 90.

1. Before servicing the vehicle, refer to the precautions section.

2. Disconnect the negative battery cable.

3. Remove the air duct (inlet) and air duct (between air cleaner case and electric throttle control actuator). Refer to Air Cleaner Removal & Installation.

4. Remove the reservoir tank.

5. Remove the oil level gauge. Cover the oil level gauge guide openings to avoid entry of foreign materials.

6. Remove electric throttle control actuator and position aside. Handle carefully and avoid impacts. Never disassemble or adjust electric throttle control actuator.

7. Disconnect the harness connector and EVAP hose from the EVAP canister purge volume control solenoid valve.

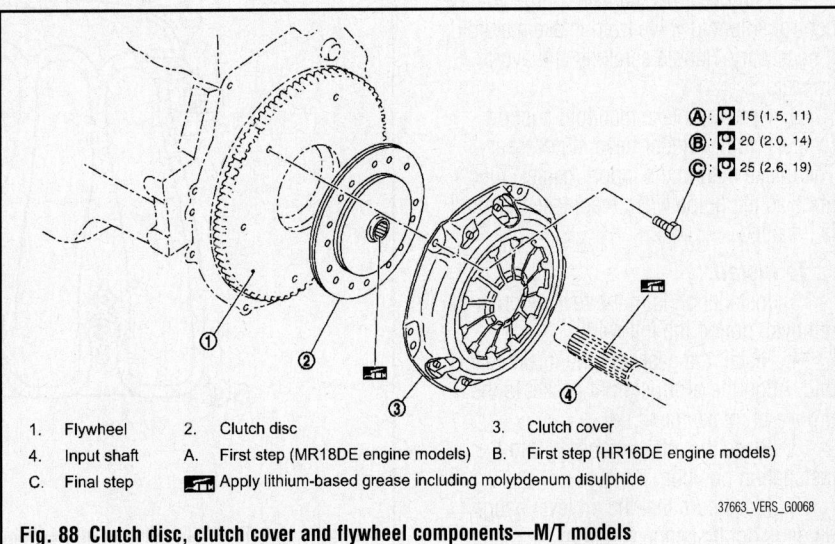

1.	Flywheel	2.	Clutch disc	3.	Clutch cover
4.	Input shaft	A.	First step (MR18DE engine models)	B.	First step (HR16DE engine models)
C.	Final step		Apply lithium-based grease including molybdenum disulphide		

Ⓐ : ⟳ 15 (1.5, 11)
Ⓑ : ⟳ 20 (2.0, 14)
Ⓒ : ⟳ 25 (2.6, 19)

Fig. 88 Clutch disc, clutch cover and flywheel components—M/T models

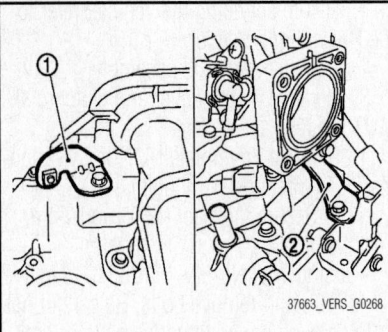

Fig. 89 Remove front intake manifold support bracket (1) and bolt from rear (2)

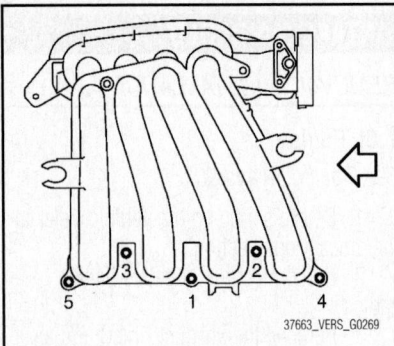

Fig. 90 Intake manifold bolt installation sequence—reverse for removal

8. Disconnect vacuum hose for brake booster from intake manifold.

9. Remove the front intake manifold support bracket (1) and the bolt from the rear (2). Rear bracket (2) is not removed. Remove bolt from intake manifold.

10. Remove intake manifold.

➡**Loosen bolts in the reverse of the order shown.**

11. Remove EVAP canister purge volume control solenoid valve from intake manifold, if necessary. Handle carefully and avoid impacts.

12. Remove intake manifold support (center) from cylinder head, if necessary. The intake manifold support (center) functions as the guide when the intake manifold is installed.

To install:

13. Installation is in the reverse order of removal, noting the following:

14. Install the gasket to the intake manifold. Align the protrusion of gasket to the groove of intake manifold.

15. Place the intake manifold into the installation position.

16. Make sure that the oil level gauge guide is not disconnected from the fixing

clip of water inlet due to interference with intake manifold.

17. Tighten bolts in the numerical order shown above.

18. Install intake manifold support (front and rear).

1.8L Engine

See Figures 91 and 92.

1. Before servicing the vehicle, refer to the Precautions Section.

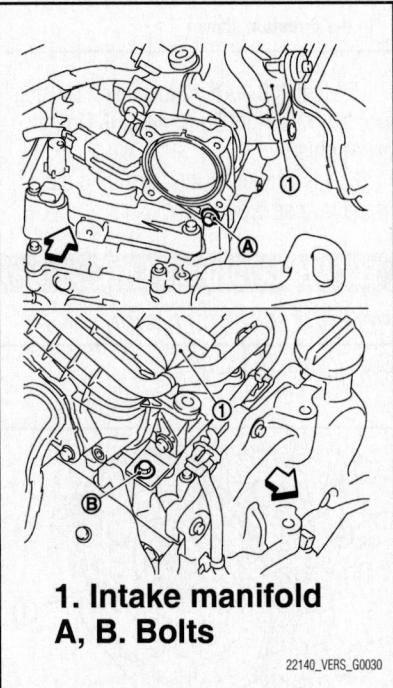

1. Intake manifold
A, B. Bolts

Fig. 91 Remove intake manifold bolts

2. Disconnect the negative battery cable.

3. Remove engine cover.

4. Remove the air duct, air duct (inlet) and air duct (front). Refer to Air Cleaner Removal & Installation.

5. Disconnect the EVAP canister purge volume control solenoid valve.

6. Partially drain engine coolant. Perform this step when engine is cold. This step is unnecessary when putting plugs to water hoses (to electric throttle control actuator).

7. Disconnect water hoses from the electronic throttle control actuator and remove the electronic throttle control actuator. Do not disassemble.

8. Remove the PCV hose and the vacuum hose.

9. Remove oil level gauge. Cover the oil level gauge guide openings to avoid entry of foreign materials.

10. Loosen and remove intake manifold bolts as shown.

11. Loosen bolts in reverse order as shown. Cover engine openings to avoid entry of foreign materials.

12. Remove intake manifold.

To install:

13. Install intake manifold. Be sure the intake manifold gasket is seated correctly in groove of intake manifold.

14. Tighten bolts in numerical order as shown above.

15. Tighten intake manifold bolts as shown above, Tighten in alphabetical order to 14 ft. lbs. (20 Nm).

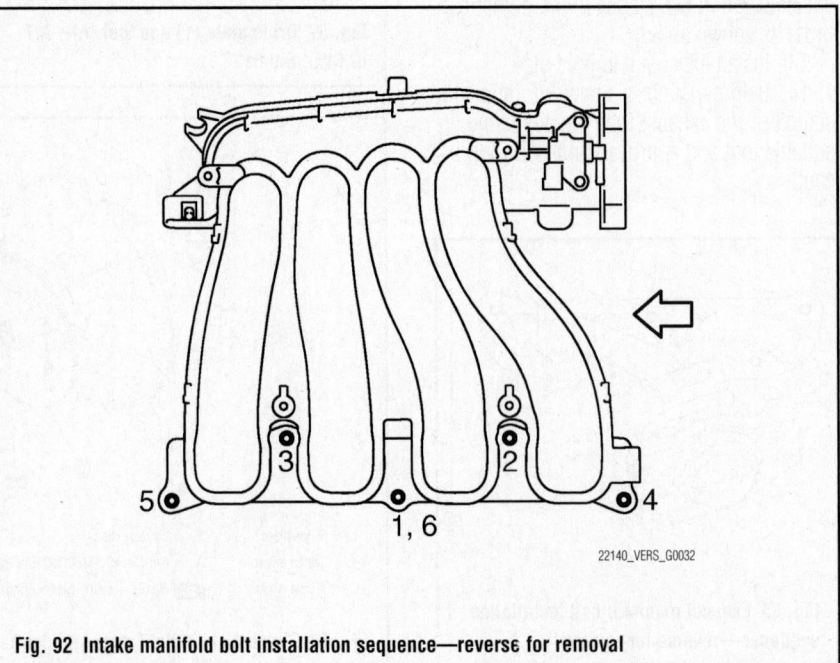

Fig. 92 Intake manifold bolt installation sequence—reverse for removal

16. Install electronic throttle control actuator.

17. Install the water hoses to electronic throttle control actuator.

18. Installation of the remaining components is in the reverse order of removal.

OIL PAN

REMOVAL & INSTALLATION

1.6L Engine

See Figures 93 and 94.

1. Before servicing the vehicle, refer to the Precautions Section.

2. Drain engine oil.

3. Loosen bolts in the reverse of the order shown.

4. Insert the tool No. KV10111100 (J-37228) between oil pan (upper) and oil pan (lower).

❊❊ WARNING

Be careful not to damage the mating surface. A more adhesive liquid gasket is applied compared to previous types when shipped, so it should not be forced off using a flat- bladed screwdriver, etc.

5. Remove the lower oil pan.

To install:

6. Use scraper to carefully remove old liquid gasket from mating surfaces. Also remove the old liquid gasket from mating surface of oil pan (upper). Remove old liquid gasket from the bolt holes and threads.

7. Apply a continuous bead of liquid gasket with tool No. WS39930000 to areas shown. Bead width:
- A: 0.295–0.374 in. (7.5–9.5 mm)
- B: 0.157–0.196 in. (4.0–5.0 mm)

➡**Attaching should be done within 5 minutes after sealant application.**

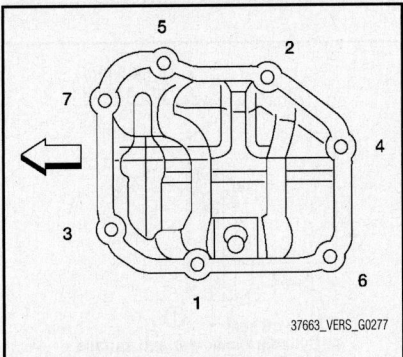

Fig. 93 Lower oil pan bolt installation sequence—reverse for removal

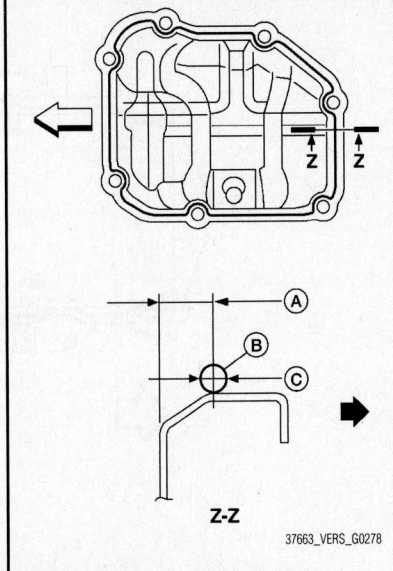

Fig. 94 Apply a continuous bead of liquid gasket (B) to areas shown

8. Tighten bolts in the numerical order shown above.

❊❊ WARNING

Do not pour engine oil until at least 30 minutes after oil pan (lower) is installed.

9. Refill the engine oil level, check and adjust as necessary.

10. Start engine, and check there are no leaks of engine oil.

11. Stop engine and wait for 10 minutes.

12. Check the engine oil level again

1.8L Engine

See Figures 95 through 102.

1. Before servicing the vehicle, refer to the Precautions Section.

2. Drain engine oil.

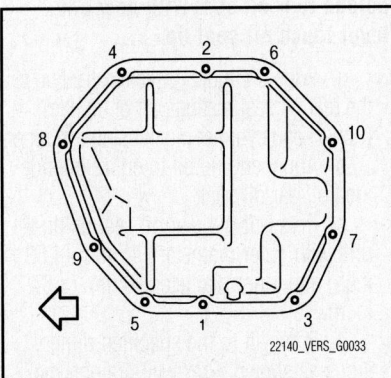

Fig. 95 Oil pan bolt installation sequence—reverse for removal

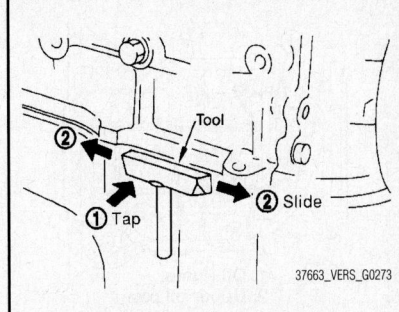

Fig. 96 Slide (2) the Tool by tapping (1) its side with a hammer to remove the oil pan (lower) from the oil pan (upper)

3. Remove engine and transaxle assembly. Refer to Engine Assembly Removal & Installation.

4. Remove flywheel (M/T models) or drive plate (CVT or A/T models).

5. Remove oil filter using tool No. KV10115801. When removing, prepare a shop cloth to absorb any engine oil leakage or spillage.

6. Remove lower oil pan bolts in reverse order as shown.

7. After removing the bolts and nuts, separate the mating surface and remove the sealant using tool No. KV10111100 (J-37228). Be careful not to damage the mating surfaces.

   a. Slide the tool by tapping its side with a hammer to remove the oil pan (lower) from the oil pan (upper).

8. Remove the front cover, timing chain, oil pump drive chain.

9. Remove oil pump. Loosen bolts in reverse order as shown.

10. Remove oil pan (upper) bolts in reverse order as shown.

11. Insert a screwdriver into the area indicated by the arrows and open up a crack between the upper oil pan and cylinder block.

➡**A more adhesive liquid gasket is applied compared to previous types when shipped, so it should not be forced off the position not specified.**

12. After removing the bolts, separate the mating surface and remove the sealant using tool No. KV10111100 (J-37228).

   a. Slide the tool by tapping its side with a hammer to remove the oil pan (upper) from the cylinder block. Be careful not to damage the mating surfaces.

13. Remove O-ring between cylinder block and upper oil pan.

To install:

14. Use a scraper to remove old liquid gasket from mating surfaces. Remove the old liquid gasket from the mating surface of

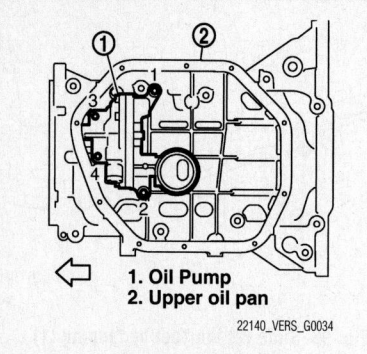

1. Oil Pump
2. Upper oil pan

22140_VERS_G0034

Fig. 97 Oil pump bolt installation sequence, oil pump (1) and oil pan (upper) (2)—reverse for removal

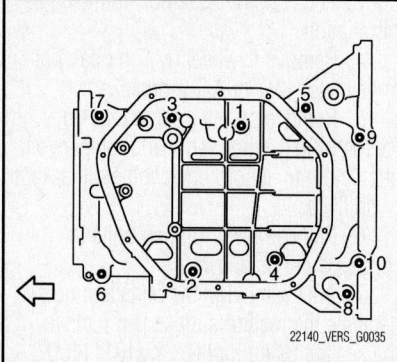

22140_VERS_G0035

Fig. 98 Upper oil pan bolt installation sequence—reverse for removal

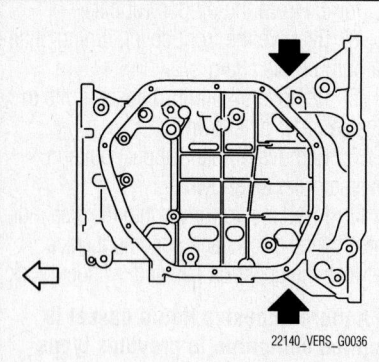

22140_VERS_G0036

Fig. 99 Pry locations between the upper oil pan and cylinder block

cylinder block and from the bolt holes and threads without damaging the area.

15. Apply the sealant (Genuine Silicone RTV Sealant or equivalent) without breaks to the specified location using tool No. WS39930000. Apply liquid gasket to outside of bolt hole for the positions shown by triangle marks.

16. Install new O-ring at cylinder block side.

17. Tighten bolts in numerical order as shown above.

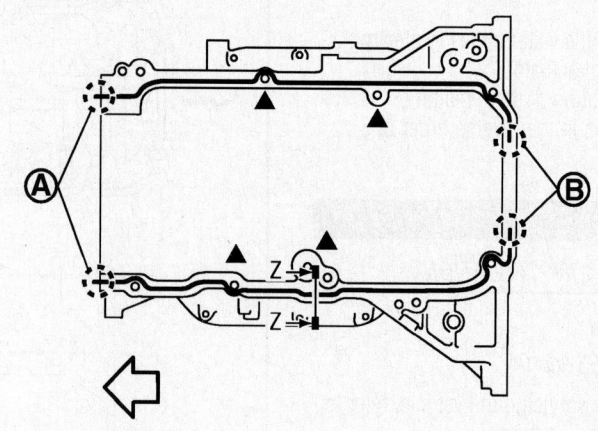

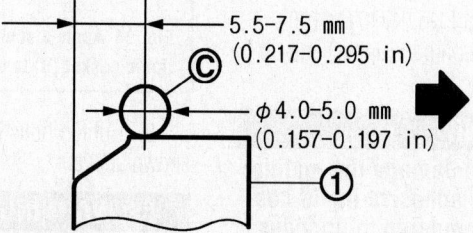

1: Upper oil pan   Z–Z
A: 2 mm protruded to outside
B: 2 mm protruded to rear oil seal mounting side
White arrow: Engine front
Black arrow: Engine outside
Black triangle positions: Apply liquid gasket to outside of bolt hole

22140_VERS_G0059

Fig. 100 Sealant application

18. Install rear oil seal with the following procedure:

➡ **The installation of rear oil seal should be completed within 5 minutes after installing oil pan (upper). Always replace rear oil seal with new one. Never touch oil seal lip.**

   a. Wipe off liquid gasket protruding to the rear oil seal mating part of oil pan (upper) and cylinder block using a scraper.

   b. Apply engine oil to entire outside area of rear oil seal.

   c. Press-fit the rear oil seal using a drift with outer diameter 115 mm (4.53 inch) and inner diameter 90 mm (3.54 inch).

   d. Press-fit to the specified dimensions as shown. Press-fit straight, making sure that rear oil seal does not curl or tilt.

※※ WARNING

Never touch the grease applied to the oil seal lip. Be careful not to damage the rear oil seal mounting part of oil pan (upper) and cylinder block or the crankshaft.

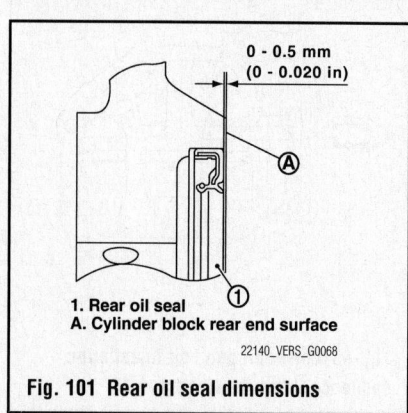

1. Rear oil seal
A. Cylinder block rear end surface

22140_VERS_G0068

Fig. 101 Rear oil seal dimensions

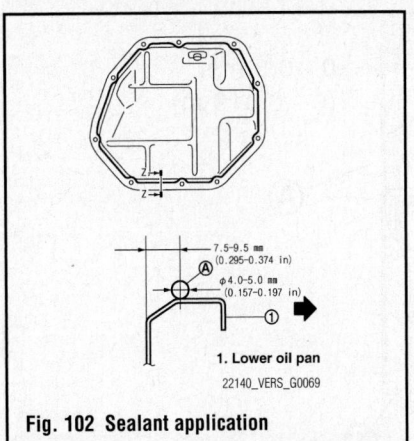

Fig. 102 Sealant application

➡ **The standard surface of the dimension is the rear end surface of cylinder block.**

19. Install oil pump and tighten

bolts in numerical order as shown above.

20. Install oil pump sprocket, oil pump drive chain, and other related parts if removed.

21. Use a scraper to remove old liquid gasket from mating surfaces. Also remove old liquid gasket from mating surface of oil pan (upper)and bolt holes and threads.

22. Apply the sealant (Genuine Silicone RTV Sealant or equivalent) without breaks to the specified location using tool No. WS39930000.

23. Tighten bolts in numerical order as shown above.

24. Install oil filter with the following procedure:

　a. Remove foreign materials adhering to the oil filter installation surface.

　b. Apply new engine oil to the oil seal contact surface of new oil filter.

　c. Screw oil filter manually until it touches the installation surface, and then tighten it by ⅔ turn. Or tighten to 13 ft. lbs. (18 Nm).

25. The remainder of installation is the reverse of removal.

OIL PUMP

REMOVAL & INSTALLATION

See Figure 103.

1. Before servicing the vehicle, refer to the Precautions Section.

2. For 1.6L engines, the oil pump is not serviced separately, it is serviced as part of the oil pan. For 1.8L engines, remove and install the oil pump as follows.

3. Remove the timing chain and oil pump drive chain. Refer to Timing Chain & Sprockets Removal & Installation.

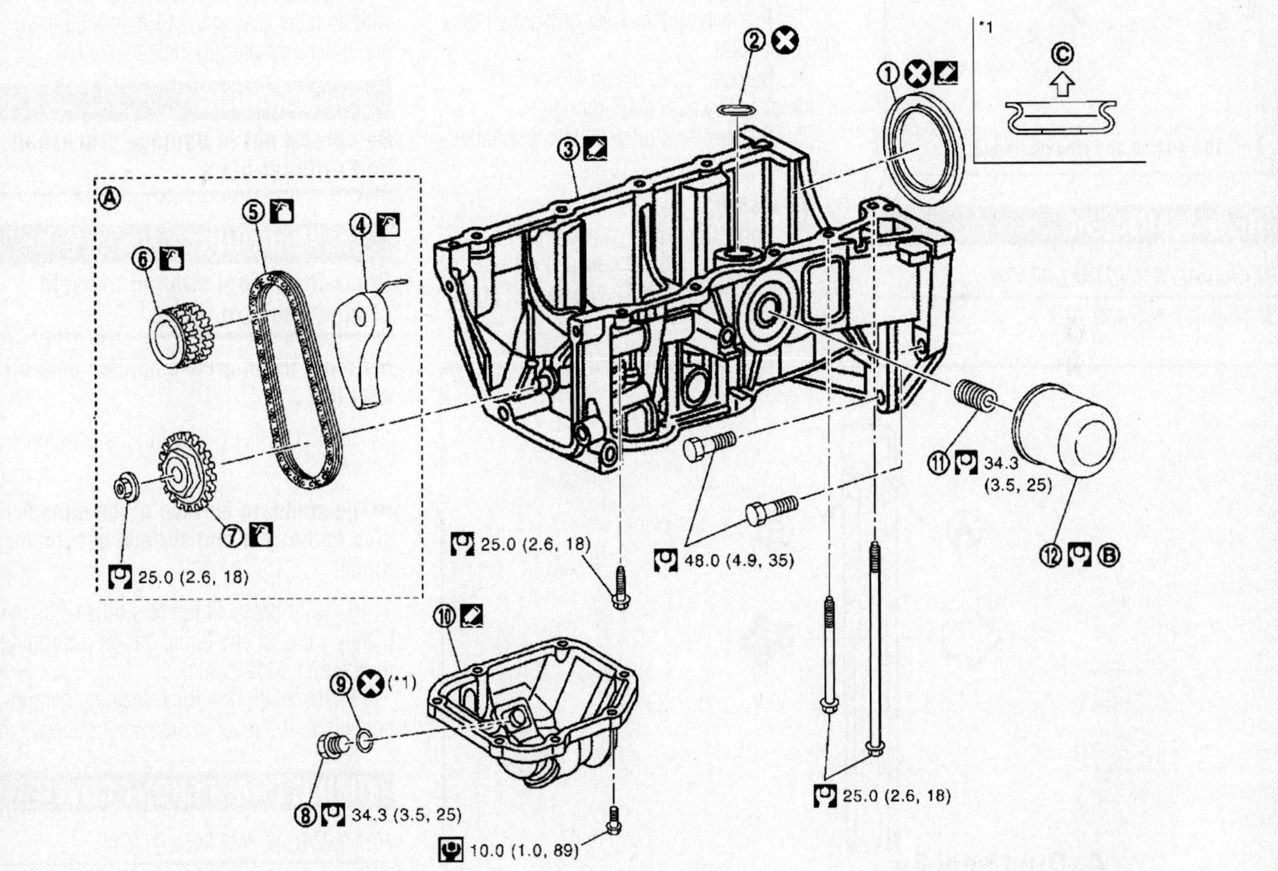

1.	Rear oil seal	2.	O-ring	3.	Oil pan (upper)
4.	Chain tensioner	5.	Oil pump drive chain	6.	Crankshaft sprocket
7.	Oil pump sprocket	8.	Oil pan drain plug	9.	Washer
10.	Oil pan (lower)	11.	Oil filter stud bolt	12.	Oil filter

Fig. 103 Oil pump (1) bolt installation sequence, oil pan (2) shown—reverse for removal

4. Remove oil pump. Loosen bolts in reverse order as shown.

To install:

5. Install oil pump. Tighten bolts in numerical order as shown above.

6. Install the timing chain and oil pump drive chain.

PISTON AND RING

POSITIONING

See Figure 104.

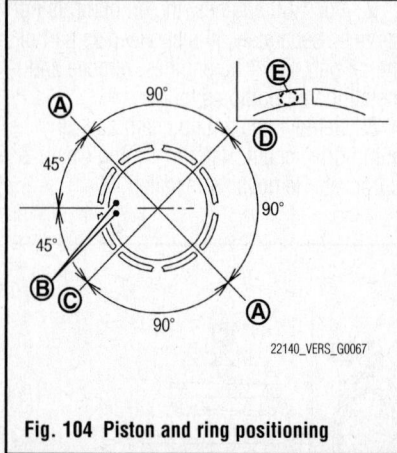

Fig. 104 Piston and ring positioning

REAR MAIN SEAL

REMOVAL & INSTALLATION

See Figures 105 and 106.

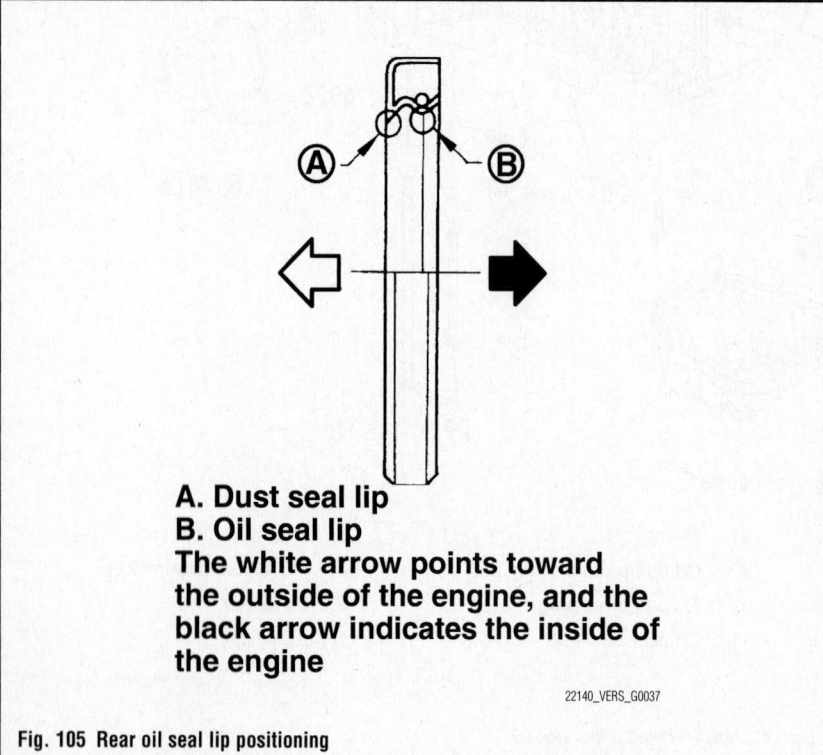

A. Dust seal lip
B. Oil seal lip
The white arrow points toward the outside of the engine, and the black arrow indicates the inside of the engine

22140_VERS_G0037

Fig. 105 Rear oil seal lip positioning

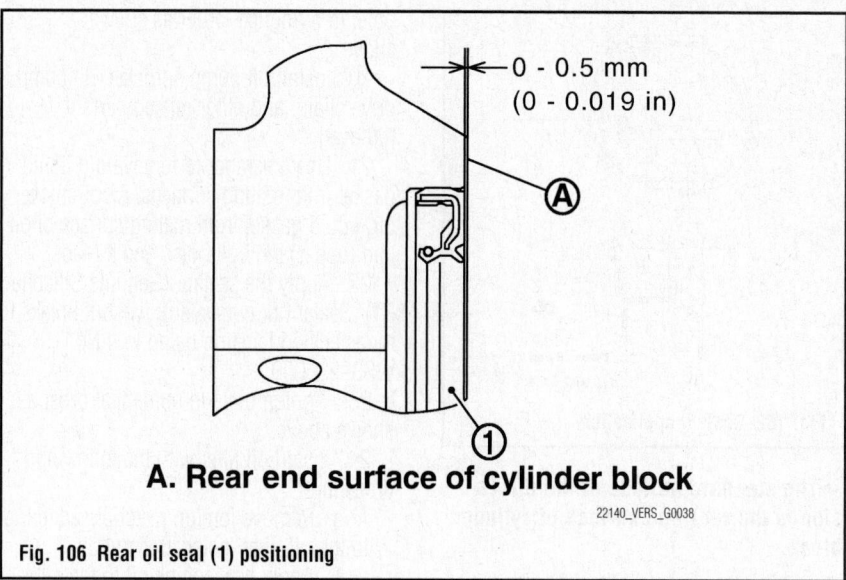

A. Rear end surface of cylinder block

22140_VERS_G0038

Fig. 106 Rear oil seal (1) positioning

1. Before servicing the vehicle, refer to the Precautions Section.

2. Remove transaxle assembly.

3. Remove clutch cover and clutch disk (M/T models).

4. Remove drive plate (A/T or CVT models) or flywheel (M/T models).

5. Remove rear oil seal with a suitable tool.

To install:

6. Apply the liquid gasket lightly to entire outside area of new rear oil seal. Use Genuine Silicone RTV Sealant or equivalent.

7. Install rear oil seal so that each seal lip is oriented as shown.

8. Install rear oil seal with a suitable tool with an outer diameter 115 mm (4.53 inch) and inner diameter 90 mm (3.54 inch).

✳✳ WARNING

Be careful not to damage crankshaft and cylinder block.

✳✳ WARNING

Press-fit oil seal straight to avoid causing burrs or tilting.

➡**Do not touch grease applied onto oil seal lip.**

9. Install rear oil seal to the position as shown.

➡**The standard surface of the dimension is the rear end surface of cylinder block.**

10. After press-fitting rear oil seal, completely wipe off any liquid gasket protruding to rear end surface side.

11. Installation of the remaining components is in the reverse order of removal.

TIMING CHAIN FRONT COVER

REMOVAL & INSTALLATION

See Timing Chain & Sprockets, Removal & Installation.

TIMING CHAIN & SPROCKETS

REMOVAL & INSTALLATION

1.6L Engine

See Figures 107 through 120.

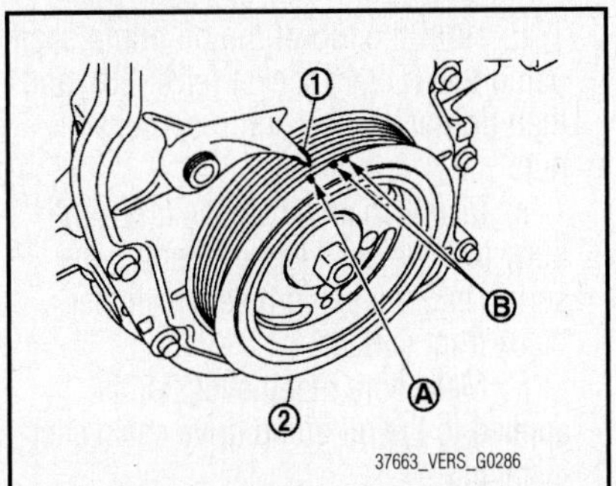
1
B
A
2
37663_VERS_G0286

Fig. 107 Rotate crankshaft pulley (2) clockwise and align TDC mark (A) to timing indicator (1) on front cover

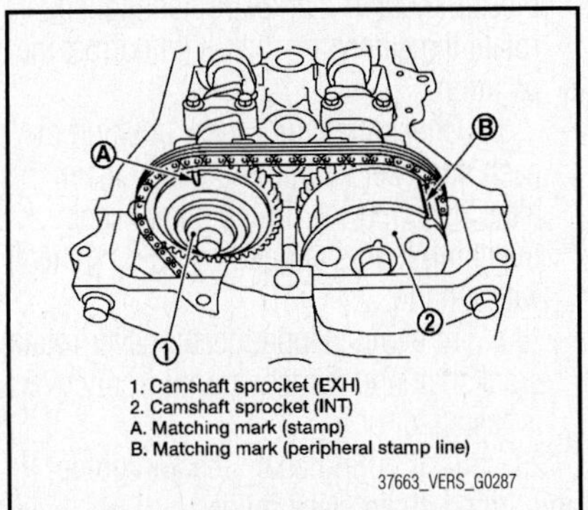
A
B
1
2
1. Camshaft sprocket (EXH)
2. Camshaft sprocket (INT)
A. Matching mark (stamp)
B. Matching mark (peripheral stamp line)
37663_VERS_G0287

Fig. 108 Camshaft sprocket matching mark positioning

1. Before servicing the vehicle, refer to the Precautions Section.
2. Disconnect the negative battery cable.
3. Drain engine oil.
4. Remove the right front wheel.
5. Remove the right front fender protector.
6. Remove the intake manifold. Refer to Intake Manifold Removal & Installation.
7. Remove the drive belt. Refer to Accessory Drive Belts Removal & Installation.
8. Remove the water pump pulley.
9. Remove the ground cable (RH).
10. Support the bottom surface of engine using a transmission jack, and remove the right-hand engine bracket and insulator.
11. Remove the rocker cover.
12. Set No. 1 cylinder at TDC on its compression stroke with the following procedure:
    a. Rotate crankshaft pulley clockwise and align TDC mark (no paint) to timing indicator on front cover.
    b. Make sure the matching marks on each camshaft sprocket are positioned as shown. If not, rotate crankshaft pulley one more turn to align matching marks to the positions.
13. Remove crankshaft pulley with the following procedure:

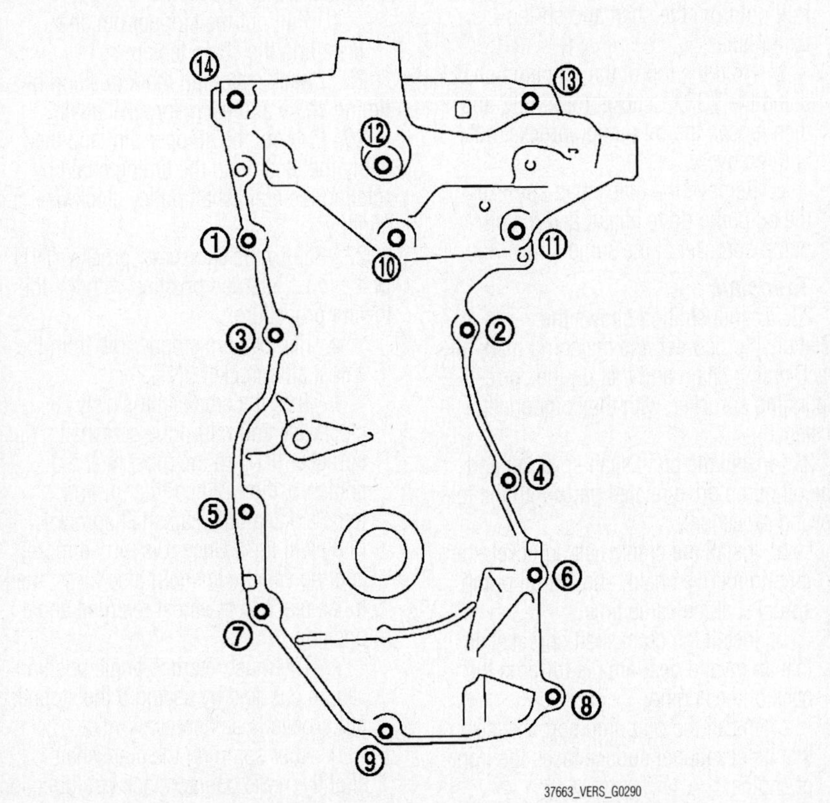
14
13
12
1
11
10
3
2
4
5
6
7
8
9
37663_VERS_G0290

Fig. 109 Timing chain cover bolt installation sequence—reverse for removal

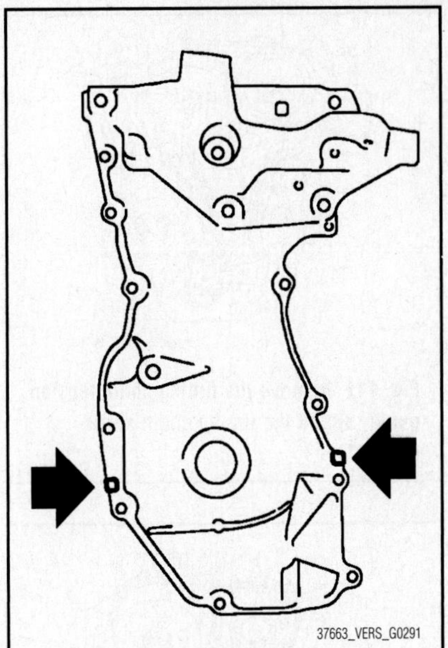
37663_VERS_G0291

Fig. 110 Cut liquid gasket by prying in the position shown

    a. Secure crankshaft pulley using a suitable tool.
    b. Loosen and pull out crankshaft pulley bolts.

⁂ WARNING

Never remove the bolts as they are used as a supporting point for the pulley puller.

    c. Attach tool (tool No. KV11103000) in the M 6 thread hole on crankshaft pulley, and remove crankshaft pulley.
14. Remove front cover with the following procedure:
    a. Loosen bolts in the reverse of the order shown.
    b. Cut liquid gasket by prying in the position shown, and then remove the front cover.
15. Remove front oil seal from front

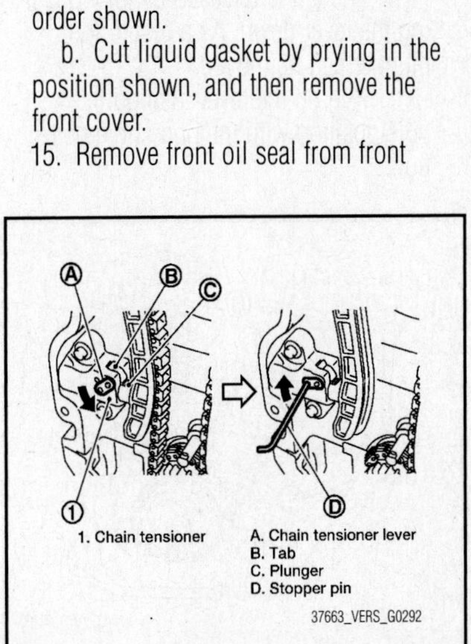
A
B
C
1
D
1. Chain tensioner
A. Chain tensioner lever
B. Tab
C. Plunger
D. Stopper pin
37663_VERS_G0292

Fig. 111 Timing chain tensioner retaining position

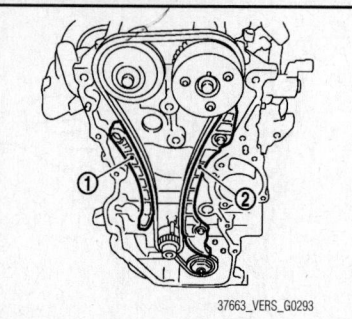

Fig. 112 Remove the timing chain tension guide (2) and the timing chain slack guide (1)

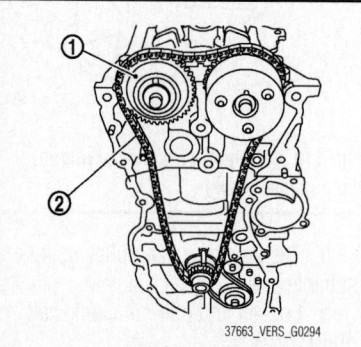

Fig. 113 Start the timing chain (2) removal from camshaft sprocket (EXH) (1) side

cover. Remove by lifting it up using a suitable tool. Be careful not to damage the front cover.

16. Remove chain tensioner with the following procedure:

   a. Fully push down the chain tensioner lever, and then push the plunger into the inside of tensioner.

   b. The tab is released by fully pushing the lever down. As a result, the plunger can be moved.

   c. Pull up the lever to align its hole position with the body hole position.

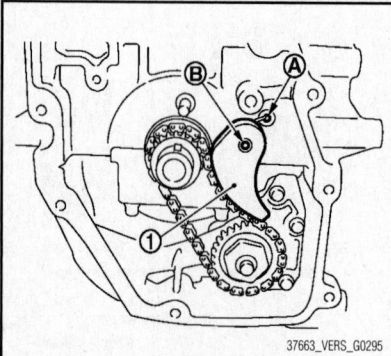

Fig. 114 Pull chain tensioner (1) out from the shaft (B) and spring fixing holes (A)

   d. When the lever hole is aligned with the body hole position, the plunger is fixed.

   e. When the protrusion parts of the plunger ratchet and the tab face each other, both hole positions are not aligned. At that time, correctly engage them and align these hole positions by slightly moving the plunger.

   f. Insert the stopper pin into the body hole through the lever hole, and then fix the lever at the upper position.

   g. Remove chain tensioner.

17. Remove the timing chain tension guide and the timing chain slack guide.

18. Remove the timing chain. Pull the looseness of timing chain toward the camshaft sprocket (EXH), and then remove the timing chain and start the removal from camshaft sprocket (EXH) side.

❋❋ WARNING

Never rotate crankshaft or camshaft while timing chain is removed. It causes interference between valve and piston.

19. Remove the crankshaft sprocket and the oil pump drive related parts with the following procedure:

   a. Remove chain tensioner. Pull out from the shaft and spring fixing holes.

   b. Hold the top of the oil pump shaft using the TORX® socket (size: E8), and then loosen the oil pump sprocket nut and remove it.

   c. Remove the crankshaft sprocket, the oil pump drive chain, and the oil pump sprocket at the same time.

To install:

20. The illustration shows the relationship between the matching mark on each timing chain and that on the corresponding sprocket, with the components installed.

21. Install the crankshaft sprocket and the oil pump drive related parts with the following procedure:

   a. Install the crankshaft sprocket, the oil pump drive chain, and the oil pump sprocket at the same time.

   b. Install the crankshaft sprocket so that its invalid gear area is towards the back of the engine.

   c. Install the oil pump sprocket so that its hexagonal surface faces the front of engine.

➠There is no matching mark in the oil pump drive related parts.

   d. Hold the top of the oil pump shaft using the TORX® socket (size: E8), and then tighten the oil pump sprocket nuts.

   e. Install chain tensioner. Insert the body into the shaft while inserting the spring into the fixing hole of cylinder block front surface.

   f. Make sure that the tension is applied to the oil pump drive chain after installing.

22. Install timing chain with the following procedure:

   a. Install by aligning matching marks on each sprocket and timing chain. If these matching marks are not aligned, rotate the camshaft slightly to correct the position.

   b. Check matching mark position of each sprocket and timing chain again after installing the timing chain, keep matching marks aligned by holding them with a hand.

   c. To avoid skipped teeth, never rotate crankshaft and camshaft until front cover is installed.

23. Install timing chain tension guide and timing chain slack guide.

24. Install chain tensioner:

   a. Secure the plunger at the most compressed position using a stopper pin, and then install it.

   b. Pull out the stopper pin after installing the chain tensioner.

25. Check matching mark position of timing chain and each sprocket again.

26. Pull out the stopper pin, and then apply the tension to the timing chain by rotating the crankshaft pulley clockwise slightly.

27. Return the camshaft sprocket (INT) in the most retarded position with the following procedure:

   a. Remove the stopper pin from the camshaft sprocket (INT).

   b. Turn the crankshaft slowly clockwise and return the camshaft sprocket (INT) to the most retarded angle position. When first turning the crankshaft the camshaft sprocket (INT) will turn. Once it is turned more, and the vane (camshaft) also turns, then it has reached the most retarded angle position.

   c. The most retarded angle position can be checked by seeing if the stopper pin groove is shifted clockwise.

   d. After spinning the crankshaft slightly in the counterclockwise direction, you can make sure the lock pin has joined by seeing if the vane (camshaft) and the sprocket move together.

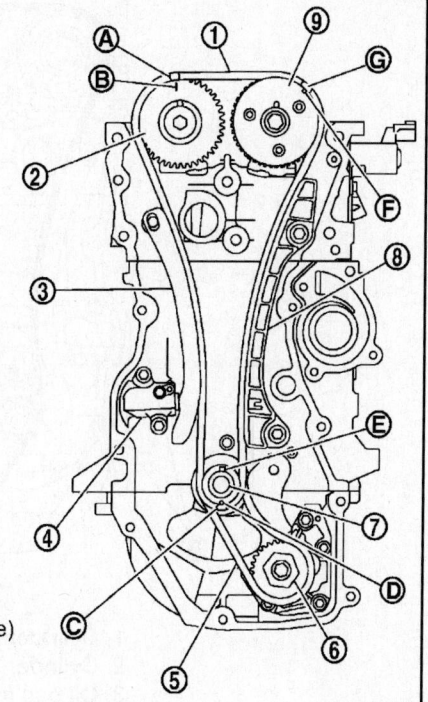

1 : Timing chain
2 : Camshaft sprocket (EXH)
3 : Timing chain slack guide
4 : Chain tensioner
5 : Oil pump drive chain
6 : Oil pump sprocket
7 : Crankshaft sprocket
8 : Timing chain tension guide
9 : Camshaft sprocket (INT)
A : Dark blue link
B : Matching mark (stamp)
C : Orange link
D : Matching mark (stamp)
E : Crankshaft key (point straight up)
F : Matching mark (peripheral stamp line)
G : Dark blue link

37663_VERS_G0297

Fig. 115 Timing chain positioning

28. Install the front oil seal to the front cover.

29. Install front cover with the following procedure:

a. Apply a continuous bead of liquid gasket [0.12–0.16 in. (3.0–4.0 mm)] wide to front of engine as shown.

b. Apply a continuous bead of liquid gasket [0.12–0.16 in. (3.0–4.0 mm)] wide to front cover as shown. Use Genuine Silicone RTV Sealant or equivalent.

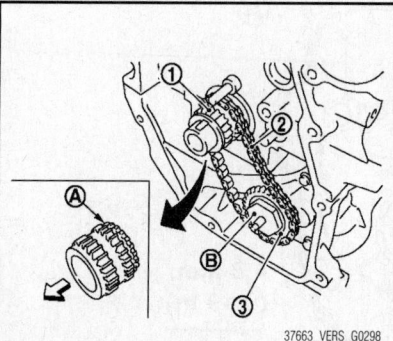

37663_VERS_G0298

Fig. 116 Crankshaft sprocket (1), with gear (A) facing engine back; oil pump sprocket (3), with hexagonal surface facing (B) engine front; and oil pump drive chain (2)

c. Tighten bolts in the numerical order shown above.

d. After all bolts are tightened, retighten them to specified torque in numerical order as shown above. Be sure to wipe off any excessive liquid gasket leaking to surface.

30. Insert crankshaft pulley by aligning with crankshaft key.

a. When inserting crankshaft pulley with a plastic hammer, tap on its center portion (not circumference).

✳✳ WARNING

Never damage front oil seal lip section.

31. Tighten crankshaft pulley bolt with the following procedure:

a. Secure crankshaft pulley with a suitable tool, and tighten crankshaft pulley bolt.

b. Apply new engine oil to thread and seat surfaces of crankshaft pulley bolt.

c. Tighten crankshaft pulley bolt to 26 ft. lbs. (35 Nm).

d. Put a paint mark on crankshaft

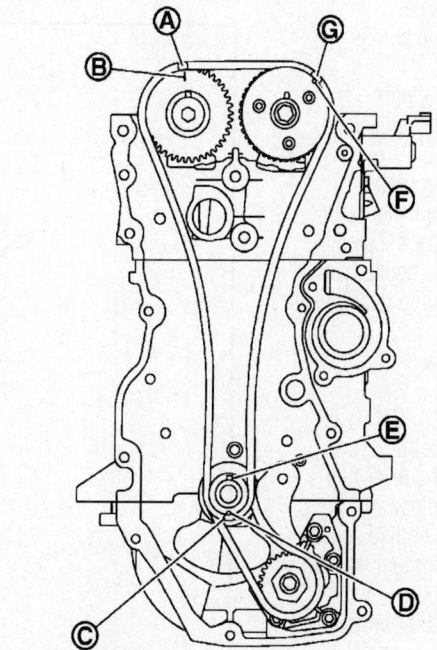

A. Dark blue link
B. Matching mark (stamp)
C. Orange link
D. Matching mark (stamp)
E. Crankshaft key (point straight up)
F. Matching mark (peripheral stamp line)
G. Dark blue link

37663_VERS_G0299

Fig. 117 Install timing chain

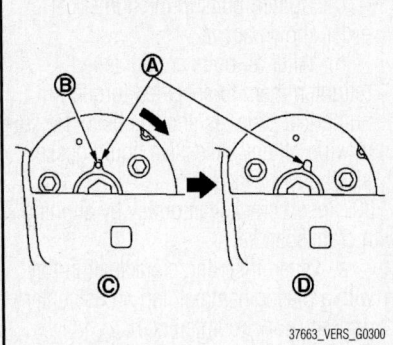

Fig. 118 Stopper pin hole (B), Stopper pin groove (A); Most advanced angle (C), Most retarded angle (lock pin engaged) (D)

pulley, mating with any one of six easy to recognize angle marks on crankshaft bolt flange.

    e. Turn another 60 degrees clockwise (angle tightening). Check the tightening angle with movement of one angle mark.

32. Make sure that crankshaft turns smoothly by rotating by hand clockwise.

33. Install the rocker cover.

34. Install the engine mounting bracket and insulator (RH).

35. Install the ground cable.

36. Install the water pump pulley.

37. Install the drive belt.

38. Install the intake manifold.

39. Install the front fender protector (RH).

40. Install the front wheel (RH).

41. Connect the negative battery cable.

42. Before starting engine, check oil/fluid levels including engine coolant and engine oil. If less than required quantity, fill to the specified level.

43. Use procedure below to check for fuel leakage:

    a. Turn ignition switch "ON" (with engine stopped). With fuel pressure applied to fuel piping, check for fuel leakage at connection points.

    b. Start engine. With engine speed increased, check again for fuel leakage at connection points.

44. Run engine to check for unusual noise and vibration.

➡**If hydraulic pressure inside chain tensioner drops after removal/installation, slack in guide may generate a pounding noise during and just after the engine start. However, this does not indicate an unusualness. Noise will stop after hydraulic pressure rises.**

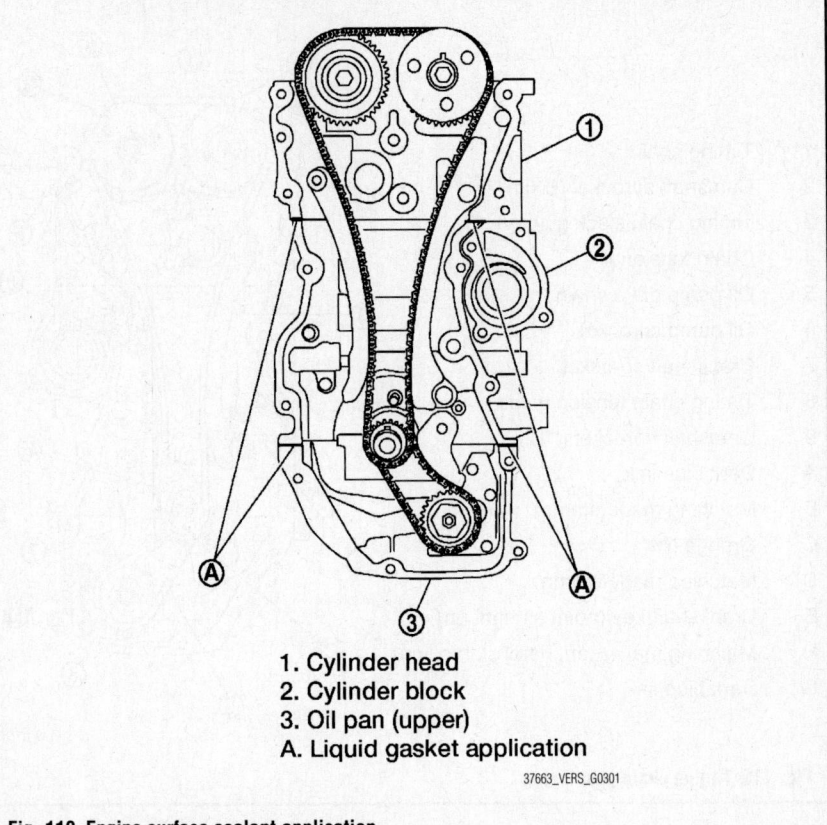

1. Cylinder head
2. Cylinder block
3. Oil pan (upper)
A. Liquid gasket application

Fig. 119 Engine surface sealant application

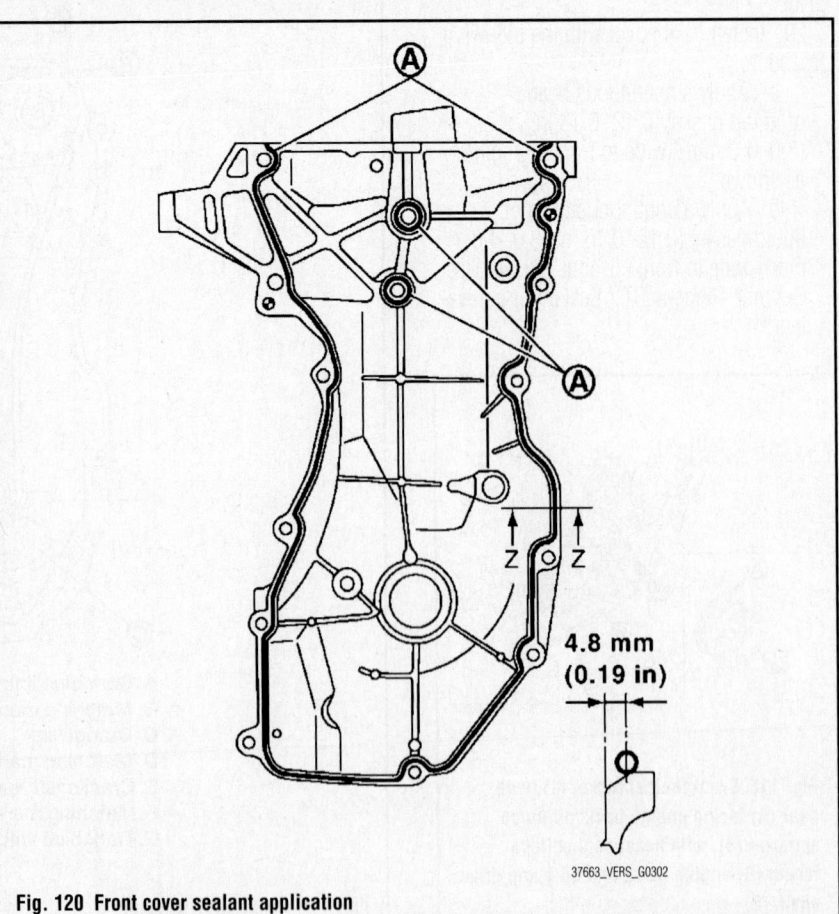

4.8 mm (0.19 in)

Fig. 120 Front cover sealant application

45. Warm up engine thoroughly to make sure there is no leakage of fuel, or any oil/fluids including engine oil and engine coolant.

46. Bleed air from lines and hoses of applicable lines, such as in cooling system.

47. After cooling down engine, again check oil/fluid levels including engine oil and engine coolant. Refill to the specified level, if necessary.

1.8L Engine

See Figures 121 through 132.

1. Before servicing the vehicle, refer to the Precautions Section.
2. Disconnect the negative battery cable.
3. Drain engine oil.
4. Partially drain engine coolant from the radiator.
5. Remove the right front wheel.
6. Remove the right front fender protector.
7. Remove the rocker cover.
8. Remove the drive belt. Refer to Accessory Drive Belts Removal & Installation.
9. Remove the water pump pulley.
10. Remove the ground cable (between

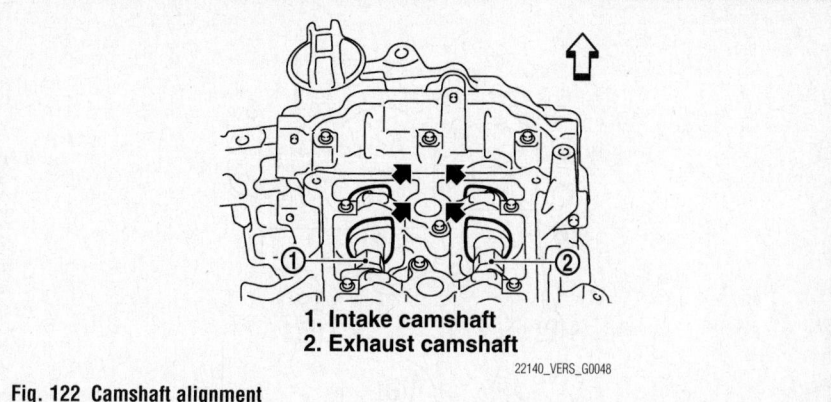

1. Intake camshaft
2. Exhaust camshaft

22140_VERS_G0048

Fig. 122 Camshaft alignment

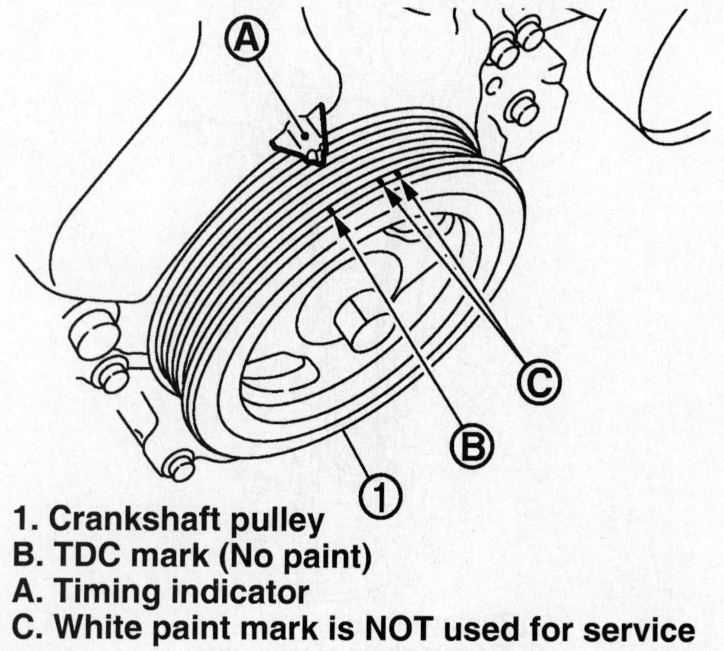

engine bracket (RH) and radiator core support).

11. Support the bottom surface of engine using a transmission jack, and remove the right-hand engine bracket and insulator.

12. Set No. 1 cylinder at TDC on its compression stroke with the following procedure:

   a. Rotate crankshaft pulley clockwise and align TDC mark (no paint) to timing indicator on front cover.

   b. At the same time, make sure that the cam noses of the No.1 cylinder are located as shown. If not, rotate crankshaft

pulley one revolution (360 degrees) and align as shown.

13. Hold crankshaft pulley using suitable tool, and loosen crankshaft pulley bolt. Locate bolt seating surface at 10 mm (0.39 inch) from its original position. Never remove the crankshaft pulley bolt as it will be used as a supporting point for the pulley puller.

14. Attach a pulley puller in the M6 thread hole on crankshaft pulley, and remove crankshaft pulley.

15. Remove the lower oil pan. When crankshaft sprocket, oil pump sprocket and other related parts are not removed, this step is unnecessary.

16. Remove the intake valve timing control solenoid valve.

17. Remove drive belt auto-tensioner.

18. Loosen the front cover bolts in reverse order as shown.

19. Cut liquid gasket by prying, and then remove the front cover. Be careful not to damage the mating surface. A more adhesive liquid gasket is applied compared to previous types when shipped, so it should not be forced off the position not specified.

20. Remove front oil seal from front cover. Lift up front oil seal using a suitable tool. Be careful not to damage front cover.

21. Push in timing chain tensioner plunger.

22. Insert a stopper pin into the body hole to retain the plunger in collapsed position. Use approximately 1.5 mm (0.059 inch) diameter hard metal pin as a stopper pin.

23. Remove timing chain tensioner.

24. Remove timing chain slack guide, timing chain tension guide and timing chain.

⁂ WARNING

Never rotate each crankshaft and camshaft individually while timing chain is removed. It causes interference between valve and piston.

1. **Crankshaft pulley**
B. **TDC mark (No paint)**
A. **Timing indicator**
C. **White paint mark is NOT used for service**

22140_VERS_G0047

Fig. 121 Rotate crankshaft pulley and align TDC mark to timing indicator

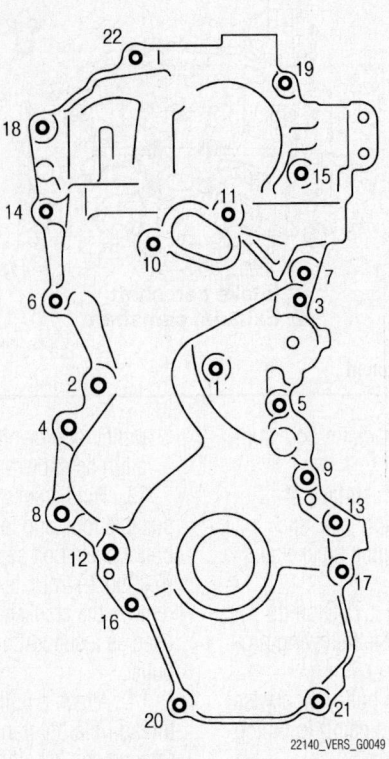

Fig. 123 Timing chain cover bolt installation sequence—reverse for removal

28. Hold the WAF part of oil pump shaft, and then loosen the oil pump sprocket bolt and remove them.

   a. Secure the oil pump shaft with the WAF part. Never loosen the oil pump sprocket bolt by tightening the oil pump drive chain.

29. Remove crankshaft sprocket, oil pump sprocket and oil pump drive chain as a set, if necessary.

30. Remove timing chain tension guide (front cover side) from front cover if necessary.

To install:

31. Make sure that crankshaft key points are aligned.

   a. There are two outer grooves in the intake camshaft sprocket. The wider one is a matchmark.

32. If the timing chain tension guide (front cover side) is removed, install it to the front cover. Check the joint condition by sound or feeling.

33. Install crankshaft sprocket, oil pump sprocket and oil pump drive chain, as follows:

   a. Install by aligning matchmarks on each sprocket and oil pump drive chain.

   b. If these matchmarks are not

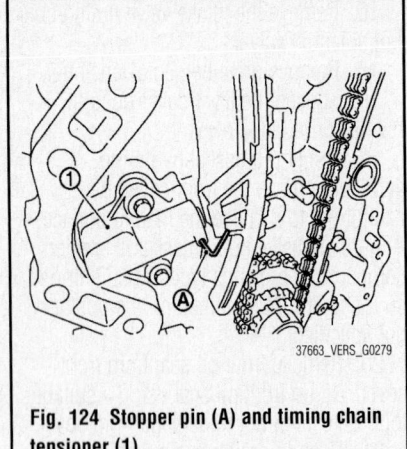

Fig. 124 Stopper pin (A) and timing chain tensioner (1)

25. Press stopper tab in the direction shown to push the timing chain slack guide toward timing chain tensioner (for oil pump). The slack guide is released by pressing the stopper tab. As a result, the slack guide can be moved.

26. Insert stopper pin into tensioner body hole to secure timing chain slack guide. Use a hard metal pin with a diameter of approximately 0.047 in. (1.2 mm) as a stopper pin.

27. Remove timing chain tensioner (for oil pump), if necessary.

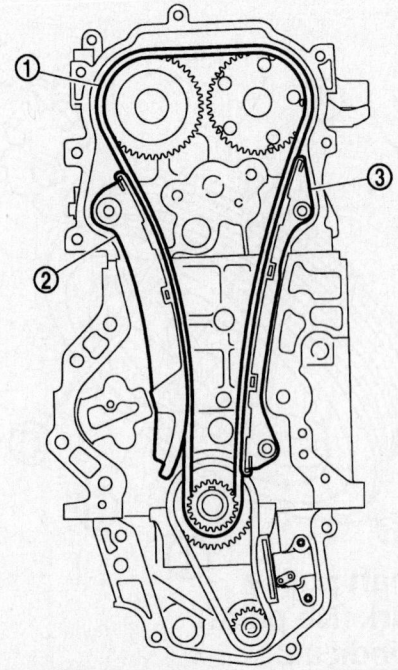

1. Timing chain
2. Timing chain slack guide
3. Timing chain tension guide

Fig. 125 Remove timing chain guides

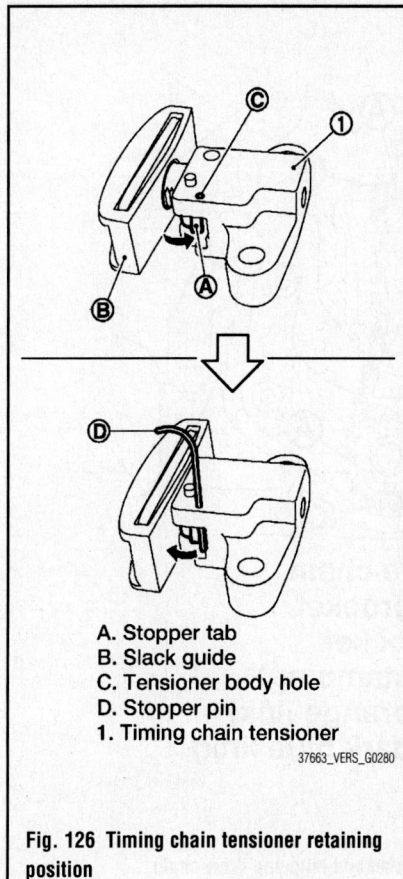

A. Stopper tab
B. Slack guide
C. Tensioner body hole
D. Stopper pin
1. Timing chain tensioner

37663_VERS_G0280

Fig. 126 Timing chain tensioner retaining position

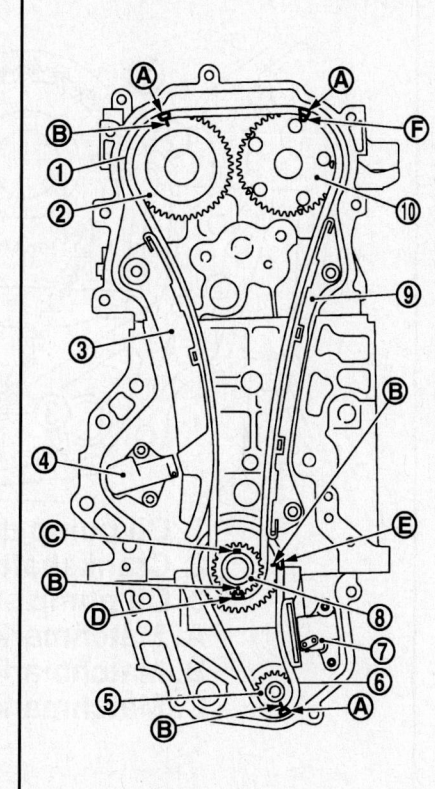

1. Timing chain
2. Exhaust camshaft sprocket
3. Timing chain slack guide
4. Timing chain tensioner
5. Oil pump sprocket
6. Oil pump drive chain
7. Chain tensioner (for oil pump)
8. Crankshaft sprocket
9. Timing chain tension guide
10. Intake camshaft sprocket
A. Matchmark (dark blue link)
B. Matchmark (stamping)
C. Crankshaft key position (straight up)
D. Matchmark (gold link)
E. Matchmark (orange link)
F. Matchmark (outer groove)

22140_VERS_G0052

Fig. 128 Timing chain positioning

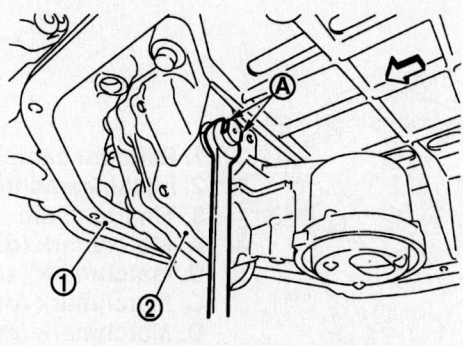

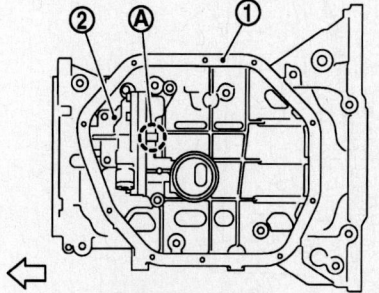

A. Oil pump shaft
1. Upper oil pan
2. Oil pump

22140_VERS_G0051

Fig. 127 Oil pump shaft and sprocket bolt removal

aligned, rotate the oil pump shaft slightly to correct the position.

c. Check matchmark position of each sprocket after installing the oil pump drive chain.

34. Hold the WAF part of oil pump shaft, and then tighten the oil pump sprocket bolt.

35. Secure the oil pump shaft with the WAF part. Never loosen the oil pump sprocket bolt by tightening the oil pump drive chain.

36. Install chain tensioner (for oil pump), as follows:

a. Secure the plunger at the most compressed position using a stopper pin, and install it.

b. Securely pull out the stopper pin after installing the chain tensioner (for oil pump).

c. Check matchmark position of oil pump drive chain and each sprocket again.

37. Align the matchmarks of each sprocket with the matchmarks of timing chain. There are two outer grooves in the intake camshaft sprocket. The wider one is a matchmark. If these matchmarks are not aligned, rotate the camshaft slightly by holding the hexagonal portion to correct the position.

38. Check matchmark position of each sprocket and timing chain again after installing the timing chain.

39. Install the timing chain tension guide and the timing chain slack guide.

40. Install timing chain tensioner, as follows:

    a. Secure the plunger at the most compressed position using a stopper pin, and install it.

    b. Securely pull out the stopper pin after installing the timing chain tensioner.

41. Check matchmark position of timing chain and each sprocket again.

42. Apply new engine oil to new front oil seal joint surface.

43. Using a suitable tool install front oil seal so that each seal lip is oriented properly. Press-fit front oil seal until it is flush with front end surface of front cover as shown below with a suitable tool. Within 0.012 in. (0.3 mm) toward engine front, within 0.020 in. (0.5 mm) toward engine rear. Press-fit oil seal straight to avoid causing burrs or tilting.

❋❋ WARNING

Be careful not to damage front cover and crankshaft. Never touch grease applied onto oil seal lip.

44. Install new O-ring to cylinder block.

45. Apply the sealant without breaks to the specified location using tool No. WS39930000. Use Genuine Silicone RTV Sealant or equivalent.

46. Make sure that matching marks of timing chain and each sprocket are still aligned.

47. Install front cover, and tighten bolts in numerical order as shown.

➡ **Attaching should be done within 5 minutes after liquid gasket application.**

48. Bolt installation positions:
- M6 bolts: No. 1
- M10 bolts: No. 6, 7, 10, 11, 14
- M12 bolts: No. 2, 4, 8, 12
- M8 bolts: Except the above

49. Tighten all bolts are in two stages to specified torque in numerical order as shown above. Be sure to wipe off any excessive liquid gasket leaking.

50. Install crankshaft pulley as follows:

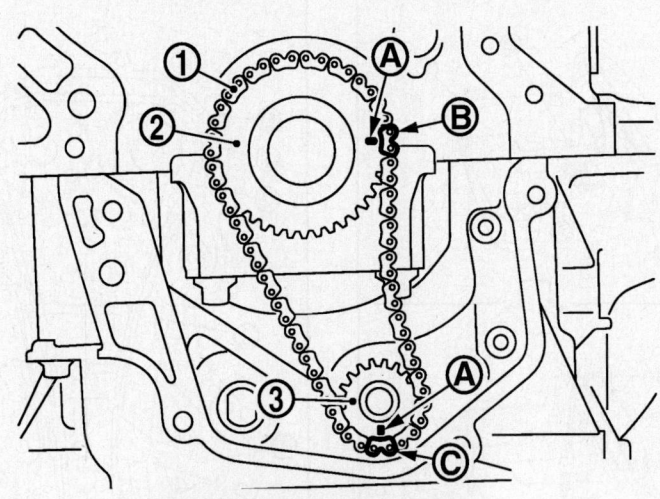

1. Oil pump drive chain
2. Crankshaft sprocket
3. Oil pump sprocket
A. Matchmark (stamping)
B. Matchmark (orange link)
C. Matchmark (dark blue link)

22140_VERS_G0053

Fig. 129 Install crankshaft sprocket, oil pump sprocket and oil pump drive chain

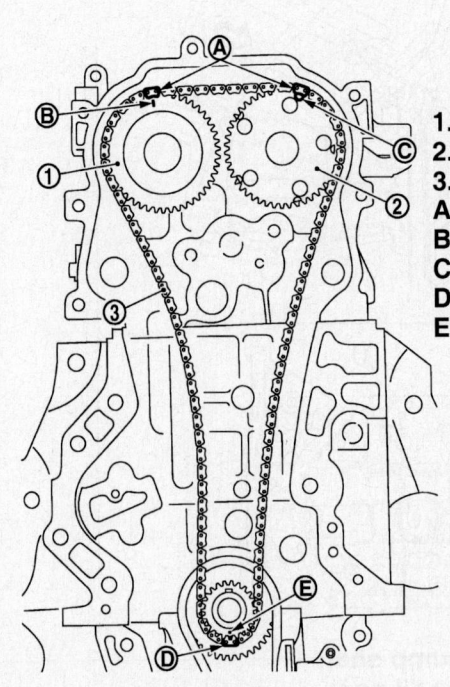

1. Exhaust camshaft sprocket
2. Intake camshaft sprocket
3. Timing chain
A. Matchmark (dark blue link)
B. Matchmark (stamping)
C. Matchmark (outer groove*)
D. Matchmark (gold link)
E. Matchmark (stamping)

22140_VERS_G0054

Fig. 130 Timing chain alignment

Never damage front oil seal lip section. If needed use a plastic hammer, tap on its center portion (not circumference) to seat crankshaft pulley.

a. Apply new engine oil to thread and seat surfaces of crankshaft pulley bolt.

b. Secure crankshaft pulley using tool No. KV10109300.

c. Tighten crankshaft pulley bolt in two steps:
- Step 1: 22 ft. lbs. (29 Nm)
- Step 2: 60° clockwise

d. For angle tightening, put a paint mark on crankshaft pulley matching with any one of six easy to recognize angle marks on crankshaft pulley bolt flange.

e. Turn 60 degrees clockwise (angle tightening).

f. Check the tightening angle with movement of one angle mark.

g. Make sure that crankshaft rotates clockwise smoothly.

51. Install the oil pan (lower).

52. Install the engine mounting bracket and insulator (RH).

53. Install the ground cable.

54. Install the water pump pulley.

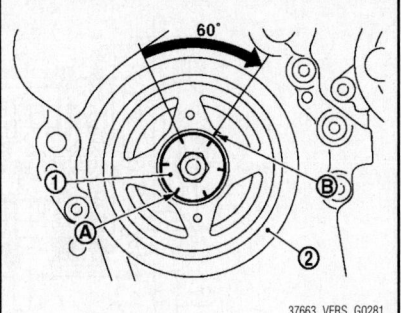

Fig. 132 Crankshaft pulley angle tightening

55. Install the drive belt.

56. Install the rocker cover.

57. Install the front fender protector (RH).

58. Install the front wheel (RH).

59. Connect the negative battery cable.

60. Before starting engine, check oil/fluid levels including engine coolant and engine oil. If less than required quantity, fill to the specified level.

61. Use procedure below to check for fuel leakage:

a. Turn ignition switch "ON" (with engine stopped). With fuel pressure applied to fuel piping, check for fuel leakage at connection points.

b. Start engine. With engine speed increased, check again for fuel leakage at connection points.

62. Run engine to check for unusual noise and vibration.

➡**If hydraulic pressure inside chain tensioner drops after removal/installation, slack in guide may generate a pounding noise during and just after the engine start. However, this does not indicate an unusualness. Noise will stop after hydraulic pressure rises.**

63. Warm up engine thoroughly to make sure there is no leakage of fuel, or any oil/fluids including engine oil and engine coolant.

64. Bleed air from lines and hoses of applicable lines, such as in cooling system.

65. After cooling down engine, again check oil/fluid levels including engine oil and engine coolant. Refill to the specified level, if necessary.

VALVE LASH

ADJUSTMENT

See Figures 133 through 136.

1. Before servicing the vehicle, refer to the Precautions Section.

2. Measure the valve clearance with the following procedure:

3. Set No. 1 cylinder at TDC of its compression stroke by rotating the crankshaft pulley clockwise and align TDC mark (no paint) to timing indicator on front cover. At the same time, make sure that both intake and exhaust cam noses of No. 1 cylinder face inside. If they do not, rotate crankshaft pulley once more (360 degrees) and align.

4. Use a feeler gauge, measure the clearance between valve lifter and camshaft.

5. Intake valve clearance:
- Cold: 0.010–0.013 inches (0.26–0.34 mm)
- Hot: 0.012–0.016 inches (0.304–0.416 mm)

6. Exhaust valve clearance:
- Cold: 0.011–0.015 inches (0.29–0.37 mm)
- Hot: 0.012–0.017 inches (0.308–0.432 mm)

7. Set No. 4 cylinder at TDC of its compression stroke by rotating crankshaft pulley one revolution (360 degrees) and align TDC

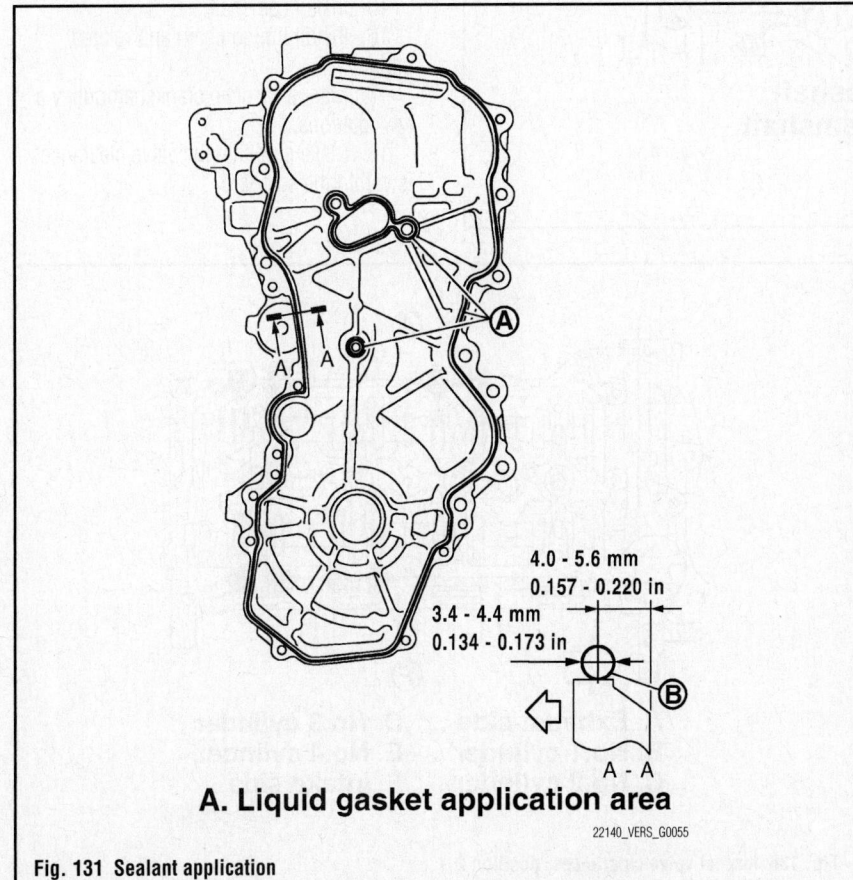

4.0 - 5.6 mm
0.157 - 0.220 in

3.4 - 4.4 mm
0.134 - 0.173 in

A. Liquid gasket application area

22140_VERS_G0055

Fig. 131 Sealant application

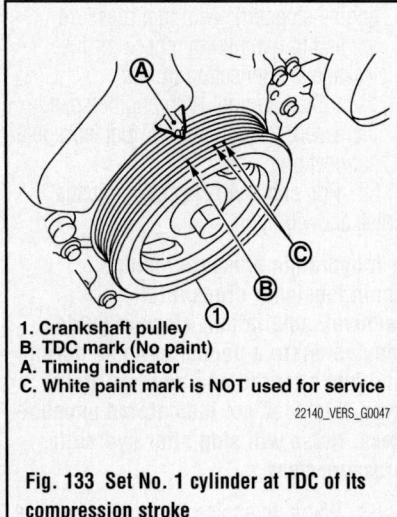

1. Crankshaft pulley
B. TDC mark (No paint)
A. Timing indicator
C. White paint mark is NOT used for service

22140_VERS_G0047

Fig. 133 Set No. 1 cylinder at TDC of its compression stroke

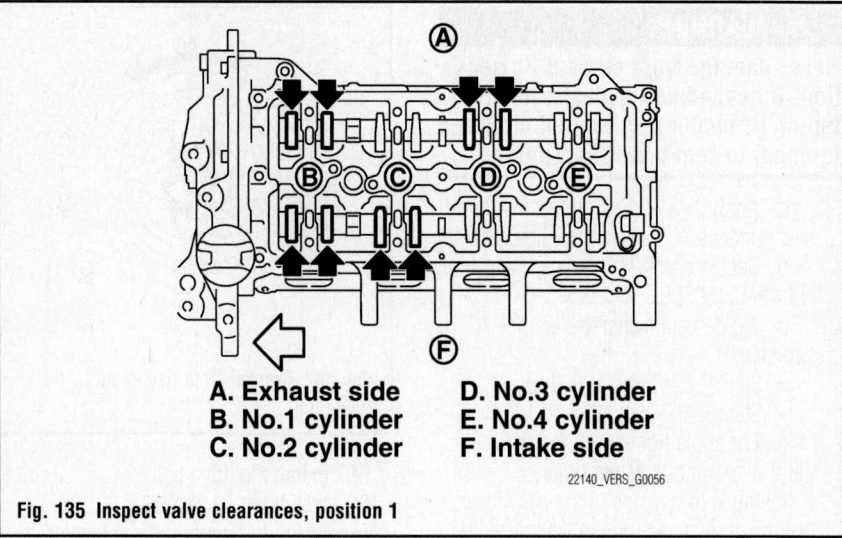

A. Exhaust side
B. No.1 cylinder
C. No.2 cylinder
D. No.3 cylinder
E. No.4 cylinder
F. Intake side

22140_VERS_G0056

Fig. 135 Inspect valve clearances, position 1

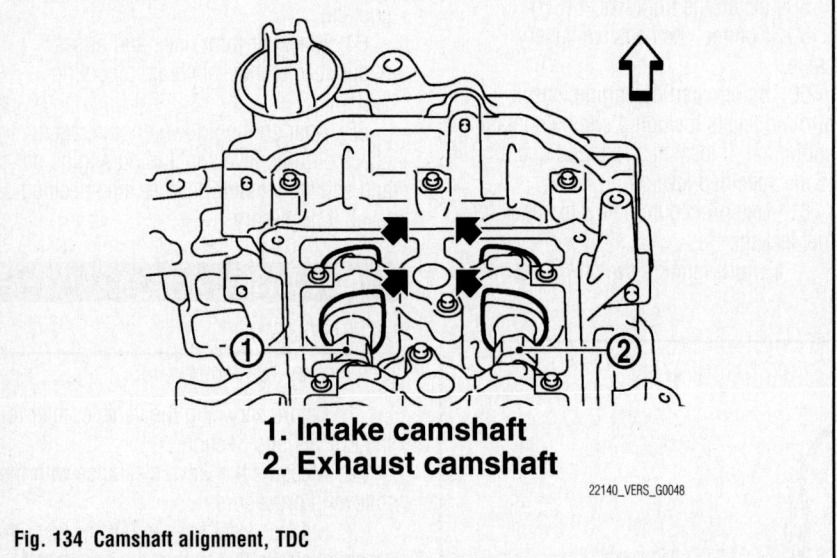

1. Intake camshaft
2. Exhaust camshaft

22140_VERS_G0048

Fig. 134 Camshaft alignment, TDC

- t1=Removed valve lifter thickness
- C1=Measured valve clearance
- C2=Standard valve clearance: Intake: 0.30 mm (0.012 inch), Exhaust: 0.33 mm (0.013 inch).
- Available thickness of valve lifter: 26 sizes range 3.00–3.50 mm (0.1181–0.1378 inch) in steps of 0.02 mm (0.0008 inch) (when manufactured at factory)

14. Install the selected valve lifter.

15. Install camshaft.

16. Install timing chain and related parts.

17. Manually rotate crankshaft pulley a few rotations.

18. Make sure that the valve clearances is within the standard.

mark (no paint) to timing indicator on front cover.

8. Use a feeler gauge, measure the clearance between valve lifter and camshaft.

9. If out of standard, perform adjustment. Perform adjustment depending on selected head thickness of valve lifter.

10. Remove camshaft.

11. Remove valve lifters at the locations that are out of the standard.

12. Measure the center thickness of the removed valve lifters with a micrometer.

13. Use the equation below to calculate valve lifter thickness for replacement:
- Valve lifter thickness calculation: $t = t1 + (C1 - C2)$
- t=Valve lifter thickness to be replaced

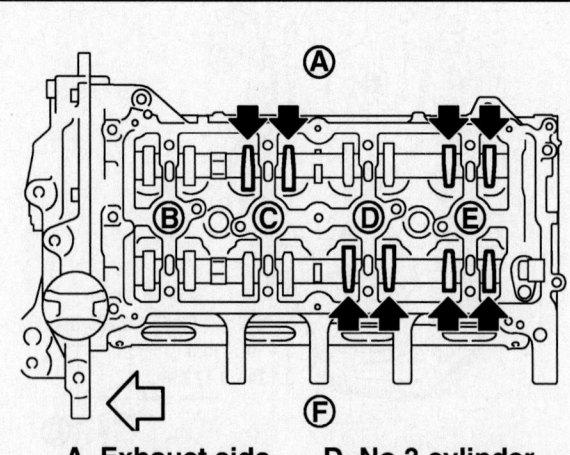

A. Exhaust side
B. No.1 cylinder
C. No.2 cylinder
D. No.3 cylinder
E. No.4 cylinder
F. Intake side

22140_VERS_G0057

Fig. 136 Inspect valve clearances, position 2

ENGINE PERFORMANCE & EMISSION CONTROLS

CAMSHAFT POSITION (CMP) SENSOR

LOCATION

See Figures 137 and 138.

Refer to the accompanying illustrations.

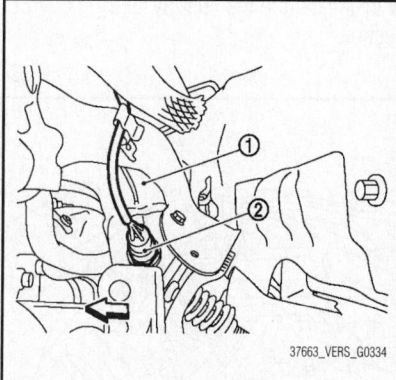

Fig. 137 Camshaft Position (CMP) sensor (4)—1.6L engine

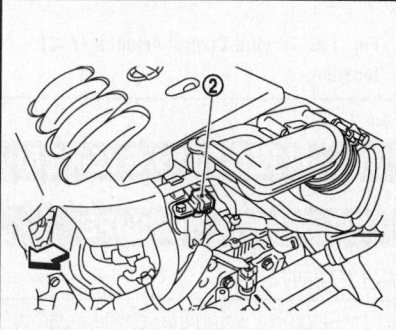

Fig. 138 Camshaft Position (CMP) sensor—1.8L engine

CRANKSHAFT POSITION (CKP) SENSOR

LOCATION

See Figure 139.

Refer to the accompanying illustration.

ELECTRONIC CONTROL MODULE (ECM)

LOCATION

See Figure 140.

Refer to the accompanying illustration.

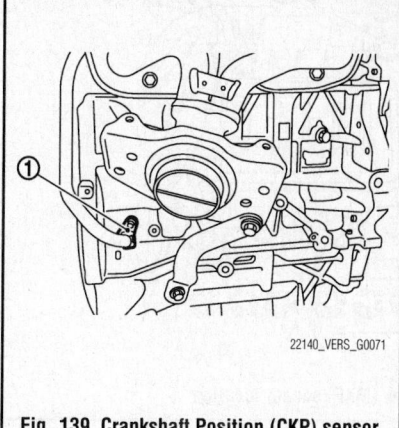

Fig. 139 Crankshaft Position (CKP) sensor

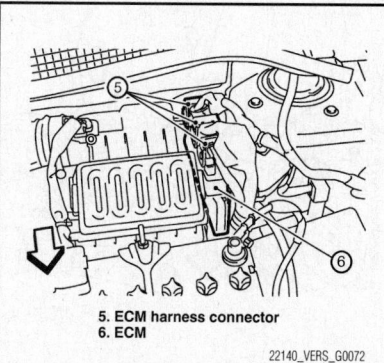

5. ECM harness connector
6. ECM

Fig. 140 Electronic Control Module (ECM) and harness location

RESET

The CONSULT-III scan tool is required.

ENGINE COOLANT TEMPERATURE (ECT) SENSOR

LOCATION

See Figure 141.

Refer to the accompanying illustration.

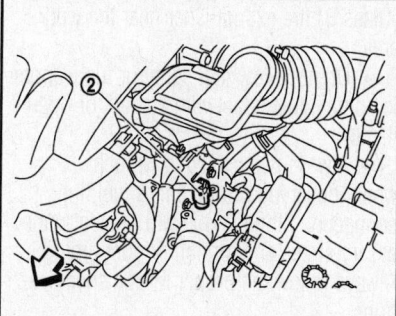

Fig. 141 Engine Coolant Temperature (ECT) sensor location

HEATED OXYGEN (HO2S) SENSOR

LOCATION

See Figure 142.

Refer to the accompanying illustration.

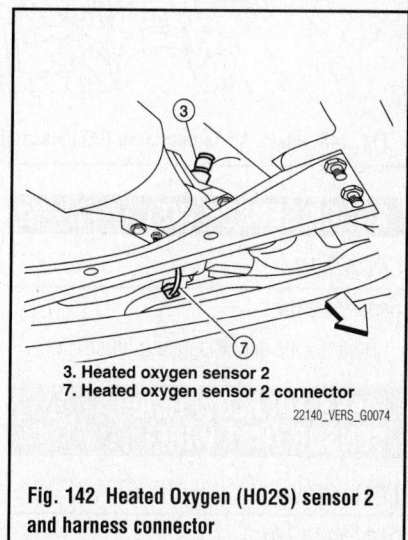

3. Heated oxygen sensor 2
7. Heated oxygen sensor 2 connector

Fig. 142 Heated Oxygen (HO2S) sensor 2 and harness connector

INTAKE AIR TEMPERATURE (IAT) SENSOR

LOCATION

See Figure 143.

Refer to the accompanying illustration.

REMOVAL & INSTALLATION

The Intake Air Temperature (IAT) sensor is located in the Mass Air Flow (MAF) sensor.

1. Before servicing the vehicle, refer to the Precautions Section.
2. Remove battery.
3. Remove the engine cover, if applicable.
4. Disconnect harness connector from Mass Air Flow (MAF) sensor.
5. Remove the air cleaner. Refer Air Cleaner Removal & Installation in the Engine Mechanical section.
6. Remove the MAF sensor from the air cleaner case.

To install:

7. Installation is the reverse of removal.

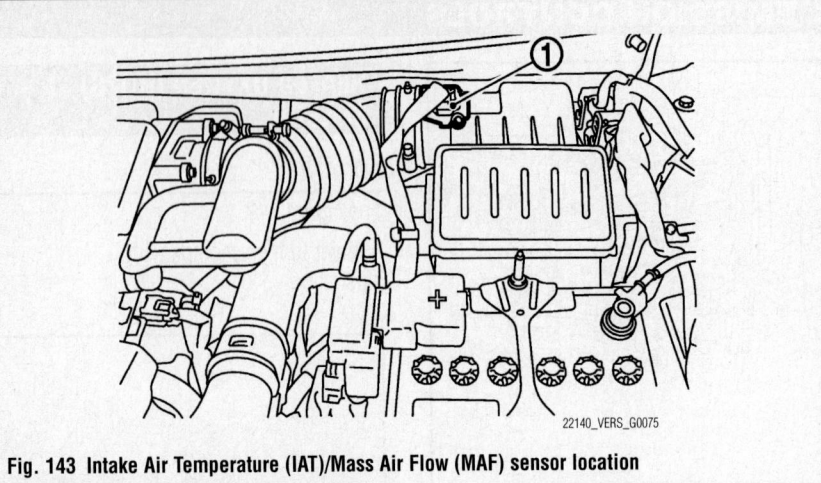

Fig. 143 Intake Air Temperature (IAT)/Mass Air Flow (MAF) sensor location

KNOCK SENSOR (KS)

LOCATION

See Figure 144.

Refer to the accompanying illustration.

MASS AIR FLOW (MAF) SENSOR (HOT WIRE)

LOCATION

See Figure 143.

Refer to Intake Air Temperature (IAT) Sensor.

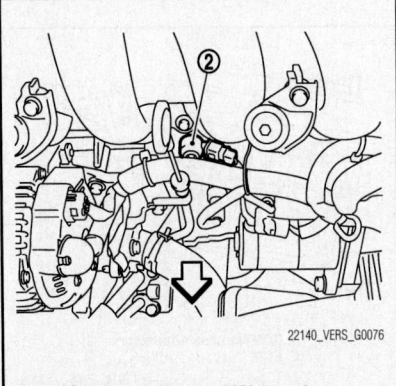

Fig. 144 Knock Sensor (KS) location

THROTTLE CONTROL ACTUATOR (TAC)

LOCATION

See Figure 145.

Refer to the accompanying illustration.

REMOVAL & INSTALLATION

See Throttle Body Removal & Installation in the Fuel System section.

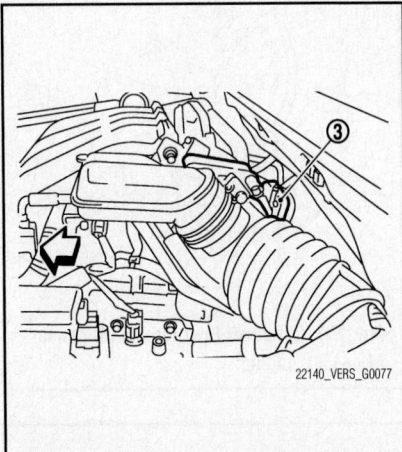

Fig. 145 Throttle Control Actuator (TAC) location

FUEL   **GASOLINE FUEL INJECTION SYSTEM**

FUEL SYSTEM SERVICE PRECAUTIONS

Safety is the most important factor when performing not only fuel system mainte-nance, but any type of maintenance. Failure to conduct maintenance and repairs in a safe manner may result in serious personal injury or death. Work on a vehicle's fuel system components can be accomplished safely and effectively by adhering to the fol-lowing rules and guidelines.

• To avoid the possibility of fire and per-sonal injury, always disconnect the negative battery cable unless the repair or test proce-dure requires that battery voltage be applied.

• Always relieve the fuel system pres-sure prior to disconnecting any fuel system component (injector, fuel rail, pressure reg-ulator, etc.) fitting or fuel line connection. Exercise extreme caution whenever relieving fuel system pressure to avoid exposing skin, face and eyes to fuel spray. Please be advised that fuel under pressure may pene-trate the skin or any part of the body that it contacts.

• Always place a shop towel or cloth around the fitting or connection prior to loosening to absorb any excess fuel due to spillage. Ensure that all fuel spillage is quickly removed from engine surfaces. Ensure that all fuel-soaked cloths or towels are deposited into a flame-proof waste con-tainer with a lid.

• Always keep a dry chemical (Class B) fire extinguisher near the work area.

• Do not allow fuel spray or fuel vapors to come into contact with a spark or open flame.

• Always use a second wrench when loosening or tightening fuel line connection fittings. This will prevent unnec-essary stress and torsion on fuel piping. Always follow the proper torque specifica-tions.

• Always replace worn fuel fitting O-rings with new ones. Do not substitute fuel hose where rigid pipe is installed.

FUEL SYSTEM PRESSURE

RELIEVING

1. Before servicing the vehicle, refer to the Precautions Section.
2. Remove fuel pump fuse.
3. Start engine.
4. After engine stalls, crank it two or three times to release all fuel pressure.
5. Turn ignition switch OFF.

FUEL FILTER

REMOVAL & INSTALLATION

The fuel filter is part of the fuel pump module. See Fuel Pump Module Removal & Installation.

FUEL LEVEL SENDING UNIT

LOCATION

See Figure 146.

The fuel level sensor unit is part of the fuel pump module.

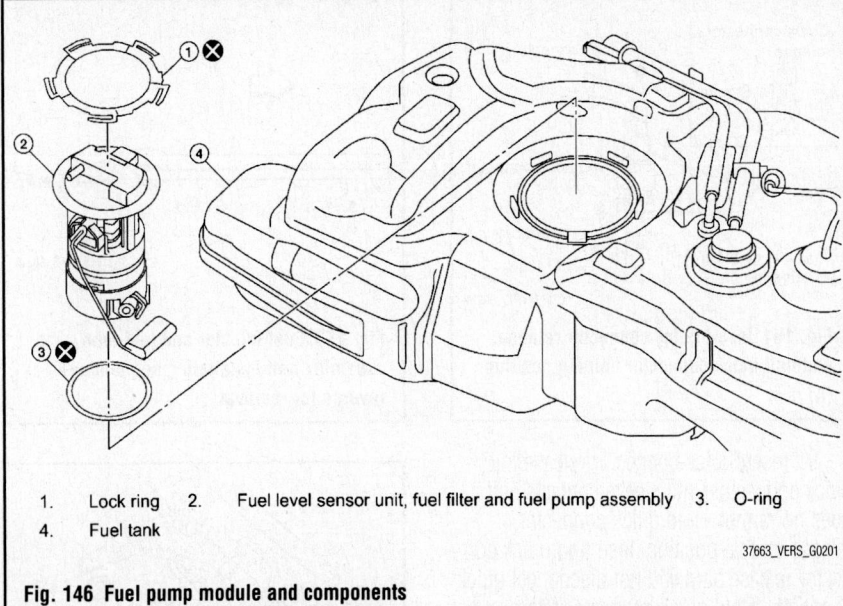

1. Lock ring 2. Fuel level sensor unit, fuel filter and fuel pump assembly 3. O-ring
4. Fuel tank

37663_VERS_G0201

Fig. 146 Fuel pump module and components

FUEL PUMP MODULE

REMOVAL & INSTALLATION

See Figures 147 through 150.

1. Before servicing the vehicle, refer to the Precautions Section.

2. Check fuel level with the vehicle on a level surface. If the fuel gauge indicates more than 7/8 full, drain fuel from the fuel tank until the fuel gauge indicates the correct level.

3. Open fuel door and unscrew the fuel filler cap to release the pressure inside the fuel tank.

4. Release the fuel pressure from the fuel lines.

5. Remove rear seat bottom.

6. Turn the four retainers 90° in a clockwise direction and remove the fuel pump inspection hole cover.

7. Disconnect electrical connector and fuel feed hose quick connector.

8. Disconnect the quick connector using the following procedure:

   a. Hold the sides of the connector, push in tabs and pull out the tube.

   b. If quick connector and tube on fuel pump assembly are stuck, push and pull several times until they move. Disconnect them by pulling.

➡ The tube can be removed when the tabs are completely depressed. Do not twist it more than necessary. Do not use any tools to remove the quick connector. Keep resin tube away from heat. Be especially careful when welding near the resin tube. Prevent acid liquid such as battery electrolyte, from getting on resin tube. Do not bend or twist resin tube during installation and removal. To keep the connecting portion clean, free of foreign materials and to avoid damage, cover them completely with plastic bags or something similar. Do not insert plug to prevent damage to O-ring in quick connector.

9. Remove the lock ring using tool No. KV991J0090 (J-46214).

10. Remove fuel level sensor unit, fuel filter and fuel pump assembly.

※※ WARNING

Do not bend float arm during removal. Do not allow foreign materials to fall into fuel tank. Use a lint free cloth when handling components. Avoid impacts such as dropping when handling components.

To install:

11. Installation is in the reverse order of removal, noting the following:

   a. Install O-ring to fuel tank without twisting.

   b. Install fuel level sensor unit with the matchmarks aligned on the fuel tank and fuel level sensor unit.

   c. Turn the lock ring until the lock ring is fully rotated into the fuel tank lock tabs.

   d. Connect fuel feed tube quick connector by aligning the connector

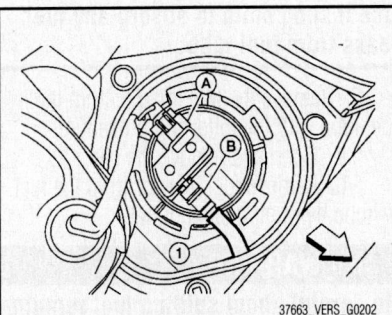

37663_VERS_G0202

Fig. 147 Electrical connector (A), fuel feed hose (1), and fuel feed hose quick connector (B)

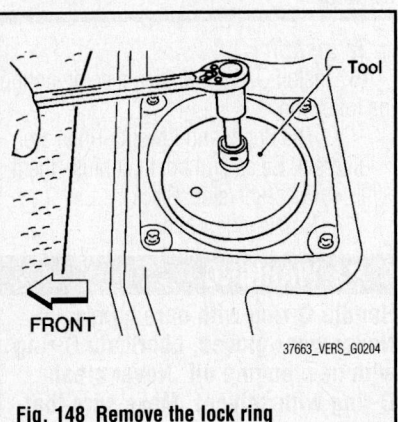

37663_VERS_G0204

Fig. 148 Remove the lock ring

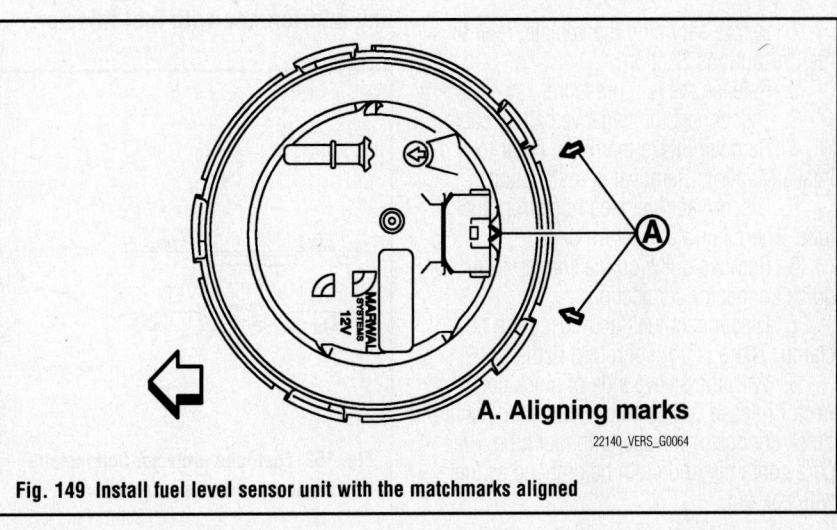

A. Aligning marks

22140_VERS_G0064

Fig. 149 Install fuel level sensor unit with the matchmarks aligned

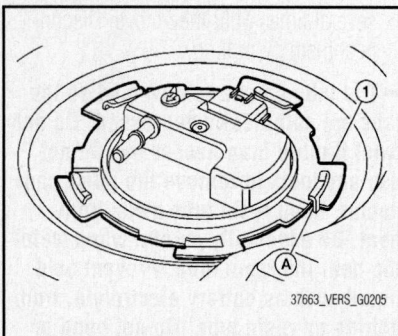

Fig. 150 Turn the lock ring (1) until the lock ring is fully rotated into the fuel tank lock tabs (A)

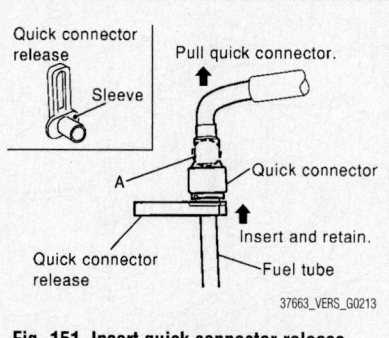

Fig. 151 Insert quick connector release, and pull quick connector holding position (A)

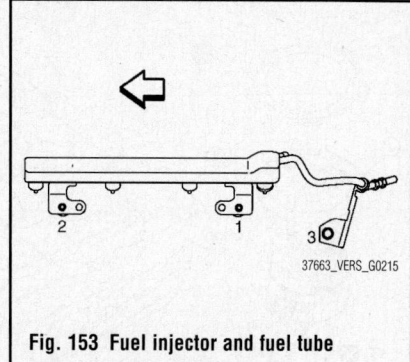

Fig. 153 Fuel injector and fuel tube assembly bolt installation sequence—reverse for removal

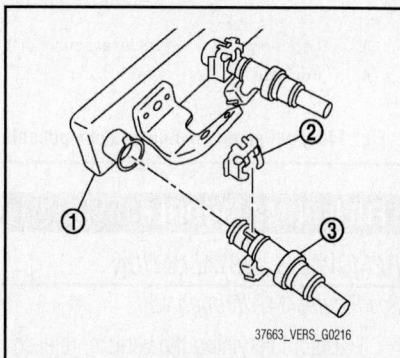

Fig. 154 Remove the clip (2), and remove the fuel injector (3) from the fuel tube (1)

with the tube, then inserting the connector straight into the tube until it clicks.

e. After connecting, visually confirm that the two retainer tabs are secured to the connector and pull the tube and the connector to make sure they are securely connected.

12. Connect electrical harness connector.

13. Before installing inspection hole cover, confirm that there are no fuel leaks, install inspection hole cover with the front mark (arrow) facing front of vehicle, and lock the clips by turning counterclockwise.

14. Check for fuel leaks:

a. Turn ignition switch "ON" (without starting the engine), to check the connections for fuel leaks with the electric fuel pump applying pressure to the fuel piping.

b. Start the engine and let it idle to check that there are no fuel leaks at the fuel system connections.

FUEL RAIL AND INJECTOR

REMOVAL & INSTALLATION

1.8L Engine

See Figures 151 through 155.

1. Before servicing the vehicle, refer to the Precautions Section.

2. Release the fuel pressure.

3. Disconnect the negative battery cable.

4. Remove intake manifold. Refer to Intake Manifold Removal & Installation.

5. Disconnect fuel feed hose from fuel tube. There is no fuel return path.

6. Remove quick connector cap from quick connector connection.

7. Disconnect fuel feed hose from hose clamp. There is no fuel return path.

8. With the sleeve side of quick connector release facing quick connector, install quick connector release onto fuel tube. Prepare container and cloth beforehand as fuel will leak out.

9. Insert quick connector release into quick connector until sleeve contacts and goes no further. Hold quick connector release on that position. Inserting quick connector release hard will not disconnect quick connector. Hold quick connector release where it contacts and goes no further.

10. Draw and pull out quick connector straight from fuel tube.

➡Do not pull with lateral force applied; O-ring may be damaged. Do not bend or twist connection between quick connector and fuel feed hose during installation/removal.

11. To keep clean the connecting portion and to avoid damage and foreign materials, cover them completely with plastic bags or something similar.

12. Disconnect harness connector from fuel injector.

13. Remove fuel tube protector. Loosen bolts in the reverse of the order shown.

14. Remove the fuel injector and fuel tube assembly. Loosen bolts in the reverse of the order shown.

❄❄ CAUTION

When removing, be careful to avoid any interference with fuel injector.

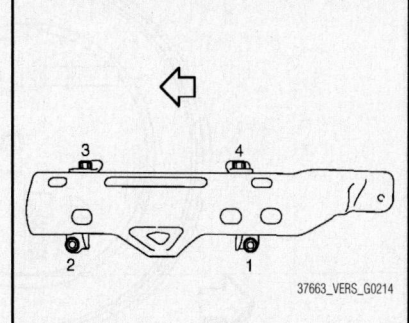

Fig. 152 Fuel tube protector bolt installation sequence—reverse for removal

Use a shop cloth to absorb any fuel leaks from fuel tube.

15. Remove the fuel injector from the fuel tube with the following procedure:

a. Open and remove the clip.

b. Remove fuel injector from the fuel tube by pulling straight out.

❄❄ WARNING

Be careful about spilling fuel remaining in fuel tube. Be careful not to damage the fuel injector nozzle during removal. Never bump or drop fuel injector. Never disassemble fuel injector.

To install:

16. Install O-rings to fuel injector, noting the following:

a. The upper and lower O-rings are different. Be careful not to confuse them.
- Fuel tube side: Black
- Nozzle side: Green

❄❄ WARNING

Handle O-ring with bare hands. Never wear gloves. Lubricate O-ring with new engine oil. Never clean O-ring with solvent. Make sure that

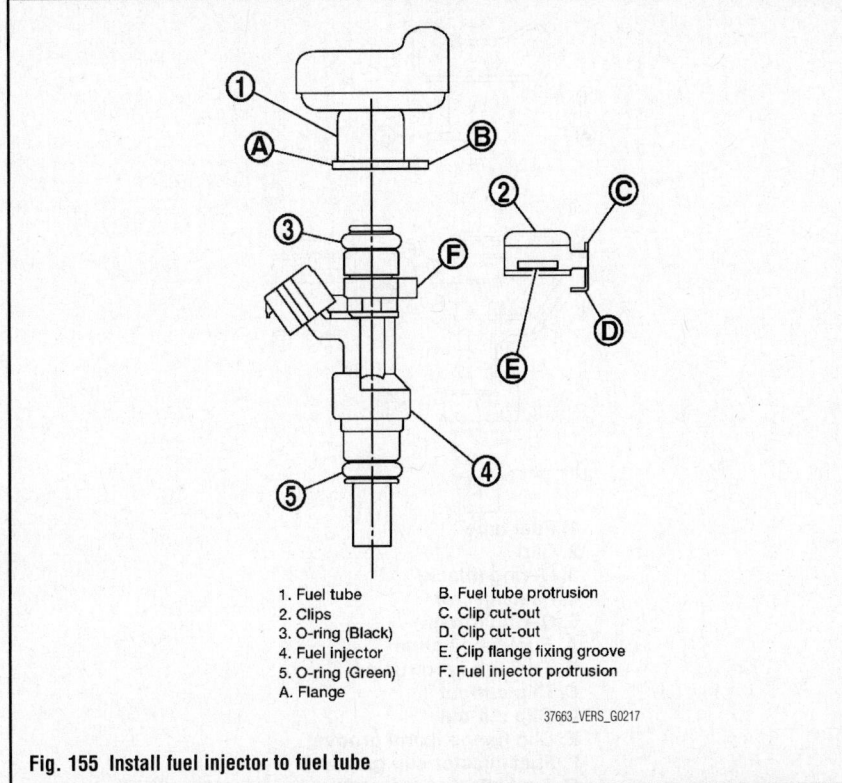

1. Fuel tube
2. Clips
3. O-ring (Black)
4. Fuel injector
5. O-ring (Green)
A. Flange

B. Fuel tube protrusion
C. Clip cut-out
D. Clip cut-out
E. Clip flange fixing groove
F. Fuel injector protrusion

37663_VERS_G0217

Fig. 155 Install fuel injector to fuel tube

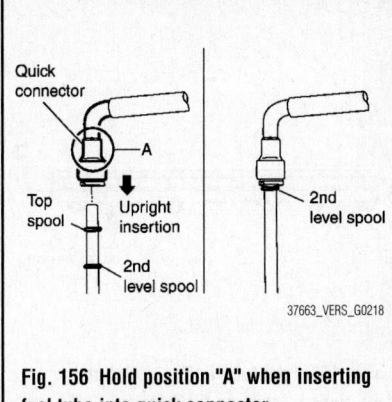

37663_VERS_G0218

Fig. 156 Hold position "A" when inserting fuel tube into quick connector

O-ring and its mating part are free of foreign material. When installing O-ring, be careful not to scratch it with tool or fingernails. Also be careful not to twist or stretch O-ring. If O-ring was stretched while it was being attached, never insert it quickly into fuel tube. Insert O-ring straight into fuel tube. Never twist it.

17. Install the fuel injector onto the fuel tube with the following procedure:

a. Insert the clips into the clip mounting grooves on the fuel injector.

b. Insert clip cut-out into fuel injector protrusion. Always replace clip with new one.

c. Make sure that the clip does not interfere with the O-ring. If interference occurs, replace the O-ring.

d. Insert the fuel injector into the fuel tube with clip attached. Make sure that the axis is lined up when inserting.

e. Insert clip cut-out into fuel tube protrusion.

f. Make sure that the flange on the fuel tube fits securely in the clip flange fixing groove.

g. Make sure that installation is complete by checking that fuel injector does not rotate or come off.

18. Install fuel tube and injector assembly onto cylinder head. Tighten bolts in numerical order as shown above.

✳✳ WARNING

Be careful not to let tip of injector nozzle interfere with other parts.

19. Install fuel tube protector. Tighten bolts in numerical order as shown above.

20. Connect harness connector to fuel injector.

21. Connect fuel feed hose with the following procedure:

a. Check for damage or foreign material on the fuel tube and quick connector.

b. Apply new engine oil lightly to area around the top of fuel tube.

c. Align center to insert quick connector straightly into fuel tube.

d. Insert quick connector to fuel tube until the top spool on fuel tube is inserted completely and the 2nd level spool is positioned slightly below quick connector bottom end.

e. Carefully align center to avoid inclined insertion to prevent damage to O-ring inside quick connector.

f. Insert until you hear a "click" sound and actually feel the engagement.

g. To avoid misidentification of engagement with a similar sound, be sure to perform the next step.

h. Before clamping fuel feed hose with hose clamp, pull quick connector hard by hand holding "A" position. Make

sure it is completely engaged (connected) so that it does not come out from fuel tube.

i. Install quick connector cap to quick connector connection. Install quick connector cap with the side arrow facing quick connector side (fuel feed hose side).

j. Make sure that the quick connector and fuel tube are securely engaged with the quick connector cap mounting groove.

k. If quick connector cap cannot be installed easily, the quick connector may not be connected correctly. Remove the quick connector cap, and then check the connection of quick connector again.

l. Install fuel feed hose to hose clamp.

22. Installation of the remaining components is in the reverse order of removal.

23. After installation is complete, check for fuel leaks:

a. Turn ignition switch "ON" (without starting the engine), to check the connections for fuel leaks with the electric fuel pump applying pressure to the fuel piping.

b. Start the engine and let it idle to check that there are no fuel leaks at the fuel system connections.

1.8L Engine

See Figures 151, 157 through 159.

1. Before servicing the vehicle, refer to the Precautions Section.

2. Release the fuel pressure.

3. Disconnect the negative battery cable.

4. Disconnect fuel feed hose from hose clamp. There is no fuel return path.

5. Remove quick connector cap from quick connector connection.

6. With the sleeve side of quick connector release facing quick connector, install quick connector release onto fuel tube. Prepare container and cloth beforehand as fuel will leak out.

7. Insert quick connector release into quick connector until sleeve contacts and

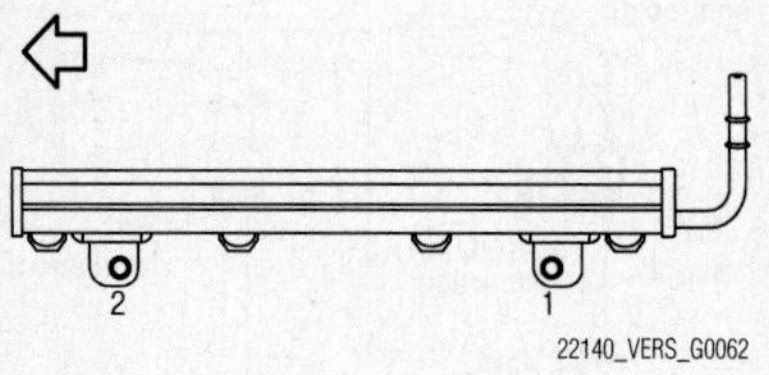
2
1
22140_VERS_G0062

Fig. 157 Fuel tube bolt installation sequence—reverse for removal

goes no further. Hold quick connector release on that position. Inserting quick connector release hard will not disconnect quick connector. Hold quick connector release where it contacts and goes no further.

8. Draw and pull out quick connector straight from fuel tube.

➡Do not pull with lateral force applied; O-ring may be damaged. Do not bend or twist connection between quick connector and fuel feed hose during installation/removal.

9. To keep clean the connecting portion and to avoid damage and foreign materials, cover them completely with plastic bags or something similar.

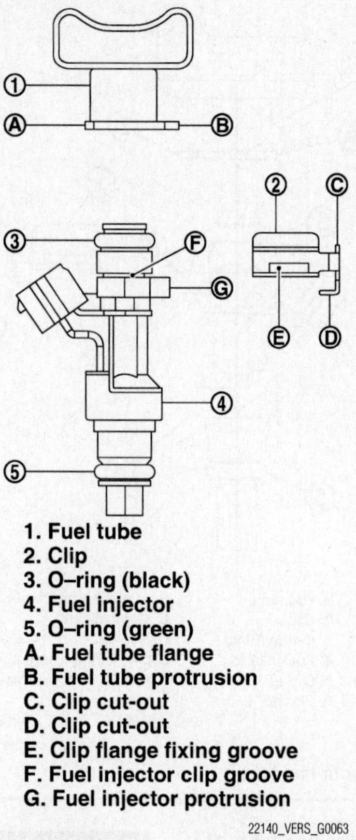
1. Fuel tube
2. Clip
3. O–ring (black)
4. Fuel injector
5. O–ring (green)
A. Fuel tube flange
B. Fuel tube protrusion
C. Clip cut-out
D. Clip cut-out
E. Clip flange fixing groove
F. Fuel injector clip groove
G. Fuel injector protrusion
22140_VERS_G0063

Fig. 159 Install fuel injector to fuel tube

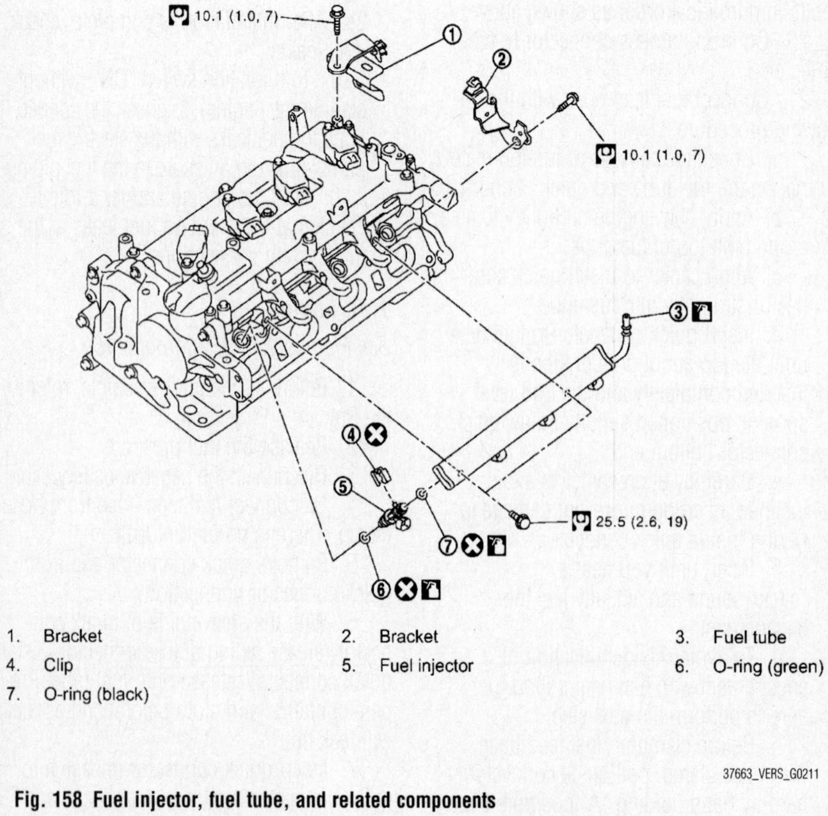
10.1 (1.0, 7)
10.1 (1.0, 7)
25.5 (2.6, 19)
1. Bracket
2. Bracket
3. Fuel tube
4. Clip
5. Fuel injector
6. O-ring (green)
7. O-ring (black)
37663_VERS_G0211

Fig. 158 Fuel injector, fuel tube, and related components

10. Remove intake manifold. Refer to Intake Manifold Removal & Installation.

11. Remove fuel tube. Loosen bolts in reverse order as shown.

12. Remove the fuel tube and fuel injector assembly.

➡When removing, be careful to avoid any interference with fuel injector. Use a shop cloth to absorb any fuel leaks from fuel tube.

13. Remove fuel injector from fuel tube with the following procedure:

a. Open and remove clip.

b. Remove fuel injector from fuel tube by pulling straight.

CAUTION

Be careful with remaining fuel that may go out from fuel tube.

WARNING

Be careful not to damage fuel injector nozzle during removal. Never bump or drop fuel injector. Never disassemble fuel injector.

To install:

14. Install O-rings to fuel injector, noting the following:

a. The upper and lower O-rings are different. Be careful not to confuse them.
- Fuel tube side: Black
- Nozzle side: Green

✳✳ WARNING

Handle O-ring with bare hands. Never wear gloves. Lubricate O-ring with new engine oil. Never clean O-ring with solvent. Make sure that O-ring and its mating part are free of foreign material. When installing O-ring, be careful not to scratch it with tool or fingernails. Also be careful not to twist or stretch O-ring. If O-ring was stretched while it was being attached, never insert it quickly into fuel tube. Insert O-ring straight into fuel tube. Never twist it.

15. Install fuel injector to fuel tube with the following procedure:

a. Insert clips into clip groove on fuel injector. Insert clip so that protrusion of fuel injector matches cutout of clip.

✳✳ WARNING

Never reuse clip. Replace it with a new one. Be careful to keep clip from interfering with O-ring. If interference occurs, replace O-ring.

b. Insert fuel injector into fuel tube with clip attached:
- Insert it while matching it to the axial center
- Insert fuel injector so that protrusion of fuel tube matches cut-out of clip
- Make sure that fuel tube flange is securely fixed in flange fixing groove on clip

16. Make sure that installation is complete by making sure that fuel injector does not rotate or come off.

17. Set fuel tube and fuel injector assembly at its position for installation on cylinder head. Be careful not to interfere with fuel injector nozzle.

18. Tighten bolts in numerical order as shown above.

19. Installation of the remaining components is in the reverse order of removal.

20. After installation is complete, check for fuel leaks:

a. Turn ignition switch "ON" (without starting the engine), to check the connections for fuel leaks with the electric fuel pump applying pressure to the fuel piping.

b. Start the engine and let it idle to check that there are no fuel leaks at the fuel system connections.

FUEL TANK

REMOVAL & INSTALLATION

See Figures 160 through 163.

1. Before servicing the vehicle, refer to the Precautions Section.

2. Drain fuel from fuel tank if necessary.

✳✳ CAUTION

Because fuel tank becomes unstable when installing/removing, fuel should be drained if the level exceeds specification.

3. Place vehicle on a flat and solid surface.

4. Open fuel door and unscrew the fuel filler cap to release the pressure inside the fuel tank.

5. Release the fuel pressure from the fuel lines.

6. Remove rear seat bottom.

7. Turn the four retainers 90° in a clockwise direction and remove the fuel pump inspection hole cover.

8. Disconnect electrical connector and fuel feed hose quick connector. For quick connector disconnecting instructions, refer to Fuel Pump Module Removal & Installation.

9. Remove center exhaust tube.

10. Remove exhaust heat shields.

11. Disconnect parking brake cables from the lower surface of fuel tank and axle and position the parking brake cables out of the way.

12. Remove brake tube protector from rear axle.

13. Loosen fuel filler hose clamp and remove fuel filler hose from fuel tank. When removing fuel filler hose at the fuel filler tube, mark components for alignment.

14. Remove vent hose and EVAP hose at rear of fuel tank.

15. Disconnect vent hose and EVAP hose quick connectors using the following procedures:

a. Pinch retaining tabs of vent hose quick connector and remove vent hose.

b. Slide sleeve of EVAP hose quick connector and remove EVAP hose.

c. If hoses are stuck, push and pull several times until they move freely, and disconnect.

16. Support the center of the fuel tank with a transmission jack. Securely support the fuel tank with a suitable tool.

17. Remove fuel tank bands.

18. Lower transmission jack carefully to remove fuel tank while supporting it by hand. Fuel tank may be in an unstable position because of the shape of fuel tank bottom. Be sure to support tank securely.

To install:

19. Installation is in the reverse order of removal, noting the following:

20. Check the EVAP canister connection for damage or any foreign materials, and align the connector with the tube, then insert the connector straight into the tube until it clicks. After connecting, make sure that the connection is secure by pulling the tube and the connector.

21. Install the fuel tank bands in the proper position by referring to the identification stamp mark "R" and "L" on the end.

a. While supporting the fuel tank, install bolts 1, 3 and 4 to support the tank, but do not fully tighten.

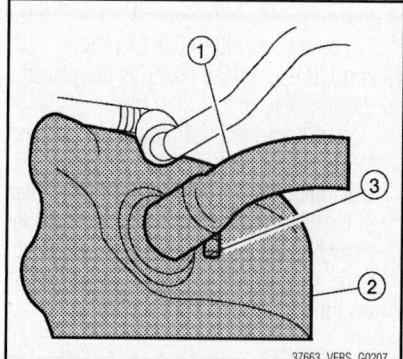

37663_VERS_G0207

Fig. 160 Loosen fuel filler hose clamp (3) and remove fuel filler hose (1) from fuel tank (2)

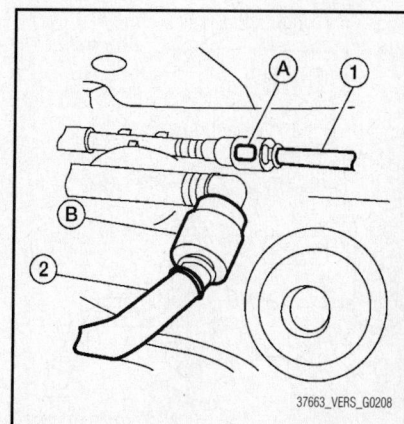

37663_VERS_G0208

Fig. 161 Vent hose (1), EVAP hose (2), retaining tabs (A), and sleeve (B)

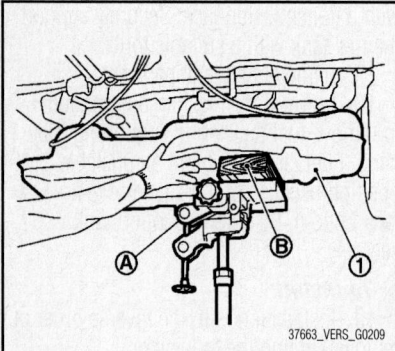

Fig. 162 Support center of fuel tank (1) with transmission jack (A), and support the fuel tank with a suitable tool (B)

b. Install bolt 2 while positioning the fuel tank toward the front of the vehicle. Tighten bolt 2 to specified torque.

c. Tighten bolts 1, 3 and 4 to specified torque.

✳✳ CAUTION

Do not allow fuel filler tube to contact the suspension during installation.

22. Insert fuel filler hose to 1.38 in. (35mm). Be sure hose clamp is not placed on swelled area of fuel filler tube.

a. Check the EVAP hose connections for damage or foreign material, align the matching quick connector with the center of EVAP hose, and insert quick connector straight until it clicks. Make sure connections are secure by pulling on quick connector and EVAP hose by hand.

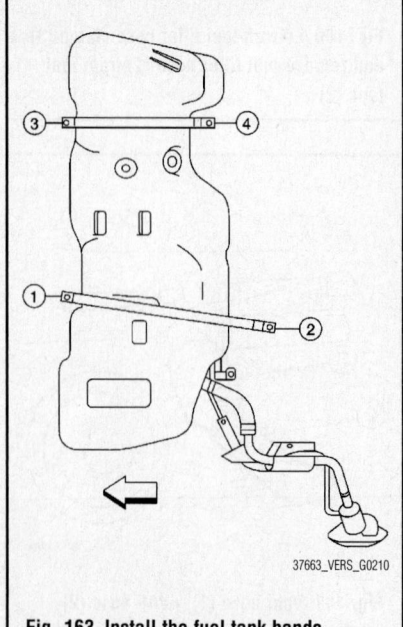

Fig. 163 Install the fuel tank bands

23. Check for fuel leaks:

a. Turn ignition switch "ON" (without starting the engine), to check the connections for fuel leaks with the electric fuel pump applying pressure to the fuel piping.

b. Start the engine and let it idle to check that there are no fuel leaks at the fuel system connections.

IDLE SPEED

ADJUSTMENT

➡**Idle speed is maintained by the ECM. No adjustment is necessary or possible.**

THROTTLE BODY

REMOVAL & INSTALLATION

See Figures 164 through 166.

1. Before servicing the vehicle, refer to the Precautions Section.
2. Disconnect the negative battery cable.
3. Remove engine cover.
4. Drain the engine coolant, or when water hoses are disconnected, attach plug to prevent engine coolant leakage. Perform this step when the engine is cold.
5. Disconnect the water hoses the electric throttle control actuator. When engine coolant is not drained from the radiator, attach plug to water hoses to prevent engine coolant leakage.
6. Disconnect the electric throttle control actuator harness connector.
7. Remove electronic throttle control actuator. Handle carefully to avoid any shock to electric throttle control actuator. Never disassemble.

To install:

8. Install electronic throttle control actuator. Tighten bolts of electric throttle control actuator equally and diagonally in several steps.
9. If applicable, install water hoses to electronic throttle control actuator as shown.
10. Add engine coolant and check for leaks.
11. Connect the negative battery cable.
12. Perform "Throttle Valve Closed Position Learning" after repair when removing harness connector of the electric throttle control actuator.
13. Perform "Throttle Valve Closed Position Learning" and "Idle Air Volume Learning" after repair when replacing electric throttle control actuator.

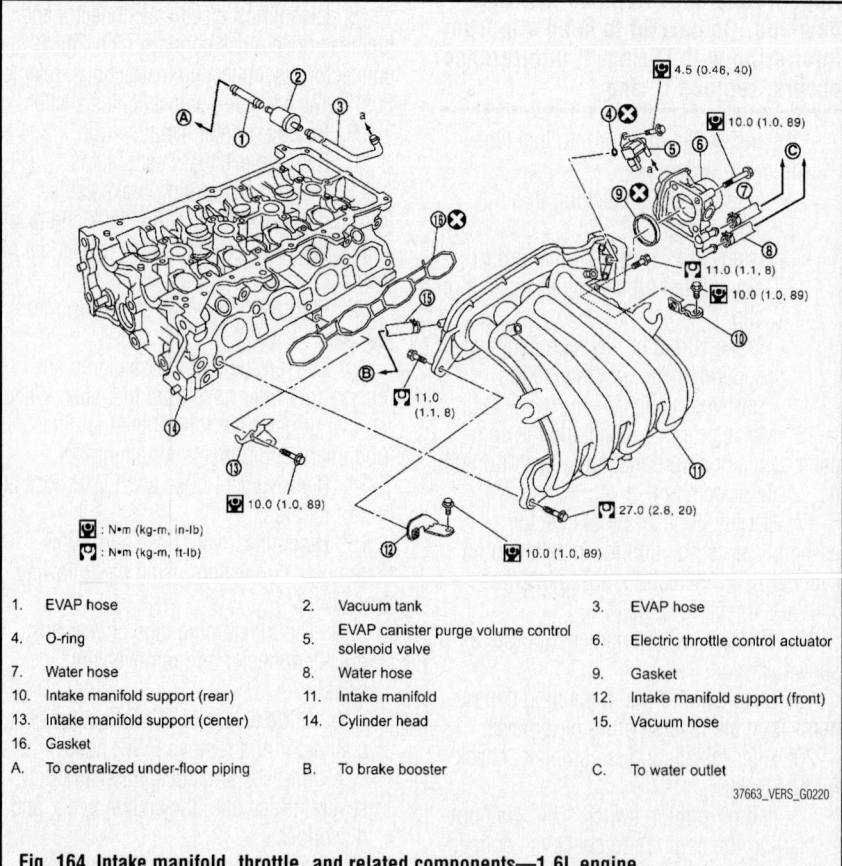

1.	EVAP hose	
2.	Vacuum tank	
3.	EVAP hose	
4.	O-ring	
5.	EVAP canister purge volume control solenoid valve	
6.	Electric throttle control actuator	
7.	Water hose	
8.	Water hose	
9.	Gasket	
10.	Intake manifold support (rear)	
11.	Intake manifold	
12.	Intake manifold support (front)	
13.	Intake manifold support (center)	
14.	Cylinder head	
15.	Vacuum hose	
16.	Gasket	
A.	To centralized under-floor piping	
B.	To brake booster	
C.	To water outlet	

🔧 : N•m (kg-m, in-lb)
🔧 : N•m (kg-m, ft-lb)

Fig. 164 Intake manifold, throttle, and related components—1.6L engine

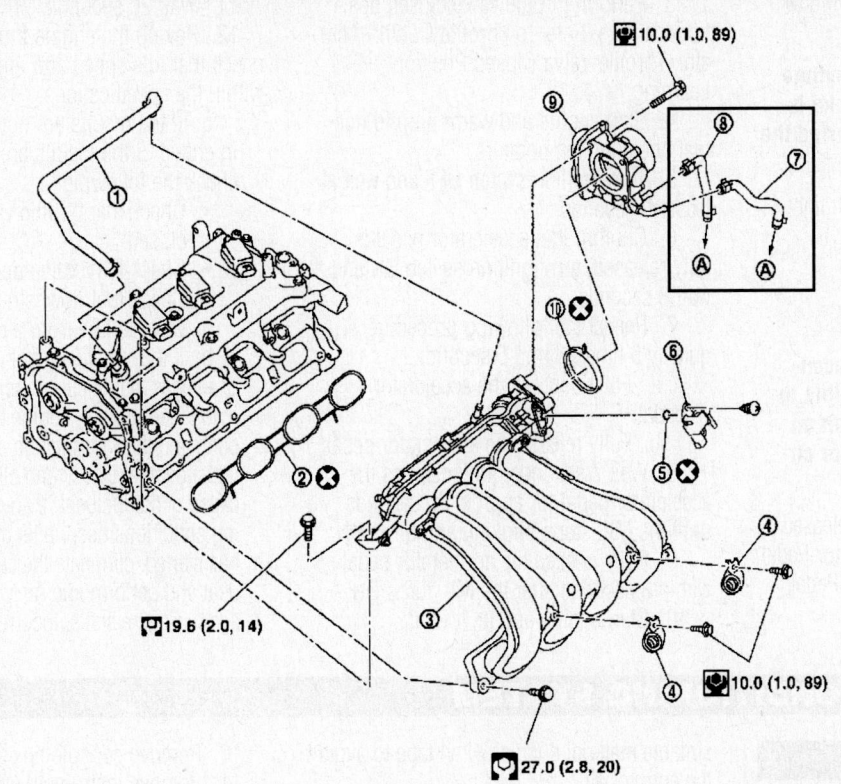

1. PCV hose
2. Gasket
3. Intake manifold
4. Bracket
5. O-ring
6. EVAP canister purge volume control solenoid valve
7. Water hose
8. Water hose
9. Electric throttle control actuator
10. Gasket
A. To water outlet

37663_VERS_G0219

Fig. 165 Intake manifold, throttle, and related components—1.8L engine

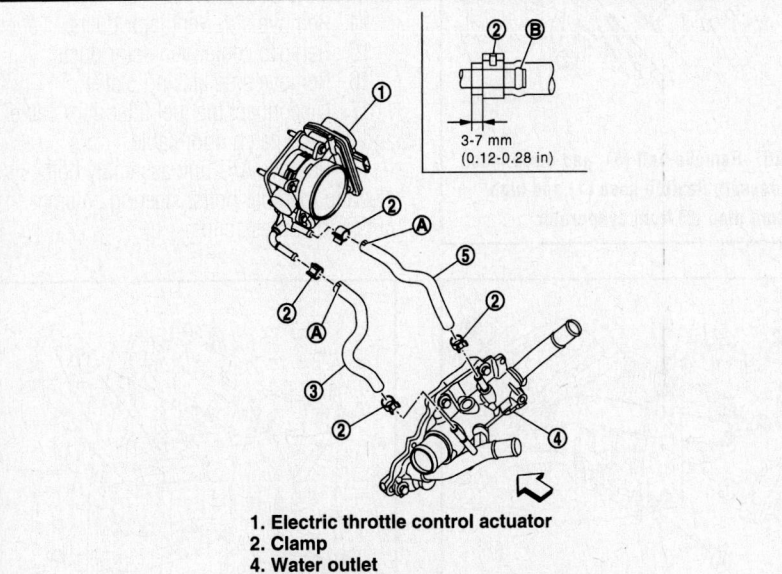

1. Electric throttle control actuator
2. Clamp
4. Water outlet
A. Paint Mark
B. The clamp shall not interfere with the bulged section

22140_VERS_G0065

Fig. 166 Install water hoses to electronic throttle control actuator

THROTTLE VALVE CLOSED POSITION LEARNING

1. Check that accelerator pedal is fully released.
2. Turn ignition switch ON.
3. Turn ignition switch OFF and wait at least 10 seconds.
4. Check that throttle valve moves during the above 10 seconds by confirming the operating sound.

IDLE AIR VOLUME LEARNING

1. Check that all of the following conditions are satisfied. Learning will be cancelled if any of the following conditions are missed for even a moment:
- Battery voltage: More than 12.9 V (At idle)
- Engine coolant temperature: 158–212°F (70–100°C)
- Selector lever position: P or N (A/T), Neutral (M/T)
- Electric load switch: OFF (Air con-

ditioner, head lamp, rear window defogger)

→On vehicles equipped with daytime light systems, if the parking brake is applied before the engine is started the head lamp will not illuminate.

- Steering wheel: Neutral (Straight-ahead position)
- Vehicle speed: Stopped
- Transmission: Warmed-up
- Drive vehicle for 10 minutes

→It is better to count the time accurately with a clock. It is impossible to switch the diagnostic mode when an accelerator pedal position sensor circuit has a malfunction.

2. Perform accelerator pedal released position learning. Refer to Accelerator Pedal Position (APP) sensor, Accelerator Pedal Released Position Learning.

3. Perform throttle valve closed position learning. Refer to Throttle Control Actuator, Throttle Valve Closed Position Learning.

4. Start engine and warm it up to normal operating temperature.

5. Turn ignition switch OFF and wait at least 10 seconds.

6. Confirm that accelerator pedal is fully released, turn ignition switch ON and wait 3 seconds.

7. Repeat the following procedure quickly 5 times within 5 seconds.

a. Fully depress the accelerator pedal.

b. Fully release the accelerator pedal.

8. Wait 7 seconds, fully depress the accelerator pedal for approx. 20 seconds until the MIL stops blinking and turns ON.

9. Fully release the accelerator pedal within 3 seconds after the MIL turns ON.

10. Start engine and let it idle.

11. Wait 20 seconds.

12. Rev up the engine 2 or 3 times and check that idle speed and ignition timing are within the specifications.

a. If the results are normal, inspection in ended. If the results are not normal, check the following:

- Check that throttle valve is fully closed
- Check PCV valve operation
- Check that downstream of throttle valve is free from air leakage

b. If the results are not normal, repair or replace malfunctioning part. If the results are normal, engine component parts and their installation condition are questionable. Check and eliminate the cause of the incident. If any of the following conditions occur after the engine has started, eliminate the cause of the incident and perform Idle Air Volume Learning again: Engine stalls, Incorrect idle.

HEATING & AIR CONDITIONING SYSTEM

A/C UNIT

REMOVAL & INSTALLATION

See Figures 167 through 170.

❋❋ CAUTION

Before servicing components near or affected by the SRS (air bag) system, read and observe all SRS Service Precautions. Refer to Supplemental Restraint System (SRS), in the Chassis Electrical section. Failure to observe all precautions may result in accidental airbag deployment, personal injury, or death.

1. Before servicing the vehicle, refer to the Precautions Section.

2. Disconnect the negative battery cable and wait at least 3 minutes for the SRS memory to drain.

3. Use a recovery/recycling equipment (for HFC-134a) to discharge refrigerant.

4. Drain coolant from cooling system.

5. Remove cowl top cover. Refer to Cowl Top Removal & Installation in the Body Exterior section.

6. Remove upper clip and position the lower dash insulator aside.

7. Remove bolt, and disconnect low-pressure flexible hose and high-pressure pipe from evaporator.

8. Cap or wrap the joint of the pipe with

suitable material such as vinyl tape to avoid the entry of air.

9. Remove clamps and disconnect heater hoses from heater core.

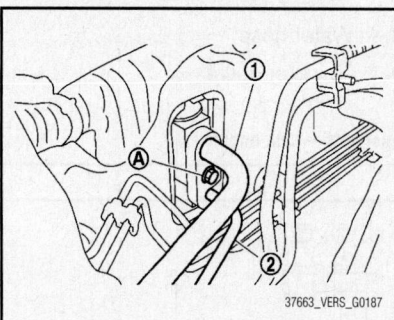

37663_VERS_G0187

Fig. 167 Remove bolt (A), and disconnect low-pressure flexible hose (1) and high-pressure pipe (2) from evaporator

37663_VERS_G0188

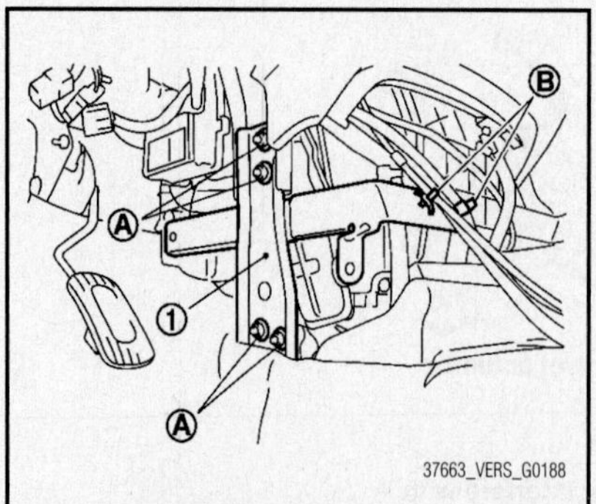

Fig. 168 Remove instrument stay nuts (A) and harness clamps (B), and then remove instrument stay (1)

10. Remove console box assembly.

11. Remove instrument stay nuts and harness clamps, and then remove instrument stay.

12. Disconnect thermo control amp. connector for MR18DE-TYPE 2 systems as shown. Disconnect the thermo control amp. short connector for MR18DE-TYPE 1 and HR16DE systems.

13. Remove instrument panel and pad. Refer to Instrument Panel Removal & Installation in the Body Interior section.

14. Remove side ventilator ducts.

15. Remove center ventilator ducts.

16. Remove side kicking plates.

17. Disconnect the fuel filler door cable and the rear hatch door cable.

18. Remove A/C unit assembly bolts, steering member bolts, steering column nuts and harness clips.

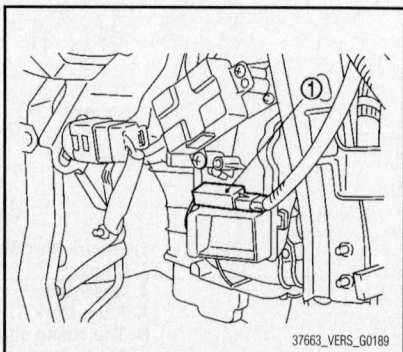

37663_VERS_G0189

Fig. 169 Disconnect thermo control amp. connector (1) for MR18DE-TYPE 2 systems

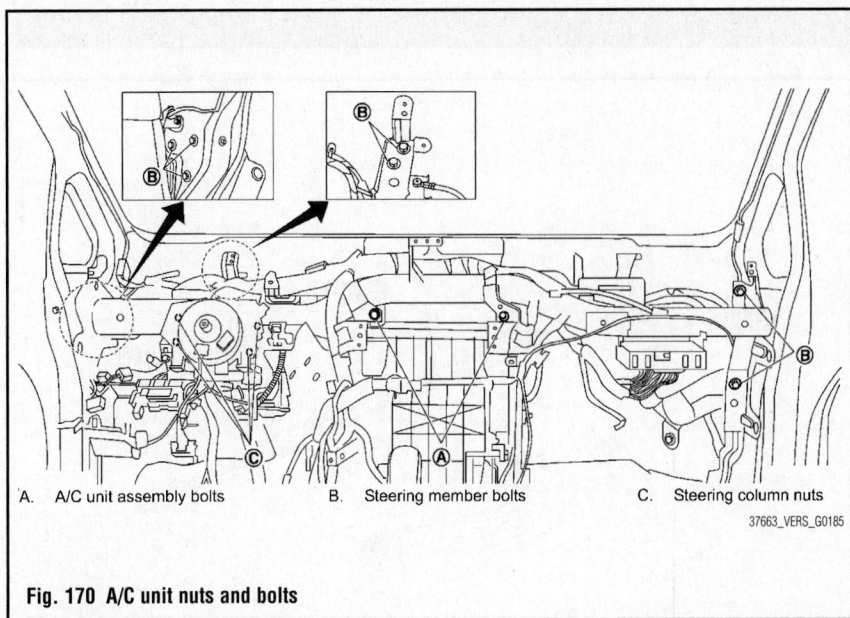

A. A/C unit assembly bolts   B. Steering member bolts   C. Steering column nuts

37663_VERS_G0185

Fig. 170 A/C unit nuts and bolts

19. Remove steering member, and then remove A/C unit assembly.

To install:
20. Installation is in the reverse order of removal, noting the following:

    a. Tighten the A/C unit assembly bolt to 61 inch lbs. (7 Nm).

    b. Tighten the steering member bolt to 9 ft. lbs. (12 Nm).

    c. Recharge the refrigerant.

    d. Replace O-rings for A/C piping with new ones, and apply compressor oil when installing.

    e. When recharging refrigerant, check for leaks.

BLOWER MOTOR

REMOVAL & INSTALLATION
See Figure 171.

✳✳ CAUTION
Before servicing components near or affected by the SRS (air bag) system, read and observe all SRS Service Precautions. Refer to Supplemental Restraint System (SRS), in the Chassis Electrical section. Failure to observe all precautions may result in accidental airbag deployment, personal injury, or death.

1. Before servicing the vehicle, refer to the Precautions Section.

2. Disconnect the negative battery cable and wait at least 3 minutes for the SRS memory to drain.

3. Remove the instrument panel and pad. Refer to Instrument Panel

Removal & Installation in the Body Interior section.

4. Remove the right side ventilator duct.

5. Disconnect the blower motor connector.

6. Push the flange holding hook toward the blower motor, then rotate the blower motor clockwise and remove it from the A/C unit assembly.

7. When blower fan and blower motor are assembled, the balance is adjusted, do not disassemble to replace the individual parts.

To install:
8. Installation is the reverse of removal, noting the following:

    a. Rotate the blower motor until the blower motor flange holding hook locks securely in A/C unit assembly.

HEATER CORE

REMOVAL & INSTALLATION
See Figure 172.

1. Before servicing the vehicle, refer to the Precautions Section.

2. Remove the A/C unit assembly.

3. Remove the left foot duct.

4. Remove the heater pipe cover screw, and then remove the heater pipe cover.

5. Remove the heater pipe clip screw, and then remove the heater pipe clip.

6. Slide the heater core out of the A/C unit assembly.

To install:
7. Installation is the reverse of removal.

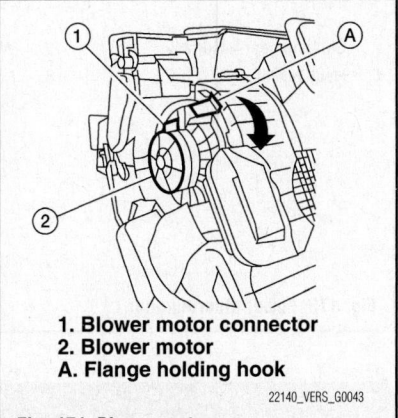

1. Blower motor connector
2. Blower motor
A. Flange holding hook

22140_VERS_G0043

Fig. 171 Blower motor

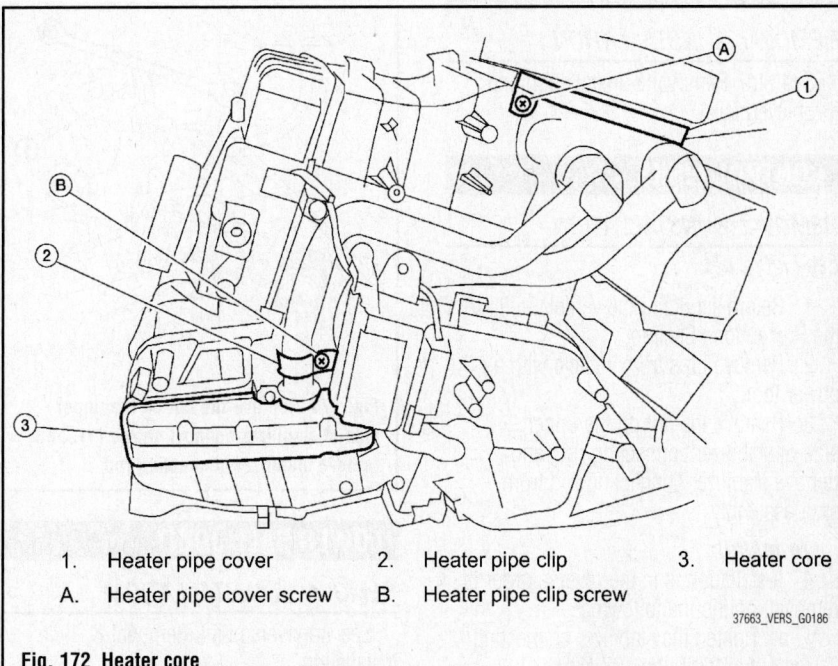

1. Heater pipe cover   2. Heater pipe clip   3. Heater core
A. Heater pipe cover screw   B. Heater pipe clip screw

37663_VERS_G0186

Fig. 172 Heater core

STEERING

POWER RACK & PINION STEERING GEAR

REMOVAL & INSTALLATION

See Figure 173.

1. Before servicing the vehicle, refer to the Precautions Section.

✳ WARNING

Spiral cable may be cut if steering wheel turns while separating steering column assembly and steering gear assembly. Be sure to secure steering wheel using string to avoid turning.

2. Remove the front suspension member. Refer to Front Suspension Assembly Removal & Installation in the Suspension section.

3. Remove mounting bolts and nuts of steering gear assembly.

To install:

4. Installation is in the reverse order of removal, noting the following:

a. Clean mounting surface on the body side of fire wall seal when installing steering gear assembly.

b. Check wheel alignment under unladen conditions with tires on level ground.

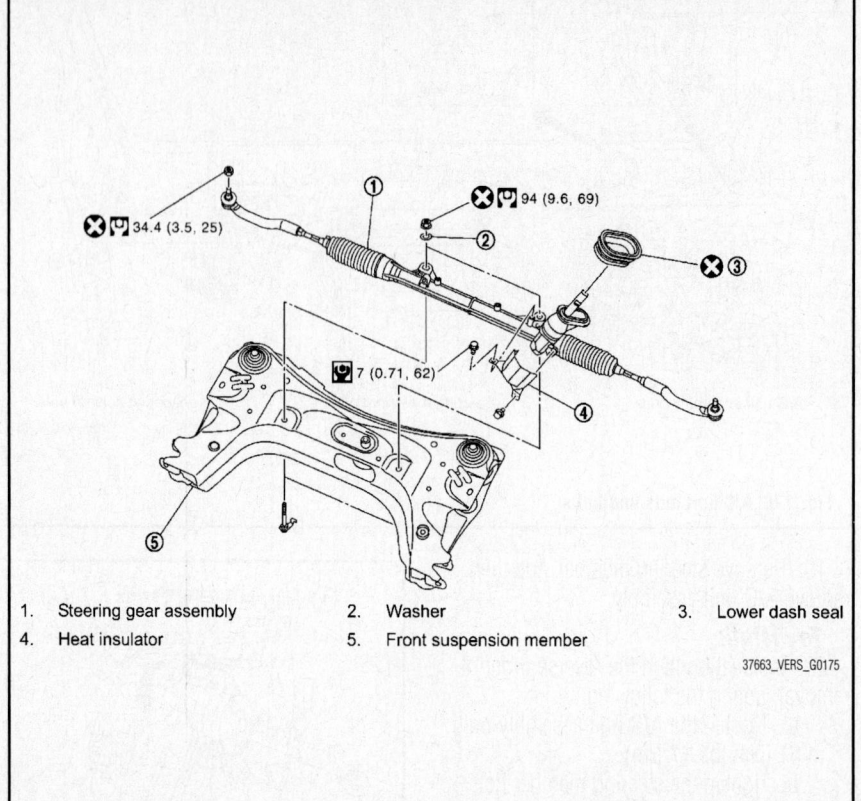

| 1. | Steering gear assembly | 2. | Washer | 3. | Lower dash seal |
| 4. | Heat insulator | 5. | Front suspension member | | |

37663_VERS_G0175

Fig. 173 Power steering gear

SUSPENSION

COIL SPRING

REMOVAL & INSTALLATION

See Strut Removal & Installation and Strut Overhaul.

CONTROL LINKS

REMOVAL & INSTALLATION

See Figure 174.

1. Before servicing the vehicle, refer to the Precautions Section.

2. Remove tires from vehicle with a power tool.

3. Remove the nut on the upper side of stabilizer connecting rod, and remove stabilizer connecting rod from strut assembly.

To install:

4. Installation is in the reverse order of removal, noting the following:

a. Tighten the stabilizer connecting rod nut to 27 ft. lbs. (37 Nm).

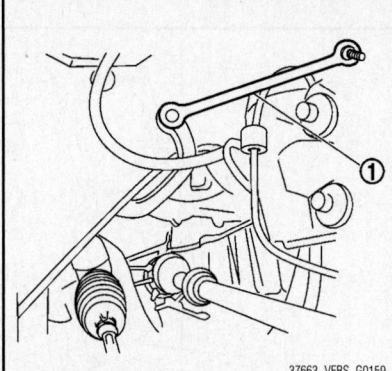

37663_VERS_G0159

Fig. 174 Remove the nut on the upper side of stabilizer connecting rod (1), and remove stabilizer connecting rod

LOWER BALL JOINT

REMOVAL & INSTALLATION

See Transverse Link Removal & Installation.

FRONT SUSPENSION

LOWER CONTROL ARM

REMOVAL & INSTALLATION

See Transverse Link Removal & Installation.

STEERING KNUCKLE

REMOVAL & INSTALLATION

See Wheel Hub & Bearing (sealed unit).

STRUT

REMOVAL & INSTALLATION

See Figures 175 and 176.

1. Before servicing the vehicle, refer to the Precautions Section.

2. Remove the cowl top panel. Refer to Cowl Top Removal & Installation in the Body Exterior section.

3. Remove the wheels and tires.

4. Remove the wheel sensor harness from the strut assembly. Do not pull on the wheel sensor harness.

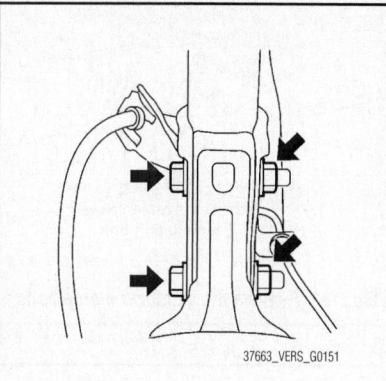

Fig. 175 Remove the strut mounting insulator bolts

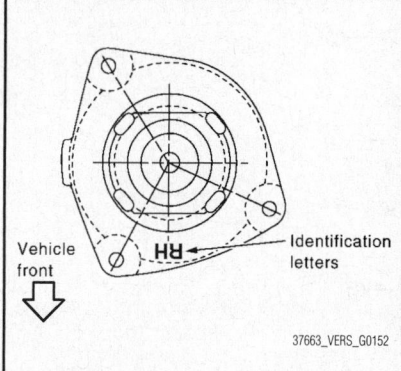

Vehicle front

RH

Identification letters

37663_VERS_G0152

Fig. 176 Attach strut mounting insulator as shown

5. Remove the brake hose lock plate.

6. Remove the nut on the upper side of the stabilizer connecting rod, and remove the stabilizer connecting rod from the strut assembly.

7. Remove the nuts and bolts, and then remove steering knuckle from the strut assembly.

8. Remove the strut mounting insulator bolts, and remove the strut assembly from vehicle.

To install:

9. Installation is in the reverse order of removal, noting the following:

a. Perform final tightening of bolts and nuts at the strut assembly lower side (rubber bushing) under unladen conditions with tires on level ground. Check wheel alignment.

b. Check wheel sensor harness for proper connection.

c. Attach strut mounting insulator as shown.

STABILIZER BAR

REMOVAL & INSTALLATION

See Figures 177 and 178.

---

1. Before servicing the vehicle, refer to the Precautions Section.

2. Separate intermediate shaft from steering gear pinion shaft.

3. Remove wheels and tires.

4. Remove the nut on the lower side of the stabilizer connecting rod, and remove the stabilizer connecting rod from the stabilizer bar.

5. If necessary remove the stabilizer connecting rod upper nut. Separate the stabilizer connecting rod and the strut.

6. Loosen the steering outer socket nut.

7. Using the ball joint remover (tool No. HT72520000 (J-25730-A) or suitable tool, remove the steering outer socket from the steering knuckle so as not to damage the ball joint boot.

8. Temporarily tighten the nut to prevent damage to threads and to prevent the ball joint remover (suitable tool) from suddenly coming off.

9. Remove the rear torque rod.

10. Set a jack under the front suspension member.

11. Remove the bolts of the member stay, and remove member stay from vehicle.

12. Gradually lower the front suspension member in order to remove the stabilizer bolts. Be careful not to lower it too far (Do not over load the links).

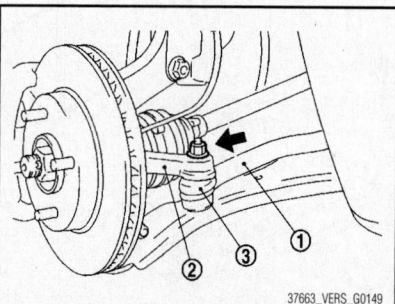

Fig. 177 Remove steering outer socket (1) from steering knuckle (2) so as not to damage ball joint boot (3)

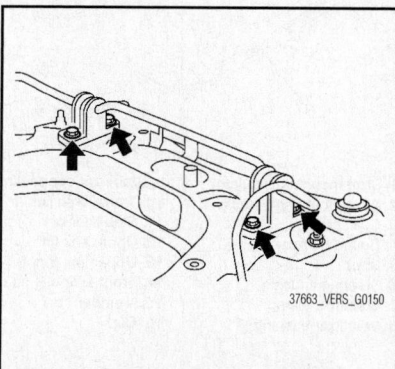

Fig. 178 Remove the stabilizer clamp bolts

---

13. Remove the bolts of the stabilizer clamp, and remove the stabilizer clamp and stabilizer bushing from vehicle.

14. Remove the stabilizer bar from the vehicle.

To install:

15. Installation is the reverse of removal, noting the following:

a. Install the stabilizer connecting rod and strut and tighten the nut to 27 ft. lbs. (37 Nm).

b. Install the stabilizer clamp and tighten the bolts to 21 ft. lbs. (28 Nm).

SUSPENSION ASSEMBLY

REMOVAL & INSTALLATION

See Figures 177, 179 and 180.

1. Before servicing the vehicle, refer to the Precautions Section.

2. Separate intermediate shaft from steering gear pinion shaft.

3. Remove wheels and tires.

4. Remove the wheel sensor from the steering knuckle. Do not pull on wheel sensor harness.

5. Remove the nut on the upper side of stabilizer connecting rod, and remove stabilizer connecting rod from strut assembly.

6. Loosen the steering outer socket nut.

7. Using the ball joint remover (tool No. HT72520000 (J-25730-A) or suitable tool, remove the steering outer socket from the steering knuckle so as not to damage the ball joint boot.

8. Temporarily tighten the nut to prevent damage to threads and to prevent the ball joint remover (suitable tool) from suddenly coming off.

9. Remove the rear torque rod.

10. Remove transverse link ball joint nut and bolt, and remove transverse link from steering knuckle.

11. Set a jack under the front suspension member.

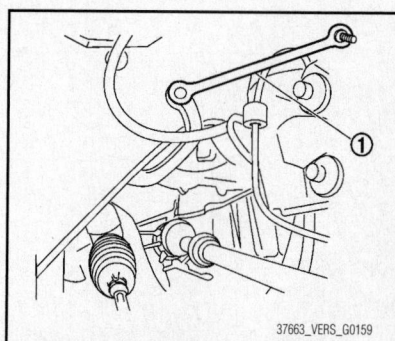

Fig. 179 Remove the nut on the upper side of stabilizer connecting rod (1), and remove stabilizer connecting rod

12. Remove upper side bolts of upper link.

13. Remove the bolts of member stay, and then remove member stay from vehicle.

14. Gradually lower a jack to remove front suspension assembly.

To install:

15. Installation is the reverse of removal, noting the following:

　a. Perform final tightening of each of parts (rubber bushing), under unladen conditions.

　b. Check wheel sensor harness for proper connection.

TRANSVERSE LINK

REMOVAL & INSTALLATION

See Figures 180 and 181.

1. Before servicing the vehicle, refer to the Precautions Section.

2. Remove wheels and tires.

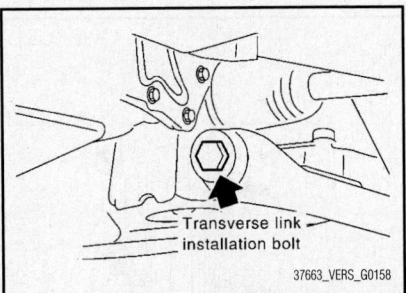

Fig. 181 Remove the stabilizer clamp bolts

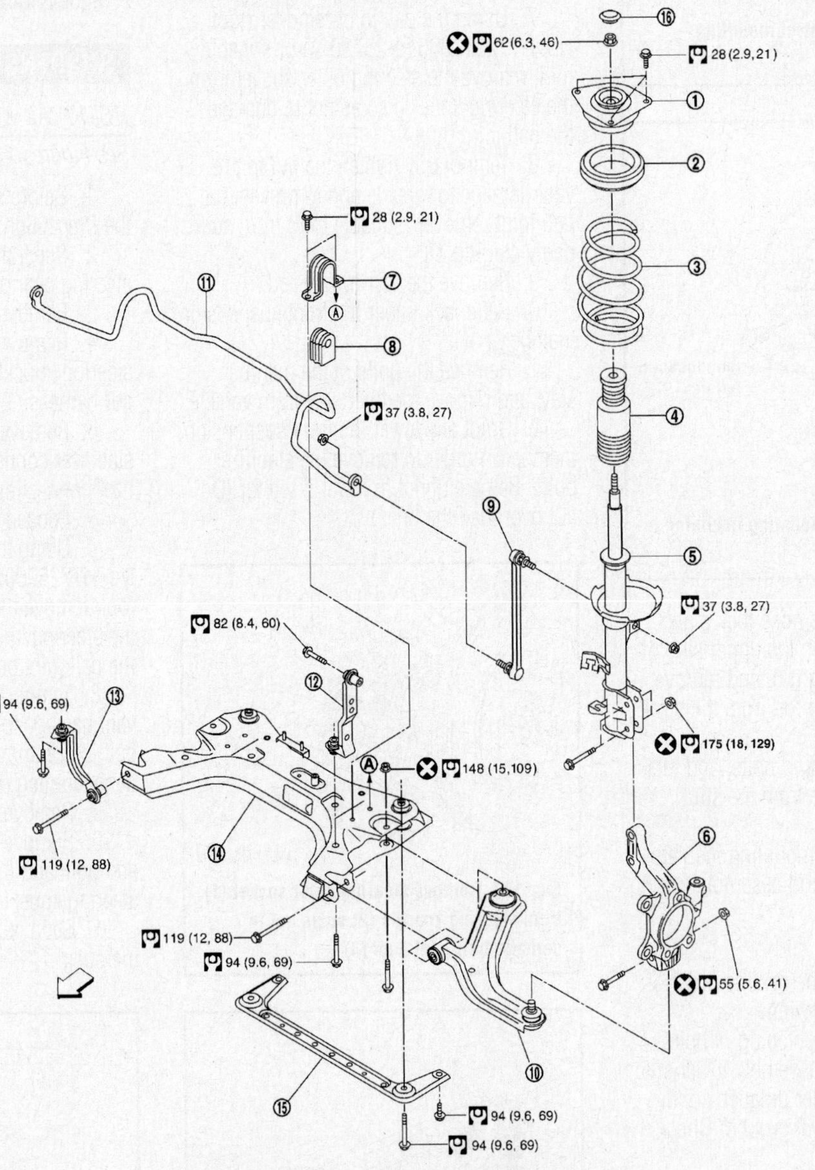

1. Strut mounting insulator
2. Strut mounting bearing
3. Coil spring
4. Bound bumper
5. Strut
6. Steering knuckle
7. Stabilizer clamp
8. Stabilizer bushing
9. Stabilizer connecting rod
10. Transverse link
11. Stabilizer bar
12. Upper link (left)
13. Upper link (right)
14. Front suspension member
15. Member stay
16. Cap

Fig. 180 Front suspension—exploded view

37663_VERS_G0148

3. Remove transverse link ball joint nut and bolt, and remove transverse link from steering knuckle.

4. Remove transverse link nuts and bolts, and remove transverse link from front suspension member.

➡**When removing the left-hand transverse link it may be necessary to lower the suspension member in order to remove bolts to avoid contact with the transaxle.**

   a. Set jack under front suspension member.

   b. Loosen right-hand upper link bolts, left-hand upper link bolt (front suspension member side), front suspension member bolts (left/right). Lower the front suspension member in order to remove transverse link bolts.

5. Remove transverse link.

To install:

6. Installation is the reverse of removal, noting the following:

   a. Perform the final tightening of each of parts under unladen conditions, which were removed when removing wheel hub and bearing assembly and steering knuckle. Check the wheel alignment.

WHEEL HUB & BEARING

REMOVAL & INSTALLATION

See Figures 182 through 185.

1. Before servicing the vehicle, refer to the Precautions Section.

2. Remove wheel and tire.

3. Without disassembling the

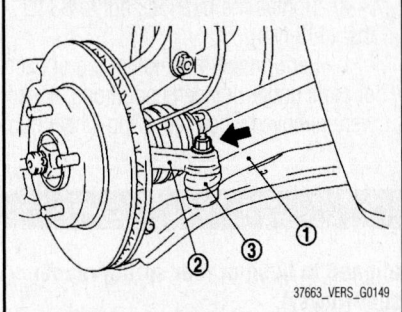

Fig. 183 Remove steering outer socket (1) from steering knuckle (2) so as not to damage ball joint boot (3)

hydraulic lines, remove the torque member bolts. Then reposition the torque member and brake caliper assembly aside with wire.

➡**Do not depress brake pedal while brake caliper is removed.**

4. Put alignment marks on disc rotor and wheel hub and bearing assembly, and remove disc rotor.

5. Remove wheel sensor from steering knuckle. Do not pull on wheel sensor harness.

6. Loosen steering outer socket nut.

7. Remove steering outer socket from steering knuckle so as not to damage ball joint boot using ball joint remover or suitable tool. Temporarily leave the outer socket nut installed to prevent damage to threads and to prevent the ball joint remover or suitable tool from suddenly coming off.

8. Remove transverse link ball joint nut and bolt, and remove transverse link from steering knuckle.

9. Remove cotter pin, and loosen hub lock nut. Temporarily leave the hub

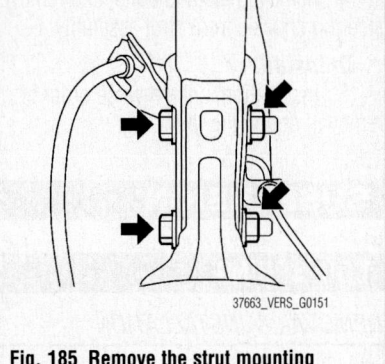

Fig. 185 Remove the strut mounting insulator bolts

lock nut installed to prevent damage to threads.

10. Separate the drive shaft from the wheel hub and bearing assembly by lightly tapping the end of the drive shaft using a hammer or suitable tool, and remove hub lock nut.

11. If wheel hub and bearing assembly and drive shaft cannot be separated after performing the above procedure, use a suitable puller.

12. Remove the drive shaft from the wheel hub and bearing assembly and support the drive shaft.

➡**Do not apply an excessive angle to drive shaft joint when removing from the wheel hub and bearing assembly. Do not excessively extend slide joint. Do not allow drive shaft to hang down. Support the entire drive shaft.**

13. Remove wheel hub and bearing assembly bolts, and then remove splash guard and wheel hub and bearing assembly from steering knuckle.

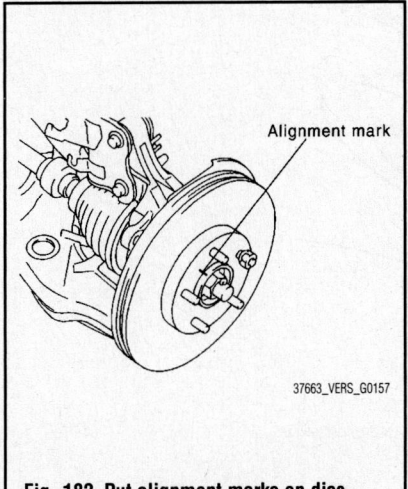

Fig. 182 Put alignment marks on disc rotor and wheel hub and bearing assembly

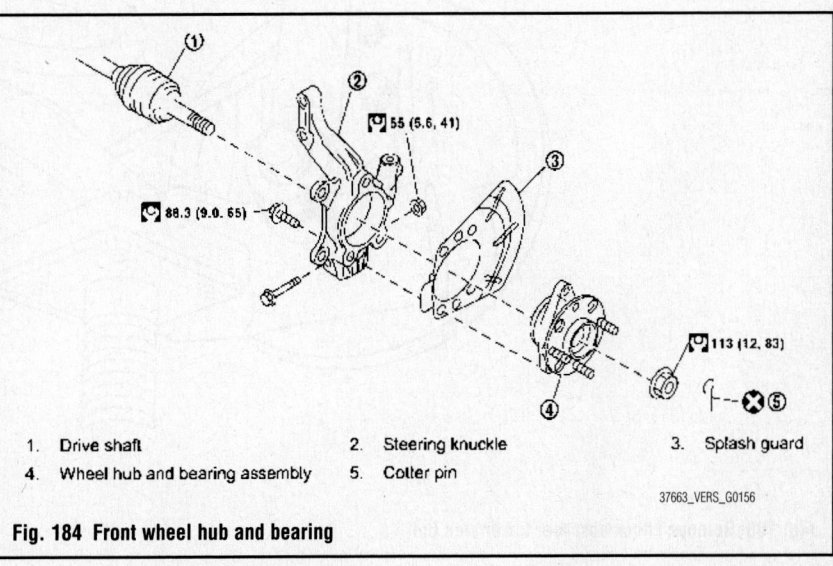

1. Drive shaft
2. Steering knuckle
3. Splash guard
4. Wheel hub and bearing assembly
5. Cotter pin

Fig. 184 Front wheel hub and bearing

14. Remove nuts and bolts, and remove steering knuckle from strut assembly.

To install:

15. Installation is the reverse order of removal, noting the following:

a. Tighten the hub lock nut to 83 ft. lbs. (113 Nm).

b. Perform the final tightening of each of parts under unladen conditions, which were removed when removing wheel hub

and bearing assembly and steering knuckle. Check the wheel alignment.

c. When installing disc rotor on wheel hub and bearing assembly, align the marks.

SUSPENSION
REAR SUSPENSION

COIL SPRING

REMOVAL & INSTALLATION

See Figures 186 and 187.

1. Before servicing the vehicle, refer to the Precautions Section.
2. Remove rear wheels and tires.
3. Remove wheel sensor from wheel hub and bearing assembly. Do not pull on wheel sensor harness.
4. Separate brake tube from wheel cylinder.
5. Set a jack under rear suspension beam.
6. Remove shock absorber lower side bolt.
7. Gradually lower the jack, and then remove coil spring and rear spring rubber seat (upper and lower).

To install:

8. Installation is the reverse of removal, noting the following:

a. Tighten the lower shock absorber side bolt to 91 ft. lbs. (124 Nm).

➡**When installing spring, be sure to securely install the spring end position**

aligned to flush of rear spring rubber seat (lower).

SHOCK ABSORBER

REMOVAL & INSTALLATION

See Figures 186 and 187.

1. Before servicing the vehicle, refer to the Precautions Section.
2. Remove rear wheels and tires.
3. Remove wheel sensor from wheel hub and bearing assembly and rear suspension beam. Do not pull on wheel sensor harness.
4. Remove shock absorber mask from trunk side finisher using a flat-bladed screwdriver with its tip taped.
5. Set jack under rear suspension beam.
6. Remove upper nut of the shock absorber, and then remove washer (upper), bushing (upper) from shock absorber.
7. Remove shock absorber lower side bolt.
8. Gradually lower the jack, and

remove the bushing (lower), washer (lower), distance tube, bound bumper cover, bound bumper and shock absorber from vehicle.

To install:

9. Installation is the reverse of removal, noting the following:

a. Tighten the lower shock absorber side bolt to 91 ft. lbs. (124 Nm).

b. Tighten the upper shock absorber nut to 15 ft. lbs. (20 Nm).

c. When installing body side bushing (upper), install the projection to the body side hole securely.

TESTING

1. Check the following:
   - Shock absorber for deformation, cracks or damage, and replace if necessary
   - Piston rod for damage, uneven wear or distortion, and replace if necessary
   - Check bound bumper and bushing for cracks, deformation or other damage, and replace applicable parts if necessary

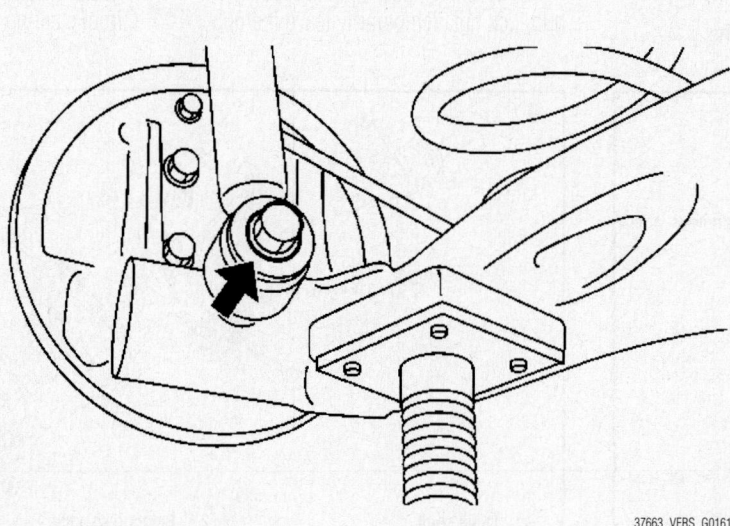

37663_VERS_G0161

Fig. 186 Remove shock absorber lower side bolt

SUSPENSION BEAM

REMOVAL & INSTALLATION

See Figure 187.

1. Before servicing the vehicle, refer to the Precautions Section.

2. Remove rear wheels and tires.

3. Separate parking brake rear cable from rear drum brake and rear suspension beam.

4. Remove wheel sensor and wheel sensor harness from wheel hub and bearing assembly and rear suspension beam.

5. Remove lock plate and separate brake tube from brake hose.

6. Remove wheel hub and bearing assembly and back plate.

7. Set jack under rear suspension beam.

8. Remove coil spring (left/right).

9. Remove bolts between body and rear suspension beam bracket.

10. Gradually lower the jack, and then remove rear suspension beam from vehicle.

11. Remove the rear suspension beam bracket bolt and nut, and then remove rear

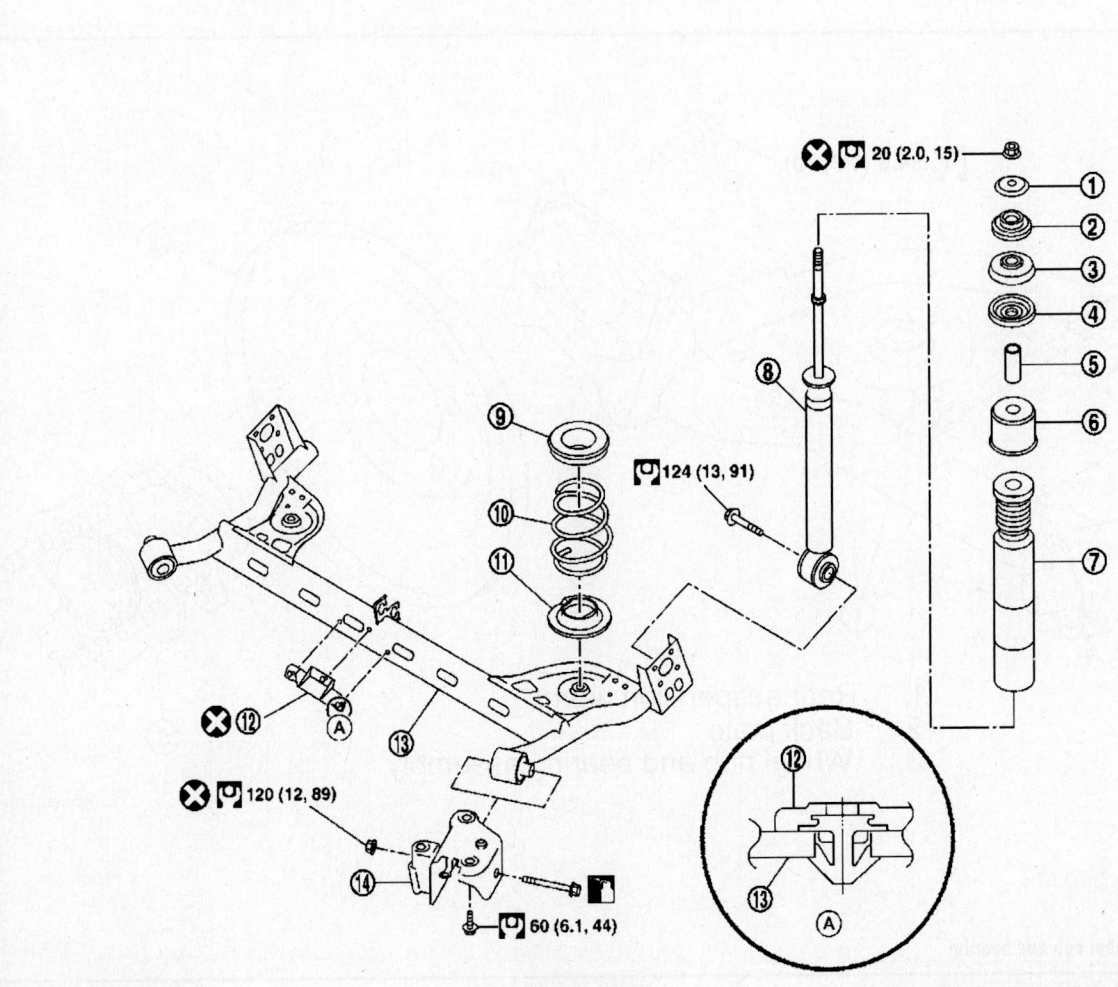

1.	Washer (upper)	2.	Bushing (upper)	
4.	Washer (lower)	5.	Distance tube	
7.	Bound bumper	8.	Shock absorber	
10.	Coil spring	11.	Rear spring rubber seat (lower)	
13.	Rear suspension beam	14.	Rear suspension beam bracket	

3.	Bushing (lower)
6.	Bound bumper cover
9.	Rear spring rubber seat (upper)
12.	Brake tube protector
A.	View of brake tube protector clip

37663_VERS_G0160

Fig. 187 Rear suspension—exploded view

suspension beam bracket from rear suspension beam.

12. Remove brake tube protector from rear suspension beam.

To install:

13. Installation is in the reverse order of removal.

14. Refill with new brake fluid and bleed air.

15. Check the following after finishing work:

- Parking brake operation (stroke)
- Wheel sensor harness for proper connection

16. Perform final tightening of rear suspension beam and rear suspension beam bracket (rubber bushing) under unladen conditions with tires on level ground.

WHEEL HUB & BEARING

REMOVAL & INSTALLATION

See Figure 188.

1. Before servicing the vehicle, refer to the Precautions Section.

2. Remove tires.

3. Remove wheel sensor from wheel hub and bearing assembly. Do not pull on wheel sensor harness.

4. Remove the drum brake assembly.

5. Remove wheel hub and bearing assembly bolts, and then remove wheel hub and bearing assembly from vehicle.

To install:

6. Installation is the reverse of removal.

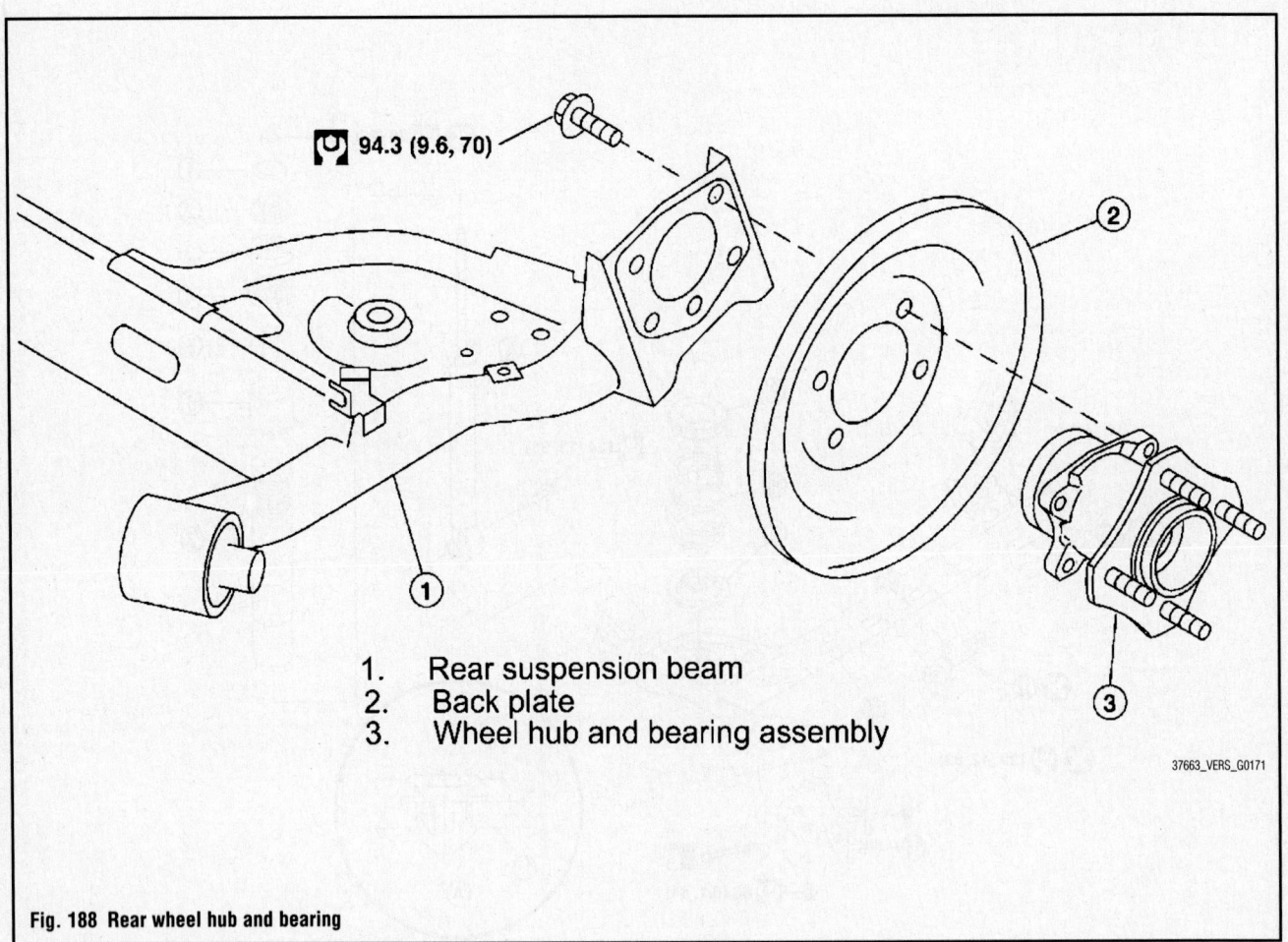

94.3 (9.6, 70)

1. Rear suspension beam
2. Back plate
3. Wheel hub and bearing assembly

37663_VERS_G0171

Fig. 188 Rear wheel hub and bearing

SPECIFICATIONS AND MAINTENANCE CHARTS

ENGINE AND VEHICLE IDENTIFICATION

ID/Code ①	Liters (cc)	Cu. In.	Cyl.	Fuel Sys.	Engine Type	Eng. Mfg.	Code ②	Year
			Engine				Model Year	
VQ40DE	4.0 (3954)	241	6	MFI	DOHC	Nissan	9	2009
							A	2010

MFI: Multi-port Fuel Injection

SOHC: Single Overhead Camshaft

DOHC: Double Overhead Camshafts

① 4th digit of the Vehicle Identification Number (VIN)

② 10th digit of the Vehicle Identification Number (VIN)

37663_XTER_C0001

GENERAL ENGINE SPECIFICATIONS

Year	Model	Engine Displacement Liters	Engine ID	Net Horsepower @ rpm	Net Torque @ rpm (ft. lbs.)	Bore x Stroke (in.)	Compression Ratio	Oil Pressure @ rpm
2009	Xterra	4.0	VQ40DE	265@5600	284@4000	3.76X3.62	9.7:1	43@2000
2010	Xterra	4.0	VQ40DE	265@5600	284@4000	3.76X3.62	9.7:1	43@2000

37663_XTER_C0002

ENGINE TUNE-UP SPECIFICATIONS

Year	Engine Displ. Liters	Engine ID	Spark Plug Gap (in.)	Ignition Timing (deg.) MT	AT	Fuel Pump (psi)	Idle Speed (rpm) MT	AT ①	Valve Clearance (in.) In.	Ex.
2009	4.0	VQ40DE	0.043	10-20B	10-20B	51 ②	575-675	650-750	③	④
2010	4.0	VQ40DE	0.043	10-20B	10-20B	51 ②	575-675	650-750	③	④

NOTE: The Vehicle Emission Control Information label often reflects specification changes made during production. The label figures must be used if they differ from those in this chart.

B: Before top dead center

HYD: Hydraulic

① Automatic transmission in Neutral

② At idle

③ 0.010-0.013 cold

   0.012-0.016 hot

④ 0.011-0.015 cold

   0.012-0.017 hot

37663_XTER_C0003

CAPACITIES

Year	Model	Engine Displacement Liters	Engine ID	Engine Oil with Filter (qts.)	Transmission (pts.)		Transfer Case (pts.)	Drive Axle		Fuel Tank (gal.)	Cooling System (qts.)
					Manual	Auto.		Front (pts.)	Rear (pts.)		
2009	Xterra	4.0	VQ40DE	5.10	①	21.50	2.1	1.75	②	21.2	11.0
2010	Xterra	4.0	VQ40DE	5.50	①	21.50	2.1	1.75	②	21.2	11.0

NOTE: All capacities are approximate. Add fluid gradually and check to be sure a proper fluid level is obtained.

① 2WD: 8.3; 4WD: 8.9

② C200: 3.3; M226: 4.25

37663_XTER_C0004

VALVE SPECIFICATIONS

Year	Engine Displacement Liters	Engine ID	Seat Angle (deg.)	Face Angle (deg.)	Spring Test Pressure (lbs. @ in.)	Spring Installed Height (in.)	Stem-to-Guide Clearance (in.)		Stem Diameter (in.)	
							Intake	Exhaust	Intake	Exhaust
2009	4.0	VQ40DE	①	②	③	1.456	0.0008-0.0021	0.0012-0.0025	0.2348-0.2354	0.2344-0.2350
2010	4.0	VQ40DE	①	②	③	1.456	0.0008-0.0021	0.0012-0.0025	0.2348-0.2354	0.2344-0.2350

① 44 degrees 22 minutes to 45 degrees 8 minutes

② 45 degrees 15 minutes to 45 degrees 45 minutes

③ Installation: 37-42@1.457

Valve open: 84-95@1.071

37663_XTER_C0005

CAMSHAFT SPECIFICATIONS CHART

All measurements are given in inches.

Year	Engine Displ. Liters	Engine VIN	Journal Dia.	Brg. Oil Clearance	Shaft End-play	Runout	Lobe Height	
							Intake	Exhaust
2009	4.0	VQ40DE	①	②	0.0045-0.0074	0.0010	1.7900-1.7921	1.7746-1.7821
2010	4.0	VQ40DE	①	②	0.0045-0.0074	0.0010	1.7900-1.7921	1.7746-1.7821

① No.1: 1.0211-1.0218. No's. 2, 3, 4: 0.9230-0.9238

② No.1: 1.0018-0.0034. No's. 2, 3, 4: 0.0014-0.0030

37663_XTER_C0006

CRANKSHAFT AND CONNECTING ROD SPECIFICATIONS

All measurements are given in inches.

Year	Engine Displacement Liters	Engine ID	Crankshaft				Connecting Rod		
			Main Brg. Journal Dia.	Main Brg. Oil Clearance	Shaft End-play	Thrust on No.	Journal Diameter	Oil Clearance	Side Clearance
2009	4.0	VQ40DE	②	0.0014-0.0018	0.0039-0.0098	NA	2.2441-2.2446	0.0013-0.0023	0.0079-0.0138
2010	4.0	VQ40DE	②	0.0014-0.0018	0.0039-0.0098	NA	2.2441-2.2446	0.0013-0.0023	0.0079-0.0138

NA: Not Available

① There are 24 different grades, ranging from A (2.1645) to 7 (2.1636)

② There are 24 different grades, ranging from A (2.7549) to 7 (2.7540)

37663_XTER_C0007

PISTON AND RING SPECIFICATIONS

All measurements are given in inches.

Year	Engine Displacement Liters	Engine ID	Piston Clearance	Ring Gap			Ring Side Clearance		
				Top Comp.	Bottom Comp.	Oil Control	Top Comp.	Bottom Comp.	Oil Control
2009	4.0	VQ40DE	0.0004-0.0012	0.0091-0.0130	0.0130-0.0189	0.0079-0.0197	0.0018-0.0031	0.0012-0.0028	0.0026-0.0053
2010	4.0	VQ40DE	0.0004-0.0012	0.0091-0.0130	0.0130-0.0189	0.0079-0.0197	0.0018-0.0031	0.0012-0.0028	0.0026-0.0053

37663_XTER_C0008

TORQUE SPECIFICATIONS

All readings in ft. lbs.

Year	Engine Displacement Liters	Engine ID	Cylinder Head Bolts	Main Bearing Bolts	Rod Bearing Bolts	Crankshaft Damper Bolts	Flywheel Bolts	Manifold Intake	Manifold Exhaust	Spark Plugs	Oil Pan Drain Plug
2009	4.0	VQ40DE	①	②	14	③	65	④	⑤	14-22	25
2010	4.0	VQ40DE	①	②	14	③	65	④	⑤	14-22	25

① Step 1: 72 ft. lbs.

    Step 2: loosen completely to 0 ft. lbs.

    Step 3: 29 ft. lbs.

    Step 4: Plus 90 degrees clockwise

    Step 5: Plus 90 degrees clockwise

② Bolts: 17-24 (M8) 16 ft. lbs.

    Install rear main seal

    Bolts: 1-16 (M10) 26 ft. lbs.

    Bolts: 1-16 (M10) Plus 90 degrees clockwise

③ Step 1: 33 ft. lbs.

    Step 2: Plus 84-90 degrees clockwise

④ Intake manifold collector:

    Bolts and nuts: 8 ft. lbs.

    Stud bolts: 61 inch lbs.

⑤ Intake manifold:

    Bolts and nuts: 5 ft. lbs. and

    than to 21 ft. lbs.

    Studs: 8 ft. lbs.

37663_XTER_C0009

WHEEL ALIGNMENT

Year	Model		Caster Range (+/-Deg.)	Caster Preferred Setting (Deg.)	Camber Range (+/-Deg.)	Camber Preferred Setting (Deg.)	Toe-in (in.)
2009	Xterra	2WD	①	①	②	②	③
		4WD	④	④	⑤	⑤	③
2010	Xterra	2WD	①	①	②	②	③
		4WD	④	④	⑤	⑤	③

NOTE: On 2009-2010 vehicles, fuel, coolant and engine oil must be full. Spare tire, jack, hand tools and mats must be in place.

Some 2009-2010 vehicles may be equipped with non adjustable lower link bolts and washers. In order to adjust caster and camber on these vehicles,

first replace these bolts with adjustable cam bolts and washers.

① Minimum: 2 degrees 15' (2.25 degrees)

    Nominal: 3 degrees 0' (3.00 degrees)

    Maximum: 3 degrees 45' (3.75 degrees)

② Minimum: 2 degrees 0' (2.00 degrees)

    Nominal: 0 degrees 15' (0.25 degrees)

    Maximum: 1 degrees 0' (1.00 degrees)

③ Minimum: 0.12 inches

    Nominal 0.16 inches

    Maximum 0.20 inches

④ Minimum: 0 degrees 15' (-0.25 degrees)

    Nominal: 2 degrees 45' (2.75 degrees)

    Maximum: 3 degrees 30' (3.50 degrees)

⑤ Minimum: 0 degrees 15' (-0.25 degrees)

    Nominal: 0 degrees 30' (0.50 degrees)

    Maximum: 1 degrees 15' (1.25 degrees)

37663_XTER_C0010

TIRE, WHEEL AND BALL JOINT SPECIFICATIONS

Year	Model	OEM Tires Standard	OEM Tires Optional	Tire Pressures (psi) Front	Tire Pressures (psi) Rear	Wheel Size	Ball Joint Inspection	Lug Nut Torque (ft. lbs.)
2009	Xterra S	P265/70R16	None	①	①	7J	②	98
	Xterra S-O/R	P265/75R16	None	①	①	7J	②	98
	Xterra SE	P255/65R17	None	①	①	7.5J	②	98
2010	Xterra X	P265/70R16	None	①	①	7J	②	98
	Xterra S	P265/70R16	None	①	①	7J	②	98
	Xterra S-O/R	P265/75R16	None	①	①	7J	②	98
	Xterra SE	P255/65R17	None	①	①	7.5J	②	98

OEM: Original Equipment Manufacturer

PSI: Pounds Per Square Inch

① See placard on vehicle

② Replace if any measurable movement is found.

37663_XTER_C0011

BRAKE SPECIFICATIONS

All measurements in inches unless noted

Year	Model	Brake Disc Original Thickness	Brake Disc Minimum Thickness	Brake Disc Maximum Runout	Brake Drum Diameter Original Inside Diameter	Brake Drum Diameter Max. Wear Limit	Minimum Lining Thickness Front	Minimum Lining Thickness Rear	Brake Caliper Bracket Bolts (ft. lbs.)	Brake Caliper Mounting Bolts (ft. lbs.)
2009	Xterra	1.102	1.024	0.002	④	①	0.079	0.079	②	③
2010	Xterra	1.102	1.024	0.002	④	①	0.079	0.079	②	③

① Rear disc brakes: 0.630

② Front: 136 ft. lbs.

   Rear: 76 ft. lbs.

③ Front: 32 ft. lbs.

   Rear: 19 ft. lbs.

④ Rear disc brakes: 0.709

37663_XTER_C0012

SCHEDULED MAINTENANCE INTERVALS
Nissan Xterra

TO BE SERVICED	TYPE OF SERVICE	VEHICLE MILEAGE INTERVAL (x1000)												
		7.5	15	22.5	30	37.5	45	52.5	60	67.5	75	82.5	90	97.5
Engine oil & filter	R	✓	✓	✓	✓	✓	✓	✓	✓	✓	✓	✓	✓	✓
Brake lines & cables	S/I		✓		✓		✓		✓		✓		✓	
Brake pads, discs, drums & linings	S/I		✓		✓		✓		✓		✓		✓	
Driveshaft boots & propeller shaft	S/I				✓				✓				✓	
Front wheel bearings (4X2)	S/I				✓				✓				✓	
Front wheel bearings (4X4)	S/I				✓				✓				✓	
Automatic & manual transmission, transfer & differential gear oil ①	S/I		✓		✓		✓		✓		✓		✓	
Air cleaner filter	R				✓				✓				✓	
Engine coolant	R								✓				✓	
Spark plugs (platinum)	R	replace every 105,000 miles												
Drive belt(s)	S/I				✓				✓				✓	
Exhaust system	S/I				✓				✓				✓	
Fuel lines	S/I				✓				✓				✓	
Steering gear (box) & linkage, axle & suspension parts	S/I				✓				✓				✓	
Vapor lines	S/I				✓				✓				✓	
Tires (rotate)	S/I	✓	✓	✓	✓	✓	✓	✓	✓	✓	✓	✓	✓	✓
Timing belt ②	R													

R: Replace   S/I: Service or Inspect

① Differential (w/limited-slip differential) oil: replace oil every 30,000 miles, 2007-2008 vehicles.

② Timing belt: replace at 105,000 miles.

FREQUENT OPERATION MAINTENANCE (SEVERE SERVICE)

If a vehicle is operated under any of the following conditions it is considered severe service:

- Extremely dusty areas.

- 50% or more of the vehicle operation is in 32°C (90°F) or higher temperatures, or constant operation in temperatures below 0°C (32°F).

- Prolonged idling (vehicle operation in stop and go traffic).

- Frequent short running periods (engine does not warm to normal operating temperatures).

- Police, taxi, delivery usage or trailer towing usage.

Oil & oil filter: replace every 3750 miles.

Brake pads, discs, drums & linings: service or inspect every 7500 miles.

Driveshaft boots & propeller shaft: service or inspect every 7500 miles.

Exhaust system: service or inspect every 7500 miles.

Steering gear (box) & linkage, (steering damper-4X4), axle & suspension parts: service or inspect every 7500 miles.

Steering linkage ball joints & front suspension ball joints: service or inspect every 7500 miles.

37663_XTER_C0013

BRAKES | **INFORMATION AND PRECAUTIONS**

ANTI-LOCK SYSTEMS

• Certain components within the ABS system are not intended to be serviced or repaired individually.

• Do not use rubber hoses or other parts not specifically specified for and ABS system. When using repair kits, replace all parts included in the kit. Partial or incorrect repair may lead to functional problems and require the replacement of components.

• Lubricate rubber parts with clean, fresh brake fluid to ease assembly. Do not use shop air to clean parts; damage to rubber components may result.

• Use only DOT 3 brake fluid from an unopened container.

• If any hydraulic component or line is removed or replaced, it may be necessary to bleed the entire system.

• A clean repair area is essential. Always clean the reservoir and cap thoroughly before removing the cap. The slightest amount of dirt in the fluid may plug an orifice and impair the system function. Perform repairs after components have been thoroughly cleaned; use only denatured alcohol to clean components. Do not allow ABS components to come into contact with any substance containing mineral oil; this includes used shop rags.

• The Anti-Lock control unit is a microprocessor similar to other computer units in the vehicle. Ensure that the ignition switch is **OFF** before removing or installing controller harnesses. Avoid static electricity discharge at or near the controller.

• If any arc welding is to be done on the vehicle, the control unit should be unplugged before welding operations begin.

DISC AND DRUM SYSTEMS

✳✳ CAUTION

Dust and dirt accumulating on brake parts during normal use may contain asbestos fibers from production or aftermarket brake linings.

Breathing excessive concentrations of asbestos fibers can cause serious bodily harm. Exercise care when servicing brake parts. Do not sand or grind brake lining unless equipment used is designed to contain the dust residue. Do not clean brake parts with compressed air or by dry brushing. Cleaning should be done by dampening the brake components with a fine mist of water, then wiping the brake components clean with a dampened cloth. Dispose of cloth and all residue containing asbestos fibers in an impermeable container with the appropriate label. Follow practices prescribed by the Occupational Safety and Health Administration (OSHA) and the Environmental Protection Agency (EPA) for the handling, processing, and disposing of dust or debris that may contain asbestos fibers.

BRAKES | **BLEEDING THE BRAKE SYSTEM**

BLEEDING PROCEDURE

BLEEDING PROCEDURE

1. Before servicing the vehicle, refer to the Precautions Section.

✳✳ CAUTION

While bleeding the brake system, pay attention to the master cylinder fluid level.

2. Disconnect the negative battery cable.
3. Raise and safely support the vehicle.
4. Attach a vinyl tube to the right, rear bleeder valve.
5. Depress the brake pedal fully 4 or 5 times.
6. With the brake pedal depressed, loosen the bleeder valve to let the air out, then tighten it immediately.

7. Repeat steps 3 and 4 until no more air comes out.
8. Tighten the bleeder valve.
9. Fill the master cylinder reservoir.
10. Repeat the above steps for the left front, left rear, and the right front calipers, in that order.

BLEEDING THE ABS SYSTEM

See Bleeding Procedure.

WHEEL SPEED SENSORS

REMOVAL & INSTALLATION

See Figure 1.

1. Before servicing the vehicle, refer to the Precautions Section.
2. Disconnect the negative battery cable.
3. Raise and support the vehicle safely.
4. Remove the tire and wheel assembly.
5. Remove the rotor.
6. Remove the wheel speed sensor mounting bolts and clip.
7. Pull the sensor out, being careful to turn it as little as possible. Do not pull on the sensor harness.

To install:

8. Before installing the sensor be sure no foreign materials such as iron fragments are adhered to the pick-up part of the sensor or to the inside of the sensor mounting hole or on the rotor mounting surface.
9. Be sure the harness is not twisted when installed.
10. Continue the installation in the reverse order of the removal procedure.

WHEEL SPEED SENSOR RINGS (TOOTHED RINGS)

REMOVAL & INSTALLATION

Front Wheel

The front sensor rotor cannot be disassembled. It is integrated with the hub and bearing assembly. To replace the sensor rotor, replace the front hub and knuckle assembly. The rear sensor rotor cannot be disassembled. It is integrated with the hub and bearing assembly. To replace the sensor rotor, replace the rear hub and bearing assembly.

Rear Wheel

C200 Axle

1. Disconnect the negative battery cable.
2. Raise and support the vehicle safely.
3. Remove the axle shaft.
4. Remove the sensor rotor using tool J25852-B or equivalent.

To install:

5. Installation is the reverse of the removal procedure.
6. Pay attention to the direction of the sensor rotor.

M226 Axle

1. Disconnect the negative battery cable.
2. Raise and support the vehicle safely.
3. Remove the axle shaft.

➡**It is necessary to disassemble the rear axle to replace the sensor rotor.**

4. Pull the sensor off of the axle shaft using tool ST30031000, or equivalent, and a press.

To install:

5. Installation is the reverse of the removal procedure.
6. Make sure that the sensor rotor is fully seated.

➡**Never reuse the old sensor rotor. Never reuse the old oil seal.**

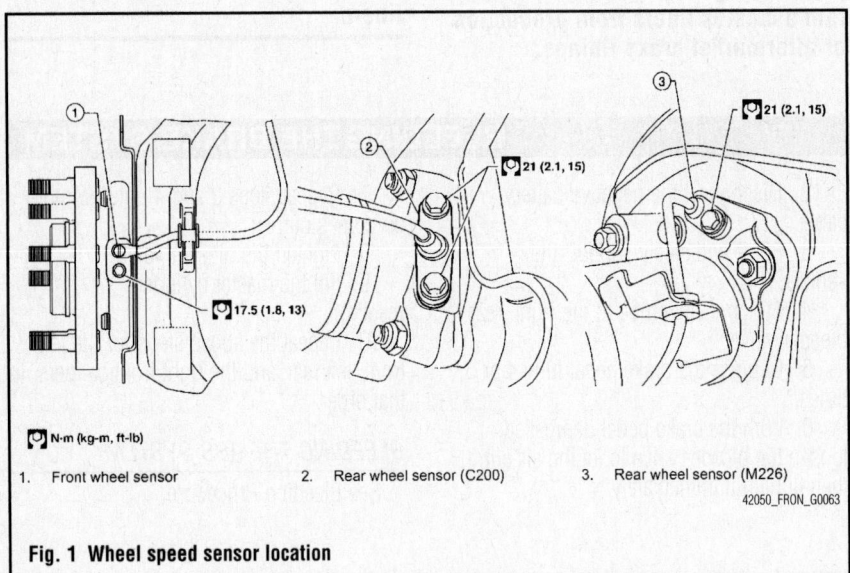

Fig. 1 Wheel speed sensor location

1. Front wheel sensor
2. Rear wheel sensor (C200)
3. Rear wheel sensor (M226)

42050_FRON_G0063

BRAKES   **FRONT DISC BRAKES**

BRAKE CALIPER

REMOVAL & INSTALLATION

See Figure 2.

1. Before servicing the vehicle, refer to the Precautions Section.

2. Drain the brake fluid, as necessary.

3. Raise the vehicle and support safely.

4. Remove the tire and wheel assembly.

5. Remove the bolt attaching the brake hose to the caliper. Plug the brake hose to prevent brake fluid loss.

6. Remove the caliper support mounting bolts and lift the caliper assembly from the knuckle.

To install:

7. Position the caliper assembly onto the knuckle and install the bolts. Make sure the rotor fits between the brake pads. Torque the bolts to specification.

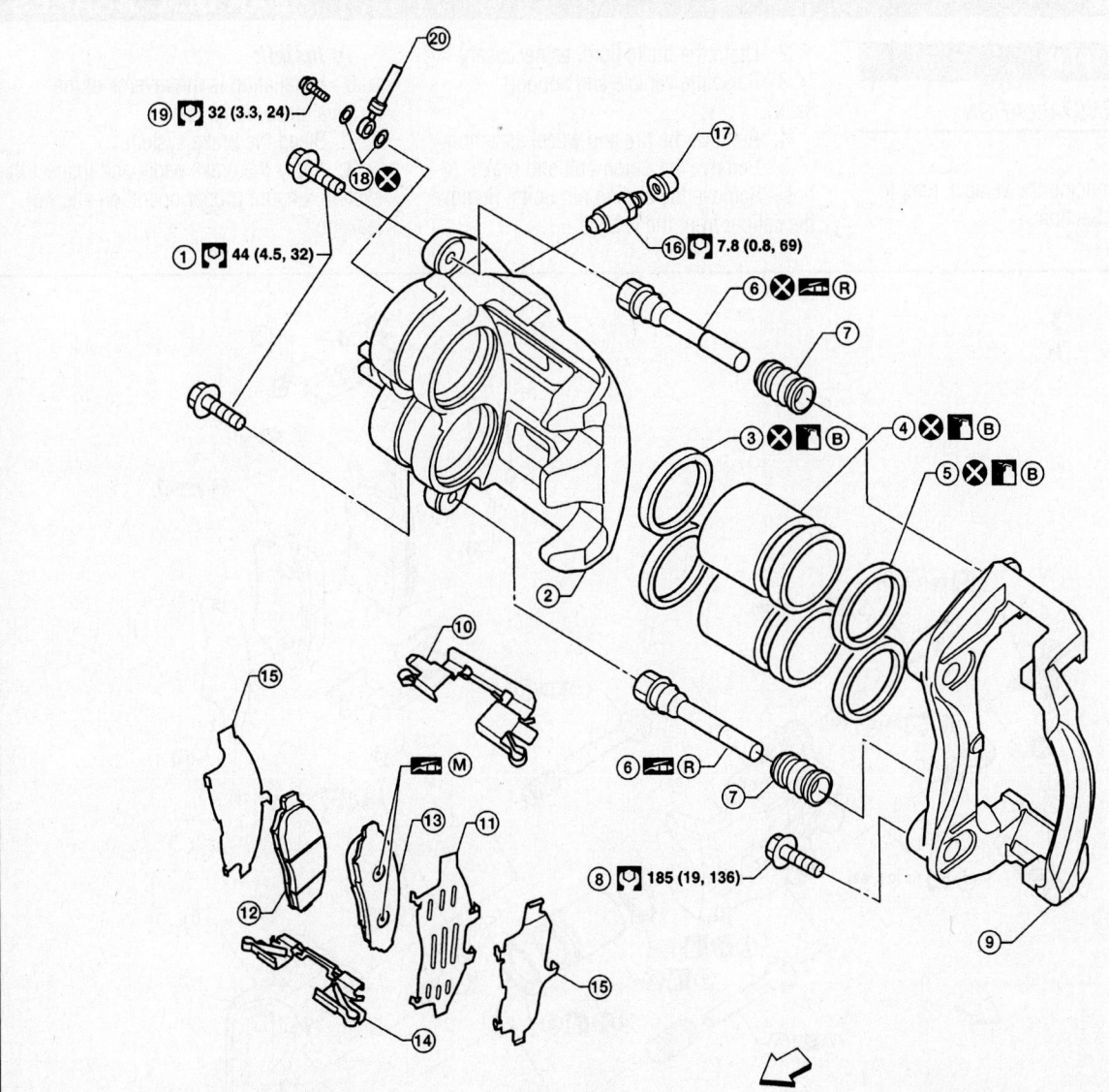

1.	Sliding pin bolt	2.	Cylinder body	3.	Piston seal
4.	Piston	5.	Piston boot	6.	Sliding pin
7.	Sliding pin boot	8.	Torque member bolt	9.	Torque member
10.	Pad retainer	11.	Inner shim	12.	Inner brake pad
13.	Outer brake pad	14.	Pad retainer	15.	Outer shim
16.	Bleed valve	17.	Cap	18.	Copper washers
19.	Union bolt	20.	Brake hose		

09482_FRON_G0121

Fig. 2 Front disc brake and related components

8. Using new copper washers, connect the brake hose to the caliper. Torque the brake hose attaching bolt to specification.

9. Bleed the brake system.

10. Apply the brake pedal and inspect the system. Ensure proper operation and no leakage.

11. Install tire and wheel assembly. Lower the vehicle and road test.

BRAKES

BRAKE CALIPER

REMOVAL & INSTALLATION

See Figure 3.

1. Before servicing the vehicle, refer to the Precautions Section.

DISC BRAKE PADS

REMOVAL & INSTALLATION

1. Before servicing the vehicle, refer to the Precautions Section.

2. Drain the brake fluid, as necessary.

3. Raise the vehicle and support safely.

4. Remove the bottom pin from the

2. Drain the brake fluid, as necessary.

3. Raise the vehicle and support safely.

4. Remove the tire and wheel assembly.

5. Remove the union bolt and brake hose. Remove the sliding pin bolts. Remove the caliper from the vehicle.

caliper and swing the caliper cylinder body upward; support the caliper with a wire.

5. Remove the brake pad retainers, shims and the pads.

To install:

6. Compress the piston of the disc brake caliper.

7. Install the brake pads and caliper assembly.

REAR DISC BRAKES

To install:

6. Installation is the reverse of the removal procedure.

7. Bleed the brake system.

8. Apply the brake pedal and inspect the system. Ensure proper operation and no leakage.

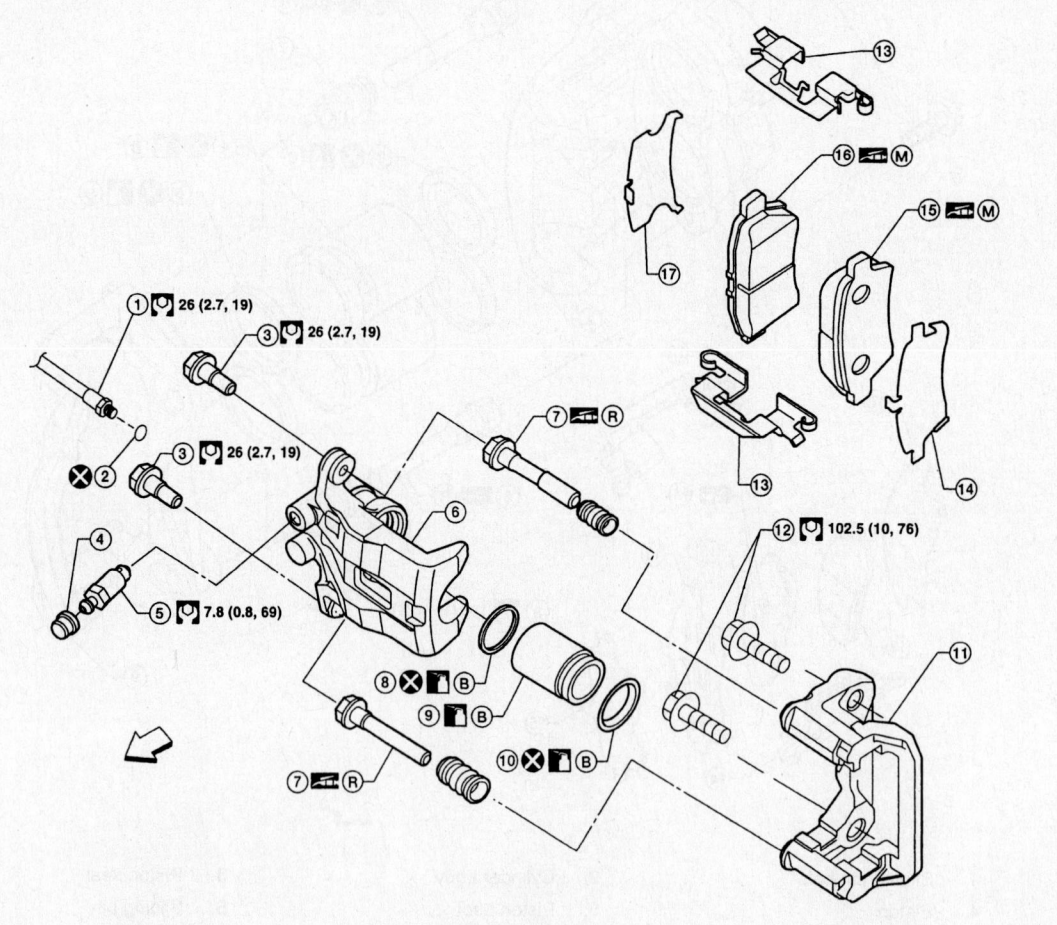

No.	Part	No.	Part	No.	Part
1.	Brake hose	2.	Copper washer	3.	Sliding pin bolt
4.	Cap	5.	Bleed valve	6.	Cylinder body
7.	Sliding pin	8.	Piston seal	9.	Piston
10.	Piston boot	11.	Torque member	12.	Torque member bolt
13.	Pad retainer	14.	Outer shim	15.	Outer brake pad
16.	Inner brake pad	17.	Inner shim	⇐:	Front

09482_FRON_G0123

Fig. 3 Rear disc brake and related components—Xterra

9. Install tire and wheel assembly. Lower the vehicle and road test.

DISC BRAKE PADS

REMOVAL & INSTALLATION

1. Before servicing the vehicle, refer to the Precautions Section.

BRAKES

PARKING BRAKE CABLES

ADJUSTMENT

1. Remove rear half of the center console.

2. Rotate adjusting nut and loosen cable until tension is sufficiently released.

3. Remove the wheel and tire using power tool.

4. Remove the rotor and measure inner diameter at widest point.

5. Transfer measurement less 0.6 mm to the parking brake shoes and adjust accordingly.

6. Using wheel nuts, secure the disc to the hub to prevent it from tilting.

7. Rotate disc rotor to make sure there is no drag.

8. Operate parking brake lever 10 or more times with a force of 110 lbs. (490 N).

9. Rotate adjusting nut to adjust lever stroke to 6–8 notches with 44 lbs. (196 N) force.

10. With parking brake lever completely disengaged, make sure there is no drag on the parking brake.

PARKING BRAKE SHOES

REMOVAL & INSTALLATION

See Figures 4 and 5.

1. Before servicing the vehicle, refer to the Precautions Section.

➡ **Clean the brakes with a vacuum dust collector to minimize the hazard of air-borne particles or other materials.**

➡ **Remove the disc rotor with the parking brake completely disengaged.**

2. Raise and safely support the vehicle.

3. Release the parking brake.
4. Remove the rear wheels.
5. Remove the rotor.
6. Remove the return springs.
7. Remove the adjuster.

---

2. Drain the brake fluid, as necessary.

3. Raise the vehicle and support safely.

4. Remove the tire and wheel assembly.

5. Remove the top bolt from the caliper.

6. Swing the caliper open and remove the pads.

To install:

7. Compress the piston of the disc brake caliper.

8. Install the brake pads and caliper assembly. Tighten caliper bolts to 24 ft. lbs. (32 Nm).

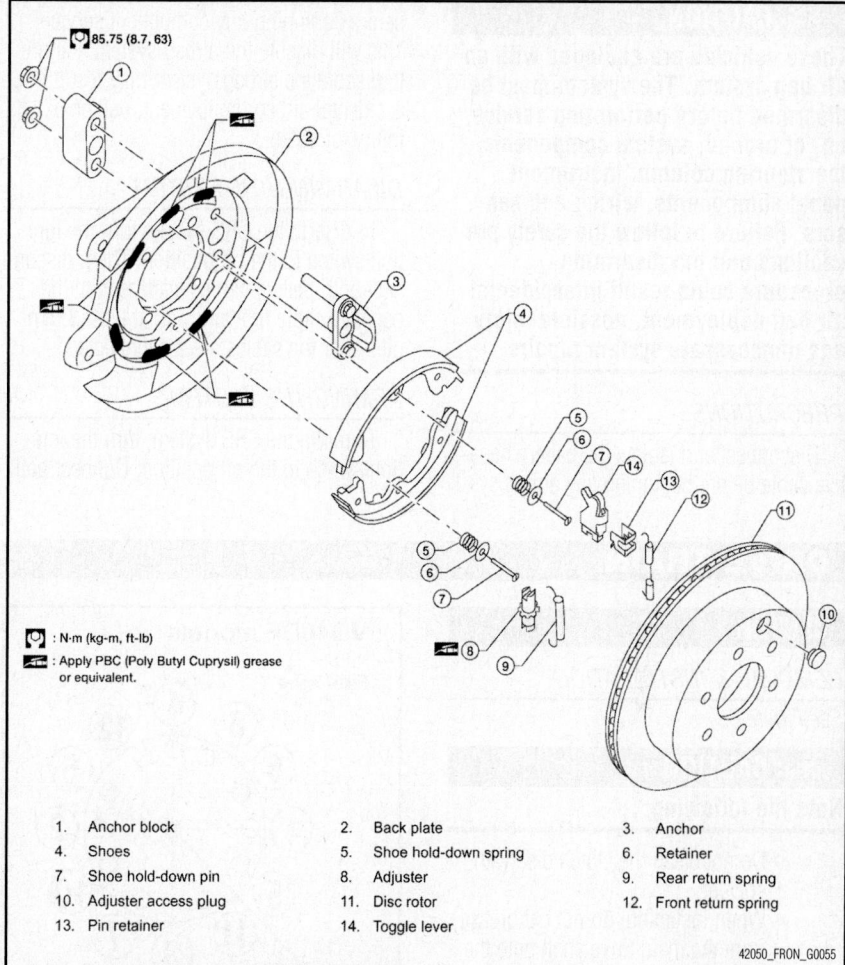

: N·m (kg-m, ft-lb)

: Apply PBC (Poly Butyl Cuprysil) grease or equivalent.

1.	Anchor block	2.	Back plate	3.	Anchor
4.	Shoes	5.	Shoe hold-down spring	6.	Retainer
7.	Shoe hold-down pin	8.	Adjuster	9.	Rear return spring
10.	Adjuster access plug	11.	Disc rotor	12.	Front return spring
13.	Pin retainer	14.	Toggle lever		

42050_FRON_G0055

Fig. 4 Parking brake and related components

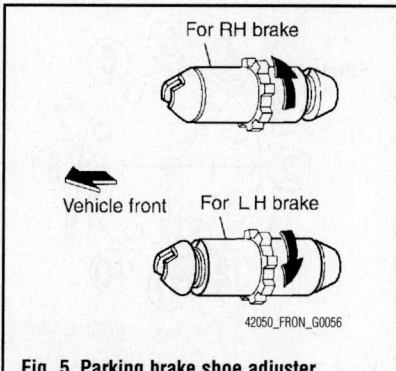

For RH brake

Vehicle front   For L H brake

42050_FRON_G0056

Fig. 5 Parking brake shoe adjuster identification

---

8. Remove the, retainers, anti-rattle pins and shoes.

9. Remove the pin retainer. Disconnect the parking brake cable from the toggle lever.

10. Remove the back plate.

To install:

11. Installation is the reverse of the removal procedure.

12. Assemble the adjuster so that the threaded part expands when rotating it in the direction shown by the arrow. Shorten the adjuster by rotating it in the opposite direction shown by the arrow.

13. Perform the parking brake break-in operation as follows: Safely, drive forward at approximately 25 mph (40 km/h) with the parking brake set with a force of approx. 45 lbs. (200 N) for about 30 seconds.

14. After the break-in operation, check the pedal stroke of parking brake. Readjust if necessary.

➡ **To prevent lining from getting too hot, allow a cool off period of approximately 5 minutes after every break-in operation.**

15. Check and adjust the parking brake pedal stroke. Correct as required.

CHASSIS ELECTRICAL

GENERAL INFORMATION

❋ CAUTION

These vehicles are equipped with an air bag system. The system must be disarmed before performing service on, or around, system components, the steering column, instrument panel components, wiring and sensors. Failure to follow the safety precautions and the disarming procedure could result in accidental air bag deployment, possible injury and unnecessary system repairs.

PRECAUTIONS

Disconnect and isolate the battery negative cable before beginning any airbag system component diagnosis, testing, removal, or installation procedures. Allow system capacitor to discharge for two minutes before beginning any component service. This will disable the airbag system. Failure to disable the airbag system may result in accidental airbag deployment, personal injury, or death.

DISARMING THE SYSTEM

To disarm the SRS system turn the ignition switch to the off position. Then, disconnect both battery cables starting with the negative cable first and wait at least 3 minutes after the cables are disconnected.

ARMING THE SYSTEM

To rearm the SRS system, turn the ignition switch to the off position. Connect both

AIR BAG (SUPPLEMENTAL RESTRAINT SYSTEM)

battery cables starting with the positive cable first.

CLOCKSPRING CENTERING

1. Be sure to align the spiral cable correctly when installing the steering wheel. Make sure that the spiral cable is in the neutral position.

➡ **The neutral position is detected by turning to the left 2.6 revolutions from the right end position and ending with the knob at the top. The spiral cable may snap due to steering operation if the cable is installed incorrectly. Also, with the steering linkage disconnected the cable may snap by turning the steering wheel beyond the limited number of turns (2.6 from the neutral position to both the left and right).**

DRIVE TRAIN

CLUTCH

REMOVAL & INSTALLATION

See Figure 6.

❋ CAUTION

Note the following:

- Do not clean the clutch disc with solvent.
- When installing, do not get grease from the main drive shaft onto the clutch disc friction surface.
- If the flywheel is removed, align the dowel pin with the smallest hole of flywheel.

1. Remove the manual transmission from the vehicle.

2. Remove the clutch cover bolts using power tool. Remove the clutch cover and clutch disc.

To install:

3. Apply recommended grease to clutch disc and main drive shaft spline.

❋ CAUTION

Do not allow grease to contaminate the clutch facing.

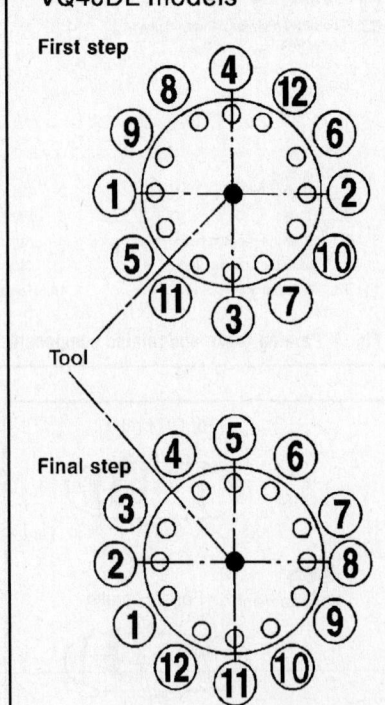

Fig. 6 Tightening order of the clutch disc and cover—Tool number ST20630000 (J-26366)

4. Install clutch disc and clutch cover. Pre-tighten the bolts and install Tool. Then tighten the clutch cover bolts evenly in two steps to the specified torque in the order shown in Figure 6.

5. Install the manual transmission.

HYDRAULIC SYSTEM BLEEDING

BLEEDING PROCEDURE

1. Before servicing the vehicle, refer to the Precautions Section.

➡ **Do not use a vacuum assist or any other type of power bleeder on this system. Use of a vacuum assist or power bleeder will not purge all the air from the system.**

2. Fill the system with the proper grade and type fluid.

3. Have an assistant pump the clutch pedal slowly several times and hold it depressed.

4. Open the slave cylinder bleeder screw and allow air to escape.

5. Close the bleeder screw before releasing the clutch pedal.

6. Repeat until all air is purged from the clutch hydraulic system.

7. Refill the reservoir to the full mark.

❋❋ CAUTION

Do not spill brake fluid on painted surfaces. If it spills, wipe up immediately and wash the affected area with water.

➡ **Do not use a vacuum assist or any other type of power bleeder on this system. Use of a vacuum assist or power bleeder will not purge all the air from the system. Monitor the fluid level in the reservoir tank to make sure it does not empty.**

8. Top off reservoir with new brake fluid.

9. Connect a transparent vinyl tube and container to the air bleeder valve on the clutch operating cylinder.

10. Fully depress the clutch pedal several times.

11. With the clutch pedal depressed, open the bleeder valve to release the air.

12. Close the bleeder valve.

13. Repeat steps 3 to 5 until clear brake fluid comes out of the air bleeder valve.

14. Tighten the air bleeder to 70 inch lbs. (7.9 Nm).

FRONT AXLE SHAFT, BEARING & SEAL

REMOVAL & INSTALLATION

1. Before servicing the vehicle, refer to the Precautions Section.

2. Raise and support the vehicle safely. Remove the tire and wheel assembly.

3. Remove the rear engine cover.

4. Remove the wheel sensor harness from the mount on the knuckle. Disconnect the harness connector.

➡ **Do not pull on the wheel sensor harness.**

5. Remove the wheel hub and bearing assembly.

➡ **It is not necessary to remove the wheel speed sensor from the wheel hub when the wheel hub is not being replaced. Carefully feed the sensor harness through the hole in the splash shield.**

6. Separate the upper link ball joint stud from the steering knuckle using tool ST29020001 (J-24319-01) or equivalent.

7. Remove the halfshaft assembly from the vehicle by prying the halfshaft from the front final drive using the proper tool.

To install:

8. Installation is the reverse of the removal procedure.

9. Be sure to use a new differential side oil seal.

FRONT CV-JOINT

OVERHAUL

Inner CV-Joint

1. Before servicing the vehicle, refer to the Precautions Section.

2. Remove the axle halfshaft from the vehicle.

3. Remove the CV-joint boot clamps and push the boot away from the joint.

4. Match mark the housing and the shaft before separation.

5. Remove the stopper ring and pull housing off.

6. Remove the snap ring using a suitable tool, then remove the ball cage, steel ball, and inner race assembly from the shaft.

7. Remove the boot from the shaft. Remove circlip and dust cover from the housing. Clean the old grease off of the housing using paper towels.

To install:

➡ **Use new circlips and boot clamps for assembly.**

8. Wrap the serrated part of the shaft with tape. Install the boot band and boot to shaft.

9. Remove the tape wound around the serrated part of the shaft.

10. Install the ball cage, steel ball, and inner race assembly on the shaft, and secure them using the snap ring.

➡ **Use new snap ring.**

11. Pack the joint with 4.23–4.94 oz. grease.

➡ **Ensure boot mounting surfaces do not have any grease.**

12. Install the stopper ring onto the housing.

13. Install the boot securely into the grooves.

14. Check that the overall boot installation length is 6.45–6.47 inches. Insert a flat-tip screwdriver or similar tool into the large end of the boot. Bleed air from boot to prevent boot deformation.

➡ **Do not to touch the tip of the screwdriver to the inside of the boot.**

15. Install the boot clamps.

16. Install the axle halfshaft to the vehicle.

Outer CV-Joint

1. Before servicing the vehicle, refer to the Precautions Section.

2. Remove the axle halfshaft from the vehicle.

3. Remove the boot bands and slide the boot back.

4. Screw a sliding hammer or suitable tool 30 mm (1.18 in) or more into threaded part of joint sub-assembly. Pull joint sub-assembly off of shaft.

5. Remove boot from the shaft. Remove circlip from the shaft.

To install:

➡ **Use new snap rings and boots for assembly.**

6. Insert the Genuine NISSAN Grease or equivalent, into the joint sub-assembly serration hole until the grease begins to ooze from the ball groove and serration hole.

7. Wrap the serrated part of the shaft with tape. Install the boot band and boot onto the shaft. Do not damage the boot.

8. Remove the tape wound around the serrated part of the shaft.

9. Attach the circlip to the shaft. The circlip must fit securely into the shaft groove. Attach the nut to the joint sub-assembly. Use a soft hammer to press-fit the circlip.

➡ **Use new circlip.**

10. Pack the joint with 4.01–4.76 oz. grease.

➡ **Ensure boot mounting surfaces do not have any grease.**

11. Install the boot securely into the grooves.

12. Check that the overall boot installation length is 5.32 inches. Insert a flat-tip screwdriver or similar tool into the large end of the boot. Bleed air from boot to prevent boot deformation.

➡ **Do not to touch the tip of the screwdriver to the inside of the boot.**

13. Install the boot clamps.

14. Install the axle halfshaft to the vehicle.

FRONT DRIVESHAFT

REMOVAL & INSTALLATION

See Figure 7.

1. Remove wheel and tire using power tool.
2. Remove rear engine under cover using power tool.
3. Remove wheel sensor harness from mount on knuckle, then disconnect wheel sensor harness connector.

✸✸ CAUTION

Do not pull on wheel sensor harness.

4. Remove wheel hub and bearing assembly.
   - It is not necessary to remove wheel sensor from wheel hub when wheel hub is not being replaced.
   - Carefully feed wheel sensor harness through hole in splash shield.
5. Separate upper link ball joint stud from steering knuckle using Tool.
   - Support lower link with jack.
6. Remove drive shaft assembly.
   - Pry drive shaft front final drive using suitable tool.
   - Remove differential side oil seal.

To install:

7. Install drive shaft assembly.
   - Install differential side oil seal.
   - Pry drive shaft front final drive using suitable tool.
8. Separate upper link ball joint stud from steering knuckle using Tool.
   - Support lower link with jack.
9. Install wheel hub and bearing assembly.
   - Carefully feed wheel sensor harness through hole in splash shield.

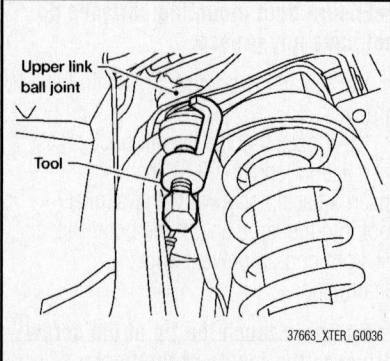

Fig. 7 Separate upper link ball joint stud from steering knuckle using tool— ST29020001 (J-24319-01)

37663_XTER_G0036

- It is not necessary to remove wheel sensor from wheel hub when wheel hub is not being replaced.

✸✸ CAUTION

Do not pull on wheel sensor harness.

10. Install wheel sensor harness from mount on knuckle, then connect wheel sensor harness connector.
11. Install rear engine under cover using power tool.
12. Install wheel and tire using power tool.

REAR AXLE SHAFT, BEARING & SEAL

REMOVAL & INSTALLATION

See Figures 8 through 10.

1. Before servicing the vehicle, refer to the Precautions Section.
2. Remove or disconnect the following:
   - Rear wheel and tire assembly
   - Wheel speed sensor
   - Brake rotor
   - Brake caliper assembly
   - Parking brake cable
   - Brake fluid line
   - Bearing cage and backing plate bolts
   - Axle shaft assembly
   - Axle seal
   - Wheel speed sensor rotor
   - Snap ring and shim washer
   - Bearing ring retainer
   - back plate and torque member
   - Axle bearing studs
   - Wheel bearing
   - Grease catcher

To install:

➡**Use new seals, bearings, circlips and snap rings for assembly.**

3. Install grease catcher
4. Install the wheel studs through the grease catcher into the axle shaft using a suitable press.

➡**All six wheel studs must be pressed on at the same time and are flush with the grease catcher when installed.**

5. Position the axle bearing on the back plate and torque member.
6. Install the axle bearing studs using a suitable press to attach the axle bearing to the back plate and torque member.

➡**Always replace the axle bearing with a new one.**

7. Install the back plate and torque member, new axle bearing and new bearing ring retainer on the axle shaft using a suitable press. Do not exceed 11 tons force.
8. Press the new bearing ring retainer on the axle shaft with the taper side positioned toward press.

➡**Always replace the bearing ring retainer with a new one.**

9. Select the correct size shim washer. Select the size of shim washer so that the installed snap ring to shim washer clearance is 0.008 inch or less.
10. Install a new snap ring on the axle shaft.
11. Do not over spread the snap ring when installing, measure the outer diameter of the snap ring after installation and replace if the snap ring outer diameter exceeds 1.87 inch maximum.
12. Check the snap ring to shim washer clearance. Repeat previous steps as necessary.
13. Perform break-in rotation of the wheel bearing.
   - Rotate the wheel bearing in the forward direction for a minimum of 10 revolutions at 50–70 RPM.
   - Rotate the wheel bearing in the reverse direction for a minimum of 10 revolutions at 50–70 RPM.
14. Measure the rotational torque of the wheel bearing. Rotational torque should be 16 inch lbs. (1.8 Nm) at 8–12 RPM.
15. Inspect that the wheel bearing is free from axial play relative to the axle shaft.
16. Install a new ABS sensor rotor on the axle shaft with notch side away from press using a suitable press.

➡**Always replace the ABS sensor rotor with a new one.**

17. Install new axle seal in housing.
18. Apply multi-purpose grease to the recess of axle case end as shown in illustration.
19. Insert tool J-34296 into the new axle oil seal as a guide. Ensure tool ends do not overlap.
20. Insert the axle shaft assembly. Tighten the axle shaft nuts evenly in a criss-cross pattern to specification. Remove the tool when the axle shaft assembly is approximately 90 percent inserted to protect the new axle oil seal.
21. Install parking brake assembly, rear caliper assembly and ABS wheel sensor.

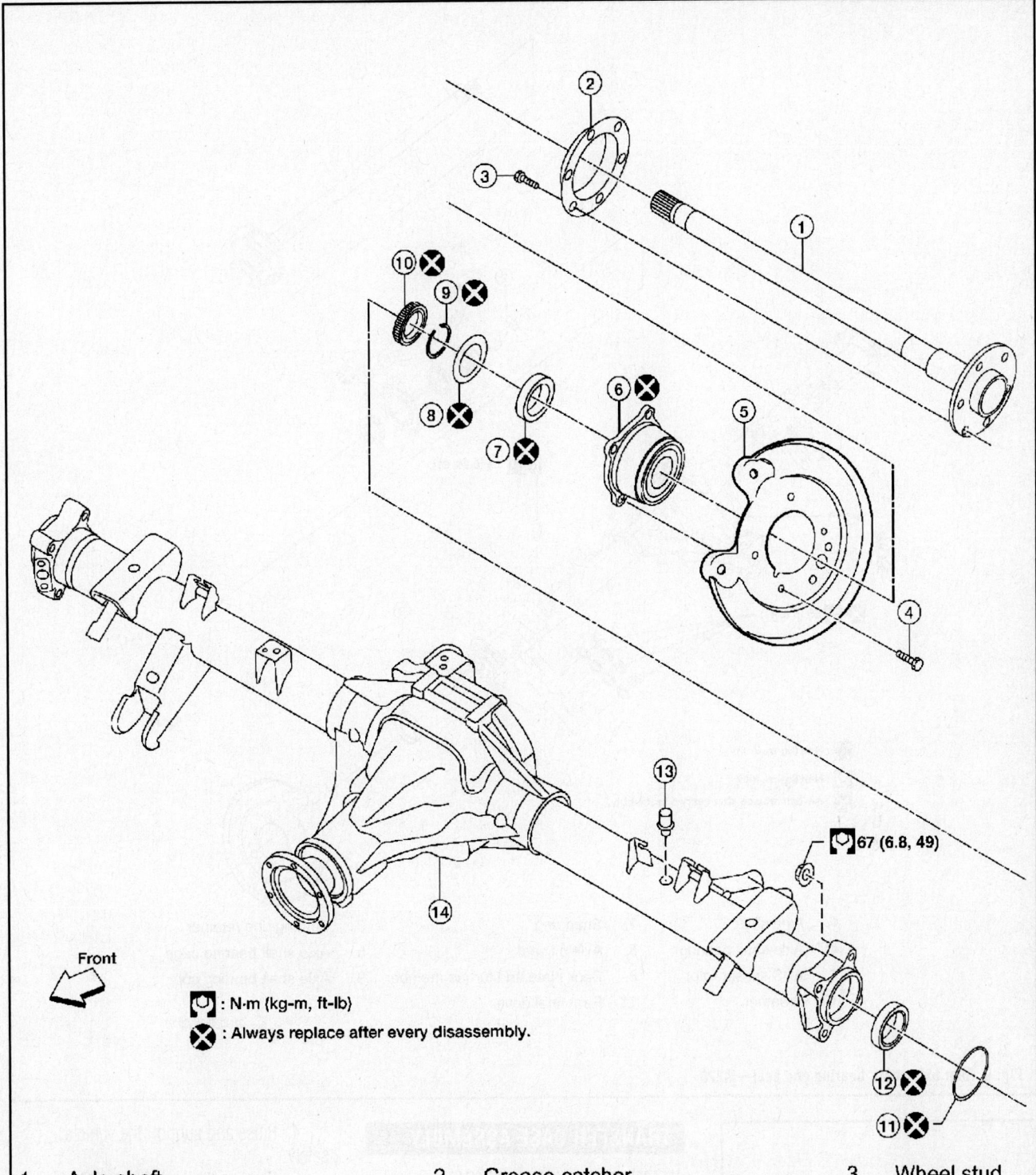

: N·m (kg-m, ft-lb)

: Always replace after every disassembly.

Front

1. Axle shaft	2. Grease catcher	3. Wheel stud
4. Axle bearing stud	5. Back plate and torque member	6. Axle bearing
7. Bearing ring retainer	8. Shim washer	9. Snap ring
10. ABS sensor rotor	11. O-ring	12. Axle oil seal
13. Breather	14. Rear final drive	

22140_FRON_G0001

Fig. 8 Rear axle shaft, bearing and seal—C200

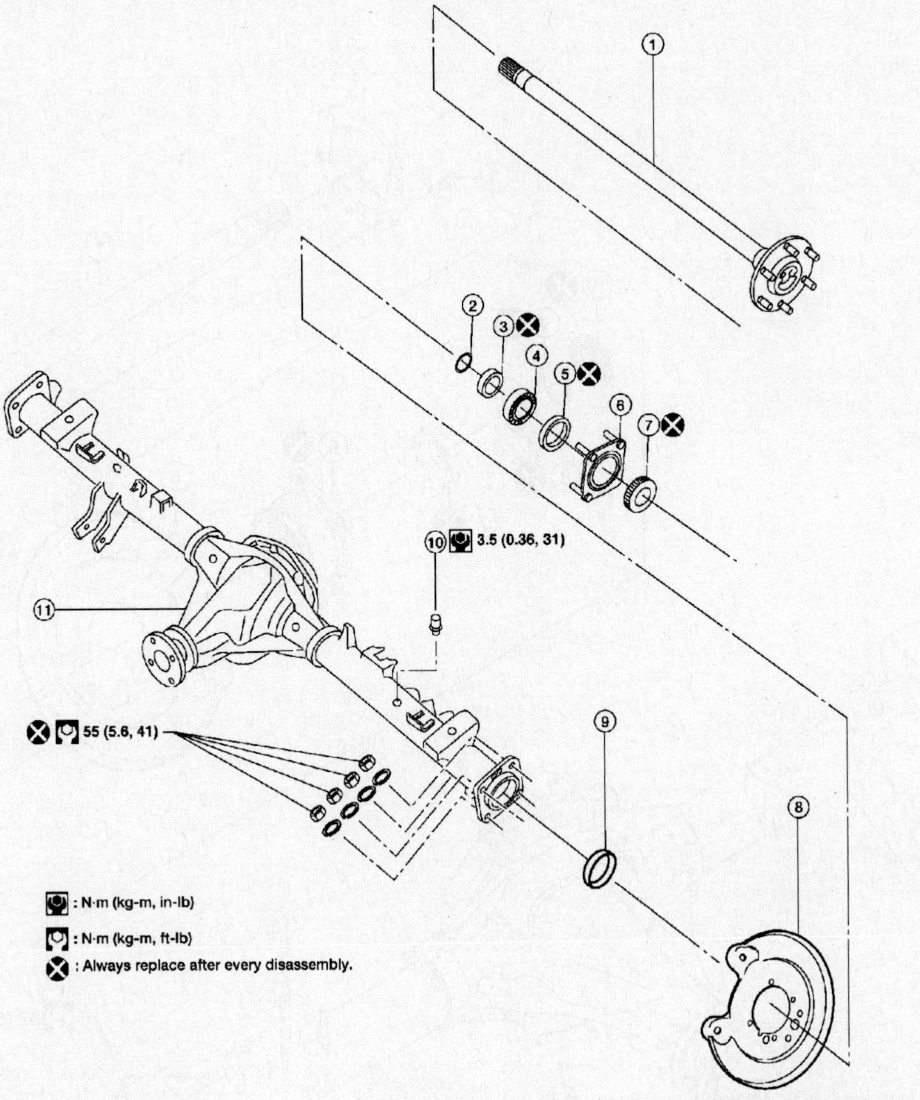

10 3.5 (0.36, 31)

55 (5.6, 41)

: N·m (kg-m, in-lb)

: N·m (kg-m, ft-lb)

: Always replace after every disassembly.

1.	Axle shaft	2.	Snap ring	3.	Bearing ring retainer
4.	Axle shaft bearing	5.	Axle oil seal	6.	Axle shaft bearing cage
7.	ABS sensor rotor	8.	Back plate and torque member	9.	Axle shaft bearing cup
10.	Breather	11.	Rear final drive		

22140_FRON_G0002

Fig. 9 Rear axle shaft, bearing and seal—M226

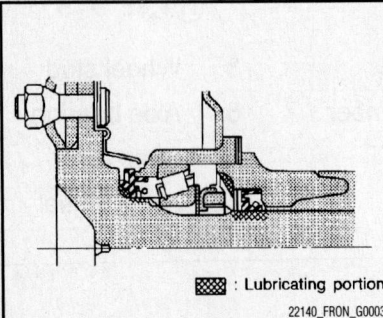

: Lubricating portion

22140_FRON_G0003

Fig. 10 Locating areas to apply grease

TRANSFER CASE ASSEMBLY

REMOVAL & INSTALLATION

TX15B

See Figure 11.

1. Before servicing the vehicle, refer to the Precautions Section.

2. Disconnect the negative battery cable.

3. Switch the 4WD switch to 2WD. Set the transfer case to 2WD.

4. Raise and support the vehicle safely.

5. Remove the undercovers. Drain the transfer case fluid.

6. Remove the center exhaust tube and main muffler.

7. Remove the front and rear driveshafts. Install plug in rear oil seal.

8. Remove the transmission-to-cross-member bolts. Properly support the transmission and transfer case assembly, using a suitable jack.

9. Remove the transmission crossmember.

➡ **Support the transmission and transfer case using two suitable jacks while removing the transmission crossmember.**

10. Disconnect the ATP electrical connector, the 4LO switch connector, the wait detection switch and the transfer control device.

11. Disconnect each air breather hose from the transfer control device and the breather tube.

12. Remove the transfer case to transmission retaining bolts.

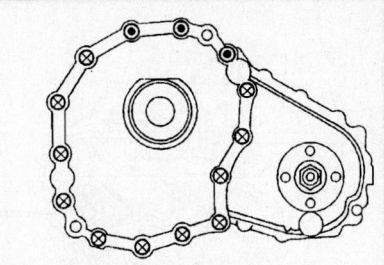

: Transfer ➝ Transmission
: Transmission ➝ Transfer

09482_FRON_G0111

Fig. 11 Transfer case bolt tightening sequence

➡ **Support the transmission and transfer case, using a suitable jack.**

13. Remove the transfer case from the vehicle.

To install:

14. Installation is the reverse of the removal procedure.

15. Tighten the transfer case to transmission retaining bolts to specification and in the proper sequence. Specification is 27 ft. lbs. (36.6 Nm).

16. Start the engine and check for leaks, correct as required.

ENGINE COOLING

ENGINE FAN

REMOVAL & INSTALLATION

Electric

1. Remove air dam using power tool.
2. Remove engine front undercover using power tool.
3. Partially drain engine coolant from radiator.
4. Release the radiator shroud (lower) from the radiator shroud (upper) and position aside. Release the tabs, pull radiator shroud (lower) rearwards and down.
5. Remove air duct.
6. Remove reservoir tank hose from shroud.
7. Remove the radiator shroud (upper) bolts and remove the radiator shroud (upper).
8. Disconnect fan motor harness connector.
9. Remove bolt and fan grille and motor assembly.

To install:

10. Installation is the reverse of the removal procedure.
11. Start the engine and check for leaks, correct as required.

Mechanical (Belt Driven)

1. Remove air dam using power tool.
2. Remove engine front undercover using power tool.
3. Partially drain engine coolant from radiator.

✳ CAUTION

Perform this step when engine is cold. Do not spill engine coolant on drive belts.

4. Remove air duct.

5. Remove reservoir tank hose from shroud.
6. Removal radiator hose (upper) from radiator.

✳ CAUTION

Be careful not to allow engine coolant to contact drive belts.

7. Release the radiator shroud (lower) from the radiator shroud (upper) and position aside. Release the tabs, pull radiator shroud (lower) rearwards and down.
8. Remove the radiator shroud (upper) bolts and remove the radiator shroud (upper).
9. Remove the drive belt.
10. Remove the engine cooling fan.

To install:

11. Installation is the reverse of the removal procedure.
12. Start the engine and check for leaks, correct as required.

RADIATOR

REMOVAL & INSTALLATION

1. Remove air dam using power tool.
2. Remove engine undercover using power tool.
3. Drain engine coolant from radiator.

✳ CAUTION

Perform this step when engine is cold. Do not spill engine coolant on drive belts.

4. Remove air duct and air cleaner case assembly.
5. Remove reservoir tank hose.
6. Remove radiator hoses (upper and lower).

✳ CAUTION

Be careful not to allow engine coolant to contact drive belts.

7. Disconnect A/T fluid cooler hoses and plug.
8. Remove radiator shroud (lower).
9. Remove radiator shroud (upper).
10. Remove radiator cooling fan assembly.
11. Remove the radiator mounting bracket bolts.
12. Remove the two A/C condenser bolts.
13. Remove radiator as follows:

✳ CAUTION

Do not damage or scratch A/C condenser and radiator core when removing.

- With lifting and pulling radiator in a rear direction, disassemble mounting rubber (lower) from radiator core support center
- Because A/C condenser is attached to the front-lower portion of radiator, moving it in the rear direction should be at a minimum
- Lift A/C condenser up and remove radiator after disengaging the fitting as front-bottom surface

✳ CAUTION

Lifting A/C condenser should be minimum to prevent a load to A/C piping.

- After removing radiator, put A/C condenser on radiator core support center to prevent a load to A/C piping, and temporarily tie it in place

To install:

14. Installation is the reverse of the removal procedure.

15. Start the engine and check for coolant and transmission fluid leaks, correct as required.

THERMOSTAT

REMOVAL & INSTALLATION

See Figures 12 and 13.

➡**Never remove the radiator cap when the engine is hot. Serious burns could occur from high-pressure engine coolant escaping from the radiator.**

1. Be sure the engine is cold.
2. Disconnect the negative battery cable.
3. Drain the coolant. Properly disposed of used coolant.
4. Remove the air duct and air cleaner case.

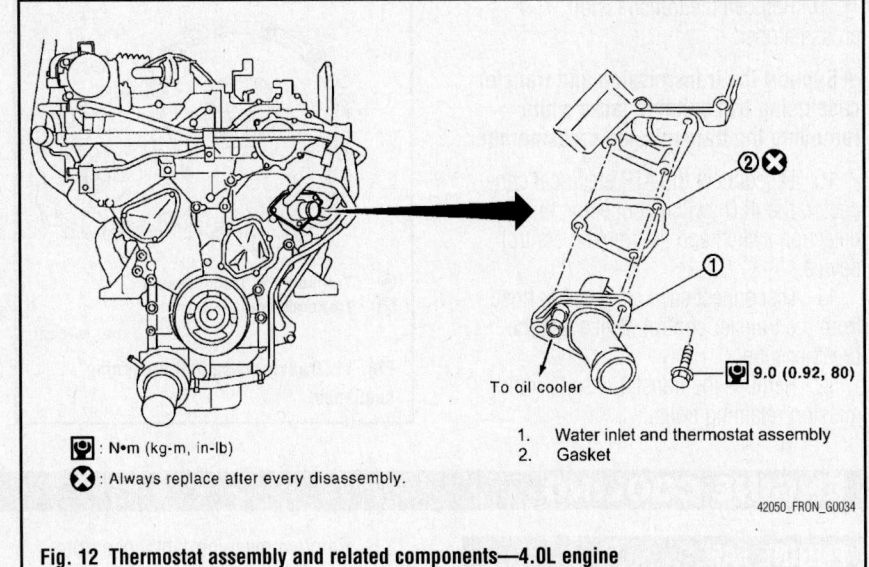

: N•m (kg-m, in-lb)

: Always replace after every disassembly.

To oil cooler

9.0 (0.92, 80)

1. Water inlet and thermostat assembly
2. Gasket

42050_FRON_G0034

Fig. 12 Thermostat assembly and related components—4.0L engine

To heater

4   1   24.5 (2.5, 18)

2   24.5 (2.5, 18)

24.5 (2.5, 18)

To electric throttle control actuator

To electric throttle control actuator

22 (2.2, 16)

22.0 (2.2, 16)

To heater

9.0 (0.92, 80)

1. Engine coolant temperature sensor	2. Washer	3. Water outlet
4. Heater hose	5. O-ring	6. Water pipe
7. Water hose	8. Radiator hose (upper)	9. Water hose
10. Water hose	11. Gasket	12. Heater pipe
13. Heater hose	14. Gasket	

: Always replace after every disassembly.

: N•m (kg-m, in-lb)

: N•m (kg-m, ft-lb)

42050_FRON_G0035

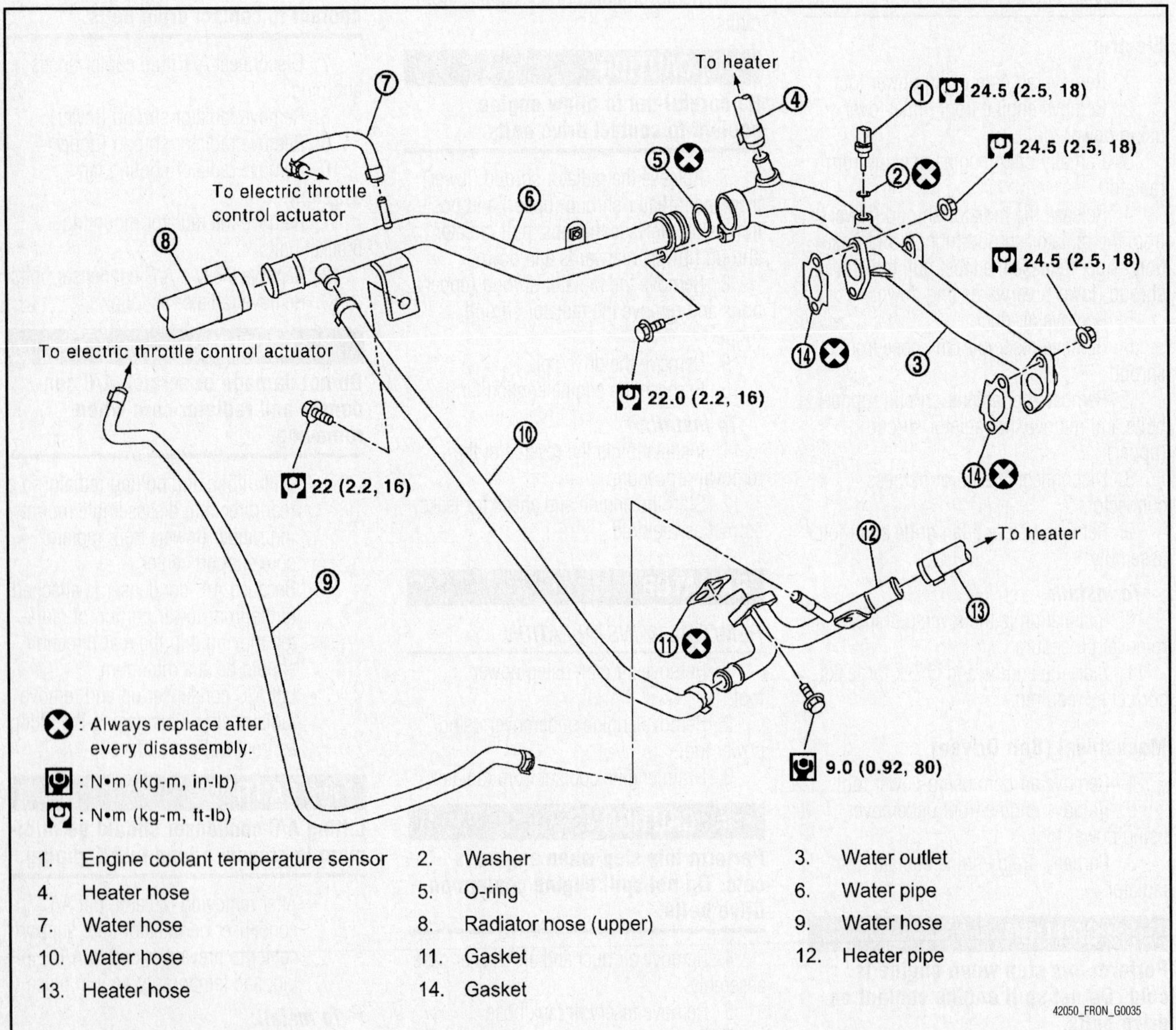

Fig. 13 Water outlet and related components—4.0L engine

5. Disconnect radiator hose (lower) and oil cooler hose from water inlet and thermostat assembly.

6. Remove the water inlet retaining bolts.

7. Remove the water inlet and the thermostat.

➡**Do not disassemble the water inlet and thermostat assembly. Replace them as a unit, if required.**

To install:

8. Installation is the reverse of the removal procedure.

9. Be sure to refill the cooling using the proper grade and type engine coolant.

10. Start the engine and check for leaks.

11. Start the engine and allow it to reach operation temperature. Recheck the coolant level, fill as required.

WATER PUMP

REMOVAL & INSTALLATION

See Figures 14 and 15.

1. Before servicing the vehicle, refer to the Precautions Section.

2. Disconnect the negative battery cable. Drain the cooling system.

3. Remove the air dam and undercover.

4. Remove air duct and resonator.

5. Remove the drive belts.

6. Remove the radiator upper hose. Remove the cooling fan.

7. Remove the chain tensioner cover and water pump cover from the front timing case, using tool KV10111100 (J-37228) or equivalent.

8. To remove the timing chain tensioner (primary), loosen the clip of the timing chain tensioner (primary) and release the plunger stopper. Insert the plunger into the tensioner body by pressing the slack guide. Keep the slack guide pressed and hold the plunger in by pushing the stopper pin through the tensioner body hole and plunger groove. Turn the crankshaft pulley clockwise so that the timing chain on the timing chain tensioner (primary) side is loose. Remove the bolts and remove the timing chain tensioner (primary).

➡**Be careful not to drop the bolts inside the timing chain case.**

9. Remove the three water pump retaining bolts. Secure a gap between the water pump gear and the timing chain, by turning the crankshaft pulley counterclockwise until timing chain looseness on the water pump sprocket is at its maximum point.

10. Screw M8 bolts approximately 1.97 inch in length into the water pumps upper and lower bolt holes until they reach the timing chain case.

11. Alternately tighten each bolt for a half turn and pull out the water pump.

➡**Pull the pump straight out while preventing the vane from contacting the socket in the installation area. Remove the pump without causing the sprocket to contact the timing chain.**

12. Remove the M8 bolts. Remove and discard the O-rings.

To install:

13. Installation is the reverse of the removal procedure.

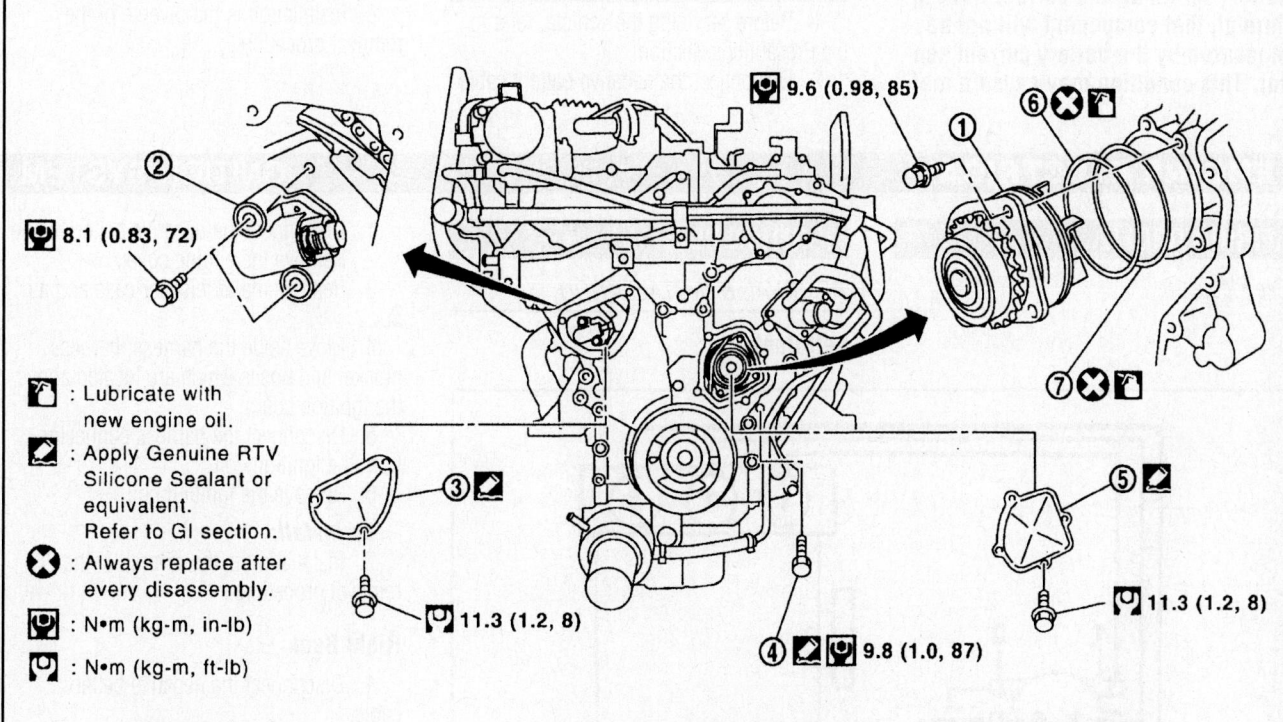

: Lubricate with new engine oil.

: Apply Genuine RTV Silicone Sealant or equivalent. Refer to GI section.

: Always replace after every disassembly.

: N•m (kg-m, in-lb)

: N•m (kg-m, ft-lb)

8.1 (0.83, 72)

9.6 (0.98, 85)

11.3 (1.2, 8)

9.8 (1.0, 87)

11.3 (1.2, 8)

1. Water pump	2. Timing chain tensioner (primary)	3. Chain tensioner cover
4. Water drain plug (front)	5. Water pump cover	6. O-ring
7. O-ring		

09482_FRON_G0016

Fig. 14 Water pump and related components—4.0L engine

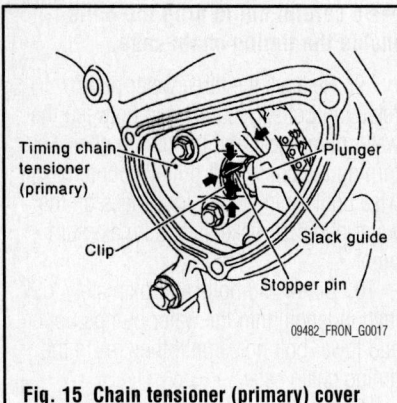

Fig. 15 Chain tensioner (primary) cover removal—4.0L engine

14. Be sure to use new gaskets and O-rings, as required. Apply engine oil to new O-rings before installation. Locate O-ring with white paint mark in forward groove.

15. When installing the water pump make sure that the timing chain and water pump sprocket are engaged. Tighten the bolts alternately and evenly to specification.

16. Before installing the chain tensioner cover and the water pump cover be sure to apply a continuous bead of sealant to the mating surfaces of the covers.

➡ **Do not allow the sealant to set for more than five minutes before installing the covers.**

17. Be sure to fill the cooling system with the proper grade and type engine coolant.

18. Start the engine and check for leaks.

19. Let the engine idle for about three minutes than rev it up to 3,000 rpm's under a no load condition to purge air from the high pressure chamber of the chain tensioner. The engine may produce a rattling noise. This indicates that air still remains in the chamber and is not a matter of concern.

ENGINE ELECTRICAL

✳✳ CAUTION

For this model, the battery current sensor that is installed to the negative battery cable measures the charging/discharging current of the battery and performs various engine controls. If an electrical component is connected directly to the negative battery terminal, the current flowing through that component will not be measured by the battery current sensor. This condition may cause a mal- function of the engine control system and battery discharge may occur. Do not connect an electrical component or ground wire directly to the battery terminal.

ALTERNATOR

REMOVAL & INSTALLATION

1. Before servicing the vehicle, refer to the Precautions Section.
2. Disconnect the negative battery cable.

CHARGING SYSTEM

3. Remove the drive belt.
4. Remove alternator stay.
5. Remove the upper alternator mounting bolt.
6. Disconnect the alternator harness electrical connectors.
7. Remove the alternator from the vehicle.

To install:

8. Installation is the reverse of the removal procedure.

ENGINE ELECTRICAL

FIRING ORDER

See Figure 16.

IGNITION COIL

REMOVAL & INSTALLATION

Left Bank

IGNITION SYSTEM

1. Disconnect the negative battery cable.
2. Remove the engine cover.
3. Remove the air cleaner case and air duct.
4. Move aside the harness, harness bracket and hoses which are located above the ignition coil.
5. Disconnect the harness connector from the ignition coil.
6. Remove the ignition coil.

To install:

7. Installation is the reverse of the removal procedure.

Right Bank

1. Disconnect the negative battery cable.
2. Remove the intake manifold collector.
3. Move aside the harness, harness bracket and hoses which are located above the ignition coil.
4. Disconnect the harness connector from the ignition coil.
5. Remove the ignition coil.

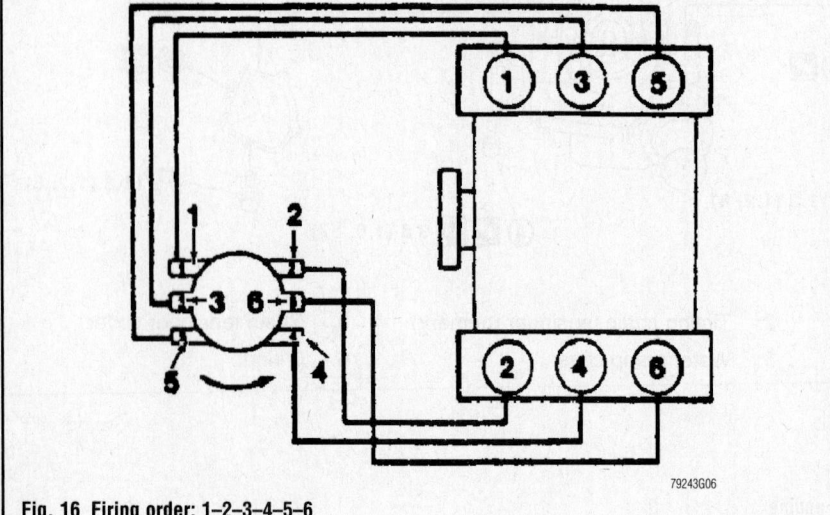

Fig. 16 Firing order: 1–2–3–4–5–6

To install:

6. Installation is the reverse of the removal procedure.

IGNITION TIMING

INSPECTION

1. Before servicing the vehicle, refer to the Precautions Section.

2. Remove the number one ignition coil.

3. Connect the number one ignition coil and spark plug with a suitable high tension wire.

4. Attach the timing light clamp to the wire.

5. Check the ignition timing.

SPARK PLUGS

REMOVAL & INSTALLATION

See Figure 17.

1. Disconnect the negative battery cable.
2. Remove the intake manifold collector.
3. Remove the ignition coil.
4. Remove the spark plug using a spark plug socket and wrench.

To install:

5. Installation is the reverse of the removal procedure.

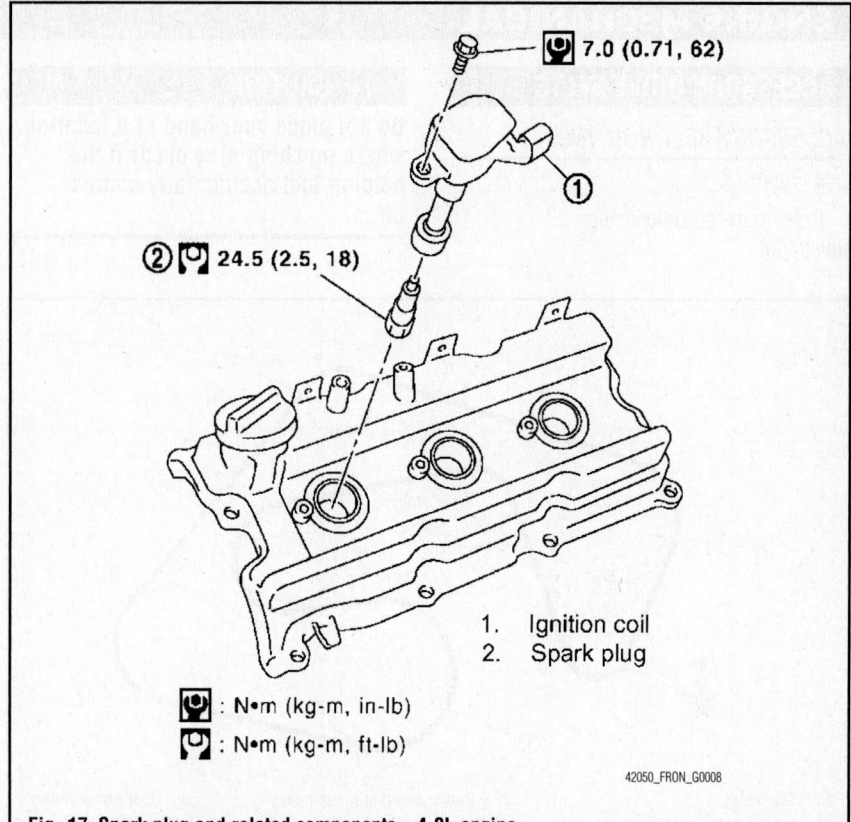

1. Ignition coil
2. Spark plug

🔧 : N•m (kg-m, in-lb)
🔧 : N•m (kg-m, ft-lb)

42050_FRON_G0008

Fig. 17 Spark plug and related components—4.0L engine

ENGINE ELECTRICAL

STARTER

REMOVAL & INSTALLATION

1. Before servicing the vehicle, refer to the Precautions Section.

2. Remove or disconnect the following:
   - Negative battery cable
   - Engine under cover

- On vehicles with 4.0L engine, remove the exhaust manifold cover to gain access to the starter retaining bolts
- Starter harness connectors
- Starter bolts
- Starter motor

To install:

3. Install or connect the following:

STARTING SYSTEM

- Starter motor
- air cleaner cover and the air cleaner to intake manifold collector duct, if equipped
- Exhaust manifold cover, if equipped
- Starter harness connectors
- Engine under cover
- Negative battery cable

ENGINE MECHANICAL

ACCESSORY DRIVE BELTS

ACCESSORY BELT ROUTING

See Figure 18.

Refer to the accompanying illustration.

※※ CAUTION

Do not place your hand in a location where pinching may occur if the holding tool accidentally comes off.

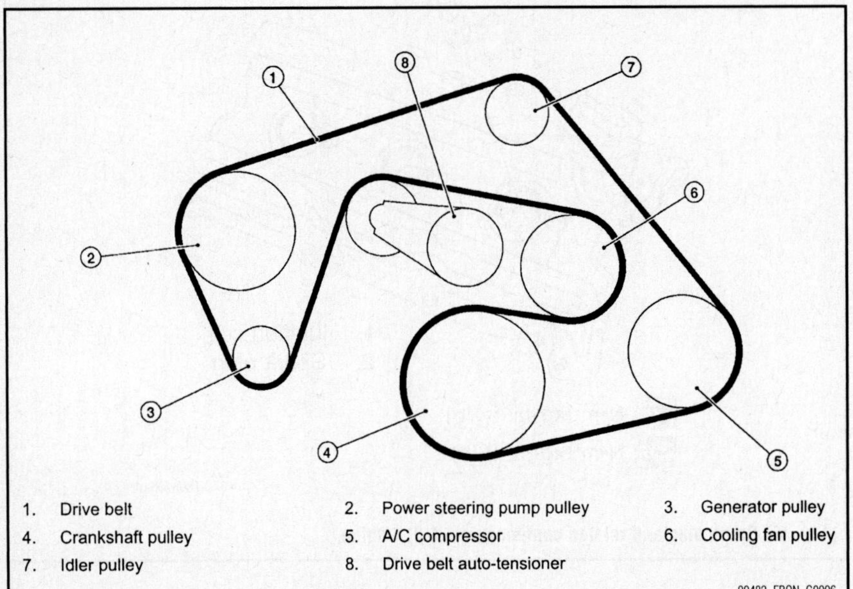

5. Remove the drive belt.

To install:

6. Installation is the reverse of the removal procedure.

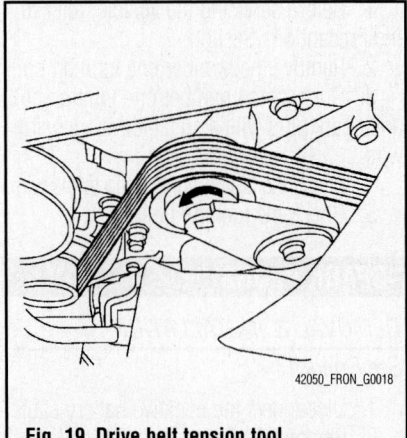

42050_FRON_G0018

Fig. 19 Drive belt tension tool installation and removal direction—4.0L engine

1. Drive belt
2. Power steering pump pulley
3. Generator pulley
4. Crankshaft pulley
5. A/C compressor
6. Cooling fan pulley
7. Idler pulley
8. Drive belt auto-tensioner

09482_FRON_G0006

Fig. 18 Accessory drive belt routing—4.0L engine

CAMSHAFT, BEARINGS & LIFTERS

REMOVAL & INSTALLATION

See Figures 21 through 29.

INSPECTION

Inspect the drive belt for signs of glazing or cracking. A glazed belt will be perfectly smooth from slippage, while a good belt will have a slight texture of fabric visible. Cracks will usually start at the inner edge of the belt and run outward. All worn or damaged drive belts should be replaced immediately.

ADJUSTMENT

Belt tensioning is not necessary, as it is automatically adjusted by the drive belt auto tensioner.

REMOVAL & INSTALLATION

See Figures 19 and 20.

1. Before servicing the vehicle, refer to the Precautions section.

2. Disconnect the negative battery cable.

3. Remove the air duct and resonator assembly (inlet).

4. Rotate the drive belt auto tensioner in the direction shown in the illustration.

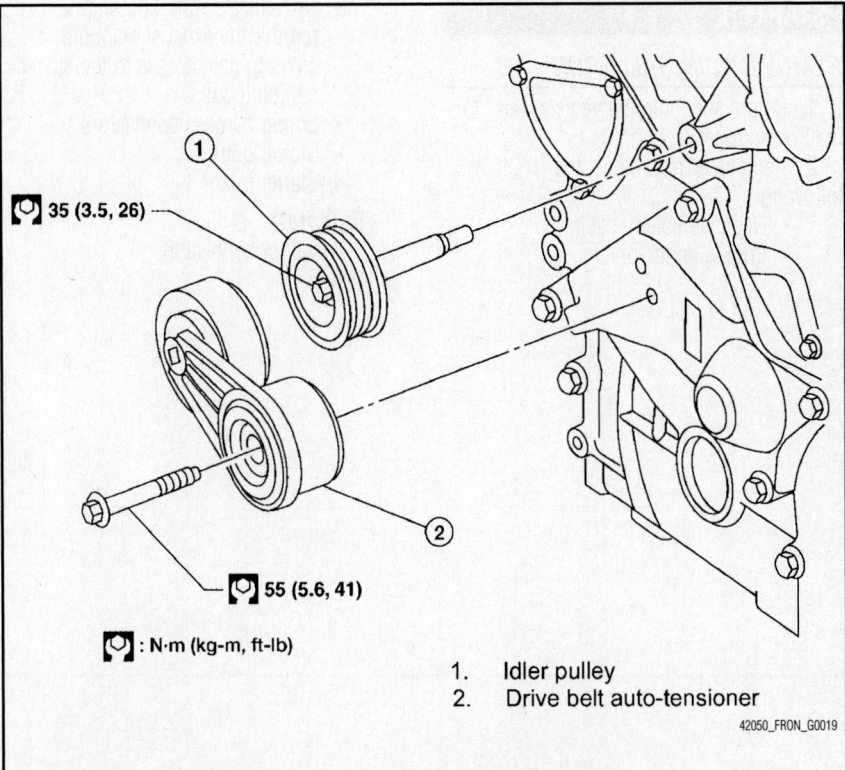

35 (3.5, 26)

55 (5.6, 41)

: N·m (kg-m, ft-lb)

1. Idler pulley
2. Drive belt auto-tensioner

42050_FRON_G0019

Fig. 20 Drive belt auto tensioner and related components—4.0L engine

1. Before servicing the vehicle, refer to the Precautions Section.

2. Properly relieve the fuel system pressure.

3. Disconnect the negative battery cable. Remove the engine cover.

4. Remove the front timing chain case, camshaft sprocket, timing chain and rear timing chain case.

5. Remove the camshaft position sensor (PHASE) from the cylinder head back side.

➡**Handle carefully to avoid dropping and shocks. Do not disassemble. Do not place in a location where the sensor can be exposed to magnetism.**

6. Remove the intake manifold collector.

7. Separate the engine harness and remove their brackets from the rocker covers. Remove the harness bracket from the cylinder head, if necessary.

8. Remove the ignition coil. Remove the PCV hoses. Remove the oil filler cap, if necessary.

9. Loosen the rocker cover retaining bolts, in the reverse order of the tightening sequence.

10. Remove the rocker covers from the engine.

11. Remove the intake valve timing control solenoid valves. Discard the gaskets.

12. Mark the camshaft brackets and bolts for reinstallation. Remove the camshaft bracket bolts. Be sure to remove the bolts by reversing the order of the tightening torque sequence and in several steps.

13. Remove the camshafts.

14. If required, remove the valve lifters. Identify them for reinstallation in their original locations.

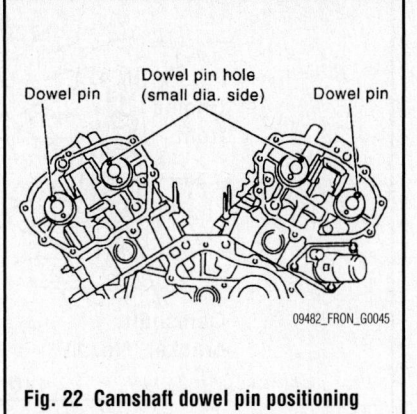

Fig. 22 Camshaft dowel pin positioning

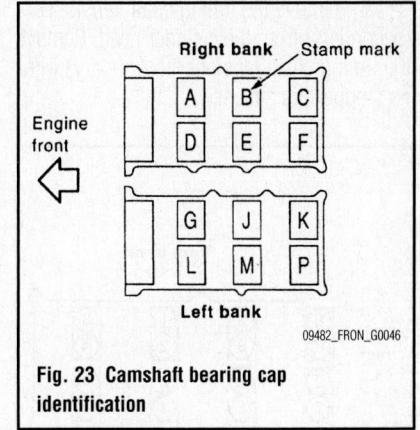

Fig. 23 Camshaft bearing cap identification

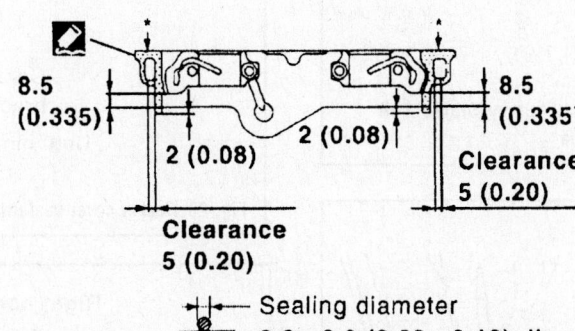

Camshaft bracket (No. 1)

8.5 (0.335)

2 (0.08)

2 (0.08)

8.5 (0.335)

Clearance 5 (0.20)

Clearance 5 (0.20)

Sealing diameter 2.0 - 3.0 (0.08 - 0.12) dia.

* : Remove the protruding liquid gasket from front face. (Remove the hardened liquid gasket from surface only.)

▨ : Apply Genuine RTV Silicone Sealant or equivalent. Refer to GI section.

Unit: mm (in)

Fig. 24 Camshaft sealant application and location

Bank	INT/EXH	Dowel pin	Paint marks		Identification mark
			M1	M2	
RH	INT	No	Green	No	RE
	EXH	Yes	No	White	RE
LH	INT	No	Green	No	LH
	EXH	Yes	No	White	LH

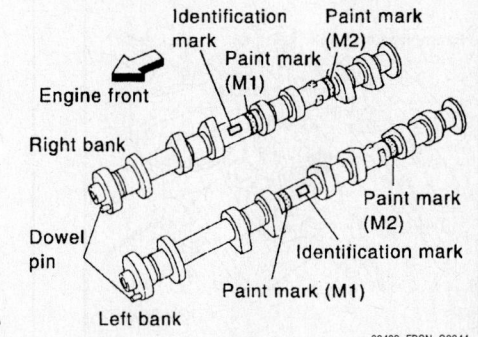

Fig. 21 Camshaft identification

15. Remove the timing chain tensioner (secondary) from the cylinder head. Remove the timing chain tensioner (secondary) with its stopper pin attached.

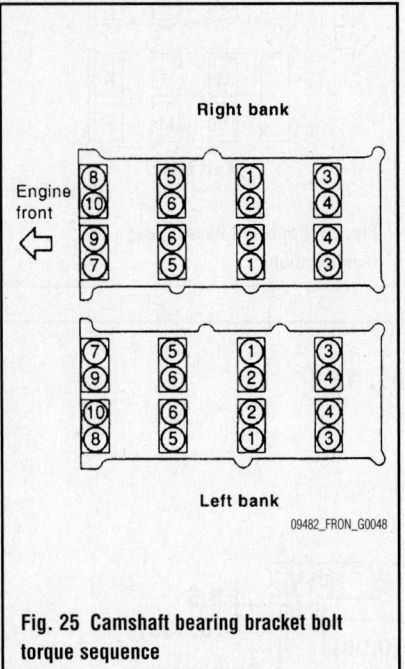

Fig. 25 Camshaft bearing bracket bolt torque sequence

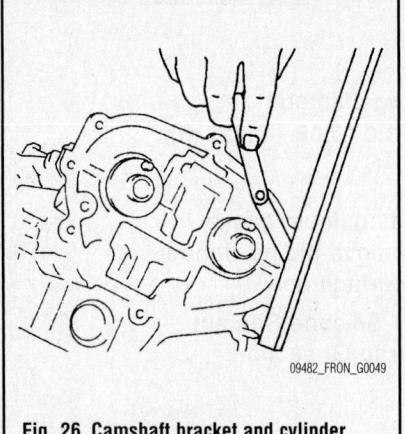

Fig. 26 Camshaft bracket and cylinder head measurement

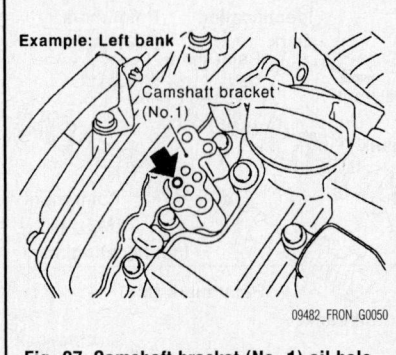

Fig. 27 Camshaft bracket (No. 1) oil hole location

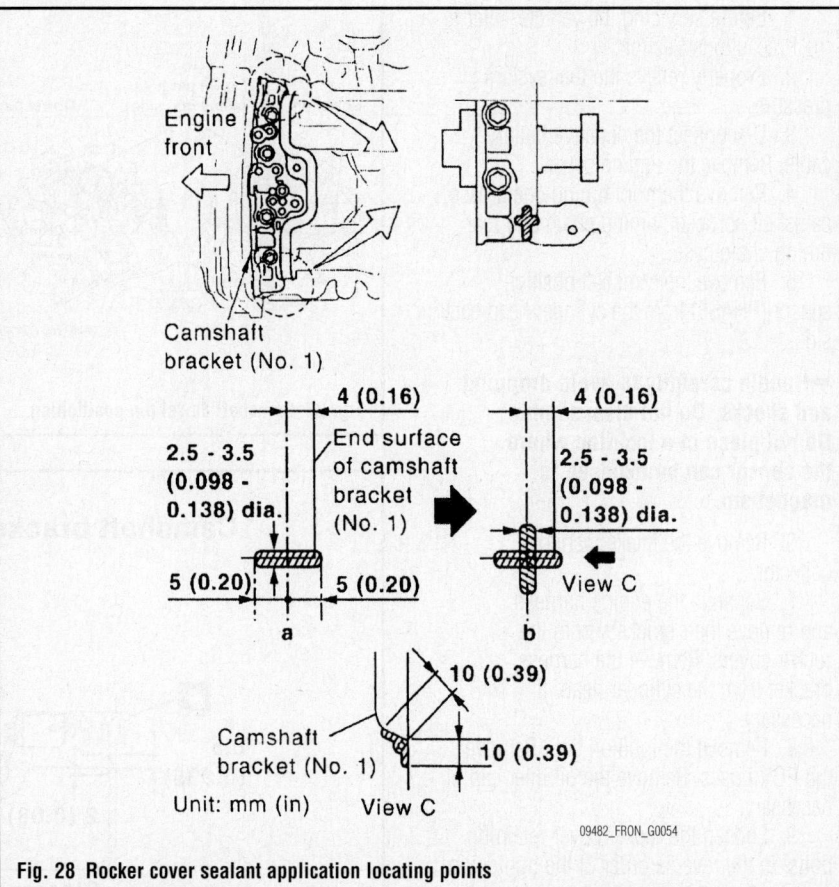

Fig. 28 Rocker cover sealant application locating points

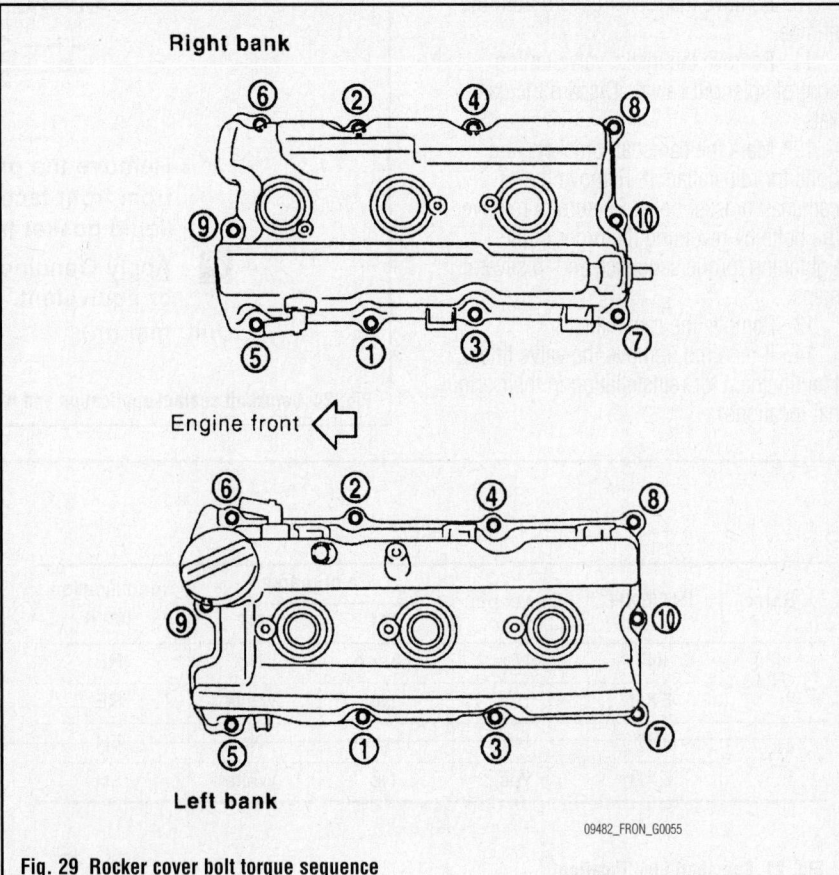

Fig. 29 Rocker cover bolt torque sequence

To install:

16. Inspect the camshafts, replace as required.

17. Install the timing chain tensioners (secondary) on both sides of the cylinder head. Be sure to use new O-rings.

➡ **Install the tensioner with its stopper pin attached. Install the tensioner with the sliding part facing downward on the right cylinder head and with the sliding part facing upward on the left cylinder head.**

18. Install the valve lifters, in their original bores.

19. Install the camshafts, with the dowel pin attached to its front end face on the exhaust side.

➡ **Follow the identification marks for proper placement and direction.**

20. Install the camshaft so that the dowel pin hole and dowel pin on the front end face are positioned as shown in the illustration (No. 1 piston at TDC on its compression stroke).

➡ **Large and small pin holes are located on the front end face of the camshaft (INT), at intervals of 180 degrees. Face small diameter side pin hole upward (in cylinder head upper face direction).**

➡ **Though the camshaft does not stop at the portion as shown, for placement of the cam nose, it is generally accepted that the camshaft is placed for the same direction as shown.**

21. Install the camshaft brackets in the same position that they were removed. Install brackets No. 2–No. 4 aligning the stamp marks as indicated in the illustration.

➡ **There are no identification marks indicating left or right for camshaft bracket No. 1.**

22. Apply liquid gasket to the mating surfaces of camshaft bracket No. 1 as shown in the illustration on both the left and right cylinder heads. Be sure to use genuine RTV sealant or equivalent.

23. Tighten the camshaft bracket bolts in the proper sequence and to specification:

    a. Bolts 7–10 to 17 inch lbs. (1.13 to 1.92 Nm)

    b. Bolts 1–6 to 17 inch lbs. (0.67 to 1.92 Nm)

    c. All bolts to 52 inch lbs. (5.88 Nm)

    d. All bolts to 92 inch lbs. (10.39 Nm)

24. Measure the difference in levels between the front end faces of the camshaft bracket No. 1 and the cylinder head. Speci-

fication should be -0.0055–0.0055 inch. If not within specification, reinstall camshaft bracket No. 1.

➡ **Measure two positions (both intake and exhaust side) for a single bank.**

25. Check and adjust valve clearance, as required.

26. Apply liquid gasket, be sure to use genuine RTV silicone sealant or equivalent, to the positions shown in the illustration. Refer to figure "a" to apply liquid gasket to joint part of camshaft bracket No. 1 and cylinder head. Refer to figure "b" to apply liquid gasket to the figure "a" squarely.

27. Install the rocker cover. Torque the retaining bolts to 17 inch lbs. and then to 74 inch lbs., in the proper sequence.

28. Continue the installation in the reverse order of the removal procedure.

29. Inspect the camshaft sprocket (INT) oil groove.

➡ **Perform this inspection only when DTC P0011 or DTC P0021 are detected in self diagnostic results of CONSULT-II.**

30. Be sure the engine is cold. Check and adjust oil level, as required.

31. Properly release the fuel system pressure. Disconnect the ignition coil and injector harness connectors.

➡ **This is being done to prevent the engine from unintentionally being started while checking.**

32. Remove the intake valve timing control solenoid valve.

33. Crank the engine, and then make sure that engine oil comes out from camshaft bracket (No. 1) oil hole.

❄❄❄ WARNING

Be careful not to touch rotating parts, (drive belt, idler pulley, crankshaft pulley, etc) as injury could result.

➡ **Oil may squirt from the intake valve timing control solenoid valve installation hole during engine cranking. Use a shop towel to prevent oil from squirting on engine components.**

34. Clean the oil groove between the oil strainer and the intake timing control solenoid valve if engine oil does not come out from camshaft bracket (No. 1) oil hole.

35. Remove the components between the intake valve timing control solenoid valve and the camshaft sprocket (INT). Check each oil groove for clogging.

36. After inspection install any removed components.

CRANKSHAFT DAMPER (BALANCER)

REMOVAL & INSTALLATION

1. Before servicing the vehicle, refer to the Precautions Section.

2. Disconnect the negative battery cable.

3. Remove the engine undercover.

4. Remove the drive belts.

5. Remove the cooling fan.

6. Loosen the crankshaft pulley retaining bolt and locate the bolt seating surface, which is about 0.39 inch from its original position.

➡ **Do not remove the crankshaft pulley bolt. Keep the loosened pulley bolt in place to protect the removed crankshaft pulley from dropping.**

7. Pull the pulley with both hands and remove it from its mounting. Remove the bolt and pulley from the engine.

To install:

8. Installation is the reverse order of the removal procedure.

CRANKSHAFT FRONT SEAL

REMOVAL & INSTALLATION

1. Before servicing the vehicle, refer to the Precautions Section.

2. Disconnect the negative battery cable.

3. Remove the engine undercover.

4. Remove the drive belts.

5. Remove the cooling fan.

6. Loosen the crankshaft pulley retaining bolt and locate the bolt seating surface, which is about 0.39 inch from its original position.

➡ **Do not remove the crankshaft pulley bolt. Keep the loosened pulley bolt in place to protect the removed crankshaft pulley from dropping.**

7. Pull the pulley with both hands and remove it from its mounting. Remove the bolt and pulley from the engine.

8. Using a seal removal tool, remove the oil seal from its mounting.

➡ **Be careful not to damage the front cover and/or the crankshaft.**

To install:

9. Installation is the reverse order of the removal procedure.

10. Press fit until the height of the front oil seal is level with the mounting surface, using the proper tools.

CYLINDER HEAD

REMOVAL & INSTALLATION

See Figures 30 through 33.

1. Before servicing the vehicle, refer to the Precautions Section.
2. Properly relieve the fuel system pressure.
3. Disconnect the negative battery cable. Drain the cooling system.
4. Remove the camshaft.
5. Remove the intake manifold.
6. Remove the exhaust manifold.
7. Remove the water inlet and thermostat assembly.
8. Remove the water outlet, water pipe and heater pipe.
9. Remove the cylinder head retaining bolts. Be sure to remove the bolts by reversing the order of the tightening torque sequence.
10. Remove the cylinder head from the engine. Discard the gasket.

To install:

11. Installation is the reverse of the removal procedure.
12. Be sure to inspect the cylinder head bolts. Replace as required.

➡**Head bolts are tightened by plastic zone tightening method. Whenever the size difference between "d1" and "d2" exceeds the limit, replace the bolt. "d1"-"d2" limit is 0.0043. If reduction of the outer diameter appears in a position other than "d2", use it the "d2" point.**

13. Install the new cylinder head gasket. Turn the crankshaft until the number one piston is at TDC.

➡**The crankshaft key should line up with the right bank center line, see illustration.**

14. Torque the cylinder head bolts to specification and in the proper sequence.

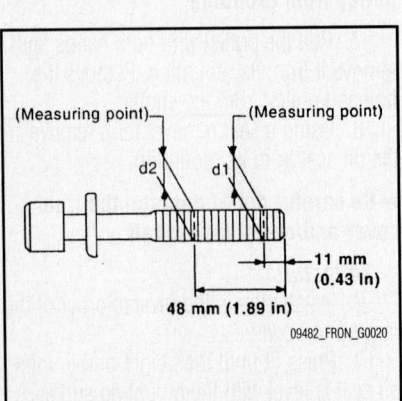

Fig. 30 Cylinder head bolt measurement

(Measuring point)   (Measuring point)

d2   d1

11 mm (0.43 In)

48 mm (1.89 in)

09482_FRON_G0020

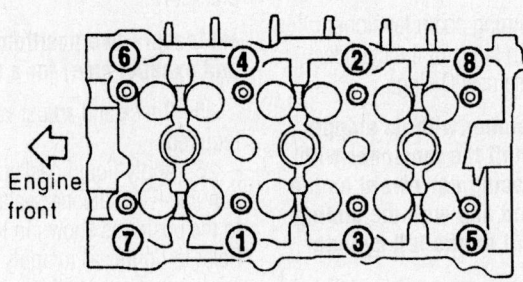

Right bank

Engine front

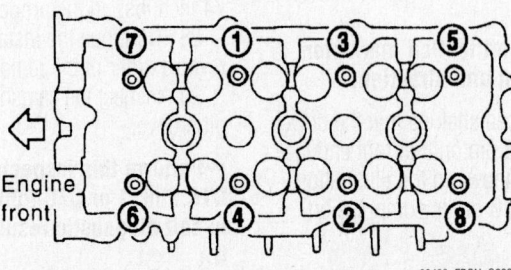

Left bank

Engine front

09482_FRON_G0021

Fig. 31 Cylinder head bolt torque sequence

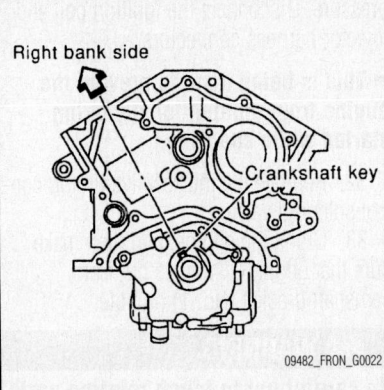

Right bank side

Crankshaft key

09482_FRON_G0022

Fig. 32 Cylinder head and crankshaft key alignment

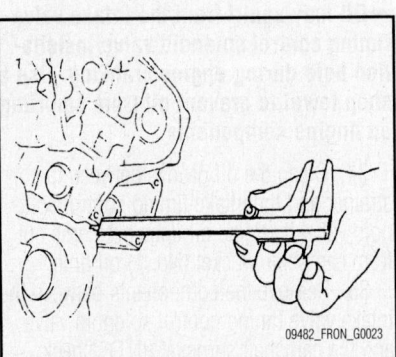

09482_FRON_G0023

Fig. 33 Cylinder head to cylinder block installation measurement

15. Measure the distance between the front end faces of the cylinder block and the cylinder head on both the left and right banks. If the measured value is not within specification reinstall the cylinder head. Specification is 0.555–0.587 inch.
16. Be sure to fill the cooling system with the proper grade and type engine coolant.
17. Start the engine and check for leaks.

EXHAUST MANIFOLD

REMOVAL & INSTALLATION

Left Side

See Figure 34.

1. Before servicing the vehicle, refer to the Precautions Section.
2. Disconnect the negative battery cable.
3. Remove air duct, PCV hose (between air duct and rocker cover) and electric throttle control actuator.
4. Disconnect the harness connector and remove the heated oxygen sensor.

➡**Be careful not to damage the air fuel ratio sensor. Discard the sensor if it has been dropped from a height of more than 19.7 inches on to a hard surface.**

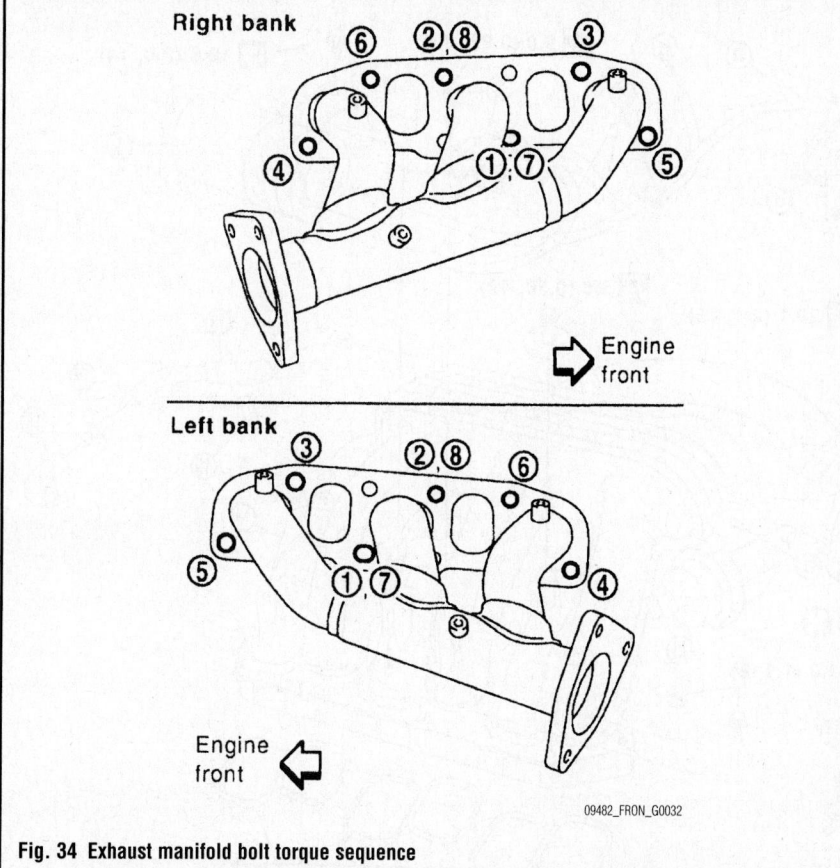

Right bank

Engine front

Left bank

Engine front

09482_FRON_G0032

Fig. 34 Exhaust manifold bolt torque sequence

5. Remove the front exhaust tube.

6. Remove the exhaust manifold cover.

7. Remove bracket between exhaust manifold–three way catalyst assembly and transmission assembly.

8. Remove the exhaust manifold retaining bolts. Be sure to remove the bolts by reversing the order of the tightening torque sequence.

9. Remove the exhaust manifold from the engine. Discard the gasket.

To install:

10. Installation is the reverse of the removal procedure.

11. Be sure to use new gaskets.

12. Be sure to tighten the exhaust manifold retaining bolts to specification and in the proper sequence.

➡**Before installing a new air fuel sensor and heated oxygen sensor, clean threads and apply anti seize lubricant to the threads. Do not over torque the sensor, doing so may cause damage to the sensor resulting in the MIL light coming on.**

Right Side

1. Before servicing the vehicle, refer to the Precautions Section.

2. Disconnect the negative battery cable.

3. Remove the engine from the vehicle.

Position the assembly in a suitable holding fixture.

4. Remove the exhaust manifold retaining bolts. Be sure to remove the bolts by reversing the order of the tightening torque sequence.

➡**Disregard the numerical order of No.7 and No.8 in the removal process.**

5. Discard the gaskets.

To install:

6. Installation is the reverse of the removal procedure.

7. Be sure to use new gaskets.

8. Be sure to tighten the exhaust manifold retaining bolts to specification and in the proper sequence.

➡**Before installing a new air fuel sensor and heated oxygen sensor apply anti seize lubricant to the threads. Do not over torque the sensor, doing so may cause damage to the sensor resulting in the MIL light coming on.**

INTAKE MANIFOLD

REMOVAL & INSTALLATION

See Figures 35 through 38.

➡**Upper intake manifold is also referred to as intake manifold collector.**

1. Before servicing the vehicle, refer to the Precautions Section.

2. Properly relieve the fuel system pressure.

3. Disconnect the negative battery cable. Drain the cooling system.

4. Remove the engine cover. Remove the air cleaner case (upper) with the mass air flow sensor and air duct assembly.

5. Disconnect the water hoses from the electric throttle control actuator. Disconnect the harness connector.

6. Remove the electric throttle control actuator retaining bolts. Be sure to remove the bolts by reversing the order of the tightening torque sequence.

7. Remove the electric throttle control actuator.

8. Remove the brake booster vacuum hose and the PCV hose. Remove the intake manifold collector support.

9. Disconnect the EVAP hoses and harness connector from the EVAP canister purge volume control solenoid valve. Remove the EVAP canister purge volume control solenoid valve.

10. Remove the VIAS control solenoid valve and vacuum tank.

11. Remove the intake manifold collector retaining bolts. Be sure to remove the bolts by reversing the order of the tightening torque sequence.

12. Remove the intake manifold collector from the engine.

13. Remove the fuel tube and fuel injector assembly.

14. Remove the intake manifold retaining bolts. Be sure to remove the bolts by reversing the order of the tightening torque sequence.

15. Remove the intake manifold from the engine.

To install:

16. Installation is the reverse of the removal procedure.

17. Be sure to use new gaskets.

18. Be sure to tighten the intake manifold retaining bolts to specification and in the proper sequence in two or more steps.

19. Be sure to tighten the intake manifold collector retaining bolts to specification and in the proper sequence.

20. Be sure to tighten the electric throttle control actuator retaining bolts to specification and in the proper sequence.

➡**See throttle valve closed position learning and idle air volume learning procedures, for relearning information.**

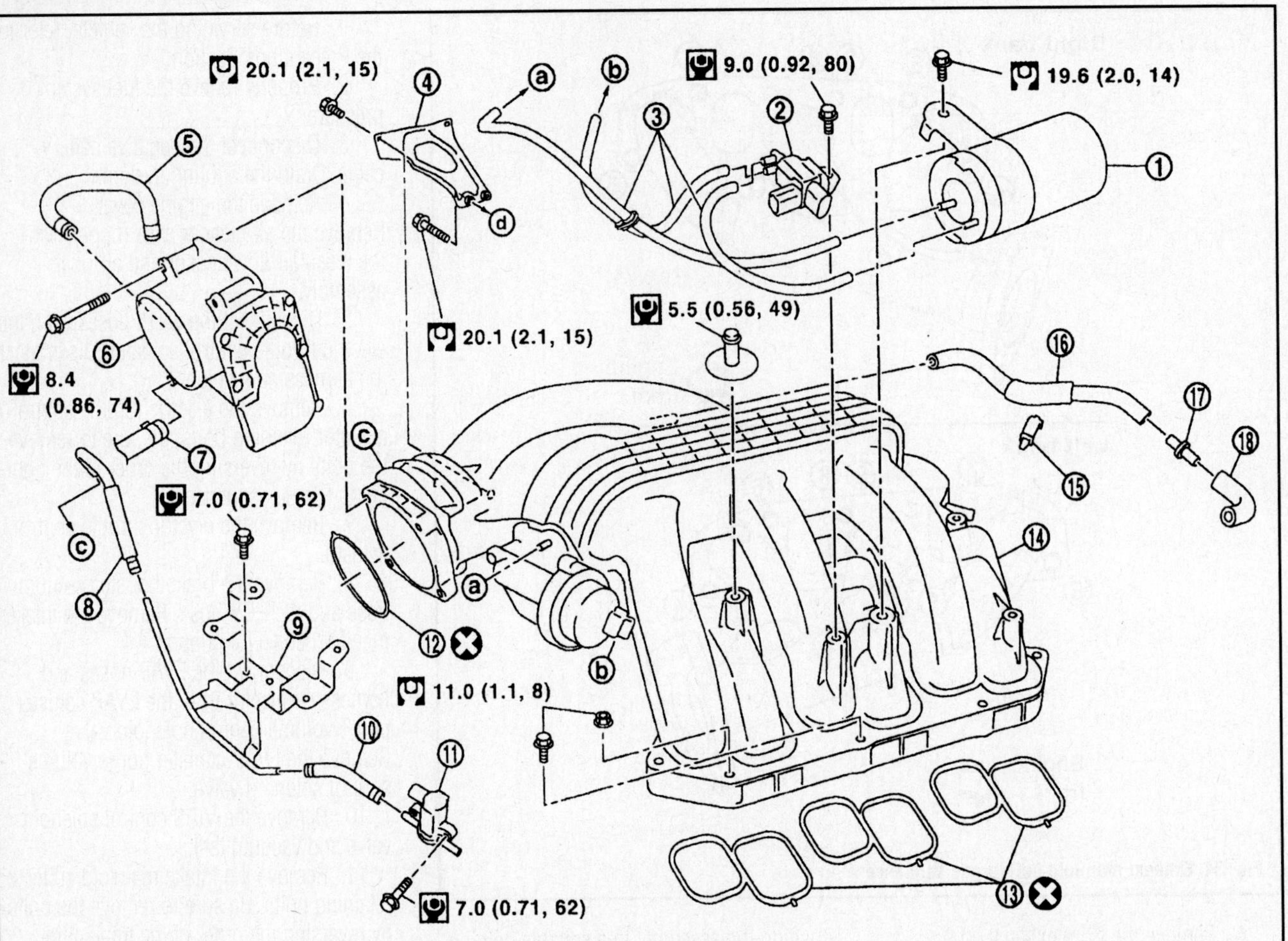

[Figure 35 diagram with torque specifications and numbered callouts]

20.1 (2.1, 15)

9.0 (0.92, 80)

19.6 (2.0, 14)

20.1 (2.1, 15)

5.5 (0.56, 49)

8.4 (0.86, 74)

7.0 (0.71, 62)

11.0 (1.1, 8)

7.0 (0.71, 62)

1. Vacuum tank
2. VIAS control solenoid valve
3. Vacuum hose
4. Intake manifold collector support
5. Water hose
6. Electric throttle control actuator
7. Water hose
8. EVAP hose
9. Bracket
10. EVAP hose
11. EVAP canister purge volume control solenoid valve
12. Gasket
13. Gasket
14. Intake manifold collector
15. Clip
16. PCV hose
17. Connector
18. PCV hose
a. To intake manifol collector
b. To power valve
c. To throttle body
d. To cylinder head (RH bank)

09482_FRON_G0024

Fig. 35 Intake manifold collector and related components

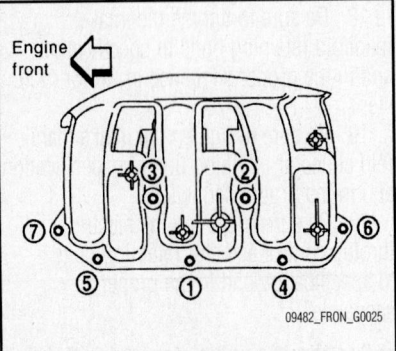

09482_FRON_G0025

Fig. 36 Intake manifold collector bolt torque sequence

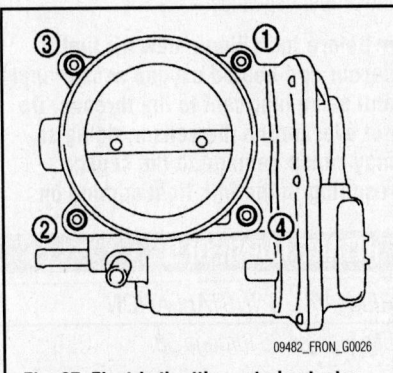

09482_FRON_G0026

Fig. 37 Electric throttle control actuator bolt torque sequence

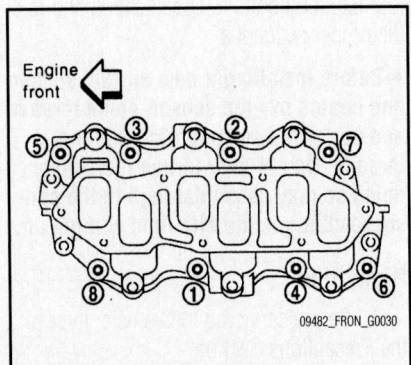

09482_FRON_G0030

Fig. 38 Intake manifold bolt torque sequence

OIL PAN

REMOVAL & INSTALLATION

Lower

See Figure 39.

1. Before servicing the vehicle, refer to the Precautions Section.
2. Disconnect the negative battery cable.
3. Remove the engine under cover. Drain the engine oil.
4. Loosen the oil pan retaining bolts, in the reverse order of the installation sequence.
5. Insert a seal cutter tool between the oil pan and the cylinder block, and slide it by tapping on the side of the tool with a hammer.
6. Remove the oil pan from the engine.

To install:

7. Be sure to clean all the oil gasket material from both the oil pan and the cylinder block surfaces, using the proper tools.
8. Apply a continuous bead of sealant 0.138–0.177 inches (3.5–4.5 mm) to the oil pan mating surface. Be sure to use genuine RTV sealant or equivalent.
9. Install the oil pan to the cylinder block. This must be done within 5 minutes after applying the liquid gasket.
10. Torque the bolts to specification and in the proper sequence.
11. Continue the installation in the reverse order of the removal procedure.

➡**Wait 30 minutes after installation of the oil pan to allow the sealant to cure before adding oil.**

12. Fill the crankcase to the correct level.
13. Start the engine and check for leaks.

Upper

See Figures 40 and 41.

1. Before servicing the vehicle, refer to the Precautions Section.
2. Disconnect the negative battery cable.
3. Remove the air duct. Remove the engine under cover.
4. Drain the engine oil. Drain the engine coolant.
5. Remove the final drive, if equipped with 4WD.
6. Disconnect the steering gear lower shaft joint bolt and steering gear nuts and bolts, position the assembly out of the way.
7. Remove the starter.
8. Disconnect the automatic transmission fluid cooler brackets, if equipped and position them out of the way.
9. Remove the oil filter, as necessary. Remove the oil cooler.
10. Remove the lower oil pan. Remove the oil strainer.
11. Remove the transmission joint bolts which pierce the oil pan.
12. Remove the rear cover plate.
13. Loosen the upper oil pan retaining bolts, in the reverse order of the installation sequence.

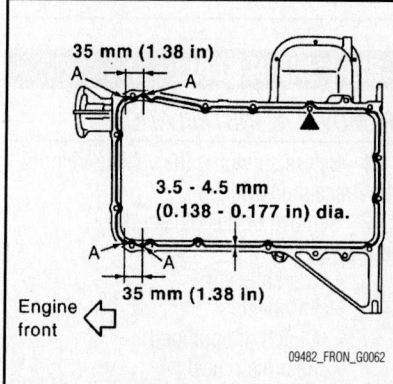

Fig. 40 Upper oil pan sealant application

14. Insert a seal cutter tool between the oil pan and the cylinder block, and slide it by tapping on the side of the tool with a hammer.
15. Remove the oil pan from the engine. Remove the O-rings from the bottom lower cylinder block and oil pump.

To install:

16. Be sure to clean all the oil gasket material from both the oil pan and the cylinder block surfaces, using the proper tools.
17. Install new O-rings on the bottom lower cylinder block and oil pump.
18. Apply a continuous bead of sealant 0.138–0.177 inches (3.5–4.5 mm) to the lower cylinder block mating surfaces of the upper oil pan. Be sure to use genuine RTV sealant or equivalent.

➡**For bolt holes marked with a solid black triangle, apply liquid gasket outside the hole. Apply a bead of sealant (0.177–0.217 inch diameter) to area "A".**

19. Install the upper oil pan. This must be done within 5 minutes after applying the liquid gasket.
20. Torque the bolts to specification and in the proper sequence. There are two types of bolts M8X100 mm (3.97 inch) bolts 7, 11, 12, 13 and M8X25 mm (0.98 inch) except 7, 11, 12 and 13.
21. Tighten the transmission joint bolts.
22. Install the oil strainer to the upper oil pan.
23. Continue the installation in the reverse order of the removal procedure.

➡**Wait 30 minutes after installation of the oil pan to allow the sealant to cure before adding oil.**

24. Fill the crankcase to the correct level.
25. Start the engine and check for leaks.

OIL PUMP

REMOVAL & INSTALLATION

See Figure 42.

1. Before servicing the vehicle, refer to the Precautions Section.
2. Disconnect the negative battery cable. Drain the engine oil. Drain the engine coolant.
3. Remove the lower oil pan.
4. Remove the upper oil pan.
5. Remove the front timing chain case and timing chain (primary).
6. Remove the oil pump from the engine.

To install:

7. Installation is the reverse of the removal procedure.

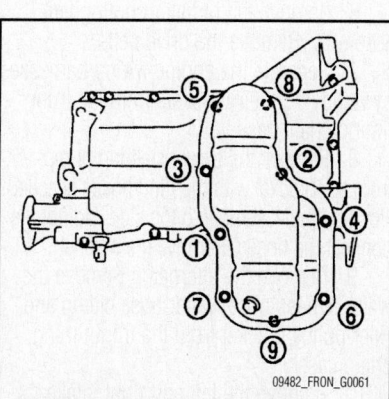

Fig. 39 Lower oil pan bolt torque sequence

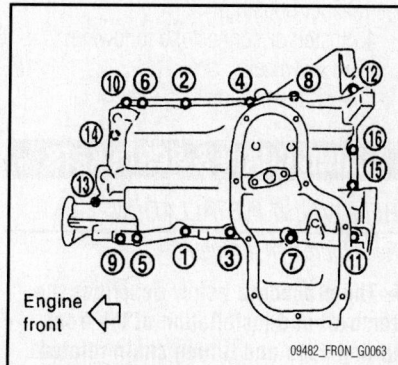

Fig. 41 Upper oil pan bolt torque sequence

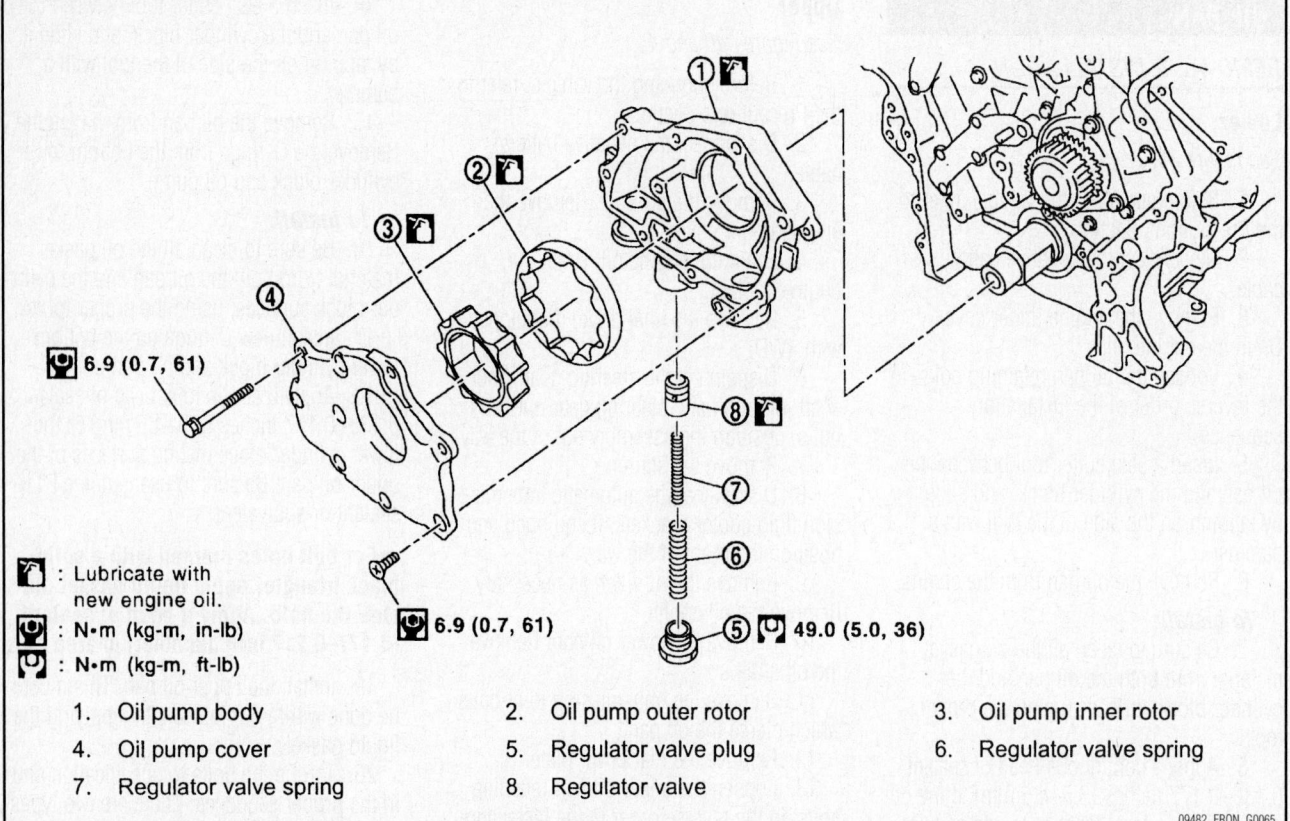

: : Lubricate with
new engine oil.

: N•m (kg-m, in-lb)

: N•m (kg-m, ft-lb)

1. Oil pump body
4. Oil pump cover
7. Regulator valve spring

2. Oil pump outer rotor
5. Regulator valve plug
8. Regulator valve

3. Oil pump inner rotor
6. Regulator valve spring

6.9 (0.7, 61)

6.9 (0.7, 61)

49.0 (5.0, 36)

09482_FRON_G0065

Fig. 42 Oil pump and related components

➡Wait 30 minutes after installation of the oil pan to allow the sealant to cure before adding oil.

8. Fill the crankcase to the correct level.

9. Start the engine and check for leaks.

PISTON AND RING

POSITIONING

See Figure 43.

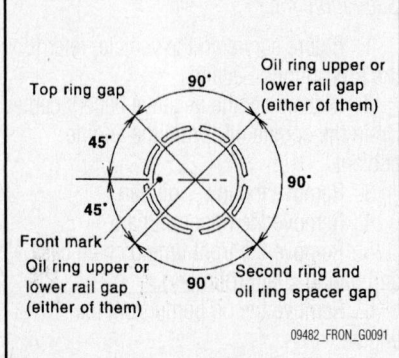

09482_FRON_G0091

Fig. 43 Piston ring positioning—4.0L engine

REAR MAIN SEAL

REMOVAL & INSTALLATION

1. Before servicing the vehicle, refer to the Precautions Section.

2. Remove or disconnect the following:
   • Transmission
   • Flywheel
   • Clutch, if equipped
   • Rear main seal

To install:

3. Install the seal so that it is flush with the retainer housing.

4. Install or connect the following:
   • Flywheel.
   • Transmission

TIMING CHAIN & SPROCKETS

REMOVAL & INSTALLATION

See Figures 44 through 59.

➡The procedure below describes the removal and installation of the front timing case and timing chain related parts and rear timing chain case, when the upper oil pan needs to be removed or installed. When only the timing chain

(primary) is being removed it is not necessary to remove the rocker covers.

1. Before servicing the vehicle, refer to the Precautions Section.

2. Properly relieve the fuel system pressure.

3. Disconnect the negative battery cable.

4. Remove the engine cover. Drain the engine oil. Drain the engine coolant.

5. Remove the upper and lower oil pans.

6. Remove the radiator cooling fan assembly. Remove the drive belts.

7. Separate the engine wiring harnesses by removing their brackets from the front timing chain case.

8. Remove the power steering pump from the bracket with the fluid hoses attached. Position the assembly to the side. Do not disconnect the hoses. Remove the bracket.

9. Remove the alternator. Remove the water bypass hose, water hose clamp and idler pulley bracket from the front timing chain case.

10. Remove the left and right intake valve timing control covers. Loosen the bolts in the reverse order of the tightening

sequence. Use tool KV10111100 (J-37228) or equivalent to cut the liquid gasket seal.

➡ **The shaft is internally jointed with the camshaft sprocket (INT) center hole. When removing, keep it horizontal until it is completely disconnected.**

11. Remove the collared O-rings from the front timing chain case on both the left and right side.

12. Remove the intake manifold collector.

13. Separate the engine harness and remove their brackets from the rocker covers. Remove the harness bracket from the cylinder head, if necessary.

14. Remove the ignition coil. Remove the PCV hoses. Remove the oil filler cap, if necessary.

15. Loosen the rocker cover retaining bolts, in the reverse order of the tightening sequence.

16. Remove the rocker covers from the engine.

➡ **When only the timing chain (primary) is being removed it is not necessary to remove the rocker covers.**

17. Set the No. 1 cylinder at TDC of its compression stroke by rotating the crankshaft pulley clockwise to align the timing mark (grooved line without color) with the timing indicator. Make sure that the intake and exhaust cam noses on No. 1 cylinder (engine front side on right bank) are in alignment as shown in the illustration. If not, rotate the crankshaft in the clockwise direction 360 degrees.

➡ **When only the timing chain (primary) is removed, the rocker cover does not need to be removed. To be sure that the No. 1 cylinder is set at TDC on the compression stroke, remove the front timing chain case cover first, then check the mating marks on the camshaft sprockets.**

18. Remove the starter. Position tool KV10117700 (J-44716) or equivalent.

19. Loosen the crankshaft pulley retaining bolt and locate the bolt seating surface, which is about 0.39 inch from its original position.

➡ **Do not remove the crankshaft pulley bolt. Keep the loosened pulley bolt in place to protect the removed crankshaft pulley from dropping.**

20. Pull the pulley with both hands and remove it from its mounting. Remove the bolt and pulley from the engine.

21. Loosen and remove the two bolts of the upper oil pan.

22. Loosen the front timing chain cover retaining bolts in the reverse order of the tightening sequence.

23. Insert a suitable tool in the notch at the top of the front timing chain case and pry off the case by moving the tool as shown in the illustration. Use tool KV10111100 (J-37228) or equivalent to cut the liquid gasket seal.

➡ **Do not use a screwdriver or something similar. After removal handle the front timing chain cover case carefully so it does not tilt, cant or warp under a load.**

24. Remove the O-rings from the rear timing chain case.

25. Remove the water pump cover and chain tensioner cover from the front timing chain case cover, as required.

26. Remove the oil seal from the front timing chain case cover, as required.

27. Remove the timing chain tensioner (primary) by loosening the clip of the timing chain tensioner (primary) and release the plunger stopper. Insert the plunger into the tensioner body by pressing the slack guide. Keep the slack guide pressed and hold the plunger in by pushing the stopper pin through the tensioner body hole and the plunger groove. Remove the bolts and remove the timing chain tensioner (primary).

28. Remove the internal chain guide, tension guide and slack guide.

➡ **The tension guide can be removed after removing the timing chain (primary).**

29. Remove the timing chain (primary) and the crankshaft sprocket.

➡ **After removing the timing chain (primary), do not turn the crankshaft and camshaft separately or the valves will strike the piston heads.**

30. To remove the timing chain (secondary) and camshaft sprockets, attach a suitable stopper pin to the right and left timing chain tensioner (secondary).

➡ **Use a 0.02 inch (approximate) metal pin as a stopper pin.**

31. Remove the camshafts. Remove the valve lifters. Identify them for reinstallation in their original locations.

32. Remove the camshaft sprocket (INT and EXH) bolts. Secure the hexagonal portion of the camshaft using a wrench to loosen the bolts.

➡ **Do not loosen the bolts with securing anything other than the camshaft hexagonal portion or with tensioning the timing chain.**

33. To remove the timing chain (secondary) together with the camshaft sprockets, turn the crankshaft slightly to secure slackness of the timing chain on the timing chain tensioner (secondary) side.

34. Insert a 0.020 inch thick metal or resin plate between the timing chain and timing chain plunger (guide). Remove the timing chain (secondary) together with the camshaft sprockets with the timing chain loose from the guide groove.

❄❄ CAUTION

Be careful of the plunger coming off when removing the timing chain (secondary). This is because the plunger of the timing chain tensioner (secondary) moves during operation, leading to coming off its fixed stopper pin.

➡ **The camshaft sprocket (INT) is a one piece integrated design sprocket for the timing chain (primary) and for the timing chain (secondary). When handling the sprocket avoid shock to the sprocket. Do not disassemble or loosen bolt "A", as shown in the illustration.**

35. Remove the water pump.

36. Remove the rear timing chain case cover bolts, in the reverse order of the tightening sequence. Using the proper tool, cut the liquid gasket sealant seal. Remove the cover.

➡ **Do not remove the metal cover of the oil passage. After removal, handle the case carefully so it does not tilt, or warp under a load.**

37. Remove the O-rings from the cylinder head and No. 1 camshaft bracket. Remove the O-rings from the cylinder block.

38. If necessary, remove the timing chain tensioners (secondary) from the cylinder head by first removing the No. 1 camshaft bracket. Remove the timing chain tensioners (secondary) with the stopper pin attached.

To install:

39. Check the chain for cracks and excessive wear, replace as required.

40. Be sure to remove all old gasket material from bolts and bolt holes.

41. If removed install the timing chain tensioners (secondary) to the cylinder head.

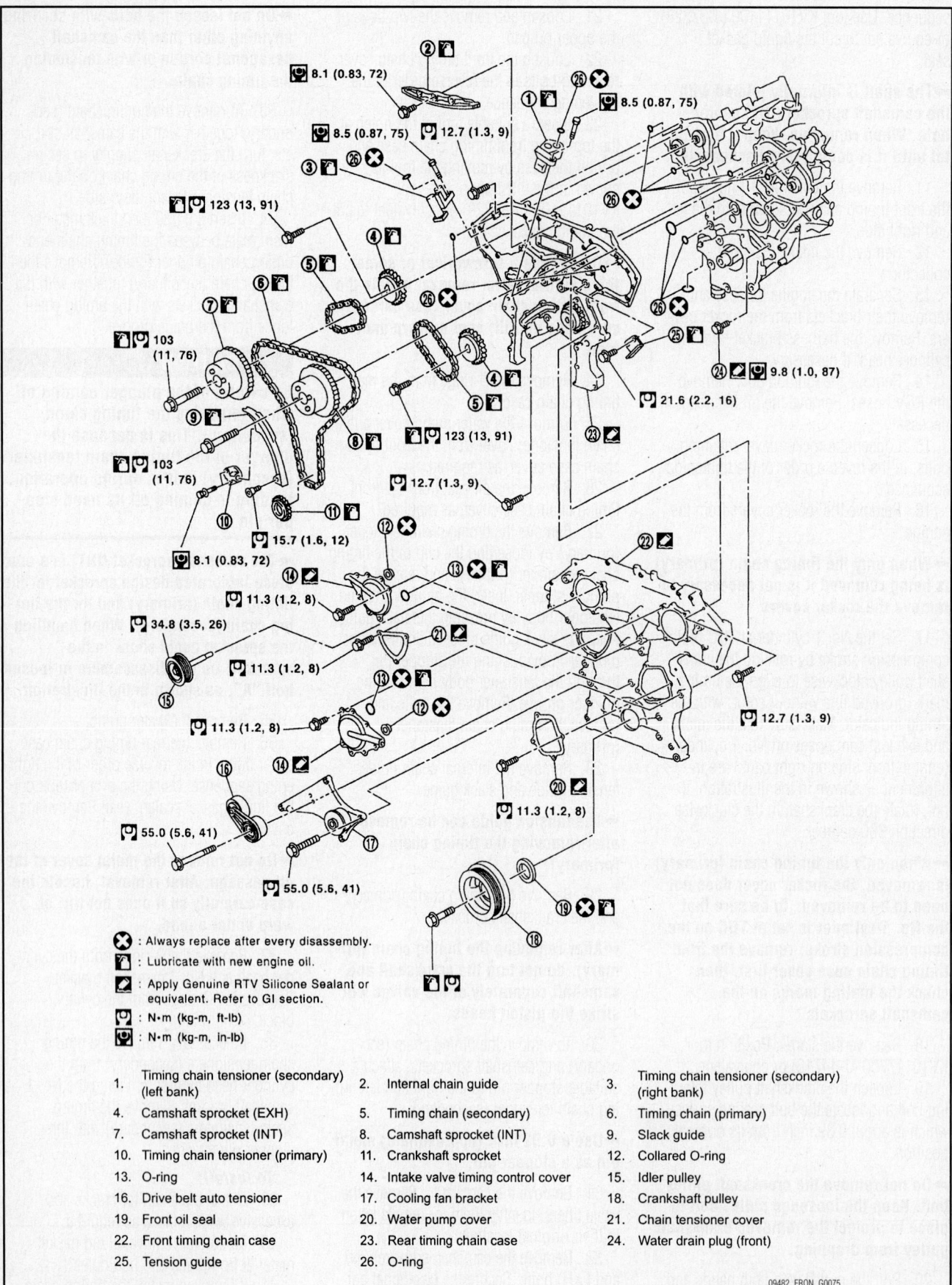

8.1 (0.83, 72)

8.5 (0.87, 75)

12.7 (1.3, 9)

8.5 (0.87, 75)

123 (13, 91)

103 (11, 76)

103 (11, 76)

15.7 (1.6, 12)

8.1 (0.83, 72)

11.3 (1.2, 8)

34.8 (3.5, 26)

11.3 (1.2, 8)

11.3 (1.2, 8)

55.0 (5.6, 41)

55.0 (5.6, 41)

123 (13, 91)

12.7 (1.3, 9)

21.6 (2.2, 16)

9.8 (1.0, 87)

12.7 (1.3, 9)

11.3 (1.2, 8)

❌ : Always replace after every disassembly.

🛢 : Lubricate with new engine oil.

✏ : Apply Genuine RTV Silicone Sealant or equivalent. Refer to GI section.

🔧 : N·m (kg-m, ft-lb)

🔧 : N·m (kg-m, in-lb)

1. Timing chain tensioner (secondary) (left bank)	2. Internal chain guide	3. Timing chain tensioner (secondary) (right bank)
4. Camshaft sprocket (EXH)	5. Timing chain (secondary)	6. Timing chain (primary)
7. Camshaft sprocket (INT)	8. Camshaft sprocket (INT)	9. Slack guide
10. Timing chain tensioner (primary)	11. Crankshaft sprocket	12. Collared O-ring
13. O-ring	14. Intake valve timing control cover	15. Idler pulley
16. Drive belt auto tensioner	17. Cooling fan bracket	18. Crankshaft pulley
19. Front oil seal	20. Water pump cover	21. Chain tensioner cover
22. Front timing chain case	23. Rear timing chain case	24. Water drain plug (front)
25. Tension guide	26. O-ring	

09482_FRON_G0075

Fig. 44 Timing chain and related components

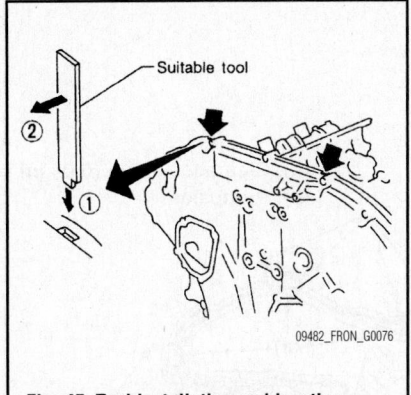

Fig. 45 Tool installation and location

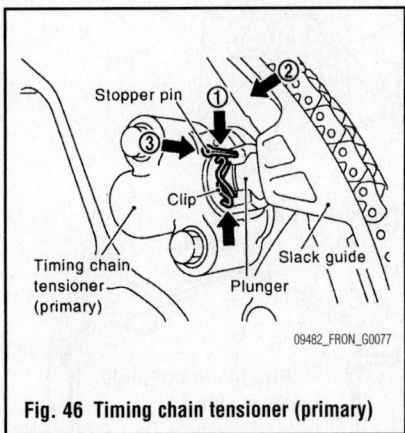

Fig. 46 Timing chain tensioner (primary)

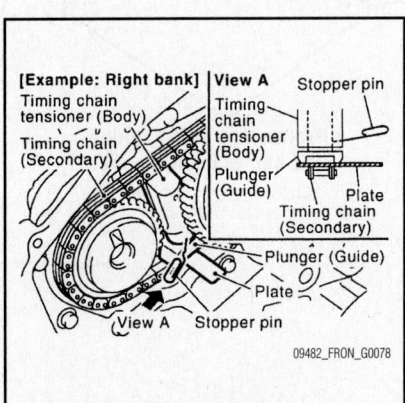

Fig. 47 Resin plate installation location

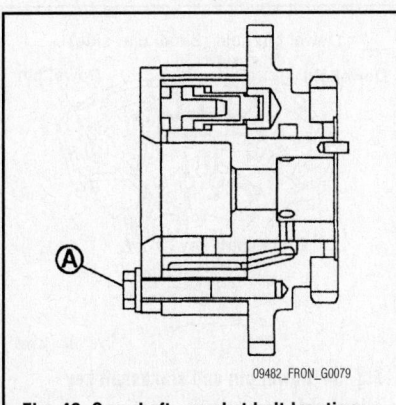

Fig. 48 Camshaft sprocket bolt location

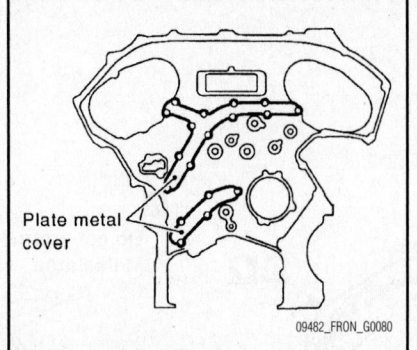

Fig. 49 Metal cover plate location on rear timing case cover

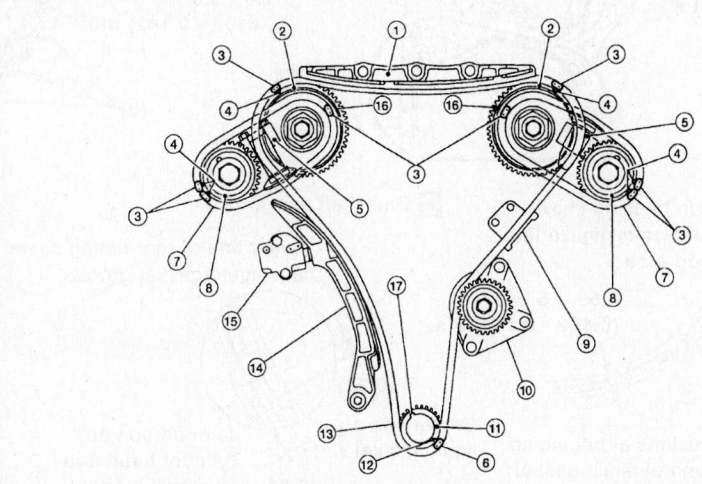

1. Internal chain guide
2. Camshaft sprocket (intake)
3. Mating mark (copper link)
4. Mating mark (punched)
5. Secondary timing chain tensioner
6. Mating mark (yellow link)
7. Secondary timing chain
8. Camshaft sprocket (exhaust)
9. Tensioner guide
10. Water pump
11. Crankshaft sprocket
12. Mating mark (notched)
13. Primary timing chain
14. Slack guide
15. Primary timing chain tensioner
16. Mating mark (back side)
17. Crankshaft key

Fig. 50 Timing chain alignment

Install the rear timing chain case. Make sure that the O-rings stay in place during installation to the cylinder block, cylinder head and camshaft bracket No. 1.

46. Tighten the bolts to specification and in the proper sequence.

 a. Bolt length: 0.79 inch. Bolt position: 1,2,3,6,7,8,9,and 10.

 b. Bolt length: 0.63 inch. Bolt position: except 1,2,3,6,7,8,9,and 10.

 c. Torque bolts to 9 ft. lbs.

 d. After all bolts are tightened, retighten them to specification and in the proper sequence

42. Install camshaft brackets No. 1.

43. To install the rear timing chain case cover, first install new O-rings to the cylinder block, Install new O-rings to the cylinder head and camshaft bracket No. 1.

44. Apply liquid gasket sealant to the rear timing chain case back side, as shown in the illustration. Be sure to use genuine RTV sealant, or equivalent.

➡**For "A" in the figure, completely wipe out excessive liquid gasket extended on a portion touching at engine coolant. Apply liquid gasket on the installation position of the water pump and cylinder head very completely.**

45. Align the rear timing case with dowel pins (right and left) on the cylinder block.

➡**Be sure to wipe off any excess liquid gasket leaking to the surface for installing the oil pan.**

47. After installing the rear timing case, check the surface height deference between the rear timing chain case and the lower cylinder block. Specification should be -0.0094–0.0055 inch. If not within specification, repeat the installation procedure.

48. Install the water pump, using new O-rings.

49. Make sure that the dowel pin hole, dowel pin of camshaft and crankshaft key are located with number one piston at TDC on the compression stroke.

➡**Though the camshaft does not stop at the position, as shown in the illustration, for placement of the cam nose it**

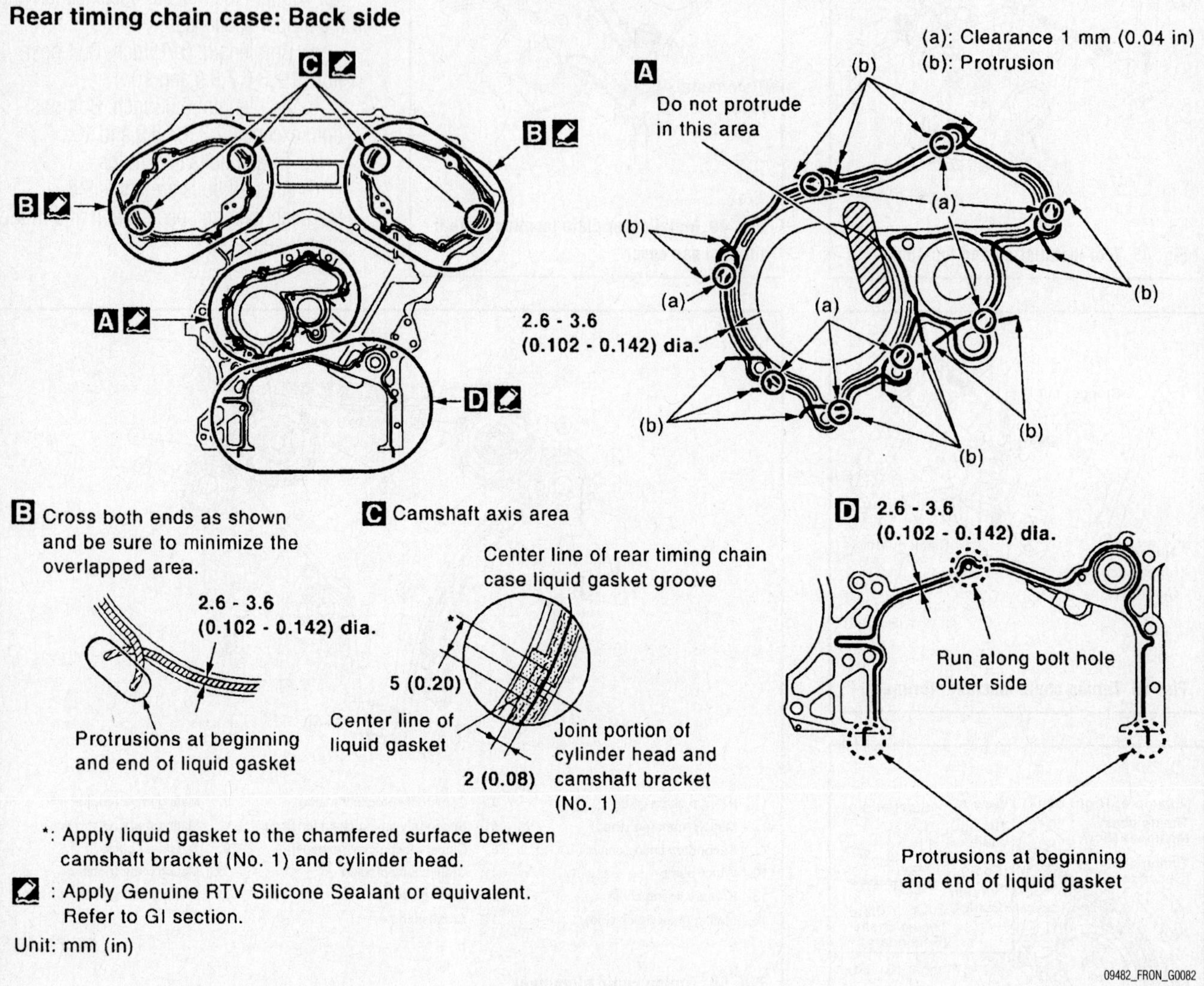

Rear timing chain case: Back side

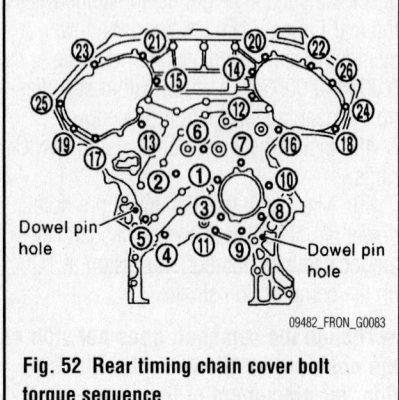

(a): Clearance 1 mm (0.04 in)
(b): Protrusion

A Do not protrude in this area

2.6 - 3.6 (0.102 - 0.142) dia.

B Cross both ends as shown and be sure to minimize the overlapped area.

2.6 - 3.6 (0.102 - 0.142) dia.

Protrusions at beginning and end of liquid gasket

C Camshaft axis area

Center line of rear timing chain case liquid gasket groove

5 (0.20)

Center line of liquid gasket

2 (0.08)

Joint portion of cylinder head and camshaft bracket (No. 1)

D 2.6 - 3.6 (0.102 - 0.142) dia.

Run along bolt hole outer side

Protrusions at beginning and end of liquid gasket

*: Apply liquid gasket to the chamfered surface between camshaft bracket (No. 1) and cylinder head.

: Apply Genuine RTV Silicone Sealant or equivalent. Refer to GI section.

Unit: mm (in)

09482_FRON_G0082

Fig. 51 Rear timing chain cover sealant application

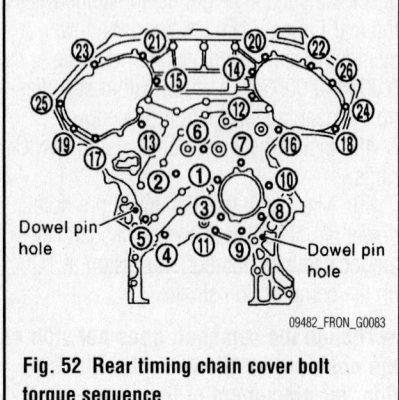

Dowel pin hole

Dowel pin hole

09482_FRON_G0083

Fig. 52 Rear timing chain cover bolt torque sequence

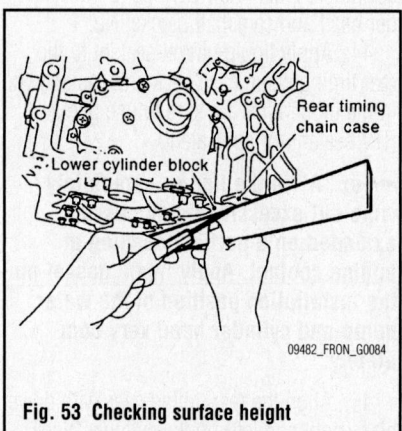

Rear timing chain case

Lower cylinder block

09482_FRON_G0084

Fig. 53 Checking surface height

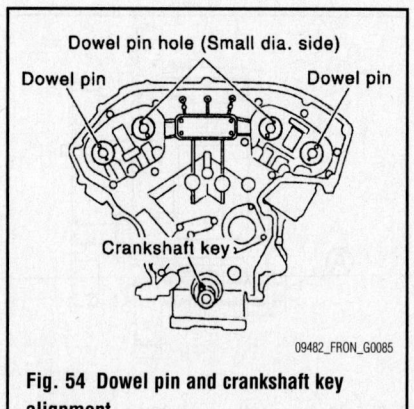

Dowel pin hole (Small dia. side)

Dowel pin

Dowel pin

Crankshaft key

09482_FRON_G0085

Fig. 54 Dowel pin and crankshaft key alignment

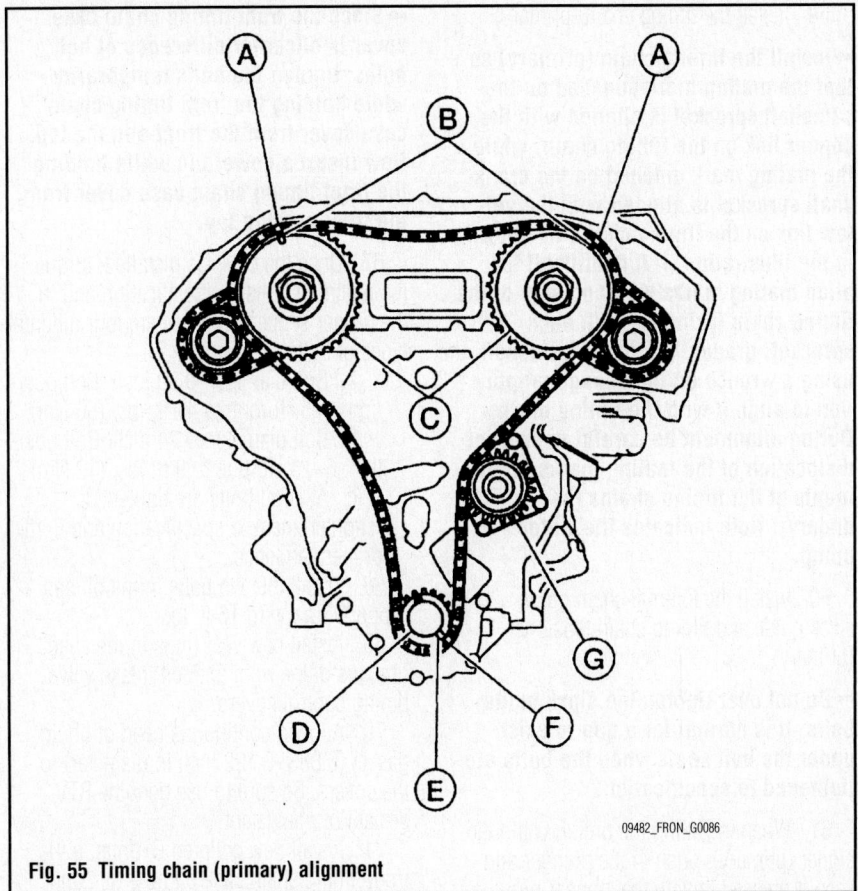

Fig. 55 Timing chain (primary) alignment

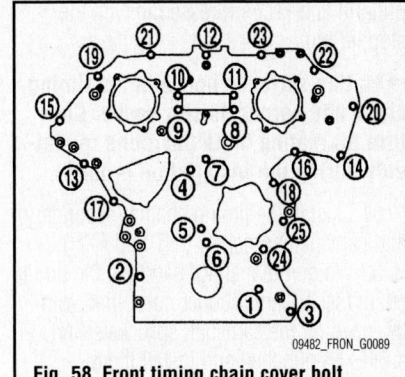

Fig. 58 Front timing chain cover bolt torque sequence

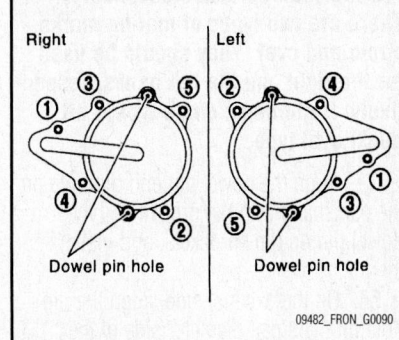

Fig. 59 Right and left intake valve timing control cover bolt torque sequence

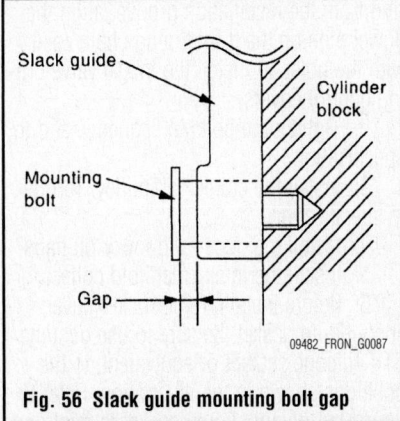

Fig. 56 Slack guide mounting bolt gap

is generally accepted that the camshaft is placed for the same direction as the illustration. Camshaft dowel pin hole (intake side): at the cylinder head upper face side in each bank. Camshaft dowel pin hole (exhaust side): at the cylinder head upper face side in each bank. Crankshaft key: at the cylinder head side of the right bank. Hole on the small diameter side must be used for the intake side dowel pin hole.

50. To install the timing chains (secondary) and camshaft sprockets, push the plunger of the timing chain tensioner (sec-

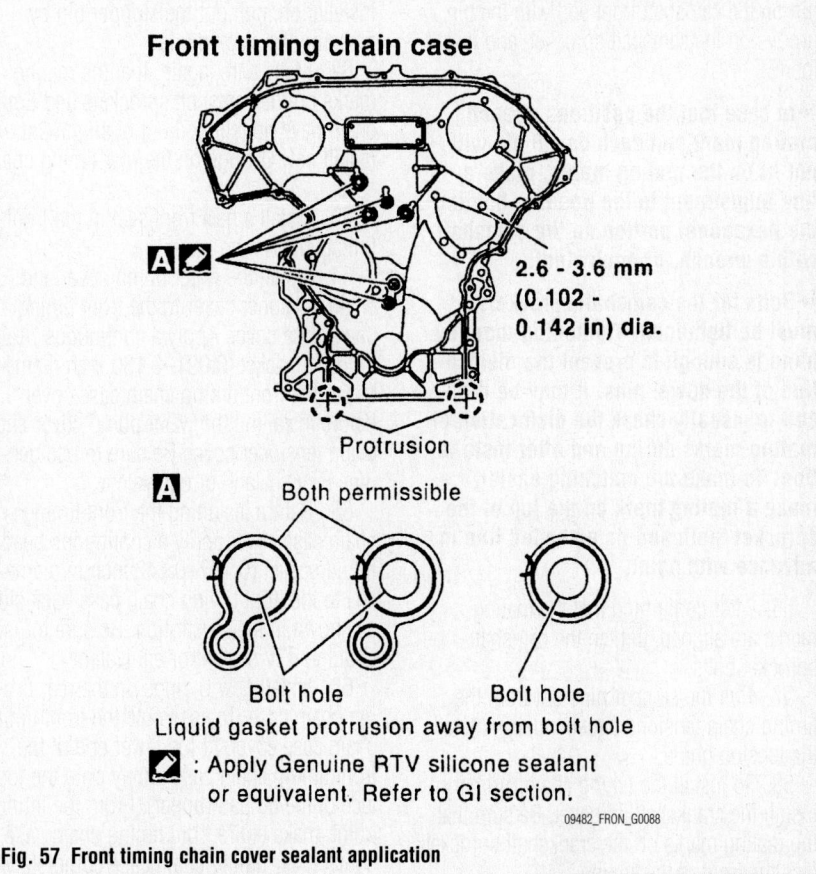

Front timing chain case

2.6 - 3.6 mm (0.102 - 0.142 in) dia.

Protrusion

A

Both permissible

Bolt hole   Bolt hole

Liquid gasket protrusion away from bolt hole

: Apply Genuine RTV silicone sealant or equivalent. Refer to GI section.

Fig. 57 Front timing chain cover sealant application

ondary) and keep it pressed in with the stopper pin.

➡ **Mating surfaces between the timing chain and sprockets slip easily. Confirm all mating mark positions repeatedly during the installation process.**

51. Install the timing chains (secondary) and camshaft sprockets (INT and EXH).

52. Align the mating marks on the timing chain (secondary) cooper color link, with the ones on the camshaft sprockets (INT and EXH) punched and install them.

➡ **Mating marks for the camshaft sprocket (INT) are on the back side of the camshaft sprocket (secondary). There are two types of mating marks, circle and oval. They should be used for the right and the left banks, respectively. Right bank: circle type. Left bank: oval type.**

53. Align the dowel pin and pin hole on the camshafts with the groove and the dowel pin on the sprockets, and install them.

54. On the exhaust side, align the pin hole on the small diameter side of the camshaft front end with the dowel pin on the back side of the camshaft sprocket, and install them.

55. On the exhaust side, align the dowel pin on the camshaft front end with the pin groove on the camshaft sprocket, and install them.

➡ **In case that the positions of each mating mark and each dowel pin will not fit on the mating marks, make a fine adjustment to the position holding the hexagonal portion on the camshaft with a wrench, or equivalent.**

➡ **Bolts for the camshaft sprockets must be tightened. Tightening them by hand is enough to prevent the dislocation of the dowel pins. It may be difficult to visually check the dislocation of mating marks during and after installation. To make the matching easier, make a mating mark on the top of the sprocket teeth and its extended line in advance with paint.**

56. After confirming that the mating marks are aligned, tighten the camshaft sprocket bolts.

57. Pull the stopper pins out from the timing chain tensioners (secondary). Install the tension guide.

58. To install the timing chain (primary), install the crankshaft sprocket. Be sure that the mating marks on the crankshaft sprocket face the front of the engine.

59. Install the timing chain (primary).

➡ **Install the timing chain (primary) so that the mating mark punched on the camshaft sprocket is aligned with the copper link on the timing chain, while the mating mark notched on the crankshaft sprocket is aligned with the yellow link on the timing chain, as shown in the illustration. If it is difficult to align mating marks with and with of the timing chain (primary) with each sprocket, gradually turn the camshaft using a wrench on the hexagonal portion to align it with the timing marks. During alignment be careful to prevent dislocation of the mating marks alignments of the timing chains (secondary). Note indicates the water pump.**

60. Install the internal chain guide, slack guide and timing chain tensioner (primary).

➡ **Do not over tighten the slack guide bolts. It is normal for a gap to exist under the bolt seats when the bolts are tightened to specification.**

61. When installing the timing chain tensioner (primary), push in the plunger and keep it pressed in with the stopper pin. Remove any dirt on the surfaces. After installation, pull out the stopper pin by pressing the slack guide.

62. Make sure, again, that the mating marks on the camshaft sprockets and timing chain have not slipped out of alignment. Install new O-rings on the rear timing chain case.

63. Install a new front seal in the front timing chain case cover.

64. Install the water pump cover and chain tensioner cover to the front timing chain case cover. Apply a continuous bead of liquid gasket (0.091–0.130 inch diameter) to the front timing chain case cover before installing the water pump cover and chain tensioner cover. Be sure to use genuine RTV sealant, or equivalent.

65. Before installing the front timing chain case cover apply a continuous bead of liquid gasket (0.102–0.142 inch in diameter) to the front timing chain case back side, as shown in the illustration. Be sure to use genuine RTV sealant or equivalent.

66. Install new O-rings on the rear timing chain case. To assemble the front timing chain case cover, fit the lower end of the front timing chain case tightly onto the top face of the oil pan (upper). From the fitting point, make entire front timing chain case contact rear timing chain case completely.

➡ **Since the front timing chain case cover is offset for difference of holt holes; tighten the bolts temporarily while holding the front timing chain case cover from the front and the top. Now insert a dowel pin while holding the front timing chain case cover from the front and the top.**

67. Once the cover is installed, torque the retaining bolts to specification and in the proper sequence. There are four different types of bolts:

   a. Bolt diameter: 0.39 inch. Bolt position: 1–5. Torque to 41 ft. lbs. (56 Nm).

   b. Bolt diameter: 0.24 inch. Bolt position: 6–25. Torque to 9 ft. lbs. (12 Nm).

   c. After all bolts are tightened, retighten them to specification and in the proper sequence.

68. Install the two bolts in the oil pan (upper). Torque to 16 ft. lbs.

69. Install new seal rings in the shaft grooves of the right and left intake valve timing control covers.

70. Apply a continuous bead of liquid gasket (0.083–0.122 inch in diameter) to the covers. Be sure to use genuine RTV sealant or equivalent.

71. Install new collared O-rings in the front timing chain case oil hole (left and right sides). Be careful not to move the seal ring from the installation groove, align the dowel pins on the front timing chain case with the holes to install the intake valve timing control covers.

72. Tighten the bolts in sequence and to specification.

73. Install the crankshaft pulley. Torque to specification.

74. Install the upper and lower oil pans.

75. Install the intake manifold collector.

76. Before installing the rocker cover, apply liquid gasket, be sure to use genuine RTV silicone sealant or equivalent, to the positions shown in the illustration. Refer to figure "a" to apply liquid gasket to joint part of camshaft bracket No. 1 and cylinder head. Refer to figure "b" to apply liquid gasket to the figure "a" squarely.

77. Install the rocker cover. Torque the retaining bolts to 17 inch lbs. and then to 74 inch lbs., in the proper sequence.

78. Continue the installation in the reverse order of the removal procedure.

➡ **If hydraulic pressure inside the timing chain tensioner drops after removal/installation, slack in the guide may generate a pounding noise during and just after engine start. This is normal the noise will stop after hydraulic pressure rises.**

VALVE LASH

ADJUSTMENT

See Figures 60 through 63.

1. Before servicing the vehicle, refer to the Precautions Section.

2. Disconnect the negative battery cable. Remove the engine under cover.

3. Remove the intake manifold collector.

4. Separate the engine harness and remove their brackets from the rocker covers. Remove the harness bracket from the cylinder head, if necessary.

5. Remove the ignition coil. Remove the PCV hoses. Remove the oil filler cap, if necessary.

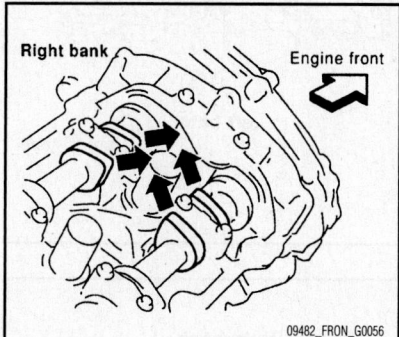

Fig. 60 No. 1 cylinder at TDC (compression stroke)—4.0L engine

6. Loosen the rocker cover retaining bolts, in the reverse order of the tightening sequence.

7. Remove the rocker covers from the engine.

8. Set the No. 1 cylinder at TDC of its compression stroke by rotating the crankshaft pulley clockwise to align the timing mark (grooved line without color) with the timing indicator. Make sure that the intake and exhaust cam noses on No. 1 cylinder (engine front side on right bank) are in alignment as shown in the illustration. If not, rotate the crankshaft in the clockwise direction 360 degrees.

9. Use a feeler gauge and measure the clearance between the valve lifter and the camshaft.

10. With the No. 1 piston at TDC, refer to the illustration and measure the valve clearances at the locations marked with an "X". The "X" locations are indicated in the illustration with an arrow.

11. Rotate the crankshaft pulley clockwise 240 degrees (when viewed from the engine front) to align No. 3 cylinder at TDC on the compression stroke.

➡**The crankshaft pulley bolt flange has a stamped line every 60 degrees, which can be used as a guide to rotation angle.**

12. With the No. 3 piston at TDC, refer to the illustration and measure the valve clearances at the locations marked with an "X". The "X" locations are indicated in the illustration with an arrow.

13. Rotate the crankshaft pulley clockwise 240 degrees (when viewed from the engine front) to align No. 5 cylinder at TDC on the compression stroke.

➡**The crankshaft pulley bolt flange has a stamped line every 60 degrees, which can be used as a guide to rotation angle.**

14. With the No. 5 piston at TDC, refer to the illustration and measure the valve clearances at the locations marked with an "X". The "X" locations are indicated in the illustration with an arrow.

15. If measurements are not within specification, proceed to the next step.

16. Remove the camshaft. Remove the valve lifters that are not within specification.

17. Measure the center thickness of the removed lifters, using a micrometer.

18. Use the equation (t=t1+(C1-C2) to calculate valve lifter thickness for replacement.

➡**t= valve lifter thickness to be replaced. t1= removed valve lifter thickness. C1= measured valve clearance. C2= standard valve clearance.**

Measuring position (right bank)		No. 1 CYL.	No. 3 CYL.	No. 5 CYL.
No. 1 cylinder at compression TDC	EXH		×	
	INT	×		
Measuring position (left bank)		No. 2 CYL.	No. 4 CYL.	No. 6 CYL.
No. 1 cylinder at compression TDC	INT			×
	EXH	×		

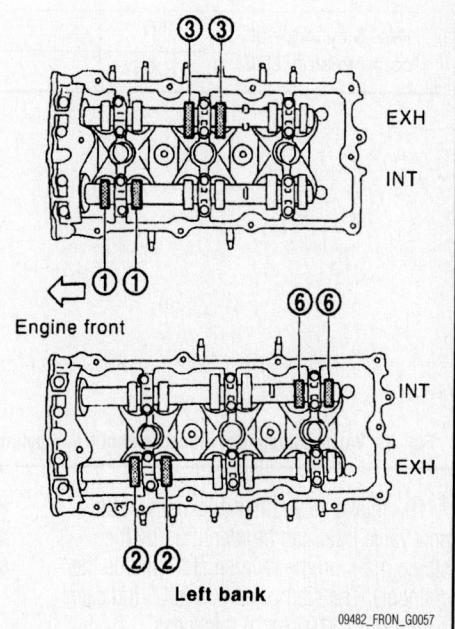

Fig. 61 Valve adjustment measurement No. 1 cylinder at TDC (compression stroke)—4.0L engine

Measuring position (right bank)		No. 1 CYL.	No. 3 CYL.	No. 5 CYL.
No. 3 cylinder at compression TDC	EXH			×
	INT		×	
Measuring position (left bank)		No. 2 CYL.	No. 4 CYL.	No. 6 CYL.
No. 3 cylinder at compression TDC	INT	×		
	EXH		×	

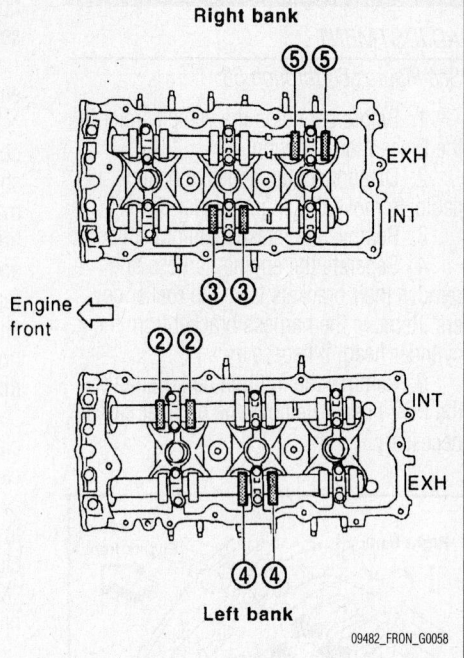

Fig. 62 Valve adjustment measurement No. 3 cylinder at TDC (compression stroke)—4.0L engine

Measuring position (right bank)		No. 1 CYL.	No. 3 CYL.	No. 5 CYL.
No. 5 cylinder at compression TDC	EXH	×		
	INT			×
Measuring position (left bank)		No. 2 CYL.	No. 4 CYL.	No. 6 CYL.
No. 5 cylinder at compression TDC	INT		×	
	EXH			×

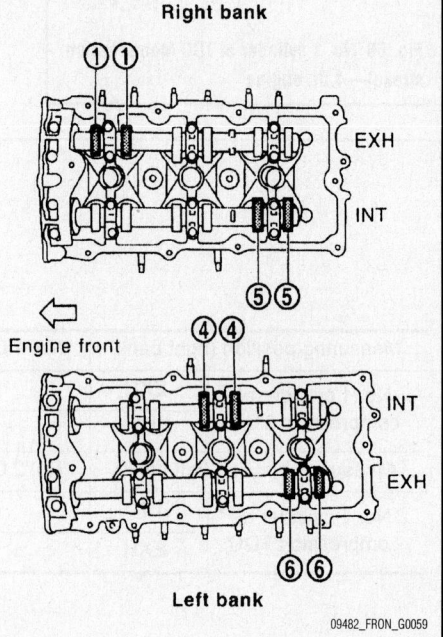

Fig. 63 Valve adjustment measurement No. 5 cylinder at TDC (compression stroke)—4.0L engine

19. Intake valve lifter thickness of the new valve lifter can be identified by the stamp mark on the reverse side (inside the cylinder). The stamp mark "788U" indicates 7.88 mm (0.3102 inch) thickness.

➡**Available thickness of a valve lifter ranges from 7.88–8.40 mm**

(0.3102–0.3307 inch) in steps of 0.02 mm (0.0008 inch). There are 27 different sizes.

20. Exhaust valve lifter thickness of the new valve lifter can be identified by the stamp mark on the reverse side (inside the cylinder). The stamp mark

"N788" indicates 7.88 mm (0.3102 inch) thickness.

➡**Available thickness of a valve lifter ranges from 7.88–8.36 mm (0.3102–0.3291 inch) in steps of 0.02 mm (0.0008 inch). There are 25 different sizes.**

21. Install the selected valve lifters.
22. Install the camshaft.
23. Manually rotate the crankshaft pulley in the clockwise direction a few rotations.
24. Check the valve clearance and be sure it is within specification.
25. When installing the rocker cover,

apply liquid gasket, be sure to use genuine RTV silicone sealant or equivalent, to the positions shown in the illustration. Refer to figure "a" to apply liquid gasket to joint part of camshaft bracket No. 1 and cylinder head. Refer to figure "b" to apply liquid gasket to the figure "a" squarely.

26. Install the rocker cover. Torque the retaining bolts to 17 inch lbs. and then to 74 inch lbs., in the proper sequence.
27. Continue the installation in the reverse of the removal procedure.

ENGINE PERFORMANCE & EMISSION CONTROLS

ACCELERATOR PEDAL POSITION (APP) SENSOR

LOCATION

The Accelerator Pedal Position (APP) sensor is installed on the upper end of the accelerator pedal assembly.

REMOVAL & INSTALLATION

1. Disconnect the negative battery terminal.
2. Disconnect the accelerator position sensor electrical connector.
   - Pull the connector lock back to unlock the connector from the Accelerator Pedal Position sensor
   - Pull up on the connector to disconnect it from the APP sensor
3. Remove the two upper and one lower accelerator pedal nuts
4. Remove the accelerator pedal assembly

> **❊❊ CAUTION**
>
> **Do not disassemble the accelerator pedal assembly. Do not remove the APP sensor from the accelerator pedal bracket. Avoid damage from dropping the accelerator pedal assembly during handling. Keep the accelerator pedal assembly away from water.**

To install:

5. Installation is in the reverse order of removal.
Inspection after installation:
- Check that the accelerator pedal moves smoothly within the specified range
- Check that the accelerator pedal smoothly returns to the original position
- Perform an electrical inspection of the APP sensor

> **❊❊ CAUTION**
>
> **When the harness connector of the APP sensor is disconnected, perform**

Accelerator Pedal Released Position Learning, Accelerator Pedal Released Position Learning and Accelerator Pedal Released Position Learning.

CAMSHAFT POSITION (CMP) SENSOR

LOCATION

See Figure 64.

Refer to the accompanying illustration.

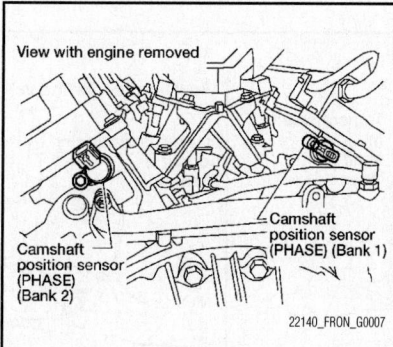

Fig. 64 Locating Camshaft Position (CMP) sensor—4.0L engine

REMOVAL & INSTALLATION

1. Loosen the fixing bolt of the sensor.
2. Disconnect CMP sensor (PHASE) harness connector.
3. Remove the sensor.

To install:

4. Installation is the reverse of the removal procedure.

CRANKSHAFT POSITION (CKP) SENSOR

LOCATION

See Figure 65.

Refer to the accompanying illustration.

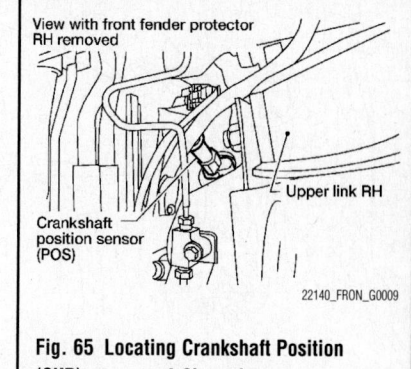

Fig. 65 Locating Crankshaft Position (CKP) sensor—4.0L engine

REMOVAL & INSTALLATION

1. Loosen the fixing bolt of the sensor.
2. Disconnect CKP sensor (POS) harness connector.
3. Remove the sensor.

To install:

4. Installation is the reverse of the removal procedure.

ELECTRONIC CONTROL MODULE (ECM)

LOCATION

See Figure 66.

Refer to the accompanying illustration.

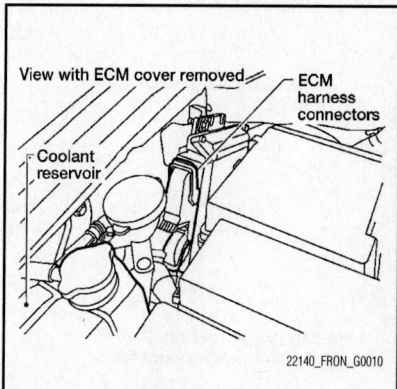

Fig. 66 Locating Electronic Control Module

HEATED OXYGEN (HO2S) SENSOR

LOCATION

See Figures 67 and 68.

Refer to the accompanying illustrations.

REMOVAL & INSTALLATION

> **✴✴ CAUTION**
>
> **Perform the operation with the exhaust system fully cooled. The system will be hot just after the engine stops.**

1. Disconnect sensor harness connector.
2. Remove the sensor using heated oxygen sensor wrench KV10114400 (J-38365) or equivalent.

To install:

3. Installation is the reverse of the removal procedure.

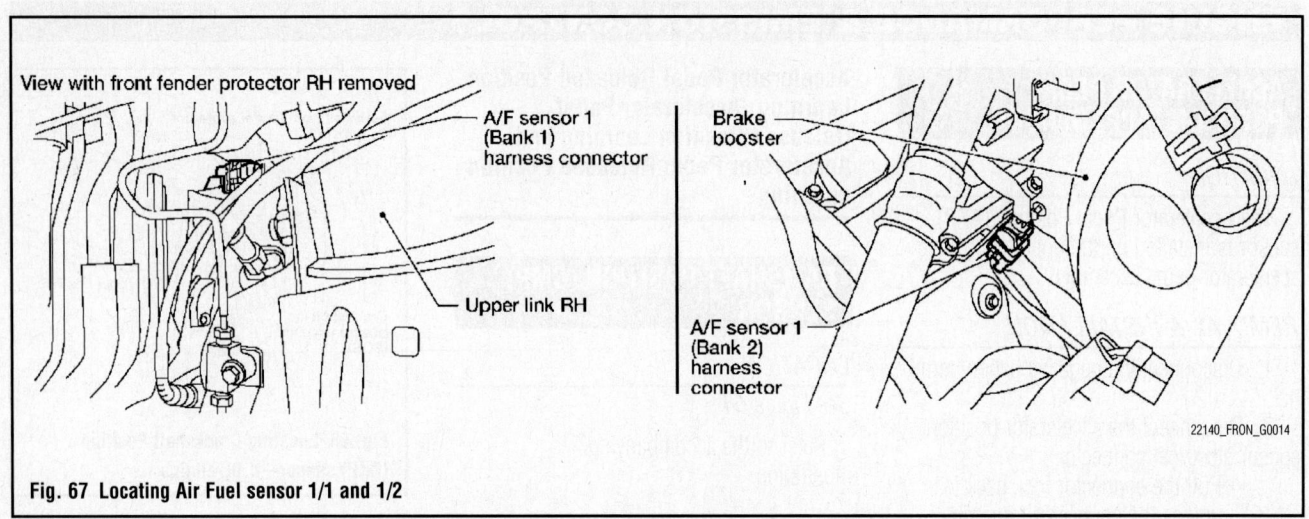

Fig. 67 Locating Air Fuel sensor 1/1 and 1/2

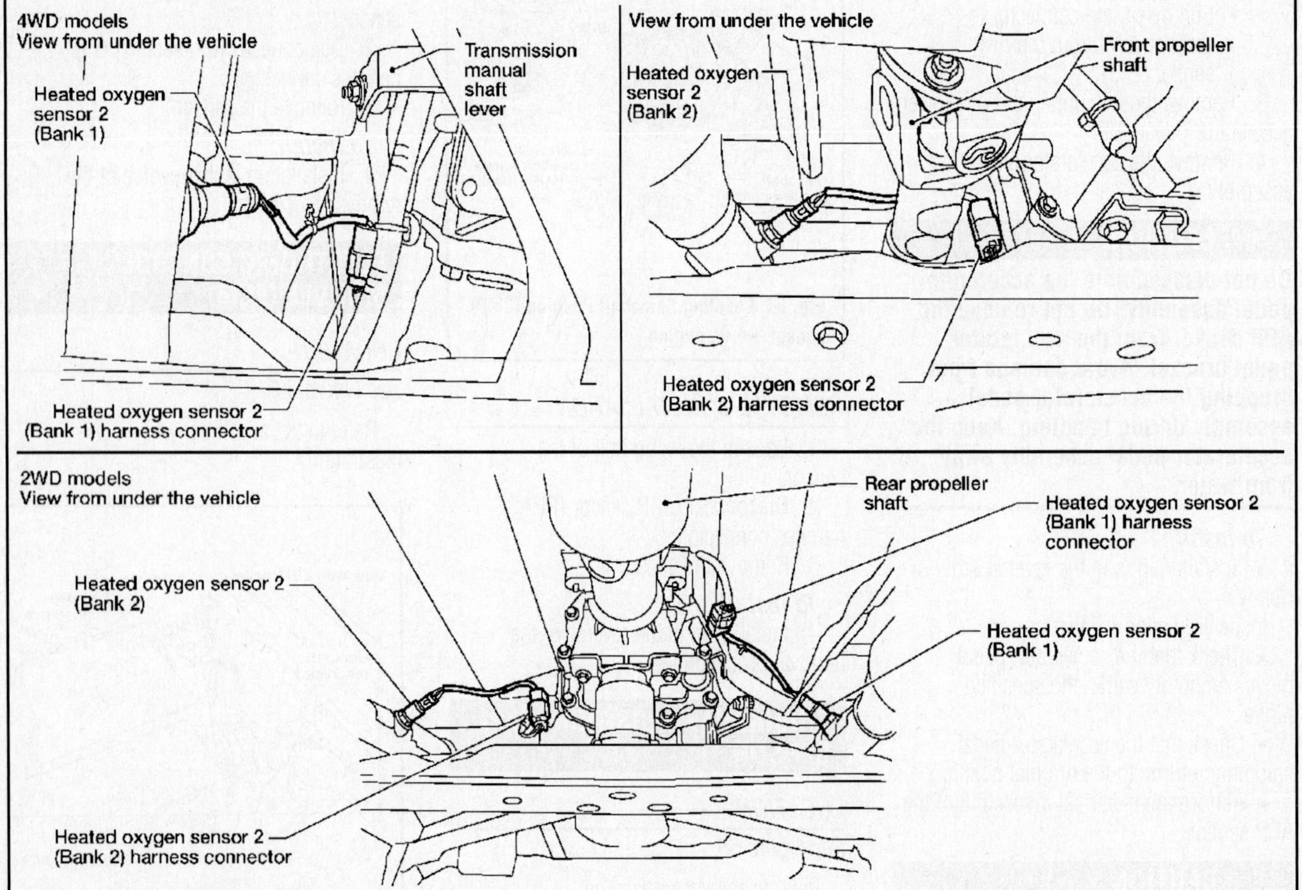

Fig. 68 Locating Heated Oxygen Sensor (HO2S) 2/1 and 2/2

➡Clean exhaust system threads before installing sensor.

INTAKE AIR TEMPERATURE (IAT) SENSOR

LOCATION

The Intake Air Temperature (IAT) sensor is built into Mass Air Flow (MAF) sensor.

REMOVAL & INSTALLATION

See Mass Air Flow (MAF) sensor.

KNOCK SENSOR (KS)

LOCATION

See Figure 69.

Refer to the accompanying illustration.

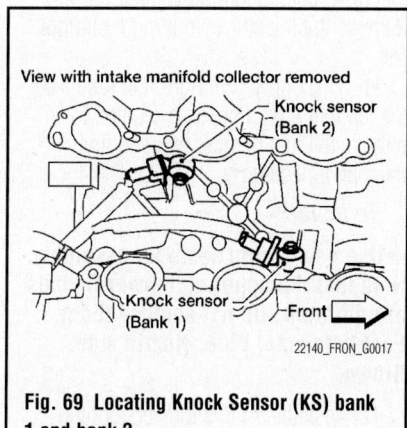

View with intake manifold collector removed

Fig. 69 Locating Knock Sensor (KS) bank 1 and bank 2

REMOVAL & INSTALLATION

1. Remove intake collector.
2. Remove sensor harness. Remove sensor.

   To install:

3. Installation is the reverse of the removal procedure.

➡Use care when installing sensor. Do not use any knock sensors that have been dropped or physically damaged. Use only new ones. Torque sensor properly.

MASS AIR FLOW (MAF) SENSOR

LOCATION

See Figure 70.

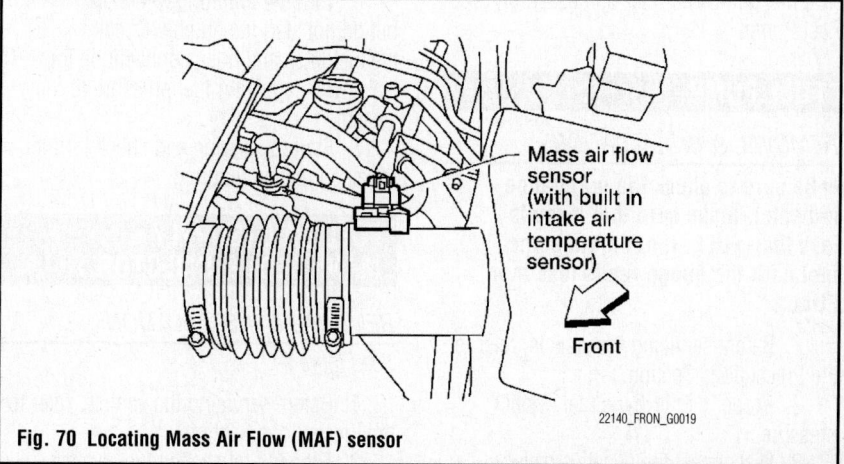

Fig. 70 Locating Mass Air Flow (MAF) sensor

Refer to the accompanying illustration.

THROTTLE POSITION SENSOR (TPS)

LOCATION

Electric Throttle Control actuator consists of Throttle Control motor, Throttle Position (TPS) sensor, etc. The TPS responds to the throttle valve movement. The TPS has two sensors.

REMOVAL & INSTALLATION

The Electric Throttle Control actuator must be replaced as a unit. See Intake Manifold.

FUEL   GASOLINE FUEL INJECTION SYSTEM

FUEL SYSTEM SERVICE PRECAUTIONS

Safety is the most important factor when performing not only fuel system maintenance but any type of maintenance. Failure to conduct maintenance and repairs in a safe manner may result in serious personal injury or death. Maintenance and testing of the vehicle's fuel system components can be accomplished safely and effectively by adhering to the following rules and guidelines.

• To avoid the possibility of fire and personal injury, always disconnect the negative battery cable unless the repair or test procedure requires that battery voltage be applied.

• Always relieve the fuel system pressure prior to disconnecting any fuel system component (injector, fuel rail, pressure reg-

ulator, etc.), fitting or fuel line connection. Exercise extreme caution whenever relieving fuel system pressure to avoid exposing skin, face and eyes to fuel spray. Please be advised that fuel under pressure may penetrate the skin or any part of the body that it contacts.

• Always place a shop towel or cloth around the fitting or connection prior to loosening to absorb any excess fuel due to spillage. Ensure that all fuel spillage (should it occur) is quickly removed from engine surfaces. Ensure that all fuel soaked cloths or towels are deposited into a suitable waste container.

• Always keep a dry chemical (Class B) fire extinguisher near the work area.

• Do not allow fuel spray or fuel vapors to come into contact with a spark or open flame.

• Always use a back-up wrench when loosening and tightening fuel line connection fittings. This will prevent unnecessary stress and torsion to fuel line piping.

• Always replace worn fuel fitting O-rings with new Do not substitute fuel hose or equivalent where fuel pipe is installed.

Before servicing the vehicle, make sure to also refer to the precautions in the beginning of this section as well.

FUEL SYSTEM PRESSURE

RELIEVING

1. Before servicing the vehicle, refer to the Precautions Section.
2. Remove the fuel pump fuse from the panel.

3. Start the engine and allow it to run until it stalls. Crank the engine for a few seconds to relieve additional fuel pressure.

4. Turn ignition switch off.

5. When repairs are complete, replace the fuel pump fuse and connect the negative battery cable.

FUEL FILTER

REMOVAL & INSTALLATION

Fuel Filter is serviced with fuel pump and sending unit assembly. See Fuel Pump.

FUEL LEVEL SENDING UNIT

REMOVAL & INSTALLATION

The fuel level sending unit is serviced with fuel pump and filter unit assembly. See Fuel Pump.

FUEL PUMP

REMOVAL & INSTALLATION

➡ **Be sure to check the fuel gauge indicator. Make sure that it reads less than FULL. If not drain some fuel until the gauge reads less than FULL.**

1. Before servicing the vehicle, refer to the Precautions Section.

2. Properly relieve the fuel system pressure.

3. Disconnect the negative battery cable.

4. Remove the fuel filler cap. Remove the left rear tire and wheel assembly.

5. Disconnect the lower fuel filler hose from the fuel tank, the EVAP hose and the vent pipe quick connector.

➡ **Disconnect the fuel feed hose from the molder clip in the side of the fuel tank.**

6. Remove the four tank shield retaining bolts. Remove the tank shield.

7. Remove the driveshaft.

8. Properly support the fuel tank. Remove the three fuel tank retaining strap bolts. Remove the fuel tank straps.

9. Lower the fuel tank to gain access to the top of the fuel pump assembly.

➡ **Be careful not to lower the tank too much as you do not want to damage**

the fuel feed hose and the fuel pump assembly.

10. Disconnect the fuel pump assembly electrical connector, and the fuel feed hose.

11. Disconnect the quick connector.

12. Lower the fuel tank and remove it from the vehicle. Disconnect the EVAP hose from the fuel pump and remove the EVAP hose from the molded clip in the top of the fuel tank.

13. Remove the fuel pump assembly lockring. Remove the fuel pump assembly. Discard the O-ring.

To install:

14. Installation is the reverse of the removal procedure.

15. Be sure to use a new O-ring upon installation.

16. Turn the ignition switch ON, but do not start the engine. Check the fuel lines and hose connections for leaks while applying fuel pressure to the system.

17. Start the engine and check for fuel leaks, correct as required.

FUEL RAIL (SUPPLY MANIFOLD) & INJECTOR

REMOVAL & INSTALLATION

See Figure 71.

1. Before servicing the vehicle, refer to the Precautions Section.

2. Properly relieve the fuel system pressure.

3. Disconnect the negative battery cable. Remove the fuel filler cap.

4. Remove the intake manifold collector.

5. Remove the quick connector cap (engine side). With the sleeve side of the quick connector release facing the quick connector, install the quick connector release on to the tube. Insert the quick connector release into the quick connector until the sleeve contacts and goes no further. Hold the quick connector release in that position.

➡ **Disconnect the quick connector using tool J-45488, or equivalent, not by picking out the retainer tabs. Inserting the quick connector hard will not disconnect the quick connector. Hold the quick connector release where it contacts and goes no further.**

6. Draw and pull out the quick connector straight from the fuel tube. Grasp the

quick connector holding "A" in the illustration. Do not pull with lateral force applied and the O-ring inside the quick connector could be damaged.

➡ **Have a cloth ready, as fuel will leak out. Avoid fire and sparks. Keep parts away from heat. Do not bend or twist the connection between the quick connector and the fuel feed hose. Cover the openings with a plastic bag.**

7. Remove the PCV hose between the rocker covers.

8. Disconnect the harness for the fuel injector.

9. Loosen the retaining bolts. Remove the fuel tube and fuel injector assembly. Remove the bolts which connect the left and right fuel tubes.

10. To remove the fuel injectors from the fuel tube, open and remove the clip. Remove the injector by pulling it straight out.

11. Disconnect the right fuel tube from the left fuel tube. Loosen the bolts, to remove the fuel damper cap and fuel damper, if necessary.

To install:

➡ **Use new O-ring seals for assembly. Note that the upper and lower O-rings are different. Do not confuse them. Fuel tube side: Blue. Nozzle side: Brown.**

12. Installation is the reverse of the removal procedure.

13. When installing the fuel feed tube be sure to torque the retaining bolts to 7 ft. lbs and then to 16 ft. lbs. in an alternating order.

14. Turn the ignition switch ON, but do not start the engine. Check the fuel lines and hose connections for leaks while applying fuel pressure to the system.

15. Start the engine and check for fuel leaks, correct as required.

FUEL TANK

REMOVAL & INSTALLATION

See Fuel Pump.

IDLE SPEED

ADJUSTMENT

Idle speed is maintained by the Powertrain Control Module (PCM). No adjustment is necessary or possible.

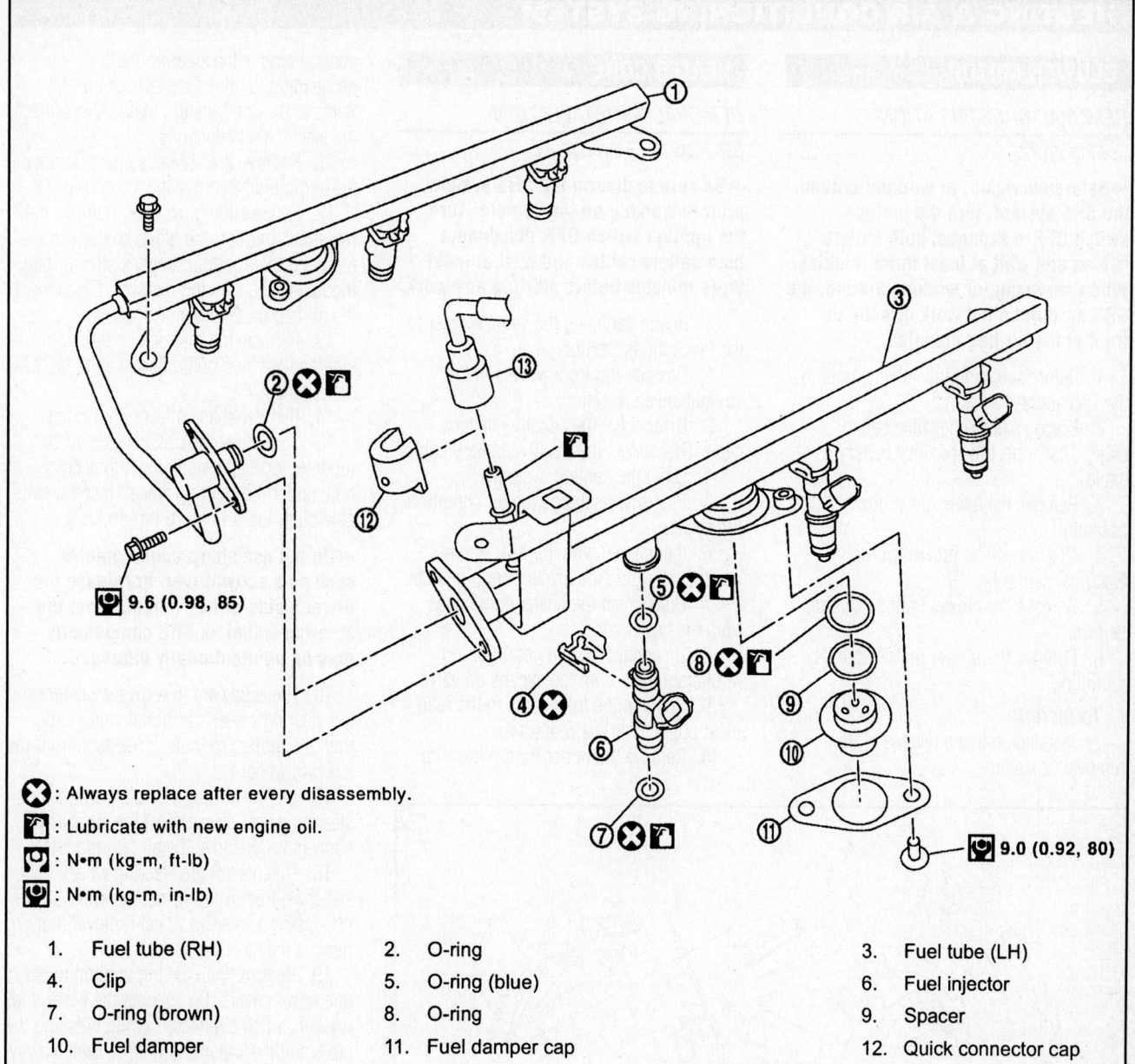

: Always replace after every disassembly.

: Lubricate with new engine oil.

: N•m (kg-m, ft-lb)

: N•m (kg-m, in-lb)

1. Fuel tube (RH)	2. O-ring	3. Fuel tube (LH)
4. Clip	5. O-ring (blue)	6. Fuel injector
7. O-ring (brown)	8. O-ring	9. Spacer
10. Fuel damper	11. Fuel damper cap	12. Quick connector cap
13. Fuel feed hose		

09482_FRON_G0097

Fig. 71 Fuel injector tube and related components

HEATING & AIR CONDITIONING SYSTEM

BLOWER MOTOR

REMOVAL & INSTALLATION
See Figure 72.

➡ **Before servicing, or working around, the SRS system, turn the ignition switch OFF, disconnect both battery cables and wait at least three minutes. When servicing, or working around, the SRS system do not work directly in front of the air bag module.**

1. Before servicing the vehicle, refer to the Precautions Section.
2. Disconnect the negative battery cable. Disconnect the positive battery cable.
3. Remove the lower glove box assembly.
4. Disconnect the blower motor electrical connector.
5. Remove the blower motor retaining screws.
6. Remove the blower motor from its mounting.

To install:
7. Installation is the reverse of the removal procedure.

HEATER CORE

REMOVAL & INSTALLATION
See Figures 73 through 79.

➡ **Be sure to disarm the SRS system, prior to working on the vehicle. Turn the ignition switch OFF, disconnect both battery cables and wait at least three minutes before starting any work.**

1. Before servicing the vehicle, refer to the Precautions Section.
2. Position the front wheels in the straight ahead direction.
3. Disconnect the negative battery cable. Disconnect the positive battery cable.
4. Drain the cooling system.
5. Properly discharge the air conditioning system.
6. If equipped with the 4.0L engine, remove the right side heater core pipe nuts.
7. Disconnect the heater core hoses from the heater core.
8. Disconnect the air conditioning refrigerant lines from the expansion valve.
9. Position the front seats in the rearmost position on the seat tracks.
10. Remove the upper front pillar trim panel. Remove the steering lock escutcheon. Remove the cluster lid "A". Remove the combination meter. Disconnect the electrical connections.
11. Remove the optical sensor. Remove the audio unit. Remove the cluster lid "D".
12. Remove the glove box. Remove the two bolts, through the glove box opening, retaining the front passenger's side air bag module to the steering member. Disconnect the air bag module connectors.
13. Remove the instrument stay right side and left side bolts. Remove the instrument panel.
14. Remove the two front floor ducts.
15. To remove the driver's side air bag module, locate the retaining clip access hole under the steering wheel. Insert a suitable blunt tool (4 mm–6 mm in size).

➡ **Do not use sharp edged objects, such as a screwdriver, to release the driver's side airbag module from the steering wheel as SRS components may be unintentionally damaged.**

16. Press upward, toward the center of the steering wheel, on the retaining clip until the air bag module is released from the steering wheel.
17. Lift the air bag module from the steering wheel. Disconnect the electrical connectors. Remove the air bag module.
18. Disconnect the steering wheel switches. Remove the steering wheel center nut. Using a steering wheel removal tool, remove the steering wheel.
19. Remove the steering column upper and lower covers. Disconnect the wiper and washer switch connector. While pressing the tabs, pull the wiper and washer switch away from the spiral cable to remove it.
20. Disconnect the light and turn signal switch connector. While pressing the tabs, pull the light and turn signal switch toward the driver's door to remove it.
21. Remove the screws. While pressing the tab, pull the spiral cable away from the steering column assembly. Disconnect the electrical connectors.

➡ **With the steering linkage disconnected, the spiral cable may snap by turning the steering wheel beyond the limited number of turns. The spiral cable can be turned counterclockwise about 2.5 turns from the neutral position.**

22. Remove the lower knee protector.
23. Remove the locknut and bolt from the upper joint and then separate the upper joint from the upper shaft.

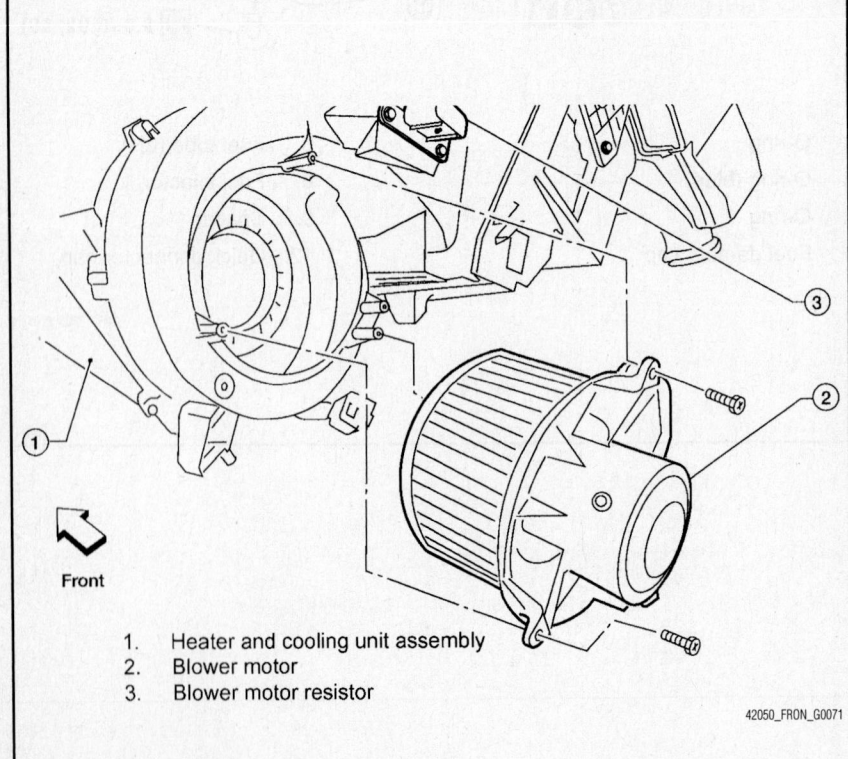

Front

1. Heater and cooling unit assembly
2. Blower motor
3. Blower motor resistor

42050_FRON_G0071

Fig. 72 Blower motor an related components

24. Remove the three nuts and bolt from the steering column and then remove the steering column assembly from the steering member.

25. Remove the hole cover seal and clamp. Remove the hole cover nuts, remove the hole cover from the dash panel.

26. Remove the bolt from the lower joint of the lower joint shaft and remove the lower joint shaft from the vehicle.

27. Disconnect the instrument panel wire harness at the right and left in-line connector brackets, and the fuse block (SMJ) electrical connectors.

28. Remove the covers and then remove the three steering member bolts from each side to disconnect the steering member from the vehicle body.

29. Remove the heater/evaporator case assembly with it attached to the steering member from the vehicle.

30. Separate the steering member from the heater/evaporator unit.

31. Remove the heater cover retaining screws. Remove the cover.

32. Remove the heater core and the evaporator pipe bracket. Remove the heater core.

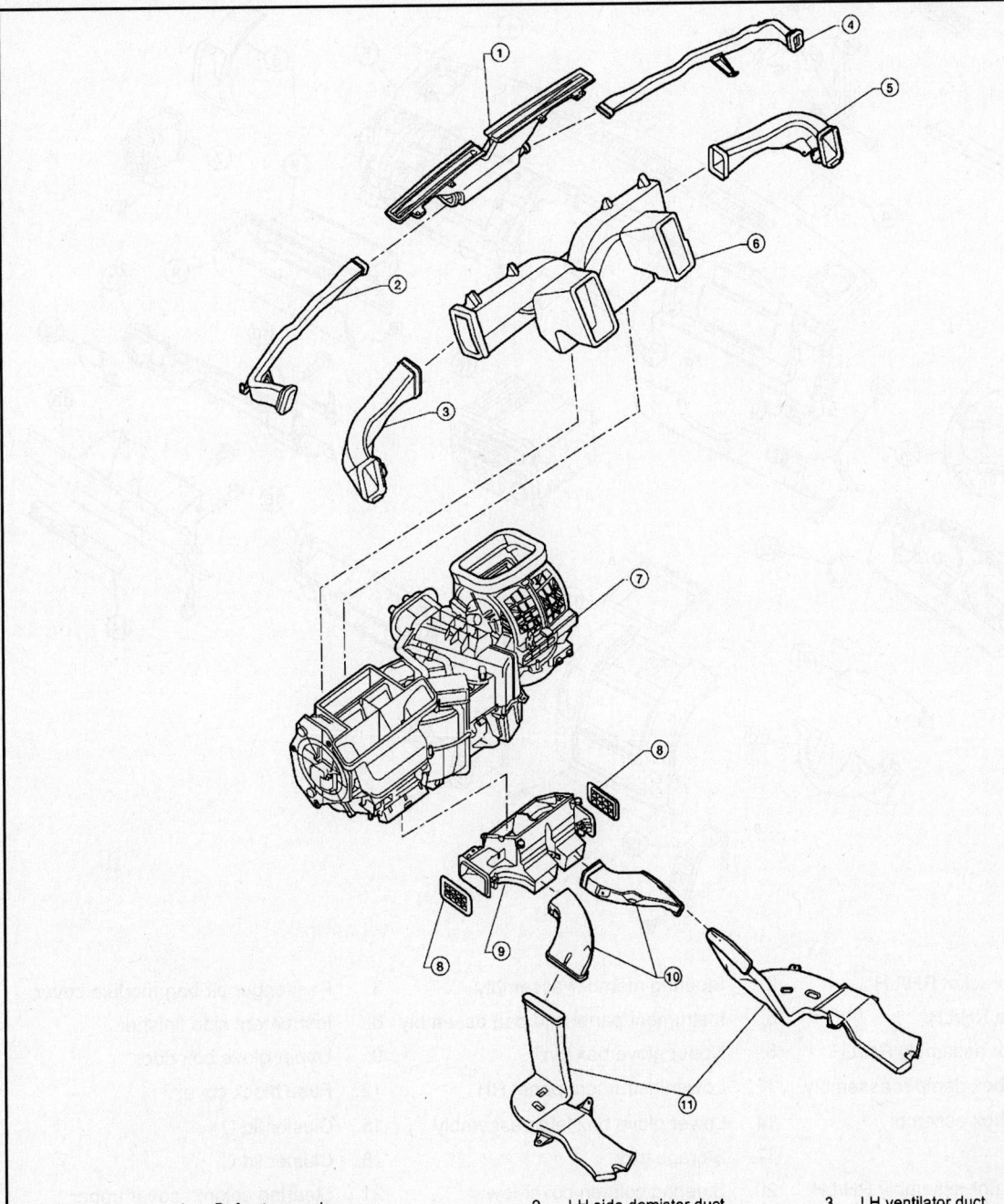

1. Defroster nozzle
4. RH side demister duct
7. Front heater and cooling unit assembly
10. Front floor ducts
2. LH side demister duct
5. RH ventilator duct
8. Floor connector duct grilles
11. Rear floor ducts
3. LH ventilator duct
6. Center ventilator duct
9. Floor connector duct

42050_FRON_G0072

Fig. 73 Duct work surrounding the heater/cooling unit

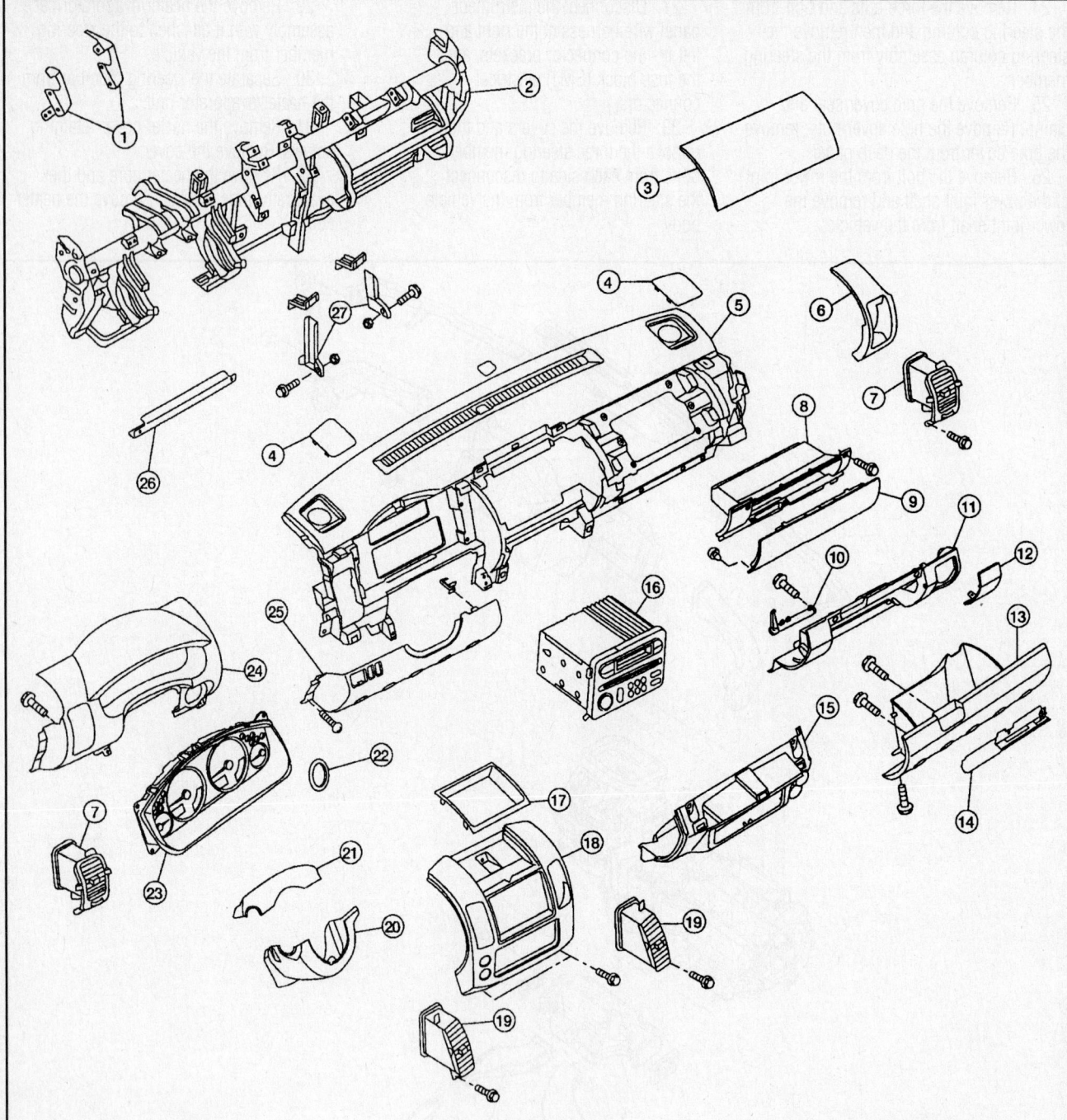

1. Display unit bracket RH/LH	2. Steering member assembly	3. Passenger air bag module cover
4. Speaker grille RH/LH	5. Instrument panel and pad assembly	6. Instrument side finisher
7. Side ventilator assembly RH/LH	8. Upper glove box bin	9. Upper glove box door
10. Lower glove box damper assembly	11. Lower instrument panel RH	12. Fuse block cover
13. Lower glove box assembly	14. Lower glove box latch assembly	15. Cluster lid D
16. Audio unit	17. Storage tray	18. Cluster lid C
19. Center ventilator assembly RH/LH	20. Steering column cover lower	21. Steering column cover upper
22. Steering lock escutcheon	23. Combination meter	24. Cluster lid A
25. Lower instrument panel LH	26. Knee protector brace	27. Instrument stay RH/LH

09482_FRON_G0009

Fig. 74 Instrument panel and related components

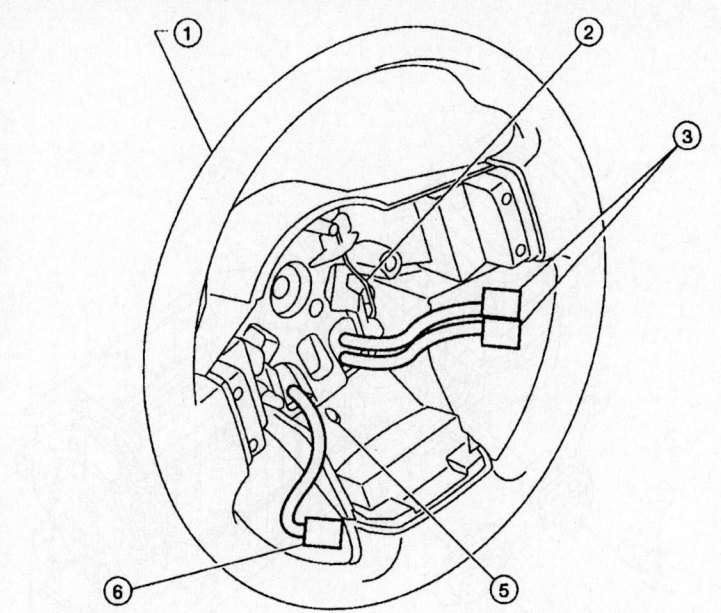

1. Steering wheel
2. Retaining clip
3. Driver air bag module connectors
4. Driver air bag module
5. Retaining clip access hole
6. Horn connector

09482_FRON_G0010

Fig. 75 Driver's side air bag module and related components

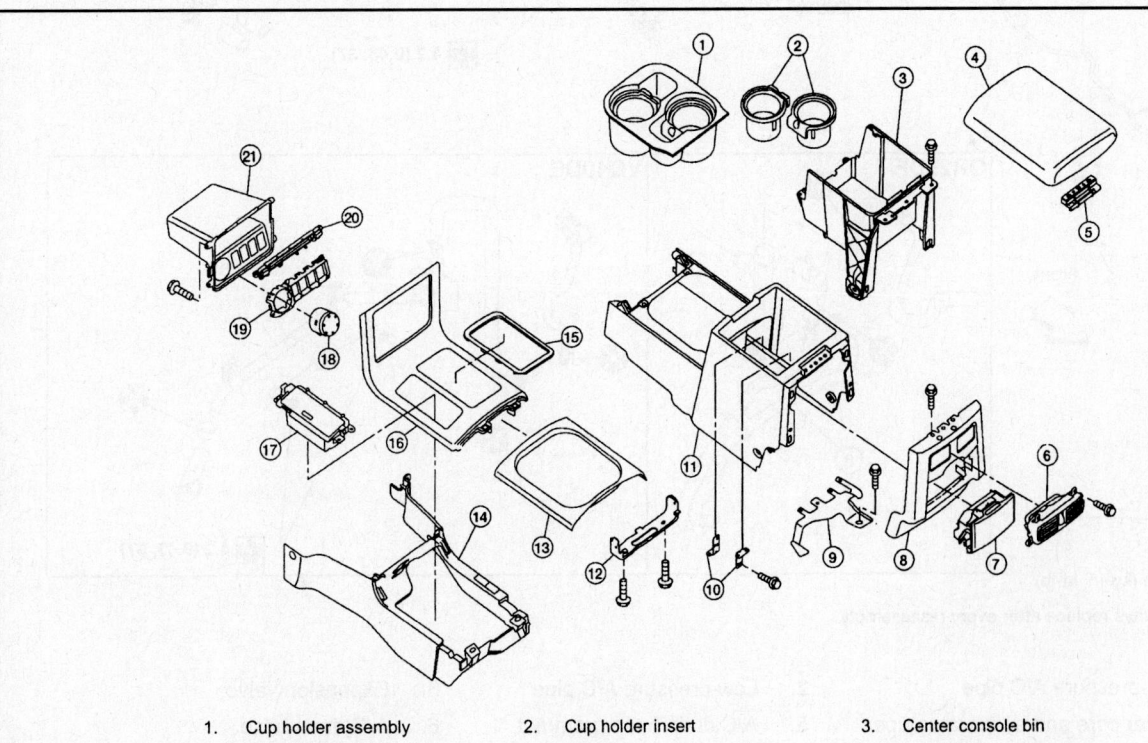

1. Cup holder assembly
2. Cup holder insert
3. Center console bin
4. Center console lid
5. Hinge
6. Ventilator console grille
7. Rear cup holder assembly
8. Rear finisher assembly
9. Wire harness bracket
10. Bracket DVD
11. Center console rear base
12. Bracket
13. Cup holder finisher
14. Center console front base
15. A/T finisher bezel
16. A/T finisher
17. Ash tray
18. Switch assembly
19. Switch finisher
20. CD changer door
21. Console bin

09482_FRON_G0011

Fig. 76 Center console and related components

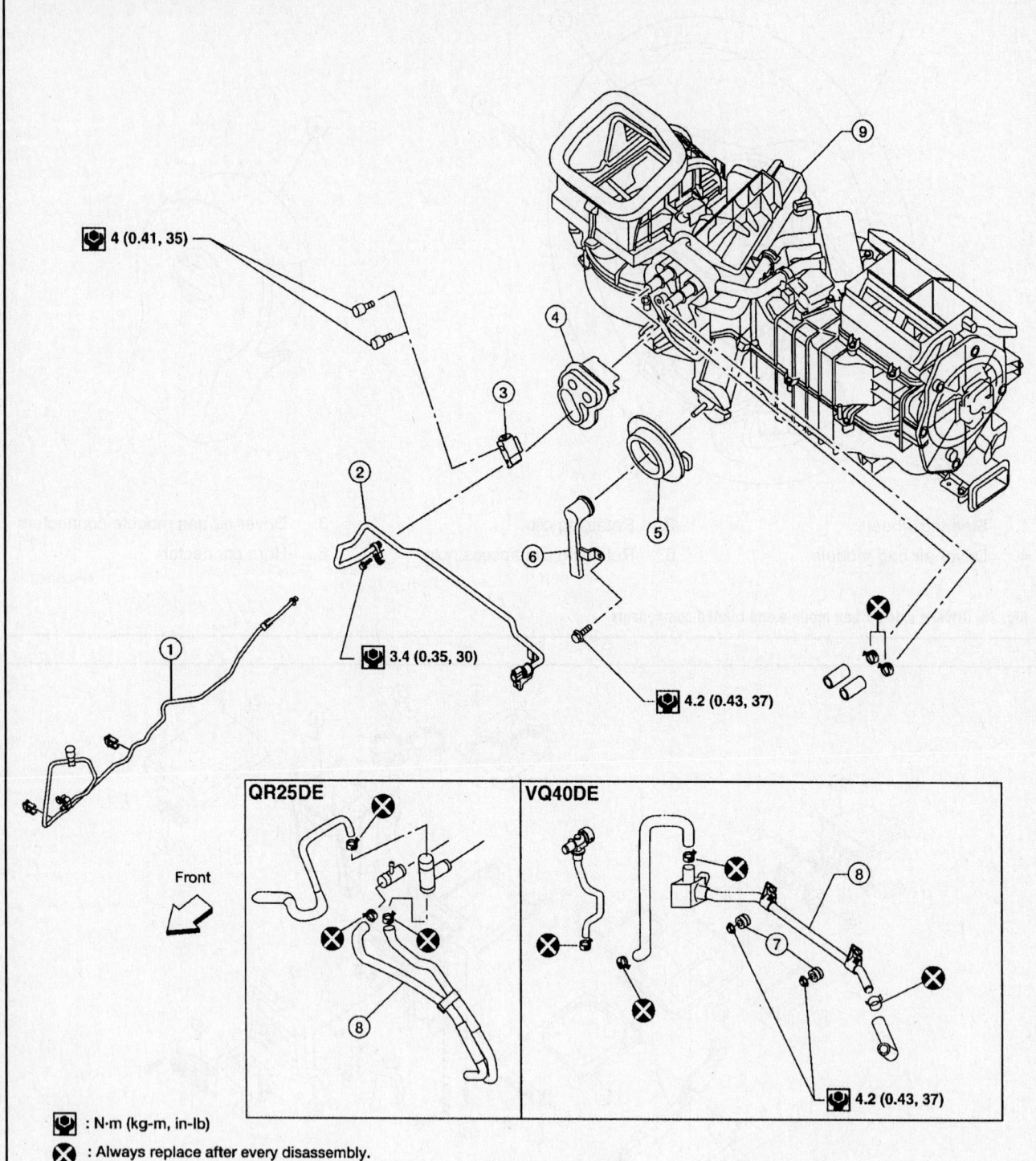

4 (0.41, 35)

3.4 (0.35, 30)

4.2 (0.43, 37)

QR25DE

VQ40DE

Front

4.2 (0.43, 37)

: N·m (kg-m, in-lb)

: Always replace after every disassembly.

1. High-pressure A/C pipe
2. Low-pressure A/C pipe
3. Expansion valve
4. Heater core and evaporator pipes grommet
5. A/C drain hose grommet
6. A/C drain hose
7. Heater core pipe mounts
8. Heater core pipes
9. Heater and cooling unit assembly

09482_FRON_G0012

Fig. 77 Heater/evaporator core and related components

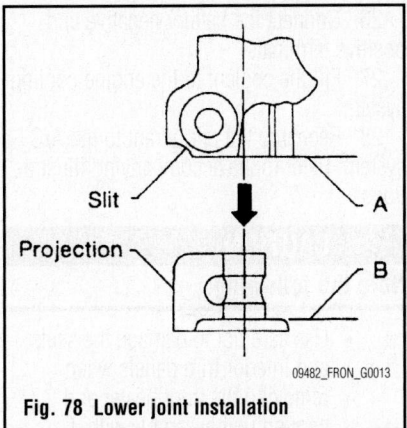

Fig. 78 Lower joint installation

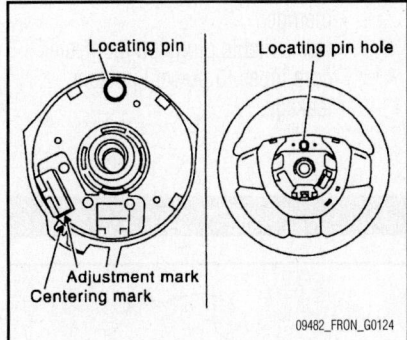

Fig. 79 Spiral cable installation and locating point

To install:

33. Installation is the reverse of the removal procedure.

➡**If the in-cabin microfilters are contaminated with coolant, replace them.**

34. Be sure to use new steering column retaining bolts and pinch bolt, as required.

➡**When installing the steering column, finger tighten all of the lower bracket and joint bolts and then tighten them to specification. Do not apply undue stress to the steering column.**

35. With the wheels in the straight ahead position align the slit of the lower joint with the projection on the dust cover. Insert the joint until surface "A" contacts surface "B".

36. Be sure to align the spiral cable correctly when installing the steering wheel. Make sure that the cable is in the neutral position. The neutral position is detected by turning left 2.6 revolutions from the right end position and ending with the locating pin at the top.

37. To adjust the steering angle sensor neutral position, position the steering wheel in the straight ahead position and rive the vehicle at 10 mph or more for ten minutes. When the procedure is complete, the SLP indicator lamp and the VDC OFF indicator lamp will turn off.

38. Be sure to fill the cooling system with the proper grade and type coolant.

39. Be sure to recharge the air conditioning system.

40. Check and adjust the front end alignment, as necessary.

HEATER & COOLING UNIT ASSEMBLY

REMOVAL & INSTALLATION

See Figures 80 through 82.

1. Discharge the refrigerant from the A/C system. Refer to the accompanying illustration.

2. Drain the coolant from the engine cooling system.

3. Disconnect the battery negative and positive terminals.

4. Remove the front heater core pipes RH nut.

5. Disconnect the front heater core hoses from the front heater core.

6. Disconnect the high- and low-pressure A/C pipes from the front expansion valve.

7. Move the two front seats to the rear most position on the seat track.

8. Remove the instrument panel and console panel.

9. Remove the two front floor ducts.

10. Remove the steering column.

11. Disconnect the instrument panel wire harness at the RH and LH in-line connector brackets, and the fuse block (SMJ) electrical connectors.

12. Remove the covers then remove the three steering member bolts from each side to disconnect the steering member from the vehicle body.

13. Remove the front heater and cooling unit assembly with it attached to the steering member, from the vehicle.

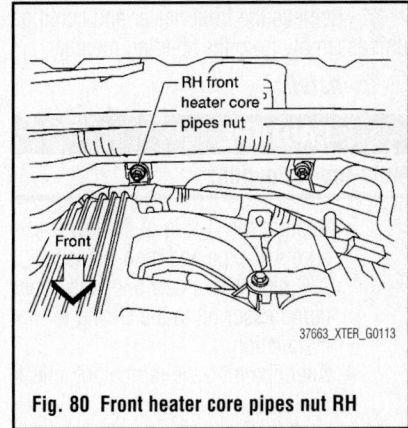

Fig. 80 Front heater core pipes nut RH

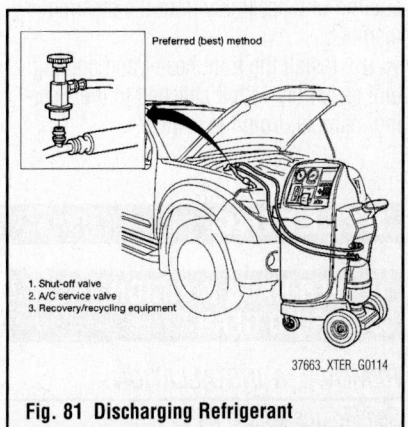

Fig. 81 Discharging Refrigerant

✳✳ CAUTION
Note the following:

- Use care not to damage the seats and interior trim panels when removing the front heater and cooling unit assembly with it attached to the steering member.
- Use suitable plugs on the heater core pipes to prevent coolant leakage.

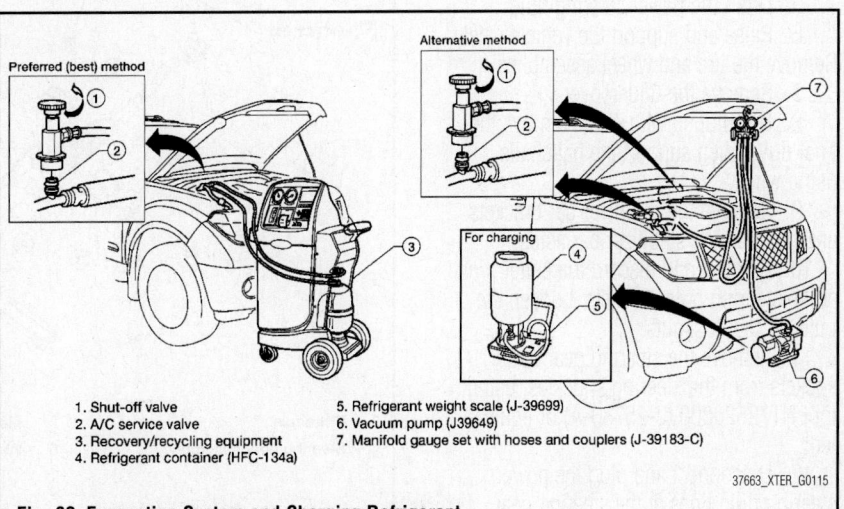

1. Shut-off valve
2. A/C service valve
3. Recovery/recycling equipment
4. Refrigerant container (HFC-134a)
5. Refrigerant weight scale (J-39699)
6. Vacuum pump (J39649)
7. Manifold gauge set with hoses and couplers (J-39183-C)

Fig. 82 Evacuating System and Charging Refrigerant

14. Remove the front heater and cooling unit assembly from the steering member.

To install:

- Replace the O-ring of the low-pressure A/C pipe and high-pressure A/C pipe with a new one, and apply compressor oil to the O-ring for installation.
- After charging the refrigerant, check for leaks.

15. Install the front heater and cooling unit assembly from the steering member.

16. Install the front heater and cooling unit assembly with it attached to the steering member, from the vehicle.

17. Install the covers then install the three steering member bolts from each side to disconnect the steering member from the vehicle body.

18. Connect the instrument panel wire harness at the RH and LH in-line connector brackets, and the fuse block (SMJ) electrical connectors.

19. Install the steering column.

20. Install the two front floor ducts.

21. Install the instrument panel and console panel.

22. Move the two front seats to the rear most position on the seat track.

23. Connect the high- and low-pressure A/C pipes from the front expansion valve.

24. Connect the front heater core hoses from the front heater core.

25. Install the front heater core pipes RH nut.

26. Connect the battery negative and positive terminals.

27. Fill the coolant to the engine cooling system.

28. Recharge the refrigerant to the A/C system. Refer to the accompanying illustrations.

- Use care not to damage the seats and interior trim panels when removing the front heater and cooling unit assembly with it attached to the steering member.
- Use suitable plugs on the heater core pipes to prevent coolant leakage.

STEERING

POWER RACK & PINION STEERING GEAR

REMOVAL & INSTALLATION

See Figures 83 and 84.

➡ **The spiral cable may snap due to steering operation if the steering column is separated from the steering gear assembly. Be sure to secure the steering wheel to avoid turning.**

1. Before servicing the vehicle, refer to the Precautions Section.

2. Position the front wheels in the straight ahead position.

3. Disarm the SRS system.

4. Disconnect the negative battery cable.

5. Drain the power steering fluid.

6. Raise and support the vehicle safely. Remove the tire and wheel assemblies.

7. Remove the undercover.

8. If equipped with 4WD, remove the final drive, then support the halfshafts, using wire.

9. Remove the stabilizer bar brackets, and position the stabilizer bar aside.

10. Remove and discard the cotter pins at the steering outer sockets. Loosen the outer socket locknuts.

11. Remove the steering gear outer sockets from the steering knuckles, using tool HT72520000 (J-25730-A) or equivalent.

12. Disconnect and plug the power steering fluid lines at the steering gear.

13. Remove the bolt from the lower joint of the lower joint assembly. Separate the lower joint from the steering gear assembly. Be careful not to damage the lower joint.

14. Remove the steering gear retaining nuts and bolts. Remove the steering gear from the vehicle.

To install:

15. With the steering wheel in the straight ahead position, align the slit of the lower joint with the projection on the dust cover. Insert the joint until both surfaces contact each other.

16. Continue the installation in the reverse order of the removal procedure.

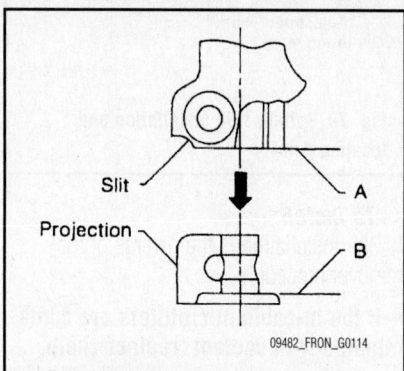

09482_FRON_G0114

Fig. 84 Power steering gear lower joint installation alignment

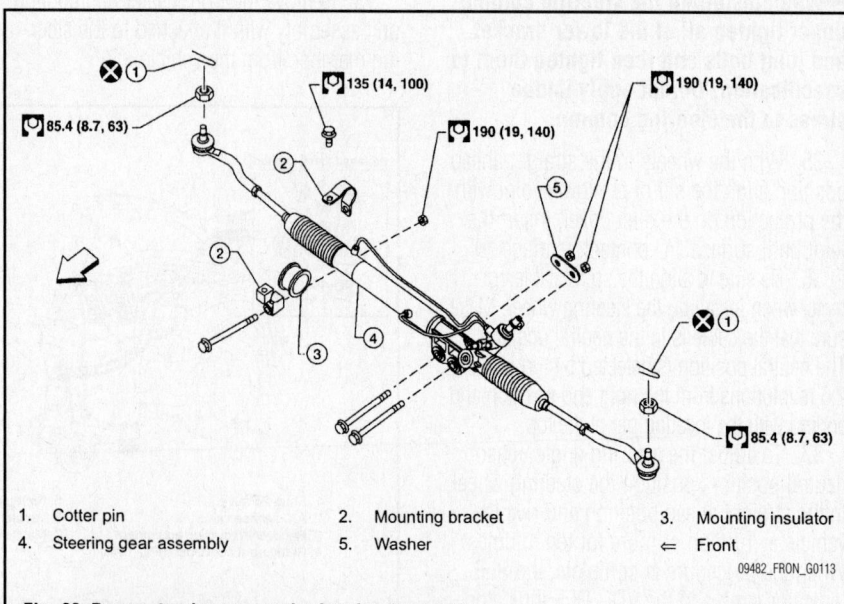

1. Cotter pin
2. Mounting bracket
3. Mounting insulator
4. Steering gear assembly
5. Washer
⇐ Front

09482_FRON_G0113

Fig. 83 Power steering gear and related components

17. Check and adjust the front alignment, as required.

18. Bleed the power steering system.

19. Fill the power steering pump with the proper grade and type fluid.

➡ **After removing/installing or replacing steering and suspension components which effect wheel alignment or after adjusting wheel alignment, or the steering angle sensor or the ABS actuator electrical unit be sure to adjust the neutral position of the steering angle sensor before running the vehicle.**

20. Position the steering wheel in the straight ahead position.

21. When this procedure is complete the SLP indicator lamp and the VDC OFF indicator lamp will turn off.

POWER STEERING PUMP

REMOVAL & INSTALLATION

See Figures 85 and 86.

1. Before servicing the vehicle, refer to the Precautions section.

2. Disconnect the negative battery cable.

3. Drain the power steering fluid from the reservoir tank. Properly dispose of used fluid.

4. On the 4.0L engine, remove the engine cover.

5. Remove the air duct assembly.

6. Remove the drive belt.

7. Disconnect the pressure sensor electrical connector.

8. Disconnect and plug the fluid lines.

9. Remove the pump retaining bolts.

10. Remove the pump from the vehicle.

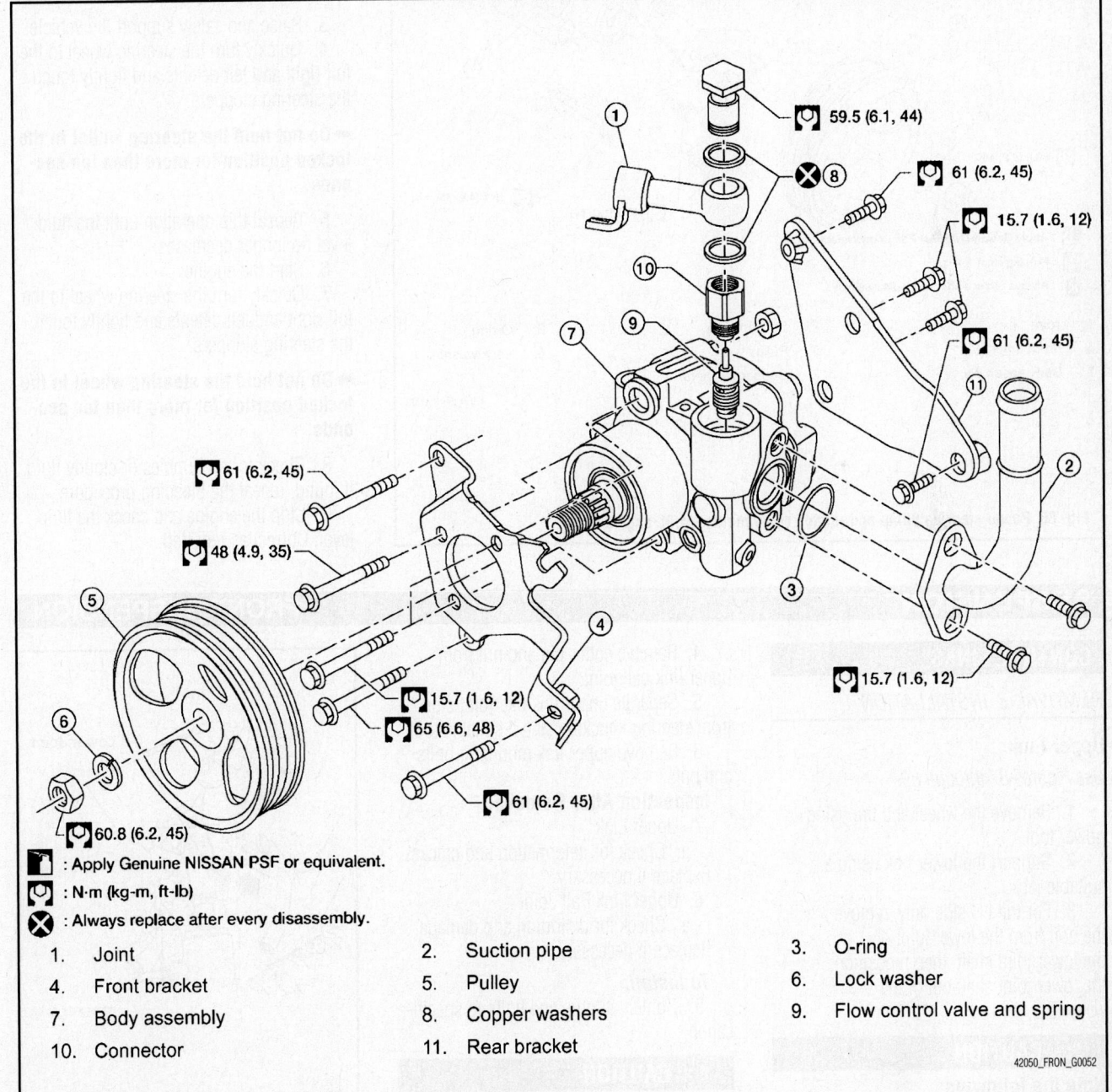

59.5 (6.1, 44)
61 (6.2, 45)
15.7 (1.6, 12)
61 (6.2, 45)
15.7 (1.6, 12)
61 (6.2, 45)
48 (4.9, 35)
15.7 (1.6, 12)
65 (6.6, 48)
61 (6.2, 45)
60.8 (6.2, 45)

🛢 : Apply Genuine NISSAN PSF or equivalent.

🔧 : N·m (kg-m, ft-lb)

✖ : Always replace after every disassembly.

1.	Joint	2.	Suction pipe	3.	O-ring
4.	Front bracket	5.	Pulley	6.	Lock washer
7.	Body assembly	8.	Copper washers	9.	Flow control valve and spring
10.	Connector	11.	Rear bracket		

42050_FRON_G0052

Fig. 85 Power steering pump and related components—2.5L engine

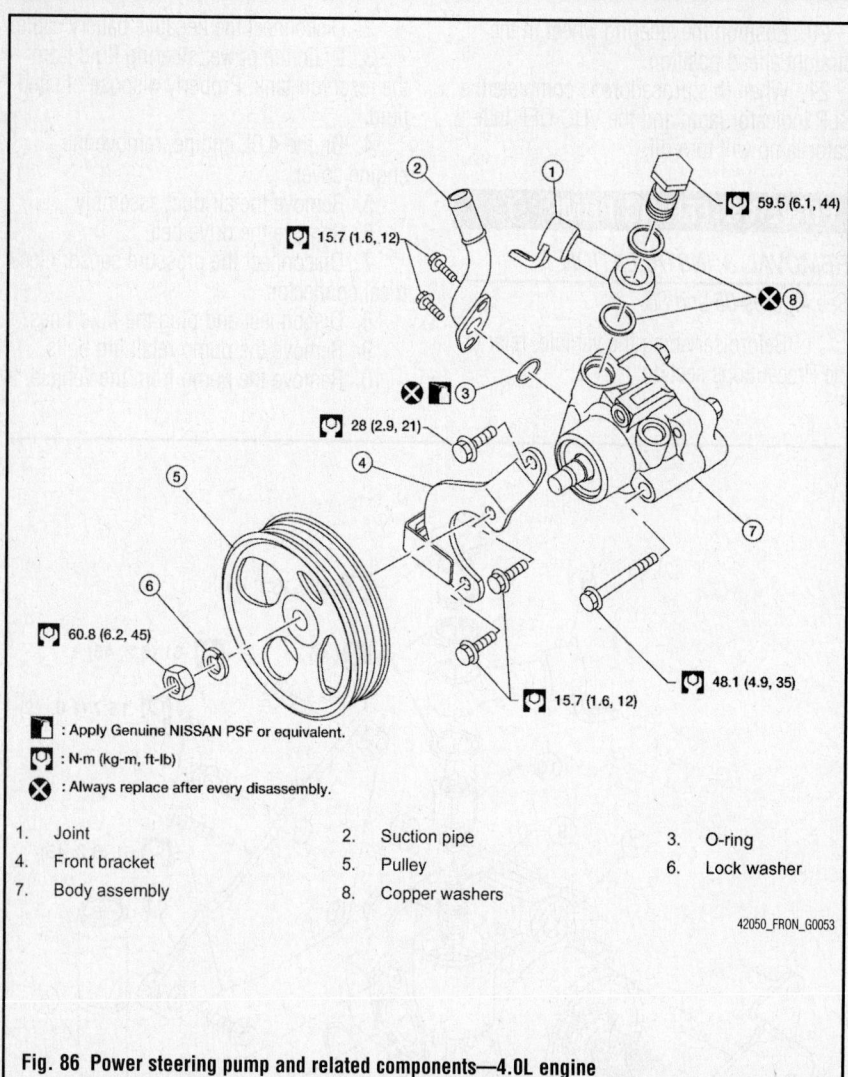

15.7 (1.6, 12)

59.5 (6.1, 44)

28 (2.9, 21)

60.8 (6.2, 45)

15.7 (1.6, 12)

48.1 (4.9, 35)

■ : Apply Genuine NISSAN PSF or equivalent.

◘ : N·m (kg-m, ft-lb)

✖ : Always replace after every disassembly.

1. Joint
2. Suction pipe
3. O-ring
4. Front bracket
5. Pulley
6. Lock washer
7. Body assembly
8. Copper washers

42050_FRON_G0053

Fig. 86 Power steering pump and related components—4.0L engine

To install:

11. Installation is the reverse of the removal procedure.

12. Bleed the power steering system.

BLEEDING

1. Before servicing the vehicle, refer to the Precautions Section.

2. Fill the power steering system with the proper grade and type steering fluid.

➡**Do not allow the fluid level in the reservoir tank to go below the MIN level line. Check and add fluid as needed.**

3. Raise and safely support the vehicle.

4. Quickly turn the steering wheel to the full right and left detents and lightly touch the steering stoppers.

➡**Do not hold the steering wheel in the locked position for more than ten seconds.**

5. Repeat this operation until the fluid level no longer decreases.

6. Start the engine.

7. Quickly turn the steering wheel to the full right and left detents and lightly touch the steering stoppers.

➡**Do not hold the steering wheel in the locked position for more than ten seconds.**

8. Check for air bubbles or cloudy fluid. If found, repeat the bleeding procedure.

9. Stop the engine and check the fluid level. Correct as required.

SUSPENSION

FRONT SUSPENSION

CONTROL LINKS

REMOVAL & INSTALLATION

Upper Link

See Figures 87 through 89.

1. Remove the wheel and tire using power tool.

2. Support the lower link using a suitable jack.

3. For the LH side only, remove the bolt from the lower joint of the lower joint shaft, then reposition the lower joint shaft out of the way.

✳✳ CAUTION

Note the following:

• Do not damage the lower joint.

4. Remove cotter pin and nut from upper link ball joint.

5. Separate upper link ball joint stud from steering knuckle using a suitable tool.

6. Remove upper link mounting bolts and nuts.

Inspection After Removal

7. Upper Link
   a. Check for deformation and cracks. Replace if necessary.

8. Upper Link Ball Joint
   a. Check for distortion and damage. Replace if necessary.

To install:

9. Tighten all nuts and bolts to specification.

✳✳ CAUTION

Always replace drive shaft lock nut and cotter pin.

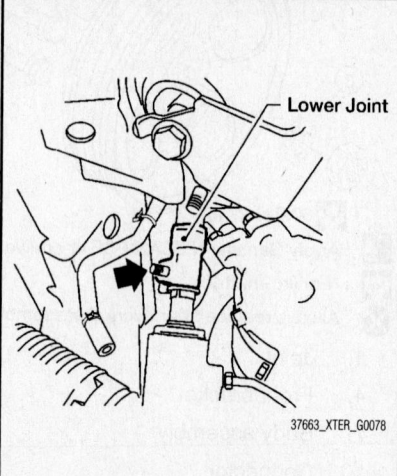

Lower Joint

37663_XTER_G0078

Fig. 87 For the left-hand side only, remove the bolt from the lower joint of the lower joint shaft

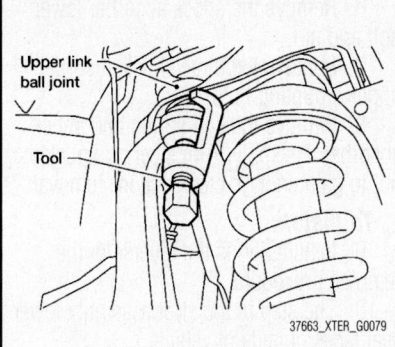

Fig. 88 Removal of the cotter pin and nut from upper link ball joint. Separate upper link ball joint stud from steering knuckle using tool number: ST290200001 (J-24319-01)

Fig. 89 Remove upper link mounting bolts and nuts

- After installation, check that the front wheel alignment is within specification.

10. Install upper link mounting bolts and nuts.

11. Install upper link ball joint stud to steering knuckle using Tool.

12. Install cotter pin and nut from upper link ball joint.

✳✳ CAUTION

Note the following:

- Do not damage the lower joint.

13. For the LH side only, install the bolt from the lower joint of the lower joint shaft, then reposition the lower joint shaft.

14. Support the lower link using a suitable jack.

15. Install the wheel and tire using power tool.

Lower Link

See Figures 90 through 92.

1. Remove the wheel and tire using power tool.

---

2. Remove lower shock absorber bolt.

3. Remove stabilizer bar connecting rod lower nut using power tool, then separate connecting rod from lower link.

4. On 4WD models, remove the drive shaft. Refer to the Drive Train Section.

5. Remove pinch bolt from steering knuckle using power tool, then separate lower link ball joint from steering knuckle.

6. Remove lower link adjusting bolts and nuts, then the lower link.

➡**Some vehicles may be equipped with straight (non-adjustable) lower link bolts and washers. In order to adjust camber and caster on these vehicles, first replace the lower link bolts and washers with adjustable (cam) bolts and washers.**

7. Remove the jounce bumper from the lower link.

Inspection After Removal

8. Lower Link
   a. Check for deformation and cracks. Replace if necessary.

9. Lower Link Bushing
   a. Check for distortion and damage. Replace if necessary.

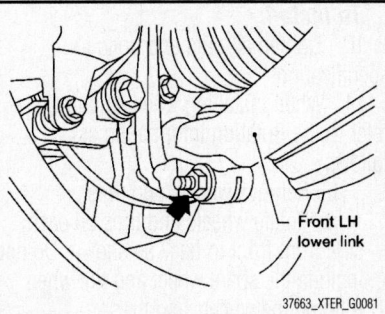

Fig. 90 Remove pinch bolt from steering knuckle using power tool, then separate lower link ball joint from steering knuckle

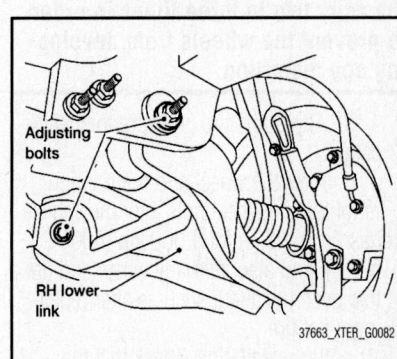

Fig. 91 Remove lower link adjusting bolts and nuts, then the lower link

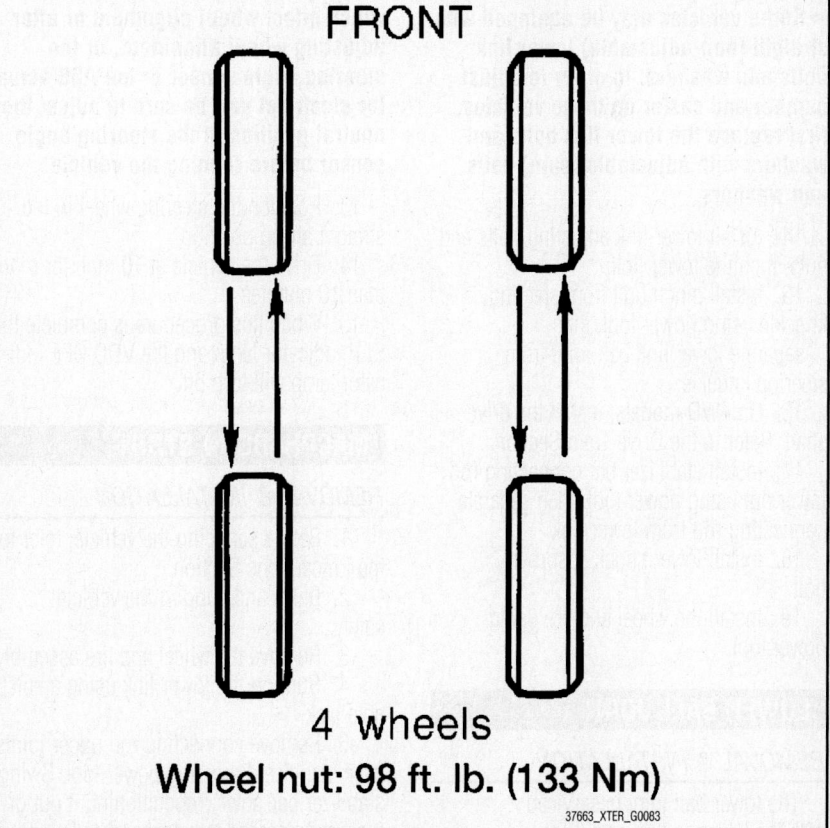

FRONT

4 wheels
Wheel nut: 98 ft. lb. (133 Nm)

Fig. 92 Tire rotation and Wheel nut tightening specifications

To install:

10. Tighten all nuts and bolts to specification.

11. When installing wheel and tire, refer to the Rotation information as follows:

    b. Remove wheels and tires.

    a. Rotate wheels and tires on each side from front to back as shown. Do not include the spare wheel and tire when rotating the wheels and tires.

※※ CAUTION

When installing wheels and tires, tighten them diagonally by dividing the work two to three times in order to prevent the wheels from developing any distortion.

    b. Adjust the tire pressure to specification.

    c. After the wheel and tire rotation, retighten the wheel nuts after the vehicle has been driven for 1,000 km (600 miles), and also after any wheel and tire has been installed, such as after repairing a flat tire.

12. After installation, check that the front wheel alignment is within specification.

13. Install the jounce bumper from the lower link.

➡**Some vehicles may be equipped with straight (non-adjustable) lower link bolts and washers. In order to adjust camber and caster on these vehicles, first replace the lower link bolts and washers with adjustable (cam) bolts and washers.**

14. Install lower link adjusting bolts and nuts, then the lower link.

15. Install pinch bolt from steering knuckle using power tool, then
    separate lower link ball joint from steering knuckle.

16. On 4WD models, install the drive shaft. Refer to the Drive Train Section.

17. Install stabilizer bar connecting rod lower nut using power tool, then separate connecting rod from lower link.

18. Install lower shock absorber bolt.

19. Install the wheel and tire using power tool.

LOWER BALL JOINT

REMOVAL & INSTALLATION

The lower ball joint is serviced with the lower control arm as an assembly.

LOWER CONTROL ARM

REMOVAL & INSTALLATION

1. Before servicing the vehicle, refer to the Precautions Section.

2. Raise and support the vehicle safely.

3. Remove the tire and wheel assembly.

4. Remove the lower strut bolt.

5. Remove the stabilizer bar connecting rod lower nut. Separate the connecting rod from the lower link.

6. If equipped with 4WD, remove the halfshaft.

7. Remove the pinch bolt from the steering knuckle. Separate the lower control arm ball joint stud from the steering knuckle, using the proper tool.

8. Remove the lower control arm adjusting bolts and nuts. Lower the control arm and remove it from the vehicle.

9. Remove the jounce bumper from the lower control arm.

To install:

10. Installation is the reverse of the removal procedure.

11. Be sure to replace all wearable components, as required.

12. Check and adjust the front alignment, as required.

➡**After removing/installing or replacing steering and suspension components which effect wheel alignment or after adjusting wheel alignment, or the steering angle sensor or the ABS actuator electrical unit be sure to adjust the neutral position of the steering angle sensor before running the vehicle.**

13. Position the steering wheel in the straight ahead position.

14. Drive the vehicle at 10 mph for more than 10 minutes.

15. When this procedure is complete the SLP indicator lamp and the VDC OFF indicator lamp will turn off.

MACPHERSON STRUT

REMOVAL & INSTALLATION

1. Before servicing the vehicle, refer to the Precautions Section.

2. Raise and support the vehicle safely.

3. Remove the wheel and tire assembly.

4. Support the lower link using a suitable jack.

5. Remove connecting rod upper joints from stabilizer bar using power tool. Swing stabilizer bar down, repositioning it out of the way to access shock absorber lower mount.

6. Remove the shock absorber lower bolt and nut.

7. Remove the three shock absorber upper mounting nuts.

8. Remove the coil spring and shock absorber assembly. Turn steering knuckle out to gain enough clearance for removal.

To install:

9. Installation is the reverse of the removal procedure.

10. The step in the strut assembly lower seat faces outside of vehicle.

OVERHAUL

Assembly

1. When installing coil spring on strut, it must be positioned so spring ends align with proper spots on seats.

2. Install the shock absorber mounting insulator in line with lower shock mount and step in lower seat. The step in the strut assembly lower seat faces outside of vehicle.

3. Tighten the new piston rod lock nut to 30 ft. lbs. (41 Nm).

4. Remove commercial service tool.

Disassembly

1. Set the shock absorber in a vise, then loosen (without removing) the piston rod lock nut.

※※ CAUTION

Do not remove piston rod lock nut at this time.

2. Compress the spring using commercial service tool until the shock absorber mounting insulator can be turned by hand.

※※ WARNING

Make sure that the pawls of the two spring compressors are firmly hooked on the spring. The spring compressors must be tightened alternately and evenly so as not to tilt the spring.

3. Remove the piston rod lock nut. Discard the piston rod lock nut, use a new nut for assembly.

4. Remove the components from the shock absorber. Keep the spring compressed in the commercial service tool if reusing it for assembly.

Inspection After Disassembly

1. Shock Absorber Assembly:
    • Check for smooth operation through a full stroke, both compression and extension

- Check for oil leakage on welded or gland packing portions
- Check piston rod for cracks, deformation or other damage and replace if necessary

2. Mounting Insulator and Rubber Parts:
- Check cemented rubber-to-metal portion for separation or cracks. Check rubber parts for deterioration and replace if necessary

3. Coil Spring:
- Check for cracks, deformation or other damage and replace if necessary
- Check the free spring height. 2WD: 345.4 mm (13.6 inches), 4WD: 356 mm (14.0 inches).

STEERING KNUCKLE

REMOVAL & INSTALLATION

1. Before servicing the vehicle, refer to the Precautions Section.
2. Raise and support the vehicle safely.
3. Remove the wheel and tire assembly.
4. Without disassembling the hydraulic lines, remove brake caliper. Reposition it aside with wire.

➡**Do not press the brake pedal while the brake caliper is removed.**

5. Put alignment marks on disc rotor and wheel hub and bearing assembly, then remove disc rotor.
6. Disconnect wheel sensor and remove bracket from steering knuckle.
7. On 4WD models, remove cotter pin, then remove lock nut from drive shaft.
8. Remove steering outer socket cotter pin at steering knuckle, then loosen mounting nut.
9. Remove the pinch bolt from the steering knuckle. Separate the lower control arm ball joint stud from the steering knuckle, using the proper tool.
10. Remove wheel hub and bearing assembly bolts.
11. Remove splash guard and wheel hub and bearing assembly from steering knuckle.
12. Remove cotter pin and nut from upper link ball joint.
13. Separate upper link ball joint from steering knuckle using Tool ST29020001 (J-24319-01).
14. Remove pinch bolt from steering knuckle, then separate lower link ball joint from steering knuckle.
15. Remove steering knuckle from vehicle.

To install:

16. Installation is the reverse of the removal procedure.

STABILIZER BAR (SWAY BAR) & LINKS

REMOVAL & INSTALLATION

1. Before servicing the vehicle, refer to the Precautions Section.
2. Remove the front valance center.
3. Raise and support the vehicle safely.
4. Remove the engine undercover.
5. Remove the connecting rod nuts.
6. Loosen the top bolts for the stabilizer bar mounting brackets. Remove the lower bolts from the mounting brackets.
7. Remove the stabilizer bar from the vehicle.
8. Remove the bushings from the stabilizer bar.

To install:

9. Installation is the reverse of the removal procedure.

UPPER BALL JOINT

REMOVAL & INSTALLATION

The upper ball joint is serviced with the upper control arm as an assembly.

UPPER CONTROL ARM

REMOVAL & INSTALLATION

1. Before servicing the vehicle, refer to the Precautions Section.
2. Raise and support the vehicle safely.
3. Remove the tire and wheel assembly.
4. Using a suitable jack, support the lower control arm.
5. If working on the left side, remove the bolt from the lower joint of the lower joint shaft, then reposition the lower joint shaft out of the way. Do not damage the lower joint.
6. Remove the cotter pin and nut from the upper control arm ball joint.
7. Separate the upper control arm ball joint stud from the steering knuckle, using tool ST29020001 (J-24319-01) or equivalent.
8. Remove the upper control arm retaining bolts and nuts.
9. Remove the upper control arm from the vehicle.

To install:

10. Installation is the reverse of the removal procedure.
11. Be sure to replace all wearable components, as required.

12. Check and adjust the front alignment, as required.

➡**After removing/installing or replacing steering and suspension components which effect wheel alignment or after adjusting wheel alignment, or the steering angle sensor or the ABS actuator electrical unit be sure to adjust the neutral position of the steering angle sensor before running the vehicle.**

13. Position the steering wheel in the straight ahead position.
14. Drive the vehicle at 10 mph for more than 10 minutes.
15. When this procedure is complete the SLP indicator lamp and the VDC OFF indicator lamp will turn off.

WHEEL BEARINGS

REMOVAL & INSTALLATION

See Figure 93.

1. Before servicing the vehicle, refer to the Precautions Section.
2. Raise and support the vehicle safely.
3. Remove the tire and wheel assembly.
4. Remove the caliper and position it to the side with wire. Do not disconnect the brake fluid line.

➡**Do not press the brake pedal while the brake caliper is removed.**

5. Matchmark the brake rotor and the wheel hub. Remove the brake rotor.
6. Remove the cotter pin. Remove the lock nut.
7. Remove the halfshaft from the wheel hub and bearing assembly.
8. Remove the wheel sensor from the hub and bearing assembly. Do not pull on the wheel sensor harness.
9. Remove the wheel hub and bearing assembly bolts.
10. Remove the splash guard. Remove the wheel hub and bearing assembly from the steering knuckle.

➡**Carefully remove the wheel sensor and harness through the hole in the splash guard.**

To install:

11. Inspect the wheel sensor O-ring, replace the wheel speed sensor assembly, as required.
12. Installation is the reverse of the removal procedure.
13. Be sure to use new bolts when installing the wheel hub and bearing assembly.

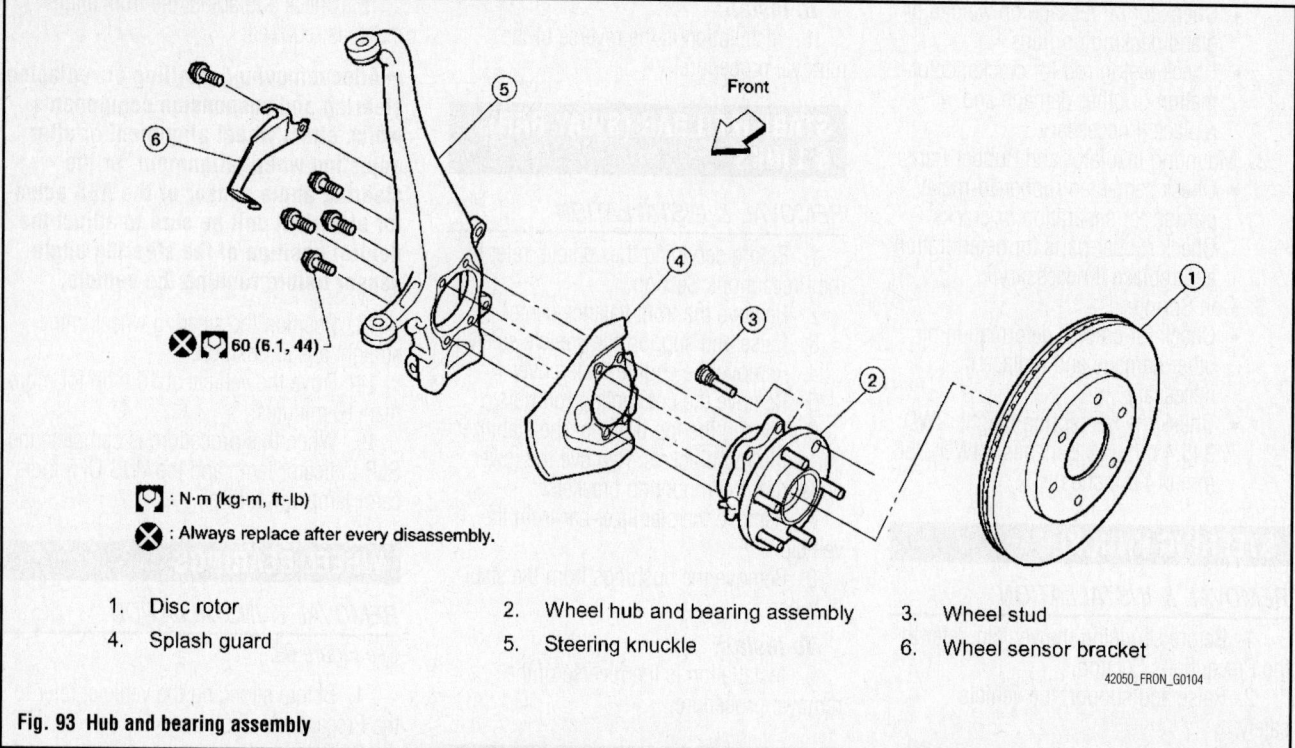

60 (6.1, 44)

⊡ : N·m (kg-m, ft-lb)

⊗ : Always replace after every disassembly.

1. Disc rotor
2. Wheel hub and bearing assembly
3. Wheel stud
4. Splash guard
5. Steering knuckle
6. Wheel sensor bracket

42050_FRON_G0104

Fig. 93 Hub and bearing assembly

SUSPENSION

LEAF SPRING

REMOVAL & INSTALLATION

1. Support rear axle assembly. Do not allow weight to be unsupported, but do not compress spring.
2. Remove lower end of shock absorber.
3. Remove u-bolt nuts and spring pad.
4. Remove rear shackle and bushings.
5. Remove front spring nut and bolt.
6. Remove spring.

To install:

7. Install front spring nut and bolt and rear shackle. Tighten finger tight.
8. Install U-bolts and bumper on axle. Install spring pad and nuts. Tighten nuts evenly to 53 ft. lbs. (72.5 Nm).
9. Install lower shock nut and bolt finger tight.

10. Lower vehicle and bounce to settle suspension. Tighten fasteners with vehicle on suspension.
11. Tighten lower shock absorber to 148 ft. lbs. (200 Nm).
12. Tighten front spring nut and bolt to 165 ft. lbs. (190 Nm).
13. Tighten rear spring shackle nuts to 77 ft. lbs. (105 Nm)

SHOCK ABSORBER

REMOVAL & INSTALLATION

1. Support rear axle assembly.
2. Remove fasteners and shock.

To install:

3. Install shock absorber.
4. Install upper nut and tighten to 33 ft. lbs. (45 Nm).

REAR SUSPENSION

5. Install lower nut and bolt. Tighten to 148 ft. lbs. (200 Nm).
6. Remove support.

STABILIZER BAR (SWAY BAR) & LINKS

REMOVAL & INSTALLATION

1. Disconnect the stabilizer bar ends from the connecting rods using power tool.
2. Remove the stabilizer bar clamps using power tool, and remove the bushings.
3. Remove the stabilizer bar.

To install:

4. Installation is the reverse of the removal procedure.
5. Install the stabilizer bar clamp and bushing so they are positioned outside of the crimp ring on the stabilizer bar.

NISSAN

Diagnostic Trouble Codes

DIAGNOSTIC TROUBLE CODES

OBD II VEHICLE APPLICATIONS

NISSAN

350Z
2009
- 3.5L V6 .... ID VQ35HR

370Z
2009–2010
- 3.7L V6 .... ID VQ37VHR

Altima
2009–2010
- 2.5L I4 .... ID QR25DE
- 3.5L V6 .... ID VQ35DE

Altima Hybrid
2009–2010
- 2.5L I4 .... ID QR25DE

Armada
2009–2010
- 5.6L V8 .... ID VK56DE

Cube
2009–2010
- 1.8L I4 .... ID MR18DE

Frontier
2009–2010
- 2.5L I4 .... ID QR25DE
- 4.0L V8 .... ID VQ40DE

Maxima
2009–2010
- 3.5L V6 .... ID VQ35DE

Murano
2009–2010
- 3.5L V6 .... ID VQ35DE

Pathfinder
2009–2010
- 4.0L V8 .... ID VQ40DE
- 5.6L V8 .... ID VQ56DE

Quest
2009–2010
- 3.5L V6 .... ID VQ35DE

Rogue
2009–2010
- 2.5L I4 .... ID QR25DE

Sentra
2009–2010
- 2.0L I4 .... ID MR20DE
- 2.5L I4 .... ID QR25DE

Titan
2009–2010
- 5.6L V8 .... ID VK56DE

Versa
2007–2008
- 1.6L I4 .... ID HR16DE
- 1.8L I4 .... ID MR18DE

Xterra
2009–2010
- 4.0L V8 .... ID VQ40DE

OBD II Trouble Code List (P0XXX Codes)

DTC	Trouble Code Title, Conditions & Possible Causes
DTC: P0011 **1T ECM, MIL: Yes** **Year:** 2009, 2010 **Model:** Armada, Titan **Engine:** 5.6L V8 **Transmission:** All	**Intake Valve Timing Control Performance (Bank 1):** NOTE: If DTC P0011 is displayed with DTC P0075, P0081, P1140 or P1145, first perform the trouble diagnosis for any other DTC before proceeding with P0011. Condition A: The alignment of the intake valve timing control has been misresistered. Condition B: There is a gap between angle of target and phase-control angle degree. **Possible Causes:** • Harness or connectors • Intake valve timing control solenoid valve circuit is open or shorted. • Intake valve timing control position sensor circuit is open or shorted. • Intake valve timing control solenoid valve • Intake valve timing control position sensor • Crankshaft position sensor (POS) • Camshaft position sensor (PHASE) • Accumulation of debris to the signal pick-up portion of the camshaft sprocket • Timing chain installation • Foreign matter caught in the oil groove for intake valve timing control
DTC: P0011 **1T ECM, MIL: Yes** **Year:** 2009, 2010 **Model:** 350Z, 370Z, Altima, Cube, Frontier, Maxima, Murano, Pathfinder, Quest, Rogue, Sentra, Versa, Xterra **Engine:** 1.6L L4, 1.8L L4, 2.0L L4, 2.5L L4, 3.5L V6, 3.5L V6 VIN A, 3.7L V6, 4.0L V6, 5.6L V8 **Transmission:** All	**Intake Valve Timing Control Performance (Includes Hybrid Models):** HYBRID MODELS CAUTION: Hybrid systems use very high-voltage battery systems. Before starting any service work involving the battery system, turn the ignition switch OFF and then remove the service plug from pocket in the trunk. After removing the service plug, wait 10 minutes before touching any of the high-voltage connectors and terminals. NOTE: On V6 and V8 engines, this applies to Bank 1. NOTE: When the malfunction is detected, the ECM enters fail-safe mode. NOTE: If DTC P0011 is displayed with DTC P0075, P0081, or P0524, perform the appropriate trouble diagnosis first. • There is a gap between angle of target and phase-control angle degree. **Possible Causes:** • Crankshaft position sensor (POS) • Camshaft position sensor (PHASE) • Intake valve control solenoid valve • Accumulation of debris to the signal pick-up portion of the camshaft • Timing chain installation • Foreign matter caught in the oil groove for intake valve timing control
DTC: P0014 **1T ECM, MIL: Yes** **Year:** 2009, 2010 **Model:** Maxima **Engine:** 3.5L V6 **Transmission:** All	**Exhaust Valve Timing (EVT) Control Performance (Bank 1):** NOTE: If DTC P0014 or P0024 is displayed with DTC P0078, P0084, P1078 or P1084 first perform trouble diagnosis for respective DTC before proceeding with P0014. • There is a gap between angle of target and phase-control angle degree. **Possible Causes:** • Crankshaft position sensor • Camshaft position sensor • EVT control position sensor • EVT control magnet retarder • Accumulation of debris to the signal pick-up portion of the camshaft • Timing chain installation • EVT control pulley assembly
DTC: P0021 **1T ECM, MIL: Yes** **Year:** 2009, 2010 **Model:** 370Z **Engine:** 3.7L V6 **Transmission:** All	**Intake Valve Timing Control Performance (Bank 2):** NOTE: If DTC P0011 or P0021 is displayed with DTC P0075, P0081 or P0524, first perform the trouble diagnosis for the applicable DTC before proceeding with P0011. • There is a gap between angle of target and phase-control angle degree. **Possible Causes:** • Crankshaft position sensor (POS) • Camshaft position sensor (PHASE) • Intake valve control solenoid valve • Accumulation of debris to the signal pick-up portion of the camshaft • Timing chain installation • Foreign matter caught in the oil groove for intake valve timing control

DTC	Trouble Code Title, Conditions & Possible Causes
DTC: P0021 **1T ECM, MIL: Yes** **Year:** 2009, 2010 **Model:** 350Z, Altima, Maxima, Murano, Pathfinder, Quest **Engine:** 2.5L L4, 3.5L V6, 3.5L V6 VIN A, 4.0L V6, 5.6L V8 **Transmission:** All	**Intake Valve Timing Control Performance (Bank 2):** • If DTC P0011 or P0021 is displayed with DTC P0075 or P0081, first perform trouble diagnosis for DTC P0075 or P0081. • There is a gap between angle of target and phase-control angle degree. • When the malfunction is detected, the ECM enters fail-safe mode. **Possible Causes:** • Crankshaft position sensor (POS) • Camshaft position sensor (PHASE) • Intake valve timing control solenoid valve • Accumulation of debris to the signal pick-up portion of the camshaft • Timing chain installation • Foreign matter caught in the oil groove for intake valve timing control
DTC: P0021 **1T ECM, MIL: Yes** **Year:** 2009, 2010 **Model:** Armada, Titan **Engine:** 5.6L V8 **Transmission:** All	**Intake Valve Timing Control Performance (Bank 2):** **NOTE: If DTC P0021 is displayed with DTC P0075, P0081, P1140 or P1145, first perform the trouble diagnosis for any other DTC before proceeding with P0021.** Condition A: The alignment of the intake valve timing control has been misresisted. Condition B: There is a gap between angle of target and phase-control angle degree. **Possible Causes:** • Harness or connectors • Intake valve timing control solenoid valve circuit is open or shorted. • Intake valve timing control position sensor circuit is open or shorted. • Intake valve timing control solenoid valve • Intake valve timing control position sensor • Crankshaft position sensor (POS) • Camshaft position sensor (PHASE) • Accumulation of debris to the signal pick-up portion of the camshaft sprocket • Timing chain installation • Foreign matter caught in the oil groove for intake valve timing control
DTC: P0024 **1T ECM, MIL: Yes** **Year:** 2009, 2010 **Model:** Maxima **Engine:** 3.5L V6 **Transmission:** All	**Exhaust Valve Timing (EVT) Control Performance (Bank 2):** **NOTE: If DTC P0014 or P0024 is displayed with DTC P0078, P0084, P1078 or P1084, first perform trouble diagnosis for the respective DTC before proceeding with P0024.** • There is a gap between angle of target and phase-control angle degree. **Possible Causes:** • Crankshaft position sensor • Camshaft position sensor • EVT control position sensor • EVT control magnet retarder • Accumulation of debris to the signal pick-up portion of the camshaft • Timing chain installation • EVT control pulley assembly
DTC: P0031 **1T ECM, MIL: Yes** **Year:** 2009, 2010 **Model:** 350Z, 370Z, Altima, Armada, Frontier, Maxima, Murano, Pathfinder, Quest, Rogue, Sentra, Titan, Versa, Xterra **Engine:** 1.6L L4, 1.8L L4, 2.0L L4, 2.5L L4, 3.5L V6, 3.5L V6 VIN A, 3.7L V6, 4.0L V6, 5.6L V8 **Transmission:** All	**Air Fuel Ratio (A/F) Sensor 1 Heater Control Circuit Low (Includes Hybrid Models):** HYBRID MODELS CAUTION: Hybrid systems use very high-voltage battery systems. Before starting any service work involving the battery system, turn the ignition switch OFF and then remove the service plug from pocket in the trunk. After removing the service plug, wait 10 minutes before touching any of the high-voltage connectors and terminals. **NOTE: On V6 and V8 engines, this applies to Bank 1.** • The ECM performs ON/OFF duty control of the A/F sensor 1 heater corresponding to the engine operating condition to keep the temperature of A/F sensor 1 element at the specified range. • The current amperage in the A/F sensor 1 heater circuit is out of the normal range. (An excessively low voltage signal is sent to ECM through the A/F sensor heater.) **Possible Causes:** • Harness or connectors • The A/F sensor 1 heater circuit is open or shorted. • A/F sensor 1 heater
DTC: P0031 **1T ECM, MIL: Yes** **Year:** 2009, 2010 **Model:** Cube **Engine:** 1.8L L4 **Transmission:** All	**Air Fuel (A/F) Ratio Sensor 1 Heater Control Circuit Low:** • The current amperage in the A/F sensor 1 heater circuit is out of the normal range. (An excessively low voltage signal is sent to ECM through the A/F sensor 1 heater.) **Possible Causes:** • Harness or connectors • The A/F sensor 1 heater circuit is open or shorted. • A/F sensor 1 heater

DTC	Trouble Code Title, Conditions & Possible Causes
DTC: P0032 **1T ECM, MIL: Yes** **Year:** 2009, 2010 **Model:** 350Z, 370Z, Altima, Armada, Frontier, Maxima, Murano, Pathfinder, Quest, Rogue, Sentra, Titan, Versa, Xterra **Engine:** 1.6L L4, 1.8L L4, 2.0L L4, 2.5L L4, 3.5L V6, 3.5L V6 VIN A, 3.7L V6, 4.0L V6, 5.6L V8 **Transmission:** All	**Air Fuel Ratio (A/F) Sensor 1 Heater (Bank 1) Control Circuit High (Includes Hybrid Models):** HYBRID MODELS CAUTION: Hybrid systems use very high-voltage battery systems. Before starting any service work involving the battery system, turn the ignition switch OFF and then remove the service plug from pocket in the trunk. After removing the service plug, wait 10 minutes before touching any of the high-voltage connectors and terminals. **NOTE: On V6 and V8 engines, this applies to Bank 1.** • The ECM performs ON/OFF duty control of the A/F sensor 1 heater corresponding to the engine operating condition to keep the temperature of A/F sensor 1 element at the specified range. • The current amperage in the A/F sensor 1 heater circuit is out of the normal range. (An excessively high voltage signal is sent to ECM through the A/F sensor heater.) **Possible Causes:** • Harness or connectors • A/F sensor 1 heater circuit is shorted. • A/F sensor 1 heater
DTC: P0032 **1T ECM, MIL: Yes** **Year:** 2009, 2010 **Model:** Cube **Engine:** 1.8L L4 **Transmission:** All	**Air Fuel (A/F) Ratio Sensor 1 Heater Control Circuit High:** • The current amperage in the A/F sensor 1 heater circuit is out of the normal range. (An excessively high voltage signal is sent to ECM through the A/F sensor 1 heater.) **Possible Causes:** • Harness or connectors • The A/F sensor 1 heater circuit is shorted. • A/F sensor 1 heater
DTC: P0037 **1T ECM, MIL: Yes** **Year:** 2009, 2010 **Model:** 350Z, 370Z, Altima, Armada, Cube, Frontier, Maxima, Murano, Pathfinder, Quest, Rogue, Titan, Versa, Xterra **Engine:** 1.6L L4, 1.8L L4, 2.5L L4, 3.5L V6, 3.5L V6 VIN A, 3.7L V6, 4.0L V6, 5.6L V8 **Transmission:** All	**Heated Oxygen Sensor 2 Heater Control Circuit Low (Includes Hybrid Models):** HYBRID MODEL CAUTION: Hybrid systems use very high-voltage battery systems. Before starting any service work involving the battery system, turn the ignition switch OFF and then remove the service plug from pocket in the trunk. After removing the service plug, wait 10 minutes before touching any of the high-voltage connectors and terminals. **NOTE: On V6 and V8 engines, this applies to Bank 1.** • The current amperage in the heated oxygen sensor 2 heater circuit is out of the normal range. (An excessively low voltage signal is sent to ECM via the heated oxygen sensor 2 heater.) **Possible Causes:** • Harness or connectors • The heated oxygen sensor 2 heater circuit is open or shorted. • Heated oxygen sensor 2 heater
DTC: P0037 **1T ECM, MIL: Yes** **Year:** 2010 **Model:** Sentra **Engine:** 2.0L L4, 2.5L L4 **Transmission:** All	**Heated Oxygen Sensor 2 Heater:** • Above 3,600 rpm heater is off. • Below 3,600 rpm after the following conditions are met: - Engine warmed up - Engine speed is between 3,500 and 4,000 rpm for 1 minute and at idle for 1 minute under no load • The current amperage in the heated oxygen sensor 2 heater circuit is out of the normal range. (An excessively low voltage signal is sent to ECM through the heated oxygen sensor 2 heater.) **Possible Causes:** • Harness or connectors (Heated oxygen sensor 2 heater circuit is open or shorted.) • Heated oxygen sensor 2 heater has failed
DTC: P0038 **1T ECM, MIL: Yes** **Year:** 2009, 2010 **Model:** Cube **Engine:** 1.8L L4 **Transmission:** All	**Heated Oxygen Sensor 2 Heater Control Circuit High:** The current amperage in the heated oxygen sensor 2 heater circuit is out of the normal range. (An excessively high voltage signal is sent to ECM through the heated oxygen sensor 2 heater.) **Possible Causes:** • Harness or connectors • The heated oxygen sensor 2 heater circuit is shorted. • Heated oxygen sensor 2 heater
DTC: P0038 **1T ECM, MIL: Yes** **Year:** 2009, 2010 **Model:** 350Z, 370Z, Altima, Armada, Frontier, Maxima, Murano, Pathfinder, Quest, Rogue, Sentra, Titan, Versa, Xterra **Engine:** 1.6L L4, 1.8L L4, 2.0L L4, 2.5L L4, 3.5L V6, 3.5L V6 VIN A, 3.7L V6, 4.0L V6, 5.6L V8 **Transmission:** All	**Heated Oxygen Sensor 2 Heater Control Circuit High (Includes Hybrid Models):** HYBRID MODELS CAUTION: Hybrid systems use very high-voltage battery systems. Before starting any service work involving the battery system, turn the ignition switch OFF and then remove the service plug from pocket in the trunk. After removing the service plug, wait 10 minutes before touching any of the high-voltage connectors and terminals. **NOTE: On V6 and V8 engines, this applies to Bank 1.** • The current amperage in the heated oxygen sensor 2 heater circuit is out of the normal range. (An excessively high voltage signal is sent to ECM via the heated oxygen sensor 2 heater.) **Possible Causes:** • Harness or connectors • The heated oxygen sensor 2 heater circuit is shorted. • Heated oxygen sensor 2 heater

DTC	Trouble Code Title, Conditions & Possible Causes
DTC: P0043 **1T ECM, MIL: Yes** **Year:** 2009, 2010 **Model:** Altima, Rogue **Engine:** 2.5L L4 **Transmission:** All	**Heated Oxygen Sensor 3 Heater Control Circuit Low (Includes Hybrid):** HYBRID MODELS CAUTION: Hybrid systems use very high-voltage battery systems. Before starting any service work involving the battery system, turn the ignition switch OFF and then remove the service plug from pocket in the trunk. After removing the service plug, wait 10 minutes before touching any of the high-voltage connectors and terminals. • The current amperage in the heated oxygen sensor 3 heater circuit is out of the normal range. (An excessively low voltage signal is sent to ECM through the heated oxygen sensor 3 heater.) **Possible Causes:** • Harness or connectors • The heated oxygen sensor 3 heater circuit is open or shorted. • Heated oxygen sensor 3 heater
DTC: P0044 **1T ECM, MIL: Yes** **Year:** 2009, 2010 **Model:** Altima, Rogue **Engine:** 2.5L L4 **Transmission:** All	**Heated Oxygen Sensor 3 Heater Control Circuit High (Includes Hybrid Models):** HYBRID MODELS CAUTION: Hybrid systems use very high-voltage battery systems. Before starting any service work involving the battery system, turn the ignition switch OFF and then remove the service plug from pocket in the trunk. After removing the service plug, wait 10 minutes before touching any of the high-voltage connectors and terminals. • The current amperage in the heated oxygen sensor 3 heater circuit is out of the normal range. (An excessively high voltage signal is sent to ECM through the heated oxygen sensor 3 heater.) **Possible Causes:** • Harness or connectors • The heated oxygen sensor 3 heater circuit is shorted. • Heated oxygen sensor 3 heater
DTC: P0051 **1T ECM, MIL: Yes** **Year:** 2009, 2010 **Model:** 350Z, 370Z, Altima, Armada, Maxima, Murano, Pathfinder, Quest, Titan, Xterra **Engine:** 2.5L L4, 3.5L V6, 3.5L V6 VIN A, 3.7L V6, 4.0L V6, 5.6L V8 **Transmission:** All	**Air Fuel Ratio (A/F) Sensor 1 (Bank 2) Heater Control Circuit Low:** **NOTE: On V6 and V8 engines, this applies to Bank 2.** • The ECM performs ON/OFF duty control of the A/F sensor 1 heater corresponding to the engine operating condition to keep the temperature of A/F sensor 1 element at the specified range. • The current amperage in the A/F sensor 1 heater circuit is out of the normal range. (An excessively low voltage signal is sent to ECM through the A/F sensor heater.) **Possible Causes:** • Harness or connectors • The A/F sensor 1 heater circuit is open or shorted. • A/F sensor 1 heater
DTC: P0052 **1T ECM, MIL: Yes** **Year:** 2009, 2010 **Model:** 350Z, 370Z, Altima, Armada, Maxima, Murano, Pathfinder, Quest, Titan, Xterra **Engine:** 2.5L L4, 3.5L V6, 3.5L V6 VIN A, 3.7L V6, 4.0L V6, 5.6L V8 **Transmission:** All	**Air Fuel Ratio (A/F) Sensor 1 (Bank 2) Heater Control Circuit High:** **NOTE: On V6 and V8 engines, this applies to Bank 2.** • The ECM performs ON/OFF duty control of the A/F sensor 1 heater corresponding to the engine operating condition to keep the temperature of A/F sensor 1 element at the specified range. • The current amperage in the A/F sensor 1 heater circuit is out of the normal range. (An excessively high voltage signal is sent to ECM through the A/F sensor heater.) **Possible Causes:** • Harness or connectors (The A/F sensor 1 heater circuit is shorted.) • A/F sensor 1 heater
DTC: P0057 **1T ECM, MIL: Yes** **Year:** 2009, 2010 **Model:** 350Z, 370Z, Altima, Armada, Maxima, Murano, Pathfinder, Quest, Titan, Xterra **Engine:** 2.5L L4, 3.5L V6, 3.5L V6 VIN A, 3.7L V6, 4.0L V6, 5.6L V8 **Transmission:** All	**Heated Oxygen Sensor 2 (Bank 2) Heater Control Circuit Low:** **NOTE: On V6 and V8 engines, this applies to Bank 2.** • The current amperage in the heated oxygen sensor 2 heater circuit is out of the normal range. (An excessively low voltage signal is sent to ECM via the heated oxygen sensor 2 heater.) **Possible Causes:** • Harness or connectors • The heated oxygen sensor 2 heater circuit is open or shorted. • Heated oxygen sensor 2 heater
DTC: P0058 **1T ECM, MIL: Yes** **Year:** 2009, 2010 **Model:** 350Z, 370Z, Altima, Armada, Maxima, Murano, Pathfinder, Quest, Titan, Xterra **Engine:** 2.5L L4, 3.5L V6, 3.5L V6 VIN A, 3.7L V6, 4.0L V6, 5.6L V8 **Transmission:** All	**Heated Oxygen Sensor 2 (Bank 2) Heater Control Circuit High:** **NOTE: On V6 and V8 engines, this applies to Bank 2.** • The current amperage in the heated oxygen sensor 2 heater circuit is out of the normal range. (An excessively high voltage signal is sent to ECM via the heated oxygen sensor 2 heater.) **Possible Causes:** • Harness or connectors • The heated oxygen sensor 2 heater circuit is shorted. • Heated oxygen sensor 2 heater

DTC	Trouble Code Title, Conditions & Possible Causes
DTC: P0075 **1T ECM, MIL:** Yes **Year:** 2009, 2010 **Model:** Cube **Engine:** 1.8L L4 **Transmission:** All	**Intake Valve Timing Control Solenoid Valve Circuit:** • An improper voltage is sent to the ECM through intake valve timing control solenoid valve. **Possible Causes:** • Harness or connectors • Intake valve timing control solenoid valve circuit is open or shorted. • Intake valve timing control solenoid valve
DTC: P0075 **1T ECM, MIL:** Yes **Year:** 2009, 2010 **Model:** 350Z, Altima, Armada, Frontier, Maxima, Murano, Pathfinder, Quest, Rogue, Titan, Versa, Xterra **Engine:** 1.6L L4, 1.8L L4, 2.5L L4, 3.5L V6, 3.5L V6 VIN A, 4.0L V6, 5.6L V8 **Transmission:** All	**Intake Valve Timing Control Solenoid Valve Circuit (Includes Hybrid Models):** HYBRID MODELS CAUTION: Hybrid systems use very high-voltage battery systems. Before starting any service work involving the battery system, turn the ignition switch OFF and then remove the service plug from pocket in the trunk. After removing the service plug, wait 10 minutes before touching any of the high-voltage connectors and terminals. **NOTE: On V6 and V8, this IVT is on bank 1.** • An improper voltage is sent to the ECM via the intake valve timing control solenoid valve. **Possible Causes:** • Harness or connectors. • Intake valve timing control solenoid valve circuit is open or shorted. • Intake valve timing control solenoid valve has failed.
DTC: P0075 **1T ECM, MIL:** Yes **Year:** 2010 **Model:** Sentra **Engine:** 2.0L L4, 2.5L L4 **Transmission:** All	**Intake Valve Timing Control Solenoid Valve Circuit:** • Engine is warmed up. • Air conditioner switch is OFF. • Shift lever is in P or N (CVT) or Neutral (M/T). • No load. • At idle, circuit is 0–2% specification. • When revving engine up to 2,000 rpm quickly, circuit is 0–90% specification. • DTC is set when an improper voltage is sent to the ECM through intake valve timing control solenoid valve. **Possible Causes:** • Harness or connectors. (Intake valve timing control solenoid valve • circuit is open or shorted.) • Intake valve timing control solenoid valve has failed.
DTC: P0078 **1T ECM, MIL:** Yes **Year:** 2009, 2010 **Model:** Maxima **Engine:** 3.5L V6 **Transmission:** All	**Exhaust Valve Timing (EVT) Control Magnet Retarder (Bank 1) Circuit:** • An improper voltage is sent to the ECM via the exhaust valve timing control magnet retarder. **Possible Causes:** • Harness or connectors • Exhaust valve timing control magnet retarder circuit is open or shorted. • Exhaust valve timing control magnet retarder
DTC: P0081 **1T ECM, MIL:** Yes **Year:** 2009, 2010 **Model:** 350Z, 370Z, Altima, Armada, Maxima, Murano, Pathfinder, Quest, Titan, Xterra **Engine:** 3.5L V6, 3.5L V6 VIN A, 3.7L V6, 4.0L V6, 5.6L V8 **Transmission:** All	**Intake Valve Timing Control Solenoid Valve Circuit (Bank 2):** **NOTE: Applies to V6 and V8 engines only.** • An improper voltage is sent to the ECM via the intake valve timing control solenoid valve. **Possible Causes:** • Harness or connectors. • Intake valve timing control solenoid valve circuit is open or shorted. • Intake valve timing control solenoid valve has failed.
DTC: P0084 **1T ECM, MIL:** Yes **Year:** 2009, 2010 **Model:** Maxima **Engine:** 3.5L V6 **Transmission:** All	**Exhaust Valve Timing (EVT) Timing Control (Magnet Retarder) Solenoid (Bank 2) Circuit:** • An improper voltage is sent to the ECM via the exhaust valve timing control (magnet retarder). **Possible Causes:** • Harness or connectors • Exhaust valve timing control magnet retarder circuit is open or shorted. • Exhaust valve timing control magnet retarder

DTC	Trouble Code Title, Conditions & Possible Causes
DTC: P0101 **1T ECM, MIL: Yes** **Year:** 2009, 2010 **Model:** Sentra **Engine:** 2.0L L4, 2.5L L4 **Transmission:** All	**Mass Air Flow Sensor Circuit Range/Performance - Low Voltage:** • Engine: After warming up • Shift lever: P or N (CVT), Neutral (M/T) • Air conditioner switch: OFF • No load • A low voltage from the sensor is sent to ECM under heavy load driving condition. **Possible Causes:** • Harness or connectors (Mass air flow sensor circuit is open or shorted.) • Intake air leaks • Mass air flow sensor has failed • EVAP control system pressure sensor • Intake air temperature sensor
DTC: P0101 **1T ECM, MIL: Yes** **Year:** 2009, 2010 **Model:** 350Z, 370Z, Altima, Armada, Cube, Frontier, Maxima, Murano, Pathfinder, Quest, Rogue, Titan, Versa, Xterra **Engine:** 1.6L L4, 1.8L L4, 2.5L L4, 3.5L V6, 3.5L V6 VIN A, 3.7L V6, 4.0L V6, 5.6L V8 **Transmission:** All	**Mass Air Flow Sensor Circuit Range/Performance (Includes Hybrid Models):** HYBRID MODELS CAUTION: Hybrid systems use very high-voltage battery systems. Before starting any service work involving the battery system, turn the ignition switch OFF and then remove the service plug from pocket in the trunk. After removing the service plug, wait 10 minutes before touching any of the high-voltage connectors and terminals. • Condition A: A high voltage from the sensor is sent to ECM under light load driving condition. • Condition B: A low voltage from the sensor is sent to ECM under heavy load driving condition. **Possible Causes:** • Condition A: • Harness or connectors • The sensor circuit is open or shorted. • Mass air flow sensor • EVAP control system pressure sensor • Intake air temperature sensor • Condition B: • Harness or connectors • The sensor circuit is open or shorted. • Intake air leakage • Mass air flow sensor • EVAP control system pressure sensor • Intake air temperature sensor
DTC: P0102 **1T ECM, MIL: Yes** **Year:** 2009, 2010 **Model:** Sentra **Engine:** 2.0L L4, 2.5L L4 **Transmission:** All	**Mass Air Flow Circuit Low Input:** • Engine: After warming up • Shift lever: P or N (CVT), Neutral (M/T) • Air conditioner switch: OFF • No load An excessively low voltage from the sensor is sent to ECM. **Possible Causes:** • Harness or connectors (Mass air flow sensor circuit is open or shorted.) • Intake air leaks • Mass air flow sensor has failed
DTC: P0102 **1T ECM, MIL: Yes** **Year:** 2009, 2010 **Model:** Cube **Engine:** 1.8L L4 **Transmission:** All	**Mass Air Flow Sensor Circuit Low Input:** • An excessively low voltage from the sensor is sent to ECM. **Possible Causes:** • Harness or connectors • The sensor circuit is open or shorted. • Intake air leaks • Mass air flow sensor
DTC: P0102 **1T ECM, MIL: Yes** **Year:** 2009, 2010 **Model:** 350Z, 370Z, Altima, Armada, Frontier, Maxima, Murano, Pathfinder, Quest, Rogue, Titan, Versa, Xterra **Engine:** 1.6L L4, 1.8L L4, 2.5L L4, 3.5L V6, 3.5L V6 VIN A, 3.7L V6, 4.0L V6, 5.6L V8 **Transmission:** All	**Mass Air Flow Circuit Low Input (Includes Hybrid Models):** HYBRID MODELS CAUTION: Hybrid systems use very high-voltage battery systems. Before starting any service work involving the battery system, turn the ignition switch OFF and then remove the service plug from pocket in the trunk. After removing the service plug, wait 10 minutes before touching any of the high-voltage connectors and terminals. **NOTE: On V6 and V8 engines, this DTC is for Bank 1.** • An excessively low voltage from the sensor is sent to ECM. **Possible Causes:** • Harness or connectors • The sensor circuit is open or shorted. • Intake air leakage • Mass air flow sensor

DTC	Trouble Code Title, Conditions & Possible Causes
DTC: P0103 **1T ECM, MIL: Yes** **Year:** 2009, 2010 **Model:** 350Z, 370Z, Altima, Armada, Cube, Frontier, Maxima, Murano, Pathfinder, Quest, Rogue, Sentra, Titan, Versa, Xterra **Engine:** 1.6L L4, 1.8L L4, 2.0L L4, 2.5L L4, 3.5L V6, 3.5L V6 VIN A, 3.7L V6, 4.0L V6, 5.6L V8 **Transmission:** All	**Mass Air Flow Sensor Circuit High Input (Includes Hybrid Models):** HYBRID MODELS CAUTION: Hybrid systems use very high-voltage battery systems. Before starting any service work involving the battery system, turn the ignition switch OFF and then remove the service plug from pocket in the trunk. After removing the service plug, wait 10 minutes before touching any of the high-voltage connectors and terminals. **NOTE: On V6 and V8 engines, this DTC is for Bank 1.** • An excessively high voltage from the sensor is sent to ECM. **Possible Causes:** • Harness or connectors • The sensor circuit is open or shorted. • Mass air flow sensor
DTC: P010A **1T ECM, MIL: Yes** **Year:** 2009, 2010 **Model:** 370Z **Engine:** 3.7L V6 **Transmission:** All	**Manifold Absolute Pressure Sensor Circuit:** **NOTE: If DTC P010A is displayed with DTC P0643, first perform the trouble diagnosis for DTC P0643.** • An excessively low voltage from the sensor is sent to ECM. • An excessively high voltage from the sensor is sent to ECM. **Possible Causes:** • Harness or connectors • The sensor circuit is open or shorted. • Manifold absolute pressure (MAP) sensor
DTC: P010B **1T ECM, MIL: Yes** **Year:** 2009, 2010 **Model:** 370Z **Engine:** 3.7L V6 **Transmission:** All	**Mass Air Flow Sensor (Bank 2) Circuit/Range Performance:** • Condition A: A high voltage from the sensor is sent to ECM under light load driving condition. • Condition B: A low voltage from the sensor is sent to ECM under heavy load driving condition. **Possible Causes:** • Condition A: • Harness or connectors • The sensor circuit is open or shorted. • Mass air flow sensor • EVAP control system pressure sensor • Condition B: • Harness or connectors • The sensor circuit is open or shorted. • Intake air leaks • Mass air flow sensor • EVAP control system pressure sensor • Intake air temperature sensor
DTC: P010C **1T ECM, MIL: Yes** **Year:** 2009, 2010 **Model:** 370Z **Engine:** 3.7L V6 **Transmission:** All	**Mass Air Flow Sensor (Bank 2) Circuit Low Input:** • An excessively low voltage from the sensor is sent to ECM. **Possible Causes:** • Harness or connectors • The sensor circuit is open or shorted. • Intake air leaks • Mass air flow sensor
DTC: P010D **1T ECM, MIL: Yes** **Year:** 2009, 2010 **Model:** 370Z **Engine:** 3.7L V6 **Transmission:** All	**Mass Air Flow Sensor (Bank 2) Circuit High Input:** An excessively high voltage from the sensor is sent to ECM. **Possible Causes:** • Harness or connectors • The sensor circuit is open or shorted. • Mass air flow sensor
DTC: P0112 **1T ECM, MIL: Yes** **Year:** 2009, 2010 **Model:** 350Z, 370Z, Altima, Armada, Cube, Frontier, Maxima, Murano, Pathfinder, Quest, Rogue, Sentra, Titan, Versa, Xterra **Engine:** 1.6L L4, 1.8L L4, 2.0L L4, 2.5L L4, 3.5L V6, 3.5L V6 VIN A, 3.7L V6, 4.0L V6, 5.6L V8 **Transmission:** All	**Intake Air Temperature Sensor Circuit Low Input (Includes Hybrid Models):** HYBRID MODELS CAUTION: Hybrid systems use very high-voltage battery systems. Before starting any service work involving the battery system, turn the ignition switch OFF and then remove the service plug from pocket in the trunk. After removing the service plug, wait 10 minutes before touching any of the high-voltage connectors and terminals. **NOTE: On V6 and V8 engines, this DTC is for Bank 1.** • An excessively low voltage from the sensor is sent to ECM. **Possible Causes:** • Harness or connectors • Intake air temperature sensor circuit is open or shorted. • Intake air temperature sensor

DTC	Trouble Code Title, Conditions & Possible Causes

DTC: P0113
1T ECM, MIL: Yes
Year: 2009, 2010
Model: 350Z, 370Z, Altima, Armada, Cube, Frontier, Maxima, Murano, Pathfinder, Quest, Rogue, Sentra, Titan, Versa, Xterra
Engine: 1.6L L4, 1.8L L4, 2.0L L4, 2.5L L4, 3.5L V6, 3.5L V6 VIN A, 3.7L V6, 4.0L V6, 5.6L V8
Transmission: All
	Intake Air Temperature Sensor Circuit High Input (Includes Hybrid Models):
	HYBRID MODELS CAUTION: Hybrid systems use very high-voltage battery systems. Before starting any service work involving the battery system, turn the ignition switch OFF and then remove the service plug from pocket in the trunk. After removing the service plug, wait 10 minutes before touching any of the high-voltage connectors and terminals.
	NOTE: On V6 and V8 engines, this DTC is for Bank 1.
	    • An excessively high voltage from the sensor is sent to ECM.
	Possible Causes:
	    • Harness or connectors
	    • Intake air temperature sensor circuit is open or shorted.
	    • Intake air temperature sensor has failed

DTC: P0116
1T ECM, MIL: Yes
Year: 2009, 2010
Model: 350Z, 370Z, Altima, Armada, Frontier, Maxima, Murano, Pathfinder, Quest, Rogue, Titan, Versa, Xterra
Engine: 1.6L L4, 1.8L L4, 2.5L L4, 3.5L V6, 3.5L V6 VIN A, 3.7L V6, 4.0L V6, 5.6L V8
Transmission: All
	Engine Coolant Temperature Sensor Circuit Range/Performance (Includes Hybrid Models):
	HYBRID MODELS CAUTION: Hybrid systems use very high-voltage battery systems. Before starting any service work involving the battery system, turn the ignition switch OFF and then remove the service plug from pocket in the trunk. After removing the service plug, wait 10 minutes before touching any of the high-voltage connectors and terminals.
	NOTE: If DTC P0116 is displayed with P0117 or P0118, first perform the trouble diagnosis for DTC P0117, P0118.
	    • Engine coolant temperature signal from engine coolant temperature sensor does not fluctuate, even when some time has passed after starting the engine with pre-warming up condition.
	Possible Causes:
	    • Harness or connectors
	    • High or low resistance in the circuit.
	    • Engine coolant temperature sensor

DTC: P0117
1T ECM, MIL: Yes
Year: 2009, 2010
Model: 350Z, 370Z, Altima, Armada, Cube, Frontier, Maxima, Murano, Pathfinder, Quest, Rogue, Sentra, Titan, Versa, Xterra
Engine: 1.6L L4, 1.8L L4, 2.0L L4, 2.5L L4, 3.5L V6, 3.5L V6 VIN A, 3.7L V6, 4.0L V6, 5.6L V8
Transmission: All
	Engine Coolant Temperature Circuit Low Input (Includes Hybrid Models):
	HYBRID MODELS CAUTION: Hybrid systems use very high-voltage battery systems. Before starting any service work involving the battery system, turn the ignition switch OFF and then remove the service plug from pocket in the trunk. After removing the service plug, wait 10 minutes before touching any of the high-voltage connectors and terminals.
	    • An excessively low voltage from the sensor is sent to ECM.
	Possible Causes:
	    • Harness or connectors
	    • Engine coolant temperature sensor circuit is open or shorted.
	    • Engine coolant temperature sensor has failed

DTC: P0118
1T ECM, MIL: Yes
Year: 2009, 2010
Model: 350Z, 370Z, Altima, Armada, Cube, Frontier, Maxima, Murano, Pathfinder, Quest, Rogue, Sentra, Titan, Versa, Xterra
Engine: 1.6L L4, 1.8L L4, 2.0L L4, 2.5L L4, 3.5L V6, 3.5L V6 VIN A, 3.7L V6, 4.0L V6, 5.6L V8
Transmission: All
	Engine Coolant Temperature Sensor Circuit High Input (Includes Hybrid Models):
	HYBRID MODELS CAUTION: Hybrid systems use very high-voltage battery systems. Before starting any service work involving the battery system, turn the ignition switch OFF and then remove the service plug from pocket in the trunk. After removing the service plug, wait 10 minutes before touching any of the high-voltage connectors and terminals.
	    • An excessively high voltage from the sensor is sent to ECM.
	Possible Causes:
	    • Harness or connectors
	    • Engine coolant temperature sensor circuit is open or shorted.
	    • Engine coolant temperature sensor has failed

DTC: P0122
1T ECM, MIL: Yes
Year: 2009, 2010
Model: Sentra
Engine: 2.0L L4, 2.5L L4
Transmission: All
	Throttle position sensor 2 circuit low input:
	    • Ignition switch: ON (Engine stopped)
	    • Shift lever: D (CVT), 1st (M/T)
	    * An excessively low voltage from the TP sensor 2 is sent to ECM.
	Possible Causes:
	    • Harness or connectors (TP sensor 2 circuit is open or shorted.)
	    • Electric throttle control actuator (TP sensor 2) has failed

DTC	Trouble Code Title, Conditions & Possible Causes
DTC: P0122 **1T ECM, MIL: Yes** **Year:** 2009, 2010 **Model:** 350Z, 370Z, Altima, Armada, Cube, Frontier, Maxima, Murano, Pathfinder, Quest, Rogue, Titan, Versa, Xterra **Engine:** 1.6L L4, 1.8L L4, 2.5L L4, 3.5L V6, 3.5L V6 VIN A, 3.7L V6, 4.0L V6, 5.6L V8 **Transmission:** All	**Throttle Position (TP) Sensor 2 Circuit Low Input (Includes Hybrid Models):** HYBRID MODELS CAUTION: Hybrid systems use very high-voltage battery systems. Before starting any service work involving the battery system, turn the ignition switch OFF and then remove the service plug from pocket in the trunk. After removing the service plug, wait 10 minutes before touching any of the high-voltage connectors and terminals. **NOTE: If this DTC is displayed with DTC P0643, first perform the trouble diagnosis for DTC P0643.** **NOTE: On V6 and V8 engines, this DTC is for Bank 1.** • An excessively low voltage from the TP sensor 2 is sent to ECM. **Possible Causes:** • Harness or connectors • TP sensor 2 circuit is open or shorted. • Electric throttle control actuator (TP sensor 2) • Accelerator pedal position sensor (APP sensor 2)
DTC: P0123 **1T ECM, MIL: Yes** **Year:** 2009, 2010 **Model:** 350Z, 370Z, Altima, Armada, Cube, Frontier, Maxima, Murano, Pathfinder, Quest, Rogue, Sentra, Titan, Versa, Xterra **Engine:** 1.6L L4, 1.8L L4, 2.0L L4, 2.5L L4, 3.5L V6, 3.5L V6 VIN A, 3.7L V6, 4.0L V6, 5.6L V8 **Transmission:** All	**Throttle Position Sensor 2 Circuit High Input (Includes Hybrid Models):** HYBRID MODELS CAUTION: Hybrid systems use very high-voltage battery systems. Before starting any service work involving the battery system, turn the ignition switch OFF and then remove the service plug from pocket in the trunk. After removing the service plug, wait 10 minutes before touching any of the high-voltage connectors and terminals. **NOTE: If DTC P0122 or P0123 is displayed with DTC P0643, first perform the trouble diagnosis for DTC P0643.** **NOTE: On 3.7L engine, this DTC is for Bank 1.** • An excessively high voltage from the TP sensor 2 is sent to ECM. • When the malfunction is detected, ECM enters fail-safe mode and the MIL lights up. **Possible Causes:** • Harness or connectors • TP sensor 2 circuit is open or shorted. • Electric throttle control actuator (TP sensor 2) • Accelerator pedal position sensor (APP sensor 2)
DTC: P0125 **1T ECM, MIL: Yes** **Year:** 2009, 2010 **Model:** 350Z, 370Z, Altima, Armada, Cube, Frontier, Maxima, Murano, Pathfinder, Quest, Rogue, Sentra, Titan, Versa, Xterra **Engine:** 1.6L L4, 1.8L L4, 2.0L L4, 2.5L L4, 3.5L V6, 3.5L V6 VIN A, 3.7L V6, 4.0L V6, 5.6L V8 **Transmission:** All	**Insufficient Engine Coolant Temperature for Closed Loop Fuel Control (Includes Hybrid Models):** HYBRID MODELS CAUTION: Hybrid systems use very high-voltage battery systems. Before starting any service work involving the battery system, turn the ignition switch OFF and then remove the service plug from pocket in the trunk. After removing the service plug, wait 10 minutes before touching any of the high-voltage connectors and terminals. • If DTC P0125 is displayed with P0116, P0117 or P0118, first perform the trouble diagnosis for the appropriate DTC, then proceed with P0125. • Voltage sent to ECM from the sensor is not practical, even when some time has passed after starting the engine. • Engine coolant temperature is insufficient for closed loop fuel control. **Possible Causes:** • Harness or connectors • High resistance in the circuit • Engine coolant temperature sensor • Thermostat
DTC: P0127 **1T ECM, MIL: Yes** **Year:** 2009, 2010 **Model:** 350Z, 370Z, Altima, Armada, Cube, Frontier, Maxima, Murano, Pathfinder, Quest, Rogue, Sentra, Titan, Versa, Xterra **Engine:** 1.6L L4, 1.8L L4, 2.0L L4, 2.5L L4, 3.5L V6, 3.5L V6 VIN A, 3.7L V6, 4.0L V6, 5.6L V8 **Transmission:** All	**Intake Air Temperature Too High (Includes Hybrid Models):** HYBRID MODELS CAUTION: Hybrid systems use very high-voltage battery systems. Before starting any service work involving the battery system, turn the ignition switch OFF and then remove the service plug from pocket in the trunk. After removing the service plug, wait 10 minutes before touching any of the high-voltage connectors and terminals. • Rationally incorrect voltage from the sensor is sent to ECM, compared with the voltage signal from engine coolant temperature sensor. **Possible Causes:** • Harness or connectors • Intake temperature sensor circuit is open or shorted • Intake air temperature sensor
DTC: P0128 **1T ECM, MIL: Yes** **Year:** 2009, 2010 **Model:** 350Z, 370Z, Altima, Armada, Cube, Frontier, Murano, Pathfinder, Quest, Rogue, Sentra, Titan, Versa, Xterra **Engine:** 1.6L L4, 1.8L L4, 2.0L L4, 2.5L L4, 3.5L V6, 3.5L V6 VIN A, 3.7L V6, 4.0L V6, 5.6L V8 **Transmission:** All	**Thermostat Function :** NOTE: If DTC P0128 is displayed with DTC P0300, P0301, P0302, P0303, P0304, P0305, P0306, P0307 or P0308 first perform the trouble diagnosis for this DTC before continuing with P0128 diagnosis. • The engine coolant temperature does not reach to specified temperature even though the engine has run long enough. **Possible Causes:** • Thermostat • Leakage from sealing portion of thermostat • Engine coolant temperature sensor

DTC	Trouble Code Title, Conditions & Possible Causes

DTC: P0128
1T ECM, MIL: Yes
Year: 2009, 2010
Model: Altima
Engine: 2.5L L4
Transmission: All
	Thermostat Function (Hybrid):
	HYBRID MODELS CAUTION: Hybrid systems use very high-voltage battery systems. Before starting any service work involving the battery system, turn the ignition switch OFF and then remove the service plug from pocket in the trunk. After removing the service plug, wait 10 minutes before touching any of the high-voltage connectors and terminals.
	    • The engine coolant temperature does not reach to specified temperature even though the engine has run long enough.
	Possible Causes:
	    • Thermostat
	    • Leakage from sealing portion of thermostat
	    • Engine coolant temperature sensor

DTC: P0130
1T ECM, MIL: Yes
Year: 2009, 2010
Model: 350Z, 370Z, Altima, Armada, Cube, Frontier, Maxima, Murano, Pathfinder, Quest, Rogue, Sentra, Titan, Versa, Xterra
Engine: 1.6L L4, 1.8L L4, 2.0L L4, 2.5L L4, 3.5L V6, 3.5L V6 VIN A, 3.7L V6, 4.0L V6, 5.6L V8
Transmission: All
	Air Fuel Ratio (A/F) Sensor 1 Circuit (Includes Hybrid Models):
	HYBRID MODELS CAUTION: Hybrid systems use very high-voltage battery systems. Before starting any service work involving the battery system, turn the ignition switch OFF and then remove the service plug from pocket in the trunk. After removing the service plug, wait 10 minutes before touching any of the high-voltage connectors and terminals.
	NOTE: On V6 and V8 engines, this applies to Bank 1.
	    • Condition A: The A/F signal computed by ECM from the A/F sensor 1 signal is constantly in the range other than approx. 2.2V.
	    • Condition B: The A/F signal computed by ECM from the A/F sensor 1 signal is constantly approx. 2.2V.
	Possible Causes:
	    • Harness or connectors
	    • Air fuel ratio (A/F) sensor 1 circuit is open or shorted.
	    • Air fuel ratio (A/F) sensor 1

DTC: P0131
1T ECM, MIL: Yes
Year: 2009, 2010
Model: 350Z, 370Z, Altima, Armada, Cube, Frontier, Maxima, Murano, Pathfinder, Quest, Rogue, Sentra, Titan, Versa, Xterra
Engine: 1.6L L4, 1.8L L4, 2.0L L4, 2.5L L4, 3.5L V6, 3.5L V6 VIN A, 3.7L V6, 4.0L V6, 5.6L V8
Transmission: All
	Air Fuel Ratio (A/F) Sensor 1 Circuit Low Voltage (Includes Hybrid Models):
	HYBRID MODELS CAUTION: Hybrid systems use very high-voltage battery systems. Before starting any service work involving the battery system, turn the ignition switch OFF and then remove the service plug from pocket in the trunk. After removing the service plug, wait 10 minutes before touching any of the high-voltage connectors and terminals.
	NOTE: On V6 and V8 engines, this applies to Bank 1.
	    • To judge the malfunction, the diagnosis checks that the A/F signal computed by ECM from the A/F sensor 1 signal is not inordinately low.
	    • The A/F signal computed by ECM from the A/F sensor 1 signal is constantly approx. 0V.
	Possible Causes:
	    • Harness or connectors
	    • Air fuel ratio (A/F) sensor circuit is open or shorted.
	    • Air fuel ratio (A/F) sensor 1

DTC: P0132
1T ECM, MIL: Yes
Year: 2009, 2010
Model: 350Z, 370Z, Altima, Armada, Cube, Frontier, Maxima, Murano, Pathfinder, Quest, Rogue, Sentra, Titan, Versa, Xterra
Engine: 1.6L L4, 1.8L L4, 2.0L L4, 2.5L L4, 3.5L V6, 3.5L V6 VIN A, 3.7L V6, 4.0L V6, 5.6L V8
Transmission: All
	Air Fuel Ratio (A/F) Sensor 1 Circuit High Voltage (Includes Hybrid Models):
	HYBRID MODELS CAUTION: Hybrid systems use very high-voltage battery systems. Before starting any service work involving the battery system, turn the ignition switch OFF and then remove the service plug from pocket in the trunk. After removing the service plug, wait 10 minutes before touching any of the high-voltage connectors and terminals.
	NOTE: On V6 and V8 engines, this applies to Bank 1.
	    • To judge the malfunction, the diagnosis checks that the A/F signal computed by ECM from the A/F sensor 1 signal is not inordinately low.
	    • The A/F signal computed by ECM from the A/F sensor 1 signal is constantly approx. 5V.
	Possible Causes:
	    • Harness or connectors
	    • Air fuel ratio (A/F) sensor circuit is open or shorted.
	    • Air fuel ratio (A/F) sensor 1

DTC	Trouble Code Title, Conditions & Possible Causes
DTC: P0133 **1T ECM, MIL: Yes** **Year:** 2009, 2010 **Model:** 350Z, 370Z, Altima, Armada, Frontier, Maxima, Murano, Pathfinder, Quest, Rogue, Sentra, Titan, Versa, Xterra **Engine:** 1.6L L4, 1.8L L4, 2.0L L4, 2.5L L4, 3.5L V6, 3.5L V6 VIN A, 3.7L V6, 4.0L V6, 5.6L V8 **Transmission:** All	**Air Fuel Ratio (A/F) Sensor 1 Circuit Slow Response (Includes Hybrid Models):** HYBRID MODELS CAUTION: Hybrid systems use very high-voltage battery systems. Before starting any service work involving the battery system, turn the ignition switch OFF and then remove the service plug from pocket in the trunk. After removing the service plug, wait 10 minutes before touching any of the high-voltage connectors and terminals. **NOTE: On V6 and V8 engines, this applies to Bank 1.** • To judge the malfunction of A/F sensor 1, this diagnosis measures response time of the A/F signal computed by ECM from the A/F sensor 1 signal. The time is compensated by engine operating (speed and load), fuel feedback control constant, and the A/F sensor 1 temperature index. Judgment is based on whether the compensated time (the A/F signal cycling time index) is inordinately long or not. • The response of the A/F signal computed by ECM from A/F sensor 1 signal takes more than the specified time. **Possible Causes:** • Harness or connectors • Air fuel ratio (A/F) sensor circuit is open or shorted. • Air fuel ratio (A/F) sensor 1 • Air fuel ratio (A/F) sensor heater 1 • Fuel pressure • Fuel injector • Intake air leaks • Exhaust gas leaks • PCV valve • Mass air flow sensor
DTC: P0137 **1T ECM, MIL: Yes** **Year:** 2009, 2010 **Model:** 350Z, 370Z, Altima, Armada, Cube, Frontier, Maxima, Murano, Pathfinder, Quest, Rogue, Sentra, Titan, Versa, Xterra **Engine:** 1.6L L4, 1.8L L4, 2.0L L4, 2.5L L4, 3.5L V6, 3.5L V6 VIN A, 3.7L V6, 4.0L V6, 5.6L V8 **Transmission:** All	**Heated Oxygen Sensor 2 Circuit Low Voltage (Includes Hybrid Models):** HYBRID MODELS CAUTION: Hybrid systems use very high-voltage battery systems. Before starting any service work involving the battery system, turn the ignition switch OFF and then remove the service plug from pocket in the trunk. After removing the service plug, wait 10 minutes before touching any of the high-voltage connectors and terminals. **NOTE: On V6 and V8 engines, this applies to Bank 1.** • The maximum voltage from the sensor is not reached to the specified voltage **Possible Causes:** • Harness or connectors • Heated oxygen sensor 2 circuit open or shorted. • Heated oxygen sensor 2 • Fuel pressure • Fuel injector • Intake air leaks
DTC: P0138 **1T ECM, MIL: Yes** **Year:** 2009, 2010 **Model:** 350Z, 370Z, Altima, Armada, Cube, Frontier, Maxima, Murano, Pathfinder, Quest, Rogue, Sentra, Titan, Versa, Xterra **Engine:** 1.6L L4, 1.8L L4, 2.0L L4, 2.5L L4, 3.5L V6, 3.5L V6 VIN A, 3.7L V6, 4.0L V6, 5.6L V8 **Transmission:** All	**Heated Oxygen Sensor 2 Circuit High Voltage (Includes Hybrid Models):** HYBRID MODELS CAUTION: Hybrid systems use very high-voltage battery systems. Before starting any service work involving the battery system, turn the ignition switch OFF and then remove the service plug from pocket in the trunk. After removing the service plug, wait 10 minutes before touching any of the high-voltage connectors and terminals. **NOTE: On V6 and V8 engines, this applies to Bank 1.** • Condition A: An excessively high voltage from the sensor is sent to ECM, or, • Condition B: The minimum voltage from the sensor is not reached to the specified voltage. **Possible Causes:** • Condition A: • Harness or connectors • Heated oxygen sensor 2 circuit is open or shorted. • Heated oxygen sensor 2 • Condition B: • Harness or connectors • Heated oxygen sensor circuit is open or shorted. • Heated oxygen sensor 2 • Fuel pressure • Fuel injector

DTC	Trouble Code Title, Conditions & Possible Causes
DTC: P0139 **1T ECM, MIL: Yes** **Year:** 2009, 2010 **Model:** 350Z, 370Z, Altima, Armada, Rogue, Sentra, Titan, Versa, Xterra **Engine:** 1.6L L4, 1.8L L4, 2.0L L4, 2.5L L4, 3.5L V6, 3.7L V6, 4.0L V6, 5.6L V8 **Transmission:** All	**Heated Oxygen Sensor 2 Circuit Slow Response (Includes Hybrid Models):** HYBRID MODELS CAUTION: Hybrid systems use very high-voltage battery systems. Before starting any service work involving the battery system, turn the ignition switch OFF and then remove the service plug from pocket in the trunk. After removing the service plug, wait 10 minutes before touching any of the high-voltage connectors and terminals. **NOTE: On 2.0L & 2.5L, Sensor 2; on 3.5L, Sensor 2 Bank 1** • It takes more time for the sensor to respond between rich and lean than the specified time. **Possible Causes:** • Harness or connectors • Heated oxygen sensor circuit is open or shorted. • Heated oxygen sensor 2 • Fuel pressure • Fuel injector • Intake air leaks
DTC: P0143 **1T ECM, MIL: Yes** **Year:** 2009, 2010 **Model:** Altima, Maxima, Murano, Pathfinder, Quest, Rogue, Sentra **Engine:** 2.5L L4, 3.5L V6, 3.5L V6 VIN A, 4.0L V6, 5.6L V8 **Transmission:** All	**Heated Oxygen Sensor 3 Circuit High Voltage (Includes Hybrid Models):** HYBRID MODELS CAUTION: Hybrid systems use very high-voltage battery systems. Before starting any service work involving the battery system, turn the ignition switch OFF and then remove the service plug from pocket in the trunk. After removing the service plug, wait 10 minutes before touching any of the high-voltage connectors and terminals. The minimum voltage from the sensor is not reached to the specified voltage. **Possible Causes:** • Harness or connectors • The sensor circuit is open or shorted • Heated oxygen sensor 3 • Fuel pressure • Fuel injector
DTC: P0144 **1T ECM, MIL: Yes** **Year:** 2009, 2010 **Model:** Altima, Armada, Frontier, Maxima, Murano, Pathfinder, Quest, Rogue **Engine:** 2.5L L4, 3.5L V6, 3.5L V6 VIN A, 4.0L V6, 5.6L V8 **Transmission:** All	**Heated Oxygen Sensor 3 Circuit Low Voltage (Includes Hybrid Models) :** HYBRID MODELS CAUTION: Hybrid systems use very high-voltage battery systems. Before starting any service work involving the battery system, turn the ignition switch OFF and then remove the service plug from pocket in the trunk. After removing the service plug, wait 10 minutes before touching any of the high-voltage connectors and terminals. • The maximum voltage from the sensor is not reached to the specified voltage. **Possible Causes:** • Harness or connectors • The sensor circuit is open or shorted • Heated oxygen sensor 3 • Fuel pressure • Fuel injector • Intake air leaks
DTC: P0145 **1T ECM, MIL: Yes** **Year:** 2009, 2010 **Model:** Altima, Armada, Frontier, Maxima, Pathfinder, Quest, Rogue **Engine:** 2.5L L4, 3.5L V6, 4.0L V6, 5.6L V8 **Transmission:** All	**Heated Oxygen Sensor 3 Circuit Slow Response (Includes Hybrid Models):** HYBRID MODELS CAUTION: Hybrid systems use very high-voltage battery systems. Before starting any service work involving the battery system, turn the ignition switch OFF and then remove the service plug from pocket in the trunk. After removing the service plug, wait 10 minutes before touching any of the high-voltage connectors and terminals. • It takes more time for the sensor to respond between rich and lean than the specified time. **Possible Causes:** • Harness or connectors • The sensor circuit is open or shorted • Heated oxygen sensor 3 • Fuel pressure • Fuel injector • Intake air leaks
DTC: P0146 **1T ECM, MIL: Yes** **Year:** 2009, 2010 **Model:** Altima, Armada, Frontier, Maxima, Murano, Pathfinder, Quest, Rogue **Engine:** 2.5L L4, 3.5L V6, 3.5L V6 VIN A, 4.0L V6, 5.6L V8 **Transmission:** All	**Heated Oxygen Sensor 3 Circuit No Response Detected (Includes Hybrid Models):** HYBRID MODELS CAUTION: Hybrid systems use very high-voltage battery systems. Before starting any service work involving the battery system, turn the ignition switch OFF and then remove the service plug from pocket in the trunk. After removing the service plug, wait 10 minutes before touching any of the high-voltage connectors and terminals. • An excessively high voltage from the sensor is sent to ECM. **Possible Causes:** • Harness or connectors • The sensor circuit is open or shorted • Heated oxygen sensor 3

DTC	Trouble Code Title, Conditions & Possible Causes

DTC: P014C
1T ECM, MIL: Yes
Year: 2009, 2010
Model: Altima, Versa
Engine: 1.8L L4, 2.5L L4, 3.5L V6
Transmission: All
	Air Fuel Ratio (A/F) Sensor 1 Circuit Slow Response:
	To judge the malfunction of A/F sensor 1, this diagnosis measures response time of the A/F signal computed by ECM from the A/F sensor 1 signal. The time is compensated by engine operating (speed and load), fuel feedback control constant, and the A/F sensor 1 temperature index. Judgment is based on whether the compensated time (the A/F signal cycling time index) is inordinately long or not.
	   • The response of the A/F signal computed by ECM from A/F sensor 1 signal takes more than the specified time.
	Possible Causes:
	   • Harness or connectors (The A/F sensor 1 circuit is open or shorted.)
	   • A/F sensor 1
	   • A/F sensor 1 heater
	   • Fuel pressure
	   • Fuel injector
	   • Intake air leaks
	   • Exhaust gas leaks
	   • PCV
	   • Mass air flow sensor

DTC: P014D
1T ECM, MIL: Yes
Year: 2009, 2010
Model: Altima, Versa
Engine: 1.8L L4, 2.5L L4, 3.5L V6
Transmission: All
	Air Fuel Ratio (A/F) Sensor 1 Circuit Slow Response:
	To judge the malfunction of A/F sensor 1, this diagnosis measures response time of the A/F signal computed by ECM from the A/F sensor 1 signal. The time is compensated by engine operating (speed and load), fuel feedback control constant, and the A/F sensor 1 temperature index. Judgment is based on whether the compensated time (the A/F signal cycling time index) is inordinately long or not.
	   • The response of the A/F signal computed by ECM from A/F sensor 1 signal takes more than the specified time.
	Possible Causes:
	   • Harness or connectors (The A/F sensor 1 circuit is open or shorted.)
	   • A/F sensor 1
	   • A/F sensor 1 heater
	   • Fuel pressure
	   • Fuel injector
	   • Intake air leaks
	   • Exhaust gas leaks
	   • PCV
	   • Mass air flow sensor

DTC: P0150
1T ECM, MIL: Yes
Year: 2009, 2010
Model: 350Z, 370Z, Altima, Armada, Maxima, Murano, Pathfinder, Quest, Titan, Xterra
Engine: 3.5L V6, 3.5L V6 VIN A, 3.7L V6, 4.0L V6, 5.6L V8
Transmission: All
	Air Fuel Ratio (A/F) Sensor 1 Bank 2 Circuit:
	   • The A/F signal computed by ECM from the A/F sensor 1 bank 2 signal is constantly in the range other than approx. 2.2V.
	   • The A/F signal computed by ECM from the A/F sensor 1 bank 2 signal is constantly approx. 2.2V.
	Possible Causes:
	   • Harness or connectors
	   • The A/F sensor 1 circuit is open or shorted.
	   • A/F sensor 1 bank 2

DTC: P0151
1T ECM, MIL: Yes
Year: 2009, 2010
Model: 350Z, 370Z, Altima, Armada, Maxima, Murano, Pathfinder, Quest, Titan, Xterra
Engine: 3.5L V6, 3.5L V6 VIN A, 3.7L V6, 4.0L V6, 5.6L V8
Transmission: All
	Air Fuel Ratio (A/F) Sensor 1 Bank 2 Circuit Low Voltage:
	   • To judge the malfunction, the diagnosis checks that the A/F signal computed by ECM from the A/F sensor 1 signal is not inordinately low.
	   • The A/F signal computed by ECM from the A/F sensor 1 signal is constantly approx. 0V.
	Possible Causes:
	   • Harness or connectors
	   • Air fuel ratio (A/F) sensor circuit is open or shorted.
	   • Air fuel ratio (A/F) sensor 1

DTC: P0152
1T ECM, MIL: Yes
Year: 2009, 2010
Model: 350Z, 370Z, Altima, Armada, Murano, Pathfinder, Quest, Titan, Xterra
Engine: 2.5L L4, 3.5L V6, 3.5L V6 VIN A, 3.7L V6, 4.0L V6, 5.6L V8
Transmission: All
	Air Fuel Ratio (A/F) Sensor 1 (Bank 2) Circuit High Voltage:
	NOTE: On V6 and V8 engines, this applies to Bank 2.
	   • Engine: After warming up
	   • Maintaining engine speed at 2,000 rpm
	   • The A/F signal computed by ECM from the A/F sensor 1 signal is constantly approx. 5V.
	Possible Causes:
	   • Harness or connectors
	   • Air fuel ratio (A/F) sensor circuit is open or shorted.
	   • Air fuel ratio (A/F) sensor 1

DTC	Trouble Code Title, Conditions & Possible Causes
DTC: P0153 **1T ECM, MIL: Yes** **Year:** 2009, 2010 **Model:** 350Z, 370Z, Altima, Armada, Maxima, Murano, Pathfinder, Quest, Titan, Xterra **Engine:** 3.5L V6, 3.5L V6 VIN A, 3.7L V6, 4.0L V6, 5.6L V8 **Transmission:** All	**Air Fuel Ratio (A/F) Sensor 1 Bank 2 Circuit Slow Response:** • The response of the A/F signal computed by ECM from A/F sensor 1 signal takes more than the specified time. **Possible Causes:** • Harness or connectors • Air fuel ratio (A/F) sensor circuit is open or shorted. • Air fuel ratio (A/F) sensor 1 • Air fuel ratio (A/F) sensor heater 1 • Fuel pressure • Fuel injector • Intake air leaks • Exhaust gas leaks • PCV valve • Mass air flow sensor
DTC: P0157 **1T ECM, MIL: Yes** **Year:** 2009, 2010 **Model:** 350Z, 370Z, Altima, Armada, Maxima, Murano, Pathfinder, Quest, Titan, Xterra **Engine:** 3.5L V6, 3.5L V6 VIN A, 3.7L V6, 4.0L V6, 5.6L V8 **Transmission:** All	**Heated Oxygen Sensor 2 Bank 2 Circuit Low Voltage:** • The maximum voltage from the sensor does not reach the specified voltage. **Possible Causes:** • Harness or connectors • Sensor circuit is open or shorted • Heated oxygen sensor 2 • Fuel pressure • Fuel injector • Intake air leakage
DTC: P0158 **1T ECM, MIL: Yes** **Year:** 2009, 2010 **Model:** 350Z, 370Z, Altima, Armada, Maxima, Murano, Pathfinder, Quest, Titan, Xterra **Engine:** 3.5L V6, 3.5L V6 VIN A, 3.7L V6, 4.0L V6, 5.6L V8 **Transmission:** All	**Heated Oxygen Sensor 2 Bank 2 Circuit High Voltage:** • Condition A: An excessively high voltage from the sensor is sent to ECM, or, • Condition B: The minimum voltage from the sensor is not reached to the specified voltage. **Possible Causes:** • Condition A: • Harness or connectors (Heated oxygen sensor 2 circuit is open or shorted.) • Heated oxygen sensor 2 • Condition B: • Harness or connectors (Heated oxygen sensor circuit is open or shorted.) • Heated oxygen sensor 2 • Fuel pressure • Fuel injector
DTC: P0159 **1T ECM, MIL: Yes** **Year:** 2009, 2010 **Model:** 350Z, 370Z, Altima, Armada, Maxima, Murano, Pathfinder, Quest, Titan, Xterra **Engine:** 3.5L V6, 3.5L V6 VIN A, 3.7L V6, 4.0L V6, 5.6L V8 **Transmission:** All	**Heated Oxygen Sensor 2 Bank 2 Circuit Slow Response:** • It takes more time for the sensor to respond between rich and lean than the specified time. **Possible Causes:** • Harness or connectors • Heated oxygen sensor circuit is open or shorted. • Heated oxygen sensor 2 • Fuel pressure • Fuel injector • Intake air leaks
DTC: P015A **T ECM, MIL: Yes** **Year:** 2010 **Model:** Versa **Engine:** 1.8L L4 **Transmission:** All	**Air Fuel (A/F) Ration Sensor 1 (Bank 2) Circuit Delayed Response:** • The response time of the A/F sensor 1 signal delays more than the specified time computed by the ECM. **Possible Causes:** • Harness or connectors • The A/F sensor 1 circuit is open or shorted • A/F sensor
DTC: P015B **T ECM, MIL: Yes** **Year:** 2010 **Model:** Versa **Engine:** 1.8L L4 **Transmission:** All	**Air Fuel (A/F) Ration Sensor 1 (Bank 2) Circuit Delayed Response:** • The response time of the A/F sensor 1 signal delays more than the specified time computed by the ECM. **Possible Causes:** • Harness or connectors • The A/F sensor 1 circuit is open or shorted • A/F sensor

DTC	Trouble Code Title, Conditions & Possible Causes
DTC: P0171 **1T ECM, MIL: Yes** **Year:** 2009, 2010 **Model:** 350Z, 370Z, Altima, Armada, Cube, Frontier, Maxima, Murano, Pathfinder, Quest, Rogue, Sentra, Titan, Versa, Xterra **Engine:** 1.6L L4, 1.8L L4, 2.0L L4, 2.5L L4, 3.5L V6, 3.5L V6 VIN A, 3.7L V6, 4.0L V6, 5.6L V8 **Transmission:** All	**Fuel Injection System Too Lean (Includes Hybrid Models):** HYBRID MODELS CAUTION: Hybrid systems use very high-voltage battery systems. Before starting any service work involving the battery system, turn the ignition switch OFF and then remove the service plug from pocket in the trunk. After removing the service plug, wait 10 minutes before touching any of the high-voltage connectors and terminals. **NOTE: On V6 and V8 engines, this applies to Bank 1.** • Fuel injection system does not operate properly. • The amount of mixture ratio compensation is too large. (The mixture ratio is too lean.) **Possible Causes:** • Intake air leaks • Air fuel ratio (A/F) sensor 1 • Fuel injector • Exhaust gas leaks • Incorrect fuel pressure • Lack of fuel • Mass air flow sensor • Incorrect PCV hose connection
DTC: P0172 **1T ECM, MIL: Yes** **Year:** 2009, 2010 **Model:** 350Z, 370Z, Altima, Armada, Cube, Frontier, Maxima, Murano, Pathfinder, Quest, Rogue, Sentra, Titan, Versa, Xterra **Engine:** 1.6L L4, 1.8L L4, 2.0L L4, 2.5L L4, 3.5L V6, 3.5L V6 VIN A, 3.7L V6, 4.0L V6, 5.6L V8 **Transmission:** All	**Fuel Injection System Too Rich (Includes Hybrid Models):** HYBRID MODELS CAUTION: Hybrid systems use very high-voltage battery systems. Before starting any service work involving the battery system, turn the ignition switch OFF and then remove the service plug from pocket in the trunk. After removing the service plug, wait 10 minutes before touching any of the high-voltage connectors and terminals. **NOTE: On V6 and V8 engines, this applies to Bank 1.** • Fuel injection system does not operate properly. • The amount of mixture ratio compensation is too large. (The mixture ratio is too rich.) **Possible Causes:** • Air fuel ratio (A/F) sensor 1 • Fuel injector • Exhaust gas leaks • Incorrect fuel pressure • Mass air flow sensor
DTC: P0174 **1T ECM, MIL: Yes** **Year:** 2009, 2010 **Model:** 350Z, 370Z, Altima, Armada, Maxima, Murano, Pathfinder, Quest, Titan, Xterra **Engine:** 3.5L V6, 3.5L V6 VIN A, 3.7L V6, 4.0L V6, 5.6L V8 **Transmission:** All	**Fuel Injection System Too Lean (Bank 2):** • Fuel injection system does not operate properly. • The amount of mixture ratio compensation is too large. (The mixture ratio is too lean.) **Possible Causes:** • Intake air leaks • Air fuel ratio (A/F) sensor 1 • Fuel injector • Exhaust gas leaks • Incorrect fuel pressure • Lack of fuel • Mass air flow sensor • Incorrect PCV hose connection
DTC: P0175 **1T ECM, MIL: Yes** **Year:** 2009, 2010 **Model:** 350Z, 370Z, Altima, Armada, Maxima, Murano, Pathfinder, Quest, Titan, Xterra **Engine:** 3.5L V6, 3.5L V6 VIN A, 3.7L V6, 4.0L V6, 5.6L V8 **Transmission:** All	**Fuel Injection System Too Rich (Bank 2):** • Fuel injection system does not operate properly. • The amount of mixture ratio compensation is too large. (The mixture ratio is too rich.) **Possible Causes:** • Air fuel ratio (A/F) sensor 1 • Fuel injector • Exhaust gas leaks • Incorrect fuel pressure • Mass air flow sensor
DTC: P0181 **1T ECM, MIL: Yes** **Year:** 2009, 2010 **Model:** 350Z, 370Z, Altima, Armada, Cube, Frontier, Maxima, Murano, Pathfinder, Quest, Rogue, Sentra, Titan, Versa, Xterra **Engine:** 1.6L L4, 1.8L L4, 2.0L L4, 2.5L L4, 3.5L V6, 3.5L V6 VIN A, 3.7L V6, 4.0L V6, 5.6L V8 **Transmission:** All	**Fuel Tank Temperature Sensor Circuit Range/Performance (Includes Hybrid Models):** HYBRID MODELS CAUTION: Hybrid systems use very high-voltage battery systems. Before starting any service work involving the battery system, turn the ignition switch OFF and then remove the service plug from pocket in the trunk. After removing the service plug, wait 10 minutes before touching any of the high-voltage connectors and terminals. • Rationally incorrect voltage from the sensor is sent to ECM, compared with the voltage signals from engine coolant temperature sensor and intake air temperature sensor. **Possible Causes:** • Harness or connectors • Fuel tank temperature sensor circuit is open or shorted • Fuel tank temperature sensor

DTC	Trouble Code Title, Conditions & Possible Causes
DTC: P0182 **1T ECM, MIL: Yes** **Year:** 2009, 2010 **Model:** 350Z, 370Z, Altima, Armada, Cube, Frontier, Maxima, Murano, Pathfinder, Quest, Rogue, Sentra, Titan, Versa, Xterra **Engine:** 1.6L L4, 1.8L L4, 2.0L L4, 2.5L L4, 3.5L V6, 3.5L V6 VIN A, 3.7L V6, 4.0L V6, 5.6L V8 **Transmission:** All	**Fuel Tank Temperature Sensor Circuit Low Input (Includes Hybrid Models):** HYBRID MODELS CAUTION: Hybrid systems use very high-voltage battery systems. Before starting any service work involving the battery system, turn the ignition switch OFF and then remove the service plug from pocket in the trunk. After removing the service plug, wait 10 minutes before touching any of the high-voltage connectors and terminals. • An excessively low voltage from the sensor is sent to ECM. **Possible Causes:** • Harness or connectors • Fuel tank temperature sensor circuit is open or shorted. • Fuel tank temperature sensor
DTC: P0183 **1T ECM, MIL: Yes** **Year:** 2009, 2010 **Model:** 350Z, 370Z, Altima, Armada, Cube, Frontier, Maxima, Murano, Pathfinder, Quest, Rogue, Sentra, Titan, Xterra **Engine:** 1.8L L4, 2.0L L4, 2.5L L4, 3.5L V6, 3.5L V6 VIN A, 3.7L V6, 4.0L V6, 5.6L V8 **Transmission:** All	**Fuel Tank Temperature Sensor Circuit High Input (Includes Hybrid Models):** HYBRID MODELS CAUTION: Hybrid systems use very high-voltage battery systems. Before starting any service work involving the battery system, turn the ignition switch OFF and then remove the service plug from pocket in the trunk. After removing the service plug, wait 10 minutes before touching any of the high-voltage connectors and terminals. • An excessively high voltage from the sensor is sent to ECM. **Possible Causes:** • Harness or connectors • Fuel tank temperature sensor circuit is open or shorted. • Fuel tank temperature sensor
DTC: P0196 **1T ECM, MIL: Yes** **Year:** 2009, 2010 **Model:** 350Z, 370Z, Altima, Maxima, Murano, Sentra **Engine:** 2.0L L4, 2.5L L4, 3.5L V6, 3.5L V6 VIN A, 3.7L V6 **Transmission:** All	**Engine Oil Temperature (EOT) Sensor Range/Performance:** NOTE: If DTC P0196 is displayed with P0197 or P0198, first perform the trouble diagnosis for DTC P0197, P0198. • Rationally incorrect voltage from the sensor is sent to ECM, compared with the voltage signals from engine coolant temperature sensor and intake air temperature sensor. **Possible Causes:** • Harness or connectors (The sensor circuit is open or shorted) • Engine oil temperature sensor
DTC: P0197 **1T ECM, MIL: Yes** **Year:** 2009, 2010 **Model:** 350Z, 370Z, Altima, Maxima, Murano, Sentra **Engine:** 2.0L L4, 2.5L L4, 3.5L V6, 3.5L V6 VIN A, 3.7L V6 **Transmission:** All	**Engine Oil Temperature (EOT) Sensor Circuit Low Input:** • An excessively low voltage from the sensor is sent to ECM. **Possible Causes:** • Harness or connectors (Engine oil temperature sensor circuit is open or shorted.) • Engine oil temperature sensor.
DTC: P0198 **1T ECM, MIL: Yes** **Year:** 2009, 2010 **Model:** 350Z, 370Z, Altima, Maxima, Murano, Sentra **Engine:** 2.0L L4, 2.5L L4, 3.5L V6, 3.5L V6 VIN A, 3.7L V6 **Transmission:** All	**Engine Oil Temperature (EOT) Sensor Circuit High Input:** • An excessively high voltage from the sensor is sent to ECM. **Possible Causes:** • Harness or connectors (Engine oil temperature sensor circuit is open or shorted.) • Engine oil temperature sensor
DTC: P0201 **T ECM, MIL: Yes** **Year:** 2009, 2010 **Model:** Altima **Engine:** 2.5L L4 **Transmission:** All	**No. 1 Cylinder Fuel Injector Circuit Open (Hybrid):** HYBRID MODELS CAUTION: Hybrid systems use very high-voltage battery systems. Before starting any service work involving the battery system, turn the ignition switch OFF and then remove the service plug from pocket in the trunk. After removing the service plug, wait 10 minutes before touching any of the high-voltage connectors and terminals. • An excessively low voltage signal is sent to ECM through the No. 1 fuel injector **Possible Causes:** • Harness or connectors • No. 1 fuel injector circuit is open or shorted. • No. 1 fuel injector

DTC	Trouble Code Title, Conditions & Possible Causes

DTC: P0202
T ECM, MIL: Yes
Year: 2009, 2010
Model: Altima
Engine: 2.5L L4
Transmission: All
	No. 2 Cylinder Fuel Injector Circuit Open (Hybrid):
	HYBRID MODELS CAUTION: Hybrid systems use very high-voltage battery systems. Before starting any service work involving the battery system, turn the ignition switch OFF and then remove the service plug from pocket in the trunk. After removing the service plug, wait 10 minutes before touching any of the high-voltage connectors and terminals.
	    • No. 2 cylinder fuel injector circuit open
	Possible Causes:
	    • Harness or connectors
	    • No. 2 fuel injector circuit is open or shorted.
	    • No. 2 fuel injector

DTC: P0203
T ECM, MIL: Yes
Year: 2009, 2010
Model: Altima
Engine: 2.5L L4
Transmission: All
	No. 3 Cylinder Fuel Injector Circuit Open (Hybrid):
	HYBRID MODELS CAUTION: Hybrid systems use very high-voltage battery systems. Before starting any service work involving the battery system, turn the ignition switch OFF and then remove the service plug from pocket in the trunk. After removing the service plug, wait 10 minutes before touching any of the high-voltage connectors and terminals.
	An excessively low voltage signal is sent to ECM through the No. 3 fuel injector
	Possible Causes:
	    • Harness or connectors
	    • No. 3 fuel injector circuit is open or shorted.
	    • No. 3 fuel injector

DTC: P0204
T ECM, MIL: Yes
Year: 2009, 2010
Model: Altima
Engine: 2.5L L4
Transmission: All
	No. 4 Cylinder Fuel Injector Circuit Open (Hybrid):
	HYBRID MODELS CAUTION: Hybrid systems use very high-voltage battery systems. Before starting any service work involving the battery system, turn the ignition switch OFF and then remove the service plug from pocket in the trunk. After removing the service plug, wait 10 minutes before touching any of the high-voltage connectors and terminals.
	An excessively low voltage signal is sent to ECM through the No. 4 fuel injector
	Possible Causes:
	    • Harness or connectors
	    • No. 4 fuel injector circuit is open or shorted.
	    • No. 4 fuel injector

DTC: P0222
1T ECM, MIL: Yes
Year: 2009, 2010
Model: 350Z, 370Z, Altima, Armada, Cube, Frontier, Maxima, Murano, Pathfinder, Quest, Rogue, Sentra, Titan, Versa, Xterra
Engine: 1.6L L4, 1.8L L4, 2.0L L4, 2.5L L4, 3.5L V6, 3.5L V6 VIN A, 3.7L V6, 4.0L V6, 5.6L V8
Transmission: All
	Throttle Position (TP) Sensor 1 Circuit Low Input (Includes Hybrid Models):
	HYBRID MODELS CAUTION: Hybrid systems use very high-voltage battery systems. Before starting any service work involving the battery system, turn the ignition switch OFF and then remove the service plug from pocket in the trunk. After removing the service plug, wait 10 minutes before touching any of the high-voltage connectors and terminals.
	NOTE: If DTC P0222 or P0223 is displayed with DTC P0643, first perform the trouble diagnosis for DTC P0643.
	NOTE: On V6 and V8 engines, this DTC is for Bank 1.
	    • An excessively low voltage from the TP sensor 1 is sent to ECM.
	Possible Causes:
	    • Harness or connectors
	    • TP sensor 1 circuit is open or shorted.
	    • APP sensor circuit is shorted (if equipped)
	    • Electric throttle control actuator (TP sensor 1)
	    • Accelerator pedal position sensor (APP sensor 2)

DTC: P0223
1T ECM, MIL: Yes
Year: 2009, 2010
Model: 350Z, 370Z, Altima, Armada, Cube, Frontier, Maxima, Murano, Pathfinder, Quest, Rogue, Sentra, Titan, Versa, Xterra
Engine: 1.6L L4, 1.8L L4, 2.0L L4, 2.5L L4, 3.5L V6, 3.5L V6 VIN A, 3.7L V6, 4.0L V6, 5.6L V8
Transmission: All
	Throttle Position (TP) Sensor 1 Circuit High Input (Includes Hybrid Models):
	HYBRID MODELS CAUTION: Hybrid systems use very high-voltage battery systems. Before starting any service work involving the battery system, turn the ignition switch OFF and then remove the service plug from pocket in the trunk. After removing the service plug, wait 10 minutes before touching any of the high-voltage connectors and terminals.
	NOTE: If DTC P0222 or P0223 is displayed with DTC P0643, first perform the trouble diagnosis for DTC P0643.
	NOTE: On V6 and V8 engines, this DTC is for Bank 1.
	    • An excessively high voltage from the TP sensor 1 is sent to ECM.
	Possible Causes:
	    • Harness or connectors
	    • TP sensor 1 circuit is open or shorted.
	    • APP sensor 2 circuit is shorted.
	    • Electric throttle control actuator (TP sensor 1)
	    • Accelerator pedal position sensor (APP sensor 2)

DTC: P0227
1T ECM, MIL: Yes
Year: 2009, 2010
Model: 370Z
Engine: 3.7L V6
Transmission: All
	Throttle Position Sensor 2 (Bank 2) Circuit Low Input:
	NOTE: If DTC P0122, P0123, P0227 or P0228 is displayed with DTC P0643, first perform the trouble diagnosis for DTC P0643.
	    • An excessively low voltage from the TP sensor 2 is sent to ECM.
	Possible Causes:
	    • Harness or connectors
	    • TP sensor 2 circuit is open or shorted.
	    • Electric throttle control actuator (TP sensor 2)

DTC	Trouble Code Title, Conditions & Possible Causes

DTC: P0228
1T ECM, MIL: Yes
Year: 2009, 2010
Model: 370Z
Engine: 3.7L V6
Transmission: All
	Throttle Position Sensor 2 (Bank 2) Circuit High Input:
	NOTE: If DTC P0122, P0123, P0227 or P0228 is displayed with DTC P0643, first perform the trouble diagnosis for DTC P0643.
	    • An excessively high voltage from the TP sensor 2 is sent to ECM.
	Possible Causes:
	    • Harness or connectors
	    • TP sensor 2 circuit is open or shorted.
	    • Electric throttle control actuator (TP sensor 2)

DTC: P0300
1T ECM, MIL: Yes
Year: 2009, 2010
Model: 350Z, 370Z, Altima, Armada, Cube, Frontier, Maxima, Murano, Pathfinder, Quest, Rogue, Sentra, Titan, Versa, Xterra
Engine: 1.6L L4, 1.8L L4, 2.0L L4, 2.5L L4, 3.5L V6, 3.5L V6 VIN A, 3.7L V6, 4.0L V6, 5.6L V8
Transmission: All
	Multiple Cylinder Misfire Detected (Includes Hybrid Models):
	HYBRID MODELS CAUTION: Hybrid systems use very high-voltage battery systems. Before starting any service work involving the battery system, turn the ignition switch OFF and then remove the service plug from pocket in the trunk. After removing the service plug, wait 10 minutes before touching any of the high-voltage connectors and terminals.
	    • Multiple cylinder misfire.
	    • One Trip Detection Logic (Three Way Catalyst Damage)
	NOTE: On the 1st trip, when a misfire condition occurs that can damage the three way catalyst (TWC) due to overheating, the MIL will blink.
	    • When a misfire condition occurs, the ECM monitors the CKP sensor (POS) signal every 200 engine revolutions for a change.
	    • When the misfire condition decreases to a level that will not damage the TWC, the MIL will turn off.
	    • If another misfire condition occurs that can damage the TWC on a second trip, the MIL will blink.
	    • Two Trip Detection Logic (Exhaust quality deterioration)
	NOTE: For misfire conditions that will not damage the TWC (but will affect vehicle emissions), the MIL will only light when the misfire is detected on a second trip. During this condition, the ECM monitors the CKP sensor signal every 1,000 engine revolutions.
	    • A misfire malfunction can be detected in any one cylinder or in multiple cylinders.
	Possible Causes:
	    • Improper spark plug
	    • Insufficient compression
	    • Incorrect fuel pressure
	    • Fuel injector circuit is open or shorted
	    • Fuel injector
	    • Intake air leak
	    • The ignition signal circuit is open or shorted
	    • Lack of fuel
	    • Drive plate or flywheel
	    • Air fuel ratio (A/F) sensor 1
	    • Incorrect PCV hose connection

DTC	Trouble Code Title, Conditions & Possible Causes

DTC: P0301
1T ECM, MIL: Yes
Year: 2009, 2010
Model: 350Z, 370Z, Altima, Armada, Cube, Frontier, Maxima, Murano, Pathfinder, Quest, Rogue, Sentra, Titan, Versa, Xterra
Engine: 1.6L L4, 1.8L L4, 2.0L L4, 2.5L L4, 3.5L V6, 3.5L V6 VIN A, 3.7L V6, 4.0L V6, 5.6L V8
Transmission: All
	No.1 Cylinder Misfire Detected (Includes Hybrid Models):
	HYBRID MODELS CAUTION: Hybrid systems use very high-voltage battery systems. Before starting any service work involving the battery system, turn the ignition switch OFF and then remove the service plug from pocket in the trunk. After removing the service plug, wait 10 minutes before touching any of the high-voltage connectors and terminals.
	   • No. 1 cylinder misfires.
	1. One Trip Detection Logic (Three Way Catalyst Damage)
	   • On the 1st trip, when a misfire condition occurs that can damage the three way catalyst (TWC) due to overheating, the MIL will blink.
	   • When a misfire condition occurs, the ECM monitors the CKP sensor (POS) signal every 200 engine revolutions for a change.
	   • When the misfire condition decreases to a level that will not damage the TWC, the MIL will turn off.
	2. Two Trip Detection Logic (Exhaust quality deterioration)
	   • For misfire conditions that will not damage the TWC (but will affect vehicle emissions), the MIL will only light when the misfire is detected on a second trip.
	   • During this condition, the ECM monitors the CKP sensor signal every 1,000 engine revolutions.
	   • A misfire malfunction can be detected on any one cylinder or on multiple cylinders.
	   • If another misfire condition occurs that can damage the TWC on a second trip, the MIL will blink.
	Possible Causes:
	   • Improper spark plug
	   • Insufficient compression
	   • Incorrect fuel pressure
	   • Fuel injector circuit is open or shorted
	   • Fuel injector
	   • Intake air leak
	   • The ignition signal circuit is open or shorted
	   • Lack of fuel
	   • Drive plate or flywheel
	   • Air fuel ratio (A/F) sensor 1
	   • Incorrect PCV hose connection

DTC: P0302
1T ECM, MIL: Yes
Year: 2009, 2010
Model: 350Z, 370Z, Altima, Armada, Cube, Frontier, Maxima, Murano, Pathfinder, Quest, Rogue, Sentra, Titan, Versa, Xterra
Engine: 1.6L L4, 1.8L L4, 2.0L L4, 2.5L L4, 3.5L V6, 3.5L V6 VIN A, 3.7L V6, 4.0L V6, 5.6L V8
Transmission: All
	No. 2 Cylinder Misfire Detected (Includes Hybrid Models):
	HYBRID MODELS CAUTION: Hybrid systems use very high-voltage battery systems. Before starting any service work involving the battery system, turn the ignition switch OFF and then remove the service plug from pocket in the trunk. After removing the service plug, wait 10 minutes before touching any of the high-voltage connectors and terminals.
	   • No. 2 cylinder misfires.
	1. One Trip Detection Logic (Three Way Catalyst Damage)
	   • On the 1st trip, when a misfire condition occurs that can damage the three way catalyst (TWC) due to overheating, the MIL will blink.
	   • When a misfire condition occurs, the ECM monitors the CKP sensor (POS) signal every 200 engine revolutions for a change.
	   • When the misfire condition decreases to a level that will not damage the TWC, the MIL will turn off.
	2. Two Trip Detection Logic (Exhaust quality deterioration)
	   • For misfire conditions that will not damage the TWC (but will affect vehicle emissions), the MIL will only light when the misfire is detected on a second trip.
	   • During this condition, the ECM monitors the CKP sensor signal every 1,000 engine revolutions.
	   • A misfire malfunction can be detected on any one cylinder or on multiple cylinders.
	   • If another misfire condition occurs that can damage the TWC on a second trip, the MIL will blink.
	Possible Causes:
	   • Improper spark plug
	   • Insufficient compression
	   • Incorrect fuel pressure
	   • Fuel injector circuit is open or shorted
	   • Fuel injector
	   • Intake air leak
	   • The ignition signal circuit is open or shorted
	   • Lack of fuel
	   • Drive plate or flywheel
	   • Air fuel ratio (A/F) sensor 1
	   • Incorrect PCV hose connection

DTC	Trouble Code Title, Conditions & Possible Causes
DTC: P0303 **1T ECM, MIL: Yes** **Year:** 2009, 2010 **Model:** 350Z, 370Z, Altima, Armada, Cube, Frontier, Maxima, Murano, Pathfinder, Quest, Rogue, Sentra, Titan, Versa, Xterra **Engine:** 1.6L L4, 1.8L L4, 2.0L L4, 2.5L L4, 3.5L V6, 3.5L V6 VIN A, 3.7L V6, 4.0L V6, 5.6L V8 **Transmission:** All	**No. 3 Cylinder Misfire Detected (Includes Hybrid Models):** HYBRID MODELS CAUTION: Hybrid systems use very high-voltage battery systems. Before starting any service work involving the battery system, turn the ignition switch OFF and then remove the service plug from pocket in the trunk. After removing the service plug, wait 10 minutes before touching any of the high-voltage connectors and terminals. • No. 3 cylinder misfires. 1. One Trip Detection Logic (Three Way Catalyst Damage) • On the 1st trip, when a misfire condition occurs that can damage the three way catalyst (TWC) due to overheating, the MIL will blink. • When a misfire condition occurs, the ECM monitors the CKP sensor (POS) signal every 200 engine revolutions for a change. • When the misfire condition decreases to a level that will not damage the TWC, the MIL will turn off. 2. Two Trip Detection Logic (Exhaust quality deterioration) • For misfire conditions that will not damage the TWC (but will affect vehicle emissions), the MIL will only light when the misfire is detected on a second trip. • During this condition, the ECM monitors the CKP sensor signal every 1,000 engine revolutions. • A misfire malfunction can be detected on any one cylinder or on multiple cylinders. • If another misfire condition occurs that can damage the TWC on a second trip, the MIL will blink. **Possible Causes:** • Improper spark plug • Insufficient compression • Incorrect fuel pressure • Fuel injector circuit is open or shorted • Fuel injector • Intake air leak • The ignition signal circuit is open or shorted • Lack of fuel • Drive plate or flywheel • Air fuel ratio (A/F) sensor 1 • Incorrect PCV hose connection
DTC: P0304 **1T ECM, MIL: Yes** **Year:** 2009, 2010 **Model:** 350Z, 370Z, Altima, Armada, Cube, Frontier, Maxima, Murano, Pathfinder, Quest, Rogue, Sentra, Titan, Versa, Xterra **Engine:** 1.6L L4, 1.8L L4, 2.0L L4, 2.5L L4, 3.5L V6, 3.5L V6 VIN A, 3.7L V6, 4.0L V6, 5.6L V8 **Transmission:** All	**No. 4 Cylinder Misfire Detected (Includes Hybrid Models):** HYBRID MODELS CAUTION: Hybrid systems use very high-voltage battery systems. Before starting any service work involving the battery system, turn the ignition switch OFF and then remove the service plug from pocket in the trunk. After removing the service plug, wait 10 minutes before touching any of the high-voltage connectors and terminals. • No. 4 cylinder misfires. • The misfire detection logic consists of the following two conditions. 1. One Trip Detection Logic (Three Way Catalyst Damage) • On the 1st trip, when a misfire condition occurs that can damage the three way catalyst (TWC) due to overheating, the MIL will blink. • When a misfire condition occurs, the ECM monitors the CKP sensor (POS) signal every 200 engine revolutions for a change. • When the misfire condition decreases to a level that will not damage the TWC, the MIL will turn off. • If another misfire condition occurs that can damage the TWC on a second trip, the MIL will blink. When the misfire condition decreases to a level that will not damage the TWC, the MIL will remain on. If another misfire condition occurs that can damage the TWC, the MIL will begin to blink again. 2. Two Trip Detection Logic (Exhaust quality deterioration) • For misfire conditions that will not damage the TWC (but will affect vehicle emissions), the MIL will only light when the misfire is detected on a second trip. • During this condition, the ECM monitors the CKP sensor signal every 1,000 engine revolutions. • A misfire malfunction can be detected on any one cylinder or on multiple cylinders. **Possible Causes:** • Improper spark plug • Insufficient compression • Incorrect fuel pressure • Fuel injector circuit is open or shorted • Fuel injector • Intake air leak • The ignition signal circuit is open or shorted • Lack of fuel • Drive plate or flywheel • Air fuel ratio (A/F) sensor 1 • Incorrect PCV hose connection

DTC	Trouble Code Title, Conditions & Possible Causes

DTC: P0305
1T ECM, MIL: Yes
Year: 2009, 2010
Model: 350Z, 370Z, Altima, Armada, Frontier, Maxima, Murano, Pathfinder, Quest, Titan, Xterra
Engine: 3.5L V6, 3.5L V6 VIN A, 3.7L V6, 4.0L V6, 5.6L V8
Transmission: All
	No. 5 Cylinder Misfire Detected:
	• No. 5 cylinder misfires.
	• The misfire detection logic consists of the following two conditions.
	1. One Trip Detection Logic (Three Way Catalyst Damage):
	  - On the first trip, when a misfire condition occurs that can damage the three way catalyst (TWC) due to overheating, the MIL will blink.
	  - When a misfire condition occurs, the ECM monitors the CKP sensor signal every 200 engine revolutions for a change.
	  - When the misfire condition decreases to a level that will not damage the TWC, the MIL will turn off.
	  - If another misfire condition occurs that can damage the TWC on a second trip, the MIL will blink.
	  - When the misfire condition decreases to a level that will not damage the TWC, the MIL will remain on.
	  - If another misfire condition occurs that can damage the TWC, the MIL will begin to blink again.
	2. Two Trip Detection Logic (Exhaust quality deterioration):
	  - For misfire conditions that will not damage the TWC (but will affect vehicle emissions), the MIL will only light when the misfire is detected on a second trip.
	  - During this condition, the ECM monitors the CKP sensor signal every 1,000 engine revolutions.
	  - A misfire malfunction can be detected in any one cylinder or in multiple cylinders.
	Possible Causes:
	• Improper spark plug
	• Insufficient compression
	• Incorrect fuel pressure
	• The fuel injector circuit is open or shorted
	• Fuel injector
	• Intake air leakage
	• The ignition signal circuit is open or shorted
	• Lack of fuel
	• Signal plate
	• A/F sensor 1
	• Incorrect PCV hose connection

DTC: P0306
1T ECM, MIL: Yes
Year: 2009, 2010
Model: 350Z, 370Z, Altima, Armada, Frontier, Maxima, Murano, Pathfinder, Quest, Titan, Xterra
Engine: 3.5L V6, 3.5L V6 VIN A, 3.7L V6, 4.0L V6, 5.6L V8
Transmission: All
	No. 6 Cylinder Misfire Detected:
	• No. 6 cylinder misfires.
	• The misfire detection logic consists of the following two conditions.
	1. One Trip Detection Logic (Three Way Catalyst Damage):
	  - On the first trip, when a misfire condition occurs that can damage the three way catalyst (TWC) due to overheating, the MIL will blink.
	  - When a misfire condition occurs, the ECM monitors the CKP sensor signal every 200 engine revolutions for a change.
	  - When the misfire condition decreases to a level that will not damage the TWC, the MIL will turn off.
	  - If another misfire condition occurs that can damage the TWC on a second trip, the MIL will blink.
	  - When the misfire condition decreases to a level that will not damage the TWC, the MIL will remain on.
	  - If another misfire condition occurs that can damage the TWC, the MIL will begin to blink again.
	2. Two Trip Detection Logic (Exhaust quality deterioration):
	  - For misfire conditions that will not damage the TWC (but will affect vehicle emissions), the MIL will only light when the misfire is detected on a second trip.
	  - During this condition, the ECM monitors the CKP sensor signal every 1,000 engine revolutions.
	  - A misfire malfunction can be detected in any one cylinder or in multiple cylinders
	Possible Causes:
	• Improper spark plug
	• Insufficient compression
	• Incorrect fuel pressure
	• The fuel injector circuit is open or shorted
	• Fuel injector
	• Intake air leakage
	• The ignition signal circuit is open or shorted
	• Lack of fuel
	• Signal plate
	• A/F sensor 1
	• Incorrect PCV hose connection

DTC	Trouble Code Title, Conditions & Possible Causes
DTC: P0307 **1T ECM, MIL: Yes** **Year:** 2009, 2010 **Model:** Armada, Pathfinder, Titan **Engine:** 5.6L V8 **Transmission:** All	**No. 7 Cylinder Misfire Detected:** The misfire detection logic consists of the following two conditions: <ul><li>One Trip Detection Logic (Three Way Catalyst Damage)</li></ul>- On the 1st trip that a misfire condition occurs that can damage the three way catalyst (TWC) due to overheating, the MIL will blink. - When a misfire condition occurs, the ECM monitors the CKP sensor signal every 200 engine revolutions for a change. - When the misfire condition decreases to a level that will not damage the TWC, the MIL will turn off. - If another misfire condition occurs that can damage the TWC on a second trip, the MIL will blink. - When the misfire condition decreases to a level that will not damage the TWC, the MIL will remain on. - If another misfire condition occurs that can damage the TWC, the MIL will begin to blink again. <ul><li>Two Trip Detection Logic (Exhaust quality deterioration)</li></ul>- For misfire conditions that will not damage the TWC (but will affect vehicle emissions), the MIL will only light when the misfire is detected on a second trip. - During this condition, the ECM monitors the CKP sensor signal every 1,000 engine revolutions. - A misfire malfunction can be detected on any one cylinder or on multiple cylinders. <ul><li>No. 7 cylinder misfires.</li></ul>**Possible Causes:** <ul><li>Improper spark plug</li><li>Insufficient compression</li><li>Incorrect fuel pressure</li><li>The fuel injector circuit is open or shorted</li><li>Fuel injector</li><li>Intake air leak</li><li>The ignition signal circuit is open or shorted</li><li>Lack of fuel</li><li>Signal plate</li><li>Air fuel ratio (A/F) sensor 1</li><li>Incorrect PCV hose connection</li></ul>
DTC: P0308 **1T ECM, MIL: Yes** **Year:** 2009, 2010 **Model:** Armada, Pathfinder, Titan **Engine:** 5.6L V8 **Transmission:** All	**No. 8 Cylinder Misfire Detected:** The misfire detection logic consists of the following two conditions: <ul><li>One Trip Detection Logic (Three Way Catalyst Damage)</li></ul>- On the 1st trip that a misfire condition occurs that can damage the three way catalyst (TWC) due to overheating, the MIL will blink. - When a misfire condition occurs, the ECM monitors the CKP sensor signal every 200 engine revolutions for a change. - When the misfire condition decreases to a level that will not damage the TWC, the MIL will turn off. - If another misfire condition occurs that can damage the TWC on a second trip, the MIL will blink. - When the misfire condition decreases to a level that will not damage the TWC, the MIL will remain on. - If another misfire condition occurs that can damage the TWC, the MIL will begin to blink again. <ul><li>Two Trip Detection Logic (Exhaust quality deterioration)</li></ul>- For misfire conditions that will not damage the TWC (but will affect vehicle emissions), the MIL will only light when the misfire is detected on a second trip. - During this condition, the ECM monitors the CKP sensor signal every 1,000 engine revolutions. - A misfire malfunction can be detected on any one cylinder or on multiple cylinders. <ul><li>No. 8 cylinder misfires.</li></ul>**Possible Causes:** <ul><li>Improper spark plug</li><li>Insufficient compression</li><li>Incorrect fuel pressure</li><li>The fuel injector circuit is open or shorted</li><li>Fuel injector</li><li>Intake air leak</li><li>The ignition signal circuit is open or shorted</li><li>Lack of fuel</li><li>Signal plate</li><li>Air fuel ratio (A/F) sensor 1</li><li>Incorrect PCV hose connection</li></ul>

DTC	Trouble Code Title, Conditions & Possible Causes

DTC: P0327
1T ECM, MIL: Yes
Year: 2009, 2010
Model: 350Z, 370Z, Altima, Armada, Cube, Frontier, Maxima, Murano, Pathfinder, Quest, Rogue, Sentra, Titan, Versa, Xterra
Engine: 1.6L L4, 1.8L L4, 2.0L L4, 2.5L L4, 3.5L V6, 3.5L V6 VIN A, 3.7L V6, 4.0L V6, 5.6L V8
Transmission: All
	Knock Sensor Circuit Low Input (Includes Hybrid Models):
	HYBRID MODELS CAUTION: Hybrid systems use very high-voltage battery systems. Before starting any service work involving the battery system, turn the ignition switch OFF and then remove the service plug from pocket in the trunk. After removing the service plug, wait 10 minutes before touching any of the high-voltage connectors and terminals.
	NOTE: On V6 and V8 engines, this applies to Bank 1.
	    • An excessively low voltage from the sensor is sent to ECM.
	Possible Causes:
	    • Harness or connectors
	    • Knock sensor circuit is open or shorted.
	    • Knock sensor has failed

DTC: P0328
1T ECM, MIL: Yes
Year: 2009, 2010
Model: 350Z, 370Z, Altima, Armada, Cube, Frontier, Maxima, Murano, Pathfinder, Quest, Rogue, Sentra, Titan, Versa, Xterra
Engine: 1.6L L4, 1.8L L4, 2.0L L4, 2.5L L4, 3.5L V6, 3.5L V6 VIN A, 3.7L V6, 4.0L V6, 5.6L V8
Transmission: All
	Knock Sensor Circuit High Input (Includes Hybrid Models):
	HYBRID MODELS CAUTION: Hybrid systems use very high-voltage battery systems. Before starting any service work involving the battery system, turn the ignition switch OFF and then remove the service plug from pocket in the trunk. After removing the service plug, wait 10 minutes before touching any of the high-voltage connectors and terminals.
	NOTE: On V6 and V8 engines, this applies to Bank 1.
	    • An excessively high voltage from the sensor is sent to ECM.
	Possible Causes:
	    • Harness or connectors
	    • Knock sensor circuit is open or shorted.
	    • Knock sensor

DTC: P0332
1T ECM, MIL: Yes
Year: 2009, 2010
Model: 350Z, 370Z, Altima, Armada, Maxima, Murano, Pathfinder, Titan, Xterra
Engine: 3.5L V6, 3.5L V6 VIN A, 3.7L V6, 4.0L V6, 5.6L V8
Transmission: All
	Knock Sensor (KS) Bank 2 Sensor Circuit Low Input:
	    • An excessively low voltage from the sensor is sent to ECM.
	Possible Causes:
	    • Harness or connectors
	    • The sensor circuit is open or shorted.
	    • Knock sensor

DTC: P0333
1T ECM, MIL: Yes
Year: 2009, 2010
Model: Altima, Armada, Pathfinder
Engine: 3.5L V6, 4.0L V6, 5.6L V8
Transmission: All
	Knock Sensor (Bank 2) Circuit High Input:
	    • An excessively high voltage from the sensor is sent to ECM.
	Possible Causes:
	    • Harness or connectors
	    • The sensor circuit is open or shorted.
	    • Knock sensor

DTC: P0333
1T ECM, MIL: Yes
Year: 2009, 2010
Model: 350Z, 370Z, Altima, Armada, Frontier, Maxima, Murano, Pathfinder, Quest, Titan, Xterra
Engine: 3.5L V6, 3.5L V6 VIN A, 3.7L V6, 4.0L V6, 5.6L V8
Transmission: All
	Knock Sensor (Bank 2) Circuit High Input:
	    • An excessively high voltage from the sensor is sent to ECM.
	Possible Causes:
	    • Harness or connectors
	    • The sensor circuit is open or shorted.
	    • Knock sensor

DTC	Trouble Code Title, Conditions & Possible Causes

DTC: P0335
1T ECM, MIL: Yes
Year: 2009, 2010
Model: 350Z, 370Z, Altima, Armada, Cube, Frontier, Maxima, Murano, Pathfinder, Quest, Rogue, Sentra, Titan, Versa, Xterra
Engine: 1.6L L4, 1.8L L4, 2.0L L4, 2.5L L4, 3.5L V6, 3.5L V6 VIN A, 3.7L V6, 4.0L V6, 5.6L V8
Transmission: All
	Crankshaft Position Sensor (POS) Circuit (Includes Hybrid Models):
	HYBRID MODELS CAUTION: Hybrid systems use very high-voltage battery systems. Before starting any service work involving the battery system, turn the ignition switch OFF and then remove the service plug from pocket in the trunk. After removing the service plug, wait 10 minutes before touching any of the high-voltage connectors and terminals.
	  • The crankshaft position sensor (POS) signal is not detected by the ECM during the first few seconds of engine cranking.
	  • The proper pulse signal from the crankshaft position sensor (POS) is not sent to ECM while the engine is running.
	  • The crankshaft position sensor (POS) signal is not in the normal pattern during engine running.
	Possible Causes:
	  • Harness or connectors
	  • CKP sensor (POS) circuit is open or shorted.
	  • CMP sensor (PHASE) (bank 2) circuit is shorted.
	  • EVT control position sensor (bank 2) circuit is shorted.
	  • Battery current sensor circuit is shorted.
	  • APP sensor 2 circuit is shorted.
	  • Crankshaft position sensor (POS) circuit is open or shorted.
	  • Accelerator pedal position sensor circuit is shorted.
	  • Refrigerant pressure sensor circuit is shorted.
	  • EVAP control system pressure sensor circuit is sorted.
	  • Tumble control valve position sensor circuit is shorted.
	  • Crankshaft position sensor (POS)
	  • Accelerator pedal position sensor
	  • Camshaft position sensor (PHASE) (bank 2)
	  • Exhaust valve timing control position sensor (bank 2)
	  • Battery current sensor
	  • Refrigerant pressure sensor
	  • EVAP control system pressure sensor
	  • Tumble control valve position sensor
	  • Signal plate
	  • Battery current sensor

DTC: P0340
1T ECM, MIL: Yes
Year: 2009, 2010
Model: 350Z, 370Z, Altima, Armada, Cube, Frontier, Maxima, Murano, Pathfinder, Quest, Rogue, Sentra, Titan, Versa, Xterra
Engine: 1.6L L4, 1.8L L4, 2.0L L4, 2.5L L4, 3.5L V6, 3.5L V6 VIN A, 3.7L V6, 4.0L V6, 5.6L V8
Transmission: All
	Camshaft Position Sensor Circuit (Includes Hybrid Models):
	HYBRID MODELS CAUTION: Hybrid systems use very high-voltage battery systems. Before starting any service work involving the battery system, turn the ignition switch OFF and then remove the service plug from pocket in the trunk. After removing the service plug, wait 10 minutes before touching any of the high-voltage connectors and terminals.
	NOTE: On V6 and V8 engines, this applies to Bank 1.
	NOTE: If DTC P0340 is displayed with DTC P0643, first perform the trouble diagnosis for DTC P0643.
	  • The cylinder No. signal is not sent to ECM for the first few seconds during engine cranking.
	  • The cylinder No. signal is not set to ECM during engine running.
	  • The cylinder No. signal is not in the normal pattern during engine running.
	Possible Causes:
	  • Harness or connectors
	  • CMP sensor (PHASE) (bank 1) circuit is open or shorted.
	  • Camshaft position sensor (PHASE) (bank 1)
	  • Camshaft (INT)
	  • Starter motor
	  • Starting system circuit
	  • Dead (Weak) battery

DTC: P0345
1T ECM, MIL: Yes
Year: 2009, 2010
Model: 350Z, Altima, Murano, Pathfinder, Quest
Engine: 3.5L V6, 3.5L V6 VIN A, 4.0L V6, 5.6L V8
Transmission: All
	Camshaft Position Sensor (PHASE) (Bank 2) Circuit:
	NOTE: If DTC P0340 or P0345 is displayed with DTC P0643, first perform the trouble diagnosis for DTC P0643.
	  • The cylinder No. signal is not sent to ECM for the first few seconds during engine cranking.
	  • The cylinder No. signal is not sent to ECM during engine running.
	  • The cylinder No. signal is not in the normal pattern during engine running.
	Possible Causes:
	  • Harness or connectors
	  • The sensor circuit is open or shorted.
	  • Camshaft position sensor (PHASE)
	  • Camshaft (INT)
	  • Starter motor
	  • Starting system circuit
	  • Dead (weak) battery

DTC	Trouble Code Title, Conditions & Possible Causes

DTC: P0345
1T ECM, MIL: Yes
Year: 2009, 2010
Model: 350Z, 370Z, Altima, Armada, Frontier, Maxima, Pathfinder, Quest, Xterra
Engine: 3.5L V6, 3.7L V6, 4.0L V6, 5.6L V8
Transmission: All
	Camshaft Position (CMP) Sensor Bank 2 Circuit:
	NOTE: If DTC P0340 or P0345 is displayed with DTC P0643, first perform the trouble diagnosis for DTC P0643.
	    • The cylinder No. signal is not sent to ECM for the first few seconds during engine cranking.
	    • The cylinder No. signal is not sent to ECM during engine running.
	    • The cylinder No. signal is not in the normal pattern during engine running.
	Possible Causes:
	    • Harness or connectors
	    • CMP sensor (PHASE) (bank 2) circuit is open or shorted.
	    • CKP sensor (POS) circuit is shorted.
	    • EVT control position sensor (bank 2) circuit is shorted.
	    • Battery current sensor circuit is shorted.
	    • APP sensor 2 circuit is shorted.
	    • EVAP control system pressure sensor circuit is shorted.
	    • Refrigerant pressure sensor circuit is shorted.
	    • Camshaft position sensor (PHASE)
	    • (bank 2)
	    • Crankshaft position sensor (POS)
	    • Exhaust valve timing control position sensor (bank 2)
	    • Battery current sensor
	    • Accelerator pedal position sensor
	    • EVAP control system pressure sensor
	    • Refrigerant pressure sensor
	    • Camshaft (INT)
	    • Starter motor
	    • Starting system circuit
	    • Dead (Weak) battery

DTC: P0400
1T ECM, MIL: Yes
Year: 2009
Model: Quest
Engine: 3.5L V6
Transmission: All
	EGR Function (Close):
	    • If the absence of EGR flow is detected by EGR temperature sensor under the condition that calls for EGR, a low-flow malfunction is diagnosed.
	    • No EGR flow is detected under the condition that calls for EGR.
	Possible Causes:
	    • Harness or connectors
	    • The EGR volume control valve circuit is open or shorted.
	    • EGR volume control valve stuck closed
	    • Dead (Weak) battery
	    • EGR passage clogged
	    • EGR temperature sensor and circuit
	    • Exhaust gas leaks

DTC: P0403
1T ECM, MIL: Yes
Year: 2009
Model: Quest
Engine: 3.5L V6
Transmission: All
	EGR Volume Control Valve Circuit:
	    • An improper voltage signal is sent to ECM through the valve
	Possible Causes:
	    • Harness or connectors
	    • The EGR volume control valve circuit is open or shorted.
	    • EGR volume control valve

DTC: P0405
1T ECM, MIL: Yes
Year: 2009
Model: Quest
Engine: 3.5L V6
Transmission: All
	EGR Temperature Sensor Circuit Low Input:
	    • An excessively low voltage from the EGR temperature sensor is sent to ECM even when engine coolant temperature is low.
	Possible Causes:
	    • Harness or connectors
	    • The EGR temperature sensor circuit is shorted.
	    • EGR temperature sensor
	    • Malfunction of EGR function

DTC: P0406
1T ECM, MIL: Yes
Year: 2009
Model: Quest
Engine: 3.5L V6
Transmission: All
	EGR Temperature Sensor Circuit High Input:
	    • An excessively high voltage from the EGR temperature sensor is sent to ECM even when engine coolant temperature is high.
	Possible Causes:
	    • Harness or connectors
	    • The EGR temperature sensor circuit is open.
	    • EGR temperature sensor
	    • Malfunction of EGR function

DTC	Trouble Code Title, Conditions & Possible Causes
DTC: P0420 **1T ECM, MIL: Yes** **Year:** 2009, 2010 **Model:** 350Z, 370Z, Altima, Armada, Cube, Frontier, Maxima, Murano, Pathfinder, Quest, Rogue, Sentra, Titan, Versa, Xterra **Engine:** 1.6L L4, 1.8L L4, 2.0L L4, 2.5L L4, 3.5L V6, 3.5L V6 VIN A, 3.7L V6, 4.0L V6, 5.6L V8 **Transmission:** All	**Catalyst System Efficiency Below Threshold (Includes Hybrid Models):** HYBRID MODELS CAUTION: Hybrid systems use very high-voltage battery systems. Before starting any service work involving the battery system, turn the ignition switch OFF and then remove the service plug from pocket in the trunk. After removing the service plug, wait 10 minutes before touching any of the high-voltage connectors and terminals. **NOTE: On models with dual exhaust, this DTC refers to Bank 1.** • Three way catalyst (manifold) does not operate properly. • Three way catalyst (manifold) does not have enough oxygen storage capacity. **Possible Causes:** • Three way catalyst (manifold) • Exhaust tube • Intake air leaks • Fuel injector • Fuel injector leaks • Spark plug • Improper ignition timing
DTC: P0430 **1T ECM, MIL: Yes** **Year:** 2009, 2010 **Model:** 350Z, 370Z, Altima, Armada, Maxima, Murano, Pathfinder, Quest, Sentra, Titan, Xterra **Engine:** 2.5L L4, 3.5L V6, 3.5L V6 VIN A, 3.7L V6, 4.0L V6, 5.6L V8 **Transmission:** All	**Catalyst System Efficiency Below Threshold (Bank 2):** • Three way catalyst (manifold) does not operate properly. • Three way catalyst (manifold) does not have enough oxygen storage capacity. **Possible Causes:** • Three way catalyst (manifold) • Exhaust tube • Intake air leaks • Fuel injector • Fuel injector leaks • Spark plug • Improper ignition timing
DTC: P0441 **1T ECM, MIL: Yes** **Year:** 2009, 2010 **Model:** 350Z, 370Z, Altima, Armada, Cube, Frontier, Maxima, Murano, Pathfinder, Quest, Rogue, Sentra, Titan, Versa, Xterra **Engine:** 1.6L L4, 1.8L L4, 2.0L L4, 2.5L L4, 3.5L V6, 3.5L V6 VIN A, 3.7L V6, 4.0L V6, 5.6L V8 **Transmission:** All	**EVAP Control System Incorrect Purge Flow (Includes Hybrid Models):** HYBRID MODELS CAUTION: Hybrid systems use very high-voltage battery systems. Before starting any service work involving the battery system, turn the ignition switch OFF and then remove the service plug from pocket in the trunk. After removing the service plug, wait 10 minutes before touching any of the high-voltage connectors and terminals. **NOTE: If DTC P0441 is displayed with other DTC such as P2122, P2123 P2127, P2128, P2138, first perform trouble diagnosis for other DTC.** • Under normal conditions (non-closed throttle), sensor output voltage indicates if pressure drop and purge flow are adequate. If not, a malfunction is determined. • EVAP control system does not operate properly – EVAP control system has a leak between intake manifold and EVAP control system pressure sensor. **Possible Causes:** • EVAP canister purge volume control solenoid valve stuck closed • EVAP control system pressure sensor and the circuit • Loose, disconnected or improper connection of rubber tube • Blocked rubber tube • Cracked EVAP canister • EVAP canister purge volume control solenoid valve circuit • Accelerator pedal position sensor • Blocked purge port • EVAP canister vent control valve • Drain filter

DTC	Trouble Code Title, Conditions & Possible Causes
DTC: P0442 **1T ECM, MIL: Yes** **Year:** 2009, 2010 **Model:** 350Z, 370Z, Altima, Armada, Cube, Frontier, Maxima, Murano, Pathfinder, Quest, Rogue, Sentra, Titan, Versa, Xterra **Engine:** 1.6L L4, 1.8L L4, 2.0L L4, 2.5L L4, 3.5L V6, 3.5L V6 VIN A, 3.7L V6, 4.0L V6, 5.6L V8 **Transmission:** All	**EVAP Control System Small Leak Detected (Negative Pressure) (Includes Hybrid Models):** **NOTE: If DTC P0442 is displayed with DTC P0456, first perform the trouble diagnosis for DTC P0456.** • EVAP control system has a leak, EVAP control system does not operate properly. **Possible Causes:** • Incorrect fuel tank vacuum relief valve • Incorrect fuel filler cap used • Fuel filler cap remains open or fails to close. • Foreign matter caught in fuel filler cap. • Leak is in line between intake manifold and EVAP canister purge volume control solenoid valve. • Foreign matter caught in EVAP canister vent control valve. • EVAP canister or fuel tank leaks • EVAP purge line (pipe and rubber tube) leaks • EVAP purge line rubber tube bent • Loose or disconnected rubber tube • EVAP canister vent control valve and the circuit • EVAP canister purge volume control solenoid valve and the circuit • Fuel tank temperature sensor • O-ring of EVAP canister vent control valve is missing or damaged • Drain filter • EVAP canister is saturated with water • EVAP control system pressure sensor • Fuel level sensor and the circuit • Refueling EVAP vapor cut valve • ORVR system leaks

DTC: P0443
1T ECM, MIL: Yes
Year: 2009, 2010
Model: 350Z, 370Z, Altima, Armada, Cube, Pathfinder, Quest, Sentra, Titan, Versa, Xterra
Engine: 1.6L L4, 1.8L L4, 2.0L L4, 2.5L L4, 3.5L V6, 3.7L V6, 4.0L V6, 5.6L V8
Transmission: All
	EVAP Canister Purge Volume Control Solenoid Valve (Includes Hybrid Models):
	HYBRID MODELS CAUTION: Hybrid systems use very high-voltage battery systems. Before starting any service work involving the battery system, turn the ignition switch OFF and then remove the service plug from pocket in the trunk. After removing the service plug, wait 10 minutes before touching any of the high-voltage connectors and terminals.
	    • Condition A: The canister purge flow is detected during the vehicle is stopped while the engine is running, even when EVAP canister purge volume control solenoid valve is completely closed.
	    • Condition B: The canister purge flow is detected during the specified driving conditions, even when EVAP canister purge volume control solenoid valve is completely closed.
	Possible Causes:
	• EVAP control system pressure sensor
	• EVAP canister purge volume control solenoid valve (EVAP canister purge volume control solenoid valve is stuck open.)
	• EVAP canister vent control valve
	• Drain filter
	• EVAP canister
	• Hoses (Hoses are connected incorrectly or clogged.)

DTC: P0443 **1T ECM, MIL: Yes** **Year:** 2009, 2010 **Model:** Frontier, Maxima, Murano, Rogue **Engine:** 2.5L L4, 3.5L V6, 3.5L V6 VIN A, 4.0L V6 **Transmission:** All	**EVAP Canister Purge Volume Control Solenoid Valve:** • The canister purge flow is detected during the specified driving conditions, even when EVAP canister purge volume control solenoid valve is completely closed. **Possible Causes:** • EVAP control system pressure sensor • EVAP canister purge volume control solenoid valve (valve is stuck open) • EVAP canister vent control valve • EVAP canister • Hoses are connected incorrectly or clogged
DTC: P0444 **1T ECM, MIL: Yes** **Year:** 2009, 2010 **Model:** 350Z, 370Z, Altima, Armada, Cube, Frontier, Maxima, Murano, Pathfinder, Quest, Rogue, Sentra, Titan, Versa, Xterra **Engine:** 1.6L L4, 1.8L L4, 2.0L L4, 2.5L L4, 3.5L V6, 3.5L V6 VIN A, 3.7L V6, 4.0L V6, 5.6L V8 **Transmission:** All	**EVAP Canister Purge Volume Control Solenoid Valve Circuit Open (Includes Hybrid Models):** HYBRID MODELS CAUTION: Hybrid systems use very high-voltage battery systems. Before starting any service work involving the battery system, turn the ignition switch OFF and then remove the service plug from pocket in the trunk. After removing the service plug, wait 10 minutes before touching any of the high-voltage connectors and terminals. • An excessively low voltage signal is sent to ECM through the valve **Possible Causes:** • Harness or connectors • EVAP canister purge volume control solenoid valve circuit is open or shorted. • EVAP canister purge volume control solenoid valve

DTC	Trouble Code Title, Conditions & Possible Causes

DTC: P0445
1T ECM, MIL: Yes
Year: 2009, 2010
Model: 350Z, 370Z, Altima, Armada, Cube, Frontier, Maxima, Murano, Pathfinder, Quest, Rogue, Sentra, Titan, Versa, Xterra
Engine: 1.6L L4, 1.8L L4, 2.0L L4, 2.5L L4, 3.5L V6, 3.5L V6 VIN A, 3.7L V6, 4.0L V6, 5.6L V8
Transmission: All
	EVAP Canister Purge Volume Control Solenoid Valve Circuit Shorted (Includes Hybrid Models):
	HYBRID MODELS CAUTION: Hybrid systems use very high-voltage battery systems. Before starting any service work involving the battery system, turn the ignition switch OFF and then remove the service plug from pocket in the trunk. After removing the service plug, wait 10 minutes before touching any of the high-voltage connectors and terminals.
	    • An excessively high voltage signal is sent to ECM through the valve
	Possible Causes:
	    • Harness or connectors
	    • EVAP canister purge volume control solenoid valve circuit is shorted.
	    • EVAP canister purge volume control solenoid valve

DTC: P0447
1T ECM, MIL: Yes
Year: 2009, 2010
Model: 350Z, 370Z, Altima, Armada, Cube, Frontier, Maxima, Murano, Pathfinder, Quest, Rogue, Sentra, Titan, Versa, Xterra
Engine: 1.6L L4, 1.8L L4, 2.0L L4, 2.5L L4, 3.5L V6, 3.5L V6 VIN A, 3.7L V6, 4.0L V6, 5.6L V8
Transmission: All
	EVAP Canister Vent Control Valve Circuit Open (Includes Hybrid Models):
	HYBRID MODELS CAUTION: Hybrid systems use very high-voltage battery systems. Before starting any service work involving the battery system, turn the ignition switch OFF and then remove the service plug from pocket in the trunk. After removing the service plug, wait 10 minutes before touching any of the high-voltage connectors and terminals.
	    • An improper voltage signal is sent to ECM through EVAP canister vent control valve.
	Possible Causes:
	    • Harness or connectors
	    • EVAP canister vent control valve circuit is open or shorted.
	    • EVAP canister vent control valve
	    • Drain filter (if equipped)

DTC: P0448
1T ECM, MIL: Yes
Year: 2009, 2010
Model: 350Z, 370Z, Altima, Armada, Cube, Frontier, Maxima, Murano, Pathfinder, Quest, Rogue, Sentra, Titan, Versa, Xterra
Engine: 1.6L L4, 1.8L L4, 2.0L L4, 2.5L L4, 3.5L V6, 3.5L V6 VIN A, 3.7L V6, 4.0L V6, 5.6L V8
Transmission: All
	EVAP Canister Vent Control Valve Closed (Includes Hybrid Models):
	HYBRID MODELS CAUTION: Hybrid systems use very high-voltage battery systems. Before starting any service work involving the battery system, turn the ignition switch OFF and then remove the service plug from pocket in the trunk. After removing the service plug, wait 10 minutes before touching any of the high-voltage connectors and terminals.
	    • EVAP canister vent control valve remains closed under specified driving conditions.
	Possible Causes:
	    • EVAP canister vent control valve
	    • EVAP control system pressure sensor and the circuit
	    • Blocked rubber tube to EVAP canister vent control valve
	    • EVAP canister is saturated with water
	    • Drain filter (if equipped)

DTC: P0451
1T ECM, MIL: Yes
Year: 2009, 2010
Model: 350Z, 370Z, Altima, Armada, Cube, Frontier, Maxima, Murano, Pathfinder, Quest, Rogue, Sentra, Titan, Versa, Xterra
Engine: 1.6L L4, 1.8L L4, 2.0L L4, 2.5L L4, 3.5L V6, 3.5L V6 VIN A, 3.7L V6, 4.0L V6, 5.6L V8
Transmission: All
	EVAP Control System Pressure Sensor Performance (Includes Hybrid Models):
	HYBRID MODELS CAUTION: Hybrid systems use very high-voltage battery systems. Before starting any service work involving the battery system, turn the ignition switch OFF and then remove the service plug from pocket in the trunk. After removing the service plug, wait 10 minutes before touching any of the high-voltage connectors and terminals.
	    • ECM detects a sloshing signal from the EVAP control system pressure sensor
	Possible Causes:
	    • Harness or connectors
	    • EVAP control system pressure sensor circuit is shorted.
	    • CKP sensor (POS) circuit is shorted.
	    • CMP sensor (PHASE) (bank 2) circuit is shorted.
	    • EVT control position sensor (bank 2) circuit is shorted.
	    • Battery current sensor circuit is shorted.
	    • APP sensor 2 circuit is shorted.
	    • Refrigerant pressure sensor circuit is shorted.
	    • EVAP control system pressure sensor
	    • Crankshaft position sensor (POS)
	    • Camshaft position sensor (PHASE) (bank 2)
	    • Exhaust valve timing control position sensor (bank 2)
	    • Battery current sensor
	    • Accelerator pedal position sensor
	    • Refrigerant pressure sensor

DTC	Trouble Code Title, Conditions & Possible Causes

DTC: P0452
1T ECM, MIL: Yes
Year: 2009, 2010
Model: 350Z, 370Z, Altima, Armada, Cube, Frontier, Maxima, Murano, Pathfinder, Quest, Rogue, Sentra, Titan, Versa, Xterra
Engine: 1.6L L4, 1.8L L4, 2.0L L4, 2.5L L4, 3.5L V6, 3.5L V6 VIN A, 3.7L V6, 4.0L V6, 5.6L V8
Transmission: All
	EVAP Control System Pressure Sensor Low Input (Includes Hybrid Models):
	HYBRID MODELS CAUTION: Hybrid systems use very high-voltage battery systems. Before starting any service work involving the battery system, turn the ignition switch OFF and then remove the service plug from pocket in the trunk. After removing the service plug, wait 10 minutes before touching any of the high-voltage connectors and terminals.
	    • An excessively low voltage from the sensor is sent to ECM.
	Possible Causes:
	    • Harness or connectors
	    • EVAP control system pressure sensor circuit is shorted.
	    • CKP sensor (POS) circuit is shorted.
	    • CMP sensor (PHASE) (bank 2) circuit is shorted.
	    • EVT control position sensor (bank 2) circuit is shorted.
	    • Battery current sensor circuit is shorted.
	    • APP sensor 2 circuit is shorted.
	    • Refrigerant pressure sensor circuit is shorted.
	    • EVAP control system pressure sensor
	    • Crankshaft position sensor (POS)
	    • Camshaft position sensor (PHASE) (bank 2)
	    • Exhaust valve timing control position sensor (bank 2)
	    • Battery current sensor
	    • Accelerator pedal position sensor
	    • Refrigerant pressure sensor

DTC: P0453
1T ECM, MIL: Yes
Year: 2009, 2010
Model: 350Z, 370Z, Altima, Armada, Cube, Frontier, Maxima, Murano, Pathfinder, Quest, Rogue, Sentra, Titan, Versa, Xterra
Engine: 1.6L L4, 1.8L L4, 2.0L L4, 2.5L L4, 3.5L V6, 3.5L V6 VIN A, 3.7L V6, 4.0L V6, 5.6L V8
Transmission: All
	EVAP Control System Pressure Sensor High Input (Includes Hybrid Models):
	HYBRID MODELS CAUTION: Hybrid systems use very high-voltage battery systems. Before starting any service work involving the battery system, turn the ignition switch OFF and then remove the service plug from pocket in the trunk. After removing the service plug, wait 10 minutes before touching any of the high-voltage connectors and terminals.
	    • An excessively high voltage from the sensor is sent to ECM.
	Possible Causes:
	    • Harness or connectors
	    • EVAP control system pressure sensor circuit is open or shorted.
	    • CKP sensor (POS) circuit is shorted.
	    • CMP sensor (PHASE) (bank 2) circuit is shorted.
	    • EVT control position sensor (bank 2) circuit is shorted.
	    • Battery current sensor circuit is shorted.
	    • APP sensor 2 circuit is shorted.
	    • Refrigerant pressure sensor circuit is shorted.
	    • EVAP control system pressure sensor
	    • Crankshaft position sensor (POS)
	    • Camshaft position sensor (PHASE)
	    • (bank 2)
	    • Exhaust valve timing control position sensor (bank 2)
	    • Battery current sensor
	    • Accelerator pedal position sensor
	    • Refrigerant pressure sensor
	    • EVAP canister vent control valve
	    • EVAP canister
	    • Rubber hose from EVAP canister vent control valve to vehicle frame

DTC	Trouble Code Title, Conditions & Possible Causes

DTC: P0455
1T ECM, MIL: Yes
Year: 2009, 2010
Model: 350Z, 370Z, Altima, Armada, Cube, Frontier, Maxima, Murano, Pathfinder, Quest, Rogue, Sentra, Titan, Versa, Xterra
Engine: 1.6L L4, 1.8L L4, 2.0L L4, 2.5L L4, 3.5L V6, 3.5L V6 VIN A, 3.7L V6, 4.0L V6, 5.6L V8
Transmission: All
	EVAP Control System Gross Leak Detected (Includes Hybrid Models):
	  • EVAP control system has a very large leak, such as fuel filler cap fell off.
	  • EVAP control system does not operate properly.
	CAUTION: Never remove fuel filler cap during the DTC Confirmation Procedure.
	Possible Causes:
	  • Fuel filler cap remains open or fails to close.
	  • Incorrect fuel tank vacuum relief valve
	  • Incorrect fuel filler cap used
	  • Foreign matter caught in fuel filler cap
	  • Leak is in line between intake manifold and EVAP canister purge volume control solenoid valve
	  • Foreign matter caught in EVAP canister vent control valve.
	  • EVAP canister or fuel tank leaks
	  • EVAP purge line (pipe and rubber tube) leaks
	  • EVAP purge line rubber tube bent.
	  • Loose or disconnected rubber tube
	  • EVAP canister vent control valve and the circuit
	  • Drain filter
	  • EVAP canister purge volume control solenoid valve and the circuit
	  • Fuel tank temperature sensor
	  • O-ring of EVAP canister vent control valve is missing or damaged.
	  • EVAP control system pressure sensor
	  • Refueling EVAP vapor cut valve
	  • ORVR system leaks

DTC: P0456
1T ECM, MIL: Yes
Year: 2009, 2010
Model: 350Z, 370Z, Altima, Armada, Cube, Frontier, Maxima, Murano, Pathfinder, Quest, Rogue, Sentra, Titan, Versa, Xterra
Engine: 1.6L L4, 1.8L L4, 2.0L L4, 2.5L L4, 3.5L V6, 3.5L V6 VIN A, 3.7L V6, 4.0L V6, 5.6L V8
Transmission: All
	Evaporative Emission Control System Very Small Leak (Negative Pressure Check) (Includes Hybrid Models):
	HYBRID MODELS CAUTION: Hybrid systems use very high-voltage battery systems. Before starting any service work involving the battery system, turn the ignition switch OFF and then remove the service plug from pocket in the trunk. After removing the service plug, wait 10 minutes before touching any of the high-voltage connectors and terminals.
	NOTE: If ECM judges a leak which corresponds to a very small leak, the very small leak P0456 will be detected.
	NOTE: If ECM judges a leak equivalent to a small leak, EVAP small leak P0442 will be detected.
	NOTE: If ECM judges there are no leaks, the diagnosis will be OK.
	  • If DTC P0456 is displayed with DTC P0442, first perform the trouble diagnosis for DTC P0456.
	  • This diagnosis detects very small leakage in the EVAP line between fuel tank and EVAP canister purge volume control solenoid valve, using the intake manifold vacuum in the same way as conventional EVAP small leakage diagnosis.
	NOTE: If ECM judges a leakage which corresponds to a very small leakage, the very small leakage P0456 will be detected.
	  • If ECM judges a leakage equivalent to a small leakage, EVAP small leakage P0442 will be detected.
	  • If ECM judges that there are no leakage, the diagnosis will be OK.
	  • EVAP system has a very small leak.
	  • EVAP system does not operate properly.
	Possible Causes:
	  • Incorrect fuel tank vacuum relief valve
	  • Incorrect fuel filler cap used
	  • Fuel filler cap remains open or fails to close.
	  • Foreign matter caught in fuel filler cap
	  • Leak is in line between intake manifold and EVAP canister purge volume control solenoid valve.
	  • Foreign matter caught in EVAP canister vent control valve
	  • EVAP canister or fuel tank leaks
	  • EVAP purge line (pipe and rubber tube) leaks
	  • EVAP purge line rubber tube bent
	  • Loose or disconnected rubber tube
	  • EVAP canister vent control valve and the circuit
	  • EVAP canister purge volume control solenoid valve and the circuit
	  • Fuel tank temperature sensor
	  • Drain filter
	  • O-ring of EVAP canister vent control valve is missing or damaged
	  • EVAP canister is saturated with water
	  • EVAP control system pressure sensor
	  • Refueling EVAP vapor cut valve
	  • ORVR system leaks
	  • Fuel level sensor and the circuit
	  • Foreign matter caught in EVAP canister purge volume control solenoid valve

DTC	Trouble Code Title, Conditions & Possible Causes

DTC: P0460
1T ECM, MIL: Yes
Year: 2009, 2010
Model: 350Z, 370Z, Altima, Armada, Cube, Frontier, Maxima, Murano, Pathfinder, Quest, Rogue, Sentra, Titan, Versa, Xterra
Engine: 1.6L L4, 1.8L L4, 2.0L L4, 2.5L L4, 3.5L V6, 3.5L V6 VIN A, 3.7L V6, 4.0L V6, 5.6L V8
Transmission: All
	Fuel Level Sensor Circuit Noise (Includes Hybrid Models):
	HYBRID MODELS CAUTION: Hybrid systems use very high-voltage battery systems. Before starting any service work involving the battery system, turn the ignition switch OFF and then remove the service plug from pocket in the trunk. After removing the service plug, wait 10 minutes before touching any of the high-voltage connectors and terminals.
	NOTE: If DTC P0461 is displayed with DTC UXXXX, first perform the trouble diagnosis for DTC UXXXX.
	NOTE: If DTC P0460 is displayed with DTC P0607, first perform the trouble diagnosis for DTC P0607.
	    • When the vehicle is parked, naturally the fuel level in the fuel tank is stable. It means that output signal of the fuel level sensor does not change. If ECM senses sloshing signal from the sensor, fuel level sensor malfunction is detected.
	    • Even though the vehicle is parked, a signal being varied is sent from the fuel level sensor to ECM.
	Possible Causes:
	    • Harness or connectors
	    • CAN communication line is open or shorted
	    • Fuel level sensor circuit is open or shorted
	    • Combination meter (unified meter and A/C amp.)
	    • Fuel level sensor

DTC: P0461
1T ECM, MIL: Yes
Year: 2009, 2010
Model: 350Z, 370Z, Altima, Armada, Cube, Frontier, Maxima, Murano, Pathfinder, Quest, Rogue, Sentra, Titan, Versa, Xterra
Engine: 1.6L L4, 1.8L L4, 2.0L L4, 2.5L L4, 3.5L V6, 3.5L V6 VIN A, 3.7L V6, 4.0L V6, 5.6L V8
Transmission: All
	Fuel Level Sensor Circuit Range/Performance (Includes Hybrid Models):
	HYBRID MODELS CAUTION: Hybrid systems use very high-voltage battery systems. Before starting any service work involving the battery system, turn the ignition switch OFF and then remove the service plug from pocket in the trunk. After removing the service plug, wait 10 minutes before touching any of the high-voltage connectors and terminals.
	NOTE: If DTC P0461 is displayed with DTC U1000 or U1001, first perform the trouble diagnosis for appropriate "U" code.
	NOTE: If DTC P0461 is displayed with DTC P0607, first perform the trouble diagnosis for DTC P0607.
	    • This diagnosis detects the fuel gauge malfunction of the gauge not moving even after a long distance has been driven. Driving long distances naturally affect fuel gauge level.
	    • The output signal of the fuel level sensor does not change within the specified range even though the vehicle has been driven a long distance.
	Possible Causes:
	    • Harness or connectors
	    • CAN communication line is open or shorted
	    • Fuel level sensor circuit is open or shorted
	    • Combination meter (unified meter and A/C amp.)
	    • Fuel level sensor

DTC: P0462
1T ECM, MIL: Yes
Year: 2009, 2010
Model: 350Z, 370Z, Altima, Armada, Cube, Frontier, Maxima, Murano, Pathfinder, Quest, Rogue, Sentra, Titan, Versa, Xterra
Engine: 1.6L L4, 1.8L L4, 2.0L L4, 2.5L L4, 3.5L V6, 3.5L V6 VIN A, 3.7L V6, 4.0L V6, 5.6L V8
Transmission: All
	Fuel Level Sensor Circuit Low Input (Includes Hybrid Models):
	HYBRID MODELS CAUTION: Hybrid systems use very high-voltage battery systems. Before starting any service work involving the battery system, turn the ignition switch OFF and then remove the service plug from pocket in the trunk. After removing the service plug, wait 10 minutes before touching any of the high-voltage connectors and terminals.
	NOTE: If DTC P0462 or P0463 is displayed with DTC UXXXX, first perform the trouble diagnosis for DTC UXXXX.
	NOTE: If DTC P0462 or P0463 is displayed with DTC P0607, first perform the trouble diagnosis for DTC P0607.
	    • An excessively low voltage from the sensor is sent to ECM.
	Possible Causes:
	    • Harness or connectors
	    • CAN communication line is open or shorted
	    • Fuel level sensor circuit is open or shorted
	    • Combination meter (unified meter and A/C amp.)
	    • Fuel level sensor

DTC: P0463
1T ECM, MIL: Yes
Year: 2009, 2010
Model: 350Z, 370Z, Altima, Armada, Cube, Frontier, Maxima, Murano, Pathfinder, Quest, Rogue, Sentra, Titan, Versa, Xterra
Engine: 1.6L L4, 1.8L L4, 2.0L L4, 2.5L L4, 3.5L V6, 3.5L V6 VIN A, 3.7L V6, 4.0L V6, 5.6L V8
Transmission: All
	Fuel Level Sensor Circuit High Input (Includes Hybrid Models):
	HYBRID MODELS CAUTION: Hybrid systems use very high-voltage battery systems. Before starting any service work involving the battery system, turn the ignition switch OFF and then remove the service plug from pocket in the trunk. After removing the service plug, wait 10 minutes before touching any of the high-voltage connectors and terminals.
	NOTE: If DTC P0462 or P0463 is displayed with DTC UXXXX, first perform the trouble diagnosis for DTC UXXXX.
	NOTE: If DTC P0462 or P0463 is displayed with DTC P0607, first perform the trouble diagnosis for DTC P0607.
	    • An excessively high voltage from the sensor is sent to ECM.
	Possible Causes:
	    • Harness or connectors
	    • CAN communication line is open or shorted
	    • Fuel level sensor circuit is open or shorted
	    • Combination meter (unified meter and A/C amp.)
	    • Fuel level sensor

DTC	Trouble Code Title, Conditions & Possible Causes

DTC: P0500
1T ECM, MIL: Yes
Year: 2009, 2010
Model: 350Z, 370Z, Altima, Armada, Cube, Frontier, Maxima, Murano, Pathfinder, Quest, Rogue, Sentra, Titan, Versa, Xterra
Engine: 1.6L L4, 1.8L L4, 2.0L L4, 2.5L L4, 3.5L V6, 3.5L V6 VIN A, 3.7L V6, 4.0L V6, 5.6L V8
Transmission: All
	Vehicle Speed Sensor Input (Includes Hybrid Models):
	HYBRID MODELS CAUTION: Hybrid systems use very high-voltage battery systems. Before starting any service work involving the battery system, turn the ignition switch OFF and then remove the service plug from pocket in the trunk. After removing the service plug, wait 10 minutes before touching any of the high-voltage connectors and terminals.
	NOTE: If DTC P0500 is displayed with DTC UXXXX, first perform the trouble diagnosis for DTC UXXXX.
	NOTE: If DTC P0500 is displayed with DTC P0607, first perform the trouble diagnosis for DTC P0607.
	  • When in fail-safe mode, the cooling fan operates (High) while engine is running.
	  • The vehicle speed signal sent to ECM is almost 0 km/h (0 MPH) even when vehicle is being driven.
	Possible Causes:
	  • Harness or connectors
	  • The CAN communication line is open or shorted
	  • The vehicle speed signal circuit is open or shorted
	  • Wheel sensor
	  • Unified meter and A/C amp. (combination meter)
	  • ABS actuator and electric unit (control unit)

DTC: P0506
1T ECM, MIL: Yes
Year: 2009, 2010
Model: 350Z, 370Z, Altima, Armada, Cube, Frontier, Maxima, Murano, Pathfinder, Quest, Rogue, Sentra, Titan, Versa, Xterra
Engine: 1.6L L4, 1.8L L4, 2.0L L4, 2.5L L4, 3.5L V6, 3.5L V6 VIN A, 3.7L V6, 4.0L V6, 5.6L V8
Transmission: All
	Idle Speed Control System RPM Lower Than Expected (Includes Hybrid Models):
	HYBRID MODELS CAUTION: Hybrid systems use very high-voltage battery systems. Before starting any service work involving the battery system, turn the ignition switch OFF and then remove the service plug from pocket in the trunk. After removing the service plug, wait 10 minutes before touching any of the high-voltage connectors and terminals.
	NOTE: If DTC P0506 is displayed with other DTC, first perform the trouble diagnosis for the other DTC.
	  • The idle speed is less than the target idle speed by 100 rpm or more.
	Possible Causes:
	  • Electric throttle control actuator
	  • Intake air leak

DTC: P0507
1T ECM, MIL: Yes
Year: 2009, 2010
Model: 350Z, Altima, Armada, Cube, Frontier, Maxima, Murano, Pathfinder, Quest, Rogue, Sentra, Titan, Versa, Xterra
Engine: 1.6L L4, 1.8L L4, 2.0L L4, 2.5L L4, 3.5L V6, 3.5L V6 VIN A, 4.0L V6, 5.6L V8
Transmission: All
	Idle Speed Control System RPM Higher Than Expected (Includes Hybrid Models):
	HYBRID MODELS CAUTION: Hybrid systems use very high-voltage battery systems. Before starting any service work involving the battery system, turn the ignition switch OFF and then remove the service plug from pocket in the trunk. After removing the service plug, wait 10 minutes before touching any of the high-voltage connectors and terminals.
	NOTE: If DTC P0507 is displayed with other DTC, first perform the trouble diagnosis for the other DTC.
	  • The idle speed is more than the target idle speed by 200 rpm or more.
	Possible Causes:
	  • Electric throttle control actuator
	  • Intake air leak
	  • PCV system

DTC: P050A
1T ECM, MIL: Yes
Year: 2010
Model: Rogue
Engine: 2.5L L4
Transmission: All
	Cold Start Idle Air Control System Performance:
	  • ECM does not control engine idle speed properly when engine is started with pre-warm up condition.
	Possible Causes:
	  • Lack of intake air volume
	  • Fuel injection system
	  • ECM

DTC: P050B
1T ECM, MIL: Yes
Year: 2010
Model: Rogue
Engine: 2.5L L4
Transmission: All
	Cold Start Ignition Timing Performance:
	NOTE: If DTC P050B is displayed with other DTCs, perform the trouble diagnosis for the other DTCs first.
	  • ECM does not control ignition timing properly when engine is started with pre-warming up condition.
	Possible Causes:
	  • Lack of intake air volume
	  • Fuel injection system
	  • ECM

DTC: P050E
1T ECM, MIL: Yes
Year: 2009, 2010
Model: Murano, Rogue
Engine: 2.5L L4, 3.5L V6 VIN A
Transmission: All
	Cold Start Engine Exhaust Temperature Too Low:
	  • The temperature of the catalyst inlet does not rise to the proper temperature when the engine is started with pre-warming up condition.
	Possible Causes:
	  • Lack of intake air volume
	  • Fuel injection system
	  • ECM

DTC	Trouble Code Title, Conditions & Possible Causes
DTC: P0524 **1T ECM, MIL: Yes** **Year:** 2009, 2010 **Model:** 370Z **Engine:** 3.7L V6 **Transmission:** All	**Engine Oil Pressure Too Low:** Engine oil pressure is low because there is a gap between angle of target and phase-control angle. **Possible Causes:** • Engine oil pressure or level too low • Crankshaft position sensor (POS) • Camshaft position sensor (PHASE) • Intake valve control solenoid valve • Accumulation of debris to the signal pick-up portion of the camshaft • Timing chain installation • Foreign matter caught in the oil groove for intake valve timing control
DTC: P0550 **1T ECM** **Year:** 2009, 2010 **Model:** 350Z, 370Z, Armada, Frontier, Maxima, Murano, Pathfinder, Quest, Titan, Xterra **Engine:** 2.5L L4, 3.5L V6, 3.5L V6 VIN A, 3.7L V6, 4.0L V6, 5.6L V8 **Transmission:** All	**Power Steering Pressure Sensor Circuit:** • The MIL will not illuminate for this diagnosis. **NOTE: If DTC P0550 is displayed with DTC P0643, first perform the trouble diagnosis for DTC P0643.** • An excessively low or high voltage from the sensor is sent to ECM. **Possible Causes:** • Harness or connectors • The sensor circuit is open or shorted • Power steering pressure sensor
DTC: P0555 **T ECM, MIL: Yes** **Year:** 2009, 2010 **Model:** 370Z **Engine:** 3.7L V6 **Transmission:** All	**Brake Booster Pressure Sensor Circuit:** • An excessively low voltage from the sensor is sent to ECM. • An excessively high voltage from the sensor is sent to ECM. **Possible Causes:** • Harness or connectors • The sensor circuit is open or shorted. • CKP sensor (POS) circuit is shorted. • APP sensor 2 circuit is shorted. • EVAP control system pressure sensor circuit is shorted. • Refrigerant pressure sensor circuit is shorted. • Gear lever position sensor circuit is shorted. • Brake booster pressure sensor • Crankshaft position sensor (POS) • Accelerator pedal position sensor • EVAP control system pressure sensor • Refrigerant pressure sensor • Gear lever position sensor
DTC: P0603 **1T ECM, MIL: Yes** **Year:** 2009, 2010 **Model:** 350Z, 370Z, Altima, Armada, Frontier, Maxima, Murano, Pathfinder, Quest, Rogue, Sentra, Versa, Xterra **Engine:** 1.6L L4, 1.8L L4, 2.0L L4, 2.5L L4, 3.5L V6, 3.5L V6 VIN A, 3.7L V6, 4.0L V6, 5.6L V8 **Transmission:** All	**ECM Power Supply Circuit (Includes Hybrid Models):** HYBRID MODELS CAUTION: Hybrid systems use very high-voltage battery systems. Before starting any service work involving the battery system, turn the ignition switch OFF and then remove the service plug from pocket in the trunk. After removing the service plug, wait 10 minutes before touching any of the high-voltage connectors and terminals. • ECM back-up RAM system does not function properly. **Possible Causes:** • Harness or connectors • The ECM power supply (back-up) circuit is open or shorted. • ECM
DTC: P0605 **1T ECM, MIL: Yes** **Year:** 2009, 2010 **Model:** 350Z, 370Z, Altima, Armada, Cube, Frontier, Maxima, Murano, Pathfinder, Quest, Rogue, Titan, Versa, Xterra **Engine:** 1.6L L4, 1.8L L4, 2.5L L4, 3.5L V6, 3.5L V6 VIN A, 3.7L V6, 4.0L V6, 5.6L V8 **Transmission:** All	**Engine Control Module (ECM) (Includes Hybrid Models):** HYBRID MODELS CAUTION: Hybrid systems use very high-voltage battery systems. Before starting any service work involving the battery system, turn the ignition switch OFF and then remove the service plug from pocket in the trunk. After removing the service plug, wait 10 minutes before touching any of the high-voltage connectors and terminals. A. ECM calculation function is malfunctioning. B. ECM EEP-ROM system is malfunctioning. C. ECM self shut-off function is malfunctioning. **Possible Causes:** • ECM

DTC	Trouble Code Title, Conditions & Possible Causes
DTC: P0607 **1T ECM, MIL: Yes** **Year:** 2009, 2010 **Model:** 350Z, 370Z, Altima, Armada, Cube, Frontier, Maxima, Murano, Pathfinder, Quest, Rogue, Sentra, Titan, Versa, Xterra **Engine:** 1.6L L4, 1.8L L4, 2.0L L4, 2.5L L4, 3.5L V6, 3.5L V6 VIN A, 3.7L V6, 4.0L V6, 5.6L V8 **Transmission:** All	**CAN Communication Bus (Includes Hybrid Models):** HYBRID MODELS CAUTION: Hybrid systems use very high-voltage battery systems. Before starting any service work involving the battery system, turn the ignition switch OFF and then remove the service plug from pocket in the trunk. After removing the service plug, wait 10 minutes before touching any of the high-voltage connectors and terminals. • When detecting error during the initial diagnosis of CAN controller of ECM. **Possible Causes:** • ECM has failed
DTC: P0615 **T TCM, TCIL: Yes** **Year:** 2009, 2010 **Model:** 350Z, 370Z, Altima, Armada, Frontier, Maxima, Murano, Pathfinder, Titan, Versa, Xterra **Engine:** 1.6L L4, 1.8L L4, 2.5L L4, 3.5L V6, 3.5L V6 VIN A, 3.7L V6, 4.0L V6, 5.6L V8 **Transmission:** All	**Starter Relay Circuit:** • This is not an OBD-II self-diagnostic item • This DTC will set if the starter monitor value is OFF when the ignition switch is ON at the "P" and "N" positions. **Possible Causes:** • Harness or connectors • Starter relay and TCM circuit is open or shorted. • Starter relay circuit
DTC: P0643 **1T ECM, MIL: Yes** **Year:** 2009, 2010 **Model:** 350Z, 370Z, Altima, Armada, Cube, Frontier, Maxima, Murano, Pathfinder, Quest, Rogue, Sentra, Titan, Versa, Xterra **Engine:** 1.6L L4, 1.8L L4, 2.0L L4, 2.5L L4, 3.5L V6, 3.5L V6 VIN A, 3.7L V6, 4.0L V6, 5.6L V8 **Transmission:** All	**Sensor Power Supply Circuit Short (Includes Hybrid Models):** HYBRID MODELS CAUTION: Hybrid systems use very high-voltage battery systems. Before starting any service work involving the battery system, turn the ignition switch OFF and then remove the service plug from pocket in the trunk. After removing the service plug, wait 10 minutes before touching any of the high-voltage connectors and terminals. • ECM detects a voltage of power source for sensor is excessively low or high. **NOTE: When the malfunction is detected, ECM enters fail-safe mode and the MIL illuminates.** • ECM stops the electric throttle control actuator control, throttle valve is maintained at a fixed opening (approx. 5 degrees) by the return spring. **Possible Causes:** • Harness or connectors • APP sensor 1 circuit is shorted. • TP sensor circuit is shorted. • CMP sensor (PHASE) (bank 1) circuit is shorted. • EVT control position sensor (bank 1) circuit is shorted. • PSP sensor circuit is shorted. • Accelerator pedal position sensor • Throttle position sensor • Camshaft position sensor (PHASE) (bank 1) • Exhaust valve timing control position sensor (bank 1) • Power steering pressure sensor
DTC: P0700 **T TCM, MIL: Yes, TCIL: Yes** **Year:** 2009, 2010 **Model:** 350Z, Armada, Frontier, Pathfinder, Titan, Xterra **Engine:** 2.5L L4, 3.5L V6, 4.0L V6, 5.6L V8 **Transmission:** All	**TCM:** • This is an OBD-II self-diagnostic item. • Diagnostic trouble code P0700 is detected when the TCM is malfunctioning. **Possible Causes:** • TCM
DTC: P0700 **T TCM, TCIL: Yes** **Year:** 2009, 2010 **Model:** Frontier **Engine:** 2.5L L4, 4.0L V6 **Transmission:** All	**Transmission Range Switch A:** • This is an OBD-II self-diagnostic item. • Diagnostic trouble code "P0705" with CONSULT-III or 9th judgment flicker without CONSULT-III is detected under the following conditions: - When TCM does not receive the correct voltage signal from the transmission range switch 1, 2, 3, 4 based on the gear position. - When no other position but "P" position is detected from "N" positions. **Possible Causes:** • Harness or connectors • The transmission range switch 1, 2, 3, 4 and TCM circuit is open or shorted. • Transmission range switch 1, 2, 3, 4

DTC	Trouble Code Title, Conditions & Possible Causes

DTC: P0703
1T TCM, TCIL: Yes
Year: 2009, 2010
Model: Altima, Sentra, Versa
Engine: 1.6L L4, 1.8L L4, 2.0L L4, 2.5L L4, 3.5L V6
Transmission: All
	Brake Lamp Switch Circuit:
	  * ON, OFF status of the stop lamp switch is sent via the CAN communication from the combination meter to TCM using the signal.
	  • This is not an OBD-II self-diagnostic item.
	  • Diagnostic trouble code P0703 is detected when the stop lamp switch does not switch to ON and OFF.
	  * The stop lamp switch does not switch to ON and OFF.
	Possible Causes:
	  • Harness or connectors: Stop lamp switch, and combination meter circuit are open or shorted and/or CAN communication line is open or shorted.
	  • Stop lamp switch

DTC: P0703
T BCM, TCIL: Yes
Year: 2009, 2010
Model: Maxima, Murano, Rogue
Engine: 2.5L L4, 3.5L V6, 3.5L V6 VIN A
Transmission: All
	Brake Switch B Circuit:
	BCM detects ON/OFF state of the stop lamp switch and transmits the data to the TCM via CAN communication by converting the data to a signal.
	  • When the brake switch does not switch to ON or OFF, DTC is detected.
	Possible Causes:
	  • Harness or connectors
	  • Stop lamp switch, and BCM circuit are open or shorted.
	  • CAN communication line is open or shorted.
	  • Stop lamp switch

DTC: P0703
T TCM, TCIL: Yes
Year: 2009, 2010
Model: Cube
Engine: 1.8L L4
Transmission: All
	Brake Switch B Circuit:
	  • TCM detects malfunction in CAN communication between BCM
	  • TCM detects a state that ON/OFF of stop lamp switch signal is not switched
	Possible Causes:
	  • Harness or connectors
	  • CAN communication line is open or shorted.
	  • Stop lamp switch circuit is open or shorted.
	  • Stop lamp switch
	  • BCM

DTC: P0705
T TCM, TCIL: Yes
Year: 2009, 2010
Model: Maxima, Murano, Rogue, Versa
Engine: 1.6L L4, 1.8L L4, 2.5L L4, 3.5L V6, 3.5L V6 VIN A
Transmission: All
	Transmission Range Sensor A Circuit (PRNDL Input):
	  • TCM does not receive the correct voltage signal (based on the gear position) from the switch.
	Possible Causes:
	  • Harness or connectors
	  • Transmission range switches circuit is open or shorted.
	  • Transmission range switch

DTC: P0705
1T TCM, TCIL: Yes
Year: 2009, 2010
Model: 350Z, Altima, Armada, Pathfinder, Quest, Sentra, Titan, Versa, Xterra
Engine: 1.6L L4, 1.8L L4, 2.0L L4, 2.5L L4, 3.5L V6, 4.0L V6, 5.6L V8
Transmission: All
	Park/Neutral Position Switch Circuit:
	NOTE: If DTC U1000 is displayed with this DTC, first perform the trouble diagnosis for DTC U1000.
	  • The PNP switch assembly includes a transaxle range switch.
	  • The transaxle range switch detects the selector lever position and sends a signal to the TCM.
	  • This is an OBD-II self-diagnostic item.
	  • Diagnostic trouble code P0705 is detected under the following conditions:
	  - When TCM does not receive the correct voltage signal from the PNP switches 1, 2, 3 and 4 based on the gear position.
	  - When the signal from monitor terminal of PNP switch 3 is different from PNP switch 3.
	  - When no other position but "P" position is detected from "N" positions.
	Possible Causes:
	  • Harness or connectors
	  • PNP switches 1, 2, 3, 4 and TCM circuit is open or shorted.
	  • PNP switches 1, 2, 3, 4
	  • PNP switch 3 monitor terminal is open or shorted

DTC: P0705
T TCM, TCIL: Yes
Year: 2009, 2010
Model: 370Z
Engine: 3.7L V6
Transmission: All
	Transmission Range Switch A:
	  • Transmission range switch 1 – 4 signals input with impossible pattern.
	  • "P" position is detected from "N" position without any other position being detected in between.
	Possible Causes:
	  • Harness or connectors
	  • Transmission range switches 1, 2, 3, 4 and TCM circuit is open or shorted.
	  • Transmission range switches 1, 2, 3 and 4

DTC	Trouble Code Title, Conditions & Possible Causes

DTC: P0705
T TCM, TCIL: Yes
Year: 2009, 2010
Model: Cube
Engine: 1.8L L4
Transmission: All
	Transmission Range Sensor A Circuit (PRNDL Input):
	  • Range signal is not transmitted to TCM
	  • 2 or more range signals are transmitted to TCM
	Possible Causes:
	  • Harness or connectors
	  • Transmission range switch circuit is open or shorted.
	  • Transmission range switch

DTC: P0710
T TCM, TCIL: Yes
Year: 2009, 2010
Model: 370Z, Cube
Engine: 1.8L L4, 3.7L V6
Transmission: All
	Transmission Fluid Temperature Sensor A Circuit:
	  • CVT fluid temperature does not rise to the specified temperature after driving for a certain period of time with the TCM-received oil temperature sensor value between −39°C (−38.2°F) and 20°C (−68°F)
	  • CVT fluid temperature sensor value that TCM receives is more than 180°C (356°F)
	  • TCM-received CVT fluid temperature sensor value while driving is less than −40°C (−40°F)
	Possible Causes:
	  • Harness or connectors
	  • Sensor circuit is open or shorted.
	  • Fluid temperature sensor

DTC: P0710
T TCM, TCIL: Yes
Year: 2009
Model: Quest
Engine: 3.5L V6
Transmission: All
	ATF Temperature Sensor Circuit:
	NOTE: If DTC U1000 is displayed with this DTC, first perform the trouble diagnosis for DTC U1000.
	  • This is an OBD-II self-diagnostic item.
	  • Diagnostic trouble code "ATF TEMP SEN/CIRC" with CONSULT-III or P0710 without CONSULT-III is detected under the following conditions:
	  - When normal voltage not applied to ATF temperature sensor due to open, short, and so on.
	  - When during running, the ATF temperature sensor signal voltage is excessively high or low.
	Possible Causes:
	  • Harness or connectors
	  • The sensor circuit is open or shorted.
	  • A/T fluid temperature sensor

DTC: P0710
1T TCM, TCIL: Yes
Year: 2009, 2010
Model: Altima, Sentra, Versa
Engine: 1.6L L4, 1.8L L4, 2.0L L4, 2.5L L4, 3.5L V6
Transmission: All
	CVT Fluid Temperature Sensor Circuit:
	  • The CVT fluid temperature sensor is included in the control valve assembly.
	  • The CVT fluid temperature sensor detects the CVT fluid temperature and sends a signal to the TCM.
	  • This is an OBD-II self-diagnostic item.
	  • Diagnostic trouble code P0710 is detected when TCM receives an excessively low or high voltage from the sensor.
	Possible Causes:
	  • Harness or connectors (Sensor circuit is open or shorted.)
	  • CVT fluid temperature sensor

DTC: P0710
T TCM, TCIL: Yes
Year: 2009, 2010
Model: Maxima, Murano, Rogue
Engine: 2.5L L4, 3.5L V6, 3.5L V6 VIN A
Transmission: All
	CVT Transmission Fluid Temperature Sensor A Circuit:
	  • During running, the CVT fluid temperature sensor signal voltage is excessively high or low.
	Possible Causes:
	  • Harness or connectors
	  • Sensor circuit is open or shorted.
	  • CVT fluid temperature sensor

DTC: P0711
T TCM, TCIL: Yes
Year: 2009
Model: Quest
Engine: 3.5L V6
Transmission: All
	ATF Temperature Sensor Performance:
	NOTE: If DTC U1000 is displayed with this DTC, first perform the trouble diagnosis for DTC U1000.
	  • This is an OBD-II self-diagnostic item.
	  • Diagnostic trouble code "FLUID TEMP SEN" with CONSULT-III or P0711 without CONSULT-III is detected when ATF temperature signal does not change.
	Possible Causes:
	  • Harness or connectors
	  • The sensor circuit is open or shorted.
	  • A/T fluid temperature sensor

DTC: P0715
T TCM, TCIL: Yes
Year: 2009, 2010
Model: Maxima, Murano, Rogue
Engine: 2.5L L4, 3.5L V6, 3.5L V6 VIN A
Transmission: All
	Input/Turbine Speed Sensor A Circuit:
	  • Input speed sensor (primary speed sensor) signal is not input due to an open circuit.
	  • An unexpected signal is input when vehicle is being driven.
	Possible Causes:
	  • Harness or connectors
	  • Sensor circuit is open or shorted.
	  • Primary speed sensor

DTC	Trouble Code Title, Conditions & Possible Causes
DTC: P0715 **T TCM, TCIL: Yes** **Year:** 2010 **Model:** Versa **Engine:** 1.6L L4, 1.8L L4 **Transmission:** All	**Input Speed Sensor A:** • This is an OBD-II self-diagnostic item. • Diagnostic trouble code P0715 is detected when TCM does not receive the proper signal from the sensor. **Possible Causes:** • Harness or connectors • Sensor circuit is open or shorted. • Input speed sensor (Primary speed sensor)
DTC: P0715 **T TCM, TCIL: Yes** **Year:** 2009, 2010 **Model:** Cube **Engine:** 1.8L L4 **Transmission:** All	**Input/Turbine Speed Sensor A Circuit:** • Primary speed sensor signal is not transmitted to TCM • Primary speed sensor value is less than 150 rpm while secondary pulley speed is more than 500 rpm **Possible Causes:** • Harness or connectors • Primary speed sensor circuit is open or shorted. • Primary speed sensor
DTC: P0715 **1T TCM, TCIL: Yes** **Year:** 2009, 2010 **Model:** Altima, Sentra, Versa **Engine:** 1.6L L4, 1.8L L4, 2.0L L4, 2.5L L4, 3.5L V6 **Transmission:** All	**Input Speed Sensor Circuit (PRI Speed Sensor):** • The input speed sensor (primary speed sensor) detects the primary pulley revolution speed and sends a signal to the TCM. • This is an OBD-II self-diagnostic item. • Diagnostic trouble code P0715 is detected when TCM does not receive the proper signal from the sensor. • The engine speed signal is determined when the engine running at any speed. The scan tool display value should closely match the tachometer reading. • The primary (PRI) speed signal is indicated during driving, with the Lock-Up ON. The display value should closely match engine speed. **Possible Causes:** • Harness or connectors (Sensor circuit is open or shorted.) • Input speed sensor (Primary speed sensor)
DTC: P0717 **T TCM, TCIL: Yes** **Year:** 2009, 2010 **Model:** 370Z **Engine:** 3.7L V6 **Transmission:** All	**Input/Turbine Speed Sensor A Circuit No Signal:** The revolution of input speed sensor 1 and/or 2 is 270 rpm or less. **Possible Causes:** • Harness or connectors • Sensor circuit is open. • Input speed sensor 1 and/or 2
DTC: P0717 **1T TCM, TCIL: Yes** **Year:** 2009, 2010 **Model:** 350Z, Armada, Frontier, Pathfinder, Quest, Titan, Xterra **Engine:** 2.5L L4, 3.5L V6, 4.0L V6, 5.6L V8 **Transmission:** All	**Input Speed Sensor A (Turbine Revolution Sensor):** • The input speed sensor detects input shaft rpm (revolutions per minute). It is located on the input side of the automatic transmission. Monitors revolution of sensor 1 and sensor 2 for non-standard conditions. • This is an OBD-II self-diagnostic item. • Diagnostic trouble code P0717 is detected under the following conditions: - When TCM does not receive the proper voltage signal from the sensor. - When TCM detects an irregularity only at position of 4th gear for input speed sensor 2. **Possible Causes:** • Harness or connectors • The sensor circuit is open or shorted. • Input speed sensor 1, 2
DTC: P0720 **1T TCM, TCIL: Yes** **Year:** 2009, 2010 **Model:** Altima, Versa **Engine:** 1.6L L4, 1.8L L4, 2.5L L4, 3.5L V6 **Transmission:** All	**Vehicle Speed Sensor CVT (Secondary Speed Sensor):** • The vehicle speed sensor CVT [output speed sensor (secondary speed sensor)] detects the revolution of the CVT output shaft and emits a pulse signal. The pulse signal is sent to the TCM, which converts it into vehicle speed. • This is an OBD-II self-diagnostic item. • Diagnostic trouble code P0720 is detected TCM does not receive the proper signal from the sensor. • The VSS sensor signal during driving should closely match the speedometer. **Possible Causes:** • Harness or connectors (Sensor circuit is open or shorted.) • Output speed sensor (Secondary speed sensor)

DTC	Trouble Code Title, Conditions & Possible Causes
DTC: P0720 **T TCM, TCIL: Yes** **Year:** 2009, 2010 **Model:** Maxima, Murano, Rogue **Engine:** 2.5L L4, 3.5L V6, 3.5L V6 VIN A **Transmission:** All	**Output Speed Sensor Circuit:** • Signal from vehicle speed sensor CVT [output speed sensor (secondary speed sensor)] is not input due to open or short circuit. • An unexpected signal is input during running. • After ignition switch is turned ON, unexpected signal input from vehicle speed signal before the vehicle starts moving. **Possible Causes:** • Harness or connectors • Sensor circuit is open or shorted. • Secondary (output) speed sensor • Vehicle speed signal
DTC: P0720 **T TCM, TCIL: Yes** **Year:** 2009, 2010 **Model:** Cube **Engine:** 1.8L L4 **Transmission:** All	**Output Speed Sensor Circuit:** • Secondary speed sensor signal is not transmitted to TCM • Secondary speed sensor value is less than 150 rpm while primary pulley speed is more than 1,000 rpm **Possible Causes:** • Harness or connectors • Secondary speed sensor circuit is open or shorted. • Secondary speed sensor
DTC: P0720 **1T TCM, TCIL: Yes** **Year:** 2009, 2010 **Model:** 350Z, Armada, Frontier, Pathfinder, Titan, Versa, Xterra **Engine:** 1.6L L4, 1.8L L4, 2.5L L4, 3.5L V6, 4.0L V6, 5.6L V8 **Transmission:** All	**Vehicle Speed Sensor A/T (Revolution Sensor/Output Speed Sensor):** • This is an OBD-II self-diagnostic item. • Diagnostic trouble code P0720 is detected under the following conditions: - When TCM does not receive the proper voltage signal from the sensor. - After ignition switch is turned "ON", irregular signal input from vehicle speed sensor MTR before the vehicle starts moving. **Possible Causes:** • Harness or connectors • The sensor circuit is open or shorted. • Revolution sensor • Vehicle speed sensor MTR
DTC: P0720 **T TCM, TCIL: Yes** **Year:** 2009, 2010 **Model:** 370Z **Engine:** 3.7L V6 **Transmission:** All	**Output Speed Sensor Circuit:** • The output speed sensor recognizes that the vehicle speed is 5 km/h (3 MPH) or less even if the vehicle speed signal recognizes that the vehicle speed is 20 km/h (12 MPH) or more. (Only when starts after the ignition switch is turned ON.) • The vehicle speed recognized by the output speed sensor decelerates 36 km/h (23 MPH) or more during 60 msec when the output speed sensor recognizes that the vehicle speed is 36 km/h (23 MPH) or more and the vehicle speed signal recognizes that the vehicle speed is 24 km/h (15 MPH) or less. • The vehicle speed of output speed sensor decelerates 36 km/h (23 MPH) or more even if the vehicle speed of vehicle speed signal accelerates or decelerates 24 km/h (15 MPH) or less during 60 msec when the output speed sensor recognizes that the vehicle speed is 36 km/h (23 MPH) or more. **Possible Causes:** • Harness or connectors • Sensor circuit is open. • Output speed sensor
DTC: P0722 **T TCM, TCIL: Yes** **Year:** 2009 **Model:** Quest **Engine:** 3.5L V6 **Transmission:** All	**Vehicle Speed Sensor (Revolution Sensor) Circuit:** • This is an OBD-II self-diagnostic item. • Diagnostic trouble code "VHCL SPEED SEN-A/T" with CONSULT-III or P0722 without CONSULT-III is detected under the following conditions: - When signal from revolution sensor does not input due to open, short, and so on. - When unexpected signal input during running. **Possible Causes:** • Harness or connectors • The sensor circuit is open or shorted. • Revolution sensor

DTC	Trouble Code Title, Conditions & Possible Causes
DTC: P0725 **1T TCM, TCIL: Yes** **Year:** 2009, 2010 **Model:** Altima, Armada, Frontier, Maxima, Murano, Pathfinder, Rogue, Sentra, Titan, Versa **Engine:** 1.6L L4, 1.8L L4, 2.0L L4, 2.5L L4, 3.5L V6, 3.5L V6 VIN A, 4.0L V6, 5.6L V8 **Transmission:** All	**Engine Speed Signal:** • The engine speed signal is sent from the ECM to the TCM. • This is not an OBD-II self-diagnostic item. • Diagnostic trouble code P0725 is detected when TCM does not receive the engine speed signal or ignition signal (input by CAN communication) from ECM. • The engine speed signal is sent with the engine running and should closely match the tachometer reading. **Possible Causes:** • Harness or connectors • The ECM to the TCM circuit is open or shorted.
DTC: P0725 **T TCM, TCIL: Yes** **Year:** 2009, 2010 **Model:** 350Z, 370Z, Xterra **Engine:** 3.5L V6, 3.7L V6, 4.0L V6 **Transmission:** All	**Engine Speed Input Circuit:** • TCM does not receive the CAN communication (ignition) signal from the ECM. • The engine speed is more less 150 rpm even if the vehicle speed is more than 10 km/h (7 MPH). **Possible Causes:** • Harness or connectors • ECM to TCM circuit is open or shorted.
DTC: P0725 **T TCM, TCIL: Yes** **Year:** 2009, 2010 **Model:** Cube **Engine:** 1.8L L4 **Transmission:** All	**Engine Speed Input Circuit:** • TCM detects a malfunction in CAN communication between TCM and ECM • When primary pulley speed is more than 1,000 rpm, engine speed (CAN signal) is less than 450 rpm **Possible Causes:** • Harness or connectors • CAN communication line is open or shorted. • Engine speed signal circuit is open or shorted. • ECM
DTC: P0726 **T TCM, TCIL: Yes** **Year:** 2009 **Model:** Quest **Engine:** 3.5L V6 **Transmission:** All	**Engine Speed Input Circuit Performance:** **NOTE: If DTC U1000 is displayed with this DTC, first perform the trouble diagnosis for DTC U1000.** • This is not an OBD-II self-diagnostic item. • Diagnostic trouble code "ENG SPD INP PERFOR" with CONSULT-III or 14th judgment flicker without CONSULT-III is detected when malfunction is detected in engine speed signal, actual engine torque signal or torque reduction signal that is output from ECM through CAN communication. **Possible Causes:** • Harness or connectors • The signal circuit is open or shorted. • ECM
DTC: P0729 **T TCM, TCIL: Yes** **Year:** 2009, 2010 **Model:** 370Z **Engine:** 3.7L V6 **Transmission:** All	**Gear 6 Incorrect Ratio:** The gear ratio is: • 0.914 or more • 0.813 or less **Possible Causes:** • Input clutch solenoid valve • Direct clutch solenoid valve • High and low reverse clutch solenoid valve • Front brake solenoid valve • Low brake solenoid valve • 2346 brake solenoid valve • Anti-interlock solenoid valve • Each clutch and brake • Output speed sensor • Input speed sensor 1, 2 • Hydraulic control circuit
DTC: P0730 **1T TCM, TCIL: Yes** **Year:** 2009, 2010 **Model:** Altima, Maxima, Murano, Rogue, Versa **Engine:** 1.6L L4, 1.8L L4, 2.5L L4, 3.5L V6, 3.5L V6 VIN A **Transmission:** All	**Incorrect Gear Ratio:** • This is not an OBD-II self-diagnostic item. • TCM calculates the actual gear ratio with primary speed sensor and secondary speed sensor. • TCM receives an unexpected gear ratio signal. **Possible Causes:** • Transaxle assembly

DTC	Trouble Code Title, Conditions & Possible Causes
DTC: P0730 **T TCM, TCIL: Yes** **Year:** 2009, 2010 **Model:** 370Z **Engine:** 3.7L V6 **Transmission:** All	**Incorrect Gear Ratio:** • The revolution of under drive sun gear is 8,000 rpm or more. **NOTE: Not detected when in "P" or "N" position and during a shift to "P" or "N" position.** **Possible Causes:** • 2346 brake solenoid valve • Front brake solenoid valve • Input speed sensor 1, 2
DTC: P0730 **1T TCM, TCIL: Yes** **Year:** 2009, 2010 **Model:** Sentra, Versa **Engine:** 1.6L L4, 1.8L L4, 2.0L L4, 2.5L L4 **Transmission:** All	**Belt Damage:** TCM selects the gear ratio using the engine load (throttle position), the primary pulley revolution speed, and the secondary pulley revolution speed as input signal. Then it changes the operating pressure of the primary pulley and the secondary pulley and changes the groove width of the pulley. • This is not an OBD-II self-diagnostic item. • TCM calculates the actual gear ratio with input speed sensor (primary speed sensor) and output speed sensor (secondary speed sensor). • Diagnostic trouble code P0730 is detected, when TCM receives an unexpected gear ratio signal. **Possible Causes:** • Transaxle assembly
DTC: P0731 **T TCM, TCIL: Yes** **Year:** 2009, 2010 **Model:** 350Z, 370Z, Armada, Frontier, Pathfinder, Titan, Xterra **Engine:** 2.5L L4, 3.5L V6, 3.7L V6, 4.0L V6, 5.6L V8 **Transmission:** All	**A/T 1st Gear Function:** • This is an OBD-II self-diagnostic item. • Diagnostic trouble code P0731 is detected when TCM detects any inconsistency in the actual gear ratio. **Possible Causes:** • Harness or connectors • Solenoid circuits are open or shorted. • Input clutch solenoid valve • Direct clutch solenoid valve • High and low reverse clutch solenoid valve • Front brake solenoid valve • Low brake solenoid valve • 2346 brake solenoid valve • Anti-interlock solenoid valve • Each clutch and brake • Output speed sensor • Input speed sensor 1, 2 • Hydraulic control circuit
DTC: P0731 **T TCM, TCIL: Yes** **Year:** 2009 **Model:** Quest **Engine:** 3.5L V6 **Transmission:** All	**A/T 1st Gear Function:** **NOTE: If DTC U1000 is displayed with this DTC, first perform the trouble diagnosis for DTC U1000.** • This is an OBD-II self-diagnostic item. • Diagnostic trouble code P0731 is detected when A/T cannot be shifted to the 1st gear position even if electrical circuit is good. **Possible Causes:** • Shift solenoid valve A (off stick) • 2nd brake • 2nd coast brake • One-way clutch No.1 • One-way clutch No.2 • Hydraulic control circuit

DTC	Trouble Code Title, Conditions & Possible Causes
DTC: P0732 **T TCM, TCIL: Yes** **Year:** 2009 **Model:** Quest **Engine:** 3.5L V6 **Transmission:** All	**A/T 2nd Gear Function:** **NOTE: If DTC U1000 is displayed with this DTC, first perform the trouble diagnosis for DTC U1000.** • This is an OBD-II self-diagnostic item. • Diagnostic trouble code P0732 is detected when A/T cannot be shifted to the 2nd gear position even if electrical circuit is good. **Possible Causes:** • Shift solenoid valve A (On stick) • Shift solenoid valve B (On stick) • Shift solenoid valve C (Off stick) • Shift solenoid valve D (On stick) • Pressure control solenoid valve A (On stick) • Pressure control solenoid valve C (On stick) • U/D brake • 2nd coast brake • 2nd brake • One-way clutch No.1 • One-way clutch No.2 • B5 brake • Hydraulic control circuit
DTC: P0732 **T TCM, TCIL: Yes** **Year:** 2009, 2010 **Model:** 350Z, 370Z, Armada, Frontier, Pathfinder, Titan, Xterra **Engine:** 2.5L L4, 3.5L V6, 3.7L V6, 4.0L V6, 5.6L V8 **Transmission:** All	**A/T 2nd Gear Function:** • This malfunction is detected when the A/T does not shift into 2GR position as instructed by TCM. This is not only caused by electrical malfunction (circuits open or shorted) but mechanical malfunction such as control valve sticking, improper solenoid valve operation. • This is an OBD-II self-diagnostic item. • Diagnostic trouble code P0732 is detected when TCM detects any inconsistency in the actual gear ratio. **Possible Causes:** • Harness or connectors • Solenoid circuits are open or shorted. • Input clutch solenoid valve • Direct clutch solenoid valve • High and low reverse clutch solenoid valve • Front brake solenoid valve • Low brake solenoid valve • 2346 brake solenoid valve • Anti-interlock solenoid valve • Each clutch and brake • Output speed sensor • Input speed sensor 1, 2 • Hydraulic control circuit
DTC: P0733 **T TCM, TCIL: Yes** **Year:** 2009, 2010 **Model:** 350Z, 370Z, Armada, Frontier, Pathfinder, Titan, Xterra **Engine:** 2.5L L4, 3.5L V6, 3.7L V6, 4.0L V6, 5.6L V8 **Transmission:** All	**A/T 3rd Gear Function:** • This malfunction is detected when the A/T does not shift into 3GR position as instructed by TCM. This is not only caused by electrical malfunction (circuits open or shorted) but mechanical malfunction such as control valve sticking, improper solenoid valve operation. • This is an OBD-II self-diagnostic item. • Diagnostic trouble code P0733 is detected when TCM detects any inconsistency in the actual gear ratio. **Possible Causes:** • Harness or connectors • Solenoid circuits are open or shorted. • Input clutch solenoid valve • Direct clutch solenoid valve • High and low reverse clutch solenoid valve • Front brake solenoid valve • Low brake solenoid valve • 2346 brake solenoid valve • Anti-interlock solenoid valve • Each clutch and brake • Output speed sensor • Input speed sensor 1, 2 • Hydraulic control circuit

DTC	Trouble Code Title, Conditions & Possible Causes
DTC: P0733 **T TCM, TCIL: Yes** **Year:** 2009 **Model:** Quest **Engine:** 3.5L V6 **Transmission:** All	**A/T 3rd Gear Function:** **NOTE: If DTC U1000 is displayed with this DTC, first perform the trouble diagnosis for DTC U1000.** • This is an OBD-II self-diagnostic item. • Diagnostic trouble code P0733 is detected when A/T cannot be shifted to the 3rd gear position even if electrical circuit is good. **Possible Causes:** • Shift solenoid valve A (On stick) • Shift solenoid valve B (On stick) • Shift solenoid valve C (Off stick) • Shift solenoid valve D (Off stick) • Pressure control solenoid valve A (On stick) • B5 brake • U/D clutch • U/D brake • Hydraulic control circuit
DTC: P0734 **T TCM, TCIL: Yes** **Year:** 2009 **Model:** Quest **Engine:** 3.5L V6 **Transmission:** All	**A/T 4th Gear Function:** **NOTE: If DTC U1000 is displayed with this DTC, first perform the trouble diagnosis for DTC U1000.** • This is an OBD-II self-diagnostic item. • Diagnostic trouble code P0734 is detected when A/T cannot be shifted to the 4th gear position even if electrical circuit is good. **Possible Causes:** • Shift solenoid valve A (On stick.) • Shift solenoid valve B (On stick.) • Shift solenoid valve C (On stick.) • Pressure control solenoid valve A (On stick.) • Forward and direct clutch assembly • U/D clutch • U/D brake • 2nd coast brake • One-way clutch No.1 • Hydraulic control circuit
DTC: P0734 **T TCM, TCIL: Yes** **Year:** 2009, 2010 **Model:** 350Z, 370Z, Armada, Frontier, Pathfinder, Titan, Xterra **Engine:** 2.5L L4, 3.5L V6, 3.7L V6, 4.0L V6, 5.6L V8 **Transmission:** All	**4th Gear Incorrect Ratio:** • This malfunction is detected when the A/T does not shift into 4GR position as instructed by TCM. This is not only caused by electrical malfunction (circuits open or shorted) but mechanical malfunction such as control valve sticking, improper solenoid valve operation. • This is an OBD-II self-diagnostic item. • P0734 is detected when TCM detects any inconsistency in the actual gear ratio. **Possible Causes:** • Input clutch solenoid valve • Direct clutch solenoid valve • High and low reverse clutch solenoid valve • Front brake solenoid valve • Low brake solenoid valve • 2346 brake solenoid valve • Anti-interlock solenoid valve • Each clutch and brake • Output speed sensor • Input speed sensor 1, 2 • Hydraulic control circuit

DTC	Trouble Code Title, Conditions & Possible Causes

DTC: P0735
T TCM, TCIL: Yes
Year: 2009, 2010
Model: 350Z, 370Z, Armada, Frontier, Pathfinder, Titan, Xterra
Engine: 2.5L L4, 3.5L V6, 3.7L V6, 4.0L V6, 5.6L V8
Transmission: All
	5th Gear Incorrect Ratio:
	This malfunction is detected when the A/T does not shift into 5GR position as instructed by TCM. This is not only caused by electrical malfunction (circuits open or shorted) but mechanical malfunction such as control valve sticking, improper solenoid valve operation.
	  • This is an OBD-II self-diagnostic item.
	  • Diagnostic trouble code P0735 is detected when TCM detects any inconsistency in the actual gear ratio.
	Possible Causes:
	  • Input clutch solenoid valve
	  • Direct clutch solenoid valve
	  • High and low reverse clutch solenoid valve
	  • Front brake solenoid valve
	  • Low brake solenoid valve
	  • 2346 brake solenoid valve
	  • Anti-interlock solenoid valve
	  • Each clutch and brake
	  • Output speed sensor
	  • Input speed sensor 1, 2
	  • Hydraulic control circuit

DTC: P0735
T TCM, TCIL: Yes
Year: 2009
Model: Quest
Engine: 3.5L V6
Transmission: All
	A/T 5th Gear Function:
	NOTE: If DTC U1000 is displayed with this DTC, first perform the trouble diagnosis for DTC U1000.
	  • This is an OBD-II self-diagnostic item.
	  • Diagnostic trouble code P0735 is detected when A/T cannot be shifted to the 5th gear position even if electrical circuit is good.
	Possible Causes:
	  • Shift solenoid valve B (Off stick.)
	  • Shift solenoid valve C (On stick.)
	  • Shift solenoid valve E (On stick.)
	  • Pressure control solenoid valve A (On stick.)
	  • Pressure control solenoid valve B (On stick.)
	  • Forward and direct clutch assembly
	  • Direct clutch
	  • 2no coast brake
	  • One-way clutch No.1
	  • Hydraulic control circuit

DTC: P0740
1T TCM, TCIL: Yes
Year: 2009, 2010
Model: 350Z, Altima, Armada, Frontier, Pathfinder, Sentra, Titan, Versa, Xterra
Engine: 1.6L L4, 1.8L L4, 2.0L L4, 2.5L L4, 3.5L V6, 4.0L V6, 5.6L V8
Transmission: All
	Torque Converter Clutch Solenoid Valve:
	  • Diagnostic trouble code P0740 is detected under the following conditions:
	  - TCM detects an improper voltage drop when it tries to operate the solenoid valve.
	  - When TCM detects as irregular by comparing target value with monitor value.
	Possible Causes:
	  • Torque converter clutch solenoid valve
	  • Harness or connectors
	  • Solenoid circuit is open or shorted.

DTC: P0740
T TCM, TCIL: Yes
Year: 2009, 2010
Model: Maxima, Murano, Rogue
Engine: 2.5L L4, 3.5L V6, 3.5L V6 VIN A
Transmission: All
	Torque Converter Clutch (TCC) Circuit Open/Short:
	  • Normal voltage is not applied to solenoid due to open or short circuit.
	Possible Causes:
	  • Torque converter clutch solenoid valve
	  • Harness or connectors
	  • Solenoid circuit is open or shorted.

DTC: P0740
T TCM, TCIL: Yes
Year: 2009, 2010
Model: Cube
Engine: 1.8L L4
Transmission: All
	Torque Converter Clutch Circuit - Open:
	  • Torque converter clutch solenoid valve monitor voltage value of TCM is less than 70% of torque converter clutch solenoid valve target voltage value.
	  • Torque converter clutch solenoid valve current command value of TCM and torque converter clutch solenoid valve current monitor value is deviated.
	Possible Causes:
	  • Harness or connectors
	  • Torque converter clutch solenoid valve circuit is open or shorted.
	  • Torque converter clutch solenoid valve

DTC	Trouble Code Title, Conditions & Possible Causes
DTC: P0740 **T TCM, TCIL: Yes** **Year:** 2009, 2010 **Model:** 370Z **Engine:** 3.7L V6 **Transmission:** All	**Torque Converter Clutch Circuit - Open:** • The torque converter clutch solenoid valve monitor value is 0.4 A or less when the torque converter clutch solenoid valve command value is more than 0.75 A. **Possible Causes:** • Harness or connectors • Solenoid valve circuit is open or shorted. • Torque converter clutch solenoid valve
DTC: P0744 **T TCM, TCIL: Yes** **Year:** 2009, 2010 **Model:** Cube **Engine:** 1.8L L4 **Transmission:** All	**Torque Converter Clutch Circuit Intermittent:** Torque converter slip speed is more than a certain value (40 rpm + vehicle speed/2) while TCM is in lock-up command state. **Possible Causes:** • Hydraulic control circuit • Torque converter clutch solenoid valve • Lock-up select solenoid valve
DTC: P0744 **T TCM, TCIL: Yes** **Year:** 2009 **Model:** Quest **Engine:** 3.5L V6 **Transmission:** All	**A/T TCC S/V Function (Lock-Up):** • This malfunction will not be detected while the O/D OFF indicator lamp is indicating another self-diagnosis malfunction. • This malfunction is detected when the torque converter clutch does not lock up as instructed by the TCM. This is not caused by electrical malfunction (circuits open or shorted) but by mechanical malfunction such as control valve sticking, improper solenoid valve operation, malfunctioning oil pump or torque converter clutch, etc. • This is an OBD-II self-diagnostic item. • Diagnostic trouble code is detected when A/T cannot perform lock-up even if electrical circuit is good. **Possible Causes:** • Shift solenoid valve D (Off stick.) • Pressure control solenoid valve C (Off stick.) • Torque converter clutch • Hydraulic control circuit
DTC: P0744 **T TCM, TCIL: Yes** **Year:** 2009, 2010 **Model:** 350Z, Altima, Armada, Frontier, Pathfinder, Sentra, Titan, Versa, Xterra **Engine:** 1.6L L4, 1.8L L4, 2.0L L4, 2.5L L4, 3.5L V6, 4.0L V6, 5.6L V8 **Transmission:** All	**A/T TCC S/V Function (Lock-Up):** This malfunction is detected when the A/T does not lock-up or does not shift to 5th gear. This is not only caused by electrical malfunction (circuits open or shorted) but also by mechanical malfunction such as control valve sticking, improper solenoid valve operation, etc. • This is an OBD-II self-diagnostic item. • Diagnostic trouble code P0744 is detected under the following conditions: - When A/T cannot perform lock-up even if electrical circuit is good. - When TCM detects as irregular by comparing difference value with slip rotation. **Possible Causes:** • Harness or connectors • The solenoid circuit is open or shorted. • Torque converter clutch solenoid valve • Hydraulic control circuit
DTC: P0744 **T TCM, TCIL: Yes** **Year:** 2009, 2010 **Model:** Maxima, Murano, Rogue **Engine:** 2.5L L4, 3.5L V6, 3.5L V6 VIN A **Transmission:** All	**Torque Converter Clutch Circuit Intermittent:** • Transmission cannot perform lock-up even if electrical circuit is good. • TCM detects as irregular by comparing difference value with slip rotation. • There is a big difference between engine speed and primary speed sensor when TCM lock-up signal is on. **Possible Causes:** • Torque converter clutch solenoid valve • Hydraulic control circuit
DTC: P0744 **T TCM, TCIL: Yes** **Year:** 2009, 2010 **Model:** 370Z **Engine:** 3.7L V6 **Transmission:** All	**Torque Converter Clutch Circuit Intermittent:** • The lock-up is not performed properly within the lock-up area. • When A/T cannot perform lock-up even if electrical circuit is good. • When TCM detects as irregular by comparing difference value with slip rotation. **Possible Causes:** • Harness or connectors • Torque converter clutch solenoid valve • Torque converter • Input speed sensor 1, 2 • Hydraulic control circuit

DTC	Trouble Code Title, Conditions & Possible Causes
DTC: P0745 **T TCM, TCIL: Yes** **Year:** 2009, 2010 **Model:** Maxima, Murano, Rogue **Engine:** 2.5L L4, 3.5L V6, 3.5L V6 VIN A **Transmission:** All	**Pressure Control Solenoid A:** • Normal voltage is not applied to solenoid due to open or short circuit. • TCM detects as irregular by comparing target value with monitor value. **Possible Causes:** • Harness or connectors • Solenoid circuit is open or shorted. • Line pressure solenoid valve
DTC: P0745 **1T TCM, TCIL: Yes** **Year:** 2009, 2010 **Model:** Altima, Sentra **Engine:** 2.0L L4, 2.5L L4, 3.5L V6 **Transmission:** All	**Pressure Control Solenoid Valve A:** • The pressure control solenoid valve A (line pressure solenoid valve) regulates the oil pump discharge pressure to suit the driving condition in response to a signal sent from the TCM. • This is an OBD-II self-diagnostic item. • Diagnostic trouble code P0745 is detected under the following conditions: - TCM detects an improper voltage drop when it tries to operate the solenoid valve. - When TCM compares target value with monitor value and detects an irregularity. **Possible Causes:** • Harness or connectors (Solenoid circuit is open or shorted.) • Pressure control solenoid valve A (Line pressure solenoid valve)
DTC: P0745 **T TCM, TCIL: Yes** **Year:** 2009 **Model:** Quest **Engine:** 3.5L V6 **Transmission:** All	**A/T Pressure Control Solenoid Valve A:** • This is an OBD-II self-diagnostic item. • Diagnostic trouble code "PC SOL A(L/PRESS)" with CONSULT-III or P0745 without CONSULT-III is detected under the following conditions: - When normal voltage is not applied to solenoid due to open, short, and so on. - When TCM detects as irregular by comparing target value with monitor value. **Possible Causes:** • Harness or connectors • The solenoid circuit is open or shorted. • Pressure control solenoid valve A
DTC: P0745 **T TCM, TCIL: Yes** **Year:** 2009, 2010 **Model:** 370Z **Engine:** 3.7L V6 **Transmission:** All	**Pressure Control Solenoid A:** The line pressure solenoid valve monitor value is 0.4 A or less when the line pressure solenoid valve command value is more than 0.75 A. **Possible Causes:** • Harness or connectors • Solenoid valve circuit is open or shorted. • Line pressure solenoid valve
DTC: P0745 **T TCM, TCIL: Yes** **Year:** 2009, 2010 **Model:** 350Z, Armada, Frontier, Pathfinder, Titan, Versa, Xterra **Engine:** 1.6L L4, 1.8L L4, 2.5L L4, 3.5L V6, 4.0L V6, 5.6L V8 **Transmission:** All	**Pressure Control Solenoid A:** The line pressure solenoid valve regulates the oil pump discharge pressure to suit the driving condition in response to a signal sent from the TCM. • This is an OBD-II self-diagnostic item. • Diagnostic trouble code P0745 is detected under the following conditions: - When TCM detects an improper voltage drop when it tries to operate the solenoid valve. - When TCM detects as irregular by comparing target value with monitor value. **Possible Causes:** • Harness or connectors • The solenoid circuit is open or shorted. • Line pressure solenoid valve
DTC: P0745 **T TCM, TCIL: Yes** **Year:** 2009, 2010 **Model:** Cube **Engine:** 1.8L L4 **Transmission:** All	**Pressure Control Solenoid A:** • Monitor voltage value of TCM line pressure solenoid valve is less than 70% of the target voltage value of line pressure solenoid valve. • Current monitor value of the Line pressure solenoid valve differs from the TCM current command value of line pressure solenoid valve. **Possible Causes:** • Harness or connectors • Line pressure solenoid valve circuit is open or shorted. • Line pressure solenoid valve

DTC	Trouble Code Title, Conditions & Possible Causes
DTC: P0746 **T TCM, TCIL: Yes** **Year:** 2009, 2010 **Model:** Cube **Engine:** 1.8L L4 **Transmission:** All	**Pressure Control Solenoid A Performance - Stuck Off:** TCM detects a state that gear ratio is more than 2.9 **Possible Causes:** • Line pressure control system • Line pressure solenoid valve • Primary speed sensor • Secondary speed sensor
DTC: P0746 **1T TCM, TCIL: Yes** **Year:** 2009, 2010 **Model:** Altima, Maxima, Murano, Rogue, Sentra, Versa **Engine:** 1.6L L4, 1.8L L4, 2.0L L4, 2.5L L4, 3.5L V6, 3.5L V6 VIN A **Transmission:** All	**Pressure Control Solenoid A Performance (Stuck Off):** • The pressure control solenoid valve A (line pressure solenoid valve) regulates the oil pump discharge pressure to suit the driving condition in response to a signal sent from the TCM. • This is an OBD-II self-diagnostic item. • Diagnostic trouble code P0746 is detected under the following conditions: - Unexpected gear ratio was detected in the LOW side due to excessively low line pressure. **Possible Causes:** • Line pressure control system • Output speed sensor (Secondary speed sensor) • Input speed sensor (Primary speed sensor)
DTC: P0750 **T TCM, TCIL: Yes** **Year:** 2009, 2010 **Model:** 370Z **Engine:** 3.7L V6 **Transmission:** All	**Shift Solenoid A:** • The anti-interlock solenoid valve monitor value is ON when the anti-interlock solenoid valve command value is OFF. • The anti-interlock solenoid valve monitor value is OFF when the anti-interlock solenoid valve command value is ON. **Possible Causes:** • Harness or connectors • Solenoid valve circuit is open or shorted. • Anti-interlock solenoid valve
DTC: P0750 **T TCM, TCIL: Yes** **Year:** 2009 **Model:** Quest **Engine:** 3.5L V6 **Transmission:** All	**Shift Solenoid Valve A:** **NOTE: If DTC U1000 is displayed with this DTC, first perform the trouble diagnosis for DTC U1000.** • Shift solenoid valves are installed directly in control valve body. The shift solenoid valves operates of ON and OFF by the control signal from TCM. Combinations of 5 shift solenoid valves, A, B, C, D and E, shifts gear positions. • The shift solenoid valve A is a normally open, ON-OFF type solenoid. • This is an OBD-II self-diagnostic item. • Diagnostic trouble code "SHIFT SOL A" with CONSULT-III or P0750 without CONSULT-III is detected under the following conditions: - When normal voltage is not applied to solenoid due to open, short, and so on. - When TCM detects as irregular by comparing target value with monitor value. **Possible Causes:** • Harness or connectors • The solenoid circuit is open or shorted. • Shift solenoid valve A
DTC: P0755 **T TCM, TCIL: Yes** **Year:** 2009 **Model:** Quest **Engine:** 3.5L V6 **Transmission:** All	**Shift Solenoid Valve B:** **NOTE: If DTC U1000 is displayed with this DTC, first perform the trouble diagnosis for DTC U1000.** • Shift solenoid valves are installed directly in control valve body. The shift solenoid valves operates of ON and OFF by the control signal from TCM. Combinations of 5 shift solenoid valves, A, B, C, D and E, shifts gear positions. • The shift solenoid valve B is a normally closed, ON-OFF type solenoid. • This is an OBD-II self-diagnostic item. • Diagnostic trouble code "SHIFT SOL B" with CONSULT-III or P0755 without CONSULT-III is detected under the following conditions: - When normal voltage is not applied to solenoid due to open, short, and so on. - When TCM detects as irregular by comparing target value with monitor value. **Possible Causes:** • Harness or connectors • The solenoid circuit is open or shorted. • Shift solenoid valve B

DTC	Trouble Code Title, Conditions & Possible Causes
DTC: P0760 **T TCM, TCIL: Yes** **Year:** 2009 **Model:** Quest **Engine:** 3.5L V6 **Transmission:** All	**Shift Solenoid Valve C:** **NOTE: If DTC U1000 is displayed with this DTC, first perform the trouble diagnosis for DTC U1000.** • Shift solenoid valves are installed directly in control valve body. The shift solenoid valves operates of ON and OFF by the control signal from TCM. Combinations of 5 shift solenoid valves, A, B, C, D and E, shifts gear positions. • The shift solenoid valve C is a normally closed, ON-OFF type solenoid. • This is an OBD-II self-diagnostic item. • Diagnostic trouble code "SHIFT SOL C" with CONSULT-III or P0760 without CONSULT-III is detected under the following conditions: - When normal voltage is not applied to solenoid due to open, short, and so on. - When TCM detects as irregular by comparing target value with monitor value. **Possible Causes:** • Harness or connectors • The solenoid circuit is open or shorted. • Shift solenoid valve C
DTC: P0762 **T TCM, TCIL: Yes** **Year:** 2009 **Model:** Quest **Engine:** 3.5L V6 **Transmission:** All	**Shift Solenoid Valve C Stuck On:** **NOTE: If DTC U1000 is displayed with this DTC, first perform the trouble diagnosis for DTC U1000.** • This malfunction will not be detected while the O/D OFF indicator lamp is indicating another self-diagnosis malfunction. • This is not caused by electrical malfunction (circuits open or shorted) but by mechanical malfunction such as control valve sticking, improper solenoid valve operation, etc. • Shift solenoid valves are installed directly in control valve body. The shift solenoid valves operates of ON and OFF by the control signal from TCM. Combinations of 5 shift solenoid valves, A, B, C, D and E, shifts gear positions. • The shift solenoid valve C is a normally closed, ON-OFF type solenoid. • This is an OBD-II self-diagnostic item. • Diagnostic trouble code "SFT SOL C STUCK ON" with CONSULT-III or P0762 without CONSULT-III is detected when condition of shift solenoid valve C is different from monitor value, and relation between gear position and actual gear ratio is irregular. **Possible Causes:** • Shift solenoid valve C (On stick.) • Hydraulic control circuit
DTC: P0765 **T TCM, TCIL: Yes** **Year:** 2009 **Model:** Quest **Engine:** 3.5L V6 **Transmission:** All	**Shift Solenoid Valve D:** **NOTE: If DTC U1000 is displayed with this DTC, first perform the trouble diagnosis for DTC U1000.** • Shift solenoid valves are installed directly in control valve body. The shift solenoid valves operates of ON and OFF by the control signal from TCM. Combinations of 5 shift solenoid valves, A, B, C, D and E, shifts gear positions. • The shift solenoid valve D is a normally open, ON-OFF type solenoid. • This is an OBD-II self-diagnostic item. • Diagnostic trouble code "SHIFT SOL D" with CONSULT-III or P0765 without CONSULT-III is detected under the following conditions: - When normal voltage is not applied to solenoid due to open, short, and so on. - When TCM detects as irregular by comparing target value with monitor value. **Possible Causes:** • Harness or connectors • The solenoid circuit is open or shorted. • Shift solenoid valve D
DTC: P0770 **T TCM, TCIL: Yes** **Year:** 2009 **Model:** Quest **Engine:** 3.5L V6 **Transmission:** All	**Shift Solenoid Valve E:** **NOTE: If DTC U1000 is displayed with this DTC, first perform the trouble diagnosis for DTC U1000.** • Shift solenoid valves are installed directly in control valve body. The shift solenoid valves operates of ON and OFF by the control signal from TCM. Combinations of 5 shift solenoid valves, A, B, C, D and E, shifts gear positions. • The shift solenoid valve E is a normally closed, ON-OFF type solenoid. • This is an OBD-II self-diagnostic item. • Diagnostic trouble code "SHIFT SOL E" with CONSULT-III or P0770 without CONSULT-III is detected under the following conditions: - When normal voltage is not applied to solenoid due to open, short, and so on. - When TCM detects as irregular by comparing target value with monitor value. **Possible Causes:** • Harness or connectors • The solenoid circuit is open or shorted. • Shift solenoid valve E

DTC	Trouble Code Title, Conditions & Possible Causes
DTC: P0775 **T TCM, TCIL: Yes** **Year:** 2009, 2010 **Model:** 370Z **Engine:** 3.7L V6 **Transmission:** All	**Pressure Control Solenoid B:** The input clutch solenoid valve monitor value is 0.4 A or less when the input clutch solenoid valve command value is more than 0.75 A. **Possible Causes:** • Harness or connectors • Solenoid valve circuit is open or shorted. • Input clutch solenoid valve
DTC: P0775 **T TCM, TCIL: Yes** **Year:** 2009 **Model:** Quest **Engine:** 3.5L V6 **Transmission:** All	**Pressure Control Solenoid Valve B (Shift Pressure):** **NOTE: If DTC U1000 is displayed with this DTC, first perform the trouble diagnosis for DTC U1000.** • The pressure control solenoid valve B is normally high, 3-port linear pressure control solenoid. • The pressure control solenoid valve B controls linear shift pressure by control signal from TCM and controls 2nd coast brake directly under 2nd, 3rd, 4th and direct clutch directly under 5th and reverse. • This is an OBD-II self-diagnostic item. • Diagnostic trouble code "PC SOL B(SFT/PRS)" with CONSULT-III or P0775 without CONSULT-III is detected under the following conditions: - When normal voltage is not applied to solenoid due to open, short, and so on. - When TCM detects as irregular by comparing target value with monitor value. **Possible Causes:** • Harness or connectors • The solenoid circuit is open or shorted. • Pressure control solenoid valve B
DTC: P0776 **T TCM, TCIL: Yes** **Year:** 2009, 2010 **Model:** Cube **Engine:** 1.8L L4 **Transmission:** All	**Pressure Control Solenoid B Performance - Stuck Off:** Difference of secondary pressure target value of TCM and secondary pressure actual value is more than 1.2 MPa **Possible Causes:** • Secondary pressure solenoid valve system • Line pressure control system • Secondary pressure solenoid valve • Secondary pressure sensor
DTC: P0776 **1T TCM, TCIL: Yes** **Year:** 2009, 2010 **Model:** Altima, Maxima, Murano, Rogue, Sentra, Versa **Engine:** 1.6L L4, 1.8L L4, 2.0L L4, 2.5L L4, 3.5L V6, 3.5L V6 VIN A **Transmission:** All	**Pressure Control Solenoid B Performance (Stuck Off):** • The pressure control solenoid valve B (secondary pressure solenoid valve) regulates the secondary pressure to suit the driving condition in response to a signal sent from the TCM. • This is an OBD-II self-diagnostic item. • Diagnostic trouble code P0776 is detected when secondary pressure is too high or too low compared with the commanded value while driving. **Possible Causes:** • Harness or connectors • Solenoid circuit is open or shorted. • Pressure control solenoid valve B (Secondary pressure solenoid valve system) • Transmission fluid pressure sensor A (Secondary pressure sensor) • Line pressure control system
DTC: P0778 **1T TCM, TCIL: Yes** **Year:** 2009, 2010 **Model:** Altima, Maxima, Murano, Rogue, Sentra, Versa **Engine:** 1.6L L4, 1.8L L4, 2.0L L4, 2.5L L4, 3.5L V6, 3.5L V6 VIN A **Transmission:** All	**Pressure Control Solenoid B Electrical:** • The pressure control solenoid valve B (secondary pressure solenoid valve) regulates the oil pump discharge pressure to suit the driving condition in response to a signal sent from the TCM. • This is an OBD-II self-diagnostic item. • Diagnostic trouble code P0778 is detected under the following conditions: - TCM detects an improper voltage drop when it tries to operate the solenoid valve. - When TCM compares target value with monitor value and detects an irregularity. • Normal voltage not applied to solenoid due to cut line, short, or the like. • TCM detects as irregular by comparing target value with monitor value. **Possible Causes:** • Harness or connectors (Solenoid circuit is open or shorted.) • Pressure control solenoid valve B (Secondary pressure solenoid valve)
DTC: P0778 **T TCM, TCIL: Yes** **Year:** 2009, 2010 **Model:** Cube **Engine:** 1.8L L4 **Transmission:** All	**Pressure Control Solenoid B Electrical:** • Current monitor value of the secondary pressure solenoid valve differs from the TCM current command value of secondary pressure solenoid valve. • Secondary pressure solenoid valve current command value of TCM and secondary pressure solenoid valve current monitor value is deviated. **Possible Causes:** • Harness or connectors • Secondary pressure solenoid valve circuit is open or shorted. • Secondary pressure solenoid valve

DTC	Trouble Code Title, Conditions & Possible Causes
DTC: P0780 **T TCM, TCIL: Yes** **Year:** 2009 **Model:** Quest **Engine:** 3.5L V6 **Transmission:** All	**Shift Malfunction:** **NOTE: If DTC U1000 is displayed with this DTC, first perform the trouble diagnosis for DTC U1000.** • This malfunction will not be detected while the O/D OFF indicator lamp is indicating another self-diagnosis malfunction. • This malfunction is detected when the A/T does not shift as instructed by the TCM. This is not caused by electrical malfunction (circuits open or shorted) but by mechanical malfunction such as control valve sticking, improper solenoid valve operation, etc. • This is an OBD-II self-diagnostic item. • Diagnostic trouble code "SHIFT" with CONSULT-III or P0780 without CONSULT-III is detected under the following conditions: - When no rotation change occurs between input (turbine revolution sensor) and output (revolution sensor) and shifting time is long. - When shifting ends immediately. - When engine revs up unusually during shifting. **Possible Causes:** • Shift solenoid valve D (Off error) • Shift solenoid valve E (Off error) • Pressure control solenoid valve A (On/Off error) • Pressure control solenoid valve B (On/Off error) • Pressure control solenoid valve C (On/Off error) • Hydraulic control circuit
DTC: P0780 **T TCM, TCIL: Yes** **Year:** 2009, 2010 **Model:** 370Z **Engine:** 3.7L V6 **Transmission:** All	**Shift Error:** • When shifting from 3GR to 4GR with the selector lever in "D" position, the gear ratio does not shift to 1.412 (gear ratio of 4GR). • When shifting from 5GR to 6GR or 6GR to 7GR, the engine speed exceeds the prescribed speed. • The shift change time from 4GR to 3GR is 0.2 second or less. **Possible Causes:** • Anti-interlock solenoid valve • Low brake solenoid valve • Hydraulic control circuit
DTC: P0795 **T TCM, TCIL: Yes** **Year:** 2009 **Model:** Quest **Engine:** 3.5L V6 **Transmission:** All	**Pressure Control Solenoid Valve C (TCC & Shift Pressure):** **NOTE: If DTC U1000 is displayed with this DTC, first perform the trouble diagnosis for DTC U1000.** • The pressure control solenoid valve C is normally low, 3-port linear pressure control solenoid. • The pressure control solenoid valve C is activated to control the apply and release of the 2nd brake and 1st and reverse brake, and torque converter clutch. • Lock-up operation, however, is prohibited when A/T fluid temperature is too low. • When the accelerator pedal is depressed (less than 1/8) in lock-up condition, the engine speed should not change abruptly. If there is a big jump in engine speed, there is no lock-up. • This is an OBD-II self-diagnostic item. • Diagnostic trouble code "PC SOL C(TCC&SFT)" with CONSULT-III or P0795 without CONSULT-III is detected under the following conditions: - When normal voltage is not applied to solenoid due to open, short, and so on. - When TCM detects as irregular by comparing target value with monitor value. **Possible Causes:** • Harness or connectors • The solenoid circuit is open or shorted. • Pressure control solenoid valve C
DTC: P0795 **T TCM, TCIL: Yes** **Year:** 2009, 2010 **Model:** 370Z **Engine:** 3.7L V6 **Transmission:** All	**Pressure Control Solenoid C:** The front brake solenoid valve monitor value is 0.4 A or less when the front brake solenoid valve command value is more than 0.75 A. **Possible Causes:** • Harness or connectors • Solenoid valve circuit is open or shorted. • Front brake solenoid valve

DTC	Trouble Code Title, Conditions & Possible Causes
DTC: P0797 **T TCM, TCIL: Yes** **Year:** 2009 **Model:** Quest **Engine:** 3.5L V6 **Transmission:** All	**Pressure Control Solenoid Valve C (Stuck On):** **NOTE: If DTC U1000 is displayed with this DTC, first perform the trouble diagnosis for DTC U1000.** • This malfunction will not be detected while the O/D OFF indicator lamp is indicating another self-diagnosis malfunction. • This is not caused by electrical malfunction (circuits open or shorted) but by mechanical malfunction such as control valve sticking, improper solenoid valve operation, etc. • The pressure control solenoid valve C is normally low, 3-port linear pressure control solenoid. • The pressure control solenoid valve C is activated to control the apply and release of the 2nd brake and 1st and reverse brake, and torque converter clutch. • Lock-up operation, however, is prohibited when A/T fluid temperature is too low. • When the accelerator pedal is depressed (less than 1/8) in lock-up condition, the engine speed should not change abruptly. If there is a big jump in engine speed, there is no lock-up. • This is an OBD-II self-diagnostic item. • Diagnostic trouble code "PC SOL C STC ON" with CONSULT-III or P0797 without CONSULT-III is detected when condition of pressure control solenoid valve C is different from monitor value, and relation between gear position and actual gear ratio or lock-up status is irregular. **Possible Causes:** • Pressure control solenoid valve C (On stick) • Hydraulic control circuit
DTC: P0820 **T ECM, MIL: Yes** **Year:** 2009, 2010 **Model:** 370Z **Engine:** 3.7L V6 **Transmission:** All	**Gear Lever Position Sensor Circuit:** Condition A: • An excessively low voltage from the sensor is sent to ECM. • An excessively high voltage from the sensor is sent to ECM. Condition B: • There is a difference between target engine speed calculated by ECM and actual engine speed. **Possible Causes:** • Harness or connectors • Gear lever position sensor circuit is open or shorted. • CKP sensor (POS) circuit is shorted. • APP sensor 2 circuit is shorted. • EVAP control system pressure sensor circuit is shorted. • Refrigerant pressure sensor circuit is shorted. • (Brake booster pressure sensor circuit is shorted. • Gear lever position sensor • Crankshaft position sensor (POS) • Accelerator pedal position sensor • EVAP control system pressure • Refrigerant pressure sensor • Brake booster pressure sensor • Transmission
DTC: P0825 **T TCM, TCIL: Yes** **Year:** 2009 **Model:** Quest **Engine:** 3.5L V6 **Transmission:** All	**Lever Switch Circuit:** **NOTE: If DTC U1000 is displayed with this DTC, first perform the trouble diagnosis for DTC U1000.** • Lever switch is installed in A/T device. It sends lever switch position (ON or OFF) signals to TCM. • This is not an OBD-II self-diagnostic item. • Diagnostic trouble code "GEAR LEVER SWITCH" with CONSULT-III is detected when TCM monitors lever switch signal, and judges as irregular when impossible input pattern occurs. **Possible Causes:** • Harness or connectors • Lever switch circuit is open or shorted. • Lever switch (built into A/T device)

DTC	Trouble Code Title, Conditions & Possible Causes
DTC: P0826 **1T TCM, TCIL: Yes** **Year:** 2009, 2010 **Model:** Altima, Sentra **Engine:** 2.0L L4, 2.5L L4, 3.5L V6 **Transmission:** All	**Manual Mode Switch (Up/Down Shift Switch) Circuit:** TCM sends the switch signals to combination meter via CAN communication line. Then manual mode switch position is indicated on the CVT position indicator. • This is not an OBD-II self-diagnostic item. • Diagnostic trouble code P0826 is detected when TCM monitors Manual mode, Non manual mode, Up or Down switch signal, and then detects irregular with impossible input pattern for 1 second or more. • When an impossible pattern of switch signals is detected, a malfunction is detected. **Possible Causes:** • Harness, switches or connectors are open or shorted: • TCM • Combination meter circuit • CAN communication line • Manual mode select switch. • Manual mode position select switch.
DTC: P0826 **T TCM, TCIL: Yes** **Year:** 2009, 2010 **Model:** Maxima, Murano, Rogue **Engine:** 2.5L L4, 3.5L V6, 3.5L V6 VIN A **Transmission:** All	**Up & Down Shift Switch Circuit:** • When an impossible pattern of switch signals is detected, a malfunction is detected. • When shift up/down signal of paddle shifter continuously remains ON for 60 seconds. **Possible Causes:** • Harness or connectors • The circuit of these switches are open or shorted. • TCM, and combination meter circuit are open or shorted. • CAN communication line is open or shorted. • Manual mode select switch (Built into CVT shift selector) • Manual mode position select switch (Built into CVT shift selector) • Paddle shifter
DTC: P0830 **T ECM, MIL: Yes** **Year:** 2009, 2010 **Model:** 370Z **Engine:** 3.7L V6 **Transmission:** All	**Clutch Interlock Switch Circuit:** Condition A: ON signals from the clutch interlock switch and the clutch pedal position switch are sent to the ECM at the same time. Condition B: Clutch interlock switch ON signal is not sent to ECM for extremely long time. **Possible Causes:** • Harness or connectors • Clutch interlock switch circuit is open or shorted. • Clutch pedal position switch circuit is open or shorted. • Clutch interlock switch • Clutch pedal position switch • Incorrect clutch interlock switch installation • Incorrect clutch pedal position switch installation
DTC: P0833 **T ECM, MIL: Yes** **Year:** 2009, 2010 **Model:** 370Z **Engine:** 3.7L V6 **Transmission:** All	**Clutch Pedal Position Switch Circuit:** Condition A: ON signals from the clutch pedal position switch and the clutch interlock switch are sent to the ECM at the same time. Condition B: Clutch pedal position switch ON signal is not sent to ECM for extremely long time. **Possible Causes:** • Harness or connectors • Clutch pedal position switch circuit is open or shorted. • Clutch interlock switch circuit is open or shorted. • Clutch pedal position switch • Clutch interlock switch • Incorrect clutch pedal position switch installation • Incorrect clutch interlock switch installation
DTC: P0840 **T TCM, TCIL: Yes** **Year:** 2009, 2010 **Model:** Cube **Engine:** 1.8L L4 **Transmission:** All	**Transmission Fluid Pressure Sensor/Switch A Circuit:** • Secondary pressure sensor voltage that TCM receives is more than 4.7 V • Secondary pressure sensor voltage that TCM receives is less than 0.9 V **Possible Causes:** • Harness or connectors • Secondary pressure sensor circuit is open or shorted. • Secondary pressure sensor

DTC	Trouble Code Title, Conditions & Possible Causes
DTC: P0840 **1T TCM, TCIL: Yes** **Year:** 2009, 2010 **Model:** Altima, Maxima, Murano, Rogue, Sentra, Versa **Engine:** 1.6L L4, 1.8L L4, 2.0L L4, 2.5L L4, 3.5L V6, 3.5L V6 VIN A **Transmission:** All	**Transmission Fluid Pressure Sensor/Switch A Circuit:** • The transmission fluid pressure sensor A (secondary pressure sensor) detects secondary pressure of CVT and sends TCM the signal. • This is an OBD-II self-diagnostic item. • Diagnostic trouble code P0840 is detected when TCM detects an improper voltage drop when it receives the sensor signal. • Signal voltage of the secondary pressure sensor is too high or too low while driving. **Possible Causes:** • Transmission fluid pressure sensor A (Secondary pressure sensor) • Harness or connectors (Switch circuit is open or shorted.)
DTC: P0841 **T TCM, TCIL: Yes** **Year:** 2009, 2010 **Model:** Cube, Rogue **Engine:** 1.8L L4, 2.5L L4 **Transmission:** All	**Transmission Fluid Pressure Sensor/Switch A Circuit Range/Performance:** • Secondary pressure sensor value exceeds line pressure value. **Possible Causes:** • Harness or connectors • Secondary pressure sensor circuit is open or shorted. • Secondary pressure sensor
DTC: P0841 **1T TCM, TCIL: Yes** **Year:** 2009, 2010 **Model:** Altima, Maxima, Murano, Sentra, Versa **Engine:** 1.6L L4, 1.8L L4, 2.0L L4, 2.5L L4, 3.5L V6, 3.5L V6 VIN A **Transmission:** All	**Transmission Fluid Pressure Sensor/Switch A Circuit Range/Performance:** Using the engine load (throttle position), the primary pulley revolution speed, and the secondary pulley revolution speed as input signal, TCM changes the operating pressure of the primary pulley and the secondary pulley and changes the groove width of the pulley to control the gear ratio. • This is not an OBD-II self-diagnostic item. • Diagnostic trouble code P0841 is detected when correlation between the values of the secondary pressure sensor and the primary pressure sensor is out of specification. • Correlation between the values of the secondary pressure sensor and the primary pressure sensor is out of specification. **Possible Causes:** • Transmission fluid pressure sensor A (Secondary pressure sensor) • Harness or connectors (Sensor circuit is open or shorted.)
DTC: P0845 **1T TCM, TCIL: Yes** **Year:** 2009, 2010 **Model:** Altima, Maxima, Murano **Engine:** 2.5L L4, 3.5L V6, 3.5L V6 VIN A **Transmission:** All	**Transmission Fluid Pressure Sensor/Switch B Circuit:** • Signal voltage of the primary pressure sensor is too high or too low while driving. **Possible Causes:** • Harness or connectors • Sensor circuit is open or shorted. • Primary pressure sensor
DTC: P0850 **1T TCM, TCIL: Yes** **Year:** 2009, 2010 **Model:** 350Z, 370Z, Frontier, Maxima, Xterra **Engine:** 2.5L L4, 3.5L V6, 3.7L V6, 4.0L V6 **Transmission:** All	**Park/Neutral Position (PNP) Switch:** • The signal of the park/neutral position (PNP) does not change during driving after the engine is started. **Possible Causes:** • Harness or connectors • The park/neutral position (PNP) signal circuit is open or shorted. • Park/neutral position (PNP) switch (M/T) • Transmission range switch (A/T) • Combination meter • TCM (A/T)
DTC: P0850 **1T ECM, MIL: Yes** **Year:** 2009, 2010 **Model:** Altima, Armada, Cube, Murano, Pathfinder, Rogue, Titan, Versa **Engine:** 1.6L L4, 1.8L L4, 2.5L L4, 3.5L V6, 3.5L V6 VIN A, 4.0L V6, 5.6L V8 **Transmission:** All	**Park/Neutral Position Switch:** • CVT & M/T: When the shift lever position is P or N (CVT), Neutral (M/T), park/neutral position (PNP) switch is ON. ECM detects the position because the continuity of the line (the ON signal) exists. • A/T: The signal of the park/neutral position (PNP) switch does not change in the process of engine starting and driving. **Possible Causes:** • Harness or connectors • Park/neutral position (PNP) switch circuit is open or shorted. • Park/neutral position (PNP) switch

DTC	Trouble Code Title, Conditions & Possible Causes
DTC: P0868 **T TCM, TCIL: Yes** **Year:** 2009, 2010 **Model:** Versa **Engine:** 1.6L L4, 1.8L L4 **Transmission:** All	**Transmission Fluid Secondary Pressure Down:** • This is not an OBD-II self-diagnostic item. • Diagnostic trouble code P0868 is detected when secondary fluid pressure is too low compared with the commanded value while driving. **Possible Causes:** • Harness or connectors • Solenoid circuit is open or shorted. • Pressure control solenoid valve B (Secondary pressure solenoid valve) system • Transmission fluid pressure sensor A (Secondary pressure sensor) • Line pressure control system
DTC: P0868 **1T TCM, TCIL: Yes** **Year:** 2009, 2010 **Model:** Altima, Sentra **Engine:** 2.0L L4, 2.5L L4, 3.5L V6 **Transmission:** All	**Secondary Pressure Down (Fluid Pressure Low):** • The pressure control solenoid valve B (secondary pressure solenoid valve) regulates the secondary pressure to suit the driving condition in response to a signal sent from the TCM. • This is not an OBD-II self-diagnostic item. • Diagnostic trouble code P0868 is detected when secondary fluid pressure is too low compared with the commanded value while driving. **Possible Causes:** • Harness or connectors • Solenoid circuit is open or shorted. • Pressure control solenoid valve B (Secondary pressure solenoid valve) system • Transmission fluid pressure sensor A (Secondary pressure sensor) • Line pressure control system
DTC: P0868 **T TCM, TCIL: Yes** **Year:** 2009, 2010 **Model:** Cube, Maxima, Murano, Rogue **Engine:** 1.8L L4, 2.5L L4, 3.5L V6, 3.5L V6 VIN A **Transmission:** All	**Transmission Fluid Pressure Low:** Secondary fluid pressure is too low compared with the commanded value while driving. **Possible Causes:** • Harness or connectors • Solenoid circuit is open or shorted. • Secondary pressure solenoid valve system • Secondary pressure sensor • Line pressure control system
DTC: P0882 **T TCM, TCIL: Yes** **Year:** 2009 **Model:** Quest **Engine:** 3.5L V6 **Transmission:** All	**TCM Power Input Signal:** **NOTE: If DTC U1000 is displayed with this DTC, first perform the trouble diagnosis for DTC U1000.** When the power supply to the TCM is cut "OFF", for example because the battery is removed, and the self diagnostics memory function stops, malfunction is detected. • This is an OBD-II self-diagnostic item. • Diagnostic trouble code "TCM POWER INPT SIG" with CONSULT-III or P0882 without CONSULT-III is detected when voltage supplied to TCM is too low. **Possible Causes:** • Harness or connectors • Battery or ignition switch and TCM circuit is open or shorted. • A/T PV IGN relay
DTC: P0AC4 **T ECM, MIL: Yes** **Year:** 2009, 2010 **Model:** Altima **Engine:** 2.5L L4 **Transmission:** All	**HV MIL On Request (Hybrid):** HYBRID MODELS CAUTION: Hybrid systems use very high-voltage battery systems. Before starting any service work involving the battery system, turn the ignition switch OFF and then remove the service plug from pocket in the trunk. After removing the service plug, wait 10 minutes before touching any of the high-voltage connectors and terminals. This DTC is displayed when a malfunction is detected by HV ECU. Check DTC for HV ECU and perform the trouble diagnosis. **Possible Causes:** • Note: See indicated HV DTC for causes.

OBD II Trouble Code List (P1XXX Codes)

DTC	Trouble Code Title, Conditions & Possible Causes

DTC: P100A
1T ECM, MIL: Yes
Year: 2009, 2010
Model: 370Z
Engine: 3.7L V6
Transmission: All
	Variable Valve Event & Lift (VVEL) Response Malfunction (Bank 1):
	NOTE: If DTC P100A or P100B is displayed with DTC P1090 or P1093, first perform the trouble diagnosis for DTC P1090 or P1093.
	   • Actual event response to target is poor.
	Possible Causes:
	   • Harness or connectors
	   • VVEL actuator motor circuit is open or shorted.
	   • VVEL actuator motor
	   • VVEL actuator sub assembly
	   • VVEL ladder assembly
	   • VVEL control module

DTC: P100B
1T ECM, MIL: Yes
Year: 2009, 2010
Model: 370Z
Engine: 3.7L V6
Transmission: All
	Variable Valve Event & Lift (VVEL) Response Malfunction (Bank 2):
	NOTE: If DTC P100A or P100B is displayed with DTC P1090 or P1093, first perform the trouble diagnosis for DTC P1090 or P1093.
	   • Actual event response to target is poor.
	Possible Causes:
	   • Harness or connectors
	   • VVEL actuator motor circuit is open or shorted.
	   • VVEL actuator motor
	   • VVEL actuator sub assembly
	   • VVEL ladder assembly
	   • VVEL control module

DTC: P1078
1T ECM, MIL: Yes
Year: 2009, 2010
Model: Maxima
Engine: 3.5L V6
Transmission: All
	Exhaust Valve Timing Control Position Sensor (Bank 1) Circuit:
	NOTE: If this DTC is displayed with DTC P0643, first perform the trouble diagnosis for DTC P0643.
	   • An excessively high or low voltage from the sensor is sent to ECM.
	Possible Causes:
	   • Harness or connectors
	   • The sensor circuit is open or shorted.
	   • Exhaust valve timing (EVT) control position sensor
	   • Crankshaft position sensor (POS)
	   • Camshaft position sensor (PHASE) (bank 1)
	   • Accumulation of debris to the signal pick-up portion of the camshaft

DTC: P1084
1T ECM, MIL: Yes
Year: 2009, 2010
Model: Maxima
Engine: 3.5L V6
Transmission: All
	Exhaust Valve Timing Control Position Sensor (Bank 2) Circuit:
	NOTE: If this DTC is displayed with DTC P0643, first perform the trouble diagnosis for DTC P0643.
	   • An excessively high or low voltage from the sensor is sent to ECM.
	Possible Causes:
	   • Harness or connectors
	   • EVT control position sensor (bank 2) circuit is open or shorted
	   • CKP sensor (POS) circuit is shorted.
	   • CMP sensor (PHASE) (bank 2) circuit is shorted.
	   • Battery current sensor circuit is shorted.
	   • APP sensor 2 circuit is shorted.
	   • EVAP control system pressure sensor circuit is shorted.
	   • Refrigerant pressure sensor circuit is shorted.
	   • Exhaust valve timing control position sensor (bank 2)
	   • Crankshaft position sensor (POS)
	   • Camshaft position sensor (PHASE) (bank 2)
	   • Battery current sensor
	   • Accelerator pedal position sensor
	   • EVAP control system pressure sensor
	   • Refrigerant pressure sensor
	   • Accumulation of debris to the signal pick-up portion of the camshaft

DTC: P1087
T ECM, MIL: Yes
Year: 2009, 2010
Model: 370Z
Engine: 3.7L V6
Transmission: All
	Variable Valve Event & Lift (VVEL) Small Event Angle Malfunction (Bank 1):
	NOTE: If DTC P1087 or P1088 is displayed with DTC P1090 or P1093, perform the diagnosis for P1090 or P1093 first.
	   • The event angle of VVEL control shaft is always small.
	Possible Causes:
	   • Harness or connectors
	   • VVEL actuator motor circuit is open or shorted.
	   • VVEL actuator motor
	   • VVEL actuator sub assembly
	   • VVEL ladder assembly
	   • VVEL control module

DTC	Trouble Code Title, Conditions & Possible Causes

DTC: P1088
T ECM, **MIL:** Yes
Year: 2009, 2010
Model: 370Z
Engine: 3.7L V6
Transmission: All
	Variable Valve Event & Lift (VVEL) Small Event Angle Malfunction (Bank 2):
	NOTE: If DTC P1087 or P1088 is displayed with DTC P1090 or P1093, perform the diagnosis for P1090 or P1093 first.
	    • The event angle of VVEL control shaft is always small.
	Possible Causes:
	    • Harness or connectors
	    • VVEL actuator motor circuit is open or shorted.
	    • VVEL actuator motor
	    • VVEL actuator sub assembly
	    • VVEL ladder assembly
	    • VVEL control module

DTC: P1089
T , **MIL:** Yes
Year: 2009, 2010
Model: 370Z
Engine: 3.7L V6
Transmission: All
	Variable Valve Event & Lift (VVEL) Control Shaft Position Sensor (Bank 1) Circuit:
	NOTE: If DTC P1089 or P1092 is displayed with DTC P1608, first perform the trouble diagnosis for DTC P1608.
	    • An excessively low voltage from the sensor is sent to VVEL control module.
	    • An excessively high voltage from the sensor is sent to VVEL control module.
	    • Rationally incorrect voltage is sent to VVEL control module compared with the signals from VVEL control shaft position sensor 1 and VVEL control shaft position sensor 2.
	Possible Causes:
	    • Harness or connectors
	    • VVEL control shaft position sensor circuit is open or shorted.
	    • VVEL control shaft position sensor
	    • VVEL control module

DTC: P1090
T , **MIL:** Yes
Year: 2009, 2010
Model: 370Z
Engine: 3.7L V6
Transmission: All
	Variable Valve Event & Lift (VVEL) System Performance (Bank 1) :
	NOTE: If DTC P1090 or P1093 is displayed with DTC P1091, first perform the trouble diagnosis for DTC P1091.
	    • Event angle difference between the actual and the target is detected.
	    • Abnormal current is sent to VVEL actuator motor.
	Possible Causes:
	    • Harness or connectors
	    • VVEL actuator motor circuit is open or shorted.
	    • VVEL actuator motor
	    • VVEL actuator sub assembly
	    • VVEL ladder assembly
	    • VVEL control module

DTC: P1091
T ECM, **MIL:** Yes
Year: 2009, 2010
Model: 370Z
Engine: 3.7L V6
Transmission: All
	Variable Valve Event & Lift (VVEL) Actuator Motor Relay Circuit:
	    • VVEL control module detects the VVEL actuator motor relay is stuck OFF.
	    • VVEL control module detects the VVEL actuator motor relay is stuck ON.
	Possible Causes:
	    • Harness or connectors
	    • VVEL actuator motor relay circuit is open or shorted.
	    • Abort circuit is open or shorted.
	    • VVEL actuator motor relay
	    • VVEL control module
	    • ECM

DTC: P1092
T ECM, **MIL:** Yes
Year: 2009, 2010
Model: 370Z
Engine: 3.7L V6
Transmission: All
	Variable Valve Event & Lift (VVEL) Control Shaft Position Sensor (Bank 2) Circuit:
	NOTE: If DTC P1089 or P1092 is displayed with DTC P1608, first perform the trouble diagnosis for DTC P1608.
	    • An excessively low voltage from the sensor is sent to VVEL control module.
	    • An excessively high voltage from the sensor is sent to VVEL control module.
	    • Rationally incorrect voltage is sent to VVEL control module compared with the signals from VVEL control shaft position sensor 1 and VVEL control shaft position sensor 2.
	Possible Causes:
	    • Harness or connectors
	    • VVEL control shaft position sensor circuit is open or shorted.
	    • VVEL control shaft position sensor
	    • VVEL control module

DTC	Trouble Code Title, Conditions & Possible Causes
DTC: P1093 **T ECM, MIL: Yes** **Year:** 2009, 2010 **Model:** 370Z **Engine:** 3.7L V6 **Transmission:** All	**Variable Valve Event & Lift (VVEL) System Performance (Bank 2) :** **NOTE: If DTC P1090 or P1093 is displayed with DTC P1091, first perform the trouble diagnosis for DTC P1091** • Event angle difference between the actual and the target is detected. • Abnormal current is sent to VVEL actuator motor. **Possible Causes:** • Harness or connectors • VVEL actuator motor circuit is open or shorted. • VVEL actuator motor • VVEL actuator sub assembly • VVEL ladder assembly • VVEL control module
DTC: P1140 **1T ECM, MIL: Yes** **Year:** 2009, 2010 **Model:** Armada, Titan **Engine:** 5.6L V8 **Transmission:** All	**Intake Valve Timing Control Position Sensor Circuit (Bank 1):** • An excessively high or low voltage from the sensor is sent to ECM. **Possible Causes:** • Harness or connectors • Intake valve timing control position sensor circuit is open or shorted • Intake valve timing control position sensor • Crankshaft position sensor (POS) • Camshaft position sensor (PHASE) • Accumulation of debris to the signal pick-up portion of the camshaft sprocket
DTC: P1145 **1T ECM, MIL: Yes** **Year:** 2009, 2010 **Model:** Armada, Titan **Engine:** 5.6L V8 **Transmission:** All	**Intake Valve Timing Control Position Sensor Circuit (Bank 2):** An excessively high or low voltage from the sensor is sent to ECM. **Possible Causes:** • Harness or connectors • Intake valve timing control position sensor circuit is open or shorted • Intake valve timing control position sensor • Crankshaft position sensor (POS) • Camshaft position sensor (PHASE) • Accumulation of debris to the signal pick-up portion of the camshaft sprocket
DTC: P1148 **1T ECM, MIL: Yes** **Year:** 2009, 2010 **Model:** 350Z, 370Z, Altima, Armada, Cube, Frontier, Maxima, Murano, Pathfinder, Quest, Rogue, Titan, Versa, Xterra **Engine:** 1.6L L4, 1.8L L4, 2.5L L4, 3.5L V6, 3.5L V6 VIN A, 3.7L V6, 4.0L V6, 5.6L V8 **Transmission:** All	**Closed Loop Control Function (Includes Hybrid Models):** HYBRID MODELS CAUTION: Hybrid systems use very high-voltage battery systems. Before starting any service work involving the battery system, turn the ignition switch OFF and then remove the service plug from pocket in the trunk. After removing the service plug, wait 10 minutes before touching any of the high-voltage connectors and **NOTE: On V6 and V8 engines, this applies to Bank 1, except 3.7L which is Bank 2.** **NOTE: DTC P1148 or P1168 is displayed with another DTC for A/F sensor 1. Perform the trouble diagnosis for the corresponding DTC.** • The closed loop control function for bank 1 does not operate even when vehicle is being driven in the specified condition. **Possible Causes:** • Harness or connectors • Air fuel ratio (A/F) sensor (sensor 1 on V6 and V8) circuit is open or shorted. • Air fuel ratio (A/F) sensor (sensor 1 on V6 and V8) • Air fuel ratio (A/F) sensor (sensor 1 on V6 and V8) heater
DTC: P1168 **1T ECM, MIL: Yes** **Year:** 2009, 2010 **Model:** 350Z, Altima, Armada, Maxima, Murano, Pathfinder, Quest, Titan, Xterra **Engine:** 3.5L V6, 3.5L V6 VIN A, 4.0L V6, 5.6L V8 **Transmission:** All	**Closed Loop Control Function (Bank 2):** **NOTE: If DTC P1148 or P1168 is displayed with another DTC for air fuel ratio (A/F) sensor 2. Perform the trouble diagnosis for the corresponding DTC.** • The closed loop control function for bank 2 does not operate even when vehicle is being driven in the specified condition. **Possible Causes:** • Harness or connectors • Air fuel ratio (A/F) sensor 1 or 2 circuit is open or shorted. • Air fuel ratio (A/F) sensor 1 or 2 • Air fuel ratio (A/F) sensor 1 or 2 heater

DTC	Trouble Code Title, Conditions & Possible Causes

DTC: P1195
T ECM, MIL: Yes
Year: 2009, 2010
Model: Altima
Engine: 2.5L L4
Transmission: All
	Engine Does Not Start (Hybrid):
	HYBRID MODELS CAUTION: Hybrid systems use very high-voltage battery systems. Before starting any service work involving the battery system, turn the ignition switch OFF and then remove the service plug from pocket in the trunk. After removing the service plug, wait 10 minutes before touching any of the high-voltage connectors and terminals.
	   • If DTC P1195 is displayed with DTC P0201, P0202, P0203, P0204, first perform the trouble diagnosis for DTC P0201, P0202, P0203, P0204.
	   • If DTC P1195 is displayed with DTC P0335, first perform the trouble diagnosis for DTC P0335.
	   • If DTC P1195 is displayed with DTC P0340, first perform the trouble diagnosis for DTC P0340.
	   • If DTC P1195 is displayed with DTC P0605, first perform the trouble diagnosis for DTC P0605.
	   • When the engine is abnormal, and the engine does not start.
	Possible Causes:
	   • Intake air leaks
	   • Incorrect PCV hose connection
	   • Mass air flow sensor
	   • Electric throttle control actuator
	   • Fuel injector
	   • Fuel run out
	   • Incorrect fuel pressure
	   • Spark plug
	   • Ignition coil
	   • Ignition signal circuit is open or shorted
	   • Insufficient compression

DTC: P1196
T ECM, MIL: Yes
Year: 2009, 2010
Model: Altima
Engine: 2.5L L4
Transmission: All
	Engine Lacks Power (Hybrid):
	HYBRID MODELS CAUTION: Hybrid systems use very high-voltage battery systems. Before starting any service work involving the battery system, turn the ignition switch OFF and then remove the service plug from pocket in the trunk. After removing the service plug, wait 10 minutes before touching any of the high-voltage connectors and terminals.
	   • If DTC P1196 is displayed with DTC P0201, P0202, P0203, P0204, first perform the trouble diagnosis for DTC P0201, P0202, P0203, P0204.
	   • If DTC P1196 is displayed with DTC P0335, first perform the trouble diagnosis for DTC P0335.
	   • If DTC P1196 is displayed with DTC P0340, first perform the trouble diagnosis for DTC P0340.
	   • If DTC P1196 is displayed with DTC P0605, first perform the trouble diagnosis for DTC P0605.
	   • The estimated torque is excessively low compared with the target torque
	Possible Causes:
	   • Intake air leaks
	   • Incorrect PCV hose connection
	   • Mass air flow sensor
	   • Electric throttle control actuator
	   • Fuel injector
	   • Fuel run out
	   • Incorrect fuel pressure
	   • Spark plug
	   • Ignition coil
	   • Ignition signal circuit is open or shorted
	   • Insufficient compression

DTC: P1197
T ECM, MIL: Yes
Year: 2009, 2010
Model: Altima
Engine: 2.5L L4
Transmission: All
	Fuel Run Out (Hybrid):
	HYBRID MODELS CAUTION: Hybrid systems use very high-voltage battery systems. Before starting any service work involving the battery system, turn the ignition switch OFF and then remove the service plug from pocket in the trunk. After removing the service plug, wait 10 minutes before touching any of the high-voltage connectors and terminals.
	NOTE: This DTC may be detected if the vehicle continues turning counterclockwise over a certain speed for a length of time.
	   • Detecting condition for P1195 or P1196 is satisfied and low voltage from the fuel level sensor is sent to ECM.
	Possible Causes:
	   • Out of fuel

DTC: P1211
1T ECM
Year: 2009, 2010
Model: 350Z, 370Z, Armada, Frontier, Maxima, Pathfinder, Quest, Titan, Versa, Xterra
Engine: 1.8L L4, 3.5L V6, 3.7L V6, 4.0L V6, 5.6L V8
Transmission: All
	TCS Control Unit:
	   • Freeze frame data is not stored in the ECM for this self-diagnosis.
	   • The MIL will not illuminate for this self-diagnosis.
	   • ECM receives malfunction information from "ABS actuator and electric unit (control unit)".
	Possible Causes:
	   • ABS actuator and electric unit (control unit)
	   • TCS related parts

DTC	Trouble Code Title, Conditions & Possible Causes
DTC: P1212 **1T ECM, MIL:** Yes **Year:** 2009, 2010 **Model:** 350Z, 370Z, Armada, Cube, Maxima, Murano, Pathfinder, Quest, Rogue, Titan, Versa, Xterra **Engine:** 1.8L L4, 2.5L L4, 3.5L V6, 3.5L V6 VIN A, 3.7L V6, 4.0L V6, 5.6L V8 **Transmission:** All	**TCS Communication Line:** NOTE: If DTC P1212 is displayed with DTC UXXXX, first perform the trouble diagnosis for DTC UXXXX. NOTE: If DTC P1212 is displayed with DTC P0607, first perform the trouble diagnosis for DTC P0607. NOTE: Be sure to erase the malfunction information such as DTC not only for "ABS actuator and electric unit (control unit)" but also for ECM after TCS related repair. • Freeze frame data is not stored in the ECM for this self-diagnosis. • The MIL will not illuminate for this self-diagnosis. • ECM cannot receive the information from "ABS actuator and electric unit (control unit)". **Possible Causes:** • Harness or connectors • The CAN communication line is open or shorted. • ABS actuator and electric unit (control unit) • Dead (Weak) battery
DTC: P1217 **1T ECM, MIL:** Yes **Year:** 2009, 2010 **Model:** 350Z, 370Z, Altima, Armada, Cube, Frontier, Maxima, Murano, Pathfinder, Quest, Rogue, Titan **Engine:** 1.8L L4, 2.5L L4, 3.5L V6, 3.5L V6 VIN A, 3.7L V6, 4.0L V6, 5.6L V8 **Transmission:** All	**Engine Over Temperature (Overheat) (Includes Hybrid Models):** HYBRID MODELS CAUTION: Hybrid systems use very high-voltage battery systems. Before starting any service work involving the battery system, turn the ignition switch OFF and then remove the service plug from pocket in the trunk. After removing the service plug, wait 10 minutes before touching any of the high-voltage connectors and terminals. NOTE: If DTC P1217 is displayed with DTC UXXXX, first perform the trouble diagnosis for DTC UXXXX. NOTE: If DTC P1217 is displayed with DTC P0607, first perform the trouble diagnosis for DTC P0607. • The ECM controls cooling fan relays through CAN communication line. • Cooling fan does not operate properly (overheat). • Cooling fan system does not operate properly (overheat). • Engine coolant was not added to the system using the proper filling method. • Engine coolant is not within the specified range. **Possible Causes:** • Harness or connectors • Cooling fan circuit is open or shorted. • Cooling fan motor • IPDM E/R (cooling fan relays-1, -2 and -3) • Cooling fan relays-4 and -5 • Radiator hose • Radiator • Reservoir tank • Radiator cap • Water pump • Thermostat
DTC: P1220 **1T ECM, MIL:** Yes **Year:** 2009, 2010 **Model:** Armada, Titan **Engine:** 5.6L V8 **Transmission:** All	**Fuel Pump Control Module (FPCM):** • An improper voltage signal from the FPCM, which is supplied to a point between the fuel pump and the dropping resistor, is detected by ECM. • During engine cranking, the signal voltage of the FPCM to the ECM is too low. **Possible Causes:** • Harness or connectors • FPCM circuit is shorted • Dropping resistor • FPCM
DTC: P1225 **1T ECM, MIL:** Yes **Year:** 2009, 2010 **Model:** 350Z, 370Z, Altima, Armada, Cube, Frontier, Maxima, Murano, Pathfinder, Quest, Rogue, Titan, Versa, Xterra **Engine:** 1.6L L4, 1.8L L4, 2.5L L4, 3.5L V6, 3.5L V6 VIN A, 3.7L V6, 4.0L V6, 5.6L V8 **Transmission:** All	**Closed Throttle Position Learning Performance (Includes Hybrid Models):** HYBRID MODELS CAUTION: Hybrid systems use very high-voltage battery systems. Before starting any service work involving the battery system, turn the ignition switch OFF and then remove the service plug from pocket in the trunk. After removing the service plug, wait 10 minutes before touching any of the high-voltage connectors and terminals. NOTE: For V6 and V8, this DTC is for Bank 1. • Closed throttle position learning value is excessively low. **Possible Causes:** • Electric throttle control actuator (TP sensor 1 and 2)

DTC	Trouble Code Title, Conditions & Possible Causes
DTC: P1226 **1T ECM, MIL: Yes** **Year:** 2009, 2010 **Model:** 350Z, 370Z, Altima, Armada, Cube, Frontier, Maxima, Murano, Pathfinder, Quest, Rogue, Titan, Versa, Xterra **Engine:** 1.6L L4, 1.8L L4, 2.5L L4, 3.5L V6, 3.5L V6 VIN A, 3.7L V6, 4.0L V6, 5.6L V8 **Transmission:** All	**Closed Throttle Position Learning Performance (Includes Hybrid Models):** HYBRID MODELS CAUTION: Hybrid systems use very high-voltage battery systems. Before starting any service work involving the battery system, turn the ignition switch OFF and then remove the service plug from pocket in the trunk. After removing the service plug, wait 10 minutes before touching any of the high-voltage connectors and terminals. **NOTE: On V6 and V8, this DTC is for Bank 1.** • Closed throttle position learning is not performed successfully, repeatedly. **Possible Causes:** • Electric throttle control actuator (TP sensor 1 and 2)
DTC: P1233 **T ECM, MIL: Yes** **Year:** 2009, 2010 **Model:** 370Z **Engine:** 3.7L V6 **Transmission:** All	**Electric Throttle Control Performance (Bank 2):** **NOTE: If DTC P1233 or P2101 is displayed with DTC P1238, P1290, P2100 or 2119, first perform the trouble diagnosis for DTC P1238, P2119 or P1290, P2100.** • Electric throttle control function does not operate properly **Possible Causes:** • Harness or connectors • Throttle control motor circuit is open or shorted • Electric throttle control actuator
DTC: P1234 **1T ECM, MIL: Yes** **Year:** 2009, 2010 **Model:** 370Z **Engine:** 3.7L V6 **Transmission:** All	**Closed Throttle Position Learning Performance (Bank 2):** • Closed throttle position learning value is excessively low. **Possible Causes:** • Electric throttle control actuator (TP sensor 1 and 2)
DTC: P1235 **1T ECM, MIL: Yes** **Year:** 2009, 2010 **Model:** 370Z **Engine:** 3.7L V6 **Transmission:** All	**Closed Throttle Position Learning Performance (Bank 2):** • Closed throttle position learning is not performed successfully, repeatedly. **Possible Causes:** • Electric throttle control actuator (TP sensor 1 and 2)
DTC: P1236 **T ECM, MIL: Yes** **Year:** 2009, 2010 **Model:** 370Z **Engine:** 3.7L V6 **Transmission:** All	**Throttle Control Motor (Bank 2) Circuit Short:** ECM detects short in both circuits between ECM and throttle control motor. **Possible Causes:** • Harness or connectors • (Throttle control motor circuit is shorted. • Electric throttle control actuator (Throttle control motor)
DTC: P1238 **T ECM, MIL: Yes** **Year:** 2009, 2010 **Model:** 370Z **Engine:** 3.7L V6 **Transmission:** All	**Electrical Throttle Control Actuator (Bank 2):** Condition A: Electric throttle control actuator does not function properly due to the return spring malfunction. Condition B: Throttle valve opening angle in fail-safe mode is not in specified range. Condition CL ECM detect the throttle valve is stuck open. **Possible Causes:** • Electric throttle control actuator
DTC: P1239 **T ECM, MIL: Yes** **Year:** 2009, 2010 **Model:** 370Z **Engine:** 3.7L V6 **Transmission:** All	**Throttle Position Sensor (Bank 2) Circuit Range/Performance:** • Rationally incorrect voltage is sent to ECM compared with the signals from TP sensor 1 and TP sensor 2. **Possible Causes:** • Harness or connector • TP sensor 1 and 2 circuit is open or shorted. • Electric throttle control actuator (TP sensor 1 and 2)
DTC: P1290 **T ECM, MIL: Yes** **Year:** 2009, 2010 **Model:** 370Z **Engine:** 3.7L V6 **Transmission:** All	**Throttle Control Motor Relay Circuit Open (Bank 2):** • ECM detects a voltage of power source for throttle control motor is excessively low. **Possible Causes:** • Harness or connectors • Throttle control motor relay circuit is open • Throttle control motor relay

DTC	Trouble Code Title, Conditions & Possible Causes

DTC: P1402
1T ECM, MIL: Yes
Year: 2009
Model: Quest
Engine: 3.5L V6
Transmission: All
	EGR Function (Open):
	NOTE: If the EGR temperature sensor detects EGR flow under the condition that does not call for EGR, a high-flow malfunction is diagnosed.
	    • EGR flow is detected under the condition that does not call for EGR.
	Possible Causes:
	    • Harness or connectors
	    • The EGR volume control valve circuit is open or shorted.
	    • EGR volume control valve leaking or stuck open
	    • EGR temperature sensor

DTC: P1421
1T ECM, MIL: Yes
Year: 2009, 2010
Model: 350Z, 370Z, Altima, Armada, Cube, Frontier, Maxima, Murano, Pathfinder, Quest, Rogue, Titan, Versa, Xterra
Engine: 1.6L L4, 1.8L L4, 2.5L L4, 3.5L V6, 3.5L V6 VIN A, 3.7L V6, 4.0L V6, 5.6L V8
Transmission: All
	Cold Start Emission Reduction Strategy Monitoring (Includes Hybrid Models):
	HYBRID MODELS CAUTION: Hybrid systems use very high-voltage battery systems. Before starting any service work involving the battery system, turn the ignition switch OFF and then remove the service plug from pocket in the trunk. After removing the service plug, wait 10 minutes before touching any of the high-voltage connectors and terminals.
	NOTE: If DTC P1421 is displayed with other DTC, first perform the trouble diagnosis for other DTC.
	    • ECM does not control ignition timing and engine idle speed properly when engine is started with pre-warming up condition.
	Possible Causes:
	    • Lack of intake air volume
	    • Fuel injection system
	    • ECM

DTC: P1550
1T ECM, MIL: Yes
Year: 2009, 2010
Model: Cube
Engine: 1.8L L4
Transmission: All
	Battery Current Sensor Circuit Range/Performance:
	The output voltage of the battery current sensor remains within the specified range while engine is running.
	Possible Causes:
	    • Harness or connectors
	    • Battery current sensor circuit is open or shorted.
	    • Crankshaft position sensor (POS) circuit is shorted.
	    • EVAP control system pressure sensor circuit is shorted.
	    • Accelerator pedal position sensor circuit is shorted.
	    • Refrigerant pressure sensor circuit is shorted.
	    • Battery current sensor
	    • Crankshaft position sensor (POS)
	    • EVAP control system pressure sensor
	    • Accelerator pedal position sensor
	    • Refrigerant pressure sensor

DTC: P1550
1T ECM, MIL: Yes
Year: 2009, 2010
Model: 350Z, Armada, Frontier, Maxima, Murano, Pathfinder, Titan, Xterra
Engine: 2.5L L4, 3.5L V6, 3.5L V6 VIN A, 4.0L V6, 5.6L V8
Transmission: All
	Battery Current Sensor Circuit Range/Performance:
	    • The MIL will not illuminate for this diagnosis.
	NOTE: If DTC P1550 is displayed with DTC P0643, first perform the trouble diagnosis for DTC P0643.
	    • The output voltage of the battery current sensor remains within the specified range while engine is running.
	Possible Causes:
	    • Harness or connectors
	    • Battery current sensor circuit is open or shorted.
	    • CKP sensor (POS) circuit is shorted.
	    • CMP sensor (PHASE) (bank 2) circuit is shorted.
	    • EVT control position sensor (bank 2) circuit is shorted.
	    • APP sensor 2 circuit is shorted.
	    • EVAP control system pressure sensor circuit is shorted.
	    • Refrigerant pressure sensor circuit is shorted.
	    • Battery current sensor
	    • Crankshaft position sensor (POS)
	    • Camshaft position sensor (PHASE) (bank 2)
	    • Exhaust valve timing control position sensor (bank 2)
	    • Accelerator pedal position sensor
	    • EVAP control system pressure sensor
	    • Refrigerant pressure sensor

DTC	Trouble Code Title, Conditions & Possible Causes
DTC: P1551 **1T ECM, MIL: Yes** **Year:** 2009, 2010 **Model:** Cube **Engine:** 1.8L L4 **Transmission:** All	**Battery Current Sensor Circuit Low Input:** • An excessively low voltage from the sensor is sent to ECM. **Possible Causes:** • Harness or connectors • Battery current sensor circuit is open or shorted. • Crankshaft position sensor (POS) circuit is shorted. • Power steering pressure sensor circuit is shorted. • Accelerator pedal position sensor circuit is shorted. • Refrigerant pressure sensor circuit is shorted. • Battery current sensor • Crankshaft position sensor (POS) • Power steering pressure sensor • Accelerator pedal position sensor • Refrigerant pressure sensor
DTC: P1551 **1T ECM, MIL: Yes** **Year:** 2009, 2010 **Model:** 350Z, Armada, Frontier, Maxima, Murano, Pathfinder, Titan, Xterra **Engine:** 2.5L L4, 3.5L V6, 3.5L V6 VIN A, 4.0L V6, 5.6L V8 **Transmission:** All	**Battery Current Sensor Circuit Low Input:** • The MIL will not illuminate for this diagnosis. **NOTE: If DTC P1551 or P1552 is displayed with DTC P0643, first perform the trouble diagnosis for DTC P0643.** • An excessively low voltage from the sensor is sent to ECM. **Possible Causes:** • Harness or connectors • Battery current sensor circuit is open or shorted. • CKP sensor (POS) circuit is shorted. • CMP sensor (PHASE) (bank 2) circuit is shorted. • EVT control position sensor (bank 2) circuit is shorted. • APP sensor 2 circuit is shorted. • EVAP control system pressure sensor circuit is shorted. • Refrigerant pressure sensor circuit is shorted. • Battery current sensor • Crankshaft position sensor (POS) • Camshaft position sensor (PHASE) • (bank 2) • Exhaust valve timing control position • sensor (bank 2) • Accelerator pedal position sensor • EVAP control system pressure sensor • Refrigerant pressure sensor
DTC: P1552 **1T ECM, MIL: Yes** **Year:** 2009, 2010 **Model:** Cube **Engine:** 1.8L L4 **Transmission:** All	**Battery Current Sensor Circuit High Input:** An excessively high voltage from the sensor is sent to ECM. **Possible Causes:** • Harness or connectors • Battery current sensor circuit is open or shorted. • Crankshaft position sensor (POS) circuit is shorted. • Power steering pressure sensor circuit is shorted. • Accelerator pedal position sensor circuit is shorted. • Refrigerant pressure sensor circuit is shorted. • Battery current sensor • Crankshaft position sensor (POS) • Power steering pressure sensor • Accelerator pedal position sensor • Refrigerant pressure sensor

DTC	Trouble Code Title, Conditions & Possible Causes
DTC: P1552 **1T ECM, MIL: Yes** **Year:** 2009, 2010 **Model:** 350Z, Armada, Frontier, Maxima, Murano, Pathfinder, Titan, Xterra **Engine:** 2.5L L4, 3.5L V6, 3.5L V6 VIN A, 4.0L V6, 5.6L V8 **Transmission:** All	**Battery Current Sensor Circuit High Input:** • The MIL will not illuminate for this diagnosis. **NOTE: If DTC P1551 or P1552 is displayed with DTC P0643, first perform the trouble diagnosis for DTC P0643.** • An excessively high voltage from the sensor is sent to ECM. **Possible Causes:** • Harness or connectors • Battery current sensor circuit is open or shorted. • CKP sensor (POS) circuit is shorted. • CMP sensor (PHASE) (bank 2) circuit is shorted. • EVT control position sensor (bank 2) circuit is shorted. • APP sensor 2 circuit is shorted. • EVAP control system pressure sensor circuit is shorted. • Refrigerant pressure sensor circuit is shorted. • Battery current sensor • Crankshaft position sensor (POS) • Camshaft position sensor (PHASE) • (bank 2) • Exhaust valve timing control position • sensor (bank 2) • Accelerator pedal position sensor • EVAP control system pressure sensor • Refrigerant pressure sensor
DTC: P1553 **1T ECM, MIL: Yes** **Year:** 2009, 2010 **Model:** 350Z, Armada, Frontier, Murano, Pathfinder, Titan, Xterra **Engine:** 2.5L L4, 3.5L V6, 3.5L V6 VIN A, 4.0L V6, 5.6L V8 **Transmission:** All	**Battery Current Sensor Performance:** • The MIL will not illuminate for this diagnosis. **NOTE: If DTC P1553 is displayed with DTC P0643, first perform the trouble diagnosis for DTC P0643.** • The signal voltage transmitted from the sensor to ECM is higher than the amount of the maximum power generation. **Possible Causes:** • Harness or connectors • Battery current sensor circuit is open or shorted. • CKP sensor (POS) circuit is shorted. • CMP sensor (PHASE) (bank 2) circuit is shorted. • EVT control position sensor (bank 2) circuit is shorted. • APP sensor 2 circuit is shorted. • EVAP control system pressure sensor circuit is shorted. • Refrigerant pressure sensor circuit is shorted. • Battery current sensor • Crankshaft position sensor (POS) • Camshaft position sensor (PHASE) (bank 2) • Exhaust valve timing control position sensor (bank 2) • Accelerator pedal position sensor • EVAP control system pressure sensor • Refrigerant pressure sensor
DTC: P1553 **1T ECM, MIL: Yes** **Year:** 2009, 2010 **Model:** Cube **Engine:** 1.8L L4 **Transmission:** All	**Battery Current Sensor Performance:** The signal voltage transmitted from the sensor to ECM is higher than the amount of the maximum power generation. **Possible Causes:** • Harness or connectors • Battery current sensor circuit is open or shorted. • Crankshaft position sensor (POS) circuit is shorted. • Power steering pressure sensor circuit is shorted. • Accelerator pedal position sensor circuit is shorted. • Refrigerant pressure sensor circuit is shorted. • Battery current sensor • Crankshaft position sensor (POS) • Power steering pressure sensor • Accelerator pedal position sensor • Refrigerant pressure sensor

DTC	Trouble Code Title, Conditions & Possible Causes
DTC: P1554 **1T ECM, MIL: Yes** **Year:** 2009, 2010 **Model:** Cube **Engine:** 1.8L L4 **Transmission:** All	**Battery Current Sensor Performance:** The output voltage of the battery current sensor is lower than the specified value while the battery voltage is high enough. **Possible Causes:** • Harness or connectors • Battery current sensor circuit is open or shorted. • Crankshaft position sensor (POS) circuit is shorted. • Power steering pressure sensor circuit is shorted. • Accelerator pedal position sensor circuit is shorted. • Refrigerant pressure sensor circuit is shorted. • Battery current sensor • Crankshaft position sensor (POS) • Power steering pressure sensor • Accelerator pedal position sensor • Refrigerant pressure sensor
DTC: P1554 **1T ECM, MIL: Yes** **Year:** 2009, 2010 **Model:** 350Z, Armada, Frontier, Maxima, Murano, Pathfinder, Titan, Xterra **Engine:** 2.5L L4, 3.5L V6, 3.5L V6 VIN A, 4.0L V6, 5.6L V8 **Transmission:** All	**Battery Current Sensor Performance:** • The MIL will not illuminate for this diagnosis. **NOTE: If DTC P1554 is displayed with DTC P0643, first perform the trouble diagnosis for DTC P0643.** • The output voltage of the battery current sensor is lower than the specified value while the battery voltage is high enough. **Possible Causes:** • Harness or connectors • Battery current sensor circuit is open or shorted. • CKP sensor (POS) circuit is shorted. • CMP sensor (PHASE) (bank 2) circuit is shorted. • EVT control position sensor (bank 2) circuit is shorted. • APP sensor 2 circuit is shorted. • EVAP control system pressure sensor circuit is shorted. • Refrigerant pressure sensor circuit is shorted. • Battery current sensor • Crankshaft position sensor (POS) • Camshaft position sensor (PHASE) (bank 2) • Exhaust valve timing control position sensor (bank 2) • Accelerator pedal position sensor • EVAP control system pressure sensor • Refrigerant pressure sensor
DTC: P1564 **1T ECM** **Year:** 2009, 2010 **Model:** 350Z, 370Z, Altima, Armada, Cube, Frontier, Murano, Pathfinder, Quest, Rogue, Titan, Versa, Xterra **Engine:** 1.8L L4, 2.5L L4, 3.5L V6, 3.5L V6 VIN A, 3.7L V6, 4.0L V6, 5.6L V8 **Transmission:** All	**ASCD Steering Switch Malfunction (Includes Hybrid Models):** HYBRID MODELS CAUTION: Hybrid systems use very high-voltage battery systems. Before starting any service work involving the battery system, turn the ignition switch OFF and then remove the service plug from pocket in the trunk. After removing the service plug, wait 10 minutes before touching any of the high-voltage connectors and terminals. • This self-diagnosis has the one trip detection logic. • The MIL will not illuminate for this self-diagnosis. **NOTE: If DTC P1564 is displayed with DTC P0605, first perform the trouble diagnosis for DTC P0605.** • An excessively high voltage signal from the ASCD steering switch is sent to ECM. • ECM detects that input signal from the ASCD steering switch is out of the specified range. • ECM detects that the ASCD steering switch is stuck ON. **Possible Causes:** • Harness or connectors • ASCD switch circuit is open or shorted. • ASCD steering switch • ECM

DTC	Trouble Code Title, Conditions & Possible Causes
DTC: P1572 **1T ECM, MIL: Yes** **Year:** 2009, 2010 **Model:** 350Z, 370Z, Altima, Armada, Cube, Frontier, Maxima, Murano, Pathfinder, Quest, Rogue, Titan, Versa, Xterra **Engine:** 1.8L L4, 2.5L L4, 3.5L V6, 3.5L V6 VIN A, 3.7L V6, 4.0L V6, 5.6L V8 **Transmission:** All	**ACSD Brake Switch Malfunction (Includes Hybrid Models):** HYBRID MODELS CAUTION: Hybrid systems use very high-voltage battery systems. Before starting any service work involving the battery system, turn the ignition switch OFF and then remove the service plug from pocket in the trunk. After removing the service plug, wait 10 minutes before touching any of the high-voltage connectors and terminals.<ul><li>This self-diagnosis has the one trip detection logic.</li><li>The MIL will not illuminate for this self-diagnosis.</li></ul>**NOTE: If DTC P1572 is displayed with DTC P0605, first perform the trouble diagnosis for DTC P0605.**<ul><li>This self-diagnosis has the one trip detection logic.</li></ul>**NOTE: When malfunction A is detected, the DTC is not stored in ECM memory. And in that case, 1st trip DTC and 1st trip freeze frame data are displayed. 1st trip DTC is erased when ignition switch is turned OFF. And even when Malfunction A is detected in two consecutive trips, DTC is not stored in ECM memory.**<ul><li>Malfunction A: When the vehicle speed is above 19 MPH, ON signals from the stop lamp switch and the ASCD brake switch are sent to ECM at the same time.</li><li>Malfunction B: ASCD brake switch signal is not sent to ECM for extremely long time while the vehicle is being driven.</li></ul>**Possible Causes:**<ul><li>Stop lamp switch circuit is shorted.</li><li>ASCD brake switch circuit is shorted.</li><li>ASCD clutch switch circuit is shorted (M/T).</li><li>Stop lamp switch</li><li>ASCD brake switch</li><li>ASCD clutch switch (M/T)</li><li>Incorrect stop lamp switch installation</li><li>Incorrect ASCD brake switch installation</li><li>Incorrect ASCD clutch switch installation (M/T)</li><li>ECM</li></ul>
DTC: P1574 **1T ECM, MIL: Yes** **Year:** 2009, 2010 **Model:** 350Z, 370Z, Altima, Armada, Cube, Frontier, Maxima, Murano, Pathfinder, Quest, Rogue, Titan, Versa, Xterra **Engine:** 1.8L L4, 2.5L L4, 3.5L V6, 3.5L V6 VIN A, 3.7L V6, 4.0L V6, 5.6L V8 **Transmission:** All	**ASCD Vehicle Speed Sensor Malfunction (Includes Hybrid Models):** HYBRID MODELS CAUTION: Hybrid systems use very high-voltage battery systems. Before starting any service work involving the battery system, turn the ignition switch OFF and then remove the service plug from pocket in the trunk. After removing the service plug, wait 10 minutes before touching any of the high-voltage connectors and terminals.<ul><li>The MIL will not illuminate for this self-diagnosis.</li></ul>**NOTE: If DTC P1574 is displayed with DTC UXXXX, first perform the trouble diagnosis for DTC UXXXX.** **NOTE: If DTC P1574 is displayed with DTC P0500, P0605 and/or P0607, first perform the trouble diagnosis for these DTCs before continuing with DTC P1574.**<ul><li>ECM detects a difference between two vehicle speed signals is out of the specified range.</li></ul>**Possible Causes:**<ul><li>Harness or connectors</li><li>CAN communication line is open or shorted.</li><li>Combination meter circuit is open or shorted.</li><li>TCM (CVT models)</li><li>Unified meter and A/C amp.</li><li>Combination meter</li><li>ABS actuator and electric unit (control unit)</li><li>ECM</li></ul>
DTC: P1604 **1T BCM, MIL: Yes** **Year:** 2009, 2010 **Model:** Sentra **Engine:** 2.0L L4, 2.5L L4 **Transmission:** All	**Steering Wheel/Column Torque Sensor:**<ul><li>Steering wheel: Not steering properly (lack of or too much steering force).</li></ul>**Possible Causes:**<ul><li>Torque sensor value is out of specification</li><li>Harness and connector</li><li>EPS control unit</li><li>EPS control unit pin terminals damaged or loose connections</li><li>Torque sensor malfunction</li></ul>
DTC: P1604 **T BCM** **Year:** 2009, 2010 **Model:** Sentra **Engine:** 2.0L L4, 2.5L L4 **Transmission:** All	**Steering Wheel Torque Sensor:** Perform self-diagnosis. Steering wheel: Not steering (there is no steering force). **Possible Causes:**<ul><li>Value of torque sensor is not within parameters.</li><li>Voltage between EPS control unit harness connector M87 terminals 11, 12, 13, 14 and ground is not correct.</li><li>EPS control unit input/output signal is out of specifications.</li><li>EPS control unit pin terminals has damage or loose connection with harness connector.</li><li>Torque sensor is malfunctioning.</li></ul>

DTC	Trouble Code Title, Conditions & Possible Causes

DTC: P1604
T BCM
Year: 2009, 2010
Model: Sentra
Engine: 2.0L L4, 2.5L L4
Transmission: All
	Steering Wheel (Column) Torque Sensor:
	   • Steering wheel: Not steering (there is no steering force)
	Possible Causes:
	   • EPS control unit input/output signal is out of range.

DTC: P1606
T ECM, MIL: Yes
Year: 2009, 2010
Model: 370Z
Engine: 3.7L V6
Transmission: All
	Variable Valve Event & Lift (VVEL) Control Module:
	   • VVEL control module calculation function is malfunctioning.
	   • VVEL EEPROM system is malfunctioning.
	Possible Causes:
	   • VVEL control module

DTC: P1607
T ECM, MIL: Yes
Year: 2009, 2010
Model: 370Z
Engine: 3.7L V6
Transmission: All
	Variable Valve Event & Lift (VVEL) Control Module Circuit:
	   • The internal circuit of the VVEL control module is malfunctioning.
	Possible Causes:
	   • VVEL control module

DTC: P1608
T ECM, MIL: Yes
Year: 2009, 2010
Model: 370Z
Engine: 3.7L V6
Transmission: All
	Variable Valve Event & Lift (VVEL) Sensor Power Supply Circuit:
	VVEL control module detects a voltage of power source for sensor is excessively low or high.
	Possible Causes:
	   • Harness or connectors
	   • VVEL control shaft position sensor power supply circuit is open or shorted.
	   • VVEL control shaft position sensor
	   • VVEL control module

DTC: P1610 **T BCM, MIL: Yes** **Year:** 2009 **Model:** Quest **Engine:** 3.5L V6 **Transmission:** All	**Nissan Vehicle Immobilizer System (NVIS):** Lock mode **Possible Causes:** • Replace ECM.

DTC: P1611
T ECM, MIL: Yes
Year: 2009
Model: Quest
Engine: 3.5L V6
Transmission: All
	ID Discord (IMM-ECM):
	System initialization has not yet been completed.
	Possible Causes:
	   • Re-attempt initialization
	   • Replace ECM

DTC: P1612
T ECM, MIL: Yes
Year: 2009
Model: Quest
Engine: 3.5L V6
Transmission: All
	Chain of ECM-IMMU:
	   • In rare cases, "CHAIN OF ECM-IMMU" might be stored during key registration procedure, even if the system is not malfunctioning.
	Possible Causes:
	   • Open circuit in battery voltage line of BCM circuit
	   • Open circuit in ignition line of BCM circuit
	   • Open circuit in ground line of BCM circuit
	   • Open or short circuit between BCM and ECM communication line
	   • ECM
	   • BCM

DTC: P1614
T ECM, MIL: Yes
Year: 2009
Model: Quest
Engine: 3.5L V6
Transmission: All
	Chain of IMMU-EMC:
	In rare case, "CHAIN OF ECM-IMMU" might be stored during key registration procedure, even if the system is not malfunctioning.
	Possible Causes:
	   • Open circuit in battery voltage line of BCM circuit
	   • Open circuit in ignition line of BCM circuit
	   • Open circuit in ground line of BCM circuit
	   • Open or short circuit between BCM and ECM communication line
	   • ECM
	   • BCM

DTC	Trouble Code Title, Conditions & Possible Causes
DTC: P1615 **T ECM** **Year:** 2009 **Model:** Quest **Engine:** 3.5L V6 **Transmission:** All	**Difference of Key:** Unregistered key **Possible Causes:** • BCM
DTC: P1700 **T TCM, TCIL: Yes** **Year:** 2009, 2010 **Model:** Cube, Murano, Sentra **Engine:** 1.8L L4, 2.0L L4, 2.5L L4, 3.5L V6 VIN A **Transmission:** All	**CVT Control System:** This DTC is displayed with other DTC regarding TCM. Perform the trouble diagnosis for corresponding DTC. **Possible Causes:** • See other displayed DTCs.
DTC: P1701 **T TCM, TCIL: Yes** **Year:** 2009, 2010 **Model:** Maxima, Murano, Rogue, Versa **Engine:** 1.6L L4, 1.8L L4, 2.5L L4, 3.5L V6, 3.5L V6 VIN A **Transmission:** All	**Power Supply Circuit:** • When the power supply to the TCM is cut off, for example because the battery is removed, and the self-diagnosis memory function stops. • This is not a malfunction message (Whenever shutting off a power supply to the TCM, this message appears on the screen). **Possible Causes:** • Harness or connectors • Battery or ignition switch and TCM circuit is open or shorted.
DTC: P1701 **T TCM, TCIL: Yes** **Year:** 2009, 2010 **Model:** Cube **Engine:** 1.8L L4 **Transmission:** All	**Power Supply Circuit:** CAUTION: Immediately after TCM is replaced or after control valve or transaxle assembly is replaced (after TCM initialization is complete), self-diagnosis result of "P1701" may be displayed. In this case, erase self-diagnosis result using CONSULT-III. After erasing self-diagnosis, perform reproduction procedures of DTC P1701 and check that a malfunction is not detected. Power supply (backup) of TCM is not supplied and learning function stops **Possible Causes:** • Harness or connectors • TCM power source circuit is open or shorted.
DTC: P1701 **1T TCM, TCIL: Yes** **Year:** 2009, 2010 **Model:** Altima, Sentra **Engine:** 2.0L L4, 2.5L L4, 3.5L V6 **Transmission:** All	**Transmission Control Module (Power Supply):** When the power supply to the TCM is cut OFF, for example because the battery is removed, and the self-diagnosis memory function stops, malfunction is detected. **NOTE: Since "P1701 TCM-POWER SUPPLY" will be indicated when replacing TCM, perform diagnosis after erasing "SELF-DIAG RESULTS"** • This is not an OBD-II self-diagnostic item. • Diagnostic trouble code "P1701 TCM-POWER SUPPLY" is detected when TCM does not receive the voltage signal from the battery power supply. • This is not a malfunction message. (Whenever shutting OFF a power supply to the TCM, this message appears on the screen.) **Possible Causes:** • Harness or connectors (Battery or ignition switch and TCM circuit is open or shorted.)
DTC: P1701 **1T ECM** **Year:** 2009, 2010 **Model:** Maxima **Engine:** 3.5L V6 **Transmission:** All	**VIAS Control Solenoid Valve 2 Circuit:** • An excessively low or high voltage signal is sent to ECM via the VIAS control solenoid valve 2. **Possible Causes:** • Harness or connectors • The solenoid valve 2 circuit is open or shorted. • VIAS control solenoid valve 2
DTC: P1705 **T TCM, TCIL: Yes** **Year:** 2009, 2010 **Model:** 370Z **Engine:** 3.7L V6 **Transmission:** All	**Accelerator Pedal Position (APP) Sensor Signal Circuit:** TCM detects improper accelerator pedal position signals received from ECM via CAN communication. **Possible Causes:** • Harness or connectors • Sensor circuit is open or shorted.
DTC: P1705 **T TCM, TCIL: Yes** **Year:** 2009, 2010 **Model:** Cube **Engine:** 1.8L L4 **Transmission:** All	**Accelerator Pedal Position Sensor Signal Circuit:** • TCM detects that difference between 2 throttle opening signals (CAN communication) from ECM is 1/8 or more. **Possible Causes:** • Harness or connectors • CAN communication line is open or shorted. • Accelerator pedal position signal circuit is open or shorted. • ECM

DTC	Trouble Code Title, Conditions & Possible Causes
DTC: P1705 **1T TCM, TCIL: Yes** **Year:** 2009, 2010 **Model:** Altima, Sentra **Engine:** 2.0L L4, 2.5L L4, 3.5L V6 **Transmission:** All	**Throttle Position/Accelerator Pedal Position Sensor Circuit:** Electric throttle control actuator consists of throttle control motor, accelerator pedal position sensor, throttle position sensor etc. The actuator sends a signal to the ECM, and ECM sends the signal to TCM with CAN communication. • This is not an OBD-II self-diagnostic item. • Diagnostic trouble code P1705 is detected when TCM does not receive the proper accelerator pedal position signals (input by CAN communication) from ECM. **Possible Causes:** • ECM • Harness or connectors • CAN communication line is open or shorted.
DTC: P1705 **T TCM, TCIL: Yes** **Year:** 2009, 2010 **Model:** 350Z, Armada, Frontier, Maxima, Murano, Pathfinder, Rogue, Titan, Versa, Xterra **Engine:** 1.6L L4, 1.8L L4, 2.5L L4, 3.5L V6, 3.5L V6 VIN A, 4.0L V6, 5.6L V8 **Transmission:** All	**Throttle Position (TP) Sensor/Accelerator Pedal Position (APP) Sensor:** Electric throttle control actuator consists of throttle control motor, accelerator pedal position sensor, throttle position sensor, etc. The actuator sends a signal to the ECM, and ECM sends signals to TCM with CAN communication. • This is not an OBD-II self-diagnostic item. • Diagnostic trouble code P1705 is detected when TCM does not receive the proper accelerator pedal position signals (input by CAN communication) from ECM. **Possible Causes:** • Harness or connectors • The sensor circuit is open or shorted.
DTC: P1710 **T TCM, TCIL: Yes** **Year:** 2009, 2010 **Model:** 350Z, Armada, Frontier, Pathfinder, Xterra **Engine:** 2.5L L4, 3.5L V6, 4.0L V6, 5.6L V8 **Transmission:** All	**A/T Fluid Temperature Sensor Circuit:** • This is an OBD-II self-diagnostic item. • Diagnostic trouble code P1710 will be detected when TCM receives an excessively low or high voltage from the sensor. • A/T fluid temperature does not rise to the specified temperature while driving. **Possible Causes:** • Harness or connectors • The sensor circuit is open or shorted. • A/T fluid temperature sensor 1
DTC: P1715 **1T ECM, TCIL: Yes** **Year:** 2009, 2010 **Model:** Pathfinder **Engine:** 4.0L V6, 5.6L V8 **Transmission:** All	**Turbine Revolution Sensor:** • This is an OBD-II self-diagnostic item. • Diagnostic trouble code P0717 is detected under the following conditions: - When TCM does not receive the proper voltage signal from the sensor. - When TCM detects an irregularity only at position of 4th gear for turbine revolution sensor 2. **Possible Causes:** • Harness or connectors • The sensor circuit is open or shorted. • Turbine revolution sensor 1, 2
DTC: P1715 **1T TCM** **Year:** 2009, 2010 **Model:** 350Z, 370Z, Altima, Cube, Frontier, Maxima, Murano, Pathfinder, Versa, Xterra **Engine:** 1.6L L4, 1.8L L4, 2.5L L4, 3.5L V6, 3.5L V6 VIN A, 3.7L V6, 4.0L V6, 5.6L V8 **Transmission:** All	**Input Speed Sensor (Primary Speed Sensor/TCM Output):** • The MIL will not illuminate for this self-diagnosis. **NOTE: If DTC P1715 is displayed with DTC UXXXX, first perform the trouble diagnosis for DTC UXXXX.** **NOTE: If DTC P1715 is displayed with DTC P0335, P0340, P0605 and/or P0607, first perform the trouble diagnosis for the appropriate DTC before proceeding with P1715 diagnosis.** • Sensor signal is different from the theoretical value calculated by ECM from secondary speed sensor signal and engine rpm signal. **Possible Causes:** • Harness or connectors • CAN communication line is open or shorted • Primary speed sensor circuit is open or shorted • TCM

DTC	Trouble Code Title, Conditions & Possible Causes

DTC: P1720
1T ECM, MIL: Yes
Year: 2009, 2010
Model: Maxima, Murano
Engine: 3.5L V6, 3.5L V6 VIN A
Transmission: All
	Vehicle Speed Sensor (TCM Output):
	NOTE: If DTC P1720 is displayed with DTC UXXXX first perform the trouble diagnosis for DTC UXXXX.
	NOTE: If DTC P1720 is displayed with DTC P0607, first perform the trouble diagnosis for DTC P0607.
	  • The difference between two vehicle speed signals is out of the specified range.
	Possible Causes:
	  • Harness or connectors
	  • Output speed sensor circuit is open or shorted.
	  • Wheel sensor circuit is open or shorted.
	  • TCM
	  • Output speed sensor
	  • ABS actuator and electric unit (control unit)
	  • Wheel sensor
	  • Combination meter

DTC: P1721
T TCM, TCIL: Yes
Year: 2009, 2010
Model: 370Z
Engine: 3.7L V6
Transmission: All
	Vehicle Speed Signal Circuit:
	  • The vehicle speed signal recognizes that the vehicle speed is 5 km/h (3 MPH) or less even if the output speed sensor recognizes that the vehicle speed is 20 km/h (12 MPH) or more. (Only when starts after the ignition switch is turned ON.)
	  • The vehicle speed recognized by the vehicle speed signal decelerates 36 km/h (23 MPH) or more during 60 msec when the vehicle speed signal recognizes that the vehicle speed is 36 km/h (23 MPH) or more and the output speed sensor recognizes that the vehicle speed is 24 km/h (15 MPH) or less.
	  • The vehicle speed of vehicle speed signal decelerates 36 km/h (23 MPH) or more even if the vehicle speed of output speed sensor accelerates or decelerates 24 km/h (15 MPH) or less during 60 msec when the vehicle speed sensor recognizes that the vehicle speed is 36 km/h (23 MPH) or more.
	Possible Causes:
	  • Harness or connectors
	  • Sensor circuit is open or shorted.

DTC: P1721
T TCM, TCIL: Yes
Year: 2009, 2010
Model: 350Z, Armada, Frontier, Pathfinder, Titan, Xterra
Engine: 2.5L L4, 3.5L V6, 4.0L V6, 5.6L V8
Transmission: All
	Vehicle Speed Sensor MTR:
	  • This is not an OBD-II self-diagnostic item.
	  • Diagnostic trouble code P1721 is detected when TCM does not receive the proper vehicle speed sensor MTR signal (input by CAN communication) from combination meter (unified meter and A/C amp).
	Possible Causes:
	  • Harness or connectors
	  • The sensor circuit is open or shorted.

DTC: P1722
T TCM, TCIL: Yes
Year: 2009, 2010
Model: Cube
Engine: 1.8L L4
Transmission: All
	Vehicle Speed Signal Circuit:
	  • TCM detects a malfunction of CAN communication between ABS actuator and electric unit (control unit)
	  • When vehicle speed that TCM detects is 10 km/h (7 MPH) or more, vehicle speed signal (CAN signal) that is received from ABS actuator and electric unit (control unit) is less than 2 km/h (1 MPH)
	  • Change of vehicle speed signal (CAN communication) that TCM receives is large
	Possible Causes:
	  • Harness or connectors
	  • CAN communication line is open or shorted.
	  • Vehicle speed signal circuit is open or shorted.
	  • ABS actuator and electric unit (control unit)

DTC: P1722
1T TCM, TCIL: Yes
Year: 2009, 2010
Model: Sentra
Engine: 2.0L L4, 2.5L L4
Transmission: All
	ESTM Vehicle Speed Signal:
	The vehicle speed signal is transmitted from ABS actuator and electric unit (control unit) to TCM by CAN communication line.
	  • This is not an OBD-II self-diagnostic item.
	  • Diagnostic trouble code "P1722 ESTM VEH SPD SIG", with scan tool, is detected when TCM does not receive the proper vehicle speed signal (input by CAN communication) from ABS actuator and electric unit (control unit).
	Possible Causes:
	  • Harness or connectors (Sensor circuit is open or shorted.)
	  • ABS actuator and electric unit (control unit)

DTC: P1722
1T TCM, TCIL: Yes
Year: 2009, 2010
Model: Altima, Maxima, Murano, Rogue, Versa
Engine: 1.6L L4, 1.8L L4, 2.5L L4, 3.5L V6, 3.5L V6 VIN A
Transmission: All
	Vehicle Speed Signal Circuit:
	  • CAN communication with the ABS actuator and the electric unit (control unit) is malfunctioning.
	  • There is a great difference between the vehicle speed signal from the ABS actuator and the electric unit (control unit), and the vehicle speed sensor signal.
	Possible Causes:
	  • Harness or connectors (Sensor circuit is open or shorted.)
	  • ABS actuator and electric unit (control unit)

DTC	Trouble Code Title, Conditions & Possible Causes
DTC: P1723 **T TCM, TCIL: Yes** **Year:** 2009, 2010 **Model:** Cube **Engine:** 1.8L L4 **Transmission:** All	**Speed Sensor:** When noise (pulse) that is generated because of connection malfunction caused by primary speed sensor and secondary speed sensor harness and others is detected, it is judged that a malfunction occurs. TCM detects that high frequency elements that are extracted from primary pulley speed and secondary pulley speed exceed a certain value **Possible Causes:** • Harness or connectors • Primary speed sensor circuit is open or shorted. • Secondary speed sensor circuit is open or shorted.
DTC: P1723 **1T TCM, TCIL: Yes** **Year:** 2009, 2010 **Model:** Altima, Rogue, Sentra, Versa **Engine:** 1.6L L4, 1.8L L4, 2.0L L4, 2.5L L4, 3.5L V6 **Transmission:** All	**CVT Speed Sensor Circuit:** • The vehicle speed sensor CVT [output speed sensor (secondary speed sensor)] detects the revolution of the parking gear and generates a pulse signal. The pulse signal is sent to the TCM, which converts it into vehicle speed. • The input speed sensor (primary speed sensor) detects the primary pulley revolution speed and sends a signal to the TCM. • Diagnostic trouble code P1723 is detected when there is a great difference between the vehicle speed signal and the secondary speed sensor signal. **NOTE: P0720, P0715 or P0725 is displayed with the DTC at the same time.** **Possible Causes:** • Harness or connectors (Sensor circuit is open or shorted.) • Output speed sensor (Secondary speed sensor) • Input speed sensor (Primary speed sensor) • Engine speed signal system
DTC: P1726 **T TCM, TCIL: Yes** **Year:** 2009, 2010 **Model:** Altima, Maxima, Murano, Rogue, Sentra, Versa **Engine:** 1.6L L4, 1.8L L4, 2.0L L4, 2.5L L4, 3.5L V6, 3.5L V6 VIN A **Transmission:** All	**Throttle Control Signal Circuit:** The electronically controlled throttle for ECM is malfunctioning. **Possible Causes:** • Harness or connectors • Sensor circuit is open or shorted.
DTC: P1726 **T TCM, TCIL: Yes** **Year:** 2009, 2010 **Model:** Cube **Engine:** 1.8L L4 **Transmission:** All	**Throttle Control Signal Circuit:** TCM receives a malfunction signal of engine system from ECM **Possible Causes:** • Harness or connectors • Electric throttle sensor signal circuit is open or shorted.
DTC: P1726 **T TCM, TCIL: Yes** **Year:** 2009 **Model:** Quest **Engine:** 3.5L V6 **Transmission:** All	**Lock-Up Operation & Learning Control:** This DTC is displayed with other DTCs regarding ECM. Perform the trouble diagnosis for other DTCs displayed. When this DTC is detected, lock-up operation and learning control are canceled. **Possible Causes:** • Perform the trouble diagnosis for other DTCs displayed.
DTC: P1730 **T TCM, TCIL: Yes** **Year:** 2009, 2010 **Model:** 370Z **Engine:** 3.7L V6 **Transmission:** All	**Interlock:** • The output sensor detects the deceleration of 12 km/h (7 MPH) or more for 1 second. **Possible Causes:** • Harness or connectors • Solenoid valve circuit is open or shorted. • Input clutch solenoid valve • Direct clutch solenoid valve • High and low reverse clutch solenoid valve • Front brake solenoid valve • Low brake solenoid valve • 2346 brake solenoid valve • Anti-interlock solenoid valve • Each clutch and brake • Hydraulic control circuit

DTC	Trouble Code Title, Conditions & Possible Causes
DTC: P1730 **T TCM, TCIL: Yes** **Year:** 2009, 2010 **Model:** 350Z, Armada, Frontier, Pathfinder, Titan, Xterra **Engine:** 2.5L L4, 3.5L V6, 4.0L V6, 5.6L V8 **Transmission:** All	**A/T Interlock:** • This is an OBD-II self-diagnostic item. • Diagnostic trouble code P1730 is detected when TCM does not receive the proper voltage signal from the sensor and switch. • TCM monitors and compares gear position and conditions of each ATF pressure switch when gear is steady. **Possible Causes:** • Harness or connectors • The solenoid and switch circuit is open or shorted. • Low coast brake solenoid valve • ATF pressure switch 2
DTC: P1731 **T TCM, TCIL: Yes** **Year:** 2009, 2010 **Model:** 350Z, Armada, Frontier, Pathfinder, Titan, Xterra **Engine:** 2.5L L4, 3.5L V6, 4.0L V6, 5.6L V8 **Transmission:** All	**A/T 1st Engine Braking:** • This is not an OBD-II self-diagnostic item. • Diagnostic trouble code P1731 is detected under the following conditions. - When TCM does not receive the proper voltage signal from the sensor. - When TCM monitors each ATF pressure switch and solenoid monitor value, and detects as irregular when engine brake of 1st gear acts other than at "1" position. **Possible Causes:** • Harness or connectors • The sensor circuit is open or shorted. • Low coast brake solenoid valve • ATF pressure switch 2
DTC: P1734 **T TCM, TCIL: Yes** **Year:** 2009, 2010 **Model:** 370Z **Engine:** 3.7L V6 **Transmission:** All	**7th Gear Incorrect Ratio:** • DTC is set when any inconsistency is recognized in gear ratio. **Possible Causes:** • Input clutch solenoid valve • Direct clutch solenoid valve • High and low reverse clutch solenoid valve • Front brake solenoid valve • Low brake solenoid valve • 2346 brake solenoid valve • Anti-interlock solenoid valve • Each clutch and brake • Output speed sensor • Input speed sensor 1, 2 • Hydraulic control circuit
DTC: P1740 **1T TCM, TCIL: Yes** **Year:** 2009, 2010 **Model:** Sentra **Engine:** 2.0L L4, 2.5L L4 **Transmission:** All	**Lock-Up Select Solenoid Valve Circuit:** • The lock-up select solenoid valve controls lock-up clutch pressure or forward clutch pressure (reverse brake pressure). • When controlling lock-up clutch, the valve is turned OFF. When controlling forward clutch, it is turned ON. • This is an OBD-II self-diagnostic item. • Diagnostic trouble code P1740 is detected under the following conditions: - When TCM compares target value with monitor value and detects an irregularity. **Possible Causes:** • Lock-up select solenoid valve • Harness or connectors (Solenoid circuit is open or shorted.)
DTC: P1740 **1T TCM, TCIL: Yes** **Year:** 2009, 2010 **Model:** Altima, Maxima, Murano, Rogue, Versa **Engine:** 1.6L L4, 1.8L L4, 2.5L L4, 3.5L V6, 3.5L V6 VIN A **Transmission:** All	**Lock-Up Select Solenoid Valve:** • Normal voltage not applied to solenoid due to cut line, short, or the like. • TCM detects as irregular by comparing target value with monitor value. **Possible Causes:** • Harness or connectors (Solenoid circuit is open or shorted.) • Lock-up select solenoid valve
DTC: P1740 **T TCM, TCIL: Yes** **Year:** 2009, 2010 **Model:** Cube **Engine:** 1.8L L4 **Transmission:** All	**Lock-Up Select Solenoid Valve Circuit:** • Lock-up select solenoid valve monitor value is OFF when lock-up select solenoid valve command value of TCM is ON • Lock-up select solenoid valve monitor value is ON when lock-up select solenoid valve command value of TCM is OFF **Possible Causes:** • Harness or connectors • Lock-up select solenoid valve circuit is open or shorted. • Lock-up select solenoid valve

DTC	Trouble Code Title, Conditions & Possible Causes
DTC: P1745 **1T TCM, TCIL: Yes** **Year:** 2009, 2010 **Model:** Altima, Maxima, Rogue, Sentra, Versa **Engine:** 1.6L L4, 1.8L L4, 2.0L L4, 2.5L L4, 3.5L V6 **Transmission:** All	**Line Pressure Control:** • The pressure control solenoid valve A (line pressure solenoid valve) regulates the oil pump discharge pressure to suit the driving condition in response to a signal sent from the TCM. • This is not an OBD-II self-diagnostic item. • Diagnostic trouble code P1745 is detected when TCM detects the unexpected line pressure. **Possible Causes:** • TCM has failed

DTC: P1752
T TCM, TCIL: Yes
Year: 2009, 2010
Model: 350Z, Armada, Frontier, Pathfinder, Titan, Xterra
Engine: 2.5L L4, 3.5L V6, 4.0L V6, 5.6L V8
Transmission: All
	Input Clutch Solenoid Valve:
	• This is an OBD-II self-diagnostic item.
	• Diagnostic trouble code P1752 is detected under the following conditions:
	  - When TCM detects an improper voltage drop when it tries to operate the solenoid valve.
	  - When TCM detects as irregular by comparing target value with monitor value.
	Possible Causes:
	• Harness or connectors
	• The solenoid circuit is open or shorted.
	• Input clutch solenoid valve

DTC: P1757
T TCM, TCIL: Yes
Year: 2009, 2010
Model: 350Z, Armada, Frontier, Pathfinder, Titan, Xterra
Engine: 2.5L L4, 3.5L V6, 4.0L V6, 5.6L V8
Transmission: All
	Front Brake Solenoid Valve:
	• This is an OBD-II self-diagnostic item.
	• Diagnostic trouble code P1757 is detected under the following conditions:
	  - When TCM detects an improper voltage drop when it tries to operate the solenoid valve.
	  - When TCM detects as irregular by comparing target value with monitor value.
	Possible Causes:
	• Harness or connectors
	• The solenoid circuit is open or shorted.
	• Front brake solenoid valve

DTC: P1762
T TCM, TCIL: Yes
Year: 2009, 2010
Model: 350Z, Armada, Frontier, Pathfinder, Titan, Xterra
Engine: 2.5L L4, 3.5L V6, 4.0L V6, 5.6L V8
Transmission: All
	Direct Clutch Solenoid Valve:
	• This is an OBD-II self-diagnostic item.
	• Diagnostic trouble code P1762 will be detected under the following conditions:
	  - When TCM detects an improper voltage drop when it tries to operate the solenoid valve.
	  - When TCM detects as irregular by comparing target value with monitor value.
	Possible Causes:
	• Harness or connectors
	• The solenoid circuit is open or shorted.
	• Direct clutch solenoid valve

DTC: P1767
T TCM, TCIL: Yes
Year: 2009, 2010
Model: 350Z, Armada, Frontier, Pathfinder, Titan, Xterra
Engine: 2.5L L4, 3.5L V6, 4.0L V6, 5.6L V8
Transmission: All
	High & Low Reverse Clutch Solenoid Valve:
	• This is an OBD-II self-diagnostic item.
	• Diagnostic trouble code P1767 will be detected under the following conditions:
	  - When TCM detects an improper voltage drop when it tries to operate the solenoid valve.
	  - When TCM detects as irregular by comparing target value with monitor value.
	Possible Causes:
	• Harness or connectors
	• The solenoid circuit is open or shorted.
	• High and low reverse clutch solenoid valve

DTC: P1772 **T TCM, TCIL: Yes** **Year:** 2009, 2010 **Model:** 350Z, Armada, Frontier, Pathfinder, Titan, Xterra **Engine:** 2.5L L4, 3.5L V6, 4.0L V6, 5.6L V8 **Transmission:** All	**Low Coast Brake Solenoid Valve:** • This is an OBD-II self-diagnostic item. • Diagnostic trouble code P1772 will be set when the TCM detects an improper voltage drop when it tries to operate the solenoid valve. **Possible Causes:** • Harness or connectors • The solenoid circuit is open or shorted. • Low coast brake solenoid valve

DTC	Trouble Code Title, Conditions & Possible Causes
DTC: P1774 **T TCM, TCIL: Yes** **Year:** 2009, 2010 **Model:** 350Z, Armada, Frontier, Pathfinder, Titan, Xterra **Engine:** 2.5L L4, 3.5L V6, 4.0L V6, 5.6L V8 **Transmission:** All	**Low Coast Brake Solenoid Valve Function:** • This is an OBD-II self-diagnostic item. • Diagnostic trouble code P1774 will be detected under the following conditions: - TCM detects an improper voltage drop when it tries to operate the solenoid valve. - When TCM detects that actual gear ratio is irregular, and relation between gear position and condition of ATF pressure switch 2 is irregular during depressing accelerator pedal (other than during shift change). - When TCM detects that relation between gear position and condition of ATF pressure switch 2 is irregular during releasing accelerator pedal. (Other than during shift change) **Possible Causes:** • Harness or connectors • The solenoid and switch circuits are open or shorted. • Low coast brake solenoid valve • ATF pressure switch 2
DTC: P1777 **1T TCM, TCIL: Yes** **Year:** 2009, 2010 **Model:** Altima, Sentra, Versa **Engine:** 1.6L L4, 1.8L L4, 2.0L L4, 2.5L L4, 3.5L V6 **Transmission:** All	**Step Motor Circuit:** • The step motor changes the step with turning 4 coils ON/OFF according to the signal from TCM. As a result, the flow of line pressure to primary pulley is changed and pulley ratio is controlled. • This is an OBD-II self-diagnostic item. • Diagnostic trouble code P1777 is detected under the following conditions: - When operating step motor ON and OFF, there is no proper change in the voltage of TCM terminal which corresponds to it. **Possible Causes:** • Step motor • Harness or connectors (Step motor circuit is open or shorted.)
DTC: P1777 **T TCM, TCIL: Yes** **Year:** 2009, 2010 **Model:** Maxima, Murano, Rogue **Engine:** 2.5L L4, 3.5L V6, 3.5L V6 VIN A **Transmission:** All	**Step Motor Circuit:** Each coil of the step motor is not energized properly due to an open or a short. **Possible Causes:** • Harness or connectors • Step motor circuit is open or shorted. • Step motor
DTC: P1777 **T TCM, TCIL: Yes** **Year:** 2009, 2010 **Model:** Cube **Engine:** 1.8L L4 **Transmission:** All	**Step Motor Circuit:** • Step motor monitor value is OFF when step motor command value of TCM is ON • Step motor monitor value is ON when step motor command value of TCM is OFF **Possible Causes:** • Harness or connectors • Step motor circuit is open or shorted. • Step motor
DTC: P1778 **1T TCM, TCIL: Yes** **Year:** 2009, 2010 **Model:** Sentra **Engine:** 2.0L L4, 2.5L L4 **Transmission:** All	**Step Motor Function:** • The step motor's 4 aspects of ON/OFF change according to the signal from TCM. As a result, the flow of line pressure to primary pulley is changed and pulley ratio is controlled. • This diagnosis item is detected when electrical system is OK, but mechanical system is NG. • This diagnosis item is detected when the state of the changing the speed mechanism in unit does not operate normally. • This is an OBD-II self-diagnostic item. • Diagnostic trouble code P1778 is detected under the following conditions: - When not changing the pulley ratio according to the instruction of TCM. **Possible Causes:** • Step motor
DTC: P1778 **T , TCIL: Yes** **Year:** 2009, 2010 **Model:** Cube **Engine:** 1.8L L4 **Transmission:** All	**Step Motor Circuit Intermittent:** • This DTC is not caused by an electrical malfunction (circuit open or short) but is caused by a mechanical malfunction (control valve clogging, solenoid valve sticking, and others). • TCM detects that primary speed sensor value and primary pulley speed estimated from secondary speed sensor are in a deviated state, and target pulley ratio and actual pulley ratio are in a deviated state **Possible Causes:** • Step motor

DTC	Trouble Code Title, Conditions & Possible Causes
DTC: P1778 **1T TCM, TCIL: Yes** **Year:** 2009, 2010 **Model:** Altima, Maxima, Murano, Rogue, Versa **Engine:** 1.6L L4, 1.8L L4, 2.5L L4, 3.5L V6, 3.5L V6 VIN A **Transmission:** All	**Step Motor Circuit Malfunction:** There is a great difference between the number of steps for the stepping motor and for the actual gear ratio, when not changing the pulley ratio according to the instruction of TCM. **Possible Causes:** • Step motor
DTC: P1800 **1T ECM** **Year:** 2009, 2010 **Model:** 350Z, Frontier, Maxima, Murano, Pathfinder, Quest, Xterra **Engine:** 3.5L V6, 3.5L V6 VIN A, 4.0L V6, 5.6L V8 **Transmission:** All	**VIAS Control Solenoid Valve Circuit (Valve 1):** **NOTE: On engines with two valves, this DTC relates to Valve 1.** • The MIL will not illuminate for this self-diagnosis. • An excessively low or high voltage signal is sent to ECM through the valve. **Possible Causes:** • Harness or connectors • The solenoid valve circuit is open or shorted. • VIAS control solenoid valve
DTC: P1801 **T TCM, TCIL: Yes** **Year:** 2009, 2010 **Model:** Murano **Engine:** 3.5L V6 VIN A **Transmission:** All	**VIAS Control Solenoid Valve 2 Circuit:** An excessively low or high voltage signal is sent to ECM through the VIAS control solenoid valve 2. **Possible Causes:** • Harness or connectors • The solenoid valve 2 circuit is open or shorted. • VIAS control solenoid valve 2
DTC: P1805 **1T ECM** **Year:** 2009, 2010 **Model:** 350Z, 370Z, Altima, Armada, Cube, Frontier, Maxima, Murano, Pathfinder, Quest, Rogue, Titan, Versa, Xterra **Engine:** 1.6L L4, 1.8L L4, 2.5L L4, 3.5L V6, 3.5L V6 VIN A, 3.7L V6, 4.0L V6, 5.6L V8 **Transmission:** All	**Brake Switch Signal Malfunction (Includes Hybrid Models):** HYBRID MODELS CAUTION: Hybrid systems use very high-voltage battery systems. Before starting any service work involving the battery system, turn the ignition switch OFF and then remove the service plug from pocket in the trunk. After removing the service plug, wait 10 minutes before touching any of the high-voltage connectors and terminals. • The MIL may not illuminate for this self-diagnosis. • A brake switch signal is not sent to ECM for extremely long time while the vehicle is being driven. **Possible Causes:** • Harness or connectors • Stop lamp switch circuit is open or shorted. • Stop lamp switch
DTC: P1815 **T TCM, TCIL: Yes** **Year:** 2009, 2010 **Model:** 370Z, Pathfinder, Titan **Engine:** 3.7L V6, 4.0L V6, 5.6L V8 **Transmission:** All	**Manual Mode Switch Circuit:** • TCM monitors manual mode, non manual mode, up or down switch signal, and detects as irregular when impossible input pattern occurs 2 seconds or more. • Shift up/down signal of paddle shifter continuously remains ON for 60 seconds. **Possible Causes:** • Harness or connectors • These switches circuit is open or shorted. • Manual mode switch (into A/T shift selector) • Manual mode shift-up switch (into A/T shift selector) • Manual mode shift-down switch (into A/T shift selector) • Paddle shifter

OBD II Trouble Code List (P2XXX Codes)

DTC	Trouble Code Title, Conditions & Possible Causes
DTC: P2004 **1T ECM, MIL: Yes** **Year:** 2009, 2010 **Model:** Rogue **Engine:** 2.5L L4 **Transmission:** All	**Tumble Control Valve Stuck:** • The target angle of tumble control valve controlled by ECM and the input signal from tumble control valve position sensor is not in the normal range. **Possible Causes:** • Harness or connectors • Tumble control valve motor circuit is open or shorted. • Tumble control valve position sensor circuit is open or shorted. • Accelerator pedal position sensor 2 circuit is shorted. • Crankshaft position sensor (POS) circuit is shorted. • EVAP control system pressure sensor circuit is shorted. • Refrigerant pressure sensor circuit is shorted. • Tumble control valve actuator • Tumble control valve motor • Tumble control valve position sensor • Accelerator pedal position sensor (APP sensor 2) • Crankshaft position sensor (POS) • EVAP control system pressure sensor • Refrigerant pressure sensor
DTC: P2014 **1T ECM, MIL: Yes** **Year:** 2009, 2010 **Model:** Rogue **Engine:** 2.5L L4 **Transmission:** All	**Tumble Control Valve Position Sensor Circuit:** An excessively low or high voltage from the sensor is sent to ECM. **Possible Causes:** • Harness or connectors • Tumble control valve position sensor circuit is open or shorted. • Accelerator pedal position sensor 2 circuit is shorted. • Crankshaft position sensor (POS) circuit is shorted. • EVAP control system pressure sensor circuit is shorted. • Refrigerant pressure sensor circuit is shorted. • Tumble control valve actuator • Tumble control valve position sensor • Accelerator pedal position sensor (APP sensor 2) • Crankshaft position sensor (POS) • EVAP control system pressure sensor • Refrigerant pressure sensor
DTC: P2100 **1T ECM, MIL: Yes** **Year:** 2009, 2010 **Model:** 350Z, 370Z, Altima, Armada, Cube, Frontier, Maxima, Murano, Pathfinder, Quest, Rogue, Titan, Versa **Engine:** 1.6L L4, 1.8L L4, 2.5L L4, 3.5L V6, 3.5L V6 VIN A, 3.7L V6, 4.0L V6, 5.6L V8 **Transmission:** All	**Throttle Control Motor Relay Circuit is Open (Includes Hybrid Models):** HYBRID MODELS CAUTION: Hybrid systems use very high-voltage battery systems. Before starting any service work involving the battery system, turn the ignition switch OFF and then remove the service plug from pocket in the trunk. After removing the service plug, wait 10 minutes before touching any of the high-voltage connectors and terminals. **NOTE: On V6 and V8, this DTC is for Bank 1.** • These self-diagnoses have the one trip detection logic. • ECM detects that the voltage of power source for throttle control motor is excessively low. **Possible Causes:** • Harness or connectors • Throttle control motor relay circuit is open. • Throttle control motor relay
DTC: P2101 **1T ECM, MIL: Yes** **Year:** 2009, 2010 **Model:** 350Z, 370Z, Altima, Armada, Cube, Frontier, Maxima, Murano, Pathfinder, Rogue, Titan, Versa, Xterra **Engine:** 1.6L L4, 1.8L L4, 2.5L L4, 3.5L V6, 3.5L V6 VIN A, 3.7L V6, 4.0L V6, 5.6L V8 **Transmission:** All	**Electric Throttle Control Performance (Includes Hybrid Models):** HYBRID MODELS CAUTION: Hybrid systems use very high-voltage battery systems. Before starting any service work involving the battery system, turn the ignition switch OFF and then remove the service plug from pocket in the trunk. After removing the service plug, wait 10 minutes before touching any of the high-voltage connectors and terminals. **NOTE: On V6 and V8, this DTC refers to Bank 1.** **NOTE: If DTC P1233 or P2101 is displayed with DTC P1238, P1290, P2100 or 2119, first perform the trouble diagnosis for DTC P1238, P2119 or P1290, P2100.** • Electric throttle control function does not operate properly. **Possible Causes:** • Harness or connectors • Throttle control motor circuit is open or shorted • Electric throttle control actuator

DTC	Trouble Code Title, Conditions & Possible Causes

DTC: P2103
1T ECM, MIL: Yes
Year: 2009, 2010
Model: 350Z, 370Z, Altima, Armada, Cube, Frontier, Maxima, Murano, Pathfinder, Quest, Rogue, Titan, Versa, Xterra
Engine: 1.6L L4, 1.8L L4, 2.5L L4, 3.5L V6, 3.5L V6 VIN A, 3.7L V6, 4.0L V6, 5.6L V8
Transmission: All
	Throttle Control Motor Relay Circuit is Short (Includes Hybrid Models):
	HYBRID MODELS CAUTION: Hybrid systems use very high-voltage battery systems. Before starting any service work involving the battery system, turn the ignition switch OFF and then remove the service plug from pocket in the trunk. After removing the service plug, wait 10 minutes before touching any of the high-voltage connectors and terminals.
	    • ECM detects the throttle control motor relay is stuck ON.
	Possible Causes:
	    • Harness or connectors
	    • Throttle control motor relay circuit is shorted
	    • Throttle control motor relay

DTC: P2118
1T ECM, MIL: Yes
Year: 2009, 2010
Model: 350Z, 370Z, Altima, Armada, Cube, Frontier, Maxima, Murano, Pathfinder, Quest, Rogue, Titan, Versa, Xterra
Engine: 1.6L L4, 1.8L L4, 2.5L L4, 3.5L V6, 3.5L V6 VIN A, 3.7L V6, 4.0L V6, 5.6L V8
Transmission: All
	Throttle Control Motor Circuit Short (Includes Hybrid Models):
	HYBRID MODELS CAUTION: Hybrid systems use very high-voltage battery systems. Before starting any service work involving the battery system, turn the ignition switch OFF and then remove the service plug from pocket in the trunk. After removing the service plug, wait 10 minutes before touching any of the high-voltage connectors and terminals.
	NOTE: On V6 and V8, this DTC is for Bank 1.
	    • ECM detects short in both circuits between ECM and throttle control motor.
	Possible Causes:
	    • Harness or connectors
	    • Throttle control motor circuit is shorted.
	    • Electric throttle control actuator (Throttle control motor)

DTC: P2119
1T ECM, MIL: Yes
Year: 2009, 2010
Model: 350Z, 370Z, Altima, Armada, Cube, Frontier, Maxima, Murano, Pathfinder, Quest, Rogue, Titan, Versa, Xterra
Engine: 1.6L L4, 1.8L L4, 2.5L L4, 3.5L V6, 3.5L V6 VIN A, 3.7L V6, 4.0L V6, 5.6L V8
Transmission: All
	Electric Throttle Control Actuator (Includes Hybrid Models):
	HYBRID MODELS CAUTION: Hybrid systems use very high-voltage battery systems. Before starting any service work involving the battery system, turn the ignition switch OFF and then remove the service plug from pocket in the trunk. After removing the service plug, wait 10 minutes before touching any of the high-voltage connectors and terminals.
	NOTE: When the malfunction is detected, ECM enters fail-safe mode and the MIL illuminates.
	NOTE: On V6 and V8, this DTC is for Bank 1.
	    • Malfunction A: Electric throttle control actuator does not function properly due to the return spring malfunction. ECM controls the electric throttle actuator by regulating the throttle opening around the idle position. The engine speed will not rise more than 2,000 rpm.
	    • Malfunction B: Throttle valve opening angle in fail-safe mode is not in specified range. ECM controls the electric throttle control actuator by regulating the throttle opening to 20 degrees or less.
	    • Malfunction C: ECM detects the throttle valve is stuck open. While the vehicle is driving, it slows down gradually by fuel cut. After the vehicle stops, the engine stalls.
	    • The engine can restart in N or P position (CVT), neutral (M/T), and engine speed will not exceed 1,000 rpm or more.
	Possible Causes:
	    • Electric throttle control actuator

DTC: P2122
1T ECM, MIL: Yes
Year: 2009, 2010
Model: 350Z, 370Z, Altima, Armada, Cube, Frontier, Maxima, Murano, Pathfinder, Quest, Rogue, Titan, Versa, Xterra
Engine: 1.6L L4, 1.8L L4, 2.5L L4, 3.5L V6, 3.5L V6 VIN A, 3.7L V6, 4.0L V6, 5.6L V8
Transmission: All
	Accelerator Pedal Position Sensor 1 Circuit Low Input:
	NOTE: If DTC P2122 or P2123 is displayed with DTC P0643, first perform the trouble diagnosis for DTC P0643.
	    • An excessively low voltage from the APP sensor 1 is sent to ECM.
	Possible Causes:
	    • Harness or connectors
	    • APP sensor 1 circuit is open or shorted.
	    • Accelerator pedal position sensor 1 (APP sensor 1)

DTC: P2123
1T ECM, MIL: Yes
Year: 2009, 2010
Model: 350Z, 370Z, Altima, Armada, Cube, Frontier, Maxima, Murano, Pathfinder, Quest, Rogue, Titan, Versa, Xterra
Engine: 1.6L L4, 1.8L L4, 2.5L L4, 3.5L V6, 3.5L V6 VIN A, 3.7L V6, 4.0L V6, 5.6L V8
Transmission: All
	Accelerator Pedal Position Sensor 1 Circuit High Input:
	NOTE: If DTC P2122 or P2123 is displayed with DTC P0643, first perform the trouble diagnosis for DTC P0643.
	    • An excessively high voltage from the APP sensor 1 is sent to ECM.
	    • When the malfunction is detected, ECM enters fail-safe mode and the MIL illuminates.
	Possible Causes:
	    • Harness or connectors
	    • APP sensor 1 circuit is open or shorted.
	    • Accelerator pedal position sensor 1 (APP sensor 1)

DTC	Trouble Code Title, Conditions & Possible Causes
DTC: P2127 **1T ECM, MIL: Yes** **Year:** 2009, 2010 **Model:** 350Z, 370Z, Altima, Armada, Cube, Frontier, Maxima, Murano, Pathfinder, Quest, Rogue, Titan, Versa, Xterra **Engine:** 1.6L L4, 1.8L L4, 2.5L L4, 3.5L V6, 3.5L V6 VIN A, 3.7L V6, 4.0L V6, 5.6L V8 **Transmission:** All	**Accelerator Pedal Position Sensor 2 Circuit Low Input:** • An excessively low voltage from the APP sensor 2 is sent to ECM. • When the malfunction is detected, ECM enters fail-safe mode and the MIL illuminates. **Possible Causes:** • Harness or connectors • APP sensor 2 circuit is open or shorted. • Crankshaft position sensor (POS) circuit is shorted. • Refrigerant pressure sensor circuit is shorted. • EVAP control system pressure sensor circuit is shorted. • Battery current sensor circuit is shorted. • Accelerator pedal position sensor (APP sensor 2) • Electric throttle control actuator (TP sensor) • Crankshaft position sensor (POS) • Refrigerant pressure sensor • EVAP control system pressure sensor • Battery current sensor
DTC: P2128 **1T ECM, MIL: Yes** **Year:** 2009, 2010 **Model:** 350Z, 370Z, Altima, Armada, Cube, Frontier, Maxima, Murano, Quest, Rogue, Titan, Versa, Xterra **Engine:** 1.6L L4, 1.8L L4, 2.5L L4, 3.5L V6, 3.5L V6 VIN A, 3.7L V6, 4.0L V6, 5.6L V8 **Transmission:** All	**Accelerator Pedal Position Sensor 2 Circuit High Input:** • An excessively high voltage from the APP sensor 2 is sent to ECM. • When the malfunction is detected, ECM enters fail-safe mode and the MIL illuminates. **Possible Causes:** • Harness or connectors • APP sensor 2 circuit is open or shorted. • Crankshaft position sensor (POS) circuit is shorted. • Refrigerant pressure sensor circuit is shorted. • EVAP control system pressure sensor circuit is shorted. • Battery current sensor circuit is shorted. • Accelerator pedal position sensor (APP sensor 2) • Electric throttle control actuator (TP sensor) • Crankshaft position sensor (POS) • Refrigerant pressure sensor • EVAP control system pressure sensor • Battery current sensor
DTC: P2132 **1T ECM, MIL: Yes** **Year:** 2009, 2010 **Model:** 370Z **Engine:** 3.7L V6 **Transmission:** All	**Throttle Position Sensor 1 (Bank 2) Circuit Low Input:** An excessively low voltage from the TP sensor 1 is sent to ECM. **Possible Causes:** • Harness or connectors • TP sensor 1 circuit is open or shorted. • Electric throttle control actuator (TP sensor 1)
DTC: P2133 **1T ECM, MIL: Yes** **Year:** 2009, 2010 **Model:** 370Z **Engine:** 3.7L V6 **Transmission:** All	**Throttle Position Sensor 1 (Bank 2) Circuit High Input:** An excessively high voltage from the TP sensor 1 is sent to ECM. **Possible Causes:** • Harness or connectors • TP sensor 1 circuit is open or shorted. • Electric throttle control actuator (TP sensor 1)
DTC: P2135 **1T ECM, MIL: Yes** **Year:** 2009, 2010 **Model:** 350Z, 370Z, Altima, Armada, Cube, Frontier, Maxima, Murano, Pathfinder, Quest, Rogue, Titan, Versa, Xterra **Engine:** 1.6L L4, 1.8L L4, 2.5L L4, 3.5L V6, 3.5L V6 VIN A, 3.7L V6, 4.0L V6, 5.6L V8 **Transmission:** All	**Throttle Position Sensor Circuit Range/Performance (Includes Hybrid Models):** HYBRID MODELS CAUTION: Hybrid systems use very high-voltage battery systems. Before starting any service work involving the battery system, turn the ignition switch OFF and then remove the service plug from pocket in the trunk. After removing the service plug, wait 10 minutes before touching any of the high-voltage connectors and terminals. **NOTE: If DTC P2135 is displayed with DTC P0643, first perform the trouble diagnosis for DTC P0643.** **NOTE: On V6 and V8, this DTC refers to Bank 1.** • Rationally incorrect voltage is sent to ECM compared with the signals from TP sensor 1 and TP sensor 2. • When the malfunction is detected, the ECM enters fail-safe mode and the MIL illuminates. **Possible Causes:** • Harness or connectors • TP sensor 1 or 2 circuit is open or shorted. • Electric throttle control actuator (TP sensor 1 or 2) • Accelerator pedal position sensor (APP sensor 2)

DTC	Trouble Code Title, Conditions & Possible Causes

DTC: P2138
1T ECM, MIL: Yes
Year: 2009, 2010
Model: 350Z, 370Z, Altima, Armada, Frontier, Maxima, Murano, Pathfinder, Quest, Rogue, Titan, Versa, Xterra
Engine: 1.6L L4, 1.8L L4, 2.5L L4, 3.5L V6, 3.5L V6 VIN A, 3.7L V6, 4.0L V6, 5.6L V8
Transmission: All
	Accelerator Pedal Position Sensor Circuit Range/Performance:
	NOTE: If DTC P2138 is displayed with DTC P0643, first perform the trouble diagnosis for DTC P0643.
	  • Rationally incorrect voltage is sent to ECM compared with the signals from APP sensor 1 and APP sensor 2.
	  • When the malfunction is detected, ECM enters fail-safe mode and the MIL illuminates.
	Possible Causes:
	  • Harness or connector
	  • APP sensor 1 or 2 circuit is open or shorted.
	  • Crankshaft position sensor (POS) circuit is shorted.
	  • Refrigerant pressure sensor circuit is shorted.
	  • EVAP control system sensor circuit is shorted.
	  • Battery current sensor circuit is shorted.
	  • Accelerator pedal position sensor (APP sensor 1 or 2)
	  • Electric throttle control actuator (TP sensor)
	  • Crankshaft position sensor (POS)
	  • Camshaft position sensor (PHASE) (bank 2)
	  • Exhaust valve timing control position sensor (bank 2)
	  • Refrigerant pressure sensor
	  • EVAP control system pressure sensor
	  • Battery current sensor

DTC: P2423
1T ECM, MIL: Yes
Year: 2009, 2010
Model: Altima, Rogue
Engine: 2.5L L4
Transmission: All
	HC Adsorption Catalyst Efficiency Below Threshold (Hybrid):
	HYBRID MODELS CAUTION: Hybrid systems use very high-voltage battery systems. Before starting any service work involving the battery system, turn the ignition switch OFF and then remove the service plug from pocket in the trunk. After removing the service plug, wait 10 minutes before touching any of the high-voltage connectors and terminals.
	  • HC adsorption catalyst (under floor) does not operate properly.
	  • HC adsorption catalyst (under floor) does not have enough oxygen storage capacity.
	Possible Causes:
	  • HC adsorption catalyst (under floor)
	  • Exhaust tube
	  • Intake air leaks
	  • Fuel injector
	  • Fuel injector leaks
	  • Spark plug
	  • Improper ignition timing

DTC: P2713
T TCM, TCIL: Yes
Year: 2009, 2010
Model: 370Z
Engine: 3.7L V6
Transmission: All
	Pressure Control Solenoid D:
	The high and low reverse clutch solenoid valve monitor value is 0.4 A or less when the high and low reverse clutch solenoid valve command value is more than 0.75 A.
	Possible Causes:
	  • Harness or connectors
	  • Solenoid valve circuit is open or shorted.
	  • High and low reverse clutch solenoid valve

DTC: P2722
T TCM, TCIL: Yes
Year: 2009, 2010
Model: 370Z
Engine: 3.7L V6
Transmission: All
	Pressure Control Solenoid E:
	The low brake solenoid valve monitor value is 0.4 A or less when the low brake solenoid valve command value is more than 0.75 A.
	Possible Causes:
	  • Harness or connectors
	  • Solenoid valve circuit is open or shorted.
	  • Low brake solenoid valve

DTC: P2731
T TCM, TCIL: Yes
Year: 2009, 2010
Model: 370Z
Engine: 3.7L V6
Transmission: All
	Pressure Control Solenoid F:
	The 2346 brake solenoid valve monitor value is 0.4 A or less when the 2346 brake solenoid valve command value is more than 0.75 A.
	Possible Causes:
	  • Harness or connectors
	  • Solenoid valve circuit is open or shorted.
	  • 2346 brake solenoid valve

DTC: P2765
T ECM, MIL: Yes
Year: 2009, 2010
Model: 370Z
Engine: 3.7L V6
Transmission: All
	Input Speed Sensor Circuit:
	NOTE: If DTC P2765 is displayed with DTC P0335, P0340 or P0345, first perform the trouble diagnosis for DTC P0335, P0340 or P0345.
	  • There is a difference between engine speed signal calculated by ECM and input shaft speed sensor signal.
	Possible Causes:
	  • Harness or connectors
	  • Input speed sensor circuit is open or shorted.
	  • Input speed sensor

DTC	Trouble Code Title, Conditions & Possible Causes
DTC: P2807 **T TCM, TCIL: Yes** **Year:** 2009, 2010 **Model:** 370Z **Engine:** 3.7L V6 **Transmission:** All	**Pressure Control Solenoid G:** The direct clutch solenoid valve monitor value is 0.4 A or less when the direct clutch solenoid valve command value is more than 0.75 A. **Possible Causes:** • Harness or connectors • Solenoid valve circuit is open or shorted. • Direct clutch solenoid valve
DTC: P2A00 **1T ECM, MIL: Yes** **Year:** 2009, 2010 **Model:** 350Z, 370Z, Altima, Armada, Cube, Frontier, Maxima, Murano, Pathfinder, Quest, Rogue, Titan, Versa, Xterra **Engine:** 1.6L L4, 1.8L L4, 2.5L L4, 3.5L V6, 3.5L V6 VIN A, 3.7L V6, 4.0L V6, 5.6L V8 **Transmission:** All	**Air Fuel (A/F) Sensor 1 Circuit Range/Performance (Includes Hybrid Models):** HYBRID MODELS CAUTION: Hybrid systems use very high-voltage battery systems. Before starting any service work involving the battery system, turn the ignition switch OFF and then remove the service plug from pocket in the trunk. After removing the service plug, wait 10 minutes before touching any of the high-voltage connectors and terminals. **NOTE: On V6 and V8 engines, this applies to Bank 1.** • To judge the malfunction, the A/F signal computed by ECM from the A/F sensor 1 signal is monitored not to be shifted to LEAN side or RICH side. • The output voltage computed by ECM from the A/F sensor 1 signal is shifted to the lean side for a specified period. • The A/F signal computed by ECM from the A/F sensor 1 signal is shifted to the rich side for a specified period. **Possible Causes:** • Air fuel ratio (A/F) sensor 1 • Air fuel ratio (A/F) sensor 1 heater • Fuel pressure • Fuel injector • Intake air leaks
DTC: P2A03 **1T ECM, MIL: Yes** **Year:** 2009, 2010 **Model:** 350Z, 370Z, Altima, Armada, Frontier, Maxima, Murano, Pathfinder, Quest, Titan, Xterra **Engine:** 2.5L L4, 3.5L V6, 3.5L V6 VIN A, 3.7L V6, 4.0L V6, 5.6L V8 **Transmission:** All	**Air Fuel (A/F) Ratio Sensor 1 Circuit Range/Performance:** NOTE: On V6 and V8 engines, this applies to Bank 2. • The output voltage computed by ECM from the A/F sensor 1 signal is shifted to the lean side for a specified period. • The A/F signal computed by ECM from the A/F sensor 1 signal is shifted to the rich side for a specified period. • To judge the malfunction, the A/F signal computed by ECM from the A/F sensor 1 signal is monitored so it will not shift to LEAN side or RICH side. **Possible Causes:** • Air fuel ratio (A/F) sensor 1 • Air fuel ratio (A/F) sensor 1 heater • Fuel pressure • Fuel injector • Intake air leaks

OBD II Trouble Code List (U0XXX Codes)

DTC	Trouble Code Title, Conditions & Possible Causes
DTC: U0101 **1T ECM, MIL: Yes** **Year:** 2009, 2010 **Model:** 350Z, 370Z, Altima, Cube, Frontier, Maxima, Murano, Quest, Rogue, Titan, Versa, Xterra **Engine:** 1.6L L4, 1.8L L4, 2.5L L4, 3.5L V6, 3.5L V6 VIN A, 3.7L V6, 4.0L V6, 5.6L V8 **Transmission:** All	**Lost communication with TCM:** When ECM is not transmitting or receiving CAN communication signal of OBD (emission-related diagnosis) with TCM for 2 seconds or more. **Possible Causes:** • CAN communication line between TCM and ECM • CAN communication line is open or shorted
DTC: U0113 **T ECM, MIL: Yes** **Year:** 2009, 2010 **Model:** 370Z **Engine:** 3.7L V6 **Transmission:** All	**Lost Communication with Variable Valve Event & Lift (VVEL) Control Module:** NOTE: If DTC U0113 or U1003 is displayed with DTC P0607, first perform the trouble diagnosis for DTC P0607. • CAN communication signal of OBD (emission related diagnosis) is not received VVEL control module and ECM for 2 seconds or more. **Possible Causes:** • Harness or connectors • VVEL CAN communication line is open or shorted • ECM • VVEL control module

DTC	Trouble Code Title, Conditions & Possible Causes
DTC: U0129 **1T ECM, MIL: Yes** **Year:** 2009, 2010 **Model:** Altima **Engine:** 2.5L L4, 3.5L V6 **Transmission:** All	**Lost Communication with Brake ECU (Hybrid):** CAUTION: Before servicing the hybrid vehicle, turn the ignition switch OFF and then remove the service plug from pocket in the trunk. After removing the service plug, wait 10 minutes before touching any of the high-voltage connectors and terminals. CAN communication signal of OBD (emission related diagnosis) is not received **Possible Causes:** • HEV SYSTEM CAN communication line between brake ECU and ECM • HEV SYSTEM CAN communication line short
DTC: U0140 **1T ECM, MIL: Yes** **Year:** 2009, 2010 **Model:** Altima, Cube, Frontier, Rogue, Versa **Engine:** 1.6L L4, 1.8L L4, 2.5L L4, 3.5L V6, 4.0L V6 **Transmission:** All	**Lost Communication with BCM:** • When ECM is not transmitting or receiving CAN communication signal of OBD (emission related diagnosis) with BCM for 2 seconds or more. **Possible Causes:** • CAN communication line between BCM and ECM • CAN communication line is open or shorted
DTC: U0164 **T ECM, MIL: Yes** **Year:** 2009, 2010 **Model:** 370Z, Maxima, Murano **Engine:** 3.5L V6, 3.5L V6 VIN A, 3.7L V6 **Transmission:** All	**Lost Communication with Unified Meter & A/C Amp or Combination Meter:** • When ECM is not transmitting or receiving CAN communication signal of OBD (emission related diagnosis) with unified meter and A/C amp. or the combination meter for 2 seconds or more. **Possible Causes:** • CAN communication line between Unified meter and A/C amp. and ECM • CAN communication line open or shorted
DTC: U0164 **1T ECM, MIL: Yes** **Year:** 2009, 2010 **Model:** Altima, Quest **Engine:** 2.5L L4, 3.5L V6 **Transmission:** All	**Lost Communication with Controller (Auto. Amp.) (Hybrid):** CAUTION: Hybrid systems use very high-voltage battery systems. Before starting any service work involving the battery system, turn the ignition switch OFF and then remove the service plug from pocket in the trunk. After removing the service plug, wait 10 minutes before touching any of the high-voltage connectors and terminals. • When ECM is not transmitting or receiving CAN communication signal of OBD (emission related diagnosis) with Controller (auto amp.) for 2 seconds or more. **Possible Causes:** • CAN communication line between Controller (auto amp.) and ECM • CAN communication line is open or shorted
DTC: U0293 **1T ECM, MIL: Yes** **Year:** 2009, 2010 **Model:** Altima **Engine:** 2.5L L4 **Transmission:** All	**Lost Communication with Hybrid Vehicle Control System (Hybrid):** CAUTION: Hybrid systems use very high-voltage battery systems. Before starting any service work involving the battery system, turn the ignition switch OFF and then remove the service plug from pocket in the trunk. After removing the service plug, wait 10 minutes before touching any of the high-voltage connectors and terminals. • CAN communication signal of OBD (emission related diagnosis) is not received between HV ECU and ECM for 1 second or more. **Possible Causes:** • HEV SYSTEM CAN communication line between HV ECU and ECM • HEV SYSTEM CAN communication line short
DTC: U0400 **1T ECM, MIL: Yes** **Year:** 2009, 2010 **Model:** Altima **Engine:** 2.5L L4 **Transmission:** All	**Invalid Data Received from Hybrid vehicle Control ECU (Hybrid):** CAUTION: Hybrid systems use very high-voltage battery systems. Before starting any service work involving the battery system, turn the ignition switch OFF and then remove the service plug from pocket in the trunk. After removing the service plug, wait 10 minutes before touching any of the high-voltage connectors and terminals. • SUM data on CAN signal of OBD (emission related diagnosis) from HV ECU is different from SUM data calculated by ECM. **Possible Causes:** • Harness or connectors • HEV SYSTEM CAN communication line is open or shorted • ECM • HV ECU
DTC: U0415 **1T BCM, MIL: Yes, TCIL: Yes** **Year:** 2009, 2010 **Model:** Altima **Engine:** 2.5L L4, 3.5L V6 **Transmission:** All	**Vehicle Speed Signal:** When the vehicle speed signal received from the ABS actuator and electric unit (control unit) remains abnormal for 2 seconds or more. **Possible Causes:** • ABS actuator and electric unit (control unit) • BCM

DTC	Trouble Code Title, Conditions & Possible Causes
DTC: U0418 **1T ECM, MIL: Yes** **Year:** 2009, 2010 **Model:** Altima **Engine:** 2.5L L4 **Transmission:** All	**Invalid Data Received from Brake ECU (Hybrid):** CAUTION: Hybrid systems use very high-voltage battery systems. Before starting any service work involving the battery system, turn the ignition switch OFF and then remove the service plug from pocket in the trunk. After removing the service plug, wait 10 minutes before touching any of the high-voltage connectors and terminals. • SUM data on CAN signal of OBD (emission related diagnosis) from brake ECU is different from SUM data calculated by ECM. **Possible Causes:** • Harness or connectors • HEV SYSTEM CAN communication line is open or shorted • ECM • Brake ECU

OBD II Trouble Code List (U1XXX Codes)

DTC	Trouble Code Title, Conditions & Possible Causes
DTC: U1000 **1T TCM, TCIL: Yes** **Year:** 2009, 2010 **Model:** 350Z, 370Z, Altima, Armada, Cube, Frontier, Maxima, Murano, Pathfinder, Quest, Rogue, Sentra, Titan, Versa, Xterra **Engine:** 1.6L L4, 1.8L L4, 2.0L L4, 2.5L L4, 3.5L V6, 3.5L V6 VIN A, 3.7L V6, 4.0L V6, 5.6L V8 **Transmission:** All	**CAN Communication Line:** • This is an OBD-II self-diagnostic item. • Diagnostic trouble code U1000 is detected when ECM or TCM cannot communicate to other control units for 2 seconds or more. **Possible Causes:** • Harness or connectors • CAN communication line is open or shorted. • TCM
DTC: U1001 **1T ECM, MIL: Yes, TCIL: Yes** **Year:** 2009, 2010 **Model:** 350Z, 370Z, Altima, Cube, Frontier, Maxima, Murano, Quest, Rogue, Titan, Versa, Xterra **Engine:** 1.6L L4, 1.8L L4, 2.5L L4, 3.5L V6, 3.5L V6 VIN A, 3.7L V6, 4.0L V6, 5.6L V8 **Transmission:** All	**CAN Communication Line (Includes Hybrid models):** HYBRID CAUTION: Hybrid systems use very high-voltage battery systems. Before starting any service work involving the battery system, turn the ignition switch OFF and then remove the service plug from pocket in the trunk. After removing the service plug, wait 10 minutes before touching any of the high-voltage connectors and terminals. • When ECM is not transmitting or receiving CAN communication signal other than OBD (emission related diagnosis) for 2 seconds or more. **Possible Causes:** • Harness or connectors • CAN communication line is open or shorted
DTC: U1003 **T ECM, MIL: Yes** **Year:** 2009, 2010 **Model:** 370Z **Engine:** 3.7L V6 **Transmission:** All	**Lost Communication with Variable Valve Event & Lift (VVEL) Control Module:** NOTE: If DTC U0113 or U1003 is displayed with DTC P0607, first perform the trouble diagnosis for DTC P0607. • CAN communication signal other than OBD (emission related diagnosis) is not received between VVEL control module and ECM for 2 seconds or more. **Possible Causes:** • Harness or connectors • VVEL CAN communication line is open or shorted • ECM • VVEL control module
DTC: U1010 **T TCM, TCIL: Yes** **Year:** 2009, 2010 **Model:** Versa **Engine:** 1.6L L4, 1.8L L4 **Transmission:** All	**Control Unit (CAN):** • This is an OBD-II self-diagnostic item. • Diagnostic trouble code U1010 is detected when TCM cannot communicate to other control units. **Possible Causes:** • Harness or connectors • CAN communication line is open or shorted.
DTC: U1010 **T TCM, TCIL: Yes** **Year:** 2009, 2010 **Model:** Cube **Engine:** 1.8L L4 **Transmission:** All	**TCM Communication Malfunction:** TCM detects a malfunction in CAN communication initial diagnosis (control unit malfunction) **Possible Causes:** • TCM

DTC	Trouble Code Title, Conditions & Possible Causes
DTC: U1010 **1T TCM, TCIL: Yes** **Year:** 2009, 2010 **Model:** Altima, Rogue **Engine:** 2.5L L4, 3.5L V6 **Transmission:** All	**TCM or BCM Communication Malfunction:** When detecting error during the initial diagnosis of CAN controller to TCM or BCM. **Possible Causes:** • Harness or connectors (CAN communication line is open or shorted.) • TCM • BCM
DTC: U1020 **1T ECM, MIL: Yes** **Year:** 2009, 2010 **Model:** Altima **Engine:** 2.5L L4 **Transmission:** All	**Lost Communication with Hybrid Vehicle (HV) Control System (Hybrid):** CAUTION: Hybrid systems use very high-voltage battery systems. Before starting any service work involving the battery system, turn the ignition switch OFF and then remove the service plug from pocket in the trunk. After removing the service plug, wait 10 minutes before touching any of the high-voltage connectors and terminals. • CAN communication signal other than OBD (emission related diagnosis) is not received between HV ECU and ECM for 1 second or more. **Possible Causes:** • HEV SYSTEM CAN communication line between HV ECU and ECM • HEV SYSTEM CAN communication line short
DTC: U1021 **1T ECM, MIL: Yes** **Year:** 2009, 2010 **Model:** Altima **Engine:** 2.5L L4 **Transmission:** All	**Invalid Data Received from Hybrid Vehicle Control ECU (Hybrid):** CAUTION: Hybrid systems use very high-voltage battery systems. Before starting any service work involving the battery system, turn the ignition switch OFF and then remove the service plug from pocket in the trunk. After removing the service plug, wait 10 minutes before touching any of the high-voltage connectors and terminals. • SUM data on CAN signal other than OBD (emission related diagnosis) from HV ECU is different from SUM data calculated by ECM. **Possible Causes:** • Harness or connectors • HEV SYSTEM CAN communication line is open or shorted • ECM • HV ECU
DTC: U1022 **1T ECM, MIL: Yes** **Year:** 2009, 2010 **Model:** Altima **Engine:** 2.5L L4, 3.5L V6 **Transmission:** All	**Lost Communication with Brake ECU:** CAUTION: Hybrid systems use very high-voltage battery systems. Before starting any service work involving the battery system, turn the ignition switch OFF and then remove the service plug from pocket in the trunk. After removing the service plug, wait 10 minutes before touching any of the high-voltage connectors and terminals. CAN communication signal other than OBD (emission related diagnosis) is not received between brake ECU and ECM for 1 second or more. **Possible Causes:** • HEV SYSTEM CAN communication line between brake ECU and ECM • HEV SYSTEM CAN communication line short
DTC: U1023 **T ECM, MIL: Yes** **Year:** 2009, 2010 **Model:** Altima **Engine:** 2.5L L4 **Transmission:** All	**Invalid Data Received from Brake ECU:** CAUTION: Hybrid systems use very high-voltage battery systems. Before starting any service work involving the battery system, turn the ignition switch OFF and then remove the service plug from pocket in the trunk. After removing the service plug, wait 10 minutes before touching any of the high-voltage connectors and terminals. • SUM data on CAN signal other than OBD (emission related diagnosis) from brake ECU is different from SUM data calculated by ECM. **Possible Causes:** • Harness or connectors • HEV SYSTEM CAN communication line is open or shorted • ECM • Brake ECU
DTC: U1024 **T ECM, MIL: Yes** **Year:** 2009, 2010 **Model:** 370Z **Engine:** 3.7L V6 **Transmission:** All	**Variable Valve Event & Lift (VVEL) CAN Communication:** • When VVEL control module cannot transmit/receive can communication signal from ECM. • When detecting error during the initial diagnosis of CAN controller of VVEL control module. **Possible Causes:** • Harness or connectors • CAN communication line is open or shorted • ECM • VVEL control module
DTC: U1200 **1T ECM, MIL: Yes** **Year:** 2009, 2010 **Model:** Sentra **Engine:** 2.0L L4, 2.5L L4 **Transmission:** All	**CAN Vehicle Speed (ABS):** Vehicle speed sensor display not consistent with speedometer indicated speed (driving or stopped) **Possible Causes:** • Harness or connectors • EPS control unit • ABS actuator • ECBM control module

Commonly Used Abbreviations

2

2WD	Two Wheel Drive

4

4WD	Four Wheel Drive

A

A/C	Air Conditioning
ABDC	After Bottom Dead Center
ABS	Anti-lock Brakes
AC	Alternating Current
ACL	Air cleaner
ACT	Air Charge Temperature
AIR	Secondary Air Injection
ALCL	Assembly Line Communications Link
ALDL	Assembly Line Diagnostic Link
AT	Automatic Transaxle/Transmission
ATDC	After Top Dead Center
ATF	Automatic Transmission Fluid
ATS	Air Temperature Sensor
AWD	All Wheel Drive

B

BAP	Barometric Absolute Pressure
BARO	Barometric Pressure
BBDC	Before Bottom Dead Center
BCM	Body Control Module
BDC	Bottom Dead Center
BPT	Backpressure Transducer
BTDC	Before Top Dead Center
BVSV	Bimetallic Vacuum Switching Valve

C

CAC	Charge Air Cooler
CARB	California Air Resources Board
CAT	Catalytic Converter
CCC	Computer Command Control
CCCC	Computer Controlled Catalytic Converter
CCCI	Computer Controlled Coil Ignition
CCD	Computer Controlled Dwell
CDI	Capacitor Discharge Ignition
CEC	Computerized Engine Control
CFI	Continuous Fuel Injection
CIS	Continuous Injection System
CIS-E	Continuous Injection System - Electronic
CKP	Crankshaft Position
CL	Closed Loop
CMP	Camshaft Position
CPP	Clutch Pedal Position
CTOX	Continuous Trap Oxidizer System
CTP	Closed Throttle Position
CVC	Constant Vacuum Control
CYL	Cylinder

D

DBC	Dual Bed Catalyst
DC	Direct Current
DFI	Direct Fuel Injection
DIS	Distributorless Ignition System
DLC	Data Link Connector
DMM	Digital Multimeter
DOHC	Double Overhead Camshaft
DRB	Diagnostic Readout Box
DTC	Diagnostic Trouble Code
DTM	Diagnostic Test Mode
DVOM	Digital Volt/Ohmmeter

E

EBCM	Electronic Brake Control Module
ECM	Engine Control Module
ECT	Engine Coolant Temperature
ECU	Engine Control Unit or Electronic Control Unit
EDIS	Electronic Distributorless Ignition System
EEC	Electronic Engine Control
EEPROM	Electrically Erasable Programmable Read Only Memory
EFE	Early Fuel Evaporation
EGR	Exhaust Gas Recirculation
EGRT	Exhaust Gas Recirculation Temperature
EGRVC	EGR Valve Control
EPROM	Erasable Programmable Read Only Memory
EVAP	Evaporative Emissions
EVP	EGR Valve Position

F

FBC	Feedback Carburetor
FEEPROM	Flash Electrically Erasable Programmable Read Only Memory
FF	Flexible Fuel
FI	Fuel Injection
FT	Fuel Trim
FWD	Front Wheel Drive

G

GND	Ground

H

HAC	High Altitude Compensation
HEGO	Heated Exhaust Gas Oxygen sensor
HEI	High Energy Ignition
HO2 Sensor	Heated Oxygen Sensor

I

IAC	Idle Air Control
IAT	Intake Air Temperature
ICM	Ignition Control Module
IFI	Indirect Fuel Injection
IFS	Inertia Fuel Shutoff
ISC	Idle Speed Control
IVSV	Idle Vacuum Switching Valve

Commonly Used Abbreviations

K

KOEO	Key On, Engine Off
KOER	Key ON, Engine Running
KS	Knock Sensor

M

MAF	Mass Air Flow
MAP	Manifold Absolute Pressure
MAT	Manifold Air Temperature
MC	Mixture Control
MDP	Manifold Differential Pressure
MFI	Multiport Fuel Injection
MIL	Malfunction Indicator Lamp or Maintenance
MST	Manifold Surface Temperature
MVZ	Manifold Vacuum Zone

N

NVRAM	Nonvolatile Random Access Memory

O

O2 Sensor	Oxygen Sensor
OBD	On-Board Diagnostic
OC	Oxidation Catalyst
OHC	Overhead Camshaft
OL	Open Loop

P

P/S	Power Steering
PAIR	Pulsed Secondary Air Injection
PCM	Powertrain Control Module
PCS	Purge Control Solenoid
PCV	Positive Crankcase Ventilation
PIP	Profile Ignition Pick-up
PNP	Park/Neutral Position
PROM	Programmable Read Only Memory
PSP	Power Steering Pressure
PTO	Power Take-Off
PTOX	Periodic Trap Oxidizer System

R

RABS	Rear Anti-lock Brake System
RAM	Random Access Memory
ROM	Read Only Memory
RPM	Revolutions Per Minute
RWAL	Rear Wheel Anti-lock Brakes
RWD	Rear Wheel Drive

S

SBC	Single Bed Converter
SBEC	Single Board Engine Controller
SC	Supercharger
SCB	Supercharger Bypass
SFI	Sequential Multiport Fuel Injection
SIR	Supplemental Inflatible Restraint
SOHC	Single Overhead Camshaft
SPL	Smoke Puff Limiter
SPOUT	Spark Output
SRI	Service Reminder Indicator
SRS	Supplemental Restraint System
SRT	System Readiness Test
SSI	Solid State Ignition
ST	Scan Tool
STO	Self-Test Output

T

TAC	Thermostatic Air Cleaner
TBI	Throttle Body Fuel Injection
TC	Turbocharger
TCC	Torque Converter Clutch
TCM	Transmission Control Module
TDC	Top Dead Center
TFI	Thick Film Ignition
TP	Throttle Position
TR Sensor	Transaxle/Transmission Range Sensor
TVV	Thermal Vacuum Valve
TWC	Three-way Catalytic Converter

V

VAF	Volume Air Flow, or Vane Air Flow
VAPS	Variable Assist Power Steering
VRV	Vacuum Regulator Valve
VSS	Vehicle Speed Sensor
VSV	Vacuum Switching Valve

W

WOT	Wide Open Throttle
WU-TWC	Warm Up Three-way Catalytic Converter